How to Use This Edition

The second edition of the *CRC Concise Encyclopedia of Mathematics* has been designed with the user in mind and for ease of accessibility. Listed below are various changes in the new edition that will make the book easier for the reader to use while navigating to different areas of interest

Alphabetization

All entries are listed in alphabetical order. There is a separate section appearing before the A's to cover the entries that are numerals. The alphabetizing of letters is not affected by dashes, apostrophes, or any other punctuation falling within a word. For example, you will find A-Integrable listed in the Ai section of the book. Following the same logic, all entries for Abel will precede entries for Abel's.

Cross-References

In many cases, a particular entry of interest can be located from a cross-reference. Cross-references are indicated in SMALL CAPS typeface in the text. In addition, for some main listings, you will be re-directed to a different entry (or multiple entries) as indicated in small caps underneath the main listing. For example,

> ABEL'S TEST
> ABEL'S UNIFORM CONVERGENCE TEST

Finally, most articles are followed by a *"See also"* list of related entries.

References

All Reference listings follow the text of the corresponding entry. Note that in this reference style, page ranges may be abbreviated. Accordingly, a page range of 132-136 will be indicated as 132-36. Another example of this is a page range of 96-100 that is indicated by 96-00.

Entries

Many new entries have been added for the user to the new edition. However, because this is a work in progress, some of the new entries have not been completed with appropriate definitions or textual description. Following many of these kinds of entries, the reader is referred to other items of interest that are closely related or similar to the article in question.

Introduction to the First Edition

The *CRC Concise Encyclopedia of Mathematics* is a compendium of mathematical definitions, formulas, figures, tabulations, and references. It is written in an informal style intended to make it accessible to a broad spectrum of readers with a wide range of mathematical backgrounds and interests. Although mathematics is a fascinating subject, it all too frequently is clothed in specialized jargon and dry formal exposition that make many interesting and useful mathematical results inaccessible to laypeople. This problem is often further compounded by the difficulty in locating concrete and easily understood examples. To give perspective to a subject, I find it helpful to learn why it is useful, how it is connected to other areas of mathematics and science, and how it is actually implemented. While a picture may be worth a thousand words, explicit examples are worth at least a few hundred! This work attempts to provide enough details to give the reader a flavor for a subject without getting lost in minutiae. While absolute rigor may suffer somewhat, I hope the improvement in usefulness and readability will more than make up for the deficiencies of this approach.

The format of this work is somewhere between a handbook, a dictionary, and an encyclopedia. It differs from existing dictionaries of mathematics in a number of important ways. The entries are extensively cross-referenced, not only to related entries but also to many external sites on the Internet. This makes locating information very convenient. It also provides a highly efficient way to "navigate" from one related concept to another. Standard mathematical references, combined with a few popular ones, are also given at the end of most entries to facilitate additional reading and exploration. In the interests of offering abundant examples, this work also contains a large number of explicit, formulas and derivations, providing a ready place to locate a particular formula, as well as including the framework for understanding where it comes from.

The selection of topics in this work is more extensive than in most mathematical dictionaries (e.g., Borowski and Borwein's *HarperCollins Dictionary of Mathematics* and Jeans and Jeans' *Mathematics Dictionary*). At the same time, the descriptions are more accessible than in "technical" mathematical encyclopedias (e.g., Hazewinkel's *Encyclopaedia of Mathematics* and Iyanaga's *Encyclopedic Dictionary of Mathematics*). While the latter remain models of accuracy and rigor, they are not terribly useful to the undergraduate, research scientist, or recreational mathematician. In this work, the most useful, interesting, and entertaining (at least to my mind) aspects of topics are discussed in addition to their technical definitions. For example, in my entry for pi (π), the definition in terms of the diameter and circumference of a circle is supplemented by a great many formulas and series for pi, including some of the amazing discoveries of Ramanujan. These formulas are comprehensible to readers with only minimal mathematical background, and are interesting to both those with and without formal mathematics training. However, they have not previously been collected in a single convenient location. For this reason, I hope that, in addition to serving as a reference source, this work has some of the same flavor and appeal of Martin Gardner's delightful *Scientific American* columns.

Everything in this work has been compiled by me alone. I am an astronomer by training, but have picked up a fair bit of mathematics along the way. It never ceases to amaze me how mathematical connections weave their way through the physical sciences. It frequently transpires that some piece of recently acquired knowledge turns out to be just what I need to solve some apparently unrelated problem. I have therefore developed the habit of picking up and storing away odd bits of information for future use. This work has provided a mechanism for organizing what has turned out to be a fairly large collection of mathematics. I have also found it very difficult to find clear yet accessible explanations of technical mathematics unless I already have some familiarity with the subject. I hope this encyclopedia will provide jumping-off points for people who are interested in the subjects listed here but who, like me, are not necessarily experts.

The encyclopedia has been compiled over the last 11 years or so, beginning in my college years and continuing during graduate school: The initial document was written in *Microsoft Word*® on a Mac Plus® computer, and had reached about 200 pages by the time I started graduate school in 1990. When Andrew Treverrow made his OzTEX program available for the Mac; I began the task of converting all my documents to TEX resulting in a vast improvement in readability. While undertaking the *Word* to TEX conversion, I also began cross-referencing entries, anticipating that eventually I would be able to convert the entire document to hypertext. This hope was realized beginning in 1995, when the Internet explosion was in full swing and I learned of Nikos Drakos's excellent TEX to HTML converter, LATEX2HTML. After some additional effort, I was able to post an HTML version of my encyclopedia to the World Wide Web.

The selection of topics included in this compendium is not based on any fixed set of criteria, but rather reflects my own random walk through mathematics. In truth, there is no good way of selecting topics in such a work. The mathematician James Sylvester may have summed up the situation most aptly. According to

Sylvester (as quoted in the introduction to Ian Stewart's book *From Here to Infinity*), "Mathematics is not a book confined within a cover and bound between brazen clasps, whose contents it needs only patience to ransack; it is not a mine, whose treasures may take long to reduce into possession, but which fill only a limited number of veins and lodes; it is not a soil, whose fertility can be exhausted by the yield of successive harvests; it is not a continent or an ocean, whose area can be mapped out and its contour defined; it is as limitless as that space which it finds too narrow for its aspiration; its possibilities are as infinite as the worlds which are forever crowding in and multiplying upon the astronomer's gaze; it is as incapable of being restricted within assigned boundaries or being reduced to definitions of permanent validity, as the consciousness of life."

Several of Sylvester's points apply particularly to this undertaking: As he points out, mathematics itself cannot be confined to the pages of a book. The results of mathematics, however, are shared and passed on primarily through the printed (and now electronic) medium. While there is no danger of mathematical results being lost through lack of dissemination, many people miss out on fascinating and useful mathematical results simply because they are not aware of them. Not only does collecting many results in one place provide a single starting point for mathematical exploration, but it should also lessen the aggravation of encountering explanations for new concepts which themselves use unfamiliar terminology. In this work, the reader is only a cross-reference away from the necessary background material. As to Sylvester's second point, the very fact that the quantity of mathematics is so great means that any attempt to catalog it with any degree of completeness is doomed to failure. This certainly does not mean that it's not worth trying. Strangely, except for relatively small works usually on particular subjects, there do not appear to have been any substantial attempts to collect and display in a place of prominence the treasure trove of mathematical results that have been discovered (invented?) over the years (one notable exception being Sloane and Plouffe's *Encyclopedia of Integer Sequences*). This work, the product of the "gazing" of a single astronomer, attempts to fill that omission.

Finally, a few words about logistics. Because of the alphabetical listing of entries in the encyclopedia, neither table of contents nor index are included. In many cases, a particular entry of interest can be located from a cross-reference (indicated in SMALL CAPS TYPEFACE in the text) in a related article. In addition, most articles are followed by a "see also" list of related entries for quick navigation. This can be particularly useful if you are looking for a specific entry (say, "Zeno's Paradoxes"), but have forgotten the exact name. By examining the "see also" list at bottom of the entry for "Paradox," you will likely recognize Zeno's name and thus quickly locate the desired entry.

In cases where the same word is applied in different contexts, the context is indicated in parentheses or appended to the end. Examples of the first type are "Crossing Number (Graph)" and "Crossing Number (Link)." Examples of the second type are "Convergent Sequence" and "Convergent Series." In the case of an entry like "Euler Theorem," which may describe one of three or four different formulas, I have taken the liberty of adding descriptive words ("Euler's Something Theorem") to all variations, or kept the standard name for the most commonly used variant and added descriptive words for the others. In cases where specific examples are derived from a general concept, em dashes (—) are used (for example, "Fourier Series," "Fourier Series — Power Series," "Fourier Series — Square Wave," "Fourier Series — Triangle"). The decision to put a possessive 's at the end of a name or to use a lone trailing apostrophe is based on whether the final "s" is pronounced. "Gauss's Theorem" is therefore written out, whereas "Archimedes' Recurrence Formula" is not. Finally, given the absence of a definitive stylistic convention, plurals of numerals are written without an apostrophe (e.g., 1990s instead of 1990's).

In an endeavor of this magnitude, errors and typographical mistakes are inevitable. The blame for these lies with me alone. Although the current length makes extensive additions in a printed version problematic, I plan to continue updating, correcting, and improving the work..

Eric Weisstein
Charlottesville, Virginia
August 8, 1998

Preface to the New Edition

The long awaited second edition of this Encyclopedia is finished, and it is now more complete than ever. Heavily revised by the author Eric Weisstein over the past three years, it contains well over 3,000 pages. Mr. Weisstein has updated all of the original material, added approximately 3,600 new entries and many illustrations, and updated the bibliographies that follow each entry to include the most recent references. As yet another enhancement, this edition integrates the use of the Mathematica software into many of its entries, presenting the precise commands that allow you to implement the formulas presented, perform many different calculations, construct graphical displays of your results, and generate remarkable mathematical illustrations. This is a unique touch and to our knowledge, a first for an encyclopedia.

With definitions, formulas, and facts presented in clear, engaging prose along with a multitude of illustrations, extensive cross-references, and even links to the Internet, this new and improved edition remains one of the most readable and accessible references in mathematics. This is truly a unique book written by an individual who is clearly dedicated to the study and field of mathematics. Users of the first edition of the Encyclopedia have described it as "extraordinary," "impressive," and "fascinating, " and report spending hours browsing its pages simply for pleasure. We hope you will do the same.

Acknowledgments

Although I alone have compiled and typeset this work, many people have contributed indirectly and directly to its creation. I have not yet had the good fortune to meet Donald Knuth of Stanford University, but he is unquestionably the person most directly responsible for making this work possible. Before his mathematical typesetting program TEX, it would have been impossible for a single individual to compile such a work as this. Had Prof. Bateman owned a personal computer equipped with TEX, perhaps his shoe box of notes would not have had to await the labors of Erdelyi, Magnus, and Oberhettinger to become a three-volume work on mathematical functions. Andrew Trevorrow's shareware implementation of TEX for the Macintosh, OzTEX (www.kagi.com/authors/akt/oztex.html), was also of fundamental importance. Nikos Drakos and Ross Moore have provided another building block for this work by developing the LATEX2HTML program (www-dsed.llnl.gov/files/programs/unix/latex2html/manual/manual.html), which has allowed me to easily maintain and update an on-line version of the encyclopedia long before it existed in book form.

I would like to thank Steven Finch of MathSoft, Inc., for his interesting on-line essays about mathematical constants (www.mathsoft.com/asolve/constant/constant.html), and also for his kind permission to reproduce excerpts from some of these essays. I hope that Steven will someday publish his detailed essays in book form. Thanks also to Neil Sloane and Simon Plouffe for compiling and making available the printed and on-line (www.research.att.com/~njas/sequences/) versions of the Encyclopedia of Integer Sequences, an immensely valuable compilation of useful information which represents a truly mind-boggling investment of labor.

Thanks to Robert Dickau, Simon Plouffe, and Richard Schroeppel for reading portions of the manuscript and providing a number of helpful suggestions and additions. Thanks also to algebraic topologist Ryan Budney for sharing some of his expertise, to Charles Walkden for his helpful comments about dynamical systems theory, and to Lambros Lambrou for his contributions. Thanks to David W. Wilson for a number of helpful comments and corrections. Thanks to Dale Rolfsen, compiler James Bailey, and artist Ali Roth for permission to reproduce their beautiful knot and link diagrams. Thanks to Gavin Theobald for providing diagrams of his masterful polygonal dissections. Thanks to Wolfram Research, not only for creating an indispensable mathematical tool in *Mathematica*®, but also for permission to include figures from the *Mathematica*® book and *MathSource* repository for the braid, conical spiral, double helix, Enneper's surfaces, Hadamard matrix, helicoid, helix, Henneberg's minimal surface, hyperbolic polyhedra, Klein bottle, Maeder's "owl" minimal surface, Penrose tiles, polyhedron, and Scherk's minimal surfaces entries.

Sincere thanks to Judy Schroeder for her skill and diligence in the monumental task of proofreading the entire document for syntax. Thanks also to Bob Stern, my executive editor from CRC Press, for his encouragement, and to Mimi Williams of CRC Press for her careful reading of the manuscript for typographical and formatting errors. As this encyclopedia's entry on PROOFREADING MISTAKES shows, the number of mistakes that are expected to remain after three independent proofreadings is much lower than the original number, but unfortunately still nonzero. Many thanks to the library staff at the University of Virginia, who have provided invaluable assistance in tracking down many an obscure citation. Finally, I would like to thank the hundreds of people who took the time to e-mail me comments and suggestions while this work was in its formative stages. Your continued comments and feedback are very welcome.

Numerals

(−1, 0, 1)-Matrix

The number of distinct $(-1, 0, 1)$-$n \times n$ matrices (counting row and column permutations, the transpose, and multiplication by -1 as equivalent) having $2n$ different row and column sums for $n = 2, 4, 6, \ldots$ are 1, 4, 39, 2260, 1338614, ... (Kleber). For example, the 2×2 matrix is given by

$$\begin{bmatrix} -1 & -1 \\ 0 & 1 \end{bmatrix},$$

To get the total number from these counts (assuming that 0 is not the missing sum, which is true for $n \leq 10$), multiply by $(2n!)^2$. In general, if an -matrix which has different column and row sums (collectively called line sums), then

1. n is even,
2. The number in $\{-n, 1-n, 2-n, \ldots, n\}$ that does not appear as a line sum is either $-n$ or , and
3. Of the largest line sums, half are column sums and half are row sums

(Bodendiek and Burosch 1995, F. Galvin).

See also ALTERNATING SIGN MATRIX, C-MATRIX, INTEGER MATRIX

References

Bodendiek, R. and Burosch, G. "Solution to the Antimagic 0, 1, −1 Matrix Problem." Aufgabe 5.30 in *Streifzüge durch die Kombinatorik: Aufgaben und Lösungen aus dem Schatz der Mathematik-Olympiaden.* Heidelberg, Germany: Spektrum Akademischer Verlag, pp. 250–253, 1995.

(−1, 1)-Matrix

See also HADAMARD MATRIX, INTEGER MATRIX

References

Kahn, J.; Komlós, J.; and Szemeredi, E. "On the Probability that a Random ±1 Matrix is Singular." *J. Amer. Math. Soc.* **8**, 223–240, 1995.

0-Free

ZEROFREE

0

DIVISION BY ZERO, FALLACY, NAUGHT, ZERO, ZERO DIVISOR, ZERO-FORM, ZERO MATRIX, ZERO-SUM GAME, ZEROFREE

0 = 1

FALLACY

(0, 1)-Matrix

A $(0, 1)$-INTEGER MATRIX, i.e., a matrix each of whose elements is 0 or 1, also called a binary matrix.

The numbers of binary matrices with no adjacent 1s (in either columns or rows) for $n = 1, 2, \ldots$, are given by 2, 7, 63, 1234, ... (Sloane's A006506). For example, the binary matrices with no adjacent 1s are

$$\begin{bmatrix} 0 & 0 \\ 0 & 0 \end{bmatrix}, \begin{bmatrix} 0 & 0 \\ 0 & 1 \end{bmatrix}, \begin{bmatrix} 0 & 0 \\ 1 & 0 \end{bmatrix}, \begin{bmatrix} 0 & 1 \\ 0 & 0 \end{bmatrix}$$

$$\begin{bmatrix} 0 & 1 \\ 1 & 0 \end{bmatrix}, \begin{bmatrix} 1 & 0 \\ 0 & 0 \end{bmatrix}, \begin{bmatrix} 1 & 0 \\ 0 & 1 \end{bmatrix},$$

These numbers are closely related to the HARD SQUARE ENTROPY CONSTANT. The numbers of binary matrices with no *three* adjacent 1s for , 2, ..., are given by 2, 16, 265, 16561, ... (Sloane's A050974).

Wilf (1997) considers the complexity of transforming an $m \times n$ binary matrix A into a TRIANGULAR MATRIX by permutations of the rows and columns of , and concludes that the problem falls in difficulty between a known easy case and a known hard case of the general NP-COMPLETE PROBLEM.

See also ADJACENCY MATRIX, FROBENIUS-KÖNIG THEOREM, GALE-RYSER THEOREM, HADAMARD'S MAXIMUM DETERMINANT PROBLEM, HARD SQUARE ENTROPY CONSTANT, IDENTITY MATRIX, INCIDENCE MATRIX, INTEGER MATRIX, LAM'S PROBLEM, s-CLUSTER, s-RUN

References

Brualdi, R. A. "Discrepancy of Matrices of Zeros and Ones." *Electronic J. Combinatorics* **6**, No. 1, R15, 1–12, 1999. http://www.combinatorics.org/Volume_6/v6i1toc.html.
Ehrlich, H. "Determinantenabschätzungen für binäre Matrizen." *Math. Z.* **83**, 123–132, 1964.
Ehrlich, H. and Zeller, K. "Binäre Matrizen." *Z. angew. Math. Mechanik* **42**, T20–21, 1962.
Komlós, J. "On the Determinant of -Matrices." *Studia Math. Hungarica* **2**, 7–21 1967.
Metropolis, N. and Stein, P. R. "On a Class of Matrices with Vanishing Determinants." *J. Combin Th.* **3**, 191–198, 1967.
Ryser, H. J. "Combinatorial Properties of Matrices of Zeros and Ones." *Canad. J. Math.* **9**, 371–377, 1957.
Sloane, N. J. A. Sequences A006506/M1816 and A050974 in "An On-Line Version of the Encyclopedia of Integer Sequences." http://www.research.att.com/~njas/sequences/eisonline.html.
Wilf, H. "On Crossing Numbers, and Some Unsolved Problems." In *Combinatorics, Geometry, and Probability: A Tribute to Paul Erdos. Papers from the Conference in Honor of Erdos' 80th Birthday Held at Trinity College, Cambridge, March 1993* (Ed. B. Bollobás and A. Thomason). Cambridge, England: Cambridge University Press, pp. 557–562, 1997.
Williamson, J. "Determinants Whose Elements Are 0 and 1." *Amer. Math. Monthly* **53**, 427–434, 1946.

1

The number one (1), also called "unity" is the first POSITIVE INTEGER. It is an ODD NUMBER. Although the number 1 used to be considered a PRIME NUMBER, it requires special treatment in so many definitions and applications involving primes greater than or equal to 2 that it is usually placed into a class of its own (Wells 1986, p. 31). The number 1 is sometimes also called "unity," so the th roots of 1 are often called the th ROOTS OF UNITY. FRACTIONS having 1 as a NUMERATOR are called UNIT FRACTIONS. If only one root, solution, etc., exists to a given problem, the solution is called UNIQUE.

The GENERATING FUNCTION having all COEFFICIENTS 1 is given by

$$\frac{1}{1-x} = 1 + x + x^2 + x^3 + x^4 + \cdots.$$

See also FALLACY, ONE-FORM, ONE-MOUTH THEOREM, ONE-NINTH CONSTANT, ONE-SHEETED HYPERBOLOID, ONE-TO-ONE, ONE-WAY FUNCTION, 2, 3, COMPLEXITY (NUMBER), EXACTLY ONE, ROOT OF UNITY, UNIQUE, UNIT FRACTION, ZERO

References

Wells, D. *The Penguin Dictionary of Curious and Interesting Numbers.* Middlesex, England: Penguin Books, pp. 30–32, 1986.

2

The number two (2) is the second POSITIVE INTEGER and the first PRIME NUMBER. It is EVEN, and is the only EVEN PRIME (the PRIMES other than 2 are called the ODD PRIMES). The number 2 is also equal to its FACTORIAL since $2! = 2$. A quantity taken to the POWER 2 is said to be SQUARED. The number of times k a given BINARY number $b_n \cdots b_2 b_1 b_0$ is divisible by 2 is given by the position of the first $b_k = 1$, counting from the right. For example, $12 = 1100$ is divisible by 2 twice, and $13 = 1101$ is divisible by 2 zero times.

The only known solutions to the CONGRUENCE

$$2^n \equiv 3 \pmod{n}$$

are $n = 4700063497$ (Sloane's A050259; Guy 1994) and

6313070745113443598938014005986613883062336144748427477099906755

(P.-L. Montgomery 1999). In general, the least satisfying

$$2^n \equiv k \pmod{n}$$

for $k = 2, 3, \ldots$ are $n = 3, 4700063497, 6, 19147, 10669, 25, 9, 2228071, \ldots$ (Sloane's A036236).

See also 1, BINARY, 3, RULER FUNCTION, SQUARED, TWO-EARS THEOREM, TWO-FORM, TWO-GRAPH, TWO-SCALE EXPANSION, TWO-SHEETED HYPERBOLOID, ZERO

References

Daiev, V. "Problem 636: Greatest Divisors of Even Integers." *Math. Mag.* **40**, 164–165, 1967.
Guy, R. K. "Residues of Powers of Two." §F10 in *Unsolved Problems in Number Theory, 2nd ed.* New York: Springer-Verlag, p. 250, 1994.
Montgomery, P.-L. "New solution to 2^n == 3 (mod n)." NMBRTHRY@listserv.nodak.edu posting, 24 Jun 1999.
Sloane, N. J. A. Sequences A036236 and A050259 in "An On-Line Version of the Encyclopedia of Integer Sequences." http://www.research.att.com/~njas/sequences/eisonline.html.
Wells, D. *The Penguin Dictionary of Curious and Interesting Numbers.* Middlesex, England: Penguin Books, pp. 41–44, 1986.

2x mod 1 Map

Let x_0 be a RATIONAL NUMBER in the CLOSED INTERVAL $[0, 1]$, and generate a SEQUENCE using the MAP

$$x_{n+1} \equiv 2x_n \pmod{1}. \tag{1}$$

Then the number of periodic ORBITS of period p (for PRIME) is given by

$$N_p = \frac{2^p - 2}{p} \tag{2}$$

(i.e, the number of period- repeating bit strings, modulo shifts). Since a typical ORBIT visits each point with equal probability, the NATURAL INVARIANT is given by

$$\rho(x) = 1. \tag{3}$$

See also TENT MAP

References

Ott, E. *Chaos in Dynamical Systems.* Cambridge, England: Cambridge University Press, pp. 26–31, 1993.

3

3 is the only INTEGER which is the sum of the preceding POSITIVE INTEGERS $(1 + 2 = 3)$ and the only number which is the sum of the FACTORIALS of the preceding POSITIVE INTEGERS $(1! + 2! = 3)$. It is also the first ODD PRIME. A quantity taken to the POWER 3 is said to be CUBED.

The sequence 1, 31, 331, 3331, 33331, ... (Sloane's A033175) consisting of $n = 0, 1, \ldots$ 3s followed by a 1. The th tern is given by

$$a(n) = \frac{10^{n+1} - 7}{3}.$$

The result is prime for , 2, 3, 4, 5, 6, 7, 17, 39, ... (Sloane's A055520); i.e., for 3, 31, 331, 3331, 33331, 333331, 3333331, 33333331, ... (Sloane's A051200), a fact which Gardner (1997) calls "a remarkable pattern that is entirely accidental and leads nowhere."

3x+1 Mapping

See also 1, 2, 3X+1 MAPPING, CUBED, PERIOD THREE THEOREM, TERNARY, THREE-CHOICE POLYGON, THREE-CHOICE WALK, THREE-COLORABLE, THREE CONICS THEOREM, THREE JUG PROBLEM, THREE-VALUED LOGIC, TREFOIL KNOT, WIGNER 3*J*-SYMBOL, ZERO

References

Gardner, M. *The Last Recreations: Hydras, Eggs, and Other Mathematical Mystifications.* New York: Springer-Verlag, p. 194, 1997.

Sloane, N. J. A. Sequences A033175, A051200, and A055520 in "An On-Line Version of the Encyclopedia of Integer Sequences." http://www.research.att.com/~njas/sequences/eisonline.html.

Smarandache, F. *Properties of Numbers.* University of Craiova, 1973.

Wells, D. *The Penguin Dictionary of Curious and Interesting Numbers.* Middlesex, England: Penguin Books, pp. 46–48, 1986.

3x+1 Mapping

COLLATZ PROBLEM

4

See also FOUR COINS PROBLEM, FOUR-COLOR THEOREM, FOUR CONICS THEOREM, FOUR EXPONENTIALS CONJECTURE, FOUR TRAVELERS PROBLEM, FOUR-VECTOR, FOUR-VERTEX THEOREM, LAGRANGE'S FOUR-SQUARE THEOREM

References

Wells, D. *The Penguin Dictionary of Curious and Interesting Numbers.* Middlesex, England: Penguin Books, pp. 55–58, 1986.

4-D Geometry

4-DIMENSIONAL GEOMETRY

4-Dimensional Geometry

4-dimensional geometry is Euclidean geometry extended into one additional DIMENSION. The prefix "hyper-" is usually used to refer to the 4- (and higher-) dimensional analogs of 3-dimensional objects, e.g. HYPERCUBE, HYPERPLANE, HYPERSPHERE. -dimensional POLYHEDRA are called POLYTOPES. the 4-dimensional cases of general -dimensional objects are often given special names, such as those summarized in the following table.

2-D	3-D	4-D	General
CIRCLE	SPHERE	GLOME	HYPERSPHERE
SQUARE	CUBE	TESSERACT	HYPERCUBE
EQUILATERAL TRIANGLE	TETRAHEDRON	PENTATOPE	SIMPLEX
POLYGON	POLYHEDRON	POLYCHORON	POLYTOPE
LINE SEGMENT	PLANE	HYPERPLANE	HYPERPLANE
SQUARE	OCTAHEDRON	16-CELL	CROSS POLYTOPE
EDGE	FACE	FACET	FACET
AREA	VOLUME	CONTENT	CONTENT

The SURFACE AREA of a HYPERSPHERE in -D is given by

$$S_n = \frac{2\pi^{n/2}}{\Gamma\left(\frac{1}{2}n\right)},$$

and the VOLUME by

$$V_n = \frac{\pi^{n/2}R^n}{\Gamma\left(1+\frac{1}{2}n\right)},$$

where $\Gamma(n)$ is the GAMMA FUNCTION.

See also DIMENSION, HYPERCUBE, HYPERSPHERE

References

Hinton, C. H. *The Fourth Dimension.* Pomeroy, WA: Health Research, 1993.

Manning, H. *The Fourth Dimension Simply Explained.* Magnolia, MA: Peter Smith, 1990.

Manning, H. *Geometry of Four Dimensions.* New York: Dover, 1956.

Neville, E. H. *The Fourth Dimension.* Cambridge, England: Cambridge University Press, 1921.

Rucker, R. von Bitter. *The Fourth Dimension: A Guided Tour of the Higher Universes.* Boston, MA: Houghton Mifflin, 1984.

Sommerville, D. M. Y. *An Introduction to the Geometry of Dimensions.* New York: Dover, 1958.

5

See also FIVE DISKS PROBLEM, MIQUEL FIVE CIRCLES THEOREM, PENTAGON, PENTAGRAM, PENTAHEDRON, TETRAHEDRON 5-COMPOUND

References

Wells, D. *The Penguin Dictionary of Curious and Interesting Numbers.* Middlesex, England: Penguin Books, pp. 58–67, 1986.

5-Cell

PENTATOPE

6

See also 6-SPHERE COORDINATES, HEXAGON, HEXAHE-

DRON, SIX CIRCLES THEOREM, SIX-COLOR THEOREM, SIX EXPONENTIALS THEOREM, WIGNER 6J-SYMBOL

References

Wells, D. *The Penguin Dictionary of Curious and Interesting Numbers.* Middlesex, England: Penguin Books, pp. 67–69, 1986.

6-Sphere Coordinates

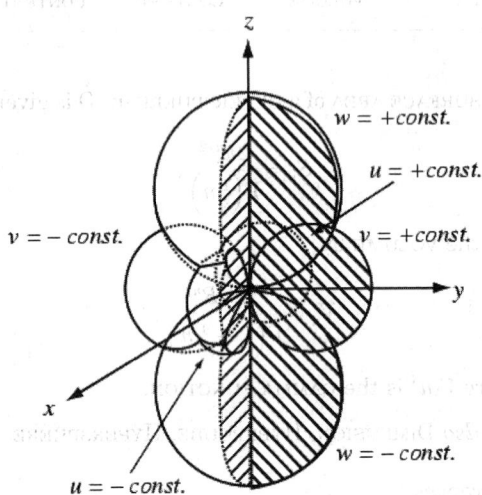

The coordinate system obtained by INVERSION of CARTESIAN COORDINATES, with $u, v, w \in (-\infty, \infty)$. The transformation equations are

$$x = \frac{u}{u^2 + v^2 + w^2} \tag{1}$$

$$y = \frac{v}{u^2 + v^2 + w^2} \tag{2}$$

$$z = \frac{w}{u^2 + v^2 + w^2}. \tag{3}$$

The equations of the surfaces of constant coordinates are given by

$$\left(x - \frac{1}{2u}\right)^2 + y^2 + z^2 = \frac{1}{4u^2}, \tag{4}$$

which gives spheres tangent to the yz-plane at the origin for u constant,

$$x^2 + \left(y - \frac{1}{2v}\right)^2 + z^2 = \frac{1}{4v^2}, \tag{5}$$

which gives spheres tangent to xz-plane at the origin for v constant, and

$$x^2 + y^2 + \left(z - \frac{1}{2w}\right)^2 = \frac{1}{4w^2}. \tag{6}$$

which gives spheres tangent to the xy-plane at the origin for w constant.

The metric coefficients are

$$g_{uu} = g_{vv} = g_{ww} = \frac{1}{\left(u^2 + v^2 + w^2\right)^2}. \tag{7}$$

See also CARTESIAN COORDINATES, INVERSION

References

Moon, P. and Spencer, D. E. "6-Sphere Coordinates (u, v, w)." Fig. 4.07 in *Field Theory Handbook, Including Coordinate Systems, Differential Equations, and Their Solutions, 2nd ed.* New York: Springer-Verlag, pp. 122–123, 1988.

7

See also SEVEN CIRCLES THEOREM

References

Wells, D. *The Penguin Dictionary of Curious and Interesting Numbers.* Middlesex, England: Penguin Books, pp. 70–71, 1986.

8

See also EIGHT CURVE, EIGHT-POINT CIRCLE THEOREM, EIGHT SURFACE

References

Wells, D. *The Penguin Dictionary of Curious and Interesting Numbers.* Middlesex, England: Penguin Books, pp. 71–73, 1986.

8-Cell

TESSERACT

9

See also NINE-POINT CENTER, NINE-POINT CIRCLE, NINE-POINT CONIC, WIGNER 9J-SYMBOL

References

Wells, D. *The Penguin Dictionary of Curious and Interesting Numbers.* Middlesex, England: Penguin Books, pp. 73–76, 1986.

10

The number 10 (ten) is the basis for the DECIMAL system of notation. In this system, each "decimal place" consists of a DIGIT 0–9 arranged such that each DIGIT is multiplied by a POWER of 10, decreasing from left to right, and with a decimal place indicating the $10^0 = 1$s place. For example, the number 1234.56 specifies

$$1 \times 10^3 + 2 \times 10^2 + 3 \times 10^1 + 4 \times 10^0 + 5 \times 10^{-1}$$
$$+6 \times 10^{-2}.$$

The decimal places to the left of the decimal point are 1, 10, 100, 1000, 10000, 100000, 1000000, 10000000, 100000000, ... (Sloane's A011557), called one, ten, HUNDRED, THOUSAND, ten thousand, hundred thousand, MILLION, 10 million, 100 million, and so on. The names of subsequent decimal places for LARGE NUMBERS differ depending on country.

Any POWER of 10 which can be written as the PRODUCT of two numbers not containing 0s must be OF THE FORM $2^n \cdot 5^n = 10^n$ for an INTEGER such that neither 2^n nor 5^n contains any ZEROS. The largest known such number is

$$10^{23} = 2^{33} \cdot 5^{33} = 8,589,934,592$$
$$\cdot 116,415,321,826,934,814,453,125.$$

A complete list of known such numbers is

$$10^1 = 2^1 \cdot 5^1$$
$$10^2 = 2^2 \cdot 5^2$$
$$10^3 = 2^3 \cdot 5^3$$
$$10^4 = 2^4 \cdot 5^4$$
$$10^5 = 2^5 \cdot 5^5$$
$$10^6 = 2^6 \cdot 5^6$$
$$10^7 = 2^7 \cdot 5^7$$
$$10^9 = 2^9 \cdot 5^9$$
$$10^{18} = 2^{18} \cdot 5^{18}$$
$$10^{33} = 2^{33} \cdot 5^{33}$$

(Madachy 1979). Since all POWERS of 2 with exponents $86 < n \leq 4.6 \times 10^7$ contain at least one ZERO (M. Cook), no other POWER of ten less than 46 million can be written as the PRODUCT of two numbers not containing 0s.

See also BILLION, DECIMAL, HUNDRED, LARGE NUMBER, MILLIARD, MILLION, THOUSAND, TRILLION, ZERO

References

Madachy, J. S. *Madachy's Mathematical Recreations.* New York: Dover, pp. 127–128, 1979.
Pickover, C. A. *Keys to Infinity.* New York: Wiley, p. 135, 1995.
Sloane, N. J. A. Sequences A011557 in "An On-Line Version of the Encyclopedia of Integer Sequences." http://www.research.att.com/~njas/sequences/eisonline.html.
Wells, D. *The Penguin Dictionary of Curious and Interesting Numbers.* Middlesex, England: Penguin Books, pp. 76–82, 1986.

11

References

Wells, D. *The Penguin Dictionary of Curious and Interesting Numbers.* Middlesex, England: Penguin Books, 1986.

12

One DOZEN, or a twelfth of a GROSS.

See also DOZEN, GROSS

References

Wells, D. *The Penguin Dictionary of Curious and Interesting Numbers.* Middlesex, England: Penguin Books, 1986.

13

A NUMBER traditionally associated with bad luck. A so-called BAKER'S DOZEN is equal to 13. Fear of the number 13 is called TRISKAIDEKAPHOBIA. There are 13 ARCHIMEDEAN SOLIDS. Mazur and Tate (1973/74) proved that there is no ELLIPTIC CURVE over the rationals $\mathbb{Q}$ having a RATIONAL POINT of order 13.

See also BAKER'S DOZEN, TRISKAIDEKAPHOBIA

References

Mazur, B. and Tate, J. "Points of Order 13 on Elliptic Curves." *Invent. Math.* **22**, 41–49, 1973/74.
Wells, D. *The Penguin Dictionary of Curious and Interesting Numbers.* Middlesex, England: Penguin Books, 1986.

14

References

Wells, D. *The Penguin Dictionary of Curious and Interesting Numbers.* Middlesex, England: Penguin Books, 1986.

15

See also 15 PUZZLE, FIFTEEN THEOREM

15 Puzzle

1	2	3	4
5	6	7	8
9	10	11	12
13	14	15	

A puzzle introduced by Sam Loyd in 1878. It consists of 15 squares numbered from 1 to 15 which are placed in a 4×4 box leaving one position out of the 16 empty. The goal is to reposition the squares from a given arbitrary starting arrangement by sliding them one at a time into the configuration shown above. For some initial arrangements, this rearrangement is possible, but for others, it is not.

To address the solubility of a given initial arrangement, proceed as follows. If the SQUARE containing the number i appears "before" (reading the squares in the box from left to right and top to bottom) numbers which are less than , then call it an inversion of order , and denote it n_i. Then define

$$N \equiv \sum_{i=1}^{15} n_i = \sum_{i=2}^{15} n_i,$$

where the sum need run only from 2 to 15 rather than 1 to 15 since there are no numbers less than 1 (so n_1 must equal 0). If N is EVEN, the position is possible, otherwise it is not. This can be formally proved using ALTERNATING GROUPS. For example, in the following arrangement

2	1	3	4
5	6	7	8
9	10	11	12
13	14	15	

$n_2 = 1$ (2 precedes 1) and all other $n_i = 0$, so $N = 1$ and the puzzle cannot be solved.

Johnson (1879) proved that odd permutations of the puzzle are impossible, which Story (1879) proved that all even permutations are possible. While Herstein and Kaplansky (1978) wrote that "no really easy proof seems to be known," Archer (1999) presented a simple proof. A more general result due to Wilson (1974) showed that for any CONNECTED GRAPH on nodes, with the exception of CYCLE GRAPHS C_n and the THETA-0 GRAPH, either exactly half or all of the $n!$ possible labelings are obtainable by sliding labels, depending on whether the graph is BIPARTITE (Archer 1999). θ_0 has six inequivalent labelings, which has $(n-2)!$ inequivalent labelings.

Reversing the order of the "8 Puzzle" made on a 3×3 board can be proved to require at least 26 moves, although the best solution requires 30 moves (Gardner 1984, pp. 200 and 206–207). The number of distinct solutions in 28, 30, 32, ... moves are 0, 10, 112, 512, ... (Sloane's A046164), giving 634 solutions better than the 36-move solution given by Dudeney (1949).

References

Archer, A. F. "A Modern Treatment of the 15 Puzzle." *Amer. Math. Monthly* **106**, 793–799, 1999.

Ball, W. W. R. and Coxeter, H. S. M. *Mathematical Recreations and Essays, 13th ed.* New York: Dover, pp. 312–316, 1987.

Beasley, J. D. *The Mathematics of Games.* Oxford, England: Oxford University Press, pp. 80–81, 1990.

Bogomolny, A. "Sam Loyd's Fifteen." http://www.cut-the-knot.com/pythagoras/fifteen.html.

Bogomolny, A. "Sam Loyd's Fifteen [History]." http://www.cut-the-knot.com/pythagoras/history15.html.

Davies, A. L. "Rotating the 15 Puzzle." *Math. Gaz.* **54**, 237–240, 1970.

Dudeney, H. E. Problem 253 in *The Canterbury Puzzles and Other Curious Problems, 7th ed.* London: Thomas Nelson and Sons, 1949.

Gardner, M. *The Sixth Book of Mathematical Games from Scientific American.* Chicago, IL: University of Chicago Press, pp. 64–65, 200–201, and 206–207, 1984.

Herstein, I. N. and Kaplansky, I. *Matters Mathematical, 2nd ed.* New York: Chelsea, pp. 114–115, 1978.

Hurd, S. and Trautman, D. "The Knight's Tour on the 15-Puzzle." *Math. Mag.* **66**, 159–166, 1993.

Johnson, W. W. "Notes on the '15 Puzzle. I.'" *Amer. J. Math.* **2**, 397–399, 1879.

Kasner, E. and Newman, J. R. *Mathematics and the Imagination.* Redmond, WA: Tempus Books, pp. 177–180, 1989.

Kraitchik, M. "The 15 Puzzle." §12.2.1 in *Mathematical Recreations.* New York: W. W. Norton, pp. 302–308, 1942.

Liebeck, H. "Some Generalizations of the 14–15 Puzzle." *Math. Mag.* **44**, 185–189, 1971.

Loyd, S. *Mathematical Puzzles of Sam Loyd, Vol. 1.* New York: Dover, pp. 19–20, 1959.

Loyd, S. Jr. *Sam Loyd's Cyclopedia of 5,000 Puzzles, Tricks, and Conundrums.* Lamb Pub., 1993.

Mallison, H. V. "An Array of Squares." *Math. Gaz.* **24**, 119–121, 1940.

Sloane, N. J. A. Sequences A046164 in "An On-Line Version of the Encyclopedia of Integer Sequences." http://www.research.att.com/~njas/sequences/eisonline.html.

Spitznagel, E. L. Jr. *Selected Topics in Mathematics.* New York: Holt, Rinehart and Winston, pp. 143–148, 1971.

Spitznagel, E. L. Jr. "A New Look at the Fifteen Puzzle." *Math. Mag.* **40**, 171–174, 1967.

Steinhaus, H. *Mathematical Snapshots, 3rd ed.* New York: Dover, pp. 14–16, 1999.

Story, W. E. "Notes on the '15 Puzzle. II.'" *Amer. J. Math.* **2**, 399–404, 1879.

Whipple, F. J. W. "The Sign of a Term in the Expansion of a Determinant." *Math. Gaz.* **13**, 126, 1926.

Wilson, R. M. "Graph Puzzles, Homotopy, and the Alternating Group." *J. Combin. Th. Ser. B* **16**, 86–96, 1974.

15 Schoolgirl Problem

KIRKMAN'S SCHOOLGIRL PROBLEM

16-Cell

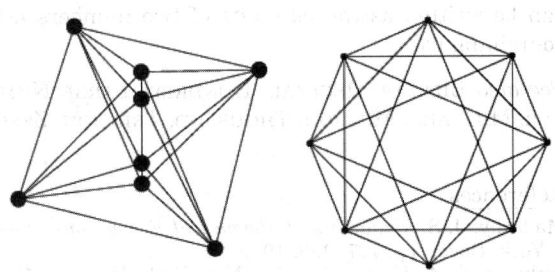

The finite regular 4-D CROSS POLYTOPE with SCHLÄFLI SYMBOL $\{3, 3, 4\}$ and VERTICES which are the PERMUTATIONS of $(, 0, 0, 0)$. The 16-cell is the dual of the TESSERACT. Its graph is isomorphic to the CIRCULANT GRAPH $Ci_{1,2,3}(8)$.

See also 24-CELL, 120-CELL, 600-CELL, CELL, CROSS POLYTOPE, HYPERCUBE, PENTATOPE, POLYCHORON, POLYTOPE, TESSERACT

References

Wells, D. *The Penguin Dictionary of Curious and Interesting Geometry.* London: Penguin, p. 210, 1991.

17

17 is a FERMAT PRIME which means that the 17-sided REGULAR POLYGON (the HEPTADECAGON) is CONSTRUCTIBLE using COMPASS and STRAIGHTEDGE (as proved by Gauss).

See also CONSTRUCTIBLE POLYGON , FERMAT PRIME, HEPTADECAGON

References
Lefevre, V. "Properties of 17." http://www.ens-lyon.fr/~vlefevre/d17_eng.html.

17-gon
HEPTADECAGON

18-Point Problem

Place a point somewhere on a LINE SEGMENT. Now place a second point and number it 2 so that each of the points is in a different half of the LINE SEGMENT. Continue, placing every th point so that all points are on different $(1/N)$th of the LINE SEGMENT. Formally, for a given , does there exist a sequence of real numbers x_1, x_2, ..., x_N such that for every $n \in \{1, ..., N\}$ and every $k \in \{1, ..., n\}$, the inequality

$$\frac{k-1}{n} \le x_i < \frac{k}{n}$$

holds for some $i \in \{1, ..., n\}$? Surprisingly, it is only possible to place 17 points in this manner (Berlekamp and Graham 1970, Warmus 1976).

Steinhaus (1979) gives a 14-point solution (0.06, 0.55, 0.77, 0.39, 0.96, 0.28, 0.64, 0.13, 0.88, 0.48, 0.19, 0.71, 0.35, 0.82), and Warmus (1976) gives the 17-point solution

$$\tfrac{4}{7} \le x_1 < \tfrac{7}{12}, \ \tfrac{2}{7} \le x_2 < \tfrac{5}{17}, \ \tfrac{16}{17} \le x_3 < 1, \ \tfrac{1}{14} \le x_4 < \tfrac{1}{13},$$

$$\tfrac{8}{11} \le x_5 < \tfrac{11}{15}, \ \tfrac{5}{11} \le x_6 < \tfrac{6}{13}, \ \tfrac{1}{7} \le x_7 < \tfrac{2}{13}, \ \tfrac{14}{17} \le x_8 < \tfrac{5}{6},$$

$$\tfrac{3}{8} \le x_9 < \tfrac{5}{13}, \ \tfrac{11}{17} \le x_{10} < \tfrac{2}{3}, \ \tfrac{3}{14} \le x_{11} < \tfrac{3}{13},$$

$$\tfrac{15}{17} \le x_{12} < \tfrac{11}{12}, \ \tfrac{1}{2} \le x_{12} < \tfrac{9}{17}, \ 0 \le x_{14} < \tfrac{1}{17},$$

$$\tfrac{13}{17} \le x_{15} < \tfrac{4}{5}, \ \tfrac{5}{16} \le x_{16} < \tfrac{6}{17}, \ \tfrac{10}{17} \le x_{17} < \tfrac{11}{17},$$

Warmus (1976) states that there are 768 patterns of 17-point solutions (counting reversals as equivalent).

See also DISCREPANCY THEOREM, POINT PICKING

References
Berlekamp, E. R. and Graham, R. L. "Irregularities in the Distributions of Finite Sequences." *J. Number Th.* **2**, 152–161, 1970.
Gardner, M. *The Last Recreations: Hydras, Eggs, and Other Mathematical Mystifications.* New York: Springer-Verlag, pp. 34–36, 1997.
Steinhaus, H. "Distribution on Numbers" and "Generalization." Problems 6 and 7 in *One Hundred Problems in Elementary Mathematics.* New York: Dover, pp. 12–13, 1979.
Warmus, M. "A Supplementary Note on the Irregularities of Distributions." *J. Number Th.* **8**, 260–263, 1976.

24-Cell

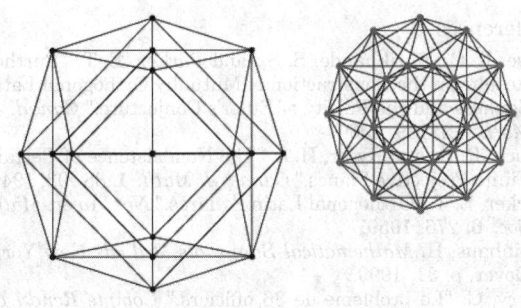

A finite regular 4-D POLYTOPE with SCHLÄFLI SYMBOL {3, 4, 3}. Coxeter (1969) gives a list of the VERTEX positions. The EVEN coefficients of the D_4 lattice are 1, 24, 24, 96, ... (Sloane's A004011), and the 24 shortest vectors in this lattice form the 24-cell (Coxeter 1973, Conway and Sloane 1993, Sloane and Plouffe 1995). The 24-cell is self-dual, and is the unique regular convex POLYCHORON which has no direct 3-D analog.

One construction for the 24-cell evokes comparison with the RHOMBIC DODECAHEDRON. Given two equal cubes, we construct this dodecahedron by cutting one cube into six congruent square pyramids, and attaching these to the six squares bounding the other cube. Similarly, given two equal tesseracts, we can construct the 24-cell by cutting one tesseract into eight congruent cubic pyramids, and attaching these to the eight cubes bounding the other tesseract (Towle).

See also 16-CELL, 120-CELL, 600-CELL, CELL, HYPERCUBE, PENTATOPE, POLYCHORON, POLYTOPE

References
Conway, J. H. and Sloane, N. J. A. *Sphere-Packings, Lattices and Groups, 2nd ed.* New York: Springer-Verlag, 1993.
Coxeter, H. S. M. *Introduction to Geometry, 2nd ed.* New York: Wiley, p. 404, 1969.
Coxeter, H. S. M. *Regular Polytopes, 3rd ed.* New York: Dover, 1973.
Sloane, N. J. A. Sequences A004011/M5140 in "An On-Line Version of the Encyclopedia of Integer Sequences." http://www.research.att.com/~njas/sequences/eisonline.html.
Sloane, N. J. A. and Plouffe, S. Figure M5150 in *The Encyclopedia of Integer Sequences.* San Diego: Academic Press, 1995.
Wells, D. *The Penguin Dictionary of Curious and Interesting Geometry.* London: Penguin, p. 210, 1991.

36 Officer Problem

How can a delegation of six regiments, each of which sends a colonel, a lieutenant-colonel, and major, a captain, a lieutenant, and a sub-lieutenant be arranged in a regular 6×6 array such that no row or column duplicates a rank *or* a regiment? The answer is that no such arrangement is possible.

See also EULER'S GRAECO-ROMAN SQUARES CONJECTURE, LATIN SQUARE

References

Bose, R. C.; Shrikhande, S. S.; and Parker, E. T. "Further Results on the Construction of Mutually Orthogonal Latin Squares and the Falsity of Euler's Conjecture." *Canad. J. Math.* **12**, 189, 1960.
Bruck, R. H. and Ryser, H. J. "The Nonexistence of Certain Finite Projective Planes." *Canad. J. Math.* **1**, 88–93, 1949.
Parker, E. T. "Orthogonal Latin Squares." *Not. Amer. Math. Soc.* **6**, 276, 1959.
Steinhaus, H. *Mathematical Snapshots, 3rd ed.* New York: Dover, p. 31, 1999.
Tarry, G. "Le problème de 36 officiers." *Compte Rendu de l'Assoc. Français Avanc. Sci. Naturel* **1**, 122–123, 1900.
Tarry, G. "Le problème de 36 officiers." *Compte Rendu de l'Assoc. Français Avanc. Sci. Naturel* **2**, 170–203, 1901.

42

According to Adams (1997), 42 is the ultimate answer to life, the universe, and everything, although it is left as an exercise to the reader to determine the actual question leading to this result.

References

Adams, D. *The Hitchhiker's Guide to the Galaxy.* New York: Ballantine Books, 1997.

72 Rule

RULE OF 72

120-Cell

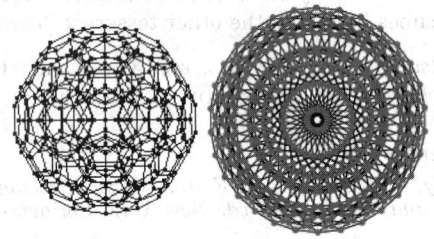

A finite regular 4-D POLYTOPE with SCHLÄFLI SYMBOL $\{5, 3, 3\}$. The 120-cell has 600 vertices (Coxeter 1969), and consists of 120 DODECAHEDRA and 720 PENTAGONS (Coxeter 1973, p. 264). In the plate following p. 176, Coxeter (1973) illustrates the polytope. The dual of the 120-cell is the 600-CELL.

See also 16-CELL, 24-CELL, 600-CELL, CELL, HYPERCUBE, PENTATOPE, POLYCHORON, POLYTOPE, SIMPLEX

References

Coxeter, H. S. M. *Introduction to Geometry, 2nd ed.* New York: Wiley, p. 404, 1969.
Coxeter, H. S. M. "Stellating ." §14.2 in *Regular Polytopes, 3rd ed.* New York: Dover, pp. 136–137, 157, 264–267, and 292, 1973.

Wells, D. *The Penguin Dictionary of Curious and Interesting Geometry.* London: Penguin, p. 210, 1991.

144

A DOZEN DOZEN, also called a GROSS. 144 is a SQUARE NUMBER and a SUM-PRODUCT NUMBER.

See also DOZEN

163

The number 163 is very important in number theory, since $d = 163$ is the largest number such that the IMAGINARY QUADRATIC FIELD $\mathbb{Q}\left(-\sqrt{d}\right)$ has CLASS NUMBER $h(-d) = 1$. It also satisfies the curious identities

$$163 \sum_{i=0}^{4} \binom{8}{i} \tag{1}$$

$$\frac{1}{2}\left[4^4 + \binom{8}{4}\right] \tag{2}$$

$$\frac{1}{2}\left[4^4 + \sum_{i=0}^{4} \binom{4}{i}^2\right], \tag{3}$$

where $\binom{n}{k}$ is a BINOMIAL COEFFICIENT (Stoschek). An approximation due to Stoschek is given by

$$\pi \approx \frac{2^9}{163} = \frac{512}{163} \approx 3.1411043, \tag{4}$$

which is good to 3 digits.

See also RAMANUJAN CONSTANT

References

Stoschek, E. "Modul 33: Algames with Numbers." http://marvin.sn.schule.de/~inftreff/modul33/task33.htm.

196-Algorithm

Take any POSITIVE INTEGER of two DIGITS or more, reverse the DIGITS, and add to the original number. Now repeat the procedure with the SUM so obtained. This procedure quickly produces PALINDROMIC NUMBERS for most INTEGERS. For example, starting with the number 5280 produces (5280, 6105, 11121, 23232). The end results of applying the algorithm to 1, 2, 3, ... are 1, 2, 3, 4, 5, 6, 7, 8, 9, 11, 11, 33, 44, 55, 66, 77, 88, 99, 121, ... (Sloane's A033865). The value for 89 is especially large, being 8813200023188.

The first few numbers not known to produce PALINDROMES are 196, 887, 1675, 7436, 13783, ... (Sloane's A006960), which are simply the numbers obtained by iteratively applying the algorithm to the number 196. This number therefore lends itself to the name of the ALGORITHM. In 1990, John Walker computed 2,415,836 iterations of the algorithm on 196 and obtained a number having 1,000,000 digits. This was extended in 1995 by Tim Irvin, who obtained a number having 2,000,000 digits. The `rec.puzzles` archive states that a 3,924,257-digit nonpalindromic number is obtained after 9,480,000 iterations.

The number of terms $a(n)$ in the iteration sequence required to produce a PALINDROMIC NUMBER from (i.e., $a(n) = 1$ for a PALINDROMIC NUMBER, $a(n) = 2$ if a PALINDROMIC NUMBER is produced after a single iteration of the 196-algorithm, etc.) for , 2, ... are 1, 1, 1, 1, 1, 1, 1, 1, 1, 2, 1, 2, 2, 2, 2, 2, 2, 3, 2, 2, 1, ... (Sloane's A030547). The smallest numbers which require, 1, 2, ... iterations to reach a palindrome are 0, 10, 19, 59, 69, 166, 79, 188, ... (Sloane's A023109).

The 196-algorithm can be implemented in *Mathematica* as

```
PalindromicQ[n_Integer?Positive]: = Module[
  {sn = ToString[n]},
  sn == StringReverse[sn]
                                   ]
Algorithm196[n_Integer?PalindromicQ, it_:0]:-
= {n}         Algorithm196[n_Integer?Positive,
it_:Infinity]: =
  FixedPointList[# + ToExpression[StringRe-
verse[ToString[#]]]&,
  n, it, SameTest-> (PalindromicQ[#2]&)
]
```

M. Sofroniou gives an efficient *Mathematica* implementation which has complexity $\mathcal{O}(k^2)$ for steps, requiring approximately 10.6 hours on a 450 MHz Pentium II to compute 250,000 iterations. Extrapolating the timing data suggests that approximately 42 days would be needed on this same machine to match Walker's 2,415,836 iterations.

See also ADDITIVE PERSISTENCE, DIGITADDITION, MULTIPLICATIVE PERSISTENCE, PALINDROMIC NUMBER, PALINDROMIC NUMBER CONJECTURE, RATS SEQUENCE, RECURRING DIGITAL INVARIANT

References

Brown, K. S. "Digit Reversal Sums Leading to Palindromes." http://www.seanet.com/~ksbrown/kmath004.htm.

De Geest, P. "Websources about '196' Becoming Palindromic by Using Reversal Sums." http://www.ping.be/~ping6758/weblinks.htm.

Eddins, S. "The Palindromic Order of a Number." *IMSA Math. J.* **4**, Spring 1996. http://www.imsa.edu/edu/math/journal/volume4/webver/palinord.html.

Gardner, M. *Mathematical Circus: More Puzzles, Games, Paradoxes and Other Mathematical Entertainments from Scientific American.* New York: Knopf, pp. 242–245, 1979.

Gruenberger, F. "How to Handle Numbers with Thousands of Digits, and Why One Might Want to." *Sci. Amer.* **250**, 19–26, Apr. 1984.

Irving, T. "About Two Months of Computing, or, An Addendum to Mr. Walker's Three Years of Computing" http://www.fourmilab.ch/documents/threeyears/two_months_more.html.

Math Forum. "Ask Dr. Math: Making Numbers into Palindromic Numbers." http://forum.swarthmore.edu/dr.math/problems/barnes10.11.html.

Peters, I. J. "Search for the Biggest Numeric Palindrome." http://www.floot.demon.co.uk/palindromes.html.

rec.puzzles archive. 1996. ftp://rtfm.mit.edu/pub/usenet/news.answers/puzzles/archive/arithmetic/part1.

Safroniou, M. "Palindromic Numbers: The 196-Algorithm." MATHEMATICA NOTEBOOK Algorithm196.NB.

Sloane, N. J. A. Sequences A006960/M5410, A023109, A030547, and A033865 in "An On-Line Version of the Encyclopedia of Integer Sequences." http://www.research.att.com/~njas/sequences/eisonline.html.

Walker, J. "Three Years of Computing: Final Report on the Palindrome Quest." http://www.fourmilab.ch/documents/threeyears/threeyears.html.

Weisstein, E. W. "Integer Sequences." MATHEMATICA NOTEBOOK IntegerSequences.m.

239

Some interesting properties (as well as a few arcane ones not reiterated here) of the number 239 are discussed in Beeler *et al.* (1972, Item 63). 239 appears in MACHIN'S FORMULA

$$\frac{1}{4}\pi = 4\tan^{-1}\left(\frac{1}{5}\right) - \tan^{-1}\left(\frac{1}{239}\right),$$

which is related to the fact that

$$2 \cdot 13^4 - 1 = 239^2,$$

which is why 239/169 is the 7th CONVERGENT of $\sqrt{2}$. Another pair of INVERSE TANGENT FORMULAS involving 239 is

$$\tan^{-1}\left(\frac{1}{239}\right)\tan^{-1}\left(\frac{1}{70}\right) - \tan^{-1}\left(\frac{1}{99}\right)$$

$$\tan^{-1}\left(\frac{1}{408}\right) + \tan^{-1}\left(\frac{1}{577}\right).$$

239 needs 4 SQUARES (the maximum) to express it, 9 CUBES (the maximum, shared only with 23) to express it, and 19 fourth POWERS (the maximum) to express it (see WARING'S PROBLEM). However, 239 doesn't need the maximum number of fifth POWERS (Beeler *et al.* 1972, Item 63).

References

Schroeppel, R. Item 63 in Beeler, M.; Gosper, R. W.; and Schroeppel, R. *HAKMEM.* Cambridge, MA: MIT Artificial Intelligence Laboratory, Memo AIM-239, p. 24, Feb. 1972.

243

Feynman (1997) noticed the curious fact that the decimal expansion

$$\frac{1}{243} = 0.004115226337448559\dots$$

repeats pairs of the digits 0, 1, 2, 3, ... separated by the digits 4, 5, 6, 7, ... Just after this point, the pattern breaks, since the fraction is given exactly by the repeating decimal

$$\frac{1}{243} = 0.\overline{004115226337448559670781893}.$$

This pattern is related to the fact that

$$\frac{1}{9} = 0.\overline{1}$$

and

$$\tfrac{1}{81} = 0.\overline{0123456789}.$$

References

Feynman, R. P. and Leighton, R. *'Surely You're Joking, Mr. Feynman!': Adventures of a Curious Character.* New York: W. W. Norton, p. 99, 1997.

257-gon

257 is a FERMAT PRIME, and the 257-gon is therefore a CONSTRUCTIBLE POLYGON using COMPASS and STRAIGHTEDGE, as proved by Gauss. An illustration of the 257-gon is not included here, since its 257 segments so closely resemble a CIRCLE. Richelot and Schwendenwein found constructions for the 257-gon in 1832 (Coxeter 1969). De Temple (1991) gives a construction using 150 CIRCLES (24 of which are CARLYLE CIRCLES) which has GEOMETROGRAPHY symbol $94S_1 + 47S_2 + 275C_1 + 0C_2 + 150C_3$ and SIMPLICITY 566.

See also 65537-GON, CONSTRUCTIBLE POLYGON, FERMAT PRIME, HEPTADECAGON, PENTAGON

References

Bachmann, P. *Die Lehre von der Kreistheilung und ihre Beziehungen zur Zahlentheorie.* Leipzig, Germany: Teubner, 1872.
Bold, B. *Famous Problems of Geometry and How to Solve Them.* New York: Dover, p. 70, 1982.
Coxeter, H. S. M. *Introduction to Geometry, 2nd ed.* New York: Wiley, 1969.
De Temple, D. W. "Carlyle Circles and the Lemoine Simplicity of Polygonal Constructions." *Amer. Math. Monthly* **98**, 97–108, 1991.
Dickson, L. E. "Constructions with Ruler and Compasses; Regular Polygons." Ch. 8 in *Monographs on Topics of Modern Mathematics Relevant to the Elementary Field* (Ed. J. W. A. Young). New York: Dover, pp. 352–386, 1955.
Dixon, R. *Mathographics.* New York: Dover, p. 53, 1991.
Klein, F. "The Construction of the Regular Polygon of 17 Sides." Part I, Ch. 4 in "Famous Problems of Elementary Geometry: The Duplication of the Cube, the Trisection of the Angle, and the Quadrature of the Circle." In *Famous Problems and Other Monographs.* New York: Chelsea, pp. 24–41, 1980.
Pascal, E. "Sulla costruzione del poligono regolare di 257 lati." *Rendiconto dell Accad. della scienze fisiche e matemat. sezione della Soc. a reale di Napoli, Ser. 2* **1**, 33–39, 1887.
Rademacher, H. *Lectures on Elementary Number Theory.* New York: Blaisdell, 1964.
Richelot, F. J. "De resolutione algebraica aequationis $X^{257} = 1$, sive de divisione circuli per bisectionem anguli septies repetitam in partes 257 inter se aequales commentatio coronata." *J. reine angew. Math.* **9**, 1–26, 146–161, 209–230, and 337–358, 1832.
Trott, M. " $\cos(2\pi/257)$ à la Gauss." *Mathematica Educ. Res.* **4**, 31–36, 1995.

600-Cell

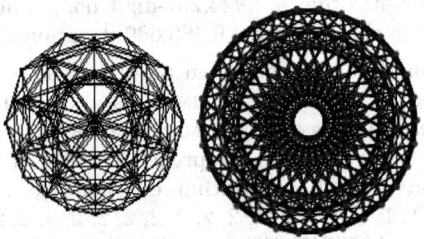

A finite regular 4-D POLYTOPE with SCHLÄFLI SYMBOL $\{3, 3, 5\}$. The 600-cell has 120 VERTICES (Coxeter 1969). In the plate following p. 160, Coxeter (1973) gives two illustrations of the polytope.

The dual of the 600-cell is the 120-CELL.

See also 16-CELL, 24-CELL, 120-CELL, CELL, HYPERCUBE, PENTATOPE, POLYCHORON, POLYTOPE, SIMPLEX

References

Coxeter, H. S. M. *Introduction to Geometry, 2nd ed.* New York: Wiley, p. 404, 1969.
Coxeter, H. S. M. "Gosset's Construction for . §8.5 in *Regular Polytopes, 3rd ed.* New York: Dover, pp. 136–137, 153–154, and 157, 1973.
Wells, D. *The Penguin Dictionary of Curious and Interesting Geometry.* London: Penguin, p. 210, 1991.

666

A number known as the BEAST NUMBER appearing in the *Bible* and ascribed various numerological properties.

See also APOCALYPTIC NUMBER, BEAST NUMBER, LEVIATHAN NUMBER

References

De Geest, P. "The Number of the Best 666." http://www.ping.be/~ping6758/weblinks.htm.
Hardy, G. H. *A Mathematician's Apology, reprinted with a foreword by C. P. Snow.* New York: Cambridge University Press, p. 96, 1993.

1729

1729 is sometimes called the HARDY-RAMANUJAN NUMBER. It is the smallest TAXICAB NUMBER, i.e., the smallest number which can be expressed as the sum of two cubes in two different ways:

$$1729 = 1^3 + 12^3 = 9^3 + 10^3.$$

See also HARDY-RAMANUJAN NUMBER, TAXICAB NUMBER

2187

The digits in the number 2187 form the two VAMPIRE NUMBERS: $21 \times 87 = 1827$ and $2187 = 27 \times 81$. 2187 is also given by 3^7.

See also VAMPIRE NUMBER

References

Gardner, M. "Lucky Numbers and 2187." *Math. Intell.* **19**, 26–29, Spring 1997.

65537-gon

65537 is the largest known FERMAT PRIME, and the 65537-gon is therefore a CONSTRUCTIBLE POLYGON using COMPASS and STRAIGHTEDGE, as proved by Gauss. The 65537-gon has so many sides that it is, for all intents and purposes, indistinguishable from a CIRCLE using any reasonable printing or display methods.

Hermes spent 10 years on the construction of the 65537-gon at Königsberg around (1900). After the Second World War, his manuscripts were moved to the Mathematical Institute in Göttingen, where they can now be viewed (Coxeter 1969).

De Temple (1991) notes that a GEOMETRIC CONSTRUCTION can be done using 1332 or fewer CARLYLE CIRCLES.

See also 257-GON, CONSTRUCTIBLE POLYGON, HEPTADECAGON, PENTAGON

References

Bold, B. *Famous Problems of Geometry and How to Solve Them.* New York: Dover, p. 70, 1982.

Coxeter, H. S. M. *Introduction to Geometry, 2nd ed.* New York: Wiley, 1969.

De Temple, D. W. "Carlyle Circles and the Lemoine Simplicity of Polygonal Constructions." *Amer. Math. Monthly* **98**, 97–108, 1991.

Dickson, L. E. "Constructions with Ruler and Compasses; Regular Polygons." Ch. 8 in *Monographs on Topics of Modern Mathematics Relevant to the Elementary Field* (Ed. J. W. A. Young). New York: Dover, pp. 352–386, 1955.

Dixon, R. *Mathographics.* New York: Dover, p. 53, 1991.

Hermes, J. "Ueber die Teilung des Kreises in 65537 gleiche Teile." *Nachr. Königl. Gesellsch. Wissensch. Göttingen, Math.-Phys. Klasse*, pp. 170–186, 1894.

A

AAA Theorem

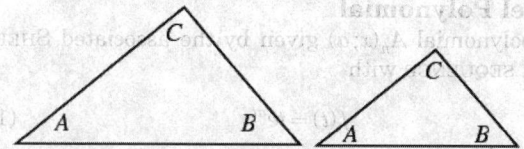

Specifying three ANGLES A, B, and C does not uniquely define a TRIANGLE, but any two TRIANGLES with the same ANGLES are SIMILAR. Specifying two ANGLES of a TRIANGLE automatically gives the third since the sum of ANGLES in a TRIANGLE sums to 180° (π RADIANS), i.e.,

$$C = \pi - A - B.$$

See also AAS THEOREM, ASA THEOREM, ASS THEOREM, SAS THEOREM, SSS THEOREM, TRIANGLE

AAS Theorem

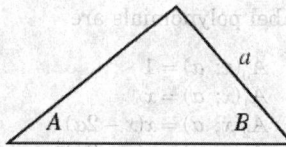

Specifying two angles A and B and a side a uniquely determines a TRIANGLE with AREA

$$K = \frac{a^2 \sin B \sin C}{2 \sin A} = \frac{a^2 \sin B \sin(\pi - A - B)}{2 \sin A}. \quad (1)$$

The third angle is given by

$$C = \pi - A - B, \quad (2)$$

since the sum of angles of a TRIANGLE is 180° (π RADIANS). Solving the LAW OF SINES

$$\frac{a}{\sin A} = \frac{b}{\sin B} \quad (3)$$

for b gives

$$b = a \frac{\sin B}{\sin A}. \quad (4)$$

Finally,

$$c = b \cos A + a \cos B = a(\sin B \cot A + \cos B) \quad (5)$$

$$= a \sin B(\cot A + \cot B). \quad (6)$$

See also AAA THEOREM, ASA THEOREM, ASS THEOREM, SAS THEOREM, SSS THEOREM, TRIANGLE

Abacus

A mechanical counting device consisting of a frame holding a series of parallel rods on each of which beads are strung. Each bead represents a counting unit, and each rod a place value. The primary purpose of the abacus is not to perform actual computations, but to provide a quick means of storing numbers during a calculation. Abaci were used by the Japanese and Chinese, as well as the Romans.

See also ROMAN NUMERAL, SLIDE RULE

References

Boyer, C. B. and Merzbach, U. C. "The Abacus and Decimal Fractions." *A History of Mathematics, 2nd ed.* New York: Wiley, pp. 199–01, 1991.

Fernandes, L. "The Abacus: The Art of Calculating with Beads." http://www.ee.ryerson.ca/~elf/abacus/.

Gardner, M. "The Abacus." Ch. 18 in *Mathematical Circus: More Puzzles, Games, Paradoxes and Other Mathematical Entertainments from Scientific American.* New York: Knopf, pp. 232–41, 1979.

Pappas, T. "The Abacus." In *The Joy of Mathematics.* San Carlos, CA: Wide World Publ./Tetra, p. 209, 1989.

Pullan, J. M. *The History of the Abacus.* New York: Prager, 1968.

Smith, D. E. "Mechanical Aids to Calculation: The Abacus." Ch. 3 §1 in *History of Mathematics, Vol. 2.* New York: Dover, pp. 156–96, 1958.

Yoshino, Y. *The Japanese Abacus Explained.* New York: Dover, 1963.

abc Conjecture

A CONJECTURE due to J. Oesterlé and D. W. Masser. It states that, for any INFINITESIMAL $\epsilon > 0$, there exists a CONSTANT C_ϵ such that for any three RELATIVELY PRIME INTEGERS a, b, c satisfying

$$a + b = c, \quad (1)$$

the INEQUALITY

$$\max(|a|, |b|, |c|) \leq C_\epsilon \prod_{p|abc} p^{1+\epsilon} \quad (2)$$

holds, where $p|abc$ indicates that the PRODUCT is over PRIMES p which DIVIDE the PRODUCT abc. If this CONJECTURE were true, it would imply FERMAT'S LAST THEOREM for sufficiently large POWERS (Goldfeld 1996). This is related to the fact that the abc conjecture implies that there are at least $C \ln x$ WIEFERICH PRIMES $\leq x$ for some constant C (Silverman 1988, Vardi 1991).

The conjecture can also be stated by defining the height and radical of the sum $P: a + b = c$ as

$$h(P) = \max\{\ln|a|, \ln|b|, \ln|c|\} \qquad (3)$$

$$r(P) = \sum_{p|abc} \ln p, \qquad (4)$$

where p runs over all prime divisors of a, b, and c. Then the abc conjecture states that for all $\epsilon > 0$, there exists a constant K such that for all $P: a + b + c$,

$$h(P) \leq r(P) + \epsilon h(P) + K \qquad (5)$$

(van Frankenhuysen 2000). van Frankenhuysen (2000) has shown that there exists an infinite sequence of sums $P: a + b = c$ or RATIONAL INTEGERS with large height compared to the radical,

$$h(p) \geq r(P) + 4K_l \frac{\sqrt{h(P)}}{\ln[h(P)]}, \qquad (6)$$

with

$$K_l = 2^{l/2}\left(\frac{2\pi}{e}\right)^{1/4} > 1.517 \qquad (7)$$

for $l = 0.5990$, improving a result of Stewart and Tijdeman (1986).

See also FERMAT'S LAST THEOREM, MASON'S THEOREM, MORDELL CONJECTURE, ROTH'S THEOREM, WIEFERICH PRIME

References

Cox, D. A. "Introduction to Fermat's Last Theorem." *Amer. Math. Monthly* **101**, 3–4, 1994.
Elkies, N. D. "ABC Implies Mordell." *Internat. Math. Res. Not.* **7**, 99–09, 1991.
Goldfeld, D. "Beyond the Last Theorem." *The Sciences* **36**, 34–0, March/April 1996.
Goldfeld, D. "Beyond the Last Theorem." *Math. Horizons*, 26–1 and 24, Sept. 1996.
Guy, R. K. *Unsolved Problems in Number Theory, 2nd ed.* New York: Springer-Verlag, pp. 75–6, 1994.
Lang, S. "Old and New Conjectures in Diophantine Inequalities." *Bull. Amer. Math. Soc.* **23**, 37–5, 1990.
Lang, S. *Number Theory III: Diophantine Geometry.* New York: Springer-Verlag, pp. 63–7, 1991.
Mason, R. C. *Diophantine Equations over Functions Fields.* Cambridge, England: Cambridge University Press, 1984.
Mauldin, R. D. "A Generalization of Fermat's Last Theorem: The Beal Conjecture and Prize Problem." *Not. Amer. Math. Soc.* **44**, 1436–437, 1997.
Nitaq, A. "The abc Conjecture Home Page." http://www.math.unicaen.fr/~nitaj/abc.html.
Silverman, J. "Wieferich's Criterion and the abc Conjecture." *J. Number Th.* **30**, 226–37, 1988.
Stewart, C. L. and Tijdeman, R. "On the Oesterlé-Masser Conjecture." *Mh. Math.* **102**, 251–57, 1986.
Stewart, C. L. and Yu, K. "On the ABC Conjecture." *Math. Ann.* **291**, 225–30, 1991.
van Frankenhuysen, M. "The ABC Conjecture Implies Roth's Theorem and Mordell's Conjecture." *Mat. Contemp.* **16**, 45–2, 1999.
van Frankenhuysen, M. "A Lower Bound in the *abc* Conjecture." *J. Number Th.* **82**, 91–5, 2000.
Vardi, I. *Computational Recreations in Mathematica.* Reading, MA: Addison-Wesley, p. 66, 1991.
Vojta, P. *Diophantine Approximations and Value Distribution Theory.* Berlin: Springer-Verlag, p. 84, 1987.

Abel Polynomial

A polynomial $A_n(x; a)$ given by the associated SHEFFER SEQUENCE with

$$f(t) = te^{at}, \qquad (1)$$

given by

$$A_n(x;\ a) = x(x - an)^{n-1}. \qquad (2)$$

The GENERATING FUNCTION is

$$\sum_{k=0}^{\infty} \frac{A_k(x;\ a)}{k!} t^k = e^{xW(at)/a}, \qquad (3)$$

where $W(x)$ is LAMBERT'S W-FUNCTION. The associated BINOMIAL IDENTITY is

$$(x + y)(x + y - an)^{n-1}$$
$$= \sum_{k=0}^{n} \binom{n}{k} xy(x - ak)^{k-1}[y - a(n-k)]^{n-k-1}, \qquad (4)$$

where $\binom{n}{k}$ is a BINOMIAL COEFFICIENT, a formula originally due to Abel (Riordan 1979, p. 18; Roman 1984, pp. 30 and 73).

The first few Abel polynomials are

$$A_0(x;\ a) = 1$$
$$A_1(x;\ a) = x$$
$$A_2(x;\ a) = x(x - 2a)$$
$$A_3(x;\ a) = x(x - 3a)^2$$
$$A_4(x;\ a) = x(x - 4a)^3.$$

References

Riordan, J. *Combinatorial Identities.* New York: Wiley, p. 18, 1979.
Roman, S. "The Abel Polynomials." §4.1.5 in *The Umbral Calculus.* New York: Academic Press, pp. 29–0 and 72–5, 1984.

Abel Transform

The following INTEGRAL TRANSFORM relationship, known as the Abel transform, exists between two functions $f(x)$ and $g(t)$ for $0 < \alpha < 1$,

$$f(x) = \int_0^x \frac{g(t)\, dt}{(x - t)^\alpha} \qquad (1)$$

$$g(t) = -\frac{\sin(\pi\alpha)}{\pi} \frac{d}{dt} \int_0^t \frac{f(x)\, dx}{(x - t)^{1-\alpha}} \qquad (2)$$

$$= -\frac{\sin(\pi\alpha)}{\pi} \left[\int_0^t \frac{df}{dx} \frac{dx}{(t - x)^{1-\alpha}} + \frac{f(0)}{t^{1-\alpha}} \right]. \qquad (3)$$

The Abel transform is used in calculating the radial

mass distribution of galaxies (Binney and Tremaine 1987) and inverting planetary radio occultation data to obtain atmospheric information as a function of height.

Bracewell (1999, p. 262) defines a slightly different form of the Abel transform given by

$$g(x) = \mathscr{A}[f(r)] = 2\int_x^\infty \frac{f(r)r\,dr}{\sqrt{r^2 - x^2}}. \qquad (4)$$

The following table gives a number of common Abel transform pairs (Bracewell 1999, p. 264). Here,

$$\Pi_a(x) \equiv \Pi\left(\frac{x}{2a} - \frac{1}{2}\right) = \begin{cases} 1 & \text{for } 0 < x < 0 \\ 0 & \text{otherwise} \end{cases} \qquad (5)$$

where $\Pi(x)$ is the RECTANGLE FUNCTION, and

$$M(x) = 2\pi\left[x^{-3}\int_0^x J_0(x)\,dx - x^{-2}J_0(x)\right] \qquad (6)$$

$$= \frac{\pi^2}{x^2}[J_1(x)\mathscr{H}_0(x) - J_0(x)\mathscr{H}_1(x)], \qquad (7)$$

where $J_n(x)$ is a BESSEL FUNCTION OF THE FIRST KIND and $\mathscr{H}_n(x)$ is a STRUVE FUNCTION.

$f(r)$	$g(x)$	conditions
$\Pi_a(r)$	$2\sqrt{a^2 - x^2}$	$a^2 > x^2$
$(a^2 - r^2)^{-1/2}\Pi_a(r)$	π	$a^2 > x^2$
$\sqrt{a^2 - r^2}\,\Pi_a(r)$	$\frac{1}{2}\pi(a^2 - x^2)$	$a^2 > x^2$
$(a^2 - r^2)\Pi_a(r)$	$\frac{4}{3}(a^2 - x^2)^{3/2}$	$a^2 > x^2$
$(a^2 - r^2)^{3/2}\Pi_a(r)$	$\frac{3}{8}\pi(a^2 - x^2)^2$	$a^2 > x^2$
$(a - r)\Pi_a(r)$	$a\sqrt{a^2 - x^2} - x^2\cosh^{-1}\left(\frac{a}{x}\right)$	
$\frac{1}{\pi}\cosh^{-1}\left(\frac{a}{r}\right)$	$a - x$	
$\delta(r - a)$	$\frac{2a}{\sqrt{a^2 - x^2}}\,\Pi_a(x)$	
e^{-r^2/σ^2}	$\sigma\sqrt{\pi}e^{-x^2/\sigma^2}$	$\sigma > 0$
$r^2 e^{-r^2/\sigma^2}$	$\sigma(x^2 + \frac{1}{2}\sigma^2)\sqrt{\pi}e^{-x^2/\sigma^2}$	$\sigma > 0$
$\frac{e^{-r^2/\sigma^2}}{\sigma\sqrt{\pi}}(r^2 - \frac{1}{2}\sigma^2)$	$x^2 e^{-x^2/\sigma^2}$	$\sigma > 0$
$\frac{1}{b^2 + r^2}$	$\frac{\pi}{\sqrt{b^2 + r^2}}$	$b^2 + x^2 > 0$
$J_0(\omega r)$	$\frac{2\cos(\omega x)}{\omega}$	$\omega > 0$
$M(r)$	$\frac{8\pi^4}{\omega^2 x^2}\sin^2\left(\frac{x\omega}{2\pi}\right)$	$\omega > 0$

See also FOURIER TRANSFORM, HILBERT TRANSFORM, INTEGRAL EQUATION

References

Abel, N. H. *Oeuvres Completes* (Ed. L. Sylow and S. Lie). New York: Johnson Reprint Corp., pp. 11 and 97, 1988.

Arfken, G. *Mathematical Methods for Physicists, 3rd ed.* Orlando, FL: Academic Press, pp. 875–76, 1985.

Binney, J. and Tremaine, S. *Galactic Dynamics.* Princeton, NJ: Princeton University Press, p. 651, 1987.

Bracewell, R. *The Fourier Transform and Its Applications, 3rd ed.* New York: McGraw-Hill, pp. 262–66, 1999.

Hilfer, R. (Ed.). *Applications of Fractional Calculus in Physics.* Singapore: World Scientific, pp. 3–, 2000.

Liouville, J. "Memoire sur quelques quéstions de géométrie et de mécanique, et sur un nouveau genre pour réspondre ces quéstions." *J. École Polytech.* **13**, 1–9, 1832.

Lützen, J. *Joseph Liouville, 1809–882. Master of Pure and Applied Mathematics.* New York: Springer-Verlag, p. 314, 1990.

Whittaker, E. T. and Robinson, G. *The Calculus of Observations: A Treatise on Numerical Mathematics, 4th ed.* New York: Dover, pp. 376–77, 1967.

Abel's Binomial Theorem

The identity

$$\sum_{y=0}^m \binom{m}{y}(w - y)^{m-y-1}(z + y)^y = w^{-1}(z + w + m)^m$$

(Bhatnagar 1995, p. 51). There are a host of other such BINOMIAL IDENTITIES.

See also BINOMIAL IDENTITY, Q-ABEL'S THEOREM

References

Abel, N. H. "Beweis eines Ausdrucks, von welchem die Binomial-Formel ein einzelner Fall ist." *J. reine angew. Math.* **1**, 159–60, 1826. Reprinted in *Euvres Complètes, 2nd ed., Vol. 1.* pp. 102–03, 1881.

Bhatnagar, G. *Inverse Relations, Generalized Bibasic Series, and their $U(n)$ Extensions.* Ph.D. thesis. Ohio State University, p. 51, 1995.

Riordan, J. *Combinatorial Identities.* New York: Wiley, p. 18, 1979.

Abel's Convergence Theorem

Given a TAYLOR SERIES

$$f(z) = \sum_{n=0}^\infty C_n z^n = \sum_{n=0}^\infty C_n r^n e^{in\theta}, \qquad (1)$$

where the COMPLEX NUMBER z has been written in the polar form $z = re^{i\theta}$, examine the REAL and IMAGINARY PARTS

$$u(r,\,\theta) = \sum_{n=0}^\infty C_n r^n\,\cos(n\theta) \qquad (2)$$

$$v(r,\,\theta) = \sum_{n=0}^\infty C_n r^n\,\sin(n\theta). \qquad (3)$$

Abel's theorem states that, if $u(1,\,\theta)$ and $v(1,\,\theta)$ are CONVERGENT, then

$$u(1,\,\theta) + iv(1,\,\theta) = \lim_{r\to 1} f(re^{i\theta}). \qquad (4)$$

Stated in words, Abel's theorem guarantees that, if a REAL POWER SERIES CONVERGES for some POSITIVE value of the argument, the DOMAIN of UNIFORM CONVERGENCE extends at least up to and including this point. Furthermore, the continuity of the sum function extends at least up to and including this point.

References
Arfken, G. *Mathematical Methods for Physicists, 3rd ed.* Orlando, FL: Academic Press, p. 773, 1985.

Abel's Curve Theorem

The sum of the values of an INTEGRAL of the "first" or "second" sort

$$\int_{x_0, y_0}^{x_1, y_1} \frac{P \, dx}{Q} + \ldots + \int_{x_0, y_0}^{x_N, y_N} \frac{P \, dx}{Q} = F(z)$$

and

$$\frac{P(x_1, y_1)}{Q(x_1, y_1)} \frac{dx_1}{dz} + \ldots + \frac{P(x_N, y_N)}{Q(x_N, y_N)} \frac{dx_N}{dz} = \frac{dF}{dz},$$

from a FIXED POINT to the points of intersection with a curve depending rationally upon any number of parameters is a RATIONAL FUNCTION of those parameters.

References
Coolidge, J. L. *A Treatise on Algebraic Plane Curves.* New York: Dover, p. 277, 1959.

Abel's Differential Equation

The Abel equation of the first kind is given by

$$y' = f_0(x) + f_1(x)y + f_2(x)y^2 + f_3(x)y^3 + \ldots$$

(Murphy 1960, p. 23; Zwillinger 1997, p. 120), and the Abel equation of the second kind by

$$[g_0(x) + g_1(x)y]y' = f_0(x) + f_1(x)y + f_2(x)y^2 + f_3(x)y^3$$

(Murphy 1960, p. 25; Zwillinger 1997, p. 120).

References
Murphy, G. M. *Ordinary Differential Equations and Their Solution.* Princeton, NJ: Van Nostrand, 1960.
Zwillinger, D. *Handbook of Differential Equations, 3rd ed.* Boston, MA: Academic Press, p. 120, 1997.

Abel's Differential Equation Identity

Given a homogeneous linear SECOND-ORDER ORDINARY DIFFERENTIAL EQUATION,

$$y'' + P(x)y' + Q(x)y = 0, \tag{1}$$

call the two linearly independent solutions $y_1(x)$ and $y_2(x)$. Then

$$y_1'' + P(x)y_1' + Q(x)y_1 = 0 \tag{2}$$

$$y_2'' + P(x)y_2' + Q(x)y_2 = 0. \tag{3}$$

Now, take $y_1 \times$ (3) minus $y_2 \times$ (2),

$$y_1[y_2'' + P(x)y_2' + Q(x)y_2] - y_2[y_1'' + P(x)y_1' + Q(x)y_1] = 0 \tag{4}$$

$$(y_1 y_2'' - y_2 y_1'') + P(y_1 y_2' - y_1' y_2) + Q(y_1 y_2 - y_1 y_2) = 0 \tag{5}$$

$$(y_1 y_2'' - y_2 y_1'') + P(y_1 y_2' - y_1' y_2) = 0. \tag{6}$$

Now, use the definition of the WRONSKIAN and take its DERIVATIVE,

$$W \equiv y_1 y_2' + y_1' y_2 \tag{7}$$

$$W' = (y' y_2' + y_1 y_2'') - (y_1' y_2' + y_1'' y_2)$$

$$y_1 y_2'' - y_1'' y_2. \tag{8}$$

Plugging W and W' into (6) gives

$$W' + PW = 0. \tag{9}$$

This can be rearranged to yield

$$\frac{dW}{W} = -P(x) \, dx \tag{10}$$

which can then be directly integrated to

$$\ln\left[\frac{W(x)}{W_0}\right] = -\int P(x) \, dx, \tag{11}$$

where $\ln x$ is the NATURAL LOGARITHM. Exponentiating then yields Abel's identity

$$W(x) = W_0 e^{-\int P(x) \, dx}, \tag{12}$$

where W_0 is a constant of integration.

See also ORDINARY DIFFERENTIAL EQUATION–SECOND-ORDER

References
Boyce, W. E. and DiPrima, R. C. *Elementary Differential Equations and Boundary Value Problems, 4th ed.* New York: Wiley, pp. 118, 262, 277, and 355, 1986.

Abel's Duplication Formula

The duplication formula for ROGERS L-FUNCTION follows from ABEL'S FUNCTIONAL EQUATION and is given by

$$\tfrac{1}{2}L(x^2) = L(x) - L\left(\frac{x}{1+x}\right).$$

See also ABEL'S FUNCTIONAL EQUATION, DILOGARITHM

References
Gordon, B. and McIntosh, R. J. "Algebraic Dilogarithm Identities." *Ramanujan J.* **1**, 431–48, 1997.

Abel's Functional Equation

Let $L(x)$ denote the ROGERS L-FUNCTION defined in terms of the usual DILOGARITHM by

$$L(x) = \frac{6}{\pi^2}\left[\text{Li}_2(x) + \tfrac{1}{2}\ln x \ln(1-x)\right]$$

$$= \frac{6}{\pi^2}\left[\sum_{n=1}^{\infty}\frac{x^n}{n^2} + \tfrac{1}{2}\ln x \ln(1-x)\right],$$

then $L(x)$ satisfies the functional equation

$$L(x) + L(y) = L(xy) + L\left(\frac{x(1-y)}{1-xy}\right) + L\left(\frac{y(1-x)}{1-xy}\right).$$

ABEL'S DUPLICATION FORMULA follows from this identity.

See also ABEL'S DUPLICATION FORMULA, DILOGARITHM, FUNCTIONAL EQUATION, POLYLOGARITHM, RIEMANN ZETA FUNCTION, ROGERS L-FUNCTION

References

Abel, N. H. *Oeuvres Completes, Vol. 2* (Ed. L. Sylow and S. Lie). New York: Johnson Reprint Corp., pp. 189–92, 1988.

Bytsko, A. G. Two-Term Dilogarithm Identities Related to Conformal Field Theory. 9 Nov 1999. http://xxx.lanl.gov/abs/math-ph/9911012/.

Gordon, B. and McIntosh, R. J. "Algebraic Dilogarithm Identities." *Ramanujan J.* **1**, 431–48, 1997.

Hardy, G. H. *Ramanujan: Twelve Lectures on Subjects Suggested by His Life and Work, 3rd ed.* New York: Chelsea, pp. 14 and 21, 1999.

Rogers, L. J. "On Function Sum Theorems Connected with the Series $\Sigma_1^\infty x^n/n^2$." *Proc. London Math. Soc.* **4**, 169–89, 1907.

Abel's Impossibility Theorem

In general, POLYNOMIAL equations higher than fourth degree are incapable of algebraic solution in terms of a finite number of ADDITIONS, SUBTRACTIONS, MULTIPLICATIONS, DIVISIONS, and ROOT EXTRACTIONS. This was also shown by Ruffini in 1813 (Wells 1986, p. 59).

See also CUBIC EQUATION, GALOIS'S THEOREM, POLYNOMIAL, QUADRATIC EQUATION, QUARTIC EQUATION, QUINTIC EQUATION

References

Abel, N. H. "Beweis der Unmöglichkeit, algebraische Gleichungen von höheren Graden als dem vierten allgemein aufzulösen." *J. reine angew. Math.* **1**, 65, 1826. Reprinted in Abel, N. H. *Oeuvres Completes* (Ed. L. Sylow and S. Lie). New York: Johnson Reprint Corp., pp. 66–7, 1988.

Artin, E. *Galois Theory, 2nd ed.* Notre Dame, IN: Edwards Brothers, 1944.

Faucette, W. M. "A Geometric Interpretation of the Solution of the General Quartic Polynomial." *Amer. Math. Monthly* **103**, 51–7, 1996.

Fraleigh, J. B. *A First Course in Abstract Algebra.* Reading, MA: Addison-Wesley, 1982.

Herstein, I. N. *Topics in Algebra, 2nd ed.* New York: Wiley, 1975.

Hungerford, T. W. *Algebra.* New York: Springer-Verlag, 1980.

van der Waerden, B. L. *A History of Algebra: From al-Khwarizmi to Emmy Noether.* New York: Springer-Verlag, pp. 85–8, 1985.

Wells, D. *The Penguin Dictionary of Curious and Interesting Numbers.* Middlesex, England: Penguin Books, p. 59, 1986.

Abel's Inequality

Let $\{f_n\}$ and $\{a_n\}$ be SEQUENCES with $f_n \geq f_{n+1} > 0$ for $n = 1, 2, ...,$ then

$$\left|\sum_{n=1}^{m} a_n f_n\right| \leq A f_1,$$

where

$$A = \max\{|a_1|, |a_1 + a_2|, ..., |a_1 + a_2 + ... + a_m|\}.$$

Abel's Irreducibility Theorem

If one ROOT of the equation $f(x) = 0$, which is irreducible over a FIELD K, is also a ROOT of the equation $F(x) = 0$ in K, then all the ROOTS of the irreducible equation $f(x) = 0$ are ROOTS of $F(x) = 0$. Equivalently, $F(x)$ can be divided by $f(x)$ without a REMAINDER,

$$F(x) = f(x)F_1(x),$$

where $F_1(x)$ is also a POLYNOMIAL over K.

See also ABEL'S LEMMA, KRONECKER'S POLYNOMIAL THEOREM, SCHÖNEMANN'S THEOREM

References

Abel, N. H. "Mémoire sur une classe particulière d'équations résolubles algébriquement." *J. reine angew. Math.* **4**, 1829.

Dörrie, H. *100 Great Problems of Elementary Mathematics: Their History and Solutions.* New York: Dover, p. 120, 1965.

Abel's Lemma

The pure equation

$$x^p = C$$

of PRIME degree p is irreducible over a FIELD when C is a number of the FIELD but not the pth POWER of an element of the FIELD.

Jeffreys and Jeffreys (1988) use the term "Abel's lemma" for another LEMMA related to ABEL'S UNIFORM CONVERGENCE TEST.

See also ABEL'S IRREDUCIBILITY THEOREM, GAUSS'S POLYNOMIAL THEOREM, KRONECKER'S POLYNOMIAL THEOREM, SCHÖNEMANN'S THEOREM

References

Dörrie, H. *100 Great Problems of Elementary Mathematics: Their History and Solutions.* New York: Dover, p. 118, 1965.

Jeffreys, H. and Jeffreys, B. S. "Abel's Lemma." §1.1153 in *Methods of Mathematical Physics, 3rd ed.* Cambridge, England: Cambridge University Press, pp. 41–2, 1988.

Abel's Test

ABEL'S UNIFORM CONVERGENCE TEST

Abel's Theorem

ABEL'S BINOMIAL THEOREM, ABEL'S CONVERGENCE THEOREM, ABEL'S CURVE THEOREM, ABEL'S IMPOSSIBILITY THEOREM, ABEL'S IRREDUCIBILITY THEOREM, ABELIAN THEOREM, q-ABEL'S THEOREM

Abel's Uniform Convergence Test

Let $\{u_n(x)\}$ be a SEQUENCE of functions. If

1. $u_n(x)$ can be written $u_n(x) = a_n f_n(x)$,
2. Σa_n is CONVERGENT,
3. $f_n(x)$ is a MONOTONIC DECREASING SEQUENCE (i.e., $f_{n+1}(x) \le f_n(x)$) for all n, and
4. $f_n(x)$ is BOUNDED in some region (i.e., $0 \le f_n(x) \le M$ for all $x \in [a, b]$)

then, for all $x \in [a, b]$, the SERIES $\Sigma u_n(x)$ CONVERGES UNIFORMLY.

See also CONVERGENCE TESTS, CONVERGENT SERIES, UNIFORM CONVERGENCE

References

Bromwich, T. J. I'a. and MacRobert, T. M. *An Introduction to the Theory of Infinite Series, 3rd ed.* New York: Chelsea, p. 59, 1991.
Jeffreys, H. and Jeffreys, B. S. "Abel's Lemma" and "Abel's Test." §1.1153–.1154 in *Methods of Mathematical Physics, 3rd ed.* Cambridge, England: Cambridge University Press, pp. 41–2, 1988.
Whittaker, E. T. and Watson, G. N. *A Course in Modern Analysis, 4th ed.* Cambridge, England: Cambridge University Press, p. 17, 1990.

Abelian

A group or other algebraic object is said to be Abelian is the law of commutativity always holds. If an algebraic object is not Abelian, it is said to be NON-ABELIAN.

See also ABELIAN CATEGORY, ABELIAN DIFFERENTIAL, ABELIAN FUNCTION, ABELIAN GROUP, ABELIAN INTEGRAL, ABELIAN VARIETY, COMMUTATIVE, NON-ABELIAN

Abelian Category

An Abelian category is an abstract mathematical CATEGORY which displays some of the characteristic properties of the CATEGORY of all ABELIAN GROUPS.

See also ABELIAN GROUP, CATEGORY

References

Freyd, P. *Abelian Categories: An Introduction to the Theory of Functors.* New York: Harper & Row, 1964.
Grothendieck, A. "Sur quelques points d'algèbre homologique." *Tôhoku Math. J.* **9**, 119–21, 1957.
Mac Lane, S. and Gehring, F. W. *Categories for the Working Mathematician, 2nd ed.* New York: Springer-Verlag, 1998.

Abelian Differential

An Abelian differential is an ANALYTIC or MEROMORPHIC DIFFERENTIAL on a COMPACT or closed RIEMANN SURFACE.

Abelian Extension

This entry contributed by NICOLAS BRAY

If F is an ALGEBRAIC GALOIS EXTENSION of K such that the GALOIS GROUP of the extension is ABELIAN, then F is said to be an Abelian extension of K.

See also ALGEBRAIC EXTENSION, GALOIS EXTENSION, GALOIS GROUP

Abelian Function

An INVERSE FUNCTION of an ABELIAN INTEGRAL. Abelian functions have two variables and four periods, and can be defined by

$$\Theta\left(v, \ \tau; \ \frac{q'}{q}\right) = \sum_{\lambda = -\infty}^{\infty} 2^{2\pi i v(\lambda + q') + \pi i \tau(\lambda + q')^2 + 2\pi i q(\lambda + q')}$$

Baker (1907, p. 21). Abelian functions are a generalization of ELLIPTIC FUNCTIONS, and are also called hyperelliptic functions.

See also ABELIAN INTEGRAL, ELLIPTIC FUNCTION, THETA FUNCTIONS

References

Baker, H. F. *Abelian Functions: Abel's Theorem and the Allied Theory, Including the Theory of the Theta Functions.* New York: Cambridge University Press, 1995.
Baker, H. F. *An Introduction to the Theory of Multiply Periodic Functions.* London: Cambridge University Press, 1907.
Weisstein, E. W. "Books about Abelian Functions." http://www.treasure-troves.com/books/AbelianFunctions.html.

Abelian Group

N.B. A detailed online essay by S. Finch was the starting point for this entry.

A GROUP for which the elements COMMUTE (i.e., $AB = BA$ for all elements A and B) is called an Abelian group. All CYCLIC GROUPS are Abelian, but an Abelian group is not necessarily CYCLIC. All SUBGROUPS of an Abelian group are NORMAL. In an Abelian group, each element is in a CONJUGACY CLASS by itself, and the CHARACTER TABLE involves POWERS of a single element known as a GENERATOR.

No general formula is known for giving the number of nonisomorphic FINITE GROUPS of a given ORDER. However, the number of nonisomorphic Abelian FINITE GROUPS $a(n)$ of any given ORDER n is given by writing n as

$$n = \prod_i p_i^{\alpha_i}, \tag{1}$$

where the p_i are distinct PRIME FACTORS, then

$$a(n) = \prod_i P(\alpha_i), \tag{2}$$

where $P(k)$ is the PARTITION FUNCTION. This gives 1, 1, 1, 2, 1, 1, 1, 3, 2, ... (Sloane's A000688). The smallest orders for which $n = 1, 2, 3, ...$ nonisomorphic Abelian groups exist are 1, 4, 8, 36, 16, 72, 32, 900, 216, 144, 64, 1800, 0, 288, 128, ... (Sloane's A046056), where 0 denotes an impossible number (i.e., not a product of partition numbers) of nonisomorphic Abelian, groups. The "missing" values are 13, 17, 19, 23, 26, 29, 31, 34, 37, 38, 39, 41, 43, 46, ... (Sloane's A046064). The incrementally largest numbers of Abelian groups as a function of order are 1, 2, 3, 5, 7, 11, 15, 22, 30, 42, 56, 77, 101, ... (Sloane's A046054), which occur for orders 1, 4, 8, 16, 32, 64, 128, 256, 512, 1024, 2048, 4096, 8192, ... (Sloane's A046055).

The KRONECKER DECOMPOSITION THEOREM states that every FINITE Abelian group can be written as a GROUP DIRECT PRODUCT of CYCLIC GROUPS of PRIME POWER ORDER. If the ORDER of a FINITE GROUP is a PRIME p, then there exists a single Abelian group of order p (denoted $\mathbb{Z}_p$) and no non-Abelian groups. If the ORDER is a prime squared p^2 then there are two Abelian groups (denoted $\mathbb{Z}_{p^2}$ and $\mathbb{Z}_p \times \mathbb{Z}_p$. If the ORDER is a prime cubed p^3, then there are three Abelian groups (denoted $\mathbb{Z}_p \times \mathbb{Z}_p \times \mathbb{Z}_p$, $\mathbb{Z}_p \times \mathbb{Z}_{p^2}$, and $\mathbb{Z}_{p^3}$), and five groups total. If the order is a PRODUCT of two primes p and q, then there exists exactly one Abelian group of ORDER pq (denoted $\mathbb{Z}_p \times \mathbb{Z}_q$).

Another interesting result is that if $a(n)$ denotes the number of nonisomorphic Abelian groups of ORDER n, then

$$\sum_{n=1}^{\infty} a(n) n^{-s} = \zeta(s)\zeta(2s)\zeta(3s)\cdots, \tag{3}$$

where $\zeta(s)$ is the RIEMANN ZETA FUNCTION. Srinivasan (1973) has also shown that

$$\sum_{n=1}^{N} a(n) = A_1 N + A_2 N^{1/2} + A_3 N^{1/3}$$
$$+ \mathcal{O}[x^{105/407}(\ln x)^2], \tag{4}$$

where

$$A_k \equiv \prod_{\substack{j=1 \\ j \neq k}} \zeta\left(\frac{j}{k}\right) = \begin{cases} 2.294856591\ldots & \text{for } k=1 \\ -14.6475663\ldots & \text{for } k=2 \\ 118.6924619\ldots & \text{for } k=3, \end{cases} \tag{5}$$

and $\zeta(s)$ is again the RIEMANN ZETA FUNCTION. [Richert (1952) incorrectly gave $A_3 = 114$.] DeKoninck and Ivic (1980) showed that

$$\sum_{n=1}^{N} \frac{1}{a(n)} = BN + \mathcal{O}[\sqrt{N}(\ln N)^{-1/2}], \tag{6}$$

where

$$B \equiv \prod \left\{ 1 - \sum_{k=2}^{\infty} \left[\frac{1}{P(k-2)} - \frac{1}{P(k)} \right] \frac{1}{p^k} \right\} = 0.752\ldots \tag{7}$$

is a product over PRIMES. Bounds for the number of nonisomorphic non-Abelian groups are given by Neumann (1969) and Pyber (1993).

See also FINITE GROUP, GROUP THEORY, KRONECKER DECOMPOSITION THEOREM, PARTITION FUNCTION P, RING

References

Arnold, D. M. and Rangaswamy, K. M. (Eds.). *Abelian Groups and Modules.* New York: Dekker, 1996.

DeKoninck, J.-M. and Ivic, A. *Topics in Arithmetical Functions: Asymptotic Formulae for Sums of Reciprocals of Arithmetical Functions and Related Fields.* Amsterdam, Netherlands: North-Holland, 1980.

Erdos, P. and Szekeres, G. "Über die Anzahl abelscher Gruppen gegebener Ordnung und über ein verwandtes zahlentheoretisches Problem." *Acta Sci. Math. (Szeged)* **7**, 95–02, 1935.

Finch, S. "Favorite Mathematical Constants." http://www.mathsoft.com/asolve/constant/abel/abel.html.

Fuchs, L. and Göbel, R. (Eds.). *Abelian Groups.* New York: Dekker, 1993.

Kendall, D. G. and Rankin, R. A. "On the Number of Abelian Groups of a Given Order." *Quart. J. Oxford* **18**, 197–08, 1947.

Kolesnik, G. "On the Number of Abelian Groups of a Given Order." *J. reine angew. Math.* **329**, 164–75, 1981.

Neumann, P. M. "An Enumeration Theorem for Finite Groups." *Quart. J. Math. Ser. 2* **20**, 395–01, 1969.

Pyber, L. "Enumerating Finite Groups of Given Order." *Ann. Math.* **137**, 203–20, 1993.

Richert, H.-E. "Über die Anzahl abelscher Gruppen gegebener Ordnung I." *Math. Zeitschr.* **56**, 21–2, 1952.

Sloane, N. J. A. Sequences A000688/M0064 in "An On-Line Version of the Encyclopedia of Integer Sequences." http://www.research.att.com/~njas/sequences/eisonline.html.

Srinivasan, B. R. "On the Number of Abelian Groups of a Given Order." *Acta Arith.* **23**, 195–05, 1973.

Abelian Integral

An INTEGRAL OF THE FORM

$$\int_0^x \frac{dt}{\sqrt{R(t)}},$$

where $R(t)$ is a POLYNOMIAL of degree > 4. They are also called HYPERELLIPTIC INTEGRALS.

See also ABELIAN FUNCTION, ELLIPTIC INTEGRAL

References

Siegel, C. L. *Topics in Complex Function Theory, Vol. 2: Automorphic Functions and Abelian Integrals.* New York: Wiley, 1988.

Abelian Theorem

A theorem which asserts that if a sequence or function behaves regularly, then some average of it behaves regularly. For example,

$$A(x) \sim x$$

implies

$$A_1(x) = \int_0^x A(t)\, dt \sim \tfrac{1}{2}x^2$$

for any $A(x)$. The converse is false, but can be made into a correct TAUBERIAN THEOREM if $A(x)$ is subjected to an appropriate additional condition (Hardy 1999, p. 46).

See also TAUBERIAN THEOREM

References

Hardy, G. H. *Ramanujan: Twelve Lectures on Subjects Suggested by His Life and Work, 3rd ed.* New York: Chelsea, p. 46, 1999.

Abelian Variety

An Abelian variety is an algebraic GROUP which is a complete ALGEBRAIC VARIETY. An Abelian variety of DIMENSION 1 is an ELLIPTIC CURVE.

See also ALBANESE VARIETY

References

Murty, V. K. *Introduction to Abelian Varieties.* Providence, RI: Amer. Math. Soc., 1993.
Shimura, G. *Abelian Varieties With Complex Multiplication and Modular Functions.* Princeton, NJ: Princeton University Press, 1999.
Shimura, G. and Taniyama, Y. *Complex Multiplication of Abelian Varieties and Its Applications to Number Theory.* Tokyo: Mathematical Society of Japan, 1961.

Abelianization

In general, groups are not ABELIAN. However, there is always a GROUP HOMOMORPHISM $h : G \to G'$ to an ABELIAN GROUP, and this homomorphism is called Abelianization. The homomorphism is abstractly described by its kernel, the COMMUTATOR SUBGROUP $[G, G]$. So $G' = G/[G, G]$. Roughly speaking, in any expression, every product becomes commutative after Abelianization. As a consequence, some previously unequal expressions may become equal, or even represent the IDENTITY ELEMENT.

For example, in the eight-element QUATERNION GROUP $G = \{\pm 1, \pm i, \pm j, \pm k\}$, the COMMUTATOR SUBGROUP is $\{\pm 1\}$. The Abelianization of G is a copy of $\mathbb{Z}_2 \times \mathbb{Z}_2$, and for instance, $i'j' = j'i'$ in the Abelianization.

See also ABELIAN, GROUP, HOMOMORPHISM

Abel-Plana Formula
This entry contributed by DAVID ANDERSON

The Abel-Plana formula gives an expression for the difference between a discrete sum and the corresponding integral. The formula can be derived from the ARGUMENT PRINCIPLE

$$\oint_\gamma f(z)\frac{g'(z)}{g(z)}\, dz = \sum_n f(\mu_n) - \sum_m f(v_m), \qquad (1)$$

where μ_n are the zeros of $g(z)$ and v_m are the poles contained within the CONTOUR γ. An appropriate choice of g and γ then yields

$$\sum_{n=0}^\infty f(n) - \int_0^\infty f(x)\, dx$$
$$= \tfrac{1}{2}f(0) - \tfrac{1}{2}\int_0^\infty [f(it) - f(-it)][\cot(\pi i t) + i]\, dt, \qquad (2)$$

or equivalently

$$\sum_{n=0}^\infty f(n) - \int_0^\infty f(x)\, dx$$
$$= \tfrac{1}{2}f(0) + i\int_0^\infty \frac{f(it) - f(-it)}{e^{2\pi t} - 1}\, dt. \qquad (3)$$

The formula is particularly useful in Casimir effect calculations involving differences between quantized modes and free modes.

See also ARGUMENT PRINCIPLE

References

Mostepanenko, V. M. and Trunov, N. N. §2.2 in *The Casimir Effect and Its Applications.* Oxford, England: Clarendon Press, 1997.
Saharian, A. A. "The Generalized Abel-Plana Formula. Applications to Bessel Functions and Casimir Effect." http://www.ictp.trieste.it/~pub_off/preprints-sources/2000/IC2000014P.pdf.

Abhyankar's Conjecture

For a FINITE GROUP G, let $p(G)$ be the SUBGROUP generated by all the SYLOW p-SUBGROUPS of G. If X is a projective curve in characteristic $p > 0$, and if $x_0, ..., x_t$ are points of X (for $t > 0$), then a NECESSARY and SUFFICIENT condition that G occur as the GALOIS GROUP of a finite covering Y of X, branched only at the points $x_0, ..., x_t$, is that the QUOTIENT GROUP $G/p(G)$ has $2g + t$ generators.

Raynaud (1994) solved the Abhyankar problem in the crucial case of the affine line (i.e., the projective line with a point deleted), and Harbater (1994) proved the full Abhyankar conjecture by building upon this special solution.

See also FINITE GROUP, GALOIS GROUP, QUOTIENT GROUP, SYLOW P-SUBGROUP

References
Abhyankar, S. "Coverings of Algebraic Curves." *Amer. J. Math.* **79**, 825–56, 1957.
American Mathematical Society. "Notices of the AMS, April 1995, 1995 Frank Nelson Cole Prize in Algebra." http://www.ams.org/notices/199504/prize-cole.pdf.
Harbater, D. "Abhyankar's Conjecture on Galois Groups Over Curves." *Invent. Math.* **117**, 1–5, 1994.
Raynaud, M. "Revêtements de la droite affine en caractéristique $p > 0$ et conjecture d'Abhyankar." *Invent. Math.* **116**, 425–62, 1994.

Ablowitz-Ramani-Segur Conjecture
The Ablowitz-Ramani-Segur conjecture states that a nonlinear PARTIAL DIFFERENTIAL EQUATION is solvable by the INVERSE SCATTERING METHOD only if every nonlinear ORDINARY DIFFERENTIAL EQUATION obtained by exact reduction has the PAINLEVÉ PROPERTY.

See also INVERSE SCATTERING METHOD

References
Tabor, M. *Chaos and Integrability in Nonlinear Dynamics: An Introduction.* New York: Wiley, p. 351, 1989.

Abnormal Number
A hypothetical number which can be factored into primes in more than one way. Hardy and Wright (1979) prove the FUNDAMENTAL THEOREM OF ARITHMETIC by showing that no abnormal numbers exist.

See also FUNDAMENTAL THEOREM OF ARITHMETIC

References
Hardy, G. H. and Wright, E. M. *An Introduction to the Theory of Numbers, 5th ed.* Oxford, England: Clarendon Press, p. 21, 1979.

Abs
ABSOLUTE VALUE

Abscissa
The x- (horizontal) coordinate of a point in a two dimensional coordinate system. Physicists and astronomers sometimes use the term to refer to the axis itself instead of the distance along it.

See also AXIS, ORDINATE, REAL LINE, x-AXIS, y-AXIS, z-AXIS

Absolute Convergence
A SERIES $\Sigma_n u_n$ is said to CONVERGE absolutely if the SERIES $\Sigma_n |u_n|$ CONVERGES, where $|u_n|$ denotes the ABSOLUTE VALUE. If a SERIES is absolutely convergent, then the sum is independent of the order in which terms are summed. Furthermore, if the SERIES is multiplied by another absolutely convergent series, the product series will also converge absolutely.

See also CONDITIONAL CONVERGENCE, CONVERGENT SERIES, RIEMANN SERIES THEOREM

References
Bromwich, T. J. I'a. and MacRobert, T. M. "Absolute Convergence." Ch. 4 in *An Introduction to the Theory of Infinite Series, 3rd ed.* New York: Chelsea, pp. 69–7, 1991.
Jeffreys, H. and Jeffreys, B. S. "Absolute Convergence." §1.051 in *Methods of Mathematical Physics, 3rd ed.* Cambridge, England: Cambridge University Press, p. 16, 1988.

Absolute Deviation
Let $\bar{u}$ denote the MEAN of a SET of quantities u_i, then the absolute deviation is defined by

$$\Delta u_i \equiv |u_i - \bar{u}|.$$

See also DEVIATION, MEAN DEVIATION, SIGNED DEVIATION, STANDARD DEVIATION

Absolute Error
The DIFFERENCE between the measured or inferred value of a quantity x_0 and its actual value x, given by

$$\Delta x \equiv x_0 - x$$

(sometimes with the ABSOLUTE VALUE taken) is called the absolute error. The absolute error of the SUM or DIFFERENCE of a number of quantities is less than or equal to the SUM of their absolute errors.

See also ERROR PROPAGATION, PERCENTAGE ERROR, RELATIVE ERROR

References
Abramowitz, M. and Stegun, C. A. (Eds.). *Handbook of Mathematical Functions with Formulas, Graphs, and Mathematical Tables, 9th printing.* New York: Dover, p. 14, 1972.

Absolute Frequency
The number of data points which fall within a given CLASS in a FREQUENCY DISTRIBUTION.

See also CUMULATIVE FREQUENCY, FREQUENCY DISTRIBUTION, RELATIVE FREQUENCY, RELATIVE CUMULATIVE FREQUENCY

References
Kenney, J. F. and Keeping, E. S. "Frequency Distributions." §1.8 in *Mathematics of Statistics, Pt. 1, 3rd ed.* Princeton, NJ: Van Nostrand, pp. 12–9, 1962.

Absolute Geometry
GEOMETRY which depends only on the first four of EUCLID'S POSTULATES and not on the PARALLEL POSTULATE. Euclid himself used only the first four

postulates for the first 28 propositions of the ELE-MENTS, but was forced to invoke the PARALLEL POSTULATE on the 29th.

See also AFFINE GEOMETRY, *ELEMENTS*, EUCLID'S POSTULATES, GEOMETRY, ORDERED GEOMETRY, PARALLEL POSTULATE

References

Hofstadter, D. R. *Gödel, Escher, Bach: An Eternal Golden Braid.* New York: Vintage Books, pp. 90–1, 1989.

Absolute Moment

The absolute moment of M_n of a probability function $P(x)$ taken about a point a is defined by

$$M_n = \int |x - a|^n P(x)\, dx.$$

See also CENTRAL MOMENT, MOMENT, RAW MOMENT

References

Papoulis, A. *Probability, Random Variables, and Stochastic Processes, 2nd ed.* New York: McGraw-Hill, p. 146, 1984.

Absolute Monotonic Sequence

See also ABSOLUTELY MONOTONIC SEQUENCE

References

Feller, W. *An Introduction to Probability Theory and Its Applications, Vol. 2, 3rd ed.* New York: Wiley, p. 224, 1971.

Absolute Pseudoprime

CARMICHAEL NUMBER

Absolute Square

Also known as the squared norm. The absolute square of a COMPLEX NUMBER z is written $|z|^2$, where $|z|$ is the MODULUS and is defined as

$$|z|^2 \equiv z\bar{z}, \tag{1}$$

where $\bar{z}$ denotes the COMPLEX CONJUGATE of z. For a REAL NUMBER, (1) simplifies to

$$|z|^2 = z^2. \tag{2}$$

If the COMPLEX NUMBER is written $z = x + iy$, then the absolute square can be written

$$|x + iy|^2 = x^2 + y^2. \tag{3}$$

An absolute square can be computed in terms of x and y using the *Mathematica* command ComplexExpand[Abs[z]2, TargetFunctions-> {Conjugate}].

An important identity involving the absolute square is given by

$$\left| a \pm be^{-i\delta} \right|^2 = (a \pm be^{-i\delta})(a \pm be^{i\delta})$$
$$= a^2 + b^2 \pm ab(e^{i\delta} + e^{-i\delta}) = a^2 + b^2 \pm 2ab \cos \delta. \tag{4}$$

If $a = 1$, then (4) becomes

$$\left| 1 \pm be^{-i\delta} \right|^2 = 1 + b^2 \pm 2b \cos \delta$$
$$= (1 \pm b)^2 \mp 4b \sin^2(\tfrac{1}{2}\delta). \tag{5}$$

If $a = 1$, and $b = 1$, then

$$\left| 1 - e^{-i\delta} \right|^2 = 4 \sin^2(\tfrac{1}{2}\delta). \tag{6}$$

Finally,

$$|e^{i\phi_1} + e^{i\phi_2}|^2 = (e^{i\phi_1} + e^{i\phi_2})(e^{-i\phi_1} + e^{-i\phi_2})$$
$$= 2[1 + \cos(\phi_2 - \phi_1)]$$
$$= 4 \cos^2[\tfrac{1}{2}(\phi_2 - \phi_1)]. \tag{7}$$

See also ARGUMENT (COMPLEX NUMBER), COMPLEX NUMBER, MODULUS (COMPLEX NUMBER)

Absolute Value

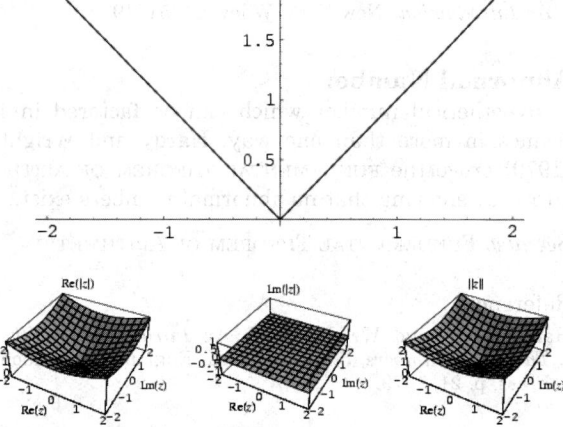

The absolute value of a REAL NUMBER x is denoted $|x|$ and given by the "unsigned" portion of x,

$$|x| = x \operatorname{sgn}(x) = \begin{cases} -x & \text{for } x \leq 0 \\ x & \text{for } x \geq 0, \end{cases}$$

where sgn x is the sign function SGN. The absolute value is therefore always greater than or equal to 0. The same notation is used to denote the MODULUS of a COMPLEX NUMBER $z = x + iy$, $|z| \equiv \sqrt{x^2 + y^2}$, a P-ADIC NORM, or a general VALUATION. The NORM of a VECTOR **x** is also denoted $|x|$, although $\|x\|$ is more commonly used.

Other NOTATIONS similar to the absolute value are the FLOOR FUNCTION $\lfloor x \rfloor$, NINT function $[x]$, and CEILING FUNCTION $\lceil x \rceil$.

The integral of the absolute value of the different of two variables is given by

$$\int_0^1 \int_0^1 |x-y|^n \, dx \, dy = \frac{2}{(n+1)(n+2)},$$

which has values 1/3, 1/6, 1/10, 1/15, 1/21, ... for $n = 1$, 2, ..., i.e., the inverses of the TRIANGULAR NUMBERS (Sloane's A000217).

See also ABSOLUTE SQUARE, CEILING FUNCTION, FLOOR FUNCTION, MODULUS (COMPLEX NUMBER), NINT, RECTANGLE FUNCTION, SGN, TRIANGLE FUNCTION, VALUATION

References

Sloane, N. J. A. Sequences A000217/M2535 in "An On-Line Version of the Encyclopedia of Integer Sequences." http://www.research.att.com/~njas/sequences/eisonline.html.

Absolutely Continuous

A MEASURE λ is absolutely continuous with respect to another measure μ if $\lambda(E) = 0$ for every set with $\mu(E) = 0$. This makes sense as long as μ is a POSITIVE MEASURE, such as LEBESGUE MEASURE, but λ can be any measure, possibly a COMPLEX MEASURE.

By the RADON-NIKODYM THEOREM, this is equivalent to saying that

$$\lambda(E) = \int_E f \, d\mu$$

where the integral is the LEBESGUE INTEGRAL, for some INTEGRABLE function f. The function f is like a derivative, and is called the RADON-NIKODYM DERIVATIVE $d\lambda/d\mu$.

The measure supported at 0 ($\mu^{(E)} = 1$ iff $0 \in E$) is not absolutely continuous with respect to LEBESGUE MEASURE, and is a SINGULAR MEASURE.

See also COMPLEX MEASURE, CONCENTRATED, HAAR MEASURE, LEBESGUE DECOMPOSITION (MEASURE), LEBESGUE MEASURE, MUTUALLY SINGULAR, POLAR REPRESENTATION (MEASURE), SINGULAR MEASURE

References

Rudin, W. *Functional Analysis, 2nd ed.* New York: McGraw-Hill, pp. 121–25, 1991.

Absolutely Fair

A sequence of random variates X_0, X_1, ... is called absolutely fair if for $n = 1, 2, ...$,

$$(X_1) = 0$$

and

$$(X_{n+1}|X_1, ..., X_n) = 0$$

(Feller 1971, p. 210).

See also MARTINGALE

References

Feller, W. *An Introduction to Probability Theory and Its Applications, Vol. 2, 3rd ed.* New York: Wiley, 1971.

Absolutely Monotonic Function

This entry contributed by RONALD M. AARTS

A function $f(x)$ is absolutely monotonic in the interval $a < x < b$ if it has nonnegative derivatives of all orders in the region, i.e.,

$$f^{(k)}(x) \geq 0 \qquad (1)$$

for $a < x < b$ and $k = 0, 1, 2, ...$ For example, the functions

$$f(x) = -\ln(-x) \quad (-1 \leq x < 0) \qquad (2)$$

and

$$f(x) = \sin^{-1} x \quad (0 \leq x \leq 1) \qquad (3)$$

are absolutely monotonic functions (Widder 1941).

See also ABSOLUTELY MONOTONIC SEQUENCE

References

Widder, D. V. Ch. 4 in *The Laplace Transform.* Princeton, NJ: Princeton University Press, 1941.

Absolutely Monotonic Sequence

See also ABSOLUTE MONOTONIC SEQUENCE, ABSOLUTELY MONOTONIC FUNCTION

References

Feller, W. *An Introduction to Probability Theory and Its Applications, Vol. 2, 3rd ed.* New York: Wiley, p. 224, 1971.

Absorption Law

The law appearing in the definition of a BOOLEAN ALGEBRA which states

$$a \wedge (a \vee b) = a \vee (a \wedge b) = a$$

for binary operators $\vee$ and $\wedge$ (which most commonly are logical OR and logical AND).

See also BOOLEAN ALGEBRA, LATTICE

References

Birkhoff, G. and Mac Lane, S. *A Survey of Modern Algebra, 5th ed.* New York: Macmillian, p. 317, 1996.

Abstract Algebra

That portion of ALGEBRA dealing with theoretical as opposed to applied topics. Ash (1998) includes the following areas in his definite of abstract algebra: logic and foundations, counting, elementary NUMBER THEORY, informal SET THEORY, LINEAR ALGEBRA, and the theory of linear operators.

See also ALGEBRA

References

Ash, R. B. *A Primer of Abstract Mathematics.* Washington, DC: Math. Assoc. Amer., 1998.

Abstract Manifold

An abstract manifold is a MANIFOLD in the context of an abstract space with no particular embedding, or representation in mind. It is a TOPOLOGICAL SPACE with an ATLAS of COORDINATE CHARTS.

For example, the SPHERE $\mathbb{S}^2$ can be considered a SUBMANIFOLD of $\mathbb{R}^3$ or a QUOTIENT SPACE $O(3)/O(2)$. But as an abstract manifold, it is just a MANIFOLD, which can be covered by two coordinate charts $\phi_1 : \mathbb{R}^2 \to \mathbb{S}^2$ and $\phi_2 : \mathbb{R}^2 \to \mathbb{S}^2$, with the single TRANSITION FUNCTION,

$$\phi_2^{-1} \circ \phi_1 : \mathbb{R}^2 - (0,\ 0) \to \mathbb{R}^2 - (0,\ 0)$$

defined by

$$\phi_2^{-1} \circ \phi_1(x,\ y) = (x/r^2,\ y/r^2)$$

where $r^2 = x^2 + y^2$. It can also be thought of as two disks glued together at their boundary.

See also ALGEBRAIC MANIFOLD, HOMOGENEOUS SPACE, MANIFOLD, SUBMANIFOLD, TOPOLOGICAL SPACE

Abstract Mathematics
ABSTRACT ALGEBRA

Abstract Simplicial Complex

An abstract simplicial complex is a collection S of finite nonempty sets such that if A is an element of S, then so is every nonempty subset of A (Munkres 1993, p. 15).

See also SIMPLICIAL COMPLEX

References
Munkres, J. R. *Elements of Algebraic Topology.* Perseus Press, 1993.

Abstract Vector Space

See also QUOTIENT VECTOR SPACE, VECTOR SPACE

Abstraction Operator
LAMBDA CALCULUS

Abundance

The abundance of a number n is the quantity

$$A(n) \equiv \sigma(n) - 2n,$$

where $\sigma(n)$ is the DIVISOR FUNCTION. Kravitz has conjectured that no numbers exist whose abundance is an ODD SQUARE (Guy 1994).

The following table lists special classifications given to a number n based on the value of $A(n)$.

$A(n)$	Number
< 0	DEFICIENT NUMBER
-1	ALMOST PERFECT NUMBER
0	PERFECT NUMBER
1	QUASIPERFECT NUMBER
> 0	ABUNDANT NUMBER

See also ABUNDANCY, DEFICIENCY

References
Guy, R. K. *Unsolved Problems in Number Theory, 2nd ed.* New York: Springer-Verlag, pp. 45–6, 1994.

Abundancy

The ratio $\sigma(n)/n$, where $\sigma(n)$ is the DIVISOR FUNCTION.

See also ABUNDANCE, ABUNDANT NUMBER

References
Guy, R. K. "The Second Strong Law of Small Numbers." *Math. Mag.* **63**, 3–0, 1990.

Abundant Number

An abundant number is an INTEGER n which is not a PERFECT NUMBER and for which

$$s(n) \equiv \sigma(n) - n > n, \tag{1}$$

where $\sigma(n)$ is the DIVISOR FUNCTION. The quantity $\sigma(n) - 2n$ is sometimes called the ABUNDANCE. The first few abundant numbers are 12, 18, 20, 24, 30, 36, ... (Sloane's A005101). Abundant numbers are sometimes called EXCESSIVE NUMBERS.

There are only 21 abundant numbers less than 100, and they are all EVEN. The first ODD abundant number is

$$945 = 3^3 \cdot 7 \cdot 5. \tag{2}$$

That 945 is abundant can be seen by computing

$$s(945) = 975 > 945. \tag{3}$$

Any multiple of a PERFECT NUMBER or an abundant number is also abundant. Every number greater than 20161 can be expressed as a sum of two abundant numbers.

Define the density function

$$A(x) \equiv \lim_{n \to \infty} \frac{|\{n : \sigma(n) \geq xn\}|}{n} \tag{4}$$

for a POSITIVE REAL NUMBER x, then Davenport (1933) proved that $A(x)$ exists and is continuous for all x, and Erdos (1934) gave a simplified proof (Finch). Wall (1971) and Wall *et al.* (1977) showed that

$$0.2441 < A(2) < 0.2909, \tag{5}$$

and Deléglise (1998) showed that

$$0.2474 < A(2) < 0.2480. \tag{6}$$

A number which is abundant but for which all its PROPER DIVISORS are DEFICIENT is called a PRIMITIVE ABUNDANT NUMBER (Guy 1994, p. 46).

See also ALIQUOT SEQUENCE, DEFICIENT NUMBER, HIGHLY ABUNDANT NUMBER, MULTIAMICABLE NUMBERS, PERFECT NUMBER, PRACTICAL NUMBER, PRIMITIVE ABUNDANT NUMBER, WEIRD NUMBER

References

Deléglise, M. "Bounds for the Density of Abundant Integers." *Exp. Math.* **7**, 137–43, 1998.

Dickson, L. E. *History of the Theory of Numbers, Vol. 1: Divisibility and Primality.* New York: Chelsea, pp. 3–3, 1952.

Erdos, P. "On the Density of the Abundant Numbers." *J. London Math. Soc.* **9**, 278–82, 1934.

Finch, S. "Favorite Mathematical Constants." http://www.mathsoft.com/asolve/constant/abund/abund.html.

Guy, R. K. *Unsolved Problems in Number Theory, 2nd ed.* New York: Springer-Verlag, pp. 45–6, 1994.

Singh, S. *Fermat's Enigma: The Epic Quest to Solve the World's Greatest Mathematical Problem.* New York: Walker, pp. 11 and 13, 1997.

Sloane, N. J. A. Sequences A005101/M4825 in "An On-Line Version of the Encyclopedia of Integer Sequences." http://www.research.att.com/~njas/sequences/eisonline.html.

Souissi, M. *Un Texte Manuscrit d'Ibn Al-Banna' Al-Marrakusi sur les Nombres Parfaits, Abondants, Deficients, et Amiables.* Karachi, Pakistan: Hamdard Nat. Found., 1975.

Wall, C. R. "Density Bounds for the Sum of Divisors Function." In *The Theory of Arithmetic Functions: Proceedings of the Conference at Western Michigan University, April 29-May 1, 1971.* (Ed. A. A. Gioia and D. L. Goldsmith). New York: Springer-Verlag, pp. 283–87, 1971.

Wall, C. R.; Crews, P. L.; and Johnson, D. B. "Density Bounds for the Sum of Divisors Function." *Math. Comput.* **26**, 773–77, 1972.

Wall, C. R.; Crews, P. L.; and Johnson, D. B. "Density Bounds for the Sum of Divisors Function." *Math. Comput.* **31**, 616, 1977.

Acceleration

Let a particle travel a distance $s(t)$ as a function of time t (here, s can be thought of as the ARC LENGTH of the curve traced out by the particle). The SPEED (the SCALAR NORM of the VECTOR VELOCITY) is then given by

$$\frac{ds}{dt} = \sqrt{\left(\frac{dx}{dt}\right)^2 + \left(\frac{dy}{dt}\right)^2 + \left(\frac{dz}{dt}\right)^2}. \tag{1}$$

The acceleration is defined as the time DERIVATIVE of the VELOCITY, so the SCALAR acceleration is given by

$$a \equiv \frac{dv}{dt} \tag{2}$$

$$= \frac{d^2s}{dt^2} \tag{3}$$

$$= \frac{\frac{dx}{dt}\frac{d^2x}{dt^2} + \frac{dy}{dt}\frac{d^2y}{dt^2} + \frac{dz}{dt}\frac{d^2z}{dt^2}}{\sqrt{\left(\frac{dx}{dt}\right)^2 + \left(\frac{dy}{dt}\right)^2 + \left(\frac{dz}{dt}\right)^2}} \tag{4}$$

$$= \frac{dx}{ds}\frac{d^2x}{dt^2} + \frac{dy}{ds}\frac{d^2y}{dt^2} + \frac{dz}{ds}\frac{d^2z}{dt^2} \tag{5}$$

$$= \frac{d\mathbf{r}}{ds} \cdot \frac{d^2\mathbf{r}}{dt^2}. \tag{6}$$

The VECTOR acceleration is given by

$$\mathbf{a} \equiv \frac{d\mathbf{v}}{dt} = \frac{d^2\mathbf{r}}{dt^2} = \frac{d^2s}{dt^2}\hat{\mathbf{T}} + \kappa\left(\frac{ds}{dt}\right)^2 \hat{\mathbf{N}}, \tag{7}$$

where $\hat{\mathbf{T}}$ is the UNIT TANGENT VECTOR, κ the CURVATURE, s the ARC LENGTH, and $\hat{\mathbf{N}}$ the UNIT NORMAL VECTOR.

Let a particle move along a straight LINE so that the positions at times t_1, t_2, and t_3 are s_1, s_2, and s_3, respectively. Then the particle is uniformly accelerated with acceleration a IFF

$$a \equiv 2\left[\frac{(s_2 - s_3)t_1 + (s_3 - s_1)t_2 + (s_1 - s_2)t_3}{(t_1 - t_2)(t_2 - t_3)(t_3 - t_1)}\right] \tag{8}$$

is a constant (Klamkin 1995, 1996).

Consider the measurement of acceleration in a rotating reference frame. Apply the ROTATION OPERATOR

$$\tilde{R} \equiv \left(\frac{d}{dt}\right)_{body} + \omega \times \tag{9}$$

twice to the RADIUS VECTOR $\mathbf{r}$ and suppress the *body* notation,

$$\mathbf{a}_{space} = \tilde{R}^2\mathbf{r} = \left(\frac{d}{dt} + \omega\times\right)^2 \mathbf{r}$$

$$= \left(\frac{d}{dt} + \omega\times\right)\left(\frac{d\mathbf{r}}{dt} + \omega \times \mathbf{r}\right)$$

$$= \frac{d^2\mathbf{r}}{dt^2} + \frac{d}{dt}(\omega \times \mathbf{r}) + \omega \times \frac{d\mathbf{r}}{dt} + \omega \times (\omega \times \mathbf{r})$$

$$= \frac{d^2\mathbf{r}}{dt^2} + \omega \times \frac{d\mathbf{r}}{dt} + \mathbf{r} \times \frac{d\omega}{dt} + \omega \times \frac{d\mathbf{r}}{dt}$$

$$+ \omega \times (\omega \times \mathbf{r}). \tag{10}$$

Grouping terms and using the definitions of the VELOCITY $\mathbf{v} \equiv d\mathbf{r}/dt$ and ANGULAR VELOCITY $\boldsymbol{\alpha} \equiv d\omega/dt$ give the expression

$$\mathbf{a}_{\text{space}} = \frac{d^2\mathbf{r}}{dt^2} + 2\omega \times \mathbf{v} + \omega \times (\omega \times \mathbf{r}) + \mathbf{r} \times \boldsymbol{\alpha}. \quad (11)$$

Now, we can identify the expression as consisting of three terms

$$\mathbf{a}_{\text{body}} \equiv \frac{d^2\mathbf{r}}{dt^2}, \quad (12)$$

$$\mathbf{a}_{\text{Coriolis}} \equiv 2\omega \times \mathbf{v}, \quad (13)$$

$$\mathbf{a}_{\text{centrifugal}} \equiv \omega \times (\omega \times \mathbf{r}), \quad (14)$$

a "body" acceleration, centrifugal acceleration, and Coriolis acceleration. Using these definitions finally gives

$$\mathbf{a}_{\text{space}} = \mathbf{a}_{\text{body}} + \mathbf{a}_{\text{Coriolis}} + \mathbf{a}_{\text{centrifugal}} + \mathbf{r} \times \boldsymbol{\alpha}, \quad (15)$$

where the fourth term will vanish in a uniformly rotating frame of reference (i.e., $\alpha = 0$). The centrifugal acceleration is familiar to riders of merry-go-rounds, and the Coriolis acceleration is responsible for the motions of hurricanes on Earth and necessitates large trajectory corrections for intercontinental ballistic missiles.

See also ANGULAR ACCELERATION, ARC LENGTH, JERK, VELOCITY

References
Klamkin, M. S. "Problem 1481." *Math. Mag.* **68**, 307, 1995.
Klamkin, M. S. "A Characteristic of Constant Acceleration." Solution to Problem 1481. *Math. Mag.* **69**, 308, 1996.

Accidental Cancellation
ANOMALOUS CANCELLATION

Accretion
CUMULATION

Accumulation Point
An accumulation point is a POINT which is the limit of a SEQUENCE, also called a LIMIT POINT. For some MAPS, periodic orbits give way to CHAOTIC ones beyond a point known as the accumulation point.

See also BOLZANO-WEIERSTRASS THEOREMBolzano-Weierstrass Theorem, CANTOR'S INTERSECTION THEOREM, CHAOS, FRACTIONAL PART, HEINE-BOREL THEOREM, LIMIT POINT, LOGISTIC MAP, MODE LOCKING, PERIOD DOUBLING, PISOT-VIJAYARAGHAVAN CONSTANT

Achilles and the Tortoise Paradox
ZENO'S PARADOXES

Achiral
AMPHICHIRAL

Ackermann Function
The Ackermann function is the simplest example of a WELL DEFINED TOTAL FUNCTION which is COMPUTABLE but not PRIMITIVE RECURSIVE, providing a counterexample to the belief in the early 1900s that every COMPUTABLE FUNCTION was also PRIMITIVE RECURSIVE (Dötzel 1991). It grows faster than an exponential function, or even a multiple exponential function. The Ackermann function $A(x, y)$ is defined by

$$A(x, y) \equiv \begin{cases} y+1 & \text{if } x=0 \\ A(x-1, 1) & \text{if } y=0 \\ A(x-1, A(x, y-1)) & \text{otherwise.} \end{cases} \quad (1)$$

Special values for INTEGER x include

$$A(0, y) = y+1 \quad (2)$$

$$A(1, y) = y+2 \quad (3)$$

$$A(2, y) = 2y+3 \quad (4)$$

$$A(3, y) = 2^{y+3} - 3 \quad (5)$$

$$A(4, y) = \underbrace{2^{2^{\cdot^{\cdot^2}}}}_{y+3} - 3. \quad (6)$$

Expressions of the latter form are sometimes called POWER TOWERS. $A(0, y)$ follows trivially from the definition. $A(1, y)$ can be derived as follows,

$$A(1, y) = A(0, A(1, y-1)) = A(1, y-1) + 1$$
$$= A(0, A(1, y-2)) + 1 = A(1, y-2) + 2$$
$$= \ldots = A(1, 0) + y = A(0, 1) + y = y + 2. \quad (7)$$

$A(2, y)$ has a similar derivation,

$$A(2, y) = A(1, A(2, y-1)) = A(2, y-1) + 2$$
$$= A(1, A(2, y-2)) + 2 = A(2, y-2) + 4 = \ldots$$
$$= A(2, 0) + 2y = A(1, 1) + 2y = 2y + 3. \quad (8)$$

Buck (1963) defines a related function using the same fundamental RECURRENCE RELATION (with arguments flipped from Buck's convention)

$$F(x, y) = F(x-1, F(x, y-1)), \quad (9)$$

but with the slightly different boundary values

$$F(0, y) = y+1 \quad (10)$$

$$F(1, 0) = 2 \quad (11)$$

$$F(2, 0) = 2 \quad (12)$$

$$F(x, 0) = 1 \quad \text{for } x = 3, 4, \ldots. \quad (13)$$

Buck's recurrence gives

$$F(1, y) = 2 + y \quad (14)$$

$$F(2, y) = 2y \quad (15)$$

$$F(3, y) = 2^y \quad (16)$$

$$F(4, \ y) = \underbrace{2^{2^{\cdot^{\cdot^{\cdot^2}}}}}_{y}. \tag{17}$$

Taking $F(4, \ n)$ gives the sequence 1, 2, 4, 16, 65536, 2^{65536}, ... (Sloane's A006263). Defining $ah(x) = F(x, \ x)$ for $x = 0, \ 1, \ ...$ then gives 1, 3, 4, 8, 65536, $\underbrace{2^{2^{\cdot^{\cdot^2}}}}_{m}$, ...

(Sloane's A001695), where $m = \underbrace{2^{\cdot^{\cdot^{\cdot^2}}}}_{65536}$, a truly huge number!

See also ACKERMANN NUMBER, COMPUTABLE FUNC-
TION, GOODSTEIN SEQUENCE, POWER TOWER, PRIMI-
TIVE RECURSIVE FUNCTION, TAK FUNCTION, TOTAL
FUNCTION

References

Buck, R. C. "Mathematical Induction and Recursive Defini-
tions." *Amer. Math. Monthly* **70**, 128–35, 1963.
Dötzel, G. "A Function to End All Functions." *Algorithm:
Recreational Programming* **2.4**, 16–7, 1991.
Kleene, S. C. *Introduction to Metamathematics.* New York:
Elsevier, 1971.
Péter, R. *Rekursive Funktionen.* Budapest: Akad. Kiado,
1951.
Reingold, E. H. and Shen, X. "More Nearly Optimal Algo-
rithms for Unbounded Searching, Part I: The Finite Case."
SIAM J. Comput. **20**, 156–83, 1991.
Rose, H. E. *Subrecursion, Functions, and Hierarchies.* New
York: Clarendon Press, 1988.
Sloane, N. J. A. Sequences A001695/M2352 and A006263/
M1310 in "An On-Line Version of the Encyclopedia of
Integer Sequences." http://www.research.att.com/~njas/
sequences/eisonline.html.
Smith, H. J. "Ackermann's Function." http://pweb.netcom.-
com/~hjsmith/Ackerman.html.
Spencer, J. "Large Numbers and Unprovable Theorems."
Amer. Math. Monthly **90**, 669–75, 1983.
Tarjan, R. E. *Data Structures and Network Algorithms.*
Philadelphia PA: SIAM, 1983.
Vardi, I. *Computational Recreations in Mathematica.* Red-
wood City, CA: Addison-Wesley, pp. 11, 227, and 232,
1991.

Ackermann Number

A number OF THE FORM $\underbrace{n \uparrow \cdots \uparrow n}_{n}$, where ARROW

NOTATION has been used. The first few Ackermann numbers are $1 \uparrow 1 = 1$, $2 \uparrow\uparrow 2 = 4$, and $3 \uparrow\uparrow\uparrow 3 = \underbrace{3^{3^{\cdot^{\cdot^3}}}}_{7,625,507,484,987}$

See also ACKERMANN FUNCTION, ARROW NOTATION,
POWER TOWER

References

Ackermann, W. "Zum hilbertschen Aufbau der reellen
Zahlen." *Math. Ann.* **99**, 118–33, 1928.
Conway, J. H. and Guy, R. K. *The Book of Numbers.* New
York: Springer-Verlag, pp. 60–1, 1996.
Crandall, R. E. "The Challenge of Large Numbers." *Sci.
Amer.* **276**, 74–9, Feb. 1997.
Vardi, I. *Computational Recreations in Mathematica.* Red-
wood City, CA: Addison-Wesley, pp. 11, 227, and 232,
1991.

Acnode

Another name for an ISOLATED POINT.

See also CRUNODE, SPINODE, TACNODE

Acoptic Polyhedron

A term invented by B. Grünbaum in an attempt to promote concrete and precise POLYHEDRON terminol-ogy. The word "coptic" derives from the Greek for "to cut," and acoptic polyhedra are defined as POLYHEDRA for which the FACES do not intersect (cut) themselves, making them 2-MANIFOLDS.

See also HONEYCOMB, NOLID, POLYHEDRON, SPONGE

Action

Let $M(X)$ denote the GROUP of all invertible MAPS $X \to X$ and let G be any GROUP. A HOMOMORPHISM $\theta : G \to M(X)$ is called an action of G on X. Therefore, θ satisfies

1. For each $g \in G$, $\theta(g)$ is a MAP $X \to X : x \mapsto \theta(g)x$,
2. $\theta(gh)x = \theta(g)(\theta(h)x)$,
3. $\theta(e)x = x$, where e is the group identity in G,
4. $\theta(g^{-1})x = \theta(g)^{-1}x$.

See also CASCADE, FLOW, SEMIDIRECT PRODUCT,
SEMIFLOW

Actuarial Polynomial

The polynomials $a_n^{(\beta)}(x)$ given by the SHEFFER SE-
QUENCE with

$$g(t) = (1 - t)^{-\beta} \tag{1}$$

$$f(t) = \ln(1 - t), \tag{2}$$

giving GENERATING FUNCTION

$$\sum_{k=0}^{\infty} \frac{a_n^{(\beta)}}{k!} t^k = e^{x(1-e^t) + \beta t}. \tag{3}$$

The Sheffer identity is

$$a_n^{(\beta)}(x + y) = \sum_{k=0}^{n} \binom{n}{k} a_k^{(\beta)}(y) \phi_{n-k}(-x), \tag{4}$$

where $\phi_n(x)$ is an EXPONENTIAL POLYNOMIAL. The actuarial polynomials are given in terms of the EXPONENTIAL POLYNOMIALS $\phi_n(x)$ by

$$a_n^{(\beta)}(x) = (1 - t)^{\beta} \phi_n(-x) \tag{5}$$

$$= \sum_{k=0}^{n} \binom{\beta}{k} \phi_n^{(k)}(-x). \tag{6}$$

They are related to the STIRLING NUMBERS OF THE SECOND KIND $S(n, \ m)$ by

$$a_n^{(\beta)}(x) = \sum_{k=0}^{n} \binom{\beta}{k} \sum_{j=k}^{n} S(n, \ j)(j)_k (-x)^{j-k}, \tag{7}$$

where $\binom{n}{k}$ is a BINOMIAL COEFFICIENT and $(x)_n$ is a FALLING FACTORIAL. The actuarial polynomials also satisfy the identity

$$a_n^{(\beta)}(-x) = e^{-x} \sum_{k=0}^{\infty} \frac{(k+\beta)^n}{k!} x^k \qquad (8)$$

(Roman 1984, p. 125; Whittaker and Watson 1990, p. 336).

The first few polynomials are

$$a_0^{(\beta)}(x) = 1$$

$$a_1^{(\beta)}(x) = -x + \beta$$

$$a_2^{(\beta)}(x) = x^2 - x(1 + 2\beta) + \beta^2$$

$$a_3^{(\beta)}(x) = -x^3 + 3x^2(\beta + 1) - x(3\beta^2 + 3\beta + 1) + \beta^3.$$

See also SHEFFER SEQUENCE

References

Boas, R. P. and Buck, R. C. *Polynomial Expansions of Analytic Functions, 2nd print., corr.* New York: Academic Press, p. 42, 1964.
Erdélyi, A.; Magnus, W.; Oberhettinger, F.; and Tricomi, F. G. *Higher Transcendental Functions, Vol. 3.* New York: Krieger, p. 254, 1981.
Roman, S. "The Actuarial Polynomial." §4.3.4 in *The Umbral Calculus.* New York: Academic Press, pp. 123–25, 1984.
Whittaker, E. T. and Watson, G. N. *A Course in Modern Analysis, 4th ed.* Cambridge, England: Cambridge University Press, 1990.

Acute Angle

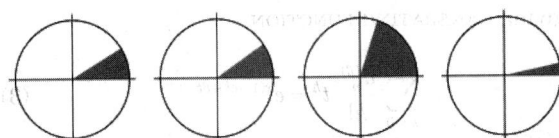

An ANGLE of less than $\pi/2$ RADIANS (90°) is called an acute angle.

See also ACUTE TRIANGLE, ANGLE, FULL ANGLE, OBTUSE ANGLE, REFLEX ANGLE, RIGHT ANGLE, STRAIGHT ANGLE

Acute Triangle

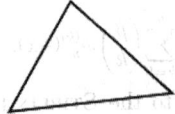

A TRIANGLE in which all three ANGLES are ACUTE ANGLES. A TRIANGLE which is neither acute nor a RIGHT TRIANGLE (i.e., it has an OBTUSE ANGLE) is called an OBTUSE TRIANGLE. From the LAW OF CO-

SINES, for a triangle with side lengths a, b, and c,

$$\cos C = \frac{a^2 + b^2 - c^2}{2ab},$$

with C the angle opposite side C. For an angle to be acute, $\cos C > 0$. Therefore, an acute triangle satisfies $a^2 + b^2 > c^2$, $b^2 + c^2 > a^2$, and $c^2 + a^2 > b^2$.

The smallest number of acute triangles into which an arbitrary OBTUSE TRIANGLE can be dissected is seven if $B > 90°$, $B - A$, $B - C < 90°$, and otherwise eight (Manheimer 1960, Gardner 1981, Wells 1991). A SQUARE can be dissected into as few as 9 acute triangles (Gardner 1981, Wells 1991).

See also OBTUSE TRIANGLE, ONO INEQUALITY, RIGHT TRIANGLE

References

Gardner, M. "Mathematical Games: A Fifth Collection of 'Brain-Teasers.'" *Sci. Amer.* **202**, 150–54, Feb. 1960.
Gardner, M. "Mathematical Games: The Games and Puzzles of Lewis Carroll and the Answers to February's Problems." *Sci. Amer.* **202**, 172–82, Mar. 1960.
Gardner, M. "Mathematical Games: The Inspired Geometrical Symmetries of Scott Kim." *Sci. Amer.* **244**, 22–1, Jun. 1981.
Goldberg, G. "Problem E1406." *Amer. Math. Monthly* **67**, 923, 1960.
Hoggatt, V. E. Jr. "Acute Isosceles Dissection of an Obtuse Triangle." *Amer. Math. Monthly* **68**, 912–13, 1961.
Johnson, R. S. "Problem 256 [1977: 155]." *Crux Math.* **4**, 53–4, 1978.
Nelson, H. L. "Solution to Problem 256." *Crux Math.* **4**, 102–04, 1978.
Wells, D. *The Penguin Dictionary of Curious and Interesting Geometry.* London: Penguin, pp. 1–, 1991.

Acyclic Digraph

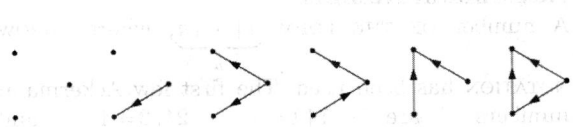

An acyclic digraph is a DIRECTED GRAPH containing no directed cycles, also known as a directed acyclic graph or a "DAG." Every acyclic digraph has at least one node of OUTDEGREE 0. The numbers of acyclic digraphs on $n = 1$, 2, ... vertices are 1, 2, 6, 31, 302, 5984, ... (Sloane's A003087).

See also DIRECTED GRAPH, FOREST

References

Harary, F. *Graph Theory.* Reading, MA: Addison-Wesley, p. 200, 1994.
Robinson, R. W. "Counting Unlabeled Acyclic Digraphs." In *Combinatorial Mathematics V (Melbourne 1976).* Providence, RI: Amer. Math. Soc., pp. 28–3, 1976.

Skiena, S. *Implementing Discrete Mathematics: Combinatorics and Graph Theory with Mathematica.* Reading, MA: Addison-Wesley, p. 190, 1990.

Sloane, N. J. A. Sequences A003087/M1696 in "An On-Line Version of the Encyclopedia of Integer Sequences." http://www.research.att.com/~njas/sequences/eisonline.html.

Acyclic Graph

FOREST

Ad

ADJOINT REPRESENTATION, ADJOINT REPRESENTATION (LIE GROUP)

Adams' Circle

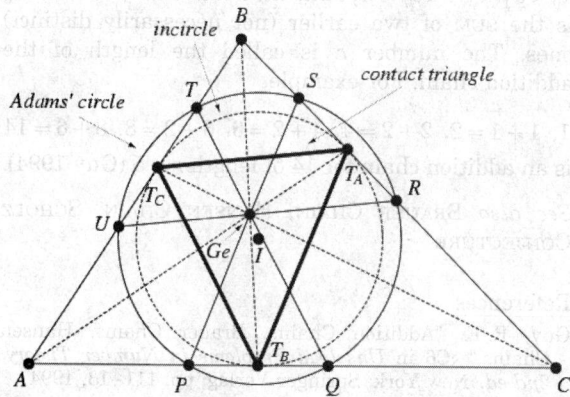

Given a TRIANGLE ΔABC, construct the CONTACT TRIANGLE $\Delta T_A T_B T_C$. Now extend lines parallel to the sides of the CONTACT TRIANGLE from the GERGONNE POINT. These intersect the triangle ΔABC in the six points P, Q, R, S, T, and U. As C. Adams proved in 1843, these points are CONCYCLIC in a CIRCLE now known as Adams' circle. Moreover, Adams' circle is concentric with the INCIRCLE of ΔABC (Honsberger 1995, pp. 62–4).

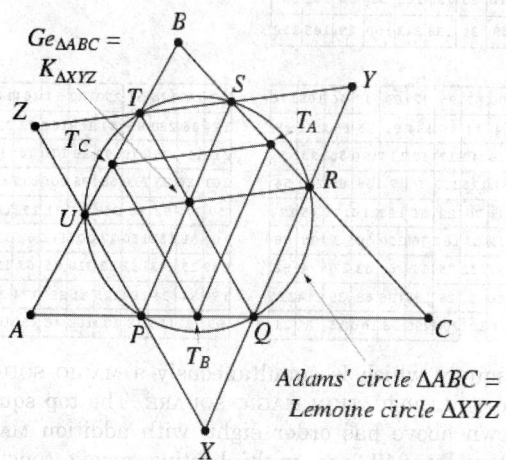

Adams' circle ΔABC = Lemoine circle ΔXYZ

Extend the segments UP, TS, and RQ to form a TRIANGLE ΔXYZ. Then the GERGONNE POINT of ΔABC is the SYMMEDIAN POINT of ΔXYZ, and Adams' circle of

ΔABC is the LEMOINE CIRCLE of ΔXYZ (Honsberger 1995, p. 98).

See also CONTACT TRIANGLE, GERGONNE POINT

References

Honsberger, R. "A Real Gem." §7.4 (v) in *Episodes in Nineteenth and Twentieth Century Euclidean Geometry.* Washington, DC: Math. Assoc. Amer., pp. 62–4 and 98, 1995.

Adams' Method

Adams' method is a numerical METHOD for solving linear FIRST-ORDER ORDINARY DIFFERENTIAL EQUATIONS OF THE FORM

$$\frac{dy}{dx} = f(x, y). \tag{1}$$

Let

$$h = x_{n+1} - x_n \tag{2}$$

be the step interval, and consider the MACLAURIN SERIES of y about x_n,

$$y_{n+1} = y_n + \left(\frac{dy}{dx}\right)_n (x - x_n) + \frac{1}{2}\left(\frac{d^2 y}{dx^2}\right)_n (x - x_n)^2 + \dots \tag{3}$$

$$\left(\frac{dy}{dx}\right)_{n+1} = \left(\frac{dy}{dx}\right)_n + \left(\frac{d^2 y}{dx^2}\right)_n (x - x_n)^2 + \dots. \tag{4}$$

Here, the DERIVATIVES of y are given by the BACKWARD DIFFERENCES

$$q_n \equiv \left(\frac{dy}{dx}\right)_n = \frac{\Delta y_n}{x_{n+1} - x_n} = \frac{y_{n+1} - y_n}{h} \tag{5}$$

$$\nabla q_n \equiv \left(\frac{d^2 y}{dx^2}\right)_n = q_n - q_{n-1} \tag{6}$$

$$\nabla^2 q_n \equiv \left(\frac{d^3 y}{dx^3}\right)_n = \nabla q_n - \nabla q_{n-1}, \tag{7}$$

etc. Note that by (1), q_n is just the value of $f(x_n, y_n)$. For first-order interpolation, the method proceeds by iterating the expression

$$y_{n+1} = y_n + q_n h \tag{8}$$

where $q_n \equiv f(x_n, y_n)$. The method can then be extended to arbitrary order using the finite difference integration formula from Beyer (1987)

$$\int_0^1 f_p\, dp =$$

$$\left(1 + \frac{1}{2}\nabla + \frac{5}{12}\nabla^2 + \frac{3}{8}\nabla^3 + \frac{251}{720}\nabla^4 + \frac{95}{288}\nabla^5 + \frac{19087}{60480}\nabla^6 + \dots\right) f_p \tag{9}$$

to obtain

$$y_{n+1} - y_n = h(q_n + \tfrac{1}{2}\nabla q_{n-1} + \tfrac{5}{12}\nabla^2 q_{n-2} + \tfrac{3}{8}\nabla^3 q_{n-3}$$
$$+ \tfrac{251}{720}\nabla^4 q_{n-4} + \tfrac{95}{288}\nabla^5 q_{n-5} + \ldots). \qquad (10)$$

Note that von Kármán and Biot (1940) confusingly use the symbol normally used for FORWARD DIFFERENCES δ to denote BACKWARD DIFFERENCES ∇.

See also GILL'S METHOD, MILNE'S METHOD, PREDICTOR-CORRECTOR METHODS, RUNGE-KUTTA METHOD

References

Abramowitz, M. and Stegun, C. A. (Eds.). *Handbook of Mathematical Functions with Formulas, Graphs, and Mathematical Tables, 9th printing.* New York: Dover, p. 896, 1972.

Bashforth, F. and Adams, J. C. *Theories of Capillary Action.* London: Cambridge University Press, 1883.

Beyer, W. H. *CRC Standard Mathematical Tables, 28th ed.* Boca Raton, FL: CRC Press, p. 455, 1987.

Jeffreys, H. and Jeffreys, B. S. "The Adams-Bashforth Method." §9.11 in *Methods of Mathematical Physics, 3rd ed.* Cambridge, England: Cambridge University Press, pp. 292–93, 1988.

Kármán, T. von and Biot, M. A. *Mathematical Methods in Engineering: An Introduction to the Mathematical Treatment of Engineering Problems.* New York: McGraw-Hill, pp. 14–0, 1940.

Press, W. H.; Flannery, B. P.; Teukolsky, S. A.; and Vetterling, W. T. *Numerical Recipes in FORTRAN: The Art of Scientific Computing, 2nd ed.* Cambridge, England: Cambridge University Press, p. 741, 1992.

Whittaker, E. T. and Robinson, G. "The Numerical Solution of Differential Equations." Ch. 14 in *The Calculus of Observations: A Treatise on Numerical Mathematics, 4th ed.* New York: Dover, pp. 363–67, 1967.

Adams-Bashforth-Moulton Method

ADAMS' METHOD

Addend

A quantity to be ADDED to another, also called a SUMMAND. For example, in the expression $a + b + c$, a, b, and c are all addends. The first of several addends, or "the one to which the others are added" (a in the previous example), is sometimes called the AUGEND.

See also ADDITION, AUGEND, PLUS, RADICAND

Addition

```
      1 1   ◀━ carries
    1 5 8   ◀━ addend 1
  + 2 4 9   ◀━ addend 2
  ───────
    4 0 7   ◀━ sum
```

The combining of two or more quantities using the PLUS operator. The individual numbers being combined are called ADDENDS, and the total is called the SUM. The first of several ADDENDS, or "the one to which the others are added," is sometimes called the AUGEND. The opposite of addition is SUBTRACTION. While the usual form of adding two n-digit INTEGERS (which consists of summing over the columns right to left and "CARRYING" a 1 to the next column if the sum exceeds 9) requires n operations (plus carries), two n-digit INTEGERS can be added in about $2\lg n$ steps by n processors using carry-lookahead addition (McGeoch 1993). Here, $\lg x$ is the LG function, the LOGARITHM to the base 2.

See also ADDEND, AMENABLE NUMBER, AUGEND, CARRY, DIFFERENCE, DIVISION, MULTIPLICATION, PLUS, SUBTRACTION, SUM

References

McGeoch, C. C. "Parallel Addition." *Amer. Math. Monthly* **100**, 867–71, 1993.

Addition Chain

An addition chain for a number n is a SEQUENCE $1 = a_0 < a_1 < \ldots < a_r = n$, such that each member after a_0 is the SUM of two earlier (not necessarily distinct) ones. The number r is called the length of the addition chain. For example,

$$1,\ 1+1=2,\ 2+2=4,\ 4+2=6,\ 6+2=8,\ 8+6=14$$

is an addition chain for 14 of length $r = 5$ (Guy 1994).

See also BRAUER CHAIN, HANSEN CHAIN, SCHOLZ CONJECTURE

References

Guy, R. K. "Addition Chains. Brauer Chains. Hansen Chains." §C6 in *Unsolved Problems in Number Theory, 2nd ed.* New York: Springer-Verlag, pp. 111–13, 1994.

Addition-Multiplication Magic Square

46	81	117	102	15	76	200	203
19	60	232	175	54	69	153	78
216	161	17	52	171	90	58	75
135	114	50	87	184	189	13	68
150	261	45	38	91	136	92	27
119	104	108	23	174	225	57	30
116	25	133	120	51	26	162	207
39	34	138	243	100	29	105	152

200	87	95	42	99	1	46	108	170
14	44	10	184	81	85	150	261	19
138	243	17	50	116	190	56	33	5
57	125	232	9	7	66	68	230	54
4	70	22	51	115	216	171	25	174
153	23	162	76	250	58	3	35	88
145	152	75	11	6	63	270	34	92
110	2	28	135	136	69	29	114	225
27	102	207	290	38	100	55	8	21

17	171	126	54	230	100	93	264	145
124	66	290	85	57	168	162	23	225
216	115	75	279	198	29	170	76	42
261	186	33	210	68	38	200	135	69
50	270	92	87	248	165	21	153	114
105	51	152	150	27	207	116	62	330
138	25	243	132	58	310	95	63	136
190	84	34	184	125	81	297	174	31
99	232	155	19	189	102	46	250	108

A square which is simultaneously a MAGIC SQUARE and MULTIPLICATION MAGIC SQUARE. The top square shown above has order eight, with addition MAGIC CONSTANT 840 and multiplicative magic constant 2,058,068,231,856,000 (Horner 1955, Hunter and Madachy 1975). The bottom two squares have order nine with addition MAGIC CONSTANTS 848 and

1200 and multiplicative magic constants 5,804,807,833,440,000 and 1,619,541,385,529,760, 000, respectively (Hunter and Madachy 1975, Madachy 1979).

1	−3	2
−4	12	−8
3	−9	6

L. Sallows has constructed an interesting 3×3 magic square in which the products of corresponding pairs of 2×2 diagonals are 12, 24, 36, and 72, while the products of the numbers in the pair of 3×3 diagonals also give 72.

See also MAGIC SQUARE

References

Horner, W. W. "Addition-Multiplication Magic Square of Order 8." *Scripta Math.* **21**, 23–7, 1955.
Hunter, J. A. H. and Madachy, J. S. "Mystic Arrays." Ch. 3 in *Mathematical Diversions.* New York: Dover, pp. 30–1, 1975.
Madachy, J. S. *Madachy's Mathematical Recreations.* New York: Dover, pp. 89–1, 1979.

Additive Number Theory

The portion of NUMBER THEORY concerned with expressing an integer as a sum of integers from some given set.

See also CIRCLE METHOD, MULTIPLICATIVE NUMBER THEORY, NUMBER THEORY

Additive Persistence

Consider the process of taking a number, adding its DIGITS, then adding the DIGITS of the number derived from it, etc., until the remaining number has only one DIGIT. The number of additions required to obtain a single DIGIT from a number n is called the additive persistence of n, and the DIGIT obtained is called the DIGITAL ROOT of n.

For example, the sequence obtained from the starting number 9876 is (9876, 30, 3), so 9876 has an additive persistence of 2 and a DIGITAL ROOT of 3. The additive persistences of the first few positive integers are 0, 0, 0, 0, 0, 0, 0, 0, 0, 1, 1, 1, 1, 1, 1, 1, 1, 1, 2, 1, ... (Sloane's A031286). The smallest numbers of additive persistence n for $n = 0$, 1, ... are 0, 10, 19, 199, 19999999999999999999999, ... (Sloane's A006050).

See also ADDITIVE PERSISTENCE, DIGITADDITION, DIGITAL ROOT, MULTIPLICATIVE PERSISTENCE, NARCISSISTIC NUMBER, RECURRING DIGITAL INVARIANT

References

Hinden, H. J. "The Additive Persistence of a Number." *J. Recr. Math.* **7**, 134–35, 1974.
Sloane, N. J. A. Sequences A006050/M4683 and A031286 in "An On-Line Version of the Encyclopedia of Integer Sequences." http://www.research.att.com/~njas/sequences/eisonline.html.
Sloane, N. J. A. "The Persistence of a Number." *J. Recr. Math.* **6**, 97–8, 1973.
Weisstein, E. W. "Integer Sequences." MATHEMATICA NOTEBOOK INTEGERSEQUENCES.M.

Adéle

An element of an ADÉLE GROUP, sometimes called a REPARTITION in older literature (e.g., Chevalley 1951, p. 25). Adéles arise in both NUMBER FIELDS and FUNCTION FIELDS. The adéles of a NUMBER FIELD are the additive SUBGROUPS of all elements in $\prod k_v$, where v is the PLACE, whose ABSOLUTE VALUE is < 1 at all but finitely many vs.

Let F be a FUNCTION FIELD of algebraic functions of one variable. Then a MAP r which assigns to every PLACE P of F an element $r(P)$ of F such that there are only a finite number of PLACES P for which $vp(r(P)) < 0$ is called an adéle (Chevalley 1951, p. 1951).

See also FUNCTION FIELD, IDELE

References

Chevalley, C. C. *Introduction to the Theory of Algebraic Functions of One Variable.* Providence, RI: Amer. Math. Soc., p. 25, 1951.
Knapp, A. W. "Group Representations and Harmonic Analysis, Part II." *Not. Amer. Math. Soc.* **43**, 537–49, 1996.

Adéle Group

The restricted topological GROUP DIRECT PRODUCT of the GROUP G_{k_v} with distinct invariant open subgroups G_{0_v}.

References

Weil, A. *Adéles and Algebraic Groups.* Princeton, NJ: Princeton University Press, 1961.

Adem Relations

Relations in the definition of a STEENROD ALGEBRA which state that, for $i < 2j$,

$$Sq^i \circ Sq^j(x) = \sum_{k=0}^{\lfloor i \rfloor} \binom{j-k-1}{i-2k} Sq^{i+j-k} \circ Sq^k(x),$$

where $f \circ g$ denotes function COMPOSITION and $\lfloor i \rfloor$ is the FLOOR FUNCTION.

See also STEENROD ALGEBRA

Adequate Knot

A class of KNOTS containing the class of ALTERNATING KNOTS. Let $c(K)$ be the CROSSING NUMBER. Then for KNOT SUM $K_1 \# K_2$ which is an adequate knot,

$$c(K_1 \# K_2) = c(K_1) + c(K_2).$$

This relationship is postulated to hold true for all KNOTS.

See also ALTERNATING KNOT, CROSSING NUMBER (LINK)

Adiabatic Invariant

A property of motion which is conserved to exponential accuracy in the small parameter representing the typical rate of change of the gross properties of the body.

See also ALGEBRAIC INVARIANT, LYAPUNOV CHARACTERISTIC NUMBER

Adjacency List

The adjacency list representation of a GRAPH consists of n lists one for each vertex v_i, $1 \le i \le n$, which gives the vertices to which v_i is adjacent. The adjacency lists of a graph g may be computed using `ToAdjacencyLists[g]` in the *Mathematica* add-on package `DiscreteMath`Combinatorica`` (which can be loaded with the command `<<DiscreteMath`). A graph may be constructed from adjacency lists using `FromAdjacencyLists[e]`.

See also ADJACENCY MATRIX

References

Skiena, S. "Adjacency Lists." §3.1.2 in *Implementing Discrete Mathematics: Combinatorics and Graph Theory with Mathematica*. Reading, MA: Addison-Wesley, pp. 86–7, 1990.

Adjacency Matrix

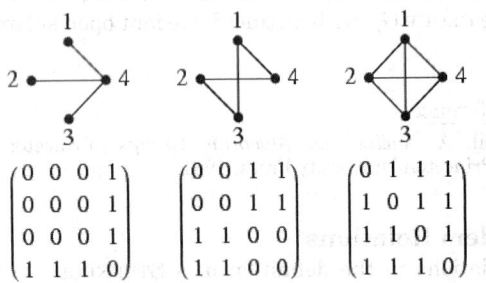

$$\begin{pmatrix} 0 & 0 & 0 & 1 \\ 0 & 0 & 0 & 1 \\ 0 & 0 & 0 & 1 \\ 1 & 1 & 1 & 0 \end{pmatrix} \quad \begin{pmatrix} 0 & 0 & 1 & 1 \\ 0 & 0 & 1 & 1 \\ 1 & 1 & 0 & 0 \\ 1 & 1 & 0 & 0 \end{pmatrix} \quad \begin{pmatrix} 0 & 1 & 1 & 1 \\ 1 & 0 & 1 & 1 \\ 1 & 1 & 0 & 1 \\ 1 & 1 & 1 & 0 \end{pmatrix}$$

The adjacency matrix of a simple GRAPH is a MATRIX with rows and columns labeled by VERTICES, with a 1 or 0 in position (v_i, v_j) according to whether v_i and v_j are ADJACENT or not. For a simple graph with no self-loops, the adjacency matrix must have 0s on the diagonal. For an undirected graph, the adjacency matrix is symmetrical. The adjacency matrix of a graph can be computed using `Edges[g]` in the *Mathematica* add-on package `DiscreteMath`Combinatorica`` (which can be loaded with the command `<<DiscreteMath`).

See also ADJACENCY LIST, INCIDENCE MATRIX, INTEGER MATRIX

References

Chartrand, G. *Introductory Graph Theory*. New York: Dover, p. 218, 1985.
Skiena, S. "Adjacency Matrices." §3.1.1 in *Implementing Discrete Mathematics: Combinatorics and Graph Theory with Mathematica*. Reading, MA: Addison-Wesley, pp. 81–5, 1990.

Adjacency Relation

The SET E of EDGES of a GRAPH (V, E), being a set of unordered pairs of elements of V, constitutes a RELATION on V. Formally, an adjacency relation is any RELATION which is IRREFLEXIVE and SYMMETRIC.

See also IRREFLEXIVE, RELATION, SYMMETRIC

Adjacent Fraction

Two FRACTIONS are said to be adjacent if their difference has a unit NUMERATOR. For example, 1/3 and 1/4 are adjacent since $1/3 - 1/4 = 1/12$, but 1/2 and 1/5 are not since $1/2 - 1/5 = 3/10$. Adjacent fractions can be adjacent in a FAREY SEQUENCE.

See also FAREY SEQUENCE, FORD CIRCLE, FRACTION, NUMERATOR

References

Pickover, C. A. *Keys to Infinity*. New York: Wiley, p. 119, 1995.

Adjacent Value

The value nearest to but still inside an inner FENCE.

References

Tukey, J. W. *Explanatory Data Analysis*. Reading, MA: Addison-Wesley, p. 667, 1977.

Adjacent Vertices

In a GRAPH G, two VERTICES are adjacent if they are joined by an EDGE.

See also EDGE (GRAPH), GRAPH, VERTEX (GRAPH)

Adjoint

Given a SECOND-ORDER ORDINARY DIFFERENTIAL EQUATION

$$\mathcal{L}u(x) \equiv p_0 \frac{d^2 u}{dx^2} + p_1 \frac{du}{dx} + p_2 u, \tag{1}$$

where $p_i \equiv p_i(x)$ and $u \equiv u(x)$, the adjoint operator $\tilde{\mathcal{L}}^*$ is defined by

$$\tilde{\mathcal{L}}^* u \equiv \frac{d}{dx^2}(p_0 u) - \frac{d}{dx}(p_1 u) + p_2 u$$

$$= p_0 \frac{d^2 u}{dx^2} + (2p_0' - p_1)\frac{du}{dx} + (p_0'' - p_1' + p_2)u. \tag{2}$$

Write the two LINEARLY INDEPENDENT solutions as $y_1(x)$ and $y_2(x)$. Then the adjoint operator can also be written

$$\tilde{\mathscr{L}}^*u = \int(y_2\tilde{\mathscr{L}}y_1 - y_1\tilde{\mathscr{L}}y_2)dx = \left[\frac{p_1}{p_0}(y_2'y_2 - y_1y_2')\right]. \quad (3)$$

In general, given two adjoint operators $\tilde{A}$ and $\tilde{B}$,

$$(\tilde{A}\tilde{B})^* = \tilde{B}^*\tilde{A}^*, \quad (4)$$

which can be generalized to

$$(\tilde{A}\tilde{B}\cdots\tilde{Z})^* = \tilde{Z}^*\cdots\tilde{B}^*\tilde{A}^*. \quad (5)$$

Note that many older physics text use the a DAGGER notation $A^\dagger$ to denote the adjoint (Arfken 1985). For example, (Dirac 1982, p. 26) denotes the adjoint of the BRA vector $\langle P|\alpha$ as $\alpha^\dagger|P\rangle$, or $\bar{\alpha}|P\rangle$. The term Hermitian conjugate is sometimes also used instead of adjoint (Griffiths 1987, p. 22)

See also ADJOINT CURVE, ADJOINT MATRIX, DAGGER, HERMITIAN OPERATOR, SELF-ADJOINT, STURM-LIOU-VILLE THEORY

References

Arfken, G. *Mathematical Methods for Physicists, 3rd ed.* Orlando, FL: Academic Press, 1985.
Dirac, P. A. M. "Conjugate Relations." §8 in *Principles of Quantum Mechanics, 4th ed.* Oxford, England: Oxford University Press, pp. 26–9, 1982.
Griffiths, D. J. *Introduction to Elementary Particles.* New York: Wiley, p. 220, 1987.

Adjoint Curve

A curve which has at least multiplicity $r_i - 1$ at each point where a given curve (having only ordinary singular points and cusps) has a multiplicity r_i is called the adjoint to the given curve. When the adjoint curve is of order $n - 3$, it is called a special adjoint curve.

References

Coolidge, J. L. *A Treatise on Algebraic Plane Curves.* New York: Dover, p. 30, 1959.

Adjoint Matrix

The adjoint matrix, sometimes also called the adjugate matrix or conjugate transpose (Golub and van Loan 1996, p. 14), of an $m \times n$ MATRIX A is the $n \times m$ matrix defined by

$$A^* \equiv \bar{A}^T, \quad (1)$$

where the ADJOINT operator is denoted with a star, T denotes the TRANSPOSE, and $\bar{A}$ denotes the CONJU-GATE MATRIX. Unfortunately, several different notations are in use. Older physics text commonly use $A^\dagger$ (Arfken 1985, p. 210), mathematicians commonly use A^* (Courant and Hilbert 1989, p. 9), and computer scientists sometimes use A^H (Golub and van Loan 1996, p. 14). In this work, a star is used to denote the adjoint operator, so care must be taken not to confuse

this with the star used in older physics and engineering texts to denote the COMPLEX CONJUGATE.

If a MATRIX is SELF-ADJOINT, it is said to be HERMI-TIAN. The adjoint matrix of a MATRIX product is given by

$$(ab)^*_{ij} \equiv \overline{[(ab)^T]}_{ij}. \quad (2)$$

Using the identity for the product of TRANSPOSE gives

$$\overline{[(ab)^T]}_{ij} = \overline{[b^Ta^T]}_{ij} = \overline{b_{ik}^Ta_{kj}^T} = [\overline{b}^T]_{ik}[\overline{a}^T]_{kj} = b^*_{ik}a^*_{kj}$$
$$= [b^*a^*]_{ij}, \quad (3)$$

where EINSTEIN SUMMATION has been used here to sum over repeated indices, it follows that

$$(AB)^* = B^*A^*. \quad (4)$$

See also ADJOINT, COMPLEX CONJUGATE, DAGGER, HERMITIAN MATRIX, SCHUR DECOMPOSITION, TRANS-POSE

References

Arfken, G. *Mathematical Methods for Physicists, 3rd ed.* Orlando, FL: Academic Press, p. 210, 1985.
Ayres, F. Jr. *Theory and Problems of Matrices.* New York: Schaum, p. 49, 1962.
Courant, R. and Hilbert, D. *Methods of Mathematical Physics, Vol. 1.* New York: Wiley, 1989.
Golub, G. H. and van Loan, C. F. *Matrix Computations, 3rd ed.* Baltimore, MD: Johns Hopkins University Press, p. 14, 1996.

Adjoint Operator

Given a SECOND-ORDER ORDINARY DIFFERENTIAL EQUATION

$$p_i \equiv p_i(x) \quad (1)$$

where $u \equiv u(x)$ and $\tilde{\mathscr{L}}^*$, the adjoint operator $\tilde{\mathscr{L}}^*u$ (denoted by a DAGGER), is defined by

$$\frac{d}{dx^2}(p_0u) - \frac{d}{dx}(p_1u) + p_2u(y_1y_2'' - y_2y_1'') + P(y_1y_2' - y_1'y_2)$$

$$+Q(y_1y_2 - y_1y_2) = 0^{p0}$$

$$p_0\frac{d^2u}{dx^2} + (2p_0' - p_1)\frac{du}{dx} + (p_0'' - p_1' + p_2)u = \tilde{\mathscr{L}}^*u$$

$$= \int(y_2\tilde{\mathscr{L}}y_1 - y_1\tilde{\mathscr{L}}y_2)\,dx = \left[\frac{p_1}{p_0}(y_1'y_2 - y_1y_2')\right]. \quad (2)$$

Write the two LINEARLY INDEPENDENT solutions as $y' = f_0(x) + f_1(x)y + f_0(x)y^2 + f_3(x)y^3 + \ldots$ and $[g_0(x) + g_1(x)y]y' = f_0(x) + f_1(x)y + f_2(x)y^2 + f_3(x)y^3$. Then the adjoint operator can also be written

$$\tilde{A}. \quad (3)$$

In general, given two adjoint operators $\tilde{B}$ and $(\tilde{A}\tilde{B})^* = \tilde{B}^*\tilde{A}^*$,

$$(\tilde{A}\tilde{B}\cdots\tilde{Z})^* = \tilde{Z}^*\cdots\tilde{B}^*\tilde{A}^*. \qquad (4)$$

which can be generalized to

$$A^\dagger \qquad (5)$$

The adjoint of the BRA vector $\langle P|\alpha$ is denoted $\alpha^\dagger|P\rangle$, or $\bar{\alpha}|P\rangle$ (Dirac 1982, p. 26). The term Hermitian conjugate is sometimes also used (Griffiths 1987, p. 22)

See also ADJOINT MATRIX, DAGGER, HERMITIAN OPERATOR, SELF-ADJOINT OPERATOR, STURM-LIOU-VILLE THEORY

References

Dirac, P. A. M. "Conjugate Relations." §8 in *Principles of Quantum Mechanics, 4th ed.* Oxford, England: Oxford University Press, pp. 26–9, 1982.

Griffiths, D. J. *Introduction to Elementary Particles.* New York: Wiley, p. 220, 1987.

Adjoint Representation

A LIE ALGEBRA is a VECTOR SPACE g with a LIE BRACKET $[X, Y]$, satisfying the JACOBI IDENTITY. Hence any element X gives a linear transformation given by

$$\mathrm{ad}(X)(Y) = [X, Y], \qquad (1)$$

which is called the adjoint representation of g. It is a LIE ALGEBRA REPRESENTATION because of the JACOBI IDENTITY,

$$[\mathrm{ad}(X_1), \ \mathrm{ad}(X_2)](Y) = [X_1, [X_2, Y]] - [X_2, [X_1, Y]]$$

$$= [[X_1, X_2], Y] = \mathrm{ad}([X_1, X_2])(Y). \qquad (2)$$

A REPRESENTATION is given by matrices. The simplest LIE ALGEBRA is gl_n the set of matrices. Consider the adjoint representation of gl_2, which has four dimensions and so will be a four dimensional representation. The matrices

$$e_1 = \begin{bmatrix} 1 & 0 \\ 0 & 0 \end{bmatrix} \qquad (3)$$

$$e_2 = \begin{bmatrix} 0 & 1 \\ 0 & 0 \end{bmatrix} \qquad (4)$$

$$e_3 = \begin{bmatrix} 0 & 0 \\ 1 & 0 \end{bmatrix} \qquad (5)$$

$$e_4 = \begin{bmatrix} 0 & 0 \\ 0 & 1 \end{bmatrix} \qquad (6)$$

give a basis for gl_2. Using this basis, the adjoint representation is described by the following matrices,

$$\mathrm{ad}\, e_1 = \begin{bmatrix} 0 & 0 & 0 & 0 \\ 0 & 1 & 0 & 0 \\ 0 & 0 & -1 & 0 \\ 0 & 0 & 0 & 0 \end{bmatrix} \qquad (7)$$

$$\mathrm{ad}\, e_2 = \begin{bmatrix} 0 & 0 & 1 & 0 \\ -1 & 0 & 0 & 1 \\ 0 & 0 & 0 & 0 \\ 0 & 0 & -1 & 0 \end{bmatrix} \qquad (8)$$

$$\mathrm{ad}\, e_3 = \begin{bmatrix} 0 & -1 & 0 & 0 \\ 0 & 0 & 0 & 0 \\ 1 & 0 & 0 & -1 \\ 0 & 1 & 0 & 0 \end{bmatrix} \qquad (9)$$

$$\mathrm{ad}\, e_4 = \begin{bmatrix} 0 & 0 & 0 & 0 \\ 0 & -1 & 0 & 0 \\ 0 & 0 & 1 & 0 \\ 0 & 0 & 0 & 0 \end{bmatrix}. \qquad (10)$$

The following *Mathematica* function gives the adjoint representation of the matrix m in the Lie algebra, given by a basis, the list of matrices g.

```
ad[g_List,  m_List?MatrixQ]:=  Transpose[Li-
nearSolve[Transpose[Flatten/@g],
    Flatten[m.#1-#1.m]]&/@g]
```

See also COMMUTATOR, LIE ALGEBRA, LIE GROUP, LIE BRACKET, NILPOTENT LIE ALGEBRA, REPRESENTATION, SEMISIMPLE LIE ALGEBRA

References

Fulton, W. and Harris, J. *Representation Theory.* New York: Springer-Verlag, 1991.

Jacobson, N. *Lie Algebras.* New York: Dover, 1979.

Knapp, A. *Lie Groups Beyond an Introduction.* Boston, MA: Birkhäuser, 1996.

Adjugate Matrix

ADJOINT MATRIX

Adjunction

If a is an element of a FIELD F over the PRIME FIELD P, then the set of all RATIONAL FUNCTIONS of a with COEFFICIENTS in P is a FIELD derived from P by adjunction of a.

Adleman-Pomerance-Rumely Primality Test

A modified MILLER'S PRIMALITY TEST which gives a guarantee of PRIMALITY or COMPOSITENESS. The ALGORITHM's running time for a number n has been proved to be as $\mathcal{O}((\ln n)^{c \ln \ln \ln n})$ for some $c > 0$. It was simplified by Cohen and Lenstra (1984), implemented by Cohen and Lenstra (1987), and subsequently optimized by Bosma and van der Hulst (1990).

References

Adleman, L. M.; Pomerance, C.; and Rumely, R. S. "On Distinguishing Prime Numbers from Composite Number." *Ann. Math.* **117**, 173–06, 1983.

Bosma, W. and van der Hulst, M.-P. "Faster Primality Testing." In *Advances in Cryptology, Proc. Eurocrypt '89, Houthalen, April 10–3, 1989* (Ed. J.-J. Quisquater). New York: Springer-Verlag, 652–56, 1990.

Brillhart, J.; Lehmer, D. H.; Selfridge, J.; Wagstaff, S. S. Jr.; and Tuckerman, B. *Factorizations of $b^n \pm 1$, $b = 2$, 3, 5, 6, 7, 10, 11, 12 Up to High Powers*, rev. ed. Providence, RI: Amer. Math. Soc., pp. lxxxiv-lxxxv, 1988.

Cohen, H. and Lenstra, A. K. "Primality Testing and Jacobi Sums." *Math. Comput.* **42**, 297–30, 1984.

Cohen, H. and Lenstra, A. K. "Implementation of a New Primality Test." *Math. Comput.* **48**, 103–21, 1987.

Mihailescu, P. "A Primality Test Using Cyclotomic Extensions." In *Applied Algebra, Algebraic Algorithms and Error-Correcting Codes* (Proc. AAECC-6, Rome, July 1988). New York: Springer-Verlag, pp. 310–23, 1989.

Adleman-Rumely Primality Test

ADLEMAN-POMERANCE-RUMELY PRIMALITY TEST

Admissible

A string or word is said to be admissible if that word appears in a given SEQUENCE. For example, in the SEQUENCE *aabaabaabaabaab* ..., *a*, *aa*, *baab* are all admissible, but *bb* is inadmissible.

See also BLOCK GROWTH

Ado's Theorem

Every finite-dimensional LIE ALGEBRA of characteristic $p = 0$ has a FAITHFUL finite-dimensional representation.

See also IWASAWA'S THEOREM, LIE ALGEBRA

References

Jacobson, N. *Lie Algebras.* New York: Dover, pp. 202–03, 1979.

Affine Complex Plane

The set $\mathbb{A}^2$ of all ORDERED PAIRS of COMPLEX NUMBERS.

See also AFFINE CONNECTION, AFFINE EQUATION, AFFINE GEOMETRY, AFFINE GROUP, AFFINE HULL, AFFINE PLANE, AFFINE SPACE, AFFINE TRANSFORMATION, AFFINITY, COMPLEX PLANE, COMPLEX PROJECTIVE PLANE

Affine Connection

CONNECTION COEFFICIENT

Affine Equation

A nonhomogeneous LINEAR EQUATION or system of nonhomogeneous LINEAR EQUATIONS is said to be affine.

See also AFFINE COMPLEX PLANE, AFFINE CONNECTION, AFFINE GEOMETRY, AFFINE GROUP, AFFINE HULL, AFFINE PLANE, AFFINE SPACE, AFFINE TRANSFORMATION, AFFINITY

Affine Geometry

A GEOMETRY in which properties are preserved by PARALLEL PROJECTION from one PLANE to another. In an affine geometry, the third and fourth of EUCLID'S POSTULATES become meaningless. This type of GEOMETRY was first studied by Euler.

See also ABSOLUTE GEOMETRY, AFFINE COMPLEX PLANE, AFFINE CONNECTION, AFFINE EQUATION, AFFINE GROUP, AFFINE HULL, AFFINE PLANE, AFFINE SPACE, AFFINE TRANSFORMATION, AFFINITY, ORDERED GEOMETRY

References

Birkhoff, G. and Mac Lane, S. "Affine Geometry." §9.13 in *A Survey of Modern Algebra, 5th ed.* New York: Macmillan, pp. 268–75, 1996.

Graustein, W. C. *Introduction to Higher Geometry.* New York: Macmillan, pp. 179–82, 1930.

Leichtweiß, K. *Affine Geometry of Convex Bodies.* Heidelberg, Germany: Barth Verlag, 1998.

Affine Group

The set of all nonsingular AFFINE TRANSFORMATIONS of a TRANSLATION in SPACE constitutes a GROUP known as the affine group. The affine group contains the full linear group and the group of TRANSLATIONS as SUBGROUPS.

See also AFFINE COMPLEX PLANE, AFFINE CONNECTION, AFFINE EQUATION, AFFINE GEOMETRY, AFFINE HULL, AFFINE PLANE, AFFINE SPACE, AFFINE TRANSFORMATION, AFFINITY

References

Birkhoff, G. and Mac Lane, S. *A Survey of Modern Algebra, 5th ed.* New York: Macmillan, p. 237, 1996.

Affine Hull

The IDEAL generated by a SET in a VECTOR SPACE.

See also AFFINE COMPLEX PLANE, AFFINE CONNECTION, AFFINE EQUATION, AFFINE GEOMETRY, AFFINE GROUP, AFFINE PLANE, AFFINE SPACE, AFFINE TRANSFORMATION, AFFINITY, CONVEX HULL, HULL

Affine Plane

A 2-D AFFINE GEOMETRY constructed over a FINITE FIELD. For a FIELD F of size n, the affine plane consists of the set of points which are ordered pairs of elements in F and a set of lines which are themselves a set of points. Adding a POINT AT INFINITY and LINE AT INFINITY allows a PROJECTIVE PLANE to be constructed from an affine plane. An affine plane of order n is a BLOCK DESIGN OF THE FORM $(n^2, n, 1)$. An affine plane of order n exists IFF a PROJECTIVE PLANE of order n exists.

See also AFFINE COMPLEX PLANE, AFFINE CONNECTION, AFFINE EQUATION, AFFINE GEOMETRY, AFFINE

Group, Affine Hull, Affine Space, Affine Transformation, Affinity, Projective Plane

References

Lindner, C. C. and Rodger, C. A. *Design Theory.* Boca Raton, FL: CRC Press, 1997.

Affine Scheme

Let P be the set of PRIME IDEALS of a COMMUTATIVE RING A. Then an affine scheme is a technical mathematical object defined as the SPECTRUM $\sigma(A)$ of P, regarded as a local-ringed space with a structure sheaf. A local-ringed space that is locally isomorphic to an affine scheme is called a SCHEME (Itô 1986, p. 69).

See also PRIME IDEAL, SCHEME, SPECTRUM (RING)

References

Itô, K. (Ed.). "Schemes." §16D in *Encyclopedic Dictionary of Mathematics, 2nd ed., Vol. 1.* Cambridge, MA: MIT Press, p. 69, 1986.

Affine Space

Let V be a VECTOR SPACE over a FIELD K, and let A be a nonempty SET. Now define addition $p + \mathbf{a} \in A$ for any VECTOR $\mathbf{a} \in V$ and element $p \in A$ subject to the conditions

1. $p + \mathbf{0} = p$,
2. $(p + \mathbf{a}) + \mathbf{b} = p + (\mathbf{a} + \mathbf{b})$,
3. For any $q \in A$, there EXISTS a unique VECTOR $\mathbf{a} \in V$ such that $q = p + \mathbf{a}$.

Here, $\mathbf{a}, \mathbf{b} \in V$. Note that (1) is implied by (2) and (3). Then A is an affine space and K is called the COEFFICIENT FIELD.

In an affine space, it is possible to fix a point and coordinate axis such that every point in the SPACE can be REPRESENTED AS an n-tuple of its coordinates. Every ordered pair of points A and B in an affine space is then associated with a VECTOR AB.

See also AFFINE COMPLEX PLANE, AFFINE CONNECTION, AFFINE EQUATION, AFFINE GEOMETRY, AFFINE GROUP, AFFINE HULL, AFFINE PLANE, AFFINE SPACE, AFFINE TRANSFORMATION, AFFINITY

Affine Transformation

Any TRANSFORMATION preserving COLLINEARITY (i.e., all points lying on a LINE initially still lie on a LINE after TRANSFORMATION) and ratios of distances (e.g., the midpoint of a line segment remains the midpoint after transformation). An affine transformation may also be thought of as a shearing transformation (Croft *et al.* 1991). An affine transformation is also called an AFFINITY.

An affine transformation of $\mathbb{R}^n$ is a MAP $F : \mathbb{R}^n \rightarrow \mathbb{R}^n$ OF THE FORM

$$F(\mathbf{p}) = A\mathbf{p} + \mathbf{q} \tag{1}$$

for all $p \in \mathbb{R}^n$, where A is a linear transformation of $\mathbb{R}^n$. If $\det(A) = 1$, the transformation is ORIENTATION-PRESERVING; if $\det(A) = -1$, it is ORIENTATION-REVERSING.

CONTRACTION, EXPANSION, DILATION, REFLECTION, SIMILARITY TRANSFORMATIONS, SPIRAL SIMILARITIES, ROTATION, and TRANSLATION are all affine transformations, as are their combinations. A particular example combining ROTATION and EXPANSION is the rotation-enlargement transformation

$$\begin{bmatrix} x' \\ y' \end{bmatrix} = s \begin{bmatrix} \cos\alpha & \sin\alpha \\ -\sin\alpha & \cos\alpha \end{bmatrix} \begin{bmatrix} x - x_0 \\ y - y_0 \end{bmatrix}$$
$$= s \begin{bmatrix} \cos\alpha(x - x_0) + \sin\alpha(y - y_0) \\ -\sin\alpha(x - x_0) + \cos\alpha(y - y_0) \end{bmatrix}. \tag{2}$$

Separating the equations,

$$x' = (s\cos\alpha)x + (s\sin\alpha)y - s(x_0\cos\alpha + y_0\sin\alpha) \tag{3}$$

$$y' = (-s\sin\alpha)x + (s\cos\alpha)y + s(x_0\sin\alpha - y_0\cos\alpha). \tag{4}$$

This can be also written as

$$x' = ax + by + c \tag{5}$$

$$y' = bx + ay + d, \tag{6}$$

where

$$a = s\cos\alpha \tag{7}$$

$$b = -s\sin\alpha. \tag{8}$$

The scale factor s is then defined by

$$s \equiv \sqrt{a^2 + b^2}, \tag{9}$$

and the rotation ANGLE by

$$\alpha = \tan^{-1}\left(-\frac{b}{a}\right). \tag{10}$$

See also AFFINE COMPLEX PLANE, AFFINE CONNECTION, AFFINE EQUATION, AFFINE GEOMETRY, AFFINE GROUP, AFFINE HULL, AFFINE PLANE, AFFINE SPACE, AFFINE TRANSFORMATION, AFFINITY, EQUIAFFINITY, EUCLIDEAN MOTION

References

Croft, H. T.; Falconer, K. J.; and Guy, R. K. *Unsolved Problems in Geometry.* New York: Springer-Verlag, p. 3, 1991.

Gray, A. *Modern Differential Geometry of Curves and Surfaces with Mathematica, 2nd ed.* Boca Raton, FL: CRC Press, p. 130, 1997.

Zwillinger, D. (Ed.). "Affine Transformations." §4.3.2 in *CRC Standard Mathematical Tables and Formulae.* Boca Raton, FL: CRC Press, pp. 265–66, 1995.

Affine Variety

An affine variety V is a VARIETY contained in AFFINE SPACE. For example,

$$\{(x,\,y,\,z): x^2 + y^2 - z^2 = 0\} \tag{1}$$

is the CONE, and

$$\{(x,\,y,\,z): x^2 + y^2 - z^2 = 0,\ ax + by + cz = 0\} \tag{2}$$

is a CONIC SECTION, which is a SUBVARIETY of the cone. The cone can be written $V(x^2 + y^2 - z^2)$ to indicate that it is the variety corresponding to $x^2 + y^2 - z^2 = 0$. Naturally, many other polynomials vanish on $V(x^2 + y^2 - z^2)$, in fact all polynomials in $I(C) = \{x^2 + y^2 - z^2\}$. The set $I(C)$ is an IDEAL in the POLYNOMIAL RING $\mathbb{C}[x,\,y,\,z]$. Note also, that the ideal of polynomials vanishing on the conic section is the IDEAL generated by $x^2 + y^2 - z^2$ and $ax + by + cz$.

A MORPHISM between two affine varieties is given by polynomial coordinate functions. For example, the map $\phi(x,\,y,\,z) = (x^2,\,y^2,\,z^2)$ is a MORPHISM from $X = V(x^2 + y^2 + z^2)$ to $Y = V(x + y + z)$. Two affine varieties are ISOMORPHIC if there is a MORPHISM which has an inverse morphism. For example, the affine variety $V(x^2 + y^2 + z^2)$ is isomorphic to the cone $V(x^2 + y^2 - z^2)$ via the coordinate change $\phi(x,\,y,\,z) = (x,\,y,\,iz)$.

Many polynomials f may be factored, for instance $f = x^2 + y^2 = (x + iy)(x - iy)$, and then $V(f) = V(x + iy) \cup V(x - iy)$. Consequently, only IRREDUCIBLE POLYNOMIALS, and more generally only PRIME IDEALS p are used in the definition of a variety. An affine variety V is the set of common zeros of a collection of polynomials $p_1, ..., p_k$, i.e.,

$$V = \{x = (x_1,\,\ldots,\,x_n): p_1(x) = \ldots = p_k(x) = 0\} \tag{3}$$

as long as the IDEAL $I = (p_1,\,\ldots,\,p_k)$ is a PRIME IDEAL. More classically, an affine variety is defined by any set of polynomials, i.e., what is now called an ALGEBRAIC SET. Most points in V will have dimension $n - k$, but V may have singular points like the origin in the cone.

When V is one-dimensional generically (at almost all points), which typically occurs when $k = n - 1$, then V is called a curve. When V is two-dimensional, it is called a surface. In the case of COMPLEX affine space, a curve is a RIEMANN SURFACE, possibly with some singularities.

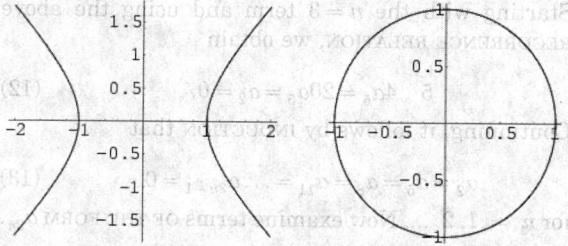

Mathematica has a built-in function `ImplicitPlot` in the *Mathematica* add-on package `Graphics`Im-`

`plicitPlot`` (which can be loaded with the command `<<Graphics`)` that will graph affine varieties in the real affine plane. For example, the following graphs a hyperbola and a circle.

```
<<Graphics`;
Show[GraphicsArray[{
    ImplicitPlot[x^2 - y^2 == 1, {x, -2, 2},
DisplayFunction -> Identity],
    ImplicitPlot[x^2 + y^2 == 1, {x, -2, 2},
DisplayFunction -> Identity]
}]]
```

An extension to this function called `Implicit-Plot3D` can be downloaded from MathSource and used to plot affine varieties in three-dimensional space.

See also ALGEBRAIC SET, CATEGORY THEORY, COMMUTATIVE ALGEBRA, CONIC SECTION, GROEBNER BASIS, PROJECTIVE VARIETY, SCHEME, STACK (MODULI SPACE), INTRINSIC VARIETY, ZARISKI TOPOLOGY

References
Bump, D. *Algebraic Geometry.* Singapore: World Scientific, pp. 1–, 1998.

Cox, D.; Little, J.; and O'Shea, D. *Ideals, Varieties, and Algorithms.* New York: Springer-Verlag, pp. 5–9, 1997.

Hartshorne, R. *Algebraic Geometry.* New York: Springer-Verlag, 1977.

Affinity
AFFINE TRANSFORMATION

Affix
In the archaic terminology of Whittaker and Watson (1990), the COMPLEX NUMBER z representing $x + iy$.

References
Whittaker, E. T. and Watson, G. N. *A Course in Modern Analysis, 4th ed.* Cambridge, England: Cambridge University Press, 1990.

Aggregate
An archaic word for infinite SETS such as those considered by Georg Cantor.

See also CLASS (SET), SET

AGM
ARITHMETIC-GEOMETRIC MEAN

Agnesi's Witch
WITCH OF AGNESI

Agnésienne
WITCH OF AGNESI

Agonic Lines
SKEW LINES

Ahlfors Five Island Theorem

Let $f(z)$ be a TRANSCENDENTAL MEROMORPHIC FUNCTION, and let $D_1, D_2, ..., D_5$ be five SIMPLY CONNECTED domains in $\mathbb{C}$ with disjoint closures (Ahlfors 1932). Then there exists $j \in \{1, 2, ..., 5\}$ and, for any $R > 0$, a SIMPLY CONNECTED domain $G \subset \{z \in \mathbb{C} : |z| > R\}$ such that $f(z)$ is a CONFORMAL MAP of G onto D_j. If $f(z)$ has only finitely many POLES, then "five" may be replaced by "three" (Ahlfors 1933).

See also MEROMORPHIC FUNCTION, TRANSCENDENTAL FUNCTION

References

Ahlfors, L. "Sur les fonctions inverses des fonctions mér-omorphes." *C. R. Acad. Sci.* **194**, 1145–147, 1932. Reprinted in *Lars Valerian Ahlfors: Collected Papers Volume 1, 1929–955* (Ed. R. M. Shortt). Boston, MA: Birkhäuser, 149–51, 1982.

Ahlfors, L. "Über die Kreise die von einer Riemannschen Fläche schlicht überdeckt werden." *Comm. Math. Helv.* **5**, 28–8, 1933. Reprinted in *Lars Valerian Ahlfors: Collected Papers Volume 1, 1929–955* (Ed. R. M. Shortt). Boston, MA: Birkhäuser, 163–73, 1982.

Bergweiler, W. "Iteration of Meromorphic Functions." *Bull. Amer. Math. Soc. (N. S.)* **29**, 151–88, 1993.

Hayman, W. K. *Meromorphic Functions.* Oxford, England: Oxford University Press, 1964.

Nevanlinna, R. *Analytic Functions.* New York: Springer-Verlag, 1970.

Ahlfors-Bers Theorem

The RIEMANN'S MODULI SPACE gives the solution to RIEMANN'S MODULI PROBLEM, which requires an ANALYTIC parameterization of the compact RIEMANN SURFACES in a fixed HOMEOMORPHISM.

A-Integrable

A generalization of the LEBESGUE INTEGRAL. A MEASURABLE FUNCTION $f(x)$ is called A-integrable over the CLOSED INTERVAL $[a, b]$ if

$$m\{x : |f(x)| > n\} = \mathcal{O}(n^{-1}), \tag{1}$$

where m is the LEBESGUE MEASURE, and

$$I = \lim_{n \to \infty} \int_a^b [f(x)]_n \, dx \tag{2}$$

exists, where

$$[f(x)]_n = \begin{cases} f(x) & \text{if } |f(x)| \leq n \\ 0 & \text{if } |f(x)| > n. \end{cases} \tag{3}$$

References

Titchmarsh, E. C. "On Conjugate Functions." *Proc. London Math. Soc.* **29**, 49–0, 1928.

Airy Differential Equation

Some authors define a general Airy differential equation as

$$y'' \pm k^2 x y = 0. \tag{1}$$

This equation can be solved by series solution using the expansions

$$y = \sum_{n=0}^{\infty} a_n x^n \tag{2}$$

$$y' = \sum_{n=0}^{\infty} n a_n x^{n-1} = \sum_{n=1}^{\infty} n a_n x^{n-1}$$

$$= \sum_{n=0}^{\infty} (n+1) a_{n+1} x^n \tag{3}$$

$$y'' = \sum_{n=0}^{\infty} (n+1) n a_{n+1} x^{n-1} = \sum_{n=1}^{\infty} (n+1) n a_{n+1} x^{n-1}$$

$$= \sum_{n=0}^{\infty} (n+2)(n+1) a_{n+2} x^n. \tag{4}$$

Specializing to the "conventional" Airy differential equation occurs by taking the MINUS SIGN and setting $k^2 = 1$. Then plug (4) into

$$y'' - xy = 0 \tag{5}$$

to obtain

$$\sum_{n=0}^{\infty} (n+2)(n+1) a_{n+2} x^n - x \sum_{n=0}^{\infty} a_n x^n = 0 \tag{6}$$

$$\sum_{n=0}^{\infty} (n+2)(n+1) a_{n+2} x^n - \sum_{n=0}^{\infty} a_n x^{n+1} = 0 \tag{7}$$

$$2a_2 + \sum_{n=1}^{\infty} (n+2)(n+1) a_{n+2} x^n - \sum_{n=1}^{\infty} a_{n-1} x^n = 0 \tag{8}$$

$$2a_2 + \sum_{n=1}^{\infty} [(n+2)(n+1) a_{n+2} - a_{n-1}] x^n = 0. \tag{9}$$

In order for this equality to hold for all x, each term must separately be 0. Therefore,

$$a_2 = 0 \tag{10}$$

$$(n+2)(n+1) a_{n+2} = a_{n-1}. \tag{11}$$

Starting with the $n = 3$ term and using the above RECURRENCE RELATION, we obtain

$$5 \cdot 4 a_5 = 20 a_5 = a_2 = 0. \tag{12}$$

Continuing, it follows by INDUCTION that

$$a_2 = a_5 = a_8 = a_{11} = \ldots a_{3n-1} = 0 \tag{13}$$

for $n = 1, 2, \ldots$. Now examine terms OF THE FORM a_{3n}.

$$a_3 = \frac{a_0}{3 \cdot 2} \tag{14}$$

$$a_6 = \frac{a_3}{6 \cdot 5} = \frac{a_0}{(6 \cdot 5)(3 \cdot 2)} \qquad (15)$$

$$a_9 = \frac{a_6}{9 \cdot 8} = \frac{a_0}{(9 \cdot 8)(6 \cdot 5)(3 \cdot 2)}. \qquad (16)$$

Again by INDUCTION,

$$a_{3n} = \frac{a_0}{[(3n)(3n-1)][(3n-3)(3n-4)] \cdots [6 \cdot 5][3 \cdot 2]} \qquad (17)$$

for $n = 1, 2, \ldots$. Finally, look at terms OF THE FORM a_{3n+1},

$$a_4 = \frac{a_1}{4 \cdot 3} \qquad (18)$$

$$a_7 = \frac{a_4}{7 \cdot 6} = \frac{a_1}{(7 \cdot 6)(4 \cdot 3)} \qquad (19)$$

$$a_{10} = \frac{a_7}{10 \cdot 9} = \frac{a_1}{(10 \cdot 9)(7 \cdot 6)(4 \cdot 3)}. \qquad (20)$$

By INDUCTION,

$$a_{3n+1}$$
$$= \frac{a_1}{[(3n+1)(3n)][(3n-2)(3n-3)] \cdots [7 \cdot 6][4 \cdot 3]} \qquad (21)$$

for $n = 1, 2, \ldots$. The general solution is therefore

$$y = a_0 \left[1 + \sum_{n=1}^{\infty} \frac{x^{3n}}{(3n)(3n-1)(3n-3)(3n-4) \cdots 3 \cdot 2} \right]$$
$$+ a_1 \left[x + \sum_{n=1}^{\infty} \frac{x^{3n+1}}{(3n+1)(3n)(3n-2)(3n-3) \cdots 4 \cdot 3} \right]. \qquad (22)$$

For a general k^2 with a MINUS SIGN, equation (1) is

$$y'' - k^2 x y = 0, \qquad (23)$$

and the solution is

$$y(x) = \tfrac{1}{3}\sqrt{x}[AI_{-1/3}(\tfrac{2}{3}kx^{3/2}) - BI_{1/3}(\tfrac{2}{3}kx^{3/2})], \qquad (24)$$

where I is a MODIFIED BESSEL FUNCTION OF THE FIRST KIND. This is usually expressed in terms of the AIRY FUNCTIONS Ai(x) and Bi(x)

$$y(x) = A' \,\text{Ai}(k^{2/3}x) + B' \text{Bi}(k^{2/3}x). \qquad (25)$$

If the PLUS SIGN is present instead, then

$$y'' + k^2 x y = 0 \qquad (26)$$

and the solutions are

$$y(x) = \tfrac{1}{3}\sqrt{x}[AJ_{-1/3}(\tfrac{2}{3}kx^{3/2}) + BJ_{1/3}(\tfrac{2}{3}kx^{3/2})], \qquad (27)$$

where $J(z)$ is a BESSEL FUNCTION OF THE FIRST KIND.

A generalization of the Airy differential equation is given by

$$y''' - 4xy' - 2y = 0, \qquad (28)$$

which has solutions

$$y = C_1[\text{Ai}(x)]^2 + C_2\text{Ai}(x)\,\text{Bi}(x) + C_3[\text{Bi}(x)]^2 \qquad (29)$$

(Abramowitz and Stegun 1972, p. 448; Zwillinger 1997, p. 128).

See also AIRY-FOCK FUNCTIONS, AIRY FUNCTIONS, BESSEL FUNCTION OF THE FIRST KIND, MODIFIED BESSEL FUNCTION OF THE FIRST KIND

References

Abramowitz, M. and Stegun, C. A. (Eds.). "Airy Functions." §10.4.1 in *Handbook of Mathematical Functions with Formulas, Graphs, and Mathematical Tables, 9th printing.* New York: Dover, pp. 446–52, 1972.

Zwillinger, D. (Ed.). *CRC Standard Mathematical Tables and Formulae.* Boca Raton, FL: CRC Press, p. 413, 1995.

Zwillinger, D. *Handbook of Differential Equations, 3rd ed.* Boston, MA: Academic Press, p. 121, 1997.

Airy Functions

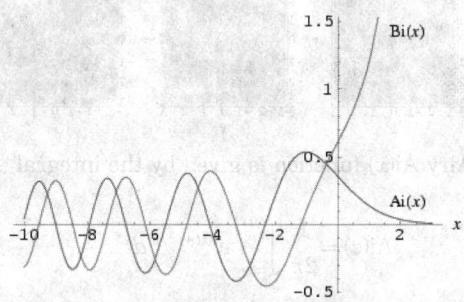

The Ai(x) and Bi(x) functions are defined as the two LINEARLY INDEPENDENT solutions to

$$y'' - yz = 0. \qquad (1)$$

(Abramowitz and Stegun 1972, pp. 446–47; illustrated above), written in the form

$$y(z) = A\,\text{Ai}(z) + B\,\text{Bi}(z), \qquad (2)$$

where

$$\text{Ai}(z) = \tfrac{1}{3}\sqrt{x}\left[I_{-1/3}\left(\tfrac{2}{3}z^{3/2}\right) - I_{1/3}\left(\tfrac{2}{3}z^{3/2}\right) \right]$$
$$= \sqrt{\frac{z}{3\pi}}K_{1/3}\left(\tfrac{2}{3}z^{3/2}\right) \qquad (3)$$

$$\text{Bi}(z) = \sqrt{\frac{z}{3}}\left[I_{-1/3}\left(\tfrac{2}{3}z^{3/2}\right) + I_{1/3}\left(\tfrac{2}{3}z^{3/2}\right) \right], \qquad (4)$$

where $I(z)$ is a MODIFIED BESSEL FUNCTION OF THE FIRST KIND and $K(z)$ is a MODIFIED BESSEL FUNCTION OF THE SECOND KIND. The functions are implemented in *Mathematica* as AiryAi[z] and AiryBi[z]. Their derivatives are implemented as AiryAiPrime[z] and

`AiryBiPrime[z]`.

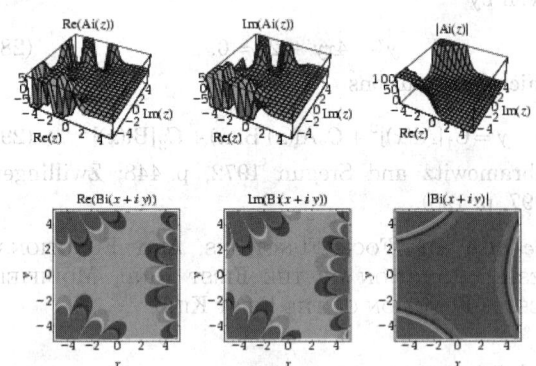

Plots of Ai(z) in the COMPLEX PLANE are illustrated above, and Bi(z) is illustrated below.

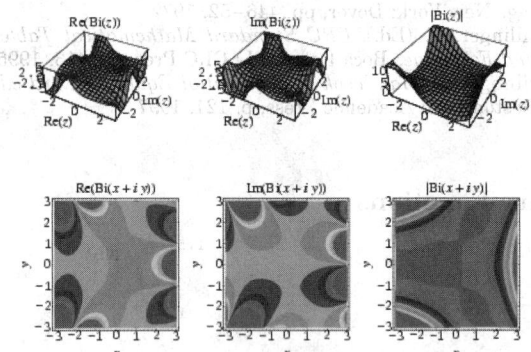

The Airy Ai(x) function is given by the integral

$$\mathrm{Ai}(z) = \frac{1}{2\pi} \int_{-\infty}^{\infty} e^{i(zt+t^3/3)} dt \tag{5}$$

and the INFINITE SERIES

$$\mathrm{Ai}(x) = \frac{1}{3^{2/3}\pi} \sum_{n=0}^{\infty} \frac{\Gamma\left(\frac{1}{3}(n+1)\right)}{n!}$$
$$\times (3^{1/3}x)^n \sin\left[\frac{2(n+1)\pi}{3}\right] \tag{6}$$

(Banderier *et al.*). A generalization of the Airy function has been constructed by Hardy. For $z = 0$,

$$\mathrm{Ai}(0) = \frac{1}{3^{2/3}\Gamma(\frac{2}{3})} \tag{7}$$

$$\mathrm{Bi}(0) = \frac{1}{3^{1/6}\Gamma(\frac{2}{3})}, \tag{8}$$

where $\Gamma(z)$ is the GAMMA FUNCTION.

The ASYMPTOTIC SERIES of Ai(z) has a different form in different QUADRANTS of the COMPLEX PLANE, a fact known as the STOKES PHENOMENON.

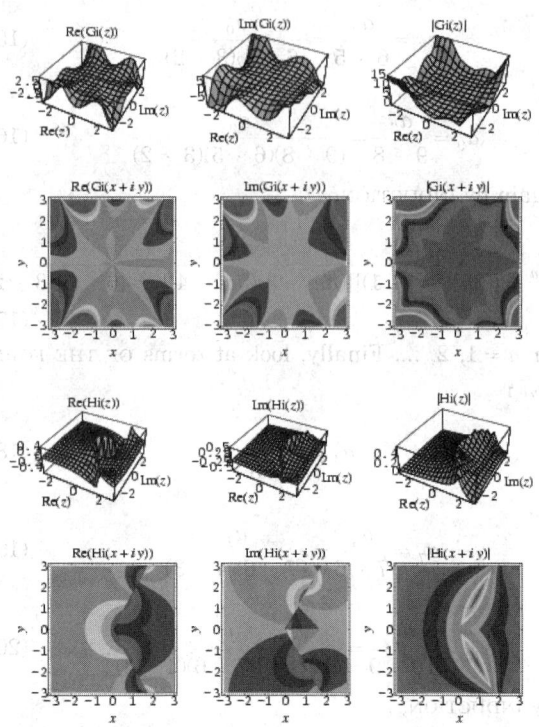

Functions related to the Airy functions have been defined as

$$\mathrm{Gi}(z) \equiv \frac{1}{\pi} \int_0^{\infty} \sin(\tfrac{1}{3}t^3 + zt) dt \tag{9}$$

$$\mathrm{Hi}(z) \equiv \frac{1}{\pi} \int_0^{\infty} \exp\left(-\tfrac{1}{3}t^3 + zt\right) dt, \tag{10}$$

where Gi(z) is defined for $\Im[z] \neq 0$ and Hi(z) for $\Re[z] \geq 0$. The can be expressed in terms of the Airy functions by

$$\mathrm{Gi}(z) = -\frac{z^2}{2\pi} {}_1F_4\left(1: \tfrac{2}{3}, \tfrac{5}{6}, \tfrac{7}{6}, \tfrac{4}{3}; \tfrac{1}{1296} z^6\right)$$
$$+ \frac{[\mathrm{sgn}(z)]^6}{360\pi z^6} {}_1F_4\left(1: \tfrac{7}{6}, \tfrac{4}{3}, \tfrac{5}{3}, \tfrac{11}{6}; \tfrac{1}{1296} z^6\right) + \frac{\overline{z^6}}{6|z|^6}$$
$$\times [\mathrm{Bi}(-|z|) + \mathrm{Bi}(|z|)] - \frac{i\sqrt{3}|z|^3}{6z^4}[\mathrm{Ai}(-|z|) - \mathrm{Ai}(|z|)]$$
$$+ \frac{1}{6z^4|z|^6}\{\Im[z] + \Re[z][\mathrm{Bi}(|z|) - \mathrm{Bi}(-|z|)]\} \tag{11}$$

$$\mathrm{Hi}(z) = \tfrac{2}{3}\sqrt{-\tfrac{z}{3}}\left[J_{-1/3}\left(\tfrac{2}{3}(-z)^{3/2}\right) - J_{1/3}\left(\tfrac{2}{3}(-z)^{3/2}\right)\right]$$
$$+ \frac{z^2}{2\pi} {}_1F_2\left(1: \tfrac{4}{3}, \tfrac{5}{3}; \tfrac{1}{9}z^3\right), \tag{12}$$

where pF_q is a GENERALIZED HYPERGEOMETRIC FUNCTION, SGN is the sign function, $|z|$ is the MODULUS of z, $\Re[z]$ is the REAL PART, $\Im[z]$ is the IMAGINARY PART, and $J_n(z)$ is a BESSEL FUNCTION OF THE FIRST KIND.

Watson (1966, pp. 188–90) gives a slightly more general definition of the Airy function as the solution to the AIRY DIFFERENTIAL EQUATION

$$\Phi'' \pm k^2 \Phi x = 0 \tag{13}$$

which is FINITE at the ORIGIN, where Φ' denotes the DERIVATIVE $d\Phi/dx$, $k^2 = 1/3$, and either SIGN is permitted. Call these solutions $(1/\pi)\Phi(\pm k^2, x)$, then

$$\frac{1}{\pi}\Phi\left(\pm\frac{1}{3}; x\right) \equiv \int_0^\infty \cos(t^3 \pm xt)\, dt \tag{14}$$

$$\Phi\left(\frac{1}{3}; x\right) = \frac{1}{3}\pi\sqrt{\frac{x}{3}}\left[J_{-1/3}\left(\frac{2x^{3/2}}{3^{3/2}}\right) + J_{1/3}\left(\frac{2x^{3/2}}{3^{3/2}}\right)\right] \tag{15}$$

$$\Phi\left(-\frac{1}{3}; x\right) = \frac{1}{3}\pi\sqrt{\frac{x}{3}}\left[I_{-1/3}\left(\frac{2x^{3/2}}{3^{3/2}}\right) - I_{1/3}\left(\frac{2x^{3/2}}{3^{3/2}}\right)\right], \tag{16}$$

where $J(z)$ is a BESSEL FUNCTION OF THE FIRST KIND. Using the identity

$$K_n(x) = \frac{\pi}{2}\frac{I_{-n}(x) - I_n(x)}{\sin(n\pi)}, \tag{17}$$

where $K(z)$ is a MODIFIED BESSEL FUNCTION OF THE SECOND KIND, the second case can be re-expressed

$$\Phi(-\tfrac{1}{3}; x) = \frac{1}{3}\pi\sqrt{\frac{x}{3}}\frac{2}{\pi}\sin\left(\frac{1}{3}\pi\right)K_{1/3}\left(\frac{2x^{3/2}}{3^{3/2}}\right) \tag{18}$$

$$= \frac{\pi}{3}\sqrt{\frac{x}{3}}\frac{2}{\pi}\frac{\sqrt{3}}{2}K_{1/3}\left(\frac{2x^{3/2}}{3^{3/2}}\right) \tag{19}$$

$$= \frac{1}{3}\sqrt{x}K_{1/3}\left(\frac{2x^{3/2}}{3^{3/2}}\right). \tag{20}$$

See also AIRY-FOCK FUNCTIONS, BESSEL FUNCTION OF THE FIRST KIND, MAP-AIRY DISTRIBUTION, MODIFIED BESSEL FUNCTION OF THE FIRST KIND, MODIFIED BESSEL FUNCTION OF THE SECOND KIND

References

Abramowitz, M. and Stegun, C. A. (Eds.). "Airy Functions." §10.4 in *Handbook of Mathematical Functions with Formulas, Graphs, and Mathematical Tables, 9th printing.* New York: Dover, pp. 446–52, 1972.
Banderier, C.; Flajolet, P.; Schaeffer, G.; and Soria, M. "Planar Maps and Airy Phenomena." Preprint.
Press, W. H.; Flannery, B. P.; Teukolsky, S. A.; and Vetterling, W. T. "Bessel Functions of Fractional Order, Airy Functions, Spherical Bessel Functions." §6.7 in *Numerical Recipes in FORTRAN: The Art of Scientific Computing, 2nd ed.* Cambridge, England: Cambridge University Press, pp. 234–45, 1992.
Spanier, J. and Oldham, K. B. "The Airy Functions Ai(x) and Bi(x)." Ch. 56 in *An Atlas of Functions.* Washington, DC: Hemisphere, pp. 555–62, 1987.
Watson, G. N. *A Treatise on the Theory of Bessel Functions, 2nd ed.* Cambridge, England: Cambridge University Press, 1966.

Airy Projection

A MAP PROJECTION. The inverse equations for ϕ are computed by iteration. Let the ANGLE of the projection plane be θ_b. Define

$$a = \begin{cases} 0 & \text{for } \theta_b = \frac{1}{2}\pi \\ \dfrac{\ln[\frac{1}{2}\cos(\frac{1}{2}\pi - \theta_b)]}{\tan[\frac{1}{2}(\frac{1}{2}\pi - \theta_b)]} & \text{otherwise.} \end{cases} \tag{1}$$

For proper convergence, let $x_i = \pi/6$ and compute the initial point by checking

$$x_i = |\exp[-(\sqrt{x^2 + y^2} + a\tan x_i)\tan x_i]|. \tag{2}$$

As long as $x_i > 1$, take $x_{i+1} = x_i/2$ and iterate again. The first value for which $x_i < 1$ is then the starting point. Then compute

$$x_i = \cos^{-1}\{\exp[-(\sqrt{x^2 + y^2} + a\tan x_i)\tan x_i]\} \tag{3}$$

until the change in x_i between evaluations is smaller than the acceptable tolerance. The (inverse) equations are then given by

$$\phi = \frac{1}{2}\pi - 2x_i \tag{4}$$

$$\lambda = \tan^{-1}\left(-\frac{x}{y}\right). \tag{5}$$

AiryAi
AIRY FUNCTIONS

AiryAiPrime
AIRY FUNCTIONS

AiryBi
AIRY FUNCTIONS

AiryBiPrime
AIRY FUNCTIONS

Airy-Fock Functions

The three Airy-Fock functions are

$$v(z) = \frac{1}{2}\sqrt{\pi}\,\text{Ai}(z) \tag{1}$$

$$w_1(z) = 2e^{i\pi/6}v(\omega z) \tag{2}$$

$$w_2(z) = 2e^{-i\pi/6}v(\omega^{-1}z), \tag{3}$$

where Ai(z) is an AIRY FUNCTION. These functions satisfy

$$v(z) = \frac{\omega_1(z) - \omega_2(z)}{2i} \qquad (4)$$

$$\overline{w_1(z)} = w_2(\bar{z}), \qquad (5)$$

where $\bar{z}$ is the COMPLEX CONJUGATE of z.

See also AIRY FUNCTIONS

References

Hazewinkel, M. (Managing Ed.). *Encyclopaedia of Mathematics: An Updated and Annotated Translation of the Soviet "Mathematical Encyclopaedia."* Dordrecht, Netherlands: Reidel, p. 65, 1988.

Aitken Interpolation

An algorithm similar to NEVILLE'S ALGORITHM for constructing the LAGRANGE INTERPOLATING POLYNOMIAL. Let $f(x|x_0, x_1, \ldots, x_k)$ be the unique POLYNOMIAL of kth ORDER coinciding with $f(x)$ at $x_0, \ldots, x_k$. Then

$$f(x|x_0, x_1) = \frac{1}{x_1 - x_0} \begin{vmatrix} f_0 & x_0 - x \\ f_1 & x_1 - x \end{vmatrix}$$

$$f(x|x_0, x_2) = \frac{1}{x_2 - x_0} \begin{vmatrix} f_0 & x_0 - x \\ f_2 & x_2 - x \end{vmatrix}$$

$$f(x|x_0, x_1, x_2) = \frac{1}{x_2 - x_1} \begin{vmatrix} f(x|x_0, x_1) & x_1 - x \\ f(x|x_0, x_2) & x_2 - x \end{vmatrix}$$

$$f(x|x_0, x_1, x_2, x_3) = \frac{1}{x_3 - x_2} \begin{vmatrix} f(x|x_0, x_1)x_2 - x \\ f(x|x_0, x_1)x_3 - x \end{vmatrix}.$$

See also LAGRANGE INTERPOLATING POLYNOMIAL

References

Abramowitz, M. and Stegun, C. A. (Eds.). *Handbook of Mathematical Functions with Formulas, Graphs, and Mathematical Tables, 9th printing.* New York: Dover, p. 879, 1972.
Acton, F. S. *Numerical Methods That Work, 2nd printing.* Washington, DC: Math. Assoc. Amer., pp. 93–4, 1990.
Press, W. H.; Flannery, B. P.; Teukolsky, S. A.; and Vetterling, W. T. *Numerical Recipes in FORTRAN: The Art of Scientific Computing, 2nd ed.* Cambridge, England: Cambridge University Press, p. 102, 1992.

Aitken's Delta Squared Process

An ALGORITHM which extrapolates the partial sums s_n of a SERIES $\sum_n a_n$ whose CONVERGENCE is approximately geometric and accelerates its rate of CONVERGENCE. The extrapolated partial sum is given by

$$s_n' \equiv s_{n+1} - \frac{(s_{n+1} - s_n)^2}{s_{n+1} - 2s_n + s_{n-1}}.$$

See also EULER'S SERIES TRANSFORMATION

References

Abramowitz, M. and Stegun, C. A. (Eds.). *Handbook of Mathematical Functions with Formulas, Graphs, and Mathematical Tables, 9th printing.* New York: Dover, p. 18, 1972.
Press, W. H.; Flannery, B. P.; Teukolsky, S. A.; and Vetterling, W. T. *Numerical Recipes in FORTRAN: The Art of Scientific Computing, 2nd ed.* Cambridge, England: Cambridge University Press, p. 160, 1992.

Ajima-Malfatti Points

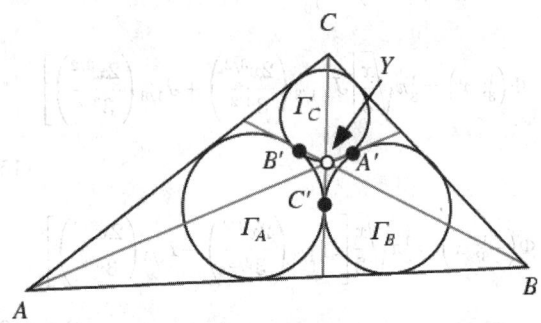

The lines connecting the vertices and corresponding circle-circle intersections in MALFATTI'S TANGENT TRIANGLE PROBLEM coincide in a point Y called the first Ajima-Malfatti point (Kimberling and MacDonald 1990, Kimberling 1994). Similarly, letting A'', B'', and C'' be the excenters of ABC, then the lines $A'A''$, $B'B''$, and $C'C''$ are coincident in another point called the second Ajima-Malfatti point. The points are sometimes simply called the malfatti points (Kimberling 1994).

References

Kimberling, C. "Central Points and Central Lines in the Plane of a Triangle." *Math. Mag.* **67**, 163–87, 1994.
Kimberling, C. "1st and 2nd Ajima-Malfatti Points." http://cedar.evansville.edu/~ck6/tcenters/recent/ajmalf.html.
Kimberling, C. and MacDonald, I. G. "Problem E 3251 and Solution." *Amer. Math. Monthly* **97**, 612–13, 1990.

Akinetor

Moon, P. and Spencer, D. E. *Theory of Holors: A Generalization of Tensors.* Cambridge, England: Cambridge University Press, 1986.

Akisation

CUMULATION

Albanese Variety

An ABELIAN VARIETY which is canonically attached to an ALGEBRAIC VARIETY which is the solution to a certain universal problem. The Albanese variety is dual to the PICARD VARIETY.

References

Hazewinkel, M. (Managing Ed.). *Encyclopaedia of Mathematics: An Updated and Annotated Translation of the Soviet "Mathematical Encyclopaedia."* Dordrecht, Netherlands: Reidel, pp. 67–8, 1988.

Albers Conic Projection
ALBERS EQUAL-AREA CONIC PROJECTION

Albers Equal-Area Conic Projection

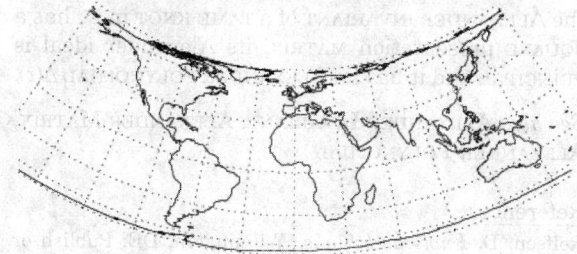

An EQUAL-AREA PROJECTION. Let ϕ_0 be the LATITUDE for the origin of the CARTESIAN COORDINATES and λ_0 its LONGITUDE. Let ϕ_1 and ϕ_2 be the standard parallels. Then

$$x = \rho \sin \theta \tag{1}$$

$$y = \rho_0 - \rho \cos \theta, \tag{2}$$

where

$$\rho = \frac{\sqrt{C - 2n \sin \phi}}{n} \tag{3}$$

$$\theta = n(\lambda - \lambda_0) \tag{4}$$

$$\rho_0 = \frac{\sqrt{C - 2n \sin \phi_0}}{n} \tag{5}$$

$$C = \cos^2 \phi_1 + 2n \sin \phi_1 \tag{6}$$

$$n = \tfrac{1}{2}(\sin \phi_1 + \sin \phi_2). \tag{7}$$

The inverse FORMULAS are

$$\phi = \sin^{-1}\left(\frac{C - \rho^2 n^2}{2n}\right) \tag{8}$$

$$\lambda = \lambda_0 + \frac{\theta}{n}, \tag{9}$$

where

$$\rho = \sqrt{x^2 + (\rho_0 - y)^2} \tag{10}$$

$$\theta = \tan^{-1}\left(\frac{x}{\rho_0 - y}\right). \tag{11}$$

See also EQUAL-AREA PROJECTION

References
Snyder, J. P. *Map Projections--A Working Manual.* U. S. Geological Survey Professional Paper 1395. Washington, DC: U. S. Government Printing Office, pp. 98–03, 1987.

Alcuin's Sequence
The INTEGER SEQUENCE 1, 0, 1, 1, 2, 1, 3, 2, 4, 3, 5, 4, 7, 5, 8, 7, 10, 8, 12, 10, 14, 12, 16, 14, 19, 16, 21, 19, ... (Sloane's A005044) given by the COEFFICIENTS of the MACLAURIN SERIES for $1/(1-x^2)(1-x^3)(1-x^4)$. The number of different TRIANGLES which have INTEGRAL sides and PERIMETER n is given by

$$T(n) = P_3(n) = \sum_{1 \le j \le \lfloor n/2 \rfloor} P_2(j) \tag{1}$$

$$= \left[\frac{n^2}{12}\right] - \left\lfloor \frac{n}{4} \right\rfloor \left\lfloor \frac{n+2}{4} \right\rfloor \tag{2}$$

$$= \begin{cases} \left[\dfrac{n^2}{48}\right] & \text{for } n \text{ even} \\[2ex] \left[\dfrac{(n+3)^2}{48}\right] & \text{for } n \text{ odd.} \end{cases} \tag{3}$$

where $P_2(n)$ and $P_3(n)$ are PARTITION FUNCTIONS, with $P_k(n)$ giving the number of ways of writing n as a sum of k terms, $[x]$ is the NINT function, and $\lfloor x \rfloor$ is the FLOOR FUNCTION (Jordan *et al.* 1979, Andrews 1979, Honsberger 1985). Strangely enough, $T(n)$ for $n = 3$, 4, ... is precisely Alcuin's sequence.

See also PARTITION FUNCTION P, TRIANGLE

References
Andrews, G. "A Note on Partitions and Triangles with Integer Sides." *Amer. Math. Monthly* **86**, 477, 1979.
Honsberger, R. *Mathematical Gems III.* Washington, DC: Math. Assoc. Amer., pp. 39–7, 1985.
Jordan, J. H.; Walch, R.; and Wisner, R. J. "Triangles with Integer Sides." *Amer. Math. Monthly* **86**, 686–89, 1979.
Sloane, N. J. A. Sequences A005044/M0146 in "An On-Line Version of the Encyclopedia of Integer Sequences." http://www.research.att.com/~njas/sequences/eisonline.html.

Aleksandrov's Uniqueness Theorem
A convex body in EUCLIDEAN n-space that is centrally symmetric with center at the ORIGIN is determined among all such bodies by its brightness function (the VOLUME of each projection).

See also TOMOGRAPHY

References
Gardner, R. J. "Geometric Tomography." *Not. Amer. Math. Soc.* **42**, 422–29, 1995.

Aleksandrov-Cech Cohomology
A theory which satisfies all the EILENBERG-STEENROD AXIOMS with the possible exception of the LONG EXACT SEQUENCE OF A PAIR AXIOM, as well as a certain additional continuity CONDITION.

References
Hazewinkel, M. (Managing Ed.). *Encyclopaedia of Mathematics: An Updated and Annotated Translation of the*

Soviet "Mathematical Encyclopaedia." Dordrecht, Netherlands: Reidel, p. 68, 1988.

Aleph

The SET THEORY symbol ($\aleph$) for the CARDINALITY of an INFINITE SET.

See also ALEPH-0, ALEPH-1, COUNTABLE SET, COUNTABLY INFINITE, FINITE, INFINITE, TRANSFINITE NUMBER, UNCOUNTABLY INFINITE

Aleph-0

The SET THEORY symbol $\aleph_0$ for a SET having the same CARDINAL NUMBER as the "small" INFINITE SET of INTEGERS. The ALGEBRAIC NUMBERS also belong to $\aleph_0$. Rather surprising properties satisfied by $\aleph_0$ include

$$\aleph_0^r = \aleph_0 \tag{1}$$

$$r\aleph_0 = \aleph_0 \tag{2}$$

$$\aleph_0 + f = \aleph_0, \tag{3}$$

where f is any FINITE SET. However,

$$\aleph_0^{\aleph_0} = C, \tag{4}$$

where C is the CONTINUUM.

See also ALEPH-1, CARDINAL NUMBER, CONTINUUM, CONTINUUM HYPOTHESIS, COUNTABLY INFINITE, FINITE, INFINITE, TRANSFINITE NUMBER, UNCOUNTABLY INFINITE

Aleph-1

The SET THEORY symbol $\aleph_1$ for the smallest INFINITE SET larger than ALEPH-0, and equal to the CARDINALITY of the set of countable ORDINAL NUMBERS.

The CONTINUUM HYPOTHESIS asserts that $\aleph_1 = c$, where c is the CARDINALITY of the "large" INFINITE SET of REAL NUMBERS (called the CONTINUUM in SET THEORY). However, the truth of the CONTINUUM HYPOTHESIS depends on the version of SET THEORY you are using and so is UNDECIDABLE.

Curiously enough, n-D SPACE has the same number of points (c) as 1-D SPACE, or any FINITE INTERVAL of 1-D SPACE (a LINE SEGMENT), as was first recognized by Georg Cantor.

See also ALEPH-0, CARDINALITY, CONTINUUM, CONTINUUM HYPOTHESIS, COUNTABLY INFINITE, FINITE, INFINITE, ORDINAL NUMBER, TRANSFINITE NUMBER, UNCOUNTABLY INFINITE

Alethic

A term in LOGIC meaning pertaining to TRUTH and FALSEHOOD.

See also FALSE, PREDICATE, TRUE

Alexander Ideal

The order IDEAL in Λ, the RING of integral LAURENT POLYNOMIALS, associated with an ALEXANDER MATRIX for a KNOT K. Any generator of a principal Alexander ideal is called an ALEXANDER POLYNOMIAL. Because the ALEXANDER INVARIANT of a TAME KNOT in $\mathbb{S}^3$ has a SQUARE presentation MATRIX, its Alexander ideal is PRINCIPAL and it has an ALEXANDER POLYNOMIAL $\Delta(t)$.

See also ALEXANDER INVARIANT, ALEXANDER MATRIX, ALEXANDER POLYNOMIAL

References
Rolfsen, D. *Knots and Links.* Wilmington, DE: Publish or Perish Press, pp. 206–07, 1976.

Alexander Invariant

The Alexander invariant $H_*(\hat{X})$ of a KNOT K is the HOMOLOGY of the INFINITE cyclic cover of the complement of K, considered as a MODULE over Λ, the RING of integral LAURENT POLYNOMIALS. The Alexander invariant for a classical TAME KNOT is finitely presentable, and only H_1 is significant.

For any KNOT K^n in $\mathbb{S}^{n+2}$ whose complement has the homotopy type of a FINITE COMPLEX, the Alexander invariant is finitely generated and therefore finitely presentable. Because the Alexander invariant of a TAME KNOT in $\mathbb{S}^3$ has a SQUARE presentation MATRIX, its ALEXANDER IDEAL is PRINCIPAL and it has an ALEXANDER POLYNOMIAL denoted $\Delta(t)$.

See also ALEXANDER IDEAL, ALEXANDER MATRIX, ALEXANDER POLYNOMIAL

References
Rolfsen, D. *Knots and Links.* Wilmington, DE: Publish or Perish Press, pp. 206–07, 1976.

Alexander Matrix

A presentation matrix for the ALEXANDER INVARIANT $H_1(\hat{X})$ of a KNOT K. If V is a SEIFERT MATRIX for a TAME KNOT K in $\mathbb{S}^3$, then $V^T - tV$ and $V^T - tV^T$ are Alexander matrices for K, where V^T denotes the MATRIX TRANSPOSE.

See also ALEXANDER IDEAL, ALEXANDER INVARIANT, ALEXANDER POLYNOMIAL, SEIFERT MATRIX

References
Rolfsen, D. *Knots and Links.* Wilmington, DE: Publish or Perish Press, pp. 206–07, 1976.

Alexander Polynomial

A POLYNOMIAL invariant of a KNOT discovered in 1923 by J. W. Alexander (Alexander 1928). In technical language, the Alexander polynomial arises from the HOMOLOGY of the infinitely cyclic cover of a KNOT's complement. Any generator of a PRINCIPAL ALEXANDER IDEAL is called an Alexander polynomial (Rolfsen

1976). Because the ALEXANDER INVARIANT of a TAME KNOT in $\mathbb{S}^3$ has a SQUARE presentation MATRIX, its ALEXANDER IDEAL is PRINCIPAL and it has an Alexander polynomial denoted $\Delta(t)$.

Let Ψ be the MATRIX PRODUCT of BRAID WORDS of a KNOT, then

$$\frac{\det(1-\Psi)}{1+t+\ldots+t^{n-1}} = \Delta_L, \tag{1}$$

where Δ_L is the Alexander polynomial and det is the DETERMINANT. The Alexander polynomial of a TAME KNOT in $\mathbb{S}^3$ satisfies

$$\Delta(t) = \det(V^T - tV). \tag{2}$$

where V is a SEIFERT MATRIX, det is the DETERMINANT, and V^T denotes the MATRIX TRANSPOSE. The Alexander polynomial also satisfies

$$\Delta(1) = \pm 1. \tag{3}$$

The Alexander polynomial of a splittable link is always 0. Surprisingly, there are known examples of nontrivial KNOTS with Alexander polynomial 1. An example is the $(-3, 5, 7)$ PRETZEL KNOT.

The Alexander polynomial remained the *only* known KNOT POLYNOMIAL until the JONES POLYNOMIAL was discovered in 1984. Unlike the Alexander polynomial, the more powerful JONES POLYNOMIAL *does,* in most cases, distinguish HANDEDNESS. A normalized form of the Alexander polynomial symmetric in t and t^{-1} and satisfying

$$\Delta(\text{unknot}) = 1 \tag{4}$$

was formulated by J. H. Conway and is sometimes denoted ∇_L. The NOTATION $[a+b+c+\ldots$ is an abbreviation for the Conway-normalized Alexander polynomial of a KNOT

$$a + b(x + x^{-1}) + c(x^2 + x^{-2}) + \ldots \tag{5}$$

For a description of the NOTATION for LINKS, see Rolfsen (1976, p. 389). Examples of the Conway-Alexander polynomials for common KNOTS include

$$\nabla_{\text{TK}} = [1-1 = -x^{-1} + 1 - x \tag{6}$$

$$\nabla_{\text{FEK}} = [3-1 = -x^{-1} + 3 - x, \tag{7}$$

$$\nabla_{\text{SSK}} = [1-1+1 = x^{-2} - x^{-1} + 1 - x + x^2 \tag{8}$$

for the TREFOIL KNOT, FIGURE-OF-EIGHT KNOT, and SOLOMON'S SEAL KNOT, respectively. Multiplying through to clear the NEGATIVE POWERS gives the usual Alexander polynomial, where the final SIGN is determined by convention.

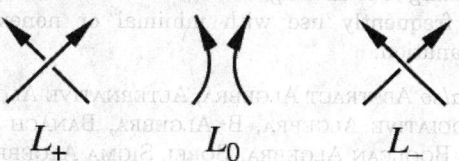

Let an Alexander polynomial be denoted Δ, then there exists a SKEIN RELATIONSHIP (discovered by J. H. Conway)

$$\Delta_{L_+}(t) - \Delta_{L_-}(t) + (t^{-1/2} - t^{1/2})\Delta_{L_0}(t) = 0 \tag{9}$$

corresponding to the above LINK DIAGRAMS (Adams 1994). A slightly different SKEIN RELATIONSHIP convention used by Doll and Hoste (1991) is

$$\nabla_{L_+} - \nabla_{L_-} = z\nabla_{L_0}. \tag{10}$$

These relations allow Alexander polynomials to be constructed for arbitrary knots by building them up as a sequence of over- and undercrossings.

For a KNOT,

$$\Delta_K(-1) \equiv \begin{cases} 1(\text{mod } 8) & \text{if } \text{Arf}(K) = 0, \\ 5(\text{mod } 8) & \text{if } \text{Arf}(K) = 1, \end{cases} \tag{11}$$

where Arf is the ARF INVARIANT (Jones 1985). If K is a KNOT and

$$|\Delta_K(i)| > 3. \tag{12}$$

then K cannot be REPRESENTED AS a closed 3-BRAID. Also, if

$$\Delta_K(e^{2\pi i/5}) > \frac{13}{2}, \tag{13}$$

then K cannot be REPRESENTED AS a closed 4-braid (Jones 1985).

The HOMFLY POLYNOMIAL $P(a, z)$ generalizes the Alexander polynomial (as well at the JONES POLYNOMIAL) with

$$\nabla(z) = P(1, z) \tag{14}$$

(Doll and Hoste 1991).

Rolfsen (1976) gives a tabulation of Alexander polynomials for KNOTS up to 10 CROSSINGS and LINKS up to 9 CROSSINGS.

See also BRAID GROUP, JONES POLYNOMIAL, KNOT, KNOT DETERMINANT, LINK, SKEIN RELATIONSHIP

References

Adams, C. C. *The Knot Book: An Elementary Introduction to the Mathematical Theory of Knots.* New York: W. H. Freeman, pp. 165–69, 1994.

Alexander, J. W. "Topological Invariants of Knots and Links." *Trans. Amer. Math. Soc.* **30**, 275–06, 1928.

Alexander, J. W. "A Lemma on a System of Knotted Curves." *Proc. Nat. Acad. Sci. USA* **9**, 93–5, 1923.

Casti, J. L. "The Alexander Polynomial." Ch. 1 in *Five More Golden Rules: Knots, Codes, Chaos, and Other Great Theories of 20th-Century Mathematics.* New York: Wiley, pp. 1–4, 2000.

Doll, H. and Hoste, J. "A Tabulation of Oriented Links." *Math. Comput.* **57**, 747–61, 1991.

Jones, V. "A Polynomial Invariant for Knots via von Neumann Algebras." *Bull. Amer. Math. Soc.* **12**, 103–11, 1985.

Murasugi, K. and Kurpita, B. I. *A Study of Braids.* Dordrecht, Netherlands: Kluwer, 1999.

Rolfsen, D. "Table of Knots and Links." Appendix C in *Knots and Links.* Wilmington, DE: Publish or Perish Press, pp. 280–87, 1976.
Stoimenow, A. "Alexander Polynomials." http://guests.mpim-bonn.mpg.de/alex/ptab/a10.html.
Stoimenow, A. "Conway Polynomials." http://guests.mpim-bonn.mpg.de/alex/ptab/c10.html.

Alexander's Horned Sphere

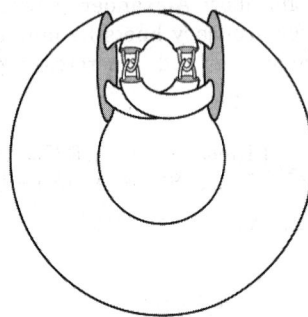

The above solid, composed of a countable UNION of COMPACT SETS, is called Alexander's horned sphere. It is HOMEOMORPHIC with the BALL $\mathbb{B}^3$, and its boundary is therefore a SPHERE. It is therefore an example of a wild embedding in $\mathbb{E}^3$. The outer complement of the solid is not SIMPLY CONNECTED, and its fundamental GROUP is not finitely generated. Furthermore, the set of nonlocally flat ("bad") points of Alexander's horned sphere is a CANTOR SET.

The complement in $\mathbb{R}^3$ of the bad points for Alexander's horned sphere is SIMPLY CONNECTED, making it inequivalent to ANTOINE'S HORNED SPHERE. Alexander's horned sphere has an uncountable infinity of WILD POINTS, which are the limits of the sequences of the horned sphere's branch points (roughly, the "ends" of the horns), since any NEIGHBORHOOD of a limit contains a horned complex.

A humorous drawing by Simon Frazer (Guy 1983, Schroeder 1991, Albers 1994) depicts mathematician John H. Conway with Alexander's horned sphere growing from his head.

See also ANTOINE'S HORNED SPHERE

References
Albers, D. J. Illustration accompanying "The Game of 'Life'." *Math Horizons,* p. 9, Spring 1994.
Guy, R. "Conway's Prime Producing Machine." *Math. Mag.* **56**, 26–3, 1983.
Hocking, J. G. and Young, G. S. *Topology.* New York: Dover, 1988.
Rolfsen, D. *Knots and Links.* Wilmington, DE: Publish or Perish Press, pp. 80–1, 1976.
Schroeder, M. *Fractals, Chaos, Power Law: Minutes from an Infinite Paradise.* New York: W. H. Freeman, p. 58, 1991.

Alexander's Theorem
Any LINK can be represented by a closed BRAID.

Alexander-Conway Polynomial
CONWAY POLYNOMIAL

Alexander-Spanier Cohomology
A fundamental result of DE RHAM COHOMOLOGY is that the kth DE RHAM COHOMOLOGY VECTOR SPACE of a MANIFOLD M is canonically isomorphic to the Alexander-Spanier cohomology VECTOR SPACE $H^k(M;\mathbb{R})$ (also called cohomology with compact support). In the case that M is COMPACT, Alexander-Spanier cohomology is exactly "singular" COHOMOLOGY.

Algebra
The branch of mathematics dealing with such topics as GROUP THEORY, invariant theory, and COHOMOLOGY which studies number systems and operations within them. The word "algebra" is a distortion of the Arabic title of a treatise by al-Khwarizmi about algebraic methods. Note that mathematicians refer to the "school algebra" generally taught in middle and high school as "ARITHMETIC," reserving the word "algebra" for the more advanced aspects of the subject.

Formally, an algebra is a VECTOR SPACE V, over a FIELD F with a MULTIPLICATION which turns it into a RING defined such that, if $f \in F$ and $x, y \in V$, then

$$f(xy) = (fx)y = x(fy).$$

In addition to the usual algebra of REAL NUMBERS, there are ≈ 1151 additional CONSISTENT algebras which can be formulated by weakening the FIELD AXIOMS, at least 200 of which have been *rigorously* proven to be self-CONSISTENT (Bell 1945).

Algebras which have been investigated and found to be of interest are usually named after one or more of their investigators. This practice leads to exotic-sounding (but unenlightening) names which algebraists frequently use with minimal or nonexistent explanation.

See also ABSTRACT ALGEBRA, ALTERNATIVE ALGEBRA, ASSOCIATIVE ALGEBRA, B*-ALGEBRA, BANACH ALGEBRA, BOOLEAN ALGEBRA, BOREL SIGMA ALGEBRA, C*-

ALGEBRA, CAYLEY ALGEBRA, CLIFFORD ALGEBRA, COMMUTATIVE ALGEBRA, DERIVATION ALGEBRA, EXTERIOR ALGEBRA, FUNDAMENTAL THEOREM OF ALGEBRA, GRADED ALGEBRA, GRASSMANN ALGEBRA, HECKE ALGEBRA, HEYTING ALGEBRA, HOMOLOGICAL ALGEBRA, HOPF ALGEBRA, JORDAN ALGEBRA, LIE ALGEBRA, LINEAR ALGEBRA, MEASURE ALGEBRA, NONASSOCIATIVE ALGEBRA, POWER ASSOCIATIVE ALGEBRA, QUATERNION, ROBBINS ALGEBRA, SCHUR ALGEBRA, SEMISIMPLE ALGEBRA, SIGMA ALGEBRA, SIMPLE ALGEBRA, STEENROD ALGEBRA, UMBRAL ALGEBRA, VON NEUMANN ALGEBRA

References

Artin, M. *Algebra.* Englewood Cliffs, NJ: Prentice-Hall, 1991.

Bell, E. T. *The Development of Mathematics, 2nd ed.* New York: McGraw-Hill, pp. 35–6, 1945.

Bhattacharya, P. B.; Jain, S. K.; and Nagpu, S. R. (Eds.). *Basic Algebra, 2nd ed.* New York: Cambridge University Press, 1994.

Birkhoff, G. and Mac Lane, S. *A Survey of Modern Algebra, 5th ed.* New York: Macmillan, 1996.

Brown, K. S. "Algebra." http://www.seanet.com/~ksbrown/ialgebra.htm.

Cardano, G. *Ars Magna or The Rules of Algebra.* New York: Dover, 1993.

Chevalley, C. C. *Introduction to the Theory of Algebraic Functions of One Variable.* Providence, RI: Amer. Math. Soc., 1951.

Chrystal, G. *Textbook of Algebra, 2 vols.* New York: Dover, 1961.

Connell, E. H. *Elements of Abstract and Linear Algebra.* http://www.cs.miami.edu/~ec/book/.

Dickson, L. E. *Algebras and Their Arithmetics.* Chicago, IL: University of Chicago Press, 1923.

Dickson, L. E. *Modern Algebraic Theories.* Chicago, IL: H. Sanborn, 1926.

Dummit, D. S. and Foote, R. M. *Abstract Algebra, 2nd ed.* Englewood Cliffs, NJ: Prentice-Hall, 1998.

Edwards, H. M. *Galois Theory, corrected 2nd printing.* New York: Springer-Verlag, 1993.

Euler, L. *Elements of Algebra.* New York: Springer-Verlag, 1984.

Gallian, J. A. *Contemporary Abstract Algebra, 3rd ed.* Lexington, MA: D. C. Heath, 1994.

Grove, L. *Algebra.* New York: Academic Press, 1983.

Hall, H. S. and Knight, S. R. *Higher Algebra, A Sequel to Elementary Algebra for Schools.* London: Macmillan, 1960.

Harrison, M. A. "The Number of Isomorphism Types of Finite Algebras." *Proc. Amer. Math. Soc.* **17**, 735–37, 1966.

Herstein, I. N. *Noncommutative Rings.* Washington, DC: Math. Assoc. Amer., 1996.

Herstein, I. N. *Topics in Algebra, 2nd ed.* New York: Wiley, 1975.

Jacobson, N. *Basic Algebra II, 2nd ed.* New York: W. H. Freeman, 1989.

Kaplansky, I. *Fields and Rings, 2nd ed.* Chicago, IL: University of Chicago Press, 1995.

Lang, S. *Undergraduate Algebra, 2nd ed.* New York: Springer-Verlag, 1990.

Spiegel, M. R. *Schaum's Outline of Theory and Problems of College Algebra, 2nd ed.* New York: McGraw-Hill, 1997.

Uspensky, J. V. *Theory of Equations.* New York: McGraw-Hill, 1948.

van der Waerden, B. L. *Algebra, Vol. 2.* New York: Springer-Verlag, 1991.

van der Waerden, B. L. *Geometry and Algebra in Ancient Civilizations.* New York: Springer-Verlag, 1983.

van der Waerden, B. L. *A History of Algebra: From al-Khwarizmi to Emmy Noether.* New York: Springer-Verlag, 1985.

Varadarajan, V. S. *Algebra in Ancient and Modern Times.* Providence, RI: Amer. Math. Soc., 1998.

Weisstein, E. W. "Books about Algebra." http://www.treasure-troves.com/books/Algebra.html.

Algebraic Closure

The FIELD $\bar{F}$ is called an algebraic closure of F if $\bar{F}$ is algebraic over F and if every polynomial $f(x) \in F[x]$ SPLITS completely over $\bar{F}$, so that $\bar{F}$ can be said to contain all the elements that are algebraic over F.

For example, the FIELD of COMPLEX NUMBERS $\mathbb{C}$ is the algebraic closure of the FIELD of REALS $\mathbb{R}$.

See also ALGEBRAICALLY CLOSED, SPLITTING FIELD

References

Dummit, D. S. and Foote, R. M. *Abstract Algebra, 2nd ed.* Englewood Cliffs, NJ: Prentice-Hall, p. 455, 1998.

Algebraic Coding Theory

CODING THEORY

Algebraic Combinatorics

The use of techniques from algebra, topology, and geometry in the solution of combinatorial problems, or the use of combinatorial methods to attack problems in these areas (Billera *et al.* 1999, p. ix).

See also COMBINATORICS

References

Billera, L. J.; Björner, A.; Greene, C.; Simion, R. E.; and Stanley, R. P. (Eds.). *New Perspectives in Algebraic Combinatorics.* Cambridge, England: Cambridge University Press, 1999.

Algebraic Congruence

A CONGRUENCE OF THE FORM

$$f(x) \equiv 0 \pmod{n}$$

where $f(x)$ is an INTEGER POLYNOMIAL (Nagell 1951, p. 73).

See also CONGRUENCE, FUNCTIONAL CONGRUENCE

References

Nagell, T. "Algebraic Congruences and Functional Congruences," "Algebraic Congruences to a Prime Modulus," "Algebraic Congruences to a Composite Modulus," "Algebraic Congruences to a Prime-Power Modulus," and "Numerical Examples of Solution of Algebraic Congruences." §22, 24, and 26–8 in *Introduction to Number Theory.* New York: Wiley, pp. 73–6, 79–1, and 83–3, 1951.

Algebraic Connectivity

The second smallest EIGENVALUE of the LAPLACIAN MATRIX of a graph G. This eigenvalue is greater than 0 IFF G is a CONNECTED GRAPH.

See also CONNECTED GRAPH, FIEDLER VECTOR, LAPLACIAN MATRIX

References

Chung, F. R. K. *Spectral Graph Theory.* Providence, RI: Amer. Math. Soc., 1997.
Demmel, J. "CS 267: Notes for Lecture 23, April 9, 1999. Graph Partitioning, Part 2." http://www.cs.berkeley.edu/~demmel/cs267/lecture20/lecture20.html.

Algebraic Curve

An algebraic curve over a FIELD K is an equation $f(X, Y) = 0$, where $f(X, Y)$ is a POLYNOMIAL in X and Y with COEFFICIENTS in K. A nonsingular algebraic curve is an algebraic curve over K which has no SINGULAR POINTS over K. A point on an algebraic curve is simply a solution of the equation of the curve. A K-RATIONAL POINT is a point (X, Y) on the curve, where X and Y are in the FIELD K.

See also ALGEBRAIC GEOMETRY, ALGEBRAIC VARIETY, CURVE

References

Griffiths, P. A. *Introduction to Algebraic Curves.* Providence, RI: Amer. Math. Soc., 1989.

Algebraic Expression

An algebraic expression in variables $\{x_1, \ldots, x_n\}$ is an expression constructed with the variables and ALGEBRAIC NUMBERS using addition, multiplication, and rational powers.

References

Strzebonski, A. "Solving Algebraic Inequalities." *Mathematica J.* **7**, 525–41, 2000.

Algebraic Extension

This entry contributed by NICOLAS BRAY

An extension F of a FIELD K is said to be algebraic if every element of F is algebraic over K (i.e., is the root of a nonzero polynomial with coefficients in K).

See also GALOIS EXTENSION

Algebraic Function

A function which can be constructed using only a finite number of ELEMENTARY OPERATIONS together with the INVERSES of functions capable of being so constructed. Nonalgebraic functions are called TRANSCENDENTAL FUNCTIONS.

See also ELEMENTARY FUNCTION, ELEMENTARY OPERATION, TRANSCENDENTAL FUNCTION

References

Knopp, K. "Algebraic Functions." Ch. 5 in *Theory of Functions Parts I and II, Two Volumes Bound as One, Part II.* New York: Dover, pp. 119–34, 1996.
Koch, H. "Algebraic Functions of One Variable." Ch. 6 in *Number Theory: Algebraic Numbers and Functions.* Providence, RI: Amer. Math. Soc., pp. 141–70, 2000.

Algebraic Function Field

FUNCTION FIELD

Algebraic Geometry

Algebraic geometry is the study of geometries that come from algebra, in particular, from RINGS. In CLASSICAL ALGEBRAIC GEOMETRY, the algebra is the RING of POLYNOMIALS, and the geometry is the set of zeros of polynomials, called an ALGEBRAIC VARIETY. For instance, the UNIT CIRCLE is the set of zeros of $x^2 + y^2 = 1$ and is an ALGEBRAIC VARIETY, as are all of the CONIC SECTIONS.

In the twentieth century, it was discovered that the basic ideas of classical algebraic geometry can be applied to any COMMUTATIVE RING with a unit, such as the INTEGERS. The geometry of such a ring is determined by its algebraic structure, in particular its PRIME IDEALS. Grothendieck defined SCHEMES as the basic geometric objects, which have the same relationship to the geometry of a ring as a MANIFOLD to a COORDINATE CHART. The language of CATEGORY THEORY evolved at around the same time, largely in response to the needs of the increasing abstraction in algebraic geometry.

As a consequence, algebraic geometry became very useful in other areas of mathematics, most notably in ALGEBRAIC NUMBER THEORY. For instance, Deligne used it to prove a variant of the RIEMANN HYPOTHESIS. Also, Andrew Wiles' proof of FERMAT'S LAST THEOREM used the tools developed in algebraic geometry.

In the latter part of the twentieth century, researchers have tried to extend the relationship between algebra and geometry to arbitrary NONCOMMUTATIVE RINGS. The study of geometries associated to noncommutative rings is called NONCOMMUTATIVE GEOMETRY.

See also ALGEBRAIC CURVE, ALGEBRAIC NUMBER THEORY, ALGEBRAIC VARIETY, CATEGORY THEORY, COMMUTATIVE ALGEBRA, CONIC SECTION, DIFFERENTIAL GEOMETRY, GEOMETRY, NONCOMMUTATIVE GEOMETRY, PLANE CURVE, SCHEME, SPACE CURVE, ZARISKI TOPOLOGY

References

Abhyankar, S. S. *Algebraic Geometry for Scientists and Engineers.* Providence, RI: Amer. Math. Soc., 1990.
Bump, D. *Algebraic Geometry.* Singapore: World Scientific, 1998.

Cox, D.; Little, J.; and O'Shea, D. *Ideals, Varieties, and Algorithms: An Introduction to Algebraic Geometry and Commutative Algebra, 2nd ed.* New York: Springer-Verlag, 1996.

Eisenbud, D. *Commutative Algebra with a View Toward Algebraic Geometry.* New York: Springer-Verlag, 1995.

Eisenbud, D. (Ed.). *Commutative Algebra, Algebraic Geometry, and Computational Methods.* Singapore: Springer-Verlag, 1999.

Griffiths, P. and Harris, J. *Principles of Algebraic Geometry.* New York: Wiley, 1978.

Greuel, G.-M. Computer Algebra and Algebraic Geometry--Achievements and Perspectives. 29 Feb 2000. http://xxx.lanl.gov/abs/math.AG/0002247/.

Harris, J. *Algebraic Geometry: A First Course.* New York: Springer-Verlag, 1992.

Hartshorne, R. *Algebraic Geometry, rev. ed.* New York: Springer-Verlag, 1997.

Hulek, K.; Catanese, F.; Peters, C.; and Reid, M. (Eds.). *New Trends in Algebraic Geometry: EuroConference on Algebraic Geometry, Warwick, July 1996.* Cambridge, England: Cambridge University Press, 1999.

Lang, S. *Introduction to Algebraic Geometry.* New York: Interscience, 1958.

Newstead, P. E. (Ed.). *Algebraic Geometry.* New York: Dekker, 1999.

Pedoe, D. and Hodge, W. V. *Methods of Algebraic Geometry, Vol. 1.* Cambridge, England: Cambridge University Press, 1994.

Pedoe, D. and Hodge, W. V. *Methods of Algebraic Geometry, Vol. 2.* Cambridge, England: Cambridge University Press, 1994.

Pedoe, D. and Hodge, W. V. *Methods of Algebraic Geometry, Vol. 3.* Cambridge, England: Cambridge University Press, 1994.

Pragacz, P.; Szurek, M.; and Wisniewski, J. *Algebraic Geometry: Hirzenbruch 70.* Providence, RI: Amer. Math. Soc., 1999.

Seidenberg, A. (Ed.). *Studies in Algebraic Geometry.* Washington, DC: Math. Assoc. Amer., 1980.

Sertöz, S. (Ed.). *Algebraic Geometry.* New York: Dekker, 1998.

van Oystaeyen, F. *Algebraic Geometry for Associative Algebras.* New York: Dekker, 2000.

Weil, A. *Foundations of Algebraic Geometry, enl. ed.* Providence, RI: Amer. Math. Soc., 1962.

Weisstein, E. W. "Books about Algebraic Geometry." http://www.treasure-troves.com/books/AlgebraicGeometry.html.

Yang, K. *Complex Algebraic Geometry: An Introduction to Curves and Surfaces, 2nd ed.* New York: Dekker, 1999.

Algebraic Integer

If r is a ROOT of the POLYNOMIAL equation

$$x^n + a_{n-1}x^{n-1} + \cdots + a_1 x + a_0 = 0,$$

where the a_is are INTEGERS and r satisfies no similar equation of degree $< n$, then r is called an algebraic integer of degree n. An algebraic integer is a special case of an ALGEBRAIC NUMBER (for which the leading COEFFICIENT a_n need not equal 1). RADICAL INTEGERS are a SUBRING of the algebraic integers.

A SUM or PRODUCT of algebraic integers is again an algebraic integer. However, ABEL'S IMPOSSIBILITY THEOREM shows that there are algebraic integers of degree ≥ 5 which are not expressible in terms of ADDITION, SUBTRACTION, MULTIPLICATION, DIVISION, and ROOT EXTRACTION (the ELEMENTARY OPERATIONS)

on COMPLEX NUMBERS. In fact, if ELEMENTARY OPERATIONS are allowed on real numbers only, then there are real numbers which are algebraic integers of degree 3 which cannot be so expressed.

The GAUSSIAN INTEGERS are algebraic integers of $\mathbb{Q}(\sqrt{-1})$, since $a + bi$ are roots of

$$z^2 - 2az + a^2 + b^2 = 0.$$

See also ALGEBRAIC NUMBER, CASUS IRREDUCIBILUS, ELEMENTARY OPERATION, EUCLIDEAN NUMBER, RADICAL INTEGER

References

Ferreirós, J. "Algebraic Integers." §3.3.2 in *Labyrinth of Thought: A History of Set Theory and Its Role in Modern Mathematics.* Basel, Switzerland: Birkhäuser, pp. 97–9, 1999.

Hancock, H. *Foundations of the Theory of Algebraic Numbers, Vol. 1: Introduction to the General Theory.* New York: Macmillan, 1931.

Hancock, H. *Foundations of the Theory of Algebraic Numbers, Vol. 2: The General Theory.* New York: Macmillan, 1932.

Pohst, M. and Zassenhaus, H. *Algorithmic Algebraic Number Theory.* Cambridge, England: Cambridge University Press, 1989.

Wagon, S. "Algebraic Numbers." §10.5 in *Mathematica in Action.* New York: W. H. Freeman, pp. 347–53, 1991.

Algebraic Invariant

A quantity such as a DISCRIMINANT which remains unchanged under a given class of algebraic transformations. Such invariants were originally called HYPERDETERMINANTS by Cayley.

See also DISCRIMINANT (POLYNOMIAL), INVARIANT, QUADRATIC INVARIANT

References

Grace, J. H. and Young, A. *The Algebra of Invariants.* New York: Chelsea, 1965.

Gurevich, G. B. *Foundations of the Theory of Algebraic Invariants.* Groningen, Netherlands: P. Noordhoff, 1964.

Hermann, R. and Ackerman, M. *Hilbert's Invariant Theory Papers.* Brookline, MA: Math Sci Press, 1978.

Hilbert, D. *Theory of Algebraic Invariants.* Cambridge, England: Cambridge University Press, 1993.

Mumford, D.; Fogarty, J.; and Kirwan, F. *Geometric Invariant Theory, 3rd enl. ed.* New York: Springer-Verlag, 1994.

Weisstein, E. W. "Books about Invariants." http://www.treasure-troves.com/books/Invariants.html.

Algebraic Knot

A single component ALGEBRAIC LINK. Most knots up to 11 crossings are algebraic, but they quickly become outnumbered by nonalgebraic knots for more crossings (Hoste *et al.* 1998).

See also ALGEBRAIC LINK, KNOT, LINK

References

Bonahon, F. and Siebermann, L. "The Classification of Algebraic Links." Unpublished manuscript. Hoste, J.; Thistlethwaite, M.; and Weeks, J. "The First 1,701,936 Knots." *Math. Intell.* **20**, 33–8, Fall 1998.

Algebraic K-Theory

K-THEORY

Algebraic Language

Let X be an alphabet (i.e., a finite and nonempty set), and call its member letters. A word on X is a finite sequence of letters $a_1 \ldots a_n$, where $a_1, \ldots, a_n \in X$. Denote the empty word by e, and the set of all words in X by X^*. Define the concatenation (also called product) of a word $u = a_1 \ldots a_n$ with a word $v = b_1 \ldots b_m$ as $uv = a_1 \ldots a_n b_1 \ldots b_m$. In general, concatenation is not commutative. Use the notation $|u|_a$ to mean the number of letters a in the word u. A language $\mathfrak{L}$ is then a subset of X^*, and $\mathfrak{L}$ is said to be algebraic when a set of rewriting rules, applied recursively, forms all the words of $\mathfrak{L}$ and no others.

See also DYCK LANGUAGE

References

Bousquet-Mélou, M. "Convex Polyominoes and Algebraic Languages." *J. Phys. A: Math. Gen.* **25**, 1935–944, 1992.
Delest, M.-P. and Viennot, G. "Algebraic Languages and Polyominoes [sic] Enumeration." *Theoret. Comput. Sci.* **34**, 169–06, 1984.

Algebraic Link

A class of fibered knots and links which arises in ALGEBRAIC GEOMETRY. An algebraic link is formed by connecting the NW and NE strings and the SW and SE strings of an ALGEBRAIC TANGLE (Adams 1994).

See also ALGEBRAIC KNOT, ALGEBRAIC TANGLE, FIBRATION, TANGLE

References

Adams, C. C. *The Knot Book: An Elementary Introduction to the Mathematical Theory of Knots.* New York: W. H. Freeman, pp. 48–9, 1994.
Bonahon, F. and Siebermann, L. "The Classification of Algebraic Links." Unpublished manuscript. Rolfsen, D. *Knots and Links.* Wilmington, DE: Publish or Perish Press, p. 335, 1976.

Algebraic Manifold

An algebraic manifold is another name for a smooth ALGEBRAIC VARIETY. It can be covered by COORDINATE CHARTS so that the TRANSITION FUNCTIONS are given by RATIONAL FUNCTIONS. Technically speaking, the coordinate charts should be to all of affine space $\mathbb{C}^n$.

For example, the SPHERE is an algebraic manifold, with a chart given by STEREOGRAPHIC PROJECTION to $\mathbb{C}$, and another chart at ∞, with the TRANSITION FUNCTION given by $1/z$. In this setting, it is called the RIEMANN SPHERE. The TORUS is also an algebraic manifold, in this setting called an ELLIPTIC CURVE, with charts given by ELLIPTIC FUNCTIONS such as the WEIERSTRASS ELLIPTIC FUNCTION.

See also ABSTRACT MANIFOLD, ALGEBRAIC GEOMETRY, ALGEBRAIC VARIETY, ELLIPTIC CURVE, MANIFOLD

Algebraic Number

If r is a ROOT of the POLYNOMIAL equation

$$a_0 x^n + a_1 x^{n-1} + \cdots + a_{n-1} x + a_n = 0, \qquad (1)$$

where the a_is are INTEGERS and r satisfies no similar equation of degree $< n$, then r is an algebraic number of degree n. If r is an algebraic number and $a_0 = 1$, then it is called an ALGEBRAIC INTEGER. It is also true that if the c_is in

$$a_0 x^n + c_1 x^{n-1} + \cdots + c_{n-1} x + c_n = 0 \qquad (2)$$

are algebraic numbers, then any ROOT of this equation is also an algebraic number.

If α is an algebraic number of degree n satisfying the POLYNOMIAL

$$a(x - \alpha)(x - \beta)(x - \gamma) \ldots, \qquad (3)$$

then there are $n - 1$ other algebraic numbers β, γ, ... called the conjugates of α. Furthermore, if α satisfies any other algebraic equation, then its conjugates also satisfy the same equation (Conway and Guy 1996).

Any number which is not algebraic is said to be TRANSCENDENTAL. The set of algebraic numbers is denoted $\mathbb{A}$ (*Mathematica*), or sometimes $\bar{\mathbb{Q}}$ (Nesterenko 1999), and is implemented in *Mathematica* as Algebraics. A number x can then be tested to see if it is algebraic using the command Element[x, Algebraics].

See also ALGEBRAIC INTEGER, EUCLIDEAN NUMBER, HERMITE-LINDEMANN THEOREM, RADICAL INTEGER, Q-BAR, TRANSCENDENTAL NUMBER

References

Conway, J. H. and Guy, R. K. "Algebraic Numbers." In *The Book of Numbers.* New York: Springer-Verlag, pp. 189–90, 1996.
Courant, R. and Robbins, H. "Algebraic and Transcendental Numbers." §2.6 in *What is Mathematics?: An Elementary Approach to Ideas and Methods, 2nd ed.* Oxford, England: Oxford University Press, pp. 103–07, 1996.
Ferreirós, J. "The Emergence of Algebraic Number Theory." §3.3 in *Labyrinth of Thought: A History of Set Theory and Its Role in Modern Mathematics.* Basel, Switzerland: Birkhäuser, pp. 94–9, 1999.
Hancock, H. *Foundations of the Theory of Algebraic Numbers. Vol. 1: Introduction to the General Theory.* New York: Macmillan, 1931.
Hancock, H. *Foundations of the Theory of Algebraic Numbers. Vol. 2: The General Theory.* New York: Macmillan, 1932.
Koch, H. *Number Theory: Algebraic Numbers and Functions.* Providence, RI: Amer. Math. Soc., 2000.

Nagell, T. *Introduction to Number Theory.* New York: Wiley, p. 35, 1951.

Narkiewicz, W. *Elementary and Analytic Number Theory of Algebraic Numbers.* Warsaw: Polish Scientific Publishers, 1974.

Nesterenko, Yu. V. *A Course on Algebraic Independence: Lectures at IHP 1999.* http://www.math.jussieu.fr/~nesteren/.

Wagon, S. "Algebraic Numbers." §10.5 in *Mathematica in Action.* New York: W. H. Freeman, pp. 347–53, 1991.

Algebraic Number Field
NUMBER FIELD

Algebraic Number Theory
NUMBER THEORY

Algebraic Projective Geometry
PROJECTIVE GEOMETRY

Algebraic Set

An algebraic set is the locus of zeros of a collection of POLYNOMIALS. For example, the circle is the set of zeros of $x^2 + y^2 - 1$ and the point at (a, b) is the set of zeros of x and y. The algebraic set $\{(x, 0)\} \cup \{(0, y)\}$ is the set of solutions to $xy = 0$. It decomposes into two irreducible algebraic sets, called ALGEBRAIC VARIETIES. In general, an algebraic set can be written uniquely as the finite union of ALGEBRAIC VARIETIES.

The intersection of two algebraic sets is an algebraic set corresponding to the union of the polynomials. For example, $x = 0$ and $y = 0$ intersect at $(0, 0)$, i.e., where $x = 0$ and $y = 0$. In fact, the intersection of an arbitrary number of algebraic sets is itself an algebraic set. However, only a finite union of algebraic sets is algebraic. If X is the set of solutions to $f_i = 0$ and Y is the set of solutions to $g_j = 0$, then $X \cup Y$ is the set of solutions to $f_i g_j = 0$. Consequently, the algebraic sets are the closed sets in a TOPOLOGY, called the ZARISKI TOPOLOGY.

The set of polynomials vanishing on an algebraic set X is an IDEAL in the POLYNOMIAL RING. Conversely, any IDEAL defines an algebraic set since it is a collection of polynomials. HILBERT'S NULLSTELLENSATZ describes the precise relationship between IDEALS and algebraic sets.

See also ALGEBRAIC VARIETY, CATEGORY THEORY, COMMUTATIVE ALGEBRA, CONIC SECTION, HILBERT'S NULLSTELLENSATZ, IDEAL, PRIME IDEAL, PROJECTIVE VARIETY, SCHEME, ZARISKI TOPOLOGY

References
Bump, D. *Algebraic Geometry.* Singapore: World Scientific, pp. 1–, 1998.

Hartshorne, R. *Algebraic Geometry.* New York: Springer-Verlag, 1977.

Algebraic Surface

The set of ROOTS of a POLYNOMIAL $f(x, y, z) = 0$. An algebraic surface is said to be of degree $n = \max(i + j + k)$, where n is the maximum sum of powers of all terms $a_m x^{i_m} y^{j_m} z^{k_m}$. The following table lists the names of algebraic surfaces of a given degree.

Order	Surface
3	CUBIC SURFACE
4	QUARTIC SURFACE
5	QUINTIC SURFACE
6	SEXTIC SURFACE
7	HEPTIC SURFACE
8	OCTIC SURFACE
9	NONIC SURFACE
10	DECIC SURFACE
12	DODECIC SURFACE

See also BARTH DECIC, BARTH SEXTIC, BOY SURFACE, CAYLEY CUBIC, CHAIR, CLEBSCH DIAGONAL CUBIC, CUSHION, DERVISH, ENDRASS OCTIC, HEART SURFACE, HENNEBERG'S MINIMAL SURFACE, KUMMER SURFACE, ORDER (ALGEBRAIC SURFACE), ROMAN SURFACE, SARTI DODECIC, SURFACE, TOGLIATTI SURFACE

References
Banchoff, T. F. "Computer Graphics Tools for Rendering Algebraic Surfaces and for Geometry of Order." In *Geometric Analysis and Computer Graphics: Proceedings of a Workshop Held May 23–5, 1988* (Eds. P. Concus, R. Finn, D. A. Hoffman). New York: Springer-Verlag, pp. 31–7, 1991.

Fischer, G. (Ed.). *Mathematical Models from the Collections of Universities and Museums.* Braunschweig, Germany: Vieweg, p. 7, 1986.

Algebraic Tangle

Any TANGLE obtained by additions and multiplications of rational TANGLES (Adams 1994).

See also ALGEBRAIC LINK, TANGLE

References
Adams, C. C. *The Knot Book: An Elementary Introduction to the Mathematical Theory of Knots.* New York: W. H. Freeman, pp. 41–1, 1994.

Algebraic Topology

The study of intrinsic qualitative aspects of spatial objects (e.g., SURFACES, SPHERES, TORI, CIRCLES, KNOTS, LINKS, configuration spaces, etc.) that remain invariant under both-directions continuous ONE-TO-ONE (HOMEOMORPHIC) transformations. The discipline of algebraic topology is popularly known as "RUBBER-SHEET GEOMETRY" and can also be viewed as the study of DISCONNECTIVITIES. Algebraic topology has a great deal of mathematical machinery for studying different kinds of HOLE structures, and it gets the prefix "algebraic" since many HOLE structures are represented best by algebraic objects like GROUPS and RINGS.

A technical way of saying this is that algebraic topology is concerned with FUNCTORS from the topological CATEGORY of GROUPS and HOMOMORPHISMS. Here, the FUNCTORS are a kind of filter, and given an "input" SPACE, they spit out something else in return. The returned object (usually a GROUP or RING) is then a representation of the HOLE structure of the SPACE, in the sense that this algebraic object is a vestige of what the original SPACE was like (i.e., much information is lost, but some sort of "shadow" of the SPACE is retained–just enough of a shadow to understand some aspect of its HOLE-structure, but no more). The idea is that FUNCTORS give much simpler objects to deal with. Because SPACES by themselves are very complicated, they are unmanageable without looking at particular aspects.

COMBINATORIAL TOPOLOGY is a special type of algebraic topology that uses COMBINATORIAL methods.

See also CATEGORY, COMBINATORIAL TOPOLOGY, DIFFERENTIAL TOPOLOGY, FUNCTOR, HOMOTOPY THEORY, TOPOLOGY

References
Dieudonné, J. *A History of Algebraic and Differential Topology: 1900–960.* Boston, MA: Birkhäuser, 1989.

Dodson, C. T. J. and Parker, P. E. *A User's Guide to Algebraic Topology.* Dordrecht, Netherlands: Kluwer, 1997.

Massey, W. S. *A Basic Course in Algebraic Topology.* New York: Springer-Verlag, 1991.

Maunder, C. R. F. *Algebraic Topology.* New York: Dover, 1997.

May, J. P. *A Concise Course on Algebraic Topology.* Chicago, IL: University of Chicago Press, 1999.

May, J. P. *Simplicial Objects in Algebraic Topology.* Chicago, IL: University of Chicago Press, 1982.

Munkres, J. R. *Elements of Algebraic Topology.* Perseus Press, 1993.

Sato, H. *Algebraic Topology: An Intuitive Approach.* Providence, RI: Amer. Math. Soc., 1999.

Weisstein, E. W. "Books about Topology." http://www.treasure-troves.com/books/Topology.html.

Algebraic Unknotting Number

The algebraic unknotting number of a knot K in $\mathbb{S}^3$ is defined as the algebraic unknotting number of the S-equivalence class of a SEIFERT MATRIX of K. The algebraic unknotting number of an element in an S-equivalent class is defined as the minimum number of algebraic unknotting operations necessary to transform the element to the S-equivalence class of the zero matrix (Saeki 1999).

See also SEIFERT MATRIX, UNKNOTTING NUMBER

References
Fogel, M. "Knots with Algebraic Unknotting Number One." *Pacific J. Math.* **163**, 277–95, 1994.

Murakami, H. "Algebraic Unknotting Operation, Q&A." *Gen. Topology* **8**, 283–92, 1990.

Saeki, O. "On Algebraic Unknotting Numbers of Knots." *Tokyo J. Math.* **22**, 425–43, 1999.

Algebraic Variety

A generalization to n-D of ALGEBRAIC CURVES. More technically, an algebraic variety is a reduced SCHEME of FINITE type over a FIELD K. An algebraic variety V is defined as the SET of points in the REALS $\mathbb{R}^n$ (or the COMPLEX NUMBERS $\mathbb{C}^n$) satisfying a system of POLYNOMIAL equations $f_i(x_1, \ldots, x_n) = 0$ for $i = 1, 2, \ldots$. According to the HILBERT BASIS THEOREM, a FINITE number of equations suffices.

A variety is the set of common zeros to a collection of POLYNOMIALS. In classical algebraic geometry, the polynomials have COMPLEX NUMBERS for coefficients. Because of the FUNDAMENTAL THEOREM OF ALGEBRA, such polynomials always have zeros. For example,

$$\{(x, y, z) : x^2 + y^2 - z^2\}$$

is the CONE, and

$$\{(x, y, z) : x^2 + y^2 - z^2, \ ax + by + cz = 0\}$$

is a CONIC SECTION, which is a SUBVARIETY of the cone.

Actually, the cone and the conic section are examples of AFFINE VARIETIES because they are in AFFINE SPACE. A general variety is comprised of affine varieties glued together, like the COORDINATE CHARTS of a MANIFOLD. The FIELD of coefficients can be any ALGEBRAICALLY CLOSED field. When a variety is embedded in projective space, it is a PROJECTIVE ALGEBRAIC VARIETY. Also, an INTRINSIC VARIETY can be thought of as an abstract object, like a MANIFOLD, independent of any particular embedding. A SCHEME is a generalization of a variety, which includes the possibility of replacing $\mathbb{C}[x, y, z]$ by any COMMUTATIVE RING with a unit. A further generalization is a STACK.

See also ABELIAN VARIETY, AFFINE VARIETY, ALBANESE VARIETY, ALGEBRAIC NUMBER THEORY, BRAUER-SEVERI VARIETY, CATEGORY THEORY, CHOW VARIETY, COMMUTATIVE ALGEBRA, CONIC SECTION, INTRINSIC VARIETY, PICARD VARIETY, PROJECTIVE ALGEBRAIC VARIETY, SCHEME, STACK (MODULI SPACE), ZARISKI TOPOLOGY

References
Bump, D. *Algebraic Geometry.* Singapore: World Scientific, pp. 79–6, 1998.

Ciliberto, C.; Laura, E.; and Somese, A. J. (Eds.). *Classification of Algebraic Varieties.* Providence, RI: Amer. Math. Soc., 1994.

Hartshorne, R. *Algebraic Geometry.* New York: Springer-Verlag, 1977.

Algebraically Closed

A FIELD K is said to be algebraically closed if every POLYNOMIAL with coefficients in K has a ROOT in K.

See also ALGEBRAIC CLOSURE, FIELD

References

Dummit, D. S. and Foote, R. M. *Abstract Algebra, 2nd ed.* Englewood Cliffs, NJ: Prentice-Hall, p. 455, 1998.

Algebraically Independent

This entry contributed by JOHNNY CHEN

Let K be a FIELD, and A a K-algebra. Elements $y_1, ..., y_n$ are algebraically independent over K if the natural surjection $K[Y_1, ..., Y_n] \to K[y_1, ...y_n]$ is an isomorphism. In other words, there are no polynomial relations $F(y_1, ..., y_n) = 0$ with coefficients in K.

References

Reid, M. *Undergraduate Commutative Algebra.* Cambridge, England: Cambridge University Press, 1995.

See also IRRATIONAL NUMBER, LINDEMANN-WEIERSTRASS THEOREM, SCHANUEL'S CONJECTURE, SHIDLOVSKII THEOREM, TRANSCENDENTAL NUMBER

Algebraics

ALGEBRAIC NUMBER

Algebroidal Function

An ANALYTIC FUNCTION $f(z)$ satisfying the irreducible algebraic equation

$$A_0(z)f^k + A_1(z)f^{k-1} + \cdots + A_k(z) = 0$$

with single-valued MEROMORPHIC FUNCTIONS $A_j(z)$ in a COMPLEX DOMAIN G is called a k-algebroidal function in G.

See also MEROMORPHIC FUNCTION

References

Iyanaga, S. and Kawada, Y. (Eds.). "Algebroidal Functions." §19 in *Encyclopedic Dictionary of Mathematics.* Cambridge, MA: MIT Press, pp. 86–8, 1980.

Algorithm

A specific set of instructions for carrying out a procedure or solving a problem, usually with the requirement that the procedure terminate at some point. Specific algorithms sometimes also go by the name METHOD, PROCEDURE, or TECHNIQUE. The word "algorithm" is a distortion of al-Khwarizmi, an Arab mathematician who wrote an influential treatise about algebraic methods.

See also 196-ALGORITHM, ALGORITHMIC COMPLEXITY, ARCHIMEDES ALGORITHM, BHASKARA-BROUCKNER ALGORITHM, BORCHARDT-PFAFF ALGORITHM, BRELAZ'S HEURISTIC ALGORITHM, BUCHBERGER'S ALGORITHM, BULIRSCH-STOER ALGORITHM, BUMPING ALGORITHM, COMPUTABLE FUNCTION, CONTINUED FRACTION FACTORIZATION ALGORITHM, DECISION PROBLEM, DIJKSTRA'S ALGORITHM, EUCLIDEAN ALGORITHM, FERGUSON-FORCADE ALGORITHM, FERMAT'S ALGORITHM, FLOYD'S ALGORITHM, GAUSSIAN APPROXIMATION ALGORITHM, GENETIC ALGORITHM, GOSPER'S ALGORITHM, GREEDY ALGORITHM, HASSE'S ALGORITHM, HJLS ALGORITHM, JACOBI ALGORITHM, KRUSKAL'S ALGORITHM, LEVINE-O'SULLIVAN GREEDY ALGORITHM, LLL ALGORITHM, MARKOV ALGORITHM, MILLER'S ALGORITHM, NEVILLE'S ALGORITHM, NEWTON'S METHOD, PRIME FACTORIZATION ALGORITHMS, PRIMITIVE RECURSIVE FUNCTION, PROGRAM, PSLQ ALGORITHM, PSOS ALGORITHM, QUOTIENT-DIFFERENCE ALGORITHM, RISCH ALGORITHM, SCHRAGE'S ALGORITHM, SHANKS' ALGORITHM, SPIGOT ALGORITHM, SYRACUSE ALGORITHM, TOTAL FUNCTION, TURING MACHINE, ZASSENHAUS-BERLEKAMP ALGORITHM, ZEILBERGER'S ALGORITHM

References

Aho, A. V.; Hopcroft, J. E.; and Ullman, J. D. *The Design and Analysis of Computer Algorithms.* Reading, MA: Addison-Wesley, 1974.

Atallah, M. J. *Algorithms and Theory of Computation Handbook.* Boca Raton, FL: CRC Press, 1998.

Baase, S. *Computer Algorithms.* Reading, MA: Addison-Wesley, 1988.

Bellman, R. E.; Cooke, K. L.; and Lockett, J. A. *Algorithms, Graphs, and Computers.* New York: Academic Press, 1970.

Brassard, G. and Bratley, P. *Fundamentals of Algorithmics.* Englewood Cliffs, NJ: Prentice-Hall, 1995.

Chabert, J.-L. (Ed.). *A History of Algorithms: From the Pebble to the Microchip.* New York: Springer-Verlag, 1999.

Collberg, C. "AlgoVista." http://www.algovista.com/.

Cormen, T. H.; Leiserson, C. E.; and Rivest, R. L. *Introduction to Algorithms.* Cambridge, MA: MIT Press, 1990.

Greene, D. H. and Knuth, D. E. *Mathematics for the Analysis of Algorithms, 3rd ed.* Boston, MA: Birkhäuser, 1990.

Harel, D. *Algorithmics: The Spirit of Computing, 2nd ed.* Reading, MA: Addison-Wesley, 1992.

Knuth, D. E. *The Art of Computer Programming, Vol. 1: Fundamental Algorithms, 3rd ed.* Reading, MA: Addison-Wesley, 1997.

Knuth, D. E. *The Art of Computer Programming, Vol. 2: Seminumerical Algorithms, 3rd ed.* Reading, MA: Addison-Wesley, 1998.

Knuth, D. E. *The Art of Computer Programming, Vol. 3: Sorting and Searching, 2nd ed.* Reading, MA: Addison-Wesley, 1998.

Kozen, D. C. *Design and Analysis and Algorithms.* New York: Springer-Verlag, 1991.

Nijenhuis, A. and Wilf, H. *Combinatorial Algorithms for Computers and Calculators, 2nd ed.* New York: Academic Press, 1978.

Sedgewick, R. *Algorithms in C, 3rd ed.* Reading, MA: Addison-Wesley, 1998.

Sedgewick, R. and Flajolet, P. *An Introduction to the Analysis of Algorithms.* Reading, MA: Addison-Wesley, 1996.

Skiena, S. S. *The Algorithm Design Manual.* New York: Springer-Verlag, 1997.

Skiena, S. *Implementing Discrete Mathematics: Combinatorics and Graph Theory with Mathematica.* Reading, MA: Addison-Wesley, 1990.

Skiena, S. S. "The Stony Brook Algorithm Repository." http://www.cs.sunysb.edu/~algorith/.

Wilf, H. *Algorithms and Complexity.* Englewood Cliffs, NJ: Prentice Hall, 1986. http://www.cis.upenn.edu/~wilf/AlgComp2.html.

Algorithmic Complexity

BIT COMPLEXITY, KOLMOGOROV COMPLEXITY

Alhazen's Billiard Problem

In a given CIRCLE, find an ISOSCELES TRIANGLE whose LEGS pass through two given POINTS inside the CIRCLE. This can be restated as: from two POINTS in the PLANE of a CIRCLE, draw LINES meeting at the POINT of the CIRCUMFERENCE and making equal ANGLES with the NORMAL at that POINT.

The problem is called the billiard problem because it corresponds to finding the POINT on the edge of a circular "BILLIARD" table at which a cue ball at a given POINT must be aimed in order to carom once off the edge of the table and strike another ball at a second given POINT. The solution leads to a BIQUADRATIC EQUATION OF THE FORM

$$H(x^2 - y^2) - 2Kxy + (x^2 + y^2)(hy - kx) = 0.$$

The problem is equivalent to the determination of the point on a spherical mirror where a ray of light will reflect in order to pass from a given source to an observer. It is also equivalent to the problem of finding, given two points and a CIRCLE such that the points are both inside or outside the CIRCLE, the ELLIPSE whose FOCI are the two points and which is tangent to the given CIRCLE.

The problem was first formulated by Ptolemy in 150 AD, and was named after the Arab scholar Alhazen, who discussed it in his work on optics. It was not until 1997 that Neumann proved the problem to be insoluble using a COMPASS and RULER construction because the solution requires extraction of a CUBE ROOT (Neumann 1998). This is the same reason that the CUBE DUPLICATION problem is insoluble.

See also BILLIARDS, BILLIARD TABLE PROBLEM, CUBE DUPLICATION

References

Dörrie, H. "Alhazen's Billiard Problem." §41 in *100 Great Problems of Elementary Mathematics: Their History and Solutions.* New York: Dover, pp. 197–00, 1965.

Hogendijk, J. P. "Al-Mutaman's Simplified Lemmas for Solving 'Alhazen's Problem'." *From Baghdad to Barcelona / De Bagdad à Barcelona, Vol. I, II (Zaragoza, 1993),* pp. 59–01, Anu. Filol. Univ. Barc., XIX B-2, Univ. Barcelona, Barcelona, 1996.

Lohne, J. A. "Alhazens Spiegelproblem." *Nordisk Mat. Tidskr.* **18**, 5–5, 1970.

Neumann, P. M. " Reflections on Reflection in a Spherical Mirror." *Amer. Math. Monthly* **105**, 523–28, 1998.

Riede, H. "Reflexion am Kugelspiegel. Oder: das Problem des Alhazen." *Praxis Math.* **31**, 65–0, 1989.

Sabra, A. I. "ibn al-Haytham's Lemmas for Solving 'Alhazen's Problem'." *Arch. Hist. Exact Sci.* **26**, 299–24, 1982.

Alhazen's Problem

ALHAZEN'S BILLIARD PROBLEM

Alias Transformation

A transformation in which the coordinate system is changed, leaving vectors in the original coordinate system "fixed" while changing their representation in the new coordinate system. In contrast, a transformation in which vectors are transformed in a fixed coordinate system is called an ALIBI TRANSFORMATION.

See also ALIBI TRANSFORMATION, ROTATION FORMULA

Aliasing

Given a power spectrum (a plot of power vs. frequency), aliasing is a false translation of power falling in some frequency range $(-f_c, f_c)$ outside the range. Aliasing can be caused by discrete sampling below the NYQUIST FREQUENCY. The sidelobes of any INSTRUMENT FUNCTION (including the simple SINC SQUARED function obtained simply from FINITE sampling) are also a form of aliasing. Although sidelobe contribution at large offsets can be minimized with the use of an APODIZATION FUNCTION, the tradeoff is a widening of the response (i.e., a lowering of the resolution).

See also APODIZATION FUNCTION, NYQUIST FREQUENCY

Alibi Transformation

A transformation in which vectors are transformed in a fixed coordinate system. In contrast, a transformation in which the coordinate system is changed, leaving vectors in the original coordinate system "fixed" while changing their representation in the new coordinate system, is called an ALIAS TRANSFORMATION.

See also ALIAS TRANSFORMATION, ROTATION FORMULA

Aliquant Divisor

A number which does not DIVIDE another exactly. For instance, 4 and 5 are aliquant divisors of 6. A number which is not an aliquant divisor (i.e., one that *does* DIVIDE another exactly) is said to be an ALIQUOT DIVISOR.

See also ALIQUOT DIVISOR, DIVISOR, PROPER DIVISOR

Aliquot Cycle

ALIQUOT SEQUENCE, SOCIABLE NUMBERS

Aliquot Divisor

A number which DIVIDES another exactly. For instance, 1, 2, 3, and 6 are aliquot divisors of 6. A number which is not an aliquot divisor is said to be an ALIQUANT DIVISOR. The term "aliquot" is frequently used to specifically mean a PROPER DIVISOR, i.e., a DIVISOR of a number other than the number itself.

See also ALIQUANT DIVISOR, DIVISOR, PROPER DIVISOR

Aliquot Sequence

Let

$$s(n) \equiv \sigma(n) - n$$

where $\sigma(n)$ is the DIVISOR FUNCTION and $s(n)$ is the RESTRICTED DIVISOR FUNCTION. Then the SEQUENCE of numbers

$$s^0(n) \equiv n, \ s^1(n) = s(n), \ s^2(n) = s(s(n)), \cdots$$

is called an aliquot sequence. If the SEQUENCE for a given n is bounded, it either ends at $s(1) = 0$ or becomes periodic.

1. If the SEQUENCE reaches a constant, the constant is known as a PERFECT NUMBER.
2. If the SEQUENCE reaches an alternating pair, it is called an AMICABLE PAIR.
3. If, after k iterations, the SEQUENCE yields a cycle of minimum length t OF THE FORM $s^{k+1}(n)$, $s^{k+2}(n)$, ..., $s^{k+1}(n)$, then these numbers form a group of SOCIABLE NUMBERS of order t.

It has not been proven that all aliquot sequences eventually terminate and become period. The smallest number whose fate is not known is 276, which has been computed up to $s^{628}(276)$ (Guy 1994). There are five such sequences less than 1000, namely 276, 552, 564, 660, and 966, sometimes called the "Lehmer five." Furthermore, there are 934 open sequences $\leq 10^5$, and 9710 open sequences $\leq 10^6$ (Creyaufmüller).

See also 196-ALGORITHM, ADDITIVE PERSISTENCE, AMICABLE NUMBERS, CATALAN'S ALIQUOT SEQUENCE CONJECTURE, MULTIAMICABLE NUMBERS, MULTIPERFECT NUMBER, MULTIPLICATIVE PERSISTENCE, PERFECT NUMBER, SOCIABLE NUMBERS, UNITARY ALIQUOT SEQUENCE

References
Creyaufmüller, W. "Aliquot Sequences." http://home.t-online.de/home/Wolfgang.Creyaufmueller/aliquote.htm.
Guy, R. K. "Aliquot Sequences." §B6 in *Unsolved Problems in Number Theory, 2nd ed.* New York: Springer-Verlag, pp. 60-2, 1994.
Guy, R. K. and Selfridge, J. L. "What Drives Aliquot Sequences." *Math. Comput.* **29**, 101-07, 1975.
Sloane, N. J. A. Sequences A003023/M0062 in "An On-Line Version of the Encyclopedia of Integer Sequences." http://www.research.att.com/~njas/sequences/eisonline.html.

Sloane, N. J. A. and Plouffe, S. Figure M0062 in *The Encyclopedia of Integer Sequences.* San Diego: Academic Press, 1995.

Alladi-Grinstead Constant

N.B. A detailed online essay by S. Finch was the starting point for this entry.

Let $N(n)$ be the number of ways in which the FACTORIAL $n!$ can be decomposed into n FACTORS of the form $P_k^{b_k}$ arranged in nondecreasing order. Also define

$$m(n) \equiv \max(p_1^{b_1}), \tag{1}$$

i.e., $m(n)$ is the LEAST PRIME FACTOR raised to its appropriate POWER in the factorization. Then define

$$\alpha(n) \equiv \frac{\ln m(n)}{\ln n} \tag{2}$$

where $\ln(x)$ is the NATURAL LOGARITHM. For instance,

$$
\begin{aligned}
9! &= 2 \cdot 2 \cdot 2 \cdot 2 \cdot 2 \cdot 2^2 \cdot 5 \cdot 7 \cdot 3^4 \\
&= 2 \cdot 2 \cdot 2 \cdot 2 \cdot 3 \cdot 5 \cdot 7 \cdot 2^3 \cdot 3^3 \\
&= 2 \cdot 2 \cdot 2 \cdot 2 \cdot 5 \cdot 7 \cdot 2^3 \cdot 3^2 \cdot 3^2 \\
&= 2 \cdot 2 \cdot 2 \cdot 3 \cdot 2^2 \cdot 2^2 \cdot 5 \cdot 7 \cdot 3^3 \\
&= 2 \cdot 2 \cdot 2 \cdot 2^2 \cdot 2^2 \cdot 5 \cdot 7 \cdot 3^2 \cdot 3^2 \\
&= 2 \cdot 2 \cdot 2 \cdot 3 \cdot 3 \cdot 5 \cdot 7 \cdot 3^2 \cdot 2^4 \\
&= 2 \cdot 2 \cdot 3 \cdot 3 \cdot 2^2 \cdot 5 \cdot 7 \cdot 2^3 \cdot 3^2 \\
&= 2 \cdot 2 \cdot 3 \cdot 3 \cdot 3 \cdot 3 \cdot 5 \cdot 7 \cdot 2^5 \\
&= 2 \cdot 3 \cdot 3 \cdot 2^2 \cdot 2^2 \cdot 2^2 \cdot 5 \cdot 7 \cdot 3^2 \\
&= 2 \cdot 3 \cdot 3 \cdot 3 \cdot 3 \cdot 2^2 \cdot 5 \cdot 7 \cdot 2^4 \\
&= 2 \cdot 3 \cdot 3 \cdot 3 \cdot 3 \cdot 5 \cdot 7 \cdot 2^3 \cdot 2^3 \\
&= 3 \cdot 3 \cdot 3 \cdot 3 \cdot 2^2 \cdot 2^2 \cdot 5 \cdot 7 \cdot 2^3, \tag{3}
\end{aligned}
$$

so

$$\alpha(9) = \frac{\ln 3}{\ln 9} = \frac{\ln 3}{2\ln 3} = \frac{1}{2}. \tag{4}$$

For large n,

$$\lim_{n \to \infty} \alpha(n) = e^{c-1} = 0.809394020534..., \tag{5}$$

where

$$c \equiv \sum_{k=2}^{\infty} \frac{1}{k} \ln\left(\frac{k}{k-1}\right). \tag{6}$$

References
Alladi, K. and Grinstead, C. "On the Decomposition of $n!$ into Prime Powers." *J. Number Th.* **9**, 452-58, 1977.
Finch, S. "Favorite Mathematical Constants." http://www.mathsoft.com/asolve/constant/aldgrns/aldgrns.html.
Guy, R. K. "Factorial n as the Product of n Large Factors." §B22 in *Unsolved Problems in Number Theory, 2nd ed.* New York: Springer-Verlag, p. 79, 1994.

Allais Paradox

Choose between the following two alternatives:

1. 90% chance of an unknown amount x and a 10% chance of $1 million, or
2. 89% chance of the same unknown amount x, 10% chance of $2.5 million, and 1% chance of nothing.

The PARADOX is to determine which choice has the larger EXPECTATION VALUE, $0.9x+\$100,000$ or $0.89x+\$250,000$. However, the best choice depends on the unknown amount, even though it is the same in both cases! This appears to violate the INDEPENDENCE AXIOM.

See also INDEPENDENCE AXIOM, MONTY HALL PROBLEM, NEWCOMB'S PARADOX

References

Allais, M. "Le comportement de l'homme rationnel devant le risque: Critique des postulats et axiomes de l'école américaine." *Econometrica* **21**, 503–46, 1953.
Kreps, D. M. *Notes on the Theory of Choice.* Boulder, CO: Westview Press, p. 192, 1988.
Fishburn, P. C. *Utility Theory for Decision Making.* New York: Wiley, 1970.
Savage, L. J. *The Foundations of Statistics, 2nd ed.* New York: Dover, 1972.

Allegory

A technical mathematical object which bears the same resemblance to binary relations as CATEGORIES do to FUNCTIONS and SETS.

See also CATEGORY

References

Freyd, P. J. and Scedrov, A. *Categories, Allegories.* Amsterdam, Netherlands: North-Holland, 1990.

Allometric

Mathematical growth in which one population grows at a rate PROPORTIONAL to the POWER of another population.

References

Coffey, W. J. *Geography Towards a General Spatial Systems Approach.* London: Routledge, Chapman & Hall, 1981.

All-Pairs Shortest Path

The shortest distance between any pair of vertices in the shortest-path spanning tree, as long as the path giving the shortest path does not pass through the root of the spanning tree (Skiena 1990, p. 228). The problem can be solved using n applications of DIJKSTRA'S ALGORITHM or FLOYD'S ALGORITHM. The latter also works in the case of a weighted graph where the edges have negative weights.

See also FLOYD'S ALGORITHM, DIJKSTRA'S ALGORITHM, GRAPH GEODESIC

References

Skiena, S. "All Pairs Shortest Paths." §6.1.2 in *Implementing Discrete Mathematics: Combinatorics and Graph Theory with Mathematica.* Reading, MA: Addison-Wesley, pp. 228–29, 1990.

All-Poles Model

MAXIMUM ENTROPY METHOD

All-to-All Communication

GOSSIPING

Almost All

Given a property P, if $P(x) \sim x$ as $x \to \infty$ (so the number of numbers less than x not satisfying the property P is $\sigma(x)$), then P is said to hold true for almost all numbers. For example, almost all positive integers are COMPOSITE NUMBERS (which is not in conflict with the second of EUCLID'S THEOREMS that there are an infinite number of PRIMES).

See also FOR ALL, NORMAL ORDER

References

Hardy, G. H. *Ramanujan: Twelve Lectures on Subjects Suggested by His Life and Work, 3rd ed.* New York: Chelsea, p. 50, 1999.
Hardy, G. H. and Wright, E. M. *An Introduction to the Theory of Numbers, 5th ed.* Oxford, England: Clarendon Press, p. 8, 1979.

Almost Alternating Knot

An ALMOST ALTERNATING LINK with a single component.

See also ALMOST ALTERNATING LINK

Almost Alternating Link

Call a projection of a LINK an almost alternating projection if one crossing change in the projection makes it an alternating projection. Then an almost alternating link is a LINK with an almost alternating projection, but no alternating projection. Every ALTERNATING KNOT has an almost alternating projection. A PRIME KNOT which is almost alternating is either a TORUS KNOT or a HYPERBOLIC KNOT. Therefore, no SATELLITE KNOT is an almost alternating knot.

All nonalternating 9-crossing PRIME KNOTS are almost alternating. Of the 393 nonalternating knots and links with 11 or fewer crossings, all but five are known to be almost alternating (and 3 of these have 11 crossings). The fate of the remaining five is not known. The $(q, 2)$, $(4, 3)$, and $(5, 3)$-TORUS KNOTS are almost alternating (Adams 1994, p. 142).

See also ALTERNATING KNOT, LINK

References

Adams, C. C. *The Knot Book: An Elementary Introduction to the Mathematical Theory of Knots.* New York: W. H. Freeman, pp. 139–46, 1994.

Almost Everywhere

A property of X is said to hold almost everywhere if the SET of points in X where this property fails has MEASURE ZERO.

See also ALMOST EVERYWHERE CONVERGENCE, MEASURE ZERO

References

Jeffreys, H. and Jeffreys, B. S. "Measure Zero': 'Almost Everywhere'." §1.1013 in *Methods of Mathematical Physics, 3rd ed.* Cambridge, England: Cambridge University Press, pp. 29–0, 1988.
Sansone, G. *Orthogonal Functions, rev. English ed.* New York: Dover, p. 1, 1991.

Almost Everywhere Convergence

A weakened version of POINTWISE CONVERGENCE hypothesis which states that, for X a MEASURE SPACE, $f_n(x) \to f(x)$ for all $x \in Y$, where Y is a measurable subset of X such that $\mu(X \setminus Y) = 0$.

See also POINTWISE CONVERGENCE

References

Browder, A. *Mathematical Analysis: An Introduction.* New York: Springer-Verlag, 1996.

Almost Integer

A number which is very close to an INTEGER. One surprising example involving both E and PI is

$$e^\pi - \pi = 19.999099979\ldots \tag{1}$$

which can also be written as

$$(\pi + 20)^i = -0.9999999992 - 0.0000388927i \approx -1 \tag{2}$$

$$\cos(\ln(\pi + 20)) \approx -0.9999999992. \tag{3}$$

Applying COSINE a few more times gives

$$\cos(\pi \cos(\pi \cos(\ln(\pi + 20))))$$
$$\approx -1 + 3.9321609261 \times 10^{-35}. \tag{4}$$

This curious near-identity was apparently noticed almost simultaneously around 1988 by N. J. A. Sloane, J. H. Conway, and S. Plouffe, but no satisfying explanation as to "why" it has been true has yet been discovered.

An interesting near-identity is given by

$$\frac{1}{4}\left[\cos\left(\tfrac{1}{10}\right) + \cosh\left(\tfrac{1}{10}\right) + 2\cos\left(\tfrac{1}{20}\sqrt{2}\right)\cosh\left(\tfrac{1}{20}\sqrt{2}\right)\right]$$
$$= 1 + 2.480\ldots \times 10^{-13} \tag{5}$$

(W. Dubuque). Other remarkable near-identities are given by

$$\frac{5(1+\sqrt{5})[\Gamma\left(\tfrac{3}{4}\right)]^2}{e^{5x/6}\sqrt{\pi}} = 1 + 4.5422\ldots \times 10^{-14} \tag{6}$$

where $\Gamma(z)$ is the GAMMA FUNCTION (S. Plouffe),

$$e^6 - \pi^4 - \pi^5 = 0.000017673\ldots \tag{7}$$

(D. Wilson),

$$r\left(\frac{160}{\pi}\right)^{1/13} \approx 0.9999996766, \tag{8}$$

where $r \approx 0.739085$ is the root of $x = \cos x$ (L. A. Broukhis),

$$\ln 2 + \log_{10} 2 = 0.994177\ldots \tag{9}$$

(D. Davis),

$$\frac{163}{\ln 163} = 31.9999983738\ldots \tag{10}$$

(posted to `sci.math`; origin unknown),

$$eK^{5/7 - \gamma}\pi^{-(2/7+\gamma)} \approx 1.00014678 \tag{11}$$

$$\frac{K^{\gamma - 19/7}\pi^{2/7 + \gamma}}{2\phi} \approx 1.00105 \tag{12}$$

$$e\gamma\phi(K\pi)^{-(2/7+\gamma)} \approx 1.01979, \tag{13}$$

where K is CATALAN'S CONSTANT, γ is the EULER-MASCHERONI CONSTANT, and ϕ is the GOLDEN RATIO (D. Barron), and

$$163(\pi - e) = 68.999664\ldots \tag{14}$$

$$\frac{53453}{\ln 53453} = 4910.00000122\ldots \tag{15}$$

$$\left[(2-1)^2 + \frac{(5^2-1)^2}{6^2+1}\right]e - \left[(2+1)^2 + \frac{(5^2+1)^2}{6^2-1}\right]^{-1}$$
$$= \tfrac{613}{37}e - \tfrac{35}{991} = 44.99999999993962\ldots \tag{16}$$

(Stoschek). Stoschek also gives an interesting near-identity involving the fine structure constant α and FEIGENBAUM CONSTANT δ,

$$(28 - \delta^{-1})(\alpha^{-1} - 137) \approx 0.999998. \tag{17}$$

The near identity

$$3\sqrt{2}(\sqrt{5} - 2) = 1.0015516\ldots \tag{18}$$

arises by noting that the stellation ratio $3(\sqrt{5} - 2)$ in the CUMULATION of the DODECAHEDRON to form the GREAT DODECAHEDRON is approximately equal to $\sqrt{2}$.

A set of almost integers due to D. Hickerson are those OF THE FORM

$$h_n = \frac{n!}{2(\ln 2)^{n+1}}. \tag{19}$$

for $1 \le n \le 15$, as summarized in the following table.

n	h_n
0	0.72135
1	1.04068
2	3.00278
3	12.99629
4	74.99874
5	541.00152
6	4683.00125
7	47292.99873
8	545834.99791
9	7087261.00162
10	102247563.00527
11	1622632572.99755
12	28091567594.98157
13	526858348381.00125
14	10641342970443.08453
15	230283190977853.03744
16	5315654681981354.51308
17	130370767029135900.45799

These numbers are close to integers due to the fact that the quotient is the dominant term in an infinite series for the number of possible outcomes of a race between n people (with ties are allowed). Calling this number $f(n)$, it follows that

$$f(n) = \sum_{k=1}^{n} \binom{n}{k} f(n-k) \quad (20)$$

for $n \geq 1$, where $\binom{n}{k}$ is a BINOMIAL COEFFICIENT. From this, we obtain the exponential generating function for f

$$\sum_{n=0}^{\infty} \frac{f(n)}{n!} z^n = \frac{1}{2 - e^z}, \quad (21)$$

and then by CONTOUR INTEGRATION it can be shown that

$$f(n) = \tfrac{1}{2} n! \sum_{k=-\infty}^{\infty} \frac{1}{(\ln 2 + 2\pi i k)^{n+1}} \quad (22)$$

for $n \geq 1$, where i is the square root of -1 and the sum is over all integers k (here, the imaginary parts of the terms for k and $-k$ cancel each other, so this sum is real.) The $k = 0$ term dominates, so $f(n)$ is asymptotic to $n!/(2(\ln 2)^{n+1})$. In fact, the other terms are quite small for n from 1 to 15, so $f(n)$ is the nearest integer to $n!/(2(\ln 2)^{n+1})$ for these values (Hickerson), given by the sequence 1, 3, 13 75, 541, 4683, ... (Sloane's A034172).

A large class of IRRATIONAL "almost integers" can be found using the theory of MODULAR FUNCTIONS, and a few rather spectacular examples are given by Ramanujan (1913–4). Such approximations were also studied by Hermite (1859), Kronecker (1863), and Smith (1965). They can be generated using some amazing (and very deep) properties of the J-FUNCTION. Some of the numbers which are closest approximations to INTEGERS are $e^{\pi\sqrt{163}}$ (sometimes known as the RAMANUJAN CONSTANT and which corresponds to the field $\mathbb{Q}(\sqrt{-163})$ which has CLASS NUMBER 1 and is the IMAGINARY QUADRATIC FIELD of maximal discriminant), $e^{\pi\sqrt{22}}, e^{\pi\sqrt{37}}$, and $e^{\pi\sqrt{58}}$, the last three of which have CLASS NUMBER 2 and are due to Ramanujan (Berndt 1994, Waldschmidt 1988).

The properties of the J-FUNCTION also give rise to the spectacular identity

$$\left[\frac{\ln(640320^3 + 744)}{\pi} \right]^2 = 163 + 2.32167\ldots \times 10^{-29} \quad (23)$$

(Le Lionnais 1983, p. 152).

The list below gives numbers OF THE FORM $x \equiv e^{\pi\sqrt{n}}$ for $n \leq 1000$ for which $[x] - x \leq 0.01$.

$e^{\pi\sqrt{6}} = 2,\ 197.990869543\ldots$

$e^{\pi\sqrt{17}} = 422,\ 150.997675680\ldots$

$e^{\pi\sqrt{18}} = 614,\ 551.992885619\ldots$

$e^{\pi\sqrt{22}} = 2,\ 508,\ 951.998257424\ldots$

$e^{\pi\sqrt{25}} = 6,\ 635,\ 623.999341134\ldots$

$e^{\pi\sqrt{37}} = 199,\ 148,\ 647.999978046551\ldots$

$e^{\pi\sqrt{43}} = 884,\ 736,\ 743.999777466\ldots$

$e^{\pi\sqrt{58}} = 24,\ 591,\ 257,\ 751.999999822213\ldots$

$e^{\pi\sqrt{59}} = 30,\ 197,\ 683,\ 486.993182260\ldots$

$e^{\pi\sqrt{67}} = 147,\ 197,\ 952,\ 743.999998662454\ldots$

$e^{\pi\sqrt{74}} = 545,\ 518,\ 122,\ 089.999174678853\ldots$

$e^{\pi\sqrt{149}} = 45,\ 116,\ 546,\ 012,\ 289,\ 599.991830287\ldots$

$e^{\pi\sqrt{163}} = 262,\ 537,\ 412,\ 640,\ 768,\ 743.999999999999250072\ldots$

$e^{\pi\sqrt{177}} = 1,\ 418,\ 556,\ 986,\ 635,\ 586,\ 485.996179355\ldots$

$e^{\pi\sqrt{232}} = 604,\ 729,\ 957,\ 825,\ 300,\ 084,\ 759.999992171526\ldots$

$e^{\pi\sqrt{267}} = 19,\ 683,\ 091,\ 854,\ 079,\ 461,\ 001,\ 445.992737040\ldots$

$e^{\pi\sqrt{326}} = 4,\ 309,\ 793,\ 301,\ 730,\ 386,\ 363,\ 005,\ 719.996011651\ldots$

$e^{\pi\sqrt{386}} = 639,\ 355,\ 180,\ 631,\ 208,\ 421,\ 212,\ 174,\ 016.997669832\ldots$

$e^{\pi\sqrt{522}} = 14,\ 871,\ 070,\ 263,\ 238,\ 043,\ 663,\ 567,\ldots$
$\ldots 627,\ 879,\ 007.999848726\ldots$

$e^{\pi\sqrt{566}} = 288,\ 099,\ 755,\ 064,\ 053,\ 264,\ 917,\ 867,\ldots$
$\ldots 975,\ 825,\ 573.993898311\ldots$

$e^{\pi\sqrt{638}} = 28,\ 994,\ 858,\ 898,\ 043,\ 231,\ 996,\ 779,\ \ldots$
$\ldots 771,\ 804,\ 797,\ 161.992372939\ldots$

$e^{\pi\sqrt{719}} = 3,\ 842,\ 614,\ 373,\ 539,\ 548,\ 891,\ 490,\ \ldots$
$\ldots 294,\ 277,\ 805,\ 829,\ 192.999987249\ldots$

$e^{\pi\sqrt{790}} = 223,\ 070,\ 667,\ 213,\ 077,\ 889,\ 794,\ 379,\ \ldots$
$\qquad\qquad \ldots 623,\ 183,\ 838,\ 336,\ 437.992055117\ldots$

$e^{\pi\sqrt{792}} = 249,\ 433,\ 117,\ 287,\ 892,\ 229,\ 255,\ 125,\ \ldots$
$\qquad\qquad \ldots 388,\ 685,\ 911,\ 710,\ 805.996097323\ldots$

$e^{\pi\sqrt{928}} = 365,\ 698,\ 321,\ 891,\ 389,\ 219,\ 219,\ 142,\ \ldots$
$\qquad\qquad \ldots 531,\ 076,\ 638,\ 716,\ 362,\ 775.998259747\ldots$

$e^{\pi\sqrt{986}} = 6,\ 954,\ 830,\ 200,\ 814,\ 801,\ 770,\ 418,\ 837,\ \ldots$
$\qquad\qquad \ldots 940,\ 281,\ 460,\ 320,\ 666,\ 108.994649611\ldots$

Gosper noted that the expression

$$1 - 262537412640768744e^{-\pi\sqrt{163}} - 196884e^{-2\pi\sqrt{163}}$$

$$+ 103378831900730205293632e^{-3\pi\sqrt{163}}. \qquad (24)$$

differs from an INTEGER by a mere 10^{-59}.

See also CLASS NUMBER, *J*-FUNCTION, PI, PISOT-VIJAYARAGHAVAN CONSTANT

References

Berndt, B. C. *Ramanujan's Notebooks, Part IV.* New York: Springer-Verlag, pp. 90–1, 1994.

Cohen, H. In *From Number Theory to Physics* (Ed. M. Waldschmidt, P. Moussa, J.-M. Luck, and C. Itzykson). New York: Springer-Verlag, 1992.

Hermite, C. "Sur la théorie des équations modulaires." *C. R. Acad. Sci. (Paris)* **48**, 1079–084 and 1095–102, 1859.

Hermite, C. "Sur la théorie des équations modulaires." *C. R. Acad. Sci. (Paris)* **49**, 16–4, 110–18, and 141–44, 1859.

Kronecker, L. "Über die Klassenzahl der aus Werzeln der Einheit gebildeten komplexen Zahlen." *Monatsber. K. Preuss. Akad. Wiss. Berlin*, 340–45. 1863.

Le Lionnais, F. *Les nombres remarquables.* Paris: Hermann, 1983.

Ramanujan, S. "Modular Equations and Approximations to π." *Quart. J. Pure Appl. Math.* **45**, 350–72, 1913–914.

Roberts, J. *The Lure of the Integers.* Washington, DC: Math. Assoc. Amer., 1992.

Sloane, N. J. A. Sequences A034172 in "An On-Line Version of the Encyclopedia of Integer Sequences." http://www.research.att.com/~njas/sequences/eisonline.html.

Smith, H. J. S. *Report on the Theory of Numbers.* New York: Chelsea, 1965.

Stoschek, E. "Modul 33: Algames with Numbers." http://marvin.sn.schule.de/~inftreff/modul33/task33.htm.

Waldschmidt, M. "Some Transcendental Aspects of Ramanujan's Work." In *Ramanujan Revisited: Proceedings of the Centenary Conference* (Ed. G. E. Andrews, B. C. Berndt, and R. A. Rankin). New York: Academic Press, pp. 57–6, 1988.

Waldschmidt, M. In *Ramanujan Centennial International Conference* (Ed. R. Balakrishnan, K. S. Padmanabhan, and V. Thangaraj). Ramanujan Math. Soc., 1988.

Almost Perfect Number

A number n for which the DIVISOR FUNCTION satisfies $\sigma(n) = 2n - 1$ is called almost perfect. The only known almost perfect numbers are the POWERS of 2, namely 1, 2, 4, 8, 16, 32, ... (Sloane's A000079). Singh (1997) calls almost perfect numbers SLIGHTLY DEFECTIVE.

See also QUASIPERFECT NUMBER

References

Guy, R. K. "Almost Perfect, Quasi-Perfect, Pseudoperfect, Harmonic, Weird, Multiperfect and Hyperperfect Numbers." §B2 in *Unsolved Problems in Number Theory, 2nd ed.* New York: Springer-Verlag, pp. 16 and 45–3, 1994.

Singh, S. *Fermat's Enigma: The Epic Quest to Solve the World's Greatest Mathematical Problem.* New York: Walker, p. 13, 1997.

Sloane, N. J. A. Sequences A000079/M1129 in "An On-Line Version of the Encyclopedia of Integer Sequences." http://www.research.att.com/~njas/sequences/eisonline.html.

Almost Periodic Function

This entry contributed by RONALD M. AARTS

A function representable as a generalized Fourier series. Let $\mathscr{R}$ be a METRIC SPACE with metric $\rho(x, y)$. Following Bohr (1947), a CONTINUOUS FUNCTION $x(t)$ for $(-\infty < t < \infty)$ with values in $\mathscr{R}$ is called an almost periodic function if, for every $\epsilon > 0$, there exists $\ell = \ell(\epsilon) > 0$ such that every interval $[t_0, t_0 + \ell(\epsilon)]$ contains at least one number τ for which

$$\rho[x(t), x(t + \tau)] < \varepsilon \quad (-\infty < t < \infty). \qquad (1)$$

Another formal description can be found in Krasnosel'skii *et al.* (1973).

Every almost periodic function is bounded and uniformly continuous on the entire REAL LINE. In addition, the range of an almost period function is compact in $\mathscr{R}$.

See also FOURIER SERIES, PERIODIC FUNCTION

References

Bohr, H. *Almost Periodic Functions.* New York: Chelsea, 1947.

Besicovitch, A. S. *Almost Periodic Functions.* New York: Dover, 1954.

Corduneanu, C. *Almost Periodic Functions.* New York: Wiley Interscience, 1961.

Krasnosel'skii, M. A.; Burd, V. Sh.; and Kolesov, Yu. S. *Nonlinear Almost Periodic Oscillations.* New York: Wiley, 1973.

Levitan, B. M. *Almost-Periodic Functions.* Moscow, 1953.

Almost Prime

A number n with prime factorization

$$n = \prod_{i=1}^{r} p_i^{a_i}$$

is called k-almost prime when the sum of the POWERS $\Sigma_{i=1}^{r} a_i = k$. The set of k-almost primes is denoted P_k.

The PRIMES correspond to the "1-almost prime" numbers 2, 3, 5, 7, 11, ... (Sloane's A000040). The 2-almost prime numbers correspond to SEMIPRIMES 4, 6, 9, 10, 14, 15, 21, 22, ... (Sloane's A001358). The first few 3-almost primes are 8, 12, 18, 20, 27, 28, 30, 42, 44, 45, 50, 52, 63, 66, 68, 70, 75, 76, 78, 92, 98, 99, ... (Sloane's A014612). The first few 4-almost primes are 16, 24, 36, 40, 54, 56, 60, 81, 84, 88, 90, 100, ... (Sloane's A014613). The first few 5-almost primes are 32, 48, 72, 80, ... (Sloane's A014614).

See also CHEN'S THEOREM, PRIME NUMBER, SEMI-PRIME

References

Sloane, N. J. A. Sequences A000040/M0652, A001358/ M3274, A014612, A014613, and A014614 in "An On-Line Version of the Encyclopedia of Integer Sequences." http:// www.research.att.com/~njas/sequences/eisonline.html.

Almost Unit

An almost unit is a nonunit in the INTEGRAL DOMAIN of FORMAL POWER SERIES with a nonzero first coefficient, $P = a_1 x + z_2 x^2 + \ldots$, where $a_1 \neq 0$. Under the operation of composition, the almost units in the INTEGRAL DOMAIN of FORMAL POWER SERIES over a FIELD F form a GROUP (Henrici 1988, p. 45).

See also SCHUR-JABOTINSKY THEOREM

References

Henrici, P. *Applied and Computational Complex Analysis, Vol. 1: Power Series-Integration-Conformal Mapping-Location of Zeros.* New York: Wiley, p. 45, 1988.

Alon-Tarsi Conjecture

See also LATIN SQUARE

References

Drisko, A. A. "Proof of the Alon-Tarsi Conjecture for $n = 2^r p$." *Electronic J. Combinatorics* **5**, No. 1, R28, 1–, 1998. http://www.combinatorics.org/Volume_5/ v5i1toc.html.

Alpha

Alpha is the name for the first letter in the Greek alphabet: α.

In finance, alpha is a financial measure giving the difference between a fund's actual return and its expected level of performance, given its level of risk (as measured by BETA). A POSITIVE alpha indicates that a fund has performed better than expected based on its BETA, whereas a NEGATIVE alpha indicates poorer performance.

See also ALPHA FUNCTION, ALPHA-TEST, ALPHA VALUE, BETA, SHARPE RATIO

Alpha Function

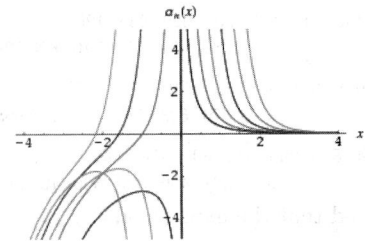

$$\alpha_n(z) \equiv \int_1^\infty t^n e^{-zt}\, dt = n! z^{-(n+1)} e^{-z} \sum_{k=0}^n \frac{z^k}{k!}.$$

It is equivalent to

$$\alpha_n(z) = \mathbf{E}_{-n}(z),$$

where $\mathbf{E}_n(z)$ is the E_N-FUNCTION.

See also BETA EXPONENTIAL FUNCTION, E_N-FUNCTION

Alpha Value

An alpha value is a number $0 \leq \alpha \leq 1$ such that $P(z \geq z_{\text{observed}}) \leq \alpha$ is considered "SIGNIFICANT," where P is a P-VALUE.

See also CONFIDENCE INTERVAL, P-VALUE, SIGNIFICANCE

Alphabet

A SET (usually of letters) from which a SUBSET is drawn. A sequence of letters is called a WORD, and a set of WORDS is called a CODE.

See also CODE, STRING, WORD

Alpha-Beta Conjecture

MANN'S THEOREM

Alphamagic Square

A MAGIC SQUARE for which the number of letters in the word for each number generates another MAGIC SQUARE. This definition depends, of course, on the language being used. In English, for example,

5	22	18		4	9	8
28	15	2		11	7	3
12	8	25		6	5	10

where the MAGIC SQUARE on the right corresponds to the number of letters in

five	twenty-two	eighteen
twenty-eight	fifteen	two
twelve	eight	twenty-five

References

Sallows, L. C. F. "Alphamagic Squares." *Abacus* **4**, 28–5, 1986.
Sallows, L. C. F. "Alphamagic Squares. 2." *Abacus* **4**, 20–9 and 43, 1987.
Sallows, L. C. F. "Alpha Magic Squares." In *The Lighter Side of Mathematics* (Ed. R. K. Guy and R. E. Woodrow). Washington, DC: Math. Assoc. Amer., 1994.

Alphametic

A CRYPTARITHM in which the letters used to represent distinct DIGITS are derived from related words or meaningful phrases. The term was coined by Hunter in 1955 (Madachy 1979, p. 178).

References

Brooke, M. *One Hundred & Fifty Puzzles in Crypt-Arithmetic.* New York: Dover, 1963.
Hunter, J. A. H. and Madachy, J. S. "Alphametics and the Like." Ch. 9 in *Mathematical Diversions.* New York: Dover, pp. 90–5, 1975.
Madachy, J. S. "Alphametics." Ch. 7 in *Madachy's Mathematical Recreations.* New York: Dover, pp. 178–00, 1979.

Alpha-Test

For some constant α_0, $\alpha(f, z) < \alpha_0$ implies that z is an APPROXIMATE ZERO of f, where

$$\alpha(f, z) = \frac{|f(z)|}{|f'(z)|} \sup_{k>1} \left| \frac{f^{(k)}(z)}{k! f'(z)} \right|^{1/(k-1)}$$

Smale (1986) found a constant $\alpha \approx 0.130707$ for the test, and this value was subsequently improved to $\alpha_0 = 3 - 2\sqrt{2} \approx 0.171573$ by Wang and Han (1989), and further improved by Wang and Zhao (1995; Petkovic *et al.* 1997, p. 2).

See also APPROXIMATE ZERO, NEWTON'S METHOD, POINT ESTIMATION THEORY

References

Kim, M. Ph.D. thesis. New York: City University of New York, 1985.
Petkovic, M. S.; Herceg, D. D.; and Ilic, S. M. *Point Estimation Theory and Its Applications.* Novi Sad, Yugoslavia: Institute of Mathematics, 1997.
Smale, S. "Newton's Method Estimates from Data at One Point." In *The Merging of Disciplines: New Directions in Pure, Applied, and Computational Mathematics* (Ed. R. E. Ewing, K. I. Gross, and C. F. Martin). New York: Springer-Verlag, pp. 185–96, 1986.
Wang, X. and Han, D. "On Dominating Sequence Method in the Point Estimate and Smale's Theorem." *Scientia Sinica Ser. A*, 905–13, 1989.
Wang, D. and Zhao, F. "The Theory of Smale's Point Estimation and Its Application." *J. Comput. Appl. Math.* **60**, 253–69, 1995.

Alternating Algebra

EXTERIOR ALGEBRA

Alternating Group

A PERMUTATION GROUP of an even number of permutations on a set of length n, denoted A_n or Alt(n) (Scott 1987, p. 267). An alternating group is a NORMAL SUBGROUP of the PERMUTATION GROUP, and has ORDER $n!/2$, the first few values of which for $n = 2, 3, \ldots$ are 1, 3, 12, 60, 360, 2520, ... (Sloane's A001710). Alternating groups are FINITE analogs of the families of simple LIE GROUPS.

Alternating groups with $n \geq 5$ are non-ABELIAN SIMPLE GROUPS (Scott 1987, p. 295). The number of conjugacy classes in the alternating groups A_n for $n = 2, 3, \ldots$ are 1, 3, 4, 5, 7, 9, ... (Sloane's A000702).

See also 15 PUZZLE, FINITE GROUP, GROUP, JORDAN'S SYMMETRY GROUP THEOREM, LIE GROUP, PERMUTATION GROUP, SIMPLE GROUP, SYMMETRIC GROUP

References

Scott, W. R. *Group Theory.* New York: Dover, pp. 267 and 295, 1987.
Sloane, N. J. A. Sequences A000702/M2307 and A001710/M2933 in "An On-Line Version of the Encyclopedia of Integer Sequences." http://www.research.att.com/~njas/sequences/eisonline.html.
Wilson, R. A. "ATLAS of Finite Group Representation." http://for.mat.bham.ac.uk/atlas/html/contents.html#alt.

Alternating Knot

An alternating knot is a KNOT which possesses a knot diagram in which crossings alternate between under- and overpasses. Not all knot diagrams of alternating knots need be alternating diagrams.

The TREFOIL KNOT and FIGURE-OF-EIGHT KNOT are alternating knots. The number of PRIME alternating and nonalternating knots of n crossings are summarized in the following table.

type	Sloane	counts
alternating	A002864	0, 0, 1, 1, 2, 3, 7, 18, 41, 123, 367, 1288, 4878, 19536, 85263, 379799, ...
nonalternating	A051763	0, 0, 0, 0, 0, 0, 0, 3, 8, 42, 185, 888, 5110, 27436, 168030, 1008906, ...

The 3 nonalternating knots of eight crossings are 08–19, 08–20, and 08–21, illustrated below (Wells 1991).

Thistlethwaite, M. "A Spanning Tree Expansion for the Jones Polynomial." *Topology* **26**, 297–09, 1987.
Wells, D. *The Penguin Dictionary of Curious and Interesting Geometry.* London: Penguin, p. 160, 1991.

One of TAIT'S KNOT CONJECTURES states that the number of crossings is the same for any diagram of a reduced alternating knot. Furthermore, a reduced alternating projection of a knot has the least number of crossings for any projection of that knot. Both of these facts were proved true by Kauffman (1988), Thistlethwaite (1987), and Murasugi (1987). FLYPE moves are sufficient to pass between all minimal diagrams of a given alternating knot (Hoste *et al.* 1998).

If K has a reduced alternating projection of n crossings, then the SPAN of K is An. Let $c(K)$ be the CROSSING NUMBER. Then an alternating knot $K_1 \# K_2$ (a KNOT SUM) satisfies

$$c(K_1 \# K_2) = c(K_1) + c(K_2).$$

In fact, this is true as well for the larger class of ADEQUATE KNOTS and postulated for all KNOTS.

It is conjectured that the proportion of knots which are alternating tends exponentially to zero with increasing crossing number (Hoste *et al.* 1998), a statement which has been proved true for alternating *links*.

See also ADEQUATE KNOT, ALMOST ALTERNATING LINK, ALTERNATING LINK, FLYPING CONJECTURE, TAIT'S KNOT CONJECTURES

References

Adams, C. C. *The Knot Book: An Elementary Introduction to the Mathematical Theory of Knots.* New York: W. H. Freeman, pp. 159–64, 1994.
Arnold, B.; Au, M.; Candy, C.; Erdener, K.; Fan, J.; Flynn, R.; Muir, J.; Wu, D.; and Hoste, J. "Tabulating Alternating Knots through 14 Crossings." ftp://chs.cusd.claremont.edu/pub/knot/paper.TeX.txt.
Arnold, B.; Au, M.; Candy, C.; Erdener, K.; Fan, J.; Flynn, R.; Muir, J.; Wu, D.; and Hoste, J. ftp://chs.cusd.claremont.edu/pub/knot/AltKnots/.
Erdener, K. and Flynn, R. "Rolfsen's Table of all Alternating Diagrams through 9 Crossings." ftp://chs.cusd.claremont.edu/pub/knot/Rolfsen_table.final.
Hoste, J.; Thistlethwaite, M.; and Weeks, J. "The First 1,701,936 Knots." *Math. Intell.* **20**, 33–8, Fall 1998.
Kauffman, L. "New Invariants in the Theory of Knots." *Amer. Math. Monthly* **95**, 195–42, 1988.
Little, C. N. "Non Alternate $\pm$ Knots of Orders Eight and Nine." *Trans. Roy. Soc. Edinburgh* **35**, 663–64, 1889.
Little, C. N. "Alternate $\pm$ Knots of Order 11." *Trans. Roy. Soc. Edinburgh* **36**, 253–55, 1890.
Little, C. N. "Non-Alternate $\pm$ Knots." *Trans. Roy. Soc. Edinburgh* **39**, 771–78, 1900.
Murasugi, K. "Jones Polynomials and Classical Conjectures in Knot Theory." *Topology* **26**, 297–07, 1987.
Sloane, N. J. A. Sequences A002864/M0847 and A051763 in "An On-Line Version of the Encyclopedia of Integer Sequences." http://www.research.att.com/~njas/sequences/eisonline.html.

Alternating Knot Diagram

A KNOT DIAGRAM which has alternating under- and overcrossings as the KNOT projection is traversed. The first KNOT which does not have an alternating diagram has 8 crossings.

Alternating Link

A LINK which has a LINK DIAGRAM with alternating underpasses and overpasses.

The proportion of links which are alternating tends exponentially to zero with increasing crossing number (Sundberg and Thistlethwaite 1998, Thistlethwaite 1998).

See also ALMOST ALTERNATING LINK, ALTERNATING KNOT

References

Hoste, J.; Thistlethwaite, M.; and Weeks, J. "The First 1,701,936 Knots." *Math. Intell.* **20**, 33–8, Fall 1998.
Menasco, W. and Thistlethwaite, M. "The Classification of Alternating Links." *Ann. Math.* **138**, 113–71, 1993.
Sundberg, C. and Thistlethwaite, M. "The Rate of Growth of the Number of Prime Alternating Links and Tangles." *Pacific J. Math.* **182**, 329–58, 1998.
Thistlethwaite, M. "On the Structure and Scarcity of Alternating Links and Tangles." *J. Knot Th. Ramifications* **7**, 981–004, 1998.

Alternating Multilinear Form

An alternating multilinear form on a REAL VECTOR SPACE V is a MULTILINEAR FORM

$$F : V \otimes \cdots \otimes V \to \mathbb{R} \qquad (1)$$

such that

$$F(x_1, \ldots, x_i, x_{i+1}, \ldots, x_n)$$
$$= -F(x_1, \ldots, x_{i+1}, x_i, \ldots, x_n) \qquad (2)$$

for any index i. For example,

$$F((a_1, a_2, a_3), (b_1, b_2, b_3), (c_1, c_2, c_3))$$
$$= a_1 b_2 c_3 - a_1 b_3 c_2 + a_2 b_3 c_1 - a_2 b_1 c_3 + a_3 b_1 c_2$$
$$- a_3 b_2 c_1 \qquad (3)$$

is an alternating form on $\mathbb{R}^3$.

An alternating multilinear form is defined on a MODULE in a similar way, by replacing $\mathbb{R}$ with the RING.

See also DUAL SPACE, EXTERIOR ALGEBRA, MODULE, MULTILINEAR FORM, VECTOR SPACE

Alternating Permutation

An arrangement of the elements $c_1, ..., c_n$ such that no element c_i has a magnitude between c_{i-1} and c_{i+1} is called an alternating (or ZIGZAG) permutation. The determination of the number of alternating permutations for the set of the first n INTEGERS $\{1, 2, ..., n\}$ is known as ANDRÉ'S PROBLEM. An example of an alternating permutation is $(1, 3, 2, 5, 4)$.

As many alternating permutations among n elements begin by rising as by falling. The magnitude of the c_ns does not matter; only the number of them. Let the number of alternating permutations be given by $Z_n = 2A_n$. This quantity can then be computed from

$$2na_n = \sum a_r a_s, \qquad (1)$$

where r and s pass through all INTEGRAL numbers such that

$$r + s = n - 1, \qquad (2)$$

$a_0 = a_1 = 1$, and

$$A_n = n! a_n. \qquad (3)$$

The numbers A_n are sometimes called the EULER ZIGZAG NUMBERS, and the first few are given by 1, 1, 1, 2, 5, 16, 61, 272, ... (Sloane's A000111). The EVEN-numbered A_ns are called EULER NUMBERS, SECANT NUMBERS, or ZIG NUMBERS, and the ODD-numbered ones are sometimes called TANGENT NUMBERS or ZAG NUMBERS.

Curiously enough, the SECANT and TANGENT MACLAURIN SERIES can be written in terms of the A_ns as

$$\sec x = A_0 + A_2 \frac{x^2}{2!} + A_4 \frac{x^4}{4!} + ... \qquad (4)$$

$$\tan x = A_1 x + A_3 \frac{x^3}{3!} + A_5 \frac{x^5}{5!} + ..., \qquad (5)$$

or combining them,

$$\sec x + \tan x$$
$$= A_0 + A_1 x + A_2 \frac{x^2}{2!} + A_3 \frac{x^3}{3!} + A_4 \frac{x^4}{4!} + A_5 \frac{x^5}{5!}$$
$$+ .... \qquad (6)$$

See also ENTRINGER NUMBER, EULER NUMBER, EULER ZIGZAG NUMBER, SECANT NUMBER, SEIDEL-ENTRINGER-ARNOLD TRIANGLE, TANGENT NUMBER

References

André, D. "Developments de sec x et tan x." *C. R. Acad. Sci. Paris* **88**, 965–67, 1879.

André, D. "Memoire sur les permutations alternées." *J. Math.* **7**, 167–84, 1881.

Arnold, V. I. "Bernoulli-Euler Updown Numbers Associated with Function Singularities, Their Combinatorics and Arithmetics." *Duke Math. J.* **63**, 537–55, 1991.

Arnold, V. I. "Snake Calculus and Combinatorics of Bernoulli, Euler, and Springer Numbers for Coxeter Groups." *Russian Math. Surveys* **47**, 3–5, 1992.

Bauslaugh, B. and Ruskey, F. "Generating Alternating Permutations Lexicographically." *BIT* **30**, 17–6, 1990.

Conway, J. H. and Guy, R. K. In *The Book of Numbers.* New York: Springer-Verlag, pp. 110–11, 1996.

Dörrie, H. "André's Deviation of the Secant and Tangent Series." §16 in *100 Great Problems of Elementary Mathematics: Their History and Solutions.* New York: Dover, pp. 64–9, 1965.

Honsberger, R. *Mathematical Gems III.* Washington, DC: Math. Assoc. Amer., pp. 69–5, 1985.

Knuth, D. E. and Buckholtz, T. J. "Computation of Tangent, Euler, and Bernoulli Numbers." *Math. Comput.* **21**, 663–88, 1967.

Millar, J.; Sloane, N. J. A.; and Young, N. E. "A New Operation on Sequences: The Boustrophedon Transform." *J. Combin. Th. Ser. A* **76**, 44–4, 1996.

Ruskey, F. "Information of Alternating Permutations." http://www.theory.csc.uvic.ca/~cos/inf/perm/Alternating.html.

Sloane, N. J. A. Sequences A000111/M1492 in "An On-Line Version of the Encyclopedia of Integer Sequences." http://www.research.att.com/~njas/sequences/eisonline.html.

Alternating Representation

See also REPRESENTATION

Alternating Series

A SERIES OF THE FORM

$$\sum_{k=1}^{\infty} (-1)^{k+1} a_k \qquad (1)$$

or

$$\sum_{k=1}^{\infty} (-1)^k a_k. \qquad (2)$$

Rather surprisingly, the alternating series

$$\sum_{k=1}^{\infty} \frac{(-1)^{k-1}}{k} = \ln 2 \qquad (3)$$

converges to the natural logarithm of 2.

See also SERIES

References

Arfken, G. "Alternating Series." §5.3 in *Mathematical Methods for Physicists, 3rd ed.* Orlando, FL: Academic Press, pp. 293–94, 1985.

Bromwich, T. J. I'a. and MacRobert, T. M. "Alternating Series." §19 in *An Introduction to the Theory of Infinite Series, 3rd ed.* New York: Chelsea, pp. 55–7, 1991.

Gardner, M. *The Sixth Book of Mathematical Games from Scientific American.* Chicago, IL: University of Chicago Press, p. 170, 1984.

Hoffman, P. *The Man Who Loved Only Numbers: The Story of Paul Erdos and the Search for Mathematical Truth.* New York: Hyperion, p. 218, 1998.

Pinsky, M. A. "Averaging an Alternating Series." *Math. Mag.* **51**, 235–37, 1978.

Alternating Series Test

Also known as the LEIBNIZ CRITERION. An ALTERNATING SERIES CONVERGES if $a_1 \geq a_2 \geq \ldots$ and

$$\lim_{k \to \infty} a_k = 0.$$

See also CONVERGENCE TESTS

Alternating Sign Matrix

A MATRIX of 0s, 1s, and -1s in which the entries in each row or column sum to 1 and the nonzero entries in each row and column alternate in sign. The number of $n \times n$ alternating sign matrices for $n = 1$, 2, ... are 1, 2, 21, 1344, 628080, ...(Sloane's A050204), illustrated below:

$$A_1' = [1] \tag{1}$$

$$A_2' = \begin{bmatrix} 1 & 0 \\ 0 & 1 \end{bmatrix}, \begin{bmatrix} 0 & 1 \\ 1 & 0 \end{bmatrix} \tag{2}$$

$$A_3' = \begin{bmatrix} -1 & 1 & 1 \\ 1 & -1 & 1 \\ 1 & 1 & -1 \end{bmatrix}, \begin{bmatrix} -1 & 1 & 1 \\ 1 & 0 & 0 \\ 1 & 0 & 0 \end{bmatrix}, \begin{bmatrix} -1 & 1 & 1 \\ 1 & 1 & -1 \\ 1 & -1 & 1 \end{bmatrix}$$

$$\begin{bmatrix} 0 & 0 & 1 \\ 0 & 0 & 1 \\ 1 & 1 & -1 \end{bmatrix}, \begin{bmatrix} 0 & 0 & 1 \\ 0 & 1 & 0 \\ 1 & 0 & 0 \end{bmatrix}, \begin{bmatrix} 0 & 0 & 1 \\ 1 & 0 & 0 \\ 0 & 1 & 0 \end{bmatrix}, \ldots . \tag{3}$$

If the additional restriction is added that any -1s in a row or column must have a +1 "outside" it (i.e., all -1s are "bordered" by +1s), then the number of these "Robins and Rumsey" $n \times n$ alternating sign matrices A_n are given by 1, 2, 7, 42, 429, 7436, 218348, ... (Sloane's A005130). The single A_1 and two A_2s are identical to A_1' and A_2', but only seven of the 21 A_3's are A_3s:

$$A_3 = \begin{bmatrix} 0 & 0 & 1 \\ 0 & 1 & 0 \\ 1 & 0 & 0 \end{bmatrix}, \begin{bmatrix} 0 & 0 & 1 \\ 1 & 0 & 0 \\ 0 & 1 & 0 \end{bmatrix}, \begin{bmatrix} 0 & 1 & 0 \\ 0 & 0 & 1 \\ 1 & 0 & 0 \end{bmatrix}, \begin{bmatrix} 0 & 1 & 0 \\ 1 & -1 & 1 \\ 0 & 1 & 0 \end{bmatrix},$$
$$\tag{4}$$

$$\begin{bmatrix} 0 & 1 & 0 \\ 1 & 0 & 0 \\ 0 & 0 & 1 \end{bmatrix}, \begin{bmatrix} 1 & 0 & 0 \\ 0 & 0 & 1 \\ 0 & 1 & 0 \end{bmatrix}, \begin{bmatrix} 1 & 0 & 0 \\ 0 & 1 & 0 \\ 0 & 0 & 1 \end{bmatrix} \tag{5}$$

The conjecture that the number A_n of A_n is explicitly given by the formula

$$A_n \prod_{j=0}^{n-1} \frac{(3j+1)!}{(n+j)!}, \tag{6}$$

now proven to be true, was known as the ALTERNATING SIGN MATRIX CONJECTURE. Let $A(n, k)$ be the number of $n \times n$ alternating sign matrices with one in the top row occurring in the kth position. Then

$$A_n = \sum_{k=1}^{n} A(n, k). \tag{7}$$

The result

$$\frac{A(n, k+1)}{A(n, k)} = \frac{(n-k)(n+k-1)}{k(2n-k-1)} \tag{8}$$

for $0 < k < n$ implies (7) (Mills *et al.* 1983).

Making a triangular array of the number of A_n' with a 1 at the top of column k gives

$$1$$
$$1 \quad 1$$
$$2 \quad 3 \quad 2$$
$$7 \quad 14 \quad 14 \quad 7$$
$$42 \quad 105 \quad 135 \quad 105 \quad 42$$

(Sloane's A048601), and taking the ratios of adjacent terms gives the array

$$2/2$$
$$2/3 \quad 3/2$$
$$2/4 \quad 5/5 \quad 4/2$$
$$2/5 \quad 7/9 \quad 9/7 \quad 5/2$$

(Sloane's A029656 and A029638). The fact that these numerators and denominators are respectively the numbers in the (2, 1)- and (1, 2)-Pascal triangles which are different from 1 is known as the REFINED ALTERNATING SIGN MATRIX CONJECTURE.

See also ALTERNATING SIGN MATRIX CONJECTURE, CONDENSATION, DESCENDING PLANE PARTITION, INTEGER MATRIX, PERMUTATION MATRIX

References

Andrews, G. E. "Plane Partitions (III): The Weak Macdonald Conjecture." *Invent. Math.* **53**, 193–25, 1979.

Bressoud, D. *Proofs and Confirmations: The Story of the Alternating Sign Matrix Conjecture.* Cambridge, England: Cambridge University Press, 1999.

Bressoud, D. and Propp, J. "How the Alternating Sign Matrix Conjecture was Solved." *Not. Amer. Math. Soc.* **46**, 637–46.

Kuperberg, G. "Another Proof of the Alternating-Sign Matrix Conjecture." *Internat. Math. Res. Notes*, No. 3, 139–50, 1996.

Mills, W. H.; Robbins, D. P.; and Rumsey, H. Jr. "Proof of the Macdonald Conjecture." *Invent. Math.* **66**, 73–7, 1982.

Mills, W. H.; Robbins, D. P.; and Rumsey, H. Jr. "Alternating Sign Matrices and Descending Plane Partitions." *J. Combin. Th. Ser. A* **34**, 340–59, 1983.

Robbins, D. P. "The Story of 1, 2, 7, 42, 429, 7436, ..." *Math. Intell.* **13**, 12–9, 1991.

Robbins, D. P. and Rumsey, H. Jr. "Determinants and Alternating Sign Matrices." *Adv. Math.* **62**, 169–84, 1986.

Sloane, N. J. A. Sequences A005130/M1808, A029638, A029656, A048601, and A050204 in "An On-Line Version of the Encyclopedia of Integer Sequences." http://www.re-search.att.com/~njas/sequences/eisonline.html.

Stanley, R. P. "A Baker's Dozen of Conjectures Concerning Plane Partitions." In *Combinatoire Énumérative. Proceedings of the colloquium held at the Université du Québec, Montreal, May 28-June 1, 1985* (Ed. G. Labelle and P. Leroux). New York: Springer-Verlag, pp. 285–93, 1986.

Zeilberger, D. "Proof of the Alternating Sign Matrix Conjecture." *Electronic J. Combinatorics* **3**, No. 2, R13, 1–4, 1996. http://www.combinatorics.org/Volume_3/volume3_2.html.

Zeilberger, D. "Proof of the Refined Alternating Sign Matrix Conjecture." *New York J. Math.* **2**, 59–8, 1996.

Zeilberger, D. "A Constant Term Identity Featuring the Ubiquitous (and Mysterious) Andrews-Mills-Robbins-Rumsey numbers 1, 2, 7, 42, 429, ..." *J. Combin. Theory A* **66**, 17–7, 1994.

Alternating Sign Matrix Conjecture

The conjecture that the number of ALTERNATING SIGN MATRICES "bordered" by +1s A_n is explicitly given by the formula

$$A_n = \prod_{j=0}^{n-1} \frac{(3j+1)!}{(n+j)!}.$$

This conjecture was proved by Doron Zeilberger in 1995 (Zeilberger 1996a). This proof enlisted the aid of an army of 88 referees together with extensive computer calculations. A beautiful, shorter proof was given later that year by Kuperberg (Kuperberg 1996), and the REFINED ALTERNATING SIGN MATRIX CONJECTURE was subsequently proved by Zeilberger (Zeilberger 1996b) using Kuperberg's method together with techniques from q-calculus and orthogonal polynomials.

See also ALTERNATING SIGN MATRIX, REFINED ALTERNATING SIGN MATRIX CONJECTURE

References

Bressoud, D. *Proofs and Confirmations: The Story of the Alternating Sign Matrix Conjecture.* Cambridge, England: Cambridge University Press, 1999.

Bressoud, D. and Propp, J. "How the Alternating Sign Matrix Conjecture was Solved." *Not. Amer. Math. Soc.* **46**, 637–46.

Kuperberg, G. "Another Proof of the Alternating-Sign Matrix Conjecture." *Internat. Math. Res. Notes*, No. 3, 139–50, 1996. Zeilberger, D. "A Constant Term Identity Featuring the Ubiquitous (and Mysterious) Andrews-Mills-Robbins-Rumsey numbers 1, 2, 7, 42, 429, ..." *J. Combin. Theory A* **66**, 17–7, 1994.

Zeilberger, D. "Proof of the Alternating Sign Matrix Conjecture." *Electronic J. Combinatorics* **3**, No. 2, R13, 1–4, 1996a. http://www.combinatorics.org/Volume_3/volume3_2.html.

Zeilberger, D. "Proof of the Refined Alternating Sign Matrix Conjecture." *New York J. Math.* **2**, 59–8, 1996b.

Alternating Tensor

ANTISYMMETRIC TENSOR

Alternative Algebra

Let A denote an $\mathbb{R}$-ALGEBRA, so that A is a VECTOR SPACE over R and

$$A \times A \to A \tag{1}$$

$$(x, y) \mapsto x \cdot y. \tag{2}$$

Then A is said to be alternative if, for all $x, y \in A$

$$(x \cdot y) \cdot y = x \cdot (y \cdot y) \tag{3}$$

$$(x \cdot x) \cdot y = x \cdot (x \cdot y). \tag{4}$$

Here, VECTOR MULTIPLICATION $x \cdot y$ is assumed to be BILINEAR.

The ASSOCIATOR (x, y, z) is an alternating function, and the SUBALGEBRA generated by two elements is associative.

See also ASSOCIATOR

References

Finch, S. "Zero Structures in Real Algebras." http://www.mathsoft.com/asolve/zerodiv/zerodiv.html.

Schafer, R. D. *An Introduction to Non-Associative Algebras.* New York: Dover, p. 5, 1995.

Alternative Denial

The term used in PROPOSITIONAL CALCULUS for the NAND CONNECTIVE. The notation $A|B$ is used for this connective, a most unfortunate choice in light of modern usage of $A|B$ or $A\|B$ to denote OR.

See also JOINT DENIAL, NAND

References

Mendelson, E. *Introduction to Mathematical Logic, 4th ed.* London: Chapman & Hall, p. 26, 1997.

Alternative Link

A category of LINK encompassing both ALTERNATING KNOTS and TORUS KNOTS.

See also ALTERNATING KNOT, LINK, TORUS KNOT

References

Kauffman, L. "Combinatorics and Knot Theory." *Contemp. Math.* **20**, 181–00, 1983.

Altitude

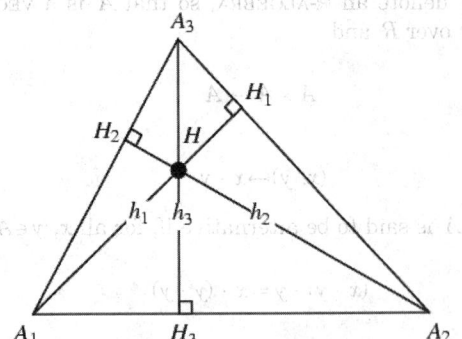

The altitudes of a TRIANGLE are the CEVIANS A_iH_i which are PERPENDICULAR to the LEGS A_jA_k opposite A_i. The three altitudes of any TRIANGLE are CONCURRENT at the ORTHOCENTER H (Durell 1928). This fundamental fact did not appear anywhere in Euclid's *ELEMENTS*.

The altitudes have lengths $h_i \equiv \overline{A_iH_i}$ given by

$$h_i = a_{i+1} \sin \alpha_{i+2} = a_{i+2} \sin \alpha_{i+1} \tag{1}$$

$$h_1 = \frac{2\sqrt{s(s - a_1)(s - a_2)(s - a_3)}}{a_1}, \tag{2}$$

where s is the SEMIPERIMETER and $a_i \equiv \overline{A_jA_k}$. Another pair of interesting FORMULAS are

$$s_h = \frac{\Delta}{R} \tag{3}$$

where Δ is the AREA of the TRIANGLE $\Delta A_1A_2A_3$ and s_h is the SEMIPERIMETER of the ALTITUDE TRIANGLE $\Delta H_1H_2H_3$, and

$$h_1 h_2 h_3 = 2s_h \Delta = \frac{2\Delta^2}{R}, \tag{4}$$

where R is the CIRCUMRADIUS of $\Delta A_1A_2A_3$ (Johnson 1929, p. 191).

Other formulas satisfied by the altitude include

$$\frac{1}{h_1} + \frac{1}{h_2} + \frac{1}{h_3} = \frac{1}{r} \tag{5}$$

$$\frac{1}{r_1} = \frac{1}{h_2} + \frac{1}{h_3} + \frac{1}{h_1} \tag{6}$$

$$\frac{1}{r_2} + \frac{1}{r_3} = \frac{1}{r} + \frac{1}{r_1} = \frac{2}{h_1}, \tag{7}$$

where r is the INRADIUS and r_i are the EXRADII (Johnson 1929, p. 189). In addition,

$$HA_1 \cdot HH_1 = HA_2 \cdot HH_2 = HA_3 \cdot HH_3 \tag{8}$$

$$HA_1 \cdot HH_1 = \tfrac{1}{2}(a_1^2 + a_2^2 + a_3^2) - 4R^2, \tag{9}$$

where R is the CIRCUMRADIUS.

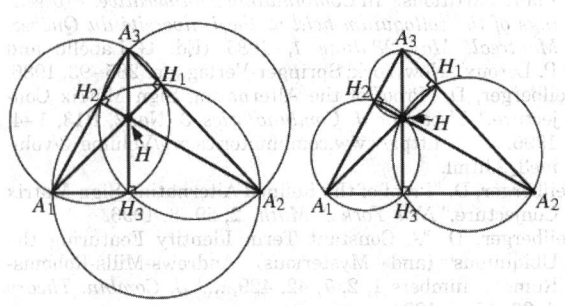

The points A_1, A_3, H_1, and H_3 (and their permutations with respect to indices) all lie on a CIRCLE, as do the points A_3, H_3, H, and H_1 (and their permutations with respect to indices). TRIANGLES $\Delta A_1A_2A_3$ and $\Delta A_1H_2H_3$ are inversely similar.

The triangle $H_1H_2H_3$ has the minimum PERIMETER of any TRIANGLE inscribed in a given ACUTE TRIANGLE (Johnson 1929, pp. 161–65). Additional properties involving the FEET of the altitudes are given by Johnson (1929, pp. 261–62). The line joining the feet to two altitudes of a triangle is ANTIPARALLEL to the third side (Johnson 1929, p. 172).

See also CEVIAN, FOOT, MALTITUDE, ORTHOCENTER, PERPENDICULAR, PERPENDICULAR FOOT, TAYLOR CIRCLE

References

Coxeter, H. S. M. and Greitzer, S. L. "More on the Altitude and Orthocentric Triangle." §2.4 in *Geometry Revisited.* Washington, DC: Math. Assoc. Amer., pp. 9 and 36–0, 1967.

Durell, C. V. *Modern Geometry: The Straight Line and Circle.* London: Macmillan, p. 20, 1928.

Johnson, R. A. *Modern Geometry: An Elementary Treatise on the Geometry of the Triangle and the Circle.* Boston, MA: Houghton Mifflin, 1929.

Altitude Plane

The plane through an edge of a TRIHEDRAL ANGLE drawn perpendicularly to the opposite face. The term was first used by J. Neuberg (Altshiller-Court 1979, p. 298).

References

Altshiller-Court, N. *Modern Pure Solid Geometry.* New York: Chelsea, p. 27, 1979.

Altitude Triangle

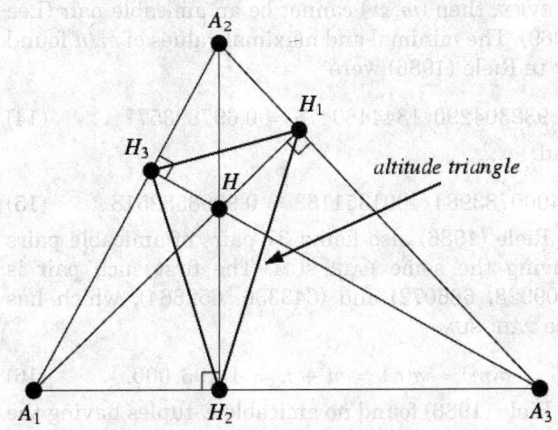

altitude triangle

The TRIANGLE $\Delta H_1 H_2 H_3$ formed by connecting the three feet H_1, H_2, and H_3 of the altitudes of a given triangle $\Delta A_1 A_2 A_3$.

See also ALTITUDE

Alysoid

CATENARY

Ambient Isotopy

An ambient isotopy from an embedding of a MANIFOLD M in N to another is a HOMOTOPY of self DIFFEOMORPHISMS (or ISOMORPHISMS, or piecewise-linear transformations, etc.) of N, starting at the IDENTITY MAP, such that the "last" DIFFEOMORPHISM compounded with the first embedding of M is the second embedding of M. In other words, an ambient isotopy is like an ISOTOPY except that instead of distorting the embedding, the whole ambient SPACE is being stretched and distorted and the embedding is just "coming along for the ride." For SMOOTH MANIFOLDS, a MAP is ISOTOPIC IFF it is ambiently isotopic.

For KNOTS, the equivalence of MANIFOLDS under continuous deformation is independent of the embedding SPACE. KNOTS of opposite CHIRALITY have ambient isotopy, but not REGULAR ISOTOPY.

See also ISOTOPY, REGULAR ISOTOPY

References

Hirsch, M. W. *Differential Topology.* New York: Springer-Verlag, 1988.
Hoste, J.; Thistlethwaite, M.; and Weeks, J. "The First 1,701,936 Knots." *Math. Intell.* **20**, 33–8, Fall 1998.

Ambiguous

An expression is said to be ambiguous (or poorly defined) if its definition does not assign it a unique interpretation or value. An expression which is *not* ambiguous is said to be WELL DEFINED.

See also ILL DEFINED, WELL DEFINED

Ambiguous Rectangle

FAULT-FREE RECTANGLE

Ambrose-Kakutani Theorem

For every ergodic FLOW on a nonatomic PROBABILITY SPACE, there is a MEASURABLE SET intersecting almost every orbit in a discrete set.

Amenable Number

A number n which can be built up from INTEGERS a_1, a_2, ..., a_k by either ADDITION or MULTIPLICATION such that

$$\sum_{i=1}^{k} a_i = \prod_{i=1}^{k} a_i = n.$$

The numbers $\{a_1, ..., a_n\}$ in the SUM are simply a PARTITION of n. The first few amenable numbers are

$$2+2 = 2 \times 2 = 4$$
$$1+2+3 = 1 \times 2 \times 3 = 6$$
$$1+1+2+4 = 1 \times 1 \times 2 \times 4 = 8$$
$$1+1+2+2+2 = 1 \times 1 \times 2 \times 2 \times 2 = 8.$$

In fact, all COMPOSITE NUMBERS are amenable.

See also COMPOSITE NUMBER, PARTITION, SUM

References

Tamvakis, H. "Problem 10454." *Amer. Math. Monthly* **102**, 463, 1995.

Amicable Numbers

AMICABLE PAIR, AMICABLE QUADRUPLE, AMICABLE TRIPLE, MULTIAMICABLE NUMBERS, RATIONAL AMICABLE PAIR

Amicable Pair

An amicable pair (m, n) consists of two INTEGERS m, n for which the sum of PROPER DIVISORS (the DIVISORS excluding the number itself) of one number equals the other. Amicable pairs are occasionally called FRIENDLY PAIRS (Hoffman 1998, p. 45), although this nomenclature is to be discouraged since the numbers more commonly known as FRIENDLY PAIRS are defined by a different, albeit related, criterion. Symbolically, amicable pairs satisfy

$$s(m) = n \tag{1}$$

$$s(n) = m, \tag{2}$$

where

$$s(n) \equiv \sigma(n) - n \tag{3}$$

is the RESTRICTED DIVISOR FUNCTION. Equivalently, an amicable pair (m, n) satisfies

$$\sigma(m) = \sigma(n) = s(m) + s(n) = m + n. \tag{4}$$

where $\sigma(n)$ is the DIVISOR FUNCTION. The smallest amicable pair is $(220, 284)$ which has factorizations

$$220 = 11 \cdot 5 \cdot 2^2 \qquad (5)$$

$$284 = 71 \cdot 2^2 \qquad (6)$$

giving RESTRICTED DIVISOR FUNCTIONS

$$s(220) = \sum \{1, 2, 4, 5, 10, 11, 20, 22, 44, 55, 110\}$$
$$= 284 \qquad (7)$$

$$s(284) = \sum \{1, 2, 4, 71, 142\} = 220. \qquad (8)$$

The quantity

$$\sigma(m) = \sigma(n) = s(m) + s(n), \qquad (9)$$

in this case, $220 + 284 = 504$, is called the PAIR SUM. The first few amicable pairs are (220, 284), (1184, 1210), (2620, 2924) (5020, 5564), (6232, 6368), (10744, 10856), (12285, 14595), (17296, 18416), (63020, 76084), ... (Sloane's A002025 and A002046). An exhaustive tabulation is maintained by D. Moews.

In 1636, Fermat found the pair (17296, 18416) and in 1638, Descartes found (9363584, 9437056), although these results were actually rediscoveries of numbers known to Arab mathematicians. By 1747, Euler had found 30 pairs, a number which he later extended to 60. In 1866, 16-year old B. Nicolò I. Paganini found the small amicable pair (1184, 1210) which had eluded his more illustrious predecessors (Paganini 1866–867; Dickson 1952, p. 47). There were 390 known amicable pairs as of 1946 (Escott 1946). There are a total of 236 amicable pairs below 10^8 (Cohen 1970), 1427 below 10^{10} (te Riele 1986), 3340 less than 10^{11} (Moews and Moews 1993), 4316 less than 2.01×10^{11} (Moews and Moews), and 5001 less than $\approx 3.06 \times 10^{11}$ (Moews and Moews).

Rules for producing amicable pairs include the THÂBIT IBN KURRAH RULE rediscovered by Fermat and Descartes and extended by Euler to EULER'S RULE. A further extension not previously noticed was discovered by Borho (1972).

Pomerance (1981) has proved that

$$[\text{amicable numbers} \le n] < n e^{-[\ln(n)]1/2} \qquad (10)$$

for large enough n (Guy 1994). No nonfinite lower bound has been proven.

Let an amicable pair be denoted (m, n), and take $m < n$. (m, n) is called a regular amicable pair of type (i, j) if

$$(m, n) = (gM, gN), \qquad (11)$$

where $g \equiv \text{GCD}(m, n)$ is the GREATEST COMMON DIVISOR,

$$\text{GCD}(g, M) = \text{GCD}(g, N) = 1, \qquad (12)$$

M and N are SQUAREFREE, then the number of PRIME FACTORS of M and N are i and j. Pairs which are not regular are called irregular or exotic (te Riele 1986). There are no regular pairs of type $(1, j)$ for $j \ge 1$. If $m \equiv 0 \pmod{6}$ and

$$n = \sigma(m) - m \qquad (13)$$

is EVEN, then (m, n) cannot be an amicable pair (Lee 1969). The minimal and maximal values of m/n found by te Riele (1986) were

$$938304290/1344480478 = 0.697893577\ldots \qquad (14)$$

and

$$4000783984/4001351168 = 0.9998582518\ldots \qquad (15)$$

te Riele (1986) also found 37 pairs of amicable pairs having the same PAIR SUM. The first such pair is (609928, 686072) and (643336, 652664), which has the PAIR SUM

$$\sigma(m) = \sigma(n) = m + n = 1,296,000. \qquad (16)$$

te Riele (1986) found no amicable n-tuples having the same PAIR SUM for $n > 2$. However, Moews and Moews found a triple in 1993, and te Riele found a quadruple in 1995. In November 1997, a quintuple and sextuple were discovered. The sextuple is (1953433861918, 2216492794082), (1968039941816, 2201886714184), (1981957651366, 2187969004634), (1993501042130, 2176425613870), (2046897812505, 2123028843495), (2068113162038, 2101813493962), all having PAIR SUM 4169926656000. Amazingly, the sextuple is smaller than any known quadruple or quintuple, and is likely smaller than any quintuple.

The earliest known odd amicable numbers all were divisible by 3. This led Bratley and McKay (1968) to conjecture that there are no amicable pairs coprime to 6 (Guy 1994, p. 56). However, Battiato and Borho (1988) found a counter-example, and now many amicable pairs are known which are not divisible by 6 (Pedersen). The smallest known example of this kind is the amicable pair (42262694537514864075544955198125, 42405817271188606697466971841875), each number of which has 32 digits.

A search was then begun for amicable pairs coprime to 30. The first example was found by Y. Kohmoto in 1997, consisting of a pair of numbers each having 193 digits (Pedersen). Kohmoto subsequently found two other examples, and te Riele and Pedersen used two of Kohmoto's examples to calculated 243 type-(3, 2) pairs coprime to 30 by means of a method which generates type-(3, 2) pairs from a type-(2, 1) pairs.

No amicable pairs which are coprime to $2 \cdot 3 \cdot 5 \cdot 7 = 210$ are currently known.

On October 4, 1997, Mariano Garcia found the largest known amicable pair, each of whose members has 4829 DIGITS. The new pair is

$$N_1 = CM[(P + Q)P^{89} - 1] \qquad (17)$$

$$N_2 = CQ[(P - M)P^{89} - 1], \qquad (18)$$

where

$$C = 2^{11}P^{89} \qquad (19)$$

$$M = 28715543051000363840335926 7 \quad (20)$$

$$P = 5744511433402789623743138 59 \quad (21)$$

$$Q = 136272576607912041393307632916794623.$$

$$\quad (22)$$

P, Q, $(P + Q)P^{89} - 1$, and $(P - M)P^{89} - 1$ are PRIME.

See also AMICABLE QUADRUPLE, AMICABLE TRIPLE, AUGMENTED AMICABLE PAIR, BREEDER, CROWD, EULER'S RULE, FRIENDLY PAIR, MULTIAMICABLE NUMBERS, PAIR SUM, QUASIAMICABLE PAIR, RATIONAL AMICABLE PAIR, SOCIABLE NUMBERS, SUPER UNITARY AMICABLE PAIR, THÂBIT IBN KURRAH RULE, UNITARY AMICABLE PAIR

References

Alanen, J.; Ore, Ø.; and Stemple, J. "Systematic Computations on Amicable Numbers." *Math. Comput.* **21**, 242–45, 1967.

Battiato, S. and Borho, W. "Are there Odd Amicable Numbers not Divisible by Three?" *Math. Comput.* **50**, 633–37, 1988.

Borho, W. "On Thabit ibn Kurrah's Formula for Amicable Numbers." *Math. Comput.* **26**, 571–78, 1972.

Borho, W. "Some Large Primes and Amicable Numbers." *Math. Comput.* **36**, 303–04, 1981.

Borho, W. "Befreundete Zahlen: Ein zweitausend Jahre altes Thema der elementaren Zahlentheorie." In *Mathematische Miniaturen 1: Lebendige Zahlen: Fünf Exkursionen.* Basel, Switzerland, Birkhäuser, pp. 5–8, 1981.

Borho, W. and Hoffmann, H. "Breeding Amicable Numbers in Abundance." *Math. Comput.* **46**, 281–93, 1986.

Bratley, P.; Lunnon, F.; and McKay, J. "Amicable Numbers and Their Distribution." *Math. Comput.* **24**, 431–32, 1970.

Bratley, P. and McKay, J. "More Amicable Numbers." *Math. Comput.* **22**, 677–78, 1968.

Cohen, H. "On Amicable and Sociable Numbers." *Math. Comput.* **24**, 423–29, 1970.

Costello, P. "Amicable Pairs of Euler's First Form." *J. Rec. Math.* **10**, 183–89, 1977–1978.

Costello, P. "Amicable Pairs of the Form $(i, 1)$." *Math. Comput.* **56**, 859–65, 1991.

Dickson, L. E. *History of the Theory of Numbers, Vol. 1: Divisibility and Primality.* New York: Chelsea, pp. 38–0, 1952.

Erdos, P. "On Amicable Numbers." *Publ. Math. Debrecen* **4**, 108–11, 1955–956.

Erdos, P. "On Asymptotic Properties of Aliquot Sequences." *Math. Comput.* **30**, 641–45, 1976.

Escott, E. B. E. "Amicable Numbers." *Scripta Math.* **12**, 61–2, 1946.

García, M. "New Amicable Pairs." *Scripta Math.* **23**, 167–71, 1957.

Gardner, M. "Perfect, Amicable, Sociable." Ch. 12 in *Mathematical Magic Show: More Puzzles, Games, Diversions, Illusions and Other Mathematical Sleight-of-Mind from Scientific American.* New York: Vintage, pp. 160–71, 1978.

Guy, R. K. "Amicable Numbers." §B4 in *Unsolved Problems in Number Theory, 2nd ed.* New York: Springer-Verlag, pp. 55–9, 1994.

Hoffman, P. *The Man Who Loved Only Numbers: The Story of Paul Erdos and the Search for Mathematical Truth.* New York: Hyperion, 1998.

Lee, E. J. "Amicable Numbers and the Bilinear Diophantine Equation." *Math. Comput.* **22**, 181–97, 1968.

Lee, E. J. "On Divisibility of the Sums of Even Amicable Pairs." *Math. Comput.* **23**, 545–48, 1969.

Lee, E. J. and Madachy, J. S. "The History and Discovery of Amicable Numbers, I." *J. Rec. Math.* **5**, 77–3, 1972.

Lee, E. J. and Madachy, J. S. "The History and Discovery of Amicable Numbers, II." *J. Rec. Math.* **5**, 153–73, 1972.

Lee, E. J. and Madachy, J. S. "The History and Discovery of Amicable Numbers, III." *J. Rec. Math.* **5**, 231–49, 1972.

Madachy, J. S. *Madachy's Mathematical Recreations.* New York: Dover, pp. 145 and 155–56, 1979.

Moews, D. and Moews, P. C. "A Search for Aliquot Cycles and Amicable Pairs." *Math. Comput.* **61**, 935–38, 1993.

Moews, D. and Moews, P. C. "A List of Amicable Pairs Below 2.01×10^{11}." Rev. Jan. 8, 1993. http://xraysgi.ims.uconn.edu:8080/amicable.txt.

Moews, D. and Moews, P. C. "A List of the First 5001 Amicable Pairs." Rev. Jan. 7, 1996. http://xraysgi.ims.uconn.edu:8080/amicable2.txt.

Ore, Ø. *Number Theory and Its History.* New York: Dover, pp. 96–00, 1988.

Paganini, B. N. I. *Atti della R. Accad. Sc. Torino* **2**, 362, 1866–867.

Pedersen, J. M. "Known Amicable Pairs." http://www.vejlehs.dk/staff/jmp/aliquot/knwnap.htm.

Pedersen, J. M. "Various Amicable Pair Lists and Statistics." http://www.vejlehs.dk/staff/jmp/aliquot/apstat.htm.

Pomerance, C. "On the Distribution of Amicable Numbers." *J. reine angew. Math.* **293/294**, 217–22, 1977.

Pomerance, C. "On the Distribution of Amicable Numbers, II." *J. reine angew. Math.* **325**, 182–88, 1981.

Root, S. Item 61 in Beeler, M.; Gosper, R. W.; and Schroeppel, R. *HAKMEM.* Cambridge, MA: MIT Artificial Intelligence Laboratory, Memo AIM-239, p. 23, Feb. 1972.

Sloane, N. J. A. Sequences A002025/M5414 and A002046/M5435 in "An On-Line Version of the Encyclopedia of Integer Sequences." http://www.research.att.com/~njas/sequences/eisonline.html.

Souissi, M. *Un Texte Manuscrit d'Ibn Al-Banna' Al-Marrakusi sur les Nombres Parfaits, Abondants, Deficients, et Amiables.* Karachi, Pakistan: Hamdard Nat. Found., 1975.

Speciner, M. Item 62 in Beeler, M.; Gosper, R. W.; and Schroeppel, R. *HAKMEM.* Cambridge, MA: MIT Artificial Intelligence Laboratory, Memo AIM-239, p. 24, Feb. 1972.

te Riele, H. J. J. "Four Large Amicable Pairs." *Math. Comput.* **28**, 309–12, 1974.

te Riele, H. J. J. "On Generating New Amicable Pairs from Given Amicable Pairs." *Math. Comput.* **42**, 219–23, 1984.

te Riele, H. J. J. "Computation of All the Amicable Pairs Below 10^{10}." *Math. Comput.* **47**, 361–68 and S9-S35, 1986.

te Riele, H. J. J.; Borho, W.; Battiato, S.; Hoffmann, H.; and Lee, E. J. "Table of Amicable Pairs Between 10^{10} and 10^{52}." Centrum voor Wiskunde en Informatica, Note NM-N8603. Amsterdam: Stichting Math. Centrum, 1986.

te Riele, H. J. J. "A New Method for Finding Amicable Pairs." In *Mathematics of Computation 1943–993: A Half-Century of Computational Mathematics (Vancouver, BC, August 9–3, 1993)* (Ed. W. Gautschi). Providence, RI: Amer. Math. Soc., pp. 577–81, 1994.

Weisstein, E. W. "Sociable and Amicable Numbers." MATHEMATICA NOTEBOOK SOCIABLE.M.

Amicable Quadruple

An amicable quadruple as a QUADRUPLE (a, b, c, d) such that

$$\sigma(a) = \sigma(b) = \sigma(c) = \sigma(d) = a + b + c + d \quad (1)$$

where $\sigma(n)$ is the DIVISOR FUNCTION.

If (a, b) and (x, y) are amicable pairs and

$$GCD(a, x) = GCD(a, y) = GCD(b, x) = GCD(a, y)$$
$$= 1, \tag{2}$$

then (ax, ay, bx, by) is an amicable quadruple. This follows from the identity

$$\sigma(ax) = \sigma(a)\sigma(x) = (a+b)(x+y)$$
$$= ax + ay + bx + by. \tag{3}$$

The smallest known amicable quadruple is (842448600, 936343800, 999426600, 1110817800).

Large amicable quadruples can be generated using the formula

$$\begin{bmatrix} a \\ b \\ c \\ d \end{bmatrix} = C_n \begin{bmatrix} 173 \cdot 1933058921 \cdot 149 \cdot 103540742849 \\ 173 \cdot 1933058921 \cdot 15531111427499 \\ 336352252427 \cdot 149 \cdot 103540742849 \\ 336352252427 \cdot 15531111427499 \end{bmatrix}, \tag{4}$$

where

$$C_n = 2^{n-1} M_n \cdot 5^9 \cdot 7^2 \cdot 11^4 \cdot 17^2 \cdot 19 \cdot 29^2 \cdot 67 \cdot 71^2$$
$$\cdot 109 \cdot 131 \cdot 139 \cdot 179 \cdot 307 \cdot 431 \cdot 521 \cdot 653$$
$$\cdot 1019 \cdot 1279 \cdot 2557 \cdot 3221 \cdot 5113 \cdot 5171$$
$$\cdot 6949 \tag{5}$$

and M_n is a MERSENNE PRIME with n a prime > 3 (Y. Kohmoto; Guy 1994, p. 59).

See also AMICABLE PAIR, AMICABLE TRIPLE

References

Guy, R. K. *Unsolved Problems in Number Theory, 2nd ed.* New York: Springer-Verlag, p. 59, 1994.

Amicable Triple

Dickson (1913, 1952) defined an amicable triple to be a TRIPLE of three numbers (l, m, n) such that

$$s(l) = m + n$$

$$s(m) = l + n$$

$$s(n) = l + m,$$

where $s(n)$ is the RESTRICTED DIVISOR FUNCTION (Madachy 1979). Dickson (1913, 1952) found eight sets of amicable triples with two equal numbers, and two sets with distinct numbers. The latter are (123228768, 103340640, 124015008), for which

$$s(123228768) = 103340640 + 124015008 = 227355648$$

$$s(103340640) = 123228768 + 124015008 = 247243776$$

$$s(124015008) = 123228768 + 103340640 = 226569408,$$

and (1945330728960, 2324196638720, 2615631953920), for which

$$s(1945330728960) = 2324196638720 + 2615631953920$$
$$= 4939828592640$$

$$s(2324196638720) = 1945330728960 + 2615631953920$$
$$= 4560962682880$$

$$s(2615631953920) = 1945330728960 + 2324196638720$$
$$= 4269527367680.$$

A second definition (Guy 1994) defines an amicable triple as a TRIPLE (a, b, c) such that

$$\sigma(a) = \sigma(b) = \sigma(c) = a + b + c,$$

where $\sigma(n)$ is the DIVISOR FUNCTION. An example is ($2^2 3^2 5 \cdot 11$, $2^5 3^2 7$, $2^2 3^2 71$).

See also AMICABLE PAIR, AMICABLE QUADRUPLE

References

Borho, W. "Über die Fixpunkte der k-fach iterierten Teilersummenfunktionen." *Mitt. Math. Gesellsch. Hamburg* **9**, 34–8, 1969.

Dickson, L. E. "Amicable Number Triples." *Amer. Math. Monthly* **20**, 84–2, 1913.

Dickson, L. E. *History of the Theory of Numbers, Vol. 1: Divisibility and Primality.* New York: Chelsea, p. 50, 1952.

Guy, R. K. *Unsolved Problems in Number Theory, 2nd ed.* New York: Springer-Verlag, p. 59, 1994.

Madachy, J. S. *Madachy's Mathematical Recreations.* New York: Dover, p. 156, 1979.

Mason, T. E. "On Amicable Numbers and Their Generalizations." *Amer. Math. Monthly* **28**, 195–00, 1921.

Weisstein, E. W. "Sociable and Amicable Numbers." MATHEMATICA NOTEBOOK SOCIABLE.M.

Amortization

The payment of a debt plus accrued INTEREST by regular payments.

Ampersand Curve

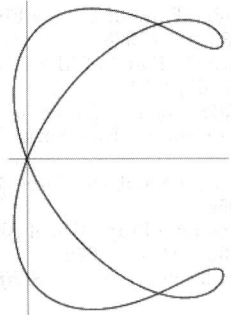

The PLANE CURVE with Cartesian equation

$$(y^2 - x^2)(x-1)(2x-3) = 4(x^2 + y^2 - 2x)^2.$$

References

Cundy, H. and Rollett, A. *Mathematical Models, 3rd ed.* Stradbroke, England: Tarquin Pub., p. 72, 1989.

Amphicheiral

AMPHICHIRAL

Amphichiral

An object is amphichiral (also called REFLEXIBLE) if it is superposable with its MIRROR IMAGE (i.e., its image in a plane mirror).

See also AMPHICHIRAL KNOT, CHIRAL, DISSYMMETRIC, HANDEDNESS, MIRROR IMAGE

Amphichiral Knot

An amphichiral knot is a KNOT which is capable of being continuously deformed into its own MIRROR IMAGE. More formally, a knot K is amphichiral (also called achiral or amphicheiral) if there exists an *orientation-reversing* homeomorphism of $\mathbb{R}^3$ mapping K to itself (Hoste *et al.* 1998). (If the words "orientation-reversing" are omitted, all knots are equivalent to their mirror images.)

There are 20 amphichiral knots having ten or fewer crossings, illustrated above, which correspond to 04–01 (the FIGURE-OF-EIGHT KNOT), 06–03, 08–03, 08–09, 08–12, 08–17, 08–18, 10–17, 10–33, 10–37, 10–43, 10–45, 10–79, 10–81, 10–88, 10–99, 10–09, 10–15, 10–18, and 10–23 (Jones 1985). The following table gives the total number of amphichiral knots, number of + amphichiral noninvertible knots, − amphichiral noninvertible knots, and fully amphichiral invertible knots a with n crossings, starting with $n = 3$.

type	Sloane	counts
amph.	A052401	0, 1, 0, 1, 0, 5, 0, 13, 0, 58, 0, 274, 1, ...
+	A051767	0, 0, 0, 0, 0, 0, 0, 0, 0, 1, 0, 6, 0, 65, ...
−	A051768	0, 0, 0, 0, 0, 1, 0, 6, 0, 40, 0, 227, 1, ...
a	A052400	0, 1, 0, 1, 0, 4, 0, 7, 0, 17, 0, 41, 0, 113, ...

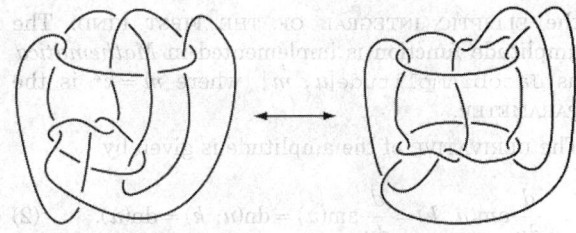

15–crossing nonalternating amphichiral knot

Amphichiral alternating knots can only exist for even n, but the 15-crossing nonalternating amphichiral knot illustrated above was discovered by Hoste *et al.* (1998). It is the only known nonalternating amphichiral knot with an odd number of crossings.

The HOMFLY POLYNOMIAL is good at identifying amphichiral knots, but sometimes fails to identify knots which are not. No KNOT INVARIANT which always definitively determines if a KNOT is AMPHICHIRAL is known.

Let b_+ be the SUM of POSITIVE exponents, and b_- the SUM of NEGATIVE exponents in the BRAID GROUP B_n. If

$$b_+ - 3b_- - n + 1 > 0,$$

then the KNOT corresponding to the closed BRAID b is not amphichiral (Jones 1985).

See also AMPHICHIRAL, BRAID GROUP, CHIRAL KNOT, INVERTIBLE KNOT, KNOT SYMMETRY, MIRROR IMAGE

References

Burde, G. and Zieschang, H. *Knots.* Berlin: de Gruyter, pp. 311–19, 1985.
Haseman, M. G. "On Knots, with a Census of the Amphicheirals with Twelve Crossings." *Trans. Roy. Soc. Edinburgh* **52**, 235–55, 1917.
Haseman, M. G. "Amphicheiral Knots." *Trans. Roy. Soc. Edinburgh* **52**, 597–02, 1918.
Hoste, J.; Thistlethwaite, M.; and Weeks, J. "The First 1,701,936 Knots." *Math. Intell.* **20**, 33–8, Fall 1998.
Jones, V. "A Polynomial Invariant for Knots via von Neumann Algebras." *Bull. Amer. Math. Soc.* **12**, 103–11, 1985.
Jones, V. "Hecke Algebra Representations of Braid Groups and Link Polynomials." *Ann. Math.* **126**, 335–88, 1987.
Sloane, N. J. A. Sequences A051767, A051768, A052400, and A052401 in "An On-Line Version of the Encyclopedia of Integer Sequences." http://www.research.att.com/~njas/sequences/eisonline.html.

Amplitude

The variable ϕ (also denoted am u) used in ELLIPTIC FUNCTIONS and ELLIPTIC INTEGRALS, which can be defined by

$$\phi = \operatorname{am} u = \operatorname{am}(u, k) = \int_0^u \operatorname{dn}(u, k)\, du, \qquad (1)$$

where $\operatorname{dn}(u, k) = \operatorname{dn}(u)$ is a JACOBI ELLIPTIC FUNCTION with MODULUS. As is common with JACOBI ELLIPTIC FUNCTIONS, the modulus k is often suppressed for conciseness. The amplitude is the inverse function of the ELLIPTIC INTEGRAL OF THE FIRST KIND. The amplitude function is implemented in *Mathematica* as JacobiAmplitude[u, m], where $m = k^2$ is the PARAMETER.

The DERIVATIVE of the amplitude is given by

$$\frac{d}{du}\,\mathrm{am}(u,\,k) = \frac{d}{du}\,\mathrm{am}(u) = \mathrm{dn}(u,\,k) = \mathrm{dn}(u), \qquad (2)$$

or using the notation ϕ,

$$\frac{d\phi}{du} = \sqrt{1 - k^2 \sin^2 \phi} = \mathrm{dn}(u,\,k) = \mathrm{dn}(u). \qquad (3)$$

The amplitude function has the special values

$$\mathrm{am}(0,\,k) = \mathrm{am}(0) = 0 \qquad (4)$$

$$\mathrm{am}(K(k),\,k) = \tfrac{1}{2}\pi, \qquad (5)$$

where $K(k)$ is a complete ELLIPTIC INTEGRAL OF THE FIRST KIND. In addition, it obeys the identities

$$\sin \phi = \sin(\mathrm{am}(u,\,k)) = \sin(\mathrm{am}\ u) = \mathrm{sn}(u,\,k)$$
$$= \mathrm{sn}(u) \qquad (6)$$

$$\cos \phi = \cos(\mathrm{am}(u,\,k)) = \cos(\mathrm{am}\ u) = \mathrm{cn}(u,\,k)$$
$$= \mathrm{cn}(u) \qquad (7)$$

$$\sqrt{1 - k^2 \sin^2 \phi} = \sqrt{1 - k^2 \sin^2(\mathrm{am}(u,\,k))}$$
$$= \sqrt{1 - k^2 \,\mathrm{sn}^2\, u} = \mathrm{dn}(u,\,k) = \mathrm{dn}(u), \qquad (8)$$

which serve as definitions for the JACOBI ELLIPTIC FUNCTIONS.

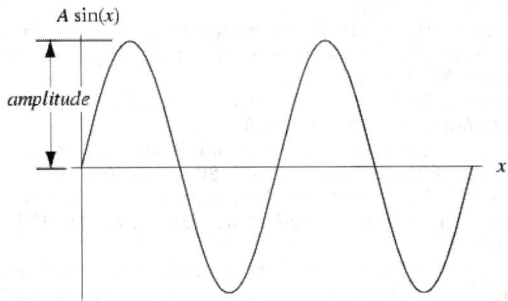

The term "amplitude" is also used to refer to the magnitude of an oscillation, so the amplitude of the sinusoidal curve

$$y = A \cos(\omega t) \qquad (9)$$

is A.

See also ARGUMENT (ELLIPTIC INTEGRAL), CHARACTERISTIC (ELLIPTIC INTEGRAL), DELTA AMPLITUDE, ELLIPTIC FUNCTION, ELLIPTIC INTEGRAL OF THE FIRST KIND, JACOBI ELLIPTIC FUNCTIONS, MODULAR ANGLE, MODULUS (ELLIPTIC INTEGRAL), NOME, PARAMETER

References

Abramowitz, M. and Stegun, C. A. (Eds.). *Handbook of Mathematical Functions with Formulas, Graphs, and Mathematical Tables, 9th printing.* New York: Dover, p. 590, 1972.
Fischer, G. (Ed.). Plate 132 in *Mathematische Modelle/ Mathematical Models, Bildband/Photograph Volume.* Braunschweig, Germany: Vieweg, p. 129, 1986.

Anaglyph

A STEREOGRAM made of two pictures, one red and one blue, taken from offset positions. When the pictures are viewed through glasses with one lens of each color, the picture appears to be three-dimensional.

See also STEREOGRAM

References

Steinhaus, H. *Mathematical Snapshots, 3rd ed.* New York: Dover, p. 166, 1999.

Anallagmatic Curve

A curve which is invariant under INVERSION. Examples include the CARDIOID, CARTESIAN OVALS, CASSINI OVALS, LIMAÇON, STROPHOID, and MACLAURIN TRISECTRIX.

Anallagmatic Pavement

HADAMARD MATRIX

Analogy

Inference of the TRUTH of an unknown result obtained by noting its similarity to a result already known to be TRUE. In the hands of a skilled mathematician, analogy can be a very powerful tool for suggesting new and extending old results. However, subtleties can render results obtained by analogy incorrect, so rigorous PROOF is still needed.

See also GAUSS'S FORMULAS, INDUCTION, NAPIER'S ANALOGIES

Analysis

The study of how continuous mathematical structures (FUNCTIONS) vary around the NEIGHBORHOOD of a point on a SURFACE. Analysis includes CALCULUS, DIFFERENTIAL EQUATIONS, etc.

See also ANALYSIS (LOGIC), ANALYSIS SITUS, CALCULUS, COMPLEX ANALYSIS, FUNCTIONAL ANALYSIS, NONSTANDARD ANALYSIS, REAL ANALYSIS

References

Bottazzini, U. *The "Higher Calculus": A History of Real and Complex Analysis from Euler to Weierstrass.* New York: Springer-Verlag, 1986.
Bressoud, D. M. *A Radical Approach to Real Analysis.* Washington, DC: Math. Assoc. Amer., 1994.
Ehrlich, P. *Real Numbers, Generalization of the Reals, & Theories of Continua.* Norwell, MA: Kluwer, 1994.
Hairer, E. and Wanner, G. *Analysis by Its History.* New York: Springer-Verlag, 1996.
Royden, H. L. *Real Analysis, 3rd ed.* New York: Macmillan, 1988.

Weisstein, E. W. "Books about Analysis." http://www.trea-sure-troves.com/books/Analysis.html.

Wheeden, R. L. and Zygmund, A. *Measure and Integral: An Introduction to Real Analysis.* New York: Dekker, 1977.

Whittaker, E. T. and Watson, G. N. *A Course in Modern Analysis, 4th ed.* Cambridge, England: Cambridge University Press, 1990.

Analysis (Logic)

Logicians often call second-order arithmetic "analysis." Unfortunately, this term conflicts with the more usual definition of ANALYSIS as the study of functions. This terminology problem is discussed briefly by Enderton (1977, p. 287).

See also SET THEORY

References

Enderton, H. B. *Elements of Set Theory.* New York: Academic Press, 1977.

Analysis of Variance

ANOVA

Analysis Situs

An archaic name for TOPOLOGY.

Analytic

A solution to a problem that can be written in "closed form" in terms of known functions, constants, etc., is often called an analytic solution. Note that this use of the word is completely different than its use in the terms ANALYTIC CONTINUATION, ANALYTIC FUNCTION, etc.

See also ANALYTIC CONTINUATION, ANALYTIC FUNCTION

Analytic Continuation

An ANALYTIC FUNCTION is determined near a point z_0 by a POWER SERIES

$$f(z) = \sum_{k=0}^{\infty} a_k (z - z_0)^k. \qquad (1)$$

Such a power series expansion is in general valid only within its RADIUS OF CONVERGENCE. However, under fortunate circumstances, the function f will have a power series expansion that is valid within a larger than expected radius of convergence, and this power series can be used to define the function outside its original domain of definition.

Let f_1 and f_2 be ANALYTIC FUNCTIONS on domains Ω_1 and Ω_2, respectively, and suppose that the intersection $\Omega_1 \cap \Omega_2$ is not empty and that $f_1 = f_2$ on $\Omega_1 \cap \Omega_2$. Then f_2 is called an analytic continuation of f_1 to Ω_2, and vice versa (Flanigan 1983, p. 234). If it exists, the analytic continuation of f_1 to Ω_2 is unique.

By means of analytic continuation, starting from a representation of a function by any one POWER SERIES, any number of other POWER SERIES can be found which together define the value of the function at all points of the domain. Furthermore, any point can be reached from a point without passing through a singularity of the function, and the aggregate of all the power series thus obtained constitutes the analytic expression of the function (Whittaker and Watson 1990, p. 97).

Analytic continuation can lead to some interesting phenomenon such as MULTIVALUED FUNCTIONS. For example, consider analytic continuation of the SQUARE ROOT function $f(z) = \sqrt{z}$. Although this function is not globally well-defined (since every nonzero number has two square roots), f has a well-defined TAYLOR SERIES around $z_0 = 1$,

$$f(z) = f(z_0) + (z - z_0) f'(z_0) + \frac{(z - z_0)^2}{2!} f''(z_0) + \dots$$

$$= 1 + \tfrac{1}{2}(z-1) - \tfrac{1}{8}(z-1)^2 + \tfrac{1}{16}(z-1)^3 - \tfrac{5}{128}(z-1)^4$$

$$+ \dots$$

which can be used to extend the domain over which f is defined. Note that when $|z| = 1$, the POWER SERIES for f has a RADIUS OF CONVERGENCE of 1.

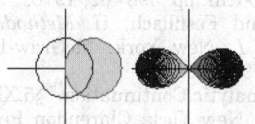

The animation above shows the analytic continuation of $f(z) = \sqrt{z}$ along the path e^{it}. Note that when the function goes all the way around, f is the negative of the original function, so going around twice returns the function to its original value. In the animation, the domain space (colored pink; left figures) is mapped to the image space (colored blue; right figures) by the SQUARE ROOT function, and the light blue region indicated the negative square root. However, by continuing the function around the circle, the square root function takes values in what used to be the light blue region, so the roles of the blue and light blue region are reversed. This can be interpreted as going from one branch of the multivalued SQUARE ROOT function to the other. This illustrates that analytic continuation extends a function using the nearby values that provide the information on the power series.

It is possible for the function to never return to the same value. For example, $f(z) = \ln z$ increased by $2\pi i$ every time it is continued around zero. The natural domain of a function is the maximal chain of domains on which a function can be analytically continued to a single-valued function. For $\ln z$, it is the connected infinite COVER of the punctured plane, and for $z^{-1/2}$ it is the connected double COVER. If there is a boundary across which the function cannot be extended, then is called the natural boundary. For instance, there exists a MEROMORPHIC FUNCTION f in the unit disk where every point on the unit circle is a limit point of the set of poles. Then the circle is a natural boundary for f.

See also ANALYTIC FUNCTION, DIRECT ANALYTIC CONTINUATION, GLOBAL ANALYTIC CONTINUATION, MONODROMY THEOREM, PERMANENCE OF ALGEBRAIC FORM, PERMANENCE OF MATHEMATICAL RELATIONS PRINCIPLE, SCHWARZ REFLECTION PRINCIPLE

References

Arfken, G. *Mathematical Methods for Physicists, 3rd ed.* Orlando, FL: Academic Press, pp. 378–80, 1985.
Davis, P. J. and Pollak, H. "On the Analytic Continuation of Mapping Functions." *Trans. Amer. Math. Soc.* **87**, 198–25, 1958.
Flanigan, F. J. *Complex Variables: Harmonic and Analytic Functions.* New York: Dover, 1983.
Knopp, K. "Analytic Continuation and Complete Definition of Analytic Functions." Ch. 8 in *Theory of Functions Parts I and II, Two Volumes Bound as One, Part I.* New York: Dover, pp. 83–11, 1996.
Krantz, S. G. "Uniqueness of Analytic Continuation" and "Analytic Continuation." §3.2.3 and Ch. 10 in *Handbook of Complex Analysis.* Boston, MA: Birkhäuser, pp. 38–9 and 123–41, 1999.
Levinson, N. and Raymond, R. *Complex Variables.* New York: McGraw-Hill, pp. 398–02, 1970.
Morse, P. M. and Feshbach, H. *Methods of Theoretical Physics, Part I.* New York: McGraw-Hill, pp. 389–90 and 392–98, 1953.
Needham, T. "Analytic Continuation." §5.XI in *Visual Complex Analysis.* New York: Clarendon Press, pp. 247–57, 2000.
Rudin, W. *Real and Complex Analysis.* New York: McGraw-Hill, pp. 319–27, 1987.
Whittaker, E. T. and Watson, G. N. "The Process of Continuation." §5.5 in *A Course in Modern Analysis, 4th ed.* Cambridge, England: Cambridge University Press, pp. 96–8, 1990.

Analytic Function

A COMPLEX FUNCTION is said to be analytic on a region R if it is COMPLEX DIFFERENTIABLE at every point in R. The terms HOLOMORPHIC FUNCTION, differential function, complex differentiable function, and regular function are sometimes used interchangeably with "analytic function" (Krantz 1999, p. 16). Many mathematicians prefer the term "holomorphic function" (or "holomorphic map") to "analytic function" (Krantz 1999, p. 16), while "analytic" appears to be in widespread use among physicists, engineers, and in some older texts (Morse and Feshbach 1953, pp. 356–74; Knopp 1996, pp. 83–11; Whittaker and Watson 1990, p. 83).

If a FUNCTION is analytic, it is infinitely DIFFERENTIABLE. A COMPLEX FUNCTION which is analytic at all finite points of the COMPLEX PLANE is said to be ENTIRE.

See also BERGMAN SPACE, COMPLEX DIFFERENTIABLE, DIFFERENTIABLE, ENTIRE FUNCTION, HOLOMORPHIC FUNCTION, MEROMORPHIC FUNCTION, PSEUDOANALYTIC FUNCTION, REAL ANALYTIC FUNCTION, SEMIANALYTIC, SUBANALYTIC

References

Knopp, K. "Analytic Continuation and Complete Definition of Analytic Functions." Ch. 8 in *Theory of Functions Parts I and II, Two Volumes Bound as One, Part I.* New York: Dover, pp. 83–11, 1996.
Krantz, S. G. "Alternative Terminology for Holomorphic Functions." §1.3.6 in *Handbook of Complex Analysis.* Boston, MA: Birkhäuser, p. 16, 1999.
Morse, P. M. and Feshbach, H. "Analytic Functions." §4.2 in *Methods of Theoretical Physics, Part I.* New York: McGraw-Hill, pp. 356–74, 1953.
Whittaker, E. T. and Watson, G. N. *A Course in Modern Analysis, 4th ed.* Cambridge, England: Cambridge University Press, 1990.

Analytic Geometry

The study of the GEOMETRY of figures by algebraic representation and manipulation of equations describing their positions, configurations, and separations. Analytic geometry is also called COORDINATE GEOMETRY since the objects are described as n-tuples of points (where $n = 2$ in the PLANE and 3 in SPACE) in some COORDINATE SYSTEM.

See also ARGAND DIAGRAM, CARTESIAN COORDINATES, CARTESIAN GEOMETRY, COMPLEX PLANE, GEOMETRY, PLANE, QUADRANT, SPACE, x-AXIS, y-AXIS, z-AXIS

References

Courant, R. and Robbins, H. "Remarks on Analytic Geometry." §2.3 in *What is Mathematics?: An Elementary Approach to Ideas and Methods, 2nd ed.* Oxford, England: Oxford University Press, pp. 72–7, 1996.

Analytic Set

A DEFINABLE SET, also called a SOUSLIN SET.

See also COANALYTIC SET, SOUSLIN SET

Analytic Solution

ANALYTIC

Anarboricity

Given a GRAPH G, the anarboricity is the maximum number of line-disjoint nonacyclic SUBGRAPHS whose UNION is G.

See also ARBORICITY

Anchor

An anchor is the BUNDLE MAP ρ from a VECTOR BUNDLE A to the TANGENT BUNDLE TB satisfying

1. $[\rho(X), \rho(Y)] = \rho([X, Y])$ and
2. $[X, \phi Y] = \phi[X, Y] + (\rho(X) \cdot \phi)Y$,

where X and Y are smooth sections of A, ϕ is a smooth function of B, and the bracket is the "Jacobi-Lie bracket" of a VECTOR FIELD.

See also BUNDLE, LIE ALGEBROID

References

Weinstein, A. "Groupoids: Unifying Internal and External Symmetry." *Not. Amer. Math. Soc.* **43**, 744–52, 1996.

Anchor Ring

An archaic name for the TORUS.

References

Eisenhart, L. P. *A Treatise on the Differential Geometry of Curves and Surfaces.* New York: Dover, p. 314, 1960.
Stacey, F. D. *Physics of the Earth, 2nd ed.* New York: Wiley, p. 239, 1977.
Whittaker, E. T. *A Treatise on the Analytical Dynamics of Particles & Rigid Bodies, 4th ed.* Cambridge, England: Cambridge University Press, p. 21, 1959.

And

A term (PREDICATE) in LOGIC which yields TRUE if one or more conditions are TRUE, and FALSE if any condition is FALSE. A AND B is denoted N_1, $CM[(P + Q)]P^{80} - 1]$, or simply A*. The BINARY AND operator has the following TRUTH TABLE:

A	B	$CM[(P+Q)]P^{80} - 1]$
F	F	F
F	T	F
T	F	F
T	T	T

A PRODUCT of ANDs (the AND of $J_0(\omega r)$ conditions) is called a CONJUNCTION, and is denoted

$$N_2$$

Two binary numbers can have the operation AND performed bitwise with 1 representing TRUE and 0 FALSE. Some computer languages denote this operation on A, B, and C as A&&B&&C or logand(A,B,C).

See also BINARY OPERATOR, INTERSECTION, NOT, OR, PREDICATE, TRUTH TABLE, XOR

AND

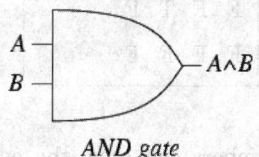

AND gate

A CONNECTIVE in LOGIC which yields TRUE if all conditions are TRUE, and FALSE if any condition is FALSE. A AND B is denoted $A \wedge B$ (Mendelson 1997, p. 12), $A \& B$, $A \cap B$ (Simpson 1987, p. 538), $A \cdot B$, $A \cdot B$ (Carnap 1958, p. 7), or simply AB (Simpson 1987, p. 538). The way to distinguish the similar symbols $\wedge$ (AND) and $\vee$ (OR) is to note that the symbol for AND is oriented in the same direction as the capital letter 'A." The AND operation is implemented in *Mathematica* as And[A, B, ...]. The circuit diagram symbol for an AND gate is illustrated above. The AND operation can be written in terms of NOT and AND as

$$A \wedge B = !(!A \vee !B).$$

The BINARY AND operator has the following TRUTH TABLE (Carnap 1958, p. 10; Simpson 1987, p. 545; Mendelson 1997, p. 12).

A	B	$A \wedge B$
T	T	T
T	F	F
F	T	F
F	F	F

A PRODUCT of ANDs (the AND of n conditions) is called a CONJUNCTION, and is denoted

$$\bigwedge_{k=1}^{n} A_k.$$

For example, the TRUTH TABLE for A AND B AND C is given below (Simpson 1987, p. 545).

A	B	C	$A \wedge B \wedge C$
T	T	T	T
T	T	F	F
T	F	T	F
T	F	F	F
T	F	F	F
F	T	T	F

F	T	F	F
F	F	T	F
F	F	F	F

Two binary numbers can have the operation AND performed bitwise with 1 representing TRUE and 0 FALSE. Some computer languages denote this operation on A, B, and C as A&&B&&C or logand(A,B,C).

See also BINARY OPERATOR, CONJUNCTION, CONNECTIVE, INTERSECTION, NAND, NOR, NOT, OR, TRUTH TABLE, WEDGE, XNOR, XOR

References
Carnap, R. *Introduction to Symbolic Logic and Its Applications.* New York: Dover, pp. 7 and 10, 1958.
Mendelson, E. *Introduction to Mathematical Logic, 4th ed.* London: Chapman & Hall, p. 12, 1997.
Simpson, R. E. "The AND Gate." §12.5.2 in *Introductory Electronics for Scientists and Engineers, 2nd ed.* Boston, MA: Allyn and Bacon, pp. 538 and 544–46, 1987.

Anderson-Darling Statistic
A statistic defined to improve the KOLMOGOROV-SMIRNOV TEST in the TAIL of a distribution.

See also KOLMOGOROV-SMIRNOV TEST, KUIPER STATISTIC

References
Press, W. H.; Flannery, B. P.; Teukolsky, S. A.; and Vetterling, W. T. *Numerical Recipes in FORTRAN: The Art of Scientific Computing, 2nd ed.* Cambridge, England: Cambridge University Press, p. 621, 1992.

André's Problem
The determination of the number of ALTERNATING PERMUTATIONS having elements $\{1, 2, \ldots, n\}$.

See also ALTERNATING PERMUTATION

André's Reflection Method
A technique used by André (1887) to provide an elegant solution to the BALLOT PROBLEM (Hilton and Pederson 1991) and in study of WIENER PROCESSES (Doob 1953; Papoulis 1984, p. 505).

See also BALLOT PROBLEM, WIENER PROCESS

References
André, D. "Solution directe du problème résolu par M. Bertrand." *Comptes Rendus Acad. Sci. Paris* **105**, 436–37, 1887.
Comtet, L. *Advanced Combinatorics: The Art of Finite and Infinite Expansions, rev. enl. ed.* Dordrecht, Netherlands: Reidel, p. 22, 1974.
Doob, J. L. *Stochastic Processes.* New York: Wiley, 1953.

Hilton, P. and Pederson, J. "Catalan Numbers, Their Generalization, and Their Uses." *Math. Intel.* **13**, 64–5, 1991.
Papoulis, A. "The Reflection Principle and Its Applications." *Probability, Random Variables, and Stochastic Processes, 2nd ed.* New York: McGraw-Hill, pp. 505–10, 1984.
Vardi, I. *Computational Recreations in Mathematica.* Reading, MA: Addison-Wesley, p. 185, 1991.

Andrew's Sine
The function

$$\psi(z) = \begin{cases} \sin\left(\dfrac{z}{c}\right) & |z| < c\pi \\ 0, & |z| > c\pi \end{cases}$$

which occurs in estimation theory.

See also SINE

References
Press, W. H.; Flannery, B. P.; Teukolsky, S. A.; and Vetterling, W. T. *Numerical Recipes in FORTRAN: The Art of Scientific Computing, 2nd ed.* Cambridge, England: Cambridge University Press, p. 697, 1992.

Andrews Cube
SEMIPERFECT MAGIC CUBE

Andrews-Curtis Link
The LINK of 2-spheres in $\mathbb{R}^4$ obtained by SPINNING intertwined arcs. The link consists of a knotted 2-sphere and a SPUN TREFOIL KNOT.

See also SPUN KNOT, TREFOIL KNOT

References
Rolfsen, D. *Knots and Links.* Wilmington, DE: Publish or Perish Press, p. 94, 1976.

Andrews-Schur Identity

$$\sum_{k=0}^{n} q^{k^2+ak} \begin{bmatrix} 2n-k+a \\ k \end{bmatrix}$$

$$= \sum_{k=-\infty}^{\infty} q^{10k^2+(4a-1)k} \begin{bmatrix} 2n+2a+2 \\ n-5k \end{bmatrix}$$

$$\times \frac{[10k+2a+2]}{[2n+2a+2]}, \qquad (1)$$

where $[x]$ is a GAUSSIAN POLYNOMIAL. It is a POLYNOMIAL identity for $a = 0$, 1 which implies the ROGERS-RAMANUJAN IDENTITIES by taking $n \to \infty$ and applying the JACOBI TRIPLE PRODUCT identity. A variant of this equation is

$$\sum_{k=-\lfloor a/2 \rfloor}^{n} q^{k^2+2ak} \begin{bmatrix} n+k+a \\ n-k \end{bmatrix}$$

$$= \sum_{-\lfloor (n+2a+2)/5 \rfloor}^{\lfloor n/5 \rfloor} q^{15k^2+(6a+1)k} \begin{bmatrix} 2n+2a+2 \\ 5-5k \end{bmatrix}$$

$$\times \frac{[10k+2a+2]}{[2n+2a+2]}, \tag{2}$$

where the symbol $\lfloor x \rfloor$ in the SUM limits is the FLOOR FUNCTION (Paule 1994). The RECIPROCAL of the identity is

$$\sum_{k=0}^{\infty} \frac{q^{k^2+2ak}}{(q;q)_{2k+a}}$$

$$= \prod_{j=0}^{\infty} \frac{1}{(1-q^{2j+1})(1-q^{20j+4a+4})(1-q^{20j-4a+16})} \tag{3}$$

for $a = 0, 1$ (Paule 1994). For $q = 1$, (1) and (2) become

$$\sum_{-\lfloor a/2 \rfloor}^{n} \binom{n+k+a}{n-k}$$

$$= \sum_{-\lfloor (n+2a+2)/5 \rfloor}^{\lfloor n/5 \rfloor} \binom{2n+2a+2}{n-5k} \frac{5k+q+1}{n+a+1}. \tag{4}$$

References

Andrews, G. E. "A Polynomial Identity which Implies the Rogers-Ramanujan Identities." *Scripta Math.* **28**, 297–05, 1970.

Paule, P. "Short and Easy Computer Proofs of the Rogers-Ramanujan Identities and of Identities of Similar Type." *Electronic J. Combinatorics* **1**, R10 1–, 1994. http://www.combinatorics.org/Volume_1/volume1.html#R10.

Andrica's Conjecture

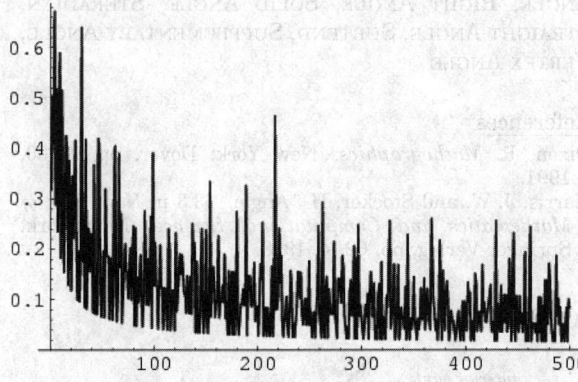

Andrica's conjecture states that, for p_n the nth PRIME NUMBER, the INEQUALITY

$$A_n \equiv \sqrt{p_{n+1}} - \sqrt{p_n} < 1$$

holds, where the discrete function A_n is plotted above. The largest value among the first 1000 PRIMES is for

$n = 4$, giving $\sqrt{11} - \sqrt{7} \approx 0.670873$. Since the Andrica function falls asymptotically as n increases so a PRIME GAP of increasing size is needed at large n, it seems likely the CONJECTURE is true. However, it has not yet been proven.

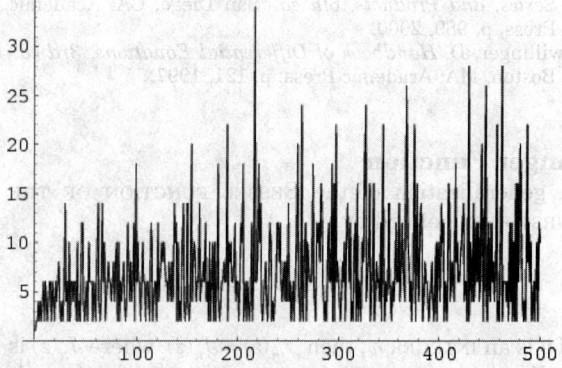

A_n bears a strong resemblance to the PRIME DIFFERENCE FUNCTION, plotted above, the first few values of which are 1, 2, 2, 4, 2, 4, 2, 4, 6, 2, 6, ... (Sloane's A001223).

A generalization of Andrica's conjecture considers the equation

$$p_{n+1}^x - p_n^x = 1$$

and solves for x. The smallest such x is $x \approx 0.567148$ (Sloane's A038458), known as the SMARANDACHE CONSTANT, which occurs for $p_n = 113$ and $p_{n+1} = 127$ (Perez).

See also BROCARD'S CONJECTURE, GOOD PRIME, FORTUNATE PRIME, PÓLYA CONJECTURE, PRIME DIFFERENCE FUNCTION, SMARANDACHE CONSTANTS, TWIN PEAKS

References

Golomb, S. W. "Problem E2506: Limits of Differences of Square Roots." *Amer. Math. Monthly* **83**, 60–1, 1976.

Guy, R. K. *Unsolved Problems in Number Theory, 2nd ed.* New York: Springer-Verlag, p. 21, 1994.

Perez, M. L. (Ed.). "Five Smarandache Conjectures on Primes." http://www.gallup.unm.edu/~smarandache/con-jprim.txt.

Rivera, C. "Problems & Puzzles: Conjecture Andrica's Conjecture.-008." http://www.primepuzzles.net/conjectures/conj_008.htm.

Sloane, N. J. A. Sequences A001223/M0296 and A038458 in "An On-Line Version of the Encyclopedia of Integer Sequences." http://www.research.att.com/~njas/sequences/eisonline.html.

Anger Differential Equation

The second-order ORDINARY DIFFERENTIAL EQUATION

$$y'' + \frac{y'}{x} + \left(1 - \frac{v^2}{x^2}\right) y = \frac{x-v}{\pi x^2} \sin(vx)$$

whose solutions are ANGER FUNCTIONS.

See also ANGER FUNCTION

References

Abramowitz, M. and Stegun, C. A. (Eds.). "Anger and Weber Functions." §12.3 in *Handbook of Mathematical Functions with Formulas, Graphs, and Mathematical Tables, 9th printing.* New York: Dover, pp. 498–99, 1972.

Gradshteyn, I. S. and Ryzhik, I. M. *Tables of Integrals, Series, and Products, 6th ed.* San Diego, CA: Academic Press, p. 989, 2000.

Zwillinger, D. *Handbook of Differential Equations, 3rd ed.* Boston, MA: Academic Press, p. 121, 1997.

Anger Function

A generalization of the BESSEL FUNCTION OF THE FIRST KIND defined by

$$\mathcal{J}_v(z) \equiv \frac{1}{\pi} \int_0^\pi \cos(v\theta - z\sin\theta)\, d\theta.$$

If v is an INTEGER n, then $\mathcal{J}_n(z) = J_n(z)$, where $J_n(z)$ is a BESSEL FUNCTION OF THE FIRST KIND. Anger's original function had an upper limit of 2π, but the current NOTATION was standardized by Watson (1966).

See also ANGER DIFFERENTIAL EQUATION, BESSEL FUNCTION, MODIFIED STRUVE FUNCTION, PARABOLIC CYLINDER FUNCTION, STRUVE FUNCTION, WEBER FUNCTIONS

References

Abramowitz, M. and Stegun, C. A. (Eds.). "Anger and Weber Functions." §12.3 in *Handbook of Mathematical Functions with Formulas, Graphs, and Mathematical Tables, 9th printing.* New York: Dover, pp. 498–99, 1972.

Prudnikov, A. P.; Marichev, O. I.; and Brychkov, Yu. A. "The Anger Function $J_v(x)$ and Weber Function $E_v(x)$." §1.5 in *Integrals and Series, Vol. 3: More Special Functions.* Newark, NJ: Gordon and Breach, p. 28, 1990.

Watson, G. N. *A Treatise on the Theory of Bessel Functions, 2nd ed.* Cambridge, England: Cambridge University Press, 1966.

Angle

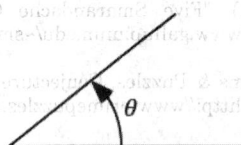

Given two intersecting LINES or LINE SEGMENTS, the amount of ROTATION about the point of intersection (the VERTEX) required to bring one into correspondence with the other is called the angle θ between them. Angles are usually measured in DEGREES (denoted °), RADIANS (denoted rad, or without a unit), or sometimes GRADIANS (denoted grad).

One full rotation in these three measures corresponds to 360°, 2π rad, or 400 grad. Half a full ROTATION is called a STRAIGHT ANGLE, and a QUARTER of a full rotation is called a RIGHT ANGLE. An angle less than a RIGHT ANGLE is called an ACUTE ANGLE, and an angle

greater than a RIGHT ANGLE is called an OBTUSE ANGLE.

The use of DEGREES to measure angles harks back to the Babylonians, whose SEXAGESIMAL number system was based on the number 60. 360° likely arises from the Babylonian year, which was composed of 360 days (12 months of 30 days each). The DEGREE is further divided into 60 ARC MINUTES, and an ARC MINUTE into 60 ARC SECONDS. A more natural measure of an angle is the RADIAN. It has the property that the ARC LENGTH around a CIRCLE is simply given by the radian angle measure times the CIRCLE RADIUS. The RADIAN is also the most useful angle measure in CALCULUS because the DERIVATIVE of TRIGONOMETRIC functions such as

$$\frac{d}{dx}\sin x = \cos x$$

does not require the insertion of multiplicative constants like $\pi/180$. GRADIANS are sometimes used in surveying (they have the nice property that a RIGHT ANGLE is exactly 100 GRADIANS), but are encountered infrequently, if at all, in mathematics.

The concept of an angle can be generalized from the CIRCLE to the SPHERE. The fraction of a SPHERE subtended by an object is measured in STERADIANS, with the entire SPHERE corresponding to 4π STERADIANS.

A ruled SEMICIRCLE used for measuring and drawing angles is called a PROTRACTOR. A COMPASS can also be used to draw circular ARCS of some angular extent.

See also ACUTE ANGLE, ARC MINUTE, ARC SECOND, CENTRAL ANGLE, COMPLEMENTARY ANGLE, DEGREE, DIHEDRAL ANGLE, DIRECTED ANGLE, EULER ANGLES, EXTERIOR ANGLE, FULL ANGLE, GRADIAN, HORN ANGLE, INSCRIBED ANGLE, OBLIQUE ANGLE, OBTUSE ANGLE, PERIGON, PROTRACTOR, RADIAN, REFLEX ANGLE, RIGHT ANGLE, SOLID ANGLE, STERADIAN, STRAIGHT ANGLE, SUBTEND, SUPPLEMENTARY ANGLE, VERTEX ANGLE

References

Dixon, R. *Mathographics.* New York: Dover, pp. 99–00, 1991.

Harris, J. W. and Stocker, H. "Angle." §3.3 in *Handbook of Mathematics and Computational Science.* New York: Springer-Verlag, pp. 62–4, 1998.

Angle Bisector

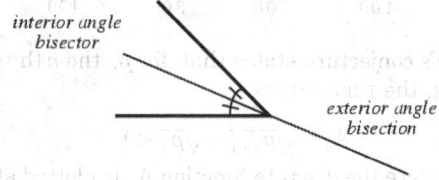

interior angle bisector

exterior angle bisection

The (interior) bisector of an ANGLE is the LINE or LINE

SEGMENT which cuts it into two equal ANGLES on the same "side" as the ANGLE.

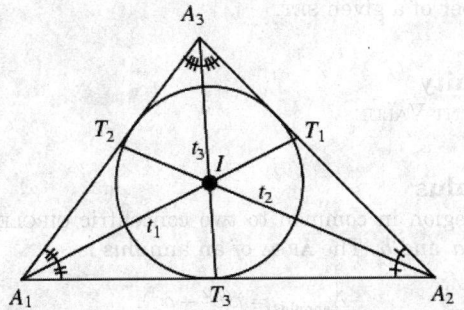

The length of the bisector of ANGLE A_1 in the above TRIANGLE $\Delta A_1 A_2 A_3$ is given by

$$t_1^2 = a_2 a_3 \left[1 - \frac{a_1^2}{(a_2 + a_3)^2} \right],$$

where $t_i \equiv \overline{A_i T_i}$ and $a_i \equiv \overline{A_j A_k}$. The angle bisectors meet at the INCENTER I, which has TRILINEAR COORDINATES 1:1:1.

See also ANGLE BISECTOR THEOREM, CYCLIC QUADRANGLE, EXTERIOR ANGLE BISECTOR, ISODYNAMIC POINTS, ORTHOCENTRIC SYSTEM, STEINER-LEHMUS THEOREM, TRISECTION

References

Coxeter, H. S. M. and Greitzer, S. L. *Geometry Revisited.* Washington, DC: Math. Assoc. Amer., pp. 9–0, 1967.
Dixon, R. *Mathographics.* New York: Dover, p. 19, 1991.
Mackay, J. S. "Properties Concerned with the Angular Bisectors of a Triangle." *Proc. Edinburgh Math. Soc.* **13**, 37–02, 1895.

Angle Bisector Theorem

The ANGLE BISECTOR of an ANGLE in a TRIANGLE divides the opposite side in the same RATIO as the sides adjacent to the ANGLE.

Angle Bracket

The combination of a BRA and KET (bra + ket = bracket) which represents the INNER PRODUCT of two functions or vectors,

$$\langle f | g \rangle = \int f(x) g(x)\, dx$$

$$\langle \mathbf{v} | \mathbf{w} \rangle = \mathbf{v} \cdot \mathbf{w}.$$

By itself, the BRA is a COVARIANT 1-VECTOR, and the KET is a CONTRAVARIANT ONE-FORM. These terms are commonly used in quantum mechanics.

See also BRA, BRACE, DIFFERENTIAL k-FORM, KET, ONE-FORM, PARENTHESIS, SQUARE BRACKET

References

Bringhurst, R. *The Elements of Typographic Style, 2nd ed.* Point Roberts, WA: Hartley and Marks, p. 271, 1997.

Angle of Parallelism

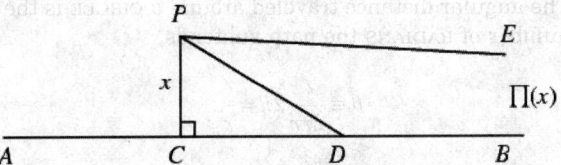

Given a point P and a LINE AB, draw the PERPENDICULAR through P and call it PC. Let PD be any other line from P which meets CB in D. In a HYPERBOLIC GEOMETRY, as D moves off to infinity along CB, then the line PD approaches the limiting line PE, which is said to be parallel to CB at P. The angle $\angle CPE$ which PE makes with PC is then called the angle of parallelism for perpendicular distance x, and is given by

$$\prod(x) = 2 \tan^{-1}(e^{-x}).$$

This is known as LOBACHEVSKY'S FORMULA.

See also HYPERBOLIC GEOMETRY, LOBACHEVSKY'S FORMULA

References

Coxeter, H. S. M. "The Angle of Parallelism." §16.3 in *Introduction to Geometry, 2nd ed.* New York: Wiley, pp. 291–95, 1969.
Manning, H. P. *Introductory Non-Euclidean Geometry.* New York: Dover, pp. 31–2 and 58, 1963.

Angle Trisection

TRISECTION

Angle-Preserving Transformation

CONFORMAL MAPPING

Angular Acceleration

The angular acceleration α is defined as the time DERIVATIVE of the ANGULAR VELOCITY ω,

$$\alpha \equiv \frac{d\omega}{dt} = \frac{d^2\theta}{dt^2}\hat{\mathbf{z}} = \frac{\mathbf{a}}{r}.$$

See also ACCELERATION, ANGULAR DISTANCE, ANGULAR VELOCITY

Angular Defect

The DIFFERENCE between the SUM of face ANGLES A_i at a VERTEX of a POLYHEDRON and 2π,

$$\delta = 2\pi - \sum_i A_i.$$

See also DESCARTES TOTAL ANGULAR DEFECT, JUMP ANGLE, SPHERICAL DEFECT

Angular Distance

The angular distance traveled around a CIRCLE is the number of RADIANS the path subtends,

$$\theta \equiv \frac{\ell}{2\pi r} 2\pi = \frac{\ell}{r}.$$

See also ANGULAR ACCELERATION, ANGULAR VELOCITY

Angular Velocity

The angular velocity ω is the time DERIVATIVE of the ANGULAR DISTANCE θ with direction $\hat{z}$ PERPENDICULAR to the plane of angular motion,

$$\omega \equiv \frac{d\theta}{dt} \hat{z} = \frac{\mathbf{v}}{r}.$$

See also ANGULAR ACCELERATION, ANGULAR DISTANCE

Anharmonic Ratio

CROSS-RATIO

Animal

1. A FIXED POLYOMINO.
2. The set of points obtained by taking the centers of a FIXED POLYOMINO.

See also POLYOMINO

References

Delest, M.-P. and Viennot, G. "Algebraic Languages and Polyominoes [sic] Enumeration." *Theoret. Comput. Sci.* **34**, 169–06, 1984.
Read, R. C. "Contributions to the Cell Growth Problem." *Canad. J. Math.* **14**, 1–0, 1962.

Anisohedral Tiling

A k-anisohedral tiling is a tiling which permits no n-ISOHEDRAL TILING with $n < k$.

References

Berglund, J. "Is There a k-Anisohedral Tile for $k \geq 5$?" *Amer. Math. Monthly* **100**, 585–88, 1993.
Klee, V. and Wagon, S. *Old and New Unsolved Problems in Plane Geometry and Number Theory.* Washington, DC: Math. Assoc. Amer., 1991.

Annealing

SIMULATED ANNEALING

Annihilator

The term annihilator is used in several different ways in various aspects of mathematics. It is most com-monly used to mean the SET of all functions satisfying a given set of conditions which is zero on every member of a given SET.

Annuity

PRESENT VALUE

Annulus

The region in common to two concentric CIRCLES of RADII a and b. The AREA of an annulus is

$$A_{\text{annulus}} = \pi(b^2 - a^2).$$

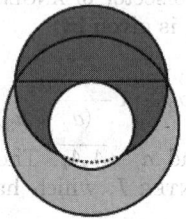

In the above figure, the area of the circle whose diameter is tangent to the inner circle and has endpoints at the outer circle is equal to the area of the annulus.

See also ANNULUS THEOREM, BULLSEYE ILLUSION, CHORD, CIRCLE, CONCENTRIC CIRCLES, LUNE, SPHERICAL SHELL

References

Harris, J. W. and Stocker, H. "Annulus, Circular Ring." §3.8.3 in *Handbook of Mathematics and Computational Science.* New York: Springer-Verlag, p. 91, 1998.
Pappas, T. "The Amazing Trick." *The Joy of Mathematics.* San Carlos, CA: Wide World Publ./Tetra, p. 69, 1989.

Annulus Conjecture

ANNULUS THEOREM

Annulus Theorem

Let K_1^n and K_2^n be disjoint bicollared KNOTS in $\mathbb{R}^{n+1}$ or $\mathbb{S}^{n+1}$ and let U denote the open region between them. Then the closure of U is a closed annulus $\mathbb{S}^n \times [0, 1]$. Except for the case $n = 3$, the theorem was proved by Kirby (1969).

References

Kirby, R. C. "Stable Homeomorphisms and the Annulus Conjecture." *Ann. Math.* **89**, 575–82, 1969.
Rolfsen, D. *Knots and Links.* Wilmington, DE: Publish or Perish Press, p. 38, 1976.

Anomalous Cancellation

The simplification of a FRACTION a/b which gives a correct answer by "canceling" DIGITS of a and b. There are only four such cases for NUMERATOR and DENOMINATORS of two DIGITS in base 10: $64/16 =$

$4/1 = 4$, $98/49 = 8/4 = 2$, $95/19 = 5/1 = 5$,   and $65/26 = 5/2$ (Boas 1979).

The concept of anomalous cancellation can be extended to arbitrary bases. PRIME bases have no solutions, but there is a solution corresponding to each PROPER DIVISOR of a COMPOSITE b. When $b - 1$ is PRIME, this type of solution is the only one. For base 4, for example, the only solution is $32_4/13_4 = 2_4$. Boas gives a table of solutions for $b \leq 39$. The number of solutions is EVEN unless b is an EVEN SQUARE.

b	N	b	N
4	1	26	4
6	2	27	6
8	2	28	10
9	2	30	6
10	4	32	4
12	4	34	6
14	2	35	6
15	6	36	21
16	7	38	2
18	4	39	6
20	4		
21	10		
22	6		
24	6		

See also FRACTION, PRINTER'S ERRORS, REDUCED FRACTION

References

Boas, R. P. "Anomalous Cancellation." Ch. 6 in *Mathematical Plums* (Ed. R. Honsberger). Washington, DC: Math. Assoc. Amer., pp. 113–29, 1979.
Moessner, A. *Scripta Math.* **19**.
Moessner, A. *Scripta Math.* **20**.
Ogilvy, C. S. and Anderson, J. T. *Excursions in Number Theory.* New York: Dover, pp. 86–7, 1988.
Wells, D. *The Penguin Dictionary of Curious and Interesting Numbers.* Middlesex, England: Penguin Books, pp. 26–7, 1986.

Anomalous Number

BENFORD'S LAW

Anonymous

A term in SOCIAL CHOICE THEORY meaning invariance of a result under permutation of voters.

See also DUAL VOTING, MONOTONIC VOTING

Anosov Automorphism

A HYPERBOLIC linear map $\mathbb{R}^n \to \mathbb{R}^n$ with INTEGER entries in the transformation MATRIX and DETERMINANT ± 1 is an ANOSOV DIFFEOMORPHISM of the n-TORUS, called an Anosov automorphism (or HYPERBOLIC AUTOMORPHISM). Here, the term automorphism is used in the GROUP THEORY sense.

Anosov Diffeomorphism

An Anosov diffeomorphism is a C^1 DIFFEOMORPHISM ϕ such that the MANIFOLD M is HYPERBOLIC with respect to ϕ. Very few classes of Anosov diffeomorphisms are known. The best known is ARNOLD'S CAT MAP.

A HYPERBOLIC linear map $\mathbb{R}^n \to \mathbb{R}^n$ with INTEGER entries in the transformation MATRIX and DETERMINANT ± 1 is an Anosov diffeomorphism of the n-TORUS. Not every MANIFOLD admits an Anosov diffeomorphism. Anosov diffeomorphisms are EXPANSIVE, and there are no Anosov diffeomorphisms on the CIRCLE.

It is conjectured that if $\phi : M \to M$ is an Anosov diffeomorphism on a COMPACT RIEMANNIAN MANIFOLD and the NONWANDERING SET $\Omega(\phi)$ of ϕ is M, then ϕ is TOPOLOGICALLY CONJUGATE to a FINITE-TO-ONE FACTOR of an ANOSOV AUTOMORPHISM of a NILMANIFOLD. It has been proved that any Anosov diffeomorphism on the n-TORUS is TOPOLOGICALLY CONJUGATE to an ANOSOV AUTOMORPHISM, and also that Anosov diffeomorphisms are C^1 STRUCTURALLY STABLE.

See also ANOSOV AUTOMORPHISM, AXIOM A DIFFEOMORPHISM, DYNAMICAL SYSTEM

References

Anosov, D. V. "Geodesic Flow on Closed Riemannian Manifolds of Negative Curvature." *Trudy Mat. Inst. Steklov* **90**, 1–09, 1970.
Smale, S. "Differentiable Dynamical Systems." *Bull. Amer. Math. Soc.* **73**, 747–17, 1967.

Anosov Flow

A FLOW defined analogously to the ANOSOV DIFFEOMORPHISM, except that instead of splitting the TANGENT BUNDLE into two invariant sub-BUNDLES, they are split into three (one exponentially contracting, one expanding, and one which is 1-dimensional and tangential to the flow direction).

See also DYNAMICAL SYSTEM

Anosov Map

An important example of a ANOSOV DIFFEOMORPHISM.

$$\begin{bmatrix} x_{n+1} \\ y_{n+1} \end{bmatrix} = \begin{bmatrix} 2 & 1 \\ 1 & 1 \end{bmatrix} \begin{bmatrix} x_n \\ y_n \end{bmatrix},$$

where x_{n+1}, y_{n+1} are computed mod 1.

See also ARNOLD'S CAT MAP

ANOVA

"Analysis of Variance." A STATISTICAL TEST for hetero-geneity of MEANS by analysis of group VARIANCES. To apply the test, assume random sampling of a variate y with equal VARIANCES, independent errors, and a NORMAL DISTRIBUTION. Let n be the number of REPLICATES (sets of identical observations) within each of K FACTOR LEVELS (treatment groups), and y_{ij} be the jth observation within FACTOR LEVEL i. Also assume that the ANOVA is "balanced" by restricting n to be the same for each FACTOR LEVEL.

Now define the sum of square terms

$$\text{SST} \equiv \sum_{i=1}^{k} \sum_{j=1}^{n} (y_{ij} - \tilde{y})^2 \tag{1}$$

$$= \sum_{i=1}^{k} \sum_{j=1}^{n} y_{ij}^2 - \frac{\left(\sum_{i=1}^{k} \sum_{j=1}^{n} y_{ij}\right)^2}{Kn} \tag{2}$$

$$\text{SSA} \equiv \frac{1}{n} \sum_{i=1}^{k} \left(\sum_{j=1}^{n} y_{ij}\right)^2 - \frac{1}{Kn} \left(\sum_{i=1}^{k} \sum_{j=1}^{n} y_{ij}\right)^2 \tag{3}$$

$$\text{SSE} \equiv \sum_{i=1}^{k} \sum_{j=1}^{n} (y_{ij} - \ddot{y}_i)^2 \tag{4}$$

$$= \text{SST} - \text{SSA}, \tag{5}$$

which are the total, treatment, and error sums of squares. Here, $\ddot{y}_i$ is the mean of observations within FACTOR LEVEL i, and $\tilde{y}$ is the "group" mean (i.e., mean of means). Compute the entries in the following table, obtaining the P-VALUE corresponding to the calculated F-RATIO of the mean squared values

$$F = \frac{\text{MSA}}{\text{MSE}}. \tag{6}$$

Category	SS	°Freedom	Mean Squared	F-RATIO
Treatment	SSA	$K-1$	$\text{MSA} \equiv \dfrac{\text{SSA}}{K-1}$	$\dfrac{\text{MSA}}{\text{MSE}}$
Error	SSE	$K(n-1)$	$\text{MSE} \equiv \dfrac{\text{SSE}}{K(n-1)}$	
Total	SST	$Kn-1$	$\text{MST} \equiv \dfrac{\text{SST}}{Kn-1}$	

If the P-VALUE is small, reject the NULL HYPOTHESIS that all MEANS are the same for the different groups.

See also FACTOR LEVEL, MANOVA, REPLICATE, VARIANCE

References
Miller, R. G. *Beyond ANOVA: Basics of Applied Statistics.* Boca Raton, FL: Chapman & Hall, 1997.

Anthropomorphic Polygon
A SIMPLE POLYGON with precisely two EARS and one MOUTH.

References
Toussaint, G. "Anthropomorphic Polygons." *Amer. Math. Monthly* **122**, 31–5, 1991.

Anthyphairetic Ratio
An archaic term for a CONTINUED FRACTION.

References
Fowler, D. H. *The Mathematics of Plato's Academy: A New Reconstruction, 2nd ed.* New York: Oxford University Press, 1987.

Antiautomorphism
If a MAP $f : G \to G'$ from a GROUP G to a GROUP G' satisfies $f(ab) = f(a)f(b)$ for all a, $b \in G$, then f is said to be an antiautomorphism.

See also AUTOMORPHISM

Anticenter

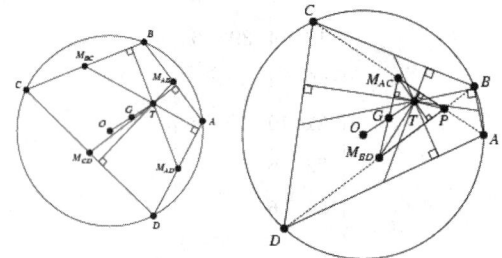

The point of concurrence of the three MALTITUDES of a CYCLIC QUADRILATERAL. Let M_{AC} and M_{BD} be the MIDPOINTS of the diagonals of a CYCLIC QUADRILATERAL $ABCD$, and let P be the intersection of the diagonals. Then the ORTHOCENTER of TRIANGLE $\Delta PM_{AC}M_{BD}$ is the anticenter T of $ABCD$ (Honsberger 1995, p. 39).

See also CYCLIC QUADRILATERAL, MALTITUDE

References
Honsberger, R. *Episodes in Nineteenth and Twentieth Century Euclidean Geometry.* Washington, DC: Math. Assoc. Amer., pp. 36–7, 1995.

Anticevian Triangle
Given a center $\alpha : \beta : \gamma$, the anticevian triangle is defined as the TRIANGLE with VERTICES $-\alpha : \beta : \gamma$, $\alpha : -\beta : \gamma$, and $\alpha : \beta : -\gamma$. If $A'B'C'$ is the CEVIAN TRIANGLE of X and $A''B''C''$ is an anticevian triangle, then X and A'' are HARMONIC CONJUGATE POINTS with respect to A and A'.

See also CEVIAN TRIANGLE

References

Kimberling, C. "Central Points and Central Lines in the Plane of a Triangle." *Math. Mag.* **67**, 163–87, 1994.

Antichain

Let P be a finite PARTIALLY ORDERED SET. An antichain in P is a set of pairwise incomparable elements (e.g., a family of SUBSETS such that, for any two of them, neither is a SUBSET of the other). Antichains are also called Sperner systems in older literature (Comtet 1974).

The following table gives the antichains on n-set $\{1, 2, \ldots, n\}$ for small n.

n	antichains
1	$\emptyset, \{(1)\}$
2	$\emptyset, \{\{1\}\}, \{\{2\}\}, \{\{1\}, \{2\}\}, \{\{1, 2\}\}$
3	$\emptyset, \{\{1\}\}, \{\{2\}\}, \{\{3\}\}, \{\{1, 2\}\},$
	$\{\{1, 3\}\}, \{\{2, 3\}\}, \{\{1\}, \{2\}\}, \{\{1\}, \{3\}\},$
	$\{\{2\}, \{3\}\}, \{\{1, 2, 3\}\}, \{\{1\}, \{2, 3\}\}, \{\{1, 2\}, \{2, 3\}\},$
	$\{\{1, 2\}, \{1, 3\}\}, \{\{1, 2\}, \{3\}\}, \{\{2\}, \{1, 3\}\}, \{\{2, 3\}, \{1,3\}\},$
	$\{\{1\}, \{2\}, \{3\}\}, \{\{1, 2\}, \{2, 3\}, \{1, 3\}\}$

The number of antichains on the n-set $\{1, 2, \ldots, n\}$ for $n = 1, 2, \ldots$, are 1, 2, 5, 19, 167, ... (Sloane's A014466). If the EMPTY SET is not considered a valid antichain, then these reduce to 0, 1, 4, 18, 166, ... (Sloane's A007153; Comtet 1974, p. 273). The numbers obtained by adding one to Sloane's A014466, 2, 3, 6, 20, 168, 7581, 7828354, ... (Sloane's A000372), are also frequently encountered (Speciner 1972).

The number of antichains on the n-set are equal to the number of monotonic increasing Boolean functions of n variables, and also the number of free distributive lattices with n generators (Comtet 1974, p. 273). Determining these numbers is known as DEDEKIND'S PROBLEM, and the numbers in each of these sequences are sometimes called Dedekind numbers (Sloane).

The WIDTH of P is the maximum CARDINALITY of an ANTICHAIN in P. For a PARTIAL ORDER, the size of the longest ANTICHAIN is called the WIDTH $w(P)$. Sperner (1928) proved that the maximum width of an antichain containing n elements is

$$w_{\max(n)} = \binom{n}{\lfloor n/2 \rfloor},$$

where $\binom{n}{k}$ is a BINOMIAL COEFFICIENT and $\lfloor n \rfloor$ is the FLOOR FUNCTION.

See also BOOLEAN FUNCTION, CHAIN, DILWORTH'S LEMMA, PARTIALLY ORDERED SET, WIDTH (PARTIAL ORDER)

References

Agnew, R. P. "Minimax Functions, Configuration Functions, and Partitions." *J. Indian Math. Soc.* **24**, 1–1, 1961.

Anderson, I. *Combinatorics of Finite Sets.* Oxford, England: Oxford University Press, p. 38, 1987.

Arocha, J. L. "Antichains in Ordered Sets" [Spanish]. *Anales del Instituto de Matematicas de la Universidad Nacional Autonoma de Mexico* **27**, 1–1, 1987.

Berman, J. "Free Spectra of 3-Element Algebras." In *Universal Algebra and Lattice Theory (Puebla, 1982)* (Ed. R. S. Freese and O. C. Garcia). New York: Springer-Verlag, 1983.

Berman, J. and Koehler, P. "Cardinalities of Finite Distributive Lattices." *Mitteilungen aus dem Mathematischen Seminar Giessen* **121**, 103–24, 1976.

Birkhoff, G. *Lattice Theory, 3rd ed.* Providence, RI: Amer. Math. Soc., p. 63, 1967.

Church, R. "Numerical Analysis of Certain Free Distributive Structures." *Duke Math. J.* **6**, 732–33, 1940.

Church. "Enumeration by Rank of the Elements of the Free Distributive Lattice with Seven Generators." *Not. Amer. Math. Soc.* **12**, 724, 1965.

Comtet, L. "Sperner Systems." §7.2 in *Advanced Combinatorics: The Art of Finite and Infinite Expansions, rev. enl. ed.* Dordrecht, Netherlands: Reidel, pp. 271–73, 1974.

Dedekind, R. "Über Zerlegungen von Zahlen durch ihre grössten gemeinsammen Teiler." In *Gesammelte Werke, Bd. 1.* pp. 103–48, 1897.

Erdos, P.; Ko, Chao; and Rado, R. "Intersection Theorems for Systems of Finite Sets." *Quart. J. Math. Oxford* **12**, 313–20, 1961.

Gilbert, E. N. "Lattice Theoretic Properties of Frontal Switching Networks." *J. Math. Phys.* **33**, 57–7, 1954.

Hansel, G. "Problèmes de dénombrement et d'évaluation de bornes concernant les éléments du trellis distributif libre." *Publ. Inst. Statist. Univ. Paris* **16**, 163–94, 1967.

Harrison, M. A. *Introduction to Switching and Automata Theory.* New York: McGraw-Hill, p. 188, 1965.

Hilton, A. J. W. and Milner, E. C. "Some Intersection Theorems of Systems of Finite Sets." *Quart. J. Math. Oxford* **18**, 369–84, 1967.

Katona, G. "On a Conjecture of Erdos and a Stronger Form of Sperner's Theorem." *Studia Sci. Math. Hung.* **1**, 59–3, 1966.

Katona, G. "A Theorem of Finite Sets." In *Theory of Graphs, Proceedings of the Colloquium Held at Tihany, Hungary* (Ed. P. Erdos and G. Katona). New York: Academic Press, pp. 187–07, 1968.

Kleitman, D. "A Conjecture of Erdos-Katona on Commensurable Pairs Among Subsets of a n-Set." In *Theory of Graphs, Proceedings of the Colloquium Held at Tihany, Hungary* (Ed. P. Erdos and G. Katona). New York: Academic Press, pp. 215–18, 1968.

Kleitman, D. "On Dedekind's Problem: The Number of Monotone Boolean Functions." *Proc. Amer. Math. Soc.* **21**, 677–82, 1969.

Kleitman, D. and Markowsky, G. "On Dedekind's Problem: The Number of Isotone Boolean Functions. II." *Trans. Amer. Math. Soc.* **213**, 373–90, 1975.

Lunnon, W. F. "The IU Function: The Size of a Free Distributive Lattice." In *Combinatorial Mathematics and Its Applications* (Ed. D. J. A. Welsh). New York: Academic Press, pp. 173–81, 1971.

Mesalkin, L. D. "A Generalization of Sperner's Theorem on the Number of Subsets of a Finite Set." *Theory Prob.* **8**, 203–04, 1963.

Milner, E. C. "A Combinatorial Theorem on Systems of Sets." *J. London Math. Soc.* **43**, 204–06, 1968.

Muroga, S. *Threshold Logic and Its Applications.* New York: Wiley, p. 38 and 214, 1971.

Rivière, N. M. "Recursive Formulas on Free Distributive Lattices." *J. Combin. Th.* **5**, 229–34, 1968.

Shapiro. "On the Counting Problem for Monotone Boolean Functions." *Comm. Pure Appl. Math.* **23**, 299–12, 1970.

Skiena, S. *Implementing Discrete Mathematics: Combinatorics and Graph Theory with Mathematica.* Reading, MA: Addison-Wesley, p. 241, 1990.

Sloane, N. J. A. Sequences A006826/M2469, A007153/M3551, and A014466 in "An On-Line Version of the Encyclopedia of Integer Sequences." http://www.research.-att.com/~njas/sequences/eisonline.html.

Speciner, M. Item 18 in Beeler, M.; Gosper, R. W.; and Schroeppel, R. *HAKMEM.* Cambridge, MA: MIT Artificial Intelligence Laboratory, Memo AIM-239, p. 10, Feb. 1972.

Sperner, E. "Ein Satz über Untermengen einer endlichen Menge." *Math. Z.* **27**, 544–48, 1928.

Ward, M. "Note on the Order of the Free Distributive Lattice." *Bull. Amer. Math. Soc.* **52**, 423, 1946.

Yamamoto, K. "Logarithmic Order of Free Distributive Lattice." *J. Math. Soc. Japan* **6**, 343–53, 1954.

Anticlastic

When the GAUSSIAN CURVATURE K is everywhere NEGATIVE, a SURFACE is called anticlastic and is saddle-shaped. A SURFACE on which K is everywhere POSITIVE is called SYNCLASTIC. A point at which the GAUSSIAN CURVATURE is NEGATIVE is called a HYPERBOLIC POINT.

See also ELLIPTIC POINT, GAUSSIAN QUADRATURE, HYPERBOLIC POINT, PARABOLIC POINT, PLANAR POINT, SYNCLASTIC

Anticommutative

An OPERATOR $*$ for which $a * b = -b * a$ is said to be anticommutative.

See also COMMUTATIVE

Anticommutator

For OPERATORS $\hat{A}$ and $\hat{B}$, the anticommutator is defined by

$$\{\hat{A}, \hat{B}\} \equiv \hat{A}\hat{B} + \hat{B}\hat{A}.$$

See also COMMUTATOR, JORDAN ALGEBRA, JORDAN PRODUCT

Anticomplementary Triangle

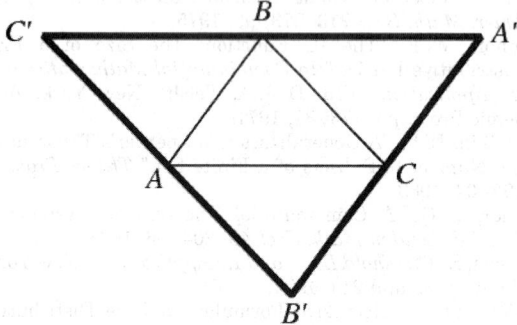

A TRIANGLE $\Delta A'B'C'$ which has a given TRIANGLE ΔABC as its MEDIAL TRIANGLE. The TRILINEAR CO-ORDINATES of the anticomplementary triangle are

$$A' = -a^{-1} : b^{-1} : c^{-1}$$

$$B' = a^{-1} : -b^{-1} : c^{-1}$$

$$C' = a^{-1} : b^{-1} : -c^{-1}.$$

See also MEDIAL TRIANGLE

Anticross-Stitch Curve

BOX FRACTAL

Antiderivative

INTEGRAL

Antidifferentiation

INTEGRATION

Antigonal Points

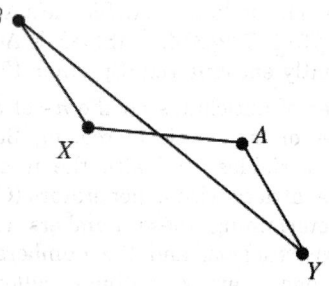

Given $\angle AXB + \angle AYB = \pi$ RADIANS in the above figure, then X and Y are said to be antigonal points with respect to A and B.

Antihomography

A CIRCLE-preserving TRANSFORMATION composed of an ODD number of INVERSIONS.

See also HOMOGRAPHY

Antihomologous Points

Two points which are COLLINEAR with respect to a SIMILITUDE CENTER but are not HOMOLOGOUS POINTS. Four interesting theorems from Johnson (1929) follow.

1. Two pairs of antihomologous points form inversely similar triangles with the HOMOTHETIC CENTER.

2. The PRODUCT of distances from a HOMOTHETIC CENTER to two antihomologous points is a constant.

3. Any two pairs of points which are antihomologous with respect to a SIMILITUDE CENTER lie on a CIRCLE.

4. The tangents to two CIRCLES at antihomologous points make equal ANGLES with the LINE through the points.

See also HOMOLOGOUS POINTS, HOMOTHETIC CENTER, SIMILITUDE CENTER

References

Johnson, R. A. *Modern Geometry: An Elementary Treatise on the Geometry of the Triangle and the Circle.* Boston, MA: Houghton Mifflin, pp. 19–1, 1929.

Antilaplacian

The antilaplacian of u with respect to x is a function whose LAPLACIAN with respect to x equals u. The antilaplacian is never unique.

See also LAPLACIAN

Antilinear

An antilinear OPERATOR $\tilde{A}$ satisfies the following two properties:

$$\tilde{A}[f_1(x) + f_2(x)] = \tilde{A}f_1(x) + \tilde{A}f_2(x)$$

$$\tilde{A}cf(x) = \tilde{c}\tilde{A}f(x),$$

where $\tilde{c}$ is the COMPLEX CONJUGATE of c.

See also ANTIUNITARY, LINEAR OPERATOR

References

Sakurai, J. J. *Modern Quantum Mechanics.* Menlo Park, CA: Benjamin/Cummings, 1985.

Antilinear Operator

An antilinear OPERATOR

$$\mathscr{L}^*u = \int (y_2\tilde{\mathscr{L}}y_1 - y_1\tilde{\mathscr{L}}y_2)\, dx = \left[\frac{p_1}{p_0}(y_1'y_2 - y_1y_2')\right]$$

satisfies the following two properties:

$$PD = CB$$

$$D = PE$$

where $\angle CPE$ is the COMPLEX CONJUGATE of C_ϵ.

See also ANTIUNITARY OPERATOR, LINEAR OPERATOR

References

Sakurai, J. J. *Modern Quantum Mechanics.* Menlo Park, CA: Benjamin/Cummings, 1985.

Antilogarithm

The INVERSE FUNCTION of the LOGARITHM, defined such that

$$\log_b(\text{antilog}_b z) = z = \text{antilog}_b(\log_b z).$$

The antilogarithm in base b of z is therefore b^z.

See also COLOGARITHM, LOGARITHM, POWER

Antimagic Graph

A GRAPH with e EDGES labeled with distinct elements $\{1, 2 \ldots, c\}$ so that the SUM of the EDGE labels at each VERTEX differ.

See also LABELED GRAPH, MAGIC GRAPH

References

Hartsfield, N. and Ringel, G. *Pearls in Graph Theory: A Comprehensive Introduction.* San Diego, CA: Academic Press, 1990.

Antimagic Square

15	2	12	4
1	14	10	5
8	9	3	16
11	13	6	7

21	18	6	17	4
7	3	13	16	24
5	20	23	11	1
15	8	19	2	25
14	12	9	22	10

10	25	32	13	16	9
22	7	3	24	21	30
20	27	18	26	11	6
1	31	23	33	17	8
19	5	36	12	15	29
34	14	2	4	35	28

14	3	34	21	47	29	22
43	16	13	25	6	26	44
30	48	24	8	12	9	45
10	5	11	38	49	46	19
4	41	37	36	33	27	1
39	17	40	20	7	35	23
31	42	18	32	28	2	15

49	16	50	10	19	28	24	56
42	43	11	15	44	38	55	5
25	21	48	46	9	37	6	63
29	47	8	40	51	30	52	1
45	22	54	23	20	34	2	62
14	59	18	33	41	26	61	13
36	12	58	32	27	64	3	35
17	39	7	57	53	4	60	31

52	19	81	22	29	15	42	31	76
61	10	67	23	54	79	25	33	16
57	9	71	24	38	1	51	47	75
26	78	7	69	66	77	13	27	12
39	21	74	20	37	17	49	55	64
8	65	4	62	50	34	73	41	40
56	68	2	63	14	72	35	44	6
53	30	60	32	36	3	46	43	58
11	70	5	59	48	80	28	45	18

An antimagic square is an $n \times n$ ARRAY of integers from 1 to n^2 such that each row, column, and main diagonal produces a different sum such that these sums form a SEQUENCE of consecutive integers. It is therefore a special case of a HETEROSQUARE. Antimagic squares of orders 4– are illustrated above (Madachy 1979). For the 4×4 square, the sums are 30, 31, 32, ..., 39; for the 5×5 square they are 59, 60, 61, ..., 70; and so on.

Let an antimagic square of order n have entries 0, 1, ..., $n^2 - 2$, $n^2 - 1$, and let

$$M(n) \equiv \tfrac{1}{2}n(n^2 + 1)$$

be the magic constant. Then if and antimagic square of order n exists, it is either positive with sums $[M(n) - n, M(n) + n + 1]$, or negative with sums $[M(n) - n - 1, M(n) + n]$ (Madachy 1979).

Antimagic squares of orders one, two, and three are impossible. In the case of the 3×3 square, there is no known method of proof of this fact except by case analysis or enumeration by computer. There are 18 families of antimagic squares of order four. The total number of antimagic squares of orders 1, 2, ... modulo the full group of symmetries (reflection, rotation, complementation, and exchanges) are 0, 0, 0, 299710, ... (Sloane's A050257; Cormie).

Abe (1994) and Madachy (1979) ask for methods of constructing antimagic squares of every order. Recently, J. Cormie and V. Linek have developed general constructions for squares of order n for all $n > 3$, as well as for bordering antimagic squares.

See also HETEROSQUARE, MAGIC SQUARE, TALISMAN SQUARE

References
Abe, G. "Unsolved Problems on Magic Squares." *Disc. Math.* **127**, 3–3, 1994.
Cormie, J. "The Anti-Magic Square Project." http://www.u-winnipeg.ca/~jcormie/.
Madachy, J. S. "Magic and Antimagic Squares." Ch. 4 in *Madachy's Mathematical Recreations.* New York: Dover, pp. 103–13, 1979.
Sloane, N. J. A. Sequences A050257 in "An On-Line Version of the Encyclopedia of Integer Sequences." http://www.research.att.com/~njas/sequences/eisonline.html.
Weisstein, E. W. "Magic Squares." MATHEMATICA NOTEBOOK MAGICSQUARES.M.

Antimorph
A number which can be represented both in the form $x_0^2 - Dy_0^2$ and in the form $Dx_1^2 - y_1^2$. This is only possible when the PELL EQUATION

$$x^2 - Dy^2 = -1$$

is solvable. Then

$$x^2 - Dy^2 = -(x_0 - Dy_0^2)(x_n^2 - Dy_n^2)$$
$$= D(x_0 y_n - y_0 x_n)^2 - (x_0 x_n - Dy_0 y_n)^2.$$

See also IDONEAL NUMBER, POLYMORPH

References
Beiler, A. H. *Recreations in the Theory of Numbers: The Queen of Mathematical Entertains.* New York: Dover, 1964.

Antimorphic Number
ANTIMORPH

Antinomy
A PARADOX or contradiction.

Antiparallel

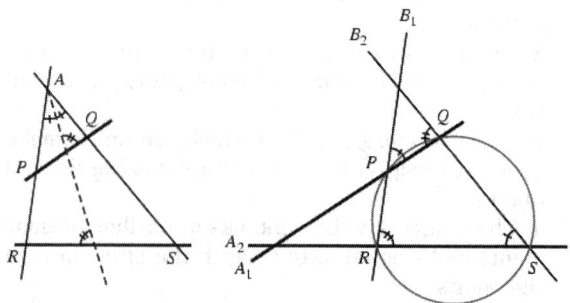

Two lines PQ and RS are said to be antiparallel with respect to the sides of an ANGLE A if they make the same angle in the opposite senses with the BISECTOR of that angle. If PQ and RS are antiparallel with respect to PR and QS, then the latter are also antiparallel with respect to the former. Furthermore, if PQ and RS are antiparallel, then the points P, Q, R, and S are CONCYCLIC (Johnson 1929, p. 172; Honsberger 1995, pp. 87–8).

There are a number of fundamental relationships involving a triangle and antiparallel lines (Johnson 1929, pp. 172–73).

1. The line joining the feet to two ALTITUDES of a triangle is antiparallel to the third side.
2. The tangent to a triangle's CIRCUMCIRCLE at a vertex is antiparallel to the opposite side.
3. The radius of the CIRCUMCIRCLE at a vertex is perpendicular to all lines antiparallel to the opposite sides.

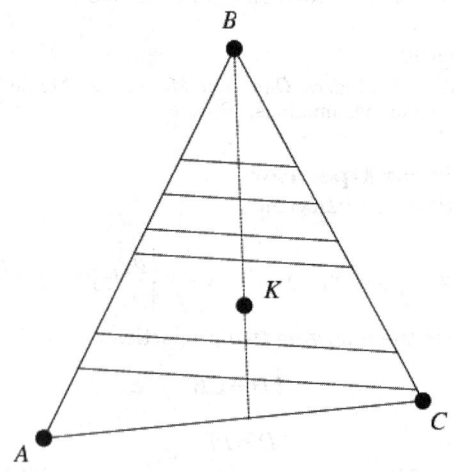

In a TRIANGLE $\triangle ABC$, a SYMMEDIAN BK bisects all segments antiparallel to a given side AC (Honsberger 1995, p. 88). Furthermore, every antiparallel to BC in $\triangle ABC$ is PARALLEL to the tangent to the CIRCUMCIRCLE of $\triangle ABC$ at A (Honsberger 1995, p. 98).

See also ANGLE, CONCYCLIC, COSINE CIRCLE, COSINE HEXAGON, HYPERPARALLEL, LEMOINE CIRCLE, LEMOINE HEXAGON, PARALLEL, TUCKER CIRCLES, TUCKER HEXAGON

References

Casey, J. "Theory of Isogonal and Isotomic Points, and of Antiparallel and Symmedian Lines." Supp. Ch. §1 in *A Sequel to the First Six Books of the Elements of Euclid, Containing an Easy Introduction to Modern Geometry with Numerous Examples, 5th ed., rev. enl.* Dublin: Hodges, Figgis, & Co., pp. 165–73, 1888.

Coolidge, J. L. *A Treatise on the Geometry of the Circle and Sphere.* New York: Chelsea, p. 65, 1971.

Honsberger, R. "Parallels and Antiparallels." §9.1 in *Episodes in Nineteenth and Twentieth Century Euclidean Geometry.* Washington, DC: Math. Assoc. Amer., pp. 87–8, 1995.

Johnson, R. A. *Modern Geometry: An Elementary Treatise on the Geometry of the Triangle and the Circle.* Boston, MA: Houghton Mifflin, p. 172, 1929.

Lachlan, R. §113 in *An Elementary Treatise on Modern Pure Geometry.* London: Macmillian, p. 63, 1893.

Phillips, A. W. and Fisher, I. *Elements of Geometry.* New York: American Book Co., 1896.

Antipedal Triangle

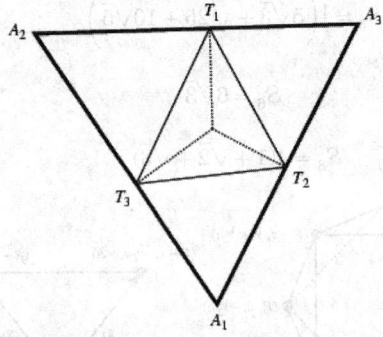

The antipedal triangle A of a given TRIANGLE T is the TRIANGLE of which T is the PEDAL TRIANGLE. For a TRIANGLE with TRILINEAR COORDINATES $\alpha : \beta : \gamma$ and ANGLES A, B, and C, the antipedal triangle has VERTICES with TRILINEAR COORDINATES

$$-(\beta + \alpha \cos C)(\gamma + \alpha \cos B) : (\gamma + \alpha \cos B)(\alpha + \beta \cos C) :$$

$$(\beta + \alpha \cos C)(\alpha + \gamma \cos B)$$

$$(\gamma + \beta \cos A)(\beta + \alpha \cos C) : -(\gamma + \beta \cos A)(\alpha + \beta \cos C) :$$

$$(\alpha + \beta \cos C)(\beta + \gamma \cos A)$$

$$(\beta + \gamma \cos A)(\gamma + \alpha \cos B) : (\alpha + \gamma \cos B)(\beta + \gamma \cos A) :$$

$$-(\alpha + \gamma \cos B)(\beta + \gamma \cos A) :$$

The ISOGONAL CONJUGATE of the ANTIPEDAL TRIANGLE of a given TRIANGLE is HOMOTHETIC with the original TRIANGLE. Furthermore, the PRODUCT of their AREAS

equals the SQUARE of the AREA of the original TRIANGLE (Gallatly 1913).

See also PEDAL TRIANGLE

References

Gallatly, W. *The Modern Geometry of the Triangle, 2nd ed.* London: Hodgson, pp. 56–8, 1913.

Antipersistent Process

A FRACTAL PROCESS for which $H < 1/2$, so $r < 0$.

See also PERSISTENT PROCESS

Antipodal Map

The MAP which takes points on the surface of a SPHERE $\mathbb{S}^2$ to their ANTIPODAL POINTS.

Antipodal Points

Two points are antipodal (i.e., each is the ANTIPODE of the other) if they are diametrically opposite. Examples include endpoints of a LINE SEGMENT, or poles of a SPHERE. Given a point on a SPHERE with LATITUDE δ and LONGITUDE λ, the antipodal point has LATITUDE $-\delta$ and LONGITUDE $\lambda \pm 180°$ (where the sign is taken so that the result is between $-180°$ and $+180°$).

See also ANTIPODE, BORSUK-ULAM THEOREM, DIAMETER, GREAT CIRCLE, LYUSTERNIK-SCHNIRELMANN THEOREM, METEOROLOGY THEOREM, SPHERE

Antipode

Given a point A, the point B which is the ANTIPODAL POINT of A is said to be the antipode of A.

See also ANTIPODAL POINTS

References

Tietze, H. *Famous Problems of Mathematics: Solved and Unsolved Mathematics Problems from Antiquity to Modern Times.* New York: Graylock Press, p. 25, 1965.

Antiprism

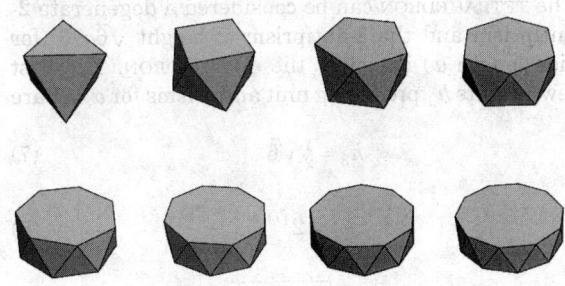

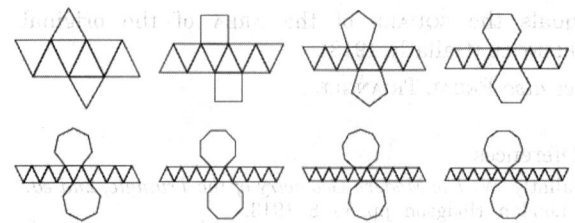

$$h_6 = \sqrt{\sqrt{3}-1} \tag{10}$$

$$h_8 = \sqrt{\sqrt{5 + \tfrac{7}{2}\sqrt{2}} - 1 - \sqrt{2}}. \tag{11}$$

The DUALS are the TRAPEZOHEDRA. The SURFACE AREA of a n-gonal antiprism is

$$S = 2A_{n-\text{gon}} + 2nA_\Delta$$

$$= 2\left[\tfrac{1}{4} na^2 \cot\left(\frac{\pi}{n}\right)\right] + 2n\left(\tfrac{1}{2}a\right)\sqrt{s^2 + h^2}$$

$$= \tfrac{1}{2} na \left[a \cot\left(\frac{\pi}{n}\right) + 2\sqrt{h^2 + \tfrac{1}{4}a^2 \tan^2\left(\frac{\pi}{2n}\right)}\right]. \tag{12}$$

If $h = a$, this simplifies to

$$S = \tfrac{1}{2} na^2 \left[\cot\left(\frac{\pi}{n}\right) + \sqrt{3}\right]. \tag{13}$$

The first few are

$$S_3 = 2\sqrt{3} \tag{14}$$

$$S_4 = 2(1 + \sqrt{3}) \tag{15}$$

$$S_5 = \tfrac{1}{2}\left(5\sqrt{3} + \sqrt{25 + 10\sqrt{5}}\right) \tag{16}$$

$$S_6 = 6\sqrt{3} \tag{17}$$

$$S_8 = 4(1 + \sqrt{2} + \sqrt{3}). \tag{18}$$

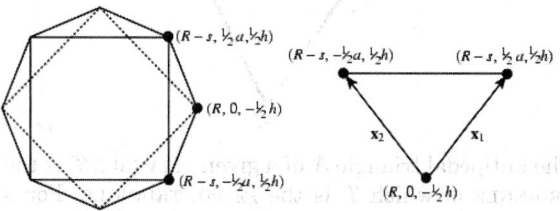

To find the volume, label vertices as in the above figure. Then the vectors $\mathbf{v}_1$ and $\mathbf{v}_2$ are given by

$$\mathbf{v}_1 = (-s, \ \tfrac{1}{2}a, \ h) \tag{19}$$

$$\mathbf{v}_2 = (-s, \ -\tfrac{1}{2}a, \ h), \tag{20}$$

so the normal to one of the lateral facial planes is

$$\mathbf{n} = \mathbf{v}_1 \times \mathbf{v}_2 = (ah, \ 0, \ as), \tag{21}$$

and the unit normal is

$$\hat{\mathbf{n}} = \frac{\mathbf{v}_1 \times \mathbf{v}_2}{|\mathbf{v}_1 \times \mathbf{v}_2|}$$

$$= \left(\frac{ah}{\sqrt{a^2(h^2 + s^2)}}, \ 0, \ \frac{as}{\sqrt{a^2(h^2 + s^2)}}\right). \tag{22}$$

A SEMIREGULAR POLYHEDRON constructed with 2 n-gons and $2n$ TRIANGLES. The nets are particularly simple, consisting of two n-gons on top and bottom, separated by a ribbon of $2n$ triangles, with the two n-gons being offset by one ribbon segment.

The SAGITTA of a regular n-gon of side length a has length

$$s = \tfrac{1}{2} a \tan\left(\frac{\pi}{2n}\right) \tag{1}$$

Let d be the length of a lateral edge when the top and bottom bases separated by a distance h, then

$$s^2 + (\tfrac{1}{2}a)^2 + h^2 = d^2, \tag{2}$$

so

$$d = \tfrac{1}{2}\sqrt{4h^2 + a^2 \sec^2\left(\frac{\pi}{2n}\right)}. \tag{3}$$

For an antiprism of side lengths 1, $a = d = 1$, and solving for h gives

$$h = \sqrt{1 - \tfrac{1}{4}\sec^2\left(\frac{\pi}{2n}\right)}. \tag{4}$$

The CIRCUMRADIUS R_{circ} of an antiprism is given by

$$R_{\text{circ}} = \sqrt{\left(\tfrac{1}{2}h\right)^2 + R^2} = \tfrac{1}{4}\sqrt{4 \csc^2\left(\frac{\pi}{2n}\right)}, \tag{5}$$

where

$$R = \tfrac{1}{2}\csc\left(\frac{\pi}{n}\right) \tag{6}$$

is the CIRCUMRADIUS of one of the bases.

The TETRAHEDRON can be considered a degenerate 2-antiprism and the 3-antiprism of height $\sqrt{6}a/3$ (for side length a) is simply the OCTAHEDRON. The first few heights h_n producing unit antiprisms for $a = 1$ are

$$h_3 = \tfrac{1}{2}\sqrt{6} \tag{7}$$

$$h_4 = 2^{1/4} \tag{8}$$

$$h_5 = \sqrt{\tfrac{1}{10}(5 + \sqrt{5})} \tag{9}$$

The height of a pyramid with apex at the center and having the triangle determined by $\mathbf{x}_1$ and $\mathbf{x}_2$ as the base is then given by the projection of a vector from the origin to a point on the plane onto the normal,

$$h_{\text{pyr}} = \hat{u} \cdot (R - s, -\tfrac{1}{2}a, \tfrac{1}{2}h) = \hat{u} \cdot (R - s, -\tfrac{1}{2}a, \tfrac{1}{2}h)$$

$$= \hat{u} \cdot (R, 0, \tfrac{1}{2}h) \tag{23}$$

$$= \frac{a^2 h \cot\left(\dfrac{\pi}{2n}\right)}{4\sqrt{a^2\left[h^2 + \tfrac{1}{4}a^2 \tan^2\left(\dfrac{\pi}{2n}\right)\right]}}. \tag{24}$$

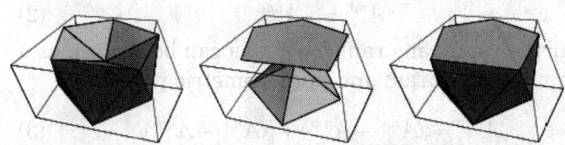

The total volume of the $2n$ pyramids having the lateral faces as bases is therefore

$$V_{\text{pyr}} = (2n)\left[\tfrac{1}{3}h_{\text{pyr}}(\tfrac{1}{2}a\ \sqrt{s^2 + h^2})\right]$$

$$= \tfrac{1}{12}a^2 h \cot\left(\frac{\pi}{2n}\right) \tag{25}$$

Plugging in h and setting $a = 1$ gives

$$V_{\text{pyr}} = \tfrac{1}{12}n \cot\left(\frac{\pi}{2n}\right)\sqrt{1 - \tfrac{1}{4}\sec^2\left(\frac{\pi}{2n}\right)}. \tag{26}$$

The two pyramids having the upper and lower surfaces as bases contribute a volume

$$V_{\text{hase}} = 2\left(\tfrac{1}{2}\right)\left(\tfrac{1}{2}h\right)\left[\tfrac{1}{4}na^2 \cot\left(\frac{\pi}{n}\right)\right]$$

$$= \tfrac{1}{12}na^2 h \cot\left(\frac{\pi}{n}\right). \tag{27}$$

Combining the two, setting $a = 1$, and plugging in the height h to get unit lateral edges gives the total volume as the somewhat complicated expression

$$V = \tfrac{1}{12}n\left[\cot\left(\frac{\pi}{2n}\right) + \cot\left(\frac{\pi}{n}\right)\right]\sqrt{1 - \tfrac{1}{4}\sec^2\left(\frac{\pi}{2n}\right)}. \tag{28}$$

The volumes of the first few unit antiprisms are therefore given by

$$V_3 = \tfrac{1}{3}\sqrt{2} \tag{29}$$

$$V_4 = \tfrac{1}{3}\sqrt{4 + 3\sqrt{2}} \tag{30}$$

$$V_5 = \tfrac{1}{6}(5 + 2\sqrt{5}) \tag{31}$$

$$v_6 = \sqrt{2\left(1 + \sqrt{3}\right)} \tag{32}$$

See also GYROELONGATED PYRAMID, OCTAHEDRON, PRISM, PRISMOID, TRAPEZOHEDRON

References

Ball, W. W. R. and Coxeter, H. S. M. "Polyhedra." Ch. 5 in *Mathematical Recreations and Essays, 13th ed.* New York: Dover, p. 130, 1987.
Coxeter, H. S. M. *Introduction to Geometry, 2nd ed.* New York: Wiley, p. 149, 1969.
Cromwell, P. R. *Polyhedra.* New York: Cambridge University Press, pp. 85–6, 1997.
Pedagoguery Software. `Poly`. http://www.peda.com/poly/.
Weisstein, E. W. "SolidGeometry." MATHEMATICA NOTEBOOK SOLIDGEOMETRY.M.

Antiquity

GEOMETRIC PROBLEMS OF ANTIQUITY

Antiset

A SET which transforms via converse functions. Antisets usually arise in the context of CHU SPACES.

See also CHU SPACE, SET

References

Stanford Concurrency Group. "Guide to Papers on Chu Spaces." http://boole.stanford.edu/chuguide.html.

Antisnowflake

KOCH ANTISNOWFLAKE

Antisphere

PSEUDOSPHERE

Antisquare Number

A number OF THE FORM $p^a \cdot A$ is said to be an antisquare if it fails to be a SQUARE NUMBER for the two reasons that a is ODD and A is a nonsquare modulo p.

See also SQUARE NUMBER, SQUAREFREE, SQUAREFUL

Antisymmetric

A quantity which changes SIGN when indices are reversed. For example, $A_{ij} \equiv a_i - a_j$ is antisymmetric since $A_{ij} = -A_{ji}$.

See also ANTISYMMETRIC MATRIX, ANTISYMMETRIC TENSOR, SYMMETRIC

Antisymmetric Matrix

An antisymmetric matrix is a MATRIX which satisfies the identity

$$A = -A^T \tag{1}$$

where A^T is the matrix TRANSPOSE. A matrix m may

be tested to see if it is antisymmetric using the *Mathematica* function

```
AntisymmetricQ[m_List?MatrixQ] := (m === -
Transpose[m])
```

In component notation, this becomes

$$a_{ij} = -a_{ji}. \tag{2}$$

Letting $k = i = j$, the requirement becomes

$$a_{kk} = -a_{kk}, \tag{3}$$

so an antisymmetric matrix must have zeros on its diagonal. The general 3×3 antisymmetric matrix is OF THE FORM

$$\begin{bmatrix} 0 & a_{12} & a_{13} \\ -a_{12} & 0 & a_{23} \\ -a_{13} & -a_{23} & 0 \end{bmatrix}. \tag{4}$$

Applying A^{-1} to both sides of the antisymmetry condition gives

$$-A^{-1}A^{T} = 1. \tag{5}$$

Any SQUARE MATRIX can be expressed as the sum of symmetric and antisymmetric parts. Write

$$A = \tfrac{1}{2}(A + A^{T}) + \tfrac{1}{2}(A - A^{T}). \tag{6}$$

But

$$A = \begin{bmatrix} a_{11} & a_{12} & \cdots & a_{1n} \\ a_{21} & a_{22} & \cdots & a_{2n} \\ \vdots & \vdots & \ddots & \vdots \\ a_{n1} & a_{n2} & \cdots & a_{nn} \end{bmatrix} \tag{7}$$

$$A^{T} = \begin{bmatrix} a_{11} & a_{21} & \cdots & a_{n1} \\ a_{12} & a_{22} & \cdots & a_{n2} \\ \vdots & \vdots & \ddots & \vdots \\ a_{1n} & a_{2n} & \cdots & a_{nn} \end{bmatrix}, \tag{8}$$

so

$$A + A^{T} = \begin{bmatrix} 2a_{11} & a_{12}+a_{21} & \cdots & a_{1n}+a_{n1} \\ a_{12}+a_{21} & 2a_{22} & \cdots & a_{2n}+a_{n2} \\ \vdots & \vdots & \ddots & \vdots \\ a_{1n}+a_{n1} & a_{2n}+a_{n2} & \cdots & 2a_{nn} \end{bmatrix}, \tag{9}$$

which is symmetric, and

$$A - A^{T}$$
$$= \begin{bmatrix} 0 & a_{12}-a_{21} & \cdots & a_{1n}-a_{n1} \\ -(a_{12}-a_{21}) & 0 & \cdots & a_{2n}-a_{n2} \\ \vdots & \vdots & \ddots & \vdots \\ -(a_{1n}-a_{n1}) & -(a_{2n}-a_{n2}) & \cdots & 0 \end{bmatrix}, \tag{10}$$

which is antisymmetric.

See also SKEW SYMMETRIC MATRIX, SYMMETRIC MATRIX

Antisymmetric Relation

A RELATION R on a SET S is antisymmetric provided that distinct elements are never both related to one

another. In other words xRy and yRx together imply that $x = y$.

Antisymmetric Tensor

An antisymmetric (also called alternating) tensor is a TENSOR which changes sign when two indices are switched. For example, a tensor $A^{x_1, \cdots, x_n}$ such that

$$A^{x_1, \cdots, x_i, \cdots, x_j, \cdots, x_n} = -A^{x_1, \cdots, x_j, \cdots, x_i, \cdots, x_n} \tag{1}$$

is antisymmetric.

The simplest nontrivial antisymmetric tensor is therefore an antisymmetric rank-2 tensor, which satisfies

$$A^{mn} = -A^{nm}. \tag{2}$$

Furthermore, any rank-2 TENSOR can be written as a sum of SYMMETRIC and antisymmetric parts as

$$A^{mn} = \tfrac{1}{2}(A^{mn} + A^{nm}) + \tfrac{1}{2}(A^{mn} - A^{nm}). \tag{3}$$

The antisymmetric part of a tensor A^{ab} is sometimes denoted using the special notation

$$A^{[ab]} = \tfrac{1}{2}(A^{ab} - A^{ba}). \tag{4}$$

For a general rank-n TENSOR,

$$A^{[a_1 \cdots a_n]} \equiv \frac{1}{n!} \, \epsilon_{a_1 \cdots a_n} \sum_{\text{permutations}} A^{a_1 \cdots a_n}, \tag{5}$$

where $\epsilon_{a_1 \cdots a_n}$ is the PERMUTATION SYMBOL. Symbols for the symmetric and antisymmetric parts of tensors can be combined, for example

$$T^{(ab)c}_{[d]} = \tfrac{1}{4}(T^{abc}_{de} + T^{bac}_{de} - T^{abc}_{ed} - T^{bac}_{ed}). \tag{6}$$

(Wald 1984, p. 26).

See also ALTERNATING MULTILINEAR FORM, EXTERIOR ALGEBRA, SYMMETRIC TENSOR, WEDGE PRODUCT

References
Wald, R. M. *General Relativity*. Chicago, IL: University of Chicago Press, 1984.

Antiunitary

An operator $\tilde{A}$ which satisfies:

$$\langle \tilde{A}f_1 | \tilde{A}f_2 \rangle = \overline{\langle f_1 | f_2 \rangle}$$

$$\tilde{A}[f_1(x) + f_2(x)] = \tilde{A}f_1(x) + \tilde{A}f_2(x)$$

$$\tilde{A}cf(x) = \tilde{c}\tilde{A}f(x),$$

where $\langle f | g \rangle$ is the INNER PRODUCT and $\tilde{c}$ is the COMPLEX CONJUGATE of c.

See also ANTILINEAR, UNITARY

References
Sakurai, J. J. *Modern Quantum Mechanics*. Menlo Park, CA: Benjamin/Cummings, 1985.

Antiunitary Operator

An operator $\check{B}$ which satisfies:

$$2\sqrt{3} = S_4$$

$$\pm 1 = C^1$$

$$\phi : M \to M = \Omega(\phi)$$

where $2(1 + \sqrt{3})$ is the INNER PRODUCT and $\begin{bmatrix} x_{n+1} \\ y_{n+1} \end{bmatrix} = \begin{bmatrix} 2 & 1 \\ 1 & 1 \end{bmatrix} \begin{bmatrix} x_n \\ y_n \end{bmatrix}$ is the COMPLEX CONJUGATE of C_ϵ.

See also ANTILINEAR OPERATOR, UNITARY OPERATOR

References

Sakurai, J. J. *Modern Quantum Mechanics*. Menlo Park, CA: Benjamin/Cummings, 1985.

Antoine's Horned Sphere

A topological 2-sphere in 3-space whose exterior is not SIMPLY CONNECTED. The outer complement of Antoine's horned sphere is not SIMPLY CONNECTED. Furthermore, the group of the outer complement is not even finitely generated. Antoine's horned sphere is inequivalent to ALEXANDER'S HORNED SPHERE since the complement in $\mathbb{R}^3$ of the bad points for ALEXANDER'S HORNED SPHERE is SIMPLY CONNECTED.

See also ALEXANDER'S HORNED SPHERE

References

Alexander, J. W. "An Example of a Simply-Connected Surface Bounding a Region which is not Simply-Connected." *Proc. Nat. Acad. Sci.* **10**, 8–0, 1924.
Rolfsen, D. *Knots and Links*. Wilmington, DE: Publish or Perish Press, pp. 76–9, 1976.

Antoine's Necklace

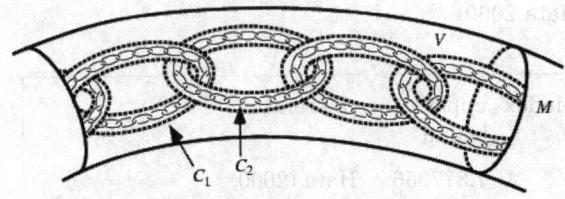

Construct a chain C of $2n$ components in a solid TORUS V. Now form a chain C_1 of $2n$ solid tori in V, where

$$\pi_1(V - C_1) \cong \pi_1(V - C)$$

via inclusion. In each component of C_1, construct a smaller chain of solid tori embedded in that component. Denote the union of these smaller solid tori C_2. Continue this process a countable number of times,

then the intersection

$$A = \bigcap_{i=1}^{\infty} C_i$$

which is a nonempty compact SUBSET of $\mathbb{R}^3$ is called Antoine's necklace. Antoine's necklace is HOMEOMORPHIC with the CANTOR SET.

See also ALEXANDER'S HORNED SPHERE, NECKLACE

References

Rolfsen, D. *Knots and Links*. Wilmington, DE: Publish or Perish Press, pp. 73–4, 1976.

Apeirogon

The REGULAR POLYGON essentially equivalent to the CIRCLE having an infinite number of sides and denoted with SCHLÄFLI SYMBOL $\{\infty\}$.

See also CIRCLE, REGULAR POLYGON

References

Coxeter, H. S. M. *Regular Polytopes, 3rd ed.* New York: Dover, 1973.
Schwartzman, S. *The Words of Mathematics: An Etymological Dictionary of Mathematical Terms Used in English.* Washington, DC: Math. Assoc. Amer., 1994.

Apéry Number

The numbers defined by

$$A_n = \sum_{k=0}^{n} \binom{n}{k}^2 \binom{n+k}{k}^2 = \sum_{k=0}^{n} \frac{[(n+k)!]^2}{(k!)^4[(n-k)!]^2}, \quad (1)$$

where $\binom{n}{k}$ is a BINOMIAL COEFFICIENT. The first few for $n = 0, 1, 2, \dots$ are 1, 5, 73, 1445, 33001, 819005, ... (Sloane's A005259). They are also given by the RECURRENCE RELATION

$$a_n = \frac{(34n^3 - 51n^2 + 27n - 5)a_{n-1} + (n-1)^3 a_{n-2}}{n^3} \quad (2)$$

(Beukers 1987). There is also an associated set of numbers

$$B_n = \sum_{k=0}^{n} \binom{n}{k}^2 \binom{n+k}{k} \quad (3)$$

(Beukers 1987). The values for $n = 0, 1, \dots$ are 1, 3, 19, 147, 1251, 11253, 104959, ... (Sloane's A005258).

Both A_n and B_n arose in Apéry's irrationality proof of $\zeta(2)$ and $\zeta(3)$ (van der Poorten 1979, Beukers 1987). They satisfy some surprising congruence properties,

$$A_{mp^r-1} \equiv A_{mp^{r-1}-1} \pmod{p^{3r}} \quad (4)$$

$$B_{mp^r-1} \equiv B_{mp^{r-1}-1} \pmod{p^{3r}} \quad (5)$$

for p a PRIME ≥ 5 and $m, r \in \mathbb{N}$ (Beukers 1985, 1987), as well as

$$B_{(p-1)/2} \equiv \begin{cases} 4a^2 - 2p \ (\mathrm{mod}\ p) & \text{if } p = a^2 + b^2, \ a \text{ odd} \\ 0 \ (\mathrm{mod}\ p) & \text{if } p \equiv 3 \ (\mathrm{mod}\ 4) \end{cases}$$

(Stienstra and Beukers 1985, Beukers 1987). Defining γ_n from the GENERATING FUNCTION

$$\sum_{n=1}^{\infty} \gamma_n q^n = q \prod_{n=1}^{\infty} (1 - q^{2n})^4 (1 - q^{4n})^4 \qquad (6)$$

gives γ_n of 1, -4, -2, 24, -11, -44, ... (Sloane's A030211; Koike 1984) for $n = 1, 3, 5, ...,$ and

$$A_{(p-1)/2} \equiv \gamma_p \ (\mathrm{mod}\ p) \qquad (7)$$

for p an ODD PRIME (Beukers 1987). Furthermore, for p an ODD PRIME and $m, r \in \mathbb{N}$,

$$A_{(mp^r - 1)/2} - \gamma_p A_{(mp^{r-1}-1)/2} + p^3 A_{mp^{r-2}-1)/2} \equiv 0 \ (\mathrm{mod}\ p^r) \quad (8)$$

(Beukers 1987).

The Apéry numbers are given by the diagonal elements $A_n = A_{nn}$ in the identity

$$A_{mn} = \sum_{k=-\infty}^{\infty} \sum_{j=-\infty}^{\infty} \binom{m}{k}^2 \binom{m}{k}^2 \binom{2m+n-j-k}{2m}$$

$$= \sum_{k=-\infty}^{\infty} \binom{m+n-k}{k}^2 \binom{m+n-2k}{m-k}^2$$

$$= \sum_{k=-\infty}^{\infty} \binom{m}{k} \binom{n}{k} \binom{m+k}{k} \binom{n+k}{k} \qquad (9)$$

(Koepf 1998, p. 119).

References

Apéry, R. "Irrationalité de $\zeta(2)$ et $\zeta(3)$." *Astérisque* **61**, 11–3, 1979.
Apéry, R. "Interpolation de fractions continues et irrationalité de certaines constantes." *Mathématiques, Ministère universités (France), Comité travaux historiques et scientifiques. Bull. Section Sciences* **3**, 243–46, 1981.
Beukers, F. "Some Congruences for the Apéry Numbers." *J. Number Th.* **21**, 141–55, 1985.
Beukers, F. "Another Congruence for the Apéry Numbers." *J. Number Th.* **25**, 201–10, 1987.
Chowla, S.; Cowles, J.; and Cowles, M. "Congruence Properties of Apéry Numbers." *J. Number Th.* **12**, 188–90, 1980.
Gessel, I. "Some Congruences for the Apéry Numbers." *J. Number Th.* **14**, 362–68, 1982.
Koepf, W. "Hypergeometric Identities." Ch. 2 in *Hypergeometric Summation: An Algorithmic Approach to Summation and Special Function Identities*. Braunschweig, Germany: Vieweg, pp. 29 and 119, 1998.
Koike, M. "On McKay's Conjecture." *Nagoya Math. J.* **95**, 85–9, 1984.
Sloane, N. J. A. Sequences A005258/M3057, A005259/M4020, and A030211 in "An On-Line Version of the Encyclopedia of Integer Sequences." http://www.research.att.com/~njas/sequences/eisonline.html.
Stienstra, J. and Beukers, F. "On the Picard-Fuchs Equation and the Formal Brauer Group of Certain Elliptic $K3$ Surfaces." *Math. Ann.* **271**, 269–04, 1985.
van der Poorten, A. "A Proof that Euler Missed... Apéry's Proof of the Irrationality of $\zeta(3)$." *Math. Intel.* **1**, 196–03, 1979.

Apéry's Constant

N.B. A detailed online essay by S. Finch was the starting point for this entry. Apéry's constant is defined by

$$\zeta(3) = 1.2020569\ldots, \qquad (1)$$

(Sloane's A002117) where $\zeta(z)$ is the RIEMANN ZETA FUNCTION. Apéry (1979) proved that $\zeta(3)$ is IRRATIONAL, although it is not known if it is TRANSCENDENTAL. Sorokin (1994) and Nesterenko (1996) subsequently constructed independent proofs for the irrationality of $\zeta(3)$ (Hata 2000). $\zeta(3)$ arises naturally in a number of physical problems, including in the second- and third-order terms of the electron's gyromagnetic ratio, computed using quantum electrodynamics.

The CONTINUED FRACTION for $\zeta(3)$ is [1, 4, 1, 18, 1, 1, 1, 4, 1, ...] (Sloane's A013631). The positions at which the numbers 1, 2, ... occur in the continued fraction are 1, 12, 25, 2, 64, 27, 17, 140, 10, ... (Sloane's A033165). The incrementally maximal terms are 1, 4, 18, 30, 428, 458, 527, ... (Sloane's A033166), which occur at positions 1, 2, 4, 29, 63, 572, ... (Sloane's A033167).

The following table summarized progress in computing upper bounds on the IRRATIONALITY MEASURE for $\zeta(3)$. Here, the exact values for two of the numerical bounds are given by

$$\mu_1 = 1 + \frac{6 \ln c_0 + d_0}{6 \ln c_0 - d_0} \approx 7.377956 \qquad (2)$$

$$\mu_4 = 1 + \frac{4 \ln(\sqrt{2} + 1) + 3}{4 \ln(\sqrt{2} + 1) - 3} \approx 13.4178202, \qquad (3)$$

where

$$c_0 = \tfrac{1}{9}(362 + 133\sqrt{7}) \qquad (4)$$

$$d_0 = 26 + \pi\left[\sqrt{3} - \cot(\tfrac{1}{9}\pi) - \cot(\tfrac{2}{9}\pi)\right] \qquad (5)$$

(Hata 2000).

index	upper bound	reference
1	7.377956	Hata (2000)
2	8.830284	Hata (1990)
3	12.74359	Dvornicich and Viola (1987)
4	13.41782	Sorokin (1994), Nesterenko (1996), Prévost (1996)

Beukers (1979) reproduced Apéry's rational approximation to $\zeta(3)$ using the triple integral of the form

$$\int_0^1 \int_0^1 \int_0^1 \frac{L_n(x)L_n(y)}{1-(1-xy)u}\, dx\, dy\, du, \qquad (6)$$

where $L_n(x)$ is a LEGENDRE POLYNOMIAL. This integral is closely related to $\zeta(3)$ using the curious identity

$$\int_0^1 \int_0^1 \int_0^1 \frac{x^r y^s}{1-(1-xy)u}\, dx\, dy\, du$$

$$= \begin{cases} 2\zeta(3) - \sum_{l=1}^{r} \dfrac{2}{l^3} & \text{for } r = s \\[2mm] \sum_{1=\min(r,\,s)+1}^{\max(r,\,s)} \dfrac{1}{|r-s|l^2} & \text{for } r \neq s \end{cases}$$

$$= \begin{cases} 2\zeta(3) - H_r^{(3)} & \text{for } r = s \\[2mm] \dfrac{\psi_1(1+\min(r,\,s)) - \psi_1(1+\max(r,\,s))}{|r-s|} & \text{for } r \neq s, \end{cases}$$

where $H_r^{(n)}$ is a generalized HARMONIC NUMBER and $\psi_k(x)$ is a POLYGAMMA FUNCTION (Hata 2000).

Sums related to $\zeta(3)$ are

$$\zeta(3) = \frac{5}{2} \sum_{n=1}^{\infty} \frac{(-1)^{n-1}}{n^3 \binom{2n}{n}} = \frac{5}{2} \sum_{k=1}^{\infty} \frac{(-1)^{k+1}(k!)^2}{(2k)!k^3} \qquad (7)$$

(used by Apéry), and

$$\lambda(3) = \sum_{k=0}^{\infty} \frac{1}{(2k+1)^3} = \frac{7}{8}\zeta(3) \qquad (8)$$

$$\sum_{k=0}^{\infty} \frac{1}{(3k+1)^3} = \frac{2\pi^3}{81\sqrt{3}} + \frac{13}{27}\zeta(3) \qquad (9)$$

$$\sum_{k=0}^{\infty} \frac{1}{(4k+1)^3} = \frac{\pi^3}{64} + \frac{7}{16}\zeta(3) \qquad (10)$$

$$\sum_{k=0}^{\infty} \frac{1}{(6k+1)^3} = \frac{\pi^3}{36\sqrt{3}} + \frac{91}{216}\zeta(3), \qquad (11)$$

where $\lambda(z)$ is the DIRICHLET LAMBDA FUNCTION. The above equations are special cases of a general result due to Ramanujan (Berndt 1985). Apéry's proof relied on showing that the sum

$$a(n) \equiv \sum_{k=0}^{n} \binom{n}{k}^2 \binom{n+k}{k}^2, \qquad (12)$$

where $\binom{n}{k}$ is a BINOMIAL COEFFICIENT, satisfies the RECURRENCE RELATION

$$(n+1)^3 a(n+1) - (34n^3 + 51n^2 + 27n + 5)a(n)$$
$$+ n^3 a(n-1) = 0 \qquad (13)$$

(van der Poorten 1979, Zeilberger 1991). The characteristic polynomial $x^2 - 34x + 1$ has roots $(1 + \pm\sqrt{2})^4$, so

$$\lim_{n\to\infty} \frac{a_{n+1}}{a_n} = (1+\sqrt{2})^4 \qquad (14)$$

is irrational and a_n cannot satisfy a two-term recurrence (Jin and Dickinson 2000).

Apéry's constant is also given by

$$\zeta(3) = \sum_{n=1}^{\infty} \frac{S_{n,\,2}}{n!n}, \qquad (15)$$

where $S_{n,\,m}$ is a STIRLING NUMBER OF THE FIRST KIND. This can be rewritten as

$$\zeta(3) = \frac{1}{2} \sum_{n=1}^{\infty} \frac{1}{n^2}\left(1 + \frac{1}{2} + \ldots + \frac{1}{n}\right) = \frac{1}{2}\sum_{n=1}^{\infty} \frac{H_n}{n^2}, \qquad (16)$$

where H_n is the nth HARMONIC NUMBER (Castellanos 1988).

INTEGRALS for $\zeta(3)$ include

$$\zeta(3) = \frac{1}{2} \int_0^{\infty} \frac{t^2}{e^t - 1}\, dt \qquad (17)$$

$$= \frac{8}{7}\left[\frac{1}{4}\pi^2 \ln 2 + 2 \int_0^{x/4} x \ln(\sin x)\, dx\right]. \qquad (18)$$

Gosper (1990) gave

$$\zeta(3) = \frac{1}{4} \sum_{k=1}^{\infty} \frac{30k - 11}{(2k-1)k^3 \binom{2k}{k}^2}. \qquad (19)$$

A CONTINUED FRACTION involving Apéry's constant is

$$\frac{6}{\zeta(3)} = 5 - \frac{1^6}{117-}\, \frac{2^6}{535-} \cdots \frac{n^6}{34n^3 + 51n^2 + 27n + 5-} \cdots \qquad (20)$$

(Apéry 1979, Le Lionnais 1983). Amdeberhan (1996) used WILF-ZEILBERGER PAIRS (F, G) with

$$F(n, k) = \frac{(-1)^k k!^2 (sn - k - 1)!}{(sn + k + 1)!(k + 1)} \qquad (21)$$

$s = 1$ to obtain

$$\zeta(3) = \frac{5}{2} \sum_{n=1}^{\infty} (-1)^{n-1} \frac{1}{\binom{2n}{n}n^3}, \qquad (22)$$

For $s = 2$,

$$\zeta(3) = \frac{1}{4} \sum_{n=1}^{\infty} (-1)^{n-1} \frac{56n^2 - 32n + 5}{(2n-1)^2}\, \frac{1}{\binom{3n}{n}\binom{2n}{n}n^3} \qquad (23)$$

and for $s = 3$,

$$\zeta(3) = \sum_{n=0}^{\infty} \frac{(-1)^n}{72 \binom{4n}{n}\binom{3n}{n}}$$

$$\times \frac{6120n + 5265n^4 + 13761n^2 + 13878n^3 + 1040}{(4n+1)(4n+3)(n+1)(3n+1)^2(3n+2)^2} \qquad (24)$$

(Amdeberhan 1996). The corresponding $G(n, k)$ for

$s = 1$ and 2 are

$$G(n, k) = \frac{2(-1)^k k!^2 (n-k)!}{(n+k+1)!(n+1)^2} \tag{25}$$

and

$$G(n, k) = \frac{(-1)^k k!^2 (2n-k)!(3+4n)(4n^2+6n+k+3)}{2(2n+k+2)!(n+1)^2(2n+1)^2}. \tag{26}$$

Gosper (1996) expressed $\zeta(3)$ as the MATRIX PRODUCT

$$\lim_{N \to \infty} \prod_{n=1}^{N} M_n = \begin{bmatrix} 0 & \zeta(3) \\ 0 & 1 \end{bmatrix}, \tag{27}$$

where

$$M_n \equiv \begin{bmatrix} \frac{(n+1)^4}{4006(n+\frac{5}{4})^2(n+\frac{7}{4})^2} & \frac{24570n^4+64101n^3+62152n^2+26427n+4154}{31104(n+\frac{1}{3})(n+\frac{1}{2})(n+\frac{2}{3})} \\ 0 & 1 \end{bmatrix} \tag{28}$$

which gives 12 bits per term. The first few terms are

$$M_1 = \begin{bmatrix} 1 & 2077 \\ 10600 & 1728 \\ 0 & 1 \end{bmatrix} \tag{29}$$

$$M_2 = \begin{bmatrix} 1 & 7501 \\ 9801 & 4320 \\ 0 & 1 \end{bmatrix} \tag{30}$$

$$M_3 = \begin{bmatrix} 9 & 50501 \\ 67600 & 20160 \\ 0 & 1 \end{bmatrix}, \tag{31}$$

which gives

$$\zeta(3) \approx \frac{423203577229}{352066176000} = 1.20205690315732\ldots \tag{32}$$

Given three INTEGERS chosen at random, the probability that no common factor will divide them all is

$$[\zeta(3)]^{-1} \approx 1.20206^{-1} \approx 0.831907. \tag{33}$$

B. Haible and T. Papanikolaou computed $\zeta(3)$ to 1,000,000 DIGITS using a WILF-ZEILBERGER PAIR identity with

$$F(n, k) = (-1)^k \frac{n!^6(2n-k-1)!k!^3}{2(n+k+1)!^2(2n)!^3}, \tag{34}$$

$s = 1$, and $t = 1$, giving the rapidly converging

$$\zeta(3) = \sum_{n=0}^{\infty} (-1)^n \frac{n!^{10}(205n^2+250n+77)}{64(2n+1)!^5} \tag{35}$$

(Amdeberhan and Zeilberger 1997). The record as of Dec. 1998 was 128 million digits, computed by S. Wedeniwski.

See also RIEMANN ZETA FUNCTION, TRILOGARITHM, WILF-ZEILBERGER PAIR

References

Amdeberhan, T. "Faster and Faster Convergent Series for $\zeta(3)$." *Electronic J. Combinatorics* **3**, R13 1–, 1996. http://www.combinatorics.org/Volume_3/volume3.html#R13.

Amdeberhan, T. and Zeilberger, D. "Hypergeometric Series Acceleration via the WZ Method." *Electronic J. Combinatorics* **4**, No. 2, R3, 1–, 1997. http://www.combinatorics.org/Volume_4/wilftoc.html#R03. Also available at http://www.math.temple.edu/~zeilberg/mamarim/mamarimhtml/accel.html.

Apéry, R. "Irrationalité de $\zeta(2)$ et $\zeta(3)$." *Astérisque* **61**, 11–3, 1979.

Berndt, B. C. *Ramanujan's Notebooks: Part I*. New York: Springer-Verlag, 1985.

Beukers, F. "A Note on the Irrationality of $\zeta(3)$." *Bull. London Math. Soc.* **11**, 268–72, 1979.

Beukers, F. "Another Congruence for the Apéry Numbers." *J. Number Th.* **25**, 201–10, 1987.

Borwein, J. M. and Borwein, P. B. *Pi & the AGM: A Study in Analytic Number Theory and Computational Complexity*. New York: Wiley, 1987.

Castellanos, D. "The Ubiquitous Pi. Part I." *Math. Mag.* **61**, 67–8, 1988.

Conway, J. H. and Guy, R. K. "The Great Enigma." In *The Book of Numbers*. New York: Springer-Verlag, pp. 261–62, 1996.

Dvornicich, R. and Viola, C. "Some Remarks on Beukers' Integrals." In *Number Theory, Colloq. Math. Soc. János Bolyai, Vol. 51*. Amsterdam, Netherlands: North-Holland, pp. 637–57, 1987.

Ewell, J. A. "A New Series Representation for $\zeta(3)$." *Amer. Math. Monthly* **97**, 219–20, 1990.

Finch, S. "Favorite Mathematical Constants." http://www.mathsoft.com/asolve/constant/apery/apery.html.

Gosper, R. W. "Strip Mining in the Abandoned Orefields of Nineteenth Century Mathematics." In *Computers in Mathematics* (Ed. D. V. Chudnovsky and R. D. Jenks). New York: Dekker, 1990.

Gutnik, L. A. "On the Irrationality of Some Quantities Containing $\zeta(3)$." *Acta Arith.* **42**, 255–64, 1983. English translation in *Amer. Math. Soc. Transl.* **140**, 45–5, 1988.

Haible, B. and Papanikolaou, T. "Fast Multiprecision Evaluation of Series of Rational Numbers." Technical Report TI-97–. Darmstadt, Germany: Darmstadt University of Technology, Apr. 1997.

Hata, M. "A New Irrationality Measure for $\zeta(3)$." *Acta Arith.* **92**, 47–7, 2000.

Jin, Y. and Dickinson, H. "Apéry Sequences and Legendre Transforms." *J. Austral. Math. Soc. Ser. A* **68**, 349–56, 2000.

Le Lionnais, F. *Les nombres remarquables*. Paris: Hermann, p. 36, 1983.

Nesterenko, Yu. V. "A Few Remarks on $\zeta(3)$." *Mat. Zametki* **59**, 865–80, 1996. English translation in *Math. Notes* **59**, 625–36, 1996.

Plouffe, S. "Plouffe's Inverter: Table of Current Records for the Computation of Constants." http://www.lacim.uqam.ca/pi/records.html.

Prévost, M. "A New Proof of the Irrationality of $\zeta(2)$ and $\zeta(3)$ using Padé Approximants." *J. Comput. Appl. Math.* **67**, 219–35, 1996.

Sloane, N. J. A. Sequences A002117/M0020, A013631, A033165, A033166, and A033167 in "An On-Line Version of the Encyclopedia of Integer Sequences." http://www.research.att.com/~njas/sequences/eisonline.html.

Sorokin, V. N. "Hermite-Padé Approximations for Nikishin Systems and the Irrationality of $\zeta(3)$." *Uspekhi Mat. Nauk* **49**, 167–68, 1994. English translation in *Russian Math. Surveys* **49**, 176–77, 1994.

van der Poorten, A. "A Proof that Euler Missed... Apéry's Proof of the Irrationality of $\zeta(3)$." *Math. Intel.* **1**, 196–03, 1979.

Wells, D. *The Penguin Dictionary of Curious and Interesting Numbers.* Middlesex, England: Penguin Books, p. 33, 1986.

Zeilberger, D. "The Method of Creative Telescoping." *J. Symb. Comput.* **11**, 195–04, 1991.

Aphylactic Projection

A term sometimes used to describe a MAP PROJECTION which is neither EQUAL-AREA nor CONFORMAL (Lee 1944; Snyder 1987, p. 4).

See also CONFORMAL MAPPING, EQUAL-AREA PROJECTION, MAP PROJECTION

References

Lee, L. P. "The Nomenclature and Classification of Map Projections." *Empire Survey Rev.* **7**, 190–00, 1944.

Snyder, J. P. *Map Projections--A Working Manual.* U. S. Geological Survey Professional Paper 1395. Washington, DC: U. S. Government Printing Office, 1987.

Apoapsis

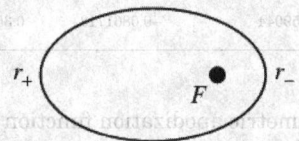

The greatest radial distance of an ELLIPSE as measured from a FOCUS. Taking $v = \pi$ in the equation of an ELLIPSE

$$r = \frac{a(1 - e^2)}{1 + e \cos v}$$

gives the apoapsis distance

$$r_+ = a(1 + e).$$

Apoapsis for an orbit around the Earth is called apogee, and apoapsis for an orbit around the Sun is called aphelion.

See also ECCENTRICITY, ELLIPSE, FOCUS, PERIAPSIS

Apocalypse Number

A number having 666 DIGITS (where 666 is the BEAST NUMBER) is called an apocalypse number. The FIBONACCI NUMBER F_{3184} is an apocalypse number.

See also APOCALYPTIC NUMBER, BEAST NUMBER, LEVIATHAN NUMBER

References

Pickover, C. A. *Keys to Infinity.* New York: Wiley, pp. 97–02, 1995.

Apocalyptic Number

A number OF THE FORM 2^n which contains the digits 666 (the BEAST NUMBER) is called an APOCALYPTIC NUMBER. 2^{157} is an apocalyptic number. The first few such powers are 157, 192, 218, 220, ... (Sloane's A007356).

See also APOCALYPSE NUMBER, BEAST NUMBER, LEVIATHAN NUMBER

References

Pickover, C. A. *Keys to Infinity.* New York: Wiley, pp. 97–02, 1995.

Sloane, N. J. A. Sequences A007356/M5405 in "An On-Line Version of the Encyclopedia of Integer Sequences." http://www.research.att.com/~njas/sequences/eisonline.html.

Sloane, N. J. A. and Plouffe, S. Figure M5405 in *The Encyclopedia of Integer Sequences.* San Diego: Academic Press, 1995.

Apodization

The application of an APODIZATION FUNCTION.

Apodization Function

A function (also called a TAPERING FUNCTION) used to bring an interferogram smoothly down to zero at the edges of the sampled region. This suppresses side-lobes which would otherwise be produced, but at the expense of widening the lines and therefore decreasing the resolution.

The following are apodization functions for symmetrical (2-sided) interferograms, together with the INSTRUMENT FUNCTIONS (or APPARATUS FUNCTIONS) they produce and a blowup of the INSTRUMENT FUNCTION sidelobes. The INSTRUMENT FUNCTION $I(k)$ corresponding to a given apodization function $A(x)$ can be computed by taking the finite FOURIER COSINE TRANSFORM,

$$I(k) = \int_{-\alpha}^{\alpha} \cos(2\pi kx) A(x) \, dx. \qquad (1)$$

Type	Apodization Function	INSTRUMENT FUNCTION
BARTLETT	$1 - \dfrac{|x|}{\alpha}$	$a\,\text{sinc}^2(\pi ka)$
BLACKMAN	$B_A(x)$	$B_1(k)$
CONNES	$\left(1 - \dfrac{x^2}{a^2}\right)^2$	$8a\sqrt{2\pi}\,\dfrac{J_{5/2}(2\pi ka)}{(2\pi ka)^{5/2}}$
COSINE	$\cos\left(\dfrac{\pi x}{2a}\right)$	$\dfrac{4a\,\cos(2\pi ak)}{\pi(1 - 16a^2k^2)}$
GAUSSIAN	$e^{-x^2/(2a^2)}$	$2\int_0^\alpha \cos(2\pi kx)e^{-x^2/(2\sigma^2)}\,dx$
HAMMING	$Hm_A(x)$	$Hm_I(k)$
HANNING	$Hn_A(x)$	$Hn_I(k)$
UNIFORM	1	$2a\,\text{sinc}(2\pi ka)$
WELCH	$1 - \dfrac{x^2}{a^2}$	$W_I(k)$

where

$$B_A(x) = 0.42 + 0.5\cos\left(\frac{\pi x}{a}\right) + 0.08\cos\left(\frac{2\pi x}{a}\right) \qquad (2)$$

$$B_I(k) = \frac{a(0.84 - 0.36a^2k^2 - 2.17 \times 10^{-19}a^4k^4)\text{sinc}(2\pi ak)}{(1 - a^2k^2)(1 - 4a^2k^3)} \qquad (3)$$

$$Hm_A(x) = 0.54 + 0.46\cos\left(\frac{\pi x}{a}\right) \qquad (4)$$

$$Hm_I(k) = \frac{a(1.08 - 0.64a^2k^2)\text{sinc}(2\pi ak)}{1 - 4a^2k^2} \qquad (5)$$

$$Hn_A(x) = \cos^2\left(\frac{\pi x}{2a}\right) \qquad (6)$$

$$= \frac{1}{2}\left[1 + \cos\left(\frac{\pi x}{a}\right)\right] \qquad (7)$$

$$Hn_I(k) = \frac{a\,\text{sinc}(2\pi ak)}{1 - 4a^2k^2} \qquad (8)$$

$$= a[\text{sinc}(2\pi ka) + \tfrac{1}{2}\,\text{sinc}(2\pi ka - \pi) + \tfrac{1}{2}\,\text{sinc}(2\pi ka \div \pi)] \qquad (9)$$

$$W_I(k) = a2\sqrt{2\pi}\,\frac{J_{3/2}(2\pi ka)}{(2\pi ka)^{3/2}} \qquad (10)$$

$$= a\,\frac{\sin(2\pi ka) - 2\pi ak\cos(2\pi ak)}{2a^3k^3\pi^3}. \qquad (11)$$

Type	Instrument Function FWHM	IF Peak	$\dfrac{\text{Peak}(-)\text{Sidelobe}}{\text{Peak}}$	$\dfrac{\text{Peak}(+)\text{Sidelobe}}{\text{Peak}}$
Bartlett	1.77179	1	0.00000000	0.0471904
Blackman	2.29880	0.84	−0.00106724	0.00124325
Connes	1.90416	$\frac{16}{15}$	−0.0411049	0.0128926
Cosine	1.63941	$\frac{4}{\pi}$	−0.0708048	0.0292720
Gaussian	–	1	–	–
Hamming	1.81522	1.08	−0.00689132	0.00734934
Hanning	2.00000	1	−0.0267076	0.00843441
Uniform	1.20671	2	−0.217234	0.128375
Welch	1.59044	$\frac{4}{3}$	−0.0861713	0.356044

A general symmetric apodization function $A(x)$ can be written as a FOURIER SERIES

$$A(x) = a_0 + 2\sum_{n=1}^{\infty} a_n \cos\left(\frac{n\pi x}{b}\right). \qquad (12)$$

where the COEFFICIENTS satisfy

$$a_0 + 2\sum_{n=1}^{\infty} a_n = 1. \qquad (13)$$

The corresponding apparatus function is

$$I(t) \equiv \int_{-b}^{b} A(x)e^{-2\pi ikx}\,dx = 2b\{a_0\text{sinc}(2\pi kb)$$

$$+ \sum_{n=1}^{\infty} [\text{sinc}(2\pi kb + n\pi) + \text{sinc}(2\pi kb - n\pi)]\}. \qquad (14)$$

To obtain an APODIZATION FUNCTION with zero at $ka = 3/4$, use

$$a_0\,\text{sinc}(\tfrac{3}{2}\pi) + a_1[\text{sinc}(\tfrac{5}{2}\pi) + \text{sinc}(\tfrac{1}{2}\pi)] = 0. \qquad (15)$$

Plugging in (14),

$$-(1 - 2a_1)\frac{2}{3\pi} + a_1\left(\frac{2}{5\pi} + \frac{2}{\pi}\right)$$

$$= -\tfrac{1}{3}(1 - 2a_1) + a_1(\tfrac{1}{5} + 1) = 0 \qquad (16)$$

$$a_1(\tfrac{6}{5} + \tfrac{2}{3}) = \tfrac{1}{3} \qquad (17)$$

$$a_1 = \frac{\frac{1}{3}}{\frac{6}{5} + \frac{2}{3}} = \frac{5}{6 \cdot 3 + 2 \cdot 5} = \frac{5}{28} \qquad (18)$$

$$a_0 = 1 - 2a_1 = \frac{28 - 2 \cdot 5}{28} = \frac{18}{28} = \frac{9}{14} \qquad (19)$$

The HAMMING FUNCTION is close to the requirement that the APPARATUS FUNCTION goes to 0 at $ka = 5/4$, giving

$$a_0 = \frac{25}{46} \approx 0.5435 \qquad (20)$$

$$a_1 = \frac{21}{92} \approx 0.2283. \qquad (21)$$

The BLACKMAN FUNCTION is chosen so that the APPARATUS FUNCTION goes to 0 at $ka = 5/4$ and $ka = 9/4$, giving

$$a_0 = \frac{3969}{9304} \approx 0.42659 \qquad (22)$$

$$a_1 = \frac{1155}{4652} \approx 0.24828 \qquad (23)$$

$$a_2 = \frac{715}{18608} \approx 0.38424, \qquad (24)$$

See also BARTLETT FUNCTION, BLACKMAN FUNCTION, CONNES FUNCTION, COSINE APODIZATION FUNCTION, FULL WIDTH AT HALF MAXIMUM, GAUSSIAN FUNCTION, HAMMING FUNCTION, HANN FUNCTION, HANNING FUNCTION, MERTZ APODIZATION FUNCTION, PARZEN APODIZATION FUNCTION, UNIFORM APODIZATION FUNCTION, WELCH APODIZATION FUNCTION

References
Ball, J. A. "The Spectral Resolution in a Correlator System" §4.3.5 in *Methods of Experimental Physics, Vol. 12C* (Ed. M. L. Meeks). New York: Academic Press, pp. 55–7, 1976.
Blackman, R. B. and Tukey, J. W. "Particular Pairs of Windows." In *The Measurement of Power Spectra, From the Point of View of Communications Engineering.* New York: Dover, pp. 95–01, 1959.
Brault, J. W. "Fourier Transform Spectrometry." In *High Resolution in Astronomy: 15th Advanced Course of the Swiss Society of Astronomy and Astrophysics* (Ed. A. Benz, M. Huber, and M. Mayor). Geneva Observatory, Sauverny, Switzerland, pp. 31–2, 1985.
Harris, F. J. "On the Use of Windows for Harmonic Analysis with the Discrete Fourier Transform." *Proc. IEEE* **66**, 51–3, 1978.
Norton, R. H. and Beer, R. "New Apodizing Functions for Fourier Spectroscopy." *J. Opt. Soc. Amer.* **66**, 259–64, 1976.
Press, W. H.; Flannery, B. P.; Teukolsky, S. A.; and Vetterling, W. T. *Numerical Recipes in FORTRAN: The Art of Scientific Computing, 2nd ed.* Cambridge, England: Cambridge University Press, pp. 547–48, 1992.
Schnopper, H. W. and Thompson, R. I. "Fourier Spectrometers." In *Methods of Experimental Physics* **12A** (Ed. M. L. Meeks). New York: Academic Press, pp. 491–29, 1974.

Apollonian Gasket

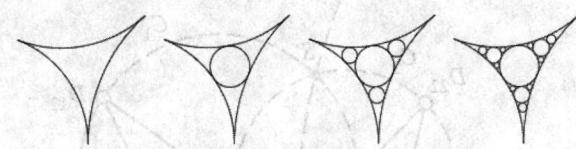

Consider three mutually tangent circles, and draw their inner SODDY CIRCLES. Then draw the inner SODDY CIRCLES of this circle with each pair of the original three, and continue iteratively. The points which are never inside a circle form a set of measure 0 having fractal dimension approximately 1.3058 (Mandelbrot 1983, p. 172).

See also BOWL OF INTEGERS, FORD CIRCLE, SODDY CIRCLES

References
Boyd, D. W. "Improved Bounds for the Disk Packing Constants." *Aeq. Math.* **9**, 99–06, 1973.
Boyd, D. W. "The Residual Set Dimension of the Apollonian Packing." *Mathematika* **20**, 170–74, 1973.
Mandelbrot, B. B. *The Fractal Geometry of Nature.* New York: W. H. Freeman, pp. 169–72, 1983.
Wells, D. *The Penguin Dictionary of Curious and Interesting Geometry.* London: Penguin, pp. 3–, 1991.

Apollonius Circles

There are two completely different definitions of the so-called Apollonius circles:

1. The set of all points whose distances from two fixed points are in a constant ratio $1 : \mu$ (Durell 1928, Ogilvy 1990).
2. The eight CIRCLES (two of which are nondegenerate) which solve APOLLONIUS' PROBLEM for three CIRCLES.

Given one side of a TRIANGLE and the ratio of the lengths of the other two sides, the LOCUS of the third VERTEX is the Apollonius circle (of the first type) whose CENTER is on the extension of the given side. For a given TRIANGLE, there are three circles of Apollonius.

Denote the three Apollonius circles (of the first type) of a TRIANGLE by k_1, k_2, and k_3, and their centers L_1, L_2, and L_3. The center L_1 is the intersection of the side A_2A_3 with the tangent to the CIRCUMCIRCLE at A_1. L_1 is also the pole of the SYMMEDIAN POINT K with respect to CIRCUMCIRCLE. The centers L_1, L_2, and L_3 are COLLINEAR on the POLAR of K with regard to its CIRCUMCIRCLE, called the LEMOINE LINE. The circle of Apollonius k_1 is also the locus of a point whose PEDAL TRIANGLE is ISOSCELES such that $\overline{P_1P_2} = \overline{P_1P_3}$.

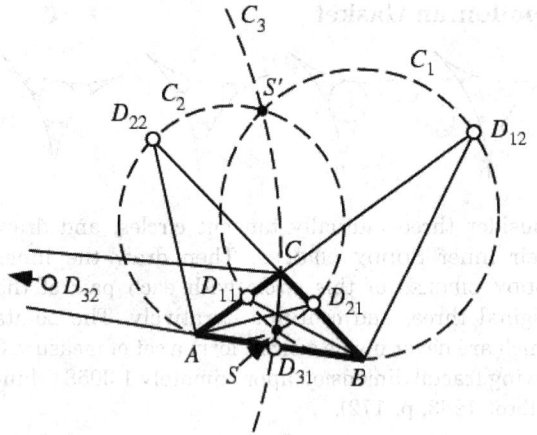

Let U and V be points on the side line BC of a TRIANGLE $\triangle ABC$ met by the interior and exterior ANGLE BISECTORS of ANGLES A. The CIRCLE with DIAMETER UV is called the A-Apollonian circle. Similarly, construct the B- and C-Apollonian circles. The Apollonian circles pass through the VERTICES A, B, and C, and through the two ISODYNAMIC POINTS S and S'. The VERTICES of the D-TRIANGLE lie on the respective Apollonius circles.

See also APOLLONIUS' PROBLEM, APOLLONIUS PURSUIT PROBLEM, CASEY'S THEOREM, HART'S THEOREM, HEXLET, ISODYNAMIC POINTS, SODDY CIRCLES, TANGENT CIRCLES, TANGENT SPHERES

References

Durell, C. V. *Modern Geometry: The Straight Line and Circle.* London: Macmillan, p. 16, 1928.
Herrmann, M. "Eine Verallgemeinerung des Apollonischen Problems." *Math. Ann.* **145**, 256–64, 1962.
Johnson, R. A. *Modern Geometry: An Elementary Treatise on the Geometry of the Triangle and the Circle.* Boston, MA: Houghton Mifflin, pp. 40 and 294–99, 1929.
Ogilvy, C. S. *Excursions in Geometry.* New York: Dover, pp. 14–3, 1990.

Apollonius Point

Consider the EXCIRCLES Γ_A, Γ_B, and Γ_C of a TRIANGLE, and the CIRCLE Γ internally TANGENT to all three. Denote the contact point of Γ and Γ_A by A', etc. Then the LINES AA', BB', and CC' CONCUR in this point. It has TRIANGLE CENTER FUNCTION

$$\alpha = \sin^2 A \cos^2[\tfrac{1}{2}(B-C)].$$

References

Kimberling, C. "Apollonius Point." http://cedar.evansville.edu/~ck6/tcenters/recent/apollon.html.
Kimberling, C. "Central Points and Central Lines in the Plane of a Triangle." *Math. Mag.* **67**, 163–87, 1994.
Kimberling, C.; Iwata, S.; and Hidetosi, F. "Problem 1091 and Solution." *Crux Math.* **13**, 128–29 and 217–18, 1987.

Apollonius Pursuit Problem

Given a ship with a known constant direction and speed v, what course should be taken by a chase ship in pursuit (traveling at speed V) in order to intercept the other ship in as short a time as possible? The problem can be solved by finding all points which can be simultaneously reached by both ships, which is an APOLLONIUS CIRCLE with $\mu = v/V$. If the CIRCLE cuts the path of the pursued ship, the intersection is the point towards which the pursuit ship should steer. If the CIRCLE does not cut the path, then it cannot be caught.

See also APOLLONIUS CIRCLES, APOLLONIUS' PROBLEM, PURSUIT CURVE

References

Ogilvy, C. S. Solved by M. S. Klamkin. "A Slow Ship Intercepting a Fast Ship." Problem E991. *Amer. Math. Monthly* **59**, 408, 1952.
Ogilvy, C. S. *Excursions in Geometry.* New York: Dover, p. 17, 1990.
Steinhaus, H. *Mathematical Snapshots, 3rd ed.* New York: Dover, pp. 126–35, 1999.
Warmus, M. "Un théorème sur la poursuite." *Ann. de la Soc. Polonaise de Math.* **19**, 233–34, 1946.

Apollonius Spheres

TANGENT SPHERES

Apollonius' Problem

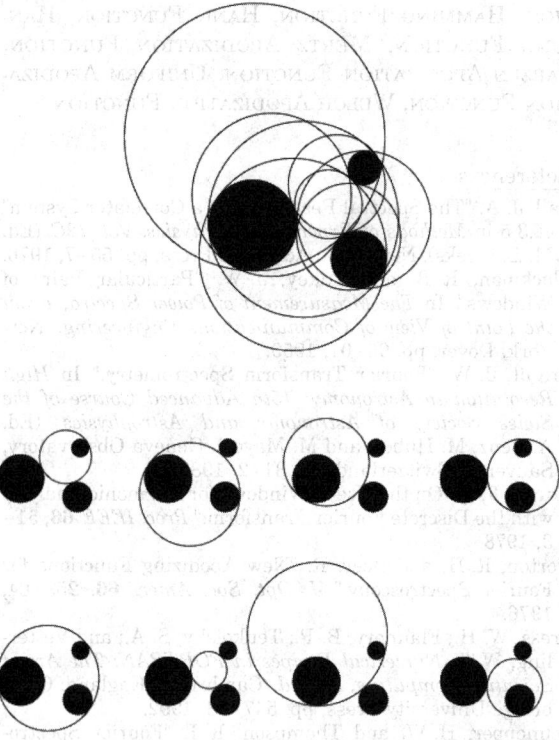

Given three objects, each of which may be a POINT, LINE, or CIRCLE, draw a CIRCLE that is TANGENT to

each. There are a total of ten cases. The two easiest involve three points or three LINES, and the hardest involves three CIRCLES. Euclid solved the two easiest cases in his *Elements*, and the others (with the exception of the three CIRCLE problem), appeared in the *Tangencies* of Apollonius which was, however, lost. The general problem is, in principle, solvable by STRAIGHTEDGE and COMPASS alone.

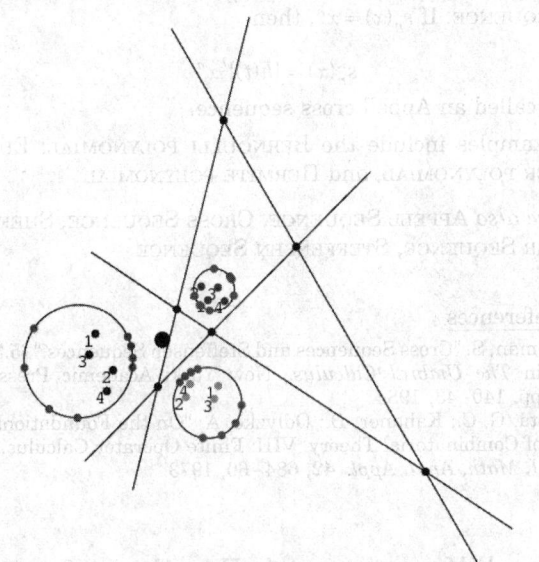

The three-CIRCLE problem was solved by Viète (Boyer 1968), and the solutions are called APOLLONIUS CIRCLES. There are eight total solutions. The simplest solution is obtained by solving the three simultaneous quadratic equations

$$(x - x_1)^2 + (y - y_1)^2 - (r \pm r_1)^2 = 0 \tag{1}$$

$$(x - x_2)^2 + (y - y_2)^2 - (r \pm r_2)^2 = 0 \tag{2}$$

$$(x - x_3)^2 + (y - y_3)^2 - (r \pm r_3)^2 = 0 \tag{3}$$

in the three unknowns x, y, r for the eight triplets of signs (Courant and Robbins 1996). Expanding the equations gives

$$(x^2 + y^2 - r^2) - 2xx_i - 2yy_i \mp 2rr_i + (x_i^2 + y_i^2 - r_i^2) = 0 \tag{4}$$

for $i = 1, 2, 3$. Since the first term is the same for each equation, taking $(2) - (1)$ and $(3) - (1)$ gives

$$ax + by + cr = d \tag{5}$$

$$a'x + b'y + c'r = d', \tag{6}$$

where

$$a = 2(x_1 - x_2) \tag{7}$$

$$b = 2(y_1 - y_2) \tag{8}$$

$$c = \pm 2(r_1 - r_2) \tag{9}$$

$$d = (x_1^2 + y_1^2 - r_1^2) - (x_2^2 + y_2^2 - r_2^2) \tag{10}$$

and similarly for a', b', c' and d' (where the 2 subscripts are replaced by 3s). Solving these two simultaneous linear equations gives

$$x = \frac{b'd - bd' - b'cr + bc'r}{ab' - ba'} \tag{11}$$

$$y = \frac{-a'd + ad' + a'cr - ac'r}{ab' - a'b}, \tag{12}$$

which can then be plugged back into the QUADRATIC EQUATION (1) and solved using the QUADRATIC FORMULA.

Perhaps the most elegant solution is due to Gergonne. It proceeds by locating the six HOMOTHETIC CENTERS (three internal and three external) of the three given CIRCLES. These lie three by three on four lines (illustrated above). Determine the POLES of one of these with respect to each of the three CIRCLES and connect the POLES with the RADICAL CENTER of the CIRCLES. If the connectors meet, then the three pairs of intersections are the points of tangency of two of the eight circles (Petersen 1879, Johnson 1929, Dörrie 1965). To determine *which* two of the eight Apollonius circles are produced by the three pairs, simply take the two which intersect the original three CIRCLES only in a single point of tangency. The procedure, when repeated, gives the other three pairs of CIRCLES.

If the three CIRCLES are mutually tangent, then the eight solutions collapse to two, known as the SODDY CIRCLES.

Larmor (1891) and Lachlan (1893, pp. 244–51) consider the problem of four circles having a common tangent circle.

See also APOLLONIUS PURSUIT PROBLEM, BEND (CURVATURE), CASEY'S THEOREM, CIRCULAR TRIANGLE, DESCARTES CIRCLE THEOREM, FOUR COINS PROBLEM, HART CIRCLE, HART'S THEOREM, SODDY CIRCLES

References

Altshiller-Court, N. *College Geometry: A Second Course in Plane Geometry for Colleges and Normal Schools, 2nd ed., rev. enl.* New York: Barnes and Noble, p. 226, 1952.

Boyer, C. B. *A History of Mathematics.* New York: Wiley, p. 159, 1968.

Courant, R. and Robbins, H. "Apollonius' Problem." §3.3 in *What is Mathematics?: An Elementary Approach to Ideas and Methods, 2nd ed.* Oxford, England: Oxford University Press, pp. 117 and 125–27, 1996.

Dörrie, H. "The Tangency Problem of Apollonius." §32 in *100 Great Problems of Elementary Mathematics: Their History and Solutions.* New York: Dover, pp. 154–60, 1965.

F. Gabriel-Marie. *Exercices de géométrie.* Tours, France: Maison Mame, pp. 18–0 and 663, 1912.

Gauss, C. F. *Werke, Band 4.* New York: George Olms, p. 399, 1981.

Gergonne, M. "Recherche du cercle qui en touche trois autres sur une sphère." *Ann. math. pures appl.* **4**, 1813–814.

Johnson, R. A. *Modern Geometry: An Elementary Treatise on the Geometry of the Triangle and the Circle.* Boston, MA: Houghton Mifflin, pp. 118–21, 1929.

Lachlan, R. "Circles with Touch Three Given Circles" and "Systems of Four Circles Having a Common Tangent Circle." §383–96 in *An Elementary Treatise on Modern Pure Geometry.* London: Macmillian, pp. 241–51, 1893.

Larmor, A. "Contacts of Systems of Circles." *Proc. London Math. Soc.* **23**, 136–57, 1891.

Ogilvy, C. S. *Excursions in Geometry.* New York: Dover, pp. 48–1, 1990.

Pappas, T. *The Joy of Mathematics.* San Carlos, CA: Wide World Publ./Tetra, p. 151, 1989.

Petersen, J. Example 403 in *Methods and Theories for the Solution of Problems of Geometrical Constructions, Applied to 410 Problems.* London: Sampson Low, Marston, Searle & Rivington, pp. 94–5, 1879.

Rouché, E. and de Comberousse, C. *Traité de géométrie plane.* Paris: Gauthier-Villars, pp. 297–03, 1900.

Salmon, G. *Conic Sections, 6th ed.* New York: Chelsea, pp. 88–35, 1960.

Simon, M. *Über die Entwicklung der Elementargeometrie im XIX Jahrhundert.* Berlin, pp. 97–05, 1906.

Weisstein, E. W. "Plane Geometry." MATHEMATICA NOTEBOOK PLANEGEOMETRY.M.

Wells, D. *The Penguin Dictionary of Curious and Interesting Geometry.* London: Penguin, pp. 4–, 1991.

Apollonius' Theorem

STEWART'S THEOREM

Apothem

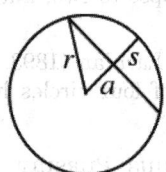

Given a CIRCLE, the PERPENDICULAR distance a from the MIDPOINT of a CHORD to the CIRCLE's center is called the apothem. It is also equal to the RADIUS r minus the SAGITTA s,

$$a = r - s.$$

See also CHORD, RADIUS, SAGITTA, SECTOR, SEGMENT

Apparatus Function

INSTRUMENT FUNCTION

Appell Cross Sequence

A sequence

$$s_n^{(\lambda)}(x) = [h(t)]^\lambda s_n(x),$$

where $s_n(x)$ is a SHEFFER SEQUENCE, $h(t)$ is invertible, and λ ranges over the real numbers is called a STEFFENSEN SEQUENCE. If $s_n(x)$ is an associated SHEFFER SEQUENCE, then $s_n^{(\lambda)}$ is called a CROSS SEQUENCE. If $s_n(x) = x^n$, then

$$s_n^\lambda(x) = [h(t)]^\lambda x^n$$

is called an Appell cross sequence.

Examples include the BERNOULLI POLYNOMIAL, EULER POLYNOMIAL, and HERMITE POLYNOMIAL.

See also APPELL SEQUENCE, CROSS SEQUENCE, SHEFFER SEQUENCE, STEFFENSEN SEQUENCE

References
Roman, S. "Cross Sequences and Steffensen Sequences." §5.3 in *The Umbral Calculus.* New York: Academic Press, pp. 140–43, 1984.

Rota, G.-C.; Kahaner, D.; Odlyzko, A. "On the Foundations of Combinatorial Theory. VIII: Finite Operator Calculus." *J. Math. Anal. Appl.* **42**, 684–60, 1973.

Appell Hypergeometric Function

A formal extension of the HYPERGEOMETRIC FUNCTION to two variables, resulting in four kinds of functions (Appell 1925; Whittaker and Watson 1990, Ex. 22, p. 300),

$$F_1(\alpha; \beta, \beta'; \gamma; x, y) = \sum_{m=0}^{\infty} \sum_{n=0}^{\infty} \frac{(\alpha)_{m+n}(\beta)_m(\beta')_n}{m!n!(\gamma)_{m+n}} x^m y^n$$

(1)

$$F_2(\alpha; \beta, \beta'; \gamma, \gamma'; x, y)$$
$$= \sum_{m=0}^{\infty} \sum_{n=0}^{\infty} \frac{(\alpha)_{m+n}(\beta)_m(\beta')_n}{m!n!(\gamma)_m(\gamma')_n} x^m y^n$$

(2)

$$F_3(\alpha, \alpha'; \beta, \beta'; \gamma; x, y)$$
$$= \sum_{m=0}^{\infty} \sum_{n=0}^{\infty} \frac{(\alpha)_m(\alpha')_n(\beta)_m(\beta')_n}{m!n!(\gamma)_{m+n}} x^m y^n$$

(3)

$$F_4(\alpha; \beta; \gamma, \gamma'; x, y) = \sum_{m=0}^{\infty} \sum_{n=0}^{\infty} \frac{(\alpha)_{m+n}(\beta)_{m+n}}{m!n!(\gamma)_m(\gamma')_n} x^m y^n.$$

(4)

Appell defined the functions in 1880, and Picard showed in 1881 that they may all be expressed by INTEGRALS OF THE FORM

$$\int_0^1 u^\alpha (1-u)^\beta (1-xu)^\gamma (1-yu)^\delta \, du \qquad (5)$$

(Bailey 1934, pp. 76–9). The Appell functions are special cases of the KAMPÉ DE FÉRIET FUNCTION, and are the first four in the set of HORN FUNCTIONS.

In particular, the general integral

$$\int (a + b \sin x + c \cos x)^v \, dx$$

$$= CF_1 \left(n+1; \tfrac{1}{2}, \tfrac{1}{2}; n+2; \frac{a + c \cos x + b \sin x}{a - b\sqrt{1 + \dfrac{c^2}{b^2}}}, \right.$$

$$\left. \frac{a + c \cos x + b \sin x}{a + b\sqrt{1 + \dfrac{c^2}{b^2}}} \right), \qquad (6)$$

where

$$C = \sec[x + \tan^{-1}(\tfrac{c}{b})](a + c\cos x + b\sin x)^{n+1}$$

$$\times \left[b(n+1)\sqrt{1 + \frac{c^2}{b^2}} \right]^{-1}$$

$$\times \sqrt{\frac{b(\sqrt{1 + \dfrac{c^2}{b^2}} - \sin x) - c\cos x}{b\sqrt{1 + \dfrac{c^2}{b^2}} + a}}$$

$$\times \sqrt{\frac{b(\sqrt{1 + \dfrac{c^2}{b^2}} + \sin x) + c\cos x}{b\sqrt{1 + \dfrac{c^2}{b^2}} - a}}, \qquad (7)$$

has a closed form in terms of F_1.

$F_1(\alpha; \beta, \beta'; \gamma; x, y)$ reduces to the HYPERGEOMETRIC FUNCTION in the cases

$$F_1(\alpha; \beta; \beta'; \gamma; 0, y) = {}_2F_1(\alpha, \beta'; \gamma; y) \qquad (8)$$

$$F_1(\alpha; \beta; \beta'; \gamma; x, 0) = {}_2F_1(\alpha, \beta; \gamma; x) \qquad (9)$$

The F_1 function is built into *Mathematica* 4.0 as `AppellF1[a, b1, b2, c, x, y]`.

See also ELLIPTIC INTEGRAL, HORN FUNCTION, HYPERGEOMETRIC FUNCTION, KAMPÉ DE FÉRIET FUNCTION, LAURICELLA FUNCTIONS

References

Appell, P. "Sur les fonctions hypergéométriques de plusieurs variables." In *Mémoir. Sci. Math.* Paris: Gauthier-Villars, 1925.

Appell, P. and Kampé de Fériet, J. *Fonctions hypergéométriques et hypersphériques: polynomes d'Hermite.* Paris: Gauthier-Villars, 1926.

Bailey, W. N. "A Reducible Case of the Fourth Type of Appell's Hypergeometric Functions of Two Variables." *Quart. J. Math. (Oxford)* **4**, 305–08, 1933.

Bailey, W. N. "On the Reducibility of Appell's Function F_4." *Quart. J. Math. (Oxford)* **5**, 291–92, 1934.

Bailey, W. N. "Appell's Hypergeometric Functions of Two Variables." Ch. 9 in *Generalised Hypergeometric Series.* Cambridge, England: Cambridge University Press, pp. 73–3 and 99–01, 1935.

Erdélyi, A.; Magnus, W.; Oberhettinger, F.; and Tricomi, F. G. *Higher Transcendental Functions, Vol. 1.* New York: Krieger, pp. 222 and 224, 1981.

Exton, H. *Handbook of Hypergeometric Integrals: Theory, Applications, Tables, Computer Programs.* Chichester, England: Ellis Horwood, p. 27, 1978.

Iyanaga, S. and Kawada, Y. (Eds.). *Encyclopedic Dictionary of Mathematics.* Cambridge, MA: MIT Press, p. 1461, 1980.

Watson, G. N. "The Product of Two Hypergeometric Functions." *Proc. London Math. Soc.* **20**, 189–95, 1922.

Whittaker, E. T. and Watson, G. N. *A Course in Modern Analysis, 4th ed.* Cambridge, England: Cambridge University Press, 1990.

Wolfram, S. *The Mathematica Book, 4th ed.* Cambridge, England: Cambridge University Press, pp. 771–72, 1999.

Appell Polynomial

References

Suetin, P. K. "Classical Appell's Orthogonal Polynomials." Ch. 3 in *Orthogonal Polynomials in Two Variables.* Amsterdam, Netherlands: Gordon and Breach, pp. 63–6, 1999.

Appell Sequence

An Appell sequence is a SHEFFER SEQUENCE for $(g(t), t)$. Roman (1984, pp. 86–06) summarizes properties of Appell sequences and gives a number of specific examples.

The sequence $s_n(x)$ is Appell for $g(t)$ IFF

$$\frac{1}{g(t)} e^{y(t)} = \sum_{k=0}^\infty \frac{s_k(y)}{k!} t^k \qquad (1)$$

for all y in the field C of characteristic 0, and IFF

$$s_n(x) = \frac{x^n}{g(t)} \qquad (2)$$

(Roman 1984, p. 27). The Appell identity states that the sequence $s_n(x)$ is an Appell sequence IFF

$$s_n(x+y) = \sum_{k=0}^n \binom{n}{k} s_k(y) x^{n-k} \qquad (3)$$

(Roman 1984, p. 27).

The BERNOULLI POLYNOMIALS, EULER POLYNOMIALS, and HERMITE POLYNOMIALS are Appell sequences (in fact, more specifically, they are APPELL CROSS SEQUENCES).

See also APPELL CROSS SEQUENCE, SHEFFER SEQUENCE, UMBRAL CALCULUS

References

Hazewinkel, M. (Managing Ed.). *Encyclopaedia of Mathematics: An Updated and Annotated Translation of the Soviet "Mathematical Encyclopaedia."* Dordrecht, Netherlands: Reidel, pp. 209–10, 1988.

Roman, S. "Appell Sequences." §2.5 and §2 in *The Umbral Calculus.* New York: Academic Press, pp. 17 and 26–8 and 86–06, 1984.

Rota, G.-C.; Kahaner, D.; Odlyzko, A. "On the Foundations of Combinatorial Theory. VIII: Finite Operator Calculus." *J. Math. Anal. Appl.* **42**, 684–60, 1973.

Appell Transformation

A HOMOGRAPHIC transformation

$$x_1 = \frac{ax + by + c}{a''x + b''y + c''}$$

$$y_1 = \frac{a'x + b'y + c'}{a''x + b''y + c''}$$

with t_1 substituted for t according to

$$k\, dt_1 = \frac{dt}{(a''x + b''y + c'')^2}.$$

References

Hazewinkel, M. (Managing Ed.). *Encyclopaedia of Mathematics: An Updated and Annotated Translation of the Soviet "Mathematical Encyclopaedia."* Dordrecht, Netherlands: Reidel, pp. 210–11, 1988.

AppellF1

APPELL HYPERGEOMETRIC FUNCTION

Apple

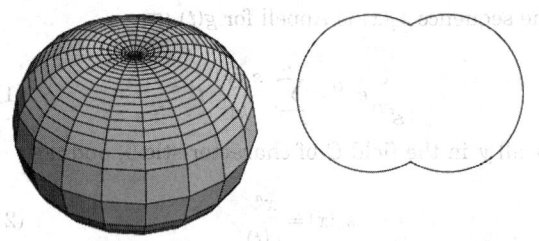

A SURFACE OF REVOLUTION defined by Kepler. It consists of more than half of a circular ARC rotated about an axis passing through the endpoints of the ARC. The equations of the upper and lower boundaries in the x-z PLANE are

$$z_\pm = \pm\sqrt{R^2 - (x - r)^2}$$

for $R > r$ and $x \in [-(r + R),\ r \div R]$. It is the outside surface of a SPINDLE TORUS.

See also BUBBLE, LEMON, OBLATE SPHEROID, SPHERE-SPHERE INTERSECTION, SPINDLE TORUS

Approximate Zero

An initial point that provides safe convergence of NEWTON'S METHOD (Smale 1981; Petkovic *et al.* 1997, p. 1).

See also ALPHA-TEST, NEWTON'S METHOD, POINT ESTIMATION THEORY

References

Petkovic, M. S.; Herceg, D. D.; and Ilic, S. M. *Point Estimation Theory and Its Applications.* Novi Sad, Yugoslavia: Institute of Mathematics, 1997.

Smale, S. "The Fundamental Theorem of Algebra and Complexity Theory." *Bull. Amer. Math. Soc.* **4**, 1–5, 1981.

Approximately Equal

If two quantities A and B are approximately equal, this is written $A \approx B$.

See also DEFINED, EQUAL

Approximately Equal To

APPROXIMATELY EQUAL

Approximation Theory

The mathematical study of how given quantities can be approximated by other (usually simpler) ones under appropriate conditions. Approximation theory also studies the size and properties of the ERROR introduced by approximation. Approximations are often obtained by POWER SERIES expansions in which the higher order terms are dropped.

See also LAGRANGE REMAINDER

References

Achieser, N. I. *Theory of Approximation.* New York: Dover, 1992.

Cheney, E. W. *Introduction to Approximation Theory, 2nd ed.* New York: Chelsea, 1982.

Golomb, M. *Lectures on Theory of Approximation.* Argonne, IL: Argonne National Laboratory, 1962.

Jackson, D. *The Theory of Approximation.* New York: Amer. Math. Soc., 1930.

Natanson, I. P. *Constructive Function Theory, Vol. 1: Uniform Approximation.* New York: Ungar, 1964.

Petrushev, P. P. and Popov, V. A. *Rational Approximation of Real Functions.* New York: Cambridge University Press, 1987.

Rivlin, T. J. *An Introduction to the Approximation of Functions.* New York: Dover, 1981.

Timan, A. F. *Theory of Approximation of Functions of a Real Variable.* New York: Dover, 1994.

Weisstein, E. W. "Books about Approximation Theory." http://www.treasure-troves.com/books/ApproximationTheory.html.

Arakelov Theory

A formal mathematical theory which introduces "components at infinity" by defining a new type of divisor class group of INTEGERS of a NUMBER FIELD.

The divisor class group is called an "arithmetic surface."

See also ARITHMETIC GEOMETRY

Arbelos

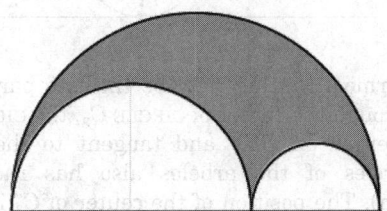

The term "arbelos" means SHOEMAKER'S KNIFE in Greek, and this term is applied to the shaded AREA in the above figure which resembles the blade of a knife used by ancient cobblers (Gardner 1979). Archimedes himself is believed to have been the first mathematician to study the mathematical properties of this figure. The position of the central notch is arbitrary and can be located anywhere along the DIAMETER.

The arbelos satisfies a number of unexpected identities (Gardner 1979, Schoch).

1. Call the diameters of the left and right SEMICIRCLES $r < 1$ and $1 - r$, respectively, so the diameter of the enclosing SEMICIRCLE is 1. Then the arc length along the bottom of the arbelos is

$$L = \pi r + \pi(1 - r) = \pi_1$$

so the arc length along the enclosing semicircle is the same as the arc length along the two smaller semicircles.

2. Draw the PERPENDICULAR BD from the tangent of the two SEMICIRCLES to the edge of the large CIRCLE. Then the AREA of the arbelos is the same as the AREA of the CIRCLE with DIAMETER BD. Let $AC = 1$ and $r = AB$, then simultaneously solve the equations

$$r^2 + h^2 = x^2 \tag{1}$$

$$(1 - r)^2 + h^2 = y^2 \tag{2}$$

$$x^2 + y^2 = 1^2 \tag{3}$$

for the sides

$$x = AD = \sqrt{r} \tag{4}$$

$$y = CD = \sqrt{1 - r} \tag{5}$$

$$h = BD = \sqrt{r(1 - r)}. \tag{6}$$

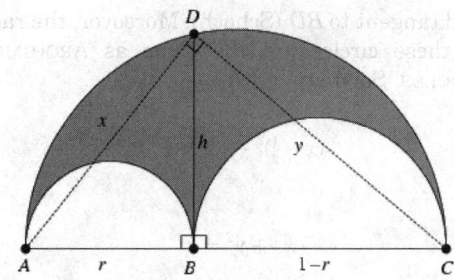

3. The CIRCLES C_1 and C_1' inscribed on each half of BD on the arbelos (called ARCHIMEDES' CIRCLES) each have DIAMETER $(AB)(BC)/(AC)$.

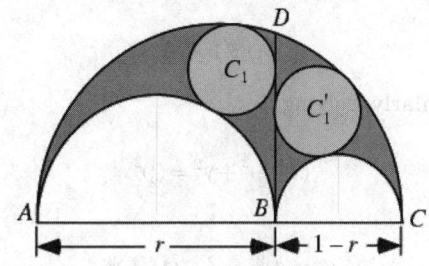

If $AC = 1$ and $AB = r$, then the radius of the Archimedes' circles is

$$R = \tfrac{1}{2}r(1 - r). \tag{7}$$

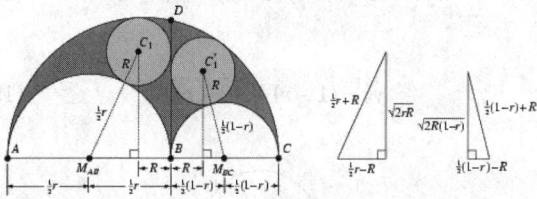

The positions of the circles can be found using the triangles shown above. The lengths of the horizonal legs and hypotenuses are known as indicated, so the vertical legs can be found using the PYTHAGOREAN THEOREM. This then gives the centers of the circles as

$$x_1 = r - R = \tfrac{1}{2}r(1 + r) \tag{8}$$

$$y_1 = \sqrt{2rR} = r\sqrt{1 - r} \tag{9}$$

and

$$x_1' = r + R = \tfrac{1}{2}r(3 - r) \tag{10}$$

$$y_1' = \sqrt{2R(1 - r)} = (1 - r)\sqrt{r}. \tag{11}$$

4. Let A' be the point at which the CIRCLE centered at A and of RADIUS $r = AB$ intersects the enclosing SEMICIRCLE, and let C' be the point at which the CIRCLE centered at C of RADIUS $1 - r = BC$ intersects the enclosing SEMICIRCLE. Then the smallest CIRCLE C_2 passing through A' and tangent to BD is equal to the smallest CIRCLE C_2' passing through C'

and tangent to BD (Schoch). Moreover, the radii R of these circles are the same as ARCHIMEDES' CIRCLES. Solving

$$(x - \tfrac{1}{2})^2 + y^2 = (\tfrac{1}{2})^2 \tag{12}$$

$$x^2 + y^2 = r^2 \tag{13}$$

gives $(x, y) = (r^2, r\sqrt{1-r^2})$, so the center of C_2 is

$$x_2 = r^2 + \tfrac{1}{2}r(1-r) = \tfrac{1}{2}r(r+1) \tag{14}$$

$$y_2 = r\sqrt{1-r^2}. \tag{15}$$

Similarly, solving

$$(x - \tfrac{1}{2})^2 + y^2 = (\tfrac{1}{2})^2 \tag{16}$$

$$(x - 1)^2 + y^2 = (1-r)^2 \tag{17}$$

gives $(x, y) = (r(2-r), (1-r)\sqrt{r(2-r)})$, so the center of C_2' is

$$x_2' = r(2-r) - \tfrac{1}{2}r(1-r) = \tfrac{1}{2}r(r-3) \tag{18}$$

$$y_2' = (1-r)\sqrt{r(2-r)}. \tag{19}$$

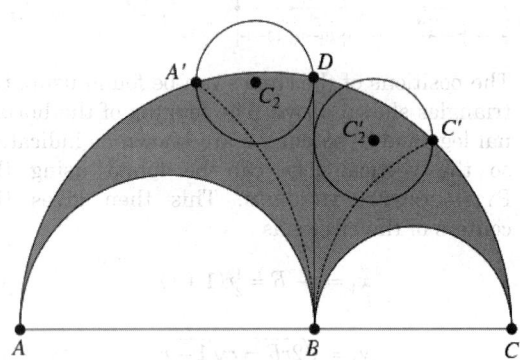

5. The APOLLONIUS CIRCLE C_3 of the circles with arcs BA', BC', and $AA'DC'C$ is located at a position

$$x = \tfrac{1}{2}r(1 + 3r - 2r^2) \tag{20}$$

$$y = r(1-r)\sqrt{(2-r)(1+r)} \tag{21}$$

and has radius R equal to that of ARCHIMEDES' CIRCLES (Schoch), as does the smallest circle C_3' passing through B and tangent to C_3.

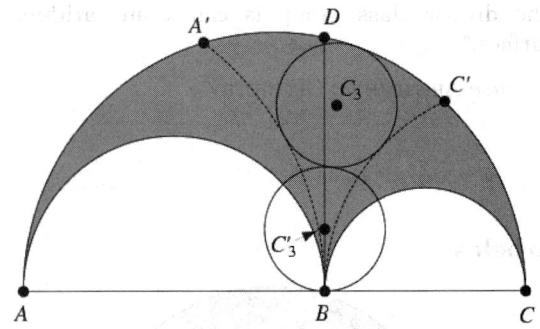

Furthermore, letting $B'D'$ be the line parallel to BD through the center of CIRCLE C_3, the CIRCLE C_3'' with center on $B'D'$ and tangent to the small semicircles of the arbelos also has radius R (Schoch). The position of the center of C_3'' is given by

$$x_3'' = x = \tfrac{1}{2}r(1 + 3r - 2r^2) \tag{22}$$

$$y_3'' = \sqrt{(\tfrac{1}{2}r + R) - (x - \tfrac{1}{2}r)^2}$$

$$= r(1-r)\sqrt{1+r-r^2}. \tag{23}$$

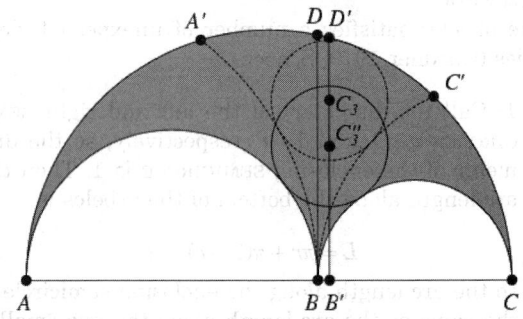

The vertical h' position of D' is

$$h' = \sqrt{\tfrac{1}{4} - \tfrac{1}{4}(2r^3 - 3r^2 - r + 1)^2}$$

$$= \tfrac{1}{2}\sqrt{r(1-r)(2r^2 - 3r - 1)(2r^2 - r - 2)}. \tag{24}$$

6. Let P be the MIDPOINT of AB, and let Q be the MIDPOINT of BC. Then draw the SEMICIRCLE having PQ as a DIAMETER with center M. This CIRCLE has RADIUS

$$R_{PQ} = \tfrac{1}{2}\{1 - \tfrac{1}{2}[r + (1-r)]\} = \tfrac{1}{4}. \tag{25}$$

The smallest circle C_4 through D' touching arc PQ then has radius R (Schoch). Using similar triangles, the center of this circle is at

$$x_4 = \frac{r(2r^4 - 5r^3 + 3r + 1)}{1 + 4r - 4r^2} \tag{26}$$

$$y_4 = \frac{2r^2 - 2r - 1}{2(4r^2 - 4r - 1)}$$

$$\times \sqrt{r(1-r)(2r^2-3r-1)(2r^2-r-2)}. \qquad (27)$$

Similarly, let U be the point of intersection of $B'D'$ and the SEMICIRCLE PQ, then the CIRCLE through B, B', and U also has RADIUS R (Schoch). The center of this CIRCLE is at

$$x'_4 = \tfrac{1}{4}r(3+3r-2r^2) \qquad (28)$$

$$y'_4 = \tfrac{1}{4}r(1-r) \qquad (29)$$

$$\times \sqrt{(2r+1)(3-2r)}.$$

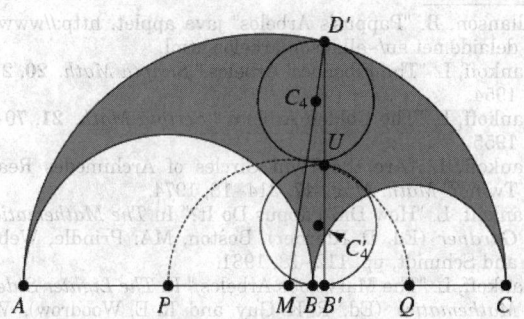

Consider the circle X of RADIUS r_X which is tangent to the two interior semicircles. Its position and radius are obtained by solving the simultaneous equations

$$h^2+z^2=(\tfrac{1}{2}r+r_X)^2 \qquad (30)$$

$$h^2+(\tfrac{1}{2}-z)^2=[\tfrac{1}{2}(1-r)+r_X]^2 \qquad (31)$$

$$(\tfrac{1}{2}r+r_X)^2+[\tfrac{1}{2}(1-r)+r_X]^2=(\tfrac{1}{2})^2. \qquad (32)$$

giving

$$z=\tfrac{1}{4}+\tfrac{1}{4}(2r-1)\sqrt{1+4r-4r^2} \qquad (33)$$

$$h=r(1-r) \qquad (34)$$

$$r_X=\tfrac{1}{4}(\sqrt{1+4r-4r^2}-1). \qquad (35)$$

Letting C''_4 be the smallest CIRCLE through X and tangent to ABC, the radius of C''_4 is therefore $h/2 = r(1-r)/2 = R$ (Schoch), and its center is located at

$$x''_4 = \tfrac{1}{4}+\tfrac{1}{2}r+\tfrac{1}{4}(2r-1)\sqrt{1+4r-4r^2} \qquad (36)$$

$$y''_4 = \tfrac{1}{2}r(1-r). \qquad (37)$$

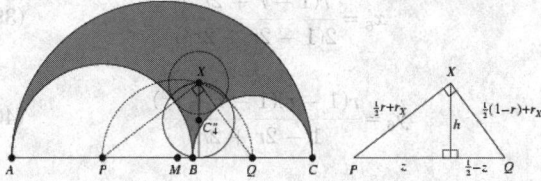

7. Within each small semicircle of an arbelos, construct arbeloses similar to the original. Then the circles C_5 and C'_5 are congruent and have radius R (Schoch). Moreover, connect the midpoints of the arcs and their cusp points to form the RECTANGLES $\square$ $EFGH$ and $\square$ $E'F'G'H'$. Then these rectangles are similar with respect to the point C''_5 (Schoch). This point lies on the line $B'D'$, and the circle with center C''_5 and radius C''_5B' also has radius R, so C''_5 has coordinates $(\tfrac{1}{2}r(1+3r-2r^2), \tfrac{1}{2}r(1-r))$. The following tables summarized the positions of the rectangle vertices.

X	Coordinates	X'	Coordinates
E	$(\tfrac{1}{2}r, \tfrac{1}{2}r)$	E'	$(r(2-r), 0)$
F	$(\tfrac{1}{2}r(1+r), \tfrac{1}{2}r(1-r))$	F'	$(\tfrac{1}{2}r(3-r), \tfrac{1}{2}r(1-r))$
G	$(r^2, 0)$	G'	$(\tfrac{1}{2}(1+r), \tfrac{1}{2}(1-r))$
H	$(\tfrac{1}{2}r^2, \tfrac{1}{2}r^2)$	H'	$(\tfrac{1}{2}(1+2r-r^2), \tfrac{1}{2}(1-r)^2)$

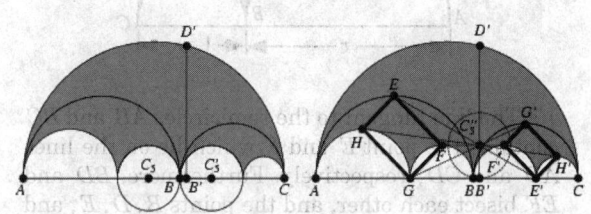

8. Let MM' be the PERPENDICULAR BISECTOR of AC, let B be the cusp of the arbelos and D lie above it, let E and G' be the tops of the large and small semicircles, respectively. Let EG' intersect the lines MM' and BD in points I and J, respectively. Then the smallest circle C_6 passing through I and tangent to arc AC at M', the smallest circle C'_6 through J and tangent to the outside semicircle at P_C, and the circle C''_6 with diameter JB are all equal to the Archimedean circles (Schoch). The circle C''_6 is called the BANKOFF CIRCLE, and is also the CIRCUMCIRCLE of the point B and tangent points P_A and P_C of the first Pappus circle. The centers of the circles C_6, C'_6, and C''_6 are given by

$$x_6 = \tfrac{1}{2}$$

$$y_6 = \tfrac{1}{2}(1-r+r^2) \qquad (38)$$

$$x_6' = \frac{r(1 - r + 2r^2)}{2(1 - 2r + 2r^2)} \quad (39)$$

$$y_6' = \frac{r(1 - r)(1 - r + r^2)}{1 - 2r + 2r^2} \quad (40)$$

$$x_6'' = r \quad (41)$$

$$y_6'' = \tfrac{1}{2}r(1 - r). \quad (42)$$

Rather amazingly, the points E, M, B, G', P_C, D, and M' are CONCYCLIC (Schoch) in a circle with center $((1 + 2r)/4, 1/4)$ and radius

$$R_{EMBG'P_CDM'} = \tfrac{1}{4}\sqrt{2(1 - 2r + 2r^2)}. \quad (43)$$

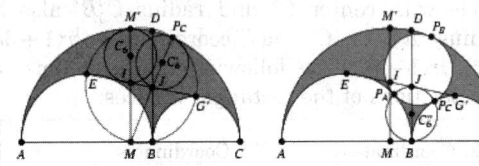

9. The smallest CIRCUMCIRCLE of the Archimedean circles has an area equal to that of the arbelos.

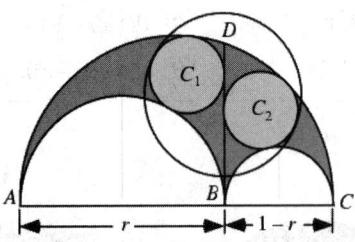

10. The line tangent to the semicircles AB and BC contains the point E and F which lie on the lines AD and CD, respectively. Furthermore, BD and EF bisect each other, and the points B, D, E, and F are CONCYCLIC.

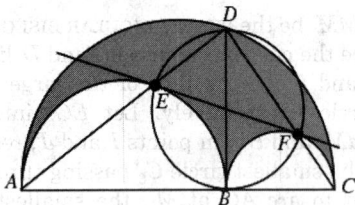

11. Construct a chain of TANGENT CIRCLES starting with the CIRCLE TANGENT to the two small ones and large one (a so-called PAPPUS CHAIN). The centers of the CIRCLES lie on an ELLIPSE, and the DIAMETER of the nth CIRCLE C_n is $((1/n))$th PERPENDICULAR distance to the base of the SEMICIRCLE. This result is most easily proven using INVERSION, but was known to Pappus, who referred to it as an ancient theorem (Hood 1961, Cadwell 1966, Gardner 1979, Bankoff 1981).

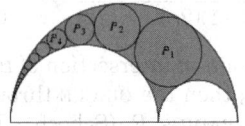 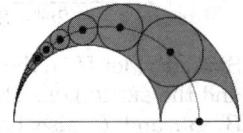

12. If B divides AC in the GOLDEN RATIO ϕ, then the circles in the chain satisfy a number of other special properties (Bankoff 1955).

See also ARCHIMEDES' CIRCLES, BANKOFF CIRCLE, COXETER'S LOXODROMIC SEQUENCE OF TANGENT CIRCLES, GOLDEN RATIO, INVERSION, PAPPUS CHAIN, STEINER CHAIN

References

Allanson, B. "Pappus's Arbelos" java applet. http://www.a-delaide.net.au/~allanson/arbelos.html.

Bankoff, L. "The Fibonacci Arbelos." *Scripta Math.* **20**, 218, 1954.

Bankoff, L. "The Golden Arbelos." *Scripta Math.* **21**, 70–6, 1955.

Bankoff, L. "Are the Twin Circles of Archimedes Really Twins?" *Math. Mag.* **47**, 214–18, 1974.

Bankoff, L. "How Did Pappus Do It?" In *The Mathematical Gardner* (Ed. D. Klarner). Boston, MA: Prindle, Weber, and Schmidt, pp. 112–18, 1981.

Bankoff, L. "The Marvelous Arbelos." In *The Lighter Side of Mathematics* (Ed. R. K. Guy and R. E. Woodrow). Washington, DC: Math. Assoc. Amer., 1994.

Cadwell, J. H. *Topics in Recreational Mathematics.* Cambridge, England: Cambridge University Press, 1966.

Coolidge, J. L. *A Treatise on the Geometry of the Circle and Sphere.* New York: Chelsea, pp. 35–6, 1971.

Dodge, C. W.; Schoch, T.; Woo, P. Y.; and Yiu, P. "Those Ubiquitous Archimedean Circles." *Math. Mag.* **72**, 202–13, 1999.

Gaba, M. G. "On a Generalization of the Arbelos." *Amer. Math. Monthly* **47**, 19–4, 1940.

Gardner, M. "Mathematical Games: The Diverse Pleasures of Circles that Are Tangent to One Another." *Sci. Amer.* **240**, 18–8, Jan. 1979.

Heath, T. L. *The Works of Archimedes with the Method of Archimedes.* New York: Dover, p. 307, 1953.

Hood, R. T. "A Chain of Circles." *Math. Teacher* **54**, 134–37, 1961.

Johnson, R. A. *Modern Geometry: An Elementary Treatise on the Geometry of the Triangle and the Circle.* Boston, MA: Houghton Mifflin, pp. 116–17, 1929.

Ogilvy, C. S. *Excursions in Geometry.* New York: Dover, pp. 54–5, 1990.

Schoch, T. "A Dozen More Arbelos Twins." http://www.bio-la.edu/academics/undergrad/math/woopy/arbel2.htm.

Soddy, F. "The Bowl of Integers and the Hexlet." *Nature* **139**, 77–9, 1937.

Wells, D. *The Penguin Dictionary of Curious and Interesting Geometry.* London: Penguin, pp. 5–, 1991.

Woo, P. "The Arbelos." http://www.biola.edu/academics/undergrad/math/woopy/arbelos.htm.

Yiu, P. "The Archimedean Circles in the Shoemaker's Knife." Lecture at the 31st Annual Meeting of the Florida Section of the Math. Assoc. Amer., Boca Raton, FL, March 6–, 1998.

Arborescence

A DIRECTED GRAPH is called an arborescence if, from a given node x known as the ROOT NODE, there is exactly one elementary path from x to every other node y.

See also ARBORICITY, DIRECTED GRAPH, ROOT NODE

Arboricity

Given a GRAPH G, the arboricity is the MINIMUM number of line-disjoint acyclic SUBGRAPHS whose UNION is G.

See also ANARBORICITY

Arc

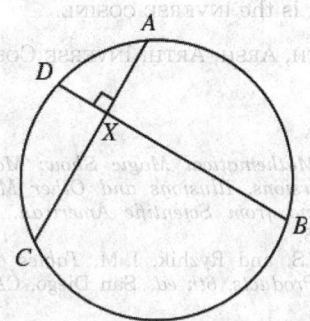

In general, any smooth curve joining two points. In particular, any portion (other than the entire curve) of a CIRCLE or ELLIPSE. As Archimedes proved, for CHORDS AC and BD which are PERPENDICULAR to each other,

$$\text{arc } AB + \text{arc } CD = \text{arc } BC + \text{arc } DA$$

(Wells 1991).

The prefix "arc" is also used to denote the INVERSE FUNCTIONS of TRIGONOMETRIC FUNCTIONS and HYPERBOLIC FUNCTIONS. Finally, any path through a graph which passes through no vertex twice is called an arc (Gardner 1984, p. 96).

See also APPLE, ARC LENGTH, CHORD, CIRCLE-CIRCLE INTERSECTION, CIRCULAR TRIANGLE, FIVE DISKS PROBLEM, FLOWER OF LIFE, LEMON, LENS, PIECEWISE CIRCULAR CURVE, REULEAUX POLYGON, REULEAUX TRIANGLE, SALINON, SEED OF LIFE, TRIANGLE ARCS, VENN DIAGRAM, YIN-YANG

References
Gardner, M. *The Sixth Book of Mathematical Games from Scientific American.* Chicago, IL: University of Chicago Press, 1984.
Wells, D. *The Penguin Dictionary of Curious and Interesting Geometry.* London: Penguin, p. 118, 1991.

Arc Length

Arc length is defined as the length along a curve,

$$s \equiv \int_a^b |dl|. \tag{1}$$

Defining the line element $ds^2 \equiv |dl|^2$, parameterizing the curve in terms of a parameter t, and noting that

ds/dt is simply the magnitude of the VELOCITY with which the end of the RADIUS VECTOR $\mathbf{r}$ moves gives

$$s = \int_a^b ds = \int_a^b \frac{ds}{dt} dt = \int_a^b |\mathbf{r}'(t)| dt. \tag{2}$$

In POLAR COORDINATES,

$$d\ell = \hat{\mathbf{r}}\, dr + r\hat{\theta}\, d\theta = \left(\frac{dr}{d\theta}\hat{\mathbf{r}} + r\hat{\theta}\right) d\theta, \tag{3}$$

so

$$ds = |d\ell| = \sqrt{r^2 + \left(\frac{dr}{d\theta}\right)^2}\, d\theta \tag{4}$$

$$s = \int |d\ell| = \int_{\theta_1}^{\theta_2} \sqrt{r^2 + \left(\frac{dr}{d\theta}\right)^2}\, d\theta. \tag{5}$$

In CARTESIAN COORDINATES,

$$d\ell = dy\hat{\mathbf{x}} + dy\hat{\mathbf{y}} \tag{6}$$

$$ds = |d\ell.d\ell| = \sqrt{dx^2 + dy^2} = \sqrt{\left(\frac{dy}{dx}\right)^2 + 1}\, dx. \tag{7}$$

Therefore, if the curve is written

$$\mathbf{r}(x) = x\hat{\mathbf{x}} + f(x)\hat{\mathbf{y}}, \tag{8}$$

then

$$s = \int_a^b \sqrt{1 + f'^2(x)}\, dx. \tag{9}$$

If the curve is instead written

$$\mathbf{r}(t) = x(t)\hat{\mathbf{x}} + y(t)\hat{\mathbf{y}}, \tag{10}$$

then

$$s = \int_a^b \sqrt{x'^2(t) + y'^2(t)}\, dt. \tag{11}$$

Or, in three dimensions,

$$\mathbf{r}(t) = x(t)\hat{\mathbf{x}} + y(t)\hat{\mathbf{y}} + z(t)\hat{\mathbf{z}}, \tag{12}$$

so

$$s = \int_a^b \sqrt{x'^2(t) + y'^2(t) + z'^2(t)}\, dt. \tag{13}$$

See also CURVATURE, GEODESIC, NORMAL VECTOR, RADIUS OF CURVATURE, RADIUS OF TORSION, SPEED, SURFACE AREA, TANGENTIAL ANGLE, TANGENT VECTOR, TORSION (DIFFERENTIAL GEOMETRY), VELOCITY

Arc Minute

A unit of ANGULAR measure equal to 60 ARC SECONDS, or 1/60 of a DEGREE. The arc minute is denoted ' (not to be confused with the symbol for feet).

See also ARC SECOND, DEGREE

Arc Second

A unit of ANGULAR measure equal to 1/60 of an ARC MINUTE, or 1/3600 of a DEGREE. The arc second is denoted (not to be confused with the symbol for inches).

See also ARC MINUTE, DEGREE

Arccos

INVERSE COSINE

ArcCos

INVERSE COSINE

Arccosecant

INVERSE COSECANT

ArcCosh

INVERSE HYPERBOLIC COSINE

Arccosine

INVERSE COSINE

ArcCot

INVERSE COTANGENT

Arccot

INVERSE COTANGENT

Arccotangent

INVERSE COTANGENT

Arccoth

INVERSE HYPERBOLIC COTANGENT

ArcCoth

INVERSE HYPERBOLIC COTANGENT

ArcCsc

INVERSE COSECANT

Arccsc

INVERSE COSECANT

Arccsch

INVERSE HYPERBOLIC COSECANT

ArcCsch

INVERSE HYPERBOLIC COSECANT

Arch

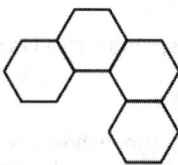

A 4-POLYHEX (Gardner 1978, p. 147).
The term is also used by Gradshteyn and Ryzhik (2000, p. xxx) to denote

$$\text{Arch } z = i \cos^{-1} z,$$

where $\cos^{-1} z$ is the INVERSE COSINE.

See also ARCTH, ARSH, ARTH, INVERSE COSINE

References

Gardner, M. *Mathematical Magic Show: More Puzzles, Games, Diversions, Illusions and Other Mathematical Sleight-of-Mind from Scientific American.* New York: Vintage, 1978.
Gradshteyn, I. S. and Ryzhik, I. M. *Tables of Integrals, Series, and Products, 6th ed.* San Diego, CA: Academic Press, 2000.

Archimedean Dual

The DUALS of the ARCHIMEDEAN SOLIDS, sometimes called the CATALAN SOLIDS, are given in the following table. Hume (1986) gives exact solutions for the side lengths, angles, and DIHEDRAL ANGLES of the Archimedean duals.

n	ARCHIMEDEAN SOLID	DUAL
1	CUBOCTAHEDRON	RHOMBIC DODECAHEDRON
2	GREAT RHOMBICOSIDODECA-HEDRON	DISDYAKIS TRIACONTAHE-DRON
3	GREAT RHOMBICUBOCTAHE-DRON	DISDYAKIS DODECAHEDRON
4	ICOSIDODECAHEDRON	RHOMBIC TRIACONTAHEDRON
5	SMALL RHOMBICOSIDODECA-HEDRON	DELTOIDAL HEXECONTAHE-DRON
6	SMALL RHOMBICUBOCTAHE-DRON	DELTOIDAL ICOSITETRAHE-DRON
7	SNUB CUBE (laevo)	PENTAGONAL ICOSITETRAHE-DRON (dextro)
8	SNUB DODECAHEDRON (lae-vo)	PENTAGONAL HEXECONTAHE-DRON (dextro)
9	TRUNCATED CUBE	SMALL TRIAKIS OCTAHEDRON
10	TRUNCATED DODECAHEDRON	TRIAKIS ICOSAHEDRON
11	TRUNCATED ICOSAHEDRON	PENTAKIS DODECAHEDRON
12	TRUNCATED OCTAHEDRON	TETRAKIS HEXAHEDRON
13	TRUNCATED TETRAHEDRON	TRIAKIS TETRAHEDRON

Here are the Archimedean DUALS (Pearce 1978, Holden 1991) displayed in the order listed above (left to right, then continuing to the next row).

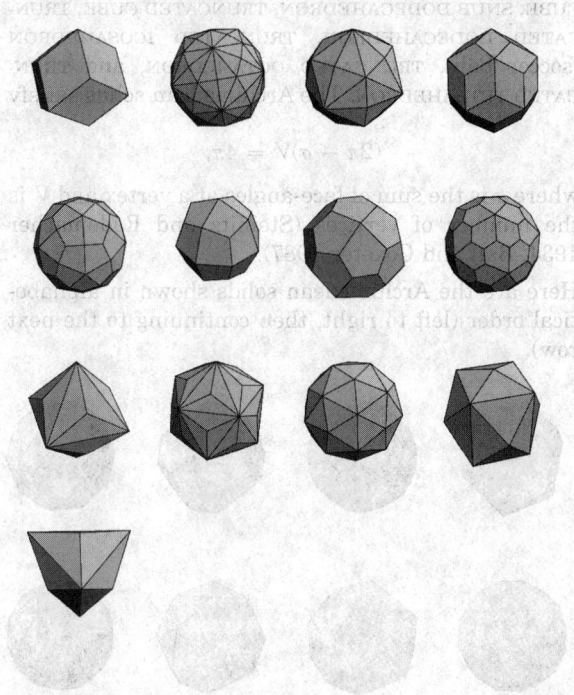

Here are the Archimedean solids paired with their DUALS.

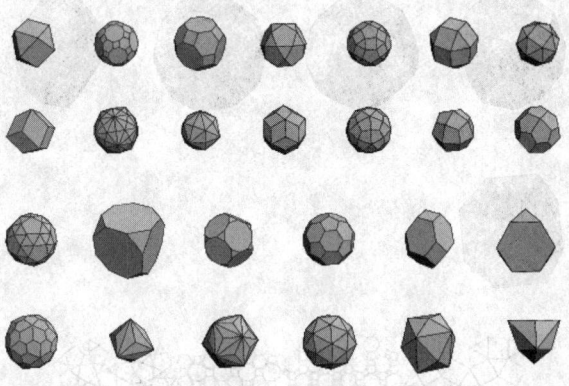

See also ARCHIMEDEAN SOLID, CATALAN SOLID

References

Holden, A. *Shapes, Space, and Symmetry.* New York: Dover, p. 54, 1991.

Hume, A. "Exact Descriptions of Regular and Semi-Regular Polyhedra and Their Duals." *Computing Science Tech. Rep.*, No. 130. Murray Hill, NJ: AT&T Bell Laboratories, 1986.

Pearce, P. *Structure in Nature Is a Strategy for Design.* Cambridge, MA: MIT Press, pp. 34–5, 1978.

Archimedean Solid

The Archimedean solids are convex POLYHEDRA which have a similar arrangement of nonintersecting regular plane CONVEX POLYGONS of two or more different types arranged in the same way about each VERTEX with all sides the same length (Cromwell 1997, pp. 91–2). The Archimedean solids are distinguished from the regular PRISMS and ANTIPRISMS by having very high symmetry, thus excluding solids belonging to a DIHEDRAL GROUP of symmetries (e.g., prisms and antiprisms with unit side lengths) and the ELONGATED SQUARE GYROBICUPOLA (because that surface's symmetry-breaking twist allows vertices "near the equator" and those "in the polar regions" to be distinguished; Cromwell 1997, p. 92). The Archimedean solids are sometimes also referred to as the SEMIREGULAR POLYHEDRA.

Nine of the Archimedean solids can be obtained by TRUNCATION of a PLATONIC SOLID, and two further can be obtained by a second truncation. The remaining two solids, the SNUB CUBE and SNUB DODECAHEDRON, are obtained by moving the faces of a CUBE and DODECAHEDRON outward while giving each face a twist. The resulting spaces are then filled with ribbons of EQUILATERAL TRIANGLES (Wells 1991).

Pugh (1976, p. 25) points out the Archimedean solids are all capable of being circumscribed by a regular TETRAHEDRON so that four of their faces lie on the faces of that TETRAHEDRON. A method of constructing the Archimedean solids using a method known as "expansion" has been enumerated by Stott (Stott 1910; Ball and Coxeter 1987, pp. 139–40).

Let the cyclic sequence $S = (p_1, p_2, \ldots p_q)$ represent the degrees of the faces surrounding a vertex (i.e., S is a list of the number of sides of all polygons surrounding any vertex). Then the definition of an Archimedean solid requires that the sequence must be the same for each vertex to within ROTATION and REFLECTION. Walsh (1972) demonstrates that S represents the degrees of the faces surrounding each vertex of a semiregular convex polyhedron or TESSELLATION of the plane IFF

1. $q \geq 3$ and every member of S is at least 3,
2. $\sum_{i=1}^{q} \frac{1}{pi} \geq \frac{1}{2} q - 1$, with equality in the case of a plane TESSELLATION, and
3. for every ODD NUMBER $p \in S$, S contains a subsequence (b, p, b).

Condition (1) simply says that the figure consists of two or more polygons, each having at least three sides. Condition (2) requires that the sum of interior angles at a vertex must be equal to a full rotation for the figure to lie in the plane, and less than a full rotation for a solid figure to be convex.

The usual way of enumerating the semiregular polyhedra is to eliminate solutions of conditions (1) and (2) using several classes of arguments and then prove that the solutions left are, in fact, semiregular (Kepler 1864, pp. 116–26; Catalan 1865, pp. 25–2; Coxeter 1940, p. 394; Coxeter *et al.* 1954; Lines 1965, pp. 202–03; Walsh 1972). The following table gives all possible regular and semiregular polyhedra and tessellations. In the table, 'P' denotes PLATONIC SOLID, 'M' denotes a PRISM or ANTIPRISM, 'A' denotes an Archimedean solid, and 'T' a plane tessellation.

S	Figure	Solid	SCHLÄFLI SYMBOL
(3, 3, 3)	P	TETRAHEDRON	$\{3, 3\}$
(3, 4, 4)	M	Triangular PRISM	$t^{\{2, 3\}}$
(3, 6, 6)	A	TRUNCATED TETRAHEDRON	$t\{3, 3\}$
(3, 8, 8)	A	TRUNCATED CUBE	$t^{\{4, 3\}}$
(3, 10, 10)	A	TRUNCATED DODECAHE-DRON	$t^{\{5, 3\}}$
(3, 12, 12)	T	(Plane TESSELLATION)	$t^{\{6, 3\}}$
(4, 4, n)	M	n-gonal PRISM	$t^{\{2, n\}}$
(4, 4, 4)	P	CUBE	$\{4, 3\}$
(4, 6, 6)	A	TRUNCATED OCTAHEDRON	$t^{\{3, 4\}}$
(4, 6, 8)	A	GREAT RHOMBICUBOCTA-HEDRON	$t\left\{{3 \atop 4}\right\}$
(4, 6, 10)	A	GREAT RHOMBICOSIDODE-CAHEDRON	$t\left\{{3 \atop 5}\right\}$
(4, 6, 12)	T	(Plane TESSELLATION)	$t\left\{{3 \atop 6}\right\}$
(4, 8, 8)	T	(Plane TESSELLATION)	$t^{\{4, 4\}}$
(5, 5, 5)	P	DODECAHEDRON	$\{5, 3\}$
(5, 6, 6)	A	TRUNCATED ICOSAHEDRON	$t^{\{3, 5\}}$
(6, 6, 6)	T	(Plane TESSELLATION)	$\{6, 3\}$
(3, 3, 3, n)	M	n-gonal ANTIPRISM	$s\left\{{2 \atop n}\right\}$
(3, 3, 3, 3)	P	OCTAHEDRON	$\{3, 4\}$
(3, 4, 3, 4)	A	CUBOCTAHEDRON	$\left\{{3 \atop 4}\right\}$
(3, 5, 3, 5)	A	ICOSIDODECAHEDRON	$\left\{{3 \atop 5}\right\}$
(3, 6, 3, 6)	T	(Plane TESSELLATION)	$\left\{{3 \atop 6}\right\}$
(3, 4, 4, 4)	A	SMALL RHOMBICUBOCTA-HEDRON	$r\left\{{3 \atop 4}\right\}$
(3, 4, 5, 4)	A	SMALL RHOMBICOSIDODE-CAHEDRON	$r\left\{{3 \atop 5}\right\}$
(3, 4, 6, 4)	T	(Plane TESSELLATION)	$r\left\{{3 \atop 6}\right\}$
(4, 4, 4, 4)	T	(Plane TESSELLATION)	$\{4, 4\}$
(3, 3, 3, 3, 3)	P	ICOSAHEDRON	$\{3, 5\}$
(3, 3, 3, 3, 4)	A	SNUB CUBE	$s\left\{{3 \atop 4}\right\}$
(3, 3, 3, 3, 5)	A	SNUB DODECAHEDRON	$s\left\{{3 \atop 5}\right\}$
(3, 3, 3, 3, 6)	T	(Plane TESSELLATION)	$s\left\{{3 \atop 6}\right\}$
(3, 3, 3, 4, 4)	T	(Plane TESSELLATION)	—
(3, 3, 4, 3, 4)	T	(Plane TESSELLATION)	$s\left\{{3 \atop 4}\right\}$
(3, 3, 3, 3, 3)	T	(Plane TESSELLATION)	$\{3, 6\}$

As shown in the above table, there are exactly 13 Archimedean solids (Walsh 1972, Ball and Coxeter 1987). They are called the CUBOCTAHEDRON, GREAT

RHOMBICOSIDODECAHEDRON, GREAT RHOMBICUBOCTA-HEDRON, ICOSIDODECAHEDRON, SMALL RHOMBICOSIDO-DECAHEDRON, SMALL RHOMBICUBOCTAHEDRON, SNUB CUBE, SNUB DODECAHEDRON, TRUNCATED CUBE, TRUN-CATED DODECAHEDRON, TRUNCATED ICOSAHEDRON (soccer ball), TRUNCATED OCTAHEDRON, and TRUN-CATED TETRAHEDRON. The Archimedean solids satisfy

$$(2\pi - \sigma)V = 4\pi,$$

where σ is the sum of face-angles at a vertex and V is the number of vertices (Steinitz and Rademacher 1934, Ball and Coxeter 1987).

Here are the Archimedean solids shown in alphabetical order (left to right, then continuing to the next row).

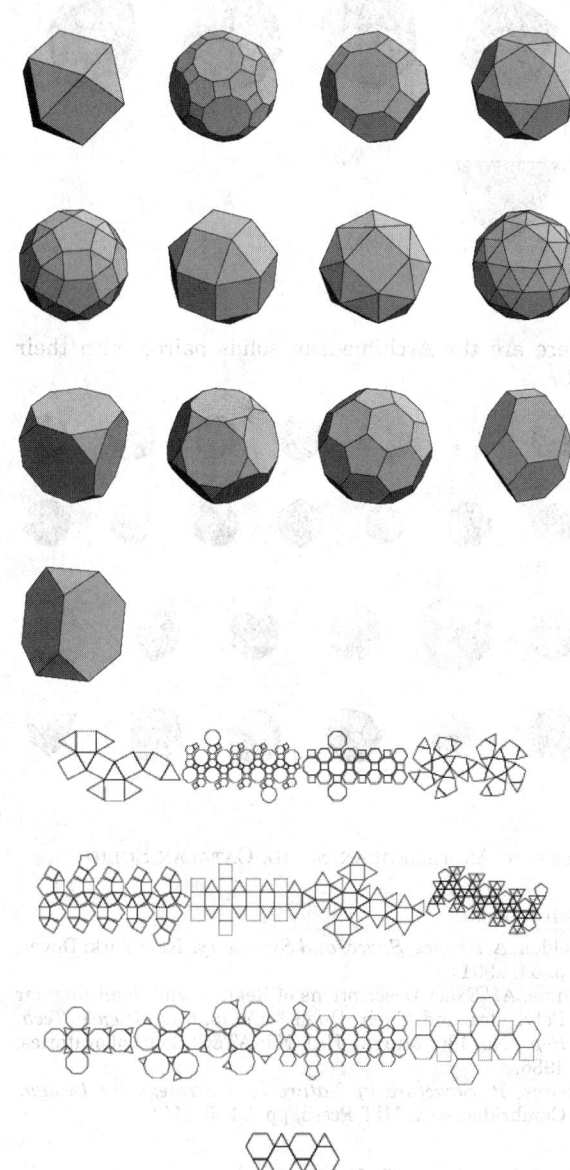

The following table lists the symbols for the Archimedean solids (Wenninger 1989, p. 9).

n	Solid	SCHLÄFLI SYMBOL	WYTHOFF SYMBOL	C&R Symbol
1	CUBOCTAHEDRON	$\left\{{3 \atop 4}\right\}$	2 2|3 4 3 4	$(3.4)^2$
2	GREAT RHOMBICOSIDODECAHEDRON	$t\left\{{3 \atop 5}\right\}$	2 3 5 2|34	
3	GREAT RHOMBICUBOCTAHEDRON	$t\left\{{3 \atop 4}\right\}$	2 3 4 2|34	
4	ICOSIDODECAHEDRON	$\left\{{3 \atop 5}\right\}$	2 2|3 4 3 5	$(3.5)^2$
5	SMALL RHOMBICOSIDODECAHEDRON	$t\left\{{3 \atop 5}\right\}$	3 5 2|34 2	3.4.5.4
6	SMALL RHOMBICUBOCTAHEDRON	$r\left\{{3 \atop 4}\right\}$	3 4 2|34 2	3.4^3
7	SNUB CUBE	$s\left\{{3 \atop 4}\right\}$	2|34 2 3 4	$3^4.4$
8	SNUB DODECAHEDRON	$s\left\{{3 \atop 5}\right\}$	2|34 2 3 5	$3^4.5$
9	TRUNCATED CUBE	$t\{4, 3\}$	2 3 2|34 4	3.8^2
10	TRUNCATED DODECAHEDRON	$t\{5, 3\}$	2 3 2|34 5	3.10^2
11	TRUNCATED ICOSAHEDRON	$t\{3, 5\}$	2 5 2|34 3	5.6^2
12	TRUNCATED OCTAHEDRON	$t\{3, 4\}$	2 4 2|34 3	4.6^2
13	TRUNCATED TETRAHEDRON	$t\{3, 3\}$	2 3 2|34 3	3.6^2

The following table gives the number of vertices v, edges e, and faces f, together with the number of n-gonal faces f_n for the Archimedean solids.

n	Solid	v	e	f	f_3	f_4	f_5	f_6	f_8	f_{10}
1	CUBOCTAHEDRON	12	24	14	8	6				
2	GREAT RHOMBICOSIDODECAHEDRON	120	180	62		30		20		12
3	GREAT RHOMBICUBOCTAHEDRON	48	72	26		12		8	6	
4	ICOSIDODECAHEDRON	30	60	32	20		12			
5	SMALL RHOMBICOSIDODECAHEDRON	60	120	62	20	30	12			
6	SMALL RHOMBICUBOCTAHEDRON	24	48	26	8	18				
7	SNUB CUBE	24	60	38	32	6				
8	SNUB DODECAHEDRON	60	150	92	80		12			
9	TRUNCATED CUBE	24	36	14	8				6	
10	TRUNCATED DODECAHEDRON	60	90	32	20					12
11	TRUNCATED ICOSAHEDRON	60	90	32			12	20		
12	TRUNCATED OCTAHEDRON	24	36	14		6		8		
13	TRUNCATED TETRAHEDRON	12	18	8	4			4		

Let r be the INRADIUS of the dual polyhedron (corresponding to the INSPHERE, which touches the faces of the dual solid), ρ be the MIDRADIUS of both the polyhedron and its dual (corresponding to the MID-SPHERE, which touches the edges of both the polyhedron and its duals), and R the CIRCUMRADIUS

(corresponding to the CIRCUMSPHERE of the solid which touches the vertices of the solid). Since the CIRCUMSPHERE and INSPHERE are dual to each other, they obey the relationship

$$R_r = \rho^2 \qquad (1)$$

(Cundy and Rollett 1989, Table II following p. 144). The following tables give the analytic and numerical values of r, ρ, and R for the Archimedean solids with EDGES of unit length (Coxeter *et al.* 1954; Cundy and Rollett 1989, Table II following p. 144). Hume (1986) gives approximate expressions for the DIHEDRAL ANGLES of the Archimedean solid (and exact expressions for their duals).

n	Solid	r	ρ	R
1	CUBOCTAHEDRON	$\frac{3}{4}$	$\frac{1}{2}\sqrt{3}$	1
2	GREAT RHOMBICOSIDODECAHEDRON	$\frac{1}{241}(105 + 6\sqrt{5}) \times \sqrt{31 + 12\sqrt{5}}$	$\frac{1}{2}\sqrt{30 + 12\sqrt{5}}$	$\frac{1}{2}\sqrt{31 + 12\sqrt{5}}$
3	GREAT RHOMBICUBOCTAHEDRON	$\frac{3}{97}(14 + \sqrt{2}) \times \sqrt{13 + 6\sqrt{2}}$	$\frac{1}{2}\sqrt{12 + 6\sqrt{2}}$	$\frac{1}{2}\sqrt{13 + 6\sqrt{2}}$
4	ICOSIDODECAHEDRON	$\frac{1}{8}(5 + 3\sqrt{5})$	$\frac{1}{2}\sqrt{5 + 2\sqrt{5}}$	$\frac{1}{4}(1 + \sqrt{5})$
5	SMALL RHOMBICOSIDODECAHEDRON	$\frac{1}{41}(15 + 2\sqrt{5}) \times \sqrt{11 + 4\sqrt{5}}$	$\frac{1}{2}\sqrt{10 + 4\sqrt{5}}$	$\frac{1}{2}\sqrt{11 + 4\sqrt{5}}$
6	SMALL RHOMBICUBOCTAHEDRON	$\frac{1}{17}(6 + \sqrt{2}) \times \sqrt{5 + 2\sqrt{2}}$	$\frac{1}{2}\sqrt{4 + 2\sqrt{2}}$	$\frac{1}{2}\sqrt{5 + 2\sqrt{2}}$
7	SNUB CUBE	*	*	*
8	SNUB DODECAHEDRON	*	*	*
9	TRUNCATED CUBE	$\frac{1}{17}(5 + 2\sqrt{2}) \times \sqrt{7 + 4\sqrt{2}}$	$\frac{1}{2}(2 + \sqrt{2})$	$\frac{1}{2}\sqrt{7 + 4\sqrt{2}}$
10	TRUNCATED DODECAHEDRON	$\frac{5}{488}(17\sqrt{2} + 3\sqrt{10}) \times \sqrt{37 + 15\sqrt{5}}$	$\frac{1}{4}(5 + 3\sqrt{5})$	$\frac{1}{4}\sqrt{74 + 30\sqrt{5}}$
11	TRUNCATED ICOSAHEDRON	$\frac{9}{872}(21 + \sqrt{5}) \times \sqrt{58 + 18\sqrt{5}}$	$\frac{3}{4}(1 + \sqrt{5})$	$\frac{1}{4}\sqrt{58 + 18\sqrt{5}}$
12	TRUNCATED OCTAHEDRON	$\frac{9}{20}\sqrt{10}$	$\frac{3}{2}$	$\frac{1}{2}\sqrt{10}$
13	TRUNCATED TETRAHEDRON	$\frac{9}{44}\sqrt{22}$	$\frac{3}{4}\sqrt{2}$	$\frac{1}{4}\sqrt{22}$

*The complicated analytic expressions for the CIRCUMRADII of these solids are given in the entries for the SNUB CUBE and SNUB DODECAHEDRON.

n	Solid	r	ρ	R
1	CUBOCTAHEDRON	0.75	0.86603	1
2	GREAT RHOMBICOSIDODECAHEDRON	3.73665	3.76938	3.80239
3	GREAT RHOMBICUBOCTAHEDRON	2.20974	2.26303	2.31761
4	ICOSIDODECAHEDRON	1.46353	1.53884	1.61803
5	SMALL RHOMBICOSIDODECAHEDRON	2.12099	2.17625	2.23295
6	SMALL RHOMBICUBOCTAHEDRON	1.22026	1.30656	1.39897
7	SNUB CUBE	1.15763	1.24719	1.34371
8	SNUB DODECAHEDRON	2.03969	2.09688	2.15583

9	TRUNCATED CUBE	1.63828	1.70711	1.77882
10	TRUNCATED DODECAHEDRON	2.88526	2.92705	2.96945
11	TRUNCATED ICOSAHEDRON	2.37713	2.42705	2.47802
12	TRUNCATED OCTAHEDRON	1.42302	1.5	1.58114
13	TRUNCATED TETRAHEDRON	0.95940	1.06066	1.17260

The Archimedean solids and their DUALS are all CANONICAL POLYHEDRA. Since the Archimedean solids of convex, the CONVEX HULL of each Archimedean solid is the solid itself.

See also ARCHIMEDEAN SOLID STELLATION, CATALAN SOLID, DELTAHEDRON, ISOHEDRON, JOHNSON SOLID, KEPLER-POINSOT SOLID, PLATONIC SOLID, QUASIREGULAR POLYHEDRON, SEMIREGULAR POLYHEDRON, UNIFORM POLYHEDRON

References

Ball, W. W. R. and Coxeter, H. S. M. *Mathematical Recreations and Essays, 13th ed.* New York: Dover, p. 136, 1987.

Behnke, H.; Bachman, F.; Fladt, K.; and Kunle, H. (Eds.). *Fundamentals of Mathematics, Vol. 2: Geometry.* Cambridge, MA: MIT Press, pp. 269–86, 1974.

Catalan, E. "Mémoire sur la Théorie des Polyèdres." *J. l'École Polytechnique (Paris)* **41**, 1–1, 1865.

Coxeter, H. S. M. "The Pure Archimedean Polytopes in Six and Seven Dimensions." *Proc. Cambridge Phil. Soc.* **24**, 1–, 1928.

Coxeter, H. S. M. "Regular and Semi-Regular Polytopes I." *Math. Z.* **46**, 380–07, 1940.

Coxeter, H. S. M. *Regular Polytopes, 3rd ed.* New York: Dover, 1973.

Coxeter, H. S. M.; Longuet-Higgins, M. S.; and Miller, J. C. P. "Uniform Polyhedra." *Phil. Trans. Roy. Soc. London Ser. A* **246**, 401–50, 1954.

Critchlow, K. *Order in Space: A Design Source Book.* New York: Viking Press, 1970.

Cromwell, P. R. *Polyhedra.* New York: Cambridge University Press, pp. 79–6, 1997.

Cundy, H. and Rollett, A. "Stellated Archimedean Polyhedra." §3.9 in *Mathematical Models, 3rd ed.* Stradbroke, England: Tarquin Pub., pp. 123–28 and Table II following p. 144, 1989.

Fejes Tóth, L. Ch. 4 in *Regular Figures.* Oxford, England: Pergamon Press, 1964.

Holden, A. *Shapes, Space, and Symmetry.* New York: Dover, p. 54, 1991.

Hume, A. "Exact Descriptions of Regular and Semi-Regular Polyhedra and Their Duals." *Computing Science Tech. Rep.*, No. 130. Murray Hill, NJ: AT&T Bell Laboratories, 1986.

Kepler, J. "Harmonice Mundi." *Opera Omnia, Vol. 5.* Frankfurt, pp. 75–34, 1864.

Kraitchik, M. *Mathematical Recreations.* New York: W. W. Norton, pp. 199–07, 1942.

Le, Ha. "Archimedean Solids." http://daisy.uwaterloo.ca/~hqle/Polyhedra/archimedean.html.

Lines, L. *Solid Geometry.* New York: Dover, 1965.

Maehara, H. "On the Sphericity of the Graphs of Semi-Regular Polyhedra." *Discr. Math.* **58**, 311–15, 1986.

Nooshin, H.; Disney, P. L.; and Champion, O. C. "Properties of Platonic and Archimedean Polyhedra." Table 12.1 in "Computer-Aided Processing of Polyhedric Configurations." Ch. 12 in *Beyond the Cube: The Architecture of Space Frames and Polyhedra* (Ed. J. F. Gabriel). New York: Wiley, pp. 360–61, 1997.

Pearce, P. *Structure in Nature Is a Strategy for Design.* Cambridge, MA: MIT Press, pp. 34–5, 1978.

Pedagoguery Software. Poly. http://www.peda.com/poly/.

Pugh, A. *Polyhedra: A Visual Approach.* Berkeley: University of California Press, p. 25, 1976.

Rawles, B. A. "Platonic and Archimedean Solids--Faces, Edges, Areas, Vertices, Angles, Volumes, Sphere Ratios." http://www.intent.com/sg/polyhedra.html.

Robertson, S. A. and Carter, S. "On the Platonic and Archimedean Solids." *J. London Math. Soc.* **2**, 125–32, 1970.

Rorres, C. "Archimedean Solids: Pappus." http://www.mcs.drexel.edu/~crorres/Archimedes/Solids/Pappus.html.

Steinitz, E. and Rademacher, H. *Vorlesungen über die Theorie der Polyheder.* Berlin, p. 11, 1934.

Stott, A. B. "Geometrical Deduction of Semiregular from Regular Polytopes and Space Fillings." *Verhandelingen der Koninklijke Akad. Wetenschappen Amsterdam* **11**, 3–4, 1910.

Vichera, M. "Archimedean Polyhedra." http://alpha.ujep.cz/~vicher/puzzle/telesa/telesa.htm.

Walsh, T. R. S. "Characterizing the Vertex Neighbourhoods of Semi-Regular Polyhedra." *Geometriae Dedicata* **1**, 117–13, 1972.

Weisstein, E. W. "Archimedean Solids with Analytic Vertices." MATHEMATICA NOTEBOOK ARCHIMEDEAN.M.

Wells, D. *The Penguin Dictionary of Curious and Interesting Geometry.* London: Penguin, pp. 6–, 1991.

Wenninger, M. J. "The Thirteen Semiregular Convex Polyhedra and Their Duals." Ch. 2 in *Dual Models.* Cambridge, England: Cambridge University Press, pp. 14–5, 1983.

Wenninger, M. J. *Polyhedron Models.* New York: Cambridge University Press, 1989.

Archimedean Solid Stellation

A large class of POLYHEDRA which includes the DODECADODECAHEDRON and GREAT ICOSIDODECAHEDRON. No complete enumeration (even with restrictive uniqueness conditions) has been worked out. There are at least four stellations of the CUBOCTAHEDRON (Wenninger 1989), although the exact number depends on what type of cells formed by plane intersections are allowed.

There are also many stellations of the Archimedean solid duals. The RHOMBIC DODECAHEDRON has three stellations (Wells 1991, pp. 216–17).

See also ARCHIMEDEAN SOLID, CATALAN SOLID

References

Coxeter, H. S. M.; Longuet-Higgins, M. S.; and Miller, J. C. P. "Uniform Polyhedra." *Phil. Trans. Roy. Soc. London Ser. A* **246**, 401–50, 1954.

Wells, D. *The Penguin Dictionary of Curious and Interesting Geometry.* London: Penguin, 1991.

Wenninger, M. J. "Commentary on the Stellation of the Archimedean Solids." In *Polyhedron Models.* New York: Cambridge University Press, pp. 66–2, 1989.

Archimedean Spiral

A SPIRAL with POLAR equation

$$r = a\theta^{1/n}, \tag{1}$$

where r is the radial distance, θ is the polar angle, and n is a constant which determines how tightly the spiral is "wrapped." The CURVATURE of an Archimedean spiral is given by

$$\kappa = \frac{n\theta^{1-1/n}(1 + n + n^2\theta^2)}{a(1 + n^2\theta^2)^{3/2}}, \tag{2}$$

and the ARC LENGTH by

$$s = a\theta^{1/n}\,_2F_1((2n)^{-1}, -\tfrac{1}{2}; 1 + (2n)^{-1}; -n^2\theta^2), \tag{3}$$

where $_2F_1(a, b; c; x)$ is a HYPERGEOMETRIC FUNCTION. Various special cases are given in the following table.

Name	n
LITUUS	-2
HYPERBOLIC SPIRAL	-1
ARCHIMEDES' SPIRAL	1
FERMAT'S SPIRAL	2

If a fly crawls radially outward along a uniformly spinning disk, the curve it traces with respect to a reference frame in which the disk is at rest is an Archimedean spiral (Steinhaus 1999, p. 137). Furthermore, a heart-shaped frame composed of two arcs of an Archimedean spiral which is fixed to a rotating disk converts uniform rotational motion to uniform back-and-forth motion (Steinhaus 1999, pp. 136–37).

See also ARCHIMEDES' SPIRAL, DAISY, FERMAT'S SPIRAL, HYPERBOLIC SPIRAL, LITUUS, SPIRAL

References

Gray, A. *Modern Differential Geometry of Curves and Surfaces with Mathematica, 2nd ed.* Boca Raton, FL: CRC Press, pp. 90–2, 1997.
Lauwerier, H. *Fractals: Endlessly Repeated Geometric Figures.* Princeton, NJ: Princeton University Press, pp. 59–0, 1991.
Lawrence, J. D. *A Catalog of Special Plane Curves.* New York: Dover, pp. 186 and 189, 1972.
Lockwood, E. H. *A Book of Curves.* Cambridge, England: Cambridge University Press, p. 175, 1967.
MacTutor History of Mathematics Archive. "Spiral of Archimedes." http://www-groups.dcs.st-and.ac.uk/~history/Curves/Spiral.html.
Pappas, T. "The Spiral of Archimedes." *The Joy of Mathematics.* San Carlos, CA: Wide World Publ./Tetra, p. 149, 1989.
Steinhaus, H. *Mathematical Snapshots, 3rd ed.* New York: Dover, pp. 136–37, 1999.

Wells, D. *The Penguin Dictionary of Curious and Interesting Geometry.* London: Penguin, pp. 8–, 1991.

Archimedean Spiral Inverse Curve

The INVERSE CURVE of the ARCHIMEDEAN SPIRAL

$$r = a\theta^{1/n}$$

with INVERSION CENTER at the origin and inversion RADIUS k is the ARCHIMEDEAN SPIRAL

$$r = ka\theta^{1/n}.$$

Archimedean Tessellation

TESSELLATION

Archimedean Valuation

A VALUATION for which $|x| \leq 1$ IMPLIES $|1 + x| \leq C$ for the constant $C = 1$ (independent of x). Such a VALUATION *does not* satisfy the strong TRIANGLE INEQUALITY

$$|x + y| \leq \max(|x|, |y|).$$

Archimedes Algorithm

Successive application of ARCHIMEDES' RECURRENCE FORMULA gives the Archimedes algorithm, which can be used to provide successive approximations to π (PI). The algorithm is also called the BORCHARDT-PFAFF ALGORITHM. Archimedes obtained the first rigorous approximation of π by CIRCUMSCRIBING and INSCRIBING $n = G \cdot 2^k$-gons on a CIRCLE. From ARCHIMEDES' RECURRENCE FORMULA, the CIRCUMFERENCES a and b of the circumscribed and inscribed POLYGONS are

$$a(n) = 2n \tan\left(\frac{\pi}{n}\right) \tag{1}$$

$$b(n) = 2n \sin\left(\frac{\pi}{n}\right), \tag{2}$$

where

$$b(n) < C = 2\pi r = 2\pi \cdot 1 = 2\pi < a(n). \tag{3}$$

For a HEXAGON, $n = 6$ and

$$a_0 \equiv a(6) = 4\sqrt{3} \tag{4}$$

$$b_0 \equiv b(6) = 6, \tag{5}$$

where $a_k \equiv a(6 \cdot 2^k)$. The first iteration of ARCHIMEDES' RECURRENCE FORMULA then gives

$$a_1 = \frac{2 \cdot 6 \cdot 4\sqrt{3}}{6 + 4\sqrt{3}} = \frac{24\sqrt{3}}{3 + 2\sqrt{3}} = 24\left(2 - \sqrt{3}\right) \tag{6}$$

$$b_1 = \sqrt{24\left(2 - \sqrt{3}\right) \cdot 6} = 12\sqrt{2 - \sqrt{3}}$$

$$= 6\left(\sqrt{6} - \sqrt{2}\right). \tag{7}$$

Additional iterations do not have simple closed forms, but the numerical approximations for $k = 0, 1, 2, 3, 4$ (corresponding to 6-, 12-, 24-, 48-, and 96-gons) are

$$3.00000 < \pi < 3.46410 \tag{8}$$

$$3.10583 < \pi < 3.21539 \tag{9}$$

$$3.13263 < \pi < 3.15966 \tag{10}$$

$$3.13935 < \pi < 3.14609 \tag{11}$$

$$3.14103 < \pi < 3.14271. \tag{12}$$

By taking $k = 4$ (a 96-gon) and using strict inequalities to convert irrational bounds to rational bounds at each step, Archimedes obtained the slightly looser result

$$\tfrac{223}{71} = 3.14084\ldots < \pi < \tfrac{22}{7} = 3.14285\ldots. \tag{13}$$

See also Pi

References

Miel, G. "Of Calculations Past and Present: The Archimedean Algorithm." *Amer. Math. Monthly* **90**, 17–5, 1983.
Phillips, G. M. "Archimedes in the Complex Plane." *Amer. Math. Monthly* **91**, 108–14, 1984.

Archimedes' Axiom

An AXIOM actually attributed to Eudoxus (Boyer and Merzbach 1991, pp. 89–0) which states that

$$\frac{a}{b} = \frac{c}{d}$$

IFF the appropriate one of following conditions is satisfied for INTEGERS m and n:

1. If $ma < nb$, then $mc < nd$.
2. If $ma = nb$, then $mc = nd$.
3. If $ma > nb$, then $mc > nd$.

Also known as the continuity axiom or Archimedes' lemma, this axiom survives in the writings of Eudoxus (Boyer and Merzbach 1991). It states that, given two magnitudes having a ratio, one can find a multiple of either which will exceed the other. This principle was the basis for the EXHAUSTION METHOD which Archimedes invented to solve problems of AREA and VOLUME.

Formally, Archimedes' axiom states that if AB and CD are two line segments, then there exist a finite number of points $A_1, A_2, \ldots, A_n$ on $A \cup B$ such that

$$CD \equiv AA_1 \equiv AA_2 \equiv \ldots \equiv A_{n-1}A_n,$$

and B is between A and A_n (Itô 1986, p. 611). A geometry in which Archimedes' lemma does not hold is called a NON-ARCHIMEDEAN GEOMETRY.

See also CONTINUITY AXIOMS, FRACTION, INEQUALITY, NON-ARCHIMEDEAN GEOMETRY

References

Boyer, C. B. and Merzbach, U. C. "The Abacus and Decimal Fractions." *A History of Mathematics, 2nd ed.* New York: Wiley, p. 100, 1991.
Itô, K. (Ed.). §155B and 155D in *Encyclopedic Dictionary of Mathematics, 2nd ed., Vol. 2.* Cambridge, MA: MIT Press, p. 611, 1986.

Archimedes' Cattle Problem

Also called the BOVINUM PROBLEMA. It is stated as follows: "The sun god had a herd of cattle consisting of bulls and cows, one part of which was white, a second black, a third spotted, and a fourth brown. Among the bulls, the number of white ones was one half plus one third the number of the black greater than the brown; the number of the black, one quarter plus one fifth the number of the spotted greater than the brown; the number of the spotted, one sixth and one seventh the number of the white greater than the brown. Among the cows, the number of white ones was one third plus one quarter of the total black cattle; the number of the black, one quarter plus one fifth the total of the spotted cattle; the number of spotted, one fifth plus one sixth the total of the brown cattle; the number of the brown, one sixth plus one seventh the total of the white cattle. What was the composition of the herd?"

Solution consists of solving the simultaneous DIOPHANTINE EQUATIONS in INTEGERS W, X, Y, Z (the number of white, black, spotted, and brown bulls) and w, x, y, z (the number of white, black, spotted, and brown cows),

$$W = \tfrac{5}{6}X + Z \tag{1}$$

$$X = \tfrac{9}{20}Y + Z \tag{2}$$

$$Y = \tfrac{13}{42}W + Z \tag{3}$$

$$w = \tfrac{7}{12}(X + x) \tag{4}$$

$$x = \tfrac{9}{20}(Y + y) \tag{5}$$

$$y = \tfrac{11}{30}(Z + z) \tag{6}$$

$$z = \tfrac{13}{42}(W + w). \tag{7}$$

The smallest solution in INTEGERS is

$$W = 10,366,482 \tag{8}$$

$$X = 7,460,514 \tag{9}$$

$$Y = 7,358,060 \tag{10}$$

$$Z = 4,149,387 \tag{11}$$

$$w = 7,206,360 \qquad (12)$$

$$x = 4,893,246 \qquad (13)$$

$$y = 3,515,820 \qquad (14)$$

$$z = 5,439,213. \qquad (15)$$

A more complicated version of the problem requires that $W + X$ be a SQUARE NUMBER and $Y + Z$ a TRIANGULAR NUMBER. The solution to this PROBLEM are numbers with 206544 or 206545 digits.

References

Amthor, A. and Krumbiegel B. "Das Problema bovinum des Archimedes." *Z. Math. Phys.* **25**, 121–71, 1880.

Archibald, R. C. "Cattle Problem of Archimedes." *Amer. Math. Monthly* **25**, 411–14, 1918.

Beiler, A. H. *Recreations in the Theory of Numbers: The Queen of Mathematics Entertains.* New York: Dover, pp. 249–52, 1966.

Bell, A. H. "Solution to the Celebrated Indeterminate Equation $x^2 - ng^2 = 1$." *Amer. Math. Monthly* **1**, 240, 1894.

Bell, A. H. "'Cattle Problem.' By Archimedes 251 BC." *Amer. Math. Monthly* **2**, 140, 1895.

Bell, A. H. "Cattle Problem of Archimedes." *Math. Mag.* **1**, 163, 1882–884.

Burton, D. M. *Elementary Number Theory, 4th ed.* Boston, MA: Allyn and Bacon, p. 391, 1989.

Calkins, K. G. "Archimedes' *Problema Bovinum.*" http://www2.andrews.edu/~calkins/profess/cattle.htm.

Dickson, L. E. *History of the Theory of Numbers, Vol. 2: Diophantine Analysis.* New York: Chelsea, pp. 342–45, 1952.

Dörrie, H. "Archimedes' *Problema Bovinum*." §1 in *100 Great Problems of Elementary Mathematics: Their History and Solutions.* New York: Dover, pp. 3–, 1965.

Grosjean, C. C. and de Meyer, H. E. "A New Contribution to the Mathematical Study of the Cattle-Problem of Archimedes." In *Constantin Carathéodory: An International Tribute, Vols. 1 and 2* (Ed. T. M. Rassias). Teaneck, NJ: World Scientific, pp. 404–53, 1991.

Merriman, M. "Cattle Problem of Archimedes." *Pop. Sci. Monthly* **67**, 660–65, 1905.

Rorres, C. "The Cattle Problem." http://www.mcs.drexel.edu/~crorres/Archimedes/Cattle/Statement.html.

Stewart, I. "Mathematical Recreations: Counting the Cattle of the Sun." *Sci. Amer.* **282**, 112–13, Apr. 2000.

Vardi, I. "Archimedes' Cattle Problem." *Amer. Math. Monthly* **105**, 305–19, 1998.

Archimedes' Circles

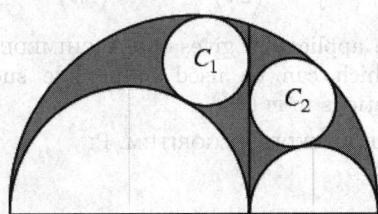

Draw the PERPENDICULAR LINE from the intersection of the two small SEMICIRCLES in the ARBELOS. The two CIRCLES C_1 and C_2 TANGENT to this line, the large

SEMICIRCLE, and each of the two SEMICIRCLES are then congruent and known as Archimedes' circles.

See also ARBELOS, BANKOFF CIRCLE, SEMICIRCLE

Archimedes' Constant

PI

Archimedes' Hat-Box Theorem

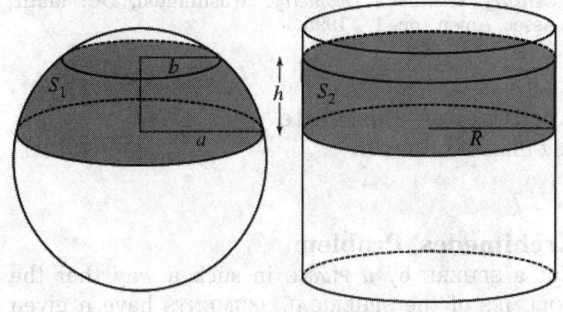

Enclose a SPHERE in a CYLINDER and cut out a SPHERICAL SEGMENT by slicing twice PERPENDICULARLY to the CYLINDER's axis. Then the *lateral* SURFACE AREA of the SPHERICAL SEGMENT S_1 is equal to the *lateral* SURFACE AREA cut out of the CYLINDER S_2 by the same slicing planes, i.e.,

$$S \equiv S_1 = S_2 = 2\pi R h,$$

where R is the RADIUS of the CYLINDER (and tangent SPHERE) and h is the height of the cylindrical (and spherical) segment.

See also ARCHIMEDES' PROBLEM, CYLINDER, SPHERE, SPHERICAL SEGMENT

References

Cundy, H. and Rollett, A. "Sphere and Cylinder--Archimedes' Theorem." §4.3.4 in *Mathematical Models, 3rd ed.* Stradbroke, England: Tarquin Pub., pp. 172–73, 1989.

Archimedes' Lemma

ARCHIMEDES' AXIOM

Archimedes' Midpoint Theorem

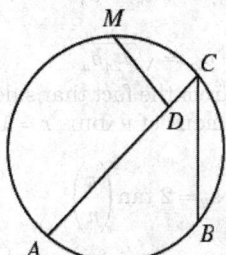

Let M be the MIDPOINT of the ARC AMB. Pick C at random and pick D such that $MD \perp AC$ (where $\perp$

denotes PERPENDICULAR). Then

$$AD = DC + BC.$$

See also MIDPOINT

References

Honsberger, R. *More Mathematical Morsels*. Washington, DC: Math. Assoc. Amer., pp. 31–2, 1991.

Honsberger, R. *Episodes in Nineteenth and Twentieth Century Euclidean Geometry*. Washington, DC: Math. Assoc. Amer., pp. 1–, 1995.

Archimedes' Postulate
ARCHIMEDES' LEMMA

Archimedes' Problem

Cut a SPHERE by a PLANE in such a way that the VOLUMES of the SPHERICAL SEGMENTS have a given RATIO.

See also ARCHIMEDES' HAT-BOX THEOREM, SPHERICAL SEGMENT

Archimedes' Recurrence Formula

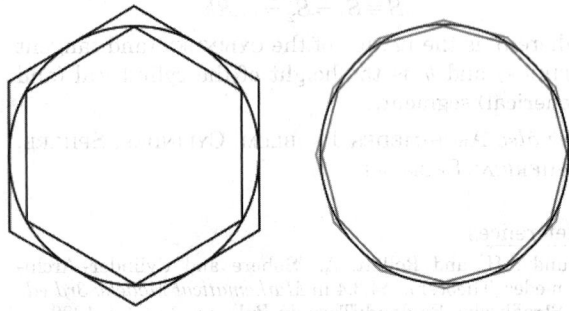

Let a_n and b_n be the PERIMETERS of the CIRCUMSCRIBED and INSCRIBED n-gon and a_{2n} and b_{2n} the PERIMETERS of the CIRCUMSCRIBED and INSCRIBED $2n$-gon. Then

$$a_{2n} = \frac{2a_n b_n}{a_n + b_n} \tag{1}$$

$$b_{2n} = \sqrt{a_{2n} b_n}. \tag{2}$$

The first follows from the fact that side lengths of the POLYGONS on a CIRCLE of RADIUS $r = 1$ are

$$s_R = 2 \tan\left(\frac{\pi}{n}\right) \tag{3}$$

$$s_r = 2 \sin\left(\frac{\pi}{n}\right), \tag{4}$$

so

$$a_n = 2n \tan\left(\frac{\pi}{n}\right) \tag{5}$$

$$b_n = 2n \sin\left(\frac{\pi}{n}\right). \tag{6}$$

But

$$\frac{2a_n b_n}{a_n + b_n} = \frac{2 \cdot 2n \tan\left(\frac{\pi}{n}\right) \cdot 2n \sin\left(\frac{\pi}{n}\right)}{2n \tan\left(\frac{\pi}{n}\right) + 2n \sin\left(\frac{\pi}{n}\right)}$$

$$= 4n \frac{\tan\left(\frac{\pi}{n}\right) \sin\left(\frac{\pi}{n}\right)}{\tan\left(\frac{\pi}{n}\right) + \sin\left(\frac{\pi}{n}\right)}. \tag{7}$$

Using the identity

$$\tan\left(\tfrac{1}{2}x\right) = \frac{\tan x \sin x}{\tan x + \sin x} \tag{8}$$

then gives

$$\frac{2a_n b_n}{a_n + b_n} = 4n \tan\left(\tfrac{\pi}{2n}\right) = a_{2n}. \tag{9}$$

The second follows from

$$\sqrt{a_{2n} b_n} = \sqrt{4n \tan\left(\frac{\pi}{2n}\right) \cdot 2n \sin\left(\frac{\pi}{n}\right)} \tag{10}$$

Using the identity

$$\sin x = 2 \sin\left(\tfrac{1}{2}x\right) \cos\left(\tfrac{1}{2}x\right) \tag{11}$$

gives

$$\sqrt{a_{2n} b_n} = 2n \sqrt{2 \tan\left(\frac{x}{2n}\right) \cdot 2 \sin\left(\frac{\pi}{2n}\right) \cos\left(\frac{\pi}{2n}\right)}$$

$$= 4n \sqrt{\sin^2\left(\frac{\pi}{2n}\right)} = 4n \sin\left(\frac{\pi}{2n}\right) = b_{2n}. \tag{12}$$

Successive application gives the ARCHIMEDES ALGORITHM, which can be used to provide successive approximations to PI (π).

See also ARCHIMEDES ALGORITHM, PI

References

Dörrie, H. *100 Great Problems of Elementary Mathematics: Their History and Solutions*. New York: Dover, p. 186, 1965.

Archimedes' Spiral

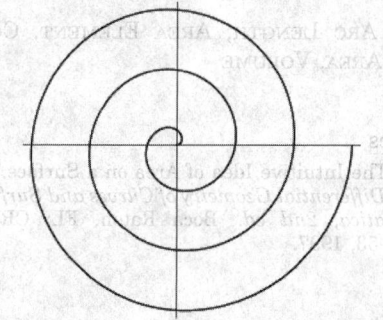

An ARCHIMEDEAN SPIRAL with POLAR equation

$$r = a\theta.$$

This spiral was studied by Conon, and later by Archimedes in *On Spirals* about 225 BC. Archimedes was able to work out the lengths of various tangents to the spiral.

Archimedes' spiral can be used for COMPASS and STRAIGHTEDGE division of an ANGLE into n parts (including ANGLE TRISECTION) and can also be used for CIRCLE SQUARING. In addition, the curve can be used as a cam to convert uniform circular motion into uniform linear motion (Steinhaus 1983, p. 137; Brown). The cam consists of one arch of the spiral above the x-AXIS together with its reflection in the x-AXIS. Rotating this with uniform angular velocity about its center will result in uniform linear motion of the point where it crosses the Y-AXIS.

See also ARCHIMEDEAN SPIRAL

References

Beyer, W. H. *CRC Standard Mathematical Tables, 28th ed.* Boca Raton, FL: CRC Press, p. 225, 1987.

Brown, H. T. *507 Mouvements mécaniques.* Liège, Belgium: Desoer, p. 28, 1923.

Gardner, M. *The Unexpected Hanging and Other Mathematical Diversions.* Chicago, IL: Chicago University Press, pp. 106–07, 1991.

Gray, A. *Modern Differential Geometry of Curves and Surfaces with Mathematica, 2nd ed.* Boca Raton, FL: CRC Press, pp. 90–2, 1997.

Lawrence, J. D. *A Catalog of Special Plane Curves.* New York: Dover, pp. 186–87, 1972.

Lockwood, E. H. *A Book of Curves.* Cambridge, England: Cambridge University Press, pp. 173–64, 1967.

Steinhaus, H. *Mathematical Snapshots, 3rd ed.* New York: Dover, p. 137, 1999.

Archimedes' Spiral Inverse

Taking the ORIGIN as the INVERSION CENTER, ARCHIMEDES' SPIRAL $r = a\theta$ inverts to the HYPERBOLIC SPIRAL $r = a/\theta$.

ArcSec

INVERSE SECANT

Arcsec

INVERSE SECANT

Arcsecant

INVERSE SECANT

ArcSech

INVERSE HYPERBOLIC SECANT

Arcsech

INVERSE HYPERBOLIC SECANT

ArcSin

INVERSE SINE

Arcsin

INVERSE SINE

Arcsine

INVERSE SINE

Arcsinh

INVERSE HYPERBOLIC SINE

ArcSinh

INVERSE HYPERBOLIC SINE

Arctan

INVERSE TANGENT

ArcTan

INVERSE TANGENT

Arctangent

INVERSE TANGENT

Arctangent Integral

INVERSE TANGENT INTEGRAL

Arctanh

INVERSE HYPERBOLIC TANGENT

ArcTanh

INVERSE HYPERBOLIC TANGENT

Arcth

$$\text{Arcth } z = \frac{1}{i} \cot^{-1}(-iz),$$

where $\cot^{-1} z$ is the INVERSE COTANGENT.

See also ARCH, ARSH, ARTH, INVERSE COTANGENT

References

Gradshteyn, I. S. and Ryzhik, I. M. *Tables of Integrals, Series, and Products, 6th ed.* San Diego, CA: Academic Press, p. xxx, 2000.

Arcwise-Connected

See also CONNECTED SET, LOCALLY PATHWISE-CONNECTED, PATH-CONNECTED, PATHWISE-CONNECTED

Arcwise-Connected Set

See also CONNECTED SET, PATH-CONNECTED SET

Area

The AREA of a SURFACE is the amount of material needed to "cover" it completely. The AREA of a TRIANGLE is given by

$$A_\Delta = \tfrac{1}{2}\, lh, \tag{1}$$

where l is the base length and h is the height, or by HERON'S FORMULA

$$A_\Delta = \sqrt{s(s-a)(s-b)(s-c)}, \tag{2}$$

where the side lengths are a, b, and c and s the SEMIPERIMETER. The AREA of a RECTANGLE is given by

$$A_{\text{rectangle}} = ab, \tag{3}$$

where the sides are length a and b. This gives the special case of

$$A_{\text{square}} = a^2 \tag{4}$$

for the SQUARE. The AREA of a REGULAR POLYGON with n sides and side length s is given by

$$A_{n-\text{gon}} = \tfrac{1}{4}\, ns^2 \cot\left(\frac{\pi}{n}\right). \tag{5}$$

CALCULUS and, in particular, the INTEGRAL, are powerful tools for computing the AREA between a curve $f(x)$ and the x-AXIS over an INTERVAL $[a, b]$, giving

$$A = \int_a^b f(x)\, dx. \tag{6}$$

The AREA of a POLAR curve with equation $r = r(\theta)$ is

$$A = \tfrac{1}{2}\int r^2\, d\theta. \tag{7}$$

Written in CARTESIAN COORDINATES, this becomes

$$A = \frac{1}{2}\int \left(x\, \frac{dy}{dt} - y\, \frac{dx}{dt}\right) dt \tag{8}$$

$$= \frac{1}{2}\int (x\, dy - y\, dx). \tag{9}$$

For the AREA of special surfaces or regions, see the entry for that region. The generalization of AREA to 3-

D is called VOLUME, and to higher DIMENSIONS is called CONTENT.

See also ARC LENGTH, AREA ELEMENT, CONTENT, SURFACE AREA, VOLUME

References

Gray, A. "The Intuitive Idea of Area on a Surface." §15.3 in *Modern Differential Geometry of Curves and Surfaces with Mathematica, 2nd ed.* Boca Raton, FL: CRC Press, pp. 351–53, 1997.

Area Element

The area element for a SURFACE with RIEMANNIAN METRIC

$$ds^2 = E\, du^2 + 2F\, du\, dv + G\, dv^2$$

is

$$dA = \sqrt{EG - F^2}\, du \wedge dv,$$

where $du \wedge dv$ is the WEDGE PRODUCT.

See also AREA, LINE ELEMENT, RIEMANNIAN METRIC, VOLUME ELEMENT

References

Gray, A. "The Intuitive Idea of Area on a Surface." §15.3 in *Modern Differential Geometry of Curves and Surfaces with Mathematica, 2nd ed.* Boca Raton, FL: CRC Press, pp. 351–53, 1997.

Area Integral

A double integral over three coordinates giving the AREA within some region R,

$$A = \int\!\!\int_R dx\, dy.$$

If a plane curve is given by $y = f(x)$, then the area between the curve and the x-AXIS from $x = a$ to $x = b$ is given by

$$A = \int_a^b f(x)dx.$$

See also INTEGRAL, LINE INTEGRAL, LUSIN AREA INTEGRAL, MULTIPLE INTEGRAL, SURFACE INTEGRAL, VOLUME INTEGRAL

Area Principle

There are at least two results known as "the area principle."

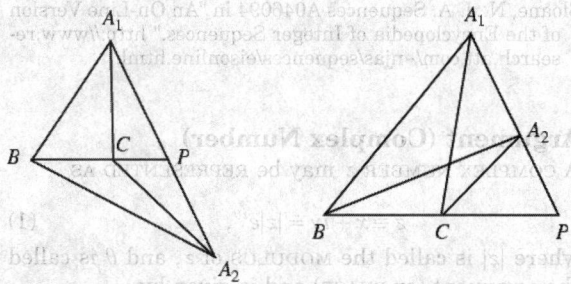

The geometric area principle states that

$$\frac{|A_1P|}{|A_2P|} = \frac{|A_1BC|}{|A_2BC|}. \tag{1}$$

This can also be written in the form

$$\left[\frac{|A_1P|}{|A_2P|}\right] = \left[\frac{|A_1BC|}{|A_2BC|}\right], \tag{2}$$

where

$$\left[\frac{AB}{CD}\right] \tag{3}$$

is the ratio of the lengths [A, B] and [C, D] for AB∥CD with a PLUS or MINUS SIGN depending on if these segments have the same or opposite directions, and

$$\left[\frac{ABC}{DEF}\right] \tag{4}$$

is the RATIO of signed AREAS of the TRIANGLES. Grünbaum and Shepard (1995) show that CEVA'S THEOREM, HOEHN'S THEOREM, and MENELAUS' THEOREM are the consequences of this result.

The area principle of complex analysis states that if f is a SCHLICHT FUNCTION and if

$$h(z) = \frac{1}{f(z)} = \frac{1}{z} + \sum_{j=0}^{\infty} b_j z^j, \tag{5}$$

then

$$\sum_{j=1}^{\infty} j|b_j|^2 \le 1 \tag{6}$$

(Krantz 1999, p. 150).

See also CEVA'S THEOREM, HOEHN'S THEOREM, ME-NELAUS' THEOREM, SCHLICHT FUNCTION, SELF-TRANS-VERSALITY THEOREM

References

Grünbaum, B. and Shepard, G. C. "Ceva, Menelaus, and the Area Principle." *Math. Mag.* **68**, 254–68, 1995.

Krantz, S. G. "Schlicht Functions." §12.1.1 in *Handbook of Complex Analysis.* Boston, MA: Birkhäuser, p. 149, 1999.

Areal Coordinates

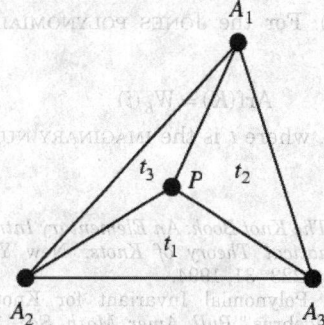

BARYCENTRIC COORDINATES (t_1, t_2, t_3) normalized so that they become the AREAS of the TRIANGLES PA_1A_2, PA_1A_3, and PA_2A_3, where P is the point whose coordinates have been specified, *normalized by the area of the original triangle* $\Delta A_1A_2A_3$. This is equivalent to application of the normalization relation

$$t_1 + t_2 + t_3 = 1$$

(Coxeter 1969, p. 218).

See also BARYCENTRIC COORDINATES, TRILINEAR CO-ORDINATES

References

Coxeter, H. S. M. *Introduction to Geometry, 2nd ed.* New York: Wiley, p. 218, 1969.

Area-Preserving Map

A MAP F from $\mathbb{R}^n$ to $\mathbb{R}^n$ is AREA-preserving if

$$m(F(A)) = m(A)$$

for every subregion A of $\mathbb{R}^n$, where $m(A)$ is the n-D MEASURE of A. A linear transformation is AREA-preserving if its corresponding DETERMINANT is equal to 1.

See also CONFORMAL MAP, SYMPLECTIC MAP

Arf Invariant

A LINK invariant which always has the value 0 or 1. A KNOT has ARF INVARIANT 0 if the KNOT is "pass equivalent" to the UNKNOT and 1 if it is pass equivalent to the TREFOIL KNOT. If K_+, K_-, and L are projections which are identical outside the region of the crossing diagram, and K_+ and K_- are KNOTS while l is a 2-component LINK with a nonintersecting crossing diagram where the two left and right strands belong to the different LINKS, then

$$a(K_+) = a(K_-) + l(L_1, L_2), \tag{1}$$

where l is the LINKING NUMBER of L_1 and L_2. The Arf invariant can be determined from the ALEXANDER POLYNOMIAL or JONES POLYNOMIAL for a KNOT. For Δ_K the ALEXANDER POLYNOMIAL of K, the Arf invariant is given by

$$\Delta_K(-1) \equiv \begin{cases} 1(\bmod 8) & \text{if } \mathrm{Arf}(K) = 0 \\ 5(\bmod 8) & \text{if } \mathrm{Arf}(K) = 1 \end{cases} \qquad (2)$$

(Jones 1985). For the JONES POLYNOMIAL W_K of a KNOT K,

$$\mathrm{Arf}(K) = W_K(i) \qquad (3)$$

(Jones 1985), where i is the IMAGINARY NUMBER.

References

Adams, C. C. *The Knot Book: An Elementary Introduction to the Mathematical Theory of Knots.* New York: W. H. Freeman, pp. 223–31, 1994.

Jones, V. "A Polynomial Invariant for Knots via von Neumann Algebras." *Bull. Amer. Math. Soc.* **12**, 103–11, 1985.

Weisstein, E. W. "Knots." MATHEMATICA NOTEBOOK KNOTS.M.

Arg

ARGUMENT (COMPLEX NUMBER)

Argand Diagram

A plot of COMPLEX NUMBERS as points

$$z = x + iy$$

using the X-AXIS as the REAL AXIS and Y-AXIS as the IMAGINARY AXIS. An Argand diagram is also called the COMPLEX PLANE or ARGAND PLANE. The Argand plane was described by C. Wessel prior to Argand.

See also COMPLEX PLANE, IMAGINARY NUMBER, REAL NUMBER

References

Argand, R. *Essai sur une manière de représenter les quantités imaginaires dans les constructions géométriques.* Paris: Albert Blanchard, 1971. Reprint of the 2nd ed., published by G. J. Hoel in 1874. First edition published Paris, 1806.

Argand Plane

ARGAND DIAGRAM

Argoh's Conjecture

Let B_k be the kth BERNOULLI NUMBER. Then does

$$nB_{n-1} \equiv -1 \ (\bmod n)$$

IFF n is PRIME? For example, for $n = 1, 2, ..., nB_{n-1}$ $(\bmod n)$ is 0, -1, -1, 0, -1, 0, -1, 0, -3, 0, -1, ... (Sloane's A046094). There are no counterexamples less than $n = 5, 600$. Any counterexample to Argoh's conjecture would be a contradiction to GIUGA'S CONJECTURE, and vice versa.

See also BERNOULLI NUMBER, GIUGA'S CONJECTURE

References

Borwein, D.; Borwein, J. M.; Borwein, P. B.; and Girgensohn, R. "Giuga's Conjecture on Primality." *Amer. Math. Monthly* **103**, 40–0, 1996.

Sloane, N. J. A. Sequences A046094 in "An On-Line Version of the Encyclopedia of Integer Sequences." http://www.research.att.com/~njas/sequences/eisonline.html.

Argument (Complex Number)

A COMPLEX NUMBER z may be REPRESENTED AS

$$z \equiv x + iy = |z|e^{i\theta}, \qquad (1)$$

where $|z|$ is called the MODULUS of z, and θ is called the argument (or PHASE) and is given by

$$\arg(x + iy) \equiv \tan^{-1}\left(\frac{y}{x}\right). \qquad (2)$$

Here, θ, sometimes also denoted ϕ, corresponds to the counterclockwise ANGLE from the POSITIVE REAL AXIS, i.e., the value of θ such that $x = \cos\theta$ and $y = \sin\theta$. The special kind of INVERSE TANGENT used here takes into account the quadrant in which z lies and is returned by the FORTRAN command ATAN2(X,Y) and the *Mathematica* command ArcTan[x, y], and is often restricted to the range $-\pi < \theta \le \pi$. In the degenerate case when $x = 0$,

$$\phi = \begin{cases} -\frac{1}{2}\pi & \text{if } y < 0 \\ \text{undefined} & \text{if } y = 0 \\ \frac{1}{2}\pi & \text{if } y > 0. \end{cases} \qquad (3)$$

From the definition of the argument,

$$\arg(zw) = \arg(|z|e^{i\theta_z}|w|e^{i\theta_w}) = \arg(e^{i\theta_z}e^{i\theta_w})$$
$$= \arg\left[e^{i(\theta_z + \theta_w)}\right] = \arg(z) + \arg(w). \qquad (4)$$

Extending this procedure gives

$$\arg(z^n) = n\arg(z). \qquad (5)$$

The argument of a COMPLEX NUMBER is sometimes called the PHASE.

See also AFFIX, COMPLEX NUMBER, DE MOIVRE'S IDENTITY, EULER FORMULA, IMAGINARY PART, INVERSE TANGENT, MODULUS (COMPLEX NUMBER), PHASE, PHASOR, REAL PART

References

Abramowitz, M. and Stegun, C. A. (Eds.). *Handbook of Mathematical Functions with Formulas, Graphs, and Mathematical Tables, 9th printing.* New York: Dover, p. 16, 1972.

Krantz, S. G. "The Argument of a Complex Number." §1.2.6 n *Handbook of Complex Analysis.* Boston, MA: Birkhäuser, p. 11, 1999.

Silverman, R. A. *Introductory Complex Analysis.* New York: Dover, 1984.

Argument (Elliptic Integral)

Given an AMPLITUDE ϕ in an ELLIPTIC INTEGRAL, the argument u is defined by the relation

$$\phi \equiv \mathrm{am}\ u.$$

See also AMPLITUDE, ELLIPTIC INTEGRAL

Argument (Function)

An argument of a FUNCTION $f(x_1, \ldots, x_n)$ is one of the n parameters on which the function's value depends. For example, the SINE $\sin x$ is a one-argument function, the BINOMIAL COEFFICIENT $\binom{n}{m}$ is a two-argument function, and the HYPERGEOMETRIC FUNCTION $2F_1(a, b; c; z)$ is a four-argument function.

Argument Addition Relation

A mathematical relationship relating $f(x+y)$ to $f(x)$ and $f(y)$.

See also ARGUMENT MULTIPLICATION RELATION, RECURRENCE RELATION, REFLECTION RELATION, TRANSLATION RELATION

Argument Multiplication Relation

A mathematical relationship relating $f(nx)$ to $f(x)$ for INTEGER n.

See also ARGUMENT ADDITION RELATION, RECURRENCE RELATION, REFLECTION RELATION, TRANSLATION RELATION

Argument Principle

If $f(z)$ is MEROMORPHIC in a region R enclosed by a CONTOUR γ, let N be the number of COMPLEX ROOTS of $f(z)$ in γ, and P be the number of POLES in γ, then

$$N - P = \frac{1}{2\pi i} \int_\gamma \frac{f'(z)\, dz}{f(z)}$$

Defining $w \equiv f(z)$ and $\sigma \equiv f(\gamma)$ gives

$$N - P = \frac{1}{2\pi i} \int_\sigma \frac{dw}{w}.$$

See also CAUCHY INTEGRAL FORMULA, CAUCHY INTEGRAL THEOREM, HURWITZ'S ROOT THEOREM, MEROMORPHIC FUNCTION, POLE, ROOT, ROUCHÉ'S THEOREM, VARIATION OF ARGUMENT

References
Duren, P.; Hengartner, W.; and Laugessen, R. S. "The Argument Principle for Harmonic Functions." *Math. Mag.* **103**, 411–15, 1996.
Knopp, K. *Theory of Functions, Parts I and II.* New York: Dover, pp. 132–34, 1996.
Krantz, S. G. "The Argument Principle." Ch. 5 in *Handbook of Complex Analysis.* Boston, MA: Birkhäuser, pp. 69–8, 1999.

Argument Variation

VARIATION OF ARGUMENT

Aristotle's Wheel Paradox

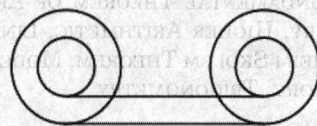

A PARADOX mentioned in the Greek work *Mechanica*, dubiously attributed to Aristotle. Consider the above diagram depicting a wheel consisting of two concentric CIRCLES of different DIAMETERS (a wheel within a wheel). there is a 1:1 correspondence of points on the large CIRCLE with points on the small CIRCLE, so the wheel should travel the same distance regardless of whether it is rolled from left to right on the top straight line or on the bottom one. this seems to imply that the two CIRCUMFERENCES of different sized CIRCLES are equal, which is impossible. The fallacy lies in the assumption that a 1:1 correspondence of points means that two curves must have the same length. In fact, the CARDINALITIES of points in a LINE SEGMENT of any length (or even an INFINITE LINE, a PLANE, a 3-D SPACE, or an infinite dimensional EUCLIDEAN SPACE) are all the same: $\aleph_1$ (ALEPH-1), so the points of any of these can be put in a ONE-TO-ONE correspondence with those of any other.

See also ZENO'S PARADOXES

References
Ballew, D. "The Wheel of Aristotle." *Math. Teacher* **65**, 507–09, 1972.
Costabel, P. "The Wheel of Aristotle and French Consideration of Galileo's Arguments." *Math. Teacher* **61**, 527–34, 1968.
Drabkin, I. "Aristotle's Wheel: Notes on the History of the Paradox." *Osiris* **9**, 162–98, 1950.
Gardner, M. *Wheels, Life, and other Mathematical Amusements.* New York: W. H. Freeman, pp. 2–, 1983.
Pappas, T. "The Wheel of Paradox Aristotle." *The Joy of Mathematics.* San Carlos, CA: Wide World Publ./Tetra, p. 202, 1989.
vos Savant, M. *The World's Most Famous Math Problem.* New York: St. Martin's Press, pp. 48–0, 1993.

Arithmetic

The branch of mathematics dealing with INTEGERS or, more generally, numerical computation. Arithmetical operations include ADDITION, CONGRUENCE calculation, DIVISION, FACTORIZATION, MULTIPLICATION, POWER computation, ROOT EXTRACTION, and SUBTRACTION. Arithmetic was part of the QUADRIVIUM taught in medieval universities.

The FUNDAMENTAL THEOREM OF ARITHMETIC, also called the UNIQUE FACTORIZATION THEOREM, states that any POSITIVE INTEGER can be represented in exactly one way as a PRODUCT of PRIMES.

The LÖWENHEIM-SKOLEM THEOREM, which is a fundamental result in MODEL THEORY, establishes the existence of "nonstandard" models of arithmetic.

See also ALGEBRA, CALCULUS, FLOATING-POINT AR-
ITHMETIC, FUNDAMENTAL THEOREM OF ARITHMETIC,
GROUP THEORY, HIGHER ARITHMETIC, LINEAR ALGE-
BRA, LÖWENHEIM-SKOLEM THEOREM, MODEL THEORY,
NUMBER THEORY, TRIGONOMETRY

References

Karpinski, L. C. *The History of Arithmetic*. Chicago, IL:
    Rand, McNally, & Co., 1925.
Maxfield, J. E. and Maxfield, M. W. *Abstract Algebra and
    Solution by Radicals*. Philadelphia, PA: Saunders, 1992.
Thompson, J. E. *Arithmetic for the Practical Man*. New
    York: Van Nostrand Reinhold, 1973.
Weisstein, E. W. "Books about Arithmetic." http://www.trea-
    sure-troves.com/books/Arithmetic.html.

Arithmetic Function

A function $\psi(n)$ such that

$$\psi(n + m) = \psi(\psi(n) + \psi(m))$$

and

$$\psi(n, m) = \psi(\psi(n)\psi(m)).$$

See also ARITHMETICAL FUNCTION

References

Atanassov, K. *Bull. Number Th.* **9**, 18, 1985.
Trott, M. "Numerical Computations." §1.2.1 in *The Mathe-
    matica Guidebook, Vol. 1: Programming in Mathematica*.
    New York: Springer-Verlag, 2000.

Arithmetic Geometry

A vaguely defined branch of mathematics dealing
with VARIETIES, the MORDELL CONJECTURE, ARAKE-
LOV THEORY, and ELLIPTIC CURVES.

References

Cornell, G. and Silverman, J. H. (Eds.). *Arithmetic Geome-
    try*. New York: Springer-Verlag, 1986.
Lorenzini, D. *An Invitation to Arithmetic Geometry*. Provi-
    dence, RI: Amer. Math. Soc., 1996.

Arithmetic Mean

For a CONTINUOUS DISTRIBUTION FUNCTION, the ar-
ithmetic mean of the population, denoted μ, $\tilde{x}$, $\langle x \rangle$, or
$A(x)$, is given by

$$\mu = \langle f(x) \rangle \equiv \int_{-\infty}^{\infty} P(x) f(x)\, dx, \qquad (1)$$

where $\langle x \rangle$ is the EXPECTATION VALUE. For a DISCRETE
DISTRIBUTION,

$$\mu = \langle f(x) \rangle \equiv \frac{\sum_{n=0}^{N} P(x_n) f(x_n)}{\sum_{n=0}^{N} P(x_n)} = \sum_{n=0}^{N} P(x_n) f(x_n). \qquad (2)$$

The population mean satisfies

$$\langle f(x) + g(x) \rangle = \langle f(x) \rangle + \langle g(x) \rangle \qquad (3)$$

$$\langle cf(x) \rangle = c \langle f(x) \rangle, \qquad (4)$$

and

$$\langle f(x) g(y) \rangle = \langle f(x) \rangle \, \langle g(y) \rangle \qquad (5)$$

if x and y are INDEPENDENT STATISTICS. The "sample
mean," which is the mean estimated from a statistical
sample, is an UNBIASED ESTIMATOR for the population
mean.

For small samples, the mean is more efficient than
the MEDIAN and approximately $\pi/2$ less (Kenney and
Keeping 1962, p. 211). A general expression which
often holds approximately is

$$\text{mean} - \text{mode} \approx 3(\text{mean} - \text{median}). \qquad (6)$$

Given a set of samples $\{x_i\}$, the arithmetic mean is

$$A(x) \equiv \tilde{x} \equiv \mu \equiv \langle x \rangle = \frac{1}{N} \sum_{i=1}^{N} x_i. \qquad (7)$$

Hoehn and Niven (1985) show that

$$\begin{aligned} A(a_1 + c,\ a_2 + c,\ \ldots,\ a_n + c) \\ = c + A(a_1, a_2, \ldots, a_n) \end{aligned} \qquad (8)$$

for any POSITIVE constant c. For positive arguments,
the arithmetic mean satisfies

$$A \geq G \geq H, \qquad (9)$$

where G is the GEOMETRIC MEAN and H is the
HARMONIC MEAN (Hardy *et al.* 1952; Mitrinovic
1970; Beckenbach and Bellman 1983; Bullen *et al.*
1988; Mitrinovic *et al.* 1993; Alzer 1996). This can be
shown as follows. For $a, b > 0$,

$$\left(\frac{1}{\sqrt{a}} - \frac{1}{\sqrt{b}} \right)^2 \geq 0 \qquad (10)$$

$$\frac{1}{a} - \frac{2}{\sqrt{ab}} + \frac{1}{b} \geq 0 \qquad (11)$$

$$\frac{1}{a} + \frac{1}{b} \geq \frac{2}{\sqrt{ab}} \qquad (12)$$

$$\sqrt{ab} \geq \frac{2}{\frac{1}{a} + \frac{1}{b}} \qquad (13)$$

$$G \geq H, \qquad (14)$$

with equality IFF $b = a$. To show the second part of
the inequality,

$$(\sqrt{a} - \sqrt{b})^2 = a - 2\sqrt{ab} + b \geq 0 \qquad (15)$$

$$\frac{a + b}{2} \geq \sqrt{ab} \qquad (16)$$

$$A \geq G, \qquad (17)$$

with equality IFF $a = b$. Combining (14) and (17) then
gives (9).

Given n independent random GAUSSIAN DISTRIBUTED variates x_i, each with population mean $\mu_i = \mu$ and VARIANCE $\sigma_i^2 = \sigma^2$,

$$\tilde{x} \equiv \frac{1}{N} \sum_{i=1}^{N} x_i \tag{18}$$

$$\langle x \rangle = \frac{1}{N} \left\langle \sum_{i=1}^{N} x_i \right\rangle = \frac{1}{N} \sum_{i=1}^{N} \langle x_i \rangle$$

$$= \frac{1}{N} \sum_{i=1}^{N} \mu = \frac{1}{N}(N\mu) = \mu, \tag{19}$$

so the sample mean is an UNBIASED ESTIMATOR of population mean. However, the distribution of $\tilde{x}$ depends on the sample size. For large samples, $\tilde{x}$ is approximately NORMAL. For small samples, STUDENT'S T-DISTRIBUTION should be used.

The VARIANCE of the sample mean is independent of the distribution.

$$\operatorname{var}(\tilde{x}) = \operatorname{var}\left(\frac{1}{n} \sum_{i=1}^{N} x_i\right) = \frac{1}{N^2} \operatorname{var}\left(\sum_{i=1}^{N} x_i\right)$$

$$= \frac{1}{N^2} \sum_{i=1}^{n} \operatorname{var}(x_i) = \left(\frac{1}{N^2}\right) \sum_{i=1}^{N} \sigma^2 = \frac{\sigma^2}{N}. \tag{20}$$

From K-STATISTIC for a GAUSSIAN DISTRIBUTION, the UNBIASED ESTIMATOR for the VARIANCE is given by

$$\sigma^2 = \frac{N}{N-1} s^2, \tag{21}$$

where

$$s \equiv \frac{1}{N} \sum_{i=1}^{N} (x_i - \bar{x})^2, \tag{22}$$

so

$$\operatorname{var}(\tilde{x}) = \frac{s^2}{N-1}. \tag{23}$$

The SQUARE ROOT of this,

$$\sigma_x = \frac{s}{\sqrt{N-1}}, \tag{24}$$

is called the STANDARD ERROR.

$$\operatorname{var}(\tilde{x}) \equiv \langle \tilde{x}^2 \rangle - \langle \tilde{x} \rangle^2, \tag{25}$$

so

$$\langle \tilde{x}^2 \rangle = \operatorname{var}(\tilde{x}) + (\tilde{x})^2 = \frac{\sigma^2}{N} + \mu^2. \tag{26}$$

See also ARITHMETIC-GEOMETRIC MEAN, ARITHMETIC-HARMONIC MEAN, CARLEMAN'S INEQUALITY, CUMU-LANT, GENERALIZED MEAN, GEOMETRIC MEAN, HARMONIC MEAN, HARMONIC-GEOMETRIC MEAN, KURTOSIS, MEAN, MEAN DEVIATION, MEDIAN (STATISTICS), MODE, MOMENT, QUADRATIC MEAN, ROOT-MEAN-SQUARE, SAMPLE VARIANCE, SKEWNESS, STANDARD DEVIATION, TRIMEAN, VARIANCE

References

Abramowitz, M. and Stegun, C. A. (Eds.). *Handbook of Mathematical Functions with Formulas, Graphs, and Mathematical Tables, 9th printing.* New York: Dover, p. 10, 1972.
Alzer, H. "A Proof of the Arithmetic Mean-Geometric Mean Inequality." *Amer. Math. Monthly* **103**, 585, 1996.
Beckenbach, E. F. and Bellman, R. *Inequalities.* New York: Springer-Verlag, 1983.
Beyer, W. H. *CRC Standard Mathematical Tables, 28th ed.* Boca Raton, FL: CRC Press, p. 471, 1987.
Bullen, P. S.; Mitrinovic, D. S.; and Vasic, P. M. *Means & Their Inequalities.* Dordrecht, Netherlands: Reidel, 1988.
Hardy, G. H.; Littlewood, J. E.; and Pólya, G. *Inequalities.* Cambridge, England: Cambridge University Press, 1952.
Hoehn, L. and Niven, I. "Averages on the Move." *Math. Mag.* **58**, 151–56, 1985.
Kenney, J. F. and Keeping, E. S. *Mathematics of Statistics, Pt. 1, 3rd ed.* Princeton, NJ: Van Nostrand, 1962.
Mitrinovic, D. S. *Analytic Inequalities.* New York: Springer-Verlag, 1970.
Mitrinovic, D. S.; Pecaric, J. E.; and Fink, A. M. *Classical and New Inequalities in Analysis.* Dordrecht, Netherlands: Kluwer, 1993.
Zwillinger, D. (Ed.). *CRC Standard Mathematical Tables and Formulae.* Boca Raton, FL: CRC Press, p. 601, 1995.

Arithmetic Progression

ARITHMETIC SEQUENCE

Arithmetic Sequence

A SEQUENCE of n numbers $\{d_0 + kd\}_{k=0}^{n-1}$ such that the differences between successive terms is a constant d.

See also ARITHMETIC SERIES, BAUDET'S CONJECTURE, NONARITHMETIC PROGRESSION SEQUENCE, SEQUENCE, SZEMERÉDI'S THEOREM

Arithmetic Series

An arithmetic series is the SUM of a SEQUENCE $\{a_k\}$, $k = 1, 2, \ldots$, in which each term is computed from the previous one by adding (or subtracting) a constant d. Therefore, for $k > 1$,

$$a_k = a_{k-1} + d = a_{k-2} + 2d = \ldots = a_1 + d(k-1). \tag{1}$$

The sum of the sequence of the first n terms is then given by

$$S_n \equiv \sum_{k=1}^{n} a_k = \sum_{k=1}^{n} [a_1 + (k-1)d] = na_1 + d \sum_{k=1}^{n}(k-1)$$

$$= na_1 + d \sum_{k=2}^{n}(k-1)$$

$$= na_1 + d \sum_{k=1}^{n-1} k \qquad (2)$$

Using the SUM identity

$$\sum_{k=1}^{n} k = \tfrac{1}{2}n(n+1) \qquad (3)$$

then gives

$$S_n = na_1 + \tfrac{1}{2}dn(n-1) = \tfrac{1}{2}n[2a_i + d(n-1)]. \qquad (4)$$

Note, however, that

$$a_1 + an = a_1 + [a_1 + d(n-1)] = 2a_1 + d(n-1), \qquad (5)$$

so

$$S_n = \tfrac{1}{2}n(a_1 + a_n), \qquad (6)$$

or n times the AVERAGE of the first and last terms! This is the trick Gauss used as a schoolboy to solve the problem of summing the INTEGERS from 1 to 100 given as busy-work by his teacher. While his classmates toiled away doing the ADDITION longhand, Gauss wrote a single number, the correct answer

$$\tfrac{1}{2}(100)(1 + 100) = 50 \cdot 101 = 5050 \qquad (7)$$

on his slate (Burton 1989, pp. 80–1; Hoffman 1998, p. 207). When the answers were examined, Gauss's proved to be the only correct one.

See also ARITHMETIC SEQUENCE, GEOMETRIC SERIES, HARMONIC SERIES, PRIME ARITHMETIC PROGRESSION

References

Abramowitz, M. and Stegun, C. A. (Eds.). *Handbook of Mathematical Functions with Formulas, Graphs, and Mathematical Tables, 9th printing.* New York: Dover, p. 10, 1972.
Beyer, W. H. (Ed.). *CRC Standard Mathematical Tables, 28th ed.* Boca Raton, FL: CRC Press, p. 8, 1987.
Burton, D. M. *Elementary Number Theory, 4th ed.* Boston, MA: Allyn and Bacon, 1989.
Courant, R. and Robbins, H. "The Arithmetical Progression." §1.2.2 in *What is Mathematics?: An Elementary Approach to Ideas and Methods, 2nd ed.* Oxford, England: Oxford University Press, pp. 12–3, 1996.
Hoffman, P. *The Man Who Loved Only Numbers: The Story of Paul Erdos and the Search for Mathematical Truth.* New York: Hyperion, 1998.
Pappas, T. *The Joy of Mathematics.* San Carlos, CA: Wide World Publ./Tetra, p. 164, 1989.

Arithmetical Function

INTEGER FUNCTION

Arithmetic-Geometric Mean

The arithmetic-geometric mean (often abbreviated AGM) $M(a, b)$ of two numbers a and b is defined by starting with $a_0 \equiv a$ and $b_0 \equiv b$, then iterating

$$a_{n+1} = \tfrac{1}{2}(a_n + b_n) \qquad (1)$$

$$b_{n+1} = \sqrt{a_n b_n} \qquad (2)$$

until $a_n = b_n$. a_n and b_n converge towards each other since

$$a_{n+1} - b_{n+1} = \tfrac{1}{2}(a_n + b_n) - \sqrt{a_n b_n}$$
$$= \frac{a_n - 2\sqrt{a_n b_n} + b_n}{2}. \qquad (3)$$

But $\sqrt{b_n} < \sqrt{a_n}$, so

$$2b_n < 2\sqrt{a_n b_n}. \qquad (4)$$

Now, add $a_n - b_n - 2\sqrt{a_n b_n}$ to each side

$$a_n + b_n - 2\sqrt{a_n b_n} < a_n - b_n, \qquad (5)$$

so

$$a_{n+1} - b_{n+1} < \tfrac{1}{2}(a_n - b_n). \qquad (6)$$

The AGM is very useful in computing the values of complete ELLIPTIC INTEGRALS and can also be used for finding the INVERSE TANGENT. In terms of the complete ELLIPTIC INTEGRAL OF THE FIRST KIND $K(k)$,

$$M(a, b) = \frac{(a + b)\pi}{4K\left(\dfrac{a - b}{a + b}\right)}. \qquad (7)$$

The special value $1/M(1, \sqrt{2})$ is called GAUSS'S CONSTANT.

The AGM has the properties

$$\lambda M(a, b) = M(\lambda a, \lambda b) \qquad (8)$$

$$M(a, b) = M\left(\tfrac{1}{2}(a + b), \sqrt{ab}\right) \qquad (9)$$

$$M(1, \sqrt{1 - x^2}) = M(1 + x, 1 - x) \qquad (10)$$

$$M(1, b) = \frac{1 + b}{2} M\left(1, \frac{2\sqrt{b}}{1 + b}\right). \qquad (11)$$

The Legendre form is given by

$$M(1, x) = \prod_{n=0}^{\infty} \tfrac{1}{2}(1 + k_n), \qquad (12)$$

where $k_0 \equiv x$ and

$$k_{n+1} \equiv \frac{2\sqrt{k_n}}{1 + k_n}. \qquad (13)$$

Solutions to the differential equation

$$(x^3 - x)\frac{d^2 y}{dx^2} + (3x^2 - 1)\frac{dy}{dx} + xy = 0 \qquad (14)$$

are given by $[M(1 + x, 1 - x)]^{-1}$ and $[M(1, x)]^{-1}$.

A generalization of the ARITHMETIC-GEOMETRIC MEAN is

$$I_p(a,\ b) = \int_0^\infty \frac{x^{p-2}\ dx}{(x^p + a^p)^{1/p}(x^p + b^p)^{(p-1)/p}} \quad (15)$$

which is related to solutions of the differential equation

$$x(1 - x^p)Y'' + [1 - (p+1)x^p]Y' - (p-1)x^{p-1}Y = 0. \quad (16)$$

When $p = 2$ or $p = 3$, there is a modular transformation for the solutions of (16) that are bounded as $x \to 0$. Letting $J_p(x)$ be one of these solutions, the transformation takes the form

$$J_p(\lambda) = \mu J_p(x), \quad (17)$$

where

$$\lambda = \frac{1 - u}{1 + (p-1)u} \quad (18)$$

$$\mu = \frac{1 + (p-1)u}{p} \quad (19)$$

and

$$x^p + u^p = 1. \quad (20)$$

The case $p = 2$ gives the ARITHMETIC-GEOMETRIC MEAN, and $p = 3$ gives a cubic relative discussed by Borwein and Borwein (1990, 1991) and Borwein (1996) in which, for $a,\ b > 0$ and $I(a,\ b)$ defined by

$$I(a,\ b) = \int_0^\infty \frac{t\ dt}{[(a^3 + t^3)(b^3 + t^3)^2]^{1/3}}, \quad (21)$$

$$I(a,\ b) = I\left(\frac{a + 2b}{3}, \left[\frac{b}{3}(a^2 + ab + b^2)\right]\right) \quad (22)$$

For iteration with $a_0 = a$ and $b_0 = b$ and

$$a_{n+1} = \frac{a_n + 2b_n}{3} \quad (23)$$

$$b_{n+1} = \frac{b_n}{3}(a_n^2 + a_n b_n + b_n^2), \quad (24)$$

$$\lim_{n \to \infty} a_n = \lim_{n \to \infty} b_n = \frac{I(1,\ 1)}{I(a,\ b)}. \quad (25)$$

Modular transformations are known when $p = 4$ and $p = 6$, but they do not give identities for $p = 6$ (Borwein 1996).

See also ARITHMETIC-HARMONIC MEAN

References

Abramowitz, M. and Stegun, C. A. (Eds.). "The Process of the Arithmetic-Geometric Mean." §17.6 in *Handbook of Mathematical Functions with Formulas, Graphs, and Mathematical Tables, 9th printing.* New York: Dover, pp. 571 ad 598–99, 1972.

Borwein, J. M. Problem 10281. "A Cubic Relative of the AGM." *Amer. Math. Monthly* **103**, 181–83, 1996.

Borwein, J. M. and Borwein, P. B. "A Remarkable Cubic Iteration." In *Computational Method & Function Theory:*

Proc. Conference Held in Valparaiso, Chile, March 13–8, 1989 (Ed. A. Dold, B. Eckmann, F. Takens, E. B Saff, S. Ruscheweyh, L. C. Salinas, L. C., and R. S. Varga). New York: Springer-Verlag, 1990.

Borwein, J. M. and Borwein, P. B. "A Cubic Counterpart of Jacobi's Identity and the AGM." *Trans. Amer. Math. Soc.* **323**, 691–01, 1991.

Press, W. H.; Flannery, B. P.; Teukolsky, S. A.; and Vetterling, W. T. *Numerical Recipes in FORTRAN: The Art of Scientific Computing, 2nd ed.* Cambridge, England: Cambridge University Press, pp. 906–07, 1992.

Arithmetic-Harmonic Mean

Let

$$a_{n+1} = \tfrac{1}{2}(a_n + b_n) \quad (1)$$

$$b_{n+1} = \frac{2a_n b_n}{a_n + b_n}. \quad (2)$$

Then

$$A(a_0,\ b_0) = \lim_{n \to \infty} a_n = \lim_{n \to \infty} b_n \sqrt{a_0 b_0}, \quad (3)$$

which is just the GEOMETRIC MEAN.

Arithmetic-Logarithmic-Geometric Mean Inequality

$$\frac{a + b}{2} > \frac{b - a}{\ln b - \ln a} > \sqrt{ab}.$$

See also NAPIER'S INEQUALITY

References

Nelson, R. B. "Proof without Words: The Arithmetic-Logarithmic-Geometric Mean Inequality." *Math. Mag.* **68**, 305, 1995.

Armstrong Number

The n-digit numbers equal to sum of nth powers of their digits (a finite sequence), also called plus perfect numbers. They first few are given by 1, 2, 3, 4, 5, 6, 7, 8, 9, 153, 370, 371, 407, 1634, 8208, 9474, 54748, ... (Sloane's A005188).

See also HARSHAD NUMBER, NARCISSISTIC NUMBER

References

Sloane, N. J. A. Sequences A005188/M0488 in "An On-Line Version of the Encyclopedia of Integer Sequences." http://www.research.att.com/~njas/sequences/eisonline.html.

Arnold Diffusion

The nonconservation of ADIABATIC INVARIANTS which arises in systems with three or more DEGREES OF FREEDOM.

References

Lichtenberg, A. and Lieberman, M. *Regular and Stochastic Motion, 2nd ed.* New York: Springer-Verlag, 1994.

Rasband, S. N. "Arnold Diffusion." §8.6 in *Chaotic Dynamics of Nonlinear Systems.* New York: Wiley, pp. 179–81, 1990.
Tabor, M. *Chaos and Integrability in Nonlinear Dynamics: An Introduction.* New York: Wiley, p. 74, 1989.

Arnold Tongue

Consider the CIRCLE MAP. If K is NONZERO, then the motion is periodic in some FINITE region surrounding each rational Ω. This execution of periodic motion in response to an irrational forcing is known as MODE LOCKING. If a plot is made of K versus Ω with the regions of periodic MODE-LOCKED parameter space plotted around rational Ω values (the WINDING NUMBERS), then the regions are seen to widen upward from 0 at $K = 0$ to some FINITE width at $K = 1$. The region surrounding each RATIONAL NUMBER is known as an ARNOLD TONGUE.

At $K = 0$, the Arnold tongues are an isolated set of MEASURE zero. At $K = 1$, they form a general CANTOR SET of dimension $d = 0.8700 \pm 3.7 \times 10^{-4}$ (Rasband 1990, p. 131). In general, an Arnold tongue is defined as a resonance zone emanating out from RATIONAL NUMBERS in a two-dimensional parameter space of variables.

See also CIRCLE MAP, DEVIL'S STAIRCASE

References

Rasband, S. N. *Chaotic Dynamics of Nonlinear Systems.* New York: Wiley, pp. 130–31, 1990.

Arnold's Cat Map

The best known example of an ANOSOV DIFFEOMORPHISM. It is given by the TRANSFORMATION

$$\begin{bmatrix} x_n + 1 \\ y_n + 1 \end{bmatrix} = \begin{bmatrix} 1 & 1 \\ 1 & 2 \end{bmatrix}\begin{bmatrix} x_n \\ y_n \end{bmatrix}, \tag{1}$$

where x_{n+1} and y_{n+1} are computed mod 1. The Arnold cat mapping is non-Hamiltonian, nonanalytic, and mixing. However, it is AREA-PRESERVING since the DETERMINANT is 1. The LYAPUNOV CHARACTERISTIC EXPONENTS are given by

$$\begin{vmatrix} 1-\sigma & 1 \\ 1 & 2-\sigma \end{vmatrix} = \sigma^2 - 3\sigma + 1 = 0, \tag{2}$$

so

$$\sigma_\pm = \tfrac{1}{2}(3 \pm \sqrt{5}). \tag{3}$$

The EIGENVECTORS are found by plugging $\sigma_\pm$ into the MATRIX EQUATION

$$\begin{bmatrix} 1-\sigma_\pm & 1 \\ 1 & 2-\sigma_\pm \end{bmatrix}\begin{bmatrix} x \\ y \end{bmatrix} = \begin{bmatrix} 0 \\ 0 \end{bmatrix}. \tag{4}$$

For σ_+, the solution is

$$y = \tfrac{1}{2}(1 + \sqrt{5})x \equiv \phi x, \tag{5}$$

where ϕ is the GOLDEN RATIO, so the unstable

(normalized) EIGENVECTOR is

$$\xi_+ = \tfrac{1}{10}\sqrt{50 - 10\sqrt{5}}\begin{bmatrix} 1 \\ \tfrac{1}{2}(1 + \sqrt{5}) \end{bmatrix}. \tag{6}$$

Similarly, for σ_-, the solution is

$$y = -\tfrac{1}{2}(\sqrt{5} - 1)x \equiv \phi^{-1}x, \tag{7}$$

so the stable (normalized) EIGENVECTOR is

$$\xi_- = \tfrac{1}{10}\sqrt{50 + 10\sqrt{5}}\begin{bmatrix} 1 \\ \tfrac{1}{2}(1 - \sqrt{5}) \end{bmatrix}. \tag{8}$$

See also ANOSOV MAP

Aronhold Process

The process used to generate an expression for a covariant in the first degree of any one of the equivalent sets of COEFFICIENTS for a curve.

See also CLEBSCH-ARONHOLD NOTATION, JOACHIMSTHAL'S EQUATION

References

Coolidge, J. L. *A Treatise on Algebraic Plane Curves.* New York: Dover, p. 74, 1959.

Aronson's Sequence

The sequence whose definition is: "t is the first, fourth, eleventh, ... letter of this sentence." The first few values are 1, 4, 11, 16, 24, 29, 33, 35, 39, ... (Sloane's A005224).

References

Hofstadter, D. R. *Metamagical Themas: Questing of Mind and Pattern.* New York: BasicBooks, p. 44, 1985.
Sloane, N. J. A. Sequences A005224/M3406 in "An On-Line Version of the Encyclopedia of Integer Sequences." http://www.research.att.com/~njas/sequences/eisonline.html.

Arrangement

In general, an arrangement of objects is simply a grouping of them. The number of "arrangements" of n items is given either by a COMBINATION (order is ignored) or PERMUTATION (order is significant).

The division of SPACE into cells by a collection of HYPERPLANES (Agarwal and Sharir 2000) is also called an arrangement.

See also COMBINATION, CONFIGURATION, CUTTING, HYPERPLANE, ORDERING, PERMUTATION

References

Agarwal, P. K. and Sharir, M. "Arrangements and Their Applications." Ch. 2 in *Handbook of Computational Geometry* (Ed. J.-R. Sack and J. Urrutia). Amsterdam, Netherlands: North-Holland, pp. 49–19, 2000.

Arrangement Number
PERMUTATION

Array
An array is a "list of lists" with the length of each level of list the same. The size (sometimes called the "shape") of a d-dimensional array is then indicated as $\underbrace{m \times n \times x \cdots \times p}_{d}$. The most common type of array encountered is the 2-D $m \times n$ rectangular array having m columns and n rows. If $m = n$, a square array results. Sometimes, the order of the elements in an array is significant (as in a MATRIX), whereas at other times, arrays which are equivalent modulo reflections (and rotations, in the case of a square array) are considered identical (as in a MAGIC SQUARE or PRIME ARRAY).

In order to exhaustively list the number of distinct arrays of a given shape with each element being one of k possible choices, the naive algorithm of running through each case and checking to see whether it's equivalent to an earlier one is already just about as efficient as can be. The running time must be at least the number of answers, and this is so close to $k^{mn \cdots p}$ that the difference isn't significant.

However, finding the *number* of possible arrays of a given shape is much easier, and an exact formula can be obtained using the POLYA ENUMERATION THEOREM. For the simple case of an $m \times n$ array, even this proves unnecessary since there are only a few possible symmetry types, allowing the possibilities to be counted explicitly. For example, consider the case of m and n EVEN and distinct, so only reflections need be included. To take a specific case, let $m = 6$ and $n = 4$ so the array looks like

$$
\begin{array}{ccc:ccc}
a & b & c & d & e & f \\
g & h & i & j & k & l \\
\cdots & \cdots & \cdots & \cdots & \cdots & \cdots \\
m & n & o & p & q & r \\
s & t & u & v & w & x
\end{array}
$$

where each a, b, ..., x can take a value from 1 to k. The total number of possible arrangements is k^{24} (k^{mn} in general). The number of arrangements which are equivalent to their left-right mirror images is k^{12} (in general, $k^{mn/2}$), as is the number equal to their up-down mirror images, or their rotations through $180°$. There are also k^6 arrangements (in general, $k^{mn/4}$) with full symmetry.

In general, it is therefore true that

$$
\begin{cases}
k^{mn/4} & \text{with full symmetry} \\
k^{mn/2} - k^{mn/4} & \text{with } only \text{ left-right reflection} \\
k^{mn/2} - k^{mn/4} & \text{with } only \text{ up-down reflection} \\
k^{mn/2} - k^{mn/4} & \text{with } only \text{ } 180° \text{ rotation,}
\end{cases}
$$

so there are

$$
k^{mn} - 3k^{mn/2} + 2k^{mn/4}
$$

arrangements with no symmetry. Now dividing by the number of images of each type, the result, for $m \neq n$ with m, n EVEN, is

$$
\begin{aligned}
N(m, &\, n, k) \\
&= \tfrac{1}{4} k^{mn} + (\tfrac{1}{2})(3)(k^{mn/2} - k^{mn/4}) \\
&\quad + \tfrac{1}{4}(k^{mn} - 3k^{mn/2} + 2k^{mn/4}) \\
&= \tfrac{1}{4} k^{mn} + \tfrac{3}{4} k^{mn/2} + \tfrac{1}{2} k^{mn/4}.
\end{aligned}
$$

The number is therefore of order $\mathcal{O}(k^{mn}/4)$, with "correction" terms of much smaller order.

See also ANTIMAGIC SQUARE, EULER SQUARE, KIRKMAN'S SCHOOLGIRL PROBLEM, LATIN RECTANGLE, LATIN SQUARE, MAGIC SQUARE, MATRIX, MRS. PERKINS' QUILT, MULTIPLICATION TABLE, ORTHOGONAL ARRAY, PERFECT SQUARE, PRIME ARRAY, QUOTIENT-DIFFERENCE TABLE, ROOM SQUARE, STOLARSKY ARRAY, TRUTH TABLE, WYTHOFF ARRAY

Arrow Notation
A NOTATION invented by Knuth (1976) to represent LARGE NUMBERS in which evaluation proceeds from the right (Conway and Guy 1996, p. 60).

$$
\begin{array}{ll}
m \uparrow n & m \cdot m \cdots m \\
m \uparrow\uparrow n & m \uparrow m \uparrow \cdots \uparrow m \\
m \uparrow\uparrow\uparrow n & \underbrace{m \uparrow\uparrow m \uparrow\uparrow \cdots \uparrow\uparrow m}_{n}
\end{array}
$$

For example,

$$
m \uparrow n = m^n \tag{1}
$$

$$
m \uparrow\uparrow n = \underbrace{m \uparrow \cdots \uparrow m}_{n} = \underbrace{m^{m \cdot^{\cdot^{\cdot m}}}}_{n}
$$

$$
m \uparrow\uparrow 2 = \underbrace{m \uparrow m}_{2} = m \uparrow m = m^m \tag{2}
$$

$$
m \uparrow\uparrow 3 = \underbrace{m \uparrow m \uparrow m}_{3} = m \uparrow (m \uparrow m)
$$

$$
= m \uparrow m^m = m^{m^m} \tag{3}
$$

$$
m \uparrow\uparrow\uparrow 2 = \underbrace{m \uparrow\uparrow m}_{2} = m \uparrow\uparrow m = \underbrace{m^{m \cdot^{\cdot^{\cdot m}}}}_{m} \tag{4}
$$

$$
m \uparrow\uparrow\uparrow 3 = \underbrace{m \uparrow\uparrow m m \uparrow\uparrow m}_{3} = m \uparrow\uparrow m = \underbrace{m^{m \cdot^{\cdot^{\cdot m}}}}_{m}
$$

$$= \underbrace{m \uparrow \cdots \uparrow m}_{\underbrace{m^{m \cdot \cdot m}}_{m}} = \underbrace{m^{m \cdot \cdot m}}_{\underbrace{m^{m \cdot \cdot m}}_{m}} \qquad (5)$$

$m \uparrow\uparrow m$ is sometimes called a POWER TOWER. The values $n \uparrow \cdots \uparrow n$ are called ACKERMANN NUMBERS.

See also ACKERMANN NUMBER, CHAINED ARROW NOTATION, DOWN ARROW NOTATION, LARGE NUMBER, POWER TOWER, STEINHAUS-MOSER NOTATION

References
Conway, J. H. and Guy, R. K. *The Book of Numbers.* New York: Springer-Verlag, pp. 59–2, 1996.
Guy, R. K. and Selfridge, J. L. "The Nesting and Roosting Habits of the Laddered Parenthesis." *Amer. Math. Monthly* **80**, 868–76, 1973.
Knuth, D. E. "Mathematics and Computer Science: Coping with Finiteness. Advances in Our Ability to Compute are Bringing Us Substantially Closer to Ultimate Limitations." *Science* **194**, 1235–242, 1976.
Vardi, I. *Computational Recreations in Mathematica.* Redwood City, CA: Addison-Wesley, pp. 11 and 226–29, 1991.

Arrow's Paradox

Perfect democratic VOTING is, not just in practice but *in principle,* impossible.

See also SOCIAL CHOICE THEORY, VOTING

References
Erickson, G. W. and Fossa, J. A. *Dictionary of Paradox.* Lanham, MD: University Press of America, pp. 13–5, 1998.
Gardner, M. *Time Travel and Other Mathematical Bewilderments.* New York: W. H. Freeman, p. 56, 1988.

Arrowhead Curve

SIERPINSKI ARROWHEAD CURVE

Arsh

$$\text{Arsh } z = \frac{1}{i} \sin^{-1}(iz),$$

where $\sin^{-1} z$ the INVERSE SINE.

See also ARCH, ARCTH, ARTH, INVERSE SINE

References
Gradshteyn, I. S. and Ryzhik, I. M. *Tables of Integrals, Series, and Products, 6th ed.* San Diego, CA: Academic Press, p. xxx, 2000.

Art Gallery Theorem

Also called Chvátal's art gallery theorem. If the walls of an art gallery are made up of n straight LINE SEGMENTS, then the entire gallery can always be supervised by $\lfloor n/3 \rfloor$ watchmen placed in corners, where $\lfloor x \rfloor$ is the FLOOR FUNCTION. This theorem was proved by Chvátal (1975). It was conjectured that an art gallery with n walls and h HOLES requires $\lfloor (n+h)/3 \rfloor$ watchmen, which has now been proven by Bjorling-Sachs and Souvaine (1991, 1995) and Hoffman *et al.* (1991).

See also ILLUMINATION PROBLEM, TRIANGULATION, VORONOI DIAGRAM

References
Bjorling-Sachs, I. and Souvaine, D. L. "A Tight Bound for Guarding Polygons with Holes." Report LCSR-TR-165. New Brunswick, NJ: Lab. Comput. Sci. Res., Rutgers Univ., 1991.
Bjorling-Sachs, I. and Souvaine, D. L. "An Efficient Algorithm for Guard Placement in Polygons with Holes." *Disc. Comput. Geom.* **13**, 77–09, 1995.
Chvátal, V. "A Combinatorial Theorem in Plane Geometry." *J. Combin. Th.* **18**, 39–1, 1975.
de Berg, M.; van Kreveld, M.; Overmans, M.; and Schwarzkopf, O. *Computational Geometry: Algorithms and Applications, 2nd rev. ed.* Berlin: Springer-Verlag, pp. 48 and 59, 2000.
Fisk, S. "A Short Proof of Chvátal's Watchman Theorem." *J. Combin. Th. Ser. B* **24**, 374, 1978.
Fournier, A. and Montuno, D. Y. "Triangulating Simple Polygons and Equivalent Problems." *ACM Trans. Graphics* **3**, 153–74, 1984.
Garey, M. R.; Johnson, D. S.; Preparata, F. P.; and Tarjan, R. E. "Triangulating a Simple Polygon." *Inform. Process. Lett.* **7**, 175–79, 1978.
Hoffmann, F.; Kaufmann, M.; and Kriegel, K. "The Art Gallery Theorem for Polygons with Holes." *Proc. 32nd Annual IEEE Sympos. Found. Comput. Sci.*, 39–8, 1991.
Honsberger, R. "Chvátal's Art Gallery Theorem." Ch. 11 in *Mathematical Gems II.* Washington, DC: Math. Assoc. Amer., pp. 104–10, 1976.
Kahn, J.; Klawe, M.; and Kleitman, D. "Traditional Galleries Require Fewer Watchmen." *SIAM J. Alg. Disc. Math.* **4**, 194–06, 1993.
Klee, V. "On the Complexity of d-Dimensional Voronoi Diagrams." *Archiv. Math.* **34**, 75–0, 1980.
O'Rourke, J. *Art Gallery Theorems and Algorithms.* New York: Oxford University Press, 1987.
O'Rourke, J. §2.3 in *Computational Geometry in C, 2nd ed.* Cambridge, England: Cambridge University Press, 1998.
Stewart, I. "How Many Guards in the Gallery?" *Sci. Amer.* **270**, 118–20, May 1994.
Tucker, A. "The Art Gallery Problem." *Math Horizons,* pp. 24–6, Spring 1994.
Urrutia, J. "Art Gallery and Illumination Problems." Ch. 22 in *Handbook of Computational Geometry* (Ed. J.-R. Sack and J. Urrutia). Amsterdam, Netherlands: North-Holland, pp. 973–027, 2000.
Wagon, S. "The Art Gallery Theorem." §10.3 in *Mathematica in Action.* New York: W. H. Freeman, pp. 333–45, 1991.
Wells, D. *The Penguin Dictionary of Curious and Interesting Geometry.* London: Penguin, p. 9, 1991.

Arth

$$\text{Arth } z = \frac{1}{i} \tan^{-1}(iz).$$

where $\tan^{-1} z$ is the INVERSE TANGENT.

See also ARCH, ARSH, ARCTH, INVERSE TANGENT

Articulation Vertex

References

Gradshteyn, I. S. and Ryzhik, I. M. *Tables of Integrals, Series, and Products, 6th ed.* San Diego, CA: Academic Press, p. xxx, 2000.

Articulation Vertex

An articulation of a CONNECTED GRAPH is a node whose removal will disconnect the graph (Chartrand 1985). In general, an articulation vertex is node of a GRAPH whose removal increases the number of components (Harary 1994, p. 26). Articulation vertices are also called cut-vertices or "cutpoints" (Harary 1994, p. 26).

A GRAPH with no articulation vertices is called a BICONNECTED GRAPH.

See also BICONNECTED GRAPH, BLOCK, BRIDGE, CUT SET, NONSEPARABLE GRAPH, VERTEX (GRAPH)

References

Chartrand, G. "Cut-Vertices and Bridges." §2.4 in *Introductory Graph Theory.* New York: Dover, pp. 45–9, 1985.
Harary, F. *Graph Theory.* Reading, MA: Addison-Wesley, 1994.
Skiena, S. *Implementing Discrete Mathematics: Combinatorics and Graph Theory with Mathematica.* Reading, MA: Addison-Wesley, p. 175, 1990.

Artin Braid Group

BRAID GROUP

Artin L-Function

An Artin L-function over the RATIONALS $\mathbb{Q}$ encodes in a GENERATING FUNCTION information about how an irreducible MONIC POLYNOMIAL over $\mathbb{Z}$ factors when reduced modulo each PRIME. For the POLYNOMIAL $x^2 + 1$, the Artin L-function is

$$L(s,\ \mathbb{Q}(i)/\mathbb{Q},\ \text{sgn}) = \prod_{p\ \text{odd prime}} \frac{1}{1 - \left(\dfrac{-1}{p}\right)p^{-s}},$$

where $(-1/p)$ is a LEGENDRE SYMBOL, which is equivalent to the EULER L-FUNCTION. The definition over arbitrary POLYNOMIALS generalizes the above expression.

See also LANGLANDS RECIPROCITY

References

Knapp, A. W. "Group Representations and Harmonic Analysis, Part II." *Not. Amer. Math. Soc.* **43**, 537–49, 1996.

Artin Reciprocity

ARTIN'S RECIPROCITY THEOREM

Artin's Conjecture

There are at least two statements which go by the name of Artin's conjecture. The first is the RIEMANN HYPOTHESIS.

The second states that every INTEGER not equal to -1 or a SQUARE NUMBER is a primitive root modulo p for infinitely many p and proposes a density for the set of such p which are always rational multiples of a constant known as ARTIN'S CONSTANT. There is an analogous theorem for functions instead of numbers which has been proved by Billharz (Shanks 1993, p. 147).

See also ARTIN'S CONSTANT, RIEMANN HYPOTHESIS

References

Matthews, K. R. "A Generalization of Artin's Conjecture for Primitive Roots." *Acta Arith.* **29**, 113–46, 1976.
Moree, P. "A Note on Artin's Conjecture." *Simon Stevin* **67**, 255–57, 1993.
Ram Murty, M. "Artin's Conjecture for Primitive Roots." *Math. Intell.* **10**, 59–7, 1988.
Shanks, D. *Solved and Unsolved Problems in Number Theory, 4th ed.* New York: Chelsea, pp. 31, 80–3, and 147, 1993.

Artin's Constant

If $n \neq -1$ and n is not a PERFECT SQUARE, then Artin conjectured that the SET $S(n)$ of all PRIMES for which n is a PRIMITIVE ROOT is infinite. Under the assumption of the EXTENDED RIEMANN HYPOTHESIS, Artin's conjecture was solved by Hooley (1967).

If, in addition, n is not an rth POWER for any $r > 1$ then let n' be the SQUAREFREE PART of n and suppose that $n' \equiv 1 \pmod 4$. Then Artin conjectured that the density of $S(n)$ relative to the PRIMES is given by C_{Artin}, where

$$C_{\text{Artin}} = \prod_{k=1}^{\infty} \left[1 - \frac{1}{p_k(p_k - 1)}\right] = 0.3739558136\ldots, \quad (1)$$

and p_k is the kth PRIME, independently of the choice of n.

C_{Artin} is connected with the PRIME ZETA FUNCTION $P(n)$ by

$$\ln C_{\text{Artin}} = -\sum_{n=2}^{\infty} \frac{(u_n - 1)P(n)}{n} \quad (2)$$

where

$$u_n = u_{n-1} + u_{n-2} \quad (3)$$

with $u_1 = 1$, $u_2 = 3$ (Ribenboim 1998, Gourdon and Sebah). Wrench (1961) gave 45 digits of C_{Artin}, and Gourdon and Sebah give 60.

If $n' \equiv 1 \pmod 4$ and n is still restricted not to be an rth power, then the density is not C_{Artin} itself, but a rational multiple thereof. The explicit formula for computing the density in this case is conjectured to be

$$C'_{\text{Artin}} = \left[1 - \mu(n') \prod_{\substack{\text{prime}\ q \\ q|n'}} \frac{1}{q^2 - q - 1}\right] C_{\text{Artin}} \quad (4)$$

(Finch, Matthews 1976), where $\mu(n)$ is the MÖBIUS

FUNCTION. Special cases can be written down explicitly for $n' = p$ a PRIME,

$$C'_{\text{Artin}} = \left(1 + \frac{1}{p^2 - p - 1}\right) C_{\text{Artin}} \qquad (5)$$

or $n' = pq$, where p, q are both PRIMES with u, $v \equiv 1 \pmod 4$,

$$C'_{\text{Artin}} = \left(1 + \frac{1}{p^2 - p - 1} \frac{1}{q^2 - q - 1}\right) C_{\text{Artin}}, \qquad (6)$$

If n is a perfect cube (which is not a perfect square), a perfect fifth power (which is not a perfect square or perfect cube), etc., other formulas apply (Hooley 1967, Western and Miller 1968).

The significance of Artin's constant is more easily seen by describing it as the fraction of PRIMES p for which $1/p$ has a maximal DECIMAL EXPANSION, i.e., p is a FULL REPTEND PRIME, (Conway and Guy 1996).

See also ARTIN'S CONJECTURE, DECIMAL EXPANSION, FULL REPTEND PRIME, PRIMITIVE ROOT, STEPHENS' CONSTANT

References

Artin, E. *Collected Papers* (Ed. S. Lang and J. T. Tate). New York: Springer-Verlag, pp. viii-ix, 1965.
Conway, J. H. and Guy, R. K. *The Book of Numbers.* New York: Springer-Verlag, p. 169, 1996.
Finch, S. "Favorite Mathematical Constants." http://www.mathsoft.com/asolve/constant/artin/artin.html.
Finch, S. "Correction Factors for Artin's Constant." http://www.mathsoft.com/asolve/constant/artin/factor.html.
Gourdon, X. and Sebah, P. "Some Constants from Number Theory." http://xavier.gourdon.free.fr/Constants/Miscellaneous/constantsNumTheory.html.
Hooley, C. "On Artin's Conjecture." *J. reine angew. Math.* **225**, 209-20, 1967.
Hooley, C. *Applications of Sieve Methods to the Theory of Numbers.* Cambridge, England: Cambridge University Press, 1976.
Ireland, K. and Rosen, M. *A Classical Introduction to Modern Number Theory, 2nd ed.* New York: Springer-Verlag, 1990.
Lehmer, D. H. and Lehmer, E. "Heuristics Anyone?" In *Studies in Mathematical Analysis and Related Topics: Essays in Honor of George Pólya* (Ed. G. Szego, C. Loewner, S. Bergman, M. M. Schiffer, J. Neyman, D. Gilbarg, and H. Solomon). Stanford, CA: Stanford University Press, 1962.
Lenstra, H. W. Jr. "On Artin's Conjecture and Euclid's Algorithm in Global Fields." *Invent. Math.* **42**, 201-24, 1977.
Matthews, K. R. "A Generalization of Artin's Conjecture for Primitive Roots." *Acta Arith.* **29**, 113-46, 1976.
Plouffe, S. "Artin's Constant." http://www.lacim.uqam.ca/piDATA/artin.txt.
Ram Murty, M. "Artin's Conjecture for Primitive Roots." *Math. Intell.* **10**, 59-7, 1988.
Ribenboim, P. *The New Book of Prime Number Records.* New York: Springer-Verlag, 1996.
Shanks, D. *Solved and Unsolved Problems in Number Theory, 4th ed.* New York: Chelsea, pp. 80-3, 1993.
Western, A. E. and Miller, J. C. P. *Tables of Indices and Primitive Roots.* Cambridge, England: Cambridge University Press, pp. xxxvii-xlii, 1968.
Wrench, J. W. "Evaluation of Artin's Constant and the Twin Prime Constant." *Math. Comput.* **15**, 396-98, 1961.

Artin's Reciprocity Theorem

A general RECIPROCITY THEOREM for all orders which covered all other known reciprocity theorems when proved by E. Artin in 1927. If R is a NUMBER FIELD and R' a finite integral extension, then there is a SURJECTION from the group of fractional IDEALS prime to the discriminant, given by the Artin symbol. For some cycle c, the kernel of this SURJECTION contains each PRINCIPAL fractional IDEAL generated by an element congruent to 1 mod c.

See also LANGLANDS PROGRAM

Artinian Group

A GROUP in which any decreasing CHAIN of distinct SUBGROUPS terminates after a FINITE number.

Artinian Ring

A noncommutative SEMISIMPLE RING satisfying the "descending chain condition."

See also GORENSTEIN RING, SEMISIMPLE RING

References

Artin, E. "Zur Theorie der hyperkomplexer Zahlen." *Hamb. Abh.* **5**, 251-60, 1928.
Artin, E. "Zur Arithmetik hyperkomplexer Zahlen." *Hamb. Abh.* **5**, 261-89, 1928.

Artistic Sequence

A SERIES is called artistic if every three consecutive terms have a common three-way ratio

$$P[a_i, a_{i+1}, a_{i+2}] = \frac{(a_i + a_{i+1} + a_{i+2})a_{i+1}}{a_i a_{i+2}}.$$

A SERIES is also artistic IFF its BIAS is a constant. A GEOMETRIC SERIES with RATIO $r > 0$ is an artistic series with

$$P = \frac{1}{r} + 1 + r \geq 3.$$

See also BIAS (SERIES), GEOMETRIC SERIES, MELODIC SEQUENCE

References

Duffin, R. J. "On Seeing Progressions of Constant Cross Ratio." *Amer. Math. Monthly* **100**, 38-7, 1993.

ASA Theorem

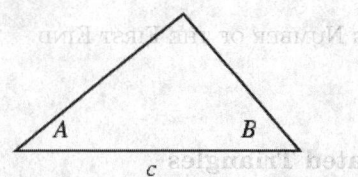

Specifying two adjacent ANGLES A and B and the side between them c uniquely determines a TRIANGLE with AREA

$$K = \frac{c^2}{2(\cot A + \cot B)} \qquad (1)$$

The angle C is given in terms of A and B by

$$C = \pi - A - B, \qquad (2)$$

and the sides a and b can be determined by using the LAW OF SINES

$$\frac{a}{\sin A} = \frac{b}{\sin B} = \frac{c}{\sin C} \qquad (3)$$

to obtain

$$a = \frac{\sin A}{\sin(\pi - A - B)} c \qquad (4)$$

$$b = \frac{\sin B}{\sin(\pi - A - B)} c. \qquad (5)$$

See also AAA THEOREM, AAS THEOREM, ASS THEOREM, SAS THEOREM, SSS THEOREM, TRIANGLE

Aschbacher's Component Theorem

Suppose that $E(G)$ (the commuting product of all components of G) is SIMPLE and G contains a semisimple INVOLUTION. Then there is some semisimple INVOLUTION x such that $C_G(x)$ has a NORMAL SUBGROUP K which is either QUASISIMPLE or ISOMORPHIC to $O + (4, q)'$ and such that $Q = C_G(K)$ is TIGHTLY EMBEDDED.

See also INVOLUTION (GROUP), ISOMORPHIC GROUPS, NORMAL SUBGROUP, QUASISIMPLE GROUP, SIMPLE GROUP, TIGHTLY EMBEDDED

A-Sequence

N.B. A detailed online essay by S. Finch was the starting point for this entry.

An INFINITE SEQUENCE of POSITIVE INTEGERS a^iS satisfying

$$1 \le a_1 < a_2 < a_3 < \ldots \qquad (1)$$

is an A-sequence if no a_k is the SUM of two or more distinct earlier terms (Guy 1994). Such sequences are sometimes also known as sum-free sets.

Erdos (1962) proved

$$S(A) \equiv \sup_{\text{all A sequences}} \sum_{k=1}^{\infty} \frac{1}{a_k} < 103. \qquad (2)$$

Any A-sequence satisfies the CHI INEQUALITY (Levine and O'Sullivan 1977), which gives $S(A) < 3.9998$. Abbott (1987) and Zhang (1992) have given a bound from below, so the best result to date is

$$2.0649 < S(A) < 3.9998. \qquad (3)$$

Levine and O'Sullivan (1977) conjectured that the sum of RECIPROCALS of an A-sequence satisfies

$$S(A) \le \sum_{k=1}^{\infty} \frac{1}{\chi_k} = 3.01\ldots, \qquad (4)$$

where χ_i are given by the LEVINE-O'SULLIVAN GREEDY ALGORITHM.

See also B2-SEQUENCE, MIAN-CHOWLA SEQUENCE, SUM-FREE SET

References

Abbott, H. L. "On Sum-Free Sequences." *Acta Arith.* **48**, 93–6, 1987.

Erdos, P. "Remarks on Number Theory III. Some Problems in Additive Number Theory." *Mat. Lapok* **13**, 28–8, 1962.

Finch, S. "Favorite Mathematical Constants." http://www.mathsoft.com/asolve/constant/erdos/erdos.html.

Guy, R. K. "B_2-Sequences." §E28 in *Unsolved Problems in Number Theory, 2nd ed.* New York: Springer-Verlag, pp. 228–29, 1994.

Levine, E. and O'Sullivan, J. "An Upper Estimate for the Reciprocal Sum of a Sum-Free Sequence." *Acta Arith.* **34**, 9–4, 1977.

Zhang, Z. X. "A Sum-Free Sequence with Larger Reciprocal Sum." Unpublished manuscript, 1992.

ASS Theorem

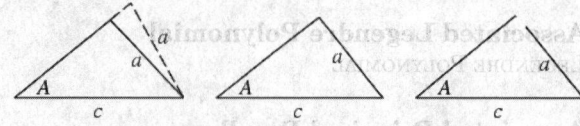

Specifying two adjacent side lengths a and c of a TRIANGLE (with $a < c$) and one ACUTE ANGLE A opposite a does not, in general, uniquely determine a triangle. If $\sin A < a/c$, there are two possible TRIANGLES satisfying the given conditions. If $\sin A = a/c$, there is one possible TRIANGLE. If $\sin A > a/c$, there are no possible TRIANGLES. Remember: don't try to prove congruence with the ASS theorem or you will make an ASS out of yourself.

See also AAA THEOREM, AAS THEOREM, SAS THEOREM, SSS THEOREM, TRIANGLE

Associate

Let p be an ODD PRIME, a a positive number such that $p|a$ (i.e., p does not DIVIDE a), and let x be one of the numbers 1, 2, 3, ..., $p - 1$. Then there is a unique x', called the associate of x, such that

$$xx' \equiv a \pmod{p}$$

with $0 < x' < p$ (Hardy and Wright 1979, p. 67). If $x' = x$, then a is called a QUADRATIC RESIDUE of p.

See also QUADRATIC RESIDUE

References

Hardy, G. H. and Wright, E. M. *An Introduction to the Theory of Numbers, 5th ed.* Oxford, England: Clarendon Press, p. 67, 1979.

Associated Fiber Bundle

Given a GROUP ACTION $G \times F \to F$ and a PRINCIPAL BUNDLE $\pi : A \to M$, the associated fiber bundle on M is

$$\tilde{\pi} : A \times F/G \to M. \tag{1}$$

In particular, it is the QUOTIENT SPACE $A \times F/G$ where $(a, x) \sim (ga, g^{-1}x)$.

For example, the torus $\mathbb{T} = \{(e^{is}, e^{it})$ has a $\mathbb{S}^1$ action given by

$$\phi(e^{i\theta})(e^{is}, e^{it}) = (e^{i(s+\theta)}, e^{i(t+\theta)}) \tag{2}$$

and the frame bundle on the sphere,

$$\pi : SO(3) \to \mathbb{S}^2, \tag{3}$$

is a principal $\mathbb{S}^1$ bundle. The associated fiber bundle is a fiber bundle on the sphere, with fiber the torus. It is an example of a four-dimensional MANIFOLD.

See also BUNDLE, FIBER BUNDLE, GROUP ACTION, PRINCIPAL BUNDLE, QUOTIENT SPACE

Associated Laguerre Polynomial

LAGUERRE POLYNOMIAL

Associated Legendre Polynomial

LEGENDRE POLYNOMIAL

Associated Principal Bundle

See also BUNDLE

Associated Sequence

A SHEFFER SEQUENCE for $(1, f(t))$ is called the associated sequence for $f(t)$, and a sequence $s_n(x)$ of polynomials satisfying the orthogonality conditions

$$\left\langle [f(t)]^k | s_n(x) \right\rangle = n! \delta_{nk},$$

where δ_{nk} is the DELTA FUNCTION, is said to be associated to $f(t)$.

See also SHEFFER SEQUENCE

References

Roman, S. *The Umbral Calculus.* New York: Academic Press, 1984.

Associated Stirling Number of the First Kind

STIRLING NUMBER OF THE FIRST KIND

Associated Triangles

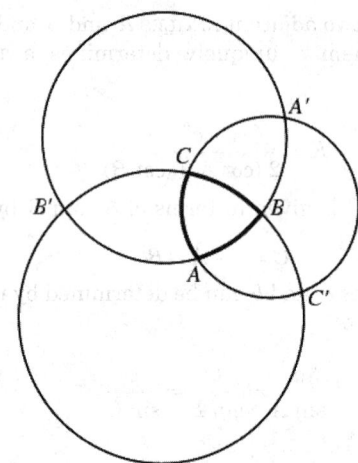

The three CIRCULAR TRIANGLES $A'B'C'$, $AB'C'$, $A'BC'$, and $A'B'C$ obtained by extending the arcs of a CIRCULAR TRIANGLE ABC into complete circles.

See also CIRCULAR TRIANGLE

References

Lachlan, R. *An Elementary Treatise on Modern Pure Geometry.* London: Macmillian, pp. 251–52, 1893.

Associated Vector Bundle

Given a PRINCIPAL BUNDLE $\pi : A \to M$, with fiber a LIE GROUP G and BASE MANIFOLD M, and a REPRESENTATION of G, say $\phi : G \times V \to V$, then the associated vector bundle is

$$\tilde{\pi} : A \times V/G \to M. \tag{1}$$

In particular, it is the QUOTIENT SPACE $A \times V/G$ where $(a, v) \sim (ga, g^{-1}v)$.

This construction has many uses. For instance, any REPRESENTATION of the ORTHOGONAL GROUP gives rise to a BUNDLE of TENSORS on a RIEMANNIAN MANIFOLD as the vector bundle associated to the FRAME BUNDLE.

For example, $\pi : SO(3) \to \mathbb{S}^2$ is the frame bundle on $\mathbb{S}^2$, where

$$\pi \left(\begin{bmatrix} w_1 \\ w_2 \\ w_3 \end{bmatrix} \right) = w_1, \tag{2}$$

writing the special orthogonal matrix with rows w_i. It is a $SO(2)$ bundle with the action defined by

$$\begin{bmatrix} \cos\theta & -\sin\theta \\ \sin\theta & \cos\theta \end{bmatrix} \cdot A = \begin{bmatrix} 1 & 0 & 0 \\ 0 & \cos\theta & -\sin\theta \\ 0 & \sin\theta & \cos\theta \end{bmatrix} A, \quad (3)$$

which preserves the map π.

The TANGENT BUNDLE is the associated vector bundle with the standard REPRESENTATION of $SO(2)$ on $V = \mathbb{R}^2$ given by pairs (v, A), with $v = (a, b) \in \mathbb{R}^2$ and $A \in SO(3)$. Two pairs (v_1, A_1) and (v_2, A_2) represent the same tangent vector IFF there is a $g \in SO(2)$ such that $v_2 = gv_1$ and $A_1 = g \cdot A_2$.

See also ASSOCIATED FIBER BUNDLE, FRAME BUNDLE, GROUP ACTION, LIE GROUP, PRINCIPAL BUNDLE, REPRESENTATION, QUOTIENT SPACE

Associative

Three elements x, y and z of a set S are said to be associative under a binary operation * if they satisfy

$$x*(y*z) = (x*y)*z.$$

Real numbers are associative under addition

$$x + (y + z) = (x + y) + z$$

and multiplication

$$x \cdot (y \cdot z) = (x \cdot y) \cdot z.$$

See also ASSOCIATIVE ALGEBRA, COMMUTATIVE, DISTRIBUTIVE, TRANSITIVE

Associative Algebra

In simple terms, let x, y, and z be members of an ALGEBRA. Then the ALGEBRA is said to be associative if

$$x \cdot (y \cdot z) = (x \cdot y) \cdot z, \quad (1)$$

where $\cdot$ denotes MULTIPLICATION. More formally, let A denote an $\mathbb{R}$-algebra, so that A is a VECTOR SPACE over $\mathbb{R}$ and

$$A \times A \to A \quad (2)$$

$$(x, y) \to x \cdot y. \quad (3)$$

Then A is said to be m-associative if there exists an m-dimensional SUBSPACE S of A such that

$$(y \cdot x) \cdot z = y \cdot (x \cdot z) \quad (4)$$

for all $y, z \in A$ and $x \in S$. Here, VECTOR MULTIPLICATION $x \cdot y$ is assumed to be BILINEAR. An n-dimensional n-associative ALGEBRA is simply said to be "associative."

See also ASSOCIATIVE

References
Finch, S. "Zero Structures in Real Algebras." http://www.mathsoft.com/asolve/zerodiv/zerodiv.html.

Associative Magic Square

1	15	24	8	17
23	7	16	5	14
20	4	13	22	6
12	21	10	19	3
9	18	2	11	25

An $n \times n$ MAGIC SQUARE for which every pair of numbers symmetrically opposite the center sum to $n^2 + 1$. The LO SHU is associative but not PANMAGIC. Order four squares can be PANMAGIC or associative, but not both. Order five squares are the smallest which can be both associative and PANMAGIC, and 16 distinct associative PANMAGIC SQUARES exist, one of which is illustrated above (Gardner 1988).

See also MAGIC SQUARE, PANMAGIC SQUARE

References
Gardner, M. "Magic Squares and Cubes." Ch. 17 in *Time Travel and Other Mathematical Bewilderments*. New York: W. H. Freeman, pp. 213-25, 1988.

Associator

For an ALGEBRA A, the associator is the trilinear map $A \times A \times A \to A$ given by

$$(x, y, z) = (xy)z - x(yz).$$

The associator is identically zero IFF A is associative.

See also ALTERNATIVE ALGEBRA, COMMUTATOR, POWER ASSOCIATIVE ALGEBRA

References
Schafer, R. D. *An Introduction to Nonassociative Algebras.* New York: Dover, p. 13, 1996.

Asterisk

STAR

Astroid

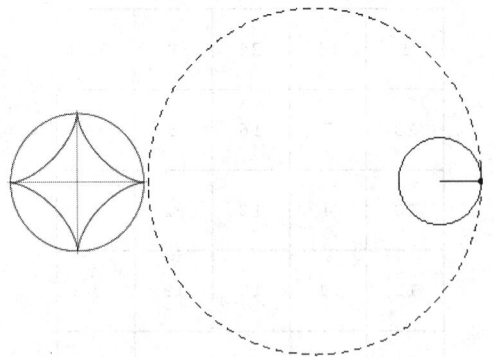

A 4-cusped HYPOCYCLOID which is sometimes also called a TETRACUSPID, CUBOCYCLOID, or PARACYCLE. The PARAMETRIC EQUATIONS of the astroid can be obtained by plugging in $n \equiv a/b = 4$ or $4/3$ into the equations for a general HYPOCYCLOID, giving

$$x = 3b \cos\phi + b \cos(3\phi) = 4b \cos^3\phi = a \cos^3\phi \quad (1)$$

$$y = 3b \sin\phi - b \sin(3\phi) = 4b \sin^3\phi = a \sin^3\phi. \quad (2)$$

In CARTESIAN COORDINATES,

$$x^{2/3} + y^{2/3} = a^{2/3}. \quad (3)$$

In PEDAL COORDINATES with the PEDAL POINT at the center, the equation is

$$r^2 + 3p^2 = a^2. \quad (4)$$

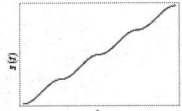 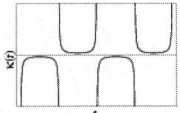 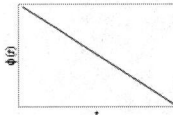

The ARC LENGTH, CURVATURE, and TANGENTIAL ANGLE are

$$s(t) = \frac{3}{2} \int_0^t |\sin(2t')| \, dt' = \frac{3}{2} \sin^2 t \quad (5)$$

$$\kappa(t) = -\frac{2}{3} \csc(2t) \quad (6)$$

$$\phi(t) = -t. \quad (7)$$

As usual, care must be taken in the evaluation of $s(t)$ for $t > \pi/2$. Since (5) comes from an integral involving the ABSOLUTE VALUE of a function, it must be monotonic increasing. Each QUADRANT can be treated correctly by defining

$$n = \left\lfloor \frac{2t}{\pi} \right\rfloor + 1, \quad (8)$$

where $\lfloor x \rfloor$ is the FLOOR FUNCTION, giving the formula

$$s(t) = (-1)^{1 + 1[n(\bmod 2)]} \frac{3}{2} \sin^2 t + 3[\tfrac{1}{2} n]. \quad (9)$$

The overall ARC LENGTH of the astroid can be computed from the general HYPOCYCLOID formula

$$s_n = \frac{Sa(n-1)}{n} \quad (10)$$

with $n = 4$,

$$s_4 = 6a. \quad (11)$$

The AREA is given by

$$A_n = \frac{(n-1)(n-2)}{n^2} \pi a^2 \quad (12)$$

with $n = 4$,

$$A_4 = \frac{3}{8} \pi a^2. \quad (13)$$

The EVOLUTE of an ELLIPSE is a stretched HYPOCYCLOID. The gradient of the TANGENT T from the point with parameter p is $-\tan p$. The equation of this TANGENT T is

$$x \sin p + y \cos p = \tfrac{1}{2} a \sin(2p) \quad (14)$$

(MacTutor Archive). Let T cut the X-AXIS and the Y-AXIS at X and Y, respectively. Then the length XY is a constant and is equal to a.

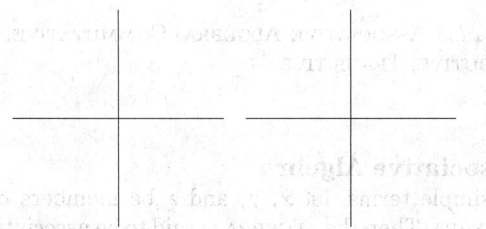

The astroid can also be formed as the ENVELOPE produced when a LINE SEGMENT is moved with each end on one of a pair of PERPENDICULAR axes (e.g., it is the curve enveloped by a ladder sliding against a wall or a garage door with the top corner moving along a vertical track; left figure above). The astroid is therefore a GLISSETTE. To see this, note that for a ladder of length L, the points of contact with the wall and floor are $(x_0, 0)$ and $(0, \sqrt{L^2 - x_0^2})$, respectively. The equation of the LINE made by the ladder with its foot at $(x_0, 0)$ is therefore

$$y - 0 = \frac{\sqrt{L^2 - x_0^2}}{-x_0} (x - x_0) \quad (15)$$

which can be written

$$U(x, y, x_0) = y + \frac{\sqrt{L^2 - x_0^2}}{x_0} (x - x_0). \quad (16)$$

The equation of the ENVELOPE is given by the simultaneous solution of

$$\begin{cases} U(x,\, y,\, x_0)=y+\dfrac{\sqrt{L^2-x_0^2}}{x_0}(x-x_0)=0 \\ \dfrac{\partial U}{\partial x_0}=\dfrac{x_0^2-L^2x}{x_0^2\sqrt{L^2-x_0^2}}=0, \end{cases} \quad (17)$$

which is

$$x=\frac{x_0^3}{L^2} \qquad (18)$$

$$y=\frac{(L^2-x_0^2)^{3/2}}{L^2} \qquad (19)$$

Noting that

$$x^{2/3}=\frac{x_0^2}{L^{4/3}} \qquad (20)$$

$$y^{2/3}=\frac{L^2-x_0^2}{L^{4/3}} \qquad (21)$$

allows this to be written implicitly as

$$x^{2/3}+y^{2/3}=L^{2/3}, \qquad (22)$$

the equation of the astroid, as promised.

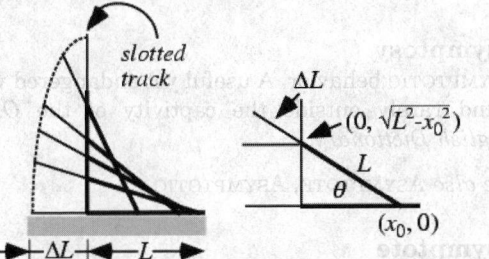

The related problem obtained by having the "garage door" of length L with an "extension" of length ΔL move up and down a slotted track also gives a surprising answer. In this case, the position of the "extended" end for the foot of the door at horizontal position x_0 and ANGLE θ is given by

$$x=-\Delta L\cos\theta \qquad (23)$$

$$y=\sqrt{L^2-x_0^2}+\Delta L\sin\theta. \qquad (24)$$

Using

$$x_0=L\cos\theta \qquad (25)$$

then gives

$$x=-\frac{\Delta L}{L}x_0 \qquad (26)$$

$$y=\sqrt{L^2-x_0^2}\left(1+\frac{\Delta L}{L}\right) \qquad (27)$$

Solving (26) for x_0, plugging into (27) and squaring then gives

$$y^2=L^2-\frac{L^2x^2}{(\Delta L)^2}\left(1+\frac{\Delta L}{L}\right)^2. \qquad (28)$$

Rearranging produces the equation

$$\frac{x^2}{(\Delta L)^2}+\frac{y^2}{(L+\Delta L)^2}=1, \qquad (29)$$

the equation of a (QUADRANT of an) ELLIPSE with SEMIMAJOR and SEMIMINOR AXES of lengths δl and $l+\delta l$.

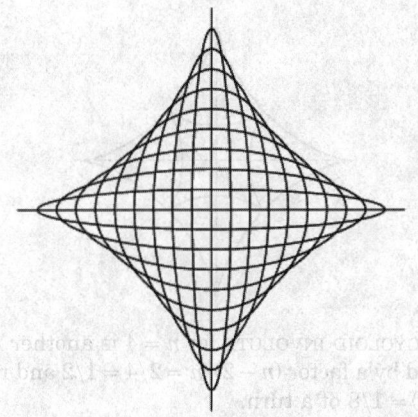

the astroid is also the ENVELOPE of the family of ELLIPSES

$$\frac{x^2}{c^2}+\frac{y^2}{(1-c)^2}-1=0, \qquad (30)$$

illustrated above (Wells 1991).

See also DELTOID, ELLIPSE ENVELOPE, LAMÉ CURVE, NEPHROID, RANUNCULOID

References

Beyer, W. H. *CRC Standard Mathematical Tables, 28th ed.* Boca Raton, FL: CRC Press, p. 219, 1987.

Lawrence, J. D. *A Catalog of Special Plane Curves.* New York: Dover, pp. 172–75, 1972.

Lockwood, E. H. "The Astroid." Ch. 6 in *A Book of Curves.* Cambridge, England: Cambridge University Press, pp. 52–1, 1967.

MacTutor History of Mathematics Archive. "Astroid." http://www-groups.dcs.st-and.ac.uk/~history/Curves/Astroid.html.

Steinhaus, H. *Mathematical Snapshots, 3rd ed.* New York: Dover, pp. 146–47, 1999.

Wells, D. *The Penguin Dictionary of Curious and Interesting Geometry.* London: Penguin, pp. 10–1, 1991.

Yates, R. C. "Astroid." *A Handbook on Curves and Their Properties.* Ann Arbor, MI: J. W. Edwards, pp. 1–, 1952.

Astroid Evolute

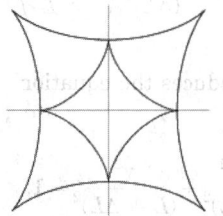

A HYPOCYCLOID EVOLUTE for $n = 4$ is another ASTROID scaled by a factor $n/(n-2) = 4/2 = 2$ and rotated $1/(2 \cdot 4) = 1/8$ of a turn.

Astroid Involute

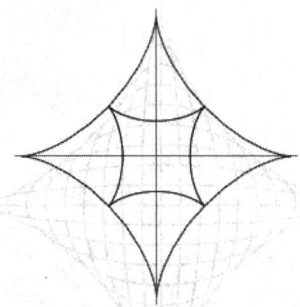

A HYPOCYCLOID INVOLUTE for $n = 4$ is another ASTROID scaled by a factor $(n-2)/n = 2/4 = 1/2$ and rotated $1/(2 \cdot 4) = 1/8$ of a turn.

Astroid Pedal Curve

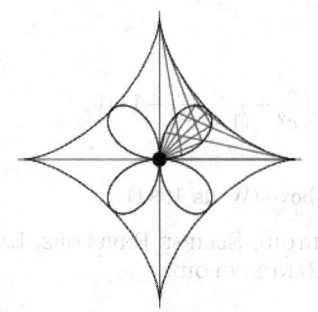

The PEDAL CURVE of an ASTROID with PEDAL POINT at the center is a QUADRIFOLIUM.

Astroid Radial Curve

The QUADRIFOLIUM

$$x = x_0 + 3a \cos t - 3a \cos(3t)$$

$$y = y_0 + 3a \sin t + 3 \sin(3t).$$

Astroidal Ellipsoid

The surface which is the inverse of the ELLIPSOID in the sense that it "goes in" where the ELLIPSOID "goes out." It is given by the PARAMETRIC EQUATIONS

$$x = (a \cos u \cos v)^3$$

$$y = (b \sin u \cos v)^3$$

$$z = (c \sin v)^3$$

for $u \in [-\pi/2, \pi/2]$ and $v \in [-\pi, \pi]$. The special case $a = b = c = 1$ corresponds to the HYPERBOLIC OCTAHEDRON.

See also ELLIPSOID, HYPERBOLIC OCTAHEDRON

References
Nordstrand, T. "Astroidal Ellipsoid." http://www.uib.no/people/nfytn/asttxt.htm.

Asymptosy

ASYMPTOTIC behavior. A useful yet endangered word, found rarely outside the captivity of the *Oxford English Dictionary.*

See also ASYMPTOTE, ASYMPTOTIC

Asymptote

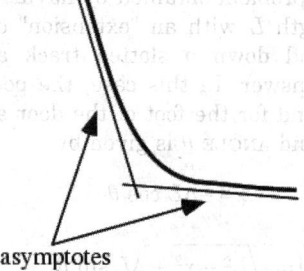

asymptotes

A curve approaching a given curve arbitrarily closely, as illustrated in the above diagram.

See also ASYMPTOSY, ASYMPTOTIC, ASYMPTOTIC CURVE

References
Giblin, P. J. "What is an Asymptote?" *Math. Gaz.* **56**, 274–84, 1972.

Asymptotic

Approaching a value or curve arbitrarily closely (i.e., as some sort of LIMIT is taken). A CURVE A which is asymptotic to given CURVE C is called the ASYMPTOTE of C. Hardy and Wright (1979, p. 7) use the symbol $\asymp$ to denote that one quantity is asymptotic to another. If $f \asymp \phi$, then Hardy and Wright say that f and ϕ are of the same ORDER OF MAGNITUDE.

See also ASYMPTOSY, ASYMPTOTE, ASYMPTOTIC CURVE, ASYMPTOTIC DIRECTION, ASYMPTOTIC NOTATION, ASYMPTOTIC SERIES, LANDAU SYMBOL, LIMIT, ORDER OF MAGNITUDE

References

Hardy, G. H. and Wright, E. M. *An Introduction to the Theory of Numbers*, 5th ed. Oxford, England: Clarendon Press, 1979.

Asymptotic Curve

Given a REGULAR SURFACE M, an asymptotic curve is formally defined as a curve $\mathbf{x}(t)$ on M such that the NORMAL CURVATURE is 0 in the direction $\mathbf{x}'(t)$ for all t in the domain of $\mathbf{x}$. The differential equation for the parametric representation of an asymptotic curve is

$$eu'^2 + 2fu'v' + gv'^2 = 0, \tag{1}$$

where e, f, and g are coefficients of the SECOND FUNDAMENTAL FORM. The differential equation for asymptotic curves on a MONGE PATCH $(u, v, h(u, v))$ is

$$h_{uu}u'^2 + 2h_{uu}u'v' + h_{vv}v'^2 = 0, \tag{2}$$

and on a polar patch $(r \cos\theta, r \sin\theta, h(r))$ is

$$h''(r)r'^2 + h'(r)r\theta'^2 = 0. \tag{3}$$

The images below show asymptotic curves for the ELLIPTIC HELICOID, FUNNEL, HYPERBOLIC PARABOLOID, and MONKEY SADDLE.

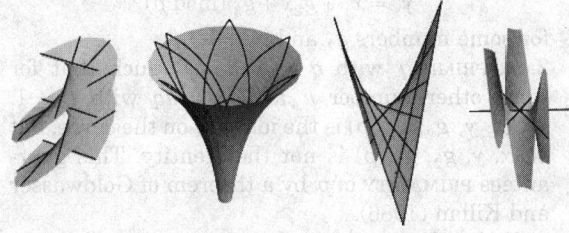

See also RULED SURFACE

References

Gray, A. "Asymptotic Curves," "Examples of Asymptotic Curves," and "Using Mathematica to Find Asymptotic Curves." §18.1, 18.2, and 18.3 in *Modern Differential Geometry of Curves and Surfaces with Mathematica*, 2nd ed. Boca Raton, FL: CRC Press, pp. 417–29, 1997.

Asymptotic Direction

An asymptotic direction at a point $\mathbf{p}$ of a REGULAR SURFACE $M \in \mathbb{R}^3$ is a direction in which the NORMAL CURVATURE of M vanishes.

1. There are no asymptotic directions at an ELLIPTIC POINT.
2. There are exactly two asymptotic directions at a HYPERBOLIC POINT.
3. There is exactly one asymptotic direction at a PARABOLIC POINT.
4. Every direction is asymptotic at a PLANAR POINT.

See also ASYMPTOTIC CURVE

References

Gray, A. *Modern Differential Geometry of Curves and Surfaces with Mathematica*, 2nd ed. Boca Raton, FL: CRC Press, pp. 364 and 418, 1997.

Asymptotic Equipartition Property

This entry contributed by ERIK G. MILLER

A theorem from INFORMATION THEORY that is a simple consequence of the WEAK LAW OF LARGE NUMBERS. It states that if a set of values $X_1, X_2, ..., X_n$ is drawn independently from a random variable X distributed according to $P(x)$ then the joint probability $P(X_1, ..., X_n)$ satisfies

$$-\frac{1}{n} \ln P(X_1, X_2, ..., X_n) \to H(X),$$

where $H(X)$ is the ENTROPY of the random variable X.

See also ENTROPY

References

Cover, T. M. and Thomas, J. A. *Elements of Information Theory*. New York: Wiley, 1991.

Asymptotic Expansion

ASYMPTOTIC SERIES

Asymptotic Notation

Let n be a integer variable which tends to infinity and let x be a continuous variable tending to some limit. Also, let $\phi(n)$ or $\phi(x)$ be a positive function and $f(n)$ or $f(x)$ any function. Then Hardy and Wright (1979) define

1. $f = \mathcal{O}(\phi)$ to mean that $|f| < A\phi$ for some constant A and all values of n and x,
2. $f = o(\phi)$ to mean that $f/\phi \to 0$,
3. $f \sim \phi$ to mean that $f/\phi \to 1$,
4. $f \prec \phi$ to mean the same as $f = o(\phi)$,
5. $f \succ \phi$ to mean $f/\phi \to \infty$, and
6. $f \asymp \phi$ to mean $A_1\phi < f < A_2\phi$ for some positive constants A_1 and A_2.

$f = o(\phi)$ implies and is stronger than $f = \mathcal{O}(\phi)$.

The term LANDAU SYMBOL is sometimes used to indicate the notation $o(\phi)$, and in general, $\mathcal{O}(x)$ and $o(x)$ are read as "is of order x."

See also LANDAU SYMBOL

References
Hardy, G. H. and Wright, E. M. "Some Notations." §1.6 in *An Introduction to the Theory of Numbers, 5th ed.* Oxford, England: Clarendon Press, pp. 7-, 1979.

Jeffreys, H. and Jeffreys, B. S. "Increasing and Decreasing Functions." §1.065 in *Methods of Mathematical Physics, 3rd ed.* Cambridge, England: Cambridge University Press, p. 22, 1988.

Asymptotic Series

An asymptotic series is a SERIES EXPANSION of a FUNCTION in a variable x which may converge or diverge (Erdélyi 1987, p. 1), but whose partial sums can be made an arbitrarily good approximation to a given function for large enough x. To form an asymptotic series $R(x)$ of

$$f(x) \sim R(x), \tag{1}$$

take

$$x^n R_n(x) = x^n [f(x) - S_n(x)], \tag{2}$$

where

$$S_n(x) \equiv a_0 + \frac{a_1}{x} + \frac{a_2}{x^2} + \cdots + \frac{a_n}{x^n}. \tag{3}$$

The asymptotic series is defined to have the properties

$$\lim_{x \to \infty} x^n R_n(x) = 0 \quad \text{for fixed } n \tag{4}$$

$$\lim_{x \to \infty} x^n R_n(x) = \infty \quad \text{for fixed } x \tag{5}$$

Therefore,

$$f(x) \approx \sum_{n=0}^{\infty} a_n x^{-n} \tag{6}$$

in the limit $x \to \infty$. If a function has an asymptotic expansion, the expansion is unique. The symbol $\sim$ is also used to mean directly SIMILAR.

See also HYPERASYMPTOTIC SERIES, SUPERASYMPTOTIC SERIES

References
Abramowitz, M. and Stegun, C. A. (Eds.). *Handbook of Mathematical Functions with Formulas, Graphs, and Mathematical Tables, 9th printing.* New York: Dover, p. 15, 1972.

Arfken, G. "Asymptotic of Semiconvergent Series." §5.10 in *Mathematical Methods for Physicists, 3rd ed.* Orlando, FL: Academic Press, pp. 339–46, 1985.

Bleistein, N. and Handelsman, R. A. *Asymptotic Expansions of Integrals.* New York: Dover, 1986.

Boyd, J. P. "The Devil's Invention: Asymptotic, Superasymptotic and Hyperasymptotic Series." *Acta Appl. Math.* **56**, 1–8, 1999.

Copson, E. T. *Asymptotic Expansions.* Cambridge, England: Cambridge University Press, 1965.

de Bruijn, N. G. *Asymptotic Methods in Analysis.* New York: Dover, 1982.

Dingle, R. B. *Asymptotic Expansions: Their Derivation and Interpretation.* London: Academic Press, 1973.

Erdélyi, A. *Asymptotic Expansions.* New York: Dover, 1987.

Morse, P. M. and Feshbach, H. "Asymptotic Series; Method of Steepest Descent." §4.6 in *Methods of Theoretical Physics, Part I.* New York: McGraw-Hill, pp. 434–43, 1953.

Olver, F. W. J. *Asymptotics and Special Functions.* New York: Academic Press, 1974.

Wasow, W. R. *Asymptotic Expansions for Ordinary Differential Equations.* New York: Dover, 1987.

Weisstein, E. W. "Books about Asymptotic Series." http://www.treasure-troves.com/books/AsymptoticSeries.html.

Atiyah-Singer Index Theorem

A theorem which states that the analytic and topological "indices" are equal for any elliptic differential operator on an n-D COMPACT DIFFERENTIABLE C^∞ boundaryless MANIFOLD.

See also COMPACT MANIFOLD, DIFFERENTIABLE MANIFOLD

References
Atiyah, M. F. and Singer, I. M. "The Index of Elliptic Operators on Compact Manifolds." *Bull. Amer. Math. Soc.* **69**, 322–33, 1963.

Atiyah, M. F. and Singer, I. M. "The Index of Elliptic Operators I, II, III." *Ann. Math.* **87**, 484–04, 1968.

Petkovsek, M.; Wilf, H. S.; and Zeilberger, D. *A = B.* Wellesley, MA: A. K. Peters, p. 4, 1996.

Atkin-Goldwasser-Kilian-Morain Certificate

A recursive PRIMALITY CERTIFICATE for a PRIME p. The certificate consists of a list of

1. A point on an ELLIPTIC CURVE C

$$y^2 = x^3 + g_2 x + g_3 \pmod{p}$$

for some numbers g_2 and g_3.

2. A PRIME q with $q > (p^{1/4} + 1)^2$ such that for some other number k and $m = kq$ with $k \neq 1$, $mC(x, y, g_2, g_3, p)$ is the identity on the curve, but $kC(x, y, g_2, g_3, p)$ is not the identity. This guarantees PRIMALITY of p by a theorem of Goldwasser and Kilian (1986).

3. Each q has its recursive certificate following it. So if the smallest q is known to be PRIME, all the numbers are certified PRIME up the chain.

A PRATT CERTIFICATE is quicker to generate for small numbers. The *Mathematica* task `ProvablePrimeQ[n]` in the *Mathematica* add-on package `NumberTheory`PrimeQ`` (which can be loaded with the command `<<NumberTheory`` `) therefore generates an Atkin-Goldwasser-Kilian-Morain certificate only for numbers above a certain limit (10^{10} by default), and a PRATT CERTIFICATE for smaller numbers.

See also ELLIPTIC CURVE PRIMALITY PROVING, ELLIPTIC PSEUDOPRIME, PRATT CERTIFICATE, PRIMALITY CERTIFICATE, WITNESS

References

Atkin, A. O. L. and Morain, F. "Elliptic Curves and Primality Proving." *Math. Comput.* **61**, 29–8, 1993.

Bressoud, D. M. *Factorization and Prime Testing.* New York: Springer-Verlag, 1989.

Goldwasser, S. and Kilian, J. "Almost All Primes Can Be Quickly Certified." *Proc. 18th STOC.* pp. 316–29, 1986.

Morain, F. "Implementation of the Atkin-Goldwasser-Kilian Primality Testing Algorithm." Rapport de Recherche 911, INRIA, Octobre 1988.

Schoof, R. "Elliptic Curves over Finite Fields and the Computation of Square Roots mod *p*." *Math. Comput.* **44**, 483–94, 1985.

Wunderlich, M. C. "A Performance Analysis of a Simple Prime-Testing Algorithm." *Math. Comput.* **40**, 709–14, 1983.

Atlas

An atlas is a collection of consistent COORDINATE CHARTS on a MANIFOLD, where "consistent" most commonly means that the TRANSITION FUNCTIONS of the charts are SMOOTH. As the name suggests, an atlas corresponds to a collection of maps, each of which shows a piece of a MANIFOLD and looks like flat two-dimensional Euclidean space. To use an atlas, one needs to know how the maps overlap. To be useful, the maps must not be too different on these overlapping areas.

The overlapping maps from one chart to another are called transition functions. They represent the transition from one chart's point of view to that of another. Let the open unit ball in $\mathbb{R}^n$ be denoted B_1. Then if $\phi : U \to B_1$ and $\psi : V \to B_1$ are two coordinate charts, the composition $\phi \circ \psi^{-1}$ is a function defined on $\psi(U \cap V)$. That is, it is a function from an open subset of B_1 to B_1, and given such a function from $\mathbb{R}^n$ to $\mathbb{R}^n$, there are conditions for it to be smooth or have k smooth derivatives (i.e., it is a C-K FUNCTION). Furthermore, when $\mathbb{R}^{2n}$ is isomorphic to $\mathbb{C}^n$ (in the even DIMENSIONAL case), a function can be HOLOMORPHIC.

A smooth atlas has transition functions that are C-INFINITY smooth (i.e., infinitely differentiable). The consequence is that a smooth function on one chart is smooth in any other chart (by the CHAIN RULE for higher derivatives). Similarly, one could have an atlas in class C^k, where the transition functions are in class C-K.

In the even-dimensional case, one may ask whether the transition functions are HOLOMORPHIC. In this case, one has a holomorphic atlas, and by the chain rule, it makes sense to ask if a function on the manifold is holomorphic.

It is possible for two atlases to be compatible, meaning the union is also an atlas. By ZORN'S LEMMA, there always exists a maximal atlas, where a maximal atlas is an atlas not contained in any other atlas. However, in typical applications, it is not necessary to use a maximal atlas and any sufficiently refined atlas will do.

See also COORDINATE CHART, HOLOMORPHIC FUNCTION, MANIFOLD, SMOOTH FUNCTION, TRANSITION FUNCTION, ZORN'S LEMMA

Atom

ATOMIC STATEMENT, URELEMENT

Atomic Statement

In LOGIC, a statement which cannot be broken down into smaller statements.

Attraction Basin

BASIN OF ATTRACTION

Attractor

An attractor is a SET of states (points in the PHASE SPACE), invariant under the dynamics, towards which neighboring states in a given BASIN OF ATTRACTION asymptotically approach in the course of dynamic evolution. An attractor is defined as the smallest unit which cannot be itself decomposed into two or more attractors with distinct BASINS OF ATTRACTION. This restriction is necessary since a DYNAMICAL SYSTEM may have multiple attractors, each with its own BASIN OF ATTRACTION.

Conservative systems do not have attractors, since the motion is periodic. For dissipative DYNAMICAL SYSTEMS, however, volumes shrink exponentially so attractors have 0 volume in n-D phase space.

A stable FIXED POINT surrounded by a dissipative region is an attractor known as a SINK. Regular attractors (corresponding to 0 LYAPUNOV CHARACTERISTIC EXPONENTS) act as LIMIT CYCLES, in which trajectories circle around a limiting trajectory which they asymptotically approach, but never reach. STRANGE ATTRACTORS are bounded regions of PHASE SPACE (corresponding to POSITIVE LYAPUNOV CHARACTERISTIC EXPONENTS) having zero MEASURE in the embedding PHASE SPACE and a FRACTAL DIMENSION. Trajectories within a STRANGE ATTRACTOR appear to skip around randomly.

See also BARNSLEY'S FERN, BASIN OF ATTRACTION, CHAOS GAME, FRACTAL DIMENSION, LIMIT CYCLE, LYAPUNOV CHARACTERISTIC EXPONENT, MEASURE, SINK (MAP), STRANGE ATTRACTOR

Aubel's Theorem

VON AUBEL'S THEOREM

Auction

A type of sale in which members of a group of buyers offer ever increasing amounts. The bidder making the last bid (for which no higher bid is subsequently made within a specified time limit: "going once, going twice, sold") must then purchase the item in question at this price. Variants of simple bidding are also possible, as in a VICKREY AUCTION.

See also VICKREY AUCTION

Augend

The first of several ADDENDS, or "the one to which the others are added," is sometimes called the augend. Therefore, while a, b, and c are ADDENDS in $a + b + c$, a is the augend.

See also ADDEND, ADDITION

Augmented Amicable Pair

A PAIR of numbers m and n such that

$$\sigma(m) = \sigma(n) = m + n - 1,$$

where $\sigma(m)$ is the DIVISOR FUNCTION. Beck and Najar (1977) found 11 augmented amicable pairs.

See also AMICABLE PAIR, DIVISOR FUNCTION, QUASIAMICABLE PAIR

References
Beck, W. E. and Najar, R. M. "More Reduced Amicable Pairs." *Fib. Quart.* **15**, 331–32, 1977.
Guy, R. K. *Unsolved Problems in Number Theory, 2nd ed.* New York: Springer-Verlag, p. 59, 1994.

Augmented Dodecahedron

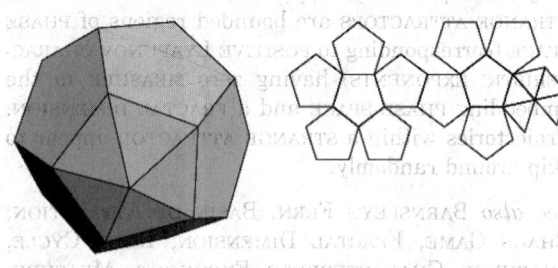

JOHNSON SOLID J_{58}.

References
Weisstein, E. W. "Johnson Solids." MATHEMATICA NOTEBOOK JOHNSONSOLIDS.M.
Weisstein, E. W. "Johnson Solid Netlib Database." MATHEMATICA NOTEBOOK JOHNSONSOLIDS.DAT.

Augmented Hexagonal Prism

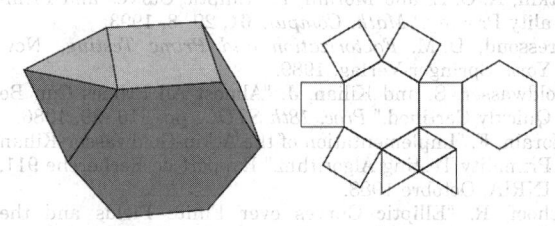

JOHNSON SOLID J_{54}.

References
Weisstein, E. W. "Johnson Solids." MATHEMATICA NOTEBOOK JOHNSONSOLIDS.M.
Weisstein, E. W. "Johnson Solid Netlib Database." MATHEMATICA NOTEBOOK JOHNSONSOLIDS.DAT.

Augmented Pentagonal Prism

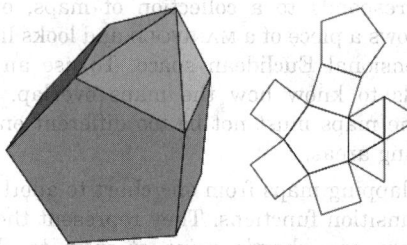

JOHNSON SOLID J_{52}.

References
Weisstein, E. W. "Johnson Solids." MATHEMATICA NOTEBOOK JOHNSONSOLIDS.M.
Weisstein, E. W. "Johnson Solid Netlib Database." MATHEMATICA NOTEBOOK JOHNSONSOLIDS.DAT.

Augmented Polyhedron

A UNIFORM POLYHEDRON with one or more other solids adjoined.

Augmented Sphenocorona

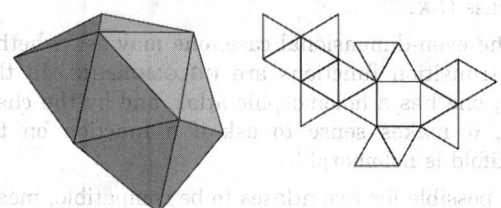

JOHNSON SOLID J_{87}.

References

Bressi, H. "Augmented Factorization in Approximate Numbers and Computers [...]

Weisstein, E. W. "Johnson Solids." MATHEMATICA NOTEBOOK JOHNSONSOLIDS.M.

Weisstein, E. W. "Johnson Solid Netlib Database." MATHEMATICA NOTEBOOK JOHNSONSOLIDS.DAT.

Augmented Truncated Cube

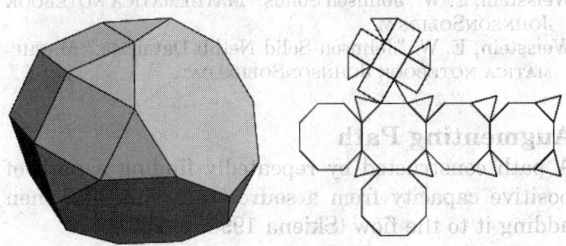

JOHNSON SOLID J_{66}.

References

Weisstein, E. W. "Johnson Solids." MATHEMATICA NOTEBOOK JOHNSONSOLIDS.M.

Weisstein, E. W. "Johnson Solid Netlib Database." MATHEMATICA NOTEBOOK JOHNSONSOLIDS.DAT.

Augmented Triangular Prism

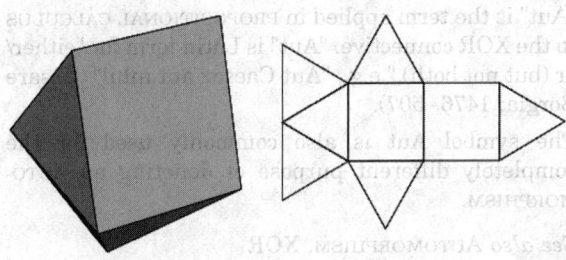

JOHNSON SOLID J_{49}.

References

Weisstein, E. W. "Johnson Solids." MATHEMATICA NOTEBOOK JOHNSONSOLIDS.M.

Weisstein, E. W. "Johnson Solid Netlib Database." MATHEMATICA NOTEBOOK JOHNSONSOLIDS.DAT.

Augmented Truncated Dodecahedron

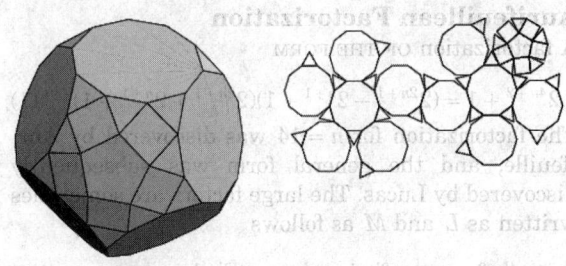

JOHNSON SOLID J_{68}.

References

Weisstein, E. W. "Johnson Solids." MATHEMATICA NOTEBOOK JOHNSONSOLIDS.M.

Weisstein, E. W. "Johnson Solid Netlib Database." MATHEMATICA NOTEBOOK JOHNSONSOLIDS.DAT.

Augmented Tridiminished Icosahedron

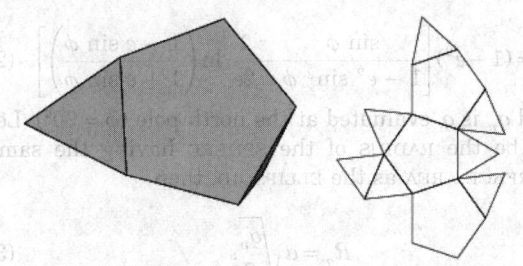

JOHNSON SOLID J_{64}.

References

Weisstein, E. W. "Johnson Solids." MATHEMATICA NOTEBOOK JOHNSONSOLIDS.M.

Weisstein, E. W. "Johnson Solid Netlib Database." MATHEMATICA NOTEBOOK JOHNSONSOLIDS.DAT.

Augmented Truncated Tetrahedron

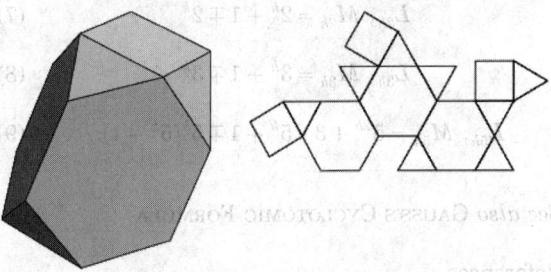

JOHNSON SOLID J_{65}.

References

Weisstein, E. W. "Johnson Solids." MATHEMATICA NOTEBOOK JOHNSONSOLIDS.M.

Weisstein, E. W. "Johnson Solid Netlib Database." MATHEMATICA NOTEBOOK JOHNSONSOLIDS.DAT.

Augmenting Path

A path constructed by repeatedly finding a path of positive capacity from a source to a sink and then adding it to the flow (Skiena 1990, p. 237).

See also BERGE'S THEOREM

References

Ford, L. R. and Fulkerson, D. R. *Flows in Networks.* Princeton, NJ: Princeton University Press, 1962.

Skiena, S. *Implementing Discrete Mathematics: Combinatorics and Graph Theory with Mathematica.* Reading, MA: Addison-Wesley, 1990.

Aureum Theorema

Gauss's name for the QUADRATIC RECIPROCITY THEOREM.

Aurifeuillean Factorization

A factorization OF THE FORM

$$2^{4n+2} + 1 = (2^{2n+1} - 2^{n+1} + 1)(2^{2n+1} + 2^{n+1} + 1). \quad (1)$$

The factorization for $n = 14$ was discovered by Aurifeuille, and the general form was subsequently discovered by Lucas. The large factors are sometimes written as L and M as follows

$$2^{4k-2} + 1 = (2^{2k-1} - 2^k + 1)(2^{2k-1} + 2^k + 1) \quad (2)$$

$$3^{6k-3} + 1 = (3^{2k-1} + 1)(3^{2k-1} - 3^k + 1)$$
$$\times (3^{2k-1} + 3^k + 1), \quad (3)$$

which can be written

$$2^{2h} + 1 = L_{2h} M_{2h} \quad (4)$$

$$3^{3h} + 1 = (3^h + 1) L_{3h} M_{3h} \quad (5)$$

$$5^{5k} - 1 = (5^h + 1) L_{5h} M_{5h}, \quad (6)$$

where $h = 2k - 1$ and

$$L_{2h}, M_{2h} = 2^h + 1 \mp 2^k \quad (7)$$

$$L_{3h}, M_{3h} = 3^h + 1 \mp 3^k \quad (8)$$

$$L_{5h}, M_{5h} = 5^{2h} + 3 \cdot 5^h + 1 \mp 5^k(5^k + 1). \quad (9)$$

See also GAUSS'S CYCLOTOMIC FORMULA

References

Brillhart, J.; Lehmer, D. H.; Selfridge, J.; Wagstaff, S. S. Jr.; and Tuckerman, B. *Factorizations of $b^n \pm 1$, $b = 2$, 3, 5, 6, 7, 10, 11, 12 Up to High Powers,* rev. ed. Providence, RI: Amer. Math. Soc., pp. lxviii-lxxii, 1988.

Riesel, H. "Aurifeullian Factorization" in Appendix 6. *Prime Numbers and Computer Methods for Factorization, 2nd ed.* Boston, MA: Birkhäuser, pp. 309–15, 1994.

Wagstaff, S. S. Jr. "Aurifeullian Factorizations and the Period of the Bell Numbers Modulo a Prime." *Math. Comput.* **65**, 383–91, 1996.

Ausdehnungslehre

EXTERIOR ALGEBRA

Aut

"Aut" is the term applied in PROPOSITIONAL CALCULUS to the XOR connective. "Aut" is Latin form for "either/ or (but not both)," e.g., "Aut Caesar aut nihil" (Cesare Borgia; 1476–507).

The symbol Aut is also commonly used for the completely different purpose of denoting an AUTOMORPHISM.

See also AUTOMORPHISM, XOR

References

Oxford University Press. *The Oxford Dictionary of Quotations, 3rd ed.* Oxford, England: Oxford University Press, p. 89, 1980.

Authalic Latitude

An AUXILIARY LATITUDE which gives a SPHERE equal SURFACE AREA relative to an ELLIPSOID. The authalic latitude is defined by

$$\beta = \sin^{-1}\left(\frac{q}{q_p}\right), \quad (1)$$

where

$$q = (1 - e^2)\left[\frac{\sin\phi}{1 - e^2 \sin^2\phi} - \frac{1}{2e}\ln\left(\frac{1 - e\sin\phi}{1 + e\sin\phi}\right)\right] \quad (2)$$

and q_p is q evaluated at the north pole ($\phi = 90°$). Let R_q be the RADIUS of the SPHERE having the same SURFACE AREA as the ELLIPSOID, then

$$R_q = a\sqrt{\frac{q_p}{2}}. \quad (3)$$

The series for β is

$$\beta = \phi - (\tfrac{1}{3}e^2 + \tfrac{31}{180}e^4 + \tfrac{59}{560}e^6 + \ldots)\sin(\phi)$$

$$+ (\tfrac{17}{360}e^4 + \tfrac{61}{1260}e^6 + \ldots)\sin(4\phi)$$

$$- (\tfrac{383}{45360}e^6 + \ldots)\sin(6\phi) + \ldots. \quad (4)$$

The inverse FORMULA is found from

$$\Delta\phi = \frac{(1 - e^2 \sin^2 \phi)^2}{2\cos\phi}$$
$$\times \left[\frac{q}{1-e^2} - \frac{\sin\phi}{1 - e^2\sin^2\phi} + \frac{1}{2e}\ln\left(\frac{1 - e\sin\phi}{1 + e\sin\phi}\right)\right], \tag{5}$$

where

$$q = q_p \sin\beta \tag{6}$$

and $\phi_0 = \sin^{-1}(q/2)$. This can be written in series form as

$$\phi = \beta + (\tfrac{1}{3}e^2 + \tfrac{31}{180}e^4 + \tfrac{517}{5040}e^6 + \ldots)\sin(2\beta).$$

$$+ (\tfrac{23}{360}e^4 + \tfrac{251}{3780}e^6 + \ldots)\sin(4\beta)$$

$$+ (\tfrac{761}{45360}e^6 + \ldots)\sin(6\beta) + \ldots. \tag{7}$$

See also LATITUDE

References

Adams, O. S. "Latitude Developments Connected with Geodesy and Cartography with Tables, Including a Table for Lambert Equal-Area Meridional Projections." Spec. Pub. No. 67. U. S. Coast and Geodetic Survey, 1921.

Snyder, J. P. *Map Projections—A Working Manual.* U. S. Geological Survey Professional Paper 1395. Washington, DC: U. S. Government Printing Office, p. 16, 1987.

Authalic Projection

Lee (1944) defines an authalic MAP PROJECTION to be one in which at any point the scales in two orthogonal directions are inversely proportional.

See also EQUAL-AREA PROJECTION

References

Lee, L. P. "The Nomenclature and Classification of Map Projections." *Empire Survey Review* **7**, 190–00, 1944.

Autocorrelation

The autocorrelation function $R_f(t)$ of a real function $f(t)$ is defined by

$$R_f(t) \equiv \lim_{T\to\infty} \frac{1}{2T} \int_{-T}^{T} f(\tau)f(T + \tau)\, d\tau \tag{1}$$

(Papoulis 1962, p. 241). For a complex function, the autocorrelation $\rho_f(t)$ is defined by

$$\rho_f(t) \equiv f \star f = \bar{f}(-t) * f(t) = \int_{-\infty}^{\infty} f(t + \tau)\bar{f}(\tau)\, d\tau. \tag{2}$$

where $*$ denotes CONVOLUTION, $\star$ denotes CROSS-CORRELATION, and $\bar{f}$ is the COMPLEX CONJUGATE (Papoulis 1962, pp. 241–42). The autocorrelation discards phase information, returning only the power, and is therefore an irreversible operation.

There is also a somewhat surprising and extremely important relationship between the autocorrelation and the FOURIER TRANSFORM known as the WIENER-KHINTCHINE THEOREM. Let $\mathscr{F}[f(x)] = F(k)$, and $\bar{F}$ denote the COMPLEX CONJUGATE of F, then the FOURIER TRANSFORM of the ABSOLUTE SQUARE of $F(k)$ is given by

$$\mathscr{F}[|F(k)|^2] = \int_{-\infty}^{\infty} \bar{f}(\tau)f(\tau + x)\, d\tau. \tag{3}$$

The autocorrelation is a HERMITIAN OPERATOR since $\rho_f(-t) = \bar{\rho}_f(t)$.

$f \star f$ is MAXIMUM at the ORIGIN; in other words,

$$\int_{-\infty}^{\infty} f(u)f(u+x)\, du \le \int_{-\infty}^{\infty} f^2(u)\, du. \tag{4}$$

To see this, let ϵ be a REAL NUMBER. Then

$$\int_{-\infty}^{\infty} [f(u) + \epsilon f(u+x)]^2\, du > 0 \tag{5}$$

$$\int_{-\infty}^{\infty} f^2(u)\, du + 2\epsilon\int_{-\infty}^{\infty} f(u)f(u+x)\, du$$
$$+ \epsilon^2 \int_{-\infty}^{\infty} f^2(u+x)\, du > 0 \tag{6}$$

$$\int_{-\infty}^{\infty} f^2(u)\, du + 2\epsilon\int_{-\infty}^{\infty} f(u)f(u+x)\, du$$
$$+ \epsilon^2 \int_{-\infty}^{\infty} f^2(u+x)\, du > 0. \tag{7}$$

Define

$$a \equiv \int_{-\infty}^{\infty} f^2(u)\, du \tag{8}$$

$$b \equiv 2\int_{-\infty}^{\infty} f(u)f(u+x)\, du. \tag{9}$$

Then plugging into above, we have $a\epsilon^2 + b\epsilon + c > 0$. This QUADRATIC EQUATION does not have any REAL ROOT, so $b^2 - 4ac \le 0$, i.e., $b/2 \le a$. It follows that

$$\int_{-\infty}^{\infty} f(u)f(u+x)\, du \le \int_{-\infty}^{\infty} f^2(u)\, du, \tag{10}$$

with the equality at $x = 0$. This proves that $f \star f$ is MAXIMUM at the ORIGIN.

See also AVERAGE POWER, CONVOLUTION, CROSS-CORRELATION, QUANTIZATION EFFICIENCY, WIENER-KHINTCHINE THEOREM

References

Bracewell, R. "The Autocorrelation Function." *The Fourier Transform and Its Applications,* 3rd ed. New York: McGraw-Hill, pp. 40–5, 1999.

Papoulis, A. *The Fourier Integral and Its Applications.* New York: McGraw-Hill, 1962.

Press, W. H.; Flannery, B. P.; Teukolsky, S. A.; and Vetterling, W. T. "Correlation and Autocorrelation Using the

FFT." §13.2 in *Numerical Recipes in FORTRAN: The Art of Scientific Computing, 2nd ed.* Cambridge, England: Cambridge University Press, pp. 538–39, 1992.

Autogonal Projection

CONFORMAL PROJECTION

Automata Theory

The mathematical study of abstract computing machines (especially TURING MACHINES) and the analysis of algorithms used by such machines.

See also CELLULAR AUTOMATON, TURING MACHINE

References

Harrison, M. A. *Introduction to Switching and Automata Theory.* New York: McGraw-Hill, p. 188, 1965.
Simon, M. *Automata Theory.* Singapore: World Scientific, 1999.
Wolfram, S. *A New Kind of Science.* Champaign, IL: Wolfram Media, 2001.

Automatic Set

A k-automatic set is a set of integers whose base-k representations form a regular language, i.e., a language accepted by a finite automaton or state machine. If bases a and b are incompatible (do not have a common power) and if an a-automatic set S_a and b-automatic set S_b are both of density 0 over the integers, then it is believed that $S_a \cap S_b$ is finite. However, this problem has not been settled.

Some automatic sets, such as the 2-automatic consisting of numbers whose BINARY representations contain at most two 1s: 1, 2, 3, 4, 5, 6, 8, 9, 10, 12, 16, 17, 18, ... (Sloane's A048645) have a simple arithmetic expression. However, this is not the case for general k-automatic sets.

See also TURING MACHINE

References

Cobham, A. "On the Base-Dependence of Sets of Numbers Recognizable by Finite Automata." *Math. Systems Th.* **3**, 186–92, 1969.
Cobham, A. "Uniform Tag Sequences." *Math. Systems Th.* **6**, 164–92, 1972.
Sloane, N. J. A. Sequences A048645 in "An On-Line Version of the Encyclopedia of Integer Sequences." http://www.research.att.com/~njas/sequences/eisonline.html.

Automaton

AUTOMATIC SET, CELLULAR AUTOMATON, TURING MACHINE

Automorphic Form

See also AUTOMORPHIC FUNCTION, LANGLANDS PROGRAM

Automorphic Function

An automorphic function $f(z)$ of a COMPLEX variable z is one which is analytic (except for POLES) in a domain D and which is invariant under a DENUMERABLY INFINITE group of LINEAR FRACTIONAL TRANSFORMATIONS (also known as MÖBIUS TRANSFORMATIONS)

$$z' = \frac{az + b}{cz + d}.$$

Automorphic functions are generalizations of TRIGONOMETRIC FUNCTIONS and ELLIPTIC FUNCTIONS.

See also AUTOMORPHIC FORM, MODULAR FUNCTION, MÖBIUS TRANSFORMATIONS, ZETA FUCHSIAN

References

Hadamard, J.; Gray, J. J.; and Shenitzer, A. *Non-Euclidean Geometry in the Theory of Automorphic Forms.* Providence, RI: Amer. Math. Soc., 1999.
Shimura, G. *Introduction to the Arithmetic Theory of Automorphic Functions.* Princeton, NJ: Princeton University Press, 1971.
Siegel, C. L. *Topics in Complex Function Theory, Vol. 2: Automorphic Functions and Abelian Integrals.* New York: Wiley, 1988.

Automorphic Number

A number k such that nk^2 has its last digits equal to k is called n-automorphic. For example, $1 \cdot 5^2 \equiv 25$ (Wells 1986, pp. 58–9) and $1 \cdot 6^2 \equiv 36$ (Wells 1986, p. 68) are 1-automorphic and $2 \cdot 8^2 \equiv 128$ and $2 \cdot 88^2 = 15488$ are 2-automorphic. de Guerre and Fairbairn (1968) give a history of automorphic numbers.

The first few 1-automorphic numbers are 1, 5, 6, 25, 76, 376, 625, 9376, 90625, ... (Sloane's A003226, Wells 1986, p. 130). There are two 1-automorphic numbers with a given number of digits, one ending in 5 and one in 6 (except that the 1-digit automorphic numbers include 1), and each of these contains the previous number with a digit prepended. Using this fact, it is possible to construct automorphic numbers having more than 25,000 digits (Madachy 1979). The first few 1-automorphic numbers ending with 5 are 5, 25, 625, 0625, 90625, ... (Sloane's A007185), and the first few ending with 6 are 6, 76, 376, 9376, 09376, ... (Sloane's A016090). The 1-automorphic numbers $a(n)$ ending in 5 are IDEMPOTENT (mod 10^n) since

$$[a(n)]^2 \equiv a(n) (\text{mod } 10^n)$$

(Sloane and Plouffe 1995).

The following table gives the 10-digit n-automorphic numbers.

n	n-Automorphic Numbers	Sloane
1	0000000001, 8212890625, 1787109376	–, A007185, A016090

2	0893554688	A030984
3	6666666667, 7262369792, 9404296875	–, A030985, A030986
4	0446777344	A030987
5	3642578125	A030988
6	3631184896	A030989
7	7142857143, 4548984375, 1683872768	A030990, A030991, A030992
8	0223388672	A030993
9	5754123264, 3134765625, 8888888889	A030994, A030995, –

The infinite 1-automorphic number ending in 5 is given by ...56259918212890625 (Sloane's A018247), while the infinite 1-automorphic number ending in 6 is given by ...740081787109376 (Sloane's A018248).

See also IDEMPOTENT, NARCISSISTIC NUMBER, NUMBER PYRAMID, TRIMORPHIC NUMBER

References

Fairbairn, R. A. "More on Automorphic Numbers." *J. Recr. Math.* **2**, 170–74, 1969.

Fairbairn, R. A. Erratum to "More on Automorphic Numbers." *J. Recr. Math.* **2**, 245, 1969.

de Guerre, V. and Fairbairn, R. A. "Automorphic Numbers." *J. Recr. Math.* **1**, 173–79, 1968.

Hunter, J. A. H. "Two Very Special Numbers." *Fib. Quart.* **2**, 230, 1964.

Hunter, J. A. H. "Some Polyautomorphic Numbers." *J. Recr. Math.* **5**, 27, 1972.

Kraitchik, M. "Automorphic Numbers." §3.8 in *Mathematical Recreations.* New York: W. W. Norton, pp. 77–8, 1942.

Madachy, J. S. *Madachy's Mathematical Recreations.* New York: Dover, pp. 34–4 and 175–76, 1979.

Schroeppel, R. Item 59 in Beeler, M.; Gosper, R. W.; and Schroeppel, R. *HAKMEM.* Cambridge, MA: MIT Artificial Intelligence Laboratory, Memo AIM-239, p. 23, Feb. 1972.

Sloane, N. J. A. Sequences A003226/M3752, A007185/M3940, A016090, A018247, and A018248 in "An On-Line Version of the Encyclopedia of Integer Sequences." http://www.research.att.com/~njas/sequences/eisonline.html.

Wells, D. *The Penguin Dictionary of Curious and Interesting Numbers.* Middlesex, England: Penguin Books, pp. 59 and 171, 178, 191–92, 1986.

Automorphism

An ISOMORPHISM of a system of objects onto itself. The term derives from the Greek prefix $\alpha\upsilon\tau o$ (*auto*) "self" and $\mu o\rho\phi\omega\sigma\iota\varsigma$ (*morphosis*) "to form" or "to shape."

The automorphisms of a GRAPH always describe a GROUP (Skiena 1990, p. 19).

An automorphism of a region of the COMPLEX PLANE is a conformal SELF-MAP (Krantz 1999, p. 81).

See also ANOSOV AUTOMORPHISM, GRAPH AUTOMORPHISM

References

Krantz, S. G. *Handbook of Complex Analysis.* Boston, MA: Birkhäuser, p. 81, 1999.

Skiena, S. *Implementing Discrete Mathematics: Combinatorics and Graph Theory with Mathematica.* Reading, MA: Addison-Wesley, 1990.

Automorphism Group

The GROUP of functions from an object G to itself which preserve the structure of the object, denoted $Aut(G)$. The automorphism group of a GROUP preserves the MULTIPLICATION table, the automorphism group of a GRAPH the INCIDENCE MATRICES, and that of a FIELD the ADDITION and MULTIPLICATION tables.

Autonomous

A differential equation or system of ORDINARY DIFFERENTIAL EQUATIONS is said to be autonomous if it does not explicitly contain the independent variable (usually denoted t). A second-order autonomous differential equation is OF THE FORM $F(y, y', y'') = 0$, where $y' \equiv dy/dt \equiv v$. By the CHAIN RULE, y'' can be expressed as

$$y'' = v' = \frac{dv}{dt} = \frac{dv}{dy}\frac{dy}{dt} = \frac{dv}{dy}v.$$

For an autonomous ODE, the solution is independent of the time at which the initial conditions are applied. This means that all particles pass through a given point in phase space. A nonautonomous system of n first-order ODEs can be written as an autonomous system of $n + 1$ ODEs by letting $t \equiv x_{n+1}$ and increasing the dimension of the system by 1 by adding the equation

$$\frac{dx_{n+1}}{dt} = 1.$$

Autoregressive Model

MAXIMUM ENTROPY METHOD

Auxiliary Circle

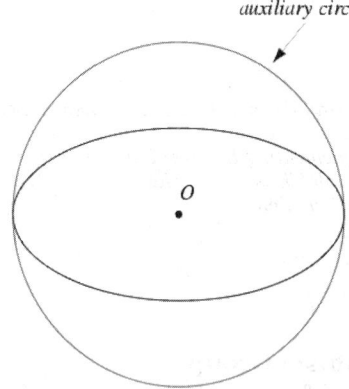

auxiliary circle

The CIRCUMCIRCLE of an ELLIPSE, i.e., the CIRCLE whose CENTER concurs with that of the ELLIPSE and whose RADIUS is equal to the ELLIPSE's SEMIMAJOR AXIS.

See also CIRCLE, ECCENTRIC ANGLE, ELLIPSE

References

Montenbruck, O. and Pfleger, T. *Astronomy on the Personal Computer, 4th ed.* Berlin: Springer-Verlag, p. 62, 2000.

Auxiliary Latitude

AUTHALIC LATITUDE, CONFORMAL LATITUDE, GEOCENTRIC LATITUDE, ISOMETRIC LATITUDE, LATITUDE, PARAMETRIC LATITUDE, RECTIFYING LATITUDE, REDUCED LATITUDE

Auxiliary Triangle

MEDIAL TRIANGLE

Average

MEAN

Average Absolute Deviation

$$\alpha = \frac{1}{N}\sum_{i=1}^{N}|x_i - \mu| = \langle|x_i - \mu|\rangle.$$

See also ABSOLUTE DEVIATION, DEVIATION, STANDARD DEVIATION, VARIANCE

Average Function

If f is CONTINUOUS on a CLOSED INTERVAL $[a, b]$, then there is at least one number x^* in $[a, b]$ such that

$$\int_a^b f(x)dx = f(x'')(b-a).$$

The average value of the FUNCTION (f^-) on this interval is then given by $f(x*)$.

See also MEAN-VALUE THEOREM

Average Power

The average power of a complex signal $f(t)$ as a function of time t is defined as

$$\langle f^2(t)\rangle = \lim_{T\to\infty}\frac{1}{2T}\int_{-T}^{T}|f(t)^2 dt|,$$

where $|z|$ is the MODULUS (Papoulis 1962, p. 240).

See also AUTOCORRELATION

References

Papoulis, A. *The Fourier Integral and Its Applications.* New York: McGraw-Hill, 1962.

Average Seek Time

POINT-POINT DISTANCE–1-D

Avoided Pattern

A pattern $\tau = (\tau_1, \ldots, \tau_n)$ is said to avoid $\alpha = (\alpha_1, \ldots, \alpha_k)$ if α is not CONTAINED in τ. In other words, τ avoids α IFF no K-SUBSET of τ is ORDER ISOMORPHIC to α.

See also CONTAINED PATTERN, ORDER ISOMORPHIC, PERMUTATION PATTERN, WILF CLASS, WILF EQUIVALENT

References

Mansour, T. Permutations Avoiding a Pattern from S_k and at Least Two Patterns from S_3. 31 Jul 2000. http://xxx.lanl.gov/abs/math.CO/0007194/.

Axial Vector

PSEUDOVECTOR

Axiom

A PROPOSITION regarded as self-evidently TRUE without PROOF. The word "axiom" is a slightly archaic synonym for POSTULATE. Compare CONJECTURE or HYPOTHESIS, both of which connote apparently TRUE *but not self-evident* statements.

See also ARCHIMEDES' AXIOM, AXIOM OF CHOICE, AXIOMATIC SYSTEM, CANTOR-DEDEKIND AXIOM, CONGRUENCE AXIOMS, CONJECTURE, CONTINUITY AXIOMS, COUNTABLE ADDITIVITY PROBABILITY AXIOM, DEDEKIND'S AXIOM, DIMENSION AXIOM, EILENBERG-STEENROD AXIOMS, EUCLID'S AXIOMS, EXCISION AXIOM, FANO'S AXIOM, FIELD AXIOMS, HAUSDORFF AXIOMS, HILBERT'S AXIOMS, HOMOTOPY AXIOM, INACCESSIBLE CARDINALS AXIOM, INCIDENCE AXIOMS, INDEPENDENCE AXIOM, INDUCTION AXIOM, LAW, LEMMA, LONG EXACT SEQUENCE OF A PAIR AXIOM, ORDERING AXIOMS, PARALLEL AXIOM, PASCH'S AXIOM, PEANO'S AXIOMS, PLAYFAIR'S AXIOM, PORISM, POSTULATE, PROBABILITY AXIOMS, PROCLUS' AXIOM, RULE, T2-SEPARATION AXIOM, THEOREM, ZERMELO'S AXIOM OF CHOICE, ZERMELO-FRAENKEL AXIOMS

Axiom A Diffeomorphism

Let $\phi : M \to M$ be a C^1 DIFFEOMORPHISM on a compact RIEMANNIAN MANIFOLD M. Then ϕ satisfies Axiom A if the NONWANDERING set $\Omega(\phi)$ of ϕ is hyperbolic and the PERIODIC POINTS of ϕ are DENSE in $\omega(\phi)$. although it was conjectured that the first of these conditions implies the second, they were shown to be independent in or around 1977. examples include the ANOSOV DIFFEOMORPHISMS and SMALE HORSESHOE MAP.

In some cases, Axiom A can be replaced by the condition that the DIFFEOMORPHISM is a hyperbolic diffeomorphism on a hyperbolic set (Bowen 1975, Parry and Pollicott 1990).

See also ANOSOV DIFFEOMORPHISM, AXIOM A FLOW, DIFFEOMORPHISM, DYNAMICAL SYSTEM, RIEMANNIAN MANIFOLD, SMALE HORSESHOE MAP

References

Bowen, R. *Equilibrium States and the Ergodic Theory of Anosov Diffeomorphisms.* New York: Springer-Verlag, 1975.
Ott, E. *Chaos in Dynamical Systems.* New York: Cambridge University Press, p. 143, 1993.
Parry, W. and Pollicott, M. "Zeta Functions and the Periodic Orbit Structure of Hyperbolic Dynamics." *Astérisque* No. 187–88, 1990.
Smale, S. "Differentiable Dynamical Systems." *Bull. Amer. Math. Soc.* **73**, 747–17, 1967.

Axiom A Flow

A FLOW defined analogously to the AXIOM A DIFFEOMORPHISM, except that instead of splitting the TANGENT BUNDLE into two invariant sub-BUNDLES, they are split into three (one exponentially contracting, one expanding, and one which is 1-dimensional and tangential to the flow direction).

See also DYNAMICAL SYSTEM

Axiom of Choice

An important and fundamental axiom in SET THEORY sometimes called ZERMELO'S AXIOM OF CHOICE. It was formulated by Zermelo in 1904 and states that, given any SET of mutually exclusive nonempty SETS, there exists at least one SET that contains exactly one element in common with each of the nonempty SETS. The axiom of choice is related to the first of HILBERT'S PROBLEMS.

In ZERMELO-FRAENKEL SET THEORY (in the form omitting the axiom of choice), the ZORN'S LEMMA, TRICHOTOMY LAW, and the WELL ORDERING PRINCIPLE are equivalent to the axiom of choice (Mendelson 1997, p. 275). In contexts sensitive to the axiom of choice, the notation "ZF" is often used to denote Zermelo-Fraenkel *without* the axiom of choice, while "ZFC" is used if the axiom of choice is included.

In 1940, Gödel proved that the axiom of choice is CONSISTENT with the axioms of VON NEUMANN-BERNAYS-GÖDEL SET THEORY (a conservative extension of ZERMELO-FRAENKEL SET THEORY). However, in 1963, Cohen (1963) unexpectedly demonstrated that the axiom of choice is also *independent* of ZERMELO-FRAENKEL SET THEORY (Mendelson 1997; Boyer and Merzbacher 1991, pp. 610–11).

See also HILBERT'S PROBLEMS, SET THEORY, VON NEUMANN-BERNAYS-GÖDEL SET THEORY, WELL ORDERED SET, WELL ORDERING PRINCIPLE, ZERMELO-FRAENKEL AXIOMS, ZERMELO-FRAENKEL SET THEORY, ZORN'S LEMMA

References

Boyer, C. B. and Merzbacher, U. C. *A History of Mathematics, 2nd ed.* New York: Wiley, 1991.
Carnap, R. *Introduction to Symbolic Logic and Its Applications.* New York: Dover, pp. 178–79, 1958.
Cohen, P. J. "The Independence of the Continuum Hypothesis." *Proc. Nat. Acad. Sci. U. S. A.* **50**, 1143–148, 1963.
Cohen, P. J. "The Independence of the Continuum Hypothesis. II." *Proc. Nat. Acad. Sci. U. S. A.* **51**, 105–10, 1964.
Conway, J. H. and Guy, R. K. *The Book of Numbers.* New York: Springer-Verlag, pp. 274–76, 1996.
Mendelson, E. *Introduction to Mathematical Logic, 4th ed.* London: Chapman & Hall, 1997.
Moore, G. H. *Zermelo's Axiom of Choice: Its Origin, Development, and Influence.* New York: Springer-Verlag, 1982.

Axiom of Comprehension

AXIOM OF SEPARATION

Axiom of Extensionality

The axiom of ZERMELO-FRAENKEL SET THEORY which asserts that sets formed by the same elements are equal,

$$\forall x(x \in a \equiv x \in b) \Rightarrow a = b.$$

Using the notation $a \subset b$ (a is a SUBSET of b) for $x \in a(x \in b)$, the axiom can be rewritten

$$a \subset b \wedge b \subset a \Rightarrow a = b.$$

See also ZERMELO-FRAENKEL SET THEORY

References

Itô, K. (Ed.). "Zermelo-Fraenkel Set Theory." §33B in *Encyclopedic Dictionary of Mathematics, 2nd ed., Vol. 1.* Cambridge, MA: MIT Press, pp. 146–48, 1986.

Axiom of Foundation

One of the ZERMELO-FRAENKEL axioms, also known the axiom of regularity (Rubin 1967, Suppes 1972). In the formal language of SET THEORY, it states that

$$x \neq 0 \Rightarrow \exists y(y \in x \wedge y \bigcap x = \phi),$$

where $\Rightarrow$ means IMPLIES, $\exists$ means EXISTS, $\wedge$ means AND, $\cap$ denotes INTERSECTION, and ϕ is the EMPTY

SET (Mendelson 1997, p. 288). More descriptively, "every nonempty set is disjoint from one of its elements."

The axiom of foundation can also be stated as "A set contains no infinitely descending (membership) sequence," or "A set contains a (membership) minimal element," i.e., there is an element of the set that shares no member with the set (Ciesielski 1997, p. 37; Moore 1982, p. 269; Rubin 1967, p. 81; Suppes 1972, p. 53).

Mendelson (1958) proved that the equivalence of these two statements necessarily relies on the AXIOM OF CHOICE. The dual expression is called ϵ-induction, and is equivalent to the axiom itself (Itô 1986, p. 147).

See also AXIOM OF CHOICE, ZERMELO-FRAENKEL AXIOMS

References

Ciesielski, K. *Set Theory for the Working Mathematician.* Cambridge, England: Cambridge University Press, 1997.

Dauben, J. W. *Georg Cantor: His Mathematics and Philosophy of the Infinite.* Princeton, NJ: Princeton University Press, 1990.

Itô, K. (Ed.). "Zermelo-Fraenkel Set Theory." §33B in *Encyclopedic Dictionary of Mathematics, 2nd ed., Vol. 1.* Cambridge, MA: MIT Press, pp. 146–48, 1986.

Mendelson, E. "The Axiom of Fundierung and the Axiom of Choice." *Archiv für math. Logik und Grundlagenfors.* **4**, 67–0, 1958.

Mendelson, E. *Introduction to Mathematical Logic, 4th ed.* London: Chapman & Hall, 1997.

Mirimanoff, D. "Les antinomies de Russell et de Burali-Forti et le problème fondamental de la théorie des ensembles." *Enseign. math.* **19**, 37–2, 1917.

Moore, G. H. *Zermelo's Axiom of Choice: Its Origin, Development, and Influence.* New York: Springer-Verlag, 1982.

Neumann, J. von. "Über eine Widerspruchsfreiheitsfrage in der axiomatischen Mengenlehre." *J. reine angew. Math.* **160**, 227–41, 1929.

Neumann, J. von. "Eine Axiomatisierung der Mengenlehre." *J. reine angew. Math.* **154**, 219–40, 1925.

Rubin, J. E. *Set Theory for the Mathematician.* New York: Holden-Day, 1967.

Suppes, P. *Axiomatic Set Theory.* New York: Dover, 1972.

Zermelo, E. "Über Grenzzahlen und Mengenbereiche." *Fund. Math.* **16**, 29–7, 1930.

Axiom of Infinity

The axiom of ZERMELO-FRAENKEL SET THEORY which asserts the existence of a set containing all the natural numbers,

$$\exists r(\varnothing \in x \lor \forall y \in x(y' \in x)).$$

Here, following von Neumann, $0 = \phi$, $1 = 0' = \{0\}$, $2 = 1' = \{0, 1\}$, $3 = 2' = \{0, 1, 2\}$, ...

See also ZERMELO-FRAENKEL SET THEORY

References

Itô, K. (Ed.). "Zermelo-Fraenkel Set Theory." §33B in *Encyclopedic Dictionary of Mathematics, 2nd ed., Vol. 1.* Cambridge, MA: MIT Press, pp. 146–48, 1986.

Axiom of Regularity
AXIOM OF FOUNDATION

Axiom of Replacement

One of the ZERMELO-FRAENKEL AXIOMS which asserts the existence for any set a of a set x such that, for any y of a, if there exists a z satisfying $A(y, z)$, then such z exists in x. This axiom was introduced by Fraenkel.

See also ZERMELO-FRAENKEL AXIOMS

References

Itô, K. (Ed.). "Zermelo-Fraenkel Set Theory." §33B in *Encyclopedic Dictionary of Mathematics, 2nd ed., Vol. 1.* Cambridge, MA: MIT Press, pp. 146–48, 1986.

Axiom of Separation

The axiom of ZERMELO-FRAENKEL SET THEORY which asserts the existence for any set a and a formula $A(y)$ of a set x consisting of all elements of a satisfying $A(y)$,

$$\exists x \, \forall y(y \in x \equiv y \in a \land A(y)).$$

This axiom is also called the axiom of comprehension or axiom of subsets, and was introduced by Zermelo.

See also ZERMELO-FRAENKEL SET THEORY

References

Itô, K. (Ed.). "Zermelo-Fraenkel Set Theory." §33B in *Encyclopedic Dictionary of Mathematics, 2nd ed., Vol. 1.* Cambridge, MA: MIT Press, pp. 146–48, 1986.

Axiom of the Empty Set

One of the ZERMELO-FRAENKEL AXIOMS which asserts the existence of the EMPTY SET ϕ. The axiom may be stated symbolically as

$$\exists x \, \forall y(!y \in x).$$

See also ZERMELO-FRAENKEL AXIOMS

References

Itô, K. (Ed.). "Zermelo-Fraenkel Set Theory." §33B in *Encyclopedic Dictionary of Mathematics, 2nd ed., Vol. 1.* Cambridge, MA: MIT Press, pp. 146–48, 1986.

Axiom of the Power Set

One of the ZERMELO-FRAENKEL AXIOMS which asserts the existence for any set a of the POWER SET x consisting of all the SUBSETS of a. The axiom may be stated symbolically as

$$\forall x \, \exists y(y \in x \equiv \forall z \in y(z \in a)).$$

See also POWER SET, ZERMELO-FRAENKEL AXIOMS

Axiom of the Sum Set

References

Itô, K. (Ed.). "Zermelo-Fraenkel Set Theory." §33B in *Encyclopedic Dictionary of Mathematics, 2nd ed., Vol. 1.* Cambridge, MA: MIT Press, pp. 146–48, 1986.

Axiom of the Sum Set

The axiom of ZERMELO-FRAENKEL SET THEORY which asserts the existence for any set a of the sum (union) x of all sets that are elements of a. The axiom may be stated symbolically as

$$\exists\, x \,\forall\, y(y \in x \equiv \exists\, z \in a(y \in z)).$$

See also ZERMELO-FRAENKEL SET THEORY

References

Itô, K. (Ed.). "Zermelo-Fraenkel Set Theory." §33B in *Encyclopedic Dictionary of Mathematics, 2nd ed., Vol. 1.* Cambridge, MA: MIT Press, pp. 146–48, 1986.

Axiom of the Unordered Pair

The axiom of ZERMELO-FRAENKEL SET THEORY which asserts the existence for any sets a and b of a set x having a and b as its only elements. x is called the unordered pair of a and b, denoted $\{a, b\}$. The axiom may be stated symbolically as

$$\exists\, x \,\forall\, y(y \in x \equiv y = a \,\vee\, y = b).$$

See also ZERMELO-FRAENKEL SET THEORY

References

Itô, K. (Ed.). "Zermelo-Fraenkel Set Theory." §33B in *Encyclopedic Dictionary of Mathematics, 2nd ed., Vol. 1.* Cambridge, MA: MIT Press, pp. 146–48, 1986.

Axiomatic Set Theory

A version of SET THEORY in which axioms are taken as uninterpreted rather than as formalizations of pre-existing truths.

See also AXIOMATIC SYSTEM, COMPLETE AXIOMATIC THEORY, NAIVE SET THEORY, SET THEORY

References

Curry, H. B. *Foundations of Mathematical Logic.* New York: Dover, pp. 22–3, 1977.

Axiomatic System

A logical system which possesses an explicitly stated SET of AXIOMS from which THEOREMS can be derived.

See also AXIOMATIC SET THEORY, COMPLETE AXIOMATIC THEORY, CONSISTENCY, MODEL THEORY, THEOREM

Axioms of Subsets

This entry contributed by NICOLAS BRAY

For any set theoretic formula $f(x, t_1, t_2, \ldots, t_n)$,

$$(\forall t_1)(\forall t_2)\cdots(\forall t_n)(\forall A)\exists B)(\forall x).$$

$$(x \in B \Leftrightarrow x \in A \,\wedge\, f(x, t_1, \ldots, t_n))$$

In other words, for any formula and set A there is a SUBSET of A consisting exactly of those elements which satisfy the formula.

Axis

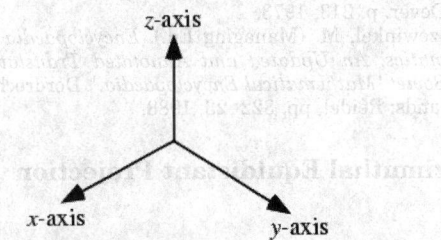

A LINE with respect to which a curve or figure is drawn, measured, rotated, etc.

The term is also used to refer to a LINE through a SHEAF OF PLANES (Woods 1961; Altshiller-Court 1979, p. 12).

See also ABSCISSA, BROCARD AXIS, HOMOLOGY AXIS, LEMOINE AXIS, LINE, MAJOR AXIS, MEDIAL AXIS, MINOR AXIS, ORDINATE, ORTHIC AXIS, PERSPECTIVE AXIS, RADICAL AXIS, REAL AXIS, SEMIMAJOR AXIS, SEMIMINOR AXIS, SHEAF OF PLANES, SIMILARITY AXIS, x-AXIS, y-AXIS, z-AXIS

References

Altshiller-Court, N. *Modern Pure Solid Geometry.* New York: Chelsea, 1979.
Woods, F. S. *Higher Geometry: An Introduction to Advanced Methods in Analytic Geometry.* New York: Dover, p. 8, 1961.

Ax-Kochen Isomorphism Theorem

Let P be the SET of PRIMES, and let $\mathbb{Q}_p$ and $\mathbb{Z}_p(t)$ be the FIELDS of P-ADIC NUMBERS and formal POWER SERIES over $\mathbb{Z}_p = (0, 1, \ldots, p-1)$. Further, suppose that D is a "nonprincipal maximal filter" on P. Then $\prod_{p \in p} \mathbb{Q}_p/D$ and $\prod_{p \in q} \mathbb{Z}_p(t)/D$ are ISOMORPHIC.

See also HYPERREAL NUMBER, NONSTANDARD ANALYSIS

Axonometry

A METHOD for mapping 3-D figures onto the PLANE.

See also CROSS SECTION, MAP PROJECTION, POHLKE'S THEOREM, PROJECTION, STEREOLOGY

References

Coxeter, H. S. M. *Regular Polytopes, 3rd ed.* New York: Dover, p. 313, 1973.

Hazewinkel, M. (Managing Ed.). *Encyclopaedia of Mathematics: An Updated and Annotated Translation of the Soviet "Mathematical Encyclopaedia."* Dordrecht, Netherlands: Reidel, pp. 322–23, 1988.

Azimuthal Equidistant Projection

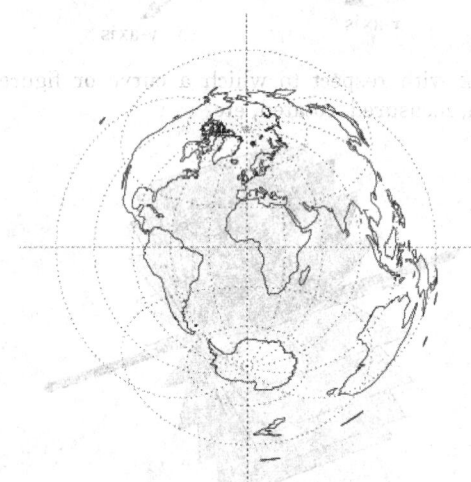

An AZIMUTHAL PROJECTION which is neither EQUAL-AREA nor CONFORMAL. Let ϕ_1 and λ_0 be the LATITUDE and LONGITUDE of the center of the projection, then the transformation equations are given by

$$x = k' \cos \phi \sin(\lambda - \lambda_0) \qquad (1)$$

$$y = k'[\cos \phi_1 \sin \phi - \sin \phi_1 \cos \phi \cos(\lambda - \lambda_0)]. \qquad (2)$$

Here,

$$k' = \frac{c}{\sin c} \qquad (3)$$

and

$$\cos c = \sin \phi_1 \sin \phi + \cos \phi_1 \cos \phi \cos(\lambda - \lambda_0), \qquad (4)$$

where c is the angular distance from the center. The inverse FORMULAS are

$$\phi = \sin^{-1}\left(\cos c \sin \phi_1 + \frac{y \sin c \cos \phi_1}{c} \right) \qquad (5)$$

and

$$\lambda = \begin{cases} \lambda_0 + \tan^{-1}\left(\dfrac{x \sin c}{c \cos \phi_1 \cos c - y \sin \phi_1 \sin c} \right) & \text{for } \phi_1 \neq \pm 90^\circ \\[2mm] \lambda_0 + \tan^{-1}\left(-\dfrac{x}{y} \right) & \text{for } \phi_1 = 90^\circ \\[2mm] \lambda_0 + \tan^{-1}\left(\dfrac{x}{y} \right) & \text{for } \phi_1 = -90^\circ. \end{cases} \qquad (6)$$

with the angular distance from the center given by

$$c = \sqrt{x^2 + y^2}. \qquad (7)$$

See also AZIMUTHAL PROJECTION, EQUIDISTANT PROJECTION

References

Snyder, J. P. *Map Projections--A Working Manual.* U. S. Geological Survey Professional Paper 1395. Washington, DC: U. S. Government Printing Office, pp. 191–02, 1987.

Azimuthal Projection

A MAP PROJECTION on which the azimuths of all points are shown correctly with respect to the center (Snyder 1987, p. 4). A plane tangent to one of the Earth's poles is the basis for polar azimuthal projection. The term "zenithal" is an older one for azimuthal projections (Hinks 1921, Lee 1944).

See also

AZIMUTHAL EQUIDISTANT PROJECTION, LAMBERT AZIMUTHAL EQUAL-AREA PROJECTION, ORTHOGRAPHIC PROJECTION, STEREOGRAPHIC PROJECTION

References

Hinks, A. R. *Map Projections, 2nd rev. ed.* Cambridge, England: Cambridge University Press, 1921.

Lee, L. P. "The Nomenclature and Classification of Map Projections." *Empire Survey Rev.* **7**, 190–00, 1944.

Snyder, J. P. *Map Projections--A Working Manual.* U. S. Geological Survey Professional Paper 1395. Washington, DC: U. S. Government Printing Office, 1987.

B

B2-Sequence

N.B. A detailed online essay by S. Finch was the starting point for this entry.

Also called a SIDON SEQUENCE. An INFINITE SEQUENCE of POSITIVE INTEGERS

$$1 \le b_1 < b_2 < b_3 < \dots \tag{1}$$

such that all pairwise sums

$$b_i + b_j \tag{2}$$

for $i \le j$ are distinct (Guy 1994). An example is 1, 2, 4, 8, 13, 21, 31, 45, 66, 81, 97, 123, 148, 182, 204, 252, 290, 361, ... (Sloane's A005282).

Zhang (1993, 1994) showed that

$$S(B2) \equiv \underset{\text{all B2 sequences}}{\text{SUP}} \sum_{k=1}^{\infty} \frac{1}{b_k} > 2.1597, \tag{3}$$

which has been increased to $S(B2) > 2.16086$ by R. Lewis using the sequence 1, 2, 4, 8, 13, 21, 31, 45, 66, 81, 97, 123, 148, 182, 204, 252, 291, 324, ... (Sloane's A046185). The definition can be extended to B_n-sequences (Guy 1994).

See also A-SEQUENCE, MIAN-CHOWLA SEQUENCE

References
Finch, S. "Favorite Mathematical Constants." http://www.mathsoft.com/asolve/constant/erdos/erdos.html.

Guy, R. K. "Packing Sums of Pairs," "Three-Subsets with Distinct Sums," and "B_2-Sequences," and B_2-Sequences Formed by the Greedy Algorithm." §C9, C11, E28, and E32 in *Unsolved Problems in Number Theory, 2nd ed.* New York: Springer-Verlag, pp. 115–118, 121–123, 228–229, and 232–233, 1994.

Mian, A. M. and Chowla, S. D. "On the B_2-Sequences of Sidon." *Proc. Nat. Acad. Sci. India* **A14**, 3–4, 1944.

Sloane, N. J. A. Sequences A005282/M1094 and A046185 in "An On-Line Version of the Encyclopedia of Integer Sequences." http://www.research.att.com/~njas/sequences/eisonline.html.

Zhang, Z. X. "A B2-Sequence with Larger Reciprocal Sum." *Math. Comput.* **60**, 835–839, 1993.

Zhang, Z. X. "Finding Finite B2-Sequences with Larger $m - a_m^{1/2}$." *Math. Comput.* **63**, 403–414, 1994.

Baby Monster Group

Also known as FISCHER'S BABY MONSTER GROUP. The SPORADIC FINITE GROUP B. It has ORDER

$$2^{41} \cdot 3^{13} \cdot 5^6 \cdot 7^2 \cdot 11 \cdot 13 \cdot 17 \cdot 19 \cdot 23 \cdot 31 \cdot 47.$$

See also FINITE GROUP, MONSTER GROUP

References
Wilson, R. A. "ATLAS of Finite Group Representation." http://for.mat.bham.ac.uk/atlas/html/BM.html.

BAC-CAB Identity

The VECTOR TRIPLE PRODUCT identity

$$\mathbf{A} \times (\mathbf{B} \times \mathbf{C}) = \mathbf{B}(\mathbf{A} \cdot \mathbf{C}) - \mathbf{C}(\mathbf{A} \cdot \mathbf{B}).$$

This identity can be generalized to n-D

$$\mathbf{a}_2 \times \cdots \times \mathbf{a}_{n-1} \times (\mathbf{b}_1 \times \cdots \times \mathbf{b}_{n-1})$$

$$= (-1)^{n+1} \begin{vmatrix} \mathbf{b}_1 & \cdots & \mathbf{b}_{n-1} \\ \mathbf{a}_2 \cdot \mathbf{b}_1 & \cdots & \mathbf{a}_2 \cdot \mathbf{b}_{n-1} \\ \vdots & \ddots & \vdots \\ \mathbf{a}_{n-1} \cdot \mathbf{b}_1 & \cdots & \mathbf{a}_{n-1} \cdot \mathbf{b}_{n-1} \end{vmatrix}.$$

See also LAGRANGE'S IDENTITY

BAC-CAB Rule

BAC-CAB IDENTITY

Bachelier Function

BROWN FUNCTION

Bachet Equation

The DIOPHANTINE EQUATION

$$x^2 + k = y^3,$$

which is also an ELLIPTIC CURVE. The general equation is still the focus of ongoing study.

Bachet's Conjecture

LAGRANGE'S FOUR-SQUARE THEOREM

Bachet's Theorem

LAGRANGE'S FOUR-SQUARE THEOREM

Backhouse's Constant

Let $P(x)$ be defined as the POWER SERIES whose nth term has a COEFFICIENT equal to the nth PRIME,

$$P(x) \equiv \sum_{k=0}^{\infty} p_k x^k = 1 + 2x + 3x^2 + 5x^3 + 7x^4 + 11x^5 + \dots,$$

and let $Q(x)$ be defined by

$$Q(x) = \frac{1}{P(x)} = \sum_{k=0}^{\infty} q_k x^k.$$

Then N. Backhouse conjectured that

$$\lim_{n \to \infty} \left| \frac{q_{n+1}}{q_n} \right| = 1.4560749485826896713995953511116\dots.$$

This list was subsequently shown to exist by P. Flajolet.

References
Finch, S. "Favorite Mathematical Constants." http://www.mathsoft.com/asolve/constant/backhous/backhous.html.

Bäcklund Transformation

A method for solving classes of nonlinear PARTIAL DIFFERENTIAL EQUATIONS.

See also INVERSE SCATTERING METHOD, SOLITON

References
Anderson, R. L. and Ibragimov, N. H. *Lie-Bäcklund Transformation in Applications.* Philadelphia, PA: SIAM, 1979.
Dodd, R. K.; Eilbeck, J. C.; and Morris, H. C. *Solitons and Nonlinear Equations.* London: Academic Press, 1984.
Infeld, E. and Rowlands, G. "Bäcklund Transformations." §7.5 in *Nonlinear Waves, Solitons, and Chaos, 2nd ed.* Cambridge, England: Cambridge University Press, pp. 175–77, 2000.
Lamb, G. L. Jr. *Elements of Soliton Theory.* New York: Wiley, 1980.
Miura, R. M. (Ed.). *Bäcklund Transformations, the Inverse Scattering Method, Solitons, and Their Applications.* New York: Springer-Verlag, 1974.
Olver, P. J. *Applications of Lie Groups to Differential Equations.* New York: Springer-Verlag, 1986.
Rogers, C. and Shadwick, W. F. *Bäcklund Transformations and Their Applications.* New York: Academic Press, 1982.
Whitham, G. B. *Linear and Nonlinear Waves.* New York: Wiley, pp. 609–11, 1974.
Zwillinger, D. "Bäcklund Transformations." §87 in *Handbook of Differential Equations, 3rd ed.* Boston, MA: Academic Press, pp. 321–24, 1997.

Backtracking

A method of solving combinatorial problems by means of an algorithm which is allowed to run forward until a dead end is reached, at which point previous steps are retraced and the algorithm is allowed to run forward again. Backtracking can greatly reduce the amount of work in an exhaustive search. Backtracking is implemented as Backtrack[s, *partialQ*, *solutionQ*] in the *Mathematica* add-on package DiscreteMath`Combinatorica` (which can be loaded with the command < <DiscreteMath`).

Backtracking also refers to a method of drawing FRACTALS by appropriate numbering of the corresponding tree diagram which does not require storage of intermediate results (Lauwerier 1991).

References
Baumert, L. D. and Golomb, S. W. "Backtrack Programming." *J. Ass. Comp. Machinery* **12**, 516–24, 1965.
Lauwerier, H. A. *Fractals: Endlessly Repeated Geometrical Figures.* Princeton, NJ: Princeton University Press, 1991.
Skiena, S. "Backtracking and Distinct Permutations." §1.1.5 in *Implementing Discrete Mathematics: Combinatorics and Graph Theory with Mathematica.* Reading, MA: Addison-Wesley, pp. 12–4, 1990.
Wilf, H. "Backtrack: An $\iota(1)$ Expected Time Algorithm for the Graph Coloring Problem." *Info. Proc. Let.* **18**, 119–21, 1984.

Backus-Gilbert Method

A method which can be used to solve some classes of INTEGRAL EQUATIONS and is especially useful in implementing certain types of data inversion. It has been applied to invert seismic data to obtain density profiles in the Earth.

References
Backus, G. and Gilbert, F. "The Resolving Power of Growth Earth Data." *Geophys. J. Roy. Astron. Soc.* **16**, 169–05, 1968.
Backus, G. E. and Gilbert, F. "Uniqueness in the Inversion of Inaccurate Gross Earth Data." *Phil. Trans. Roy. Soc. London Ser. A* **266**, 123–92, 1970.
Loredo, T. J. and Epstein, R. I. "Analyzing Gamma-Ray Burst Spectral Data." *Astrophys. J.* **336**, 896–19, 1989.
Parker, R. L. "Understanding Inverse Theory." *Ann. Rev. Earth Planet. Sci.* **5**, 35–4, 1977.
Press, W. H.; Flannery, B. P.; Teukolsky, S. A.; and Vetterling, W. T. "Backus-Gilbert Method." §18.6 in *Numerical Recipes in FORTRAN: The Art of Scientific Computing, 2nd ed.* Cambridge, England: Cambridge University Press, pp. 806–09, 1992.

Backward Difference

The backward difference is a FINITE DIFFERENCE defined by

$$\nabla_p \equiv \nabla f_p \equiv f_p - f_{p-1}. \tag{1}$$

Higher order differences are obtained by repeated operations of the backward difference operator, so

$$\nabla_p^2 = \nabla(\nabla p) = \nabla(f_p - f_{p-1}) = \nabla f_p - \nabla f_{p-1} \tag{2}$$

$$= (f_p - f_{p-1}) - (f_{p-1} - f_{p-2}) $$

$$= f_p - 2f_{p-1} + f_{p-2} \tag{3}$$

In general,

$$\nabla_p^k \equiv \nabla^k f_p \equiv \sum_{m=0}^{k} (-1)^m \binom{k}{m} f_{p-m}, \tag{4}$$

where $\binom{k}{m}$ is a BINOMIAL COEFFICIENT. NEWTON'S BACKWARD DIFFERENCE FORMULA expresses f_p as the sum of the nth backward differences

$$f_p = f_0 + p\nabla_0 + \frac{1}{2!}\, p(p+1)\nabla_0^2 + \frac{1}{3!}\, p(p+1)(p+2)\nabla_0^3 + \cdots, \tag{5}$$

where ∇_0^n is the first nth difference computed from the difference table.

See also ADAMS' METHOD, DIFFERENCE EQUATION, DIVIDED DIFFERENCE, FINITE DIFFERENCE, FORWARD DIFFERENCE, NEWTON'S BACKWARD DIFFERENCE FORMULA, RECIPROCAL DIFFERENCE

References
Beyer, W. H. *CRC Standard Mathematical Tables, 28th ed.* Boca Raton, FL: CRC Press, pp. 429 and 433, 1987.

Backward Stability

The property of certain algorithms that accurate answers are returned for well-conditioned problems, and the inaccuracy of the answers returned for ill-conditioned problems is proportional to the sensitivity.

Bader-Deuflhard Method

A generalization of the BULIRSCH-STOER ALGORITHM for solving ORDINARY DIFFERENTIAL EQUATIONS.

References

Bader, G. and Deuflhard, P. "A Semi-Implicit Mid-Point Rule for Stiff Systems of Ordinary Differential Equations." *Numer. Math.* **41**, 373–98, 1983.

Press, W. H.; Flannery, B. P.; Teukolsky, S. A.; and Vetterling, W. T. *Numerical Recipes in FORTRAN: The Art of Scientific Computing, 2nd ed.* Cambridge, England: Cambridge University Press, p. 730, 1992.

Baer Differential Equation

The Baer differential equation is given by

$$(x - a_1)(x - a_2)y'' + \tfrac{1}{2}[2x - (a_1 + a_2)]y' - (p^2 x + q^2)y = 0,$$

while the Baer "wave equation" is

$$(x - a_1)(x - a_2)y'' + \tfrac{1}{2}[2x - (a_1 + a_2)]y' - (k^2 x^2 - p^2 x + q^2)y = 0$$

(Moon and Spencer 1961, pp. 156–57; Zwillinger 1997, p. 121).

References

Moon, P. and Spencer, D. E. *Field Theory for Engineers.* New York: Van Nostrand, 1961.

Zwillinger, D. *Handbook of Differential Equations, 3rd ed.* Boston, MA: Academic Press, p. 121, 1997.

Bagging

See also RESAMPLING STATISTICS

Baguenaudier

A PUZZLE involving disentangling a set of rings from a looped double rod, originally used by French peasants to lock chests (Steinhaus 1983). The word "baguenaudier" means "time-waster" in French, and the puzzle is also called the Chinese rings or Devil's needle puzzle. ("Bague" also means "ring," but this appears to be an etymological coincidence. Interestingly, the bladder-senna tree is also known as "baguenaudier" in French.) Culin (1965) attributes the puzzle to Chinese general Hung Ming (A.D. 181–34), who gave it to his wife as a present to occupy her while he was away at the wars.

The solution of the baguenaudier is intimately related to the theory of GRAY CODES.

The minimum number of moves $a(n)$ needed for n rings is

$$a(n) = \left[\tfrac{2}{3}(2^n - 1)\right] = \begin{cases} \tfrac{1}{3}(2^{n+1} - 2) & n \text{ even} \\ \tfrac{1}{3}(2^{n+1} - 1) & n \text{ odd,} \end{cases} \quad (1)$$

where $\lceil x \rceil$ is the CEILING FUNCTION, giving 1, 2, 5, 10, 21, 42, 85, 170, 341, 682, ... (Sloane's A000975). The GENERATING FUNCTION for these numbers is

$$\frac{1}{(1 - 2x)(1 - x^2)} = 1 + 2x + 5x^2 + 10x^3 + 21x^4 + \dots. \quad (2)$$

They are also given by the RECURRENCE RELATION

$$a(n) = a(n - 1) + 2a(n - 2) + 1 \quad (3)$$

with $a(1) = 1$ and $a(2) = 2$.

By simultaneously moving the two end rings, the number of moves for n rings can be reduced to

$$b(n) = \begin{cases} 2^{n-1} - 1 & n \text{ even} \\ 2^{n-1} & n \text{ odd,} \end{cases} \quad (4)$$

giving 1, 1, 4, 7, 16, 31, 64, 127, 256, 511, ... (Sloane's A051049).

Defining the complexity of a solution as the minimal number of times the ring passes through the arc from the last ring to the base of the puzzle, the minimal complexity of a solution if 2^{n-1}, as conjectured by Kauffman (1996) and proved by Przytycki and Sikora (2000).

See also GRAY CODE, HABIRO MOVE

References

Culin, S. "Ryou-Kaik-Tjyo--Dilguel Guest Instrument (Ring Puzzle)." §20 in *Games of the Orient: Korea, China, Japan.* Rutland, VT: Charles E. Tuttle, pp. 31–2, 1965.

Dubrovsky, V. "Nesting Puzzles, Part II: Chinese Rings Produce a Chinese Monster." *Quantum* **6**, 61–5 (Mar.) and 58–9 (Apr.), 1996.

Gardner, M. "The Binary Gray Code." In *Knotted Doughnuts and Other Mathematical Entertainments.* New York: W. H. Freeman, pp. 15–7, 1986.

Kauffman, L. H. "Tangle Complexity and the Topology of the Chinese Rings." In *Mathematical Approaches to Biomolecular Structure and Dynamics.* New York: Springer-Verlag, pp. 1–0, 1996.

Kraitchik, M. "Chinese Rings." §3.12.3 in *Mathematical Recreations.* New York: W. W. Norton, pp. 89–1, 1942.

Przytycki, J. H. and Sikora, A. S. Topological Insights from the Chinese Rings. 21 Jul 2000. http://xxx.lanl.gov/abs/math.GT/0007134/.

Sloane, N. J. A. Sequences A000975 and A051049 in "An On-Line Version of the Encyclopedia of Integer Se-

quences." http://www.research.att.com/~njas/sequences/
eisonline.html.

Slocum, J. and Botermans, J. *Puzzles Old and New: How to Make and Solve Them.* Seattle, WA: University of Washington Press, p. 105, 1988.

Steinhaus, H. *Mathematical Snapshots, 3rd ed.* New York: Dover, pp. 268–69, 1999.

University of Waterloo. "Wire and RIng Puzzles." http://www.ahs.uwaterloo.ca/~museum/vexhibit/puzzles/wire/wire.html.

Bailey's Lemma

If, for $n \geq 0$,

$$\beta_n = \sum_{r=0}^{n} \frac{\alpha_r}{(q;\ q)_{n-r}(aq;\ q)_{n+r}}, \tag{1}$$

then

$$\beta_n' = \sum_{r=0}^{n} \frac{\alpha_r'}{(q;\ q)_{n-r}(aq;\ q)_{n+r}}, \tag{2}$$

where

$$\alpha_r' = \frac{(\rho_1;\ q)_r(\rho_2;\ q)_r(aq/\rho_1\rho_2)^r \alpha_r}{(aq/\rho_1;\ q)_r(aq/\rho_2;\ q)_r} \tag{3}$$

$$\beta_n' = \sum_{j \geq 0} \frac{(\rho_1;\ q)_j(\rho_2;\ q)_j(aq/\rho_1\rho_2;\ q)_{n-j}(aq/\rho_1\rho_2)^j \beta_j}{(q;\ q)_{n-j}(aq/\rho_1;\ q)_n(aq/\rho_2;\ q)_n}. \tag{4}$$

References

Andrews, G. E. "Multiple Series Rogers-Ramanujan Type Identities." *Pacific J. Math.* **114**, 267–83, 1984.

Andrews, G. E. "Bailey's Lemma" and "Bailey's Lemma in Computer Algebra." §3.4 and 10.4 in *q-Series: Their Development and Application in Analysis, Number Theory, Combinatorics, Physics, and Computer Algebra.* Providence, RI: Amer. Math. Soc., pp. 25–7 and 99–00, 1986.

Bailey, W. N. "Identities of the Rogers-Ramanujan Type." *Proc. London Math. Soc.* **50**, 1–0, 1949.

Bailey's Method

LAMBERT'S METHOD

Bailey's Theorem

Let $\Gamma(z)$ be the GAMMA FUNCTION, then

$$\left[\frac{\Gamma(m+\tfrac{1}{2})}{\Gamma(m)}\right]^2$$

$$\times \underbrace{\left[\frac{1}{m} + \left(\frac{1}{2}\right)^2 \frac{1}{m+1} + \left(\frac{1 \cdot 3}{2 \cdot 4}\right)^2 \frac{1}{m+2} + \cdots\right]}_{n}$$

$$= \left[\frac{\Gamma(n+\tfrac{1}{2})}{\Gamma(n)}\right]^2$$

$$\times \underbrace{\left[\frac{1}{n} + \left(\frac{1}{2}\right)^2 \frac{1}{n+1} + \left(\frac{1 \cdot 3}{2 \cdot 4}\right)^2 \frac{1}{n+2} + \cdots\right]}_{m}.$$

Writing the sums explicitly, Bailey's theorem states

$$\left[\frac{\Gamma(m+\tfrac{1}{2})}{\Gamma(m)}\right]^2 \sum_{k=0}^{n-1} \frac{1}{m+k}\left[\frac{(2k-1)!!}{(2k)!!}\right]^2$$

$$\left[\frac{\Gamma(n+\tfrac{1}{2})}{\Gamma(n)}\right]^2 \sum_{k=0}^{m-1} \frac{1}{n+k}\left[\frac{(2k-1)!!}{(2k)!!}\right]^2.$$

See also GAMMA FUNCTION

References

Bailey, W. N. "The Partial Sum of the Coefficients of the Hypergeometric Series." *J. London Math. Soc.* **6**, 40–1, 1931.

Bailey, W. N. "On One of Ramanujan's Theorems." *J. London Math. Soc.* **7**, 34–6, 1932.

Darling, H. B. C. "On a Proof of One of Ramanujan's Theorems." *J. London Math. Soc.* **5**, 8–, 1930.

Hardy, G. H. *Ramanujan: Twelve Lectures on Subjects Suggested by His Life and Work, 3rd ed.* New York: Chelsea, pp. 106–07 and 112, 1999.

Hodgkinson, J. "Note on One of Ramanujan's Theorems." *J. London Math. Soc.* **6**, 42–3, 1931.

Watson, G. N. "Theorems Stated by Ramanujan (VIII): Theorems on Divergent Series." *J. London Math. Soc.* **4**, 82–6, 1929.

Watson, G. N. *Quart. J. Math. (Oxford)* **1**, 310–18, 1930.

Whipple, F. J. W. "The Sum of the Coefficients of a Hypergeometric Series." *J. London Math. Soc.* **5**, 192, 1930.

Bailey's Transformation

The very general transformation

$${}_9F_8\left[\begin{matrix} a, & 1+\tfrac{1}{2}a, & b, & c, & d \\ & \tfrac{1}{2}a & 1+a-b, & 1+a-c, & 1+a-d. \end{matrix}\right.$$

$$\left.\begin{matrix} e, & f, & g, & -m; \\ 1+a-e, & 1+a-f, & 1+a-g, & 1+a+m \end{matrix}\right]$$

$$= \frac{(1+a)_m(1+k-e)_m(1+k-f)_m(1+k-g)_m}{(1+k)_m(1+a-e)_m(1+a-f)_m(1+a-g)_m}$$

$$\times_9F_8 \begin{bmatrix} k, & 1+\tfrac{1}{2}k, & k+b-a, & k+c-a, & k+d-a, \\ & \tfrac{1}{2}k, & 1+a-b, & a+a-c, & 1+a-d, \\ e, & f, & g, & -m; \\ 1+k-e, & 1+k-f, & 1+k-g, & 1+k+m \end{bmatrix},$$

where $k = 1 + 2a - b - c - d$, and the parameters are subject to the restriction

$$b+c+d+e+f+g-m = 2+3a$$

(Bailey 1935, p. 27).

Bhatnagar (1995, pp. 17–8) defines the Bailey transform as follows. Let $(a;\ q)_n$ be the q-POCHHAMMER SYMBOL, and let a be an indeterminate, and let the LOWER TRIANGULAR MATRICES $F = (F(n,\ k))$ and $F = (G(n,\ k))$ be defined as

$$F(n,\ k) = \frac{1}{(q;\ q)_{n-k}(aq;\ q)_{n+k}}$$

and

$$G(n,\ k) = \frac{(1 - aq^{2n})(a;\ q)_{n+k}}{(1-a)(q;\ q)_{n-k}}(-1)^{n-k}q^{\binom{n-k}{2}}$$

Then F and G are MATRIX INVERSES.

See also DOUGALL-RAMANUJAN IDENTITY, GENERALIZED HYPERGEOMETRIC FUNCTION

References

Bailey, W. N. "Some Identities Involving Generalized Hypergeometric Series." *Proc. London Math. Soc.* **29**, 503–16, 1929.
Bailey, W. N. *Generalised Hypergeometric Series.* Cambridge, England: University Press, 1935.
Bhatnagar, G. *Inverse Relations, Generalized Bibasic Series, and their U(n) Extensions.* Ph.D. thesis. Ohio State University, 1995.
Milne, S. C. and Lilly, G. M. "The A_ℓ and C_ℓ Bailey Transform and Lemma." *Bull. Amer. Math. Soc.* **26**, 258–63, 1992.

Bailey-Borwein-Plouffe Algorithm

The DIGIT-EXTRACTION ALGORITHM for calculating the digits of PI given by the formula

$$\pi = \sum_{n=0}^{\infty} \left(\frac{4}{8n+1} - \frac{2}{8n+4} - \frac{1}{8n+5} - \frac{1}{8n+6} \right) \left(\frac{1}{16} \right)^n.$$

See also PI, PI FORMULAS

References

Adamchik, V. and Wagon, S. "A Simple Formula for π." *Amer. Math. Monthly* **104**, 852–55, 1997.
Adamchik, V. and Wagon, S. "Pi: A 2000-Year Search Changes Direction." http://members.wri.com/victor/articles/pi.html.
Bailey, D.; Borwein, P.; and Plouffe, S. "On the Rapid Computation of Various Polylogarithmic Constants." http://www.cecm.sfu.ca/~pborwein/PAPERS/P123.ps.

Finch, S. "Unsolved Mathematics Problems: The Miraculous Bailey-Borwein-Plouffe Pi Algorithm." http://www.mathsoft.com/asolve/plouffe/plouffe.html.

Baire Category Theorem

A nonempty complete METRIC SPACE cannot be REPRESENTED AS the UNION of a COUNTABLE family of NOWHERE DENSE SUBSETS.

See also COUNTABLE SET, METRIC SPACE, NOWHERE DENSE

Baire Function

References

Feller, W. *An Introduction to Probability Theory and Its Applications, Vol. 2, 3rd ed.* New York: Wiley, pp. 104–06, 1971.

Baire Space

A TOPOLOGICAL SPACE X in which each SUBSET of X of the "first category" has an empty interior. A TOPOLOGICAL SPACE which is HOMEOMORPHIC to a complete METRIC SPACE is a Baire space.

Bairstow's Method

A procedure for finding the quadratic factors for the COMPLEX CONJUGATE ROOTS of a POLYNOMIAL $P(x)$ with REAL COEFFICIENTS.

$$[x-(a-ib)][x-(a-ib)] = x^2 + 2ax + (a^2+b^2)$$
$$\equiv x^2 + Bx + C. \tag{1}$$

Now write the original POLYNOMIAL as

$$P(x) = (x^2 + Bx + C)Q(x) + Rx + S \tag{2}$$

$$R(B+\delta B,\ C+\delta C) \approx R(B,\ C) + \frac{\partial R}{\partial B}\ dB + \frac{\partial R}{\partial C}\ dC \tag{3}$$

$$S(B+\delta B,\ C+\delta C) \approx S(B,\ C) + \frac{\partial S}{\partial B}\ dB + \frac{\partial S}{\partial C}\ dC \tag{4}$$

$$\frac{\partial P}{\partial C} = 0 = (x^2+Bx+C)\frac{\partial Q}{\partial C} + Q(x) + \frac{\partial R}{\partial C} + \frac{\partial S}{\partial C} \tag{5}$$

$$-Q(x) = (x^2+Bx+C)\frac{\partial Q}{\partial C} + \frac{\partial R}{\partial C} + \frac{\partial S}{\partial C} \tag{6}$$

$$\frac{\partial P}{\partial B} = 0 = (x^2+Bx+C)\frac{\partial Q}{\partial B} + xQ(x) + \frac{\partial R}{\partial B} + \frac{\partial S}{\partial B} \tag{7}$$

$$-xQ(x) = (x^2+Bx+C)\frac{\partial Q}{\partial B} + \frac{\partial R}{\partial B} + \frac{\partial S}{\partial B}. \tag{8}$$

Now use the 2-D NEWTON'S METHOD to find the simultaneous solutions.

References

Press, W. H.; Flannery, B. P.; Teukolsky, S. A.; and Vetterling, W. T. *Numerical Recipes in C: The Art of Scientific Computing.* Cambridge, England: Cambridge University Press, pp. 277 and 283–84, 1989.

Baker's Dozen

The number 13.

See also 13, DOZEN

Baker's Map

The MAP

$$x_{n+1} = 2\mu x_n, \tag{1}$$

where x is computed modulo 1. A generalized Baker's map can be defined as

$$x_{n+1} = \begin{cases} \lambda_a x_n & y_n < \alpha \\ (1 - \lambda_b) + \lambda_b x_n & y_n > \alpha \end{cases} \tag{2}$$

$$y_{n+1} = \begin{cases} \dfrac{y_n}{\alpha} & y_n < \alpha \\ \dfrac{y_n - \alpha}{\beta} & y_n > \alpha, \end{cases} \tag{3}$$

where $\beta \equiv 1 - \alpha$, $\lambda_a + \lambda_b \leq 1$, and x and y are computed mod 1. The $q = 1$ Q-DIMENSION is

$$D_1 = 1 + \frac{\alpha \ln\left(\dfrac{1}{\alpha}\right) + \beta \ln\left(\dfrac{1}{\beta}\right)}{\alpha \ln\left(\dfrac{1}{\gamma_a}\right) + \beta \ln\left(\dfrac{1}{\gamma_b}\right)}. \tag{4}$$

If $\lambda_a = \lambda_b$, then the general Q-DIMENSION is

$$D_q = 1 + \frac{1}{q - 1} \frac{\ln(\alpha^q + \beta^q)}{\ln \lambda_a}. \tag{5}$$

References

Lichtenberg, A. and Lieberman, M. *Regular and Stochastic Motion.* New York: Springer-Verlag, p. 60, 1983.
Ott, E. *Chaos in Dynamical Systems.* Cambridge, England: Cambridge University Press, pp. 81–2, 1993.
Rasband, S. N. *Chaotic Dynamics of Nonlinear Systems.* New York: Wiley, p. 32, 1990.

Bakos' Compound

CUBE 4-COMPOUND

Balanced ANOVA

An ANOVA in which the number of REPLICATES (sets of identical observations) is restricted to be the same for each FACTOR LEVEL (treatment group).

See also ANOVA

Balanced Binomial Coefficient

An integer n is p-balanced for p a prime if, among all nonzero binomial coefficients $\binom{n}{k}$; for $k = 0, \ldots, n$ (mod p), there are equal numbers of quadratic residues and nonresidues (mod p). Let T_p be the set of integers n, $0 \leq n \leq p - 1$, that are p-balanced. Among all the primes $< 1,000,000$, only those with $p = 2, 3$, and 11 have $T_p = \varnothing$.

p	T_p
2	$\varnothing$
3	$\varnothing$
5	$\{3\}$
7	$\{3\}$
11	$\varnothing$
13	$\{7, 11\}$
17	$\{3, 15\}$

See also BINOMIAL COEFFICIENT

References

Garfield, R. and Wilf, H. S. "The Distribution of the Binomial Coefficients Modulo p." *J. Number Th.* **41**, 1, 1992.
Wilf, H. "On Crossing Numbers, and Some Unsolved Problems." In *Combinatorics, Geometry, and Probability: A Tribute to Paul Erdos. Papers from the Conference in Honor of Erdos' 80th Birthday Held at Trinity College, Cambridge, March 1993* (Ed. B. Bollobás and A. Thomason). Cambridge, England: Cambridge University Press, pp. 557–62, 1997.

Balanced Incomplete Block Design

BLOCK DESIGN

Ball

The n-ball, denoted $\mathbb{B}^n$, is the interior of a SPHERE $\mathbb{S}^{n-1}$, and sometimes also called the n-DISK. (Although physicists often use the term "SPHERE" to mean the solid ball, mathematicians definitely do not!) Let $\mathrm{Vol}(\mathbb{B}^n)$ denote the volume of an n-D ball of RADIUS r. Then

$$\sum_{n=0}^{\infty} \mathrm{Vol}(B^n) = e^{\pi r^2}[1 + \mathrm{erf}(r\sqrt{\pi})],$$

where $\mathrm{erf}(x)$ is the ERF function.

See also ALEXANDER'S HORNED SPHERE, BALL LINE PICKING, BALL TRIANGLE PICKING, BANACH-TARSKI PARADOX, BING'S THEOREM, BISHOP'S INEQUALITY, BOUNDED SET, DISK, HYPERSPHERE, SPHERE, WILD POINT

References

Freden, E. Problem 10207. "Summing a Series of Volumes." *Amer. Math. Monthly* **100**, 882, 1993.

Ball Line Picking

Given an n-ball $\mathbb{B}^n$ of radius R, find the distribution of the lengths s of the lines determined by two points chosen at random within the ball. The probability distribution of lengths is given by

$$P_n(s) = n\, \frac{s^{n-1}}{R^n}\, I_x(\tfrac{1}{2}(n+1), \tfrac{1}{2}), \tag{1}$$

where

$$x \equiv 1 - \frac{s^2}{4R^2} \tag{2}$$

and

$$I_x(p, q) = \frac{B(x;\ p,\ q)}{B(p,\ q)} \tag{3}$$

is a REGULARIZED BETA FUNCTION, with $B(x;\ p,\ q)$ is an INCOMPLETE BETA FUNCTION and $B(p,\ q)$ is a BETA FUNCTION (Tu and Fischbach 2000). The first few are

$$P_1(s) = \frac{1}{R} - \frac{s}{2R} \tag{4}$$

$$P_2(s) = \frac{4s}{\pi R^2} \cos^{-1}\left(\frac{s}{2R}\right) - \frac{2s^2}{\pi R^3} \sqrt{1 - \frac{s^2}{4R^2}} \tag{5}$$

$$P_3(s) = \frac{3s^2}{R^3} - \frac{9s^3}{4R^4} + \frac{3s^5}{16R^6} \tag{6}$$

$$P_4(s) = \frac{8s^3}{\pi R^4} \cos^{-1}\left(\frac{s}{2R}\right) - \frac{8s^4}{3\pi R^5}$$
$$\times \left(1 - \frac{s^2}{4R^2}\right)^{3/2} - \frac{4s^4}{\pi R^5} \sqrt{1 - \frac{s^2}{4R^2}}. \tag{7}$$

The average lengths are given by

$$\bar{s}_1 = \frac{2R}{3} \tag{8}$$

$$\bar{s}_2 = \frac{128R}{45\pi} \tag{9}$$

$$\bar{s}_3 = \frac{36R}{35} \tag{10}$$

$$\bar{s}_4 = \frac{16384R}{4725\pi}. \tag{11}$$

See also BALL POINT PICKING, SPHERE LINE PICKING

References

Kendall, M. G. and Moran, P. A. P. *Geometrical Probability.* New York: Hafner, 1963.

Santaló, L. A. *Integral Geometry and Geometric Probability.* Reading, MA: Addison-Wesley, 1976.

Tu, S.-J. and Fischbach, E. A New Geometric Probability Technique for an *N.*-Dimensional Sphere and Its Applications 17 Apr 2000. http://xxx.lanl.gov/abs/math-ph/0004021/.

Ball Point Picking

See also BALL LINE PICKING, DISK POINT PICKING, NOISE SPHERE, SPHERE POINT PICKING

Ball Tetrahedron Picking

The mean volume of a TETRAHEDRON formed by four random points in a UNIT SPHERE is $\bar{V} = 12\pi/715$ (Hostinsky 1925; Solomon 1978, p. 124).

See also SPHERE TETRAHEDRON PICKING

References

Hostinsky, B. "Sur les probabilités géométriques." *Publ. Fac. Sci. Univ. Masaryk*, No. 50. Brno, Czechoslovakia, 1925.

Solomon, H. *Geometric Probability.* Philadelphia, PA: SIAM, 1978.

Ball Triangle Picking

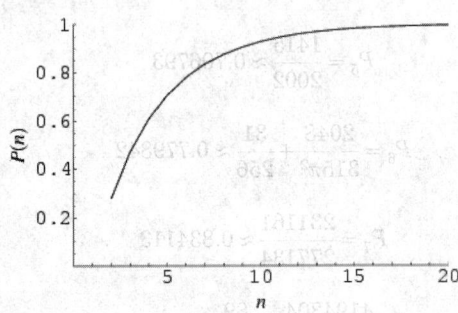

The determination of the probability for obtaining an OBTUSE TRIANGLE by picking three points at random in the unit DISK was generalized by Hall (1982) to the n-dimensional BALL. Buchta (1986) subsequently gave closed form evaluations for Hall's integrals. Let P_n be the probability that that three points chosen independently and uniformly from the n-BALL form an ACUTE TRIANGLE, then

$$P_{2m+1} = -\frac{1}{2} - 2^{2m-1} \frac{\binom{2m}{m}\binom{4m}{2m}}{\binom{4m}{m}\binom{6m+1}{2m}} + m\binom{2m}{m}^2 2^{2m}$$

$$\times \sum_{k=0}^{m} \frac{\binom{2k}{k}}{\binom{2m+k}{m}\binom{4m+2k}{2m+k}}$$

$$\times \frac{3m+k+1}{(m+k)(3m+2k+1)} \quad (1)$$

$$P_{2m+2} = \frac{1}{4} - \frac{3}{2^{2m+4}} \frac{\binom{4m+4}{m+1}}{\binom{2m+2}{m+1}} + \frac{2^{4m}}{\binom{2m}{m}\pi^2}$$

$$\times \left[\frac{1}{(2m+1)^2\binom{2m}{m}} \right.$$

$$\left. + \sum_{k=0}^{m} \frac{2^{2k}(3m+k-3)}{(2k+1)\binom{2k}{k}\binom{2m+k}{m}\binom{2m+k+2}{m}} \right], \quad (2)$$

the first few being

$$P_2 = \frac{4}{\pi^2} - \frac{1}{8} \approx 0.280285 \quad (3)$$

$$P_3 = \tfrac{33}{70} \approx 0.471429 \quad (4)$$

$$P_4 = \frac{256}{45\pi^2} + \frac{1}{32} \approx 0.607655 \quad (5)$$

$$P_5 = \frac{1415}{2002} \approx 0.706793 \quad (6)$$

$$P_6 = \frac{2048}{315\pi^2} + \frac{31}{256} \approx 0.779842 \quad (7)$$

$$P_7 = \frac{231161}{277134} \approx 0.834113 \quad (8)$$

$$P_8 = \frac{4194304}{606375\pi^2} + \frac{89}{512} \approx 0.874668 \quad (9)$$

$$P_9 = \frac{9615369}{10623470} \approx 0.905106. \quad (10)$$

The case P_2 corresponds to DISK TRIANGLE PICKING case.

See also CUBE TRIANGLE PICKING, OBTUSE TRIANGLE, SPHERE POINT PICKING

References

Buchta, C. "A Note on the Volume of a Random Polytope in a Tetrahedron." *Ill. J. Math.* **30**, 653–59, 1986.
Hall, G. R. "Acute Triangles in the *n*-Ball." *J. Appl. Prob.* **19**, 712–15, 1982.

Ballantine

BORROMEAN RINGS

Ballieu's Theorem

Let the CHARACTERISTIC POLYNOMIAL of an $n \times n$ COMPLEX MATRIX A be written in the form

$$P(\lambda) = |\lambda 1 - A| = \lambda^n + b_1 \lambda^{n-1} + b_2 \lambda^{n-2} + \ldots + b_{n-1}\lambda + b_n.$$

Then for any set $\boldsymbol{\mu} = (\mu_1, \mu_2, \ldots, \mu_n)$ of POSITIVE numbers with $\mu_0 = 0$ and

$$M^- = \max_{0 \le k \le n-1} \frac{\mu_k + \mu_n |b_{n-k}|}{\mu_{k+1}},$$

all the EIGENVALUES λ_i (for $i = 1, \ldots, n$) lie on the CLOSED DISK $|z| \le M^-$ in the COMPLEX PLANE.

References

Gradshteyn, I. S. and Ryzhik, I. M. *Tables of Integrals, Series, and Products, 6th ed.* San Diego, CA: Academic Press, p. 1153, 2000.

Ballot Problem

Suppose A and B are candidates for office and there are $2n$ voters, n voting for A and n for B. In how many ways can the ballots be counted so that A is always ahead of or tied with B? The solution is a CATALAN NUMBER C_n.

A related problem also called "the" ballot problem is to let A receive a votes and B b votes with $a > b$. This version of the ballot problem then asks for the probability that A stays ahead of B as the votes are counted (Vardi 1991). The solution is $(a-b)/(a+b)$, as first shown by M. Bertrand (Hilton and Pedersen 1991). Another elegant solution was provided by André (1887) using the so-called ANDRÉ'S REFLECTION METHOD.

The problem can also be generalized (Hilton and Pedersen 1991). Furthermore, the TAK FUNCTION is connected with the ballot problem (Vardi 1991).

See also ANDRÉ'S REFLECTION METHOD, CATALAN NUMBER, STAIRCASE WALK, TAK FUNCTION

References

André, D. "Solution directe du problème résolu par M. Bertrand." *Comptes Rendus Acad. Sci. Paris* **105**, 436–37, 1887.
Ball, W. W. R. and Coxeter, H. S. M. *Mathematical Recreations and Essays, 13th ed.* New York: Dover, p. 49, 1987.
Carlitz, L. "Solution of Certain Recurrences." *SIAM J. Appl. Math.* **17**, 251–59, 1969.

Comtet, L. *Advanced Combinatorics: The Art of Finite and Infinite Expansions, rev. enl. ed.* Dordrecht, Netherlands: Reidel, p. 22, 1974.

Feller, W. *An Introduction to Probability Theory and Its Applications, Vol. 1, 3rd ed.* New York: Wiley, pp. 67–7, 1968.

Hilton, P. and Pedersen, J. "The Ballot Problem and Catalan Numbers." *Nieuw Archief voor Wiskunde* **8**, 209–16, 1990.

Hilton, P. and Pedersen, J. "Catalan Numbers, Their Generalization, and Their Uses." *Math. Intel.* **13**, 64–5, 1991.

Kraitchik, M. "The Ballot-Box Problem." §6.13 in *Mathematical Recreations.* New York: W. W. Norton, p. 132, 1942.

Motzkin, T. "Relations Between Hypersurface Cross Ratios, and a Combinatorial Formula for Partitions of a Polygon, for Permanent Preponderance, and for Non-Associative Products." *Bull. Amer. Math. Soc.* **54**, 352–60, 1948.

Vardi, I. *Computational Recreations in Mathematica.* Redwood City, CA: Addison-Wesley, pp. 185–87, 1991.

Balthasart Projection

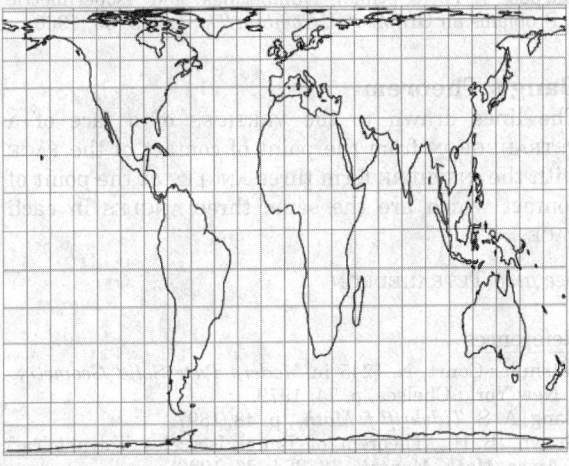

A CYLINDRICAL EQUAL-AREA PROJECTION which uses a standard parallel of $\phi_s = 50°$.

See also CYLINDRICAL EQUAL-AREA PROJECTION, BEHRMANN CYLINDRICAL EQUAL-AREA PROJECTION, GALL ORTHOGRAPHIC PROJECTION, LAMBERT AZIMUTHAL EQUAL-AREA PROJECTION, PETERS PROJECTION, TRISTAN EDWARDS PROJECTION

Banach Algebra

A Banach algebra is an ALGEBRA B over a FIELD F endowed with a NORM $\|\cdot\|$ such that B is a BANACH SPACE under the norm $\|\cdot\|$ and multiplication is continuous in the sense that if x, $y \in B$ then $\|xy\| \le \|x\|\|y\|$. Continuity of multiplication is the most important property.

F is frequently taken to be the COMPLEX NUMBERS in order to assure that the SPECTRUM fully characterizes an OPERATOR (i.e., the spectral theorems for normal or compact normal operators do not, in general, hold in the SPECTRUM over the REAL NUMBERS).

If B has a unit, then $x \in B$ is invertible if and only if $\hat{x}(\phi) \ne 0$ for all ϕ, where $x \mapsto \hat{x}$ is the GELFAND TRANSFORM.

See also B*-ALGEBRA, BANACH SPACE, GELFAND TRANSFORM

References

Helemskii, A. Ya. *Banach and Locally Convex Algebras.* Oxford, England: Oxford University Press, 1993.

Katznelson, Y. *An Introduction to Harmonic Analysis.* New York: Dover, 1976.

Rudin, W. *Real and Complex Analysis, 3rd ed.* New York: McGraw-Hill, 1987.

Banach Fixed Point Theorem

Let f be a contraction mapping from a closed SUBSET F of a BANACH SPACE E into F. Then there exists a unique $z \in F$ such that $f(z) = z$.

See also FIXED POINT THEOREM

References

Debnath, L. and Mikusinski, P. *Introduction to Hilbert Spaces with Applications.* San Diego, CA: Academic Press, 1990.

Banach Measure

An "AREA" which can be defined for every set–even those without a true geometric AREA–which is rigid and finitely additive.

Banach Space

A Banach space is a COMPLETE VECTOR SPACE B with a norm $\|v\|$. Its topology is determined by its norm, and the vector space operations of addition and scalar multiplication are required to be continuous. Two norms $\langle v \rangle_1$ and $\langle v \rangle_2$ are called equivalent if they give the same TOPOLOGY, which is equivalent to the existence of constants c and C such that

$$c\langle v \rangle_1 \le \langle v \rangle_2 \le C\langle v \rangle_1 \qquad (1)$$

holds for all v. In the finite dimensional case, all norms are equivalent. An infinite dimensional space can have many different norms.

A basic example is n dimensional EUCLIDEAN SPACE with the Euclidean norm. Usually, the notion of Banach space is only used in the infinite dimensional setting, typically as a VECTOR SPACE of functions. For example, the set of continuous functions on the real line with the norm of a function f given by

$$\|f\| = sup_{x \in \mathbb{R}} |f(x)| \qquad (2)$$

is a Banach space, where sup denotes the SUPREMUM.

On the other hand, the set of continuous functions on the unit interval [0, 1] with the norm of a function f given by

$$\|f\| = \int_0^1 |f(x)|\, dx \qquad (3)$$

is not a Banach space because it is not complete. For instance, the CAUCHY SEQUENCE of functions

$$f_n \begin{cases} 1 & \text{for } x \leq 1/2 \\ \frac{1}{2}n + 1 - nx & \text{for } x \leq 1/2 + 1/n \\ 0 & \text{for } x > 1/2 + 1/n \end{cases} \qquad (4)$$

does not converge to a continuous function.

HILBERT SPACES with their norm given by the inner product are examples of Banach spaces. While a HILBERT SPACE is always a Banach space, the converse need not hold. Therefore, it is possible for a Banach space not to have a norm given by an inner product. For instance, the supremum norm cannot be given by an INNER PRODUCT.

See also BESOV SPACE, COMPLETE SPACE, HILBERT SPACE, SCHAUDER FIXED POINT THEOREM, VECTOR SPACE

Banach-Hausdorff-Tarski Paradox

BANACH-TARSKI PARADOX

Banach-Steinhaus Theorem

UNIFORM BOUNDEDNESS PRINCIPLE

Banach-Tarski Paradox

First stated in 1924, the Banach-Tarski paradox states that it is possible to dissect a BALL into six pieces which can be reassembled by rigid motions to form two balls of the same size as the original. The number of pieces was subsequently reduced to five by R. M. Robinson in 1944, although the pieces are extremely complicated. (Actually, four pieces are sufficient as long as the single point at the center is neglected.) A generalization of this theorem is that any two bodies in $\mathbb{R}^3$ which do not extend to infinity and each containing a ball of arbitrary size can be dissected into each other (i.e., they are EQUIDECOMPOSABLE).

See also BALL, CIRCLE SQUARING, DISSECTION, EQUIDECOMPOSABLE

References

Banach, S. and Tarski, A. "Sur la décomposition des ensembles de points en parties respectivement congruentes." *Fund. Math.* **6**, 244–77, 1924.
Erickson, G. W. and Fossa, J. A. *Dictionary of Paradox.* Lanham, MD: University Press of America, pp. 16–7, 1998.
Gardner, M. *The Sixth Book of Mathematical Games from Scientific American.* Chicago, IL: University of Chicago Press, p. 48, 1984.
Hertel, E. "On the Set-Theoretical Circle-Squaring Problem." http://www.minet.uni-jena.de/Math-Net/reports/sources/2000/00-6report.ps.
Stromberg, K. "The Banach-Tarski Paradox." *Amer. Math. Monthly* **86**, 3, 1979.

Wagon, S. "A Hyperbolic Interpretation of the Banach-Tarski Paradox." *Mathematica J.* **3**, 58–0, 1993.
Wagon, S. *The Banach-Tarski Paradox.* New York: Cambridge University Press, 1993.

Bandwidth

The bandwidth of a MATRIX $\mathsf{M} = (m_{ij})$ is the maximum value of $|i - j|$ such that m_{ij} is nonzero.

The bandwidth of a GRAPH G is the minimum bandwidth among ADJACENCY MATRICES of GRAPHS isomorphic to G. Bounds for the bandwidth of a graph have been considered by (Harper 1964), and the bandwidth of the k-cube was determined by Harper (1966).

References

Chvátalová, J. "Optimal Labelling of a Product of Two Paths." *Disc. Math.* **11**, 249–53, 1975.
Harper, L. H. "Optimal Assignments of Numbers to Vertices." *J. Soc. Indust. Appl. Math.* **12**, 131–35, 1964.
Harper, L. H. "Optimal Numberings and Isoperimetric Problems on Graphs." *J. Combin. Th.* **1**, 385–93, 1966.

Bang's Theorem

The lines drawn to the VERTICES of a face of a TETRAHEDRON from the point of contact of the FACE with the INSPHERE form three ANGLES at the point of contact which are the same three ANGLES in each FACE.

See also TETRAHEDRON

References

Altshiller-Court, N. §245 in *Modern Pure Solid Geometry.* New York: Chelsea, p. 74, 1979.
Bang, A. S. *Tidskrift f. Math.*, p. 48, 1897.
Brown, B. H. "Theorem of Bang. Isosceles Tetrahedra." *Amer. Math. Monthly* **33**, 224–26, 1926.
Honsberger, R. *Mathematical Gems II.* Washington, DC: Math. Assoc. Amer., p. 93, 1976.
Wells, D. *The Penguin Dictionary of Curious and Interesting Geometry.* London: Penguin, p. 13, 1991.
White, H. S. "Two Tetrahedron Theorems." *Nouvelles Ann. de Math* **14**, 220–22, 1907–908.

Bankoff Circle

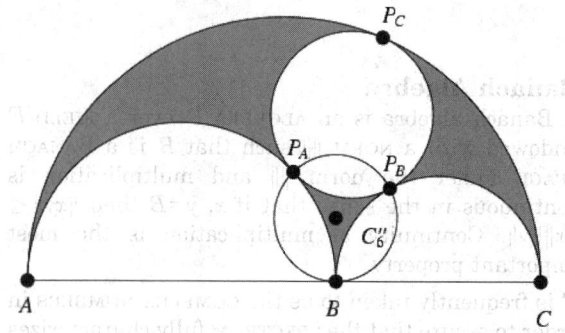

The circle through the cusp of the ARBELOS and the tangent points of the first Pappus circle, which is congruent to the two ARCHIMEDES' CIRCLES. If $AB = r$

and $AC = 1$, then the radius of the Bankoff circle is

$$R = \tfrac{1}{2}r(1 - r).$$

See also ARCHIMEDES' CIRCLES, ARBELOS, PAPPUS CHAIN

References

Bankoff, L. "Are the Twin Circles of Archimedes Really Twins?" *Math. Mag.* **47**, 214–18, 1974.

Gardner, M. "Mathematical Games: The Diverse Pleasures of Circles that Are Tangent to One Another." *Sci. Amer.* **240**, 18–8, Jan. 1979.

Banzhaf Power Index

The number of ways in which a group of n with weights $\sum_{i=1}^{n} w_i = 1$ can change a losing coalition (one with $\sum w_i < 1/2$)) to a winning one, or vice versa. It was proposed by the lawyer J. F. Banzhaf in 1965.

References

Paulos, J. A. *A Mathematician Reads the Newspaper.* New York: BasicBooks, pp. 9–0, 1995.

Bar

A bar (also called an overbar) is a horizontal line written above a mathematical symbol to give it some special meaning. If the bar is placed over a single symbol, as in $\bar{x}$ (voiced "x-bar"), it is sometimes called a MACRON. If placed over multiple symbols (especially in the context of a RADICAL), it is known as a VINCULUM. Common uses of the bar symbol include the following.

1. The MEAN

$$\bar{x} \equiv \frac{1}{n} \sum_{i=1}^{n} x_i$$

of a set $\{x_i\}_{i=1}^{n}$.
2. The COMPLEX CONJUGATE

$$\bar{z} \equiv x - iy$$

for $z = x + iy$.
3. The COMPLEMENT $\bar{F}$ of a set F.
4. A SET stripped of any structure besides order, hence the ORDER TYPE of the set.

In conventional typography, "bar" refers to a vertical (instead a horizontal) bar, such as those used to denote ABSOLUTE VALUE ($|x|$) (Bringhurst 1997, p. 271).

See also DOUBLE BAR, HAT, MACRON, VINCULUM

References

Bringhurst, R. *The Elements of Typographic Style, 2nd ed.* Point Roberts, WA: Hartley and Marks, p. 271, 1997.

Bar (Edge)

The term in rigidity theory for the EDGES of a GRAPH.

See also CONFIGURATION, FRAMEWORK

Bar Chart

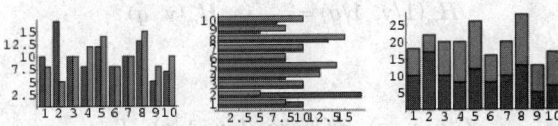

A bar graph is any plot of a set of data such that the number of data elements falling within one or more categories is indicated using a rectangle whose height or width is a function of the number of elements.

See also HISTOGRAM, PIE CHART

References

Kenney, J. F. and Keeping, E. S. *Mathematics of Statistics, Pt. 1, 3rd ed.* Princeton, NJ: Van Nostrand, p. 23, 1962.

Bar Graph

BAR CHART

Bar Graph Polygon

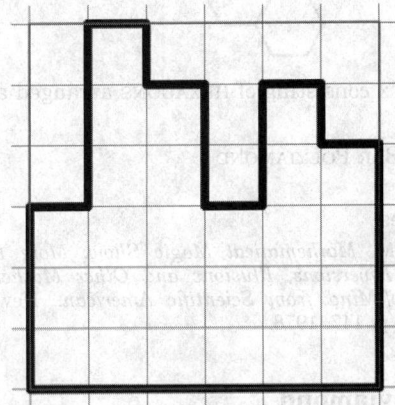

A column-convex SELF-AVOIDING POLYGON which contains the bottom edge of its minimal bounding rectangle. The anisotropic perimeter and area generating function

$$G(x,\, y,\, q) = \sum m \ge 1 \sum_{n \ge 1} \sum_{a \ge a} C(m,\, n,\, a) x^m y^n q^a,$$

where $C(m, n, a)$ is the number of polygons with $2m$ horizonal bonds, $2n$ vertical bonds, and area a, has been computed exactly for the bar graph polygons (Bousquet-Mélou 1996, Bousquet-Mélou *et al.* 1999). The anisotropic area and perimeter generating function $G(x,\, y,\, q)$ and partial generating functions

$H_m(y, q)$, connected by

$$G(x, y, q) = \sum_{m \geq 1} H_m(y, q)x^m,$$

satisfy the self-reciprocity and inversion relations

$$H_m(1/y, 1/q) = \frac{(-1)^m}{yq^m} H_m(y, q)$$

and

$$G(x, y, q) - yG(-xq, 1/y, 1/q) = 0$$

(Bousquet-Mélou *et al.* 1999).

See also LATTICE POLYGON, SELF-AVOIDING POLYGON

References

Bousquet-Mélou, M. "A Method for Enumeration of Various Classes of Column-Convex Polygons." *Disc. Math.* **154**, 1–5, 1996.
Bousquet-Mélou, M.; Guttmann, A. J.; Orrick, W. P.; and Rechnitzer, A. Inversion Relations, Reciprocity and Polyominoes. 23 Aug 1999. http://xxx.lanl.gov/abs/math.CO/9908123/.

Bar Polyhex

A POLYHEX consisting of HEXAGONS arranged along a line.

See also BAR POLYIAMOND

References

Gardner, M. *Mathematical Magic Show: More Puzzles, Games, Diversions, Illusions and Other Mathematical Sleight-of-Mind from Scientific American.* New York: Vintage, p. 147, 1978.

Bar Polyiamond

A POLYIAMOND consisting of EQUILATERAL TRIANGLES arranged along a line.

See also BAR POLYHEX

References

Golomb, S. W. *Polyominoes: Puzzles, Patterns, Problems, and Packings, 2nd ed.* Princeton, NJ: Princeton University Press, p. 92, 1994.

Barber Paradox

A man of Seville is shaved by the Barber of Seville IFF the man does not shave himself. Does the barber shave himself? This PSEUDOPARADOX was proposed by Bertrand Russell.

See also PSEUDOPARADOX, RUSSELL'S PARADOX

References

Curry, H. B. *Foundations of Mathematical Logic.* New York: Dover, pp. 4–, 1977.
Erickson, G. W. and Fossa, J. A. *Dictionary of Paradox.* Lanham, MD: University Press of America, pp. 17–8, 1998.
Hoffman, P. *The Man Who Loved Only Numbers: The Story of Paul Erdos and the Search for Mathematical Truth.* New York: Hyperion, p. 116, 1998.

Barbier's Theorem

All CURVES OF CONSTANT WIDTH of width w have the same PERIMETER πw.

Bare Angle Center

The TRIANGLE CENTER with TRIANGLE CENTER FUNCTION

$$\alpha = A.$$

References

Kimberling, C. "Major Centers of Triangles." *Amer. Math. Monthly* **104**, 431–38, 1997.

Barlow Packing

A face-centered cubic SPHERE PACKING obtained by placing layers of spheres one on top of another. Because there are two distinct ways to place each layer on top of the previous one, there are an infinite number of such packings as the number of layers is increased.

See also KEPLER CONJECTURE, SPHERE PACKING

References

Barlow, W. "Probable Nature of the Internal Symmetry of Crystals." *Nature* **29**, 186–88, 1883.
Sloane, N. J. A. "Kepler's Conjecture Confirmed." *Nature* **395**, 435–36, 1998.

Barnes' G-Function

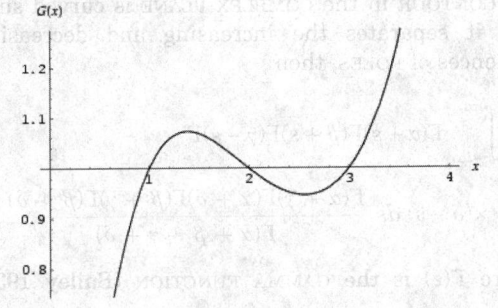

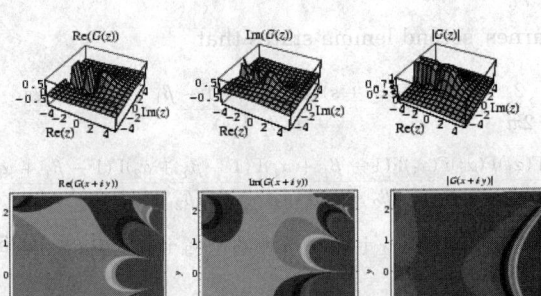

Barnes' G-function is defined by

$$G(z+1)$$

$$\equiv (2\pi)^{z/2} e^{-[z(z+1)+\gamma z^2]/2} \prod_{n=1}^{\infty} \left[\left(1 + \frac{z}{n} \right)^n e^{-z+z^2/(2n)} \right] \quad (1)$$

where γ is the EULER-MASCHERONI CONSTANT (Whittaker and Watson 1990, p. 264; Voros 1987). It is an ENTIRE FUNCTION analogous to $1/\Gamma(z)$, where $\Gamma(z)$ is the GAMMA FUNCTION, except that it has order 2 instead of 1.

This is an ANALYTIC CONTINUATION of the G-function defined in the construction of the GLAISHER-KINKELIN CONSTANT

$$G(n) \equiv \frac{[\Gamma(n)]^{n-1}}{K_n}, \quad (2)$$

where

$$K_n \equiv 0^0 1^1 2^2 3^3 \cdots (n-1)^{n-1}, \quad (3)$$

which has the special values

$$G(n) = \begin{cases} 0 & \text{if } n = 0, -1, -2, \ldots \\ 1 & \text{if } n = 1 \\ 0!1!2!\cdots(n-2)! & \text{if } n = 2, 3, 4 \ldots \end{cases} \quad (4)$$

for INTEGER n. This function is what Sloane and Plouffe (1995) call the SUPERFACTORIAL, and the first few values for $n = 1, 2, \ldots$ are 1, 1, 1, 2, 12, 288, 34560, 24883200, 125411328000, 5056584744960000, ... (Sloane's A000178).

Barnes' G-function satisfies the functional equation

$$G(z+1) = \Gamma(z)G(z), \quad (5)$$

and has the TAYLOR SERIES

$$\ln G(1+z) = \frac{1}{2}[\ln(2\pi) - 1]z - (1 + \gamma)\frac{z^2}{2}$$

$$+ \sum_{n=3}^{\infty} (-1)^{n-1} \zeta(n-1) \frac{z^n}{n} \quad (6)$$

in $|z| < 1$. It also gives an analytic solution to the finite product

$$\prod_{i=1}^{n} \Gamma(k+i) = \frac{G(n+k+1)}{G(k+1)}, \quad (7)$$

has the identities

$$\frac{[\Gamma(n)]^n}{G(n)} = K(n), \quad (8)$$

where $K(n)$ is the K-FUNCTION, and the equivalent reflection formulas

$$\frac{G'(z+1)}{G(z+1)} = \frac{1}{2}\ln(2\pi) - \frac{1}{2} - z + z\frac{\Gamma'(z)}{\Gamma(z)} \quad (9)$$

$$\ln\left[\frac{G(1-z)}{G(1+z)}\right] = \pi \int_0^z z \cot(\pi z)\, dz - z\ln(2\pi) \quad (10)$$

$$\frac{G(\frac{1}{2}+z)}{(\frac{1}{2}-z)} = \frac{(2\pi)^2}{\Gamma(\frac{1}{2}+z)} \sqrt{\frac{\pi}{\cos(\pi z)}} \exp\left[\pi \int_0^z \tan(\pi z)\, dz\right] \quad (11)$$

(Voros 1987; Whittaker and Watson 1990, p. 264). A Stirling-like ASYMPTOTIC SERIES as $z \to \infty$ is given by

$$\ln G(1+z) \sim z^2 \left(\frac{1}{2}\ln z - \frac{3}{4}\right) + \frac{1}{2}\ln(2\pi)z - \frac{1}{12}\ln z - \ln A$$

$$+ \mathcal{O}\left(\frac{1}{z}\right) \quad (12)$$

(Voros 1987).

$G(n)$ has the special values

$$G(\tfrac{1}{2}) = \pi^{-1/4} \exp\left[\frac{1}{24}\ln 2 + \frac{3}{2}\zeta'(-1)\right] \quad (13)$$

$$= A^{-3/2} \pi^{-1/4} e^{1/8} 2^{1/24} \quad (14)$$

$$G(\tfrac{3}{2}) = A^{-3/2} \pi^{1/4} e^{1/8} 2^{1/24}, \quad (15)$$

and so on, where $\zeta'(-1)$ is the derivative of the RIEMANN ZETA FUNCTION evaluated at -1 and the GLAISHER-KINKELIN CONSTANT A is defined by

$$A = \exp[\tfrac{1}{12} - \zeta'(-1)] = 1.28242712\ldots \quad (16)$$

(Voros 1987). *Mathematica* 4.0 implements the constant A as Glaisher. In general, for odd $n = 2k+1$,

$$G(\tfrac{1}{2}(2k+1)) = c_k \frac{A^{-3/2}\pi^{-(2k-3)/4}e^{1/8}2^{1/24}}{2^{(k-1)(k-2)/2}}, \quad (17)$$

where

$$c_k = \prod_{i=1}^{k-2} \frac{2^i \Gamma(\tfrac{1}{2}+i)}{\sqrt{\pi}} \quad (18)$$

for $k > 1$, of which the first few terms are 1, 1, 1, 3, 45, 4725 4465125, ... (Sloane's A057863).

Barnes' G-function can arise in spectral functions in mathematical physics (Voros 1987).

Another G-FUNCTION is defined by Erdélyi *et al.* (1981, p. 20) as

$$G(z) \equiv \psi_0\left(\tfrac{1}{2} + hz\right) - \psi_0(\tfrac{1}{2}z), \quad (19)$$

where $\psi_0(z)$ is the DIGAMMA FUNCTION. An unrelated pair of functions are denoted g_n and G_n and are known as RAMANUJAN g- AND G-FUNCTIONS.

See also EULER-MASCHERONI CONSTANT, G-FUNCTION, GLAISHER-KINKELIN CONSTANT, K-FUNCTION, MEIJER'S G-FUNCTION, RAMANUJAN g- AND G-FUNCTIONS, SUPERFACTORIAL

References

Barnes, E. W. "The Theory of the G-Function." *Quart. J. Pure Appl. Math.* **31**, 264–14, 1900.
Dyson, F. J. "Fredholm Determinants and Inverse Scattering Problems." *Commun. Math. Phys.* **47**, 171–83, 1976.
Glaisher, J. W. L. "On a Numerical Continued Product." *Messenger Math.* **6**, 71–6, 1877.
Glaisher, J. W. L. "On the Product $1^1 2^2 3^3 \cdots n^n$." *Messenger Math.* **7**, 43–7, 1878.
Glaisher, J. W. L. "On Certain Numerical Products." *Messenger Math.* **23**, 145–75, 1893.
Glaisher, J. W. L. "On the Constant Which Occurs in the Formula for $1^1 2^2 3^3 \cdots n^n$." *Messenger Math.* **24**, 1–6, 1894.
Kinkelin. "Über eine mit der Gammafunktion verwandte Transcendente und deren Anwendung auf die Integralrechnung." *J. reine angew. Math.* **57**, 122–58, 1860.
Lenard, A. "Some Remarks on Large Toeplitz Matrices." *Pacific J. Math.* **42**, 137–45, 1972.
McCoy, B. and Wu, T. T. *The Two-Dimensional Ising Model.* Cambridge, MA: Harvard University Press, p. 264 and Appendix B, 1973.
Sloane, N. J. A. Sequences A000178/M2049 and A057863 in "An On-Line Version of the Encyclopedia of Integer Sequences." http://www.research.att.com/~njas/sequences/eisonline.html.
Sloane, N. J. A. and Plouffe, S. *The Encyclopedia of Integer Sequences.* San Diego: Academic Press, 1995.
Voros, A. "Spectral Functions, Special Functions and the Selberg Zeta Function." *Commun. Math. Phys.* **110**, 439–65, 1987.
Whittaker, E. T. and Watson, G. N. *A Course in Modern Analysis, 4th ed.* Cambridge, England: Cambridge University Press, p. 264, 1990.
Widom, H. "The Strong Szego Limit Theorem for Circular Arcs." *Indiana Univ. Math. J.* **21**, 277–83, 1971.
Widom, H. "Toeplitz Determinants with Singular Generating Functions." *Amer. J. Math.* **95**, 333–83, 1973.

Barnes' Lemma

If a CONTOUR in the COMPLEX PLANE is curved such that it separates the increasing and decreasing sequences of POLES, then

$$\frac{1}{2\pi i}\int_{-i\infty}^{i\infty} \Gamma(\alpha+s)\Gamma(\beta+s)\Gamma(\gamma-s)\Gamma$$
$$\times(\delta-s)\,ds \; \frac{\Gamma(\alpha+\gamma)\Gamma(\alpha+\delta)\Gamma(\beta+\gamma)\Gamma(\beta+\delta)}{\Gamma(\alpha+\beta+\gamma+\delta)},$$

where $\Gamma(z)$ is the GAMMA FUNCTION (Bailey 1935, p. 7).

Barnes' second lemma states that

$$\int \frac{2}{2\pi i}\, \frac{\Gamma(\alpha_1+s)\Gamma(\alpha_2+s)\Gamma(\alpha_3+s)\Gamma(1-\beta_1-s)\Gamma(-s)\,ds}{\Gamma(\beta_2+s)}$$
$$= \frac{\Gamma(\alpha_1)\Gamma(\alpha_2)\Gamma(\alpha_3)\Gamma(1-\beta_1+\alpha_1)\Gamma(1-\beta_1+\alpha_2)\Gamma(1-\beta_1+\alpha_3)}{\Gamma(\beta_2-\alpha_1)\Gamma(\beta_2-\alpha_2)\Gamma(\beta_2-\alpha_3)}$$

provided that $\beta_1 + \beta_2 = \alpha_1 + \alpha_2 + \alpha_3 + 1$ (Bailey 1935, pp. 42–3).

References

Bailey, W. N. "Barnes' Lemma" and "Barnes' Second Lemma." §1.7 and 6.2 in *Generalised Hypergeometric Series.* Cambridge, England: University Press, pp. 7 and 42–3, 1935.
Barnes, E. W. "A New Development in the Theory of the Hypergeometric Functions." *Proc. London Math. Soc.* **6**, 141–77, 1908.

Barnes-Wall Lattice

A lattice which can be constructed from the LEECH LATTICE Λ_{24}.

See also COXETER-TODD LATTICE, LATTICE POINT, LEECH LATTICE

References

Barnes, E. S. and Wall, G. E. "Some Extreme Forms Defined in Terms of Abelian Groups." *J. Austral. Math. Soc.* **1**, 47–3, 1959.
Conway, J. H. and Sloane, N. J. A. "The 16-Dimensional Barnes-Wall Lattice Λ_{16}." §4.10 in *Sphere Packings, Lattices, and Groups, 2nd ed.* New York: Springer-Verlag, pp. 127–29, 1993.

Barnette's Conjecture

The conjecture that every 3-connected BIPARTITE CUBIC PLANAR GRAPH is HAMILTONIAN.

See also BIPARTITE GRAPH, CUBIC GRAPH, HAMILTONIAN GRAPH

References

Barnette, D. Conjecture 5 in *Recent Progress in Combinatorics* (Ed. W. T. Tutte). New York: Academic Press, 1969.
Owens, P. J. "Bipartite Cubic Graphs and a Shortness Exponent." *Disc. Math.* **44**, 327–30, 1983.

Barnsley's Fern

The ATTRACTOR of the ITERATED FUNCTION SYSTEM given by the set of "fern functions"

$$f_1(x, y) = \begin{bmatrix} 0.85 & 0.04 \\ -0.04 & 0.85 \end{bmatrix} \begin{bmatrix} x \\ y \end{bmatrix} + \begin{bmatrix} 0.00 \\ 1.60 \end{bmatrix} \quad (1)$$

$$f_2(x, y) = \begin{bmatrix} -0.15 & 0.28 \\ 0.26 & 0.24 \end{bmatrix} \begin{bmatrix} x \\ y \end{bmatrix} + \begin{bmatrix} 0.00 \\ 0.44 \end{bmatrix} \quad (2)$$

$$f_3(x, y) = \begin{bmatrix} 0.20 & -0.26 \\ 0.23 & 0.22 \end{bmatrix} \begin{bmatrix} x \\ y \end{bmatrix} + \begin{bmatrix} 0.00 \\ 1.60 \end{bmatrix} \quad (3)$$

$$f_4(x, y) = \begin{bmatrix} 0.00 & 0.00 \\ 0.00 & 0.16 \end{bmatrix} \begin{bmatrix} x \\ y \end{bmatrix} \quad (4)$$

(Barnsley 1993, p. 86; Wagon 1991). These AFFINE TRANSFORMATIONS are contractions. The tip of the fern (which resembles the black spleenwort variety of fern) is the fixed point of f_1, and the tips of the lowest two branches are the images of the main tip under f_2 and f_3 (Wagon 1991).

See also DYNAMICAL SYSTEM, FRACTAL, ITERATED FUNCTION SYSTEM

References

Barnsley, M. *Fractals Everywhere, 2nd ed.* Boston, MA: Academic Press, pp. 86, 90, 102 and Plate 2, 1993.
Gleick, J. *Chaos: Making a New Science.* New York: Penguin Books, p. 238, 1988.
Wagon, S. "Biasing the Chaos Game: Barnsley's Fern." §5.3 in *Mathematica in Action.* New York: W. H. Freeman, pp. 156–63, 1991.
Weisstein, E. W. "Fractals." MATHEMATICA NOTEBOOK FRACTAL.M.

Barrel

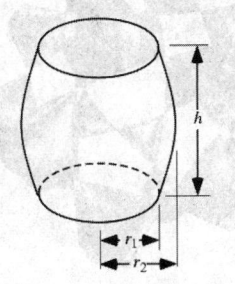

A SOLID OF REVOLUTION composed of parallel circular top and bottom with a common axis and a side formed by a smooth curve symmetrical about the midplane.

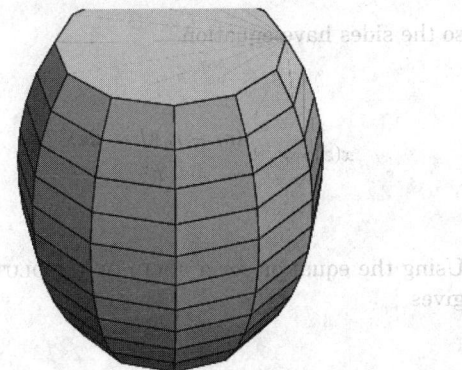

For sides consisting of an arc of an ELLIPSE, the equation of the side is given by

$$x(z) = r_2 \sqrt{1 - \frac{(z - \frac{1}{2}h)^2}{a^2}}, \quad (1)$$

with $x(0) = r_1$. Solving for a gives

$$a = \frac{hr_2}{2\sqrt{r_2^2 - r_1^2}}, \quad (2)$$

so the sides have equation

$$x(z) = \sqrt{r_2^2 + \frac{(r_1 - r_2)(r_1 + r_2)(h - 2z)^2}{h^2}}. \quad (3)$$

Using the equation for a SOLID OF REVOLUTION then gives

$$V = \pi \int_0^h [x(z)]^2 dx = \frac{1}{3}\pi h(2r_2^2 + r_1^2). \quad (4)$$

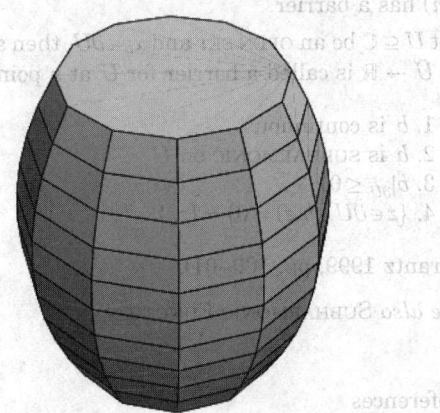

For sides consisting of a PARABOLIC SEGMENT, the equation of the side is given by

$$x(z) = r_2 + a(z - \frac{1}{2}h)^2 \quad (5)$$

with $x(0) = r_1$. Solving for a gives

$$a = \frac{4(r_1 - r_2)}{h^2}, \qquad (6)$$

so the sides have equation

$$x(z) = r_2 + \frac{(r_1 - r_2)(h - 2z)^2}{h^2}. \qquad (7)$$

Using the equation for a SOLID OF REVOLUTION then gives

$$V = \pi \int_0^h [x(z)]^2 \, dx = \tfrac{1}{15}\, \pi h (3r_1^2 + 4r_1 r_2 + 8r_2^2). \qquad (8)$$

See also CYLINDER

References

Harris, J. W. and Stocker, H. "Barrel." §4.10.4 in *Handbook of Mathematics and Computational Science.* New York: Springer-Verlag, p. 112, 1998.

Barrier

A number n is called a barrier of a number-theoretic function $f(m)$ if, for all $m < n$, $m + f(m) \le n$. Neither the TOTIENT FUNCTION $\phi(n)$ nor the DIVISOR FUNCTION $\sigma(n)$ has a barrier.

Let $U \subseteq \mathbb{C}$ be an OPEN SET and $x_0 \in \partial U$, then a function $b : \bar{U} \to \mathbb{R}$ is called a barrier for U at a point x_0 if

1. b is continuous,
2. b is SUBHARMONIC on U,
3. $b|_{\partial U} \le 0$,
4. $\{z \in \partial U : b(z) = 0\} = \{z_0\}$

(Krantz 1999, pp. 100–01).

See also SUBHARMONIC FUNCTION

References

Guy, R. K. *Unsolved Problems in Number Theory, 2nd ed.* New York: Springer-Verlag, pp. 64–5, 1994.
Krantz, S. G. "The Concept of a Barrier." §7.7.9 in *Handbook of Complex Analysis.* Boston, MA: Birkhäuser, pp. 100–01, 1999.

Barth Decic

The Barth decic is a DECIC SURFACE in complex three-dimensional projective space having the maximum possible number of ORDINARY DOUBLE POINTS (345). It is given by the implicit equation

$$8(x^2 - \phi^4 y^2)(y^2 - \phi^4 z^2)(z^2 - \phi^4 x^2)$$
$$\times (x^4 + y^4 + z^4 - 2x^2 y^2 - 2x^2 z^2 - 2y^2 z^2) + (3 + 5\phi)$$
$$\times (x^2 + y^2 + z^2 - w^2)^2 [x^2 + y^2 + z^2 - (2 - \phi)w^2]^2 w^2$$
$$= 0,$$

where ϕ is the GOLDEN MEAN and w is a parameter (Endraß, Nordstrand), taken as $w = 1$ in the above plot. The Barth decic is invariant under the ICOSAHEDRAL GROUP.

See also ALGEBRAIC SURFACE, BARTH SEXTIC, DECIC SURFACE, ORDINARY DOUBLE POINT

References

Barth, W. "Two Projective Surfaces with Many Nodes Admitting the Symmetries of the Icosahedron." *J. Alg. Geom.* **5**, 173–86, 1996.
Endraß, S. "Flächen mit vielen Doppelpunkten." *DMV-Mitteilungen* **4**, 17–0, 4/1995.
Endraß, S. "Barth's Decic." http://enriques.mathematik.uni-mainz.de/kon/docs/Ebarthdecic.shtml.
Nordstrand, T. "Batch Decic." http://www.uib.no/people/nfytn/bdectxt.htm.

Barth Sextic

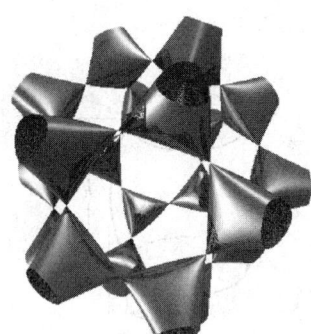

The Barth-sextic is a SEXTIC SURFACE in complex three-dimensional projective space having the maximum possible number of ORDINARY DOUBLE POINTS (65). Of these, 20 nodes are at the vertices of a regular DODECAHEDRON of side length $2/\phi$, and 30 are at the midpoints of the edges of a concentric DODECAHEDRON of side length $2/\phi^2$, where ϕ is the GOLDEN RATIO. The surface was discovered by W. Barth in 1994, and is given by the implicit equation

$$4(\phi^2 x^2 - y^2)(\phi^2 y^2 - z^2)(\phi^2 z^2 - x^2) - (1+2\phi)$$

$$\times (x^2 + y^2 + z^2 - w^2)^2 w^2 = 0,$$

where ϕ is the GOLDEN MEAN, and w is a parameter (Endraß, Nordstrand), taken as $w = 1$ in the above plot.

The Barth sextic is invariant under the ICOSAHEDRAL GROUP. Under the map

$$(x, y, z, w) \to (x^2, y^2, z^2, w^2),$$

the surface is the eightfold cover of the CAYLEY CUBIC (Endraß).

See also ALGEBRAIC SURFACE, BARTH DECIC, CAYLEY CUBIC, ORDINARY DOUBLE POINT, SEXTIC SURFACE

References

Barth, W. "Two Projective Surfaces with Many Nodes Admitting the Symmetries of the Icosahedron." *J. Alg. Geom.* **5**, 173–86, 1996.
Dominici, P. "Flight Through Barth's Sextic." http://www.mi.uni-erlangen.de/~bauerth/flight/.
Endraß, S. "Flächen mit vielen Doppelpunkten." *DMV-Mitteilungen* **4**, 17–0, 4/1995.
Endraß, S. "Barth's Sextic." http://enriques.mathematik.uni-mainz.de/kon/docs/Ebarthsextic.shtml.
Knapp, A. W. (Ed.). *Notices Amer. Math. Soc.* **46**, cover and p. 318, 1999.
Nordstrand, T. "Barth Sextic." http://www.uib.no/people/nfytn/sexttxt.htm.

Bartlett Function

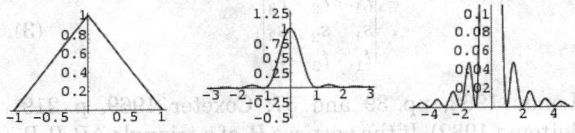

The APODIZATION FUNCTION

$$f(x) = 1 - \frac{|x|}{a} \tag{1}$$

which is a generalization of the one-argument TRIANGLE FUNCTION. Its FULL WIDTH AT HALF MAXIMUM is a.

It has INSTRUMENT FUNCTION

$$I(x) = \int_{-a}^{a} e^{-2\pi i k x}\left(1 - \frac{|x|}{a}\right) dx$$

$$= \int_{-a}^{0} e^{-2\pi i k x}\left(1 + \frac{x}{a}\right) dx$$

$$+ \int_{0}^{a} e^{-2\pi i k x}\left(1 - \frac{x}{a}\right) dx. \tag{2}$$

Letting $x' \equiv -x$ in the first part therefore gives

$$\int_{-a}^{0} e^{-2\pi i k x}\left(1 + \frac{x}{a}\right) dx = \int_{a}^{0} e^{-2\pi i k x'}\left(1 - \frac{x'}{a}\right)(-dx')$$

$$= \int_{0}^{a} e^{-2\pi i k x}\left(1 - \frac{x}{a}\right) dx. \tag{3}$$

Rewriting (2) using (3) gives

$$I(x) = (e^{2\pi i k x} + e^{-2\pi i k x})\left(1 - \frac{x}{a}\right) dx$$

$$= 2\int_{0}^{a} \cos(2\pi k x)\left(1 - \frac{x}{a}\right) dx. \tag{4}$$

Integrating the first part and using the integral

$$\int x \cos(bx)\, dx = \frac{1}{b^2}\cos(bx) + \frac{x}{b}\sin(bx) \tag{5}$$

for the second part gives

$$I(x) = 2\left[\frac{\sin(2\pi k x)}{2\pi k} - \frac{1}{a}\left\{\frac{1}{4\pi^2 k^2}\cos(2\pi k x) + \frac{x}{2\pi k}\sin(2\pi k x)\right\}\right]_0^a$$

$$= 2\left\{\left[\frac{\sin(2\pi k a)}{2\pi k} - 0\right] - \frac{1}{a}\left[\frac{\cos(2\pi k a) - 1}{4\pi^2 k^2} + \frac{a\sin(2\pi k a)}{2\pi k}\right]\right\}$$

$$= \frac{1}{2\pi^2 a k^2}[\cos(2\pi k a) - 1] = a\frac{\sin^2(\pi k a)}{\pi^2 k^2 a^2} = a\,\mathrm{sinc}^2(\pi k a) \tag{6}$$

where sinc x is the SINC FUNCTION. The peak (in units of a) is 1. The function $I(x)$ is always positive, so there are no NEGATIVE sidelobes. The extrema are given by letting $\beta \equiv \pi k a$ and solving

$$\frac{d}{d\beta}\left(\frac{\sin\beta}{\beta}\right)^2 = 2\frac{\sin\beta}{\beta}\frac{\sin\beta - \beta\cos\beta}{\beta^2} = 0 \tag{7}$$

$$\sin\beta(\sin\beta - \beta\cos\beta) = 0 \tag{8}$$

$$\sin\beta - \beta\cos\beta = 0 \tag{9}$$

$$\tan\beta = \beta. \tag{10}$$

Solving this numerically gives $\beta = 4.49341$ for the first maximum, and the peak POSITIVE sidelobe is 0.047190. The full width at half maximum is given by setting $x \equiv \pi k a$ and solving

$$\text{sinc}^2 x = \tfrac{1}{2} \tag{11}$$

for $x_{1/2}$, yielding

$$x_{1/2} = \pi k_{1/2} a = 1.39156. \tag{12}$$

Therefore, with $L \equiv 2a$,

$$\text{FWHM} = 2k_{1/2} = \frac{0.885895}{a} = \frac{1.77179}{L}. \tag{13}$$

See also APODIZATION FUNCTION, PARZEN APODIZATION FUNCTION, TRIANGLE FUNCTION

References

Bartlett, M. S. "Periodogram Analysis and Continuous Spectra." *Biometrika* **37**, 1–6, 1950.
Blackman, R. B. and Tukey, J. W. *The Measurement of Power Spectra, From the Point of View of Communications Engineering.* New York: Dover, pp. 98–9, 1959.

Barycentric Coordinates

Barycentric coordinates are triples of numbers (t_1, t_2, t_3) corresponding to masses placed at the vertices of a reference triangle $\Delta A_1 A_2 A_3$. These masses then determine a point P, which is the centroid of the three masses, and is identified with coordinates (t_1, t_2, t_3). The vertices of the triangle are given by $(1, 0, 0)$, $(0, 1, 0)$, and $(0, 0, 1)$. Barycentric coordinates were discovered by Möbius in 1827 (Coxeter 1969, p. 217; Fauvel *et al.* 1993).

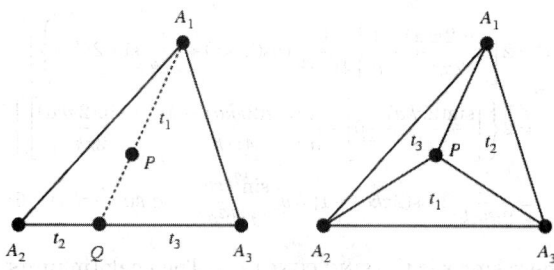

To find the barycentric coordinates for an arbitrary point P, find t_2 and t_3 from the point Q at the intersection of the line $A_1 P$ with the side $A_2 A_3$, and then determine t_1 as the mass at A_n that will balance a mass $t_2 + t_3$ at Q, thus making P the centroid (left figure). Furthermore, the areas of the triangles $\Delta A_1 A_2 P$, $\Delta A_1 A_3 P$, and $\Delta A_2 A_3 P$ are proportional to the barycentric coordinates t_3, t_2, and t_1 of P (right figure; Coxeter 1969, p. 217).

Barycentric coordinates are homogeneous, so

$$(t_1, t_2, t_3) = (\mu t_1, \mu t_2, \mu t_3) \tag{1}$$

for $\mu \neq 0$. Barycentric coordinates normalized so that they become the actual areas of the subtriangles are called homogeneous barycentric coordinates, and barycentric coordinates normalized so that

$$t_1 + t_2 + t_3 = 1, \tag{2}$$

so that the coordinates give the areas of the subtriangles *normalized by the area of the original triangle* are called AREAL COORDINATES (Coxeter 1969, p. 218). Barycentric and areal coordinates can provide particular elegant proofs of geometric theorems such as ROUTH'S THEOREM, CEVA'S THEOREM, and MENELAUS' THEOREM (Coxeter 1969, pp. 219–21).

The homogeneous barycentric coordinates corresponding to TRILINEAR COORDINATES $\alpha : \beta : \gamma$ are $(a\alpha, b\beta, c\gamma)$, and the TRILINEAR COORDINATES corresponding to homogeneous barycentric coordinates (t_1, t_2, t_3) are $t_1/a : t_2/b : t_3/c$. The homogeneous barycentric coordinates for some common triangle centers are summarized in the following table, where $s = (a + b + c)/2$ is the SEMIPERIMETER.

triangle center	homogeneous barycentric coordinates
CENTROID (TRIANGLE)	$(1, 1, 1)$
CIRCUMCENTER	$(a^2(b^2 + c^2 - a^2), b^2(c^2 + a^2 - b^2), c^3(a^2 + b^2 - c^2))$
EXCENTERS	$(-a, b, c)$
	$(a, -b, c)$
	$(a, b, -c)$
GERGONNE POINT	$((s-b)(s-c), (s-c)(s-a), (s-a)(s-b))$
INCENTER	(a, b, c)
NAGEL POINT	$(s-a, s-b, s-c)$
ORTHOCENTER	$((a^2 + b^2 - c^2)(c^2 + a^2 - b^2), (b^2 + c^2 - a^2)(a^2 + b^2 - c^2)),$ $(c^2 + a^2 - b^2)(b^2 + c^2 - a^2))$
SYMMEDIAN POINT	(a^2, b^2, c^2)

In barycentric coordinates, a line has a linear homogeneous equation. In particular, the line joining points (r_1, r_2, r_3) and (s_1, s_2, s_3) has equation

$$\begin{vmatrix} r_1 & r_2 & r_3 \\ s_1 & s_2 & s_3 \\ t_1 & t_2 & t_3 \end{vmatrix} \tag{3}$$

(Loney 1962, pp. 39 and 57; Coxeter 1969, p. 219; Bottema 1982). If the vertices P_i of a triangle $\Delta P_1 P_2 P_3$ have barycentric coordinates (x_i, y_i, z_i), then the area of the triangle is

$$\Delta P_1 P_2 P_3 = \begin{vmatrix} x_1 & y_1 & z_1 \\ x_2 & y_2 & z_2 \\ x_3 & y_3 & z_3 \end{vmatrix} \Delta ABC \tag{4}$$

(Bottema 1982, Yiu 2000).

See also AREAL COORDINATES, TRILINEAR COORDINATES

References

Bottema, O. "On the Area of a Triangle in Barycentric Coordinates." *Crux. Math.* **8**, 228–31, 1982.

Coxeter, H. S. M. "Barycentric Coordinates." §13.7 in *Introduction to Geometry, 2nd ed.* New York: Wiley, pp. 216–21, 1969.

Fauvel, J.; Flood, R.; and Wilson, R. J. (Eds.) *Möbius and his Band: Mathematics and Astronomy in Nineteenth-Century Germany.* Oxford, England: Oxford University Press, 1993.

Loney, S. L. *The Elements of Coordinate Geometry, 2 vols. in 1. Part II: Trilinear Coordinates.* London: Macmillan, 1962.

Yiu, P. "The Uses of Homogeneous Barycentric Coordinates in Plane Euclidean Geometry." *Int. J. Educ. Math. Sci. Tech.* 2000.

Base (Logarithm)

The number used to define the number system in which a LOGARITHM is computed. In general, the logarithm of a number x in base b is written $\log_b x$. The symbol $\log x$ is an abbreviation regrettably used both for the COMMON LOGARITHM $\log_{10} x$ (by engineers and physicists and indicated on pocket calculators) and for the NATURAL LOGARITHM $\log_e x$ (by mathematicians). $\ln x$ denotes the NATURAL LOGARITHM $\log_e x$ (as used by engineers and physicists and indicated on pocket calculators), and $\lg x$ denotes $\log_2 x$. In this work, the notations $\log x = \log_{10} x$ and $\ln x = \log_e x$ are used.

To convert between logarithms in different bases, the formula

$$\log_b x = \frac{\ln x}{\ln b}$$

can be used.

See also COMMON LOGARITHM, *E*, LG, LN, LOGARITHM, NAPIERIAN LOGARITHM, NATURAL LOGARITHM, BASE (NUMBER)

Base (Neighborhood System)

A base for a neighborhood system of a point x is a collection N of OPEN SETS such that x belongs to every member of N, and any OPEN SET containing x also contains a member of N as a SUBSET.

Base (Number)

A REAL NUMBER x can be represented using any INTEGER number b as a base (sometimes also called a RADIX or SCALE). The choice of a base yields to a representation of numbers known as a NUMBER SYSTEM. In base b, the DIGITS $0, 1, ..., b-1$ are used (where, by convention, for bases larger than 10, the symbols A, B, C, ...are generally used as symbols representing the DECIMAL numbers 10, 11, 12, ...).

Base	Name
2	BINARY
3	TERNARY
4	QUATERNARY
5	Quinary
6	Senary
7	Septenary
8	OCTAL
9	Nonary
10	DECIMAL
11	Undenary
12	DUODECIMAL
16	HEXADECIMAL
20	VIGESIMAL
60	SEXAGESIMAL

Let the base b representation of a number x be written

$$(a_n a_{n-1} \ldots a_0 . a_{-1} \ldots)_b, \tag{1}$$

(e.g., 123.456_{10}), then the index of the leading DIGIT needed to represent the number is

$$n \equiv \lfloor \log_b x \rfloor, \tag{2}$$

where $\lfloor x \rfloor$ is the FLOOR FUNCTION. Now, recursively compute the successive DIGITS

$$a_i = \left\lfloor \frac{r_i}{b^i} \right\rfloor, \tag{3}$$

where $r_n \equiv x$ and

$$r_{i-1} = r_i - a_i b^i \tag{4}$$

for $i = n, n-1, ..., 1, 0, ...$ This gives the base b representation of x. Note that if x is an INTEGER, then i need only run through 0, and that if x has a fractional part, then the expansion may or may not terminate. For example, the HEXADECIMAL representation of 0.1 (which terminates in DECIMAL notation) is the infinite expression $0.19999\ldots_h$.

Some number systems use a mixture of bases for counting. Examples include the Mayan calendar and the old British monetary system (in which ha'pennies, pennies, threepence, sixpence, shillings, half crowns, pounds, and guineas corresponded to units of 1/2, 1, 3, 6, 12, 30, 240, and 252, respectively).

Knuth (1998) has considered using TRANSCENDENTAL bases. This leads to some rather unfamiliar results, such as equating π to 1 in "base π," $\pi = 10_{\pi}$.

See also BINARY, DECIMAL, DUODECIMAL, HEREDI-

TARY REPRESENTATION, HEXADECIMAL, OCTAL, QUATERNARY, SEXAGESIMAL, TERNARY, VIGESIMAL

References

Abramowitz, M. and Stegun, C. A. (Eds.). *Handbook of Mathematical Functions with Formulas, Graphs, and Mathematical Tables, 9th printing.* New York: Dover, p. 28, 1972.

Bogomolny, A. "Base Converter." http://www.cut-the-knot.com/binary.html.

Knuth, D. E. "Positional Number Systems." §4.1 in *The Art of Computer Programming, Vol. 2: Seminumerical Algorithms, 3rd ed.* Reading, MA: Addison-Wesley, pp. 195–13, 1998.

Lauwerier, H. *Fractals: Endlessly Repeated Geometric Figures.* Princeton, NJ: Princeton University Press, pp. 6–1, 1991.

Weisstein, E. W. "Bases." MATHEMATICA NOTEBOOK BASES.M.

Base Curve

DIRECTRIX (RULED SURFACE)

Base Manifold

The base manifold in a BUNDLE is analogous to the domain for a set of functions. In fact, a bundle, by definition, comes with a map to the base manifold, often called π or projection.

For example, the base manifold to the TANGENT BUNDLE of a MANIFOLD M is the MANIFOLD M. A VECTOR FIELD is a function from the manifold to the TANGENT BUNDLE, with the restriction that every point gets mapped to a vector at that point. In general, a BUNDLE has SECTIONS, at least locally, which are maps from the base manifold to the BUNDLE.

See also BUNDLE, MANIFOLD, SECTION (BUNDLE), TANGENT BUNDLE, VECTOR BUNDLE

Base Space

The SPACE B of a FIBER BUNDLE given by the MAP $f : E \rightarrow B$, where E is the TOTAL SPACE of the FIBER BUNDLE.

See also FIBER BUNDLE, TOTAL SPACE

Baseball

The numbers three and four appear prominently in the game of baseball. There are three strikes for an out, and three outs per inning, $3 \cdot 3 = 9$ innings in a game, giving $3^3 = 27$ outs per game (assuming no extra innings). In addition, there are 3×3 players per team. Four balls are needed for a walk. The number of bases can either be regarded as three (excluding HOME PLATE) or four (including it).

See also BASEBALL COVER, HOME PLATE

Baseball Cover

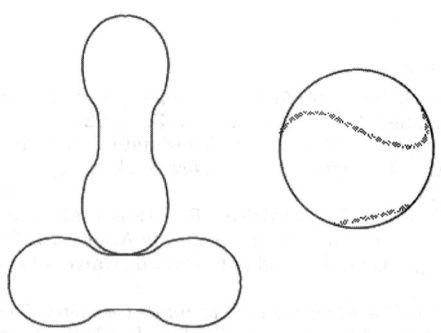

A pair of identical plane regions (mirror symmetric about two perpendicular lines through the center) which can be stitched together to form a baseball (or tennis ball). A baseball has a CIRCUMFERENCE of 9 1/8 inches. The practical consideration of separating the regions far enough to allow the pitcher a good grip requires that the "neck" distance be about 1 3/16 inches. The baseball cover was invented by Elias Drake as a boy in the 1840s. (Thompson's attribution of the current design to trial and error development by C. H. Jackson in the 1860s is apparently unsubstantiated, as discovered by George Bart.)

One way to produce a baseball cover is to draw the regions on a SPHERE, then cut them out. However, it is difficult to produce two identical regions in this manner. Thompson (1996) gives mathematical expressions giving baseball cover curves both in the plane and in 3-D. J. H. Conway has humorously proposed the following "baseball curve conjecture:" no two definitions of "the" baseball curve will give the same answer unless their equivalence was obvious from the start.

See also BASEBALL, HOME PLATE, TENNIS BALL THEOREM, YIN-YANG

References

Thompson, R. B. "Designing a Baseball Cover. 1860's: Patience, Trial, and Error. 1990's: Geometry, Calculus, and Computation." http://www.mathsoft.com/asolve/baseball/baseball.html. Rev. March 5, 1996.

Basepoint

See also LOOP

Basic Polynomial Sequence

A POLYNOMIAL SEQUENCE $p_n(x)$ is called the basic polynomial sequence for a DELTA OPERATOR Q if

1. $p_0(x) = 1$,
2. $p_n(0) = 0$ for all $n > 0$,
3. $Qp_n(x) = np_{n-1}(x)$.

If $p_n(x)$ is a basic polynomial sequence for some DELTA OPERATOR Q, then it is a BINOMIAL-TYPE SEQUENCE of polynomials. Furthermore, if $p_n(x)$ is a BINOMIAL-TYPE SEQUENCE of polynomials, then it is a basic polynomial sequence for some DELTA OPERATOR.

See also BINOMIAL-TYPE SEQUENCE, DELTA OPERATOR, POLYNOMIAL SEQUENCE, UMBRAL OPERATOR

References

Rota, G.-C.; Kahaner, D.; Odlyzko, A. "On the Foundations of Combinatorial Theory. VIII: Finite Operator Calculus." *J. Math. Anal. Appl.* **42**, 684–60, 1973.

Basin of Attraction

The set of points in the space of system variables such that initial conditions chosen in this set dynamically evolve to a particular ATTRACTOR.

See also WADA BASIN

Basis

The word basis can arise in several different contexts. Speaking in general terms, an object is "generated" by a basis in whatever manner is appropriate. For example, a VECTOR SPACE can have a BASIS which SPANS the vector space by finite LINEAR COMBINATIONS.

See also BASIS POINT, BASIS (TOPOLOGY), BASIS (VECTOR SPACE), HAMEL BASIS, HILBERT BASIS, ORTHONORMAL BASIS, VECTOR BASIS

Basis (Topology)

If X is a SET, a basis for a TOPOLOGY on X is a collection B of SUBSETS of X (called basis elements) satisfying the following properties.

1. For each $x \in X$, there is at least one basis element B containing X.
2. If x belongs to the intersection of two basis elements B_1 and B_2, then there is a basis element B_3 containing x such that $B_3 \subset B_1 \cap B_2$.

References

Munkres, J. R. *Topology: A First Course.* Englewood Cliffs, NJ: Prentice-Hall, 1975.

Basis (Vector Space)

A basis of a VECTOR SPACE V is defined as a subset $v_1, \ldots, v_n$ of vectors in V that are LINEARLY INDEPENDENT and SPAN V. Consequently, if $(v_1, v_2, \ldots, v_n)$ is a list of vectors in V, then these vectors form a basis if and only if every $v \in V$ can be uniquely written as

$$v = a_1 b_1 + a_2 b_2 + \ldots + a_n b_n,$$

where $a_1, \ldots, a_\pi$ are elements of $\mathbb{R}$ or $\mathbb{C}$. A VECTOR SPACE V will have many different bases, but there are always the same number of basis vectors in each of them. The number of basis vectors in V is called the DIMENSION of V. Every spanning list in a vector space can be reduced to a basis of the vector space.

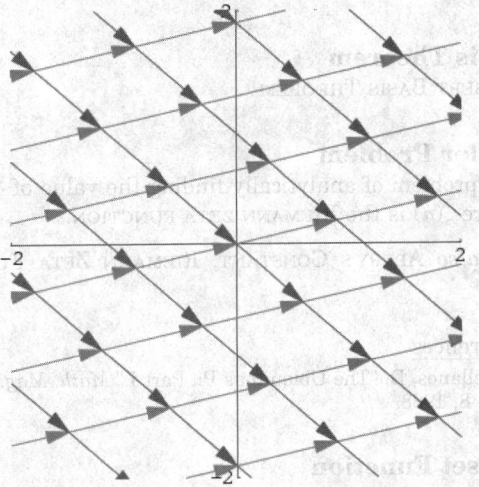

The simplest example of a basis is the standard basis in $\mathbb{R}^n$ consisting of the coordinate axes. For example, in $\mathbb{R}^2$, the standard basis consists of two VECTORS $e_1 = (1, 0)$ and $e_2 = (0, 1)$. Any VECTOR $w = (a, b)$ can be written uniquely as the LINEAR COMBINATION $w = ae_1 + be_2$. Indeed, a vector is defined by its coordinates. The VECTORS $v_1 = (3, 2)$ and $v_2 = (2, 1)$ are also a basis for $\mathbb{R}^2$ because any VECTOR $w = (a, b)$ can be uniquely written as $w = (-a + 2b)v_1 + (2a - 3b)v_2$. The above figure shows $(0.6, -0.5)n + (0.9, .02)m$, which are linear combinations of the basis $\{(0.6, -0.5), (0.9, 0.2)\}$.

Here is a *Mathematica* function which will return the coefficients a_i given a basis v_i.

```
LinearCombination[v_List?MatrixQ, w_] :=
    LinearSolve[Transpose[v], w]
```

For example, `LinearCombo[{{1, 2}, {0, 1}}, {-3, 4}]` yields $\{3, -2\}$, since $3(-1, 2) - 2(0, 1) = (-3, 4)$.

When a VECTOR SPACE is infinite dimensional, then a basis exists, as long as one assumes the AXIOM OF CHOICE. A subset of the basis which is linearly independent and whose span is DENSE is called a complete set, and is similar to a basis. When V is a HILBERT SPACE, a complete set is called a HILBERT BASIS.

See also BASIS, DIMENSION, HILBERT BASIS, LINEAR COMBINATION, ORTHONORMAL BASIS, SPAN (VECTOR SPACE), VECTOR SPACE

Basis Element

A collection B of subsets of a set X forming a topological BASIS.

See also BASIS (TOPOLOGY)

Basis Point

One basis point is defined to be 0.01 PERCENTAGE POINTS. Therefore, a change of 0.21% could also be expressed as a change by 21 "basis points."

See also PERCENTAGE POINT

Basis Theorem

HILBERT BASIS THEOREM

Basler Problem

The problem of analytically finding the value of $\zeta(2)$, where $\zeta(n)$ is the RIEMANN ZETA FUNCTION.

See also APÉRY'S CONSTANT, RIEMANN ZETA FUNCTION

References
Castellanos, D. "The Ubiquitous Pi. Part I." *Math. Mag.* **61**, 67–8, 1988.

Basset Function

MODIFIED BESSEL FUNCTION OF THE SECOND KIND

Bat

CHEVRON

Batch

A set of values of similar meaning obtained in any manner.

References
Tukey, J. W. *Explanatory Data Analysis*. Reading, MA: Addison-Wesley, p. 667, 1977.

Bateman Equation

References
Fairlie, D. B. and Leznov, A. N. The Complex Bateman Equation in a Space of Arbitrary Dimension. 16 Sep 1999. http://xxx.lanl.gov/abs/solv-int/9909013/.

Bateman Function

$$k_v(x) \equiv \frac{e^{-x}}{\Gamma(1 + \frac{1}{2}v)} \, U(-\tfrac{1}{2}v, \, 0, \, 2x)$$

for $x > 0$, where U is a CONFLUENT HYPERGEOMETRIC FUNCTION OF THE SECOND KIND.

See also CONFLUENT HYPERGEOMETRIC DIFFERENTIAL EQUATION, HYPERGEOMETRIC FUNCTION

References
Bateman, H. "The k-Function, a Particular Case of the Confluent Hypergeometric Function." *Trans. Amer. Math. Soc.* **33**, 817–31, 1931.
Koepf, W. *Hypergeometric Summation: An Algorithmic Approach to Summation and Special Function Identities.* Braunschweig, Germany: Vieweg, p. 179, 1998.
Koepf, W. and Schmersau, D. "Bounded Nonvanishing Functions are Bateman Functions." *Complex Variables* **25**, 237–59, 1994.

Batrachion

A class of CURVE defined at INTEGER values which hops from one value to another. Their name derives from the Greek word $\beta\alpha\tau\rho\alpha\chi\iota\upsilon$ *batrachion*, which means "small frog." Many batrachions are FRACTAL. Examples include the BLANCMANGE FUNCTION, HOFSTADTER-CONWAY $10,000 SEQUENCE, HOFSTADTER'S Q-SEQUENCE, and MALLOWS' SEQUENCE.

References
Pickover, C. A. "The Crying of Fractal Batrachion 1,489." Ch. 25 in *Keys to Infinity*. New York: W. H. Freeman, pp. 183–91, 1995.

Baudet's Conjecture

If $C_1, C_2, \ldots, f > C_r$ are sets of positive integers and

$$\bigcup_{i=1}^{r} C_i = \mathbb{N},$$

where $\mathbb{N}$ is the set of positive integers, then some C_i contains arbitrarily long ARITHMETIC SEQUENCES. The conjecture was proved in 1928 by B. L. van der Waerden.

See also ARITHMETIC SEQUENCE, VAN DER WAERDEN'S THEOREM

References
van der Waerden, B. L."How the Proof of Baudet's Conjecture Was Found." *Studies in Pure Mathematics (Presented to Richard Rado).* London: Academic Press, pp. 251–60, 1971.

Bauer's Identical Congruence

Let $T(m)$ denote the set of the $\phi(m)$ numbers less than and RELATIVELY PRIME to m, where $\phi(n)$ is the TOTIENT FUNCTION. Define

$$f_m(x) = \prod_{t \in T(m)} (x - t). \tag{1}$$

Then a theorem of Lagrange states that

$$f_p(x) \equiv x^{\phi(p)} - 1 \pmod{p} \tag{2}$$

for p an ODD PRIME (Hardy and Wright 1979, p. 98).

This can be generalized as follows. Let p be an ODD PRIME DIVISOR of m and p^a the highest POWER which divides m, then

$$f_m(x) \equiv (x^{p-1} - 1)^{\phi(m)/(p-1)} \pmod{p^a} \tag{3}$$

and, in particular,

$$f_{p^a}(x) \equiv (x^{p-1} - 1)^{p^{a-1}} \pmod{p^a}. \tag{4}$$

Now, if $m > 2$ is EVEN and 2^a is the highest POWER of 2 that divides m, then

$$f_m(x) \equiv (x^2 - 1)^{\phi(m)/2} \pmod{2^a} \tag{5}$$

and, in particular,

$$f_{2^a}(x) \equiv (x^2 - 1)^{2^{a-2}} \pmod{2^a}. \tag{6}$$

See also CONGRUENCE, LEUDESDORF THEOREM

References

Bauer. *Nouvelles annales* **2**, 256–64, 1902.

Hardy, G. H. and Wright, E. M. *J. London Math. Soc.* **9**, 38–1 and 240, 1934.

Hardy, G. H. and Wright, E. M. "Bauer's Identical Congruence." §8.5 in *An Introduction to the Theory of Numbers, 5th ed.* Oxford, England: Clarendon Press, pp. 98–00, 1979.

Bauer's Theorem

Let $m \geq 3$ be an integer and let

$$f(x) = \sum_{k=0}^{n} a_k x^{n-k}$$

be an INTEGER POLYNOMIAL that has at least one real zero. Then $f(x)$ has infinitely many PRIME DIVISORS that are not congruent to 1 $(\mathrm{mod}\ m)$ (Nagell 1951, p. 168).

See also BAUER'S IDENTICAL CONGRUENCE, PRIME DIVISOR

References

Nagell, T. "A Theorem of Bauer on the Prime Divisors of Certain Polynomials." §49 in *Introduction to Number Theory.* New York: Wiley, pp. 168–69, 1951.

Bauer-Muir Transformation

A transformation formula for CONTINUED FRACTIONS (Lorentzen and Waadeland 1992) which can, for example, be used to prove identities such as

$$1 + \cfrac{1}{1 + \cfrac{2+q}{1 + \cfrac{2+q^2}{1 + \cfrac{2+q^3}{1+\cdots}}}} = \cfrac{1}{2 + \cfrac{q}{2 + q + \cfrac{q^2}{2 + q^2 + \cfrac{q^3}{2 + q^3 + \cdots}}}}$$

(Berndt *et al.*).

See also CONTINUED FRACTION

References

Berndt, B. C.; Huang, S.-S.; Sohn, J.; and Son, S. H. "Some Theorems on the Rogers-Ramanujan Continued Fraction in Ramanujan's Lost Notebook." To appears in *Trans. Amer. Math. Soc.*

Lorentzen, L. and Waadeland, H. *Continued Fractions with Applications.* Amsterdam, Netherlands: North-Holland, p. 76, 1992.

Bauspiel

A construction for the RHOMBIC DODECAHEDRON.

References

Coxeter, H. S. M. *Regular Polytopes, 3rd ed.* New York: Dover, pp. 26 and 50, 1973.

Baxter-Hickerson Function

In April 1999, Ed Pegg conjectured on `sci.math` that there were only finitely many ZEROFREE cubes, to which D. Hickerson responded with a counterexample. A few days later, Lew Baxter posted the slightly simpler example

$$f(n) = \tfrac{1}{3}(2 \cdot 10^{5n} - 10^{4n} + 2 \cdot 10^{3n} + 10^{2n} + 10^{n} + 1),$$

which produces numbers whose cubes lack zeros. The first few terms for $n = 0, 1, \ldots$ are 2, 64037, 6634003367, 666334000333667, ... (Sloane's A052-427). Primes occur for $n = 0, 1, 7, 133, \ldots$ (Sloane's A051832) with no others ≤ 470 (Weisstein, Dec. 15, 1999), corresponding to 2, 64037, ... (Sloane's A051833).

See also NUMBER PATTERN, ZEROFREE

References

Pegg, E. Jr. "Fun with Numbers." http://www.mathpuzzle.com/numbers.html.

Sloane, N. J. A. Sequences A051832, A051833, and A052427 in "An On-Line Version of the Encyclopedia of Integer Sequences." http://www.research.att.com/~njas/sequences/eisonline.html.

Bayes' Formula

BAYES' THEOREM

Bayes' Theorem

Let A and B_j be SETS. CONDITIONAL PROBABILITY requires that

$$P(A \cap B_j) = P(A)P(B_j|A), \tag{1}$$

where $\cap$ denotes INTERSECTION ("and"), and also that

$$P(A \cap B_j) = P(B_j \cap A) = P(B_j)P(A|B_j). \tag{2}$$

Therefore,

$$P(B_j|A) = \frac{P(B_j)P(A|B_j)}{P(A)}. \tag{3}$$

Now, let

$$S \equiv \bigcup_{i=1}^{N} A_i, \tag{4}$$

so A_i is an event in S and $A_i \cap A_j = \emptyset$ for $i \neq j$, then

$$A = A \cap S = A \cap \left(\bigcup_{i=1}^{N} A_i \right) = \bigcup_{i=1}^{N} (A \cap A_i) \qquad (5)$$

$$P(A) = P\left(\bigcup_{i=1}^{N} (A \cap A_i) \right) = \sum_{i=1}^{N} P(A \cap A_i). \qquad (6)$$

But this can be written

$$P(A) = \sum_{i=1}^{N} P(A_i) P(A|A_i), \qquad (7)$$

so

$$P(A_i|A) = \frac{P(A_i) P(A|A_i)}{\sum\limits_{j=1}^{N} P(A_j) P(A|A_j)} \qquad (8)$$

(Papoulis 1984, pp. 38–9).

See also CONDITIONAL PROBABILITY, INCLUSION-EX-CLUSION PRINCIPLE, INDEPENDENT STATISTICS, TOTAL PROBABILITY THEOREM

References

Papoulis, A. "Bayes' Theorem in Statistics" and "Bayes' Theorem in Statistics (Reexamined)." §3– and 4– in *Probability, Random Variables, and Stochastic Processes, 2nd ed.* New York: McGraw-Hill, pp. 38–9, 78–1, and 112–14, 1984.
Press, W. H.; Flannery, B. P.; Teukolsky, S. A.; and Vetterling, W. T. *Numerical Recipes in FORTRAN: The Art of Scientific Computing, 2nd ed.* Cambridge, England: Cambridge University Press, p. 810, 1992.

Bayesian Analysis

A statistical procedure which endeavors to estimate parameters of an underlying distribution based on the observed distribution. Begin with a "PRIOR DISTRIBUTION" which may be based on anything, including an assessment of the relative likelihoods of parameters or the results of non-Bayesian observations. In practice, it is common to assume a UNIFORM DISTRIBUTION over the appropriate range of values for the PRIOR DISTRIBUTION.

Given the PRIOR DISTRIBUTION, collect data to obtain the observed distribution. Then calculate the LIKE-LIHOOD of the observed distribution as a function of parameter values, multiply this likelihood function by the PRIOR DISTRIBUTION, and normalize to obtain a unit probability over all possible values. This is called the POSTERIOR DISTRIBUTION. The MODE of the distribution is then the parameter estimate, and "probability intervals" (the Bayesian analog of CONFIDENCE INTERVALS) can be calculated using the standard procedure. Bayesian analysis is somewhat controversial because the validity of the result depends on how valid the PRIOR DISTRIBUTION is, and this cannot be assessed statistically.

See also MAXIMUM LIKELIHOOD, PRIOR DISTRIBUTION, UNIFORM DISTRIBUTION

References

Gelman, A.; Carlin, J.; Stern, H.; and Rubin, D. *Bayesian Data Analysis*. Boca Raton, FL: Chapman & Hall, 1995.
Hoel, P. G.; Port, S. C.; and Stone, C. J. *Introduction to Statistical Theory*. New York: Houghton Mifflin, pp. 36–2, 1971.
Iversen, G. R. *Bayesian Statistical Inference*. Thousand Oaks, CA: Sage Pub., 1984.
Press, W. H.; Flannery, B. P.; Teukolsky, S. A.; and Vetterling, W. T. *Numerical Recipes in FORTRAN: The Art of Scientific Computing, 2nd ed.* Cambridge, England: Cambridge University Press, pp. 799–06, 1992.
Sivia, D. S. *Data Analysis: A Bayesian Tutorial*. New York: Oxford University Press, 1996.

Bays' Shuffle

A shuffling algorithm used in a class of RANDOM NUMBER generators.

References

Knuth, D. E. §3.2 and 3.3 in *The Art of Computer Programming, Vol. 2: Seminumerical Algorithms, 2nd ed.* Reading, MA: Addison-Wesley, 1981.
Press, W. H.; Flannery, B. P.; Teukolsky, S. A.; and Vetterling, W. T. *Numerical Recipes in FORTRAN: The Art of Scientific Computing, 2nd ed.* Cambridge, England: Cambridge University Press, pp. 270–71, 1992.

Beal's Conjecture

A generalization of FERMAT'S LAST THEOREM which states that if $a^x + b^y = c^z$, where a, b, c, x, y, and z are POSITIVE INTEGERS and x, y, $z > 2$, then a, b, and c have a common factor. The conjecture was announced in Mauldin (1997), and a cash prize of $75,000 has been offered for its proof or a counterexample.

See also ABC CONJECTURE, FERMAT'S LAST THEOREM

References

Brun, V. "Über hypothesesenbildungen." *Arc. Math. Naturvidenskab* **34**, 1–4, 1914.
Darmon, H. and Granville, A. "On the Equations $z^m = F(x, y)$ and $Ax^p + By^q = cZ^r$." *Bull. London Math. Soc.* **27**, 513–43, 1995.
Mauldin, R. D. "A Generalization of Fermat's Last Theorem: The Beal Conjecture and Prize Problem." *Not. Amer. Math. Soc.* **44**, 1436–437, 1997.
Mauldin, R. D. "The Beal Conjecture and Prize." http://www.math.unt.edu/~mauldin/beal.html.

Beam Detector

N.B. A detailed online essay by S. Finch was the starting point for this entry.

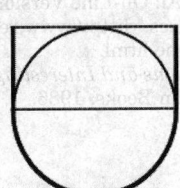 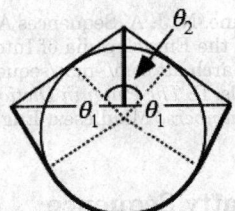

A "beam detector" for a given curve C is defined as a curve (or set of curves) through which every LINE tangent to or intersecting C passes. The shortest 1-arc beam detector, illustrated in the upper left figure, has length $L_1 = \pi + 2$. The shortest known 2-arc beam detector, illustrated in the right figure, has angles

$$\theta_1 \approx 1.286 \text{ rad} \tag{1}$$

$$\theta_2 \approx 1.191 \text{ rad}, \tag{2}$$

given by solving the simultaneous equations

$$2\cos\theta_1 - \sin(\tfrac{1}{2}\theta_2) = 0 \tag{3}$$

$$\tan(\tfrac{1}{2}\theta_1)\cos(\tfrac{1}{2}\theta_2) + \sin(\tfrac{1}{2}\theta_2)[\sec^2(\tfrac{1}{2}\theta_2)+1] = 2. \tag{4}$$

The corresponding length is

$$L_2 = 2\pi - 2\theta_1 - \theta_2 + 2\tan\left(\tfrac{1}{2}\theta_1\right) + \sec\left(\tfrac{1}{2}\theta_2\right) - \cos\left(\tfrac{1}{2}\theta_2\right)$$
$$+ \tan\left(\tfrac{1}{2}\theta_1\right)\sin\left(\tfrac{1}{2}\theta_2\right)$$
$$= 4.8189264563\ldots. \tag{5}$$

A more complicated expression gives the shortest known 3-arc length $L_3 = 4.799891547\ldots$. Finch defines

$$L = \inf_{n \geq 1} L_n \tag{6}$$

as the beam detection constant, or the TRENCH DIGGERS' CONSTANT. It is known that $L \geq \pi$.

References

Croft, H. T.; Falconer, K. J.; and Guy, R. K. §A30 in *Unsolved Problems in Geometry.* New York: Springer-Verlag, 1991.
Faber, V.; Mycielski, J.; and Pedersen, P. "On the Shortest Curve which Meets All Lines which Meet a Circle." *Ann. Polon. Math.* **44**, 249–66, 1984.
Faber, V. and Mycielski, J. "The Shortest Curve that Meets All Lines that Meet a Convex Body." *Amer. Math. Monthly* **93**, 796–01, 1986.
Finch, S. "Favorite Mathematical Constants." http://www.mathsoft.com/asolve/constant/beam/beam.html.
Makai, E. "On a Dual of Tarski's Plank Problem." In *Diskrete Geometrie.* 2 Kolloq., Inst. Math. Univ. Salzburg, 127–32, 1980.
Stewart, I. "The Great Drain Robbery." *Sci. Amer.* **273**, 206–07, Sep. 1995.
Stewart, I. *Sci. Amer.* **273**, 106, Dec. 1995.
Stewart, I. *Sci. Amer.* **274**, 125, Feb. 1996.

Bean Curve

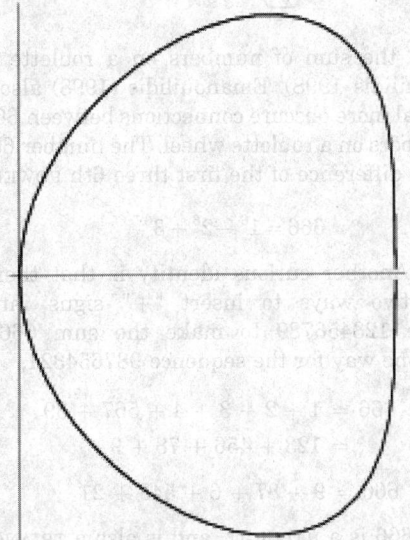

The PLANE CURVE given by the Cartesian equation

$$x^4 + x^2 y^2 + y^4 = x(x^2 + y^2).$$

References

Cundy, H. and Rollett, A. *Mathematical Models, 3rd ed.* Stradbroke, England: Tarquin Pub., 1989.

Beast Number

The occult "number of the beast" associated in the Bible with the Antichrist. It has figured in many numerological studies. It is mentioned in Revelation 13:18: "Here is wisdom. Let him that hath understanding count the number of the beast: for it is the number of a man; and his number is 666." The origin of this number is not entirely clear, although it may be as simple as the number containing the concatenation of one symbol of each type (exclude $M = 1000$) in ROMAN NUMERALS: $DCLXVI = 666$ (Wells 1986).

The first few numbers containing the beast number in their digits are 666, 1666, 2666, 3666, 4666, 5666, 6660, ... (Sloane's A051003).

The beast number has several interesting properties which numerologists may find particularly interesting (Keith 1982–3). In particular, the beast number is equal to the sum of the squares of the first 7 PRIMES

$$2^2 + 3^2 + 5^2 + 7^2 + 11^2 + 13^2 + 17^2 = 666, \tag{1}$$

satisfies the identity

$$\phi(666) = 6 \cdot 6 \cdot 6, \tag{2}$$

where ϕ is the TOTIENT FUNCTION, as well as the sum

$$\sum_{i=1}^{6 \cdot 6} i = 666 \qquad (3)$$

which is the sum of numbers on a roulette wheel (Emanouilidis 1998). Emanouilidis (1998) also gives additional more obscure connections between 666 and the numbers on a roulette wheel. The number 666 is a sum and difference of the first three 6th POWERS,

$$666 = 1^6 - 2^6 + 3^6 \qquad (4)$$

(Keith). Another curious identity is that there are exactly two ways to insert "+" signs into the sequence 123456789 to make the sum 666, and exactly one way for the sequence 987654321,

$$666 = 1 + 2 + 3 + 4 + 567 + 89$$
$$= 123 + 456 + 78 + 9 \qquad (5)$$
$$666 = 9 + 87 + 6 + 543 + 21 \qquad (6)$$

(Keith). 666 is a REPDIGIT, and is also a TRIANGULAR NUMBER

$$T_{6 \cdot 6} = T_{36} = 666. \qquad (7)$$

In fact, it is the largest REPDIGIT TRIANGULAR NUMBER (Bellew and Weger 1975–6). 666 is also a SMITH NUMBER. The first 144 DIGITS of $\pi - 3$, where π is PI, add to 666. In addition $144 = (6+6) \times (6+6)$ (Blatner 1997). Finally,

$$\sum_{i=0}^{5} 2048^i \equiv 691 \pmod{666}. \qquad (8)$$

A number OF THE FORM 2^i which contains the digits of the beast number "666" is called an APOCALYPTIC NUMBER, and a number having 666 digits is called an APOCALYPSE NUMBER.

See also APOCALYPSE NUMBER, APOCALYPTIC NUMBER, BIMONSTER, MONSTER GROUP, ROMAN NUMERAL

References
Bellew, D. W. and Weger, R. C. "Repdigit Triangular Numbers." *J. Recr. Math.* **8**, 96–7, 1975–6.
Blatner, D. *The Joy of Pi.* New York: Walker, back jacket, 1997.
Castellanos, D. "The Ubiquitous π." *Math. Mag.* **61**, 153–54, 1988.
Eco, U. *Foucault's Pendulum.* San Diego: Harcourt Brace Jovanovich, p. 31, 1989.
Emanouilidis, E. "Roulette and the Beastly Number." *J. Recr. Math.* **29**, 246–47, 1998.
Gardner, M. "Mathematical Games: A Fanciful Dialogue About the Wonders of Numerology." *Sci. Amer.* **202**, 150–56, Feb. 1960.
Hardy, G. H. *A Mathematician's Apology, reprinted with a foreword by C. P. Snow.* New York: Cambridge University Press, p. 96, 1993.
Keith, M. "The Number of the Beast." http://member.aol.com/s6sj7gt/mike666.htm.
Keith, M. "The Number 666." *J. Recr. Math.* **15**, 85–7, 1982–983.

Sloane, N. J. A. Sequences A051003 in "An On-Line Version of the Encyclopedia of Integer Sequences." http://www.research.att.com/~njas/sequences/eisonline.html.
Wells, D. *The Penguin Dictionary of Curious and Interesting Numbers.* Middlesex, England: Penguin Books, 1986.

Beatty Sequence

The Beatty sequence is a SPECTRUM SEQUENCE with an IRRATIONAL base. In other words, the Beatty sequence corresponding to an IRRATIONAL NUMBER θ is given by $\lfloor \theta \rfloor$, $\lfloor 2\theta \rfloor$, $\lfloor 3\theta \rfloor$, ..., where $\lfloor x \rfloor$ is the FLOOR FUNCTION. If α and β are POSITIVE IRRATIONAL NUMBERS such that

$$\frac{1}{\alpha} + \frac{1}{\beta} = 1,$$

then the Beatty sequences $\lfloor \alpha \rfloor$, $\lfloor 2\alpha \rfloor$, ... and $\lfloor \beta \rfloor$, $\lfloor 2\beta \rfloor$, ... together contain all the POSITIVE INTEGERS without repetition.

The sequences for particular values of α and β are given in the following table (Sprague 1963; Wells 1986, pp. 35 and 40), where ϕ is the GOLDEN RATIO.

parameter	Sloane	sequence
$\alpha = \sqrt{2}$	A001951	1, 2, 4, 5, 7, 8, 9, 11, 12, ...
$\beta = 2 + \sqrt{2}$	A001952	3, 6, 10, 13, 17, 20, 23, 27, 30, ...
$\alpha = \sqrt{3}$	A022838	1, 3, 5, 6, 8, 10, 12, 13, 15, 17, ...
$\beta = \frac{1}{2}(3 + \sqrt{3})$	A054406	2, 4, 7, 9, 11, 14, 16, 18, 21, 23, 26, ...
$\alpha = e$	A022843	2, 5, 8, 10, 13, 16, 19, 21, 24, 27, 29, ...
$\beta = e/(e-1)$	A054385	1, 3, 4, 6, 7, 9, 11, 12, 14, 15, 17, 18, ...
$\alpha = \pi$	A022844	3, 6, 9, 12, 15, 18, 21, 25, 28, 31, 34, ...
$\beta = \pi/(\pi-1)$	A054386	1, 2, 4, 5, 7, 8, 10, 11, 13, 14, 16, 17, 19,
$\alpha = \phi$	A000201	1, 3, 4, 6, 8, 9, 11, 12, 14, 16, 17, 19, 21, ...
$\beta = \phi^2$	A001950	2, 5, 7, 10, 13, 15, 18, 20, 23, 26, 28, 31, 34, ...

See also FRACTIONAL PART, WYTHOFF ARRAY, WYTHOFF'S GAME

References
Gardner, M. *Penrose Tiles and Trapdoor Ciphers...and the Return of Dr. Matrix, reissue ed.* New York: W. H. Freeman, p. 21, 1989.
Graham, R. L.; Lin, S.; and Lin, C.-S. "Spectra of Numbers." *Math. Mag.* **51**, 174–76, 1978.
Guy, R. K. *Unsolved Problems in Number Theory, 2nd ed.* New York: Springer-Verlag, p. 227, 1994.
Sloane, N. J. A. *A Handbook of Integer Sequences.* Boston, MA: Academic Press, pp. 29–0, 1973.

Sloane, N. J. A. and Plouffe, S. *The Encyclopedia of Integer Sequences.* San Diego, CA: Academic Press, p. 18, 1995.

Sprague, R. *Recreations in Mathematics: Some Novel Puzzles.* London: Blackie and Sons, 1963.

Sloane, N. J. A. Sequences A000201/M2322, A001950/M1332, A001951/M0955, A001952/M2534, A022838, A022843, A022844, A054406, A054385, and A054386 in "An On-Line Version of the Encyclopedia of Integer Sequences." http://www.research.att.com/~njas/sequences/eisonline.html.

Wells, D. *The Penguin Dictionary of Curious and Interesting Numbers.* Middlesex, England: Penguin Books, p. 35, 1986.

Beauzamy and Dégot's Identity

For P, Q, R, and S POLYNOMIALS in n variables

$$[P \cdot Q, R \cdot S] = \sum_{i_1, \ldots, i_n \geq 0} \frac{A}{i_1! \cdots i_n!},$$

where

$$A \equiv [R^{(i_1, \ldots, i_n)}(D_1, \ldots, D_n)Q(x_1, \ldots, x_n)$$
$$\times P^{(i_1, \ldots, i_n)}(D_1, \ldots, D_n)S(x_1, \ldots, x_n)],$$

$D_i = \partial/\partial x_i$ is the DIFFERENTIAL OPERATOR, $[X, Y]$ is the BOMBIERI INNER PRODUCT, and

$$P^{(i_1, \ldots, i_n)} = D_1^{i_1} \cdots D_n^{i_n} P.$$

See also REZNIK'S IDENTITY

Bed-of-Nails Function

SHAH FUNCTION

Bee

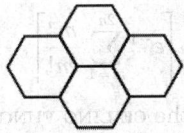

A 4-POLYHEX.

References

Gardner, M. *Mathematical Magic Show: More Puzzles, Games, Diversions, Illusions and Other Mathematical Sleight-of-Mind from Scientific American.* New York: Vintage, p. 147, 1978.

Behrens-Fisher Test

FISHER-BEHRENS PROBLEM

Behrmann Cylindrical Equal-Area Projection

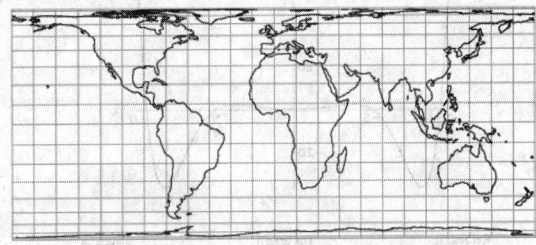

A CYLINDRICAL EQUAL-AREA PROJECTION which uses a standard parallel of $\phi_s = 30°$.

See also BALTHASART PROJECTION, CYLINDRICAL EQUAL-AREA PROJECTION, EQUAL-AREA PROJECTION, GALL ORTHOGRAPHIC PROJECTION, LAMBERT AZIMUTHAL EQUAL-AREA PROJECTION, PETERS PROJECTION, TRISTAN EDWARDS PROJECTION

References

Dana, P. H. "Map Projections." http://www.colorado.edu/geography/gcraft/notes/mapproj/mapproj_f.html.

Bei

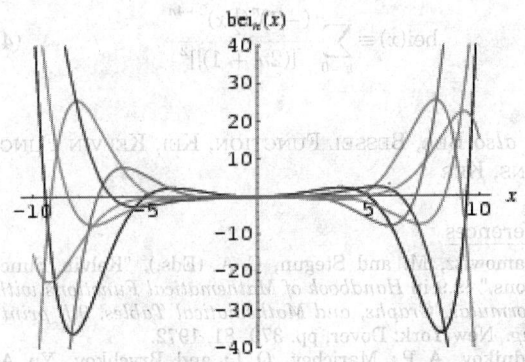

The IMAGINARY PART of

$$J_\nu(xe^{3\pi i/4}) = \mathrm{ber}_\nu(x) + i\,\mathrm{bei}_\nu(x). \tag{1}$$

The function $\mathrm{bei}_\nu(x)$ has the series expansion

$$\mathrm{bei}_\nu(x) = (\tfrac{1}{2}x)^\nu \sum_{k=0}^{\infty} \frac{\sin[(\tfrac{3}{4}\nu + \tfrac{1}{2}k)\pi]}{k!\,\Gamma(\nu + k + 1)} (\tfrac{1}{4}x^2)^k, \tag{2}$$

where $\Gamma(x)$ is the GAMMA FUNCTION (Abramowitz and

Stegun 1972, p. 379).

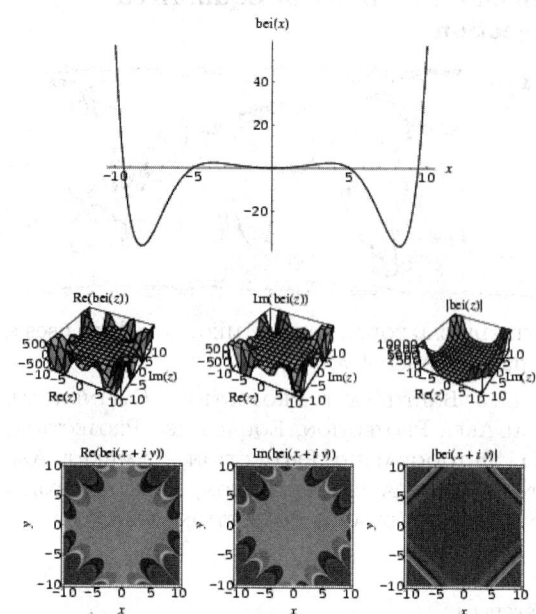

The special case $v = 0$ gives

$$J_0\left(i\sqrt{i}x\right) \equiv \mathrm{ber}(x) + i\,\mathrm{bei}(x), \tag{3}$$

where $J_0(x)$ is the zeroth order BESSEL FUNCTION OF THE FIRST KIND. The function $\mathrm{bei}_0(x) \equiv \mathrm{bei}(x)$ has the series expansion

$$\mathrm{bei}(x) \equiv \sum_{n=0}^{\infty} \frac{(-1)^n (\frac{1}{2}x)^{2+4n}}{[(2n+1)!]^2}. \tag{4}$$

See also BER, BESSEL FUNCTION, KEI, KELVIN FUNCTIONS, KER

References

Abramowitz, M. and Stegun, C. A. (Eds.). "Kelvin Functions." §9.9 in *Handbook of Mathematical Functions with Formulas, Graphs, and Mathematical Tables, 9th printing.* New York: Dover, pp. 379–81, 1972.

Prudnikov, A. P.; Marichev, O. I.; and Brychkov, Yu. A. "The Kelvin Functions $\mathrm{ber}_v(x)$, $\mathrm{bei}\,v(x)$, $\mathrm{ker}_v(x)$ and $\mathrm{kei}_v(x)$." §1.7 in *Integrals and Series, Vol. 3: More Special Functions.* Newark, NJ: Gordon and Breach, pp. 29–0, 1990.

Spanier, J. and Oldham, K. B. "The Kelvin Functions." Ch. 55 in *An Atlas of Functions.* Washington, DC: Hemisphere, pp. 543–54, 1987.

Bell Curve

GAUSSIAN DISTRIBUTION, NORMAL DISTRIBUTION

Bell Number

The number of ways a SET of n elements can be PARTITIONED into nonempty SUBSETS is called a BELL NUMBER and is denoted B_n. For example, there are five ways the numbers $\{1, 2, 3\}$ can be partitioned:

$\{\{1\},\{2\},\{3\}\}$, $\{\{1, 2\},\{3\}\}$, $\{\{1, 3\},\{2\}\}$, $\{\{1\}, \{2, 3\}\}$, and $\{\{1, 2, 3\}\}$, so $B_3 = 5$. $B_0 = 1$ and the first few Bell numbers for $n = 1, 2, \ldots$ are 1, 2, 5, 15, 52, 203, 877, 4140, 21147, 115975, ... (Sloane's A000110).

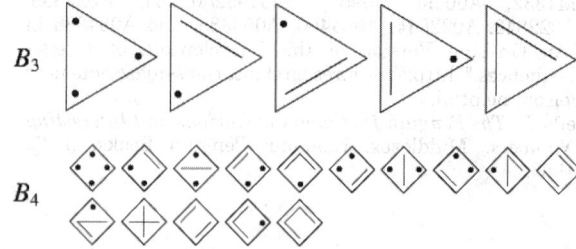

Bell numbers are closely related to CATALAN NUMBERS. The diagram above shows the constructions giving $B_3 = 5$ and $B_4 = 15$, with line segments representing elements in the same SUBSET and dots representing subsets containing a single element (Dickau). The INTEGERS B_n can be defined by the sum

$$B_n = \sum_{k=1}^{n} S(n, k), \tag{1}$$

where $S(n, k)$ is a STIRLING NUMBER OF THE SECOND KIND, i.e., as the STIRLING TRANSFORM of the sequence $1, 1, 1, \ldots$

The Bell number are given by the EXPONENTIAL GENERATING FUNCTION

$$e^{e^n - 1} = \sum_{n=0}^{\infty} \frac{B_n}{n!} x^n. \tag{2}$$

The Bell numbers can also be generated using the BELL TRIANGLE, using the RECURRENCE RELATION

$$B_{n+1} = \sum_{k=0}^{n} B_k \binom{n}{k}, \tag{3}$$

where $\binom{a}{b}$ is a BINOMIAL COEFFICIENT, or using the formula of Comtet (1974)

$$B_n = \left\lceil e^{-1} \sum_{m=1}^{2n} \frac{m^n}{m!} \right\rceil, \tag{4}$$

where $\lceil x \rceil$ denotes the CEILING FUNCTION.

The Bell number B_n is also equal to $\phi_n(1)$, where $\phi_n(x)$ is an EXPONENTIAL POLYNOMIAL. DOBINSKI'S FORMULA gives the nth Bell number

$$B_n = \frac{1}{e} \sum_{k=0}^{\infty} \frac{k^n}{k!}. \tag{5}$$

Lovász (1993) showed that this formula gives the asymptotic limit

$$B_n \sim n^{-1/2} [\lambda(n)]^{n+1/2} e^{\lambda(n) - n - 1}, \tag{6}$$

where $\lambda(n)$ is defined implicitly by the equation

$$\lambda(n)\log[\lambda(n)] = n. \qquad (7)$$

A variation of DOBINSKI'S FORMULA gives

$$B_n = \sum_{k=1}^{n} \frac{k^n}{k!} \sum_{j=0}^{n-k} \frac{(-1)^j}{j!} \qquad (8)$$

(Pitman 1997). de Bruijn (1958) gave the asymptotic formula

$$\frac{\ln B_n}{n} = \ln n - \ln\ln n - 1 + \frac{\ln\ln n}{\ln n} + \frac{1}{\ln n}$$

$$+ \frac{1}{2}\left(\frac{\ln\ln n}{\ln n}\right)^2 + \mathcal{O}\left[\frac{\ln\ln n}{(\ln n)^2}\right] \qquad (9)$$

TOUCHARD'S CONGRUENCE states

$$B_{p+k} \equiv B_k + B_{k+1} \pmod{p}, \qquad (10)$$

when p is PRIME. The only PRIME Bell numbers for $n \le 1000$ are B_2, B_3, B_7, B_{13}, B_{42}, and B_{55}. The Bell numbers also have the curious property that

$$\begin{vmatrix} B_0 & B_1 & B_2 & \cdots & B_n \\ B_1 & B_2 & B_3 & \cdots & B_{n+1} \\ \vdots & \vdots & \vdots & \ddots & \vdots \\ B_n & B_{n+1} & B_{n+2} & \cdots & B_{2n} \end{vmatrix} = \prod_{i=1}^{n} i! \qquad (11)$$

(Lenard 1986), where the product is simply a SUPER-FACTORIAL, the first few of which for $n = 0, 1, 2, \ldots$ are 1, 1, 2, 12, 288, 34560, 24883200, ... (Sloane's A000178).

See also BELL TRIANGLE, DOBINSKI'S FORMULA, EXPONENTIAL POLYNOMIAL, STIRLING NUMBER OF THE SECOND KIND, TOUCHARD'S CONGRUENCE

References

Bell, E. T. "Exponential Numbers." *Amer. Math. Monthly* **41**, 411–19, 1934.

Comtet, L. *Advanced Combinatorics: The Art of Finite and Infinite Expansions, rev. enl. ed.* Dordrecht, Netherlands: Reidel, 1974.

Conway, J. H. and Guy, R. K. In *The Book of Numbers.* New York: Springer-Verlag, pp. 91–4, 1996.

de Bruijn, N. G. *Asymptotic Methods in Analysis.* New York: Dover, pp. 102–09, 1958.

Dickau, R. M. "Bell Number Diagrams." http://forum.s-warthmore.edu/advanced/robertd/bell.html.

Dickau, R. "Visualizing Combinatorial Enumeration." *Mathematica in Educ. Res.* **8**, 11–8, 1999.

Gardner, M. "The Tinkly Temple Bells." Ch. 2 in *Fractal Music, Hypercards, and More Mathematical Recreations from Scientific American Magazine.* New York: W. H. Freeman, pp. 24–8, 1992.

Gould, H. W. *Bell & Catalan Numbers: Research Bibliography of Two Special Number Sequences, 6th ed.* Morgantown, WV: Math Monongliae, 1985.

Lenard, A. In *Fractal Music, Hypercards, and More Mathematical Recreations from Scientific American Magazine.* (M. Gardner). New York: W. H. Freeman, pp. 35–6, 1992.

Levine, J. and Dalton, R. E. "Minimum Periods, Modulo p, of First Order Bell Exponential Integrals." *Math. Comput.* **16**, 416–23, 1962.

Lovász, L. *Combinatorial Problems and Exercises, 2nd ed.* Amsterdam, Netherlands: North-Holland, 1993.

Pitman, J. "Some Probabilistic Aspects of Set Partitions." *Amer. Math. Monthly* **104**, 201–09, 1997.

Rota, G.-C. "The Number of Partitions of a Set." *Amer. Math. Monthly* **71**, 498–04, 1964.

Sloane, N. J. A. Sequences A000110/M1484 and A000178/M2049 in "An On-Line Version of the Encyclopedia of Integer Sequences." http://www.research.att.com/~njas/sequences/eisonline.html.

Bell Polynomial

The Bell polynomial are defined by

$$B_{n,k}(x_1, x_2, \ldots) = \sum_{\substack{j_1+j_2+\cdots=k \\ j_1+2j_2+\cdots=n}} \frac{n!}{j_1! j_2! \cdots} \left(\frac{x_1}{1!}\right)^{j_1} \left(\frac{x_2}{2!}\right)^{j_2} \cdots.$$

They have GENERATING FUNCTION

$$\sum_{k=0}^{\infty} \frac{b_k(x; x_1, x_2, \ldots)}{k!} t^k = e^x\left(\sum_{k=1}^{\infty} \frac{x_k}{k!} t^k\right).$$

See also EXPONENTIAL POLYNOMIAL, IDEMPOTENT NUMBER, LAH NUMBER

References

Comtet, L. *Advanced Combinatorics: The Art of Finite and Infinite Expansions, rev. enl. ed.* Dordrecht, Netherlands: Reidel, p. 133, 1974.

Roman, S. "The Bell Polynomials." §4.1.8 in *The Umbral Calculus.* New York: Academic Press, pp. 82–6, 1984.

Bell Triangle

$$
\begin{array}{ccccccc}
1 & 2 & 5 & 15 & 52 & 203 & 877 \cdots \\
& 1 & 3 & 10 & 37 & 151 & 674 \\
& & 2 & 7 & 27 & 114 & 523 \\
& & & 5 & 20 & 87 & 409 \\
& & & & 15 & 67 & 322 \\
& & & & & 52 & 255 \\
& & & & & & 203
\end{array}
$$

A triangle of numbers which allow the BELL NUMBERS to be computed using the RECURRENCE RELATION

$$B_{n+1} = \sum_{k=0}^{n} B_k \binom{n}{k}.$$

See also BELL NUMBER, CLARK'S TRIANGLE, LEIBNIZ HARMONIC TRIANGLE, LOSSNITSCH'S TRIANGLE, NUMBER TRIANGLE, PASCAL'S TRIANGLE, SEIDEL-ENTRINGER-ARNOLD TRIANGLE

Bellows Conjecture

The conjecture proposed by Dennis Sullivan that all FLEXIBLE POLYHEDRA keep a constant VOLUME as they are flexed (Cromwell 1997). This conjecture was proven by Connelly *et al.* (1997).

See also FLEXIBLE POLYHEDRON

References

Connelly, R.; Sabitov, I.; and Walz, A. "The Bellows Conjecture." *Contrib. Algebra Geom.* **38**, 1–0, 1997.
Cromwell, P. R. *Polyhedra.* New York: Cambridge University Press, pp. 245 and 247, 1997.
Mackenzie, D. "Polyhedra Can Bend But Not Breathe." *Science* **279**, 1637, 1998.

Beltrami Differential Equation

For a MEASURABLE FUNCTION μ, the Beltrami differential equation is given by

$$f_{\bar{z}} = \mu f_z,$$

where f_z is a PARTIAL DERIVATIVE and $\bar{z}$ denotes the COMPLEX CONJUGATE of z.

See also QUASICONFORMAL MAP

References

Iyanaga, S. and Kawada, Y. (Eds.). *Encyclopedic Dictionary of Mathematics.* Cambridge, MA: MIT Press, p. 1087, 1980.
Zwillinger, D. *Handbook of Differential Equations, 3rd ed.* Boston, MA: Academic Press, p. 137, 1997.

Beltrami Field

A VECTOR FIELD **u** satisfying the vector identity

$$\mathbf{u} \times (\nabla \times \mathbf{u}) = \mathbf{0}$$

where $\mathbf{A} \times \mathbf{B}$ is the CROSS PRODUCT and $\nabla \times \mathbf{A}$ is the CURL is said to be a Beltrami field.

See also DIVERGENCELESS FIELD, IRROTATIONAL FIELD, SOLENOIDAL FIELD

Beltrami Identity

An identity in CALCULUS OF VARIATIONS discovered in 1868 by Beltrami. The EULER-LAGRANGE DIFFERENTIAL EQUATION is

$$\frac{\partial f}{\partial y} - \frac{d}{dx}\left(\frac{\partial f}{\partial y_x}\right) = 0. \tag{1}$$

Now, examine the DERIVATIVE of f with respect to x

$$\frac{df}{dx} = \frac{\partial f}{\partial y}\,y_x + \frac{\partial f}{\partial y_x}\,y_{xx} + \frac{\partial f}{\partial x}. \tag{2}$$

Solving for the $\partial f / \partial y$ term gives

$$\frac{\partial f}{\partial y}\,y_x = \frac{df}{dx} - \frac{\partial f}{\partial y_x}\,y_{xx} - \frac{\partial f}{\partial x}. \tag{3}$$

Now, multiplying (1) by y_x gives

$$y_x\frac{\partial f}{\partial y} - y_x\frac{d}{dx}\left(\frac{\partial f}{\partial y_x}\right) = 0. \tag{4}$$

Substituting (3) into (4) then gives

$$\frac{df}{dx} - \frac{\partial f}{\partial y_x}\,y_{xx} - \frac{\partial f}{\partial x} - y_x\frac{d}{dx}\left(\frac{\partial f}{\partial y_x}\right) = 0 \tag{5}$$

$$-\frac{\partial f}{\partial x} + \frac{d}{dx}\left(f - y_x\frac{\partial f}{\partial y_x}\right) = 0. \tag{6}$$

This form is especially useful if $f_x = 0$, since in that case

$$\frac{d}{dx}\left(f - y_x\frac{\partial f}{\partial y_x}\right) = 0, \tag{7}$$

which immediately gives

$$f - y_x\frac{\partial f}{\partial y_x} = C, \tag{8}$$

where C is a constant of integration (Weinstock 1974, pp. 24–5; Arfken 1985, pp. 928–29; Fox 1988, pp. 8–).

The Beltrami identity greatly simplifies the solution for the minimal AREA SURFACE OF REVOLUTION about a given axis between two specified points. It also allows straightforward solution of the BRACHISTOCHRONE PROBLEM.

See also BRACHISTOCHRONE PROBLEM, CALCULUS OF VARIATIONS, EULER-LAGRANGE DIFFERENTIAL EQUATION, SURFACE OF REVOLUTION

References

Arfken, G. *Mathematical Methods for Physicists, 3rd ed.* Orlando, FL: Academic Press, 1985.
Fox, C. *An Introduction to the Calculus of Variations.* New York: Dover, 1988.
Weinstock, R. *Calculus of Variations, with Applications to Physics and Engineering.* New York: Dover, 1974.

Beltrami's Theorem

Let $f : M \to N$ be a GEODESIC MAPPING. If either M or N has constant curvature, then both surfaces have constant curvature (Ambartzumian 1982, p. 26; Kreyszig 1991).

See also GEODESIC MAPPING

References

Ambartzumian, R. V. *Combinatorial Integral Geometry.* Chichester, England: Wiley, 1982.
Kreyszig, E. §91 in *Differential Geometry.* New York: Dover, 1991.

Bend (Curvature)

The bend of a circle C mutually tangent to three other circles is defined as the signed CURVATURE of C. If the contacts are all external, the signs of the bends of all four circles are taken as POSITIVE, whereas if one circle surrounds the other three, the sign of this circle is taken as NEGATIVE (Coxeter 1969). Bends can also be defined for spheres.

See also CURVATURE, DESCARTES CIRCLE THEOREM, SODDY CIRCLES

References

Coxeter, H. S. M. *Introduction to Geometry, 2nd ed.* New York: Wiley, pp. 13–4, 1969.

Bend (Knot)

A KNOT used to join the ends of two ropes together to form a longer length.

References

Owen, P. *Knots*. Philadelphia, PA: Courage, p. 49, 1993.

Benford's Law

A phenomenological law also called the first digit law, first digit phenomenon, or leading digit phenomenon. Benford's law states that in listings, tables of statistics, etc., the DIGIT 1 tends to occur with PROBABILITY ~30%, much greater than the expected 10% (i.e., one digit out of 10). Benford's law can be observed, for instance, by examining tables of LOGARITHMS and noting that the first pages are much more worn and smudged than later pages (Newcomb 1881). While Benford's law unquestionably applies to many situations in the real world, a satisfactory explanation has been given only recently through the work of Hill (1996).

Benford's law applies to data that are *not* dimensionless, so the numerical values of the data depend on the units. If there exists a universal probability distribution $P(x)$ over such numbers, then it must be invariant under a change of scale, so

$$P(kx) = f(k)P(x). \qquad (1)$$

If $\int P(x)\,dx = 1$, then $\int P(kx)\,dx = 1/k$, and normalization implies $f(k) = 1/k$. Differentiating with respect to k and setting $k = 1$ gives

$$xP'(x) = -P(x), \qquad (2)$$

having solution $P(x) = 1/x$. Although this is not a proper probability distribution (since it diverges), both the laws of physics and human convention impose cutoffs. For example, if street addresses are distributed uniformly over the range of 1 to some maximum cutoff value, then they'll obey something close to Benford's law.

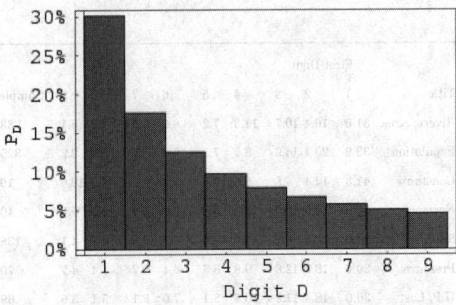

If many powers of 10 lie between the cutoffs, then the probability that the first (decimal) digit is D is given by the LOGARITHMIC DISTRIBUTION

$$P_D = \frac{\int_{D/10}^{D+1} P(x)\,dx}{\int_1^{10} P(x)\,dx} = \frac{\ln\left(\dfrac{D+1}{D}\right)}{\ln 10} = \frac{\ln(D+1) - \ln(D)}{\ln 10} \qquad (3)$$

for $D = 1, \ldots, 9$, illustrated above and tabulated below.

D	P_D	D	P_D
1	0.30103	6	0.0669468
2	0.176091	7	0.0579919
3	0.124939	8	0.0511525
4	0.09691	9	0.0457575
5	0.0791812		

However, Benford's law applies not only to scale-invariant data, but also to numbers chosen from a variety of different sources. Explaining this fact requires a more rigorous investigation of CENTRAL LIMIT-like theorems for the MANTISSAS of random variables under MULTIPLICATION. As the number of variables increases, the density function approaches that of a LOGARITHMIC DISTRIBUTION. Hill (1996) rigorously demonstrated that the "distribution of distributions" given by random samples taken from a variety of different distributions is, in fact, Benford's law (Matthews 1999).

One striking example of Benford's law is given by the 54 million real constants in Plouffe's "Inverse Symbolic Calculator" database, 30% of which begin with the DIGIT 1. Taking data from several disparate sources, the table below, shows the distribution of first digits as compiles by Benford (1938) in his original paper.

Col.	Title	First Digit 1	2	3	4	5	6	7	8	9	Samples
A	Rivers, Area	31.0	16.4	10.7	11.3	7.2	8.6	5.5	4.2	5.1	335
B	Population	33.9	20.4	14.2	8.1	7.2	6.2	4.1	3.7	2.2	3259
C	Constants	41.3	14.4	4.8	8.6	10.6	5.8	1.0	2.9	10.6	104
D	Newspapers	30.0	18.0	12.0	10.0	8.0	6.0	6.0	5.0	5.0	100
E	Specific Heat	24.0	18.4	16.2	14.6	10.6	4.1	3.2	4.8	4.1	1389
F	Pressure	29.6	18.3	12.8	9.8	8.3	6.4	5.7	4.4	4.7	703
G	H.P. Lost	30.0	18.4	11.9	10.8	8.1	7.0	5.1	5.1	3.6	690
H	Mol. Wgt.	26.7	25.2	15.4	10.8	6.7	5.1	4.1	2.8	3.2	1800
I	Drainage	27.1	23.9	13.8	12.6	8.2	5.0	5.0	2.5	1.9	159
J	Atomic Wgt.	47.2	18.7	5.5	4.4	6.6	4.4	3.3	4.4	5.5	91
K	$n^{-1}, \sqrt{n}$	25.7	20.3	9.7	6.8	6.6	6.8	7.2	8.0	8.9	5000
L	Design	26.8	14.8	14.3	7.5	8.3	8.4	7.0	7.3	5.6	560
M	*Reader's Digest*	33.4	18.5	12.4	7.5	7.1	6.5	5.5	4.9	4.2	308
N	Cost Data	32.4	18.8	10.1	10.1	9.8	5.5	4.7	5.5	3.1	741
O	X-Ray Volts	27.9	17.5	14.4	9.0	8.1	7.4	5.1	5.8	4.8	707
P	Am. League	32.7	17.6	12.6	9.8	7.4	6.4	4.9	5.6	3.0	1458
Q	Blackbody	31.0	17.3	14.1	8.7	6.6	7.0	5.2	4.7	5.4	1165
R	Addresses	28.9	19.2	12.6	8.8	8.5	6.4	5.6	5.0	5.0	342
S	$n^1, n^2 \cdots n!$	25.3	16.0	12.0	10.0	8.5	8.8	6.8	7.1	5.5	900
T	Death Rate	27.0	18.6	15.7	9.4	6.7	6.5	7.2	4.8	4.1	418
	Average	30.6	18.5	12.4	9.4	8.0	6.4	5.1	4.9	4.7	1011
	Probable Error	± 0.8	± 0.4	± 0.4	± 0.3	± 0.2	± 0.2	± 0.2	± 0.3		

The following table gives the distribution of the first digit of the mantissa following Benford's Law using a number of different methods.

method	Sloane	sequence
Sainte-Lague	A055439	1, 2, 3, 1, 4, 5, 6, 1, 2, 7, 8, 9, ...
d'Hondt	A055440	1, 2, 1, 3, 1, 4, 2, 5, 1, 6, 3, 1, ...
largest remainder, Hare quotas	A055441	1, 2, 3, 4, 1, 5, 6, 7, 1, 2, 8, 1, ...
largest remainder, Droop quotas	A055442	1, 2, 3, 1, 4, 5, 6, 1, 2, 7, 8, 1, ...

References

Barlow, J. L. and Bareiss, E. H. "On Roundoff Error Distributions in Floating Point and Logarithmic Arithmetic." *Computing* **34**, 325–47, 1985.

Benford, F. "The Law of Anomalous Numbers." *Proc. Amer. Phil. Soc.* **78**, 551–72, 1938.

Bogomolny, A. "Benford's Law and Zipf's Law." http://www.cut-the-knot.com/do_you_know/zipfLaw.html.

Boyle, J. "An Application of Fourier Series to the Most Significant Digit Problem." *Amer. Math. Monthly* **101**, 879–86, 1994.

Flehinger, B. J. "On the Probability that a Random Integer Has Initial Digit *A*." *Amer. Math. Monthly* **73**, 1056–061, 1966.

Franel, J. *Naturforschende Gesellschaft, Vierteljahrsschrift (Zürich)* **62**, 286–95, 1917.

Hill, T. P. "Base-Invariance Implies Benford's Law." *Proc. Amer. Math. Soc.* **12**, 887–95, 1995.

Hill, T. P. "The Significant-Digit Phenomenon." *Amer. Math. Monthly* **102**, 322–27, 1995.

Hill, T. P. "A Statistical Derivation of the Significant-Digit Law." *Stat. Sci.* **10**, 354–63, 1996.

Hill, T. P. "The First Digit Phenomenon." *Amer. Sci.* **86**, 358–63, 1998.

Knuth, D. E. "The Fraction Parts." §4.2.4B in *The Art of Computer Programming, Vol. 2: Seminumerical Algorithms, 3rd ed.* Reading, MA: Addison-Wesley, pp. 254–62, 1998.

Ley, E. "On the Peculiar Distribution of the U.S. Stock Indices Digits." *Amer. Stat.* **50**, 311–13, 1996.

Matthews, R. "The Power of One." http://www.newscientist.com/ns/19990710/thepowerof.html.

Newcomb, S. "Note on the Frequency of the Use of Digits in Natural Numbers." *Amer. J. Math.* **4**, 39–0, 1881.

Nigrini, M. "A Taxpayer Compliance Application of Benford's Law." *J. Amer. Tax. Assoc.* **18**, 72–1, 1996.

Nigrini, M. "I've Got Your Number." *J. Accountancy*, pp. 79–3, May 1999.

Plouffe, S. "Graph of the Number of Entries in Plouffe's Inverter." http://www.lacim.uqam.ca/plouffe/statistics.html.

Raimi, R. A. "The Peculiar Distribution of First Digits." *Sci. Amer.* **221**, 109–19, Dec. 1969.

Raimi, R. A. "On the Distribution of First Significant Digits." *Amer. Math. Monthly* **76**, 342–48, 1969.

Raimi, R. A. "The First Digit Phenomenon." *Amer. Math. Monthly* **83**, 521–38, 1976.

Schatte, P. "Zur Verteilung der Mantisse in der Gleitkommadarstellung einer Zufallsgröße." *Z. Angew. Math. Mech.* **53**, 553–65, 1973.

Schatte, P. "On Mantissa Distributions in Computing and Benford's Law." *J. Inform. Process. Cybernet.* **24**, 443–55, 1988.

Sloane, N. J. A. Sequences A055439, A055440, A055441, and A055442 in "An On-Line Version of the Encyclopedia of Integer Sequences." http://www.research.att.com/~njas/sequences/eisonline.html.

Benham's Wheel

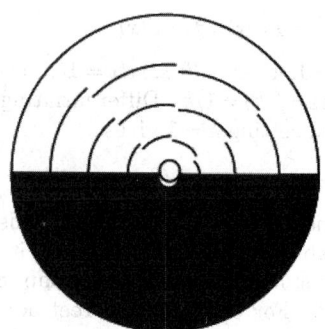

An optical ILLUSION consisting of a spinnable top marked in black with the pattern shown above. When the wheel is spun (especially slowly), the black broken lines appear as green, blue, and red colored bands!

References

Cohen, J. and Gordon, D. A. "The Prevost-Fechner-Benham Subjective Colors." *Psycholog. Bull.* **46**, 97–36, 1949.

Festinger, L.; Allyn, M. R.; and White, C. W. "The Perception of Color with Achromatic Stimulation." *Vision Res.* **11**, 591–12, 1971.

Fineman, M. *The Nature of Visual Illusion.* New York: Dover, pp. 148–51, 1996.

Trolland, T. L. "The Enigma of Color Vision." *Amer. J. Physiology* **2**, 23–8, 1921.

Benjamin-Bona-Mahony Equation

The PARTIAL DIFFERENTIAL EQUATION

$$u_t - u_{xxx} + uu_x = 0$$

(Arvin and Goldstein 1985; Zwillinger 1997, p. 130). A generalized version is given by

$$u_t - \nabla^2 u_t + \div(\phi(u)) = 0$$

(Goldstein and Wichnoski 1980; Zwillinger 1997, p. 132).

References

Arvin, J. and Goldstein, J. A. "Global Existence for the Benjamin-Bona-Mahony Equation in Arbitrary Dimensions." *Nonlinear Anal.* **9**, 861–65, 1985.

Goldstein, J. A. and Wichnoski, B. J. "On the Benjamin-Bona-Mahony Equation in Higher Dimensions." *Nonlinear Anal.* **4**, 665–75, 1980.

Zwillinger, D. *Handbook of Differential Equations, 3rd ed.* Boston, MA: Academic Press, pp. 130 and 132, 1997.

Bennequin's Conjecture

A BRAID with M strands and R components with P positive crossings and N negative crossings satisfies

$$|P - N| \le 2U + M - R \le P + N,$$

where U is the UNKNOTTING NUMBER. While the second part of the INEQUALITY was already known to be true (Boileau and Weber, 1983, 1984) at the time the conjecture was proposed, the proof of the entire conjecture was completed using results of Kronheimer and Mrowka on MILNOR'S CONJECTURE (and, independently, using MENASCO'S THEOREM).

See also BRAID, MENASCO'S THEOREM, MILNOR'S CONJECTURE, UNKNOTTING NUMBER

References

Bennequin, D. "L'instanton gordien (d'après P. B. Kronheimer et T. S. Mrowka)." *Astérisque* **216**, 233–77, 1993.

Birman, J. S. and Menasco, W. W. "Studying Links via Closed Braids. II. On a Theorem of Bennequin." *Topology Appl.* **40**, 71–2, 1991.

Boileau, M. and Weber, C. "Le problème de J. Milnor sur le nombre gordien des nœuds algébriques." *Enseign. Math.* **30**, 173–22, 1984.

Boileau, M. and Weber, C. "Le problème de J. Milnor sur le nombre gordien des nœuds algébriques." In *Knots, Braids and Singularities (Plans-sur-Bex, 1982).* Geneva, Switzerland: Monograph. Enseign. Math. Vol. 31, pp. 49–8, 1983.

Cipra, B. *What's Happening in the Mathematical Sciences, Vol. 2.* Providence, RI: Amer. Math. Soc., pp. 8–3, 1994.

Kronheimer, P. B. "The Genus-Minimizing Property of Algebraic Curves." *Bull. Amer. Math. Soc.* **29**, 63–9, 1993.

Kronheimer, P. B. and Mrowka, T. S. "Gauge Theory for Embedded Surfaces. I." *Topology* **32**, 773–26, 1993.

Kronheimer, P. B. and Mrowka, T. S. "Recurrence Relations and Asymptotics for Four-Manifold Invariants." *Bull. Amer. Math. Soc.* **30**, 215–21, 1994.

Menasco, W. W. "The Bennequin-Milnor Unknotting Conjectures." *C. R. Acad. Sci. Paris Sér. I Math.* **318**, 831–36, 1994.

Benson's Formula

An equation for a LATTICE SUM with $n = 3$

$$-b_3(1) = \sum_{i,\,j,\,k=-\infty}^{\infty}{}' \frac{(-1)^{i+j+k+1}}{\sqrt{i^2 + j^2 + k^2}}$$

$$= 12\pi \sum_{m,\,n=1,\,3,\,\dots}^{\infty} \operatorname{sech}^2(\tfrac{1}{2}\pi\sqrt{m^2 + n^2}).$$

Here, the prime denotes that summation over $(0, 0, 0)$ is excluded. The sum is numerically equal to $-1.74756\dots$, a value known as "the" MADELUNG CONSTANT.

See also MADELUNG CONSTANTS

References

Borwein, J. M. and Borwein, P. B. *Pi & the AGM: A Study in Analytic Number Theory and Computational Complexity.* New York: Wiley, p. 301, 1987.

Finch, S. "Favorite Mathematical Constants." http://www.mathsoft.com/asolve/constant/mdlung/mdlung.html.

Ber

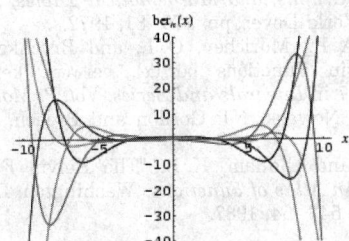

The REAL PART of

$$J_\nu(xe^{3\pi i/4}) = \operatorname{ber}_\nu(x) + i\operatorname{bei}_\nu(x). \tag{1}$$

The function $\operatorname{ber}_\nu(x)$ has the series expansion

$$\operatorname{ber}_\nu(x) = (\tfrac{1}{2}x)^\nu \sum_{k=0}^{\infty} \frac{\cos[(\tfrac{3}{4}\nu + \tfrac{1}{2}k)\pi]}{k!\,\Gamma(\nu + k + 1)} (\tfrac{1}{4}x^2)^k, \tag{2}$$

where $\Gamma(x)$ is the GAMMA FUNCTION (Abramowitz and Stegun 1972, p. 379).

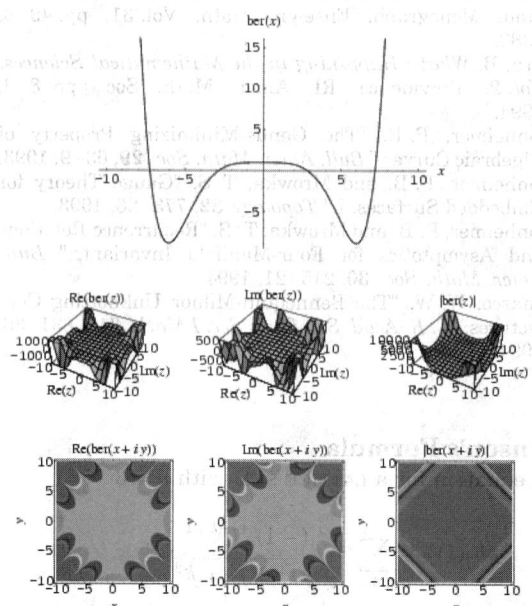

The special case $\nu = 0$ gives

$$J_0\left(i\sqrt{i}\,x\right) \equiv \mathrm{ber}(x) + i\,\mathrm{bei}(x), \tag{3}$$

where $J_0(x)$ is the zeroth order BESSEL FUNCTION OF THE FIRST KIND. The function $\mathrm{ber}_0(x) \equiv \mathrm{ber}(x)$ has the series expansion

$$\mathrm{ber}(x) \equiv \sum_{n=0}^{\infty} \frac{(-1)^n (\frac{1}{2}x)^{4n}}{[(2n)!]^2}. \tag{4}$$

See also BEI, BESSEL FUNCTION, KEI, KELVIN FUNCTIONS, KER

References

Abramowitz, M. and Stegun, C. A. (Eds.). "Kelvin Functions." §9.9 in *Handbook of Mathematical Functions with Formulas, Graphs, and Mathematical Tables, 9th printing.* New York: Dover, pp. 379–81, 1972.
Prudnikov, A. P.; Marichev, O. I.; and Brychkov, Yu. A. "The Kelvin Functions $\mathrm{ber}_\nu(x)$, $\mathrm{bei}\,\nu(x)$, $\mathrm{ker}_\nu(x)$ and $\mathrm{kei}_\nu(x)$." §1.7 in *Integrals and Series, Vol. 3: More Special Functions.* Newark, NJ: Gordon and Breach, pp. 29–0, 1990.
Spanier, J. and Oldham, K. B. "The Kelvin Functions." Ch. 55 in *An Atlas of Functions.* Washington, DC: Hemisphere, pp. 543–54, 1987.

Beraha Constants

The nth Beraha constant (or number) is given by

$$B(n) \equiv 2 + 2\cos\left(\frac{2\pi}{n}\right).$$

They appear to be ROOTS of the CHROMATIC POLY-

NOMIALS of planar triangular GRAPHS. $B(5)$ is $\phi + 1$, where ϕ is the GOLDEN RATIO, $B(7)$ is the SILVER CONSTANT, and $B(10) = \phi + 2$. The following table summarizes the first few Beraha numbers.

n	$B(n)$	Approx.
1	4	
2	0	
3	1	
4	2	
5	$\frac{1}{2}(3+\sqrt{5})$	2.618
6	3	
7	$2+2\cos(\frac{2}{7}\pi)$	3.247
8	$2+\sqrt{2}$	3.414
9	$2+2\cos(\frac{2}{9}\pi)$	3.532
10	$\frac{1}{2}(5+\sqrt{5})$	3.618

See also CHROMATIC POLYNOMIAL, GOLDEN RATIO, SILVER CONSTANT

References

Beraha, S. Ph.D. thesis. Baltimore, MD: Johns Hopkins University, 1974.
Le Lionnais, F. *Les nombres remarquables.* Paris: Hermann, p. 143, 1983.
Saaty, T. L. and Kainen, P. C. *The Four-Color Problem: Assaults and Conquest.* New York: Dover, pp. 160–63, 1986.
Tutte, W. T. "Chromials." University of Waterloo, 1971.
Tutte, W. T. "More about Chromatic Polynomials and the Golden Ratio." In *Combinatorial Structures and their Applications.* New York: Gordon and Breach, p. 439, 1969.
Tutte, W. T. "Chromatic Sums for Planar Triangulations I: The Case $\lambda = 1$." Research Report COPR 72–, University of Waterloo, 1972a.
Tutte, W. T. "Chromatic Sums for Planar Triangulations IV: The Case $\lambda = \infty$." Research Report COPR 72–, University of Waterloo, 1972b.

Berezin Transform

The operator $\tilde{B}$ defined by

$$\tilde{B}f(x) = \int_D \frac{(1-|z|^2)^2}{|1-z\bar{w}|^4} f(w)\,dA(w)$$

for $z \in \mathbb{D}$, where $\mathbb{D}$ is the unit open disk and $\bar{w}$ is the COMPLEX CONJUGATE (Hedenmalm *et al.* 2000, p. 29).

References

Hedenmalm, H.; Korenblum, B.; and Zhu, K. "The Berezin Transform." Ch. 2 in *Theory of Bergman Spaces.* New York: Springer-Verlag, pp. 28–1, 2000.

Berge's Theorem

A MATCHING is maximal IFF it contains no AUGMENTING PATH.

See also MATCHING

References

Berge, C. "Two Theorems in Graph Theory." *Proc. Nat. Acad. Sci. USA* **43**, 842–44, 1957.

Skiena, S. *Implementing Discrete Mathematics: Combinatorics and Graph Theory with Mathematica.* Reading, MA: Addison-Wesley, 1990.

Berger-Kazdan Comparison Theorem

Let M be a compact n-D MANIFOLD with INJECTIVITY radius inj(M). Then

$$\text{Vol}(M) \geq \frac{c_n \, \text{inj}(M)}{\pi},$$

with equality IFF M is ISOMETRIC to the standard round SPHERE $\mathbb{S}^n$ with RADIUS inj(M), where $c_n(r)$ is the VOLUME of the standard n-HYPERSPHERE of RADIUS r.

See also BLASCHKE CONJECTURE, HYPERSPHERE, INJECTIVE, ISOMETRY

References

Chavel, I. *Riemannian Geometry: A Modern Introduction.* New York: Cambridge University Press, 1994.

Bergman Kernel

A Bergman kernel is a function of a COMPLEX VARIABLE with the "reproducing kernel" property defined for any DOMAIN in which there exist NONZERO ANALYTIC FUNCTIONS of class $l_2(d)$ with respect to the LEBESGUE MEASURE dv.

References

HazewinKel, M. (Managing Ed.). *Encyclopaedia of Mathematics: An Updated and Annotated Translation of the Soviet "Mathematical Encyclopaedia."* Dordrecht, Netherlands: Reidel, pp. 356–57, 1988.

Bergman Space

Let G be an open subset of the COMPLEX PLANE $\mathbb{C}$, and let $L_a^2(G)$ denote the collection of all ANALYTIC FUNCTIONS $f : G \to C$ whose MODULUS is square integrable with respect to AREA measure. Then $L_a^2(G)$, sometimes also denoted $A^2(G)$, is called the Bergman space for G. Thus, the Bergman space consists of all the ANALYTIC FUNCTIONS in $L^2(G)$. The Bergman space can also be generalized to $L_a^p(G)$, where $0 < p < \infty$.

See also HARDY SPACE

References

Hedenmalm, H.; Korenblum, B.; and Zhu, K. *Theory of Bergman Spaces.* New York: Springer-Verlag, 2000.

Shields, A. L. "Weighted Shift Operators and Analytic Function Theory." In *Topics in Operator Theory.* Providence, RI: Amer. Math. Soc., pp. 49–28, 1974.

Zhu, K. *Operator Theory in Function Spaces.* New York: Dekker, 1990.

Berlekamp-Massey Algorithm

If a sequence takes only a small number of different values, then by regarding the values as the elements of a FINITE FIELD, the Berlekamp-Massey algorithm is an efficient procedure for finding the shortest linear recurrence from the field that will generate the sequence.

See also REED-SLOANE ALGORITHM

References

Berlekamp, E. R. Ch. 7 in *Algorithmic Coding Theory.* New York: McGraw-Hill, 1968.

Berlekamp, E. R.; Fredricksen, H. M.; and Proto, R. C. "Minimum Conditions for Uniquely Determining the Generator of a Linear Sequence." *Util. Math.* **5**, 305–15, 1974.

Brent, R. P.; Gustavson, F. G.; and Yun, D. Y. Y. "Fast Solution of Toeplitz Systems of Equations and Computation of Padé Approximants." *J. Algorithms* **1**, 259–95, 1980.

Dickinson, B. W.; Morf, M.; and Kailath, T. "A Minimal Realization Algorithm for Matrix Sequences." *IEEE Trans. Automatic Control* **18**, 31–8, 1974.

Gustavson, F. G. "Analysis of the Berlekamp-Massey Linear Feedback Shift-Register Synthesis Algorithm." *IBM J. Res. Dev.* **20**, 204–12, 1976.

MacWilliams, F. J. and Sloane, N. J. A. Ch. 9 in *The Theory of Error-Correcting Codes.* New York: Elsevier, 1978.

Massey, J. L. "Shift-Register Synthesis and BCH Decoding." *IEEE Trans. Information Th.* **15**, 122–27, 1969.

McEliece, R. J. *The Theory of Information Coding.* Reading, MA: Addison-Wesley, 1977.

Mills, W. H. "Continued Fractions and Linear Recurrences." *Math. Comput.* **29**, 173–80, 1975.

Sloane, N. J. A. and Plouffe, S. *The Encyclopedia of Integer Sequences.* San Diego, CA: Academic Press, pp. 25–6, 1995.

Berlekamp-Zassenhaus Algorithm

An algorithm that can be used to find subsets S of a set for which the product of elements of S of a set of monic irreducible polynomials in $\mathbb{Z}_P$ for which the product of the elements of S has integer coefficients (van Hoeij 2000).

References

van Hoeij, M. "Factoring Polynomials and the Knapsack Problem." Preprint. http://www.math.fsu.edu/~aluffi/archive/paper124.ps.gz.

Zassenhaus, H. "On Hensel Factorization, I." *J. Number Th.* **1**, 291–11, 1969.

Bernays-Gödel Set Theory

VON NEUMANN-BERNAYS-GÖDEL SET THEORY

Bernoulli Differential Equation

$$\frac{dy}{dx} + p(x)y = q(x)y^n. \tag{1}$$

Let $v \equiv y^{1-n}$ for $n \neq 1$, then

$$\frac{dv}{dx} = (1-n)y^{-n}\frac{dy}{dx}. \tag{2}$$

Rewriting (1) gives

$$y^{-n}\frac{dy}{dx} = q(x) - p(x)y^{1-n} = q(x) - vp(x). \tag{3}$$

Plugging (3) into (2),

$$\frac{dv}{dx} = (1-n)[q(x) - vp(x)]. \tag{4}$$

Now, this is a linear FIRST-ORDER ORDINARY DIFFERENTIAL EQUATION OF THE FORM

$$\frac{dv}{dx} + vP(x) = Q(x), \tag{5}$$

where $P(x) \equiv (1-n)p(x)$ and $Q(x) \equiv (1-n)q(x)$. It can therefore be solved analytically using an INTEGRATING FACTOR

$$v = \frac{\int e^{\int P(x)\,dx}Q(x)\,dx + C}{e^{\int P(x)\,dx}}$$

$$= \frac{(1-n)\int e^{(1-n)\int p(x)\,dx}q(x)\,dx + C}{e^{(1-n)\int p(x)\,dx}}, \tag{6}$$

where C is a constant of integration. If $n = 1$, then equation (1) becomes

$$\frac{dy}{dx} = y(q-p) \tag{7}$$

$$\frac{dy}{y} = (q-p)\,dx \tag{8}$$

$$y = C_2 e^{\int [q(x)-p(x)]\,dx}. \tag{9}$$

The general solution is then, with C_1 and C_2 con-

stants,

$$y = \begin{cases} \left[\dfrac{(1-n)\int e^{(1-n)\int p(x)\,dx}q(x)\,dx + C_1}{e^{(1-n)\int p(x)\,dx}}\right]^{1/(1-n)} & \text{for } n \neq 1 \\[4mm] C_2 e^{\int [(q(x)-p(x)]\,dx} & \text{for } n = 1. \end{cases} \tag{10}$$

References

Boyce, W. E. and DiPrima, R. C. *Elementary Differential Equations and Boundary Value Problems, 5th ed.* New York: Wiley, p. 28, 1992.

Ince, E. L. *Ordinary Differential Equations.* New York: Dover, p. 22, 1956.

Rainville, E. D. and Bedient, P. E. *Elementary Differential Equations.* New York: Macmillian, pp. 69–1, 1964.

Simmons, G. F. *Differential Equations, With Applications and Historical Notes.* New York: McGraw-Hill, p. 49, 1972.

Zwillinger, D. (Ed.). *CRC Standard Mathematical Tables and Formulae.* Boca Raton, FL: CRC Press, p. 413, 1995.

Zwillinger, D. "Bernoulli Equation." §II.A.37 in *Handbook of Differential Equations, 3rd ed.* Boston, MA: Academic Press, pp. 120 and 157–58, 1997.

Bernoulli Distribution

A STATISTICAL DISTRIBUTION given by

$$P(n) = \begin{cases} q \equiv 1-p & \text{for } n = 0 \\ p & \text{for } n = 1 \end{cases} \tag{1}$$

$$= p^n(1-p)^{1-n} \qquad \text{for } n = 0, 1. \tag{2}$$

The distribution of heads and tails in COIN TOSSING is a Bernoulli distribution with $p = q = 1/2$. The MOMENT-GENERATING FUNCTION of the Bernoulli distribution is

$$M(t) = \langle e^{tn} \rangle = \sum_{n=0}^{1} e^{tn}p^n(1-p)^{1-n} = e^0(1-p) + e^t p, \tag{3}$$

so

$$M(t) = (1-p) + pe^t \tag{4}$$

$$M'(t) = pe^t \tag{5}$$

$$M''(t) = pe^t \tag{6}$$

$$M^{(n)}(t) = pe^t, \tag{7}$$

and the MOMENTS about 0 are

$$\mu_1' = \mu = M'(0) = p \tag{8}$$

$$\mu_2' = M''(0) = p \tag{9}$$

$$\mu_n' = M^{(n)}(0) = p. \tag{10}$$

The MOMENTS about the MEAN are

$$\mu_2 = \mu_2' - (\mu_1')^2 = p - p^2 = p(1-p) \tag{11}$$

$$\mu_3 = \mu_3' - 3\mu_2'\mu_1' + 2(\mu_1')^3 = p - 3p^2 + 2p^3$$
$$= p(1-p)(1-2p) \tag{12}$$

$$\mu_4 = \mu_4' - 4\mu_3'\mu_1' + 6\mu_2'(\mu_1')^2 - 3(\mu_1')^4$$
$$= p - 4p^2 + 6p^3 - 3p^4 = p(1-p)(3p^2 - 3p + 1). \tag{13}$$

The MEAN, VARIANCE, SKEWNESS, and KURTOSIS are then

$$\mu = p \tag{14}$$

$$\sigma^2 = \mu_2 = p(1-p) \tag{15}$$

$$\gamma_1 = \frac{\mu_3}{\sigma_3} = \frac{p(1-p)(1-2p)}{[p(1-p)]^{3/2}} = \frac{1-2p}{\sqrt{p(1-p)}} \tag{16}$$

$$\gamma_2 = \frac{\mu_4}{\sigma_4} - 3 = \frac{p(1-2p)(2p^2 - 2p + 1)}{p^2(1-p)^2} - 3$$
$$= \frac{6p^2 - 6p + 1}{p(1-p)}. \tag{17}$$

To find an estimator $\hat{p}$ for the mean of a Bernoulli population with actual mean p, let N trials be made and suppose n successes are obtained. Assume an estimator given by

$$\hat{} \equiv \frac{n}{N}, \tag{18}$$

so that the probability of obtaining the observed n successes in N trials is then

$$\binom{N}{n} p^n (1-p)^{N-n}. \tag{19}$$

The expectation value of the estimator $\hat{p}$ is therefore given by

$$\langle \hat{p} \rangle \cong \sum_{n=0}^{N} p \binom{N}{n} p^n (1-p)^{N-n}$$
$$= (1-p)^N \left(\frac{1}{1-p} \right)^N p = p, \tag{20}$$

so $\langle p \rangle$ is indeed an UNBIASED ESTIMATOR for the population mean p.

See also BERNOULLI TRIAL, BINOMIAL DISTRIBUTION, COIN TOSSING, RUN

References

Evans, M.; Hastings, N.; and Peacock, B. "Bernoulli Distribution." Ch. 4 in *Statistical Distributions, 3rd ed.* New York: Wiley, pp. 31–3, 2000.

Bernoulli Function

BERNOULLI POLYNOMIAL

Bernoulli Inequality

$$(1+x)^n > 1 + nx, \tag{1}$$

where $x > -1 \neq 0$ is a REAL NUMBER and $n > 1$ an INTEGER. This inequality can be proven by taking a MACLAURIN SERIES of $(1+x)^n$,

$$(1+x)^n = 1 + nx + \tfrac{1}{2}n(n-1)x^2 + \tfrac{1}{6}n(n-1)(n-2)x^3 + \cdots. \tag{2}$$

Since the series terminates after a finite number of terms for INTEGRAL n, the Bernoulli inequality for $x > 0$ is obtained by truncating after the first-order term.

When $-1 < x < 0$, slightly more finesse is needed. In this case, let $y = |x| = -x > 0$ so that $0 < y < 1$, and take

$$(1-y)^n = 1 - ny + \tfrac{1}{2}n(n-1)y^2 - \tfrac{1}{6}n(n-1)(n-2)y^3 + \cdots. \tag{3}$$

Since each POWER of y multiplies by a number < 1 and since the ABSOLUTE VALUE of the COEFFICIENT of each subsequent term is smaller than the last, it follows that the sum of the third order and subsequent terms is a POSITIVE number. Therefore,

$$(1-y)^n > 1 - ny, \tag{4}$$

or

$$(1+x)^n > 1 + nx, \quad \text{for } -1 < x < 0, \tag{5}$$

completing the proof of the INEQUALITY over all ranges of parameters.

For $x > -1 \neq 0$, the following generalizations of Bernoulli inequality are valid for real exponents:

$$(1+x)^a > 1 + ax \quad \text{if } a > 1 \text{ or } a < 0, \tag{6}$$

and

$$(1+x)^a < 1 + ax \quad \text{if } 0 < a < 1 \tag{7}$$

(Mitrinovic 1970).

References

Mitrinovic, D. S. *Analytic Inequalities.* New York: Springer-Verlag, 1970.

Bernoulli Lemniscate

LEMNISCATE

Bernoulli Number

There are two definitions for the Bernoulli numbers. In modern usage, the Bernoulli numbers are written B_n, while the Bernoulli numbers encountered in older literature (where they are confusingly also denoted B_n) are distinguished by writing them as B_n^*. In each case, the Bernoulli numbers are a special case of the BERNOULLI POLYNOMIALS $B_n(x)$ or $B_n^*(x)$ with $B_n = B_n(0)$ and $B_n^* = B_n^*(0)$.

The older definition of the Bernoulli numbers, no longer in widespread use, defines B_n^* using the equations

$$\frac{x}{e^x - 1} + \frac{x}{2} - 1 \equiv \sum_{n=1}^{\infty} \frac{(-1)^{n-1} B_n^* x^{2n}}{(2n)!}$$

$$= \frac{B_1^* x^2}{2!} - \frac{B_2^* x^4}{4!} + \frac{B_3^* x^6}{6!} + \cdots \quad (1)$$

for $|x| < 2\pi$, or

$$1 - \frac{x}{2} \cot\left(\frac{x}{2}\right) \equiv \sum_{n=1}^{\infty} \frac{B_n^* x^{2n}}{(2n)!}$$

$$= \frac{B_1^* x^2}{2!} - \frac{B_2^* x^4}{4!} + \frac{B_3^* x^6}{6!} + \cdots \quad (2)$$

for $|x| < \pi$ (Whittaker and Watson 1990, p. 125). Gradshteyn and Ryzhik (2000) denote these numbers B_n^*, while Bernoulli numbers defined by the newer (National Bureau of Standards) definition are denoted B_n. The B_n^* Bernoulli numbers may be calculated from the integral

$$B_n^* = 4n \int_0^{\infty} \frac{t^{2n-1} \, dt}{e^{2\pi t} - 1}, \quad (3)$$

and analytically from

$$B_n^* = \frac{2(2n)!}{(2\pi)^{2n}} \sum_{p=1}^{\infty} p^{-2n} = \frac{2(2n)!}{(2\pi)^{2n}} \zeta(2n) \quad (4)$$

for $n = 1, 2, \ldots$, where $\zeta(z)$ is the RIEMANN ZETA FUNCTION.

The first few Bernoulli numbers b_n^* are

$$B_1^* = \frac{1}{6}$$

$$B_2^* = \frac{1}{30}$$

$$B_3^* = \frac{1}{42}$$

$$B_4^* = \frac{1}{30}$$

$$B_5^* = \frac{5}{66}$$

$$B_6^* = \frac{691}{2,730}$$

$$B_7^* = \frac{7}{6}$$

$$B_8^* = \frac{3,617}{510}$$

$$B_9^* = \frac{43,867}{798}$$

$$B_{10}^* = \frac{174,611}{330}$$

$$B_{11}^* = \frac{854,513}{138}.$$

Bernoulli numbers defined by the modern definition are denoted B_n and sometimes called "even-index" Bernoulli numbers. These are the Bernoulli numbers returned, by example, by the *Mathematica* function BernoulliB[n]. The first few are

$$B_0 = 1$$

$$B_1 = -\frac{1}{2}$$

$$B_2 = \frac{1}{6}$$

$$B_4 = -\frac{1}{30}$$

$$B_6 = \frac{1}{42}$$

$$B_8 = -\frac{1}{30}$$

$$B_{10} = \frac{5}{66}$$

$$B_{12} = -\frac{691}{2,730}$$

$$B_{14} = \frac{7}{6}$$

$$B_{16} = -\frac{3,617}{510}$$

$$B_{18} = \frac{43,867}{798}$$

$$B_{20} = -\frac{174,611}{330}$$

$$B_{22} = \frac{854,513}{138}$$

(Sloane's A000367 and A002445), with

$$B_{2n+1} = 0 \quad (5)$$

for $n = 1, 2, \ldots$ The Bernoulli numbers B_n are a superset of the archaic ones B_n^* since

$$B_n \equiv \begin{cases} 1 & \text{for } n = 0 \\ -\frac{1}{2} & \text{for } n = 1 \\ (-1)^{(n/2)-1} B_{n/2}^* & \text{for } n \text{ even} \\ 0 & \text{for } n \text{ odd.} \end{cases} \quad (6)$$

The B_n can be defined by the identity

$$\frac{x}{e^x - 1} \equiv \sum_{n=0}^{\infty} \frac{B_n x^n}{n!}. \quad (7)$$

These relationships can be derived using the generating function

$$F(x, t) = \sum_{n=0}^{\infty} \frac{B_n(x) t^n}{n!}, \quad (8)$$

which converges uniformly for $|t| < 2\pi$ and all x (Castellanos 1988). Taking the partial derivative gives

$$\frac{\partial F(x, t)}{\partial x} = \sum_{n=0}^{\infty} \frac{B_{n-1}(x) t^n}{(n-1)!} = t \sum_{n=0}^{\infty} \frac{B_n(x) t^n}{n!} = t F(x, t). \quad (9)$$

The solution to this differential equation can be found

using SEPARATION OF VARIABLES as

$$F(x, t) = T(t)e^{xt}, \qquad (10)$$

so integrating gives

$$\int_0^1 F(x, t)\, dx = T(t) \int_0^1 e^{xt}\, dx = T(t)\, \frac{e^t - 1}{t}. \qquad (11)$$

But integrating (11) explicitly gives

$$\int_0^1 F(x, t)\, dx = \sum_{n=0}^{\infty} \frac{t^n}{n!} \int_0^1 B_n(x)\, dx$$

$$= 1 + \sum_{n=0}^{\infty} \frac{t^n}{n!} \int_0^1 B_n(x)\, dx = 1, \qquad (12)$$

so

$$T(t)\, \frac{e^t - 1}{t} = 1. \qquad (13)$$

Solving for $T(t)$ and plugging back into (10) then gives

$$\frac{te^{xt}}{e^t - 1} = \sum_{n=0}^{\infty} \frac{B_n(x)t^n}{n!}. \qquad (14)$$

(Castellanos 1988). Setting $x = 0$ and adding $t/2$ to both sides then gives

$$\tfrac{1}{2}t \coth(\tfrac{1}{2}t) = \sum_{n=0}^{\infty} \frac{B_{2n}t^{2n}}{(2n)!}. \qquad (15)$$

Letting $t = 2ix$ then gives

$$x \cot x = \sum_{n=0}^{\infty} (-1)^n B_{2n} \frac{(2x)^{2n}}{(2n)!} \qquad (16)$$

for $x \in [-\pi, \pi]$. The Bernoulli numbers may also be calculated from the integral

$$B_n = \frac{n!}{2\pi i} \int \frac{z}{e^z - 1} \frac{dz}{z^{n+1}}, \qquad (17)$$

or from

$$B_n = \lim_{x \to 0} \frac{d^n}{dx^n} \frac{x}{e^x - 1}. \qquad (18)$$

The Bernoulli numbers satisfy the identity

$$\binom{k+1}{1} B_k + \binom{k+1}{2} B_{k-1} + \cdots + \binom{k+1}{k} B_1 + B_0 = 0, \qquad (19)$$

where $\binom{n}{k}$ is a BINOMIAL COEFFICIENT. They also satisfy the nice sum identity

$$\sum_{i=0}^{n} \frac{(1 - 2^{1-i})(1 - 2^{i-n+1})B_{n-i}B_i}{(n-i)!\, i!} = \frac{(1-n)B_n}{n!} \qquad (20)$$

(Gosper).

An ASYMPTOTIC SERIES for the even Bernoulli numbers is

$$B_{2n} \sim (-1)^{n-1} 4\sqrt{\pi n} \left(\frac{n}{\pi e}\right)^{2n}. \qquad (21)$$

Bernoulli numbers appear in expressions OF THE FORM $\Sigma_{k=1}^n k^p$, where $p = 1, 2, \ldots$ Bernoulli numbers also appear in the series expansions of functions involving $\tan x$, $\cot x$, $\csc x$, $\ln|\sin x|$, $\ln|\cos x|$, $\ln|\tan x|$, $\tanh x$, $\coth x$, and $\operatorname{csch} x$. An analytic solution exists for EVEN orders,

$$B_{2n} = \frac{(-1)^{n-1}2(2n)!}{(2\pi)^{2n}} \sum_{p=1}^{\infty} p^{-2n} = \frac{(-1)^{n-1}2(2n)!}{(2\pi)^{2n}} \zeta(2n) \qquad (22)$$

for $n = 1, 2, \ldots$, where $\zeta(2n)$ is the RIEMANN ZETA FUNCTION. Another intimate connection with the RIEMANN ZETA FUNCTION is provided by the identity

$$B_n = (-1)^{n+1} n \zeta(1-n). \qquad (23)$$

The DENOMINATOR of B_{2k} is given by the VON STAUDT-CLAUSEN THEOREM

$$\operatorname{denom}(B_{2k}) = \prod_{\substack{p \text{ prime} \\ (p-1)|2k}}^{2k+1} p, \qquad (24)$$

which also implies that the DENOMINATOR of B_{2k} is SQUAREFREE (Hardy and Wright 1979). Another curious property is that the fraction part of B_n in DECIMAL has a DECIMAL PERIOD which divides n, and there is a single digit before that period (Conway 1996).

Bernoulli first used the Bernoulli numbers while computing $\Sigma_{k=1}^n k^p$. He used the property of the FIGURATE NUMBER TRIANGLE that

$$\sum_{i=0}^{n} a_{ij} = \frac{(n+1)a_{nj}}{j+1}, \qquad (25)$$

along with a form for a_{nj} which he derived inductively to compute the sums up to $n = 10$ (Boyer 1968, p. 85). For $p \in \mathbb{Z} > 0$, the sum is given by

$$\sum_{k=1}^{n} k^p = \frac{(B + n + 1)^{[p+1]} - B^{p+1}}{p+1}, \qquad (26)$$

where the NOTATION $B^{[k]}$ means the quantity in question is raised to the appropriate POWER k, and all terms OF THE FORM B^m are replaced with the corresponding Bernoulli numbers B_m. Written explicitly in terms of a sum of POWERS,

$$\sum_{k=1}^{n} k^p = n^p + \sum_{k=0}^{p} \frac{B_k p!}{k!(p-k+1)!} n^{p-k+1}. \qquad (27)$$

It is also true that the COEFFICIENTS of the terms in

such an expansion sum to 1 (which Bernoulli stated without proof). Ramanujan gave a number of curious infinite sum identities involving Bernoulli numbers (Berndt 1994).

G. J. Fee and S. Plouffe have computed $B_{200,000}$, which has $\sim 800,000$ DIGITS (Plouffe). Plouffe and collaborators have also calculated B_n for n up to 72,000.

See also ARGOH'S CONJECTURE, BERNOULLI FUNCTION, BERNOULLI NUMBER OF THE SECOND KIND, BERNOULLI POLYNOMIAL, DEBYE FUNCTIONS, EULER-MACLAURIN INTEGRATION FORMULAS, EULER NUMBER, FIGURATE NUMBER TRIANGLE, GENOCCHI NUMBER, MODIFIED BERNOULLI NUMBER, PASCAL'S TRIANGLE, RIEMANN ZETA FUNCTION, VON STAUDT-CLAUSEN THEOREM

References

Abramowitz, M. and Stegun, C. A. (Eds.). "Bernoulli and Euler Polynomials and the Euler-Maclaurin Formula." §23.1 in *Handbook of Mathematical Functions with Formulas, Graphs, and Mathematical Tables, 9th printing.* New York: Dover, pp. 804–06, 1972.

Arfken, G. "Bernoulli Numbers, Euler-Maclaurin Formula." §5.9 in *Mathematical Methods for Physicists, 3rd ed.* Orlando, FL: Academic Press, pp. 327–38, 1985.

Ball, W. W. R. and Coxeter, H. S. M. *Mathematical Recreations and Essays, 13th ed.* New York: Dover, p. 71, 1987.

Berndt, B. C. *Ramanujan's Notebooks, Part IV.* New York: Springer-Verlag, pp. 81–5, 1994.

Boyer, C. B. *A History of Mathematics.* New York: Wiley, 1968.

Castellanos, D. "The Ubiquitous Pi. Part I." *Math. Mag.* **61**, 67–8, 1988.

Conway, J. H. and Guy, R. K. In *The Book of Numbers.* New York: Springer-Verlag, pp. 107–10, 1996.

Gradshteyn, I. S. and Ryzhik, I. M. *Tables of Integrals, Series, and Products, 6th ed.* San Diego, CA: Academic Press, 2000.

Graham, R. L.; Knuth, D. E.; and Patashnik, O. "Bernoulli Numbers." §6.5 in *Concrete Mathematics: A Foundation for Computer Science, 2nd ed.* Reading, MA: Addison-Wesley, pp. 283–90, 1994.

Hardy, G. H. and Wright, W. M. *An Introduction to the Theory of Numbers, 5th ed.* Oxford, England: Oxford University Press, pp. 91–3, 1979.

Hauss, M. *Verallgemeinerte Stirling, Bernoulli und Euler Zahlen, deren Anwendungen und schnell konvergente Reihen für Zeta Funktionen.* Aachen, Germany: Verlag Shaker, 1995.

Ireland, K. and Rosen, M. "Bernoulli Numbers." Ch. 15 in *A Classical Introduction to Modern Number Theory, 2nd ed.* New York: Springer-Verlag, pp. 228–48, 1990.

Knuth, D. E. and Buckholtz, T. J. "Computation of Tangent, Euler, and Bernoulli Numbers." *Math. Comput.* **21**, 663–88, 1967.

Nielsen, N. *Traité élémentaire des nombres de Bernoulli.* Paris: Gauthier-Villars, 1923.

Plouffe, S. "Plouffe's Inverter: Table of Current Records for the Computation of Constants." http://www.lacim.uqam.ca/pi/records.html.

Ramanujan, S. "Some Properties of Bernoulli's Numbers." *J. Indian Math. Soc.* **3**, 219–34, 1911.

Roman, S. *The Umbral Calculus.* New York: Academic Press, p. 31, 1984.

Sloane, N. J. A. Sequences A000367/M4039 and A002445/M4189 in "An On-Line Version of the Encyclopedia of Integer Sequences." http://www.research.att.com/~njas/sequences/eisonline.html.

Spanier, J. and Oldham, K. B. "The Bernoulli Numbers, B_n." Ch. 4 in *An Atlas of Functions.* Washington, DC: Hemisphere, pp. 35–8, 1987.

Wagstaff, S. S. Jr. "Ramanujan's Paper on Bernoulli Numbers." *J. Indian Math. Soc.* **45**, 49–5, 1981.

Whittaker, E. T. and Watson, G. N. *A Course in Modern Analysis, 4th ed.* Cambridge, England: Cambridge University Press, 1990.

Woon, S C. Generalization of a Relation Between the Riemann Zeta Function and Bernoulli Numbers. 24 Dec 1998. http://xxx.lanl.gov/abs/math.NT/9812143/.

Young, P. T. "Congruences for Bernoulli, Euler, and Stirling Numbers." *J. Number Th.* **78**, 204–27, 1999.

Bernoulli Number of the Second Kind

A number defined by $b_n = b_n(0)$, where $b_n(x)$ is a BERNOULLI POLYNOMIAL OF THE SECOND KIND (Roman 1974, p. 294), also called Cauchy numbers of the first kind. The first few for $n = 0, 1, 2, \ldots$ are 1, 1/2, $-1/6$, 1/4, $-19/30$, 9/4, $\ldots$ (Sloane's A006232 and A006233). They are given by

$$b_n = \int_0^1 (x)_n \, dx,$$

where $(x)_n$ is a FALLING FACTORIAL, and have EXPONENTIAL GENERATING FUNCTION

$$E(x) = \frac{x}{\ln(1+x)} = 1 + \frac{1!}{2} x - \frac{2!}{6} x^2 + \frac{3!}{4} x^3 + \cdots.$$

See also BERNOULLI NUMBER, BERNOULLI POLYNOMIAL OF THE SECOND KIND

References

Comtet, L. *Advanced Combinatorics: The Art of Finite and Infinite Expansions, rev. enl. ed.* Dordrecht, Netherlands: Reidel, p. 294, 1974.

Jeffreys, H. and Jeffreys, B. S. *Methods of Mathematical Physics, 3rd ed.* Cambridge, England: Cambridge University Press, p. 259, 1988.

Roman, S. *The Umbral Calculus.* New York: Academic Press, p. 114, 1984.

Sloane, N. J. A. Sequences A006232/M5067 and A006233/M1558 in "An On-Line Version of the Encyclopedia of Integer Sequences." http://www.research.att.com/~njas/sequences/eisonline.html.

Bernoulli Polynomial

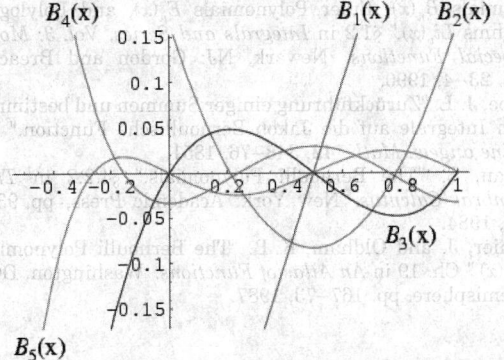

$B_4(x)$ $B_1(x)$ $B_2(x)$

$B_3(x)$

$B_5(x)$

There are two definitions of Bernoulli polynomials in use. The nth Bernoulli polynomial is denoted here by $B_n(x)$ (Abramowitz and Stegun 1972), and the archaic form of the Bernoulli polynomial by $B_n^*(x)$ (or sometimes $\phi_n(x)$). When evaluated at zero, these definitions correspond to the BERNOULLI NUMBERS,

$$B_n \equiv B_n(0) \tag{1}$$

$$B_n^* \equiv B_n^*(0). \tag{2}$$

The Bernoulli polynomials are an APPELL SEQUENCE with

$$g(t) = \frac{e^t - 1}{t} \tag{3}$$

(Roman 1984, p. 31), giving the GENERATING FUNCTION

$$\frac{te^{tx}}{e^t - 1} \equiv \sum_{n=0}^{\infty} B_n(x)\frac{t^n}{n!} \tag{4}$$

(Abramowitz and Stegun 1972, p. 804), first obtained by Euler (1738). The first few Bernoulli polynomials are

$$B_0(x) = 1$$

$$B_1(x) = x - \tfrac{1}{2}$$

$$B_2(x) = x^2 - x + \tfrac{1}{6}$$

$$B_3(x) = x^3 - \tfrac{3}{2}x^2 + \tfrac{1}{2}x$$

$$B_4(x) = x^4 - 2x^3 + x^2 - \tfrac{1}{30}$$

$$B_5(x) = x^5 - \tfrac{5}{2}x^4 + \tfrac{5}{3}x^3 - \tfrac{1}{6}x$$

$$B_6(x) = x^6 - 3x^5 + \tfrac{5}{2}x^4 - \tfrac{1}{2}x^2 + \tfrac{1}{42}.$$

Whittaker and Watson (1990, p. 126) define an older type of "Bernoulli polynomial" by writing

$$t\frac{e^{zt} - 1}{e^t - 1} = \sum_{n=1}^{\infty} \frac{\phi_n(z)t^n}{n!} \tag{5}$$

instead of (5). This gives the polynomials

$$\phi_n(x) = B_n(x) - B_n, \tag{6}$$

where B_n is a BERNOULLI NUMBER, the first few of which are

$$\phi_1(x) = x$$

$$\phi_2(x) = x^2 - x$$

$$\phi_3(x) = x^3 - \tfrac{3}{2}x^2 + \tfrac{1}{2}x$$

$$\phi_4(x) = x^4 - 2x^3 + x^2$$

$$\phi_5(x) = x^5 - \tfrac{5}{2}x^4 + \tfrac{5}{3}x^3 - \tfrac{1}{6}x.$$

The Bernoulli polynomials also satisfy

$$B_n(1) = (-1)^n B_n(0) \tag{7}$$

and

$$B_n(1 - x) = (-1)^n B_n(x) \tag{8}$$

(Lehmer 1988), as well as the relation

$$B_n(x + 1) - B_n(x) = nx^{n-1} \tag{9}$$

(Whittaker and Watson 1990, p. 127). Bernoulli (1713) defined the polynomials in terms of sums of the POWERS of consecutive integers,

$$\sum_{k=0}^{m-1} k^{n-1} = \frac{1}{n}[B_n(m) - B_n(0)]. \tag{10}$$

The Bernoulli polynomials satisfy the RECURRENCE RELATION

$$\frac{dB_n}{dx} = nB_{n-1}(x) \tag{11}$$

(Appell 1882), and obey the identity

$$B_n(x) = (B + x)^n, \tag{12}$$

where B^k is interpreted here as $B_k(x)$. Hurwitz gave the FOURIER SERIES

$$B_n(x) = -\frac{n!}{(2\pi i)^n} \sum_{k=-\infty}^{\infty}{}' k^{-n} e^{2\pi i k x}, \tag{13}$$

for $0 < x < 1$, where the prime in the summation indicates that the term $k = 0$ is omitted. Performing the sum gives

$$B_n(x) = -\frac{n!}{(2\pi i)^n}[(-1)^n \mathrm{Li}_n(e^{-2\pi i x}) + \mathrm{Li}_n(e^{2\pi i x})], \tag{14}$$

where $\mathrm{Li}_n(x)$ is the POLYLOGARITHM function. Raabe (1851) found

$$\frac{1}{m}\sum_{k=0}^{m-1} B_n\left(x + \frac{k}{m}\right) = m^{-n} B_n(mx). \tag{15}$$

A sum identity involving the Bernoulli polynomials is

$$\sum_{k=0}^{m} \binom{m}{k} B_k(\alpha) B_{m-k}(\beta)$$
$$= -(m-1)B_m(\alpha+\beta) + m(\alpha+\beta-1)B_{m-1}(\alpha+\beta) \quad (16)$$

for m an INTEGER. A sum identity due to S. M. Ruiz is

$$\sum_{k=0}^{n} (-1)^{k+n} \binom{n}{k} B_n(k) = n!, \quad (17)$$

where $\binom{n}{k}$ is a BINOMIAL COEFFICIENT. The Bernoulli polynomials are also given by the formula

$$B_n(x) = B_n(0) + \sum_{k=1}^{n} \frac{n}{k} S(n-1, k-1)(x)_k, \quad (18)$$

where $S(n, m)$ is a STIRLING NUMBER OF THE SECOND KIND and $(x)_k$ is a FALLING FACTORIAL (Roman 1984, p. 94). A general identity is given by

$$(n)_m x^{n-m} = \sum_{k=m}^{n} \frac{(n)_k}{(k-m+1)!} B_{n-k}(x), \quad (19)$$

which simplifies to

$$nx^{n-1} = \sum_{k=1}^{n} \binom{n}{k} B_{n-k}(x) \quad (20)$$

(Roman 1984, p. 97). Gosper gave the identity

$$\sum_{j=0}^{i} \frac{[2(i-j)-1]3^{2f}(2^{(2f+1)+1})B_{2(i-j)}B_{2j+1}(\tfrac{1}{3})}{[2(i-j)]!(2j+1)!}$$
$$= \frac{2 \cdot 3^{2(i-1)}(2^{2i-1}+1)B_{2i-1}(\tfrac{1}{3}) - (i-\tfrac{1}{2})B_{2i}}{(2i)!}. \quad (21)$$

Roman (1984, p. 93) defines a generalization $B_n^{(\alpha)}(x)$ of the Bernoulli numbers with an additional free parameter such that $B_n(x) = B_n^{(1)}(x)$.

See also BERNOULLI NUMBER, BERNOULLI POLYNO-MIAL OF THE SECOND KIND, EULER-MACLAURIN INTEGRATION FORMULAS, EULER POLYNOMIAL

References

Abramowitz, M. and Stegun, C. A. (Eds.). "Bernoulli and Euler Polynomials and the Euler-Maclaurin Formula." §23.1 in *Handbook of Mathematical Functions with Formulas, Graphs, and Mathematical Tables, 9th printing.* New York: Dover, pp. 804–06, 1972.

Appell, P. E. "Sur une classe de polynomes." *Annales d'École Normal Superieur, Ser. 2* **9**, 119–44, 1882.

Arfken, G. *Mathematical Methods for Physicists, 3rd ed.* Orlando, FL: Academic Press, p. 330, 1985.

Bernoulli, J. *Ars conjectandi.* Basel, Switzerland, p. 97, 1713. Published posthumously.

Euler, L. "Methodus generalis summandi progressiones." *Comment. Acad. Sci. Petropol.* **6**, 68–7, 1738.

Lehmer, D. H. "A New Approach to Bernoulli Polynomials." *Amer. Math. Monthly.* **95**, 905–11, 1988.

Lucas, E. Ch. 14 in *Théorie des Nombres.* Paris, 1891.

Prudnikov, A. P.; Marichev, O. I.; and Brychkov, Yu. A. "The Generalized Zeta Function $\zeta(s, x)$, Bernoulli Polynomials $B_n(x)$, Euler Polynomials $E_n(x)$, and Polylogarithms $Li_v(x)$." §1.2 in *Integrals and Series, Vol. 3: More Special Functions.* Newark, NJ: Gordon and Breach, pp. 23–4, 1990.

Raabe, J. L. "Zurückführung einiger Summen und bestimmten Integrale auf die Jakob Bernoullische Function." *J. reine angew. Math.* **42**, 348–76, 1851.

Roman, S. "The Bernoulli Polynomials." §4.2.2 in *The Umbral Calculus.* New York: Academic Press, pp. 93–00, 1984.

Spanier, J. and Oldham, K. B. "The Bernoulli Polynomial $B_n(x)$." Ch. 19 in *An Atlas of Functions.* Washington, DC: Hemisphere, pp. 167–73, 1987.

Bernoulli Polynomial of the Second Kind

Polynomials $b_n(x)$ which form a SHEFFER SEQUENCE with

$$g(t) = \frac{t}{e^t - 1} \quad (1)$$

$$f(t) = e^t - 1, \quad (2)$$

giving GENERATING FUNCTION

$$\sum_{k=0}^{\infty} \frac{b_k(x)}{k!} t^k = \frac{t(t+1)^x}{\ln(1+t)}. \quad (3)$$

Roman (1984) defines BERNOULLI NUMBERS OF THE SECOND KIND as $b_n = b_n(0)$. They are related to the STIRLING NUMBERS OF THE FIRST KIND $s(n, m)$ by

$$b_n(x) = b_n(0) + \sum_{k=1}^{n} \frac{n}{k} s(n-1, k-1)x^k \quad (4)$$

(Roman 1984, p. 115), and obey the reflection formula

$$b_n(\tfrac{1}{2}n - 1 - x) = (-1)^n b_n(\tfrac{1}{2}n - 1 + x) \quad (5)$$

(Roman 1984, p. 119).

The first few Bernoulli polynomials of the second kind are

$$b_0(x) = 1$$
$$b_1(x) = \tfrac{1}{2}(2x + 1)$$
$$b_2(x) = \tfrac{1}{6}(6x^2 - 1)$$
$$b_3(x) = \tfrac{1}{4}(4x^3 - 6x^2 + 1)$$
$$b_4(x) = \tfrac{1}{30}(30x^4 - 120x^3 + 120x^2 - 19).$$

See also BERNOULLI NUMBER OF THE SECOND KIND, BERNOULLI POLYNOMIAL, SHEFFER SEQUENCE, STIRLING NUMBER OF THE FIRST KIND

References

Roman, S. "The Bernoulli Polynomials of the Second Kind." §5.3.2 in *The Umbral Calculus.* New York: Academic Press, pp. 113–19, 1984.

Bernoulli Scheme

References

Petersen, K. *Ergodic Theory*. Cambridge, England: Cambridge University Press, 1983.

Bernoulli Trial

An experiment in which s TRIALS are made of an event, with probability p of success in any given TRIAL.

See also BERNOULLI DISTRIBUTION, COIN TOSSING, RUN

References

Papoulis, A. "Bernoulli Trials." §3 – in *Probability, Random Variables, and Stochastic Processes, 2nd ed.* New York: McGraw-Hill, pp. 57–3, 1984.

Bernoulli's Method

In order to find a root of a polynomial equation

$$a_0 x^n + a_1 x^{n-1} + \cdots + a_n = 0, \qquad (1)$$

consider the difference equation

$$a_0 y(t+n) + a_1 y(t+n-1) + \cdots + a_n y(t),$$

which is known to have solution

$$y(t) = w_1 x_1^t + w_2 x_2^t + \cdots + w_n x_n^t + \cdots, \qquad (2)$$

where $w_1, w_2, \ldots$, are arbitrary functions of t with period 1, and $x_1, \ldots, x_n$ are roots of (1). In order to find the absolutely greatest root (1), take any arbitrary values for $y(0), y(1), \ldots, y(n-1)$. By repeated application of (2), calculate in succession the values $y(n), y(n+1), y(n+2), \ldots$ Then the ratio of two successive members of this sequence tends in general to a limit, which is the absolutely greatest root of (1).

See also ROOT

References

Whittaker, E. T. and Robinson, G. "A Method of Daniel Bernoulli." §52 in *The Calculus of Observations: A Treatise on Numerical Mathematics, 4th ed.* New York: Dover, pp. 98–9, 1967.

Bernoulli's Paradox

Suppose the HARMONIC SERIES converges to h:

$$\sum_{k=1}^{\infty} \frac{1}{k} = h.$$

Then rearranging the terms in the sum gives

$$h - 1 = h,$$

which is a contradiction.

See also HARMONIC SERIES

References

Boas, R. P. "Some Remarkable Sequences of Integers." Ch. 3 in *Mathematical Plums* (Ed. R. Honsberger). Washington, DC: Math. Assoc. Amer., pp. 39–0, 1979.

Bernoulli's Theorem

WEAK LAW OF LARGE NUMBERS

BernoulliB

BERNOULLI NUMBER, BERNOULLI POLYNOMIAL

Bernstein Minimal Surface Theorem

If a MINIMAL SURFACE is given by the equation $z = f(x, y)$ and f has CONTINUOUS first and second PARTIAL DERIVATIVES for all REAL x and y, then f is a PLANE.

See also MINIMAL SURFACE

References

Hazewinkel, M. (Managing Ed.). *Encyclopaedia of Mathematics: An Updated and Annotated Translation of the Soviet "Mathematical Encyclopaedia."* Dordrecht, Netherlands: Reidel, p. 369, 1988.
Osserman, R. "Bernstein's Theorem." §5 in *A Survey of Minimal Surfaces*. New York: Dover, pp. 34–2, 1986.

Bernstein Polynomial

The POLYNOMIALS defined by

$$B_{i,\,n}(t) = \binom{n}{i} t^i (1-t)^{n-i},$$

where $\binom{n}{k}$ is a BINOMIAL COEFFICIENT. The Bernstein polynomials of degree n form a basis for the POWER - POLYNOMIALS of degree n.

Another form of Bernstein polynomials is given by

$$B_n(f,\,x) = \sum_{j=0}^{n} \binom{n}{j} x^j (1-x)^{n-j} f\left(\frac{j}{n}\right)$$

(Gzyl and Palacios 1997, Mathé 1999).

See also BÉZIER CURVE

References

Bernstein, S. "Démonstration du théorème de Weierstrass fondée sur le calcul des probabilités." *Comm. Soc. Math. Kharkov* **13**, 1–, 1912.
Feller, W. *An Introduction to Probability Theory and Its Applications, Vol. 2, 3rd ed.* New York: Wiley, p. 222, 1971.
Gzyl, H. and Palacios, J. L. "The Weierstrass Approximation Theorem and Large Deviations." *Amer. Math. Monthly* **104**, 650–53, 1997.
Kac, M. "Une remarque sur les polynomes de M. S. Bernstein." *Studia Math.* **7**, 49–1, 1938.
Kac, M. "Reconnaissance de priorité relative à ma note, 'Une remarque sur les polynomes de M. S. Bernstein.'" *Studia Math.* **8**, 170, 1939.
Lorentz, G. G. *Bernstein Polynomials*. Toronto: University of Toronto Press, 1953.

Mathé, P. "Approximation of Hölder Continuous Functions by Bernstein Polynomials." *Amer. Math. Monthly* **106**, 568–74, 1999.

Widder, D. V. *The Laplace Transform.* Princeton, NJ: Princeton University Press, p. 101, 1941.

Bernstein's Constant

N.B. A detailed online essay by S. Finch was the starting point for this entry.

Let $E_n(f)$ be the error of the best uniform approximation to a REAL function $f(x)$ on the INTERVAL $[-1, 1]$ by REAL POLYNOMIALS of degree at most n. If

$$\alpha(x) = |x|, \tag{1}$$

then Bernstein showed that

$$0.267\ldots < \lim_{n \to \infty} 2nE_{2n}(\alpha) < 0.286. \tag{2}$$

He conjectured that the lower limit (β) was $\beta = 1/(2 \times \sqrt{\pi})$. However, this was disproven by Varga and Carpenter (1987) and Varga (1990), who computed

$$\beta = 0.2801694990\ldots. \tag{3}$$

For rational approximations $p(x)/q(x)$ for p and q of degree m and n, D. J. Newman (1964) proved

$$\tfrac{1}{2}e^{-9\sqrt{n}} \le E_{n,n}(\alpha) \le 3e^{-\sqrt{n}} \tag{4}$$

for $n \ge 4$. Gonchar (1967) and Bulanov (1975) improved the lower bound to

$$e^{-\pi\sqrt{n+1}} \le E_{n,n}(\alpha) \le 3e^{-\sqrt{n}}. \tag{5}$$

Vjacheslavo (1975) proved the existence of POSITIVE constants m and M such that

$$m \le e^{\pi\sqrt{n}}E_{n,n}(\alpha) < M \tag{6}$$

(Petrushev 1987, pp. 105–06). Varga *et al.* (1993) conjectured and Stahl (1993) proved that

$$\lim_{n \to \infty} e^{\pi\sqrt{2n}}E_{2n,2n}(\alpha) = 8. \tag{7}$$

References

Bulanov, A. P. "Asymptotics for the Best Rational Approximation of the Function Sign x." *Mat. Sbornik* **96**, 171–78, 1975.

Finch, S. "Favorite Mathematical Constants." http://www.mathsoft.com/asolve/constant/brnstn/brnstn.html.

Gonchar, A. A. "Estimates for the Growth of Rational Functions and their Applications." *Mat. Sbornik* **72**, 489–03, 1967.

Newman, D. J. "Rational Approximation to $|x|$." *Michigan Math. J.* **11**, 11–4, 1964.

Petrushev, P. P. and Popov, V. A. *Rational Approximation of Real Functions.* New York: Cambridge University Press, 1987.

Stahl, H. "Best Uniform Rational Approximation of $|x|$ on $[-1, 1]$." *Russian Acad. Sci. Sb. Math.* **76**, 461–87, 1993.

Varga, R. S. *Scientific Computations on Mathematical Problems and Conjectures.* Philadelphia, PA: SIAM, 1990.

Varga, R. S. and Carpenter, A. J. "On a Conjecture of S. Bernstein in Approximation Theory." *Math. USSR Sbornik* **57**, 547–60, 1987.

Varga, R. S.; Ruttan, A.; and Carpenter, A. J. "Numerical Results on Best Uniform Rational Approximations to $|x|$ on $[-1, +1]$." *Math. USSR Sbornik* **74**, 271–90, 1993.

Vjacheslavo, N. S. "On the Uniform Approximation of $|x|$ by Rational Functions." *Dokl. Akad. Nauk SSSR* **220**, 512–15, 1975.

Bernstein's Inequality

Let P be a POLYNOMIAL of degree n with derivative P'. Then

$$\|P'\|_\infty \le n\|P\|_\infty,$$

where

$$\|P\|_\infty \equiv \max_{|z|=1} |P(z)|.$$

Bernstein's Polynomial Theorem

If $g(\theta)$ is a trigonometric POLYNOMIAL of degree m satisfying the condition $|g(\theta)| \le 1$ where θ is arbitrary and real, then $g'(\theta) \le m$.

References

Szego, G. *Orthogonal Polynomials, 4th ed.* Providence, RI: Amer. Math. Soc., p. 5, 1975.

Bernstein-Bézier Curve

BÉZIER CURVE

Bernstein-Szego Polynomials

The POLYNOMIALS on the interval $[-1, 1]$ associated with the WEIGHT FUNCTIONS

$$w(x) = (1 - x^2)^{-1/2}$$

$$w(x) = (1 - x^2)^{1/2}$$

$$w(x) = \sqrt{\frac{1-x}{1+x}},$$

also called BERNSTEIN POLYNOMIALS.

References

Szego, G. *Orthogonal Polynomials, 4th ed.* Providence, RI: Amer. Math. Soc., pp. 31–3, 1975.

Berry Conjecture

The longstanding conjecture that the nonimaginary solutions E_n of

$$\zeta(\tfrac{1}{2} + iE_n) = 0,$$

where $\zeta(z)$ is the RIEMANN ZETA FUNCTION, are the EIGENVALUES of an "appropriate" HERMITIAN OPERATOR H. Berry and Keating (1999) further conjecture that this operator is

$$H = xp = -i\left(x\,\frac{d}{dx} + \frac{1}{2} \right),$$

where x and p are the position and conjugate momentum operators, respectively.

See also RIEMANN HYPOTHESIS, RIEMANN ZETA FUNCTION

References

Berry, M. V. and Keating, J. P. "$H = xp$ and the Riemann Zeros." In *Supersymmetry and Trace Formulae: Chaos and Disorder* (Ed. I. V. Lerner, J. P. Keating, and D. E. Khmelnitskii). New York: Kluwer, pp. 355–67, 1999.

Berry Paradox

There are several versions of the Berry paradox, the original version of which was published by Bertrand Russell and attributed to Oxford University librarian Mr. G. Berry. In one form, the paradox notes that the number "one million, one hundred thousand, one hundred and twenty one" can be named by the description: "the first number not nameable in under ten words." However, this latter expression has only nine words, so the number *can* be named in under ten words, so there is an inconsistency in naming it in this manner!

References

Chaitin, G. J. "The Berry Paradox." *Complexity* **1**, 26–0, 1995.
Curry, H. B. *Foundations of Mathematical Logic.* New York: Dover, p. 6, 1977.
Erickson, G. W. and Fossa, J. A. *Dictionary of Paradox.* Lanham, MD: University Press of America, pp. 20–1, 1998.
Whitehead, A. N. and Russell, B. *Principia Mathematica.* New York: Cambridge University Press, p. 60, 1927.

Berry-Esséen Theorem

If $F(x)$ is a probability distribution with zero mean and

$$\rho = \int_{-\infty}^{\infty} |x|^3\, dF(x) < \infty, \tag{1}$$

where the above integral is a STIELTJES INTEGRAL, then for all x and n,

$$|F_n(x) - \Phi(x) - \tfrac{1}{2}| < \frac{33}{4}\,\frac{\rho}{\sigma^3 \sqrt{n}}, \tag{2}$$

where $\Phi(x)$ is the NORMAL DISTRIBUTION FUNCTION, $\Phi(x) + 1/2 = \Re(x)$ in Feller's notation, and

$$F_n(x) = F^{n*}(x\sigma\sqrt{n}) \tag{3}$$

is the normalized n-fold CONVOLUTION of $F(x)$ (Wallace 1958, Feller 1971).

See also CENTRAL LIMIT THEOREM

References

Bergström, H. "On the Central Limit Theorem." *Skand. Aktuarietidskr.* **27**, 139–53, 1944.
Bergström, H. "On the Central Limit Theorem in the Space R_k, $k > 1$." *Skand. Aktuarietidskr.* **28**, 106–27, 1945.
Bergström, H. "On the Central Limit Theorem in the Case of not Equally Distributed Random Variables." *Skand. Aktuarietidskr.* **32**, 37–2, 1949.
Berry, A. C. "The Accuracy of the Gaussian Approximation to the Sum of Independent Variates." *Trans. Amer. Math. Soc.* **49**, 122–36 1941.
Esseen, C. G. "On the Liapounoff Limit of Error in the Theory of Probability." *Ark. Mat. Astr. och Fys.* **28A**, No. 9, 1–9, 1942.
Esseen, C. G. "Fourier Analysis of Distribution Functions." *Acta Math.* **77**, 1–25, 1945.
Esseen, C. G. "A Moment Inequality with an Application to the Central Limit Theorem." *Skand. Aktuarietidskr.* **39**, 160–70, 1956.
Feller, W. "The Berry-Esséen Theorem." §16.5 in *An Introduction to Probability Theory and Its Applications, Vol. 2, 3rd ed.* New York: Wiley, pp. 542–46, 1971.
Hazewinkel, M. (Managing Ed.). *Encyclopaedia of Mathematics: An Updated and Annotated Translation of the Soviet "Mathematical Encyclopaedia."* Dordrecht, Netherlands: Reidel, p. 369, 1988.
Hsu, P. L. "The Approximate Distribution of the Mean and Variance of a Sample of Independent Variables." *Ann. Math. Stat.* **16**, 1–9, 1945.
Wallace, D. L. "Asymptotic Approximations to Distributions." *Ann. Math. Stat.* **29**, 635–54, 1958.

Bertelsen's Number

An erroneous value of $\pi(10^9)$, where $\pi(x)$ is the PRIME COUNTING FUNCTION. Bertelsen's value of $50{,}847{,}478$ is 56 lower than the correct value of $50{,}847{,}534$.

See also PRIME COUNTING FUNCTION

References

Brown, K. S. "Bertelsen's Number." http://www.seanet.com/~ksbrown/kmath049.htm.

Bertini's Theorem

The general curve of a system which is LINEARLY INDEPENDENT on a certain number of given irreducible curves will not have a singular point which is not fixed for all the curves of the system.

References

Coolidge, J. L. *A Treatise on Algebraic Plane Curves.* New York: Dover, p. 115, 1959.

Bertrand Curves

Two curves which, at any point, have a common principal NORMAL VECTOR are called Bertrand curves. The product of the TORSIONS of Bertrand curves is a constant.

Bertrand's Paradox

BERTRAND'S PROBLEM

Bertrand's Postulate

If $n > 3$, there is always at least one PRIME between n and $2n - 2$. Equivalently, if $n > 1$, then there is always at least one PRIME between n and $2n$. The conjecture was first made by Bertrand in 1845 (Nagell 1951, p. 67). It was proved in 1850–1 by Chebyshev, and is therefore sometimes known as CHEBYSHEV'S THEOREM. An extension of this result is that if $n > k$, then there is a number containing a PRIME divisor $> k$ in the sequence $n, n + 1, \ldots, n + k - 1$. (The case $n = k + 1$ then corresponds to Bertrand's postulate.) This was first proved by Sylvester, independently by Schur, and a simple proof was given by Erdos (Hoffman 1998, p. 37).

A related problem is to find the least value of θ so that there exists at least one PRIME between n and $n + \mathbb{O}(n^\theta)$ for sufficiently large n (Berndt 1994). The smallest known value is $\theta = 6/11 + \epsilon$ (Lou and Yao 1992).

See also CHOQUET THEORY, DE POLIGNAC'S CONJECTURE, PRIME NUMBER

References

Berndt, B. C. *Ramanujan's Notebooks, Part IV.* New York: Springer-Verlag, p. 135, 1994.
Erdos, P. "Ramanujan and I." In *Proceedings of the International Ramanujan Centenary Conference held at Anna University, Madras, Dec. 21, 1987.* (Ed. K. Alladi). New York: Springer-Verlag, pp. 1–0, 1989.
Hoffman, P. *The Man Who Loved Only Numbers: The Story of Paul Erdos and the Search for Mathematical Truth.* New York: Hyperion, 1998.
Lou, S. and Yau, Q. "A Chebyshev's Type of Prime Number Theorem in a Short Interval (II)." *Hardy-Ramanujan J.* **15**, 1–3, 1992.
Nagell, T. *Introduction to Number Theory.* New York: Wiley, p. 70, 1951.
Séroul, R. *Programming for Mathematicians.* Berlin: Springer-Verlag, pp. 7–, 2000.

Bertrand's Problem

What is the PROBABILITY that a CHORD drawn at random on a CIRCLE of RADIUS r (i.e., CIRCLE LINE PICKING) has length $\geq r$ (or sometimes greater than or equal to the side length of an inscribed equilateral triangle; Solomon 1978, p. 2)? The answer depends on the interpretation of "two points drawn at random," or more specifically on the "natural" measure for the problem.

In the most commonly considered measure, the ANGLES θ_1 and θ_2 are picked at random on the CIRCUMFERENCE of the circle. Without loss of generality, this can be formulated as the probability that the chord length of a single point at random angle θ measured from the x-AXIS on the unit circle. Since the length as a function of θ (CIRCLE LINE PICKING) is given by

$$s(\theta) = 2 \left| \sin(\tfrac{1}{2}\theta) \right|, \tag{1}$$

solving for $s(\theta) = 1$ gives $\pi/3$, so the fraction of the top unit semicircle having chord length greater than 1 is

$$P = \frac{\pi - \dfrac{\pi}{3}}{\pi} = \frac{2}{3}. \tag{2}$$

However, if a point is instead placed at random on a RADIUS of the CIRCLE and a CHORD drawn PERPENDICULAR to it, then

$$P = \frac{\dfrac{\sqrt{3}}{2}r}{r} = \frac{\sqrt{3}}{2}. \tag{3}$$

The latter interpretation is more satisfactory in the sense that the result remains the same for a rotated CIRCLE, a slightly smaller CIRCLE INSCRIBED in the first, or for a CIRCLE of the same size but with its center slightly offset. Jaynes (1983) shows that the interpretation of "random" as a continuous UNIFORM DISTRIBUTION over the RADIUS is the only one possessing all these three invariances.

See also CHORD, CIRCLE LINE PICKING, GEOMETRIC PROBABILITY

References

Bogomolny, A. "Bertrand's Paradox." http://www.cut-the-knot.com/bertrand.html.
Erickson, G. W. and Fossa, J. A. *Dictionary of Paradox.* Lanham, MD: University Press of America, pp. 21–3, 1998.
Isaac, R. *The Pleasures of Probability.* New York: Springer-Verlag, 1995.
Jaynes, E. T. *Papers on Probability, Statistics, and Statistical Physics.* Dordrecht, Netherlands: Reidel, 1983.
Pickover, C. A. *Keys to Infinity.* New York: Wiley, pp. 42–5, 1995.
Papoulis, A. *Probability, Random Variables, and Stochastic Processes, 2nd ed.* New York: McGraw-Hill, pp. 11–2, 1984.
Solomon, H. *Geometric Probability.* Philadelphia, PA: SIAM, p. 2, 1978.

Bertrand's Test

A CONVERGENCE TEST also called DE MORGAN'S AND BERTRAND'S TEST. If the ratio of terms of a SERIES $\{a_n\}_{n=1}^\infty$ can be written in the form

$$\frac{a_n}{a_{n+1}} = 1 + \frac{1}{n} + \frac{\rho_n}{n \ln n},$$

then the series converges if $\lim_{n \to \infty} \rho_n > 1$ and diverges if $\overline{\lim}_{n \to \infty} \rho_n < 1$, where $\underline{\lim}_{n \to \infty}$ is the LOWER LIMIT and $\overline{\lim}_{n \to \infty}$ is the UPPER LIMIT.

See also KUMMER'S TEST

References

Bromwich, T. J. I'a and MacRobert, T. M. *An Introduction to the Theory of Infinite Series, 3rd ed.* New York: Chelsea, p. 40, 1991.

Bertrand's Theorem

BERTRAND'S POSTULATE

Besov Space

A type of abstract SPACE which occurs in SPLINE and RATIONAL FUNCTION approximations. The Besov space $B_{p,q}^\alpha$ is a complete quasinormed space which is a BANACH SPACE when $1 \le p, q \le \infty$ (Petrushev and Popov 1987).

See also BANACH SPACE

References

Bergh, J. and Löfström, J. *Interpolation Spaces.* New York: Springer-Verlag, 1976.
Peetre, J. *New Thoughts on Besov Spaces.* Durham, NC: Duke University Press, 1976.
Petrushev, P. P. and Popov, V. A. "Besov Spaces." §7.2 in *Rational Approximation of Real Functions.* New York: Cambridge University Press, pp. 201–03, 1987.
Triebel, H. *Interpolation Theory, Function Spaces, Differential Operators.* New York: Wiley, 1998.

Bessel Differential Equation

$$x^2 \frac{d^2 y}{dx^2} + x \frac{dy}{dx} + (x^2 - m^2) y = 0. \tag{1}$$

Equivalently, dividing through by x^2,

$$\frac{d^2 y}{dx^2} + \frac{1}{x} \frac{dy}{dx} + \left(1 - \frac{m^2}{x^2}\right) y = 0, \tag{2}$$

The solutions to this equation define the BESSEL FUNCTIONS. The equation has a regular SINGULARITY at 0 and an irregular SINGULARITY at ∞.

A transformed version of the Bessel differential equation given by Bowman (1958) is

$$x^2 \frac{d^2 y}{dx^2} + (2p+1)x \frac{dy}{dx} + (a^2 x^{2r} + \beta^2) y = 0. \tag{3}$$

The solution is

$$y = x^{-p} \left[C_1 J_{q/r} \left(\frac{\alpha}{r} x^r\right) + C_2 Y_{q/r} \left(\frac{\alpha}{r} x^r\right) \right], \tag{4}$$

where

$$q \equiv \sqrt{p^2 - \beta^2}, \tag{5}$$

$J_n(x)$ and $Y_n(x)$ are the BESSEL FUNCTIONS OF THE FIRST and SECOND KINDS, and C_1 and C_2 are constants. Another form is given by letting $y = x^\alpha J_n(\beta x^\gamma)$, $\eta = yx^{-\alpha}$, and $\xi = \beta x^\gamma$ (Bowman 1958, p. 117), then

$$\frac{d^2 y}{dx^2} - \frac{2\alpha - 1}{x} \frac{dy}{dx} + \left(\beta^2 \gamma^2 x^{2\gamma - 2} + \frac{\alpha^2 - n^2 \gamma^2}{x^2}\right) y = 0. \tag{6}$$

The solution is

$$y = \begin{cases} x^\alpha [A J_n(\beta x^\gamma) + B Y_n(\beta x^\gamma)] & \text{for integer } n \\ A J_n(\beta x^\gamma) + B J_{-n}(\beta x^\gamma) & \text{for noninteger } n. \end{cases} \tag{7}$$

See also AIRY FUNCTIONS, ANGER FUNCTION, BEI, BER, BESSEL FUNCTION, BOURGET'S HYPOTHESIS, CATALAN INTEGRALS, CYLINDRICAL FUNCTION, DINI EXPANSION, HANKEL FUNCTION, HANKEL'S INTEGRAL, HEMISPHERICAL FUNCTION, KAPTEYN SERIES, LIPSCHITZ'S INTEGRAL, LOMMEL DIFFERENTIAL EQUATION, LOMMEL FUNCTION, LOMMEL'S INTEGRALS, NEUMANN SERIES (BESSEL FUNCTION), PARSEVAL'S INTEGRAL, POISSON INTEGRAL, RAMANUJAN'S INTEGRAL, RICCATI DIFFERENTIAL EQUATION, SONINE'S INTEGRAL, STRUVE FUNCTION, WEBER FUNCTIONS, WEBER'S DISCONTINUOUS INTEGRALS

References

Abramowitz, M. and Stegun, C. A. (Eds.). §9.1.1 in *Handbook of Mathematical Functions with Formulas, Graphs, and Mathematical Tables, 9th printing.* New York: Dover, 1972.
Bowman, F. *Introduction to Bessel Functions.* New York: Dover, 1958.
Morse, P. M. and Feshbach, H. *Methods of Theoretical Physics, Part I.* New York: McGraw-Hill, p. 550, 1953.
Zwillinger, D. (Ed.). *CRC Standard Mathematical Tables and Formulae.* Boca Raton, FL: CRC Press, p. 413, 1995.
Zwillinger, D. *Handbook of Differential Equations, 3rd ed.* Boston, MA: Academic Press, p. 121, 1997.

Bessel Function

A function $Z_n(x)$ defined by the RECURRENCE RELATIONS

$$Z_{n+1} + Z_{n-1} = \frac{2n}{x} Z_n$$

and

$$Z_{n+1} - Z_{n-1} = -2 \frac{dZ_n}{dx}.$$

The Bessel functions are more frequently defined as solutions to the DIFFERENTIAL EQUATION

$$x^2 \frac{d^2 y}{dx^2} + x \frac{dy}{dx} + (x^2 - n^2) y = 0.$$

There are two classes of solution, called the BESSEL FUNCTION OF THE FIRST KIND $J_n(x)$ and BESSEL FUNCTION OF THE SECOND KIND $Y_n(x)$. (A BESSEL FUNCTION OF THE THIRD KIND is a special combination of the first and second kinds.) Several related functions are also defined by slightly modifying the defining equations.

See also BESSEL FUNCTION OF THE FIRST KIND, BESSEL FUNCTION OF THE SECOND KIND, BESSEL FUNCTION OF THE THIRD KIND, CYLINDER FUNCTION, HEMICYLINDRICAL FUNCTION, MODIFIED BESSEL FUNCTION OF THE FIRST KIND, MODIFIED BESSEL

FUNCTION OF THE SECOND KIND, SPHERICAL BESSEL FUNCTION OF THE FIRST KIND, SPHERICAL BESSEL FUNCTION OF THE SECOND KIND

References

Abramowitz, M. and Stegun, C. A. (Eds.). "Bessel Functions of Integer Order," "Bessel Functions of Fractional Order," and "Integrals of Bessel Functions." Chs. 9–1 in *Handbook of Mathematical Functions with Formulas, Graphs, and Mathematical Tables, 9th printing.* New York: Dover, pp. 355–89, 435–56, and 480–91, 1972.

Adamchik, V. "The Evaluation of Integrals of Bessel Functions via *G*-Function Identities." *J. Comput. Appl. Math.* **64**, 283–90, 1995.

Arfken, G. "Bessel Functions." Ch. 11 in *Mathematical Methods for Physicists, 3rd ed.* Orlando, FL: Academic Press, pp. 573–36, 1985.

Bickley, W. G. *Bessel Functions and Formulae.* Cambridge, England: Cambridge University Press, 1957.

Bowman, F. *Introduction to Bessel Functions.* New York: Dover, 1958.

Byerly, W. E. "Cylindrical Harmonics (Bessel's Functions)." Ch. 7 in *An Elementary Treatise on Fourier's Series, and Spherical, Cylindrical, and Ellipsoidal Harmonics, with Applications to Problems in Mathematical Physics.* New York: Dover, pp. 219–37, 1959.

Gray, A. and Mathews, G. B. *A Treatise on Bessel Functions and Their Applications to Physics, 2nd ed.* New York: Dover, 1966.

Luke, Y. L. *Integrals of Bessel Functions.* New York: McGraw-Hill, 1962.

McLachlan, N. W. *Bessel Functions for Engineers, 2nd ed. with corrections.* Oxford, England: Clarendon Press, 1961.

Press, W. H.; Flannery, B. P.; Teukolsky, S. A.; and Vetterling, W. T. "Bessel Functions of Integral Order" and "Bessel Functions of Fractional Order, Airy Functions, Spherical Bessel Functions." §6.5 and 6.7 in *Numerical Recipes in FORTRAN: The Art of Scientific Computing, 2nd ed.* Cambridge, England: Cambridge University Press, pp. 223–29 and 234–45, 1992.

Watson, G. N. *A Treatise on the Theory of Bessel Functions, 2nd ed.* Cambridge, England: Cambridge University Press, 1966.

Weisstein, E. W. "Books about Bessel Functions." http://www.treasure-troves.com/books/BesselFunctions.html.

Bessel Function Fourier Expansion

Let $n \geq 1/2$ and α_1, α_2, ... be the POSITIVE ROOTS of $J_n(x) = 0$. An expansion of a function in the interval $(0, 1)$ in terms of BESSEL FUNCTIONS OF THE FIRST KIND

$$f(x) = \sum_{l=1}^{\infty} A_r J_n(x\alpha_r), \tag{1}$$

has COEFFICIENTS found as follows:

$$\int_0^1 xf(x)J_n(xa_l)\,dx = \sum_{r=1}^{\infty} A_r \int_0^1 xJ_n(x\alpha_r)J_n(x\alpha_l)\,dx. \tag{2}$$

But ORTHOGONALITY of BESSEL FUNCTION ROOTS gives

$$\int_0^1 xJ_n(x\alpha_l)J_n(x\alpha_r)\,dx = \tfrac{1}{2}\delta_{l,r}J_{n+1}^2(\alpha_r) \tag{3}$$

(Bowman 1958, p. 108), so

$$\int_0^1 xf(x)J_n(xa_l)\,dx = \tfrac{1}{2}\sum_{r=1}^{\infty} A_r\delta_{l,r}J_{n+1}^2(x\alpha_r)$$

$$= \tfrac{1}{2}A_l J_{n+1}^2(\alpha_l), \tag{4}$$

and the COEFFICIENTS are given by

$$A_l = \frac{2}{J_{n+1}^2(\alpha_l)}\int_0^1 xf(x)J_n(x\alpha_l)\,dx. \tag{5}$$

References

Bowman, F. *Introduction to Bessel Functions.* New York: Dover, 1958.

Bessel Function of the First Kind

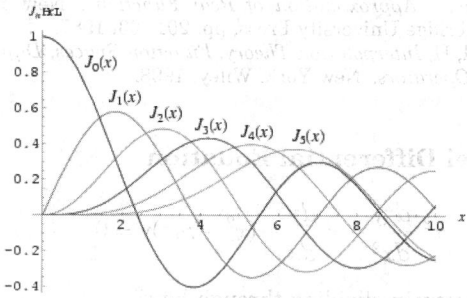

The Bessel functions of the first kind $J_n(x)$ are defined as the solutions to the BESSEL DIFFERENTIAL EQUATION

$$x^2\frac{d^2y}{dx^2} + x\frac{dy}{dx} + (x^2 - m^2)y = 0 \tag{1}$$

which are nonsingular at the origin. They are sometimes also called CYLINDER FUNCTIONS or CYLINDRICAL HARMONICS. The above plot shows $J_n(x)$ for $n = 1$, 2, ..., 5.

To solve the differential equation, apply FROBENIUS METHOD using a series solution OF THE FORM

$$y = x^k\sum_{n=0}^{\infty} a_n x^n = \sum_{n=0}^{\infty} a_n x^{n+k}. \tag{2}$$

Plugging into (1) yields

$$x^2\sum_{n=0}^{\infty}(k+n)(k+n-1)a_n x^{k+n-2}$$

$$+ x\sum_{n=0}^{\infty}(k+n)a_n x^{k+n-1}$$

$$+ x^2\sum_{n=0}^{\infty}a_n x^{k+n} - m^2\sum_{n=0}^{\infty}a_n x^{n+k} = 0 \tag{3}$$

$$\sum_{n=0}^{\infty}(k+n)(k+n-1)a_n x^{k+n} + \sum_{n=0}^{\infty}(k+n)a_n x^{k+n}$$

$$+\sum_{n=2}^{\infty} a_{n-2}x^{k+n} - m^2 \sum_{n=0}^{\infty} a_n x^{n+k} = 0. \qquad (4)$$

The INDICIAL EQUATION, obtained by setting $n=0$, is

$$a_0[k(k-1)+k-m^2] = a_0(k^2-m^2) = 0. \qquad (5)$$

Since a_0 is defined as the first NONZERO term, $k^2 - m^2 = 0$, so $k = \pm m$. Now, if $k = m$,

$$\sum_{n=0}^{\infty}[(m+n)(m+n-1)+(m+n)-m^2]$$

$$\times a_n x^{m+n} + \sum_{n=2}^{\infty} a_{n-2}x^{m+n} = 0 \qquad (6)$$

$$\sum_{n=0}^{\infty}[(m+n)^2 - m^2]a_n x^{m+n} + \sum_{n=2}^{\infty} a_{n-2}x^{m+n} = 0 \qquad (7)$$

$$\sum_{n=0}^{\infty} n(2m+n)a_n x^{m+n} + \sum_{n=2}^{\infty} a_{n-2}x^{m+n} = 0 \qquad (8)$$

$$a_1(2m+1) + \sum_{n=2}^{\infty}[a_n n(2m+n)+a_{n-2}]x^{m+n} = 0. \qquad (9)$$

First, look at the special case $m = -1/2$, then (9) becomes

$$\sum_{n=2}^{\infty}[a_n n(n-1)+a_{n-2}]x^{m+n} = 0, \qquad (10)$$

so

$$a_n = -\frac{1}{n(n-1)}a_{n-2}. \qquad (11)$$

Now let $n \equiv 2l$, where $l = 1, 2, \ldots$

$$a_{2l} = -\frac{1}{2l(2l-1)}a_{2l-2}$$

$$= \frac{(-1)^l}{[2l(2l-1)][2(l-1)(2l-3)]\cdots[2\cdot 1\cdot 1]}a_0$$

$$= \frac{(-1)^l}{2^l l!(2l-1)!!}a_0, \qquad (12)$$

which, using the identity $2^l l!(2l-1)!! = (2l)!$, gives

$$a_{2l} = \frac{(-1)^l}{(2l)!}a_0. \qquad (13)$$

Similarly, letting $n \equiv 2l+1$,

$$a_{2l+1} = -\frac{1}{(2l+1)(2l)}a_{2l-1}$$

$$= \frac{(-1)^l}{[2l(2l+1)][2(l-1)(2l-1)]\cdots[2\cdot 1\cdot 3][1]}a_1, \qquad (14)$$

which, using the identity $2^l l!(2l+1)!! = (2l+1)!$, gives

$$a_{2l+1} = \frac{(-1)^l}{2^l l!(2l+1)!!}a_1 = \frac{(-1)^l}{(2l+1)!}a_1. \qquad (15)$$

Plugging back into (2) with $k = m = -1/2$ gives

$$y = x^{-1/2}\sum_{n=0}^{\infty} a_n x^n$$

$$= x^{-1/2}\left[\sum_{n=1,3,5,\ldots}^{\infty} a_n x^n + \sum_{n=0,2,4,\ldots}^{\infty} a_n x^n\right]$$

$$= x^{-1/2}\left[\sum_{l=0}^{\infty} a_{2l}x^{2l} + \sum_{l=0}^{\infty} a_{2l+1}x^{2l+1}\right]$$

$$= x^{-1/2}\left[a_0\sum_{l=0}^{\infty}\frac{(-1)^l}{(2l)!}x^{2l} + a_1\sum_{l=0}^{\infty}\frac{(-1)^l}{(2l+1)!}x^{2l+1}\right]$$

$$= x^{-1/2}(a_0\cos x + a_1\sin x). \qquad (16)$$

The BESSEL FUNCTIONS of order $\pm 1/2$ are therefore defined as

$$J_{-1/2}(x) \equiv \sqrt{\frac{2}{\pi x}}\cos x \qquad (17)$$

$$J_{1/2}(x) \equiv \sqrt{\frac{2}{\pi x}}\sin x, \qquad (18)$$

so the general solution for $m = \pm 1/2$ is

$$y = a_0' J_{-1/2}(x) + a_1' J_{1/2}(x). \qquad (19)$$

Now, consider a general $m \neq -1/2$. Equation (9) requires

$$a_1(2m+1) = 0 \qquad (20)$$

$$[a_n n(2m+n)+a_{n-2}]x^{m+n} = 0 \qquad (21)$$

for $n = 2, 3, \ldots$, so

$$a_1 = 0 \qquad (22)$$

$$a_n = -\frac{1}{n(2m+n)}a_{n-2} \qquad (23)$$

for $n = 2, 3, \ldots$ Let $n \equiv 2l+1$, where $l = 1, 2, \ldots$, then

$$a_{2l+1} = -\frac{1}{(2l+1)[2(m+1)+1]}a_{2l-1} = \cdots$$

$$= \cdots = f(n, m)a_1 = 0, \qquad (24)$$

where $f(n, m)$ is the function of l and m obtained by iterating the recursion relationship down to a_1. Now let $n \equiv 2l$, where $l = 1, 2, \ldots$, so

$$a_{2l} = -\frac{1}{2l(2m+2l)}a_{2l-2} = -\frac{1}{4l(m+l)}a_{2l-2}$$

$$= \frac{(-1)^l}{[4l(m+l)][4(l-1)(m+l-1)]\cdots[4\cdot(m+1)]}a_0. \qquad (25)$$

Plugging back into (9),

$$y = \sum_{n=0}^{\infty} a_n x^{n+m} = \sum_{n=1,\,3,\,5,\,\dots}^{\infty} a_n x^{n+m} + \sum_{n=0,\,2,\,4,\,\dots}^{\infty} a_n x^{n+m}$$

$$= \sum_{l=0}^{\infty} a_{2l+1} x^{2l+m+1} + \sum_{l=0}^{\infty} a_{2l} x^{2l+m}$$

$$= a_0 \sum_{l=0}^{\infty} \frac{(-1)^l}{[4l(m+l)][4(l-1)(m+l-1)]\cdots[4(m+1)]} x^{2l+m}$$

$$= a_0 \sum_{l=0}^{\infty} \frac{[(-1)^l m(m-1)\cdots 1]x^{2l+m}}{[4l(m+l)][4(l-1)(m+l-1)]\cdots[4(m+1)m(m-1)\cdots 1]}$$

$$= a_0 \sum_{l=0}^{\infty} \frac{(-1)^l m!}{2^{2l} l!(m+l)!} x^{2l+m}, \qquad (26)$$

Now define

$$J_m(x) \equiv \sum_{l=0}^{\infty} \frac{(-1)^l}{2^{2l+m} l!(m+l)!} x^{2l+m}, \qquad (27)$$

where the factorials can be generalized to GAMMA FUNCTIONS for nonintegral m. The above equation then becomes

$$y = a_0 2^m m! J_m(x) = a_0' J_m(x). \qquad (28)$$

Returning to equation (5) and examining the case $k = -m$,

$$a_1(1-2m) + \sum_{n=2}^{\infty} [a_n n(n-2m) + a_{n-2}]x^{n-m} = 0. \qquad (29)$$

However, the sign of m is arbitrary, so the solutions must be the same for $+m$ and $-m$. We are therefore free to replace $-m$ with $-|m|$, so

$$a_1(1+2|m|) + \sum_{n=2}^{\infty} [a_n n(n+2|m|) + a_{n-2}]x^{|m|+n} = 0, \qquad (30)$$

and we obtain the same solutions as before, but with m replaced by $|m|$.

$$J_m(x) = \begin{cases} \displaystyle\sum_{l=0}^{\infty} \frac{(-1)^l}{2^{2l+|m|} l!(|m|+l)l} x^{2l+|m|} & \text{for } |m| \neq -\tfrac{1}{2} \\[2ex] \sqrt{\dfrac{2}{\pi x}} \cos x & \text{for } m = -\tfrac{1}{2} \\[2ex] \sqrt{\dfrac{2}{\pi x}} \sin x & \text{for } m = \tfrac{1}{2}. \end{cases} \qquad (31)$$

We can relate J_m and J_{-m} (when m is an INTEGER) by writing

$$J_{-m}(x) = \sum_{l=0}^{\infty} \frac{(-1)^l}{2^{2l-m} l!(l-m)!} x^{2l-m}. \qquad (32)$$

Now let $l \equiv l' + m$. Then

$$J_{-m}(x) = \sum_{l'+m=0}^{\infty} \frac{(-1)^{l'+m}}{2^{2l'+m}(l'+m)!l!} x^{2l'+m}$$

$$= \sum_{l'=-m}^{-1} \frac{(-1)^{l'+m}}{2^{2l'+m} l'!(l'+m)!} x^{2l'+m}$$

$$+ \sum_{l'=0}^{\infty} \frac{(-1)^{l'+m}}{2^{2l'+m} l'!(l'+m)!} x^{2l'+m}. \qquad (33)$$

But $l'! = \infty$ for $l' = -m, \dots, -1$, so the DENOMINATOR is infinite and the terms on the right are zero. We therefore have

$$J_{-m}(x) = \sum_{l=0}^{\infty} \frac{(-1)^{l+m}}{2^{2l+m} l!(l+m)!} x^{2l+m} = (-1)^m J_m(x). \qquad (34)$$

Note that the BESSEL DIFFERENTIAL EQUATION is second-order, so there must be two linearly independent solutions. We have found both only for $|m|=1/2$. For a general nonintegral order, the independent solutions are J_m and J_{-m}. When m is an INTEGER, the general (real) solution is OF THE FORM

$$Z_m \equiv C_1 J_m(x) + C_2 Y_m(x), \qquad (35)$$

where J_m is a Bessel function of the first kind, Y_m (a.k.a. N_m) is the BESSEL FUNCTION OF THE SECOND KIND (a.k.a. NEUMANN FUNCTION or WEBER FUNCTION), and C_1 and C_2 are constants. Complex solutions are given by the HANKEL FUNCTIONS (a.k.a. BESSEL FUNCTIONS OF THE THIRD KIND).

The Bessel functions are ORTHOGONAL in [0, 1] with respect to the weight factor x. Except when $2n$ is a NEGATIVE INTEGER,

$$J_m(z) = \frac{z^{-1/2}}{2^{2m+1/2} i^{m+1/2} \Gamma(m+1)} M_{0,\,m}(2iz), \qquad (36)$$

where $\Gamma(x)$ is the GAMMA FUNCTION and $M_{0,\,m}$ is a WHITTAKER FUNCTION. In terms of a CONFLUENT HYPERGEOMETRIC FUNCTION OF THE FIRST KIND, the Bessel function is written

$$J_\nu(z) = \frac{(\tfrac{1}{2}z)^\nu}{\Gamma(\nu+1)} \, {}_0F_1(\nu+1; \, -\tfrac{1}{4}z^2). \qquad (37)$$

A derivative identity for expressing higher order Bessel functions in terms of $J_0(x)$ is

$$J_n(x) = i^n T_n\!\left(i\frac{d}{dx}\right) J_0(x), \qquad (38)$$

where $T_n(x)$ is a CHEBYSHEV POLYNOMIAL OF THE FIRST KIND. Asymptotic forms for the Bessel functions are

$$J_m(x) \approx \frac{1}{\Gamma(m+1)} \left(\frac{x}{2}\right)^m \qquad (39)$$

for $x \ll 1$ and

$$J_m(x) \approx \sqrt{\frac{2}{\pi x}} \cos\left(x - \frac{m\pi}{2} - \frac{\pi}{4}\right) \qquad (40)$$

for $x \gg 1$.

A derivative identity is

$$\frac{d}{dx}[x^m J_m(x)] = x^m J_{m-1}(x). \qquad (41)$$

An integral identity is

$$\int_0^u u' J_0(u')\, du' = u J_1(u). \qquad (42)$$

Some sum identities are

$$1 = [J_0(x)]^2 + 2 \sum_{k=1}^{\infty} [J_k(x)]^2 \qquad (43)$$

(Abramowitz and Stegun 1972, p. 363),

$$1 = J_0(x) + 2 \sum_{k=1}^{\infty} J_{2k}(x) \qquad (44)$$

(Abramowitz and Stegun 1972, p. 361),

$$0 = \sum_{k=0}^{2n} (-1)^k J_k(z) J_{2n-k}(z) + 2 \sum_{k=1}^{\infty} J_k(z) J_{2n+k}(z) \qquad (45)$$

for $n \geq 1$ (Abramowitz and Stegun 1972, p. 361),

$$J_n(2z) = \sum_{k=0}^{n} J_k(z) J_{n-k}(z)$$

$$+ 2 \sum_{k=1}^{\infty} (-1)^k J_k(z) J_{n+k}(z) \qquad (46)$$

(Abramowitz and Stegun 1972, p. 361), and the JACOBI-ANGER EXPANSION

$$e^{iz \cos \theta} = \sum_{n=-\infty}^{\infty} i^n J_n(z) e^{in\theta}, \qquad (47)$$

which can also be written

$$e^{iz \cos \theta} = J_0(z) + 2 \sum_{n=1}^{\infty} i^n J_n(z) \cos(n\theta). \qquad (48)$$

The Bessel function addition theorem states

$$J_n(y+z) = \sum_{m=-\infty}^{\infty} J_m(y) J_{n-m}(z). \qquad (49)$$

The first k roots $x_1, \ldots, x_k$ of the Bessel function $J_n(x)$ can be found in *Mathematica* (Wolfram Research, Urbana, IL) using the command `BesselJZeros[n, k]` in the *Mathematica* add-on package `Numerical-Math`BesselZeros`` (which can be loaded with the command `<<NumericalMath`). ROOTS of the FUNCTION $J_n(x)$ are given in the following table.

zero	$J_0(x)$	$J_1(x)$	$J_2(x)$	$J_3(x)$	$J_4(x)$	$J_5(x)$
1	2.4048	3.8317	5.1336	6.3802	7.5883	8.7715
2	5.5201	7.0156	8.4172	9.7610	11.0647	12.3386
3	8.6537	10.1735	11.6198	13.0152	14.3725	15.7002
4	11.7915	13.3237	14.7960	16.2235	17.6160	18.9801
5	14.9309	16.4706	17.9598	19.4094	20.8269	22.2178

The first k roots $x_1, \ldots, x_k$ of the derivative of the Bessel function $J_n'(x)$ can be found in *Mathematica* using the command `BesselJPrimeZeros[n, k]` in the *Mathematica* add-on package `NumericalMath`-`BesselZeros`` (which can be loaded with the command $<<$ `NumericalMath`). The first few such ROOTS are given in the following table.

zero	$J_0'(x)$	$J_1'(x)$	$J_2'(x)$	$J_3'(x)$	$J_4'(x)$	$J_5'(x)$
1	3.8317	1.8412	3.0542	4.2012	5.3175	6.4156
2	7.0156	5.3314	6.7061	8.0152	9.2824	10.5199
3	10.1735	8.5363	9.9695	11.3459	12.6819	13.9872
4	13.3237	11.7060	13.1704	14.5858	15.9641	17.3128
5	16.4706	14.8636	16.3475	17.7887	19.1960	20.5755

Various integrals can be expressed in terms of Bessel functions

$$J_n(z) = \frac{1}{\pi} \int_0^{\pi} \cos(z \sin \theta - n\theta)\, d\theta, \qquad (50)$$

which is BESSEL'S FIRST INTEGRAL,

$$J_n(z) = \frac{i^{-n}}{\pi} \int_0^{\pi} e^{iz \cos \theta} \cos(n\theta)\, d\theta \qquad (51)$$

$$J_n(z) = \frac{1}{2\pi i^n} \int_0^{2\pi} e^{iz \cos \phi} e^{in\phi}\, d\phi \qquad (52)$$

for $n = 1, 2, \ldots$,

$$J_n(z) = \frac{2}{\pi} \frac{x^n}{(2rn-1)!!} \int_0^{\pi/2} \sin^{2n} u \cos(x \cos u)\, du \qquad (53)$$

for $n = 1, 2, \ldots$,

$$J_n(x) = \frac{1}{2\pi i} \int_\gamma e^{(x/2)(z-1/z)} z^{-n-1}\, dz \qquad (54)$$

for $n > -1/2$. The Bessel functions are normalized so that

$$\int_0^{\infty} J_n(x)\, dx = 1 \qquad (55)$$

for positive integral (and real) n. Integrals involving $J_1(x)$ include

$$\int_0^\infty \left[\frac{J_1(x)}{x}\right]^2 dx = \frac{4}{3\pi} \tag{56}$$

$$\int_0^\infty \left[\frac{J_1(x)}{x}\right]^2 x \, dx = \frac{1}{2}. \tag{57}$$

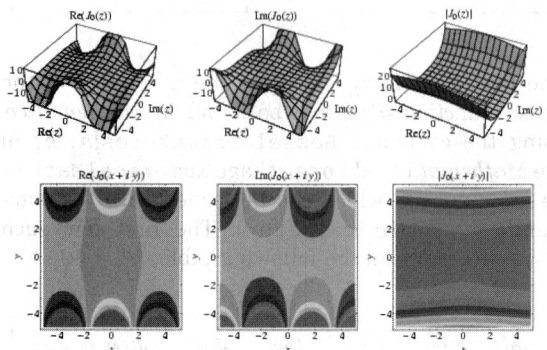

The special case of $n = 0$ gives $J_0(z)$ as the series

$$J_0(z) = \sum_{k=0}^\infty (-1)^k \frac{\left(\frac{1}{4}z^2\right)^k}{(k!)^2} \tag{58}$$

(Abramowitz and Stegun 1972, p. 360), or the integral

$$J_0(z) = \frac{1}{\pi} \int_0^\pi e^{iz\cos\theta} \, d\theta. \tag{59}$$

See also Bessel Function of the Second Kind, Debye's Asymptotic Representation, Dixon-Ferrar Formula, Hansen-Bessel Formula, Kapteyn Series, Kneser-Sommerfeld Formula, Mehler's Bessel Function Formula, Nicholson's Formula, Poisson's Bessel Function Formula, Rayleigh Function, Schläfli's Formula, Schlömilch's Series, Sonine-Schafheitlin Formula, Watson's Formula, Watson-Nicholson Formula, Weber's Discontinuous Integrals, Weber's Formula, Weber-Sonine Formula, Weyrich's Formula

References

Abramowitz, M. and Stegun, C. A. (Eds.). "Bessel Functions J and Y." §9.1 in *Handbook of Mathematical Functions with Formulas, Graphs, and Mathematical Tables, 9th printing.* New York: Dover, pp. 358–64, 1972.

Arfken, G. "Bessel Functions of the First Kind, $J_\nu(x)$" and "Orthogonality." §11.1 and 11.2 in *Mathematical Methods for Physicists, 3rd ed.* Orlando, FL: Academic Press, pp. 573–91 and 591–96, 1985.

Lehmer, D. H. "Arithmetical Periodicities of Bessel Functions." *Ann. Math.* **33**, 143–50, 1932.

Le Lionnais, F. *Les nombres remarquables.* Paris: Hermann, 1983.

Morse, P. M. and Feshbach, H. *Methods of Theoretical Physics, Part I.* New York: McGraw-Hill, pp. 619–22, 1953.

Spanier, J. and Oldham, K. B. "The Bessel Coefficients $J_0(x)$ and $J_1(x)$" and "The Bessel Function $J_\nu(x)$." Chs. 52–3 in *An Atlas of Functions.* Washington, DC: Hemisphere, pp. 509–20 and 521–32, 1987.

Watson, G. N. *A Treatise on the Theory of Bessel Functions, 2nd ed.* Cambridge, England: Cambridge University Press, 1966.

Bessel Function of the Second Kind

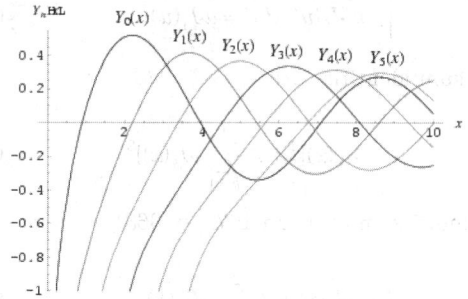

A Bessel function of the second kind $Y_n(x)$ is a solution to the Bessel differential equation which is singular at the origin. Bessel functions of the second kind are also called Neumann functions or Weber functions. The above plot shows $Y_n(x)$ for $n = 1, 2, \ldots, 5$.

Let $v \equiv J_m(x)$ be the first solution and u be the other one (since the Bessel differential equation is second-order, there are two linearly independent solutions). Then

$$xu'' + u' + xu = 0 \tag{1}$$

$$xv'' + v' + xv = 0. \tag{2}$$

Take $v \times$ (1) minus $u \times$ (2),

$$x(u''v - uv'') + u'v - uv' = 0 \tag{3}$$

$$\frac{d}{dx}[x(u'v - uv')] = 0, \tag{4}$$

so $x(u'v - uv') = B$, where B is a constant. Divide by xv^2,

$$\frac{u'v - uv'}{v^2} = \frac{d}{dx}\left(\frac{u}{v}\right) = \frac{B}{xv^2} \tag{5}$$

$$\frac{u}{v} = A + B \int \frac{dx}{xv^2}. \tag{6}$$

Rearranging and using $v \equiv J_m(x)$ gives

$$u = AJ_m(x) + BJ_m(x) \int \frac{dx}{xJ_m^2(x)}$$

$$\equiv A'J_m(x) + B'Y_m(x), \tag{7}$$

where Y_m is the so-called Bessel function of the second kind.

$Y_\nu(z)$ can be defined by

$$Y_\nu(z) = \frac{J_v(z)\cos(\nu\pi) - J_{-\nu}(z)}{\sin(\nu\pi)} \qquad (8)$$

(Abramowitz and Stegun 1972, p. 358), where $J_\nu(z)$ is a BESSEL FUNCTION OF THE FIRST KIND and, for ν an integer n by the SERIES

$$Y_n(z) = -\frac{(\tfrac{1}{2}z)^{-n}}{\pi} \sum_{k=0}^{n-1} \frac{(n-k-1)!}{k!}(\tfrac{1}{4}z^2)^k + \frac{2}{\pi}\ln(\tfrac{1}{2}z)J_n(z)$$

$$-\frac{(\tfrac{1}{2}z)^n}{\pi} \sum_{k=0}^{\infty} [\psi_0(k+1) + \psi_0(n+k+1)]\frac{(-\tfrac{1}{4}z^2)^k}{k!(n+k)!}, \qquad (9)$$

where $\psi_0(x)$ is the DIGAMMA FUNCTION (Abramowitz and Stegun 1972, p. 360).

The function has the integral representations

$$Y_\nu(z) = \frac{1}{\pi}1\nu t_0^\pi \sin(z\sin\theta - \nu\theta)\,d\theta$$

$$-\frac{1}{\pi}1\nu t_0^\infty [e^{\nu t} + e^{-\nu t}(-1)^\nu]e^{-z\sin ht}\,dt. \qquad (10)$$

$$= -\frac{2(\tfrac{1}{2}x)^{-v}}{\sqrt{\pi}\Gamma(\tfrac{1}{2}-v)}\int_1^\infty \frac{\cos(xt)\,dt}{(t^2-1)^{v+1/2}} \qquad (11)$$

(Abramowitz and Stegun 1972, p. 360).

ASYMPTOTIC SERIES are

$$Y_m(x) \sim \begin{cases} \dfrac{2}{\pi}[\ln(\tfrac{1}{2}x) + \gamma] & m = 0,\ x \ll 1 \\[2mm] -\dfrac{\Gamma(m)}{\pi}\left(\dfrac{2}{x}\right)^m & m \neq 0,\ x \ll 1 \end{cases} \qquad (12)$$

$$Y_m(x) \sim \sqrt{\frac{2}{\pi x}}\sin\left(x - \frac{m\pi}{2} - \frac{\pi}{4}\right) \quad x \gg 1, \qquad (13)$$

where $\Gamma(z)$ is a GAMMA FUNCTION.

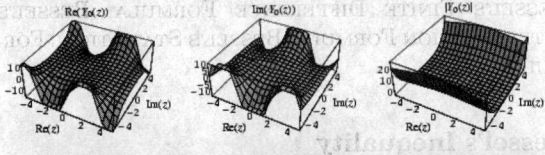

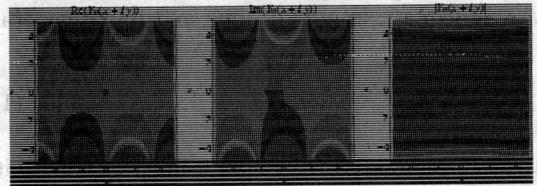

For the special case $n = 0$, $Y_0(x)$ is given by the series

$$Y_0(z)$$

$$= \frac{2}{\pi}\left\{[\ln(\tfrac{1}{2}z) + \gamma]J_0(z) + \sum_{k=1}^{\infty}(-1)^{k+1}H_k\frac{(\tfrac{1}{4}z^2)^k}{(k!)^2}\right\}, \qquad (14)$$

(Abramowitz and Stegun 1972, p. 360), where γ is the EULER-MASCHERONI CONSTANT and H_n is a HARMONIC NUMBER.

See also BESSEL FUNCTION OF THE FIRST KIND, BOURGET'S HYPOTHESIS, HANKEL FUNCTION

References

Abramowitz, M. and Stegun, C. A. (Eds.). "Bessel Functions J and Y." §9.1 in *Handbook of Mathematical Functions with Formulas, Graphs, and Mathematical Tables, 9th printing.* New York: Dover, pp. 358–64, 1972.

Arfken, G. "Neumann Functions, Bessel Functions of the Second Kind, $N_\nu(x)$." §11.3 in *Mathematical Methods for Physicists, 3rd ed.* Orlando, FL: Academic Press, pp. 596–04, 1985.

Morse, P. M. and Feshbach, H. *Methods of Theoretical Physics, Part I.* New York: McGraw-Hill, pp. 625–27, 1953.

Spanier, J. and Oldham, K. B. "The Neumann Function $Y_\nu(x)$." Ch. 54 in *An Atlas of Functions.* Washington, DC: Hemisphere, pp. 533–42, 1987.

Watson, G. N. *A Treatise on the Theory of Bessel Functions, 2nd ed.* Cambridge, England: Cambridge University Press, 1966.

Bessel Function of the Third Kind

HANKEL FUNCTION

Bessel Polynomial

Krall and Find (1948) defined the Bessel polynomials as the function

$$y_n(x) = \sum_{k=0}^{n} \frac{(n+k)!}{(n-k)!k!}\left(\frac{x}{2}\right)^k \qquad (1)$$

which satisfies the differential equation

$$x^2 y'' + (2x+2)y' + n(n+1)y = 0. \qquad (2)$$

Carlitz (1957) subsequently considered the related polynomials

$$p_n(x) = x^n y_{n-1}\left(\frac{1}{x}\right).$$

This polynomial forms an associated SHEFFER SEQUENCE with

$$f(t) = t - \tfrac{1}{2}t^2. \qquad (3)$$

This gives the GENERATING FUNCTION

$$\sum_{k=0}^{\infty} \frac{p_k(x)}{k!}t^k = e^{x(1-\sqrt{1-2t})}. \qquad (4)$$

The explicit formula is

$$p_n(x) = \sum_{k=1}^{\infty} \frac{(2n-k-1)!}{2^{n-k}(k-1)!(n-k)!}x^k. \qquad (5)$$

The polynomials satisfy the recurrence formula

$$p_n''(x) - 2p_n'(x) + 2np_{n-1}(x) = 0. \tag{6}$$

The first few polynomials are

$$\begin{aligned}
p_0(x) &= 1 \\
p_1(x) &= x \\
p_2(x) &= x^2 + x \\
p_3(x) &= x^3 + 3x^2 + 3x \\
p_4(x) &= x^4 + 6x^3 + 15x^2 + 15x.
\end{aligned}$$

See also BESSEL FUNCTION, SHEFFER SEQUENCE

References

Carlitz, L. "A Note on the Bessel Polynomials." *Duke Math. J.* **24**, 151–62, 1957.

Grosswald, E. *Bessel Polynomials.* New York: Springer-Verlag, 1978.

Krall, H. L. and Fink, O. "A New Class of Orthogonal Polynomials: The Bessel Polynomials." *Trans. Amer. Math. Soc.* **65**, 100–15, 1948.

Roman, S. "The Bessel Polynomials." §4.1.7 in *The Umbral Calculus.* New York: Academic Press, pp. 78–2, 1984.

Bessel Transform

HANKEL TRANSFORM

Bessel's Correction

The factor $(N-1)/N$ in the relationship between the VARIANCE σ and the EXPECTATION VALUES of the SAMPLE VARIANCE,

$$\langle s^2 \rangle = \frac{N-1}{N}\, \sigma^2, \tag{1}$$

where

$$s^2 \equiv \langle x^2 \rangle - \langle x \rangle^2. \tag{2}$$

For two samples,

$$\hat{\sigma}^2 = \frac{N_1 s_1^2 + N_2 s_2^2}{N_1 + N_2 - 2}. \tag{3}$$

See also SAMPLE VARIANCE, VARIANCE

References

Kenney, J. F. and Keeping, E. S. *Mathematics of Statistics, Pt. 2, 2nd ed.* Princeton, NJ: Van Nostrand, p. 161, 1951.

Bessel's Finite Difference Formula

An INTERPOLATION formula also sometimes known as

$$\begin{aligned}
f_p &= f_0 + p\delta_{1/2} + B_2(\delta_0^2 + \delta_1^2) + B_3\delta_{1/2}^3 + B_4(\delta_0^4 + \delta_1^4) \\
&\quad + B_5\delta_{1/2}^5 + \cdots,
\end{aligned} \tag{1}$$

for $p \in [0, 1]$, where δ is the CENTRAL DIFFERENCE and

$$B_{2n} \equiv \tfrac{1}{2}\, G_{2n} \equiv \tfrac{1}{2}\,(E_{2n} + F_{2n}) \tag{2}$$

$$B_{2n+1} \equiv G_{2n+1} - \tfrac{1}{2}\, G_{2n} \equiv \tfrac{1}{2}(F_{2n} - E_{2n}) \tag{3}$$

$$E_{2n} \equiv G_{2n} - G_{2n+1} \equiv B_{2n} - B_{2n+1} \tag{4}$$

$$F_{2n} \equiv G_{2n+1} \equiv B_{2n} - B_{2n+1}, \tag{5}$$

where G_k are the COEFFICIENTS from GAUSS'S BACKWARD FORMULA and GAUSS'S FORWARD FORMULA and E_k and F_k are the COEFFICIENTS from EVERETT'S FORMULA. The B_ks also satisfy

$$B_{2n}(p) = B_{2n}(q) \tag{6}$$

$$B_{2n+1}(p) = -B_{2n+1}(q), \tag{7}$$

for

$$q \equiv 1 - p. \tag{8}$$

See also EVERETT'S FORMULA

References

Abramowitz, M. and Stegun, C. A. (Eds.). *Handbook of Mathematical Functions with Formulas, Graphs, and Mathematical Tables, 9th printing.* New York: Dover, p. 880, 1972.

Acton, F. S. *Numerical Methods That Work, 2nd printing.* Washington, DC: Math. Assoc. Amer., pp. 90–1, 1990.

Beyer, W. H. *CRC Standard Mathematical Tables, 28th ed.* Boca Raton, FL: CRC Press, p. 433, 1987.

Whittaker, E. T. and Robinson, G. "The Newton-Bessel Formula." §24 in *The Calculus of Observations: A Treatise on Numerical Mathematics, 4th ed.* New York: Dover, pp. 39–0, 1967.

Bessel's First Integral

$$J_n(x) = \frac{1}{\pi} \int_0^\pi \cos(n\theta - x\sin\theta)\, d\theta,$$

where $J_n(x)$ is a BESSEL FUNCTION OF THE FIRST KIND.

Bessel's Formula

BESSEL'S FINITE DIFFERENCE FORMULA, BESSEL'S INTERPOLATION FORMULA, BESSEL'S STATISTICAL FORMULA

Bessel's Inequality

If $f(x)$ is PIECEWISE CONTINUOUS and has a general FOURIER SERIES

$$\sum_i a_i \phi_i(x) \tag{1}$$

with WEIGHTING FUNCTION $w(x)$, it must be true that

$$\int \left[f(x) - \sum_i a_i \phi_i(x) \right]^2 w(x)\, dx \geq 0 \tag{2}$$

$$\int f^2(x) w(x)\, dx - 2 \sum_i a_i \int f(x) \phi_i(x) w(x)\, dx$$

$$+\sum_i a_i^2 f \phi_i^2(x) w(x)\, dx \geq 0. \tag{3}$$

But the COEFFICIENT of the generalized FOURIER SERIES is given by

$$a_m \equiv \int f(x) \phi_m(x) w(x)\, dx, \tag{4}$$

so

$$\int f^2(x) w(x)\, dx - 2 \sum_i a_i^2 + \sum_i a_i^2 \geq 0 \tag{5}$$

$$\int f^2(x) w(x)\, dx \geq \sum_i a_i^2. \tag{6}$$

Equation (6) is an inequality if the functions ϕ_i are not COMPLETE. If they are COMPLETE, then the inequality (2) becomes an equality, so (6) becomes an equality and is known as PARSEVAL'S THEOREM. If $f(x)$ has a simple FOURIER SERIES expansion with COEFFICIENTS a_0, a_1, a_n, a_π and $b_1, \ldots, b_n$, then

$$\frac{1}{2} a_0^2 + \sum_{k=1}^{\infty} (a_k^2 + b_k^2) \leq \frac{1}{\pi} \int_{-\pi}^{\pi} [f(x)]^2\, dx. \tag{7}$$

The inequality can also be derived from SCHWARZ'S INEQUALITY

$$|\langle f|g \rangle|^2 \leq \langle f|f \rangle \langle g|g \rangle \tag{8}$$

by expanding g in a superposition of EIGENFUNCTIONS of f, $g = \sum_i a_i f_i$. Then

$$\langle f|g \rangle = \sum_i a_i \langle f|f_i \rangle \leq \sum_i a_i \tag{9}$$

$$|\langle f|g \rangle|^2 \leq \left| \sum_i a_i \right|^2 = \left(\sum_i a_i \right) \left(\sum_i \bar{a}_i \right) = \sum_i a_i \bar{a}_i$$

$$\leq \langle f|f \rangle \langle g|g \rangle, \tag{10}$$

where $\bar{f}$ is the COMPLEX CONJUGATE. If g is normalized, then $\langle g|g \rangle = 1$ and

$$\langle f|f \rangle \geq \sum_i a_i \bar{a}_i \tag{11}$$

See also SCHWARZ'S INEQUALITY, TRIANGLE INEQUALITY

References

Arfken, G. *Mathematical Methods for Physicists, 3rd ed.* Orlando, FL: Academic Press, pp. 526–27, 1985.
Gradshteyn, I. S. and Ryzhik, I. M. *Tables of Integrals, Series, and Products, 6th ed.* San Diego, CA: Academic Press, p. 1102, 2000.

Bessel's Interpolation Formula
BESSEL'S FINITE DIFFERENCE FORMULA

Bessel's Second Integral
POISSON INTEGRAL

Bessel's Statistical Formula

Let $\bar{x}_1$ and s_1^2 be the observed mean and variance of a sample of N_1 drawn from a normal universe with unknown mean $\mu_{(1)}$ and let $\bar{x}_2$ and s_2^2 be the observed mean and variance of a sample of N_2 drawn from a normal universe with unknown mean $\mu_{(2)}$. Assume the two universes have a common variance σ^2, and define

$$\bar{w} \equiv \hat{x}_1 - \bar{x}_2 \tag{1}$$

$$\omega \equiv \mu_{(1)} - \mu_{(2)} \tag{2}$$

$$N \equiv N_1 + N_2 \tag{3}$$

Then

$$t = \frac{\bar{w} - \omega}{\sigma_w / \sqrt{N}} = \frac{\bar{w} - \omega}{\sqrt{\dfrac{\sum_{i=1}^{n}(w_i - \bar{w})^2}{N(N-1)}}} \tag{4}$$

is distributed as STUDENT'S T-DISTRIBUTION $f_n(t)$ with $n = N - 2$.

See also STUDENT'S T-DISTRIBUTION

References

Kenney, J. F. and Keeping, E. S. *Mathematics of Statistics, Pt. 2, 2nd ed.* Princeton, NJ: Van Nostrand, p. 186, 1951.

BesselI
MODIFIED BESSEL FUNCTION OF THE FIRST KIND

BesselJ
BESSEL FUNCTION OF THE FIRST KIND

BesselK
MODIFIED BESSEL FUNCTION OF THE SECOND KIND

BesselY
BESSEL FUNCTION OF THE SECOND KIND

Beta

A financial measure of a fund's sensitivity to market movements which measures the relationship between a fund's excess return over Treasury Bills and the excess return of a benchmark index (which, by definition, has $\beta = 1$). A fund with a beta of β has performed $r \equiv (\beta - 1) \times 100\%$ better (or $|r|$ worse if $r < 0$) than its benchmark index (after deducting the T-bill rate) in up markets and $|r|$ worse (or $|r|$ better if $r < 0$) in down markets.

See also ALPHA, BETA DISTRIBUTION, BETA FUNCTION, BETA INTEGRAL, SHARPE RATIO

Beta Distribution

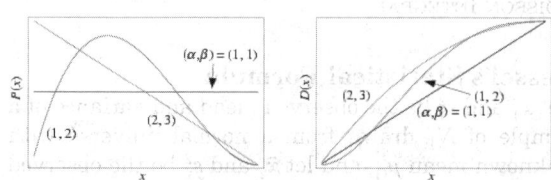

A general type of STATISTICAL DISTRIBUTION which is related to the GAMMA DISTRIBUTION. Beta distributions have two free parameters, which are labeled according to one of two notational conventions. The usual definition calls these α and β, and the other uses $\beta' \equiv \beta - 1$ and $\alpha' \equiv \alpha - 1$ (Beyer 1987, p. 534). The above plots are for various values of (α, β). The domain is $[0, 1]$, and the probability function $P(x)$ and DISTRIBUTION FUNCTION $D(x)$ are given by

$$P(x) = \frac{(1-x)^{\beta-1}x^{\alpha-1}}{B(\alpha, \beta)} = \frac{\Gamma(\alpha+\beta)}{\Gamma(\alpha)\Gamma(\beta)}(1-x)^{\beta-1}x^{\alpha-1} \quad (1)$$

$$D(x) = I(x; a, b), \quad (2)$$

where $B(a, b)$ is the BETA FUNCTION, $I(x; a, b)$ is the REGULARIZED BETA FUNCTION, and $\alpha, \beta > 0$. The distribution is normalized since

$$\int_0^1 P(x)\,dx = \frac{\Gamma(\alpha+\beta)}{\Gamma(\alpha)\Gamma(\beta)} \int_0^1 x^{\alpha-1}(1-x)^{\beta-1}\,dx \quad (3)$$

$$= \frac{\Gamma(\alpha+\beta)}{\Gamma(\alpha)\Gamma(\beta)} B(\alpha, \beta) = 1. \quad (4)$$

The CHARACTERISTIC FUNCTION is

$$\phi(t) = \mathscr{F}\left\{\frac{x^{a-1}(1-x)^{b-1}}{\beta(a, b)}\left[\tfrac{1}{2}\operatorname{sgn}(1-x) + \operatorname{sgn}x\right]\right\}$$

$$= {}_1F_1(a, a+b, it), \quad (5)$$

where $\mathscr{F}[f]$ is a FOURIER TRANSFORM with parameters $a = b = 1$ and ${}_1F_1(a; b; z)$ is a CONFLUENT HYPERGEOMETRIC FUNCTION.
The MEAN is

$$\mu = \frac{\Gamma(\alpha+\beta)}{\Gamma(\alpha)\Gamma(\beta)} \int_0^1 x^{\alpha-1}(1-x)^{\beta-1}x\,dx$$

$$= \frac{\Gamma(\alpha+\beta)}{\Gamma(\alpha)\Gamma(\beta)} B(\alpha+1, \beta) = \frac{\Gamma(\alpha+\beta)}{\Gamma(\alpha)\Gamma(\beta)} \frac{\Gamma(\alpha+1)\Gamma(\beta)}{\Gamma(\alpha+\beta+1)}$$

$$= \frac{\alpha}{\alpha+\beta}. \quad (6)$$

The RAW MOMENTS are given by

$$\mu_r' = \int_0^1 P(x)(x-\mu)^r\,dx = \frac{\Gamma(\alpha+\beta)\Gamma(\alpha+r)}{\Gamma(\alpha+\beta+r)\Gamma(\alpha)} \quad (7)$$

(Papoulis 1984, p. 147), and the CENTRAL MOMENTS by

$$\mu_r = \left(-\frac{\alpha}{\alpha+\beta}\right)^r {}_2F_1\left(-r, \alpha; \alpha+\beta; \frac{\alpha+\beta}{\alpha}\right), \quad (8)$$

where ${}_2F_1(a, b; c; x)$ is a HYPERGEOMETRIC FUNCTION. The VARIANCE, SKEWNESS, and KURTOSIS are therefore given by

$$\sigma^2 = \frac{\alpha\beta}{(\alpha+\beta)^2(\alpha+\beta+1)} \quad (9)$$

$$\gamma_1 = \frac{2(\beta-\alpha)\sqrt{1+\alpha+\beta}}{\sqrt{\alpha\beta}(2+\alpha+\beta)} \quad (10)$$

$$\gamma_2 = \frac{6[\alpha^3 + \alpha^2(1-2\beta) + \beta^2(1+\beta) - 2\alpha\beta(2+\beta)]}{\alpha\beta(\alpha+\beta+2)(\alpha+\beta+3)}. \quad (11)$$

The MODE of a variate distributed as $\beta(\alpha, \beta)$ is

$$\hat{x} = \frac{\alpha-1}{\alpha+\beta-2}. \quad (12)$$

See also GAMMA DISTRIBUTION

References

Abramowitz, M. and Stegun, C. A. (Eds.). *Handbook of Mathematical Functions with Formulas, Graphs, and Mathematical Tables, 9th printing.* New York: Dover, pp. 944–45, 1972.
Beyer, W. H. *CRC Standard Mathematical Tables, 28th ed.* Boca Raton, FL: CRC Press, pp. 534–35, 1987.
Jambunathan, M. V. "Some Properties of Beta and Gamma Distributions." *Ann. Math. Stat.* **25**, 401–05, 1954.
Kolarski, I. "On Groups of n Independent Random Variables whose Product Follows the Beta Distribution." *Colloq. Math. IX Fasc.* **2**, 325–32, 1962.
Krysicki, W. "On Some New Properties of the Beta Distribution." *Stat. Prob. Let.* **42**, 131–37, 1999.

Beta Exponential Function

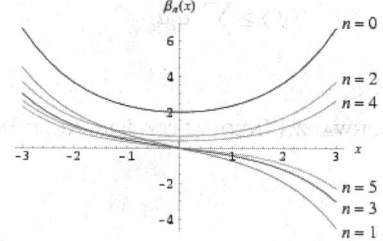

Another "BETA FUNCTION" defined in terms of an integral is the "exponential" beta function, given by

$$\beta_n(z) \equiv \int_{-1}^1 t^n e^{-zt}\,dt \quad (1)$$

$$= n!z^{-(n+1)}\left[e^z\sum_{k=0}^n\frac{(-1)^kz^k}{k!} - e^{-z}\sum_{k=0}^n\frac{z^k}{k!}\right]. \quad (2)$$

If n is an integer, then

$$\beta_n(z) = (-1)^{n+1}\mathrm{E}_{-n}(-z) - \mathrm{E}_{-n}(z), \qquad (3)$$

where $\mathrm{E}_n(z)$ is the *En-Function*. The exponential beta function satisfies the RECURRENCE RELATION

$$z\beta_n(z) = (-1)^n e^z - e^{-z} + n\beta_{n-1}(z). \qquad (4)$$

The values for $n = 0$, 1, and 2 are

$$\beta_0(z) = \frac{2\sinh z}{z} \qquad (5)$$

$$\beta_1(z) = \frac{2(\sinh z - z\cosh z)}{z^2}. \qquad (6)$$

$$\beta_2(z) = \frac{2(2 + z^2)\sinh z - 4z\cosh z}{z^3}. \qquad (7)$$

See also ALPHA FUNCTION, *En*-FUNCTION

Beta Function

The beta function is the name used by Legendre and Whittaker and Watson (1990) for the BETA INTEGRAL (also called the Eulerian integral of the first kind). To derive the integral representation of the beta function, write the product of two FACTORIALS as

$$m!n! = \int_0^\infty e^{-u}u^m\,du \int_0^\infty e^{-v}v^n\,dv. \qquad (1)$$

Now, let $u \equiv x^2$, $v \equiv y^2$, so

$$m!n! = 4\int_0^\infty e^{-x^2}x^{2m+1}\,dx \int_0^\infty e^{-y^2}y^{2n+1}\,dy$$

$$= 4\int_{-\infty}^\infty\int_{-\infty}^\infty e^{-(x^2+y^2)}x^{2m+1}y^{2n+1}\,dx\,dy. \qquad (2)$$

Transforming to POLAR COORDINATES with $x = r\cos\theta$, $y = r\sin\theta$

$$m!n! = 4\int_0^{\pi/2}\int_0^\infty e^{-r^2}(r\cos\theta)^{2m+1}(r\sin\theta)^{2n+1}r\,dr\,d\theta$$

$$= 4\int_0^\infty e^{-r^2}r^{2m+2n+3}\,dr\int_0^{\pi/2}\cos^{2m+1}\theta\sin^{2n+1}\theta\,d\theta$$

$$= 2(m+n+1)!\int_0^{\pi/2}\cos^{2m+1}\theta\sin^{2n+1}\theta\,d\theta. \qquad (3)$$

The beta function is then defined by

$$B(m+1,\,n+1) = B(n+1,\,m+1)$$

$$\equiv 2\int_0^{\pi/2}\cos^{2m+1}\theta\sin^{2n+1}\theta\,d\theta = \frac{m!n!}{(m+n+1)!}. \qquad (4)$$

Rewriting the arguments,

$$B(p,\,q) = \frac{\Gamma(p)\Gamma(q)}{\Gamma(p+q)} = \frac{(p-1)!(q-1)!}{(p+q-1)!}. \qquad (5)$$

The general trigonometric form is

$$\int_0^{\pi/2}\sin^n x\cos^m x\,dx = \tfrac{1}{2}B(\tfrac{1}{2}(n+1),\,\tfrac{1}{2}(m+1)). \qquad (6)$$

Equation (6) can be transformed to an integral over POLYNOMIALS by letting $u \equiv \cos^2\theta$,

$$B(m+1,\,n+1) \equiv \frac{m!n!}{(m+n+1)!} = \int_0^1 u^m(1-u)^n\,du \quad (7)$$

$$B(m,\,n) \equiv \frac{\Gamma(m)\Gamma(n)}{\Gamma(m+n)} = \int_0^1 u^{m-1}(1-u)^{n-1}\,du. \qquad (8)$$

The beta function is implemented in *Mathematica* as `Beta[a, b]`.

For any z_1, z_2 with $\Re[z_1]$, $\Re[z_2] > 0$,

$$B(z_1,\,z_2) = B(z_2,\,z_1) \qquad (9)$$

(Krantz 1999, p. 158).

The INCOMPLETE BETA FUNCTION $B(z;\,a,\,b)$, implemented in *Mathematica* as `Beta[z, a, b]`, is defined by the integral in (8) with an upper limit of z instead of 1. The REGULARIZED BETA FUNCTION $I(z;\,a,\,b)$, implemented in *Mathematica* as `BetaRegularized[z, a, b]` is defined by

$$I(z;\,a,\,b) = \frac{B(z;\,a,\,b)}{B(a,\,b)}. \qquad (10)$$

To put it in a form which can be used to derive the LEGENDRE DUPLICATION FORMULA, let $x \equiv \sqrt{u}$, so $u = x^2$ and $du = 2x\,dx$, and

$$B(m,\,n) = \int_0^1 x^{2(m-1)}(1-x^2)^{n-1}(2x\,dx)$$

$$= 2\int_0^1 x^{2m-1}(1-x^2)^{n-1}\,dx. \qquad (11)$$

To put it in a form which can be used to develop integral representations of the BESSEL FUNCTIONS and HYPERGEOMETRIC FUNCTION, let $u \equiv x/(1+x)$, so

$$B(m+1,\,n+1) = \int_0^\infty \frac{u^m\,du}{(1+u)^{m+n+2}}. \qquad (12)$$

Derivatives of the beta function are given by

$$\frac{d}{da}B(a,\,b) = B(a,\,b)[\psi_0(a) - \psi_0(a+b)] \qquad (13)$$

$$\frac{d}{db}B(a,\,b) = B(a,\,b)[\psi_0(b) - \psi_0(a+b)] \qquad (14)$$

$$\frac{d^2}{da^2}B(a,\,b) = B(a,\,b)$$

$$\times \{[\psi_0(a) - \psi_0(a+b)]^2 + \psi_1(a) - \psi_1(a+b)\}, \quad (15)$$

$$\frac{d^2}{db^2} B(a, \ b) = B(a, \ b)$$

$$\times \{[\psi_0(b) - \psi_0(a+b)]^2 + \psi_1(b) - \psi_1(a+b)\}, \quad (16)$$

$$\frac{d^2}{da\, db} B(a, \ b)$$

$$= B(a, \ b)\{[\psi_0(a) - \psi_0(a+b)][\psi_0(b) - \psi_0(a+b)]$$

$$-\psi_1(a+b)\} \quad (17)$$

where $\psi_n(x)$ is the POLYGAMMA FUNCTION.

Various identities can be derived using the GAUSS MULTIPLICATION FORMULA

$$B(np, \ nq) = \frac{\Gamma(np)\Gamma(nq)}{\Gamma[n(p+q)]}$$

$$= n^{-nq} \frac{B(p, \ q)B\left(p + \dfrac{1}{n}, \ q\right) \cdots B\left(p + \dfrac{n-1}{n}, \ q\right)}{B(q, \ q)B(2q, \ q) \cdots B([n-1]q, \ q)}.$$

$$(18)$$

Additional identities include

$$B(p, \ q+1) = \frac{\Gamma(p)\Gamma(q+1)}{\Gamma(p+q+1)} = \frac{q}{p} \frac{\Gamma(p+1)\Gamma(q)}{\Gamma([p+1]q)}$$

$$= \frac{q}{p} B(p+1, \ q) \quad (19)$$

$$B(p, \ q) = B(p+1, \ q) + B(p, \ q+1) \quad (20)$$

$$B(p, \ q+1) = \frac{q}{p+q} B(p, \ q). \quad (21)$$

If n is a POSITIVE INTEGER, then

$$B(p, \ n+1) = \frac{1 \cdot 2 \cdots n}{p(p+1) \cdots (p+n)} \quad (22)$$

$$B(p, \ p)B(p + \tfrac{1}{2}, \ p + \tfrac{1}{2}) = \frac{\pi}{2^{4p-1}p} \quad (23)$$

$$B(p+q)B(p+q, \ r) = B(q, \ r)B(q+r, \ p). \quad (24)$$

Gosper gives the general formulas

$$\prod_{i=0}^{2n} B\left(\frac{i}{2n+1} + a, \ \frac{i}{2n+1} + b\right)$$

$$= \frac{(2n+1)^{(2n+1)/2} \pi^n B(n, \ \tfrac{1}{2}[(b+a)(2n+1)+1])B(a(2n+1), \ b(2n+1))}{(n-1)!}$$

$$(25)$$

for ODD n, and

$$\prod_{i=0}^{2n-1} B\left(\frac{i}{2n} + a, \ \frac{i}{2n} + b\right)$$

$$= \frac{n^n \pi^n B(n, \ 2(a+b)n)B(2an, \ 2bn)}{2^{2(a+b)n = n = 1}(n-1)!B((a+b)n, \ (a+b+1)n)},$$

$$(26)$$

which are an immediate consequence of the analogous identities for GAMMA FUNCTIONS. Plugging $n = 1$ and $n = 2$ into the above give the special cases

$$B(a, \ b)B(a + \tfrac{1}{3}, \ b + \tfrac{1}{3})B(a + \tfrac{2}{3}, \ b + \tfrac{2}{3})$$

$$= \frac{6\pi\sqrt{3}B(3a, \ 3b)}{1 + 3(a+b)} \quad (27)$$

$$B(a, \ b)B(a + \tfrac{1}{4}, \ b + \tfrac{1}{4})B(a + \tfrac{1}{2}, \ b + \tfrac{1}{2})B(a + \tfrac{3}{4}, \ b + \tfrac{3}{4})$$

$$= \frac{2^{3-4(a+b)}\pi^2 B(4a, \ 4b)}{(a+b)[1 + 4(a+b)]B(2(a+b), \ 2(a+b+1)}. \quad (28)$$

See also BETA INTEGRAL, CENTRAL BETA FUNCTION, DIRICHLET INTEGRALS, GAMMA FUNCTION, INCOMPLETE BETA FUNCTION, REGULARIZED BETA FUNCTION

References

Abramowitz, M. and Stegun, C. A. (Eds.). "Beta Function" and "Incomplete Beta Function." §6.2 and 6.6 in *Handbook of Mathematical Functions with Formulas, Graphs, and Mathematical Tables, 9th printing.* New York: Dover, pp. 258 and 263, 1972.

Arfken, G. "The Beta Function." §10.4 in *Mathematical Methods for Physicists, 3rd ed.* Orlando, FL: Academic Press, pp. 560–65, 1985.

Erdélyi, A.; Magnus, W.; Oberhettinger, F.; and Tricomi, F. G. "The Beta Function." §1.5 in *Higher Transcendental Functions, Vol. 1.* New York: Krieger, pp. 9–3, 1981.

Jeffreys, H. and Jeffreys, B. S. "The Beta Function." §15.02 in *Methods of Mathematical Physics, 3rd ed.* Cambridge, England: Cambridge University Press, pp. 463–64, 1988.

Koepf, W. *Hypergeometric Summation: An Algorithmic Approach to Summation and Special Function Identities.* Braunschweig, Germany: Vieweg, pp. 6–, 1998.

Krantz, S. G. "The Beta Function." §13.1.11 in *Handbook of Complex Analysis.* Boston, MA: Birkhäuser, pp. 157–58, 1999.

Morse, P. M. and Feshbach, H. *Methods of Theoretical Physics, Part I.* New York: McGraw-Hill, p. 425, 1953.

Press, W. H.; Flannery, B. P.; Teukolsky, S. A.; and Vetterling, W. T. "Gamma Function, Beta Function, Factorials, Binomial Coefficients" and "Incomplete Beta Function, Student's Distribution, F-Distribution, Cumulative Binomial Distribution." §6.1 and 6.2 in *Numerical Recipes in FORTRAN: The Art of Scientific Computing, 2nd ed.* Cambridge, England: Cambridge University Press, pp. 206–09 and 219–23, 1992.

Spanier, J. and Oldham, K. B. "The Incomplete Beta Function $B(v; \ \mu; \ x)$." Ch. 58 in *An Atlas of Functions.* Washington, DC: Hemisphere, pp. 573–80, 1987.

Whittaker, E. T. and Watson, G. N. *A Course of Modern Analysis, 4th ed.* Cambridge, England: Cambridge University Press, 1990.

Beta Function (Exponential)

$$\mu_r = \left(-\frac{\alpha}{\alpha + \beta}\right)^r {}_2F_1\left(-r, \; \alpha; \; \alpha + \beta; \; \frac{\alpha + \beta}{\alpha}\right),$$

Another "BETA FUNCTION" defined in terms of an integral is the "exponential" beta function, given by

$$_2F_1(a, \; b; \; c; \; x)\theta_2 \; \frac{\alpha\beta}{(\alpha + \beta)^2(\alpha + \beta + 1)} \tag{1}$$

$$= \frac{2(\beta - \alpha)\sqrt{1 + \alpha + \beta}}{\sqrt{\alpha\beta}(2 + \alpha + \beta)} \tag{2}$$

The exponential beta function satisfies the RECURRENCE RELATION

$$\frac{6[\alpha^3 + \alpha^2(1 - 2\beta) + \beta^2(1 + \beta) - 2\alpha\beta(2 + \beta)]}{\alpha\beta(\alpha + \beta + 2)(\alpha + \beta + 3)}. \tag{3}$$

The first few integral values are

$$\beta(\alpha, \; \beta) = \hat{x} = \frac{\alpha - 1}{\alpha + \beta - 2}. \tag{4}$$

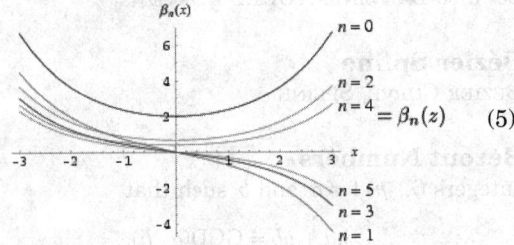

$$= \beta_n(z) \tag{5}$$

$$\int_{-1}^{1} t^n e^{-zt} \, dt$$

$$= n!z^{-(n+1)}\left[e^z \sum_{k=0}^{n} \frac{(-1)^k z^k}{k!} - e^{-z} \sum_{k=0}^{n} \frac{z^k}{k!}\right]. \tag{6}$$

See also ALPHA FUNCTION

Beta Integral

The integral

$$\int_0^1 x^p(1 - x)^q \, dx$$

called the EULERIAN INTEGRAL OF THE FIRST KIND by Legendre and Whittaker and Watson (1990). The solution is the BETA FUNCTION $B(p + 1, \; q + 1)$.

See also BETA FUNCTION, EULERIAN INTEGRAL OF THE FIRST KIND, EULERIAN INTEGRAL OF THE SECOND KIND

References

Whittaker, E. T. and Watson, G. N. *A Course in Modern Analysis,* 4th ed. Cambridge, England: Cambridge University Press, 1990.

Beta Prime Distribution

A distribution with probability function

$$P(x) = \frac{x^{\alpha-1}(1 + x)^{-\alpha-\beta}}{B(\alpha, \; \beta)},$$

where B is a BETA FUNCTION. The MODE of a variate distributed as $\beta'(\alpha, \; \beta)$ is

$$\hat{x} = \frac{\alpha - 1}{\beta + 1}.$$

If x is a $\beta'(\alpha, \; \beta)$ variate, then $1/x$ is a $\beta'(\beta, \; \alpha)$ variate. If x is a $\beta(\alpha, \; \beta)$ variate, then $(1 - x)/x$ and $x/(1 - x)$ are $\beta'(\beta, \; \alpha)$ and $\beta'(\alpha, \; \beta)$ variates. If x and y are $\gamma(\alpha_1)$ and $\gamma(\alpha_2)$ variates, then x/y is a $\beta'(\alpha_1, \; \alpha_2)$ variate. If $x^2/2$ and $y^2/2$ are $\gamma(1/2)$ variates, then $z^2 \equiv (x/y)^2$ is a $\beta'(1/2, \; 1/2)$ variate.

BetaRegularized

REGULARIZED BETA FUNCTION

Bethe Lattice

CAYLEY TREE

Betrothed Numbers

QUASIAMICABLE PAIR

Betti Group

The free part of the HOMOLOGY GROUP with a domain of COEFFICIENTS in the GROUP of INTEGERS (if this HOMOLOGY GROUP is finitely generated).

See also HOMOLOGY GROUP

References

Alexandrov, P. S. *Combinatorial Topology.* New York: Dover, 1998.
Hazewinkel, M. (Managing Ed.). *Encyclopaedia of Mathematics: An Updated and Annotated Translation of the Soviet "Mathematical Encyclopaedia."* Dordrecht, Netherlands: Reidel, p. 380, 1988.

Betti Number

Betti numbers are topological objects which were proved to be invariants by Poincaré, and used by him to extend the POLYHEDRAL FORMULA to higher dimensional spaces. Informally, the Betti number is the maximum number of cuts that can be made without dividing a surface into two separate pieces (Gardner 1984, pp. 9–0). Formally, the nth Betti number is the rank of the nth HOMOLOGY GROUP of a TOPOLOGICAL SPACE. The following table gives the Betti number of some common surfaces.

SURFACE	Betti number
CROSS-CAP	1
CYLINDER	1
KLEIN BOTTLE	2
MÖBIUS STRIP	1
plane lamina	0
PROJECTIVE PLANE	1
SPHERE	0
TORUS	2

Let p_r be the RANK of the HOMOLOGY GROUP H_r of a TOPOLOGICAL SPACE K. For a closed, orientable surface of GENUS g, the Betti numbers are $p_0 = 1$, $p_1 = 2g$, and $p_2 = 1$. For a NONORIENTABLE SURFACE with k CROSS-CAPS, the Betti numbers are $p_0 = 1$, $p_1 = k - 1$, and $p_2 = 0$.

See also CHROMATIC NUMBER, EULER CHARACTERISTIC, GENUS (SURFACE), HOMOLOGY GROUP, POINCARÉ DUALITY, TOPOLOGICAL SPACE

References

Gardner, M. *The Sixth Book of Mathematical Games from Scientific American.* Chicago, IL: University of Chicago Press, pp. 9–1 and 15–6, 1984.

Bézier Curve

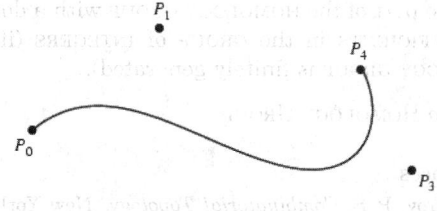

Given a set of $n + 1$ control points $P_0, P_1, \ldots, P_n$, the corresponding Bézier curve (or Bernstein-Bézier curve) is given by

$$\mathbf{C}(t) = \sum_{i=0}^{n} \mathbf{P}_i B_{i,\,n}(t),$$

where $B_{i,\,n}(t)$ is a BERNSTEIN POLYNOMIAL and $t \in [0, 1]$.
A "rational" Bézier curve is defined by

$$\mathbf{C}(t) = \frac{\sum_{i=0}^{n} B_{i,\,p}(t) w_i \mathbf{P}_i}{\sum_{i=0}^{n} B_{i,\,p}(t) w_i},$$

where p is the order, $B_{i,\,p}$ are the BERNSTEIN POLYNOMIALS, $\mathbf{P}_i$ are control points, and the weight w_i of $\mathbf{P}_i$ is the last ordinate of the homogeneous point $\mathbf{P}_i^\omega$. These curves are CLOSED under perspective transformations, and can represent CONIC SECTIONS exactly.

The Bézier curve always passes through the first and last control points and lies within the CONVEX HULL of the control points. The curve is tangent to $\mathbf{P}_1 - \mathbf{P}_0$ and $\mathbf{P}_n - \mathbf{P}_{n-1}$ at the endpoints. The "variation diminishing property" of these curves is that no line can have more intersections with a Bézier curve than with the curve obtained by joining consecutive points with straight line segments. A desirable property of these curves is that the curve can be translated and rotated by performing these operations on the control points.

Undesirable properties of Bézier curves are their numerical instability for large numbers of control points, and the fact that moving a single control point changes the global shape of the curve. The former is sometimes avoided by smoothly patching together low-order Bézier curves. A generalization of the Bézier curve is the B-SPLINE.

See also B-SPLINE, NURBS CURVE

Bézier Spline

BÉZIER CURVE, SPLINE

Bézout Numbers

Integers (λ, μ) for a and b such that

$$\lambda a + \mu b = \text{GCD}(a, b).$$

For INTEGERS $a_1, \ldots, a_\pi$, the Bézout numbers are a set of numbers $k_1, \ldots, k_n$ such that

$$k_1 a_1 + k_2 a_2 + \cdots + k_n a_n = d,$$

where d is the GREATEST COMMON DIVISOR of $a_1, \ldots, a_\pi$.

See also GREATEST COMMON DIVISOR

Bézout's Theorem

In general, two algebraic curves of degrees m and n intersect in $m \cdot n$ points and cannot meet in more than $m \cdot n$ points unless they have a component in common (i.e., the equations defining them have a common factor). This can also be stated: if P and Q are two POLYNOMIALS with no roots in common, then there exist two other POLYNOMIALS A and B such that $AP + BQ = 1$. Similarly, given N POLYNOMIAL equations of degrees $n_1, n_2, \ldots, n_N$ in N variables, there are in general $n_1 n_2 \cdots n_N$ common solutions.

Séroul (2000, p. 10) uses the term Bézout's theorem for the following two theorems.

    1. Let $a, b \in \mathbb{Z}$ be any two integers, then there exist $u, v \in \mathbb{Z}$ such that

$$au + bv = \text{GCD}(a, b).$$

2. Two integers a and b are RELATIVELY PRIME if there exist u, $v \in \mathbb{Z}$ such that

$$au + bv = 1.$$

See also BLANKINSHIP ALGORITHM, GREATEST COMMON DIVISOR, POLYNOMIAL

References

Coolidge, J. L. *A Treatise on Algebraic Plane Curves.* New York: Dover, p. 10, 1959.

Séroul, R. "The Bézout Theorem." §2.4.1 in *Programming for Mathematicians.* Berlin: Springer-Verlag, p. 10, 2000.

Shub, M. and Smale, S. "Complexity of Bézout's Theorem. I. Geometric Aspects." *J. Amer. Math. Soc.* **6**, 459–01, 1993.

Shub, M. and Smale, S. "Complexity of Bézout's Theorem. II. Volumes and Probabilities." In *Computational Algebraic Geometry (Nice, 1992).* Boston, MA: Birkhäuser, pp. 267–85, 1993.

Shub, M. and Smale, S. "Complexity of Bézout's Theorem. III. Condition Number and Packing." *J. Complexity* **9**, 4–4, 1993.

Shub, M. and Smale, S. "Complexity of Bézout's Theorem. IV. Probability of Success; Extensions." *SIAM J. Numer. Anal.* **33**, 128–48, 1996.

Shub, M. and Smale, S. "Complexity of Bézout's Theorem. V. Polynomial Time." *Theoret. Comput. Sci.* **134**, 141–64, 1994.

Bhargava's Theorem

Let the nth composition of a function $f(x)$ be denoted $f^{(n)}(x)$, such that $f^{(0)}(x) = f(x)$ and $f^{(1)}(x) = f(x)$. Denote the COMPOSITION of f and g by $f \circ g(x) = f(g(x))$, and define

$$\sum F(a, b, c)$$
$$= F(a, b, c) + F(b, c, a) + F(c, b, a). \quad (1)$$

Let

$$\mathbf{u} \equiv (a, b, c) \quad (2)$$

$$\|\mathbf{u}\| \equiv a + b + c \quad (3)$$

$$\|\mathbf{u}\| \equiv a^4 + b^4 + c^4, \quad (4)$$

and

$$f(\mathbf{u}) = (a(b-c), \ b(c-a), \ c(a-b)) \quad (5)$$

$$g(\mathbf{u}) = \left(\sum a^2 b, \ \sum ab^2, \ 3abc \right). \quad (6)$$

Then if $|\mathbf{u}| = 0$ (i.e., $c = -a - b$),

$$\|f^{(m)} \circ g^{(n)}(\mathbf{u})\| = \|g^{(n)} \circ f^{(m)}(\mathbf{u})\|$$
$$= 2(ab + bc + ca)^{2^m + 1 3^n}, \quad (7)$$

where m, $n \in \{0, 1, \ldots\}$ and COMPOSITION is done in terms of components.

See also DIOPHANTINE EQUATION–4TH POWERS, FORD'S THEOREM

References

Berndt, B. C. *Ramanujan's Notebooks, Part IV.* New York: Springer-Verlag, pp. 97–00, 1994.

Bhargava, S. "On a Family of Ramanujan's Formulas for Sums of Fourth Powers." *Ganita* **43**, 63–7, 1992.

Bhaskara-Brouckner Algorithm

SQUARE ROOT

Bialtitude

The common perpendicular to two opposite edges of a TETRAHEDRON.

See also ALTITUDE, BIMEDIAN, TETRAHEDRON

References

Altshiller-Court, N. *Modern Pure Solid Geometry.* New York: Chelsea, p. 50, 1979.

Bianchi Identities

The RIEMANN TENSOR is defined by

$$R_{\lambda\mu\nu\kappa;\ \eta} = \frac{1}{2} \frac{\partial}{\partial x^\eta}$$
$$\times \left(\frac{\partial^2 g_{\lambda\nu}}{\partial x^\kappa \partial x^\mu} - \frac{\partial^2 g_{\mu\nu}}{\partial x^\kappa \partial x^\lambda} - \frac{\partial^2 g_{\lambda\kappa}}{\partial x^\mu \partial x^\nu} + \frac{\partial^2 g_{\mu\kappa}}{\partial x^\nu \partial x^\lambda} \right). \quad (1)$$

Permuting ν, κ, and η (Weinberg 1972, pp. 146–47) gives the Bianchi identities

$$R_{\lambda\mu\nu\kappa;\ \eta} + R_{\lambda\mu\eta\nu;\ \kappa} + R_{\lambda\mu\kappa\eta;\ \nu} = 0, \quad (2)$$

which can be written concisely as

$$R^\alpha_{\ \beta[\lambda\mu;\ \nu]} = 0 \quad (3)$$

(Misner *et al.* 1973, p. 221), where $T_{[a_1 \ldots a_n]}$ denoted the ANTISYMMETRIC TENSOR part. Wald (1984, p. 39) calls

$$\nabla_{[a} R_{bc]d}^{\ \ \ \varepsilon} = 0 \quad (4)$$

the Bianchi identity, where ∇ is the COVARIANT DERIVATIVE, and $R_{abc}^{\ \ \ d}$ is the RIEMANN TENSOR.

See also BIANCHI IDENTITIES (CONTRACTED), RIEMANN TENSOR

References

Misner, C. W.; Thorne, K. S.; and Wheeler, J. A. *Gravitation.* San Francisco: W. H. Freeman, 1973.

Wald, R. M. *General Relativity.* Chicago, IL: University of Chicago Press, 1984.

Weinberg, S. *Gravitation and Cosmology: Principles and Applications of the General Theory of Relativity.* New York: Wiley, 1972.

Bianchi Identities (Contracted)

CONTRACTING λ with ν in the BIANCHI IDENTITIES

$$R_{\lambda\mu\nu\kappa;\ \eta} + R_{\lambda\mu\eta\nu;\ \kappa} + R_{\lambda\mu\kappa\eta;\ \nu} = 0 \quad (1)$$

gives

$$R_{\mu\kappa;\,\eta} - R_{\mu\eta;\,\kappa} + R^{\nu}_{\,\mu\kappa\eta;\,\nu} = 0. \tag{2}$$

CONTRACTING again,

$$R_{;\,\eta} - R^{\mu}_{\,\eta;\,\mu} - R^{\nu}_{\,\eta;\,\nu} = 0, \tag{3}$$

or

$$(R^{\mu}_{\,\eta} - \tfrac{1}{2}\delta^{\mu}_{\,\eta}R)_{;\,\mu} = 0, \tag{4}$$

or

$$(R^{\mu\nu} - \tfrac{1}{2}g^{\mu\nu}R)_{;\,\mu} = 0. \tag{5}$$

Bias (Estimator)

The bias of an ESTIMATOR $\tilde{\theta}$ is defined as

$$B(\tilde{\theta}) \equiv \langle \tilde{\theta} \rangle - \theta.$$

It is therefore true that

$$\tilde{\theta} - \theta = (\tilde{\theta} - \langle\tilde{\theta}\rangle) + (\langle\tilde{\theta}\rangle - \theta) = (\tilde{\theta} - \langle\tilde{\theta}\rangle) + B(\tilde{\theta}).$$

An ESTIMATOR for which $B = 0$ is said to be UNBIASED ESTIMATOR.

See also BIASED ESTIMATOR, ESTIMATOR, UNBIASED ESTIMATOR

Bias (Series)

The bias of a SERIES is defined as

$$Q[a_i,\, a_{i+1},\, a_{i+2}] \equiv \frac{a_i a_{i+2} - a_{i+1}^2}{a_1 a_{i+1} a_{i+2}}.$$

A SERIES is GEOMETRIC IFF $Q = 0$. A SERIES is ARTISTIC IFF the bias is constant.

See also ARTISTIC SEQUENCE, GEOMETRIC SEQUENCE

References

Duffin, R. J. "On Seeing Progressions of Constant Cross Ratio." *Amer. Math. Monthly* **100**, 38–7, 1993.

Biased Estimator

An ESTIMATOR which exhibits BIAS.

See also BIAS (ESTIMATOR), ESTIMATOR, UNBIASED ESTIMATOR

Biaugmented Pentagonal Prism

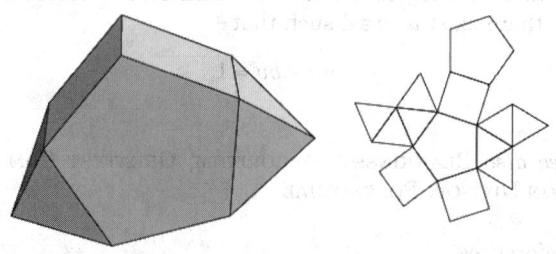

JOHNSON SOLID J_{53}.

References

Weisstein, E. W. "Johnson Solids." MATHEMATICA NOTEBOOK JOHNSONSOLIDS.M.
Weisstein, E. W. "Johnson Solid Netlib Database." MATHEMATICA NOTEBOOK JOHNSONSOLIDS.DAT.

Biaugmented Triangular Prism

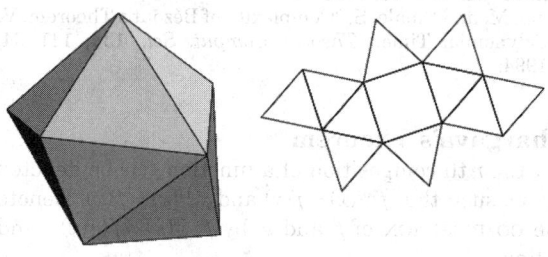

JOHNSON SOLID J_{50}.

References

Weisstein, E. W. "Johnson Solids." MATHEMATICA NOTEBOOK JOHNSONSOLIDS.M.
Weisstein, E. W. "Johnson Solid Netlib Database." MATHEMATICA NOTEBOOK JOHNSONSOLIDS.DAT.

Biaugmented Truncated Cube

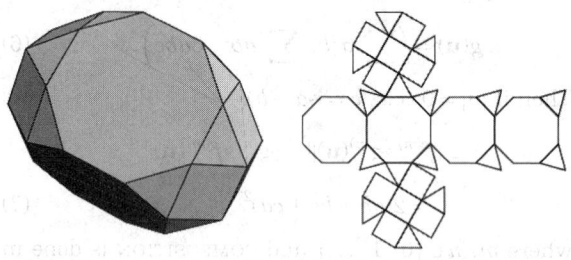

JOHNSON SOLID J_{67}.

BIBD

BLOCK DESIGN

Bicentered Tree

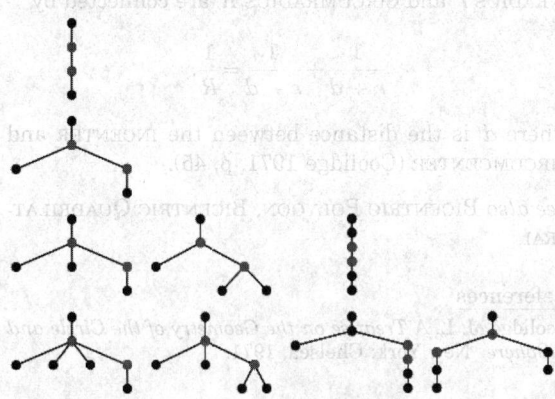

A TREE (also called a bicentral tree) having two nodes that are GRAPH CENTERS. The numbers of bicentered trees on $n = 1, 2, \ldots$ nodes are 0, 1, 0, 1, 1, 3, 4, 11, 20, 51, 108 ... (Sloane's A000677).

See also CENTERED TREE, GRAPH CENTER, TREE

References

Biggs, N. L.; Lloyd, E. K.; and Wilson, R. J. *Graph Theory 1736–936.* Oxford, England: Oxford University Press, p. 49, 1976.
Cayley, A. "On the Analytical Forms Called Trees, with Application to the Theory of Chemical Combinations." *Reports Brit. Assoc. Advance. Sci.* **45**, 237–05, 1875. Reprinted in *Math Papers, Vol. 9*, pp. 427–60.
Sloane, N. J. A. Sequences A000677/M2366 in "An On-Line Version of the Encyclopedia of Integer Sequences." http://www.research.att.com/~njas/sequences/eisonline.html.

Bicentral Tree

BICENTERED TREE

Bicentric Perspective

Bicentric perspective is the study of the projection of 3D space from a pair of fiducial points instead of a single one, the latter of which may be called "centric" or "natural" PERSPECTIVE by way of distinction.

See also PERSPECTIVE, PROJECTION

References

Koenderink, J. J. "Fundamentals of Bicentric Perspective." In *Future Tendencies in Computer Science, Control and Applied Mathematics. Proceedings of the International Conference on Research in Computer Science and Control held on the occasion of the 25th Anniversary of INRIA in Paris, December 8–1, 1992* (Ed. A. Bensoussan and J.-P. Verjus). New York: Springer-Verlag, 233–51, 1992.

Bicentric Polygon

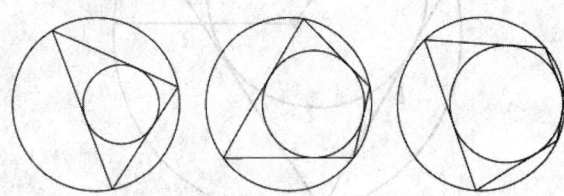

A POLYGON which has both a CIRCUMCIRCLE (which touches each vertex) and an INCIRCLE (which is tangent to each side). All TRIANGLES are bicentric with

$$R^2 - x^2 = 2Rr, \tag{1}$$

where R is the CIRCUMRADIUS, r is the INRADIUS, and x is the separation of centers. For BICENTRIC QUAD-RILATERALS (Fuss's problem), the CIRCLES satisfy

$$2r^2(R^2 + x^2) = (R^2 - x^2)^2 \tag{2}$$

(Dörrie 1965) or, in another form,

$$\frac{1}{(R-x)^2} + \frac{1}{(R+x)^2} = \frac{1}{r^2} \tag{3}$$

(Davis; Durége; Casey 1888, pp. 109–10; Johnson 1929; Dörrie 1965).

If the circles permit successive tangents around the INCIRCLE which close the POLYGON for one starting point on the CIRCUMCIRCLE, then they do so for all points on the CIRCUMCIRCLE, a result known as PONCELET'S PORISM.

See also BICENTRIC QUADRILATERAL, BICENTRIC TRI-ANGLE, CIRCUMCIRCLE, INCIRCLE, POLYGON, PONCE-LET'S PORISM, PONCELET TRANSVERSE, TANGENTIAL QUADRILATERAL, TRIANGLE, WEILL'S THEOREM

References

Beyer, W. H. (Ed.). *CRC Standard Mathematical Tables, 28th ed.* Boca Raton, FL: CRC Press, p. 124, 1987.
Casey, J. *A Sequel to the First Six Books of the Elements of Euclid, Containing an Easy Introduction to Modern Geometry with Numerous Examples, 5th ed., rev. enl.* Dublin: Hodges, Figgis, & Co., 1888.
Dörrie, H. "Fuss' Problem of the Chord-Tangent Quadrilateral." §39 in *100 Great Problems of Elementary Mathematics: Their History and Solutions.* New York: Dover, pp. 188–93, 1965.
Johnson, R. A. *Modern Geometry: An Elementary Treatise on the Geometry of the Triangle and the Circle.* Boston, MA: Houghton Mifflin, pp. 91–6, 1929.

Bicentric Quadrilateral

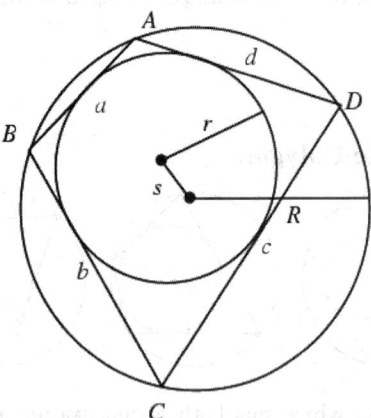

A 4-sided BICENTRIC POLYGON, also called a CYCLIC-INSCRIPTABLE QUADRILATERAL. The INRADIUS r, CIRCUMRADIUS R, and offset s are connected by the equation

$$\frac{1}{(R-s)^2}+\frac{1}{(R+s)^2}=\frac{1}{r^2} \tag{1}$$

(Davis; Durége; Casey 1888, pp. 109–10; Johnson 1929; Dörie 1965; Coolidge 1971, p. 46). In addition

$$r=\frac{\sqrt{abcd}}{s} \tag{2}$$

$$R=\tfrac{1}{4}\sqrt{\frac{(ac+bd)(ad+bc)(ad+cd)}{abcd}} \tag{3}$$

(Beyer 1987), and

$$a+c=b+d. \tag{4}$$

The AREA of a bicentric quadrilateral is

$$A=\sqrt{abcd}. \tag{5}$$

See also BICENTRIC POLYGON, BICENTRIC TRIANGLE, CYCLIC QUADRILATERAL, PONCELET'S PORISM

References

Beyer, W. H. (Ed.). *CRC Standard Mathematical Tables,* 28th ed. Boca Raton, FL: CRC Press, p. 124, 1987.

Casey, J. *A Sequel to the First Six Books of the Elements of Euclid, Containing an Easy Introduction to Modern Geometry with Numerous Examples,* 5th ed., rev. enl. Dublin: Hodges, Figgis, & Co., 1888.

Coolidge, J. L. *A Treatise on the Geometry of the Circle and Sphere.* New York: Chelsea, 1971.

Davis, M. A. *Educ. Times* **32**.

Dörrie, H. "Fuss' Problem of the Chord-Tangent Quadrilateral." §39 in *100 Great Problems of Elementary Mathematics: Their History and Solutions.* New York: Dover, pp. 188–93, 1965.

Durége, H. *Theorie der elliptischen Functionen: Versuch einer elementaren Darstellung.* Leipzig, Germany: Teubner, p. 185, 1861.

Johnson, R. A. *Modern Geometry: An Elementary Treatise on the Geometry of the Triangle and the Circle.* Boston, MA: Houghton Mifflin, pp. 91–6, 1929.

Bicentric Triangle

All triangles are bicentric, i.e., possess both an INCIRCLE and a CIRCUMCIRCLE. This is not necessarily the case for polygons with four or more sides. The INRADIUS r and CIRCUMRADIUS R are connected by

$$\frac{1}{r+d}+\frac{1}{r-d}=\frac{1}{R},$$

where d is the distance between the INCENTER and CIRCUMCENTER (Coolidge 1971, p. 45).

See also BICENTRIC POLYGON, BICENTRIC QUADRILATERAL

References

Coolidge, J. L. *A Treatise on the Geometry of the Circle and Sphere.* New York: Chelsea, 1971.

Bichromatic Graph

A GRAPH with EDGES of two possible "colors," usually identified as red and blue. For a bichromatic graph with R red EDGES and B blue EDGES,

$$R+B\geq 2.$$

See also BLUE-EMPTY GRAPH, EXTREMAL COLORING, EXTREMAL GRAPH, MONOCHROMATIC FORCED TRIANGLE, RAMSEY NUMBER

Bicollared

A SUBSET $X\subset Y$ is said to be bicollared in Y if there exists an embedding $b:X\times[-1,\,1]\to Y$ such that $b(x,\,0)=x$ when $x\in X$. The MAP b or its image is then said to be the bicollar.

References

Rolfsen, D. *Knots and Links.* Wilmington, DE: Publish or Perish Press, pp. 34–5, 1976.

Biconditional

The CONNECTIVE in $A\Leftrightarrow B$ (also denoted $A\equiv B$) that returns a true result IFF A and B are either both true or both false. The biconditional is also called an EQUIVALENCE.

See also CONDITIONAL, EQUIVALENT

References

Carnap, R. *Introduction to Symbolic Logic and Its Applications.* New York: Dover, p. 8, 1958.

Mendelson, E. *Introduction to Mathematical Logic,* 4th ed. London: Chapman & Hall, p. 14, 1997.

Bicone

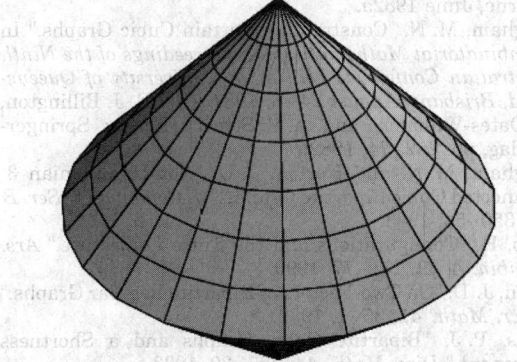

Two cones placed base-to-base.

See also DIPYRAMID, CONE, DOUBLE CONE, NAPPE, SPHERICON

Bi-Connected Component

A maximal SUBGRAPH of an undirected graph such that any two edges in the SUBGRAPH lie on a common simple cycle.

See also STRONGLY CONNECTED COMPONENT

Biconnected Component

BLOCK

Biconnected Graph

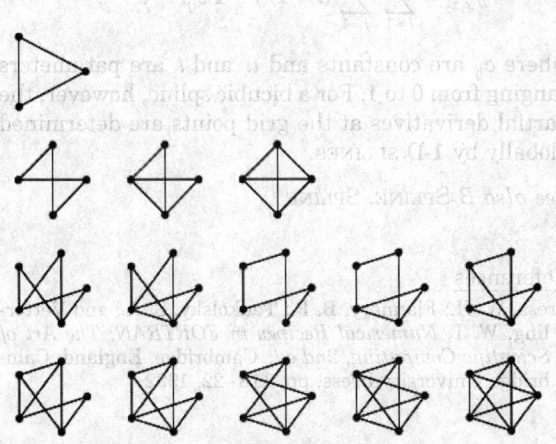

A GRAPH with no ARTICULATION VERTICES is called biconnected (Skiena 1990, p. 175), block, or "nonseparable graph" (Harary 1994, p. 26). The numbers of biconnected simple graphs on $n = 1, 2, \ldots$ nodes are 0, 1, 1, 3, 10, 56, 468, ... (Sloane's A002218). A graph can be tested for biconnectivity using `BiconnectedQ[g]` in the *Mathematica* add-on package Discrete-

`Math'Combinatorica'` (which can be loaded with the command `< <DiscreteMath'`).

Any graph containing a node of degree 1 cannot be biconnected. All HAMILTONIAN GRAPHS are biconnected (Skiena 1990, p. 177).

See also ARTICULATION VERTEX, BLOCK, CONNECTED GRAPH, K-CONNECTED GRAPH

References

Harary, F. *Graph Theory.* Reading, MA: Addison-Wesley, 1994.

Skiena, S. *Implementing Discrete Mathematics: Combinatorics and Graph Theory with Mathematica.* Reading, MA: Addison-Wesley, 1990.

Sloane, N. J. A. Sequences A002218/M2873 in "An On-Line Version of the Encyclopedia of Integer Sequences." http://www.research.att.com/~njas/sequences/eisonline.html.

Bicorn

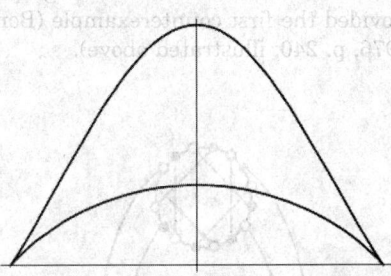

The bicorn is the name of a collection of QUARTIC CURVES studied by Sylvester in 1864 and Cayley in 1867 (MacTutor Archive). The bicorn is given by the PARAMETRIC EQUATIONS

$$x = a \sin t \tag{1}$$

$$y = \frac{a \cos^2 t (2 + \cos t)}{3 + \sin^2 t} \tag{2}$$

and Cartesian equation

$$y^2(a^2 - x^2) = (x^2 + 2ay - a^2)^2 \tag{3}$$

(Mactutor, with the final a squared instead of to the first power). The graph of the bicorn is similar to that of the COCKED HAT CURVE.

The CURVATURE is given by

$$\kappa = \frac{6\sqrt{2}(\cos t - 2)^3(3 \cos t - 2) \sec t}{a[73 - 80 \cos t + 9 \cos(2t)]^{3/2}}. \tag{4}$$

References

Lawrence, J. D. *A Catalog of Special Plane Curves.* New York: Dover, pp. 147–49, 1972.

MacTutor History of Mathematics Archive. "Bicorn." http://www-groups.dcs.st-and.ac.uk/~history/Curves/Bicorn.html.

Bicubic Graph

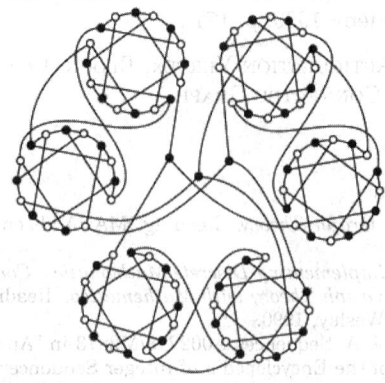

A BIPARTITE CUBIC GRAPH. Tutte (1971) conjectured that all 3-connected bicubic graphs are Hamiltonian (the TUTTE CONJECTURE). The Horton graph on 96 nodes provided the first counterexample (Bondy and Murty 1976, p. 240; illustrated above).

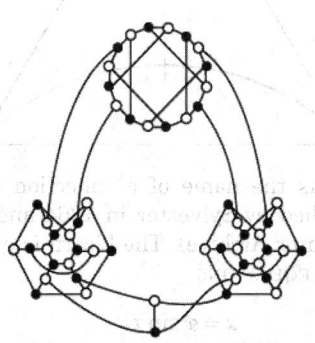

Horton subsequently found a counterexample on 92 nodes (Horton 1982). Two smaller (nonisomorphic) counterexamples on 78 nodes have since been found (Ellingham 1981, 1982b; Owens 1983). Ellingham and Horton (1983) subsequently found a nonhamiltonian 3-connected bicubic graph on 54 vertices, illustrated above.

See also BIPARTITE GRAPH, CUBIC GRAPH, TUTTE CONJECTURE

References

Bondy, J. A. and Murty, U. S. R. *Graph Theory with Applications.* New York: North Holland, pp. 61 and 240, 1976.

Ellingham, M. N. "Non-Hamiltonian 3-Connected Cubic Partite Graphs." Research Report No. 28, Dept. of Math., Univ. Melbourne, Melbourne, 1981.

Ellingham, M. N. *Cycles in 3-Connected Cubics Graphs.* M.Sc. thesis. Melbourne, Australia: University of Melbourne, June 1982a.

Ellingham, M. N. "Constructing Certain Cubic Graphs." In *Combinatorial Mathematics, IX: Proceedings of the Ninth Australian Conference held at the University of Queensland, Brisbane, August 24–8, 1981)* (Ed. E. J. Billington, S. Oates-Williams, and A. P. Street). Berlin: Springer-Verlag, pp. 252–74, 1982b.

Ellingham, M. N. and Horton, J. D. "Non-Hamiltonian 3-Connected Cubic Bipartite Graphs." *J. Combin. Th. Ser. B* **34**, 350–53, 1983.

Gropp, H. "Configurations and the Tutte Conjecture." *Ars. Combin. A* **29**, 171–77, 1990.

Horton, J. D. "On Two-Factors of Bipartite Regular Graphs." *Discr. Math.* **41**, 35–1, 1982.

Owens, P. J. "Bipartite Cubic Graphs and a Shortness Exponent." *Disc. Math.* **44**, 327–30, 1983.

Tutte, W. T. "On the 2-Factors of Bicubic Graphs." *Discr. Math.* **1**, 203–08, 1971.

Bicubic Spline

A bicubic spline is a special case of bicubic interpolation which uses an interpolation function OF THE FORM

$$y(x_1, x_2) = \sum_{i=1}^{4} \sum_{j=1}^{4} c_{ij} t^{i-1} u^{j-1}$$

$$y_{x_1}(x_1, x_2) = \sum_{i=1}^{4} \sum_{j=1}^{4} (i-1) c_{ij} t^{i-2} u^{j-1}$$

$$y_{x_2}(x_1, x_2) = \sum_{i=1}^{4} \sum_{j=1}^{4} (j-1) c_{ij} t^{i-1} u^{j-2}$$

$$y_{x_1 x_2} = \sum_{i=1}^{4} \sum_{j=1}^{4} (i-1)(j-1) c_{ij} t^{i-2} u^{j-2},$$

where c_{ij} are constants and u and t are parameters ranging from 0 to 1. For a bicubic spline, however, the partial derivatives at the grid points are determined globally by 1-D SPLINES.

See also B-SPLINE, SPLINE

References

Press, W. H.; Flannery, B. P.; Teukolsky, S. A.; and Vetterling, W. T. *Numerical Recipes in FORTRAN: The Art of Scientific Computing, 2nd ed.* Cambridge, England: Cambridge University Press, pp. 118–22, 1992.

Bicupola

Two adjoined CUPOLAS.

See also CUPOLA, ELONGATED GYROBICUPOLA, ELONGATED ORTHOBICUPOLA, GYROBICUPOLA, ORTHOBICUPOLA

Bicuspid Curve

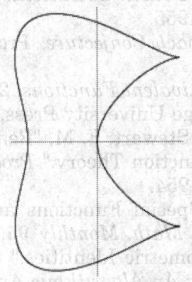

The PLANE CURVE given by the Cartesian equation

$$(x^2 - a^2)(x - a)^2 + (y^2 - a^2)^2 = 0.$$

Bi-Cyclide Coordinates

BICYCLIDE COORDINATES

Bicyclide Coordinates

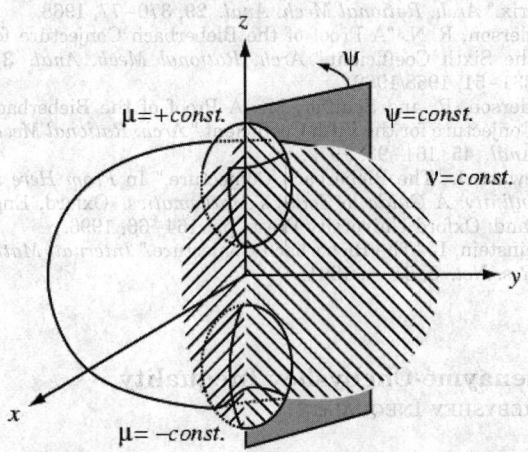

A coordinate system which is similar to BISPHERICAL COORDINATES but having fourth-degree surfaces instead of second-degree surfaces for constant μ. The coordinates are given by the transformation equations

$$x = \frac{a}{\Lambda} \, \mathrm{cn}\, \mu \, \mathrm{dn}\, \mu \, \mathrm{sn}\, \nu \, \mathrm{cn}\, \nu \cos \psi \qquad (1)$$

$$y = \frac{a}{\Lambda} \, \mathrm{cn}\, \mu \, \mathrm{dn}\, \mu \, \mathrm{sn}\, \nu \, \mathrm{cn}\, \nu \sin \psi \qquad (2)$$

$$z = \frac{a}{\Lambda} \sin \mu \, \mathrm{dn}\, \nu, \qquad (3)$$

where

$$\Lambda \equiv 1 - \mathrm{dn}^2 \, \mu \, \mathrm{sn}^2 \, \nu, \qquad (4)$$

$\mu \in [0, K]$, $\nu \in [0, K']$, $\psi \in [0, 2\pi)$, and $\mathrm{cn}\, x$, $\mathrm{dn}\, x$, and

$\mathrm{sn}\, x$ are JACOBI ELLIPTIC FUNCTIONS. Surfaces of constant μ are given by the bicyclides

$$(x^2 + y^2 + z^2)^2$$

$$+ \frac{a^2}{k^4} \frac{(1 - k^2)^2 - 2(1 - k^2)\, \mathrm{dn}^2\, \mu + (1 + k^2)\, \mathrm{dn}^4\, \mu}{\mathrm{dn}^2\, \mu \, \mathrm{cn}^2\, \mu}$$

$$\times (x^2 + y^2) - a^2 \left(\mathrm{sn}^2\, \mu + \frac{1}{k^2 \, \mathrm{sn}^2\, \mu} \right) z^2 + \frac{a^4}{k^2} = 0, \qquad (5)$$

surfaces of constant ν by the cyclides of rotation

$$\left[\frac{\mathrm{cn}^2\, \nu}{a^2 \, \mathrm{sn}^2\, \nu} (x^2 + y^2) + \frac{\mathrm{dn}^2\, \nu}{a^2} \, z^2 \right]^2 - \frac{2\, \mathrm{cn}^2\, \nu}{a^2 \, \mathrm{sn}^2\, \nu} (x^2 + y^2)$$

$$- \frac{2\, \mathrm{dn}^2\, \nu}{a^2} \, z^2 + 1 = 0, \qquad (6)$$

and surfaces of constant ψ by the half-planes

$$\tan \psi = \frac{y}{x}. \qquad (7)$$

See also BISPHERICAL COORDINATES, CAP-CYCLIDE COORDINATES, CYCLIDIC COORDINATES

References

Moon, P. and Spencer, D. E. "Bicyclide Coordinates (μ, ν, ψ)." Fig. 4.08 in *Field Theory Handbook, Including Coordinate Systems, Differential Equations, and Their Solutions,* 2nd ed. New York: Springer-Verlag, pp. 124–26, 1988.

Bicylinder

STEINMETZ SOLID

Bidiakis Cube

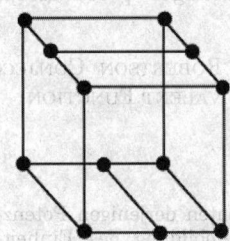 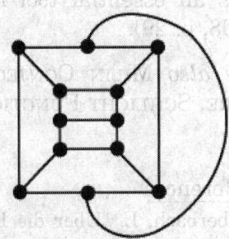

The 12-VERTEX graph consisting of a CUBE in which two opposite faces (say, top and bottom) have edges drawn across them which connect the centers of opposite sides of the faces in such a way that the orientation of the edges added on top and bottom are PERPENDICULAR to each other.

See also BISLIT CUBE, CUBE, CUBICAL GRAPH

Bieberbach Conjecture

The nth COEFFICIENT in the POWER SERIES of a UNIVALENT FUNCTION should be no greater than n. In other words, if

$$f(z) = a_0 + a_1 z + a_2 z^2 + \ldots + a_n z^n + \ldots$$

is a CONFORMAL MAP of a UNIT DISK on any domain, then $|a_n| \leq n|a_1|$. In more technical terms, "geometric extremality implies metric extremality." An alternate formulation is that $|a_j| leq j$ for any SCHLICHT FUNCTION f (Krantz 1999, p. 150).

The conjecture had been proven for the first six terms (the cases $n = 2$, 3, and 4 were done by Bieberbach, Lowner, and Garabedian and Schiffer, respectively), was known to be false for only a finite number of indices (Hayman 1954), and true for a convex or symmetric domain (Le Lionnais 1983). The general case was proved by Louis de Branges (1985). de Branges proved the MILIN CONJECTURE, which established the ROBERTSON CONJECTURE, which in turn established the Bieberbach conjecture (Stewart 1996).

author	result
Bieberbach (1916)	$|a_2| \leq 2$
Löwner (1923)	$|a_3| \leq 3$
Garabedian and Schiffer (1955)	$|a_4| \leq 4$
Pederson (1968), Ozawa (1969)	$|a_6| \leq 6$
Pederson and Schiffer (1972)	$|a_5| \leq 5$
de Branges (1985)	$|a_j| leq j$ for all j

The sum

$$\sum_{j=k}^{n} (-1)^{k+j} \binom{2j}{j-k} \binom{n+j+1}{n-j} e^{-jt}$$

was an essential tool in de Branges' proof (Koepf 1998, p. 29).

See also MILIN CONJECTURE, ROBERTSON CONJECTURE, SCHLICHT FUNCTION, UNIVALENT FUNCTION

References

Bieberbach, L. "Über die Koeffizienten derjenigen Potenzreihen, welche eine schlichte Abbildung des Einheitskreises vermitteln." *Sitzungsber. Preuss. Akad. Wiss.*, pp. 940–55, 1916.

Charzynski, Z. and Schiffer, M. "A New Proof of the Bieberbach Conjecture for the Fourth Coefficient." *Arch. Rational Mech. Anal.* **5**, 187–93, 1960.

de Branges, L. "A Proof of the Bieberbach Conjecture." *Acta Math.* **154**, 137–52, 1985.

Duren, P.; Drasin, D.; Bernstein, A.; and Marden, A. *The Bieberbach Conjecture: Proceedings of the Symposium on the Occasion of the Proof.* Providence, RI: Amer. Math. Soc., 1986.

Garabedian, P. R. "Inequalities for the Fifth Coefficient." *Comm. Pure Appl. Math.* **19**, 199–14, 1966.

Garabedian, P. R.; Ross, G. G.; and Schiffer, M. "On the Bieberbach Conjecture for Even *n*." *J. Math. Mech.* **14**, 975–89, 1965.

Garabedian, R. and Schiffer, M. "A Proof of the Bieberbach Conjecture for the Fourth Coefficient." *J. Rational Mech. Anal.* **4**, 427–65, 1955.

Gong, S. *The Bieberbach Conjecture.* Providence, RI: Amer. Math. Soc., 1999.

Hayman, W. K. *Multivalent Functions, 2nd ed.* Cambridge, England: Cambridge University Press, 1994.

Hayman, W. K. and Stewart, F. M. "Real Inequalities with Applications to Function Theory." *Proc. Cambridge Phil. Soc.* **50**, 250–60, 1954.

Kazarinoff, N. D. "Special Functions and the Bieberbach Conjecture." *Amer. Math. Monthly* **95**, 689–96, 1988.

Koepf, W. "Hypergeometric Identities." Ch. 2 in *Hypergeometric Summation: An Algorithmic Approach to Summation and Special Function Identities.* Braunschweig, Germany: Vieweg, p. 29, 1998.

Korevaar, J. "Ludwig Bieberbach's Conjecture and its Proof." *Amer. Math. Monthly* **93**, 505–13, 1986.

Krantz, S. G. "The Bieberbach Conjecture." §12.1.2 in *Handbook of Complex Analysis.* Boston, MA: Birkhäuser, pp. 149–50, 1999.

Le Lionnais, F. *Les nombres remarquables.* Paris: Hermann, p. 53, 1983.

Löwner, K. "Untersuchungen über schlichte konforme Abbildungen des Einheitskreises. I." *Math. Ann.* **89**, 103–21, 1923.

Ozawa, M. "On the Bieberbach Conjecture for the Sixth Coefficient." *Kodai Math. Sem. Rep.* **21**, 97–28, 1969.

Pederson, R. N. "On Unitary Properties of Grunsky's Matrix." *Arch. Rational Mech. Anal.* **29**, 370–77, 1968.

Pederson, R. N. "A Proof of the Bieberbach Conjecture for the Sixth Coefficient." *Arch. Rational Mech. Anal.* **31**, 331–51, 1968/1969.

Pederson, R. and Schiffer, M. "A Proof of the Bieberbach Conjecture for the Fifth Coefficient." *Arch. Rational Mech. Anal.* **45**, 161–93, 1972.

Stewart, I. "The Bieberbach Conjecture." In *From Here to Infinity: A Guide to Today's Mathematics.* Oxford, England: Oxford University Press, pp. 164–66, 1996.

Weinstein, L. "The Bieberbach Conjecture." *Internat. Math. Res. Not.* **5**, 61–4, 1991.

Bienaymé-Chebyshev Inequality

CHEBYSHEV INEQUALITY

Bifoliate

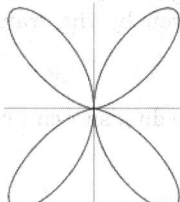

The PLANE CURVE given by the Cartesian equation

$$x^4 + y^4 = 2axy^2.$$

References

Cundy, H. and Rollett, A. *Mathematical Models, 3rd ed.* Stradbroke, England: Tarquin Pub., p. 72, 1989.

Bifolium

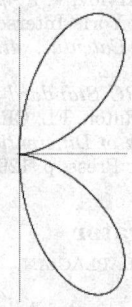

A FOLIUM with $b = 0$. The bifolium is the PEDAL CURVE of the DELTOID, where the PEDAL POINT is the MIDPOINT of one of the three curved sides. The Cartesian equation is

$$(x^2 + y^2)^2 = 4axy^2$$

and the POLAR equation is

$$r = 4a \sin^2 \theta \cos \theta.$$

See also FOLIUM, QUADRIFOLIUM, TRIFOLIUM

References

Beyer, W. H. *CRC Standard Mathematical Tables, 28th ed.* Boca Raton, FL: CRC Press, p. 214, 1987.

Lawrence, J. D. *A Catalog of Special Plane Curves.* New York: Dover, pp. 152–53, 1972.

MacTutor History of Mathematics Archive. "Double Folium." http://www-groups.dcs.st-and.ac.uk/~history/Curves/Double.html.

Bifurcation

A period doubling, quadrupling, etc., that accompanies the onset of CHAOS. It represents the sudden appearance of a qualitatively different solution for a nonlinear system as some parameter is varied. Bifurcations come in four basic varieties: FLIP BIFURCATION, FOLD BIFURCATION, PITCHFORK BIFURCATION, and TRANSCRITICAL BIFURCATION (Rasband 1990).

See also CODIMENSION, FEIGENBAUM CONSTANT, FEIGENBAUM FUNCTION, FLIP BIFURCATION, HOPF BIFURCATION, LOGISTIC MAP, PERIOD DOUBLING, PITCHFORK BIFURCATION, TANGENT BIFURCATION, TRANSCRITICAL BIFURCATION

References

Guckenheimer, J. and Holmes, P. "Local Bifurcations." Ch. 3 in *Nonlinear Oscillations, Dynamical Systems, and Bifurcations of Vector Fields, 2nd pr., rev. corr.* New York: Springer-Verlag, pp. 117–65, 1983.

Lichtenberg, A. J. and Lieberman, M. A. "Bifurcation Phenomena and Transition to Chaos in Dissipative Systems." Ch. 7 in *Regular and Chaotic Dynamics, 2nd ed.* New York: Springer-Verlag, pp. 457–69, 1992.

Rasband, S. N. "Asymptotic Sets and Bifurcations." §2.4 in *Chaotic Dynamics of Nonlinear Systems.* New York: Wiley, pp. 25–1, 1990.

Weisstein, E. W. "Books about Chaos." http://www.treasuretroves.com/books/Chaos.html.

Wiggins, S. "Local Bifurcations." Ch. 3 in *Introduction to Applied Nonlinear Dynamical Systems and Chaos.* New York: Springer-Verlag, pp. 253–19, 1990.

Bifurcation Theory

The study of the nature and properties of BIFURCATIONS.

See also CHAOS, DYNAMICAL SYSTEM

References

Chen, Z.; Chow, S.-N.; and Li, K. (Eds.) *Bifurcation Theory and Its Numerical Analysis: Proceedings of the 2nd International Conference, Xi'an China, June 29-July 3, 1998.* Singapore: Springer-Verlag, 1999.

Bigraph

BIPARTITE GRAPH

Bigyrate Diminished Rhombicosidodecahedron

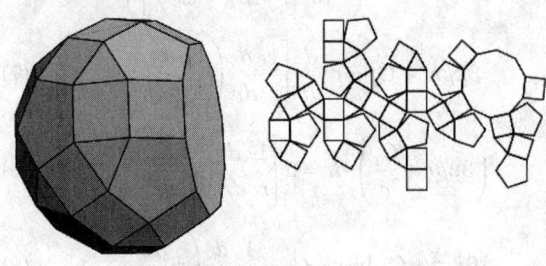

JOHNSON SOLID J_{79}.

References

Weisstein, E. W. "Johnson Solids." MATHEMATICA NOTEBOOK JOHNSONSOLIDS.M.

Weisstein, E. W. "Johnson Solid Netlib Database." MATHEMATICA NOTEBOOK JOHNSONSOLIDS.DAT.

Biharmonic Equation

The differential equation obtained by applying the BIHARMONIC OPERATOR and setting to zero.

$$\nabla^4 \phi = 0. \tag{1}$$

In CARTESIAN COORDINATES, the biharmonic equation is

$$\nabla^4 \phi = \nabla^2(\nabla^2)\phi$$
$$= \left(\frac{\partial^2}{\partial x^2} + \frac{\partial^2}{\partial y^2} + \frac{\partial^2}{\partial z^2}\right)\left(\frac{\partial^2}{\partial x^2} + \frac{\partial^2}{\partial y^2} + \frac{\partial^2}{\partial z^2}\right)\phi$$
$$= \frac{\partial^4 \phi}{\partial x^4} + \frac{\partial^4 \phi}{\partial y^4} + \frac{\partial^4 \phi}{\partial z^4} + 2\frac{\partial^4 \phi}{\partial x^2 \partial y^2} + 2\frac{\partial^4 \phi}{\partial y^2 \partial z^2} + 2\frac{\partial^4 \phi}{\partial x^2 \partial z^2}$$
$$= 0. \tag{2}$$

In POLAR COORDINATES (Kaplan 1984, p. 148)

$$\nabla^4 \phi = \phi_{rrrr} + \frac{2}{r^2}\phi_{rr\theta\theta} + \frac{1}{r^4}\phi_{\theta\theta\theta\theta} + \frac{2}{r}\phi_{rrr} - \frac{2}{r^3}\phi_{r\theta\theta}$$

$$-\frac{1}{r^2}\phi_{rr} + \frac{4}{r^4}\phi_{\theta\theta} + \frac{1}{r^3}\phi_r = 0. \qquad (3)$$

For a radial function $\phi(r)$, the biharmonic equation becomes

$$\nabla^4\phi = \frac{1}{r}\frac{d}{dr}\left\{ r\frac{d}{dr}\left[\frac{1}{r}\frac{d}{dr}\left(r\frac{d\phi}{dr}\right)\right]\right\}$$

$$= \phi_{rrrr} + \frac{2}{r}\phi_{rrr} - \frac{1}{r^2}\phi_{rr} + \frac{1}{r^3}\phi_r = 0. \qquad (4)$$

Writing the inhomogeneous equation as

$$\nabla^4\phi = 64\beta, \qquad (5)$$

we have

$$64\beta r\,dr = d\left\{ r\frac{d}{dr}\left[\frac{1}{r}\frac{d}{dr}\left(r\frac{d\phi}{dr}\right)\right]\right\} \qquad (6)$$

$$32\beta r^2 + C_1 = r\frac{d}{dr}\left[\frac{1}{r}\frac{d}{dr}\left(r\frac{d\phi}{dr}\right)\right] \qquad (7)$$

$$\left(32\beta r + \frac{C_1}{r}\right) dr = d\left[\frac{1}{r}\frac{d}{dr}\left(r\frac{d\phi}{dr}\right)\right] \qquad (8)$$

$$16\beta r^2 + C_1 \ln r + C_2 = \frac{1}{r}\frac{d}{dr}\left(r\frac{d\phi}{dr}\right) \qquad (9)$$

$$(16\beta r^3 + C_1 r \ln r + C_2 r)\,dr = d\left(r\frac{d\phi}{dr}\right). \qquad (10)$$

Now use

$$\int r \ln r\,dr = \tfrac{1}{2}r^2 \ln r - \tfrac{1}{4}r^2 \qquad (11)$$

to obtain

$$4\beta r^4 + C_1(\tfrac{1}{2}r^2 \ln r - \tfrac{1}{4}r^2) + \tfrac{1}{2}C_2 r^2 + C_3 = r\frac{d\phi}{dr} \qquad (12)$$

$$\left(4\beta r^3 + C_1' r \ln r + C_2' r + \frac{C_3}{r}\right) dr = d\phi \qquad (13)$$

$$\phi(r) = \beta r^4 + C_1'(\tfrac{1}{2}r^2 \ln r - \tfrac{1}{4}r^2) + \tfrac{1}{2}C_2' r^2 + C_3 \ln r + C_4$$

$$= \beta r^4 + ar^2 + b + (cr^2 + d)\ln\left(\frac{r}{R}\right). \qquad (14)$$

The homogeneous biharmonic equation can be separated and solved in 2-D BIPOLAR COORDINATES.

See also BIHARMONIC OPERATOR, VON KÁRMÁN EQUATIONS

References

Kantorovich, L. V. and Krylov, V. I. *Approximate Methods of Higher Analysis.* New York: Interscience, 1958.

Kaplan, W. *Advanced Calculus, 4th ed.* Reading, MA: Addison-Wesley, 1991.

Zwillinger, D. (Ed.). *CRC Standard Mathematical Tables and Formulae.* Boca Raton, FL: CRC Press, p. 417, 1995.

Zwillinger, D. *Handbook of Differential Equations, 3rd ed.* Boston, MA: Academic Press, p. 129, 1997.

Biharmonic Operator

Also known as the BILAPLACIAN.

$$\nabla^4 = (\nabla^2)^2.$$

In n-D space,

$$\nabla^4\left(\frac{1}{r}\right) = \frac{3(15 - 8n + n^2)}{r^5}.$$

See also BIHARMONIC EQUATION, D'ALEMBERTIAN, LAPLACIAN, VON KÁRMÁN EQUATIONS

Biholomorphic Function

CONFORMAL MAPPING

Biholomorphic Map

CONFORMAL MAPPING

Biholomorphic Transformation

CONFORMAL MAPPING

Bijection

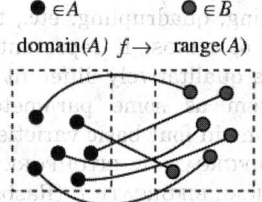

one-to-one and onto (bijection)

A transformation which is ONE-TO-ONE and ONTO.

See also DOMAIN, ONE-TO-ONE, ONTO, PERMUTATION, RANGE (IMAGE)

Bilaplacian

BIHARMONIC OPERATOR

Bilinear Basis

A bilinear basis is a BASIS, which satisfies the conditions

$$(a\mathbf{x} + b\mathbf{y})\cdot\mathbf{z} = a(\mathbf{x}\cdot\mathbf{z}) + b(\mathbf{y}\cdot\mathbf{z})$$

$$\mathbf{z} \cdot (a\mathbf{x} + b\mathbf{y}) = a(\mathbf{z} \cdot \mathbf{x}) + b(\mathbf{z} \cdot \mathbf{y}),$$

See also BASIS, BILINEAR FUNCTION, MULTILINEAR BASIS

Bilinear Form

A bilinear form on a REAL VECTOR SPACE is a function

$$b : V \times V \to \mathbb{R}$$

that satisfies the following axioms for any scalar α and any choice of vectors v, w, v_1, v_2, w_1 and w_2.

1. $b(\alpha v, \ w) = b(v, \ \alpha w) = \alpha b(v, \ w)$
2. $b(v_1 + v_2, \ w) = b(v_1, \ w) + b(v_2, \ w)$
3. $b(v, \ w_1 + w_2) = b(v, \ w_1) = +b(v, \ w_2)$.

For example, the function $b((x_1, \ x_2), \ (y_1, \ y_2)) = x_1 y_2 + x_2 y_1$ is a bilinear form on $\mathbb{R}^2$.

On a COMPLEX VECTOR SPACE, a bilinear form takes values in the COMPLEX NUMBERS. In fact, a bilinear form can take values in any VECTOR SPACE, since the axioms make sense as long as VECTOR ADDITION and SCALAR MULTIPLICATION are defined.

See also BILINEAR FUNCTION, MULTILINEAR FORM, SYMMETRIC BILINEAR FORM, VECTOR SPACE

Bilinear Function

A function of two variables is bilinear if it is linear with respect to each of its variables. The simplest example is $f(x, \ y) = xy$.

See also BILINEAR BASIS, LINEAR FUNCTION, SYMMETRIC BILINEAR FORM

Billiard Table Problem

BILLIARDS

Billiards

The game of billiards is played on a RECTANGULAR table (known as a billiard table) upon which balls are placed. One ball (the "cue ball") is then struck with the end of a "cue" stick, causing it to bounce into other balls and REFLECT off the sides of the table. Real billiards can involve spinning the ball so that it does not travel in a straight LINE, but the mathematical study of billiards generally consists of REFLECTIONS in which the reflection and incidence angles are the same. However, strange table shapes such as CIRCLES and ELLIPSES are often considered.

Many interesting problems can arise in the detailed study of billiards trajectories. For example, any smooth plane convex set has at least two DOUBLE NORMALS, so there are always two distinct "to and fro" paths for any smoothly curved table. More amazingly, there are always $\phi(k)$ distinct k-gonal periodic orbits on smooth billiard table, where $\phi(k)$ is the TOTIENT FUNCTION (Croft *et al.* 1991, p. 16). This gives Steinhaus's result that there are always two distinct periodic triangular orbits (Croft and Swinnerton-Dyer 1963) as a special case. Analysis of billiards path can involve sophisticated use of ERGODIC THEORY and DYNAMICAL SYSTEMS.

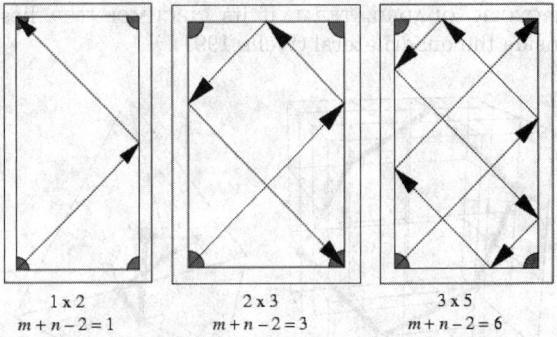

| 1 x 2 | 2 x 3 | 3 x 5 |
| $m+n-2=1$ | $m+n-2=3$ | $m+n-2=6$ |

Given a rectangular billiard table with only corner pockets and sides of INTEGER lengths m and n (with m and n RELATIVELY PRIME), a ball sent at a 45° angle from a corner will be pocketed in another corner after $m + n - 2$ bounces (Steinhaus 1983, p. 63; Gardner 1984, pp. 211–14). Steinhaus (1983, p. 64) also gives a method for determining how to hit a billiard ball such that it caroms off all four sides before hitting a second ball (Knaster and Steinhaus 1946, Steinhaus 1948).

ALHAZEN'S BILLIARD PROBLEM seeks to find the point at the edge of a circular "billiards" table at which a cue ball at a given point must be aimed in order to carom once off the edge of the table and strike another ball at a second given point. It was not until 1997 that Neumann proved that the problem is insoluble using a COMPASS and RULER construction.

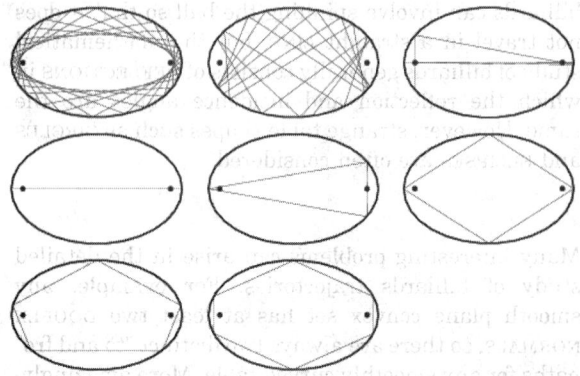

On an ELLIPTICAL billiard table, the ENVELOPE of a trajectory is a smaller ELLIPSE, a HYPERBOLA, a LINE through the FOCI of the ELLIPSE, or a closed polygon (Steinhaus 1983, pp. 239 and 241; Wagon 1991). The closed polygon case is related to PONCELET'S PORISM.

The only closed billiard path of a single circuit in an ACUTE TRIANGLE is the PEDAL TRIANGLE. There are an infinite number of multiple-circuit paths, but all segments are parallel to the sides of the PEDAL TRIANGLE. There exists a closed billiard path inside a CYCLIC QUADRILATERAL if its CIRCUMCENTER lies inside the quadrilateral (Wells 1991).

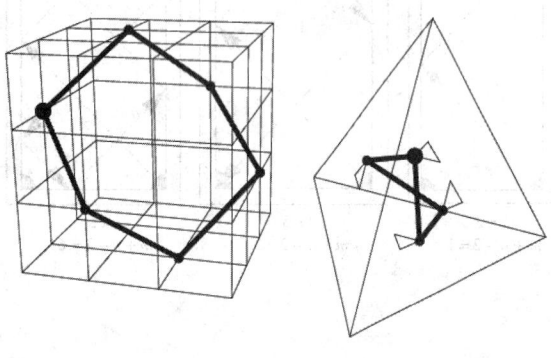

There are four identical closed billiard paths inside and touching each face of a CUBE such that each leg on the path has the same length (Hayward 1962; Steinhaus 1979; Steinhaus 1983; Gardner 1984, pp. 33–5; Wells 1991). This path is in the form of a chair-shaped hexagon, and each leg has length $\sqrt{3}/3$. For a unit cube, one such path has vertices (0, 2/3, 2/3), (1/3, 1, 1/3), (2/3, 2/3, 0), (1, 1/3, 1/3), (2/3, 0, 2/3), (1/3, 1/3, 1). Lewis Carroll (Charles Dodgson) also considered this problem (Weaver 1954).

There are three identical closed billiard paths inside and touching each face of a TETRAHEDRON such that each leg of the path has the same length (Gardner 1984, pp. 35–6; Wells 1991). These were discovered by J. H. Conway and independently by Hayward (1962). The vertices of the path are appropriately chosen vertices of equilateral triangles in each facial

plane which are scaled by a factor of 1/10. For a tetrahedron with unit side lengths, each leg has length $\sqrt{10}/10$. For a tetrahedron with vertices (0, 0, 0), (0, $\sqrt{2}/2$, $\sqrt{2}/2$), ($\sqrt{2}/2$, 0, $\sqrt{2}/2$), ($\sqrt{2}/2$, $\sqrt{2}/2$, 0), the vertices of one such path are ($3\sqrt{2}/20$, $7\sqrt{2}/20$, $\sqrt{2}/5$), ($3\sqrt{2}/20$, $3\sqrt{2}/20$, $3\sqrt{2}/10$), ($7\sqrt{2}/20$, $3\sqrt{2}/20$, $\sqrt{2}/5$), ($7\sqrt{2}/20$, $7\sqrt{2}/20$, $3\sqrt{2}/10$).

Conway has shown that period orbits exist in all TETRAHEDRA, but it is not known if there are periodic orbits in every POLYHEDRON (Croft *et al.* 1991, p. 16).

See also ALHAZEN'S BILLIARD PROBLEM, BILLIARD TABLE PROBLEM, PONCELET'S PORISM, REFLECTION PROPERTY, SALMON'S THEOREM

References

Altshiller Court, N. "Pouring Problems: The Robot Method." *Mathematics in Fun and Earnest.* New York: Dial Press, pp. 223–31, 1958.

Bakst, A. *Mathematical Puzzles and Pastimes.* New York: Van Nostrand, pp. 10–1, 1954.

Bellman, R. E.; Cooke, K. L.; and Lockett, J. A. Ch. 5 in *Algorithms, Graphs, and Computers.* New York: Academic Press, 1970.

Boldrighini, C.; Keane, M.; and Marchetti, F. "Billiards in Polygons." *Ann. Probab.* **6**, 532–40, 1978.

Coxeter, H. S. M. and Greitzer, S. L. *Geometry Revisited.* Washington, DC: Math. Assoc. Amer., pp. 89–3, 1967.

Croft, H. T.; Falconer, K. J.; and Guy, R. K. "Billiard Ball Trajectories in Convex Regions." §A4 in *Unsolved Problems in Geometry.* New York: Springer-Verlag, pp. 15–8, 1991.

Croft, H. T. and Swinnerton, H. P. F. "On the Steinhaus Billiard Table Problem." *Proc. Cambridge Philos. Soc.* **59**, 37–1, 1963.

Davis, D.; Ewing, C.; He, Z.; and Shen, T. "The Billiards Simulation." http://serendip.brynmawr.edu/chaos/home.html.

De Temple, D. W. and Robertson, J. M. "A Billiard Path Characterization of Regular Polygons." *Math. Mag.* **54**, 73–5, 1981.

De Temple, D. E. and Robertson, J. M. "Convex Curves with Periodic Billiard Polygons." *Math. Mag.* **58**, 40–2, 1985.

Dullin, H. R.; Richter, P. H.; and Wittek, A. "A Two-Parameter Study of the Extent of Chaos in a Billiard System." *Chaos* **6**, 43–8, 1996.

Gardner, M. "Bouncing Balls in Polygons and Polyhedrons." Ch. 4 in *The Sixth Book of Mathematical Games from Scientific American.* Chicago, IL: University of Chicago Press, pp. 29–8 and 211–14, 1984.

Gutkin, E. "Billiards in Polygons." *Physica D* **19**, 311–33, 1986.

Halpern, B. "Strange Billiard Tables." *Trans. Amer. Math. Soc.* **232**, 297–05, 1977.

Hayward, R. "The Bouncing Billiard Ball." *Recr. Math. Mag.*, No. 9, 16–8, June 1962.

Klamkin, M. S. "Problem 116." *Pi Mu Epsilon J.* **3**, 410–11, Spring 1963.

Knaster, B. and Steinhaus, H. *Ann. de la Soc. Polonaise de Math.* **19**, 228–31, 1946.

Knuth, D. E. "Billiard Balls in an Equilateral Triangle." *Recr. Math. Mag.* **14**, 20–3, Jan. 1964.

Madachy, J. S. "Bouncing Billiard Balls." In *Madachy's Mathematical Recreations.* New York: Dover, pp. 231–41, 1979.

Marlow, W. C. *The Physics of Pocket Billiards.* Philadelphia, PA: AIP, 1995.

Mauldin, R. D. (Ed.). Problem 147 in *The Scottish Book: Math at the Scottish Cafe.* Boston, MA: Birkhäuser, 1982.

Neumann, P. Submitted to *Amer. Math. Monthly.*

O'Beirne, T. H. Ch. 4 in *Puzzles and Paradoxes: Fascinating Excursions in Recreational Mathematics.* New York: Dover, 1984.

Pappas, T. "Mathematics of the Billiard Table." *The Joy of Mathematics.* San Carlos, CA: Wide World Publ./Tetra, p. 43, 1989.

Peterson, I. "Billiards in the Round." http://www.science-news.org/sn_arc97/3_1_97/mathland.htm.

Sine, R. and Krei͡novic, V. "Remarks on Billiards." *Amer. Math. Monthly* **86**, 204–06, 1979.

Steinhaus, H. *Econometrica* **16**, 101–04, 1948.

Steinhaus, H. "Problems P.175, P.176, and P.181." *Colloq. Math.* **4**, 243 and 262, 1957.

Steinhaus, H. Problem 33 in *One Hundred Problems in Elementary Mathematics.* New York: Dover, 1979.

Steinhaus, H. *Mathematical Snapshots, 3rd ed.* New York: Dover, 1999.

Tabachnikov, S. *Billiards.* Providence, RI: Amer. Math. Soc., 1995.

Turner, P. H. "Convex Caustics for Billiards in $\mathbb{R}^2$ and $\mathbb{R}^3$." In *Conference on Convexity and Related Combinatorial Geometry, Oklahoma, 1980* (Ed. D. C. Kay and M. Breen). New York: Dekker, 1982.

Tweedie, M. C. K. "A Graphical Method of Solving Tartaglian Measuring Problems." *Math. Gaz.* **23**, 278–82, 1939.

Wagon, S. "Billiard Paths on Elliptical Tables." §10.2 in *Mathematica in Action.* New York: W. H. Freeman, pp. 330–33, 1991.

Weaver, W. "The Mathematical Manuscripts of Lewis Carroll." *Proc. Amer. Philosoph. Soc.* **98**, 377–81, 1954.

Wells, D. *The Penguin Dictionary of Curious and Interesting Geometry.* London: Penguin, pp. 13–5, 1991.

Billion

The word billion denotes different numbers in American and British usage. In the American system, one billion equals 10^9. In the British, French, and German systems, one billion equals 10^{12}. Fortunately, in recent years, the "American" system has become common in both the United States and Britain.

See also LARGE NUMBER, MILLIARD, MILLION, TRILLION

Bilunabirotunda

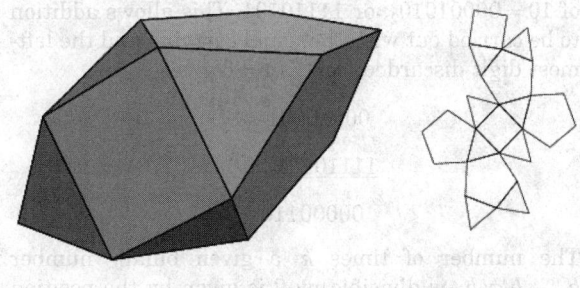

JOHNSON SOLID J_{91}

References

Weisstein, E. W. "Johnson Solids." MATHEMATICA NOTEBOOK JOHNSONSOLIDS.M.

Weisstein, E. W. "Johnson Solid Netlib Database." MATHEMATICA NOTEBOOK JOHNSONSOLIDS.DAT.

Bimagic Cube

A bimagic cube of order 25 is known.

See also MAGIC CUBE

References

Hendricks, J. R. *A Bimagic Cube: Order 25.* Published by the author, 2000.

Bimagic Square

16	41	36	5	27	62	55	18
26	63	54	19	13	44	33	8
1	40	45	12	22	51	58	31
23	50	59	30	4	37	48	9
38	3	10	47	49	24	29	60
52	21	32	57	39	2	11	46
43	14	7	34	64	25	20	53
61	28	17	56	42	15	6	35

If replacing each number by its square in a MAGIC SQUARE produces another MAGIC SQUARE, the square is said to be a bimagic square. Bimagic squares are also called DOUBLY MAGIC SQUARES, and are 2-MULTIMAGIC SQUARES.

The first known bimagic square (shown above) has order 8 with magic constant 260 for addition and 11,180 after squaring. It is believed that no bimagic squares of order less than 8 exists (Benson and Jacoby 1976), and Hendricks (1998) shows that a bimagic square of order 3 is impossible for *any* set of numbers except the trivial case of using the same number 9 times.

See also MAGIC SQUARE, MULTIMAGIC SQUARE, TRIMAGIC SQUARE

References

Ball, W. W. R. and Coxeter, H. S. M. *Mathematical Recreations and Essays, 13th ed.* New York: Dover, p. 212, 1987.

Benson, W. H. and Jacoby, O. *New Recreations with Magic Squares.* New York: Dover, 1976.

Hendricks, J. R. "Note on the Bimagic Square of Order 3." *J. Recr. Math.* **29**, 265–67, 1998.

Hunter, J. A. H. and Madachy, J. S. "Mystic Arrays." Ch. 3 in *Mathematical Diversions.* New York: Dover, p. 31, 1975.

Kraitchik, M. "Multimagic Squares." §7.10 in *Mathematical Recreations.* New York: W. W. Norton, pp. 143 and 176–78, 1942.

Bimedian

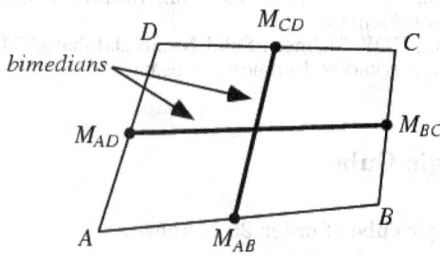

A LINE SEGMENT joining the MIDPOINTS of opposite sides of a QUADRILATERAL or TETRAHEDRON.

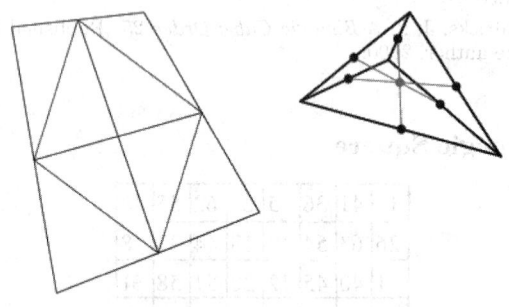

VARIGNON'S THEOREM states that the bimedians of a QUADRILATERAL bisect each other (left figure). In addition, the three bimedians of a tetrahedron are CONCURRENT and bisect each other (right figure; Altshiller-Court 1979, p. 48).

See also COMMANDINO'S THEOREM, MEDIAN (TRIANGLE), VARIGNON'S THEOREM

References

Altshiller-Court, N. *Modern Pure Solid Geometry.* New York: Chelsea, 1979.
Neuberg, J. "Notes Mathématiques: 49. Probléme sur les tétraèdres." *Mathesis* **38**, 446–48, 1924.

Bimodal Distribution

A STATISTICAL DISTRIBUTION having two separated peaks.

See also UNIMODAL DISTRIBUTION

Bimonster

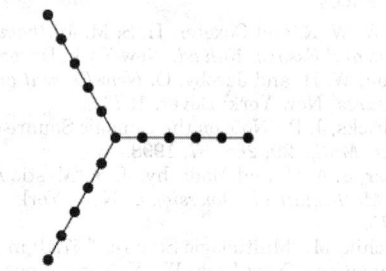

The wreathed product of the MONSTER GROUP by $\mathbb{Z}_2$.

The bimonster is a quotient of the COXETER GROUP with the above COXETER-DYNKIN DIAGRAM. This had been conjectured by Conway, but was proven around 1990 by Ivanov and Norton. If the parameters p, q, r in Coxeter's NOTATION $[3^{p,\,q,\,r}]$ are written side by side, the bimonster can be denoted by the BEAST NUMBER 666.

Bin

An interval into which a given data point does or does not fall.

See also BIN-PACKING PROBLEM, HISTOGRAM

Binary

The BASE 2 method of counting in which only the digits 0 and 1 are used. In this BASE, the number 1011 equals $1 \cdot 2^0 + 1 \cdot 2^1 + 0 \cdot 2^2 + 1 \cdot 2^3 = 11$. This BASE is used in computers, since all numbers can be simply REPRESENTED AS a string of electrically pulsed ons and offs. The following table gives the binary equivalents of the first few decimal numbers.

1	1	11	1011	21	10101
2	10	12	1100	22	10110
3	11	13	1101	23	10111
4	100	14	1110	24	11000
5	101	15	1111	25	11001
6	110	16	10000	26	11010
7	111	17	10001	27	11011
8	1000	18	10010	28	11100
9	1001	19	10011	29	11101
10	1010	20	10100	30	11110

A NEGATIVE $-n$ is most commonly REPRESENTED AS the complement of the POSITIVE number $n-1$, so $-11 = 00001011_2$ would be written as the complement of $10 = 00001010_2$, or 11110101. This allows addition to be carried out with the usual carrying and the leftmost digit discarded, so $17 - 1 = 6$ gives

$$
\begin{array}{ll}
00010001 & 17 \\
\underline{11110101} & \underline{-11} \\
00000110 & 6
\end{array}
$$

The number of times k a given binary number $b_n \ldots b_2 b_1 b_0$ is divisible by 2 is given by the position of the first $b_k = 1$ counting from the right. For example, $12 = 1100$ is divisible by 2 twice, and $13 = 1101$ is divisible by 2 0 times.

The number of 1s $N(1, n)$ in the binary representation of a number is given by

$$N(1, n) = n - \text{gde}(n!, 2) = n - \sum_{k=1}^{\lfloor \log_2 n \rfloor} \left\lfloor \frac{n}{2^k} \right\rfloor, \quad (1)$$

where $\text{gde}(n!, 2)$ is the GREATEST DIVIDING EXPONENT of 2 with respect to $n!$. This is a special application of the general result that the POWER of a PRIME p dividing a FACTORIAL (Graham *et al.* 1990, Vardi 1991). Writing $a(n)$ for $N(1, n)$, the number of 1s is also given by the RECURRENCE RELATION

$$a(2n) = a(n) \quad (2)$$

$$a(2n + 1) = a(n) + 1, \quad (3)$$

with $a(0) = 0$, and by

$$N(1, n) = 2n - \log_2(d), \quad (4)$$

where d is the DENOMINATOR of

$$\frac{1}{n!} \left[\frac{d^n}{dx^n} (1 - x)^{-1/2} \right]_{x=0}. \quad (5)$$

For $n = 1, 2, \ldots$, the first few values are 1, 1, 2, 1, 2, 2, 3, 1, 2, 2, 3, ... (Sloane's A000120; Smith 1966, Graham 1970, McIlroy 1974).

Unfortunately, the storage of binary numbers in computers is not entirely standardized. Because computers store information in 8-bit bytes (where a bit is a single binary digit), depending on the "word size" of the machine, numbers requiring more than 8 bits must be stored in multiple bytes. The usual FORTRAN77 integer size is 4 bytes long. However, a number REPRESENTED AS (byte1 byte2 byte3 byte4) in a VAX would be read and interpreted as (byte4 byte3 byte2 byte1) on a Sun. The situation is even worse for floating point (real) numbers, which are represented in binary as a MANTISSA and CHARACTERISTIC, and worse still for long (8-byte) reals!

Binary multiplication of single bit numbers (0 or 1) is equivalent to the AND operation, as can be seen in the following MULTIPLICATION TABLE.

$\times$	0	1
0	0	0
1	0	1

See also BASE (NUMBER), BINARY CARRY SEQUENCE, DECIMAL, FACTORIAL, HEXADECIMAL, MOSER-DE BRUIJN SEQUENCE, NEGABINARY, OCTAL, QUATERNARY, RUDIN-SHAPIRO SEQUENCE, STOLARSKY-HARBORTH CONSTANT, TERNARY

References

Graham, R. L. "On Primitive Graphs and Optimal Vertex Assignments." *Ann. New York Acad. Sci.* **175**, 170–86, 1970.

Graham, R. L.; Knuth, D. E.; and Patashnik, O. "Factorial Factors." §4.4 in *Concrete Mathematics: A Foundation for Computer Science, 2nd ed.* Reading, MA: Addison-Wesley, pp. 111–115, 1994.

Heath, F. G. "Origin of the Binary Code." *Sci. Amer.*, Aug. 1972.

Lauwerier, H. *Fractals: Endlessly Repeated Geometric Figures.* Princeton, NJ: Princeton University Press, pp. 6–, 1991.

McIlroy, M. D. "The Number of 1's in Binary Integers: Bounds and Extremal Properties." *SIAM J. Comput.* **3**, 255–61, 1974.

Pappas, T. "Computers, Counting, & Electricity." *The Joy of Mathematics.* San Carlos, CA: Wide World Publ./Tetra, pp. 24–5, 1989.

Press, W. H.; Flannery, B. P.; Teukolsky, S. A.; and Vetterling, W. T. "Error, Accuracy, and Stability" and "Diagnosing Machine Parameters." §1.2 and §20.1 in *Numerical Recipes in FORTRAN: The Art of Scientific Computing, 2nd ed.* Cambridge, England: Cambridge University Press, pp. 18–1, 276, and 881–86, 1992.

Sloane, N. J. A. Sequences A000120/M0105 in "An On-Line Version of the Encyclopedia of Integer Sequences." http://www.research.att.com/~njas/sequences/eisonline.html.

Smith, N. "Problem B-82." *Fib. Quart.* **4**, 374–65, 1966.

Vardi, I. *Computational Recreations in Mathematica.* Reading, MA: Addison-Wesley, p. 67, 1991.

Weisstein, E. W. "Bases." MATHEMATICA NOTEBOOK BASES.M.

Wells, D. *The Penguin Dictionary of Curious and Interesting Numbers.* Middlesex, England: Penguin Books, pp. 42–4, 1986.

Binary Bracketing

A binary bracketing is a BRACKETING built up entirely of binary operations. The number of binary bracketings of n letters (CATALAN'S PROBLEM) are given by the CATALAN NUMBERS C_{n-1}, where

$$C_n \equiv \frac{1}{n+1} \binom{2n}{n} = \frac{1}{n+1} \frac{(2n)!}{n!^2} = \frac{(2n)!}{(n+1)!n!},$$

where $\binom{2n}{n}$ denotes a BINOMIAL COEFFICIENT and $n!$ is the usual FACTORIAL, as first shown by Catalan in 1838. For example, for the four letters a, b, c, and d there are five possibilities: $((ab)c)d$, $(a(bc))d$, $(ab)(cd)$, $a((bc)d$, and $a(b(cd))$, written in shorthand as $((xx)x)x$, $(x(xx))x$, $(xx)(xx)$, $x((xx)x$, and $x(x(xx))$.

See also BRACKETING, CATALAN NUMBER, CATALAN'S PROBLEM

References

Schröder, E. "Vier combinatorische Probleme." *Z. Math. Physik* **15**, 361–76, 1870.

Sloane, N. J. A. Sequences A000108/M1459 in "An On-Line Version of the Encyclopedia of Integer Sequences." http://www.research.att.com/~njas/sequences/eisonline.html.

Sloane, N. J. A. and Plouffe, S. Figure M1459 in *The Encyclopedia of Integer Sequences.* San Diego: Academic Press, 1995.

Stanley, R. P. "Hipparchus, Plutarch, Schröder, and Hough." *Amer. Math. Monthly* **104**, 344–50, 1997.

Binary Carry Sequence

The sequence $a(n)$ given by the exponents of the highest power of 2 dividing n, i.e., the number of trailing 0s in the BINARY representation of n. For $n = 1, 2, \ldots$, the first few are 0, 1, 0, 2, 0, 1, 0, 3, 0, 1, 0, 2, ... (Sloane's A007814). Amazingly, this corresponds to one less than the number of disk to be moved at nth step of optimal solution to TOWERS OF HANOI problem, 1, 2, 1, 3, 1, 2, 1, 4, 1, 2, 1, ... (Sloane's A001511).

The anti-PARITY of this sequence is given by 1, 0, 1, 1, 1, 0, 1, 0, 1, 0, 1, 1, ... (Sloane's A035263) which, amazingly, also corresponds to the ACCUMULATION POINT of 2^n cycles through successive bifurcations.

See also DOUBLE-FREE SET, TOWERS OF HANOI

References

Atanassov, K. "On the 37th and the 38th Smarandache Problems. *Notes on Number Theory and Discrete Mathematics, Sophia, Bulgaria* **5**, 83–5, 1999.

Atanassov, K. *On Some of the Smarandache's Problems.* Lupton, AZ: American Research Press, pp. 16–1, 1999.

Derrida, B.; Gervois, A.; and Pomeau, Y. "Iteration of Endomorphisms on the Real Axis and Representation of Number." *Ann. Inst. Henri Poincaré, Section A: Physique Théorique* **29**, 305–56, 1978.

Karamanos, K. and Nicolis, G. "Symbolic Dynamics and Entropy Analysis of Feigenbaum Limit Sets." *Chaos, Solitons, Fractals* **10**, 1135–150, 1999.

Metropolis, M.; Stein, M. L.; and Stein, P R. "On Finite Limit Sets for Transformations on the Unit Interval." *J. Combin. Th. A* **15**, 25–4, 1973.

Sloane, N. J. A. Sequences A001511/M0127, A007814, and A035263 in "An On-Line Version of the Encyclopedia of Integer Sequences." http://www.research.att.com/~njas/sequences/eisonline.html.

Smarandache, F. *Only Problems, Not Solutions!*, 4th ed. Phoenix, AZ: Xiquan, 1993.

Vitanyi, P. M. B. " An Optimal Simulation of Counter Machines." *SIAM J. Comput.* **14**, 1–3, 1985.

Binary Goldbach Conjecture

GOLDBACH CONJECTURE

Binary Heap

HEAP

Binary Matrix

(0,1)-MATRIX

Binary Operation

This entry contributed by J. BRAD WEATHERLY

A binary operation on a nonempty set A is a map $f : A \times A \to A$, such that f is defined for every element in A and the image of f is unique. Examples of binary operations on A from $A \times A$ to A include + and -.

See also BINARY OPERATOR

Binary Operator

An OPERATOR defined on a set S which takes two elements from S as inputs and returns a single element of S. Binary operators are called compositions by Rosenfeld (1968). Sets possessing a binary multiplication operation include the GROUP, GROUPOID, MONOID, QUASIGROUP, and SEMIGROUP. Sets possessing both a binary multiplication and a binary addition operation include the DIVISION ALGEBRA, FIELD, RING, RINGOID, SEMIRING, and UNIT RING.

See also AND, BINARY OPERATION, BOOLEAN ALGEBRA, CLOSURE (SET), CONNECTIVE, DIVISION ALGEBRA, FIELD, GROUP, GROUPOID, MONOID, OPERATOR, OR, MONOID, NOT, QUASIGROUP, RING, RINGOID, SEMIGROUP, SEMIRING, XNOR, XOR, UNIT RING

References

Rosenfeld, A. *An Introduction to Algebraic Structures.* New York: Holden-Day, 1968.

Binary Quadratic Form

A QUADRATIC FORM in two variables having the form

$$Q(x, y) = a_{11}x^2 + 2a_{12}xy + a_{22}y^2. \quad (1)$$

Consider a binary quadratic form with real coefficients a_{11}, a_{12}, and a_{22}, determinant

$$D \equiv a_{11}a_{22} - a_{12}^2 = 1, \quad (2)$$

and $a_{11} > 0$. Then $Q(x, y)$ is POSITIVE DEFINITE. An important result states that exist two integers x and y not both 0 such that

$$Q(x, y) \leq \frac{2}{\sqrt{3}} \quad (3)$$

for all values of a_{ij} satisfying the above constraint (Hilbert and Cohn-Vossen 1999, p. 39).

See also PELL EQUATION, POSITIVE DEFINITE QUADRATIC FORM, QUADRATIC FORM, QUADRATIC INVARIANT

References

Hilbert, D. and Cohn-Vossen, S. "The Minimum Value of Quadratic Forms." §6.2 in *Geometry and the Imagination.* New York: Chelsea, pp. 39–1, 1999.

Binary Relation

Given a set of objects S, a binary relation is a subset of the CARTESIAN PRODUCT $S \otimes S$.

See also RELATION

References

Skiena, S. *Implementing Discrete Mathematics: Combinatorics and Graph Theory with Mathematica.* Reading, MA: Addison-Wesley, p. 161, 1990.

Binary Remainder Method

An ALGORITHM for computing a UNIT FRACTION (Stewart 1992).

References

Eppstein, D. Egypt.ma Mathematica notebook. http://www.ics.uci.edu/~eppstein/numth/egypt/egypt.ma.

Stewart, I. "The Riddle of the Vanishing Camel." *Sci. Amer.* **266**, 122–24, June 1992.

Binary Search

A SEARCHING algorithm which works on a sorted table by testing the middle of an interval, eliminating the half of the table in which the key cannot lie, and then repeating the procedure iteratively.

See also SEARCHING

References

Lewis, G. N.; Boynton, N. J.; and Burton, F. W. "Expected Complexity of Fast Search with Uniformly Distributed Data." *Inform. Proc. Let.* **13**, 4–, 1981.

Skiena, S. "Backtracking and Distinct Permutations." §1.1.5 in *Implementing Discrete Mathematics: Combinatorics and Graph Theory with Mathematica*. Reading, MA: Addison-Wesley, pp. 12–4, 1990.

Binary Splitting

References

Borwein, J. M. and Borwein, P. B. *Pi & the AGM: A Study in Analytic Number Theory and Computational Complexity.* New York: Wiley, 1987.

Brent, R. P. "The Complexity of Multiple-Precision Arithmetic." *Complexity of Computational Problem Solving* (Ed. R. S. Andressen and R. P. Brent). Brisbane, Australia: University of Queensland Press, 1976.

Gourdon, X. and Sebah, P. "Binary Splitting Method." http://xavier.gourdon.free.fr/Constants/Algorithms/splitting.html.

Haible, B. and Papanikolaou, T. "Fast Multiprecision Evaluation of Series of Rational Numbers." Report TI-97–. TH Darmstadt.

Binary Tree

A TREE with two BRANCHES at each FORK and with one or two LEAVES at the end of each BRANCH. (This definition corresponds to what is sometimes known as an "extended" binary tree.) The height of a binary tree is the number of levels within the TREE. For a binary tree of height H with n nodes,

$$H \leq n \leq 2^H - 1.$$

These extremes correspond to a balanced tree (each node except the LEAVES has a left and right CHILD, and all LEAVES are at the same level) and a degenerate tree (each node has only one outgoing BRANCH), respectively. For a search of data organized into a binary tree, the number of search steps $S(n)$ needed to find an item is bounded by

$$\lg n \leq S(n) \leq n.$$

Partial balancing of an arbitrary tree into a so-called AVL binary search tree can improve search speed.

The number of binary trees with n internal nodes is the CATALAN NUMBER C_n (Sloane's A000108), and the number of binary trees of height b is given by Sloane's A001699. The numbers of binary trees on $n = 1, 2, \ldots$ nodes (i.e., n-node trees having VERTEX DEGREE either 1 or 3; also called 3-Cayley trees, 3-valent trees, or boron trees) are 1, 1, 0, 1, 0, 1, 0, 1, 0, 2, 0, 2, 0 ,4, 0, 6, 0, 11, ... (Sloane's A052120).

See also B-TREE, CAYLEY TREE, COMPLETE BINARY TREE, EXTENDED BINARY TREE, HEAP, QUADTREE, QUATERNARY TREE, RAMUS TREE, RED-BLACK TREE, SPLAY TREE, STERN-BROCOT TREE, WEAKLY BINARY TREE

References

Lucas, J.; Roelants van Baronaigien, D.; and Ruskey, F. "Generating Binary Trees by Rotations." *J. Algorithms* **15**, 343–66, 1993.

Ranum, D. L. "On Some Applications of Fibonacci Numbers." *Amer. Math. Monthly* **102**, 640–45, 1995.

Ruskey, F. "Information on Binary Trees." http://www.theory.csc.uvic.ca/~cos/inf/tree/BinaryTrees.html.

Ruskey, F. and Proskurowski, A. "Generating Binary Trees by Transpositions." *J. Algorithms* **11**, 68–4, 1990.

Skiena, S. S. *The Algorithm Design Manual.* New York: Springer-Verlag, pp. 177–78, 1997.

Sloane, N. J. A. Sequences A000108/M1459, A001699/M3087, and A052120 in "An On-Line Version of the Encyclopedia of Integer Sequences." http://www.research.att.com/~njas/sequences/eisonline.html.

Binet Forms

The two RECURRENCE SEQUENCES

$$U_n = mU_{n-1} + U_{n-2} \qquad (1)$$

$$V_n = mV_{n-1} + V_{n-2} \qquad (2)$$

with $U_0 = 0$, $U_1 = 1$ and $V_0 = 2$, $V_1 = m$, can be solved for the individual U_n and V_n. They are given by

$$U_n = \frac{\alpha^n - \beta^n}{\Delta} \qquad (3)$$

$$V_n = \alpha^n + \beta^n, \qquad (4)$$

where

$$\Delta \equiv \sqrt{m^2 + 4} \qquad (5)$$

$$\alpha \equiv \frac{m + \Delta}{2} \qquad (6)$$

$$\beta \equiv \frac{m - \Delta}{2}. \qquad (7)$$

A useful related identity is

$$U_{n-1} + U_{n+1} = V_n. \tag{8}$$

BINET'S FIBONACCI NUMBER FORMULA is a special case of the Binet form for U_n corresponding to $m = 1$.

See also BINET'S FIBONACCI NUMBER FORMULA, FIBONACCI Q-MATRIX

Binet's Fibonacci Number Formula

A special case of the U_n BINET FORM with $m = 1$, corresponding to the nth FIBONACCI NUMBER,

$$F_n = \frac{(1 + \sqrt{5})^n - (1 - \sqrt{5})^n}{2^n \sqrt{5}}.$$

It was derived by Binet in 1843, although the result was known to Euler and to Daniel Bernoulli more than a century earlier.

See also BINET FORMS, FIBONACCI NUMBER

References

Séroul, R. *Programming for Mathematicians.* Berlin: Springer-Verlag, p. 21, 2000.
Wells, D. *The Penguin Dictionary of Curious and Interesting Numbers.* Middlesex, England: Penguin Books, p. 62, 1986.

Binet's Log Gamma Formulas

Binet's first formula for $\ln \Gamma(z)$, where $\Gamma(z)$ is a GAMMA FUNCTION, is given by

$$\ln \Gamma(z) = (z - \tfrac{1}{2}) \ln z - z + \tfrac{1}{2}\ln(2\pi)$$
$$+ \int_0^\infty [(e^t - 1)^{-1} - t^{-1} + \tfrac{1}{2}]t^{-1}e^{-tz}\, dt$$

for $\Re[z] > 0$ (Erdélyi *et al.* 1981, p. 21). Binet's second formula is

$$\ln \Gamma(z) = \left(z - \tfrac{1}{2}\right)\ln z - z + \tfrac{1}{2}\ln(2\pi) + 2\int_0^\infty \frac{\tan\left(\frac{t}{2}\right)}{e^{2\pi t} - 1}\, dt$$

for $\Re[z] > 0$ (Erdélyi *et al.* 1981, p. 22; Whittaker and Watson 1990, p. 251).

See also GAMMA FUNCTION, MALMSTÉN'S FORMULA

References

Erdélyi, A.; Magnus, W.; Oberhettinger, F.; and Tricomi, F. G. *Higher Transcendental Functions, Vol. 1.* New York: Krieger, 1981.
Whittaker, E. T. and Watson, G. N. *A Course in Modern Analysis, 4th ed.* Cambridge, England: Cambridge University Press, 1990.

Binet-Cauchy Identity

The algebraic identity

$$\left(\sum_{i=1}^n a_i c_i\right)\left(\sum_{i=1}^n b_i d_i\right) - \left(\sum_{i=1}^n a_i d_i\right)\left(\sum_{i=1}^n b_i c_i\right)$$

$$= \sum_{1 \le i \le j \le n} (a_i b_j - a_j b_i)(c_i d_j - c_j d_i). \tag{1}$$

Letting $c_i = a_i$ and $d_i = b_i$ gives LAGRANGE'S IDENTITY.

The identity can be coded in *Mathematica* as follows.

```
< <DiscreteMath`Combinatorica`;
  BinetCauchyId[n_] :=  Module[{
    aa = Array[a, n], bb = Array[b, n],
    cc = Array[c, n], dd = Array[d, n]
    },
    aa.cc bb.dd - aa.dd bb.cc ==
              Plus   @@   ((a[#1]b[#2]  -
a[#2]b[#1])(c[#1]d[#2]  -  c[#2]d[#1]) & @@@
KSubsets[Range[n], 2])
  ]
```

The $n = 2$ case then gives

$$(a_1 c_1 + a_2 c_2)(b_1 d_1 + b_2 d_2) - (b_1 c_1 + b_2 c_2)(a_1 d_1 + a_2 d_2)$$

$$= (a_1 b_2 - a_2 b_1)(c_1 d_2 - c_2 d_1). \tag{2}$$

The $n = 3$ case is equivalent to the vector identity

$$(\mathbf{A} \times \mathbf{B}) \cdot (\mathbf{C} \times \mathbf{D}) = (\mathbf{A} \cdot \mathbf{C})(\mathbf{B} \cdot \mathbf{D}) - (\mathbf{A} \cdot \mathbf{D})(\mathbf{B} \cdot \mathbf{C}), \tag{3}$$

where $\mathbf{A} \cdot \mathbf{B}$ is the DOT PRODUCT and $\mathbf{A} \times \mathbf{B}$ is the CROSS PRODUCT. Note that this identity itself is sometimes known as LAGRANGE'S IDENTITY.

See also LAGRANGE'S IDENTITY

References

Mitrinovic, D. S. *Analytic Inequalities.* New York: Springer-Verlag, p. 42, 1970.

Bing's Theorem

If M^3 is a closed oriented connected 3-MANIFOLD such that every simple closed curve in M lies interior to a BALL in M, then M is HOMEOMORPHIC with the HYPERSPHERE, $\mathbb{S}^3$.

See also BALL, HYPERSPHERE

References

Bing, R. H. "Necessary and Sufficient Conditions that a 3-Manifold be S^3." *Ann. Math.* **68**, 17–7, 1958.
Rolfsen, D. *Knots and Links.* Wilmington, DE: Publish or Perish Press, pp. 251–57, 1976.

Binomial

A POLYNOMIAL with 2 terms.

See also BINOMIAL COEFFICIENT, MONOMIAL, POLYNOMIAL, TRINOMIAL

Binomial Coefficient

The number of ways of picking n unordered outcomes from N possibilities, also known as a COMBINATION or combinatorial number. The symbols $_N C_n$ and $\binom{N}{n}$ are used to denote a binomial coefficient, and are sometimes read as "N CHOOSE n." The value of the binomial coefficient is given by

$$ {}_N C_n \equiv \binom{N}{n} \equiv \frac{N!}{(N-n)!\,n!}, \qquad (1) $$

where $n!$ denotes a FACTORIAL. Writing the FACTORIAL as a GAMMA FUNCTION $n! = \Gamma(n+1)$ allows the binomial coefficient to be generalized to non-integral arguments.

The binomial coefficients form the rows of PASCAL'S TRIANGLE, and the number of LATTICE PATHS from the ORIGIN $(0, 0)$ to a point (a, b) is the binomial coefficient $\binom{a+b}{a}$ (Hilton and Pedersen 1991).

For a POSITIVE INTEGER n, the BINOMIAL THEOREM gives

$$ (x+a)^n = \sum_{k=0}^{n} \binom{n}{k} x^k a^{n-k}. \qquad (2) $$

The FINITE DIFFERENCE analog of this identity is known as the CHU-VANDERMONDE IDENTITY. A similar formula holds for NEGATIVE INTEGERS,

$$ (x+a)^{-n} = \sum_{k=0}^{\infty} \binom{-n}{k} x^k a^{-n-k}. \qquad (3) $$

There are a number of elegant BINOMIAL SUMS.

The binomial coefficients satisfy the identities

$$ \binom{n}{0} = \binom{n}{n} = 1 \qquad (4) $$

$$ \binom{n}{k} = \binom{n}{n-k} = (-1)^k \binom{k-n-1}{k} \qquad (5) $$

$$ \binom{n+1}{k} = \binom{n}{k} + \binom{n}{k-1}. \qquad (6) $$

As shown by Kummer in 1852, if p^k is the largest power of a PRIME p that divides $\binom{n+k}{k}$, where n and k are nonnegative integers, then k is the number of carries that occur when k is added to n in base p (Graham *et al.* 1989, Exercise 5.36, p. 245; Ribenboim 1989; Vardi 1991, p. 68). Kummer's result can also be stated in the form that the exponent of a PRIME p dividing $\binom{n}{m}$ is given by the number of integers $j \geq 0$ for which

$$ \mathrm{frac}(m/p^j) > \mathrm{frac}(n/p^j), \qquad (7) $$

where $\mathrm{frac}(x)$ denotes the FRACTIONAL PART of x. This inequality may be reduced to the study of the exponential sums $\Sigma_n \Lambda(n) e(x/n)$, where $\Lambda(n)$ is the MANGOLDT FUNCTION. Estimates of these sums are given by Jutila (1974, 1975), but recent improvements have been made by Granville and Ramare (1996).

R. W. Gosper showed that

$$ f(n) = \binom{n-1}{\frac{1}{2}(n-1)} \equiv (-1)^{(n-1)/2} \pmod{n} \qquad (8) $$

for all PRIMES, and conjectured that it holds *only* for

PRIMES. This was disproved when Skiena (1990) found it also holds for the COMPOSITE NUMBER $n = 3 \times 11 \times 179$. Vardi (1991, p. 63) subsequently showed that $n = p^2$ is a solution whenever p is a WIEFERICH PRIME and that if $n = p^k$ with $k > 3$ is a solution, then so is $n = p^{k-1}$. This allowed him to show that the only solutions for COMPOSITE $n < 1.3 \times 10^7$ are 5907, 1093^2, and 3511^2, where 1093 and 3511 are WIEFERICH PRIMES.

Consider the binomial coefficients $f(n) = \binom{2n-1}{n}$, the first few of which are 1, 3, 10, 35, 126, ... (Sloane's A001700). The GENERATING FUNCTION is

$$ \frac{1}{2}\left[\frac{1}{\sqrt{1-4x}} - 1 \right] = x + 3x^2 + 10x^3 + 35x^4 + \dots. \qquad (9) $$

These numbers are SQUAREFREE only for $n = 2, 3, 4, 6, 9, 10, 12, 36, \dots$ (Sloane's A046097), with no others known. It turns out that $f(n)$ is divisible by 4 unless n belongs to a 2-AUTOMATIC SET S_2, which happens to be the set of numbers whose BINARY representations contain at most two 1s: 1, 2, 3, 4, 5, 6, 8, 9, 10, 12, 16, 17, 18, ... (Sloane's A048645). Similarly, $f(n)$ is divisible by 9 unless n belongs to a 3-AUTOMATIC SET S_3, consisting of numbers n for which the representation of $2n$ in TERNARY consists entirely of 0s and 2s (except possibly for a pair of adjacent 1s; D. Wilson, A. Karttunen). The initial elements of S3 are 1, 2, 3, 4, 6, 7, 9, 10, 11, 12, 13, 18, 19, 21, 22, 27, ... (Sloane's A051382). If $f(n)$ is squarefree, then n must belong to $S = S_2 \cap S_3$. It is very probable that S is finite, but no proof is known. Now, squares larger than 4 and 9 might also divide $f(n)$, but by eliminating these two alone, the only possible n for $n \leq 2^6 4$ are 1, 2, 3, 4, 6, 9, 10, 12, 18, 33, 34, 36, 40, 64, 66, 192, 256, 264, 272, 513, 514, 516, 576 768, 1026, 1056, 2304, 16392, 65664, 81920, 532480, and 545259520. All of these but the last have been checked (D. Wilson), establishing that there are no other n such that $f(n)$ is squarefree for $n \leq 545, 259, 520$.

Erdos showed that the binomial coefficient $\binom{n}{k}$, with $3 \leq k \leq n/2$ is a POWER of an INTEGER for the single case $\binom{50}{3} = 140^2$ (Le Lionnais 1983, p. 48). Binomial coefficients $T_{n-1} = \binom{n}{2}$ are squares a^2 when a^2 is a TRIANGULAR NUMBER, which occur for $a = 1, 6, 35, 204, 1189, 6930, \dots$ (Sloane's A001109). These values of a have the corresponding values $n = 2, 9, 50, 289, 1682, 9801, \dots$ (Sloane's A052436).

The binomial coefficients $\binom{n}{\lfloor n/2 \rfloor}$ are called CENTRAL BINOMIAL COEFFICIENTS, where $\lfloor x \rfloor$ is the FLOOR FUNCTION, although the subset of coefficients $\binom{2n}{n}$ is sometimes also given this name. Erdos and Graham (1980, p. 71) conjectured that the CENTRAL BINOMIAL COEFFICIENT $\binom{2n}{n}$ is *never* SQUAREFREE for $n > 4$, and this is sometimes known as the ERDOS SQUAREFREE CONJECTURE. SÁRKOZY'S THEOREM (Sárkozy 1985) provides a partial solution which states that the BINOMIAL COEFFICIENT $\binom{2n}{n}$ is never SQUAREFREE for all sufficiently large $n \geq n_0$ (Vardi 1991). Granville and Ramare (1996) proved that the *only* SQUAREFREE values are $n = 2$ and 4. Sander (1992) subsequently showed that $\binom{2n+d}{n}$ are also never SQUAREFREE for sufficiently large n as long as d is not "too big."

For p, q, and r distinct PRIMES, then the function (8) satisfies

$$f(pqr)f(p)f(q)f(r) \equiv f(pq)f(pr)f(qr) \pmod{pqr} \quad (10)$$

(Vardi 1991, p. 66).

Most binomial coefficients $\binom{n}{k}$ with $n \geq 2k$ have a prime factor $p \leq n/k$, and Lacampagne *et al.* (1993) conjecture that this inequality is true for all $n > 17.125k$, or more strongly that any such binomial coefficient has LEAST PRIME FACTOR $p \leq n/k$ or $p \leq 17$ with the exceptions $\binom{62}{6}$, $\binom{959}{56}$, $\binom{474}{66}$, $\binom{284}{28}$ for which $p = 19, 19, 23, 29$ (Guy 1994, p. 84).

The binomial coefficient $\binom{m}{n}$ (mod 2) can be computed using the XOR operation n XOR m, making PASCAL'S TRIANGLE mod 2 very easy to construct.

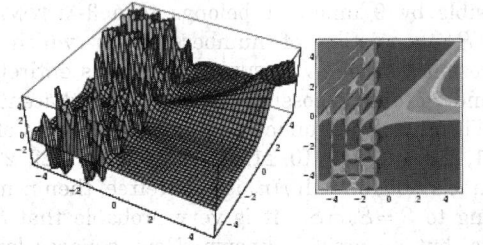

The binomial coefficient "function" can be defined as

$$C(x, y) \equiv \frac{x!}{y!(x - y)!} \quad (11)$$

(Fowler 1996), shown above. It has a very complicated GRAPH for NEGATIVE x and y which is difficult to render using standard plotting programs.

See also APÉRY NUMBER, BALANCED BINOMIAL COEFFICIENT, BALLOT PROBLEM, BINOMIAL DISTRIBUTION, BINOMIAL IDENTITY, BINOMIAL SUMS, BINOMIAL THEOREM, CENTRAL BINOMIAL COEFFICIENT, CHOOSE, CHU-VANDERMONDE IDENTITY, COMBINATION, DEFICIENCY, ERDOS SQUAREFREE CONJECTURE, EXCEPTIONAL BINOMIAL COEFFICIENT, FACTORIAL, GAMMA FUNCTION, GAUSSIAN COEFFICIENT, GAUSSIAN POLYNOMIAL, GOOD BINOMIAL COEFFICIENT, KINGS PROBLEM, KLEE'S IDENTITY, LAH NUMBER, MULTICHOOSE, MULTINOMIAL COEFFICIENT, PERMUTATION, ROMAN COEFFICIENT, SÁRKOZY'S THEOREM, STANLEY'S IDENTITY, STAR OF DAVID THEOREM, STOLARSKY-HARBORTH CONSTANT, STREHL IDENTITIES, SZÉKELY IDENTITY, WOLSTENHOLME'S THEOREM

References

Abramowitz, M. and Stegun, C. A. (Eds.). "Binomial Coefficients." §24.1.1 in *Handbook of Mathematical Functions with Formulas, Graphs, and Mathematical Tables, 9th printing.* New York: Dover, pp. 10 and 822–23, 1972.

Comtet, L. *Advanced Combinatorics: The Art of Finite and Infinite Expansions, rev. enl. ed.* Dordrecht, Netherlands: Reidel, 1974.

Conway, J. H. and Guy, R. K. In *The Book of Numbers.* New York: Springer-Verlag, pp. 66–4, 1996.

Erdos, P.; Graham, R. L.; Nathanson, M. B.; and Jia, X. *Old and New Problems and Results in Combinatorial Number Theory.* New York: Springer-Verlag, 1998.

Erdos, P.; Lacampagne, C. B.; and Selfridge, J. L. "Estimates of the Least Prime Factor of a Binomial Coefficient." *Math. Comput.* **61**, 215–24, 1993.

Feller, W. "Binomial Coefficients" and "Problems and Identities Involving Binomial Coefficients." §2.8 and 2.12 in *An Introduction to Probability Theory and Its Applications, Vol. 1, 3rd ed.* New York: Wiley, pp. 48–0 and 61–4, 1968.

Fowler, D. "The Binomial Coefficient Function." *Amer. Math. Monthly* **103**, 1–7, 1996.

Graham, R. L.; Knuth, D. E.; and Patashnik, O. "Binomial Coefficients." Ch. 5 in *Concrete Mathematics: A Foundation for Computer Science, 2nd ed.* Reading, MA: Addison-Wesley, pp. 153–42, 1994.

Granville, A. and Ramaré, O. "Explicit Bounds on Exponential Sums and the Scarcity of Squarefree Binomial Coefficients." *Mathematika* **43**, 73–07, 1996.

Guy, R. K. "Binomial Coefficients," "Largest Divisor of a Binomial Coefficient," and "Series Associated with the ς-Function." §B31, B33, and F17 in *Unsolved Problems in Number Theory, 2nd ed.* New York: Springer-Verlag, pp. 84–5, 87–9, and 257–58, 1994.

Harborth, H. "Number of Odd Binomial Coefficients." *Not. Amer. Math. Soc.* **23**, 4, 1976.

Hilton, P. and Pedersen, J. "Catalan Numbers, Their Generalization, and Their Uses." *Math. Intel.* **13**, 64–5, 1991.

Jutila, M. "On Numbers with a Large Prime Factor." *J. Indian Math. Soc.* **37**, 43–3, 1973.

Jutila, M. "On Numbers with a Large Prime Factor. II." *J. Indian Math. Soc.* **38**, 125–30, 1974.

Le Lionnais, F. *Les nombres remarquables.* Paris: Hermann, 1983.

Ogilvy, C. S. "The Binomial Coefficients." *Amer. Math. Monthly* **57**, 551–52, 1950.

Press, W. H.; Flannery, B. P.; Teukolsky, S. A.; and Vetterling, W. T. "Gamma Function, Beta Function, Factorials, Binomial Coefficients." §6.1 in *Numerical Recipes in FORTRAN: The Art of Scientific Computing, 2nd ed.* Cambridge, England: Cambridge University Press, pp. 206–09, 1992.

Prudnikov, A. P.; Marichev, O. I.; and Brychkow, Yu. A. Formula 41 in *Integrals and Series, Vol. 1: Elementary Functions.* Newark, NJ: Gordon & Breach, p. 611, 1986.

Ribenboim, P. *The Book of Prime Number Records, 2nd ed.* New York: Springer-Verlag, pp. 23–4, 1989.

Riordan, J. "Inverse Relations and Combinatorial Identities." *Amer. Math. Monthly* **71**, 485–98, 1964.

Sander, J. W. "On Prime Divisors of Binomial Coefficients." *Bull. London Math. Soc.* **24**, 140–42, 1992.

Sárkozy, A. "On the Divisors of Binomial Coefficients, I." *J. Number Th.* **20**, 70–0, 1985.

Skiena, S. *Implementing Discrete Mathematics: Combinatorics and Graph Theory with Mathematica.* Reading, MA: Addison-Wesley, p. 262, 1990.

Sloane, N. J. A. Sequences A001109/M4217, A001700/M2848, A046097, A048645, A051382, and A052436, in "An On-Line Version of the Encyclopedia of Integer

Sequences." http://www.research.att.com/~njas/sequences/eisonline.html.

Spanier, J. and Oldham, K. B. "The Binomial Coefficients $\binom{v}{m}$." Ch. 6 in *An Atlas of Functions*. Washington, DC: Hemisphere, pp. 43–2, 1987.

Sved, M. "Counting and Recounting." *Math. Intel.* **5**, 21–6, 1983.

Vardi, I. "Application to Binomial Coefficients," "Binomial Coefficients," "A Class of Solutions," "Computing Binomial Coefficients," and "Binomials Modulo an Integer." §2.2, 4.1, 4.2, 4.3, and 4.4 in *Computational Recreations in Mathematica*. Redwood City, CA: Addison-Wesley, pp. 25–8 and 63–1, 1991.

Wolfram, S. "Geometry of Binomial Coefficients." *Amer. Math. Monthly* **91**, 566–71, 1984.

Binomial Differential Equation

The ORDINARY DIFFERENTIAL EQUATION

$$(y')^m = f(x, y)$$

(Hille 1969, p. 675; Zwillinger 1997, p. 120).

References

Hille, E. *Lectures on Ordinary Differential Equations*. Reading, MA: Addison-Wesley, 1969.

Zwillinger, D. *Handbook of Differential Equations, 3rd ed.* Boston, MA: Academic Press, p. 120, 1997.

Binomial Distribution

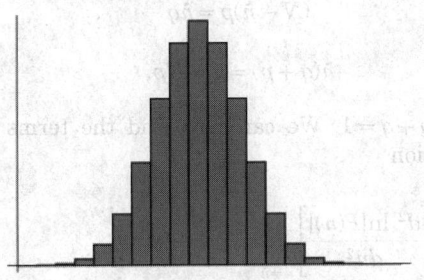

The binomial distribution gives the probability distribution $P_p(n|N)$ of obtaining exactly n successes out of N BERNOULLI TRIALS (where the result of each BERNOULLI TRIAL is true with probability p and false with probability $q = 1 - p$). The binomial distribution is therefore given by

$$P_p(n|N) = \binom{N}{n} p^n (1-p)^{N-n} = \frac{N!}{n!(N-n)!} p^n q^{N-n}. \quad (1)$$

The above plot shows the distribution of n successes out of $N = 20$ trials with $p = q = 1/2$. Steinhaus (1983, pp. 25–8) considers the expected number of squares

$S(n, N, s)$ containing a given number of grains n on board of size s after random distribution of N of grains,

$$S(n, N, s) = sP_{1/s}(n|N). \quad (2)$$

Taking $N = s = 64$ gives the results summarized in the following table.

S	n
0	23.3591
1	23.7299
2	11.8650
3	3.89221
4	0.942162
5	0.179459
6	0.0280109
7	0.0036840
8	4.16639×10^{-4}
9	4.11495×10^{-5}
10	3.59242×10^{-6}

The probability of obtaining *more* successes than the n observed in a binomial distribution is

$$P = \sum_{k=n+1}^{N} \binom{N}{k} p^k (1-p)^{N-k} = I_p(n+1, N-n), \quad (3)$$

where

$$I_x(a, b) \equiv \frac{B(x; a, b)}{B(a, b)}, \quad (4)$$

$B(a, b)$ is the BETA FUNCTION, and $B(x; a, b)$ is the incomplete BETA FUNCTION.

The CHARACTERISTIC FUNCTION for the binomial distribution is

$$\phi(t) = (q + pe^{it})^n \quad (5)$$

(Papoulis 1984, p. 154). The MOMENT-GENERATING FUNCTION M for the distribution is

$$M(t) = \langle e^{tn} \rangle = \sum_{n=0}^{N} e^{tn} \binom{N}{n} p^n q^{N-n}$$

$$= \sum_{n=0}^{N} \binom{N}{n} (pe^t)(1-p)^{N-n} = [pe^t + (1-p)]^N \quad (6)$$

$$M'(t) = N[pe^t + (1-p)]^{N-1}(pe^t) \quad (7)$$

$$M''(t) = N(N-1)[pe^t + (1-p)]^{N-2}(pe^t)^2$$

$$+N[pe^t + (1-p)]^{N-1}(pe^t). \qquad (8)$$

The MEAN is

$$\mu = M'(0) = N(p+1-p)p = Np. \qquad (9)$$

The MOMENTS about 0 are

$$\mu_1' = \mu = Np \qquad (10)$$

$$\mu_2' = Np(1-p+Np) \qquad (11)$$

$$\mu_3' = Np(1 - 3p + 3Np + 2p^2 - 3Np^2 + N^2p^2) \qquad (12)$$

$$\mu_4' = Np(1 - 7p + 7Np + 12p^2 - 18Np^2 + 6N^2p^2 - 6p^3 + 11Np^3 - 6N^2p^3 + N^3p^3), \qquad (13)$$

so the MOMENTS about the MEAN are

$$\mu_2 = \sigma^2 = [N(N-1)p^2 + Np] - (Np)^2$$
$$= N^2p^2 - Np^2 + Np - N^2p^2 = Np(1-p) = Npq \quad (14)$$

$$\mu_3 = \mu_3' - 3\mu_2'\mu_1' + 2(\mu_1')^3 = Np(1-p)(1-2p) \qquad (15)$$

$$\mu_4 = \mu_4' - 4\mu_3'\mu_1' + 6\mu_2'(\mu_1')^2 - 3(\mu_1')^4$$
$$= Np(1-p)[3p^2(2-N) + 3p(N-2) + 1]. \qquad (16)$$

The SKEWNESS and KURTOSIS are

$$\gamma_1 = \frac{\mu_3}{\sigma^3} = \frac{Np(1-p)(1-2p)}{[Np(1-p)]^{3/2}} = \frac{1-2p}{\sqrt{Np(1-p)}}$$
$$= \frac{q-p}{\sqrt{Npq}} \qquad (17)$$

$$\gamma_2 = \frac{\mu_4}{\sigma^4} - 3 = \frac{6p^2 - 6p + 1}{Np(1-p)} = \frac{1 - 6pq}{Npq}. \qquad (18)$$

An approximation to the Bernoulli distribution for large N can be obtained by expanding about the value $\tilde{n}$ where $P(n)$ is a maximum, i.e., where $dP/dn = 0$. Since the LOGARITHM function is MONOTONIC, we can instead choose to expand the LOGARITHM. Let $n \equiv \tilde{n} + \eta$, then

$$\ln[P(n)] = \ln[P(\tilde{n})] + B_1\eta + \tfrac{1}{2}B_2\eta^2 + \tfrac{1}{3!}B_3\eta^3 + \dots, \quad (19)$$

where

$$B_k = \left[\frac{d^k \ln[P(n)]}{dn^k}\right]_{n=\tilde{n}}. \qquad (20)$$

But we are expanding about the maximum, so, by definition,

$$B_1 = \left[\frac{d \ln[P(n)]}{dn}\right]_{n=\tilde{n}} = 0. \qquad (21)$$

This also means that B_2 is negative, so we can write $B_2 = -|B_2|$. Now, taking the LOGARITHM of (1) gives

$$\ln[P(n)] = \ln N! - \ln n! - \ln(N-n)! + n \ln p$$
$$+ (N-n) \ln q. \qquad (22)$$

For large n and $N-n$ we can use STIRLING'S APPROXIMATION

$$\ln(n!) \approx n \ln n - n, \qquad (23)$$

so

$$\frac{d[\ln(n!)]}{dn} \approx (\ln n + 1) - 1 = \ln n \qquad (24)$$

$$\frac{d[\ln(N-n)!]}{dn} \approx \frac{d}{dn}[(N-n)\ln(N-n) - (N-n)]$$
$$= \left[-\ln(N-n) + (N-n)\frac{-1}{N-n} + 1\right]$$
$$= -\ln(N-n), \qquad (25)$$

and

$$\frac{d \ln[P(n)]}{dn} \approx -\ln n + \ln(N-n)\ln p - \ln q. \qquad (26)$$

To find $\tilde{n}$, set this expression to 0 and solve for n,

$$\ln\left(\frac{N-\tilde{n}}{\tilde{n}}\frac{p}{q}\right) = 0 \qquad (27)$$

$$\frac{N-\tilde{n}}{\tilde{n}}\frac{p}{q} = 1 \qquad (28)$$

$$(N-\tilde{n})p = \tilde{n}q \qquad (29)$$

$$\tilde{n}(q+p) = \tilde{n} = Np, \qquad (30)$$

since $p + q = 1$. We can now find the terms in the expansion

$$B_2 = \left[\frac{d^2 \ln[P(n)]}{dn^2}\right]_{n=\tilde{n}} = -\frac{1}{\tilde{n}} - \frac{1}{N-\tilde{n}}$$
$$= -\frac{1}{Np} - \frac{1}{N(1-p)} = -\frac{1}{N}\left(\frac{1}{p} + \frac{1}{q}\right) = -\frac{1}{N}\left(\frac{p+q}{pq}\right)$$
$$= -\frac{1}{Npq} = -\frac{1}{N(1-p)} \qquad (31)$$

$$B_3 \equiv \left[\frac{d^3 \ln[P(n)]}{dn^3}\right]_{n=\tilde{n}} = -\frac{1}{\tilde{n}^2} - \frac{1}{(N-\tilde{n})^2} = \frac{1}{N^2p^2} - \frac{1}{N^2q^2}$$
$$= \frac{q^2 - p^2}{N^2p^2q^2} = \frac{(1 - 2p + p^2) - p^2}{N^2p^2(1-p)^2}$$
$$= \frac{1 - 2p}{N^2p^2(1-p)^2} \qquad (32)$$

$$B_4 \equiv \left[\frac{d^4 \ln[P(n)]}{dn^4}\right]_{n=\tilde{n}} = -\frac{2}{\tilde{n}^3} - \frac{2}{(n-\tilde{n})^3}$$

$$= -2\left(\frac{1}{N^3 p^3} + \frac{1}{N^3 q^3}\right) = \frac{2(p^3 + q^3)}{N^3 p^3 q^3}$$

$$= \frac{2(p^2 - pq + q^2)}{N^3 p^3 q^3}$$

$$= \frac{2[p^2 - p(1-p) + (1 - 2p + p^2)]}{N^3 p^3 (1-p^3)}$$

$$= \frac{2(3p^2 - 3p + 1)}{N^3 p^3 (1-p^3)}. \tag{33}$$

Now, treating the distribution as continuous,

$$\lim_{N \to \infty} \sum_{n=0}^{N} P(n) \approx \int P(n)\, dn = \int_{-\infty}^{\infty} P(\tilde{n} + \eta)\, d\eta = 1. \tag{34}$$

Since each term is of order $1/N \sim 1/\sigma^2$ smaller than the previous, we can ignore terms higher than B_2, so

$$P(n) = P(\tilde{n}) e^{-|B_2|\eta^2/2}. \tag{35}$$

The probability must be normalized, so

$$\int_{-\infty}^{\infty} P(\tilde{n})\, e^{-|B_2|\eta^2/2}\, d\eta = P(\tilde{n})\sqrt{\frac{2\pi}{|B_2|}} = 1, \tag{36}$$

and

$$P(n) = \sqrt{\frac{|B_2|}{2\pi}}\, e^{-|B_2|(n-\tilde{n})^2/2}$$

$$= \frac{1}{\sqrt{2\pi Npq}} \exp\left[-\frac{(n - Np)^2}{2Npq}\right]. \tag{37}$$

Defining $\sigma^2 \equiv Npq$,

$$P(n) = \frac{1}{\sigma \sqrt{2\pi}} \exp\left[-\frac{(n-\tilde{n})^2}{2\sigma^2}\right], \tag{38}$$

which is a GAUSSIAN DISTRIBUTION. For $p \ll 1$, a different approximation procedure shows that the binomial distribution approaches the POISSON DISTRIBUTION. The first CUMULANT is

$$\kappa_1 = np, \tag{39}$$

and subsequent CUMULANTS are given by the RECURRENCE RELATION

$$\kappa_{r+1} = pq\, \frac{d\kappa_r}{dp}. \tag{40}$$

Let x and y be independent binomial RANDOM VARIABLES characterized by parameters n, p and m, p. The CONDITIONAL PROBABILITY of x given that $x + y = k$ is

$$P(x = i | x + y = k) = \frac{P(x = i,\ x + y = k)}{P(x + y = k)}$$

$$= \frac{P(x = i,\ y = k - i)}{P(x + y = k)} = \frac{P(x = i)P(y = k - i)}{P(x + y = k)}$$

$$= \frac{\binom{n}{i} p^i (1-p)^{n-i} \binom{m}{k-i} p^{k-i} (1-p)^{m-(k-i)}}{\binom{n+m}{k} p^k (1-p)^{n+m-k}}$$

$$= \frac{\binom{n}{i}\binom{m}{k-i}}{\binom{n+m}{k}}. \tag{41}$$

Note that this is a HYPERGEOMETRIC DISTRIBUTION.

See also DE MOIVRE-LAPLACE THEOREM, HYPERGEOMETRIC DISTRIBUTION, NEGATIVE BINOMIAL DISTRIBUTION

References

Beyer, W. H. *CRC Standard Mathematical Tables, 28th ed.* Boca Raton, FL: CRC Press, p. 531, 1987.

Papoulis, A. *Probability, Random Variables, and Stochastic Processes, 2nd ed.* New York: McGraw-Hill, pp. 102–03, 1984.

Press, W. H.; Flannery, B. P.; Teukolsky, S. A.; and Vetterling, W. T. "Incomplete Beta Function, Student's Distribution, F-Distribution, Cumulative Binomial Distribution." §6.2 in *Numerical Recipes in FORTRAN: The Art of Scientific Computing, 2nd ed.* Cambridge, England: Cambridge University Press, pp. 219–23, 1992.

Spiegel, M. R. *Theory and Problems of Probability and Statistics.* New York: McGraw-Hill, pp. 108–09, 1992.

Steinhaus, H. *Mathematical Snapshots, 3rd ed.* New York: Dover, 1999.

Binomial Expansion

BINOMIAL SERIES

Binomial Formula

BINOMIAL SERIES, BINOMIAL THEOREM

Binomial Identity

Roman (1984, p. 26) defines "the" binomial identity as the equation

$$p_n(x + y) = \sum_{k=0}^{n} \binom{n}{k} p_k(y) p_{n-k}(x). \tag{1}$$

IFF the sequence $p_n(x)$ satisfies this identity for all y in a FIELD C of characteristic 0, then $p_n(x)$ is an ASSOCIATED SEQUENCE known as a BINOMIAL-TYPE SEQUENCE.

In general, a binomial identity is a formula expressing products of factors as a sum over terms, each including a BINOMIAL COEFFICIENT $\binom{n}{k}$. The prototypical example is the BINOMIAL THEOREM

$$(x+a)^n = \sum_{k=0}^{n} \binom{n}{k} x^k a^{n-k} \qquad (2)$$

for $n > 0$. Abel (1826) gave a host of such identities (Riordan 1979, Roman 1984), some of which include

$$(x+y)(x+y-an)^{n-1}$$
$$= \sum_{k=0}^{n} \binom{n}{k} xy(x-ak)^{k-1}[y-a(n-k)]^{n-k-1}, \qquad (3)$$

$$x^{-1}(x+y-na)^n$$
$$= \sum_{k=0}^{n} \sum_{k=0}^{n} \binom{n}{k}(x-ak)^{k-1}[y-a(n-k)]^{n-k} \qquad (4)$$

(Abel 1826, Riordan 1979, p. 18; Roman 1984, pp. 30 and 73), and

$$x^{-1}(x+y)^n = \sum_{k=0}^{n} \binom{n}{k}(x-ak)^{k-1}(y+ak)^{n-k} \qquad (5)$$

(Saslaw 1989).

See also ABEL'S BINOMIAL THEOREM, ABEL POLYNOMIAL, BINOMIAL COEFFICIENT, DILCHER'S FORMULA, Q-ABEL'S THEOREM

References

Abel, N. H. "Beweis eines Ausdrucks, von welchem die Binomial-Formel ein einzelner Fall ist." *J. reine angew. Math.* **1**, 159–60, 1826. Reprinted in (*Euvres Complètes*, 2nd ed., Vol. 1. pp. 102–03, 1881.

Bhatnagar, G. *Inverse Relations, Generalized Bibasic Series, and their $U(n)$ Extensions.* Ph.D. thesis. Ohio State University, p. 61, 1995.

Comtet, L. *Advanced Combinatorics: The Art of Finite and Infinite Expansions, rev. enl. ed.* Dordrecht, Netherlands: Reidel, p. 128, 1974.

Ekhad, S. B. and Majewicz, J. E. "A Short WZ-Style Proof of Abel's Identity." *Electronic J. Combinatorics* **3**, No. 2, R16, 1, 1996. http://www.combinatorics.org/Volume_3/volume3_2.html.

Foata, D. "Enumerating k-Trees." *Discr. Math.* **1**, 181–86, 1971.

Riordan, J. *Combinatorial Identities.* New York: Wiley, p. 18, 1979.

Roman, S. "The Abel Polynomials." §4.1.5 in *The Umbral Calculus.* New York: Academic Press, pp. 29–0 and 72–5, 1984.

Saslaw, W. C. "Some Properties of a Statistical Distribution Function for Galaxy Clustering." *Astrophys. J.* **341**, 588–98, 1989.

Strehl, V. "Binomial Sums and Identities." *Maple Technical Newsletter* **10**, 37–9, 1993.

Strehl, V. "Binomial Identities--Combinatorial and Algorithmic Aspects." *Discrete Math.* **136**, 309–46, 1994.

Binomial Number

A number OF THE FORM $a^n \pm b^n$, where a, b, and n are INTEGERS. They can be factored algebraically

$$a^n - b^n = (a-b)(a^{n-1} + a^{n-2}b + \ldots + ab^{n-2} + b^{n-1}) \qquad (1)$$

for all n,

$$a^n + b^n = (a+b)(a^{n-1} - a^{n-2}b + \ldots - ab^{n-2} + b^{n-1}) \qquad (2)$$

for n not a power of 2, and

$$a^{nm} - b^{nm} = (a^m - b^m)$$
$$\times [a^{m(n-1)} + a^{m(n-2)}b^m + \ldots + b^{m(n-1)}]. \qquad (3)$$

for all positive integers m, n. For example,

$$a^2 - b^2 = (a-b)(a+b) \qquad (4)$$
$$a^3 - b^3 = (a-b)(a^2 + ab + b^2) \qquad (5)$$
$$a^4 - b^4 = (a-b)(a+b)(a^2 + b^2) \qquad (6)$$
$$a^5 - b^5 = (a-b)(a^4 + a^3b + a^2b^2 + ab^3 + b^4) \qquad (7)$$
$$a^6 - b^6 = (a-b)(a+b)(a^2 - ab + b^2)(a^2 + ab + b^2) \qquad (8)$$
$$a^7 - b^7 = (a-b)(a^6 + a^5b + a^4b^2 + a^3b^3 + a^2b^4 + ab^5 + b^6) \qquad (9)$$
$$a^8 - b^8 = (a-b)(a+b)(a^2 + b^2)(a^4 + b^4) \qquad (10)$$
$$a^9 - b^9 = (a-b)(a^2 + ab + b^2)(a^6 + a^3b^3 + b^6) \qquad (11)$$
$$a^{10} - b^{10} = (a-b)(a+b)(a^4 - a^3b + a^2b^2 - ab^3 + b^4)$$
$$\times (a^4 + a^3b + a^2b^2 + ab^3 + b^4) \qquad (12)$$

and

$$a^2 + b^2 = a^2 + b^2 \qquad (13)$$
$$a^3 + b^3 = (a+b)(a^2 - ab + b^2) \qquad (14)$$
$$a^4 + b^4 = a^4 + b^4 \qquad (15)$$
$$a^5 + b^5 = (a+b)(a^4 - a^3b + a^2b^2 - ab^3 + b^4) \qquad (16)$$
$$a^6 + b^6 = (a^2 + b^2)(a^4 - a^2b^2 + b^4) \qquad (17)$$
$$a^7 + b^7 = (a+b)(a^6 - a^5b + a^4b^2 - a^3b^3 + a^2b^4 - ab^5 + b^6) \qquad (18)$$
$$a^8 + b^8 = a^8 + b^8 \qquad (19)$$
$$a^9 + b^9 = (a+b)(a^2 - ab + b^2)(a^6 - a^3b^3 + b^6) \qquad (20)$$
$$a^{10} + b^{10} = (a^2 + b^2)(a^8 - a^6b^2 + a^4b^4 - a^2b^6 + b^8). \qquad (21)$$

In 1770, Euler proved that if $(a, b) = 1$, then every FACTOR of

$$a^{2^n} + b^{2^n} \qquad (22)$$

is either 2 or OF THE FORM $2^{n+1}K + 1$. (A number OF THE FORM $2^{2^n} + 1$ is called a FERMAT NUMBER.)

If p and q are PRIMES, then

$$\frac{(a^{pq} - 1)(a - 1)}{(a^p - 1)(a^q - 1)} - 1 \qquad (23)$$

is DIVISIBLE by every PRIME FACTOR of a^{p-1} not dividing a^{q-1}.

See also CUNNINGHAM NUMBER, FERMAT NUMBER, MERSENNE NUMBER, RIESEL NUMBER, SIERPINSKI NUMBER OF THE SECOND KIND

References

Guy, R. K. "When Does $2^a - 2^b$ Divide $n^a - n^b$." §B47 in *Unsolved Problems in Number Theory, 2nd ed.* New York: Springer-Verlag, p. 102, 1994.

Qi, S and Ming-Zhi, Z. "Pairs where $2^a - 2^b$ Divides $n^a - n^b$ for All n." *Proc. Amer. Math. Soc.* **93**, 218-20, 1985.

Schinzel, A. "On Primitive Prime Factors of $a^n - b^n$." *Proc. Cambridge Phil. Soc.* **58**, 555-62, 1962.

Binomial Polynomial

FALLING FACTORIAL

Binomial Series

For $|x| < 1$,

$$(1+x)^n = \sum_{k=0}^{n} \binom{n}{k} x^k \qquad (1)$$

$$= \binom{n}{0}x^0 + \binom{n}{1}x^1 + \binom{n}{2}x^2 + \cdots \qquad (2)$$

$$= 1 + \frac{n!}{1!(n-1)!} x + \frac{n!}{(n-2)!2!} x^2 + \cdots \qquad (3)$$

$$= 1 + nx + \frac{n(n-1)}{2} x^2 + \cdots . \qquad (4)$$

The binomial series also has the CONTINUED FRACTION representation

$$(1+x)^n = \cfrac{1}{1 - \cfrac{nx}{1 + \cfrac{1 \cdot (1+n)}{1 \cdot 2}x}{1 + \cfrac{\frac{1 \cdot (1-n)}{2 \cdot 3}x}{1 + \cfrac{\frac{2(2+n)}{3 \cdot 4}x}{1 + \cfrac{\frac{2(2-n)}{4 \cdot 5}x}{1 + \cfrac{\frac{3(3+n)}{5 \cdot 6}x}{1 + \cdots}}}}}} . \qquad (5)$$

See also BINOMIAL IDENTITY, BINOMIAL THEOREM, MULTINOMIAL SERIES, NEGATIVE BINOMIAL SERIES

References

Abramowitz, M. and Stegun, C. A. (Eds.). *Handbook of Mathematical Functions with Formulas, Graphs, and Mathematical Tables, 9th printing.* New York: Dover, pp. 14-5, 1972.

Pappas, T. "Pascal's Triangle, the Fibonacci Sequence & Binomial Formula." *The Joy of Mathematics.* San Carlos, CA: Wide World Publ./Tetra, pp. 40-1, 1989.

Binomial Sums

The important BINOMIAL THEOREM states that

$$\sum_{k=0}^{n} \binom{n}{k} r^k = (1+r)^n. \qquad (1)$$

Sums of powers of BINOMIAL COEFFICIENTS

$$a_r(n) = \sum_{k=0}^{n} \binom{n}{k}^r \qquad (2)$$

are given by

$$a_1(n) = 2^n \qquad (3)$$

$$a_2(n) = \binom{2n}{n} \qquad (4)$$

$a_1(n)$ and $a_2(n)$ obey the RECURRENCE RELATION

$$a_1(n+1) - 2a_1(n) = 0 \qquad (5)$$

$$(n+1)a_2(n+1) - (4n+2)a_2(n) = 0. \qquad (6)$$

Franel (1894, 1895) was the first to obtain recurrences for $a_3 n$ (Riordan 1948, p. 193) and $a_4(n)$,

$$(n+1)^2 a_3(n+1) - (7n^2 + 7n + 2)a_3(n) - 8n^2 a_3(n-1)$$
$$= 0 \qquad (7)$$

(Barrucand 1975, Cusick 1989, Jin and Dickinson 2000)

$$(n+1)^3 a_4(n+1) - 2(2n+1)(3n^2 + 3n + 1)a_4(n)$$

$$-4n(4n+1)(4n-1)a_4(n-1) = 0. \qquad (8)$$

(Jin and Dickinson 2000). Therefore, $a_3 n$ are sometimes called FRANEL NUMBERS. The sequence for $a_3 n$ cannot be expressed as a fixed number of hypergeometric terms (Petkovsek *et al.* 1996, p. 160), and therefore has no closed-form hypergeometric expression. Perlstadt (1987) found recurrences of length 4 for $r = 5$ and 6, while Schmidt and Yuan (1995) showed that the give recurrences for $r = 3$, 4, 5, and 6 are minimal, are the minimal lengths for $r > 6$ are at least 3. The following table summarizes the first few values of $a_r(n)$ for small r.

k	Sloane	$a_k(n)$
1	A000079	1, 2, 4, 8, 16, 32, 54, ...
2	A000984	1, 2, 6, 20, 70, 252, 924, ...
3	A000172	1, 2, 10, 56, 346, 2252, ...
4	A005260	1, 2, 18, 164, 1810, 21252, ...
5	A005260	1, 2, 34, 488, 9826, 206252, ...

The corresponding alternating series is

$$b_r \equiv \sum_{k=0}^{n}(-1)^k \binom{n}{k}^k = 0. \qquad (9)$$

The first few values are

$$b_1(n) = 0 \qquad (10)$$

$$b_2(n) = \frac{2^n \sqrt{\pi}}{\Gamma(\frac{1}{2}-\frac{1}{2}n)\Gamma(1+\frac{1}{2}n)}, \qquad (11)$$

$$= \begin{cases} 0 & \text{for } n=2k \\ (-1)^k \binom{n}{k} & \text{for } n=2k-1 \end{cases} \qquad (12)$$

$$b_3(n) = \frac{2^n \sqrt{\pi}\,\Gamma(1+\frac{3}{2}n)}{n!\,\Gamma(\frac{1}{2}(1-n))\Gamma(1+\frac{1}{2}n)^2} \qquad (13)$$

$$= \begin{cases} 0 & \text{for } n=2k-1 \\ \dfrac{(-1)^k(3k)!}{(k!)^3} & \text{for } n=2k, \end{cases} \qquad (14)$$

where $\Gamma(z)$ is the GAMMA FUNCTION, and the odd terms of $b_3(n)$ are given by de Bruijn's $s(3, n)$ with alternating signs.

de Bruijn (1982) has considered the sum

$$s(m, n) = \sum_{k=0}^{2n}(-1)^{k+n}\binom{2n}{k}^m \qquad (15)$$

for $m, n \geq 1$. This sum has closed form for $m=1, 2,$ and 3,

$$s(1,n) = 0 \qquad (16)$$

$$s(2, n) = \frac{(2n)!}{(n!)^2}, \qquad (17)$$

the CENTRAL BINOMIAL COEFFICIENT, giving 1, 2, 6, 20, 70, 252, 924, ... (Sloane's A000984), and

$$s(3, n) = \frac{(3n)!}{(n!)^3}, \qquad (18)$$

giving 1, 6, 90, 1680, 36450, 756756, ... (Sloane's A006480; Aizenberg and Yuzhakov 1984). However, there is no similar formula for $m \geq 4$ (Finch). The first few terms of $s(4, n)$ are 1, 14, 786, 61340, 5562130, ... (Sloane's A050983), and for $s(5,n)$ are 1, 30, 5730, 1696800, 613591650, ... (Sloane's A050984).

An interesting generalization of $b_1(n)$ was found by Ruiz (1996),

$$\sum_{k=0}^{\infty}(-1)^k \binom{n}{k}(x-k)^n = n! \qquad (19)$$

and

$$\sum_{k=0}^{n}(-1)^k \binom{n}{k}(x-k)^n = n! \qquad (20)$$

for positive integer n and all x.

The infinite sum of inverse binomial coefficients has the analytic form

$$\sum_{k=0}^{\infty}\frac{1}{\binom{n}{k}} = {}_2F_1(1, 1; -n; -1) \qquad (21)$$

$$= -(n+1)\int_0^1 \frac{dx}{(1-x)^{n+2}(x+1)}, \qquad (22)$$

where ${}_2F_1(a, b; c; x)$ is a HYPERGEOMETRIC FUNCTION. In fact, in general,

$$\sum_{k=0}^{\infty}\frac{1}{\binom{n}{k}^p} = {}_{p+1}F_p(\underbrace{1, \ldots, 1;}_{p+1} \underbrace{-n, \ldots, -n;}_{p} (-1)^k) \qquad (23)$$

and

$$\sum_{k=0}^{\infty}\frac{(-1)^k}{\binom{n}{k}^p} = {}_{p+1}F_p(\underbrace{1, \ldots, 1);}_{p+1} \underbrace{-n, \ldots, -n;}_{p} (-1)^{k+1}). \qquad (24)$$

A fascinating series of identities involving inverse central binomial coefficients times small powers are given by

$$\sum_{n=1}^{\infty}\frac{1}{\binom{2n}{n}} = \tfrac{1}{27}(2\pi\sqrt{3}+9) = 0.7363998587\ldots \qquad (25)$$

$$\sum_{n=1}^{\infty}\frac{1}{n\binom{2n}{n}} = \tfrac{1}{9}\pi\sqrt{3} = 0.6045997881\ldots \qquad (26)$$

$$\sum_{n=1}^{\infty}\frac{1}{n^2\binom{2n}{n}} = \tfrac{1}{3}\zeta(2) = \tfrac{1}{8}\pi^2 \qquad (27)$$

$$\sum_{n=1}^{\infty}\frac{1}{n^4\binom{2n}{n}} = \tfrac{17}{36}\zeta(4) = \tfrac{17}{3240}\pi^4 \qquad (28)$$

(Comtet 1974, p. 89; Le Lionnais 1983, pp. 29, 30, 41, 36), which follow from the beautiful formula

$$\sum_{n=1}^{\infty}\frac{1}{n^k\binom{2n}{n}} = \tfrac{1}{2}\,{}_{k+1}F_k(\underbrace{1, \ldots, 1;}_{k+1} \tfrac{3}{2}, \underbrace{2, \ldots, 2;}_{k-1} \tfrac{1}{4}) \qquad (29)$$

for $k \geq 1$, where $mF_n(a_1, \ldots, a_m; b_1, \ldots, b_n; x)$ is a GENERALIZED HYPERGEOMETRIC FUNCTION. Additional sums of this type include

$$\sum_{n=1}^{\infty} \frac{1}{n^3 \binom{2n}{n}} = \frac{1}{18}\pi \sqrt{3}[\psi_1(\tfrac{1}{3}) - \psi_1(\tfrac{2}{3})] - \frac{4}{3}\zeta(3) \qquad (30)$$

$$\sum_{n=1}^{\infty} \frac{1}{n^5 \binom{2n}{n}}$$

$$= \frac{1}{432}\pi \sqrt{3}[\psi_3(\tfrac{1}{3}) - \psi_3(\tfrac{2}{3})] - \frac{19}{3}\zeta(5) + \frac{1}{9}\zeta(3)\pi^2 \qquad (31)$$

$$\sum_{n=1}^{\infty} \frac{1}{n^7 \binom{2n}{n}} = \frac{11}{311040}\pi \sqrt{3}[\psi_5(\tfrac{1}{3}) - \psi_5(\tfrac{2}{3})] - \frac{493}{24}\zeta(7) + \frac{1}{3}\zeta(5)\pi^2$$
$$+ \frac{17}{1620}\zeta(3)\pi^4, \qquad (32)$$

where $\psi_n(x)$ is the POLYGAMMA FUNCTION and $\zeta(x)$ is the RIEMANN ZETA FUNCTION (Plouffe 1998).

SUMS OF THE FORM

$$\sum_{n=1}^{\infty} \frac{(-1)^{n+1}}{n^k \binom{2n}{n}} = \frac{1}{2} \ _{k+1}F_k(\underbrace{1, \ldots, 1}_{k+1}; \tfrac{3}{2}, \underbrace{2, \ldots, 2}_{k-1}; -\tfrac{1}{4}) \qquad (33)$$

can also be simplified (Plouffe) to give the special cases

$$\sum_{n=1}^{\infty} \frac{(-1)^{n-1}}{n \binom{2n}{n}} = \frac{2}{5}\sqrt{5}\ \sinh^{-1}(\tfrac{1}{2}) \qquad (34)$$

$$\sum_{n=1}^{\infty} \frac{(-1)^{n-1}}{n^2 \binom{2n}{n}} = 2[\sinh^{-1}(\tfrac{1}{2})]^2 \qquad (35)$$

$$\sum_{n=1}^{\infty} \frac{(-1)^{n-1}}{n^3 \binom{2n}{n}} = \frac{2}{5}\zeta(3). \qquad (36)$$

Other general identities include

$$\frac{(a+b)^n}{a} = \sum_{k=0}^{n} \binom{n}{k}(a - kc)^{k-1}(b + kc)^{n-k} \qquad (37)$$

(Prudnikov *et al.* 1986), which gives the BINOMIAL THEOREM as a special case with $c = 0$, and

$$\sum_{n=0}^{\infty} \binom{2n+s}{n}x^n = \ _2F_1(\tfrac{1}{2}(s+1), \ \tfrac{1}{2}(s+2); \ s+1, \ 4x)$$

$$= \frac{2'}{(\sqrt{1-4x}+1)'\sqrt{1-4x}}, \qquad (38)$$

where $_2F_1(a, b; c; z)$ is a HYPERGEOMETRIC FUNCTION (Abramowitz and Stegun 1972, p. 555; Graham *et al.* 1994, p. 203).

For NONNEGATIVE INTEGERS n and r with $r \le n + 1$,

$$\sum_{k=0}^{n} \frac{(-1)^k}{k+1} \binom{n}{k} \left[\sum_{j=0}^{r-1}(-1)^j \binom{n}{j}(r-j)^{n-k} \right.$$

$$\left. + \sum_{j=0}^{n-r}(-1)^j \binom{n}{j}(n+1-r-j)^{n-k} \right] = n! . \qquad (39)$$

Taking $n = 2r - 1$ gives

$$\sum_{k=0}^{n} \frac{(-1)^k}{K-1} \binom{n}{k} \sum_{j=0}^{r-1} \binom{n}{j}(r-j)^{n-k} = \frac{1}{2}n! . \qquad (40)$$

Other identities are

$$\sum_{k=0}^{n} \binom{n+k}{k}[x^{n+1}(1-x)^k + (1-x)^{n+1}x^k] = 1 \qquad (41)$$

(Gosper 1972) and

$$\sum_i \binom{n_i}{2} + \sum_{i>j} n_i n_j = \binom{n}{2}, \qquad (42)$$

where

$$n \equiv \sum_i n_i . \qquad (43)$$

The latter is the umbral analog of the multinomial theorem for n^2

$$\frac{(a+b+c)^2}{2} = \frac{a^2}{2} + \frac{b^2}{2} + \frac{c^2}{2} + ab + ac + bc \qquad (44)$$

using the lower-factorial polynomial $(n)_2 = n(n-1)/2$, giving

$$\binom{a+b+c}{2} = \binom{a}{2} + \binom{b}{2} + \binom{c}{2} + ab + ac + bc. \qquad (45)$$

The identity holds true not only for $(n)_2$ and $n^2/2$, but also for any quadratic polynomial OF THE FORM $n(n + a)/2$ (Dubuque).

See also APÉRY NUMBER, BINOMIAL COEFFICIENT, CENTRAL BINOMIAL COEFFICIENT, HYPERGEOMETRIC IDENTITY, HYPERGEOMETRIC SERIES, IDEMPOTENT NUMBER, JONAH FORMULA KLEE'S IDENTITY, LUCAS CORRESPONDENCE THEOREM, MARRIED COUPLES PROBLEM, MORLEY'S FORMULA, NEXUS NUMBER, STANLEY'S IDENTITY, STREHL IDENTITIES, SZÉKELY IDENTITY, WARING FORMULA, WORPITZKY'S IDENTITY

References

Aizenberg, I. A. and Yuzhakov, A. P. *Integral Representations and Residues in Multidimensional Complex Analysis.* Providence, RI: Amer. Math. Soc., p. 194, 1984.

Barrucand, P. "Problem 75–: A Combinatorial Identity." *SIAM Rev.* **17**, 168, 1975.

Beukers, F. "Another Congruence for the Apéry Numbers." *J. Number Th.* **25**, 201–10, 1987.

Cusick, T. W. "Recurrences for Sums of Powers of Binomial Coefficients." *J. Combin. Th. Ser. A* **52**, 77–3, 1989.

de Bruijn, N. G. *Asymptotic Methods in Analysis.* New York: Dover, 1982.

Egorychev, G. P. *Integral Representation and the Computation of Combinatorial Sums.* Providence, RI: Amer. Math. Soc., 1984.

Finch, S. "Favorite Mathematical Constants." http://www.mathsoft.com/asolve/constant/nielram/nielram.html.

Franel, J. "On a Question of Laisant." *L'intermédiaire des mathématiciens* **1**, 45–7, 1894.

Franel, J. "On a Question of J. Franel." *L'intermédiaire des mathématiciens* **2**, 33–5, 1895.

Gosper, R. W. Item 42 in Beeler, M.; Gosper, R. W.; and Schroeppel, R. *HAKMEM.* Cambridge, MA: MIT Artificial Intelligence Laboratory, Memo AIM-239, p. 16, Feb. 1972.

Graham, R. L.; Knuth, D. E.; and Patashnik, O. "Binomial Coefficients." Ch. 5 in *Concrete Mathematics: A Foundation for Computer Science, 2nd ed.* Reading, MA: Addison-Wesley, pp. 153–42, 1994.

Jin, Y. and Dickinson, H. "Apéry Sequences and Legendre Transforms." *J. Austral. Math. Soc. Ser. A* **68**, 349–56, 2000.

MacMahon P. A. "The Sums of the Powers of the Binomial Coefficients." *Quart. J. Math.* **33**, 274–88, 1902.

McIntosh, R. J. "Recurrences for Alternating Sums of Powers of Binomial Coefficients." *J. Combin. Th. A* **63**, 223–33, 1993.

Perlstadt, M. A. "Some Recurrences for Sums of Powers of Binomial Coefficients." *J. Number Th.* **27**, 304–09, 1987.

Petkovsek, M.; Wilf, H. S.; and Zeilberger, D. *A = B.* Wellesley, MA: A. K. Peters, 1996.

Plouffe, S. "The Art of Inspired Guessing." Aug. 7, 1998. http://www.lacim.uqam.ca/plouffe/inspired.html.

Riordan, J. *An Introduction to Combinatorial Analysis.* New York: Wiley, 1980.

Ruiz, S. *Math. Gaz.* **80**, 579–82, Nov. 1996.

Schmidt, A. L. and Yuan, J. "On Recurrences for Sums of Powers of Binomial Coefficients." Tech. Rep., 1995.

Shanks, E. B. "Iterated Sums of Powers of the Binomial Coefficients." *Amer. Math. Monthly* **58**, 404–07, 1951.

Sloane, N. J. A. Sequences A000079/M1129, A000172/M1971, A000984/M1645, A005260/M2110, A005261/M2156, A006480/M4284, A050983, and A050984 in "An On-Line Version of the Encyclopedia of Integer Sequences." http://www.research.att.com/~njas/sequences/eisonline.html.

Strehl, V. "Binomial Identities--Combinatorial and Algorithmic Aspects. Trends in Discrete Mathematics." *Disc. Math.* **136**, 309–46, 1994.

Binomial Theorem

The theorem that, for POSITIVE INTEGERS n,

$$(x+a)^n = \sum_{k=0}^{n} \frac{n!}{k!(n-k)!} x^k a^{n-k} = \sum_{k=0}^{n} \binom{n}{k} x^k a^{n-k},$$

the so-called BINOMIAL SERIES, where $\binom{n}{k}$ are BINOMIAL COEFFICIENTS. The theorem was known for the case $n=2$ by Euclid around 300 BC, and stated in its modern form by Pascal in a posthumous pamphlet published in 1665. Newton (1676) showed that a similar formula (with INFINITE upper limit) holds for NEGATIVE INTEGERS n,

$$(x+a)^{-n} = \sum_{k=0}^{\infty} \binom{-n}{k} x^k a^{-n-k},$$

the so-called NEGATIVE BINOMIAL SERIES, which converges for $|x| > |a|$.

See also BINOMIAL COEFFICIENT, BINOMIAL IDENTITY, BINOMIAL SERIES, CAUCHY BINOMIAL THEOREM, CHU-VANDERMONDE IDENTITY, LOGARITHMIC BINOMIAL

FORMULA, NEGATIVE BINOMIAL SERIES, Q-BINOMIAL THEOREM, RANDOM WALK

References
Abramowitz, M. and Stegun, C. A. (Eds.). *Handbook of Mathematical Functions with Formulas, Graphs, and Mathematical Tables, 9th printing.* New York: Dover, p. 10, 1972.

Arfken, G. *Mathematical Methods for Physicists, 3rd ed.* Orlando, FL: Academic Press, pp. 307–08, 1985.

Boyer, C. B. and Merzbach, U. C. "The Binomial Theorem." *A History of Mathematics, 2nd ed.* New York: Wiley, pp. 393–94, 1991.

Conway, J. H. and Guy, R. K. "Choice Numbers Are Binomial Coefficients." In *The Book of Numbers.* New York: Springer-Verlag, pp. 72–4, 1996.

Coolidge, J. L. "The Story of the Binomial Theorem." *Amer. Math. Monthly* **56**, 147–57, 1949.

Courant, R. and Robbins, H. "The Binomial Theorem." §1.6 in *What is Mathematics?: An Elementary Approach to Ideas and Methods, 2nd ed.* Oxford, England: Oxford University Press, pp. 16–18, 1996.

Pascal, B. *Traite du Triangle Arithmetic.* 1665.

Whittaker, E. T. and Robinson, G. "The Binomial Theorem." §10 in *The Calculus of Observations: A Treatise on Numerical Mathematics, 4th ed.* New York: Dover, pp. 15–9, 1967.

Binomial Transform

The binomial transform takes the sequence $a_0, a_1, a_2, \ldots$ to the sequence $b_0, b_1, b_2, \ldots$ via the transformation

$$b_n = \sum_{k=0}^{n} (-1)^{n-k} \binom{n}{k} a_k.$$

The inverse transform is

$$a_n = \sum_{k=0}^{n} \binom{n}{k} b_k.$$

(Sloane and Plouffe 1995, pp. 13 and 22). The inverse binomial transform of $b_n = 1$ for prime n and $b_n = 0$ for composite n is 0, 1, 3, 6, 11, 20, 37, 70, ... (Sloane's A052467). The inverse binomial transform of $b_n = 1$ for even n and $b_n = 0$ for odd n is 0, 1, 2, 4, 8, 16, 32, 64, ... (Sloane's A000079). Similarly, the inverse binomial transform of $b_n = 1$ for odd n and $b_n = 0$ for even n is 1, 2, 4, 8, 16, 32, 64, ... (Sloane's A000079). The inverse binomial transform of the BELL NUMBERS 1, 1, 2, 5, 15, 52, 203, ... (Sloane's A000110) is a shifted version of the same numbers: 1, 2, 5, 15, 52, 203, ... (Bernstein and Sloane 1995, Sloane and Plouffe 1995, p. 22).

The CENTRAL and RAW MOMENTS of statistical distributions are also related by the binomial transform.

See also CENTRAL MOMENT, EULER TRANSFORM, EXPONENTIAL TRANSFORM, MÖBIUS TRANSFORM, RAW MOMENT

Binomial Triangle

References

Bernstein, M. and Sloane, N. J. A. "Some Canonical Sequences of Integers." *Linear Algebra Appl.* **226//228**, 57–2, 1995.

Sloane, N. J. A. Sequences A000079/M1129, A000110/M1484, and A052467 in "An On-Line Version of the Encyclopedia of Integer Sequences." http://www.research.-att.com/~njas/sequences/eisonline.html.

Sloane, N. J. A. and Plouffe, S. *The Encyclopedia of Integer Sequences.* San Diego, CA: Academic Press, 1995.

Binomial Triangle
PASCAL'S TRIANGLE

Binomial-Type Sequence

A sequence of POLYNOMIALS p_n satisfying the identities

$$p_n(x+y) = \sum_{k \geq 0} \binom{n}{k} p_k(x) p_{n-k}(y).$$

See also BINOMIAL IDENTITY, SHEFFER SEQUENCE, UMBRAL CALCULUS

References

Rota, G.-C.; Kahaner, D.; Odlyzko, A. "On the Foundations of Combinatorial Theory. VIII: Finite Operator Calculus." *J. Math. Anal. Appl.* **42**, 684–60, 1973.

Binormal Developable

A RULED SURFACE M is said to be a binormal developable of a curve $\mathbf{y}$ if M can be parameterized by $\mathbf{x}(u,\ v) = \mathbf{y}(u) + v\hat{\mathbf{B}}(u)$, where $\mathbf{B}$ is the BINORMAL VECTOR.

See also NORMAL DEVELOPABLE, TANGENT DEVELOPABLE

References

Gray, A. "Developables." §17.6 in *Modern Differential Geometry of Curves and Surfaces with Mathematica.* Boca Raton, FL: CRC Press, pp. 352–54, 1993.

Binormal Vector

$$\tilde{\mathbf{B}} \equiv \hat{\mathbf{T}} \times \hat{\mathbf{N}} \tag{1}$$

$$= \frac{\mathbf{r}' \times \mathbf{r}''}{|\mathbf{r}' \times \mathbf{r}''|}, \tag{2}$$

where the unit TANGENT VECTOR $\mathbf{T}$ and unit "principal" NORMAL VECTOR $\mathbf{N}$ are defined by

$$\hat{\mathbf{T}} \equiv \frac{\mathbf{r}'(s)}{|\hat{\mathbf{r}}(s)|} \tag{3}$$

$$\hat{\mathbf{N}} \equiv \frac{\mathbf{r}''(s)}{|\mathbf{r}''(s)|} \tag{4}$$

Here, $\mathbf{r}$ is the RADIUS VECTOR, s is the ARC LENGTH, τ

is the TORSION, and κ is the CURVATURE. The binormal vector satisfies the remarkable identity

$$[\dot{\mathbf{B}}, \ddot{\mathbf{B}}, \dddot{\mathbf{B}}] = \tau^5 \frac{d}{ds}\left(\frac{\kappa}{\tau}\right). \tag{5}$$

See also FRENET FORMULAS, NORMAL VECTOR, TANGENT VECTOR

References

Kreyszig, E. "Binormal. Moving Trihedron of a Curve." §13 in *Differential Geometry.* New York: Dover, pp. 36–7, 1991.

Bin-Packing Problem

The problem of packing a set of items into a number of bins such that the total weight, volume, etc. does not exceed some maximum value. A simple algorithm (the first-fit algorithm) takes items in the order they come an places them in the first bin in which they fit. In 1973, J. Ullman proved that this algorithm can differ from an optimal packing by as much at 70% (Hoffman 1998, p. 171). An alternative strategy first orders the items from largest to smallest, then places them sequentially in the first bin in which they fit. In 1973, D. Johnson showed that this strategy is never suboptimal by more than 22%, and furthermore that *no* efficient bin-packing algorithm can be guaranteed to do better than 22% (Hoffman 1998, p. 172).

There exist arrangements of items such that applying the packing algorithm after *removing* an item results in *one more* bin being required than the number obtained if the item is *included* (Hoffman 1998, pp. 172–73).

See also COOKIE-CUTTER PROBLEM, TILING PROBLEM

References

Hoffman, P. *The Man Who Loved Only Numbers: The Story of Paul Erdos and the Search for Mathematical Truth.* New York: Hyperion, 1998.

Bioche's Theorem

If two complementary PLÜCKER CHARACTERISTICS are equal, then each characteristic is equal to its complement except in four cases where the sum of order and class is 9.

References

Coolidge, J. L. *A Treatise on Algebraic Plane Curves.* New York: Dover, p. 101, 1959.

Biotic Potential
LOGISTIC EQUATION

Bipartite Graph

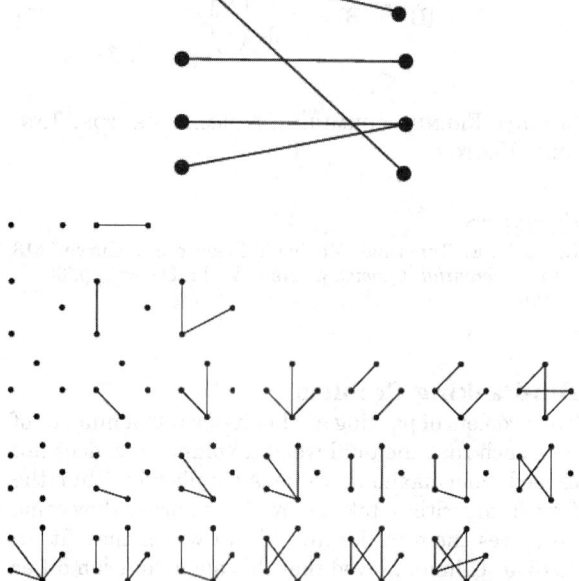

A set of VERTICES decomposed into two disjoint sets such that no two VERTICES within the same set are adjacent. A bigraph is a special case of a K-PARTITE GRAPH with $k = 2$. Bipartite graphs are equivalent to two-colorable graphs, and a graph is bipartite IFF all its cycles are of even length (Skiena 1990, p. 213). The numbers of bipartite graphs on $n = 1, 2, \ldots$ nodes are 1, 2, 3, 7, 13, 35, 88, 303, ... (Sloane's A033995). A graph can be tested for bipartiteness using `BipartiteQ[g]` in the *Mathematica* add-on package `DiscreteMath`Combinatorica`` (which can be loaded with the command `< <DiscreteMath`).

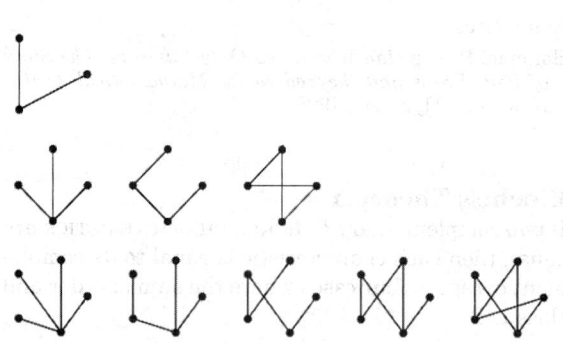

The numbers of CONNECTED bipartite graphs on $n = 1, 2 \ldots$ nodes are 1, 1, 1, 3, 5, 17, 44, 182, ... (Sloane's A005142).
All TREES are bipartite (Skiena 1990, p. 213).

See also BICUBIC GRAPH, COMPLETE BIPARTITE GRAPH, K-PARTITE GRAPH, KÖNIG-EGEVÁRY THEOREM

References

Chartrand, G. *Introductory Graph Theory.* New York: Dover, p. 116, 1985.
Read, R. C. and Wilson, R. J. *An Atlas of Graphs.* Oxford, England: Oxford University Press, 1998.
Saaty, T. L. and Kainen, P. C. *The Four-Color Problem: Assaults and Conquest.* New York: Dover, p. 12, 1986.
Skiena, S. "Coloring Bipartite Graphs." §5.5.2 in *Implementing Discrete Mathematics: Combinatorics and Graph Theory with Mathematica.* Reading, MA: Addison-Wesley, p. 213, 1990.
Sloane, N. J. A. Sequences A033995 in "An On-Line Version of the Encyclopedia of Integer Sequences." http://www.research.att.com/~njas/sequences/eisonline.html.
Steinbach, P. *Field Guide to Simple Graphs.* Albuquerque, NM: Design Lab, 1990.

Biplanar Double Point

ISOLATED SINGULARITY

Bipolar Coordinates

Bipolar coordinates are a 2-D system of coordinates. There are two commonly defined types of bipolar coordinates, the first of which is defined by

$$x = \frac{a \sinh v}{\cosh v - \cos u} \tag{1}$$

$$y = \frac{a \sin u}{\cosh v - \cos u}, \tag{2}$$

where $u \in [0, 2\pi)$, $v \in (-\infty, \infty)$. The following identities show that curves of constant u and v are CIRCLES in xy-space.

$$x^2 + (y - a \cot u)^2 = a^2 \csc^2 u \tag{3}$$

$$(x - a \coth v)^2 + y^2 = a^2 \operatorname{csch}^2 v. \tag{4}$$

The SCALE FACTORS are

$$h_u = \frac{a}{\cosh v - \cos u} \tag{5}$$

$$h_v = \frac{a}{\cosh v - \cos u} \tag{6}$$

The LAPLACIAN is

$$\nabla^2 = \frac{(\cosh v - \cos u)^2}{a^2} \left(\frac{\partial^2}{\partial u^2} + \frac{\partial^2}{\partial v^2} \right). \tag{7}$$

LAPLACE'S EQUATION is separable.

Two-center bipolar coordinates are two coordinates giving the distances from two fixed centers r_1 and r_2, sometimes denoted r and r'. For two-center bipolar coordinates with centers at $(\pm c, 0)$,

$$r_1^2 = (x + c)^2 + y^2 \tag{8}$$

$$r_2^2 = (x - c)^2 + y^2. \tag{9}$$

Combining (8) and (9) gives

$$r_1^2 - r_2^2 = 4cx. \tag{10}$$

Solving for CARTESIAN COORDINATES x and y gives

$$x = \frac{r_1^2 - r_2^2}{4c} \tag{11}$$

$$y = \pm \frac{1}{4c} \sqrt{16c^2 r_1^2 - (r_1^2 - r_2^2 + 4c^2)^2}. \tag{12}$$

Solving for POLAR COORDINATES gives

$$r = \sqrt{\frac{r_1^2 + r_2^2 - 2c^2}{2}} \tag{13}$$

$$\theta = \tan^{-1} \left[\frac{\sqrt{r_2^4 - 2(4c^2 + r_1^2) r_2^2 - (4c^2 - r_1^2)^2}}{r_1^2 - r_2^2} \right]. \tag{14}$$

See also BIPOLAR CYLINDRICAL COORDINATES, POLAR COORDINATES

References

Lockwood, E. H. "Bipolar Coordinates." Ch. 25 in *A Book of Curves.* Cambridge, England: Cambridge University Press, pp. 186–90, 1967.

Bipolar Cylindrical Coordinates

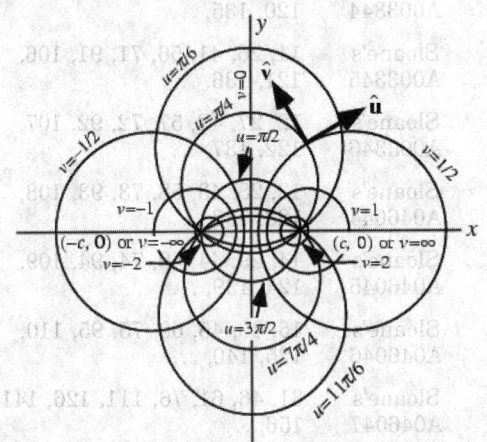

A set of CURVILINEAR COORDINATES defined by

$$x = \frac{a \sinh v}{\cosh v - \cos u} \tag{1}$$

$$y = \frac{a \sin u}{\cosh v - \cos u} \tag{2}$$

$$z = z, \tag{3}$$

where $u \in [0, 2\pi)$, $v \in (-\infty, \infty)$, and $z \in (-\infty, \infty)$. There are several notational conventions, and whereas (u, v, z) is used in this work, Arfken (1970) prefers (η, ξ, z). The following identities show that

curves of constant u and v are CIRCLES in xy-space.

$$x^2 + (y - a \cot u)^2 = a^2 \csc^2 u \tag{4}$$

$$(x - a \coth v)^2 + y^2 = a^2 \operatorname{csch}^2 v. \tag{5}$$

The SCALE FACTORS are

$$h_u = \frac{a}{\cosh v - \cos u} \tag{6}$$

$$h_v = \frac{a}{\cosh v - \cos u} \tag{7}$$

$$h_z = 1. \tag{8}$$

The LAPLACIAN is

$$\nabla^2 = \frac{(\cosh v - \cos u)^2}{a^2} \left(\frac{\partial^2}{\partial u^2} + \frac{\partial^2}{\partial v^2} \right) + \frac{\partial^2}{\partial z^2}. \tag{9}$$

LAPLACE'S EQUATION is not separable in BIPOLAR CYLINDRICAL COORDINATES, but it is in 2-D BIPOLAR COORDINATES.

See also BIPOLAR COORDINATES, POLAR COORDINATES

References

Arfken, G. "Bipolar Coordinates (ξ, η, z)." §2.9 in *Mathematical Methods for Physicists, 2nd ed.* Orlando, FL: Academic Press, pp. 97–02, 1970.

Bipolyhedral Group

The image of $A_5 \times A_5$ in the SPECIAL ORTHOGONAL GROUP $SO(4)$, where A_5 is the ICOSAHEDRAL GROUP.

See also ICOSAHEDRAL GROUP, SPECIAL ORTHOGONAL GROUP

References

Endraß, S. "The Sarti Surface." http://enriques.mathematik.uni-mainz.de/kon/docs/Esarti.shtml.

Biprism

Two slant triangular PRISMS fused together.

See also PRISM, SCHMITT-CONWAY BIPRISM

Bipyramid

DIPYRAMID

Biquadratefree

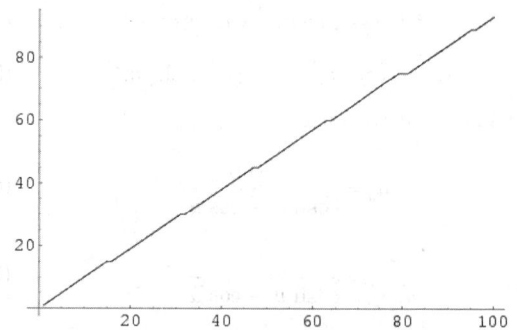

A number is said to be biquadratefree (or quarticfree) if its PRIME FACTORIZATION contains no quadrupled factors. All PRIMES and PRIME POWERS p^n with $n \leq 3$ are therefore trivially biquadratefree. The biquadratefree numbers are 1, 2, 3, 4, 5, 6, 7, 8, 9, 10, 11, 12, 13, 14, 15, 17, ... (Sloane's A046100). The biquadrateful numbers (i.e., those that contain at least one biquadrate) are 16, 32, 48, 64, 80, 81, 96, ... (Sloane's A046101). The number of biquadratefree numbers less than 10, 100, 1000, ... are 10, 93, 925, 9240, 92395, 923939, ..., and their asymptotic density is $1/\zeta(4) = 90/\pi^4 \approx 0.923938$, where $\zeta(n)$ is the RIEMANN ZETA FUNCTION.

See also CUBEFREE, PRIME NUMBER, RIEMANN ZETA FUNCTION, SQUAREFREE

References

Sloane, N. J. A. Sequences A046100 and A046101 in "An On-Line Version of the Encyclopedia of Integer Sequences." http://www.research.att.com/~njas/sequences/eisonline.html.

Biquadratic Equation

QUARTIC EQUATION

Biquadratic Number

A biquadratic number is a fourth POWER, n^4. The first few biquadratic numbers are 1, 16, 81, 256, 625, ... (Sloane's A000583). The minimum number of biquadratic numbers needed to represent the numbers 1, 2, 3, ... are 1, 2, 3, 4, 5, 6, 7, 8, 9, 10, 11, 12, 13, 14, 15, 1, 2, 3, 4, 5, ... (Sloane's A002377), and the number of distinct ways to represent the numbers 1, 2, 3, ... in terms of biquadratic numbers are 1, 1, 1, 1, 1, 1, 1, 1, 1, 1, 1, 1, 1, 1, 1, 2, 2, 2, 2, 2, ... A brute-force algorithm for enumerating the biquadratic permutations of n is repeated application of the GREEDY ALGORITHM.

Every POSITIVE integer is expressible as a SUM of (at most) $g(4) = 19$ biquadratic numbers (WARING'S PROBLEM). Davenport (1939) showed that $G(4) = 16$, meaning that all sufficiently large integers require only 16 biquadratic numbers. It is also known that every integer is a sum of at most 10 signed biquadrates ($eg(4) \leq 10$; although it is not known if 10 can

be reduced to 9). The following table gives the first few numbers which require 1, 2, 3, ..., 19 biquadratic numbers to represent them as a sum, with the sequences for 17, 18, and 19 being finite.

#	Sloane	Numbers
1	Sloane's A000290	1, 16, 81, 256, 625, 1296, 2401, 4096, ...
2	Sloane's A003336	2, 17, 32, 82, 97, 162, 257, 272, ...
3	Sloane's A003337	3, 18, 33, 48, 83, 98, 113, 163, ...
4	Sloane's A003338	4, 19, 34, 49, 64, 84, 99, 114, 129, ...
5	Sloane's A003339	5, 20, 35, 50, 65, 80, 85, 100, 115, ...
6	Sloane's A003340	6, 21, 36, 51, 66, 86, 96, 101, 116, ...
7	Sloane's A003341	7, 22, 37, 52, 67, 87, 102, 112, 117, ...
8	Sloane's A003342	8, 23, 38, 53, 68, 88, 103, 118, 128, ...
9	Sloane's A003343	9, 24, 39, 54, 69, 89, 104, 119, 134, ...
10	Sloane's A003344	10, 25, 40, 55, 70, 90, 105, 120, 135, ...
11	Sloane's A003345	11, 26, 41, 56, 71, 91, 106, 121, 136, ...
12	Sloane's A003346	12, 27, 42, 57, 72, 92, 107, 122, 137, ...
13	Sloane's A046044	13, 28, 43, 58, 73, 93, 108, 123, 138, ...
14	Sloane's A046045	14, 29, 44, 59, 74, 94, 109, 124, 139, ...
15	Sloane's A046046	15, 30, 45, 60, 75, 95, 110, 125, 140, ...
16	Sloane's A046047	31, 46, 61, 76, 111, 126, 141, 156, ...
17	Sloane's A046048	47, 62, 77, 127, 142, 157, 207, 222, ...
18	Sloane's A046049	63, 78, 143, 158, 223, 238, 303, 318, ...
19	Sloane's A046050	79, 159, 239, 319, 399

The following table gives the numbers which can be represented in n different ways as a sum of k biquadrates.

k	n	Sloane	Numbers
1	1	Sloane's A000290	1, 16, 81, 256, 625, 1296, 2401, 4096, ...
2	2	Sloane's A018786	635318657, 3262811042, 8657437697, ...

The numbers 2, 3, 4, 5, 6, 7, 8, 9, 10, 11, 12, 13, 14, 15, 18, 19, 20, 21, ... (Sloane's A046039) cannot be represented using distinct biquadrates.

See also CUBIC NUMBER, PARTITION, SQUARE NUMBER, WARING'S PROBLEM

References
Davenport, H. "On Waring's Problem for Fourth Powers." *Ann. Math.* **40**, 731–47, 1939.

Hardy, G. H. and Wright, E. M. "The Representation of a Number by Two or Four Squares." Ch. 20 in *An Introduction to the Theory of Numbers, 5th ed.* Oxford, England: Clarendon Press, pp. 297–16, 1979.

Sloane, N. J. A. Sequences A000290, A000583/M5004, A002377, A003336, A003337, A003338, A003339, A003340, A003341, A003342, A003343, A003344, A003345, A003346, A018786, and A046039 in "An On-Line Version of the Encyclopedia of Integer Sequences." http://www.research.att.com/~njas/sequences/eisonli-ne.html.

Biquadratic Reciprocity Theorem
Gauss stated the reciprocity theorem for the case $n = 4$

$$x^4 \equiv q \pmod{p} \tag{1}$$

can be solved using the GAUSSIAN INTEGERS as

$$\left(\frac{\pi}{\sigma}\right)_4 \left(\frac{\sigma}{\pi}\right)_4 = (-1)^{[(N(\pi)-1)/4][(N(\sigma)-1)/4]} \tag{2}$$

Here, π and σ are distinct GAUSSIAN INTEGER PRIMES, and

$$N(a + bi) = \sqrt{a^2 + b^2} \tag{3}$$

is the norm. The symbol $\left(\frac{\alpha}{\pi}\right)$ means

$$\left(\frac{\alpha}{\pi}\right)_4 = \begin{cases} 1 & \text{if } x^4 \equiv \alpha \pmod{\pi} \text{ is solvable} \\ -1, i, \text{ or } -i & \text{otherwise} \end{cases} \tag{4}$$

where "solvable" means solvable in terms of GAUSSIAN INTEGERS.

2 is a quartic residue (mod p) IFF there are integers x, y such that

$$x^2 + 64y^2 = p. \tag{5}$$

This is a generalization of the GENUS THEOREM.

See also BIQUADRATIC RESIDUE, GENUS THEOREM, RECIPROCITY THEOREM

References
Ireland, K. and Rosen, M. "Cubic and Biquadratic Reciprocity." Ch. 9 in *A Classical Introduction to Modern Number Theory, 2nd ed.* New York: Springer-Verlag, pp. 108–37, 1990.

Biquadratic Residue
If there is an INTEGER x such that

$$x^4 \equiv q \pmod{p}, \tag{1}$$

then q is said to be a biquadratic residue (mod p). If not, q is said to be a biquadratic nonresidue (mod p).

See also BIQUADRATIC RECIPROCITY THEOREM, CUBIC RESIDUE, QUADRATIC RESIDUE

References
Nagell, T. *Introduction to Number Theory.* New York: Wiley, p. 115, 1951.

Biquaternion
A QUATERNION with COMPLEX coefficients. The ALGEBRA of biquaternions is isomorphic to a full matrix ring over the complex number field (van der Waerden 1985).

See also QUATERNION

References
Clifford, W. K. "Preliminary Sketch of Biquaternions." *Proc. London Math. Soc.* **4**, 381–95, 1873.

Hamilton, W. R. *Lectures on Quaternions: Containing a Systematic Statement of a New Mathematical Method.* Dublin: Hodges and Smith, 1853.

Study, E. "Von den Bewegung und Umlegungen." *Math. Ann.* **39**, 441–66, 1891.

van der Waerden, B. L. *A History of Algebra from al-Khwarizmi to Emmy Noether.* New York: Springer-Verlag, pp. 188–89, 1985.

Birational Transformation
A transformation in which coordinates in two SPACES are expressed rationally in terms of those in another.

See also RIEMANN CURVE THEOREM, WEBER'S THEOREM

Birch Conjecture
SWINNERTON-DYER CONJECTURE

Birch-Swinnerton-Dyer Conjecture
SWINNERTON-DYER CONJECTURE

Birkhoff's Ergodic Theorem

Let T be an ergodic ENDOMORPHISM of the PROBABIL-ITY SPACE X and let $f : X \to \mathbb{R}$ be a real-valued MEASURABLE FUNCTION. Then for ALMOST EVERY $x \in X$, we have

$$\frac{1}{n} \sum_{j=1}^{n} f \circ T^j(x) \to \int f \, dm \qquad (1)$$

as $n \to \infty$. To illustrate this, take f to be the characteristic function of some SUBSET A of X so that

$$f(x) = \begin{cases} 1 & \text{if } x \in A \\ 0 & \text{if } x \notin A. \end{cases} \qquad (2)$$

The left-hand side of (1) just says how often the orbit of x (that is, the points $x, Tx, T^2x, \ldots$) lies in A, and the right-hand side is just the MEASURE of A. Thus, for an ergodic ENDOMORPHISM, "space-averages = time-averages almost everywhere." Moreover, if T is continuous and uniquely ergodic with BOREL PROBABILITY MEASURE m and f is continuous, then we can replace the ALMOST EVERYWHERE convergence in (1) with "everywhere."

See also BIRKHOFF'S THEOREM, ERGODIC THEORY

References
Cornfeld, I.; Fomin, S.; and Sinai, Ya. G. Appendix 3 in *Ergodic Theory*. New York: Springer-Verlag, 1982.

Birkhoff-Khinchin Ergodic Theorem

BIRKHOFF'S ERGODIC THEOREM

Birkhoff-Witt Theorem

POINCARÉ-BIRKHOFF-WITT THEOREM

Birotunda

Two adjoined ROTUNDAS.

See also BILUNABIROTUNDA, CUPOLAROTUNDA, ELON-GATED GYROCUPOLAROTUNDA, ELONGATED ORTHOCU-POLAROTUNDA, ELONGATED ORTHOBIROTUNDA, GYROCUPOLAROTUNDA, GYROELONGATED ROTUNDA, ORTHOBIROTUNDA, TRIANGULAR HEBESPHENOROTUN-DA

Birthday Attack

Birthday attacks are a class of brute-force techniques used in an attempt to solve a class of CRYPTOGRAPHIC HASH FUNCTION problems. These methods take advantage of functions which, when supplied with a random input, return one of k equally likely values. By repeatedly evaluating the function for different inputs, the same output is expected to be obtained after about $1.2\sqrt{k}$ evaluations.

See also BIRTHDAY PROBLEM, CRYPTOGRAPHIC HASH FUNCTION

References
RSA Laboratories. "Question 95. What is a Birthday Attack" and "Question 96. How Does the Length of a Hash Value Affect Security?" http://www.rsasecurity.com/rsalabs/faq/.
van Oorschot, P. and Wiener, M. "A Known Plaintext Attack on Two-Key Triple Encryption." In *Advances in Cryptology--Eurocrypt '90*. New York: Springer-Verlag, pp. 366–77, 1991.
Yuval, G. "How to Swindle Rabin." *Cryptologia* **3**, 187–89, Jul. 1979.

Birthday Problem

Consider the probability $Q_1(n, d)$ that *no two people* out of a group of n will have matching birthdays out of d equally possible birthdays. Start with an arbitrary person's birthday, then note that the probability that the second person's birthday is different is $(d-1)/d$, that the third person's birthday is different from the first two is $[(d-1)/d][(d-2)/d]$, and so on, up through the nth person. Explicitly,

$$Q_1(n, d) = \frac{d-1}{d} \frac{d-2}{d} \cdots \frac{d-(n-1)}{d}$$
$$= \frac{(d-1)(d-2)\cdots[d-(n-1)]}{d^{n-1}}. \qquad (1)$$

But this can be written in terms of FACTORIALS as

$$Q_1(n, d) = \frac{d!}{(d-n)! d^n}, \qquad (2)$$

so the probability $P_2(n, 365)$ that two people out of a group of n *do have the same birthday* is therefore

$$P_2(n, d) = 1 - Q_1(n, d) = 1 - \frac{d!}{(d-n)! d^n}. \qquad (3)$$

If 365-day years have been assumed, i.e., the existence of leap days is ignored, then the number of people needed for there to be at least a 50% chance that two share birthdays is the smallest n such that $P_2(n, 365) \geq 1/2$. This is given by $n = 23$, since

$P_2(23, 365)$

$$= \frac{38093904702297390785243708291056390518886454060947}{50918832685153501254262074252223147563269805908207}$$

$$\approx 0.507297. \qquad (4)$$

The number n of people needed to obtain $P_2(n, d) \geq 1/2$ for $d = 1, 2, \ldots$, are 2, 2, 3, 3, 3, 4, 4, 4, 4, 5, $\ldots$ (Sloane's A033810).

The probability $P_2(n, d)$ can be estimated as

$$P_2(n, d) \approx 1 - e^{-n(n-1)/2d} \qquad (5)$$

$$\approx 1 - \left(1 - \frac{n}{2d}\right)^{n-1}, \qquad (6)$$

where the latter has error

$$\epsilon < \frac{n^3}{6(d - n + 1)^2} \qquad (7)$$

(Sayrafiezadeh 1994).

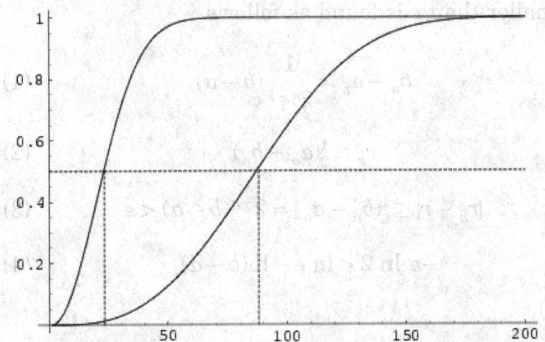

In general, let $Q_i(n, d)$ denote the probability that a birthday is shared by exactly i (and no more) people out of a group of n people. Then the probability that a birthday is shared by k *or more* people is given by

$$P_k(n, d) = 1 - \sum_{i=1}^{k-1} Q_i(n, d). \qquad (8)$$

Q_2 can be computed explicitly as

$$Q_2(n, d) = \frac{n!}{d^n} \sum_{i=2}^{\lfloor n/2 \rfloor} \frac{1}{2^i} \binom{d}{i}\binom{d-i}{n-2i}$$

$$= \frac{n!}{d^n} \sum_{i=1}^{\lfloor n/2 \rfloor} \frac{d!}{2^i i!(n - 2i)!(d - n + i)!}$$

$$= \frac{(-1)^n}{d^n}\left[2^{-n/2}\Gamma(1 + n)P_n^{(-d)}(\tfrac{1}{2}\sqrt{2}) \right.$$

$$\left. - \frac{\Gamma(1 + d)}{\Gamma(1 + d - n)} \right], \qquad (9)$$

where $\binom{n}{m}$ is a BINOMIAL COEFFICIENT, $\Gamma(n)$ is a GAMMA FUNCTION, and $P_n^{(\lambda)}(x)$ is an ULTRASPHERICAL POLYNOMIAL. This gives the explicit formula for $P_3(n, d)$ as

$$P_3(n, d) = 1 - Q_1(n, d) - Q_2(n, d)$$

$$= 1 + \frac{(-1)^{n+1}\Gamma(n + 1)P_n^{(-d)}(2^{-1/2})}{2^{n/2}d^n}. \qquad (10)$$

$Q_3(n, d)$ cannot be computed in entirely closed form, but a partially reduced form is

$$Q_3(n, d) = \frac{\Gamma(d + 1)}{d^n}\left[\frac{(-1)^n F(\tfrac{9}{8}) - F(-\tfrac{9}{8})}{\Gamma(d - n + 1)} + (-1)^n \Gamma \right.$$

$$\left. \times (1 + n) \sum_{i=1}^{\lfloor n/3 \rfloor} \frac{(-3)^{-i} 2^{(i-n)/2} P_{n-3i}^{(i-d)}(\tfrac{1}{2}\sqrt{2})}{\Gamma(d - i + 1)\Gamma(i + 1)} \right], \qquad (11)$$

where

$$F = F(n, d, a) \equiv 1 - {}_3F_2\left[\begin{array}{c} \tfrac{1}{3}(1 - n), \tfrac{1}{3}(2 - n), -\tfrac{1}{3} \\ \tfrac{1}{2}(d - n + 1), \tfrac{1}{2}(d - n + 2) \end{array}; a \right] \qquad (12)$$

and ${}_3F_2(a, b, c; d, e; z)$ is a GENERALIZED HYPERGEOMETRIC FUNCTION.

In general, $Q_k(n, d)$ can be computed using the RECURRENCE RELATION

$$Q_k(n, d) = \sum_{i=1}^{\lfloor n/k \rfloor}\left[\frac{n!d!}{d^{ik}i!(k!)^i(n - ik)!(d - i)!} \right.$$

$$\left. \times \sum_{j=1}^{k-1} Q_j(n-k, d-i)\frac{(d - i)^{n-ik}}{d^{n-ik}} \right] \qquad (13)$$

(Finch). However, the time to compute this recursive function grows exponentially with k and so rapidly becomes unwieldy. The minimal number of people to give a 50% probability of having at least n coincident birthdays is 1, 23, 88, 187, 313, 460, 623, 798, 985, 1181, 1385, 1596, 1813, ... (Sloane's A014088; Diaconis and Mosteller 1989).

A good approximation to the number of people n such that $p = P_k(n, d)$ is some given value can be given by solving the equation

$$ne^{-n/(dk)} = \left[d^{k-1}k! \ln\left(\frac{1}{1 - p}\right)\left(1 - \frac{n}{d(k + 1)}\right) \right]^{1/k} \qquad (14)$$

for n and taking $\lceil n \rceil$, where $\lceil n \rceil$ is the CEILING FUNCTION (Diaconis and Mosteller 1989). For $p = 0.5$ and $k = 1, 2, 3, ...$, this formula gives $n = 1, 23, 88, 187, 313, 459, 622, 797, 983, 1179, 1382, 1592, 1809, ...$ (Sloane's A050255), which differ from the true values by from 0 to 4. A much simpler but also poorer approximation for n such that $p = 0.5$ for $k < 20$ is given by

$$n = 47(k - 1.5)^{3/2} \qquad (15)$$

(Diaconis and Mosteller 1989), which gives 86, 185, 307, 448, 606, 778, 965, 1164, 1376, 1599, 1832, ... for $k = 3, 4, ...$ (Sloane's A050256).

The "almost" birthday problem, which asks the number of people needed such that two have a birthday within a day of each other, was considered by Abramson and Moser (1970), who showed that 14 people suffice. An approximation for the minimum number of people needed to get a 50–0 chance that two have a match within k days out of d possible is given by

$$n(k, d) = 1.2\sqrt{\frac{d}{2k + 1}} \qquad (16)$$

(Sevast'yanov 1972, Diaconis and Mosteller 1989).

See also BIRTHDAY ATTACK, COINCIDENCE, SMALL
WORLD PROBLEM, SULTAN'S DOWRY PROBLEM

References

Abramson, M. and Moser, W. O. J. "More Birthday Sur-
prises." *Amer. Math. Monthly* **77**, 856–58, 1970.
Ball, W. W. R. and Coxeter, H. S. M. *Mathematical Recrea-
tions and Essays, 13th ed.* New York: Dover, pp. 45–6,
1987.
Bloom, D. M. "A Birthday Problem." *Amer. Math. Monthly*
80, 1141–1142, 1973.
Bogomolny, A. "Coincidence." http://www.cut-the-knot.com/
do_you_know/coincidence.html.
Clevenson, M. L. and Watkins, W. "Majorization and the
Birthday Inequality." *Math. Mag.* **64**, 183–88, 1991.
Diaconis, P. and Mosteller, F. "Methods of Studying Coin-
cidences." *J. Amer. Statist. Assoc.* **84**, 853–61, 1989.
Feller, W. *An Introduction to Probability Theory and Its
Applications, Vol. 1, 3rd ed.* New York: Wiley, pp. 31–2,
1968.
Finch, S. "Puzzle #28 [June 1997]: Coincident Birthdays."
http://www.mathsoft.com/mathcad/library/puzzle/soln28/
soln28.html.
Gehan, E. A. "Note on the 'Birthday Problem.'" *Amer. Stat.*
22, 28, Apr. 1968.
Heuer, G. A. "Estimation in a Certain Probability Problem."
Amer. Math. Monthly **66**, 704–06, 1959.
Hocking, R. L. and Schwertman, N. C. "An Extension of the
Birthday Problem to Exactly *k* Matches." *College Math. J.*
17, 315–21, 1986.
Hunter, J. A. H. and Madachy, J. S. *Mathematical Diver-
sions.* New York: Dover, pp. 102–03, 1975.
Klamkin, M. S. and Newman, D. J. "Extensions of the
Birthday Surprise." *J. Combin. Th.* **3**, 279–82, 1967.
Levin, B. "A Representation for Multinomial Cumulative
Distribution Functions." *Ann. Statistics* **9**, 1123–126,
1981.
McKinney, E. H. "Generalized Birthday Problem." *Amer.
Math. Monthly* **73**, 385–87, 1966.
Mises, R. von. "Über Aufteilungs--und Besetzungs-
Wahrscheinlichkeiten." *Revue de la Faculté des Sciences
de l'Université d'Istanbul, N. S.* **4**, 145–63, 1939. Rep-
rinted in *Selected Papers of Richard von Mises, Vol. 2* (Ed.
P. Frank, S. Goldstein, M. Kac, W. Prager, G. Szego, and
G. Birkhoff). Providence, RI: Amer. Math. Soc., pp. 313–
34, 1964.
Riesel, H. *Prime Numbers and Computer Methods for
Factorization, 2nd ed.* Boston, MA: Birkhäuser, pp. 179–
80, 1994.
Sayrafiezadeh, M. "The Birthday Problem Revisited." *Math.
Mag.* **67**, 220–23, 1994.
Sevast'yanov, B. A. "Poisson Limit Law for a Scheme of
Sums of Dependent Random Variables." *Th. Prob. Appl.*
17, 695–99, 1972.
Sloane, N. J. A. Sequences A014088, A033810, A050255,
and A050256 in "An On-Line Version of the Encyclopedia
of Integer Sequences." http://www.research.att.com/~njas/
sequences/eisonline.html.
Stewart, I. "What a Coincidence!" *Sci. Amer.* **278**, 95–6,
June 1998.
Tesler, L. "Not a Coincidence!" http://www.nomodes.com/
coincidence.html.

Bisected Perimeter Point

NAGEL POINT

Bisection Procedure

A simple procedure for iteratively converging on a
solution which is known to lie inside some interval $[a,
b]$. Let a_π and b_n be the endpoints at the nth iteration
and r_n be the nth approximate solution. Then, the
number of iterations required to obtain an error
smaller than ϵ is found as follows.

$$b_n - a_n = \frac{1}{2^{n-1}}(b - a) \tag{1}$$

$$r_n \equiv \tfrac{1}{2}(a_n + b_n) \tag{2}$$

$$|r_n - r| \le \tfrac{1}{2}(b_n - a_n) = 2^{-n}(b - a) < \epsilon \tag{3}$$

$$-n \ln 2 < \ln \epsilon - \ln(b - a), \tag{4}$$

so

$$n > \frac{\ln(b - a) - \ln \epsilon}{\ln 2}. \tag{5}$$

See also ROOT

References

Arfken, G. *Mathematical Methods for Physicists, 3rd ed.*
Orlando, FL: Academic Press, pp. 964–65, 1985.
Press, W. H.; Flannery, B. P.; Teukolsky, S. A.; and Vetter-
ling, W. T. "Bracketing and Bisection." §9.1 in *Numerical
Recipes in FORTRAN: The Art of Scientific Computing,
2nd ed.* Cambridge, England: Cambridge University
Press, pp. 343–47, 1992.

Bisector

Bisection is the division of a given curve or figure into
two equal parts (halves).

See also ANGLE BISECTOR, BISECTION PROCEDURE,
EXTERIOR ANGLE BISECTOR, HALF, HEMISPHERE, LINE
BISECTOR, PERPENDICULAR BISECTOR, TRISECTION

Bishop's Inequality

Let $V(r)$ be the volume of a BALL of radius r in a
complete n-D RIEMANNIAN MANIFOLD with RICCI
CURVATURE $\ge (n-1)\kappa$. Then $V(r) \ge V_\kappa(r)$, where V_κ
is the volume of a BALL in a space having constant
SECTIONAL CURVATURE. In addition, if equality holds
for some BALL, then this BALL is ISOMETRIC to the
BALL of radius r in the space of constant SECTIONAL
CURVATURE κ.

See also BALL, ISOMETRY

References

Chavel, I. *Riemannian Geometry: A Modern Introduction.*
New York: Cambridge University Press, 1994.

Bishops Problem

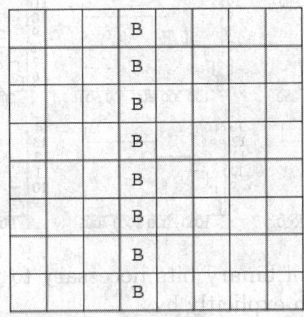

Find the maximum number of bishops $B(n)$ which can be placed on an $n \times n$ CHESSBOARD such that no two attack each other. The answer is $2n - 2$ (Dudeney 1970, Madachy 1979), giving the sequence 2, 4, 6, 8, ... (the EVEN NUMBERS) for $n = 2$, 3, ... One maximal solution for $n = 8$ is illustrated above. The number of distinct maximal arrangements of bishops for $n = 1$, 2, ... are 1, 4, 26, 260, 3368, ... (Sloane's A002465). The number of rotationally and reflectively distinct solutions on an $n \times n$ board for $n \geq 2$ is

$$B(n) = \begin{cases} 2^{(n-4)/2}[2^{(n-2)/2}+1] & \text{for } n \text{ even} \\ 2^{(n-3)/2}[2^{(n-3)/2}+1] & \text{for } n \text{ odd} \end{cases}$$

(Dudeney 1970, p. 96; Madachy 1979, p. 45; Pickover 1995). An equivalent formula is

$$B(n) = 2^{n-3} + 2^{\lfloor (n-1)/2 \rfloor - 1},$$

where $\lfloor n \rfloor$ is the FLOOR FUNCTION, giving the sequence for $n = 1$, 2, ... as 1, 1, 2, 3, 6, 10, 20, 36, ... (Sloane's A005418).

The minimum number of bishops needed to occupy or attack all squares on an $n \times n$ CHESSBOARD is n, arranged as illustrated above.

See also CHESS, KINGS PROBLEM, KNIGHTS PROBLEM, QUEENS PROBLEM, ROOKS PROBLEM

References

Ahrens, W. *Mathematische Unterhaltungen und Spiele, Vol. 1, 3rd ed.* Leipzig, Germany: Teubner, p. 271, 1921.
Dudeney, H. E. "Bishops--Unguarded" and "Bishops--Guarded." §297 and 298 in *Amusements in Mathematics.* New York: Dover, pp. 88–9, 1970.
Guy, R. K. "The n Queens Problem." §C18 in *Unsolved Problems in Number Theory, 2nd ed.* New York: Springer-Verlag, pp. 133–35, 1994.
Madachy, J. *Madachy's Mathematical Recreations.* New York: Dover, pp. 36–6, 1979.
Pickover, C. A. *Keys to Infinity.* New York: Wiley, pp. 74–5, 1995.
Sloane, N. J. A. Sequences A002465/M3616 and A005418/M0771 in "An On-Line Version of the Encyclopedia of Integer Sequences." http://www.research.att.com/~njas/sequences/eisonline.html.

Bislit Cube

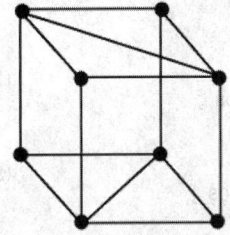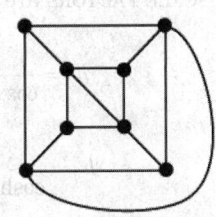

The 8-VERTEX graph consisting of a CUBE in which two opposite faces have DIAGONALS oriented PERPENDICULAR to each other.

See also BIDIAKIS CUBE, CUBE, CUBICAL GRAPH

Bispherical Coordinates

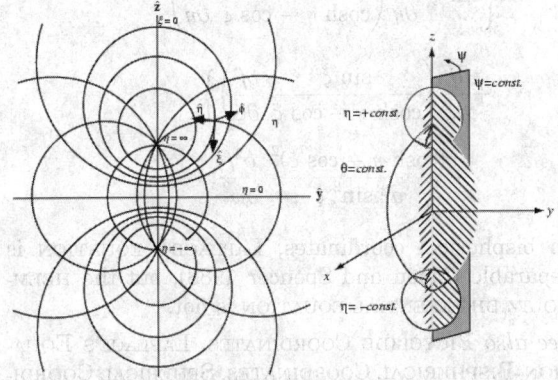

A system of CURVILINEAR COORDINATES variously denoted (ξ, η, ϕ) (Arfken 1970) or (θ, η, ψ) (Moon and Spencer 1988). Using the notation of Arfken, the bispherical coordinates are defined by

$$x = \frac{a \sin \xi \cos \phi}{\cosh \eta - \cos \xi} \tag{1}$$

$$y = \frac{a \sin \xi \sin \phi}{\cosh \eta - \cos \xi} \tag{2}$$

$$z = \frac{a \sinh \eta}{\cosh \eta - \cos \xi}. \tag{3}$$

Surfaces of constant η are given by the spheres

$$x^2 + y^2 + (z - a \coth \eta)^2 = \frac{a^2}{\sinh^2 \eta}, \qquad (4)$$

surfaces of constant ξ by the APPLES ($\xi < \pi/2$) or LEMONS ($\xi > \pi/2$)

$$x^2 + y^2 + z^2 - 2a\sqrt{x^2 + y^2} \cot \xi = a^2, \qquad (5)$$

and surface of constant ψ by the half-planes

$$\tan \phi = y/x. \qquad (6)$$

The SCALE FACTORS are

$$h_\xi = \frac{a}{\cos \eta - \cos \xi} \qquad (7)$$

$$h_\eta = \frac{a}{\cosh \eta - \cos \xi} \qquad (8)$$

$$h_\phi = \frac{a \sin \xi}{\cosh \eta - \cos \xi}. \qquad (9)$$

The LAPLACIAN is given by

$$\nabla^2 f = \frac{(\cosh \eta - \cos \xi)^2}{a^2 \sin \xi}$$

$$\times \left\{ \sin \xi \frac{\partial}{\partial \eta} \left(\frac{1}{\cosh \eta - \cos \xi} \frac{\partial f}{\partial \eta} \right) \right.$$

$$+ \frac{\partial}{\partial \xi} \left(\frac{\sin \xi}{\cosh \eta - \cos \xi} \frac{\partial f}{\partial \xi} \right) \Bigg\}$$

$$+ \frac{(\cosh \eta - \cos \xi)^2}{a^2 \sin^2 \xi} \frac{\partial^2 f}{\partial \phi^2}.$$

In bispherical coordinates, LAPLACE'S EQUATION is separable (Moon and Spencer 1988), but the HELMHOLTZ DIFFERENTIAL EQUATION is not.

See also BICYCLIDE COORDINATES, LAPLACE'S EQUATION–BISPHERICAL COORDINATES, SPHERICAL COORDINATES, TOROIDAL COORDINATES

References

Arfken, G. "Bispherical Coordinates (ξ, η, ϕ)." §2.14 in *Mathematical Methods for Physicists, 2nd ed.* Orlando, FL: Academic Press, pp. 115–17, 1970.
Moon, P. and Spencer, D. E. "Bispherical Coordinates (η, θ, ψ)." Fig. 4.03 in *Field Theory Handbook, Including Coordinate Systems, Differential Equations, and Their Solutions, 2nd ed.* New York: Springer-Verlag, pp. 110–12, 1988.
Morse, P. M. and Feshbach, H. *Methods of Theoretical Physics, Part I.* New York: McGraw-Hill, pp. 665–66, 1953.

Bisymmetric Matrix

A SQUARE MATRIX is called bisymmetric if it is both CENTROSYMMETRIC and either SYMMETRIC or SKEW SYMMETRIC (Muir 1960, p. 19).

See also CENTROSYMMETRIC MATRIX, SKEW SYMMETRIC MATRIX, SYMMETRIC MATRIX

References

Muir, T. *A Treatise on the Theory of Determinants.* New York: Dover, 1960.

Bit Complexity

The number of single operations (of ADDITION, SUBTRACTION, and MULTIPLICATION) required to complete an algorithm.

See also STRASSEN FORMULAS

References

Borodin, A. and Munro, I. *The Computational Complexity of Algebraic and Numeric Problems.* New York: American Elsevier, 1975.

Bit Length

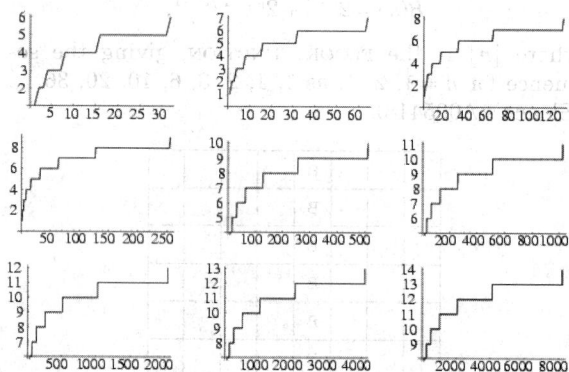

The number of binary bits necessary to represent a number, given explicitly by

$$BL(n) = \lceil \lg n \rceil,$$

where $\lceil x \rceil$ is the CEILING FUNCTION and $\lg n$ is LG, the LOGARITHM to base 2. For $n = 0, 1, 2, \ldots$, the first few values are 0, 1, 2, 2, 3, 3, 3, 3, 4, 4, ... (Sloane's A036377). The function is given by the *Mathematica* 4.0 function `BitLength[n]` in the `Developer` context.

References

Sloane, N. J. A. Sequences A036377 in "An On-Line Version of the Encyclopedia of Integer Sequences." http://www.research.att.com/~njas/sequences/eisonline.html.

Bitangent

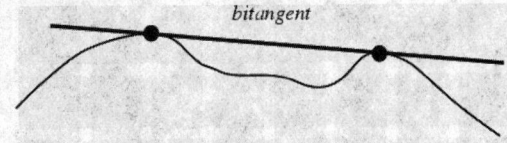

A LINE which is TANGENT to a curve at two distinct points.

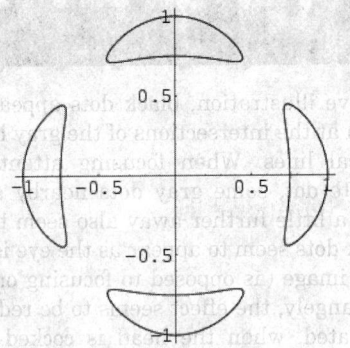

There exist plane QUARTIC CURVES

$$\sum_{i+j \leq 4} a_{ij} x^i y^j = 0$$

that have 28 real bitangents (Shioda 1995, Trott 1997), for example

$$12^2(x^4 + y^4) - 15^2(x^2 + y^2) + 350x^2y^2 + 81 = 0$$

(Trott 1997), illustrated above.

See also KLEIN'S EQUATION, PLÜCKER CHARACTERISTICS, SECANT LINE, SOLOMON'S SEAL LINES, TANGENT LINE

References
Shioda, F. *Comm. Math. Univ. Sancti Pauli* **44**, 109, 1995.
Trott, M. "Applying GroebnerBasis to Three Problems in Geometry." *Mathematica Educ. Res.* **6**, 15–8, 1997.

Bitwin Chain
A bitwin chain of length one consists of two pairs of TWIN PRIMES with the property that they are related by being of the form:

$$(n-1, n+1) \text{ and } (2n-1, 2n+1).$$

In general a chain of length i consists of $i+1$ pairs of TWIN PRIMES,

$$(n-1, n+1), \ (2n-1, 2n+1), \ ..., \ (2^i \cdot n - 1, 2^i \cdot n + 1).$$

Bitwin chains can also be viewed as consisting of two related CUNNINGHAM CHAINS of the first and second kinds,

$$(n-1, 2n-1, 4n-1, ...) \text{ and}$$

$$(n+1, 2n+1, 4n+1, ...).$$

P. Jobling (1999) found the largest known chain of length six,

$$337190719854678690 \cdot 2^n \pm 1,$$

where $n = 0$ to 6.

See also CUNNINGHAM CHAIN, TWIN PRIMES

References
Jobling, P. "A BiTwin chain of length 6 discovered." NMBRTHRY@listserv.nodak.edu posting, 4 Oct 1999.

Biunitary Divisor
A divisor d of a positive integer n is biunitary if the greatest common unitary divisor of d and n/d is 1. For a prime power p^y, the biunitary divisors are the powers $1, p, p^2, ..., p^y$, except for $p^{y/2}$ when y is EVEN (Cohen 1990).

See also DIVISOR, K-ARY DIVISOR, UNITARY DIVISOR

References
Cohen, G. L. "On an Integer's Infinary Divisors." *Math. Comput.* **54**, 395–11, 1990.
Suryanarayana, D. "The Number of Bi-Unitary Divisors of an Integer." *The Theory of Arithmetic Functions (Proc. Conf., Western Michigan Univ., Kalamazoo, Mich., 1971.* New York: Springer-Verlag, pp. 273–82, 1972.
Suryanarayana, D. and Rao, R. S. R. C. "The Number of Bi-Unitary Divisors of an Integer. II." *J. Indian Math. Soc.* **39**, 261–80, 1975.

Bivalent
Capable of taking on one out of two possible values.

See also EXCLUDED MIDDLE LAW, UNIVALENT

Bivalent Range
If the CROSS-RATIO κ of $\{AB, CD\}$ satisfy

$$\kappa^2 - \kappa + 1 = 0, \tag{1}$$

then the points are said to form a bivalent range, and

$$\{AB, CD\} = \{AC, DB\} = \{AD, BC\} = \kappa \tag{2}$$

$$\{AC, BD\} = \{AD, BC\} = \{AB, DC\} = -\kappa^2. \tag{3}$$

See also HARMONIC RANGE

References
Lachlan, R. *An Elementary Treatise on Modern Pure Geometry.* London: Macmillian, p. 268, 1893.

Bivariate Distribution

See also GAUSSIAN BIVARIATE DISTRIBUTION

Bivariate Normal Distribution
GAUSSIAN BIVARIATE DISTRIBUTION

Bivector
An antisymmetric TENSOR of second RANK (a.k.a. 2-form).

$$\vec{X} = X_{ab}\,\omega^a \wedge \omega^b,$$

where $\wedge$ is the WEDGE PRODUCT (or OUTER PRODUCT).

See also TENSOR, VECTOR

Biweight
TUKEY'S BIWEIGHT

Björling Curve
Let $\alpha(z)$, $\gamma(z) : (a, b) \to \mathbb{R}^3$ be curves such that $\|\gamma\| = 1$ and $\alpha \cdot \gamma = 0$, and suppose that α and γ have holomorphic extensions α, $\gamma : (a, b) \times (c, d) \to \mathbb{C}^3$ such that $\|\gamma\| = 1$ and $\alpha \cdot \gamma = 0$ also for $z \in (a, b) \times (c, d)$. Fix $z_0 \in (a, b) \times (c, d)$. Then the Björling curve, defined by

$$B(z) = \alpha(z) - i \int_{z_0}^{z} \gamma(z) \times \alpha'(z)\,dz,$$

is a minimal curve (Gray 1997, p. 762).

References
Björling, E. G. "In integrationem aequationis derivatarum partialum superficiei, cujus in puncto, unoquoque principales ambo radii curvedinis aequales sunt signoque contrario." *Arch. Math. Phys.* **4**, 290–15, 1844.

Dierkes, U.; Hildebrand, S.; Küster, A.; and Wohlrab, O. *Minimal Surfaces, 2 vols.* New York: Springer-Verlag, pp. 120–35, 1992.

Gray, A. "Minimal Surfaces via Björling's Formula." Ch. 33 in *Modern Differential Geometry of Curves and Surfaces with Mathematica, 2nd ed.* Boca Raton, FL: CRC Press, pp. 761–72, 1997.

Nitsche, J. C. C. *Lectures on Minimal Surfaces, Vol. 1: Introduction, Fundamentals, Geometry and Basic Boundary Value Problems.* Cambridge, England: Cambridge University Press, pp. 139–45, 1989.

Schwarz, H. A. *Gesammelte Mathematische Abhandlungen, Vols. 1–.* New York: Chelsea, pp. 179–89, 1972.

Black Dot Illusion

In the above illustration, black dots appear to form and vanish at the intersections of the gray horizontal and vertical lines. When focusing attention on a single white dot, some gray dots nearby and some black dots a little further away also seem to appear. More black dots seem to appear as the eye is scanned across the image (as opposed to focusing on a single point). Strangely, the effect seems to be reduced, but not eliminated, when the head is cocked at a 45° angle. The effect seems to exist only at intermediate distances; if the eye is moved very close to or very far away from the figure, the phantom black dots do not appear.

See also ILLUSION

References
Gephart, J. "Find the Black Dot." http://udel.edu/~jgephart/fun2.htm.

Black Spleenwort Fern
BARNSLEY'S FERN

Blackboard Bold
DOUBLESTRUCK

Blackman Function

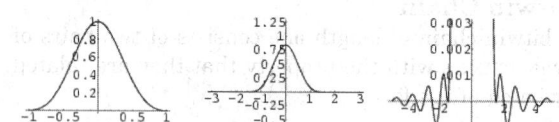

An APODIZATION FUNCTION given by

$$A(x) = 0.42 + 0.5\cos\left(\frac{\pi x}{a}\right) + 0.08\cos\left(\frac{2\pi x}{a}\right). \quad (1)$$

Its FULL WIDTH AT HALF MAXIMUM is $0.810957a$. The APPARATUS FUNCTION is

$$I(k) = \frac{a(0.84 - 0.36a^2k^2 - 2.17 \times 10^{-19}a^4k^4)\sin(2\pi ak)}{(1 - a^2k^2)(1 - 4a^2k^2)}. \quad (2)$$

The COEFFICIENTS are approximations in the general

expansion

$$A(x) = a_0 + 2 \sum_{n=1}^{\infty} a_n \cos\left(\frac{n\pi x}{b}\right), \qquad (3)$$

to

$$a_0 = \frac{3969}{9304} \approx 0.42659 \qquad (4)$$

$$a_1 = \frac{1155}{4652} \approx 0.24828 \qquad (5)$$

$$a_2 = \frac{715}{18608} \approx 0.38424, \qquad (6)$$

which produce zeros of $I(k)$ at $ka = 7/4$ and $ka = 9/4$.

See also APODIZATION FUNCTION

References

Blackman, R. B. and Tukey, J. W. "Particular Pairs of Windows." In *The Measurement of Power Spectra, From the Point of View of Communications Engineering.* New York: Dover, pp. 98–9, 1959.

Black-Scholes Theory

The theory underlying financial derivatives which involves "stochastic calculus" and assumes an uncorrelated LOG NORMAL DISTRIBUTION of continuously varying prices. A simplified "binomial" version of the theory was subsequently developed by Sharpe *et al.* (1995) and Cox *et al.* (1979). It reproduces many results of the full-blown theory, and allows approximation of options for which analytic solutions are not known (Price 1996).

See also GARMAN-KOHLHAGEN FORMULA

References

Black, F. and Scholes, M. S. "The Pricing of Options and Corporate Liabilities." *J. Political Econ.* **81**, 637–59, 1973.
Cox, J. C.; Ross, A.; and Rubenstein, M. "Option Pricing: A Simplified Approach." *J. Financial Economics* **7**, 229–63, 1979.
Price, J. F. "Optional Mathematics is Not Optional." *Not. Amer. Math. Soc.* **43**, 964–71, 1996.
Sharpe, W. F.; Alexander, G. J.; Bailey, J. V.; and Sharpe, W. C. *Investments, 6th ed.* Englewood Cliffs, NJ: Prentice-Hall, 1998.

Blanche's Dissection

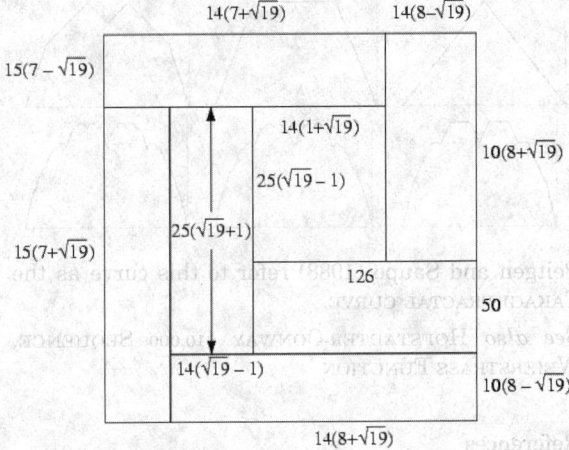

The simplest dissection of a SQUARE into rectangles of the same AREAS but different shapes, composed of the seven pieces illustrated above. The square is 210 units on a side, and each RECTANGLE has AREA $210^2/7 = 6300$.

See also PERFECT SQUARE DISSECTION, RECTANGLE

References

Descartes, B. "Division of a Square into Rectangles." *Eureka,* No. 34, 31–5, 1971.
Wells, D. *The Penguin Dictionary of Curious and Interesting Geometry.* London: Penguin, pp. 14–5, 1991.

Blancmange Function

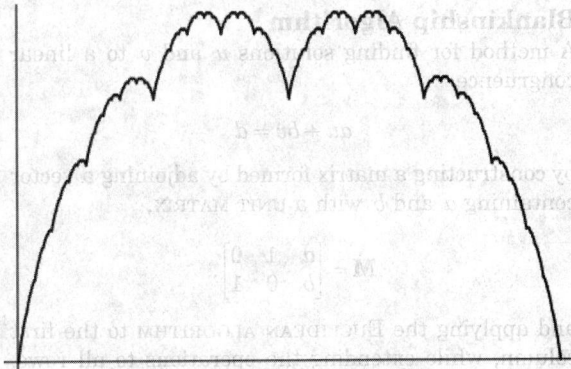

A CONTINUOUS FUNCTION which is nowhere DIFFERENTIABLE. The iterations towards the continuous function are BATRACHIONS resembling the HOFSTADTER-CONWAY \$10,000 SEQUENCE. The first six iterations are illustrated below. The dth iteration contains $N + 1$ points, where $N = 2^d$, and can be obtained by setting $b(0) = b(N) = 0$, letting

$$b(m + 2^{n-1}) = 2^n + \tfrac{1}{2}[b(m) + b(m + 2^n)],$$

and looping over $n = d$ to 1 by steps of -1 and $m = 0$

to $N - 1$ by steps of 2^n.

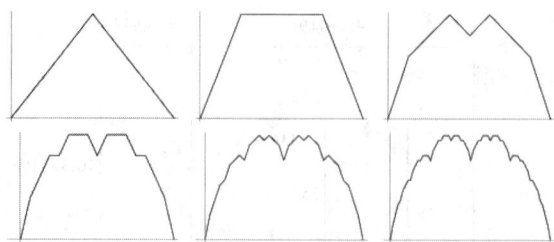

Peitgen and Saupe (1988) refer to this curve as the TAKAGI FRACTAL CURVE.

See also HOFSTADTER-CONWAY $10,000 SEQUENCE, WEIERSTRASS FUNCTION

References

Dixon, R. *Mathographics.* New York: Dover, pp. 175–76 and 210, 1991.

Peitgen, H.-O. and Saupe, D. (Eds.). "Midpoint Displacement and Systematic Fractals: The Takagi Fractal Curve, Its Kin, and the Related Systems." §A.1.2 in *The Science of Fractal Images.* New York: Springer-Verlag, pp. 246–48, 1988.

Takagi, T. "A Simple Example of the Continuous Function without Derivative." *Proc. Phys. Math. Japan* **1**, 176–77, 1903.

Tall, D. O. "The Blancmange Function, Continuous Everywhere but Differentiable Nowhere." *Math. Gaz.* **66**, 11–2, 1982.

Tall, D. "The Gradient of a Graph." *Math. Teaching* **111**, 48–2, 1985.

Wells, D. *The Penguin Dictionary of Curious and Interesting Geometry.* London: Penguin, pp. 16–7, 1991.

Blankinship Algorithm

A method for finding solutions u and v to a linear congruence

$$au + bv = d$$

by constructing a matrix formed by adjoining a vector containing a and b with a UNIT MATRIX,

$$\mathbf{M} = \begin{bmatrix} a & 1 & 0 \\ b & 0 & 1 \end{bmatrix},$$

and applying the EUCLIDEAN ALGORITHM to the first column, while extending the operations to all rows. The algorithm terminates when the first column contains the GREATEST COMMON DIVISOR $\mathrm{GCD}(a, b)$.

See also EUCLIDEAN ALGORITHM, GREATEST COMMON DIVISOR

References

Blankinship, W. A. "A New Version of the Euclidean Algorithm." *Amer. Math. Monthly* **70**, 742–45, 1963.

Séroul, R. "The Blankinship Algorithm." §8.2 in *Programming for Mathematicians.* Berlin: Springer-Verlag, pp. 161–63, 2000.

Blaschke Condition

If $\{a_j\} \subseteq D(0, 1)$ (with possible repetitions) satisfies

$$\sum_{j=1}^{\infty} (1 - |a_j|) \le \infty,$$

where $D(0, 1)$ is the unit open disk, and no $a_j = 0$, then there is a bounded ANALYTIC FUNCTION on $D(0, 1)$ which has ZERO SET consisting precisely of the a_js, counted according to their MULTIPLICITIES. More specifically, the INFINITE PRODUCT

$$\prod_{j=1}^{\infty} -\frac{\bar{a}_j}{|a_j|} B_{a_j}(z),$$

where $B_{a_j}(z)$ is a BLASCHKE FACTOR and $\tilde{z}$ is the COMPLEX CONJUGATE, converges uniformly on compact subsets of $D(0, 1)$ to a bounded analytic function $B(z)$.

See also BLASCHKE FACTOR, BLASCHKE FACTORIZATION, BLASCHKE PRODUCT

References

Krantz, S. G. "The Blaschke Condition." §9.1.5 in *Handbook of Complex Analysis.* Boston, MA: Birkhäuser, pp. 118–19, 1999.

Blaschke Conjecture

The only WIEDERSEHEN MANIFOLDS are the standard round spheres. The conjecture has been proven by combining the BERGER-KAZDAN COMPARISON THEOREM with A. Weinstein's results for n EVEN and C. T. Yang's for n ODD.

See also WIEDERSEHEN MANIFOLD

References

Chavel, I. *Riemannian Geometry: A Modern Introduction.* New York: Cambridge University Press, 1994.

Blaschke Factor

If a is a point in the open UNIT DISK, then the Blaschke factor is defined by

$$B_a(z) = \frac{z - a}{1 - \bar{a}z},$$

where $\bar{a}$ is the COMPLEX CONJUGATE of a. Blaschke factors allow the manipulation of the zeros of a HOLOMORPHIC FUNCTION analogously to factors of $(z - a)$ for complex polynomials (Krantz 1999, p. 117).

See also BLASCHKE CONDITION, BLASCHKE FACTORIZATION

References

Krantz, S. G. "Blaschke Factors." §9.1.1 in *Handbook of Complex Analysis.* Boston, MA: Birkhäuser, p. 117, 1999.

Blaschke Factorization

Let f be a bounded ANALYTIC FUNCTION on $D(0, 1)$ vanishing to order $m \geq 0$ at 0 and let $\{a_j\}$ be its other zeros, listed with multiplicities. Then

$$f(z) = z^m F(z) \prod_{j=1}^{\infty} -\frac{\bar{a}_j}{|a_j|} B_{a_j}(z),$$

where F is a bounded ANALYTIC FUNCTION on $D(0, 1)$, F is zerofree, $\tilde{z}$ is the COMPLEX CONJUGATE, and

$$\sup_{z \in D(0,\,1)} |f(z)| = \sup_{z \in D(0,\,1)} |F(z)|.$$

See also BLASCHKE FACTOR

References
Krantz, S. G. "Blaschke Factorization." §9.1.7 in *Handbook of Complex Analysis*. Boston, MA: Birkhäuser, p. 119, 1999.

Blaschke Product

A Blaschke product is an expression of the form

$$B(z) = z^m \prod_{j=1}^{\infty} -\frac{\bar{a}_j}{|a_j|} B_{a_j}(z),$$

where m is a nonnegative integer and $\tilde{z}$ is the COMPLEX CONJUGATE.

See also BLASCHKE FACTOR

References
Krantz, S. G. "Blaschke Products." §9.1.6 in *Handbook of Complex Analysis*. Boston, MA: Birkhäuser, p. 119, 1999.

Blaschke's Theorem

A convex planar domain in which the minimal GENERALIZED DIAMETER is >1 always contains a CIRCLE of RADIUS 1/3.

See also GENERALIZED DIAMETER

References
Le Lionnais, F. *Les nombres remarquables*. Paris: Hermann, p. 25, 1983.
Wells, D. *The Penguin Dictionary of Curious and Interesting Geometry*. London: Penguin, pp. 17–8, 1991.

Blasius Differential Equation

The third-order ORDINARY DIFFERENTIAL EQUATION

$$2y''' + yy'' = 0.$$

This equation arises in the theory of fluid boundary layers, and must be solved numerically (Rosenhead 1963; Schlichting 1979; Tritton 1989, p. 129). The velocity profile produced by this differential equation is known as the Blasius profile.

References
Meyer, G. H. *Initial Value Methods for Boundary Value Problems: Theory and Application of Invariant Imbedding*. New York: Academic Press, 1973.
Rosenhead, L. (Ed.). *Laminar Boundary Layers*. Oxford, England: Oxford University Press, 1963.
Schlichting, H. *Boundary Layer Theory*, 7th ed. New York: McGraw-Hill, 1979.
Tritton, D. J. *Physical Fluid Dynamics*, 2nd ed. Oxford, England: Clarendon Press, p. 129, 1989.
Zwillinger, D. *Handbook of Differential Equations*, 3rd ed. Boston, MA: Academic Press, p. 128, 1997.

Blecksmith-Brillhart-Gerst Theorem

A generalization of SCHRÖTER'S FORMULA.

References
Berndt, B. C. *Ramanujan's Notebooks, Part III*. New York: Springer-Verlag, p. 73, 1985.

Blichfeldt's Lemma

BLICHFELDT'S THEOREM

Blichfeldt's Theorem

Any bounded planar region with POSITIVE AREA $> A$ placed in any position of the UNIT SQUARE LATTICE can be TRANSLATED so that the number of LATTICE POINTS inside the region will be at least $A + 1$ (Blichfeldt 1914, Steinhaus 1983) The theorem can be generalized to n-D.

See also LATTICE POINT, MINKOWSKI CONVEX BODY THEOREM, PICK'S THEOREM

References
Blichfeldt, H. F. "A New Principle in the Geometry of Numbers, with Some Applications." *Trans. Amer. Math. Soc.* **15**, 227–35, 1914.
Steinhaus, H. *Mathematical Snapshots*, 3rd ed. New York: Dover, pp. 97–9, 1999.

B-Line

A line which simultaneously bisects a triangle's perimeter and area.

See also CLEAVER, SPLITTER

References
Todd, A. "Bisecting a Triangle." *Pi Mu Epsilon J.* **11**, 31–7, Fall 1999.
Todd, A. "Bisecting a Triangle." http://www.math.colostate.edu/~todd/triangle.html.

BLM/Ho Polynomial

A 1-variable unoriented KNOT POLYNOMIAL $Q(x)$. It satisfies

$$Q_{\text{unknot}} = 1 \tag{1}$$

and the SKEIN RELATIONSHIP

$$Q_{L_+} + Q_{L_-} = x(Q_{L_0} + Q_{L_\infty}). \tag{2}$$

It also satisfies

$$Q_{L_1 \# L_2} = Q_{L_1} Q_{L_2}, \qquad (3)$$

where is the KNOT SUM and

$$Q_{L^*} = Q_L, \qquad (4)$$

where L^* is the MIRROR IMAGE of L. The BLM/Ho polynomials of MUTANT KNOTS are also identical. Brandt *et al.* (1986) give a number of interesting properties. For any LINK L with ≥ 2 components, $Q_L - 1$ is divisible by $2(x - 1)$. If L has c components, then the lowest POWER of x in $Q_L(x)$ is $1 - c$, and

$$\lim_{x \to 0} x^{c-1} Q_L(x) = \lim_{(\ell, m) \to (1, 0)} (-m)^{c-1} P_L(\ell, m), \qquad (5)$$

where P_L is the HOMFLY POLYNOMIAL. Also, the degree of Q_L is less than the CROSSING NUMBER of L. If L is a 2-BRIDGE KNOT, then

$$Q_L(z) = 2z^{-1} V_L(t) V_L(t^{-1} + 1 - 2z^{-1}), \qquad (6)$$

where $z \equiv -t - t^{-1}$ (Kanenobu and Sumi 1993).

The POLYNOMIAL was subsequently extended to the 2-variable KAUFFMAN POLYNOMIAL F, which satisfies

$$Q(x) = F(1, x). \qquad (7)$$

Brandt *et al.* (1986) give a listing of Q POLYNOMIALS for KNOTS up to 8 crossings and links up to 6 crossings.

References

Brandt, R. D.; Lickorish, W. B. R.; and Millett, K. C. "A Polynomial Invariant for Unoriented Knots and Links." *Invent. Math.* **84**, 563–73, 1986.

Ho, C. F. "A New Polynomial for Knots and Links--Preliminary Report." *Abstracts Amer. Math. Soc.* **6**, 300, 1985.

Kanenobu, T. and Sumi, T. "Polynomial Invariants of 2-Bridge Knots through 22-Crossings." *Math. Comput.* **60**, 771–78 and S17-S28, 1993.

Stoimenow, A. "Brandt-Lickorish-Millett-Ho Polynomials." http://guests.mpim-bonn.mpg.de/alex/ptab/blmh10.html.

Weisstein, E. W. "Knots." MATHEMATICA NOTEBOOK KNOTS.M.

Bloch Constant

N.B. A detailed online essay by S. Finch was the starting point for this entry. Let F be the set of COMPLEX ANALYTIC FUNCTIONS f defined on an open region containing the CLOSURE of the UNIT DISK $D = \{z : |z| < 1\}$ satisfying $f(0) = 0$ and $df/dz(0) = 1$. For each f in F, let $b(f)$ be the SUPREMUM of all numbers r such that there is a disk S in D on which f is ONE-TO-ONE and such that $f(S)$ contains a disk of radius r. In 1925, Bloch (Conway 1978) showed that $b(f) \geq 1/72$. Define Bloch's constant by

$$B \equiv \inf\{b(f) : f \in F\}.$$

Ahlfors and Grunsky (1937) derived

$$0.433012701\ldots = \tfrac{1}{4}\sqrt{3} \leq B < \frac{1}{\sqrt{1 + \sqrt{3}}} \frac{\Gamma(\tfrac{1}{3})\Gamma(\tfrac{11}{12})}{\Gamma(\tfrac{1}{4})} < 0.4718617.$$

They also conjectured that the upper limit is actually the value of B,

$$B = \frac{1}{\sqrt{1 + \sqrt{3}}} \frac{\Gamma(\tfrac{1}{3})\Gamma(\tfrac{11}{12})}{\Gamma(\tfrac{1}{4})}$$

$$= \sqrt{\pi} 2^{1/4} \frac{\Gamma(\tfrac{1}{3})}{\Gamma(\tfrac{1}{4})} \sqrt{\frac{\Gamma(\tfrac{11}{12})}{\Gamma(\tfrac{1}{12})}}$$

$$= 0.4718617\ldots$$

(Le Lionnais 1983).

See also LANDAU CONSTANT

References

Conway, J. B. *Functions of One Complex Variable I, 2nd ed.* New York: Springer-Verlag, 1989.

Finch, S. "Favorite Mathematical Constants." http://www.mathsoft.com/asolve/constant/bloch/bloch.html.

Le Lionnais, F. *Les nombres remarquables.* Paris: Hermann, p. 25, 1983.

Minda, C. D. "Bloch Constants." *J. d'Analyse Math.* **41**, 54–4, 1982.

Bloch-Landau Constant

LANDAU CONSTANT

Block

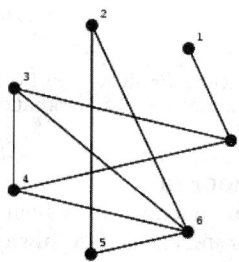

A maximal BICONNECTED SUBGRAPH of a given GRAPH G. In the illustration above, the blocks are $\{2, 5, 6\}$, $\{3, 4, 6, 7\}$, and $\{1, 7\}$.

If a graph G is biconnected, then G itself is called a block (Harary 1994, p. 26) or a BICONNECTED GRAPH (Skiena 1990, p. 175).

See also BICONNECTED GRAPH, BLOCK DESIGN, DIGIT BLOCK, SQUARE POLYOMINO

References

Aho, A. V.; Hopcroft, J. E.; and Ullman, J. D. *The Design and Analysis of Computer Algorithms.* Reading, MA: Addison-Wesley, 1974.

Harary, F. *Graph Theory.* Reading, MA: Addison-Wesley, 1994.

Skiena, S. "Biconnected Components." §5.1.4 in *Implementing Discrete Mathematics: Combinatorics and Graph Theory with Mathematica.* Reading, MA: Addison-Wesley, pp. 175–77, 1990.

Block (Group Action)

A GROUP ACTION $G \times \Omega \to \Omega$ might preserve a special kind of PARTITION of Ω called a system of blocks. A block is a SUBSET Δ of Ω such that for any group element g either

     1. g preserves Δ, i.e., $g\Delta = \Delta$, or
     2. g translates everything in Δ out of Δ, i.e., $g\Delta \cap \Delta = \phi$.

For example, the GENERAL LINEAR GROUP $GL(2, \mathbb{R})$ acts on the plane minus the origin, $\mathbb{R}^2 - (0, 0)$. The lines $A = \{(at, bt)\}$ are blocks because either a line is mapped to itself, or to another line. Of course, the points on the line may be rescaled, so the lines in A are minimal blocks.

In fact, if two blocks intersect then their intersection is also a block. Hence, the minimal blocks form a PARTITION of Ω. It is important to avoid confusion with the notion of a block in a BLOCK DESIGN, which is different.

See also GROUP, PRIMITIVE (GROUP ACTION), STEINER SYSTEM

References

Dixon, J. and Mortimer, B. *Permutation Groups.* New York: Springer-Verlag, 1996.

Block (Set)

One of the disjoint SUBSETS making up a SET PARTITION. A block containing n elements is called an n-block. The partitioning of sets into blocks can be denoted using a RESTRICTED GROWTH STRING.

See also BLOCK DESIGN, RESTRICTED GROWTH STRING, SET PARTITION

Block Design

An incidence system (v, k, λ, r, b) in which a set X of v points is partitioned into a family A of b subsets (blocks) in such a way that any two points determine λ blocks with k points in each block, and each point is contained in r different blocks. It is also generally required that $k < v$, which is where the "incomplete" comes from in the formal term most often encountered for block designs, BALANCED INCOMPLETE BLOCK DESIGNS (BIBD). The five parameters are not independent, but satisfy the two relations

$$vr = bk \tag{1}$$

$$\lambda(v-1) = r(k-1). \tag{2}$$

A BIBD is therefore commonly written as simply (v, k, λ), since b and r are given in terms of v, k, and λ by

$$b = \frac{v(v-1)\lambda}{k(k-1)} \tag{3}$$

$$r = \frac{\lambda(v-1)}{k-1}. \tag{4}$$

A BIBD is called SYMMETRIC if $b = v$ (or, equivalently, $r = k$).

Writing $X = \{x_i\}_{i=1}^{v}$ and $A = \{A_j\}_{j=1}^{b}$, then the INCIDENCE MATRIX of the BIBD is given by the $v \times b$ MATRIX **M** defined by

$$m_{ij} = \begin{cases} 1 & \text{if } x_i \in A \\ 0 & \text{otherwise.} \end{cases} \tag{5}$$

This matrix satisfies the equation

$$\mathsf{M}\mathsf{M}^{\mathsf{T}} = (r - \lambda)\mathsf{I} + \lambda\mathsf{J}, \tag{6}$$

where I is a $v \times v$ IDENTITY MATRIX and J is the $v \times v$ UNIT MATRIX (Dinitz and Stinson 1992).

Examples of BIBDs are given in the following table.

Block Design	(v, k, λ)
AFFINE PLANE	$(n^2, n, 1)$
FANO PLANE	$(7, 3, 1)$
HADAMARD DESIGN	SYMMETRIC $(4n + 3, 2n + 1, n)$
PROJECTIVE PLANE	SYMMETRIC $(n^2 + n + 1, n + 1, 1)$
STEINER TRIPLE SYSTEM	$(v, 3, 1)$
UNITAL	$(q^3 + 1, q + 1, 1)$

See also AFFINE PLANE, DESIGN, FANO PLANE, HADAMARD DESIGN, PARALLEL CLASS, PROJECTIVE PLANE, RESOLUTION, RESOLVABLE, STEINER TRIPLE SYSTEM, SYMMETRIC BLOCK DESIGN, UNITAL

References

Dinitz, J. H. and Stinson, D. R. "A Brief Introduction to Design Theory." Ch. 1 in *Contemporary Design Theory: A Collection of Surveys* (Ed. J. H. Dinitz and D. R. Stinson). New York: Wiley, pp. 1–2, 1992.
Ryser, H. J. "The (b, v, r, k, λ)-Configuration." §8.1 in *Combinatorial Mathematics.* Buffalo, NY: Math. Assoc. Amer., pp. 96–02, 1963.

Block Diagonal Matrix

A block diagonal matrix, also called a diagonal block matrix, is a SQUARE DIAGONAL MATRIX in which the diagonal elements are SQUARE MATRICES of any size (possibly even 1×1), and the off-diagonal elements are 0. A block diagonal matrix is therefore a BLOCK MATRIX in which the blocks off the diagonal are the ZERO MATRICES, and the diagonal matrices are SQUARE.

Block diagonal matrices can be constructed in *Mathematica* using the following code snippet.

```
< <LinearAlgebra`MatrixManipulation`
BlockDiagonal[a_List] :=
Module[{n = Length[a], lens = Length/@a, i, k, tmp},
  k = Outer[List, lens, lens];
tmp = Map[ZeroMatrix[#1[[1]], #1[[2]]]&, k, {2}];
  BlockMatrix@
ReplacePart[tmp, a, Table[{i, i}, {i, Length[a]}],
  Table[{i}, {i, Length[a]}]]]]
```

See also BLOCK MATRIX, CAYLEY-HAMILTON THEO-REM, DIAGONAL MATRIX, DIRECT SUM, JORDAN CANONICAL FORM, LINEAR TRANSFORMATION, MATRIX, MATRIX DIRECT SUM

Block Growth

Let $(x_0 x_1 x_2 \ldots)$ be a sequence over a finite ALPHABET A (all the entries are elements of A). Define the block growth function $B(n)$ of a sequence to be the number of ADMISSIBLE words of length n. For example, in the sequence $aabaabaabaabaab \ldots$, the following words are ADMISSIBLE

Length	Admissible Words
1	a, b
2	aa, ab, ba
3	aab, aba, baa
4	$aaba\ abaa,\ baab$

so $B(1) = 2$, $B(2) = 3$, $B(3) = 3$, $B(4) = 3$, and so on. Notice that $B(n) \le B(n+1)$, so the block growth function is always nondecreasing. This is because any ADMISSIBLE word of length n can be extended rightwards to produce an ADMISSIBLE word of length $n+1$. Moreover, suppose $B(n) = B(n+1)$ for some n. Then each admissible word of length n extends to a *unique* ADMISSIBLE word of length $n+1$.

For a SEQUENCE in which each substring of length n uniquely determines the next symbol in the SE-QUENCE, there are only finitely many strings of length n, so the process must eventually cycle and the SEQUENCE must be eventually periodic. This gives us the following theorems:

1. If the SEQUENCE is eventually periodic, with least period p, then $B(n)$ is strictly increasing until it reaches p, and $B(n)$ is constant thereafter.
2. If the SEQUENCE is not eventually periodic, then $B(n)$ is strictly increasing and so $B(n) \ge n+1$ for

all n. If a SEQUENCE has the property that $B(n) = n+1$ for all n, then it is said to have minimal block growth, and the SEQUENCE is called a STURMIAN SEQUENCE.

The block growth is also called the GROWTH FUNCTION or the COMPLEXITY of a SEQUENCE.

Block Matrix

A block matrix is a MATRIX that is defined using smaller matrices, called blocks. For example,

$$\begin{bmatrix} A & B \\ C & D \end{bmatrix}, \tag{1}$$

where A, B, C, and D are themselves matrices, is a block matrix. In the specific example

$$A = \begin{bmatrix} 0 & 2 \\ 2 & 0 \end{bmatrix} \tag{2}$$

$$B = \begin{bmatrix} 3 & 3 & 3 \\ 3 & 3 & 3 \end{bmatrix} \tag{3}$$

$$C = \begin{bmatrix} 4 & 4 \\ 4 & 4 \\ 4 & 4 \end{bmatrix} \tag{4}$$

$$D = \begin{bmatrix} 5 & 0 & 5 \\ 0 & 5 & 0 \\ 5 & 0 & 5 \end{bmatrix}, \tag{5}$$

it is the matrix

$$\begin{bmatrix} 0 & 2 & 3 & 3 & 3 \\ 2 & 0 & 3 & 3 & 3 \\ 4 & 4 & 5 & 0 & 5 \\ 4 & 4 & 0 & 5 & 0 \\ 4 & 4 & 5 & 0 & 5 \end{bmatrix}. \tag{6}$$

Block matrices can be created using BlockMa-trix[*blocks*] in the *Mathematica* add-on package LinearAlgebra`MatrixMultiplication` (which can be loaded with the command < <LinearAlgebra`).

When two block matrices have the same shape and their diagonal blocks are square matrices, then they multiply similarly to MATRIX MULTIPLICATION. For example,

$$\begin{bmatrix} A_1 & B_1 \\ C_1 & D_1 \end{bmatrix} \begin{bmatrix} A_2 & B_2 \\ C_2 & D_2 \end{bmatrix}$$
$$= \begin{bmatrix} A_1 A_2 + B_1 C_2 & A_1 B_2 \\ C_1 A_2 + D_1 C_2 & C_1 B_2 + D_1 D_2 \end{bmatrix}. \tag{7}$$

When the blocks are SQUARE MATRICES, the set of invertible block matrices form a group, which is a special case of the GENERAL LINEAR GROUP. In this case, it is $GL_2(R^*)$, the invertible two by two matrices with entries in the UNITS of a RING R, where here R is the ring of square matrices.

See also BLOCK DIAGONAL MATRIX, CAYLEY-HAMILTON THEOREM, MATRIX, RING

Blow-Up

A common mechanism which generates SINGULARITIES from smooth initial conditions.

See also BLOW-UP LEMMA

Blow-Up Lemma

The blow-up lemma essentially says that regular pairs in SZEMERÉDI'S REGULARITY LEMMA behave like COMPLETE BIPARTITE GRAPHS from the point of view of embedding bounded degree subgraphs.

In particular, given a graph R of order r, minimal VERTEX DEGREE δ and maximal VERTEX DEGREE Δ, then there exists an $\epsilon > 0$ such that the following holds. Let N be an arbitrary positive integer, and replace the vertices of R with pairwise disjoint N-sets $V_1, V_2, ..., V_r$ (blowing up). Now construct two graphs on the same vertex set $V = \cup V_i$. The graph $R(N)$ is obtained by replacing all edges of R with copies of the complete bipartite graph $K_{N, N}$, and construct a sparser graph by replacing the edges of R with some (ϵ, δ)-superregular pair. If a graph H with $\Delta(H) \leq \Delta$ is embeddable into $R(N)$, then it is already embeddable into G (Komlós *et al.* 1998).

See also SZEMERÉDI'S REGULARITY LEMMA

References

Komlós, J.; Sárkozy, G. N.; and Szemerédi, E. "Blow-Up Lemma." *Combinatorica* **17**, 109–23, 1997.
Komlós, J.; Sárkozy, G. N.; and Szemerédi, E. "Proof of the Seymour Conjecture for Large Graphs." *Ann. Comb.* **2**, 43–0, 1998.

Blue-Empty Coloring

BLUE-EMPTY GRAPH

Blue-Empty Graph

An EXTREMAL GRAPH in which the forced TRIANGLES are all the same color. Call R the number of red MONOCHROMATIC FORCED TRIANGLES and B the number of blue MONOCHROMATIC FORCED TRIANGLES, then a blue-empty graph is an EXTREMAL GRAPH with $B = 0$. For EVEN n, a blue-empty graph can be achieved by coloring red two COMPLETE SUBGRAPHS of $n/2$ points (the RED NET method). There is no blue-empty coloring for ODD n except for $n = 7$ (Lorden 1962).

See also COMPLETE GRAPH, EXTREMAL GRAPH, MONOCHROMATIC FORCED TRIANGLE, RED NET

References

Lorden, G. "Blue-Empty Chromatic Graphs." *Amer. Math. Monthly* **69**, 114–20, 1962.
Sauvé, L. "On Chromatic Graphs." *Amer. Math. Monthly* **68**, 107–11, 1961.

Board

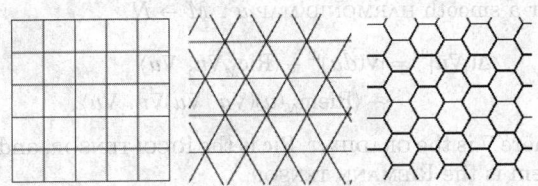

A board is a subset of the polygons determined by a number of (usually regularly spaced and oriented) lines. These polygons form the spaces on which "pieces" can be placed and move in many games (called board games). The simplest division the plane is into equal squares. The 3×3 square board is used in TIC-TAC-TOE. The 8×8 square board is used in CHECKERS and CHESS. Hexagonal boards are used in some games. Chinese checkers uses a board in the space of a pentagram with spaces at the vertices of a regular triangular tiling.

See also CHECKERS, CHESS, CHESSBOARD, GRID, ROOK NUMBER, TIC-TAC-TOE

References

Bell, R. C. *Board and Table Games from Many Civilizations.* New York: Dover, 1980.
Gardner, M. "Four Unusual Board Games." Ch. 5 in *The Sixth Book of Mathematical Games from Scientific American.* Chicago, IL: University of Chicago Press, pp. 39–7, 1984.
Murray, H. J. R. *A History of Board-Games Other than Chess.* New York: Oxford University Press, 1952.
Parlett, D. *The Oxford History of Board Games.* Oxford, England: Oxford University Press, 1999.
Steinhaus, H. *Mathematical Snapshots, 3rd ed.* New York: Dover, p. 10, 1999.

Boatman's Knot

CLOVE HITCH

Bôcher Equation

A second-order ORDINARY DIFFERENTIAL EQUATION OF THE FORM

$$y'' + \frac{1}{2}\left[\frac{m_1}{x - a_1} + \ldots + \frac{m_{n-1}}{x - a_{n-1}}\right]y'$$

$$+ \frac{1}{4}\left[\frac{A_0 + A_1 x + \ldots + A_1 x^1}{(x - a_1)^{m_1}(x - a_2)^{m_2}\ldots(x - a_{n-1})^{m_{n-1}}}\right]y = 0.$$

References

Moon, P. and Spencer, D. E. "Differential Equations." §6 in *Field Theory Handbook, Including Coordinate Systems, Differential Equations, and Their Solutions, 2nd ed.* New York: Springer-Verlag, pp. 144–62, 1988.
Zwillinger, D. (Ed.). *CRC Standard Mathematical Tables and Formulae.* Boca Raton, FL: CRC Press, p. 413, 1995.

Bochner Identity

For a smooth HARMONIC MAP $u : M \to N$,

$$\Delta(|\nabla u|^2) = |\nabla(du)|^2 + \langle \mathrm{Ric}_M \nabla u, \ \nabla u \rangle$$
$$- \langle \mathrm{Riem}_N(u)(\nabla u, \ \nabla u)\nabla u, \ \nabla u \rangle,$$

where ∇ is the GRADIENT, Ric is the RICCI TENSOR, and Riem is the RIEMANN TENSOR.

References

Eels, J. and Lemaire, L. "A Report on Harmonic Maps." *Bull. London Math. Soc.* **10**, 1–8, 1978.

Bochner's Theorem

Among the continuous functions on $\mathbb{R}^n$, the POSITIVE DEFINITE FUNCTIONS are those functions which are the FOURIER TRANSFORMS of finite measures.

Bode's Rule

Let the values of a function $f(x)$ be tabulated at points x_i equally spaced by $h = x_{i+1} - x_i$, so $f_1 = f(x_1)$, $f_2 = f(x_2)$, ..., $f_5 = f(x_5)$. Then Bode's rule approximating the integral of $f(x)$ is given by the NEWTON-COTES-like formula

$$\int_{x_1}^{x_5} f(x) \, dx = \frac{2}{45}h(7f_1 + 32f_2 + 12f_3 + 32f_4 + 7f_5)$$
$$- \frac{8}{945}h^7 f^{(6)}(\xi).$$

See also HARDY'S RULE, NEWTON-COTES FORMULAS, SIMPSON'S 3/8 RULE, SIMPSON'S RULE, TRAPEZOIDAL RULE, WEDDLE'S RULE

References

Abramowitz, M. and Stegun, C. A. (Eds.). *Handbook of Mathematical Functions with Formulas, Graphs, and Mathematical Tables, 9th printing.* New York: Dover, p. 886, 1972.

Bogdanov Map

A 2-D MAP which is conjugate to the HÉNON MAP in its nondissipative limit. It is given by

$$x' = x + y'$$
$$y' = y + \epsilon y + kx(x - 1) + \mu xy.$$

See also HÉNON MAP

References

Arrowsmith, D. K.; Cartwright, J. H. E.; Lansbury, A. N.; and Place, C. M. "The Bogdanov Map: Bifurcations, Mode Locking, and Chaos in a Dissipative System." *Int. J. Bifurcation Chaos* **3**, 803–42, 1993.
Bogdanov, R. "Bifurcations of a Limit Cycle for a Family of Vector Fields on the Plane." *Selecta Math. Soviet* **1**, 373–88, 1981.

Bogomolov-Miyaoka-Yau Inequality

Relates invariants of a curve defined over the INTEGERS. If this inequality were proven true, then FERMAT'S LAST THEOREM would follow for sufficiently large exponents. Miyaoka claimed to have proven this inequality in 1988, but the proof contained an error.

See also FERMAT'S LAST THEOREM

References

Cox, D. A. "Introduction to Fermat's Last Theorem." *Amer. Math. Monthly* **101**, 3–4, 1994.

Bohemian Dome

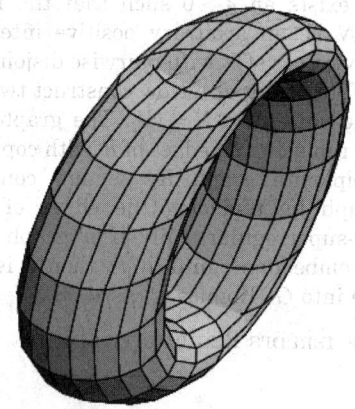

A QUARTIC SURFACE which can be constructed as follows. Given a CIRCLE C and PLANE E PERPENDICULAR to the PLANE of C, move a second CIRCLE K of the same RADIUS as C through space so that its CENTER always lies on C and it remains PARALLEL to E. Then K sweeps out the Bohemian dome. It can be given by the PARAMETRIC EQUATIONS

$$x = a \cos u$$
$$y = b \cos v + a \sin u$$
$$z = c \sin v$$

where u, $v \in [0, \ 2\pi)$. In the above plot, $a = 0.5$, $b = 1.5$, and $c = 1$.

See also QUARTIC SURFACE

References

Fischer, G. (Ed.). *Mathematical Models from the Collections of Universities and Museums.* Braunschweig, Germany: Vieweg, pp. 19–0, 1986.
Fischer, G. (Ed.). Plate 50 in *Mathematische Modelle/ Mathematical Models, Bildband/Photograph Volume.* Braunschweig, Germany: Vieweg, p. 50, 1986.
Gray, A. *Modern Differential Geometry of Curves and Surfaces with Mathematica, 2nd ed.* Boca Raton, FL: CRC Press, p. 389, 1997.
Nordstrand, T. "Bohemian Dome." http://www.uib.no/people/nfytn/bodtxt.htm.

Bohr Matrix

A finite or infinite SQUARE MATRIX with RATIONAL entries. (If the matrix is infinite, all but a finite number of entries in each row must be 0.) The sum or product of two Bohr matrices is another Bohr matrix.

References

Apostol, T. M. "Bohr Matrices." §8.4 in *Modular Functions and Dirichlet Series in Number Theory*, 2nd ed. New York: Springer-Verlag, pp. 167–68, 1997.

Bohr-Favard Inequalities

If f has no spectrum in $[-\lambda, \lambda]$, then

$$\|f\|_\infty \le \frac{\pi}{2\lambda} \|f'\|_\infty$$

(Bohr 1935). A related inequality states that if A_k is the class of functions such that

$$f(x) = f(x + 2\pi), f(x), f'(x), \ldots, f^{(k-1)}(x)$$

are absolutely continuous and $\int_0^{2\pi} f(x)\,dx = 0$, then

$$\|f\|_\infty \le \frac{4}{\pi} \sum_{v=0}^\infty \frac{(-1)^{v(k+1)}}{(2v+1)^{k+1}} \|f^{(k)}(x)\|_\infty$$

(Northcott 1939). Further, for each value of k, there is always a function $f(x)$ belonging to A_k and not identically zero, for which the above inequality becomes an equality (Favard 1936). These inequalities are discussed in Mitrinovic *et al.* (1991).

References

Bohr, H. "Ein allgemeiner Satz über die Integration eines trigonometrischen Polynoms." *Prace Matem.-Fiz.* **43**, 1935.

Favard, J. "Application de la formule sommatoire d'Euler à la démonstration de quelques propriétés extrémales des intégrale des fonctions périodiques ou presquepériodiques." *Mat. Tidsskr. B*, 81–4, 1936. Reviewed in *Zentralblatt f. Math.* **16**, 58–9, 1939.

Mitrinovic, D. S.; Pecaric, J. E.; and Fink, A. M. *Inequalities Involving Functions and Their Integrals and Derivatives*. Dordrecht, Netherlands: Kluwer, pp. 71–2, 1991.

Northcott, D. G. "Some Inequalities Between Periodic Functions and Their Derivatives." *J. London Math. Soc.* **14**, 198–02, 1939.

Tikhomirov, V. M. "Approximation Theory." In *Analysis II. Convex Analysis and Approximation Theory* (Ed. R. V. Gamkrelidze). New York: Springer-Verlag, pp. 93–55, 1990.

Bohr-Mollerup Theorem

If a function $\varphi : (0, \infty) \to (0, \infty)$ satisfies

1. $\ln[\varphi(x)]$ is convex,
2. $\varphi(x+1) = x\varphi(x)$ for all $x > 0$, and
3. $\varphi(1) = 1$,

then $\varphi(x)$ is the GAMMA FUNCTION $\Gamma(x)$. Therefore, by ANALYTIC CONTINUATION, $\Gamma(z)$ is the only MEROMORPHIC FUNCTION on $\mathbb{C}$ satisfying the functional equation

$$z\Gamma(z) = \Gamma(z+1)$$

with $\Gamma(1) = 1$ and which is logarithmically convex on the positive REAL AXIS.

See also GAMMA FUNCTION

References

Krantz, S. G. "The Bohr-Mollerup Theorem." §13.1.10 in *Handbook of Complex Analysis*. Boston, MA: Birkhäuser, p. 157, 1999.

Bolyai-Gerwein Theorem

WALLACE-BOLYAI-GERWEIN THEOREM

Bolza Problem

Given the functional

$$U = \int_{t_0}^{t_1} f(y_1, \ldots, y_n; y'_1, \ldots, y'_n)\,dt$$
$$+ G(y_{10}, \ldots, y_{nr}; y_{11}, \ldots, y_{n1}), \quad (1)$$

find in a class of arcs satisfying p differential and q finite equations

$$\phi_\alpha(y_1, \ldots, y_n; y'_1, \ldots, y'_n) = 0 \quad \text{for } \alpha = 1, \ldots, p \quad (2)$$

$$\psi_\beta(y_1, \ldots, y_n) = 0 \quad \text{for } \beta = 1, \ldots, q \quad (3)$$

as well as the r equations on the endpoints

$$\chi_\gamma(y_{10}, \ldots, y_{nr}; y_{11}, \ldots, y_{n1}) = 0$$
$$\text{for } \gamma = 1, \ldots, r, \quad (4)$$

one which renders U a minimum.

References

Goldstine, H. H. *A History of the Calculus of Variations from the 17th through the 19th Century*. New York: Springer-Verlag, p. 374, 1980.

Bolzano Theorem

BOLZANO-WEIERSTRASS THEOREM

Bolzano-Weierstrass Theorem

Every BOUNDED infinite set in $\mathbb{R}^n$ has an ACCUMULATION POINT.

For $n = 1$, an infinite subset of a closed bounded set S has an ACCUMULATION POINT in S. For instance, given a bounded SEQUENCE a_n, with $-C \le a_n \le C$ for all n, it must have a MONOTONIC subsequence a_{n_k}. The SUBSEQUENCE a_{n_k} must converge because it is monotonic and bounded. Because S is closed, it contains the limit of a_{n_k}.

The Bolzano-Weierstrass theorem is closely related to the HEINE-BOREL THEOREM and CANTOR'S INTERSECTION THEOREM, each of which can be easily derived from either of the other two.

See also ACCUMULATION POINT, CANTOR'S INTERSECTION THEOREM, HEINE-BOREL THEOREM, INTERMEDIATE VALUE THEOREM

References

Jeffreys, H. and Jeffreys, B. S. §1.034 in *Methods of Mathematical Physics, 3rd ed.* Cambridge, England: Cambridge University Press, pp. 9–0, 1988.

Knopp, K. *Theory of Functions Parts I and II, Two Volumes Bound as One, Part I.* New York: Dover, p. 7, 1996.

Bombieri Inner Product

For HOMOGENEOUS POLYNOMIALS P and Q of degree n,

$$[P,\,Q] \equiv \sum_{i_1,\,\ldots,\,i_n \geq 0} (i_1! \ldots i_n!)(a_{i,\,\ldots,\,i_n} b_{i_1,\,\ldots,\,i_n}).$$

Bombieri Norm

This entry contributed by KEVIN O'BRYANT

The Bombieri p-norm of a polynomial

$$Q(x) = \sum_{i=0}^{n} a_i x^i \qquad (1)$$

is defined by

$$[Q]_p \equiv \left[\sum_{i=0}^{n} \binom{n}{i}^{1-p} |a_i|^p \right]^{1/p}, \qquad (2)$$

where $\binom{n}{k}$ is a BINOMIAL COEFFICIENT. The most remarkable feature of Bombieri's n norm is that given polynomials R and S such that $RS = Q$, then BOMBIERI'S INEQUALITY

$$[R]_2 [S]_2 \leq \binom{n}{m}^{1/2} [Q]_2 \qquad (3)$$

holds, where n is the degree of Q, and m is the degree of either R or S. This theorem captures the heuristic that if R and S have big coefficients, then so does RS, i.e., there can't be too much cancellation.

See also NORM, BOMBIERI'S INEQUALITY, POLYNOMIAL NORM

References

Beauzamy, B.; Bombieri, E.; Enflo, P.; and Montgomery, H. L. "Products of Polynomials in Many Variables." *J. Number Th.* **36**, 219–45, 1990.

Borwein, P. and Erdélyi, T. "Bombieri's Norm." §5.3.E.7 in *Polynomials and Polynomial Inequalities.* New York: Springer-Verlag, p. 274, 1995.

Reznick, B. "An Inequality for Products of Polynomials." *Proc. Amer. Math. Soc.* **117**, 1063–073, 1993.

Bombieri's Inequality

For HOMOGENEOUS POLYNOMIALS P and Q of degree m and n, then

$$[P \cdot Q]_2 \geq \sqrt{\frac{m! \, n!}{(m+n)!}} [P]_2 [Q]_2,$$

where $[P \cdot Q]_2$ is the BOMBIERI NORM. If $m = n$, this becomes

$$[P \cdot Q]_2 \geq [P]_2 [Q]_2,$$

See also BOMBIERI NORM, BEAUZAMY AND DÉGOT'S IDENTITY, REZNIK'S IDENTITY

References

Borwein, P. and Erdélyi, T. "Bombieri's Norm." §5.3.E.7 in *Polynomials and Polynomial Inequalities.* New York: Springer-Verlag, p. 274, 1995.

Bombieri's Theorem

Define

$$E(x;\,q,\,a) \equiv \psi(x;\,q,\,a) - \frac{x}{\phi(q)}, \qquad (1)$$

where

$$\psi(x;\,q,\,a) = \sum_{\substack{n \leq x \\ n \equiv a \pmod q}} \Lambda(n) \qquad (2)$$

(Davenport 1980, p. 121), $\Lambda(n)$ is the MANGOLDT FUNCTION, and $\phi(q)$ is the TOTIENT FUNCTION. Now define

$$E(x;\,q) = \max_{(a,\,q)=1} |E(x;\,q,\,a)| \qquad (3)$$

where the sum is over a RELATIVELY PRIME to q, $(a,\,q) = 1$, and

$$E^*(x,\,q) = \max_{y \leq x} E(y,\,q). \qquad (4)$$

Bombieri's theorem then says that for fixed $A > 0$,

$$\sum_{q \leq Q} E^*(x,\,q) \ll \sqrt{x} Q (\ln x)^5, \qquad (5)$$

provided that $\sqrt{x}(\ln x)^{-4} < Q < \sqrt{x}$.

References

Bombieri, E. "On the Large Sieve." *Mathematika* **12**, 201–25, 1965.

Davenport, H. "Bombieri's Theorem." Ch. 28 in *Multiplicative Number Theory, 2nd ed.* New York: Springer-Verlag, pp. 161–68, 1980.

Bond Percolation

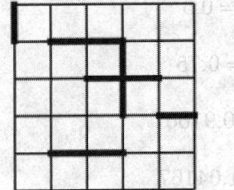 

bond percolation   *site percolation*

A PERCOLATION which considers the lattice edges as the relevant entities (left figure).

See also PERCOLATION THEORY, SITE PERCOLATION

Bonferroni Correction

The Bonferroni correction is a multiple-comparison correction used when several independent STATISTICAL TESTS are being performed simultaneously (since while a given ALPHA VALUE α may be appropriate for each individual comparison, it is not for the set of *all* comparisons). In order to avoid a lot of spurious positives, the ALPHA VALUE needs to be lowered to account for the number of comparisons being performed.

The simplest and most conservative approach is the Bonferroni correction, which sets the ALPHA VALUE for the entire *set* of n comparisons equal to α by taking the ALPHA VALUE for *each* comparison equal to α/n. Explicitly, given n tests T_i for hypotheses H_i ($1 \le i \le n$) under the assumption H_0 that all hypotheses H_i are false, and if the individual test critical values are $\le \alpha/n$, then the experiment-wide critical value is $\le \alpha$. In equation form, if

$$P(T_i \text{ passes } |H_0) \le \frac{\alpha}{n}$$

for $1 \le i \le n$, then

$$P(\text{some } T_i \text{ passes } |H_0) \le \alpha,$$

which follows from BONFERRONI'S INEQUALITIES.

Another correction instead uses $1 - (1 - \alpha)^{1/n}$. While this choice is applicable for two-sided hypotheses, multivariate normal statistics, and positive orthant dependent statistics, it is not, in general, correct (Shaffer 1995).

See also ALPHA VALUE, HYPOTHESIS TESTING, STATISTICAL TEST

References

Bonferroni, C. E. "Il calcolo delle assicurazioni su gruppi di teste." In *Studi in Onore del Professore Salvatore Ortu Carboni.* Rome: Italy, pp. 13–0, 1935.
Bonferroni, C. E. "Teoria statistica delle classi e calcolo delle probabilità." *Pubblicazioni del R Istituto Superiore di Scienze Economiche e Commerciali di Firenze* **8**, 3–2, 1936.

Dewey, M. "Carlo Emilio Bonferroni: Life and Works." http://www.nottingham.ac.uk/~mhzmd/life.html.
Miller, R. G. Jr. *Simultaneous Statistical Inference.* New York: Springer-Verlag, 1991.
Perneger, T. V. "What's Wrong with Bonferroni Adjustments." *Brit. Med. J.* **316**, 1236–238, 1998.
Shaffer, J. P. "Multiple Hypothesis Testing." *Ann. Rev. Psych.* **46**, 561–84, 1995.

Bonferroni Test

BONFERRONI CORRECTION

Bonferroni's Inequalities

Let $P(E_i)$ be the probability that E_i is true, and $P(\cup_{i=1}^n E_i)$ be the probability that at least one of E_1, E_2, ..., E_n is true. Then

$$P\left(\bigcup_{i=1}^n E_i\right) \le \sum_{i=1}^n P(E_i).$$

A slightly wider class of inequalities are also known as "Bonferroni inequalities."

References

Comtet, L. "Bonferroni Inequalities." §4.7 in *Advanced Combinatorics: The Art of Finite and Infinite Expansions,* rev. enl. ed. Dordrecht, Netherlands: Reidel, pp. 193–94, 1974.
Galambos, J.; and Simonelli, I. *Bonferroni-Type Inequalities with Applications.* New York: Springer-Verlag, 1996.

Bonne Projection

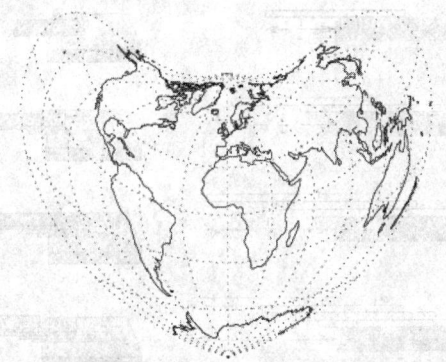

A MAP PROJECTION which resembles the shape of a heart. Let ϕ_1 be the standard parallel, λ_0 the central meridian, ϕ be the LATITUDE, and λ the LONGITUDE on a UNIT SPHERE. Then

$$x = \rho \sin E \tag{1}$$

$$y = \cot \phi_1 - \rho \cos E, \tag{2}$$

where

$$\rho = \cot \phi_1 + \phi_1 - \phi \tag{3}$$

$$E = \frac{(\lambda - \lambda_0) \cos \phi}{\rho}. \tag{4}$$

The inverse FORMULAS are

$$\phi = \cot \phi_1 + \phi_1 - \rho \qquad (5)$$

$$\lambda = \lambda_0 + \frac{\rho}{\cos \phi} \tan^{-1}\left(\frac{x}{\cot \phi_1 - y}\right), \qquad (6)$$

where

$$\rho = \pm\sqrt{x^2 + (\cot \phi_1 - y)^2}. \qquad (7)$$

The WERNER PROJECTION is a special case of the Bonne projection.

See also MAP PROJECTION, WERNER PROJECTION

References

MathWorks. "Mapping Toolbox: Bonne Projection." http://www.mathworks.com/access/helpdesk/help/toolbox/map/bonneprojection.shtml.
Snyder, J. P. *Map Projections--A Working Manual.* U. S. Geological Survey Professional Paper 1395. Washington, DC: U. S. Government Printing Office, pp. 138–40, 1987.

Book Stacking Problem

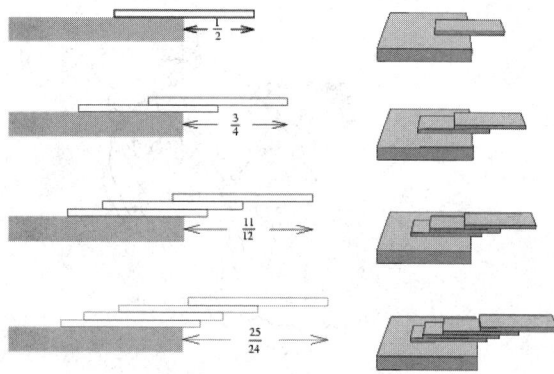

How far can a stack of n books protrude over the edge of a table without the stack falling over? It turns out that the maximum overhang possible d_n for n books (in terms of book lengths) is half the nth partial sum of the HARMONIC SERIES, given explicitly by

$$d_n = \frac{1}{2}\sum_{k=1}^{n}\frac{1}{k} = \tfrac{1}{2}[\gamma + \Psi(1+n)]$$

where $\Psi(z)$ is the DIGAMMA FUNCTION and γ is the EULER-MASCHERONI CONSTANT. The first few values

are

$$d_1 = \tfrac{1}{2} = 0.5$$

$$d_2 = \tfrac{3}{4} = 0.75$$

$$d_3 = \tfrac{11}{12} \approx 0.91667$$

$$d_4 = \tfrac{25}{24} \approx 1.04167,$$

(Sloane's A001008 and A002805).

In order to find the number of stacked books required to obtain d book-lengths of overhang, solve the d_n equation for d, and take the CEILING FUNCTION. For $n = 1, 2, ...$ book-lengths of overhang, 4, 31, 227, 1674, 12367, 91380, 675214, 4989191, 36865412, 272400600, ... (Sloane's A014537) books are needed.

References

Dickau, R. M. "The Book-Stacking Problem." http://www.prairienet.org/~pops/BookStacking.html.
Eisner, L. "Leaning Tower of the Physical Review." *Amer. J. Phys.* **27**, 121, 1959.
Gamow, G. and Stern, M. *Puzzle Math.* New York: Viking, 1958.
Gardner, M. *Martin Gardner's Sixth Book of Mathematical Games from Scientific American.* New York: Scribner's, pp. 167–69, 1971.
Graham, R. L.; Knuth, D. E.; and Patashnik, O. *Concrete Mathematics: A Foundation for Computer Science.* Reading, MA: Addison-Wesley, pp. 272–74, 1990.
Johnson, P. B. "Leaning Tower of Lire." *Amer. J. Phys.* **23**, 240, 1955.
Sharp, R. T. "Problem 52." *Pi Mu Epsilon J.* **1**, 322, 1953.
Sharp, R. T. "Problem 52." *Pi Mu Epsilon J.* **2**, 411, 1954.
Sloane, N. J. A. Sequences A001008/M2885, A002805/M1589, and A014537 in "An On-Line Version of the Encyclopedia of Integer Sequences." http://www.research.att.com/~njas/sequences/eisonline.html.

Boole

IVERSON BRACKET

Boole Polynomial

Polynomials $s_k(x; \lambda)$ which form a SHEFFER SEQUENCE with

$$g(t) = 1 + e^{\lambda t} \qquad (1)$$

$$f(t) = e^t - 1 \qquad (2)$$

and have GENERATING FUNCTION

$$\sum_{k=0}^{\infty}\frac{s_k(x; \lambda)}{\kappa!}\,t^k = \frac{(1+t)^x}{1 + (1+t)^\lambda}. \qquad (3)$$

The first few are

$$s_0(x; \lambda) = \tfrac{1}{2}$$
$$s_1(x; \lambda) = \tfrac{1}{4}(2x - \lambda)t$$
$$x_2(x; \lambda) = \tfrac{1}{4}[2x(x - \lambda - 1) + \lambda].$$

Jordan (1950) considers the related polynomials $r_n(x)$

which form a SHEFFER SEQUENCE with

$$g(t) = \tfrac{1}{2}(1 + e^t) \tag{4}$$

$$f(t) = e^t - 1. \tag{5}$$

These polynomials have GENERATING FUNCTION

$$\sum_{k=0}^{\infty} \frac{r_n(x)}{k!} t^k = \frac{2(1 + t)^x}{2 + t}. \tag{6}$$

The first few are

$$r_0(x) = 1$$
$$r_1(x) = \tfrac{1}{2}(2x - 1)$$
$$r_2(x) = \tfrac{1}{2}(2x^2 - 4x + 1)$$
$$r_3(x) = \tfrac{1}{4}(4x^3 - 18x^2 + 20x - 3).$$

The PETERS POLYNOMIALS are a generalization of the Boole polynomials.

See also PETERS POLYNOMIAL

References

Boas, R. P. and Buck, R. C. *Polynomial Expansions of Analytic Functions, 2nd print., corr.* New York: Academic Press, p. 37, 1964.
Jordan, C. *Calculus of Finite Differences, 3rd ed.* New York: Chelsea, 1965.
Roman, S. *The Umbral Calculus.* New York: Academic Press, 1984.

Boole's Inequality

Let $P(E_i)$ be the probability of an event E_i occurring. Then

$$P\left(\bigcup_{i=1}^{N} E_i\right) \le \sum_{i=1}^{N} P(E_i),$$

where $\cup$ denotes the UNION. If E_i and E_j are DISJOINT SETS for all i and j, then the INEQUALITY becomes an equality.

See also DISJOINT SETS, UNION

Boolean Algebra

A mathematical structure which is similar to a BOOLEAN RING, but which is defined using the meet and join operators instead of the usual addition and multiplication operators. Explicitly, a Boolean algebra is the PARTIAL ORDER on subsets defined by inclusion (Skiena 1990, p. 207), i.e., the Boolean algebra $b(A)$ of a set A is the set of subsets of A that can be obtained by means of a finite number of the set operations UNION (OR), INTERSECTION (AND), and COMPLEMENTATION (NOT) (Comtet 1974, p. 185). A Boolean algebra also forms a LATTICE (Skiena 1990, p. 170), and each of the elements of $b(A)$ is called a BOOLEAN FUNCTION. There are 2^{2^n} BOOLEAN FUNCTIONS in a Boolean algebra of order n (Comtet 1974, p. 186).

In 1938, Shannon proved that a two-valued Boolean algebra (whose members are most commonly denoted 0 and 1, or false and true) can describe the operation of two-valued electrical switching circuits. In modern times, Boolean algebra and BOOLEAN FUNCTIONS are therefore indispensable in the design of computer chips and integrated circuits.

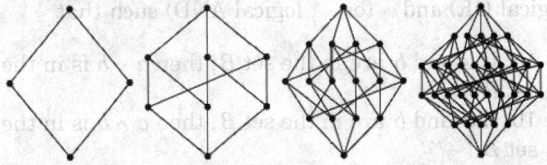

Boolean algebras have a recursive structure apparent in the HASSE DIAGRAMS illustrated above for Boolean algebras of orders $n = 2$, 3, 4, and 5. These figures illustrate the partition between left and right halves of the lattice, each of which is the Boolean algebra on $n - 1$ elements (Skiena 1990, pp. 169–70).

A Boolean algebra can be formally defined as a SET B of elements a, b, ... with the following properties:

1. B has two binary operations, $\wedge$ (logical AND, or "WEDGE") and $\vee$ (logical OR, or "VEE"), which satisfy the IDEMPOTENT laws

$$a \wedge a = a \vee a = a, \tag{1}$$

the COMMUTATIVE laws

$$a \wedge b = b \wedge a, \tag{2}$$

$$a \vee b = b \vee a, \tag{3}$$

and the ASSOCIATIVE laws

$$a \wedge (b \wedge c) = (a \wedge b) \wedge c \tag{4}$$

$$a \vee (b \vee c) = (a \vee b) \vee c. \tag{5}$$

2. The operations satisfy the ABSORPTION LAW

$$a \wedge (a \vee b) = a \vee (a \wedge b) = a. \tag{6}$$

3. The operations are mutually distributive

$$a \wedge (b \vee c) = (a \wedge b) \wedge (a \wedge c) \tag{7}$$

$$a \vee (b \wedge c) = (a \vee b) \wedge (a \wedge c). \tag{8}$$

4. B contains universal bounds $\varnothing$ and I which satisfy

$$\varnothing \wedge a = \varnothing \tag{9}$$

$$\varnothing \vee a = a \tag{10}$$

$$I \wedge a = a \tag{11}$$

$$I \vee a = I. \tag{12}$$

5. B has a unary operation $a \to a'$ of complementation which obeys the laws

$$a \wedge a' = \varnothing \tag{13}$$

$$a \vee a' = I \qquad (14)$$

(Birkhoff and Mac Lane 1965).

In the slightly archaic terminology of (Bell 1937, p. 444), a Boolean algebra can be defined as a set B of elements a, b, ... with BINARY OPERATORS $\vee$ (or +; logical OR) and $\wedge$ (or .; logical AND) such that

1a. If a and b are in the set B, then $a \vee b$ is in the set B.
1b. If a and b are in the set B, then $a \wedge b$ is in the set B.
2a. There is an element Z (zero) such that $a \vee Z = a$ for every element a.
2b. There is an element U (unity) such that $a \wedge U = a$ for every element a.
3a. $a \vee b = b \vee a$.
3b. $a \wedge b = b \wedge a$.
4a. $a \vee b \wedge c = (a \vee b) \wedge (a \vee c)$.
4b. $a \wedge (b \vee c) = (a \wedge b) \vee (a \wedge c)$.
5. For every element a there is an element a' such that $a \vee a' = U$ and $a \wedge a' = Z$.
6. There are at least two distinct elements in the set B.

Huntington (1933ab) presented the following basis for Boolean algebra:

1. Commutativity. $x \vee y = y \vee x$.
2. Associativity. $(x \vee y) \vee z = x \vee (y \vee z)$.
3. HUNTINGTON AXIOM. $!(!x \vee y) \vee !(!x \vee !y) = x$.

H. Robbins then conjectured that the HUNTINGTON AXIOM could be replaced with the simpler ROBBINS AXIOM,

$$!(!(x \vee y) \vee !(x \vee !y)) = x \qquad (15)$$

The ALGEBRA defined by commutativity, associativity, and the ROBBINS AXIOM is called ROBBINS ALGEBRA. Computer theorem proving demonstrated that every ROBBINS ALGEBRA satisfies the second WINKLER CONDITION, from which it follows immediately that all ROBBINS ALGEBRAS are Boolean (McCune, Kolata 1996).

See also BOOLEAN FUNCTION, BOOLEANS, HUNTINGTON AXIOM, MAXIMAL IDEAL THEOREM, ROBBINS ALGEBRA, ROBBINS AXIOM, WINKLER CONDITIONS, WOLFRAM AXIOM

References

Bell, E. T. *Men of Mathematics.* New York: Simon and Schuster, 1986.
Birkhoff, G. and Mac Lane, S. *A Survey of Modern Algebra, 5th ed.* New York: Macmillian, p. 317, 1996.
Comtet, L. "Boolean Algebra Generated by a System of Subsets." §4.4 in *Advanced Combinatorics: The Art of Finite and Infinite Expansions, rev. enl. ed.* Dordrecht, Netherlands: Reidel, pp. 185–89, 1974.
Halmos, P. *Lectures on Boolean Algebras.* Princeton, NJ: Van Nostrand, 1963.
Huntington, E. V. "New Sets of Independent Postulates for the Algebra of Logic." *Trans. Amer. Math. Soc.* **35**, 274–04, 1933a.
Huntington, E. V. "Boolean Algebras: A Correction." *Trans. Amer. Math. Soc.* **35**, 557–58, 1933.
Kolata, G. "Computer Math Proof Shows Reasoning Power." *New York Times*, Dec. 10, 1996.
McCune, W. "Robbins Algebras are Boolean." http://www-unix.mcs.anl.gov/~mccune/papers/robbins/.
Mendelson, E. *Introduction to Boolean Algebra and Switching Circuits.* New York: McGraw-Hill, 1973.
Sikorski, R. *Boolean Algebra, 3rd ed.* New York: Springer-Verlag, 1969.
Skiena, S. *Implementing Discrete Mathematics: Combinatorics and Graph Theory with Mathematica.* Reading, MA: Addison-Wesley, 1990.
Wells, C. F. "Boolean Expression Manipulation." http://www.mathsource.com/cgi-bin/msitem?0204-69.

Boolean Connective

One of the LOGIC operators AND $\wedge$, OR $\vee$, and NOT $\neg$.

See also QUANTIFIER

Boolean Function

Consider a Boolean algebra of subsets $b(A)$ generated by a set A, which is the set of subsets of A that can be obtained by means of a finite number of the set operations union, intersection, and complementation. Then each of the elements of $b(A)$ is called a Boolean function generated by A (Comtet 1974, p. 185). Each Boolean function has a unique representation (up to order) as a union of COMPLETE PRODUCTS. It follows that there are 2^{2^p} inequivalent Boolean functions for a set A with cardinality p (Comtet 1974, p. 187).

In 1938, Shannon proved that a two-valued Boolean algebra (whose members are most commonly denoted 0 and 1, or false and true) can describe the operation of two-valued electrical switching circuits. The following table gives the TRUTH TABLE for the $2^{2^2} = 16$ possible Boolean functions of two binary variables.

A	B	F_0	F_1	F_2	F_3	F_4	F_5	F_6	F_7
0	0	0	0	0	0	0	0	0	0
0	1	0	0	0	0	1	1	1	1
1	0	0	0	1	1	0	0	1	1
1	1	0	1	0	1	0	1	0	1

A	B	F_8	F_9	F_{10}	F_{11}	F_{12}	F_{13}	F_{14}	F_{15}
0	0	1	1	1	1	1	1	1	1
0	1	0	0	0	0	1	1	1	1

$$
\begin{array}{cccccccccccc}
1 & 0 & 0 & 0 & 1 & 1 & 0 & 0 & 1 & 1 \\
1 & 1 & 0 & 1 & 0 & 1 & 0 & 1 & 0 & 1
\end{array}
$$

The names and symbols for these functions are given in the following table (Simpson 1987, p. 539).

operation	symbol	name
F_0	0	FALSE
F_1	$A \wedge B$	AND
F_2	$A \wedge !B$	A AND NOT B
F_3	A	A
F_4	$!A \wedge B$	NOT A AND B
F_5	B	B
F_6	$A \underline{\vee} B$	XOR
F_7	$A \vee B$	OR
F_8	$A \underline{\triangledown} B$	NOR
F_9	A XNOR B	XNOR
F_{10}	$!B$	NOT B
F_{11}	$A \vee !B$	A OR NOT B
F_{12}	$!A$	NOT A
F_{13}	$!A \vee B$	NOT A OR B
F_{14}	$A \overline{\wedge} B$	NAND
F_{15}	1	TRUE

Determining the number of *monotone* Boolean functions of n variables is known as DEDEKIND'S PROBLEM and is equivalent to the number of ANTICHAINS on the n-set $\{1, 2, \ldots, n\}$. Boolean functions can also be thought of as colorings of a Boolean n-cube. The numbers of inequivalent monotone Boolean functions in $n = 1, 2, \ldots$ variables are given by 2, 3, 5, 10, 30, ...(Sloane's A003182).

Let $M(n, k)$ denote the number of distinct monotone Boolean functions of n variables with k MINCUTS. Then

$$M(n, 0) = 1$$

$$M(n, 1) = 2^n$$

$$M(n, 2) = 2^{n-1}(2^n - 1) - 3^n + 2^n$$

$$M(n, 3) = \tfrac{1}{6}(2^n)(2^n - 1)(2^n - 2) - 6^n + 5^n + 4^n - 3^n.$$

See also ANTICHAIN, BOOLEAN ALGEBRA, BOOLEANS, COMPLETE PRODUCT, CONJUNCTION, DEDEKIND'S PROBLEM, MINCUT, MONOTONE FUNCTION

References

Comtet, L. "Boolean Algebra Generated by a System of Subsets." §4.4 in *Advanced Combinatorics: The Art of Finite and Infinite Expansions, rev. enl. ed.* Dordrecht, Netherlands: Reidel, pp. 185–89, 1974.
Shapiro. "On the Counting Problem for Monotone Boolean Functions." *Comm. Pure Appl. Math.* **23**, 299–12, 1970.
Simpson, R. E. *Introductory Electronics for Scientists and Engineers, 2nd ed.* Boston, MA: Allyn and Bacon, 1987.
Sloane, N. J. A. Sequences A003182/M0729 in "An On-Line Version of the Encyclopedia of Integer Sequences." http://www.research.att.com/~njas/sequences/eisonline.html.

Boolean Representation Theorem

Every BOOLEAN ALGEBRA is isomorphic to the BOOLEAN ALGEBRA of sets. It is equivalent to the MAXIMAL IDEAL THEOREM, which can be proved without using the AXIOM OF CHOICE (Mendelson 1997, p. 121).

See also BOOLEAN ALGEBRA, MAXIMAL IDEAL THEOREM

References

Mendelson, E. *Introduction to Mathematical Logic, 4th ed.* London: Chapman & Hall, p. 121, 1997.
Stone, M. "The Representation Theorem for Boolean Algebras." *Trans. Amer. Math. Soc.* **40**, 37–11, 1936.

Boolean Ring

A RING with a unit element in which every element is IDEMPOTENT.

See also BOOLEAN ALGEBRA

Booleans

The domain of Booleans, sometimes denoted $\mathbb{B}$, consisting of the elements TRUE and FALSE, implemented in *Mathematica* as `Booleans`. In *Mathematica*, a quantity can be tested to determine if it is in the domain of Booleans using `Element[e, Booleans]`.

See also BOOLEAN ALGEBRA, BOOLEAN FUNCTION, FALSE, TRUE

Boomeron Equation

The system of PARTIAL DIFFERENTIAL EQUATIONS

$$u_t = \mathbf{b} \cdot \mathbf{v}_x$$

$$\mathbf{b}_{xt} = u_{xx}\mathbf{b} + \mathbf{a} \times \mathbf{v}_x - 2\mathbf{v} \times (\mathbf{v} \times \mathbf{b}).$$

References

Calogero, F. and Degasperis, A. *Spectral Transform and Solitons: Tools to Solve and Investigate Nonlinear Evolution Equations.* New York: North-Holland, p. 57, 1982.
Zwillinger, D. *Handbook of Differential Equations, 3rd ed.* Boston, MA: Academic Press, p. 137, 1997.

Boosting

See also RESAMPLING STATISTICS

Bootstrap Methods

A set of methods that are generally superior to ANOVA for small data sets or where sample distributions are non-normal.

See also ANOVA, JACKKNIFE, PERMUTATION TESTS, RESAMPLING STATISTICS

References

Chernick, M. R. *Bootstrap Methods: A Practitioner's Guide.* New York: Wiley, 1999.
Davison, A. C. and Hinkley, D. V. *Bootstrap Methods and Their Application.* Cambridge, England: Cambridge University Press, 1997.
Efron, B. and Tibshirani, R. J. *An Introduction to the Bootstrap.* Boca Raton, FL: CRC Press, 1994.
Mooney, C. Z. and Duval, R. D. *Bootstrapping: A Nonparametric Approach to Statistical Inference.* Sage, 1993.

Borchardt-Pfaff Algorithm

ARCHIMEDES ALGORITHM

Border Square

40	1	2	3	42	41	46
38	31	13	14	32	35	12
39	30	26	21	28	20	11
43	33	27	25	23	17	7
6	16	22	29	24	34	44
5	15	37	36	18	19	45
4	49	48	47	8	9	10

31	13	14	32	35
30	26	21	28	20
33	27	25	23	17
16	22	29	24	34
15	37	36	18	19

26	21	28
27	25	23
22	29	24

A MAGIC SQUARE that remains magic when its border is removed. A nested magic square remains magic after the border is successively removed one ring at a time. An example of a nested magic square is the order 7 square illustrated above (i.e., the order 7, 5, and 3 squares obtained from it are all magic).

See also MAGIC SQUARE

References

Chabert, J.-L. (Ed.). "Squares with Borders" and "Arnauld's Borders Method." §2.1 and 2.4 in *A History of Algorithms: From the Pebble to the Microchip.* New York: Springer-Verlag, pp. 53–8 and 70–0, 1999.
Kraitchik, M. "Border Squares." §7.7 in *Mathematical Recreations.* New York: W. W. Norton, pp. 167–70, 1942.

Bordism

A relation between COMPACT boundaryless MANIFOLDS (also called closed MANIFOLDS). Two closed MANIFOLDS are bordant IFF their disjoint union is the boundary of a compact $(n + 1)$-MANIFOLD. Roughly, two MANIFOLDS are bordant if together they form the boundary of a MANIFOLD. The word bordism is now used in place of the original term COBORDISM.

References

Budney, R. "The Bordism Project." http://www.math.cornell.edu/~rybu/bordism/bordism.html.

Bordism Group

There are bordism groups, also called COBORDISM GROUPS or COBORDISM RINGS, and there are singular bordism groups. The bordism groups give a framework for getting a grip on the question, "When is a compact boundaryless MANIFOLD the boundary of another MANIFOLD?" The answer is, precisely when all of its STIEFEL-WHITNEY CLASSES are zero. Singular bordism groups give insight into STEENROD'S REALIZATION PROBLEM: "When can homology classes be realized as the image of fundamental classes of manifolds?" That answer is known, too.

The machinery of the bordism group winds up being important for HOMOTOPY THEORY as well.

References

Budney, R. "The Bordism Project." http://www.math.cornell.edu/~rybu/bordism/bordism.html.

Borel Algebra

See also BOREL SIGMA ALGEBRA, BOREL SUBALGEBRA

Borel Determinacy Theorem

Let T be a TREE defined on a metric over a set of paths such that the distance between paths p and q is $1/n$, where n is the number of nodes shared by p and q. Let A be a BOREL SET of paths in the topology induced by this metric. Suppose two players play a game by choosing a path down the tree, so that they alternate and each time choose an immediate successor of the previously chosen point. The first player wins if the chosen path is in A. Then one of the players has a winning STRATEGY in this GAME.

See also GAME THEORY, TREE

Borel Field

If a FIELD has the property that, if the sets $A_n, ..., A_n,$ ... belong to it, then so do the sets $A_1 + ... + A_n + ...$ and $A_1 ... A_n ...$, then the field is called a Borel field (Papoulis 1984, p. 29).

See also FIELD

References

Papoulis, A. *Probability, Random Variables, and Stochastic Processes, 2nd ed.* New York: McGraw-Hill, 1984.

Borel Measure

If F is the BOREL SIGMA ALGEBRA on some TOPOLOGICAL SPACE, then a MEASURE $m : F \to \mathbb{R}$ is said to be a Borel measure (or BOREL PROBABILITY MEASURE). For a Borel measure, all continuous functions are MEASURABLE.

Borel Probability Measure

BOREL MEASURE

Borel Set

A Borel set is an element of a BOREL SIGMA ALGEBRA. Roughly speaking, Borel sets are the sets that can be constructed from open or closed sets by repeatedly taking countable unions and intersections. Formally, the class B of Borel sets in Euclidean $\mathbb{R}^n$ is the smallest collection of sets that includes the open and closed sets such that if $E, E_1, E_2, \ldots$ are in B, then so are $\cup_{i=1}^{\infty} E_i$, $\cap_{i=1}^{\infty} E_i$, and $\mathbb{R}^n \backslash E$, where $F \backslash E$ is a SET DIFFERENCE (Croft *et al.* 19991).

The set of rational numbers is a Borel set, as is the CANTOR SET.

See also CLOSED SET, OPEN SET, STANDARD SPACE

References

Croft, H. T.; Falconer, K. J.; and Guy, R. K. *Unsolved Problems in Geometry.* New York: Springer-Verlag, p. 3, 1991.

Borel Sigma Algebra

A SIGMA ALGEBRA which is related to the TOPOLOGY of a SET. The Borel σ-algebra is defined to be the SIGMA ALGEBRA generated by the OPEN SETS (or equivalently, by the CLOSED SETS).

See also BOREL ALGEBRA, BOREL MEASURE, BOREL SUBALGEBRA

Borel Space

A SET equipped with a SIGMA ALGEBRA of SUBSETS.

Borel Subalgebra

See also BOREL ALGEBRA, BOREL SIGMA ALGEBRA

Borel's Expansion

Let $\phi(t) = \Sigma_{n=0}^{\infty} A_n t^n$ be any function for which the integral

$$I(x) \equiv \int_0^{\infty} e^{-tx} t^p \phi(t) \, dt$$

converges. Then the expansion

$$I(x) \frac{\Gamma(p+1)}{x^{p+1}} \left[A_0 + (p+1) \frac{A_1}{x} + (p+1)(p+2) \frac{A_2}{x^2} + \ldots \right],$$

where $\Gamma(z)$ is the GAMMA FUNCTION, is usually an ASYMPTOTIC SERIES for $I(x)$.

Borel-Cantelli Lemma

Let $\{A_n\}_{n=0}^{\infty}$ be a SEQUENCE of events occurring with a certain probability distribution, and let A be the event consisting of the occurrence of a finite number of events A_n, $n = 1, \ldots$. Then if

$$\sum_{n=1}^{\infty} P(A_n) < \infty,$$

then

$$P(A) = 1.$$

References

Hazewinkel, M. (Managing Ed.). *Encyclopaedia of Mathematics: An Updated and Annotated Translation of the Soviet "Mathematical Encyclopaedia."* Dordrecht, Netherlands: Reidel, pp. 435–36, 1988.

Borel-Weyl Theorem

Let $G = SL(n, \mathbb{C})$. If $\lambda \in Z^n$ is the highest weight of an irreducible holomorphic representation V of G, (i.e., λ is a dominant integral weight), then the G-map $\phi :$ $V^* \to \Gamma(\lambda)$ defined by $\alpha \mapsto F_\alpha$, where $F_\alpha(g) = \langle \alpha, gv \rangle$, is an ISOMORPHISM. Thus, $V \cong \Gamma(\lambda)^*$.

References

Huang, J.-S. "The Borel-Weyl Theorem." §8.7 in *Lectures on Representation Theory.* Singapore: World Scientific, pp. 105–07, 1999.

Born-Infeld Equation

The PARTIAL DIFFERENTIAL EQUATION

$$(1 - u_t^2) u_{xx} + 2 u_x u_t u_{xt} - (1 + u_x^2) u_{tt} = 0.$$

References

Whitham, G. B. *Linear and Nonlinear Waves.* New York: Wiley, p. 617, 1974.
Zwillinger, D. *Handbook of Differential Equations, 3rd ed.* Boston, MA: Academic Press, p. 132, 1997.

Boron Tree

BINARY TREE

Borromean Rings

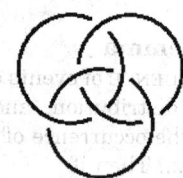

Three mutually interlocked rings, named after the Italian Renaissance family who used them on their coat of arms. The configuration of rings is also known as a "Ballantine," and a brand of beer (illustrated above) has been brewed under this name. In the Borromean rings, no two rings are linked, so if any one of the rings is cut, all three rings fall apart. Any number of rings can be linked in an analogous manner (Steinhaus 1983, Wells 1991).

The Borromean rings have LINK symbol 06-3-2, BRAID WORD $\sigma_1^{-1}\sigma_2\sigma_1^{-1}\sigma_2\sigma_1^{-1}\sigma_2$, and are also the simplest BRUNNIAN LINK.

See also BRUNNIAN LINK, CIRCLE-CIRCLE INTERSECTION, TRIQUETRA, VENN DIAGRAM

References
Cundy, H. and Rollett, A. *Mathematical Models, 3rd ed.* Stradbroke, England: Tarquin Pub., pp. 58–9, 1989.
Gardner, M. *The Unexpected Hanging and Other Mathematical Diversions.* Chicago, IL: University of Chicago Press, 1991.
Jablan, S. "Borromean Triangles." http://members.tripod.com/~modularity/links.htm.
Pappas, T. "Trinity of Rings--A Topological Model." *The Joy of Mathematics.* San Carlos, CA: Wide World Publ./Tetra, p. 31, 1989.
Steinhaus, H. *Mathematical Snapshots, 3rd ed.* New York: Dover, pp. 266–67, 1999.
Wells, D. *The Penguin Dictionary of Curious and Interesting Geometry.* London: Penguin, p. 18, 1991.

Borrow

$$
\begin{array}{r}
1\ 2\ 3\ 4 \\
-\ 7\ 8\ 9 \\
\hline
4\ 4\ 5
\end{array}
\qquad
\begin{array}{r}
\text{borrows} \\
1\ ^1\!2\ ^1\!1\ ^1\!4 \\
1\ \cancel{2}\ \cancel{3}\ 4 \\
-\ 7\ 8\ 9 \\
\hline
4\ 4\ 5
\end{array}
$$

The procedure used in SUBTRACTION to "borrow" 10 from the next higher DIGIT column in order to obtain a POSITIVE DIFFERENCE in the column in question.

See also CARRY

Borsuk's Conjecture

Borsuk conjectured that it is possible to cut an n-D shape of GENERALIZED DIAMETER 1 into $n+1$ pieces each with diameter smaller than the original. It is true for $n=2, 3$ and when the boundary is "smooth." However, the minimum number of pieces required has been shown to increase as $\sim 1.1^{\sqrt{n}}$. Since $1.1^{\sqrt{n}} > n+1$ at $n=9162$, the conjecture becomes false at high dimensions. In fact, the conjecture is false for every $n > 561$.

See also GENERALIZED DIAMETER, KELLER'S CONJECTURE, LEBESGUE MINIMAL PROBLEM

References
Borsuk, K. "Über die Zerlegung einer Euklidischen n-dimensionalen Vollkugel in n Mengen." *Verh. Internat. Math.-Kongr. Zürich* **2**, 192, 1932.
Borsuk, K. "Drei Sätze über die n-dimensionale euklidische Sphäre." *Fund. Math.* **20**, 177–90, 1933.
Cipra, B. "If You Can't See It, Don't Believe It...." *Science* **259**, 26–7, 1993.
Cipra, B. *What's Happening in the Mathematical Sciences, Vol. 1.* Providence, RI: Amer. Math. Soc., pp. 21–5, 1993.
Grünbaum, B. "Borsuk's Problem and Related Questions." In *Convexity: Proceedings of the Seventh Symposium in Pure Mathematics of the American Mathematical Society, Held at the University of Washington, Seattle, June 13–5, 1961.* Providence, RI: Amer. Math. Soc., pp. 271–84, 1963.
Kalai, J. K. G. "A Counterexample to Borsuk's Conjecture." *Bull. Amer. Math. Soc.* **329**, 60–2, 1993. Lyusternik, L. and Schnirel'mann, L. *Topological Methods in Variational Problems.* Moscow, 1930.
Lyusternik, L. and Schnirel'mann, L. "Topological Methods in Variational Problems and Their Application to the Differential Geometry of Surfaces." *Uspehi Matem. Nauk (N.S.)* **2**, 166–17, 1947.

Borsuk-Ulam Theorem

Every continuous map $f : \mathbb{S}^n \to \mathbb{R}^n$ must identify a pair of ANTIPODAL POINTS.

References
Dodson, C. T. J. and Parker, P. E. *A User's Guide to Algebraic Topology.* Dordrecht, Netherlands: Kluwer, pp. 121 and 284, 1997.

Borwein Conjectures

Use the definition of the Q-SERIES

$$(a;\, q)_n \equiv \prod_{j=0}^{n-1}(1 - aq^j) \qquad (1)$$

and define

$$\begin{bmatrix} N \\ M \end{bmatrix} \equiv \frac{(q^{N-M+1};\, q)_M}{(q;\, q)_m}. \qquad (2)$$

Then P. Borwein has conjectured that (1) the POLYNOMIALS $A_n(q)$, $B_n(q)$, and $C_n(q)$ defined by

$$(q;\, q^3)_n (q^2;\, q^3)_n = A_n(q^3) - qB_n(q^3) - q^2C_n(q^3) \qquad (3)$$

have NONNEGATIVE COEFFICIENTS, (2) the POLYNOMIALS $A_n^*(q)$, $B_n^*(q)$, and $C_n^*(q)$ defined by

$$(q;\, q^3)_n^2 (q^2;\, q^3)_n^2 = A_n^*(q^3) - qB_n^*(q^3) - q^2C_n^*(q^3) \qquad (4)$$

have NONNEGATIVE COEFFICIENTS, (3) the POLYNOMIALS $A_n^*(q)$, $B_n^*(q)$, $C_n^*(q)$, $D_n^*(q)$, and $E_n^*(q)$ defined by

$$(q;\, q^5)_n (q^2;\, q^5)_n (q^3;\, q^5)_n (q^4;\, q^5)_n =$$

$$A_n^*(q^5) - qB_n^*(q^5) - q^2C_n^*(q^5) - q^3D_n^*(q^5) - q^4E_n^*(q^5) \qquad (5)$$

have NONNEGATIVE COEFFICIENTS, (4) the POLYNOMIALS $A_n^\dagger(m, n, t, q)$, $B_n^\dagger(m, n, t, q)$, and $C_n^\dagger(m, n, t, q)$ defined by

$$(q;\, q^3)_m (q^2;\, q^3)_m (zq;\, q^3)_n (zq^2;\, q^3)_n$$

$$= \sum_{t=0}^{2m} z^t [A^\dagger(m, n, t, q^3) - qB^\dagger(m, n, t, q^3)$$

$$- q^2C^\dagger(m, n, t, q^3)] \qquad (6)$$

have NONNEGATIVE COEFFICIENTS, (5) for k ODD and $1 \le a \le k/2$, consider the expansion

$$(q^a;\, q^k)_m (q^{k-a};\, q^k)_n = \sum_{v=(1-k)/2}^{(k-1)/2} (-1)^v q^{k(v^2+v)/2-av} F_v(q^k) \qquad (7)$$

with

$$F_v(q) = \sum_{j=-\infty}^{\infty} (-1)^j q^{j(k^2j+2kv+k-2a)/2} \begin{bmatrix} m+n \\ m+v+kj \end{bmatrix}, \qquad (8)$$

then if a is RELATIVELY PRIME to k and $m = n$, the COEFFICIENTS of $F_v(q)$ are NONNEGATIVE, and (6) given $\alpha + \beta < 2K$ and $-K + \beta \le n - m \le K - \alpha$, consider

$$G(\alpha, \beta, K; q)$$

$$= \sum_q (-1)^j q^{j[K(\alpha+\beta)j + K(\alpha+\beta)]/2} \begin{bmatrix} m+n \\ m+Kj \end{bmatrix}, \qquad (9)$$

the GENERATING FUNCTION for partitions inside an $m \times n$ rectangle with hook difference conditions specified by α, β, and K. Let α and β be POSITIVE RATIONAL NUMBERS and $k > 1$ an INTEGER such that

αk and βk are integers. then if $1 \le \alpha + \beta \le 2k - 1$ (with strict inequalities for $k = 2$) and $-k + \beta \le n - m \le k - \alpha$, then $g(\alpha, \beta, k;\, q)$ has NONNEGATIVE COEFFICIENTS.

See also Q-SERIES

References

Andrews, G. E. *et al.* "Partitions with Prescribed Hook Differences." *Europ. J. Combin.* **8**, 341–50, 1987.

Bressoud, D. M. "The Borwein Conjecture and Partitions with Prescribed Hook Differences." *Electronic J. Combinatorics* **3**, No. 2, R4, 1–4, 1996. http://www.combinatorics.org/Volume_3/volume3_2.html#R4.

Bott Periodicity Theorem

Define

$$O = \lim_{\to} O(n),\quad F = \mathbb{R} \qquad (1)$$

$$U = \lim_{\to} U(n),\quad F = \mathbb{C} \qquad (2)$$

$$Sp = \lim_{\to} Sp(n),\quad F = \mathbb{H}. \qquad (3)$$

Then

$$\Omega^2 BU \cong BU \times \mathbb{Z} \qquad (4)$$

$$\Omega^4 BO \cong BSp \times \mathbb{Z} \qquad (5)$$

$$\Omega^4 BSp \cong BO \times \mathbb{Z}. \qquad (6)$$

References

Atiyah, M. F. *K-Theory.* New York: Benjamin, 1967.

Bott, R. "The Stable Homotopy of the Classical Groups." *Ann. Math.* **70**, 313–37, 1959.

Dodson, C. T. J. and Parker, P. E. *A User's Guide to Algebraic Topology.* Dordrecht, Netherlands: Kluwer, p. 229, 1997.

Milnor, J. W. *Morse Theory.* Princeton, NJ: Princeton University Press, 1963.

Bottle Imp Paradox

In Robert Louis Stevenson's "bottle imp paradox," you are offered the opportunity to buy, for whatever price you wish, a bottle containing a genie who will fulfill your every desire. The only catch is that the bottle must thereafter be resold for a price smaller than what you paid for it, or you will be condemned to live out the rest of your days in excruciating torment. Obviously, no one would buy the bottle for 1c since he would have to give the bottle away, but no one would accept the bottle knowing he would be unable to get rid of it. Similarly, no one would buy it for 2c, and so on. However, for some reasonably large amount, it will always be possible to find a next buyer, so the bottle will be bought (Paulos 1995).

See also UNEXPECTED HANGING PARADOX

References

Erickson, G. W. and Fossa, J. A. *Dictionary of Paradox.* Lanham, MD: University Press of America, pp. 25–7, 1998.

Paulos, J. A. *A Mathematician Reads the Newspaper.* New York: BasicBooks, p. 97, 1995.

Bouligand Dimension

MINKOWSKI-BOULIGAND DIMENSION

Bound

GREATEST LOWER BOUND, INFIMUM, LEAST UPPER BOUND, SUPREMUM

Bound Variable

An occurrence of a variable in a LOGIC which is not FREE. Bound variables are also called DUMMY VARIABLES.

See also DUMMY VARIABLE, SENTENCE

References

Comtet, L. "Bound Variables." §1.11 in *Advanced Combinatorics: The Art of Finite and Infinite Expansions, rev. enl. ed.* Dordrecht, Netherlands: Reidel, pp. 30–4, 1974.

Boundary

The set of points, known as BOUNDARY POINTS, which are members of the CLOSURE of a given set S and the CLOSURE of its complement set. The boundary is sometimes called the FRONTIER.

See also BOUNDARY CONDITIONS, BOUNDARY MAP, BOUNDARY POINT, BOUNDARY SET, NATURAL BOUNDARY, SURGERY

Boundary Conditions

There are several types of boundary conditions commonly encountered in the solution of PARTIAL DIFFERENTIAL EQUATIONS.

1. DIRICHLET BOUNDARY CONDITIONS specify the value of the function on a surface $T = f(\mathbf{r}, t)$.
2. NEUMANN BOUNDARY CONDITIONS specify the normal derivative of the function on a surface,

$$\frac{\partial T}{\partial n} = \hat{\mathbf{n}} \cdot \nabla T = f(\mathbf{r}, y).$$

3. CAUCHY BOUNDARY CONDITIONS specify a weighted average of first and second kinds.
4. ROBIN BOUNDARY CONDITIONS. For an elliptic partial differential equation in a region Ω, Robin boundary conditions specify the sum of αu and the normal derivative of $u = f$ at all points of the boundary of Ω, with α and f being prescribed.

See also BOUNDARY VALUE PROBLEM, DIRICHLET BOUNDARY CONDITIONS, GOURSAT PROBLEM, INITIAL VALUE PROBLEM, NEUMANN BOUNDARY CONDITIONS, PARTIAL DIFFERENTIAL EQUATION, ROBIN BOUNDARY CONDITIONS

References

Arfken, G. *Mathematical Methods for Physicists, 3rd ed.* Orlando, FL: Academic Press, pp. 502–04, 1985.

Morse, P. M. and Feshbach, H. "Boundary Conditions and Eigenfunctions." Ch. 6 in *Methods of Theoretical Physics, Part I.* New York: McGraw-Hill, pp. 495–98 and 676–90, 1953.

Boundary Map

The MAP $H_n(X, A) \to H_{n-1}(A)$ appearing in the LONG EXACT SEQUENCE OF A PAIR AXIOM.

See also LONG EXACT SEQUENCE OF A PAIR AXIOM

Boundary Point

A point which is a member of the CLOSURE of a given set S and the CLOSURE of its complement set. If A is a subset of $\mathbb{R}^n$, then a point $\mathbf{x} \in \mathbb{R}^n$ is a boundary point of A if every NEIGHBORHOOD of $\mathbf{x}$ contains at least one point in A and at least one point not in A.

See also BOUNDARY

Boundary Set

A (symmetrical) boundary set of RADIUS r and center $\mathbf{x}_0$ is the set of all points $\mathbf{x}$ such that

$$|\mathbf{x} - \mathbf{x}_0| = r.$$

Let $\mathbf{x}_0$ be the ORIGIN. In $\mathbb{R}^1$, the boundary set is then the pair of points $x = r$ and $x = -r$. In $\mathbb{R}^2$, the boundary set is a CIRCLE. In $\mathbb{R}^3$, the boundary set is a SPHERE.

See also CIRCLE, COMPACT SET, DISK, OPEN SET, SPHERE

References

Croft, H. T.; Falconer, K. J.; and Guy, R. K. *Unsolved Problems in Geometry.* New York: Springer-Verlag, p. 2, 1991.

Boundary Value Problem

A boundary value problem is a problem, typically an ORDINARY DIFFERENTIAL EQUATION or a PARTIAL DIFFERENTIAL EQUATION, which has values assigned on the physical boundary of the DOMAIN in which the problem is specified. For example,

$$\begin{cases} \dfrac{\partial^2 u}{\partial t^2} - \nabla^2 u = f & \text{in } \Omega \\ u(0, t) = u_1 & \text{on } \partial\Omega \\ \dfrac{\partial u}{\partial t}(0, t) = u_2 & \text{on } \partial\Omega, \end{cases}$$

where $\partial\Omega$ denotes the boundary of Ω, is a boundary problem.

See also BOUNDARY CONDITIONS, INITIAL VALUE PROBLEM

References

Eriksson, K.; Estep, D.; Hansbo, P.; and Johnson, C. *Computational Differential Equations*. Lund: Studentlitteratur, 1996.

Powers, D. L. *Boundary Value Problems, 4th ed.* San Diego, CA: Academic Press, 1999.

Press, W. H.; Flannery, B. P.; Teukolsky, S. A.; and Vetterling, W. T. "Two Point Boundary Value Problems." Ch. 17 in *Numerical Recipes in FORTRAN: The Art of Scientific Computing, 2nd ed.* Cambridge, England: Cambridge University Press, pp. 745–78, 1992.

Bounded

A mathematical object (such as a set or function) is said to bounded if it possesses a BOUND, i.e., a value which all members of the set, functions, etc., are less than.

See also BOUNDED SET

Bounded Set

A SET in a METRIC SPACE (X, d) is bounded if it has a FINITE GENERALIZED DIAMETER, i.e., there is an $R < \infty$ such that $d(x, y) \leq R$ for all $x, y \in X$. A SET in $\mathbb{R}^n$ is bounded if it is contained inside some BALL $x_1^2 + \ldots + x_n^2 \leq R^2$ of FINITE RADIUS R (Adams 1994).

See also BOUND, FINITE

References

Adams, R. A. *Calculus: A Complete Course*. Reading, MA: Addison-Wesley, p. 707, 1994.

Croft, H. T.; Falconer, K. J.; and Guy, R. K. *Unsolved Problems in Geometry*. New York: Springer-Verlag, p. 2, 1991.

Jeffreys, H. and Jeffreys, B. S. "Bounded, Unbounded, Convergent, Oscillatory." §1.041 in *Methods of Mathematical Physics, 3rd ed.* Cambridge, England: Cambridge University Press, pp. 11–2, 1988.

Bounded Variation

A FUNCTION $f(x)$ is said to have bounded variation if, over the CLOSED INTERVAL $x \in [a, b]$, there exists an M such that

$$|f(x_i) - f(a)| + |f(x_2) - f(x_1)| + \ldots + |f(b) - f(x_n - 1)|$$
$$\leq M \tag{1}$$

for all $a < x_1 < x_2 < \ldots < x_{n-1} < b$.

The space of functions of bounded variation is denoted "BV," and has the SEMINORM

$$\Phi(f) = \sup \int f \, \frac{d\phi}{dx}, \tag{2}$$

where ϕ ranges over all COMPACTLY SUPPORTED functions bounded by -1 and 1. The seminorm is equal to the SUPREMUM over all sums above, and is also equal to $\int |df/dx| \, dx$ (when this expression

makes sense).

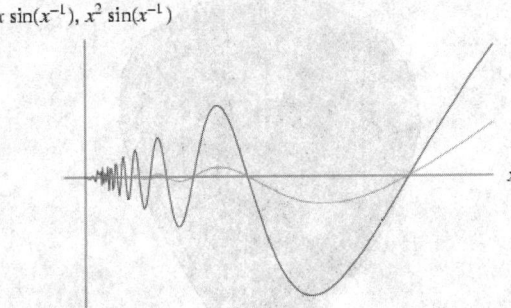

$x \sin(x^{-1})$, $x^2 \sin(x^{-1})$

On the interval $[0, 1]$, the function $x^2 \sin(1/x)$ (purple) is of bounded variation, but $x \sin 1/x$ (red) is not. More generally, a function f is locally of bounded variation in a domain U if f is LOCALLY INTEGRABLE, $f \in L^1_{\text{loc}}$, and for all open subsets W, with COMPACT CLOSURE in U, and all SMOOTH VECTOR FIELDS g COMPACTLY SUPPORTED in W,

$$\int_W f \, \text{div} \, g \, dx \leq c(W) \sup|g|, \tag{3}$$

div denotes DIVERGENCE and c is a constant which only depends on the choice of W and f.

Such functions form the space $BV_{\text{loc}}(U)$. They may not be DIFFERENTIABLE, but by the RIESZ REPRESENTATION THEOREM, the derivative of a BV_{loc}-function f is a REGULAR BOREL MEASURE Df. Functions of bounded variation also satisfy a compactness theorem.

Given a sequence f_n of functions in $BV_{\text{loc}}(U)$, such that

$$\sup_n \left(\|f_n\|_{L^1(W)} + \int_W |Df_n| \, dx \right) < \infty,$$

that is the TOTAL VARIATION of the functions is bounded, in any COMPACTLY SUPPORTED open subset W, there is a SUBSEQUENCE f_{n_k} which converges to a function $f \in BV_{\text{loc}}$ in the topology of L^1_{loc}. Moreover, the limit satisfies

$$\int_W |Df| \, dx \leq \liminf \int_W |Df_{n_k}| \, dx. \tag{4}$$

They also satisfy a version of POINCARÉ'S LEMMA.

See also DIFFERENTIABLE, WEAKLY DIFFERENTIABLE

References

Jeffreys, H. and Jeffreys, B. S. "Functions of Bounded Variation." §1.09 in *Methods of Mathematical Physics, 3rd ed.* Cambridge, England: Cambridge University Press, pp. 24–6, 1988.

Simon, L. §2.6 in *Lectures on Geometric Measure Theory* Canberra: Centre for Mathematical Analysis, Australian National University, 1984.

Bour's Minimal Surface

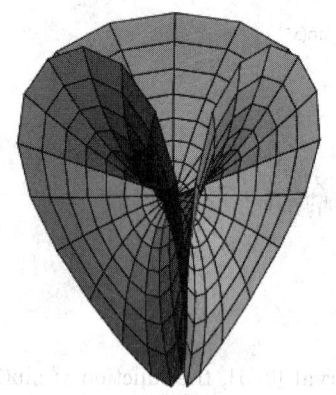

Gray (1997) defines Bour's minimal curve over complex z by

$$x' = \frac{z^{m-1}}{m-1} - \frac{z^{m+1}}{m+1} \tag{1}$$

$$y' = i\left(\frac{z^{m-1}}{m-1} + \frac{z^{m+1}}{m+1}\right) \tag{2}$$

$$z' = \frac{2z^m}{m}, \tag{3}$$

and then derives a family of MINIMAL SURFACES.

The order three Bour surface resembles a CROSS-CAP and is given using ENNEPER-WEIERSTRASS PARAMETERIZATION by

$$f = 1 \tag{4}$$

$$g = \sqrt{z} \tag{5}$$

or explicitly by the PARAMETRIC EQUATIONS

$$x = r \cos\theta - \tfrac{1}{2}r^2 \cos(2\theta) \tag{6}$$

$$y = -r \sin\theta - \tfrac{1}{2}r^2 \sin(2\theta), \tag{7}$$

$$z = \tfrac{4}{3} r^{3/2} \cos(\tfrac{3}{2}\theta) \tag{8}$$

(Maeder 1997). The coefficients of the FIRST FUNDAMENTAL FORM are given by

$$E = 1 + r^2 \tag{9}$$

$$F = 0 \tag{10}$$

$$G = r^2(r^2 + 1) \tag{11}$$

and the coefficients of the SECOND FUNDAMENTAL FORM by

$$e = -r^{-1/2} \cos(\tfrac{3}{2}\phi) \tag{12}$$

$$f = \sqrt{r} \sin(\tfrac{3}{2}\phi) \tag{13}$$

$$g = r^{3/2} \cos(\tfrac{3}{2}\phi). \tag{14}$$

The AREA ELEMENT is

$$dA = r(r+1)^2 \, dr \wedge d\phi. \tag{15}$$

The GAUSSIAN and MEAN CURVATURES are given by

$$K = -\frac{1}{r(r+1)^4} \tag{16}$$

$$H = 0. \tag{17}$$

See also CROSS-CAP, ENNEPER-WEIERSTRASS PARAMETERIZATION, MINIMAL SURFACE

References

Gray, A. *Modern Differential Geometry of Curves and Surfaces with Mathematica, 2nd ed.* Boca Raton, FL: CRC Press, pp. 732–33, 1997.
Maeder, R. *Programming in Mathematica, 3rd ed.* Reading, MA: Addison-Wesley, pp. 29–0, 1997.

Bourget Function

The function defined by the CONTOUR INTEGRAL

$$J_{n,\,k}(z)$$

$$= \frac{1}{2\pi i} \int_{}^{(0+)} t^{-n-1} \left(t + \frac{1}{t}\right)^k \exp\left[\tfrac{1}{2}z\left(t - \frac{1}{t}\right)\right] dt,$$

where $\int_{(0+)}$ denotes the CONTOUR encircling the point $z = 0$ once in a counterclockwise direction. It is equal to

$$J_{n,\,k}(z) = \frac{1}{\pi} \int_0^\pi (2\cos\theta)^k \cos(n\theta - z\sin\theta) \, d\theta$$

(Watson 1966, p. 326).

See also BESSEL FUNCTION OF THE FIRST KIND

References

Bourget, J. "Mémoire sue les nombres de Cauchy et leur application à divers problèmes de mécanique céleste." *J. de Math.* **6**, 33–4, 1861.
Giuliani, G. "Alcune osservazioni sopra le funzioni spheriche di ordine superiore al secondo e sopra altre funzioni che se ne possono dedurre (April, 1888)." *Giornale di Mat.* **26**, 155–71, 1888.
Hazewinkel, M. (Managing Ed.). *Encyclopaedia of Mathematics: An Updated and Annotated Translation of the Soviet "Mathematical Encyclopaedia."* Dordrecht, Netherlands: Reidel, p. 465, 1988.
Watson, G. N. "The Functions of Bourget and Giuliani." §10.31 in *A Treatise on the Theory of Bessel Functions, 2nd ed.* Cambridge, England: Cambridge University Press, pp. 326–27, 1966.

Bourget's Hypothesis

When n is an INTEGER ≥ 0, then $J_n(z)$ and $J_{n+m}(z)$ have no common zeros other than at $z = 0$ for m an

INTEGER ≥ 1, where $J_n(z)$ is a BESSEL FUNCTION OF THE FIRST KIND. The theorem has been proved true for $m = 1$ 2, 3, and 4.

References

Watson, G. N. *A Treatise on the Theory of Bessel Functions, 2nd ed.* Cambridge, England: Cambridge University Press, 1966.

Bourque-Ligh Conjecture

Bourque and Ligh (1992) conjectured that the LEAST COMMON MULTIPLE MATRIX on a GCD-CLOSED SET S is nonsingular. This conjecture was shown to be false by Hong (1999).

See also GCD-CLOSED SET, LEAST COMMON MULTIPLE MATRIX

References

Bourque, K. and Ligh, S. "On GCD and LCM Matrices." *Linear Algebra Appl.* **174**, 65–4, 1992.
Hong, S. "On the Bourque-Ligh Conjecture of Least Common Multiple Matrices." *J. Algebra* **218**, 216–28, 1999.

Boussinesq Equation

The linear Boussinesq equation is the PARTIAL DIFFERENTIAL EQUATION

$$u_{tt} - \alpha^2 u_{xx} = \beta^2 u_{xxtt} \qquad (1)$$

(Whitham 1974, p. 9; Zwillinger 1997, p. 129). The nonlinear Boussinesq equation is

$$u_{tt} - u_{xx} - u_{xxxx} + 3(u^2)_{xx} = 0 \qquad (2)$$

(Calogero and Degasperis 1982; Zwillinger 1997, p. 130). The modified Boussinesq equation is

$$\tfrac{1}{3} u_{tt} - u_t u_{xx} - \tfrac{3}{2} u_x^2 u_{xx} + u_{xxxx} = 0 \qquad (3)$$

(Clarkson 1986; Zwillinger 1997, p. 132).

References

Calogero, F. and Degasperis, A. *Spectral Transform and Solitons: Tools to Solve and Investigate Nonlinear Evolution Equations.* New York: North-Holland, 1982.
Clarkson, P. A. "The Painlevé Property, a Modified Boussinesq Equation and a Modified Kadomtsev-Petviashvili Equation." *Physica D* **19**, 447–50, 1986.
Whitham, G. B. *Linear and Nonlinear Waves.* New York: Wiley, 1974.
Zwillinger, D. *Handbook of Differential Equations, 3rd ed.* Boston, MA: Academic Press, pp. 129–30, 1997.

Boustrophedon Transform

The boustrophedon ("ox-plowing") transform **b** of a sequence **a** is given by

$$b_n = \sum_{k=0}^{n} \binom{n}{k} a_k E_{n-k} \qquad (1)$$

$$a_n = \sum_{k=0}^{n} (-1)^{n-k} \binom{n}{k} b_k E_{n-k} \qquad (2)$$

for $n \geq 0$, where E_n is a SECANT NUMBER or TANGENT NUMBER defined by

$$\sum_{n=0}^{\infty} E_n \frac{x^n}{n!} = \sec x + \tan x. \qquad (3)$$

The exponential generating functions of **a** and **b** are related by

$$\mathscr{B}(x) = (\sec x + \tan x)\mathscr{A}(x), \qquad (4)$$

where the exponential generating function is defined by

$$\mathscr{A}(x) = \sum_{n=0}^{\infty} A_n \frac{x^n}{n!}. \qquad (5)$$

See also ALTERNATING PERMUTATION, ENTRINGER NUMBER, SECANT NUMBER, SEIDEL-ENTRINGER-ARNOLD TRIANGLE, TANGENT NUMBER

References

Millar, J.; Sloane, N. J. A.; and Young, N. E. "A New Operation on Sequences: The Boustrophedon Transform." *J. Combin. Th. Ser. A* **76**, 44–4, 1996.

Bovinum Problema

ARCHIMEDES' CATTLE PROBLEM

Bow

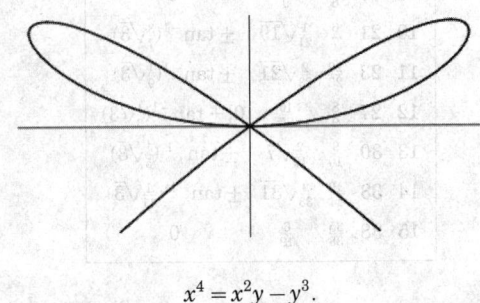

$$x^4 = x^2 y - y^3.$$

References

Cundy, H. and Rollett, A. *Mathematical Models, 3rd ed.* Stradbroke, England: Tarquin Pub., p. 72, 1989.

Bowditch Curve

LISSAJOUS CURVE

Bowl of Integers

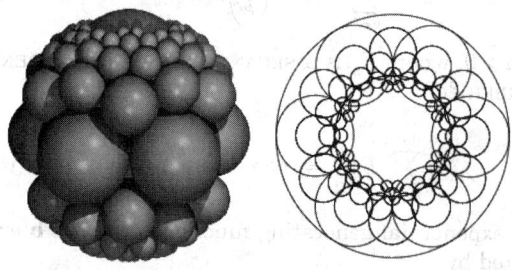

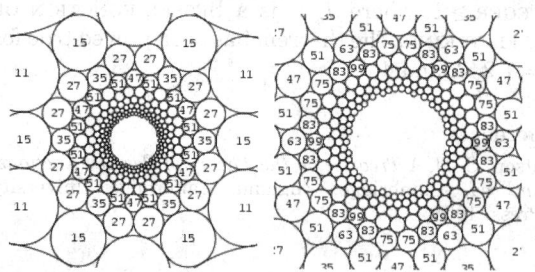

Place two solid spheres of radius 1/2 inside a hollow sphere of radius 1 so that the two smaller circles touch each other at the center of the large circle and are tangent to the large circle on the extremities of one of its diameters. This arrangement is called the "bowl of integers" (Soddy 1937) since the BEND of each of the infinite chain of spheres that can be packed into it such that each successive sphere is tangent to its neighbors is an integer. The first few bends are then $-1, 2, 5, 6, 9, 11, 14, 15, 18, 21, 23, \ldots$ (Sloane's A046160). The sizes and positions of the first few rings of spheres are given in the table below.

Spheres can also be packed along the plane tangent to the two spheres of radius 2 (Soddy 1937). The sequence of integers for can be found using the equation of five TANGENT SPHERES. Letting $\kappa_3 = \kappa_4 = 2$ gives

$$\kappa(\kappa_1, \kappa_2)$$

$$= \tfrac{1}{2}(4 + \kappa_1 + \kappa_2 + \sqrt{3[\kappa_2(8 - \kappa_2) + 2\kappa_1(\kappa_2 + 4) - 3\kappa_1^2]}).$$

For example, $\kappa(3, 3) = 11$, $\kappa(3, 11) = 15$, $\kappa(11, 15) = 27$, $\kappa(15, 27) = 35$, $\kappa(27, 27)47$, and so on, giving the sequence -1, 2, 3, 11, 15, 27, 35, 47, 51, 63, 75, 83, ... (Sloane's A046159). The sizes and positions of the first few rings of spheres are given in the table below.

n	κ_n	z_n	R_n	ϕ_n
1	-1	0	0	–
2	2	$\frac{1}{2}$	0	–
3	5	$\frac{2}{5}$	$\frac{2}{5}\sqrt{3}$	$\frac{1}{6}\pi$
4	6	$\frac{1}{2}$	$\frac{2}{3}$	0
5	9	$\frac{2}{3}$	$\frac{2}{9}\sqrt{7}$	$\pm\tan^{-1}(\frac{1}{2}\sqrt{3})$
6	11	$\frac{8}{11}$	$\frac{6}{11}$	0
7	14	$\frac{11}{14}$	$\frac{2}{7}\sqrt{3}$	$\frac{1}{6}\pi$
8	15	$\frac{4}{5}$	$\frac{2}{15}\sqrt{13}$	$\pm\tan^{-1}(2\sqrt{3})$
9	18	$\frac{5}{6}$	$\frac{4}{9}$	0
10	21	$\frac{6}{7}$	$\frac{1}{21}\sqrt{19}$	$\pm\tan^{-1}(\frac{3}{7}\sqrt{3})$
11	23	$\frac{20}{23}$	$\frac{2}{23}\sqrt{21}$	$\pm\tan^{-1}(\frac{1}{9}\sqrt{3})$
12	27	$\frac{8}{9}$	$\frac{10}{27}$	$0, \pm\tan^{-1}(\frac{1}{3}\sqrt{3})$
13	30	$\frac{9}{10}$	$\frac{2}{15}\sqrt{7}$	$\pm\tan^{-1}(\frac{1}{5}\sqrt{3})$
14	33	$\frac{10}{11}$	$\frac{2}{33}\sqrt{31}$	$\pm\tan^{-1}(\frac{1}{11}\sqrt{3})$
15	38	$\frac{35}{38}$	$\frac{6}{19}$	0

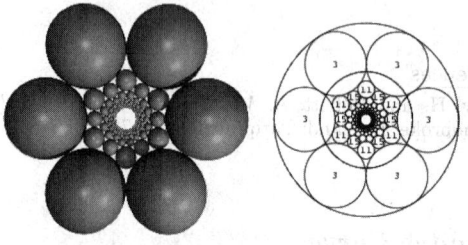

n	κ_n	R_n	ϕ_n
1	-1	0	–
2	2	0	–
3	3	$\frac{2}{3}$	0
4	11	$\frac{2}{11}\sqrt{3}$	$\frac{1}{6}\pi$
5	15	$\frac{4}{15}$	0
6	27	$\frac{2}{27}\sqrt{7}$	$\pm\tan^{-1}(3\sqrt{3})$
7	35	$\frac{6}{35}$	0
8	47	$\frac{4}{47}\sqrt{3}$	$\frac{1}{6}\pi$
9	51	$\frac{2}{51}\sqrt{13}$	$\pm\tan^{-1}(\frac{3}{5}\sqrt{3})$
10	63	$\frac{8}{63}$	0
11	75	$\frac{2}{75}\sqrt{19}$	$\pm\tan^{-1}(5\sqrt{3})$
12	83	$\frac{2}{83}\sqrt{21}$	$\pm\tan^{-1}(\frac{5}{3}\sqrt{3})$
13	99	$\frac{10}{99}$	0
14	107	$\frac{6}{107}\sqrt{3}$	$\frac{1}{6}\pi$
15	111	$\frac{4}{111}\sqrt{7}$	$\pm\tan^{-1}(\frac{1}{2}\sqrt{3})$
16	123	$\frac{2}{123}\sqrt{31}$	$\pm\tan^{-1}(\frac{5}{7}\sqrt{3})$
17	143	$\frac{12}{143}$	0
18	147	$\frac{2}{147}\sqrt{37}$	$\pm\tan^{-1}(7\sqrt{3})$
19	155	$\frac{2}{155}\sqrt{39}$	$\pm\tan^{-1}(\frac{1}{6}\sqrt{3})$
20	171	$\frac{2}{171}\sqrt{43}$	$\pm\tan^{-1}(\frac{7}{5}\sqrt{3})$

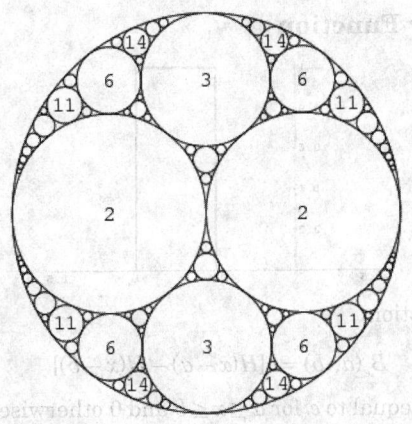

The analogous problem of placing two circles of bend 2 inside a circle of bend -1 and then constructing chains of mutually tangent circles was considered by B. L. Galebach and A. R. Wilks. The circle have integral bends given by -1, 2, 3, 6, 11, 14, 15, 18, 23, 26, 27, 30, 35, 38, ... (Sloane's A042944). Of these, the only known numbers congruent to 2, 3, 6, 11 (mod 12) missing from this sequence are 78, 159, 207, 243, 246, 342, ... (Sloane's A042945), a sequence which is conjectured to be finite.

See also APOLLONIAN GASKET, BEND (CURVATURE), COXETER'S LOXODROMIC SEQUENCE OF TANGENT CIRCLES, HEXLET, SPHERE, TANGENT SPHERES

References

Borkovec, M.; de Paris, W.; and Peikert, R. "The Fractal Dimension of the Apollonian Sphere Packing." *Fractals* **2**, 521–26, 1994.

Sloane, N. J. A. Sequences A042944, A042945, A046159, and A046160 in "An On-Line Version of the Encyclopedia of Integer Sequences." http://www.research.att.com/~njas/sequences/eisonline.html.

Soddy, F. "The Bowl of Integers and the Hexlet." *Nature* **139**, 77–9, 1937.

Bowley Index

The statistical INDEX

$$P_B \equiv \tfrac{1}{2}(P_L + P_P),$$

where P_L is LASPEYRES' INDEX and P_P is PAASCHE'S INDEX.

See also INDEX

References

Kenney, J. F. and Keeping, E. S. *Mathematics of Statistics, Pt. 1, 3rd ed.* Princeton, NJ: Van Nostrand, p. 66, 1962.

Bowley Skewness

Also known as QUARTILE SKEWNESS COEFFICIENT,

$$\frac{(Q_3 - Q_2) - (Q_2 - Q_1)}{Q_3 - Q_1} = \frac{Q_1 - 2Q_2 + Q_3}{Q_3 - Q_1},$$

where the Qs denote the INTERQUARTILE RANGES.

See also INTERQUARTILE RANGE, SKEWNESS

References

Kenney, J. F. and Keeping, E. S. *Mathematics of Statistics, Pt. 1, 3rd ed.* Princeton, NJ: Van Nostrand, p. 102, 1962.

Bowling

Bowling is a game played by rolling a heavy ball down a long narrow track and attempting to knock down ten pins arranged in the form of a TRIANGLE with its vertex oriented towards the bowler. The number 10 is, in fact, the TRIANGULAR NUMBER $T_4 = 4(4+1)/2 = 10$.

Two "bowls" are allowed per "frame." If all the pins are knocked down in the two bowls, the score for that frame is the number of pins knocked down. If some or none of the pins are knocked down on the first bowl, then all the pins knocked down on the second, it is called a "spare," and the number of points tallied is 10 plus the number of pins knocked down on the bowl of the next frame. If all of the pins are knocked down on the first bowl, the number of points tallied is 10 plus the number of pins knocked down on the next two bowls. Ten frames are bowled, unless the last frame is a strike or spare, in which case an additional bowl is awarded.

The maximum number of points possible, corresponding to knocking down all 10 pins on every bowl, is 300.

References

Cooper, C. N. and Kennedy, R. E. "A Generating Function for the Distribution of the Scores of All Possible Bowling Games." In *The Lighter Side of Mathematics* (Ed. R. K. Guy and R. E. Woodrow). Washington, DC: Math. Assoc. Amer., 1994.

Cooper, C. N. and Kennedy, R. E. "Is the Mean Bowling Score Awful?" In *The Lighter Side of Mathematics* (Ed. R. K. Guy and R. E. Woodrow). Washington, DC: Math. Assoc. Amer., 1994.

Box

CUBOID

Box Counting Dimension

CAPACITY DIMENSION

Box Fractal

A FRACTAL also called the anticross-stitch curve which can be constructed using STRING REWRITING by creating a matrix with 3 times as many entries as

the current matrix using the rules

line 1: "$*$" → "$* *$", "$\ $" → "$\ \ $"
line 2: "$*$" → "$\ *$", "$\ $" → "$\ \ $"
line 3: "$*$" → "$* *$", "$\ $" → "$\ \ $"

Let N_n be the number of black boxes, L_n the length of a side of a white box, and A_n the fractional AREA of black boxes after the nth iteration.

$$N_n = 5^n \tag{1}$$

$$L_n = \left(\tfrac{1}{3}\right)^n = 3^{-n} \tag{2}$$

$$A_n = L_n^2 \, N_n = \left(\tfrac{5}{9}\right)^n. \tag{3}$$

The CAPACITY DIMENSION is therefore

$$d_{\mathrm{cap}} = -\lim_{n\to\infty} \frac{\ln N_n}{\ln L_n} = -\lim_{n\to\infty} \frac{\ln(5^n)}{\ln(3^{-n})}$$

$$= \frac{\ln 5}{\ln 3} = 1.464973521\ldots. \tag{4}$$

See also CANTOR DUST, CROSS-STITCH CURVE, SIERPINSKI CARPET, SIERPINSKI SIEVE

References

Weisstein, E. W. "Fractals." MATHEMATICA NOTEBOOK FRACTAL.M.

Box-and-Whisker Plot

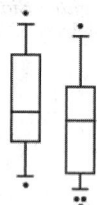

A HISTOGRAM-like method of displaying data invented by J. Tukey (1977). Draw a box with ends at the QUARTILES Q_1 and Q_3. Draw the MEDIAN as a horizontal line in the box. Extend the "whiskers" to the farthest points. For every point that is more than 3/2 times the INTERQUARTILE RANGE from the end of a box, draw a dot on the corresponding top or bottom of the whisker. If two dots have the same value, draw them side by side.

References

Tukey, J. W. *Explanatory Data Analysis.* Reading, MA: Addison-Wesley, pp. 39–1, 1977.

Boxcar Function

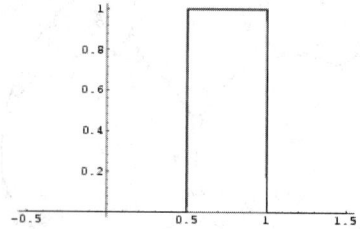

The function

$$B_c(a,\ b) = c[H(x-a) - H(x-b)]$$

which is equal to c for $a \le x \le b$ and 0 otherwise. Here $H(x)$ is the HEAVISIDE STEP FUNCTION. The special case $B_1(-1/2,\ 1/2)$ gives the unit RECTANGLE FUNCTION.

See also HEAVISIDE STEP FUNCTION, RECTANGLE FUNCTION

References

von Seggern, D. *CRC Standard Curves and Surfaces.* Boca Raton, FL: CRC Press, p. 324, 1993.

Boxcars

 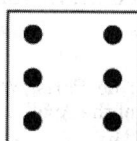

A roll of two 6s (the highest roll possible) on a pair of 6-sided DICE. The probability of rolling boxcars in a single roll of two dice is 1/36, or 2.777...%. In order to have a 50% chance of obtaining at least one boxcars in n rolls of two dice, it must be true that

$$1\left(\frac{35}{36}\right)^n = \frac{1}{2}, \tag{1}$$

so solving for n gives

$$n = \frac{\ln 2}{\ln 36 - \ln 35} = 24.605\ldots. \tag{2}$$

In fact, rolling two dice 25 times gives a probability of

$$1 - \left(\frac{35}{36}\right)^{25} \approx 0.505532 \tag{3}$$

that at least once boxcars will occur.

See also DICE, DE MÉRÉ'S PROBLEM, SNAKE EYES

Box-Counting Dimension
CAPACITY DIMENSION

Box-Muller Transformation

A transformation which transforms from a 2-D continuous UNIFORM DISTRIBUTION to a 2-D GAUSSIAN BIVARIATE DISTRIBUTION (or COMPLEX GAUSSIAN DISTRIBUTION). If x_1 and x_2 are uniformly and independently distributed between 0 and 1, then z_1 and z_2 as defined below have a GAUSSIAN DISTRIBUTION with MEAN $\mu = 0$ and VARIANCE $\sigma^2 = 1$.

$$z_1 = \sqrt{-2 \ln x_1} \, \cos(2\pi x_2) \tag{1}$$

$$z_2 = \sqrt{-2 \ln x_1} \, \sin(2\pi x_2). \tag{2}$$

This can be verified by solving for x_1 and x_2,

$$x_1 = e^{-(z_1^2 + z_2^2)/2} \tag{3}$$

$$x_2 = \frac{1}{2\pi} \tan^{-1}\left(\frac{z_2}{z_1}\right). \tag{4}$$

Taking the JACOBIAN yields

$$\frac{\partial(x_1, x_2)}{\partial(z_1, z_2)} = \begin{vmatrix} \dfrac{\partial x_1}{\partial z_1} & \dfrac{\partial x_1}{\partial z_2} \\ \dfrac{\partial x_2}{\partial z_1} & \dfrac{\partial x_2}{\partial z_2} \end{vmatrix}$$

$$= -\left[\frac{1}{\sqrt{2\pi}} e^{-z_1^2/2}\right]\left[\frac{1}{\sqrt{2\pi}} e^{-z_2^2/2}\right]. \tag{5}$$

See also GAUSSIAN BIVARIATE DISTRIBUTION, GAUSSIAN DISTRIBUTION, NORMAL DEVIATES

References

Box, G. E. P. and Muller, M. E. "A Note on the Generation of Random Normal Deviates." *Ann. Math. Stat.* **28**, 610–611, 1958.

Box-Packing Theorem

The number of "prime" boxes is always finite, where a set of boxes is prime if it cannot be built up from one or more given configurations of boxes.

See also CONWAY PUZZLE, CUBOID, DE BRUIJN'S THEOREM, KLARNER'S THEOREM, SLOTHOUBER-GRAATSMA PUZZLE

References

Honsberger, R. *Mathematical Gems II*. Washington, DC: Math. Assoc. Amer., p. 74, 1976.

Boy Surface

A NONORIENTABLE SURFACE which is one of the three possible SURFACES obtained by sewing a MÖBIUS STRIP to the edge of a DISK. The other two are the CROSS-CAP and ROMAN SURFACE. The Boy surface is a model of the PROJECTIVE PLANE without singularities and is a SEXTIC SURFACE. The Boy surface can be described using the general method for NONORIENTA-

BLE SURFACES, but this was not known until the analytic equations were found by Apéry (1986). Based on the fact that it had been proven impossible to describe the surface using quadratic polynomials, Hopf had conjectured that quartic polynomials were also insufficient (Pinkall 1986). Apéry's IMMERSION proved this conjecture wrong, giving the equations explicitly in terms of the standard form for a NONORIENTABLE SURFACE,

$$f_1(x, y, z) = \tfrac{1}{2}[(2x^2 - y^2 - z^2)(x^2 + y^2 + z^2) + 2yz(y^2 - z^2) + zx(x^1 - z^2) + xy(y^2 - x^2)] \tag{1}$$

$$f_2(x, y, z) = \tfrac{1}{2}\sqrt{3}[(y^2 - z^2)(x^2 + y^2 + z^2) + zx(z^2 - x^2) + xy(y^2 - x^2)] \tag{2}$$

$$f_3(x, y, z) = \tfrac{1}{8}(x + y + z) \times [(x + y + z)^3 + 4(y - x)(z - y)(x - z)]. \tag{3}$$

Plugging in

$$x = \cos u \sin v \tag{4}$$

$$y = \sin u \sin v \tag{5}$$

$$z = \cos v \tag{6}$$

and letting $u \in [0, \pi]$ and $v \in [0, \pi]$ then gives the Boy surface, three views of which are shown above.

The $\mathbb{R}^3$ parameterization can also be written as

$$x = \frac{\sqrt{2} \cos^2 v \cos(2u) + \cos u \sin(2v)}{2 - \sqrt{2} \sin(3u) \sin(2v)} \tag{7}$$

$$y = \frac{\sqrt{2} \cos^2 v \cos(2u) + \cos u \sin(2v)}{2 - \sqrt{2} \sin(3u) \sin(2v)} \tag{8}$$

$$z = \frac{2 \cos^2 v}{2 - \sqrt{2} \sin(3u) \sin(2v)} \tag{9}$$

(Nordstrand) for $u \in [-\pi/2, \pi/2]$ and $v \in [0, \pi]$.

Three views of the surface obtained using this parameterization are shown above.

In fact, a HOMOTOPY (smooth deformation) between the ROMAN SURFACE and Boy surface is given by the equations

$$x(u, v) = \frac{\sqrt{2} \cos(2u) \cos^2 v + \cos u \sin(2v)}{2 - \alpha\sqrt{2} \sin(3u) \sin(2v)} \quad (10)$$

$$y(u, v) = \frac{\sqrt{2} \sin(2u) \cos^2 v - \sin u \sin(2v)}{2 - \alpha\sqrt{2} \sin(3u) \sin(2v)} \quad (11)$$

$$z(u, v) = \frac{3 \cos^2 v}{2 - \alpha\sqrt{2} \sin(3u) \sin(2v)} \quad (12)$$

as α varies from 0 to 1, where $\alpha = 0$ corresponds to the ROMAN SURFACE and $\alpha = 1$ to the Boy surface (Wang), shown below.

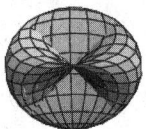  

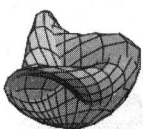  

In $\mathbb{R}^4$, the parametric representation is

$$x_0 = 3[(u^2 + v^2 + w^2)(u^2 + v^2) - \sqrt{2}\, vw(3u^2 - v^2)] \quad (13)$$

$$x_1 = \sqrt{2}(u^2 + v^2)(u^2 - v^2 + \sqrt{2}\, uw) \quad (14)$$

$$x_2 = \sqrt{2}(u^2 + v^2)(2uv - \sqrt{2}\, vw) \quad (15)$$

$$x_3 = 3(u^2 + v^2)^2, \quad (16)$$

and the algebraic equation is

$$64(x_0 - x_3)^3 x_3^3 - 48(x_0 - x_3)^2 x_3^2 (3x_1^2 + 3x_2^2 + 2x_3^2)$$
$$+ 12(x_0 - x_3)x_3[27(x_1^2 + x_2^2)^2 - 24x_3^2(x_1^2 + x_2^2)$$
$$+ 36\sqrt{2} x_2 x_3(x_2^2 - 3x_1^2) + x_3^4] + (9x_1^2 + 9x_2^2 - 2x_3^2)$$
$$\times [-81(x_1^2 + x_2^2)^2 - 72x_3^2(x_1^2 + x_2^2)$$
$$+ 108\sqrt{2} x_1 x_3(x_1^2 - 3x_2^2) + 4x_3^4] = 0 \quad (17)$$

(Apéry 1986). Letting

$$x_0 = 1 \quad (18)$$

$$x_1 = x \quad (19)$$

$$x_2 = y \quad (20)$$

$$x_3 = z \quad (21)$$

gives another version of the surface in $\mathbb{R}^3$.

See also CROSS-CAP, IMMERSION, MÖBIUS STRIP, NONORIENTABLE SURFACE, REAL PROJECTIVE PLANE, ROMAN SURFACE, SEXTIC SURFACE

References

Apéry, F. "The Boy Surface." *Adv. Math.* **61**, 185–266, 1986.

Apéry, F. *Models of the Real Projective Plane: Computer Graphics of Steiner and Boy Surfaces.* Braunschweig, Germany: Vieweg, 1987.

Boy, W. "Über die Curvatura integra und die Topologie geschlossener Flächen." *Math. Ann* **57**, 151–184, 1903.

Brehm, U. "How to Build Minimal Polyhedral Models of the Boy Surface." *Math. Intell.* **12**, 51–56, 1990.

Carter, J. S. "On Generalizing Boy Surface--Constructing a Generator of the 3rd Stable Stem." *Trans. Amer. Math. Soc.* **298**, 103–122, 1986.

Fischer, G. (Ed.). Plates 115–120 in *Mathematische Modelle/Mathematical Models, Bildband/Photograph Volume.* Braunschweig, Germany: Vieweg, pp. 110–115, 1986.

Hilbert, D. and Cohn-Vossen, S. §46–47 in *Geometry and the Imagination.* New York: Chelsea, 1999.

Nordstrand, T. "Boy's Surface." http://www.uib.no/people/nfytn/boytxt.htm.

Petit, J.-P. and Souriau, J. "Une représentation analytique de la surface de Boy." *C. R. Acad. Sci. Paris Sér. 1 Math* **293**, 269–272, 1981.

Pinkall, U. *Mathematical Models from the Collections of Universities and Museums* (Ed. G. Fischer). Braunschweig, Germany: Vieweg, pp. 64–65, 1986.

Stewart, I. *Game, Set and Math.* New York: Viking Penguin, 1991.

Bp-Theorem

If $O_{p'}(G) = 1$ and if x is a p-element of G, then

$$L_{p'}(C_G(x)) \leq E(C_G(x)),$$

where $L_{p'}$ is the P-LAYER.

Bra

A (COVARIANT) 1-VECTOR denoted $\langle \psi |$. The bra is DUAL to the CONTRAVARIANT KET, denoted $| \psi \rangle$. Taken together, the bra and KET form an ANGLE BRACKET (bra + ket = bracket). The bra is commonly encountered in quantum mechanics.

See also ANGLE BRACKET, BRACKET PRODUCT, COVARIANT VECTOR, DIFFERENTIAL K-FORM, KET, ONE-FORM

References

Dirac, P. A. M. "Bra and Ket Vectors." §6 in *Principles of Quantum Mechanics, 4th ed.* Oxford, England: Oxford University Press, pp. 18–22, 1982.

Brace

One of the symbols { and } used in many different contexts in mathematics. Braces are used

1. To denote grouping of mathematical terms, usually as the outermost delimiter in a complex expression such as $\{a + b[c + d(e + f)]\}$,

2. To delineate a SET, as in $\{a_1, \ldots, a_n\}$,

3. Using a left bracket only, to denote different cases for an expression, such as

$$p(n) = \begin{cases} 1 & \text{for } n \text{ even} \\ 0 & \text{for } n \text{ odd}, \end{cases}$$

4. Using a single horizontal underbrace, to indicate the number of items in a list with not all elements shown explicitly, as in $\underbrace{1, 1, \ldots, 1}$.

5. As an alternate notation to the FRACTIONAL PART function, $\{x\} = \text{frac } x$.

See also ANGLE BRACKET, PARENTHESIS, SQUARE BRACKET

References

Bringhurst, R. *The Elements of Typographic Style, 2nd ed.* Point Roberts, WA: Hartley and Marks, p. 273, 1997.

Braced Square

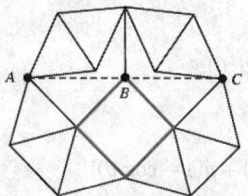 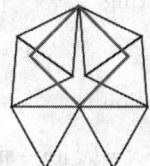

27 nonoverlapping rods   *21 overlapping rods*

The braced square problem asks: given a hinged SQUARE composed of four equal rods (indicated by the thick lines above), how many more hinged rods must be added in the same plane (with no two rods crossing) so that the original square is rigid in the plane. The best solution known, illustrated in the left figure above, uses a total of 27 rods, where A, B, and C are COLLINEAR. If rods are allowed to cross, the best known solution, discovered by E. Friedman in Jan. 2000, requires 21 rods, as illustrated in the right figure above.

Friedman has also considered the minimum number of rods needed to construct RIGID regular n-gons (with overlapping permitted). The best known solutions for $n = 3, 4, \ldots$ are 3, 21, 69, 11, 45, 99, 51, ...

See also HINGED TESSELLATION, RIGID GRAPH, SQUARE

References

Friedman, E. "Problem of the Month (January 2000)." http://www.stetson.edu/~efriedma/mathmagic/0100.html.
Gardner, M. "The Rigid Square." §6.1 in *The Sixth Book of Mathematical Games from Scientific American*. Chicago, IL: University of Chicago Press, pp. 48–49 and 54–55, 1984.
Wells, D. *The Penguin Dictionary of Curious and Interesting Geometry*. London: Penguin, p. 19, 1991.

Brachistochrone Problem

Find the shape of the CURVE down which a bead sliding from rest and ACCELERATED by gravity will slip (without friction) from one point to another in the least time. The term derives from the Greek βραχιστος (*brachistos*) "the shortest" and χρονος (*chronos*) "time, delay."

The brachistochrone problem was one of the earliest problems posed in the CALCULUS OF VARIATIONS. The solution, a segment of a CYCLOID, was found by Leibniz, L'Hospital, Newton, and the two Bernoullis. Johann Bernoulli solved the problem using the analogous one of considering the path of light refracted by transparent layers of varying density (Mach 1893, Gardner 1984, Courant and Robbins 1996). Note that bead may actually travel uphill along the cycloid for a distance, but the path is nonetheless faster than a straight line or any other line.

The time to travel from a point P_1 to another point P_2 is given by the INTEGRAL

$$t_{12} = \int_1^2 \frac{ds}{v}, \tag{1}$$

The VELOCITY at any point is given by a simple application of energy conservation equating kinetic energy to gravitational potential energy,

$$\tfrac{1}{2}mv^2 = mgy, \tag{2}$$

so

$$v = \sqrt{2gy}. \tag{3}$$

Plugging this into (1) then gives

$$t_{12} = \int_1^2 \frac{\sqrt{1+y'^2}}{\sqrt{2gy}} \, dx = \int_1^2 \sqrt{\frac{1+y'^2}{2gy}} \, dx. \tag{4}$$

The function to be varied is thus

$$f = (1+y'^2)^{1/2}(2gy)^{-1/2}, \tag{5}$$

To proceed, one would normally have to apply the full-blown EULER-LAGRANGE DIFFERENTIAL EQUATION

$$\frac{\partial f}{\partial y} - \frac{d}{dx}\left(\frac{\partial f}{\partial y'}\right) = 0. \tag{6}$$

However, the function $f(y, y', x)$ is particularly nice since x does not appear explicitly. Therefore, $\partial f/\partial x = 0$, and we can immediately use the BELTRAMI IDENTITY

$$f - y' \frac{\partial f}{\partial y'} = C. \tag{7}$$

Computing

$$\frac{\partial f}{\partial y'} = y'(1+y'^2)^{-1/2}(2\,gy)^{-1/2}, \tag{8}$$

subtracting $y'(\partial f/\partial y')$ from f, and simplifying then gives

$$\frac{1}{\sqrt{2\,gy}}\frac{1}{\sqrt{1+y'^2}} = C. \tag{9}$$

Squaring both sides and rearranging slightly results in

$$\left[1+\left(\frac{dy}{dx}\right)^2\right]y = \frac{1}{2g\,C^2} = k^2, \tag{10}$$

where the square of the old constant C has been expressed in terms of a new (POSITIVE) constant k^2. This equation is solved by the PARAMETRIC EQUATIONS

$$x = \tfrac{1}{2}k^2(\theta - \sin\theta) \tag{11}$$

$$y = \tfrac{1}{2}k^2(1 - \cos\theta), \tag{12}$$

which are–lo and behold–the equations of a CYCLOID.

If kinetic friction is included, the problem can also be solved analytically, although the solution is significantly messier. In that case, terms corresponding to the normal component of weight and the normal component of the ACCELERATION (present because of path CURVATURE) must be included. Including both terms requires a constrained variational technique (Ashby *et al.* 1975), but including the normal component of weight only gives an elementary solution. The TANGENT and NORMAL VECTORS are

$$\mathbf{T} = \frac{dx}{ds}\,\hat{\mathbf{x}} + \frac{dy}{ds}\,\hat{\mathbf{y}} \tag{13}$$

$$\mathbf{N} = -\frac{dy}{ds}\,\hat{\mathbf{x}} + \frac{dx}{ds}\,\hat{\mathbf{y}}, \tag{14}$$

gravity and friction are then

$$\mathbf{F}_{\text{gravity}} = mg\hat{\mathbf{y}} \tag{15}$$

$$\mathbf{F}_{\text{friction}} = -\mu(\mathbf{F}_{\text{gravity}}\dot{\mathbf{N}})\mathbf{T} = -\mu mg\,\frac{dx}{ds}\,\mathbf{T}, \tag{16}$$

and the components along the curve are

$$\mathbf{F}_{\text{gravity}}\dot{\mathbf{T}} = mg\,\frac{dy}{ds} \tag{17}$$

$$\mathbf{F}_{\text{friction}}\dot{\mathbf{T}} = -\mu mg\,\frac{dx}{ds}, \tag{18}$$

so Newton's Second Law gives

$$m\,\frac{dv}{dt} = mg\,\frac{dy}{ds} - \mu mg\,\frac{dx}{ds}. \tag{19}$$

But

$$\frac{dv}{dt} = v\,\frac{dv}{ds} = \frac{1}{2}\frac{d}{ds}(v^2) \tag{20}$$

$$\tfrac{1}{2}v^2 = g(y - \mu x) \tag{21}$$

$$v = \sqrt{2g(y - \mu x)}, \tag{22}$$

so

$$t = \int \sqrt{\frac{1+(y')^2}{2g(y-\mu x)}}\,dx. \tag{23}$$

Using the EULER-LAGRANGE DIFFERENTIAL EQUATION gives

$$[1+y'^2](1+\mu y') + 2(y-\mu x)y'' = 0. \tag{24}$$

This can be reduced to

$$\frac{1+(y')^2}{(1+\mu y')^2} = \frac{C}{y-\mu x}. \tag{25}$$

Now letting

$$y' = \cot(\tfrac{1}{2}\theta), \tag{26}$$

the solution is

$$x = \tfrac{1}{2}k^2[(\theta - \sin\theta) + \mu(1 - \cos\theta)] \tag{27}$$

$$y = \tfrac{1}{2}k^2[(1 - \cos\theta) + \mu(\theta - \sin\theta)]. \tag{28}$$

See also CALCULUS OF VARIATIONS, CYCLOID, TAUTO-CHRONE PROBLEM

References

Ashby, N.; Brittin, W. E.; Love, W. F.; and Wyss, W. "Brachistochrone with Coulomb Friction." *Amer. J. Phys.* **43**, 902–905, 1975.

Courant, R. and Robbins, H. *What is Mathematics?: An Elementary Approach to Ideas and Methods, 2nd ed.* Oxford, England: Oxford University Press, 1996.

Gardner, M. *The Sixth Book of Mathematical Games from Scientific American.* Chicago, IL: University of Chicago Press, pp. 130–131, 1984.

Haws, L. and Kiser, T. "Exploring the Brachistochrone Problem." *Amer. Math. Monthly* **102**, 328–336, 1995.

Mach, E. *The Science of Mechanics.* Chicago, IL: Open Court, 1893.

Phillips, J. P. "Brachistochrone, Tautochrone, Cycloid--Apple of Discord." *Math. Teacher* **60**, 506–508, 1967.

Steinhaus, H. *Mathematical Snapshots, 3rd ed.* New York: Dover, pp. 148–149, 1999.

Wagon, S. *Mathematica in Action.* New York: W. H. Freeman, pp. 60–66 and 385–389, 1991.

Wells, D. *The Penguin Dictionary of Curious and Interesting Geometry.* London: Penguin, p. 46, 1991.

Bracket

Mathematicians often use the term "bracket" to mean "COMMUTATOR," which is denoted using SQUARE BRACKETS.

See also ANGLE BRACKET, BRA, BRACE, BRACKET POLYNOMIAL, BRACKET PRODUCT, IVERSON BRACKET,

KET, LAGRANGE BRACKET, POISSON BRACKET, SQUARE BRACKET

Bracket Polynomial

A one-variable KNOT POLYNOMIAL related to the JONES POLYNOMIAL. The bracket polynomial, however, is *not* a topological invariant, since it is changed by type I REIDEMEISTER MOVES. However, the SPAN of the bracket polynomial is a knot invariant. The bracket polynomial is occasionally given the grandiose name REGULAR ISOTOPY INVARIANT. It is defined by

$$\langle L \rangle (A, B, d) \equiv \sum_{\sigma} \langle L | \sigma \rangle d^{\|\sigma\|}, \tag{1}$$

where A and B are the "splitting variables," σ runs through all "states" of L obtained by SPLITTING the LINK, $\langle L | \sigma \rangle$ is the product of "splitting labels" corresponding to σ, and

$$\|\sigma\| \equiv N_L - 1, \tag{2}$$

where N_L is the number of loops in σ. Letting

$$B = A^{-1} \tag{3}$$

$$d = -A^2 - A^{-2} \tag{4}$$

gives a KNOT POLYNOMIAL which is invariant under REGULAR ISOTOPY, and normalizing gives the KAUFFMAN POLYNOMIAL X which is invariant under AMBIENT ISOTOPY. The bracket POLYNOMIAL of the UNKNOT is 1. The bracket POLYNOMIAL of the MIRROR IMAGE K^* is the same as for K but with A replaced by A^{-1}. In terms of the one-variable KAUFFMAN POLYNOMIAL X, the two-variable KAUFFMAN POLYNOMIAL F and the JONES POLYNOMIAL V,

$$X(A) = (-A^3)^{-w(L)} \langle L \rangle, \tag{5}$$

$$\langle L \rangle (A) = F(-A^3, A + A^{-1}) \tag{6}$$

$$\langle L \rangle (A) = V(A^{-4}), \tag{7}$$

where $w(L)$ is the WRITHE of L.

See also JONES POLYNOMIAL, SQUARE BRACKET POLYNOMIAL

References
Adams, C. C. *The Knot Book: An Elementary Introduction to the Mathematical Theory of Knots.* New York: W. H. Freeman, pp. 148–155, 1994.
Kauffman, L. "New Invariants in the Theory of Knots." *Amer. Math. Monthly* **95**, 195–242, 1988.
Kauffman, L. *Knots and Physics.* Teaneck, NJ: World Scientific, pp. 26–29, 1991.
Weisstein, E. W. "Knots and Links." MATHEMATICA NOTEBOOK KNOTS.M.

Bracket Product
$L2$-INNER PRODUCT

Bracketing

Take x itself to be a bracketing, then recursively define a bracketing as a sequence $B = (B_1, \ldots, B_k)$ where $k \geq 2$ and each B_i is a bracketing. A bracketing can be REPRESENTED AS a parenthesized string of xs, with parentheses removed from any single letter x for clarity of notation (Stanley 1997). Bracketings built up of binary operations only are called BINARY BRACKETINGS. For example, four letters have 11 possible bracketings:

$xxxx$	$(xx)xx$	$x(xx)x$	$xx(xx)$
$(xxx)x$	$x(xxx)$	$((xx)x)x$	$(x(xx))x$
$(xx)(xx)$	$x((xx)x)$	$x(x(xx))$,	

the last five of which are binary.

The number of bracketings on n letters is given by the GENERATING FUNCTION

$$\tfrac{1}{4}(1 + x - \sqrt{1 - 6x + x^2}) = x + x^2 + 3x^3 + 11x^4 + 45x^5$$

(Schröder 1870, Stanley 1997) and the RECURRENCE RELATION

$$s_n = \frac{3(2n - 3)s_{n-1} - (n - 3)s_{n-2}}{n}$$

(Sloane), giving the sequence for s_n as 1, 1, 3, 11, 45, 197, 903, ... (Sloane's A001003). The numbers are also given by

$$s_n = \sum_{i_1 + \ldots + i_k = n} s(i_1) \cdots s(i_k)$$

for $n \geq 2$ (Stanley 1997).

The first PLUTARCH NUMBER 103,049 is equal to s_{10} (Stanley 1997), suggesting that Plutarch's problem of ten compound propositions is equivalent to the number of bracketings. In addition, Plutarch's second number 310,954 is given by $(s_{10} + s_{11})/2 = 310{,}954$ (Habsieger *et al.* 1998).

See also BINARY BRACKETING, PLUTARCH NUMBERS

References
Comtet, L. "Bracketing Problems." §1.15 in *Advanced Combinatorics: The Art of Finite and Infinite Expansions, rev. enl. ed.* Dordrecht, Netherlands: Reidel, pp. 52–57, 1974.
Habsieger, L.; Kazarian, M.; and Lando, S. "On the Second Number of Plutarch." *Amer. Math. Monthly* **105**, 446, 1998.
Schröder, E. "Vier combinatorische Probleme." *Z. Math. Physik* **15**, 361–376, 1870.
Sloane, N. J. A. Sequences A001003/M2898 in "An On-Line Version of the Encyclopedia of Integer Sequences." http://www.research.att.com/~njas/sequences/eisonline.html.
Stanley, R. P. "Hipparchus, Plutarch, Schröder, and Hough." *Amer. Math. Monthly* **104**, 344–350, 1997.

Bradley's Theorem

Let

$$S(\alpha, \beta, m; z)$$

$$\equiv m \sum_{j=0}^{\infty} \frac{\Gamma(m+j(z+1))\Gamma(\beta+1+jz)}{\Gamma(m+jz+1)\Gamma(\alpha+\beta+1+j(z+1))} \frac{(\alpha)_j}{j!},$$

where $(\alpha)_j$ is a POCHHAMMER SYMBOL, and let α be a NEGATIVE INTEGER. Then

$$S(\alpha, \beta, m; z) = \frac{\Gamma(\beta+1-m)}{\Gamma(\alpha+\beta+1-m)},$$

where $\Gamma(z)$ is the GAMMA FUNCTION.

References

Berndt, B. C. *Ramanujan's Notebooks, Part IV.* New York: Springer-Verlag, pp. 346–348, 1994.

Bradley, D. "On a Claim by Ramanujan about Certain Hypergeometric Series." *Proc. Amer. Math. Soc.* **121**, 1145–1149, 1994.

Brahmagupta Identity

Let

$$\beta \equiv \det B = x^2 - ty^2,$$

where B is the BRAHMAGUPTA MATRIX, then

$$\det[B(x_1, y_1)B(x_2, y_2)] = \det[B(x_1, y_1)] \det[B(x_2, y_2)]$$
$$= \beta_1 \beta_2.$$

References

Suryanarayan, E. R. "The Brahmagupta Polynomials." *Fib. Quart.* **34**, 30–39, 1996.

Brahmagupta Matrix

$$B(x, y) = \begin{bmatrix} x & y \\ \pm ty & \pm x \end{bmatrix}.$$

It satisfies

$$B(x_1, y_1)B(x_2, y_2) = B(x_1 x_2 \pm ty_1 y_2, \; x_1 y_2 \pm y_1 x_2).$$

Powers of the matrix are defined by

$$B^n = \begin{bmatrix} x & y \\ ty & x \end{bmatrix}^n = \begin{bmatrix} x_n & y_n \\ ty_n & x_n \end{bmatrix} \equiv B_n.$$

The x_n and y_n are called BRAHMAGUPTA POLYNOMIALS. The Brahmagupta matrices can be extended to NEGATIVE INTEGERS

$$B^{-n} = \begin{bmatrix} x & y \\ ty & x \end{bmatrix}^{-n} = \begin{bmatrix} x_{-n} & y_{-n} \\ ty_{-n} & x_{-n} \end{bmatrix} \equiv B_{-n}.$$

See also BRAHMAGUPTA IDENTITY

References

Suryanarayan, E. R. "The Brahmagupta Polynomials." *Fib. Quart.* **34**, 30–39, 1996.

Brahmagupta Polynomial

One of the POLYNOMIALS obtained by taking POWERS of the BRAHMAGUPTA MATRIX. They satisfy the RECURRENCE RELATION

$$x_{n+1} = xx_n + tyy_n \tag{1}$$

$$y_{n+1} = xy_n + yx_n. \tag{2}$$

A list of many others is given by Suryanarayan (1996). Explicitly,

$$x_n = x^n + t\binom{n}{2}x^{n-2}y^2 + t^2\binom{n}{4}x^{n-4}y^4 + \ldots \tag{3}$$

$$y_n = nx^{n-1}y + t\binom{n}{3}x^{n-3}y^3 + t^2\binom{n}{5}x^{n-5}y^5 + \ldots \tag{4}$$

The Brahmagupta POLYNOMIALS satisfy

$$\frac{\partial x_n}{\partial x} = \frac{\partial y_n}{\partial y} = nx_{n-1} \tag{5}$$

$$\frac{\partial x_n}{\partial y} = t\frac{\partial y_n}{\partial y} = nty_{n-1}. \tag{6}$$

The first few POLYNOMIALS are

$$x_0 = 0$$
$$x_1 = x$$
$$x_2 = x^2 + ty^2$$
$$x_3 = x^3 + 3txy^2$$
$$x_4 = x^4 + 6tx^2y^2 + t^2y^4$$

and

$$y_0 = 0$$
$$y_1 = y$$
$$y_2 = 2xy$$
$$y_3 = 3x^2y + ty^3$$
$$y_4 = 4x^3y + 4txy^3.$$

Taking $x = y = 1$ and $t = 2$ gives y_n equal to the PELL NUMBERS and x_n equal to half the Pell-Lucas numbers. The Brahmagupta POLYNOMIALS are related to the MORGAN-VOYCE POLYNOMIALS, but the relationship given by Suryanarayan (1996) is incorrect.

References

Suryanarayan, E. R. "The Brahmagupta Polynomials." *Fib. Quart.* **34**, 30–39, 1996.

Brahmagupta's Formula

For a QUADRILATERAL with sides of length a, b, c, and d, the AREA K is given by

$$K = \sqrt{(s-a)(s-b)(s-c)(s-d) - abcd\,\cos^2[\tfrac{1}{2}(A+B)]},$$

(1)

where

$$s \equiv \tfrac{1}{2}(a+b+c+d)$$

(2)

is the SEMIPERIMETER, A is the ANGLE between a and d, and B is the ANGLE between b and c. For a CYCLIC QUADRILATERAL (i.e., a QUADRILATERAL inscribed in a CIRCLE), $A+B=\pi$, so

$$K = \sqrt{(s-a)(s-b)(s-c)(s-d)}$$

(3)

$$= \frac{\sqrt{(bc+ad)(ac+bd)(ab+cd)}}{4R},$$

(4)

where R is the RADIUS of the CIRCUMCIRCLE. If the QUADRILATERAL is INSCRIBED in one CIRCLE and CIRCUMSCRIBED on another, then the AREA FORMULA simplifies to

$$K = \sqrt{abcd}.$$

(5)

See also BRETSCHNEIDER'S FORMULA, HERON'S FORMULA, QUADRILATERAL

References

Brown, K. S. "Heron's FOrmula and Brahmagupta's Generalization." http://www.seanet.com/~ksbrown/kmath19 6.htm.

Coxeter, H. S. M. and Greitzer, S. L. "Cyclic Quadrangles; Brahmagupta's Formula." §3.2 in *Geometry Revisited*. Washington, DC: Math. Assoc. Amer., pp. 56–60, 1967.

Johnson, R. A. *Modern Geometry: An Elementary Treatise on the Geometry of the Triangle and the Circle*. Boston, MA: Houghton Mifflin, pp. 81–82, 1929.

Brahmagupta's Problem

Solve the PELL EQUATION

$$x^2 - 92y^2 = 1$$

in INTEGERS. The smallest solution is $x = 1151$, $y = 120$.

See also DIOPHANTINE EQUATION, PELL EQUATION

Brahmagupta's Theorem

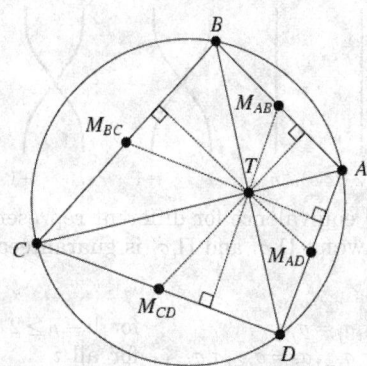

In a CYCLIC QUADRILATERAL $ABCD$ having perpendicular diagonals $AC \perp BD$, the perpendiculars to the sides through point T of intersection of the diagonals (the ANTICENTER) always bisects the opposite side (so M_{AB}, M_{BC}, M_{CD}, and M_{DA} are the MIDPOINTS of the corresponding sides of the QUADRILATERAL).

See also ANTICENTER, CYCLIC QUADRILATERAL, MIDPOINT

References

Honsberger, R. *Episodes in Nineteenth and Twentieth Century Euclidean Geometry*. Washington, DC: Math. Assoc. Amer., p. 37, 1995.

Braid

An intertwining of strings attached to top and bottom "bars" such that each string never "turns back up." In other words, the path of each string in a braid could be traced out by a falling object if acted upon only by gravity and horizontal forces.

See also BRAID GROUP

References

Christy, J. "Braids." http://www.mathsource.com/cgi-bin/ msitem?0202–228.

Murasugi, K. and Kurpita, B. I. *A Study of Braids*. Dordrecht, Netherlands: Kluwer, 1999.

Braid Group

Also called ARTIN BRAID GROUPS. Consider n strings, each oriented vertically from a lower to an upper "bar." If this is the least number of strings needed to make a closed braid representation of a LINK, n is called the BRAID INDEX. Now enumerate the possible braids in a group, denoted B_n. A general n-braid is constructed by iteratively applying the σ_i ($i = 1, \ldots, n-1$) operator, which switches the lower endpoints of the ith and $(i+1)$th strings–keeping the upper endpoints fixed–with the $(i+1)$th string brought *above* the ith string. If the $(i+1)$th string passes *below* the ith string, it is denoted σ_i^{-1}.

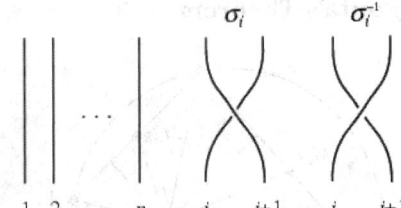

Topological equivalence for different representations of a BRAID WORD $\Pi_i \sigma_i$ and $\Pi_i \sigma_i'$ is guaranteed by the conditions

$$\begin{cases} \sigma_i \sigma_j = \sigma_j \sigma_i & \text{for } |i - j| \geq 2 \\ \sigma_i \sigma_{i+1} \sigma_i = \sigma_{i+1} \sigma_i \sigma_{i+1} & \text{for all } i \end{cases}$$

as first proved by E. Artin. Any n-braid is expressed as a BRAID WORD, e.g., $\sigma_1 \sigma_2 \sigma_3 \sigma_2^{-1} \sigma_1$ is a BRAID WORD for the braid group B_3. When the opposite ends of the braids are connected by nonintersecting lines, KNOTS are formed which are identified by their braid group and BRAID WORD. The BURAU REPRESENTATION gives a matrix representation of the braid groups.

References

Birman, J. S. "Braids, Links, and the Mapping Class Groups." *Ann. Math. Studies*, No. 82. Princeton, NJ: Princeton University Press, 1976.

Birman, J. S. "Recent Developments in Braid and Link Theory." *Math. Intell.* **13**, 52–60, 1991.

Christy, J. "Braids." http://www.mathsource.com/cgi-bin/msitem?0202-228.

Jones, V. F. R. "Hecke Algebra Representations of Braid Groups and Link Polynomials." *Ann. Math.* **126**, 335–388, 1987.

Murasugi, K. and Kurpita, B. I. *A Study of Braids*. Dordrecht, Netherlands: Kluwer, 1999.

Weisstein, E. W. "Knots and Links." MATHEMATICA NOTEBOOK KNOTS.M.

Braid Index

The least number of strings needed to make a closed braid representation of a LINK. The braid index is equal to the least number of SEIFERT CIRCLES in any projection of a KNOT (Yamada 1987). Also, for a nonsplittable LINK with CROSSING NUMBER $c(L)$ and braid index $i(L)$,

$$c(L) \geq 2[i(L) - 1]$$

(Ohyama 1993). Let E be the largest and e the smallest POWER of ℓ in the HOMFLY POLYNOMIAL of an oriented LINK, and i be the braid index. Then the MORTON-FRANKS-WILLIAMS INEQUALITY holds,

$$i \geq \tfrac{1}{2}(E - e) + 1$$

(Franks and Williams 1987). The inequality is sharp for all PRIME KNOTS up to 10 crossings with the exceptions of 09–042, 09–049, 10–132, 10–150, and 10–156.

References

Franks, J. and Williams, R. F. "Braids and the Jones Polynomial." *Trans. Amer. Math. Soc.* **303**, 97–108, 1987.

Jones, V. F. R. "Hecke Algebra Representations of Braid Groups and Link Polynomials." *Ann. Math.* **126**, 335–388, 1987.

Ohyama, Y. "On the Minimal Crossing Number and the Brad Index of Links." *Canad. J. Math.* **45**, 117–131, 1993.

Yamada, S. "The Minimal Number of Seifert Circles Equals the Braid Index of a Link." *Invent. Math.* **89**, 347–356, 1987.

Braid Word

Any n-braid is expressed as a braid word, e.g., $\sigma_1 \sigma_2 \sigma_3 \sigma_2^{-1} \sigma_1$ is a braid word for the BRAID GROUP B_3. By ALEXANDER'S THEOREM, any LINK is representable by a closed braid, but there is no general procedure for reducing a braid word to its simplest form. However, MARKOV'S THEOREM gives a procedure for identifying different braid words which represent the same LINK.

Let b_+ be the sum of POSITIVE exponents, and b_- the sum of NEGATIVE exponents in the BRAID GROUP B_n. If

$$b_+ - 3b_- \geq n,$$

then the closed braid b is not AMPHICHIRAL (Jones 1985).

See also BRAID GROUP

References

Jones, V. F. R. "A Polynomial Invariant for Knots via von Neumann Algebras." *Bull. Amer. Math. Soc.* **12**, 103–111, 1985.

Jones, V. F. R. "Hecke Algebra Representations of Braid Groups and Link Polynomials." *Ann. Math.* **126**, 335–388, 1987.

Murasugi, K. and Kurpita, B. I. *A Study of Braids*. Dordrecht, Netherlands: Kluwer, 1999.

Braikenridge-Maclaurin Construction

Let $A_n, B_2, C_1, A_2,$ and B_1 be five points determining a CONIC. Then the CONIC is the LOCUS of the point

$$C_2 = A_1(L \cdot C_1 A_2) \cdot B_1(L \cdot C_1 B_2),$$

where L is a line through the point $A_1 B_2 \cdot B_1 A_2$.

See also BRAIKENRIDGE-MACLAURIN THEOREM, CONIC SECTION

Braikenridge-Maclaurin Theorem

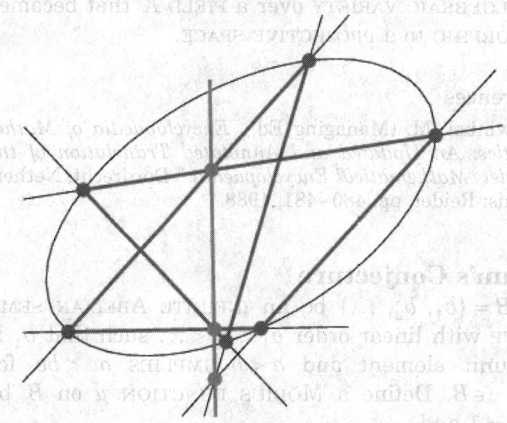

The converse of PASCAL'S THEOREM, which states that if the three pairs of opposite sides of (an irregular) HEXAGON meet at three COLLINEAR points, then the six vertices lie on a conic, which may degenerate into a pair of lines (Coxeter and Greitzer 1967, p. 76).

See also BRAIKENRIDGE-MACLAURIN CONSTRUCTION, CONIC SECTION, PASCAL'S THEOREM

References

Coxeter, H. S. M. *Projective Geometry, 2nd ed.* New York: Springer-Verlag, p. 85, 1987.
Coxeter, H. S. M. and Greitzer, S. L. *Geometry Revisited.* Washington, DC: Math. Assoc. Amer., p. 76, 1967.

Branch

A branch at a point u in a TREE is a maximal SUBTREE containing u as an ENDPOINT (Harary 1994, p. 35).

See also FORK, LEAF (TREE), LIMB, TREE

References

Harary, F. *Graph Theory.* Reading, MA: Addison-Wesley, 1994.
Lu, T. "The Enumeration of Trees with and without Given Limbs." *Disc. Math.* **154**, 153–165, 1996.
Schwenk, A. "Almost All Trees are Cospectral." In *New Directions in the Theory of Graphs* (Ed. F. Harary). New York: Academic Press, pp. 275–307, 1973.

Branch Cut

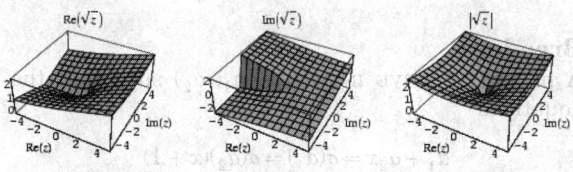

A line in the COMPLEX PLANE across which a MULTI-

VALUED FUNCTION is discontinuous. Some functions have a relatively simple branch cut structure, but branch cuts for some functions are extremely complicated. The illustrations above show the single branch cut present in the definition of the square root function in the complex plane. In general, branch cuts are not unique, but are chosen by convention to give simple analytic properties. An alternative to branch cuts is the use of RIEMANN SURFACES.

function	branch cut(s)
$\cos^{-1} z$	$(-\infty, -1)$ and $(1, \infty)$
$\cosh^{-1}$	$(-\infty, 1)$
$\cot^{-1} z$	$(-i, i)$
$\coth^{-1}$	$[-1, 1]$
$\csc^{-1} z$	$(-1, 1)$
csch^{-1}	$(-i, i)$
$\ln z$	$(-\infty, 0]$
$\sec^{-1} z$	$(-1, 1)$
sech^{-1}	$(\infty, 0]$ and $(1, \infty)$
$\sin^{-1} z$	$(-\infty, -1)$ and $(1, \infty)$
$\sinh^{-1}$	$(-i\infty, -i)$ and $(i, i\infty)$
$\sqrt{z}$	$(-\infty, 0)$
$\tan^{-1} z$	$(-i\infty, -i)$ and $(i, i\infty)$
$\tanh^{-1}$	$(-\infty, -1]$ and $[1, \infty)$
$z^n, \; n \notin \mathbb{Z}$	$(-\infty, 0)$ for $\Re[n] \leq 0$; $(-\infty, 0]$ for $\Re[n] > 0$

See also BRANCH POINT, CUT, MULTIVALUED FUNCTION, RIEMANN SURFACE

References

Kahan, W. "Branch Cuts for Complex Elementary Functions, or Much Ado About Nothing's Sign Bit." In *The State of the Art in Numerical Analysis: Proceedings of the Joint IMA/SIAM Conference on the State of the Art in Numerical Analysis Held at the UN* (Ed. A. Iserles and M. J. D. Powell). New York: Clarendon Press, pp. 165–211, 1987.
Morse, P. M. and Feshbach, H. *Methods of Theoretical Physics, Part I.* New York: McGraw-Hill, pp. 399–401, 1953.

Branch Line

BRANCH CUT

Branch Point

An argument at which identical points in the COMPLEX PLANE are mapped to different points. For example, consider

$$f(z) = z^a.$$

Then $f(e^{0i}) = f(1) = 1$, but $f(e^{2\pi i}) = e^{2\pi ia}$, despite the fact that $e^{i0} = e^{2\pi i}$. PINCH POINTS are also called branch points.

See also BRANCH CUT, PINCH POINT

References
Arfken, G. *Mathematical Methods for Physicists, 3rd ed.* Orlando, FL: Academic Press, pp. 397–399, 1985.
Morse, P. M. and Feshbach, H. *Methods of Theoretical Physics, Part I.* New York: McGraw-Hill, pp. 391–392 and 399–401, 1953.

Brauer Chain
A Brauer chain is an ADDITION CHAIN in which each member uses the previous member as a summand. A number n for which a shortest chain exists which is a Brauer chain is called a BRAUER NUMBER.

See also ADDITION CHAIN, BRAUER NUMBER, HANSEN CHAIN

References
Guy, R. K. "Addition Chains. Brauer Chains. Hansen Chains." §C6 in *Unsolved Problems in Number Theory, 2nd ed.* New York: Springer-Verlag, pp. 111–113, 1994.

Brauer Group
The GROUP of classes of finite dimensional central simple ALGEBRAS over k with respect to a certain equivalence.

References
Hazewinkel, M. (Managing Ed.). *Encyclopaedia of Mathematics: An Updated and Annotated Translation of the Soviet "Mathematical Encyclopaedia."* Dordrecht, Netherlands: Reidel, p. 479, 1988.

Brauer Number
A number n for which a shortest chain exists which is a BRAUER CHAIN is called a Brauer number. There are infinitely many non-Brauer numbers.

See also BRAUER CHAIN, HANSEN NUMBER

References
Guy, R. K. "Addition Chains. Brauer Chains. Hansen Chains." §C6 in *Unsolved Problems in Number Theory, 2nd ed.* New York: Springer-Verlag, pp. 111–113, 1994.

Brauer's Theorem
If, in the GERSGORIN CIRCLE THEOREM for a given m,

$$|a_{jj} - a_{mm}| > \Lambda_j + \Lambda_m$$

for all $j \neq m$, then exactly one EIGENVALUE of A lies in the DISK Γ_m.

References
Gradshteyn, I. S. and Ryzhik, I. M. *Tables of Integrals, Series, and Products, 6th ed.* San Diego, CA: Academic Press, p. 1121, 2000.

Brauer-Severi Variety
An ALGEBRAIC VARIETY over a FIELD K that becomes ISOMORPHIC to a PROJECTIVE SPACE.

References
Hazewinkel, M. (Managing Ed.). *Encyclopaedia of Mathematics: An Updated and Annotated Translation of the Soviet "Mathematical Encyclopaedia."* Dordrecht, Netherlands: Reidel, pp. 480–481, 1988.

Braun's Conjecture
Let $B = \{b_1, b_2, \ldots\}$ be an INFINITE ABELIAN SEMIGROUP with linear order $b_1 < b_2 < \ldots$ such that b_1 is the unit element and $a < b$ IMPLIES $ac < bc$ for $a, b, c \in B$. Define a MÖBIUS FUNCTION μ on B by $\mu(b_1) = 1$ and

$$\sum_{b_d | b_n} \mu(b_d) = 0$$

for $n = 2, 3, \ldots$. Further suppose that $\mu(b_n) = \mu(n)$ (the true MÖBIUS FUNCTION) for all $n \geq 1$. Then Braun's conjecture states that

$$b_{mn} = b_m \, b_n$$

for all $m, n \geq 1$.

See also MÖBIUS PROBLEM

References
Flath, A. and Zulauf, A. "Does the Möbius Function Determine Multiplicative Arithmetic?" *Amer. Math. Monthly* **102**, 354–256, 1995.

Breadth-First Traversal
A search algorithm of a GRAPH which explores all nodes adjacent to the current node before moving on. For cyclic graphs, care must be taken to make sure that no nodes are repeated. When properly implemented, all nodes in a given connected component are explored.

See also DEPTH-FIRST TRAVERSAL

References
Skiena, S. "Breadth-First and Depth-First Search." §3.2.5 in *Implementing Discrete Mathematics: Combinatorics and Graph Theory with Mathematica.* Reading, MA: Addison-Wesley, pp. 95–97, 1990.

Breeder
A pair of POSITIVE INTEGERS (a_1, a_2) such that the equations

$$a_1 + a_2 x = \sigma(a_1) = \sigma(a_2)(x + 1)$$

have a POSITIVE INTEGER solution x, where $\sigma(n)$ is the DIVISOR FUNCTION. If x is PRIME, then $(a_1, a_2 x)$ is an AMICABLE PAIR (te Riele 1986). (a_1, a_2) is a "special" breeder if

$$a_1 = au$$

$$a_2 = a,$$

where a and u are RELATIVELY PRIME, $(a, u) = 1$. If regular amicable pairs of type $(i, 1)$ with $i \geq 2$ are OF THE FORM (au, ap) with p PRIME, then (au, a) are special breeders (te Riele 1986).

See also AMICABLE PAIR

References

te Riele, H. J. J. "Computation of All the Amicable Pairs Below 10^{10}." *Math. Comput.* **47**, 361–368 and S9-S35, 1986.

Brelaz's Heuristic Algorithm

An ALGORITHM which can be used to find a good, but not necessarily minimal, EDGE or VERTEX COLORING for a GRAPH. However, the algorithm does minimally color COMPLETE K-PARTITE GRAPH.

See also CHROMATIC NUMBER, EDGE COLORING, VERTEX COLORING

References

Brelaz, D. "New Methods to Color the Vertices of a Graph." *Comm. ACM* **22**, 251–256, 1979.
Skiena, S. "Finding a Vertex Coloring." §5.5.3 in *Implementing Discrete Mathematics: Combinatorics and Graph Theory with Mathematica.* Reading, MA: Addison-Wesley, pp. 214–215, 1990.

Brent's Factorization Method

A modification of the POLLARD RHO FACTORIZATION METHOD which uses

$$x_{i+1} = x_i^2 - c \pmod{n}.$$

References

Brent, R. "An Improved Monte Carlo Factorization Algorithm." *Nordisk Tidskrift for Informationsbehandling (BIT)* **20**, 176–184, 1980.

Brent's Method

A ROOT-finding ALGORITHM which combines root bracketing, bisection, and INVERSE QUADRATIC INTERPOLATION. It is sometimes known as the VAN WIJNGAARDEN-DEKER-BRENT METHOD.

Brent's method uses a LAGRANGE INTERPOLATING POLYNOMIAL of degree 2. Brent (1973) claims that this method will always converge as long as the values of the function are computable within a given region containing a ROOT. Given three points x_1, x_2, and x_3, Brent's method fits x as a quadratic function of y, then uses the interpolation formula

$$x = \frac{[y - f(x_1)][y - f(x_2)]x_3}{[f(x_3) - f(x_1)][f(x_3) - f(x_2)]}$$
$$+ \frac{[y - f(x_2)][y - f(x_3)]x_1}{[f(x_1) - f(x_2)][f(x_1) - f(x_3)]}$$
$$+ \frac{[y - f(x_3)][y - f(x_1)]x_2}{[f(x_2) - f(x_3)][f(x_2) - f(x_1)]}. \tag{1}$$

Subsequent root estimates are obtained by setting $y = 0$, giving

$$x = x_2 + \frac{P}{Q}, \tag{2}$$

where

$$P = S[R(R - T)(x_3 - x_2) - (1 - R)(x_2 - x_1)] \tag{3}$$

$$Q = (T - 1)(R - 1)(S - 1) \tag{4}$$

with

$$R \equiv \frac{f(x_2)}{f(x_3)} \tag{5}$$

$$S \equiv \frac{f(x_2)}{f(x_1)} \tag{6}$$

$$T \equiv \frac{f(x_1)}{f(x_3)} \tag{7}$$

(Press *et al.* 1992).

References

Brent, R. P. Ch. 3–4 in *Algorithms for Minimization Without Derivatives.* Englewood Cliffs, NJ: Prentice-Hall, 1973.
Forsythe, G. E.; Malcolm, M. A.; and Moler, C. B. §7.2 in *Computer Methods for Mathematical Computations.* Englewood Cliffs, NJ: Prentice-Hall, 1977.
Press, W. H.; Flannery, B. P.; Teukolsky, S. A.; and Vetterling, W. T. "Van Wijngaarden-Dekker-Brent Method." §9.3 in *Numerical Recipes in FORTRAN: The Art of Scientific Computing, 2nd ed.* Cambridge, England: Cambridge University Press, pp. 352–355, 1992.

Brent-Salamin Formula

A formula which uses the ARITHMETIC-GEOMETRIC MEAN to compute PI. It has quadratic convergence and is also called the GAUSS-SALAMIN FORMULA and SALAMIN FORMULA. Let

$$a_{n+1} = \tfrac{1}{2}(a_n + b_n) \tag{1}$$

$$b_{n+1} = \sqrt{a_n b_n} \tag{2}$$

$$c_{n+1} = \tfrac{1}{2}(a_n - b_n) \tag{3}$$

$$d_n \equiv a_n^2 - b_n^2, \tag{4}$$

and define the initial conditions to be $a_0 = 1$, $b_0 = 1/\sqrt{2}$. Then iterating a_n and b_n gives the ARITHMETIC-

GEOMETRIC MEAN, and π is given by

$$\pi = \frac{4[M(1,\ 2^{-1/2})]^2}{1 - \sum_{j=1}^{\infty}\ 2^{j+1}d_j} \tag{5}$$

$$= \frac{4[M(1,\ 2^{-1/2})]^2}{1 - \sum_{j=1}^{\infty}\ 2^{j+1}c_j^2}. \tag{6}$$

King (1924) showed that this formula and the LEGENDRE RELATION are equivalent and that either may be derived from the other.

See also ARITHMETIC-GEOMETRIC MEAN, PI

References

Borwein, J. M. and Borwein, P. B. *Pi & the AGM: A Study in Analytic Number Theory and Computational Complexity.* New York: Wiley, pp. 48–51, 1987.
Castellanos, D. "The Ubiquitous Pi. Part II." *Math. Mag.* **61**, 148–163, 1988.
King, L. V. *On the Direct Numerical Calculation of Elliptic Functions and Integrals.* Cambridge, England: Cambridge University Press, 1924.
Lord, N. J. "Recent Calculations of π: The Gauss-Salamin Algorithm." *Math. Gaz.* **76**, 231–242, 1992.
Salamin, E. "Computation of π Using Arithmetic-Geometric Mean." *Math. Comput.* **30**, 565–570, 1976.

Bretschneider's Formula

Given a general QUADRILATERAL with sides of lengths a, b, c, and d (Beyer 1987), the AREA is given by

$$A_{\text{quadrilateral}} = \tfrac{1}{4}\sqrt{4p^2q^2 - (b^2+d^2-a^2-c^2)^2},$$

where p and q are the diagonal lengths.

See also BRAHMAGUPTA'S FORMULA, HERON'S FORMULA

References

Beyer, W. H. (Ed.). *CRC Standard Mathematical Tables, 28th ed.* Boca Raton, FL: CRC Press, p. 123, 1987.

Brianchon Point

The point of CONCURRENCE of the joins of the VERTICES of a TRIANGLE and the points of contact of a CONIC SECTION INSCRIBED in the TRIANGLE. A CONIC INSCRIBED in a TRIANGLE has an equation OF THE FORM

$$\frac{f}{u}+\frac{g}{v}+\frac{h}{w}=0,$$

so its Brianchon point has TRILINEAR COORDINATES $(1/f, 1/g, 1/h)$. For KIEPERT'S PARABOLA, the Branchion point has TRIANGLE CENTER FUNCTION

$$\alpha = \frac{1}{a(b^2 - c^2)},$$

which is the STEINER POINT.

See also HEPTAGON THEOREM, KIEPERT'S PARABOLA, STEINER POINTS

References

Evelyn, C. J. A.; Money-Coutts, G. B.; and Tyrrell, J. A. "The Heptagon Theorem." §2.1 in *The Seven Circles Theorem and Other New Theorems.* London: Stacey International, pp. 8–11, 1974.

Brianchon's Theorem

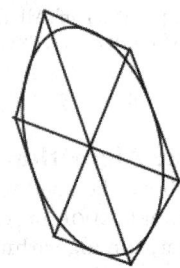

The DUAL of PASCAL'S THEOREM (Casey 1888, p. 146). It states that, given a HEXAGON CIRCUMSCRIBED on a CONIC SECTION, the lines joining opposite VERTICES (DIAGONALS) meet in a single point.

See also DUALITY PRINCIPLE, PASCAL'S THEOREM

References

Casey, J. *A Sequel to the First Six Books of the Elements of Euclid, Containing an Easy Introduction to Modern Geometry with Numerous Examples, 5th ed., rev. enl.* Dublin: Hodges, Figgis, & Co., pp. 146–147, 1888.
Coxeter, H. S. M. and Greitzer, S. L. "Brianchon's Theorem." §3.9 in *Geometry Revisited.* Washington, DC: Math. Assoc. Amer., pp. 77–79, 1967.
Evelyn, C. J. A.; Money-Coutts, G. B.; and Tyrrell, J. A. "Extensions of Pascal's and Brianchon's Theorems." Ch. 2 in *The Seven Circles Theorem and Other New Theorems.* London: Stacey International, pp. 8–30, 1974.
Graustein, W. C. *Introduction to Higher Geometry.* New York: Macmillan, p. 261, 1930.
Johnson, R. A. §387 in *Modern Geometry: An Elementary Treatise on the Geometry of the Triangle and the Circle.* Boston, MA: Houghton Mifflin, p. 237, 1929.
Ogilvy, C. S. *Excursions in Geometry.* New York: Dover, p. 110, 1990.
Smogorzhevskii, A. S. *The Ruler in Geometrical Constructions.* New York: Blaisdell, pp. 33–34, 1961.
Wells, D. *The Penguin Dictionary of Curious and Interesting Geometry.* London: Penguin, pp. 20–21, 1991.

Brick

A RECTANGULAR PARALLELEPIPED.

See also CANONICAL BRICK, EULER BRICK, HARMONIC BRICK, RECTANGULAR PARALLELEPIPED

Bride's Chair

One name for the figure used by Euclid to prove the PYTHAGOREAN THEOREM.

See also PEACOCK'S TAIL, WINDMILL

Bridge

References
Wells, D. *The Penguin Dictionary of Curious and Interesting Geometry.* London: Penguin, p. 203, 1991.

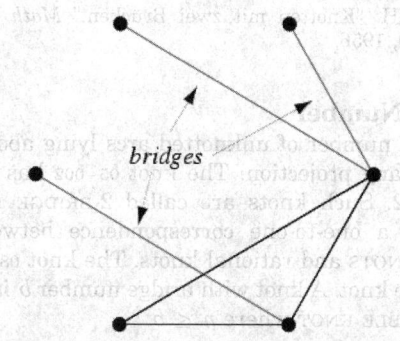

bridges

The bridges of a CONNECTED GRAPH are the EDGES whose removal disconnects the GRAPH (Chartrand 1985, p. 45; Skiena 1990, p. 177). More generally, a bridge is an edge of a GRAPH G whose removal increases the number of components of G (Harary 1994, p. 26). An edge of a CONNECTED GRAPH is a bridge IFF is does not lie on any cycle. The bridges of a graph can be found using Bridges[g] in the *Mathematica* add-on package DiscreteMath`Combinatorica` (which can be loaded with the command <<DiscreteMath`).

Every edge of a TREE is a bridge. A CUBIC GRAPH contains a bridge IFF it contains an ARTICULATION VERTEX (Skiena 1990, p. 177).

See also ARTICULATION VERTEX, BLOCK

References
Chartrand, G. "Cut-Vertices and Bridges." §2.4 in *Introductory Graph Theory.* New York: Dover, pp. 45–49, 1985.
Harary, F. *Graph Theory.* Reading, MA: Addison-Wesley, 1994.
Skiena, S. *Implementing Discrete Mathematics: Combinatorics and Graph Theory with Mathematica.* Reading, MA: Addison-Wesley, pp. 171 and 177, 1990.

Bridge Card Game

Bridge is a CARD game played with a normal deck of 52 cards. The number of possible distinct 13-card hands is

$$N = \binom{52}{13} = 635,013,559,600.$$

where $\binom{n}{k}$ is a BINOMIAL COEFFICIENT. While the chances of being dealt a hand of 13 CARDS (out of 52) of the same suit are

$$\frac{4}{\binom{52}{13}} = \frac{1}{158,753,389,900},$$

the chance that one of four players will receive a hand of a single suit is

$$\frac{1}{39,688,347,497}.$$

There are special names for specific types of hands. A ten, jack, queen, king, or ace is called an "honor." Getting the three top cards (ace, king, and queen) of three suits and the ace, king, and queen, and jack of the remaining suit is called 13 top honors. Getting all cards of the same suit is called a 13-card suit. Getting 12 cards of same suit with ace high and the 13th card *not* an ace is called 2-card suit, ace high. Getting *no* honors is called a Yarborough.

The probabilities of being dealt 13-card bridge hands of a given type are given below. As usual, for a hand with probability P, the ODDS against being dealt it are $(1/P) - 1 : 1$.

Hand	Exact Probability	Probability	ODDS
13 top honors	$\dfrac{4}{N} = \dfrac{1}{158,753,389,900}$	6.30×10^{-12}	158,753,389,899:1
13-card suit	$\dfrac{4}{N} = \dfrac{1}{158,753,389,900}$	6.30×10^{-12}	158,753,389,899:1
12-card suit, ace high	$\dfrac{4 \cdot 12 \cdot 36}{N} = \dfrac{4}{1,469,938,705}$	2.72×10^{-9}	367,484,697.8:1
Yarborough	$\dfrac{\binom{32}{13}}{N} = \dfrac{5,394}{9,860,459}$	5.47×10^{-4}	1,827.0:1
four aces	$\dfrac{\binom{48}{9}}{N} = \dfrac{11}{4,165}$	2.64×10^{-3}	377.6:1
nine honors	$\dfrac{\binom{20}{9}\binom{32}{4}}{N} = \dfrac{888,212}{93,384,347}$	9.51×10^{-3}	104.1:1

See also CARDS, POKER

References
Ball, W. W. R. and Coxeter, H. S. M. *Mathematical Recreations and Essays, 13th ed.* New York: Dover, pp. 48–49, 1987.
Kraitchik, M. "Bridge Hands." §6.3 in *Mathematical Recreations.* New York: W. W. Norton, pp. 119–121, 1942.
Reese, T. *Bridge for Bright Beginners.* New York: Dover, 1973.
Rubens, J. *The Secrets of Winning Bridge.* New York: Dover, 1981.

Bridge Index

A numerical KNOT invariant. For a TAME KNOT K, the bridge index is the least BRIDGE NUMBER of all planar representations of the KNOT. The bridge index of the UNKNOT is defined as 1.

See also BRIDGE NUMBER, CROOKEDNESS

References

Rolfsen, D. *Knots and Links.* Wilmington, DE: Publish or Perish Press, p. 114, 1976.
Schubert, H. "Über eine numerische Knotteninvariante." *Math. Z.* **61**, 245–288, 1954.

Bridge Knot

An n-bridge knot is a knot with BRIDGE NUMBER n. The set of 2-bridge knots is identical to the set of rational knots. If L is a 2-BRIDGE KNOT, then the BLM/HO POLYNOMIAL Q and JONES POLYNOMIAL V satisfy

$$Q_L(z) = 2z^{-1}V_L(t)V_L(t^{-1} + 1 - 2z^{-1}),$$

where $z \equiv -t - t^{-1}$ (Kanenobu and Sumi 1993). Kanenobu and Sumi also give a table containing the number of distinct 2-bridge knots of n crossings for $n = 10$ to 22, both not counting and counting MIRROR IMAGES as distinct.

n	K_n	$K_n + K_n^*$
3	0	0
4	0	0
5		
6		
7		
8		
9		
10	45	85
11	91	182
12	176	341
13	352	704
14	693	1365
15	1387	2774
16	2752	5461
17	5504	11008
18	10965	21845
19	21931	43862
20	43776	87381
21	87552	175104
22	174933	349525

References

Kanenobu, T. and Sumi, T. "Polynomial Invariants of 2-Bridge Links through 20 Crossings." *Adv. Studies Pure Math.* **20**, 125–145, 1992.
Kanenobu, T. and Sumi, T. "Polynomial Invariants of 2-Bridge Knots through 22-Crossings." *Math. Comput.* **60**, 771–778 and S17-S28, 1993.
Schubert, H. "Knotten mit zwei Brücken." *Math. Z.* **65**, 133–170, 1956.

Bridge Number

The least number of unknotted arcs lying above the plane in any projection. The knot 05-002 has bridge number 2. Such knots are called 2-BRIDGE KNOTS. There is a one-to-one correspondence between 2-BRIDGE KNOTS and rational knots. The knot 08-010 is a 3-bridge knot. A knot with bridge number b is an n-EMBEDDABLE KNOT where $n \leq b$.

See also BRIDGE INDEX

References

Adams, C. C. *The Knot Book: An Elementary Introduction to the Mathematical Theory of Knots.* New York: W. H. Freeman, pp. 64–67, 1994.
Rolfsen, D. *Knots and Links.* Wilmington, DE: Publish or Perish Press, p. 115, 1976.

Bridge of Königsberg

KÖNIGSBERG BRIDGE PROBLEM

Brightness

The area of the SHADOW of a body on a plane, also called the "outer quermass."

See also INNER QUERMASS, SHADOW

References

Blaschke, W. *Kreis und Kugel.* New York: Chelsea, p. 140, 1949.
Bonnesen, T. "Om Minkowski's uligheder fur konvexer legemer." *Mat. Tidsskr.* **B**, 80, 1926.
Bonnesen, R. and Fenchel, W. *Theorie der Konvexer Körper.* New York: Chelsea, p. 140, 1971.
Chakerian, G. D. "Is a Body Spherical If All Its Projections Have the Same I.Q.?" *Amer. Math. Monthly* **77**, 989–992, 1970.
Croft, H. T.; Falconer, K. J.; and Guy, R. K. *Unsolved Problems in Geometry.* New York: Springer-Verlag, p. 23, 1991.
Firey, W. J. "Blaschke Sum of Convex Bodies and Mixed Bodies." In *Proceedings of the Colloquium on Convexity* (Ed. W. Fenchel). Copenhagen, Denmark: Københavns Univ. Math. Inst., pp. 94–101, 1967.

Brill-Noether Theorem

If the total group of the canonical series is divided into two parts, the difference between the number of points in each part and the double of the dimension of the complete series to which it belongs is the same.

Bring Quintic Form

References

Coolidge, J. L. *A Treatise on Algebraic Plane Curves*. New York: Dover, p. 263, 1959.

Bring Quintic Form

A TSCHIRNHAUSEN TRANSFORMATION can be used to take a general QUINTIC EQUATION to the form

$$x^5 - x - a = 0,$$

where a may be COMPLEX.

See also BRING-JERRARD QUINTIC FORM, QUINTIC EQUATION

References

Bring, E. S. *Quart. J. Math.* **6**, 1864.
Grunert, J. A. "VIII. Miscellen von dem Herausgeber." *Archiv der Math. Phys.* **41**, 105–112, 1864.
Harley, R. "A Contribution to the History of the Problem of the Reduction of the General Equation of the Fifth Degree to a Trinomial Form." *Quart. J. Math.* **6**, 38–47, 1864.
Ruppert, W. M. "On the Bring Normal Form of a Quintic in Characteristic 5." *Arch. Math.* **58**, 44–46, 1992.
Tortolini, B. "Rivista bibliografica sopra a transformazione del Sig. Jerrard per l'equazioni di quinto grado." *Annali di Mat. pura appl.* **6**, 33–42, 1864.

Bring-Jerrard Quintic Form

A TSCHIRNHAUSEN TRANSFORMATION can be used to algebraically transform a general QUINTIC EQUATION to the form

$$z^5 + c_1 z + c_0 = 0. \qquad (1)$$

In practice, the general quintic is first reduced to the PRINCIPAL QUINTIC FORM

$$y^5 + b_2 y^2 + b_1 y + b_0 = 0 \qquad (2)$$

before the transformation is done. Then, we require that the sum of the third POWERS of the ROOTS vanishes, so $s_3(y_j) = 0$. We assume that the ROOTS z_i of the Bring-Jerrard quintic are related to the ROOTS y_i of the PRINCIPAL QUINTIC FORM by

$$z_i = \alpha y_i^4 + \beta y_i^3 + \gamma y_i^2 + \delta y_i + \epsilon. \qquad (3)$$

In a similar manner to the PRINCIPAL QUINTIC FORM transformation, we can express the COEFFICIENTS c_j in terms of the b_j.

See also BRING QUINTIC FORM, PRINCIPAL QUINTIC FORM, QUINTIC EQUATION

References

Grunert, J. A. "VIII. Miscellen von dem Herausgeber." *Archiv der Math. Phys.* **41**, 105–112, 1864.
Klein, F. "Über die Transformation der elliptischen Funktionen und die Auflösung der Gleichungen fünften Grades." *Math. Ann.* **14**, 1878/79.
Tortolini, B. "Rivista bibliografica sopra a transformazione del Sig. Jerrard per l'equazioni di quinto grado." *Annali di Mat. pura appl.* **6**, 33–42, 1864.

Brioschi Formula

For a curve with METRIC

$$ds^2 = E\, du^2 + F\, du\, dv + G\, dv^2, \qquad (1)$$

where E, F, and G is the first FUNDAMENTAL FORM, the GAUSSIAN CURVATURE is

$$K = \frac{M_1 + M_2}{(EG - F^2)^2}, \qquad (2)$$

where

$$M_1 \equiv \begin{vmatrix} -\tfrac{1}{2}E_{uv} + F_{uv} - \tfrac{1}{2}G_{uu} & \tfrac{1}{2}E_u & F_u - \tfrac{1}{2}E_v \\ F_v - \tfrac{1}{2}G_u & E & F \\ \tfrac{1}{2}G_v & F & G \end{vmatrix} \qquad (3)$$

$$M_2 \equiv \begin{vmatrix} 0 & \tfrac{1}{2}E_v & \tfrac{1}{2}G_u \\ \tfrac{1}{2}E_v & E & F \\ \tfrac{1}{2}G_u & F & G \end{vmatrix}, \qquad (4)$$

which can also be written

$$K = -\frac{1}{\sqrt{EG}}\left[\frac{\partial}{\partial u}\left(\frac{1}{\sqrt{E}}\frac{\partial\sqrt{G}}{\partial u}\right) + \frac{\partial}{\partial v}\left(\frac{1}{\sqrt{G}}\frac{\partial\sqrt{E}}{\partial v}\right)\right] \qquad (5)$$

$$= -\frac{1}{\sqrt{EG}}\left[\frac{\partial}{\partial u}\left(\frac{G_u}{\sqrt{EG}}\right) + \frac{\partial}{\partial v}\left(\frac{E_v}{\sqrt{EG}}\right)\right]. \qquad (6)$$

See also FUNDAMENTAL FORMS, GAUSSIAN CURVATURE

References

Gray, A. *Modern Differential Geometry of Curves and Surfaces with Mathematica, 2nd ed.* Boca Raton, FL: CRC Press, pp. 504–507, 1997.

Briot-Bouquet Equation

An ORDINARY DIFFERENTIAL EQUATION OF THE FORM

$$x^m y' = f(x, y),$$

where m is a POSITIVE INTEGER, f is ANALYTIC at $x = y = 0$, $f(0, 0) = 0$, and $f'_y(0, 0) \neq 0$.

Zwillinger (1997, p. 120), citing Ince (1956, p. 295), define the Briot-Bouquet equation as

$$xy' - \lambda y = a_{10}x + a_{20}x^2 + a_{11}yx + a_{02}y^2 + \cdots$$

References

Briot and Bouquet. "Propriétés des fonctions définie par des équations différentielles." *J. l'Ecole Polytechnique*, Cah. 36.
Hazewinkel, M. (Managing Ed.). *Encyclopaedia of Mathematics: An Updated and Annotated Translation of the Soviet "Mathematical Encyclopaedia."* Dordrecht, Netherlands: Reidel, pp. 481–482, 1988.

Ince, E. L. *Ordinary Differential Equations.* New York: Dover, 1956.

Zwillinger, D. *Handbook of Differential Equations, 3rd ed.* Boston, MA: Academic Press, p. 120, 1997.

Brjuno Number

Let p_n/q_n be the sequence of CONVERGENTS of the CONTINUED FRACTION of a number α. Then a Brjuno number is an IRRATIONAL NUMBER such that

$$\sum_{n=0}^{\infty} \frac{\ln q_{n+1}}{q_n} < \infty$$

(Marmi *et al.* 1999). Brjuno numbers arise in the study of one-dimensional analytic small divisors problems, and Brjuno (1971, 1972) proved that all "germs" with linear part $\lambda = e^{2\pi i \alpha}$ are linearizable if α is a Brjuno number. Yoccoz (1995) proved that this condition is also NECESSARY.

References

Brjuno, A. D. "Analytical Form of Differential Equations." *Trans. Moscow Math. Soc.* **25**, 131–288, 1971.

Brjuno, A. D. "Analytical Form of Differential Equations. II." *Trans. Moscow Math. Soc.* **26**, 199–239, 1972.

Marmi, S.; Moussa, P.; and Yoccoz, J.-C. "The Brjuno Functions and Their Regularity Properties." *Comm. Math. Phys.* **186**, 265–293, 1997.

Marmi, S.; Moussa, P.; and Yoccoz, J.-C. "Complex Brjuno Functions." Preprint. 5 Dec 1999. http://rene.ma.utexas.edu/mp_arc/index-99.html.

Moussa, P. and Marmi, S. "Diophantine Conditions and Real of Complex Brjuno Functions." Preprint. 5 Dec 1999. http://rene.ma.utexas.edu/mp_arc/index-99.html.

Siegel, C. L. "Iteration of Analytic Functions." *Ann. Math.* **43**, 807–812, 1942.

Yoccoz, J.-C. "Théorème de Siegel, nombres de Bruno et polynômes quadratiques." *Astérique* **231**, 3–88, 1995.

Broadcasting

GOSSIPING

Brocard Angle

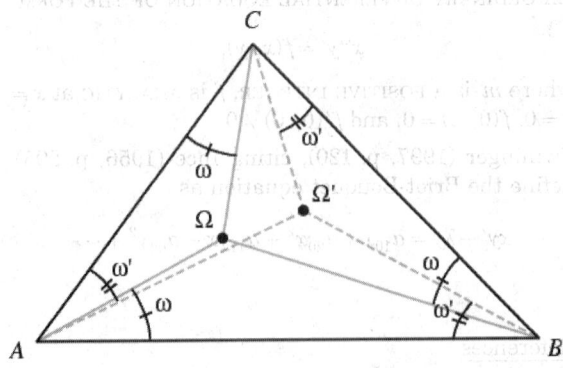

Define the first BROCARD POINT as the interior point Ω of a TRIANGLE for which the ANGLES $\angle \Omega AB$, $\angle \Omega BC$, and $\angle \Omega CA$ are equal to an angle ω. Similarly, define the second BROCARD POINT as the interior point Ω' for which the ANGLES $\angle \Omega' AC$, $\angle \Omega' CB$, and $\angle \Omega' BA$ are

equal to an angle ω'. Then $\omega = \omega'$, and this angle is called the Brocard angle.

The Brocard angle ω of a TRIANGLE $\Delta A_1 A_2 A_3$ is given by the formulas

$$\cot \omega = \cot A_1 + \cot A_2 + \cot A_3 \tag{1}$$

$$= \left(\frac{a_1^2 + a_2^2 + a_3^2}{4\Delta} \right) \tag{2}$$

$$= \frac{1 + \cos \alpha_1 \cos \alpha_2 \cos \alpha_3}{\sin \alpha_1 \sin \alpha_2 \sin \alpha_3} \tag{3}$$

$$= \frac{\sin^2 \alpha_1 + \sin^2 \alpha_2 + \sin^2 \alpha_3}{2 \sin \alpha_1 \sin \alpha_2 \sin \alpha_3} \tag{4}$$

$$= \frac{a_1 \sin \alpha_1 + a_2 \sin \alpha_2 + a_3 \sin \alpha_3}{a_1 \cos \alpha_1 + a_2 \cos \alpha_2 + a_3 \cos \alpha_3} \tag{5}$$

$$\csc^2 \omega = \csc^2 \alpha_1 + \csc^2 \alpha_2 + \csc^2 \alpha_3 \tag{6}$$

$$\sin \omega = \frac{2\Delta}{\sqrt{a_1^2 a_2^2 + a_2^2 a_3^2 + a_3^2 a_1^2}} \tag{7}$$

where Δ is the TRIANGLE AREA, A, B, and C are ANGLES, and a, b, and c are side lengths (Johnson 1929), where (6) is due to Neuberg (Tucker 1883).

If an ANGLE α of a TRIANGLE is given, the maximum possible Brocard angle is given by

$$\cot \omega = \tfrac{3}{2} \tan(\tfrac{1}{2}\alpha) + \tfrac{1}{2} \cos(\tfrac{1}{2}\alpha) \tag{8}$$

(Johnson 1929, p. 289). If ω is specified, that the largest possible value α_{max} and minimum possible value α_{min} of any possible triangle having Brocard angle ω are given by

$$\cot(\tfrac{1}{2}\alpha_{max}) = \cot \omega - \sqrt{\cot^2 - 3} \tag{9}$$

$$\cot(\tfrac{1}{2}\alpha_{min}) = \cot \omega + \sqrt{\cot^2 - 3}, \tag{10}$$

where the square rooted quantity is the radius of the corresponding NEUBERG CIRCLE (Johnson 1929, p. 288). The maximum possible Brocard angle for any triangle is $30°$ (Honsberger 1995, pp. 102–103). Let a TRIANGLE have ANGLES A, B, and C. Then

$$\sin A \sin B \sin C \leq kABC, \tag{11}$$

where

$$k = \left(\frac{3\sqrt{3}}{2\pi} \right)^3 \tag{12}$$

(Le Lionnais 1983). This can be used to prove that

$$8\omega^3 < ABC \tag{13}$$

(Abi-Khuzam 1974).

See also BROCARD CIRCLE, BROCARD LINE, EQUI-

BROCARD CENTER, FERMAT POINTS, NEUBERG CIRCLE

References

Abi-Khuzam, F. "Proof of Yff's Conjecture on the Brocard Angle of a Triangle." *Elem. Math.* **29**, 141–142, 1974.

Casey, J. *A Sequel to the First Six Books of the Elements of Euclid, Containing an Easy Introduction to Modern Geometry with Numerous Examples,* 5th ed., rev. enl. Dublin: Hodges, Figgis, & Co., p. 172, 1888.

Coolidge, J. L. *A Treatise on the Geometry of the Circle and Sphere.* New York: Chelsea, p. 61, 1971.

Emmerich, A. *Die Brocardschen Gebilde und ihre Beziehungen zu den verwandten merkwürdigen Punkten und Kreisen des Dreiecks.* Berlin: Georg Reimer, 1891.

Honsberger, R. "The Brocard Angle." §10.2 in *Episodes in Nineteenth and Twentieth Century Euclidean Geometry.* Washington, DC: Math. Assoc. Amer., pp. 101–106, 1995.

Johnson, R. A. *Modern Geometry: An Elementary Treatise on the Geometry of the Triangle and the Circle.* Boston, MA: Houghton Mifflin, pp. 263–286 and 289–294, 1929.

Lachlan, R. *An Elementary Treatise on Modern Pure Geometry.* London: Macmillian, pp. 65–66, 1893.

Le Lionnais, F. *Les nombres remarquables.* Paris: Hermann, p. 28, 1983.

Tucker, R. "The 'Triplicate Ratio' Circle." *Quart. J. Pure Appl. Math.* **19**, 342–348, 1883.

Brocard Axis

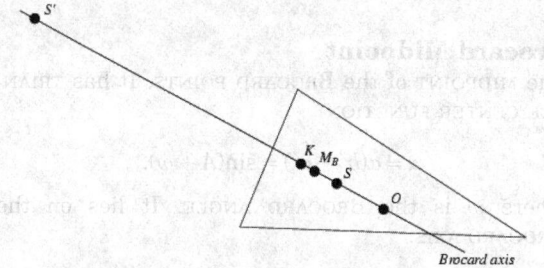

Brocard axis

The LINE KO passing through the SYMMEDIAN POINT K and CIRCUMCENTER O of a TRIANGLE. The distance $\overline{OK}$ is called the BROCARD DIAMETER. The Brocard axis is PERPENDICULAR to the LEMOINE AXIS and is the ISOGONAL CONJUGATE of KIEPERT'S HYPERBOLA. It has equations

$$\sin(B-C)\alpha + \sin(C-A)\beta + \sin(A-B)\gamma = 0$$

$$bc(b^2 - c^2)\alpha + ca(c^2 - a^2)\beta + ab(a^2 - b^2)\gamma = 0.$$

The SYMMEDIAN POINT K, CIRCUMCENTER O, ISODY-NAMIC POINTS S and S', and BROCARD MIDPOINT M_B all lie along the Brocard axis.

Note that the Brocard axis is *not* equivalent to the BROCARD LINE.

See also BROCARD CIRCLE, BROCARD DIAMETER, BROCARD LINE

Brocard Circle

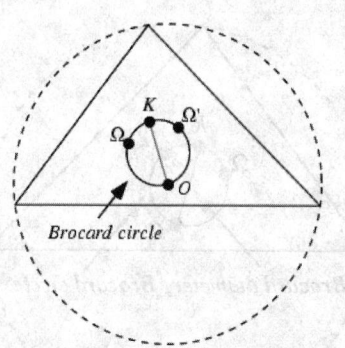

Brocard circle

The CIRCLE passing through the first and second BROCARD POINTS Ω and Ω', the LEMOINE POINT K, and the CIRCUMCENTER O of a given TRIANGLE. The BROCARD POINTS Ω and Ω' are symmetrical about the LINE $\check{K}O$, which is called the BROCARD LINE. The LINE SEGMENT $\overline{KO}$ is called the BROCARD DIAMETER, and it has length

$$\overline{OK} = \frac{\overline{O\Omega}}{\cos\omega} = \frac{R\sqrt{1 - 4\sin^2\omega}}{\cos\omega},$$

where R is the CIRCUMRADIUS and ω is the BROCARD ANGLE. The distance between either of the BROCARD POINTS and the SYMMEDIAN POINT is

$$\overline{\Omega K} = \overline{\Omega' K} = \overline{\Omega O} \tan\omega.$$

The Brocard circle and LEMOINE CIRCLE are concentric.

See also BROCARD ANGLE, BROCARD DIAMETER, BROCARD POINTS

References

Brocard, M. H. "Etude d'un nouveau cercle du plan du triangle." *Assoc. Français pour l'Academie des Sciences-Congrés d'Alger*, 1881.

Coolidge, J. L. *A Treatise on the Geometry of the Circle and Sphere.* New York: Chelsea, p. 75, 1971.

Emmerich, A. *Die Brocardschen Gebilde und ihre Beziehungen zu den verwandten merkwürdigen Punkten und Kreisen des Dreiecks.* Berlin: Georg Reimer, 1891.

Honsberger, R. "The Brocard Circle." §10.3 in *Episodes in Nineteenth and Twentieth Century Euclidean Geometry.* Washington, DC: Math. Assoc. Amer., pp. 106–110, 1995.

Johnson, R. A. *Modern Geometry: An Elementary Treatise on the Geometry of the Triangle and the Circle.* Boston, MA: Houghton Mifflin, p. 272, 1929.

Lachlan, R. "The Brocard Circle." §134–135 in *An Elementary Treatise on Modern Pure Geometry.* London: Macmillian, pp. 78–81, 1893.

Brocard Diameter

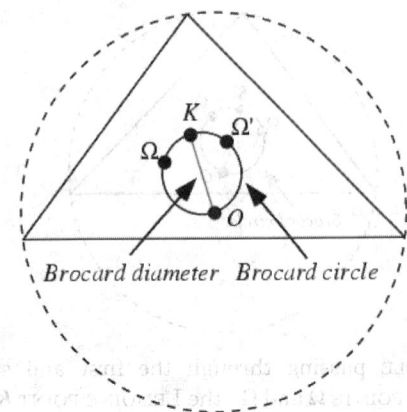

Brocard diameter   Brocard circle

The LINE SEGMENT $\overline{KO}$ joining the SYMMEDIAN POINT K and CIRCUMCENTER O of a given TRIANGLE. It is the DIAMETER of the TRIANGLE'S BROCARD CIRCLE, and lies along the BROCARD AXIS. The Brocard diameter has length

$$\overline{OK} = \frac{\overline{O\Omega}}{\cos \omega} = \frac{R\sqrt{1 - 4\sin^2 \omega}}{\cos \omega},$$

where Ω is the first BROCARD POINT, R is the CIRCUMRADIUS, and ω is the BROCARD ANGLE.

See also BROCARD AXIS, BROCARD CIRCLE, BROCARD LINE, BROCARD POINTS

Brocard Line

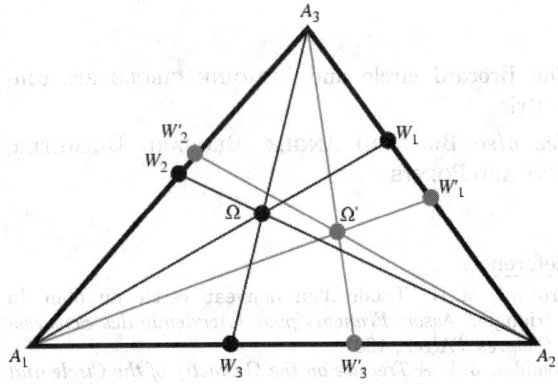

A LINE from any of the VERTICES A_i of a TRIANGLE to the first Ω or second Ω' BROCARD POINT. Let the ANGLE at a VERTEX A_i also be denoted A_i, and denote the intersections of $A_1\Omega$ and $A_1\Omega'$ with A_2A_3 as W_1 and W_2. Then the ANGLES involving these points are

$$\angle A_1\Omega W_3 = A_1 \qquad (1)$$

$$\angle W_3\Omega A_2 = A_3 \qquad (2)$$

$$\angle A_2\Omega W_1 = A_2 \qquad (3)$$

Distances involving the points W_i and W_i' are given by

$$\overline{A_2\Omega} = \frac{a_3}{\sin A_2} \sin \omega \qquad (4)$$

$$\frac{\overline{A_2\Omega}}{\overline{A_3\Omega}} = \frac{a_3^2}{a_1 a_2} = \frac{\sin(A_3 - \omega)}{\sin \omega} \qquad (5)$$

$$\frac{\overline{W_3 A_1}}{\overline{W_3 A_2}} = \frac{a_2 \sin \omega}{a_1 \sin(A_3 - \omega)} = \left(\frac{a_2}{a_3}\right)^2, \qquad (6)$$

where ω is the BROCARD ANGLE (Johnson 1929, pp. 267–268).

The Brocard line, MEDIAN M, and SYMMEDIAN POINT K are concurrent, with $A_1\Omega_1$, A_2K, and A_3M meeting at a point P. Similarly, $A_1\Omega'$, A_2M, and A_3K meet at a point which is the ISOGONAL CONJUGATE point of P (Johnson 1929, pp. 268–269).

See also BROCARD AXIS, BROCARD DIAMETER, BROCARD POINTS, ISOGONAL CONJUGATE, SYMMEDIAN POINT, MEDIAN (TRIANGLE)

References

Emmerich, A. *Die Brocardschen Gebilde und ihre Beziehungen zu den verwandten merkwürdigen Punkten und Kreisen des Dreiecks.* Berlin: Georg Reimer, 1891.

Johnson, R. A. *Modern Geometry: An Elementary Treatise on the Geometry of the Triangle and the Circle.* Boston, MA: Houghton Mifflin, pp. 263–286, 1929.

Brocard Midpoint

The MIDPOINT of the BROCARD POINTS. It has TRIANGLE CENTER FUNCTION

$$\alpha = a(b^2 + c^2) = \sin(A + \omega),$$

where ω is the BROCARD ANGLE. It lies on the BROCARD AXIS.

References

Kimberling, C. "Central Points and Central Lines in the Plane of a Triangle." *Math. Mag.* **67**, 163–187, 1994.

Brocard Points

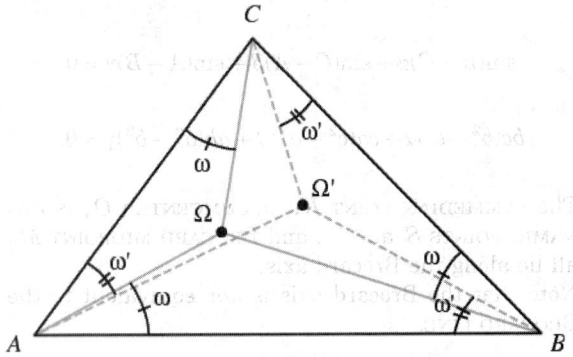

The first Brocard point is the interior point Ω (or τ_1 or

Z_1) of a TRIANGLE for which the ANGLES $\angle \Omega AB$, $\angle \Omega BC$, and $\angle \Omega CA$ are equal to an angle ω. The second Brocard point is the interior point Ω' (or τ_2 or Z_2) for which the ANGLES $\angle \Omega' AC$, $\angle \Omega' CB$, and $\angle \Omega' BA$ are equal to an angle ω'. The two angles $\omega = \omega'$ are equal, and this angle is called the BROCARD ANGLE,

$$\omega = \angle \Omega AB = \angle \Omega BC = \angle \Omega CA$$

$$= \angle \Omega' AC = \angle \Omega' CB = \angle \Omega' BA.$$

The first two Brocard points are ISOGONAL CONJUGATES (Johnson 1929, p. 266). They were described by French army officer Henri Brocard in 1875, although they had previously been investigated by Jacobi and, in 1816, Crelle (Wells 1991; Honsberger 1995, p. 98). The satisfy $\Omega O = \Omega' O$ and $\angle \Omega O \Omega' = 2\omega$, where O is the CIRCUMCENTER and ω is the BROCARD ANGLE (Honsberger 1995, p. 106).

If three dogs start at the vertices of a triangle and chase either their left or right neighbor at a constant speed, that the three will meet at either Ω or Ω' (Wells 1991).

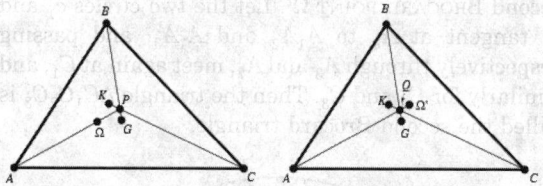

One BROCARD LINE, MEDIAN, and SYMMEDIAN (out of the three of each) are CONCURRENT, with $A\Omega$, CK, and BG meeting at a point, where G is the CENTROID and K is the SYMMEDIAN POINT. Similarly, $A\Omega'$, BG, and CK meet at a point which is the ISOGONAL CONJUGATE of the first (Johnson 1929, pp. 268–269; Honsberger 1995, pp. 121–124).

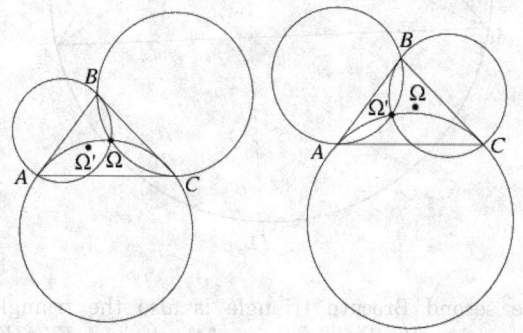

Let C_{BC} be the CIRCLE which passes through the vertices B and C and is TANGENT to the line AC at C, and similarly for C_{AB} and C_{BC}. Then the CIRCLES C_{AB}, C_{BC}, and C_{AC} intersect in the first Brocard point Ω. Similarly, let C'_{BC} be the CIRCLE which passes through the vertices B and C and is TANGENT to the line AB at B, and similarly for C'_{AB} and C'_{AC}. Then the CIRCLES C'_{AB}, C'_{BC}, and C'_{AC} intersect in the second Brocard points Ω' (Johnson 1929, pp. 264–265; Honsberger

1995, pp. 99–100).

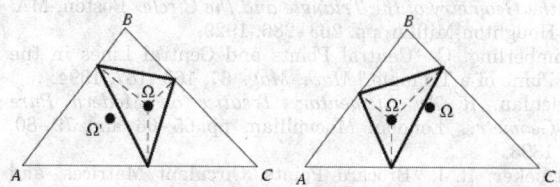

The PEDAL TRIANGLES of Ω and Ω' are congruent, and SIMILAR to the TRIANGLE ΔABC (Johnson 1929, p. 269). Lengths involving the Brocard points include

$$\overline{O\Omega} = \overline{O\Omega'} = R\sqrt{1 - 4\sin^2 \omega} \tag{1}$$

$$\overline{\Omega\Omega'} = 2R\sin \omega \sqrt{1 - 4\sin^2 \omega}. \tag{2}$$

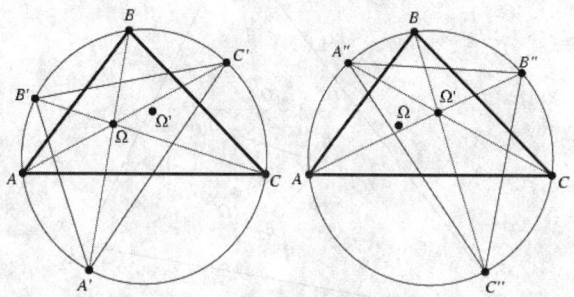

Extend the segments $A\Omega$, $B\Omega$, and $C\Omega$ to the CIRCUMCIRCLE of ΔABC to form $\Delta C'A'B'$, and the segments $A\Omega'$, $B\Omega'$, and $C\Omega'$ to form $\Delta B''C''A''$. Then $\Delta A'B'C'$ and $\Delta A''B''C''$ are congruent to ΔABC (Honsberger 1995, pp. 104–106).

Brocard's third point is related to a given TRIANGLE by the TRIANGLE CENTER FUNCTION

$$\alpha = a^{-3} \tag{3}$$

(Casey 1893, Kimberling 1994). The third Brocard point Ω'' (or τ_3 or Z_3) is COLLINEAR with the SPIEKER CENTER and the ISOTOMIC CONJUGATE POINT of its TRIANGLE'S INCENTER.

See also BROCARD ANGLE, BROCARD MIDPOINT, EQUI-BROCARD CENTER, YFF POINTS

References

Casey, J. *A Treatise on the Analytical Geometry of the Point, Line, Circle, and Conic Sections, Containing an Account of Its Most Recent Extensions, with Numerous Examples,* 2nd ed., rev. enl. Dublin: Hodges, Figgis, & Co., p. 66, 1893.

Coolidge, J. L. "The Brocard Figures." §1.5 in *A Treatise on the Geometry of the Circle and Sphere.* New York: Chelsea, pp. 60–84, 1971.

Emmerich, A. *Die Brocardschen Gebilde und ihre Beziehungen zu den verwandten merkwürdigen Punkten und Kreisen des Dreiecks.* Berlin: Georg Reimer, 1891.

Honsberger, R. "The Brocard Points." Ch. 10 in *Episodes in Nineteenth and Twentieth Century Euclidean Geometry.* Washington, DC: Math. Assoc. Amer., pp. 99–124, 1995.

Johnson, R. A. *Modern Geometry: An Elementary Treatise on the Geometry of the Triangle and the Circle.* Boston, MA: Houghton Mifflin, pp. 263–286, 1929.

Kimberling, C. "Central Points and Central Lines in the Plane of a Triangle." *Math. Mag.* **67**, 163–187, 1994.

Lachlan, R. *An Elementary Treatise on Modern Pure Geometry.* London: Macmillian, pp. 65–66 and 79–80, 1893.

Stroeker, R. J. "Brocard Points, Circulant Matrices, and Descartes' Folium." *Math. Mag.* **61**, 172–187, 1988.

Wells, D. *The Penguin Dictionary of Curious and Interesting Geometry.* London: Penguin, pp. 21–22, 1991.

Brocard Triangles

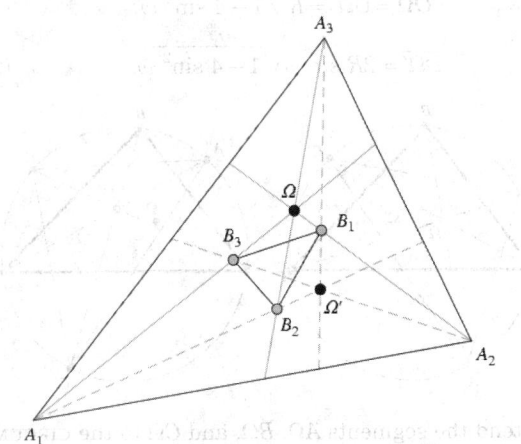

Given TRIANGLE $\Delta A_1 A_2 A_3$, let the point of intersection of $A_2\Omega$ and $A_3\Omega'$ be B_1, where Ω and Ω' are the BROCARD POINTS, and similarly define B_2 and B_3. Then $B_1 B_2 B_3$ is called the first Brocard triangle, and is INVERSELY SIMILAR to $A_1 A_2 A_3$ (Honsberger 1995, p. 112). It is inscribed in the BROCARD CIRCLE drawn with OK as the DIAMETER.

The triangles $B_1 A_2 A_3$, $B_2 A_3 A_1$, and $B_3 A_1 A_2$ are ISOSCELES TRIANGLES with base angles ω, where ω is the BROCARD ANGLE. The sum of the areas of the ISOSCELES TRIANGLES is Δ, the AREA of TRIANGLE $A_1 A_2 A_3$. The first Brocard triangle is in perspective with the given TRIANGLE, with $A_1 B_1$, $A_2 B_2$, and $A_3 B_3$ CONCURRENT. The CENTROID of the first brocard triangle is the CENTROID G of the original triangle (Honsberger 1995, pp. 112–116).

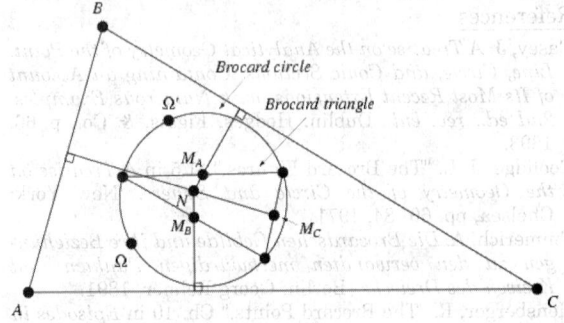

Let perpendiculars be drawn from the midpoints M_A, M_B, and M_C of each side of the first Brocard triangle

to the opposite sides of the triangle ΔABC. Then the extensions of these lines CONCUR in the NINE-POINT CENTER (Honsberger 1995, pp. 116–118).

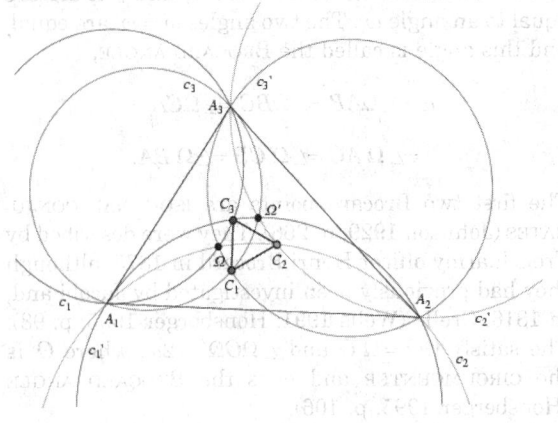

Let c_1, c_2, and c_3 be the CIRCLES through the vertices A_2 and A_3, A_n and A_3, and A_n and A_2, respectively, which intersect in the first BROCARD POINT Ω. Similarly, define c_1', c_2', and c_3' with respect to the second BROCARD POINT Ω'. Let the two circles c_1 and c_1' tangent at A_n to $A_1 A_2$ and $A_1 A_3$, and passing respectively through A_3 and A_2, meet again at C_1, and similarly for C_2 and C_3. Then the triangle $\Delta C_1 C_2 C_3$ is called the second Brocard triangle.

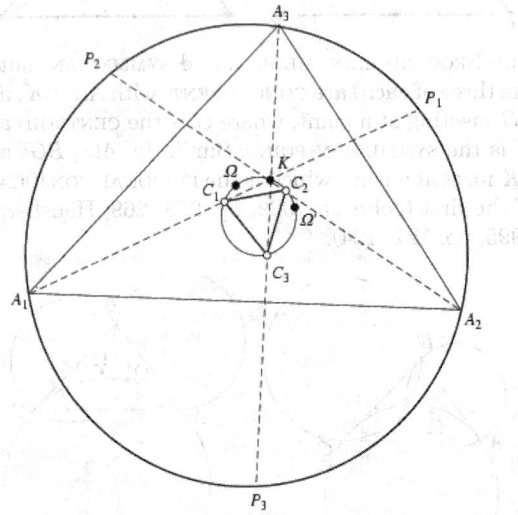

The second Brocard triangle is also the triangle obtained as the intersections of the lines $A_1 K$, $A_2 K$, and $A_3 K$ with the BROCARD CIRCLE, where K is the SYMMEDIAN POINT. Let P_1, P_2, and P_3 be the intersections of the lines $A_1 K$, $A_2 K$, and $A_3 K$ with the CIRCUMCIRCLE of $\Delta A_1 A_2 A_3$. Then C_1, C_2, and C_3 are the midpoints of $A_1 P_1$, $A_2 P_2$, and $A_3 P_3$, respectively (Lachlan 1893).

The two Brocard triangles are in PERSPECTIVE at M.

See also BROCARD CIRCLE, CIRCLE-CIRCLE INTERSEC-

TION, MCCAY CIRCLE, NINE-POINT CENTER, STEINER POINTS, TARRY POINT

References

Coolidge, J. L. *A Treatise on the Geometry of the Circle and Sphere.* New York: Chelsea, p. 75, 1971.

Emmerich, A. *Die Brocardschen Gebilde und ihre Beziehungen zu den verwandten merkwürdigen Punkten und Kreisen des Dreiecks.* Berlin: Georg Reimer, 1891.

Honsberger, R. "The Brocard Triangles." §10.4 in *Episodes in Nineteenth and Twentieth Century Euclidean Geometry.* Washington, DC: Math. Assoc. Amer., pp. 110–118, 1995.

Johnson, R. A. *Modern Geometry: An Elementary Treatise on the Geometry of the Triangle and the Circle.* Boston, MA: Houghton Mifflin, pp. 277–281, 1929.

Lachlan, R. *An Elementary Treatise on Modern Pure Geometry.* London: Macmillian, pp. 78–81, 1893.

Brocard's Conjecture

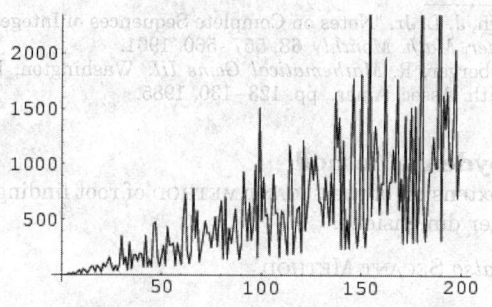

$$\pi(p_{n+1}^2) - \pi(p_n^2) \geq 4$$

for $n \geq 2$ where $\pi(n)$ is the PRIME COUNTING FUNCTION and p_n is the nth PRIME. For $n = 1, 2, \ldots$, the first few values are 2, 5, 6, 15, 9, 22, 11, 27, 47, 16, ... (Sloane's A050216).

See also ANDRICA'S CONJECTURE

References

Sloane, N. J. A. Sequences A050216 in "An On-Line Version of the Encyclopedia of Integer Sequences." http://www.research.att.com/~njas/sequences/eisonline.html.

Brocard's Problem

Find the values of n for which $n! + 1$ is a SQUARE NUMBER m^2, where $n!$ is the FACTORIAL (Brocard 1876, 1885). Pairs of numbers (m, n) are called BROWN NUMBERS. The only known solutions are $n = 4, 5$, and 7, and there are no other solutions with $n \leq 10^7$ (Wells 1986, p. 70; D. Wilson). It is virtually certain that there are no more solutions (Guy 1994). In fact, Dabrowski (1996) has shown that $n! + A = k^2$ has only finitely many solutions for general A, although this result requires assumption of a weak form of the ABC CONJECTURE if A is SQUARE).

Wilson has also computed the least k such that $n! + k^2$ is square starting at $n = 4$, giving 1, 1, 3, 1, 9, 27, 15, 18, 288, 288, 420, 464, 1856, ... (Sloane's A038202).

See also BROWN NUMBERS, FACTORIAL, SQUARE NUMBER

References

Brocard, H. Question 166. *Nouv. Corres. Math.* **2**, 287, 1876.

Brocard, H. Question 1532. *Nouv. Ann. Math.* **4**, 391, 1885.

Dabrowski, A. "On the Diophantine Equation $x! + A = y^2$." *Nieuw Arch. Wisk.* **14**, 321–324, 1996.

Erdos, P. and Obláth, R. "Über diophantische Gleichungen der Form $n! = x^p \pm y^p$ und $n! \pm m! = x^p$" *Acta Szeged* **8**, 241–255, 1937.

Gupta. *Math. Student* **3**, 71, 1935.

Guy, R. K. "Equations Involving Factorial n." §D25 in *Unsolved Problems in Number Theory, 2nd ed.* New York: Springer-Verlag, pp. 193–194, 1994.

Hardy, G. H.; Aiyar, S.; Venkatesvara, P.; and Wilson, B. M. (Eds.). *Collected Papers of Srinivasa Ramanujan.* Cambridge, England: The University Press, p. 327, 1927.

Overholt, M. "The Diophantine Equation $n! + 1 = m^2$." *Bull. London Math. Soc.* **25**, 104, 1993.

Sloane, N. J. A. Sequences A038202 in "An On-Line Version of the Encyclopedia of Integer Sequences." http://www.research.att.com/~njas/sequences/eisonline.html.

Wells, D. *The Penguin Dictionary of Curious and Interesting Numbers.* Middlesex, England: Penguin Books, p. 70, 1986.

Bromwich Integral

The inverse of the LAPLACE TRANSFORM, given by

$$F(t) = \frac{1}{2\pi i} \int_{\gamma - i\infty}^{\gamma + i\infty} e^{\pi i} f(s) \, ds,$$

where γ is a vertical CONTOUR in the COMPLEX PLANE chosen so that all singularities of $f(s)$ are to the left of it.

See also LAPLACE TRANSFORM

References

Arfken, G. "Inverse Laplace Transformation." §15.12 in *Mathematical Methods for Physicists, 3rd ed.* Orlando, FL: Academic Press, pp. 853–861, 1985.

Brooks' Theorem

The CHROMATIC NUMBER of a graph is at most the maximum VERTEX DEGREE Δ, unless the graph is COMPLETE or an odd cycle.

See also CHROMATIC NUMBER

References

Brooks, R. L. "On Coloring the Nodes of a Network." *Proc. Cambridge Philos. Soc.* **37**, 194–197, 1941.

Lovász, L. "Three Short Proofs in Graph Theory." *J. Combin. Th. Ser. B* **19**, 111–113, 1975.

Skiena, S. *Implementing Discrete Mathematics: Combinatorics and Graph Theory with Mathematica.* Reading, MA: Addison-Wesley, p. 215, 1990.

Brothers

A PAIR of consecutive numbers.

See also PAIR, SMITH BROTHERS, TWINS

Brouwer Fixed Point Theorem

Any continuous FUNCTION $G : \mathbb{B} \to \mathbb{B}^n$ has a FIXED POINT, where

$$\mathbb{B}^n = \{x \in \mathbb{R}^n : x_1^2 + \cdots + x_n^2 \le 1\}$$

is the unit n-BALL.

See also BALL, FIXED POINT THEOREM

References

Kannai, Y. "An Elementary Proof of the No Retraction Theorem." *Amer. Math. Monthly* **88**, 264–268, 1981.
Milnor, J. W. *Topology from the Differentiable Viewpoint.* Princeton, NJ: Princeton University Press, p. 14, 1965.
Munkres, J. R. *Elements of Algebraic Topology.* Perseus Press, p. 117, 1993.
Samelson, H. "On the Brouwer Fixed Point Theorem." *Portugal. Math.* **22**, 189–191, 1963.

Browkin's Theorem

For every POSITIVE INTEGER n, there exists a SQUARE in the plane with exactly n LATTICE POINTS in its interior. This was extended by Schinzel and Kulikowski to *all* plane figures of a given shape. The generalization of the SQUARE in 2-D to the CUBE in 3-D was also proved by Browkin.

See also CUBE, SCHINZEL'S THEOREM, SQUARE

References

Honsberger, R. *Mathematical Gems I.* Washington, DC: Math. Assoc. Amer., pp. 121–125, 1973.

Brown Function

For a FRACTAL PROCESS with values $y(t - \Delta t)$ and $y(t + \Delta t)$, the correlation between these two values is given by the Brown function

$$r = 2^{2H-1} - 1,$$

also known as the BACHELIER FUNCTION, LÉVY FUNCTION, or WIENER FUNCTION.

Brown Numbers

Brown numbers are PAIRS (m, n) of INTEGERS satisfying the condition of BROCARD'S PROBLEM, i.e., such that

$$n! + 1 = m^2$$

where $n!$ is the FACTORIAL and m^2 is a SQUARE NUMBER. Only three such PAIRS of numbers are known: (5, 4), (11, 5), (71, 7), and Erdos conjectured that these are the only three such PAIRS.

See also BROCARD'S PROBLEM, FACTORIAL, SQUARE NUMBER, WILSON PRIME

References

Guy, R. K. *Unsolved Problems in Number Theory, 2nd ed.* New York: Springer-Verlag, p. 193, 1994.

Pickover, C. A. *Keys to Infinity.* New York: Wiley, p. 170, 1995.

Brown's Criterion

A SEQUENCE $\{v_i\}$ of nondecreasing POSITIVE INTEGERS is COMPLETE IFF

1. $v_1 = 1$.
2. For all $k = 2, 3, \ldots$,

$$s_{k-1} = v_1 + v_2 + \cdots + v_{k-1} \ge v_k - 1.$$

A corollary states that a SEQUENCE for which $v_1 = 1$ and $v_{k+1} \le 2v_k$ is COMPLETE (Honsberger 1985).

See also COMPLETE SEQUENCE

References

Brown, J. L. Jr. "Notes on Complete Sequences of Integers." *Amer. Math. Monthly* **68**, 557–560, 1961.
Honsberger, R. *Mathematical Gems III.* Washington, DC: Math. Assoc. Amer., pp. 123–130, 1985.

Broyden's Method

An extension of the SECANT METHOD of root finding to higher dimensions.

See also SECANT METHOD

References

Broyden, C. G. "A Class of Methods for Solving Nonlinear Simultaneous Equations." *Math. Comput.* **19**, 577–593, 1965.
Press, W. H.; Flannery, B. P.; Teukolsky, S. A.; and Vetterling, W. T. *Numerical Recipes in FORTRAN: The Art of Scientific Computing, 2nd ed.* Cambridge, England: Cambridge University Press, pp. 382–385, 1992.

Bruck-Ryser Theorem

BRUCK-RYSER-CHOWLA THEOREM

Bruck-Ryser-Chowla Theorem

If $n \equiv 1, 2 \pmod 4$, and the SQUAREFREE part of n is divisible by a PRIME $p \equiv 3 \pmod 4$, then no DIFFERENCE SET of ORDER n exists. Equivalently, if a PROJECTIVE PLANE of order n exists, and $n = 1$ or 2 (mod 4), then n is the sum of two SQUARES.

Dinitz and Stinson (1992) give the theorem in the following form. If a symmetric (v, k, λ)-BLOCK DESIGN exists, then

1. If v is EVEN, then $k - \lambda$ is a SQUARE NUMBER,
2. If v is ODD, then the DIOPHANTINE EQUATION

$$x^2 = (k - \lambda)y^2 + (-1)^{(v-1)/2}\lambda z^2$$

has a solution in integers, not all of which are 0.

See also BLOCK DESIGN, DIFFERENCE SET, FISHER'S BLOCK DESIGN INEQUALITY

Bruhat Order

References

Dinitz, J. H. and Stinson, D. R. "A Brief Introduction to Design Theory." Ch. 1 in *Contemporary Design Theory: A Collection of Surveys* (Ed. J. H. Dinitz and D. R. Stinson). New York: Wiley, pp. 1–12, 1992.
Gordon, D. M. "The Prime Power Conjecture is True for $n < 2,000,000$." *Electronic J. Combinatorics* **1**, R6 1–7, 1994. http://www.combinatorics.org/Volume_1/volume 1.html#R6.
Ryser, H. J. *Combinatorial Mathematics*. Buffalo, NY: Math. Assoc. Amer., 1963.

Bruhat Order

References

Björner, A. and Wachs, M. "Bruhat Order of Coxeter Groups and Shellability." *Adv. Math.* **43**, 87–100, 1982.
Stanley, R. P. Exercise 3.75(a) in *Enumerative Combinatorics, Vol. 1.* Cambridge, England: Cambridge University Press, 1999.
Stanley, R. P. Exercises 6.47 and 7.103d in *Enumerative Combinatorics, Vol. 2.* Cambridge, England: Cambridge University Press, pp. 243 and 485, 1999.

Brun's Constant

The number obtained by adding the reciprocals of the odd TWIN PRIMES,

$$B \equiv (\tfrac{1}{3} + \tfrac{1}{5}) + (\tfrac{1}{5} + \tfrac{1}{7}) + (\tfrac{1}{11} + \tfrac{1}{13}) + (\tfrac{1}{17} + \tfrac{1}{19}) + \cdots, \quad (1)$$

By BRUN'S THEOREM, the constant converges to a definite number as $p \to \infty$. Any finite sum underestimates B. Shanks and Wrench (1974) used all the TWIN PRIMES among the first 2 million numbers. Brent (1976) calculated all TWIN PRIMES up to 100 billion and obtained (Ribenboim 1989, p. 146)

$$B \approx 1.90216054, \quad (2)$$

assuming the truth of the first HARDY-LITTLEWOOD CONJECTURE. Using TWIN PRIMES up to 10^{14}, Nicely (1996) obtained

$$B \approx 1.9021605778 \pm 2.1 \times 10^{-9} \quad (3)$$

(Cipra 1995, 1996), in the process discovering a bug in Intel's® Pentium™ microprocessor. Using TWIN PRIMES up to 2.55^{15}, Nicely subsequently obtained the result

$$B \approx 1.9021605820 \pm 2.4 \times 10^{-9}. \quad (4)$$

(Note that the value given by Le Lionnais 1983 is incorrect)

Segal (1930) proved that Brun-type sums B_d of $1/p$ over consecutive primes separated by d are finite (Halberstam and Richert 1983, p. 92). Wolf suggests that B_d is roughly equal to $4/d$ which, in the $d = 2$ case of twin primes, gives $B_2 \approx 2$ instead of 1.902.... Wolf also considers the "COUSIN PRIMES" Brun's constant B_4.

See also COUSIN PRIMES, TWIN PRIMES, TWIN PRIME CONJECTURE, TWIN PRIMES CONSTANT

References

Ball, W. W. R. and Coxeter, H. S. M. *Mathematical Recreations and Essays, 13th ed.* New York: Dover, p. 64, 1987.
Brent, R. P. "Tables Concerning Irregularities in the Distribution of Primes and Twin Primes Up to 10^{11}." *Math. Comput.* **30**, 379, 1976.
Brun, V. "La serie $1/5 + 1/7 + \cdots$ est convergente ou finie." *Bull. Sci. Math.* **43**, 124–128, 1919.
Cipra, B. "How Number Theory Got the Best of the Pentium Chip." *Science* **267**, 175, 1995.
Cipra, B. "Divide and Conquer." *What's Happening in the Mathematical Sciences, 1995–1996, Vol. 3.* Providence, RI: Amer. Math. Soc., pp. 38–47, 1996.
Finch, S. "Favorite Mathematical Constants." http://www.mathsoft.com/asolve/constant/brun/brun.html.
Halberstam, H. and Richert, H.-E. *Sieve Methods.* New York: Academic Press, 1974.
Le Lionnais, F. *Les nombres remarquables.* Paris: Hermann, p. 41, 1983.
Nagell, T. *Introduction to Number Theory.* New York: Wiley, p. 67, 1951.
Nicely, T. "Enumeration to 10^{14} of the Twin Primes and Brun's Constant." *Virginia J. Sci.* **46**, 195–204, 1996.
Ribenboim, P. *The Book of Prime Number Records, 2nd ed.* New York: Springer-Verlag, 1989.
Segal, B. "Généralisation du théorème de Brun." *Dokl. Akad. Nauk SSSR*, 501–507, 1930.
Shanks, D. and Wrench, J. W. "Brun's Constant." *Math. Comput.* **28**, 293–299, 1974.
Wells, D. *The Penguin Dictionary of Curious and Interesting Numbers.* Middlesex, England: Penguin Books, pp. 40–41, 1986.

Brun's Sieve

See also SIEVE

References

Blecksmith, R.; Erdos, P.; and Selfridge, J. L. "Cluster Primes." *Amer. Math. Monthly* **106**, 43–48, 1999.
Halberstam, H. and Richert, H.-E. *Sieve Methods.* New York: Academic Press, 1974.

Brun's Sum

BRUN'S CONSTANT

Brun's Theorem

The series producing BRUN'S CONSTANT CONVERGES even if there are an infinite number of TWIN PRIMES. Proved in 1919 by V. Brun.

Brunnian Link

A Brunnian link is a set of n linked loops such that each proper sublink is trivial, so that the removal of any component leaves a set of trivial unlinked UNKNOTS. The BORROMEAN RINGS are the simplest example and have $n = 3$.

See also BORROMEAN RINGS

References

Rolfsen, D. *Knots and Links.* Wilmington, DE: Publish or Perish Press, 1976.

Brunn-Minkowski Inequality

The nth root of the CONTENT of the set sum of two sets in Euclidean n-space is greater than or equal to the sum of the nth roots of the CONTENTS of the individual sets.

See also TOMOGRAPHY

References

Cover, T. M. "The Entropy Power Inequality and the Brunn-Minkowski Inequality" §5.10 in *Open Problems in Communications and Computation.* (Ed. T. M. Cover and B. Gopinath). New York: Springer-Verlag, p. 172, 1987.

Schneider, R. *Convex Bodies: The Brunn-Minkowski Theory.* Cambridge, England: Cambridge University Press, 1993.

Brusselator Equations

The system of ordinary differential equations

$$u' = A + u^2 v - (B+1)u \tag{1}$$

$$v' = Bu - u^2 v \tag{2}$$

(Hairer *et al.* 1987, p. 112; Zwillinger 1997, p. 136). The so-called full Brusselator equations are given by

$$u' = 1 + u^2 v - (w+1)u \tag{3}$$

$$v' = uw - u^2 v \tag{4}$$

$$w' = -uw + \alpha \tag{5}$$

(Hairer *et al.* 1987, p. 114; Zwillinger 1997, p. 136).

References

Hairer, E.; Nørsett, S. P.; and Wanner, G. *Solving Ordinary Differential Equations I.* New York: Springer-Verlag, 1987.

Zwillinger, D. *Handbook of Differential Equations, 3rd ed.* Boston, MA: Academic Press, p. 136, 1997.

Brute Force Factorization

DIRECT SEARCH FACTORIZATION

B-Spline

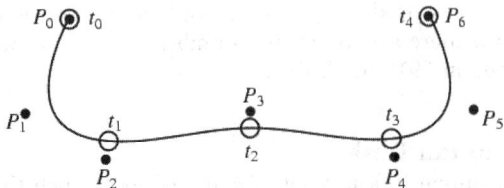

A generalization of the BÉZIER CURVE. Let a vector known as the KNOT VECTOR be defined

$$\mathbf{T} = \{t_0, t_1, \ldots, t_m\}, \tag{1}$$

where $\mathbf{T}$ is a nondecreasing SEQUENCE with $t_i \in [0, 1]$,

and define control points $\mathbf{P}_0, \ldots, \mathbf{P}_n$. Define the degree as

$$p \equiv m - n - 1. \tag{2}$$

The "knots" $t_{p+1}, \ldots, t_{m-p-1}$ are called INTERNAL KNOTS.

Define the basis functions as

$$N_{i,0}(t) = \begin{cases} 1 & \text{if } t_i \le t < t_{i+1} \text{ and } t_i < t_{i+1} \\ 0 & \text{otherwise} \end{cases} \tag{3}$$

$$N_{i,p}(t) = \frac{t - t_i}{t_{i+p} - t_i} N_{i,p-1}(t)$$
$$+ \frac{t_{i+p+1} - t}{t_{i+p+1} - t_{i+1}} N_{i+1,p-1}(t). \tag{4}$$

Then the curve defined by

$$\mathbf{C}(t) = \sum_{i=0}^{n} \mathbf{P}_i N_{i,p}(t) \tag{5}$$

is a B-spline. Specific types include the nonperiodic B-spline (first $p + 1$ knots equal 0 and last $p + 1$ equal to 1) and uniform B-spline (INTERNAL KNOTS are equally spaced). A B-spline with no INTERNAL KNOTS is a BÉZIER CURVE.

A curve is $p - k$ times differentiable at a point where k duplicate knot values occur. The knot values determine the extent of the control of the control points.

See also BÉZIER CURVE, NURBS CURVE

B-Tree

B-trees were introduced by Bayer (1972) and McCreight. They are a special m-ary balanced tree used in databases because their structure allows records to be inserted, deleted, and retrieved with guaranteed worst-case performance. An n-node B-tree has height $\mathcal{O}(1g2)$, where LG is the LOGARITHM to base 2. The Apple ® Macintosh ® (Apple Computer, Cupertino, CA) HFS filing system uses B-trees to store disk directories (Benedict 1995). A B-tree satisfies the following properties:

1. The ROOT is either a LEAF (TREE) or has at least two CHILDREN.
2. Each node (except the ROOT and LEAVES) has between $\lceil m/2 \rceil$ and m CHILDREN, where $\lceil x \rceil$ is the CEILING FUNCTION.
3. Each path from the ROOT to a LEAF (TREE) has the same length.

Every 2– TREE is a B-tree of order 3. The number of B-trees of order-3 with $n = 1, 2, \ldots$ leaves are 1, 1, 1, 1, 2, 2, 3, 4, 5, 8, 14, 23, 32, 43, 63, ... (Ruskey, Sloane's A014535). The number of order-4 B-trees with $n = 1$,

2, ... leaves are 1, 1, 1, 2, 2, 4, 5, 9, 15, 28, 45, ... (Sloane's A037026).

See also RED-BLACK TREE, TREE

References

Aho, A. V.; Hopcroft, J. E.; and Ullmann, J. D. *Data Structures and Algorithms*. Reading, MA: Addison-Wesley, pp. 369–374, 1987.

Bayer, R. and McCreight, E. "Organization and Maintenance of Large Ordered Indexes." *Acta Informatica* **1**, 173–189, 1972.

Benedict, B. *Using Norton Utilities for the Macintosh.* Indianapolis, IN: Que, pp. B-17-B-33, 1995.

Beyer, R. "Symmetric Binary *B*-Trees: Data Structures and Maintenance Algorithms." *Acta Informat.* **1**, 290–306, 1972.

Knuth, D. E. "B-Trees." *The Art of Computer Programming, Vol. 3: Sorting and Searching, 2nd ed.* Reading, MA: Addison-Wesley, pp. 482–485 and 490–491, 1998.

Ruskey, F. "Information on B-Trees." http://www.theory.cs-c.uvic.ca/~cos/inf/tree/BTrees.html.

Skiena, S. S. *The Algorithm Design Manual.* New York: Springer-Verlag, p. 178, 1997.

Sloane, N. J. A. Sequences A014535 and A037026 in "An On-Line Version of the Encyclopedia of Integer Sequences." http://www.research.att.com/~njas/sequences/eisonline.html.

Bubble

A bubble is a minimal-energy surface of the type that is formed by soap film. The simplest bubble is a single SPHERE, illustrated above (courtesy of J. M. Sullivan). More complicated forms occur when multiple bubbles are joined together. The simplest example is the DOUBLE BUBBLE, and beautiful configurations can form when three or more bubbles are conjoined (Sullivan).

An outstanding problem involving bubbles is the determination of the arrangements of bubbles with the smallest SURFACE AREA which enclose and separate n given volumes in space.

See also DOUBLE BUBBLE, PLATEAU'S LAWS, PLATEAU'S PROBLEM, SPHERE

References

Morgan, F. "Mathematicians, Including Undergraduates, Look at Soap Bubbles." *Amer. Math. Monthly* **101**, 343–351, 1994.

Pappas, T. "Mathematics & Soap Bubbles." *The Joy of Mathematics.* San Carlos, CA: Wide World Publ./Tetra, p. 219, 1989.

Steinhaus, H. *Mathematical Snapshots, 3rd ed.* New York: Dover, pp. 214–216, 1999.

Sullivan, J. M. "Generating and Rendering Four-Dimensional Polytopes." *Mathematica J.* **1**, 76–85, Winter 1991.

Sullivan, J. M. "Polytope Bubble Images." http://www.math.uiuc.edu/~jms/Images/polyt.html.

Williams, R. *The Geometrical Foundation of Natural Structure: A Source Book of Design.* New York: Dover, pp. 44–45, 1979.

Buchberger's Algorithm

The algorithm for the construction of a GRÖBNER BASIS from an arbitrary ideal basis.

See also GRÖBNER BASIS

References

Becker, T. and Weispfenning, V. *Gröbner Bases: A Computational Approach to Commutative Algebra.* New York: Springer-Verlag, pp. 213–214, 1993.

Buchberger, B. "Theoretical Basis for the Reduction of Polynomials to Canonical Forms." *SIGSAM Bull.* **39**, 19–24, Aug. 1976.

Cox, D.; Little, J.; and O'Shea, D. *Ideals, Varieties, and Algorithms: An Introduction to Algebraic Geometry and Commutative Algebra, 2nd ed.* New York: Springer-Verlag, 1996.

Buchowski Paradox

A paradox arising in the use of comparative adjectives. Suppose you have exactly two brothers, both of whom are older than you are. Then the following apparently false statement is actually true: "My younger brother is older than I am."

Buckminster Fuller Dome

GEODESIC DOME

Buffon's Needle Problem

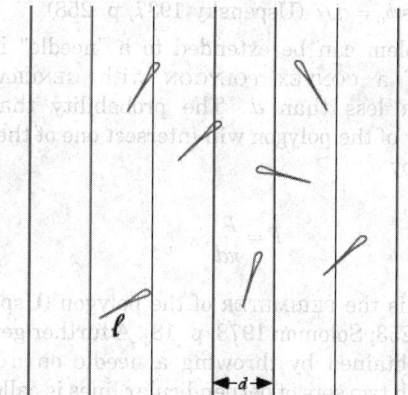

Find the probability $P(\ell, d)$ that a needle of length ℓ will land on a line, given a floor with equally spaced PARALLEL LINES a distance d apart. The problem was first posed by the French naturalist Buffon in 1733, and reproduced with the solution by Buffon in 1777. For $\ell \leq d$,

$$P(\ell,\, d) = \int_0^{2\pi} \frac{\ell|\cos\theta|}{d}\, \frac{d\theta}{2\pi} = \frac{\ell}{2\pi d}\, 4 \int_0^{\pi/2} \cos\theta\, d\theta$$

$$= \frac{2\ell}{\pi d}[\sin\theta]_0^{\pi/2} = \frac{2\ell}{\pi d}. \tag{1}$$

For $\ell \geq d$, the solution is slightly more complicated,

$$P(\ell, d) = \frac{1}{\pi d}\left\{d\left[\pi - 2\sin^{-1}\left(\frac{d}{\ell}\right)\right] + 2\ell\left(1 - \sqrt{1 - \frac{d^2}{\ell^2}}\right)\right\} \quad (2)$$

(Uspensky 1937, p. 252; Kunkel).

Several attempts have been made to experimentally determine π by needle-tossing. For a discussion of the relevant statistics and a critical analysis of one of the more accurate (and least believable) needle-tossings, see Badger (1994). Uspensky (1937, pp. 112–113) discusses experiments conducted with 2520, 3204, and 5000 trials. An asymptotically unbiased estimator for π from the needle-tossing experiment is

$$\hat{\pi} = \frac{2rn}{N}, \quad (3)$$

where $r = \ell/d$, n is the number of throws, and N is the number of line crossings, which has asymptotic variance

$$\mathrm{var}(\hat{\pi}) = \frac{\pi^2}{n}(\tfrac{1}{2}\pi - 1) \approx \frac{5.63}{n} \quad (4)$$

(Mantel 1953; Solomon 1978, p. 7).

If the needle is longer than the distance between two lines, then the probability that it intersects at least one line is

$$P(\ell) = \frac{2\ell}{\pi d}(1 - \sin\phi_0) + \frac{2\phi_0}{\pi}, \quad (5)$$

where $\cos\phi_0 = d/\ell$ (Uspensky 1937, p. 258).

The problem can be extended to a "needle" in the shape of a CONVEX POLYGON with GENERALIZED DIAMETER less than d. The probability that the boundary of the polygon will intersect one of the lines is given by

$$P = \frac{p}{\pi d}, \quad (6)$$

where p is the PERIMETER of the polygon (Uspensky 1937, p. 253; Solomon 1978, p. 18). A further generalization obtained by throwing a needle on a board ruled with two sets of perpendicular lines is called the BUFFON-LAPLACE NEEDLE PROBLEM.

See also BUFFON-LAPLACE NEEDLE PROBLEM

References

Badger, L. "Lazzarini's Lucky Approximation of π." *Math. Mag.* **67**, 83–91, 1994.
Buffon, G. *Proc. Paris Acad. Sci.* 1733.
Buffon, G. *Essai d'arithmétique morale.* Supplément a l'Histoire Naturelle, Vol. 4, 1777.
Diaconis, P. "Buffon's Needle Problem with a Long Needle." *J. Appl. Prob.* **13**, 614–618, 1976.

Dörrie, H. "Buffon's Needle Problem." §18 in *100 Great Problems of Elementary Mathematics: Their History and Solutions.* New York: Dover, pp. 73–77, 1965.
Edelman, A. and Kostlan, E. "How Many Zeros of a Random Polynomial are Real?" *Bull. Amer. Math. Soc.* **32**, 1–37, 1995.
Hoffman, P. *The Man Who Loved Only Numbers: The Story of Paul Erdos and the Search for Mathematical Truth.* New York: Hyperion, p. 209, 1998.
Isaac, R. *The Pleasures of Probability.* New York: Springer-Verlag, 1995.
Klain, Daniel A. and Rota, G.-C. *Introduction to Geometric Probability.* New York: Cambridge University Press, 1997.
Kraitchik, M. "The Needle Problem." §6.14 in *Mathematical Recreations.* New York: W. W. Norton, p. 132, 1942.
Kunkel, P. "Buffon's Needle." http://www.nas.com/~kunkel/buffon/buffon.htm.
Mantel, L. "An Extension of the Buffon Needle Problem." *Ann. Math. Stat.* **24**, 674–677, 1953.
Perlman, M. and Wichura, M. "On Sharpening Buffon's Needle." *Amer. Stat.* **20**, 157–163, 1975.
Santaló, L. A. *Integral Geometry and Geometric Probability.* Reading, MA: Addison-Wesley, 1976.
Schuster, E. F. "Buffon's Needle Experiment." *Amer. Math. Monthly* **81**, 26–29, 1974.
Solomon, H. "Buffon Needle Problem, Extensions, and Estimation of π." Ch. 1 in *Geometric Probability.* Philadelphia, PA: SIAM, pp. 1–24, 1978.
Stoka, M. "Problems of Buffon Type for Convex Test Bodies." *Conf. Semin. Mat. Univ. Bari,* No. 268, 1–17, 1969.
Uspensky, J. V. "Buffon's Needle Problem," "Extension of Buffon's Problem," and "Second Solution of Buffon's Problem." §12.14–12.16 in *Introduction to Mathematical Probability.* New York: McGraw-Hill, pp. 112–115, 251–255, and 258, 1937.
Wegert, E. and Trefethen, L. N. "From the Buffon Needle Problem to the Kreiss Matrix Theorem." *Amer. Math. Monthly* **101**, 132–139, 1994.
Wells, D. *The Penguin Dictionary of Curious and Interesting Numbers.* Middlesex, England: Penguin Books, p. 53, 1986.

Buffon-Laplace Needle Problem

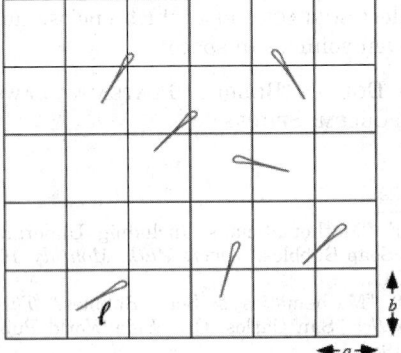

Find the probability $P(\ell, a, b)$ that a needle of length ℓ will land on a line, given a floor with a grid of equally spaced PARALLEL LINES distances a and b apart, with $\ell < a$, b. The position of the needle can be specified with points (x, y) and its orientation with coordinate ϕ. By symmetry, we can consider a single rectangle of the grid, so $0 < x < a$ and $0 < y < b$. In

addition, since opposite orientations are equivalent, we can take $-\pi/2 < \phi < \pi/2$.
The probability is given by

$$P(\ell; a, b) = 1 - \frac{\displaystyle\int_{-\pi/2}^{\pi/2} F(\phi)\, d\phi}{\pi a b}, \tag{1}$$

where

$$F(\phi) = ab - b\ell \cos\phi - \ell a|\sin\phi| + \tfrac{1}{2}\ell^2 |\sin(2\phi)| \tag{2}$$

(Uspensky 1937, p. 256; Solomon 1978, p. 4), giving

$$P(\ell; a, b) = \frac{2\ell(a + b) - \ell^2}{\pi a b}. \tag{3}$$

If the plane is instead tiled with congruent triangles with sides a, b, c, and a needle with length ℓ less than the shortest altitude is thrown, the probability that the needle is contained entirely within one of the triangles is given by

$$P = 1 + \frac{(Aa^2 + Bb^2 + Cc^2)\ell^2}{2\pi K^2}$$

$$\qquad - \frac{(4a + 4b + 4c - 3\ell)\ell}{2\pi K}, \tag{4}$$

where A, B, and C are the angles opposite a, b, and c, respectively, and K is the AREA of the triangle. For equilateral triangles, this simplifies to

$$P = 1 + \frac{2}{3}\left(\frac{\ell}{a}\right)^2 - \frac{\ell\sqrt{3}}{\pi a}\left(4 - \frac{\ell}{a}\right) \tag{5}$$

(Uspensky 1937, p. 258).

See also BUFFON'S NEEDLE PROBLEM

References

Schuster, E. F. "Buffon's Needle Experiment." *Amer. Math. Monthly* **81**, 26–29, 1974.
Solomon, H. *Geometric Probability*. Philadelphia, PA: SIAM, pp. 3–6, 1978.
Uspensky, J. V. "Laplace's Problem." §12.17 in *Introduction to Mathematical Probability*. New York: McGraw-Hill, pp. 255–257, 1937.

Bug Problem

MICE PROBLEM

Building

A highly structured geometric object used to study GROUPS which act upon them.

See also COXETER GROUP, GROUP

References

Garrett, P. *Buildings and Classical Groups*. Boca Raton, FL: Chapman and Hall, 1997.

Bulirsch-Stoer Algorithm

An algorithm which finds RATIONAL FUNCTION extrapolations OF THE FORM

$$R_{i(i+1)\dots(i+m)} = \frac{P_\mu(x)}{P_\nu(x)} = \frac{p_0 + p_1 x + \dots + p_\mu x^\mu}{q_0 + q_1 x + \dots + q_\nu x^\nu}$$

and can be used in the solution of ORDINARY DIFFERENTIAL EQUATIONS.

References

Bulirsch, R. and Stoer, J. §2.2 in *Introduction to Numerical Analysis*. New York: Springer-Verlag, 1991.
Press, W. H.; Flannery, B. P.; Teukolsky, S. A.; and Vetterling, W. T. "Richardson Extrapolation and the Bulirsch-Stoer Method." §16.4 in *Numerical Recipes in FORTRAN: The Art of Scientific Computing, 2nd ed.* Cambridge, England: Cambridge University Press, pp. 718–725, 1992.

Bullet Nose

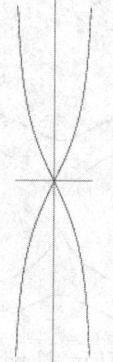

A plane curve with implicit equation

$$\frac{a^2}{x^2} - \frac{b^2}{y^2} = 1. \tag{1}$$

In parametric form,

$$x = a \cos t \tag{2}$$

$$y = b \cot t. \tag{3}$$

The CURVATURE is

$$\kappa = \frac{3ab \cot t \csc t}{(b^2 \csc^4 t + a^2 \sin^2 t)^{3/2}} \tag{4}$$

and the TANGENTIAL ANGLE is

$$\phi = \tan^{-1}\left(\frac{b \csc^3 t}{a}\right). \tag{5}$$

References

Lawrence, J. D. *A Catalog of Special Plane Curves*. New York: Dover, pp. 127–129, 1972.

Bullseye Illusion

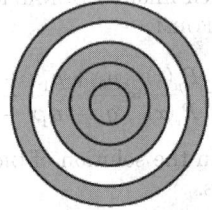

Although the inner shaded region has the same area as the outer shaded ANNULUS, it appears to be larger. Since the rings are equally spaced,

$$A_{\text{inner}} = \pi \cdot 3^2 = 9\pi$$
$$A_{\text{outer}} = \pi \cdot 5^2 - \pi \cdot 4^2 = 9\pi.$$

See also ILLUSION

References
Wells, D. *The Penguin Dictionary of Curious and Interesting Geometry.* London: Penguin, p. 87, 1991.

Bump Function

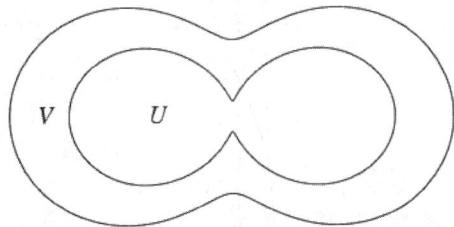

Given any OPEN SET U in $\mathbb{R}^n$ with COMPACT CLOSURE $K = \bar{U}$, there exists SMOOTH FUNCTIONS which are identically one on U and vanish arbitrarily close to U. One way to express this more precisely is that for any OPEN SET V containing K, there is a SMOOTH FUNCTION f such that

    1. $f(x) = 1$ for all $x \in U$ and
    2. $f(x) = 0$ for all $x \notin V$.

A function f that satisfies (1) and (2) is called a bump function. If $\int f = 1$ then by rescaling f, namely $f_k(x) = k^n f(kx)$, one gets a sequence of smooth functions which converges to the DELTA FUNCTION.

See also COMPACT SUPPORT, CONVOLUTION, DIRAC DISTRIBUTION, SMOOTH FUNCTION

Bumping Algorithm

Given a PERMUTATION $\{p_1, p_2, \ldots, p_n\}$ of $\{1, \ldots, n\}$, the bumping algorithm constructs a standard YOUNG TABLEAU by inserting the p_i one by one into an already constructed YOUNG TABLEAU. To apply the bumping algorithm, start with $\{\{p_1\}\}$, which is a YOUNG TABLEAU. If p_1 through p_k have already been inserted, then in order to insert p_{k+1}, start with the first line of the already constructed YOUNG TABLEAU and search for the first element of this line which is greater than p_{k+1}. If there is no such element, append p_{k+1} to the first line and stop. If there is such an element (say, p_p), exchange p_p for p_{k+1}, search the second line using p_p, and so on.

See also TABLEAU CLASS, YOUNG TABLEAU

References
Skiena, S. *Implementing Discrete Mathematics: Combinatorics and Graph Theory with Mathematica.* Reading, MA: Addison-Wesley, 1990.

Bundle

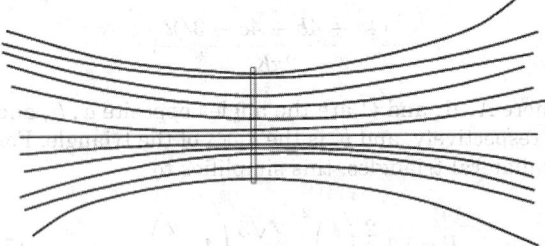

The term "bundle" is an abbreviated form of the full term FIBER BUNDLE. Depending on context, it may mean one of the special cases of FIBER BUNDLES, such as a VECTOR BUNDLE or a PRINCIPAL BUNDLE. Bundles are so named because they contain a collection of objects which, like a bundle of hay, are held together in a special way. All of the fibers line up–or at least they line up to nearby fibers.

LOCALLY, a bundle looks like a PRODUCT MANIFOLD in a TRIVIALIZATION. The graph of a function f sits inside the product as $(x, f(x))$. The SECTIONS of a bundle generalize functions in this way. It is necessary to use bundles when the range of a function only makes sense locally, as in the case of a VECTOR FIELD on the SPHERE.

Bundles are a special kind of SHEAF.

See also FIBER BUNDLE, JET BUNDLE, LINE BUNDLE, PRINCIPAL BUNDLE, SHEAF, TANGENT BUNDLE, VECTOR BUNDLE

Bundle Map

A bundle map is a map between bundles along with a compatible map between the BASE MANIFOLDS. Suppose $p : X \to M$ and $q : Y \to N$ are two BUNDLES, then

$$F : X \to Y$$

is a bundle map if there is a map $f : M \to N$ such that $q(F(x)) = f(p(x))$ for all $x \in X$. In particular, the FIBER of X over a point $m \in M$, gets mapped to the fiber of Y over $f(m) \in N$.

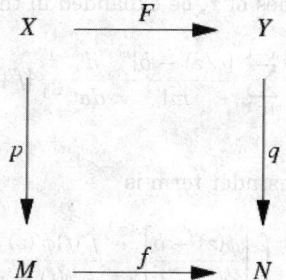

In the language of CATEGORY THEORY, the above diagram COMMUTES. To be more precise, the induced map between fibers has to be a map in the category of the fiber. For instance, in a bundle map between VECTOR BUNDLES the fiber over $m \in M$ is mapped to the fiber over $f(m) \in M$ by a LINEAR TRANSFORMATION.

For example, when $f : M \to N$ is a SMOOTH MAP between SMOOTH MANIFOLDS then $df : TM \to TN$ is the differential, which is a bundle map between the tangent bundles. Over any point in $m \in M$, the tangent vectors at m get mapped to tangent vectors at $f(m) \in N$ by the JACOBIAN.

See also BUNDLE, COMMUTATIVE DIAGRAM, FIBER (BUNDLE), JACOBIAN, PRINCIPAL BUNDLE, VECTOR BUNDLE

Buniakowsky Inequality

SCHWARZ'S INEQUALITY

Burali-Forti Paradox

In the theory of transfinite ORDINAL NUMBERS,

1. Every WELL ORDERED SET has a unique ORDINAL NUMBER,
2. Every segment of ordinals (i.e., any set of ordinals arranged in natural order which contains all the predecessors of each of its elements) has an ORDINAL NUMBER which is greater than any ordinal in the segment, and
3. The set B of all ordinals in natural order is well ordered.

Then by statements (3) and (1), B has an ordinal β. Since β is in B, it follows that $\beta < \beta$ by (2), which is a contradiction.

See also ORDINAL NUMBER

References
Copi, I. M. "The Burali-Forti Paradox." *Philos. Sci.* **25**, 281–286, 1958.
Curry, H. B. *Foundations of Mathematical Logic.* New York: Dover, p. 5, 1977.
Erickson, G. W. and Fossa, J. A. *Dictionary of Paradox.* Lanham, MD: University Press of America, pp. 29–30, 1998.
Mirimanoff, D. "Les antinomies de Russell et de Burali-Forti et le problème fondamental de la théorie des ensembles." *Enseign. math.* **19**, 37–52, 1917.

Burau Representation

Gives a MATRIX representation b_i of a BRAID GROUP in terms of $(n-1) \times (n-1)$ MATRICES. A $-t$ always appears in the (i, i) position.

$$b_1 = \begin{bmatrix} -t & 0 & 0 & \dots & 0 \\ -1 & 1 & 0 & \dots & 0 \\ 0 & 0 & 1 & \dots & 0 \\ \vdots & \vdots & \vdots & \ddots & \vdots \\ 0 & 0 & 1 & \dots & 1 \end{bmatrix} \tag{1}$$

$$b_i = \begin{bmatrix} 1 & \dots & 0 & 0 & \dots & 0 \\ \vdots & \ddots & \vdots & \vdots & & \vdots \\ 0 & \dots & -t & 0 & \dots & 0 \\ 0 & \dots & -t & 0 & \dots & 0 \\ 0 & \dots & -1 & 1 & \dots & 0 \\ 0 & \ddots & 0 & 0 & & \vdots \\ 0 & \dots & 0 & 0 & \dots & 1 \end{bmatrix} \tag{2}$$

$$b_{n-1} = \begin{bmatrix} 1 & 0 & \dots & 0 & 0 \\ 0 & 1 & \dots & 0 & 0 \\ \vdots & \vdots & \ddots & \vdots & \vdots \\ 0 & 0 & \dots & 0 & -t \\ 0 & 0 & \dots & 0 & -t \end{bmatrix} \tag{3}$$

Let Ψ be the MATRIX PRODUCT of BRAID WORDS, then

$$\frac{\det(1 - \Psi)}{1 + t + \dots + t^{n-1}} = \Delta_L, \tag{4}$$

where Δ_L is the ALEXANDER POLYNOMIAL and det is the DETERMINANT.

References
Burau, W. "Über Zopfgruppen und gleichsinnig verdrilte Verkettungen." *Abh. Math. Sem. Hanischen Univ.* **11**, 171–178, 1936.
Jones, V. "Hecke Algebra Representation of Braid Groups and Link Polynomials." *Ann. Math.* **126**, 335–388, 1987.

Burgers' Equation

The PARTIAL DIFFERENTIAL EQUATION

$$u_t + u u_x = v u_{xx}$$

(Benton and Platzman 1972; Zwillinger 1995, p. 417; Zwillinger 1997, p. 130). The so-called nonplanar Burgers equation is given by

$$u_t + u u_x + \frac{Ju}{2t} = \tfrac{1}{2} \delta u_x x$$

(Sachdev and Nair 1987; Zwillinger 1997, p. 131).

References

Benton, E. R. and Platzman, G. W. "A Table of Solutions of the of the One-Dimensional Burgers Equation." *Quart. Appl. Math.*, 195–212, Jul. 1972.

Sachdev, P. L. and Nair, K. R. C. "Generalized Burgers Equations and Euler-Painlevé Transcendents. II." *J. Math. Phys.* **28**, 997–1004, 1987.

Zwillinger, D. (Ed.). *CRC Standard Mathematical Tables and Formulae.* Boca Raton, FL: CRC Press, p. 417, 1995.

Zwillinger, D. *Handbook of Differential Equations, 3rd ed.* Boston, MA: Academic Press, p. 130, 1997.

Burkhardt Quartic

The VARIETY which is an invariant of degree four and is given by the equation

$$y_0^4 - y_0(y_1^3 + y_2^3 + y_3^3 + y_4^3) + 3y_1y_2y_3y_4 = 0.$$

See also QUARTIC EQUATION

References

Burkhardt, H. "Untersuchungen aus dem Gebiet der hyper-elliptischen Modulfunctionen. II." *Math. Ann.* **38**, 161–224, 1890.

Burkhardt, H. "Untersuchungen aus dem Gebiet der hyper-elliptischen Modulfunctionen. III." *Math. Ann.* **40**, 313–343, 1892.

Hunt, B. "The Burkhardt Quartic." Ch. 5 in *The Geometry of Some Special Arithmetic Quotients.* New York: Springer-Verlag, pp. 168–221, 1996.

Bürmann's Theorem

Bürmann's theorem deals with the expansion of functions in powers of another function. Let $\phi(z)$ be a function of z which is analytic in a closed region S, of which a is an interior point, and let $\phi(a) = b$. Suppose also that $\phi'(a) \neq 0$. Then TAYLOR'S THEOREM furnishes the expansion

$$\phi(z) - b = \phi'(a)(z - a) + \frac{\phi''(a)}{2!}(z - a)^2 + \ldots, \quad (1)$$

and if it is legitimate to revert this series, we obtain

$$z - a = \frac{\phi(z) - b}{\phi'(a)} - \frac{1}{2}\frac{\phi''(a)}{[\phi'(a)]^3}[\phi(z) - b]^2 + \ldots, \quad (2)$$

which expresses z as an ANALYTIC FUNCTION of the variable $\phi(z) - b$ for sufficiently small values of $|z - a|$. If then $f(z)$ is analytic near $z = a$, it follows that $f(z)$ is an ANALYTIC FUNCTION of $\phi(z) - b$ when $|z - a|$ is sufficiently small, and so there will be an expansion in the form

$$f(z) = f(a) + a_1[\phi(z) - b] + \frac{a_2}{2!}[\phi(z) - b]^2 + \frac{a_3}{3!}[\phi(z) - b]^3 + \ldots \quad (3)$$

The actual coefficients in the expansion are given by the following theorem, which is generally known as Bürmann's theorem. Let $\psi(z)$ be a function of z

defined by the equation

$$\psi(z) = \frac{z - a}{\phi(z) - b}. \quad (4)$$

Then an ANALYTIC FUNCTION $f(z)$ can, in a certain domain of values of z, be expanded in the form

$$f(z) = f(a) + \sum_{m=1}^{n-1} \frac{[\phi(z) - b]^m}{m!}\frac{d^{m-1}}{da^{m-1}}\{f'(a)[\psi(a)]^m\} + R_n, \quad (5)$$

where the remainder term is

$$R_n = \frac{1}{2\pi i}\int_a^x \int_\gamma \left[\frac{\phi(z) - b}{\phi(t) - b}\right]^{n-1}\frac{f'(t)\phi'(z)\, dt\, dz}{\phi(t) - \phi(z)}, \quad (6)$$

and γ is a CONTOUR in the t-plane enclosing the points a and z such that if ζ is any point inside γ, the equation $\phi(t) = \phi(\zeta)$ has no roots on or inside the CONTOUR except a simple root $t = \zeta$.

TEIXEIRA'S THEOREM is extended form of Bürmann's theorem. The LAGRANGE EXPANSION gives another such extension.

See also DARBOUX'S FORMULA, LAGRANGE EXPANSION, LAGRANGE INVERSION THEOREM, TAYLOR SERIES, TEIXEIRA'S THEOREM

References

Bürmann. "Rapport sur deux mémoirs d'analyse." *Mémoires de l'Institut National des Sci. et Arts: Sci. Math. Phys.* **2**, 13–17, 1799.

Dixon, A. C. "On Burmann's Theorem." *Proc. London Math. Soc.* **34**, 151–153, 1902.

Whittaker, E. T. and Watson, G. N. "Bürmann's Theorem" and "Teixeira's Extended Form of Bürmann's Theorem." §7.3 and 7.3.1 in *A Course in Modern Analysis, 4th ed.* Cambridge, England: Cambridge University Press, pp. 128–132, 1990.

Burnside Problem

A problem originating with W. Burnside (1902), who wrote, "A still undecided point in the theory of discontinuous groups is whether the ORDER of a GROUP may be not finite, while the order of every operation it contains is finite." This question would now be phrased as "Can a finitely generated group be infinite while every element in the group has finite order?" (Vaughan-Lee 1990). This question was answered by Golod (1964) when he constructed finitely generated infinite P-GROUP. These GROUPS, however, do not have a finite exponent.

Let F_r be the FREE GROUP of RANK r and let N be the NORMAL SUBGROUP generated by the set of nth POWERS $\{g^n | g \in F_r\}$. Then N is a normal subgroup of F_r. We define $B(r, n) = F_r/N$ to be the QUOTIENT GROUP. We call $B(r, n)$ the r-generator Burnside group of exponent n. It is the largest r-generator group of exponent n, in the sense that every other such group is a HOMOMORPHIC image of $B(r, n)$. The Burnside problem is usually stated as: "For which values of r and n is $B(r, n)$ a FINITE GROUP?"

An answer is known for the following values. For $r = 1$, $B(1, n)$ is a CYCLIC GROUP of ORDER n. For $n = 2$, $B(r, 2)$ is an elementary ABELIAN 2-group of ORDER 2^r. For $n = 3$, $B(r, 3)$ was proved to be finite by Burnside. The ORDER of the $B(r, 3)$ groups was established by Levi and van der Waerden (1933), namely 3^a where

$$a \equiv r + \binom{r}{2} + \binom{r}{3}, \tag{1}$$

where $\binom{n}{k}$ is a BINOMIAL COEFFICIENT. For $n = 4$, $B(r, 4)$ was proved to be finite by Sanov (1940). Groups of exponent four turn out to be the most complicated for which a POSITIVE solution is known. The precise nilpotency class and derived length are known, as are bounds for the ORDER. For example,

$$|B(2, 4)| = 2^{12} \tag{2}$$

$$|B(3, 4)| = 2^{69} \tag{3}$$

$$|B(4, 4)| = 2^{422} \tag{4}$$

$$|B(5, 4)| = 2^{2728}, \tag{5}$$

while for larger values of r the exact value is not yet known. For $n = 6$, $B(r, 6)$ was proved to be finite by Hall (1958) with ORDER $2^a 3^b$, where

$$a \equiv 1 + (r - 1)3^c \tag{6}$$

$$b \equiv 1 + (r - 1)2^r \tag{7}$$

$$c \equiv r + \binom{r}{2} + \binom{r}{3}. \tag{8}$$

No other Burnside groups are known to be finite. On the other hand, for $r > 2$ and $n \geq 665$, with n ODD, $B(r, n)$ is infinite (Novikov and Adjan 1968). There is a similar fact for $r > 2$ and n a large POWER of 2.

E. Zelmanov was awarded a FIELDS MEDAL in 1994 for his solution of the "restricted" Burnside problem.

See also FREE GROUP

References

Burnside, W. "On an Unsettled Question in the Theory of Discontinuous Groups." *Quart. J. Pure Appl. Math.* **33**, 230–238, 1902.
Golod, E. S. "On Nil-Algebras and Residually Finite p-Groups." *Isv. Akad. Nauk SSSR Ser. Mat.* **28**, 273–276, 1964.
Hall, M. "Solution of the Burnside Problem for Exponent Six." *Ill. J. Math.* **2**, 764–786, 1958.
Levi, F. and van der Waerden, B. L. "Über eine besondere Klasse von Gruppen." *Abh. Math. Sem. Univ. Hamburg* **9**, 154–158, 1933.
Novikov, P. S. and Adjan, S. I. "Infinite Periodic Groups I, II, III." *Izv. Akad. Nauk SSSR Ser. Mat.* **32**, 212–244, 251–524, and 709–731, 1968.

Sanov, I. N. "Solution of Burnside's problem for exponent four." *Leningrad State Univ. Ann. Math. Ser.* **10**, 166–170, 1940.
Vaughan-Lee, M. *The Restricted Burnside Problem, 2nd ed.* New York: Clarendon Press, 1993.

Burnside's Conjecture

This entry contributed by NICOLAS BRAY

In Note M, Burnside (1955) states, "The contrast that these results shew between groups of odd and of even order suggests inevitably that simple groups of odd order do not exist." Of course, SIMPLE GROUPS of prime order do exist, namely the groups $\mathbb{Z}_p$ for any prime p. Therefore, Burnside conjectured that every FINITE SIMPLE GROUP of non-prime order must have even order. The conjecture was proven true by Feit and Thompson (1963).

See also ABELIAN GROUP, FEIT-THOMPSON CONJECTURE, FEIT-THOMPSON THEOREM, SIMPLE GROUP

References

Burnside, W. *Theory of Groups of Finite Order, 2nd ed.* New York: Dover, 1955.
Feit, W. and Thompson, J. G. "Solvability of Groups of Odd Order." *Pacific J. Math.* **13**, 775–1029, 1963.

Burnside's Lemma

CAUCHY-FROBENIUS LEMMA

Buschman Transform

The INTEGRAL TRANSFORM defined by

$$(K\phi)(x) = \int_{-\infty}^{\infty} (x^2 - t^2)_+^{\lambda/2} P_\nu^\lambda \left(\frac{t}{x}\right) \phi(t) \, dt,$$

where y_+^a is the TRUNCATED POWER FUNCTION and $P_\nu^\lambda(x)$ is an associated LEGENDRE POLYNOMIAL.

References

Samko, S. G.; Kilbas, A. A.; and Marichev, O. I. *Fractional Integrals and Derivatives.* Yverdon, Switzerland: Gordon and Breach, p. 23, 1993.

Busemann-Petty Problem

If the section function of a centered convex body in Euclidean n-space ($n \geq 3$) is smaller than that of another such body, is its volume also smaller?

References

Gardner, R. J. "Geometric Tomography." *Not. Amer. Math. Soc.* **42**, 422–429, 1995.

Busy Beaver

A busy beaver is an n-state, 2-symbol, 5-tuple TURING MACHINE which writes the maximum possible number $BB(n)$ of 1s on an initially blank tape before halting. For $n = 0, 1, 2, \ldots,$ $BB(n)$ is given by 0, 1, 4, 6, 13,

$\geq 4098, \geq 136612, \ldots$ The busy beaver sequence is also known as RADO'S SIGMA FUNCTION.

See also HALTING PROBLEM, TURING MACHINE

References

Chaitin, G. J. "Computing the Busy Beaver Function." §4.4 in *Open Problems in Communication and Computation* (Ed. T. M. Cover and B. Gopinath). New York: Springer-Verlag, pp. 108–112, 1987.

Dewdney, A. K. "A Computer Trap for the Busy Beaver, the Hardest-Working Turing Machine." *Sci. Amer.* **251**, 19–23, Aug. 1984.

Marxen, H. and Buntrock, J. "Attacking the Busy Beaver 5." *Bull. EATCS* **40**, 247–251, Feb. 1990.

Sloane, N. J. A. Sequences A028444 in "An On-Line Version of the Encyclopedia of Integer Sequences." http://www.research.att.com/~njas/sequences/eisonline.html.

Butterfly Catastrophe

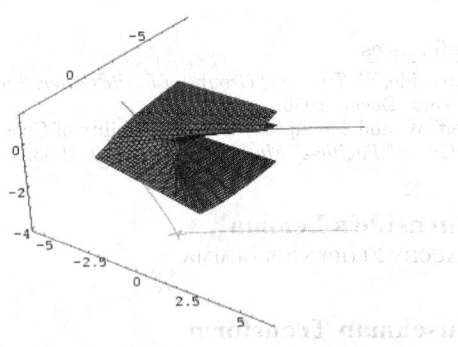

A CATASTROPHE which can occur for four control factors and one behavior axis. The butterfly catastrophe is the universal unfolding of the singularity $f(x) = x^6$ of codimension 4, i.e., with four unfolding parameters. It has the form $F(x, u, v, w, t) = x^6 + ux^4 + vx^3 + wx^2 + tx.$

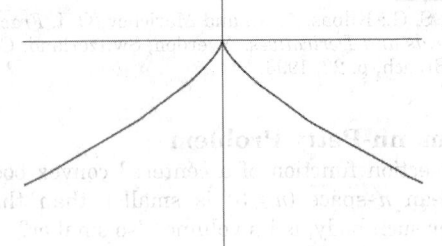

The equations

$$x = c(8at^3 + 24t^5)$$

$$y = c(-6at^2 - 15t^4)$$

display such a catastrophe (von Seggern 1993).

References

Sanns, W. *Catastrophe Theory with Mathematica: A Geometric Approach.* Germany: DAV, 2000.

von Seggern, D. *CRC Standard Curves and Surfaces.* Boca Raton, FL: CRC Press, p. 94, 1993.

Butterfly Curve

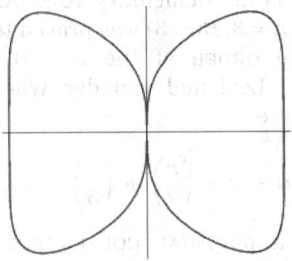

A PLANE CURVE given by the implicit equation

$$y^6 = (x^2 - x^6).$$

See also DUMBBELL CURVE, EIGHT CURVE, PIRIFORM

References

Cundy, H. and Rollett, A. *Mathematical Models, 3rd ed.* Stradbroke, England: Tarquin Pub., p. 72, 1989.

Butterfly Effect

Due to nonlinearities in weather processes, a butterfly flapping its wings in Tahiti can, in theory, produce a tornado in Kansas. This strong dependence of outcomes on very slightly differing initial conditions is a hallmark of the mathematical behavior known as CHAOS.

See also CHAOS, LORENZ SYSTEM

Butterfly Fractal

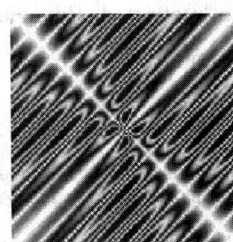

The FRACTAL-like curve generated by the 2-D function

$$f(x, y) = \frac{(x^2 - y^2)\sin\left(\dfrac{x+y}{a}\right)}{x^2 + y^2}.$$

Butterfly Polyiamond

A 6-POLYIAMOND.

References

Golomb, S. W. *Polyominoes: Puzzles, Patterns, Problems, and Packings, 2nd ed.* Princeton, NJ: Princeton University Press, p. 92, 1994.

Butterfly Theorem

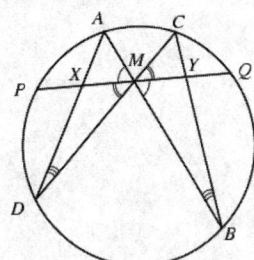 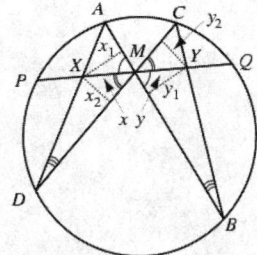

Given a CHORD PQ of a CIRCLE, draw any other two CHORDS AB and CD passing through its MIDPOINT. Call the points where AD and BC meet PQ X and Y. Then M is also the MIDPOINT of XY. There are a number of proofs of this theorem, including those by W. G. Horner, Johnson (1929, p. 78), and Coxeter (1987, pp. 78 and 144). The latter concise proof employs PROJECTIVE GEOMETRY.

The following proof is given by Coxeter and Greitzer (1967, p. 46). In the figure at right, drop perpendiculars x_1 and y_1 from X and Y to AB, and x_2 and y_2 from X and Y to CD. Write $a = PM = MQ$, $x = XM$, and $y = MY$, and then note that by SIMILAR TRIANGLES

$$\frac{x}{y} = \frac{x_1}{y_1} = \frac{x_2}{y_2} \tag{1}$$

$$\frac{x_1}{y_2} = \frac{AX}{CY} \tag{2}$$

$$\frac{x_2}{y_1} = \frac{XD}{YB}, \tag{3}$$

so

$$\frac{x^2}{y^2} = \frac{x_1}{y_1} \frac{x_2}{y_2} = \frac{x_1}{y_2} \frac{x_2}{y_1} = \frac{AX \cdot XD}{CY \cdot YB} = \frac{PX \cdot XQ}{PY \cdot YQ}$$

$$= \frac{(a-x)(a+x)}{(a+y)(a-y)} = \frac{a^2 - x^2}{a^2 - y^2} = \frac{a^2}{a^2} = 1, \tag{4}$$

so $x = y$. Q.E.D.

See also CHORD, CIRCLE, CYCLIC QUADRILATERAL, MIDPOINT, QUADRILATERAL

References

Coxeter, H. S. M. *Projective Geometry, 2nd ed.* New York: Springer-Verlag, pp. 78 and 144, 1987.
Coxeter, H. S. M. and Greitzer, S. L. "The Butterfly." §2.8 in *Geometry Revisited.* Washington, DC: Math. Assoc. Amer., pp. 45–46, 1967.
Johnson, R. A. *Modern Geometry: An Elementary Treatise on the Geometry of the Triangle and the Circle.* Boston, MA: Houghton Mifflin, p. 78, 1929.

C

Cable

TENSEGRITY

Cable Knot

Let K_1 be a TORUS KNOT. Then the SATELLITE KNOT with COMPANION KNOT K_2 is a cable knot on K_2.

See also SATELLITE KNOT

References

Adams, C. C. *The Knot Book: An Elementary Introduction to the Mathematical Theory of Knots.* New York: W. H. Freeman, p. 118, 1994.
Burde, G. and Zieschang, H. *Knots.* Berlin: de Gruyter, 1985.
Hoste, J.; Thistlethwaite, M.; and Weeks, J. "The First 1,701,936 Knots." *Math. Intell.* **20**, 33–48, Fall 1998.
Rolfsen, D. *Knots and Links.* Wilmington, DE: Publish or Perish Press, pp. 112 and 283, 1976.

Cactus Fractal

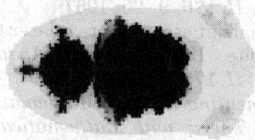

A MANDELBROT SET-like FRACTAL obtained by iterating the map

$$z_{n+1} = z_n^3 + (z_0 - 1)z_n - z_0.$$

See also FRACTAL, JULIA SET, MANDELBROT SET

Cage Graph

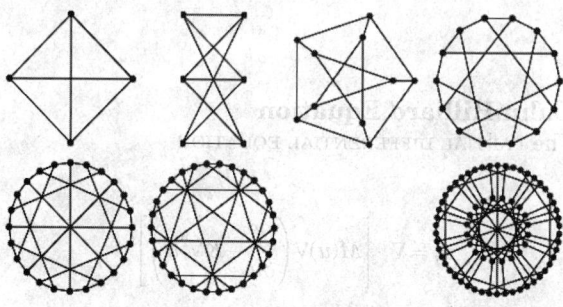

A 3-regular g-cage for $g \geq 3$ is a CUBIC GRAPH of GIRTH g with the minimum possible number of points. More generally, an (v, g)-cage graph is a smallest v-regular graph with GIRTH g. Cubic cages were first discussed by Tutte (1947), but the intensive study of cage graphs did not begin until publication of an article by Erdos and Sachs (1963). There exists a $(3, g)$-cage

for all $g \geq 3$, and the $(3, g)$-cages are unique for $g = 3$ to 8. The number of nonisomorphic $(3, g)$ cages for $g = 1, 2, \ldots$ are given by 0, 0, 1, 1, 1, 1, 1, 1, 18, 3, ... (Sloane's A052453; Gould 1988, Royle). The number of vertices in the $(3, g)$ cages for $g = 3, 4, \ldots$ are 4, 6, 10, 14, 22, 30, 46, 62, 94, ... (Sloane's A052454). A selection of known $(3, g)$-cages are illustrated above. There are a number of special cases (Wong 1982). The $(2, g)$-cage is the CYCLE GRAPH C_g, the $(v, 2)$-cage is the MULTIGRAPH of v edges on two vertices, the $(v, 3)$-cage is the COMPLETE GRAPH K_{v+1}, and the $(v, 4)$-cage is the BIPARTITE GRAPH $K_{v, v}$.

Computing the number of vertices in a (v, g)-cage is very difficult for $g \geq 5$ and $n \geq 3$ (Wong 1982). The following table summarizes known cages. A lower bound for the number of vertices $f(v, g)$ in a (v, g)-cage is given by

$$f_l(v, g) = \begin{cases} \dfrac{v(v-1)^r - 2}{v - 2} & \text{for } g = 2r+1 \\ \dfrac{2(v-1)^r - 2}{v - 2} & \text{for } g = 2r \end{cases}$$

(Tutte 1967, p. 70; Bollobás 1978, p. 105; Wong 1982). Sauer (1967ab) has obtained the best known upper bounds

$$f_u(3, g) = \begin{cases} \dfrac{4}{3} + \dfrac{29}{12} 2^{g-2} & \text{for } g \text{ odd} \\ \dfrac{2}{3} + \dfrac{29}{12} 2^{g-2} & \text{for } g \text{ even} \end{cases} \tag{1}$$

$$f_u(n, g) = \begin{cases} 2(n-1)^{g-2} & \text{for } g \text{ odd} \\ 2(n-1)^{g-3} & \text{for } g \text{ even,} \end{cases} \tag{2}$$

with $v \geq 4$ (Wong 1982).

In the table, K_n denotes a COMPLETE GRAPH, and $K_{m, n}$ a complete bipartite graph.

g	$(3, g)$	$(4, g)$	$(5, g)$	$(6, g)$	$(7, g)$-cage
3	K_4	K_5	K_6	K_7	K_8
4	$K_{3, 3}$	$K_{4, 4}$	$K_{5, 5}$	$K_{6, 6}$	$K_{7, 7}$
5	PETERSEN GRAPH	ROBERTSON GRAPH	ROBERTSON-WEGNER GRAPH		HOFFMAN-SINGLETON GRAPH
6	HEAWOOD GRAPH				
7	McGEE GRAPH				
8	LEVI GRAPH				

g	$f(3, g)$	$f(4, g)$	$f(5, g)$	$f(6, g)$	$f(7, g)$
3	4	5	6	7	8
4	6	8	10	12	14
5	10	19	30	40	50
6	14	26	42	62	90
7	24				
8	30				
9	[54, 58]				
10	70				
11	< 112				

The first (3, 9)-cage was found by Biggs and Hoare (1980), and Brinkmann *et al.* (1995) completed an exhaustive search yielding all 18 (3, 9)-cages (Royle). The three (3, 10)-cages were found by O'Keefe and Wong (1980). Computations by McKay and W. Myrvold have demonstrated that a (3, 11)-cage must have 112 vertices (Royle). The single known example was found by Balaban (1973).

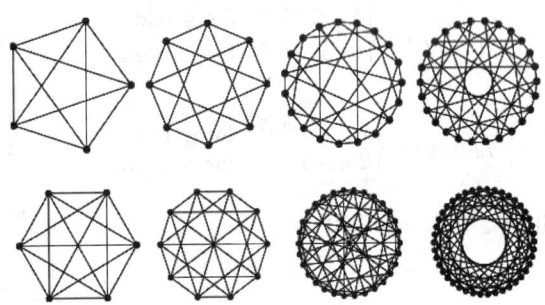

The known (4, g)- and (5, g)-cages are shown above (Wong 1982).

See also CAYLEY GRAPH, CUBIC GRAPH, EXCESS, HOFFMAN-SINGLETON GRAPH, MOORE GRAPH, REGULAR GRAPH, ROBERTSON GRAPH, ROBERTSON-WEGNER GRAPH, UNITRANSITIVE GRAPH

References

Balaban, A. T. "Trivalent Graphs of Girth Nine and Eleven and Relationships among the Cages." *Rev. Roumaine Math. Pures Appl.* **18**, 1033–1043, 1973.
Biggs, N. L. Ch. 23 in *Algebraic Graph Theory, 2nd ed.* Cambridge, England: Cambridge University Press, 1993.
Biggs, N. L. "Constructions for Cubic Graphs of Large Girth." LSE Tech Report 97–11.
Biggs, N. L. and Hoare, M. J. "A Trivalent Graph with 58 Vertices and Girth 9." *Disc. Math.* **30**, 299–301, 1980.
Bollobás, B. *Extremal Graph Theory.* New York: Academic Press, 1978.
Bondy, J. A. and Murty, U. S. R. *Graph Theory with Applications.* New York: North Holland, pp. 236–239, 1976.
Brinkmann, G.; McKay, B. D.; and Saager, C. "The Smallest Cubic Graphs of Girth Nine." *Combin., Probability, and Computing* **5**, 1–13, 1995.
Brouwer, A. E.; Cohen, A. M.; and Neumaier, A. §6.9 in *Distance Regular Graphs.* New York: Springer-Verlag, 1989.
Erdos, P. and Sachs, H. "Reguläre graphen gegebener Taillenweite mit minimaler Knotenzahl." *Wiss. Z. Uni. Halle (Math. Nat.)* **12**, 251–257, 1963.
Friedman, E. "Cages." http://www.stetson.edu/~efriedma/girth/.
Gould, R. (Ed.). *Graph Theory.* Menlo Park, CA: Benjamin-Cummings, 1988.
Harary, F. *Graph Theory.* Reading, MA: Addison-Wesley, pp. 174–175, 1994.
Holton, D A. and Sheehan, J. (Eds.). Ch. 6 in *The Petersen Graph.* Cambridge, England: Cambridge University Press, 1993.
O'Keefe, M. and Wong, P. K. "A Smallest Graph of Girth 10 and Valency 3." *J. Combin. Th. B* **29**, 91–105, 1980.
Royle, G. "Cubic Cages." http://www.cs.uwa.edu.au/~gordon/cages/.
Sauer, N. 'Extremaleigenschaften regulärer Graphen gegebener Taillenweite, I." *Österreich. Akad. Wiss. Math. Natur. Kl. S.-B. II* **176**, 9–25, 1967.
Sauer, N. 'Extremaleigenschaften regulärer Graphen gegebener Taillenweite, II." *Österreich. Akad. Wiss. Math. Natur. Kl. S.-B. II* **176**, 27–43, 1967.
Skiena, S. *Implementing Discrete Mathematics: Combinatorics and Graph Theory with Mathematica.* Reading, MA: Addison-Wesley, pp. 191 and 221, 1990.
Sloane, N. J. A. Sequences A052453 and A052454 in "An On-Line Version of the Encyclopedia of Integer Sequences." http://www.research.att.com/~njas/sequences/eisonline.html.
Tutte, W. T. "A Family of Cubical Graphs." *Proc. Cambridge Philos. Soc.*, 459–474, 1947.
Tutte, W. T. *The Connectivity of Graphs.* Toronto, Canada: Toronto University Press, pp. 71–83, 1967.
Weisstein, E. W. "Graphs." MATHEMATICA NOTEBOOK GRAPHS.M.
Wong, P. K. "Cages--A Survey." *J. Graph Th.* **6**, 1–22, 1982.

Cahn-Hilliard Equation

The PARTIAL DIFFERENTIAL EQUATION

$$u_t = \nabla \cdot \left[M(u)\nabla \left(\frac{\partial f}{\partial u} - K\nabla^2 u \right) \right].$$

References

Novick-Cohen, A. and Segal, L. A. "Nonlinear Aspects of the Cahn-Hilliard Equation." *Physica D* **10**, 277–298, 1984.
Zwillinger, D. *Handbook of Differential Equations, 3rd ed.* Boston, MA: Academic Press, p. 132, 1997.

Cairo Tessellation

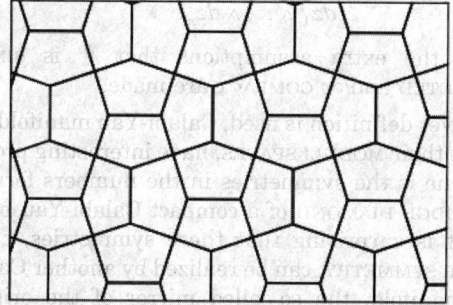

A TESSELLATION appearing in the streets of Cairo and in many Islamic decorations. Its tiles are obtained by projection of a DODECAHEDRON, and it is the DUAL TESSELLATION of the semiregular tessellation of squares and equilateral triangles.

See also DODECAHEDRON, TESSELLATION

References

Wells, D. *The Penguin Dictionary of Curious and Interesting Geometry.* London: Penguin, p. 23, 1991.

Williams, R. *The Geometrical Foundation of Natural Structure: A Source Book of Design.* New York: Dover, p. 38, 1979.

Cake Cutting

It is always possible to "fairly" divide a cake among n people using only vertical cuts. Furthermore, it is possible to cut and divide a cake such that each person believes that *everyone* has received $1/n$ of the cake according to his own measure (Steinhaus 1983, pp. 65–71). Finally, if there is some piece on which two people disagree, then there is a way of partitioning and dividing a cake such that each participant believes that he has obtained more than $1/n$ of the cake according to his own measure.

There are also similar methods of dividing collections of individually indivisible objects among two or more people when cash payments are used to even up the final division (Steinhaus 1983, pp. 67–68).

Ignoring the height of the cake, the cake-cutting problem is really a question of fairly dividing a CIRCLE into n equal AREA pieces using cuts in its plane. One method of proving fair cake cutting to always be possible relies on the FROBENIUS-KÖNIG THEOREM.

See also CIRCLE DIVISION BY CHORDS, CIRCLE DIVISION BY LINES, CYLINDER CUTTING, ENVYFREE, FROBENIUS-KÖNIG THEOREM, HAM SANDWICH THEOREM, PANCAKE THEOREM, PIZZA THEOREM, SQUARE DIVISION BY LINES, TORUS CUTTING, VOTING

References

Beck, A. "Constructing a Fair Share." *Amer. Math. Monthly* **94**, 157–162, 1987.

Brams, S. J. and Taylor, A. D. "An Envy-Free Cake Division Protocol." *Amer. Math. Monthly* **102**, 9–19, 1995.

Brams, S. J. and Taylor, A. D. *Fair Division: From Cake-Cutting to Dispute Resolution.* New York: Cambridge University Press, 1996.

Dubbins, L. "Group Decision Devices." *Amer. Math. Monthly* **84**, 350–356, 1997.

Dubbins, L. and Spanier, E. "How to Cut a Cake Fairly." *Amer. Math. Monthly* **68**, 1–17, 1961.

Gale, D. "Dividing a Cake." *Math. Intel.* **15**, 50, 1993.

Hill, T. "Determining a Fair Border." *Amer. Math. Monthly* **90**, 438–442, 1983.

Hill, T. P. "Mathematical Devices for Getting a Fair Share." *Amer. Sci.* **88**, 325–331, Jul.-Aug. 2000.

Jones, M. L. "A Note on a Cake Cutting Algorithm of Banach and Knaster." *Amer. Math. Monthly* **104**, 353–355, 1997.

Knaster, B. "Sur le problème du partage pragmatique de H. Steinhaus." *Ann. de la Soc. Polonaise de Math.* **19**, 228–230, 1946.

Rebman, K. "How to Get (At Least) a Fair Share of the Cake." In *Mathematical Plums* (Ed. R. Honsberger). Washington, DC: Math. Assoc. Amer., pp. 22–37, 1979.

Robertson, J. and Webb, W. *Cake Cutting Algorithms: Be Fair If You Can.* Natick, MA: Peters, 1998.

Steinhaus, H. "Remarques sur le partage pragmatique." *Ann. de la Soc. Polonaise de Math.* **19**, 230–231, 1946.

Steinhaus, H. "The Problem of Fair Division." *Econometrica* **16**, 101–104, 1948.

Steinhaus, H. "Sur la division pragmatique." *Ekonometrika (Supp.)* **17**, 315–319, 1949.

Steinhaus, H. *Mathematical Snapshots, 3rd ed.* New York: Dover, pp. 64–67, 1999.

Stromquist, W. "How to Cut a Cake Fairly." *Amer. Math. Monthly* **87**, 640–644, 1980.

Cal

WALSH FUNCTION

Calabi's Triangle

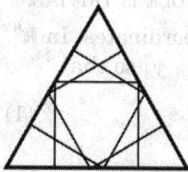 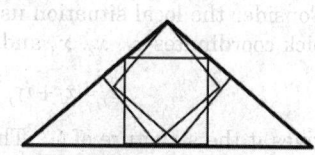

Equilateral Triangle   *Calabi's Triangle*

The one TRIANGLE, in addition to the EQUILATERAL TRIANGLE, for which the largest inscribed SQUARE can be inscribed in three different ways. The ratio of the sides to that of the base is given by $x = 1.55138752454\ldots$ (Sloane's A046095), where

$$x = \frac{1}{3} + \frac{(-23 + 3i\sqrt{237})^{1/3}}{3 \cdot 2^{2/3}} + \frac{11}{3[2(-23 + 3i\sqrt{237})]^{1/3}}$$

is the largest POSITIVE ROOT of

$$2x^3 - 2x^2 - 3x + 2 = 0,$$

which has CONTINUED FRACTION [1, 1, 1, 4, 2, 1, 2, 1, 5, 2, 1, 3, 1, 1, 390, ...] (Sloane's A046096).

See also GRAHAM'S BIGGEST LITTLE HEXAGON, TRIANGLE

References

Conway, J. H. and Guy, R. K. "Calabi's Triangle." In *The Book of Numbers.* New York: Springer-Verlag, p. 206, 1996.

Sloane, N. J. A. Sequences A046095 and A046096 in "An On-Line Version of the Encyclopedia of Integer Sequences." http://www.research.att.com/~njas/sequences/eisonline.html.

Weisstein, E. W. "Plane Geometry." MATHEMATICA NOTEBOOK PLANEGEOMETRY.M.

Calabi-Yau Manifold

CALABI-YAU SPACE

Calabi-Yau Space

Calabi-Yau spaces are important in string theory, where one model posits the geometry of the universe to consist of a ten-dimensional space OF THE FORM $M \times V$, where M is a four dimensional manifold (space-time) and V is a six dimensional COMPACT Calabi-Yau space. They are related to KUMMER SURFACES. Although the main application of Calabi-Yau spaces is in theoretical physics, they are also interesting from a purely mathematical standpoint. Consequently, they go by slightly different names, depending mostly on context, such as Calabi-Yau manifolds or Calabi-Yau varieties.

Although the definition can be generalized to any dimension, they are usually considered to have three complex dimensions. Since their COMPLEX STRUCTURE may vary, it is convenient to think of them as having six real dimensions and a fixed SMOOTH STRUCTURE.

A Calabi-Yau space is characterized by the existence of a NONVANISHING HARMONIC SPINOR ϕ. This condition implies that its CANONICAL BUNDLE is TRIVIAL.

Consider the local situation using coordinates. In $\mathbb{R}^6$, pick coordinates x_1, x_2, x_3 and y_1, y_2, y_3 so that

$$z_j = x_j + iy_j \tag{1}$$

gives it the structure of $\mathbb{C}^3$. Then

$$\phi_z = dz_1 \wedge dz_2 \wedge dz_3 \tag{2}$$

is a local section of the canonical bundle. A unitary change of coordinates $w = Az$, where A is a UNITARY MATRIX, transforms ϕ by det A, i.e.

$$\phi_w = \det A\phi_z. \tag{3}$$

If the linear transformation A has DETERMINANT 1, that is, it is a special unitary transformation, then ϕ is consistently defined as ϕ_z or as ϕ_w.

On a Calabi-Yau manifold V, such a ϕ can be defined globally, and the LIE GROUP $SU(3)$ is very important in the theory. In fact, one of the many equivalent definitions, coming from RIEMANNIAN GEOMETRY, says that a Calabi-Yau manifold is a $2n$-dimensional manifold whose HOLONOMY GROUP reduces to $SU(n)$. Another is that it is a CALIBRATED MANIFOLD with a CALIBRATION FORM ψ, which is algebraically the same

as the REAL PART of

$$dz_1 \wedge \ldots \wedge dz_n. \tag{4}$$

Often, the extra assumptions that V is SIMPLY CONNECTED and/or COMPACT are made.

Whatever definition is used, Calabi-Yau manifolds, as well as their MODULI SPACES, have interesting properties. One is the symmetries in the numbers forming the HODGE DIAMOND of a compact Calabi-Yau manifold. It is surprising that these symmetries, called MIRROR SYMMETRY, can be realized by another Calabi-Yau manifold, the so-called mirror of the original Calabi-Yau manifold. The two manifolds together form a MIRROR PAIR. Some of the symmetries of the geometry of mirror pairs have been the object of recent research.

See also CALIBRATED MANIFOLD, CANONICAL BUNDLE, COMPLEX MANIFOLD, DOLBEAULT COHOMOLOGY, HARMONIC, HODGE DIAMOND, KÄHLER FORM, LIE GROUP, MIRROR PAIR, MODULI SPACE, SPINOR, VARIETY

Calabi-Yau Variety

CALABI-YAU SPACE

Calculus

In general, "a" calculus is an abstract theory developed in a purely formal way.

"The" calculus, more properly called ANALYSIS (or REAL ANALYSIS or, in older literature, INFINITESIMAL ANALYSIS) is the branch of mathematics studying the rate of change of quantities (which can be interpreted as SLOPES of curves) and the length, AREA, and VOLUME of objects. The calculus is sometimes divided into DIFFERENTIAL and INTEGRAL CALCULUS, concerned with DERIVATIVES

$$\frac{d}{dx}f(x)$$

and INTEGRALS

$$\int f(x)\, dx,$$

respectively.

While ideas related to calculus had been known for some time (Archimedes' EXHAUSTION METHOD was a form of calculus), it was not until the independent work of Newton and Leibniz that the modern elegant tools and ideas of calculus were developed. Even so, many years elapsed until the subject was put on a mathematically rigorous footing by mathematicians such as Weierstrass.

See also ARC LENGTH, AREA, CALCULUS OF VARIATIONS, CHANGE OF VARIABLES THEOREM, DERIVATIVE, DIFFERENTIAL CALCULUS, ELLIPSOIDAL CALCULUS, EXTENSIONS CALCULUS, FLUENT, FLUXION, FRAC-

TIONAL CALCULUS, FUNCTIONAL CALCULUS, FUNDA-
MENTAL THEOREMS OF CALCULUS, HEAVISIDE CALCU-
LUS, INTEGRAL, INTEGRAL CALCULUS, JACOBIAN,
LAMBDA CALCULUS, KIRBY CALCULUS, MALLIAVIN
CALCULUS, PREDICATE CALCULUS, PROPOSITIONAL
CALCULUS, SLOPE, STOCHASTIC CALCULUS, TENSOR
CALCULUS, UMBRAL CALCULUS, VOLUME

References

Anton, H. *Calculus: A New Horizon, 6th ed.* New York: Wiley, 1999.

Apostol, T. M. *Calculus, 2nd ed., Vol. 1: One-Variable Calculus, with an Introduction to Linear Algebra.* Waltham, MA: Blaisdell, 1967.

Apostol, T. M. *Calculus, 2nd ed., Vol. 2: Multi-Variable Calculus and Linear Algebra, with Applications to Differential Equations and Probability.* Waltham, MA: Blaisdell, 1969.

Apostol, T. M.; Chrestenson, H. E.; Ogilvy, C. S.; Richmond, D. E.; and Schoonmaker, N. J. *A Century of Calculus, Part I: 1894–1968.* Washington, DC: Math. Assoc. Amer., 1992.

Apostol, T. M.; Mugler, D. H.; Scott, D. R.; Sterrett, A. Jr.; and Watkins, A. E. *A Century of Calculus, Part II: 1969–1991.* Washington, DC: Math. Assoc. Amer., 1992.

Ayres, F. Jr. and Mendelson, E. *Schaum's Outline of Theory and Problems of Differential and Integral Calculus, 3rd ed.* New York: McGraw-Hill, 1990.

Borden, R. S. *A Course in Advanced Calculus.* New York: Dover, 1998.

Boyer, C. B. *A History of the Calculus and Its Conceptual Development.* New York: Dover, 1989.

Brown, K. S. "Calculus and Differential Equations." http://www.seanet.com/~ksbrown/icalculu.htm.

Courant, R. and John, F. *Introduction to Calculus and Analysis, Vol. 1.* New York: Springer-Verlag, 1999.

Courant, R. and John, F. *Introduction to Calculus and Analysis, Vol. 2.* New York: Springer-Verlag, 1990.

Hahn, A. *Basic Calculus: From Archimedes to Newton to Its Role in Science.* New York: Springer-Verlag, 1998.

Kaplan, W. *Advanced Calculus, 4th ed.* Reading, MA: Addison-Wesley, 1992.

Marsden, J. E. and Tromba, A. J. *Vector Calculus, 4th ed.* New York: W. H. Freeman, 1996.

Mendelson, E. *3000 Solved Problems in Calculus.* New York: McGraw-Hill, 1988. Strang, G. *Calculus.* Wellesley, MA: Wellesley-Cambridge Press, 1991.

Weisstein, E. W. "Books about Calculus." http://www.treasure-troves.com/books/Calculus.html.

Calculus of Variations

A branch of mathematics which is a sort of generalization of CALCULUS. Calculus of variations seeks to find the path, curve, surface, etc., for which a given FUNCTION has a STATIONARY VALUE (which, in physical problems, is usually a MINIMUM or MAXIMUM). Mathematically, this involves finding STATIONARY VALUES of integrals OF THE FORM

$$i = \int_b^a f(y, \dot{y}, x) \, dx. \tag{1}$$

i has an extremum only if the EULER-LAGRANGE

DIFFERENTIAL EQUATION is satisfied, i.e., if

$$\frac{\partial f}{\partial y} - \frac{d}{dx}\left(\frac{\partial f}{\partial \dot{y}}\right) = 0. \tag{2}$$

the FUNDAMENTAL LEMMA OF CALCULUS OF VARIATIONS states that, if

$$\int_a^b M(x)h(x) \, dx = 0 \tag{3}$$

for all $h(x)$ with CONTINUOUS second PARTIAL DERIVATIVES, then

$$M(x) = 0 \tag{4}$$

on (a, b).

A generalization of calculus of variations known as MORSE THEORY (and sometimes called "calculus of variations in the large" uses nonlinear techniques to address variational problems.

See also BELTRAMI IDENTITY, BOLZA PROBLEM, BRACHISTOCHRONE PROBLEM, CATENARY, ENVELOPE THEOREM, EULER-LAGRANGE DIFFERENTIAL EQUATION, ISOPERIMETRIC PROBLEM, ISOVOLUME PROBLEM, LINDELOF'S THEOREM, MORSE THEORY, PLATEAU'S PROBLEM, POINT-POINT DISTANCE–2-D, POINT-POINT DISTANCE–3-D, ROULETTE, SKEW QUADRILATERAL, SPHERE WITH TUNNEL, SURFACE OF REVOLUTION, UNDULOID, WEIERSTRASS-ERDMAN CORNER CONDITION

References

Arfken, G. "Calculus of Variations." Ch. 17 in *Mathematical Methods for Physicists, 3rd ed.* Orlando, FL: Academic Press, pp. 925–962, 1985.

Bliss, G. A. *Calculus of Variations.* Chicago, IL: Open Court, 1925.

Forsyth, A. R. *Calculus of Variations.* New York: Dover, 1960.

Fox, C. *An Introduction to the Calculus of Variations.* New York: Dover, 1988.

Isenberg, C. *The Science of Soap Films and Soap Bubbles.* New York: Dover, 1992.

Jeffreys, H. and Jeffreys, B. S. "Calculus of Variations." Ch. 10 in *Methods of Mathematical Physics, 3rd ed.* Cambridge, England: Cambridge University Press, pp. 314–332, 1988.

Menger, K. "What is the Calculus of Variations and What are Its Applications?" In *The World of Mathematics* (Ed. K. Newman). Redmond, WA: Microsoft Press, pp. 886–890, 1988.

Sagan, H. *Introduction to the Calculus of Variations.* New York: Dover, 1992.

Smith, D. R. *Variational Methods in Optimization.* New York: Dover, 1998.

Todhunter, I. *History of the Calculus of Variations During the Nineteenth Century.* New York: Chelsea, 1962.

Weinstock, R. *Calculus of Variations, with Applications to Physics and Engineering.* New York: Dover, 1974.

Weisstein, E. W. "Books about Calculus of Variations." http://www.treasure-troves.com/books/CalculusofVariations.html.

Calcus

$$1 \text{ calcus} \equiv \tfrac{1}{2304}.$$

See also HALF, QUARTER, SCRUPLE, UNCIA, UNIT FRACTION

Calderón's Formula

$$f(x) = C_\psi \int_{-\infty}^{\infty} \int_{-\infty}^{\infty} \langle f, \psi^{a,\,b} \rangle \psi^{a,\,b}(x) a^{-2} \, da \, db,$$

where

$$\psi^{a,\,b}(x) = |a|^{-1/2} \psi\left(\frac{x-b}{a}\right).$$

This result was originally derived using HARMONIC ANALYSIS, but also follows from a WAVELETS viewpoint.

C*-Algebra

A special type of B*-ALGEBRA in which the INVOLUTION is the ADJOINT operator in a HILBERT SPACE.

See also B*-ALGEBRA, *K*-THEORY

References

Davidson, K. R. *C*-Algebras by Example.* Providence, RI: Amer. Math. Soc., 1996.
Wegge-Olsen, N. E. *K-Theory and C*-Algebras: A Friendly Approach.* Oxford, England: Oxford University Press, 1993.

Caliban Puzzle

A puzzle in LOGIC in which one or more facts must be inferred from a set of given facts.

Calibration Form

A calibration form on a RIEMANNIAN MANIFOLD M is a DIFFERENTIAL K-FORM ϕ such that

1. ϕ is a CLOSED FORM.
2. The COMASS of ϕ,

$$\sup_{v \in \wedge pTM, \, |v|=1} |\phi(v)| \tag{1}$$

defined as the largest value of ϕ on a p vector of p-volume one, equals 1.

A p-dimensional submanifold is calibrated when ϕ restricts to give the VOLUME FORM.

It is not hard to see that a calibrated submanifold N minimizes its volume among objects in its HOMOLOGY CLASS. By STOKES' THEOREM, if N' represents the

same homology class, then

$$\int_N \phi = \int_{N'} \phi. \tag{2}$$

Since

$$\text{vol}(N) = \int_N \phi \tag{3}$$

and

$$\text{vol}(N') \geq \int_{N'} \phi, \tag{4}$$

it follows that the volume of N is less than or equal to the volume of N'.

A simple example is dx on the plane, for which the lines $y = c$ are calibrated submanifolds. In fact, in this example, the calibrated submanifolds give a FOLIATION. On a KÄHLER MANIFOLD, the KÄHLER FORM ω is a calibration form, which is INDECOMPOSABLE. For example, on

$$\mathbb{C}^2 = \{(x_1 + y_1 i, \; x_2 + y_2 i)\}, \tag{5}$$

the Kähler form is

$$dx_1 \wedge dy_1 + dx_2 \wedge dy_2. \tag{6}$$

On a KÄHLER MANIFOLD, the calibrated submanifolds are precisely the complex submanifolds. Consequently, the complex submanifolds are locally volume minimizing.

See also KÄHLER FORM, KÄHLER MANIFOLD, VOLUME FORM

Calogero-Degasperis-Fokas Equation

The PARTIAL DIFFERENTIAL EQUATION

$$u_{xxx} - \tfrac{1}{8} u_x^3 + u_x(Ae^u + Be^{-u}) = 0.$$

References

Gerdt, V. P.; Shvachka, A. B.; and Zharkov, A. Y. "Computer Algebra Applications for Classification of Integrable Non-Linear Evolution Equations." *J. Symb. Comput.* **1**, 101–107, 1985.
Zwillinger, D. *Handbook of Differential Equations,* 3rd ed. Boston, MA: Academic Press, p. 132, 1997.

Calugareanu Theorem

Letting Lk be the LINKING NUMBER of the two components of a ribbon, Tw be the TWIST, and Wr be the WRITHE, then

$$\text{Lk}(K) = \text{Tw}(K) + \text{Wr}(K).$$

(Adams 1994, p. 187).

See also GAUSS INTEGRAL, LINKING NUMBER, TWIST, WRITHE

References

Adams, C. C. *The Knot Book: An Elementary Introduction to the Mathematical Theory of Knots.* New York: W. H. Freeman, 1994.

Calugareanu, G. "L'intégrale de Gauss et l'Analyse des nœuds tridimensionnels." *Rev. Math. Pures Appl.* **4**, 5–20, 1959.

Calugareanu, G. "Sur les classes d'isotopie des noeuds tridimensionnels et leurs invariants." *Czech. Math. J.* **11**, 588–625, 1961.

Calugareanu, G. "Sur les enlacements tridimensionnels des courbes fermées." *Comm. Acad. R. P. Romîne* **11**, 829–832, 1961.

Kaul, R. K. Topological Quantum Field Theories--A Meeting Ground for Physicists and Mathematicians. 15 Jul 1999. http://xxx.lanl.gov/abs/hep-th/9907119/.

Pohl, W. F. "The Self-Linking Number of a Closed Space Curve." *J. Math. Mech.* **17**, 975–985, 1968.

Calvary Cross

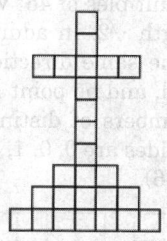

See also CROSS

Cameron's Sum-Free Set Constant

A set of POSITIVE INTEGERS S is sum-free if the equation $x + y = z$ has no solutions x, y, $z \in S$. The probability that a random sum-free set S consists entirely of ODD INTEGERS satisfies

$$0.21759 \le c \le 0.21862.$$

References

Cameron, P. J. "Cyclic Automorphisms of a Countable Graph and Random Sum-Free Sets." *Graphs and Combinatorics* **1**, 129–135, 1985.

Cameron, P. J. "Portrait of a Typical Sum-Free Set." In *Surveys in Combinatorics 1987* (Ed. C. Whitehead). New York: Cambridge University Press, 13–42, 1987.

Finch, S. "Favorite Mathematical Constants." http://www.mathsoft.com/asolve/constant/cameron/cameron.html.

Campbell's Theorem

Any n-dimensional RIEMANNIAN MANIFOLD can be locally EMBEDDED into an $(n+1)$-dimensional manifold with RICCI CURVATURE $R_{ab} = 0$. A similar version of the theorem for a PSEUDO-RIEMANNIAN MANIFOLD states that any n-dimensional PSEUDO-RIEMANNIAN MANIFOLD can be locally and isometrically embedded in an $n(n+1)/2$-dimensional PSEUDO-EUCLIDEAN SPACE.

See also EMBEDDING, PSEUDO-EUCLIDEAN SPACE, PSEUDO-RIEMANNIAN MANIFOLD, RICCI CURVATURE, RIEMANNIAN MANIFOLD

References

Eisenhart, L. P. *Riemannian Geometry.* Princeton, NJ: Princeton University Press, 1964.

Cancellation

ANOMALOUS CANCELLATION

Cancellation Law

If $bc \equiv bd \pmod{a}$ and $(b, a) = 1$ (i.e., a and b are RELATIVELY PRIME), then $c \equiv d \pmod{a}$.

See also CONGRUENCE

References

Courant, R. and Robbins, H. *What is Mathematics?: An Elementary Approach to Ideas and Methods,* 2nd ed. Oxford, England: Oxford University Press, p. 36, 1996.

Shanks, D. *Solved and Unsolved Problems in Number Theory,* 4th ed. New York: Chelsea, p. 56, 1993.

Cannonball Problem

Find a way to stack a SQUARE of cannonballs laid out on the ground into a SQUARE PYRAMID (i.e., find a SQUARE NUMBER which is also SQUARE PYRAMIDAL). This corresponds to solving the DIOPHANTINE EQUATION

$$\sum_{i=1}^{k} i^2 = \tfrac{1}{6} k(1+k)(1+2k) = N^2$$

for some pyramid height k. The only solution is $k = 24$, $N = 70$, corresponding to 4900 cannonballs (Ball and Coxeter 1987, Dickson 1952), as conjectured by Lucas (1875, 1876) and proved by Watson (1918).

See also SPHERE PACKING, SQUARE NUMBER, SQUARE PYRAMID, SQUARE PYRAMIDAL NUMBER

References

Ball, W. W. R. and Coxeter, H. S. M. *Mathematical Recreations and Essays,* 13th ed. New York: Dover, p. 59, 1987.

Dickson, L. E. *History of the Theory of Numbers, Vol. 2: Diophantine Analysis.* New York: Chelsea, p. 25, 1952.

Lucas, É. Question 1180. *Nouvelles Ann. Math. Ser. 2* **14**, 336, 1875.

Lucas, É. Solution de Question 1180. *Nouvelles Ann. Math. Ser. 2* **15**, 429–432, 1876.

Ogilvy, C. S. and Anderson, J. T. *Excursions in Number Theory.* New York: Dover, pp. 77 and 152, 1988.

Pappas, T. "Cannon Balls & Pyramids." *The Joy of Mathematics.* San Carlos, CA: Wide World Publ./Tetra, p. 93, 1989.

Watson, G. N. "The Problem of the Square Pyramid." *Messenger. Math.* **48**, 1–22, 1918.

Canonical

The word canonical is used to indicate a particular choice from of a number of possible conventions. This convention allows a mathematical object or class of objects to be uniquely identified or standardized. For example, the RIGHT-HAND RULE for the CROSS PRODUCT is a convention, which corresponds to the canonical ORIENTATION in $\mathbb{R}^3$.

See also BASIS (VECTOR SPACE), CANONICAL BRICK, CANONICAL BUNDLE, CANONICAL TRANSFORMATION, RATIONAL CANONICAL FORM

Canonical Box Matrix

JORDAN BLOCK

Canonical Brick

A $1 \times 2 \times 4$ RECTANGULAR PARALLELEPIPED.

See also BRICK

References

Gardner, M. "Mathematical Games: In Which a Mathematical Aesthetic is Applied to Modern Minimal Art." *Sci. Amer.* **239**, 22–32, Nov. 1978.

Canonical Bundle

The canonical bundle is a HOLOMORPHIC LINE BUNDLE on a COMPLEX MANIFOLD which is determined by its COMPLEX STRUCTURE. On a coordinate chart $(z_1, \ldots z_n)$, it is spanned by the nonvanishing section $dz_1 \wedge \ldots \wedge dz_n$. The TRANSITION FUNCTION between COORDINATE CHARTS is given by the determinant of the JACOBIAN of the coordinate change.

The canonical bundle is defined in a similar way to the HOLOMORPHIC TANGENT BUNDLE. In fact, it is the nth EXTERIOR POWER of the DUAL BUNDLE to the HOLOMORPHIC TANGENT BUNDLE.

Canonical Form

A clear-cut way of describing every object in a class in a ONE-TO-ONE manner.

See also NORMAL FORM, ONE-TO-ONE

References

Petkovsek, M.; Wilf, H. S.; and Zeilberger, D. *A = B.* Wellesley, MA: A. K. Peters, p. 7, 1996.

Canonical Polygon

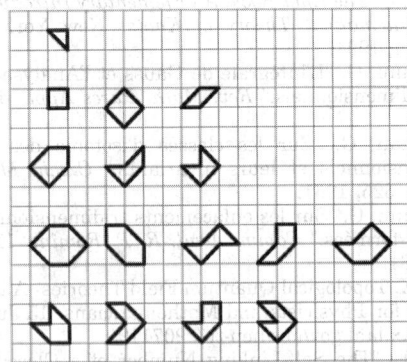

A closed polygon whose vertices lie on a POINT LATTICE and whose edges consist of vertical and horizontal steps of unit length or diagonal steps (at angles which are multiples of 45° with respect to the lattice axes) of length $\sqrt{2}$. In addition, no two steps may be taken in the same direction, no edge intersections are allowed, and no point may be a vertex of two edges. The numbers of distinct canonical polygons of $n = 1, 2, \ldots$ sides are 0, 0, 1, 3, 3, 9, 13, 48, 125, ... (Sloane's A052436).

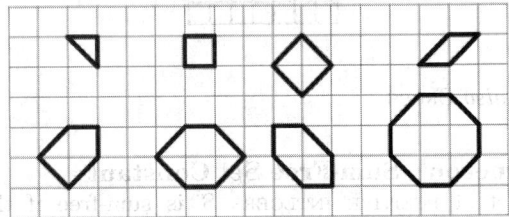

There are exactly eight distinct convex canonical polygons, illustrated above.

The concept can also be generalized to diagonals rotated with respect to the lattice axes.

See also GOLYGON, LATTICE POLYGON

References

Kyrmse, R. E. "Canonical Polygons." http://users.sti.com.br/rkyrmse/canonic-e.htm.
Sloane, N. J. A. Sequences A052436 in "An On-Line Version of the Encyclopedia of Integer Sequences." http://www.research.att.com/~njas/sequences/eisonline.html.

Canonical Polyhedron

A POLYHEDRON is said to be canonical if all its EDGES touch a SPHERE and the center of gravity of their contact points is the center of that SPHERE. Each combinatorial type of (GENUS zero) polyhedron contains just one canonical version. The ARCHIMEDEAN SOLIDS and their DUALS are all canonical.

References

Hart, G. W. "Calculating Canonical Polyhedra." *Mathematica Educ. Res.* **6**, 5–10, Summer 1997.

Hart, G. "Calculating Canonical Polyhedra." http://www.georgehart.com/canonical/canonical-supplement.html.
Hart, G. "Canonical Polyhedra." http://www.georgehart.com/virtual-polyhedra/canonical.html.

Canonical Transformation
SYMPLECTIC DIFFEOMORPHISM

Cantor Comb
CANTOR SET

Cantor Diagonal Argument
CANTOR DIAGONAL METHOD

Cantor Diagonal Method
A clever technique used by Georg Cantor to show that the INTEGERS and REALS cannot be put into a ONE-TO-ONE correspondence (i.e., the UNCOUNTABLY INFINITE set of REAL NUMBERS is "larger" than the COUNTABLY INFINITE set of INTEGERS).

It proceeds by first considering a countably infinite list of elements from a set S, each of which is an infinite set (in the case of the REALS, the decimal expansion of each REAL). A new member S' of S is then created by arranging its nth term to differ from the nth term of the nth member of S. This shows that S is not COUNTABLE, since any attempt to put it in one-to-one correspondence with the integers will fail to include some elements of S. The argument is rather subtle, and requires some care to describe clearly.

See also CARDINALITY, CONTINUUM HYPOTHESIS, COUNTABLE SET, COUNTABLY INFINITE

References
Courant, R. and Robbins, H. *What is Mathematics?: An Elementary Approach to Ideas and Methods*, 2nd ed. Oxford, England: Oxford University Press, pp. 81–83, 1996.
Hoffman, P. *The Man Who Loved Only Numbers: The Story of Paul Erdos and the Search for Mathematical Truth.* New York: Hyperion, pp. 220–223, 1998.
Penrose, R. *The Emperor's New Mind: Concerning Computers, Minds, and the Laws of Physics.* Oxford, England: Oxford University Press, pp. 84–85, 1989.

Cantor Diagonal Slash
CANTOR DIAGONAL METHOD

Cantor Dust

A FRACTAL which can be constructed using STRING REWRITING by creating a matrix three times the size of the current matrix using the rules

$$\text{line } 1: \text{"} * \text{"} \to \text{"} * \ * \text{"}, \text{" "} \to \text{" "}$$

$$\text{line } 2: \text{"} * \text{"} \to \text{" "}, \text{" "} \to \text{" "}$$

$$\text{line } 3: \text{"} * \text{"} \to \text{"} * \ * \text{"}, \text{" "} \to \text{" "}$$

Let N_n be the number of black boxes, L_n the length of a side of a box, and A_n the fractional AREA of black boxes after the nth iteration.

$$N_n = 4^n \tag{1}$$

$$L_n = (\tfrac{1}{3})^n = 3^{-n} \tag{2}$$

$$A_n = L_n^2 N_n = (\tfrac{4}{9})^n. \tag{3}$$

The CAPACITY DIMENSION is therefore

$$d_{\text{cap}} = -\lim_{n\to\infty} \frac{\ln N_n}{\ln L_n} = -\lim_{n\to\infty} \frac{\ln(4^n)}{\ln(3^{-n})} = \frac{2\ln 2}{\ln 3}$$

$$\approx 1.26186. \tag{4}$$

See also BOX FRACTAL, SIERPINSKI CARPET, SIERPINSKI SIEVE

References
Dickau, R. M. "Cantor Dust." http://forum.swarthmore.edu/advanced/robertd/cantor.html.
Ott, E. *Chaos in Dynamical Systems.* New York: Cambridge University Press, pp. 103–104, 1993.
Weisstein, E. W. "Fractals." MATHEMATICA NOTEBOOK FRACTAL.M.

Cantor Function
The function whose values are

$$\frac{1}{2}\left(\frac{c_1}{2} + \ldots + \frac{c_{m-1}}{2_{m-1}} + \frac{2}{2^m}\right)$$

for any number between

$$a \equiv \frac{c_1}{3} + \ldots + \frac{c_{m-1}}{3^{m-1}} + \frac{1}{3^m}$$

and

$$b \equiv \frac{c_1}{3} + \ldots + \frac{c_{m-1}}{3^{m-1}} + \frac{2}{3^m}.$$

Chalice (1991) shows that any real-valued function $F(x)$ on $[0, 1]$ which is MONOTONE INCREASING and satisfies

1. $F(0) = 0$,
2. $F(x/3) = F(x)/2$,
3. $F(1-x) = 1 - F(x)$

is the Cantor function.

The DEVIL'S STAIRCASE is sometimes also called the Cantor function (Devaney 1987, p. 110).

See also CANTOR SET, DEVIL'S STAIRCASE

References

Chalice, D. R. "A Characterization of the Cantor Function." *Amer. Math. Monthly* **98**, 255–258, 1991.
Devaney, R. L. *An Introduction to Chaotic Dynamical Systems.* Redwood City, CA: Addison-Wesley, 1987.
Wagon, S. "The Cantor Function" and "Complex Cantor Sets." §4.2 and 5.1 in *Mathematica in Action.* New York: W. H. Freeman, pp. 102–108 and 143–149, 1991.

Cantor Set

The Cantor set (T_∞) is given by taking the interval [0, 1] (set T_0), removing the middle third (T_1), removing the middle third of each of the two remaining pieces (T_2), and continuing this procedure ad infinitum. It is therefore the set of points in the INTERVAL [0, 1] whose ternary expansions do not contain 1, illustrated above.

This produces the SET of REAL NUMBERS $\{x\}$ such that

$$x = \frac{c_1}{3} + \ldots + \frac{c_n}{3^n} + \ldots, \qquad (1)$$

where c_n may equal 0 or 2 for each n. This is an infinite, PERFECT SET. The total length of the LINE SEGMENTS in the nth iteration is

$$\ell_n = \left(\frac{2}{3}\right)^n, \qquad (2)$$

and the number of LINE SEGMENTS is $N_n = 2^n$, so the length of each element is

$$\epsilon_n \equiv \frac{\ell}{N} = \left(\frac{1}{3}\right)^n \qquad (3)$$

and the CAPACITY DIMENSION is

$$d_{\text{cap}} \equiv -\lim_{\epsilon \to 0+} \frac{\ln N}{\ln \epsilon} = -\lim_{n \to \infty} \frac{n \ln 2}{-n \ln 3} = \frac{\ln 2}{\ln 3}$$
$$= 0.630929\ldots. \qquad (4)$$

The Cantor set is nowhere DENSE, so it has LEBESGUE MEASURE 0.

A general Cantor set is a CLOSED SET consisting entirely of BOUNDARY POINTS. Such sets are UNCOUNTABLE and may have 0 or POSITIVE LEBESGUE MEASURE. The Cantor set is the only totally disconnected, perfect, COMPACT METRIC SPACE up to a HOMEOMORPHISM (Willard 1970).

See also ALEXANDER'S HORNED SPHERE, ANTOINE'S NECKLACE, CANTOR FUNCTION, CLOSED SET, SCRAWNY CANTOR SET

References

Boas, R. P. Jr. *A Primer of Real Functions.* Washington, DC: Amer. Math. Soc., 1996.
Lauwerier, H. *Fractals: Endlessly Repeated Geometric Figures.* Princeton, NJ: Princeton University Press, pp. 15–20, 1991.
Harris, J. W. and Stocker, H. "Cantor Set." §4.11.4 in *Handbook of Mathematics and Computational Science.* New York: Springer-Verlag, p. 114, 1998.
Willard, S. §30.4 in *General Topology.* Reading, MA: Addison-Wesley, 1970.

Cantor Square Fractal

A FRACTAL which can be constructed using STRING REWRITING by creating a matrix three times the size of the current matrix using the rules

$$\text{line 1}: \ "*" \to "***", \ " \ " \to " \ \ "$$
$$\text{line 2}: \ "*" \to "* *", \ " \ " \to " \ \ "$$
$$\text{line 3}: \ "*" \to "***", \ " \ " \to " \ \ "$$

The first three steps are illustrated above.

The size of the unit element after the nth iteration is

$$L_n = \left(\frac{1}{3}\right)^n$$

and the number of elements is given by the RECURRENCE RELATION

$$N_n = 4N_{n-1} + 5(9^n)$$

where $N_1 \equiv 5$, and the first few numbers of elements are 5, 65, 665, 6305, ... Expanding out gives

$$N_n = 5 \sum_{k=0}^{n} 4^{n-k} 9^{k-1} = 9^n - 4^n.$$

The CAPACITY DIMENSION is therefore

$$D = -\lim_{n \to \infty} \frac{\ln N_n}{\ln L_n} = -\lim_{n \to \infty} \frac{\ln(9^n - 4^n)}{\ln(3^{-n})} = -\lim_{n \to \infty} \frac{\ln(9^n)}{\ln(3^{-n})}$$
$$= \frac{\ln 9}{\ln 3} = \frac{2 \ln 3}{\ln 3} = 2.$$

Since the DIMENSION of the filled part is 2 (i.e., the SQUARE is completely filled), Cantor's square fractal is not a true FRACTAL.

See also BOX FRACTAL, CANTOR DUST

References

Lauwerier, H. *Fractals: Endlessly Repeated Geometric Figures.* Princeton, NJ: Princeton University Press, pp. 82–83, 1991.

Weisstein, E. W. "Fractals." MATHEMATICA NOTEBOOK FRACTAL.M.

Cantor-Dedekind Axiom

The points on a line can be put into a ONE-TO-ONE correspondence with the REAL NUMBERS.

See also CARDINAL NUMBER, CONTINUUM HYPOTHESIS, DEDEKIND CUT

Cantor's Equation

$$\omega^\epsilon = \epsilon,$$

where ω is an ORDINAL NUMBER and ϵ is an INACCESSIBLE CARDINAL.

See also CARDINAL NUMBER, INACCESSIBLE CARDINAL, ORDINAL NUMBER

References

Conway, J. H. and Guy, R. K. *The Book of Numbers.* New York: Springer-Verlag, p. 274, 1996.

Cantor's Intersection Theorem

A theorem about (or providing an equivalent definition of)) COMPACT SETS, originally due to Georg Cantor. Given a decreasing sequence of bounded nonempty CLOSED SETS

$$C_1 \supset C_2 \supset C_3 \supset \ldots$$

in the real numbers, then Cantor's intersection theorem states that there must exist a point p in their intersection, $p \in C_n$ for all n. For example, $0 \in \cap [0, 1/n]$. It is also true in higher DIMENSIONS of EUCLIDEAN SPACE.

Note that the hypotheses stated above are crucial. The infinite intersection of open intervals may be empty, for instance $\cap (0, 1/n)$. Also, the infinite intersection of unbounded closed sets may be EMPTY, e.g., $\cap [n, \infty]$.

Cantor's intersection theorem is closely related to the HEINE-BOREL THEOREM and BOLZANO-WEIERSTRASS THEOREM, each of which can be easily derived from either of the other two. It can be used to show that the CANTOR SET is nonempty.

See also BOLZANO-WEIERSTRASS THEOREM, BOUNDED SET, CANTOR SET, CLOSED SET, COMPACT SET, HEINE-BOREL THEOREM, INTERSECTION, REAL NUMBER, TOPOLOGICAL SPACE

Cantor's Paradox

The SET of all SETS is its own POWER SET. Therefore, the CARDINALITY of the SET of all SETS must be bigger than itself.

See also CANTOR'S THEOREM, POWER SET

References

Curry, H. B. *Foundations of Mathematical Logic.* New York: Dover, p. 5, 1977.

Erickson, G. W. and Fossa, J. A. *Dictionary of Paradox.* Lanham, MD: University Press of America, pp. 32–33, 1998.

Cantor's Theorem

The CARDINAL NUMBER of any set is lower than the CARDINAL NUMBER of the set of all its subsets. A COROLLARY is that there is no highest ℵ (ALEPH).

See also CANTOR'S PARADOX

Cap

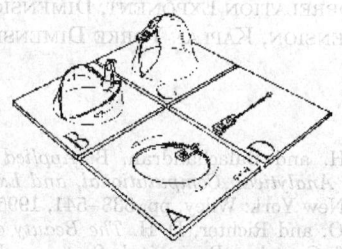

A topological object produced by puncturing a surface a single time, attaching two ZIPS around the puncture in opposite directions, distorting the hole so that the zips line up, and then zipping up. The cap is topologically trivial in the sense that a surface with a cap is topologically equivalent to a surface without one.

See also CROSS-CAP, CROSS-HANDLE, CUP, HANDLE, SPHERICAL CAP

References

Feller, W. *An Introduction to Probability Theory and Its Applications, Vol. 2, 3rd ed.* New York: Wiley, p. 104, 1971.

Francis, G. K. and Weeks, J. R. "Conway's ZIP Proof." *Amer. Math. Monthly* **106**, 393–399, 1999.

Capacity

TRANSFINITE DIAMETER

Capacity Dimension

A DIMENSION also called the FRACTAL DIMENSION, HAUSDORFF DIMENSION, and HAUSDORFF-BESICOVITCH DIMENSION in which nonintegral values are permitted. Objects whose capacity dimension is different from their TOPOLOGICAL DIMENSION are called FRACTALS. The capacity dimension of a compact METRIC SPACE X is a REAL NUMBER d_{capicity} such that if $n(\epsilon)$ denotes the minimum number of open sets of diameter less than or equal to ϵ, then $n(\epsilon)$ is proportional to ϵ^{-D} as $\epsilon \to 0$. Explicitly,

$$d_{capacity} \equiv -\lim_{\epsilon \to 0^+} \frac{\ln N}{\ln \epsilon}$$

(if the limit exists), where N is the number of elements forming a finite COVER of the relevant METRIC SPACE and ϵ is a bound on the diameter of the sets involved (informally, ϵ is the size of each element used to cover the set, which is taken to approach 0). If each element of a FRACTAL is equally likely to be visited, then $d_{capacity} = d_{information}$, where $d_{information}$ is the INFORMATION DIMENSION. The capacity dimension satisfies

$$d_{correlation} \leq d_{information} \leq d_{capacity}$$

where $d_{correlation}$ is the CORRELATION DIMENSION, and is conjectured to be equal to the LYAPUNOV DIMENSION.

See also CORRELATION EXPONENT, DIMENSION, HAUSDORFF DIMENSION, KAPLAN-YORKE DIMENSION

References

Nayfeh, A. H. and Balachandran, B. *Applied Nonlinear Dynamics: Analytical, Computational, and Experimental Methods.* New York: Wiley, pp. 538–541, 1995.

Peitgen, H.-O. and Richter, D. H. *The Beauty of Fractals: Images of Complex Dynamical Systems.* New York: Springer-Verlag, 1986.

Wheeden, R. L. and Zygmund, A. *Measure and Integral: An Introduction to Real Analysis.* New York: Dekker, 1977.

Cap-Cyclide Coordinates

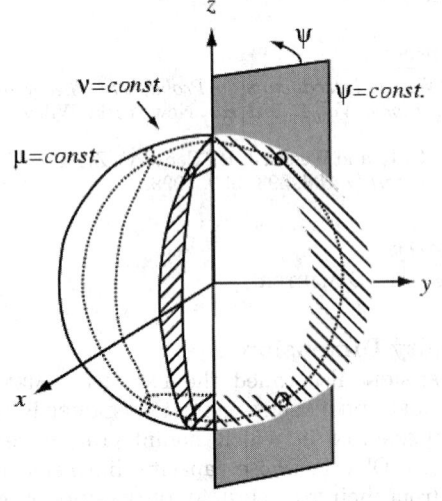

A coordinate system obtained by INVERSION of the BICYCLIDE COORDINATES. They are given by the transformation equations

$$x = \frac{\Lambda}{a\Upsilon} \operatorname{sn} \mu \operatorname{dn} v \cos \psi \tag{1}$$

$$y = \frac{\Lambda}{a\Upsilon} \operatorname{sn} \mu \operatorname{dn} v \sin \psi \tag{2}$$

$$z = \frac{\sqrt{k}\,Pi}{2a\Upsilon}, \tag{3}$$

where

$$\Lambda = 1 - \operatorname{dn}^2 \mu \operatorname{sn}^2 v \tag{4}$$

$$\Upsilon = \operatorname{sn}^2 \mu \operatorname{dn}^2 v + \left[\frac{\Lambda}{\sqrt{k}} + \operatorname{cn} \mu \operatorname{dn} \mu \operatorname{sn} v \operatorname{cn} v \right]^2 \tag{5}$$

$$\Pi = \frac{\Lambda^2}{k} - (\operatorname{sn}^2 \mu \operatorname{dn}^2 v + \operatorname{cn}^2 \mu \operatorname{dn}^2 \mu \operatorname{sn}^2 v \operatorname{cn}^2 v), \tag{6}$$

and $\operatorname{cn} x$, $\operatorname{dn} x$, and $\operatorname{sn} x$ are JACOBI ELLIPTIC FUNCTIONS. Surfaces of constant μ are ring cyclides with complicated equations (Moon and Spencer 1988, p. 133), surfaces of constant v are cap-cyclides with complicated equations (Moon and Spencer 1988, p. 133), and surfaces of constant ψ are half-planes

$$\tan \psi = \frac{y}{x}. \tag{7}$$

See also BICYCLIDE COORDINATES, CYCLIDIC COORDINATES, DISK-CYCLIDE COORDINATES, FLAT-RING CYCLIDE COORDINATES

References

Moon, P. and Spencer, D. E. "Cap-Cyclide Coordinates (μ, v, ψ)." Fig. 4.11 in *Field Theory Handbook, Including Coordinate Systems, Differential Equations, and Their Solutions, 2nd ed.* New York: Springer-Verlag, pp. 132–135, 1988.

Capping

CUMULATION

Carathéodory Derivative

A function f is Carathéodory differentiable at a if there exists a function ϕ which is CONTINUOUS at a such that

$$f(x) - f(a) = \phi(x)(x - a).$$

Every function which is Carathéodory differentiable is also FRÉCHET DIFFERENTIABLE.

See also DERIVATIVE, FRÉCHET DERIVATIVE

Carathéodory's Fundamental Theorem

Each point in the CONVEX HULL of a set S in $\mathbb{R}^n$ is in the convex combination of $n+1$ or fewer points of S.

See also CONVEX HULL, HELLY'S THEOREM

References

Eckhoff, J. "Helly, Radon, and Carathéodory Type Theorems." Ch. 2.1 in *Handbook of Convex Geometry* (Ed. P. M. Gruber and J. M. Wills). Amsterdam, Netherlands: North-Holland, pp. 389–448, 1993.

Carathéodory's Theorem

If Ω_1 and Ω_2 are bounded domains, $\partial\Omega_1$, $\partial\Omega_2$ are JORDAN CURVES, and $\varphi : \Omega_1 \to \Omega_2$ is a CONFORMAL MAPPING, then φ (respectively, φ^{-1}) extends one-to-one and continuously to $\partial\Omega_1$ (respectively, $\partial\Omega_2$).

References

Krantz, S. G. *Handbook of Complex Analysis.* Boston, MA: Birkhäuser, p. 152, 1999.

Cardano's Formula

CUBIC EQUATION

Cardinal Addition

Let A and B be any sets with empty INTERSECTION, and let $|X|$ denote the CARDINAL NUMBER of a SET X. Then

$$|A| + |B| = |A \cup B|$$

(Ciesielski 1997, p. 68; Dauben 1990, p. 173; Rubin 1967, p. 274; Suppes 1972, pp. 112–113).

It is an interesting exercise to show that cardinal addition is WELL DEFINED. The main steps are to show that for any CARDINAL NUMBERS a and b, there exist disjoint sets A and B with CARDINAL NUMBERS a and b, and to show that if A and B are disjoint and C and D disjoint with $|A| = |C|$ and $|B| = |D|$ then $|A \cup B| = |C \cup D|$. The second of these is easy. The first is a little tricky and requires an appeal to the axioms of SET THEORY. Also, one needs to restrict the definition of cardinal to guarantee if a is a cardinal, then there is a set A satisfying $|A| = a$.

See also CARDINAL MULTIPLICATION, CARDINAL EXPONENTIATION

References

Ciesielski, K. *Set Theory for the Working Mathematician.* Cambridge, England: Cambridge University Press, 1997.
Dauben, J. W. *Georg Cantor: His Mathematics and Philosophy of the Infinite.* Princeton, NJ: Princeton University Press, 1990.
Rubin, J. E. *Set Theory for the Mathematician.* New York: Holden-Day, 1967.
Suppes, P. *Axiomatic Set Theory.* New York: Dover, 1972.

Cardinal Comparison

For any sets A and B, their CARDINAL NUMBERS satisfy $|A| \le |B|$ IFF there is a one-to-one function f from A into B (Rubin 1967, p. 266; Suppes 1972, pp. 94 and 116). It is easy to show this satisfies the reflexive and transitive axioms of a PARTIAL ORDER. However, it is difficult to show the antisymmetry property, whose proof is known as the SCHRÖDER-BERNSTEIN THEOREM. To show the trichotomy property, one must use the AXIOM OF CHOICE.

Although an order type can be defined similarly, it does not seem usual to do so.

See also SCHRÖDER-BERNSTEIN THEOREM

References

Rubin, J. E. *Set Theory for the Mathematician.* New York: Holden-Day, 1967.
Suppes, P. *Axiomatic Set Theory.* New York: Dover, 1972.

Cardinal Exponentiation

Let A and B be any sets, and let $|X|$ be the CARDINAL NUMBER of a set X. Then cardinal exponentiation is defined by

$$|A|^{|B|} = |\text{set of all function from } B \text{ into } A|$$

(Ciesielski 1997, p. 68; Dauben 1990, p. 174; Moore 1982, p. 37; Rubin 1967, p. 275, Suppes 1972, p. 116).

It is easy to show that the CARDINAL NUMBER of the POWER SET of A is $2^{|A|}$, sine $|\{0, 1\}| = 2$ and there is a natural BIJECTION between the SUBSETS of A and the functions from A into $\{0, 1\}$.

See also CARDINAL ADDITION, CARDINAL MULTIPLICATION, CARDINAL NUMBER, POWER SET

References

Ciesielski, K. *Set Theory for the Working Mathematician.* Cambridge, England: Cambridge University Press, 1997.
Dauben, J. W. *Georg Cantor: His Mathematics and Philosophy of the Infinite.* Princeton, NJ: Princeton University Press, 1990.
Moore, G. H. *Zermelo's Axiom of Choice: Its Origin, Development, and Influence.* New York: Springer-Verlag, 1982.
Rubin, J. E. *Set Theory for the Mathematician.* New York: Holden-Day, 1967.
Suppes, P. *Axiomatic Set Theory.* New York: Dover, 1972.

Cardinal Multiplication

Let A and B be any sets. Then the product of $|A|$ and $|B|$ is defined as the CARTESIAN PRODUCT

$$|A| * |B| = |A \times B|$$

(Ciesielski 1997, p. 68; Dauben 1990, p. 173; Moore 1982, p. 37; Rubin 1967, p. 274; Suppes 1972, pp. 114–115).

See also CARDINAL ADDITION, CARDINAL EXPONENTIATION

References

Ciesielski, K. *Set Theory for the Working Mathematician.* Cambridge, England: Cambridge University Press, 1997.

Dauben, J. W. *Georg Cantor: His Mathematics and Philosophy of the Infinite.* Princeton, NJ: Princeton University Press, 1990.

Moore, G. H. *Zermelo's Axiom of Choice: Its Origin, Development, and Influence.* New York: Springer-Verlag, 1982.

Rubin, J. E. *Set Theory for the Mathematician.* New York: Holden-Day, 1967.

Suppes, P. *Axiomatic Set Theory.* New York: Dover, 1972.

Cardinal Number

In common usage, a cardinal number is a number used in counting (a COUNTING NUMBER), such as 1, 2, 3, ...

In formal SET THEORY, a cardinal number (also called "the cardinality") is a type of number defined in such a way that any method of counting SETS using it gives the same result. (This is not true for the ORDINAL NUMBERS.) In fact, the cardinal numbers are obtained by collecting all ORDINAL NUMBERS which are obtainable by counting a given set. A set has $\aleph_0$ (ALEPH-0) members if it can be put into a ONE-TO-ONE correspondence with the finite ORDINAL NUMBERS. The cardinality of a set is also frequently referred to as the "power" of a set (Moore 1982, Dauben 1990, Suppes 1972).

In Cantor's original notation, the symbol for a SET A annotated with a single overbar $\bar{A}$ indicated A stripped of any structure besides order, hence it represented the ORDER TYPE of the set. A double overbar $\bar{\bar{A}}$ then indicated stripping the order from the set and thus indicated the cardinal number of the set. However, in modern notation, the symbol $|A|$ is used to denote the cardinal number of set.

Cantor, the father of modern SET THEORY, noticed that while the ORDINAL NUMBERS $\omega + 1$, $\omega + 2$, ... were bigger than omega in the sense of order, they were not bigger in the sense of EQUIPOLLENCE. This led him to study what would come to be called cardinal numbers. He called the ordinals ω, $\omega + 1$, ... that are equipollent to the integers "the second number class" (as opposed to the finite ordinals, which he called the "first number class"). Cantor showed

1. The second number class is bigger than the first.
2. There is no class bigger than the first number class and smaller than the second.
3. The class of real numbers is bigger than the first number class.

One of the first serious mathematical definitions of cardinal was the one devised by Gottlob Frege and Bertrand Russell, who defined a cardinal number $|A|$ as the set of all sets EQUIPOLLENT to A. (Moore 1982, p. 153; Suppes 1972, p. 109). Unfortunately, the objects produced by this definition are not sets in the sense of ZERMELO-FRAENKEL SET THEORY, but

rather "PROPER CLASSES" in the terminology of von Neumann.

Tarski (1924) proposed to instead define a cardinal number by stating that every set A is associated with a cardinal number $|A|$, and two sets A and B have the same cardinal number IFF they are EQUIPOLLENT (Moore 1982, pp. 52 and 214; Rubin 1967, p. 266; Suppes 1972, p. 111). The problem is that this definition requires a special axiom to guarantee that cardinals exist.

A. P. Morse and Dana Scott defined cardinal number by letting A be any set, then calling $|A|$ the set of all sets EQUIPOLLENT to A and of least possible RANK (Rubin 1967, p. 270).

It is possible to associate cardinality with a specific set, but the process required either the AXIOM OF FOUNDATION or the AXIOM OF CHOICE. However, these are two of the more controversial ZERMELO-FRAENKEL AXIOMS. With the AXIOM OF CHOICE, the cardinals can be enumerated through the ordinals. In fact, the two can be put into one-to-one correspondence. The AXIOM OF CHOICE implies that every set can be WELL ORDERED and can therefore be associated with an ORDINAL NUMBER.

This leads to the definition of cardinal number for a SET A as the least ORDINAL NUMBER b such that A and b are EQUIPOLLENT. In this model, the cardinal numbers are just the INITIAL ORDINALS. This definition obviously depends on the AXIOM OF CHOICE, because if the AXIOM OF CHOICE is not true, then there are sets that cannot be well ordered. Cantor believed that every set could be well ordered and used this correspondence to define the $\aleph$s ("alephs"). For any ORDINAL NUMBER α, $\aleph_\alpha = \omega_\alpha$.

An INACCESSIBLE CARDINAL cannot be expressed in terms of a smaller number of smaller cardinals.

See also ALEPH, ALEPH-0, ALEPH-1, CANTOR-DEDEKIND AXIOM, CANTOR DIAGONAL SLASH, CARDINAL ADDITION, CARDINAL EXPONENTIATION, CARDINAL MULTIPLICATION, CONTINUUM, CONTINUUM HYPOTHESIS, EQUIPOLLENT, INACCESSIBLE CARDINALS AXIOM, INFINITY, ORDINAL NUMBER, POWER SET, SURREAL NUMBER, UNCOUNTABLE SET

References

Cantor, G. *Über unendliche, lineare Punktmannigfaltigkeiten, Arbeiten zur Mengenlehre aus dem Jahren 1872–1884.* Leipzig, Germany: Teubner, 1884.

Conway, J. H. and Guy, R. K. "Cardinal Numbers." In *The Book of Numbers.* New York: Springer-Verlag, pp. 277–282, 1996.

Courant, R. and Robbins, H. "Cantor's 'Cardinal Numbers.'" §2.4.3 in *What is Mathematics?: An Elementary Approach to Ideas and Methods, 2nd ed.* Oxford, England: Oxford University Press, pp. 83–86, 1996.

Dauben, J. W. *Georg Cantor: His Mathematics and Philosophy of the Infinite.* Princeton, NJ: Princeton University Press, 1990.

Ferreirós, J. "The Notion of Cardinality and the Continuum Hypothesis." Ch. 6 in *Labyrinth of Thought: A History of Set Theory and Its Role in Modern Mathematics.* Basel, Switzerland: Birkhäuser, pp. 171–214, 1999.

Moore, G. H. *Zermelo's Axiom of Choice: Its Origin, Development, and Influence.* New York: Springer-Verlag, 1982.

Rubin, J. E. *Set Theory for the Mathematician.* New York: Holden-Day, 1967.

Suppes, P. *Axiomatic Set Theory.* New York: Dover, 1972.

Tarski, A. "Sur quelques théorèmes qui équivalent à l'axiome du choix." *Fund. Math.* **5**, 147–154, 1924.

Cardinality

CARDINAL NUMBER

Cardioid

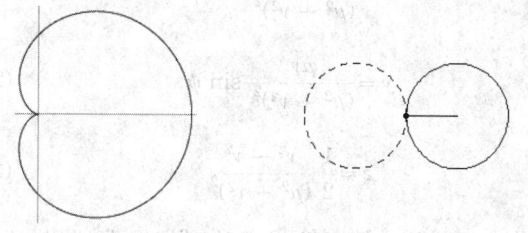

The curve given by the POLAR equation

$$r = a(1 + \cos\theta), \qquad (1)$$

sometimes also written

$$r = 2b(1 + \cos\theta), \qquad (2)$$

where $b \equiv a/2$, the CARTESIAN equation

$$(x^2 + y^2 - ax)^2 = a^2(x^2 + y^2), \qquad (3)$$

and the PARAMETRIC EQUATIONS

$$x = a\cos t(1 + \cos t) \qquad (4)$$

$$y = a\sin t(1 + \cos t). \qquad (5)$$

The cardioid is a degenerate case of the LIMAÇON. It is also a 1-CUSPED EPICYCLOID (with $r = r$) and is the CAUSTIC formed by rays originating at a point on the circumference of a CIRCLE and reflected by the CIRCLE.

the name cardioid was first used by de castillon in *philosophical transactions of the royal society* in 1741. its ARC LENGTH was found by la hire in 1708. there are exactly three PARALLEL TANGENTS to the cardioid with any given gradient. also, the TANGENTS at the ends of any CHORD through the CUSP point are at RIGHT ANGLES. The length of any CHORD through the CUSP point is $2a$.

The cardioid may also be generated as follows. Draw a CIRCLE C and fix a point A on it. Now draw a set of CIRCLES centered on the CIRCUMFERENCE of C and passing through A. The ENVELOPE of these CIRCLES is then a cardioid (Pedoe 1995). Let the CIRCLE C be centered at the origin and have RADIUS 1, and let the fixed point be $A = (1, 0)$. Then the RADIUS of a CIRCLE centered at an ANGLE θ from $(1, 0)$ is

$$r^2 = (0 - \cos\theta)^2 + (1 - \sin\theta)^2$$
$$= \cos^2\theta + 1 - 2\sin\theta + \sin^2\theta = 2(1 - \sin\theta). \qquad (6)$$

If the fixed point A is not on the circle, then the resulting ENVELOPE is a LIMAÇON instead of a cardioid.

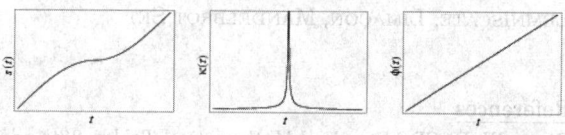

The ARC LENGTH, CURVATURE, and TANGENTIAL ANGLE are

$$s = \int_0^t 2|\cos(\tfrac{1}{2}t)|\,dt = 4a\sin(\tfrac{1}{2}\theta) \qquad (7)$$

$$\kappa = \frac{3|\sec(\tfrac{1}{2}\theta)|}{4a} \qquad (8)$$

$$\phi = \tfrac{3}{2}\theta. \qquad (9)$$

As usual, care must be taken in the evaluation of $s(t)$ for $t > \pi$. Since (7) comes from an integral involving the ABSOLUTE VALUE of a function, it must be monotonic increasing. Each QUADRANT can be treated correctly by defining

$$n = \left\lfloor \frac{t}{\pi} \right\rfloor + 1, \qquad (10)$$

where $\lfloor x \rfloor$ is the FLOOR FUNCTION, giving the formula

$$s(t) = (-1)^{1 + [n(\bmod 2)]}4\sin(\tfrac{1}{2}t) + 8\left\lfloor \tfrac{1}{2}n \right\rfloor. \qquad (11)$$

The PERIMETER of the curve is

$$L = \int_0^{2\pi} |2a \cos(\tfrac{1}{2}\theta)|\, d\theta = 4a \int_0^{\pi} \cos(\tfrac{1}{2}\theta)\, d\theta$$

$$= 4a \int_0^{\pi/2} \cos\phi(2\, d\phi) = 8a \int_0^{\pi/2} \cos\phi\, d\phi$$

$$= 8a[\sin\phi]_0^{\pi/2} = 8a. \tag{12}$$

The AREA is

$$A = \tfrac{1}{2} \int_0^{2\pi} r^2\, d\theta = \tfrac{1}{2} a^2 \int_0^{2\pi} (1 + 2\cos\theta + \cos^2\theta)\, d\theta$$

$$= \tfrac{1}{2} a^2 \int_0^{2\pi} \{1 + 2\cos\theta + \tfrac{1}{2}[1 + \cos(2\theta)]\}\, d\theta$$

$$= \tfrac{1}{2} a^2 \int_0^{2\pi} [\tfrac{3}{2} + 2\cos\theta + \tfrac{1}{2}\cos(2\theta)]\, d\theta$$

$$= \tfrac{1}{2} a^2 [\tfrac{3}{2}\theta + 2\sin\theta + \tfrac{1}{4}\sin(2\theta)]_0^{2\pi} = \tfrac{3}{2}\pi a^2. \tag{13}$$

See also CARDIOID COORDINATES, CIRCLE, CISSOID, COIN PARADOX, CONCHOID, EQUIANGULAR SPIRAL, LEMNISCATE, LIMAÇON, MANDELBROT SET

References

Beyer, W. H. *CRC Standard Mathematical Tables, 28th ed.* Boca Raton, FL: CRC Press, p. 214, 1987.

Gray, A. "Cardioids." §3.3 in *Modern Differential Geometry of Curves and Surfaces with Mathematica, 2nd ed.* Boca Raton, FL: CRC Press, pp. 54–55, 1997.

Lawrence, J. D. *A Catalog of Special Plane Curves.* New York: Dover, pp. 118–121, 1972.

Lockwood, E. H. "The Cardioid." Ch. 4 in *A Book of Curves.* Cambridge, England: Cambridge University Press, pp. 34–43, 1967.

MacTutor History of Mathematics Archive. "Cardioid." http://www-groups.dcs.st-and.ac.uk/~history/Curves/Cardioid.html.

Pedoe, D. *Circles: A Mathematical View, rev. ed.* Washington, DC: Math. Assoc. Amer., pp. xxvi–xxvii, 1995.

Wells, D. *The Penguin Dictionary of Curious and Interesting Geometry.* London: Penguin, pp. 24–25, 1991.

Yates, R. C. "The Cardioid." *Math. Teacher* **52**, 10–14, 1959.

Yates, R. C. "Cardioid." *A Handbook on Curves and Their Properties.* Ann Arbor, MI: J. W. Edwards, pp. 4–7, 1952.

Cardioid Caustic

The CATACAUSTIC of a CARDIOID for a RADIANT POINT at the CUSP is a NEPHROID. The CATACAUSTIC for PARALLEL rays crossing a CIRCLE is a CARDIOID.

Cardioid Coordinates

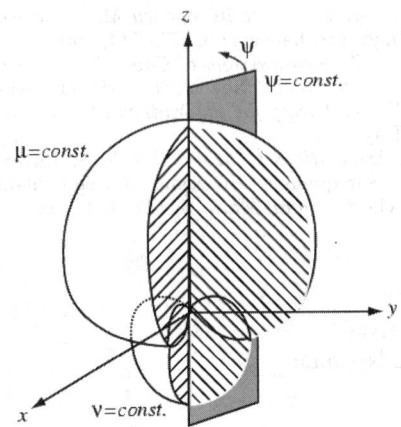

A coordinate system (μ, v, ψ) defined by the coordinate transformation

$$x = \frac{\mu v}{(\mu^2 + v^2)^2} \cos\psi \tag{1}$$

$$y = \frac{\mu v}{(\mu^2 + v^2)^2} \sin\psi \tag{2}$$

$$z = \frac{1}{2} \frac{v^2 - v^2}{(\mu^2 + v^2)^2} \tag{3}$$

with $\mu, v \leq 0$ and $\psi \in [0, 2\pi)$. Surfaces of constant μ are given by the cardioids of revolution intersecting the positive half of the z-axis

$$x^2 + y^2 + z^2 = \frac{1}{4\mu^2}[\sqrt{x^2 + y^2 + z^2} + 1], \tag{4}$$

surfaces of constant v by the cardioids of revolution intersecting the negative half of the z-axis

$$x^2 + y^2 + z^2 = \frac{1}{4v^2}[\sqrt{x^2 + y^2 + z^2} - z], \tag{5}$$

and surfaces of constant ψ by the half-planes

$$\tan\psi = \frac{y}{x}. \tag{6}$$

The metric coefficients are

$$g_{\mu\mu} = \frac{1}{(\mu^2 + v^2)^3} \tag{7}$$

$$g_{vv} = \frac{1}{(\mu^2 + v^2)^3} \tag{8}$$

$$g_{\psi\psi} = \frac{\mu^2 v^2}{(\mu^2 + v^2)^4} \tag{9}$$

See also CARDIOID

References

Moon, P. and Spencer, D. E. "Cardioid Coordinate (μ, ν, ψ)." Fig. 4.02 in *Field Theory Handbook, Including Coordinate Systems, Differential Equations, and Their Solutions, 2nd ed.* New York: Springer-Verlag, pp. 107–109, 1988.

Cardioid Evolute

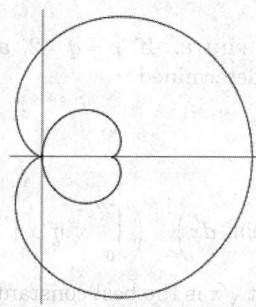

$$x = \tfrac{2}{3}a + \tfrac{1}{3}a \cos\theta(1 - \cos\theta)$$

$$y = \tfrac{1}{3}a \sin\theta(1 - \cos\theta).$$

This is a mirror-image CARDIOID with $a' = a/3$.

Cardioid Inverse Curve

If the CUSP of the cardioid is taken as the INVERSION CENTER, the cardioid inverts to a PARABOLA.

Cardioid Involute

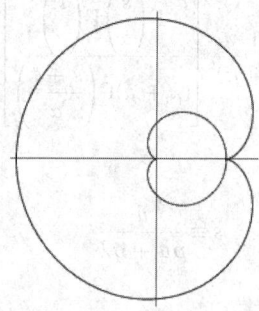

$$x = 2a + 3a \cos\theta(1 - \cos\theta)$$

$$y = 3a \sin\theta(1 - \cos\theta).$$

This is a mirror-image CARDIOID with $a' = 3a$.

Cardioid Pedal Curve

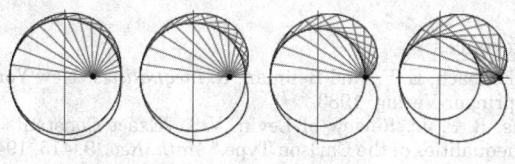

The PEDAL CURVE of the CARDIOID where the PEDAL POINT is the CUSP is CAYLEY'S SEXTIC.

Cards

Cards are a set of n rectangular pieces of cardboard with markings on one side and a uniform pattern on the other. The collection of all cards is called a "deck," and a normal deck of cards consists of 52 cards having 14 distinct values for each of four different "suits." The suits are called clubs (♣), diamonds (♢), hearts (♡), and spades (♠). Spades and clubs are colored black, while hearts and diamonds are colored red. The cards of each suit are numbered 1 through 13, where the special terms ace (1), jack (11), queen (12), and king (13) are used instead of numbers 1 and 11–13. However, in BRIDGE and a number of other games, the ace is considered the *highest* card, and so would be assigned a value of 14 instead of 1.

The randomization of the order of cards in a deck is called SHUFFLING. Cards are used in many gambling games (such as POKER), and the investigation of the probabilities of various outcomes in card games was one of the original motivations for the development of modern PROBABILITY theory.

See also BRIDGE CARD GAME, CLOCK SOLITAIRE, COIN, COIN TOSSING, CRIBBAGE, DICE, POKER, SHUFFLE

References

Chatto, W. A. *Facts and Speculations on the Origin and History of Playing Cards.* Saint Clair Shores, MI: Scholarly Press, 1977.
Hargrave, C. P. *History of Playing Cards and a Bibliography of Cards and Gaming.* New York: Dover, 1986.
Horr, N. T. *Bibliography of Card Games and of the History of Playing Cards.* Montclair, NJ: Patterson Smith, 1972.
Jessel, F. and Horr, N. T. *Bibliographies of Works on Playing Cards and Gaming.* Montclair, NJ: Patterson Smith, 1972.
Leeming, J. *Games and Fun with Playing Cards.* New York: Dover, 1980.
Parlett, D. S. *A Dictionary of Card Games.* Oxford, England: Oxford University Press, 1992.
Parlett, D. S. *The Oxford Guide to Card Games: A History of Card Games.* Oxford, England: Oxford University Press, 1991.
Parlett, D. S. *Solitaire: Aces Up and 399 Other Card Games.* New York: Pantheon, 1991.
Sackson, S. *Card Games Around the World.* New York: Dover, 1994.
University of Waterloo. "Playing Cards." http://www.ahs.uwaterloo.ca/~museum/vexhibit/plcards/plcards.html.

Caret

The symbol ∧ which is used to denote partial conjunction in symbolic logic. It also appears in several other contexts in mathematics and is sometimes called a "WEDGE". The shape of the caret is similar to that of the HAT.

See also HAT, WEDGE

References

Bringhurst, R. *The Elements of Typographic Style, 2nd ed.* Point Roberts, WA: Hartley and Marks, p. 274, 1997.

Carleman Equation

The system of PARTIAL DIFFERENTIAL EQUATIONS

$$u_t + u_x = v^2 - u^2$$
$$v_t - v_x = u^2 - v^2.$$

References

Kaper, H. G. and Leaf, G. K. "Initial Value Problems for the Carleman Equation." *Nonlinear Anal.* **4**, 343–362, 1980.
Zwillinger, D. *Handbook of Differential Equations, 3rd ed.* Boston, MA: Academic Press, p. 137, 1997.

Carleman's Inequality

Let $\{a_i\}_{i=1}^n$ be a SET of POSITIVE numbers. Then

$$\sum_{i=1}^n (a_1 a_2 \ldots a_i)^{1/i} \le e \sum_{i=1}^n a_i$$

(which is given incorrectly in Gradshteyn and Ryzhik 1994). Here, the constant e is the best possible, in the sense that counterexamples can be constructed for any stricter INEQUALITY which uses a smaller constant. The theorem is suggested by writing $a_i' = a_i^p$ in HARDY'S INEQUALITY

$$\sum_{i=1}^n \left(\frac{a_1 + \ldots + a_i}{i} \right)^p < \left(\frac{p}{p-1} \right)^p \sum_{i=1}^n a_i^p \qquad (1)$$

and letting $p \to \infty$.

See also ARITHMETIC MEAN, e, GEOMETRIC MEAN, HARDY'S INEQUALITY

References

Carleman, T. "Sur les fonctions quasi-analytiques." *Conférences faites au cinqui'eme congrès des mathématiciens scandinaves.* Helsingfors, pp. 181–196, 1923.
Gradshteyn, I. S. and Ryzhik, I. M. *Tables of Integrals, Series, and Products, 6th ed.* San Diego, CA: Academic Press, p. 1126, 2000.
Hardy, G. H.; Littlewood, J. E.; and Pólya, G. "Carleman's Inequality." §9.12 in *Inequalities, 2nd ed.* Cambridge, England: Cambridge University Press, pp. 249–250, 1988.
Kaluza, T. and Szego, G. "Über Reihen mit lauter positiven Gliedern." *J. London Math. Soc.* **2**, 266–272, 1927.
Knopp, K. "Über Reihen mit positiven Gliedern." *J. London Math. Soc.* **3**, 205–211, 1928.
Mitrinovic, D. S. *Analytic Inequalities.* New York: Springer-Verlag, p. 131, 1970.
Ostrowski, A. "Über quasi-analytischen Funktionen und Bestimmtheit asymptotischer Entwicklungen." *Acta Math.* **53**, 181–266, 1929.
Pólya, G. "Proof of an Inequality." *Proc. London Math. Soc.* **24**, lvii, 1926.
Valiron, G. §3, Appendix B in *Lectures on the General Theory of Integral Functions.* New York: Chelsea, pp. 186–187, 1949.

Carlson-Levin Constant

N.B. A detailed online essay by S. Finch was the starting point for this entry.

Assume that f is a NONNEGATIVE REAL function on $[0, \infty)$ and that the two integrals

$$\int_0^\infty x^{p-1-\lambda} [f(x)]^p \, dx \qquad (1)$$

$$\int_0^\infty x^{q-1+\mu} [f(x)]^q \, dx \qquad (2)$$

exist and are FINITE. If $p = q = 2$ and $\lambda = \mu = 1$, Carlson (1934) determined

$$\int_0^\infty f(x) \, dx$$
$$\le \sqrt{\pi} \left(\int_0^\infty [f(x)]^2 \, dx \right)^{1/4} \left(\int_0^\infty x^2 [f(x)]^2 \, dx \right)^{1/4} \qquad (3)$$

and showed that $\sqrt{\pi}$ is the best constant (in the sense that counterexamples can be constructed for any stricter INEQUALITY which uses a smaller constant). For the general case

$$\int_0^\infty f(x) \, dx$$
$$\le C \left(\int_0^\infty x^{p-1-\lambda} [f(x)]^p \, dx \right)^s \left(\int_0^\infty x^{q-1+\mu} [f(x)]^q \, dx \right)^t, \qquad (4)$$

and Levin (1948) showed that the best constant

$$C = \frac{1}{(ps)^s (qt)^t} \left[\frac{\Gamma\left(\dfrac{s}{\alpha}\right) \Gamma\left(\dfrac{t}{\alpha}\right)}{(\lambda + \mu) \Gamma\left(\dfrac{s+t}{\alpha}\right)} \right]^\alpha, \qquad (5)$$

where

$$s \equiv \frac{\mu}{p\mu + q\lambda} \qquad (6)$$

$$t \equiv \frac{\lambda}{p\mu + q\lambda} \qquad (7)$$

$$\alpha \equiv 1 - s - t \qquad (8)$$

and $\Gamma(z)$ is the GAMMA FUNCTION.

References

Beckenbach, E. F.; and Bellman, R. *Inequalities.* New York: Springer-Verlag, 1983.
Boas, R. P. Jr. Review of Levin, V. I. "Exact Constants in Inequalities of the Carlson Type." *Math. Rev.* **9**, 415, 1948.
Finch, S. "Favorite Mathematical Constants." http://www.mathsoft.com/asolve/constant/crlslvn/crlslvn.html.
Levin, V. I. "Exact Constants in Inequalities of the Carlson Type." *Doklady Akad. Nauk. SSSR (N. S.)* **59**, 635–638, 1948. English review in Boas (1948).

Mitrinovic, D. S.; Pecaric, J. E.; and Fink, A. M. *Inequalities Involving Functions and Their Integrals and Derivatives.* Amsterdam, Netherlands: Kluwer, 1991.

Carlson's Theorem

If $f(z)$ is regular and OF THE FORM $\mathcal{O}(e^{k|z|})$ where $k < \pi$, for $\Re[z] \geq 0$, and if $f(z) = 0$ for $z = 0, 1, \ldots$, then $f(z)$ is identically zero.

See also GENERALIZED HYPERGEOMETRIC FUNCTION

References

Bailey, W. N. "Carlson's Theorem." §5.3 in *Generalised Hypergeometric Series.* Cambridge, England: Cambridge University Press, pp. 36–40, 1935.
Carlson, F. "Sur une classe de séries de Taylor." Dissertation. Uppsala, Sweden, 1914.
Hardy, G. H. "On Two Theorems of F. Carlson and S. Wigert." *Acta Math.* **42**, 327–339, 1920.
Riesz, M. "Sur le principe de Phragmén-Lindelöf." *Proc. Cambridge Philos. Soc.* **20**, 205–207, 1920.
Riesz, M. Erratum to "Sur le principe de Phragmén-Lindelöf." *Proc. Cambridge Philos. Soc.* **21**, 6, 1921.
Titchmarsh, E. C. Ch. 5 in *The Theory of Functions, 2nd ed.* Oxford, England: Oxford University Press, 1960.
Wigert, S. "Sur un théorème concernant les fonctions entières." *Archiv för Mat. Astr. o Fys.* **11**, No. 22, 1916.

Carlyle Circle

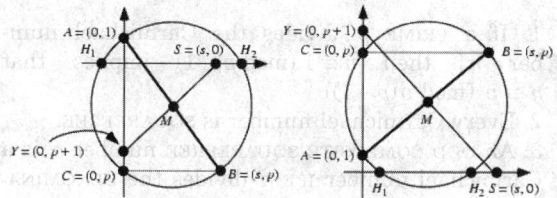

Consider a QUADRATIC EQUATION $x^2 - sx + p = 0$ where s and p denote *signed* lengths. The CIRCLE which has the points $A = (0, 1)$ and $B = (s, p)$ as a DIAMETER is then called the Carlyle circle $C_{s,p}$ of the equation. The CENTER of $C_{s,p}$ is then at the MIDPOINT of AB, $M = (s/2, (1+p)/2)$, which is also the MIDPOINT of $S = (s, 0)$ and $Y = (0, 1+p)$. Call the points at which $C_{s,p}$ crosses the X-AXIS $H_1 = (x_1, 0)$ and $H_2 = (x_2, 0)$ (with $x_1 \geq x_2$). Then

$$s = x_1 + x_2$$

$$p = x_1 x_2$$

$$(x - x_1)(x - x_2) = x^2 - sx + p,$$

so x_1 and x_2 are the ROOTS of the quadratic equation.

See also 257-GON, 65537-GON, HEPTADECAGON, PENTAGON

References

Bold, B. *Famous Problems of Geometry and How to Solve Them.* New York: Dover, pp. 4–5, 1982.
De Temple, D. W. "Carlyle Circles and the Lemoine Simplicity of Polygonal Constructions." *Amer. Math. Monthly* **98**, 97–108, 1991.

Eves, H. *An Introduction to the History of Mathematics, 6th ed.* Philadelphia, PA: Saunders, 1990.
Leslie, J. *Elements of Geometry and Plane Trigonometry with an Appendix and Very Copious Notes and Illustrations, 4th ed., improved and exp.* Edinburgh: W. & G. Tait, 1820.

Carmichael Condition

A number n satisfies the Carmichael condition IFF $(p-1)|(n/p-1)$ for all PRIME DIVISORS p of n. This is equivalent to the condition $(p-1)|(n-1)$ for all PRIME DIVISORS p of n.

See also CARMICHAEL NUMBER

References

Borwein, D.; Borwein, J. M.; Borwein, P. B.; and Girgensohn, R. "Giuga's Conjecture on Primality." *Amer. Math. Monthly* **103**, 40–50, 1996.

Carmichael Function

There are two definitions of the Carmichael function. One is the reduced totient function (also called the least universal exponent function), defined as the smallest integer m such that $k^n \equiv 1 \pmod{n}$ for all k RELATIVELY PRIME to n. The ORDER of $a \pmod{n}$ is at most $\lambda(n)$ (Ribenboim 1989). The first few values of this function, implemented in *Mathematica* 4.0 as `CarmichaelLambda[n]`, are 1, 1, 2, 2, 4, 2, 6, 2, 6, 4, 10, ... (Sloane's A002322). It can be defined recursively as

$$\lambda(n) = \begin{cases} \phi(n) & \text{for } n = p^\alpha, \ p = 2 \text{ and } \alpha \leq 2, \text{ or } p \geq 3 \\ \frac{1}{2}\phi(n) & \text{for } n = 2^\alpha \text{ and } \alpha \geq 3 \\ \text{LCM}[\lambda(p_i^{\alpha_i})]_i & \text{for } n = \prod_i p_i^{\alpha_i}. \end{cases}$$

Some special values are

$$\lambda(1) = 1$$

$$\lambda(2) = 1$$

$$\lambda(4) = 2$$

$$\lambda(2^r) = 2^{r-2}$$

for $r \geq 3$, and

$$\lambda'(p^r) = \phi(p^r)$$

for p an ODD PRIME and $r \geq 1$.

The second Carmichael's function $\lambda'(n)$ is given by the LEAST COMMON MULTIPLE (LCM) of all the FACTORS of the TOTIENT FUNCTION $\phi(n)$, except that if $8|n$, then $2^{\alpha-2}$ is a FACTOR instead of $2^{\alpha-1}$. The values of $\lambda'(n)$ for the first few n are 1, 1, 2, 2, 4, 2, 6, 4, 6, 4, 10, 2, 12, ... (Sloane's A011773).

See also MODULO MULTIPLICATION GROUP, TOTIENT FUNCTION

References

Ribenboim, P. *The Book of Prime Number Records, 2nd ed.* New York: Springer-Verlag, p. 27, 1989.

Riesel, H. "Carmichael's Function." *Prime Numbers and Computer Methods for Factorization, 2nd ed.* Boston, MA: Birkhäuser, pp. 273–275, 1994.

Sloane, N. J. A. Sequences A002322/M0298 and A011773 in "An On-Line Version of the Encyclopedia of Integer Sequences." http://www.research.att.com/~njas/sequences /eisonline.html.

Vardi, I. *Computational Recreations in Mathematica.* Redwood City, CA: Addison-Wesley, p. 226, 1991.

Carmichael Lambda

CARMICHAEL FUNCTION

Carmichael Number

A Carmichael number is an ODD COMPOSITE NUMBER n which satisfies FERMAT'S LITTLE THEOREM

$$a^{n-1} - 1 \equiv 0 \pmod{n} \tag{1}$$

for *every* choice of a satisfying $(a, n) = 1$ (i.e., a and n are RELATIVELY PRIME) with $1 < a < n$. A Carmichael number is therefore a PSEUDOPRIME to any base. Carmichael numbers therefore cannot be found to be COMPOSITE using FERMAT'S LITTLE THEOREM. However, if $(a, n) \neq 1$, the congruence of FERMAT'S LITTLE THEOREM is sometimes NONZERO, thus identifying a Carmichael number n as COMPOSITE.

Carmichael numbers are sometimes called "absolute pseudoprimes" and also satisfy KORSELT'S CRITERION. R. D. Carmichael first noted the existence of such numbers in 1910, computed 15 examples, and conjectured that there were infinitely many. In 1956, Erdos sketched a technique for constructing large Carmichael numbers (Hoffman 1998, p. 183), and a proof was given by Alford *et al.* (1994).

The first few Carmichael numbers are 561, 1105, 1729, 2465, 2821, 6601, 8911, 10585, 15841, 29341, ... (Sloane's A002997). The number of Carmichael numbers less than 10^2, 10^3, ... are 0, 1, 7, 16, 43, 105, ... (Sloane's A055553; Pinch 1993). The smallest Carmichael numbers having 3, 4, ... factors are $561 = 3 \times 11 \times 17$, $41041 = 7 \times 11 \times 13 \times 41$, 825265, 321197185, ... (Sloane's A006931).

Carmichael numbers have at least three PRIME FACTORS. For Carmichael numbers with exactly three PRIME FACTORS, once one of the PRIMES has been specified, there are only a finite number of Carmichael numbers which can be constructed. Indeed, for Carmichael numbers with k prime factors, there are only a finite number with the least $k - 2$ specified.

Numbers OF THE FORM $(6k + 1)(12k + 1)(18k + 1)$ are Carmichael numbers if each of the factors is PRIME (Korselt 1899, Ore 1988, Guy 1994). This can be seen since for

$$N \equiv (6k + 1)(12k + 1)(18k + 1)$$
$$= 1296k^3 + 396k^2 + 36k + 1, \tag{2}$$

$N - 1$ is a multiple of $36k$ and the LEAST COMMON MULTIPLE of $6k$, $12k$, and $18k$ is $36k$, so $a^{N-1} \equiv 1$ modulo each of the PRIMES $6k + 1$, $12k + 1$, and $18k + 1$, hence $a^{N-1} \equiv 1$ modulo their product. The first few such Carmichael numbers correspond to $k = 1, 6, 35, 45, 51, 55, 56, \ldots$ (Sloane's A046025) and are 1729, 294409, 56052361, 118901521, ... (Sloane's A033502). In Jan. 1999, Dubner found the largest known Carmichael of this form, having 4848 digits and index

$$k = 133752260 \cdot 3003 \cdot 10^{1604} \tag{3}$$

The prime factors of N have 1616, 1616, and 1617 digits.

Let $C(n)$ denote the number of Carmichael numbers less than n. Then, for all sufficiently large n,

$$C(n) > n^{2/7} \tag{4}$$

(Alford *et al.* 1994), which proves that there are infinitely many Carmichael numbers. The upper bound

$$C(n) < n \exp\left(\frac{\ln n \ln \ln \ln n}{\ln \ln n}\right) \tag{5}$$

has also been proved (R. G. E. Pinch).

The Carmichael numbers have the following properties:

1. If a PRIME p divides the Carmichael number n, then $n \equiv 1 \pmod{p - 1}$ implies that $n \equiv p \pmod{p(p - 1)}$.
2. Every Carmichael number is SQUAREFREE.
3. An ODD COMPOSITE SQUAREFREE number n is a Carmichael number IFF n divides the DENOMINATOR of the BERNOULLI NUMBER B_{n-1}.

The largest known Carmichael numbers having a given number of factors are summarized in the following table (Dubner 1989, Dubner 1998).

Factors	Digits	Discoverer
3	10200	Dubner
4	2467	Caldwell and Dubner
5	1015	Caldwell and Dubner
6	827	Caldwell and Dubner

See also CARMICHAEL CONDITION, PSEUDOPRIME

References

Alford, W. R.; Granville, A.; and Pomerance, C. "There are Infinitely Many Carmichael Numbers." *Ann. Math.* **139**, 703–722, 1994.

Beyer, W. H. *CRC Standard Mathematical Tables, 28th ed.* Boca Raton, FL: CRC Press, p. 87, 1987.

Carlini, A. and Hosoya, A. Carmichael Numbers on a Quantum Computer. 5 Aug 1999. http://xxx.lanl.gov/abs/ quant-ph/9908022/.

Dubner, H. "A New Method for Producing Large Carmichael Numbers." *Math. Comput.* **53**, 411–414, 1989.

Dubner, H. "Carmichael Number Record." Posting to nmbrthry@listserv.nodak.edu. Sep. 11, 1998.

Dubner, H. "3-Component Carmichael Number." Posting to nmbrthry@listserv.nodak.edu. Jan. 15, 1999.

Guy, R. K. "Carmichael Numbers." §A13 in *Unsolved Problems in Number Theory, 2nd ed.* New York: Springer-Verlag, pp. 30–32, 1994.

Hoffman, P. *The Man Who Loved Only Numbers: The Story of Paul Erdos and the Search for Mathematical Truth.* New York: Hyperion, pp. 182–183, 1998.

Korselt, A. "Problème chinois." *L'intermédiaire math.* **6**, 143–143, 1899.

Ore, Ø. *Number Theory and Its History.* New York: Dover, 1988.

Pinch, R. G. E. "The Carmichael Numbers up to 10^{15}." *Math. Comput.* **55**, 381–391, 1993.

Pinch, R. G. E. ftp://ftp.dpmms.cam.ac.uk/pub/Carmichael/.

Pomerance, C.; Selfridge, J. L.; and Wagstaff, S. S. Jr. "The Pseudoprimes to $25 \cdot 10^9$." *Math. Comput.* **35**, 1003–1026, 1980.

Ribenboim, P. *The New Book of Prime Number Records.* New York: Springer-Verlag, pp. 118–125, 1996.

Riesel, H. *Prime Numbers and Computer Methods for Factorization, 2nd ed.* Basel: Birkhäuser, pp. 89–90 and 94–95, 1994.

Shanks, D. *Solved and Unsolved Problems in Number Theory, 4th ed.* New York: Chelsea, p. 116, 1993.

Sloane, N. J. A. Sequences A002997/M5462, A006931/ M5463, A033502, A046025, and A055553 in "An On-Line Version of the Encyclopedia of Integer Sequences." http://www.research.att.com/~njas/sequences/eisonline.html.

Carmichael Sequence

A FINITE, INCREASING SEQUENCE of INTEGERS $\{a_1, \ldots, a_m\}$ such that

$$(a_i - 1) | (a_1 \ldots a_{i-1})$$

for $i = 1, \ldots, m$, where $m|n$ indicates that m DIVIDES n. A Carmichael sequence has exclusive EVEN or ODD elements. There are infinitely many Carmichael sequences for every order.

See also GIUGA SEQUENCE

References

Borwein, D.; Borwein, J. M.; Borwein, P. B.; and Girgensohn, R. "Giuga's Conjecture on Primality." *Amer. Math. Monthly* **103**, 40–50, 1996.

Carmichael's Conjecture

Carmichael's conjecture asserts that there are an INFINITE number of CARMICHAEL NUMBERS. This was proven by Alford *et al.* (1994).

See also CARMICHAEL NUMBER, CARMICHAEL'S TOTIENT FUNCTION CONJECTURE

References

Alford, W. R.; Granville, A.; and Pomerance, C. "There Are Infinitely Many Carmichael Numbers." *Ann. Math.* **139**, 703–722, 1994.

Cipra, B. *What's Happening in the Mathematical Sciences, Vol. 1.* Providence, RI: Amer. Math. Soc., 1993.

Guy, R. K. "Carmichael's Conjecture." §B39 in *Unsolved Problems in Number Theory, 2nd ed.* New York: Springer-Verlag, p. 94, 1994.

Pomerance, C.; Selfridge, J. L.; and Wagstaff, S. S. Jr. "The Pseudoprimes to $25 \cdot 10^9$." *Math. Comput.* **35**, 1003–1026, 1980.

Ribenboim, P. *The Book of Prime Number Records, 2nd ed.* New York: Springer-Verlag, pp. 29–31, 1989.

Schlafly, A. and Wagon, S. "Carmichael's Conjecture on the Euler Function is Valid Below $10^{10,000,000}$." *Math. Comput.* **63**, 415–419, 1994.

Carmichael's Theorem

If a and n are RELATIVELY PRIME so that the GREATEST COMMON DIVISOR $GCD(a, n) = 1$, then

$$a^{\lambda(n)} \equiv 1 \pmod{n}$$

where λ is the CARMICHAEL FUNCTION.

Carmichael's Totient Function Conjecture

It is thought that the TOTIENT VALENCE FUNCTION $N_\phi(n) \geq 2$, i.e., if there is an n such that $\phi(x) = n$, then there are at least two solutions x. This assertion is called Carmichael's totient function conjecture and is equivalent to the statement that there exists an $m \neq n$ such that $\phi(n) = \phi(m)$ (Ribenboim 1996, pp. 39–40).

Dickson 1952 (p. 137) states that the conjecture was proved by Carmichael (1907), who also developed a method of finding the solution (Carmichael 1909). The result also appears as in exercise in Carmichael (1914). However, Carmichael (1922) subsequently discovered an error in the proof, and the conjecture currently remain open. Any counterexample to the conjecture must have more than 10,000,000 DIGITS (Schlafly and Wagon 1994; conservatively given as 10,000 in Conway and Guy 1996, p. 155).

Ford (1998ab) showed that if there is a counterexample to Carmichael's conjecture, then a positive proportion of totients are counterexamples.

SIERPINSKI'S CONJECTURE states that all integers > 1 appear as multiplicities of the TOTIENT VALENCE FUNCTION.

See also TOTIENT FUNCTION, SIERPINSKI'S CONJECTURE, TOTIENT VALENCE FUNCTION

References

Carmichael, R. D. "On Euler's ϕ-Function." *Bull. Amer. Math. Soc.* **13**, 241–243, 1907.

Carmichael, R. D. "Notes on the Simplex Theory of Numbers." *Bull. Amer. Math. Soc.* **15**, 217–223, 1909.

Carmichael, R. D. *The Theory of Numbers.* New York: Wiley, 1914.

Carmichael, R. D. "Note on Euler's ϕ-Function." *Bull. Amer. Math. Soc.* **28**, 109–110, 1922.

Conway, J. H. and Guy, R. K. *The Book of Numbers.* New York: Springer-Verlag, 1996.

Dickson, L. E. *History of the Theory of Numbers, Vol. 1: Divisibility and Primality.* New York: Chelsea, 1952.

Ford, K. "The Distribution of Totients." *Ramanujan J.* **2**, 67–151, 1998a.

Ford, K. "The Distribution of Totients, *Electron. Res. Announc. Amer. Math. Soc.* **4**, 27–34, 1998b.

Guy, R. K. "Carmichael's Conjecture." §B39 in *Unsolved Problems in Number Theory, 2nd ed.* New York: Springer-Verlag, pp. 94–95, 1994.

Klee, V. "On a Conjecture of Carmichael." *Bull. Amer. Math. Soc.* **53**, 1183–1186, 1947.

Masai, P. and Valette, A. "A Lower Bound for a Counter-example to Carmichael's Conjecture." *Boll. Un. Mat. Ital.* **1**, 313–316, 1982.

Ribenboim, P. *The New Book of Prime Number Records.* New York: Springer-Verlag, 1996.

Schlafly, A. and Wagon, S. "Carmichael's Conjecture on the Euler Function is Valid Below $10^{10,000,000}$." *Math. Comput.* **63**, 415–419, 1994.

Carnot's Polygon Theorem

If a PLANE cuts the sides AB, BC, CD, and DA of a SKEW QUADRILATERAL $ABCD$ in points P, Q, R, and S, then

$$\frac{AP}{PB} \cdot \frac{BQ}{QC} \cdot \frac{CR}{RD} \cdot \frac{DS}{SA} = 1$$

both in magnitude and sign (Altshiller-Court 1979, p. 111).

More generally, if P_1, P_2, ..., are the VERTICES of a finite POLYGON with no "minimal sides" and the side P_iP_j meets a curve in the POINTS P_{ij1} and P_{ij2}, then

$$\frac{\prod_i \overline{P_1P_{12i}} \prod_i \overline{P_2P_{23i}} \cdots \prod_i \overline{P_NP_{N1i}}}{\prod_i \overline{P_NP_{N1i}} \cdots \prod_i \overline{P_2P_{2i1}}} = 1,$$

where $\overline{AB}$ denotes the DISTANCE from POINT A to B.

References

Altshiller-Court, N. "Carnot's Theorem." §329 in *Modern Pure Solid Geometry.* New York: Chelsea, p. 111, 1979.

Carnot, L. N. M. *Géométrie de position.* Paris: Duprat, p. 287, 1803.

Carnot, L. N. M. *Mémoir sur la relation qui existe entre les distances respectives de cinq points quelconques pris dans l'espace; suivi d'un Essai sur la théorie des transversales.* Paris: Courcier, p. 71, 1806.

Casey, J. *A Sequel to the First Six Books of the Elements of Euclid, Containing an Easy Introduction to Modern Geometry with Numerous Examples, 5th ed., rev. enl.* Dublin: Hodges, Figgis, & Co., p. 160, 1888.

Coolidge, J. L. *A Treatise on Algebraic Plane Curves.* New York: Dover, p. 190, 1959.

Carnot's Theorem

Given any TRIANGLE $A_1A_2A_3$, the signed sum of PERPENDICULAR distances from the CIRCUMCENTER O to the sides is

$$OO_1 + OO_2 + OO_3 = R + r,$$

where r is the INRADIUS and R is the CIRCUMRADIUS. The sign of the distance is chosen to be POSITIVE IFF the entire segment OO_i lies outside the TRIANGLE.

See also JAPANESE TRIANGULATION THEOREM

References

Eves, H. W. *A Survey of Geometry, rev. ed.* Boston, MA: Allyn and Bacon, pp. 256 and 262, 1972.

Honsberger, R. *Mathematical Gems III.* Washington, DC: Math. Assoc. Amer., p. 25, 1985.

Carotid-Kundalini Fractal

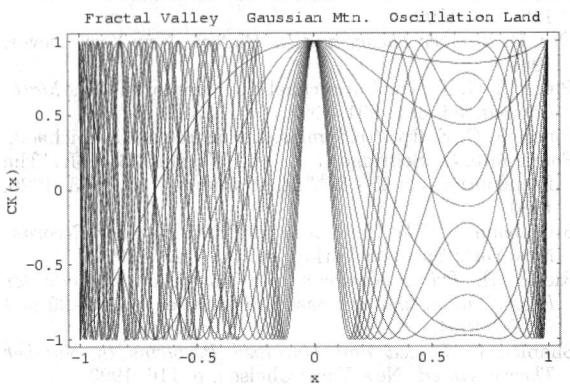

A fractal-like structure is produced for $x < 0$ by superposing plots of CAROTID-KUNDALINI FUNCTIONS ck_n of different orders n. the region $-1 < x < 0$ is called FRACTAL LAND by pickover (1995), the central region the GAUSSIAN MOUNTAIN RANGE, and the region $0 < x < 1$ OSCILLATION LAND. The plot above shows $n = 1$ to 25. Gaps in FRACTAL LAND occur whenever

$$x \cos^{-1} x = 2\pi \frac{p}{q}$$

for p and q RELATIVELY PRIME INTEGERS. At such points x, the functions assume the $\lceil (q+1)/2 \rceil$ values $\cos(2\pi r/q)$ for $r = 0$, 1, ..., $\lfloor q/2 \rfloor$, where $\lceil z \rceil$ is the CEILING FUNCTION and $\lfloor z \rfloor$ is the FLOOR FUNCTION.

References

Pickover, C. A. "Are Infinite Carotid-Kundalini Functions Fractal?" Ch. 24 in *Keys to Infinity.* New York: Wiley, pp. 179–181, 1995.

Weisstein, E. W. "Fractals." MATHEMATICA NOTEBOOK FRACTAL.M.

Carotid-Kundalini Function

The FUNCTION given by

$$CK_n(x) \equiv \cos(nx \cos^{-1} x),$$

where n is an INTEGER and $-1 < x < 1$.

See also CAROTID-KUNDALINI FRACTAL

Carry

$$\begin{array}{r} 1\ 1 \longleftarrow \text{carries} \\ 1\ 5\ 8 \longleftarrow \text{addend 1} \\ +\ 2\ 4\ 9 \longleftarrow \text{addend 2} \\ \hline 4\ 0\ 7 \longleftarrow \text{sum} \end{array}$$

The operating of shifting the leading DIGITS of an ADDITION into the next column to the left when the SUM of that column exceeds a single DIGIT (i.e., 9 in base 10).

See also ADDEND, ADDITION, BORROW

Carrying Capacity

LOGISTIC GROWTH CURVE

Cartan Decomposition

References
Huang, J.-S. "Linear Reductive Groups and Cartan Decomposition." §10.1 in *Lectures on Representation Theory*. Singapore: World Scientific, pp. 129–130, 1999.

Cartan Matrix

A Cartan matrix is a SQUARE INTEGER MATRIX who elements (A_{ij}) satisfy the following conditions.

1. A_{ij} is an integer, one of $\{-3, -2, -1, 0, 2\}$.
2. $A_{ii} = 2$ the diagonal entries are all 2.
3. $A_{ij} \leq 0$ off of the diagonal.
4. $A_{ij} = 0$ iff $A_{ji} = 0$.
5. There exists a DIAGONAL MATRIX D such that DAD^{-1} gives a SYMMETRIC and POSITIVE DEFINITE QUADRATIC FORM.

A Cartan matrix can be associated to a SEMISIMPLE LIE ALGEBRA $\mathfrak{g}$. It is a $k \times k$ SQUARE MATRIX, where k is the RANK of $\mathfrak{g}$. The SIMPLE ROOTS are the basis vectors, and A_{ij} is determined by their inner product, using the KILLING form.

$$A_{ij} = 2\langle \alpha_i, \alpha_j \rangle / \langle \alpha_j, \alpha_j \rangle \qquad (1)$$

In fact, it is more a table of values than a matrix. By reordering the basis vectors, one gets another Cartan matrix, but it is considered equivalent to the original Cartan matrix.

The Lie algebra $\mathfrak{g}$ can be reconstructed, up to ISOMORPHISM, by the $3k$ generators $\{e_j, f_i, h_i\}$ which satisfy the SERRE RELATIONS. In fact,

$$\mathfrak{g} = \mathfrak{h} \oplus \mathfrak{e} \oplus \mathfrak{f} \qquad (2)$$

where $\mathfrak{h}$, $\mathfrak{e}$, $\mathfrak{f}$ are the LIE SUBALGEBRAS generated by the generators of the same letter.

For example,

$$A = \begin{bmatrix} 2 & -1 \\ -1 & 2 \end{bmatrix} \qquad (3)$$

is a Cartan matrix. The LIE ALGEBRA $\mathfrak{g}$ has six generators $\{h_1, h_2, e_1, e_2, f_1, f_2\}$. They satisfy the following relations.

1. $[h_1, h_2] = 0$.
2. $[e_1, f_1] = h_1$ and $[e_2, f_2] = h_2$ while $[e_1, f_2] = [e_2, f_1] = 0$.
3. $[h_i, e_j] = -A_{ij} e_j$.
4. $[h_i, f_j] = -A_{ij} f_j$.
5. $e_{12} = [e_1, e_2] \neq 0$ and $f_{12} = [f_1, f_2] \neq 0$.
6. $[e_i, e_{12}] = 0$ and $[f_i, f_{12}] = 0$.

From these relations, it is not hard to see that $\mathfrak{g} = \mathfrak{sl}_3$ with the standard REPRESENTATION

$$h_1 = \begin{bmatrix} 1 & 0 & 1 \\ 0 & -1 & 0 \\ 0 & 0 & 0 \end{bmatrix} \qquad (4)$$

$$h_2 = \begin{bmatrix} 0 & 0 & 0 \\ 0 & 1 & 0 \\ 0 & 0 & -1 \end{bmatrix} \qquad (5)$$

$$e_1 = \begin{bmatrix} 0 & 1 & 0 \\ 0 & 0 & 0 \\ 0 & 0 & 0 \end{bmatrix} \qquad (6)$$

$$e_2 = \begin{bmatrix} 0 & 0 & 0 \\ 0 & 0 & 1 \\ 0 & 0 & 0 \end{bmatrix} \qquad (7)$$

$$e_{12} = \begin{bmatrix} 0 & 0 & 1 \\ 0 & 0 & 0 \\ 0 & 0 & 0 \end{bmatrix} \qquad (8)$$

$$f_1 = \begin{bmatrix} 0 & 0 & 0 \\ 1 & 0 & 0 \\ 0 & 0 & 0 \end{bmatrix} \qquad (9)$$

$$f_2 = \begin{bmatrix} 0 & 0 & 0 \\ 0 & 0 & 0 \\ 0 & 1 & 0 \end{bmatrix} \qquad (10)$$

$$f_{12} = \begin{bmatrix} 0 & 0 & 0 \\ 0 & 0 & 0 \\ -1 & 0 & 0 \end{bmatrix} \qquad (11)$$

In addition, the WEYL GROUP can be constructed directly from the Cartan matrix. Its rows determine the reflections against the simple roots. The following Mathematica command converts a Cartan matrix to a list of generators for the Weyl group, in its representation on the ROOT LATTICE. In particular, its output represents the matrices of the Weyl group as INTEGER MATRICES.

See also DYNKIN DIAGRAM, LIE ALGEBRA, ROOT (LIE ALGEBRA), ROOT SYSTEM, SEMISIMPLE LIE ALGEBRA, SPECIAL LINEAR LIE ALGEBRA, WEYL GROUP

References

Fulton, W. and Harris, J. *Representation Theory.* New York: Springer-Verlag, 1991.

Jacobson, N. "The Determination of the Cartan Matrices." §4.5 in *Lie Algebras.* New York: Dover, pp. 121 and 128–135, 1979.

Knapp, A. *Lie Groups Beyond an Introduction.* Boston, MA: Birkhäuser, 1996.

Cartan Relation

The relationship $Sq^i(x \smile y) = \Sigma_{j+k=i} Sq^j(x) \smile Sq^k(y)$ encountered in the definition of the STEENROD ALGEBRA.

Cartan Subgroup

A type of maximal ABELIAN SUBGROUP.

References

Knapp, A. W. "Group Representations and Harmonic Analysis, Part II." *Not. Amer. Math. Soc.* **43**, 537–549, 1996.

Cartan Torsion Coefficient

The ANTISYMMETRIC parts of the CHRISTOFFEL SYMBOL OF THE SECOND KIND $\Gamma^\lambda_{\mu\nu}$.

Cartesian Coordinates

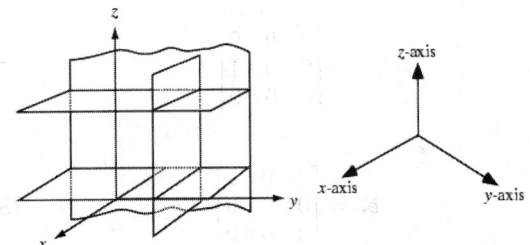

Cartesian coordinates are rectilinear 2-D or 3-D coordinates (and therefore a special case of CURVILINEAR COORDINATES) which are also called rectangular coordinates. The three axes of 3-D Cartesian coordinates, conventionally denoted the x-, y-, and z-AXES (a NOTATION due to Descartes) are chosen to be linear and mutually PERPENDICULAR. In 3-D, the coordinates x, y, and z may lie anywhere in the INTERVAL $(-\infty, \infty)$.

The INVERSION of 3-D Cartesian is called 6-SPHERE COORDINATES coordinates.

The SCALE FACTORS of Cartesian coordinates are all unity, $h_i = 1$. The LINE ELEMENT is given by

$$d\mathbf{s} = dx\,\hat{\mathbf{x}} + dy\,\hat{\mathbf{y}} + dz\,\hat{\mathbf{z}}, \tag{1}$$

and the VOLUME ELEMENT by

$$dV = dx\,dy\,dz. \tag{2}$$

The GRADIENT has a particularly simple form,

$$\nabla \equiv \hat{\mathbf{x}}\,\frac{\partial}{\partial x} + \hat{\mathbf{y}}\,\frac{\partial}{\partial y} + \hat{\mathbf{z}}\,\frac{\partial}{\partial z}, \tag{3}$$

as does the LAPLACIAN

$$\nabla^2 \equiv \frac{\partial^2}{\partial x^2} + \frac{\partial^2}{\partial y^2} + \frac{\partial^2}{\partial z^2}. \tag{4}$$

The LAPLACIAN is

$$\nabla^2 \mathbf{F} \equiv \nabla \cdot (\nabla \mathbf{F}) = \frac{\partial^2 \mathbf{F}}{\partial x^2} + \frac{\partial^2 \mathbf{F}}{\partial y^2} + \frac{\partial^2 \mathbf{F}}{\partial z^2}$$

$$= \hat{\mathbf{x}}\left(\frac{\partial^2 F_x}{\partial x^2} + \frac{\partial^2 F_x}{\partial y^2} + \frac{\partial^2 F_x}{\partial z^2}\right)$$

$$+ \hat{\mathbf{y}}\left(\frac{\partial^2 F_y}{\partial x^2} + \frac{\partial^2 F_y}{\partial y^2} + \frac{\partial^2 F_y}{\partial z^2}\right)$$

$$+ \hat{\mathbf{z}}\left(\frac{\partial^2 F_z}{\partial x^2} + \frac{\partial^2 F_z}{\partial y^2} + \frac{\partial^2 F_z}{\partial z^2}\right). \tag{5}$$

The DIVERGENCE is

$$\nabla \cdot \mathbf{F} = \frac{\partial F_x}{\partial x} + \frac{\partial F_y}{\partial y} + \frac{\partial F_z}{\partial z}, \tag{6}$$

and the CURL is

$$\nabla \times \mathbf{F} \equiv \begin{vmatrix} \hat{\mathbf{x}} & \hat{\mathbf{y}} & \hat{\mathbf{z}} \\ \frac{\partial}{\partial x} & \frac{\partial}{\partial y} & \frac{\partial}{\partial z} \\ F_x & F_y & F_z \end{vmatrix} = \left(\frac{\partial F_z}{\partial y} - \frac{\partial F_y}{\partial z}\right)\hat{\mathbf{x}} + \left(\frac{\partial F_x}{\partial z} - \frac{\partial F_z}{\partial x}\right)\hat{\mathbf{y}}$$

$$+ \left(\frac{\partial F_y}{\partial x} - \frac{\partial F_x}{\partial y}\right)\hat{\mathbf{z}}. \tag{7}$$

The GRADIENT of the DIVERGENCE is

$$\nabla(\nabla \cdot \mathbf{u}) = \begin{bmatrix} \frac{\partial}{\partial x}\left(\frac{\partial u_z}{\partial x} + \frac{\partial u_y}{\partial y} + \frac{\partial u_z}{\partial z}\right) \\ \frac{\partial}{\partial y}\left(\frac{\partial u_z}{\partial x} + \frac{\partial u_y}{\partial y} + \frac{\partial u_z}{\partial z}\right) \\ \frac{\partial}{\partial z}\left(\frac{\partial u_z}{\partial x} + \frac{\partial u_y}{\partial y} + \frac{\partial u_z}{\partial z}\right) \end{bmatrix}$$

$$= \begin{bmatrix} \frac{\partial}{\partial x} \\ \frac{\partial}{\partial y} \\ \frac{\partial}{\partial z} \end{bmatrix} \left(\frac{\partial u_x}{\partial x} + \frac{\partial u_y}{\partial y} + \frac{\partial u_z}{\partial z}\right). \tag{8}$$

LAPLACE'S EQUATION is separable in Cartesian coordinates.

See also CARTESIAN GEOMETRY, COORDINATES, HELMHOLTZ DIFFERENTIAL EQUATION–CARTESIAN COORDINATES, 6-SPHERE COORDINATES

References

Arfken, G. "Special Coordinate Systems--Rectangular Cartesian Coordinates." §2.3 in *Mathematical Methods for Physicists, 3rd ed.* Orlando, FL: Academic Press, pp. 94–95, 1985.

Moon, P. and Spencer, D. E. "Rectangular Coordinates (x, y, z)." Table 1.01 in *Field Theory Handbook, Including Coordinate Systems, Differential Equations, and Their Solutions, 2nd ed.* New York: Springer-Verlag, pp. 9–11, 1988.

Morse, P. M. and Feshbach, H. *Methods of Theoretical Physics, Part I.* New York: McGraw-Hill, p. 656, 1953.

Cartesian Geometry

The use of coordinates (such as CARTESIAN COORDINATES) in the study of GEOMETRY. Cartesian geometry is named after René Descartes (Bell 1986, p. 48), although Descartes may have been anticipated by Fermat (Coxeter and Greitzer 1967, p. 31).

See also ANALYTIC GEOMETRY, CARTESIAN COORDINATES

References

Bell, E. T. *Men of Mathematics.* New York: Simon and Schuster, p. 48, 1986.

Coxeter, H. S. M. and Greitzer, S. L. *Geometry Revisited.* Washington, DC: Math. Assoc. Amer., p. 31, 1967.

Cartesian Ovals

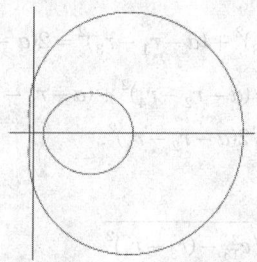

A curve consisting of two ovals which was first studied by Descartes in 1637. It is the locus of a point P whose distances from two FOCI F_1 and F_2 in two-center BIPOLAR COORDINATES satisfy

$$mr \pm nr' = k, \tag{1}$$

where m, n are POSITIVE INTEGERS, k is a POSITIVE real, and r and r' are the distances from F_1 and F_2. If $m = n$, the oval becomes an ELLIPSE. In CARTESIAN COORDINATES, the Cartesian ovals can be written

$$m \sqrt{(x-a)^2 + y^2} + n \sqrt{(x+a)^2 + y^2} = k^2 \tag{2}$$

$$(x^2 + y^2 + a^2)(m^2 - n^2) - 2ax(m^2 + n^2) - k^2$$
$$= -2n \sqrt{(x+a)^2 + y^2}, \tag{3}$$

$$[(m^2 - n^2)(x^2 + y^2 + a^2) - 2ax(m^2 + n^2)]^2$$
$$= 2(m^2 + n^2)(n^2 + y^2 + a^2) - 4ax(m^2 - n^2) - k^2. \tag{4}$$

Now define

$$b \equiv m^2 - n^2 \tag{5}$$

$$c \equiv m^2 + n^2, \tag{6}$$

and set $a = 1$. Then

$$[b(x^2 + y^2) - 2cx + b]^2 + 4bx + k^2 - 2c = 2c(x^2 + y^2). \tag{7}$$

If c' is the distance between F_1 and F_2, and the equation

$$r + mr' = a \tag{8}$$

is used instead, an alternate form is

$$[(1 - m^2)(x^2 + y^2) + 2m^2 c'x + a'^2 - m^2 c'^2]^2$$
$$= 4a'^2(x^2 + y^2). \tag{9}$$

The curves possess three FOCI. If $m = 1$, one Cartesian oval is a central CONIC, while if $m = a/c$, then the curve is a LIMAÇON and the inside oval touches the outside one. Cartesian ovals are ANALLAGMATIC CURVES.

References

Baudoin, P. *Les ovales de Descartes et le limaçon de Pascal.* Paris: Vuibert, 1938.

Cundy, H. and Rollett, A. *Mathematical Models, 3rd ed.* Stradbroke, England: Tarquin Pub., p. 35, 1989.

Lawrence, J. D. *A Catalog of Special Plane Curves.* New York: Dover, pp. 155–157, 1972.

Lockwood, E. H. *A Book of Curves.* Cambridge, England: Cambridge University Press, p. 188, 1967.

MacTutor History of Mathematics Archive. "Cartesian Oval." http://www-groups.dcs.st-and.ac.uk/~history/Curves/Cartesian.html.

Cartesian Product

The Cartesian product of two sets A and B (also called the product set, set direct product, or cross product) is defined to be the set of all points (a, b) where $a \in A$ and $b \in B$. It is denoted $A \times B$, and is called the Cartesian product since it originated in Descartes' formulation of analytic geometry. In the Cartesian view, points in the plane are specified by their vertical and horizontal coordinates, with points on a line being specified by just one coordinate. The main examples of direct products are EUCLIDEAN 3-space ($\mathbb{R} \times \mathbb{R} \times \mathbb{R}$, where $\mathbb{R}$ are the REAL NUMBERS), and the plane ($\mathbb{R} \times \mathbb{R}$).

The GRAPH PRODUCT is sometimes called the Cartesian product (Vizing 1963, Cark and Suen 2000).

See also DIRECT PRODUCT, DISJOINT UNION, EXTERNAL DIRECT PRODUCT, EXTERNAL DIRECT SUM, GRAPH PRODUCT, GROUP DIRECT PRODUCT, PRODUCT SPACE

References

Clark, W. E. and Suen, S. "An Inequality Related to Vizing's Conjecture." *Electronic J. Combinatorics* **7**, No. 1, N4, 1–3, 2000. http://www.combinatorics.org/Volume_7/v7i1toc.html#N4.

Comtet, L. "Product Sets." §1.2 in *Advanced Combinatorics: The Art of Finite and Infinite Expansions, rev. enl. ed.* Dordrecht, Netherlands: Reidel, pp. 3–4, 1974.

Papoulis, A. *Probability, Random Variables, and Stochastic Processes, 2nd ed.* New York: McGraw-Hill, pp. 49–50, 1984.

Vizing, V. G. "The Cartesian Product of Graphs." *Vycisl. Sistemy* **9**, 30–43, 1963.

Cartesian Space
EUCLIDEAN SPACE

Cartesian Trident
TRIDENT OF DESCARTES

Cartography
The study of MAP PROJECTIONS and the making of geographical maps.

See also MAP PROJECTION

Cascade
A $\mathbb{Z}$-ACTION or $\mathbb{N}$-ACTION. A cascade and a single MAP $X \to X$ are essentially the same, but the term "cascade" is preferred by many Russian authors.

See also ACTION, FLOW

Casey's Theorem
Four CIRCLES c_1, c_2, c_3, and c_4 are TANGENT to a fifth CIRCLE or a straight LINE IFF

$$T_{12}T_{34} \pm T_{13}T_{42} \pm T_{14}T_{23} = 0. \tag{1}$$

where T_{ij} is the length of a common TANGENT to CIRCLES i and j (Johnson 1929, pp. 121–122). The following cases are possible:

1. If all the Ts are direct common tangents, then c_5 has like contact with all the circles,
2. If the Ts from one circle are transverse while the other three are direct, then this one circle has contact with c_5 unlike that of the other three,
3. If the given circles can be so paired that the common tangents to the circles of each pair are direct, while the other four are transverse, then the members of each pair have like contact with c_5

(Johnson 1929, p. 125).

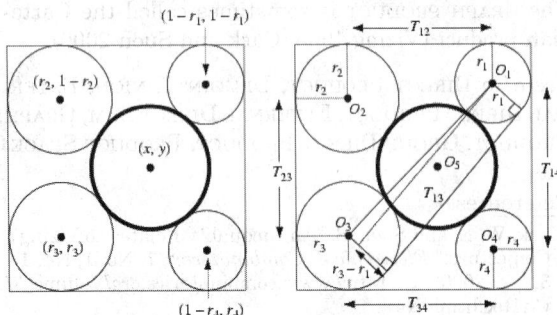

$(1-r_1, 1-r_1)$ $(r_2, 1-r_2)$ (x, y) (r_3, r_3) $(1-r_4, r_4)$

T_{12} T_{23} T_{14} T_{34} T_{13} r_1 O_1 O_2 O_3 O_4 O_5 $r_3 - r_1$

The special case of Casey's theorem shown above was given in a SANGAKU PROBLEM from 1874 in the Gumma Prefecture. In this form, a single circle is drawn inside a square, and four circles are then

drawn around it, each of which is tangent to the square on two of its sides. For a square of side length a with lower left corner at $(0, 0)$ containing a central circle of radius r with center (x, y), the radii and positions of the four circles can be found by solving

$$(1-r_4-x)^2 + (y-r_4)^2 = (r+r_4)^2 \tag{2}$$

$$(1-r_1-x)^2 + (1-r_1-y)^2 = (r+r_1)^2 \tag{3}$$

$$(x-r_3)^2 + (y-r_3)^2 + (r+r_3)^2 \tag{4}$$

$$(x-r_2)^2 + (1-r_2-y)^2 = (r+r_2)^2. \tag{5}$$

Four of the T_{ij} for the theorem are given immediately for the figure as

$$T_{12} = a - r_1 - r_2 \tag{6}$$

$$T_{34} = a - r_r - r_4 \tag{7}$$

$$T_{14} = a - r_1 - r_4 \tag{8}$$

$$T_{23} = a - r_2 - r_3. \tag{9}$$

The remaining T_{13} and T_{24} can be found as shown in the above right figure. Let c_{ij} be the distance from O_i to O_j, then

$$c_{13}^2 = (a - r_1 - r_3)^2 + (a - r_1 - r_3)^2 = 2(a - r_1 - r_3)^2 \tag{10}$$

$$c_{24}^2 = (a - r_2 - r_4)^2 + (a - r_2 - r_4)^2$$
$$= 2(a - r_2 - r_4)^2, \tag{11}$$

so

$$T_{13} = \sqrt{c_{13}^2 - (r_3 - r_1)^2}$$
$$= \sqrt{2(a - r_1 - r_3)^2 - (r_3 - r_1)^2} \tag{12}$$

$$T_{24} = \sqrt{c_{24}^2 - (r_2 - r_4)^2}$$
$$= \sqrt{2(a - r_2 - r_4)^2 - (r_2 - r_4)^2}. \tag{13}$$

Since the four circles are all externally tangent to c_5, the relevant form of Casey's theorem to use has signs $(+, -)$, so we have the equation

$$(a-r_1-r_2)(a-r_3-r_4)+(a-r_1-r_4)(a-r_2-r_3)$$
$$-\sqrt{[2(a-r_1-r_3)^2-(r_3-r_1)^2][2(a-r_2-r_4)^2-(r_2-r_4)^2]} = 0 \tag{14}$$

(Rothman 1998). Solving for a then gives the relationship

$$a = \frac{2(r_1 r_3 - r_2 r_4) + \sqrt{2(r_1 - r_2)(r_1 - r_4)(r_3 - r_2)(r_3 - r_4)}}{r_1 - r_2 + r_3 - r_4} \tag{15}$$

Durell (1928) calls the following Casey's theorem: if t is the length of a common tangent of two circles of radii a and b, t' is the length of the corresponding common tangent of their inverses with respect to any point, and a' and b' are the radii of their inverses,

then

$$\frac{t^2}{ab} = \frac{t'^2}{a'b'}. \tag{16}$$

See also PURSER'S THEOREM, TANGENT CIRCLES

References

Casey, J. *A Sequel to the First Six Books of the Elements of Euclid, Containing an Easy Introduction to Modern Geometry with Numerous Examples, 5th ed., rev. enl.* Dublin: Hodges, Figgis, & Co., p. 103, 1888.

Casey, J. *A Treatise on the Analytical Geometry of the Point, Line, Circle, and Conic Sections, Containing an Account of Its Most Recent Extensions, with Numerous Examples, 2nd ed., rev. enl.* Dublin: Hodges, Figgis, & Co., p. 125, 1893.

Coolidge, J. L. *A Treatise on the Geometry of the Circle and Sphere.* New York: Chelsea, p. 37, 1971.

Durell, C. V. *Modern Geometry: The Straight Line and Circle.* London: Macmillan, p. 117, 1928.

Fukagawa, H. and Pedoe, D. "Many Circles and Squares (Casey's Theorem)." §3.3 in *Japanese Temple Geometry Problems.* Winnipeg, Manitoba, Canada: Charles Babbage Research Foundation, pp. 41–42 and 120–1989.

Johnson, R. A. *Modern Geometry: An Elementary Treatise on the Geometry of the Triangle and the Circle.* Boston, MA: Houghton Mifflin, pp. 121–127, 1929.

Lachlan, R. *An Elementary Treatise on Modern Pure Geometry.* London: Macmillian, pp. 244–251, 1893.

Rothman, T. "Japanese Temple Geometry." *Sci. Amer.* **278**, 85–91, May 1998.

Casimir Operator

An OPERATOR

$$\Gamma = \sum_{i=1}^{m} e_i^R u^{iR}$$

on a representation R of a LIE ALGEBRA.

References

Jacobson, N. *Lie Algebras.* New York: Dover, p. 78, 1979.

Casoratian

The Casoratian of sequences $x_n^{(1)}, x_n^{(2)}, ..., x_n^{(k)}$ is defined by the $k \times k$ DETERMINANT

$$C(x_n^{(1)}, x_n^{(2)}, x_n^{(k)}) = \begin{vmatrix} x_n^{(1)} & x_n^{(2)} & \cdots & x_n^{(k)} \\ x_{n+1}^{(1)} & x_{n+1}^{(2)} & \cdots & x_{n+1}^{(k)} \\ \vdots & \vdots & \ddots & \vdots \\ x_{n+k-1}^{(1)} & x_{n+k-1}^{(2)} & \cdots & x_{n+k-1}^{(k)} \end{vmatrix}.$$

The solutions $x_n^{(1)}, x_n^{(2)}, ..., x_n^{(k)}$ of the linear difference equation

$$x_{n+k} + b_n^{(k-1)} x_{n+(k-1)} + \ldots + b_n^{(1)} x_{n+1} + b_n^{(0)} x_n = 0$$

for $n = 0, 1, ...,$ are linearly independent sequences IFF their Casoratian is nonzero for $n = 0$ (Zwillinger 1995).

See also LINEARLY DEPENDENT SEQUENCES

References

Zwillinger, D. (Ed.). *CRC Standard Mathematical Tables and Formulae.* Boca Raton, FL: CRC Press, p. 229, 1995.

Casorati-Weierstrass Theorem

WEIERSTRASS-CASORATI THEOREM

Cassini Ellipses

CASSINI OVALS

Cassini Ovals

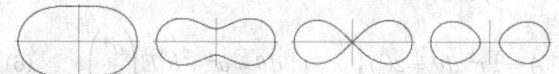

The curves, also called Cassini ellipses, described by a point such that the product of its distances from two fixed points a distance $2a$ apart is a constant b^2. The shape of the curve depends on b/a. If $a < b$, the curve is a single loop with an OVAL (left figure above) or dog bone (second figure) shape. The case $a = b$ produces a LEMNISCATE (third figure). If $a > b$, then the curve consists of two loops (right figure). Cassini ovals are ANALLAGMATIC CURVES.

The curve was first investigated by Cassini in 1680 when he was studying the relative motions of the Earth and the Sun. Cassini believed that the Sun traveled around the Earth on one of these ovals, with the Earth at one FOCUS of the oval.

The Cassini ovals are defined in two-center BIPOLAR COORDINATES by the equation

$$r_1 r_2 = b^2, \tag{1}$$

with the origin at a FOCUS. Even more incredible curves are produced by the locus of a point the product of whose distances from 3 or more fixed points is a constant.

The Cassini ovals have the CARTESIAN equation

$$[(x-a)^2 + y^2][(x+a)^2 + y^2] = b^4 \tag{2}$$

or the equivalent form

$$(x^2 + y^2 + a^2)^2 - 4a^2 x^2 = b^4 \tag{3}$$

and the polar equation

$$r^4 + a^4 - 2a^2 r^2 \cos(2\theta) = b^4. \tag{4}$$

Solving for r^2 using the QUADRATIC EQUATION gives

$$r^2 = \frac{2a^2 \cos(2\theta) \pm \sqrt{4a^4 \cos^2(2\theta) - 4(a^4 - b^4)}}{2}$$

$$= a^2 \cos(2\theta) \pm \sqrt{a^4 \cos^2(2\theta) + b^4 - a^4}$$

$$= a^2 \cos(2\theta) \pm \sqrt{a^4[\cos^2(2\theta) - 1] + b^4}$$

$$= a^2 \cos(2\theta) \pm \sqrt{b^4 - a^4 \sin^2(2\theta)}$$

$$= a^2 \left[\cos(2\theta) \pm \sqrt{\left(\frac{b}{a}\right)^4 - \sin^2(2\theta)}\right]. \qquad (5)$$

Let a TORUS of tube radius a be cut by a plane perpendicular to the plane of the torus's centroid. Call the distance of this plane from the center of the torus hole r, let $a = r$, and consider the intersection of this plane with the torus as r is varied. The resulting curves are Cassini ovals, with a LEMNISCATE occurring at $r = 1/2$ (Gosper). Cassini ovals are therefore TORIC SECTIONS.

If $a < b$, the curve has AREA

$$A = \tfrac{1}{2} r^2 \, d\theta = 2(\tfrac{1}{2}) \int_{-\pi/4}^{\pi/4} r^2 \, d\theta = a^2 + b^2 E\left(\frac{a^4}{b^4}\right), \qquad (6)$$

where the integral has been done over half the curve and then multiplied by two and $E(x)$ is the complete ELLIPTIC INTEGRAL OF THE SECOND KIND. If $a = b$, the curve becomes

$$r^2 = a^2 \left[\cos(2\theta) + \sqrt{1 - \sin^2 \theta}\right] = 2a^2 \cos(2\theta), \qquad (7)$$

which is a LEMNISCATE having AREA

$$A = 2a^2 \qquad (8)$$

(two loops of a curve $\sqrt{2}$ the linear scale of the usual lemniscate $r^2 = a^2 \cos(2\theta)$, which has area $A = a^2/2$ for each loop). If $a > b$, the curve becomes two disjoint ovals with equations

$$r = \pm a \sqrt{\cos(2\theta) \pm \sqrt{\left(\frac{b}{a}\right)^2 - \sin^2(2\theta)}}, \qquad (9)$$

where $\theta \in [-\theta_0, \, \theta_0]$ and

$$\theta_0 \equiv \tfrac{1}{2} \sin^{-1}\left[\left(\frac{b}{a}\right)^2\right]. \qquad (10)$$

See also CASSINI SURFACE, LEMNISCATE, MANDELBROT SET, OVAL, TORUS

References

Beyer, W. H. *CRC Standard Mathematical Tables, 28th ed.* Boca Raton, FL: CRC Press, p. 221, 1987.

Gray, A. "Cassinian Ovals." §4.2 in *Modern Differential Geometry of Curves and Surfaces with Mathematica, 2nd ed.* Boca Raton, FL: CRC Press, pp. 82–86, 1997.

Lawrence, J. D. *A Catalog of Special Plane Curves.* New York: Dover, pp. 153–155, 1972.

Lockwood, E. H. *A Book of Curves.* Cambridge, England: Cambridge University Press, pp. 187–188, 1967.

MacTutor History of Mathematics Archive. "Cassinian Ovals." http://www-groups.dcs.st-and.ac.uk/~history/Curves/Cassinian.html.

Piziak, R. and Turner, D. "Exploring Gerschgorin Circles and Cassini Ovals." *Mathematica Educ.* **3**, 13–21, 1994.

Wells, D. *The Penguin Dictionary of Curious and Interesting Geometry.* London: Penguin, pp. 25–26, 1991.

Yates, R. C. "Cassinian Curves." *A Handbook on Curves and Their Properties.* Ann Arbor, MI: J. W. Edwards, pp. 8–11, 1952.

Cassini Projection

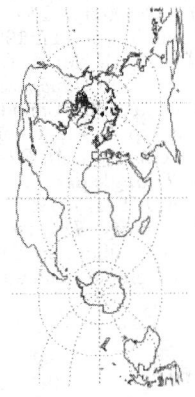

A MAP PROJECTION defined by

$$x = \sin^{-1} B \qquad (1)$$

$$y = \tan^{-1}\left[\frac{\tan \phi}{\cos(\lambda - \lambda_0)}\right], \qquad (2)$$

where

$$B = \cos \phi \, \sin(\lambda - \lambda_0). \qquad (3)$$

The inverse FORMULAS are

$$\phi = \sin^{-1}(\sin D \cos x) \qquad (4)$$

$$\lambda = \lambda_0 + \tan^{-1}\left(\frac{\tan x}{\cos D}\right), \qquad (5)$$

where

$$D = y + \phi_0. \qquad (6)$$

References

Snyder, J. P. *Map Projections--A Working Manual.* U. S. Geological Survey Professional Paper 1395. Washington, DC: U. S. Government Printing Office, pp. 92–95, 1987.

Cassini Surface

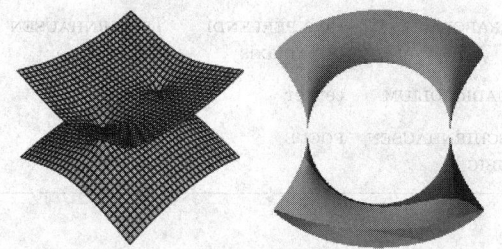

The QUARTIC SURFACE obtained by replacing the constant b in the equation of the CASSINI OVALS with $b = z$, obtaining

$$[(x-a)^2 + y^2][(x+a)^2 + y^2] = z^4. \tag{1}$$

As can be seen by letting $y = 0$ to obtain

$$(x^2 - a^2)^2 = z^4 \tag{2}$$

$$x^2 + z^2 = a^2, \tag{3}$$

the intersection of the surface with the $y = 0$ PLANE is a CIRCLE of RADIUS a.

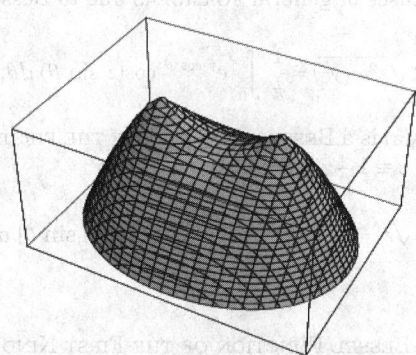

Let a TORUS of tube radius a be cut by a plane perpendicular to the plane of the torus's centroid. Call the distance of this plane from the center of the torus hole r, let $a = r$, and consider the intersection of this plane with the torus as r is varied. The resulting curves are CASSINI OVALS, and the surface having these curves as CROSS SECTIONS is the Cassini surface

$$(x + 2 + z^2 + c^2) - 4c^2 x^2 = 4c^2 r^2,$$

which has a scaled r^2 on the right side instead of z^4 (Gosper).

See also CASSINI OVALS, TORUS

References

Fischer, G. (Ed.). *Mathematical Models from the Collections of Universities and Museums.* Braunschweig, Germany: Vieweg, p. 20, 1986.
Fischer, G. (Ed.). Plate 51 in *Mathematische Modelle / Mathematical Models, Bildband / Photograph Volume.* Braunschweig, Germany: Vieweg, p. 51, 1986.

Cassini's Identity

For F_n the nth FIBONACCI NUMBER,

$$F_{n-1}F_{n+1} - F_n^2 = (-1)^n.$$

This identity was also discovered by Simson (Coxeter and Greitzer 1967, p. 41; Coxeter 1969, pp. 165–168). It is a special case of CATALAN'S IDENTITY with $r = 1$.

See also D'OCAGNE'S IDENTITY, CATALAN'S IDENTITY, FIBONACCI NUMBER

References

Coxeter, H. S. M. *Introduction to Geometry, 2nd ed.* New York: Wiley, 1969.
Coxeter, H. S. M. and Greitzer, S. L. *Geometry Revisited.* Washington, DC: Math. Assoc. Amer., p. 41, 1967.
Petkovsek, M.; Wilf, H. S.; and Zeilberger, D. *A = B.* Wellesley, MA: A. K. Peters, p. 12, 1996.

Casson Invariant

References

Akbulut, S. and McCarthy, J. *Casson's Invariant for Oriented Homology 3-Spheres--An Exposition.* Princeton, NJ: Princeton University Press, 1990.
Saveliev, N. *Lectures on the Topology of 3-Manifolds: An Introduction to the Casson Invariant.* Berlin: de Gruyter, 1999.

Castillon's Problem

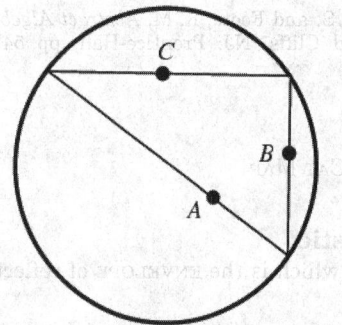

Inscribe a TRIANGLE in a CIRCLE such that the sides of the TRIANGLE pass through three given POINTS A, B, and C.

References

Dörrie, H. "Castillon's Problem." §29 in *100 Great Problems of Elementary Mathematics: Their History and Solutions.* New York: Dover, pp. 144–147, 1965.
F. Gabriel-Marie. *Exercices de géométrie.* Tours, France: Maison Mame, pp. 20–22, 1912.
Rouché, E. and de Comberousse, C. *Traité de géométrie plane.* Paris: Gauthier-Villars, pp. 310–311, 1900.

Casting Out Nines

An elementary check of a MULTIPLICATION which makes use of the CONGRUENCE $10^n \equiv 1 \pmod 9$ for $n \geq 2$. From this CONGRUENCE, a MULTIPLICATION $ab = c$ must give

$$a \equiv \sum a_i = a^*$$

$$b \equiv \sum b_i = b^*$$

$$c \equiv \sum c_i = c^*,$$

so $ab \equiv a^*b^*$ must be $\equiv c^*$ (mod 9). Casting out nines was transmitted to Europe by the Arabs, but was probably an Indian invention and is therefore sometimes also called "the Hindu check." The procedure was described by Fibonacci in his *Liber Abaci* (Wells 1986, p. 74).

References

Conway, J. H. and Guy, R. K. *The Book of Numbers.* New York: Springer-Verlag, pp. 28–29, 1996.

Hilton, P.; Holton, D.; and Pedersen, J. "Casting Out 9's and 11's: Tricks of the Trade." *Mathematical Reflections in a Room with Many Mirrors.* New York: Springer-Verlag, pp. 53–57, 1997.

Wells, D. *The Penguin Dictionary of Curious and Interesting Numbers.* Middlesex, England: Penguin Books, p. 74, 1986.

Casus Irreducibilus

If $P(x)$ is an irreducible CUBIC EQUATION all of whose roots are real, then to obtain them by radicals, you must take roots of nonreal numbers at some point.

See also ALGEBRAIC INTEGER

References

Dummit, D. S. and Foote, R. M. *Abstract Algebra, 2nd ed.* Englewood Cliffs, NJ: Prentice-Hall, pp. 547 and 551, 1998.

Cat Map

ARNOLD'S CAT MAP

Catacaustic

The curve which is the ENVELOPE of reflected rays.

CARDIOID	CUSP of CARDIOID	NEPHROID
CIRCLE	not on CIRCUMFERENCE	LIMAÇON
CIRCLE	on CIRCUMFERENCE	CARDIOID
CIRCLE	point at ∞	NEPHROID
CISSOID OF DIOCLES	FOCUS	CARDIOID
one arch of a CYCLOID	rays PERPENDICULAR axis	two arches of a CYCLOID
DELTOID	point at infinity	ASTROID
ln x	rays PARALLEL axis	CATENARY

LOGARITHMIC SPIRAL	ORIGIN	equal LOGARITHMIC SPIRAL
PARABOLA	rays PERPENDICULAR axis	TSCHIRNHAUSEN CUBIC
QUADRIFOLIUM	center	ASTROID
TSCHIRNHAUSEN CUBIC	FOCUS	SEMICUBICAL PARABOLA

See also CAUSTIC, CIRCLE CAUSTIC, DIACAUSTIC

References

Lawrence, J. D. *A Catalog of Special Plane Curves.* New York: Dover, pp. 60 and 207, 1972.

Catafusene

POLYHEX

Catalan

CATALAN'S CONSTANT

Catalan Integrals

Special cases of general FORMULAS due to Bessel.

$$J_0(\sqrt{z^2 - y^2}) = \frac{1}{\pi} \int_0^\pi e^{y \cos \theta} \cos(z \sin \theta) \, d\theta,$$

where $J_0(z)$ is a BESSEL FUNCTION OF THE FIRST KIND. Now, let $z \equiv 1 - z'$ and $y \equiv 1 + z'$. Then

$$J_0(2i\sqrt{z}) = \frac{1}{\pi} \int_0^\pi e^{(1+z) \cos \theta} \cos[(1 - z) \sin \theta] \, d\theta.$$

See also BESSEL FUNCTION OF THE FIRST KIND

Catalan Number

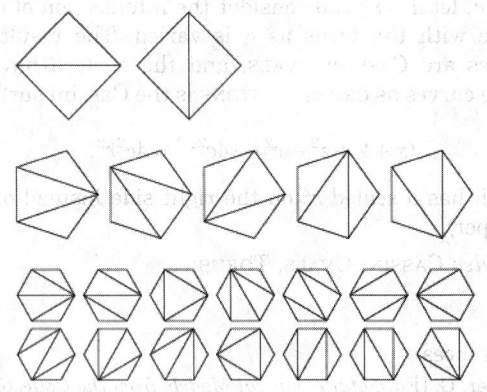

The Catalan numbers are an INTEGER SEQUENCE $\{C_n\}$ which appears in TREE enumeration problems of the type, "In how many ways can a regular n-gon be divided into $n-2$ TRIANGLES if different orientations are counted separately?" (EULER'S POLYGON DIVISION PROBLEM). The solution is the Catalan number C_{n-2} (Dörrie 1965, Honsberger 1973), as graphically illustrated above (Dickau). The first few Catalan numbers for $n = 1, 2, \ldots$ are 1, 2, 5, 14, 42, 132, 429, 1430, 4862, 16796, ... (Sloane's A000108).

The only ODD Catalan numbers are those OF THE FORM C_{2^k-1}, and the last DIGIT is five for $k = 9$ to 15. The only PRIME Catalan numbers for $n \leq 2^{15} - 1$ are $C_2 = 2$ and $C_3 = 5$.

The Catalan numbers turn up in many other related types of problems. C_{n-1} can also be defined as the number of $(-1, 1)$-sequences $\{s_1, s_2, \ldots, s_n\}$ such that $\Sigma_{i=1}^{2n} s_j = 0$ and $\Sigma_{j=1}^{i} s_j \geq 0$ for $i \leq 2n - 1$ (Mays and Wojciechowski 2000). The following table gives the first few such sequences.

n	lists
1	$\{1, -1\}$
2	$\{1, 1, -1, -1\}$
3	$\{1, 1, -1, 1, -1, -1\}, \{1, 1, 1, -1, -1, -1\}$
4	$\{1, 1, -1, 1, -1, 1, -1, -1\}$,
	$\{1, 1, -1, 1, 1, -1, -1, -1\}$,
	$\{1, 1, 1, -1, -1, 1, -1, -1\}$,
	$\{1, 1, 1, -1, 1, -1, -1, -1\}$,
	$\{1, 1, 1, 1, -1, -1, -1, -1\}$

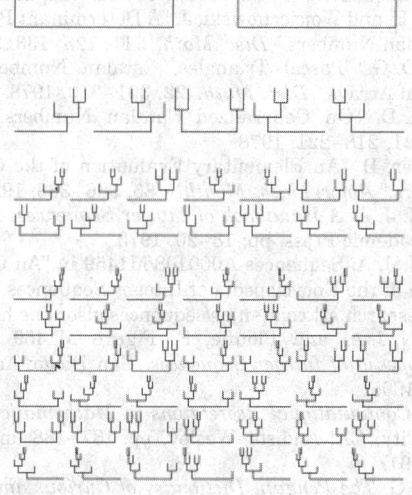

The Catalan number C_{n-1} also gives the number of BINARY BRACKETINGS of n letters (CATALAN'S PROBLEM), the solution to the BALLOT PROBLEM, the

number of trivalent PLANTED PLANAR TREES (Dickau; illustrated above), the number of states possible in an n-FLEXAGON, the number of different diagonals possible in a FRIEZE PATTERN with $n+1$ rows, the number of ways of forming an n-fold exponential, the number of rooted planar binary trees with n internal nodes, the number of rooted plane bushes with n EDGES, the number of extended BINARY TREES with n internal nodes, the number of mountains which can be drawn with n upstrokes and n downstrokes, the number of noncrossing handshakes possible across a round table between n pairs of people (Conway and Guy 1996), and the number of SEQUENCES with NONNEGATIVE PARTIAL SUMS which can be formed from n 1s and n -1s (Bailey 1996, Brualdi 1992)!

An explicit formula for C_n is given by

$$C_n \equiv \frac{1}{n+1}\binom{2n}{n} = \frac{1}{n+1}\frac{(2n)!}{n!^2} = \frac{(2n)!}{(n+1)!n!}, \quad (1)$$

where $\binom{2n}{n}$ denotes a BINOMIAL COEFFICIENT and $n!$ is the usual FACTORIAL. A RECURRENCE RELATION for C_n is obtained from

$$\frac{C_{n+1}}{C_n} = \frac{(2n+2)!}{(n+2)[(n+1)!]^2}\frac{(n+1)(n!)^2}{(2n)!}$$

$$= \frac{(2n+2)(2n+1)(n+1)}{(n+2)(n+1)^2} = \frac{2(2n+1)(n+1)^2}{(n+1)^2(n+2)}$$

$$= \frac{2(2n+1)}{n+2}, \quad (2)$$

so

$$C_{n+1} = \frac{2(2n+1)}{n+2} C_n. \quad (3)$$

Other forms include

$$C_n = \frac{2 \cdot 6 \cdot 10 \cdots (4n-2)}{(n+1)!} \quad (4)$$

$$= \frac{2^n(2n-1)!!}{(n+1)!} \quad (5)$$

$$= \frac{(2n)!}{n!(n+1)!}. \quad (6)$$

SEGNER'S RECURRENCE FORMULA, given by Segner in 1758, gives the solution to EULER'S POLYGON DIVISION PROBLEM

$$E_n = E_2 E_{n-1} + E_3 E_{n-2} + \ldots + E_{n-1} E_2. \quad (7)$$

With $E_1 = E_2 = 1$, the above RECURRENCE RELATION gives the Catalan number $C_{n-2} = E_n$.

The GENERATING FUNCTION for the Catalan numbers is given by

$$\frac{1-\sqrt{1-4x}}{2x}=\sum_{n=0}^{\infty}C_n x^n = 1 + x + 2x^2 + 5x^3 + \ldots . \quad (8)$$

The asymptotic form for the Catalan numbers is

$$C_k \sim \frac{4^k}{\sqrt{\pi}k^{3/2}} \quad (9)$$

(Vardi 1991, Graham *et al.* 1994).

A generalization of the Catalan numbers is defined by

$$_p d_k = \frac{1}{k}\binom{pk}{k-1} = \frac{1}{(p-1)k+1}\binom{pk}{k} \quad (10)$$

for $k \geq 1$ (Klarner 1970, Hilton and Pederson 1991). The usual Catalan numbers $C_k = {}_2 d_k$ are a special case with $p = 2$. $_p d_k$ gives the number of p-ary TREES with k source-nodes, the number of ways of associating k applications of a given p-ary OPERATOR, the number of ways of dividing a convex POLYGON into k disjoint $(p+1)$-gons with nonintersecting DIAGONALS, and the number of P-GOOD PATHS from $(0, -1)$ to $(k, (p-1)k-1)$ (Hilton and Pederson 1991).

A further generalization is obtained as follows. Let p be an INTEGER > 1, let $P_k = (k, (p-1)k-1)$ with $k \geq 0$, and $q \leq p-1$. Then define $_p d_{q0} = 1$ and let $_p d_{qk}$ be the number of P-GOOD PATHS from $(1, q-1)$ to P_k (Hilton and Pederson 1991). Formulas for $_p d_{qi}$ include the generalized JONAH FORMULA

$$\binom{n-q}{k-1} = \sum_{i=1}^{k} {}_p d_{qi}\binom{n-pi}{k-i} \quad (11)$$

and the explicit formula

$$_p d_{qk} = \frac{p-q}{pk-q}\binom{pk-q}{k-1} . \quad (12)$$

A RECURRENCE RELATION is given by

$$_p d_{qk} = \sum_{i,j} {}_p d_{p-r,\, i}\, {}_p d_{q+r,\, j} \quad (13)$$

where $i, j, r \geq 1$, $k \geq 1$, $q < p-r$, and $i+j = k+1$ (Hilton and Pederson 1991).

See also BALLOT PROBLEM, BINARY BRACKETING, BINARY TREE, CATALAN'S PROBLEM, CATALAN'S TRIANGLE, DELANNOY NUMBER, EULER'S POLYGON DIVISION PROBLEM, FLEXAGON, FRIEZE PATTERN, MOTZKIN NUMBER, P-GOOD PATH, PLANTED PLANAR TREE, SCHRÖDER NUMBER, STAIRCASE POLYGON, SUPER CATALAN NUMBER

References

Alter, R. "Some Remarks and Results on Catalan Numbers." *Proc. 2nd Louisiana Conf. Comb., Graph Th., and Comput.*, 109–132, 1971.

Alter, R. and Kubota, K. K. "Prime and Prime Power Divisibility of Catalan Numbers." *J. Combin. Th. A* **15**, 243–256, 1973.

Bailey, D. F. "Counting Arrangements of 1's and -1's." *Math. Mag.* **69**, 128–131, 1996.

Brualdi, R. A. *Introductory Combinatorics, 3rd ed.* New York: Elsevier, 1997.

Campbell, D. "The Computation of Catalan Numbers." *Math. Mag.* **57**, 195–208, 1984.

Chorneyko, I. Z. and Mohanty, S. G. "On the Enumeration of Certain Sets of Planted Trees." *J. Combin. Th. Ser. B* **18**, 209–221, 1975.

Chu, W. "A New Combinatorial Interpretation for Generalized Catalan Numbers." *Disc. Math.* **65**, 91–94, 1987.

Conway, J. H. and Guy, R. K. In *The Book of Numbers.* New York: Springer-Verlag, pp. 96–106, 1996.

Dershowitz, N. and Zaks, S. "Enumeration of Ordered Trees." *Disc. Math.* **31**, 9–28, 1980.

Dickau, R. M. "Catalan Numbers." http://forum.swarthmore.edu/advanced/robertd/catalan.html.

Dörrie, H. "Euler's Problem of Polygon Division." §7 in *100 Great Problems of Elementary Mathematics: Their History and Solutions.* New York: Dover, pp. 21–27, 1965.

Eggleton, R. B. and Guy, R. K. "Catalan Strikes Again! How Likely is a Function to be Convex?" *Math. Mag.* **61**, 211–219, 1988.

Gardner, M. "Catalan Numbers." Ch. 20 in *Time Travel and Other Mathematical Bewilderments.* New York: W. H. Freeman, pp. 253–266, 1988.

Gardner, M. "Catalan Numbers: An Integer Sequence that Materializes in Unexpected Places." *Sci. Amer.* **234**, 120–125, June 1976.

Gould, H. W. *Bell & Catalan Numbers: Research Bibliography of Two Special Number Sequences, 6th ed.* Morgantown, WV: Math Monongliae, 1985.

Graham, R. L.; Knuth, D. E.; and Patashnik, O. Exercise 9.8 in *Concrete Mathematics: A Foundation for Computer Science, 2nd ed.* Reading, MA: Addison-Wesley, 1994.

Guy, R. K. "Dissecting a Polygon Into Triangles." *Bull. Malayan Math. Soc.* **5**, 57–60, 1958.

Hilton, P. and Pederson, J. "Catalan Numbers, Their Generalization, and Their Uses." *Math. Int.* **13**, 64–75, 1991.

Honsberger, R. *Mathematical Gems I.* Washington, DC: Math. Assoc. Amer., pp. 130–134, 1973.

Honsberger, R. *Mathematical Gems III.* Washington, DC: Math. Assoc. Amer., pp. 146–150, 1985.

Klarner, D. A. "Correspondences Between Plane Trees and Binary Sequences." *J. Comb. Th.* **9**, 401–411, 1970.

Mays, M. E. and Wojciechowski, J. "A Determinant Property of Catalan Numbers." *Disc. Math.* **211**, 125–133, 2000.

Rogers, D. G. "Pascal Triangles, Catalan Numbers and Renewal Arrays." *Disc. Math.* **22**, 301–310, 1978.

Sands, A. D. "On Generalized Catalan Numbers." *Disc. Math.* **21**, 218–221, 1978.

Singmaster, D. "An Elementary Evaluation of the Catalan Numbers." *Amer. Math. Monthly* **85**, 366–368, 1978.

Sloane, N. J. A. *A Handbook of Integer Sequences.* Boston, MA: Academic Press, pp. 18–20, 1973.

Sloane, N. J. A. Sequences A000108/M1459 in "An On-Line Version of the Encyclopedia of Integer Sequences." http://www.research.att.com/~njas/sequences/eisonline.html.

Sloane, N. J. A. and Plouffe, S. Figure M1459 in *The Encyclopedia of Integer Sequences.* San Diego: Academic Press, 1995.

Vardi, I. *Computational Recreations in Mathematica.* Redwood City, CA: Addison-Wesley, pp. 187–188 and 198–199, 1991.

Wells, D. G. *The Penguin Dictionary of Curious and Interesting Numbers.* London: Penguin, pp. 121–122, 1986.

Catalan Solid

The DUAL POLYHEDRA of the ARCHIMEDEAN SOLIDS, given in the following table. They are known as Catalan solids in honor of the French mathematician who first published them in 1862 (Wenninger 1983, p. 1).

n	ARCHIMEDEAN SOLID	DUAL
1	CUBOCTAHEDRON	RHOMBIC DODECAHEDRON
2	GREAT RHOMBICOSIDODECAHEDRON	DISDYAKIS TRIACONTAHEDRON
3	GREAT RHOMBICUBOCTAHEDRON	DISDYAKIS DODECAHEDRON
4	ICOSIDODECAHEDRON	RHOMBIC TRIACONTAHEDRON
5	RHOMBICOSIDODECAHEDRON	DELTOIDAL HEXECONTAHEDRON
6	SMALL RHOMBICUBOCTAHEDRON	DELTOIDAL ICOSITETRAHEDRON
7	SNUB CUBE (laevo)	PENTAGONAL ICOSITETRAHEDRON (dextro)
8	SNUB DODECAHEDRON (laevo)	PENTAGONAL HEXECONTAHEDRON (dextro)
9	TRUNCATED CUBE	SMALL TRIAKIS OCTAHEDRON
10	TRUNCATED DODECAHEDRON	TRIAKIS ICOSAHEDRON
11	TRUNCATED ICOSAHEDRON	PENTAKIS DODECAHEDRON
12	TRUNCATED OCTAHEDRON	TETRAKIS HEXAHEDRON
13	TRUNCATED TETRAHEDRON	TRIAKIS TETRAHEDRON

Here are the ARCHIMEDEAN DUALS (Pearce 1978, Holden 1991) displayed in the order listed above (left to right, then continuing to the next row).

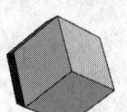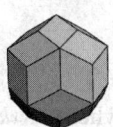

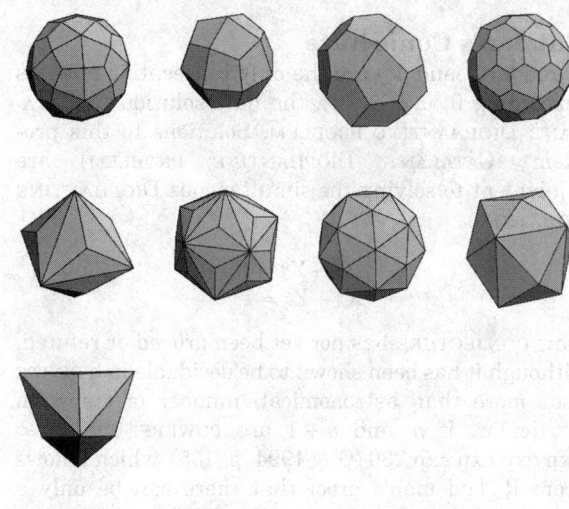

Here are the Archimedean solids paired with the corresponding Catalan solids.

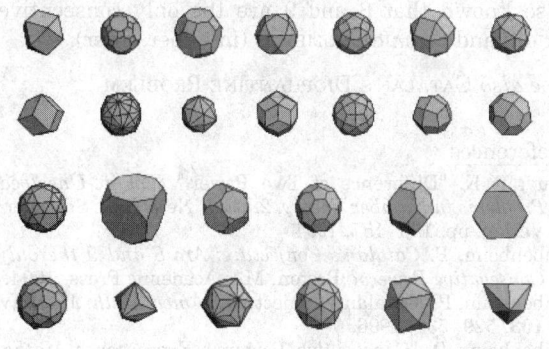

See also ARCHIMEDEAN SOLID, DUAL POLYHEDRON, SEMIREGULAR POLYHEDRON

References

Catalan, E. "Mémoire sur la Théorie des Polyèdres." *J. l'École Polytechnique (Paris)* **41**, 1–71, 1865.
Holden, A. *Shapes, Space, and Symmetry.* New York: Dover, 1991.
Pedagoguery Software. `Poly`. http://www.peda.com/poly/.
Wenninger, M. J. *Dual Models.* Cambridge, England: Cambridge University Press, 1983.

Catalan's Aliquot Sequence Conjecture

The conjecture proposed by Catalan in 1888 and extended by E. Dickson that each ALIQUOT SEQUENCE ends in a PRIME, a PERFECT NUMBER, or a set of SOCIABLE NUMBERS. The conjecture remains open to this day.

See also ALIQUOT SEQUENCE, SOCIABLE NUMBERS

References

Creyaufmüller, W. "Aliquot Sequences." http://home.t-online.de/home/Wolfgang.Creyaufmueller/aliquote.htm.

Catalan's Conjecture

8 and 9 (2^3 and 3^2) are the only consecutive POWERS (excluding 0 and 1), i.e., the only solution to CATALAN'S DIOPHANTINE PROBLEM. Solutions to this problem (CATALAN'S DIOPHANTINE PROBLEM) are equivalent to solving the simultaneous DIOPHANTINE EQUATIONS

$$X^2 - Y^3 = 1$$
$$X^3 - Y^2 = 1.$$

This CONJECTURE has not yet been proved or refuted, although it has been shown to be decidable in a FINITE (but more than astronomical) number of steps. In particular, if n and $n+1$ are POWERS, then $n <$ exp exp exp exp 730 (Guy 1994, p. 155), which follows from R. Tijdeman's proof that there can be only a FINITE number of exceptions should the CONJECTURE not hold.

Hyyro and Makowski proved that there do not exist three consecutive POWERS (Ribenboim 1996), and it is also known that 8 and 9 are the only consecutive CUBIC and SQUARE NUMBERS (in either order).

See also CATALAN'S DIOPHANTINE PROBLEM

References

Guy, R. K. "Difference of Two Power." §D9 in *Unsolved Problems in Number Theory, 2nd ed.* New York: Springer-Verlag, pp. 155–157, 1994.
Ribenboim, P. *Catalan's Conjecture: Are 8 and 9 the only Consecutive Powers?* Boston, MA: Academic Press, 1994.
Ribenboim, P. "Catalan's Conjecture." *Amer. Math. Monthly* **103**, 529–538, 1996.
Ribenboim, P. "Consecutive Powers." *Expositiones Mathematicae* **2**, 193–221, 1984.
Wells, D. *The Penguin Dictionary of Curious and Interesting Numbers.* Middlesex, England: Penguin Books, pp. 71 and 73, 1986.

Catalan's Constant

A constant which appears in estimates of combinatorial functions. It is usually denoted K, $\beta(2)$, or G. It is not known if K is IRRATIONAL. Numerically,

$$K = 0.915965594177\ldots \tag{1}$$

(Sloane's A006752). The CONTINUED FRACTION for K is [0, 1, 10, 1, 8, 1, 88, 4, 1, 1, ...] (Sloane's A014538). K can be given analytically by the following expressions,

$$K \equiv \beta(2) \tag{2}$$

$$= -i\chi_2(i) \tag{3}$$

$$= \sum_{k=0}^{\infty} \frac{(-1)^k}{(2k+1)^2} = \frac{1}{1^2} - \frac{1}{3^2} + \frac{1}{5^2} + \ldots \tag{4}$$

$$= 1 + \sum_{n=1}^{\infty} \frac{1}{(4n+1)^2} - \frac{1}{9} - \sum_{n=1}^{\infty} \frac{1}{(4n+3)^2} \tag{5}$$

$$= \int_0^1 \frac{\tan^{-1} x \, dx}{x} \tag{6}$$

$$= -\int_0^1 \frac{\ln x \, dx}{1+x^2}, \tag{7}$$

where $\beta(z)$ is the DIRICHLET BETA FUNCTION and $\chi_v(z)$ is LEGENDRE'S CHI-FUNCTION. In terms of the POLYGAMMA FUNCTION $\Psi_1(x)$,

$$K = \frac{1}{16} \Psi_1\left(\frac{1}{4}\right) - \frac{1}{16} \Psi_1\left(\frac{3}{4}\right) \tag{8}$$

$$= \frac{1}{80} \Psi_1\left(\frac{5}{12}\right) + \frac{1}{80} \Psi_1\left(\frac{1}{12}\right) - \frac{1}{10} \pi^2 \tag{9}$$

$$= \frac{1}{32} \Psi_1\left(\frac{1}{8}\right) - \frac{1}{32} \Psi_1\left(\frac{3}{8}\right) - \frac{1}{16} \sqrt{2}. \tag{10}$$

Applying CONVERGENCE IMPROVEMENT to (4) gives

$$K = \frac{1}{16} \sum_{m=1}^{\infty} (m+1) \frac{3^m - 1}{4^m} \zeta(m+2), \tag{11}$$

where $\zeta(z)$ is the RIEMANN ZETA FUNCTION and the identity

$$\frac{1}{(1-3z)^2} - \frac{1}{(1-z)^2} = \sum_{m=1}^{\infty} (m+1) \frac{3^m - 1}{4^m} z^m \tag{12}$$

has been used (Flajolet and Vardi 1996). The Flajolet and Vardi algorithm also gives

$$K = \frac{1}{\sqrt{2}} \prod_{k=1}^{\infty} \left[\left(1 - \frac{1}{2^{2^k}}\right) \frac{\zeta(2^k)}{\beta(2^k)} \right]^{1/(2^{k+1})}, \tag{13}$$

where $\beta(z)$ is the DIRICHLET BETA FUNCTION. Glaisher (1913) gave

$$K = 1 - \sum_{n=1}^{\infty} \frac{n \zeta(2n+1)}{16^n} \tag{14}$$

(Vardi 1991, p. 159). W. Gosper used the related FORMULA

$$K = \frac{1}{\sqrt{2}} \left[\frac{1}{\Psi(2) - 1} \right]^{2^{1/2}} \prod_{k=2}^{\infty} \left[\frac{1}{-\Psi(2^k) - 1} \right]^{1/(2^{k+1})}, \tag{15}$$

where

$$\Psi(m) = \frac{m \psi_{m-1}\left(\frac{1}{4}\right)}{\pi^m (2^m - 1) 4^{m-1} B_m}, \tag{16}$$

where B_n is a BERNOULLI NUMBER and $\psi(x)$ is a POLYGAMMA FUNCTION (Finch). The Catalan constant may also be defined by

$$K \equiv \frac{1}{2} \int_0^1 K(k) \, dk, \tag{17}$$

where $K(k)$ (not to be confused with Catalan's constant itself, denoted K) is a complete ELLIPTIC INTEGRAL OF THE FIRST KIND.

$$K = \frac{\pi \ln 2}{8} + \sum_{i=1}^{\infty} \frac{a_i}{2^{\lfloor (i+1)/2 \rfloor} i^2}, \qquad (18)$$

where

$$\{a_i\} = \{1, 1, 1, 0, -1, -1, -1, 0\} \qquad (19)$$

is given by the periodic sequence obtained by appending copies of $\{1, 1, 1, 0, -1, -1, -1, 0\}$ (in other words, $a_i \equiv a_{[i-1 \ (\mathrm{mod}\,8)]+1}$ for $i > 8$) and $\lfloor x \rfloor$ is the FLOOR FUNCTION (Nielsen 1909).

See also DIRICHLET BETA FUNCTION

References

Abramowitz, M. and Stegun, C. A. (Eds.). *Handbook of Mathematical Functions with Formulas, Graphs, and Mathematical Tables, 9th printing.* New York: Dover, pp. 807–808, 1972.

Adamchik, V. "Integral and Series Representations for Catalan's Constant." http://members.wri.com/victor/articles/catalan.html.

Adamchik, V. "Thirty-Three Representations of Catalan's Constant." http://library.wolfram.com/demos/v4/Catalan-Formulas.nb.

Arfken, G. *Mathematical Methods for Physicists, 3rd ed.* Orlando, FL: Academic Press, pp. 551–552, 1985.

Fee, G. J. "Computation of Catalan's Constant using Ramanujan's Formula." *ISAAC '90. Proc. Internat. Symp. Symbolic Algebraic Comp., Aug. 1990.* Reading, MA: Addison-Wesley, 1990.

Finch, S. "Favorite Mathematical Constants." http://www.mathsoft.com/asolve/constant/catalan/catalan.html.

Flajolet, P. and Vardi, I. "Zeta Function Expansions of Classical Constants." Unpublished manuscript. 1996. http://pauillac.inria.fr/algo/flajolet/Publications/landau.ps.

Glaisher, J. W. L. "Numerical Values of the Series $1 - 1/3^n + 1/5^n - 1/7^n + 1/9^n - \&c$ for $n = 2, 4, 6$." *Messenger Math.* **42**, 35–58, 1913.

Gosper, R. W. "A Calculus of Series Rearrangements." In *Algorithms and Complexity: New Directions and Recent Results* (Ed. J. F. Traub). New York: Academic Press, 1976.

Nielsen, N. *Der Eulersche Dilogarithms.* Leipzig, Germany: Halle, pp. 105 and 151, 1909.

Plouffe, S. "Plouffe's Inverter: Table of Current Records for the Computation of Constants." http://www.lacim.uqam.ca/pi/records.html.

Sloane, N. J. A. Sequences A006752/M4593 and A014538 in "An On-Line Version of the Encyclopedia of Integer Sequences." http://www.research.att.com/~njas/sequences/eisonline.html.

Srivastava, H. M. and Miller, E. A. "A Simple Reducible Case of Double Hypergeometric Series involving Catalan's Constant and Riemann's Zeta Function." *Int. J. Math. Educ. Sci. Technol.* **21**, 375–377, 1990.

Vardi, I. *Computational Recreations in Mathematica.* Reading, MA: Addison-Wesley, p. 159, 1991.

Yang, S. "Some Properties of Catalan's Constant G." *Int. J. Math. Educ. Sci. Technol.* **23**, 549–556, 1992.

Catalan's Diophantine Problem

Find consecutive POWERS, i.e., solutions to

$$a^b - c^d = 1,$$

excluding 0 and 1. CATALAN'S CONJECTURE is that the only solution is $3^2 - 2^3 = 1$, so 8 and 9 (2^3 and 3^2) are the only consecutive POWERS (again excluding 0 and 1).

See also CATALAN'S CONJECTURE

References

Cassels, J. W. S. "On the Equation $a^x - b^y = 1$. II." *Proc. Cambridge Phil. Soc.* **56**, 97–103, 1960.

Inkeri, K. "On Catalan's Problem." *Acta Arith.* **9**, 285–290, 1964.

Catalan's Identity

$$F_n^2 - F_{n+r}F_{n-r} = (-1)^{n-r}F_r^2,$$

where F_n is a FIBONACCI NUMBER. Letting $r = 1$ gives CASSINI'S IDENTITY.

See also CASSINI'S IDENTITY, D'OCAGNE'S IDENTITY, FIBONACCI NUMBER

Catalan's Problem

The problem of finding the number of different ways in which a PRODUCT of n different ordered FACTORS can be calculated by pairs (i.e., the number of BINARY BRACKETINGS of n letters). For example, for the four FACTORS a, b, c, and d, there are five possibilities: $((ab)c)d$, $(a(bc))d$, $(ab)(cd)$, $a((bc)d)$, and $a(b(cd))$. The solution was given by Catalan in 1838 as

$$C'_n = \frac{(4n-6)!!!!}{n!} = \frac{2 \cdot 6 \cdot 10 \cdots (4n-6)}{n!},$$

where $n!!!!$ is a MULTIFACTORIAL and $n!$ is the usual FACTORIAL, which is equal to the CATALAN NUMBER $C_{n-1} = C'_n$.

See also BINARY BRACKETING, CATALAN'S DIOPHANTINE PROBLEM, CATALAN NUMBER, EULER'S POLYGON DIVISION PROBLEM

References

Dörrie, H. *100 Great Problems of Elementary Mathematics: Their History and Solutions.* New York: Dover, p. 23, 1965.

Catalan's Surface

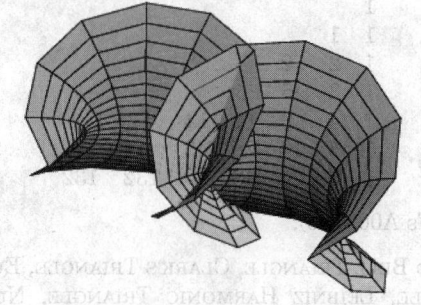

A MINIMAL SURFACE given by the PARAMETRIC EQUA-

TIONS

$$x(u, v) = u - \sin u \cosh v \qquad (1)$$

$$y(u, v) = 1 - \cos u \cosh v \qquad (2)$$

$$z(u, v) = 4 \sin(\tfrac{1}{2}u) \sinh(\tfrac{1}{2}v) \qquad (3)$$

(Gray 1997), or

$$x(r, \phi) = a \sin(2\phi) - 2a\phi + \tfrac{1}{2}av^2 \cos(2\phi) \qquad (4)$$

$$y(r, \phi) = -a \cos(2\phi) - \tfrac{1}{2}av^2 \cos(2\phi) \qquad (5)$$

$$z(r, \phi) = 2av \sin \phi, \qquad (6)$$

where

$$v = -r + \frac{1}{r} \qquad (7)$$

(do Carmo 1986).

References

Catalan, E. "Mémoire sur les surfaces dont les rayons de courbures en chaque point, sont égaux et les signes contraires." *C. R. Acad. Sci. Paris* **41**, 1019–1023, 1855.
do Carmo, M. P. "Catalan's Surface" §3.5D in *Mathematical Models from the Collections of Universities and Museums* (Ed. G. Fischer). Braunschweig, Germany: Vieweg, pp. 45–46, 1986.
Fischer, G. (Ed.). Plates 94–95 in *Mathematische Modelle / Mathematical Models, Bildband / Photograph Volume.* Braunschweig, Germany: Vieweg, pp. 90–91, 1986.
Gray, A. "Catalan's Minimal Surface." *Modern Differential Geometry of Curves and Surfaces with Mathematica, 2nd ed.* Boca Raton, FL: CRC Press, pp. 692–693, 1997.
JavaView. "Classic Surfaces from Differential Geometry: Catalan Surface." http://www-sfb288.math.tu-berlin.de/vgp/javaview/demo/surface/common/PaSurface_Catalan.html.

Catalan's Triangle

A triangle of numbers with entries given by

$$c_{nm} = \frac{(n + m)!(n - m + 1)}{m!(n + 1)!}$$

for $0 \le m \le n$, where each element is equal to the one above plus the one to the left. Furthermore, the sum of each row is equal to the last element of the next row and also equal to the CATALAN NUMBER C_n.

```
1
1  1
1  2   2
1  3   5   5
1  4   9  14  14
1  5  14  28  42   42
1  6  20  48  90  132  132
```

(Sloane's A009766).

See also BELL TRIANGLE, CLARK'S TRIANGLE, EULER'S TRIANGLE, LEIBNIZ HARMONIC TRIANGLE, NUMBER TRIANGLE, PASCAL'S TRIANGLE, PRIME TRIANGLE,

SEIDEL-ENTRINGER-ARNOLD TRIANGLE

References

Sloane, N. J. A. Sequences A009766 in "An On-Line Version of the Encyclopedia of Integer Sequences." http://www.research.att.com/~njas/sequences/eisonline.html.

Catalan's Trisectrix

TSCHIRNHAUSEN CUBIC

Catalogue Paradox

Consider a library which compiles a bibliographic catalog of all (and only those) catalogs which do not list themselves. Then does the library's catalog list itself?

See also PSEUDOPARADOX, RUSSELL'S PARADOX

References

Curry, H. B. *Foundations of Mathematical Logic.* New York: Dover, p. 5, 1977.
Gonseth, F. "La structure du paradoxe des catalogues." §106 in *Les mathématiques et la réalité: Essai sur la méthode axiomatique.* Paris: Félix Alcan, pp. 255–257, 1936.

Catastrophe

For any system that seeks to minimize a function, only seven different local forms of CATASTROPHE "typically" occur for four or fewer variables:

1. FOLD CATASTROPHE,
2. CUSP CATASTROPHE,
3. SWALLOWTAIL CATASTROPHE,
4. BUTTERFLY CATASTROPHE,
5. ELLIPTIC UMBILIC CATASTROPHE,
6. HYPERBOLIC UMBILIC CATASTROPHE, and
7. PARABOLIC UMBILIC CATASTROPHE.

More specifically, for any system with fewer than five control factors and fewer than three behavior axes, these are the only seven catastrophes possible. The following tables gives the possible catastrophes as a function of control factors and behavior axes (Goetz).

Control Factors	1 Behavior Axis	2 Behavior Axes
1	FOLD	
2	CUSP	
3	SWALLOWTAIL	HYPERBOLIC UMBILIC, ELLIPTIC UMBILIC
4	BUTTERFLY	PARABOLIC UMBILIC

The following table gives prototypical examples for equations showing each type of catastrophe.

equation	catastrophe
$x^3 + ux$	FOLD CATASTROPHE
$x^4 + ux^2 + vx$	CUSP CATASTROPHE, Riemann-Hugoniot catastrophe
$x^5 + ux^3 + vx^2 + wx$	SWALLOWTAIL CATASTROPHE
$x^3 + y^3 + uxy + vx + wy$	HYPERBOLIC UMBILIC CATASTROPHE
$x^3 - xy^2 + u(x^2 + y^2) + vx + wy$	ELLIPTIC UMBILIC CATASTROPHE
$x^6 + ux^4 + vx^3 + wx^2 + tx$	BUTTERFLY CATASTROPHE
$x^2y + y^4 + ux^2 + vy^2 + wx + ty$	PARABOLIC UMBILIC CATASTROPHE

See also BUTTERFLY CATASTROPHE, CATASTROPHE THEORY, CUSP CATASTROPHE, ELLIPTIC UMBILIC CATASTROPHE, FOLD CATASTROPHE, HYPERBOLIC UMBILIC CATASTROPHE, PARABOLIC UMBILIC CATASTROPHE, SWALLOWTAIL CATASTROPHE

References

Sanns, W. *Catastrophe Theory with Mathematica: A Geometric Approach.* Germany: DAV, 2000.

Catastrophe Theory

Catastrophe theory studies how the qualitative nature of equation solutions depends on the parameters that appear in the equations. Subspecializations include bifurcation theory, nonequilibrium thermodynamics, singularity theory, synergetics, and topological dynamics. For any system that seeks to minimize a function, only seven different local forms of CATASTROPHE "typically" occur for four or fewer variables.

See also CATASTROPHE

References

Arnold, V. I. *Catastrophe Theory, 3rd ed.* Berlin: Springer-Verlag, 1992.
Dujardin, L. "Catastrophe Teacher: An Introduction for Experimentalists." http://perso.wanadoo.fr/l.d.v.dujardin/ct/eng_index.html.
Gilmore, R. *Catastrophe Theory for Scientists and Engineers.* New York: Dover, 1993.
Goetz, P. "Phil's Good Enough Complexity Dictionary." http://www.cs.buffalo.edu/~goetz/dict.html.
Sanns, W. *Catastrophe Theory with Mathematica: A Geometric Approach.* Germany: DAV, 2000.
Saunders, P. T. *An Introduction to Catastrophe Theory.* Cambridge, England: Cambridge University Press, 1980.

Stewart, I. *The Problems of Mathematics, 2nd ed.* Oxford, England: Oxford University Press, p. 211, 1987.
Thom, R. *Structural Stability and Morphogenesis: An Outline of a General Theory of Models.* Reading, MA: Addison-Wesley, 1993.
Thompson, J. M. T. *Instabilities and Catastrophes in Science and Engineering.* New York: Wiley, 1982.
Weisstein, E. W. "Books about Catastrophe Theory." http://www.treasure-troves.com/books/CatastropheTheory.html.
Woodcock, A. E. R. and Davis, M. *Catastrophe Theory.* New York: E. P. Dutton, 1978.
Zeeman, E. C. *Catastrophe Theory--Selected Papers 1972–1977.* Reading, MA: Addison-Wesley, 1977.

Categorical Game

A GAME in which no DRAW is possible. All CATEGORICAL GAMES are unfair (Steinhaus 1983, p. 16).

See also DRAW, GAME

References

Steinhaus, H. *Mathematical Snapshots, 3rd ed.* New York: Dover, p. 16 1999.

Categorical Variable

A variable which belongs to exactly one of a finite number of CATEGORIES.

See also CATEGORY

Category

A category consists of two things: a collection of OBJECTS and, for each pair of OBJECTS, a collection of MORPHISMS (sometimes called "arrows") from one to another.

In most concrete categories over sets, an OBJECT is some mathematical structure (e.g., a GROUP, VECTOR SPACE, or DIFFERENTIABLE MANIFOLD) and a MORPHISM is a MAP between two OBJECTS. The MORPHISMS are then required to satisfy some fairly natural conditions; for instance, the IDENTITY MAP between any object and itself is always a MORPHISM, and the composition of two MORPHISMS (if defined) is always a MORPHISM.

One usually requires the MORPHISMS to preserve the mathematical structure of the objects. So if the objects are all groups, a good choice for a MORPHISM would be a group HOMOMORPHISM. Similarly, for vector spaces, one would choose linear maps, and for differentiable manifolds, one would choose differentiable maps.

In the category of TOPOLOGICAL SPACES, homomorphisms are usually continuous maps between topological spaces. However, there are also other category structures having TOPOLOGICAL SPACES as objects, but they are not nearly as important as the "standard" category of TOPOLOGICAL SPACES and continuous maps.

See also ABELIAN CATEGORY, ALLEGORY, EILENBERG-

STEENROD AXIOMS, GROUPOID, HOLONOMY, LOGOS, MONODROMY, TOPOS

References

Freyd, P. J. and Scedrov, A. *Categories, Allegories*. Amsterdam, Netherlands: North-Holland, 1990.

Getzler, E. and Kapranov, M. (Eds.). *Higher Category Theory*. Providence, RI: Amer. Math. Soc., 1998.

Lawvere, F. W. and Schanuel, S. H. *Conceptual Mathematics: A First Introduction to Categories*. Cambridge, England: Cambridge University Press, 1997.

Mac Lane, S. and Gehring, F. W. *Categories for the Working Mathematician, 2nd ed.* New York: Springer-Verlag, 1998.

Munkres, J. R. "Categories and Functors." §28 in *Elements of Algebraic Topology*. Perseus Press, pp. 154–160, 1993.

Category Theory

The branch of mathematics which formalizes a number of algebraic properties of collections of transformations between mathematical objects (such as binary relations, groups, sets, topological spaces, etc.) of the same type, subject to the constraint that the collections contain the identity mapping and are closed with respect to compositions of mappings. The objects studied in category theory are called CATEGORIES.

See also CATEGORY

Catenary

The curve a hanging flexible wire or chain assumes when supported at its ends and acted upon by a uniform gravitational force. The word catenary is derived from the Latin word for "chain." In 1669, Jungius disproved Galileo's claim that the curve of a chain hanging under gravity would be a PARABOLA (MacTutor Archive). The curve is also called the alysoid and chainette. The equation was obtained by Leibniz, Huygens, and Johann Bernoulli in 1691 in response to a challenge by Jakob Bernoulli.

Huygens was the first to use the term catenary in a letter to Leibniz in 1690, and David Gregory wrote a treatise on the catenary in 1690 (MacTutor Archive). If you roll a PARABOLA along a straight line, its FOCUS traces out a catenary. As proved by Euler in 1744, the catenary is also the curve which, when rotated, gives the surface of minimum SURFACE AREA (the CATENOID) for the given bounding CIRCLE.

The PARAMETRIC EQUATIONS for the catenary are given by

$$x(t) = t \tag{1}$$

$$y(t) = \tfrac{1}{2} a(e^{t/a} + e^{-t/a}) = a \cosh\left(\frac{t}{a}\right), \tag{2}$$

where $t = 0$ corresponds to the vertex, and the CESÀRO EQUATION is

$$(s^2 + a^2)\kappa = -a. \tag{3}$$

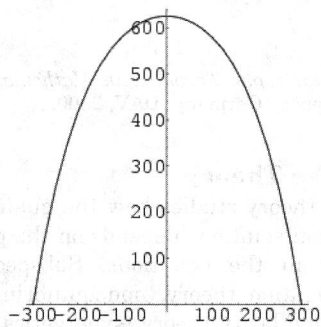

The ARC LENGTH, CURVATURE, and TANGENTIAL ANGLE are

$$s(t) = a \sinh\left(\frac{t}{a}\right), \tag{4}$$

$$\kappa(t) = -\frac{1}{a} \operatorname{sech}^2\left(\frac{t}{a}\right), \tag{5}$$

$$\phi(t) = -2 \tan^{-1}\left[\tanh\left(\frac{t}{2a}\right)\right]. \tag{6}$$

The slope is proportional to the ARC LENGTH as measured from the center of symmetry.

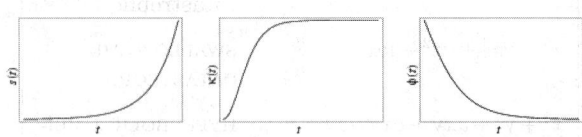

The St. Louis Arch closely approximates an inverted catenary, but it has a finite thickness and varying cross sectional area (thicker at the base; thinner at the apex). The centroid has half-length of $L = 299.2239$ feet at the base, height of 625.0925 feet, top cross sectional area 125.1406 square feet, and bottom cross sectional area 1262.6651 square feet.

The catenary also gives the shape of the road (ROULETTE) over which a regular polygonal "wheel" can travel smoothly. For a regular n-gon, the Cartesian equation of the corresponding catenary is

$$y = -A \cosh\left(\frac{x}{A}\right), \tag{7}$$

where

$$A \equiv R \cos\left(\frac{\pi}{n}\right). \tag{8}$$

See also CALCULUS OF VARIATIONS, CATENOID, LINDE-LOF'S THEOREM, ROULETTE, SURFACE OF REVOLUTION

References

Beyer, W. H. *CRC Standard Mathematical Tables, 28th ed.* Boca Raton, FL: CRC Press, p. 214, 1987.

Gray, A. "The Evolute of a Tractrix is a Catenary." §5.3 in *Modern Differential Geometry of Curves and Surfaces with Mathematica, 2nd ed.* Boca Raton, FL: CRC Press, pp. 102–103, 1997.

Lawrence, J. D. *A Catalog of Special Plane Curves.* New York: Dover, pp. 195 and 199–200, 1972.

Lockwood, E. H. "The Tractrix and Catenary." Ch. 13 in *A Book of Curves.* Cambridge, England: Cambridge University Press, pp. 118–124, 1967.

MacTutor History of Mathematics Archive. "Catenary." http://www-groups.dcs.st-and.ac.uk/~history/Curves/Catenary.html.

National Park Service. "Arch History and Architecture: Catenary Curve Equation." http://www.nps.gov/jeff/equation.htm.

Pappas, T. "The Catenary & the Parabolic Curves." *The Joy of Mathematics.* San Carlos, CA: Wide World Publ./Tetra, p. 34, 1989.

Steinhaus, H. *Mathematical Snapshots, 3rd ed.* New York: Dover, pp. 247–249, 1999.

Wells, D. *The Penguin Dictionary of Curious and Interesting Geometry.* London: Penguin, pp. 26–27, 1991.

Yates, R. C. "Catenary." *A Handbook on Curves and Their Properties.* Ann Arbor, MI: J. W. Edwards, pp. 12–14, 1952.

Catenary Evolute

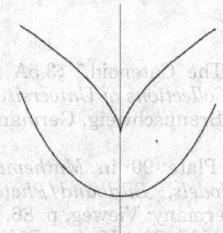

$$x = a[x - \tfrac{1}{2}\sinh(2t)]$$

$$y = 2a \cosh t.$$

Catenary Involute

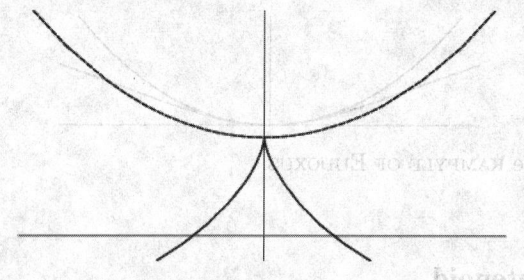

The parametric equation for a CATENARY is

$$\mathbf{r}(t) = a \begin{bmatrix} t \\ \cosh t \end{bmatrix}, \tag{1}$$

so

$$\frac{d\mathbf{r}}{dt} = a \begin{bmatrix} 1 \\ \sinh t \end{bmatrix} \tag{2}$$

$$\left|\frac{d\mathbf{r}}{dt}\right| = a\sqrt{1 + \sinh^2 t} = a \cosh t \tag{3}$$

and

$$\hat{\mathbf{T}} = \frac{\dfrac{d\mathbf{r}}{dt}}{\left|\dfrac{d\mathbf{r}}{dt}\right|} = \begin{bmatrix} \operatorname{sech} t \\ \tanh t \end{bmatrix} \tag{4}$$

$$ds^2 = |d\mathbf{r}^2| = a^2(1 + \sinh^2 t)\, dt^2 = a^2 \cosh^2 dt^2 \tag{5}$$

$$\frac{ds}{dt} = a \cosh t. \tag{6}$$

Therefore,

$$s = a \int \cosh t \, dt = a \sinh t \tag{7}$$

and the equation of the INVOLUTE is

$$x = a(t - \tanh t) \tag{8}$$

$$y = a \operatorname{sech} t. \tag{9}$$

This curve is called a TRACTRIX.

Catenary Radial Curve

The KAMPYLE OF EUDOXUS.

Catenoid

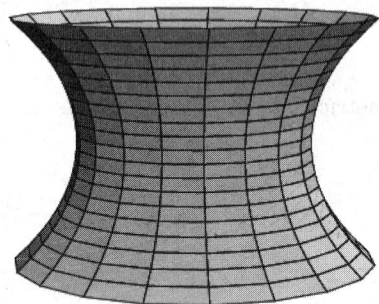

A CATENARY of REVOLUTION. The catenoid and PLANE are the only SURFACES OF REVOLUTION which are also MINIMAL SURFACES. The catenoid can be given by the PARAMETRIC EQUATIONS

$$x = c \cosh\left(\frac{v}{c}\right) \cos u \tag{1}$$

$$y = c \cosh\left(\frac{v}{c}\right) \sin u \tag{2}$$

$$z = v, \tag{3}$$

where $u \in [0, 2\pi)$. The differentials are

$$dx = \sinh\left(\frac{v}{c}\right) \cos u \, dv - \cosh\left(\frac{v}{c}\right) \sin u \, du \tag{4}$$

$$dy = \sinh\left(\frac{v}{c}\right) \sin u \, dv + \cosh\left(\frac{v}{c}\right) \cos u \, du \tag{5}$$

$$dz = du, \tag{6}$$

so the LINE ELEMENT is

$$\begin{aligned} ds^2 &= dx^2 + dy^2 + dz^2 \\ &= \left[\sinh^2\left(\frac{v}{c}\right) + 1\right] dv^2 + \cosh^2\left(\frac{v}{c}\right) du^2 \\ &= \cosh^2\left(\frac{v}{c}\right) dv^2 + \cosh^2\left(\frac{v}{c}\right) du^2. \end{aligned} \tag{7}$$

The PRINCIPAL CURVATURES are

$$\kappa_1 = -\frac{1}{c} \operatorname{sech}^2\left(\frac{v}{c}\right) \tag{8}$$

$$\kappa_2 = \frac{1}{c} \operatorname{sech}^2\left(\frac{v}{c}\right). \tag{9}$$

The MEAN CURVATURE of the catenoid is

$$H = 0 \tag{10}$$

and the GAUSSIAN CURVATURE is

$$K = -\frac{1}{c^2} \operatorname{sech}^4\left(\frac{v}{c}\right). \tag{11}$$

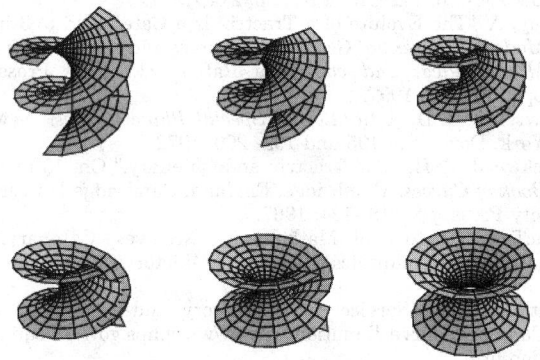

The HELICOID can be continuously deformed into a catenoid with $c = 1$ by the transformation

$$x(u, v) = \cos \alpha \sinh v \sin u + \sin \alpha \cosh v \cos u \tag{12}$$

$$y(u, v) = -\cos \alpha \sinh v \cos u + \sin \alpha \cosh v \sin u \tag{13}$$

$$z(u, v) = u \cos \alpha + v \sin \alpha, \tag{14}$$

where $\alpha = 0$ corresponds to a HELICOID and $\alpha = \pi/2$ to a catenoid.

See also CATENARY, COSTA MINIMAL SURFACE, HELICOID, MINIMAL SURFACE, SURFACE OF REVOLUTION

References

do Carmo, M. P. "The Catenoid." §3.5A in *Mathematical Models from the Collections of Universities and Museums* (Ed. G. Fischer). Braunschweig, Germany: Vieweg, p. 43, 1986.

Fischer, G. (Ed.). Plate 90 in *Mathematische Modelle/Mathematical Models, Bildband/Photograph Volume.* Braunschweig, Germany: Vieweg, p. 86, 1986.

Gray, A. "The Catenoid." §20.4 *Modern Differential Geometry of Curves and Surfaces with Mathematica, 2nd ed.* Boca Raton, FL: CRC Press, pp. 467–469, 1997.

JavaView. "Classic Surfaces from Differential Geometry: Catenoid/Helicoid." http://www-sfb288.math.tu-berlin.de/vgp/javaview/demo/surface/common/PaSurface_Catenoid-Helicoid.html.

Meusnier, J. B. "Mémoire sur la courbure des surfaces." *Mém. des savans étrangers* **10** (lu 1776), 477–510, 1785.

Ogawa, A. "Helicatenoid." *Mathematica J.* **2**, 21, 1992.

Osserman, R. *A Survey of Minimal Surfaces.* New York: Dover, p. 18 1986.

Steinhaus, H. *Mathematical Snapshots, 3rd ed.* New York: Dover, pp. 247–249, 1999.

Caterpillar Graph

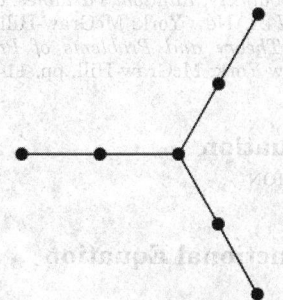

A TREE with every NODE on a central stalk or only one EDGE away from the stalk. A tree is a caterpillar graph IFF all nodes of degree ≥ 3 are surrounded by at most two nodes of degree two or greater. The number of caterpillar graphs on $n = 1, 2, \ldots$ nodes are 1, 1, 1, 2, 3, 6, 10, 20, 36, 72, 136, ... (Sloane's A005418), giving the number of noncaterpillar graphs on $n = 7, 8, \ldots$ as 1, 3, 11, 34, 99, ... (Sloane's A052471). The non-caterpillar graphs on $n \leq 9$ nodes are illustrated above.

See also TREE

References

Gardner, M. *Wheels, Life, and other Mathematical Amusements.* New York: W. H. Freeman, p. 160, 1983.

Hoffman, N. "Binary Grids and a Related Counting Problem." *Two Year Coll. Math. J.* **9**, 267–272, 1978.

Sloane, N. J. A. Sequences A005418/M0771 and A052471 in "An On-Line Version of the Encyclopedia of Integer Sequences." http://www.research.att.com/~njas/sequences/eisonline.html.

Sulanke, R. A.. "Moments of Generalized Motzkin Paths." *J. Integer Sequences* **3**, No. 00.1.1, 2000. http://www.research.att.com/~njas/sequences/JIS/SULANKE/sulanke.html.

Cattle Problem of Archimedes

ARCHIMEDES' CATTLE PROBLEM

Cauchy Binomial Theorem

$$
\begin{aligned}
\prod_{k=1}^{n} (1 + y q^k) &= \sum_{m=0}^{n} y^m q^{m(m+1)/2} \begin{bmatrix} n \\ m \end{bmatrix}_q \\
&= \sum_{m=0}^{n} y^m q^{m(m+1)/2} \frac{(q)_n}{(q)_m (q)_{n-m}},
\end{aligned}
$$

where $[nrm]_q$ is a Q-BINOMIAL COEFFICIENT.

See also Q-BINOMIAL COEFFICIENT, Q-BINOMIAL THEOREM

Cauchy Boundary Conditions

BOUNDARY CONDITIONS of a PARTIAL DIFFERENTIAL EQUATION which are a weighted AVERAGE of DIRICHLET BOUNDARY CONDITIONS (which specify the value of the function on a surface) and NEUMANN BOUNDARY CONDITIONS (which specify the normal derivative of the function on a surface).

See also BOUNDARY CONDITIONS, CAUCHY PROBLEM, DIRICHLET BOUNDARY CONDITIONS, NEUMANN BOUNDARY CONDITIONS

References

Morse, P. M. and Feshbach, H. *Methods of Theoretical Physics, Part I.* New York: McGraw-Hill, pp. 678–679, 1953.

Cauchy Condition

UNIFORMLY CAUCHY

Cauchy Criterion

A NECESSARY and SUFFICIENT condition for a SEQUENCE S_i to CONVERGE. The Cauchy criterion is satisfied when, for all $\epsilon > 0$, there is a fixed number N such that $|S_j - S_i| < \epsilon$ for all $i, j > N$.

Cauchy Distribution

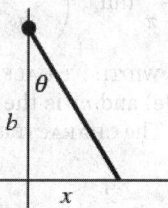

The Cauchy distribution, also called the LORENTZIAN DISTRIBUTION, is a continuous distribution describing resonance behavior. It also describes the distribution of horizontal distances at which a LINE SEGMENT tilted at a random ANGLE cuts the x-AXIS. Let θ represent the ANGLE that a line, with fixed point of rotation, makes with the vertical axis, as shown above. Then

$$
\tan \theta = \frac{x}{b} \tag{1}
$$

$$
\theta = \tan^{-1}\left(\frac{x}{b}\right) \tag{2}
$$

$$
d\theta = -\frac{1}{1 + \dfrac{x^2}{b^2}} \frac{dx}{b} = -\frac{b\,dx}{b^2 + x^2}, \tag{3}
$$

so the distribution of ANGLE θ is given by

$$
\frac{d\theta}{\pi} = -\frac{1}{\pi} \frac{b\,dx}{b^2 + x^2}. \tag{4}
$$

This is normalized over all angles, since

$$
\int_{-\pi/2}^{\pi/2} \frac{d\theta}{\pi} = 1 \tag{5}
$$

and

$$-\int_{-\infty}^{\infty} \frac{1}{\pi} \frac{b\,dx}{b^2 + x^2} = \frac{1}{\pi}\left[\tan^{-1}\left(\frac{b}{x}\right)\right]_{-\infty}^{\infty}$$

$$= \frac{1}{\pi}[\tfrac{1}{2}\pi - (-\tfrac{1}{2}\pi)] = 1. \qquad (6)$$

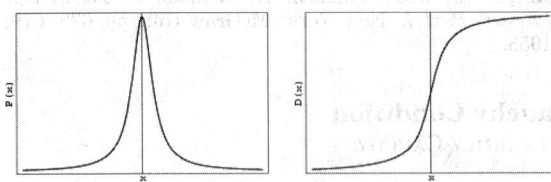

The general Cauchy distribution and its cumulative distribution can be written as

$$P(x) = \frac{1}{\pi} \frac{\tfrac{1}{2}\Gamma}{(x-m)^2 + (\tfrac{1}{2}\Gamma)^2} \qquad (7)$$

$$D(x) = \frac{1}{2} + \frac{1}{\pi}\tan^{-1}\left(\frac{x-m}{b}\right), \qquad (8)$$

where Γ is the FULL WIDTH AT HALF MAXIMUM ($\Gamma = 2b$ in the above example) and m is the MEDIAN ($m = 0$ in the above example). The CHARACTERISTIC FUNCTION is

$$\phi(t) = \frac{1}{\pi}\int_{-\infty}^{\infty} e^{itx} \frac{\tfrac{1}{2}\Gamma}{(\tfrac{1}{2}\Gamma)^2 + (x-m)^2}\,dx = e^{imt - \Gamma|t|/2}. \quad (9)$$

The MOMENTS μ_n of the distribution are undefined since the integrals

$$\mu_n = \int_{-\infty}^{\infty} \frac{\Gamma}{2\pi} \frac{x^n}{(x-m)^2 + (\tfrac{1}{2}\Gamma)^2} \qquad (10)$$

diverge for $n \geq 1$.

If X and Y are variates with a NORMAL DISTRIBUTION, then $Z \equiv X/Y$ has a Cauchy distribution with MEDIAN $m = 0$ and full width

$$\Gamma = \frac{2\sigma_y}{\sigma_x}. \qquad (11)$$

The sum of n variates each from a Cauchy distribution has itself a Cauchy distribution, as can be seen from

$$P_n(x) = \mathscr{F}^{-1}\{[\phi(t)]^n\} = \frac{(\tfrac{1}{2}n\Gamma)}{\pi[(\tfrac{1}{2}n\Gamma)^2 + (x-nm)^2]}, \qquad (12)$$

where $\phi(t)$ is the CHARACTERISTIC FUNCTION and $\mathscr{F}^{-1}|f|$ is the inverse FOURIER TRANSFORM, taken with parameters $a = b = 1$.

See also GAUSSIAN DISTRIBUTION, NORMAL DISTRIBUTION

References

Papoulis, A. *Probability, Random Variables, and Stochastic Processes, 2nd ed.* New York: McGraw-Hill, p. 104, 1984.

Spiegel, M. R. *Theory and Problems of Probability and Statistics.* New York: McGraw-Hill, pp. 114–115, 1992.

Cauchy Equation

EULER EQUATION

Cauchy Functional Equation

The fifth of HILBERT'S PROBLEMS is a generalization of this equation.

See also HILBERT'S PROBLEMS

Cauchy Integral Formula

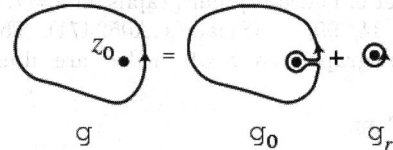

Given a CONTOUR INTEGRAL OF THE FORM

$$\oint_{\gamma} \frac{f(z)\,dz}{z - z_0}, \qquad (1)$$

define a path γ_r as an infinitesimal clockwise CIRCLE around the point z_0 (the dot in the above illustration), and define the path γ_0 as an arbitrary loop with a cut line (on which the forward and reverse contributions cancel each other out) so as to go around z_0.

The total path is then

$$\gamma = \gamma_0 + \gamma_r, \qquad (2)$$

so

$$\oint_{\gamma} \frac{f(z)\,dz}{z - z_0} = \oint_{\gamma_0} \frac{f(z)\,dz}{z - z_0} + \oint_{\gamma_r} \frac{f(z)\,dz}{z - z_0}. \qquad (3)$$

From the CAUCHY INTEGRAL THEOREM, the CONTOUR INTEGRAL along any path not enclosing a POLE is 0. Therefore, the first term in the above equation is 0 since γ_0 does not enclose the POLE, and we are left with

$$\oint_{\gamma} \frac{f(z)\,dz}{z - z_0} = \oint_{\gamma_r} \frac{f(z)\,dz}{z - z_0}. \qquad (4)$$

Now, let $z \equiv z_0 + re^{i\theta}$, so $dz = ire^{i\theta}\,d\theta$. Then

$$\oint_{\gamma} \frac{f(z)\,dz}{z - z_0} = \oint_{\gamma_r} \frac{f(z_0 + re^{i\theta})}{re^{i\theta}}\,ire^{i\theta}\,d\theta$$

$$= \oint_{\gamma_r} f(z_0 + re^{i\theta})i\,d\theta. \qquad (5)$$

But we are free to allow the radius r to shrink to 0, so

$$\oint_\gamma \frac{f(z)\,dz}{z-z_0} = \lim_{r\to 0}\oint_{\gamma_r} f(z_0 + re^{i\theta})i\,d\theta = \oint_{\gamma_r} f(z_0)i\,d\theta$$

$$= if(z_0)\oint_{\gamma_r} d\theta = 2\pi i f(z_0), \qquad (6)$$

and

$$f(z_0) = \frac{1}{2\pi i}\oint_\gamma \frac{f(z)\,dz}{z-z_0}. \qquad (7)$$

If multiple loops are made around the POLE, then equation (7) becomes

$$n(\gamma, z_0)f(z_0) = \frac{1}{2\pi i}\oint_\gamma \frac{f(z)\,dz}{z-z_0}, \qquad (8)$$

where $n(\gamma, z_0)$ is the WINDING NUMBER.

A similar formula holds for the derivatives of $f(z)$,

$$f'(z_0) = \lim_{h\to 0}\frac{f(z_0+h)-f(z_0)}{h}$$

$$= \lim_{h\to 0}\frac{1}{2\pi i h}\left[\oint_\gamma \frac{f(z)\,dz}{z-z_0-h} - \oint_\gamma \frac{f(z)\,dz}{z-z_0}\right]$$

$$= \lim_{h\to 0}\frac{1}{2\pi i h}\oint_\gamma \frac{f(z)[(z-z_0)-(z-z_0-h)]\,dz}{(z-z_0-h)(z-z_0)}$$

$$= \lim_{h\to 0}\frac{1}{2\pi i h}\oint_\gamma \frac{hf(z)\,dz}{(z-z_0-h)(z-z_0)}$$

$$= \frac{1}{2\pi i}\oint_\gamma \frac{f(z)\,dz}{(z-z_0)^2}. \qquad (9)$$

Iterating again,

$$f''(z_0) = \frac{2}{2\pi i}\oint_\gamma \frac{f(z)\,dz}{(z-z_0)^3}. \qquad (10)$$

Continuing the process and adding the WINDING NUMBER n,

$$n(\gamma, z_0)f^{(r)}(z_0) = \frac{r!}{2\pi i}\oint_\gamma \frac{f(z)\,dz}{(z-z_0)^{r+1}}. \qquad (11)$$

See also ARGUMENT PRINCIPLE, CONTOUR INTEGRAL, MORERA'S THEOREM

References

Arfken, G. "Cauchy's Integral Formula." §6.4 in *Mathematical Methods for Physicists, 3rd ed.* Orlando, FL: Academic Press, pp. 371–376, 1985.

Kaplan, W. "Cauchy's Integral Formula." §9.9 in *Advanced Calculus, 4th ed.* Reading, MA: Addison-Wesley, pp. 598–599, 1991.

Knopp, K. "Cauchy's Integral Formulas." Ch. 5 in *Theory of Functions Parts I and II, Two Volumes Bound as One, Part I.* New York: Dover, pp. 61–66, 1996.

Krantz, S. G. "The Cauchy Integral Theorem and Formula." §2.3 in *Handbook of Complex Analysis.* Boston, MA: Birkhäuser, pp. 26–29, 1999.

Morse, P. M. and Feshbach, H. *Methods of Theoretical Physics, Part I.* New York: McGraw-Hill, pp. 367–372, 1953.

Woods, F. S. "Cauchy's Theorem." §146 in *Advanced Calculus: A Course Arranged with Special Reference to the Needs of Students of Applied Mathematics.* Boston, MA: Ginn, pp. 352–353, 1926.

Cauchy Integral Test

INTEGRAL TEST

Cauchy Integral Theorem

If $f(z)$ is analytic in some simply connected region R, then

$$\oint_\gamma f(z)\,dz = 0 \qquad (1)$$

for any closed CONTOUR γ completely contained in R. Writing z as

$$z \equiv x + iy \qquad (2)$$

and $f(z)$ as

$$f(z) \equiv u + iv \qquad (3)$$

then gives

$$\oint_\gamma f(z)\,dz = \int_\gamma (u+iv)(dx+i\,dy)$$

$$= \int_\gamma u\,dx - v\,dy + i\int_\gamma v\,dx + u\,dy. \qquad (4)$$

From GREEN'S THEOREM,

$$\int_\gamma f(x,y)\,dx - g(x,y)\,dy = -\iint\left(\frac{\partial g}{\partial x}+\frac{\partial f}{\partial y}\right)dx\,dy, \qquad (5)$$

$$\int_\gamma f(x,y)\,dx + g(x,y)\,dy = \iint\left(\frac{\partial g}{\partial x}-\frac{\partial f}{\partial y}\right)dx\,dy \qquad (6)$$

so (4) becomes

$$\oint_\gamma f(z)\,dz = -\iint\left(\frac{\partial v}{\partial x}+\frac{\partial u}{\partial y}\right)dx\,dy$$

$$+ i\iint\left(\frac{\partial u}{\partial x}+\frac{\partial v}{\partial y}\right)dx\,dy. \qquad (7)$$

But the CAUCHY-RIEMANN EQUATIONS require that

$$\frac{\partial u}{\partial x} = \frac{\partial v}{\partial y} \qquad (8)$$

$$\frac{\partial u}{\partial y} = -\frac{\partial v}{\partial x}, \qquad (9)$$

so

$$\oint_\gamma f(z)\,dz = 0, \qquad (10)$$

Q.E.D.

For a MULTIPLY CONNECTED region,

$$\oint_{C_1} f(z)\,dz = \oint_{C_2} f(z)\,dz. \qquad (11)$$

See also ARGUMENT PRINCIPLE, CAUCHY INTEGRAL THEOREM, CONTOUR INTEGRAL, MORERA'S THEOREM, RESIDUE THEOREM

References
Arfken, G. "Cauchy's Integral Theorem." §6.3 in *Mathematical Methods for Physicists, 3rd ed.* Orlando, FL: Academic Press, pp. 365–371, 1985.
Kaplan, W. "Integrals of Analytic Functions. Cauchy Integral Theorem." §9.8 in *Advanced Calculus, 4th ed.* Reading, MA: Addison-Wesley, pp. 594–598, 1991.
Knopp, K. "Cauchy's Integral Theorem." Ch. 4 in *Theory of Functions Parts I and II, Two Volumes Bound as One, Part I.* New York: Dover, pp. 47–60, 1996.
Krantz, S. G. "The Cauchy Integral Theorem and Formula." §2.3 in *Handbook of Complex Analysis.* Boston, MA: Birkhäuser, pp. 26–29, 1999.
Morse, P. M. and Feshbach, H. *Methods of Theoretical Physics, Part I.* New York: McGraw-Hill, pp. 363–367, 1953.
Woods, F. S. "Integral of a Complex Function." §145 in *Advanced Calculus: A Course Arranged with Special Reference to the Needs of Students of Applied Mathematics.* Boston, MA: Ginn, pp. 351–352, 1926.

Cauchy Mean Theorem
CAUCHY'S FORMULA

Cauchy Number of the First Kind
BERNOULLI NUMBER OF THE SECOND KIND

Cauchy Principal Value

$$PV \int_{-\infty}^{\infty} f(x)\,dx \equiv \lim_{R\to\infty} \int_{-R}^{R} f(x)\,dx$$

$$PV \int_a^b f(x)\,dx \equiv \lim_{\epsilon\to 0}\left[\int_a^{c-\epsilon} f(x)\,dx + \int_{c+\epsilon}^b f(x)\,dx\right],$$

where $\epsilon > 0$ and $a \le c \le b$. Russian authors use the notation $\mathscr{P}(x)$ instead of PVx for the principal value of x.

References
Arfken, G. *Mathematical Methods for Physicists, 3rd ed.* Orlando, FL: Academic Press, pp. 401–403, 1985.
Sansone, G. *Orthogonal Functions, rev. English ed.* New York: Dover, p. 158, 1991.

Cauchy Problem
If $f(x, y)$ is an ANALYTIC FUNCTION in a NEIGHBORHOOD of the point (x_0, y_0) (i.e., it can be expanded in a series of NONNEGATIVE INTEGER POWERS of $(x - x_0)$ and $(y - y_0)$), find a solution $y(x)$ of the DIFFERENTIAL EQUATION

$$\frac{dy}{dx} = f(x),$$

with initial conditions $y = y_0$ and $x = x_0$. The existence and uniqueness of the solution were proven by Cauchy and Kovalevskaya in the CAUCHY-KOVALEVSKAYA THEOREM. The Cauchy problem amounts to determining the shape of the boundary and type of equation which yield unique and reasonable solutions for the CAUCHY BOUNDARY CONDITIONS.

See also CAUCHY BOUNDARY CONDITIONS, CAUCHY-KOVALEVSKAYA THEOREM

Cauchy Product
The Cauchy product of two sequences $f(n)$ and $g(n)$ defined for nonnegative integers n is defined by

$$(f \circ g)(n) = \sum_{k=0}^{n} f(k)g(n-k).$$

See also CONVOLUTION

References
Apostol, T. M. *Modular Functions and Dirichlet Series in Number Theory, 2nd ed.* New York: Springer-Verlag, p. 24, 1997.

Cauchy Ratio Test
RATIO TEST

Cauchy Remainder
The remainder after n terms of a TAYLOR SERIES is given by

$$R_n = \frac{(x-x^*)^n (x-x_0)^{n+1}}{n!}\, f^{(n+1)}(x^*),$$

where $x^* \in (x_0, x)$.

Note that the Cauchy remainder R_n is also sometimes taken to refer to the remainder when terms up to the $(n-1)$st power are taken in the TAYLOR SERIES, and that a notation in which $h \to x - x_0$, $x^* \to a + \theta h$, and $x - x^* \to 1 - \theta$ is sometimes used (Blumenthal 1926; Whittaker and Watson 1990, pp. 95–96).

See also LAGRANGE REMAINDER, SCHLÖMILCH REMAINDER, TAYLOR SERIES

References
Beesack, P. R. "A General Form of the Remainder in Taylor's Theorem." *Amer. Math. Monthly* **73**, 64–67, 1966.

Blumenthal, L. M. "Concerning the Remainder Term in Taylor's Formula." *Amer. Math. Monthly* **33**, 424–426, 1926.

Hamilton, H. J. "Cauchy's Form of R_n from the Iterated Integral Form." *Amer. Math. Monthly* **59**, 320, 1952.

Whittaker, E. T. and Watson, G. N. "Forms of the Remainder in Taylor's Series." §5.41 in *A Course in Modern Analysis, 4th ed.* Cambridge, England: Cambridge University Press, pp. 95–96, 1990.

Cauchy Root Test

ROOT TEST

Cauchy Sequence

A SEQUENCE $a_1, a_2, \ldots$ such that the METRIC $d(a_m, a_n)$ satisfies

$$\lim_{\min(m,\, n)\to\infty} d(a_m,\, a_n) = 0.$$

Cauchy sequences in the rationals do not necessarily CONVERGE, but they do CONVERGE in the REALS.

REAL NUMBERS can be defined using either DEDEKIND CUTS or Cauchy sequences.

See also DEDEKIND CUT

Cauchy Test

RATIO TEST

Cauchy-Davenport Theorem

Let t be a NONNEGATIVE INTEGER and let $x_1, \ldots, x_t$ be nonzero elements of $\mathbb{Z}_p$ which are not necessarily distinct. Then the number of elements of $\mathbb{Z}_p$ that can be written as the sum of some SUBSET (possibly empty) of the x_i is at least $\min\{p, t+1\}$. In particular, if $t \geq p-1$, then every element of $\mathbb{Z}_p$ can be so written.

References

Martin, G. "Dense Egyptian Fractions." *Trans. Amer. Math. Soc.* **351**, 3641–3657, 1999.

Vaughan, R. C. Lemma 2.14 in *The Hardy-Littlewood Method, 2nd ed.* Cambridge, England: Cambridge University Press, 1997.

Cauchy-Frobenius Lemma

Let J be a FINITE GROUP and the image $R(J)$ be a representation which is a HOMEOMORPHISM of J into a PERMUTATION GROUP $S(X)$, where $S(X)$ is the GROUP of all permutations of a SET X. Define the orbits of $R(J)$ as the equivalence classes under $x \sim y$, which is true if there is some permutation p in $R(J)$ such that $p(x) = y$. Define the fixed points of p as the elements x of X for which $p(x) = x$. Then the AVERAGE number of FIXED POINTS of permutations in $R(J)$ is equal to the number of orbits of $R(J)$.

The LEMMA was apparently known by Cauchy (1845) in obscure form and Frobenius (1887) prior to Burnside's (1900) rediscovery. It is sometimes also called BURNSIDE'S LEMMA, the PÓLYA-BURNSIDE LEMMA, or even "the LEMMA THAT IS NOT BURNSIDE'S!" Whatever its name, the lemma was subsequently extended and refined by Pólya (1937) for applications in COMBINATORIAL counting problems. In this form, it is known as PÓLYA ENUMERATION THEOREM.

See also PÓLYA ENUMERATION THEOREM

References

Cauchy, A. "Mémoire sur diverses propriétés remarquables des substitutions régulières ou irrégulières, et des systémes de substitutiones conjugées." *C. R. Acad. Sci. Paris* **21**, 835, 1845. Reprinted in *Œuvres Complètes d'Augustin Cauchy, Tome IX.* Paris: Gauthier-Villars, 342–360, 1896.

Frobenius, F. G. "Über die Congruenz nach einem aus zwei endlichen Gruppen gebildeten Doppelmodul." *J. reine angew. Math.* **101**, 273–299, 1887. Reprinted in *Ferdinand Georg Frobenius Gesammelte Abhandlungen, Band II.* Berlin: Springer-Verlag, pp. 304–330, 1968.

Neumann, P. M. "A Lemma that is not Burnside's." *Math. Scientist* **4**, 133–141, 1979.

Khan, M. R. "A Counting Formula for Primitive Tetrahedra in Z^3." *Amer. Math. Monthly* **106**, 525–533, 1999.

Pólya, G. "Kombinatorische Anzahlbestimmungen für Gruppen, Graphen, und chemische Verbindungen." *Acta Math.* **68**, 145–254, 1937.

Rotman, J. *A First Course in Abstract Algebra, 2nd ed.* Englewood Cliffs, NJ: Prentice-Hall, 2000.

Cauchy-Hadamard Theorem

The RADIUS OF CONVERGENCE of the TAYLOR SERIES

$$a_0 + a_1 z + a_2 z^2 + \ldots$$

is

$$r = \frac{1}{\overline{\lim_{n\to\infty}}(|a_n|)^{1/n}}.$$

See also RADIUS OF CONVERGENCE, TAYLOR SERIES

Cauchy-Kovalevskaya Theorem

The theorem which proves the existence and uniqueness of solutions to the CAUCHY PROBLEM.

See also CAUCHY PROBLEM

Cauchy-Lagrange Identity

LAGRANGE'S IDENTITY

Cauchy-Maclaurin Theorem

MACLAURIN-CAUCHY THEOREM

Cauchy-Riemann Equations

Let

$$f(x,\, y) \equiv u(x,\, y) + iv(x,\, y), \tag{1}$$

where

$$z \equiv x + iy, \tag{2}$$

so

$$dz = dx + i\, dy. \tag{3}$$

The total derivative of f with respect to z may then be computed as follows.

$$y = \frac{z - x}{i} \tag{4}$$

$$x = z - iy, \tag{5}$$

so

$$\frac{\partial y}{\partial z} = \frac{1}{i} = -i \tag{6}$$

$$\frac{\partial x}{\partial z} = 1, \tag{7}$$

and

$$\frac{df}{dz} = \frac{\partial f}{\partial x}\frac{\partial x}{\partial z} + \frac{\partial f}{\partial y}\frac{\partial y}{\partial z} = \frac{\partial f}{\partial x} - i\frac{\partial f}{\partial y}. \tag{8}$$

In terms of u and v, (8) becomes

$$\frac{df}{dz} = \left(\frac{\partial u}{\partial x} + i\frac{\partial v}{\partial x}\right) - i\left(\frac{\partial u}{\partial y} + i\frac{\partial v}{\partial y}\right)$$

$$= \left(\frac{\partial u}{\partial x} + i\frac{\partial v}{\partial x}\right) + \left(-i\frac{\partial u}{\partial y} + \frac{\partial v}{\partial y}\right). \tag{9}$$

Along the real, or X-AXIS, $\partial f/\partial y = 0$, so

$$\frac{df}{dz} = \frac{\partial u}{\partial x} + i\frac{\partial v}{\partial x}. \tag{10}$$

Along the imaginary, or Y-AXIS, $\partial f/\partial x = 0$, so

$$\frac{df}{dz} = -i\frac{\partial u}{\partial y} + \frac{\partial v}{\partial y}. \tag{11}$$

If f is COMPLEX DIFFERENTIABLE, then the value of the derivative must be the same for a given dz, regardless of its orientation. Therefore, (10) must equal (11), which requires that

$$\frac{\partial u}{\partial x} = \frac{\partial v}{\partial y} \tag{12}$$

and

$$\frac{\partial v}{\partial x} = -\frac{\partial u}{\partial y}. \tag{13}$$

These are known as the Cauchy-Riemann equations. They lead to the condition

$$\frac{\partial^2 u}{\partial x\, \partial y} = -\frac{\partial^2 v}{\partial x\, \partial y}. \tag{14}$$

The Cauchy-Riemann equations may be concisely written as

$$\frac{df}{d\bar{z}} = \frac{\partial f}{\partial x} + i\,\frac{\partial f}{\partial y} = \left(\frac{\partial u}{\partial x} + i\,\frac{\partial v}{\partial x}\right) + i\left(\frac{\partial u}{\partial y} + i\,\frac{\partial v}{\partial y}\right)$$

$$= \left(\frac{\partial u}{\partial x} - \frac{\partial v}{\partial y}\right) + i\left(\frac{\partial u}{\partial y} + \frac{\partial v}{\partial x}\right) = 0, \tag{15}$$

where $\bar{z}$ is the COMPLEX CONJUGATE.

If $z = re^{i\theta}$, then the Cauchy-Riemann equations become

$$\frac{\partial u}{\partial r} = \frac{1}{r}\frac{\partial v}{\partial \theta} \tag{16}$$

$$\frac{1}{r}\frac{\partial u}{\partial \theta} = -\frac{\partial v}{\partial r} \tag{17}$$

(Abramowitz and Stegun 1972, p. 17).

If u and v satisfy the Cauchy-Riemann equations, they also satisfy LAPLACE'S EQUATION in 2-D, since

$$\frac{\partial^2 u}{\partial x^2} + \frac{\partial^2 u}{\partial y^2} = \frac{\partial}{\partial x}\left(\frac{\partial v}{\partial y}\right) + \frac{\partial}{\partial y}\left(-\frac{\partial v}{\partial x}\right) = 0 \tag{18}$$

$$\frac{\partial^2 v}{\partial x^2} + \frac{\partial^2 v}{\partial y^2} = \frac{\partial}{\partial x}\left(-\frac{\partial u}{\partial y}\right) + \frac{\partial}{\partial y}\left(\frac{\partial u}{\partial x}\right) = 0. \tag{19}$$

By picking an arbitrary $f(z)$, solutions can be found which automatically satisfy the Cauchy-Riemann equations and LAPLACE'S EQUATION. This fact is used to use CONFORMAL MAPPINGS to find solutions to physical problems involving scalar potentials such as fluid flow and electrostatics.

See also ANALYTIC FUNCTION, CAUCHY INTEGRAL THEOREM, COMPLEX DERIVATIVE, CONFORMAL TRANSFORMATION, ENTIRE FUNCTION, MONOGENIC FUNCTION, POLYGENIC FUNCTION

References

Abramowitz, M. and Stegun, C. A. (Eds.). *Handbook of Mathematical Functions with Formulas, Graphs, and Mathematical Tables, 9th printing.* New York: Dover, p. 17, 1972.

Arfken, G. "Cauchy-Riemann Conditions." §6.2 in *Mathematical Methods for Physicists, 3rd ed.* Orlando, FL: Academic Press, pp. 3560–365, 1985.

Knopp, K. "The Cauchy-Riemann Differential Equations." §7 in *Theory of Functions Parts I and II, Two Volumes Bound as One, Part I.* New York: Dover, pp. 28–31, 1996.

Krantz, S. G. "The Cauchy-Riemann Equations." §1.3.2 in *Handbook of Complex Analysis.* Boston, MA: Birkhäuser, p. 13, 1999.

Levinson, N. and Redheffer, R. M. *Complex Variables.* San Francisco, CA: Holden-Day, 1970.

Zwillinger, D. *Handbook of Differential Equations, 3rd ed.* Boston, MA: Academic Press, p. 137, 1997.

Cauchy's Cosine Integral Formula

$$\int_{-\pi/2}^{\pi/2} \cos^{\mu+\nu-2}\theta\, e^{i\theta(\mu-\nu+2\xi)}\, d\theta$$

$$= \frac{\pi\Gamma(\mu+\nu-1)}{2^{\mu+\nu-2}\,\Gamma(\mu+\xi)\Gamma(\nu-\xi)},$$

where $\Gamma(z)$ is the GAMMA FUNCTION.

Cauchy's Determinant Theorem

Any row r and column s of a DETERMINANT being selected, if the element common to them be multiplied by its COFACTOR in the DETERMINANT, and every product of another element of the row by another element of the columns be multiplied by its COFACTOR, the sum of the results is equal to the given DETERMINANT. Symbolically,

$$\Delta = a_{rs}\frac{\partial\Delta}{\partial a_{rs}} + \sum a_{ri}a_{ks}\frac{\partial^2\Delta}{\partial a_{ri}\,\partial a_{ks}} \qquad (1)$$

$$= (-1)^{r+s}a_{rs}A_{rs} + \sum \pm a_{ri}a_{ks}A_{rk,\,is}, \qquad (2)$$

where $i,\ k = 1, 2, ..., n$; $i \neq s$; $k \neq r$; and the sign before $a_{ri}a_{ks}A_{rk,\,is}$ is determined by the formula $(-1)^{v_1+v_2}$, with v_1 the total number of PERMUTATION INVERSIONS in the suffix and $v_2 = r + i + k + s$.

See also DETERMINANT

References

Muir, T. "Cauchy's Theorem." §110 in *A Treatise on the Theory of Determinants*. New York: Dover, pp. 95–96, 1960.

Cauchy's Formula

The GEOMETRIC MEAN is smaller than the ARITHMETIC MEAN,

$$\left(\prod_{i=1}^{N} n_i\right)^{1/N} \le \frac{\sum_{i=1}^{N} n_i}{N},$$

with equality in the cases (1) $N = 1$ or (2) $n_i = n_j$ for all i, j.

See also ARITHMETIC MEAN, GEOMETRIC MEAN

Cauchy's Inequality

A special case of HÖLDER'S SUM INEQUALITY with $p = q = 2$,

$$\left(\sum_{k=1}^{n} a_k b_k\right)^2 \le \left(\sum_{k=1}^{n} a_k^2\right)\left(\sum_{k=1}^{n} b_k^2\right), \qquad (1)$$

where equality holds for $a_k = cb_k$. The inequality is sometimes also called Lagrange's inequality (Mitrinovic 1970, p. 42), and can be written in vector form as

$$|\mathbf{a}\cdot\mathbf{b}| \le |\mathbf{a}||\mathbf{b}|. \qquad (2)$$

In 2-D, it becomes

$$(a^2 + b^2)(c^2 + d^2) \ge (ac + bd)^2. \qquad (3)$$

It can be proven by writing

$$\sum_{i=1}^{n}(a_i x + b_i)^2 = \sum_{i=1}^{n} a_i^2\left(x + \frac{b_i}{a_i}\right)^2 = 0. \qquad (4)$$

If b_i/a_i is a constant c, then $x = -c$. If it is not a constant, then all terms cannot simultaneously vanish for REAL x, so the solution is COMPLEX and can be found using the QUADRATIC EQUATION

$$x = \frac{-2\sum a_i b_i \pm \sqrt{4(\sum a_i b_i)^2 - 4\sum a_i^2 \sum b_i^2}}{2\sum a_i^2}. \qquad (5)$$

In order for this to be COMPLEX, it must be true that

$$\left(\sum_i a_i b_i\right)^2 \le \left(\sum_i a_i^2\right)\left(\sum_i b_i^2\right), \qquad (6)$$

with equality when b_i/a_i is a constant. The VECTOR derivation is much simpler,

$$(\mathbf{a}\cdot\mathbf{b})^2 = a^2 b^2 \cos^2\theta \le a^2 b^2, \qquad (7)$$

where

$$a^2 \equiv \mathbf{a}\cdot\mathbf{a} = \sum_i a_i^2, \qquad (8)$$

and similarly for b.

See also CHEBYSHEV INEQUALITY, HÖLDER'S INEQUALITIES

References

Abramowitz, M. and Stegun, C. A. (Eds.). *Handbook of Mathematical Functions with Formulas, Graphs, and Mathematical Tables, 9th printing.* New York: Dover, p. 11, 1972.

Apostol, T. M. *Calculus, 2nd ed., Vol. 1: One-Variable Calculus, with an Introduction to Linear Algebra.* Waltham, MA: Blaisdell, pp. 42–43, 1967.

Cauchy, A. L. *Cours d'analyse de l'École Royale Polytechnique, 1ère partie: Analyse algébrique.* Paris: p. 373, 1821. Reprinted in *Œuvres complètes, 2e série, Vol. 3.*

Gradshteyn, I. S. and Ryzhik, I. M. *Tables of Integrals, Series, and Products, 6th ed.* San Diego, CA: Academic Press, p. 1092, 2000.

Hardy, G. H.; Littlewood, J. E.; and Pólya, G. "Cauchy's Inequality." §2.4 in *Inequalities, 2nd ed.* Cambridge, England: Cambridge University Press, pp. 16–18, 1952.

Jeffreys, H. and Jeffreys, B. S. "Cauchy's Inequality." §1.16 in *Methods of Mathematical Physics, 3rd ed.* Cambridge, England: Cambridge University Press, p. 54, 1988.

Krantz, S. G. *Handbook of Complex Analysis.* Boston, MA: Birkhäuser, p. 12, 1999.

Mitrinovic, D. S. "Cauchy's and Related Inequalities." §2.6 in *Analytic Inequalities.* New York: Springer-Verlag, pp. 41–48, 1970.

Cauchy's Rigidity Theorem

RIGIDITY THEOREM

Cauchy's Theorem

CAUCHY BINOMIAL THEOREM, CAUCHY-DAVENPORT THEOREM, CAUCHY'S DETERMINANT THEOREM, CAUCHY'S FORMULA, CAUCHY-HADAMARD THEOREM, CAUCHY INTEGRAL THEOREM, CAUCHY-KOVALEVSKAYA THEOREM, MACLAURIN-CAUCHY THEOREM, RIGIDITY THEOREM

Cauchy-Schwarz Inequality

SCHWARZ'S INEQUALITY

Cauchy-Schwarz Integral Inequality

Let a_1 and a_2 by any two REAL integrable functions in $[a, b]$, then

$$\lim_{\min(m, n) \to \infty} d(a_m, a_n) = 0.$$

with equality IFF F with k real.

References

Gradshteyn, I. S. and Ryzhik, I. M. *Tables of Integrals, Series, and Products, 6th ed.* San Diego, CA: Academic Press, p. 1099, 2000.

Cauchy-Schwarz Sum Inequality

$$p \neq 2$$

$$u_1$$

Equality holds IFF the sequences $u_2, u_8, \ldots$ and $\mu_1, \mu_2, \ldots$ are proportional.

See also FIBONACCI IDENTITY

References

Apostol, T. M. *Calculus, 2nd ed., Vol. 1: One-Variable Calculus, with an Introduction to Linear Algebra.* Waltham, MA: Blaisdell, pp. 42–43, 1967.
Gradshteyn, I. S. and Ryzhik, I. M. *Tables of Integrals, Series, and Products, 6th ed.* San Diego, CA: Academic Press, p. 1092, 2000.
Krantz, S. G. *Handbook of Complex Analysis.* Boston, MA: Birkhäuser, p. 12, 1999.

Caudrey-Dodd-Gibbon-Sawada-Kotera Equation

The PARTIAL DIFFERENTIAL EQUATION

$$u_t + u_{xxxxx} + 30uu_{xxx} + 30u_x u_{xx} + 180u^2 u_x = 0.$$

See also SAWADA-KOTERA EQUATION

References

Aiyer, R. N.; Fuchssteiner, B.; and Oevel, W. "Solitons and Discrete Eigenfunctions of the Recursion Operator of Non-Linear Evolution Equations: I. The Caudrey-Dodd-Gib-bon-Sawada-Kotera Equations." *J. Phys. A: Math. Gen.* **19**, 3755–3770, 1986.
Zwillinger, D. *Handbook of Differential Equations, 3rd ed.* Boston, MA: Academic Press, p. 132, 1997.

Caustic

The curve which is the ENVELOPE of reflected (CATACAUSTIC) or refracted (DIACAUSTIC) rays of a given curve for a light source at a given point (known as the RADIANT POINT). The caustic is the EVOLUTE of the ORTHOTOMIC.

See also CATACAUSTIC, CIRCLE CAUSTIC, DIACAUSTIC, ENVELOPE, EVOLUTE, ORTHOTOMIC, RADIANT POINT

References

Lawrence, J. D. *A Catalog of Special Plane Curves.* New York: Dover, p. 60, 1972.
Lockwood, E. H. "Caustic Curves." Ch. 24 in *A Book of Curves.* Cambridge, England: Cambridge University Press, pp. 182–185, 1967.
Wells, D. *The Penguin Dictionary of Curious and Interesting Geometry.* London: Penguin, p. 28, 1991.
Yates, R. C. "Caustics." *A Handbook on Curves and Their Properties.* Ann Arbor, MI: J. W. Edwards, pp. 15–20, 1952.

Cavalieri's Principle

1. If the lengths of every one-dimensional slice are equal for two regions, then the regions have equal AREAS.
2. If the AREAS of every two-dimensional SECTION are equal for two SOLIDS, then the SOLIDS have equal VOLUMES.

See also CROSS SECTION, PAPPUS'S CENTROID THEOREM, SECTION, VOLUME THEOREM

References

Beyer, W. H. (Ed.). *CRC Standard Mathematical Tables, 28th ed.* Boca Raton, FL: CRC Press, p. 126 and 132, 1987.
Harris, J. W. and Stocker, H. "Cavalieri's Theorem." §4.1.1 in *Handbook of Mathematics and Computational Science.* New York: Springer-Verlag, p. 95, 1998.
Kern, W. F. and Bland, J. R. "Cavalieri's Theorem" and "Proof of Cavalieri's Theorem." §11 and 49 in *Solid Mensuration with Proofs, 2nd ed.* New York: Wiley, pp. 25–27 and 145–146, 1948.

Cavalieri's Theorem

CAVALIERI'S PRINCIPLE

Cayley Algebra

The only NONASSOCIATIVE DIVISION ALGEBRA with REAL SCALARS. There is an 8-square identity corresponding to this algebra.

The elements of a Cayley algebra are called CAYLEY NUMBERS or OCTONIONS, and the MULTIPLICATION TABLE for any Cayley algebra over a FIELD F with characteristic $p \neq 2$ may be taken as shown in the following table, where u_1, u_2, ..., u_8 are a bases over F and μ_1, μ_2, and μ_3 are nonzero elements of F (Schafer 1996, pp. 5-).

	u_1	u_2	u_3	u_4	u_5	u_6	u_7	u_8
u_1	u_1	u_2	u_3	u_4	u_5	u_6	u_7	u_8
u_2	u_2	$\mu_1 u_1$	$-u_4$	$-\mu_1 u_3$	$-u_6$	$-\mu_1 u_5$	u_8	$\mu_1 u_7$
u_3	u_3	u_4	$\mu_2 u_1$	$\mu_2 u_2$	$-u_7$	$-u_8$	$-\mu_2 u_5$	$-\mu_2 u_6$
u_4	u_4	$\mu_1 u_3$	$-\mu_2 u_2$	$-\mu_1 \mu_2 u_1$	$-u_8$	$-\mu_1 u_7$	$\mu_2 u_6$	$\mu_1 \mu_2 u_5$
u_5	u_5	u_6	u_7	u_8	$\mu_3 u_1$	$\mu_3 u_2$	$\mu_3 u_3$	$\mu_3 u_4$
u_6	u_6	$\mu_1 u_5$	u_8	$\mu_1 u_7$	$-\mu_3 u_2$	$-\mu_1 \mu_3 u_1$	$-\mu_3 u_4$	$-\mu_1 \mu_2 u_3$
u_7	u_7	$-u_8$	$\mu_2 u_5$	$-\mu_2 u_6$	$-\mu_3 u_3$	$\mu_3 u_4$	$-\mu_2 \mu_3 u_1$	$\mu_2 \mu_3 u_2$
u_8	u_8	$-\mu_1 u_7$	$\mu_2 u_6$	$-\mu_1 \mu_2 u_5$	$-\mu_3 u_4$	$\mu_1 \mu_3 u_3$	$-\mu_2 \mu_3 u_2$	$\mu_1 \mu_2 \mu_3 u_1$

See also CAYLEY NUMBER, DIVISION ALGEBRA, OCTONION, NONASSOCIATIVE ALGEBRA

References

Kurosh, A. G. *General Algebra.* New York: Chelsea, pp. 226–28, 1963.

Schafer, R. D. *An Introduction to Nonassociative Algebras.* New York: Dover, pp. 5–6, 1996.

Cayley Cubic

A CUBIC RULED SURFACE (Fischer 1986) in which the director line meets the director CONIC SECTION. Cayley's surface is the unique cubic surface having four ORDINARY DOUBLE POINTS (Hunt), the maximum possible for CUBIC SURFACE (Endraß). The Cayley cubic is invariant under the TETRAHEDRAL GROUP and contains exactly nine lines, six of which connect the four nodes pairwise and the other three of which are coplanar (Endraß).

If the ORDINARY DOUBLE POINTS in projective 3-space are taken as $(1, 0, 0, 0)$, $(0, 1, 0, 0)$, $(0, 0, 1, 0)$, $(0, 0, 0, 1)$, then the equation of the surface in projective coordinates is

$$\frac{1}{x_0} + \frac{1}{x_1} + \frac{1}{x_2} + \frac{1}{x_3} = 0 \qquad (1)$$

(Hunt). Defining "affine" coordinates with plane at infinity $v = x_0 + x_1 + x_2 + 2x_3$ and

$$x = \frac{x_0}{v} \qquad (2)$$

$$y = \frac{x_1}{v} \qquad (3)$$

$$z = \frac{x_2}{v} \qquad (4)$$

then gives the equation

$$-5(x^2 y + x^2 z + y^2 x + y^2 z + z^2 y + z^2 x) + 2(xy + xz + yz) = 0 \qquad (5)$$

plotted in the left figure above (Hunt). The slightly different form

$$4(x^3 + y^3 + z^3 + w^3) - (x + y + z + w)^3 = 0 \qquad (6)$$

is given by Endraß which, when rewritten in TETRAHEDRAL COORDINATES, becomes

$$x^2 + y^2 - x^2 z + y^2 z + z^2 - 1 = 0, \qquad (7)$$

plotted in the right figure above.

The Hessian of the Cayley cubic is given by

$$0 = x_0^2(x_1 x_2 + x_1 x_3 + x_2 x_3) + x_1^2(x_0 x_2 + x_0 x_3 + x_2 x_3)$$
$$+ x_2^2(x_0 x_1 + x_0 x_3 + x_1 x_3) + x_3^2(x_0 x_1 + x_0 x_2 + x_1 x_2) \qquad (8)$$

in homogeneous coordinates x_0, x_1, x_2, and x_3. Taking the plane at infinity as $v = 5(x_0 + x_1 + x_2 + 2x_3)/2$ and setting x, y, and z as above gives the equation

$$25[x^3(y + z) + y^3(x + z) + z^3(x + y)] + 50(x^2 y^2 + x^2 z^2 + y^2 z^2)$$
$$- 125(x^2 yz + y^2 xz + z^2 xy) + 60xyz - 4(xy + xz + yz) = 0, \qquad (9)$$

plotted above (Hunt). The Hessian of the Cayley cubic has 14 ORDINARY DOUBLE POINTS, four more than a the general Hessian of a smooth CUBIC SURFACE (Hunt).

See also CAYLEY SURFACE

References

Endraß, S. "Flächen mit vielen Doppelpunkten." *DMV-Mitteilungen* **4**, 17–20, Apr. 1995.

Endraß, S. "The Cayley Cubic." http://enriques.mathematik.uni-mainz.de/kon/docs/Ecayley.shtml.

Fischer, G. (Ed.). *Mathematical Models from the Collections of Universities and Museums.* Braunschweig, Germany: Vieweg, p. 14, 1986.

Fischer, G. (Ed.). Plate 33 in *Mathematische Modelle/Mathematical Models, Bildband/Photograph Volume.* Braunschweig, Germany: Vieweg, p. 33, 1986.

Hunt, B. "Algebraic Surfaces." http://www.mathematik.uni-kl.de/~wwwagag/E/Galerie.html.

Hunt, B. *The Geometry of Some Special Arithmetic Quotients.* New York: Springer-Verlag, pp. 115–122, 1996.

Nordstrand, T. "The Cayley Cubic." http://www.uib.no/people/nfytn/cleytxt.htm.

Cayley Graph

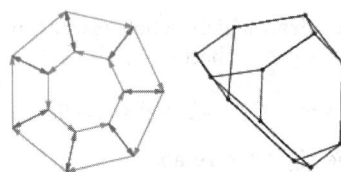

The Cayley graph of a GROUP G is a DIRECTED GRAPH determined by a set of generators $g_1, ..., g_k$. The vertices correspond to the elements of the group, and whenever $g_i a = b$, an edge is drawn between a and b. For example, the DIHEDRAL GROUP D_7 (left figure) is generated by the two elements, flips (red) and rotations (blue). The Cayley graph depends on the choice of a generating set. The right figure above illustrates the Cayley graph for the ALTERNATING GROUP A_4.

Royle has constructed all cubic Cayley graphs up to 1000 vertices, excluding those on 512 and 768 vertices.

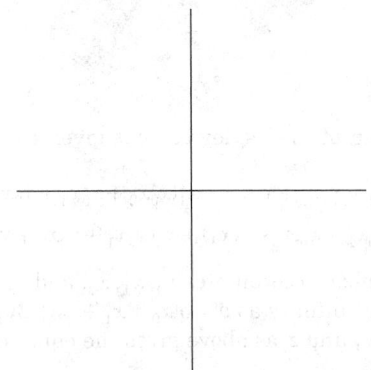

The Cayley graphs of infinite groups provide interesting geometries. For example, the Cayley graphs of the FREE GROUP on two generators are illustrated above (drawn out to successive levels), representing horizontal and vertical displacement respectively. Each new edge is drawn at half the size to give FRACTAL images.

See also CAGE GRAPH, CAYLEY TREE, DISCRETE GROUP, FREE GROUP, GRAPH, GROUP, TREE

References

Dixon, J. and Mortimer, B. *Permutation Groups.* New York: Springer-Verlag, 1996.

Grossman, I. and Magnus, W. *Groups and Their Graphs.* New York: Random House, p. 45, 1964.

Royle, G. "Cubic Cages." http://www.cs.uwa.edu.au/~gordon/cages/.

Cayley Lines

The 60 PASCAL LINES of a hexagon inscribed in a conic intersect three at a time through 20 STEINER POINTS, and also three at a time in 60 KIRKMAN POINTS. Each STEINER POINT lies together with three KIRKMAN POINTS on a total of 20 lines known as Cayley lines. The 20 Cayley lines pass four at a time though 15 points known as SALMON POINTS (Wells 1991). There is a dual relationship between the 20 Cayley lines and the 20 STEINER POINTS.

See also KIRKMAN POINTS, PASCAL LINES, PASCAL'S THEOREM, PLÜCKER LINES, SALMON POINTS, STEINER POINTS

References

Johnson, R. A. *Modern Geometry: An Elementary Treatise on the Geometry of the Triangle and the Circle.* Boston, MA: Houghton Mifflin, pp. 236–237, 1929.

Salmon, G. "Notes: Pascal's Theorem, Art. 267" in *A Treatise on Conic Sections, 6th ed.* New York: Chelsea, pp. 379–382, 1960.

Wells, D. *The Penguin Dictionary of Curious and Interesting Geometry.* London: Penguin, p. 172, 1991.

Cayley Number

There are two completely different definitions of Cayley numbers. The first and most commonly encountered type of Cayley number is the eight elements in a CAYLEY ALGEBRA, also known as octonions. The set of octonions is sometimes denoted $\mathbb{O}$. A typical Cayley number is OF THE FORM

$$a + bi_0 + ci_1 + di_2 + ei_3 + fi_4 + gi_5 + hi_6,$$

where each of the triples (i_0, i_1, i_3), (i_1, i_2, i_4), (i_2, i_3, i_5), (i_3, i_4, i_6), (i_4, i_5, i_0), (i_5, i_6, i_1), (i_6, i_0, i_2) behaves like the QUATERNIONS (i, j, k). Cayley numbers are *not* ASSOCIATIVE. They have been used in the study of 7- and 8-D space, and a general rotation in 8-D space can be written

$$x' \rightarrow ((((((xc_1)c_2)c_3)c_4)c_5)c_6)c_7.$$

A quantity which describes a DEL PEZZO SURFACE is sometimes also called a Cayley number (Coxeter 1973, p. 211).

See also COMPLEX NUMBER, DEL PEZZO SURFACE, QUATERNION, REAL NUMBER

References

Conway, J. H. and Guy, R. K. "Cayley Numbers." In *The Book of Numbers*. New York: Springer-Verlag, pp. 234–235, 1996.

Coxeter, H. S. M. *Regular Polytopes, 3rd ed.* New York: Dover, 1973.

Okubo, S. *Introduction to Octonion and Other Non-Associative Algebras in Physics*. New York: Cambridge University Press, 1995.

Cayley Surface

In affine 3-space the Cayley surface is given by

$$x_3 = x_1 x_2 - \tfrac{1}{3} x_1^3$$

(Nomizu and Sasaki 1994). The surface has been generalized by Eastwood and Ezhov (2000) to

$$\Phi_N(x_1, x_2, \ldots, x_N) \equiv \sum_{d=1}^{N} (-1)^d \sum_{i+j+\ldots+m=N} \underbrace{x_i x_j \ldots x_m}_{d} = 0.$$

This gives the first few hypersurfaces as

$$x_4 = x_1 x_3 + \tfrac{1}{2} x_2^2 - x_1^2 x_2 + \tfrac{1}{4} x_1^4$$

$$x_5 = x_1 x_4 + x_2 x_3 - x_1^2 x_3 - x_1 x_2^2 + x_1^3 x_2 - \tfrac{1}{5} x_1^5.$$

See also CAYLEY CUBIC

References

Eastwood, M. and Ezhov, V. Cayley Hypersurfaces. 25 Jan 2000. http://xxx.lanl.gov/abs/math.DG/0001134/.

Nomizu, K. and Sasaki, T. *Affine Differential Geometry: Geometry of Affine Immersions*. Cambridge, England: Cambridge University Press, 1994.

Nomizu, K. and Pinkall, U. "Cayley Surfaces in Affine Differential Geometry." *Tôhoku Math. J.* **41**, 589–596, 1989.

Cayley Transform

$$f(z) = \frac{i - z}{i + z}$$

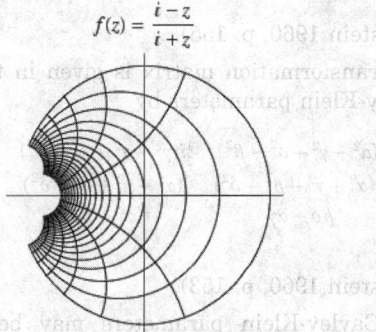

The LINEAR FRACTIONAL TRANSFORMATION

$$z \mapsto \frac{i - z}{i + z}$$

that maps the UPPER HALF-PLANE $\{z : \Im[z] > 0\}$ CONFORMALLY onto the UNIT DISK $\{z : |z| < 1\}$.

See also CONFORMAL MAPPING, LINEAR FRACTIONAL TRANSFORMATION

References

Krantz, S. G. "The Cayley Transform." §6.3.5 in *Handbook of Complex Analysis*. Boston, MA: Birkhäuser, p. 85, 1999.

Cayley Tree

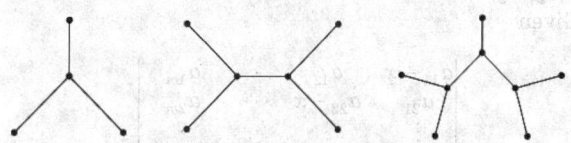

A TREE in which each non-leaf NODE has a constant number of branches n is called an n-Cayley tree. 2-Cayley trees are PATH GRAPHS. The unique n-Cayley tree on $n + 1$ nodes is the STAR GRAPH. The illustration above shows the first few 3-Cayley trees (also called trivalent trees, binary trees, or boron trees). The numbers of binary trees on $n = 1, 2, \ldots$ nodes (i.e., n-node trees having VERTEX DEGREE either 1 or 3; also called 3-Cayley trees, 3-valent trees, or boron trees) are 1, 1, 0, 1, 0, 1, 0, 1, 0, 2, 0, 2, 0 ,4, 0, 6, 0, 11, ... (Sloane's A052120).

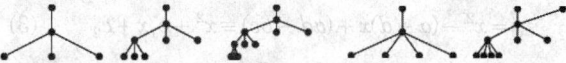

The illustrations above show the first few 4-Cayley and 5-Cayley trees.

The PERCOLATION THRESHOLD for a Cayley tree having z branches is

$$p_c = \frac{1}{z - 1}.$$

See also CAYLEY GRAPH, PATH GRAPH, STAR GRAPH, TREE

References

Sloane, N. J. A. Sequences A052120 in "An On-Line Version of the Encyclopedia of Integer Sequences." http://www.research.att.com/~njas/sequences/eisonline.html.

Cayley-Bacharach Theorem

Let $X_1, X_2 \subset \mathbb{P}^2$ be CUBIC plane curves meeting in nine points $p_1, \ldots, p_9$. If $X \subset \mathbb{P}^2$ is any CUBIC containing $p_1, \ldots, p_8$, then X contains p_9 as well. It is related to GORENSTEIN RINGS, and is a generalization of PAPPUS'S HEXAGON THEOREM and PASCAL'S THEOREM.

See also PASCAL'S THEOREM, PAPPUS'S HEXAGON THEOREM

References

Eisenbud, D.; Green, M.; and Harris, J. "Cayley-Bacharach Theorems and Conjectures." *Bull. Amer. Math. Soc.* **33**, 295–324, 1996.

Cayley-Dickson Algebra

CAYLEY ALGEBRA

Cayley-Hamilton Theorem

Given

$$\begin{vmatrix} a_{11}-x & a_{12} & \cdots & a_{1m} \\ a_{21} & a_{22}-x & \cdots & a_{2m} \\ \vdots & \vdots & \ddots & \vdots \\ a_{m1} & a_{m2} & \cdots & a_{mm}-x \end{vmatrix}$$

$$= x^m + c_{m-1}x^{m-1} + \ldots + c_0, \tag{1}$$

then

$$\mathsf{A}^m + c_{m-1}\mathsf{A}^{m-1} + \ldots + c_0 \mathsf{I} = 0, \tag{2}$$

where I is the IDENTITY MATRIX. Cayley verified this identity for $m = 2$ and 3 and postulated that it was true for all m. For $m = 2$, direct verification gives

$$\begin{vmatrix} a-x & b \\ c & d-x \end{vmatrix} = (a-x)(d-x)-bc$$

$$= x^2 - (a+d)x + (ad-bc) \equiv x^2 + c_1 x + c_2 \tag{3}$$

$$\mathsf{A} = \begin{bmatrix} a & b \\ c & d \end{bmatrix} \tag{4}$$

$$\mathsf{A}^2 = \begin{bmatrix} a & b \\ c & d \end{bmatrix}\begin{bmatrix} a & b \\ c & d \end{bmatrix} = \begin{bmatrix} a^2+bc & ab+bd \\ ac+cd & bc+d^2 \end{bmatrix} \tag{5}$$

$$-(a+d)\mathsf{A} = \begin{bmatrix} -a^2-ad & -ab-bd \\ -ac-dc & -ad-d^2 \end{bmatrix} \tag{6}$$

$$(ad-bc)\mathsf{I} = \begin{bmatrix} ad-bc & 0 \\ 0 & ad-bc \end{bmatrix}, \tag{7}$$

so

$$\mathsf{A}^2 - (a+d)\mathsf{A} + (ad-bc)\mathsf{I} = \begin{bmatrix} 0 & 0 \\ 0 & 0 \end{bmatrix}. \tag{8}$$

The Cayley-Hamilton theorem states that a $n \times n$ MATRIX A is annihilated by its CHARACTERISTIC POLYNOMIAL $\det(x\mathsf{I} - \mathsf{A})$, which is monic of degree n.

References

Ayres, F. Jr. *Theory and Problems of Matrices.* New York: Schaum, p. 181, 1962.

Gradshteyn, I. S. and Ryzhik, I. M. *Tables of Integrals, Series, and Products, 6th ed.* San Diego, CA: Academic Press, p. 1117, 2000.

Segercrantz, J. "Improving the Cayley-Hamilton Equation for Low-Rank Transformations." *Amer. Math. Monthly* **99**, 42–44, 1992.

Cayleyian Curve

The ENVELOPE of the lines connecting corresponding points on the JACOBIAN CURVE and STEINERIAN CURVE. The Cayleyian curve of a net of curves of order n has the same GENUS (CURVE) as the JACOBIAN CURVE and STEINERIAN CURVE and, in general, the class $3n(n-1)$.

References

Coolidge, J. L. *A Treatise on Algebraic Plane Curves.* New York: Dover, p. 150, 1959.

Cayley-Klein Parameters

The parameters α, β, γ, and δ which, like the three EULER ANGLES, provide a way to uniquely characterize the orientation of a solid body. These parameters satisfy the identities

$$\alpha\bar{\alpha} + \gamma\bar{\gamma} = 1 \tag{1}$$

$$\alpha\bar{\alpha} + \beta\bar{\beta} = 1 \tag{2}$$

$$\beta\bar{\beta} + \delta\bar{\delta} = 1 \tag{3}$$

$$\bar{\alpha}\beta + \bar{\gamma}\delta = 0 \tag{4}$$

$$\alpha\delta - \beta\gamma = 1 \tag{5}$$

and

$$\beta = -\bar{\gamma} \tag{6}$$

$$\delta = \bar{\alpha}, \tag{7}$$

where $\bar{z}$ denotes the COMPLEX CONJUGATE. In terms of the EULER ANGLES θ, ϕ, and ψ, the Cayley-Klein parameters are given by

$$\alpha = e^{i(\psi+\phi)/2}\cos(\tfrac{1}{2}\theta) \tag{8}$$

$$\beta = ie^{i(\psi+\phi)/2}\sin(\tfrac{1}{2}\theta) \tag{9}$$

$$\gamma = ie^{i(\psi+\phi)/2}\sin(\tfrac{1}{2}\theta) \tag{10}$$

$$\delta = e^{i(\psi+\phi)/2}\cos(\tfrac{1}{2}\theta) \tag{11}$$

(Goldstein 1960, p. 155).

The transformation matrix is given in terms of the Cayley-Klein parameters by

$$\mathsf{A} = \begin{bmatrix} \tfrac{1}{2}(\alpha^2-\gamma^2+\delta^2-\beta^2) & \tfrac{1}{2}i(\gamma^2-\alpha^2+\delta^2-\beta^2) & \gamma\delta-\alpha\beta \\ \tfrac{1}{2}i(\alpha^2+\gamma^2-\beta^2-\delta^2) & \tfrac{1}{2}(\alpha^2+\gamma^2+\beta^2+\delta^2) & -i(\alpha\beta+\gamma\delta) \\ \beta\delta-\alpha\gamma & i(\alpha\gamma+\beta\delta) & \alpha\delta+\beta\gamma \end{bmatrix} \tag{12}$$

(Goldstein 1960, p. 153).

The Cayley-Klein parameters may be viewed as parameters of a matrix (denoted Q for its close relationship with QUATERNIONS)

$$\mathsf{Q} = \begin{bmatrix} \alpha & \beta \\ \gamma & \delta \end{bmatrix} \tag{13}$$

which characterizes the transformations

$$u' = \alpha u + \beta v \qquad (14)$$

$$v' = \gamma u + \delta v. \qquad (15)$$

of a linear space having complex axes. This matrix satisfies

$$Q^*Q = QQ^* = I, \qquad (16)$$

where I is the IDENTITY MATRIX and A^* the ADJOINT MATRIX, as well as

$$|Q|^*|Q| = 1. \qquad (17)$$

In terms of the EULER PARAMETERS e_i and the PAULI MATRICES σ_i, the Q-matrix can be written as

$$Q = e_0 I + i(e_1\sigma_1 + e_2\sigma_2 + e_3\sigma_3) \qquad (18)$$

(Goldstein 1980, p. 156).

See also EULER ANGLES, EULER PARAMETERS, PAULI MATRICES, QUATERNION, ROTATION

References

Goldstein, H. "The Cayley-Klein Parameters and Related Quantities." §4–5 in *Classical Mechanics, 2nd ed.* Reading, MA: Addison-Wesley, pp. 148–158, 1980.

Varshalovich, D. A.; Moskalev, A. N.; and Khersonskii, V. K. "Description of Rotations in Terms of Unitary 2×2 Matrices. Cayley-Klein Parameters." §1.4.3 in *Quantum Theory of Angular Momentum.* Singapore: World Scientific, pp. 24–27, 1988.

Cayley-Klein-Hilbert Metric

The METRIC of Felix Klein's model for HYPERBOLIC GEOMETRY,

$$g_{11} = \frac{a^2(1 - x_2^2)}{(1 - x_1^2 - x_2^2)^2}$$

$$g_{12} = \frac{a^2 x_1 x_2}{(1 - x_1^2 - x_2^2)^2}$$

$$g_{22} = \frac{a^2(1 - x_1^2)}{(1 - x_1^2 - x_2^2)^2}.$$

See also HYPERBOLIC GEOMETRY

Cayley-Menger Determinant

This entry contributed by KAREN D. COLLINS

A DETERMINANT that gives the volume of a SIMPLEX in j dimensions. If S is a j-simplex in $\mathbb{R}^n$ with vertices $v_1, \ldots, v_j + 1$ and $B = (\beta_{ik})$ denotes the $(j+1) \times (j+1)$ matrix given by

$$\beta_{ik} = \|v_i - v_k\|_2^2, \qquad (1)$$

then the CONTENT V_j is given by

$$V_j^2(S) = \frac{(-1)^{j+1}}{2^j(j!)^2} \det(\hat{B}), \qquad (2)$$

where $\hat{B}$ is the $(j+2) \times (j+2)$ matrix obtained from B by bordering B with a top row $(0, 1, \ldots, 1)$ and a left column $(0, 1, \ldots, 1)^T$. Here, the vector L 2-NORMS $\|v_i - v_k\|_2$ are the edge lengths and the DETERMINANT in (2) is the Cayley-Menger determinant (Sommerville 1958, Gritzmann and Klee 1994). The first few coefficients for $j = 0, 1, \ldots$ are $-1, 2, -16, 288, -9216, 460800, \ldots$ (Sloane's A055546).

For $j = 2$, (2) becomes

$$-16\Delta^2 = \begin{vmatrix} 0 & 1 & 1 & 1 \\ 1 & 0 & c^2 & b^2 \\ 1 & c^2 & 0 & a^2 \\ 1 & b^2 & a^2 & 0 \end{vmatrix}, \qquad (3)$$

which gives the AREA for a plane triangle with side lengths a, b, and c, and is a form of HERON'S FORMULA.

For $j = 3$, the content of the 3-simplex (i.e., volume of the general TETRAHEDRON) is given by the determinant

$$288V^2 = \begin{vmatrix} 0 & 1 & 1 & 1 & 1 \\ 1 & 0 & d_{12}^2 & d_{13}^2 & d_{14}^2 \\ 1 & d_{21}^2 & 0 & d_{23}^2 & d_{24}^2 \\ 1 & d_{31}^2 & d_{32}^2 & 0 & d_{34}^2 \\ 1 & d_{41}^2 & d_{42}^2 & d_{43}^2 & 0 \end{vmatrix}, \qquad (4)$$

where the edge between vertices i and j has length d_{ij}. Setting the left side equal to 0 (corresponding to a TETRAHEDRON of volume 0) gives a relationship between the DISTANCES between vertices of a planar QUADRILATERAL (Uspensky 1948, p. 256).

See also HERON'S FORMULA, QUADRILATERAL, TETRAHEDRON

References

Gritzmann, P. and Klee, V. §3.6.1 in "On the Complexity of Some Basic Problems in Computational Convexity II. Volume and Mixed Volumes." In *Polytopes: Abstract, Convex and Computational* (Ed. T. Bisztriczky, P. McMullen, R. Schneider, R.; and A. W. Weiss). Dordrecht, Netherlands: Kluwer, 1994.

Sloane, N. J. A. Sequences A055546 in "An On-Line Version of the Encyclopedia of Integer Sequences." http://www.research.att.com/~njas/sequences/eisonline.html.

Sommerville, D. M. Y. *An Introduction to the Geometry of N Dimensions.* New York: Dover, p. 124, 1958.

Uspensky, J. V. *Theory of Equations.* New York: McGraw-Hill, p. 256, 1948.

Cayley's Group Theorem

Every FINITE GROUP of order n can be REPRESENTED AS A PERMUTATION GROUP on n letters, as first proved by Cayley in 1878 (Rotman 1995).

See also FINITE GROUP, PERMUTATION GROUP

References

Rotman, J. J. *An Introduction to the Theory of Groups, 4th ed.* New York: Springer-Verlag, p. 52, 1995.

Cayley's Hypergeometric Function Theorem

If

$$(1-z)^{a+b-c} {}_2F_1(2a,\ 2b;\ 2c;\ z) = \sum_{n=0}^{\infty} a_n z^n,$$

then

$$ {}_2F_1(a,\ b;\ c+\tfrac{1}{2};\ z) {}_2F_1(c-a,\ c-b;\ c\tfrac{1}{2};\ z) $$

$$ = \sum_{n=0}^{\infty} \frac{(c)_n}{(c+\tfrac{1}{2})} a_n z^n, $$

where ${}_2F_1(a,\ b;\ c;\ z)$ is a HYPERGEOMETRIC FUNCTION.

See also HYPERGEOMETRIC FUNCTION

Cayley's Ruled Surface

CAYLEY CUBIC

Cayley's Sextic

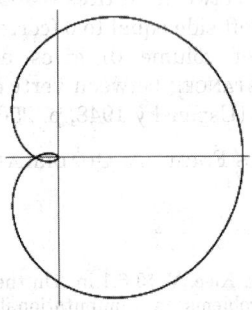

A plane curve discovered by Maclaurin but first studied in detail by Cayley. The name Cayley's sextic is due to R. C. Archibald, who attempted to classify curves in a paper published in Strasbourg in 1900 (MacTutor Archive). Cayley's sextic is given in POLAR COORDINATES by

$$r = 4a\ \cos^3(\tfrac{1}{3}\theta). \tag{1}$$

Parametric equations can be given by

$$x(t) = 4a\ \cos^4(\tfrac{1}{2}t)(2\cos t - 1) \tag{2}$$

$$y(t) = 4a\ \cos^3(\tfrac{1}{2}t)\ \sin(\tfrac{3}{2}t) \tag{3}$$

(Gray 1997, p. 119). Calculating r gives

$$r = \sqrt{x^2 + y^2} = 4\ \cos^3(\tfrac{1}{2}t), \tag{4}$$

and t is related to θ by

$$\theta = \tan^{-1}\left(\frac{y}{x}\right) = \tfrac{3}{2}t, \tag{5}$$

thus recovering (1). The CARTESIAN equation is

$$4(x^2 + y^2 - ax)^3 = 27a^2(x^2 + y^2)^2. \tag{6}$$

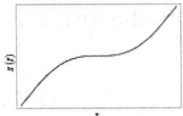 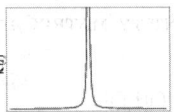 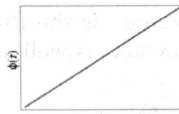

The ARC LENGTH, CURVATURE, and TANGENTIAL ANGLE for the curve with $a = 1$ are

$$s(t) = 3(t + \sin t), \tag{7}$$

$$\kappa(t) = \tfrac{1}{3}\ \sec^2(\tfrac{1}{2}t), \tag{8}$$

$$\phi(t) = 2t. \tag{9}$$

References

Gray, A. *Modern Differential Geometry of Curves and Surfaces with Mathematica, 2nd ed.* Boca Raton, FL: CRC Press, pp. 119–120, 1997.

Lawrence, J. D. *A Catalog of Special Plane Curves.* New York: Dover, pp. 178 and 180, 1972.

MacTutor History of Mathematics Archive. "Cayley's Sextic." http://www-groups.dcs.st-and.ac.uk/~history/Curves/Cayleys.html.

Cayley's Sextic Evolute

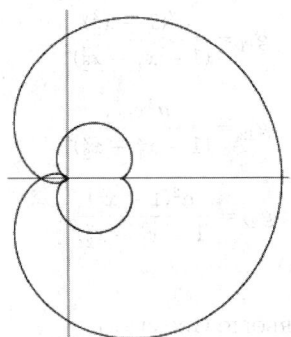

The EVOLUTE of Cayley's sextic is

$$x = \tfrac{1}{8}a + \tfrac{1}{16}a[3\cos(\tfrac{2}{3}t) - \cos(2t)]$$

$$y = \tfrac{1}{16}a[3\sin(\tfrac{2}{3}t) - \sin(2t)],$$

which is a NEPHROID.

C-Curve

LÉVY FRACTAL

C-Determinant

A DETERMINANT appearing in PADÉ APPROXIMANT identities:

$$C_{r/s} = \begin{vmatrix} a_{r-s+1} & a_{r-s+2} & \cdots & a_r \\ \vdots & \vdots & \ddots & \vdots \\ a_r & a_{r+1} & \cdots & a_{r+s-1} \end{vmatrix}.$$

See also PADÉ APPROXIMANT

Cech Cohomology

The direct limit of the COHOMOLOGY groups with COEFFICIENTS in an ABELIAN GROUP of certain coverings of a TOPOLOGICAL SPACE.

Ceiling

CEILING FUNCTION

Ceiling Function

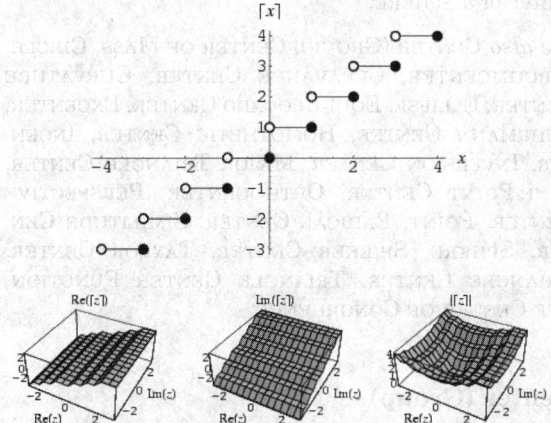

The function $\lceil x \rceil$ which gives the smallest INTEGER $\geq x$, shown as the thick curve in the above plot. Schroeder (1991) calls the ceiling function symbols the "GALLOWS" because of the similarity in appearance to the structure used for hangings. The name and symbol for the ceiling function were coined by K. E. Iverson (Graham *et al.* 1990). Although some authors used the symbol $]x[$ to denote the ceiling function (by analogy with the older notation $[x]$ for the FLOOR FUNCTION), this practice is strongly discouraged (Graham *et al.* 1990, p. 67).

Since usage concerning fractional part/value and integer part/value can be confusing, the following table gives a summary of names and notations used (D. W. Cantrell). Here, S&O indicates Spanier and Oldham (1987).

notation	name	S&O	Graham *et al.*	*Mathematica*
$\lfloor x \rfloor$	integer-value	$\mathrm{Int}(x)$	floor or integer part	`Floor[x]`
$\mathrm{sgn}(x)\lfloor \lvert x \rvert \rfloor$	integer-part	$\mathrm{Ip}(x)$	no name	`IntegerPart[x]`
$x - \lfloor x \rfloor$	fractional-value	$\mathrm{frac}(x)$	fractional part or $\{x\}$	no name
$\mathrm{sgn}(x)(\lvert x \rvert - \lfloor \lvert x \rvert \rfloor)$	fractional-part	$\mathrm{Fp}(x)$	no name	`FractionalPart[x]`

Odlyzko and Wilf (1991) have shown that the sequence $\{x_n\}$ defined by $x_0 = 1$ and

$$x_{n+1} = \left\lceil \tfrac{3}{2} x_n \right\rceil$$

satisfies

$$x_n = \left\lfloor K \left(\tfrac{3}{2} \right)^n \right\rfloor$$

for all n, where $K = 1.6222705028\ldots$ is analogous to MILLS' CONSTANT in the sense that the formula is useless unless K is known exactly ahead of time (Finch).

See also FLOOR FUNCTION, INTEGER PART, MILLS' CONSTANT, NEAREST INTEGER FUNCTION, STAIRCASE FUNCTION

References

Croft, H. T.; Falconer, K. J.; and Guy, R. K. *Unsolved Problems in Geometry.* New York: Springer-Verlag, p. 2, 1991.

Finch, S. "Powers of 3/2 Modulo One." http://www.mathsoft.com/asolve/pwrs32/pwrs32.html.

Graham, R. L.; Knuth, D. E.; and Patashnik, O. "Integer Functions." Ch. 3 in *Concrete Mathematics: A Foundation for Computer Science, 2nd ed.* Reading, MA: Addison-Wesley, pp. 67–101, 1994.

Iverson, K. E. *A Programming Language.* New York: Wiley, p. 12, 1962.

Odlyzko, A. M. and Wilf, H. S. "Functional Iteration and the Josephus Problem." *Glasgow Math. J.* **33**, 235–240, 1991.

Schroeder, M. *Fractals, Chaos, Power Laws: Minutes from an Infinite Paradise.* New York: W. H. Freeman, p. 57, 1991.

Cell

A finite regular POLYTOPE.

See also 16-CELL, 24-CELL, 120-CELL, 600-CELL

Cellular Automaton

A cellular automaton is a grid (possibly 1-D) of cells which evolves according to a set of rules based on the states of surrounding cells. von Neumann was one of the first people to consider such a model, and incorporated a cellular model into his "universal constructor." von Neumann proved that an automaton consisting of cells with four orthogonal neighbors and 29 possible states would be capable of simulating a TURING MACHINE for some configuration of about 200,000 cells (Gardner 1983, p. 227).

1-D automata called "ELEMENTARY CELLULAR AUTOMATA" are represented by a row of pixels with states either 0 or 1. These can be indexed with an 8-bit binary number, as shown by Stephen Wolfram. Wolfram further restricted the number from $2^8 = 256$ to 32 by requiring certain symmetry conditions.

The most well-known cellular automaton is Conway's game of LIFE, popularized in Martin Gardner's *Scientific American* columns. Although the computation of successive LIFE generations was originally done by hand, the computer revolution soon arrived and allowed more extensive patterns to be studied and propagated.

See also AUTOMATA THEORY, ELEMENTARY CELLULAR AUTOMATON, LIFE, LANGTON'S ANT, TOTALISTIC CELLULAR AUTOMATON, TURING MACHINE

References
Adami, C. *Artificial Life.* Cambridge, MA: MIT Press, 1998.
Buchi, J. R. and Siefkes, D. (Eds.). *Finite Automata, Their Algebras and Grammars: Towards a Theory of Formal Expressions.* New York: Springer-Verlag, 1989.
Burks, A. W. (Ed.). *Essays on Cellular Automata.* Urbana-Champaign, IL: University of Illinois Press, 1970.
Cipra, B. "Cellular Automata Offer New Outlook on Life, the Universe, and Everything." In *What's Happening in the Mathematical Sciences, 1995–1996, Vol. 3.* Providence, RI: Amer. Math. Soc., pp. 70–81, 1996.
Dewdney, A. K. *The Armchair Universe: An Exploration of Computer Worlds.* New York: W. H. Freeman, 1988.
Gardner, M. "The Game of Life, Parts I-III." Chs. 20–22 in *Wheels, Life, and Other Mathematical Amusements.* New York: W. H. Freeman, pp. 219 and 222, 1983.
Goles, E. and Martínez, S. (Eds.). *Cellular Automata and Complex Systems.* Amsterdam, Netherlands: Kluwer, 1999.
Gutowitz, H. (Ed.). *Cellular Automata: Theory and Experiment.* Cambridge, MA: MIT Press, 1991.
Hopcroft, J. E. and Ullman, J. D. *Introduction to Automata Theory, Languages, and Computation.* Reading, MA: Addison Wesley, 1979.
Hopcroft J. E. "An $n \log n$ Algorithm for Minimizing the States in a Finite Automaton." In *The Theory of Machines and Computations* (Ed. Z. Kohavi.) New York: Academic Press, pp. 189–196, 1971.
Levy, S. *Artificial Life: A Report from the Frontier Where Computers Meet Biology.* New York: Vintage, 1993.
Martin, O.; Odlyzko, A.; and Wolfram, S. "Algebraic Aspects of Cellular Automata." *Communications in Mathematical Physics* **93**, 219–258, 1984.
Preston, K. Jr. and Duff, M. J. B. *Modern Cellular Automata: Theory and Applications.* New York: Plenum, 1985.
Sigmund, K. *Games of Life: Explorations in Ecology, Evolution and Behaviour.* New York: Penguin, 1995.
Sloane, N. J. A. Sequences A006977/M2497 in "An On-Line Version of the Encyclopedia of Integer Sequences." http://www.research.att.com/~njas/sequences/eisonline.html.
Sloane, N. J. A. and Plouffe, S. Figure M2497 in *The Encyclopedia of Integer Sequences.* San Diego: Academic Press, 1995.
Toffoli, T. and Margolus, N. *Cellular Automata Machines: A New Environment for Modeling.* Cambridge, MA: MIT Press, 1987.
Weisstein, E. W. "Books about Cellular Automata." http://www.treasure-troves.com/books/CellularAutomata.html.
Wolfram, S. "Statistical Mechanics of Cellular Automata." *Rev. Mod. Phys.* **55**, 601–644, 1983.
Wolfram, S. "Twenty Problems in the Theory of Cellular Automata." *Physica Scripta* **T9**, 170–183, 1985.
Wolfram, S. (Ed.). *Theory and Application of Cellular Automata.* Reading, MA: Addison-Wesley, 1986.
Wolfram, S. *Cellular Automata and Complexity: Collected Papers.* Reading, MA: Addison-Wesley, 1994.
Wolfram, S. *A New Kind of Science.* Champaign, IL: Wolfram Media, 2001.
Wuensche, A. and Lesser, M. *The Global Dynamics of Cellular Automata: An Atlas of Basin of Attraction Fields of One-Dimensional Cellular Automata.* Reading, MA: Addison-Wesley, 1992.

Cellular Space

A HAUSDORFF SPACE which has the structure of a so-called CW-COMPLEX.

Center

A special POINT which usually has some symmetric placement with respect to points on a curve or in a SOLID. The center of a CIRCLE is equidistant from all points on the CIRCLE and is the intersection of any two distinct DIAMETERS. The same holds true for the center of a SPHERE.

See also CENTER (GROUP), CENTER OF MASS, CIRCLE, CIRCUMCENTER, CLEAVANCE CENTER, CURVATURE CENTER, ELLIPSE, EQUI-BROCARD CENTER, EXCENTER, FUHRMANN CENTER, HOMOTHETIC CENTER, INCENTER, INVERSION CENTER, MAJOR TRIANGLE CENTER, NINE-POINT CENTER, ORTHOCENTER, PERSPECTIVE CENTER, POINT, RADICAL CENTER, SIMILITUDE CENTER, SPHERE, SPIEKER CENTER, TAYLOR CENTER, TRIANGLE CENTER, TRIANGLE CENTER FUNCTION, YFF CENTER OF CONGRUENCE

Center (Group)

The center of a GROUP is the set of elements which commute with every element of the GROUP. It is equal to the intersection of the CENTRALIZERS of the GROUP elements.

See also CENTRALIZER, ISOCLINIC GROUPS, NILPOTENT GROUP

Center Function
TRIANGLE CENTER FUNCTION

Center of Gravity
CENTROID (GEOMETRIC)

Center of Mass
CENTROID (GEOMETRIC)

Center of Similitude
SIMILITUDE CENTER

Centered Cube Number

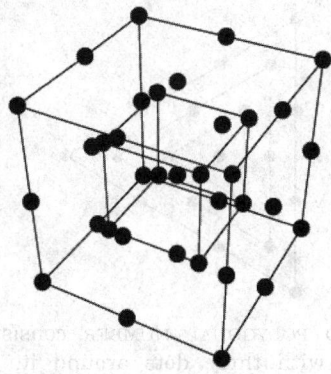

A FIGURATE NUMBER OF THE FORM,

$$CCub_n = n^3 + (n-1)^3 = (2n-1)(n^2 - n + 1).$$

The first few are 1, 9, 35, 91, 189, 341, ... (Sloane's A005898). The GENERATING FUNCTION for the centered cube numbers is

$$\frac{x(x^3 + 5x^2 + 5x + 1)}{(x-1)^4} = x + 9x^2 + 35x^3 + 91x^4 + \dots.$$

See also CUBIC NUMBER

References

Conway, J. H. and Guy, R. K. *The Book of Numbers.* New York: Springer-Verlag, p. 51, 1996.
Sloane, N. J. A. Sequences A005898/M4616 in "An On-Line Version of the Encyclopedia of Integer Sequences." http://www.research.att.com/~njas/sequences/eisonline.html.

Centered Hexagonal Number

HEX NUMBER

Centered Pentagonal Number

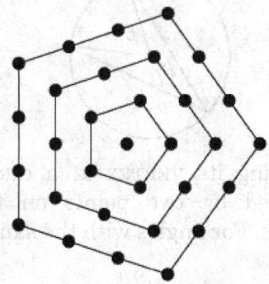

A CENTERED POLYGONAL NUMBER consisting of a central dot with five dots around it, and then additional dots in the gaps between adjacent dots. The general term is $(5n^2 - 5n + 2)/2$, and the first few such numbers are 1, 6, 16, 31, 51, 76, ... (Sloane's A005891). The GENERATING FUNCTION of the centered

pentagonal numbers is

$$\frac{x(x^2 + 3x + 1)}{(1-x)^3} = x + 6x^2 + 16x^3 + 31x^4 + \dots.$$

See also CENTERED POLYGONAL NUMBER, CENTERED SQUARE NUMBER, CENTERED TRIANGULAR NUMBER, HEX NUMBER

References

Sloane, N. J. A. Sequences A005891/M4112 in "An On-Line Version of the Encyclopedia of Integer Sequences." http://www.research.att.com/~njas/sequences/eisonline.html.

Centered Polygonal Number

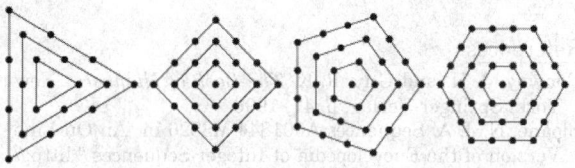

A FIGURATE NUMBER in which layers of POLYGONS are drawn centered about a point instead of with the point at a VERTEX.

See also CENTERED PENTAGONAL NUMBER, CENTERED SQUARE NUMBER, CENTERED TRIANGULAR NUMBER

References

Sloane, N. J. A. Sequences A001844/M3826 in "An On-Line Version of the Encyclopedia of Integer Sequences." http://www.research.att.com/~njas/sequences/eisonline.html.
Sloane, N. J. A. and Plouffe, S. Figure M3826 in *The Encyclopedia of Integer Sequences.* San Diego: Academic Press, 1995.

Centered Square Number

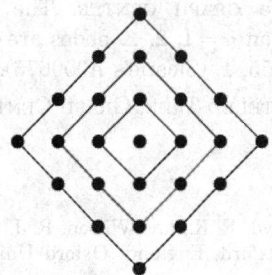

A CENTERED POLYGONAL NUMBER consisting of a central dot with four dots around it, and then additional dots in the gaps between adjacent dots. The general term is $n^2 + (n-1)^2$, and the first few

such numbers are 1, 5, 13, 25, 41, ... (Sloane's A001844). Centered square numbers are the sum of two consecutive SQUARE NUMBERS and are congruent to 1 (mod 4). The GENERATING FUNCTION giving the centered square numbers is

$$\frac{x(x+1)^2}{(1-x)^3} = x + 5x^2 + 13x^3 + 25x^4 + \dots.$$

See also CENTERED PENTAGONAL NUMBER, CENTERED POLYGONAL NUMBER, CENTERED TRIANGULAR NUMBER, SQUARE NUMBER

References

Conway, J. H. and Guy, R. K. *The Book of Numbers.* New York: Springer-Verlag, p. 41, 1996.
Sloane, N. J. A. Sequences A001844/M3826 in "An On-Line Version of the Encyclopedia of Integer Sequences." http://www.research.att.com/~njas/sequences/eisonline.html.

Centered Tree

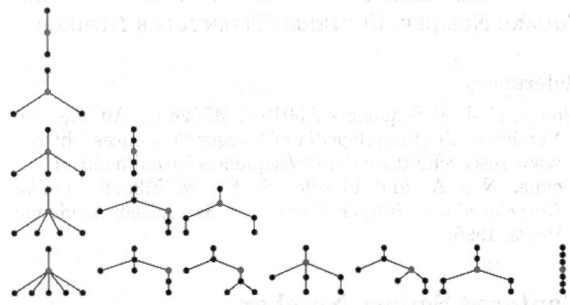

A TREE (also called a central tree) having a single node that is a GRAPH CENTER. The numbers of centered trees on $n = 1, 2, \dots$ nodes are 1, 1, 0, 1, 1, 2, 3, 7, 12, 27, 55, ... (Sloane's A000676).

See also BICENTERED TREE, GRAPH CENTER, TREE

References

Biggs, N. L.; Lloyd, E. K.; and Wilson, R. J. *Graph Theory 1736–1936.* Oxford, England: Oxford University Press, p. 49, 1976.
Cayley, A. "On the Analytical Forms Called Trees, with Application to the Theory of Chemical Combinations." *Reports Brit. Assoc. Advance. Sci.* **45**, 237–305, 1875. Reprinted in *Math Papers, Vol. 9*, pp. 427–460.
Sloane, N. J. A. Sequences A000676/M0831 in "An On-Line Version of the Encyclopedia of Integer Sequences." http://www.research.att.com/~njas/sequences/eisonline.html.

Centered Triangular Number

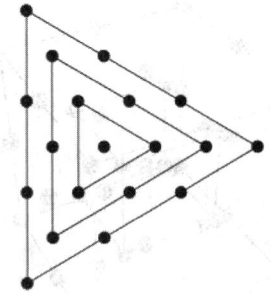

A CENTERED POLYGONAL NUMBER consisting of a central dot with three dots around it, and then additional dots in the gaps between adjacent dots. The general term is $(3n^2 - 3n + 2)/2$, and the first few such numbers are 1, 4, 10, 19, 31, 46, 64, ... (Sloane's A005448). The GENERATING FUNCTION giving the centered triangular numbers is

$$\frac{x(x^2 + x + 1)}{(1-x)^3} = x + 4x^2 + 10x^3 + 19x^4 + \dots.$$

See also CENTERED PENTAGONAL NUMBER, CENTERED SQUARE NUMBER

References

Sloane, N. J. A. Sequences A005448/M3378 in "An On-Line Version of the Encyclopedia of Integer Sequences." http://www.research.att.com/~njas/sequences/eisonline.html.

Centillion

In the American system, 10^{303}.

See also LARGE NUMBER

Central Angle

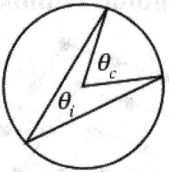

An ANGLE having its VERTEX at a CIRCLE's center which is formed by two points on the CIRCLE'S CIRCUMFERENCE. For angles with the same endpoints,

$$\theta_c = 2\theta_i,$$

where θ_i is the INSCRIBED ANGLE.

References

Pedoe, D. *Circles: A Mathematical View, rev. ed.* Washington, DC: Math. Assoc. Amer., pp. xxi-xxii, 1995.

Central Beta Function

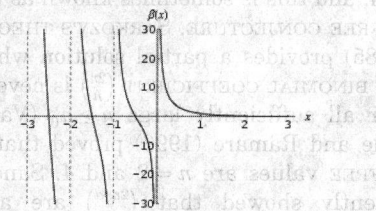

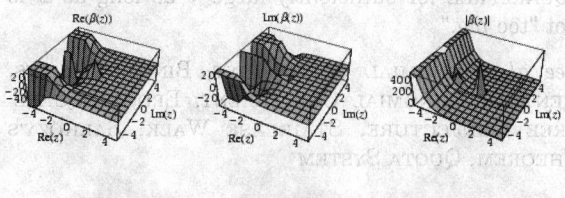

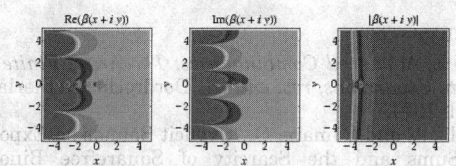

The central beta function is defined by

$$\beta(p) \equiv B(p, p), \qquad (1)$$

where $B(p, q)$ is the BETA FUNCTION. It satisfies the identities

$$\beta(p) = 2^{1-2p} B(p, \tfrac{1}{2}) \qquad (2)$$

$$= 2^{1-2p} \cos(\pi p)(\tfrac{1}{2} - p, p) \qquad (3)$$

$$= \int_0^1 \frac{t^p dt}{(1+t)^{2p}} \qquad (4)$$

$$\frac{2}{p} \prod_{n=1}^{\infty} \frac{n(n+2p)}{(n+p)(n+p)}. \qquad (5)$$

With $p = 1/2$, the latter gives the WALLIS FORMULA. When $p = a/b$,

$$b\beta(a/b) = 2^{1-2a/b} J(a, b), \qquad (6)$$

where

$$J(a, b) \equiv \int_0^1 \frac{t^{a-1} dt}{\sqrt{1-t^b}}. \qquad (7)$$

The central beta function satisfies

$$(2 + 4x)\beta(1 + x) = x\beta(x) \qquad (8)$$

$$(1 - 2x)\beta(1 - x)\beta(x) = 2\pi \cot(\pi x) \qquad (9)$$

$$\beta(\tfrac{1}{2} - x) = 2^{4x-1} \tan(\pi x)\beta(x) \qquad (10)$$

$$\beta(x)\beta(x + \tfrac{1}{2}) = 2^{4x+1} \pi \beta(2x)\beta(2x + \tfrac{1}{2}). \qquad (11)$$

For p an ODD POSITIVE INTEGER, the central beta function satisfies the identity

$$\beta(px) = \frac{1}{\sqrt{p}} \prod_{k=1}^{(p-1)/2} \frac{2x + \dfrac{2k-1}{p}}{2\pi} \prod_{k=0}^{p-1} \beta\left(x + \frac{k}{p}\right). \qquad (12)$$

See also BETA FUNCTION, REGULARIZED BETA FUNCTION

References

Borwein, J. M. and Zucker, I. J. "Elliptic Integral Evaluation of the Gamma Function at Rational Values of Small Denominators." *IMA J. Numerical Analysis* **12**, 519–526, 1992.

Central Binomial Coefficient

The nth central binomial coefficient is defined as $\binom{n}{\lfloor n/2 \rfloor}$, where $\binom{n}{k}$ is a BINOMIAL COEFFICIENT and $\lfloor n \rfloor$ is the FLOOR FUNCTION. The first few values are 1, 2, 3, 6, 10, 20, 35, 70, 126, 252, ... (Sloane's A001405). The central binomial coefficients have GENERATING FUNCTION

$$\frac{1 - 4x^2 - \sqrt{1 - 4x^2}}{2(2x^3 - x^2)} = 1 + 2x + 3x^2 + 6x^3 + 10x^4 + \dots.$$

The central binomial coefficients are SQUAREFREE only for $n = 1, 2, 3, 4, 5, 7, 8, 11, 17, 19, 23, 71, \dots$ (Sloane's A046098), with no others less than 7320.

The above coefficients are a superset of the alternative "central" binomial coefficients

$$\binom{2n}{n} = \frac{(2n)!}{(n!)^2},$$

which have GENERATING FUNCTION

$$\frac{1}{\sqrt{1 - 4x}} = 1 + 2x + 6x^2 + 20x^3 + 70x^4 + \dots.$$

The first few values are 2, 6, 20, 70, 252, 924, 3432, 12870, 48620, 184756, ... (Sloane's A000984).

A fascinating series of identities involving inverse central binomial coefficients times small powers are given by

$$\sum_{n=1}^{\infty} \frac{1}{\binom{2n}{n}} = \tfrac{1}{27}(2\pi\sqrt{3} + 9) = 0.7363998587\dots \qquad (1)$$

$$\sum_{n=1}^{\infty} \frac{1}{n\binom{2n}{n}} = \tfrac{1}{9}\pi\sqrt{3} = 0.6045997881\dots \qquad (2)$$

$$\sum_{n=1}^{\infty} \frac{1}{n^2\binom{2n}{n}} = \tfrac{1}{3}\zeta(2) = \tfrac{1}{8}\pi^2 \qquad (3)$$

$$\sum_{n=1}^{\infty} \frac{1}{n^4\binom{2n}{n}} = \frac{17}{36}\zeta(4) = \frac{17}{3240}\pi^4 \qquad (4)$$

(Comtet 1974, p. 89; Le Lionnais 1983, pp. 29, 30, 41, 36), which follow from the beautiful formula

$$\sum_{n=1}^{\infty} \frac{1}{n^k\binom{2n}{n}} = \frac{1}{2}k+1 F_k(\underbrace{1, \ldots, 1}_{k+1}; \frac{3}{2}, \underbrace{2, \ldots, 2}_{k-1}; \frac{1}{4}). \quad (5)$$

for $k \geq 1$, where $_mF_n(a_1, \ldots, a_m; b_1, \ldots, b_n; x)$ is a GENERALIZED HYPERGEOMETRIC FUNCTION. Additional sums of this type include

$$\sum_{n=1}^{\infty} \frac{1}{n^3\binom{2n}{n}} = \frac{1}{18}\pi\sqrt{3}[\psi_1(\tfrac{1}{3}) - \psi_1(\tfrac{2}{3})] - \frac{4}{3}\zeta(3) \qquad (6)$$

$$\sum_{n=1}^{\infty} \frac{1}{n^5\binom{2n}{n}}$$
$$= \frac{1}{432}\pi\sqrt{3}[\psi_3(\tfrac{1}{3}) - \psi_3(\tfrac{2}{3})] - \frac{19}{3}\zeta(5) + \frac{1}{9}\zeta(3)\pi^3, \qquad (7)$$

$$\sum_{n=1}^{\infty} \frac{1}{n^7\binom{2n}{n}} = \frac{11}{311040}\pi\sqrt{3}[\psi_5(\tfrac{1}{3}) - \psi_5(\tfrac{2}{3})] - \frac{493}{24}\zeta(7)$$
$$+ \frac{1}{3}\zeta(5)\pi^2 + \frac{17}{1620}\zeta(3)\pi^4, \qquad (8)$$

where $\psi_n(x)$ is the POLYGAMMA FUNCTION and $\zeta(x)$ is the RIEMANN ZETA FUNCTION (Plouffe 1998).

Similarly, we have

$$\sum_{n=1}^{\infty} \frac{(-1)^{n-1}}{\binom{2n}{n}} = \frac{1}{25}[5 + 4\sqrt{5}\operatorname{csch}^{-1}(2)] \qquad (9)$$

$$\sum_{n=1}^{\infty} \frac{(-1)^{n-1}}{n\binom{2n}{n}} = \frac{2}{5}\sqrt{5}\operatorname{csch}^{-1}(2) \qquad (10)$$

$$\sum_{n=1}^{\infty} \frac{(-1)^{n-1}}{n^2\binom{2n}{n}} = 2[\operatorname{csch}^{-1}(2)]^2 \qquad (11)$$

$$\sum_{n=1}^{\infty} \frac{(-1)^{n-1}}{n^3\binom{2n}{n}} = \frac{2}{5}\zeta(3) \qquad (12)$$

(Le Lionnais 1983, p. 35; Guy 1994, p. 257), where $\zeta(z)$ is the RIEMANN ZETA FUNCTION. These follow from the analogous identity

$$\sum_{n=1}^{\infty} \frac{(-1)^{n-1}}{n^k\binom{2n}{n}} = \frac{1}{2}k+1 F_k(\underbrace{1, \ldots, 1}_{k+1}; \frac{3}{2}, \underbrace{2, \ldots, 2}_{k-1}; -\frac{1}{4}). \quad (13)$$

Erdos and Graham (1980, p. 71) conjectured that the central binomial coefficient $\binom{2n}{n}$ is *never* SQUAREFREE for $n > 4$, and this is sometimes known as the ERDOS SQUAREFREE CONJECTURE. SÁRKOZY'S THEOREM (Sárkozy 1985) provides a partial solution which states that the BINOMIAL COEFFICIENT $\binom{2n}{n}$ is never SQUAREFREE for all sufficiently large $n \geq n_0$ (Vardi 1991). Granville and Ramare (1996) proved that the *only* SQUAREFREE values are $n = 2$ and 4. Sander (1992) subsequently showed that $\binom{2n\pm d}{n}$ are also never SQUAREFREE for sufficiently large n as long as d is not "too big."

See also BINOMIAL COEFFICIENT, BINOMIAL SUMS, CENTRAL TRINOMIAL COEFFICIENT, ERDOS SQUAREFREE CONJECTURE, STAIRCASE WALK, SÁRKÖZY'S THEOREM, QUOTA SYSTEM

References

Comtet, L. *Advanced Combinatorics: The Art of Finite and Infinite Expansions, rev. enl. ed.* Dordrecht, Netherlands: Reidel, 1974.

Granville, A. and Ramare, O. "Explicit Bounds on Exponential Sums and the Scarcity of Squarefree Binomial Coefficients." *Mathematika* **43**, 73–107, 1996.

Le Lionnais, F. *Les nombres remarquables.* Paris: Hermann, 1983.

Plouffe, S. "The Art of Inspired Guessing." Aug. 7, 1998. http://www.lacim.uqam.ca/plouffe/inspired.html.

Sander, J. W. "On Prime Divisors of Binomial Coefficients." *Bull. London Math. Soc.* **24**, 140–142, 1992.

Sárkozy, A. "On Divisors of Binomial Coefficients. I." *J. Number Th.* **20**, 70–80, 1985.

Sloane, N. J. A. Sequences A000984/M1645, A001405/M0769, and A046098 in "An On-Line Version of the Encyclopedia of Integer Sequences." http://www.research.att.com/~njas/sequences/eisonline.html.

Vardi, I. "Application to Binomial Coefficients," "Binomial Coefficients," "A Class of Solutions," "Computing Binomial Coefficients," and "Binomials Modulo and Integer." §2.2, 4.1, 4.2, 4.3, and 4.4 in *Computational Recreations in Mathematica.* Redwood City, CA: Addison-Wesley, pp. 25–28 and 63–71, 1991.

Central Conic

An ELLIPSE or HYPERBOLA.

See also CONIC SECTION

References

Coxeter, H. S. M. and Greitzer, S. L. *Geometry Revisited.* Washington, DC: Math. Assoc. Amer., pp. 146–150, 1967.

Ogilvy, C. S. *Excursions in Geometry.* New York: Dover, p. 77, 1990.

Central Difference

The central difference for a function tabulated at equal intervals f_n is defined by

$$\delta(f_n) = \delta_n = \delta_n^1 = f_{n+1/2} - f_{n-1/2}. \qquad (1)$$

First and higher order central differences arranged so as to involve integer indices are then given by

$$\delta_{n+1/2} = \delta_{n+1/2}^1 = f_{n+1} - f_n \qquad (2)$$

$$\delta_n^2 = \delta_{n+1/2}^1 - \delta_{n-1/2}^1 = f_{n+1} - 2f_n + f_{n-1} \qquad (3)$$

$$\delta_{n+1/2}^3 = \delta_{n+1}^2 - \delta_n^2 = f_{n+2} - 3f_{n+1} + 3f_n - f_{n-1}. \qquad (4)$$

Higher order differences may be computed for EVEN and ODD powers,

$$\delta_{n+1/2}^{2k} = \sum_{j=0}^{2k} (-1)^j \binom{2k}{j} f_{n+k-j} \qquad (5)$$

$$\delta_{n+1/2}^{2k+1} = \sum_{j=0}^{2k+1} (-1)^j \binom{2k+1}{j} f_{n+k+1-j}. \qquad (6)$$

See also BACKWARD DIFFERENCE, DIVIDED DIFFERENCE, FORWARD DIFFERENCE

References

Abramowitz, M. and Stegun, C. A. (Eds.). "Differences." §25.1 in *Handbook of Mathematical Functions with Formulas, Graphs, and Mathematical Tables, 9th printing.* New York: Dover, pp. 877–878, 1972.

Jeffreys, H. and Jeffreys, B. S. "Central Differences Formula." §9.084 in *Methods of Mathematical Physics, 3rd ed.* Cambridge, England: Cambridge University Press, pp. 284–286, 1988.

Sheppard, W. F. *Proc. London Math. Soc.* **31**, 459, 1899.

Whittaker, E. T. and Robinson, G. "Central-Difference Formulae." Ch. 3 in *The Calculus of Observations: A Treatise on Numerical Mathematics, 4th ed.* New York: Dover, pp. 35–52, 1967.

Central Dilation

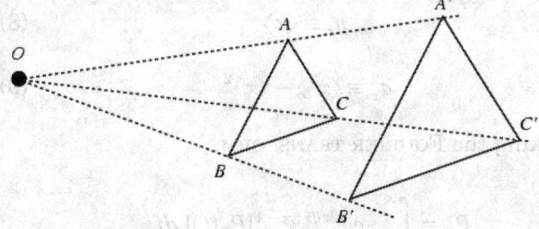

A DILATION that is not merely a TRANSLATION. Two triangles related by a central dilation are said to be PERSPECTIVE TRIANGLES because the lines joining corresponding vertices CONCUR.

See also DILATION, PERSPECTIVE TRIANGLES, SPIRAL SIMILARITY, TRANSLATION

References

Coxeter, H. S. M. and Greitzer, S. L. "Dilation." §4.7 in *Geometry Revisited.* Washington, DC: Math. Assoc. Amer., pp. 94–95, 1967.

Central Factorial

The central factorials $x^{[k]}$ form an associated SHEFFER SEQUENCE with

$$f(t) = e^{t/2} - e^{-t/2} = 2\sinh(\tfrac{1}{2}t),$$

giving the GENERATING FUNCTION

$$\sum_{k=0}^{\infty} \frac{x^{[k]}}{k!} t^k = e^{2x \sinh^{-1}(t/2)}.$$

The first central factorials are

$$x^{[0]} = 1$$
$$x^{[1]} = x$$
$$x^{[2]} = x^2$$
$$x^{[3]} = \tfrac{1}{4}(4x^3 - x) = -\tfrac{1}{4}(1 - 2x)x(1 + 2x)$$
$$x^{[4]} = x^4 - x^2 = -(1 - x)x^2(1 + x)$$
$$x^{[5]} = \tfrac{1}{16}(16x^5 - 40x^3 + 9x)$$
$$= \tfrac{1}{16}(1 - 2x)(3 - 2x)x(1 + 2x)(3 + 2x).$$

See also FACTORIAL, FALLING FACTORIAL, GOULD POLYNOMIAL, RISING FACTORIAL

References

Roman, S. *The Umbral Calculus.* New York: Academic Press, pp. 133–134, 1984.

Central Limit Theorem

Let $x_1, x_2, \ldots, x_N$ be a set of N INDEPENDENT random variates and each x_i have an *arbitrary* probability distribution $P(x_1, \ldots, x_N)$ with MEAN μ_i and a finite VARIANCE σ_i^2. Then the normal form variate

$$X_{\text{norm}} \equiv \frac{\sum_{i=1}^N x_i - \sum_{i=1}^N \mu_i}{\sqrt{\sum_{i=1}^N \sigma_i^2}} \qquad (1)$$

has a limiting cumulative distribution function which approaches a NORMAL (GAUSSIAN).

Under additional conditions on the distribution of the summand, the probability density itself is also GAUSSIAN (Feller 1971) with MEAN $\mu = 0$ and VARIANCE $\sigma^2 = 1$. If conversion to normal form is not performed, then the variate

$$X \equiv \frac{1}{N} \sum_{i=1}^N x_i \qquad (2)$$

is NORMALLY DISTRIBUTED with $\mu_X = \mu_x$ and $\sigma_X = \sigma_x / \sqrt{N}$.

Kallenberg (1997) gives a six-line proof of the central limit theorem. An elementary, but slightly more cumbersome proof of the central limit theorem, consider the INVERSE FOURIER TRANSFORM of $P_X(f)$.

$$\mathscr{F}^{-1}[P_X(f)] \equiv \int_{-\infty}^{\infty} e^{2\pi i f X} P(X)\, dX$$

$$= \int_{-\infty}^{\infty} \sum_{n=0}^{\infty} \frac{(2\pi i f X)^n}{n!}\, P(X)\, dX$$

$$= \sum_{n=0}^{\infty} \frac{(2\pi i f)^n}{n!} \int_{-\infty}^{\infty} X^n P(X)\, dX$$

$$= \sum_{n=0}^{\infty} \frac{(2\pi i f)^n}{n!} \langle X^n \rangle. \qquad (3)$$

Now write

$$\langle X^n \rangle = \langle N^{-n}(x_1 + x_2 + \ldots + x_N)^n \rangle$$

$$= \int_{-\infty}^{\infty} N^{-n}(x_1 + \ldots + x_N)^n P(x_1) \cdots P(x_N)\, dx_1 \cdots dx_N, \qquad (4)$$

so we have

$$\mathscr{F}^{-1}[P_X(f)] = \sum_{n=0}^{\infty} \frac{(2\pi i f)^n}{n!} \langle X^n \rangle$$

$$= \sum_{n=0}^{\infty} \frac{(2\pi i f)^n}{n!} \int_{-\infty}^{\infty} N^{-n}(x_1 + \ldots + x_N)^n$$
$$\times P(x_1) \cdots P(x_N)\, dx_1 \cdots dx_N$$

$$= \int_{-\infty}^{\infty} \sum_{n=0}^{\infty} \left[\frac{2\pi i f(x_1 + \ldots + x_N)}{N} \right]^n \frac{1}{n!}$$
$$\times P(x_1) \cdots P(x_N)\, dx_1 \cdots dx_N$$

$$= \int_{-\infty}^{\infty} e^{2\pi i f(x_1 + \ldots + x_N)/N} P(x_1) \cdots P(x_N)\, dx_1 \cdots dx_N$$

$$= \left[\int_{-\infty}^{\infty} e^{2\pi i f x_1/N} P(x_1)\, dx_1 \right]$$

$$\times \cdots \times \left[\int_{-\infty}^{\infty} e^{2\pi i f x_N/N} P(x_N)\, dx_N \right]$$

$$= \left[\int_{-\infty}^{\infty} e^{2\pi i f x/N} P(x)\, dx \right]^N$$

$$= \left\{ \int_{-\infty}^{\infty} \left[1 + \left(\frac{2\pi i f}{N} \right) x + \frac{1}{2} \left(\frac{2\pi i f}{N} \right)^2 x^2 + \ldots \right] P(x)\, dx \right\}^N$$

$$= \left[\int_{-\infty}^{\infty} P(x)\, dx + \frac{2\pi i f}{N} \int_{-\infty}^{\infty} x P(x)\, dx \right.$$

$$-\frac{(2\pi f)^2}{2N^2} \int_{-\infty}^{\infty} x^2 P(x)\, dx + \mathcal{O}(N^{-3}) \Big]^N$$

$$= \left[1 + \frac{2\pi i f}{N} \langle x \rangle - \frac{(2\pi f)^2}{2N^2} \langle x^2 \rangle + \mathcal{O}(N^{-3}) \right]^N$$

$$= \exp\left\{ N \ln\left[1 + \frac{2\pi i f}{N} \langle x \rangle - \frac{(2\pi f)^2}{2N^2} \langle x^2 \rangle + \mathcal{O}(N^{-3}) \right] \right\} \qquad (5)$$

Now expand

$$\ln(1 + x) = x - \tfrac{1}{2} x^2 + \tfrac{1}{3} x^3 + \ldots, \qquad (6)$$

so

$$\mathscr{F}^{-1}[P_X(f)] \approx \exp\left\{ N \left[\frac{2\pi i f}{N} \langle x \rangle - \frac{(2\pi f)^2}{2N^2} \langle x^2 \rangle \right. \right.$$

$$\left. \left. + \frac{1}{2} \frac{(2\pi i f)^2}{N^2} \langle x \rangle^2 + \mathcal{O}(N^{-3}) \right] \right\}$$

$$= \exp\left[2\pi i f \langle x \rangle - \frac{(2\pi f)^2 (\langle x^2 \rangle - \langle x \rangle^2)}{2N} + \mathcal{O}(N^{-2}) \right]$$

$$\approx \exp\left[2\pi i f \mu_x - \frac{(2\pi f)^2 \sigma_x^2}{2N} \right], \qquad (7)$$

since

$$\mu_x \equiv \langle x \rangle \qquad (8)$$

$$\sigma_x^2 \equiv \langle x^2 \rangle - \langle x \rangle^2. \qquad (9)$$

Taking the FOURIER TRANSFORM,

$$P_X \equiv \int_{-\infty}^{\infty} e^{-2\pi i f x} \mathscr{F}^{-1}[P_X(f)]\, df$$

$$= \int_{-\infty}^{\infty} e^{2\pi i f(\mu_x - x) - (2\pi f)^2 \sigma_x^2/2N}\, df. \qquad (10)$$

This is OF THE FORM

$$\int_{-\infty}^{\infty} e^{iaf - bf^2}\, df, \qquad (11)$$

where $a \equiv 2\pi(\mu_x - x)$ and $b \equiv (2\pi\sigma_x)^2/2N$. But, from Abramowitz and Stegun (1972, p. 302, equation 7.4.6),

$$\int_{-\infty}^{\infty} e^{iaf - bf^2}\, df = e^{-a^2/4b} \sqrt{\frac{\pi}{b}}. \qquad (12)$$

Therefore,

$$P_X = \sqrt{\frac{\pi}{\frac{(2\pi\sigma_z)^2}{2N}}} \exp\left\{\frac{-[2\pi(\mu_x - x)]^2}{4\frac{(2\pi\sigma_z)^2}{2N}}\right\}$$

$$= \sqrt{\frac{2\pi N}{4\pi^2\sigma_x^2}} \exp\left[-\frac{4\pi^2(\mu_x - x)^2 2N}{4 \cdot 4\pi^2\sigma_x^2}\right]$$

$$= \frac{\sqrt{N}}{\sigma_x\sqrt{2\pi}} e^{-(\mu_z - x)^2 N/2\sigma_z^2}. \tag{13}$$

But $\mu_X = \mu_x$ and $\mu_X = \mu_x$, so

$$P_X = \frac{1}{\sigma_X\sqrt{2\pi}} e^{-(\mu_X - x)^2/2\sigma_X^2}. \tag{14}$$

The "fuzzy" central limit theorem says that data which are influenced by many small and unrelated random effects are approximately NORMALLY DISTRIBUTED.

See also BERRY-ESSÉEN THEOREM, LINDEBERG CONDITION, LINDEBERG-FELLER CENTRAL LIMIT THEOREM, LYAPUNOV CONDITION

References

Abramowitz, M. and Stegun, C. A. (Eds.). *Handbook of Mathematical Functions with Formulas, Graphs, and Mathematical Tables, 9th printing.* New York: Dover, 1972.

Feller, W. "The Fundamental Limit Theorems in Probability." *Bull. Amer. Math. Soc.* **51**, 800–832, 1945.

Feller, W. *An Introduction to Probability Theory and Its Applications, Vol. 1, 3rd ed.* New York: Wiley, p. 229, 1968.

Feller, W. *An Introduction to Probability Theory and Its Applications, Vol. 2, 3rd ed.* New York: Wiley, 1971.

Kallenberg, O. *Foundations of Modern Probability.* New York: Springer-Verlag, 1997.

Lindeberg, J. W. "Eine neue Herleitung des Exponentialgesetzes in der Wahrscheinlichkeitsrechnung." *Math. Z.* **15**, 211–225, 1922.

Spiegel, M. R. *Theory and Problems of Probability and Statistics.* New York: McGraw-Hill, pp. 112–113, 1992.

Trotter, H. F. "An Elementary Proof of the Central Limit Theorem." *Arch. Math.* **10**, 226–234, 1959.

Zabell, S. L. "Alan Turing and the Central Limit Theorem." *Amer. Math. Monthly* **102**, 483–494, 1995.

Central Moment

A MOMENT μ_n of a probability function $P(x)$ taken about the mean μ,

$$\mu_n = \int (x - \mu)^n P(x)\, dx. \tag{1}$$

The central moments μ_n can be expressed as terms of the RAW MOMENTS μ_n' (i.e., those taken about zero) using the BINOMIAL TRANSFORM

$$\mu_n = \sum_{k=0}^{n} \binom{n}{k} (-1)^{n-k} \mu_k' \mu_1'^{n-k}, \tag{2}$$

with $\mu_0' = 1$ (Papoulis 1986, p. 146). The first few

values are therefore

$$\mu_1 = 0 \tag{3}$$

$$\mu_2 = -\mu_1'^2 + \mu_2' \tag{4}$$

$$\mu_3 = 2\mu_1'^3 - 3\mu_1'\mu_2' + \mu_3' \tag{5}$$

$$\mu_4 = -3\mu_1'^4 + 6\mu_1'^2\mu_2' - 4\mu_1'\mu_3' + \mu_4' \tag{6}$$

$$\mu_5 = 4\mu_1'^5 - 10\mu_1'^3\mu_2' + 10\mu_1'^2\mu_3' - 5\mu_1'\mu_4' + \mu_5'. \tag{6}$$

See also ABSOLUTE MOMENT, CUMULANT, KURTOSIS, MOMENT, PEARSON KURTOSIS, RAW MOMENT, SKEWNESS

References

Papoulis, A. *Probability, Random Variables, and Stochastic Processes, 2nd ed.* New York: McGraw-Hill, p. 146, 1984.

Kenney, J. F. and Keeping, E. S. "Moments About the Mean." §7.3 in *Mathematics of Statistics, Pt. 1, 3rd ed.* Princeton, NJ: Van Nostrand, pp. 92–93, 1962.

Central Point

A point v is a central point of a graph if the eccentricity of the point equals the GRAPH RADIUS. The set of all central points is called the GRAPH CENTER.

See also CENTROID POINT, GRAPH CENTER, GRAPH ECCENTRICITY, GRAPH RADIUS

References

Harary, F. *Graph Theory.* Reading, MA: Addison-Wesley, p. 35, 1994.

Central Tree

CENTERED TREE

Central Trinomial Coefficient

The nth central trinomial coefficient is defined as the coefficient of x_n in the expansion of $(1 + x + x^2)^n$. It is also the number of permutations of n symbols, each -1, 0, or 1, which sum to 0. For example, there are seven such permutations of three symbols: $\{-1, 0, 1\}$, $\{-1, 1, 0\}$, $\{0, -1, 1\}$, $\{0, 0, 0\}$, and $\{0, 1, -1\}$, $\{1, -1, 0\}$, $\{1, 0, -1\}$. The first few central binomial coefficients are 1, 3, 7, 19, 51, 141, 393, ... (Sloane's A002426). This sequence cannot be expressed as a fixed number of hypergeometric terms (Petkovsek *et al.* 1996, p. 160). The GENERATING FUNCTION is given by

$$f(x) = \frac{1}{\sqrt{(1+x)(1-3x)}} = 1 + x + 3x^2 + 7x^3 + \ldots.$$

See also CENTRAL BINOMIAL COEFFICIENT, TRINOMIAL COEFFICIENT

References

Petkovsek, M.; Wilf, H. S.; and Zeilberger, D. *A = B.* Wellesley, MA: A. K. Peters, 1996.

Sloane, N. J. A. Sequences A002426/M2673 in "An On-Line Version of the Encyclopedia of Integer Sequences." http://www.research.att.com/~njas/sequences/eisonline.html.

Central Value

CLASS MARK

Centralizer

The centralizer of an element z of a GROUP G is the set of elements of G which commute with z,

$$C_G(z) = \{x \in G, \; xz = zx\}.$$

Likewise, the centralizer of a SUBGROUP H of a GROUP G is the set of elements of G which commute with every element of H,

$$C_G(H) = \{x \in G, \; \forall h \in H, \; xh = hx\}.$$

The centralizer always contains the CENTER of the group and is contained in the corresponding NORMALIZER. In an ABELIAN GROUP, the centralizer is the whole group.

See also ABELIAN GROUP, CENTER (GROUP), GROUP, NORMALIZER, SUBGROUP

Centrally Symmetric Set

CENTROSYMMETRIC SET

Centric Perspective

PERSPECTIVE

Centrode

$$\mathbf{C} \equiv \tau\mathbf{T} + \kappa\mathbf{B},$$

where τ is the TORSION, κ is the CURVATURE, $\mathbf{T}$ is the TANGENT VECTOR, and $\mathbf{B}$ is the BINORMAL VECTOR.

Centroid (Function)

By analogy with the GEOMETRIC CENTROID, the centroid of an arbitrary function $f(x)$ is defined as

$$\langle x \rangle = \frac{\int_{-\infty}^{\infty} x f(x)\, dx}{\int_{-\infty}^{\infty} f(x)\, dx}.$$

References

Bracewell, R. *The Fourier Transform and Its Applications,* 3rd ed. New York: McGraw-Hill, pp. 139–140 and 156, 1999.

Centroid (Geometric)

The CENTER OF MASS of a 2-D planar LAMINA or a 3-D solid. The mass of a LAMINA with surface density function $\sigma(x, y)$ is

$$M = \iint \sigma(x, y)\, dA, \tag{1}$$

and the coordinates of the centroid (also called the CENTER OF GRAVITY) are

$$\bar{x} = \frac{\iint x\sigma(x, y)\, dA}{M} \tag{2}$$

$$\bar{y} = \frac{\iint y\sigma(x, y)\, dA}{M}. \tag{3}$$

The centroid of a lamina is the point on which it would balance when placed on a needle. The centroid of a solid is the point on which the solid would "balance."

The centroid of a set of n point masses m_i located at positions $\mathbf{x}_i$ is

$$\bar{\mathbf{x}} = \frac{\sum_{i=1}^{n} m_i \mathbf{x}_i}{\sum_{i=1}^{n} m_i}, \tag{4}$$

which, if all masses are equal, simplifies to

$$\bar{\mathbf{x}} = \frac{\sum_{i=1}^{n} \mathbf{x}_i}{n}. \tag{5}$$

The centroid of n point masses also gives the location at which a school should be built in order to minimize the distance travelled by children from n cities, located at the positions of the masses, and with m_i equal to the number of students from city i (Steinhaus 1983, pp. 113–116).

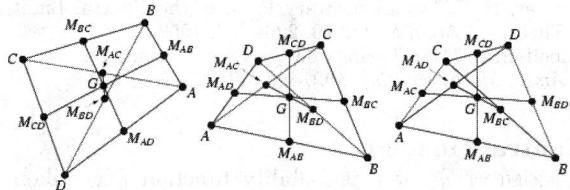

The centroid of the vertices of a quadrilateral occurs at the point of intersection of the BIMEDIANS (i.e., the lines $M_{AB}M_{CD}$ and $M_{AD}M_{BC}$ joining pairs of opposite MIDPOINTS) (Honsberger 1995, pp. 36–37). In addition, it is the MIDPOINT of the line $M_{AC}M_{BD}$ connecting the midpoints of the diagonals AC and BD (Honsberger 1995, pp. 39–40).

Given an arbitrary HEXAGON, connecting the centroids of each consecutive three sides gives the so-called CENTROID HEXAGON, a hexagon with equal and parallel sides (Wells 1991).

The centroids of several common laminas along the nonsymmetrical axis are summarized in the following table.

Figure	$\bar{y}$
PARABOLIC SEGMENT	$\frac{2}{5}h$
SEMICIRCLE	$\frac{4r}{3\pi}$

In 3-D, the mass of a solid with density function $\rho(x, y, z)$ is

$$M = \iiint \rho(x, y, z)\, dV, \qquad (6)$$

and the coordinates of the center of mass are

$$\bar{x} = \frac{\iiint x\rho(x, y, z)\, dV}{M} \qquad (7)$$

$$\bar{y} = \frac{\iiint y\rho(x, y, z)\, dV}{M} \qquad (8)$$

$$\bar{z} = \frac{\iiint z\rho(x, y, z)\, dV}{M}. \qquad (9)$$

Figure	$\bar{z}$
CONE	$\frac{1}{4}h$
CONICAL FRUSTUM	$\dfrac{h(R_1^2 + 2R_1R_2 + 3R_2^2)}{4(R_1^2 + R_1R_2 + R_2^2)}$
HEMISPHERE	$\frac{3}{8}R$
PARABOLOID	$\frac{2}{3}h$
PYRAMID	$\frac{1}{4}h$

See also CENTROID HEXAGON, PAPPUS'S CENTROID THEOREM

References

Beyer, W. H. *CRC Standard Mathematical Tables, 28th ed.* Boca Raton, FL: CRC Press, p. 132, 1987.

Honsberger, R. *Episodes in Nineteenth and Twentieth Century Euclidean Geometry.* Washington, DC: Math. Assoc. Amer., 1995.

Kern, W. F. and Bland, J. R. "Center of Gravity." §39 in *Solid Mensuration with Proofs, 2nd ed.* New York: Wiley, p. 110, 1948.

McLean, W. G. and Nelson, E. W. "First Moments and Centroids." Ch. 9 in *Schaum's Outline of Theory and Problems of Engineering Mechanics: Statics and Dynamics, 4th ed.* New York: McGraw-Hill, pp. 134–162, 1988.

Steinhaus, H. *Mathematical Snapshots, 3rd ed.* New York: Dover, 1999.

Wells, D. *The Penguin Dictionary of Curious and Interesting Geometry.* London: Penguin, pp. 53–54, 1991.

Centroid (Orthocentric System)

The centroid of the four points constituting an ORTHOCENTRIC SYSTEM is the center of the common NINE-POINT CIRCLE (Johnson 1929, p. 249). This fact automatically guarantees that the centroid of the INCENTER and EXCENTERS of a TRIANGLE is located at the CIRCUMCENTER.

References

Johnson, R. A. *Modern Geometry: An Elementary Treatise on the Geometry of the Triangle and the Circle.* Boston, MA: Houghton Mifflin, 1929.

Centroid (Triangle)

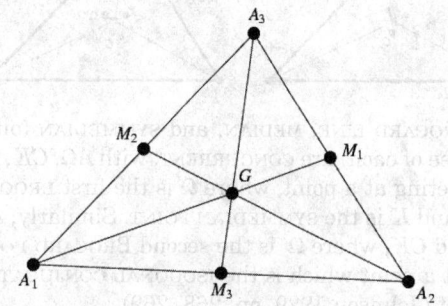

The CENTROID (CENTER OF MASS) of the VERTICES of a TRIANGLE is the point G (sometimes also denoted M) which is also the intersection of the TRIANGLE's three MEDIANS (Johnson 1929, p. 249; Wells 1991, p. 150). The point is therefore sometimes called the median point. The centroid is always in the interior of the TRIANGLE. It has TRILINEAR COORDINATES

$$\frac{1}{a} : \frac{1}{b} : \frac{1}{c}, \qquad (1)$$

or

$$\csc A\ :\ \csc B\ :\ \csc C, \qquad (2)$$

and homogeneous BARYCENTRIC COORDINATES

$(1, 1, 1)$.

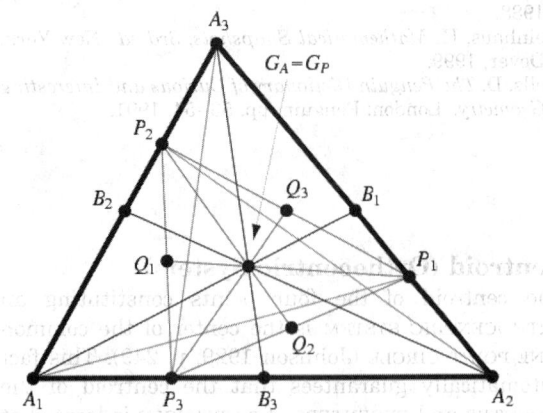

If the sides of a TRIANGLE $\Delta A_1 A_2 A_3$ are divided by points P_1, P_2, and P_3 so that

$$\frac{\overline{A_2 P_1}}{\overline{P_1 A_3}} = \frac{\overline{A_3 P_2}}{\overline{P_2 A_1}} = \frac{\overline{A_1 P_3}}{\overline{P_3 A_2}} = \frac{p}{q}, \tag{3}$$

then the centroid of the TRIANGLE $\Delta P_1 P_2 P_3$ is M, the centroid of the original triangle $\Delta A_1 A_2 A_3$ (Johnson 1929, p. 250).

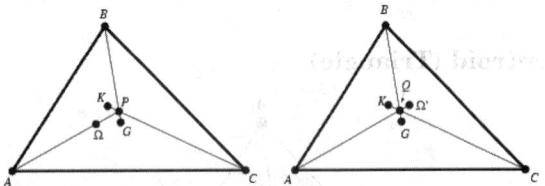

One BROCARD LINE, MEDIAN, and SYMMEDIAN (out of the three of each) are CONCURRENT, with $A\Omega$, CK, and BG meeting at a point, where Ω is the first BROCARD POINT and K is the SYMMEDIAN POINT. Similarly, $A\Omega'$, BG, and CK, where Ω' is the second BROCARD POINT, meet at a point which is the ISOGONAL CONJUGATE of the first (Johnson 1929, pp. 268–269).

Pick an interior point X. The TRIANGLES BXC, CXA, and AXB have equal areas IFF X corresponds to the centroid. The centroid is located one third of the way from each VERTEX to the MIDPOINT of the opposite side. Each median divides the triangle into two equal areas; all the medians together divide it into six equal parts, and the lines from the MEDIAN POINT to the VERTICES divide the whole into three equivalent TRIANGLES. In general, for any line in the plane of a TRIANGLE ABC,

$$d = \tfrac{1}{3}(d_A + d_B + d_C), \tag{4}$$

where d, d_A, d_B, and d_C are the distances from the centroid and VERTICES to the line.

A TRIANGLE will balance at the centroid, and along any line passing through the centroid. The TRILINEAR POLAR of the centroid is called the LEMOINE AXIS. The PERPENDICULARS from the centroid are proportional

to s_i^{-1},

$$a_1 p_2 = a_2 p_2 = a_3 p_3 = \tfrac{2}{3} \Delta, \tag{5}$$

where Δ is the AREA of the TRIANGLE. Let P be an arbitrary point, the VERTICES be A_1, A_2, and A_3, and the centroid G. Then

$$\overline{PA_1}^2 + \overline{PA_2}^2 + \overline{PA_3}^2$$
$$= \overline{GA_1}^2 + \overline{GA_2}^2 + \overline{GA_3}^2 + 3\overline{PG}^2. \tag{6}$$

If O is the CIRCUMCENTER of the triangle's centroid, then

$$\overline{OG}^2 = R^2 - \tfrac{1}{9}(a^2 + b^2 + c^2). \tag{7}$$

The centroid lies on the EULER LINE and NAGEL LINE. The centroid of the PERIMETER of a TRIANGLE is the triangle's SPIEKER CENTER (Johnson 1929, p. 249). The SYMMEDIAN POINT of a triangle is the centroid of its PEDAL TRIANGLE (Honsberger 1995, pp. 72–74).

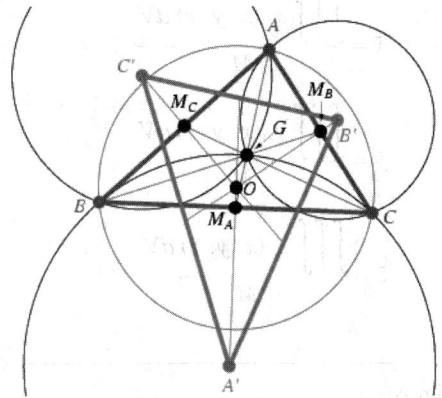

Given a triangle ΔABC, construct circles through each pair of vertices which also pass through the CENTROID G. The TRIANGLE $\Delta A'B'C'$ determined by the center of these circles then satisfies a number of interesting properties. The first is that the CIRCUMCIRCLE O and CENTROID G of ΔABC are, respectively, the CENTROID G' and SYMMEDIAN POINT K' of the triangle $\Delta A'B'C'$ (Honsberger 1995, p. 77). In addition, the MEDIANS of ΔABC and $\Delta A'B'C$ intersect in the midpoints of the sides of ΔABC.

See also CIRCUMCENTER, EULER LINE, EXMEDIAN POINT, INCENTER, NAGEL LINE, ORTHOCENTER

References

Carr, G. S. *Formulas and Theorems in Pure Mathematics,* *2nd ed.* New York: Chelsea, p. 622, 1970.

Coxeter, H. S. M. and Greitzer, S. L. *Geometry Revisited.* Washington, DC: Math. Assoc. Amer., p. 7, 1967.

Dixon, R. *Mathographics.* New York: Dover, pp. 55–57, 1991.

Honsberger, R. *Episodes in Nineteenth and Twentieth Century Euclidean Geometry.* Washington, DC: Math. Assoc. Amer., pp. 72–74 and 77, 1995.

Johnson, R. A. *Modern Geometry: An Elementary Treatise on the Geometry of the Triangle and the Circle.* Boston, MA: Houghton Mifflin, pp. 173–176 and 249, 1929.
Kimberling, C. "Central Points and Central Lines in the Plane of a Triangle." *Math. Mag.* **67**, 163–187, 1994.
Kimberling, C. "Centroid." http://cedar.evansville.edu/~ck6/tcenters/class/centroid.html.
Lachlan, R. *An Elementary Treatise on Modern Pure Geometry.* London: Macmillian, pp. 62–63, 1893.
Wells, D. *The Penguin Dictionary of Curious and Interesting Geometry.* London: Penguin, p. 150, 1991.

Centroid Hexagon

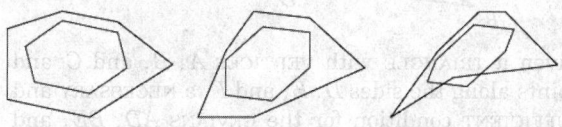

The hexagon obtained from an arbitrary HEXAGON by connecting the centroids of each consecutive three sides. This hexagon has equal and parallel sides (Wells 1991).

References

Cadwell, J. H. *Topics in Recreational Mathematics.* Cambridge, England: Cambridge University Press, 1966.
Wells, D. *The Penguin Dictionary of Curious and Interesting Geometry.* London: Penguin, pp. 53–54, 1991.

Centroid Point

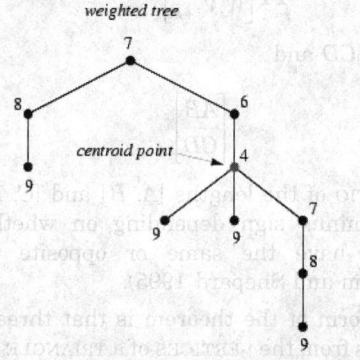

A point in a WEIGHTED TREE that has minimum weight for the tree. The set of all centroid points is called a TREE CENTROID (Harary 1994, p. 36). The largest possible values for a centroid point (i.e., the maximum minimum weight) for a tree on $n = 2, 3, \ldots$ nodes are 1, 1, 2, 2, 3, 3, ...

See also TREE CENTROID, WEIGHTED TREE

References

Harary, F. *Graph Theory.* Reading, MA: Addison-Wesley, 1994.
Weisstein, E. W. "Graphs." MATHEMATICA NOTEBOOK GRAPHS.M.

Centroidal Line

The three planes determined by the edges of a TRIHEDRON and the internal bisectors of the respec-

tively opposite faces are coaxal, and the common line of these planes is called the centroidal line.

See also TRIHEDRON

References

Altshiller-Court, N. "Centroidal Lines." §2.5 in *Modern Pure Solid Geometry.* New York: Chelsea, pp. 40–41, 1979.

Centrosymmetric Matrix

A SQUARE MATRIX is called centrosymmetric if it is symmetric with respect to the center (Muir 1960, p. 19).

See also BISYMMETRIC MATRIX, SYMMETRIC MATRIX

References

Muir, T. *A Treatise on the Theory of Determinants.* New York: Dover, 1960.

Centrosymmetric Set

A CONVEX SET K is centro-symmetric, sometimes also called centrally symmetric, if it has a center **p** that bisects every CHORD of K through **p**.

References

Croft, H. T.; Falconer, K. J.; and Guy, R. K. *Unsolved Problems in Geometry.* New York: Springer-Verlag, p. 7, 1991.

Certificate of Compositeness
COMPOSITENESS CERTIFICATE

Certificate of Primality
PRIMALITY CERTIFICATE

Cesàro Equation
An INTRINSIC EQUATION which expresses a curve in terms of its ARC LENGTH s and RADIUS OF CURVATURE R (or equivalently, the CURVATURE κ).

See also ARC LENGTH, INTRINSIC EQUATION, NATURAL EQUATION, RADIUS OF CURVATURE, WHEWELL EQUATION

References

Yates, R. C. "Intrinsic Equations." *A Handbook on Curves and Their Properties.* Ann Arbor, MI: J. W. Edwards, pp. 123–126, 1952.

Cesàro Fractal

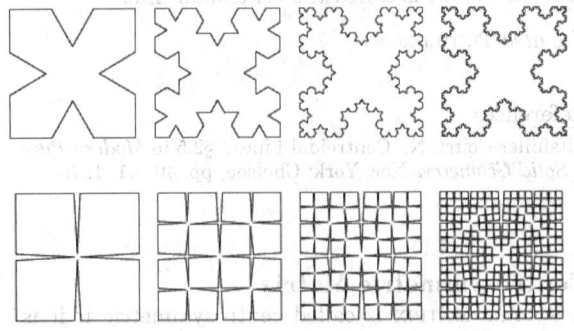

A FRACTAL also known as the TORN SQUARE FRACTAL. The base curves and motifs for the two fractals illustrated above are shown below.

See also FRACTAL, KOCH SNOWFLAKE

References

Cesàro, E. "Remarques sur la courbe de von Koch." *Atti della R. Accad. della Scienze fisiche e matem. Napoli* **12**, No. 15, 1905. Reprinted as §228 in *Opere scelte, a cura dell'Unione matematica italiana e col contributo del Consiglio nazionale delle ricerche, Vol. 2: Geometria, analisi, fisica matematica.* Rome: Edizioni Cremonese, pp. 464–479, 1964.

Lauwerier, H. *Fractals: Endlessly Repeated Geometric Figures.* Princeton, NJ: Princeton University Press, p. 43, 1991.

Pappas, T. *The Joy of Mathematics.* San Carlos, CA: Wide World Publ./Tetra, p. 79, 1989.

Weisstein, E. W. "Fractals." MATHEMATICA NOTEBOOK FRACTAL.M.

Cesàro Mean

FEJES TÓTH'S INTEGRAL

Cesàro's Theorem

The three points determined on three coplanar edges of a TETRAHEDRON by the external bisecting planes of the opposite DIHEDRAL ANGLES are COLLINEAR. Furthermore, this line belongs to the plane determined by the three points in which the remaining three (concurrent) edges of the TETRAHEDRON are met by the internal bisecting planes of the respectively opposite DIHEDRAL ANGLE.

References

Altshiller-Court, N. "Gergonne's Theorem." §235 in *Modern Pure Solid Geometry.* New York: Chelsea, p. 71, 1979.

Ceva's Theorem

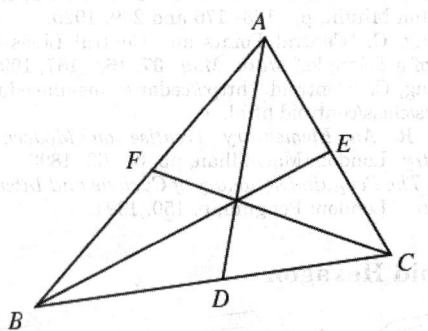

Given a TRIANGLE with VERTICES A, B, and C and points along the sides D, E, and F, a NECESSARY and SUFFICIENT condition for the CEVIANS AD, BE, and CF to be CONCURRENT (intersect in a single point) is that

$$BD \cdot CE \cdot AF = DC \cdot EA \cdot FB. \tag{1}$$

This theorem was first published by Giovanni Cevian 1678.

Let $P = [V_1, \ldots, V_n]$ be an arbitrary n-gon, C a given point, and k a POSITIVE INTEGER such that $1 \leq k \leq n/2$. For $i = 1, \ldots, n$, let W_i be the intersection of the lines CV_i and $V_{i-k}V_{i+k}$, then

$$\prod_{i=1}^{n} \left[\frac{V_{i-k}W_i}{W_iV_{i+k}} \right] = 1. \tag{2}$$

Here, $AB \| CD$ and

$$\left[\frac{AB}{CD} \right] \tag{3}$$

is the RATIO of the lengths $[A, B]$ and $[C, D]$ with a plus or minus sign depending on whether these segments have the same or opposite directions (Grünbaum and Shepard 1995).

Another form of the theorem is that three CONCURRENT lines from the VERTICES of a TRIANGLE divide the opposite sides in such fashion that the product of three nonadjacent segments equals the product of the other three (Johnson 1929, p. 147).

See also HOEHN'S THEOREM, MENELAUS' THEOREM

References

Beyer, W. H. (Ed.). *CRC Standard Mathematical Tables, 28th ed.* Boca Raton, FL: CRC Press, p. 122, 1987.

Coxeter, H. S. M. and Greitzer, S. L. "Ceva's Theorem." §1.2 in *Geometry Revisited.* Washington, DC: Math. Assoc. Amer., pp. 4–5, 1967.

Durell, C. V. *A Course of Plane Geometry for Advanced Students, Part I.* London: Macmillan, p. 54, 1909.

Durell, C. V. *Modern Geometry: The Straight Line and Circle.* London: Macmillan, pp. 40–41, 1928.

Graustein, W. C. *Introduction to Higher Geometry.* New York: Macmillan, p. 81, 1930.

Grünbaum, B. and Shepard, G. C. "Ceva, Menelaus, and the Area Principle." *Math. Mag.* **68**, 254–268, 1995.

Honsberger, R. "Ceva's Theorem." §12.1 in *Episodes in Nineteenth and Twentieth Century Euclidean Geometry.* Washington, DC: Math. Assoc. Amer., pp. 136–138, 1995.

Johnson, R. A. *Modern Geometry: An Elementary Treatise on the Geometry of the Triangle and the Circle.* Boston, MA: Houghton Mifflin, pp. 145–151, 1929.

Pedoe, D. *Circles: A Mathematical View, rev. ed.* Washington, DC: Math. Assoc. Amer., p. xx, 1995.

Wells, D. *The Penguin Dictionary of Curious and Interesting Geometry.* London: Penguin, pp. 28–29, 1991.

Cevian

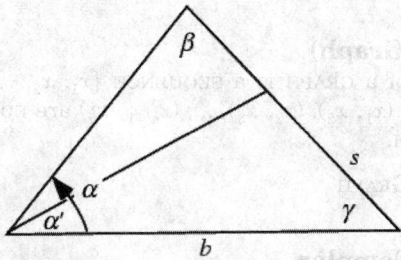

A line segment which joins a VERTEX of a TRIANGLE with a point on the opposite side (or its extension). In the above figure,

$$s = \frac{b \sin \alpha'}{\sin(\gamma + \alpha')}.$$

The condition for Cevians from the three sides of a TRIANGLE to CONCUR is known as CEVA'S THEOREM. If AD, BE, and CF are cevians of a TRIANGLE $\triangle ABC$ through an arbitrary point P inside $\triangle ABC$, then the ratios

$$\frac{AP}{PD}, \frac{BP}{PE}, \frac{CP}{PF}$$

into which P divides the Cevians have a sum ≥ 6 and a product ≥ 8 (Ramler 1958; Honsberger 1995, pp. 138–141).

See also ANGLE BISECTOR, CEVA'S THEOREM, CEVIAN CIRCLE, CEVIAN TRIANGLE, MEDIAN (TRIANGLE), PEDAL-CEVIAN POINT, ROUTH'S THEOREM, SPLITTER

References

Honsberger, R. "On Cevians." Ch. 12 in *Episodes in Nineteenth and Twentieth Century Euclidean Geometry.* Washington, DC: Math. Assoc. Amer., pp. 13 and 137–146, 1995.

Ramler, O. J. Solved by C. W. Trigg. "Problem E1043." *Amer. Math. Monthly* **65**, 421, 1958.

Thébault, V. "On the Cevians of a Triangle." *Amer. Math. Monthly* **60**, 167–173, 1953.

Cevian Circle

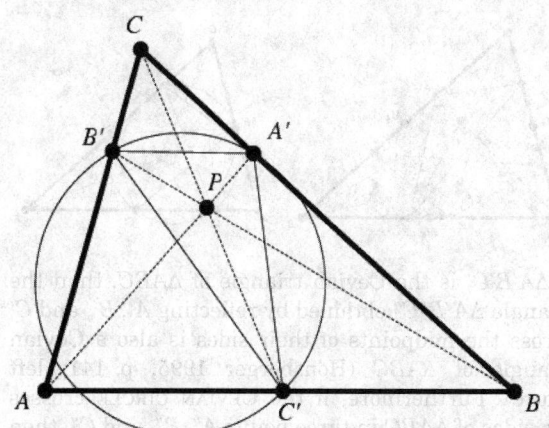

The CIRCUMCIRCLE of the CEVIAN TRIANGLE $\triangle A'B'C'$ of a given TRIANGLE $\triangle ABC$ with respect to a point P.

See also CEVIAN TRIANGLE, CIRCUMCIRCLE

Cevian Conjugate Point

ISOTOMIC CONJUGATE POINT

Cevian Transform

Vandeghen's (1965) name for the transformation taking points to their ISOTOMIC CONJUGATE POINTS.

See also ISOTOMIC CONJUGATE POINT

References

Vandeghen, A. "Some Remarks on the Isogonal and Cevian Transforms. Alignments of Remarkable Points of a Triangle." *Amer. Math. Monthly* **72**, 1091–1094, 1965.

Cevian Triangle

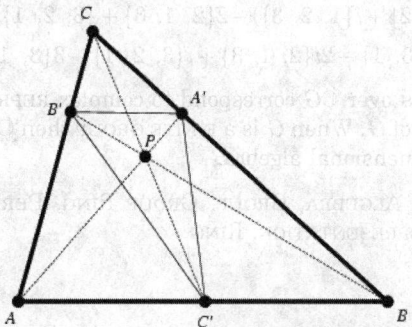

Given a point P and a TRIANGLE $\triangle ABC$, the Cevian triangle $\triangle A'B'C'$ is defined as the triangle composed of the endpoints of the CEVIANS though P. If the point P has TRILINEAR COORDINATES $\alpha:\beta:\gamma$, then the Cevian triangle has VERTICES $0:\beta:\gamma$, $\alpha:0:\gamma$, and $\alpha:\beta:0$. If $A'B'C'$ is the CEVIAN TRIANGLE of X and $A''B''C''$ is the ANTICEVIAN TRIANGLE, then X and A'' are HARMO-

NIC CONJUGATE POINTS with respect to A and A'.

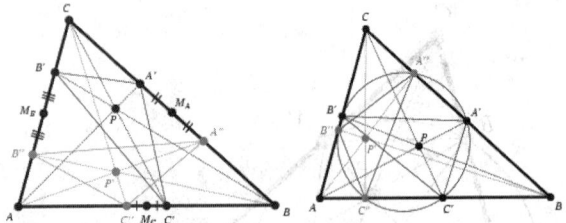

If $\triangle A'B'C'$ is the Cevian triangle of $\triangle ABC$, then the triangle $\triangle A''B''C''$ obtained by reflecting A', B', and C' across the midpoints of their sides is also a Cevian triangle of $\triangle ABC$ (Honsberger 1995, p. 141; left figure). Furthermore, if the CEVIAN CIRCLE crosses the sides of $\triangle ABC$ in three points A'', B'', and C'', then $\triangle A''B''C''$ is also a Cevian triangle of $\triangle ABC$ (Honsberger 1995, pp. 141–142; right figure).

See also ANTICEVIAN TRIANGLE, CEVIAN, CEVIAN CIRCLE

References

Honsberger, R. *Episodes in Nineteenth and Twentieth Century Euclidean Geometry.* Washington, DC: Math. Assoc. Amer., pp. 141–143, 1995.

CG

Given a GROUP G, the algebra $\mathbb{C}G$ is a VECTOR SPACE

$$\mathbb{C}G = \left\{ \sum a_i g_i \,|\, a_i \in \mathbb{C},\, g_i \in G \right\}$$

of finite sums of elements of G, with multiplication defined by $g \cdot h = gh$, the group operation. It is an example of a GROUP RING.

For example, when the group is the SYMMETRIC GROUP on three letters, S_3, the GROUP RING $\mathbb{C}S_3$ is a six-dimensional algebra. An example of the product of elements is

$$(3\{1, 3, 2\} + i\{1, 2, 3\})(-2\{2, 1, 3\} + \{3, 2, 1\})$$
$$= -6\{2, 3, 1\} - 2i\{2, 1, 3\} + i\{3, 2, 1\} + 3\{3, 1, 2\}.$$

MODULES over $\mathbb{C}G$ correspond to complex REPRESENTATIONS of G. When G is a FINITE GROUP then $\mathbb{C}G$ is a finite-dimensional algebra.

See also ALGEBRA, GROUP, GROUP RING, PERMUTATION, REPRESENTATION, RING

Ch

HYPERBOLIC COSINE

Chain

Let P be a finite PARTIALLY ORDERED SET. A chain in P is a set of pairwise comparable elements (i.e., a TOTALLY ORDERED subset). The LENGTH of P is the maximum CARDINALITY of a chain in P. For a PARTIAL

ORDER, the size of the longest chain is called the LENGTH.

See also ADDITION CHAIN, ANTICHAIN, BRAUER CHAIN, CHAIN (GRAPH), CHAIN OF CIRCLES, DILWORTH'S LEMMA, HANSEN CHAIN, LENGTH (PARTIAL ORDER), PAPPUS CHAIN, PARTIAL ORDER

References

Comtet, L. *Advanced Combinatorics: The Art of Finite and Infinite Expansions, rev. enl. ed.* Dordrecht, Netherlands: Reidel, p. 272, 1974.
Skiena, S. *Implementing Discrete Mathematics: Combinatorics and Graph Theory with Mathematica.* Reading, MA: Addison-Wesley, p. 241, 1990.

Chain (Graph)

A chain of a GRAPH is a SEQUENCE $\{x_1, x_2, \ldots, x_n\}$ such that (x_1, x_2), (x_2, x_3), ..., (x_{n-1}, x_n) are EDGES of the GRAPH.

See also GRAPH

Chain Complex

A chain complex is a sequence of maps

$$\cdots \overset{\partial_{i+1}}{\to} C_i \overset{\partial_i}{\to} C_{i-1} \overset{\partial_{i-1}}{\to} \cdots, \tag{1}$$

where the spaces C_i may be GROUPS or MODULES. The maps must satisfy $\partial_{i-1} \circ \partial_i = 0$. Making the domain implicitly understood, the maps are denoted by ∂, called the BOUNDARY OPERATOR or the differential. Chain complexes are an algebraic tool for computing or defining HOMOLOGY and have a variety of applications. A COCHAIN COMPLEX is used in the case of COHOMOLOGY.

Elements of C_p are called CHAINS. For each p, the kernel of $\partial_p : C_p \to C_{p-1}$ is called the group of cycles,

$$Z_p = \{c \in C_p : \partial(c) = 0\}. \tag{2}$$

The letter Z is short for the German word for cycle, "Zyklus." The image $\partial(C_{p+1})$ is contained in the group of cycles because $\partial \circ \partial = 0$. It is called the group of boundaries.

$$B_p = \{c \in C_p : \text{there exists } b \in C_{p+1} \text{ such that } \partial(b) = c\}. \tag{3}$$

The quotients $H_p = Z_p / B_p$ are the HOMOLOGY GROUPS of the chain.

For example, the sequence

$$\cdots \overset{\times 4}{\to} \mathbb{Z}/8\mathbb{Z} \overset{\times 4}{\to} \mathbb{Z}/8\mathbb{Z} \overset{\times 4}{\to} \cdots, \tag{4}$$

where every space is $\mathbb{Z}/8\mathbb{Z}$ and each map is given by multiplication by 4 is a chain complex. The cycles at each stage are $Z_p = \{0, 2, 4, 6\}$ and the boundaries are $B_p = \{0, 4\}$. So the homology at each stage is the group of two elements $\mathbb{Z}/2\mathbb{Z}$. A simpler example is given by a LINEAR TRANSFORMATION $\alpha : V \to W$, which can be extended to a chain complex by the zero vector space and the ZERO MAP. Then the nontrivial homology groups are ker α and $W/\text{im}(\alpha)$.

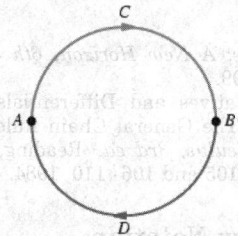

The terminology of chain complexes comes from the calculation for HOMOLOGY of geometric objects in a TOPOLOGICAL SPACE, like a MANIFOLD. For example, the figure above is the circle as a SIMPLICIAL COMPLEX. Let A and B denote the points, and C and D denote the oriented segments, which are the chains. The boundary of C is $B - A$, and the boundary of D is $A - B$.

The group C_1 is the FREE ABELIAN GROUP $\langle C, D \rangle$ and the group C_0 is the FREE ABELIAN GROUP $\langle A, B \rangle$. The BOUNDARY OPERATOR is

$$\begin{aligned} \partial(nC + mD) &= n(B-A) + m(A-B) \\ &= (m-n)A + (n-m)B. \end{aligned} \quad (5)$$

The other groups C_p are the TRIVIAL GROUP, and the other maps are the ZERO MAP. Then Z_1 is generated by $C + D$ and B_1 is the trivial subgroup. So H_1 is the rank one FREE ABELIAN GROUP isomorphic to $\mathbb{Z}$. The zero-dimensional case is slightly more interesting. Every element of C_0 has no boundary and so is in Z_0 while the boundaries B_0 are generated by $A - B$. Hence, $H_0 = Z_0/B_0$ is also isomorphic to $\mathbb{Z}$. Note that the result is not affected by how the circle is cut into pieces, or by how many cuts are used.

See also CHAIN EQUIVALENCE, CHAIN HOMOMORPHISM, CHAIN HOMOTOPY, COCHAIN COMPLEX, COHOMOLOGY, FREE ABELIAN GROUP, HOMOLOGY, HOMOLOGY (CHAIN), SIMPLICIAL HOMOLOGY

References

Hilton, P. and Stammbach, U. *A Course in Homological Algebra.* New York: Springer-Verlag, pp. 117–118, 1997.
Munkres, J. *Elements of Algebraic Topology.* Reading, MA: Addison-Wesley, pp. 58 and 71–76, 1984.

Chain Equivalence

Chain equivalences give an EQUIVALENCE RELATION on the space of CHAIN HOMOMORPHISMS. Two CHAIN COMPLEXES are chain equivalent if there are chain maps $\phi : C_* \to D_*$ and $\gamma : D_* \to C_*$ such that $\phi \circ \gamma$ is CHAIN HOMOTOPIC to the identity on D_* and $\gamma \circ \phi$ is CHAIN HOMOTOPIC to the identity on C_*.

See also CHAIN COMPLEX. CHAIN HOMOMORPHISM, CHAIN HOMOTOPY, HOMOTOPY EQUIVALENCE, SNAKE LEMMA

References

Hilton, P. and Stammbach, U. *A Course in Homological Algebra.* New York: Springer-Verlag, pp. 117–118, 1997.
Munkres, J. *Elements of Algebraic Topology.* Reading, MA: Addison-Wesley, pp. 58 and 71–76, 1984.

Chain Fraction

CONTINUED FRACTION

Chain Homomorphism

Also called a chain map. Given two CHAIN COMPLEXES C_* and D_*, a chain homomorphism is given by homomorphisms $\alpha_i : C_i \to D_i$ such that

$$\alpha \circ \partial_C = \partial_D \circ \alpha,$$

where ∂_C and ∂_D are the BOUNDARY OPERATORS.

See also CHAIN COMPLEX, CHAIN EQUIVALENCE, CHAIN HOMOTOPY, HOMOMORPHISM (MODULE)

References

Hilton, P. and Stammbach, U. *A Course in Homological Algebra.* New York: Springer-Verlag, pp. 117–118, 1997.
Munkres, J. *Elements of Algebraic Topology.* Addison-Wesley, pp. 58 and 71–76, 1984.

Chain Homotopy

Suppose $\alpha : C_* \to D_*$ and $\beta : C_* \to D_*$ are two CHAIN HOMOMORPHISMS. Then a chain homotopy is given by a sequence of maps

$$\delta_p : C_p \to D_{p+1}$$

such that

$$\partial_D \circ \delta + \delta \circ \partial_C = \alpha - \beta,$$

where ∂ denotes the BOUNDARY OPERATOR.

See also CHAIN COMPLEX, CHAIN EQUIVALENCE, CHAIN HOMOMORPHISM, HOMOTOPY, SNAKE LEMMA

References

Hilton, P. and Stammbach, U. *A Course in Homological Algebra.* New York: Springer-Verlag, p. 124, 1997.
Munkres, J. *Elements of Algebraic Topology.* Reading, MA: Addison-Wesley, pp. 58 and 71–76, 1984.

Chain Map

CHAIN HOMOMORPHISM

Chain of Circles

A sequence of circles which closes (such as a STEINER CHAIN or the circles inscribed in the ARBELOS) is called a chain.

See also ARBELOS, COXETER'S LOXODROMIC SEQUENCE OF TANGENT CIRCLES, NINE CIRCLES THEOREM, PAPPUS CHAIN, SEVEN CIRCLES THEOREM, SIX CIRCLES THEOREM, STEINER CHAIN, STEINER'S PORISM

References

Evelyn, C. J. A.; Money-Coutts, G. B.; and Tyrrell, J. A. "Chains of Circles." Ch. 3 in *The Seven Circles Theorem and Other New Theorems*. London: Stacey International, pp. 31–68, 1974.

Chain Rule

If $g(x)$ is DIFFERENTIABLE at the point x and $f(x)$ is DIFFERENTIABLE at the point $g(x)$, then $f \circ g$ is DIFFERENTIABLE at x. Furthermore, let $y = f(g(x))$ and $u = g(x)$, then

$$\frac{dy}{dx} = \frac{dy}{du} \cdot \frac{du}{dx}. \tag{1}$$

There are a number of related results which also go under the name of "chain rules." For example, if $z = f(x, y)$, $x = g(t)$, and $y = h(t)$, then

$$\frac{dz}{dt} = \frac{\partial z}{\partial x}\frac{dx}{dt} + \frac{\partial z}{\partial y}\frac{dy}{dt}. \tag{2}$$

The "general" chain rule applies to two sets of functions

$$y_1 = f_1(u_1, \ldots, u_p)$$
$$\vdots \tag{3}$$
$$y_m = f_m(u_1, \ldots, u_p)$$

and

$$u_1 = g_1(x_1, \ldots, x_n)$$
$$\vdots \tag{4}$$
$$u_p = g_p(x_1, \ldots, x_n).$$

Defining the $m \times n$ JACOBI MATRIX by

$$\left(\frac{\partial y_i}{\partial x_j}\right) = \begin{bmatrix} \dfrac{\partial y_1}{\partial x_1} & \dfrac{\partial y_1}{\partial x_2} & \cdots & \dfrac{\partial y_1}{\partial x_n} \\ \vdots & \vdots & \ddots & \vdots \\ \dfrac{\partial y_m}{\partial x_1} & \dfrac{\partial y_m}{\partial x_2} & \cdots & \dfrac{\partial y_m}{\partial x_n} \end{bmatrix}, \tag{5}$$

and similarly for $(\partial y_i / \partial u_j)$ and $(\partial u_i / \partial x_j)$ then gives

$$\left(\frac{\partial y_i}{\partial x_j}\right) = \left(\frac{\partial y_i}{\partial u_j}\right)\left(\frac{\partial u_i}{\partial x_j}\right). \tag{6}$$

In differential form, this becomes

$$dy_1 = \left(\frac{\partial y_1}{\partial u_1}\frac{\partial u_1}{\partial x_1} + \ldots + \frac{\partial y_1}{\partial u_p}\frac{\partial u_p}{\partial x_1}\right) dx_1$$
$$+ \left(\frac{\partial y_1}{\partial u_1}\frac{\partial u_1}{\partial x_2} + \ldots + \frac{\partial y_1}{\partial u_p}\frac{\partial u_p}{\partial x_2}\right) dx_2 + \ldots \tag{7}$$

(Kaplan 1984).

See also DERIVATIVE, JACOBIAN, POWER RULE, PRODUCT RULE

References

Anton, H. *Calculus: A New Horizon, 6th ed.* New York: Wiley, p. 165, 1999.
Kaplan, W. "Derivatives and Differentials of Composite Functions" and "The General Chain Rule." §2.8 and 2.9 in *Advanced Calculus, 3rd ed.* Reading, MA: Addison-Wesley, pp. 101–105 and 106–110, 1984.

Chained Arrow Notation

A NOTATION which generalizes ARROW NOTATION and is defined as

$$\underbrace{a \uparrow \cdots \uparrow b}_{c} \equiv a \to b \to c.$$

See also ARROW NOTATION

References

Conway, J. H. and Guy, R. K. *The Book of Numbers*. New York: Springer-Verlag, p. 61, 1996.

Chainette

CATENARY

Chair

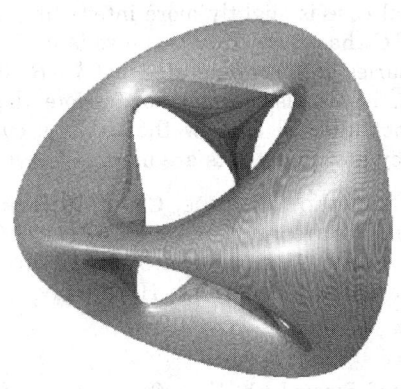

A SURFACE with tetrahedral symmetry which, according to Nordstrand, looks like an inflatable chair from the 1970s. It is given by the implicit equation

$$(x^2 + y^2 + z^2 - ak^2)^2 - b[(z-k)^2 - 2x^2][(z+k)^2 - 2y^2]$$
$$= 0.$$

The surface illustrated above has $k = 5$, $a = 0.95$, and $b = 0.8$.

See also BRIDE'S CHAIR

References

Nordstrand, T. "Chair." http://www.uib.no/people/nfytn/chairtxt.htm.

Chaitin's Constant

An IRRATIONAL NUMBER Ω which gives the probability that for any set of instructions, a UNIVERSAL TURING MACHINE will halt. The digits in Ω are random and cannot be computed ahead of time.

See also HALTING PROBLEM, TURING MACHINE, UNIVERSAL TURING MACHINE

References
Finch, S. "Favorite Mathematical Constants." http://www.mathsoft.com/asolve/constant/chaitin/chaitin.html.
Gardner, M. "The Random Number Ω Bids Fair to Hold the Mysteries of the Universe." *Sci. Amer.* **241**, 20–34, Nov. 1979.
Gardner, M. "Chaitin's Omega." Ch. 21 in *Fractal Music, Hypercards, and More Mathematical Recreations from Scientific American Magazine.* New York: W. H. Freeman, pp. 307–319, 1992.
Kobayashi, K. "Sigma(N)O-Complete Properties of Programs and Lartin-Lof Randomness." *Information Proc. Let.* **46**, 37–42, 1993.

Chaitin's Number

CHAITIN'S CONSTANT

Chaitin's Omega

CHAITIN'S CONSTANT

Champernowne Constant

Champernowne's constant 0.1234567891011... (Sloane's A033307) is the number obtained by concatenating the POSITIVE INTEGERS and interpreting them as decimal digits to the right of a decimal point. It is NORMAL in base 10. In 1961, Mahler showed it to also be TRANSCENDENTAL.

The first few terms in the CONTINUED FRACTION of the Champernowne constant are 0, 8, 9, 1, 149083, 1, 1, 1, 4, 1, 1, 1, 3, 4, 1, 1, 1, 15,

45754011139103107648364662824295611859960393 9...

71045755500066200439309026265925631493795320 7...

74712865631386412093755035520946071830899845 7...

580146986314883359241178301098 7,

6, 1, 1, 21, 1, 9, 1, 1, 2, 3, 1, 7, 2, 1, 83, 1, 156, 4, 58, 8, 54, ... (Sloane's A030167). The next term of the CONTINUED FRACTION is huge, having 2504 digits.

In fact, the coefficients eventually become unbounded, making the continued fraction difficult to calculate for too many more terms. Large terms greater than 10^5 occur at positions 5, 19, 41, 102, 163, 247, 358, 460, ... and have 6, 166, 2504, 140, 33102, 109, 2468, 136, ... digits, respectively (Plouffe). The 527th partial quotient of the continued fraction expansion has 411,100 decimal digits and the 1709th partial quotient has 4,911,098 decimal digits, as computed using *Mathematica* 4.0. This result was obtained by Mark Sofroniou and Giulia Spaletta and

presented at the conference on Foundations of Computational Mathematics in Oxford, UK, July 1999.

Interestingly, the COPELAND-ERDOS CONSTANT, which is the decimal number obtained by concatenating the PRIMES (instead of all the positive integers), has a well-behaved CONTINUED FRACTION that does not show the "large term" phenomenon.

See also COPELAND-ERDOS CONSTANT, SMARANDACHE SEQUENCES

References
Champernowne, D. G. "The Construction of Decimals Normal in the Scale of Ten." *J. London Math. Soc.* **8**, 1933.
Copeland, A. H. and Erdos, P. "Note on Normal Numbers." *Bull. Amer. Math. Soc.* **52**, 857–860, 1946.
Finch, S. "Favorite Mathematical Constants." http://www.mathsoft.com/asolve/constant/cntfrc/cntfrc.html.
Sloane, N. J. A. Sequences A030167 and A033307 in "An On-Line Version of the Encyclopedia of Integer Sequences." http://www.research.att.com/~njas/sequences/eisonline.html.
Wells, D. *The Penguin Dictionary of Curious and Interesting Numbers.* Middlesex, England: Penguin Books, p. 26, 1986.

Change of Variables Theorem

A theorem which effectively describes how lengths, areas, volumes, and generalized n-dimensional volumes (CONTENTS) are distorted by DIFFERENTIABLE FUNCTIONS. In particular, the change of variables theorem reduces the whole problem of figuring out the distortion of the content to understanding the infinitesimal distortion, i.e., the distortion of the DERIVATIVE (a linear MAP), which is given by the linear MAP's DETERMINANT. So $f : \mathbb{R}^n \to \mathbb{R}^n$ is an AREA-PRESERVING linear MAP IFF $|\det(f)|=1$, and in more generality, if S is any subset of $\mathbb{R}^n$, the CONTENT of its image is given by $|\det(f)|$ times the CONTENT of the original. The change of variables theorem takes this infinitesimal knowledge, and applies CALCULUS by breaking up the DOMAIN into small pieces and adds up the change in AREA, bit by bit.

The change of variable formula persists to the generality of DIFFERENTIAL FORMS on MANIFOLDS, giving the formula

$$\int_M (f^*\omega) = \int_W (\omega) \tag{1}$$

under the conditions that M and W are compact connected oriented MANIFOLDS with nonempty boundaries, $f : M \to W$ is a smooth map which is an orientation-preserving DIFFEOMORPHISM of the boundaries.

In 1-D, the explicit statement of the theorem for f a continuous function of y is

$$\int_s f(\phi(x)) \frac{d\phi}{dx} \, dx = \int_T f(y) \, dy, \tag{2}$$

where $y = \phi(x)$ is a differential mapping on the interval $[c, d]$ and T is the interval $[a, b]$ with $\phi(c) = a$ and $\phi(d) = b$ (Lax 1999). In 2-D, the explicit statement of the theorem is

$$\int_R f(x, y)\, dx\, dy$$
$$= \int_{R^*} f[x(u, v), y(u, v)] \left| \frac{\partial(x, y)}{\partial(u, v)} \right| du\, dv$$

and in 3-D, it is

$$\int_R f(x, y, z)\, dx\, dy\, dz$$
$$= \int_{R^*} f[x(u, v, w), y(u, v, w), z(u, v, w)]$$
$$\times \left| \frac{\partial(x, y, z)}{\partial(u, v, w)} \right| du\, dv\, dw,$$

$$\tag{3}$$

where $R = f(R^*)$ is the image of the original region R^*,

$$\left| \frac{\partial(x, y, z)}{\partial(u, v, w)} \right| \tag{4}$$

is the JACOBIAN, and f is a global orientation-preserving DIFFEOMORPHISM of R and R^* (which are open subsets of $\mathbb{R}^n$).

The change of variables theorem is a simple consequence of the CURL THEOREM and a little DE RHAM COHOMOLOGY. The generalization to n-D requires no additional assumptions other than the regularity conditions on the boundary.

See also IMPLICIT FUNCTION THEOREM, JACOBIAN

References

Jeffreys, H. and Jeffreys, B. S. "Change of Variable in an Integral." §1.1032 in *Methods of Mathematical Physics, 3rd ed.* Cambridge, England: Cambridge University Press, pp. 32–33, 1988.
Kaplan, W. "Change of Variables in Integrals." §4.6 in *Advanced Calculus, 3rd ed.* Reading, MA: Addison-Wesley, pp. 238–245, 1984.
Lax, P. D. "Change of Variables in Multiple Integrals." *Amer. Math. Monthly* **106**, 497–501, 1999.

Chaos

A DYNAMICAL SYSTEM is chaotic if it

1. Has a DENSE collection of points with periodic orbits,
2. Is sensitive to the initial condition of the system (so that initially nearby points can evolve quickly into very different states), and
3. Is TOPOLOGICALLY TRANSITIVE.

Chaotic systems exhibit irregular, unpredictable behavior (the BUTTERFLY EFFECT). The boundary between linear and chaotic behavior is often characterized by PERIOD DOUBLING, followed by quadrupling, etc., although other routes to chaos are also possible (Abarbanel *et al.* 1993; Hilborn 1994; Strogatz 1994, pp. 363–365).

An example of a simple physical system which displays chaotic behavior is the motion of a magnetic pendulum over a plane containing two or more attractive magnets. The magnet over which the pendulum ultimately comes to rest (due to frictional damping) is highly dependent on the starting position and velocity of the pendulum (Dickau). Another such system is a double pendulum (a pendulum with another pendulum attached to its end).

See also ACCUMULATION POINT, ATTRACTOR, BASIN OF ATTRACTION, BUTTERFLY EFFECT, CHAOS GAME, DYNAMICAL SYSTEM, FEIGENBAUM CONSTANT, FRACTAL DIMENSION, GINGERBREADMAN MAP, HÉNON-HEILES EQUATION, HÉNON MAP, LIMIT CYCLE, LOGISTIC EQUATION, LYAPUNOV CHARACTERISTIC EXPONENT, PERIOD THREE THEOREM, PHASE SPACE, QUANTUM CHAOS, RESONANCE OVERLAP METHOD, SARKOVSKII'S THEOREM, SHADOWING THEOREM, SINK (MAP), STRANGE ATTRACTOR

References

Abarbanel, H. D. I.; Rabinovich, M. I.; and Sushchik, M. M. *Introduction to Nonlinear Dynamics for Physicists.* Singapore: World Scientific, 1993.
Bai-Lin, H. *Chaos.* Singapore: World Scientific, 1984.
Baker, G. L. and Gollub, J. B. *Chaotic Dynamics: An Introduction, 2nd ed.* Cambridge, England: Cambridge University Press, 1996.
Smith, P. *Explaining Chaos.* Cambridge, England: Cambridge University Press, 1998.
Cvitanovic, P. *Universality in Chaos: A Reprint Selection, 2nd ed.* Bristol: Adam Hilger, 1989.
Devaney, R. L. *An Introduction to Chaotic Dynamical Systems.* Redwood City, CA: Addison-Wesley, 1987.
Dickau, R. M. "Magnetic Pendulum." http://forum.swarthmore.edu/advanced/robertd/magneticpendulum.html.
Drazin, P. G. *Nonlinear Systems.* Cambridge, England: Cambridge University Press, 1992.
Field, M. and Golubitsky, M. *Symmetry in Chaos: A Search for Pattern in Mathematics, Art and Nature.* Oxford, England: Oxford University Press, 1992.
Gleick, J. *Chaos: Making a New Science.* New York: Penguin, 1988.
Guckenheimer, J. and Holmes, P. *Nonlinear Oscillations, Dynamical Systems, and Bifurcations of Vector Fields, 3rd ed.* New York: Springer-Verlag, 1997.
Hall, N. (Ed.). *Exploring Chaos: A Guide to the New Science of Disorder.* New York: W. W. Norton, 1994.

Hilborn, R. C. *Chaos and Nonlinear Dynamics.* New York: Oxford University Press, 1994.

Kapitaniak, T. and Bishop, S. R. *The Illustrated Dictionary of Nonlinear Dynamics and Chaos.* New York: Wiley, 1998.

Lichtenberg, A. and Lieberman, M. *Regular and Stochastic Motion, 2nd ed.* New York: Springer-Verlag, 1994.

Lorenz, E. N. *The Essence of Chaos.* Seattle, WA: University of Washington Press, 1996.

Ott, E. *Chaos in Dynamical Systems.* New York: Cambridge University Press, 1993.

Ott, E.; Sauer, T.; and Yorke, J. A. *Coping with Chaos: Analysis of Chaotic Data and the Exploitation of Chaotic Systems.* New York: Wiley, 1994.

Peitgen, H.-O.; Jürgens, H.; and Saupe, D. *Chaos and Fractals: New Frontiers of Science.* New York: Springer-Verlag, 1992.

Poon, L. "Chaos at Maryland." http://www-chaos.umd.edu.

Rasband, S. N. *Chaotic Dynamics of Nonlinear Systems.* New York: Wiley, 1990.

Strogatz, S. H. *Nonlinear Dynamics and Chaos, with Applications to Physics, Biology, Chemistry, and Engineering.* Reading, MA: Addison-Wesley, 1994.

Tabor, M. *Chaos and Integrability in Nonlinear Dynamics: An Introduction.* New York: Wiley, 1989.

Tufillaro, N.; Abbott, T. R.; and Reilly, J. *An Experimental Approach to Nonlinear Dynamics and Chaos.* Redwood City, CA: Addison-Wesley, 1992.

Wiggins, S. *Global Bifurcations and Chaos: Analytical Methods.* New York: Springer-Verlag, 1988.

Wiggins, S. *Introduction to Applied Nonlinear Dynamical Systems and Chaos.* New York: Springer-Verlag, 1990.

Chaos Game

Pick a point at random inside a regular n-gon. Then draw the next point a fraction r of the distance between it and a VERTEX picked at random. Continue the process (after throwing out the first few points). The result of this "chaos game" is sometimes, but not always, a FRACTAL. The case $(n, r) = (4, 1/2)$ gives the interior of a SQUARE with all points visited with equal probability.

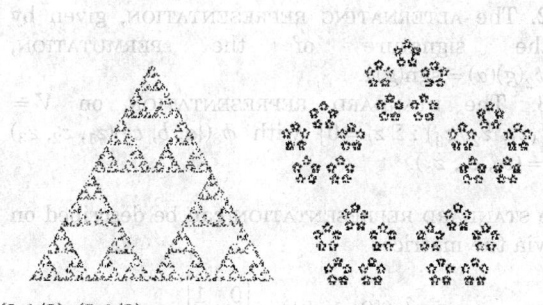

(3, 1/2) (5, 1/3)

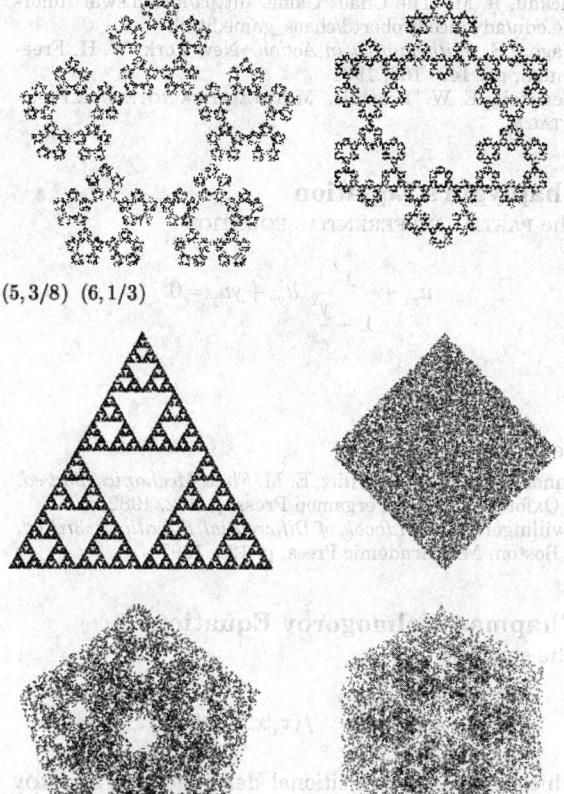

(5, 3/8) (6, 1/3)

The above plots show the chaos game for 10,000 points in the regular 3-, 4-, 5-, and 6-gons with $r = 1/2$.

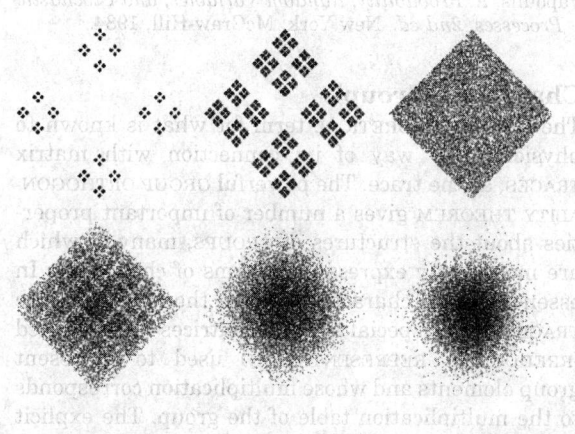

The above plots show the chaos game for 10,000 points in the square with $r = 0.25$, 0.4, 0.5, 0.6, 0.75, and 0.9.

See also BARNSLEY'S FERN

References

Barnsley, M. F. and Rising, H. *Fractals Everywhere, 2nd ed.* Boston, MA: Academic Press, 1993.

Dickau, R. M. "The Chaos Game." http://forum.swarthmore.edu/advanced/robertd/chaos_game.html.

Wagon, S. *Mathematica in Action.* New York: W. H. Freeman, pp. 149–163, 1991.

Weisstein, E. W. "Fractals." Mathematica notebook Fractal.M.

Chaplygin's Equation

The PARTIAL DIFFERENTIAL EQUATION

$$u_{xx} + \frac{y^2}{1 - \frac{y^2}{c^2}} u_{yy} + y u_y = 0.$$

References

Landau, L. D. and Lifschitz, E. M. *Fluid Mechanics, 2nd ed.* Oxford, England: Pergamon Press, p. 432, 1982.

Zwillinger, D. *Handbook of Differential Equations, 3rd ed.* Boston, MA: Academic Press, p. 129, 1997.

Chapman-Kolmogorov Equation

The equation

$$f(x_n | x_s) = \int_{-\infty}^{\infty} f(x_n | x_r) f(x_r | x_s) \, dx_r$$

which gives the transitional densities of a MARKOV SEQUENCE. Here, $n > r > s$ are any integers (Papoulis 1984, p. 531).

See also MARKOV PROCESS

References

Papoulis, A. *Probability, Random Variables, and Stochastic Processes, 2nd ed.* New York: McGraw-Hill, 1984.

Character (Group)

The GROUP THEORETICAL term for what is known to physicists, by way of its connection with matrix TRACES, as the trace. The powerful GROUP ORTHOGONALITY THEOREM gives a number of important properties about the structures of GROUPS, many of which are most easily expressed in terms of characters. In essence, group characters can be thought of as the TRACES of a special set of matrices (a so-called IRREDUCIBLE REPRESENTATION) used to represent group elements and whose multiplication corresponds to the multiplication table of the group. The explicit construction of a set of characters (CHARACTER TABLE) is illustrated for the FINITE GROUP $D3$.

All members of the same CONJUGACY CLASS in the same representation have the same character. Members of other CONJUGACY CLASSES may also have the same character, however. An (abstract) GROUP can be uniquely identified by a listing of the characters of its various representations, known as a CHARACTER TABLE. Some of the SCHÖNFLIES symbols denote different sets of symmetry operations but correspond to the same abstract GROUP and so have the same CHARACTER TABLES.

See also CHARACTER TABLE, CONJUGACY CLASS, GROUP ORTHOGONALITY THEOREM, TRACE (MATRIX)

Character (Number Theory)

A number theoretic function $\chi_k(n)$ for POSITIVE integral n is a character modulo k if

$$\chi_k(1) = 1$$

$$\chi_k(n) = \chi_k(n + k)$$

$$\chi_k(m)\chi_k(n) = \chi_k(mn)$$

for all m, n, and

$$\chi_k(n) = 0$$

if $(k, n) \neq 1$. χ_k can only assume values which are $\phi(k)$ ROOTS OF UNITY, where ϕ is the TOTIENT FUNCTION.

See also DIRICHLET L-SERIES, MULTIPLICATIVE CHARACTER, PRIMITIVE CHARACTER

Character Table

A FINITE GROUP G has a finite number of CONJUGACY CLASSES and a finite number of distinct IRREDUCIBLE REPRESENTATIONS. The CHARACTER of a REPRESENTATION is constant on a CONJUGACY CLASS. Hence, the values of the characters can be written as an array, known as a character table. Typically, the rows are given by the IRREDUCIBLE REPRESENTATIONS and the columns are given the CONJUGACY CLASSES. A character table contains enough information to uniquely identify a given abstract group and distinguish it from others.

For example, the SYMMETRIC GROUP on three letters S_3 has three CONJUGACY CLASSES, represented by the PERMUTATIONS $\{1, 2, 3\}$, $\{2, 1, 3\}$, and $\{2, 3, 1\}$. It also has three IRREDUCIBLE REPRESENTATIONS; two are one-dimensional and the third is two-dimensional:

1. The TRIVIAL REPRESENTATION $\phi_1(g)(\alpha) = \alpha$.
2. The ALTERNATING REPRESENTATION, given by the signature of the PERMUTATION, $\phi_2(g)(\alpha) = \text{sgn}(g)\alpha$.
3. The STANDARD REPRESENTATION on $V = \{(z_1, z_2, z_3) : \Sigma z_i = 0\}$ with $\phi_3(\{a, b, c\})(z_1, z_2, z_3) = (z_a, z_b, z_c)$.

The STANDARD REPRESENTATION can be described on $\mathbb{C}^2$ via the matrices

$$\tilde{\phi}_3(\{2, 1, 3\}) = \begin{bmatrix} 0 & 1 \\ 1 & 0 \end{bmatrix}$$

$$\tilde{\phi}_3(\{2, 3, 1\}) = \begin{bmatrix} 0 & -1 \\ 1 & -1 \end{bmatrix},$$

and hence the CHARACTER of the first matrix is 0 and that of the second is -1. The CHARACTER of the identity is always the dimension of the VECTOR SPACE. The trace of the alternating representation is just the SIGNATURE of the PERMUTATION. Consequently, the character table for S_3 is shown below.

S_3	1 e	2 (12)	3 (123)
trivial	1	1	1
alternating	1	-1	1
standard	2	0	-1

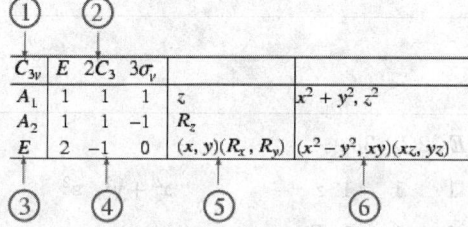

C_{3v}	E	$2C_3$	$3\sigma_v$		
A_1	1	1	1	z	x^2+y^2, z^2
A_2	1	1	-1	R_z	
E	2	-1	0	$(x,y)(R_x, R_y)$	$(x^2-y^2, xy)(xz, yz)$

Chemists and physicists use a special convention for representing character tables which is applied especially to the so-called POINT GROUPS, which are the 32 finite symmetry groups possible in a lattice. In the example above, the numbered regions contain the following contents (Cotton 1990 pp. 90–92).

1. The symbol used to represent the group in question (in this case C_{3v}).
2. The CONJUGACY CLASSES, indicated by number and symbol, where the sum of the coefficients gives the ORDER of the group.
3. MULLIKEN SYMBOLS, one for each IRREDUCIBLE REPRESENTATION.
4. An array of the CHARACTERS of the IRREDUCIBLE REPRESENTATION of the group, with one column for each CONJUGACY CLASS, and one row for each IRREDUCIBLE REPRESENTATION.
5. Combinations of the symbols x, y, z, R_x, R_y, and R_z, the first three of which represent the coordinates $x, y,$ and z, and the last three of which stand for rotations about these axes. These are related to transformation properties and basis representations of the group.
6. All square and binary products of coordinates according to their transformation properties.

The character tables for many of the POINT GROUPS are reproduced below using this notation.

C_1	E
A	1

C_s	E	σ_h		
A	1	1	x, y, R_z	x^2, y^2, z^2, xy
B	1	-1	z, R_x, R_y	yz, xz

C_i	E	i		
A_g	1	1	R_x, R_y, R_z	$x^2, y^2, z^2, xy, xz, yz$
A_u	1	-1	x, y, z	

C_2	E	C_2		
A	1	1	z, R_z	x^2, y^2, z^2, xy
B	1	-1	x, y, R_x, R_z	yz, xz

C_3	E	C_3	$C_3{}^2$			$\varepsilon=\exp(2\pi i/3)$
A	1	1	1	z, R_z		x^2, y^2, z^2, xy
E	$\left\{ \begin{smallmatrix}1\\1\end{smallmatrix}\right.$	ε^*	$\varepsilon \}$	$(x, y)(R_x, R_y)$		$(x^2-y^2, xy)(yz, xz)$

C_4	E	C_3	C_2	$C_4{}^3$		
A	1	1	1	1	z, R_z	x^2+y^2, z^2
B	1	-1	1	-1		x^2-y^2, xy
E	$\left\{ \begin{smallmatrix}1\\1\end{smallmatrix}\right.$	$-i$	1	$i \}$	$(x, y)(R_x, R_y)$	(yz, xz)

C_5	E	C_5	$C_5{}^2$	$C_5{}^3$	$C_5{}^4$		$\varepsilon=\exp(2\pi i/5)$
A	1	1	1	1	1	z, R_z	x^2+y^2, z^2
E_1	$\left\{ \begin{smallmatrix}1\\1\end{smallmatrix}\right.$ ε^*	ε^{2*}	ε^2	$\varepsilon \}$		$(x, y)(R_x, R_y)$	(yz, xz)
E_2	$\left\{ \begin{smallmatrix}1\\1\end{smallmatrix}\right.$ ε^{2*}	ε	ε^*	$\varepsilon^2 \}$			(x^2-y^2, xy)

C_6	E	C_6	C_3	C_2	$C_3{}^2$	$C_6{}^5$		$\varepsilon=\exp(2\pi i/6)$
A	1	1	1	1	1	1	z, R_z	x^2+y^2, z^2

B	1	-1	1	-1	1	-1		
E_1	$\begin{cases} 1 \\ 1 \end{cases}$	$\begin{matrix} \varepsilon^* \\ \varepsilon \end{matrix}$	$\begin{matrix} -\varepsilon \\ -\varepsilon^* \end{matrix}$	$\begin{matrix} -1 \\ -1 \end{matrix}$	$\begin{matrix} -\varepsilon^* \\ -\varepsilon \end{matrix}$	$\begin{matrix} \varepsilon \\ \varepsilon^* \end{matrix}$	(R_x, R_y)	(yz, xz)
E_2	$\begin{cases} 1 \\ 1 \end{cases}$	$\begin{matrix} -\varepsilon^* \\ -\varepsilon \end{matrix}$	$\begin{matrix} -\varepsilon^* \\ -\varepsilon \end{matrix}$	$\begin{matrix} 1 \\ 1 \end{matrix}$	$\begin{matrix} -\varepsilon \\ -\varepsilon^* \end{matrix}$	$\begin{matrix} \varepsilon^* \\ \varepsilon^* \end{matrix}$		(x^2-y^2, xy)

B_1	1	-1	1	-1	1	-1	
B_2	1	-1	1	1	-1	1	$(x, y)(R_x, R_y)$
E_1	2	1	-1	-2	0	0	(xz, yz)
E_2	2	-1	-1	2	0	0	(x^2-y^2, xy)

D_2	E	$C_2(z)$	$C_2(y)$	$C_2(x)$		
A_1	1	1	1	1		$x^2+y^2,\ z^2$
B_1	1	1	-1	-1	z, R_z	xy
B_2	1	-1	1	-1	y, R_y	xz
B_3	1	-1	-1	1	x, R_x	yz

C_{2v}	E	C_2	$\sigma_v(xz)$	$\sigma_v'(yz)$		
A_1	1	1	1	1	z	$x^2,\ y^2,\ z^2$
A_2	1	1	-1	-1	R_z	xy
B_1	1	-1	1	-1	x, R_y	xz
B_2	1	-1	-1	1	y, R_x	yz

D_3	E	$2C_3$	$3C_2$		
A_1	1	1	1		$x^2+y^2,\ z^2$
A_2	1	1	-1	z, R_z	xy
E	2	-1	0	$(x, y)(R_x, R_y)$	$(x^2-y^2, xy)(xz, yz)$

C_{3v}	E	$2C_3$	$3\sigma_v$		
A_1	1	1	1	z	$x^2+y^2,\ z^2$
A_2	1	1	-1	R_z	
E	2	-1	0	$(x, y)(R_x, R_y)$	$(x^2-y^2, xy)(xz, yz)$

D_4	E	$2C_4$	C_2	$2C_2'$	$2C_2''$		
A_1	1	1	1	1	1		$x^2+y^2,\ z^2$
A_2	1	1	1	-1	-1	z, R_z	
B_1	1	-1	1	1	-1		x^2-y^2
B_2	1	-1	1	-1	1		xy
E	2	0	-2	0	0	$(x, y)(R_x, R_y)$	(xz, yz)

C_{4v}	E	$2C_4$	C_2	$2\sigma_v$	$2\sigma_d$		
A_1	1	1	1	1	1	z	$x^2+y^2,\ z^2$
A_2	1	1	1	-1	-1	R_z	
B_1	1	-1	1	1	-1		x^2-y^2
B_2	1	-1	1	-1	1		xy
E	2	0	-2	0	0	$(x, y)(R_x, R_y)$	(xz, yz)

D_5	E	$2C_5$	$2C_5^2$	$5C_2$		
A_1	1	1	1	1		$x^2+y^2,\ z^2$
B_1	1	1	1	-1	z, R_z	
B_2	2	$2\cos 72°$	$2\cos 144°$	0	$(x, y)(R_x, R_y)$	(xz, yz)
B_3	2	$2\cos 144°$	$2\cos 72°$	0		(x^2-y^2, xy)

C_{5v}	E	$2C_5$	$2C_5^2$	$5\sigma_v$		
A_1	1	1	1	1	z	$x^2+y^2,\ z^2$
B_1	1	1	1	-1	R_z	
B_2	2	$2\cos 72°$	$2\cos 144°$	0	$(x, y)(R_x, R_y)$	(xz, yz)
B_3	2	$2\cos 144°$	$2\cos 72°$	0		(x^2-y^2, xy)

D_6	E	$2C_6$	$2C_3$	C_2	$3C_2'$	$3C_2''$		
A_1	1	1	1	1	1	1		$x^2+y^2,\ z^2$
A_2	1	1	1	1	-1	-1	z, R_z	

C_{6v}	E	$2C_6$	$2C_3$	C_2	$3\sigma_v$	$3\sigma_d$		
A_1	1	1	1	1	1	1	z	$x^2+y^2,\ z^2$
A_2	1	1	1	1	-1	-1	R_z	

B_1	1	-1	1	-1	1	-1		
B_2	1	-1	1	-1	-1	1		
E_1	2	1	-1	-2	0	0	$(x,y)(R_x,R_y)$	(xz,yz)
E_2	2	-1	-1	2	0	0		(x^2-y^2,xy)

$C_{\infty v}$	E	$C_\infty{}^\Phi$	...	$\infty\sigma_v$		
$A_1\equiv\Sigma^+$	1	1	...	1	z	x^2+y^2, z^2
$A_2\equiv\Sigma^-$	1	1	...	-1	R_z	
$E_1\equiv\Pi$	2	$2\cos\Phi$	...	0	$(x,y); (R_x,R_y)$	(xz,yz)
$E_2\equiv\Delta$	2	$2\cos 2\Phi$	...	0		(x^2-y^2,xy)
$E_3\equiv\Phi$	2	$2\cos 3\Phi$	...	0		
$\vdots$	$\vdots$	$\vdots$	$\ddots$	$\vdots$		

See also CHARACTER (GROUP), CONJUGACY CLASS, GROUP, IRREDUCIBLE REPRESENTATION, POINT GROUPS, REPRESENTATION

References

Bishop, D. M. "Character Tables." Appendix 1 in *Group Theory and Chemistry.* New York: Dover, pp. 279–288, 1993.

Cotton, F. A. "Character Tables." §4.4 in *Chemical Applications of Group Theory, 3rd ed.* New York: Wiley, pp. 90–95, 1990.

Huang, J.-S. "Characters of Representations." §2.2 in *Lectures on Representation Theory.* Singapore: World Scientific, pp. 9–11, 1999.

Iyanaga, S. and Kawada, Y. (Eds.). "Characters of Finite Groups." Appendix B, Table 5 in *Encyclopedic Dictionary of Mathematics.* Cambridge, MA: MIT Press, pp. 1496–1503, 1980.

Sosnovsky, A. and Demarco, G. L. "Character Tables of Finite Groups." *Mathematica Educ. Res.* **6**, 5–8, 1997.

Characteristic (Elliptic Integral)

A parameter n used to specify an ELLIPTIC INTEGRAL OF THE THIRD KIND $\Pi(n;\phi,k)$.

See also AMPLITUDE, ELLIPTIC INTEGRAL, MODULAR ANGLE, MODULUS (ELLIPTIC INTEGRAL), NOME, PARAMETER

References

Abramowitz, M. and Stegun, C. A. (Eds.). *Handbook of Mathematical Functions with Formulas, Graphs, and Mathematical Tables, 9th printing.* New York: Dover, p. 590, 1972.

Characteristic (Euler)

EULER CHARACTERISTIC

Characteristic (Field)

For a FIELD K with multiplicative identity 1, consider the numbers $2 = 1+1$, $3 = 1+1+1$, $4 = 1+1+1+1$, etc. Either these numbers are all different, in which case we say that K has characteristic 0, or two of them will be equal. In the latter case, it is straightforward to show that, for some number p, we have

$$\underbrace{1+1+\ldots+1}_{p \text{ times}} = 0.$$

If p is chosen to be as small as possible, then p will be a PRIME, and we say that K has characteristic p. The characteristic of a field K is sometimes denoted $\mathrm{ch}(K)$.

The FIELDS $\mathbb{Q}$ (rationals), $\mathbb{R}$ (reals), $\mathbb{C}$ (complex numbers), and the P-ADIC NUMBERS $\mathbb{Q}_p$ have characteristic 0. For p a PRIME, the FINITE FIELD $\mathrm{GF}(p^n)$ has characteristic p.

If H is a SUBFIELD of K, then H and K have the same characteristic.

See also FIELD, FINITE FIELD, SUBFIELD

References

Dummit, D. S. and Foote, R. M. *Abstract Algebra, 2nd ed.* Englewood Cliffs, NJ: Prentice-Hall, p. 422, 1998.

Characteristic (Partial Differential Equation)

Paths in a 2-D plane used to transform PARTIAL DIFFERENTIAL EQUATIONS into systems of ORDINARY DIFFERENTIAL EQUATIONS. They were invented by Riemann. For an example of the use of characteristics, consider the equation

$$u_1 - 6uu_x = 0.$$

Now let $u(s) = u(x(s), t(s))$. Since

$$\frac{du}{ds} = \frac{dx}{ds}u_x + \frac{dt}{ds}u_t,$$

it follows that $dt/ds = 1$, $dx/ds = -6u$, and $du/ds = 0$. Integrating gives $t(s) = s$, $x(s) = -6su_0(x)$, and $u(s) = u_0(x)$, where the constants of integration are 0 and $u_0(x) = u(x, 0)$.

References

Farlow, S. J. *Partial Differential Equations for Scientists and Engineers.* New York: Dover, pp. 205–212, 1993.

Landau, L. D. and Lifschitz, E. M. *Fluid Mechanics, 2nd ed.* Oxford, England: Pergamon Press, pp. 310–346, 1982.

Moon, P. and Spencer, D. E. *Partial Differential Equations.* Lexington, MA: Heath, pp. 27–29, 1969.

Whitham, G. B. *Linear and Nonlinear Waves.* New York: Wiley, pp. 113–142, 1974.

Zauderer, E. *Partial Differential Equations of Applied Mathematics, 2nd ed.* New York: Wiley, pp. 78–121, 1989.

Zwillinger, D. "Method of Characteristics." §88 in *Handbook of Differential Equations, 3rd ed.* Boston, MA: Academic Press, pp. 325–330, 1997.

Characteristic (Real Number)

For a REAL NUMBER x, $\lfloor x \rfloor = \text{int}(x)$ is called the characteristic, where $\lfloor x \rfloor$ is the FLOOR FUNCTION.

See also MANTISSA, SCIENTIFIC NOTATION

Characteristic Class

Characteristic classes are COHOMOLOGY classes in the BASE SPACE of a VECTOR BUNDLE, defined through OBSTRUCTION theory, which are (perhaps partial) obstructions to the existence of k everywhere linearly independent vector FIELDS on the VECTOR BUNDLE. The most common examples of characteristic classes are the CHERN, PONTRYAGIN, and STIEFEL-WHITNEY CLASSES.

Characteristic Equation

The equation which is solved to find a matrix's EIGENVALUES, also called the characteristic polynomial. For a general $k \times k$ MATRIX M, the characteristic equation in variable t is defined by

$$\det(\mathsf{M} - t\mathsf{I}) = 0, \tag{1}$$

where I is the IDENTITY MATRIX and $\det(\mathsf{A})$ is the DETERMINANT of the MATRIX A. Writing M out explicitly gives

$$\mathsf{M} \equiv \begin{bmatrix} a_{11} & a_{12} & \cdots & a_{1k} \\ a_{21} & a_{22} & \cdots & a_{2k} \\ \vdots & \vdots & \ddots & \vdots \\ a_{k1} & a_{k2} & \cdots & a_{kk} \end{bmatrix}, \tag{2}$$

so the characteristic equation is given by

$$\begin{vmatrix} a_{11} - t & a_{12} & \cdots & a_{1k} \\ a_{21} & a_{22} - t & \cdots & a_{2k} \\ \vdots & \vdots & \ddots & \vdots \\ a_{k1} & a_{k2} & \cdots & a_{kk} - t \end{vmatrix} = 0 \tag{3}$$

The solutions t of the characteristic equation are called EIGENVALUES, and are extremely important in the analysis of many problems in mathematics and physics.

See also BALLIEU'S THEOREM, CAYLEY-HAMILTON THEOREM, DIAGONAL MATRIX, EIGENVALUE, PARODI'S THEOREM, ROUTH-HURWITZ THEOREM

References

Gradshteyn, I. S. and Ryzhik, I. M. *Tables of Integrals, Series, and Products, 6th ed.* San Diego, CA: Academic Press, pp. 1117–1119, 2000.

Characteristic Factor

A characteristic factor is a factor in a particular factorization of the TOTIENT FUNCTION $\phi(n)$ such that the product of characteristic factors gives the representation of a corresponding abstract GROUP as a GROUP DIRECT PRODUCT. By computing the characteristic factors, any ABELIAN GROUP can be expressed as a GROUP DIRECT PRODUCT of CYCLIC SUBGROUPS, for example, the FINITE GROUP $Z_2 \otimes Z_4$ or $Z_2 \otimes Z_2 \otimes Z_2$. There is a simple algorithm for determining the characteristic factors of MODULO MULTIPLICATION GROUPS.

See also CYCLIC GROUP, GROUP DIRECT PRODUCT, MODULO MULTIPLICATION GROUP, TOTIENT FUNCTION

References

Shanks, D. *Solved and Unsolved Problems in Number Theory, 4th ed.* New York: Chelsea, p. 94, 1993.

Characteristic Function (Probability)

The characteristic function $\phi(t)$ is defined as the FOURIER TRANSFORM of the PROBABILITY DENSITY FUNCTION *using FOURIER TRANSFORM parameters* $(a, b) = (1, 1)$,

$$\phi(t) = \mathscr{F}[P(x)] = \int_{-\infty}^{\infty} e^{itx} P(x)\, dx \tag{1}$$

$$= \int_{-\infty}^{\infty} P(x)\, dx + it \int_{-\infty}^{\infty} x P(x)\, dx$$

$$+ \frac{1}{2}(it)^2 \int_{-\infty}^{\infty} x^2 P(x)\, dx + \ldots \tag{2}$$

$$= \sum_{k=0}^{\infty} \frac{(it)^k}{k!} \mu_k' \tag{3}$$

$$= 1 + it\mu_1' - \frac{1}{2} t^2 \mu_2' - \frac{1}{3!} it^3 \mu_3' + \frac{1}{4!} t^4 \mu_4' + \ldots, \tag{4}$$

where μ_n' (sometimes also denoted v_n) is the nth MOMENT about 0 and $\mu_0' \equiv 1$ (Abramowitz and Stegun 1972, p. 928). A DISTRIBUTION is *not* uniquely specified by its MOMENTS, but is uniquely specified by its characteristic function,

$$P(x) = \mathscr{F}^{-1}[\phi(t)] = \frac{1}{2\pi} \int_{-\infty}^{\infty} e^{-itx} \phi(t)\, dt \tag{5}$$

(Papoulis 1984, p. 155).

The characteristic function can therefore be used to generate RAW MOMENTS,

$$\phi^{(n)}(0) \equiv \left[\frac{d^n \phi}{dt^n} \right]_{t=0} = i^n \mu_n' \tag{6}$$

or the CUMULANTS κ_n,

$$\ln \phi(t) \equiv \sum_{n=o}^{\infty} \kappa_n \frac{(it)^n}{n!}. \tag{7}$$

See also CUMULANT, MOMENT, MOMENT-GENERATING

FUNCTION, PROBABILITY DENSITY FUNCTION

References

Abramowitz, M. and Stegun, C. A. (Eds.). *Handbook of Mathematical Functions with Formulas, Graphs, and Mathematical Tables, 9th printing.* New York: Dover, p. 928, 1972.

Kenney, J. F. and Keeping, E. S. "Moment-Generating and Characteristic Functions," "Some Examples of Moment-Generating Functions," and "Uniqueness Theorem for Characteristic Functions." §4.6–4.8 in *Mathematics of Statistics, Pt. 2, 2nd ed.* Princeton, NJ: Van Nostrand, pp. 72–77, 1951.

Papoulis, A. "Characteristic Functions." §5–5 in *Probability, Random Variables, and Stochastic Processes, 2nd ed.* New York: McGraw-Hill, pp. 153–162, 1984.

Characteristic Function (Set)

Given a SUBSET A of a larger set, the characteristic function χ_A is identically one on A, and is zero elsewhere.

These kinds of functions get their own name because they are useful tools. It is easier to say "the characteristic function of the rationals" or "the characteristic function of PRIMES" than to keep repeating the definition.

A characteristic function is a special case of a SIMPLE FUNCTION.

See also SET, SIMPLE FUNCTION

References

Lukacs, E. *Characteristic Functions.* London: Griffin, 1970.

Characteristic Polynomial

The expanded form of the CHARACTERISTIC EQUATION,

$$\det(x\mathsf{I} - \mathsf{A}),$$

where A is an $n \times n$ MATRIX and I is the IDENTITY MATRIX. The characteristic polynomial of a GRAPH G takes A as the ADJACENCY MATRIX of A.

See also CAYLEY-HAMILTON THEOREM, EIGENVALUE, SPECTRUM (MATRIX)

References

Golub, G. H. and van Loan, C. F. *Matrix Computations, 3rd ed.* Baltimore, MD: Johns Hopkins University Press, p. 310, 1996.

Hagos, E. M. "The Characteristic Polynomial of a Graph is Reconstructible from the Characteristic Polynomials of its Vertex-Deleted Subgraphs and Their Complements." *Electronic J. Combinatorics* **7**, No. 1, R12, 1–9, 2000. http://www.combinatorics.org/Volume_7/v7i1toc.html.

Characteristic Root

EIGENVALUE

Characteristic Vector

EIGENVECTOR

Charlier A-Series

CHARLIER SERIES

Charlier Differential Series

CHARLIER SERIES

Charlier Polynomial

The orthogonal polynomials defined by

$$c_n^{(\mu)}(x) = {}_2F_0(-n, -x; ; -\mu^{-1}) \tag{1}$$

$$= \frac{(-1)^n}{\mu^n}(x - n + 1)_n \, {}_1F_1(-n; x - n + 1; \mu) \tag{2}$$

$$= {}_2F_0(-n, -x; ; -1/\mu) \tag{3}$$

where $(x)_n$ is the POCHHAMMER SYMBOL (Koekoek and Swarttouw 1998). The first few are given by

$$c_0^{(\mu)}(x) = 1$$

$$c_1^{(\mu)}(x) = 1 - \frac{x}{\mu}$$

$$c_2^{(\mu)}(x) = \frac{x^2 + \mu^2 - x(1 + 2\mu)}{\mu^2}.$$

References

Koekoek, R. and Swarttouw, R. F. "Charlier." §1.12 in *The Askey-Scheme of Hypergeometric Orthogonal Polynomials and its q-Analogue.* Delft, Netherlands: Technische Universiteit Delft, Faculty of Technical Mathematics and Informatics Report 98–17, pp. 49–50, 1998. ftp://www.twi.tudelft.nl/publications/tech-reports/1998/DUT-TWI-98–17.ps.gz.

Koepf, W. *Hypergeometric Summation: An Algorithmic Approach to Summation and Special Function Identities.* Braunschweig, Germany: Vieweg, p. 115, 1998.

Charlier Series

A class of formal series expansions in derivatives of a distribution $\Psi(t)$ which may (but need not) be the NORMAL DISTRIBUTION FUNCTION

$$\Phi(t) \equiv \frac{1}{\sqrt{2\pi}} \, e^{-t^2/2}$$

and moments or other measured parameters. Edgeworth series are known as the Charlier series or Gram-Charlier series. Let $\psi(t)$ be the CHARACTERISTIC FUNCTION of the function $\Psi(t)$, and γ_r its CUMULANTS. Similarly, let $F(t)$ be the distribution to be approximated, $f(t)$ its CHARACTERISTIC FUNCTION, and κ_r its CUMULANTS. By definition, these quantities are connected by the formal series

$$f(t) = \exp\left[\sum_{r=1}^{\infty} (\kappa_r - \gamma_r) \frac{(it)^r}{r!}\right] \psi(t)$$

(Wallace 1958). Integrating by parts gives $(it)^r \psi(t)$ as the CHARACTERISTIC FUNCTION of $(-1)^r \Psi^{(r)}(x)$, so the formal identity corresponds pairwise to the identity

$$F(x) = \exp\left[\sum_{r=1}^{\infty}(\kappa_r - \gamma_r)\frac{(-D)^r}{r!}\right]\Psi(x),$$

where D is the DIFFERENTIAL OPERATOR. The most important case $\Psi(t) = \Phi(t)$ was considered by Chebyshev (1890), Charlier (1905), and Edgeworth (1905).

Expanding and collecting terms according to the order of the derivatives gives the so-called Gram-Charlier A-Series, which is identical to the formal expansion of $F - \Psi$ in Hermite polynomials. The A-series converges for functions F whose tails approach zero faster than $\Psi'^{1/2}$ (Cramér 1925, Wallace 1958, Szego 1975).

See also CORNISH-FISHER ASYMPTOTIC EXPANSION, EDGEWORTH SERIES

References

Charlier, C. V. L. "Über das Fehlergesetz." *Ark. Math. Astr. och Phys.* **2**, No. 8, 1–9, 1905–06.
Chebyshev, P. L. "Sur deux théorèmes relatifs aux probabilités." *Acta Math.* **14**, 305–315, 1890.
Cramér, H. "On Some Classes of Series Used in Mathematical Statistics." *Proceedings of the Sixth Scandinavian Congress of Mathematicians, Copenhagen.* pp. 399–425, 1925.
Edgeworth, F. Y. "The Law of Error." *Cambridge Philos. Soc.* **20**, 36–66 and 113–141, 1905.
Gram, J. P. "Über die Entwicklung reeler Funktionen in Reihen mittelst der Methode der kleinsten Quadrate." *J. reine angew. Math.* **94**, 41–73, 1883.
Szego, G. *Orthogonal Polynomials, 4th ed.* Providence, RI: Amer. Math. Soc., 1975.
Wallace, D. L. "Asymptotic Approximations to Distributions." *Ann. Math. Stat.* **29**, 635–654, 1958.

Charlier's Check

A check which can be used to verify correct computations in a table of grouped classes. For example, consider the following table with specified class limits and frequencies f. The class marks x_i are then computed as well as the rescaled frequencies u_i, which are given by

$$u_i = \frac{f_i - x_0}{c}, \tag{1}$$

where the class mark is taken as $x_0 = 74.5$ and the class interval is $c = 10$. The remaining quantities are then computed as follows.

class limits	x_i	$f_i \ (\mu)_n$	$f_i u_i$	$f_i u_i^2$	$f_i(u_i + 1)^2$
30–39	34.5	2	−4	−8	32	18
40–49	44.5	3	−3	−9	27	12
50–59	54.5	11	−2	−22	44	11
60–69	64.5	20	−1	−20	20	0
70–79	74.5	32	0	0	0	32
80–89	84.5	25	1	25	25	100
90–99	94.5	7	2	14	28	63
total		100		−20	176	236

In order to compute the VARIANCE, note that

$$s_u^2 = \frac{\sum_i f_i u_i^2}{\sum_i f_i} - \left(\frac{\sum_i f_i u_i}{\sum_i f_i}\right)^2 \tag{2}$$

$$= \frac{176}{100} - \left(\frac{-20}{100}\right)^2 = 1.72, \tag{3}$$

so the VARIANCE of the original data is

$$s_x^2 = c^2 s_u^2 = 172. \tag{4}$$

Charlier's check makes use of the additional column $f_i(u_i + 1)^2$ added to the right side of the table. By noting that the identity

$$\sum_i f_i(u_i + 1)^2 = \sum_i f_i(u_i^2 + 2u_i + 1)$$

$$= \sum_i f_i u_i^2 + 2\sum_i f_i u_i + \sum_i f_i, \tag{5}$$

connects columns five through seven, it can be checked that the computations have been done correctly. In the example above,

$$236 = 176 + 2(-20) + 100, \tag{6}$$

so the computations pass Charlier's check.

See also VARIANCE

References

Kenney, J. F. and Keeping, E. S. "Charlier Check." §6.8 in *Mathematics of Statistics, Pt. 1, 3rd ed.* Princeton, NJ: Van Nostrand, pp. 47–48, 81, 94–95, and 104, 1962.

Chart
COORDINATE CHART

Chasles-Cayley-Brill Formula

The number of coincidences of a (v, v') correspondence of value γ on a curve of GENUS p is given by

$$v + v' + 2p\gamma.$$

See also ZEUTHEN'S THEOREM

References

Coolidge, J. L. *A Treatise on Algebraic Plane Curves.* New York: Dover, p. 129, 1959.

Chasles's Contact Theorem

If a one-parameter family of curves has index N and class M, the number tangent to a curve of order n_1 and class m_1 in general position is

$$m_1 N + n_1 M.$$

References

Coolidge, J. L. *A Treatise on Algebraic Plane Curves.* New York: Dover, p. 436, 1959.

Chasles's Polars Theorem

If the TRILINEAR POLARS of the VERTICES of a TRIANGLE are distinct from the respectively opposite sides, they meet the sides in three COLLINEAR points.

See also COLLINEAR, TRIANGLE, TRILINEAR POLAR

Chasles's Theorem

If two projective PENCILS of curves of orders n and n' have no common curve, the LOCUS of the intersections of corresponding curves of the two is a curve of order $n + n'$ through all the centers of either PENCIL. Conversely, if a curve of order $n + n'$ contains all centers of a PENCIL of order n to the multiplicity demanded by NOETHER'S FUNDAMENTAL THEOREM, then it is the LOCUS of the intersections of corresponding curves of this PENCIL and one of order n' projective therewith.

See also NOETHER'S FUNDAMENTAL THEOREM, PENCIL

References

Coolidge, J. L. *A Treatise on Algebraic Plane Curves.* New York: Dover, p. 33, 1959.

Chebyshev

This entry contributed by RONALD M. AARTS

A number of spellings of "Chebyshev" (which is the spelling used exclusively in this work) are commonly found in the literature. These include Tchebicheff, Cebysev, Tschebyscheff, Chebishev, and Tschebyscheff (Clenshaw).

References

Clenshaw, C. W. *Mathematical Tables, Vol. 5: Chebyshev Series for Mathematical Functions.* Department of Scientific and Industrial Research.

Chebyshev Approximation Formula

Using a CHEBYSHEV POLYNOMIAL OF THE FIRST KIND $T(x)$, define

$$c_j \equiv \frac{2}{N} \sum_{k=1}^{N} f(x_k) T_j(x_k)$$

$$= \frac{2}{N} \sum_{k=1}^{N} f\left[\cos\left\{\frac{\pi(k - \frac{1}{2})}{N}\right\}\right] \cos\left\{\frac{\pi j(k - \frac{1}{2})}{N}\right\}.$$

Then

$$f(x) \approx \sum_{k=0}^{N-1} c_k T_k(x) - \tfrac{1}{2} c_0.$$

It is exact for the N zeros of $T_N(x)$. This type of approximation is important because, when truncated, the error is spread smoothly over $[-1, 1]$. The Chebyshev approximation formula is very close to the MINIMAX POLYNOMIAL.

References

Press, W. H.; Flannery, B. P.; Teukolsky, S. A.; and Vetterling, W. T. "Chebyshev Approximation," "Derivatives or Integrals of a Chebyshev-Approximated Function," and "Polynomial Approximation from Chebyshev Coefficients." §5.8, 5.9, and 5.10 in *Numerical Recipes in FORTRAN: The Art of Scientific Computing, 2nd ed.* Cambridge, England: Cambridge University Press, pp. 184–188, 189–190, and 191–192, 1992.

Chebyshev Constants

N.B. A detailed online essay by S. Finch was the starting point for this entry.

The constants

$$\lambda_{m, n} = \inf_{r \in R_{m, n}} \sup_{x \geq 0} |e^{-x} - r(x)|,$$

where

$$r(x) = \frac{p(x)}{q(x)},$$

p and q are mth and nth order POLYNOMIALS, and $R_{m, n}$ is the set all RATIONAL FUNCTIONS with REAL coefficients.

See also ONE-NINTH CONSTANT, RATIONAL FUNCTION

References

Finch, S. "Favorite Mathematical Constants." http://www.mathsoft.com/asolve/constant/onenin/onenin.html.
Petrushev, P. P. and Popov, V. A. *Rational Approximation of Real Functions.* New York: Cambridge University Press, 1987.
Varga, R. S. *Scientific Computations on Mathematical Problems and Conjectures.* Philadelphia, PA: SIAM, 1990. Philadelphia, PA: SIAM, 1990.

Chebyshev Deviation

$$\max_{a \leq x \leq b} \{ |f(x) - \rho(x)| w(x) \}.$$

References

Szego, G. *Orthogonal Polynomials, 4th ed.* Providence, RI: Amer. Math. Soc., p. 41, 1975.

Chebyshev Differential Equation

$$(1 - x^2) \frac{d^2 y}{dx^2} - x \frac{dy}{dx} + \alpha^2 y = 0 \tag{1}$$

for $|x| < 1$. The Chebyshev differential equation has regular SINGULARITIES at -1, 1, and ∞. It can be solved by series solution using the expansions

$$y = \sum_{n=0}^{\infty} a_n x^n \tag{2}$$

$$y' = \sum_{n=0}^{\infty} n a_n x^{n-1} = \sum_{n=1}^{\infty} n a_n x^{n-1} = \sum_{n=0}^{\infty} (n+1) a_{n+1} x^n \tag{3}$$

$$y'' = \sum_{n=0}^{\infty} (n+1) n a_{n+1} x^{n-1} = \sum_{n=1}^{\infty} (n+1) n a_{n+1} x^{n-1}$$
$$= \sum_{n=0}^{\infty} (n+2)(n+1) a_{n+2} x^n. \tag{4}$$

Now, plug (2–4) into the original equation (1) to obtain

$$(1 - x^2) \sum_{n=0}^{\infty} (n+2)(n+1) a_{n+2} x^n$$

$$-x \sum_{n=0}^{\infty} (n+1) n_{n+1} x^n + \alpha^2 \sum_{n=0}^{\infty} a_n x^n = 0 \tag{5}$$

$$\sum_{n=0}^{\infty} (n+2)(n+1) a_{n+2} x^n - \sum_{n=0}^{\infty} (n+2)(n+1) a_{n+2} x^{n+2}$$

$$-\sum_{n=0}^{\infty} (n+1) a_{n+2} x^{n+1} + \alpha^2 \sum_{n=0}^{\infty} a_n x^n = 0 \tag{6}$$

$$\sum_{n=0}^{\infty} (n+2)(n+1) a_{n+2} x^n - \sum_{n=2}^{\infty} n(n-1) a_n x^{n+2}$$

$$-\sum_{n=1}^{\infty} n a_n x^n + \alpha^2 \sum_{n=0}^{\infty} a_n x^n = 0 \tag{7}$$

$$2 \cdot 1 a_2 + 3 \cdot 2 a_3 x - 1 \cdot a x + \alpha^2 a_0 + \alpha^2 a_1 x$$

$$+\sum_{n=2}^{\infty} [(n+2)(n+1) a_{n+2} - n(n-1) a_n - n a_n + \alpha^2 a_n] x^n$$
$$= 0 \tag{8}$$

$$(2 a_2 + \alpha^2 a_0) + [(\alpha^2 - 1) a_1 + 6 a_3] x$$

$$+\sum_{n=2}^{\infty} [(n+2)(n+1) a_{n+2} + (\alpha^2 - n^2) a_n] x^n = 0, \tag{9}$$

so

$$2 a_2 + \alpha^2 a_0 = 0 \tag{10}$$

$$(\alpha^2 - 1) a_1 + 6 a_3 = 0, \tag{11}$$

and by induction,

$$a_{n+2} = \frac{n^2 - \alpha^2}{(n+1)(n+2)} a_n \tag{12}$$

for $n = 2, 3, \ldots$.

Since (10) and (11) are special cases of (12), the general RECURRENCE RELATION can be written

$$a_{n+2} = \frac{n^2 - \alpha^2}{(n+1)(n+2)} a_n \tag{13}$$

for $n = 0, 1, \ldots$. From this, we obtain for the EVEN COEFFICIENTS

$$a_2 = \frac{-\alpha^2}{2} a_0 \tag{14}$$

$$a_4 = \frac{2^2 - \alpha^2}{3 \cdot 4} a_2 = \frac{(2^2 - \alpha^2)(-\alpha^2)}{1 \cdot 2 \cdot 3 \cdot 4} a_0 \tag{15}$$

$$a_{2n} = \frac{[(2n)^2 - \alpha^2][(2n-2)^2 - \alpha^2] \cdots (-\alpha^2)}{(2n)!} a_0. \tag{16}$$

and for the ODD COEFFICIENTS

$$a_3 = \frac{1 - \alpha^2}{6} a_0 \tag{17}$$

$$a_5 = \frac{3^2 - \alpha^2}{4 \cdot 5} a_3 = \frac{(3^2 - \alpha^2)(1^2 - \alpha^2)}{5!} a_1 \qquad (18)$$

$$a_{2n-1} =$$

$$\frac{[(2n-1)^2 - \alpha^2][(2n-3)^2 - \alpha^2]\cdots[1^2 - \alpha^2]}{(2n+1)!} a_1. \quad (19)$$

The even coefficients $k = 2n$ can be given in closed form by as

$$a_{k \text{ even}} = a_0 \prod_{j=1}^{k/2} (k - 2j)^2 - \alpha^2$$

$$= \frac{2^{k-1} \pi \alpha \csc(\frac{1}{2} \pi \alpha)}{\Gamma(1 - \frac{1}{2}k - \frac{1}{2}\alpha)\Gamma(1 - \frac{1}{2}k + \frac{1}{2}\alpha)} a_0, \quad (20)$$

and the odd coefficients $k = 2n - 1$ as

$$a_{k \text{ odd}} = a_1 \prod_{j=1}^{(k-1)/2} (k - 2j)^2 - \alpha^2$$

$$= \frac{2^{k-1} \pi \alpha \sec(\frac{1}{2} \pi \alpha)}{\Gamma(1 - \frac{1}{2}k - \frac{1}{2}\alpha)\Gamma(1 - \frac{1}{2}k + \frac{1}{2}\alpha)} a_1. \quad (21)$$

The general solution is then given by summing over all indices,

$$y = a_0 \left[1 + \sum_{k=2,4\ldots}^{\infty} \frac{a_{k \text{ even}}}{k!} x^k \right] + \left[x + \sum_{k=3,5\ldots}^{\infty} \frac{a_{k \text{ odd}}}{k!} x^k \right], \quad (22)$$

which can be done in closed form as

$$y = a_0 \cos(\alpha \sin^{-1} x) + \frac{a_1}{\alpha} \sin(\alpha \sin^{-1} x). \quad (23)$$

Performing a change of variables gives the equivalent form of the solution

$$y = b_1 \cos(\alpha \cos^{-1} x) + b_2 \sin(\alpha \cos^{-1} x) \quad (24)$$

$$= b_1 T_\alpha(x) + b_2 \sqrt{1 - x^2}\, U_{\alpha-1}(x), \quad (25)$$

where $T_n(x)$ is a CHEBYSHEV POLYNOMIAL OF THE FIRST KIND and $U_n(x)$ is a CHEBYSHEV POLYNOMIAL OF THE SECOND KIND. Another equivalent form of the solution is given by

$$y = c_1 \cosh[\alpha \ln(x + \sqrt{x^2 - 1})]$$

$$+ i c_2 \sinh[\alpha \ln(x + \sqrt{x^2 - 1})]. \quad (26)$$

See also CHEBYSHEV POLYNOMIAL OF THE FIRST KIND, CHEBYSHEV POLYNOMIAL OF THE SECOND KIND

References

Arfken, G. *Mathematical Methods for Physicists, 3rd ed.* Orlando, FL: Academic Press, p. 735, 1985.

Boyce, W. E. and DiPrima, R. C. *Elementary Differential Equations and Boundary Value Problems, 4th ed.* New York: Wiley, pp. 232 and 252, 1986.

Zwillinger, D. *Handbook of Differential Equations, 3rd ed.* Boston, MA: Academic Press, p. 127, 1997.

Chebyshev Functions

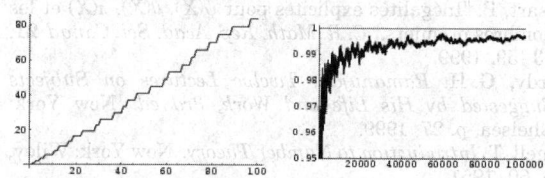

The function defined by

$$\theta(n) \equiv \sum_{i=1}^{n} \ln p_i = \ln \left(\prod_{p \leq n} p \right), \quad (1)$$

where p_i is the ith PRIME (left figure), so

$$\lim_{x \to \infty} \frac{x}{\theta(x)} = 1 \quad (2)$$

(right figure). The function has asymptotic behavior

$$\theta(n) \sim n \quad (3)$$

(Bach and Shallit 1996; Hardy 1999, p. 28). The notation $\vartheta(n)$ is also commonly used for this function (Hardy 1999, p. 27).

Chebyshev also defined the related function

$$\psi(n) \equiv \sum_{\substack{p,\, \nu \\ p^\nu \leq n}} \ln p, \quad (4)$$

which is equal to the summatory MANGOLDT FUNCTION and is given by the logarithm of the LEAST COMMON MULTIPLE of the numbers from 1 to n. The values of LCM$(1, 2, \cdots, n)$ for $n = 1, 2, \ldots$ are 1, 2, 6, 12, 60, 60, 420, 840, 2520, 2520, ... (Sloane's A003418). For example,

$$\psi(10) = \ln 2520 = 3 \ln 2 + 2 \ln 3 + \ln 5 + \ln 7. \quad (5)$$

The function has asymptotic behavior

$$\psi(n) \sim n \quad (6)$$

(Hardy 1999, p. 27).

According to Hardy (1999, p. 27), the functions $\theta(n)$ and $\psi(n)$ are in some ways more natural than the PRIME COUNTING FUNCTION $\pi(x)$ since they deal with multiplication of primes instead of the counting of them.

See also MANGOLDT FUNCTION, PRIME COUNTING FUNCTION, PRIME NUMBER THEOREM

References

Bach, E. and Shallit, J. *Algorithmic Number Theory, Vol. 1: Efficient Algorithms.* Cambridge, MA: MIT Press, pp. 206 and 233, 1996.

Costa Pereira, N. "Estimates for the Chebyshev Function $\psi(x) - \theta(x)$." *Math. Comp.* **44**, 211–221, 1985.

Costa Pereira, N. "Corrigendum: Estimates for the Chebyshev Function $\psi(x) - \theta(x)$." *Math. Comp.* **48**, 447, 1987.

Costa Pereira, N. "Elementary Estimates for the Chebyshev Function $\psi(x)$ and for the Möbius Function $M(x)$." *Acta Arith.* **52**, 307–337, 1989.

Dusart, P. "Inégalités explicites pour $\psi(X)$, $\theta(X)$, $\pi(X)$ et les nombres premiers." *C. R. Math. Rep. Acad. Sci. Canad* **21**, 53–59, 1999.

Hardy, G. H. *Ramanujan: Twelve Lectures on Subjects Suggested by His Life and Work, 3rd ed.* New York: Chelsea, p. 27, 1999.

Nagell, T. *Introduction to Number Theory.* New York: Wiley, p. 60, 1951.

Panaitopol, L. "Several Approximations of $\pi(x)$." *Math. Ineq. Appl.* **2**, 317–324, 1999.

Robin, G. "Estimation de la foction de Tchebychef θ sur le $k$ième nombre premier er grandes valeurs de la fonctions $\omega(n)$, nombre de diviseurs premiers de n." *Acta Arith.* **42**, 367–389, 1983.

Rosser, J. B. and Schoenfeld, L. "Sharper Bounds for Chebyshev Functions $\theta(x)$ and $\psi(x)$." *Math. Comput.* **29**, 243–269, 1975.

Schoenfeld, L. "Sharper Bounds for Chebyshev Functions $\theta(x)$ and $\psi(x)$, II." *Math. Comput.* **30**, 337–360, 1976.

Selmer, E. S. "On the Number of Prime Divisors of a Binomial Coefficient." *Math. Scand.* **39**, 271–281, 1976.

Sloane, N. J. A. Sequences A003418/M1590 in "An On-Line Version of the Encyclopedia of Integer Sequences." http://www.research.att.com/~njas/sequences/eisonline.html.

Chebyshev Inequality

Apply MARKOV'S INEQUALITY with $a \equiv k^2$ to obtain

$$P[(x-\mu)^2 \geq k^2] \leq \frac{\langle (x-\mu)^2 \rangle}{k^2} = \frac{\sigma^2}{k^2}. \qquad (1)$$

Therefore, if a RANDOM VARIABLE x has a finite MEAN μ and finite VARIANCE σ^2, then $\forall k \geq 0$,

$$P(|x-\mu| \geq k) \leq \frac{\sigma^2}{k^2} \qquad (2)$$

$$P(|x-\mu| \geq k\sigma) \leq \frac{1}{k^2}. \qquad (3)$$

See also CHEBYSHEV SUM INEQUALITY

References

Abramowitz, M. and Stegun, C. A. (Eds.). *Handbook of Mathematical Functions with Formulas, Graphs, and Mathematical Tables, 9th printing.* New York: Dover, p. 11, 1972.

Hardy, G. H.; Littlewood, J. E.; and Pólya, G. "Tchebychef's Inequality." §2.17 and §5.8 in *Inequalities, 2nd ed.* Cambridge, England: Cambridge University Press, pp. 43–45 and 123, 1988.

Papoulis, A. *Probability, Random Variables, and Stochastic Processes, 2nd ed.* New York: McGraw-Hill, pp. 149–151, 1984.

Chebyshev Integral

$$\int x^p (1-x)^q \, dx = \frac{x^{1+p} \, {}_2F_1(p+1, \, -q; \, p+2; \, x)}{p+1}.$$

See also CHEBYSHEV INTEGRAL INEQUALITY

Chebyshev Integral Inequality

$$\int_a^b f_1(x)\, dx \int_a^b f_2(x)\, dx \, \cdots \, \int_a^b f_n(x)\, dx$$

$$\leq (b-a)^{n-1} \int_a^b f_1(x) f_2(x) \cdots f_n(x)\, dx$$

where f_1, f_2, ..., f_n are NONNEGATIVE integrable functions on $[a, b]$ which are *all* either monotonic increasing or monotonic decreasing.

References

Gradshteyn, I. S. and Ryzhik, I. M. *Tables of Integrals, Series, and Products, 6th ed.* San Diego, CA: Academic Press, p. 1092, 2000.

Chebyshev Phenomenon

PRIME QUADRATIC EFFECT

Chebyshev Polynomial of the First Kind

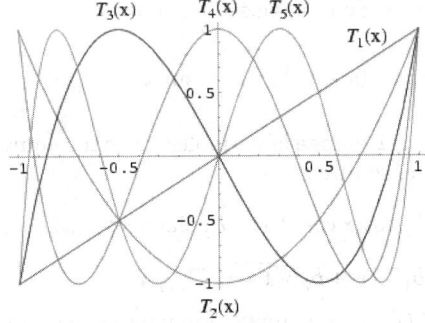

A set of ORTHOGONAL POLYNOMIALS defined as the solutions to the CHEBYSHEV DIFFERENTIAL EQUATION and denoted $T_n(x)$. They are used as an approximation to a LEAST SQUARES FIT, and are a special case of the ULTRASPHERICAL POLYNOMIAL with $\alpha = 0$. They are also intimately connected with trigonometric MULTIPLE-ANGLE FORMULAS. The Chebyshev polynomials of the first kind are denoted $T_n(x)$, and are implemented in *Mathematica* as ChebyshevT[n, x]. They are normalized such that $T_n(1) = 1$. The first few polynomials are illustrated above for $x \in [-1, 1]$ and $n = 1, 2, ..., 5$.

The Chebyshev polynomials of the first kind can be obtained from the GENERATING FUNCTIONS

$$g_1(t, x) \equiv \frac{1-t^2}{1-2xt+t^2} = T_0(x) + 2\sum_{n=1}^{\infty} T_n(x) t^n \qquad (1)$$

and

$$g_2(t, x) \equiv \frac{1 - xt}{1 - 2xt + t^2} = \sum_{n=0}^{\infty} T_n(x)t^n \qquad (2)$$

for $|x| \leq 1$ and $|t| < 1$ (Beeler *et al.* 1972, Item 15). (A closely related GENERATING FUNCTION is the basis for the definition of CHEBYSHEV POLYNOMIAL OF THE SECOND KIND.)

The polynomials can also be defined in terms of the sums

$$T_n(x) = \frac{n}{2} \sum_{r=0}^{\lfloor n/2 \rfloor} \frac{(-1)^r}{n - r} \binom{n - r}{r} (2x)^{n-2r} \qquad (3)$$

$$T_n(x) = \cos(\cos^{-1} x) = \sum_{m=0}^{\lfloor n/2 \rfloor} \binom{n}{2m} x^{n-2m}(x^2 - 1)^m, \qquad (4)$$

where $\binom{n}{k}$ is a BINOMIAL COEFFICIENT and $\lfloor x \rfloor$ is the FLOOR FUNCTION, or the product

$$T_n(x) = 2^{n-1} \prod_{k=1}^{n} \left\{ x - \cos\left[\frac{(2k-1)\pi}{2n}\right] \right\} \qquad (5)$$

(Zwillinger 1995, p. 696).

T_n also satisfy the curious DETERMINANT equation

$$T_n = \begin{vmatrix} x & 1 & 0 & 0 & \cdots & 0 & 0 \\ 1 & 2x & 1 & 0 & \ddots & 0 & 0 \\ 0 & 1 & 2x & 1 & \ddots & 0 & 0 \\ 0 & 0 & 1 & 2x & \ddots & 0 & 0 \\ 0 & 0 & 0 & 1 & \ddots & 1 & 0 \\ \vdots & & & & \ddots & & 1 \\ 0 & 0 & 0 & 0 & & 1 & 2x \end{vmatrix}. \qquad (6)$$

The Chebyshev polynomials of the first kind are a special case of the JACOBI POLYNOMIALS $P_n^{(\alpha,\beta)}$ with $\alpha = \beta = -1/2$,

$$T_n(x) = \frac{P_n^{(-1/2, -1/2)}(x)}{P_n^{(-1/2, -1/2)}(1)} = {}_2F_1(-n, -n; \tfrac{1}{2}; \tfrac{1}{2}(1-x)), \qquad (7)$$

where ${}_2F_1(a, b; c; x)$ is a HYPERGEOMETRIC FUNCTION (Koekoek and Swarttouw 1998).

Zeros occur when

$$x = \cos\left[\frac{\pi\left(k - \frac{1}{2}\right)}{n}\right] \qquad (8)$$

for $k = 1, 2, ..., n$. Extrema occur for

$$x = \cos\left(\frac{\pi k}{n}\right), \qquad (9)$$

where $k = 0, 1, \ldots, n$. At maximum, $T_n(x) = 1$, and at minimum, $T_n(x) = -1$. The Chebyshev POLYNOMIALS are ORTHONORMAL with respect to the WEIGHTING FUNCTION $(1 - x^2)^{-1/2}$

$$\int_{-1}^{1} \frac{T_m(x)T_n(x)\, dx}{\sqrt{1 - x^2}} = \begin{cases} \frac{1}{2}\pi\delta_{nm} & \text{for } m \neq 0,\ n \neq 0 \\ \pi & \text{for } m = n = 0, \end{cases} \qquad (10)$$

where δ_{mn} is the KRONECKER DELTA. Chebyshev polynomials of the first kind satisfy the additional discrete identity

$$\sum_{k=1}^{m} T_i(x_k)T_j(x_k) = \begin{cases} \frac{1}{2}m\delta_{ij} & \text{for } i \neq 0,\ j \neq 0 \\ m & \text{for } i = j = 0, \end{cases} \qquad (11)$$

where x_k for $k = 1, ..., m$ are the m zeros of $T_m(x)$. They also satisfy the RECURRENCE RELATIONS

$$T_{n+1}(x) = 2xT_n(x) - T_{n-1}(x) \qquad (12)$$

$$T_{n+1}(x) = xT_n(x) - \sqrt{(1 - x^2)\{1 - [T_n(x)]^2\}} \qquad (13)$$

for $n \geq 1$. They have a COMPLEX integral representation

$$T_n(x) = \frac{1}{4\pi i} \int_\gamma \frac{(1 - z^2)z^{-n-1}\, dz}{1 - 2xz + z^2} \qquad (14)$$

and a Rodrigues representation

$$T_n(x) = \frac{(-1)^n \sqrt{\pi}(1 - x^2)^{1/2}}{2n(n - \frac{1}{2})!} \frac{d^n}{dx^n}[(1 - x^2)^{n-1/2}]. \qquad (15)$$

Using a FAST FIBONACCI TRANSFORM with multiplication law

$$(A, B)(C, D) = (AD + BC + 2xAC, BD - AC) \qquad (16)$$

gives

$$T_{n+1}(x), -T_n(x) = (T_1(x), -T_0(x))(1, 0)^n. \qquad (17)$$

Using GRAM-SCHMIDT ORTHONORMALIZATION in the range $(-1,1)$ with WEIGHTING FUNCTION $(1 - x^2)^{(-1/2)}$ gives

$$p_0(x) = 1 \qquad (18)$$

$$p_1(x) = \left[x - \frac{\int_{-1}^{1} x(1 - x^2)^{-1/2}\, dx}{\int_{-1}^{1} (1 - x^2)^{-1/2}\, dx}\right]$$

$$= x - \frac{[-1(1 - x^2)^{1/2}]_{-1}^{1}}{[\sin^{-1} x]_{-1}^{1}} = x \qquad (19)$$

$$p_2(x) = \left[x - \frac{\int_{-1}^{1} x^3(1-x^2)^{-1/2}\,dx}{\int_{-1}^{1} x^2(1-x^2)^{-1/2}\,dx} \right] x$$

$$- \left[\frac{\int_{-1}^{1} x^2(1-x^2)^{-1/2}\,dx}{\int_{-1}^{1} (1-x^2)^{-1/2}\,dx} \right] \cdot 1$$

$$= [x - 0]x - \frac{\frac{\pi}{2}}{\pi} = x^2 - \tfrac{1}{2}, \tag{20}$$

etc. Normalizing such that $T_n(1) = 1$ gives

$$T_0(x) = 1$$

$$T_1(x) = x$$

$$T_2(x) = 2x^2 - 1$$

$$T_3(x) = 4x^3 - 3x$$

$$T_4(x) = 8x^4 - 8x^2 + 1$$

$$T_5(x) = 16x^5 - 20x^3 + 5x$$

$$T_6(x) = 32x^6 - 48x^4 + 18x^2 - 1$$

The Chebyshev polynomial of the first kind is related to the BESSEL FUNCTION OF THE FIRST KIND $J_n(x)$ and MODIFIED BESSEL FUNCTION OF THE FIRST KIND $I_n(x)$ by the relations

$$J_n(x) = i^n T_n\left(i\,\frac{d}{dx}\right) J_0(x) \tag{21}$$

$$I_n(x) = T_n\left(\frac{d}{dx}\right) I_0(x). \tag{22}$$

Letting $x \equiv \cos\theta$ allows the Chebyshev polynomials of the first kind to be written as

$$T_n(x) = \cos(n\theta) = \cos(n\,\cos^{-1} x). \tag{23}$$

The second linearly dependent solution to the transformed differential equation

$$\frac{d^2 T_n}{d\theta^2} + n^2 T_n = 0 \tag{24}$$

is then given by

$$V_n(x) = \sin(n\theta) = \sin(n\,\cos^{-1} x), \tag{25}$$

which can also be written

$$V_n(x) = \sqrt{1-x^2}\,U_{n-1}(x), \tag{26}$$

where U_n is a CHEBYSHEV POLYNOMIAL OF THE SECOND KIND. Note that $V_n(x)$ is therefore *not* a POLYNOMIAL.

The triangle of RESULTANTS $\rho(T_n(x),\ T_k(x))$ is given by $\{0\}$, $\{-1,\ 0\}$, $\{0,\ -4,\ 0\}$, $\{1,\ 16,\ 64,\ 0\}$, $\{0,\ -16,\ 0,\ 4096,\ 0\}$, ... (Sloane's A054375).

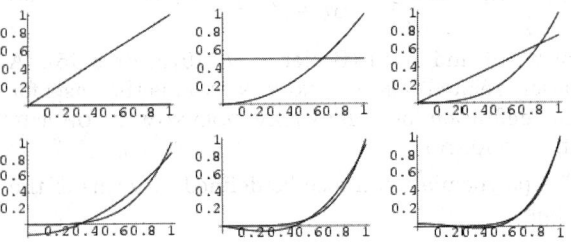

The POLYNOMIALS

$$p_n(x) = x^n - 2^{1-n} T_n(x) \tag{27}$$

of degree $n-2$, the first few of which are

$$p_1(x) = 0$$
$$p_2(x) = \tfrac{1}{2}$$
$$p_3(x) = \tfrac{3}{4}x$$
$$p_4(x) = x^2 - \tfrac{1}{8}$$
$$p_5(x) = \tfrac{5}{16}(4x^3 - x)$$

are the POLYNOMIALS of degree $< n$ which stay closest to x_n in the interval $(-1,\ 1)$. The maximum deviation is 2^{1-n} at the $n+1$ points where

$$x = \cos\left(\frac{k\pi}{n}\right), \tag{28}$$

for $k = 0,\ 1,\ ...,\ n$ (Beeler *et al.* 1972).

See also CHEBYSHEV APPROXIMATION FORMULA, CHEBYSHEV POLYNOMIAL OF THE SECOND KIND

References

Abramowitz, M. and Stegun, C. A. (Eds.). "Orthogonal Polynomials." Ch. 22 in *Handbook of Mathematical Functions with Formulas, Graphs, and Mathematical Tables, 9th printing.* New York: Dover, pp. 771–802, 1972.

Arfken, G. "Chebyshev (Tschebyscheff) Polynomials" and "Chebyshev Polynomials--Numerical Applications." §13.3 and 13.4 in *Mathematical Methods for Physicists, 3rd ed.* Orlando, FL: Academic Press, pp. 731–748, 1985.

Beeler *et al.* . Item 15 in Beeler, M.; Gosper, R. W.; and Schroeppel, R. *HAKMEM.* Cambridge, MA: MIT Artificial Intelligence Laboratory, Memo AIM-239, p. 9, Feb. 1972.

Iyanaga, S. and Kawada, Y. (Eds.). "Cebysev (Tschebyscheff) Polynomials." Appendix A, Table 20.II in *Encyclopedic Dictionary of Mathematics.* Cambridge, MA: MIT Press, pp. 1478–1479, 1980.

Koekoek, R. and Swarttouw, R. F. "Chebyshev." §1.8.2 in *The Askey-Scheme of Hypergeometric Orthogonal Polynomials and its q-Analogue.* Delft, Netherlands: Technische Universiteit Delft, Faculty of Technical Mathematics and Informatics Report 98–17, pp. 41–43, 1998. ftp://www.twi.tudelft.nl/publications/tech-reports/1998/DUT-TWI-98–17.ps.gz.

Koepf, W. "Efficient Computation of Chebyshev Polynomials." In *Computer Algebra Systems: A Practical Guide* (Ed. M. J. Wester). New York: Wiley, pp. 79–99, 1999.

Rivlin, T. J. *Chebyshev Polynomials.* New York: Wiley, 1990.

Shohat, J. *Théorie générale des polynomes orthogonaux de Tchebichef.* Paris: Gauthier-Villars, 1934.

Sloane, N. J. A. Sequences A054375 in "An On-Line Version of the Encyclopedia of Integer Sequences." http://www.research.att.com/~njas/sequences/eisonline.html.

Spanier, J. and Oldham, K. B. "The Chebyshev Polynomials $T_n(x)$ and $U_n(x)$." Ch. 22 in *An Atlas of Functions.* Washington, DC: Hemisphere, pp. 193–207, 1987.

Vasilyev, N. and Zelevinsky, A. "A Chebyshev Polyplayground: Recurrence Relations Applied to a Famous Set of Formulas." *Quantum* **10**, 20–26, Sept./Oct. 1999.

Zwillinger, D. (Ed.). *CRC Standard Mathematical Tables and Formulae.* Boca Raton, FL: CRC Press, 1995.

Chebyshev Polynomial of the Second Kind

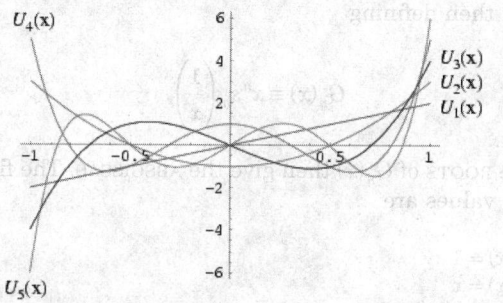

A modified set of Chebyshev POLYNOMIALS defined by a slightly different GENERATING FUNCTION. They arise in the development of four-dimensional SPHERICAL HARMONICS in angular momentum theory. They are a special case of the ULTRASPHERICAL POLYNOMIAL with $\alpha = 1$. They are also intimately connected with trigonometric MULTIPLE-ANGLE FORMULAS. The Chebyshev polynomials of the second kind are denoted $U_n(x)$, and implemented in *Mathematica* as `ChebyshevU[n, x]`. The polynomials $U_n(x)$ are illustrated above for $x \in [-1, 1]$ and $n = 1, 2, ..., 5$.

The defining GENERATING FUNCTION of the Chebyshev polynomials of the second kind is

$$g_2(t, x) = \frac{1}{1 - 2xt + t^2} = \sum_{n=0}^{\infty} U_n(x) t^n \qquad (1)$$

for $|x| < 1$ and $|t| < 1$. To see the relationship to a CHEBYSHEV POLYNOMIAL OF THE FIRST KIND $T(x)$, take $\partial g / \partial t$,

$$\frac{\partial g}{\partial t} = -(1 - 2xt + t^2)^{-2}(-2x + 2t)$$

$$= 2(t - x)(1 - 2xt + t^2)^{-2} = \sum_{n=0}^{\infty} n U_n(x) t^{n-1}. \qquad (2)$$

Multiply (2) by t,

$$(2t^2 - 2xt)(1 - 2xt + t^2)^{-2} = \sum_{n=0}^{\infty} n U_n(x) t^n \qquad (3)$$

and take (3) minus (2),

$$\frac{(2t^2 - 2tx) - (1 - 2xt + t^2)}{(1 - 2xt + t^2)^2} = \frac{t^2 - 1}{(1 - 2xt + t^2)^2}$$

$$= \sum_{n=0}^{\infty} (n - 1) U_n(x) t^n. \qquad (4)$$

The Rodrigues representation is

$$U_n(x) = \frac{(-1)^n (n+1) \sqrt{\pi}}{2^{n+1} (n + \frac{1}{2})! (1 - x^2)^{1/2}} \frac{d^n}{dx^n} [(1 - x^2)^{n+1/2}]. \quad (5)$$

The polynomials can also be defined in terms of the sums

$$U_n(x) = \sum_{r=0}^{\lfloor n/2 \rfloor} (-1)^r \binom{n-r}{r} (2x)^{n-2r}$$

$$= \sum_{m=0}^{\lceil n/2 \rceil} \binom{n+1}{2m+1} x^{n-2m} (x^2 - 1)^m, \qquad (6)$$

where $\lfloor x \rfloor$ is the FLOOR FUNCTION and $\lceil x \rceil$ is the CEILING FUNCTION, or in terms of the product

$$U_n(x) = 2^n \prod_{k=1}^{n} \left[x - \cos\left(\frac{k\pi}{n+1} \right) \right] \qquad (7)$$

(Zwillinger 1995, p. 696).

$U_n(x)$ also obey the interesting DETERMINANT identity

$$U_n = \begin{vmatrix} 2x & 1 & 0 & 0 & \cdots & 0 & 0 \\ 1 & 2x & 1 & 0 & \ddots & 0 & 0 \\ 0 & 1 & 2x & 1 & \ddots & 0 & 0 \\ 0 & 0 & 1 & 2x & \ddots & 0 & 0 \\ 0 & 0 & 0 & 1 & \ddots & 1 & 0 \\ \vdots & & & & \ddots & \ddots & 1 \\ 0 & 0 & 0 & 0 & \cdots & 1 & 2x \end{vmatrix}. \qquad (8)$$

The Chebyshev polynomials of the second kind are a special case of the JACOBI POLYNOMIALS $P_n^{(\alpha, \beta)}$ with $\alpha = \beta = 1/2$,

$$U_n(x) = (n+1) \frac{P_n^{(1/2, \, 1/2)}(x)}{P_n^{(1/2, \, 1/2)}(1)}$$

$$= {}_2F_1(-n, \, n+2; \, \tfrac{3}{2}; \, \tfrac{1}{2}(1 - x)), \qquad (9)$$

where ${}_2F_1(a, \, b; \, c; \, x)$ is a HYPERGEOMETRIC FUNCTION (Koekoek and Swarttouw 1998).

The first few POLYNOMIALS are

$$U_0(x) = 1$$

$$U_1(x) = 2x$$

$$U_2(x) = 4x^2 - 1$$

$$U_3(x) = 8x^3 - 4x$$

$$U_4(x) = 16x^4 - 12x^2 + 1$$

$$U_5(x) = 32x^5 - 32x^3 + 6x$$

$$U_6(x) = 64x^6 - 80x^4 + 24x^2 - 1.$$

Letting $x \equiv \cos\theta$ allows the Chebyshev polynomials of the second kind to be written as

$$U_n(x) = \frac{\sin[(n+1)]\theta}{\sin\theta}. \tag{10}$$

The second linearly dependent solution to the transformed differential equation is then given by

$$W_n(x) = \frac{\cos[(n+1)\theta]}{\sin\theta}, \tag{11}$$

which can also be written

$$W_n(x) = (1-x^2)^{-1/2} T_{n+1}(x), \tag{12}$$

where $T_n(x)$ is a CHEBYSHEV POLYNOMIAL OF THE FIRST KIND. Note that $W_n(x)$ is therefore *not* a POLYNOMIAL.

The triangle of RESULTANTS $\rho(U_n(x), U_k(x))$ is given by $\{0\}$, $\{-4, 0\}$, $\{0, -64, 0\}$, $\{16, 256, 4096, 0\}$, $\{0, 0, 0, 1048576, 0\}$, ... (Sloane's A054376).

See also CHEBYSHEV APPROXIMATION FORMULA, CHEBYSHEV POLYNOMIAL OF THE FIRST KIND, ULTRASPHERICAL POLYNOMIAL

References

Abramowitz, M. and Stegun, C. A. (Eds.). "Orthogonal Polynomials." Ch. 22 in *Handbook of Mathematical Functions with Formulas, Graphs, and Mathematical Tables, 9th printing.* New York: Dover, pp. 771–802, 1972.

Arfken, G. "Chebyshev (Tschebyscheff) Polynomials" and "Chebyshev Polynomials--Numerical Applications." §13.3 and 13.4 in *Mathematical Methods for Physicists, 3rd ed.* Orlando, FL: Academic Press, pp. 731–748, 1985.

Koekoek, R. and Swarttouw, R. F. "Chebyshev." §1.8.2 in *The Askey-Scheme of Hypergeometric Orthogonal Polynomials and its q-Analogue.* Delft, Netherlands: Technische Universiteit Delft, Faculty of Technical Mathematics and Informatics Report 98–17, pp. 41–43, 1998. ftp://www.twi.tudelft.nl/publications/tech-reports/1998/DUT-TWI-98–17.ps.gz.

Koepf, W. "Efficient Computation of Chebyshev Polynomials." In *Computer Algebra Systems: A Practical Guide* (Ed. M. J. Wester). New York: Wiley, pp. 79–99, 1999.

Pegg, E. Jr. "ChebyshevU." http://www.mathpuzzle.com/ChebyshevU.html.

Rivlin, T. J. *Chebyshev Polynomials.* New York: Wiley, 1990.

Sloane, N. J. A. Sequences A054376 in "An On-Line Version of the Encyclopedia of Integer Sequences." http://www.research.att.com/~njas/sequences/eisonline.html.

Spanier, J. and Oldham, K. B. "The Chebyshev Polynomials $T_n(x)$ and $U_n(x)$." Ch. 22 in *An Atlas of Functions.* Washington, DC: Hemisphere, pp. 193–207, 1987.

Vasilyev, N. and Zelevinsky, A. "A Chebyshev Polyplayground: Recurrence Relations Applied to a Famous Set of Formulas." *Quantum* **10**, 20–26, Sept./Oct. 1999.

Zwillinger, D. (Ed.). *CRC Standard Mathematical Tables and Formulae.* Boca Raton, FL: CRC Press, 1995.

Chebyshev Quadrature

A GAUSSIAN QUADRATURE-like FORMULA for numerical estimation of integrals. It uses WEIGHTING FUNCTION $W(x) = 1$ in the interval $[-1, 1]$ and forces all the weights to be equal. The general FORMULA is

$$\int_{-1}^{1} f(x)\,dx = \frac{2}{n} \sum_{i=1}^{n} f(x_i).$$

The ABSCISSAS are found by taking terms up to y^n in the MACLAURIN SERIES of

$$s_n(y) = \exp\left\{\frac{1}{2}n\left[-2 + \ln(1-y)\left(1 - \frac{1}{y}\right) + \ln(1+y)\left(1 + \frac{1}{y}\right)\right]\right\},$$

and then defining

$$G_n(x) \equiv x^n s_n\left(\frac{1}{x}\right).$$

The ROOTS of $G_n(x)$ then give the ABSCISSAS. The first few values are

$$G_0(x) = 1$$
$$G_1(x) = x$$
$$G_2(x) = \tfrac{1}{3}(3x^2 - 1)$$
$$G_3(x) = \tfrac{1}{2}(2x^3 - x)$$
$$G_4(x) = \tfrac{1}{45}(45x^4 - 30x^2 + 1)$$
$$G_5(x) = \tfrac{1}{72}(72x^5 - 60x^3 + 7x)$$
$$G_6(x) = \tfrac{1}{105}(105x^6 - 105x^4 + 21x^2 - 1)$$
$$G_7(x) = \tfrac{1}{6480}(6480x^7 - 7560x^5 + 2142x^3 - 149x)$$
$$G_8(x) = \tfrac{1}{42525}(42525x^8 - 56700x^6 + 20790x^4 - 2220x^2 - 43)$$
$$G_9(x) = \tfrac{1}{22400}(22400x^9 - 33600x^7 + 15120x^5 - 2280x^3 + 53x).$$

Because the ROOTS are all REAL for $n \leq 7$ and $n = 9$ only (Hildebrand 1956), these are the only permissible orders for Chebyshev quadrature. The error term is

$$E_n = \begin{cases} c_n \dfrac{f^{(n+1)}(\xi)}{(n+1)!} & n \text{ odd} \\[2mm] c_n \dfrac{f^{(n+2)}(\xi)}{(n+2)!} & n \text{ even}, \end{cases}$$

where

$$c_n = \begin{cases} \displaystyle\int_{-1}^{1} xG_n(x)\,dx & n \text{ odd} \\[2mm] \displaystyle\int_{-1}^{1} x^2 G_n(x)\,dx & n \text{ even}. \end{cases}$$

The first few values of c_n are 2/3, 8/45, 1/15, 32/945, 13/756, and 16/1575 (Hildebrand 1956). Beyer (1987) gives abscissas up to $n = 7$ and Hildebrand (1956) up to $n = 9$.

n	x_i
2	$\pm$ 0.57735
3	0
	$\pm$ 0.707107
4	$\pm$ 0.187592
	$\pm$ 0.794654
5	0
	$\pm$ 0.374541
	$\pm$ 0.832497
6	$\pm$ 0.266635
	$\pm$ 0.422519
	$\pm$ 0.866247
7	0
	$\pm$ 0.323912
	$\pm$ 0.529657
	$\pm$ 0.883862
9	0
	$\pm$ 0.167906
	$\pm$ 0.528762
	$\pm$ 0.601019
	$\pm$ 0.911589

The ABSCISSAS and weights can be computed analytically for small n.

n	x_i
2	$\pm\frac{1}{3}\sqrt{3}$
3	0
	$\pm\frac{1}{2}\sqrt{2}$
4	$\pm\sqrt{\dfrac{\sqrt{5}-2}{3\sqrt{5}}}$
	$\pm\sqrt{\dfrac{\sqrt{5}+2}{3\sqrt{5}}}$
5	0
	$\pm\frac{1}{2}\sqrt{\dfrac{5-\sqrt{11}}{3}}$
	$\pm\frac{1}{2}\sqrt{\dfrac{5+\sqrt{11}}{3}}$

See also GAUSSIAN QUADRATURE, LOBATTO QUADRATURE

References

Beyer, W. H. *CRC Standard Mathematical Tables, 28th ed.* Boca Raton, FL: CRC Press, p. 466, 1987.

Hildebrand, F. B. *Introduction to Numerical Analysis.* New York: McGraw-Hill, pp. 345–351, 1956.

Chebyshev Sum Inequality

If

$$a_1 \geq a_2 \geq \ldots \geq a_n$$
$$b_1 \geq b_2 \geq \ldots \geq b_n,$$

then

$$n \sum_{k=1}^{n} a_k b_k \geq \left(\sum_{k=1}^{n} a_k\right)\left(\sum_{k=1}^{n} b_k\right).$$

This is true for *any* distribution.

See also CAUCHY'S INEQUALITY, HÖLDER'S INEQUALITIES

References

Gradshteyn, I. S. and Ryzhik, I. M. *Tables of Integrals, Series, and Products, 6th ed.* San Diego, CA: Academic Press, p. 1092, 2000.

Hardy, G. H.; Littlewood, J. E.; and Pólya, G. *Inequalities, 2nd ed.* Cambridge, England: Cambridge University Press, pp. 43–44, 1988.

Chebyshev-Gauss Quadrature

Also called CHEBYSHEV QUADRATURE. A GAUSSIAN QUADRATURE over the interval $[-1, 1]$ with WEIGHTING FUNCTION $W(x) = (1-x^2)^{-1/2}$ (Abramowitz and Stegun 1972, p. 889). The ABSCISSAS for quadrature order n are given by the roots of the CHEBYSHEV POLYNOMIAL OF THE FIRST KIND $T_n(x)$, which occur symmetrically about 0. The WEIGHTS are

$$w_i = -\frac{A_{n+1}\gamma_n}{A_n T_n'(x_i)T_{n+1}(x_i)} = \frac{A_n}{A_{n-1}}\frac{\gamma_{n-1}}{T_{n-1}(x_i)T_n'(x_i)}, \quad (1)$$

where A_n is the COEFFICIENT of x_n in $T_n(x)$. For HERMITE POLYNOMIALS,

$$A_n = 2^{n-1}, \quad (2)$$

so

$$\frac{A_{n+1}}{A_n} = 2. \quad (3)$$

Additionally,

$$\gamma_n = \frac{1}{2}\pi, \quad (4)$$

so

$$w_i = -\frac{\pi}{T_{n+1}(x_i)T_n'(x_i)}. \quad (5)$$

Since

$$T_n(x) = \cos(n \cos^{-1} x), \qquad (6)$$

the ABSCISSAS are given explicitly by

$$x_i = \cos\left[\frac{(2i-1)\pi}{2n}\right]. \qquad (7)$$

Since

$$T_n'(x_i) = \frac{(-1)^{i+1} n}{\alpha_i} \qquad (8)$$

$$T_{n+1}(x_i) = (-1)^i \sin \alpha_i, \qquad (9)$$

where

$$\alpha_i = \frac{(2i-1)\pi}{2n}, \qquad (10)$$

all the WEIGHTS are

$$w_i = \frac{\pi}{n}. \qquad (11)$$

The explicit FORMULA is then

$$\int_{-1}^{1} \frac{f(x)\,dx}{\sqrt{1-x^2}}$$

$$= \frac{\pi}{n} \sum_{k=1}^{n} f\left[\cos\left(\frac{2k-1}{2n}\,\pi\right)\right] + \frac{2\pi}{2^{2n}(2n)!} f^{(2n)}(\xi). \qquad (12)$$

The following two tables give the numerical and analytic values for the first few points and weights.

n	x_i	w_i
2	$\pm$ 0.707107	1.5708
3	0	1.0472
	$\pm$ 0.866025	1.0472
4	$\pm$ 0.382683	0.785398
	$\pm$ 0.92388	0.785398
5	0	0.628319
	$\pm$ 0.587785	0.628319
	$\pm$ 0.951057	0.628319

2	$\pm\frac{1}{2}\sqrt{2}$	$\frac{1}{2}\pi$
3	0	$\frac{1}{3}\pi$
3	$\pm\frac{1}{2}\sqrt{3}$	$\frac{1}{3}\pi$
4	$\pm\frac{1}{2}\sqrt{2-\sqrt{2}}$	$\frac{1}{4}\pi$
4	$\pm\frac{1}{2}\sqrt{2+\sqrt{2}}$	$\frac{1}{4}\pi$

5	0	$\frac{1}{5}\pi$
5	$\pm\frac{1}{2}\sqrt{\frac{1}{2}(5-\sqrt{5})}$	$\frac{1}{5}\pi$
5	$\pm\frac{1}{2}\sqrt{\frac{1}{2}(5+\sqrt{5})}$	$\frac{1}{5}\pi$

References

Abramowitz, M. and Stegun, C. A. (Eds.). *Handbook of Mathematical Functions with Formulas, Graphs, and Mathematical Tables, 9th printing.* New York: Dover, p. 889, 1972.

Bronwin, B. "On the Determination of the Coefficients in Any Series of Sines and Cosines of Multiples of a Variable Angle from Particular Values of that Series." *Phil. Mag.* **34**, 260–268, 1849.

Hildebrand, F. B. *Introduction to Numerical Analysis.* New York: McGraw-Hill, pp. 330–331, 1956.

Tchebicheff, P. "Sur les quadratures." *J. de math. pures appliq.* **19**, 19–34, 1874.

Whittaker, E. T. and Robinson, G. "Chebyshef's Formulae." §79 in *The Calculus of Observations: A Treatise on Numerical Mathematics, 4th ed.* New York: Dover, pp. 158–159, 1967.

Chebyshev-Radau Quadrature

A GAUSSIAN QUADRATURE-like FORMULA over the interval $[-1, 1]$ which has WEIGHTING FUNCTION $W(x) = x$. The general FORMULA is

$$\int_{-1}^{1} x f(x)\,dx = \sum_{i=1}^{n} w_i[f(x_i) - f(-x_i)].$$

n	x_i	w_i
1	0.7745967	0.4303315
2	0.5002990	0.2393715
	0.8922365	0.2393715
3	0.4429861	0.1599145
	0.7121545	0.1599145
	0.9293066	0.1599145
4	0.3549416	0.1223363
	0.6433097	0.1223363
	0.7783202	0.1223363
	0.9481574	0.1223363

References

Beyer, W. H. *CRC Standard Mathematical Tables, 28th ed.* Boca Raton, FL: CRC Press, p. 466, 1987.

Chebyshev's Formula
CHEBYSHEV-GAUSS QUADRATURE

Chebyshev's Theorem
There are at least two theorems known as Chebyshev's theorem. The first is BERTRAND'S POSTULATE, and the second is a weak form of the PRIME NUMBER THEOREM stating that the ORDER OF MAGNITUDE of the PRIME COUNTING FUNCTION $\pi(x)$ is

$$\pi(x) \asymp \frac{x}{\ln x},$$

where $\asymp$ denotes "is ASYMPTOTIC to" (Hardy and Wright 1979, p. 9).

See also BERTRAND'S POSTULATE, PRIME COUNTING FUNCTION, PRIME NUMBER THEOREM

References
Hardy, G. H. and Wright, E. M. *An Introduction to the Theory of Numbers, 5th ed.* Oxford, England: Clarendon Press, 1979.

Chebyshev-Sylvester Constant
In 1891, Chebyshev and Sylvester showed that for sufficiently large x, there exists at least one PRIME NUMBER p satisfying

$$x < p < (1 + \alpha)x,$$

where $\alpha = 0.092\ldots$. Since the PRIME NUMBER THEOREM shows the above inequality is true for all $\alpha > 0$ for sufficiently large x, this constant is only of historical interest.

References
Le Lionnais, F. *Les nombres remarquables.* Paris: Hermann, p. 22, 1983.

ChebyshevT
CHEBYSHEV POLYNOMIAL OF THE FIRST KIND

ChebyshevU
CHEBYSHEV POLYNOMIAL OF THE SECOND KIND

Checkerboard
CHESSBOARD

Checker-Jumping Problem
Seeks the minimum number of checkers placed on a board required to allow pieces to move by a sequence of horizontal or vertical jumps (removing the piece jumped over) n rows beyond the forward-most initial checker. The first few cases are 2, 4, 8, 20. It is, however, impossible to reach level five.

See also CHECKERS

References
Honsberger, R. *Mathematical Gems II.* Washington, DC: Math. Assoc. Amer., pp. 23–28, 1976.

Checkers
Schroeppel (1972) estimated that there are about 10^{12} possible positions. However, this disagrees with the estimate of Jon Schaeffer of 5×10^{20} plausible positions, with 10^{18} reachable under the rules of the game. Because "solving" checkers may require only the SQUARE ROOT of the number of positions in the search space (i.e., 10^9), there is hope that some day checkers may be solved (i.e., it may be possible to guarantee a win for the first player to move before the game is even started; Dubuque 1996).

Depending on how they are counted, the number of EULERIAN circuits on an $n \times n$ checkerboard are either 1, 40, 793, 12800, 193721, ... (Sloane's A006240) or 1, 13, 108, 793, 5611, 39312, ... (Sloane's A006239).

See also BOARD, CHECKER-JUMPING PROBLEM, CHESSBOARD

References
Dubuque, W. "Re: number of legal chess positions." math-fun@cs.arizona.edu posting, Aug 15, 1996.
Hopper, M. *Win at Checkers.* New York: Dover, 1956.
Kraitchik, M. "Chess and Checkers" and "Checkers (Draughts)." §12.1.1 and 12.1.10 in *Mathematical Recreations.* New York: W. W. Norton, pp. 267–276 and 284–287, 1942.
Parlett, D. S. *Oxford History of Board Games.* Oxford, England: Oxford University Press, 1999.
Schaeffer, J. *One Jump Ahead: Challenging Human Supremacy in Checkers.* New York: Springer-Verlag, 1997.
Schroeppel, R. Item 93 in Beeler, M.; Gosper, R. W.; and Schroeppel, R. *HAKMEM.* Cambridge, MA: MIT Artificial Intelligence Laboratory, Memo AIM-239, p. 35, Feb. 1972.
Sloane, N. J. A. Sequences A006239/M4909 and A006240/M5271 in "An On-Line Version of the Encyclopedia of Integer Sequences." http://www.research.att.com/~njas/sequences/eisonline.html.

Checksum
A sum of the digits in a given transmission modulo some number. The simplest form of checksum is a parity bit appended on to 7-bit numbers (e.g., ASCII characters) such that the total number of 1s is always EVEN ("even parity") or ODD ("odd parity"). A significantly more sophisticated checksum is the CYCLIC REDUNDANCY CHECK (or CRC), which is based on the algebra of polynomials over the integers (mod 2). It is substantially more reliable in detecting transmission errors, and is one common error-checking protocol used in modems.

See also CYCLIC REDUNDANCY CHECK, ERROR-CORRECTING CODE

References

Press, W. H.; Flannery, B. P.; Teukolsky, S. A.; and Vetterling, W. T. "Cyclic Redundancy and Other Checksums." Ch. 20.3 in *Numerical Recipes in FORTRAN: The Art of Scientific Computing, 2nd ed.* Cambridge, England: Cambridge University Press, pp. 888–895, 1992.

Cheeger's Finiteness Theorem

Consider the set of compact n-RIEMANNIAN MANIFOLDS M with diameter$(M) \le d$, Volume$(M) \ge V$, and $|\mathcal{K}| \le \kappa$ where κ is the SECTIONAL CURVATURE. Then there is a bound on the number of DIFFEOMORPHISMS classes of this set in terms of the constants n, d, V, and κ.

References

Chavel, I. *Riemannian Geometry: A Modern Introduction.* New York: Cambridge University Press, 1994.

Chefalo Knot

A fake KNOT created by tying a SQUARE KNOT, then looping one end twice through the KNOT such that when both ends are pulled, the KNOT vanishes.

Chen's Theorem

Every "large" EVEN NUMBER may be written as $2n = p + m$ where p is a PRIME and $m \in P_2$ is the SET of SEMIPRIMES (i.e., 2-ALMOST PRIMES).

See also ALMOST PRIME, GOLDBACH CONJECTURE, PRIME NUMBER, SCHNIRELMANN'S THEOREM, SEMIPRIME

References

Chen, J. R. "On the Representation of a Large Even Integer as the Sum of a Prime and the Product of at Most Two Primes." *Kexue Tongbao* **17**, 385–386, 1966.
Chen, J. R. "On the Representation of a Large Even Integer as the Sum of a Prime and the Product of at Most Two Primes. I." *Sci. Sinica* **16**, 157–176, 1973.
Chen, J. R. "On the Representation of a Large Even Integer as the Sum of a Prime and the Product of at Most Two Primes. II." *Sci. Sinica* **16**, 421–430, 1978.
Hardy, G. H. and Wright, W. M. "Unsolved Problems Concerning Primes." Appendix §3 in *An Introduction to the Theory of Numbers, 5th ed.* Oxford, England: Oxford University Press, pp. 415–416, 1979.
Ribenboim, P. *The New Book of Prime Number Records.* New York: Springer-Verlag, p. 297, 1996.
Rivera, C. "Problems & Puzzles: Conjecture Chen's Conjecture.-002." http://www.primepuzzles.net/conjectures/conj_002.htm.
Ross, P. M. "On Chen's Theorem that Each Large Even Number has the Form $p_1 + p_2$ or $p_1 + p_2 p_3$." *J. London Math. Soc.* **10**, 500–506, 1975.

Chern Class

A GADGET defined for COMPLEX VECTOR BUNDLES. The Chern classes of a COMPLEX MANIFOLD are the Chern classes of its TANGENT BUNDLE. The ith Chern class is an OBSTRUCTION to the existence of $(n - i + 1)$ everywhere COMPLEX linearly independent VECTOR FIELDS on that VECTOR BUNDLE. The ith Chern class is in the $(2i)$th cohomology group of the base SPACE.

See also CHERN NUMBER, OBSTRUCTION, PONTRYAGIN CLASS, STIEFEL-WHITNEY CLASS

Chern Number

The Chern number is defined in terms of the CHERN CLASS of a MANIFOLD as follows. For any collection CHERN CLASSES such that their cup product has the same DIMENSION as the MANIFOLD, this cup product can be evaluated on the MANIFOLD'S FUNDAMENTAL CLASS. The resulting number is called the Chern number for that combination of Chern classes. The most important aspect of Chern numbers is that they are COBORDISM invariant.

See also CHERN CLASS, PONTRYAGIN NUMBER, STIEFEL-WHITNEY NUMBER

Chernoff Face

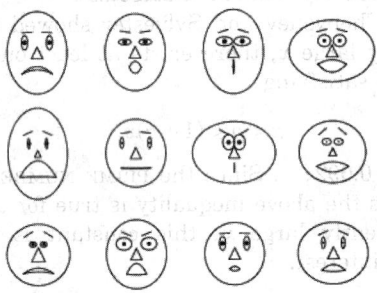

A way to display n variables on a 2-D surface. For instance, let x be eyebrow slant, y be eye size, z be nose length, etc. The above figures show faces produced using 10 characteristics–head eccentricity, eye size, eye spacing, eye eccentricity, pupil size, eyebrow slant, nose size, mouth shape, mouth size, and mouth opening)–each assigned one of 10 possible values, generated using *Mathematica* (S. Dickson).

References

Dickson, S. "Faces" Mathematica notebook. http://mathworld.wolfram.com/notebooks/ChernoffFaces.nb.
Gonick, L. and Smith, W. *The Cartoon Guide to Statistics.* New York: Harper Perennial, p. 212, 1993.

Chess

Chess is a game played on an 8×8 BOARD, called a CHESSBOARD, of alternating black and white squares. Pieces with different types of allowed moves are placed on the board, a set of black pieces in the first two rows and a set of white pieces in the last two rows. The pieces are called the bishop (2), king (1), knight (2), pawn (8), queen (1), and rook (2). The object of the game is to capture the opponent's king. It is believed that chess was played in India as early as the sixth century AD.

Hardy (1999, p. 17) estimated the number of possible games of chess as

$$10^{10^{50}}$$

In a game of 40 moves, the number of possible board positions is at least 10^{120} according to Peterson (1996). However, this value does not agree with the 10^{40} possible positions given by Beeler *et al.* (1972). This value was obtained by estimating the number of pawn positions (in the no-captures situation, this is 15^8), times all pieces in all positions, dividing by 2 for each of the (rook, knight) which are interchangeable, dividing by 2 for each pair of bishops (since half the positions will have the bishops on the same color squares). There are more positions with one or two captures, since the pawns can then switch columns (Schroeppel 1996). Shannon (1950) gave the value

$$P(40) \approx \frac{64!}{32!(8!)^2(2!)^6} \approx 10^{43}.$$

The number of chess games which end in exactly n plies (including games that mate in fewer than n plies) for $n = 1, 2, 3, \ldots$ are 20, 400, 8902, 197742, 4897256, 120921506, 3284294545, ... (K. Thompson, Sloane's A006494). Rex Stout's fictional detective Nero Wolfe quotes the number of possible games after ten moves as follows: "Wolfe grunted. One hundred and sixty-nine million, five hundred and eighteen thousand, eight hundred and twenty-nine followed by twenty-one ciphers. The number of ways the first ten moves, both sides, may be played" (Stout 1983). The number of chess positions after n moves for $n = 1, 2, \ldots$ are 20, 400, 5362, 71852, 809896?, 9132484?, ... (Schwarzkopf 1994, Sloane's A019319).

Cunningham (1889) incorrectly found 197,299 games and 71,782 positions after the fourth move. C. Flye St. Marie was the first to find the correct number of positions after four moves: 71,852. Dawson (1946) gives the source as *Intermediare des Mathematiques* (1895), but K. Fabel writes that Flye St. Marie corrected the number 71,870 (which he found in 1895) to 71,852 in 1903. The history of the determination of the chess sequences is discussed in Schwarzkopf (1994).

The analysis of chess is extremely complicated due to the many possible options at each move. Steinhaus (1983, pp. 11–14), as well as many entire books, consider clever end-game positions which may be analyzed completely.

Two problems in recreational mathematics ask

1. How many pieces of a given type can be placed on a CHESSBOARD without any two attacking.
2. What is the smallest number of pieces needed to occupy or attack every square.

The answers are given in the following table (Madachy 1979).

Piece	Max.	Min.
BISHOPS	14	8
KINGS	16	9
KNIGHTS	32	12
QUEENS	8	5
ROOKS	8	8

See also BISHOPS PROBLEM, BOARD, CHECKERBOARD, CHECKERS, FAIRY CHESS, GO, GOMORY'S THEOREM, HARD HEXAGON ENTROPY CONSTANT, KINGS PROBLEM, KNIGHT'S TOUR, MAGIC TOUR, QUEENS PROBLEM, ROOKS PROBLEM, TOUR

References

Ball, W. W. R. and Coxeter, H. S. M. *Mathematical Recreations and Essays, 13th ed.* New York: Dover, pp. 124–127, 1987.

Beeler, M. *et al.* Item 95 in Beeler, M.; Gosper, R. W.; and Schroeppel, R. *HAKMEM.* Cambridge, MA: MIT Artificial Intelligence Laboratory, Memo AIM-239, p. 35, Feb. 1972.

Culin, S. "Tjyang-keui--Chess." §82 in *Games of the Orient: Korea, China, Japan.* Rutland, VT: Charles E. Tuttle, pp. 82–91, 1965.

Dawson, T. R. "A Surprise Correction." *The Fairy Chess Review* **6**, 44, 1946.

Dickins, A. "A Guide to Fairy Chess." p. 28, 1967/1969/1971.

Fabel, K. "Nüsse." *Die Schwalbe* **84**, 196, 1934.

Fabel, K. "Weihnachtsnüsse." *Die Schwalbe* **190**, 97, 1947.

Fabel, K. "Weihnachtsnüsse." *Die Schwalbe* **195**, 14, 1948.

Fabel, K. "Eröffnungen." *Am Rande des Schachbretts*, 34–35, 1947.

Fabel, K. "Die ersten Schritte." *Rund um das Schachbrett*, 107–109, 1955.

Fabel, K. "Eröffnungen." *Schach und Zahl* **8**, 1966/1971.

Hardy, G. H. *Ramanujan: Twelve Lectures on Subjects Suggested by His Life and Work, 3rd ed.* New York: Chelsea, 1999.

Hunter, J. A. H. and Madachy, J. S. *Mathematical Diversions.* New York: Dover, pp. 86–89, 1975.

Kraitchik, M. "Chess and Checkers." §12.1.1 in *Mathematical Recreations.* New York: W. W. Norton, pp. 267–276, 1942.

Lasker, E. *Lasker's Manual of Chess.* New York: Dover, 1960.

Madachy, J. S. "Chessboard Placement Problems." Ch. 2 in *Madachy's Mathematical Recreations.* New York: Dover, pp. 34–54, 1979.

Parlett, D. S. *Oxford History of Board Games.* Oxford, England: Oxford University Press, 1999.

Peterson, I. "The Soul of a Chess Machine: Lessons Learned from a Contest Pitting Man Against Computer." *Sci. News* **149**, 200–201, Mar. 30, 1996.

Petkovic, M. *Mathematics and Chess.* New York: Dover, 1997.

Schroeppel, R. "Reprise: Number of legal chess positions." tech-news@cs.arizona.edu posting, Aug. 18, 1996.

Schwarzkopf, B. "Die ersten Züge." *Problemkiste*, 142–143, No. 92, Apr. 1994.

Shannon, C. "Programming a Computer for Playing Chess." *Phil. Mag.* **41**, 256–275, 1950.

Sloane, N. J. A. Sequences A006494, A007545/M5100, and A019319 in "An On-Line Version of the Encyclopedia of Integer Sequences." http://www.research.att.com/~njas/sequences/eisonline.html.

Steinhaus, H. *Mathematical Snapshots, 3rd ed.* New York: Dover, pp. 11–14, 1999.

Stout, R. "Gambit." In *Seven Complete Nero Wolfe Novels.* New York: Avenic Books, p. 475, 1983.

Velucchi, M. "Some On-Line PostScript MathChess Papers." http://anduin.eldar.org/~problemi/papers.html.

Chessboard

A board containing 8×8 squares alternating in color between black and white on which the game of CHESS is played. The checkerboard is identical to the chessboard except that chess's black and white squares are colored red and white in CHECKERS.

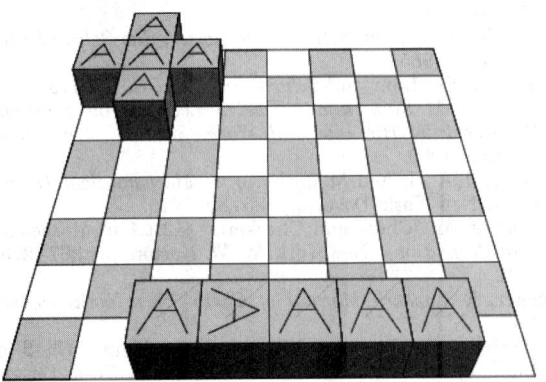

It is impossible to cover a chessboard from which two opposite corners have been removed with DOMINOES. Sprague (1963) considered the problem of "rolling" five cubes, each which an upright letter "A" on its top, on a chessboard. Here "rolling" means the cubes are moved from square to adjacent square by being tipped over along an edge (as one might move a heavy box) in a series of quarter turns. If five such cubes are initially arranged in the shape of a plus sign with the edges of the of plus sign aligned with the upper and left corners of a chessboard (top left in above figure), then it is impossible to obtain a straight row or column with all "A"s on top and oriented identically. The best that can be done is to place four out of the five "A"s in the same orientation and facing upward, with the remaining "A" also facing upward and rotated a quarter turn, illustrated above in the bottom row (Gardner 1984, pp. 75–78).

 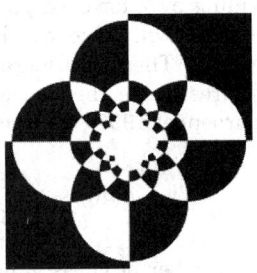

The above plot shows a chessboard centered at (0, 0) and its INVERSE about a small circle also centered at (0, 0) (Gardner 1984, pp. 244–245; Dixon 1991).

See also CHECKERS, CHESS, CIRCULAR CHESSBOARD, DOMINO, GOMORY'S THEOREM, INVERSION, KINGS PROBLEM, KNIGHTS PROBLEM, KNIGHT'S TOUR, QUEENS PROBLEM, ROOKS PROBLEM, WHEAT AND CHESSBOARD PROBLEM

References

Dixon, R. "Inverse Points and Mid-Circles." §1.6 in *Mathographics.* New York: Dover, pp. 62–73, 1991.

Gardner, M. *The Sixth Book of Mathematical Games from Scientific American.* Chicago, IL: University of Chicago Press, 1984.

Pappas, T. "The Checkerboard." *The Joy of Mathematics.* San Carlos, CA: Wide World Publ./Tetra, pp. 136 and 232, 1989.

Sprague, R. *Recreations in Mathematics: Some Novel Puzzles.* London: Blackie and Sons, 1963.

Steinhaus, H. *Mathematical Snapshots, 3rd ed.* New York: Dover, pp. 29–30, 1999.

Chevalley Groups

Finite SIMPLE GROUPS of LIE-TYPE. They include four families of linear SIMPLE GROUPS: $PSL(n, q)$, $PSU(n, q)$, $PSp(2n, q)$, or $P\Omega^\epsilon(n, q)$.

See also TWISTED CHEVALLEY GROUPS

References

Wilson, R. A. "ATLAS of Finite Group Representation."
http://for.mat.bham.ac.uk/atlas/html/contents.html#exc.

Chevalley's Theorem

Let $f(x)$ be a member of a FINITE FIELD $F[x_1, x_2 \ldots, x_n]$ and suppose $f(0, 0, \ldots, 0) = 0$ and n is greater than the degree of f, then f has at least two zeros in $A^n(F)$.

References

Chevalley, C. "Démonstration d'une hypothèse de M. Artin." *Abhand. Math. Sem. Hamburg* **11**, 73–75, 1936.
Ireland, K. and Rosen, M. "Chevalley's Theorem." §10.2 in *A Classical Introduction to Modern Number Theory, 2nd ed.* New York: Springer-Verlag, pp. 143–144, 1990.

Chevron

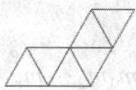

A 6-POLYIAMOND.

References

Golomb, S. W. *Polyominoes: Puzzles, Patterns, Problems, and Packings, 2nd ed.* Princeton, NJ: Princeton University Press, p. 92, 1994.

Chi

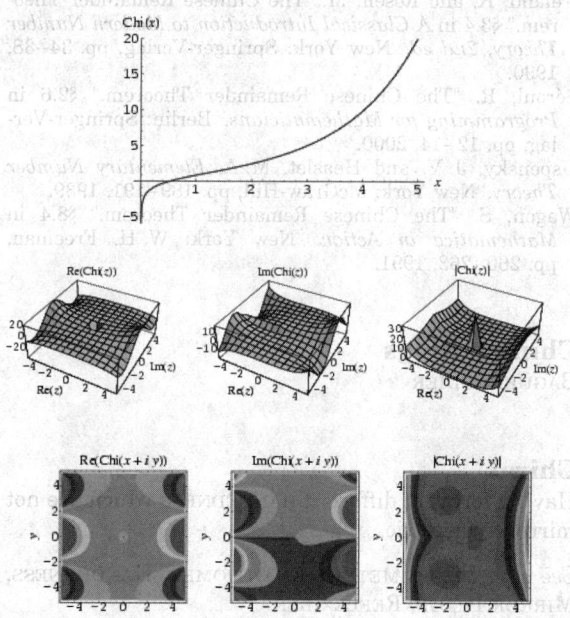

The Chi function is defined by

$$\text{Chi}(z) = \gamma + \ln z + \int_0^z \frac{\cosh t - 1}{t}\, dt,$$

where γ is the EULER-MASCHERONI CONSTANT. The function is given by the *Mathematica* command `CoshIntegral[z]`.

See also COSINE INTEGRAL, SHI, SINE INTEGRAL

References

Abramowitz, M. and Stegun, C. A. (Eds.). "Sine and Cosine Integrals." §5.2 in *Handbook of Mathematical Functions with Formulas, Graphs, and Mathematical Tables, 9th printing.* New York: Dover, pp. 231–233, 1972.

Chi Distribution

The probability density function and cumulative distribution function are

$$P_n(x) = \frac{2^{1-n/2} x^{n-1} e^{-x^2/2}}{\Gamma(\frac{1}{2} n)} \tag{1}$$

$$D_n(x) = Q(\tfrac{1}{2} n, \tfrac{1}{2} x^2), \tag{2}$$

where Q is the REGULARIZED GAMMA FUNCTION.

$$\mu = \frac{\sqrt{2}\,\Gamma(\frac{1}{2}(n+1))}{\Gamma(\frac{1}{2} n)} \tag{3}$$

$$\sigma^2 = \frac{2[\Gamma(\frac{1}{2} n)\Gamma(1+\frac{1}{2} n) - \Gamma^2(\frac{1}{2}(n+1))]}{\Gamma^2(\frac{1}{2} n)} \tag{4}$$

$$\gamma_1 = \frac{2\Gamma^3(\frac{1}{2}(n+1)) - 3\Gamma(\frac{1}{2} n)\Gamma(\frac{1}{2}(n+1))\Gamma(1+\frac{1}{2} n)}{[\Gamma(\frac{1}{2} n)\Gamma(1+\frac{1}{2} n) - \Gamma^2(\frac{1}{2}(n+1))]^{3/2}}$$
$$+ \frac{\Gamma^2(\frac{1}{2} n)\Gamma\left(\dfrac{3+n}{2}\right)}{[\Gamma(\frac{1}{2} n)\Gamma(1+\frac{1}{2} n) - \Gamma^2(\frac{1}{2}(n+1))]^{3/2}} \tag{5}$$

$$\gamma_2 = \frac{-3\Gamma^4(\frac{1}{2}(n+1)) + 6\Gamma(\frac{1}{2} n)\Gamma^2(\frac{1}{2}(n+1))\Gamma(1+\frac{1}{2} n)}{\left[\Gamma(\frac{1}{2} n)\Gamma\left(\dfrac{2+n}{2}\right) - \Gamma^2(\frac{1}{2}(n+1))\right]^2}$$
$$+ \frac{-4\Gamma^2(\frac{1}{2} n)\Gamma(\frac{1}{2}(n+1))\Gamma\left(\dfrac{3+n}{2}\right) + \Gamma^3(\frac{1}{2} n)\Gamma\left(\dfrac{4+n}{2}\right)}{\left[\Gamma(\frac{1}{2} n)\Gamma\left(\dfrac{2+n}{2}\right) - \Gamma^2(\frac{1}{2}(n+1))\right]^2}, \tag{6}$$

where μ is the MEAN, σ^2 the VARIANCE, γ_1 the SKEWNESS, and γ_2 the KURTOSIS. For $n = 1$, the χ distribution is a HALF-NORMAL DISTRIBUTION with $\theta = 1$. For $n = 2$, it is a RAYLEIGH DISTRIBUTION with $\sigma = 1$.

See also CHI-SQUARED DISTRIBUTION, HALF-NORMAL DISTRIBUTION, RAYLEIGH DISTRIBUTION

Chi Inequality

The inequality

$$(j+1)a_j + a_i \ge (j+1)i,$$

which is satisfied by all A-SEQUENCE.

References

Levine, E. and O'Sullivan, J. "An Upper Estimate for the Reciprocal Sum of a Sum-Free Sequence." *Acta Arith.* **34**, 9–24, 1977.

Child

A node which is one EDGE further away from a given node in a ROOTED TREE.

See also ROOT NODE, ROOTED TREE, SIBLING

Chinese Hypothesis

A PRIME p always satisfies the condition that $2^p - 2$ is divisible by p. However, this condition is not true *exclusively* for PRIMES (e.g., $2^{341} - 2$ is divisible by $341 = 11 \cdot 31$). COMPOSITE NUMBERS n (such as 341) for which $2^n - 2$ is divisible by n are called POULET NUMBERS, and are a special class of FERMAT PSEUDO-PRIMES. The Chinese hypothesis is a special case of FERMAT'S LITTLE THEOREM.

See also CARMICHAEL NUMBER, EULER'S THEOREM, FERMAT'S LITTLE THEOREM, FERMAT PSEUDOPRIME, POULET NUMBER, PSEUDOPRIME

References

Shanks, D. *Solved and Unsolved Problems in Number Theory, 4th ed.* New York: Chelsea, pp. 19–20, 1993.

Chinese Postman Problem

A problem asking for the shortest tour of a graph which visits each edge at least once (Kwan 1962; Skiena 1990, p. 194). For an EULERIAN GRAPH, an EULERIAN CIRCUIT is the optimal solution. In a TREE, however, the path crosses each twice.

See also EULERIAN CIRCUIT, TRAVELING SALESMAN PROBLEM

References

Edmonds, J. and Johnson, E. L. "Matching, Euler Tours, and the Chinese Postman." *Math. Programm.* **5**, 88–124, 1973.
Kwan, M. K. "Graphic Programming Using Odd or Even Points." *Chinese Math.* **1**, 273–277, 1962.
Skiena, S. *Implementing Discrete Mathematics: Combinatorics and Graph Theory with Mathematica.* Reading, MA: Addison-Wesley, 1990.

Chinese Remainder Theorem

Let r and s be POSITIVE INTEGERS which are RELATIVELY PRIME and let a and b be any two INTEGERS. Then there is an INTEGER N such that

$$N \equiv a \pmod{r} \tag{1}$$

and

$$N \equiv b \pmod{s}. \tag{2}$$

Moreover, N is uniquely determined modulo rs. An equivalent statement is that if $(r, s) = 1$, then every pair of RESIDUE CLASSES modulo r and s corresponds to a simple RESIDUE CLASS modulo rs.

The theorem can also be generalized as follows. Given a set of simultaneous CONGRUENCES

$$x \equiv a_i \pmod{m_i} \tag{3}$$

for $i = 1, \ldots, r$ and for which the m_i are pairwise RELATIVELY PRIME, the solution of the set of CONGRUENCES is

$$x = a_1 b_1 \frac{M}{m_1} + \ldots + a_r b_r \frac{M}{m_r} \pmod{M}, \tag{4}$$

where

$$M \equiv m_1 m_2 \cdots m_r \tag{5}$$

and the b_i are determined from

$$b_i \frac{M}{m_i} \equiv 1 \pmod{m_i}. \tag{6}$$

References

Ireland, K. and Rosen, M. "The Chinese Remainder Theorem." §3.4 in *A Classical Introduction to Modern Number Theory, 2nd ed.* New York: Springer-Verlag, pp. 34–38, 1990.
Séroul, R. "The Chinese Remainder Theorem." §2.6 in *Programming for Mathematicians.* Berlin: Springer-Verlag, pp. 12–14, 2000.
Uspensky, J. V. and Heaslet, M. A. *Elementary Number Theory.* New York: McGraw-Hill, pp. 189–191, 1939.
Wagon, S. "The Chinese Remainder Theorem." §8.4 in *Mathematica in Action.* New York: W. H. Freeman, pp. 260–263, 1991.

Chinese Rings

BAGUENAUDIER

Chiral

Having forms of different HANDEDNESS which are not mirror-symmetric.

See also DISSYMMETRIC, ENANTIOMER, HANDEDNESS, MIRROR IMAGE, REFLEXIBLE

Chiral Knot

A chiral knot is a KNOT which is *not* capable of being continuously deformed into its own MIRROR IMAGE.

See also AMPHICHIRAL KNOT, KNOT SYMMETRY

Chi-Squared Distribution

A χ^2 distribution is a GAMMA DISTRIBUTION with $\theta \equiv 2$ and $\alpha \equiv r/2$, where r is the number of DEGREES OF FREEDOM. If Y_i have NORMAL INDEPENDENT distributions with MEAN 0 and VARIANCE 1, then

$$\chi^2 \equiv \sum_{i=1}^{r} Y_i^2 \tag{1}$$

is distributed as χ^2 with r DEGREES OF FREEDOM. If χ_i^2 are independently distributed according to a χ^2 distribution with $r_1, r_2, \ldots, r_k$ DEGREES OF FREEDOM, then

$$\sum_{j=1}^{k} \chi_j^2 \tag{2}$$

is distributed according to χ^2 with $r \equiv \Sigma_{j=1}^{k} r_j$ DEGREES OF FREEDOM. The probability density function is

$$P_r(x) = \frac{x^{r/2-1}e^{-x/2}}{\Gamma(\frac{1}{2}r)2^{r/2}} \tag{3}$$

for $x \in [0, \infty)$. The cumulative distribution function is then

$$D_r(\chi^2) = \int_0^{\chi^2} \frac{t^{r/2-1}e^{-t/2}\,dt}{\Gamma(\frac{1}{2}r)2^{r/2}} = \frac{\gamma(\frac{1}{2}r,\,\frac{1}{2}\chi^2)}{\Gamma(\frac{1}{2}r)}$$

$$= P(\tfrac{1}{2}r, \tfrac{1}{2}\chi^2), \tag{4}$$

where $P(a, z)$ is a REGULARIZED GAMMA FUNCTION. The CONFIDENCE INTERVALS can be found by finding the value of x for which $D_r(x)$ equals a given value. The MOMENT-GENERATING FUNCTION of the χ^2 distribution is

$$M(t) = (1 - 2t)^{-r/2} \tag{5}$$

$$R(t) \equiv \ln M(t) = -\tfrac{1}{2}r\ln(1 - 2t) \tag{6}$$

$$R'(t) = \frac{r}{1 - 2t} \tag{7}$$

$$R''(t) = \frac{2r}{(1 - 2t)^2}, \tag{8}$$

so

$$\mu = R'(0) = r \tag{9}$$

$$\sigma^2 = R''(0) = 2r \tag{10}$$

$$\gamma_1 = 2\sqrt{\frac{2}{r}} \tag{11}$$

$$\gamma_2 = \frac{12}{r}. \tag{12}$$

The nth MOMENT about zero for a distribution with r DEGREES OF FREEDOM is

$$m_n' = 2^n \frac{\Gamma(n + \frac{1}{2}r)}{\Gamma(\frac{1}{2}r)} = r(r+2)\cdots(r + 2n - 2), \tag{13}$$

and the moments about the MEAN are

$$\mu_2 = 2r \tag{14}$$

$$\mu_3 = 8r \tag{15}$$

$$\mu_4 = 12r(r + 4). \tag{16}$$

The nth CUMULANT is

$$\kappa_n = 2^n \Gamma(n)(\tfrac{1}{2}r) = 2^{n-1}(n - 1)!r. \tag{17}$$

The MOMENT-GENERATING FUNCTION is

$$M(t) = e^{rt/\sqrt{2r}} \left(1 - \frac{2t}{\sqrt{2r}}\right)^{-r/2}$$

$$= \left[e^{t\sqrt{2/r}}\left(1 - \sqrt{\frac{2}{r}}\,t\right)\right]^{-r/2}$$

$$= \left[1 - \frac{t^2}{r} - \frac{1}{3}\left(\frac{2}{r}\right)^{3/2} t^3 - \ldots\right]^{-r/2}. \tag{18}$$

As $r \to \infty$,

$$\lim_{r \to \infty} M(t) = e^{t^2/2}, \tag{19}$$

so for large r,

$$\sqrt{2\chi^2} = \sqrt{\sum_i \frac{(x_i - \mu_i)^2}{\sigma_i^2}} \tag{20}$$

is approximately a GAUSSIAN DISTRIBUTION with MEAN $\sqrt{2r}$ and VARIANCE $\sigma^2 = 1$. Fisher showed that

$$\frac{\chi^2 - r}{\sqrt{2r - 1}} \tag{21}$$

is an improved estimate for moderate r. Wilson and Hilferty showed that

$$\left(\frac{\chi^2}{r}\right)^{1/3} \tag{22}$$

is a nearly GAUSSIAN DISTRIBUTION with MEAN $\mu = 1 - 2/(9r)$ and VARIANCE $\sigma^2 = 2/(9r)$.

In a GAUSSIAN DISTRIBUTION,

$$P(x)\,dx = \frac{1}{\sigma\sqrt{2\pi}}\,e^{-(x-\mu)^2/2\sigma^2}\,dx, \tag{23}$$

let

$$z \equiv (x - \mu)^2/\sigma^2. \tag{24}$$

Then

$$dz = \frac{2(x - \mu)^2}{\sigma^2} \, dx = \frac{2\sqrt{z}}{\sigma} \, dx \qquad (25)$$

so

$$dx = \frac{\sigma}{2\sqrt{z}} \, dz. \qquad (26)$$

But

$$P(z) \, dz = 2P(x) \, dx, \qquad (27)$$

so

$$P(x) \, dx = 2 \, \frac{1}{\sigma\sqrt{2\pi}} \, e^{-z/2} \, dz = \frac{1}{\sigma\sqrt{\pi}} \, e^{-z/2} \, dz. \qquad (28)$$

This is a χ^2 distribution with $r = 1$, since

$$P(z) \, dz = \frac{z^{1/2-1}e^{-z/2}}{\Gamma(\frac{1}{2})2^{1/2}} \, dz = \frac{x^{-1/2}e^{-1/2}}{\sqrt{2\pi}} \, dz. \qquad (29)$$

If X_i are independent variates with a NORMAL DISTRIBUTION having MEANS μ_i and VARIANCES σ_i^2 for $i = 1, \ldots, n$, then

$$\frac{1}{2}\chi^2 \equiv \sum_{i=1}^{n} \frac{(x_i - \mu_i)^2}{2\sigma_i^2} \qquad (30)$$

is a GAMMA DISTRIBUTION variate with $\alpha = n/2$,

$$P(\tfrac{1}{2}\chi^2)d(\tfrac{1}{2}\chi^2) = \frac{1}{\Gamma(\frac{1}{2}n)} \, e^{-\chi^2/2}(\tfrac{1}{2}\chi^2)^{(n/2)-1}d(\tfrac{1}{2}\chi^2). \qquad (31)$$

The noncentral chi-squared distribution is given by

$$P(x) = 2^{-n/2}e^{-(\lambda+x)/2}x^{n/2-1}F(\tfrac{1}{2}n, \tfrac{1}{4}\lambda x), \qquad (32)$$

where

$$F(a, z) \equiv \frac{{}_0F_1(; \; a; \; z)}{\Gamma(a)}, \qquad (33)$$

${}_0F_1$ is the CONFLUENT HYPERGEOMETRIC LIMIT FUNCTION and Γ is the GAMMA FUNCTION. The MEAN, VARIANCE, SKEWNESS, and KURTOSIS are

$$\mu = \lambda + n \qquad (34)$$

$$\sigma^2 = 2(2\lambda + n) \qquad (35)$$

$$\gamma_1 = \frac{2\sqrt{2}(3\lambda + n)}{(2\lambda + n)^{3/2}} \qquad (36)$$

$$\gamma_2 = \frac{12(4\lambda + n)}{(2\lambda + n)^2}. \qquad (37)$$

See also CHI DISTRIBUTION, SNEDECOR'S F-DISTRIBUTION, STATISTICAL DISTRIBUTION

References

Abramowitz, M. and Stegun, C. A. (Eds.). *Handbook of Mathematical Functions with Formulas, Graphs, and Mathematical Tables, 9th printing.* New York: Dover, pp. 940–943, 1972.

Beyer, W. H. *CRC Standard Mathematical Tables, 28th ed.* Boca Raton, FL: CRC Press, p. 535, 1987.

Kenney, J. F. and Keeping, E. S. "The Chi-Square Distribution." §5.3 in *Mathematics of Statistics, Pt. 2, 2nd ed.* Princeton, NJ: Van Nostrand, pp. 98–100, 1951.

Press, W. H.; Flannery, B. P.; Teukolsky, S. A.; and Vetterling, W. T. "Incomplete Gamma Function, Error Function, Chi-Square Probability Function, Cumulative Poisson Function." §6.2 in *Numerical Recipes in FORTRAN: The Art of Scientific Computing, 2nd ed.* Cambridge, England: Cambridge University Press, pp. 209–214, 1992.

Spiegel, M. R. *Theory and Problems of Probability and Statistics.* New York: McGraw-Hill, pp. 115–116, 1992.

Chi-Squared Test

Let the probabilities of various classes in a distribution be $p_1, p_2, \ldots, p_k$, with means $m_1, m_2, \ldots$. The expected frequency

$$\chi_s^2 = \sum_{i=1}^{k} \frac{(m_i - Np_i)^2}{Np_i}$$

is a measure of the deviation of a sample from expectation. Karl Pearson proved that the limiting distribution of χ_s^2 is χ^2 (Kenney and Keeping 1951, pp. 114–116).

$$\Pr(\chi^2 \geq \chi_s^2) = \int_{\chi_s^2}^{\infty} f(\chi^2) \, d(\chi^2)$$

$$= \frac{1}{2} \int_{\chi_s^2}^{\infty} \frac{\left(\dfrac{\chi^2}{2}\right)^{(k-3)/2}}{\Gamma\left(\dfrac{k-1}{2}\right)} e^{-\chi^2/2} \, d(\chi^2)$$

$$= 1 - \frac{\Gamma\left(\dfrac{1}{2}\chi_s^2, \dfrac{k-1}{2}\right)}{\Gamma\left(\dfrac{k-1}{2}\right)}$$

$$= 1 - I\left(\frac{\chi_s^2}{\sqrt{2(k-1)}}, \frac{k-3}{2}\right),$$

where $I(x, n)$ is PEARSON'S FUNCTION. There are some subtleties involved in using the χ^2 test to fit curves (Kenney and Keeping 1951, pp. 118–119).

When fitting a one-parameter solution using χ^2, the best-fit parameter value can be found by calculating χ^2 at three points, plotting against the parameter values of these points, then finding the minimum of a PARABOLA fit through the points (Cuzzi 1972, pp. 162–168).

See also CHI-SQUARED DISTRIBUTION

Chmutov Surface

References

Cuzzi, J. *The Subsurface Nature of Mercury and Mars from Thermal Microwave Emission.* Ph.D. Thesis. Pasadena, CA: California Institute of Technology, 1972.
Kenney, J. F. and Keeping, E. S. *Mathematics of Statistics, Pt. 2, 2nd ed.* Princeton, NJ: Van Nostrand, 1951.

Chmutov Surface

An ALGEBRAIC SURFACE with affine equation

$$P_d(x_1, x_2) + T_d(x_3) = 0, \tag{1}$$

where $T_d(x)$ is a CHEBYSHEV POLYNOMIAL OF THE FIRST KIND and $P_d(x_1, x_2)$ is a polynomial defined by

$$P_d(x_1, x_2) = \begin{vmatrix} x_1 & 1 & 0 & \cdots & 0 & 0 & 0 \\ 2x_2 & x_1 & 1 & \ddots & 0 & 0 & 0 \\ 3 & x_2 & x_1 & \ddots & \ddots & \vdots & \vdots \\ 0 & 1 & x_2 & \ddots & 1 & 0 & 0 \\ 0 & 0 & 1 & \ddots & x_1 & 1 & 0 \\ \vdots & \ddots & \ddots & \ddots & x_2 & x_1 & 1 \\ 0 & 0 & 0 & \cdots & 1 & x_2 & x_1 \end{vmatrix}$$

$$+ \begin{vmatrix} x_2 & 1 & 0 & \cdots & 0 & 0 & 0 \\ 2x_1 & x_2 & 1 & \ddots & 0 & 0 & 0 \\ 3 & x_1 & x_2 & \ddots & \ddots & \vdots & \vdots \\ 0 & 1 & x_1 & \ddots & 1 & 0 & 0 \\ 0 & 0 & 1 & \ddots & x_2 & 1 & 0 \\ \vdots & \ddots & \ddots & \ddots & x_1 & x_2 & 1 \\ 0 & 0 & 0 & \cdots & 1 & x_1 & x_2 \end{vmatrix}, \tag{2}$$

where the matrices have dimensions $d \times d$. These represent surfaces in $\mathbb{C}P^3$ with only ORDINARY DOUBLE POINTS as singularities. The first few surfaces are given by

$$x + y + z = 0 \tag{3}$$

$$x^2 + y^2 + 2z^2 = 1 + 2x + 2y \tag{4}$$

$$6 + x^3 + y^3 + 4z^3 = 3(2xy + z). \tag{5}$$

The dth order such surface has

$$N(d) = \begin{cases} \frac{1}{12}(5d^3 - 13d^2 + 12d) & \text{if } d \equiv 0 \pmod 6 \\ \frac{1}{12}(5d^3 - 13d^2 + 16d - 8) & \text{if } d \equiv 2, 4 \pmod 6 \\ \frac{1}{12}(5d^3 - 13d^2 + 13d - 4) & \text{if } d \equiv 1, 5 \pmod 6 \\ \frac{1}{12}(5d^3 - 14d^2 + 9d) & \text{if } d \equiv 3 \pmod 6 \end{cases}$$

singular points (Chmutov 1992), giving the sequence 0, 1, 3, 14, 28, 57, 93, 154, 216, 321, 425, 576, 732, 949, 1155, ... for $d = 1, 2, \ldots$. For a number of orders d, Chmutov surfaces have more ordinary double points than any other known equations of the same degree.

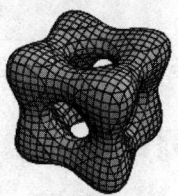

Based on Chmutov's equations, Banchoff (1991) defined the simpler set of surfaces

$$T_n(x) + T_n(y) + T_n(z) = 0, \tag{6}$$

where n is EVEN and $T_n(x)$ is again a CHEBYSHEV POLYNOMIAL OF THE FIRST KIND. For example, the surfaces illustrated above have orders 2, 4, and 6 are given by the equations

$$2(x^2 + y^2 + z^2) = 3 \tag{7}$$

$$3 + 8(x^4 + y^4 + z^4) = 8(x^2 + y^2 + z^2) \tag{8}$$

$$2[x^2(3 - 4x^2)^2 + y^2(3 - 4y^2)^2 + z^2(3 - 4z^2)^2] = 3. \tag{9}$$

See also GOURSAT'S SURFACE, ORDINARY DOUBLE POINT, SUPERELLIPSE

References

Banchoff, T. F. "Computer Graphics Tools for Rendering Algebraic Surfaces and for Geometry of Order." In *Geometric Analysis and Computer Graphics: Proceedings of a Workshop Held May 23–25, 1988* (Eds. P. Concus, R. Finn, D. A. Hoffman). New York: Springer-Verlag, pp. 31–37, 1991.
Chmutov, S. V. "Examples of Projective Surfaces with Many Singularities." *J. Algebraic Geom.* **1**, 191–196, 1992.
Hirzebruch, F. "Singularities of Algebraic Surfaces and Characteristic Numbers." In *The Lefschetz Centennial Conference, Part I: Proceedings of the Conference on Algebraic Geometry, Algebraic Topology, and Differential Equations, Held in Mexico City, December 10–14, 1984* (Ed. S. Sundararaman). Providence, RI: Amer. Math. Soc., pp. 141–155, 1986.
Trott, M. *The Mathematica Guidebook, Vol. 2: Graphics.* New York: Springer-Verlag, 2000.

Choice Axiom

AXIOM OF CHOICE

Choice Number

COMBINATION

Cholesky Decomposition

Given a symmetric POSITIVE DEFINITE MATRIX A, the Cholesky decomposition is an UPPER TRIANGULAR MATRIX U such that

$$\mathsf{A} = \mathsf{U}^\mathsf{T}\mathsf{U}.$$

Cholesky decomposition is implemented as CholeskyDecomposition[*m*] in the *Mathematica* add-on package LinearAlgebra`Cholesky` (which can be loaded with the command < <LinearAlgebra`).

See also LU DECOMPOSITION, MATRIX DECOMPOSITION, QR DECOMPOSITION

References

Gentle, J. E. "Cholesky Factorization." §3.2.2 in *Numerical Linear Algebra for Applications in Statistics.* Berlin: Springer-Verlag, pp. 93–95, 1998.

Nash, J. C. "The Choleski Decomposition." Ch. 7 in *Compact Numerical Methods for Computers: Linear Algebra and Function Minimisation, 2nd ed.* Bristol, England: Adam Hilger, pp. 84–93, 1990.

Press, W. H.; Flannery, B. P.; Teukolsky, S. A.; and Vetterling, W. T. "Cholesky Decomposition." §2.9 in *Numerical Recipes in FORTRAN: The Art of Scientific Computing, 2nd ed.* Cambridge, England: Cambridge University Press, pp. 89–91, 1992.

Choose

An alternative term for a BINOMIAL COEFFICIENT, in which $\binom{n}{k}$ is read as "n choose k." R. K. Guy suggested this pronunciation around 1950, when the notations $^{n}C_r$ and $_{n}C_r$ were commonly used. Leo Moser liked the pronunciation and he and others spread it around. It got the final seal of approval from Donald Knuth when he incorporated it into the TEX mathematical typesetting language as {n\choose k}.

See also BINOMIAL COEFFICIENT, MULTICHOOSE

Choquet Theory

Erdos proved that there exist at least one PRIME OF THE FORM $4k + 1$ and at least one PRIME OF THE FORM $4k + 3$ between n and $2n$ for all $n > 6$.

See also EQUINUMEROUS, PRIME NUMBER

Chord

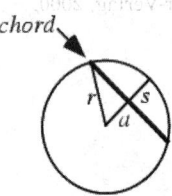

chord

The LINE SEGMENT joining two points on a curve. The term is often used to describe a LINE SEGMENT whose ends lie on a CIRCLE. In the above figure, r is the RADIUS of the CIRCLE, a is called the APOTHEM, and s the SAGITTA.

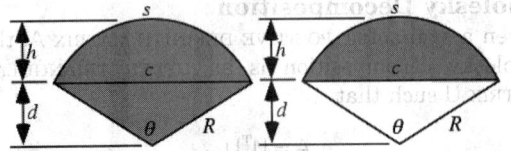

The shaded region in the left figure is called a SECTOR, and the shaded region in the right figure is called a SEGMENT.

All ANGLES inscribed in a CIRCLE and subtended by the same chord are equal. The converse is also true: The LOCUS of all points from which a given segment subtends equal ANGLES is a CIRCLE.

Given any closed convex curve, it is possible to find a point P through which three chords, inclined to one another at angles of $60°$, pass such that P is the MIDPOINT of all three (Wells 1991).

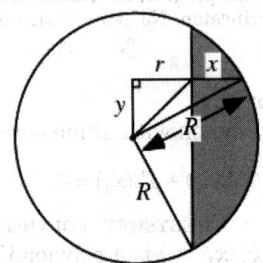

Let a CIRCLE of RADIUS R have a CHORD at distance r. The AREA enclosed by the CHORD, shown as the shaded region in the above figure, is then

$$A = 2 \int_0^{\sqrt{R^2-r^2}} x(y)\,dy. \qquad (1)$$

But

$$y^2 + (r + x)^2 = R^2, \qquad (2)$$

so

$$x(y) = \sqrt{R^2 - y^2} - r \qquad (3)$$

and

$$A = 2 \int_0^{\sqrt{R^2-r^2}} (\sqrt{R^2 - y^2} - r)\,dy \qquad (4)$$

$$= R^2 \tan^{-1}\left[\sqrt{\left(\frac{R}{r}\right)^2 - 1}\right] - r\sqrt{R^2 - r^2}. \qquad (5)$$

Checking the limits, when $r = R$, $A = 0$ and when $r \to 0$,

$$A = \tfrac{1}{2}\pi R^2, \qquad (6)$$

the expected area of the SEMICIRCLE.

See also ANNULUS, APOTHEM, BERTRAND'S PROBLEM, CONCENTRIC CIRCLES, HOLDITCH'S THEOREM, RADIUS, SAGITTA, SECTOR, SEGMENT, SEMICIRCLE

References

Wells, D. *The Penguin Dictionary of Curious and Interesting Geometry.* London: Penguin, p. 29, 1991.

Chord Diagram

See also ALGEBRA OF CHORD DIAGRAMS, KONTSEVICH INTEGRAL

Chordal

RADICAL AXIS

Chordal Theorem

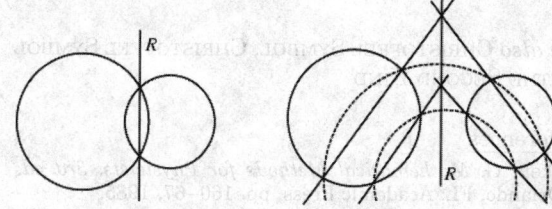

The LOCUS of the point at which two given CIRCLES possess the same POWER is a straight line PERPENDICULAR to the line joining the MIDPOINTS of the CIRCLE and is known as the chordal (or, more commonly, the RADICAL AXIS) of the two CIRCLES.

See also POWER (CIRCLE), RADICAL LINE

References

Dörrie, H. *100 Great Problems of Elementary Mathematics: Their History and Solutions*. New York: Dover, p. 153, 1965.

Chow Coordinates

A generalization of GRASSMANN COORDINATES to m-D ALGEBRAIC VARIETIES of degree d in P^n, where P^n is an n-D projective space. To define the Chow coordinates, take the intersection of an m-D ALGEBRAIC VARIETY Z of degree d by an $(n-m)$-D SUBSPACE U of P^n. Then the coordinates of the d points of intersection are algebraic functions of the GRASSMANN COORDINATES of U, and by taking a symmetric function of the algebraic functions, a HOMOGENEOUS POLYNOMIAL known as the Chow form of Z is obtained. The Chow coordinates are then the COEFFICIENTS of the Chow form. Chow coordinates can generate the smallest field of definition of a divisor.

See also CHOW RING, CHOW VARIETY

References

Chow, W.-L. and van der Waerden., B. L. "Zur algebraische Geometrie IX." *Math. Ann.* **113**, 692–704, 1937.
Wilson, W. S.; Chern, S. S.; Abhyankar, S. S.; Lang, S.; and Igusa, J.-I. "Wei-Liang Chow." *Not. Amer. Math. Soc.* **43**, 1117–1124, 1996.

Chow Ring

The intersection product for classes of rational equivalence between cycles on an ALGEBRAIC VARIETY.

See also CHOW COORDINATES, CHOW VARIETY

References

Chow, W.-L. "On Equivalence Classes of Cycles in an Algebraic Variety." *Ann. Math.* **64**, 450–479, 1956.
Wilson, W. S.; Chern, S. S.; Abhyankar, S. S.; Lang, S.; and Igusa, J.-I. "Wei-Liang Chow." *Not. Amer. Math. Soc.* **43**, 1117–1124, 1996.

Chow Variety

The set $C_{n,m,d}$ of all m-D varieties of degree d in an n-D projective space P^n into an M-D projective space P^M.

See also CHOW COORDINATES, CHOW RING

References

Wilson, W. S.; Chern, S. S.; Abhyankar, S. S.; Lang, S.; and Igusa, J.-I. "Wei-Liang Chow." *Not. Amer. Math. Soc.* **43**, 1117–1124, 1996.

Christoffel Formula

Let $\{p_n(x)\}$ be orthogonal POLYNOMIALS associated with the distribution $d\alpha(x)$ on the interval $[a, b]$. Also let

$$\rho \equiv c(x-x_1)(x-x_2)\cdots(x-x_l)$$

(for $c \neq 0$) be a POLYNOMIAL of order l which is NONNEGATIVE in this interval. Then the orthogonal polynomials $\{q(x)\}$ associated with the distribution $\rho(x)\,d\alpha(x)$ can be represented in terms of the polynomials $p_n(x)$ as

$$\rho(x)q_n(x) = \begin{vmatrix} p_n(x) & p_{n+1}(x) & \cdots & p_{n+l}(x) \\ p_n(x_1) & p_{n+1}(x_1) & \cdots & p_{n+l}(x_1) \\ \vdots & \vdots & \vdots & \vdots \\ p_n(x_l) & p_{n+1}(x_l) & \cdots & p_{n+l}(x_l) \end{vmatrix}.$$

In the case of a zero x_k of multiplicity $m > 1$, we replace the corresponding rows by the derivatives of order 0, 1, 2, ..., $m-1$ of the POLYNOMIALS $p_n(x_l)$, ..., $p_{n+l}(x_l)$ at $x = x_k$.

References

Szego, G. *Orthogonal Polynomials, 4th ed.* Providence, RI: Amer. Math. Soc., pp. 29–0, 1975.

Christoffel Number

One of the quantities λ_i appearing in the GAUSS-JACOBI MECHANICAL QUADRATURE. They satisfy

$$\lambda_1 + \lambda_2 + \ldots + \lambda_n = \int_a^b d\alpha(x) = \alpha(b) - \alpha(a) \tag{1}$$

and are given by

$$\lambda_v = \int_a^b \left[\frac{p_n(x)}{p_n'(x_v)(x-x_v)} \right]^2 d\alpha(x) \tag{2}$$

$$\lambda_v = -\frac{k_{n+1}}{k_n} \frac{1}{p_{n+1}(x_v)p_n'(x_v)} \tag{3}$$

$$= \frac{k_n}{k_{n-1}} \frac{1}{p_{n-1}(x_v)P_n'(x_v)} \tag{4}$$

$$(\lambda_v)^{-1} = [p_0(x_v)]^2 + \ldots + [p_n(x_v)]^2, \tag{5}$$

where k_n is the higher COEFFICIENT of $p_n(x)$.

See also COTES NUMBER, HERMITE'S INTERPOLATING POLYNOMIAL

References
Szego, G. *Orthogonal Polynomials, 4th ed.* Providence, RI: Amer. Math. Soc., pp. 47–8, 1975.

Christoffel Symbol

The Christoffel symbols are TENSOR-like objects derived from a RIEMANNIAN METRIC g. They are used to study the geometry of the metric and appear, for example, in the GEODESIC EQUATION. There are two closely related kinds of Christoffel symbols, the FIRST KIND $\Gamma_{i,j,k}$, and the SECOND KIND $\Gamma_{i,j}^k$.

It is always possible to pick a coordinate system on a RIEMANNIAN MANIFOLD such that the Christoffel symbol vanishes at a chosen point. In general relativity, Christoffel symbols are "gravitational forces," and the preferred coordinate system referred to above would be one attached to a body in free fall.

See also CHRISTOFFEL SYMBOL OF THE FIRST KIND, CHRISTOFFEL SYMBOL OF THE SECOND KIND, GEODESIC, LEVI-CIVITA CONNECTION, RIEMANNIAN GEOMETRY

References
Carmo, M. *Differential Geometry of Curves and Surfaces.* Englewood Cliffs, NJ: Prentice-Hall, pp. 441–42, 1976.
Sternberg, S. *Differential Geometry.* New York: Chelsea, pp. 353–54, 1983.

Christoffel Symbol of the First Kind

The first type of TENSOR derived from a RIEMANNIAN METRIC g which is used to study the geometry of the metric. Christoffel symbols of the first kind are variously denoted $[ij, k]$, $\left[{}^i{}_k{}^j\right]$, Γ_{abc}, or $\{ab, c\}$.

$$[ij,k] = g_{mk}\Gamma_{ij}^m \tag{1}$$

$$= g_{mk}\vec{e}^m \cdot \frac{\partial \vec{e}_i}{\partial q^i} \tag{2}$$

$$= \vec{e}_k \cdot \frac{\partial \vec{e}_i}{\partial q^j}, \tag{3}$$

where g_{mk} is the METRIC TENSOR, Γ_{ij}^m is a CHRISTOFFEL SYMBOL OF THE SECOND KIND, and

$$\vec{e}_i \equiv \frac{\partial \vec{r}}{\partial q^i} = h_i \hat{e}_i. \tag{4}$$

But

$$\frac{\partial g_{ij}}{\partial q^k} = \frac{\partial}{\partial q^k}(\vec{e}_i \cdot \vec{e}_j) = \frac{\partial \vec{e}_i}{\partial q^k} \cdot \vec{e}_j + \vec{e}_i \cdot \frac{\partial \vec{e}_j}{\partial q^k}$$

$$= [ik, j] + [jk, i], \tag{5}$$

so

$$[ab, c] = \tfrac{1}{2}(g_{ac,\, b} + g_{bc,\, a} - g_{ab,\, c}). \tag{6}$$

See also CHRISTOFFEL SYMBOL, CHRISTOFFEL SYMBOL OF THE SECOND KIND

References
Arfken, G. *Mathematical Methods for Physicists, 3rd ed.* Orlando, FL: Academic Press, pp. 160–67, 1985.

Christoffel Symbol of the Second Kind

The second type of TENSOR-like object derived from a RIEMANNIAN METRIC g which is used to study the geometry of the metric. Christoffel symbols of the second kind are variously denoted as $\left\{{}^m_{i\ j}\right\}$ or Γ_{ij}^m. In the latter case, they are sometimes known as connection coefficients.

$$\Gamma_{ij}^m \equiv \vec{e}^m \cdot \frac{\partial \vec{e}_i}{\partial q^j} \tag{1}$$

$$= g^{km}[ij, k] \tag{2}$$

$$= \frac{1}{2} g^{km}\left(\frac{\partial g_{ik}}{\partial q^j} + \frac{\partial g_{jk}}{\partial q^i} - \frac{\partial g_{ij}}{\partial q^k}\right), \tag{3}$$

where g^{km} is the METRIC TENSOR. The Christoffel symbol of the second kind is related to the CHRISTOFFEL SYMBOL OF THE FIRST KIND $[bc, d]$ by

$$\Gamma_{bc}^a = g^{ad}\{bc, d\}. \tag{4}$$

Christoffel symbols of the second kind can also be defined by

$$\Gamma_{\vec{e}_\beta \vec{e}_\gamma}^{\vec{e}_\alpha} \equiv \vec{e}^\alpha \cdot (\nabla_{\vec{e}_\gamma}\vec{e}_\beta) \tag{5}$$

(long form) or

$$\Gamma_{\beta\gamma}^\alpha \equiv \vec{e}^\alpha \cdot (\nabla_\gamma \vec{e}_\beta), \tag{6}$$

(abbreviated form), and satisfy

$$\nabla_{\vec{e}_\gamma}\vec{e}_\beta = \Gamma_{\vec{e}_\beta \vec{e}_\gamma}^{\vec{e}_\alpha}\vec{e}_\alpha \tag{7}$$

(long form) and

$$\nabla_\gamma \vec{e}_\beta = \Gamma_{\beta\gamma}^\alpha \vec{e}_\alpha \tag{8}$$

(abbreviated form).

Christoffel symbols of the second kind are not TENSORS, but have TENSOR-like CONTRAVARIANT and COVARIANT indices. Christoffel symbols of the second kind also do not transform as tensors. In fact, changing coordinates from $x_1, \ldots, x_n$ to $y_1, \ldots, y_n$ gives

$$\Gamma_{ij}^{k'} = \sum \frac{\partial^2 x_l}{\partial y_i \partial y_j}\frac{\partial y_k}{\partial x_l} + \sum \Gamma_{rs}^{\mathrm{T}}\frac{\partial x_r}{\partial y_i}\frac{\partial x_s}{\partial y_j}\frac{\partial y_k}{\partial x_t}. \tag{9}$$

However, a fully COVARIANT Christoffel symbol of the second kind is given by

$$\Gamma_{\alpha\beta\gamma} \equiv \tfrac{1}{2}(g_{\alpha\beta,\ \gamma} + g_{\alpha\gamma,\ \beta} + c_{\alpha\beta\gamma} + c_{\alpha\gamma\beta} - c_{\beta\gamma\alpha}), \qquad (10)$$

where the gs are the METRIC TENSORS, the cs are COMMUTATION COEFFICIENTS, and the commas indicate the COMMA DERIVATIVE. In an ORTHONORMAL BASIS, $g_{\alpha\beta,\ \gamma} = 0$ and $g_{\mu\gamma} = \delta_{\mu\gamma}$, so

$$\Gamma_{\alpha\beta\gamma} = \Gamma^{\mu}_{\alpha\beta} g_{\mu\gamma} = \Gamma^{\mu}_{\alpha\beta} = \tfrac{1}{2}(c_{\alpha\beta\gamma} + c_{\alpha\gamma\beta} - c_{\beta\gamma\alpha}) \qquad (11)$$

and

$$\Gamma_{ijk} = 0 \quad \text{for } i \neq j \neq k \qquad (12)$$

$$\Gamma_{iik} = -\frac{1}{2}\frac{\partial g_{ii}}{\partial x^k} \quad \text{for } i \neq k \qquad (13)$$

$$\Gamma_{iji} = \Gamma_{jii} = \frac{1}{2}\frac{\partial g_{ii}}{\partial x^j} \qquad (14)$$

$$\Gamma^k_{ij} = 0 \quad \text{for } i \neq j \neq k \qquad (15)$$

$$\Gamma^k_{ii} = -\frac{1}{2g_{kk}}\frac{\partial g_{ii}}{\partial x^k} \quad \text{for } i \neq k \qquad (16)$$

$$\Gamma^i_{ij} = \Gamma^i_{ji} = \frac{1}{2g_{ii}}\frac{\partial g_{ii}}{\partial x^j} = \frac{1}{2}\frac{\partial \ln g_{ii}}{\partial x^j}. \qquad (17)$$

For TENSORS of RANK 3, the Christoffel symbols of the second kind may be concisely summarized in MATRIX form:

$$\mathbf{\Gamma}^\theta \equiv \begin{bmatrix} \Gamma^\theta_{rr} & \Gamma^\theta_{r\theta} & \Gamma^\theta_{r\phi} \\ \Gamma^\theta_{\theta r} & \Gamma^\theta_{\theta\theta} & \Gamma^\theta_{\theta\phi} \\ \Gamma^\theta_{\phi r} & \Gamma^\theta_{\phi\theta} & \Gamma^\theta_{\phi\phi} \end{bmatrix}. \qquad (18)$$

The Christoffel symbols are given in terms of the coefficients of the FIRST FUNDAMENTAL FORM E, F, and G by

$$\Gamma^1_{11} = \frac{GE_u - 2FF_u + FE_v}{2(EG - F^2)} \qquad (19)$$

$$\Gamma^1_{12} = \frac{GE_v - FG_u}{2(EG - F^2)} \qquad (20)$$

$$\Gamma^1_{22} = \frac{2GF_v - GG_u - FG_v}{2(EG - F^2)} \qquad (21)$$

$$\Gamma^2_{11} = \frac{2EF_u - EE_v - FE_u}{2(EG - F^2)} \qquad (22)$$

$$\Gamma^2_{12} = \frac{EG_u - FE_v}{2(EG - F^2)} \qquad (23)$$

$$\Gamma^2_{22} = \frac{EG_v - 2FF_v + FG_u}{2(EG - F^2)}, \qquad (24)$$

and $\Gamma^1_{21} = \Gamma^1_{12}$ and $\Gamma^2_{21} = \Gamma^2_{12}$. If $F = 0$, the Christoffel symbols of the second kind simplify to

$$\Gamma^1_{11} = \frac{E_u}{2E} \qquad (25)$$

$$\Gamma^1_{12} = \frac{E_v}{2E} \qquad (26)$$

$$\Gamma^1_{22} = -\frac{G_u}{2E} \qquad (27)$$

$$\Gamma^2_{11} = -\frac{E_v}{2G} \qquad (28)$$

$$\Gamma^2_{12} = \frac{G_u}{2G} \qquad (29)$$

$$\Gamma^2_{22} = \frac{G_v}{2G} \qquad (30)$$

(Gray 1997).

The following relationships hold between the Christoffel symbols of the second kind and coefficients of the first FUNDAMENTAL FORM,

$$\Gamma^1_{11}E + \Gamma^2_{11}F = \tfrac{1}{2}E_u \qquad (31)$$

$$\Gamma^1_{12}E + \Gamma^2_{12}F = \tfrac{1}{2}E_v \qquad (32)$$

$$\Gamma^1_{22}E + \Gamma^2_{22}F = F_v - \tfrac{1}{2}G_u \qquad (33)$$

$$\Gamma^1_{11}F + \Gamma^2_{11}G = F_u - \tfrac{1}{2}E_v \qquad (34)$$

$$\Gamma^1_{12}F + \Gamma^2_{12}G = \tfrac{1}{2}G_u \qquad (35)$$

$$\Gamma^1_{22}F + \Gamma^2_{22}G = \tfrac{1}{2}G_v \qquad (36)$$

$$\Gamma^1_{11} + \Gamma^2_{12} = (\ln \sqrt{EG - F^2})_u \qquad (37)$$

$$\Gamma^1_{12} + \Gamma^2_{22} = (\ln \sqrt{EG - F^2})_v \qquad (38)$$

(Gray 1997).

For a surface given in MONGE'S FORM $z = F(x, y)$,

$$\Gamma^k_{ij} = \frac{z_{ij}z_k}{1 + z_1^2 + z_2^2}. \qquad (39)$$

Christoffel symbols of the second kind arise in the computation of GEODESICS. The GEODESIC EQUATION of free motion is

$$d\tau^2 = -\eta_{\alpha\beta}d\xi^\alpha d\xi^\beta, \qquad (40)$$

or

$$\frac{d^2\xi^\alpha}{d\tau^2} = 0. \qquad (41)$$

Expanding,

$$\frac{d}{d\tau}\left(\frac{\partial\xi^\alpha}{\partial x^\mu}\frac{dx^\mu}{d\tau}\right) = \frac{\partial\xi^\alpha}{\partial x^\mu}\frac{d^2x^\mu}{d\tau^2} + \frac{\partial^2\xi^\alpha}{\partial x^\mu \partial x^\nu}\frac{dx^\mu}{d\tau}\frac{dx^\nu}{d\tau} = 0 \quad (42)$$

$$\frac{\partial \xi^\alpha}{\partial x^\mu} \frac{d^2 x^\mu}{d\tau^2} \frac{\partial x^\lambda}{\partial \xi^\alpha} + \frac{\partial^2 \xi^\alpha}{\partial x^\mu \partial x^\nu} \frac{dx^\mu}{d\tau} \frac{dx^\nu}{d\tau} \frac{\partial x^\lambda}{\partial \xi^\alpha} = 0. \qquad (43)$$

But

$$\frac{\partial \xi^\alpha}{\partial x^\nu} \frac{\partial x^\lambda}{\partial \xi^\alpha} = \delta_\mu^\lambda, \qquad (44)$$

so

$$\delta_\mu^\lambda \frac{d^2 x^\mu}{d\tau^2} + \left(\frac{\partial^2 \xi^\alpha}{\partial x^\mu \partial x^\nu} \frac{\partial x^\lambda}{\partial \xi^\alpha} \right) \frac{dx^\mu}{d\tau} \frac{dx^\nu}{d\tau}$$

$$= \frac{d^2 x^\lambda}{d\tau^2} + \Gamma_{\mu\nu}^\lambda \frac{dx^\mu}{d\tau} \frac{dx^\nu}{d\tau}, \qquad (45)$$

where

$$\Gamma_{\mu\nu}^\lambda \equiv \frac{\partial^2 \xi^\alpha}{\partial x^\mu \partial x^\nu} \frac{\partial x^\lambda}{\partial \xi^\alpha}. \qquad (46)$$

See also CARTAN TORSION COEFFICIENT, CHRISTOFFEL SYMBOL, CHRISTOFFEL SYMBOL OF THE FIRST KIND, COMMA DERIVATIVE, COMMUTATION COEFFICIENT, CONNECTION COEFFICIENT, GAUSS EQUATIONS, SEMI-COLON DERIVATIVE, TENSOR

References
Arfken, G. *Mathematical Methods for Physicists, 3rd ed.* Orlando, FL: Academic Press, pp. 160–67, 1985.
Gray, A. "Christoffel Symbols." §22.3 in *Modern Differential Geometry of Curves and Surfaces with Mathematica, 2nd ed.* Boca Raton, FL: CRC Press, pp. 509–13, 1997.
Morse, P. M. and Feshbach, H. *Methods of Theoretical Physics, Part I.* New York: McGraw-Hill, pp. 47–8, 1953.
Sternberg, S. *Differential Geometry.* New York: Chelsea, p. 354, 1983.

Christoffel-Darboux Formula

For three consecutive orders of an ORTHOGONAL POLYNOMIAL, the following relationship holds for $n = 2, 3, \ldots$,

$$p_n(x) = (A_n x + B_n) p_{n-1}(x) - C_n p_{n-2}(x), \qquad (1)$$

where $A_n > 0$, B_n, and $C_n > 0$ are constants. Denoting the highest COEFFICIENT of $p_n(x)$ by k_n,

$$A_n = \frac{k_n}{k_{n-1}} \qquad (2)$$

$$C_n = \frac{A_n}{A_{n-1}} = \frac{k_n k_{n-2}}{k_{n-1}^2}. \qquad (3)$$

Then

$$p_0(x)p_0(y) + \ldots + p_n(x)p_n(y)$$

$$= \frac{k_n}{k_{n+1}} \frac{p_{n+1}(x)p_n(y) - p_n(x)p_{n+1}(y)}{x - y}. \qquad (4)$$

In the special case of $x = y$, (4) gives

$$[p_0(x)]^2 + \ldots + [p_n(x)]^2$$

$$= \frac{k_n}{k_{n+1}} [p_{n+1}'(x)p_n(x) - p_n'(x)p_{n+1}(x)]. \qquad (5)$$

References
Abramowitz, M. and Stegun, C. A. (Eds.). *Handbook of Mathematical Functions with Formulas, Graphs, and Mathematical Tables, 9th printing.* New York: Dover, p. 785, 1972.
Szego, G. *Orthogonal Polynomials, 4th ed.* Providence, RI: Amer. Math. Soc., pp. 42–44, 1975.

Christoffel-Darboux Identity

$$\sum_{k=0}^{\infty} \frac{\phi_k(x)\phi_k(y)}{\gamma_k}$$

$$= \frac{\phi_{m+1}(x)\phi_m(y) - \phi_m(x)\phi_{m+1}(y)}{a_m \gamma_m (x - y)}, \qquad (1)$$

where $\phi_k(x)$ are ORTHOGONAL POLYNOMIALS with WEIGHTING FUNCTION $W(x)$,

$$\gamma_m \equiv \int [\phi_m(x)]^2 W(x)\, dx, \qquad (2)$$

and

$$a_k \equiv \frac{A_{k+1}}{A_k} \qquad (3)$$

where A_k is the COEFFICIENT of x^k in $\phi_k(x)$.

References
Hildebrand, F. B. *Introduction to Numerical Analysis.* New York: McGraw-Hill, p. 322, 1956.

Chromatic Number

The fewest number of colors $\gamma(G)$ necessary to color the vertices of GRAPH or regions of a SURFACE (Skiena 1990, p. 210). The chromatic number is the smallest positive integer z such that the CHROMATIC POLYNOMIAL $\pi_G(z) > 0$. Calculating the chromatic number of a GRAPH is an NP-COMPLETE PROBLEM (Skiena 1990, pp. 211–12).

For any two positive integers g and k, there exists a graph of girth at least g and chromatic number at least k (Erdos 1961, Lovász 1968; Skiena 1990, p. 215).

The chromatic number of a surface of GENUS g is given by the HEAWOOD CONJECTURE,

$$\gamma(g) = \left\lfloor \tfrac{1}{2}(7 + \sqrt{48g + 1}) \right\rfloor,$$

where $\lfloor x \rfloor$ is the FLOOR FUNCTION. $\gamma(g)$ is sometimes also denoted $\chi(g)$ (which is unfortunate, since $\chi(g) = 2 - 2g$ commonly refers to the EULER CHARACTERISTIC). For $g = 0, 1, \ldots$, the first few values of $\chi(g)$ are 4,

7, 8, 9, 10, 11, 12, 12, 13, 13, 14, 15, 15, 16, ... (Sloane's A000934).

Erdos (1959) proved that there are graphs with arbitrarily large GIRTH and CHROMATIC NUMBER (Bollobás and West 2000).

See also BETTI NUMBER, BRELAZ'S HEURISTIC ALGORITHM, BROOKS' THEOREM, CHROMATIC POLYNOMIAL, EDGE CHROMATIC NUMBER, EDGE COLORING, EULER CHARACTERISTIC, GENUS (SURFACE), HEAWOOD CONJECTURE, MAP COLORING, PERFECT GRAPH, TORUS COLORING

References

Bollobás, B. and West, D. B. "A Note on Generalized Chromatic Number and Generalized Girth." *Discr. Math.* **213**, 29–4, 2000.
Chartrand, G. "A Scheduling Problem: An Introduction to Chromatic Numbers." §9.2 in *Introductory Graph Theory.* New York: Dover, pp. 202–09, 1985.
Eppstein, D. "The Chromatic Number of the Plane." http://www.ics.uci.edu/~eppstein/junkyard/plane-color/.
Erdos, P. "Graph Theory and Probability." *Canad. J. Math.* **11**, 34–8, 1959.
Erdos, P. "Graph Theory and Probability II." *Canad. J. Math.* **13**, 346–52, 1961.
Gardner, M. *The Sixth Book of Mathematical Games from Scientific American.* Chicago, IL: University of Chicago Press, p. 9, 1984.
Lovász, L. "On Chromatic Number of Finite Set-Systems." *Acta Math. Acad. Sci. Hungar.* **19**, 59–7, 1968.
Skiena, S. *Implementing Discrete Mathematics: Combinatorics and Graph Theory with Mathematica.* Reading, MA: Addison-Wesley, 1990.
Sloane, N. J. A. Sequences A000934/M3292 in "An On-Line Version of the Encyclopedia of Integer Sequences." http://www.research.att.com/~njas/sequences/eisonline.html.

Chromatic Polynomial

A POLYNOMIAL $\pi_G(z)$ of a GRAPH G which counts the number of ways to color g with exactly z colors. For example, the CUBICAL GRAPH has chromatic polynomial

$$\pi_G(z) = z^8 - 12z^7 + 66z^6 - 214z^5 + 441z^4 - 572z^3$$
$$+ 423z^2 - 133z, \tag{1}$$

so the number of 1-, 2-, ... colorings are 0, 2, 114, 2652, 29660, 198030, ... The chromatic polynomial of a graph g in the variable z can be determined using `ChromaticPolynomial[g, z]` in the *Mathematica* add-on package `DiscreteMath`Combinatorica` (which can be loaded with the command `<<DiscreteMath`).

The chromatic polynomial of a DISCONNECTED GRAPH is the product of the chromatic polynomials of its CONNECTED COMPONENTS. The chromatic polynomial of a graph of order n has degree n, with leading coefficient 1 and constant term 0. Furthermore, the coefficients alternate signs, and the coefficient of the $(n-1)$st term is $-e$, where e is the number of edges.

Interestingly, $\pi_G(-1)$ is equal to the number of acyclic orientations of G (Stanley 1973).

Except for special cases (such as TREES), the calculation of $\Pi_G(z)$ is exponential in the minimum number of edges in G and the COMPLEMENT GRAPH $\bar{G}$ (Skiena 1990, p. 211), and calculating the chromatic polynomial of a GRAPH is at least an NP-COMPLETE PROBLEM (Skiena 1990, pp. 211–12).

Tutte (1970) showed that the chromatic polynomial of a planar triangulation possess a ROOT close to $\phi^2 = \phi + 1 = 2.618033\ldots$, where ϕ is the GOLDEN MEAN. More precisely, if n is the number of VERTICES of G, then

$$P_G(\phi^2) \leq \phi^{5-n} \tag{2}$$

(Tutte 1970, Le Lionnais 1983).

Read (1968) conjectured that, for any chromatic polynomial

$$c_n z^n + \ldots + c_1 z, \tag{3}$$

there does not exist a $1 \leq p \leq q \leq r \leq n$ such that $|c_p| > |c_q|$ and $|c_q| < |c_r|$ (Skiena 1990, p. 221).

The CHROMATIC NUMBER of a graph gives the smallest number of colors with which a graph can be colored, and so is the smallest positive integer z such that $\pi_G(z) > 0$ (Skiena 1990, p. 211).

See also CHROMATIC NUMBER, K-COLORING

References

Berman, G. and Tutte, W. T. "The Golden Root of a Chromatic Polynomial." *J. Combin. Th.* **6**, 301–02, 1969.
Birkhoff, G. D. "A Determinant Formula for the Number of Ways of Coloring a Map." *Ann. Math.* **14**, 42–6, 1912.
Birkhoff, G. D. and Lewis, D. C. "Chromatic Polynomials." *Trans. Amer. Math. Soc.* **60**, 355–51, 1946.
Chvátal, V. "A Note on Coefficients of Chromatic Polynomials." *J. Combin. Th.* **9**, 95–6, 1970.
Erdos, P. and Hajnal, A. "On Chromatic Numbers of Graphs and Set-Systems." *Acta Math. Acad. Sci. Hungar.* **17**, 61–9, 1966.
Le Lionnais, F. *Les nombres remarquables.* Paris: Hermann, p. 46, 1983.
Read, R. C. "An Introduction to Chromatic Polynomials." *J. Combin. Th.* **4**, 52–1, 1968.
Saaty, T. L. and Kainen, P. C. "Chromatic Numbers and Chromatic Polynomials." Ch. 6 in *The Four-Color Problem: Assaults and Conquest.* New York: Dover, pp. 134–63 1986.
Skiena, S. "Chromatic Polynomials." §5.5.1 in *Implementing Discrete Mathematics: Combinatorics and Graph Theory with Mathematica.* Reading, MA: Addison-Wesley, pp. 210–12, 1990.
Stanley, R. P. "Acyclic Orientations of Graphs." *Disc. Math.* **5**, 171–78, 1973.
Tutte, W. T. "On Chromatic Polynomials and the Golden Ratio." *J. Combin. Th.* **9**, 289–96, 1970.

Chu Identity

CHU-VANDERMONDE IDENTITY

Chu Space

A Chu space is a BINARY RELATION from a SET A to an ANTISET X which is defined as a SET which transforms via converse functions.

See also ANTISET

References

Stanford Concurrency Group. "Guide to Papers on Chu Spaces." http://boole.stanford.edu/chuguide.html.

Church's Theorem

No decision procedure exists for ARITHMETIC.

Church's Thesis

CHURCH-TURING THESIS

Church-Turing Thesis

The TURING MACHINE concept defines what is meant mathematically by an algorithmic procedure. Stated another way, a function f is effectively COMPUTABLE IFF it can be computed by a TURING MACHINE.

See also ALGORITHM, COMPUTABLE FUNCTION, DECIDABLE, TURING MACHINE

References

Penrose, R. *The Emperor's New Mind: Concerning Computers, Minds, and the Laws of Physics.* Oxford, England: Oxford University Press, pp. 47–9, 1989.
Pour-El, M. B. "The Structure of Computability in Analysis and Physical Theory: An Extension of Church's Thesis." Ch. 13 in *Handbook of Computability Theory* (Ed. E. R. Griffor). Amsterdam, Netherlands: Elsevier, pp. 449–70, 1999.

Chu-Vandermonde Identity

A special case of GAUSS'S THEOREM, with a being a NEGATIVE INTEGER $-n$:

$$_2F_1(-n, b; c; 1) = \frac{(c-b)_n}{(c)_n},$$

where $_2F_1(a, b; c; z)$ is a HYPERGEOMETRIC FUNCTION and $(a)_n$ is a POCHHAMMER SYMBOL (Bailey 1935, p. 3; Koepf 1998, p. 32). The identity is sometimes also called Vandermonde's theorem.

The identity

$$(x+a)_n = \sum_{k=0}^{\infty} \binom{n}{k}(x)_k(a)_{n-k}$$

(Koepf 1998, p. 42), where $\binom{n}{k}$ is a BINOMIAL COEFFICIENT and $(a)_n \equiv a(a-1)\cdots(a-n+1)$ is the POCHHAMMER SYMBOL is sometimes also known as the Chu-Vandermonde identity. (0) can be written as

$$\binom{x+a}{n} = \sum_{k=0}^{n}\binom{x}{k}\binom{a}{n-k},$$

which is sometimes known as VANDERMONDE'S CONVOLUTION FORMULA (Roman 1984). A special case gives the identity

$$\sum_{l=0}^{\max(k,\,n)}\binom{m}{k-l}\binom{n}{l} = \binom{m+n}{k}.$$

The identities

$$\sum_{k=0}^{n}\binom{a}{k}\binom{b}{n-k} = \binom{a+b}{n} \tag{1}$$

$$\sum_{k=0}^{n}\binom{n}{k}\binom{s}{t-k} = \binom{n+s}{t} \tag{2}$$

$$\sum_{k=0}^{n}\binom{n}{k}\binom{s}{t+k} = \binom{n+s}{n+t} \tag{3}$$

are all special instances of the Chu-Vandermonde identity (Koepf 1998, p. 41).

See also BINOMIAL THEOREM, GAUSS'S HYPERGEOMETRIC THEOREM, Q-CHU-VANDERMONDE IDENTITY, UMBRAL CALCULUS

References

Bailey, W. N. *Generalised Hypergeometric Series.* Cambridge, England: Cambridge University Press, 1935.
Koepf, W. *Hypergeometric Summation: An Algorithmic Approach to Summation and Special Function Identities.* Braunschweig, Germany: Vieweg, 1998.
Petkovsek, M.; Wilf, H. S.; and Zeilberger, D. *A = B.* Wellesley, MA: A. K. Peters, pp. 130 and 181–82, 1996.
Roman, S. *The Umbral Calculus.* New York: Academic Press, p. 29, 1984.

Chvátal Graph

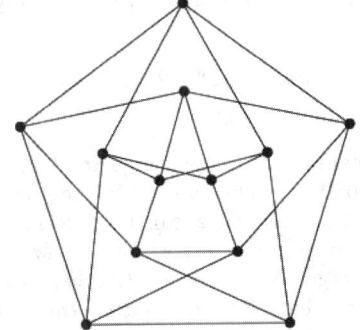

Grünbaum conjectured that for every $m > 1$, $n > 2$, there exists an m-regular, m-chromatic graph of GIRTH at least n. This result is trivial for $n = 2$ and $m = 2$, 3, but only two other such graphs are known: the Chvátal graph illustrated above, and the GRÜNBAUM GRAPH.

See also GRÜNBAUM GRAPH

References

Bondy, J. A. and Murty, U. S. R. *Graph Theory with Applications.* New York: North Holland, p. 241, 1976.

Grünbaum, B. "A Problem in Graph Coloring." *Amer. Math. Monthly* **77**, 1088–092, 1970.

Chvátal's Art Gallery Theorem

ART GALLERY THEOREM

Chvátal's Theorem

Let a GRAPH G have VERTICES with VERTEX DEGREES $d_1 \leq \cdots \leq d_m$. If for every $i < n/2$ we have either $d_i \geq i+1$ or $d_{n-i} \geq n-i$, then the GRAPH is HAMILTONIAN.

See also HAMILTONIAN GRAPH

References

Chvátal, V. "On Hamilton's Ideals." *J. Combin. Th.* **12**, 163–68, 1972.

ci

COSINE INTEGRAL

Ci

COSINE INTEGRAL

Cigarettes

It is possible to place 7 cigarettes in such a way that each touches the other if $l/d > 7\sqrt{3}/2$ (Gardner 1959, p. 115).

References

Gardner, M. *The Scientific American Book of Mathematical Puzzles & Diversions.* New York: Simon and Schuster, 1959.

Cin

COSINE INTEGRAL

C-Infinity Function

e^{2x} is smooth; $|x^3|$ is not smooth at 0

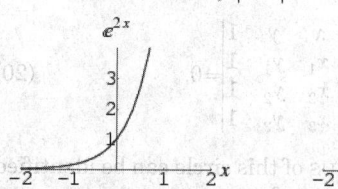 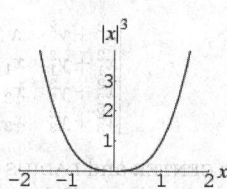

A C^∞ function is a function that is DIFFERENTIABLE for all degrees of differentiation. For instance, $f(x) = e^{2x}$ is C^∞ because its nth derivative $f^{(n)}(x) = 2^n e^{2x}$ exists and is CONTINUOUS. All polynomials are C^∞. The reason for the notation is that c^k FUNCTIONS have k continuous derivatives.

C^∞ functions are also called "smooth" because neither they nor their derivatives have "corners," which would make their graph look somewhat rough. For example, $f(x) = |x^3|$ is not smooth.

There are special C^∞ functions which are very useful in analysis and geometry. For example, there are smooth functions called BUMP FUNCTIONS, which are smooth approximations to a CHARACTERISTIC FUNCTION. Typically, these functions require some CALCULUS to show that they are indeed C^∞.

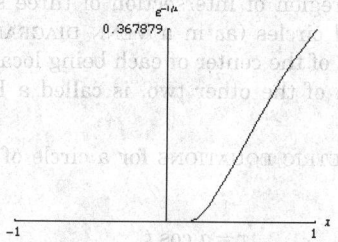

Any ANALYTIC FUNCTION is smooth. But a smooth function is not necessarily analytic. For instance, an analytic function cannot be a BUMP FUNCTION. Consider the following function, whose TAYLOR SERIES at 0 is identically zero, yet the function is not zero:

$$f(x) = \begin{cases} 0 & \text{for } x \leq 0 \\ e^{-1/x} & \text{for } x > 0. \end{cases}$$

The function f goes to zero very quickly. One property of smooth functions is that they can look very different at different scales.

The set of smooth functions cannot be made into a BANACH SPACE, which makes some problems hard, but instead has the weaker structure of a FRÉCHET SPACE.

See also C-K FUNCTION, C-INFINITY TOPOLOGY, CALCULUS, DIFFERENTIAL TOPOLOGY, FRÉCHET SPACE, PARTITION OF UNITY, SARD'S THEOREM

Circle

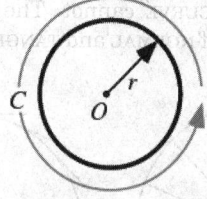

A circle is the set of points equidistant from a given point O. The distance r from the CENTER is called the RADIUS, and the point O is called the CENTER. Twice the RADIUS is known as the DIAMETER $d = 2r$. The PERIMETER C of a circle is called the CIRCUMFERENCE, and is given by

$$C = \pi d = 2\pi r. \tag{1}$$

The angle a circle subtends from its center is a FULL ANGLE, equal to $360°$ or 2π RADIANS.

The circle is a CONIC SECTION obtained by the intersection of a CONE with a PLANE PERPENDICULAR to the CONE's symmetry axis. A circle is the degenerate case of an ELLIPSE with equal semimajor and semiminor axes (i.e., with ECCENTRICITY 0). The interior of a circle is called a DISK. The generalization of a circle to 3-D is called a SPHERE, and to n-D for $n \geq 4$ a HYPERSPHERE.

The region of intersection of two circles is called a LENS. The region of intersection of three symmetrically placed circles (as in a VENN DIAGRAM), in the special case of the center of each being located at the intersection of the other two, is called a REULEAUX TRIANGLE.

The PARAMETRIC EQUATIONS for a circle of RADIUS a are

$$x = a \cos t \tag{2}$$

$$y = a \sin t. \tag{3}$$

For a body moving uniformly around the circle,

$$x' = -a \sin t \tag{4}$$

$$y' = a \cos t, \tag{5}$$

and

$$x'' = -a \cos t \tag{6}$$

$$y'' = -a \sin t. \tag{7}$$

When normalized, the former gives the equation for the unit TANGENT VECTOR of the circle, $(-\sin t, \cos t)$. The circle can also be parameterized by the rational functions

$$x = \frac{1 - t^2}{1 + t^2} \tag{8}$$

$$y = \frac{2t}{1 + t^2}, \tag{9}$$

but an ELLIPTIC CURVE cannot. The following plots show a sequence of NORMAL and TANGENT VECTORS for the circle.

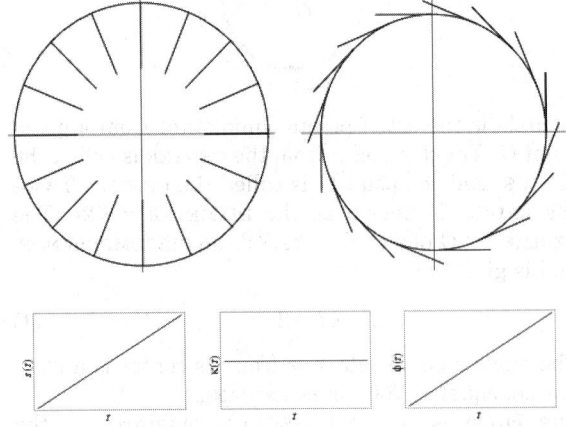

The ARC LENGTH s, CURVATURE κ, and TANGENTIAL ANGLE ϕ of the circle are

$$s(t) = \int ds = \int \sqrt{x'^2 + y'^2} \, dt = at \tag{10}$$

$$\kappa(t) = \frac{x'y'' - y'x''}{(x'^2 + y'^2)^{3/2}} = \frac{1}{a} \tag{11}$$

$$\phi(t) = \int \kappa(t) \, dt = \frac{t}{a}. \tag{12}$$

The CESÀRO EQUATION is

$$\kappa = \frac{1}{a}. \tag{13}$$

In POLAR COORDINATES, the equation of the circle has a particularly simple form.

$$r = a \tag{14}$$

is a circle of RADIUS a centered at ORIGIN,

$$r = 2a \cos \theta \tag{15}$$

is circle of RADIUS a centered at $(a, 0)$, and

$$r = 2a \sin \theta \tag{16}$$

is a circle of RADIUS a centered on $(0, a)$. In CARTESIAN COORDINATES, the equation of a circle of RADIUS a centered on (x_0, y_0) is

$$(x - x_0)^2 + (y - y_0)^2 = a^2. \tag{17}$$

In PEDAL COORDINATES with the PEDAL POINT at the center, the equation is

$$pa = r^2 \tag{18}$$

The circle having $P_1 P_2$ as a diameter is given by

$$(x - x_1)(x - x_2) + (y - y_1)(y - y_2) = 0. \tag{19}$$

The equation of a circle passing through the three points (x_i, y_i) for $i = 1, 2, 3$ (the CIRCUMCIRCLE of the TRIANGLE determined by the points) is

$$\begin{vmatrix} x^2 + y^2 & x & y & 1 \\ x_1^2 + y_1^2 & x_1 & y_1 & 1 \\ x_2^2 + y_2^2 & x_2 & y_2 & 1 \\ x_3^2 + y_3^2 & x_3 & y_3 & 1 \end{vmatrix} = 0. \tag{20}$$

The CENTER and RADIUS of this circle can be identified by assigning coefficients of a QUADRATIC CURVE

$$ax^2 + cy^2 + dx + ey + f = 0, \tag{21}$$

where $a = c$ and $b = 0$ (since there is no xy cross term). COMPLETING THE SQUARE gives

$$a\left(x + \frac{d}{2a}\right)^2 + a\left(y + \frac{e}{2a}\right)^2 + f - \frac{d^2 + e^2}{4a} = 0. \tag{22}$$

The CENTER can then be identified as

$$x_0 = -\frac{d}{2a}. \tag{23}$$

$$y_0 = -\frac{e}{2a} \tag{24}$$

and the RADIUS as

$$r = \sqrt{\frac{d^2 + e^2}{4a^2} - \frac{f}{a}}, \tag{25}$$

where

$$a = \begin{vmatrix} x_1 & y_1 & 1 \\ x_2 & y_2 & 1 \\ x_3 & y_3 & 1 \end{vmatrix} \tag{26}$$

$$d = -\begin{vmatrix} x_1^2 + y_1^2 & y_1 & 1 \\ x_2^2 + y_2^2 & y_2 & 1 \\ x_3^2 + y_3^2 & y_3 & 1 \end{vmatrix} \tag{27}$$

$$e = \begin{vmatrix} x_1^2 + y_1^2 & x_1 & 1 \\ x_2^2 + y_2^2 & x_2 & 1 \\ x_3^2 + y_3^2 & x_3 & 1 \end{vmatrix} \tag{28}$$

$$f = -\begin{vmatrix} x_1^2 + y_1^2 & x_1 & y_1 \\ x_2^2 + y_2^2 & x_2 & y_2 \\ x_3^2 + y_3^2 & x_3 & y_3 \end{vmatrix} \tag{29}$$

Four or more points which lie on a circle are said to be CONCYCLIC. Three points are trivially concyclic since three noncollinear points determine a circle.

The CIRCUMFERENCE-to-DIAMETER ratio C/d for a circle is constant as the size of the circle is changed (as it must be since scaling a plane figure by a factor s increases its PERIMETER by s), and d also scales by s. This ratio is denoted π (PI), and has been proved TRANSCENDENTAL. With d the DIAMETER and r the RADIUS,

$$C = \pi d = 2\pi r. \tag{30}$$

Knowing C/d, we can then compute the AREA of the circle either geometrically or using CALCULUS. From CALCULUS,

$$A = \int_0^{2\pi} d\theta \int_0^r r\, dr = (2\pi)\left(\tfrac{1}{2}r^2\right) = \pi r^2. \tag{31}$$

Now for a few geometrical derivations. Using concentric strips, we have

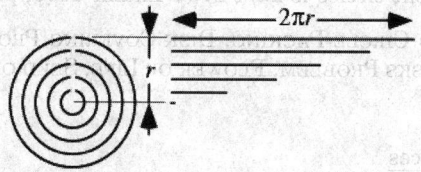

As the number of strips increases to infinity, we are left with a TRIANGLE on the right, so

$$A = \tfrac{1}{2}(2\pi r)r = \pi r^2. \tag{32}$$

This derivation was first recorded by Archimedes in *Measurement of a Circle* (ca. 225 BC). If we cut the circle instead into wedges,

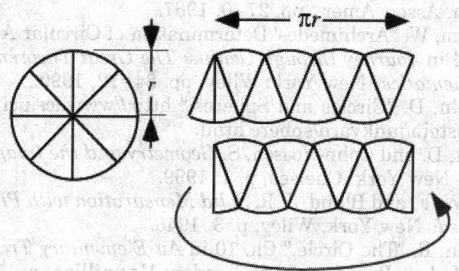

As the number of wedges increases to infinity, we are left with a RECTANGLE, so

$$A = (\pi r)r = \pi r^2. \tag{33}$$

See also ADAMS' CIRCLE, ARC, BLASCHKE'S THEOREM, BRAHMAGUPTA'S FORMULA, BROCARD CIRCLE, CASEY'S THEOREM, CEVIAN CIRCLE, CHORD, CIRCLE INSCRIBING, CIRCLE-LINE INTERSECTION, CIRCUMCIRCLE, CIRCUMFERENCE, CLIFFORD'S CIRCLE THEOREM, CLOSED DISK, CONCENTRIC CIRCLES, COSINE CIRCLE, COTES CIRCLE PROPERTY, DIAMETER, DISK, DROZ-FARNY CIRCLES, EULER TRIANGLE FORMULA, EXCIRCLE, EXCOSINE CIRCLE, EYEBALL THEOREM, FEUERBACH'S THEOREM, FIVE CIRCLES THEOREM, FIVE DISKS PROBLEM, FLOWER OF LIFE, FORD CIRCLE, FUHRMANN CIRCLE, GERSGORIN CIRCLE THEOREM, HART CIRCLE, HOPF CIRCLE, INCIRCLE, INVERSIVE DISTANCE, JOHNSON CIRCLE, KINNEY'S SET, LEMOINE CIRCLE, LENS, LESTER CIRCLE, MAGIC CIRCLES, MALFATTI CIRCLES, MCCAY CIRCLE, MIDCIRCLE, MONGE'S THEOREM, NEUBERG CIRCLE, NINE-POINT CIRCLE, OPEN DISK, *P*-CIRCLE, PARRY CIRCLE, PI, POINT CIRCLE, POLAR CIRCLE, POWER (CIRCLE), PRIME CIRCLE, PSEUDOCIRCLE, PTOLEMY'S THEOREM, PURSER'S THEOREM, RADICAL AXIS, RADIUS, REULEAUX TRIANGLE, SEED OF LIFE, SEIFERT CIRCLE, SEMICIRCLE, SEVEN CIRCLES THEOREM, SIMILITUDE CIRCLE, SIX CIRCLES THEOREM, SODDY CIRCLES, SPHERE, TAYLOR CIRCLE, TRIPLICATE-RATIO CIRCLE, TUCKER CIRCLES, UNIT CIRCLE, VENN DIAGRAM, VILLARCEAU CIRCLES, YIN-YANG

References

Beyer, W. H. *CRC Standard Mathematical Tables, 28th ed.* Boca Raton, FL: CRC Press, pp. 125 and 197, 1987.

Casey, J. "The Circle." Ch. 3 in *A Treatise on the Analytical Geometry of the Point, Line, Circle, and Conic Sections, Containing an Account of Its Most Recent Extensions, with Numerous Examples, 2nd ed., rev. enl.* Dublin: Hodges, Figgis, & Co., pp. 96–50, 1893.

Coolidge, J. L. *A Treatise on the Geometry of the Circle and Sphere.* New York: Chelsea, 1971.

Courant, R. and Robbins, H. *What is Mathematics?: An Elementary Approach to Ideas and Methods, 2nd ed.* Oxford, England: Oxford University Press, pp. 74–5, 1996.

Coxeter, H. S. M. and Greitzer, S. L. "Some Properties of Circles." Ch. 2 in *Geometry Revisited.* Washington, DC: Math. Assoc. Amer., pp. 27–0, 1967.

Dunham, W. "Archimedes' Determination of Circular Area." Ch. 4 in *Journey through Genius: The Great Theorems of Mathematics.* New York: Wiley, pp. 84–12, 1990.

Eppstein, D. "Circles and Spheres." http://www.ics.uci.edu/~eppstein/junkyard/sphere.html.

Hilbert, D. and Cohn-Vossen, S. *Geometry and the Imagination.* New York: Chelsea, p. 1, 1999.

Kern, W. F. and Bland, J. R. *Solid Mensuration with Proofs, 2nd ed.* New York: Wiley, p. 3, 1948.

Lachlan, R. "The Circle." Ch. 10 in *An Elementary Treatise on Modern Pure Geometry.* London: Macmillian, pp. 148–73, 1893.

Lawrence, J. D. *A Catalog of Special Plane Curves.* New York: Dover, pp. 65–6, 1972.

MacTutor History of Mathematics Archive. "Circle." http://www-groups.dcs.st-and.ac.uk/~history/Curves/Circle.html.

Pappas, T. "Infinity & the Circle" and "Japanese Calculus." *The Joy of Mathematics.* San Carlos, CA: Wide World Publ./Tetra, pp. 68 and 139, 1989.

Pedoe, D. *Circles: A Mathematical View, rev. ed.* Washington, DC: Math. Assoc. Amer., 1995.

Yates, R. C. "The Circle." *A Handbook on Curves and Their Properties.* Ann Arbor, MI: J. W. Edwards, pp. 21–5, 1952.

Circle Bundle

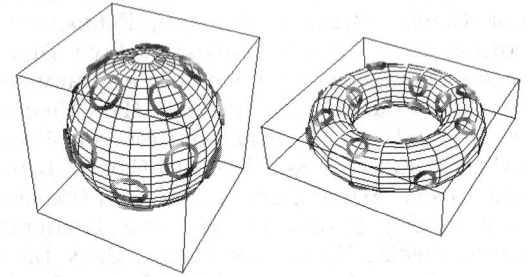

A circle bundle $\pi : E \to M$ is a FIBER BUNDLE whose FIBERS $\pi^{-1}(x)$ are circles. It may also have the structure of a PRINCIPAL BUNDLE if there is an action of $SO(2)$ that preserves the fibers, and is locally trivial. That is, if every point has a TRIVIALIZATION $U \times \mathbb{S}^1$ such that the action of $SO(2)$ on $\mathbb{S}^1$ is the usual one.

See also BUNDLE, GROUP ACTION, PRINCIPAL BUNDLE

Circle Caustic

Consider a point light source located at a point $(\mu, 0)$. The CATACAUSTIC of a unit CIRCLE for the light at $\mu = \infty$ is the NEPHROID

$$x = \tfrac{1}{4}[3 \cot t - \cos(3t)] \tag{1}$$

$$y = \tfrac{1}{4}[3 \sin t - \sin(3t)]. \tag{2}$$

The CATACAUSTIC for the light at a finite distance $\mu >$

1 is the curve

$$x = \frac{\mu(1 - 3\mu \cos t + 2\mu \cos^3 t)}{-(1 + 2\mu^2) + 3\mu \cos t} \tag{3}$$

$$y = \frac{2\mu^2 \sin^3 t}{1 + 2\mu^2 - 3\mu \cos t}, \tag{4}$$

and for the light on the CIRCUMFERENCE of the CIRCLE $\mu = 1$ is the CARDIOID

$$x = \tfrac{2}{3} \cos t(1 + \cos t) - \tfrac{1}{3} \tag{5}$$

$$y = \tfrac{2}{3} \sin t(1 + \cos t). \tag{6}$$

If the point is inside the circle, the catacaustic is a discontinuous two-part curve. These four cases are illustrated below.

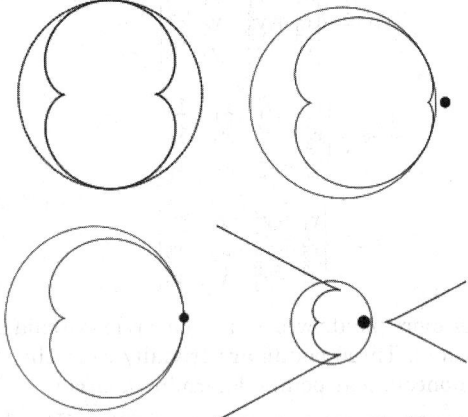

The CATACAUSTIC for PARALLEL rays crossing a CIRCLE is a CARDIOID.

See also CATACAUSTIC, CAUSTIC

Circle Chord Picking
CIRCLE LINE PICKING

Circle Covering

An arrangement of overlapping circles which cover the entire plane. A lower bound for a covering using equivalent circles is $2\pi/\sqrt{27}$ (Williams 1979, p. 51).

See also CIRCLE PACKING, DISK COVERING PROBLEM, FIVE DISKS PROBLEM, FLOWER OF LIFE, SEED OF LIFE

References

Williams, R. "Circle Coverings." §2– in *The Geometrical Foundation of Natural Structure: A Source Book of Design.* New York: Dover, pp. 51–2, 1979.

Circle Covering by Arcs

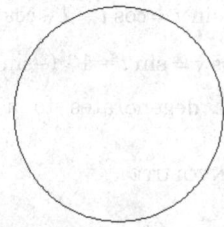

The probability $P(a, n)$ that n random arcs of angular size a cover the circumference of a circle completely (for a circle with unit circumference) is

$$P(a, n) = \sum_{k=0}^{\lfloor 1/a \rfloor} (-1)^k \binom{n}{k}(1-ka)^{n-1},$$

where $\lfloor x \rfloor$ is the FLOOR FUNCTION (Solomon 1978, p. 75). This was first given correctly by Stevens (1939), although partial results were obtains by Whitworth (1897), Baticle (1935), Garwood (1940), Darling (1953), and Shepp (1972).

The probability that n arcs leave exactly l gaps is given by

$$P_{l \text{ gaps}}(a, n) = \binom{n}{l}\sum_{j=1}^{k}(-1)^{j-l}\binom{n-l}{j-l}(1-ja)^{n-1}$$

(Stevens 1939; Solomon 1978, p. 76).

See also CIRCLE POINT PICKING, CIRCLE LINE PICKING

References

Baticle, M. "Le problème des répartitions." *C. R. Acad. Sci. Paris* **201**, 862–64, 1935.
Fisher, R. A. "Tests of Significance in Harmonic Analysis." *Proc. Roy. Soc. London Ser. A* **125**, 54–9, 1929.
Fisher, R. A. "On the Similarity of the Distributions Found for the Test of Significance in Harmonic Analysis, and in Stevens's Problem in Geometric Probability." *Eugenics* **10**, 14–7, 1940.
Darling, D. A. "On a Class of Problems Related to the Random Division of an Interval." *Ann. Math. Stat.* **24**, 239–53, 1953.
Garwood, F. "An Application to the Theory of Probability of the Operation of Vehicular-Controlled Traffic Signals." *J. Roy. Stat. Soc. Suppl.* **7**, 65–7, 1940.
Shepp, L. A. "Covering the Circle with Random Arcs." *Israel J. Math.* **11**, 328–45, 1972.
Siegel, A. F. *Random Coverage Problems in Geometric Probability with an Application to Time Series Analysis.* Ph.D. thesis. Stanford, CA: Stanford University, 1977.
Solomon, H. "Covering a Circle Circumference and a Sphere Surface." Ch. 4 in *Geometric Probability.* Philadelphia, PA: SIAM, pp. 75–6, 1978.
Stevens, W. L. "Solution to a Geometrical Problem in Probability." *Ann. Eugenics* **9**, 315–20, 1939.
Whitworth, W. A. *DCC Exercises in Choice and Chance.* 1897. Reprinted New York: Hafner, 1965.

Circle Cutting

CIRCLE DIVISION BY CHORDS, CIRCLE DIVISION BY LINES

Circle Division by Chords

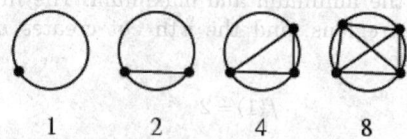

A related problem, sometimes called Moser's circle problem, is to find the number of pieces into which a CIRCLE is divided if n points on its CIRCUMFERENCE are joined by CHORDS with no three CONCURRENT. The answer is

$$g(n) = \binom{n}{4} + \binom{n}{2} + 1 \qquad (1)$$

$$= \tfrac{1}{24}(n^4 - 6n^3 + 23n^2 - 18n + 24), \qquad (2)$$

(Yaglom and Yaglom 1987, Guy 1988, Conway and Guy 1996, Noy 1996), where $\binom{n}{m}$ is a BINOMIAL COEFFICIENT. The first few values are 1, 2, 4, 8, 16, 31, 57, 99, 163, 256, ... (Sloane's A000127). This sequence demonstrates the danger in making assumptions based on limited trials. While the series starts off like 2^{n-1}, it begins differing from this GEOMETRIC SERIES at $n = 6$.

See also CAKE CUTTING, CIRCLE DIVISION BY LINES, CYLINDER CUTTING, HAM SANDWICH THEOREM, PANCAKE THEOREM, PIZZA THEOREM, PLANE DIVISION BY CIRCLES, PLANE DIVISION BY ELLIPSES, PLANE DIVISION BY LINES, SQUARE DIVISION BY LINES, TORUS CUTTING

References

Conway, J. H. and Guy, R. K. "How Many Regions." In *The Book of Numbers.* New York: Springer-Verlag, pp. 76–9, 1996.
Guy, R. K. "The Strong Law of Small Numbers." *Amer. Math. Monthly* **95**, 697–12, 1988.
Noy, M. "A Short Solution of a Problem in Combinatorial Geometry." *Math. Mag.* **69**, 52–3, 1996.
Sloane, N. J. A. Sequences A000127/M1119 in "An On-Line Version of the Encyclopedia of Integer Sequences." http://www.research.att.com/~njas/sequences/eisonline.html.
Yaglom, A. M. and Yaglom, I. M. Problem 47 in *Challenging Mathematical Problems with Elementary Solutions, Vol. 1.* New York: Dover, 1987.

Circle Division by Lines

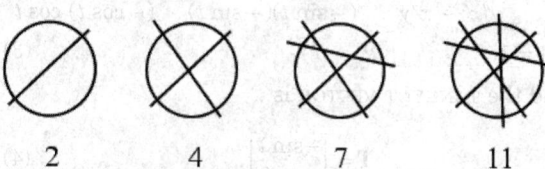

Determining the maximum number of pieces in which it is possible to divide a CIRCLE for a given number of cuts is called the circle cutting, or sometimes PANCAKE CUTTING, problem. The minimum number is always $n + 1$, where n is the number of cuts, and it is

always possible to obtain any number of pieces between the minimum and maximum. The first cut creates 2 regions, and the nth cut creates n new regions, so

$$f(1) = 2 \tag{1}$$

$$f(2) = 2 + f(1) \tag{2}$$

$$f(n) = n + f(n - 1). \tag{3}$$

Therefore,

$$f(n) = n + [(n - 1) + f(n - 2)]$$

$$= n + (n - 1) + \ldots + 2 + f(1) = f(1) + \sum_{k=2}^{n} kf(1)$$

$$= 2 + \tfrac{1}{2}(n + 2)(n - 1) = \tfrac{1}{2}(n^2 + n + 2). \tag{4}$$

Evaluating for $n = 1, 2, \ldots$ gives 2, 4, 7, 11, 16, 22, ... (Sloane's A000124). This is equivalent to the maximal number of regions into which a PLANE can be cut by n lines.

See also CIRCLE DIVISION BY CHORDS, PLANE DIVISION BY CIRCLES, SPACE DIVISION BY PLANES, SPACE DIVISION BY SPHERES, SQUARE DIVISION BY LINES

References
Sloane, N. J. A. Sequences A000124/M1041 in "An On-Line Version of the Encyclopedia of Integer Sequences." http://www.research.att.com/~njas/sequences/eisonline.html.
Sloane, N. J. A. and Plouffe, S. Figure M1041 in *The Encyclopedia of Integer Sequences.* San Diego: Academic Press, 1995.
Yaglom, A. M. and Yaglom, I. M. *Challenging Mathematical Problems with Elementary Solutions, Vol. 1.* New York: Dover, pp. 102–06, 1987.
Wells, D. *The Penguin Dictionary of Curious and Interesting Numbers.* Middlesex, England: Penguin Books, p. 31, 1986.

Circle Evolute

$$x = \cos t \quad x' = -\sin t \quad x'' = -\cos t \tag{1}$$

$$y = \sin t \quad y' = \cos t \quad y'' = -\sin t, \tag{2}$$

so the RADIUS OF CURVATURE is

$$R = \frac{(x'^2 + y'^2)^{3/2}}{y''x' - x''y'} = \frac{(\sin^2 t + \cos^2 t)^{3/2}}{(-\sin t)(-\sin t) - (-\cos t)\cos t}$$

$$= 1, \tag{3}$$

and the TANGENT VECTOR is

$$\hat{\mathbf{T}} = \begin{bmatrix} -\sin t \\ \cos t \end{bmatrix}. \tag{4}$$

Therefore,

$$\cos \tau = \hat{\mathbf{T}} \cdot \hat{\mathbf{x}} = -\sin t \tag{5}$$

$$\sin \tau = \hat{\mathbf{T}} \cdot \hat{\mathbf{y}} = \cos t, \tag{6}$$

so

$$\xi(t) = x - R \sin \tau = \cos t - 1 \cdot \cos t = 0 \tag{7}$$

$$\eta(t) = y + R \cos \tau = \sin t + 1 \cdot (-\sin t) = 0, \tag{8}$$

and the EVOLUTE degenerates to a POINT at the ORIGIN.

See also CIRCLE INVOLUTE

References
Gray, A. *Modern Differential Geometry of Curves and Surfaces with Mathematica, 2nd ed.* Boca Raton, FL: CRC Press, p. 99, 1997.
Lauwerier, H. *Fractals: Endlessly Repeated Geometric Figures.* Princeton, NJ: Princeton University Press, pp. 55–9, 1991.
Steinhaus, H. *Mathematical Snapshots, 3rd ed.* New York: Dover, p. 137, 1999.

Circle Inscribing

If r is the INRADIUS of a CIRCLE inscribed in a RIGHT TRIANGLE with sides a and b and HYPOTENUSE c, then

$$r = \tfrac{1}{2}(a + b - c).$$

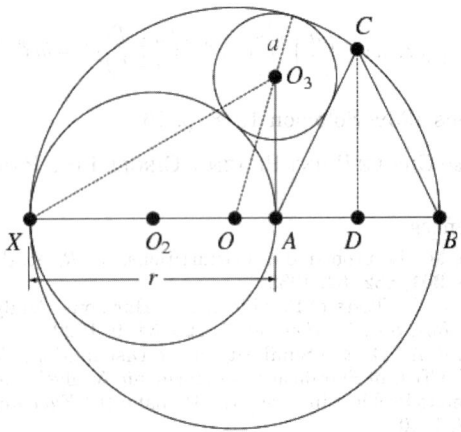

A SANGAKU PROBLEM dated 1803 from the Gumma Prefecture asks to construct the figure consisting of a circle centered at O, a second smaller circle centered at O_2 tangent to the first, and an ISOSCELES TRIANGLE whose base AB completes the diameter of the larger circle through the smaller XB. Now inscribe a third circle with center O_3 inside the large circle, outside the small one, and on the side of a leg of the triangle. It then follows that the line $O_3A \perp XB$. To find the explicit position and size of the circle, let the circle O have radius $1/2$ and be centered at $(0, 0)$ and let the circle O_2 have diameter $0 < r < 1$. Then solving the simultaneous equations

$$\left(\tfrac{1}{2}r + a\right)^2 = \left(\tfrac{1}{2}r\right)^2 + y^2 \tag{1}$$

$$\left(\tfrac{1}{2} - a\right)^2 = \left(r - \tfrac{1}{2}\right)^2 + y^2 \tag{2}$$

for a and y gives

$$a = \frac{r(1-r)}{1+r} \tag{3}$$

$$y = \frac{r\sqrt{2(1-r)}}{1+r}. \tag{4}$$

See also INCIRCLE, INSCRIBED, POLYGON

References

Rothman, T. "Japanese Temple Geometry." *Sci. Amer.* **278**, 85–1, May 1998.

Circle Involute

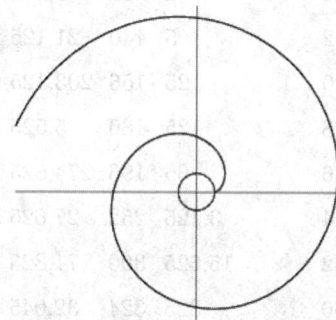

First studied by Huygens when he was considering clocks without pendula for use on ships at sea. He used the circle involute in his first pendulum clock in an attempt to force the pendulum to swing in the path of a CYCLOID. For a CIRCLE with $a = 1$, the PARAMETRIC EQUATIONS of the circle and their derivatives are given by

$$x = \cos t \quad x' = -\sin t \quad x'' = -\cos t \tag{1}$$

$$y = \sin t \quad y' = \cos t \quad y'' = -\sin t. \tag{2}$$

The TANGENT VECTOR is

$$\hat{\mathbf{T}} = \begin{bmatrix} -\sin t \\ \cos t \end{bmatrix} \tag{3}$$

and the ARC LENGTH along the circle is

$$s = \int \sqrt{x'^2 + y'^2}\, dt = \int dt = t, \tag{4}$$

so the involute is given by

$$\mathbf{r}_i = \mathbf{r} - s\hat{\mathbf{T}} = \begin{bmatrix} \cos t \\ \sin t \end{bmatrix} - t \begin{bmatrix} -\sin t \\ \cos t \end{bmatrix} = \begin{bmatrix} \cos t + t \sin t \\ \sin t - t \cos t \end{bmatrix}, \tag{5}$$

or

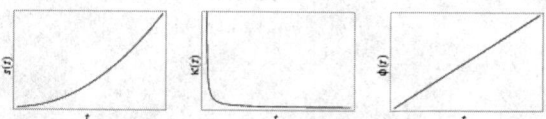

$$x = a(\cos t + t \sin t) \tag{6}$$

$$y = a(\sin t - t \cos t). \tag{7}$$

The ARC LENGTH, CURVATURE, and TANGENTIAL ANGLE are

$$s = \int ds = \int \sqrt{x'^2 + y'^2}\, dt = \tfrac{1}{2}at^2 \tag{8}$$

$$\kappa = \frac{1}{at} \tag{9}$$

$$\phi = t. \tag{10}$$

The CESÀRO EQUATION is

$$\kappa = \frac{1}{\sqrt{as}}. \tag{11}$$

See also CIRCLE, CIRCLE EVOLUTE, ELLIPSE INVOLUTE, INVOLUTE

References

Beyer, W. H. *CRC Standard Mathematical Tables, 28th ed.* Boca Raton, FL: CRC Press, p. 220, 1987.

Gray, A. *Modern Differential Geometry of Curves and Surfaces with Mathematica, 2nd ed.* Boca Raton, FL: CRC Press, p. 105, 1997.

Hilbert, D. and Cohn-Vossen, S. *Geometry and the Imagination.* New York: Chelsea, pp. 6–, 1999.

Lawrence, J. D. *A Catalog of Special Plane Curves.* New York: Dover, pp. 190–91, 1972.

MacTutor History of Mathematics Archive. "Involute of a Circle." http://www-groups.dcs.st-and.ac.uk/~history/Curves/Involute.html.

Circle Involute Pedal Curve

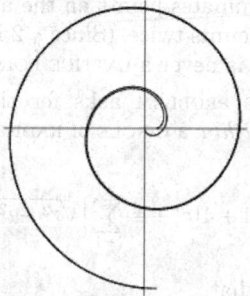

The PEDAL CURVE of CIRCLE INVOLUTE

$$f = \cos t + t \sin t$$

$$g = \sin t - t \cos t$$

with the center as the PEDAL POINT is the ARCHI-

MEDES' SPIRAL

$$x = t \sin t$$

$$y = -t \cos t.$$

Circle Lattice Points

For every POSITIVE INTEGER n, there exists a CIRCLE which contains exactly n lattice points in its interior. H. Steinhaus proved that for every POSITIVE INTEGER n, there exists a CIRCLE of AREA n which contains exactly n lattice points in its interior.

SCHINZEL'S THEOREM shows that for every POSITIVE INTEGER n, there exists a CIRCLE in the PLANE having exactly n LATTICE POINTS on its CIRCUMFERENCE. The theorem also explicitly identifies such "SCHINZEL CIRCLES" as

$$\begin{cases} \left(x - \frac{1}{2}\right)^2 + y^2 = \frac{1}{4} 5^{k-1} & \text{for } n = 2k \\ \left(x - \frac{1}{3}\right)^2 + y^2 = \frac{1}{9} 5^{2k} & \text{for } n = 2k + 1. \end{cases} \quad (1)$$

Note, however, that these solutions do not necessarily have the smallest possible RADIUS. For example, while the SCHINZEL CIRCLE centered at $(1/3, 0)$ and with RADIUS $625/3$ has nine lattice points on its CIRCUMFERENCE, so does the CIRCLE centered at $(1/3, 0)$ with RADIUS $65/3$.

Let r be the smallest INTEGER RADIUS of a CIRCLE centered at the ORIGIN $(0, 0)$ with $L(r)$ LATTICE POINTS. In order to find the number of lattice points of the CIRCLE, it is only necessary to find the number in the first octant, i.e., those with $0 \leq y \leq \lfloor r/\sqrt{2} \rfloor$, where $\lfloor z \rfloor$ is the FLOOR FUNCTION. Calling this $N(r)$, then for $r \geq 1$, $L(r) = 8N(r) - 4$, so $L(r) \equiv 4 \pmod 8$. The multiplication by eight counts all octants, and the subtraction by four eliminates points on the axes which the multiplication counts twice. (Since $\sqrt{2}$ is IRRATIONAL, a mid-arc point is never a LATTICE POINT.)

GAUSS'S CIRCLE PROBLEM asks for the number of lattice points *within* a CIRCLE of RADIUS r

$$N(r) = 1 + 4\lfloor r \rfloor + 4 \sum_{i=1}^{\lfloor r \rfloor} \left\lfloor \sqrt{r^2 - i^2} \right\rfloor. \quad (2)$$

Gauss showed that

$$N(r) = \pi r^2 + E(r), \quad (3)$$

where

$$|E(r)| \leq 2\sqrt{2}\pi r. \quad (4)$$

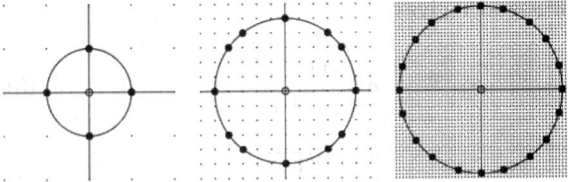

The number of lattice points on the CIRCUMFERENCE of circles centered at $(0, 0)$ with radii $0, 1, 2, \ldots$ are 1, 4, 4, 4, 4, 12, 4, 4, 4, 4, 12, 4, 4, ... (Sloane's A046109). The following table gives the smallest RADIUS $r \leq 390, 800$ for a circle centered at $(0, 0)$ having a given number of LATTICE POINTS $L(r)$ (Sloane's A046112). Note that the high-water mark radii are always multiples of five.

$L(r)$	r	$L(r)$	r
1	0	108	1,105
4	1	132	40,625
12	5	140	21,125
20	25	156	203,125
28	125	180	5,525
36	65	196	274,625
44	3,125	252	27,625
52	15,625	300	71,825
60	325	324	32,045
68	390,625	420	359,125
76	$\leq 1,953,125$	540	160,225
84	1,625		
92	$\leq 48,828,125$		
100	4,225		

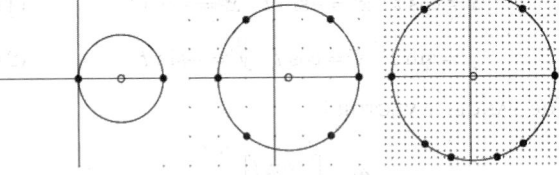

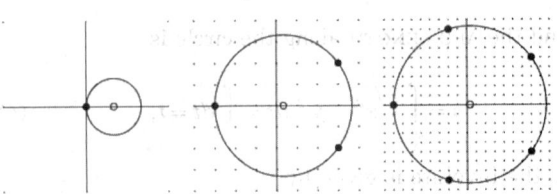

If the CIRCLE is instead centered at $(1/2, 0)$, then the CIRCLES of RADII $1/2, 3/2, 5/2, \ldots$ have 2, 2, 6, 2, 2, 2, 6,

6, 6, 2, 2, 2, 10, 2, ... (Sloane's A046110) on their CIRCUMFERENCES. If the CIRCLE is instead centered at (1/3, 0), then the number of lattice points on the CIRCUMFERENCE of the CIRCLES of RADIUS 1/3, 2/3, 4/3, 5/3, 7/3, 8/3, ... are 1, 1, 1, 3, 1, 1, 3, 1, 3, 1, 1, 3, 1, 3, 1, 1, 5, 3, ... (Sloane's A046111).

Let

1. a_n be the RADIUS of the CIRCLE centered at (0, 0) having $8n + 4$ lattice points on its CIRCUMFERENCE,
2. $b_n/2$ be the RADIUS of the CIRCLE centered at (1/2, 0) having $4n + 2$ lattice points on its CIRCUMFERENCE,
3. $c_n/3$ be the RADIUS of CIRCLE centered at (1/3, 0) having $2n + 1$ lattice points on its CIRCUMFERENCE.

Then the sequences $\{a_n\}$, $\{b_n\}$, and $\{c_n\}$ are equal, with the exception that $b_n = 0$ if $2|n$ and $c_n = 0$ if $3|n$. However, the sequences of *smallest* radii having the above numbers of lattice points are equal in the three cases and given by 1, 5, 25, 125, 65, 3125, 15625, 325, ... (Sloane's A046112).

KULIKOWSKI'S THEOREM states that for every POSITIVE INTEGER n, there exists a 3-D SPHERE which has exactly n LATTICE POINTS on its surface. The SPHERE is given by the equation

$$(x - a)^2 + (y - b)^2 + (z - \sqrt{2})^2 = c^2 + 2,$$

where a and b are the coordinates of the center of the so-called SCHINZEL CIRCLE and c is its RADIUS (Honsberger 1973).

See also CIRCLE, CIRCUMFERENCE, GAUSS'S CIRCLE PROBLEM, KULIKOWSKI'S THEOREM, LATTICE POINT, SCHINZEL CIRCLE, SCHINZEL'S THEOREM

References

Honsberger, R. "Circles, Squares, and Lattice Points." Ch. 11 in *Mathematical Gems I*. Washington, DC: Math. Assoc. Amer., pp. 117–27, 1973.
Kulikowski, T. "Sur l'existence d'une sphère passant par un nombre donné aux coordonnées entières." *L'Enseignement Math. Ser. 2* **5**, 89–0, 1959.
Schinzel, A. "Sur l'existence d'un cercle passant par un nombre donné aux points aux coordonnées entières." *L'Enseignement Math. Ser. 2* **4**, 71–2, 1958.
Sierpinski, W. "Sur quelques problèmes concernant les points aux coordonnées entières." *L'Enseignement Math. Ser. 2* **4**, 25–1, 1958.
Sierpinski, W. "Sur un problème de H. Steinhaus concernant les ensembles de points sur le plan." *Fund. Math.* **46**, 191–94, 1959.
Sierpinski, W. *A Selection of Problems in the Theory of Numbers.* New York: Pergamon Press, 1964.
Weisstein, E. W. "Circle Lattice Points." MATHEMATICA NOTEBOOK CIRCLELATTICEPOINTS.M.

Circle Lattice Theorem

GAUSS'S CIRCLE PROBLEM

Circle Line Picking

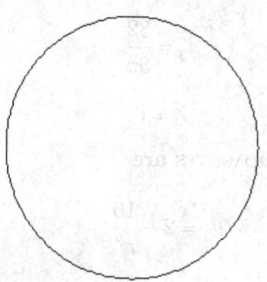

Given a UNIT CIRCLE, pick two points at random on its circumference, forming a CHORD. Without loss of generality, the first point can be taken as (1, 0), and the second by $(\cos \theta, \sin \theta)$, with $\theta \in [0, \pi]$ (by symmetry, the range can be limited to π instead of 2π). The distance s between the two points is then

$$s(\theta) = \sqrt{2 - 2\cos\theta} = 2|\sin(\tfrac{1}{2}\theta)|. \tag{1}$$

The average distance is then given by

$$\bar{s} = \frac{\int_0^\pi s(\theta)\,d\theta}{\int_0^\pi d\theta} = \frac{4}{\pi}. \tag{2}$$

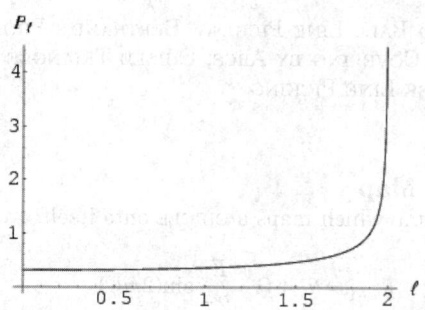

The probability function P_s is obtained from

$$P_s = \left|\frac{d\theta}{ds}\right| P_\theta = \frac{1}{\pi} \frac{1}{\sqrt{1 - (\tfrac{1}{2}s)^2}}. \tag{3}$$

The RAW MOMENTS are then

$$\mu'_n = \frac{\int_0^\pi [2\sin(\tfrac{1}{2}\theta)]^n\,d\theta}{\int_0^\pi d\theta} \tag{4}$$

$$= \int_0^2 \frac{s^n}{\pi\sqrt{1 - (\tfrac{1}{2}s)^2}} \tag{5}$$

$$= \frac{2^n \Gamma(\tfrac{1}{2}(1 + n))}{\sqrt{\pi}\,\Gamma(1 + \tfrac{1}{2}n)}, \tag{6}$$

giving the first few as

$$\mu_2' = 2 \tag{7}$$

$$\mu_3' = \frac{32}{3\pi} \tag{8}$$

$$\mu_4' = 6. \tag{9}$$

The CENTRAL MOMENTS are

$$\mu_2 = 2 - \frac{16}{\pi^2} \tag{10}$$

$$\mu_3 = \frac{8(48 - 5\pi^2)}{3\pi^3} \tag{11}$$

$$\mu_4 = 6 + \frac{64(\pi^2 - 36)}{3\pi^4}, \tag{12}$$

giving the SKEWNESS and KURTOSIS as

$$\gamma_1 = \frac{2\sqrt{2}(48 - 5\pi^2)}{3(\pi^2 - 8)^{3/2}} \tag{13}$$

$$\gamma_2 = \frac{-9\pi^4 + 320\pi^2 - 2304}{6(\pi^2 - 8)^2}. \tag{14}$$

BERTRAND'S PROBLEM asks for the PROBABILITY that a CHORD drawn at random on a CIRCLE of RADIUS r has length $\geq r$.

See also BALL LINE PICKING, BERTRAND'S PROBLEM, CIRCLE COVERING BY ARCS, CIRCLE TRIANGLE PICKING, DISK LINE PICKING

Circle Map

A 1-D MAP which maps a CIRCLE onto itself

$$\theta_{n+1} = \theta_n + \Omega - \frac{K}{2\pi} \sin(2\pi\theta_n), \tag{1}$$

where θ_{n+1} is computed mod 1 and K is a constant. Note that the circle map has two parameters: Ω and K. Ω can be interpreted as an externally applied frequency, and K as a strength of nonlinearity. The 1-D JACOBIAN is

$$\frac{\partial \theta_{n+1}}{\partial \theta_n} = 1 - K \cos(2\pi\theta_n), \tag{2}$$

so the circle map is not AREA-PRESERVING. It is related to the STANDARD MAP

$$I_{n+1} = I_n + \frac{K}{2\pi} \sin(2\pi\theta_n) \tag{3}$$

$$\theta_{n+1} = \theta_n + I_{n+1}, \tag{4}$$

for I and θ computed mod 1. Writing θ_{n+1} as

$$\theta_{n+1} = \theta_n + I_n + \frac{K}{2\pi} \sin(2\pi\theta_n) \tag{5}$$

gives the circle map with $I_n = \Omega$ and $K = -K$. The unperturbed circle map has the form

$$\theta_{n+1} = \theta_n + \Omega. \tag{6}$$

If Ω is RATIONAL, then it is known as the map WINDING NUMBER, defined by

$$\Omega = W \equiv \frac{p}{q}, \tag{7}$$

and implies a periodic trajectory, since θ_n will return to the same point (at most) every q ORBITS. If Ω is IRRATIONAL, then the motion is quasiperiodic. If K is NONZERO, then the motion may be periodic in some finite region surrounding each RATIONAL Ω. This execution of periodic motion in response to an IRRATIONAL forcing is known as MODE LOCKING.

If a plot is made of K vs. Ω with the regions of periodic MODE-LOCKED parameter space plotted around RATIONAL Ω values (WINDING NUMBERS), then the regions are seen to widen upward from 0 at $K = 0$ to some finite width at $K = 1$. The region surrounding each RATIONAL NUMBER is known as an ARNOLD TONGUE. At $K = 0$, the ARNOLD TONGUES are an isolated set of MEASURE zero. At $K = 1$, they form a CANTOR SET of DIMENSION $d \approx 0.08700$. For $K > 1$, the tongues overlap, and the circle map becomes non-invertible.

Let Ω_n be the parameter value of the circle map for a cycle with WINDING NUMBER $W_n = F_n/F_{n+1}$ passing with an angle $\theta = 0$, where F_n is a FIBONACCI NUMBER. Then the parameter values Ω_n accumulate at the rate

$$\delta \equiv \lim_{n \to \infty} \frac{\Omega_n - \Omega_{n-1}}{\Omega_{n+1} - \Omega_n} = -2.833 \tag{8}$$

(Feigenbaum *et al.* 1982).

See also ARNOLD TONGUE, DEVIL'S STAIRCASE, MODE LOCKING, WINDING NUMBER (MAP)

References

Devaney, R. L. *An Introduction to Chaotic Dynamical Systems.* Redwood City, CA: Addison-Wesley, pp. 108–11, 1987.

Feigenbaum, M. J.; Kadanoff, L. P.; and Shenker, S. J. "Quasiperiodicity in Dissipative Systems: A Renormalization Group Analysis." *Physica D* **5**, 370–86, 1982.

Rasband, S. N. "The Circle Map and the Devil's Staircase." §6.5 in *Chaotic Dynamics of Nonlinear Systems.* New York: Wiley, pp. 128–32, 1990.

Circle Method

A method employed by Hardy, Ramanujan, and Littlewood to solve many asymptotic problems in ADDITIVE NUMBER THEORY, particularly in deriving an asymptotic formula for the PARTITION FUNCTION P. The circle method proceeds by choosing a circular CONTOUR satisfying certain technical properties (Apostol 1997). The method was modified by Rademacher using a different contour in his derivative of the exact convergent formula for the PARTITION FUNCTION P.

See also PARTITION FUNCTION P

References

Apostol, T. M. "The Plan of the Proof." §5.2 in *Modular Functions and Dirichlet Series in Number Theory, 2nd ed.* New York: Springer-Verlag, pp. 95–6, 1997.

Circle Negative Pedal Curve

The NEGATIVE PEDAL CURVE of a circle is an ELLIPSE if the PEDAL POINT is inside the CIRCLE, and a HYPERBOLA if the PEDAL POINT is outside the CIRCLE.

Circle Notation

A NOTATION for LARGE NUMBERS due to Steinhaus (1983). In circle notation, $\textcircled{n}$ is defined as n in n SQUARES, where numbers written inside squares (and triangles) are interpreted in terms of STEINHAUS-MOSER NOTATION. The particular number known as the MEGA is then defined as follows (correcting the typographical error of Steinhaus).

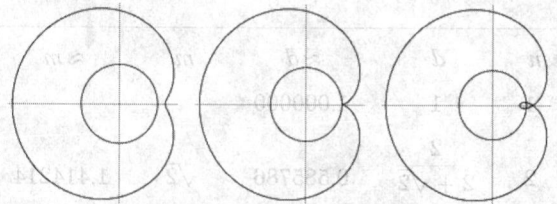

See also MEGA, MEGISTRON, STEINHAUS-MOSER NOTATION

References

Steinhaus, H. *Mathematical Snapshots, 3rd ed.* New York: Dover, pp. 28–9, 1999.

Circle Order

A POSET P is a circle order if it is ISOMORPHIC to a SET of DISKS ordered by containment.

See also ISOMORPHIC POSETS, PARTIALLY ORDERED SET

Circle Orthotomic

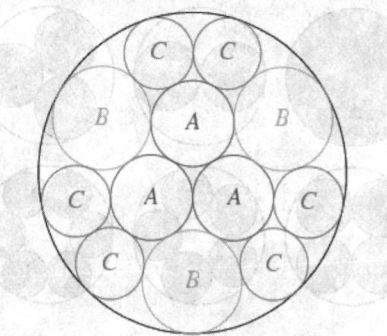

The ORTHOTOMIC of the CIRCLE represented by

$$x = \cos t \tag{1}$$
$$y = \sin t \tag{2}$$

with a source at (x, y) is

$$x = x \cos(2t) - y \sin(2t) + 2 \sin t \tag{3}$$
$$y = -x \sin(2t) - y \cos(2t) + 2 \cos t. \tag{4}$$

Circle Packing

A circle packing is an arrangement of circles inside a given boundary such that no two overlap and some (or all) of them are mutually tangent. The generalization to spheres is called a SPHERE PACKING. TESSELLATIONS of regular polygons correspond to particular circle packings (Williams 1979, pp. 35–1). There is a well developed theory of circle packing in the context of discrete conformal mapping (Stephenson).

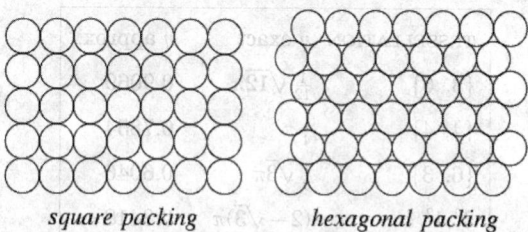

square packing   *hexagonal packing*

The densest packing of circles in the PLANE is the hexagonal lattice of the bee's honeycomb (right figure; Steinhaus 1983, p. 202), which has a PACKING DENSITY of

$$\eta_h = \tfrac{1}{6} \pi \sqrt{3} \approx 0.9068996821 \tag{1}$$

(Wells 1986, p. 30). Gauss proved that the hexagonal lattice is the densest plane *lattice* packing, and in 1940, L. Fejes Tóth proved that the hexagonal lattice is indeed the densest of *all* possible plane packings.

Wells (1991, pp. 30–1) considers the maximum size possible for n identical circles packed on the surface of a UNIT SPHERE.

Using discrete conformal mapping, the radii of the circles in the above packing inside a UNIT CIRCLE can be determined as roots of the polynomial equations

$$a^6 + 378a^5 + 3411a^4 - 8964a^3 - 10233a^2 + 3402a - 27$$
$$= 0 \tag{2}$$

$$169b^6 + 24978b^5 + 2307b^4 - 14580b^3 + 3375b^2 + 162b$$
$$-27 = 0 \tag{3}$$

$$c^6 + 438c^5 + 19077c^4 - 15840c^3 - 360c^2 + 2592c - 432$$
$$= 0 \tag{4}$$

with

$$a \approx 0.266746 \tag{5}$$

$$b \approx 0.321596 \tag{6}$$

$$c \approx 0.223138. \tag{7}$$

The following table gives the packing densities η for the circle packings corresponding to the regular and semiregular plane tessellations (Williams 1979, p. 49).

TESSELLATION	η exact	η approx.
$\{3, 6\}$	$\frac{1}{12}\sqrt{12}\pi$	0.9069
$\{4, 4\}$	$\frac{1}{4}\pi$	0.7854
$\{6, 3\}$	$\frac{1}{9}\sqrt{3}\pi$	0.6046
$3^2.4^2$	$(2-\sqrt{3})\pi$	0.8418
$3^2.4.3.4$	$(2-\sqrt{3})\pi$	0.8418
3.6.3.6	$\frac{1}{8}\sqrt{3}\pi$	0.6802
$3^4.6$	$\frac{1}{7}\sqrt{2}\pi$	0.7773
3.12^2	$(7\sqrt{3}-12)\pi$	0.3907
4.8^2	$(3-2\sqrt{2})\pi$	0.5390
3.4.6.4	$\frac{1}{3}(2\sqrt{3}-3)\pi$	0.7290
3.4.6.4	$\frac{1}{3}(2\sqrt{3}-3)\pi$	0.4860

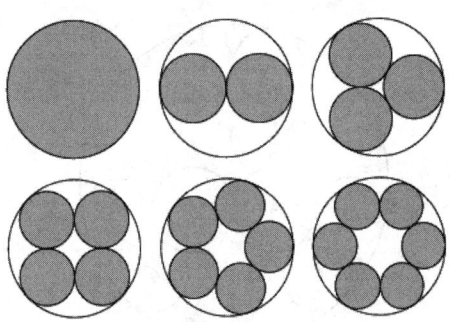

Solutions for the smallest diameter CIRCLES into which n UNIT CIRCLES can be packed have been proved optimal for $n = 1$ through 10 (Kravitz 1967). The best known results are summarized in the following table, and the first few cases are illustrated above (Friedman).

n	d exact	d approx.
1	1	1.00000
2	2	2.00000
3	$1 + \frac{2}{3}\sqrt{3}$	2.15470...
4	$1 + \sqrt{2}$	2.41421...
5	$1 + \sqrt{2(1 + 1/\sqrt{5})}$	2.70130...
6	3	3.00000
7	3	3.00000
8	$1 + \csc(\pi/7)$	3.30476...
9	$1 + \sqrt{2(2 + \sqrt{2})}$	3.61312...
10		3.82...
11		
12		4.02...

The following table gives the diameters d of circles giving the densest known packings of n equal circles packed inside a UNIT SQUARE, the first few of which are illustrated above (Friedman). All $n = 1$ to 20 solutions (in addition to all solutions $n = k^2$) have been proved optimal (Friedman). Peikert (1994) uses a normalization in which the *centers* of n circles of diameter m are packed into a square of side length 1. Friedman lets the circles have unit radius and gives the smallest square side length s. A tabulation of analytic s and diagrams for $n = 1$ to 25 circles is given by Friedman. Coordinates for optimal packings are given by Nurmela and Östergård.

n	d	$\approx d$	m	$\approx m$
1	1	1.000000		
2	$\dfrac{2}{2+\sqrt{2}}$	0.585786	$\sqrt{2}$	1.414214
3	$\dfrac{4}{4+\sqrt{2}+\sqrt{6}}$	0.508666	$\sqrt{6}-\sqrt{2}$	1.035276
4	$\frac{1}{2}$	0.500000	1	1.000000

5	$\sqrt{2}-1$	0.414214	$\frac{1}{2}\sqrt{2}$	0.707107
6	$\frac{1}{23}(6\sqrt{13}-13)$	0.375361	$\frac{1}{6}\sqrt{13}$	0.600925
7	$\frac{2}{13}(4-\sqrt{3})$	0.348915	$4-2\sqrt{3}$	0.535898
8	$\dfrac{2}{2+\sqrt{2}+\sqrt{6}}$	0.341081	$\frac{1}{2}(\sqrt{6}-\sqrt{2})$	0.517638
9	$\frac{1}{3}$	0.333333	$\frac{1}{2}$	0.500000
10		0.296408		0.421280

The smallest SQUARE into which two UNIT CIRCLES, one of which is split into two pieces by a chord, can be packed is not known (Goldberg 1968, Ogilvy 1990).

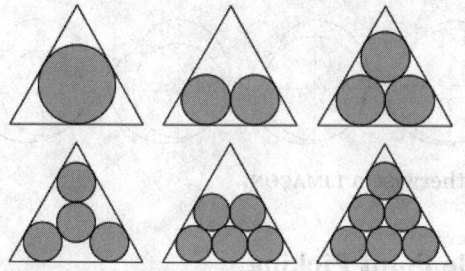

The best known packings of circles into an equilateral triangle are shown above for the first few cases (Friedman).

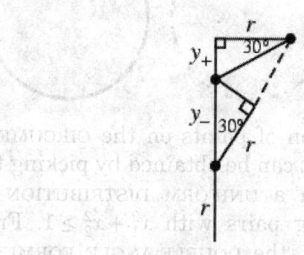

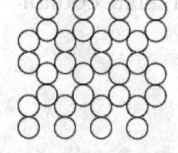 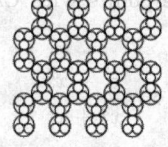 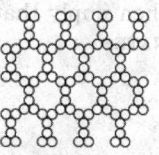

A rigid packing of circles can be obtained from a hexagonal tessellation by removing the centers of a hexagonal web, then replacing each remaining circle with three equal inscribed circles (appropriately oriented), as illustrated above (Meschkowski 1966, Wells 1991). If the original circles have unit radius, the lengths r, y_-, and y_+ can be obtained by solving

$$r = y_- \cos 30°, \tag{8}$$

$$r + y_- = 1, \tag{9}$$

$$y_+ = r \tan 30°, \tag{10}$$

giving

$$r = 2\sqrt{3}-3 \tag{11}$$

$$y_- = 4-2\sqrt{3} \tag{12}$$

$$y_+ = 2-\sqrt{3}. \tag{13}$$

The resulting circles cover a fraction

$$\eta = \eta_h\left(\frac{2}{3}\,\frac{3\pi r^2}{\pi 1^2}\right) = (7\sqrt{3}-12)\pi \approx 0.390675 \tag{14}$$

of the plane, believed to be the smallest possible for a rigid packing of circles (Wells 1991).

See also CIRCLE COVERING, DESCARTES CIRCLE THEOREM, FOUR COINS PROBLEM, HYPERSPHERE PACKING, MALFATTI'S RIGHT TRIANGLE PROBLEM, MERGELYAN-WESLER THEOREM, SANGAKU PROBLEM, SODDY CIRCLES, SPHERE PACKING, SQUARE PACKING, TANGENT CIRCLES, TRIANGLE PACKING, UNIT CELL

References

Boll, D. "Packing Results." http://www.frii.com/~dboll/packing.html.

Bowers, P. L. and Stephenson, K. "Uniformizing Dessins and Bely Maps via Circle Packing." Preprint.

Casado, L. G.and Szabó, P. G. "Equal Circle Packing in a Square." http://www.inf.u-szeged.hu/~pszabo/Packing_circles.html.

Collins, C. R. and Stephenson, K. "A Circle Packing Algorithm." Preprint.

Conway, J. H. and Sloane, N. J. A. *Sphere Packings, Lattices, and Groups, 2nd ed.* New York: Springer-Verlag, 1992.

Croft, H. T.; Falconer, K. J.; and Guy, R. K. *Unsolved Problems in Geometry.* New York: Springer-Verlag, 1991.

Donovan, J. "Packing Circles in Squares and Circles Page." http://home.att.net/~donovanhse/Packing/.

Eppstein, D. "Covering and Packing." http://www.ics.uci.edu/~eppstein/junkyard/cover.html.

Fejes Tóth, L. *Lagerungen in der Ebene auf der Kugel und im Raum.* Berlin: Springer-Verlag, 1953.

Fejes Tóth, L. "On the Stability of a Circle Packing." *Ann. Univ. Sci. Budapestinensis, Sect. Math.* **3**-, 63–6, 1960/1961.

Folkman, J. H. and Graham, R. "A Packing Inequality for Compact Convex Subsets of the Plane." *Canad. Math. Bull.* **12**, 745–52, 1969.

Friedman, E. "Circles in Circles." http://www.stetson.edu/~efriedma/cirincir/.

Friedman, E. "Squares in Circles." http://www.stetson.edu/~efriedma/squincir/.

Friedman, E. "Triangles in Circles." http://www.stetson.edu/~efriedma/triincir/.

Gardner, M. "Mathematical Games: The Diverse Pleasures of Circles that Are Tangent to One Another." *Sci. Amer.* **240**, 18–8, Jan. 1979.

Gardner, M. "Tangent Circles." Ch. 10 in *Fractal Music, Hypercards, and More Mathematical Recreations from Scientific American Magazine.* New York: W. H. Freeman, pp. 149–66, 1992.

Goldberg, M. "Problem E1924." *Amer. Math. Monthly* **75**, 195, 1968.

Goldberg, M. "The Packing of Equal Circles in a Square." *Math. Mag.* **43**, 24–0, 1970.

Goldberg, M. "Packing of 14, 16, 17, and 20 Circles in a Circle." *Math. Mag.* **44**, 134–39, 1971.

Graham, R. L. and Luboachevsky, B. D. "Repeated Patterns of Dense Packings of Equal Disks in a Square." *Electronic J. Combinatorics* **3**, R16 1–7, 1996. http://www.combinatorics.org/Volume_3/volume3.html#R16.

Graham, R. L.; Luboachevsky, B. D.; Nurmela, K. J.; and Östergård, P. R. J. "Dense Packings of Congruent Circles in a Circle." *Discrete Mat.* **181**, 139–54, 1998.

Kravitz, S. "Packing Cylinders into Cylindrical Containers." *Math. Mag.* **40**, 65–0, 1967.

Likos, C. N. and Henley, C. L. "Complex Alloy Phases for Binary Hard-Disc Mixtures." *Philos. Mag. B* **68**, 85–13, 1993.

Maranas, C. D.; Floudas, C. A.; and Pardalos, P. M. "New Results in the Packing of Equal Circles in a Square." *Disc. Math.* **142**, 287–93, 1995.

McCaughan, F. "Circle Packings." http://www.pmms.cam.a-c.uk/~gjm11/cpacking/info.html.

Meschkowski, H. *Unsolved and Unsolvable Problems in Geometry.* London: Oliver & Boyd, 1966.

Molland, M. and Payan, Charles. "A Better Packing of Ten Equal Circles in a Square." *Discrete Math.* **84**, 303–05, 1990.

Nurmela, K. J. and Östergård, P. R. J. "Packing Up to 50 Equal Circles in a Square." *Disc. Comput. Geom.* **18**, 111–20, 1997.

Nurmela, K. J. and Östergård, P. R. J. packings/square/. http://www.tcs.hut.fi/packings/square/.

Ogilvy, C. S. *Excursions in Geometry.* New York: Dover, p. 145, 1990.

Peikert, R. "Dichteste Packungen von gleichen Kreisen in einem Quadrat." *Elem. Math.* **49**, 16–6, 1994.

Peikert, R.; Würtz, D.; Monagan, M.; and de Groot, C. "Packing Circles in a Square: A Review and New Results." In *System Modelling and Optimization, Proceedings of the Fifteenth IFIP Conference Held at the University of Zürich, September 2–, 1991* (Ed. P. Kall). Berlin: Springer-Verlag, pp. 45–4, 1992.

Peikert, R. "Packing of Equal Circles in a Square." http://www.inf.ethz.ch/~peikert/personal/CirclePackings/.

Reis, G. E. "Dense Packing of Equal Circle within a Circle." *Math. Mag.* **48**, 33–7, 1975.

Schaer, J. "The Densest Packing of Nine Circles in a Square." *Can. Math. Bul.* **8**, 273–77, 1965.

Schaer, J. "The Densest Packing of Ten Equal Circles in a Square." *Math. Mag.* **44**, 139–40, 1971.

Specht, E. "The Best Known Packings of Equal Circles in the Unit Square." http://hydra.nat.uni-magdeburg.de/packing/csq.html.

Steinhaus, H. *Mathematical Snapshots, 3rd ed.* New York: Dover, p. 202, 1999.

Stephenson, K. "Circle Packing." http://www.math.utk.edu/~kens/#Packing.

Stephenson, K. "Circle Packing Bibliography as of April 1999." http://www.math.utk.edu/~kens/CP-bib.ps.

Stephenson, K. "Circle Packings in the Approximation of Conformal Mappings." *Bull. Amer. Math. Soc.* **23**, 407–16, 1990.

Stephenson, K. "A Probabilistic Proof of Thurston's Conjecture on Circle Packings." *Rend. Sem. Math. Fis. Milano* **66**, 201–91, 1998.

Valette, G. "A Better Packing of Ten Equal Circles in a Square." *Discrete Math.* **76**, 57–9, 1989.

Wells, D. *The Penguin Dictionary of Curious and Interesting Numbers.* Middlesex, England: Penguin Books, p. 30, 1986.

Wells, D. *The Penguin Dictionary of Curious and Interesting Geometry.* London: Penguin, pp. 30–1, 1991.

Williams, R. "Circle Packings, Plane Tessellations, and Networks." §2.3 in *The Geometrical Foundation of Natural Structure: A Source Book of Design.* New York: Dover, pp. 34–7, 1979.

Circle Pedal Curve

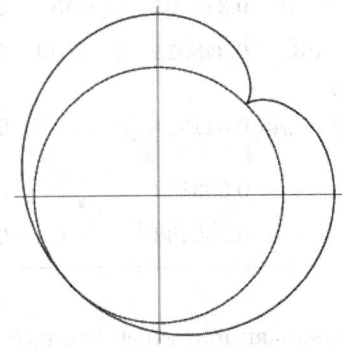

The PEDAL CURVE of a CIRCLE is a CARDIOID if the PEDAL POINT is taken on the CIRCUMFERENCE,

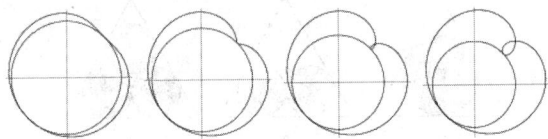

and otherwise a LIMAÇON.

Circle Point Picking

A uniform distribution of points on the CIRCUMFERENCE of a UNIT CIRCLE can be obtained by picking two numbers x_1, x_2 from a UNIFORM DISTRIBUTION on $(-1, 1)$, and rejecting pairs with $x_1^2 + x_2^2 \geq 1$. From the remaining points, the DOUBLE-ANGLE FORMULAS then imply that the points with CARTESIAN COORDINATES

$$x = \frac{x_1^2 - x_2^2}{x_1^2 + x_2^2}$$

$$y = \frac{2x_1 x_2}{x_1^2 + x_2^2}$$

have the desired distribution (von Neumann 1951, Cook 1957). This method can also be extended to SPHERE POINT PICKING (Cook 1957). The plots above show the distribution of points for 50, 100, and 500 initial points (where the counts refer to the number of points before throwing away).

See also CIRCLE COVERING BY ARCS, DISK POINT PICKING, SPHERE POINT PICKING

References

Cook, J. M. "Technical Notes and Short Papers: Rational Formulae for the Production of a Spherically Symmetric

Probability Distribution." *Math. Tables Aids Comput.* **11**, 81–2, 1957.

von Neumann, J. "Various Techniques Used in Connection with Random Digits." *NBS Appl. Math. Ser.*, No. 12. Washington, DC: U.S. Government Printing Office, pp. 36–8, 1951.

Watson, G. S. and Williams, E. J. "On the Construction of Significance Tests on the Circle and Sphere." *Biometrika* **43**, 344–52, 1956.

Circle Quadrature

CIRCLE SQUARING

Circle Radial Curve

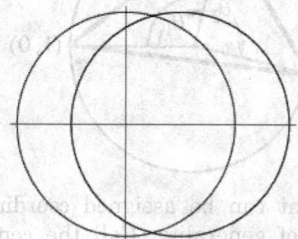

The RADIAL CURVE of a unit CIRCLE from a RADIAL POINT $(x, 0)$ is another CIRCLE with PARAMETRIC EQUATIONS

$$x(t) = x - \cos t$$

$$y(t) = -\sin t.$$

Circle Squaring

Construct a SQUARE equal in AREA to a CIRCLE using only a STRAIGHTEDGE and COMPASS. This was one of the three GEOMETRIC PROBLEMS OF ANTIQUITY, and was perhaps first attempted by Anaxagoras. It was finally proved to be an impossible problem when PI was proven to be TRANSCENDENTAL by Lindemann in 1882.'

However, approximations to circle squaring are given by constructing lengths close to $\pi = 3.1415926\ldots$. Ramanujan (1913–4), Olds (1963), Gardner (1966, pp. 92–3), and (Bold 1982, p. 45) give geometric constructions for $355/113 = 3.1415929\ldots$. Dixon (1991) gives constructions for $6/5(1 + \phi) = 3.141640\ldots$ and $\sqrt{40/3 - 2\sqrt{3}} = 3.141533\ldots$ (KOCHANSKY'S APPROXIMATION).

While the circle cannot be squared in EUCLIDEAN SPACE, it *can* in GAUSS-BOLYAI-LOBACHEVSKY SPACE (Gray 1989).

See also BANACH-TARSKI PARADOX, GEOMETRIC CONSTRUCTION, KOCHANSKY'S APPROXIMATION, QUADRATURE, SQUARING

References

Bold, B. "The Problem of Squaring the Circle." Ch. 6 in *Famous Problems of Geometry and How to Solve Them.* New York: Dover, pp. 39–8, 1982.

Conway, J. H. and Guy, R. K. *The Book of Numbers.* New York: Springer-Verlag, pp. 190–91, 1996.

Dixon, R. *Mathographics.* New York: Dover, pp. 44–9 and 52–3, 1991.

Dunham, W. "Hippocrates' Quadrature of the Lune." Ch. 1 in *Journey through Genius: The Great Theorems of Mathematics.* New York: Wiley, pp. 20–6, 1990.

Gardner, M. "The Transcendental Number Pi." Ch. 8 in *Martin Gardner's New Mathematical Diversions from Scientific American.* New York: Simon and Schuster, pp. 91–02, 1966.

Gray, J. *Ideas of Space: Euclidean, Non-Euclidean, and Relativistic,* 2nd ed. Oxford, England: Oxford University Press, 1989.

Hertel, E. "On the Set-Theoretical Circle-Squaring Problem." http://www.minet.uni-jena.de/Math-Net/reports/sources/2000/00–6report.ps.

Jesseph, D. M. *Squaring the Circle: The War Between Hobbes and Wallis.* Chicago: University of Chicago Press, 1999.

Klein, F. "Transcendental Numbers and the Quadrature of the Circle." Part II in "Famous Problems of Elementary Geometry: The Duplication of the Cube, the Trisection of the Angle, and the Quadrature of the Circle." In *Famous Problems and Other Monographs.* New York: Chelsea, pp. 49–0, 1980.

Meyers, L. F. "Update on William Wernick's 'Triangle Constructions with Three Located Points.'" *Math. Mag.* **69**, 46–9, 1996.

Olds, C. D. *Continued Fractions.* New York: Random House, pp. 59–0, 1963.

Ramanujan, S. "Modular Equations and Approximations to π." *Quart. J. Pure. Appl. Math.* **45**, 350–72, 1913–914.

Wells, D. *The Penguin Dictionary of Curious and Interesting Numbers.* Middlesex, England: Penguin Books, p. 48, 1986.

Circle Strophoid

The STROPHOID of a CIRCLE with pole at the center and fixed point on the CIRCUMFERENCE is a FREETH'S NEPHROID.

Circle Tangents

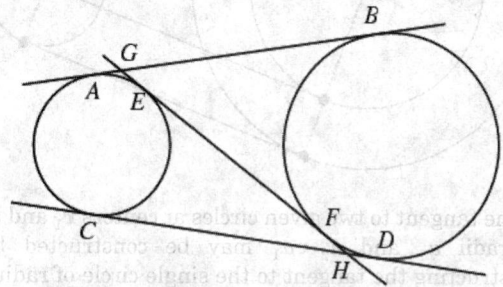

Given the above figure, $GE = FH$, since

$$AB = AG + GB = GE + GF = GE + (GE + EF)$$
$$= 2GE + EF$$

$$CD = CH + HD = EH + FH = FH + (FH + EF)$$
$$= EF + 2FH.$$

Because $AB = CD$, it follows that $GE = FH$.

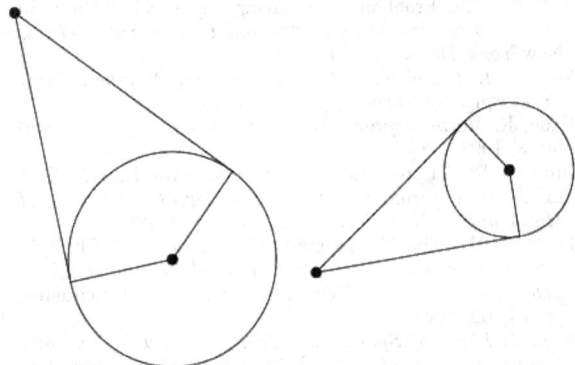

The line tangent to a CIRCLE of RADIUS a centered at (x, y)

$$x' = x + a \cos t$$

$$y' = y + a \sin t$$

through $(0, 0)$ can be found by solving the equation

$$\begin{bmatrix} x + a \cos t \\ y + a \sin t \end{bmatrix} \cdot \begin{bmatrix} a \cos t \\ a \sin t \end{bmatrix} = 0,$$

giving

$$t = \pm \cos^{-1} \left(\frac{-ax \pm y \sqrt{x^2 + y^2 - a^2}}{x^2 + y^2} \right).$$

Two of these four solutions give tangent lines, as illustrated above, and the lengths of these lines are equal (Casey 1888, p. 29).

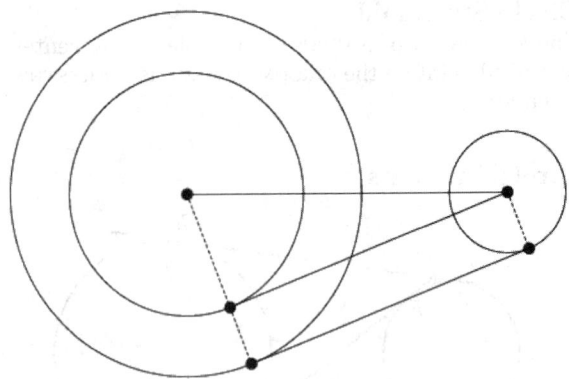

A line tangent to two given circles at centers $\mathbf{r}_1$ and $\mathbf{r}_2$ of radii a_1 and $a_2 < a_1$ may be constructed by constructing the tangent to the single circle of radius $a_1 - a_2$ centered at $\mathbf{r}_1$ and through $\mathbf{r}_2$, then translating this line along the radius through $\mathbf{r}_1$ a distance a_2 until it falls on the original two circles (Casey 1888, pp. 31–2).

See also KISSING CIRCLES PROBLEM, MIQUEL POINT, MONGE'S PROBLEM, NINE-POINT CIRCLE, PEDAL CIRCLE, TANGENT CIRCLES, TANGENT LINE, TRIANGLE

References

Casey, J. *A Sequel to the First Six Books of the Elements of Euclid, Containing an Easy Introduction to Modern Geometry with Numerous Examples*, 5th ed., rev. enl. Dublin: Hodges, Figgis, & Co., 1888.

Dixon, R. *Mathographics*. New York: Dover, p. 21, 1991.

Honsberger, R. *More Mathematical Morsels*. Washington, DC: Math. Assoc. Amer., pp. 4–, 1991.

Circle Triangle Picking

Select three points at random on a unit CIRCLE. Find the distribution of possible areas.

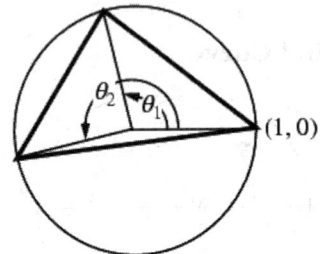

The first point can be assigned coordinates $(1, 0)$ without loss of generality. Call the central angles from the first point to the second and third θ_1 and θ_2. The range of θ_1 can be restricted to $[0, \pi]$ because of symmetry, but θ_2 can range from $[0, 2\pi)$. Then

$$A(\theta_1, \theta_2) = 2|\sin(\tfrac{1}{2}\theta_1) \sin(\tfrac{1}{2}\theta_2) \sin[\tfrac{1}{2}(\theta_1 - \theta_2)]|, \quad (1)$$

so

$$\bar{A} = \frac{\int_0^\pi \int_0^{2\pi} A(\theta_1, \theta_2) \, d\theta_2 \, d\theta_1}{C}, \quad (2)$$

where

$$C \equiv \int_0^\pi \int_0^{2\pi} d\theta_2 \, d\theta_1 = 2\pi^2. \quad (3)$$

Therefore,

$$\bar{A} = \frac{2}{2\pi^2} \int_0^\pi \int_0^{2\pi} |\sin(\tfrac{1}{2}\theta_1) \sin(\tfrac{1}{2}\theta_2) \sin[\tfrac{1}{2}(\theta_1 - \theta_2)]| \, d\theta_2 \, d\theta_1$$

$$= \frac{1}{\pi^2} \int_0^\pi \sin(\tfrac{1}{2}\theta_1) \left[\int_0^{2\pi} \sin(\tfrac{1}{2}\theta_2)|\sin[\tfrac{1}{2}(\theta_2 - \theta_1)]| \, d\theta_2 \right] d\theta_1$$

$$= \frac{1}{\pi^2} \int_0^\pi \int_{\substack{0 \\ \theta_2 - \theta_1 > 0}}^{2\pi} \sin(\tfrac{1}{2}\theta_1) \sin(\tfrac{1}{2}\theta_2) \sin[\tfrac{1}{2}(\theta_1 - \theta_2)] \, d\theta_2 \, d\theta_1$$

$$+ \frac{1}{\pi^2} \int_0^\pi \int_{\substack{0 \\ \theta_2 - \theta_1 < 0}}^{2\pi} \sin(\tfrac{1}{2}\theta_1) \sin(\tfrac{1}{2}\theta_2) \sin[\tfrac{1}{2}(\theta_1 - \theta_2)] \, d\theta_2 \, d\theta_1$$

$$= \frac{1}{\pi^2} \int_0^\pi \sin(\tfrac{1}{2}\theta_1) \left[\int_{\theta_1}^{2\pi} \sin(\tfrac{1}{2}\theta_2) \sin[\tfrac{1}{2}(\theta_2 - \theta_1)] \, d\theta_2 \right] d\theta_1$$

$$+\frac{1}{\pi^2}\int_0^\pi \sin(\tfrac{1}{2}\theta_1)$$

$$\times\left[\int_0^{\theta_1}\sin(\tfrac{1}{2}\theta_2)\sin[\tfrac{1}{2}(\theta_2-\theta_1)]\,d\theta_2\right]d\theta_1. \qquad (4)$$

But

$$\int(\tfrac{1}{2}\theta_2)\sin[\tfrac{1}{2}(\theta_2-\theta_1)]\,d\theta_2$$

$$=\int\sin(\tfrac{1}{2}\theta_2)[\sin(\tfrac{1}{2}\theta_2)\cos(\tfrac{1}{2}\theta_2)-\sin(\tfrac{1}{2}\theta_1)\cos(\tfrac{1}{2}\theta_2)]\,d\theta_2$$

$$=\cos(\tfrac{1}{2}\theta_1)\int\sin^2(\tfrac{1}{2}\theta_2)\,d\theta_2$$

$$\quad-\sin(\tfrac{1}{2}\theta_1)\int\sin(\tfrac{1}{2}\theta_1)\cos(\tfrac{1}{2}\theta_2)\,d\theta_2$$

$$=\tfrac{1}{2}\cos(\tfrac{1}{2}\theta_1)\int(1-\cos\theta_2)\,d\theta_2$$

$$\quad-\tfrac{1}{2}\sin(\tfrac{1}{2}\theta_2)\int\sin\theta_2\,d\theta_2 \qquad (5)$$

Write (4) as

$$\bar{A}=\frac{1}{\pi^2}\left[\int_0^\pi\sin(\tfrac{1}{2}\theta_1)I_1\,d\theta_1+\int_0^\pi\sin(\tfrac{1}{2}\theta_1)I_2\,d\theta_1\right], \qquad (6)$$

then

$$I_1\equiv\int_0^{2\pi}\sin(\tfrac{1}{2}\theta_2)\sin[\tfrac{1}{2}(\theta_2-\theta_1)]\,d\theta_2, \qquad (7)$$

and

$$I_2\equiv\int_0^{\theta_1}\sin(\tfrac{1}{2}\theta_2)\sin[\tfrac{1}{2}(\theta_1-\theta_2)]\,d\theta_2. \qquad (8)$$

From (6),

$$I_1=\tfrac{1}{2}\cos(\tfrac{1}{2}\theta_2)[\theta_2-\sin\theta_2]_{\theta_1}^{2\pi}+\tfrac{1}{2}\sin(\tfrac{1}{2}\theta_1)[\cos\theta_2]_{\theta_1}^{2\pi}$$

$$=\tfrac{1}{2}\cos(\tfrac{1}{2}\theta_1)(2\pi-\theta_1+\sin\theta_1)+\tfrac{1}{2}\sin(\tfrac{1}{2}\theta_1)(1-\cos\theta_1)$$

$$=\pi\cos(\tfrac{1}{2}\theta_1)-\tfrac{1}{2}\theta_1\cos(\tfrac{1}{2}\theta_1)$$

$$\quad+\tfrac{1}{2}[\cos(\tfrac{1}{2}\theta_1)\sin\theta_1-\cos\theta_1\sin(\tfrac{1}{2}\theta_1)]+\tfrac{1}{2}\sin(\tfrac{1}{2}\theta_1)$$

$$=\pi\cos(\tfrac{1}{2}\theta_1)-\tfrac{1}{2}\theta_1\cos(\tfrac{1}{2}\theta_1)+\tfrac{1}{2}+\tfrac{1}{2}\sin(\theta_1-\tfrac{1}{2}\theta_1)$$

$$\quad+\tfrac{1}{2}\sin(\tfrac{1}{2}\theta_1)$$

$$=\pi\cos(\tfrac{1}{2}\theta_1)-\tfrac{1}{2}\theta_1\cos(\tfrac{1}{2}\theta_1)+\sin(\tfrac{1}{2}\theta_1), \qquad (9)$$

so

$$\int_0^\pi I_1\sin(\tfrac{1}{2}\theta_1)\,d\theta_1=\tfrac{5}{4}\pi. \qquad (10)$$

Also,

$$I_2=\tfrac{1}{2}\cos(\tfrac{1}{2}\theta_1)[\sin\theta_2-\theta_2]_0^{\theta_1}-\tfrac{1}{2}\sin(\tfrac{1}{2}\theta_1)[\cos\theta]_0^{\theta_1}$$

$$=\tfrac{1}{2}\cos(\tfrac{1}{2}\theta_2)(\sin\theta_1-\theta_1)-\tfrac{1}{2}\sin(\tfrac{1}{2}\theta_1)(\cos\theta_1-1)$$

$$=-\tfrac{1}{2}\theta_1\cos(\tfrac{1}{2}\theta_1)+\tfrac{1}{2}[\sin\theta_1\cos(\tfrac{1}{2}\theta_1)$$

$$\quad-\cos\theta_1\sin(\tfrac{1}{2}\theta_2)]+\tfrac{1}{2}\sin(\tfrac{1}{2}\theta_1)$$

$$=-\tfrac{1}{2}\theta_1\cos(\tfrac{1}{2}\theta_1)+\sin(\tfrac{1}{2}\theta_1), \qquad (11)$$

so

$$\int_0^\pi I_2\sin(\tfrac{1}{2}\theta_1)\,d\theta_1=\tfrac{1}{4}\pi. \qquad (12)$$

Combining (10) and (12) gives

$$\bar{A}=\frac{1}{\pi^2}\left(\frac{5\pi}{4}+\frac{\pi}{4}\right)=\frac{3}{2\pi}\approx 0.4775. \qquad (13)$$

The first few moments are

$$\mu_2'=\tfrac{3}{8} \qquad (14)$$

$$\mu_3'=\frac{41}{32\pi} \qquad (15)$$

$$\mu_4'=\frac{45}{128}, \qquad (16)$$

so the VARIANCE is

$$\sigma_A^2=\langle A\rangle^2-\langle A^2\rangle=\frac{3(\pi^2-6)}{8\pi^2}\approx 0.1470. \qquad (17)$$

See also CIRCLE LINE PICKING, DISK TRIANGLE PICKING, POINT-POINT DISTANCE–1-D, SPHERE POINT PICKING

Circle-Circle Intersection

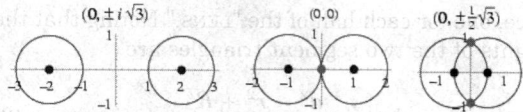

Two circles may intersect in two imaginary points, a single degenerate point, or two distinct points.

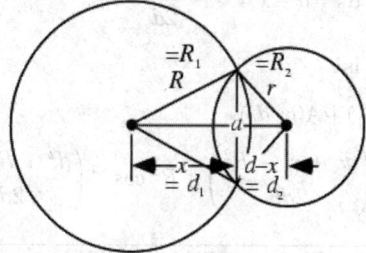

Let two CIRCLES of RADII R and r and centered at $(0, 0)$ and $(d, 0)$ intersect in a LENS-shaped region.

The equations of the two circles are

$$x^2 + y^2 = R^2 \tag{1}$$

$$(x-d)^2 + y^2 = r^2. \tag{2}$$

Combining (1) and (2) gives

$$(x-d)^2 + (R^2 + x^2) = r^2. \tag{3}$$

Multiplying through and rearranging gives

$$x^2 - 2\,dx + d^2 - x^2 = r^2 - R^2. \tag{4}$$

Solving for x results in

$$x = \frac{d^2 - r^2 + R^2}{2d}. \tag{5}$$

The line connecting the cusps of the LENS therefore has half-length given by plugging x back in to obtain

$$y^2 = R^2 - x^2 = R^2 - \left(\frac{d^2 - r^2 + R^2}{2d} \right)^2$$

$$= \frac{4d^2 R^2 - (d^2 - r^2 + R^2)^2}{4d^2}, \tag{6}$$

giving a half-height $y = a/2$ of

$$a = \frac{1}{d} \sqrt{4d^2 R^2 - (d^2 - r^2 + R^2)^2}$$

$$= \frac{1}{d} [(-d+r-R)(-d-r+R)(-d+r+R)(d+r+R)]^{1/2}. \tag{7}$$

This same formulation applies directly to the SPHERE-SPHERE INTERSECTION problem.

To find the AREA of the asymmetric "LENS" in which the CIRCLES intersect, simply use the formula for the circular SEGMENT of radius R' and triangular height d'

$$A(R', d') = R'^2 \cos^{-1}\left(\frac{d'}{R'} \right) - d'\sqrt{R'^2 - d'^2} \tag{8}$$

twice, one for each half of the "LENS." Noting that the heights of the two segment triangles are

$$d_1 = x = \frac{d^2 - r^2 + R^2}{2d} \tag{9}$$

$$d_2 = d - x = \frac{d^2 + r^2 - R^2}{2d}. \tag{10}$$

The result is

$$A = A(R, d_1) + A(r, d_2)$$

$$= r^2 \cos^{-1}\left(\frac{d^2 + r^2 - R^2}{2dr} \right) + R^2 \cos^{-1}\left(\frac{d^2 + R^2 - r^2}{2dR} \right)$$

$$- \tfrac{1}{2}\sqrt{(-d+r+R)(d+r-R)(d-r+R)(d+r+R)}. \tag{11}$$

The limiting cases of this expression can be checked

to give 0 when $d = R + r$ and

$$A = 2R^2 \cos^{-1}\left(\frac{d}{2R} \right) - \tfrac{1}{2} d\sqrt{4R^2 - d^2} \tag{12}$$

$$= 2A\left(\tfrac{1}{2} d,\ R \right) \tag{13}$$

when $r = R$, as expected. In order for half the area of two UNIT DISKS ($R = 1$) to overlap, set $A = \pi R^2/2 = \pi/2$ in the above equation

$$\tfrac{1}{2} \pi = 2 \cos^{-1}\left(\tfrac{1}{2} d \right) - \tfrac{1}{2} d\sqrt{4 - d^2} \tag{14}$$

and solve numerically, yielding $d \approx 0.807946$.

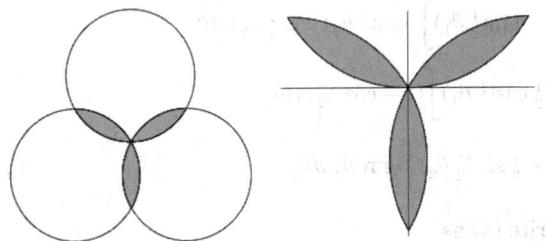

If three symmetrically placed equal circles intersect in a single point, as illustrated above, the total area of the three lens-shaped regions formed by the pairwise intersection of circles is given by

$$A = \pi - \tfrac{3}{2}\sqrt{3}. \tag{15}$$

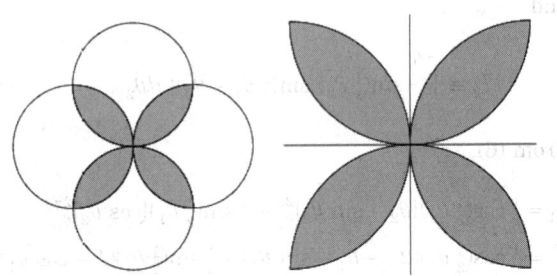

Similarly, the total area of the four lens-shaped regions formed by the pairwise intersection of circles is given by

$$A = 2(\pi - 2). \tag{16}$$

See also BORROMEAN RINGS, BROCARD TRIANGLES, CIRCLE-ELLIPSE INTERSECTION, CIRCLE-LINE INTERSECTION, CIRCULAR TRIANGLE, DOUBLE BUBBLE, GOAT PROBLEM, LENS, REULEAUX TRIANGLE, SEGMENT, SPHERE-SPHERE INTERSECTION, TRIQUETRA, VENN DIAGRAM

Circle-Ellipse Intersection

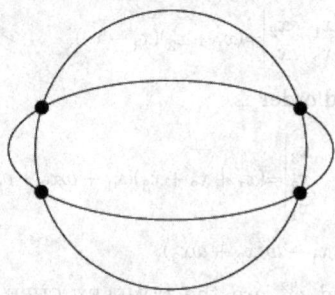

An ellipse intersects a circle in 0, 1, 2, 3, or 4 points. The points of intersection of a circle of center (x_0, y_0) and radius r with an ellipse of semi-major and semi-minor axes a and b, respectively and center (x_e, y_e) can be determined by simultaneously solving

$$(x - x_0)^2 + (y - y_0)^2 = r^2 \qquad (1)$$

$$\frac{(x - x_e)^2}{a^2} + \frac{(y - y_e)^2}{b^2} = 1. \qquad (2)$$

If $(x_0, y_0) = (x_e, y_e) = (0, 0)$, then the solution takes on the particularly simple form

$$x = \pm a \sqrt{\frac{r^2 - b^2}{a^2 - b^2}} \qquad (3)$$

$$y = \pm b \sqrt{\frac{a^2 - r^2}{a^2 - b^2}}. \qquad (4)$$

See also CIRCLE, CIRCLE-CIRCLE INTERSECTION, EL-LIPSE

Circle-Line Intersection

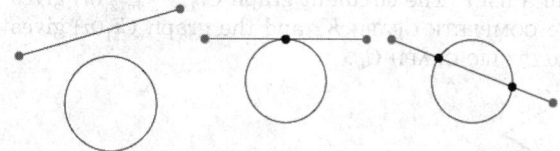

A LINE determined by two points (x_1, y_1) and (x_2, y_2) may intersect a CIRCLE of RADIUS r and center $(0, 0)$ in two imaginary points, a degenerate single point (corresponding to the line being tangent to the circle), or two real points. Defining

$$d_x = x_2 - x_1 \qquad (1)$$

$$d_y = y_2 - y_1 \qquad (2)$$

$$d_r = \sqrt{d_x^2 + d_y^2} \qquad (3)$$

$$D = \begin{vmatrix} x_1 & x_2 \\ y_1 & y_2 \end{vmatrix} = x_1 y_2 - x_2 y_1 \qquad (4)$$

gives the points of intersection as

$$x = \frac{-D d_y \pm \text{sgn}^*(d_y)\, d_x \sqrt{r^2 d_r^2 - D^2}}{d_r^2}, \qquad (5)$$

$$y = \frac{-D d_x \pm |d_y| \sqrt{r^2 d_r^2 - D^2}}{d_r^2}, \qquad (6)$$

where the function sgn* is defined as

$$\text{sgn}^*(x) \equiv \begin{cases} -1 & \text{for } x < 0 \\ 1 & \text{otherwise.} \end{cases} \qquad (7)$$

The discriminant

$$\Delta \equiv r^2 d_r^2 - D^2 \qquad (8)$$

therefore determines the incidence of the line and circle as summarized in the following table.

Δ	Incidence
$\Delta < 0$	no intersection
$\Delta = 0$	tangent
$\Delta > 0$	intersection

Circle-Point Midpoint Theorem

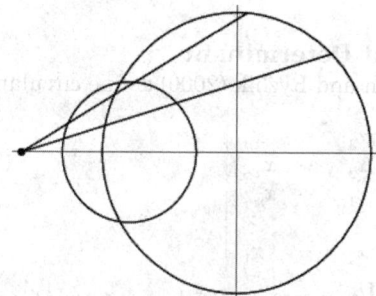

Taking the locus of MIDPOINTS from a fixed point to a circle of radius r results in a circle of radius $r/2$. This

follows trivially from

$$\mathbf{r}(\theta) = \begin{bmatrix} -x \\ 0 \end{bmatrix} + \frac{1}{2}\left(\begin{bmatrix} r\cos\theta \\ r\sin\theta \end{bmatrix} - \begin{bmatrix} -x \\ 0 \end{bmatrix}\right)$$
$$= \begin{bmatrix} \frac{1}{2}r\cos\theta - \frac{1}{2}x \\ \frac{1}{2}\sin\theta \end{bmatrix}.$$

References

Johnson, R. A. *Modern Geometry: An Elementary Treatise on the Geometry of the Triangle and the Circle.* Boston, MA: Houghton Mifflin, p. 17, 1929.

Circles-and-Squares Fractal

A FRACTAL produced by iteration of the equation

$$z_{n+1} = z_n^2 \pmod{m}$$

which results in a MOIRÉ-like pattern.

See also FRACTAL, MOIRÉ PATTERN

Circuit

GRAPH CYCLE

Circuit Rank

Also known as the CYCLOMATIC NUMBER. The circuit rank is the smallest number of EDGES γ which must be removed from a GRAPH of N EDGES and n nodes such that no CIRCUIT remains.

$$\gamma = N - n + 1.$$

Circulant Determinant

Gradshteyn and Ryzhik (2000) define circulants by

$$\begin{vmatrix} x_1 & x_2 & x_3 & \cdots & x_n \\ x_n & x_1 & x_2 & \cdots & x_{n-1} \\ x_{n-1} & x_n & x_1 & \cdots & x_{n-2} \\ \vdots & \vdots & \vdots & \ddots & \vdots \\ x_2 & x_3 & x_4 & \cdots & x_1 \end{vmatrix}$$
$$= \prod_{j=1}(x_1 + x_2\omega_j + x_3\omega_j^2 + \ldots + x_n\omega_j^{n-1}) \quad (1)$$

where ω_j is the nth ROOT OF UNITY. The second-order

circulant determinant is

$$\begin{vmatrix} x_1 & x_2 \\ x_2 & x_1 \end{vmatrix} = (x_1 + x_2)(x_1 - x_2), \quad (2)$$

and the third order is

$$\begin{vmatrix} x_1 & x_2 & x_3 \\ x_3 & x_1 & x_2 \\ x_2 & x_3 & x_1 \end{vmatrix} = (x_1 + x_2 + x_3)(x_1 + \omega x_2 + \omega^2 x_3)$$
$$\times (x_1 + \omega^2 x_2 + \omega x_3), \quad (3)$$

where ω and ω^2 are the COMPLEX CUBE ROOTS of UNITY.

The EIGENVALUES λ of the corresponding $n \times n$ CIRCULANT MATRIX are

$$\lambda_j = x_1 + x_2\omega_j + x_3\omega_j^2 + \ldots + x_n\omega_j^{n-1}. \quad (4)$$

See also CIRCULANT MATRIX

References

Gradshteyn, I. S. and Ryzhik, I. M. *Tables of Integrals, Series, and Products, 6th ed.* San Diego, CA: Academic Press, pp. 1111–112, 2000.
Vardi, I. *Computational Recreations in Mathematica.* Reading, MA: Addison-Wesley, p. 114, 1991.

Circulant Graph

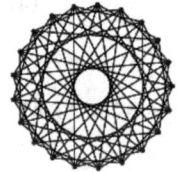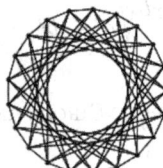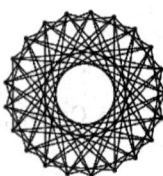

A GRAPH of n VERTICES in which the ith VERTEX is adjacent to the $(i+j)$th and $(i-j)$th VERTICES for each j in a list l. The circulant graph $Ci_{1,2,\ldots,\lfloor n/2\rfloor}(n)$ gives the COMPLETE GRAPH K_n and the graph $Ci_1(n)$ gives the CYCLIC GRAPH C_n.

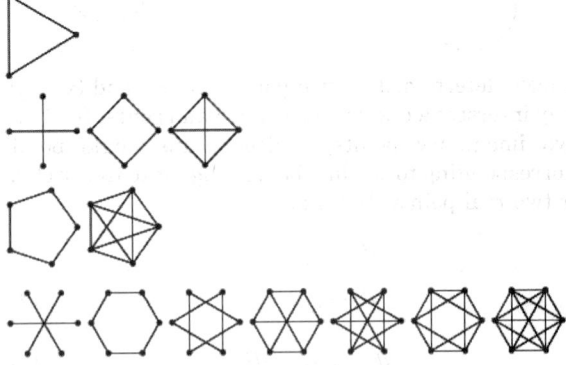

The number of circulant graphs on $n = 1, 2, \ldots$ nodes (counting empty graphs) are given by 1, 2, 2, 4, 3, 8, 4, 12, ... (Sloane's A049287). Note that these numbers cannot be counted simply by enumerating the number of nonempty subsets of $\{1, 2, \ldots, \lfloor n/2 \rfloor\}$ since, for example, $Ci_1(5) = Ci_2(5) = C_5$. There is an easy formula for prime orders, and formulas are known for squarefree and prime-squared orders.

Special cases are summarized in the table below.

Graph	Symbol
OCTAHEDRAL GRAPH	$Ci_{1,2}(6)$
16-CELL	$Ci_{1,2,3}(8)$

See also 16-CELL, OCTAHEDRAL GRAPH

References
Buckley, F. and Harary, F. *Distances in Graphs.* Redwood City, CA: Addison-Wesley, 1990.

Liskovets, V. A.; and Pöschel, R. "On the Enumeration of Circulant Graphs of Prime-Power and Square-Free Orders." Preprint. MATH-AL-8-996, TU-Dresden.

Klin, M.; Liskovets, V.; and Pöschel, R. "Analytical Enumeration of Circulant Graphs with Prime-Squared Number of Vertices." *Sém. Lothar. Combin.* **36**, Art. B36d, 1996.

Skiena, S. *Implementing Discrete Mathematics: Combinatorics and Graph Theory with Mathematica.* Reading, MA: Addison-Wesley, pp. 99 and 140, 1990.

Zhou, A. and Zhang, X. D. "Enumeration of Circulant Graphs with Order n and Degree 4 or 5" [Chinese]. *Dianzi Keji Daxue Xuebao* **25**, 272–76, 1996.

Circulant Matrix

An $n \times n$ MATRIX C defined as follows,

$$C_n = \begin{bmatrix} 1 & \binom{n}{1} & \binom{n}{2} & \cdots & \binom{n}{n-1} \\ \binom{n}{n-1} & 1 & \binom{n}{1} & \cdots & \binom{n}{n-2} \\ \vdots & \vdots & \vdots & \ddots & \vdots \\ \binom{n}{1} & \binom{n}{2} & \binom{n}{3} & \cdots & 1 \end{bmatrix},$$

where $\binom{n}{k}$ is a BINOMIAL COEFFICIENT. The DETERMINANT of C_n is given by the beautiful formula

$$C_n = \prod_{j=0}^{n-1} [(1 + \omega_j)^n - 1],$$

where $\omega_0 \equiv 1$, ω_1, ..., ω_{n-1} are the nth ROOTS OF UNITY. The determinants for $n = 1, 2, \ldots$, are given by 1, -3, 28, -375, 3751, 0, 6835648, -1343091375, 364668913756, ... (Sloane's A048954), which is 0 when $n \equiv 0 \pmod 6$.

Circulant matrices are examples of LATIN SQUARES.

See also CIRCULANT DETERMINANT

References
Davis, P. J. *Circulant Matrices, 2nd ed.* New York: Chelsea, 1994.

Sloane, N. J. A. Sequences A048954 and A049287 in "An On-Line Version of the Encyclopedia of Integer Sequences." http://www.research.att.com/~njas/sequences/eisonline.html.

Stroeker, R. J. "Brocard Points, Circulant Matrices, and Descartes' Folium." *Math. Mag.* **61**, 172–87, 1988.

Vardi, I. *Computational Recreations in Mathematica.* Reading, MA: Addison-Wesley, p. 114, 1991.

Circular Chessboard

A circular pattern obtained by superposing parallel equally spaced lines on a set of concentric circles of increasing radii, then coloring the regions in chessboard fashion. The pattern appeared on the cover of early editions of *Scripta Mathematica.*

See also CHESSBOARD

References
Gardner, M. *The Sixth Book of Mathematical Games from Scientific American.* Chicago, IL: University of Chicago Press, pp. 243–45 and 249–51, 1984.

Circular Cylinder
CYLINDER

Circular Cylindrical Coordinates
CYLINDRICAL COORDINATES

Circular Functions
The functions describing the horizontal and vertical positions of a point on a CIRCLE as a function of ANGLE (COSINE and SINE) and those functions derived from them:

$$\cot x \equiv \frac{1}{\tan x} = \frac{\cos x}{\sin x} \qquad (1)$$

$$\csc x \equiv \frac{1}{\sin x} \qquad (2)$$

$$\sec x \equiv \frac{1}{\cos x} \qquad (3)$$

$$\tan x \equiv \frac{\sin x}{\cos x}. \qquad (4)$$

Circular functions are also called TRIGONOMETRIC FUNCTIONS, and the study of circular functions is called TRIGONOMETRY.

See also COSECANT, COSINE, COTANGENT, ELLIPTIC FUNCTION, GENERALIZED HYPERBOLIC FUNCTIONS, HYPERBOLIC FUNCTIONS, SECANT, SINE, TANGENT, TRIGONOMETRIC FUNCTIONS, TRIGONOMETRY

References

Abramowitz, M. and Stegun, C. A. (Eds.). "Circular Functions." §4.3 in *Handbook of Mathematical Functions with Formulas, Graphs, and Mathematical Tables, 9th printing.* New York: Dover, pp. 71–9, 1972.

Circular Permutation

The number of ways to arrange n distinct objects along a FIXED (i.e., cannot be picked up out of the plane and turned over) CIRCLE is

$$P_n = (n-1)!.$$

The number is $(n-1)!$ instead of the usual FACTORIAL $n!$ since all CYCLIC PERMUTATIONS of objects are equivalent because the CIRCLE can be rotated.

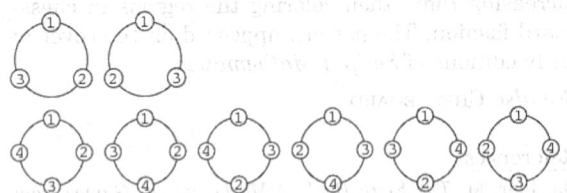

For example, of the $3! = 6$ permutations of three objects, the $(3-1)! = 2$ distinct circular permutations are $\{1, 2, 3\}$ and $\{1, 3, 2\}$. Similarly, of the $4! = 24$ permutations of four objects, the $(3-1)! = 6$ distinct circular permutations are $\{1, 2, 3, 4\}$, $\{1, 2, 4, 3\}$, $\{1, 3, 2, 4\}$, $\{1, 3, 4, 2\}$, $\{1, 4, 2, 3\}$, and $\{1, 4, 3, 2\}$. Of these, there are only three FREE permutations (i.e., inequivalent when flipping the circle is allowed): $\{1, 2, 3, 4\}$, $\{1, 2, 4, 3\}$, and $\{1, 3, 2, 4\}$. The number of free circular permutations of order n is $P'_n = 1$ for $n = 1, 2$, and

$$P'_n = \tfrac{1}{2}(n-1)!$$

for $n \geq 3$, giving the sequence 1, 1, 1, 3, 12, 60, 360, 2520, ... (Sloane's A001710).

See also CYCLIC PERMUTATION, FACTORIAL, PERMUTATION, PRIME CIRCLE

References

Sloane, N. J. A. Sequences A001710/M2933 in "An On-Line Version of the Encyclopedia of Integer Sequences." http://www.research.att.com/~njas/sequences/eisonline.html.

Circular Reciprocation

RECIPROCATION

Circular Triangle

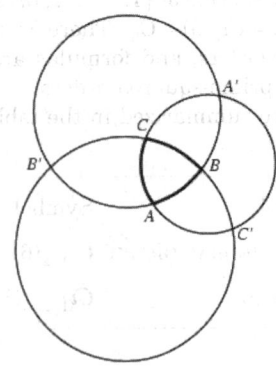

A triangle ABC formed by three circular ARCS. By extending the arcs into complete circles, the points of intersection A', B', and C' are obtained. This gives the three circular triangles, $A'B'C'$, $AB'C'$, $A'BC'$, and $A'B'C$, which are called the ASSOCIATED TRIANGLES to ABC. In addition, circular triangles $A'B'C'$, $AB'C'$, $A'BC'$, and $A'B'C$ can also be drawn.

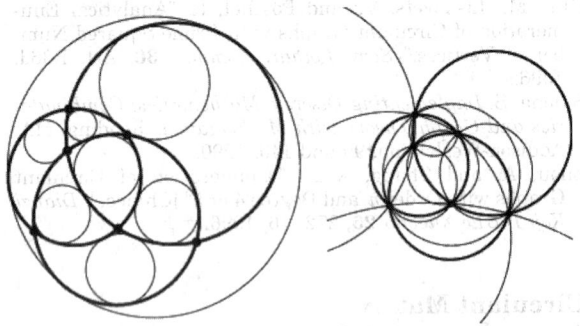

The circular triangle and its associated circles have a total of eight INCIRCLES and six CIRCUMCIRCLE. These systems of circles have some remarkable properties, including the HART CIRCLE, which is an analog of the NINE-POINT CIRCLE in FEUERBACH'S THEOREM.

See also APOLLONIUS' PROBLEM, ARC, ASSOCIATED TRIANGLES, CIRCLE-CIRCLE INTERSECTION, FEUERBACH'S THEOREM, HART CIRCLE, HARUKI'S THEOREM, NINE-POINT CIRCLE, SPHERICAL TRIANGLE, TRIQUETRA

References

Lachlan, R. "Properties of a Circular Triangle." §397–04 in *An Elementary Treatise on Modern Pure Geometry.* London: Macmillian, pp. 251–57, 1893.

Circular-Cylinder Coordinates

CYLINDRICAL COORDINATES

Circumcenter

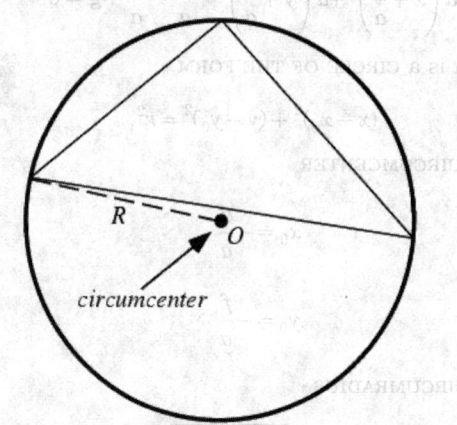

circumcenter

The center O of a TRIANGLE'S CIRCUMCIRCLE. It can be found as the intersection of the PERPENDICULAR BISECTORS. If the TRIANGLE is ACUTE, the circumcenter is in the interior of the TRIANGLE. In a RIGHT TRIANGLE, the circumcenter is the MIDPOINT of the HYPOTENUSE.

$$\overline{OO_1} + \overline{OO_2} + \overline{OO_3} = R + r, \tag{1}$$

where O_i are the MIDPOINTS of sides A_i, R is the CIRCUMRADIUS, and r is the INRADIUS (Johnson 1929, p. 190). The TRILINEAR COORDINATES of the circumcenter are

$$\cos A : \cos B : \cos C, \tag{2}$$

and the exact trilinears are therefore

$$R \cos A : R \cos B : R \cos C. \tag{3}$$

The AREAL COORDINATES are

$$(\tfrac{1}{2} a \cot A, \ \tfrac{1}{2} b \cot B, \ \tfrac{1}{2} c \cot C). \tag{4}$$

The distance between the INCENTER and circumcenter is $\sqrt{R(R - 2r)}$. Given an interior point, the distances to the VERTICES are equal IFF this point is the circumcenter. It lies on the BROCARD AXIS.

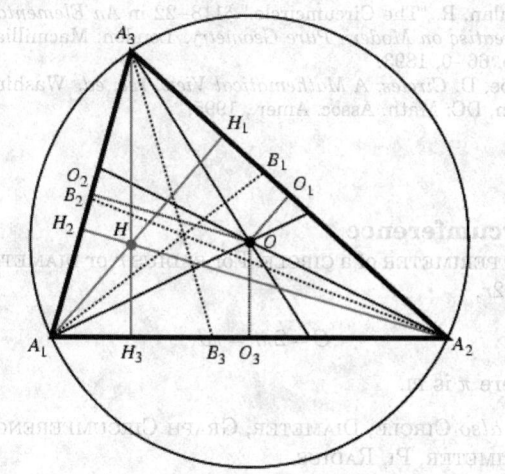

The circumcenter O and ORTHOCENTER H are ISOGONAL CONJUGATES.

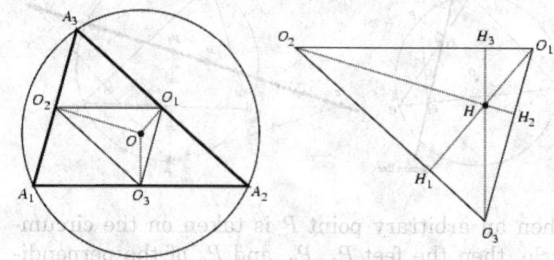

The ORTHOCENTER H of the PEDAL TRIANGLE $\Delta O_1 O_2 O_3$ formed by the CIRCUMCENTER O concurs with the circumcenter O itself, as illustrated above. The circumcenter also lies on the EULER LINE.

See also BROCARD DIAMETER, CARNOT'S THEOREM, CENTROID (TRIANGLE), CIRCLE, EULER LINE, INCENTER, LESTER CIRCLE, ORTHOCENTER

References
Carr, G. S. *Formulas and Theorems in Pure Mathematics,* 2nd ed. New York: Chelsea, p. 623, 1970.
Dixon, R. *Mathographics.* New York: Dover, p. 55, 1991.
Eppstein, D. "Circumcenters of Triangles." http://www.ics.uci.edu/~eppstein/junkyard/circumcenter.html.
Johnson, R. A. *Modern Geometry: An Elementary Treatise on the Geometry of the Triangle and the Circle.* Boston, MA: Houghton Mifflin, 1929.
Kimberling, C. "Central Points and Central Lines in the Plane of a Triangle." *Math. Mag.* **67**, 163–87, 1994.
Kimberling, C. "Circumcenter." http://cedar.evansville.edu/~ck6/tcenters/class/ccenter.html.

Circumcircle

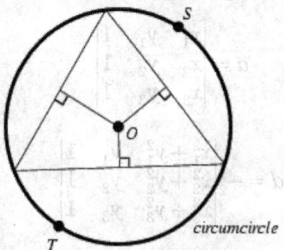

circumcircle

A TRIANGLE'S circumscribed circle. Its center O is called the CIRCUMCENTER, and its RADIUS R the CIRCUMRADIUS. The circumcircle can be specified using TRILINEAR COORDINATES as

$$\beta\gamma a + \gamma\alpha b + \alpha\beta c = 0. \tag{1}$$

The STEINER POINT S and TARRY POINT T lie on the

circumcircle.

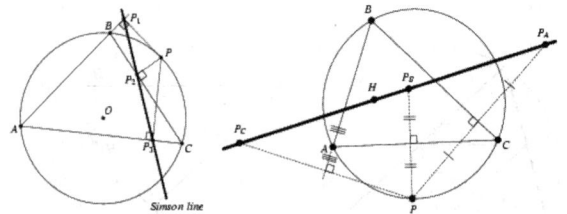

Simson line

When an arbitrary point P is taken on the circumcircle, then the feet P_1, P_2, and P_3 of the perpendiculars from P to the sides (or their extensions) of the TRIANGLE are COLLINEAR on a line called the SIMSON LINE. Furthermore, the reflections P_A, P_B, P_C of any point P on the CIRCUMCIRCLE taken with respect to the sides BC, AC, AB of the triangle are COLLINEAR, not only with each other but also with the ORTHO-CENTER H (Honsberger 1995, pp. 44-7).

The tangent to a triangle's circumcircle at a vertex is ANTIPARALLEL to the opposite side, the sides of the ORTHIC TRIANGLE are parallel to the tangents to the circumcircle at the vertices, and the radius of the circumcircle at a vertex is perpendicular to all lines ANTIPARALLEL to the opposite sides (Johnson 1929, pp. 172–73).

A GEOMETRIC CONSTRUCTION for the circumcircle is given by Pedoe (1995, pp. xii-xiii). The equation for the circumcircle of the TRIANGLE with VERTICES (x_i, y_i) for $i = 1, 2, 3$ is

$$\begin{vmatrix} x^2+y^2 & x & y & 1 \\ x_1^2+y_1^2 & x_1 & y_1 & 1 \\ x_2^2+y_2^2 & x_2 & y_2 & 1 \\ x_3^2+y_3^2 & x_3 & y_3 & 1 \end{vmatrix} = 0. \tag{2}$$

Expanding the DETERMINANT,

$$a(x^2 + y^2) + 2dx + 2fy + g = 0, \tag{3}$$

where

$$a = \begin{vmatrix} x_1 & y_1 & 1 \\ x_2 & y_2 & 1 \\ x_3 & y_3 & 1 \end{vmatrix} \tag{4}$$

$$d = -\frac{1}{2}\begin{vmatrix} x_1^2+y_1^2 & y_1 & 1 \\ x_2^2+y_2^2 & y_2 & 1 \\ x_3^2+y_3^2 & y_3 & 1 \end{vmatrix} \tag{5}$$

$$f = \frac{1}{2}\begin{vmatrix} x_1^2+y_1^2 & x_1 & 1 \\ x_2^2+y_2^2 & x_2 & 1 \\ x_3^2+y_3^2 & x_3 & 1 \end{vmatrix} \tag{6}$$

$$g = -\begin{vmatrix} x_1^2+y_1^2 & x_1 & y_1 \\ x_2^2+y_2^2 & x_2 & y_2 \\ x_3^2+y_3^2 & x_3 & y_3 \end{vmatrix}. \tag{7}$$

COMPLETING THE SQUARE gives

$$a\left(x + \frac{d}{a}\right)^2 + a\left(y + \frac{f}{a}\right)^2 - \frac{d^2}{a} - \frac{f^2}{a} + g = 0 \tag{8}$$

which is a CIRCLE OF THE FORM

$$(x - x_0)^2 + (y - y_0)^2 = r^2, \tag{9}$$

with CIRCUMCENTER

$$x_0 = -\frac{d}{a} \tag{10}$$

$$y_0 = -\frac{f}{a} \tag{11}$$

and CIRCUMRADIUS

$$r = \sqrt{\frac{f^2 + d^2}{a^2} - \frac{g}{a}}. \tag{12}$$

If a polygon with side lengths a, b, c, ... and standard trilinear equations $\alpha = 0$, $\beta = 0$, $\gamma = 0$, ... has a circumcircle, then for any point of the circle,

$$\frac{a}{\alpha} + \frac{b}{\beta} + \frac{c}{\gamma} + \ldots = 0 \tag{13}$$

(Casey 1878, 1893).

See also CIRCLE, CIRCUMCENTER, CIRCUMRADIUS, EXCIRCLE, INCIRCLE, PARRY POINT, PIVOT THEOREM, PURSER'S THEOREM, SIMSON LINE, STEINER POINTS, TARRY POINT

References

Casey, J. *Trans. Roy. Irish Acad.* **26**, 527–10, 1878.
Casey, J. *A Treatise on the Analytical Geometry of the Point, Line, Circle, and Conic Sections, Containing an Account of Its Most Recent Extensions, with Numerous Examples, 2nd ed., rev. enl.* Dublin: Hodges, Figgis, & Co., pp. 128–29, 1893.
Coxeter, H. S. M. and Greitzer, S. L. *Geometry Revisited.* Washington, DC: Math. Assoc. Amer., p. 7, 1967.
Honsberger, R. *Episodes in Nineteenth and Twentieth Century Euclidean Geometry.* Washington, DC: Math. Assoc. Amer., 1995.
Lachlan, R. "The Circumcircle." §118–22 in *An Elementary Treatise on Modern Pure Geometry.* London: Macmillian, pp. 66–0, 1893.
Pedoe, D. *Circles: A Mathematical View, rev. ed.* Washington, DC: Math. Assoc. Amer., 1995.

Circumference

The PERIMETER of a CIRCLE. For RADIUS r or DIAMETER $d = 2r$,

$$C = 2\pi r = \pi d,$$

where π is PI.

See also CIRCLE, DIAMETER, GRAPH CIRCUMFERENCE, PERIMETER, PI, RADIUS

Circumflex

HAT

Circuminscribed

Given two CLOSED CURVES, the circuminscribed curve is simultaneously INSCRIBED in the outer one and CIRCUMSCRIBED on the inner one.

See also PONCELET'S PORISM, STEINER CHAIN

Circumradius

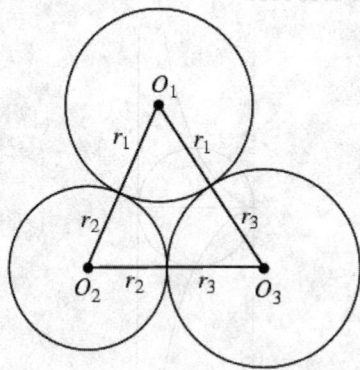

The radius of a TRIANGLE'S CIRCUMCIRCLE or of a POLYHEDRON'S CIRCUMSPHERE, denoted R. For a TRIANGLE,

$$R = \frac{abc}{\sqrt{(a+b+c)(b+c-a)(c+a-b)(a+b-c)}}$$

(1)

where the side lengths of the TRIANGLE are a, b, and c.

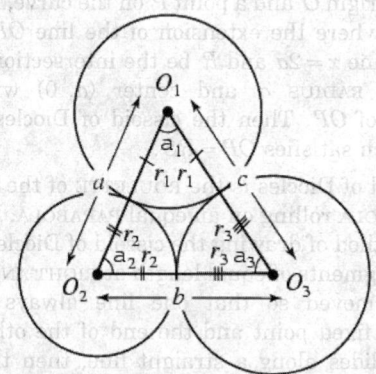

This equation can also be expressed in terms of the RADII of the three mutually tangent CIRCLES centered at the TRIANGLE'S VERTICES. Relabeling the diagram for the SODDY CIRCLES with VERTICES O_1, O_2, and O_3 and the radii r_1, r_2, and r_3, and using

$$a = r_1 + r_2 \tag{2}$$
$$b = r_2 + r_3 \tag{3}$$
$$c = r_1 + r_3 \tag{4}$$

then gives

$$R = \frac{(r_1+r_2)(r_1+r_3)(r_2+r_3)}{4\sqrt{r_1 r_2 r_3(r_1+r_2+r_3)}}.$$

(5)

If O is the CIRCUMCENTER and M is the triangle CENTROID, then

$$\overline{OM}^2 = R^2 - \tfrac{1}{9}(a^2+b^2+c^2). \tag{6}$$

$$Rr = \frac{abc}{4s} \tag{7}$$

$$\cos\alpha_1 + \cos\alpha_2 + \cos\alpha_3 = 1 + \frac{r}{R} \tag{8}$$

$$r = 2R\cos\alpha_1\cos\alpha_2\cos\alpha_3 \tag{9}$$

(Johnson 1929, pp. 189–91). Let d be the distance between INRADIUS r and circumradius R, $d = \overline{rR}$. Then

$$R^2 - d^2 = 2Rr \tag{10}$$

$$\frac{1}{R-d} + \frac{1}{R+d} = \frac{1}{r} \tag{11}$$

(Mackay 1886–7; Casey 1888, pp. 74–5). These and many other identities are given in Johnson (1929, pp. 186–90).

The HYPOTENUSE of a RIGHT TRIANGLE is a DIAMETER of the triangle's CIRCUMCIRCLE, so the circumradius is given by

$$R = \tfrac{1}{2}c, \tag{12}$$

where c is the HYPOTENUSE.

For an ARCHIMEDEAN SOLID, expressing the circumradius in terms of the INRADIUS r and MIDRADIUS ρ gives

$$R = \tfrac{1}{2}(r + \sqrt{r^2 + a^2}) \tag{13}$$

$$= \sqrt{\rho^2 + \tfrac{1}{4}a^2} \tag{14}$$

for an ARCHIMEDEAN SOLID.

See also CARNOT'S THEOREM, CIRCUMCIRCLE, CIRCUMSPHERE, INCIRCLE, INRADIUS

References

Casey, J. *A Sequel to the First Six Books of the Elements of Euclid, Containing an Easy Introduction to Modern Geometry with Numerous Examples,* 5th ed., rev. enl. Dublin: Hodges, Figgis, & Co., 1888.

Johnson, R. A. *Modern Geometry: An Elementary Treatise on the Geometry of the Triangle and the Circle.* Boston, MA: Houghton Mifflin, 1929.

Mackay, J. S. "Historical Notes on a Geometrical Theorem and its Developments [18th Century]." *Proc. Edinburgh Math. Soc.* **5**, 62–8, 1886–887.

Circumscribed

A geometric figure which touches only the vertices (or other extremities) of another figure.

See also CIRCUMCENTER, CIRCUMCIRCLE, CIRCUMIN-SCRIBED, CIRCUMRADIUS, INSCRIBED

Circumsphere

A SPHERE circumscribed in a given solid. Its radius is called the CIRCUMRADIUS. The figures above depict the circumspheres of the Platonic solids.

See also INSPHERE, MIDSPHERE

Cis

Another name for the complex exponential,

$$\text{Cis } x \equiv e^{ix} = \cos x + i \sin x.$$

See also EXPONENTIAL FUNCTION, PHASOR

Cissoid

Given two curves C_1 and C_2 and a fixed point O, let a line from O cut C_1 at Q and C_2 at R. Then the LOCUS of a point P such that $OP = QR$ is the cissoid. The word cissoid means "ivy shaped."

Curve 1	Curve 2	Pole	Cissoid
LINE	PARALLEL LINE	any point	line
LINE	CIRCLE	center	CONCHOID OF NICOMEDES
CIRCLE	tangent line	on CIRCUM-FERENCE	oblique cissoid
CIRCLE	tangent line	on CIRCUM-FERENCE opp. tangent	CISSOID OF DIOCLES
CIRCLE	radial line	on CIRCUM-FERENCE	strophoid
CIRCLE	concentric CIRCLE	center	CIRCLE
CIRCLE	same CIRCLE	$(a\sqrt{2}, 0)$	LEMNISCATE

See also CISSOID OF DIOCLES

References

Lawrence, J. D. *A Catalog of Special Plane Curves*. New York: Dover, pp. 53–6 and 205, 1972.
Lockwood, E. H. "Cissoids." Ch. 15 in *A Book of Curves*. Cambridge, England: Cambridge University Press, pp. 130–33, 1967.
Yates, R. C. "Cissoid." *A Handbook on Curves and Their Properties*. Ann Arbor, MI: J. W. Edwards, pp. 26–0, 1952.

Cissoid of Diocles

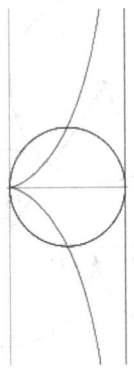

A curve invented by Diocles in about 180 BC in connection with his attempt to duplicate the cube by geometrical methods. The name "cissoid" first appears in the work of Geminus about 100 years later. Fermat and Roberval constructed the tangent in 1634. Huygens and Wallis found, in 1658, that the AREA between the curve and its asymptote was $3a$ (MacTutor Archive). From a given point there are either one or three TANGENTS to the cissoid.

Given an origin O and a point P on the curve, let S be the point where the extension of the line OP intersects the line $x = 2a$ and R be the intersection of the CIRCLE of RADIUS a and center $(a, 0)$ with the extension of OP. Then the cissoid of Diocles is the curve which satisfies $OP = RS$.

The cissoid of Diocles is the ROULETTE of the VERTEX of a PARABOLA rolling on an equal PARABOLA. Newton gave a method of drawing the cissoid of Diocles using two line segments of equal length at RIGHT ANGLES. If they are moved so that one line always passes through a fixed point and the end of the other line segment slides along a straight line, then the MIDPOINT of the sliding line segment traces out a cissoid of Diocles.

The cissoid of Diocles is given by the PARAMETRIC EQUATIONS

$$x = 2a \sin^2 \theta \qquad (1)$$

$$y = \frac{2a \sin^3 \theta}{\cos \theta}. \qquad (2)$$

Converting these to POLAR COORDINATES gives

$$r^2 = x^2 + y^2 = 4a^2 \left(\sin^4 \theta + \frac{\sin^6 \theta}{\cos^2 \theta} \right)$$

$$= 4a^2 \sin^4 \theta (1 + \tan^2 \theta) = 4a^2 \sin^4 \theta \sec^2 \theta, \quad (3)$$

so

$$r = 2a \sin^2 \theta \sec \theta = 2a \sin \theta \tan \theta. \quad (4)$$

In CARTESIAN COORDINATES,

$$\frac{x^3}{2a - x} = \frac{8a^3 \sin^6 \theta}{2a - 2a \sin^2 \theta} = 4a^2 \frac{\sin^6 \theta}{1 - \sin^2 \theta}$$

$$= 4a^2 \frac{\sin^6 \theta}{\cos^2 \theta} = y^2. \quad (5)$$

An equivalent form is

$$x(x^2 + y^2) = 2ay^2. \quad (6)$$

Using the alternative parametric form

$$x(t) = \frac{2at^2}{1 + t^2} \quad (7)$$

$$y(t) = \frac{2at^3}{1 + t^2} \quad (8)$$

(Gray 1997), gives the CURVATURE as

$$\kappa(t) = \frac{3}{a|t|(t^2 + 4)^{3/2}}. \quad (9)$$

References

Beyer, W. H. *CRC Standard Mathematical Tables, 28th ed.* Boca Raton, FL: CRC Press, p. 214, 1987.

Gray, A. "The Cissoid of Diocles." §3.5 in *Modern Differential Geometry of Curves and Surfaces with Mathematica, 2nd ed.* Boca Raton, FL: CRC Press, pp. 57–1, 1997.

Lawrence, J. D. *A Catalog of Special Plane Curves.* New York: Dover, pp. 98–00, 1972.

Lockwood, E. H. *A Book of Curves.* Cambridge, England: Cambridge University Press, pp. 130–33, 1967.

MacTutor History of Mathematics Archive. "Cissoid of Diocles." http://www-groups.dcs.st-and.ac.uk/~history/Curves/Cissoid.html.

Wells, D. *The Penguin Dictionary of Curious and Interesting Numbers.* Middlesex, England: Penguin Books, p. 34, 1986.

Yates, R. C. "Cissoid." *A Handbook on Curves and Their Properties.* Ann Arbor, MI: J. W. Edwards, pp. 26–0, 1952.

Cissoid of Diocles Caustic

The CAUSTIC of the cissoid where the RADIANT POINT is taken as $(8a, 0)$ is a CARDIOID.

Cissoid of Diocles Inverse Curve

If the cusp of the CISSOID OF DIOCLES is taken as the INVERSION CENTER, then the cissoid inverts to a PARABOLA.

Cissoid of Diocles Pedal Curve

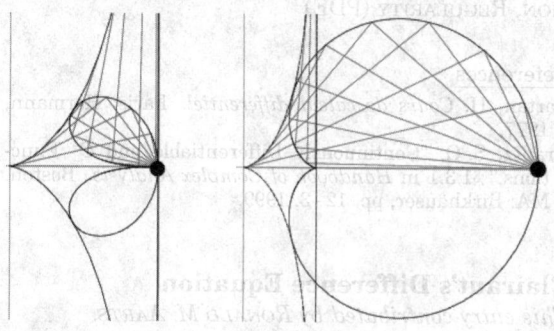

The PEDAL CURVE of the cissoid, when the PEDAL POINT is on the axis beyond the ASYMPTOTE at a distance from the cusp which is four times that of the ASYMPTOTE is a CARDIOID.

C-k Function

A function with k CONTINUOUS DERIVATIVES is called a C^k function. In order to specify a C^k function on a domain X, the notation $C^k(X)$ is used. The most common C^k space is C^0, the space of CONTINUOUS FUNCTIONS, whereas C^1 is the space of CONTINUOUSLY DIFFERENTIABLE FUNCTIONS. Cartan (1977, p. 327) writes humorously that "by 'differentiable,' we mean of class C^k, with k being as large as necessary."

Of course, any SMOOTH FUNCTION is C^k, and when $l > k$, then any C^l function is C^k. It is natural to think of a C^k function as being a little bit rough, but the graph of a C^3 function "looks" smooth.

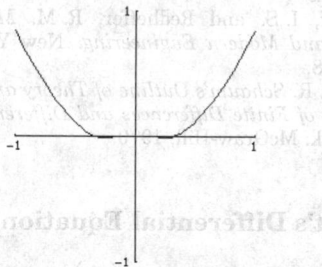

Examples of C^k functions are $|x|^{k+1}$ (for k even) and $x^{k+1} \sin(1/x)$, which do not have a $(k+1)$st derivative at 0.

The notion of C^k function may be restricted to those whose first k derivatives are BOUNDED functions. The reason for this restriction is that the set of C^k functions has a NORM which makes it a BANACH SPACE,

$$\|f\|_{C^k(X)} = \sum_{n=0}^{k} \sup_{x \in X} |f^{(n)}(x)|.$$

See also BANACH SPACE, C-INFINITY FUNCTION, CALCULUS, CONTINUOUSLY DIFFERENTIABLE FUNC-

TION, CONTINUOUS FUNCTION, DIFFERENTIAL EQUATION, REGULARITY (PDE)

References

Cartan, H. *Cours de calcul différentiel*. Paris: Hermann, 1977.

Krantz, S. G. "Continuously Differentiable and C^k Functions." §1.3.1 in *Handbook of Complex Analysis*. Boston, MA: Birkhäuser, pp. 12–3, 1999.

Clairaut's Difference Equation

This entry contributed by RONALD M. AARTS

Clairaut's difference equation is a special case of Lagrange's equation (Sokolnikoff and Redheffer 1958) defined by

$$u_k = k\Delta u_k + F(\Delta u_k),$$

or in "x notation,"

$$y = x\frac{\Delta y}{\Delta x} + F\left(\frac{\Delta y}{\Delta x}\right)$$

(Spiegel 1970). It is so named by analogy with CLAIRAUT'S DIFFERENTIAL EQUATION

$$y = x\frac{dy}{dx} + F\left(\frac{dy}{dx}\right).$$

See also CLAIRAUT'S DIFFERENTIAL EQUATION

References

Sokolnikoff, I. S. and Redheffer, R. M. *Mathematics of Physics and Modern Engineering*. New York: McGraw-Hill, 1958.

Spiegel, M. R. *Schaum's Outline of Theory and Problems of Calculus of Finite Differences and Difference Equations*. New York: McGraw-Hill, 1970.

Clairaut's Differential Equation

$$y = x\frac{dy}{dx} + f\left(\frac{dy}{dx}\right) \qquad (1)$$

or

$$y = px + f(p), \qquad (2)$$

where f is a FUNCTION of one variable and $p \equiv dy/dx$. The general solution is

$$y = cx + f(c). \qquad (3)$$

The singular solution ENVELOPES are $x = -f'(c)$ and $y = f(c) - cf'(c)$.

A PARTIAL DIFFERENTIAL EQUATION known as Clairaut's equation is given by

$$u = xu_x + yu_y + f(u_x, u_y) \qquad (4)$$

(Iyanaga and Kawada 1980, p. 1446; Zwillinger 1997, p. 132).

See also CLAIRAUT'S DIFFERENCE EQUATION, D'ALEMBERT'S EQUATION

References

Boyer, C. B. *A History of Mathematics*. New York: Wiley, p. 494, 1968.

Ford, L. R. *Differential Equations*. New York: McGraw-Hill, p. 16, 1955.

Ince, E. L. *Ordinary Differential Equations*. New York: Dover, pp. 39–0, 1956.

Iyanaga, S. and Kawada, Y. (Eds.). *Encyclopedic Dictionary of Mathematics*. Cambridge, MA: MIT Press, p. 1446, 1980.

Zwillinger, D. "Clairaut's Equation." §II.A.38 in *Handbook of Differential Equations, 3rd ed.* Boston, MA: Academic Press, pp. 120 and 158–60, 1997.

Clarity

The RATIO of a measure of the size of a "fit" to the size of a "residual."

References

Tukey, J. W. *Explanatory Data Analysis*. Reading, MA: Addison-Wesley, p. 667, 1977.

Clark's Triangle

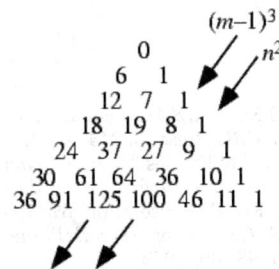

A NUMBER TRIANGLE created by setting the vertex equal to 0, filling one diagonal with 1s, the other diagonal with multiples of an INTEGER f, and filling in the remaining entries by summing the elements on either side from one row above. Call the first column $n = 0$ and the last column $m = n$ so that

$$c(m, 0) = fm \qquad (1)$$

$$c(m, m) = 1 \qquad (2)$$

then use the RECURRENCE RELATION

$$c(m, n) = c(m-1, n-1) + c(m-1, n) \qquad (3)$$

to compute the rest of the entries. For $n = 1$, we have

$$c(m, 1) = c(m-1, 0) + c(m-1, 1) \qquad (4)$$

$$c(m, 1) - c(m-1, 1) = c(m-1, 0) = f(m-1). \qquad (5)$$

For arbitrary m, the value can be computed by

SUMMING this RECURRENCE,

$$c(m,\ 1) = f\left(\sum_{k=1}^{m-1} k\right) + 1 = \tfrac{1}{2}fm(m-1) + 1. \qquad (6)$$

Now, for $n = 2$ we have

$$c(m,\ 2) = c(m-1,\ 1) + c(m-1,\ 2) \qquad (7)$$

$$c(m,\ 2) - c(m-1,\ 2) = c(m-1,\ 1)$$
$$= \tfrac{1}{2}f(m-1)m + 1, \qquad (8)$$

so SUMMING the RECURRENCE gives

$$c(m,\ 2) = \sum_{k=1}^{m-1}[\tfrac{1}{2}fk(k-1)+1] = \sum_{k=1}^{m}(\tfrac{1}{2}fk^2 - \tfrac{1}{2}fk + 1)$$
$$= \tfrac{1}{2}f[\tfrac{1}{6}m(m+1)(2m+1)] - \tfrac{1}{2}f[\tfrac{1}{2}m(m+1)] + m$$
$$= \tfrac{1}{6}(m-1)(fm^2 - 2fm + 6). \qquad (9)$$

Similarly, for $n = 3$ we have

$$c(m,\ 3) - c(m-1,\ 3) = c(m-1,\ 2)$$
$$= \tfrac{1}{6}fm^3 - fm^2 + (\tfrac{11}{6}f + 1)m - (f+2). \qquad (10)$$

Taking the SUM,

$$c(m,\ 3) = \sum_{k=2}^{m} \tfrac{1}{6}fk^3 - fk^2 + (\tfrac{11}{6}f+1)k - (f+2). \qquad (11)$$

Evaluating the SUM gives

$$c(m,\ 3) = \tfrac{1}{24}(m-1)(m-2)(fm^2 - 3fm + 12). \qquad (12)$$

So far, this has just been relatively boring ALGEBRA. But the amazing part is that if $f = 6$ is chosen as the INTEGER, then $c(m, 2)$ and $c(m, 3)$ simplify to

$$c(m,\ 2) = \tfrac{1}{6}(m-1)(6m^2 - 12m + 6) = (m-1)^3 \qquad (13)$$

$$c(m,\ 3) = \tfrac{1}{4}(m-1)^2(m-2)^2, \qquad (14)$$

which are consecutive CUBES $(m-1)^3$ and nonconsecutive SQUARES $n^2 = [(m-1)(m-2)/2]^2$.

See also BELL TRIANGLE, CATALAN'S TRIANGLE, EULER'S TRIANGLE, LEIBNIZ HARMONIC TRIANGLE, LOSSNITSCH'S TRIANGLE, NUMBER TRIANGLE, PASCAL'S TRIANGLE, SEIDEL-ENTRINGER-ARNOLD TRIANGLE, SUM

References

Clark, J. E. "Clark's Triangle." *Math. Student* **26**, No. 2, p. 4, Nov. 1978.

Class

The word "class" has many specialized meanings in mathematics in which it refers to a group of objects with some common property (e.g., CHARACTERISTIC CLASS or CONJUGACY CLASS.)

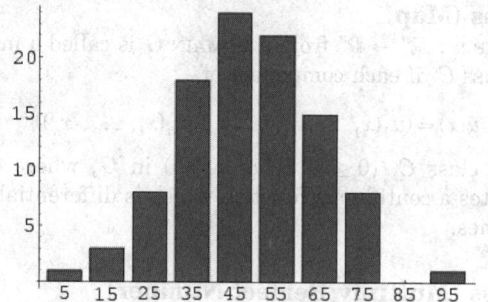

In statistics, a class is a grouping of values by which data is binned for computation of a FREQUENCY DISTRIBUTION (Kenney and Keeping 1962, p. 14). The range of values of a given class is called a CLASS INTERVAL, the boundaries of an interval are called CLASS LIMITS, and the middle of a CLASS INTERVAL is called the CLASS MARK.

class interval	class mark	absolute frequency	relative frequency	cumulative absolute frequency	relative cumulative frequency
0.00–9.99	5	1	0.01	1	0.01
10.00–9.99	15	3	0.03	4	0.04
20.00–9.99	25	8	0.08	12	0.12
30.00–9.99	35	18	0.18	30	0.30
40.00–9.99	45	24	0.24	54	0.54
50.00–9.99	55	22	0.22	76	0.76
60.00–9.99	65	15	0.15	91	0.91
70.00–9.99	75	8	0.08	99	0.99
80.00–9.99	85	0	0.00	99	0.99
90.00–9.99	95	1	0.01	100	1.00

See also CHARACTERISTIC CLASS, CLASS BOUNDARIES, CLASS GROUP FACTORIZATION METHOD, CLASS INTERVAL, CLASS LIMITS, CLASS MARK, CLASS (MULTIPLY PERFECT NUMBER), CLASS NUMBER, CLASS (SET), CONJUGACY CLASS, FREQUENCY DISTRIBUTION

Class (Group)

CONJUGACY CLASS

Class (Map)

A MAP $u : \mathbb{R}^n \to \mathbb{R}^n$ from a DOMAIN G is called a map of class C^r if each component of

$$u(x) = (u_1(x_1, \ldots, x_n), \ldots, u_m(x_1, \ldots, x_n))$$

is of class C^r ($0 \leq r \leq \infty$ or $r = \omega$) in G, where C^d denotes a continuous function which is differentiable d times.

Class (Multiply Perfect Number)

The number k in the expression $s(n) = kn$ for a MULTIPLY PERFECT NUMBER is called its class.

See also MULTIPLY PERFECT NUMBER

Class (Set)

A class is a generalized set invented to get around RUSSELL'S PARADOX while retaining the arbitrary criteria for membership which leads to difficulty for SETS. The members of classes are SETS, but it is possible to have the class C of "all SETS which are not members of themselves" without producing a PARADOX (since C is a PROPER CLASS (and not a SET), it is not a candidate for membership in C).

The distinction between classes and sets is a concept from VON NEUMANN-BERNAYS-GÖDEL SET THEORY.

See also AGGREGATE, PROPER CLASS, RUSSELL'S PARADOX, SET, TYPE, VON NEUMANN-BERNAYS-GÖDEL SET THEORY

References
Gonseth, F. "Faiblesse des idées générales de classe et d'attribut." §108 in *Les mathématiques et la réalité: Essai sur la méthode axiomatique.* Paris: Félix Alcan, pp. 259–61, 1936.

Class Boundaries

Because of rounding, the stated CLASS LIMITS do not correspond to the actual ranges of data falling in them. For example, if the CLASS LIMITS are 1.00 and 2.00, then all values between 0.95 and 2.05 would actually fall in the given CLASS, so the class boundaries are 0.95 and 2.05 (Kenney and Keeping 1962, p. 17).

See also CLASS LIMITS

References
Kenney, J. F. and Keeping, E. S. *Mathematics of Statistics, Pt. 1, 3rd ed.* Princeton, NJ: Van Nostrand, p. 17, 1962.

Class Field

See also CLASS FIELD THEORY

Class Field Theory

See also CLASS FIELD, CLASS NUMBER, RECIPROCITY LAW

References
Garbanati, D. "Class Field Theory Summarized." *Rocky Mtn. J. Math.* **11**, 195–25, 1981.
Hazewinkel, M. "Local Class Field Theory is Easy." *Adv. Math.* **18**, 148–81, 1975.

Class Group Factorization Method

A PRIME FACTORIZATION ALGORITHM.

References
Lenstra, A. K. and Lenstra, H. W. Jr. "Algorithms in Number Theory." In *Handbook of Theoretical Computer Science, Volume A: Algorithms and Complexity* (Ed. J. van Leeuwen). New York: Elsevier, pp. 673–15, 1990.

Class Interval

One of the ranges into which data in a FREQUENCY DISTRIBUTION table (or HISTOGRAM) are BINNED. The ends of a class interval are called CLASS LIMITS, and the middle of an interval is called a CLASS MARK.

See also BIN, CLASS BOUNDARIES, CLASS LIMITS, CLASS MARK, HISTOGRAM, SHEPPARD'S CORRECTION

References
Kenney, J. F. and Keeping, E. S. "Class Intervals." §1.9 in *Mathematics of Statistics, Pt. 1, 3rd ed.* Princeton, NJ: Van Nostrand, pp. 15–7, 1962.

Class Limits

The end values which specify a CLASS INTERVAL.

See also CLASS BOUNDARIES, CLASS INTERVAL

References
Kenney, J. F. and Keeping, E. S. "Class Limits and Class Boundaries." §1.10 in *Mathematics of Statistics, Pt. 1, 3rd ed.* Princeton, NJ: Van Nostrand, p. 17, 1962.

Class Mark

The average of the values of the CLASS LIMITS for a given class. A class mark is also called a midvalue or central value (Kenney and Keeping 1962, p. 14), and is commonly denoted x_c.

See also CLASS INTERVAL, CLASS LIMITS

References
Kenney, J. F. and Keeping, E. S. *Mathematics of Statistics, Pt. 1, 3rd ed.* Princeton, NJ: Van Nostrand, p. 14, 1962.

Class Number

For any IDEAL I, there is an IDEAL I_i such that

$$II_i = z, \tag{1}$$

where z is a PRINCIPAL IDEAL, (i.e., an IDEAL of rank 1). Moreover, there is a finite list of ideals I_i such that this equation may be satisfied for every I. The size of this list is known as the class number. When the class number is 1, the RING corresponding to a given IDEAL has unique factorization and, in a sense, the class number is a measure of the failure of unique factorization in the original number ring.

A finite series giving exactly the class number of a RING is known as a CLASS NUMBER FORMULA. A CLASS NUMBER FORMULA is known for the full ring of cyclotomic integers, as well as for any subring of the cyclotomic integers. Finding the class number is a computationally difficult problem.

Let $h(d)$ denote the class number of a quadratic ring, corresponding to the BINARY QUADRATIC FORM

$$ax^2 + bxy + cy^2, \tag{2}$$

with DISCRIMINANT

$$d \equiv b^2 - 4ac. \tag{3}$$

Then the class number $h(d)$ for DISCRIMINANT d gives the number of possible factorizations of $ax^2 + bxy + cy^2$ in the QUADRATIC FIELD $\mathbb{Q}(\sqrt{d})$. Here, the factors are of the form $x + y\sqrt{d}$, with x and y half INTEGERS.

Some fairly sophisticated mathematics shows that the class number for discriminant d can be given by the CLASS NUMBER FORMULA

$$h(d) \equiv \begin{cases} -\dfrac{1}{2 \ln \eta(d)} \displaystyle\sum_{r=1}^{d-1} (d/r) \ln \sin\left(\dfrac{\pi r}{d}\right) & \text{for } d > 0 \\[3mm] -\dfrac{w(d)}{2|d|} \displaystyle\sum_{r=1}^{|d|-1} (d/r) r & \text{for } d < 0, \end{cases} \tag{4}$$

where (d/r) is the KRONECKER SYMBOL, $\eta(d)$ is the FUNDAMENTAL UNIT, $w(d)$ is the number of substitutions which leave the BINARY QUADRATIC FORM unchanged

$$w(d) = \begin{cases} 6 & \text{for } d = -3 \\ 4 & \text{for } d = -4 \\ 2 & \text{otherwise,} \end{cases} \tag{5}$$

and the sums are taken over all terms where the KRONECKER SYMBOL is defined (Cohn 1980). The class number for $d > 0$ can also be written

$$\eta^{2h(d)} = \prod_{r=1}^{d-1} \sin^{-(d/r)}\left(\frac{\pi r}{d}\right) \tag{6}$$

for $d > 0$, where the PRODUCT is taken over terms for which the KRONECKER SYMBOL is defined.

The class number $h(d)$ is related to the DIRICHLET L-SERIES by

$$h(d) = \frac{L_d(1)}{\kappa(d)}, \tag{7}$$

where $\kappa(d)$ is the DIRICHLET STRUCTURE CONSTANT.

Oesterlé (1985) showed that class number $h(-d)$ satisfies the INEQUALITY

$$h(-d) > \frac{1}{7000} \prod_{p|d}{}^{*} \left(1 - \frac{\lfloor 2\sqrt{p} \rfloor}{p + 1}\right) \ln d, \tag{8}$$

for $-d < 0$, where $\lfloor x \rfloor$ is the FLOOR FUNCTION, the product is over PRIMES dividing d, and the $*$ indicates that the GREATEST PRIME FACTOR of d is omitted from the product. It is also known that if d is RELATIVELY PRIME to 5077, then the denominator 7000 in (8) can be replaced by 55.

The *Mathematica* function NumberTheory'NumberTheoryFunctions'ClassNumber[n] gives the class number $h(d)$ for d a NEGATIVE SQUAREFREE number OF THE FORM $4k + 1$.

GAUSS'S CLASS NUMBER PROBLEM asks to determine a complete list of fundamental DISCRIMINANTS $-d$ such that the CLASS NUMBER is given by $h(-d) = n$ for a given n. This problem has been solved for $n \leq 7$ and ODD $n \leq 23$. Gauss conjectured that the class number $h(-d)$ of an IMAGINARY QUADRATIC FIELD with DISCRIMINANT $-d$ tends to infinity with d, an assertion now known as GAUSS'S CLASS NUMBER CONJECTURE.

The discriminants d having $h(-d) = 1, 2, 3, 4, 5, \ldots$ are Sloane's A014602 (Cohen 1993, p. 229; Cox 1997, p. 271), Sloane's A014603 (Cohen 1993, p. 229), Sloane's A006203 (Cohen 1993, p. 504), Sloane's A013658 (Cohen 1993, p. 229), Sloane's A046002, Sloane's A046003, ... The complete set of negative discriminants having class numbers 1– and ODD 7–3 are known. Buell (1977) gives the smallest and largest fundamental class numbers for $d < 4,000,000$, partitioned into EVEN discriminants, discriminants 1 (mod 8), and discriminants 5 (mod 8). Arno *et al.* (1993) give complete lists of values of d with $h(-d) = k$ for ODD $k = 5, 7, 9, \ldots, 23$. Wagner gives complete lists of values for $k = 5, 6$, and 7.

Lists of NEGATIVE discriminants corresponding to IMAGINARY QUADRATIC FIELDS $\mathbb{Q}(\sqrt{-d(n)})$ having small class numbers $h(-d)$ are given in the table below. In the table, N is the number of "fundamental" values of $-d$ with a given class number $h(-d)$, where "fundamental" means that $-d$ is not divisible by any SQUARE NUMBER s^2 such that $h(-d/s^2) < h(-d)$. For example, although $h(-63) = 2$, -63 is not a fundamental discriminant since $63 = 3^2 \cdot 7$ and $h(-63/3^2) = h(-7) = 1 < h(-63)$. EVEN values $8 \leq h(-d) \leq 24$ have been computed by Weisstein. The number of negative discriminants having class number 1, 2, 3, ... are 9, 18, 16, 54, 25, 51, 31, ... (Sloane's A046125). The largest negative discriminants having class numbers 1, 2, 3, ... are 163, 427, 907, 1555, 2683, ... (Sloane's A038552).

The following table lists the numbers having class numbers $h \leq 25$. The search was terminated at 50000, 70000, 90000, and 90000 for class numbers 18, 20, 22, and 24, respectively. As far as I know, analytic upper bounds are not currently known for these cases.

$h(-d)$	N	Sloane	d
1	9	A014602	3, 4, 7, 8, 11, 19, 43, 67, 163
2	18	A014603	15, 20, 24, 35, 40, 51, 52, 88, 91, 115, 123, 148, 187, 232, 235, 267, 403, 427
3	16	A006203	23, 31, 59, 83, 107, 139, 211, 283, 307, 331, 379, 499, 547, 643, 883, 907
4	54	A013658	39, 55, 56, 68, 84, 120, 132, 136, 155, 168, 184, 195, 203, 219, 228, 259, 280, 291, 292, 312, 323, 328, 340, 355, 372, 388, 408, 435, 483, 520, 532, 555, 568, 595, 627, 667, 708, 715, 723, 760, 763, 772, 795, 955, 1003, 1012, 1027, 1227, 1243, 1387, 1411, 1435, 1507, 1555
5	25	A046002	47, 79, 103, 127, 131, 179, 227, 347, 443, 523, 571, 619, 683, 691, 739, 787, 947, 1051, 1123, 1723, 1747, 1867, 2203, 2347, 2683
6	51	A046003	87, 104, 116, 152, 212, 244, 247, 339, 411, 424, 436, 451, 472, 515, 628, 707, 771, 808, 835, 843, 856, 1048, 1059, 1099, 1108, 1147, 1192, 1203, 1219, 1267, 1315, 1347, 1363, 1432, 1563, 1588, 1603, 1843, 1915, 1963, 2227, 2283, 2443, 2515, 2563, 2787, 2923, 3235, 3427, 3523, 3763
7	31	A046004	71, 151, 223, 251, 463, 467, 487, 587, 811, 827, 859, 1163, 1171, 1483, 1523, 1627, 1787, 1987, 2011, 2083, 2179, 2251, 2467, 2707, 3019, 3067, 3187, 3907, 4603, 5107, 5923
8	131	A046005	95, 111, 164, 183, 248, 260, 264, 276, 295, 299, 308, 371, 376, 395, 420, 452, 456, 548, 552, 564, 579, 580, 583, 616, 632, 651, 660, 712, 820, 840, 852, 868, 904, 915, 939, 952, 979, 987, 995, 1032, 1043, 1060, 1092, 1128, 1131, 1155, 1195, 1204, 1240, 1252, 1288, 1299, 1320, 1339, 1348, 1380, 1428, 1443, 1528, 1540, 1635, 1651, 1659, 1672, 1731, 1752, 1768, 1771, 1780, 1795, 1803, 1828, 1848, 1864, 1912, 1939, 1947, 1992, 1995, 2020, 2035, 2059, 2067, 2139, 2163, 2212, 2248, 2307, 2308, 2323, 2392, 2395, 2419, 2451, 2587, 2611, 2632, 2667, 2715, 2755, 2788, 2827, 2947, 2968, 2995, 3003, 3172, 3243, 3315, 3355, 3403, 3448, 3507, 3595, 3787, 3883, 3963, 4123, 4195, 4267, 4323, 4387, 4747, 4843, 4867, 5083, 5467, 5587, 5707, 5947, 6307
9	34	A046006	199, 367, 419, 491, 563, 823, 1087, 1187, 1291, 1423, 1579, 2003, 2803, 3163, 3259, 3307, 3547, 3643, 4027, 4243, 4363, 4483, 4723, 4987, 5443, 6043, 6427, 6763, 6883, 7723, 8563, 8803, 9067, 10627
10	87	A046007	119, 143, 159, 296, 303, 319, 344, 415, 488, 611, 635, 664, 699, 724, 779, 788, 803, 851, 872, 916, 923, 1115, 1268, 1384, 1492, 1576, 1643, 1684, 1688, 1707, 1779, 1819, 1835, 1891, 1923, 2152, 2164, 2363, 2452, 2643, 2776, 2836, 2899, 3028, 3091, 3139, 3147, 3291, 3412, 3508, 3635, 3667, 3683, 3811, 3859, 3928, 4083, 4227, 4372, 4435, 4579, 4627, 4852, 4915, 5131, 5163, 5272, 5515, 5611, 5667, 5803, 6115, 6259, 6403, 6667, 7123, 7363, 7387, 7435, 7483, 7627, 8227, 8947, 9307, 10147, 10483, 13843
11	41	A046008	167, 271, 659, 967, 1283, 1303, 1307, 1459, 1531, 1699, 2027, 2267, 2539, 2731, 2851, 2971, 3203, 3347, 3499, 3739, 3931, 4051, 5179, 5683, 6163, 6547, 7027, 7507, 7603, 7867, 8443, 9283, 9403, 9643, 9787, 10987, 13003, 13267, 14107, 14683, 15667
12	206	A046009	231, 255, 327, 356, 440, 516, 543, 655, 680, 687, 696, 728, 731, 744, 755, 804, 888, 932, 948, 964, 984, 996, 1011, 1067, 1096, 1144, 1208, 1235, 1236, 1255, 1272, 1336, 1355, 1371, 1419, 1464, 1480, 1491, 1515, 1547, 1572, 1668, 1720, 1732, 1763, 1807, 1812, 1892, 1955, 1972, 2068, 2091, 2104, 2132, 2148, 2155, 2235, 2260, 2355, 2387, 2388, 2424, 2440, 2468, 2472, 2488, 2491, 2555, 2595, 2627, 2635, 2676, 2680, 2692, 2723, 2728, 2740, 2795, 2867, 2872, 2920, 2955, 3012, 3027, 3043, 3048, 3115, 3208, 3252, 3256, 3268, 3304, 3387, 3451, 3459, 3592, 3619, 3652, 3723, 3747, 3768, 3796, 3835, 3880, 3892, 3955, 3972, 4035, 4120, 4132, 4147, 4152, 4155, 4168, 4291, 4360, 4411, 4467, 4531, 4552, 4555, 4587, 4648, 4699, 4708, 4755, 4771, 4792, 4795, 4827, 4888, 4907, 4947, 4963, 5032, 5035, 5128, 5140, 5155, 5188, 5259, 5299, 5307, 5371, 5395, 5523, 5595, 5755, 5763, 5811, 5835, 6187, 6232, 6235, 6267, 6283, 6472, 6483, 6603, 6643, 6715, 6787, 6843, 6931, 6955, 6963, 6987, 7107, 7291, 7492, 7555, 7683, 7891, 7912, 8068, 8131, 8155, 8248, 8323, 8347, 8395, 8787, 8827, 9003, 9139, 9355, 9523, 9667, 9843, 10003, 10603, 10707, 10747, 10795, 10915, 11155, 11347, 11707, 11803, 12307, 12643, 14443, 15163, 15283, 16003, 17803
13	37	A046010	191, 263, 607, 631, 727, 1019, 1451, 1499, 1667, 1907, 2131, 2143, 2371, 2659, 2963, 3083, 3691, 4003, 4507, 4643, 5347, 5419, 5779, 6619, 7243, 7963, 9547, 9739, 11467, 11587, 11827, 11923, 12043, 14347, 15787, 16963, 20563
14	96	A046011	215, 287, 391, 404, 447, 511, 535, 536, 596, 692, 703, 807, 899, 1112, 1211, 1396, 1403, 1527, 1816, 1851, 1883, 2008, 2123, 2147, 2171, 2335, 2427, 2507, 2536, 2571, 2612, 2779, 2931, 2932, 3112, 3227, 3352, 3579, 3707, 3715, 3867, 3988, 4187, 4315, 4443, 4468, 4659, 4803, 4948, 5027, 5091, 5251, 5267, 5608, 5723, 5812, 5971, 6388, 6499, 6523, 6568, 6979, 7067, 7099, 7147, 7915, 8035, 8187, 8611, 8899, 9115, 9172, 9235, 9427, 10123, 10315, 10363, 10411, 11227, 12147, 12667, 12787, 13027, 13435, 13483, 13603, 14203, 16867, 18187, 18547, 18643, 20227, 21547, 23083, 23692, 30067
15	68	A046012	239, 439, 751, 971, 1259, 1327, 1427, 1567, 1619, 2243, 2647, 2699, 2843, 3331, 3571, 3803, 4099, 4219, 5003, 5227, 5323, 5563, 5827, 5987, 6067, 6091, 6211, 6571, 7219, 7459, 7547, 8467, 8707, 8779, 9043, 9907, 10243, 10267, 10459, 10651, 10723, 11083, 11971, 12163, 12763, 13147, 13963, 14323, 14827, 14851, 15187, 15643, 15907, 16603, 16843, 17467, 17923, 18043, 18523, 19387, 19867, 20707, 22003, 26203, 27883, 29947, 32323, 34483
16	322	A046013	399, 407, 471, 559, 584, 644, 663, 740, 799, 884, 895, 903, 943, 1015, 1016, 1023, 1028, 1047, 1139, 1140, 1159, 1220, 1379, 1412, 1416, 1508, 1560, 1595, 1608, 1624, 1636, 1640, 1716, 1860, 1876, 1924, 1983, 2004, 2019, 2040, 2056, 2072, 2095, 2195, 2211, 2244, 2280, 2292, 2296, 2328, 2356, 2379, 2436, 2568, 2580, 2584, 2739, 2760, 2811, 2868, 2884, 2980, 3063, 3108, 3140, 3144, 3160, 3171, 3192, 3220, 3336, 3363, 3379, 3432, 3435, 3443, 3460, 3480, 3531, 3556, 3588, 3603, 3640, 3732, 3752, 3784, 3795, 3819, 3828, 3832, 3939, 3976, 4008, 4020, 4043, 4171, 4179, 4180, 4216, 4228, 4251, 4260, 4324, 4379, 4420, 4427, 4440, 4452, 4488, 4515, 4516, 4596, 4612, 4683, 4687, 4712, 4740, 4804, 4899, 4939, 4971, 4984, 5115, 5160, 5187, 5195, 5208, 5363, 5380, 5403, 5412, 5428, 5460, 5572, 5668, 5752, 5848, 5860, 5883, 5896, 5907, 5908, 5992, 5995, 6040, 6052, 6099, 6123, 6148, 6195, 6312, 6315, 6328, 6355, 6395, 6420, 6532, 6580, 6595, 6612, 6628, 6708, 6747, 6771, 6792, 6820, 6868, 6923, 6952, 7003, 7035, 7051, 7195, 7288, 7315, 7347, 7368, 7395, 7480, 7491, 7540, 7579, 7588, 7672, 7707, 7747, 7755, 7780, 7795, 7819, 7828, 7843, 7923, 7995, 8008, 8043, 8052, 8083, 8283, 8299, 8308, 8452, 8515, 8547, 8548, 8635, 8643, 8680, 8683, 8715, 8835, 8859, 8932, 8968, 9208, 9219, 9412, 9483, 9507, 9508, 9595, 9640, 9763, 9835, 9867, 9955, 10132, 10168, 10195, 10203, 10227, 10312, 10387, 10420, 10563, 10587, 10635, 10803, 10843, 10948, 10963, 11067, 11092, 11107, 11179, 11203, 11512, 11523, 11563, 11572, 11635, 11715, 11848, 11995, 12027, 12259, 12387, 12523, 12595, 12747, 12772, 12835, 12859, 12868, 13123, 13192, 13195, 13288, 13323, 13363, 13507, 13795, 13819, 13827, 14008, 14155, 14371, 14403, 14547, 14707, 14763, 14995, 15067, 15387, 15403, 15547, 15715, 16027, 16195, 16347, 16531, 16555, 16723, 17227, 17323, 17347, 17427, 17515, 18403, 18715, 18883, 18907, 19147, 19195, 19947, 19987, 20155, 20395, 21403, 21715, 21835, 22243, 22843, 23395, 23587, 24403, 25027, 25267, 27307, 27787, 28963, 31243
17	45	A046014	383, 991, 1091, 1571, 1663, 1783, 2531, 3323, 3947, 4339, 4447, 4547, 4651, 5483, 6203, 6379, 6451, 6827, 6907, 7883, 8539, 8731, 9883, 11251, 11443, 12907, 13627, 14083, 14779, 14947, 16699, 17827, 18307, 19963, 21067, 23563, 24907, 25243, 26083, 26107, 27763, 31627, 33427, 36523, 37123
18	150	A046015	335, 519, 527, 679, 1135, 1172, 1207, 1383, 1448, 1687, 1691, 1927, 2047, 2051, 2167, 2228, 2291, 2315, 2344, 2644, 2747, 2859, 3035, 3107, 3543, 3544, 3651, 3688, 4072, 4299, 4307, 4568, 4819, 4883, 5224, 5315, 5464, 5492, 5539, 5899, 6196, 6227, 6331, 6387, 6484, 6739, 6835, 7323, 7339, 7528, 7571, 7715, 7732, 7771, 7827, 8152, 8203, 8212, 8331, 8403, 8488, 8507, 8587, 8884, 9123, 9211, 9563, 9627, 9683, 9748, 9832, 10228, 10264, 10347, 10523, 11188, 11419, 11608, 11643, 11683, 11851, 11992, 12067, 12148, 12187, 12235, 12283, 12651, 12723, 12811, 12952, 13227, 13315, 13387, 13747, 13947, 13987, 14163, 14227, 14515, 14667, 14932, 15115, 15243, 16123, 16171, 16387, 16627, 17035, 17131, 17403, 17635, 18283, 18712, 19027, 19123, 19651, 20035, 20827, 21043, 21652, 21667, 21907, 22267, 22443, 22507, 22947, 23347, 23467, 23683, 23923, 24067, 24523, 24667, 24787, 25435, 26587, 26707, 28147, 29467, 32827, 33763, 34027, 34507, 36667, 39307, 40987, 41827, 43387, 48427
19	47	A046016	311, 359, 919, 1063, 1543, 1831, 2099, 2339, 2459, 3343, 3463, 3467, 3607, 4019, 4139, 4327, 5059, 5147, 5527, 5659, 6803,

8419, 8923, 8971, 9619, 10891, 11299, 15091, 15331, 16363, 16747, 17011, 17299, 17539, 17683, 19507, 21187, 21211, 21283, 23203, 24763, 26227, 27043, 29803, 31123, 37507, 38707

20	350	A046017	455, 615, 776, 824, 836, 920, 1064, 1124, 1160, 1263, 1284, 1460, 1495, 1524, 1544, 1592, 1604, 1652, 1695, 1739, 1748, 1796, 1880, 1887, 1896, 1928, 1940, 1956, 2136, 2247, 2360, 2404, 2407, 2483, 2487, 2532, 2552, 2596, 2603, 2712, 2724, 2743, 2948, 2983, 2987, 3007, 3016, 3076, 3099, 3103, 3124, 3131, 3155, 3219, 3288, 3320, 3367, 3395, 3496, 3512, 3515, 3567, 3655, 3668, 3684, 3748, 3755, 3908, 3979, 4011, 4015, 4024, 4036, 4148, 4264, 4355, 4371, 4395, 4403, 4408, 4539, 4548, 4660, 4728, 4731, 4756, 4763, 4855, 4891, 5019, 5028, 5044, 5080, 5092, 5268, 5331, 5332, 5352, 5368, 5512, 5560, 5592, 5731, 5944, 5955, 5956, 5988, 6051, 6088, 6136, 6139, 6168, 6280, 6339, 6467, 6504, 6648, 6712, 6755, 6808, 6856, 7012, 7032, 7044, 7060, 7096, 7131, 7144, 7163, 7171, 7192, 7240, 7428, 7432, 7467, 7572, 7611, 7624, 7635, 7651, 7667, 7720, 7851, 7876, 7924, 7939, 8067, 8251, 8292, 8296, 8355, 8404, 8472, 8491, 8632, 8692, 8755, 8808, 8920, 8995, 9051, 9124, 9147, 9160, 9195, 9331, 9339, 9363, 9443, 9571, 9592, 9688, 9691, 9732, 9755, 9795, 9892, 9976, 9979, 10027, 10083, 10155, 10171, 10291, 10299, 10308, 10507, 10515, 10552, 10564, 10819, 10888, 11272, 11320, 11355, 11379, 11395, 11427, 11428, 11539, 11659, 11755, 11860, 11883, 11947, 11955, 12019, 12139, 12280, 12315, 12328, 12331, 12355, 12363, 12467, 12468, 12472, 12499, 12532, 12587, 12603, 12712, 12883, 12931, 12955, 12963, 13155, 13243, 13528, 13555, 13588, 13651, 13803, 13960, 14307, 14331, 14467, 14491, 14659, 14755, 14788, 15235, 15268, 15355, 15603, 15688, 15691, 15763, 15883, 15892, 15955, 16147, 16228, 16395, 16408, 16435, 16483, 16507, 16612, 16648, 16683, 16707, 16915, 16923, 17067, 17187, 17368, 17563, 17643, 17763, 17907, 18067, 18163, 18195, 18232, 18355, 18363, 19083, 19443, 19492, 19555, 19923, 20083, 20203, 20587, 20683, 20755, 20883, 21091, 21235, 21268, 21307, 21387, 21508, 21595, 21723, 21763, 21883, 22387, 22467, 22555, 22603, 22723, 23443, 23947, 24283, 24355, 24747, 24963, 25123, 25363, 26635, 26755, 26827, 26923, 27003, 27955, 27987, 28483, 28555, 29107, 29203, 30283, 30787, 31003, 31483, 31747, 31987, 32923, 33163, 34435, 35683, 35995, 36283, 37627, 37843, 37867, 38347, 39187, 39403, 40243, 40363, 40555, 40723, 43747, 47083, 48283, 51643, 54763, 58507
21	85	A046018	431, 503, 743, 863, 1931, 2503, 2579, 2767, 2819, 3011, 3371, 4283, 4523, 4691, 5011, 5647, 5851, 5867, 6323, 6691, 7907, 8059, 8123, 8171, 8243, 8387, 8627, 8747, 9091, 9187, 9811, 9859, 10067, 10771, 11731, 12107, 12547, 13171, 13291, 13339, 13723, 14419, 14563, 15427, 16339, 16987, 17107, 17707, 17971, 18427, 18979, 19483, 19531, 19819, 20947, 21379, 22027, 22483, 22963, 23227, 23827, 25603, 26683, 27427, 28387, 28723, 28867, 31963, 32803, 34147, 34963, 35323, 36067, 36187, 39043, 40483, 44683, 46027, 49603, 51283, 52627, 55603, 58963, 59467, 61483
22	139	A046019	591, 623, 767, 871, 879, 1076, 1111, 1167, 1304, 1556, 1591, 1639, 1903, 2215, 2216, 2263, 2435, 2623, 2648, 2815, 2863, 2935, 3032, 3151, 3316, 3563, 3587, 3827, 4084, 4115, 4163, 4328, 4456, 4504, 4667, 4811, 5383, 5416, 5603, 5716, 5739, 5972, 6019, 6127, 6243, 6616, 6772, 6819, 7179, 7235, 7403, 7763, 7768, 7899, 8023, 8143, 8371, 8659, 8728, 8851, 8907, 8915, 9267, 9304, 9496, 10435, 10579, 10708, 10851, 11035, 11283, 11363, 11668, 12091, 12115, 12403, 12867, 13672, 14019, 14059, 14179, 14548, 14587, 14635, 15208, 15563, 15832, 16243, 16251, 16283, 16291, 16459, 17147, 17587, 17779, 17947, 18115, 18267, 18835, 18987, 19243, 19315, 19672, 20108, 20392, 22579, 22587, 22987, 24243, 24427, 25387, 25507, 25843, 25963, 26323, 26548, 27619, 28267, 29227, 29635, 29827, 30235, 30867, 31315, 33643, 33667, 34003, 34387, 35347, 41083, 43723, 44923, 46363, 47587, 47923, 49723, 53827, 77683, 85507
23	68	A046020	647, 1039, 1103, 1279, 1447, 1471, 1811, 1979, 2411, 2671, 3491, 3539, 3847, 3923, 4211, 4783, 5387, 5507, 5531, 6563, 6659, 6703, 7043, 9587, 9931, 10867, 10883, 12203, 12739, 13099, 13187, 15307, 15451, 16267, 17203, 17851, 18379, 20323, 20443, 20899, 21019, 21163, 22171, 22531, 24043, 25147, 25579, 25939, 26251, 26947, 27283, 28843, 30187, 31147, 31267, 32467, 34843, 35107, 37003, 40627, 40867, 41203, 42667, 43003, 45427, 45523, 47947, 90787
24	511	A048925	695, 759, 1191, 1316, 1351, 1407, 1615, 1704, 1736, 1743, 1988, 2168, 2184, 2219, 2372, 2408, 2479, 2660, 2696, 2820, 2824, 2852, 2856, 2915, 2964, 3059, 3064, 3127, 3128, 3444, 3540, 3560, 3604, 3620, 3720, 3864, 3876, 3891, 3899, 3912, 3940, 4063, 4292, 4308, 4503, 4564, 4580, 4595, 4632, 4692, 4715, 4744, 4808, 4872, 4920, 4936, 5016, 5124, 5172, 5219, 5235, 5236, 5252, 5284, 5320, 5348, 5379, 5432, 5448, 5555, 5588, 5620, 5691, 5699, 5747, 5748, 5768, 5828, 5928, 5963, 5979, 6004, 6008, 6024, 6072, 6083, 6132, 6180, 6216, 6251, 6295, 6340, 6411, 6531, 6555, 6699, 6888, 6904, 6916, 7048, 7108,

7188, 7320, 7332, 7348, 7419, 7512, 7531, 7563, 7620, 7764, 7779, 7928, 7960, 7972, 8088, 8115, 8148, 8211, 8260, 8328, 8344, 8392, 8499, 8603, 8628, 8740, 8760, 8763, 8772, 8979, 9028, 9048, 9083, 9112, 9220, 9259, 9268, 9347, 9352, 9379, 9384, 9395, 9451, 9480, 9492, 9652, 9672, 9715, 9723, 9823, 9915, 9928, 9940, 10011, 10059, 10068, 10120, 10180, 10187, 10212, 10248, 10283, 10355, 10360, 10372, 10392, 10452, 10488, 10516, 10612, 10632, 10699, 10740, 10756, 10788, 10792, 10840, 10852, 10923, 11019, 11032, 11139, 11176, 11208, 11211, 11235, 11267, 11307, 11603, 11620, 11627, 11656, 11667, 11748, 11752, 11811, 11812, 11908, 11928, 12072, 12083, 12243, 12292, 12376, 12408, 12435, 12507, 12552, 12628, 12760, 12808, 12820, 12891, 13035, 13060, 13080, 13252, 13348, 13395, 13427, 13444, 13512, 13531, 13539, 13540, 13587, 13611, 13668, 13699, 13732, 13780, 13912, 14035, 14043, 14212, 14235, 14260, 14392, 14523, 14532, 14536, 14539, 14555, 14595, 14611, 14632, 14835, 14907, 14952, 14968, 14980, 15019, 15112, 15267, 15339, 15411, 15460, 15483, 15528, 15555, 15595, 15640, 15652, 15747, 15748, 15828, 15843, 15931, 15940, 15988, 16107, 16132, 16315, 16360, 16468, 16563, 16795, 16827, 16872, 16888, 16907, 16948, 17032, 17043, 17059, 17092, 17283, 17560, 17572, 17620, 17668, 17752, 17812, 17843, 18040, 18052, 18088, 18132, 18148, 18340, 18507, 18568, 18579, 18595, 18627, 18628, 18667, 18763, 18795, 18811, 18867, 18868, 18915, 19203, 19528, 19579, 19587, 19627, 19768, 19803, 19912, 19915, 20260, 20307, 20355, 20427, 20491, 20659, 20692, 20728, 20803, 20932, 20955, 20980, 20995, 21112, 21172, 21352, 21443, 21448, 21603, 21747, 21963, 21988, 22072, 22107, 22180, 22323, 22339, 22803, 22852, 22867, 22939, 23032, 23035, 23107, 23115, 23188, 23235, 23307, 23368, 23752, 23907, 23995, 24115, 24123, 24292, 24315, 24388, 24595, 24627, 24628, 24643, 24915, 24952, 24955, 25048, 25195, 25347, 25467, 25683, 25707, 25732, 25755, 25795, 25915, 25923, 25972, 25987, 26035, 26187, 26395, 26427, 26467, 26643, 26728, 26995, 27115, 27163, 27267, 27435, 27448, 27523, 27643, 27652, 27907, 28243, 28315, 28347, 28372, 28459, 28747, 28891, 29128, 29283, 29323, 29395, 29563, 29669, 29668, 29755, 29923, 30088, 30163, 30363, 30387, 30523, 30667, 30739, 30907, 30955, 30979, 31252, 31348, 31579, 31683, 31795, 31915, 32008, 32043, 32155, 32547, 32635, 32883, 33067, 33187, 33883, 34203, 34363, 34827, 34923, 36003, 36043, 36547, 36723, 36763, 36883, 37227, 37555, 37563, 38227, 38443, 38467, 39603, 39643, 39787, 40147, 40195, 40747, 41035, 41563, 42067, 42163, 42267, 42387, 42427, 42835, 43483, 44947, 45115, 45787, 46195, 46243, 46267, 47203, 47443, 47707, 48547, 49107, 49267, 49387, 49987, 50395, 52123, 52915, 54307, 55867, 56947, 57523, 60523, 60883, 61147, 62155, 62203, 63043, 64267, 79363, 84043, 84547, 111763

| 25 | 95 | A056987 | 479, 599, 1367, 2887, 3851, 4787, 5023, 5503, 5843, 7187, 7283, 7307, 7411, 8011, 8179, 9227, 9923, 10099, 11059, 11131, 11243, 11867, 12211, 12379, 12451, 12979, 14011, 14923, 15619, 17483, 18211, 19267, 19699, 19891, 20347, 21107, 21323, 21499, 21523, 21739, 21787, 21859, 24091, 24571, 25747, 26371, 27067, 27091, 28123, 28603, 28627, 28771, 29443, 30307, 30403, 30427, 30643, 32203, 32443, 32563, 32587, 33091, 34123, 34171, 34651, 34939, 36307, 37363, 37747, 37963, 38803, 39163, 44563, 45763, 48787, 49123, 50227, 51907, 54667, 55147, 57283, 57667, 57787, 59707, 61027, 62563, 63067, 64747, 66763, 68443, 69763, 80347, 85243, 89083, 93307 |

The table below gives lists of POSITIVE fundamental discriminants d having small class numbers $h(d)$, corresponding to REAL QUADRATIC FIELDS. All POSITIVE SQUAREFREE values of $d \leq 97$ (for which the KRONECKER SYMBOL is defined) are included.

$h(d)$	d
1	5, 13, 17, 21, 29, 37, 41, 53, 57, 61, 69, 73, 77
2	65

The POSITIVE d for which $h(d = 1)$ is given by Sloane's A014539.

See also CLASS FIELD THEORY, CLASS NUMBER

FORMULA, DIRICHLET L-SERIES, DISCRIMINANT (BINARY QUADRATIC FORM), GAUSS'S CLASS NUMBER CONJECTURE, GAUSS'S CLASS NUMBER PROBLEM, HEEGNER NUMBER, IDEAL, J-FUNCTION, RING

References

Arno, S. "The Imaginary Quadratic Fields of Class Number 4." *Acta Arith.* **40**, 321–34, 1992.

Arno, S.; Robinson, M. L.; and Wheeler, F. S. "Imaginary Quadratic Fields with Small Odd Class Number." http://www.math.uiuc.edu/Algebraic-Number-Theory/0009/.

Buell, D. A. "Small Class Numbers and Extreme Values of L-Functions of Quadratic Fields." *Math. Comput.* **139**, 786–96, 1977.

Cohen, H. *A Course in Computational Algebraic Number Theory.* New York: Springer-Verlag, 1993.

Cohn, H. *Advanced Number Theory.* New York: Dover, pp. 163 and 234, 1980.

Cox, D. A. *Primes of the Form $x^2 + ny^2$: Fermat, Class Field Theory and Complex Multiplication.* New York: Wiley, 1997.

Davenport, H. "Dirichlet's Class Number Formula." Ch. 6 in *Multiplicative Number Theory, 2nd ed.* New York: Springer-Verlag, pp. 43–3, 1980.

Himmetoglu, S. *Berechnung von Klassenzahlen Imaginaer-Quadratischer Zahlkörper.* Diplomarbeit. Heidelberg, Germany: University of Heidelberg Faculty for Mathematics, March 1986.

Iyanaga, S. and Kawada, Y. (Eds.). "Class Numbers of Algebraic Number Fields." Appendix B, Table 4 in *Encyclopedic Dictionary of Mathematics.* Cambridge, MA: MIT Press, pp. 1494–496, 1980.

Montgomery, H. and Weinberger, P. "Notes on Small Class Numbers." *Acta. Arith.* **24**, 529–42, 1974.

Müller, H. "A Calculation of Class-Numbers of Imaginary Quadratic Numberfields." *Tamkang J. Math.* **9**, 121–28, 1978.

Oesterlé, J. "Nombres de classes des corps quadratiques imaginaires." *Astérique* **121–22**, 309–23, 1985.

Sloane, N. J. A. Sequences A003657/M2332, A006203/M5131, A013658, A014539, A014602, A014603, A038552, A046002, A046003, A046125, A048925, and A056987 in "An On-Line Version of the Encyclopedia of Integer Sequences." http://www.research.att.com/~njas/sequences/eisonline.html.

Stark, H. M. "A Complete Determination of the Complex Quadratic Fields of Class Number One." *Michigan Math. J.* **14**, 1–7, 1967.

Stark, H. M. "On Complex Quadratic Fields with Class Number Two." *Math. Comput.* **29**, 289–02, 1975.

Wagner, C. "Class Number 5, 6, and 7." *Math. Comput.* **65**, 785–00, 1996.

Weisstein, E. W. "Class Numbers." MATHEMATICA NOTEBOOK CLASSNUMBERS.M.

Class Number Formula

A class number formula is a finite series giving exactly the CLASS NUMBER of a RING. For a RING of quadratic integers, the class number is denoted $h(d)$, where d is the discriminant. A class number formula is known for the full ring of cyclotomic integers, as well as for any subring of the cyclotomic integers. This formula includes the quadratic case as well as many cubic and higher-order RINGS.

See also CLASS NUMBER, RING

Class Representative

A set of class representatives is a SUBSET of X which contains exactly one element from each EQUIVALENCE CLASS.

See also EQUIVALENCE CLASS

Classical Algebraic Geometry

Classical algebraic geometry is the study of ALGEBRAIC VARIETIES, both AFFINE VARIETIES in $\mathbb{C}^n$ and PROJECTIVE VARIETIES in $\mathbb{C}\P^n$. The original motivation was to study systems of polynomials and their roots.

See also ALGEBRAIC GEOMETRY, ALGEBRAIC VARIETY, POLYNOMIAL

Classical Canonical Form

JORDAN CANONICAL FORM

Classical Groups

The four following types of GROUPS,

1. LINEAR GROUPS,
2. ORTHOGONAL GROUPS,
3. SYMPLECTIC GROUPS, and
4. UNITARY GROUPS,

which were studied before more exotic types of groups (such as the SPORADIC GROUPS) were discovered.

See also GROUP, GROUP THEORY, LINEAR GROUP, ORTHOGONAL GROUP, SIMPLE GROUP, SYMPLECTIC GROUP, UNITARY GROUP

Classification

The classification of a collection of objects generally means that a list has been constructed with exactly one member from each ISOMORPHISM type among the objects, and that tools and techniques can effectively be used to identify any combinatorially given object with its unique representative in the list. Examples of mathematical objects which have been classified include the finite SIMPLE GROUPS and 2-MANIFOLDS but not, for example, KNOTS.

See also ENUMERATION PROBLEM

Classification Theorem

CLASSIFICATION THEOREM OF FINITE GROUPS, CLASSIFICATION THEOREM OF SURFACES

Classification Theorem of Finite Groups

The classification theorem of FINITE SIMPLE GROUPS, also known as the ENORMOUS THEOREM, which states that the FINITE SIMPLE GROUPS can be classified completely into

1. CYCLIC GROUPS $\mathbb{Z}_p$ of PRIME ORDER,
2. ALTERNATING GROUPS A_n of degree at least five,
3. LIE-TYPE CHEVALLEY GROUPS $PSL(n, q)$, $PSU(n, q)$, $PsP(2n, q)$, and $P\Omega^\epsilon(n, q)$,
4. LIE-TYPE (TWISTED CHEVALLEY GROUPS or the TITS GROUP) ${}^3D_4(q)$, $E_6(q)$, $E_7(q)$, $E_8(q)$, $F_4(q)$, ${}^2F_4(2^n)'$, $G_2(q)$, ${}^2G_2(3^n)$, ${}^2B(2^n)$,
5. SPORADIC GROUPS M_{11}, M_{12}, M_{22}, M_{23}, M_{24}, $J_2 = HJ$, Suz, HS, McL, Co_3, Co_2, Co_1, He, Fi_{22}, Fi_{23}, Fi'_{24}, HN, Th, B, M, J_1, $O'N$, J_3, Ly, Ru, J_4.

The "PROOF" of this theorem is spread throughout the mathematical literature and is estimated to be approximately 15,000 pages in length.

See also FINITE GROUP, GROUP, J-FUNCTION, SIMPLE GROUP

References

Cartwright, M. "Ten Thousand Pages to Prove Simplicity." *New Scientist* **109**, 26–0, 1985.

Cipra, B. "Are Group Theorists Simpleminded?" *What's Happening in the Mathematical Sciences, 1995–996*, Vol. 3. Providence, RI: Amer. Math. Soc., pp. 82–9, 1996.

Cipra, B. "Slimming an Outsized Theorem." *Science* **267**, 794–95, 1995.

Gorenstein, D. "The Enormous Theorem." *Sci. Amer.* **253**, 104–15, Dec. 1985.

Solomon, R. "On Finite Simple Groups and Their Classification." *Not. Amer. Math. Soc.* **42**, 231–39, 1995.

Wells, D. *The Penguin Dictionary of Curious and Interesting Numbers.* Middlesex, England: Penguin Books, p. 57, 1986.

Classification Theorem of Surfaces

All closed surfaces, despite their seemingly diverse forms, are topologically equivalent to SPHERES with some number of HANDLES or CROSS-CAPS. The traditional proof follows Seifert and Threlfall (1980), but Conway's so-called "zero-irrelevancy" ("ZIP") provides a more streamlined approach (Francis and Weeks 1999).

See also CROSS-CAP, HANDLE

References

Francis, G. K. and Weeks, J. R. "Conway's ZIP Proof." *Amer. Math. Monthly* **106**, 393–99, 1999.

Seifert, H. and Threlfall, W. *A Textbook of Topology.* New York: Academic Press, 1980.

Clausen Formula

Clausen's ${}_4F_3$ identity

$${}_4F_3\left(\begin{matrix} a, b, c, d \\ e, f, g \end{matrix}; 1\right) = \frac{(2a)_{|d|}(a+b)_{|d|}(2b)_{|d|}}{(2a+2b)_{|d|}a_{|d|}b_{|d|}}, \quad (1)$$

holds for $a + b + c - d = 1/2$, $e = a + b + 1/2$, $a + f = d + 1 = b + g$, where d a nonpositive integer and $(a)_n$ is the POCHHAMMER SYMBOL (Petkovsek *et al.* 1996). Closely related identities include

$${}_4F_3 = \left[\begin{matrix} \tfrac{1}{2}a, \tfrac{1}{2}(a+1), b+n, -n \\ \tfrac{1}{2}b, \tfrac{1}{2}(b+1), a+1 \end{matrix}; 1\right] = \frac{(b-a)_n}{(b)_n} \quad (2)$$

and

$${}_4F_3\left(\begin{matrix} \tfrac{1}{2}a, \tfrac{1}{2}(a+1), b+n, -n \\ \tfrac{1}{2}(b+1), \tfrac{1}{2}(b+2), a \end{matrix}; 1\right)$$

$$= \frac{(b-a+1)_n}{(b+1)_{n-1}(b+2n)} \quad (3)$$

(Bailey 1935; Slater 1966, p. 245; Andrews and Burge 1993)

Another identity ascribed to Clausen which involves the HYPERGEOMETRIC FUNCTION ${}_2F_1(a, b; c; z)$ and the GENERALIZED HYPERGEOMETRIC FUNCTION ${}_3F_2(a, b, c; d, e; z)$ is given by

$$\left[{}_2F_1\left(\begin{matrix} a, b \\ a+b+\tfrac{1}{2} \end{matrix}; x\right)\right]^2 = {}_3F_2\left(\begin{matrix} 2a, a+b, 2b \\ a+b+\tfrac{1}{2}, 2a+2b \end{matrix}; x\right) \quad (4)$$

(Clausen 1828; Bailey 1935, p. 86; Hardy 1999, p. 106).

See also GENERALIZED HYPERGEOMETRIC FUNCTION, HYPERGEOMETRIC FUNCTION

References

Andrews, G. E. and Burge, W. H. "Determinant Identities." *Pacific J. Math.* **158**, 1–4, 1993.

Bailey, W. N. *Generalised Hypergeometric Series.* Cambridge, England: Cambridge University Press, 1935.

Clausen, T. "Ueber die Falle wenn die Reihe $y = 1 + \frac{\alpha\beta}{1\cdot\gamma}x + \ldots$ ein quadrat von der Form $x = 1 + \frac{\alpha'\beta'\gamma'}{1\cdot\delta\epsilon}x + \ldots$ hat." *J. für Math.* **3**, 89–5, 1828.

Hardy, G. H. *Ramanujan: Twelve Lectures on Subjects Suggested by His Life and Work, 3rd ed.* New York: Chelsea, 1999.

Petkovsek, M.; Wilf, H. S.; and Zeilberger, D. $A = B$. Wellesley, MA: A. K. Peters, pp. 43 and 127, 1996.

Slater, L. J. *Generalized Hypergeometric Functions.* Cambridge, England: Cambridge University Press, 1966.

Clausen Function

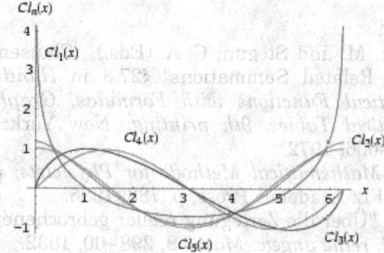

Define

$$S_n(x) \equiv \sum_{k=1}^{\infty} \frac{\sin(kx)}{k^n} \quad (1)$$

$$C_n(x) \equiv \sum_{k=1}^{\infty} \frac{\cos(kx)}{k^n}, \quad (2)$$

and write

$$Cl_n(x) \equiv \begin{cases} S_n(x) = \sum_{k=1}^{\infty} \dfrac{\sin(kx)}{k^n} & n \text{ even} \\ C_n(x) = \sum_{k=1}^{\infty} \dfrac{\cos(kx)}{k^n} & n \text{ odd}. \end{cases} \quad (3)$$

Then the Clausen function $Cl_n(x)$ can be given symbolically in terms of the POLYLOGARITHM as

$$Cl_n(x) = \begin{cases} \frac{1}{2}i[\mathrm{Li}_n(e^{-ix}) - \mathrm{Li}_n(e^{ix})] & n \text{ even} \\ \frac{1}{2}[\mathrm{Li}_n(e^{-ix}) + \mathrm{Li}_n(e^{ix})] & n \text{ odd}. \end{cases} \quad (4)$$

For $n = 1$, the function takes on the special form

$$Cl_1(x) = C_1(x) = -\ln|2 \sin(\tfrac{1}{2}x)| \quad (5)$$

and for $n = 2$, it becomes CLAUSEN'S INTEGRAL

$$Cl_2(x) = S_2(x) = -\int_0^x \ln[2 \sin(\tfrac{1}{2}t)] \, dt. \quad (6)$$

The symbolic sums of opposite parity are summable symbolically, and the first few are given by

$$C_2(x) = \tfrac{1}{6}\pi^2 - \tfrac{1}{2}\pi x + \tfrac{1}{4}x^2 \quad (7)$$

$$C_4(x) = \tfrac{1}{90} - \tfrac{1}{12}\pi^2 x^2 + \tfrac{1}{12}\pi x^3 - \tfrac{1}{48}x^4 \quad (8)$$

$$S_1(x) = \tfrac{1}{2}(\pi - x) \quad (9)$$

$$S_3(x) = \tfrac{1}{6}\pi^2 x - \tfrac{1}{4}\pi x^2 + \tfrac{1}{12}x^3 \quad (10)$$

$$S_5(x) = \tfrac{1}{90}\pi^4 x - \tfrac{1}{36}\pi^2 x^3 + \tfrac{1}{48}\pi x^4 - \tfrac{1}{240}x^5 \quad (11)$$

for $0 \le x \le 2\pi$ (Abramowitz and Stegun 1972).

See also CLAUSEN'S INTEGRAL, POLYGAMMA FUNCTION, POLYLOGARITHM

References

Abramowitz, M. and Stegun, C. A. (Eds.). "Clausen's Integral and Related Summations" §27.8 in *Handbook of Mathematical Functions with Formulas, Graphs, and Mathematical Tables, 9th printing.* New York: Dover, pp. 1005–006, 1972.

Arfken, G. *Mathematical Methods for Physicists, 3rd ed.* Orlando, FL: Academic Press, p. 783, 1985.

Clausen, R. "Über die Zerlegung reeller gebrochener Funktionen." *J. reine angew. Math.* **8**, 298–00, 1832.

Grosjean, C. C. "Formulae Concerning the Computation of the Clausen Integral $Cl_2(\alpha)$." *J. Comput. Appl. Math.* **11**, 331–42, 1984.

Jolley, L. B. W. *Summation of Series.* London: Chapman, 1925.

Lewin, L. *Dilogarithms and Associated Functions.* London: Macdonald, pp. 170–80, 1958.

Wheelon, A. D. *A Short Table of Summable Series.* Report No. SM-14642. Santa Monica, CA: Douglas Aircraft Co., 1953.

Clausen's Integral

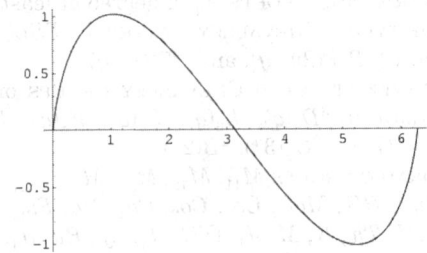

The $n = 2$ case of the S_2 CLAUSEN FUNCTION

$$Cl_2(\theta) = -\int_0^\theta \ln[2 \sin(\tfrac{1}{2}t)] \, dt.$$

See also CLAUSEN FUNCTION

References

Abramowitz, M. and Stegun, C. A. (Eds.). *Handbook of Mathematical Functions with Formulas, Graphs, and Mathematical Tables, 9th printing.* New York: Dover, pp. 1005–006, 1972.

Ashour, A. and Sabri, A. "Tabulation of the Function $\psi(\theta) = \sum_{n=1}^{\infty} \frac{\sin(n\theta)}{n^2}$." *Math. Tables Aids Comp.* **10**, 54 and 57–5, 1956.

Clausen, R. "Über die Zerlegung reeller gebrochener Funktionen." *J. reine angew. Math.* **8**, 298–00, 1832.

Lewin, L. "Clausen's Integral." Ch. 4 in *Dilogarithms and Associated Functions.* London: Macdonald, pp. 91–05, 1958.

Clausen's Product Identity

$$\begin{aligned} &{}_2F_1(\tfrac{1}{4}+a, \tfrac{1}{4}+b; \; q+a+b; \; x) \, {}_2F_1(\tfrac{1}{4}-a, \tfrac{1}{4}-b; \; 1-a \\ &\quad -b; \; x) \\ &= {}_3F_2(\tfrac{1}{2}, \tfrac{1}{2}+a-b, \tfrac{1}{2}-a+b; \; 1+a+b, 1-a \\ &\quad -b; \; x), \end{aligned}$$

where ${}_2F_1(a, b; c; x)$ is a HYPERGEOMETRIC FUNCTION.

Koepf, W. *Hypergeometric Summation: An Algorithmic Approach to Summation and Special Function Identities.* Braunschweig, Germany: Vieweg, p. 118, 1998.

Cleavance Center

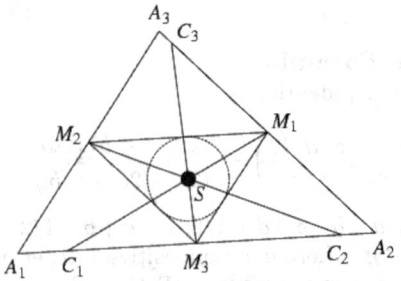

The point of concurrence S of a triangle's CLEAVERS

M_1C_1, M_2C_2, and M_3C_3, which is simply the SPIEKER CENTER, i.e., the INCENTER of the MEDIAL TRIANGLE (Honsberger 1995, p. 2).

See also CLEAVANCE CENTER, MEDIAL TRIANGLE, NAGEL POINT, SPIEKER CENTER

References

Honsberger, R. *Episodes in Nineteenth and Twentieth Century Euclidean Geometry.* Washington, DC: Math. Assoc. Amer., p. 2, 1995.

Cleaver

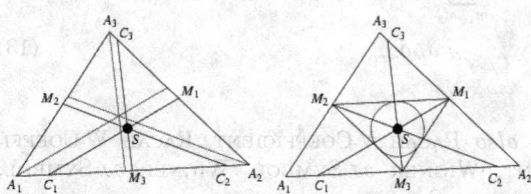

A PERIMETER-bisecting segment of a polygon originating from the MIDPOINT of one side. Each cleaver M_1C_1, M_2C_2, and M_3C_3 in a TRIANGLE $\Delta A_1A_2A_3$ is parallel to an ANGLE BISECTOR of the triangle (shown as dashed lines above). In addition, the three cleavers CONCUR in a point S known as the CLEAVANCE CENTER, which is the SPIEKER CENTER, i.e., INCENTER of the MEDIAL TRIANGLE (Honsberger 1995, p. 2).

See also B-LINE, CLEAVANCE CENTER, MEDIAL TRIANGLE, MIDPOINT, SPLITTER

References

Avishalom, D. "Perimeter-Bisectors in a Triangle" [Hebrew]. *Riveon Lematematika* **13**, 46–9, 1959.
Avishalom, D. "The Perimetric Bisection of Triangles." *Math. Mag.* **36**, 60–2, 1963.
Honsberger, R. "Cleavers and Splitters." *Episodes in Nineteenth and Twentieth Century Euclidean Geometry.* Washington, DC: Math. Assoc. Amer., pp. 1–4, 1995.
Jarden, D. "Synthetical Proof for the Theorem on the Center of Perimeter-Bisectors in a Triangle" [Hebrew]. *Riveon Lematematika* **13**, 50, 1959.

Clebsch Diagonal Cubic

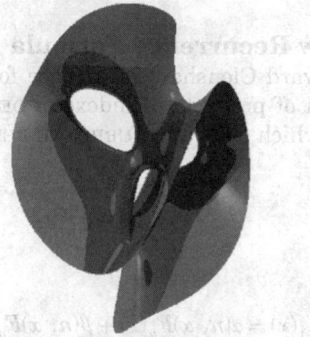

A CUBIC ALGEBRAIC SURFACE given by the equation

$$x_0^3 + x_1^3 + x_2^3 + x_3^3 + x_4^3 = 0, \tag{1}$$

with the added constraint

$$x_0 + x_1 + x_2 + x_3 + x_4 = 0. \tag{2}$$

The implicit equation obtained by taking the plane at infinity as $x_0 + x_1 + x_2 + x_3/2$ is

$$81(x^3 + y^3 + z^3) - 189(x^2y + x^2z + y^2x + y^2z + z^2x + z^2y)$$
$$+ 54xyz + 126(xy + xz + yz) - 9(x^2 + y^2 + z^2)$$
$$-9(x + y + z) + 1 = 0 \tag{3}$$

(Hunt, Nordstrand). On Clebsch's diagonal surface, all 27 of the complex lines (SOLOMON'S SEAL LINES) present on a general smooth CUBIC SURFACE are real. In addition, there are 10 points on the surface where 3 of the 27 lines meet. These points are called ECKARDT POINTS (Fischer 1986, Hunt), and the Clebsch diagonal surface is the unique CUBIC SURFACE containing 10 such points (Hunt).

If one of the variables describing Clebsch's diagonal surface is dropped, leaving the equations

$$x_0^3 + x_1^3 + x_2^3 + x_3^3 = 0, \tag{4}$$

$$x_0 + x_1 + x_2 + x_3 = 0, \tag{5}$$

the equations degenerate into two intersecting PLANES given by the equation

$$(x + y)(x + z)(y + z) = 0. \tag{6}$$

See also CUBIC SURFACE, ECKARDT POINT

References

Fischer, G. (Ed.). *Mathematical Models from the Collections of Universities and Museums.* Braunschweig, Germany: Vieweg, pp. 9–1, 1986.
Fischer, G. (Ed.). Plates 10–2 in *Mathematische Modelle/ Mathematical Models, Bildband/Photograph Volume.* Braunschweig, Germany: Vieweg, pp. 13–5, 1986.
Hunt, B. *The Geometry of Some Special Arithmetic Quotients.* New York: Springer-Verlag, pp. 122–28, 1996.
Nordstrand, T. "Clebsch Diagonal Surface." http://www.uib.no/people/nfytn/clebtxt.htm.

Clebsch-Aronhold Notation

A notation used to describe curves. The fundamental principle of Clebsch-Aronhold notation states that if each of a number of forms be replaced by a POWER of a linear form in the same number of variables equal to the order of the given form, and if a sufficient number of equivalent symbols are introduced by the ARONHOLD PROCESS so that no actual COEFFICIENT appears except to the first degree, then every identical relation holding for the new specialized forms holds for the general ones.

References

Coolidge, J. L. *A Treatise on Algebraic Plane Curves.* New York: Dover, p. 79, 1959.

ClebschGordan

CLEBSCH-GORDAN COEFFICIENT

Clebsch-Gordan Coefficient

A mathematical symbol used to integrate products of three SPHERICAL HARMONICS. Clebsch-Gordan coefficients commonly arise in applications involving the addition of angular momentum in quantum mechanics. If products of more than three SPHERICAL HARMONICS are desired, then a generalization known as WIGNER 6J-SYMBOLS or WIGNER 9J-SYMBOLS is used. The Clebsch-Gordan coefficients are written

$$C^j_{m_1 m_2} = (j_1 j_2 m_1 m_2 | j_1 j_2 jm) \tag{1}$$

and are defined by

$$\Psi_{JM} = \sum_{M=M_1+M_2} C^J_{M_1 M_2} \Psi_{M_1 M_2}, \tag{2}$$

where $J \equiv J_1 + J_2$.

The coefficients are subject to the restrictions that (j_1, j_2, j) be positive integers or half-integers, $j_1 + j_2 + j$ is an integer, (m_1, m_2, m) are positive or negative integers or half integers,

$$j_1 + j_2 - j \geq 0 \tag{3}$$

$$j_1 - j_2 + j \geq 0 \tag{4}$$

$$-j_1 + j_2 + j \geq 0, \tag{5}$$

and $-|j_1| \leq m_1 \leq |j_1|$, $-|j_2| \leq m_2 \leq |j_2|$, and $-|j| \leq m \leq |j|$ (Abramowitz and Stegun 1972, p. 1006). In addition, by use of symmetry relations, coefficients may always be put in the standard form $j_1 < j_2 < j$ and $m \geq 0$.

The Clebsch-Gordan coefficients are implemented in *Mathematica* as ClebschGordan[{j1, m1}, {j2, m2}, {j, m}] (assumed to be in standard form) and satisfy

$$(j_1 j_2 m_1 m_2 | j_1 j_2 jm) = 0 \quad \text{for } m_1 + m_2 \neq m \tag{6}$$

and are

The Clebsch-Gordan coefficients are sometimes expressed using the related RACAH V-COEFFICIENTS,

$$V(j_1 j_2 j; \; m_1 m_2 m) \tag{7}$$

or WIGNER 3J-SYMBOLS. Connections among the three are

$$(j_1 j_2 m_1 m_2 | j_1 j_2 jm)$$
$$= (-1)^{m+j_1-j_2} \sqrt{2j+1} \begin{pmatrix} j_1 & j_2 & j \\ m_1 & m_2 & -m \end{pmatrix} \tag{8}$$

$$(j_1 j_2 m_1 m_2 | j_1 j_2 jm)$$
$$= (-1)^{j+m} \sqrt{2j+1} V(j_1 j_2; \; m_1 m_2 - m) \tag{9}$$

$$V(j_1 j_2 j; \; m_1 m_2 m) = (-1)^{-j_1+j_2+j} \begin{pmatrix} j_1 & j_2 & j_1 \\ m_1 & m_2 & m_2 \end{pmatrix}. \tag{10}$$

They have the symmetry

$$(j_1 j_2 m_1 m_2 | j_1 j_2 jm) = (-1)^{j_1+j_2-j}(j_2 j_1 m_2 m_1 | j_2 j_1 jm), \tag{11}$$

and obey the orthogonality relationships

$$\sum_{j,\,m} (j_1 j_2 m_1 m_2 | j_1 j_2 jm)(j_1 j_2 jm | j_1 j_2 m'_1 m'_2)$$
$$= \delta_{m_1 m'_1} \delta_{m_2 m'_2} \tag{12}$$

$$\sum_{m_1,\,m_2} (j_1 j_2 m_1 m_2 | j_1 j_2 jm)(j_1 j_2 j'm' | j_1 j_2 m_1 m_2)$$
$$= \delta_{jj'} \delta_{mm'}. \tag{13}$$

See also RACAH V-COEFFICIENT, RACAH W-COEFFICIENT, WIGNER 3J-SYMBOL, WIGNER 6J-SYMBOL, WIGNER 9J-SYMBOL

References

Abramowitz, M. and Stegun, C. A. (Eds.). "Vector-Addition Coefficients." §27.9 in *Handbook of Mathematical Functions with Formulas, Graphs, and Mathematical Tables, 9th printing.* New York: Dover, pp. 1006–010, 1972.

Cohen-Tannoudji, C.; Diu, B.; and Laloë, F. "Clebsch-Gordan Coefficients." Complement B_X in *Quantum Mechanics, Vol. 2.* New York: Wiley, pp. 1035–047, 1977.

Condon, E. U. and Shortley, G. §3.6–.14 in *The Theory of Atomic Spectra.* Cambridge, England: Cambridge University Press, pp. 56–8, 1951.

Fano, U. and Fano, L. *Basic Physics of Atoms and Molecules.* New York: Wiley, p. 240, 1959.

Messiah, A. "Clebsch-Gordan (C.-G.) Coefficients and '3j' Symbols." Appendix C.I in *Quantum Mechanics, Vol. 2.* Amsterdam, Netherlands: North-Holland, pp. 1054–060, 1962.

Rose, M. E. *Elementary Theory of Angular Momentum.* New York: Dover, 1995.

Shore, B. W. and Menzel, D. H. "Coupling and Clebsch-Gordan Coefficients." §6.2 in *Principles of Atomic Spectra.* New York: Wiley, pp. 268–76, 1968.

Sobel'man, I. I. "Angular Momenta." Ch. 4 in *Atomic Spectra and Radiative Transitions, 2nd ed.* Berlin: Springer-Verlag, 1992.

Clement Matrix

KAC MATRIX

Clenshaw Recurrence Formula

The downward Clenshaw recurrence formula evaluates a sum of products of indexed COEFFICIENTS by functions which obey a RECURRENCE RELATION. If

$$f(x) = \sum_{k=0}^{N} c_k F_k(x)$$

and

$$F_{n+1}(x) = \alpha(n,\,x)F_n(x) + \beta(n,\,x)F_{n-1}(x),$$

where the c_ks are known, then define

$$y_{N+2} = y_{N+1} = 0$$

$$y_k = \alpha(k, x)y_{k+1} + \beta(k+1, x)y_{k+2} + c_k$$

for $k = N, N-1, \ldots$ and solve backwards to obtain y_2 and y_1.

$$c_k = y_k - \alpha(k, x)y_{k+1} - \beta(k+1, x)y_{k+2}$$

$$
\begin{aligned}
f(x) &= \sum_{k=0}^{N} c_k F_k(x) \\
&= c_0 F_0(x) + [y_1 - \alpha(1, x)y_2 - \beta(2, x)y_3]F_1(x) \\
&\quad + [y_2 - \alpha(2, x)y_3 - \beta(3, x)y_4]F_2(x) \\
&\quad + [y_3 - \alpha(3, x)y_4 - \beta(4, x)y_5]F_3(x) \\
&\quad + [y_4 - \alpha(4, x)y_5 - \beta(5, x)y_6]F_4(x) + \ldots \\
&= c_0 F_0(x) + y_1 F_1(x) + y_2 [F_2(x) - \alpha(1, x)F_1(x)] \\
&\quad + y_3[F_3(x) - \alpha(2, x)F_2(x) - \beta(2, x)] \\
&\quad + y_4[F_4(x) - \alpha(3, x)F_3(x) - \beta(3, x)] + \ldots \\
&= c_0 F_0(x) + y_2[\{\alpha(1, x)F_1(x) + \beta(1, x)F_0(x)\} \\
&\quad - \alpha(1, x)F_1(x)] + y_1 F_1(x) \\
&= c_0 F_0(x) + y_1 F_1(x) + \beta(1, x)F_0(x)y_2.
\end{aligned}
$$

The upward Clenshaw recurrence formula is

$$y_{-2} = y_{-1} = 0$$

$$y_k = \frac{1}{\beta(k+1, x)}[y_{k-2} - \alpha(k, x)y_{k-1} - c_k]$$

for $k = 0, 1, \ldots, N-1$.

$$f(x) = c_N F_N(x) - \beta(N, x)F_{N-1}(x)y_{N-1} - F_N(x)y_{N-2}.$$

References

Press, W. H.; Flannery, B. P.; Teukolsky, S. A.; and Vetterling, W. T. "Recurrence Relations and Clenshaw's Recurrence Formula." §5.5 in *Numerical Recipes in FORTRAN: The Art of Scientific Computing,* 2nd ed. Cambridge, England: Cambridge University Press, pp. 172–78, 1992.

Cliff Random Number Generator

A RANDOM NUMBER generator produced by iterating

$$X_{n+1} = |100 \ln X_n \pmod 1|$$

for a SEED $X_0 = 0.1$. This simple generator passes the NOISE SPHERE test for randomness by showing no structure.

See also RANDOM NUMBER, SEED

References

Pickover, C. A. "Computers, Randomness, Mind, and Infinity." Ch. 31 in *Keys to Infinity.* New York: W. H. Freeman, pp. 233–47, 1995.

Clifford Algebra

Let V be an n-D linear SPACE over a FIELD K, and let Q be a QUADRATIC FORM on V. A Clifford algebra is then defined over the $T(V)/I(Q)$, where $T(V)$ is the tensor algebra over V and I is a particular IDEAL of $T(V)$.

Clifford algebraists call their higher dimensional numbers HYPERCOMPLEX even though they do not share all the properties of complex numbers and no classical function theory can be constructed over them.

See also HYPERCOMPLEX NUMBER, QUATERNION

References

Ablamowicz, R. Hecke Algebra, SVD, and Other Computational Examples with CLIFFORD. 14 Oct 1999. http://xxx.lanl.gov/abs/math.RA/9910069/.

Ablamowicz, R.; Lounesto, P.; and Parra, J. M. *Clifford Algebras with Numeric and Symbolic Computations.* Boston, MA: Birkhäuser, 1996.

Huang, J.-S. "The Clifford Algebra." §6.2 in *Lectures on Representation Theory.* Singapore: World Scientific, pp. 63–5, 1999.

Iyanaga, S. and Kawada, Y. (Eds.). "Clifford Algebras." §64 in *Encyclopedic Dictionary of Mathematics.* Cambridge, MA: MIT Press, pp. 220–22, 1980.

Lounesto, P. "Counterexamples to Theorems Published and Proved in Recent Literature on Clifford Algebras, Spinors, Spin Groups, and the Exterior Algebra." http://www.hit.fi/~lounesto/counterexamples.htm.

Clifford's Circle Theorem

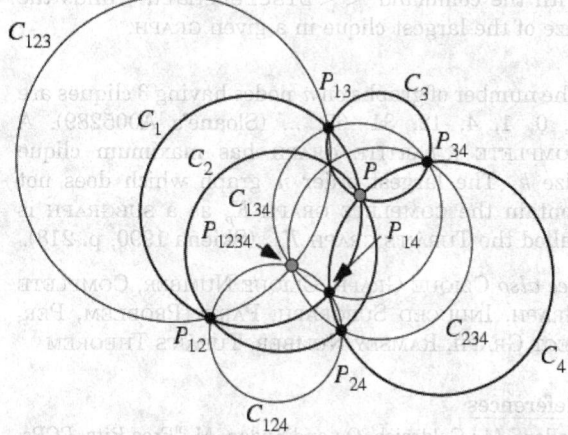

Let C_1, C_2, C_3, and C_4 be four CIRCLES of GENERAL POSITION through a point P. Let P_{ij} be the second intersection of the CIRCLES C_i and C_j. Let C_{ijk} be the CIRCLE $P_{ij}P_{ik}P_{jk}$. Then the four CIRCLES C_{234}, C_{134}, C_{124}, and C_{123} all pass through the point P_{1234}. Similarly, let C_5 be a fifth CIRCLE through P. Then the five points P_{2345}, P_{1345}, P_{1245}, P_{1235} and P_{1234} all lie on one CIRCLE C_{12345}. And so on.

See also CIRCLE, COX'S THEOREM

References

Wells, D. *The Penguin Dictionary of Curious and Interesting Geometry*. London: Penguin, pp. 32–3, 1991.

Clifford's Curve Theorem

The dimension of a special series can never exceed half its order.

References

Coolidge, J. L. *A Treatise on Algebraic Plane Curves*. New York: Dover, p. 263, 1959.

Clique

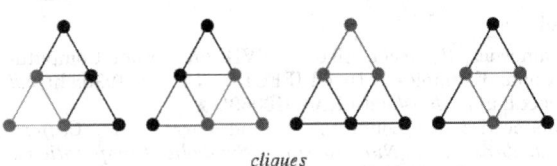

cliques

A clique of a GRAPH is its maximal COMPLETE SUBGRAPH (Harary 1994, p. 20), although some authors define a clique as any COMPLETE SUBGRAPH and then refer to "maximum cliques" (Skiena 1990, p. 217). The problem of finding the size of a clique for a given GRAPH is an NP-COMPLETE PROBLEM (Skiena 1997).

Cliques arise in a number of areas of GRAPH THEORY and combinatorics, including the theory of ERROR-CORRECTING CODES. The command MaximumClique[g] in the *Mathematica* add-on package DiscreteMath`Combinatorica` (which can be loaded with the command < <DiscreteMath`) finds the size of the largest clique in a given GRAPH.

The number of graphs on n nodes having 3 cliques are 0, 0, 1, 4, 12, 31, 67, ... (Sloane's A005289). A COMPLETE K-PARTITE GRAPH has maximum clique size k. The largest order n graph which does not contain the COMPLETE GRAPH K_p as a SUBGRAPH is called the TURÁN'S GRAPH $T_{n,p}$ (Skiena 1990, p. 218).

See also CLIQUE GRAPH, CLIQUE NUMBER, COMPLETE GRAPH, INDUCED SUBGRAPH, PARTY PROBLEM, PERFECT GRAPH, RAMSEY NUMBER, TURÁN'S THEOREM

References

Bellare, M.; Goldreich, O.; and Sudan, M. "Free Bits, PCPs, and Non-Approximability--Towards Tight Results." *SIAM J. Comput.* **27**, 804–15, 1998.
Cormen, T.; Leiserson, C.; and Rivest, R. *Introduction to Algorithms*. Cambridge, MA: MIT Press, 1990.
Harary, F. *Graph Theory*. Reading, MA: Addison-Wesley, 1994.
Karp, R. M. "Reducibility Among Combinatorial Problems." In *Complexity of Computer Calculations* (Ed. R. Miller and J. Thatcher). New York: Plenum, pp. 85–03, 1972.
Garey, M. R. and Johnson, D. S. *Computers and Intractability: A Guide to the Theory of NP-Completeness*. New York: W. H. Freeman, 1983.
Manber, U. *Introduction to Algorithms: A Creative Approach*. Reading, MA: Addison-Wesley, 1989.
Skiena, S. "Maximum Cliques." §5.6.1 in *Implementing Discrete Mathematics: Combinatorics and Graph Theory with Mathematica*. Reading, MA: Addison-Wesley, pp. 215 and 217–18, 1990.
Skiena, S. S. "Clique and Independent Set" and "Clique." §6.2.3 and 8.5.1 in *The Algorithm Design Manual*. New York: Springer-Verlag, pp. 144 and 312–14, 1997.
Sloane, N. J. A. Sequences A005289/M3440 in "An On-Line Version of the Encyclopedia of Integer Sequences." http://www.research.att.com/~njas/sequences/eisonline.html.

Clique Graph

clique graph of G

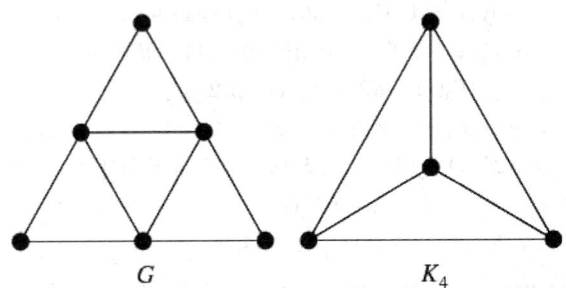

G K_4

The clique graph of a given GRAPH G is the GRAPH INTERSECTION of the family of CLIQUES of G. A GRAPH G is a clique graph IFF it contains a family F of COMPLETE SUBGRAPHS whose GRAPH UNION is G, such that whenever every pair of such complete graphs in some subfamily F' has a nonempty graph intersection, the intersection of all members of F' is not empty (Harary 1994, p. 20).

See also CLIQUE, CLIQUE NUMBER, COMPLETE GRAPH

References

Harary, F. *Graph Theory*. Reading, MA: Addison-Wesley, 1994.

Clique Number

The number of VERTICES in the largest CLIQUE of G, denoted $\omega(G)$. For an arbitrary GRAPH,

$$\omega(G) \geq \sum_{i=1}^{n} \frac{1}{n - d_i},$$

where d_i is the DEGREE of VERTEX i. The following table gives the number $N_k(n)$ of n-node graphs having clique number k for small k.

k	Sloane	$N_k(n)$
1		1, 1, 1, 1, 1, 1, ...

2	A052450	0, 1, 2, 6, 13, 37, 106, ...
3	A052451	0, 0, 1, 3, 15, 82, 578, ...
4	A052452	0, 0, 0, 1, 4, 30, 301, ...
5		0, 0, 0, 0, 1, 5, 51, ...
6		0, 0, 0, 0, 0, 1, 6, ...

See also CLIQUE, CLIQUE GRAPH

References

Aigner, M. "Turán's Graph Theorem." *Amer. Math. Monthly* **102**, 808–16, 1995.

Sloane, N. J. A. Sequences A052450, 052451, and A052452 in "An On-Line Version of the Encyclopedia of Integer Sequences." http://www.research.att.com/~njas/sequences/eisonline.html.

Clock Arithmetic

CONGRUENCE

Clock Prime

A prime number obtained by reading digits around an analog clock. In a clockwise directions, the primes are 2, 3, 5, 7, 11, 23, 67, 89, 4567, 23456789, 23456789101112123, ... (Sloane's A036342). In a counterclockwise direction, the primes are 2, 3, 5, 7, 11, 43, 109, 10987, 76543, 6543211211, 4321121110987, ... (Sloane's A036342). In either direction, the primes are 2, 3, 5, 7, 11, 23, 43, 67, 89, 109, 4567, 10987, 76543, 23456789, 6543211211, ... (Sloane's A036344).

On a 24-hour digital clock, there are 211 possible prime values: 2, 3, 5, 7, 11, 13, 17, 19, 23, 29, 31, 37, 41, 43, 47, 53, 59, 101, ... (Sloane's A050246).

References

Rivera, C. "Problems & Puzzles: Puzzle Primes on a Clock.-019." http://www.primepuzzles.net/puzzles/puzz_019.htm.

Sloane, N. J. A. Sequences A036342, A036343, A036344, and A050246 in "An On-Line Version of the Encyclopedia of Integer Sequences." http://www.research.att.com/~njas/sequences/eisonline.html.

Weisstein, E. W. "Integer Sequences." MATHEMATICA NOTE-BOOK INTEGERSEQUENCES.M.

Clock Solitaire

A solitaire game played with CARDS. The chance of winning is 1/13, and the AVERAGE number of CARDS turned up is 42.4.

References

Gardner, M. *Mathematical Magic Show: More Puzzles, Games, Diversions, Illusions and Other Mathematical Sleight-of-Mind from Scientific American.* New York: Vintage, pp. 244–47, 1978.

Knuth, D. E. *The Art of Computer Programming, Vol. 1: Fundamental Algorithms, 3rd ed.* Reading, MA: Addison-Wesley, pp. 377 and 577, 1997.

Moyse, A. Jr. *150 Ways to Play Solitaire.* Chicago: Whitman, 1950.

Close Packing

SPHERE PACKING

Closed

A mathematical structure A is said to be closed under an operation $+$ if, whenever a and b are both elements of A, then so is $a + b$.

A mathematical object taken together with its boundary is also called closed. For example, while the interior of a SPHERE is an OPEN BALL, the interior together with the sphere itself is a CLOSED BALL.

See also CLOSED BALL, CLOSED CURVE, CLOSED DISK, CLOSED FORM, CLOSURE (TOPOLOGY)

Closed Ball

The closed ball with center $\mathbf{x}$ and radius r is defined by

$$B_r(\mathbf{x}) = \{ \mathbf{y} : |\mathbf{y} - \mathbf{x}| \le r \}.$$

See also BALL, CLOSED DISK, OPEN BALL

References

Croft, H. T.; Falconer, K. J.; and Guy, R. K. *Unsolved Problems in Geometry.* New York: Springer-Verlag, p. 1, 1991.

Closed Curve

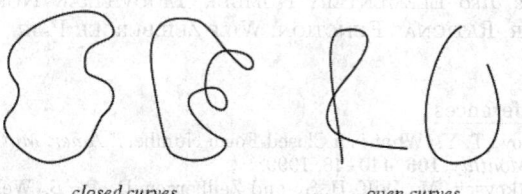

closed curves   *open curves*

In the plane, a closed curve is a CURVE with no endpoints and which completely encloses an AREA.

See also CURVE, JORDAN CURVE, SIMPLE CURVE

References

Krantz, S. G. "Closed Curves." §2.1.2 in *Handbook of Complex Analysis.* Boston, MA: Birkhäuser, pp. 19–0, 1999.

Closed Curve Problem

Find NECESSARY and SUFFICIENT conditions that determine when the integral curve of two periodic functions $\kappa(s)$ and $\tau(s)$ with the same period L is a CLOSED CURVE.

Closed Disk

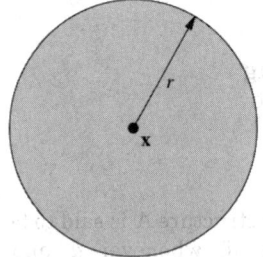

An n-D closed disk of RADIUS r is the collection of points of distance $\leq r$ from a fixed point in EUCLIDEAN n-space. Krantz (1999, p. 3) uses the symbol $\bar{D}(x, r)$ to denote the closed disk, and $\bar{D} = \bar{D}(\mathbf{0}, 1)$ to denote the unit closed disk centered at the origin

See also DISK, OPEN DISK

References
Krantz, S. G. *Handbook of Complex Analysis.* Boston, MA: Birkhäuser, p. 3, 1999.

Closed Form

A discrete FUNCTION $A(n, k)$ is called closed form (or sometimes "hypergeometric") in two variables if the ratios $A(n+1, k)/A(n, k)$ and $A(n, k+1)/A(n, k)$ are both RATIONAL FUNCTIONS. A pair of closed form functions (F, G) is said to be a WILF-ZEILBERGER PAIR if

$$F(n+1, k) - F(n, k) = G(n, k+1) - G(n, k).$$

See also ELEMENTARY NUMBER, LIOUVILLIAN NUMBER, RATIONAL FUNCTION, WILF-ZEILBERGER PAIR

References
Chow, T. Y. "What is a Closed-Form Number?" *Amer. Math. Monthly* **106**, 440–48, 1999.
Petkovsek, M.; Wilf, H. S.; and Zeilberger, D. *A = B.* Wellesley, MA: A. K. Peters, p. 141, 1996.
Zeilberger, D. "Closed Form (Pun Intended!)." *Contemporary Math.* **143**, 579–07, 1993.

Closed Graph Theorem

A linear OPERATOR between two BANACH SPACES is continuous IFF it has a "closed" graph.

See also BANACH SPACE

References
Zeidler, E. *Applied Functional Analysis: Applications to Mathematical Physics.* New York: Springer-Verlag, 1995.

Closed Interval

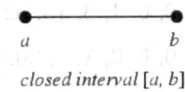

closed interval $[a, b]$

An INTERVAL which includes its LIMIT POINTS. If the endpoints of the interval are FINITE numbers a and b, then the INTERVAL is denoted $[a, b]$. If one of the endpoints is $\pm\infty$, then the interval still contains all of its LIMIT POINTS, so $[a, \infty)$ and $(-\infty, b]$ are also closed intervals.

See also CLOSED BALL, CLOSED DISK, CLOSED SET, HALF-CLOSED INTERVAL, INTERVAL, OPEN INTERVAL

References
Croft, H. T.; Falconer, K. J.; and Guy, R. K. *Unsolved Problems in Geometry.* New York: Springer-Verlag, p. 1, 1991.

Closed Manifold

A COMPACT MANIFOLD without boundary.

See also OPEN MANIFOLD

Closed Set

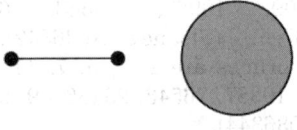

closed interval   *closed disk*

There are several equivalent definitions of a closed SET. A SET S is closed if

1. The COMPLEMENT of S is an OPEN SET,
2. S is its own CLOSURE,

3. Sequences/nets/filters in S which converge do so within S,
4. Every point outside S has a NEIGHBORHOOD disjoint from S.

The POINT-SET TOPOLOGICAL definition of a closed set is a set which contains all of its LIMIT POINTS. Therefore, a closed set C is one for which, whatever point x is picked outside of C, x can always be isolated in some OPEN SET which doesn't touch C.

The most commonly encountered closed sets are the CLOSED INTERVAL, closed path, CLOSED DISK, interior of a closed path together with the path itself, and CLOSED BALL. The CANTOR SET is an unusual closed set in the sense that it consists entirely of BOUNDARY POINTS (and is nowhere DENSE, so it has LEBESGUE MEASURE 0).

It is possible for a set to be neither OPEN nor closed, e.g., the HALF-CLOSED INTERVAL $(0, 1]$.

See also BOREL SET, BOUNDARY POINT, CANTOR SET, CLOSED BALL, CLOSED INTERVAL, CLOSED DISK, COMPACT SET, HALF-CLOSED INTERVAL, OPEN SET

References
Croft, H. T.; Falconer, K. J.; and Guy, R. K. *Unsolved Problems in Geometry.* New York: Springer-Verlag, p. 2, 1991.
Krantz, S. G. *Handbook of Complex Analysis.* Boston, MA: Birkhäuser, p. 3, 1999.

Closed Star
The CLOSURE $\overline{\text{St}}\ v$ of a STAR St v at a vertex v of a SIMPLICIAL COMPLEX K.

See also LINK (SIMPLICIAL COMPLEX), STAR

References
Munkres, J. R. *Elements of Algebraic Topology.* Perseus Press, 1993.

Closed Subgroup
A SUBSET of a TOPOLOGICAL GROUP which is CLOSED as a SUBSET and also a SUBGROUP.

See also EFFECTIVE ACTION, FREE ACTION, GROUP, ISOTROPY GROUP, MATRIX GROUP, ORBIT (GROUP), QUOTIENT SPACE (LIE GROUP), REPRESENTATION, TOPOLOGICAL GROUP, TRANSITIVE

Closure (Set)
A SET S and a BINARY OPERATOR * are said to exhibit closure if applying the BINARY OPERATOR to two elements S returns a value which is itself a member of S.

The term "closure" is also used to refer to a "closed" version of a given set. The closure of a SET can be defined in several equivalent ways, including

   1. The SET plus its LIMIT POINTS, also called "boundary" points, the union of which is also called the "frontier."
   2. The unique smallest CLOSED SET containing the given SET.
   3. The COMPLEMENT of the interior of the COMPLEMENT of the set.
   4. The collection of all points such that every NEIGHBORHOOD of these points intersects the original SET in a nonempty SET.

In topologies where the T2-SEPARATION AXIOM is assumed, the closure of a finite SET S is S itself.

See also BINARY OPERATOR, BOUNDARY SET, CLOSURE (TOPOLOGY), CONNECTED SET, EXISTENTIAL CLOSURE, REFLEXIVE CLOSURE, TIGHT CLOSURE, TRANSITIVE CLOSURE

References
Croft, H. T.; Falconer, K. J.; and Guy, R. K. *Unsolved Problems in Geometry.* New York: Springer-Verlag, p. 2, 1991.

Closure (Topology)
The closure of a set A is the smallest closed set containing A. Closed sets are CLOSED under arbitrary intersection, so it is also the intersection of all closed sets containing A. Typically, it is just A with all of its ACCUMULATION POINTS.

See also CLOSED SET, CLOSURE (SET), SEQUENCE, TOPOLOGY

Closure Relation

$$\delta(x - t) = \sum_{n=0}^{\infty} \phi_n(x)\phi_n(t),$$

where $\delta(x)$ is the DELTA FUNCTION.

Clothoid
CORNU SPIRAL

Clove Hitch

A HITCH also called the BOATMAN'S KNOT or PEG KNOT.

References
Owen, P. *Knots.* Philadelphia, PA: Courage, pp. 24–7, 1993.

Club
SPHINX

Clump
RUN

Cluster
Given a POINT LATTICE, a cluster is a group of filled cells which are all connected to their neighbors vertically or horizontally.

See also CLUSTER PERIMETER, PERCOLATION THEORY, s-CLUSTER, s-RUN

References
Stauffer, D. and Aharony, A. *Introduction to Percolation Theory, 2nd ed.* London: Taylor & Francis, 1992.

Cluster Perimeter
The number of empty neighbors of a CLUSTER.

See also PERIMETER POLYNOMIAL

Cluster Prime

An ODD PRIME p is called a cluster prime if every EVEN positive integer less than $p - 2$ can be written as a difference of two primes $q - q'$, where q, $q' \leq p$. The first 23 odd primes 3, 5, 7, ..., 89 are all cluster primes. The first few odd primes that are not cluster primes are 97, 127, 149, 191, 211, ... (Sloane's A038133).

The numbers of cluster primes less than 10^1, 10^2, ... are 23, 99, 420, 1807, ... (Sloane's A039506), and the corresponding numbers of noncluster primes are 0, 1, 68, 808, 7784, ... (Sloane's A039507). It is not known if there are infinitely many cluster primes, but Blecksmith *et al.* (1999) show that for every positive integer s, there is a bound $x_0 = x_x(s)$ such that if $x \geq x_0$, then

$$\pi_c(x) < \frac{x}{(\ln x)^s},$$

where $\pi_c(x)$ is the number of cluster primes not exceeding x. Blecksmith *et al.* (1999) also show that the sum of the reciprocals of the cluster primes is finite.

See also PRIME CONSTELLATION

References

Blecksmith, R.; Erdos, P.; and Selfridge, J. L. "Cluster Primes." *Amer. Math. Monthly* **106**, 43–8, 1999.
Sloane, N. J. A. Sequences A038133, A039506, and A039507 in "An On-Line Version of the Encyclopedia of Integer Sequences." http://www.research.att.com/~njas/sequences/eisonline.html.

C-Matrix

Any SYMMETRIC MATRIX ($C^T = C$) or SKEW SYMMETRIC MATRIX ($C^T = -C$) C_n with diagonal elements 0 and others ± 1 satisfying

$$CC^T = (n-1)I,$$

where I is the IDENTITY MATRIX, is known as a C-matrix (Ball and Coxeter 1987). There are two symmetric C-matrices of order 2,

$$\begin{bmatrix} 0 & -1 \\ -1 & 0 \end{bmatrix}, \begin{bmatrix} 0 & 1 \\ 1 & 0 \end{bmatrix}$$

and two antisymmetric C-matrices of order 2,

$$\begin{bmatrix} 0 & 1 \\ -1 & 0 \end{bmatrix}, \begin{bmatrix} 0 & 1 \\ -1 & 0 \end{bmatrix}.$$

Further examples include

$$C_4 = \begin{bmatrix} 0 & + & + & + \\ - & 0 & - & + \\ - & + & 0 & - \\ - & - & + & 0 \end{bmatrix}$$

$$C_6 = \begin{bmatrix} 0 & + & + & + & + & + \\ + & 0 & + & - & - & + \\ + & + & 0 & + & - & - \\ + & - & + & 0 & + & - \\ + & - & - & + & 0 & + \\ + & + & - & - & + & 0 \end{bmatrix}$$

There are no symmetric C-matrices of order 4 or 22 (Ball and Coxeter 1987, p. 309). The following table gives the number of C-matrices of orders $n = 1, 2, ...$.

Type	Sloane Numbers
symmetric	0, 2, 0, 0, 0, 384, 0, 0, ...
antisymmetric	0, 2, 0, 16, 0, 0, 0, 30720, ...
total	0, 4, 0, 16, 0, 384, 0, 30720, ...

A C-matrix of an odd prime power order may be constructed using a general method due to Paley (Paley 1933, Ball and Coxeter 1987).

References

Ball, W. W. R. and Coxeter, H. S. M. *Mathematical Recreations and Essays, 13th ed.* New York: Dover, pp. 308–09, 1987.
Belevitch, V. *Ann. de la Société scientifique de Bruxelles* **82**, 13–2, 1968.
Brenner, J. and Cummings, L. "The Hadamard Maximum Determinant Problem." *Amer. Math. Monthly* **79**, 626–30, 1972.
Colbourn, C. J. and Dinitz, J. H. (Eds.). *CRC Handbook of Combinatorial Designs.* Boca Raton, FL: CRC Press, p. 689, 1996.
Paley, R. E. A. C. "On Orthogonal Matrices." *J. Math. Phys.* **12**, 311–20, 1933.
Raghavarao, D. *Constructions and Combinatorial Problems in Design of Experiments.* New York: Dover, 1988.

Coanalytic Set

A DEFINABLE SET which is the complement of an ANALYTIC SET.

See also ANALYTIC SET

Coastline Paradox

Determining the length of a country's coastline is not as simple as it first appears, as first considered by L. F. Richardson (1881–953). In fact, the answer depends on the length of the RULER you use for the measurements. A shorter RULER measures more of the sinuosity of bays and inlets than a larger one, so the estimated length continues to increase as the RULER length decreases.

In fact, a coastline is an example of a FRACTAL, and plotting the length of the RULER versus the measured length of the coastline on a log-log plot gives a straight line, the slope of which is the FRACTAL

DIMENSION of the coastline (and will be a number between 1 and 2).

See also LONGIMETER

References

Lauwerier, H. *Fractals: Endlessly Repeated Geometric Figures.* Princeton, NJ: Princeton University Press, pp. 29–1, 1991.

Steinhaus, H. *Mathematical Snapshots, 3rd ed.* New York: Dover, pp. 109–10, 1999.

Coates-Wiles Theorem

In 1976, Coates and Wiles showed that ELLIPTIC CURVES with COMPLEX MULTIPLICATION having an infinite number of solutions have L-functions which are zero at the relevant fixed point. This is a special case of the SWINNERTON-DYER CONJECTURE.

References

Cipra, B. "Fermat Prover Points to Next Challenges." *Science* **271**, 1668–669, 1996.

Coaxal Circles

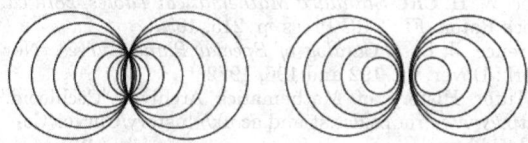

CIRCLES which share a RADICAL LINE with a given circle are said to be coaxal. The centers of coaxal circles are COLLINEAR, and the collection of all coaxal circles is called a pencil of coaxal circles (Coxeter and Greitzer 1967, p. 35). It is possible to combine the two types of coaxal systems illustrated above such that the sets are orthogonal.

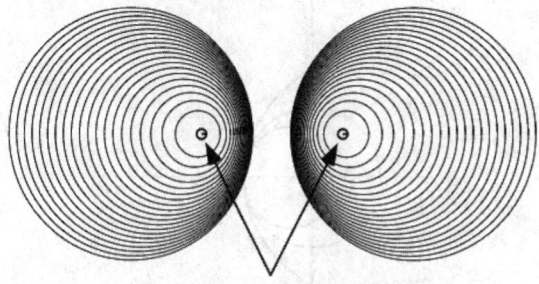

point circles (limit points)

Members of a COAXAL SYSTEM satisfy

$$x^2 + y^2 + 2\lambda x + c = (x+\lambda)^2 + y^2 + c - \lambda^2 = 0$$

for values of λ. Picking $\lambda^2 = c$ then gives the two circles

$$(x \pm \sqrt{c})^2 + y^2 = 0$$

of zero RADIUS, known as POINT CIRCLES. The two

point circles $(\pm\sqrt{c}, 0)$, real or imaginary, are called the LIMITING POINTS.

See also CIRCLE, COAXALOID SYSTEM, GAUSS-BODEN-MILLER THEOREM, LIMITING POINT, POINT CIRCLE, RADICAL LINE

References

Casey, J. "Coaxal Circles." §6.5 in *A Sequel to the First Six Books of the Elements of Euclid, Containing an Easy Introduction to Modern Geometry with Numerous Examples, 5th ed., rev. enl.* Dublin: Hodges, Figgis, & Co., pp. 113–26, 1888.

Coolidge, J. L. "Coaxal Circles." §1.7 in *A Treatise on the Geometry of the Circle and Sphere.* New York: Chelsea, pp. 95–13, 1971.

Coxeter, H. S. M. and Greitzer, S. L. "Coaxal Circles." §2.3 in *Geometry Revisited.* Washington, DC: Math. Assoc. Amer., pp. 35–6 and 122, 1967.

Dixon, R. *Mathographics.* New York: Dover, pp. 68–2, 1991.

Durell, C. V. "Coaxal Circles." Ch. 11 in *Modern Geometry: The Straight Line and Circle.* London: Macmillan, pp. 121–25, 1928.

Johnson, R. A. *Modern Geometry: An Elementary Treatise on the Geometry of the Triangle and the Circle.* Boston, MA: Houghton Mifflin, pp. 34–7, 199, and 279, 1929.

Lachlan, R. "Coaxal Circles." Ch. 13 in *An Elementary Treatise on Modern Pure Geometry.* London: Macmillian, pp. 199–17, 1893.

Steinhaus, H. *Mathematical Snapshots, 3rd ed.* New York: Dover, pp. 143–44, 1999.

Wells, D. *The Penguin Dictionary of Curious and Interesting Geometry.* London: Penguin, pp. 33–4, 1991.

Coaxal Planes

SHEAF OF PLANES

Coaxal System

A system of COAXAL CIRCLES.

See also COAXAL CIRCLES, PONCELET'S COAXAL THEOREM

Coaxaloid System

A system of circles obtained by multiplying each RADIUS in a COAXAL SYSTEM by a constant.

References

Johnson, R. A. *Modern Geometry: An Elementary Treatise on the Geometry of the Triangle and the Circle.* Boston, MA: Houghton Mifflin, pp. 276–77, 1929.

Coaxial Circles

COAXAL CIRCLES

Cobordant Manifold

Two open MANIFOLDS M and M' are cobordant if there exists a MANIFOLD with boundary W^{n+1} such that an acceptable restrictive relationship holds.

See also COBORDISM, H-COBORDISM THEOREM, MORSE THEORY

Cobordism

BORDISM, H-COBORDISM

Cobordism Group

BORDISM GROUP

Cobordism Ring

BORDISM GROUP

Cobweb Equation

This entry contributed by RONALD M. AARTS

The simple first-order DIFFERENCE EQUATION

$$y_{t+1} - Ay_t = B, \tag{1}$$

where

$$A = -\frac{m_s}{m_d} \tag{2}$$

$$B = \frac{b_d - b_s}{m_d} \tag{3}$$

and

$$D_t = -m_d p_t + b_d \tag{4}$$

$$S_{t+1} = m_s p_t + b_s \tag{5}$$

are the price-demand and price-supply curves, where $-m_d$ and b_d represent the slope and D-intercept, respectively, for the demand curve, and m_s and b_s represent the corresponding constants for the supply curve (Ezekiel 1938, Goldberg 1986).

A class of behaviors related to this equation is known as "Cobweb phenomena" in economics.

See also DIFFERENCE EQUATION

References

Ezekiel, M. "The Cobweb Theorem." *Quart. J. Econ.* **52**, 255–80, 1938.
Goldberg, S. *Introduction to Difference Equations, with Illustrative Examples from Economics, Psychology, and Sociology.* New York: Dover, 1986.

Cochleoid

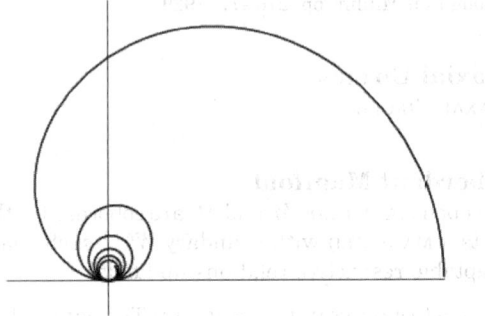

The cochleoid, whose name means "snail-form" in

Latin, was first discussed by J. Peck in 1700 (Mac-Tutor Archive). It has also been called the oui-ja board curve (Beyer 1987, p. 215). The points of contact of PARALLEL TANGENTS to the cochleoid lie on a STROPHOID.

In POLAR COORDINATES,

$$r = \frac{a \sin \theta}{\theta}. \tag{1}$$

In CARTESIAN COORDINATES,

$$(x^2 + y^2) \tan^{-1}\left(\frac{y}{x}\right) = ay. \tag{2}$$

The CURVATURE is

$$\kappa = \frac{2\sqrt{2}\theta^3[2\theta - \sin(2\theta)]}{[1 + 2\theta^2 - \cos(2\theta) - 2\theta \sin(2\theta)]^{3/2}}. \tag{3}$$

See also QUADRATRIX OF HIPPIAS

References

Beyer, W. H. *CRC Standard Mathematical Tables, 28th ed.* Boca Raton, FL: CRC Press, p. 215, 1987.
Lawrence, J. D. *A Catalog of Special Plane Curves.* New York: Dover, pp. 192 and 196, 1972.
MacTutor History of Mathematics Archive. "Cochleoid." http://www-groups.dcs.st-and.ac.uk/~history/Curves/Cochleoid.html.

Cochleoid Inverse Curve

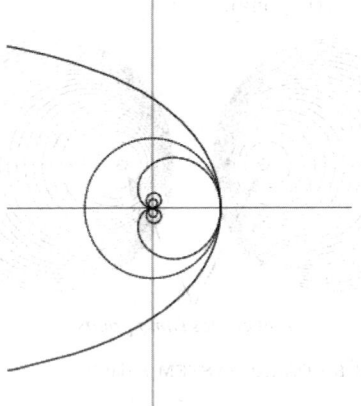

The INVERSE CURVE of the COCHLEOID

$$r = \frac{\sin \theta}{\theta} \tag{1}$$

with INVERSION CENTER at the ORIGIN and inversion

radius k is the QUADRATRIX OF HIPPIAS.

$$x = kt \cot \theta \qquad (2)$$

$$y = kt. \qquad (3)$$

Cochloid
CONCHOID OF NICOMEDES

Cochran's Theorem
The converse of FISHER'S THEOREM.

Cocked Hat Curve

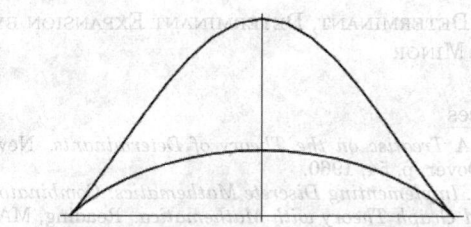

The PLANE CURVE

$$(x^2 + 2ay - a^2)^2 = y^2(a^2 - x^2),$$

which is similar to the BICORN.

References
Cundy, H. and Rollett, A. *Mathematical Models, 3rd ed.* Stradbroke, England: Tarquin Pub., p. 72, 1989.

Cocktail Party Graph

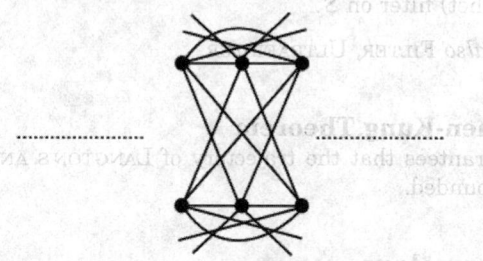

A GRAPH consisting of two rows of paired nodes in which all nodes but the paired ones are connected with an EDGE. It is the complement of the LADDER GRAPH.

See also LADDER GRAPH

Coconut
MONKEY AND COCONUT PROBLEM

Codazzi Equations
MAINARDI-CODAZZI EQUATIONS

Code
A code is a set of n-tuples of elements ("WORDS") taken from an ALPHABET.

See also ALPHABET, CODING THEORY, ENCODING, ERROR-CORRECTING CODE, GRAY CODE, HUFFMAN CODING, ISBN, LINEAR CODE, UPC, WORD

Codimension
The minimum number of parameters needed to fully describe all possible behaviors near a nonstructurally stable element.

See also BIFURCATION

Coding Theory
Coding theory, sometimes called ALGEBRAIC CODING THEORY, deals with the design of ERROR-CORRECTING CODES for the reliable transmission of information across noisy channels. It makes use of classical and modern algebraic techniques involving FINITE FIELDS, GROUP THEORY, and polynomial algebra. It has connections with other areas of DISCRETE MATHEMATICS, especially NUMBER THEORY and the theory of experimental designs.

See also ENCODING, ERROR-CORRECTING CODE, FINITE FIELD, HADAMARD MATRIX

References
Alexander, B. "At the Dawn of the Theory of Codes." *Math. Intel.* **15**, 20–6, 1993.
Berlekamp, E. R. *Algebraic Coding Theory, rev. ed.* New York: McGraw-Hill, 1968.
Golomb, S. W.; Peile, R. E.; and Scholtz, R. A. *Basic Concepts in Information Theory and Coding: The Adventures of Secret Agent 00111.* New York: Plenum, 1994.
Hill, R. *First Course in Coding Theory.* Oxford, England: Oxford University Press, 1986.
Humphreys, O. F. and Prest, M. Y. *Numbers, Groups, and Codes.* New York: Cambridge University Press, 1990.
MacWilliams, F. J. and Sloane, N. J. A. *The Theory of Error-Correcting Codes.* New York: Elsevier, 1978.
Roman, S. *Coding and Information Theory.* New York: Springer-Verlag, 1992.
Stepanov, S. A. *Codes on Algebraic Curves.* New York: Kluwer, 1999.
Vermani, L. R. *Elements of Algebraic Coding Theory.* Boca Raton, FL: CRC Press, 1996.
Weisstein, E. W. "Books about Coding Theory." http://www.treasure-troves.com/books/CodingTheory.html.

Codomain
A SET within which the values of a function lie (as opposed to the RANGE, which is the set of values that the function actually takes).

See also DOMAIN, RANGE (IMAGE)

References
Borowski, E. J. and Borwein, J. M. (Eds.). *The HarperCollins Dictionary of Mathematics.* New York: HarperCollins, p. 89, 1991.
Griffel, D. H. *Applied Functional Analysis.* New York: Wiley, p. 116, 1984.

Coefficient

A multiplicative factor (usually indexed) such as one of the constants a_i in the POLYNOMIAL $a_n x^n + a_{n-1} x^{n-1} + \ldots + a_2 x^2 + a_1 x + a_0$.

See also BINOMIAL COEFFICIENT, CARTAN TORSION COEFFICIENT, CENTRAL BINOMIAL COEFFICIENT, CLEBSCH-GORDAN COEFFICIENT, COEFFICIENT FIELD, COEFFICIENT NOTATION, COMMUTATION COEFFICIENT, CONNECTION COEFFICIENT, CORRELATION COEFFICIENT, CROSS-CORRELATION COEFFICIENT, EXCESS COEFFICIENT, GAUSSIANCOEFFICIENT, LAGRANGIAN COEFFICIENT, MULTINOMIAL COEFFICIENT, PEARSON'S SKEWNESS COEFFICIENTS, PRODUCT-MOMENT COEFFICIENT OF CORRELATION, QUARTILE SKEWNESS COEFFICIENT, QUARTILE VARIATION COEFFICIENT, RACAH V-COEFFICIENT, RACAH W-COEFFICIENT, REGRESSION COEFFICIENT, ROMAN COEFFICIENT, TRIANGLE COEFFICIENT, UNDETERMINED COEFFICIENTS METHOD, VARIATION COEFFICIENT

Coefficient Field

Let V be a VECTOR SPACE over a FIELD K, and let A be a nonempty SET. For an appropriately defined AFFINE SPACE A, K is called the coefficient field.

Coefficient Notation

Given a SERIES OF THE FORM

$$A(z) = \sum_k a_k z^k,$$

the notation $[z^k](A(z))$ is used to indicate the coefficient a_k (Sedgewick and Flajolet 1996). This corresponds to the *Mathematica* functions Coefficient$[A[z]$, z, $k]$ and SeriesCoefficient$[series, k]$.

References

Sedgewick, R. and Flajolet, P. *An Introduction to the Analysis of Algorithms*. Reading, MA: Addison-Wesley, 1996.

Coercive Functional

A bilinear FUNCTIONAL ϕ on a normed SPACE E is called coercive (or sometimes ELLIPTIC) if there exists a POSITIVE constant K such that

$$\phi(x, x) \geq K\|x\|^2$$

for all $x \in E$.

See also LAX-MILGRAM THEOREM

References

Debnath, L. and Mikusinski, P. *Introduction to Hilbert Spaces with Applications*. San Diego, CA: Academic Press, 1990.

Cofactor

The signed version C_{ij} of a MINOR M_{ij} of a MATRIX

$$C_{ij} \equiv (-1)^{i+j} M_{ij}$$

used in the computation of the matrix's DETERMINANT

$$\det(A) = \sum_i a_i C_{ij}.$$

The cofactor can be computed in *Mathematica* using

```
Cofactor[m_List,{i_Integer,j_Integer}] :=
(-1)^(i+j)Drop[Transpose[Drop[Transpose[m],
{j}]],{i}]
```

See also DETERMINANT, DETERMINANT EXPANSION BY MINORS, MINOR

References

Muir, T. *A Treatise on the Theory of Determinants*. New York: Dover, p. 54, 1960.
Skiena, S. *Implementing Discrete Mathematics: Combinatorics and Graph Theory with Mathematica*. Reading, MA: Addison-Wesley, p. 235, 1990.

Cofactor Expansion

DETERMINANT EXPANSION BY MINORS

Cofinite Filter

This entry contributed by VIKTOR BENGTSSON

If S is an infinite set, then the collection $F_S = \{A \subseteq S : S - A$ is finite$\}$ is a FILTER called the cofinite (or Fréchet) filter on S.

See also FILTER, ULTRAFILTER

Cohen-Kung Theorem

Guarantees that the trajectory of LANGTON'S ANT is unbounded.

Cohomology

Cohomology is an invariant of a TOPOLOGICAL SPACE, formally "dual" to HOMOLOGY, and so it detects "holes" in a SPACE. Cohomology has more algebraic structure than HOMOLOGY, making it into a GRADED RING (with multiplication given by the so-called "CUP PRODUCT"), whereas HOMOLOGY is just a graded ABELIAN GROUP invariant of a SPACE.

A generalized homology or cohomology theory must satisfy all of the EILENBERG-STEENROD AXIOMS with the exception of the dimension axiom.

See also ALEKSANDROV-CECH COHOMOLOGY, ALEXANDER-SPANIER COHOMOLOGY, CECH COHOMOLOGY, CUP PRODUCT, DE RHAM COHOMOLOGY, DOLBEAULT COHOMOLOGY, GRADED ALGEBRA, HOMOLOGY (TOPOLOGY)

Cohomology Class

See also INTEGRAL COHOMOLOGY CLASS

Cohomotopy Group

Cohomotopy groups are similar to HOMOTOPY GROUPS. A cohomotopy group is a GROUP related to the HOMOTOPY classes of MAPS from a SPACE X into a SPHERE $\mathbb{S}^n$.

See also HOMOTOPY GROUP

Coin

A flat disk which acts as a two-sided DIE.

See also BERNOULLI TRIAL, CARDS, COIN PARADOX, COIN TOSSING, DICE, FELLER'S COIN-TOSSING CONSTANTS, FOUR COINS PROBLEM, GAMBLER'S RUIN

References

Brooke, M. *Fun for the Money.* New York: Scribner's, 1963.

Coin Flipping

COIN TOSSING

Coin Paradox

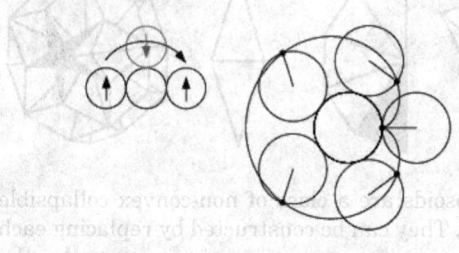

After a half rotation of the coin on the left around the central coin (of the same RADIUS), the coin undergoes a complete rotation. In other words, a coin makes two complete rotations when rolled around the boundary of an identical coin. This fact is readily apparent in the generation of the CARDIOID as one disk rolling on another.

See also CARDIOID

References

Pappas, T. "The Coin Paradox." *The Joy of Mathematics.* San Carlos, CA: Wide World Publ./Tetra, p. 220, 1989.
Steinhaus, H. *Mathematical Snapshots, 3rd ed.* New York: Dover, p. 145, 1999.

Coin Problem

Let there be $n \geq 2$ INTEGERS $0 < a_1 < \ldots < a_n$ with $(a_1, a_2, \ldots, a_n) = 1$ (all RELATIVELY PRIME). For large enough $N = \sum_{i=1}^{n} a_i x_i$, there is a solution in NONNEGATIVE INTEGERS x_i. The greatest $N = g(a_1, a_2, \ldots, a_n)$ for which there is no solution is called the coin problem. Sylvester showed

$$g(a_1, a_2) = (a_1 - 1)(a_2 - 1) - 1,$$

and an explicit solution is known for $n = 3$, but no closed form solution is known for larger N.

References

Guy, R. K. "The Money-Changing Problem." §C7 in *Unsolved Problems in Number Theory, 2nd ed.* New York: Springer-Verlag, pp. 113–14, 1994.

Coin Tossing

An idealized coin consists of a circular disk of zero thickness which, when thrown in the air and allowed to fall, will rest with either side face up ("heads" H or "tails" T) with equal probability. A coin is therefore a two-sided DIE. Despite slight differences between the sides and NONZERO thickness of actual coins, the distribution of their tosses makes a good approximation to a $p = 1/2$ BERNOULLI DISTRIBUTION.

There are, however, some rather counterintuitive properties of coin tossing. For example, it is twice as likely that the triple TTH will be encountered before THT than after it, and three times as likely that THH will precede HHT. Furthermore, it is six times as likely that HTT will be the first of HTT, TTH, and TTT to occur (Honsberger 1979). There are also strings S of Hs and Ts that have the property that the expected wait $W(S_1)$ to see string S_1 is less than the expected wait $W(S_2)$ to see S_2, but the probability of seeing S_1 before seeing S_2 is less than 1/2 (Berlekamp *et al.* 1982; Gardner 1988). Examples include

1. $THTH$ and $HTHH$, for which $W(THTH) = 20$ and $W(HTHH) = 18$, but for which the probability that $THTH$ occurs before $HTHH$ is 9/14 (Gardner 1988, p. 64),
2. $W(TTHH) = W(THHH) = 16$, $W(HHH)$, but for which the probability that $TTHH$ occurs before HHH is 7/12, and for which the probability that $THHH$ occurs before HHH is 7/8 (Penney 1969; Gardner 1988, p. 66).

More amazingly still, *spinning* a penny instead of tossing it results in heads only about 30% of the time (Paulos 1995).

The study of RUNS of two or more identical tosses is well-developed, but a detailed treatment is surprisingly complicated given the simple nature of the underlying process.

See also BERNOULLI DISTRIBUTION, BERNOULLI TRIAL, CARDS, COIN, DICE, GAMBLER'S RUIN, MARTINGALE, RUN, SAINT PETERSBURG PARADOX

References

Berlekamp, E. R.; Conway, J. H; and Guy, R. K. *Winning Ways for Your Mathematical Plays, Vol. 1: Games in General.* London: Academic Press, p. 777, 1982.

Ford, J. "How Random is a Coin Toss?" *Physics Today* **36**, 40–7, 1983.

Gardner, M. "Nontransitive Paradoxes." *Time Travel and Other Mathematical Bewilderments.* New York: W. H. Freeman, pp. 64–6, 1988.

Honsberger, R. "Some Surprises in Probability." Ch. 5 in *Mathematical Plums* (Ed. R. Honsberger). Washington, DC: Math. Assoc. Amer., pp. 100–03, 1979.

Keller, J. B. "The Probability of Heads." *Amer. Math. Monthly* **93**, 191–97, 1986.

Paulos, J. A. *A Mathematician Reads the Newspaper.* New York: BasicBooks, p. 75, 1995.

Peterson, I. *Islands of Truth: A Mathematical Mystery Cruise.* New York: W. H. Freeman, pp. 238–39, 1990.

Penney, W. "Problem 95. Penney-Ante." *J. Recr. Math.* **2**, 241, 1969.

Sloane, N. J. A. Sequences A000225/M2655 and A050227 in "An On-Line Version of the Encyclopedia of Integer Sequences." http://www.research.att.com/~njas/sequences/eisonline.html.

Spencer, J. "Combinatorics by Coin Flipping." *Coll. Math. J.*, **17**, 407–12, 1986.

Whittaker, E. T. and Robinson, G. "The Frequency Distribution of Tosses of a Coin." §90 in *The Calculus of Observations: A Treatise on Numerical Mathematics, 4th ed.* New York: Dover, pp. 176–77, 1967.

Coincidence

A coincidence is a surprising concurrence of events, perceived as meaningfully related, with no apparent causal connection (Diaconis and Mosteller 1989). Given a large number events, extremely unlikely coincidences are possible–and perhaps even common. To quote Sherlock Holmes, "Amid the action and reaction of so dense a swarm of humanity, every possible combination of events may be expected to take place, and many a little problem will be presented which may be striking and bizarre..." (Conan Doyle 1988, p. 245).

See also BIRTHDAY PROBLEM, LAW OF TRULY LARGE NUMBERS, ODDS, PROBABILITY, RANDOM NUMBER, SIGNIFICANCE

References

Bogomolny, A. "Coincidence." http://www.cut-the-knot.com/do_you_know/coincidence.html.

Conan Doyle, A. "The Adventure of the Blue Carbuncle." In *The Complete Sherlock Holmes.* New York: Doubleday, pp. 244–57, 1988.

Falk, R. "On Coincidences." *Skeptical Inquirer* **6**, 18–1, 1981–2.

Falk, R. "The Judgment of Coincidences: Mine Versus Yours." *Amer. J. Psych.* **102**, 477–93, 1989.

Falk, R. and MacGregor, D. "The Surprisingness of Coincidences." In *Analysing and Aiding Decision Processes* (Ed. P. Humphreys, O. Svenson, and A. Vári). New York: Elsevier, pp. 489–02, 1984.

Diaconis, P. and Mosteller, F. "Methods of Studying Coincidences." *J. Amer. Statist. Assoc.* **84**, 853–61, 1989.

Jung, C. G. *Synchronicity: An Acausal Connecting Principle.* Princeton, NJ: Princeton University Press, 1973.

Kammerer, P. *Das Gesetz der Serie: Eine Lehre von den Wiederholungen im Lebens--und im Weltgeschehen.* Stuttgart, Germany: Deutsche Verlags-Anstahlt, 1919.

Stewart, I. "What a Coincidence!" *Sci. Amer.* **278**, 95–6, June 1998.

Coincident

Two LINES or plane CONGRUENT geometric figures which lie on top of each other are said to be coincident.

See also CONGRUENT, HOMOTHETIC, SIMILAR

Colatitude

The polar angle on a SPHERE measured from the North Pole instead of the equator. The angle ϕ in SPHERICAL COORDINATES is the COLATITUDE. It is related to the LATITUDE δ by $\phi = 90° - \delta$.

See also LATITUDE, LONGITUDE, SPHERICAL COORDINATES

Colinear

COLLINEAR

Collapsoid

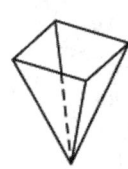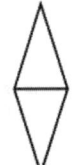

The collapsoids are a class of non-convex collapsible polyhedra. They can be constructed by replacing each edge of a DODECAHEDRON or ICOSAHEDRON by the diagonal of a pyramid (with base removed). Thirty such pyramids are then fitted together using tabs.

References

Pedersen, J. "Collapsoids." *Math. Gaz.* **59**, 81–4, 1975.

Wells, D. *The Penguin Dictionary of Curious and Interesting Geometry.* London: Penguin, p. 34, 1991.

Collatz Problem

A problem posed by L. Collatz in 1937, also called the $3X+1$ MAPPING, HASSE'S ALGORITHM, KAKUTANI'S PROBLEM, SYRACUSE ALGORITHM, SYRACUSE PROBLEM, THWAITES CONJECTURE, and ULAM'S PROBLEM (Lagarias 1985). Thwaites (1996) has offered a £1000 reward for resolving the CONJECTURE. Let a_0 be an INTEGER. Then the Collatz problem asks if iterating

$$a_n = \begin{cases} \frac{1}{2}a_{n-1} & \text{for } a_{n-1} \text{ even} \\ 3a_{n-1} + 1 & \text{for } a_{n-1} \text{ odd} \end{cases} \quad (1)$$

always returns to 1 for POSITIVE a_0. This question has been tested and found to be true for all numbers $\leq 3 \cdot 2^{53} \approx 2.702 \times 10^{16}$ (Oliveira e Silva 1999), improving the earlier results of 10^{15} (Vardi 1991, p. 129) and 5.6×10^{13} (Leavens and Vermeulen 1992). The members of the SEQUENCE produced by the Collatz are sometimes known as HAILSTONE NUMBERS. Because of the difficulty in solving this problem, Erdos commented that "mathematics is not yet ready for such problems" (Lagarias 1985). If NEGATIVE numbers are included, there are four known cycles (excluding the trivial 0 cycle): $(4, 2, 1)$, $(-2, -1)$, $(-5, -7, -10)$, and $(-17, -25, -37, -55, -82, -41, -61, -91, -136, -68, -34)$. The number of tripling steps needed to reach 1 for $n = 1, 2, \ldots$ are 0, 0, 2, 0, 1, 2, 5, 0, 6, ... (Sloane's A006667).

The Collatz problem was modified by Terras (1976, 1979), who asked if iterating

$$t_n = \begin{cases} \frac{1}{2} t_{n-1} & \text{for } t_{n-1} \text{ even} \\ \frac{1}{2}(3t_{n-1} + 1) & \text{for } t_{n-1} \text{ odd} \end{cases} \quad (2)$$

always returns to 1 for initial integer value t_0. If NEGATIVE numbers are included, there are 4 known cycles: $(1, 2)$, (-1), $(-5, -7, -10)$, and $(-17, -25, -37, -55, -82, -41, -61, -91, -136, -68, -34)$. It is a special case of the "generalized Collatz problem" with $d = 2$, $m_0 = 1$, $m_1 = 3$, $r_0 = 0$, and $r_1 = -1$. Terras (1976, 1979) also proved that the set of INTEGERS $S_k \equiv \{n : n \text{ has stopping time} \leq k\}$ has a limiting asymptotic density $F(k)$, such that if $N_x(k)$ is the number of n such that $n \leq x$ and $\sigma(n) \leq k$, then the limit

$$F(k) = \lim_{x \to \infty} \frac{N_x(k)}{x}, \quad (3)$$

exists. Furthermore, $F(k) \to 1$ as $k \to \infty$, so almost all INTEGERS have a finite stopping time. Finally, for all $k \geq 1$,

$$1 - F(k) = \lim_{x \to \infty} \frac{N_x(k)}{x} \leq 2^{-nk}, \quad (4)$$

where

$$H(x) = -x \lg x - (1 - x) \lg(1 - x) \quad (5)$$

$$\theta = \frac{1}{\lg 3} \quad (6)$$

$$\eta = 1 - H(\theta) = 0.05004\ldots \quad (7)$$

(Lagarias 1985).

Conway proved that the original Collatz problem has no nontrivial cycles of length < 400. Lagarias (1985) showed that there are no nontrivial cycles with length $< 275,000$. Conway (1972) also proved that Collatz-type problems can be formally UNDECIDABLE.

A generalization of the COLLATZ PROBLEM lets $d \geq 2$ be a POSITIVE INTEGER and $m_0, \ldots, m_{d-1}$ be NONZERO INTEGERS. Also let $r_i \in \mathbb{Z}$ satisfy

$$r_i \equiv i m_i \pmod{d}. \quad (8)$$

Then

$$T(x) = \frac{m_i x - r_i}{d} \quad (9)$$

for $x \equiv i \pmod{d}$ defines a generalized Collatz mapping. An equivalent form is

$$T(x) = \left\lfloor \frac{m_i x}{d} \right\rfloor + X_i \quad (10)$$

for $x \equiv i \pmod{d}$ where $X_0, \ldots, X_{d-1}$ are INTEGERS and $\lfloor r \rfloor$ is the FLOOR FUNCTION. The problem is connected with ERGODIC THEORY and MARKOV CHAINS (Matthews 1995). Matthews (1995) obtained the following table for the mapping

$$T_k(x) = \begin{cases} \frac{1}{2} x & \text{for } x \equiv 0 \pmod 2 \\ \frac{1}{2}(3x + k) & \text{for } x \equiv 1 \pmod 2, \end{cases} \quad (11)$$

where $k = T_{5^k}$.

k	# Cycles	Max. Cycle Length
0	5	27
1	10	34
2	13	118
3	17	118
4	19	118
5	21	165
6	23	433

Matthews and Watts (1984) proposed the following conjectures.

1. If $|m_0 \cdots m_{d-1}| < d^d$, then all trajectories $\{T^K(n)\}$ for $n \in \mathbb{Z}$ eventually cycle.
2. If $|m_0 \cdots m_{d-1}| > d^d$, then almost all trajectories $\{T^K(n)\}$ for $n \in \mathbb{Z}$ are divergent, except for an exceptional set of INTEGERS n satisfying

$$\#\{n \in S | -X \leq n < X\} = o(X).$$

3. The number of cycles is finite.
4. If the trajectory $\{T^K(n)\}$ for $n \in \mathbb{Z}$ is not eventually cyclic, then the iterates are uniformly distribution mod d^α for each $\alpha \geq 1$, with

$$\lim_{N \to \infty} \frac{1}{N + 1} \text{ card}\{K \leq N | T^K(n) \equiv j \pmod{d^\alpha}\}$$

$$= d^{-\alpha} \qquad (12)$$

for $0 \leq j \leq d^{\alpha} - 1$.

Matthews believes that the map

$$T(x) = \begin{cases} 7x + 3 & \text{for } x \equiv 0 \ (\text{mod } 3) \\ \frac{1}{3}(7x + 2) & \text{for } x \equiv 1 \ (\text{mod } 3) \\ \frac{1}{3}(x - 2) & \text{for } x \equiv 2 \ (\text{mod } 3) \end{cases} \qquad (13)$$

will either reach 0 (mod 3) or will enter one of the cycles (-1) or $(-2, -4)$, and offers a $100 (Australian?) prize for a proof.

See also HAILSTONE NUMBER

References

Applegate, D. and Lagarias, J. C. "Density Bounds for the $3x + 1$ Problem 1. Tree-Search Method." *Math. Comput.* **64**, 411–26, 1995.

Applegate, D. and Lagarias, J. C. "Density Bounds for the $3x + 1$ Problem 2. Krasikov Inequalities." *Math. Comput.* **64**, 427–38, 1995.

Burckel, S. "Functional Equations Associated with Congruential Functions." *Theor. Comp. Sci.* **123**, 397–06, 1994.

Conway, J. H. "Unpredictable Iterations." *Proc. 1972 Number Th. Conf.*, University of Colorado, Boulder, Colorado, pp. 49–2, 1972.

Crandall, R. "On the '$3x + 1$' Problem." *Math. Comput.* **32**, 1281–292, 1978.

Everett, C. "Iteration of the Number Theoretic Function $f(2n) = n$, $f(2n + 1) = f(3n + 2)$." *Adv. Math.* **25**, 42–5, 1977.

Guy, R. K. "Collatz's Sequence." §E16 in *Unsolved Problems in Number Theory, 2nd ed.* New York: Springer-Verlag, pp. 215–18, 1994.

Lagarias, J. C. "The $3x + 1$ Problem and Its Generalizations." *Amer. Math. Monthly* **92**, 3–3, 1985. http://www.cecm.sfu.ca/organics/papers/lagarias/.

Leavens, G. T. and Vermeulen, M. "$3x + 1$ Search Programs." *Comput. Math. Appl.* **24**, 79–9, 1992.

Margenstern, M. and Matiyasevich, Y. "A Binomial Representation of the $3x + 1$ Problem." *Acta Arith.* **91**, 367–78, 1999.

Matthews, K. R. "The Generalized $3x + 1$ Mapping." http://www.maths.uq.oz.au/~krm/survey.ps. Rev. Mar. 30, 1999.

Matthews, K. R. and Watts, A. M. "A Generalization of Hasses's Generalization of the Syracuse Algorithm." *Acta Arith.* **43**, 167–75, 1984.

Oliveira e Silva, T. "Maximum Excursion and Stopping Time Record-Holders for the $3x + 1$ Problem: Computational Results." *Math. Comput.* **68**, 371–84, 1999.

Schroeppel, R.; Gosper, R. W.; Henneman, W.; and Banks, R. Item 133 in Beeler, M.; Gosper, R. W.; and Schroeppel, R. *HAKMEM.* Cambridge, MA: MIT Artificial Intelligence Laboratory, Memo AIM-239, p. 64, Feb. 1972.

Sloane, N. J. A. Sequences A006667/M0019 in "An On-Line Version of the Encyclopedia of Integer Sequences." http://www.research.att.com/~njas/sequences/eisonline.html.

Terras, R. "A Stopping Time Problem on the Positive Integers." *Acta Arith.* **30**, 241–52, 1976.

Terras, R. "On the Existence of a Density." *Acta Arith.* **35**, 101–02, 1979.

Thwaites, B. "Two Conjectures, or How to Win £1100." *Math.Gaz.* **80**, 35–6, 1996.

Vardi, I. "The $3x + 1$ Problem." Ch. 7 in *Computational Recreations in Mathematica.* Redwood City, CA: Addison-Wesley, pp. 129–37, 1991.

Collinear

Three or more points P_1, P_2, P_3, ..., are said to be collinear if they lie on a single straight LINE L. A line on which points lie, especially if it is related to a geometric figure such as a TRIANGLE, is sometimes called an AXIS. Three points are collinear IFF the ratios of distances satisfy

$$x_2 - x_1 : y_2 - y_1 : z_2 - z_1 = x_3 - x_1 : y_3 - y_1 : z_3 - z_1.$$

Two points are trivially collinear since two points determine a LINE.

Let points P_1, P_2, and P_3 lie, one each, on the sides of a triangle $\Delta A_1 A_2 A_3$ or their extensions, and reflect these points about the midpoints of the triangle sides to obtain P_1', P_2', and P_3'. Then P_1', P_2', and P_3' are collinear IFF P_1, P_2, and P_3 are (Honsberger 1995).

See also AXIS, CONCYCLIC, CONFIGURATION, DIRECTED ANGLE, DROZ-FARNY THEOREM, GENERAL POSITION, LINE, N-CLUSTER, SYLVESTER'S LINE PROBLEM

References

Coxeter, H. S. M. and Greitzer, S. L. "Collinearity and Concurrence." Ch. 3 in *Geometry Revisited.* Washington, DC: Math. Assoc. Amer., pp. 51–9, 1967.

Honsberger, R. *Episodes in Nineteenth and Twentieth Century Euclidean Geometry.* Washington, DC: Math. Assoc. Amer., pp. 153–54, 1995.

Collineation

A transformation of the plane which transforms COLLINEAR points into COLLINEAR points. A projective collineation transforms every 1-D form projectively, and a perspective collineation is a collineation which leaves all lines through a point and points through a line invariant. In an ELATION, the center and axis are incident; in a HOMOLOGY they are not. For further discussion, see Coxeter (1969, p. 248).

See also AFFINITY, CORRELATION, ELATION, EQUIAFFINITY, HOMOLOGY (GEOMETRY), PERSPECTIVE COLLINEATION, PROJECTIVE COLLINEATION

References

Coxeter, H. S. M. "Collineations and Correlations." §14.6 in *Introduction to Geometry, 2nd ed.* New York: Wiley, pp. 247–51, 1969.

Collision-Free Hash Function

A function H that maps an arbitrary length message M to a fixed length message digest MD is a collision-free hash function if

1. It is a ONE-WAY HASH FUNCTION.

2. It is hard to find two distinct messages (M', M) that hash to the same result $H(M') = H(M)$. More

precisely, any efficient algorithm (solving a P-PROBLEM) succeeds in finding such a collision with negligible probability (Russell 1992).

See also HASH FUNCTION

References

Bakhtiari, S.; Safavi-Naini, R.; and Pieprzyk, J. *Cryptographic Hash Functions: A Survey.* Technical Report 95–9, Department of Computer Science, University of Wollongong, July 1995. ftp://ftp.cs.uow.edu.au/pub/papers/1995/tr-95-9.ps.Z.
Russell, A. "Necessary and Sufficient Conditions for Collision-Free Hashing." In *Abstracts of Crypto 92.* pp. 10–2–0–7. ftp://theory.lcs.mit.edu/pub/people/acr/hash.ps.

Collocation Method

A method of determining coefficients α_l in an expansion

$$y(x) = y_0(x) + \sum_{l=1}^{q} \alpha_l y_l(x)$$

so as to nullify the values of an ORDINARY DIFFERENTIAL EQUATION $L[y(x)] = 0$ at prescribed points.

References

Itô, K. (Ed.). "Methods Other than Difference Methods." §303I in *Encyclopedic Dictionary of Mathematics, 2nd ed., Vol. 2.* Cambridge, MA: MIT Press, p. 1139, 1980.

Cologarithm

The LOGARITHM of the RECIPROCAL of a number, equal to the NEGATIVE of the LOGARITHM of the number itself,

$$\operatorname{colog} x \equiv \log\left(\frac{1}{x}\right) = -\log x.$$

See also ANTILOGARITHM, LOGARITHM

Colon Product

Let **AB** and **CD** be DYADS. Their colon product is defined by

$$\mathbf{AB} : \mathbf{CD} \equiv \mathbf{C} \cdot \mathbf{AB} \cdot \mathbf{D} = (\mathbf{A} \cdot \mathbf{C})(\mathbf{B} \cdot \mathbf{D}).$$

See also DYAD

Colorable

Color each segment of a KNOT DIAGRAM using one of three colors. If

1. At any crossing, either the colors are all different or all the same, and
2. At least two colors are used,

then a KNOT is said to be colorable (or more specifically, THREE-COLORABLE). Colorability is invariant under REIDEMEISTER MOVES, and can be generalized. For instance, for five colors 0, 1, 2, 3, and 4, a KNOT is five-colorable if

1. at any crossing, three segments meet. If the overpass is numbered a and the two underpasses B and C, then $2a \equiv b + c \pmod{5}$, and
2. at least two colors are used.

Colorability cannot always distinguish HANDEDNESS. For instance, three-colorability can distinguish the mirror images of the TREFOIL KNOT but not the FIGURE-OF-EIGHT KNOT. Five-colorability, on the other hand, distinguishes the MIRROR IMAGES of the FIGURE-OF-EIGHT KNOT but not the TREFOIL KNOT.

See also COLORING, WORM

Coloring

A coloring of plane regions, LINK segments, etc., is an assignment of a distinct labeling (which could be a number, letter, color, etc.) to each component. Coloring problems generally involve TOPOLOGICAL considerations (i.e., they depend on the abstract study of the arrangement of objects), and theorems about colorings, such as the famous FOUR-COLOR THEOREM, can be extremely difficult to prove.

See also COLORABLE, EDGE COLORING, FOUR-COLOR THEOREM, κ-COLORING, LOVÁSZ NUMBER, POLYHEDRON COLORING, SIX-COLOR THEOREM, THREE-COLORABLE, VERTEX COLORING

References

Eppstein, D. "Coloring." http://www.ics.uci.edu/~eppstein/junkyard/color.html.
Saaty, T. L. and Kainen, P. C. *The Four-Color Problem: Assaults and Conquest.* New York: Dover, 1986.

Columbian Number

SELF NUMBER

Column Space

See also ROW SPACE

Column Vector

An $m \times 1$ MATRIX

$$\begin{bmatrix} a_{11} \\ a_{21} \\ \vdots \\ a_{m1} \end{bmatrix}.$$

See also MATRIX, ROW VECTOR, VECTOR

Column-Convex Polyomino

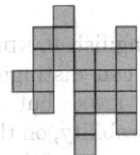

column-convex   not column-convex

A column-convex polyomino is a self-avoiding CONVEX POLYOMINO such that the intersection of any vertical line with the polyomino has at most two connected components. Column-convex polyominos are also called vertically convex polyominoes. A ROW-CONVEX POLYOMINO is similarly defined. The number $a(n)$ of column-convex n-polyominoes are given by the third-order RECURRENCE RELATION

$$a(n) = 5a(n-1) - 7a(n-2) + 4a(n-3)$$

with $a(1) = 1$, $a(2) = 2$, $a(3) = 6$, and $a(4) = 19$ (Hickerson 1999). The first few are 1, 2, 6, 19, 61, 196, 629, 2017, ... (Sloane's A001169). $a(n)$ has GENERATING FUNCTION

$$f(x)\frac{x(1-x)^3}{1 - 5x + 7x^2 - 4x^3} = x + 2x^2 + 6x^3 + 19x^4 + \ldots.$$

See also CONVEX POLYOMINO, POLYOMINO, ROW-CONVEX POLYOMINO

References

Enting, I. G. and Guttmann, A. J. "On the Area of Square Lattice Polygons." *J. Statist. Phys.* **58**, 475–84, 1990. *Phys. Rev. Ser. 2* **103**, 1–6, 1956.
Hickerson, D.. "Counting Horizontally Convex Polyominoes." *J. Integer Sequences* **2**, No. 99.1.8, 1999. http://www.research.att.com/~njas/sequences/JIS/HICK2/chcp.html.
Klarner, D. A. "Some Results Concerning Polyominoes." *Fib. Quart.* **3**, 9–0, 1965.
Klarner, D. A. "Cell Growth Problems." *Canad. J. Math.* **19**, 851–63, 1967.
Klarner, D. A. "The Number of Graded Partially Ordered Sets." *J. Combin. Th.* **6**, 12–9, 1969.
Lunnon, W. F. "Counting Polyominoes." In *Computers in Number Theory, Proc. Science Research Council Atlas Symposium No. 2 held at Oxford, from 18–3 August, 1969* (Ed. A. O. L. Atkin and B. J. Birch). London: Academic Press, pp. 347–72, 1971.

Pólya, G. "On the Number of Certain Lattice Polygons." *J. Combin. Th.* **6**, 102–05, 1969.
Sloane, N. J. A. Sequences A001169/M1636 in "An On-Line Version of the Encyclopedia of Integer Sequences." http://www.research.att.com/~njas/sequences/eisonline.html.
Stanley, R. P. "Generating Functions." In *Studies in Combinatorics* (Ed. G.-C. Rota). Washington, DC: Amer. Math. Soc., pp. 100–41, 1978.
Stanley, R. P. *Enumerative Combinatorics, Vol. 1.* Cambridge, England: Cambridge University Press, p. 259, 1999.
Temperley, H. N. V. "Combinatorial Problems Suggested By the Statistical Mechanics of Domains and of Rubber-Like Molecules."

Colunar Triangle

Given a SCHWARZ TRIANGLE (pqr), replacing each VERTEX with its antipodes gives the three colunar SPHERICAL TRIANGLES

$$(pq'r'), \ (p'qr'), \ (p'q'r)$$

where

$$\frac{1}{p} + \frac{1}{p'} = 1$$

$$\frac{1}{q} + \frac{1}{q'} = 1$$

$$\frac{1}{r} + \frac{1}{r'} = 1.$$

See also SCHWARZ TRIANGLE, SPHERICAL TRIANGLE

References

Coxeter, H. S. M. *Regular Polytopes, 3rd ed.* New York: Dover, p. 112, 1973.

Comass

The comass of a DIFFERENTIAL P-FORM ϕ is the largest value of ϕ on a p vector of p-volume one,

$$\sup_{v \in \Lambda^p TM, |v| = 1} |\phi(v)|.$$

See also CALIBRATION FORM

Comb Function

SHAH FUNCTION

Combination

The number of ways of picking k *unordered* outcomes from n possibilities. Also known as the BINOMIAL COEFFICIENT or CHOICE NUMBER and read "n choose r."

$${}_nC_k \equiv \binom{n}{k} \equiv \frac{n!}{k!(n-k)!},$$

where $n!$ is a FACTORIAL (Uspensky 1937, p. 18). For example, there are $\binom{4}{2} = 6$ combinations on $\{1, 2, 3, 4\}$, namely $\{1, 2\}$, $\{1, 3\}$, $\{1, 4\}$, $\{2, 3\}$, $\{2, 4\}$, and $\{3, 4\}$. These combinations are known as K-SUBSETS.

Muir (1960, p. 7) uses the nonstandard notations $(n)_k = \binom{n}{k}$ and $(\bar{n})_k = \binom{n-k}{k}$.

See also BINOMIAL COEFFICIENT, DERANGEMENT, FACTORIAL, K-SUBSET, PERMUTATION, SUBFACTORIAL

References

Conway, J. H. and Guy, R. K. "Choice Numbers." In *The Book of Numbers*. New York: Springer-Verlag, pp. 67–8, 1996.

Muir, T. *A Treatise on the Theory of Determinants*. New York: Dover, 1960.

Ruskey, F. "Information on Combinations of a Set." http://www.theory.csc.uvic.ca/~cos/inf/comb/CombinationsInfo.html.

Skiena, S. "Combinations." §1.5 in *Implementing Discrete Mathematics: Combinatorics and Graph Theory with Mathematica*. Reading, MA: Addison-Wesley, pp. 40–6, 1990.

Uspensky, J. V. *Introduction to Mathematical Probability*. New York: McGraw-Hill, p. 18, 1937.

Combination Lock

Let a combination of n buttons be a SEQUENCE of disjoint nonempty SUBSETS of the SET $\{1, 2, \ldots, n\}$. If the number of possible combinations is denoted a_n, then a_n satisfies the RECURRENCE RELATION

$$a_n = \sum_{i=0}^{n-1} \binom{n}{n-i} a_i, \tag{1}$$

with $a_0 = 1$. This can also be written

$$a_n = \frac{d^n}{dx^n} \left(\frac{1}{2 - e^x} \right) \Bigg|_{x=0} = \frac{1}{2} \sum_{k=0}^{\infty} \frac{k^n}{2^k}, \tag{2}$$

where the definition $0^0 = 1$ has been used. Furthermore,

$$a_n \sum_{k=1}^{n} A_{n,k} 2^{n-k} = \sum_{k=1}^{n} A_{n,k} 2^{k-1}, \tag{3}$$

where $A_{n,k}$ are EULERIAN NUMBERS. In terms of the STIRLING NUMBERS OF THE SECOND KIND $s(n, k)$,

$$a_n = \sum_{k=1}^{n} k! s(n, k). \tag{4}$$

a_n can also be given in closed form as

$$a_n = \frac{1}{2} \mathrm{Li}_{-n}(\tfrac{1}{2}), \tag{5}$$

where $\mathrm{Li}_n(z)$ is the POLYLOGARITHM. The first few values of a_n for $n = 1, 2, \ldots$ are 1, 3, 13, 75, 541, 4683, 47293, 545835, 7087261, 102247563, ... (Sloane's A000670).

The quantity

$$b_n \equiv \frac{a_n}{n!} \tag{6}$$

satisfies the inequality

$$\frac{1}{2(\ln 2)^n} \le b_n \le \frac{1}{(\ln 2)^n}. \tag{7}$$

References

Sloane, N. J. A. Sequences A000670/M2952 in "An On-Line Version of the Encyclopedia of Integer Sequences." http://www.research.att.com/~njas/sequences/eisonline.html.

Velleman, D. J. and Call, G. S. "Permutations and Combination Locks." *Math. Mag.* **68**, 243–53, 1995.

Combinatorial Composition

COMPOSITION

Combinatorial Design

References

Colbourn, C. J. and Dinitz, J. H. *CRC Handbook of Combinatorial Designs*. Boca Raton, FL: CRC Press, 1996.

Lindner, C. C. and Rodger, C. A. *Design Theory*. Boca Raton, FL: CRC Press, 1997.

Combinatorial Dual Graph

Let $m(G)$ be the cycle rank of a graph G, $m^*(G)$ be the cocycle rank, and the relative complement $G - H$ of a SUBGRAPH H of G be defined as that subgraph obtained by deleting the lines of H. Then a graph G^* is a combinatorial dual of G if there is a one-to-one correspondence between their sets of lines such that for any choice Y and Y^* of corresponding subsets of lines,

$$M^*(G - Y) = m^*(G) - m(\langle Y^* \rangle),$$

where $\langle Y^* \rangle$ is the subgraph of G^* with the line set Y^*.

Whitney showed that the GEOMETRIC DUAL GRAPH and combinatorial dual graph are equivalent (Harary 1994, p. 115), and so may simply be called "the" DUAL GRAPH. Also, a graph is PLANAR IFF it has a combinatorial dual (Harary 1994, p. 115).

See also DUAL GRAPH, GEOMETRIC DUAL GRAPH, PLANAR GRAPH

References

Harary, F. *Graph Theory*. Reading, MA: Addison-Wesley, pp. 113–15, 1994.

Combinatorial Geometry

See also MATROID

References

Friedman, E. "Erich's Combinatorial Geometry Page." http://www.stetson.edu/~efriedma/comb.html.

Pach, J. and Agarwal, P. K. *Combinatorial Geometry.* New York: Wiley, 1995.

Combinatorial Number

BINOMIAL COEFFICIENT

Combinatorial Optimization

References

Ausiello, G.; Crescenzi, P.; Gambois, G.; Kann, V.; Marchetti-Spaccamela, A.; and Protasi, M. *Complexity and Approximation: Combinatorial Optimization Problems and Their Approximability Properties.* Berlin: Springer-Verlag, 1999.

Du, D.-Z. and Pardalos, P. M. (Eds.). *Handbook of Combinatorial Optimization, Vols. 1–.* Amsterdam, Netherlands: Kluwer, 1998.

Combinatorial Species

SPECIES

Combinatorial Topology

Combinatorial topology is a special type of ALGEBRAIC TOPOLOGY that uses COMBINATORIAL methods. For example, SIMPLICIAL HOMOLOGY is a combinatorial construction in ALGEBRAIC TOPOLOGY, so it belongs to combinatorial topology.

See also ALGEBRAIC TOPOLOGY, SIMPLICIAL HOMOLOGY, TOPOLOGY

References

Alexandrov, P. S. *Combinatorial Topology.* New York: Dover, 1998.

Pontryagin, L. S. *Foundations of Combinatorial Topology.* New York: Dover, 1999.

Combinatorics

The branch of mathematics studying the enumeration, combination, and permutation of sets of elements and the mathematical relations which characterize these properties.

See also ALGEBRAIC COMBINATORICS, ANTICHAIN, CHAIN, DILWORTH'S LEMMA, DIVERSITY CONDITION, ENUMERATION PROBLEM, ERDOS-SZEKERES THEOREM, INCLUSION-EXCLUSION PRINCIPLE, KIRKMAN'S SCHOOLGIRL PROBLEM, KIRKMAN TRIPLE SYSTEM, LENGTH (PARTIAL ORDER), PARTIAL ORDER, PIGEONHOLE PRINCIPLE, RAMSEY'S THEOREM, SCHRÖDER-BERNSTEIN THEOREM, SCHUR'S LEMMA, SPERNER'S THEOREM, TOTAL ORDER, UMBRAL CALCULUS, VAN DER WAERDEN'S THEOREM, WIDTH (PARTIAL ORDER)

References

Abramowitz, M. and Stegun, C. A. (Eds.). "Combinatorial Analysis." Ch. 24 in *Handbook of Mathematical Functions with Formulas, Graphs, and Mathematical Tables, 9th printing.* New York: Dover, pp. 821–827, 1972.

Aigner, M. *Combinatorial Theory.* New York: Springer-Verlag, 1997.

Bellman, R. and Hall, M. *Combinatorial Analysis.* Amer. Math. Soc., 1979.

Berge, C. *Principles of Combinatorics.* New York: Academic Press, 1971.

Bergeron, F.; Labelle, G.; and Leroux, P. *Combinatorial Species and Tree-Like Structures.* Cambridge, England: Cambridge University Press, 1998.

Biggs, N. L. "The Roots of Combinatorics." *Historia Mathematica* **6**, 109–36, 1979.

Bose, R. C. and Manvel, B. *Introduction to Combinatorial Theory.* New York: Wiley, 1984.

Brown, K. S. "Combinatorics." http://www.seanet.com/~ksbrown/icombina.htm.

Cameron, P. J. *Combinatorics: Topics, Techniques, Algorithms.* New York: Cambridge University Press, 1994.

Cohen, D. *Basic Techniques of Combinatorial Theory.* New York: Wiley, 1978.

Cohen, D. E. *Combinatorial Group Theory: A Topological Approach.* New York: Cambridge University Press, 1989.

Colbourn, C. J. and Dinitz, J. H. *CRC Handbook of Combinatorial Designs.* Boca Raton, FL: CRC Press, 1996.

Comtet, L. *Advanced Combinatorics: The Art of Finite and Infinite Expansions, rev. enl. ed.* Dordrecht, Netherlands: Reidel, 1974.

Dinitz, J. H. and Stinson, D. R. (Eds.). *Contemporary Design Theory: A Collection of Surveys.* New York: Wiley, 1992.

Eisen, M. *Elementary Combinatorial Analysis.* New York: Gordon and Breach, 1969.

Electronic Journal of Combinatorics. http://www.combinatorics.org/previous_volumes.html.

Eppstein, D. "Combinatorial Geometry." http://www.ics.uci.edu/~eppstein/junkyard/combinatorial.html.

Erdos, P. and Spencer, J. *Probabilistic Methods in Combinatorics.* New York: Academic Press, 1974.

Erickson, M. J. *Introduction to Combinatorics.* New York: Wiley, 1996.

Fields, J. "On-Line Dictionary of Combinatorics." http://www.math.uic.edu/~fields/comb_dic/.

Gardner, M. "Combinatorial Theory." Ch. 3 in *The Sixth Book of Mathematical Games from Scientific American.* Chicago, IL: University of Chicago Press, pp. 19–8, 1984.

Godsil, C. D. "Problems in Algebraic Combinatorics." *Electronic J. Combinatorics* **2**, F1 1–0, 1995. http://www.combinatorics.org/Volume_2/volume2.html#F1.

Graham, R. L.; Grötschel, M.; and Lovász, L. (Eds.). *Handbook of Combinatorics, 2 vols.* Cambridge, MA: MIT Press, 1996.

Graham, R. L.; Knuth, D. E.; and Patashnik, O. *Concrete Mathematics: A Foundation for Computer Science, 2nd ed.* Reading, MA: Addison-Wesley, 1994.

Grimaldi, R. P. *Discrete and Combinatorial Mathematics: An Applied Introduction, 4th ed.* Longman, 1998.

Hall, M. Jr. *Combinatorial Theory, 2nd ed.* New York: Wiley, 1986.

Harary, F. *Applied Combinatorial Mathematics.* New York: Wiley, 1964.

Knuth, D. E. (Ed.). *Stable Marriage and Its Relation to Other Combinatorial Problems.* Providence, RI: Amer. Math. Soc., 1997.

Kreher, D. L. and Stinson, D. *Combinatorial Algorithms: Generation, Enumeration, and Search.* Boca Raton, FL: CRC Press, 1999.

Kucera, L. *Combinatorial Algorithms.* Bristol, England: Adam Hilger, 1989.

Liu, C. L. *Introduction to Combinatorial Mathematics.* New York: McGraw-Hill, 1968.

MacMahon, P. A. *Combinatory Analysis, 2 vols.* New York: Chelsea, 1960.

Marcus, D. *Combinatorics: A Problem Oriented Approach.* Washington, DC: Math. Assoc. Amer., 1998.

Nijenhuis, A. and Wilf, H. *Combinatorial Algorithms for Computers and Calculators, 2nd ed.* New York: Academic Press, 1978.

Petit, S. "Encyclopedia of Combinatorial Structures." http://algo.inria.fr/encyclopedia/.

Raghavarao, D. *Constructions and Combinatorial Problems in Design of Experiments.* New York: Dover, 1988.

Riordan, J. *Combinatorial Identities, reprint ed. with corrections.* Huntington, NY: Krieger, 1979.

Riordan, J. *An Introduction to Combinatorial Analysis.* New York: Wiley, 1980.

Roberts, F. S. *Applied Combinatorics.* Englewood Cliffs, NJ: Prentice-Hall, 1984.

Rosen, K. H. (Ed.). *Handbook of Discrete and Combinatorial Mathematics.* Boca Raton, FL: CRC Press, 2000.

Rota, G.-C. (Ed.). *Studies in Combinatorics.* Providence, RI: Math. Assoc. Amer., 1978.

Ruskey, F. "The (Combinatorial) Object Server." http://www.theory.csc.uvic.ca/~cos/.

Ryser, H. J. *Combinatorial Mathematics.* Buffalo, NY: Math. Assoc. Amer., 1963.

Skiena, S. *Implementing Discrete Mathematics: Combinatorics and Graph Theory with Mathematica.* Reading, MA: Addison-Wesley, 1990.

Sloane, N. J. A. "An On-Line Version of the Encyclopedia of Integer Sequences." http://www.research.att.com/~njas/sequences/eisonline.html.

Sloane, N. J. A. and Plouffe, S. *The Encyclopedia of Integer Sequences.* San Diego, CA: Academic Press, 1995.

Slomson, A. *Introduction to Combinatorics.* Boca Raton, FL: Chapman and Hall, 1997.

Stanley, R. P. *Enumerative Combinatorics, Vol. 1.* Cambridge, England: Cambridge University Press, 1999.

Stanley, R. P. *Enumerative Combinatorics, Vol. 2.* Cambridge, England: Cambridge University Press, 1999.

Street, A. P. and Wallis, W. D. *Combinatorial Theory: An Introduction.* Winnipeg, Manitoba: Charles Babbage Research Center, 1977.

Tucker, A. *Applied Combinatorics, 3rd ed.* New York: Wiley, 1995.

van Lint, J. H. and Wilson, R. M. *A Course in Combinatorics.* New York: Cambridge University Press, 1992.

Weisstein, E. W. "Books about Combinatorics." http://www.treasure-troves.com/books/Combinatorics.html.

Wilf, H. S. *Combinatorial Algorithms: An Update.* Philadelphia, PA: SIAM, 1989.

Comedian Triangles

Two triangles having the same MEDIAN are said to be comedian triangles.

See also COSYMMEDIAN TRIANGLES, MEDIAN (TRIANGLE)

References

Lachlan, R. *An Elementary Treatise on Modern Pure Geometry.* London: Macmillian, p. 63, 1893.

Comma

A typesetting symbol which has several distinct meanings in mathematics. It is used for a number of purposes.

1. To denote Boundaries between elements in a list, as in $\{1, 2, 3, \ldots\}$.
2. To delimit indices in the element of a MATRIX, as in $a_{i,j}$ (although it is frequently omitted when implied by context).
3. To indicate the COMMA DERIVATIVE of a TENSOR.
4. In place of a DECIMAL POINT in continental Europe, e.g., 3,14159.

See also COMMA DERIVATIVE, DECIMAL POINT

References

Bringhurst, R. *The Elements of Typographic Style, 2nd ed.* Point Roberts, WA: Hartley and Marks, p. 275, 1997.

Comma Derivative

For A a TENSOR,

$$A_{,k} \equiv \frac{\partial A}{\partial x^k} \equiv \partial_k A$$

$$A^k_{,k} \equiv \frac{1}{g_k} \frac{\partial A^k}{\partial x^k} \equiv \partial_k A^k.$$

Schmutzer (1968, p. 70) uses the older notation $A_{|k}$.

See also COVARIANT DERIVATIVE, TENSOR

References

Schmutzer, E. *Relativistische Physik (Klassische Theorie).* Leipzig, Germany: Akademische Verlagsgesellschaft, 1968.

Comma of Didymus

The musical interval by which four fifths exceed a seventeenth (i.e., two octaves and a major third),

$$\frac{\partial A}{\partial x^k} \equiv \partial_k A$$

also called a SYNTONIC COMMA.

See also COMMA OF PYTHAGORAS, DIESIS, SCHISMA

Comma of Pythagoras

The musical interval by which twelve fifths exceed seven octaves,

$$A^k_{,k}$$

Successive CONTINUED FRACTION CONVERGENTS to

$$\frac{1}{g_k} \frac{\partial A^k}{\partial x^k} \equiv \partial_k A^k$$

give increasingly close approximations $A_{|k}$ of m fifths by n octaves as 1, 2, 5/3, 12/7, 41/24, 53/31, 306/179, 665/389, ... (Sloane's A005664 and A046102; Jeans 1968, p. 188), shown in **bold** in the table below. All near-equalities of m fifths and n octaves having

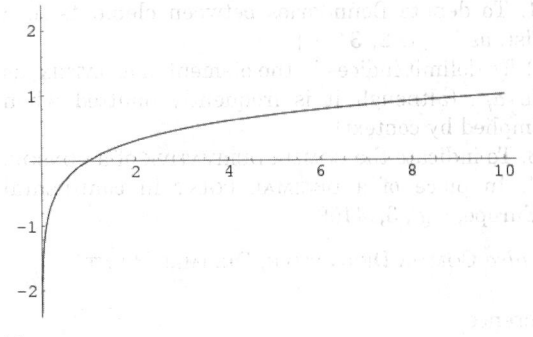

with

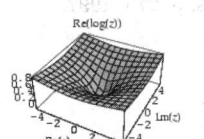 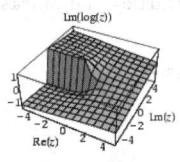 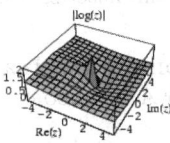

are given in the following table.

m	n	Ratio	m	n	Ratio
12	**7**	1.013643265	265	155	1.010495356
41	**24**	0.9886025477	294	172	0.9855324037
53	**31**	1.002090314	**306**	**179**	0.9989782832
65	38	1.015762098	318	186	1.012607608
94	55	0.9906690375	347	203	0.9875924759
106	62	1.004184997	359	210	1.001066462
118	69	1.017885359	371	217	1.014724276
147	86	0.9927398469	400	234	0.9896568543
159	93	1.006284059	412	241	1.003159005
188	110	0.9814251419	424	248	1.016845369
200	117	0.994814985	453	265	0.9917255479
212	124	1.008387509	465	272	1.005255922
241	141	0.9834766286	477	279	1.018970895
253	148	0.9968944607	494	289	0.9804224033

See also COMMA OF DIDYMUS, DIESIS, SCHISMA

References

Conway, J. H. and Guy, R. K. *The Book of Numbers.* New York: Springer-Verlag, p. 257, 1995.
Guy, R. K. "Small Differences Between Powers of 2 and 3." §F23 in *Unsolved Problems in Number Theory, 2nd ed.* New York: Springer-Verlag, p. 261, 1994.
Sloane, N. J. A. Sequences A005664/M1428 and A046102 in "An On-Line Version of the Encyclopedia of Integer Sequences." http://www.research.att.com/~njas/sequences/eisonline.html.

Commandino's Theorem

The four medians of a TETRAHEDRON CONCUR in a point which divides each MEDIAN in the ratio 1:3, the longer segment being on the side of the vertex of the TETRAHEDRON.

See also BIMEDIAN, MEDIAN (TETRAHEDRON), TETRAHEDRON

References

Altshiller-Court, N. "Commandino's Theorem." §170 in *Modern Pure Solid Geometry.* New York: Chelsea, pp. 51–2, 1979.
Commandino, F. Prop. 17 in *De centro gravitatis solidorum.* p. 21, 1565.

Common Cycloid

CYCLOID

Common Fraction

A FRACTION in which NUMERATOR and DENOMINATOR are both integers, as opposed to fractions. Common fractions are sometimes also called vulgar fractions.

See also COMPLEX FRACTION, FRACTION

Common Logarithm

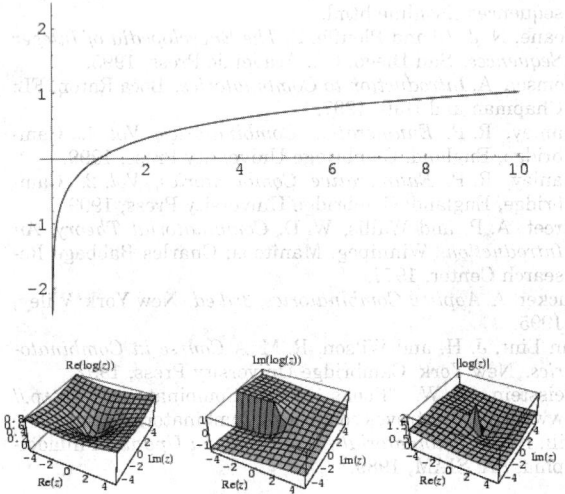

The LOGARITHM in BASE 10. The notation $\log x$ is used by physicists, engineers, and calculator keypads to denote the common logarithm. However, mathematicians generally use the same symbol to mean the NATURAL LOGARITHM LN, $\ln x$. Worse still, in Russian literature the notation $\lg x$ is used to denote a base-10 logarithm, which conflicts with the use of the symbol LG to indicate the logarithm to base 2. To avoid all ambiguity, it is best to explicitly specify $\log_{10} x$ when the logarithm to base 10 is intended. In this work, $\log x = \log_{10} x$, $\ln x = \log_e x$ is used for the NATURAL

LOGARITHM, and $\lg x = \log_2 x$ is the logarithm to the base 2.

Hardy and Wright (1979, p. 8) assert that the common logarithm has "no mathematical interest." Common and natural logarithms can be expressed in terms of each other as

$$\ln x = \frac{\log_{10} x}{\log_{10} e}$$

$$\log_{10} x = \frac{\ln x}{\ln 10}.$$

See also LG, LN, LOGARITHM, NATURAL LOGARITHM

References

Hardy, G. H. and Wright, E. M. *An Introduction to the Theory of Numbers, 5th ed.* Oxford, England: Clarendon Press, 1979.

Common Residue

The value of b, where $a \equiv b \pmod{m}$, taken to be NONNEGATIVE and smaller than m.

See also MINIMAL RESIDUE, RESIDUE (CONGRUENCE)

Commutation Coefficient

A TENSOR-like coefficient which gives the difference between PARTIAL DERIVATIVES of two coordinates with respect to the other coordinate,

$$c_{\alpha\beta}^{\mu}\vec{e}_{\mu} = [\vec{e}_{\alpha}, \vec{e}_{\beta}] = \nabla_{\alpha}\vec{e}_{\beta} - \nabla_{\beta}\vec{e}_{\alpha}.$$

See also CONNECTION COEFFICIENT, PARTIAL DERIVATIVE

Commutative

Two elements x and y of a set S are said to be commutative under a binary operation $*$ if they satisfy

$$x * y = y * x.$$

Real numbers are commutative under addition

$$x + y = y + x$$

and multiplication

$$x \cdot y = y \cdot x.$$

See also ASSOCIATIVE, COMMUTE, COMMUTATIVE ALGEBRA, COMMUTATIVE MATRICES, COMMUTATIVE RING, DISTRIBUTIVE, TRANSITIVE

Commutative Algebra

Let A denote an $\mathbb{R}$-algebra, so that A is a VECTOR SPACE over R and

$$A \times A \to A \tag{1}$$

$$(x, y) \mapsto x \cdot y. \tag{2}$$

Now define

$$Z \equiv \{x \in A : x \cdot y = 0 \text{ for some } y \in A \neq 0\}, \tag{3}$$

where $0 \in Z$. An ASSOCIATIVE $\mathbb{R}$-algebra is commutative if $x \cdot y = y \cdot x$ for all $x, y \in A$. Similarly, a RING is commutative if the MULTIPLICATION operation is commutative, and a LIE ALGEBRA is commutative if the COMMUTATOR $[A, B]$ is 0 for every A and B in the LIE ALGEBRA.

See also ABELIAN GROUP, COMMUTATIVE

References

Atiyah, M. F. and MacDonald, I. G. *Introduction to Commutative Algebra.* Reading, MA: Addison-Wesley, pp. 9–0, 1969.

Cox, D.; Little, J.; and O'Shea, D. *Ideals, Varieties, and Algorithms: An Introduction to Algebraic Geometry and Commutative Algebra, 2nd ed.* New York: Springer-Verlag, 1996.

Eisenbud, D. (Ed.). *Commutative Algebra, Algebraic Geometry, and Computational Methods.* Singapore: Springer-Verlag, 1999.

Finch, S. "Zero Structures in Real Algebras." http://www.mathsoft.com/asolve/zerodiv/zerodiv.html.

MacDonald, I. G. and Atiyah, M. F. *Introduction to Commutative Algebra.* Reading, MA: Addison-Wesley, 1969.

Samuel, P. and Zariski, O. *Commutative Algebra, Vol. 2.* New York: Springer-Verlag, 1997.

Zariski, O. and Samuel, P. *Commutative Algebra I.* New York: Springer-Verlag, 1958.

Commutative Group

ABELIAN GROUP

Commutative Matrices

COMMUTING MATRICES

Commutative Ring

A RING is commutative if the MULTIPLICATION operation is COMMUTATIVE.

See also COMMUTATIVE, RING

Commutator

Let $\tilde{A}, \tilde{B}, \ldots$ be OPERATORS. Then the commutator of $\tilde{A}$ and $\tilde{B}$ is defined as

$$[\tilde{A}, \tilde{B}] \equiv \tilde{A}\tilde{B} - \tilde{B}\tilde{A}. \tag{1}$$

Let $a, b, \ldots$ be constants. Identities include

$$[f(x), x] = 0 \tag{2}$$

$$[\tilde{A}, \tilde{A}] = 0 \tag{3}$$

$$[\tilde{A}, \tilde{B}] = -[\tilde{B}, \tilde{A}] \tag{4}$$

$$[\tilde{A}, \tilde{B}\tilde{C}] = [\tilde{A}, \tilde{B}]\tilde{C} + \tilde{B}[\tilde{A}, \tilde{C}] \tag{5}$$

$$[\tilde{A}\tilde{B}, \tilde{C}] = [\tilde{A}, \tilde{C}]\tilde{B} + \tilde{A}[\tilde{B}, \tilde{C}] \tag{6}$$

$$[a + \tilde{A}, b + \tilde{B}] = [\tilde{A}, \tilde{B}] \tag{7}$$

$$[\tilde{A} + \tilde{B}, \tilde{C} + \tilde{D}] = [\tilde{A}, \tilde{C}] + [\tilde{A}, \tilde{D}] + [\tilde{B}, \tilde{C}] + [\tilde{B}, \tilde{D}]. \tag{8}$$

Let A and B be TENSORS. Then

$$[A, B] \equiv \nabla_A B - \nabla_B A. \tag{9}$$

There is a related notion of commutator in the theory of groups. The commutator of two GROUP elements A and B is $ABA^{-1}B^{-1}$, and two elements A and B are said to COMMUTE when their commutator is the IDENTITY ELEMENT. When the group is a LIE GROUP, the LIE BRACKET in its LIE ALGEBRA is an infinitesimal version of the group commutator. For instance, let A and B be square matrices, and let $\alpha(s)$ and $\beta(t)$ be paths in the LIE GROUP of INVERTIBLE MATRICES which satisfy

$$\alpha(0) = \beta(0) = 1 \tag{10}$$

$$\left.\frac{\partial x}{\partial s}\right|_{s=0} = A \tag{11}$$

$$\left.\frac{\partial \beta}{\partial s}\right|_{s=0} = B, \tag{12}$$

then

$$\left.\frac{\partial}{\partial s}\frac{\partial}{\partial t}\alpha(s)\beta(t)\alpha^{-1}(s)\beta^{-1}(t)\right|_{(s=0,\,t=0)} = 2[A, B]. \tag{13}$$

See also AD, ad, ANTICOMMUTATOR, COMMUTATOR SUBGROUP, JACOBI IDENTITIES

References

Schafer, R. D. *An Introduction to Nonassociative Algebras.* New York: Dover, p. 13, 1996.

Commutator Series (Lie Algebra)

The commutator series of a LIE ALGEBRA $\mathfrak{g}$, sometimes called the derived series, is the sequence of subalgebras recursively defined by

$$\mathfrak{g}^{k+1} = [\mathfrak{g}^k, \mathfrak{g}^k],$$

with $\mathfrak{g}^0 = \mathfrak{g}$. The sequence of subspaces is always decreasing with respect to inclusion or dimension, and becomes stable when $\mathfrak{g}$ is finite dimensional. The notation $[\mathfrak{a}, \mathfrak{b}]$ means the linear span of elements of the form $[A, B]$, where $A \in \mathfrak{a}$ and $B \in \mathfrak{b}$.

When the commutator series ends in the zero subspace, the Lie algebra is called SOLVABLE. For example, consider the LIE ALGEBRA of strictly UPPER TRIANGULAR MATRICES, then

$$\mathfrak{g}^0 = \begin{bmatrix} 0 & a_{12} & a_{13} & a_{14} & a_{15} \\ 0 & 0 & a_{23} & a_{24} & a_{25} \\ 0 & 0 & 0 & a_{34} & a_{35} \\ 0 & 0 & 0 & 0 & a_{45} \\ 0 & 0 & 0 & 0 & 0 \end{bmatrix} \tag{1}$$

$$\mathfrak{g}^1 = \begin{bmatrix} 0 & 0 & a_{13} & a_{14} & a_{15} \\ 0 & 0 & 0 & a_{24} & a_{25} \\ 0 & 0 & 0 & 0 & a_{35} \\ 0 & 0 & 0 & 0 & 0 \\ 0 & 0 & 0 & 0 & 0 \end{bmatrix} \tag{2}$$

$$\mathfrak{g}^2 = \begin{bmatrix} 0 & 0 & 0 & 0 & a_{15} \\ 0 & 0 & 0 & 0 & 0 \\ 0 & 0 & 0 & 0 & 0 \\ 0 & 0 & 0 & 0 & 0 \\ 0 & 0 & 0 & 0 & 0 \end{bmatrix}, \tag{3}$$

and $\mathfrak{g}^3 = 0$. By definition, $\mathfrak{g}^k \subset \mathfrak{g}_k$ where $\mathfrak{g}_k$ is the term in the LOWER CENTRAL SERIES, as can be seen by the example above.

In contrast to the SOLVABLE LIE ALGEBRAS, the SEMISIMPLE LIE ALGEBRAS have a constant commutator series. Others are in between, e.g.,

$$[\mathfrak{gl}_n, \mathfrak{gl}_n] = \mathfrak{sl}_n, \tag{4}$$

which is semisimple, because the TRACE satisfies

$$\mathrm{Tr}(AB) = \mathrm{Tr}(BA). \tag{5}$$

Here, $\mathfrak{gl}_n$ is a general linear Lie algebra and $\mathfrak{sl}_n$ is the SPECIAL LINEAR LIE ALGEBRA.

Here are some *Mathematica* functions for determining the commutator series, given a list of matrices which is a basis for $\mathfrak{g}$.

```
MatrixBasis[a_List]:=
Partition[#1,Length[a[[1]]]]&/@
  LatticeReduce[Flatten/@a]
    LieCommutator[a_,b_]:=a.b-b.a
NextDerived[{}]={};
NextDerived[g_List]:=
MatrixBasis[Flatten[Outer[LieCommutator,g,g,1]
,1]]
kthDerived[g_List,k_Integer]:=
Nest[NextDerived,g,k]
```

For example,

```
g15=Flatten[Table[ReplacePart
[Table
[0,{i,5},{j,5}],1,{k,1}],{k,5},{1,5}],1];s15=
kthDerived[g15,1]
```

See also BOREL SUBALGEBRA, COMMUTATOR SERIES (GROUP), LIE ALGEBRA, LIE GROUP, NILPOTENT LIE GROUP, NILPOTENT LIE ALGEBRA, REPRESENTATION

(Lie Algebra), Representation (Solvable Lie Group), Solvable Lie Group, Split Solvable Lie Algebra

Commutator Subgroup

The commutator subgroup of a GROUP G is the SUBGROUP generated by the COMMUTATORS of its elements, and is denoted $[G, G]$. It is always a NORMAL SUBGROUP. It can range from the identity subgroup (in the case of an ABELIAN GROUP), to the whole group. For instance, in the QUATERNION group $\{\pm 1, \pm i, \pm j, \pm k\}$ with eight elements, the commutators form the subgroup $\{1, -1\}$. The commutator subgroup of the SYMMETRIC GROUP is the ALTERNATING GROUP. The commutator subgroup of the ALTERNATING GROUP A_n is the whole group A_n. When $n \geq 5$, A_n is a SIMPLE GROUP and its only nontrivial normal subgroup is itself. Since $[A_n, A_n]$ is a nontrivial normal subgroup, it must be A_n.

The first homology of a group G is the ABELIANIZATION

$$H_1(G) = G/[G, G].$$

See also ABELIAN GROUP, ABELIANIZATION, COMMUTATOR, GROUP, GROUP COHOMOLOGY, NORMAL SUBGROUP

Commute

Two algebraic objects that are COMMUTATIVE, i.e., A and B such that $A * B = B * A$ for some operation $*$, are said to commute with each other.

See also COMMUTATIVE, COMMUTATOR

Commuting Matrices

This entry contributed by RONALD M. AARTS

Two matrices A and B which satisfy

$$AB = BA$$

under MATRIX MULTIPLICATION are said to be commuting.

In general, MATRIX MULTIPLICATION is *not* COMMUTATIVE. Furthermore, in general there is no MATRIX INVERSE A^{-1} even when $A \neq 0$. Finally, AB can be zero even without $A = 0$ or $B = 0$. And when $AB = 0$, we may still have $BA \neq 0$, a simple example of which is provided by

$$A = \begin{bmatrix} 0 & 1 \\ 0 & 0 \end{bmatrix}$$

$$B = \begin{bmatrix} 1 & 0 \\ 0 & 0 \end{bmatrix},$$

for which

$$AB = 0,$$

but

$$BA = \begin{bmatrix} 0 & 1 \\ 0 & 0 \end{bmatrix} = A$$

(Taussky 1957).

See also COMMUTATIVE

References

Gantmacher, F. R. Ch. 8 in *The Theory of Matrices, Vol. 1*. Providence, RI: Amer. Math. Soc., 1998.
Taussky, O. "Commutativity in Finite Matrices." *Amer. Math. Monthly* **64**, 229–35, 1957.

Co-Monotone Approximation

COMONOTONE APPROXIMATION

Comonotone Approximation

This entry contributed by RONALD M. AARTS

The approximation of a piecewise MONOTONIC FUNCTION f by a polynomial with the same monotonicity. Such comonotonic approximations can always be accomplished with nth degree polynomials, and have an error of $A\omega(f; 1/n)$ (Passow and Raymon 1974, Passow *et al.* 1974, Newman 1979).

References

Newman, D. J. "Efficient Co-Monotone Approximation." *J. Approx. Th.* **25**, 189–92, 1979.
Passow, E. and Raymon, L. "Monotone and Comonotone Approximation." *Proc. Amer. Math. Soc.* **42**, 340–49, 1974.
Passow, E.; Raymon, L.; and Roulier, J. A. "Comonotone Polynomial Approximation." *J. Approx. Th.* **11**, 221–24, 1974.

Compact Closure

A set U has compact closure if its CLOSURE is COMPACT. Typically, compact closure is equivalent to the condition that U is BOUNDED.

See also BOUNDED, COMPACT SET, TOPOLOGY

Compact Group

COMPACT LIE GROUP

Compact Lie Group

If the parameters of a LIE GROUP vary over a CLOSED INTERVAL, them the LIE GROUP is said to be compact. Every representation of a compact group is equivalent to a UNITARY representation.

See also LIE GROUP

References

Huang, J.-S. "Compact Lie Groups." Part 3 in *Lectures on Representation Theory*. Singapore: World Scientific, pp. 71–28, 1999.

Compact Manifold

A compact manifold is a MANIFOLD which is compact as a TOPOLOGICAL SPACE. Examples are the CIRCLE (the only 1-D compact manifold) and the n-dimensional sphere and torus. Compact manifolds in two dimensions are completely classified by their orientation and the number of holes (GENUS).

For many problems in topology and geometry, it is convenient to study compact manifolds because of their "nice" behavior. Among the properties making compact manifolds "nice" are the fact that they can be covered by finitely many CHARTS, and that any continuous real-valued function is bounded on a compact manifold. However, it is an open question if the known compact manifolds in 3-D are complete, and it is not even known what a complete list in 4-D should look like. The following terse table therefore summarizes current knowledge about the number of compact manifolds $N(D)$ of D dimensions.

D	$N(D)$
1	1
2	2

See also MANIFOLD, SPHERE, TOPOLOGICAL SPACE, TORUS, TYCHONOF COMPACTNESS THEOREM

Compact Set

The SET S is compact if, from any SEQUENCE of elements X_1, X_2, ...of S, a subsequence can always be extracted which tends to some limit element X of S. Compact sets are therefore sets which are both CLOSED and BOUNDED.

See also BOUNDED SET, CLOSED SET

References
Croft, H. T.; Falconer, K. J.; and Guy, R. K. *Unsolved Problems in Geometry.* New York: Springer-Verlag, p. 2, 1991.

Compact Space

A TOPOLOGICAL SPACE is compact if every open cover of X has a finite subcover. In other words, if X is the union of a family of open sets, there is a finite subfamily whose union is X. A subset A of a TOPOLOGICAL SPACE X is compact if it is compact as a TOPOLOGICAL SPACE with the relative topology (i.e., every family of open sets of X whose union contains A has a finite subfamily whose union contains A).

Compact Support

A function has compact support if it is zero outside of a COMPACT SET. A function with compact support is only interesting in a BOUNDED domain. Alternatively, one can say that a function has compact support if its SUPPORT is a COMPACT SET. For example, the function $f : x \to x^2$ in its entire domain (i.e., $f : \mathbb{R} \to \mathbb{R}^+$) does not have compact support, while any BUMP FUNCTION does have compact support.

See also BUMP FUNCTION, COMPACT SET, SUPPORT

Compact Surface

A compact surface is a SURFACE which is also a COMPACT SET. A compact surface has a TRIANGULATION with a finite number of triangles. The SPHERE and TORUS are compact.

See also COMPACT SET, TRIANGULATION

Compactification

A compactification of a TOPOLOGICAL SPACE X is a larger space Y containing X which is also compact. The smallest compactification is the ONE-POINT COMPACTIFICATION. For example, the real line is not compact. It is contained in the circle, which is obtained by adding a point at infinity. Similarly, the plane is compactified by adding one point at infinity, giving the SPHERE.

See also COMPACT SET, STEREOGRAPHIC PROJECTION, TOPOLOGICAL SPACE

Compactness Theorem

Inside a BALL B in $\mathbb{R}^3$,

$$\{\text{rectifiable currents } S \text{ in } BL \text{ area } S \leq c, \text{ length } \partial S \leq c\}$$

is compact under the FLAT NORM.

References
Morgan, F. "What Is a Surface?" *Amer. Math. Monthly* **103**, 369–76, 1996.

Compact-Open Topology

The compact-open topology is a common topology used on FUNCTION SPACES. Suppose X and Y are TOPOLOGICAL SPACES and $C(X, Y)$ is the set of continuous maps from $f : X \to Y$. The compact-open topology on $C(X, Y)$ is generated by subsets of the following form,

$$B(K, U) = \{f | f(K) \subset U\},$$

where K is compact in X and U is open in Y. (Hence the terminology "compact-open.") It is important to note that these sets are not CLOSED under intersection, and do not form a BASIS. Instead, the sets $B(K, U)$ form a SUBBASIS for the compact-open topology. That is, the open sets in the compact-open topology are the arbitrary unions of finite intersections of $B(K, U)$.

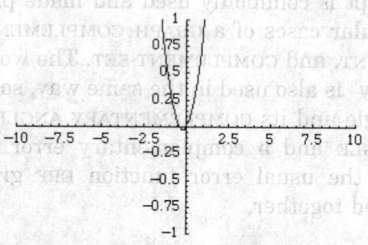

The simplest FUNCTION SPACE to compare topologies is the space of real-valued continuous functions $f : \mathbb{R} \to \mathbb{R}$. A sequence of functions f_n converges to $f = 0$ IFF for every $B(K, U)$ containing f contains all but a finite number of the f_n. Hence, for all $K > 0$ and all $\epsilon > 0$, there exists an N such that for all $n > N$,

$$|f_n(x)| < \epsilon \quad \text{for all } |x| \leq K.$$

For example, the sequence of functions $f_n = \sin(nx/2)/(n+1) + x^{2n}/e^{-n^2/2}$ converges to the zero function, although each function is unbounded.

When Y is a METRIC SPACE, the compact-open topology is the same as the topology of COMPACT CONVERGENCE. If X is a LOCALLY COMPACT HAUSDORFF space, a fairly weak condition, then the evaluation map

$$e : X \times C(X, Y) \to Y$$

defined by $e(x, f) = f(x)$ is CONTINUOUS. Similarly, $H : X \times Z \to Y$ is CONTINUOUS IFF the map $\tilde{H} : Z \to C(X, Y)$, given by $H(x, z) = \tilde{H}(z)(x)$, is CONTINUOUS. Hence, the compact-open topology is the right topology to use in HOMOTOPY theory.

See also ALGEBRAIC TOPOLOGY, COMPACT CONVERGENCE, HOMOTOPY THEORY, TOPOLOGICAL SPACE

References

Munkres, J. *Topology*. Englewood Cliffs, NJ: Prentice Hall, pp. 285–89, 1975.

Companion Knot

Let K_1 be a knot inside a TORUS. Now knot the TORUS in the shape of a second knot (called the companion knot) K_2. Then the new knot resulting from K_1 is called the SATELLITE KNOT K_3.

References

Adams, C. C. *The Knot Book: An Elementary Introduction to the Mathematical Theory of Knots*. New York: W. H. Freeman, pp. 115–18, 1994.

Companion Matrix

The companion matrix to a MONIC POLYNOMIAL

$$a(x) = a_0 + a_1 x + \ldots + a_{n-1} x^{n-1} + x^n \tag{1}$$

is the $n \times n$ SQUARE MATRIX

$$A = \begin{bmatrix} 0 & 0 & \cdots & 0 & -a_0 \\ 1 & 0 & \cdots & 0 & -a_1 \\ 0 & 1 & \cdots & 0 & -a_2 \\ \vdots & \vdots & \ddots & \vdots & \vdots \\ 0 & 0 & \cdots & 1 & -a_{n-1} \end{bmatrix} \tag{2}$$

with ones on the SUBDIAGONAL and the last column given by the coefficients of $a(x)$. Note that in the literature, the companion matrix is sometimes defined with the rows and columns switched, i.e., the TRANSPOSE of the above matrix.

When e_i is the STANDARD BASIS, a companion matrix satisfies

$$Ae_i = e_{i+1} \tag{3}$$

for $i < n$, as well as

$$Ae_n = \sum -a_i e_i, \tag{4}$$

including

$$A^n e_1 = \sum -a_i A^i e_1. \tag{5}$$

The MINIMAL POLYNOMIAL of the companion matrix is therefore $a(x)$, which is also its CHARACTERISTIC POLYNOMIAL.

Companion matrices are used to write a matrix in RATIONAL CANONICAL FORM. In fact, any $n \times n$ matrix whose MINIMAL POLYNOMIAL $p(x)$ has DEGREE n is SIMILAR to the companion matrix for $p(x)$. The RATIONAL CANONICAL FORM is more interesting when the degree of $p(x)$ is less than n.

The following *Mathematica* command gives the companion matrix for a polynomial p in the variable x.

```
CompanionMatrix[p_,x_]:=
  Module[{rnk = Exponent[p,x],
    v = CoefficientList[p,x],w},
    w = Drop[v/Last[v],-1];
If[rnk = =1,{-w},
Transpose[Append[(Prepend[#1,0]&/@IdentityMa-
trix[rnk-1]),-w]]]]
```

See also MATRIX, MINIMAL POLYNOMIAL (MATRIX), RATIONAL CANONICAL FORM

References

Dummit, D. and Foote, R. *Abstract Algebra*. Englewood Cliffs, NJ: Prentice Hall, 1991.
Herstein, I. §6.7 in *Topics in Algebra, 2nd ed.* New York: Wiley, 1975.
Jacobson, N. §3.10 in *Basic Algebra I*. New York: W. H. Freeman, 1985.

Comparability Graph

The comparability graph of a POSET $P = (X, \leq)$ is the GRAPH with vertex set X for which vertices x and y are adjacent IFF either $x \leq y$ or $y \leq x$ in P.

See also INTERVAL GRAPH, PARTIALLY ORDERED SET

Comparison Test

Let $\Sigma\, a_k$ and $\Sigma\, b_k$ be a SERIES with POSITIVE terms and suppose $a_1 \le b_1$, $a_2 \le b_2$, ...

  1. If the bigger series CONVERGES, then the smaller series also CONVERGES.
  2. If the smaller series DIVERGES, then the bigger series also DIVERGES.

See also CONVERGENCE TESTS

References
Arfken, G. *Mathematical Methods for Physicists, 3rd ed.* Orlando, FL: Academic Press, pp. 280–81, 1985.

Compass
A tool with two arms joined at their ends which can be used to draw CIRCLES. In GEOMETRIC CONSTRUCTIONS, the classical Greek rules stipulate that the compass cannot be used to mark off distances, so it must "collapse" whenever one of its arms is removed from the page. This results in significant complication in the complexity of GEOMETRIC CONSTRUCTIONS.

See also CONSTRUCTIBLE POLYGON, GEOMETRIC CONSTRUCTION, GEOMETROGRAPHY, MASCHERONI CONSTRUCTION, PLANE GEOMETRY, POLYGON, PONCELET-STEINER THEOREM, RULER, SIMPLICITY, STEINER CONSTRUCTION, STRAIGHTEDGE

References
Dixon, R. "Compass Drawings." Ch. 1 in *Mathographics.* New York: Dover, pp. 1–8, 1991.

Compatible
Let $\|A\|$ be the MATRIX NORM associated with the MATRIX A and $\|\mathbf{x}\|$ be the VECTOR NORM associated with a VECTOR $\mathbf{x}$. Let the product $A\mathbf{x}$ be defined, then $\|A\|$ and $\|\mathbf{x}\|$ are said to be compatible if

$$\|A\mathbf{x}\| \le \|A\|\|\mathbf{x}\|.$$

References
Gradshteyn, I. S. and Ryzhik, I. M. *Tables of Integrals, Series, and Products, 6th ed.* San Diego, CA: Academic Press, p. 2000, 1980.

Complement
In general, the word "complement" refers to that subset F' of some set S which excludes a given subset F. Taking F and its complement F' together then gives the whole of the original set. The notations F' and $\bar{F}$ are commonly used to denote the complement of a set F.

This concept is commonly used and made precise in the particular cases of a GRAPH COMPLEMENT, KNOT COMPLEMENT, and COMPLEMENT SET. The word "complementary" is also used in the same way, so combining an angle and its COMPLEMENTARY ANGLE gives a RIGHT ANGLE and a complementary error function ERFC and the usual error function ERF give unity when added together,

$$\operatorname{erfc} x + \operatorname{erf} x = 1.$$

See also COMPLEMENT SET, COMPLEMENTARY ANGLE, ERFC, GRAPH COMPLEMENT, KNOT COMPLEMENT

References
Papoulis, A. *Probability, Random Variables, and Stochastic Processes, 2nd ed.* New York: McGraw-Hill, p. 23, 1984.

Complement Graph
GRAPH COMPLEMENT

Complement Knot
KNOT COMPLEMENT

Complement Set
Given a set S with a subset E, the complement of E is defined as

$$E' \equiv \{F : F \in S,\ F \notin E\}. \tag{1}$$

Using SET DIFFERENCE notation, the complement is defined by

$$E' = S \backslash E. \tag{2}$$

If $E = S$, then

$$E' \equiv S' = \varnothing, \tag{3}$$

where $\varnothing$ is the EMPTY SET. The complement is implemented in *Mathematica* as Complement[l, $l1$, ...].

Given a single SET, the second PROBABILITY AXIOM gives

$$1 = P(S) = P(E \cup E'). \tag{4}$$

Using the fact that $E \cap E' = \varnothing$,

$$1 = P(E) + P(E') \tag{5}$$

$$P(E') = 1 - P(E). \tag{6}$$

This demonstrates that

$$P(S') = P(\varnothing) = 1 - P(S) = 1 - 1 = 0. \tag{7}$$

Given two SETS,

$$P(E \cap F') = P(E) - P(E \cap F) \tag{8}$$

$$P(E' \cap F') = 1 - P(E) - P(F) + P(E \cap F). \qquad (9)$$

See also INTERSECTION, SET DIFFERENCE, SYMMETRIC DIFFERENCE

References

Croft, H. T.; Falconer, K. J.; and Guy, R. K. *Unsolved Problems in Geometry.* New York: Springer-Verlag, p. 2, 1991.

Complementary Angle

Two ANGLES α and $\pi/2 - \alpha$ are said to be complementary.

See also ANGLE, RIGHT ANGLE, SUPPLEMENTARY ANGLE

Complementary Error Function

ERFC

Complementary Modulus

If k is the MODULUS of an ELLIPTIC INTEGRAL or ELLIPTIC FUNCTION, then

$$k' \equiv \sqrt{1 - k^2}$$

is called the complementary modulus. Complete elliptic integrals with respect to the complementary modulus are often denoted

$$K'(k) \equiv K(k') = K(\sqrt{1 - k^2})$$

and

$$E'(k) \equiv E(k') = E(\sqrt{1 - k^2}).$$

See also MODULUS (ELLIPTIC INTEGRAL)

References

Tölke, F. "Parameterfunktionen." Ch. 3 in *Praktische Funktionenlehre, zweiter Band: Theta-Funktionen und spezielle Weierstraßsche Funktionen.* Berlin: Springer-Verlag, pp. 83–15, 1966.

Complementation

The process of taking the COMPLEMENT of a set or truth function. In the latter case, complementation is equivalent to the NOT operation.

See also COMPLEMENT, NOT

Complete

COMPLETE AXIOMATIC THEORY, COMPLETE BIGRAPH, COMPLETE GRAPH, COMPLETE QUADRANGLE, COMPLETE QUADRILATERAL, COMPLETE SEQUENCE, COMPLETE SET OF FUNCTIONS, COMPLETE SPACE,

COMPLETENESS PROPERTY, WEAKLY COMPLETE SEQUENCE

Complete Axiomatic Theory

An axiomatic theory (such as a GEOMETRY) is said to be complete if each valid statement in the theory is capable of being proven true or false.

See also CONSISTENCY

Complete Beta Function

BETA FUNCTION, INCOMPLETE BETA FUNCTION

Complete Bigraph

COMPLETE BIPARTITE GRAPH

Complete Binary Tree

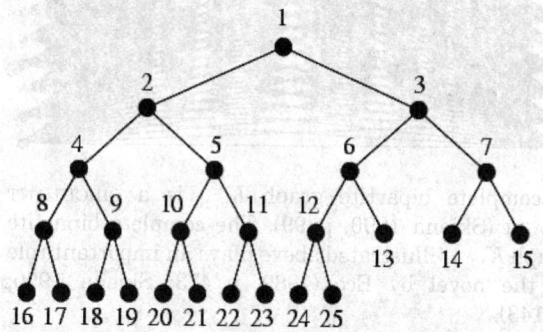

A labeled BINARY TREE containing the labels 1 to n with root 1, branches leading to nodes labeled 2 and 3, branches from these leading to 4, 5 and 6, 7, respectively, and so on (Knuth 1997, p. 401).

See also BINARY TREE, COMPLETE TREE, COMPLETE TERNARY TREE, HEAP

References

Knuth, D. E. *The Art of Computer Programming, Vol. 1: Fundamental Algorithms, 3rd ed.* Reading, MA: Addison-Wesley, 1997.
Knuth, D. E. *The Art of Computer Programming, Vol. 3: Sorting and Searching, 2nd ed.* Reading, MA: Addison-Wesley, p. 144, 1998.

Complete Bipartite Graph

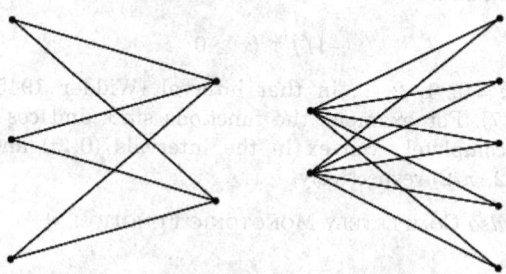

A BIPARTITE GRAPH (i.e., a set of VERTICES decomposed into two disjoint sets such that there are no two

VERTICES within the same set are adjacent) such that every pair of VERTICES in the two sets are adjacent. If there are p and q VERTICES in the two sets, the complete bipartite graph (sometimes also called a COMPLETE BIGRAPH) is denoted $K_{p, q}$. The above figures show $K_{3, 2}$ and $K_{2, 5}$. $K_{3, 3}$ is also known as the UTILITY GRAPH, and is the unique 4-CAGE GRAPH.

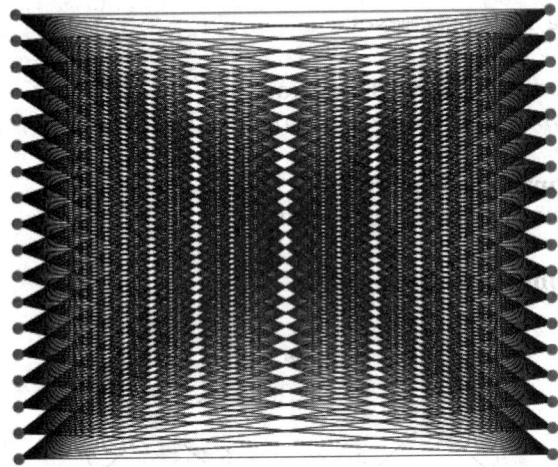

A complete bipartite graph $K_{n, n}$ is a CIRCULANT GRAPH (Skiena 1990, p. 99). The complete bipartite graph $K_{18, 18}$ illustrated above plays an important role in the novel by Eco (1989, p. 473; Skiena 1990, p. 143).

See also BIPARTITE GRAPH, CAGE GRAPH, COMPLETE GRAPH, COMPLETE K-PARTITE GRAPH, K-PARTITE GRAPH, THOMASSEN GRAPH, UTILITY GRAPH

References

Eco, U. *Foucault's Pendulum.* San Diego: Harcourt Brace Jovanovich, p. 473, 1989.
Saaty, T. L. and Kainen, P. C. *The Four-Color Problem: Assaults and Conquest.* New York: Dover, p. 12, 1986.
Skiena, S. *Implementing Discrete Mathematics: Combinatorics and Graph Theory with Mathematica.* Reading, MA: Addison-Wesley, 1990.

Complete Convex Function

This entry contributed by RONALD M. AARTS

A function $f(x)$ is completely convex in an OPEN INTERVAL (a, b) if it has DERIVATIVES of all orders there and if

$$(-1)^k f^{(2k)}(x) \geq 0$$

for $k = 0, 1, 2, \ldots$ in that interval (Widder 1945, p. 177). For example, the functions $\sin x$ and $\cos x$ are completely convex in the intervals $(0, \pi)$ and $(-\pi/2, \pi/2)$ respectively.

See also COMPLETELY MONOTONIC FUNCTION

References

Widder, D. V. *The Laplace Transform.* Princeton, NJ: Princeton University Press, 1941.

Complete Digraph

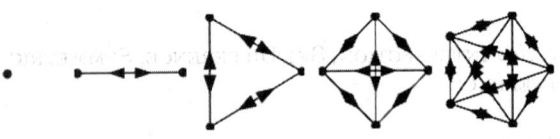

Complete digraphs are digraphs in which every pair of nodes is connected by a bidirectional edge.

See also COMPLETE GRAPH, DIGRAPH, RAMSEY'S THEOREM

Complete Direct Sum

RING DIRECT PRODUCT

Complete Functions

COMPLETE SET OF FUNCTIONS

Complete Gamma Function

GAMMA FUNCTION, INCOMPLETE GAMMA FUNCTION

Complete Graph

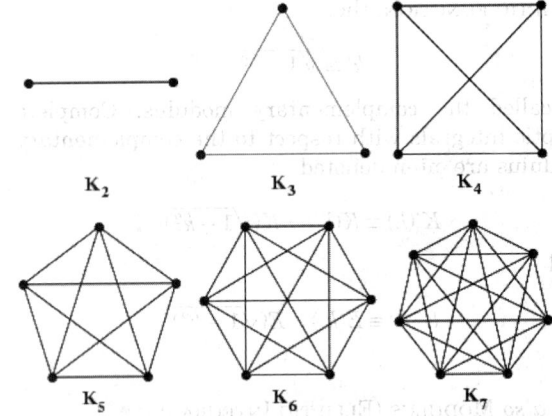

A GRAPH in which each pair of VERTICES is connected by an EDGE. The complete graph with n VERTICES is denoted K_n, and has $\binom{n}{2}$ undirected edges, where $\binom{n}{k}$ is a BINOMIAL COEFFICIENT. In older literature, complete GRAPHS are called UNIVERSAL GRAPHS.

The number of EDGES in K_v is $v(v-1)/2$ (the triangular numbers), and the GENUS is $(v-3)(v-4)/12$ for $v \geq 3$. The ADJACENCY MATRIX A of the complete graph G takes the particularly simple form of all 1s with 0s on the diagonal, i.e., the UNIT MATRIX minus the IDENTITY MATRIX,

$$A = J - I. \tag{1}$$

K_3 is the CYCLE GRAPH C_3, as well as the ODD GRAPH O_2 (Skiena 1990, p. 162). K_4 is the TETRAHEDRAL GRAPH, as well as the WHEEL GRAPH W_4, and is also a PLANAR GRAPH. K_5 is nonplanar. Conway and Gordon (1983) proved that every embedding of K_6 is INTRINSICALLY LINKED with at least one pair of linked triangles. They also showed that any embedding of K_7 contains a knotted HAMILTONIAN CYCLE.

The CHROMATIC POLYNOMIAL $\pi K_n(z)$ of K_n is given by the FALLING FACTORIAL $(z)_n$, and the CHROMATIC NUMBER by n.

It is not known in general if a set of TREES with 1, 2, ..., $n-1$ EDGES can always be packed into K_n. However, if the choice of TREES is restricted to either the path or star from each family, then the packing can always be done (Zaks and Liu 1977, Honsberger 1985).

See also CLIQUE, COMPLETE BIPARTITE GRAPH, COMPLETE DIGRAPH, COMPLETE K-PARTITE GRAPH, EMPTY GRAPH, GRAPH COMPLEMENT, ODD GRAPH

References

Chartrand, G. *Introductory Graph Theory.* New York: Dover, pp. 29–0, 1985.
Conway, J. H. and Gordon, C. M. "Knots and Links in Spatial Graphs." *J. Graph Th.* **7**, 445–53, 1983.
Honsberger, R. *Mathematical Gems III.* Washington, DC: Math. Assoc. Amer., pp. 60–3, 1985.
Saaty, T. L. and Kainen, P. C. *The Four-Color Problem: Assaults and Conquest.* New York: Dover, p. 12, 1986.
Skiena, S. "Complete Graphs." §4.2.1 in *Implementing Discrete Mathematics: Combinatorics and Graph Theory with Mathematica.* Reading, MA: Addison-Wesley, pp. 82 and 140–41, 1990.
Zaks, S. and Liu, C. L. "Decomposition of Graphs into Trees." In *Proceedings of the Eighth Southeastern Conference on Combinatorics, Graph Theory and Computing (Louisiana State Univ., Baton Rouge, La., 1977* (Ed. F. Hoffman, L. Lesniak-Foster, D. McCarthy, R. C. Mullin, K. B. Reid, and R. G. Stanton). *Congr. Numerantum* **19**, 643–54, 1977.

Complete k-Partite Graph

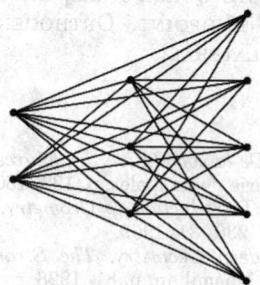

A K-PARTITE GRAPH (i.e., a set of VERTICES decomposed into k disjoint sets such that no two VERTICES within the same set are adjacent) such that every pair of VERTICES in the k sets are adjacent. If there are p, q, ..., r VERTICES in the k sets, the complete k-partite graph is denoted $K_{p,q,...,r}$. The above figure shows $K_{2,3,5}$.

See also COMPLETE GRAPH, COMPLETE K-PARTITE GRAPH, K-PARTITE GRAPH

References

Harary, F. *Graph Theory.* Reading, MA: Addison-Wesley, p. 23, 1994.
Saaty, T. L. and Kainen, P. C. *The Four-Color Problem: Assaults and Conquest.* New York: Dover, p. 12, 1986.
Skiena, S. "Complete k-Partite Graphs." §4.2.2 in *Implementing Discrete Mathematics: Combinatorics and Graph Theory with Mathematica.* Reading, MA: Addison-Wesley, pp. 142–44, 1990.

Complete Metric Space

A complete metric space is a METRIC SPACE in which every CAUCHY SEQUENCE is CONVERGENT. Examples include the REAL NUMBERS with the usual metric and the P-ADIC NUMBERS.

Complete Minimal Surface

A surface which is simultaneously COMPLETE and MINIMAL. There have been a large number of fundamental breakthroughs in the study of such surfaces in recent years, and they remain the focus of intensive current research.

Until the COSTA MINIMAL SURFACE was discovered in 1984, the only other known complete minimal embeddable surfaces in $\mathbb{R}^3$ with no self-intersections were the PLANE, CATENOID, and HELICOID. The plane is genus 0 and the catenoid and the helicoid are genus 0 with two punctures, but the Costa minimal surface is genus 1 with three punctures (Schwalbe and Wagon 1999).

See also COMPLETE SURFACE, COSTA MINIMAL SURFACE, MINIMAL SURFACE, NIRENBERG'S CONJECTURE

References

Schwalbe, D. and Wagon, S. "The Costa Surface, in Show and *Mathematica*." *Mathematica in Educ. Res.* **8**, 56–3, 1999.

Complete Permutation

DERANGEMENT

Complete Product

The complete products of a BOOLEAN ALGEBRA of subsets generated by a set $\{A_k\}_{k=1}^p$ of CARDINALITY p are the 2^p BOOLEAN FUNCTIONS

$$B_1 B_2 \cdots B_p \equiv B_1 \cap B_2 \cap \cdots \cap B_p,$$

where each B_k may equal A_k or its complement $\bar{A}_k$. For example, the $2^3 = 8$ complete products of $A = \{A_1, A_2, A_3\}$ are

$$A_1 A_2 A_3, \; A_1 A_2 \bar{A}_3, \; A_1 \bar{A}_2 A_3, \; \bar{A}_1 A_2 A_3,$$

$$A_1 \bar{A}_2 \bar{A}_3, \; \bar{A}_1 A_2 \bar{A}_3, \; \bar{A}_1 \bar{A}_2 A_3, \; \bar{A}_1 \bar{A}_2 \bar{A}_3.$$

Each BOOLEAN FUNCTION has a unique representation (up to order) as a union of complete products. For example,

$$A_1 A_2 \cup \bar{A}_3 = (A_1 A_2 A_3 \cup A_1 A_2 \bar{A}_3)$$

$$\cup (A_1 A_2 \bar{A}_3 \cup \bar{A}_1 A_2 \bar{A}_3 \cup A_1 \bar{A}_2 \bar{A}_3 \cup \bar{A}_1 \bar{A}_2 \bar{A}_3)$$

$$= A_1 A_2 A_3 \cup a_1 A_2 \bar{A}_3 \cup \bar{A}_1 A_2 \bar{A}_3 \cup A_1 \bar{A}_2 \bar{A}_3 \cup \bar{A}_1 \bar{A}_2 \bar{A}_3$$

$$= A_1 A_2 A_3 + A_1 A_2 \bar{A}_3 + \bar{A}_1 \bar{A}_2 \bar{A}_3$$

(Comtet 1974, p. 186).

See also BOOLEAN FUNCTION, CONJUNCTION

References

Comtet, L. *Advanced Combinatorics: The Art of Finite and Infinite Expansions, rev. enl. ed.* Dordrecht, Netherlands: Reidel, p. 186, 1974.

Complete Quadrangle

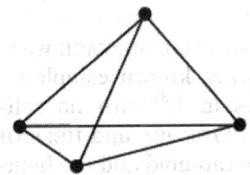

If the four points making up a QUADRILATERAL are joined pairwise by six distinct lines, a figure known as a complete quadrangle results. A complete quadrangle is therefore a set of four points, no three collinear, and the six lines which join them. Note that a complete quadrilateral is different from a COMPLETE QUADRANGLE.

The midpoints of the sides of any complete quadrangle and the three diagonal points all lie on a CONIC known as the NINE-POINT CONIC. If it is an ORTHOCENTRIC QUADRANGLE, the CONIC reduces to a CIRCLE. The ORTHOCENTERS of the four TRIANGLES of a complete quadrangle are COLLINEAR on the RADICAL LINE of the CIRCLES on the diameters of a QUADRILATERAL.

See also COMPLETE QUADRANGLE, PTOLEMY'S THEOREM

References

Coxeter, H. S. M. *Introduction to Geometry, 2nd ed.* New York: Wiley, pp. 230–31, 1969.
Demir, H. "The Compleat [sic] Cyclic Quadrilateral." *Amer. Math. Monthly* **79**, 777–78, 1972.
Durell, C. V. *Modern Geometry: The Straight Line and Circle.* London: Macmillan, p. 80, 1928.
Graustein, W. C. *Introduction to Higher Geometry.* New York: Macmillan, p. 25, 1930.
Johnson, R. A. *Modern Geometry: An Elementary Treatise on the Geometry of the Triangle and the Circle.* Boston, MA: Houghton Mifflin, pp. 61–2, 1929.
Ogilvy, C. S. *Excursions in Geometry.* New York: Dover, pp. 101–04, 1990.

Complete Quadrilateral

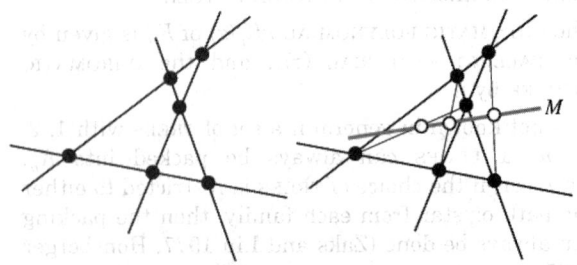

The figure determined by four lines, no three of which are concurrent, and their six points of intersection (Johnson 1929, pp. 61–2). Note that this figure is different from a COMPLETE QUADRANGLE. A complete quadrilateral has three diagonals (compared to two for an ordinary QUADRILATERAL). The MIDPOINTS of the diagonals of a complete quadrilateral are COLLINEAR on a line M (Johnson 1929, pp. 152–53).

A theorem due to Steiner (Mention 1862, Johnson 1929, Steiner 1971) states that in a complete quadrilateral, the bisectors of angles are CONCURRENT at 16 points which are the incenters and EXCENTERS of the four TRIANGLES. Furthermore, these points are the intersections of two sets of four CIRCLES each of which is a member of a conjugate coaxal system. The axes of these systems intersect at the point common to the CIRCUMCIRCLES of the quadrilateral.

Newton proved that, if a CONIC SECTION is inscribed in a complete quadrilateral, then its center lies on M (Wells 1991). In addition, the ORTHOCENTERS of the four triangles formed by a complete quadrilateral lie on a line which is perpendicular to M. Plücker proved that the circles having the three diagonals as diameters have two common points which lie on the line joining the four triangles' ORTHOCENTERS (Wells 1991).

See also COMPLETE QUADRANGLE, GAUSS-BODENMILLER THEOREM, MIDPOINT, ORTHOCENTER, POLAR CIRCLE, QUADRILATERAL

References

Carnot, L. N. M. *De la corrélation des figures de géométrie.* Paris: l'Imprimerie de Crapelet, p. 122, 1801.
Coxeter, H. S. M. *Introduction to Geometry, 2nd ed.* New York: Wiley, pp. 230–31, 1969.
Durell, C. V. *Modern Geometry: The Straight Line and Circle.* London: Macmillan, p. 81, 1928.
Graustein, W. C. *Introduction to Higher Geometry.* New York: Macmillan, p. 25, 1930.
Johnson, R. A. *Modern Geometry: An Elementary Treatise on the Geometry of the Triangle and the Circle.* Boston, MA: Houghton Mifflin, pp. 61–2, 149, 152–53, and 255–56, 1929.
Mention, M. J. "Démonstration d'un Théorème de M. Steiner." *Nouv. Ann. Math., 2nd Ser.* **1**, 16–0, 1862.
Mention, M. J. "Démonstration d'un Théorème de M. Steiner." *Nouv. Ann. Math., 2nd Ser.* **1**, 65–7, 1862.
Steiner, J. *Gesammelte Werke, 2nd ed, Vol. 1.* New York: Chelsea, p. 223, 1971.

Wells, D. *The Penguin Dictionary of Curious and Interesting Geometry*. London: Penguin, p. 35, 1991.

Complete Residue System

A set of numbers a_0, a_1, ..., a_{m-1} (mod m) form a complete set of residues, also called a covering system, if they satisfy

$$a_i \equiv i \pmod{m}$$

for $i = 0, 1, ..., m-1$. For example, a complete system of residues is formed by a base b and a modulus m if the residues r_i in $b^i \equiv r_i \pmod{m}$ for $i = 1, ..., m-1$ run through the values $1, 2, ..., m-1$.

See also CONGRUENCE, EXACT COVERING SYSTEM, HAUPT-EXPONENT, ORDER (MODULO), REDUCED RESIDUE SYSTEM, RESIDUE CLASS

References

Guy, R. K. "Covering Systems of Congruences." §F13 in *Unsolved Problems in Number Theory, 2nd ed.* New York: Springer-Verlag, pp. 251–53, 1994.
Nagell, T. "Residue Classes and Residue Systems." §20 in *Introduction to Number Theory*. New York: Wiley, pp. 69–1, 1951.

Complete Sequence

A SEQUENCE of numbers $V = \{v_n\}$ is complete if every POSITIVE INTEGER n is the sum of some subsequence of V, i.e., there exist $a_i = 0$ or 1 such that

$$n = \sum_{i=1}^{\infty} a_i v_i$$

(Honsberger 1985, pp. 123–26). The FIBONACCI NUMBERS are complete. In fact, dropping one number still leaves a complete sequence, although dropping two numbers does not (Honsberger 1985, pp. 123 and 126). The SEQUENCE of PRIMES with the element $\{1\}$ prepended,

$$\{1, 2, 3, 5, 7, 11, 13, 17, 19, 23, ...\}$$

is complete, even if any number of PRIMES each > 7 are dropped, as long as the dropped terms do not include two consecutive PRIMES (Honsberger 1985, pp. 127–28). This is a consequence of BERTRAND'S POSTULATE.

See also BERTRAND'S POSTULATE, BROWN'S CRITERION, FIBONACCI DUAL THEOREM, GREEDY ALGORITHM, WEAKLY COMPLETE SEQUENCE, ZECKENDORF'S THEOREM

References

Brown, J. L. Jr. "Unique Representations of Integers as Sums of Distinct Lucas Numbers." *Fib. Quart.* **7**, 243–52, 1969.
Hoggatt, V. E. Jr.; Cox, N.; and Bicknell, M. "A Primer for Fibonacci Numbers. XII." *Fib. Quart.* **11**, 317–31, 1973.
Honsberger, R. *Mathematical Gems III.* Washington, DC: Math. Assoc. Amer., 1985.

Complete Set of Functions

A set of ORTHONORMAL FUNCTIONS $\{\phi_n(x)\}$ is termed complete in the CLOSED INTERVAL $x \in [a, b]$ if, for every PIECEWISE CONTINUOUS function $f(x)$ in the interval, the minimum square error

$$E_n \equiv \|f - (c_1 \phi_1 + \ldots + c_n \phi_n)\|^2$$

(where $\|f\|$ denotes the $L2$-NORM with respect to a WEIGHTING FUNCTION $w(x)$) converges to zero as n becomes infinite. Symbolically, a set of functions is complete if

$$\lim_{m \to \infty} \int_a^b \left[f(x) - \sum_{n=0}^{m} a_n \phi_n(x) \right]^2 w(x)\, dx = 0,$$

where the above integral is a LEBESGUE INTEGRAL.

See also BESSEL'S INEQUALITY, HILBERT SPACE, $L2$-NORM

References

Arfken, G. "Completeness of Eigenfunctions." §9.4 in *Mathematical Methods for Physicists, 3rd ed.* Orlando, FL: Academic Press, pp. 523–38, 1985.

Complete Space

A SPACE of COMPLETE FUNCTIONS.

See also COMPLETE METRIC SPACE

Complete Surface

A surface which has no edges.

See also COMPLETE MINIMAL SURFACE, EMBEDDED SURFACE, MINIMAL SURFACE

Complete Ternary Tree

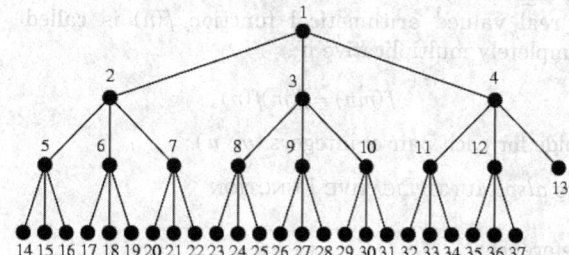

A labeled TERNARY TREE containing the labels 1 to n with root 1, branches leading to nodes labeled 2, 3, 4, branches from these leading to 5, 6, 7 and 8, 9, 10 respectively, and so on (Knuth 1997, p. 401).

See also COMPLETE BINARY TREE, COMPLETE TREE, TERNARY TREE

References

Knuth, D. E. *The Art of Computer Programming, Vol. 1: Fundamental Algorithms, 3rd ed.* Reading, MA: Addison-Wesley, 1997.

Complete Tree

See also COMPLETE BINARY TREE, COMPLETE TERNARY TREE

Complete Vector Space

A VECTOR SPACE is complete if every CAUCHY SEQUENCE in the space converges to an element in the space. For example, the rationals are not complete, whereas the real numbers are.

See also VECTOR SPACE

Completely Monotonic Function

This entry contributed by RONALD M. AARTS

A completely monotonic function is a function $f(x)$ such that

$$(-1)^{-n} f^{(n)}(x) \geq 0$$

for $n = 0, 1, 2, \ldots$. Such functions occur in areas such as probability theory (Feller 1971), numerical analysis, and elasticity (Ismail *et al.* 1986).

See also COMPLETE CONVEX FUNCTION, MONOTONIC FUNCTION

References

Feller, W. *An Introduction to Probability Theory and Its Applications, Vol. 2, 3rd ed.* New York: Wiley, 1971.
Ismail, M. E. H.; Lorch, L.; and Muldon, M. E. "Completely Monotonic Functions Associated with the Gamma Function and Its q-Analogues." *J. Math. Anal. Appl.* **116**, 1–, 1986.
Widder, D. V. *The Laplace Transform.* Princeton, NJ: Princeton University Press, 1941.

Completely Multiplicative Function

A real valued arithmetical function $f(n)$ is called completely multiplicative if

$$f(mn) = f(m)f(n)$$

holds for each pair of integers (m, n).

See also MULTIPLICATIVE FUNCTION

References

Kátai, I. and Kovács, B. "Multiplicative Functions with Nearly Integer Values." *Acta Sci. Math.* **48**, 221–25, 1985.

Completely Regular Graph

A POLYHEDRAL GRAPH is completely regular if the DUAL GRAPH is also REGULAR. There are only five types. Let ρ be the number of EDGES at each node, ρ^* the number of EDGES at each node of the DUAL GRAPH, V the number of VERTICES, E the number of EDGES, and F the number of faces in the PLATONIC SOLID corresponding to the given graph. The following table summarizes the completely regular graphs.

Type	ρ	ρ^*	V	E	F
Tetrahedral	3	3	4	6	4
Cubical	3	4	8	12	6
Dodecahedral	3	5	20	39	12
Octahedral	4	3	6	12	8
Icosahedral	5	3	12	30	20

Completeness Property

All lengths can be expressed as REAL NUMBERS.

Completing the Square

The conversion of an equation OF THE FORM $ax^2 + bx + c$ to the form

$$a\left(x + \frac{b}{2a}\right)^2 + \left(c - \frac{b^2}{4a}\right),$$

which, defining $B \equiv b/2a$ and $C \equiv c - b^2/4a$, simplifies to

$$a(x+B)^2 + C.$$

Completion

A METRIC SPACE X which is not complete has a CAUCHY SEQUENCE which does not CONVERGE. The completion of X is obtained by adding the limits to the Cauchy sequences. The completion is always COMPLETE.

For example, the rational numbers, with the distance metric, are not complete because there exist CAUCHY SEQUENCES that do not converge, e.g., 1, 1.4, 1.41, 1.414, ... does not converge because $\sqrt{2}$ is not rational. The completion of the rationals is the real numbers. Note that the completion depends on the METRIC. For instance, for any PRIME p, the rationals have a METRIC given by the P-ADIC NORM, and then the completion of the rationals is the set of P-ADIC NUMBERS. Another common example of a completion is the space of $L2$-FUNCTIONS.

Technically speaking, the completion of X is the set of CAUCHY SEQUENCES and X is contained in this set, ISOMETRICALLY, as the constant sequences.

See also CAUCHY SEQUENCE, $L2$-SPACE, LOCAL FIELD, METRIC SPACE, P-ADIC NUMBER, REAL NUMBER

Complex

CW-COMPLEX, SIMPLICIAL COMPLEX

Complex Addition

Two COMPLEX NUMBERS $z = x + iy$ and $z' = x' + iy'$ are added together componentwise,

$$z + z' = (x + x') + i(y + y').$$

In component form,

$$(x, y) + (x', y') = (x + x', y + y')$$

(Krantz 1999, p. 1).

See also COMPLEX DIVISION, COMPLEX MULTIPLICATION, COMPLEX NUMBER, VECTOR ADDITION

References

Krantz, S. G. *Handbook of Complex Analysis.* Boston, MA: Birkhäuser, p. 1, 1999.

Complex Analysis

The study of COMPLEX NUMBERS, their DERIVATIVES, manipulation, and other properties. Complex analysis is an extremely powerful tool with an unexpectedly large number of practical applications to the solution of physical problems. CONTOUR INTEGRATION, for example, provides a method of computing difficult INTEGRALS by investigating the singularities of the function in regions of the COMPLEX PLANE near and between the limits of integration.

The most fundamental result of complex analysis is the CAUCHY-RIEMANN EQUATIONS, which give the conditions a FUNCTION must satisfy in order for a complex generalization of the DERIVATIVE, the so-called COMPLEX DERIVATIVE, to exist. When the COMPLEX DERIVATIVE is defined "everywhere," the function is said to be ANALYTIC. A single example of the unexpected power of complex analysis is PICARD'S THEOREM, which states that an ANALYTIC FUNCTION assumes every COMPLEX NUMBER, with possibly one exception, infinitely often in any NEIGHBORHOOD of an ESSENTIAL SINGULARITY!

See also ANALYTIC CONTINUATION, ARGUMENT PRINCIPLE, BRANCH CUT, BRANCH POINT, CAUCHY INTEGRAL FORMULA, CAUCHY INTEGRAL THEOREM, CAUCHY PRINCIPAL VALUE, CAUCHY-RIEMANN EQUATIONS, COMPLEX NUMBER, CONFORMAL MAPPING, CONTOUR INTEGRATION, DE MOIVRE'S IDENTITY, EULER FORMULA, INSIDE-OUTSIDE THEOREM, JORDAN'S LEMMA, LAURENT SERIES, LIOUVILLE'S CONFORMALITY THEOREM, MONOGENIC FUNCTION, MORERA'S THEOREM, PERMANENCE OF ALGEBRAIC FORM, PICARD'S THEOREM, POLE, POLYGENIC FUNCTION, RESIDUE (COMPLEX ANALYSIS)

References

Arfken, G. "Functions of a Complex Variable I: Analytic Properties, Mapping" and "Functions of a Complex Variable II: Calculus of Residues." Chs. 6– in *Mathematical Methods for Physicists, 3rd ed.* Orlando, FL: Academic Press, pp. 352–95 and 396–36, 1985.

Boas, R. P. *Invitation to Complex Analysis.* New York: Random House, 1987.
Churchill, R. V. and Brown, J. W. *Complex Variables and Applications, 6th ed.* New York: McGraw-Hill, 1995.
Conway, J. B. *Functions of One Complex Variable, 2nd ed.* New York: Springer-Verlag, 1995.
Forsyth, A. R. *Theory of Functions of a Complex Variable, 3rd ed.* Cambridge, England: Cambridge University Press, 1918.
Knopp, K. *Theory of Functions Parts I and II, Two Volumes Bound as One, Part I.* New York: Dover, 1996.
Krantz, S. G. *Handbook of Complex Analysis.* Boston, MA: Birkhäuser, 1999.
Lang, S. *Complex Analysis, 3rd ed.* New York: Springer-Verlag, 1993.
Morse, P. M. and Feshbach, H. "Functions of a Complex Variable" and "Tabulation of Properties of Functions of Complex Variables." Ch. 4 in *Methods of Theoretical Physics, Part I.* New York: McGraw-Hill, pp. 348–91 and 480–85, 1953.
Needham, T. *Visual Complex Analysis.* New York: Clarendon Press, 2000.
Silverman, R. A. *Introductory Complex Analysis.* New York: Dover, 1984.
Weisstein, E. W. "Books about Complex Analysis." http://www.treasure-troves.com/books/ComplexAnalysis.html.

Complex Conjugate

The complex conjugate of a COMPLEX NUMBER $z \equiv a + bi$ is defined to be

$$\bar{z} \equiv a - bi. \tag{1}$$

Note that there are several notations in common use for the complex conjugate. Older physics and engineering texts tend to prefer z^* (Bekefi and Barrett 1987, p. 616; Arfken 1985, p. 356; Harris and Stocker 1998, p. 21; Hecht 1998, p. 18; Herkommer 1999, p. 262), while many modern math and physics texts favor $\bar{z}$ (Abramowitz and Stegun 1972, p. 16; Kaplan 1981, p. 28; Roman 1987, p. 534; Kreyszig 1988, p. 568; Kaplan 1992, p. 572; Harris and Stocker 1998, p. 21; Krantz 1999, p. 2; Anton 2000, p. 528). In the latter case, the notation z^* is then reserved to denote the ADJOINT operator, which is denoted $z^\dagger$ in many older physics texts. In this work, $\bar{z}$ is used to denote the complex conjugate, and z^* is used to denote the ADJOINT.

The CONJUGATE MATRIX of a MATRIX $A = (a_{ij})$ is the MATRIX obtained by replacing each element a_{ij} with its complex conjugate, $\bar{A} = (\bar{a}_{ij})$ (Arfken 1985, p. 210). The complex conjugate is implemented in *Mathematica* as `Conjugate[z]`.

The common notational conventions are summarized in the table below.

convention	complex conjugate	ADJOINT
mathematics	$\bar{A}$	A^*

engineering	A*	A†

By definition, the complex conjugate satisfies

$$\bar{\bar{z}} = z. \tag{2}$$

The complex conjugate is DISTRIBUTIVE under COMPLEX ADDITION,

$$\overline{z_1 + z_2} = \overline{z_1} + \overline{z_2}, \tag{3}$$

since

$$\overline{(a_1 + ib_1) + (a_2 + ib_2)} = \overline{(a_1 + a_2) + i(b_1 + b_2)}$$
$$= (a_1 + a_2) - i(b_1 + b_2) = (a_1 - ib_1) + (a_2 - ib_2)$$
$$= \overline{a_1 + ib_1} + \overline{a_2 + ib_2},$$

and DISTRIBUTIVE over COMPLEX MULTIPLICATION,

$$\overline{z_1 z_2} = \bar{z}_1 \bar{z}_2, \tag{4}$$

since

$$\overline{(a_1 + b_1 i)(a_2 + b_2 i)} = \overline{(a_1 a_2 - b_1 b_2) + i(a_1 b_2 + a_2 b_1)}$$
$$= (a_1 a_2 - b_1 b_2) - i(a_1 b_2 + a_2 b_1) = (a_1 - ib_1)(a_2 - ib_2)$$
$$= \overline{a_1 + ib_1} \, \overline{a_2 + ib_2}.$$

See also ADJOINT MATRIX, COMPLEX ANALYSIS, COMPLEX DIVISION, COMPLEX NUMBER, CONJUGATE MATRIX, MODULUS (COMPLEX NUMBER)

References

Abramowitz, M. and Stegun, C. A. (Eds.). *Handbook of Mathematical Functions with Formulas, Graphs, and Mathematical Tables, 9th printing.* New York: Dover, p. 16, 1972.
Anton, H. *Elementary Linear Algebra, 8th ed.* New York: Wiley, 2000.
Arfken, G. *Mathematical Methods for Physicists, 3rd ed.* Orlando, FL: Academic Press, pp. 355–56, 1985.
Bekefi, G. and Barrett, A. H. *Electromagnetic Vibrations, Waves, and Radiation.* Cambridge, MA: MIT Press, p. 616, 1987.
Hecht, E. *Optics, 3rd ed.* Reading, MA: Addison-Wesley, p. 18, 1998.
Herkommer, M. A. *Number Theory: A Programmer's Guide.* New York: McGraw-Hill, p. 262, 1999.
Harris, J. W. and Stocker, H. *Handbook of Mathematics and Computational Science.* New York: Springer-Verlag, p. 21, 1998.
Kaplan, W. *Advanced Calculus, 4th ed.* Reading, MA: Addison-Wesley, 1992.
Kaplan, W. *Advanced Mathematics for Engineers.* Reading, MA: Addison-Wesley, 1981.
Krantz, S. G. "Complex Conjugate." §1.1.3 in *Handbook of Complex Analysis.* Boston, MA: Birkhäuser, p. 2, 1999.
Kreyszig, E. *Advanced Engineering Mathematics, 6th ed.* New York: Wiley, p. 568, 1988.
Roman, S. "The Conjugate of a Complex Number and Complex Division." §11.2 in *College Algebra and Trigonometry.* San Diego, CA: Harcourt, Brace, Jovanovich, pp. 534–41, 1987.

Complex Derivative

A DERIVATIVE of a COMPLEX function, which must satisfy the CAUCHY-RIEMANN EQUATIONS in order to be COMPLEX DIFFERENTIABLE.

See also CAUCHY-RIEMANN EQUATIONS, COMPLEX DIFFERENTIABLE, DERIVATIVE

References

Krantz, S. G. "The Complex Derivative." §1.3.5 and 2.2.3 in *Handbook of Complex Analysis.* Boston, MA: Birkhäuser, pp. 15–6 and 24, 1999.

Complex Differentiable

Let $z = x + iy$ and $f(z) = u(x, y) + iv(x, y)$ on some region G containing the point z_0. If $f(z)$ satisfies the CAUCHY-RIEMANN EQUATIONS and has continuous first PARTIAL DERIVATIVES at z_0, then $f'(z_0)$ exists and is given by

$$f'(z_0) = \lim_{z \to z_0} \frac{f(z) - f(z_0)}{z - z_0},$$

and the function is said to be COMPLEX DIFFERENTIABLE (or, equivalently, ANALYTIC, HOLOMORPHIC, or regular).

A function $f : \mathbb{C} \to \mathbb{C}$ can be thought of as a map from the plane to the plane, $f : \mathbb{R}^2 \to \mathbb{R}^2$. Then f is complex differentiable iff its JACOBIAN is of the form

$$\begin{bmatrix} a & -b \\ b & a \end{bmatrix}$$

at every point. That is, its derivative is given by the multiplication of a COMPLEX NUMBER $a + bi$. For instance, the function $f(z) = \bar{z}$, where $\bar{z}$ is the COMPLEX CONJUGATE, is *not* complex differentiable.

See also ANALYTIC FUNCTION, CAUCHY-RIEMANN EQUATIONS, COMPLEX DERIVATIVE, DIFFERENTIABLE, ENTIRE FUNCTION, HOLOMORPHIC FUNCTION, PSEUDOANALYTIC FUNCTION

References

Krantz, S. G. "Alternative Terminology for Holomorphic Functions." §1.3.6 in *Handbook of Complex Analysis.* Boston, MA: Birkhäuser, p. 16, 1999.

Complex Division

The division of two COMPLEX NUMBERS can be accomplished by multiplying the NUMERATOR and DENOMINATOR by the COMPLEX CONJUGATE of the DENOMINATOR, for example, with $z_1 = a + bi$ and $z_2 = c + di$, $z = z_1/z_2$ is given by

$$z = \frac{a + bi}{c + di} = \frac{(a + bi)\overline{c + di}}{(c + di)\overline{c + di}} = \frac{(a + bi)(c - di)}{(c + di)(c - di)}$$
$$= \frac{(ac + bd) + i(bc - ad)}{c^2 + d^2},$$

where $\bar{z}$ denotes the COMPLEX CONJUGATE. In component notation,

$$\frac{(x, y)}{(x', y')} = \left(\frac{xx' + yy'}{\sqrt{x'^2 + y'^2}}, \frac{yx' - xy'}{\sqrt{x'^2 + y'^2}} \right).$$

See also COMPLEX ADDITION, COMPLEX MULTIPLICATION, COMPLEX NUMBER, DIVISION

Complex Form (Type)

The DIFFERENTIAL FORMS on $\mathbb{C}^n$ decompose into forms of type (p, q). For example, on $\mathbb{C}$, the EXTERIOR ALGEBRA decomposes into four types:

$$\wedge \mathbb{C} = \wedge^0 \oplus \wedge^{1,0} \oplus \wedge^{0,1} \oplus \wedge^{1,1}$$

$$= \langle 1 \rangle \oplus \langle dz \rangle \oplus \langle d\bar{z} \rangle \oplus \langle dz \wedge d\bar{z} \rangle, \quad (1)$$

where $dz = dx + i\,dy$, $d\bar{z} = dx - i\,dy$, and $\otimes$ denotes the DIRECT SUM. In general, a (p, q)-form is the sum of terms with p dzs and q $d\bar{z}$s. A k-form decomposes into a sum of (p, q)-forms, where $k = p + q$.

For example, the 2-forms on $\mathbb{C}^2$ decompose as

$$\wedge^2 \mathbb{C}^2 = \wedge^{2,0} \oplus \wedge^{1,1} \oplus \wedge^{0,2} \quad (2)$$

$$= \langle dz_1 \wedge dz_2 \rangle \oplus \langle dz_1 \wedge d\bar{z}_1, dz_1 \wedge d\bar{z}_2, dz_2 \wedge d\bar{z}_1,$$

$$dz_2 \wedge d\bar{z}_2 \rangle \oplus \langle d\bar{z}_1 \wedge d\bar{z}_2 \rangle. \quad (3)$$

The decomposition into forms of type (p, q) is preserved by HOLOMORPHIC MAPS. More precisely, when $f : X \to Y$ is holomorphic and α is a (p, q)-form on Y, then the PULLBACK $f^*\alpha$ is a (p, q)-form on X.

Recall that the EXTERIOR ALGEBRA is generated by the ONE-FORMS, by WEDGE PRODUCT and addition. Then the forms of type (p, q) are generated by

$$\wedge^p(\Lambda^{1, 0}) \wedge \wedge^q(\Lambda^{0, 1}). \quad (4)$$

The SUBSPACE $\Lambda^{1, 0}$ of the complex one-forms can be identified as the $+i$-EIGENSPACE of the ALMOST COMPLEX STRUCTURE J, which satisfies $J^2 = -I$. Similarly, the $-i$-EIGENSPACE is the SUBSPACE $\wedge^{0, 1}$. In fact, the decomposition of $TX \otimes \mathbb{C} = TX^{1, 0} \oplus TX^{0, 1}$ determines the ALMOST COMPLEX STRUCTURE J on TX.

More abstractly, the forms into type (p, q) are a REPRESENTATION of $\mathbb{C}^*$, where λ acts by multiplication by $\lambda^p \bar{\lambda}^q$.

See also ALMOST COMPLEX STRUCTURE, COMPLEX MANIFOLD, DEL BAR OPERATOR, DOLBEAULT COHOMOLOGY

References

Griffiths, P. and Harris, J. *Principles of Algebraic Geometry.* New York: Wiley, pp. 106–26, 1994.
Weil, A. *Introduction à l'étude des variétés Kähleriennes.* Publications de l'Institut de Mathématiques de l'Université de Nancago, VI, Actualites Scientifiques et Industrielles, no. 1267. Paris: Hermann, 1958.
Wells, R. O. *Differential Analysis on Complex Manifolds.* New York: Springer-Verlag, 1980.

Complex Fraction

A FRACTION in which NUMERATOR and DENOMINATOR are themselves fractions.

See also COMMON FRACTION, FRACTION

Complex Function

A FUNCTION whose RANGE is in the COMPLEX NUMBERS is said to be a complex function, or a complex-valued function.

See also REAL FUNCTION, SCALAR FUNCTION, VECTOR FUNCTION

Complex Infinity

An infinite number in the COMPLEX PLANE whose ARGUMENT is unknown.

See also C*, DIVISION BY ZERO, EXTENDED COMPLEX PLANE, INFINITY, POINT AT INFINITY, RIEMANN SPHERE

Complex Line Integral

LINE INTEGRAL

Complex Manifold

A complex manifold is a MANIFOLD M whose COORDINATE CHARTS are open subsets of $\mathbb{C}^n$ and the TRANSITION FUNCTIONS between charts are HOLOMORPHIC FUNCTIONS. Naturally, a complex manifold of dimension n also has the structure of a REAL SMOOTH MANIFOLD of dimension $2n$.

A function $f : M \to \mathbb{C}$ is HOLOMORPHIC if it is HOLOMORPHIC in every COORDINATE CHART. Similarly, a map $f : M \to N$ is HOLOMORPHIC if its restrictions to coordinate charts on N are holomorphic. Two complex manifolds M and N are considered equivalent if there is a map $f : M \to N$ which is a DIFFEOMORPHISM and whose inverse is HOLOMORPHIC.

See also ALGEBRAIC VARIETY, CONFORMAL MAPPING, HOLOMORPHIC FUNCTION, MANIFOLD, RIEMANN SURFACE, STEIN MANIFOLD

Complex Matrix

A MATRIX whose elements may contain COMPLEX NUMBERS.

The MATRIX PRODUCT of two 2×2 complex matrices is given by

$$\begin{bmatrix} x_{11} + y_{11}i & x_{12} + y_{12}i \\ x_{21} + y_{21}i & x_{22} + y_{22}i \end{bmatrix} \begin{bmatrix} u_{11} + v_{11}i & u_{12} + v_{12}i \\ u_{21} + v_{21}i & u_{22} + v_{22}i \end{bmatrix}$$

$$= \begin{bmatrix} R_{11} & R_{12} \\ R_{21} & R_{22} \end{bmatrix} + i \begin{bmatrix} I_{11} & I_{12} \\ I_{21} & I_{22} \end{bmatrix},$$

where

$$R_{11} = u_{11}x_{11} + u_{21}x_{21} - v_{11}y_{11} - v_{21}y_{12}$$

$$R_{12} = u_{12}x_{11} + u_{22}x_{12} - v_{11}y_{11} - v_{22}y_{12}$$

$$R_{21} = u_{11}x_{21} + u_{21}x_{22} - v_{11}y_{21} - v_{21}y_{22}$$

$$R_{22} = u_{12}x_{21} + u_{22}x_{22} - v_{12}y_{21} - v_{22}y_{22}$$

$$I_{11} = v_{11}x_{11} + v_{21}x_{21} + u_{11}y_{11} + u_{21}y_{12}$$

$$I_{12} = v_{12}x_{11} + v_{22}x_{12} + u_{12}y_{11} + u_{22}y_{12}$$

$$I_{21} = v_{11}x_{21} + u_{21}x_{22} + u_{11}y_{21} + u_{21}y_{22}$$

$$I_{22} = v_{12}x_{21} + v_{22}x_{22} + u_{12}y_{21} + u_{22}y_{22}.$$

Hadamard (1893) proved that the DETERMINANT of any complex $n \times n$ matrix A with entries in the closed UNIT DISK $|a_{ij}| \leq 1$ satisfies

$$|\det A| \leq n^{n/2} \qquad (1)$$

(HADAMARD'S MAXIMUM DETERMINANT PROBLEM), with equality attained by the VANDERMONDE MATRIX of the n ROOTS OF UNITY (Faddeev and Sominskii 1965, p. 331; Brenner 1972). The first few values for $n = 1, 2, \ldots$ are $1, 2, 3\sqrt{3}, 16, 25\sqrt{5}, 216, \ldots$.

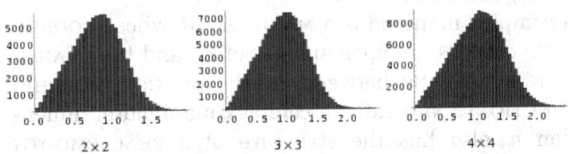

Studying the maximum possible eigenvalue norms for random complex $n \times n$ matrices is computationally intractable. Although average properties of the distribution of $|\lambda|$ can be determined, finding the maximum value corresponds to determining if the set of matrices contains a SINGULAR MATRIX, which has been proven to be an NP-COMPLETE PROBLEM (Poljak and Rohn 1993, Kaltofen 1999). The above plots show the distributions for 2×2, 3×3, and 4×4 matrix eigenvalue norms for elements uniformly distributed inside the unit disk $|z| \leq 1$. Similar plots are obtained for elements uniformly distributed inside $|\Re[z]|, |\Im[z]| \leq 1$. The exact distribution of eigenvalues for complex matrices with both real and imaginary parts distributed as independent standard normal variates is given by Ginibre (1965), Hwang (1986), and Mehta (1991).

See also COMPLEX VECTOR, HADAMARD'S MAXIMUM DETERMINANT PROBLEM, INTEGER MATRIX, k-MATRIX, MATRIX, REAL MATRIX

References

Brenner, J. and Cummings, L. "The Hadamard Maximum Determinant Problem." *Amer. Math. Monthly* **79**, 626–30, 1972.

Edelman, A. "The Probability that a Random Real Gaussian Matrix has k Real Eigenvalues, Related Distributions, and the Circular Law." *J. Multivariate Anal.* **60**, 203–32, 1997.

Faddeev, D. K. and Sominskii, I. S. *Problems in Higher Algebra.* San Francisco: W. H. Freeman, 1965.

Ginibre, J. "Statistical Ensembles of Complex, Quaternion, and Real Matrices." *J. Math. Phys.* **6**, 440–49, 1965.

Hadamard, J. "Résolution d'une question relative aux déterminants." *Bull. Sci. Math.* **17**, 30–1, 1893.

Hwang, C. R. "A Brief Survey on the Spectral Radius and the Spectral Distribution of Large Random Matrices with i.i.d. Entries." In *Random Matrices and Their Applications.* Providence, RI: Amer. Math. Soc., pp. 145–52, 1986.

Kaltofen, E. "Challenges of Symbolic Computation: My Favorite Open Problems." Submitted to *J. Symb. Comput.*

Mehta, M. L. *Random Matrices, 2nd rev. enl. ed.* New York: Academic Press, 1991.

Poljak, S. and Rohn, J. "Checking Robust Nonsingularity is NP-Hard." *Math. Control Signals Systems* **6**, 1–, 1993.

Complex Measure

A MEASURE which takes values in the COMPLEX NUMBERS. The set of complex measures on a MEASURE SPACE X forms a VECTOR SPACE. Note that this is not the case for the more common POSITIVE MEASURES. Also, the space of finite measures ($|\mu(X)| < \infty$) has a norm given by the TOTAL VARIATION MEASURE $\|\mu\| = |\mu|(X)|$, which makes it a BANACH SPACE.

Using the POLAR REPRESENTATION of μ, it is possible to define the LEBESGUE INTEGRAL using a complex measure,

$$\int f \, d\mu = \int e^{i\theta} f \, d|\mu|.$$

Sometimes, the term "complex measure" is used to indicate an arbitrary measure. The definitions for measure can be extended to measures which take values in any VECTOR SPACE. For instance in SPECTRAL THEORY, measures on $\mathbb{C}$, which take values in the bounded linear maps from a HILBERT SPACE to itself, represent the SPECTRUM of an operator.

See also BANACH SPACE, LEBESGUE INTEGRAL, MEASURE, MEASURE SPACE, POLAR REPRESENTATION (MEASURE), SPECTRAL THEORY

References

Rudin, W. *Real and Complex Analysis.* New York: McGraw-Hill, pp. 116–32, 1987.

Complex Modulus

MODULUS (COMPLEX NUMBER)

Complex Multiplication

Two COMPLEX NUMBERS $x = a + ib$ and $y = c + id$ are multiplied as follows:

$$xy = (a + ib)(c + id) = ac + ibc + iad - bd$$

$$= (ac - bd) + i(ad + bc).$$

In component form,

$$(x, y)(x', y') = (xx' - yy', xy' + yx') \qquad (1)$$

(Krantz 1999, p. 1). The special case of a COMPLEX NUMBER multiplied by a SCALAR a is then given by

$$(x, y)(x', y') = (a, 0)(x, y) = (ax, ay). \qquad (2)$$

Surprisingly, complex multiplication can be carried out using only three REAL multiplications, ac, bd, and $(a + b)(c + d)$ as

$$\Re[(a + ib)(c + id)] = ac - bd$$

$$\Im[(a + ib)(c + id)] = (a + b)(c + d) - ac - bd.$$

Complex multiplication has a special meaning for ELLIPTIC CURVES.

See also COMPLEX ADDITION, COMPLEX DIVISION, COMPLEX NUMBER, ELLIPTIC CURVE, IMAGINARY PART, MULTIPLICATION, REAL PART

References

Cox, D. A. *Primes of the Form $x^2 + ny^2$: Fermat, Class Field Theory and Complex Multiplication.* New York: Wiley, 1997.

Krantz, S. G. *Handbook of Complex Analysis.* Boston, MA: Birkhäuser, p. 1, 1999.

Complex Number

The complex numbers are the FIELD $\mathbb{C}$ of numbers OF THE FORM $x + iy$, where x and y are REAL NUMBERS and I is the IMAGINARY UNIT equal to the SQUARE ROOT of -1, $\sqrt{-1}$. When a single letter $z = x + iy$ is used to denote a complex number, it is sometimes called an "AFFIX." In component notation, $z = x + iy$ can be written (x, y). The FIELD of complex numbers includes the FIELD of REAL NUMBERS as a SUBFIELD.

The set of complex numbers is implemented in *Mathematica* as `Complexes`. A number x can then be tested to see if it is complex using the command `Element[x, Complexes]`.

Through the EULER FORMULA, a complex number

$$z = x + iy \qquad (1)$$

may be written in "PHASOR" form

$$z = |z|(\cos \theta + i \sin \theta) = |z|e^{i\theta}. \qquad (2)$$

Here, $|z|$ is known as the MODULUS and θ is known as the ARGUMENT or PHASE. The ABSOLUTE SQUARE of z is defined by $|z|^2 = z\bar{z}$, with $\bar{z}$ the COMPLEX CONJUGATE, and the argument may be computed from

$$\arg(z) = \theta = \tan^{-1}\left(\frac{y}{x}\right). \qquad (3)$$

DE MOIVRE'S IDENTITY relates POWERS of complex numbers

$$z^n = |z|^n[\cos(n\theta) + i \sin(n\theta)]. \qquad (4)$$

COMPLEX DIVISION and COMPLEX MULTIPLICATION can also be defined for complex numbers.

Finally, the REAL $\Re(z)$ and IMAGINARY PARTS $\Im(z)$ are given by

$$\Re(z) = \tfrac{1}{2}(z + \bar{z}) \qquad (5)$$

$$\Im(z) = \frac{z - \bar{z}}{2i} = -\tfrac{1}{2}i(z - \bar{z}) = \tfrac{1}{2}i(\bar{z} - z). \qquad (6)$$

The POWERS of complex numbers can be written in closed form as follows:

$$z^n = \left[x^n - \binom{n}{2}x^{n-2}y^2 + \binom{n}{4}x^{n-4}y^4 - \cdots\right]$$
$$+ i\left[\binom{n}{1}x^{n-1}y - \binom{n}{3}x^{n-3}y^3 + \cdots\right]. \qquad (7)$$

The first few are explicitly

$$z^2 = (x^2 - y^2) + i(2xy) \qquad (8)$$

$$z^3 = (x^3 - 3xy^2) + i(3x^2y - y) \qquad (9)$$

$$z^4 = (x^4 - 6x^2y^2 + y^4) + i(4x^3y - 4xy^3) \qquad (10)$$

$$z^5 = (x^5 - 10x^3y^2 + 5xy^4) + i(5x^4y - 10x^2y^3 + y^5) \qquad (11)$$

(Abramowitz and Stegun 1972).

See also ABSOLUTE SQUARE, ARGUMENT (COMPLEX NUMBER), COMPLEX DIVISION, COMPLEX MULTIPLICATION, COMPLEX PLANE, I, IMAGINARY NUMBER, MODULUS (COMPLEX NUMBER), PHASE, PHASOR, REAL NUMBER, SURREAL NUMBER

References

Abramowitz, M. and Stegun, C. A. (Eds.). *Handbook of Mathematical Functions with Formulas, Graphs, and Mathematical Tables, 9th printing.* New York: Dover, pp. 16–7, 1972.

Arfken, G. *Mathematical Methods for Physicists, 3rd ed.* Orlando, FL: Academic Press, pp. 353–57, 1985.

Bold, B. "Complex Numbers." Ch. 3 in *Famous Problems of Geometry and How to Solve Them.* New York: Dover, pp. 19–7, 1982.

Courant, R. and Robbins, H. "Complex Numbers." §2.5 in *What is Mathematics?: An Elementary Approach to Ideas and Methods, 2nd ed.* Oxford, England: Oxford University Press, pp. 88–03, 1996.

Ebbinghaus, H. D.; Hirzebruch, F.; Hermes, H.; Prestel, A; Koecher, M.; Mainzer, M.; and Remmert, R. *Numbers.* New York: Springer-Verlag, 1990.

Krantz, S. G. "Complex Arithmetic." §1.1 in *Handbook of Complex Analysis.* Boston, MA: Birkhäuser, pp. 1–, 1999.

Morse, P. M. and Feshbach, H. "Complex Numbers and Variables." §4.1 in *Methods of Theoretical Physics, Part I.* New York: McGraw-Hill, pp. 349–56, 1953.

Nahin, P. J. *An Imaginary Tale: The Story of $\sqrt{-1}$.* Princeton, NJ: Princeton University Press, 1998.

Press, W. H.; Flannery, B. P.; Teukolsky, S. A.; and Vetter-
ling, W. T. "Complex Arithmetic." §5.4 in *Numerical
Recipes in FORTRAN: The Art of Scientific Computing,
2nd ed.* Cambridge, England: Cambridge University
Press, pp. 171–72, 1992.
Wells, D. *The Penguin Dictionary of Curious and Interesting
Numbers.* Middlesex, England: Penguin Books, pp. 21–3,
1986.

Complex Plane

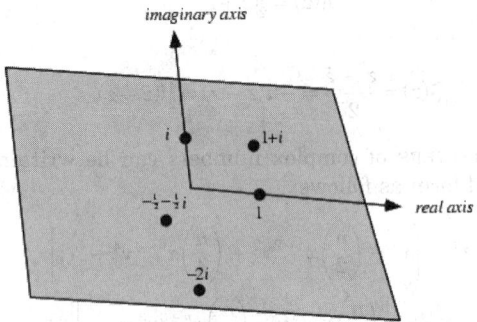

The plane of COMPLEX NUMBERS spanned by the
vectors 1 and i, where i is the IMAGINARY NUMBER.
Every COMPLEX NUMBER corresponds to a unique
POINT in the complex plane. The LINE in the plane
with $i = 0$ is the REAL LINE. The complex plane is
sometimes called the ARGAND PLANE or GAUSS PLANE,
and a plot of COMPLEX NUMBERS in the plane is
sometimes called an ARGAND DIAGRAM.

See also AFFINE COMPLEX PLANE, ARGAND DIAGRAM,
ARGAND PLANE, BERGMAN SPACE, C*, COMPLEX
PROJECTIVE PLANE, EXTENDED COMPLEX PLANE,
ISOTROPIC LINE, LEFT HALF-PLANE, LOWER HALF-
DISK, LOWER HALF-PLANE, RIGHT HALF-PLANE,
UPPER HALF-DISK, UPPER HALF-PLANE

References
Courant, R. and Robbins, H. "The Geometric Interpretation
of Complex Numbers." §5.2 in *What is Mathematics?: An
Elementary Approach to Ideas and Methods, 2nd ed.*
Oxford, England: Oxford University Press, pp. 92–7,
1996.
Krantz, S. G. "The Topology of the Complex Plane." §1.1.5 in
Handbook of Complex Analysis. Boston, MA: Birkhäuser,
pp. 3–, 1999.

Complex Projective Plane
The set $\mathbb{P}^2$ is the set of all EQUIVALENCE CLASSES
$[a, b, c]$ of ordered triples $(a, b, c) \in \mathbb{C}^3 \backslash (0, 0, 0)$ under the equivalence relation $(a, b, c) \sim (a', b', c')$ if
$(a, b, c) = (\lambda a', \lambda b', \lambda c')$ for some NONZERO COMPLEX
NUMBER λ.

See also COMPLEX PROJECTIVE PLANE

Complex Projective Space

See also COMPLEX SPACE, REAL PROJECTIVE SPACE

Complex Representation
PHASOR

Complex Space

See also COMPLEX PROJECTIVE SPACE, REAL SPACE,
TWISTOR SPACE

Complex Structure
The complex structure of a point $\mathbf{x} = x_1, x_2$ in the
PLANE is defined by the linear MAP $J : \mathbb{R}^2 \to \mathbb{R}^2$

$$J(x_1, x_2) = (-x_2, x_1),$$

and corresponds to a clockwise rotation by $\pi/2$. This
map satisfies

$$J^2 = -I$$
$$(J\mathbf{x}) \cdot (J\mathbf{y}) = \mathbf{x} \cdot \mathbf{y}$$
$$(J\mathbf{x}) \cdot \mathbf{x} = 0,$$

where I is the IDENTITY MAP.

More generally, if V is a 2-D VECTOR SPACE, a linear
map $J : V \to V$ such that $J^2 = -I$ is called a complex
structure on V. If $V = \mathbb{R}^2$, this collapses to the
previous definition.

See also MODULI SPACE

References
Gray, A. *Modern Differential Geometry of Curves and
Surfaces with Mathematica, 2nd ed.* Boca Raton, FL:
CRC Press, pp. 4 and 247, 1997.

Complex System

References
Goles, E. and Martínez, S. (Eds.). *Cellular Automata and
Complex Systems.* Amsterdam, Netherlands: Kluwer,
1999.

Complex Vector
A VECTOR whose elements are COMPLEX NUMBERS.

See also COMPLEX NUMBER, REAL VECTOR, VECTOR

Complex Vector Bundle
A complex vector bundle is a VECTOR BUNDLE $\pi : E \to
M$ whose FIBER $\pi^{-1}(x)$ is a COMPLEX VECTOR SPACE. It
is not necessarily a COMPLEX MANIFOLD, even if its
BASE MANIFOLD M is a COMPLEX MANIFOLD. If a
complex vector bundle also has the structure of a
COMPLEX MANIFOLD, and π is HOLOMORPHIC, then it is
called a HOLOMORPHIC VECTOR BUNDLE.

See also BUNDLE, COMPLEX VECTOR SPACE, HOLO-MORPHIC VECTOR BUNDLE, MANIFOLD, VECTOR SPACE

Complex Vector Space

A complex vector space is a VECTOR SPACE whose FIELD of scalars is the COMPLEX numbers. A linear transformation between complex vector spaces is given by a matrix with complex entries (i.e., a COMPLEX MATRIX).

See also BASIS (VECTOR SPACE), COMPLEX STRUCTURE, LINEAR TRANSFORMATION, REAL VECTOR SPACE, VECTOR SPACE

Complexes

COMPLEX NUMBER

Complexity (Number)

The number of 1s needed to represent an INTEGER using only additions, multiplications, and parentheses are called the integer's complexity. For example,

$$1 = 1$$
$$2 = 1 + 1$$
$$3 = 1 + 1 + 1$$
$$4 = (1+1)(1+1) = 1 + 1 + 1 + 1$$
$$5 = (1+1)(1+1) + 1 = 1 + 1 + 1 + 1 + 1$$
$$6 = (1+1)(1+1+1)$$
$$7 = (1+1)(1+1+1) + 1$$
$$8 = (1+1)(1+1)(1+1)$$
$$9 = (1+1+1)(1+1+1)$$
$$10 = (1+1+1)(1+1+1) + 1$$
$$= (1+1)(1+1+1+1+1)$$

So, for the first few n, the complexity is 1, 2, 3, 4, 5, 5, 6, 6, 6, 7, 8, 7, 8, ... (Sloane's A005245).

References

Guy, R. K. "Expressing Numbers Using Just Ones." §F26 in *Unsolved Problems in Number Theory, 2nd ed.* New York: Springer-Verlag, p. 263, 1994.

Guy, R. K. "Some Suspiciously Simple Sequences." *Amer. Math. Monthly* **93**, 186–90, 1986.

Guy, R. K. "Monthly Unsolved Problems, 1969–987." *Amer. Math. Monthly* **94**, 961–70, 1987.

Guy, R. K. "Unsolved Problems Come of Age." *Amer. Math. Monthly* **96**, 903–09, 1989.

Rawsthorne, D. A. "How Many 1's are Needed?" *Fib. Quart.* **27**, 14–7, 1989.

Sloane, N. J. A. Sequences A005245/M0457 in "An On-Line Version of the Encyclopedia of Integer Sequences." http://www.research.att.com/~njas/sequences/eisonline.html.

Complexity (Sequence)

BLOCK GROWTH

Complexity Theory

The theory of classifying problems based on how difficult they are to solve. A problem is assigned to the P-PROBLEM (polynomial time) class if the number of steps needed to solve it is bounded by some POWER of the problem's size. A problem is assigned to the NP-PROBLEM (nondeterministic polynomial time) class if it permits a nondeterministic solution and the number of steps of the solution is bounded by some power of the problem's size. The class of P-PROBLEMS is a subset of the class of NP-PROBLEMS, but there also exist problems which are not NP.

If a solution is known to an NP-PROBLEM, it can be reduced to a single period verification. A problem is NP-COMPLETE if an ALGORITHM for solving it can be translated into one for solving any other NP-PROBLEM. Examples of NP-COMPLETE PROBLEMS include the HAMILTONIAN CYCLE and TRAVELING SALESMAN PROBLEMS. LINEAR PROGRAMMING, thought to be an NP-PROBLEM, was shown to actually be a P-PROBLEM by L. Khachian in 1979. It is not known if all apparently NP-PROBLEMS are actually P-PROBLEMS.

See also BIT COMPLEXITY, NP-COMPLETE PROBLEM, NP-PROBLEM, P-PROBLEM

References

Bridges, D. S. *Computability.* New York: Springer-Verlag, 1994.

Brookshear, J. G. *Theory of Computation: Formal Languages, Automata, and Complexity.* Redwood City, CA: Benjamin/Cummings, 1989.

Cooper, S. B.; Slaman, T. A.; and Wainer, S. S. (Eds.). *Computability, Enumerability, Unsolvability: Directions in Recursion Theory.* New York: Cambridge University Press, 1996.

Davis, M. *Computability and Unsolvability.* New York: Dover, 1982.

Du, D.-Z. and Ko, K.-I. *Theory of Computational Complexity.* New York; Wiley, 2000.

Garey, M. R. and Johnson, D. S. *Computers and Intractability: A Guide to the Theory of NP-Completeness.* New York: W. H. Freeman, 1983.

Goetz, P. "Phil Goetz's Complexity Dictionary." http://www.cs.buffalo.edu/~goetz/dict.html.

Griffor, E. R. (Ed.). *Handbook of Computability Theory.* Amsterdam, Netherlands: Elsevier, 1999.

Hopcroft, J. E. and Ullman, J. D. *Introduction to Automated Theory, Languages, and Computation.* Reading, MA: Addison-Wesley, 1979.

Lewis, H. R. and Papadimitriou, C. H. *Elements of the Theory of Computation, 2nd ed.* Englewood Cliffs, NJ: Prentice-Hall, 1997.

Sudkamp, T. A. *Language and Machines: An Introduction to the Theory of Computer Science, 2nd ed.* Reading, MA: Addison-Wesley, 1996.

Weisstein, E. W. "Books about Computational Complexity." http://www.treasure-troves.com/books/Computational-Complexity.html.

Welsh, D. J. A. *Complexity: Knots, Colourings and Counting.* New York: Cambridge University Press, 1993.

Complex-Valued Function
COMPLEX FUNCTION

Component
A GROUP L is a component of H if L is a QUASISIMPLE GROUP which is a SUBNORMAL SUBGROUP of H.

See also GROUP, QUASISIMPLE GROUP, SUBGROUP, SUBNORMAL SUBGROUP

Component Graph
An n-component of a GRAPH G is a maximal n-connected SUBGRAPH.

References
Harary, F. *Graph Theory*. Reading, MA: Addison-Wesley, 1994.

Composite Knot
A KNOT which is not a PRIME KNOT. Composite knots are special cases of SATELLITE KNOTS.

See also KNOT, PRIME KNOT, SATELLITE KNOT

Composite Number
A composite number n is a POSITIVE INTEGER $n > 1$ which is not PRIME (i.e., which has FACTORS other than 1 and itself). The first few composite numbers (sometimes called "composites" for short) are 4, 6, 8, 9, 10, 12, 14, 15, 16, ... (Sloane's A002808), which can be written 2^2, $2 \cdot 3$, 2^3, 3^2, $2 \cdot 5$, $2^2 \cdot 3$, $2 \cdot 7$, $3 \cdot 5$, and 2^4, respectively. The number 1 is a special case which is considered to be neither composite nor PRIME.

A composite number C can always be written as a PRODUCT in at least two ways (since $1 \cdot C$ is always possible). Call these two products

$$C = ab = cd, \tag{1}$$

then it is obviously the case that $C|ab$ (C divides ab). Set

$$c = mn, \tag{2}$$

where m is the part of C which divides a, and n is the part of C which divides b. Then there are p and q such that

$$a = mp \tag{3}$$

$$b = nq. \tag{4}$$

Solving $ab = cd$ for d gives

$$d = \frac{ab}{c} = \frac{(mp)(nq)}{mn} = pq. \tag{5}$$

It then follows that

$$S \equiv a^2 + b^2 + c^2 + d^2 = m^2 p^2 + n^2 q^2 + m^2 n^2 + p^2 q^2$$
$$= (m^2 + q^2)(n^2 + p^2). \tag{6}$$

It therefore follows that $a^2 + b^2 + c^2 + d^2$ is never PRIME! In fact, the more general result that

$$S \equiv a^k + b^k + c^k + d^k \tag{7}$$

is never PRIME for k an INTEGER ≥ 0 also holds (Honsberger 1991).

See also AMENABLE NUMBER, GRIMM'S CONJECTURE, HIGHLY COMPOSITE NUMBER, PRIME FACTORIZATION PRIME GAPS, PRIME NUMBER, WEAKLY PRIME

References
Honsberger, R. *More Mathematical Morsels*. Washington, DC: Math. Assoc. Amer., pp. 19–0, 1991.
Sloane, N. J. A. Sequences A002808/M3272 in "An On-Line Version of the Encyclopedia of Integer Sequences." http://www.research.att.com/~njas/sequences/eisonline.html.

Composite Runs
PRIME GAPS

Compositeness Certificate
A compositeness certificate is a piece of information which guarantees that a given number p is COMPOSITE. Possible certificates consist of a FACTOR of a number (which, in general, is much quicker to check by direct division than to determine initially), or of the determination that either

$$a^{p-1} \not\equiv 1 \pmod{p},$$

(i.e., p violates FERMAT'S LITTLE THEOREM), or

$$a \neq -1, 1 \text{ and } a^2 \equiv 1 \pmod{p}.$$

A quantity a satisfying either property is said to be a WITNESS to p's compositeness.

See also ADLEMAN-POMERANCE-RUMELY PRIMALITY TEST, FERMAT'S LITTLE THEOREM, MILLER'S PRIMALITY TEST, PRIMALITY CERTIFICATE, WITNESS

Compositeness Test
A test which always identifies PRIME NUMBERS correctly, but may incorrectly identify a COMPOSITE NUMBER as a PRIME.

See also PRIMALITY TEST

Composition
The combination of two FUNCTIONS to form a single new FUNCTION. The composition of two functions f and g is denoted $f \circ g$ and is defined by

$$f \circ g = f(g(x)), \tag{1}$$

where f is a function whose domain includes the range of g. The notation

$$f \circ g(x) = f(g(x)), \tag{2}$$

is sometimes used to explicitly indicate the symbol used for the variable.

Composition is associative, so that

$$f \circ (g \circ h) = (f \circ g) \circ h. \qquad (3)$$

If the functions g is continuous at x_0 and f is continuous at $g(x_0)$, then $f \circ g$ is also continuous at x_0.

A combinatorial composition is defined as an unordered arrangement of k nonnegative integers which sum to n (Skiena 1990, p. 60). The compositions of n into k parts is given by Compositions[n, k] in the *Mathematica* add-on package DiscreteMath`Combinatorica` (which can be loaded with the command < <DiscreteMath`), and the number $C_k(n)$ of compositions of a number n of length k is given by the formula

$$C_k(n) = \binom{n+k-1}{k-1} = \frac{(n+k-1)!}{n!(k-1)!}, \qquad (4)$$

implemented as NumberOfCompositions[n, k] in the *Mathematica* add-on package DiscreteMath`-Combinatorica` (which can be loaded with the command < <DiscreteMath`). The following table gives $C_k(n)$ for $n = 1, 2, ...$ and small k.

k	Sloane	$C_k(1), C_k(2), ...$
2	Sloane's A000027	2, 3, 4, 5, 6, 7, 8, 9, 10, 11, 12, 13, 14, ...
3	Sloane's A000217	3, 6, 10, 15, 21, 28, 36, 45, 55, 66, 78, 91, 105, 120, ...
4	Sloane's A000292	4, 10, 20, 35, 56, 84, 120, 165, 220, 286, 364, 455, 560, 680, ...
5	Sloane's A000332	5, 15, 35, 70, 126, 210, 330, 495, 715, 1001, 1365, 1820, ...
6	Sloane's A000389	6, 21, 56, 126, 252, 462, 792, 1287, 2002, 3003, 4368, ...
7	Sloane's A000579	7, 28, 84, 210, 462, 924, 1716, 3003, 5005, 8008, 12376, ...
8	Sloane's A000580	8, 36, 120, 330, 792, 1716, 3432, 6435, 11440, 19448, ...
9	Sloane's A000581	9, 45, 165, 495, 1287, 3003, 6435, 12870, 24310, 43758, ...

An operation called composition is also defined on BINARY QUADRATIC FORMS. For two numbers represented by two forms, the product can then be represented by the composition. For example, the composition OF THE FORMS $2x^2 + 15y^2$ and $3x^2 + 10y^2$

is given by $6x^2 + 5y^2$, and in this case, the product of 17 and 13 would be REPRESENTED AS $((6 \cdot 36 + 5 \cdot 1 = 221))$. There are several algorithms for computing binary quadratic form composition, which is the basis for some factoring methods.

See also ADEM RELATIONS, BHARGAVA'S THEOREM, BINARY OPERATOR, BINARY QUADRATIC FORM, RANDOM COMPOSITION

References

Apostol, T. M. "Composite Functions and Continuity." §3.7 in *Calculus, 2nd ed., Vol. 1: One-Variable Calculus, with an Introduction to Linear Algebra.* Waltham, MA: Blaisdell, pp. 140–41, 1967.

Klingsberg, P. "A Gray Code for Compositions." *J. Algorithms* **3**, 41–4, 1982.

Skiena, S. "Compositions." §2.2 in *Implementing Discrete Mathematics: Combinatorics and Graph Theory with Mathematica.* Reading, MA: Addison-Wesley, pp. 60–2, 1990.

Composition Series

Every FINITE GROUP G of order greater than one possesses a finite series of SUBGROUPS, called a composition series, such that

$$I \lhd H_s \lhd ... \lhd H_2 \lhd H_1 \lhd G,$$

where H_{i+1} is a maximal subgroup of H_i and $H \lhd G$ means that H is a NORMAL SUBGROUP of G. A composition series is therefore a NORMAL SERIES without repetition whose factors are all simple (Scott 1987, p. 36).

The QUOTIENT GROUPS $G/H_1, H_1/H_2, ..., H_{s-1}/H_s, H_s$ are called composition quotient groups.

See also FINITE GROUP, INVARIANT SUBGROUP, JORDAN-HÖLDER THEOREM, NORMAL SERIES, NORMAL SUBGROUP, QUOTIENT GROUP, SUBGROUP

References

Lomont, J. S. *Applications of Finite Groups.* New York: Dover, p. 26, 1993.

Scott, W. R. "Composition Series." §2.5 in *Group Theory.* New York: Dover, pp. 36–8, 1987.

Composition Theorem

Given a QUADRATIC FORM

$$Q(x, y) \equiv x^2 + y^2,$$

then

$$Q(x, y)Q(x', y') = Q(xx' - yy', x'y + x'y),$$

since

$$(x^2 + y^2)(x'^2 + y'^2) = (xx' - yy')^2 + (xy' + x'y)^2$$
$$= x^2x'^2 + y^2y'^2 + x'^2y^2 + x^2y'^2.$$

See also GENUS THEOREM, QUADRATIC FORM

Compound Interest

Let P be the PRINCIPAL (initial investment), r be the annual compounded rate, $i^{(n)}$ the "nominal rate," n be the number of times INTEREST is compounded per year (i.e., the year is divided into n CONVERSION PERIODS), and t be the number of years (the "term"). The INTEREST rate per CONVERSION PERIOD is then

$$r \equiv \frac{i^{(n)}}{n}. \tag{1}$$

If interest is compounded n times at an annual rate of r (where, for example, 10% corresponds to $r = 0.10$), then the effective rate over $1/n$ the time (what an investor would earn if he did not redeposit his interest after each compounding) is

$$(1+r)^{1/n}. \tag{2}$$

The total amount of holdings A after a time t when interest is re-invested is then

$$A = P\left(1 + \frac{i^{(n)}}{n}\right)^{nt} = P(1+r)^{nt}. \tag{3}$$

Note that even if interest is compounded continuously, the return is still finite since

$$\lim_{n \to \infty}\left(1 + \frac{1}{n}\right)^{n} = e, \tag{4}$$

where e is the base of the NATURAL LOGARITHM.

The time required for a given PRINCIPAL to double (assuming $n = 1$ CONVERSION PERIOD) is given by solving

$$2P = P(1+r)^{t}, \tag{5}$$

or

$$t = \frac{\ln 2}{\ln(1 + r)}, \tag{6}$$

where LN is the NATURAL LOGARITHM. This function can be approximated by the so-called RULE OF 72:

$$t \approx \frac{0.72}{r}. \tag{7}$$

See also e, INTEREST, LN, NATURAL LOGARITHM, PRINCIPAL, RULE OF 72, SIMPLE INTEREST

References
Kellison, S. G. *The Theory of Interest, 2nd ed.* Burr Ridge, IL: Richard D. Irwin, pp. 14–6, 1991.
Milanfar, P. "A Persian Folk Method of Figuring Interest." *Math. Mag.* **69**, 376, 1996.

Compound Polyhedron
POLYHEDRON COMPOUND

Compressible Surface

Let L be a LINK in $\mathbb{R}^3$ and let there be a DISK $\mathbb{D}$ in the LINK COMPLEMENT $\mathbb{R}^3 - L$. Then a surface F such that $\mathbb{D}$ intersects F exactly in its boundary and its boundary does not bound another disk on F is called a compressible surface (Adams 1994, p. 86).

See also KNOT COMPLEMENT

References
Adams, C. C. *The Knot Book: An Elementary Introduction to the Mathematical Theory of Knots.* New York: W. H. Freeman, 1994.

Compression

See also INFORMATION THEORY

References
Hankerson, D.; Harris, G. A.; and Johnson, P. D. Jr. *Introduction to Information Theory and Data Compression.* Boca Raton, FL: CRC Press, 1998.

Computability
COMPLEXITY THEORY

Computable Function

Any computable function can be incorporated into a PROGRAM using while-loops (i.e., "while something is true, do something else"). For-loops (which have a fixed iteration limit) are a special case of while-loops, so computable functions could also be coded using a combination of for- and while-loops. The ACKERMANN FUNCTION is the simplest example of a WELL DEFINED TOTAL FUNCTION which is computable but not PRIMITIVE RECURSIVE, providing a counterexample to the belief in the early 1900s that every computable function was also primitive recursive (Dötzel 1991).

See also ACKERMANN FUNCTION, CHURCH'S THESIS, COMPUTABLE NUMBER, PRIMITIVE RECURSIVE FUNCTION, TURING MACHINE

References
Dötzel, G. "A Function to End All Functions." *Algorithm: Recreational Programming* **2**, 16–7, 1991.

Computable Number

A number which can be computed to any number of DIGITS desired by a TURING MACHINE. Surprisingly, most IRRATIONALS are not computable numbers!

References
Penrose, R. *The Emperor's New Mind: Concerning Computers, Minds, and the Laws of Physics.* Oxford, England: Oxford University Press, 1989.

Turing, A. M. "On Computable Numbers with an Application to the Entscheidungsproblem." *Proc. London Math. Soc.* **42**, 230–65, 1936.

Computational Complexity
COMPLEXITY THEORY

Computational Geometry

The study of efficient algorithms for solving geometric problems. Examples of problems treated by computational geometry include determination of the CONVEX HULL and VORONOI DIAGRAM for a set of points, TRIANGULATION of points in a plane or in space, and other related problems.

See also CONVEX HULL, DELAUNAY TRIANGULATION, DISCRETE GEOMETRY, GEOMETRIC PROBABILITY, HAPPY END PROBLEM, INTERSECTION DETECTION, MINKOWSKI SUM, NEAREST NEIGHBOR PROBLEM, POLYHEDRON PACKING, SPAN (GEOMETRY), SYLVESTER'S FOUR-POINT PROBLEM, TESSELLATION, TRIANGULATION, VERTEX ENUMERATION, VORONOI DIAGRAM

References

de Berg, M.; van Kreveld, M.; Overmans, M.; and Schwarzkopf, O. *Computational Geometry: Algorithms and Applications, 2nd rev. ed.* Berlin: Springer-Verlag, 2000.
Goodman, J. E. and O'Rourke, J. *Handbook of Discrete and Computational Geometry.* Boca Raton, FL: CRC Press, 1997.
O'Rourke, J. *Computational Geometry in C, 2nd ed.* Cambridge, England: Cambridge University Press, 1998.
Preparata, F. R. and Shamos, M. I. *Computational Geometry: An Introduction.* New York: Springer-Verlag, 1985.
Sack, J.-R. and Urrutia, J. (Eds.) *Handbook of Computational Geometry.* Amsterdam, Netherlands: North-Holland, 2000.
Skiena, S. S. "Computational Geometry." §8.6 in *The Algorithm Design Manual.* New York: Springer-Verlag, pp. 345–96, 1997.

Concatenated Number Sequences
CONSECUTIVE NUMBER SEQUENCES

Concatenation

The concatenation of two strings a and b is the string ab formed by joining a and b. Thus the concatenation of the strings "book" and "case" is the string "bookcase". The concatenation of two strings a and b is often denoted ab, $a\|b$, or, in *Mathematica*, $a <> b$. Concatenation is an associative operation, so that the concatenation of three or more strings, for example abc, $abcd$, etc., is WELL DEFINED.

The concatenation of two or more numbers is the number formed by concatenating their numerals. For example, the concatenation of 1, 234, and 5678 is 12345678. The value of the result depends on the numeric base, which is typically understood from context.

The formula for the concatenation of numbers p and q in base b is

$$p\|q = pb^{l(q)} + q,$$

where

$$l(q) = \lfloor \log_b q \rfloor + 1$$

is the LENGTH of q in base b and $\lfloor x \rfloor$ is the FLOOR FUNCTION.

See also CONSECUTIVE NUMBER SEQUENCES, LENGTH (NUMBER), SMARANDACHE SEQUENCES

Concave

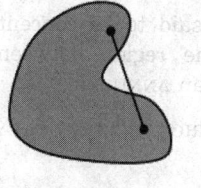 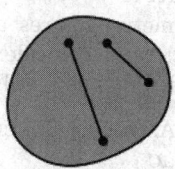

concave   convex

A SET in $\mathbb{R}^d$ is concave if it *does not* contain all the LINE SEGMENTS connecting any pair of its points. If the SET *does* contain all the LINE SEGMENTS, it is called CONVEX.

See also CONNECTED SET, CONVEX FUNCTION, CONVEX HULL, CONVEX OPTIMIZATION THEORY, CONVEX POLYGON, DELAUNAY TRIANGULATION, SIMPLY CONNECTED

Concave Function

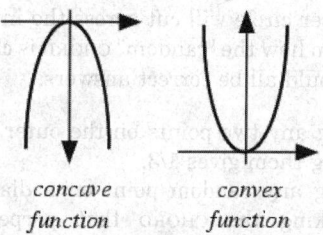

concave   convex
function   function

A function $f(x)$ is said to be concave on an interval $[a, b]$ if, for any points x_1 and x_2 in $[a, b]$, the function $-f(x)$ is CONVEX on that interval (Gradshteyn and Ryzhik 2000).

See also CONVEX FUNCTION

References

Gradshteyn, I. S. and Ryzhik, I. M. *Tables of Integrals, Series, and Products, 6th ed.* San Diego, CA: Academic Press, p. 1132, 2000.

Concentrated

Let μ be a POSITIVE MEASURE on a SIGMA ALGEBRA M, and let λ be an arbitrary (real or complex) MEASURE on M. If there is a SET $A \in M$ such that $\lambda(E) = \lambda(A \cap E)$ for every $E \in M$, then λ is said to be concentrated on A.

This is equivalent to requiring that $\lambda(E) = 0$ whenever $E \cap A = \varnothing$.

See also ABSOLUTELY CONTINUOUS, MUTUALLY SINGULAR

References
Rudin, W. *Functional Analysis, 2nd ed.* New York: McGraw-Hill, p. 121, 1991.

Concentric

Two geometric figures are said to be concentric if their CENTERS coincide. The region between two concentric CIRCLES is called an ANNULUS.

See also ANNULUS, CONCENTRIC CIRCLES, CONCYCLIC, ECCENTRIC

Concentric Circles

Concentric circles are circles with a common center. The region between two CONCENTRIC circles of different RADII is called an ANNULUS. Any two circles can be made concentric by INVERSION by picking the INVERSION CENTER as one of the LIMITING POINTS.

Given two concentric circles with RADII R and $2R$, what is the probability that a chord chosen at random from the outer circle will cut across the inner circle? Depending on how the "random" CHORD is chosen, 1/2, 1/3, or 1/4 could all be correct answers.

1. Picking any two points on the outer circle and connecting them gives 1/3.
2. Picking any random point on a diagonal and then picking the CHORD that perpendicularly bisects it gives 1/2.
3. Picking any point on the large circle, drawing a line to the center, and then drawing the perpendicularly bisected CHORD gives 1/4.

So some care is obviously needed in specifying what is meant by "random" in this problem.

Given an arbitrary CHORD BB' to the larger of two concentric CIRCLES centered on O, the distance between inner and outer intersections is equal on both sides ($AB = A'B'$). To prove this, take the PERPENDICULAR to BB' passing through O and crossing at P. By symmetry, it must be true that PA and PA' are equal. Similarly, PB and PB' must be equal. Therefore, $PB - PA = AB$ equals $PB' - PA' = A'B'$. Incidentally, this is also true for HOMEOIDS, but the proof is nontrivial.

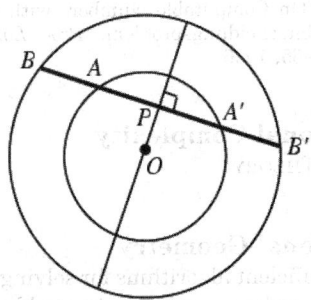

See also ANNULUS, LIMITING POINT

Conchoid

A curve whose name means "shell form." Let C be a curve and O a fixed point. Let P and P' be points on a line from O to C meeting it at Q, where $P'Q = QP = k$, with k a given constant. For example, if C is a CIRCLE and O is on C, then the conchoid is a LIMAÇON, while in the special case that k is the DIAMETER of C, then the conchoid is a CARDIOID. The equation for a parametrically represented curve $(f(t), g(t))$ with $O = (x_0, y_0)$ is

$$x = f \pm \frac{k(f - x_0)}{\sqrt{(f - x_0)^2 + (g - y_0)^2}}$$

$$y = g \pm \frac{k(g - y_0)}{\sqrt{(f - x_0)^2 + (g - y_0)^2}}.$$

See also CONCHO-SPIRAL, CONCHOID OF DE SLUZE, CONCHOID OF NICOMEDES, CONICAL SPIRAL, DÜRER'S CONCHOID

References
Lawrence, J. D. *A Catalog of Special Plane Curves.* New York: Dover, pp. 49–1, 1972.
Lockwood, E. H. "Conchoids." Ch. 14 in *A Book of Curves.* Cambridge, England: Cambridge University Press, pp. 126–29, 1967.
Wells, D. *The Penguin Dictionary of Curious and Interesting Geometry.* London: Penguin, pp. 38–9, 1991.
Yates, R. C. "Conchoid." *A Handbook on Curves and Their Properties.* Ann Arbor, MI: J. W. Edwards, pp. 31–3, 1952.

Conchoid of de Sluze

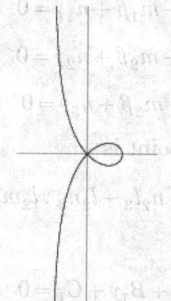

A curve first constructed by René de Sluze in 1662. In CARTESIAN COORDINATES,

$$a(x-a)(x^2+y^2) = k^2x^2,$$

and in POLAR COORDINATES,

$$r = \frac{k^2 \cos \theta}{a} + a \sec \theta.$$

The above curve has $k^2/a = 1$, $a = -0.5$.

Conchoid of Nicomedes

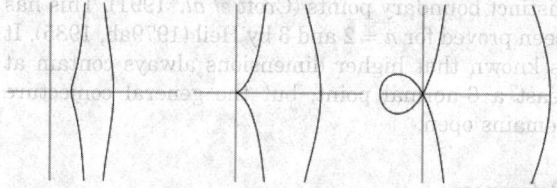

A curve studied by the Greek mathematician Nicomedes in about 200 BC, also called the COCHLOID. It is the LOCUS of points a fixed distance away from a line as measured along a line from the FOCUS point (MacTutor Archive). Nicomedes recognized the three distinct forms seen in this family. This curve was a favorite with 17th century mathematicians and could be used to solve the problems of CUBE DUPLICATION, ANGLE TRISECTION, HEPTAGON construction, and other NEUSIS CONSTRUCTIONS (Johnson 1975).
In POLAR COORDINATES,

$$r = b + a \sec \theta. \tag{1}$$

In CARTESIAN COORDINATES,

$$(x-a)^2(x^2+y^2) = b^2x^2. \tag{2}$$

The conchoid has $x = a$ as an asymptote and the AREA between either branch and the ASYMPTOTE is infinite. The AREA of the loop is

$$A = a\sqrt{b^2-a^2} - 2ab \ln\left(\frac{b+\sqrt{b^2-a^2}}{a}\right)$$

$$+ b^2 \cos^{-1}\left(\frac{a}{b}\right). \tag{3}$$

See also CONCHOID

References

Beyer, W. H. *CRC Standard Mathematical Tables, 28th ed.* Boca Raton, FL: CRC Press, p. 215, 1987.

Johnson, C. "A Construction for a Regular Heptagon." *Math. Gaz.* **59**, 17–1, 1975.

Lawrence, J. D. *A Catalog of Special Plane Curves.* New York: Dover, pp. 135–39, 1972.

MacTutor History of Mathematics Archive. "Conchoid." http://www-groups.dcs.st-and.ac.uk/~history/Curves/Conchoid.html.

Pappas, T. "Conchoid of Nicomedes." *The Joy of Mathematics.* San Carlos, CA: Wide World Publ./Tetra, pp. 94–5, 1989.

Steinhaus, H. *Mathematical Snapshots, 3rd ed.* New York: Dover, pp. 154–55, 1999.

Szmulowicz, F. "Conchoid of Nicomedes from Reflections and Refractions in a Cone." *Amer. J. Phys.* **64**, 467–71, Apr. 1996.

Wells, D. *The Penguin Dictionary of Curious and Interesting Numbers.* Middlesex, England: Penguin Books, p. 34, 1986.

Wells, D. *The Penguin Dictionary of Curious and Interesting Geometry.* London: Penguin, pp. 38–9, 1991.

Yates, R. C. "Conchoid." *A Handbook on Curves and Their Properties.* Ann Arbor, MI: J. W. Edwards, pp. 31–3, 1952.

Concho-Spiral

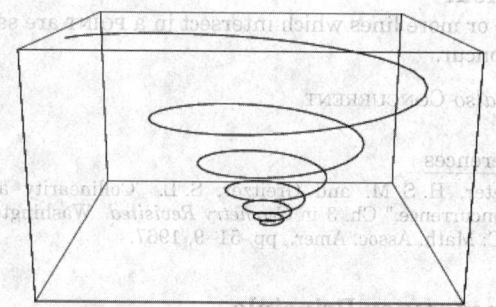

The SPACE CURVE with PARAMETRIC EQUATIONS

$$r = \mu^u a$$

$$\theta = u$$

$$z = \mu^u c.$$

See also CONICAL SPIRAL, SPIRAL

Concordant Form

A concordant form is an integer TRIPLE (a, b, N) where

$$\begin{cases} a^2 + b^2 = c^2 \\ a^2 + Nb^2 = d^2, \end{cases}$$

with c and d integers. Examples include

$$\begin{cases} 14663^2 + 111384^2 = 112345^2 \\ 14663^2 + 47 \cdot 111384^2 = 763751^2 \end{cases}$$

$$\begin{cases} 1141^2 + 13260^2 = 13309^2 \\ 1141^2 + 53 \cdot 13260^2 = 96541^2 \end{cases}$$

$$\begin{cases} 2873161^2 + 2401080^2 = 3744361^2 \\ 2873161^2 + 83 \cdot 2401080^2 = 22062761^2. \end{cases}$$

Dickson (1962) states that C. H. Brooks and S. Watson found in *The Ladies' and Gentlemen's Diary* (1857) that $x^2 + y^2$ and $x^2 + Ny^2$ can be simultaneously squares for $N < 100$ only for 1, 7, 10, 11, 17, 20, 22, 23, 24, 27, 30, 31, 34, 41, 42, 45, 49, 50, 52, 57, 58, 59, 60, 61, 68, 71, 72, 74, 76, 77, 79, 82, 85, 86, 90, 92, 93, 94, 97, 99, and 100 (which evidently omits 47, 53, and 83 from above). The list of concordant primes less than 1000 is now complete with the possible exception of the 16 primes 103, 131, 191, 223, 271, 311, 431, 439, 443, 593, 607, 641, 743, 821, 929, and 971 (Brown).

See also CONGRUUM

References

Brown, K. S. "Concordant Forms." http://www.seanet.com/~ksbrown/kmath286.htm.

Dickson, L. E. *History of the Theory of Numbers, Vol. 1: Divisibility and Primality.* New York: Chelsea, p. 475, 1952.

Concur

Two or more lines which intersect in a POINT are said to concur.

See also CONCURRENT

References

Coxeter, H. S. M. and Greitzer, S. L. "Collinearity and Concurrence." Ch. 3 in *Geometry Revisited.* Washington, DC: Math. Assoc. Amer., pp. 51–9, 1967.

Concurrency Principle

See also CONCURRENT RELATION

Concurrent

Two or more LINES are said to be concurrent if they intersect in a single point. Two LINES concur if their TRILINEAR COORDINATES satisfy

$$\begin{vmatrix} l_1 & m_1 & n_1 \\ l_2 & m_2 & n_2 \\ l_3 & m_3 & n_3 \end{vmatrix} = 0. \tag{1}$$

Three LINES concur if their TRILINEAR COORDINATES

satisfy

$$l_1 \alpha + m_1 \beta + n_1 \gamma = 0 \tag{2}$$

$$l_2 \alpha + m_2 \beta + n_2 \gamma = 0 \tag{3}$$

$$l_3 \alpha + m_3 \beta + n_3 \gamma = 0, \tag{4}$$

in which case the point is

$$m_2 n_3 - n_2 m_3 : n_2 l_3 - l_2 n_3 : l_2 m_3 - m_2 l_3. \tag{5}$$

Three lines

$$A_1 x + B_1 y + C_1 = 0 \tag{6}$$

$$A_2 x + B_2 y + C_2 = 0 \tag{7}$$

$$A_3 x + B_3 y + C_3 = 0 \tag{8}$$

are concurrent if their COEFFICIENTS satisfy

$$\begin{vmatrix} A_1 & B_1 & C_1 \\ A_2 & B_2 & C_2 \\ A_3 & B_3 & C_3 \end{vmatrix} = 0. \tag{9}$$

See also CONCYCLIC, POINT

Concurrent Normals Conjecture

It is conjectured that any convex body in Euclidean n-space has an interior lying on normals through $2n$ distinct boundary points (Croft *et al.* 1991). This has been proved for $n = 2$ and 3 by Heil (1979ab, 1985). It is known that higher dimensions always contain at least a 6-normal point, but the general conjecture remains open.

References

Coxeter, H. S. M. and Greitzer, S. L. "Collinearity and Concurrence." Ch. 3 in *Geometry Revisited.* Washington, DC: Math. Assoc. Amer., pp. 51–9, 1967.

Croft, H. T.; Falconer, K. J.; and Guy, R. K. "Concurrent Normals." §A3 in *Unsolved Problems in Geometry.* New York: Springer-Verlag, pp. 14–5, 1991.

Heil, E. "Existenz eines 6-Normalenpunktes in einem konvexen Körper." *Arch. Math. (Basel)* **32**, 412–16, 1979a.

Heil, E. "Correction to 'Existenz eines 6-Normalenpunktes in einem konvexen Körper.'" *Arch. Math. (Basel)* **33**, 496, 1979b.

Heil, E. "Concurrent Normals and Critical Points under Weak Smoothness Assumptions." In *Discrete Geometry and Convexity* (Ed. J. E. Goodman, E. Lutwak, J. Malkevitch, and R. Pollack). *Ann. New York Acad. Sci.* **440**, pp. 170–78, 1985.

Concurrent Relation

Let X and Y be sets, and let $R \subseteq X \times Y$ be a relation on $X \times Y$. Then R is a concurrent relation if and only if for any finite subset F of X, there exists a single element p of Y such that if $a \in F$, then aRp. Examples of concurrent relations include the following:

1. The relation $<$ on either the natural numbers, the integers, the rational numbers, or the real numbers.

2. The relation R between elements of an extension $\mathbb{E}$ of a field $\mathbb{F}$, defined by

$$R = \{(a, b) \in \mathbb{E} \times \mathbb{E} : b \text{ is algebraic over } \mathbb{F} \text{ and}$$

$$x \text{ is in the extension of } \mathbb{F} \text{ by } y\}.$$

3. The containment relation $\subseteq$ between open neighborhoods of a given point p of a TOPOLOGICAL SPACE X.

See also CONCURRENCY PRINCIPLE

References

Hurd, A. E. and Loeb, P. A. *An Introduction to Nonstandard Real Analysis.* Orlando, FL: Academic Press, 1985.

Robinson, A. "Germs." In *Applications of Model Theory to Algebra, Analysis and Probability (International Sympos., Pasadena, Calif., 1967).* New York: Holt, Rinehart and Winston, pp. 138–49, 1969.

Insall, M. "Hyperalgebraic Primitive Elements for Relational Algebraic and Topological Algebraic Models." *Studia Logica* **57**, 409–18, 1996.

Concyclic

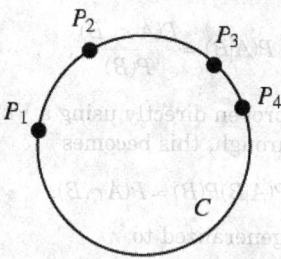

Four or more points P_1, P_2, P_3, P_4, ... which lie on a CIRCLE C are said to be concyclic. Three points are trivially concyclic since three noncollinear points determine a CIRCLE. The number of the n^2 LATTICE POINTS x, $y \in [1, n]$ which can be picked with no four concyclic is $\iota(n^{2/3} - \epsilon)$ (Guy 1994).

A theorem states that if any four consecutive points of a POLYGON are not concyclic, then its AREA can be increased by making them concyclic. This fact arises in some PROOFS that the solution to the ISOPERIMETRIC PROBLEM is the CIRCLE.

See also ANTIPARALLEL, CIRCLE, COLLINEAR, CONCENTRIC, CYCLIC HEXAGON, CYCLIC PENTAGON, CYCLIC QUADRILATERAL, ECCENTRIC, N-CLUSTER

References

Coolidge, J. L. "Concurrent Circles and Concyclic Points." §1.6 in *A Treatise on the Geometry of the Circle and Sphere.* New York: Chelsea, pp. 85–5, 1971.

Guy, R. K. "Lattice Points, No Four on a Circle." §F3 in *Unsolved Problems in Number Theory, 2nd ed.* New York: Springer-Verlag, p. 241, 1994.

Condensation

A method of computing the DETERMINANT of a SQUARE MATRIX due to Charles Dodgson (1866) (who is more famous under his pseudonym Lewis Carroll). The method is useful for hand calculations because, for an INTEGER MATRIX, all entries in submatrices computed along the way must also be integers. The method is also implemented efficiently in a parallel computation. Condensation is also known as the method of contractants (Macmillan 1955, Lotkin 1959).

Given an $n \times n$ matrix, condensation successively computes an $(n-1) \times (n-1)$ matrix, an $(n-2) \times (n-2)$ matrix, etc., until arriving at a 1×1 matrix whose only entry ends up being the DETERMINANT of the original matrix. To compute the $k \times k$ matrix $(n-1 \geq k \geq 1)$, take the k^2 2×2 connected subdeterminants of the $(k+1) \times (k+1)$ matrix and divide them by the k^2 central entries of the $(k+2) \times (k+2)$ matrix, with no divisions performed for $k = n-1$. The $k \times k$ matrices arrived at in this manner are the matrices of determinants of the $k^2(n-k+1) \times (n-k+1)$ connected submatrices of the original matrices.

For example, the first condensation of the 3×3 matrix

$$\begin{bmatrix} a & b & c \\ d & e & f \\ g & h & i \end{bmatrix}$$

yields the matrix

$$\begin{bmatrix} ae - bd & bf - ce \\ dh - eg & ei - fh \end{bmatrix},$$

and the second condensation yields

$$[((ae^2i - aefh - bdei + bdfh)$$
$$- (bdfh - befg - cdeh + ce^2g))/e]$$

which is the determinant of the original matrix. Collecting terms gives

$$(1)aei + (-1)afh + (-1)bdi + (0)bde^{-1}fh + (1)bfg$$
$$+ (1)cdh + (-1)ceg,$$

of which the nonzero terms correspond to the PERMUTATION MATRICES. In the 4×4 case, 24 nonzero terms are obtained together with 18 vanishing ones. These 42 terms correspond to the ALTERNATING SIGN MATRICES for which any -1s in a row or column must have a $+1$ "outside" it (i.e., all -1s are "bordered" by $+1$s).

See also ALTERNATING SIGN MATRIX, DETERMINANT, DETERMINANT EXPANSION BY MINORS

References

Bareiss, E. H. "Sylvester's Identity and Multistep Integer-Preserving Gaussian Elimination." *Math. Comput.* **22**, 565–78, 1968.

Bressoud, D. and Propp, J. "How the Alternating Sign Matrix Conjecture was Solved." *Not. Amer. Math. Soc.* **46**, 637–46.

Dodgson, C. L. "Condensation of Determinants, Being a New and Brief Method for Computing their Arithmetic Values." *Proc. Roy. Soc. Ser. A* **15**, 150–55, 1866.

Lotkin, M. "Note on the Method of Contractants." *Amer. Math. Soc.* **55**, 476–79, 1959.

Macmillan, R. H. A New Method for the Numerical Evaluation of Determinants." *J. Roy. Aeronaut. Soc.* **59**, 772, 1955.

Robbins, D. P. and Rumsey, H. Jr. "Determinants and Alternating Sign Matrices." *Adv. Math.* **62**, 169–84, 1986.

Condition

A requirement NECESSARY for a given statement or theorem to hold. Also called a CRITERION.

See also BOUNDARY CONDITIONS, CARMICHAEL CONDITION, CAUCHY BOUNDARY CONDITIONS, CONDITION NUMBER, DIRICHLET BOUNDARY CONDITIONS, DIVERSITY CONDITION, FELLER-LÉVY CONDITION, HÖLDER CONDITION, LICHNEROWICZ CONDITIONS, LINDEBERG CONDITION, LIPSCHITZ CONDITION, LYAPUNOV CONDITION, NEUMANN BOUNDARY CONDITIONS, ROBERTSON CONDITION, ROBIN BOUNDARY CONDITIONS, TAYLOR'S CONDITION, TRIANGLE CONDITION, WEIERSTRASS-ERDMAN CORNER CONDITION, WINKLER CONDITIONS

Condition Number

The ratio of the largest to smallest SINGULAR VALUE of a MATRIX. A system is said to be SINGULAR if the condition number is INFINITE, and ILL-CONDITIONED if it is too large. The p-norm condition number of a matrix can be computed using `MatrixCondition-Number[m, p]` in the *Mathematica* add-on package `LinearAlgebra'MatrixMultiplication'` (which can be loaded with the command `<<LinearAlgebra'`) for $p = 1$, 2, or ∞, where omitting the p is equivalent to specifying `Infinity`.

See also ILL-CONDITIONED MATRIX, SINGULAR MATRIX, SINGULAR VALUE DECOMPOSITION

Conditional

The formal term in PROPOSITIONAL CALCULUS for the CONNECTIVE IMPLIES.

See also BICONDITIONAL, IMPLIES

References

Mendelson, E. *Introduction to Mathematical Logic, 4th ed.* London: Chapman & Hall, p. 13, 1997.

Conditional Convergence

If the SERIES

$$\sum_{n=0}^{\infty} u_n$$

CONVERGES, but

$$\sum_{n=0}^{\infty} |u_n|$$

does not, where $|x|$ is the ABSOLUTE VALUE, then the SERIES is said to be conditionally CONVERGENT. The RIEMANN SERIES THEOREM states that, by a suitable rearrangement of terms, a conditionally convergent SERIES may be made to converge to any desired value, or to DIVERGE.

See also ABSOLUTE CONVERGENCE, CONVERGENCE TESTS, DIVERGENT SERIES, RIEMANN SERIES THEOREM, SERIES

References

Bromwich, T. J. I'a and MacRobert, T. M. *An Introduction to the Theory of Infinite Series, 3rd ed.* New York: Chelsea, 1991.

Gardner, M. *The Sixth Book of Mathematical Games from Scientific American.* Chicago, IL: University of Chicago Press, pp. 170–71, 1984.

Hardy, G. H. *Divergent Series.* New York: Oxford University Press, 1949.

Conditional Probability

The conditional probability of an EVENT A assuming that B has occurred, denoted $P(A|B)$, equals

$$P(A|B) = \frac{P(A \cap B)}{P(B)}, \tag{1}$$

which can be proven directly using a VENN DIAGRAM. Multiplying through, this becomes

$$P(A|B)P(B) = P(A \cap B), \tag{2}$$

which can be generalized to

$$P(A \cap B \cap C) = P(A)P(B|A)P(C|A \cap B). \tag{3}$$

Rearranging (1) gives

$$P(B|A) = \frac{P(B \cap A)}{P(A)}. \tag{4}$$

Solving (4) for $P(B \cap A) = P(A \cap B)$ and plugging in to (1) gives

$$P(A|B) = \frac{P(A)P(B|A)}{P(B)}. \tag{5}$$

See also BAYES' FORMULA, FERMAT'S PRINCIPLE OF CONJUNCTIVE PROBABILITY, TOTAL PROBABILITY THEOREM

References

Papoulis, A. "Conditional Probabilities and Independent Sets." §2– in *Probability, Random Variables, and Stochastic Processes, 2nd ed.* New York: McGraw-Hill, pp. 33–5, 1984.

Condom Problem

GLOVE PROBLEM

Condon-Shortley Phase

The $(-1)^m$ phase factor in some definitions (e.g., Arfken 1985) of the SPHERICAL HARMONICS and associated LEGENDRE POLYNOMIALS. Using the Condon-Shortley convention gives

$$Y_l^m(\theta, \phi) = (-1)^m \sqrt{\frac{2l+1}{4\pi} \frac{(l-m)!}{(l+m)!}} P_l^m(\cos\theta) e^{im\phi}.$$

The Condon-Shortley phase is not necessary in the definition of the SPHERICAL HARMONICS, but including it simplifies the treatment of angular moment in quantum mechanics. In particular, they are a consequence of the ladder operators L_- and L_+ (Arfken 1985, p. 693).

See also LEGENDRE POLYNOMIAL, SPHERICAL HARMONIC

References

Arfken, G. *Mathematical Methods for Physicists, 3rd ed.* Orlando, FL: Academic Press, pp. 682 and 692, 1985.

Condon, E. U. and Shortley, G. *The Theory of Atomic Spectra.* Cambridge, England: Cambridge University Press, 1951.

Shore, B. W. and Menzel, D. H. *Principles of Atomic Spectra.* New York: Wiley, p. 158, 1968.

Conductor

J-CONDUCTOR

Cone

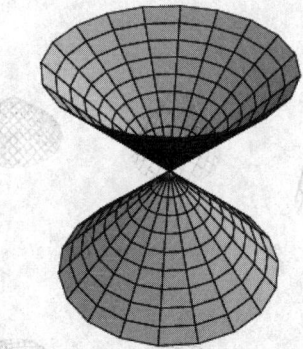

A cone is a PYRAMID with a circular CROSS SECTION, and a right cone is a cone with its vertex above the center of its base. However, in discussions of CONIC SECTIONS, the word "cone" is taken mean "DOUBLE CONE," consisting of two cones placed apex to apex. This is a QUADRATIC SURFACE, and each single cone is called a "NAPPE." The HYPERBOLA can then be defined as the intersection of a PLANE with both NAPPES of the cone.

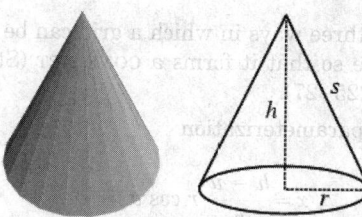

A right cone of height h can be described by the PARAMETRIC EQUATIONS

$$x = \frac{h-z}{h} r \cos\theta \tag{1}$$

$$y = \frac{h-z}{h} r \sin\theta \tag{2}$$

$$z = z \tag{3}$$

for $z \in [0, h]$ and $\theta \in [0, 2\pi)$. The VOLUME of a cone is therefore

$$V = \tfrac{1}{3} A_b h, \tag{4}$$

where A_b is the base AREA and h is the height. If the base is circular, then

$$V = \tfrac{1}{3} \pi r^2 h. \tag{5}$$

This amazing fact was first discovered by Eudoxus, and other proofs were subsequently found by Archimedes in *On the Sphere and Cylinder* (ca. 225 BC) and Euclid in Proposition XII.10 of his *ELEMENTS* (Dunham 1990).

The CENTROID can be obtained by setting $R_2 = 0$ in the equation for the centroid of the CONICAL FRUSTUM,

$$\bar{z} = \frac{\langle z \rangle}{V} = \frac{h(R_1^2 + 2R_1 R_2 + 3R_2^2)}{4(R_1^2 + R_1 R_2 + R_2^2)}, \tag{6}$$

(Eshbach 1975, p. 453; Beyer 1987, p. 133) yielding

$$\bar{z} = \tfrac{1}{4} h. \tag{7}$$

For a right circular cone, the SLANT HEIGHT s is

$$s = \sqrt{r^2 + h^2} \tag{8}$$

and the surface AREA (not including the base) is

$$S = \pi r s = \pi r \sqrt{r^2 + h^2}. \tag{9}$$

The LOCUS of the apex of a variable cone containing an ELLIPSE fixed in 3-space is a HYPERBOLA through the FOCI of the ELLIPSE. In addition, the LOCUS of the apex of a cone containing that HYPERBOLA is the original ELLIPSE. Furthermore, the ECCENTRICITIES of the ELLIPSE and HYPERBOLA are reciprocals.

There are three ways in which a grid can be mapped onto a cone so that it forms a CONE NET (Steinhaus 1983, pp. 225–27).

Using the parameterization

$$x = \frac{h - u}{h} \, r \cos v \qquad (10)$$

$$y = \frac{h - u}{h} \, r \sin v \qquad (11)$$

$$z = u \qquad (12)$$

gives coefficients of the FIRST FUNDAMENTAL FORM

$$E = 1 + \frac{r^2}{h^2} \qquad (13)$$

$$F = 0 \qquad (14)$$

$$G = \frac{r^2 (h - u)^2}{h^2}, \qquad (15)$$

SECOND FUNDAMENTAL FORM coefficients

$$e = 0 \qquad (16)$$

$$f = 0 \qquad (17)$$

$$g = \frac{r(h - u)}{\sqrt{h^2 + r^2}}, \qquad (18)$$

AREA ELEMENT

$$dS = \frac{r \sqrt{h^2 + r^2}}{h^2} (h - u), \qquad (19)$$

GAUSSIAN CURVATURE

$$K = 0, \qquad (20)$$

and MEAN CURVATURE

$$M = \frac{h^2}{\sqrt{h^2 + r^2}(2hr - 2ru)}. \qquad (21)$$

Note that writing $z = v$ instead of $z = u$ would give a HELICOID instead of a CONE.

See also BICONE, CONE NET, CONIC SECTION, CONICAL FRUSTUM, CYLINDER, DOUBLE CONE, GENERALIZED CONE, HELICOID, NAPPE, PYRAMID, SPHERE, SPHER-ICON

References

Beyer, W. H. (Ed.). *CRC Standard Mathematical Tables, 28th ed.* Boca Raton, FL: CRC Press, pp. 129 and 133, 1987.
Dunham, W. *Journey through Genius: The Great Theorems of Mathematics.* New York: Wiley, pp. 76–7, 1990.
Eshbach, O. W. *Handbook of Engineering Fundamentals.* New York: Wiley, 1975.
Harris, J. W. and Stocker, H. "Cone." §4.7 in *Handbook of Mathematics and Computational Science.* New York: Springer-Verlag, pp. 104–05, 1998.
Hilbert, D. and Cohn-Vossen, S. "The Cylinder, the Cone, the Conic Sections, and Their Surfaces of Revolution." §2 in *Geometry and the Imagination.* New York: Chelsea, pp. 7–1, 1999.
Kern, W. F. and Bland, J. R. "Cone" and "Right Circular Cone." §24–5 in *Solid Mensuration with Proofs, 2nd ed.* New York: Wiley, pp. 57–4, 1948.
Steinhaus, H. *Mathematical Snapshots, 3rd ed.* New York: Dover, 1999.
Yates, R. C. "Cones." *A Handbook on Curves and Their Properties.* Ann Arbor, MI: J. W. Edwards, pp. 34–5, 1952.

Cone (Space)

The JOIN of a TOPOLOGICAL SPACE X and a point P, $C(X) = X * P$.

References

Rolfsen, D. *Knots and Links.* Wilmington, DE: Publish or Perish Press, p. 6, 1976.

Cone Graph

A GRAPH $C_n + \overline{K_m}$, where C_n is a CYCLIC GRAPH and K_m is a COMPLETE GRAPH.

Cone Net

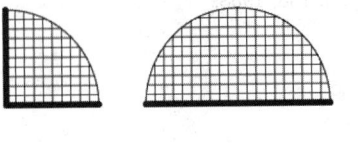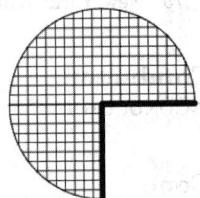

The mapping of a grid of regularly ruled squares onto a CONE with no overlap or misalignment. Cone nets are possible for vertex angles of 90°, 180°, and 270°, where the dark edges in the upper diagrams above are joined. Beautiful photographs of cone net models (lower diagrams above) are presented in Steinhaus (1983). The transformation from a point (x, y) in the grid plane to a point (x', y', z') on the cone is given by

$$x' = rn \cos\left(\frac{\theta}{n}\right) \tag{1}$$

$$y' = rn \sin\left(\frac{\theta}{n}\right) \tag{2}$$

$$z' = (1 - r)h, \tag{3}$$

where $n = 1/4$, $1/2$, or $3/4$ is the fraction of a circle forming the base, and

$$h = \sqrt{1 - n^2} \tag{4}$$

$$\theta = \tan^{-1}\left(\frac{y}{x}\right) \tag{5}$$

$$r = \sqrt{x^2 + y^2}. \tag{6}$$

See also CONE, SPHERICON

References

Steinhaus, H. *Mathematical Snapshots, 3rd ed.* New York: Dover, pp. 224–28, 1999.

Cone-Plane Intersection

CONIC SECTION

Cone-Sphere Intersection

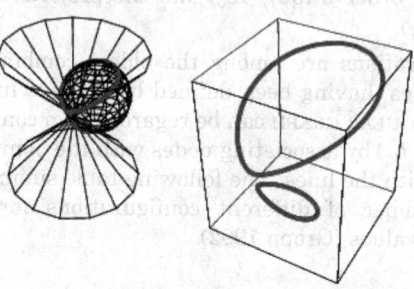

Let a CONE of opening parameter c and vertex at $(0, 0, 0)$ intersect a SPHERE of RADIUS r centered at (x_0, y_0, z_0), with the CONE oriented such that its axis does not pass through the center of the SPHERE. Then the equations of the curve of intersection are

$$\frac{x^2 + y^2}{c^2} = z^2 \tag{1}$$

$$(x - x_0)^2 + (y - y_0)^2 + (z - z_0)^2 = r^2. \tag{2}$$

Combining (1) and (2) gives

$$(x - x_0)^2 + (y - y_0)^2 + \frac{x^2 + y^2}{c^2} - \frac{2z_0}{c}\sqrt{x^2 + y^2} + z_0^2 = r^2 \tag{3}$$

$$x^2\left(1 + \frac{1}{c^2}\right) - 2x_0 x + y^2\left(1 + \frac{1}{c^2}\right) - 2y_0 y$$
$$+ (x_0^2 + y_0^2 + z_0^2 - r^2) - \frac{2z_0}{c}\sqrt{x^2 + y^2} = 0. \tag{4}$$

Therefore, x and y are connected by a complicated QUARTIC EQUATION, and x, y, and z by a QUADRATIC EQUATION.

If the CONE-SPHERE intersection is *on-axis* so that a CONE of opening parameter c and vertex at $(0, 0, z_0)$ is oriented with its AXIS along a radial of the SPHERE of radius r centered at $(0, 0, 0)$, then the equations of the curve of intersection are

$$(z - z_0)^2 = \frac{x^2 + y^2}{c^2} \tag{5}$$

$$x^2 + y^2 + z^2 = r^2. \tag{6}$$

Combining (5) and (6) gives

$$c^2(z - z_0)^2 + z^2 = r^2 \tag{7}$$

$$c^2(z^2 - 2z_0 z + z_0^2) + z^2 = r^2 \tag{8}$$

$$z^2(c^2 + 1) - 2c^2 z_0 z + (z_0^2 c^2 - r^2) = 0. \tag{9}$$

Using the QUADRATIC EQUATION gives

$$z = \frac{2c^2 z_0 \pm \sqrt{4c^4 z_0^2 - 4(c^2 + 1)(z_0^2 c^2 - r^2)}}{2(c^2 + 1)}$$
$$= \frac{c^2 z_0 \pm \sqrt{c^2(r^2 - z_0^2) + r^2}}{c^2 + 1}. \tag{10}$$

So the curve of intersection is planar. Plugging (10) into (5) shows that the curve is actually a CIRCLE, with RADIUS given by

$$a = \sqrt{r^2 - z^2}. \tag{11}$$

See also CONE, SPHERE

References

Kenison, E. and Bradley, H. C. *Descriptive Geometry.* New York: Macmillan, pp. 282–83, 1935.

Confidence Interval

The probability that a measurement will fall within a given CLOSED INTERVAL $[a, b]$. For a CONTINUOUS DISTRIBUTION,

$$\mathrm{CI}(a, b) \equiv \int_b^a P(x)\, dx, \tag{1}$$

where $P(x)$ is the PROBABILITY DISTRIBUTION FUNCTION. Usually, the confidence interval of interest is symmetrically placed around the mean, so

$$\text{CI}(x) \equiv \text{CI}(\mu - x,\ \mu + x) = \int_{\mu-x}^{\mu+x} P(x)\, dx, \qquad (2)$$

where μ is the MEAN. For a GAUSSIAN DISTRIBUTION, the probability that a measurement falls within $n\sigma$ of the mean μ is

$$\text{CI}(n\sigma) \equiv \frac{1}{\sigma\sqrt{2\pi}} \int_{\mu-n\sigma}^{\mu+n\sigma} e^{-(x-\mu)^2/2\sigma^2}\, dx$$

$$= \frac{2}{\sigma\sqrt{2\pi}} \int_{0}^{\mu+n\sigma} e^{-(x-\mu)^2/2\sigma^2}\, dx. \qquad (3)$$

Now let $u \equiv (x-\mu)/\sqrt{2}\sigma$, so $du = dx/\sqrt{2}\sigma$. Then

$$\text{CI}(n\sigma) = \frac{2}{\sigma\sqrt{2\pi}}\sqrt{2}\sigma \int_{0}^{n/\sqrt{2}} e^{-u^2}\, du = \frac{2}{\sqrt{\pi}} \int_{0}^{n/\sqrt{2}} e^{-u^2}\, du$$

$$= \text{erf}\left(\frac{n}{\sqrt{2}}\right) \qquad (4)$$

where $\text{erf}(x)$ is the so-called ERF function. The variate value producing a confidence interval CI is often denoted x_{CI}, so

$$x_{\text{CI}} = \sqrt{2}\ \text{erf}^{-1}(\text{CI}). \qquad (5)$$

range	CI
σ	0.6826895
2σ	0.9544997
3σ	0.9973002
4σ	0.9999366
5σ	0.9999994

To find the standard deviation range corresponding to a given confidence interval, solve (4) for n.

$$n = \sqrt{2}\,\text{erf}^{-1}(\text{CI}) \qquad (6)$$

CI	range
0.800	$\pm 1.28155\sigma$
0.900	$\pm 1.64485\sigma$
0.950	$\pm 1.95996\sigma$
0.990	$\pm 2.57583\sigma$
0.995	$\pm 2.80703\sigma$
0.999	$\pm 3.29053\sigma$

Configuration

The word configuration is sometimes used to describe a finite collection of points $p = (p_1, \ldots, p_n)$, $p_i \in \mathbb{R}^d$, where $\mathbb{R}^d$ is a EUCLIDEAN SPACE.

The term "configuration" also is used to describe a finite incidence structure (v_r, b_k) with the following properties (Gropp 1992).

   1. There are v points and b lines.
   2. There are k points on each line and r lines through each point.
   3. Two different lines intersect each other at most once and two different points are connected by a line at most once.

The conditions

$$vr = bk$$

$$v \geq r(k-1) + 1$$

are NECESSARY for the existence of a configuration. For $k = 3$, these conditions are also SUFFICIENT, and for $k = 4$ this is probably also the case (Gropp 1992). The necessary conditions hold, but there is no 22_5. For $k = 6$ and 7, the above conditions are not SUFFICIENT, as illustrated by the affine projective plane of order 6 (36_7, 42_6) and the projective plane (43_7, 43_7).

Configurations are among the oldest combinatorial structures, having been defined by T. Reye in 1876. An r-REGULAR GRAPH can be regarded as a configuration (v_r, b_2) by associating nodes with the points, and edges with the lines. The following table summarizes the number of different configurations for some special values (Gropp 1992).

configuration	distinct
$(12_2, 8_3)$	5
$(15_2, 10_3)$	18

A symmetric configuration $n_k = (n_k,\ n_k)$ consists of n lines and n points arranged such that k lines pass through each point and there are k points on each line. All symmetric n_3 configurations are known for $n \leq 14$. The number of 7_3, 8_3, $9_3 \ldots$ configurations are 1, 1, 3, 10, 31, 229, 2036, 21399, 245342, ..., correcting an error of von Sterneck for 12_3 (Sloane's A001403; Sterneck 1894, 1895; Wells 1991, p. 72; Colbourn and Dinitz 1996; Gropp 1997; Hilbert and Cohn-Vossen 1999).

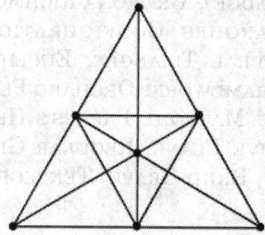

The FANO PLANE, in which the central point corresponds to the POINT AT INFINITY, is the unique 7_3 configuration. There are no 7_3 configurations using points all at finite distances (Wells 1986, p. 75).

There are no 8_3 configurations using points all at finite distances (Wells 1986, p. 75), but a single configuration exists with a POINT AT INFINITY.

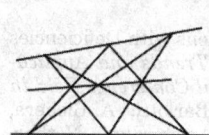 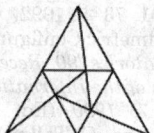 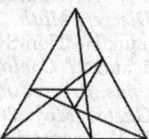

There are three 9_3 configurations, of which PAPPUS'S HEXAGON THEOREM (left figure) is one (Wells 1985, p. 75). The other two consist of embedded EQUILATERAL TRIANGLES (Wells 1991, pp. 159–60).

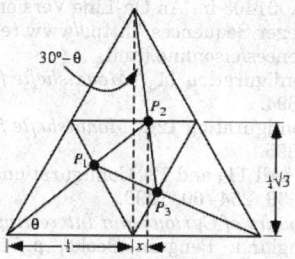 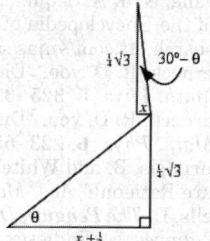

In the second 9_3 configuration, the angle θ can be computed using the above figure. For the top triangle, trigonometry gives

$$\tan(30° - \theta) = \frac{x}{\frac{1}{4}\sqrt{3}}. \tag{1}$$

Solving for x and plugging into the trigonometric equation from the bottom triangle gives

$$\tan \theta = \frac{\frac{1}{4}\sqrt{3}}{\frac{1}{2} + x} = \frac{\sqrt{3}}{2 + \sqrt{3}\tan(30° - \theta)}. \tag{2}$$

Now using the identity

$$\tan(\alpha - \beta) = \frac{\tan \alpha - \tan \beta}{1 + \tan \alpha \tan \beta} \tag{3}$$

with $\alpha = \theta$, $\beta = 30°$ gives

$$\tan(\theta - 30°) = \frac{\tan \theta - \frac{1}{\sqrt{3}}}{1 + \frac{1}{\sqrt{3}}\tan \theta} = \frac{\sqrt{3}\tan \theta - 1}{\sqrt{3} + \tan \theta}. \tag{4}$$

Plugging in gives

$$\tan \theta \left(2 + \sqrt{3}\,\frac{\sqrt{3}\tan \theta - 1}{\sqrt{3} + \tan \theta} \right) = \sqrt{3}, \tag{5}$$

which simplifies to

$$\tan^2 \theta = \sec^2 \theta - 1 = \tfrac{3}{5} \tag{6}$$

$$\sec^2 \theta = \tfrac{8}{5} \tag{7}$$

$$\cos^2 \theta = \tfrac{1}{2}[1 + \cos(2\theta)] = \tfrac{5}{8} \tag{8}$$

$$\tfrac{1}{4} = \cos(2\theta) \tag{9}$$

$$\theta = \tfrac{1}{2}\cos^{-1}\left(\tfrac{1}{4}\right) \approx 0.659058 \text{ rad.} \tag{10}$$

Some additional trigonometry then gives the positions of the three innermost EQUILATERAL TRIANGLE vertices,

$$P_1 = \left(\tfrac{1}{8}(5 - \sqrt{5}),\ \tfrac{1}{8}(\sqrt{15} - \sqrt{3}) \right) \tag{11}$$

$$P_2 = \left(\tfrac{1}{4}\sqrt{5},\ \tfrac{1}{4}\sqrt{3} \right) \tag{12}$$

$$P_3 = \left(\tfrac{1}{8}(7 - \sqrt{5}),\ \tfrac{1}{8}(3\sqrt{3} - \sqrt{15}) \right). \tag{13}$$

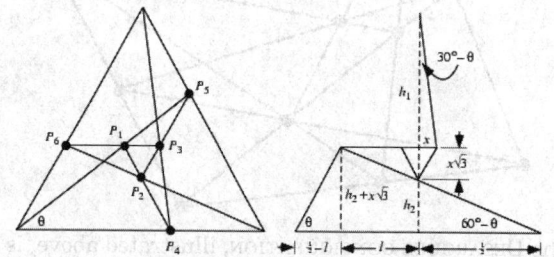

For the third 9_3 configuration, solving the five simultaneous equations

$$\tan(\theta - 30°) = \frac{x}{h_1} \tag{14}$$

$$\tan(60° - \theta) = \frac{h_2}{\frac{1}{2}} \tag{15}$$

$$h_1 + x\sqrt{3} + h_2 = \tfrac{1}{2}\sqrt{3} \tag{16}$$

$$\tan(60° - \theta) = \frac{\sqrt{3} - \tan \theta}{1 + \sqrt{3}\tan \theta} = \frac{h_2 + x\sqrt{3}}{l + \frac{1}{2}} \tag{17}$$

$$\tan 60° = \sqrt{3} = \frac{h_2 + x\sqrt{3}}{\frac{1}{2} - l} \tag{18}$$

gives

$$\theta = \tfrac{1}{2} \cos^{-1}\left(\tfrac{1}{4}\right) \tag{19}$$

$$x = \tfrac{1}{4}(7 - 3\sqrt{5}) \tag{20}$$

$$l = \tfrac{1}{4}(\sqrt{5} - 1) \tag{21}$$

$$h_1 = \tfrac{1}{4}(\sqrt{15} - \sqrt{3}) \tag{22}$$

$$h_2 = \tfrac{1}{2}(\sqrt{15} - 2\sqrt{3}). \tag{23}$$

The six points are then given by

$$P_1 = \left(\tfrac{1}{4}(3\sqrt{5} - 5),\ \tfrac{1}{4}(3\sqrt{3} - \sqrt{15})\right) \tag{24}$$

$$P_2 = \left(\tfrac{1}{2},\ \tfrac{1}{2}(\sqrt{15} - 2\sqrt{3})\right) \tag{25}$$

$$P_3 = \left(\tfrac{3}{4}(3 - \sqrt{5}),\ \tfrac{1}{4}(3\sqrt{3} - \sqrt{15})\right) \tag{26}$$

$$P_4 = \left(\tfrac{1}{2}(\sqrt{5} - 1),\ 0\right) \tag{27}$$

$$P_5 = \left(\tfrac{1}{4}(5 - \sqrt{5}),\ \tfrac{1}{4}(\sqrt{15} - \sqrt{3})\right) \tag{28}$$

$$P_6 = \left(\tfrac{1}{4}(3 - \sqrt{5}),\ \tfrac{1}{4}(3\sqrt{3} - \sqrt{15})\right). \tag{29}$$

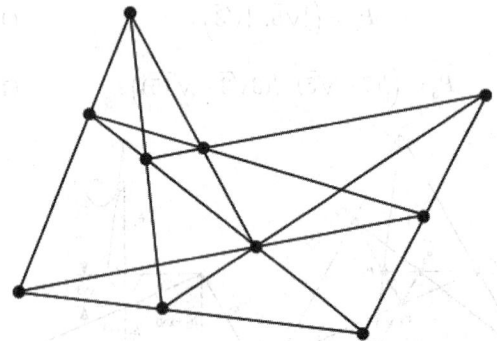

The DESARGUES CONFIGURATION, illustrated above, is one of the ten 10_3 configurations. Page and Dorwart (1984) discuss the 31 11_3 configurations (Wells 1991, p. 63).

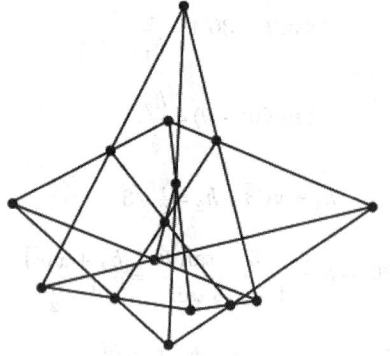

The CREMONA-RICHMOND CONFIGURATION, illustrated above, is one of the 245342 15_3 configurations.

See also BAR (EDGE), CREMONA-RICHMOND CONFIGURATION, DESARGUES CONFIGURATION, DOUBLE SIXES, EQUILATERAL TRIANGLE, EUCLIDEAN SPACE, FANO PLANE, FRAMEWORK, ORCHARD-PLANTING PROBLEM, ORIENTED MATROID, PAPPUS'S HEXAGON THEOREM, PROJECTIVE PLANE, REGULAR GRAPH, REYE'S CONFIGURATION, RIGID GRAPH, TENSEGRITY, TESSERACT

References

Bokowski, J. and Sturmfels, B. *Computational Synthetic Geometry.* Berlin: Springer-Verlag, p. 41, 1988.

Colbourn, C. J. and Dinitz, J. H. (Eds.). *CRC Handbook of Combinatorial Designs.* Boca Raton, FL: CRC Press, p. 255, 1996.

Gropp, H. "Configurations and the Tutte Conjecture." *Ars. Combin. A* **29**, 171–77, 1990.

Gropp, H. "On the History of Configurations." Conference San Sebastien (Spain). Sept. 1990.

Gropp, H. "Enumeration of Regular Graphs 100 Years Ago." *Discrete Math.* **101**, 73–5, 1992.

Gropp, H. "Non-Symmetric Configurations with Deficiencies 1 and 2." *Combinatorics '90. Recent Trends and Applications. Proceedings of the International Conference Held in Gaeta, May 20–7, 1990* (Ed. A. Barlotti, A. Bichera, P. V. Ceccherini, and G. Tallini). Amsterdam, Netherlands: North-Holland, pp. 227–39, 1992.

Gropp, H. "Configurations and Their Realization." *Discr. Math.* **174**, 137–51, 1997.

Hilbert, D. and Cohn-Vossen, S. *Geometry and the Imagination.* New York: Chelsea, 1999.

Page, W. and Dorwart, H. L. "Numerical Patterns and Geometrical Configurations." *Math. Mag.* **57**, 82–2, 1984.

Sloane, N. J. A. Sequences A001403 in "An On-Line Version of the Encyclopedia of Integer Sequences." http://www.research.att.com/~njas/sequences/eisonline.html.

Sterneck, R. D. von. "Die Configuration 11_3." *Monatshefte f. Math. Phys.* **5**, 325–31, 1894.

Sterneck, R. D. von. "Die Configuration 12_3." *Monatshefte f. Math. Phys.* **6**, 223–55, 1895.

Sturmfels, B. and White, N. "All 11_3 and 12_3 Configurations are Rational." *Aeq. Math.* **39**, 254–60, 1990.

Wells, D. *The Penguin Dictionary of Curious and Interesting Numbers.* Middlesex, England: Penguin Books, p. 75, 1986.

Wells, D. *The Penguin Dictionary of Curious and Interesting Geometry.* London: Penguin, pp. 63 and 159–60, 1991.

Confluent Hypergeometric Differential Equation

The second-order ordinary differential equation

$$xy'' + (c - x)y' - ay = 0, \tag{1}$$

sometimes also called Kummer's differential equation (Zwillinger 1997, p. 124). It has a REGULAR SINGULAR POINT at 0 and an irregular singularity at ∞. The solutions

$$y = b_1 {}_1F_1(a;\ c;\ x) + b_2 U(a,\ c,\ x) \tag{2}$$

are called CONFLUENT HYPERGEOMETRIC FUNCTION OF THE FIRST and SECOND KINDS, respectively. Note that the CONFLUENT HYPERGEOMETRIC FUNCTION OF THE FIRST KIND is also denoted $M(a,\ c,\ x)$ or $\Phi(a;\ c;\ z)$.

See also CONFLUENT HYPERGEOMETRIC FUNCTION OF

THE FIRST KIND, CONFLUENT HYPERGEOMETRIC FUNCTION OF THE SECOND KIND, GENERAL CONFLUENT HYPERGEOMETRIC DIFFERENTIAL EQUATION, HYPERGEOMETRIC DIFFERENTIAL EQUATION, WHITTAKER DIFFERENTIAL EQUATION

References

Abramowitz, M. and Stegun, C. A. (Eds.). *Handbook of Mathematical Functions with Formulas, Graphs, and Mathematical Tables, 9th printing.* New York: Dover, p. 504, 1972.

Arfken, G. "Confluent Hypergeometric Functions." §13.6 in *Mathematical Methods for Physicists, 3rd ed.* Orlando, FL: Academic Press, pp. 753–58, 1985.

Morse, P. M. and Feshbach, H. *Methods of Theoretical Physics, Part I.* New York: McGraw-Hill, pp. 551–55, 1953.

Zwillinger, D. *Handbook of Differential Equations, 3rd ed.* Boston, MA: Academic Press, pp. 123–24, 1997.

Confluent Hypergeometric Function

CONFLUENT HYPERGEOMETRIC FUNCTION OF THE FIRST KIND, CONFLUENT HYPERGEOMETRIC FUNCTION OF THE SECOND KIND, CONFLUENT HYPERGEOMETRIC LIMIT FUNCTION

Confluent Hypergeometric Function of the First Kind

The confluent hypergeometric function is a degenerate form the HYPERGEOMETRIC FUNCTION $_2F_1(a, b; c; z)$ which arises as a solution the CONFLUENT HYPERGEOMETRIC DIFFERENTIAL EQUATION. It is commonly denoted $_1F_1(a; b; z)$, $M(a, b, z)$, or $\Phi(a; b; z)$, and is also known as KUMMER'S FUNCTION of the first kind. An alternate form of the solution to the CONFLUENT HYPERGEOMETRIC DIFFERENTIAL EQUATION is known as the WHITTAKER FUNCTION.

The confluent hypergeometric function has a HYPERGEOMETRIC SERIES given by

$$_1F_1(a; b; z) = 1 + \frac{a}{b}z + \frac{a(a+1)}{b(b+1)}\frac{z^2}{2!} + \cdots$$

$$= \sum_{k=0}^{\infty} \frac{(a)_k}{(b)_k} \frac{z^k}{k!}, \tag{1}$$

where $(a)_k$ and $(b)_k$ are POCHHAMMER SYMBOLS. If a and b are INTEGERS, $a < 0$, and either $b > 0$ or $b < a$, then the series yields a POLYNOMIAL with a finite number of terms. If b is an INTEGER ≤ 0, then $_1F_1(a; b; z)$ is undefined. The confluent hypergeometric function is given in terms of the LAGUERRE POLYNOMIAL by

$$L_n^m(x) = \frac{(m+n)!}{m!n!} {}_1F_1(-n; m+1; x), \tag{2}$$

(Arfken 1985, p. 755), and also has an integral representation

$$_1F_1(a; b; z)$$

$$= \frac{\Gamma(b)}{\Gamma(b-a)\Gamma(a)} \int_0^1 e^{zt} t^{a-1}(1-t)^{b-a-1} \, dt \tag{3}$$

(Abramowitz and Stegun 1972, p. 505).

BESSEL FUNCTIONS, the ERROR FUNCTION, the incomplete GAMMA FUNCTION, HERMITE POLYNOMIAL, LAGUERRE POLYNOMIAL, as well as other are all special cases of this function (Abramowitz and Stegun 1972, p. 509). Kummer showed that

$$e^x \, _1F_1(a; b; -x) = \, _1F_1(b-a, b, x) \tag{4}$$

(Koepf 1998, p. 42).

KUMMER'S SECOND FORMULA gives

$$_1F_1\left(\frac{1}{2} + m; \; 2m + 1; \; z\right) = M_{0,m}(z)$$

$$= z^{m+1/2}\left[1 + \sum_{p=1}^{\infty} \frac{z^{2p}}{2^{4p} p!(m+1)(m+2)\cdots(m+p)}\right], \tag{5}$$

where $_1F_1(a; b; z)$ is the CONFLUENT HYPERGEOMETRIC FUNCTION and $m \neq -1/2, -1, -3/2, \ldots$.

See also CONFLUENT HYPERGEOMETRIC DIFFERENTIAL EQUATION, CONFLUENT HYPERGEOMETRIC FUNCTION OF THE SECOND KIND, CONFLUENT HYPERGEOMETRIC LIMIT FUNCTION, GENERALIZED HYPERGEOMETRIC FUNCTION, HEINE HYPERGEOMETRIC SERIES, HYPERGEOMETRIC FUNCTION, HYPERGEOMETRIC SERIES, KUMMER'S FORMULAS, WEBER-SONINE FORMULA, WHITTAKER FUNCTION

References

Abad, J. and Sesma, J. "Computation of the Regular Confluent Hypergeometric Function." *Mathematica J.* **5**, 74–6, 1995.

Abramowitz, M. and Stegun, C. A. (Eds.). "Confluent Hypergeometric Functions." Ch. 13 in *Handbook of Mathematical Functions with Formulas, Graphs, and Mathematical Tables, 9th printing.* New York: Dover, pp. 503–15, 1972.

Arfken, G. "Confluent Hypergeometric Functions." §13.6 in *Mathematical Methods for Physicists, 3rd ed.* Orlando, FL: Academic Press, pp. 753–58, 1985.

Buchholz, H. *The Confluent Hypergeometric Function with Special Emphasis on its Applications.* New York: Springer-Verlag, 1969.

Iyanaga, S. and Kawada, Y. (Eds.). "Hypergeometric Function of Confluent Type." Appendix A, Table 19.I in *Encyclopedic Dictionary of Mathematics.* Cambridge, MA: MIT Press, p. 1469, 1980.

Koepf, W. *Hypergeometric Summation: An Algorithmic Approach to Summation and Special Function Identities.* Braunschweig, Germany: Vieweg, 1998.

Morse, P. M. and Feshbach, H. *Methods of Theoretical Physics, Part I.* New York: McGraw-Hill, pp. 551–54 and 604–05, 1953.

Slater, L. J. *Confluent Hypergeometric Functions.* Cambridge, England: Cambridge University Press, 1960.

Spanier, J. and Oldham, K. B. "The Kummer Function $M(a;c;x)$." Ch. 47 in *An Atlas of Functions.* Washington, DC: Hemisphere, pp. 459–69, 1987.

Tricomi, F. G. *Fonctions hypergéométriques confluentes.* Paris: Gauthier-Villars, 1960.

Confluent Hypergeometric Function of the Second Kind

Gives the second linearly independent solution to the CONFLUENT HYPERGEOMETRIC DIFFERENTIAL EQUATION. It is also known as the KUMMER'S FUNCTION of the second kind, the TRICOMI FUNCTION, or the GORDON FUNCTION. It is denoted $U(a, b, z)$ and has an integral representation

$$U(a, b, z) = \frac{1}{\Gamma(a)} \int_0^\infty e^{-zt} t^{a-1} (1+t)^{b-a-1} \, dt$$

(Abramowitz and Stegun 1972, p. 505). The WHITTAKER FUNCTIONS give an alternative form of the solution. For small z, the function behaves as z^{1-b}.

See also BATEMAN FUNCTION, CONFLUENT HYPERGEOMETRIC FUNCTION OF THE FIRST KIND, CONFLUENT HYPERGEOMETRIC LIMIT FUNCTION, COULOMB WAVE FUNCTION, CUNNINGHAM FUNCTION, GORDON FUNCTION, HYPERGEOMETRIC FUNCTION, POISSON-CHARLIER POLYNOMIAL, TORONTO FUNCTION, WEBER FUNCTIONS, WHITTAKER FUNCTION

References
Abramowitz, M. and Stegun, C. A. (Eds.). "Confluent Hypergeometric Functions." Ch. 13 in *Handbook of Mathematical Functions with Formulas, Graphs, and Mathematical Tables, 9th printing.* New York: Dover, pp. 503–15, 1972.
Arfken, G. "Confluent Hypergeometric Functions." §13.6 in *Mathematical Methods for Physicists, 3rd ed.* Orlando, FL: Academic Press, pp. 753–58, 1985.
Buchholz, H. *The Confluent Hypergeometric Function with Special Emphasis on its Applications.* New York: Springer-Verlag, 1969.
Morse, P. M. and Feshbach, H. *Methods of Theoretical Physics, Part I.* New York: McGraw-Hill, pp. 671–72, 1953.
Spanier, J. and Oldham, K. B. "The Tricomi Function $U(a;c;x)$." Ch. 48 in *An Atlas of Functions.* Washington, DC: Hemisphere, pp. 471–77, 1987.

Confluent Hypergeometric Limit Function

$$_0F_1(; a; z) \equiv \lim_{q \to \infty} {}_1F_1\left(q; a; \frac{z}{q}\right). \tag{1}$$

It has a series expansion

$$_0F_1(; a; z) = \sum_{n=0}^\infty \frac{z^n}{(a)_n n!} \tag{2}$$

and satisfies

$$z \frac{d^2 y}{dz^2} + a \frac{dy}{dz} - y = 0. \tag{3}$$

A BESSEL FUNCTION OF THE FIRST KIND can be

expressed in terms of this function by

$$J_n(x) = \frac{\left(\frac{1}{2} x\right)^n}{n!} \, {}_0F_1(; \, n+1; \, -\tfrac{1}{4} x^2) \tag{4}$$

(Petkovsek *et al.* 1996).

See also CONFLUENT HYPERGEOMETRIC FUNCTION, GENERALIZED HYPERGEOMETRIC FUNCTION, HYPERGEOMETRIC FUNCTION

References
Petkovsek, M.; Wilf, H. S.; and Zeilberger, D. *A = B.* Wellesley, MA: A. K. Peters, p. 38, 1996.

Confocal Conics

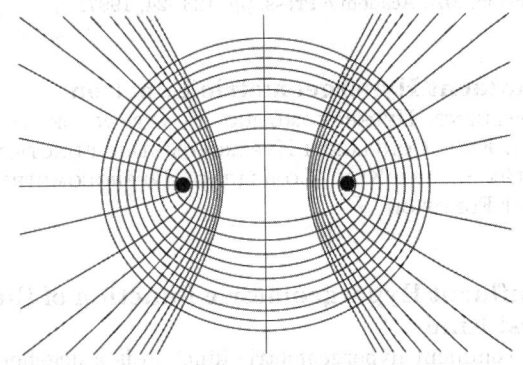

Confocal conics are CONIC SECTIONS sharing a common FOCUS. Any two confocal CENTRAL CONICS are orthogonal (Ogilvy 1990, p. 77).

See also CONFOCAL ELLIPSES, CONFOCAL ELLIPSOIDAL COORDINATES, CONFOCAL HYPERBOLAS, CONFOCAL PARABOLAS, CONFOCAL QUADRICS, CONIC SECTION, FOCUS

References
Ogilvy, C. S. *Excursions in Geometry.* New York: Dover, pp. 77–8, 1990.
Wells, D. *The Penguin Dictionary of Curious and Interesting Geometry.* London: Penguin, pp. 39–0, 1991.

Confocal Ellipses

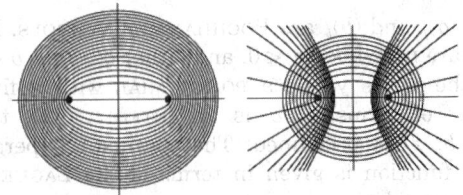

ELLIPSES sharing common FOCI (left figure). The family of confocal ellipses covers the plane simply, in the sense that there is a unique ellipse passing through each point in the plane (Hilbert and Cohn-Vossen 1999, p. 5). The figure on the right shows confocal ellipses superimposed on CONFOCAL HYPER-

BOLAS, which form an orthogonal net of curves (Hilbert and Cohn-Vossen 1999, pp. 5-).

See also CONFOCAL CONICS, CONFOCAL HYPERBOLAS, CONFOCAL PARABOLAS, ELLIPSE

References

Hilbert, D. and Cohn-Vossen, S. *Geometry and the Imagination.* New York: Chelsea, 1999.

Confocal Ellipsoidal Coordinates

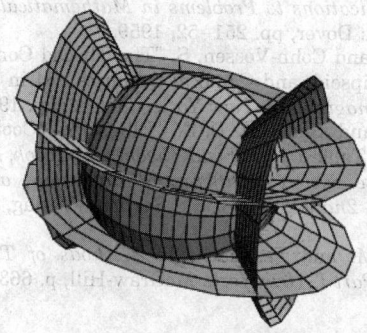

The confocal ellipsoidal coordinates, called simply "ellipsoidal coordinates" by Morse and Feshbach (1953) and "elliptic coordinates" by Hilbert and Cohn-Vossen (1999, p. 22), are given by the equations

$$\frac{x^2}{a^2 + \xi} + \frac{y^2}{b^2 + \xi} + \frac{z^2}{c^2 + \xi} = 1 \tag{1}$$

$$\frac{x^2}{a^2 + \eta} + \frac{y^2}{b^2 + \eta} + \frac{z^2}{c^2 + \eta} = 1 \tag{2}$$

$$\frac{x^2}{a^2 + \zeta} + \frac{y^2}{b^2 + \zeta} + \frac{z^2}{c^2 + \zeta} = 1, \tag{3}$$

where $-c < \xi < \infty$, $-b^2 < \eta < -c^2$, and $-a^2 < \zeta < -b^2$. These coordinates correspond to three CONFOCAL QUADRICS all sharing the same pair of foci. Surfaces of constant ζ are confocal ELLIPSOIDS, surfaces of constant η are one-sheeted HYPERBOLOIDS, and surfaces of constant ζ are two-sheeted HYPERBOLOIDS (Hilbert and Cohn-Vossen 1999, pp. 22-3). For every (x, y, z), there is a unique set of ellipsoidal coordinates. However, (ξ, η, ζ) specifies eight points symmetrically located in OCTANTS.
Solving for x, y, and z gives

$$x^2 = \frac{(a^2 + \xi)(a^2 + \eta)(a^2 + \zeta)}{(b^2 - a^2)(c^2 - a^2)} \tag{4}$$

$$y^2 = \frac{(b^2 + \xi)(b^2 + \eta)(b^2 + \zeta)}{(a^2 - b^2)(c^2 - b^2)} \tag{5}$$

$$z^2 = \frac{(c^2 + \xi)(c^2 + \eta)(c^2 + \zeta)}{(a^2 - c^2)(b^2 - c^2)}. \tag{6}$$

The LAPLACIAN is

$$\nabla^2\Psi = (\eta - \zeta)f(\xi)\frac{\partial}{\partial\xi}\left[f(\xi)\frac{\partial\Psi}{\partial\xi}\right] + (\zeta - \xi)f(\eta)\frac{\partial}{\partial\eta}$$

$$\times \left[f(\eta)\frac{\partial\Psi}{\partial\eta}\right] + (\xi - \eta)f(\zeta)\frac{\partial}{\partial\zeta}\left[f(\zeta)\frac{\partial\Psi}{\partial\zeta}\right], \tag{7}$$

where

$$f(x) \equiv \sqrt{(x + a^2)(x + b^2)(x + c^2)}. \tag{8}$$

Another definition is

$$\frac{x^2}{a^2 - \lambda} + \frac{y^2}{b^2 - \lambda} + \frac{z^2}{c^2 - \lambda} = 1 \tag{9}$$

$$\frac{x^2}{a^2 - \mu} + \frac{y^2}{b^2 - \mu} + \frac{z^2}{c^2 - \mu} = 1 \tag{10}$$

$$\frac{x^2}{a^2 - \nu} + \frac{y^2}{b^2 - \nu} + \frac{z^2}{c^2 - \nu} = 1, \tag{11}$$

where

$$\lambda < c^2 < \mu < b^2 < \nu < a^2 \tag{12}$$

(Arfken 1970, pp. 117-18). Byerly (1959, p. 251) uses a slightly different definition in which the Greek variables are replaced by their squares, and $a = 0$. Equation (9) represents an ELLIPSOID, (10) represents a one-sheeted HYPERBOLOID, and (11) represents a two-sheeted HYPERBOLOID.

In terms of CARTESIAN COORDINATES,

$$x^2 = \frac{(a^2 - \lambda)(a^2 - \mu)(a^2 - \nu)}{(a^2 - b^2)(a^2 - c^2)} \tag{13}$$

$$y^2 = \frac{(b^2 - \lambda)(b^2 - \mu)(b^2 - \nu)}{(b^2 - a^2)(b^2 - c^2)} \tag{14}$$

$$z^2 = \frac{(c^2 - \lambda)(c^2 - \mu)(c^2 - \nu)}{(c^2 - a^2)(c^2 - b^2)}. \tag{15}$$

The SCALE FACTORS are

$$h_\lambda = \sqrt{\frac{(\mu - \lambda)(\nu - \lambda)}{4(a^2 - \lambda)(b^2 - \lambda)(c^2 - \lambda)}} \tag{16}$$

$$h_\mu = \sqrt{\frac{(\nu - \mu)(\lambda - \mu)}{4(a^2 - \mu)(b^2 - \mu)(c^2 - \mu)}} \tag{17}$$

$$h_\nu = \sqrt{\frac{(\lambda - \nu)(\mu - \nu)}{4(a^2 - \nu)(b^2 - \nu)(c^2 - \nu)}}. \tag{18}$$

The LAPLACIAN is

$$\nabla^2 = 2\,\frac{a^2b^2 + a^2c^2 + b^2c^2 - 2\nu(a^2 + b^2 + c^2) + 3\nu^2}{(\mu - \nu)(\nu - \lambda)}\,\frac{\partial}{\partial \nu}$$

$$+\,\frac{4(a^2 - \nu)(b^2 - \nu)(c^2 - \nu)}{(\mu - \nu)(\nu - \lambda)}\,\frac{\partial^2}{\partial \nu^2}$$

$$+\,2\,\frac{a^2b^2 + a^2c^2 + b^2c^2 - 2\mu(a^2 + b^2 + c^2) + 3\mu^2}{(\nu - \mu)(\mu - \lambda)}\,\frac{\partial}{\partial \mu}$$

$$+\,\frac{4(a^2 - \mu)(b^2 - \mu)(c^2 - \mu)}{(\mu - \lambda)(\nu - \mu)}\,\frac{\partial^2}{\partial \mu^2}$$

$$+\,2\,\frac{-(a^2b^2 + a^2c^2 + b^2c^2) + 2\lambda(a^2 + b^2 + c^2) - 3\lambda^2}{(\mu - \lambda)(\nu - \lambda)}\,\frac{\partial}{\partial \lambda}$$

$$(19)$$

Using the NOTATION of Byerly (1959, pp. 252–53), this can be reduced to

$$\nabla^2 = (\mu^2 - \nu^2)\frac{\partial^2}{\partial \alpha^2} + (\lambda^2 - \nu^2)\frac{\partial^2}{\partial \beta^2} + (\lambda^2 - \mu^2)\frac{\partial^2}{\partial \gamma^2}, \quad (20)$$

where

$$\alpha = c\int_c^\lambda \frac{d\lambda}{\sqrt{(\lambda^2 - b^2)(\lambda^2 - c^2)}}$$

$$= F\left(\frac{b}{c}, \frac{\pi}{2}\right) - F\left(\frac{b}{c}, \sin^{-1}\left(\frac{c}{\lambda}\right)\right) \quad (21)$$

$$\beta = c\int_b^\mu \frac{d\mu}{\sqrt{(c^2 - \mu^2)(\mu^2 - b^2)}}$$

$$= F\left[\sqrt{1 - \frac{b^2}{c^2}}, \sin^{-1}\left(\sqrt{\frac{1 - \frac{b^2}{\mu^2}}{1 - \frac{b^2}{c^2}}}\right)\right] \quad (22)$$

$$\gamma = c\int_0^\nu \frac{d\nu}{\sqrt{(b^2 - \nu^2)(c^2 - \nu^2)}} = F\left(\frac{b}{c}, \sin^{-1}\left(\frac{\nu}{b}\right)\right). \quad (23)$$

Here, F is an ELLIPTIC INTEGRAL OF THE FIRST KIND. In terms of α, β, and γ,

$$\lambda = c\,\mathrm{dc}\left(\alpha, \frac{b}{c}\right) \quad (24)$$

$$\mu = b\,\mathrm{nd}\left(\beta, \sqrt{1 - \frac{b^2}{c^2}}\right) \quad (25)$$

$$\nu = b\,\mathrm{sn}\left(\gamma, \frac{b}{c}\right), \quad (26)$$

where dc, nd, and sn are JACOBI ELLIPTIC FUNCTIONS. The HELMHOLTZ DIFFERENTIAL EQUATION is separable in confocal ellipsoidal coordinates.

See also HELMHOLTZ DIFFERENTIAL EQUATION–CON-

FOCAL ELLIPSOIDAL COORDINATES

References

Abramowitz, M. and Stegun, C. A. (Eds.). "Definition of Elliptical Coordinates." §21.1 in *Handbook of Mathematical Functions with Formulas, Graphs, and Mathematical Tables, 9th printing.* New York: Dover, p. 752, 1972.

Arfken, G. "Confocal Ellipsoidal Coordinates (ξ_1, ξ_2, ξ_3)." §2.15 in *Mathematical Methods for Physicists, 2nd ed.* New York: Academic Press, pp. 117–18, 1970.

Byerly, W. E. *An Elementary Treatise on Fourier's Series, and Spherical, Cylindrical, and Ellipsoidal Harmonics, with Applications to Problems in Mathematical Physics.* New York: Dover, pp. 251–52, 1959.

Hilbert, D. and Cohn-Vossen, S. "The Thread Construction of the Ellipsoid, and Confocal Quadrics." §4 in *Geometry and the Imagination.* New York: Chelsea, pp. 19–5, 1999.

Moon, P. and Spencer, D. E. "Ellipsoidal Coordinates (η, θ, λ)." Table 1.10 in *Field Theory Handbook, Including Coordinate Systems, Differential Equations, and Their Solutions, 2nd ed.* New York: Springer-Verlag, pp. 40–4, 1988.

Morse, P. M. and Feshbach, H. *Methods of Theoretical Physics, Part I.* New York: McGraw-Hill, p. 663, 1953.

Confocal Hyperbolas

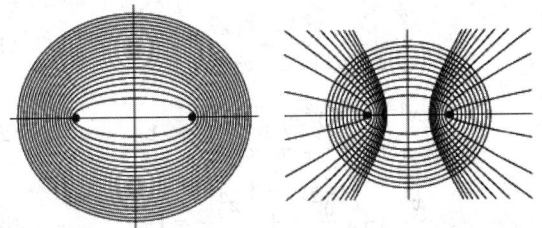

HYPERBOLAS sharing common FOCI (left figure). The family of confocal hyperbolas covers the plane simply, in the sense that there is a unique hyperbola passing through each point in the plane (Hilbert and Cohn-Vossen 1999, p. 5). The figure on the right shows confocal hyperbolas superimposed on CONFOCAL ELLIPSES, which form an orthogonal net of curves (Hilbert and Cohn-Vossen 1999, pp. 5–).

See also CONFOCAL CONICS, CONFOCAL ELLIPSES, CONFOCAL PARABOLAS, ELLIPSE

References

Hilbert, D. and Cohn-Vossen, S. *Geometry and the Imagination.* New York: Chelsea, p. 5, 1999.

Confocal Parabolas

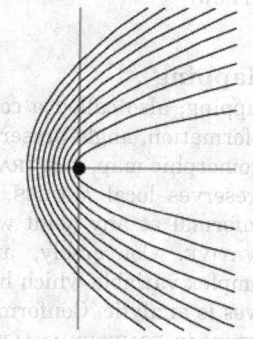

PARABOLAS sharing a common FOCUS.

See also CONFOCAL CONICS, CONFOCAL ELLIPSES, CONFOCAL HYPERBOLAS, PARABOLA

References

Hilbert, D. and Cohn-Vossen, S. *Geometry and the Imagination.* New York: Chelsea, p. 5, 1999.

Confocal Parabolic Coordinates

CONFOCAL PARABOLOIDAL COORDINATES

Confocal Paraboloidal Coordinates

$$\frac{x^2}{a^2 - \lambda} + \frac{y^2}{b^2 - \lambda} = z - \lambda \qquad (1)$$

$$\frac{x^2}{a^2 - \mu} + \frac{y^2}{b^2 - \mu} = z - \mu \qquad (2)$$

$$\frac{x^2}{a^2 - v} + \frac{y^2}{b^2 - v} = z - v, \qquad (3)$$

where $\lambda \in (-\infty, b^2)$, $\mu \in (b^2, a^2)$, and $v \in (a^2, \infty)$.

$$x^2 = \frac{(a^2 - \lambda)(a^2 - \mu)(a^2 - v)}{(b^2 - a^2)} \qquad (4)$$

$$y^2 = \frac{(b^2 - \lambda)(b^2 - \mu)(b^2 - v)}{(a^2 - b^2)} \qquad (5)$$

$$z = \lambda + \mu + v - a^2 - b^2. \qquad (6)$$

The SCALE FACTORS are

$$h_\lambda = \sqrt{\frac{(\mu - \lambda)(v - \lambda)}{4(a^2 - \lambda)(b^2 - \lambda)}} \qquad (7)$$

$$h_\mu = \sqrt{\frac{(v - \mu)(\lambda - \mu)}{4(a^2 - \mu)(b^2 - \mu)}} \qquad (8)$$

$$h_v = \sqrt{\frac{(\lambda - v)(\mu - v)}{16(a^2 - v)(b^2 - v)}}. \qquad (9)$$

The LAPLACIAN is

$$\nabla^2 = \frac{2(a^2 + b^2 - 2v)}{(\mu - v)(v - \lambda)} \frac{\partial}{\partial v} + \frac{4(a^2 - v)(v - b^2)}{(\mu - v)(v - \lambda)} \frac{\partial^2}{\partial v^2}$$

$$+ \frac{2(a^2 + b^2 - 2\mu)}{(\mu - \lambda)(v - \mu)} \frac{\partial}{\partial \mu} + \frac{4(a^2 - \mu)(\mu - b^2)}{(\mu - \lambda)(v - \mu)} \frac{\partial^2}{\partial \mu^2}$$

$$+ \frac{2(2\lambda - a^2 - b^2)}{(\mu - \lambda)(v - \lambda)} \frac{\partial}{\partial \lambda} + \frac{4(\lambda - a^2)(\lambda - b^2)}{(\mu - \lambda)(v - \lambda)} \frac{\partial^2}{\partial \lambda^2}. \qquad (10)$$

The HELMHOLTZ DIFFERENTIAL EQUATION is SEPARABLE.

See also HELMHOLTZ DIFFERENTIAL EQUATION–CONFOCAL PARABOLOIDAL COORDINATES

References

Arfken, G. "Confocal Parabolic Coordinates (ξ_1, ξ_2, ξ_3)." §2.17 in *Mathematical Methods for Physicists, 2nd ed.* Orlando, FL: Academic Press, pp. 119–20, 1970.
Moon, P. and Spencer, D. E. "Paraboloidal Coordinates (μ, v, λ)." Table 1.11 in *Field Theory Handbook, Including Coordinate Systems, Differential Equations, and Their Solutions, 2nd ed.* New York: Springer-Verlag, pp. 44–8, 1988.
Morse, P. M. and Feshbach, H. *Methods of Theoretical Physics, Part I.* New York: McGraw-Hill, p. 664, 1953.

Confocal Quadrics

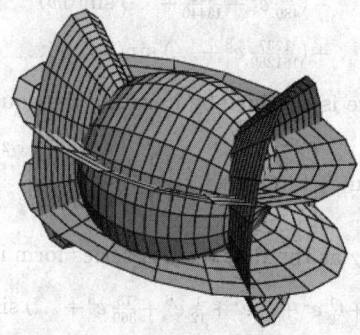

A set of QUADRATIC SURFACES which share FOCI. Ellipsoids and one- and two-sheeted hyperboloids can be confocal. These three types of surfaces can be combined to form an orthogonal coordinate system known as CONFOCAL ELLIPSOIDAL COORDINATES (Hilbert and Cohn-Vossen 1991, pp. 22–3).

The planes of symmetry of the tangent cone from any point P in space to any surface of the confocal system which does not enclose P are the tangent planes at P to the three surfaces of the system that pass through P. As a limiting case, this result means that every surface of the confocal system when viewed from a point lying on a focal curve and not enclosed by the surface looks like a circle with its center on the line of sight, provided that the line of sight is tangent to the focal curve (Hilbert and Cohn-Vossen 1999, p. 24).

See also CONFOCAL ELLIPSOIDAL COORDINATES, ELLIPSOID, HYPERBOLOID, QUADRATIC SURFACE

References

Hilbert, D. and Cohn-Vossen, S. "The Thread Construction of the Ellipsoid, and Confocal Quadrics." §4 in *Geometry and the Imagination.* New York: Chelsea, pp. 19–5, 1999.

Confoliation

A topological structure which interpolates between contact structures and codimension-one FOLIATIONS.

See also FOLIATION

References

Eliashberg, Y. M. and Thurston, W. P. *Confolations.* Providence, RI: Amer. Math. Soc., 1998.

Conformal Latitude

An AUXILIARY LATITUDE defined by

$$\chi \equiv 2\tan^{-1}\left\{\tan(\tfrac{1}{4}\pi + \tfrac{1}{2}\phi)\left[\frac{1 - e\sin\phi}{1 + e\sin\phi}\right]^{e/2}\right\} - \tfrac{1}{2}\pi$$

$$= 2\tan^{-1}\left\{\frac{1 + \sin\phi}{1 - \sin\phi}\left(\frac{1 - e\sin\phi}{1 + e\sin\phi}\right)^{e}\right\}^{1/2} - \tfrac{1}{2}\pi$$

$$= \phi - (\tfrac{1}{2}e^2 + \tfrac{5}{24}e^4 + \tfrac{3}{32}e^6 + \tfrac{281}{5760}e^8 + \ldots)\sin(2\phi)$$

$$+ (\tfrac{5}{48}e^4 + \tfrac{7}{80}e^6 + \tfrac{697}{11520}e^8 + \ldots)\sin(4\phi)$$

$$- (\tfrac{13}{480}e^6 + \tfrac{461}{13440} + \ldots)\sin(6\phi)$$

$$+ (\tfrac{1237}{161280}e^8 + \ldots)\sin(8\phi) + \ldots$$

The inverse is obtained by iterating the equation

$$\phi = 2\tan^{-1}\left[\tan(\tfrac{1}{4}\pi + \tfrac{1}{2}\chi)\left(\frac{1 + e\sin\phi}{1 - e\sin\phi}\right)^{e/2}\right] - \tfrac{1}{2}\pi$$

using $\phi = \chi$ as the first trial. A series form is

$$\phi = \chi + (\tfrac{1}{2}e^2 + \tfrac{5}{24}e^4 + \tfrac{1}{12}e^6 + \tfrac{13}{360}e^8 + \ldots)\sin(2\chi)$$

$$+ (\tfrac{7}{48}e^4 + \tfrac{29}{240}e^6 + \tfrac{811}{11520}e^8 + \ldots)\sin(4\chi)$$

$$+ (\tfrac{7}{120}e^6 + \tfrac{81}{1120}e^8 + \ldots)\sin(6\chi)$$

$$+ (\tfrac{4279}{161280}e^8 + \ldots)\sin(8\chi) + \ldots$$

The conformal latitude was called the ISOMETRIC LATITUDE by Adams (1921), but this term is now used to refer to a different quantity.

See also AUXILIARY LATITUDE, LATITUDE

References

Adams, O. S. "Latitude Developments Connected with Geodesy and Cartography with Tables, Including a Table for Lambert Equal-Area Meridianal Projections." Spec. Pub. No. 67. U. S. Coast and Geodetic Survey, pp. 18 and 84–5, 1921.

Snyder, J. P. *Map Projections--A Working Manual.* U. S. Geological Survey Professional Paper 1395. Washington, DC: U. S. Government Printing Office, pp. 15–6, 1987.

Conformal Map

CONFORMAL MAPPING

Conformal Mapping

A conformal mapping, also called a conformal map, conformal transformation, angle-preserving transformation, or biholomorphic map, is a TRANSFORMATION $w = f(z)$ that preserves local ANGLES. An ANALYTIC FUNCTION is conformal at any point where it has a NONZERO DERIVATIVE. Conversely, any conformal mapping of a complex variable which has continuous partial derivatives is analytic. Conformal mapping is extremely important in COMPLEX ANALYSIS, as well as in many areas of physics and engineering.

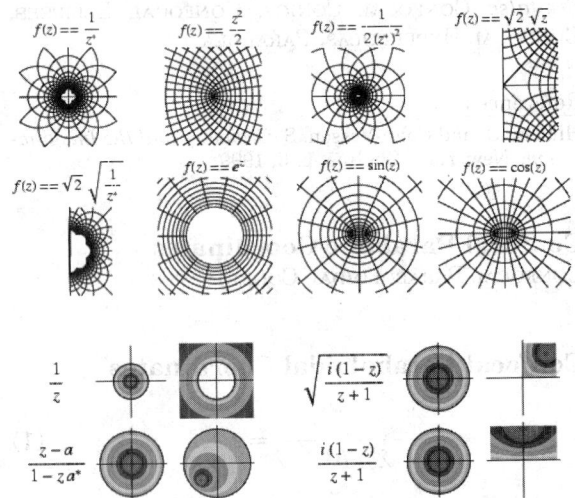

Several conformal transformations of regular grids are illustrated in the first figure above, and are implemented as `ComplexMap` in the *Mathematica* add-on package `Graphics`ComplexMap`` (which can be loaded with the command `<<Graphics``). In the second figure above, contours of constant $|z|$ are shown together with their corresponding contours after the transformation. Moon and Spencer (1988) and Krantz (1999, pp. 183–94) give tables of conformal mappings.

Let θ and ϕ be the tangents to the curves γ and $f(\gamma)$ at z_0 and w_0 in the COMPLEX PLANE,

$$w - w_0 \equiv f(z) - f(z_0) = \frac{f(z) - f(z_0)}{z - z_0}(z - z_0) \qquad (1)$$

$$\arg(w - w_0) = \arg\left[\frac{f(z) - f(z_0)}{z - z_0}\right] + \arg(z - z_0). \qquad (2)$$

Then as $w \to w_0$ and $z \to z_0$,

$$\phi = \arg f'(z_0) + \theta \qquad (3)$$

$$|w| = |f'(z_0)||z|. \qquad (4)$$

A function $f : \mathbb{C} \to \mathbb{C}$ is conformal IFF there are complex numbers $a \neq 0$ and b such that

$$f(z) = az + b \qquad (5)$$

for $z \in \mathbb{C}$ (Krantz 1999, p. 80). Furthermore, if $h : \mathbb{C} \to \mathbb{C}$ is an analytic function such that

$$\lim_{|z| \to +\infty} |h(z)| = +\infty, \qquad (6)$$

then h is a polynomial in z (Greene and Krantz 1997; Krantz 1999, p. 80).

Conformal transformations can prove extremely useful in solving physical problems. By letting $w \equiv f(z)$, the REAL and IMAGINARY PARTS of $w(z)$ must satisfy the CAUCHY-RIEMANN EQUATIONS and LAPLACE'S EQUATION, so they automatically provide a scalar POTENTIAL and a so-called stream function. If a physical problem can be found for which the solution is valid, we obtain a solution—which may have been very difficult to obtain directly—by working backwards.

For example, let

$$w(z) = Az^n = Ar^n e^{in\theta}, \qquad (7)$$

the REAL and IMAGINARY PARTS then give

$$\phi = Ar^n \cos(n\theta) \qquad (8)$$

$$\psi = Ar^n \sin(n\theta). \qquad (9)$$

For $n = -2$,

$$\phi = \frac{A}{r^2} \cos(2\theta) \qquad (10)$$

$$\psi = -\frac{A}{r^2} \sin(2\theta), \qquad (11)$$

which is a double system of LEMNISCATES (Lamb 1945, p. 69).

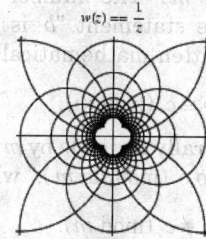

For $n = -1$,

$$\phi = \frac{A}{r} \cos \theta \qquad (12)$$

$$\psi = \frac{A}{r} \sin \theta. \qquad (13)$$

This solution consists of two systems of CIRCLES, and ϕ is the POTENTIAL FUNCTION for two PARALLEL opposite charged line charges (Feynman *et al.* 1989, §7–; Lamb 1945, p. 69).

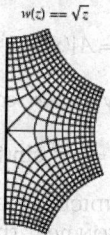

For $n = 1/2$,

$$\phi = Ar^{1/2} \cos\left(\frac{\theta}{2}\right) = A\sqrt{\frac{\sqrt{x^2 + y^2} + x}{2}} \qquad (14)$$

$$\psi = Ar^{1/2} \sin\left(\frac{\theta}{2}\right) = A\sqrt{\frac{\sqrt{x^2 + y^2} - x}{2}}. \qquad (15)$$

ϕ gives the field near the edge of a thin plate (Feynman *et al.* 1989, §7–).

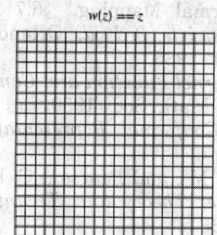

For $n = 1$,

$$\phi = Ar \cos \theta = Ax \qquad (16)$$

$$\psi = Ar \sin \theta = Ay, \qquad (17)$$

giving two straight lines (Lamb 1945, p. 68).

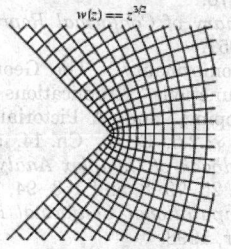

For $n = 3/2$,

$$w = Ar^{3/2} e^{3i\theta/2}. \qquad (18)$$

ϕ gives the field near the outside of a rectangular corner (Feynman *et al.* 1989, §7–).

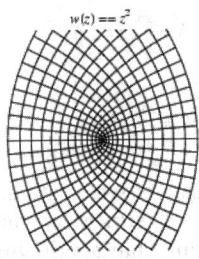

$w(z) == z^2$

For $n = 2$,

$$w = A(x + iy)^2 = A[(x^2 - y^2) + 2ixy] \tag{19}$$

$$\phi = A(x^2 - y^2) = Ar^2 \cos(2\theta) \tag{20}$$

$$\psi = 2Axy = Ar^2 \sin(2\theta). \tag{21}$$

These are two PERPENDICULAR HYPERBOLAS, and ϕ is the POTENTIAL FUNCTION near the middle of two point charges or the field on the opening side of a charged RIGHT ANGLE conductor (Feynman 1989, §7−).

See also ANALYTIC FUNCTION, CAUCHY-RIEMANN EQUATIONS, CAYLEY TRANSFORM, CONFORMAL PROJECTION, HARMONIC FUNCTION, LAPLACE'S EQUATION, MÖBIUS TRANSFORMATION, QUASICONFORMAL MAP, SCHWARZ-CHRISTOFFEL MAPPING, SIMILAR

References

Arfken, G. "Conformal Mapping." §6.7 in *Mathematical Methods for Physicists, 3rd ed.* Orlando, FL: Academic Press, pp. 392–94, 1985.

Bergman, S. *The Kernel Function and Conformal Mapping.* New York: Amer. Math. Soc., 1950.

Carathéodory, C. *Conformal Representation.* New York: Dover, 1998.

Carrier, G.; Crook, M.; and Pearson, C. E. *Functions of a Complex Variable: Theory and Technique.* New York: McGraw-Hill, 1966.

Coxeter, H. S. M. and Greitzer, S. L. *Geometry Revisited.* Washington, DC: Math. Assoc. Amer., p. 80, 1967.

Feynman, R. P.; Leighton, R. B.; and Sands, M. *The Feynman Lectures on Physics, Vol. 1.* Redwood City, CA: Addison-Wesley, 1989.

Greene, R. E. and Krantz, S. G. *Function Theory of One Complex Variable.* New York: Wiley, 1997.

Katznelson, Y. *An Introduction to Harmonic Analysis.* New York: Dover, 1976.

Kober, H. *Dictionary of Conformal Representations.* New York: Dover, 1957.

Krantz, S. G. "Conformality," "The Geometric Theory of Holomorphic Functions," "Applications That Depend on Conformal Mapping," and "A Pictorial Catalog of Conformal Maps." §2.2.5, Ch. 6, Ch. 14, and Appendix to Ch. 14 in *Handbook of Complex Analysis.* Boston, MA: Birkhäuser, pp. 25, 79–8, and 163–94, 1999.

Kythe, P. K. *Computational Conformal Mapping.* Boston, MA: Birkhäuser, 1998.

Lamb, H. *Hydrodynamics, 6th ed.* New York: Dover, 1945.

Mathews, J. "Conformal Mappings." http://www.ecs.fullerton.edu/~mathews/fofz/cmaps.html.

Moon, P. and Spencer, D. E. "Conformal Transformations." §2.01 in *Field Theory Handbook, Including Coordinate Systems, Differential Equations, and Their Solutions, 2nd ed.* New York: Springer-Verlag, pp. 49–6, 1988.

Morse, P. M. and Feshbach, H. "Conformal Mapping." §4.7 in *Methods of Theoretical Physics, Part I.* New York: McGraw-Hill, pp. 358–62 and 443–53, 1953.

Nehari, Z. *Conformal Map.* New York: Dover, 1982.

Conformal Projection

A MAP PROJECTION which is a CONFORMAL MAPPING, i.e., one for which local (infinitesimal) angles on a sphere are mapped to the same angles in the projection. On maps of an entire sphere, however, there are usually singular points at which local angles are distorted.

The term conformal was applied to map projections by Gauss in 1825, and eventually supplanted the alternative terms "orthomorphic" (Germain 1865, Lee 1944; Snyder 1987, p. 4) and "autogonal" (Tissot 1881, Lee 1944).

No projection can be both EQUAL-AREA and conform, and projections which are neither EQUAL-AREA nor conformal are sometimes called APHYLACTIC (Lee 1944; Snyder 1987, p. 4).

See also CONFORMAL MAPPING, EQUIDISTANT PROJECTION, LAMBERT CONFORMAL CONIC PROJECTION, MAP PROJECTION

References

Lee, L. P. "The Nomenclature and Classification of Map Projections." *Empire Survey Rev.* **7**, 190–00, 1944.

Snyder, J. P. *Map Projections--A Working Manual.* U. S. Geological Survey Professional Paper 1395. Washington, DC: U. S. Government Printing Office, 1987.

Thomas, P. S. *Conformal Projections in Geodesy and Cartography.* Washington, DC: U. S. Coast and Geodetic Survey Spec. Pub. 251, 1952.

Tissot, A. *Mémoir sur la représentation des surfaces et les projections des cartes géographiques.* Paris: Gauthier-Villars, 1881.

Conformal Tensor

WEYL TENSOR

Conformal Transformation

CONFORMAL MAPPING

Congruence

If two numbers b and c have the property that their difference $b - c$ is integrally divisible by a number m (i.e., $b - c/m$ is an integer), then b and c are said to be "congruent modulo m." The number m is called the MODULUS, and the statement "b is congruent to c (modulo m)" is written mathematically as

$$b \equiv c \pmod{m}. \tag{1}$$

If $b - c$ is *not* integrally divisible by m, then we say "b is *not* congruent to c (modulo m)," which is written

$$b \not\equiv c \pmod{m}. \tag{2}$$

The explicit "(mod m)" is sometimes omitted when the

MODULUS m is understood by context, so in such cases, care must be taken not to confuse the symbol $\equiv$ with the EQUIVALENCE sign.

The quantity b is sometimes called the "base," and the quantity c is called the RESIDUE or REMAINDER. There are several types of residues. The COMMON RESIDUE defined to be NONNEGATIVE and smaller than m, while the MINIMAL RESIDUE is c or $c - m$, whichever is smaller in ABSOLUTE VALUE. In many computer languages (such as FORTRAN or *Mathematica*), the COMMON RESIDUE of b (mod m) is written mod(b,m) (FORTRAN) or Mod[b,m] (*Mathematica*).

12:40   1:15

Congruence arithmetic is perhaps most familiar as a generalization of the arithmetic of the clock. Since there are 60 minutes in an hour, "minute arithmetic" uses a modulus of $m = 60$. If one starts at 40 minutes past the hour and then waits another 35 minutes, $40 + 35 \equiv 15$ (mod 60), so the current time would be 15 minutes past the (next) hour.

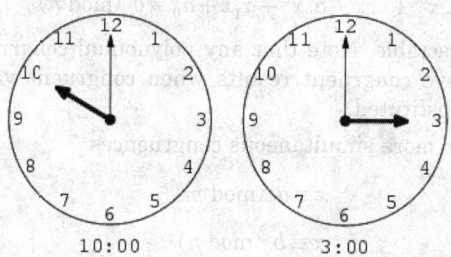

10:00   3:00

Similarly, "hour arithmetic" on a 12-hour clock uses a modulus of $m = 12$, so 10 o'clock (a.m.) plus five hours gives $10 + 5 \equiv 3$ (mod 12), or 3 o'clock (p.m.)

Congruences satisfy a number of important properties, and are extremely useful in many areas of NUMBER THEORY. Using congruences, simple DIVISIBILITY TESTS to check whether a given number is divisible by another number can sometimes be derived. For example, if the sum of a number's digits is divisible by 3 (9), then the original number is divisible by 3 (9).

Congruences also have their limitations. For example, if $a \equiv b$ and $c \equiv d$ (mod n), then it follows that $a^x \equiv b^x$, but usually not that $x^c \equiv x^d$ or $a^c \equiv b^d$. In

addition, by "rolling over," congruences discard absolute information. For example, knowing the number of minutes past the hour is useful, but knowing the hour the minutes are past is often more useful still.

Let $a \equiv a'$ (mod m) and $b \equiv b'$ (mod m), then important properties of congruences include the following, where $\Rightarrow$ means "IMPLIES":

1. Equivalence: $a \equiv b$ (mod 0) $\Rightarrow a = b$ (which can be regarded as a definition).

2. Determination: either $a \equiv b$ (mod m) or $a \not\equiv b$ (mod m).
3. Reflexivity: $a \equiv a$ (mod m).
4. Symmetry: $a \equiv b$ (mod m) $\Rightarrow b \equiv a$ (mod m).
5. Transitivity: $a \equiv b$ (mod m) and $b \equiv c$ (mod m) $\Rightarrow a \equiv c$ (mod m).
6. $a + b \equiv a' + b'$ (mod m).
7. $a - b \equiv a' - b'$ (mod m).
8. $ab \equiv a'b'$ (mod m).
9. $a \equiv b$ (mod m) $\Rightarrow ka \equiv kb$ (mod m).
10. $a \equiv b$ (mod m) $\Rightarrow a^n \equiv b^n$ (mod m).
11. $a \equiv b$ (mod m_1) and $a \equiv b$ (mod m_2) $\Rightarrow a \equiv b$ (mod[m_1, m_2]), where [m_1, m_2] is the LEAST COMMON MULTIPLE.
12. $ak \equiv bk$ (mod m) $\Rightarrow a \equiv b \left(\text{mod } \frac{m}{(k, m)}\right)$, where (k, m) is the GREATEST COMMON DIVISOR.
13. If $a \equiv b$ (mod m), then $P(a) \equiv P(b)$ (mod m), for $P(x)$ a POLYNOMIAL.

Properties (6–) can be proved simply by defining

$$a \equiv a' + rd \tag{3}$$

$$b \equiv b' + sd, \tag{4}$$

where r and s are INTEGERS. Then

$$a + b = a' + b' + (r + s)d \tag{5}$$

$$a - b = a' - b' + (r - s)d \tag{6}$$

$$ab = a'b' + (a's + b'r + rsd)d, \tag{7}$$

so the properties are true.

Congruences also apply to FRACTIONS. For example, note that

$$2 \times 4 \equiv 1 \quad 3 \times 3 \equiv 2 \quad 6 \times 6 \equiv 1 \text{ (mod 7)}, \tag{8}$$

so

$$\tfrac{1}{2} \equiv 4 \quad \tfrac{1}{4} \equiv 2 \quad \tfrac{2}{3} \equiv 3 \quad \tfrac{1}{6} \equiv 6 \text{ (mod 7)}. \tag{9}$$

To find p/q (mod m), use an ALGORITHM similar to the GREEDY ALGORITHM. Let $q_0 \equiv q$ and find

$$p_0 = \left\lceil \frac{m}{q_0} \right\rceil, \tag{10}$$

where $\lceil x \rceil$ is the CEILING FUNCTION, then compute

$$q_1 \equiv q_0 p_0 \pmod{m}. \tag{11}$$

Iterate until $q_n = 1$, then

$$\frac{p}{q} \equiv p \prod_{i=0}^{n-1} p_i \pmod{m}. \tag{12}$$

This method always works for m PRIME, and sometimes even for m COMPOSITE. However, for a COMPOSITE m, the method can fail by reaching 0 (Conway and Guy 1996). Finding a fractional congruence is equivalent to solving a corresponding LINEAR CONGRUENCE EQUATION

$$ax \equiv b \pmod{m}. \tag{13}$$

See also ALGEBRAIC CONGRUENCE, CANCELLATION LAW, CHINESE REMAINDER THEOREM, COMMON RESIDUE, CONGRUENCE AXIOMS, CONGRUENCE EQUATION, DIVISIBILITY TESTS, FUNCTIONAL CONGRUENCE, GREATEST COMMON DIVISOR, LEAST COMMON MULTIPLE, LINEAR CONGRUENCE EQUATION, MINIMAL RESIDUE, MODULUS (CONGRUENCE), QUADRATIC CONGRUENCE EQUATION, QUADRATIC RECIPROCITY LAW, RESIDUE (CONGRUENCE), RSA ENCRYPTION

References

Burton, D. M. "The Theory of Congruences." Ch. 4 in *Elementary Number Theory, 4th ed.* Boston, MA: Allyn and Bacon, pp. 80–05, 1989.
Conway, J. H. and Guy, R. K. "Arithmetic Modulo p." In *The Book of Numbers*. New York: Springer-Verlag, pp. 130–32, 1996.
Courant, R. and Robbins, H. "Congruences." §2 in Supplement to Ch. 1 in *What is Mathematics?: An Elementary Approach to Ideas and Methods, 2nd ed.* Oxford, England: Oxford University Press, pp. 31–0, 1996.
Hardy, G. H. and Wright, E. M. "Congruences and Classes of Residues," "Elementary Properties of Congruences," "Linear Congruences," "General Properties of Congruences," and "Congruences to Composite Moduli." §5.2–.4 and Chs. 7– in *An Introduction to the Theory of Numbers, 5th ed.* Oxford, England: Clarendon Press, pp. 49–2 and 82–06, 1979.
Hilton, P.; Holton, D.; and Pedersen, J. "A Far Nicer Arithmetic." Ch. 2 in *Mathematical Reflections in a Room with Many Mirrors*. New York: Springer-Verlag, pp. 25–0, 1997.
Nagell, T. "Theory of Congruences." Ch. 3 in *Introduction to Number Theory*. New York: Wiley, pp. 68–31, 1951.
Séroul, R. "Congruences." §2.5 in *Programming for Mathematicians*. Berlin: Springer-Verlag, pp. 11–2, 2000.
Shanks, D. *Solved and Unsolved Problems in Number Theory, 4th ed.* New York: Chelsea, p. 55, 1993.
Weisstein, E. W. "Fractional Congruences." MATHEMATICA NOTEBOOK MODFRACTION.M.

Congruence Arithmetic
CONGRUENCE

Congruence Axioms
The five of HILBERT'S AXIOMS which concern geometric equivalence.

See also CONGRUENCE AXIOMS, CONTINUITY AXIOMS, HILBERT'S AXIOMS, INCIDENCE AXIOMS, ORDERING AXIOMS, PARALLEL POSTULATE

References

Hilbert, D. *The Foundations of Geometry, 2nd ed.* Chicago, IL: Open Court, 1980.
Iyanaga, S. and Kawada, Y. (Eds.). "Hilbert's System of Axioms." §163B in *Encyclopedic Dictionary of Mathematics*. Cambridge, MA: MIT Press, pp. 544–45, 1980.

Congruence Equation
An equation OF THE FORM

$$f(x) \equiv b \pmod{m}, \tag{1}$$

where the values of $0 \leq x < m$ for which the equation holds are sought. Such an equation may have none, one, or many solutions. There is a general method for solving both the general LINEAR CONGRUENCE EQUATION

$$ax \equiv b \pmod{m} \tag{2}$$

and the general QUADRATIC CONGRUENCE EQUATION

$$a_2 x^2 + a_1 x + a_0 \equiv 0 \pmod{n}. \tag{3}$$

However, solution of the general polynomial congruence

$$a_m x^m + \ldots + a_2 x^2 + a_1 x + a_0 \equiv 0 \pmod{n} \tag{4}$$

is intractable. Note that any polynomial congruence will give congruent results when congruent values are substituted.

Two or more simultaneous congruences

$$x \equiv a \pmod{m} \tag{5}$$

$$x \equiv b \pmod{n} \tag{6}$$

are solvable using the CHINESE REMAINDER THEOREM.

See also CHINESE REMAINDER THEOREM, CONGRUENCE, LINEAR CONGRUENCE EQUATION, QUADRATIC CONGRUENCE EQUATION

Congruence Transformation
A transformation OF THE FORM $g = D^T \eta D$, where $\det(D) \neq 0$ and $\det(D)$ is the DETERMINANT. ISOMETRIES are also called congruence transformations.

See also SYLVESTER'S INERTIA LAW

Congruent
There are at least two meanings on the word congruent in mathematics. Two geometric figures are said to be congruent if they are equivalent to

within ROTATION and TRANSLATION (i.e., IFF one can be transformed into the other by an ISOMETRY). This relationship is written $A \cong B$. Unfortunately, the symbol $\cong$ is also used to denote an ISOMORPHISM.

A number a is said to be congruent to b modulo m if $m|a-b$ (m DIVIDES $a-b$).

See also COINCIDENT, CONGRUENCE, HOMOTHETIC, ISOMETRY, ROTATION, SIMILAR, TRANSLATION

References

Coxeter, H. S. M. and Greitzer, S. L. *Geometry Revisited.* Washington, DC: Math. Assoc. Amer., p. 80, 1967.

Congruent Incircles Point

The point Y for which TRIANGLES BYC, CYA, and AYB have congruent INCIRCLES. It is a special case of an ELKIES POINT.

References

Kimberling, C. "Central Points and Central Lines in the Plane of a Triangle." *Math. Mag.* **67**, 163–87, 1994.

Congruent Isoscelizers Point

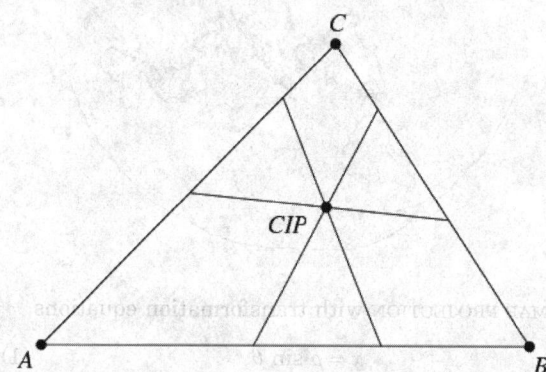

In 1989, P. Yff proved there is a unique configuration of ISOSCELIZERS for a given TRIANGLE such that all three have the same length. Furthermore, these ISOSCELIZERS meet in a point called the congruent isoscelizers point, which has TRIANGLE CENTER FUNCTION

$$\alpha = \cos(\tfrac{1}{2}B) + \cos(\tfrac{1}{2}C) - \cos(\tfrac{1}{2}A).$$

See also ISOSCELIZER

References

Kimberling, C. "Congruent Isoscelizers Point." http://cedar.-evansville.edu/~ck6/tcenters/recent/conisos.html.

Congruent Matrices

Two SQUARE MATRICES A and B are called congruent if there exists a nonsingular matrix P such that

$$B = P^{T}AP,$$

where P^{T} is the TRANSPOSE.

See also TRANSPOSE

References

Ayres, F. Jr. *Theory and Problems of Matrices.* New York: Schaum, p. 115 1962.

Congruent Numbers

A set of numbers (a, x, y, t) such that

$$\begin{cases} x^2 + ay^2 = z^2 \\ x^2 - ay^2 = t^2. \end{cases}$$

They are a generalization of the CONGRUUM PROBLEM, which is the case $y = 1$. For $a = 101$, the smallest solution is

$$x = 2015242462949760001961$$

$$y = 118171431852779451900$$

$$z = 2339148435306225006961$$

$$t = 1628124370727269996961.$$

See also CONGRUUM

References

Guy, R. K. "Congruent Number." §D76 in *Unsolved Problems in Number Theory, 2nd ed.* New York: Springer-Verlag, pp. 195–97, 1994.

Congruum

A number h which satisfies the conditions of the CONGRUUM PROBLEM:

$$x^2 + h = a^2$$

and

$$x^2 - h = b^2,$$

where x, h, a, b are integers. The list of congrua is given by 24, 96, 120, 240, 336, 384, 480, 720, ... (Sloane's A057102).

See also CONCORDANT FORM, CONGRUUM PROBLEM

References

Sloane, N. J. A. Sequences A057102 in "An On-Line Version of the Encyclopedia of Integer Sequences." http://www.re-search.att.com/~njas/sequences/eisonline.html.

Congruum Problem

Find a SQUARE NUMBER x^2 such that, when a given integer h is added or subtracted, new SQUARE NUMBERS are obtained so that

$$x^2 + h = a^2 \qquad (1)$$

and

$$x^2 - h = b^2. \tag{2}$$

This problem was posed by the mathematicians Théodore and Jean de Palerma in a mathematical tournament organized by Frederick II in Pisa in 1225. The solution (Ore 1988, pp. 188–91) is

$$x = m^2 + n^2 \tag{3}$$

$$h = 4mn(m^2 - n^2), \tag{4}$$

where m and n are INTEGERS. a and b are then given by

$$a = m^2 + 2mn - n^2 \tag{5}$$

$$b = n^2 + 2mn - m^2 \tag{6}$$

Fibonacci proved that all numbers h (the CONGRUA) are divisible by 24. FERMAT'S RIGHT TRIANGLE THEOREM is equivalent to the result that a congruum cannot be a SQUARE NUMBER.

A table for small m and n is given in Ore (1988, p. 191), and a larger one (for $h \le 1000$) by Lagrange (1977). The first

m	n	h	x	a	b
Sloane		A057103	A055096	A057104	A057105
2	1	24	5	7	1
3	1	96	10	14	2
3	2	120	13	17	7
4	1	240	17	23	7
4	2	384	20	28	4
4	3	336	25	31	17

See also CONCORDANT FORM, CONGRUENT NUMBERS, CONGRUUM, SQUARE NUMBER

References

Alter, R. and Curtz, T. B. "A Note on Congruent Numbers." *Math. Comput.* **28**, 303–05, 1974.

Alter, R.; Curtz, T. B.; and Kubota, K. K. "Remarks and Results on Congruent Numbers." In *Proc. Third Southeastern Conference on Combinatorics, Graph Theory, and Computing, 1972, Boca Raton, FL.* Boca Raton, FL: Florida Atlantic University, pp. 27–5, 1972.

Bastien, L. "Nombres congruents." *Interméd. des Math.* **22**, 231–32, 1915.

Gérardin, A. "Nombres congruents." *Interméd. des Math.* **22**, 52–3, 1915.

Lagrange, J. "Construction d'une table de nombres congruents." *Calculateurs en Math., Bull. Soc. math. France.*, Mémoire 49–0, 125–30, 1977.

Ore, Ø. *Number Theory and Its History.* New York: Dover, 1988.

Sloane, N. J. A. Sequences A055096, A057103, A057104, and A057105 in "An On-Line Version of the Encyclopedia of Integer Sequences." http://www.research.att.com/~njas/sequences/eisonline.html.

Conic

CONIC SECTION

Conic Constant

$$K \equiv -e^2,$$

where e is the ECCENTRICITY of a CONIC SECTION.

See also CONIC SECTION, ECCENTRICITY

Conic Double Point

ISOLATED SINGULARITY

Conic Equidistant Projection

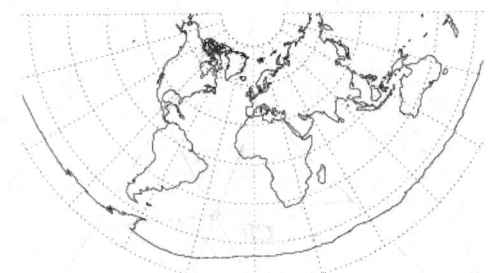

A MAP PROJECTION with transformation equations

$$x = \rho \sin \theta \tag{1}$$

$$y = \rho_0 - \rho \cos \theta, \tag{2}$$

where

$$\rho = (G - \phi) \tag{3}$$

$$\theta = n(\lambda - \lambda_0) \tag{4}$$

$$\rho_0 = (G - \theta_0) \tag{5}$$

$$G = \frac{\cos \phi_1}{n} + \phi_1 \tag{6}$$

$$n = \frac{\cos \phi_1 - \cos \phi_2}{\phi_2 - \phi_1}. \tag{7}$$

The inverse FORMULAS are given by

$$\phi = G - \rho \tag{8}$$

$$\lambda = \lambda_0 + \frac{\theta}{n}, \tag{9}$$

where

$$\rho = \mathrm{sgn}(n)\sqrt{x^2 + (\rho_0 - y)^2} \qquad (10)$$

$$\theta = \tan^{-1}\left(\frac{x}{\rho_0 - y}\right). \qquad (11)$$

See also EQUIDISTANT PROJECTION

Conic Projection

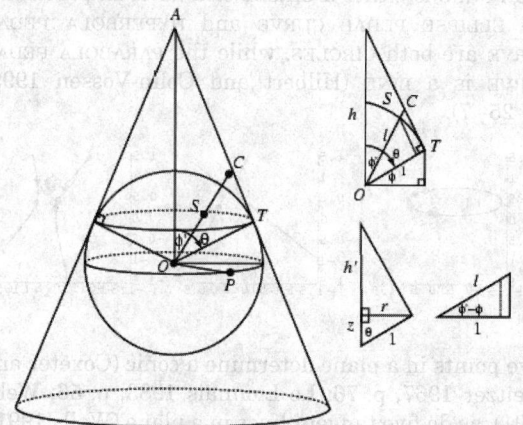

A conic projection of points on a unit sphere centered at O consists of extending the line OS for each point S until it intersects a cone with apex A which tangent to the sphere along a circle passing through a point T in a point C. For a cone with apex a height h above O, the angle from the z-AXIS at which the cone is tangent is given by

$$\theta = \sec^{-1} h, \qquad (1)$$

and the radius of the circle of tangency and height above O at which it is located are given by

$$r = \sin\theta = \frac{\sqrt{h^2 - 1}}{h} \qquad (2)$$

$$z = \cos\theta = \frac{1}{h}. \qquad (3)$$

Letting $\phi' = \pi/2 - \phi$ be the colatitude of a point S on a

sphere, the length of the vector OC along OS is

$$\begin{aligned} l &= \sec(\theta - \phi') = \sec(\sec^{-1} h - \phi') \\ &= \csc(\phi + \sec^{-1} h). \end{aligned} \qquad (4)$$

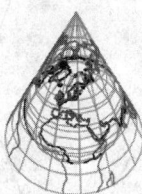

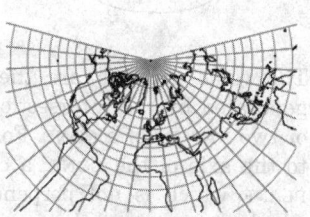

The left figure above shows the result of re-projecting onto a plane perpendicular to the z-AXIS (equivalent to looking at the cone from above the apex), while the figure on the right shows the cone cut along the solid line and flattened out. The equations transforming a point on a sphere (ϕ, λ) to a point on the flattened cone are

$$x = \csc(\sec^{-1} h + \phi)\cos\phi\sin\left(\frac{\lambda}{\sqrt{h^2 - 1}}\right) \qquad (5)$$

$$y = \csc(\sec^{-1} h + \phi)\cos\phi\cos\left(\frac{\lambda}{\sqrt{h^2 - 1}}\right). \qquad (6)$$

This form of the projection, however, is seldom used in practice, and the term "conic projection" is used instead to refer to *any* projection in which lines of latitude are mapped to equally spaced radial lines and lines of latitude (parallels) are mapped to circumferential lines with arbitrary mathematically spaced separations (Snyder 1987, p. 5).

See also ALBERS EQUAL-AREA CONIC PROJECTION, CONIC EQUIDISTANT PROJECTION, CYLINDRICAL PROJECTION, LAMBERT AZIMUTHAL EQUAL-AREA PROJECTION, POLYCONIC PROJECTION

References

Lee, L. P. "The Nomenclature and Classification of Map Projections." *Empire Survey Rev.* **7**, 190–00, 1944.
Snyder, J. P. *Map Projections--A Working Manual.* U. S. Geological Survey Professional Paper 1395. Washington, DC: U. S. Government Printing Office, p. 5, 1987.

Conic Section

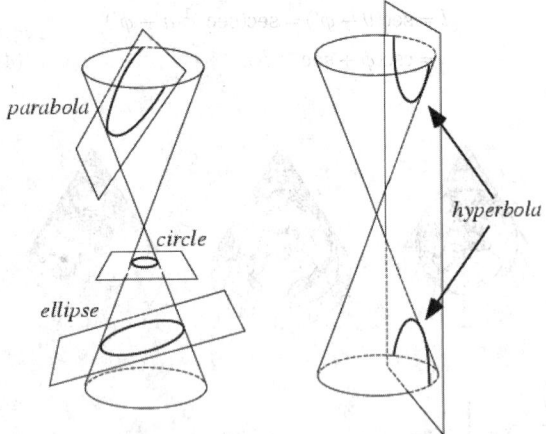

The conic sections are the nondegenerate curves generated by the intersections of a PLANE with one or two NAPPES of a CONE. For a PLANE perpendicular to the axis of the CONE, a circle is produced. For a PLANE which is not perpendicular to the axis and which intersects only a single nappe, the curve produced is either an ELLIPSE or a PARABOLA (Hilbert and Cohn-Vossen 1999, p. 8). The curve produced by a PLANE intersecting both NAPPES is a HYPERBOLA (Hilbert and Cohn-Vossen 1999, pp. 8–).
The ELLIPSE and HYPERBOLA are known as CENTRAL CONICS.

Because of this simple geometric interpretation, the conic sections were studied by the Greeks long before their application to inverse square law orbits was known. Apollonius wrote the classic ancient work on the subject entitled *On Conics*. Kepler was the first to notice that planetary orbits were ELLIPSES, and Newton was then able to derive the shape of orbits mathematically using CALCULUS, under the assumption that gravitational force goes as the inverse square of distance. Depending on the energy of the orbiting body, orbit shapes which are any of the four types of conic sections are possible.

A conic section may more formally be defined as the locus of a point P that moves in the PLANE of a fixed point F called the FOCUS and a fixed line d called the DIRECTRIX (with F not on d) such that the ratio of the distance of P from F to its distance from d is a constant e called the ECCENTRICITY. If $e = 0$, the conic is a CIRCLE, if $0 < e < 1$, the conic is an ELLIPSE, if $e = 1$, the conic is a PARABOLA, and if $e > 1$, it is a HYPERBOLA.

A conic section with DIRECTRIX at $x = 0$, focus at $(p, 0)$, and ECCENTRICITY $e > 0$ has Cartesian equation

$$y^2 + (1 - e^2)x^2 - 2px + p^2 = 0 \qquad (1)$$

(Yates 1952, p. 36), where p is called the FOCAL PARAMETER. Plugging in p for an ELLIPSE gives

$$y^2 + (1 - e^2)x^2 - \frac{2a(1 - e^2)}{e}x + \frac{a^2(1 - e^2)^2}{e^2} = 0, \qquad (2)$$

for a PARABOLA (1) simplifies to

$$y^2 = 4p(x - p), \qquad (3)$$

and for a HYPERBOLA, (1) simplifies to

$$y^2 + (1 - e^2)x^2 - \frac{2a(e^2 - 1)}{e}x + \frac{a^2(e^2 - 1)^2}{e^2} = 0. \qquad (4)$$

The polar equation of a conic section with FOCAL PARAMETER p is given by

$$r = \frac{ep}{1 + e \cos \theta}. \qquad (5)$$

The PEDAL CURVE of a conic section with PEDAL POINT at a FOCUS is either a CIRCLE or a LINE. In particular the ELLIPSE PEDAL CURVE and HYPERBOLA PEDAL CURVE are both CIRCLES, while the PARABOLA PEDAL CURVE is a LINE (Hilbert and Cohn-Vossen 1999, pp. 25–7).

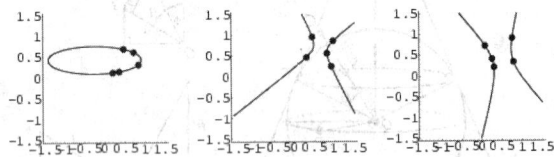

Five points in a plane determine a conic (Coxeter and Greitzer 1967, p. 76; Le Lionnais 1983, p. 56; Wells 1991), as do five tangent lines in a plane (Wells 1991). This follows from the fact that a conic section is a QUADRATIC CURVE, which has general form

$$ax^2 + 2bxy + cy^2 + dx + fy + g = 0, \qquad (6)$$

so dividing through by a to obtain

$$x^2 + 2b'xy + c'y^2 + d'x + f'y + g' = 0 \qquad (7)$$

leaves five constants. Five points, (x_i, y_i) for $i = 1, ..., 5$, therefore determine the constants uniquely. The GEOMETRIC CONSTRUCTION of a conic section from five points lying on it is called the BRAIKENRIDGE-MACLAURIN CONSTRUCTION.

Two conics that do not coincide or have an entire straight line in common cannot meet at more than four points (Hilbert and Cohn-Vossen 1999, pp. 24 and 160). There is an infinite family of conics touching four lines. However, of the eleven regions into which plane division cuts the plane, only five can contain a conic section which is tangent to all four lines. Parabolas can occur in one region only (which also contains ellipses and one branch of hyperbolas), and the only closed region contains only ellipses.

Let a polygon of $2n$ sides be inscribed in a given conic, with the sides of the polygon being termed alternately "odd" and "even" according to some definite convention. Then the $n(n-2)$ points where an odd side meet a nonadjacent even side lie on a curve of order $n-2$ (Evelyn *et al.* 1974, p. 30).

See also BRAIKENRIDGE-MACLAURIN CONSTRUCTION, BRIANCHON'S THEOREM, CENTRAL CONIC, CIRCLE, CONE, CYLINDRICAL SECTION, ECCENTRICITY, ELLIPSE, FERMAT CONIC, FOCAL PARAMETER, FOUR CONICS THEOREM, FRÉGIER'S THEOREM, HYPERBOLA, NAPPE, PARABOLA, PASCAL'S THEOREM, PLANE DIVISION BY ELLIPSES, QUADRATIC CURVE, SEYDEWITZ'S THEOREM, SKEW CONIC, STEINER'S THEOREM, THREE CONICS THEOREM

References

Besant, W. H. *Conic Sections, Treated Geometrically, 8th ed. rev.* Cambridge, England: Deighton, Bell, 1890.
Casey, J. "Special Relations of Conic Sections" and "Invariant Theory of Conics." Chs. 9 and 15 in *A Treatise on the Analytical Geometry of the Point, Line, Circle, and Conic Sections, Containing an Account of Its Most Recent Extensions, with Numerous Examples, 2nd ed., rev. enl.* Dublin: Hodges, Figgis, & Co., pp. 307–32 and 462–45, 1893.
Chasles, M. *Traité des sections coniques.* Paris, 1865.
Coolidge, J. L. *A History of the Conic Sections and Quadric Surfaces.* New York: Dover, 1968.
Coxeter, H. S. M. "Conics" §8.4 in *Introduction to Geometry, 2nd ed.* New York: Wiley, pp. 115–19, 1969.
Coxeter, H. S. M. and Greitzer, S. L. *Geometry Revisited.* Washington, DC: Math. Assoc. Amer., pp. 138–41, 1967.
Downs, J. W. *Practical Conic Sections.* Palo Alto, CA: Dale Seymour, 1993.
Evelyn, C. J. A.; Money-Coutts, G. B.; and Tyrrell, J. A. *The Seven Circles Theorem and Other New Theorems.* London: Stacey International, p. 30, 1974.
Hilbert, D. and Cohn-Vossen, S. "The Cylinder, the Cone, the Conic Sections, and Their Surfaces of Revolution." §2 in *Geometry and the Imagination.* New York: Chelsea, pp. 7–1, 1999.
Iyanaga, S. and Kawada, Y. (Eds.). "Conic Sections." §80 in *Encyclopedic Dictionary of Mathematics.* Cambridge, MA: MIT Press, pp. 271–76, 1980.
Klein, F. "Famous Problems of Elementary Geometry: The Duplication of the Cube, the Trisection of the Angle, and the Quadrature of the Circle." In *Famous Problems and Other Monographs.* New York: Chelsea, pp. 42–4, 1980.
Le Lionnais, F. *Les nombres remarquables.* Paris: Hermann, p. 56, 1983.
Lebesgue, H. *Les Coniques.* Paris: Gauthier-Villars, 1955.
Ogilvy, C. S. "The Conic Sections." Ch. 6 in *Excursions in Geometry.* New York: Dover, pp. 73–5, 1990.
Pappas, T. "Conic Sections." *The Joy of Mathematics.* San Carlos, CA: Wide World Publ./Tetra, pp. 196–97, 1989.
Salmon, G. *Conic Sections, 6th ed.* New York: Chelsea, 1960.
Smith, C. *Geometric Conics.* London: MacMillan, 1894.
Sommerville, D. M. Y. *Analytical Conics, 3rd ed.* London: G. Bell and Sons, 1961.
Steinhaus, H. *Mathematical Snapshots, 3rd ed.* New York: Dover, pp. 238–40, 1999.
Weisstein, E. W. "Books about Conic Sections." http://www.treasure-troves.com/books/ConicSections.html.
Wells, D. *The Penguin Dictionary of Curious and Interesting Geometry.* London: Penguin, p. 175, 1991.
Yates, R. C. "Conics." *A Handbook on Curves and Their Properties.* Ann Arbor, MI: J. W. Edwards, pp. 36–6, 1952.

Conic Section Tangent

Given a CONIC SECTION

$$x^2 + y^2 + 2gx + 2fy + c = 0,$$

the tangent at (x_1, y_1) is given by the equation

$$xx_1 + yy_1 + g(x+x_1) + f(y+y_1) + c = 0.$$

Conical Coordinates

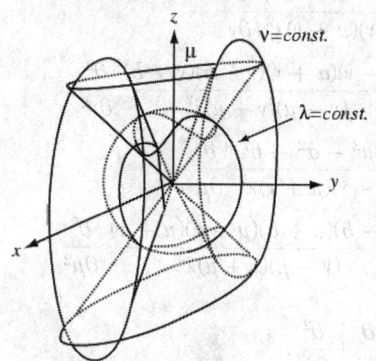

There are several different definitions of conical coordinates defined by Morse and Feshbach (1953), Byerly (1959), Arfken (1970), and Moon and Spencer (1988). The (λ, μ, ν) system defined in *Mathematica* is

$$x = \frac{\lambda\mu\nu}{ab} \tag{1}$$

$$y = \frac{\lambda}{a}\sqrt{\frac{(\mu^2-a^2)(\nu^2-a^2)}{a^2-b^2}} \tag{2}$$

$$z = \frac{\lambda}{b}\sqrt{\frac{(\mu^2-b^2)(\nu^2-b^2)}{b^2-a^2}}, \tag{3}$$

where $b^2 > \mu^2 > c^2 > \nu^2$. Byerly (1959) uses a (r, μ, ν) system which is essentially the same coordinate system as above, but replacing λ with r, a with b, and b with c. Moon and Spencer (1988) use (r, θ, λ) instead of (λ, μ, ν).

The above equations give

$$x^2 + y^2 + z^2 = \lambda^2 \tag{4}$$

$$\frac{x^2}{\mu^2} + \frac{y^2}{\mu^2 - a^2} + \frac{z^2}{\mu^2 - b^2} = 0 \tag{5}$$

$$\frac{x^2}{\nu^2} + \frac{y^2}{\nu^2 - a^2} + \frac{z^2}{\nu^2 - b^2} = 0. \tag{6}$$

The SCALE FACTORS are

$$h_\lambda = 1 \tag{7}$$

$$h_\mu = \sqrt{\frac{\lambda^2(\mu^2 - \nu^2)}{(\mu^2 - a^2)(b^2 - \mu^2)}} \tag{8}$$

$$h_\nu = \sqrt{\frac{\lambda^2(\mu^2 - \nu^2)}{(\nu^2 - a^2)(\nu^2 - b^2)}}. \tag{9}$$

The LAPLACIAN is

$$\nabla^2 = \frac{\nu(2\nu^2 - a^2 - b^2)}{(\mu - \nu)(\mu + \nu)\lambda^2} \frac{\partial}{\partial \nu}$$

$$+ \frac{(a - \nu)(a + \nu)(\nu - b)(\nu + b)}{(\nu - \mu)(\nu + \mu)\lambda^2} \frac{\partial^2}{\partial \nu^2}$$

$$+ \frac{\mu(2\mu^2 - a^2 - b^2)}{(\mu - \nu)(\mu + \nu)\lambda^2} \frac{\partial}{\partial \mu}$$

$$+ \frac{(\mu - b)(\mu + b)(\mu - a)(\mu + a)}{(\nu - \mu)(\nu + \mu)\lambda^2} \frac{\partial^2}{\partial \mu^2}$$

$$+ \frac{2}{\lambda} \frac{\partial}{\partial \lambda} + \frac{\partial^2}{\partial \lambda^2}. \tag{10}$$

The HELMHOLTZ DIFFERENTIAL EQUATION is separable in conical coordinates.

See also HELMHOLTZ DIFFERENTIAL EQUATION–CONICAL COORDINATES

References

Arfken, G. "Conical Coordinates (ξ_1, ξ_2, ξ_3)." §2.16 in *Mathematical Methods for Physicists, 2nd ed.* Orlando, FL: Academic Press, pp. 118–19, 1970.

Byerly, W. E. *An Elementary Treatise on Fourier's Series, and Spherical, Cylindrical, and Ellipsoidal Harmonics, with Applications to Problems in Mathematical Physics.* New York: Dover, p. 263, 1959.

Moon, P. and Spencer, D. E. "Conical Coordinates (r, θ, λ)." Table 1.09 in *Field Theory Handbook, Including Coordinate Systems, Differential Equations, and Their Solutions, 2nd ed.* New York: Springer-Verlag, pp. 37–0, 1988.

Morse, P. M. and Feshbach, H. *Methods of Theoretical Physics, Part I.* New York: McGraw-Hill, p. 659, 1953.

Spence, R. D. "Angular Momentum in Sphero-Conal Coordinates." *Amer. J. Phys.* **27**, 329–35, 1959.

Conical Frustum

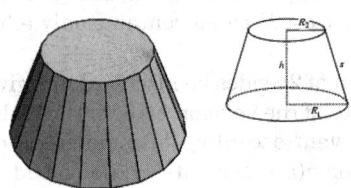

A conical frustum is a FRUSTUM created by slicing the top off a CONE (with the cut made parallel to the base). For a right circular CONE, let s be the slant height and R_1 and R_2 the top and bottom RADII. Then

$$s = \sqrt{(R_1 - R_2)^2 + h^2}. \tag{1}$$

The SURFACE AREA, not including the top and bottom CIRCLES, is

$$A = \pi(R_1 + R_2)s = \pi(R_1 + R_2)\sqrt{(R_1 - R_2)^2 + h^2}. \tag{2}$$

The VOLUME of the frustum is given by

$$V = \pi \int_0^h [r(z)]^2 \, dz. \tag{3}$$

But

$$r(z) = R_1 + (R_2 - R_1)\frac{z}{h}, \tag{4}$$

so

$$V = \pi \int_0^h [r(z)]^2 \, dz = \pi \int_0^h \left[R_1 + (R_2 - R_1)\frac{z}{h}\right]^2 dz$$

$$= \tfrac{1}{3}\pi h(R_1^2 + R_1 R_2 + R_2^2). \tag{5}$$

This formula can be generalized to any PYRAMID by letting A_i be the base AREAS of the top and bottom of the frustum. Then the VOLUME can be written as

$$V = \tfrac{1}{3}h(A_1 + A_2 + \sqrt{A_1 A_2}). \tag{6}$$

The area-weighted integral of z over the frustum is

$$\langle z \rangle = \pi \int_0^h z[r(z)]^2 \, dz = \tfrac{1}{12}\pi h^2(R_1^2 + 2R_1 R_2 + 3R_2^2), \tag{7}$$

so the CENTROID is located along the z-AXIS at a height

$$\bar{z} = \frac{\langle z \rangle}{V} = \frac{h(R_1^2 + 2R_1 R_2 + 3R_2^2)}{4(R_1^2 + R_1 R_2 + R_2^2)} \tag{8}$$

(Eshbach 1975, p. 453; Beyer 1987, p. 133; Harris and Stocker 1998, p. 105). The special case of the CONE is given by taking $R_2 = 0$, yielding $\bar{z} = h/4$.

See also CONE, FRUSTUM, PYRAMIDAL FRUSTUM, SPHERICAL SEGMENT

Conical Function

Conical Function

References

Beyer, W. H. (Ed.). *CRC Standard Mathematical Tables, 28th ed.* Boca Raton, FL: CRC Press, pp. 129–30 and 133, 1987.

Eshbach, O. W. *Handbook of Engineering Fundamentals.* New York: Wiley, 1975.

Harris, J. W. and Stocker, H. "Frustum of a Right Circular Cone." §4.7.2 in *Handbook of Mathematics and Computational Science.* New York: Springer-Verlag, p. 105, 1998.

Kern, W. F. and Bland, J. R. "Frustum of Right Circular Cone." §29 in *Solid Mensuration with Proofs, 2nd ed.* New York: Wiley, pp. 71–5, 1948.

Conical Function

Functions which can be expressed in terms of LEGENDRE FUNCTIONS OF THE FIRST and SECOND KINDS. See Abramowitz and Stegun (1972, p. 337).

$$P^{\mu}_{-1/2+ip}(\cos\theta) = 1 + \frac{4p^2 + 1^2}{2^2}\sin^2(\tfrac{1}{2}\theta)$$
$$+ \frac{(4p^2 + 1^2)(4p^2 + 3^2)}{2^2 4^2}\sin^4(\tfrac{1}{2}\theta) + \dots$$

$$= \frac{2}{\pi}\int_0^\theta \frac{\cosh(pt)dt}{\sqrt{2(\cos t - \cos\theta)}}$$

$$Q^{\mu}_{-1/2\mp ip}(\cos\theta) = \pm i\sinh(p\pi)\int_0^\infty \frac{\cos(pt)dt}{\sqrt{2(\cosh t + \cos\theta)}}$$

$$+ \int_0^\infty \frac{\cosh(pt)dt}{\sqrt{2(\cos t - \cos\theta)}}.$$

See also TOROIDAL FUNCTION

References

Abramowitz, M. and Stegun, C. A. (Eds.). "Conical Functions." §8.12 in *Handbook of Mathematical Functions with Formulas, Graphs, and Mathematical Tables, 9th printing.* New York: Dover, p. 337, 1972.

Iyanaga, S. and Kawada, Y. (Eds.). *Encyclopedic Dictionary of Mathematics.* Cambridge, MA: MIT Press, p. 1464, 1980.

Conical Projection

CONIC PROJECTION

Conical Spiral

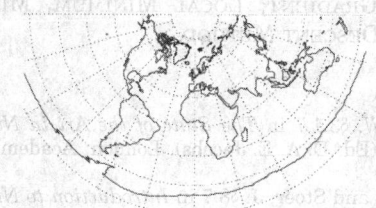

A SPACE CURVE given by the PARAMETRIC EQUATIONS

$$x = \frac{h-z}{h}\, r\cos(az)$$

$$y = \frac{h-z}{h} r\sin(az)$$

$$z = z$$

for h the height of the cone, r its radius, and a a constant.

See also CONE, SEASHELL

Conical Surface

GENERALIZED CONE

Conical Wedge

The SURFACE also called the CONOCUNEUS OF WALLIS and given by the parametric equation

$$x = u\cos v$$
$$y = u\sin v$$
$$z = c(1 - 2\cos^2 v).$$

See also CYLINDRICAL WEDGE, WEDGE

References

von Seggern, D. *CRC Standard Curves and Surfaces.* Boca Raton, FL: CRC Press, p. 302, 1993.

Conjecture

A proposition which is consistent with known data, but has neither been verified nor shown to be false. It is synonymous with HYPOTHESIS.

References

Rivera, C. "Problems & Puzzles: Conjectures." http://www.primepuzzles.net/conjectures/.

Conjugacy Class

A complete set of mutually conjugate GROUP elements. Each element in a GROUP belongs to exactly one class, and the IDENTITY ELEMENT ($I = 1$) is always in its own class. The ORDERS of all classes must be integral FACTORS of the ORDER of the GROUP. From the last two statements, a GROUP of PRIME order has one class for each element. More generally, in an ABELIAN GROUP, each element is in a conjugacy class by itself.

Two operations belong to the same class when one may be replaced by the other in a new COORDINATE SYSTEM which is accessible by a symmetry operation (Cotton 1990, p. 52). These sets correspond directly to the sets of equivalent operations.

To see how to compute conjugacy classes, consider the FINITE GROUP D_3, which has the following MULTIPLICATION TABLE.

D_3	1	A	B	C	D	E
1	1	A	B	C	D	E
A	A	1	D	E	B	C
B	B	E	1	D	C	A
C	C	D	E	1	A	B
D	D	C	A	B	E	1
E	E	B	C	A	1	D

$\{1\}$ is always in a conjugacy class of its own. To find another conjugacy class take some element, say A, and find the results of all similarity transformations $X^{-1}AX = X^{-1}(AX)$ on A. For example, for $X = A$, the product of A by A can be read of as the element at the intersection of the row containing A (the first multiplicand) with the column containing A (the second multiplicand), giving $A^{-1}AA = A^{-1}1$. Now, we want to find Z where $A^{-1}1 = Z$, so pre-multiply both sides by A to obtain $(AA^{-1})1 = 1 = AZ$, so Z is the element whose column intersects row A in 1, i.e., A. Thus, $A^{-1}AA = A$. Similarly, $B^{-1}AB = C$, and continuing the process for all elements gives

$$A^{-1}AA = A \tag{1}$$

$$B^{-1}AB = C \tag{2}$$

$$C^{-1}AC = B \tag{3}$$

$$D^{-1}AD = C \tag{4}$$

$$E^{-1}AE = B \tag{5}$$

The possible outcomes are A, B, or C, so $\{A, B, C\}$ forms a conjugacy class. To find the next conjugacy class, take one of the elements not belonging to an existing class, say D. Applying a similarity transformation gives

$$A^{-1}DA = E \tag{6}$$

$$B^{-1}DB = D, \tag{7}$$

so we need proceed no further since D and E both appear, meaning $\{D, E\}$ form a conjugacy class and we have exhausted all elements of the group.

Let G be a FINITE GROUP of ORDER $|G|$, and let s be the number of conjugacy classes of G. If $|G|$ is ODD, then

$$|G| \equiv s \pmod{16}$$

(Burnside 1955, p. 295). Furthermore, if every PRIME p_i DIVIDING $|G|$ satisfies $p_i \equiv 1 \pmod 4$, then

$$|G| \equiv s \pmod{32}$$

(Burnside 1955, p. 320). Poonen (1995) showed that if every PRIME p_i DIVIDING $|G|$ satisfies $p_i \equiv 1 \pmod m$ for $m \geq 2$, then

$$|G| \equiv s \pmod{2m^2}.$$

References

Burnside, W. *Theory of Groups of Finite Order, 2nd ed.* New York: Dover, 1955.
Cotton, F. A. *Chemical Applications of Group Theory, 3rd ed.* New York: Wiley, 1990.
Poonen, B. "Congruences Relating the Order of a Group to the Number of Conjugacy Classes." *Amer. Math. Monthly* **102**, 440–42, 1995.

Conjugate

COMPLEX CONJUGATE, CONJUGATE ELEMENT, CONJUGATE GRADIENT METHOD, CONJUGATE MATRIX, CONJUGATE POINTS, CONJUGATE SUBGROUP, CONJUGATION MOVE

Conjugate Element

Given a GROUP with elements A and X, there must be an element B which is a SIMILARITY TRANSFORMATION of A, $B = X^{-1}AX$ so A and B are conjugate with respect to X. Conjugate elements have the following properties:

1. Every element is conjugate with itself.
2. If A is conjugate with B with respect to X, then B is conjugate to A with respect to X.
3. If A is conjugate with B and C, then B and C are conjugate with each other.

See also CONJUGACY CLASS, CONJUGATE SUBGROUP

Conjugate Gradient Method

An ALGORITHM for finding the nearest LOCAL MINIMUM of a function of n variables which presupposes that the GRADIENT of the function can be computed. It uses conjugate directions instead of the local GRADIENT for going downhill. If the vicinity of the MINIMUM has the shape of a long, narrow valley, the minimum is reached in far fewer steps than would be the case using the STEEPEST DESCENT METHOD.

See also GRADIENT, LOCAL MINIMUM, MINIMUM, STEEPEST DESCENT METHOD

References

Brodie, K. W. §3.1.7 in *The State of the Art in Numerical Analysis* (Ed. D. A. E. Jacobs). London: Academic Press, 1977.
Bulirsch, R. and Stoer, J. §8.7 in *Introduction to Numerical Analysis.* New York: Springer-Verlag, 1991.
Polak, E. §2.3 in *Computational Methods in Optimization.* New York: Academic Press, 1971.

Press, W. H.; Flannery, B. P.; Teukolsky, S. A.; and Vetterling, W. T. *Numerical Recipes in FORTRAN: The Art of Scientific Computing, 2nd ed.* Cambridge, England: Cambridge University Press, pp. 413–17, 1992.

Conjugate Matrix

The matrix **Ā** obtained from a given matrix A by taking the COMPLEX CONJUGATE of each element of A (Courant and Hilbert 1989, p. 9). The notation A* is sometimes also used, which can lead to confusion since this symbol is also used to denote the ADJOINT MATRIX.

See also ADJOINT MATRIX, COMPLEX CONJUGATE

References

Arfken, G. *Mathematical Methods for Physicists, 3rd ed.* Orlando, FL: Academic Press, pp. 355–56, 1985.
Ayres, F. Jr. *Theory and Problems of Matrices.* New York: Schaum, pp. 12–3, 1962.
Courant, R. and Hilbert, D. *Methods of Mathematical Physics, Vol. 1.* New York: Wiley, 1989.

Conjugate Partition

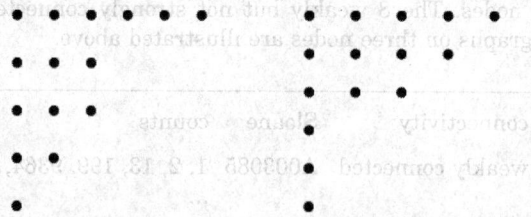

Pairs of partitions for a single number whose FERRERS DIAGRAMS transform into each other when reflected about the line $y = -x$, with the coordinates of the upper left dot taken as $(0, 0)$, are called conjugate (or transpose) partitions. For example, the conjugate partitions illustrated above correspond to the partitions $6 + 3 + 3 + 2 + 1$ and $5 + 4 + 3 + 1 + 1 + 1$ of 15. A partition that is conjugate to itself is said to be a SELF-CONJUGATE PARTITION.

The conjugate partition of a given partition l can be implemented in *Mathematica* as follows.

```
ConjugatePartition[l_List] :=
Module[{i, r = Reverse[l], n = Length[l]},
 Table[n + 1 - Position[r, _?(# > = i &),
 Infinity, 1][[1, 1]], {i, l[[1]]}
 ] ]
```

A similar implementation is given as `Transpose-Partition[`*l*`]` in the *Mathematica* add-on package `DiscreteMath`Combinatorica`` (which can be loaded with the command `< <DiscreteMath`).

See also DURFEE SQUARE, FERRERS DIAGRAM, PARTITION FUNCTION P, SELF-CONJUGATE PARTITION

References

Andrews, G. E. *The Theory of Partitions.* Cambridge, England: Cambridge University Press, pp. 7–, 1998.

Skiena, S. *Implementing Discrete Mathematics: Combinatorics and Graph Theory with Mathematica.* Reading, MA: Addison-Wesley, pp. 55–6, 1990.

Conjugate Permutation

INVERSE PERMUTATION

Conjugate Points

HARMONIC CONJUGATE POINTS, INVERSE POINTS, ISOGONAL CONJUGATE, ISOTOMIC CONJUGATE POINT

Conjugate Subgroup

A SUBGROUP H of an original GROUP G has elements h_i. Let x be a fixed element of the original GROUP G which is not a member of H. Then the transformation $x h_i x^{-1}$, $(i = 1, 2, ...)$ generates the so-called conjugate subgroup $x H x^{-1}$. If, for all x, $x H x^{-1} = H$, then H is a SELF-CONJUGATE (also called "invariant" or "normal") SUBGROUP.

All SUBGROUPS of an ABELIAN GROUP are SELF-CONJUGATE.

See also SELF-CONJUGATE SUBGROUP, SUBGROUP, SYLOW THEOREMS

Conjugate Transpose Matrix

ADJOINT MATRIX

Conjugation

The process of taking a COMPLEX CONJUGATE of a COMPLEX NUMBER, COMPLEX MATRIX, etc., or of performing a CONJUGATION MOVE on a KNOT.

See also COMPLEX CONJUGATE, COMPLEX MATRIX, COMPLEX NUMBER, CONJUGATE MATRIX, CONJUGATION MOVE

Conjugation Move

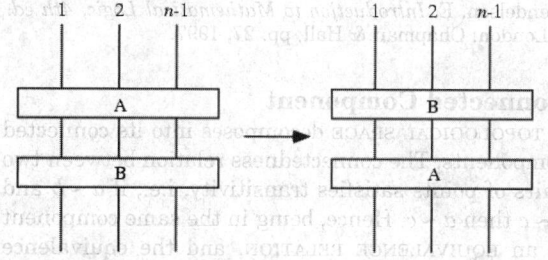

A type I MARKOV MOVE.

See also MARKOV MOVES, STABILIZATION

Conjunction

A product of ANDs, denoted

$$\bigwedge_{k=1}^{n} A_k.$$

The conjunctions of a BOOLEAN ALGEBRA A of subsets of cardinality p are the 2^p functions

$$A_\lambda = \bigcup_{i \in \lambda} A_i,$$

where $\lambda \subset \{1, 2, \ldots, p\}$. For example, the 8 conjunctions of $A = \{A_1, A_2, A_3\}$ are $\varnothing$, A_1, A_2, A_3, $A_1 A_2$, $A_2 A_3$, $A_3 A_1$, and $A_1 A_2 A_3$ (Comtet 1974, p. 186).

See also AND, BOOLEAN ALGEBRA, BOOLEAN FUNCTION, COMPLETE PRODUCT, DISJUNCTION, NOT, OR

References

Comtet, L. *Advanced Combinatorics: The Art of Finite and Infinite Expansions, rev. enl. ed.* Dordrecht, Netherlands: Reidel, p. 186, 1974.

Conjunctive Normal Form

A statement is in conjunctive normal form if it is a CONJUNCTION (sequence of ANDs) consisting of one or more conjuncts, each of which is a DISJUNCTION (OR) of one or more statement letters and negations of statement letters. Examples of disjunctive normal forms include

$$A \tag{1}$$

$$(A \vee B) \wedge (!A \vee C) \tag{2}$$

$$(A \vee B \vee !A) \wedge (C \vee !B) \wedge (A \vee !C) \tag{3}$$

$$A \vee B \tag{4}$$

$$A \wedge (B \vee C), \tag{5}$$

where $\vee$ denotes OR, $\wedge$ denotes AND, and ! denotes NOT. Every statement in logic consisting of a combination of multiple $\wedge$, $\vee$, and !s can be written in conjunctive normal form.

See also DISJUNCTIVE NORMAL FORM

References

Mendelson, E. *Introduction to Mathematical Logic, 4th ed.* London: Chapman & Hall, pp. 27, 1997.

Connected Component

A TOPOLOGICAL SPACE decomposes into its connected components. The connectedness relation between two pairs of points satisfies transitivity, i.e., if $a \sim b$ and $b \sim c$ then $a \sim c$. Hence, being in the same component is an EQUIVALENCE RELATION, and the equivalence classes are the connected components.

Using PATH-CONNECTEDNESS, the path-connected component containing $x \in X$ is the set of all y path-connected to x. That is, it is the set of y such that there is a continuous path from x to y.

Technically speaking, in some TOPOLOGICAL SPACES, path-connected is not the same as connected. A subset Y of X is connected if there is no way to write $Y =$

$U \cup V$ with U and V disjoint OPEN SETS. Every TOPOLOGICAL SPACE decomposes into a disjoint union $X = \cup Y_i$ where the Y_i are connected. The Y_i are called the connected components of X.

See also CONNECTED SET, PATH-CONNECTED, TOPOLOGICAL SPACE

Connected Digraph

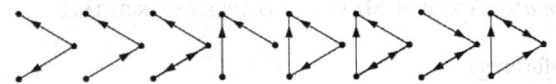

There are two distinct notions of connectivity in a DIGRAPH. A DIGRAPH is WEAKLY CONNECTED if there is an *undirected* path between any pair of vertices, and STRONGLY CONNECTED if there is a *directed* path between every pair of vertices (Skiena 1990, p. 173). The following tables summarized the number of weakly and strongly connected digraphs on $n = 1, 2,$ ... nodes. The 8 weakly but not strongly connected digraphs on three nodes are illustrated above.

connectivity	Sloane	counts
weakly connected	A003085	1, 2, 13, 199, 9364, ...
strongly connected	A035512	1, 1, 5, 83, 5048, 1047008, ...
weakly but not strongly	A056988	0, 1, 8, 116, 4316, 483835, ...

See also CONNECTED GRAPH, DIGRAPH, STRONGLY CONNECTED DIGRAPH, WEAKLY CONNECTED DIGRAPH

References

Skiena, S. "Strong and Weak Connectivity." §5.1.2 in *Implementing Discrete Mathematics: Combinatorics and Graph Theory with Mathematica.* Reading, MA: Addison-Wesley, pp. 172–74, 1990.

Sloane, N. J. A. Sequences A003085/M2067, A035512, and A056988 in "An On-Line Version of the Encyclopedia of Integer Sequences." http://www.research.att.com/~njas/sequences/eisonline.html.

Connected Graph

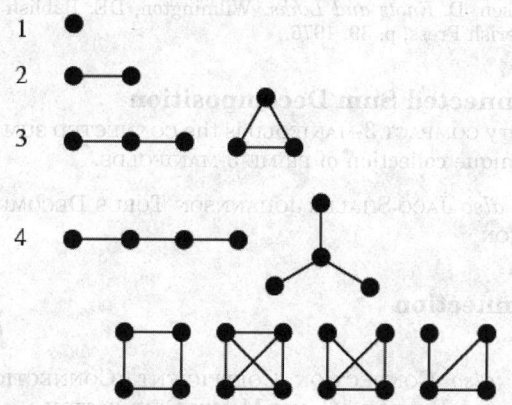

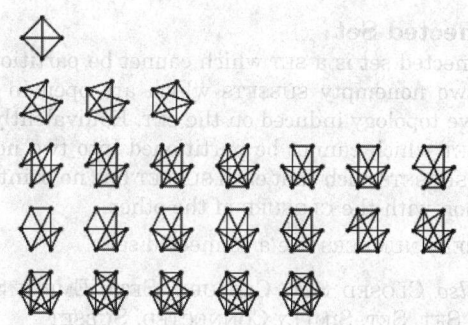

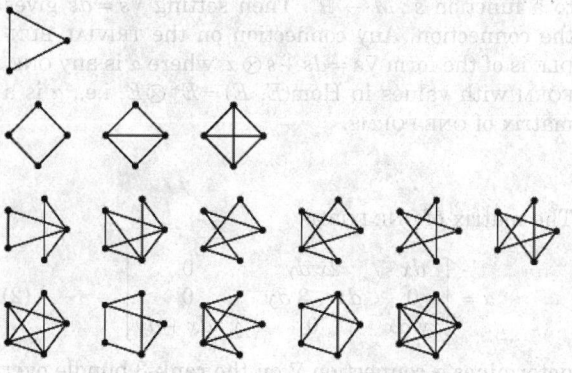

A GRAPH which is connected in the sense of a TOPOLOGICAL SPACE, i.e., there is a path from any point to any other point in the GRAPH. The number of n-node connected unlabeled graphs for $n = 1, 2, \ldots$ are 1, 1, 2, 6, 21, 112, 853, 11117, ... (Sloane's A001349). The *total* number of (not necessarily connected) unlabeled n-node graphs is given by the EULER TRANSFORM of the preceding sequence, 1, 2, 4, 11, 34, 156, 1044, 12346, ... (Sloane's A000088; Sloane and Plouffe 1995, p. 20).

The numbers of connected labeled graphs on n-nodes are 1, 1, 4, 38, 728, 26704, ... (Sloane's A001187), and the *total* number of (not necessarily connected) labeled n-node graphs is given by the EXPONENTIAL TRANSFORM of the preceding sequence: 1, 2, 8, 64, 1024, 32768, ... (Sloane's A006125; Sloane and Plouffe 1995, p. 19).

If a_n is the number of unlabeled connected graphs on n nodes satisfying some property, than the EULER TRANSFORM b_n is the total number of unlabeled graphs (connected or not) with the same property. This application of the EULER TRANSFORM is called RIDDELL'S FORMULA.

If G is DISCONNECTED, then its complement $\bar{G}$ is connected (Skiena 1990, p. 171; Bollobás 1998). However, the converse is not true, as can be seen using the example of the CYCLE GRAPH C_5 which is connected and isomorphic to its complement.

One can also speak of connected graphs in which each vertex has degree at least k (i.e., the minimum of the DEGREE SEQUENCE is $\geq k$). The usual CONNECTED GRAPH is therefore connected with minimal degree ≥ 1. The following table gives the number of connected graphs with minimal degree $\geq k$ on n vertices for small k.

k	Sloane	sequence
1	A001349	1, 1, 2, 6, 21, 112, 853, 11117, ...
2	A004108	0, 0, 1, 3, 11, 61, 507, 7442, ...
3	A007112	0, 0, 0, 1, 3, 19, 150, 2589, ...

See also ALGEBRAIC CONNECTIVITY, BICONNECTED GRAPH, DEGREE SEQUENCE, DISCONNECTED GRAPH, EULER TRANSFORM, PLANAR CONNECTED GRAPH, POLYHEDRAL GRAPH, POLYNEMA, REGULAR GRAPH, RIDDELL'S FORMULA, SEQUENTIAL GRAPH, STEINITZ'S THEOREM, TAIT'S HAMILTONIAN GRAPH CONJECTURE

References

Bollobás, B. *Modern Graph Theory*. New York: Springer-Verlag, 1998.

Cadogan, C. C. "The Möbius Function and Connected Graphs." *J. Combin. Th. B* **11**, 193–00, 1971.

Chartrand, G. "Connected Graphs." §2.3 in *Introductory Graph Theory*. New York: Dover, pp. 41–5, 1985.

Harary, F. *Graph Theory*. Reading, MA: Addison-Wesley, p. 13, 1994.

Skiena, S. "Connectivity." §5.1 in *Implementing Discrete Mathematics: Combinatorics and Graph Theory with Mathematica*. Reading, MA: Addison-Wesley, pp. 171–80, 1990.

Sloane, N. J. A. Sequences A000088/M1253, A001187/M3671, A001349/M1657, A004108/M2910, A006125/M1897, and A007112/M3059 in "An On-Line Version of the Encyclopedia of Integer Sequences." http://www.research.att.com/~njas/sequences/eisonline.html.

Sloane, N. J. A. and Plouffe, S. *The Encyclopedia of Integer Sequences*. San Diego, CA: Academic Press, 1995.

Tutte, W. T. *The Connectivity of Graphs*. Toronto, Canada: Toronto University Press, 1967.

Connected Set

A connected set is a SET which cannot be partitioned into two nonempty SUBSETS which are open in the relative topology induced on the SET. Equivalently, it is a SET which cannot be partitioned into two nonempty SUBSETS such that each SUBSET has no points in common with the CLOSURE of the other.

The REAL NUMBERS are a connected set.

See also CLOSED SET, CLOSURE (SET), EMPTY SET, OPEN SET, SET, SIMPLY CONNECTED, SUBSET

References

Croft, H. T.; Falconer, K. J.; and Guy, R. K. *Unsolved Problems in Geometry.* New York: Springer-Verlag, p. 2, 1991.
Krantz, S. G. *Handbook of Complex Analysis.* Boston, MA: Birkhäuser, p. 3, 1999.

Connected Space

A SPACE D is connected if any two points in D can be connected by a curve lying wholly within D. A SPACE is 0-connected (a.k.a. PATHWISE-CONNECTED) if every MAP from a 0-SPHERE to the SPACE extends continuously to the 1-DISK. Since the 0-SPHERE is the two endpoints of an interval (1-DISK), every two points have a path between them. A space is 1-connected (a.k.a. SIMPLY CONNECTED) if it is 0-connected and if every MAP from the 1-SPHERE to it extends continuously to a MAP from the 2-DISK. In other words, every loop in the SPACE is CONTRACTIBLE. A SPACE is n-MULTIPLY CONNECTED if it is $(n-1)$-connected and if every MAP from the n-SPHERE into it extends continuously over the $(n+1)$-DISK.

A theorem of Whitehead says that a SPACE is infinitely connected IFF it is CONTRACTIBLE.

See also CONNECTIVITY, CONTRACTIBLE, LOCALLY PATHWISE-CONNECTED, MULTIPLY CONNECTED, PATHWISE-CONNECTED, SIMPLY CONNECTED

Connected Sum

The connected sum $M_1 \# M_2$ of n-manifolds M_1 and M_2 is formed by deleting the interiors of n-BALLS b_i^n in m_i^n and attaching the resulting punctured MANIFOLDS $M_i - \dot{B}_i$ to each other by a HOMEOMORPHISM $h : \partial B_2 \rightarrow \partial B_1$, so

$$M_1 \# M_2 = (M_1 - \dot{B}_1) \underset{h}{\cup} (M_2 - \dot{B}_2).$$

B_i is required to be interior to M_i and ∂B_i bicollared in M_i to ensure that the connected sum is a MANIFOLD.

The connected sum of two KNOTS is called a KNOT SUM.

See also KNOT SUM

References

Rolfsen, D. *Knots and Links.* Wilmington, DE: Publish or Perish Press, p. 39, 1976.

Connected Sum Decomposition

Every COMPACT 3-MANIFOLD is the CONNECTED SUM of a unique collection of PRIME 3-MANIFOLDS.

See also JACO-SHALEN-JOHANNSON TORUS DECOMPOSITION

Connection

See also CONNECTION COEFFICIENT, CONNECTION (VECTOR BUNDLE), GAUSS-MANIN CONNECTION

Connection (Vector Bundle)

A connection on a VECTOR BUNDLE $\pi : E \rightarrow M$ is a way to "differentiate" SECTIONS, in a way that is analogous to the EXTERIOR DERIVATIVE df of a function f. In particular, a connection ∇ is a function from smooth sections $\Gamma(M, E)$ to smooth sections of E TENSOR with ONE-FORMS $\Gamma(M, E \otimes T^*M)$ that satisfies the following conditions.

1. $\nabla fs = s \otimes df + f \nabla s$ (Leibniz rule), and
2. $\nabla s_1 + s_2 = \nabla s_1 + \nabla s_2$.

Alternatively, a connection can be considered as a linear map from SECTIONS of $E \otimes TM$, i.e., a section of E with a VECTOR FIELD X, to sections of E, in analogy to the DIRECTIONAL DERIVATIVE. The DIRECTIONAL DERIVATIVE of a function f, in the direction of a vector field X, is given by $df(X)$. The connection, along with a vector field X, may be applied to a section s of E to get the section $\nabla_X s$. From this perspective, connections must also satisfy

$$\nabla_{fX} s = f \nabla_X s \qquad (1)$$

for any smooth function f. This property follows from the first definition.

For example, the TRIVIAL BUNDLE $E = M \times \mathbb{R}^k$ admits a FLAT CONNECTION since any SECTION s corresponds to a function $\tilde{s} : M \rightarrow \mathbb{R}^k$. Then setting $\nabla s = ds$ gives the connection. Any connection on the TRIVIAL BUNDLE is of the form $\nabla s = ds + s \otimes \alpha$, where α is any ONE-FORM with values in $\text{Hom}(E, E) = E^* \otimes E$, i.e., α is a matrix of ONE-FORMS.

The matrix of ONE-FORMS

$$\alpha = \begin{bmatrix} dx & 2x\,dy & 0 \\ 0 & dx - 3\,dy & 0 \\ xy\,dx & 0 & y^2\,dx + dy \end{bmatrix} \qquad (2)$$

determines a connection ∇ on the rank-3 bundle over

$\mathbb{R}^2$. It acts on a section $s = (s_1,\ s_2,\ s_3)$ by the following.

$$\nabla_{\partial/\partial x} s = s_x + \alpha(\partial/\partial x) s = s_x + \begin{bmatrix} 1 & 0 & 0 \\ 0 & 1 & 0 \\ xy & 0 & y^2 \end{bmatrix} s$$

$$= (\partial s_1/\partial x + s_1,\ \partial s_2/\partial x + s_2,\ \partial s_3/\partial x + xys_1 + y^2 s_3) \quad (3)$$

$$\nabla_{\partial/\partial y} s = s_y + \alpha(\partial/\partial y) s = s_y + \begin{bmatrix} 0 & 2x & 0 \\ 0 & -3 & 0 \\ 0 & 0 & 1 \end{bmatrix} s$$

$$= (\partial s_1/\partial x + 2x s_2,\ \partial s_2/\partial x - 3s_2,\ \partial s_3/\partial x + s_3). \quad (4)$$

In any TRIVIALIZATION, a connection can be described just as in the case of a TRIVIAL BUNDLE. However, if the bundle E is not TRIVIAL, then the EXTERIOR DERIVATIVE ds is not WELL DEFINED (globally) for a SECTION s. Still, the difference between any two connections must be ONE-FORMS with values in ENDOMORPHISMS of E, i.e., matrices of one forms. So the space of connections forms an AFFINE SPACE.

The CURVATURE of the bundle is given by the formula $\Omega = \nabla \circ \nabla$. In coordinates, $\Omega = \alpha \wedge \alpha$ is matrix of TWO-FORMS. For instance, in the example above,

$$\Omega = \begin{bmatrix} 0 & 2x\,dx \wedge dy & 0 \\ 0 & -3x \wedge dy & 0 \\ 0 & 2x^3 y\,dx \wedge dy & y^2\,dx \wedge dy \end{bmatrix} \quad (5)$$

is the curvature.

Another way of describing a connection is as a splitting of the TANGENT BUNDLE TE of E as $TM \oplus E$. The vertical part of TE corresponds to tangent vectors along the fibers, and is the kernel of $d\pi$: $TE \to TM$. The horizontal part is not WELL DEFINED a priori. A connection defines a subspace of $TE_{(x,\,v)}$ which is isomorphic to TM_x. It defines k FLAT SECTIONS s_i such that $\nabla s_i = 0$, which are a BASIS for the FIBERS of E, at least nearby x. These flat sections determine the horizontal part of TE near x. Also, a connection on a vector bundle can be defined by a CONNECTION on the ASSOCIATED PRINCIPAL BUNDLE.

In some settings there is a canonical connection. For example, a RIEMANNIAN MANIFOLD has the LEVI-CIVITA CONNECTION, given by the CHRISTOFFEL SYMBOLS OF THE FIRST and SECOND KINDS, which is the unique torsion-free connection compatible with the metric. A HOLOMORPHIC VECTOR BUNDLE with a HERMITIAN METRIC has a unique connection which is compatible with both metric and the COMPLEX STRUCTURE.

See also CONNECTION (PRINCIPAL BUNDLE), CURVATURE, CURVATURE (BUNDLE), HERMITIAN METRIC, LEVI-CIVITA CONNECTION, PARALLEL TRANSPORT, PRINCIPAL BUNDLE, SECOND FUNDAMENTAL FORM, SECTION (BUNDLE), TORSION (BUNDLE)

Connective

A function, or the symbol representing a function, which corresponds to English conjunctions such as "and," "or," "not," etc. that takes one or more truth values as input and returns a single truth value as output. The terms "logical connective" and "propositional connective" are also used. The following table summarizes some common connectives and their notations.

connective	symbol
AND	$A \wedge B, A \cdot B, A.B, AB, A\&B,$ $A\&\&B$
EQUIVALENT	$A \equiv B, A \Leftrightarrow B, A \rightleftharpoons B$
IMPLIES	$A \Rightarrow B, A \supset B, A \rightarrow B$
NAND	$A \bar{\wedge} B, A|B, \overline{A \cdot B}$
NONEQUIVALENT	$A \not\equiv B, A \not\Leftrightarrow B, A \not\rightleftharpoons B$
NOR	$A \bar{\vee} B, A \downarrow B, \overline{A+B}$
NOT	$!A, \neg\, A, \bar{A}, \sim A$
OR	$A \vee B, A + B, A|B, A\|B$
XNOR	$A \text{ XNOR } B$
XOR	$A \underline{\vee} B, A \oplus B$

See also AND, BINARY OPERATOR, EQUIVALENT, IMPLIES, OR, NAND, NONEQUIVALENT, NOR, NOT, PROPOSITIONAL CALCULUS, TRUTH TABLE, XNOR, XOR

References
Mendelson, E. *Introduction to Mathematical Logic, 4th ed.* London: Chapman & Hall, 1997.

Connective Constant

SELF-AVOIDING WALK CONNECTIVE CONSTANT

Connectivity

CONNECTED SPACE, EDGE CONNECTIVITY, VERTEX CONNECTIVITY

Connectivity Pair

An ordered pair (a, b) of nonnegative integers such that there is some set of a points and b edges whose removal disconnects the graph and there is no set of $a-1$ nodes and b edges or a nodes and $b-1$ edges with this property.

References
Harary, F. *Graph Theory.* Reading, MA: Addison-Wesley, 1994.

Connes Function

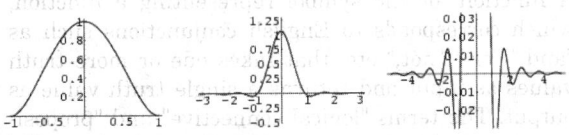

The APODIZATION FUNCTION

$$A(x) = \left(1 - \frac{x^2}{a^2}\right)^2 .$$

Its FULL WIDTH AT HALF MAXIMUM is $\sqrt{4 - 2\sqrt{2}}\,a$, and its INSTRUMENT FUNCTION is

$$I(x) = 8a\sqrt{2\pi}\, \frac{J_{5/2}(2\pi ka)}{(2\pi ka)^{5/2}},$$

where $J_n(z)$ is a BESSEL FUNCTION OF THE FIRST KIND.

See also APODIZATION FUNCTION

Conocuneus of Wallis

CONICAL WEDGE

Conoid

PLÜCKER'S CONOID, RIGHT CONOID

Consecutive Number Sequences

Consecutive number sequences are sequences constructed by concatenating numbers of a given type. Many of these sequences were considered by Smarandache, so they are sometimes known as SMARANDACHE SEQUENCES.

The nth term of the consecutive integer sequence consists of the concatenation of the first n POSITIVE INTEGERS: 1, 12, 123, 1234, ... (Sloane's A007908; Smarandache 1993, Dumitrescu and Seleacu 1994, sequence 1; Mudge 1995; Stephan 1998). This sequence gives the digits of the CHAMPERNOWNE CONSTANT and contains no PRIMES in the first 7,746 terms (Weisstein, Jan. 23, 2000). Fleuren (1999) has verified the absence of primes up to $n = 200$. This is roughly consistent with simple arguments based on the distribution of primes which suggest that only a single prime is expected in the first 15,000 or so terms. The number of digits of the n term can be computed by noticing the pattern in the following table, where $d = [\log_{10} n] + 1$ is the number of digits in n.

d	n Range	Digits
1	1–	n
2	10–9	$9 + 2(n - 9)$
3	100–99	$9 + 90 \cdot 2 + 3(n - 99)$
4	1000–999	$9 + 90 \cdot 2 + 900 \cdot 3 + 4(n - 999)$

Therefore, the number of digits $D(n)$ in the nth term can be written

$$D(n) = d(n + 1 - 10^{d-1}) + \sum_{k=1}^{d-1} 9k \cdot 10^{k-1}$$

$$= (n + 1)d - \frac{10^d - 1}{9},$$

where the second term is the REPUNIT R_d.

The nth term of the reverse integer sequence consists of the concatenation of the first n POSITIVE INTEGERS written backwards: 1, 21, 321, 4321, ... (Sloane's A000422; Smarandache 1993, Dumitrescu and Seleacu 1994, Stephan 1998). The only PRIME in the first 7,287 terms (Weisstein, Jan. 23, 2000) of this sequence is the 82nd term 828180...321 (Stephan 1998, Fleuren 1999), which has 155 digits. This is roughly consistent with simple arguments based on the distribution of prime which suggest that a single prime is expected in the first 15,000 or so terms. The terms of the reverse integer sequence have the same number of digits as do the consecutive integer sequence.

The concatenation of the first n PRIMES gives 2, 23, 235, 2357, 235711, ... (Sloane's A019518; Smith 1996, Mudge 1997). This sequence converges to the digits of the COPELAND-ERDOS CONSTANT and is PRIME for terms 1, 2, 4, 128, 174, 342, 435, 1429, ... (Sloane's A046035; Ibstedt 1998, pp. 78–9), with no others less than 4,706 (Weisstein, Jan. 23, 2000).

The concatenation of the first n ODD NUMBERS gives 1, 13, 135, 1357, 13579, ... (Sloane's A019519; Smith 1996, Marimutha 1997, Mudge 1997). This sequence is PRIME for terms 2, 10, 16, 34, 49, 2570, ... (Sloane's A046036; Weisstein, Ibstedt 1998, pp. 75–6), with no others less than 4,354 (Weisstein, Jan. 1, 2000). The 2570th term, given by 1 3 5 7...5137 5139, has 9725 digits and was discovered by Weisstein in Aug. 1998.

The concatenation of the first n EVEN NUMBERS gives 2, 24, 246, 2468, 246810, ... (Sloane's A019520; Smith 1996; Marimutha 1997; Mudge 1997; Ibstedt 1998, pp. 77–8).

The concatenation of the first n SQUARE NUMBERS gives 1, 14, 149, 14916, ... (Sloane's A019521; Marimutha 1997). The only PRIME in the first 2,822 terms is the third term, 149, (Weisstein).

The concatenation of the first n CUBIC NUMBERS gives 1, 18, 1827, 182764, ... (Sloane's A019522; Marimutha 1997). There are no PRIMES in the first 2,652 terms (Weisstein).

See also CHAMPERNOWNE CONSTANT, CONCATENATION, COPELAND-ERDOS CONSTANT, CUBIC NUMBER, DEMLO NUMBER, EVEN NUMBER, ODD NUMBER, SMARANDACHE SEQUENCES, SQUARE NUMBER

References

Dumitrescu, C. and Seleacu, V. (Eds.). *Some Notions and Questions in Number Theory.* Glendale, AZ: Erhus University Press, 1994.

Fleuren, M. "Smarandache Factors and Reverse Factors." *Smarandache Notions J.* **10**, 5–8, 1999.

Ibstedt, H. "Smarandache Concatenated Sequences." Ch. 5 in *Computer Analysis of Number Sequences.* Lupton, AZ: American Research Press, pp. 75–9, 1998.

Marimutha, H. "Smarandache Concatenate Type Sequences." *Bull. Pure Appl. Sci.* **16E**, 225–26, 1997.

Mudge, M. "Top of the Class." *Personal Computer World,* 674–75, June 1995.

Mudge, M. "Not Numerology but Numeralogy!" *Personal Computer World,* 279–80, 1997.

Rivera, C. "Problems & Puzzles: Puzzle Primes by Listing.-008." http://www.primepuzzles.net/puzzles/puzz_008.htm.

Sloane, N. J. A. Sequences A000422, A007908, A019518, A019519, A019520, A019521, A019522, A046035, and A046036 in "An On-Line Version of the Encyclopedia of Integer Sequences." http://www.research.att.com/~njas/sequences/eisonline.html.

Smarandache, F. *Only Problems, Not Solutions!, 4th ed.* Phoenix, AZ: Xiquan, 1993.

Smith, S. "A Set of Conjectures on Smarandache Sequences." *Bull. Pure Appl. Sci.* **15E**, 101–07, 1996.

Stephan, R. W. "Factors and Primes in Two Smarandache Sequences." *Smarandache Notions J.* **9**, 4–0, 1998.

Conservation of Number Principle

A generalization of Poncelet's CONTINUITY PRINCIPLE made by H. Schubert in 1874–9. The conservation of number principle asserts that the number of solutions of any determinate algebraic problem in any number of parameters under variation of the parameters is invariant in such a manner that no solutions become INFINITE. Schubert called the application of this technique the CALCULUS of ENUMERATIVE GEOMETRY.

See also CONTINUITY PRINCIPLE, DUALITY PRINCIPLE, HILBERT'S PROBLEMS

References

Bell, E. T. *The Development of Mathematics, 2nd ed.* New York: McGraw-Hill, p. 340, 1945.

Conservative Field

The following conditions are equivalent for a conservative VECTOR FIELD:

1. For any oriented simple closed curve C, the LINE INTEGRAL $\oint_C \mathbf{F} \cdot ds = 0$.
2. For any two oriented simple curves C_1 and C_2 with the same endpoints, $\int_{C_1} \mathbf{F} \cdot ds = \int_{C_2} \mathbf{F} \cdot ds$.
3. There exists a SCALAR POTENTIAL FUNCTION f such that $\mathbf{F} = \nabla f$, where ∇ is the GRADIENT.
4. The CURL $\nabla \times \mathbf{F} = 0$.

See also CURL, GRADIENT, LINE INTEGRAL, POINCARÉ'S THEOREM, POTENTIAL FUNCTION, VECTOR FIELD

Consistency

The absence of CONTRADICTION (i.e., the ability to prove that a statement and its negative are both true) in an AXIOMATIC SYSTEM is known as consistency.

See also AXIOMATIC SET THEORY, AXIOMATIC SYSTEM, COMPLETE AXIOMATIC THEORY, CONSISTENCY STRENGTH, GÖDEL'S INCOMPLETENESS THEOREM

Consistency Strength

If the CONSISTENCY of one of two propositions implies the CONSISTENCY of the other, the first is said to have greater consistency strength.

Constant

Any REAL NUMBER which is "significant" (or interesting) in some way. In this work, the term "constant" is generally reserved for REAL nonintegral numbers of interest, while "NUMBER" is reserved for interesting INTEGERS (e.g., BRUN'S CONSTANT, but BEAST NUMBER). In contexts like LINEAR COMBINATION, the term "constant" is generally used to mean "SCALAR" or "REAL NUMBER," and need not exclude integer values.

Certain constants are known to many DECIMAL DIGITS and recur throughout many diverse areas of mathematics, often in unexpected and surprising places (e.g., PI, E, and to some extent, the EULER-MASCHERONI CONSTANT γ). Other constants are more specialized and may be known to only a few DIGITS. S. Plouffe maintains a site about the computation and identification of numerical constants. Plouffe's site also contains a page giving the largest number of DIGITS computed for the most common constants. S. Finch maintains a delightful, more expository site containing detailed essays and references on constants both common and obscure.

The mathematician Glaisher remarked, "No doubt the desire to obtain the values of these quantities to a great many figures is also partly due to the fact that most of them are interesting in themselves; for e, π, γ, $\ln 2$, and many other numerical quantities occupy a curious, and some of them almost a mysterious, place in mathematics, so that there is a natural tendency to do all that can be done towards their precise determination" (Gourdon and Sebah).

See also COEFFICIENT, NUMBER, REAL NUMBER, SCALAR

References

Bailey, D. H. and Crandall, R. E. "On the Random Character of Fundamental Constant Expansions." Manuscript, Mar. 2000. http://www.nersc.gov/~dhbailey/dhbpapers/dhbpapers.html.

Borwein, J. and Borwein, P. *A Dictionary of Real Numbers.* London: Chapman & Hall, 1990.

Finch, S. "Favorite Mathematical Constants." http://www.mathsoft.com/asolve/constant/constant.html.

Gourdon, X. and Sebah, P. "Mathematical Constants and Computation." http://xavier.gourdon.free.fr/Constants/constants.html.

Le Lionnais, F. *Les nombres remarquables.* Paris: Hermann, 1983.

Plouffe, S. "Plouffe's Inverter." http://www.lacim.uqam.ca/pi/.

Plouffe, S. "Plouffe's Inverter: Table of Current Records for the Computation of Constants." http://www.lacim.u-qam.ca/pi/records.html.

Robinson, H. P. and Potter, E. *Mathematical Constants.* Report UCRL-20418. Berkeley, CA: University of California, 1971.

Wells, D. W. *The Penguin Dictionary of Curious and Interesting Numbers.* Harmondsworth, England: Penguin Books, 1986.

Constant Function

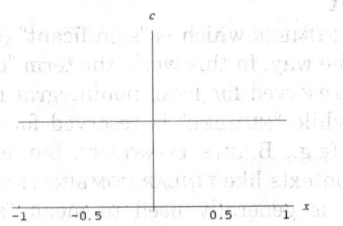

A FUNCTION $f(x) = c$ which does not change as its parameters vary. The GRAPH of a 1-D constant FUNCTION is a straight LINE. The DERIVATIVE of a constant FUNCTION c is

$$\frac{d}{dx} c = 0, \tag{1}$$

and the INTEGRAL is

$$\int c\, dx = cx. \tag{2}$$

The FOURIER transform of the constant function $f(x) = 1$ is given by

$$\mathscr{F}[1] = \int_{-\infty}^{\infty} e^{-2\pi i k x}\, dx = \delta(k), \tag{3}$$

where $\delta(k)$ is the DELTA FUNCTION.

See also FOURIER TRANSFORM–1

References
Spanier, J. and Oldham, K. B. "The Constant Function c." Ch. 1 in *An Atlas of Functions.* Washington, DC: Hemisphere, pp. 11–4, 1987.

Constant Precession Curve
CURVE OF CONSTANT PRECESSION

Constant Problem

Given an expression involving known constants, integration in finite terms, computation of limits, etc., determine if the expression is equal to ZERO. The constant problem is a very difficult unsolved problem in transcendental NUMBER THEORY. However, it is known that the problem is UNDECIDABLE if the expression involves oscillatory functions such as SINE. However, the FERGUSON-FORCADE ALGORITHM is a practical algorithm for determining if there exist integers a_i for given real numbers x_i such that

$$a_1 x_1 + a_2 x_2 + \ldots + a_n x_n = 0,$$

or else establish bounds within which no relation can exist (Bailey 1988).

See also FERGUSON-FORCADE ALGORITHM, HERMITE-LINDEMANN THEOREM, INTEGER RELATION, SCHANUEL'S CONJECTURE

References
Bailey, D. H. "Numerical Results on the Transcendence of Constants Involving π, e, and Euler's Constant." *Math. Comput.* **50**, 275–81, 1988.

Chow, T. Y. "What is a Closed-Form Number." *Amer. Math. Monthly* **106**, 440–48, 1999.

Chen, Z.-Z. and Kao, M.-Y. Reducing Randomness via Irrational Numbers. 7 Jul 1999. http://xxx.lanl.gov/abs/cs.DS/9907011/.

Richardson, D. "The Elementary Constant Problem." In *Proc. Internat. Symp. on Symbolic and Algebraic Computation, Berkeley, July 27–9, 1992* (Ed. P. S. Wang). ACM Press, 1992.

Richardson, D. "How to Recognize Zero." *J. Symb. Comp.* **24**, 627–45, 1997.

Sackell, J. "Zero-Equivalence in Function Fields Defined by Algebraic Differential Equations." *Trans. Amer. Math. Soc.* **336**, 151–71, 1993.

Constant Width Curve
CURVE OF CONSTANT WIDTH

Constructible Number

A number which can be represented by a FINITE number of ADDITIONS, SUBTRACTIONS, MULTIPLICATIONS, DIVISIONS, and FINITE SQUARE ROOT extractions of integers. Such numbers correspond to LINE SEGMENTS which can be constructed using only STRAIGHTEDGE and COMPASS.

All RATIONAL NUMBERS are constructible, and all constructible numbers are ALGEBRAIC NUMBERS (Courant and Robbins 1996, p. 133). If a CUBIC EQUATION with rational coefficients has no rational root, then none of its roots is constructible (Courant and Robbins, p. 136).

In particular, let F_0 be the FIELD of RATIONAL NUMBERS. Now construct an extension field F_1 of constructible numbers by the adjunction of $\sqrt{k_0}$, where k_0 is in F_0, but $\sqrt{k_0}$ is not, consisting of all numbers OF THE FORM $a_0 + b_0\sqrt{k_0}$, where a_0, $b_0 \in F_0$. Next, construct an extension field F_2 of F_1 by the adjunction of $\sqrt{K_1}$, defined as the numbers $a_1 + b_1\sqrt{k_1}$, where a_1, $b_1 \in F_1$, and k_1 is a number in F_1 for which $\sqrt{K_1}$ does not lie in F_1. Continue the process n times. Then constructible numbers are precisely those which can be reached by such a sequence of extension fields F_n, where n is a measure of the "complexity" of the construction (Courant and Robbins 1996).

See also ALGEBRAIC NUMBER, COMPASS, CONSTRUCTIBLE POLYGON, EUCLIDEAN NUMBER, RATIONAL NUMBER, STRAIGHTEDGE

References

Bold, B. "Achievement of the Ancient Greeks" and "An Analytic Criterion for Contractibility." Chs. 1– in *Famous Problems of Geometry and How to Solve Them.* New York: Dover, pp. 1–7, 1982.
Courant, R. and Robbins, H. "Constructible Numbers and Number Fields." §3.2 in *What is Mathematics?: An Elementary Approach to Ideas and Methods, 2nd ed.* Oxford, England: Oxford University Press, pp. 127–34, 1996.

Constructible Polygon

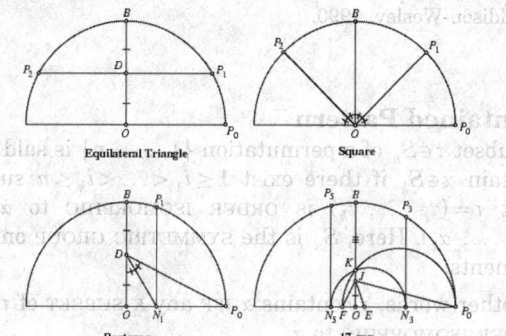

COMPASS and STRAIGHTEDGE constructions dating back to Euclid were capable of inscribing regular polygons of 3, 4, 5, 6, 8, 10, 12, 16, 20, 24, 32, 40, 48, 64, ..., sides. However, this listing is not a complete enumeration of "constructible" polygons. A regular n-gon ($n \geq 3$) can be constructed by STRAIGHTEDGE and COMPASS IFF

$$n = 2^k p_1 p_2 \cdots p_s,$$

where k is in INTEGER ≥ 0 and the p_i are distinct FERMAT PRIMES. FERMAT NUMBERS are OF THE FORM

$$F_m = 2^{2^m} + 1,$$

where m is an INTEGER ≥ 0. The only known PRIMES of this form are 3, 5, 17, 257, and 65537. The fact that this condition was SUFFICIENT was first proved by Gauss in 1796 when he was 19 years old. That this condition was also NECESSARY was not explicitly proven by Gauss, and the first proof of this fact is credited to Wantzel (1836).

See also COMPASS, CONSTRUCTIBLE NUMBER, CYCLOTOMIC POLYNOMIAL, FERMAT NUMBER, GEOMETRIC CONSTRUCTION, GEOMETROGRAPHY, HEPTADECAGON, HEXAGON, OCTAGON, PENTAGON, POLYGON, SQUARE, STRAIGHTEDGE, TRIANGLE

References

Bachmann, P. *Die Lehre von der Kreistheilung und ihre Beziehungen zur Zahlentheorie.* Leipzig, Germany: Teubner, 1872.
Ball, W. W. R. and Coxeter, H. S. M. *Mathematical Recreations and Essays, 13th ed.* New York: Dover, pp. 94–6, 1987.
Bold, B. "The Problem of Constructing Regular Polygons." Ch. 7 in *Famous Problems of Geometry and How to Solve Them.* New York: Dover, pp. 49–1, 1982.
Courant, R. and Robbins, H. *What is Mathematics?: An Elementary Approach to Ideas and Methods, 2nd ed.* Oxford, England: Oxford University Press, p. 119, 1996.
De Temple, D. W. "Carlyle Circles and the Lemoine Simplicity of Polygonal Constructions." *Amer. Math. Monthly* **98**, 97–08, 1991.
Dickson, L. E. "Constructions with Ruler and Compasses; Regular Polygons." Ch. 8 in *Monographs on Topics of Modern Mathematics Relevant to the Elementary Field* (Ed. J. W. A. Young). New York: Dover, pp. 352–86, 1955.
Dixon, R. "Compass Drawings." Ch. 1 in *Mathographics.* New York: Dover, pp. 1–8, 1991.
Gauss, C. F. §365 and 366 in *Disquisitiones Arithmeticae.* Leipzig, Germany, 1801. Translated by A. A. Clarke. New Haven, CT: Yale University Press, 1965.
Kazarinoff, N. D. "On Who First Proved the Impossibility of Constructing Certain Regular Polygons with Ruler and Compass Alone." *Amer. Math. Monthly* **75**, 647–48, 1968.
Klein, F. "The Division of the Circle into Equal Parts." Part I, Ch. 3 in "Famous Problems of Elementary Geometry: The Duplication of the Cube, the Trisection of the Angle, and the Quadrature of the Circle." In *Famous Problems and Other Monographs.* New York: Chelsea, pp. 16–3, 1980.
Ogilvy, C. S. *Excursions in Geometry.* New York: Dover, pp. 137–38, 1990.
Wantzel, M. L. "Recherches sur les moyens de reconnaître si un problème de géométrie peut se résoudre avec la règle et le compas." *J. Math. pures appliq.* **1**, 366–72, 1836.

Construction

BRAIKENRIDGE-MACLAURIN CONSTRUCTION, CONSTRUCTIBLE NUMBER, CONSTRUCTIBLE POLYGON, CONSTRUCTIVE DILEMMA, GEOMETRIC CONSTRUCTION, HAUY CONSTRUCTION, MASCHERONI CONSTRUCTION, MATCHSTICK CONSTRUCTION, NEUSIS CONSTRUCTION, PALEY CONSTRUCTION, STEINER CONSTRUCTION, WYTHOFF CONSTRUCTION

Constructive Dilemma

A formal argument in LOGIC in which it is stated that (1) $P \Rightarrow Q$ and $R \Rightarrow S$ (where $\Rightarrow$ means "IMPLIES"), and (2) either P or R is true, from which two statements it follows that either Q or S is true.

See also DESTRUCTIVE DILEMMA, DILEMMA

Contact Angle

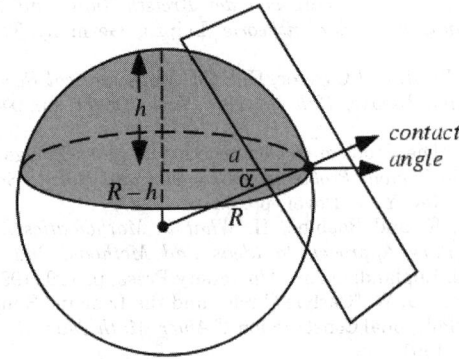

The ANGLE α between the normal vector of a SPHERE (or other geometric object) at a point where a PLANE is tangent to it and the normal vector of the plane. In the above figure,

$$\alpha = \cos^{-1}\left(\frac{a}{R}\right) = \sin^{-1}\left(\frac{R-h}{R}\right).$$

See also SPHERICAL CAP

Contact Number
KISSING NUMBER

Contact Triangle

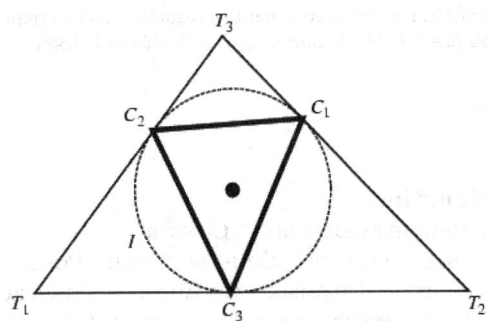

The TRIANGLE formed by the points of intersection of a TRIANGLE T's INCIRCLE with T. This is the PEDAL TRIANGLE of T with the INCENTER as the PEDAL POINT (cf., TANGENTIAL TRIANGLE). The lines from the vertices of the contact triangle to the vertices of the original triangle CONCUR in the GERGONNE POINT. Furthermore, the contact triangle and TANGENTIAL TRIANGLE are perspective from the GERGONNE POINT.

See also ADAMS' CIRCLE, GERGONNE POINT, PEDAL TRIANGLE, SEVEN CIRCLES THEOREM, TANGENTIAL TRIANGLE

References
Oldknow, A. "The Euler-Gergonne-Soddy Triangle of a Triangle." *Amer. Math. Monthly* **103**, 319–29, 1996.

Contained Partition

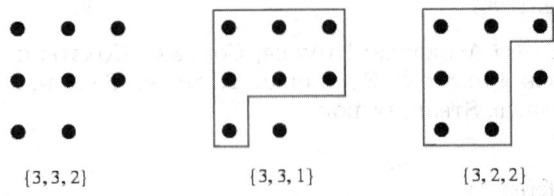

A PARTITION p is said to contain another partition q if the FERRERS DIAGRAM of p contains the FERRERS DIAGRAM of q. For example, $\{3, 3, 2\}$ (left figure) contains both $\{3, 3, 1\}$ and $\{3, 3, 2\}$ (right figures). YOUNG'S LATTICE Y_p is the PARTIAL ORDER of partitions contained within p ordered by containment (Skiena 1990, p. 77).

See also PARTITION, YOUNG'S LATTICE

References
Skiena, S. *Implementing Discrete Mathematics: Combinatorics and Graph Theory with Mathematica.* Reading, MA: Addison-Wesley, 1990.

Contained Pattern
A subset $\tau \in S_n$ of a permutation $\{1, \ldots, n\}$ is said to contain $\alpha \in S_k$ if there exist $1 \le i_1 < \ldots < i_k \le n$ such that $\tau = (\tau_i, \ldots, \tau_k)$ is ORDER ISOMORPHIC to $\alpha = (\alpha_1, \ldots, \alpha_k)$. Here, S_n is the SYMMETRIC GROUP on n elements.

In other words, τ contains α IFF any K-SUBSET of τ is ORDER ISOMORPHIC to α.

See also AVOIDED PATTERN, ORDER ISOMORPHIC, PERMUTATION PATTERN, WILF CLASS, WILF EQUIVALENT

References
Mansour, T. Permutations Avoiding a Pattern from S_k and at Least Two Patterns from S_3. 31 Jul 2000. http://xxx.lanl.gov/abs/math.CO/0007194/.

Content
The content of a POLYTOPE or other n-dimensional object is its generalized VOLUME (i.e., its "hypervolume"). Just as a three-dimensional object has VOLUME, SURFACE AREA, and GENERALIZED DIAMETER, an n-dimensional object has "measures" of order 1, 2, ..., n.

The content of an integer polynomial $P \in \mathbb{Z}(x)$, denoted cont(P), is the largest integer $k \ge 1$ such that P/k also has integer coefficients. Gauss's lemma for contents states that if P and Q are two polynomials with integer coefficients, then cont(PQ) = cont(P)cont(Q) (Séroul 2000, p. 287).

See also POLYNOMIAL, VOLUME

References

Séroul, R. *Programming for Mathematicians.* Berlin: Springer-Verlag, p. 287, 2000.

Contests

MATHEMATICS CONTESTS

Contiguous Function

A HYPERGEOMETRIC FUNCTION in which one parameter changes by $+1$ or -1 is said to be contiguous. There are 26 functions contiguous to $_2F_1(a, b; c; x)$ taking one pair at a time. There are 325 taking two or more pairs at a time. See Abramowitz and Stegun (1972, pp. 557–58).

See also HYPERGEOMETRIC FUNCTION

References

Abramowitz, M. and Stegun, C. A. (Eds.). *Handbook of Mathematical Functions with Formulas, Graphs, and Mathematical Tables, 9th printing.* New York: Dover, 1972.

Contingency

A SENTENCE is called a contingency if its TRUTH TABLE contains at least one 'T' and at least one 'F.'

See also CONTRADICTION, TAUTOLOGY, TRUTH TABLE

References

Carnap, R. *Introduction to Symbolic Logic and Its Applications.* New York: Dover, p. 13, 1958.

Continued Fraction

A "general" continued fraction representation of a REAL NUMBER x IS OF THE FORM

$$x = a_0 + \cfrac{b_1}{a_1 + \cfrac{b_2}{a_2 + \cfrac{b_3}{a_3 + \ldots}}}, \tag{1}$$

which can be written

$$x = a_0 + \frac{b_1}{a_1+} \frac{b_2}{a_2+} \ldots . \tag{2}$$

An archaic word for a continued fraction is ANTHYPHAIRETIC RATIO.

The SIMPLE CONTINUED FRACTION representation of a number x (which is usually what is meant when the term "continued fraction" is used without qualification) is given by

$$x = a_0 + \cfrac{1}{a_1 + \cfrac{1}{a_2 + \cfrac{1}{a_3 + \ldots}}}, \tag{3}$$

which can be written in a compact abbreviated

NOTATION AS

$$x = [a_0, a_1, a_2, a_3, \ldots]. \tag{4}$$

Some care is needed, since some authors begin indexing the terms at a_1 instead of a_0, causing the parity of certain fundamental results in continued fraction theory to be reversed. Starting the indexing with a_0,

$$a_0 = \lfloor x \rfloor \tag{5}$$

is the integral part of x, where $\lfloor x \rfloor$ is the FLOOR FUNCTION,

$$a_1 = \left\lfloor \frac{1}{x - a_0} \right\rfloor \tag{6}$$

is the integral part of the RECIPROCAL of $x - a_0$,

$$a_2 = \left\lfloor \cfrac{1}{\cfrac{1}{x - a_0} - a_1} \right\rfloor \tag{7}$$

is the integral part of the reciprocal of the remainder, etc. Writing the remainders according to the RECURRENCE RELATION

$$r_0 = x \tag{8}$$

$$r_n = \frac{1}{r_{n-1} - a_{n-1}} \tag{9}$$

gives the concise formula

$$a_n = \lfloor r_n \rfloor . \tag{10}$$

The quantities a_n are called PARTIAL QUOTIENTS, and the quantity obtained by including n terms of the continued fraction

$$c_n = \frac{p_n}{q_n} = [a_0, a_1, \ldots, a_n]$$

$$= a_0 + \cfrac{1}{a_1 + \cfrac{1}{a_2 + \cfrac{1}{\ldots + \cfrac{1}{a_n}}}} \tag{11}$$

is called the nth CONVERGENT. For example, consider the computation of the continued fraction of π, given by $\pi = [3, 7, 15, 1, 292, 1, 1, \ldots]$.

Term	Value	PQs		Convergent Value
a_0	$\lfloor \pi \rfloor = 3$	[3]	3	3.00000
a_1	$\left\lfloor \frac{1}{\pi - 3} \right\rfloor = 7$	[3, 7]	$\frac{22}{7}$	3.14286
a_2	$\left\lfloor \frac{1}{\frac{1}{\pi-3}-7} \right\rfloor = 15$	[3, 7, 15]	$\frac{333}{106}$	3.14151

Continued fractions provide, in some sense, a series of "best" estimates for an IRRATIONAL NUMBER. Functions can also be written as continued fractions, providing a series of better and better rational approximations. Continued fractions have also proved useful in the proof of certain properties of numbers such as E and π (PI). Because irrationals which are square roots of RATIONAL NUMBERS have periodic continued fractions, an exact representation for a tabulated numerical value (i.e., 1.414... for PYTHAGORAS'S CONSTANT, $\sqrt{2}$) can sometimes be found if it is suspected to represent an unknown QUADRATIC SURD.

Continued fractions are also useful for finding near commensurabilities between events with different periods. For example, the Metonic cycle used for calendrical purposes by the Greeks consists of 235 lunar months which very nearly equal 19 solar years, and 235/19 is the sixth CONVERGENT of the ratio of the lunar phase (synodic) period and solar period (365.2425/29.53059). Continued fractions can also be used to calculate gear ratios, and were used for this purpose by the ancient Greeks (Guy 1990).

Let P_n/Q_n be convergents of a *nonsimple* continued fraction. Then

$$P_{-1} \equiv 1 \quad Q_{-1} \equiv 0 \tag{12}$$

$$P_0 \equiv a_0 \quad Q_0 \equiv 1 \tag{13}$$

and subsequent terms are calculated from the RECURRENCE RELATIONS

$$P_j = a_j P_{j-1} + b_j P_{j-2} \tag{14}$$

$$Q_j = a_j Q_{j-1} + b_j Q_{j-2} \tag{15}$$

for $j = 1, 2, ..., n$. It is also true that

$$P_n Q_{n-1} - P_{n-1} Q_n = (-1)^{n-1} \prod_{k=1}^{n} b_k. \tag{16}$$

The error in approximating a number by a given CONVERGENT is roughly the MULTIPLICATIVE INVERSE of the square of the DENOMINATOR of the first neglected term.

A *finite simple* continued fraction representation terminates after a finite number of terms. To "round" a continued fraction, truncate the last term unless it is ± 1, in which case it should be added to the previous term (Gosper 1972, Item 101A). To take one over a continued fraction, add (or possibly delete) an initial 0 term. To negate, take the NEGATIVE of all terms, optionally using the identity

$$[-a, -b, -c, -d, \ldots]$$
$$= [-a-1, 1, b-1, c, d, \ldots]. \tag{17}$$

A particularly beautiful identity involving the terms

of the continued fraction is

$$\frac{[a_0, a_1, \ldots, a_n]}{[a_0, a_1, \ldots, a_{n-1}]} = \frac{[a_n, a_{n-1}, \ldots, a_1, a_0]}{[a_n, a_{n-1}, \ldots, a_1]}. \tag{18}$$

Finite simple fractions represent rational numbers and all rational numbers are represented by finite continued fractions. There are two possible representations for a finite simple fraction:

$$[a_0, \ldots, a_n]$$
$$= \begin{cases} [a_0, \ldots, a_{n-1}, a_n - 1, 1] & \text{for } a_n > 1 \\ [a_0, \ldots, a_{n-2}, a_{n-1} + 1] & \text{for } a_n = 1 \end{cases} \tag{19}$$

On the other hand, an infinite simple fraction represents a unique IRRATIONAL NUMBER, and each IRRATIONAL NUMBER has a unique infinite continued fraction.

Consider the CONVERGENTS $c_n = p_n/q_n$ of a *simple* continued fraction, and define

$$p_{-2} \equiv 0 \quad q_{-2} \equiv 1 \tag{20}$$

$$p_{-1} \equiv 1 \quad q_{-1} \equiv 0 \tag{21}$$

$$p_0 \equiv a_0 \quad q_0 \equiv 1. \tag{22}$$

Then subsequent terms can be calculated from the RECURRENCE RELATIONS

$$p_n = a_n p_{n-1} + p_{n-2} \tag{23}$$

$$q_n = a_n q_{n-1} + q_{n-2}. \tag{24}$$

The CONTINUED FRACTION FUNDAMENTAL RECURRENCE RELATION for *simple* continued fractions is

$$p_n q_{n-1} - p_{n-1} q_n = (-1)^{n+1}. \tag{25}$$

It is also true that if $a_0 \neq 0$,

$$\frac{p_n}{p_{n-1}} = [a_n, a_{n-1}, \ldots, a_0] \tag{26}$$

$$\frac{q_n}{q_{n-1}} = [a_n, \ldots, a_1]. \tag{27}$$

Furthermore,

$$\frac{p_n}{q_n} = \frac{p_{n+1} - p_{n-1}}{q_{n+1} - q_{n-1}}. \tag{28}$$

Also, if a convergent $c_n = p_n/q_n > 1$, then

$$\frac{q_n}{p_n} = [0, a_0, a_1, \ldots, a_n]. \tag{29}$$

Similarly, if $c_n = p_n/q_n < 1$, then $a_0 = 0$ and

$$\frac{q_n}{p_n} = [a_1, \ldots, a_n]. \tag{30}$$

The convergents $c_n = p_n/q_n$ also satisfy

$$c_n - c_{n-1} = \frac{(-1)^{n+1}}{q_n q_{n-1}} \qquad (31)$$

$$c_n - c_{n-2} = \frac{a_n (-1)^n}{q_n q_{n-2}}. \qquad (32)$$

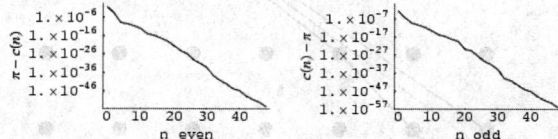

Plotted above on semilog scales are $c_n - \pi$ (n even; left figure) and $\pi - c_n$ (n odd; right figure) as a function of n for the convergents of π. In general, the EVEN convergents c_{2n+1} of an infinite simple continued fraction for a number x form an INCREASING SEQUENCE, and the ODD convergents c_{2n} form a DECREASING SEQUENCE (so any EVEN convergent is less than any ODD convergent). Summarizing,

$$c_0 < c_2 < c_4 < \cdots < c_{2n-2} < c_{2n} < \cdots < x \qquad (33)$$

$$x < \cdots < c_{2n+1} < c_{2n-1} < c_5 < c_3 < c_1. \qquad (34)$$

Furthermore, each convergent for $n \geq 3$ lies between the two preceding ones. Each convergent is nearer to the value of the infinite continued fraction than the previous one. In addition, for a number $x = [a_0, a_1, \ldots]$,

$$\frac{1}{(a_{n+1} + 2) q_n^2} < \left| x - \frac{p_n}{q_n} \right| < \frac{1}{a_{n+1} q_n^2}. \qquad (35)$$

The SQUARE ROOT of a SQUAREFREE INTEGER has a periodic continued fraction OF THE FORM

$$\sqrt{n} = [a_0, \overline{a_1, \ldots, a_n, 2a_0}] \qquad (36)$$

(Rose 1994, p. 130). Furthermore, if D is not a SQUARE NUMBER, then the terms of the continued fraction of $\sqrt{D}$ satisfy

$$0 < a_n < 2\sqrt{D}. \qquad (37)$$

In particular,

$$[\bar{a}] = \frac{a + \sqrt{a^2 + 4}}{2} \qquad (38)$$

$$[1, \bar{a}] = \frac{-1 + \sqrt{1 + 4a}}{2} \qquad (39)$$

$$[a, \overline{2a}] = \sqrt{a^2 + 1} \qquad (40)$$

$$[\overline{a, b}] = \frac{ab\sqrt{ab} + (ab(ab + 4)}{2b} \qquad (41)$$

$$[\overline{a_1, \ldots, a_n}]$$

$$= \frac{-(q_{n-1} - p_n) + \sqrt{(q_{n-1} - p_n)^2 + 4q_n p_{n-1}}}{2q_n} \qquad (42)$$

$$[a_0, \overline{b_1, \ldots, b_n}] = a_0 + \frac{1}{[\overline{b_1, \ldots, b_n}]} \qquad (43)$$

$$[\overline{b_1, \ldots, b_n}] = \frac{[\overline{b_1, \ldots, b_n}] p_n + p_{n-1}}{[\overline{b_1, \ldots, b_n}] q_n + q_{n-1}}. \qquad (44)$$

The first follows from

$$\alpha = n + \cfrac{1}{n + \cfrac{1}{n + \cfrac{1}{n + \ldots}}}$$

$$= n + \cfrac{1}{n + \left(\cfrac{1}{n + \cfrac{1}{n + \ldots}} \right)}. \qquad (45)$$

Therefore,

$$\alpha - n = \cfrac{1}{n + \cfrac{1}{n + \cfrac{1}{n + \ldots}}}, \qquad (46)$$

so plugging (46) into (45) gives

$$\alpha = n + \frac{1}{n + (\alpha - n)} = n + \frac{1}{\alpha}. \qquad (47)$$

Expanding

$$\alpha^2 - n\alpha - 1 = 0, \qquad (48)$$

and solving using the QUADRATIC FORMULA gives

$$\alpha = \frac{n + \sqrt{n^2 + 4}}{2}. \qquad (49)$$

The analog of this treatment in the general case gives

$$\alpha = \frac{\alpha p_n + p_{n-1}}{\alpha q_n + q_{n-1}}. \qquad (50)$$

The following table gives the repeating simple continued fractions for the square roots of the first few integers (excluding the trivial SQUARE NUMBERS).

N	$\alpha\sqrt{N}$	N	$\alpha\sqrt{N}$
2	$[1, \bar{2}]$	22	$[4, \overline{1, 2, 4, 2, 1, 8}]$
3	$[1, \overline{1, 2}]$	23	$[4, \overline{1, 3, 1, 8}]$
5	$[2, \bar{4}]$	24	$[4, \overline{1, 8}]$

6	[2, $\overline{2, 4}$]	26	[5, $\overline{10}$]
7	[2, $\overline{1, 1, 1, 4}$]	27	[5, $\overline{5, 10}$]
8	[2, $\overline{1, 4}$]	28	[5, $\overline{3, 2, 3, 10}$]
10	[3, $\overline{6}$]	29	[5, $\overline{2, 1, 1, 2, 10}$]
11	[3, $\overline{3, 6}$]	30	[5, $\overline{2, 10}$]
12	[3, $\overline{2, 6}$]	31	[5, $\overline{1, 1, 3, 5, 3, 1, 1, 10}$]
13	[3, $\overline{1, 1, 1, 1, 6}$]	32	[5, $\overline{1, 1, 1, 10}$]
14	[3, $\overline{1, 2, 1, 6}$]	33	[5, $\overline{1, 2, 1, 10}$]
15	[3, $\overline{1, 6}$]	34	[5, $\overline{1, 4, 1, 10}$]
17	[4, $\overline{8}$]	35	[5, $\overline{1, 10}$]
18	[4, $\overline{4, 8}$]	37	[6, $\overline{12}$]
19	[4, $\overline{2, 1, 3, 1, 2, 8}$]	38	[6, $\overline{6, 12}$]
20	[4, $\overline{2, 8}$]	39	[6, $\overline{4, 12}$]
21	[4, $\overline{1, 1, 2, 1, 1, 8}$]	40	[6, $\overline{3, 12}$]

The periods of the continued fractions of the square roots of the first few nonsquare integers 2, 3, 5, 6, 7, 8, 10, 11, 12, 13, ... (Sloane's A000037) are 1, 2, 1, 2, 4, 2, 1, 2, 2, 5, ... (Sloane's A013943; Williams 1981, Jacobson *et al.* 1995). An upper bound for the length is roughly $\mathcal{O}(\ln D \sqrt{D})$.

An even stronger result is that a continued fraction is periodic IFF it is a ROOT of a quadratic POLYNOMIAL. Calling the portion of a number x remaining after a given convergent the "tail," it must be true that the relationship between the number x and terms in its tail is OF THE FORM

$$x = \frac{ax + b}{cd + d}, \qquad (51)$$

which can only lead to a QUADRATIC EQUATION.

LOGARITHMS $\log_{b_0} b_1$ can be computed by defining b_2, ... and the POSITIVE INTEGER n_1, ...such that

$$b_1^{n_1} < b_0 < b_1^{n_1+1} \quad b_2 = \frac{b_0}{b_1^{n_1}} \qquad (52)$$

$$b_2^{n_2} < b_1 < b_2^{n_2+1} \quad b_3 = \frac{b_1}{b_2^{n_2}} \qquad (53)$$

and so on. Then

$$\log_{b_0} b_1 = [n_1, n_2, n_3, ...]. \qquad (54)$$

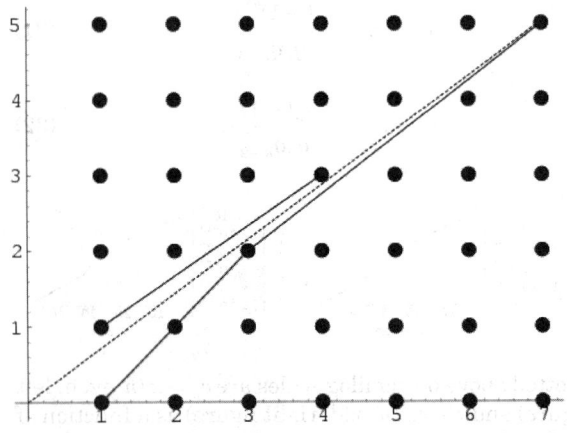

A geometric interpretation for a reduced FRACTION y/x consists of a string through a LATTICE of points with ends at $(1, 0)$ and (x, y) (Klein 1907, 1932; Steinhaus 1983, p. 40; Gardner 1984, pp. 210–11, Ball and Coxeter 1987, pp. 86–7; Davenport 1992). This interpretation is closely related to a similar one for the GREATEST COMMON DIVISOR. The pegs it presses against (x_i, y_i) give alternate CONVERGENTS y_i/x_i, while the other CONVERGENTS are obtained from the pegs it presses against with the initial end at $(0, 1)$. The above plot is for $e - 2$, which has convergents 0, 1, 2/3, 3/4, 5/7, ...

Let the continued fraction for x be written $[a_0, a_1, ..., a_n]$. Then the limiting value is *almost always* KHINTCHINE'S CONSTANT

$$K \equiv \lim_{n \to \infty} (a_1 a_2 \ldots a_n)^{1/n} = 2.68545\ldots. \qquad (55)$$

Continued fractions can be used to express the POSITIVE ROOTS of any POLYNOMIAL equation. Continued fractions can also be used to solve linear DIOPHANTINE EQUATIONS and the PELL EQUATION. Euler showed that if a CONVERGENT SERIES can be written in the form

$$c_1 + c_1 c_2 + c_1 c_2 c_3 + \ldots, \qquad (56)$$

then it is equal to the continued fraction

$$\cfrac{c_1}{1 - \cfrac{c_2}{1 + c_2 - \cfrac{c_3}{1 + c_3 - \ldots}}}. \qquad (57)$$

Gosper has invented an ALGORITHM for performing analytic ADDITION, SUBTRACTION, MULTIPLICATION, and DIVISION using continued fractions. It requires keeping track of eight INTEGERS which are conceptually arranged at the VERTICES of a CUBE. Although this ALGORITHM has not appeared in print, similar algorithms have been constructed by Vuillemin (1987) and Liardet and Stambul (1998).

Gosper's algorithm for computing the continued fraction for $(ax + b)/(cx + d)$ from the continued fraction for x is described by Gosper (1972), Knuth (1981, Exercise 4.5.3.15, pp. 360 and 601), and Fowler (1999). (In line 9 of Knuth's solution, $X_k \leftarrow \lfloor A/C \rfloor$ should be replaced by $X_k \leftarrow \min(\lfloor A/C \rfloor, \lfloor (A+B)/(C+D) \rfloor)$.) Gosper (1972) and Knuth (1981) also mention the bivariate case $(axy + bx + cy + d)/(Axy + Bx + Cy + D)$.

Ramanujan developed a number of interesting closed-form expressions for continued fractions, including

$$\cfrac{1}{1+} \cfrac{e^{-2\pi}}{1+} \cfrac{e^{-4\pi}}{1 + \ldots} = \left[\sqrt{\frac{5+\sqrt{5}}{2}} - \frac{\sqrt{5}+1}{2} \right] e^{2\pi/5} \quad (58)$$

$$\cfrac{1}{1+} \cfrac{e^{-2\pi\sqrt{5}}}{1+} \cfrac{e^{-4\pi\sqrt{5}}}{1 + \ldots}$$

$$= \left\{ \cfrac{\sqrt{5}}{1 + \left[5^{3/4} \left(\dfrac{\sqrt{5}-1}{2} \right)^{5/2} - 1 \right]} - \frac{\sqrt{5}+1}{2} \right\} e^{2\pi/\sqrt{5}} \quad (59)$$

and

$$4 \int_0^\infty \frac{x e^{-2\sqrt{5}}}{\cosh x} \, dx = \tfrac{1}{2}[\zeta(2, \tfrac{1}{4}(1+\sqrt{5})) - \zeta(2, \tfrac{1}{4}(3+\sqrt{5})]$$

$$= \cfrac{1}{1+} \cfrac{1^2}{1+} \cfrac{1^2}{1+} \cfrac{2^2}{1+} \cfrac{2^2}{1+} \cfrac{3^2}{1+} \cfrac{3^2}{1+} \quad (60)$$

(Watson 1929; Preece 1931; Watson 1931; Hardy 1999, p. 8).

See also Gaussian Brackets, Hurwitz's Irrational Number Theorem, Khintchine's Constant, Lagrange's Continued Fraction Theorem, Lamé's Theorem, Lehmer Continued Fraction, Lévy Constant, Lochs Theorem, Padé Approximant, Partial Quotient, Pi, Quadratic Irrational Number, Quotient-Difference Algorithm, Rogers-Ramanujan Continued Fraction, Segre's Theorem, Trott's Constant

References

Abramowitz, M. and Stegun, C. A. (Eds.). *Handbook of Mathematical Functions with Formulas, Graphs, and Mathematical Tables, 9th printing.* New York: Dover, p. 19, 1972.

Acton, F. S. "Power Series, Continued Fractions, and Rational Approximations." Ch. 11 in *Numerical Methods That Work, 2nd printing.* Washington, DC: Math. Assoc. Amer., 1990.

Adamchik, V. "Limits of Continued Fractions and Nested Radicals." *Mathematica J.* **2**, 54–7, 1992.

Ball, W. W. R. and Coxeter, H. S. M. *Mathematical Recreations and Essays, 13th ed.* New York: Dover, pp. 54–7 and 86–7, 1987.

Berndt, B. C. and Gesztesy, F. (Eds.). *Continued Fractions: From Analytic Number Theory to Constructive Approxi-*

mation, *A Volume in Honor of L.J. Lange.* Providence, RI: Amer. Math. Soc., 1999.

Beskin, N. M. *Fascinating Fractions.* Moscow: Mir Publishers, 1980.

Brezinski, C. *History of Continued Fractions and Padé Approximants.* New York: Springer-Verlag, 1980.

Conway, J. H. and Guy, R. K. "Continued Fractions." In *The Book of Numbers.* New York: Springer-Verlag, pp. 176–79, 1996.

Courant, R. and Robbins, H. "Continued Fractions. Diophantine Equations." §2.4 in Supplement to Ch. 1 in *What is Mathematics?: An Elementary Approach to Ideas and Methods, 2nd ed.* Oxford, England: Oxford University Press, pp. 49–1, 1996.

Davenport, H. §IV.12 in *The Higher Arithmetic: An Introduction to the Theory of Numbers, 6th ed.* New York: Cambridge University Press, 1992.

Dunne, E. and McConnell, M. "Pianos and Continued Fractions." *Math. Mag.* **72**, 104–15, 1999.

Euler, L. *Introduction to Analysis of the Infinite, Book I.* New York: Springer-Verlag, 1980.

Fowler, D. H. *The Mathematics of Plato's Academy: A New Reconstruction, 2nd ed.* Oxford, England: Oxford University Press, 1999.

Gardner, M. *The Sixth Book of Mathematical Games from Scientific American.* Chicago, IL: University of Chicago Press, pp. 210–11, 1984.

Gosper, R. W. Item 101a in Beeler, M.; Gosper, R. W.; and Schroeppel, R. *HAKMEM.* Cambridge, MA: MIT Artificial Intelligence Laboratory, Memo AIM-239, pp. 37–9, Feb. 1972.

Gosper, R. W. Item 101b in Beeler, M.; Gosper, R. W.; and Schroeppel, R. *HAKMEM.* Cambridge, MA: MIT Artificial Intelligence Laboratory, Memo AIM-239, pp. 39–4, Feb. 1972.

Graham, R. L.; Knuth, D. E.; and Patashnik, O. "Continuants." §6.7 in *Concrete Mathematics: A Foundation for Computer Science, 2nd ed.* Reading, MA: Addison-Wesley, pp. 301–09, 1994.

Guy, R. K. "Continued Fractions" §F20 in *Unsolved Problems in Number Theory, 2nd ed.* New York: Springer-Verlag, p. 259, 1994.

Hardy, G. H. *Ramanujan: Twelve Lectures on Subjects Suggested by His Life and Work, 3rd ed.* New York: Chelsea, 1999.

Jacobson, M. J. Jr.; Lukes, R. F.; and Williams, H. C. "An Investigation of Bounds for the Regulator of Quadratic Fields." *Experiment. Math.* **4**, 211–25, 1995.

Khinchin, A. Ya. *Continued Fractions.* New York: Dover, 1997.

Kimberling, C. "Continued Fractions." http://cedar.evansville.edu/~ck6/integer/contfr.html.

Klein, F. *Ausgewählte Kapitel der Zahlentheorie I.* Göttingen, Germany: n.p., 1896.

Klein, F. *Elementary Number Theory.* New York, p. 44, 1932.

Kline, M. *Mathematical Thought from Ancient to Modern Times.* New York: Oxford University Press, 1972.

Knuth, D. E. *The Art of Computer Programming, Vol. 2: Seminumerical Algorithms, 3rd ed.* Reading, MA: Addison-Wesley, p. 316, 1998.

Liardet, P. and Stambul, P. "Algebraic Computation with Continued Fractions." *J. Number Th.* **73**, 92–21, 1998.

Lorentzen, L. and Waadeland, H. *Continued Fractions with Applications.* Amsterdam, Netherlands: North-Holland, 1992.

Moore, C. D. *An Introduction to Continued Fractions.* Washington, DC: National Council of Teachers of Mathematics, 1964.

Olds, C. D. *Continued Fractions.* New York: Random House, 1963.

Perron, O. *Die Lehre von Kettenbrüchen, 3. verb. und erweiterte Aufl.* Stuttgart, Germany: Teubner, 1954–7.

Pettofrezzo, A. J. and Bykrit, D. R. *Elements of Number Theory.* Englewood Cliffs, NJ: Prentice-Hall, 1970.

Preece, C. T. "Theorems Stated by Ramanujan (X)." *J. London Math. Soc.* **6**, 22–2, 1931.

Press, W. H.; Flannery, B. P.; Teukolsky, S. A.; and Vetterling, W. T. "Evaluation of Continued Fractions." §5.2 in *Numerical Recipes in FORTRAN: The Art of Scientific Computing, 2nd ed.* Cambridge, England: Cambridge University Press, pp. 163–67, 1992.

Riesel, H. "Continued Fractions." Appendix 8 in *Prime Numbers and Computer Methods for Factorization, 2nd ed.* Boston, MA: Birkhäuser, pp. 327–42, 1994.

Rockett, A. M. and Szüsz, P. *Continued Fractions.* New York: World Scientific, 1992.

Rose, H. E. *A Course in Number Theory, 2nd ed.* Oxford, England: Oxford University Press, 1994.

Rosen, K. H. *Elementary Number Theory and Its Applications.* New York: Addison-Wesley, 1980.

Schur, I. "Ein Beitrag zur additiven Zahlentheorie und zur Theorie der Kettenbrüche." *Sitzungsber. Preuss. Akad. Wiss. Phys.-Math. Klasse*, pp. 302–21, 1917.

Sloane, N. J. A. Sequences A000037/M0613 and A013943 in "An On-Line Version of the Encyclopedia of Integer Sequences." http://www.research.att.com/~njas/sequences/eisonline.html.

Steinhaus, H. *Mathematical Snapshots, 3rd ed.* New York: Dover, pp. 39–2, 1999.

Van Tuyl, A. L. "Continued Fractions." http://www.calvin.edu/academic/math/confrac/.

Vuillemin, J. "Exact Real Computer Arithmetic with Continued Fractions." INRIA Report 760. Le Chasny, France: INRIA, Nov. 1987. http://www.inria.fr/RRRT/RR-0760.html.

Wagon, S. "Continued Fractions." §8.5 in *Mathematica in Action.* New York: W. H. Freeman, pp. 263–71, 1991.

Wall, H. S. *Analytic Theory of Continued Fractions.* New York: Chelsea, 1948.

Watson, G. N. "Theorems Stated by Ramanujan (VII): Theorems on a Continued Fraction." *J. London Math. Soc.* **4**, 39–8, 1929.

Watson, G. N. "Theorems Stated by Ramanujan (IX): Two Continued Fractions." *J. London Math. Soc.* **4**, 231–37, 1929.

Weisstein, E. W. "Books about Continued Fractions." http://www.treasure-troves.com/books/ContinuedFractions.html.

Williams, H. C. "A Numerical Investigation into the Length of the Period of the Continued Fraction Expansion of $\sqrt{D}$." *Math. Comp.* **36**, 593–01, 1981.

Continued Fraction Constant

A continued fraction with partial quotients which increase in ARITHMETIC PROGRESSION is

$$[A+D,\ A+2D,\ A+3D,\ \ldots] = \frac{I_{A/D}\left(\dfrac{2}{D}\right)}{I_{1+A/D}\left(\dfrac{2}{D}\right)},$$

where $I_n(x)$ is a MODIFIED BESSEL FUNCTION OF THE FIRST KIND (Schroeppel 1972). A special case is

$$C = 0 + \cfrac{1}{1 + \cfrac{1}{2 + \cfrac{1}{3 + \cfrac{1}{4 + \cfrac{1}{5 + \ldots}}}}},$$

which has the value

$$C = \frac{I_1(2)}{I_0(2)} = 0.697774658\ldots$$

(Lehmer 1973, Rabinowitz 1990).

See also E, GOLDEN RATIO, MODIFIED BESSEL FUNCTION OF THE FIRST KIND, PI, RABBIT CONSTANT, THUE-MORSE CONSTANT

References

Finch, S. "Favorite Mathematical Constants." http://www.mathsoft.com/asolve/constant/cntfrc/cntfrc.html.

Guy, R. K. "Review: The Mathematics of Plato's Academy." *Amer. Math. Monthly* **97**, 440–43, 1990.

Lehmer, D. H. "Continued Fractions Containing Arithmetic Progressions." *Scripta Math.* **29**, 17–4, 1973.

Rabinowitz, S. Problem E3264. "Asymptotic Estimates from Convergents of a Continued Fraction." *Amer. Math. Monthly* **97**, 157–59, 1990.

Schroeppel, R. Item 99 in Beeler, M.; Gosper, R. W.; and Schroeppel, R. *HAKMEM.* Cambridge, MA: MIT Artificial Intelligence Laboratory, Memo AIM-239, p. 36, Feb. 1972.

Continued Fraction Factorization Algorithm

A PRIME FACTORIZATION ALGORITHM which uses RESIDUES produced in the CONTINUED FRACTION of $\sqrt{mN}$ for some suitably chosen m to obtain a SQUARE NUMBER. The ALGORITHM solves

$$x^2 \equiv y^2 \pmod{n}$$

by finding an m for which $m^2 \pmod{n}$ has the smallest upper bound. The method requires (by conjecture) about $\exp(\sqrt{2 \ln n \ln \ln n})$ steps, and was the fastest PRIME FACTORIZATION ALGORITHM in use before the QUADRATIC SIEVE, which eliminates the 2 under the SQUARE ROOT (Pomerance 1996), was developed.

See also EXPONENT VECTOR, PRIME FACTORIZATION ALGORITHMS

References

Morrison, M. A. and Brillhart, J. "A Method of Factoring and the Factorization of F_7." *Math. Comput.* **29**, 183–05, 1975.

Pomerance, C. "A Tale of Two Sieves." *Not. Amer. Math. Soc.* **43**, 1473–485, 1996.

Continued Fraction Fundamental Recurrence Relation

For a SIMPLE CONTINUED FRACTION $x = [a_0, a_1, \ldots]$ with CONVERGENTS p_n/q_n, the fundamental RECUR-

RENCE RELATION is given by

$$p_n q_{n-1} - p_{n-1} q_n = (-1)^{n+1}.$$

See also SIMPLE CONTINUED FRACTION, CONTINUED FRACTION

References

Olds, C. D. *Continued Fractions.* New York: Random House, p. 27, 1963.

Continued Fraction Map

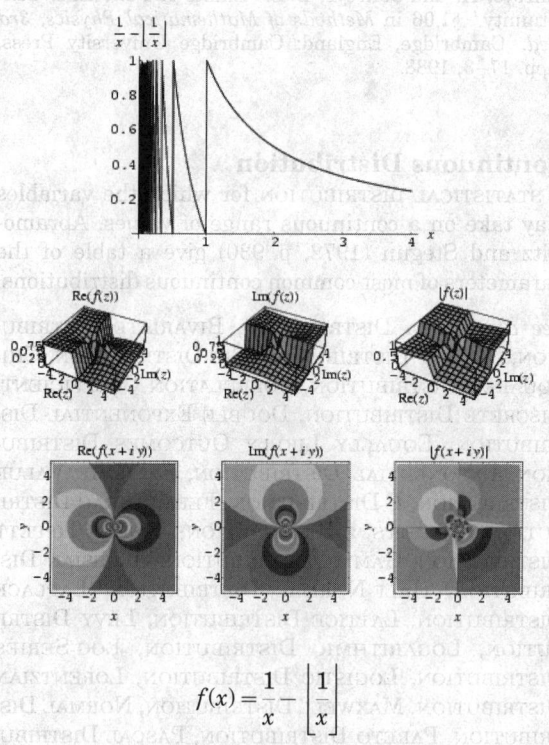

$$f(x) = \frac{1}{x} - \left\lfloor \frac{1}{x} \right\rfloor$$

for $x \in [0, 1]$, where $\lfloor x \rfloor$ is the FLOOR FUNCTION. The NATURAL INVARIANT of the map is

$$\rho(y) = \frac{1}{(1 + y) \ln 2}.$$

References

Beck, C. and Schlögl, F. *Thermodynamics of Chaotic Systems.* Cambridge, England: Cambridge University Press, pp. 194–95, 1995.

Continued Fraction Unit Fraction Algorithm

An algorithm for computing a UNIT FRACTION, called the FAREY SEQUENCE method by Bleicher (1972).

References

Bleicher, M. N. "A New Algorithm for the Expansion of Continued Fractions." *J. Number Th.* **4**, 342–82, 1972.
Eppstein, D. Egypt.ma Mathematica notebook. http://www.ics.uci.edu/~eppstein/numth/egypt/egypt.ma.

Continued Square Root

NESTED RADICAL

Continued Vector Product

VECTOR TRIPLE PRODUCT

Continuity

The property of being CONTINUOUS.

See also CONTINUITY AXIOMS, CONTINUITY CORRECTION, CONTINUITY PRINCIPLE, CONTINUOUS DISTRIBUTION, CONTINUOUS FUNCTION, CONTINUOUS SPACE, FUNDAMENTAL CONTINUITY THEOREM, LIMIT

References

Kaplan, W. "Limits and Continuity." §2.4 in *Advanced Calculus, 4th ed.* Reading, MA: Addison-Wesley, pp. 82–6, 1992.
Smith, W. K. *Limits and Continuity.* New York: Macmillan, 1964.

Continuity Axioms

"The" continuity axiom is an additional AXIOM which must be added to those of Euclid's *ELEMENTS* in order to guarantee that two equal CIRCLES of RADIUS r intersect each other if the separation of their centers is less than $2r$ (Dunham 1990). The continuity *axioms* are the three of HILBERT'S AXIOMS which concern geometric equivalence.

ARCHIMEDES' LEMMA is sometimes also known as "the continuity axiom."

See also CONGRUENCE AXIOMS, HILBERT'S AXIOMS, INCIDENCE AXIOMS, ORDERING AXIOMS, PARALLEL POSTULATE

References

Dunham, W. *Journey through Genius: The Great Theorems of Mathematics.* New York: Wiley, p. 38, 1990.
Hilbert, D. *The Foundations of Geometry.* Chicago, IL: Open Court, 1980.
Iyanaga, S. and Kawada, Y. (Eds.). "Hilbert's System of Axioms." §163B in *Encyclopedic Dictionary of Mathematics.* Cambridge, MA: MIT Press, pp. 544–45, 1980.

Continuity Correction

A correction to a discrete BINOMIAL DISTRIBUTION to approximate a continuous distribution.

$$P(a \le X \le b) \approx P\left(\frac{a - \frac{1}{2} - np}{\sqrt{np(1-p)}} \le z \le \frac{b + \frac{1}{2} - np}{\sqrt{np(1-p)}} \right),$$

where

$$z \equiv \frac{(x - \mu)}{\sigma}$$

is a continuous variate with a NORMAL DISTRIBUTION and X is a variate of a BINOMIAL DISTRIBUTION.

See also BINOMIAL DISTRIBUTION, NORMAL DISTRIBUTION

References

Gonick, L. and Smith, W. *The Cartoon Guide to Statistics.* New York: Harper Perennial, p. 87, 1993.

Continuity Principle

The metric properties discovered for a primitive figure remain applicable, without modifications other than changes of signs, to all correlative figures which can be considered to arise from the first. As stated by Lachlan (1893), the principle states that if, from the nature of a particular problem, a certain number of solutions are expected (and are, in fact, found in any one case), then there will be the same number of solutions in all cases, although some solutions may be imaginary.

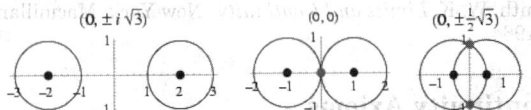

For example, two circles intersect in two points, so it can be stated that every two circles intersect in two points, although the points may be imaginary or may coincide. The principle is extremely powerful (if somewhat difficult to state precisely), and allows immediate derivation of some geometric propositions from other propositions which may appear simpler and may be substantially easier to prove.

The continuity principle was first enunciated by Kepler and thereafter enunciated by Boscovich. However, it was not generally accepted until formulated by Poncelet in 1822. Formally, it amounts to the statement that if an analytic identity in any finite number of variables holds for all real values of the variables, then it also holds by ANALYTIC CONTINUATION for all complex values (Bell 1945). This principle is also called "Poncelet's continuity principle," or sometimes the "permanence of mathematical relations principle" (Bell 1945).

See also ANALYTIC CONTINUATION, CONSERVATION OF NUMBER PRINCIPLE, DUALITY PRINCIPLE, PERMANENCE OF ALGEBRAIC FORM

References

Bell, E. T. *The Development of Mathematics, 2nd ed.* New York: McGraw-Hill, p. 340, 1945.
Lachlan, R. "The Principle of Continuity." §8 in *An Elementary Treatise on Modern Pure Geometry.* London: Macmillian, pp. 4–, 1893.
Poncelet, J.-V. *Traité des Propriétés Projectives.* 1822.

Continuous

A general mathematical property obeyed by mathematical objects in which all elements are within a NEIGHBORHOOD of nearby points. The continuous maps between TOPOLOGICAL SPACES form a CATEGORY. The designation "continuous" is sometimes used to indicate membership in this category.

See also ABSOLUTELY CONTINUOUS, CONTINUOUS DISTRIBUTION, CONTINUITY, CONTINUOUS FUNCTION, CONTINUOUS SPACE, DIFFERENTIABLE, JUMP, PIECEWISE CONTINUOUS

References

Jeffreys, H. and Jeffreys, B. S. "Limits of Functions: Continuity." §1.06 in *Methods of Mathematical Physics, 3rd ed.* Cambridge, England: Cambridge University Press, pp. 17–3, 1988.

Continuous Distribution

A STATISTICAL DISTRIBUTION for which the variables may take on a continuous range of values. Abramowitz and Stegun (1972, p. 930) give a table of the parameters of most common continuous distributions.

See also BETA DISTRIBUTION, BIVARIATE DISTRIBUTION, CAUCHY DISTRIBUTION, CHI DISTRIBUTION, CHI-SQUARED DISTRIBUTION, CORRELATION COEFFICIENT, DISCRETE DISTRIBUTION, DOUBLE EXPONENTIAL DISTRIBUTION, EQUALLY LIKELY OUTCOMES DISTRIBUTION, EXPONENTIAL DISTRIBUTION, EXTREME VALUE DISTRIBUTION, F-DISTRIBUTION, FERMI-DIRAC DISTRIBUTION, FISHER'S z-DISTRIBUTION, FISHER-TIPPETT DISTRIBUTION, GAMMA DISTRIBUTION, GAUSSIAN DISTRIBUTION, HALF-NORMAL DISTRIBUTION, LAPLACE DISTRIBUTION, LATTICE DISTRIBUTION, LÉVY DISTRIBUTION, LOGARITHMIC DISTRIBUTION, LOG-SERIES DISTRIBUTION, LOGISTIC DISTRIBUTION, LORENTZIAN DISTRIBUTION, MAXWELL DISTRIBUTION, NORMAL DISTRIBUTION, PARETO DISTRIBUTION, PASCAL DISTRIBUTION, PEARSON TYPE III DISTRIBUTION, POISSON DISTRIBUTION, PÓLYA DISTRIBUTION, RATIO DISTRIBUTION, RAYLEIGH DISTRIBUTION, RICE DISTRIBUTION, SNEDECOR'S F-DISTRIBUTION, STUDENT'S t-DISTRIBUTION, STUDENT'S z-DISTRIBUTION, UNIFORM DISTRIBUTION, WEIBULL DISTRIBUTION

References

Abramowitz, M. and Stegun, C. A. (Eds.). *Handbook of Mathematical Functions with Formulas, Graphs, and Mathematical Tables, 9th printing.* New York: Dover, pp. 927 and 930, 1972.
Evans, M.; Hastings, N.; and Peacock, B. *Statistical Distributions, 3rd ed.* New York: Wiley, 2000.
Kotz, S.; Balakrishnan, N.; and Johnson, N. L. *Continuous Multivariate Distributions, Vol. 1: Models and Applications, 2nd ed.* New York: Wiley, 2000.
McLaughlin, M. "Common Probability Distributions." http://www.geocities.com/~mikemclaughlin/math_stat/Dists/Compendium.html.

Continuous Function

There are several commonly used methods of defining the slippery, but extremely important, concept of a continuous function. The space of continuous functions is denoted C^0, and corresponds to the $k = 0$ case of a C-K FUNCTION.

A continuous function can be formally defined as a FUNCTION $f : X \to Y$ where the pre-image of every OPEN SET in Y is OPEN in X. More concretely, a function $f(x)$ in a single variable x is said to be continuous at point x_0 if

1. $f(x_0)$ is defined, so that x_0 is in the DOMAIN of f.
2. $\lim_{x \to x_0} f(x)$ exists for x in the DOMAIN of f.
3. $\lim_{x \to x_0} f(x) = f(x_0)$,

where lim denotes a LIMIT.

Many mathematicians prefer to define the continuity of a function via a so-called EPSILON-DELTA DEFINITION of a LIMIT. In this formalism, a LIMIT c of function $f(x)$ as x approaches a point x_0,

$$\lim_{x \to x_0} f(x) = c, \tag{1}$$

is defined when, given any $\epsilon > 0$, a $\delta > 0$ can be found such that for every x in some domain D and within the neighborhood of x_0 of radius δ (except possibly x_0 itself),

$$|f(x) - c| < \epsilon. \tag{2}$$

Then if x_0 is in D and

$$\lim_{x \to x_0} f(x) = f(x_0) = c, \tag{3}$$

$f(x)$ is said to be continuous at x_0.

If f is DIFFERENTIABLE at point x_0, then it is also continuous at x_0. If two functions f and g are continuous at x_0, then

1. $f + g$ is continuous at x_0.
2. $f - g$ is continuous at x_0.
3. $f \times g$ is continuous at x_0.
4. $f \div g$ is continuous at x_0 if $g(x_0) \neq 0$ and is discontinuous at x_0 if $g(x_0) = 0$.
5. $f \circ g$ is continuous at x_0, where $f \circ g$ denotes $f(g(x))$, the COMPOSITION of the functions f and g.

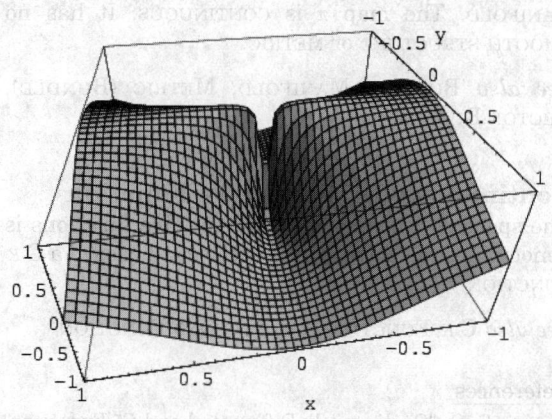

The notion of continuity for a function in two variables is slightly trickier, as illustrated above by the plot of the function

$$z = \frac{x^2 - y^2}{x^2 + y^2}. \tag{4}$$

This function is discontinuous at the origin, but has limit 0 along the line $x = y$, limit 1 along the X-AXIS, and limit -1 along the Y-AXIS (Kaplan 1992, p. 83).

See also C-K FUNCTION, CONTINUOUSLY DIFFERENTIABLE FUNCTION, CRITICAL POINT, DIFFERENTIABLE, LIMIT, NEIGHBORHOOD, PIECEWISE CONTINUOUS, STATIONARY POINT

References

Bartle, R. G. and Sherbert, D. *Introduction to Real Analysis.* New York: Wiley, p. 141, 1991.
Kaplan, W. "Limits and Continuity." §2.4 in *Advanced Calculus, 4th ed.* Reading, MA: Addison-Wesley, pp. 82–6, 1992.

Continuous Group

A GROUP having CONTINUOUS group operations. A continuous group is necessarily infinite, since an INFINITE GROUP just has to contain an infinite number of elements. But some infinite groups, such as the integers or rationals, are not continuous groups.

See also DISCRETE GROUP, FINITE GROUP, INFINITE GROUP

Continuous Space

A TOPOLOGICAL SPACE.

See also NET

Continuous Transformation

HOMEOMORPHISM

Continuous Vector Bundle

A continuous vector bundle is a VECTOR BUNDLE $\pi : E \to M$ with only the structure of a TOPOLOGICAL

MANIFOLD. The map π is CONTINUOUS. It has no SMOOTH STRUCTURE or METRIC.

See also BUNDLE, MANIFOLD, METRIC (BUNDLE), VECTOR BUNDLE

Continuously Differentiable Function

The space of continuously differentiable functions is denoted C^1, and corresponds to the $k = 1$ case of a C-K FUNCTION.

See also C-K FUNCTION, CONTINUOUS FUNCTION

References

Krantz, S. G. "Continuously Differential and C^k Functions" and "Differentiable and C^k Curves." §1.3.1 and 2.1.3 in *Handbook of Complex Analysis.* Boston, MA: Birkhäuser, pp. 12–3 and 21, 1999.

Continuum

The nondenumerable set of REAL NUMBERS, denoted C. It satisfies

$$\aleph + C = C \tag{1}$$

and

$$C^r = C \tag{2}$$

where $\aleph_0$ is ALEPH-0. It is also true that

$$\aleph_0^{\aleph_0} = C. \tag{3}$$

However,

$$C^C = F \tag{4}$$

is a SET larger than the continuum. Paradoxically, there are exactly as many points C on a LINE (or LINE SEGMENT) as in a PLANE, a 3-D SPACE, or finite HYPERSPACE, since all these SETS can be put into a ONE-TO-ONE correspondence with each other.

The CONTINUUM HYPOTHESIS, first proposed by Georg Cantor, holds that the CARDINAL NUMBER of the continuum is the same as that of ALEPH-1. The surprising truth is that this proposition is UNDECIDABLE, since neither it nor its converse contradicts the tenets of SET THEORY.

See also ALEPH-0, ALEPH-1, CONTINUUM HYPOTHESIS, DENUMERABLE SET

Continuum Hypothesis

Portions of this entry contributed by MATTHEW SZUDZIK

The proposal originally made by Georg Cantor that there is no infinite set with a CARDINAL NUMBER between that of the "small" infinite set of INTEGERS $\aleph_0$ and the "large" infinite set of REAL NUMBERS C (the "CONTINUUM"). Symbolically, the continuum hypothesis is that $\aleph_1 = C$.

Gödel showed that no CONTRADICTION would arise if the continuum hypothesis were added to conventional ZERMELO-FRAENKEL SET THEORY. However, using a technique called FORCING, Paul Cohen (1963, 1964) proved that no contradiction would arise if the *negation* of the continuum hypothesis was added to SET THEORY. Together, Gödel's and Cohen's results established that the validity of the continuum hypothesis depends on the version of SET THEORY being used, and is therefore UNDECIDABLE (assuming the ZERMELO-FRAENKEL axioms together with the AXIOM OF CHOICE).

Conway and Guy (1996, p. 282) recount a generalized version of the continuum hypothesis originally due to Hausdorff in 1908 which is also UNDECIDABLE: is $2^{\aleph_\alpha} = \aleph_{\alpha+1}$ for every α? The continuum hypothesis follows from generalized continuum hypothesis, so $ZF + GCH \vdash CH$.

In 2000, H. Woodin formulated a new plausible "axiom" whose adoption (in addition to the ZERMELO-FRAENKEL axioms and AXIOM OF CHOICE) would imply that the Continuum Hypothesis is false. Since set theoreticians have felt for some time that the Continuum Hypothesis should be false, if Woodin's axiom proves to be particularly elegant, useful, or intuitive, it may catch on. It is interesting to compare this to a situation with Euclid's PARALLEL POSTULATE more than 300 years ago, when Wallis proposed an additional axiom that would imply the PARALLEL POSTULATE (Greenberg 1994, pp. 152–53).

See also ALEPH-0, ALEPH-1, AXIOM OF CHOICE, CARDINAL NUMBER, CONTINUUM, DENUMERABLE SET, FORCING, HILBERT'S PROBLEMS, LEBESGUE MEASURABILITY PROBLEM, UNDECIDABLE, ZERMELO-FRAENKEL AXIOMS, ZERMELO-FRAENKEL SET THEORY

References

Cohen, P. J. "The Independence of the Continuum Hypothesis." *Proc. Nat. Acad. Sci. U. S. A.* **50**, 1143–148, 1963.

Cohen, P. J. "The Independence of the Continuum Hypothesis. II." *Proc. Nat. Acad. Sci. U. S. A.* **51**, 105–10, 1964.

Cohen, P. J. *Set Theory and the Continuum Hypothesis.* New York: W. A. Benjamin, 1966.

Conway, J. H. and Guy, R. K. *The Book of Numbers.* New York: Springer-Verlag, p. 282, 1996.

Ferreirós, J. "The Notion of Cardinality and the Continuum Hypothesis." Ch. 6 in *Labyrinth of Thought: A History of Set Theory and Its Role in Modern Mathematics.* Basel, Switzerland: Birkhäuser, pp. 171–14, 1999.

Gödel, K. *The Consistency of the Continuum-Hypothesis.* Princeton, NJ: Princeton University Press, 1940.

Greenberg, M. J. *Euclidean and Non-Euclidean Geometries: Development and History, 3rd ed.* San Francisco, CA: W. H. Freeman, 1994.

Hoffman, P. *The Man Who Loved Only Numbers: The Story of Paul Erdos and the Search for Mathematical Truth.* New York: Hyperion, pp. 225–26, 1998.

Jech, T. J. *Set Theory, 2nd ed.* Berlin: Springer-Verlag, 1997.

McGough, N. "The Continuum Hypothesis." http://www.ii-.com/math/ch/.

Contour

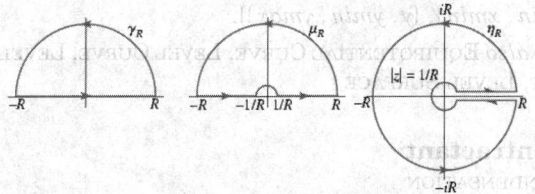

A path in the COMPLEX PLANE over which CONTOUR INTEGRATION is performed to compute a CONTOUR INTEGRAL. When choosing a contour to evaluate an integral on the REAL LINE, a contour is generally chosen based on the range of integration and the position of POLES in the COMPLEX PLANE. For example, for an integral from $-\infty$ to $+\infty$ along the real axis, the contour at left could be chosen if the function f had no POLES on the REAL LINE, and the middle contour could be chosen if it had a POLE at the origin. To perform an integral over the positive real axis from 0 to $+\infty$ for a function with a POLE at 0, the contour at right could be chosen.

See also CONTOUR INTEGRAL, CONTOUR INTEGRATION, HANKEL CONTOUR, INSIDE-OUTSIDE THEOREM, POLE, RESIDUE (COMPLEX ANALYSIS)

Contour Integral

An integral obtained by CONTOUR INTEGRATION. The particular path in the COMPLEX PLANE used to compute the integral is called a CONTOUR. Watson (1966 p. 20) uses the notation $\int^{(a+)} f(z) \, dz$ to denote the contour integral of $f(z)$ with CONTOUR encircling the point a once in a counterclockwise direction.

See also CONTOUR, CONTOUR INTEGRATION

References
Watson, G. N. *A Treatise on the Theory of Bessel Functions,* *2nd ed.* Cambridge, England: Cambridge University Press, 1966.

Contour Integration

Contour integration is the process of calculating the values of a CONTOUR INTEGRAL around a given CONTOUR in the COMPLEX PLANE. As a result of a truly amazing property of HOLOMORPHIC FUNCTIONS, such integrals can be computed easily simply by summing the values of the RESIDUES *inside* the CONTOUR.

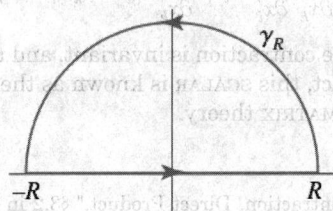

Let $P(x)$ and $Q(x)$ be POLYNOMIALS of DEGREES n and

m with COEFFICIENTS $b_n, ..., b_0$ and $c_m, ..., c_0$. Take the CONTOUR in the UPPER HALF-PLANE, replace x by z, and write $z \equiv Re^{i\theta}$. Then

$$\int_{-\infty}^{\infty} \frac{P(z) \, dz}{Q(z)} = \lim_{R \to \infty} \int_{-R}^{R} \frac{P(z) \, dz}{Q(z)}. \tag{1}$$

Define a path γ_R which is straight along the REAL axis from $-R$ to R and make a circular half-arc to connect the two ends in the upper half of the COMPLEX PLANE. The RESIDUE THEOREM then gives

$$\lim_{R \to \infty} \int_{\gamma_R} \frac{P(z) \, dz}{Q(z)}$$

$$= \lim_{R \to \infty} \int_{-R}^{R} \frac{P(z) \, dz}{Q(z)} + \lim_{R \to \infty} \int_0^{\pi} \frac{P(Re^{i\theta})}{Q(Re^{i\theta})} iRe^{i\theta} \, d\theta$$

$$= 2\pi i \sum_{\Im[z]>0} \text{Res} \left[\frac{P(z)}{Q(z)} \right], \tag{2}$$

where Res denotes the RESIDUES. Solving,

$$\lim_{R \to \infty} \int_{-R}^{R} \frac{P(z) \, dz}{Q(z)}$$

$$= 2\pi i \sum_{\Im[z]>0} \text{Res} \, \frac{P(z)}{Q(z)} - \lim_{R \to \infty} \int_0^{\pi} \frac{P(Re^{i\theta})}{Q(Re^{i\theta})} iRe^{i\theta} \, d\theta$$

Define

$$I_r \equiv \lim_{R \to \infty} \int_0^{\pi} \frac{P(Re^{i\theta})}{Q(Re^{i\theta})} iRe^{i\theta} \, d\theta$$

$$= \lim_{R \to \infty} \int_0^{\pi} \frac{b_n (Re^{i\theta})^n + b_{n-1}(Re^{i\theta})^{n-1} + \ldots + b_0}{c_m (Re^{i\theta})^m + c_{m-1}(Re^{i\theta})^{m-1} + \ldots + c_0} iR \, d\theta$$

$$= \lim_{R \to \infty} \int_0^{\pi} \frac{b_n}{c_m} (Re^{i\theta})^{n-m} iR \, d\theta$$

$$= \lim_{R \to \infty} \int_0^{\pi} \frac{b_n}{c_m} R^{n+1-m} i(e^{i\theta})^{n-m} \, d\theta \tag{3}$$

and set

$$\epsilon \equiv -(n+1-m), \tag{4}$$

then equation (3) becomes

$$I_r \equiv \lim_{R \to \infty} \frac{i}{R^{\epsilon}} \frac{b_r}{c_m} \int_0^{\pi} e^{i(n-m)\theta} \, d\theta. \tag{5}$$

Now,

$$\lim_{R \to \infty} R^{-\epsilon} = 0 \tag{6}$$

for $\epsilon > 0$. That means that for $-n - 1 + m \geq 1$, or $m \geq n + 2$, $I_R = 0$, so

$$\int_{-\infty}^{\infty} \frac{P(z) \, dz}{Q(z)} = 2\pi i \sum_{\Im[z]>0} \text{Res} \left[\frac{P(z)}{Q(z)} \right] \tag{7}$$

for $m \geq n + 2$. Apply JORDAN'S LEMMA with $f(x) \equiv P(x)/Q(x)$. We must have

$$\lim_{x \to \infty} f(x) = 0, \tag{8}$$

so we require $m \geq n + 1$. Then

$$\int_{-\infty}^{\infty} \frac{P(z)}{Q(z)} e^{iaz} \, dz = 2\pi i \sum_{\Im[z]>0} \text{Res}\left[\frac{P(z)}{Q(z)} e^{iaz}\right] \tag{9}$$

for $m \geq n + 1$.

Since this must hold separately for REAL and IMAGINARY PARTS, this result can be extended to

$$\int_{-\infty}^{\infty} \frac{P(x)}{Q(x)} \cos(ax) \, dx$$
$$= 2\pi \Re\left\{\sum_{\Im[z]>0} \text{Res}\left[\frac{P(z)}{Q(z)} e^{iaz}\right]\right\} \tag{10}$$

$$\int_{-\infty}^{\infty} \frac{P(x)}{Q(x)} \sin(ax) \, dx$$
$$= 2\pi \Im\left\{\sum_{\Im[z]>0} \text{Res}\left[\frac{P(z)}{Q(z)} e^{iaz}\right]\right\}. \tag{11}$$

It is also true that

$$\int_{-\infty}^{\infty} \frac{P(z)}{Q(z)} \ln(az) \, dz = 0. \tag{12}$$

See also CAUCHY INTEGRAL FORMULA, CAUCHY INTEGRAL THEOREM, CONTOUR, CONTOUR INTEGRAL, INSIDE-OUTSIDE THEOREM, JORDAN'S LEMMA, RESIDUE (COMPLEX ANALYSIS), SINE INTEGRAL

References

Krantz, S. G. "Applications to the Calculation of Definite Integrals and Sums." §4.5 in *Handbook of Complex Analysis*. Boston, MA: Birkhäuser, pp. 51–3, 1999.

Morse, P. M. and Feshbach, H. *Methods of Theoretical Physics, Part I.* New York: McGraw-Hill, pp. 353–56, 1953.

Contour Plot

 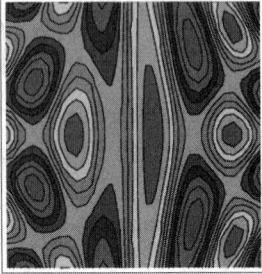

A plot of EQUIPOTENTIAL CURVES. If desired, the regions between contours can be shaded or colored to indicate their magnitude. Contour plots are implemented in *Mathematica* as `ContourPlot[f, {x, xmin, xmin}, {y, ymin, ymax}]`.

See also EQUIPOTENTIAL CURVE, LEVEL CURVE, LEVEL SET, LEVEL SURFACE

Contractant
CONDENSATION

Contracted Cycloid
CURTATE CYCLOID

Contraction (Geometry)

An AFFINE TRANSFORMATION in which the scale is reduced.

See also EXPANSION

Contraction (Graph)

The merging of nodes in a GRAPH by eliminating segments between two nodes.

Contraction (Ideal)

When $f : A \to B$ is a ring HOMOMORPHISM and $\mathfrak{b}$ is an IDEAL in B, then $f - 1(\mathfrak{b})$ is an ideal in A, called the contraction of $\mathfrak{b}$ and sometimes denoted $\mathfrak{b}^c$.

The contraction of a PRIME IDEAL is always prime. For example, consider $f : \mathbb{Z} \to \mathbb{Z}[\sqrt{2}]$. Then the contraction of $\langle \sqrt{2} \rangle$ is the ideal of even integers.

See also ALGEBRAIC NUMBER THEORY, EXTENSION (IDEAL), IDEAL, PRIME IDEAL, RING

References

Atiyah, M. F. and MacDonald, I. G. *Introduction to Commutative Algebra.* Reading, MA: Addison-Wesley, pp. 9–0, 1969.

Contraction (Tensor)

The contraction of a TENSOR is obtained by setting unlike indices equal and summing according to the EINSTEIN SUMMATION convention. Contraction reduces the RANK of a TENSOR by 2. For a second RANK TENSOR,

$$\text{contr}(B_j^{\prime i}) \equiv B_i^{\prime i}$$

$$B_i^{\prime i} = \frac{\partial x_i'}{\partial x_k} \frac{\partial x_l}{\partial x_i'} B_l^k = \frac{\partial x_l}{\partial x_k} B_l^k = \delta_k^l B_l^k = B_k^k.$$

Therefore, the contraction is invariant, and must be a SCALAR. In fact, this SCALAR is known as the TRACE of a MATRIX in MATRIX theory.

References

Arfken, G. "Contraction, Direct Product." §3.2 in *Mathematical Methods for Physicists, 3rd ed.* Orlando, FL: Academic Press, pp. 124–26, 1985.

Jeffreys, H. and Jeffreys, B. S. "Transformation of Coordinates." §3.02 in *Methods of Mathematical Physics, 3rd ed.* Cambridge, England: Cambridge University Press, pp. 86–7, 1988.

Contradiction

A SENTENCE is called a contradiction if its TRUTH TABLE contains only 'F.'

See also CONSISTENCY STRENGTH, CONTINGENCY, TAUTOLOGY, TRUTH TABLE

References

Carnap, R. *Introduction to Symbolic Logic and Its Applications.* New York: Dover, p. 13, 1958.

Contradiction Law

No *A* is not-*A*.

See also NOT

Contravariant Tensor

A contravariant tensor is a TENSOR having specific transformation properties (cf., a COVARIANT TENSOR). To examine the transformation properties of a contravariant tensor, first consider a TENSOR of RANK 1 (a VECTOR)

$$d\mathbf{r} = dx_1\hat{\mathbf{x}}_1 + dx_2\hat{\mathbf{x}}_2 + dx_3\hat{\mathbf{x}}_3, \quad (1)$$

for which

$$dx_i' = \frac{\partial x_i'}{\partial x_j} dx_j. \quad (2)$$

Now let $A_i \equiv dx_i$, then any set of quantities A_j which transform according to

$$A_i' = \frac{\partial x_i'}{\partial x_j} A_j, \quad (3)$$

or, defining

$$a_{ij} \equiv \frac{\partial x_i'}{\partial x_j}, \quad (4)$$

according to

$$A_i' = a_{ij}A_j \quad (5)$$

is a contravariant tensor. Contravariant tensors are indicated with raised indices, i.e., a^μ.

COVARIANT TENSORS are a type of TENSOR with differing transformation properties, denoted a_ν. However, in 3-D CARTESIAN COORDINATES,

$$\frac{\partial x_j}{\partial x_i'} = \frac{\partial x_i'}{\partial x_j} \equiv a_{ij} \quad (6)$$

for $i, j = 1, 2, 3$, meaning that contravariant and covariant tensors are equivalent. The two types of tensors do differ in higher dimensions, however.

Contravariant FOUR-VECTORS satisfy

$$a^\mu = \Lambda_\nu^\mu a^\nu, \quad (7)$$

where Λ is a LORENTZ TENSOR.

To turn a COVARIANT TENSOR a_ν into a contravariant tensor a^μ (INDEX RAISING), use the METRIC TENSOR $g^{\mu\nu}$ to write

$$g^{\mu\nu}a_\nu = a^\mu. \quad (8)$$

Covariant and contravariant indices can be used simultaneously in a MIXED TENSOR.

See also CONTRAVARIANT VECTOR, COVARIANT TENSOR, FOUR-VECTOR, INDEX RAISING, LORENTZ TENSOR, METRIC TENSOR, MIXED TENSOR, TENSOR

References

Arfken, G. "Noncartesian Tensors, Covariant Differentiation." §3.8 in *Mathematical Methods for Physicists, 3rd ed.* Orlando, FL: Academic Press, pp. 158–64, 1985.
Morse, P. M. and Feshbach, H. *Methods of Theoretical Physics, Part I.* New York: McGraw-Hill, pp. 44–6, 1953.

Contravariant Vector

The usual type of VECTOR, which can be viewed as a CONTRAVARIANT TENSOR ("KET") of RANK 1. Contravariant vectors are dual to ONE-FORMS ("BRAS," a.k.a. COVARIANT VECTORS).

See also BRA, COVARIANT VECTOR, CONTRAVARIANT TENSOR, KET, ONE-FORM, VECTOR

Control Theory

The mathematical study of how to manipulate the parameters affecting the behavior of a system to produce the desired or optimal outcome.

See also KALMAN FILTER, LINEAR ALGEBRA, PONTRYAGIN MAXIMUM PRINCIPLE

References

Zabczyk, J. *Mathematical Control Theory: An Introduction.* Boston, MA: Birkhäuser, 1993.

Convective Acceleration

The acceleration of an element of fluid, given by the CONVECTIVE DERIVATIVE of the VELOCITY **v**,

$$\frac{D\mathbf{v}}{Dt} = \frac{\partial \mathbf{v}}{\partial t} + \mathbf{v} \cdot \nabla\mathbf{v},$$

where ∇ is the GRADIENT operator.

See also ACCELERATION, CONVECTIVE DERIVATIVE, CONVECTIVE OPERATOR

References

Batchelor, G K. *An Introduction to Fluid Dynamics.* Cambridge, England: Cambridge University Press, p. 73, 1977.

Convective Derivative

A DERIVATIVE taken with respect to a moving coordinate system, also called a LAGRANGIAN DERIVATIVE. It is given by

$$\frac{D}{Dt} = \frac{\partial}{\partial t} + \mathbf{v} \cdot \nabla,$$

where ∇ is the GRADIENT operator and $\mathbf{v}$ is the VELOCITY of the fluid. This type of derivative is especially useful in the study of fluid mechanics. When applied to $\mathbf{v}$,

$$\frac{D\mathbf{v}}{Dt} = \frac{\partial \mathbf{v}}{\partial t} + (\nabla \times \mathbf{v}) \times \mathbf{v} + \nabla(\tfrac{1}{2}\mathbf{v}^2).$$

See also CONVECTIVE OPERATOR, DERIVATIVE, VELOCITY

References

Batchelor, G K. *An Introduction to Fluid Dynamics.* Cambridge, England: Cambridge University Press, p. 73, 1977.

Convective Operator

Defined for a VECTOR FIELD $\mathbf{A}$ by $(\mathbf{A} \cdot \nabla)$, where ∇ is the GRADIENT operator.

Applied in arbitrary orthogonal 3-D coordinates to a VECTOR FIELD $\mathbf{B}$, the convective operator becomes

$$[(\mathbf{A} \cdot \nabla)\mathbf{B}]_j = \sum_{k=1}^{3}\left[\frac{A_k}{h_k}\frac{\partial B_j}{\partial q_k} + \frac{B_k}{h_k h_j}\left(A_j \frac{\partial h_j}{\partial q_k} - A_k \frac{\partial h_k}{\partial q_j}\right)\right], \quad (1)$$

where the h_is are related to the METRIC TENSORS by $h_i = \sqrt{g_{ii}}$. In CARTESIAN COORDINATES,

$$(\mathbf{A} \cdot \nabla)\mathbf{B} = \left(A_x \frac{\partial B_x}{\partial x} + A_y \frac{\partial B_x}{\partial y} + A_z \frac{\partial B_x}{\partial z}\right)\hat{\mathbf{x}}$$
$$+ \left(A_x \frac{\partial B_y}{\partial x} + A_y \frac{\partial K_y}{\partial y} + A_z \frac{\partial B_y}{\partial z}\right)\hat{\mathbf{y}}$$
$$+ \left(A_x \frac{\partial B_z}{\partial x} + A_y \frac{\partial B_z}{\partial y} + A_z \frac{\partial B_z}{\partial z}\right)\hat{\mathbf{z}}. \quad (2)$$

In CYLINDRICAL COORDINATES,

$$(\mathbf{A} \cdot \nabla)\mathbf{B} = \left(A_r \frac{\partial B_r}{\partial r} + \frac{A_\phi}{r}\frac{\partial B_r}{\partial \phi} + A_z \frac{\partial B_r}{\partial z} - \frac{A_\phi B_\phi}{r}\right)\hat{\mathbf{r}}$$
$$+ \left(A_r \frac{\partial B_\phi}{\partial r} + \frac{A_\phi}{r}\frac{\partial B_\phi}{\partial \phi} + A_z \frac{\partial B_\phi}{\partial z} + \frac{A_\phi B_r}{r}\right)\hat{\boldsymbol{\phi}}$$
$$+ \left(A_r \frac{\partial B_z}{\partial r} + \frac{A_\phi}{r}\frac{\partial B_z}{\partial \phi} + A_z \frac{\partial B_z}{\partial z}\right)\hat{\mathbf{z}}. \quad (3)$$

In SPHERICAL COORDINATES,

$$(\mathbf{A} \cdot \nabla)\mathbf{B}$$
$$= \left(A_r \frac{\partial B_r}{\partial r} + \frac{A_\phi}{r}\frac{\partial B_r}{\partial \theta} + \frac{A_\phi}{r \sin\theta}\frac{\partial B_r}{\partial \phi} - \frac{A_\theta B_\theta + A_\phi B_\phi}{r}\right)\hat{\mathbf{r}}$$
$$+ \left(A_r \frac{\partial B_\theta}{\partial r} + \frac{A_\theta}{r}\frac{\partial B_\theta}{\partial \theta} + \frac{A_\phi}{r \sin\theta}\frac{\partial B_\theta}{\partial \phi} + \frac{A_\theta B_r}{r} - \frac{A_\phi B_\phi \cot\theta}{r}\right)\hat{\boldsymbol{\theta}}$$
$$+ \left(A_r \frac{\partial B_\phi}{\partial r} + \frac{A_\theta}{r}\frac{\partial B_\phi}{\partial \theta} + \frac{A_\phi}{r \sin\theta}\frac{\partial B_\phi}{\partial \phi} + \frac{A_\phi B_r}{r} + \frac{A_\phi B_\theta \cot\theta}{r}\right)\hat{\boldsymbol{\phi}}. \quad (4)$$

See also CONVECTIVE ACCELERATION, CONVECTIVE DERIVATIVE, CURVILINEAR COORDINATES, GRADIENT

Convergence

ALMOST EVERYWHERE CONVERGENCE, CONVERGENCE IMPROVEMENT, CONVERGENCE TESTS, CONVERGENT, CONVERGENT SEQUENCE, CONVERGENT SERIES, POINTWISE CONVERGENCE

Convergence Acceleration

CONVERGENCE IMPROVEMENT

Convergence Improvement

The improvement of the convergence properties of a SERIES, also called CONVERGENCE ACCELERATION, such that a SERIES reaches its limit to within some accuracy with fewer terms than required before. Convergence improvement can be effected by forming a LINEAR COMBINATION with a SERIES whose sum is known. Useful sums include

$$\sum_{n=1}^{\infty}\frac{1}{n(n+1)} = 1 \quad (1)$$

$$\sum_{n=1}^{\infty}\frac{1}{n(n+1)(n+2)} = \frac{1}{4} \quad (2)$$

$$\sum_{n=1}^{\infty}\frac{1}{n(n+1)(n+2)(n+3)} = \frac{1}{18} \quad (3)$$

$$\sum_{n=1}^{\infty}\frac{1}{n(n+1)\cdots(n+p)} = \frac{1}{p \cdot p!}. \quad (4)$$

Kummer's transformation takes a convergent series

$$s = \sum_{k=0}^{\infty} a_k \quad (5)$$

and another convergent series

$$c = \sum_{k=0}^{\infty} c_k \quad (6)$$

with known c such that

$$\lim_{k \to \infty} \frac{a_k}{c_k} = \lambda \neq 0. \tag{7}$$

Then a series with more rapid convergence to the same value is given by

$$s = \lambda c + \sum_{k=0}^{\infty} \left(1 - \lambda \frac{c_k}{a_k}\right) a_k \tag{8}$$

(Abramowitz and Stegun 1972).

The EULER TRANSFORM takes a convergent alternating series

$$\sum_{k=0}^{\infty} (-1)^k a_k = a_0 - a_1 + a_2 \ldots \tag{9}$$

into a series with more rapid convergence to the same value to

$$s = \sum_{k=0}^{\infty} \frac{(-1)^k \Delta^k a_0}{2^{k+1}}, \tag{10}$$

where

$$\Delta^k a_0 = \sum_{m=0}^{k} \equiv (-1)^m \binom{k}{m} a_{k-m} \tag{11}$$

(Abramowitz and Stegun 1972; Beeler *et al.* 1972).

Given a series OF THE FORM

$$S = \sum_{n=1}^{\infty} f\left(\frac{1}{n}\right) \tag{12}$$

where $f(z)$ is an ANALYTIC at 0 and on the closed unit DISK, and

$$f(z)|_{z \to 0} = \mathcal{O}(z^2), \tag{13}$$

then the series can be rearranged to

$$S = \sum_{n=1}^{\infty} \sum_{m=2}^{\infty} f_m \left(\frac{1}{n}\right)^m$$

$$= \sum_{m=2}^{\infty} \sum_{n=1}^{\infty} f_m \left(\frac{1}{n}\right)^m = \sum_{m=2}^{\infty} f_m \zeta(m), \tag{14}$$

where

$$f(z) = \sum_{m=2}^{\infty} f_m z^m \tag{15}$$

is the MACLAURIN SERIES of f and ζ is the RIEMANN ZETA FUNCTION (Flajolet and Vardi 1996). The transformed series exhibits geometric convergence. Similarly, if $f(z)$ is ANALYTIC in $|z| \leq 1/n_0$ for some POSITIVE INTEGER n_0, then

$$S = \sum_{n=1}^{n_0 - 1} f\left(\frac{1}{n}\right)$$

$$+ \sum_{m=2}^{\infty} f_m \left[\zeta(m) - \frac{1}{1^m} - \ldots - \frac{1}{(n_0 - 1)^m}\right], \tag{16}$$

which converges geometrically (Flajolet and Vardi 1996). (16) can also be used to further accelerate the convergence of series (14).

See also EULER TRANSFORM, WILF-ZEILBERGER PAIR

References

Abramowitz, M. and Stegun, C. A. (Eds.). *Handbook of Mathematical Functions with Formulas, Graphs, and Mathematical Tables, 9th printing.* New York: Dover, p. 16, 1972.

Arfken, G. *Mathematical Methods for Physicists, 3rd ed.* Orlando, FL: Academic Press, pp. 288–89, 1985.

Beeler *et al.* Item 120 in Beeler, M.; Gosper, R. W.; and Schroeppel, R. *HAKMEM.* Cambridge, MA: MIT Artificial Intelligence Laboratory, Memo AIM-239, p. 55, Feb. 1972.

Flajolet, P. and Vardi, I. "Zeta Function Expansions of Classical Constants." Unpublished manuscript. 1996. http://pauillac.inria.fr/algo/flajolet/Publications/landau.ps.

Convergence Tests

A test to determine if a given SERIES CONVERGES or DIVERGES.

See also ABEL'S UNIFORM CONVERGENCE TEST, BERTRAND'S TEST, D'ALEMBERT RATIO TEST, DIVERGENCE TESTS, ERMAKOFF'S TEST, GAUSS'S TEST, INTEGRAL TEST, KUMMER'S TEST, LIMIT COMPARISON TEST, LIMIT TEST, RAABE'S TEST, RADIUS OF CONVERGENCE, RATIO TEST, RIEMANN SERIES THEOREM, ROOT TEST

References

Arfken, G. "Convergence Tests." §5.2 in *Mathematical Methods for Physicists, 3rd ed.* Orlando, FL: Academic Press, pp. 280–93, 1985.

Bromwich, T. J. I'a and MacRobert, T. M. *An Introduction to the Theory of Infinite Series, 3rd ed.* New York: Chelsea, pp. 55–7, 1991.

Convergent

The RATIONAL NUMBER obtained by keeping only a limited number of terms in a CONTINUED FRACTION is called a convergent. For example, in the SIMPLE CONTINUED FRACTION for the GOLDEN RATIO,

$$\phi = 1 + \cfrac{1}{1 + \cfrac{1}{1 + \cfrac{1}{1 + \cdots}}},$$

the convergents are

$$1, \; 1 + \frac{1}{1} = 2, \; 1 + \frac{1}{1 + \frac{1}{1}} = \frac{3}{2}, \ldots$$

The word convergent is also used to describe a CONVERGENT SEQUENCE or CONVERGENT SERIES.

See also CONTINUED FRACTION, CONVERGENT SEQUENCE, CONVERGENT SERIES, PARTIAL QUOTIENT, SIMPLE CONTINUED FRACTION

Convergent Sequence

A SEQUENCE S_n converges to the limit S

$$\lim_{n \to \infty} S_n = S$$

if, for any $\epsilon > 0$, there exists an N such that $|S_n - S| < \epsilon$ for $n > N$. If S_n does not converge, it is said to DIVERGE. This condition can also be written as

$$\overline{\lim_{n \to \infty}} S_n = \underline{\lim_{n \to \infty}} S_n = S.$$

Every bounded MONOTONIC SEQUENCE converges. Every unbounded SEQUENCE diverges.

See also CONDITIONAL CONVERGENCE, STRONG CONVERGENCE, WEAK CONVERGENCE

References
Jeffreys, H. and Jeffreys, B. S. "Bounded, Unbounded, Convergent, Oscillatory." §1.041 in *Methods of Mathematical Physics, 3rd ed.* Cambridge, England: Cambridge University Press, pp. 11–2, 1988.

Convergent Series

The infinite SERIES $\Sigma_{n=1}^{\infty} a_n$ is convergent if the SEQUENCE of partial sums

$$S_n = \sum_{k=1}^{n} a_k$$

is convergent. Conversely, a SERIES is divergent if the SEQUENCE of partial sums is divergent. If Σu_k and Σv_k are convergent SERIES, then $\Sigma(u_k + v_k)$ and $\Sigma(u_k - v_k)$ are convergent. If $c \neq 0$, then Σu_k and $c\Sigma u_k$ both converge or both diverge. Convergence and divergence are unaffected by deleting a finite number of terms from the beginning of a series. Constant terms in the denominator of a sequence can usually be deleted without affecting convergence. All but the highest POWER terms in POLYNOMIALS can usually be deleted in both NUMERATOR and DENOMINATOR of a SERIES without affecting convergence. If a SERIES converges absolutely, then it converges.

See also CONVERGENCE TESTS, RADIUS OF CONVERGENCE

References
Bromwich, T. J. I'a. and MacRobert, T. M. *An Introduction to the Theory of Infinite Series, 3rd ed.* New York: Chelsea, 1991.

Conversion Period

The period of time between INTEREST payments.

See also COMPOUND INTEREST, INTEREST, SIMPLE INTEREST

Convex

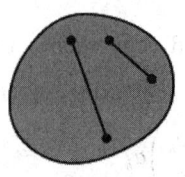 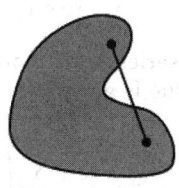

convex   *concave*

A SET in EUCLIDEAN SPACE $\mathbb{R}^d$ is a CONVEX SET if it contains *all* the LINE SEGMENTS connecting any pair of its points. If the SET does not contain all the LINE SEGMENTS, it is called CONCAVE.

See also CONNECTED SET, CONVEX FUNCTION, CONVEX HULL, CONVEX OPTIMIZATION THEORY, CONVEX POLYGON, CONVEX SET, DELAUNAY TRIANGULATION, MINKOWSKI CONVEX BODY THEOREM, SIMPLY CONNECTED

References
Benson, R. V. *Euclidean Geometry and Convexity.* New York: McGraw-Hill, 1966.
Busemann, H. *Convex Surfaces.* New York: Interscience, 1958.
Croft, H. T.; Falconer, K. J.; and Guy, R. K. "Convexity." Ch. A in *Unsolved Problems in Geometry.* New York: Springer-Verlag, pp. 6–7, 1994.
Eggleston, H. G. *Problems in Euclidean Space: Applications of Convexity.* New York: Pergamon Press, 1957.
Gruber, P. M. "Seven Small Pearls from Convexity." *Math. Intell.* **5**, 16–9, 1983.
Gruber, P. M. "Aspects of Convexity and Its Applications." *Expos. Math.* **2**, 47–3, 1984.
Guggenheimer, H. *Applicable Geometry--Global and Local Convexity.* New York: Krieger, 1977.
Kelly, P. J. and Weiss, M. L. *Geometry and Convexity: A Study of Mathematical Methods.* New York: Wiley, 1979.
Webster, R. *Convexity.* Oxford, England: Oxford University Press, 1995.

Convex Function

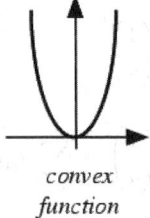

convex function

A function whose value at the MIDPOINT of every INTERVAL in its DOMAIN does not exceed the AVERAGE of its values at the ends of the INTERVAL. In other words, a function $f(x)$ is convex on an INTERVAL $[a, b]$ if for any two points x_1 and x_2 in $[a, b]$,

$$f[\tfrac{1}{2}(x_1 + x_2)] \leq \tfrac{1}{2}[f(x_1) + f(x_2)]$$

(Gradshteyn and Ryzhik 2000). If $f(x)$ has a second DERIVATIVE in $[a, b]$, then a NECESSARY and SUFFI-

CIENT condition for it to be convex on that INTERVAL is that the second DERIVATIVE $f''(x) > 0$ for all x in $[a, b]$. If the inequality above is STRICT for all x_1 and x_2, then $f(x)$ is called strictly convex. Examples of convex functions include x^p for $p \geq 1$, $x \ln x$ for $x > 0$, and $|x|$ for all x. If the sign of the inequality is reversed, the function is called CONCAVE.

See also CONCAVE FUNCTION, LOGARITHMICALLY CONVEX FUNCTION

References

Eggleton, R. B. and Guy, R. K. "Catalan Strikes Again! How Likely is a Function to be Convex?" *Math. Mag.* **61**, 211–19, 1988.

Gradshteyn, I. S. and Ryzhik, I. M. *Tables of Integrals, Series, and Products, 6th ed.* San Diego, CA: Academic Press, p. 1132, 2000.

Webster, R. *Convexity.* Oxford, England: Oxford University Press, 1995.

Convex Hull

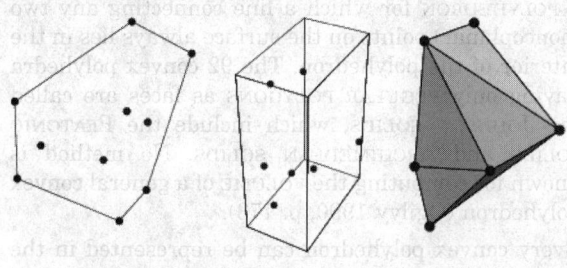

The convex hull of a set of points S in n-D is the INTERSECTION of all convex sets containing S. For N points $p_1, \ldots, p_N$, the convex hull C is then given by the expression

$$C \equiv \left\{ \sum_{j=1}^{N} \lambda_j p_j : \ \lambda_j \geq 0 \ \text{for all} \ j \ \text{and} \ \sum_{j=1}^{N} \lambda_j = 1 \right\}.$$

Computing the convex hull is a problem in COMPUTATIONAL GEOMETRY. The *indices* of the points specifying the convex hull of a set of points in two dimensions is given by the command Convex-Hull[*pts*] in the *Mathematica* add-on package DiscreteMath'ComputationalGeometry' (which can be loaded with the command < <DiscreteMath'). Future versions of *Mathematica* will support n-dimensional convex hulls.

In d dimensions, the "gift wrapping" algorithm, which has complexity $\mathcal{O}(n^{\lfloor d/2 \rfloor + 1})$, where $\lfloor x \rfloor$ is the FLOOR FUNCTION, can be used (Skiena 1997, p. 352). In 2- and 3-D, however, specialized algorithms exist with complexity $\mathcal{O}(n \ln n)$ (Skiena 1997, pp. 351–52). Yao (1981) has proved that any decision-tree algorithm for the 2-D case requires quadratic or higher-order tests, and that any algorithm using quadratic

tests (which includes all currently known algorithms) cannot be done with lower complexity than $\mathcal{O}(n \ln n)$. However, it remains an open problem whether better complexity can be obtained using higher-order polynomial tests (Yao 1981). O'Rourke (1997) gives a robust 2-D implementation as well as an $\mathcal{O}(n^2)$ 3-D implementation. Qhull works efficiently in 2 to 8 dimensions (Barber *et al.* 1997).

The DUAL POLYHEDRON of any non-convex UNIFORM POLYHEDRON is a stellated form of the CONVEX HULL of the given polyhedron (Wenninger 1983, pp. 3– and 40).

See also CARATHÉODORY'S FUNDAMENTAL THEOREM, COMPUTATIONAL GEOMETRY, CROSS POLYTOPE, GROEMER PACKING, GROEMER THEOREM, HAPPY END PROBLEM, RADON'S THEOREM, SAUSAGE CONJECTURE, SPAN (GEOMETRY), SYLVESTER'S FOUR-POINT PROBLEM, TEMPERATURE

References

Barber, C.; Dobkin, D.; and Huhdanpaa, H. "The Quickhull Algorithm for Convex Hulls." *ACM Trans. Mathematical Software* **22**, 469–83, 1997.

Croft, H. T.; Falconer, K. J.; and Guy, R. K. *Unsolved Problems in Geometry.* New York: Springer-Verlag, p. 8, 1991.

de Berg, M.; van Kreveld, M.; Overmans, M.; and Schwarzkopf, O. "Convex Hulls: Mixing Things." Ch. 11 in *Computational Geometry: Algorithms and Applications, 2nd rev. ed.* Berlin: Springer-Verlag, pp. 235–50, 2000.

Edelsbrunner, H. and Mücke, E. P. "Three-Dimensional Alpha Shapes." *ACM Trans. Graphics* **13**, 43–2, 1994.

O'Rourke, J. *Computational Geometry in C, 2nd ed.* Cambridge, England: Cambridge University Press, 1998.

Preparata, F. R. and Shamos, M. I. *Computational Geometry: An Introduction.* New York: Springer-Verlag, 1985.

Santaló, L. A. *Integral Geometry and Geometric Probability.* Reading, MA: Addison-Wesley, 1976.

Seidel, R. "Convex Hull Computations." Ch. 19 in *Handbook of Discrete and Computational Geometry* (Ed. J. E. Goodman and J. O'Rourke). Boca Raton, FL: CRC Press, pp. 361–75, 1997.

Skiena, S. S. "Convex Hull." §8.6.2 in *The Algorithm Design Manual.* New York: Springer-Verlag, pp. 351–54, 1997.

Weisstein, E. W. "Convex Hull 3D." MATHEMATICA NOTEBOOK CONVEXHULL.M.

Wenninger, M. J. *Dual Models.* Cambridge, England: Cambridge University Press, 1983.

Yao, A. C.-C. "A Lower Bound to Finding Convex Hulls." *J. ACM* **28**, 780–87, 1981.

Convex Optimization Theory

The problem of maximizing a linear function over a CONVEX POLYHEDRON, also known as OPERATIONS RESEARCH or OPTIMIZATION THEORY. The general problem of convex optimization is to find the minimum of a convex (or quasiconvex) function f on a FINITE-dimensional convex body A. Methods of solution include Levin's algorithm and the method of circumscribed ELLIPSOIDS, also called the Nemirovsky-Yudin-Shor method.

References

Tokhomirov, V. M. "The Evolution of Methods of Convex Optimization." *Amer. Math. Monthly* **103**, 65–1, 1996.

Convex Polygon

A POLYGON is CONVEX if it contains *all* the LINE SEGMENTS connecting any pair of its points. Let $f(n)$ be the smallest number such that when W is a set of more than $f(n)$ points in GENERAL POSITION (with no three points COLLINEAR) in the plane, all of the VERTICES of some convex n-gon are contained in W. The answers for $n = 2$, 3, and 4 are 2, 4, and 8. It is conjectured that $f(n) = 2^{n-2}$, but only proven that

$$2^{n-2} \leq f(n) \leq \binom{2n-4}{n-2},$$

where $\binom{n}{k}$ is a BINOMIAL COEFFICIENT.

See also CONVEX POLYOMINO, CONVEX POLYHEDRON, CONVEX POLYOMINO, CONVEX POLYTOPE, HAPPY END PROBLEM, LATTICE POLYGON, POLYGON

Convex Polyhedron

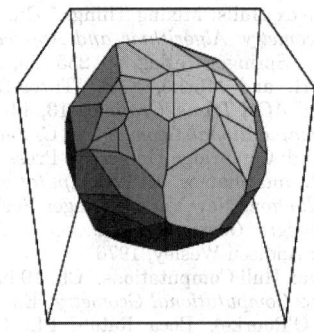

A convex polyhedron can be defined algebraically as the set of solutions to a system of linear inequalities

$$\mathsf{m}\mathbf{x} \leq \mathbf{b}, \tag{1}$$

where m is a real $s \times 3$ MATRIX and $\mathbf{b}$ is a real s-VECTOR. Although usage varies, most authors additionally require that a solution be bounded for it to define a CONVEX POLYHEDRON. An example of a convex polyhedron is illustrated above. The more simple DODECAHEDRON is given by a system with $s = 12$. Explicit examples are given in the following table.

convex polyhedron	s	m			$\mathbf{b}$
TETRAHEDRON	4	1	1	1	2
		1	−1	−1	0
		−1	1	−1	0
		−1	−1	1	0
CUBE	6	1	0	0	1
		−1	0	0	1
		0	1	0	1
		0	−1	0	1
		0	0	1	1
		0	0	−1	1
OCTAHEDRON	8	1	1	1	1
		1	1	−1	1
		1	−1	1	1
		1	−1	−1	1
		−1	1	1	1
		−1	1	−1	1
		−1	−1	1	1
		−1	−1	−1	1

In general, given the MATRICES, the VERTICES (and FACES) can be found using an algorithmic procedure known as VERTEX ENUMERATION.

Geometrically, a convex polyhedron can be defined as a POLYHEDRON for which a line connecting any two (noncoplanar) points on the surface always lies in the interior of the polyhedron. The 92 convex polyhedra having only REGULAR POLYGONS as faces are called the JOHNSON SOLIDS, which include the PLATONIC SOLIDS and ARCHIMEDEAN SOLIDS. No method is known for computing the VOLUME of a general convex polyhedron (Ogilvy 1990, p. 173).

Every convex polyhedron can be represented in the plane or on the surface of a sphere by a 3-connected PLANAR GRAPH (called a POLYHEDRAL GRAPH). Conversely, by a theorem of Steinitz as restated by Grünbaum, every 3-connected PLANAR GRAPH can be realized as a convex polyhedron (Duijvestijn and Federico 1981). The numbers of vertices V, edges E, and faces F of a convex polyhedron are related by the POLYHEDRAL FORMULA

$$V + F - E = 2.$$

See also ARCHIMEDEAN SOLID, CONVEX POLYGON, CONVEX POLYOMINO, CONVEX POLYTOPE, DELTAHEDRON, JOHNSON SOLID, KEPLER-POINSOT SOLID, PLATONIC SOLID, POLYHEDRAL FORMULA, POLYHEDRAL GRAPH, POLYHEDRON, REGULAR POLYHEDRON, VERTEX ENUMERATION

References

Duijvestijn, A. J. W. and Federico, P. J. "The Number of Polyhedral (3-Connected Planar) Graphs." *Math. Comput.* **37**, 523–32, 1981.

Ogilvy, C. S. *Excursions in Geometry.* New York: Dover, 1990.

Lyusternik, L. A. *Convex Figures and Polyhedra.* New York: Dover, 1963.

Yaglom, I. M. and Boltianskii, V. G. *Convex Figures*. New York: Holt, Rinehart and Winston, 1961.

Convex Polyomino

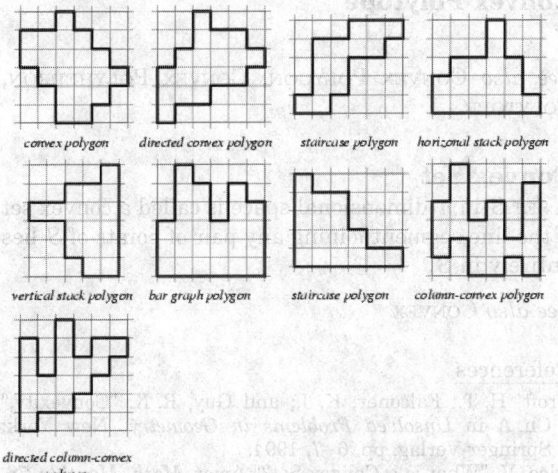

convex polygon   directed convex polygon   staircase polygon   horizontal stack polygon

vertical stack polygon   bar graph polygon   staircase polygon   column-convex polygon

directed column-convex polygon

A convex polyomino (sometimes called a "convex polygon") is a polyomino whose PERIMETER is equal to that of its minimal bounding box (Bousquet-Mélou *et al.* 1999). Furthermore, if it contains at least one corner of its minimal bounding box, it is said to be a DIRECTED CONVEX POLYOMINO. A COLUMN-CONVEX POLYOMINO is a self-avoiding polyomino such that the intersection of any vertical line with the polyomino has at most two connected components, and a ROW-CONVEX POLYOMINO is similarly defined.

The anisotropic perimeter and area generating function

$$G(x, y, q) = \sum_{m \geq 1} \sum_{n \geq 1} \sum_{a \geq 1} C(m, n, a) x^m y^n q^a, \quad (1)$$

where $C(m, n, a)$ is the number of polygons with $2m$ horizonal bonds, $2n$ vertical bonds, and area a is given by

$$G(x, y, q) = 2 \sum_{m \geq 1} \frac{y^{m+2}}{(xq)_m^2 N(xq^{m-1}) N(xq^m)}$$
$$\times [T_{m+1} S(xq^m) - y T_m S(xq^{m+1})]^2$$
$$+ \sum_{m \geq 1} \frac{xy^m q^m (T_m)^2}{(xq)_{m-1}(xq)m}, \quad (2)$$

where

$$N(x) = \sum_{n \geq 0} \frac{(-1)^n x^n q^{\binom{n+1}{2}}}{(q)_n (yq)_n} \quad (3)$$

$$S(x) = \sum_{n \geq 1} \left[\frac{x^n q^n}{(yq)_n} \sum_{j=0}^{n-1} \frac{(-1)^j q^{\binom{j}{2}}}{(q)_j (yq^{j+1})_{n-j}} \right] \quad (4)$$

and $T_n(x)$ is the polynomial RECURRENCE RELATION

$$T_n(x) = 2 T_{n-1}(x) + (xq^{n-1} - 1) T_{n-2}(x) \quad (5)$$

with $T_0(x) = 1$ and $T_1(x) = 1$ (Bousquet-Mélou 1992b). The first few of these polynomials are given by

$$T_2(x) = 1 + qx$$

$$T_3(x) = 1 + (2q + q^2)x$$

$$T_4(x) = 1 + (3q + 2q^2 + q^3)x + q^4 x^2$$

$$T_5(x) = 1 + (4q + 3q^2 + 2q^3 + q^4)x + (2q^4 + 2q^5 + q^6)x^2.$$

Expanding the generating function shows that the number of convex polyominoes having PERIMETER $2n + 8$ is given by

$$(2n + 11)4^n - 4(2n + 1)\binom{2n}{n}, \quad (6)$$

where $\binom{n}{k}$ is a BINOMIAL COEFFICIENT (Delest and Viennot 1984, Bousquet-Mélou 1992).

This function has been computed exactly for the column-convex and directed column-convex polyminoes (Bousquet-Mélou 1996, Bousquet-Mélou *et al.* 1999). $G(1, 1, q)$ is a Q-SERIES, but becomes algebraic for column-convex polyominoes. However, $G(x, y, q)$ for column-convex polyominoes again involves Q-SERIES (Temperley 1956, Bousquet-Mélou *et al.* 1999).

$G(x, y) = G(x, y, 1)$ is an algebraic function of x and y (called the "fugacities") given by

$$G(x, y) = \sum_{x \geq 1} \sum_{y \geq 1} C(m, n) x^m y^n$$
$$= \frac{R(x, y)xy}{[\Delta(x, y)]^2} - \frac{4x^2 y^2}{\Delta^{3/2}}, \quad (7)$$

where

$$R(x, y) = 1 - 3x - 3y + 3x^2 + 3y^2 + 5xy - x^3 - y^3 - x^2 y$$
$$- xy^2 - xy(x - y)^2 \quad (8)$$

$$\Delta(x, y) = 1 - 2x - 2y - 2xy + x^2 + y^2$$
$$= (1 - y)^2 \left[1 - \frac{x(2 + 2y - x)}{(1 - y)^2} \right] \quad (9)$$

(Lin and Chang 1988, Bousquet-Mélou 1992). This can be solved to explicitly give

$$C(m, n) = \frac{mn - 1}{m + n - 2}\binom{2m + 2n - 4}{2m - 2}$$
$$- 2(m + n - 2)\binom{m + n - 3}{m - 1}\binom{m + n - 3}{n - 1} \quad (10)$$

(Gessel 1990, Bousquet-Mélou 1992).

$G(x, y)$ satisfies the inversion relation

$$G(x, y) + y^3 G(x/y, 1/y) = xy - x^3 y \frac{\partial}{\partial x} \frac{1 - x + y}{\Delta(x, y)}, \quad (11)$$

where

$$\Delta(x, y) = 1 - 2x - 2y - 2xy + x^2 + y^2$$

$$= (1 - y)^2 \left[1 - \frac{x(2 + 2y - x)}{(1 - y)^2} \right] \qquad (12)$$

(Lin and Chang 1988, Bousquet-Mélou *et al.* 1999).

The half-vertical perimeter and area generating function for column-convex polyominos of width 3 is given by the special case

$$H_3(y, q) = \frac{yq^3}{(1 - yq)^4(1 - yq^2)^2(1 - yq^3)}$$

$$\times (y^6q^8 + 4y^5q^7 + 2y^5q^6 + y^4q^6 - y^4q^4$$

$$- 4y^3q^5 - 6y^3q^4 - 4y^3q^3 - y^2q^4 + y^2q^2 + 2yq^2 + 4yq + 1) \qquad (13)$$

of the general rational function (Bousquet-Mélou *et al.* 1999), which satisfies the reciprocity relation

$$H_3(1/y, 1/q) = -\frac{1}{yq^3} H_3(y, q). \qquad (14)$$

The anisotropic area and perimeter generating function $G(x, y, q)$ and partial generating functions $H_m(y, q)$, connected by

$$G(x, y, q) = \sum_{m \geq 1} H_m(y, q)x^m, \qquad (15)$$

satisfy the self-reciprocity and inversion relations

$$H_m(1/y, 1/q) = -\frac{1}{yq^m} H_m(y, q) \qquad (16)$$

and

$$G(x, y, q) + yG(xq, 1/y, 1/q) = 0$$

(Bousquet-Mélou *et al.* 1999).

See also COLUMN-CONVEX POLYOMINO, DIRECTED CONVEX POLYOMINO, POLYOMINO

References

Bousquet-Mélou, M. "Convex Polyominoes and Heaps of Segments." *J. Phys. A: Math. Gen.* **25**, 1925–934, 1992a.
Bousquet-Mélou, M. "Convex Polyominoes and Algebraic Languages." *J. Phys. A: Math. Gen.* **25**, 1935–944, 1992b.
Bousquet-Mélou, M. "A Method for Enumeration of Various Classes of Column-Convex Polygons." *Disc. Math.* **154**, 1–5, 1996.
Bousquet-Mélou, M.; Guttmann, A. J.; Orrick, W. P.; and Rechnitzer, A. Inversion Relations, Reciprocity and Polyominoes. 23 Aug 1999. http://xxx.lanl.gov/abs/math.CO/9908123/.
Delest, M.-P. and Viennot, G. "Algebraic Languages and Polyominoes [sic] Enumeration." *Theoret. Comput. Sci.* **34**, 169–06, 1984.
Gessel, I. M. "On the Number of Convex Polyominoes." Preprint. 1990.
Lin, K. Y. and Chang, S. J. "Rigorous Results for the Number of Convex Polygons on the Square and Honeycomb Lattices." *J. Phys. A: Math. Gen.* **21**, 2635–642, 1988.

Temperley, H. N. V. "Combinatorial Problems Suggested by the Statistical Mechanics of Domains and of Rubber-Like Molecules." *Phys. Rev.* **103**, 1–6, 1956.

Convex Polytope

See also CONVEX POLYGON, CONVEX POLYHEDRON, POLYTOPE

Convex Set

A SET S in n-dimensional space is called a convex set if the line segment joining any pair of points of S lies entirely in S.

See also CONVEX

References

Croft, H. T.; Falconer, K. J.; and Guy, R. K. "Convexity." Ch. A in *Unsolved Problems in Geometry.* New York: Springer-Verlag, pp. 6–7, 1994.
Klee, V. "What is a Convex Set?" *Amer. Math. Monthly* **78**, 616–31, 1971.
Lay, S. R. *Convex Sets and Their Applications.* New York: Wiley, 1979.
Valentine, F. A. *Convex Sets.* New York: McGraw-Hill, 1964.

Convolution

A convolution is an integral which expresses the amount of overlap of one function $g(t)$ as it is shifted over another function $f(t)$. It therefore "blends" one function with another. For example, in synthesis imaging, the measured dirty map is a convolution of the "true" CLEAN map with the dirty beam (the FOURIER TRANSFORM of the sampling distribution). The convolution is sometimes also known by its German name, *faltung* ("folding").

A convolution over a finite range $[0, t]$ is given by

$$f(t) * g(t) \equiv \int_0^t f(\tau)g(t - \tau) \, d\tau, \qquad (1)$$

where the symbol $f * g$ (occasionally also written as $f \otimes g$) denotes convolution of f and g. Convolution is more often taken over an infinite range,

$$f(t) * g(t) \equiv \int_{-\infty}^{\infty} f(\tau)g(t - \tau) \, d\tau$$

$$= \int_{-\infty}^{\infty} g(\tau)f(t - \tau) \, d\tau. \qquad (2)$$

Let f, g, and h be arbitrary functions and a a constant. Convolution the satisfies the following properties,

$$f * g = g * f \qquad (3)$$

$$f * (g * h) = (f * g) * h \qquad (4)$$

$$f * (g + h) = (f * g) + (f * h) \qquad (5)$$

(Bracewell 1999, p. 27), as well as

$$a(f * g) = (af) * g = f * (ag). \tag{6}$$

Taking the DERIVATIVE of a convolution gives

$$\frac{d}{dx}(f * g) = \frac{df}{dx} * g = f * \frac{dg}{dx}. \tag{7}$$

The AREA under a convolution is the product of areas under the factors,

$$\int_{-\infty}^{\infty} (f * g)\, dx = \int_{-\infty}^{\infty} \left[\int_{-\infty}^{\infty} f(u) g(x - u)\, du \right] dx$$

$$= \int_{-\infty}^{\infty} f(u) \left[\int_{-\infty}^{\infty} g(x - u)\, dx \right] du$$

$$= \left[\int_{-\infty}^{\infty} f(u)\, du \right] \left[\int_{-\infty}^{\infty} g(x)\, dx \right]. \tag{8}$$

The horizontal CENTROIDS add

$$\langle x(f * g) \rangle = \langle xf \rangle + \langle xg \rangle, \tag{9}$$

as do the VARIANCES

$$\langle x^2(f * g) \rangle = \langle x^2 f \rangle + \langle x^2 g \rangle, \tag{10}$$

where

$$\langle x^n f \rangle \equiv \frac{\displaystyle\int_{-\infty}^{\infty} x^n f(x)\, dx}{\displaystyle\int_{-\infty}^{\infty} f(x)\, dx}. \tag{11}$$

There is also a definition of the convolution which arises in probability theory and is given by

$$F(t) * G(t) = \int F(t - x)\, dG(x), \tag{12}$$

where $\int F(t - x)\, dG(x)$ is a STIELTJES INTEGRAL.

See also AUTOCORRELATION, CAUCHY PRODUCT, CONVOLUTION THEOREM, CROSS-CORRELATION, WIENER-KHINTCHINE THEOREM

References

Bracewell, R. "Convolution" and "Two-Dimensional Convolution." Ch. 3 in *The Fourier Transform and Its Applications, 3rd ed.* New York: McGraw-Hill, pp. 25–0 and 243–44, 1999.

Hirschman, I. I. and Widder, D. V. *The Convolution Transform.* Princeton, NJ: Princeton University Press, 1955.

Morse, P. M. and Feshbach, H. *Methods of Theoretical Physics, Part I.* New York: McGraw-Hill, pp. 464–65, 1953.

Press, W. H.; Flannery, B. P.; Teukolsky, S. A.; and Vetterling, W. T. "Convolution and Deconvolution Using the FFT." §13.1 in *Numerical Recipes in FORTRAN: The Art of Scientific Computing, 2nd ed.* Cambridge, England: Cambridge University Press, pp. 531–37, 1992.

Weisstein, E. W. "Books about Convolution." http://www.treasure-troves.com/books/Convolution.html.

Convolution Theorem

Let $f(t)$ and $g(t)$ be arbitrary functions of time t with FOURIER TRANSFORMS. Take

$$f(t) = \mathscr{F}^{-1}[F(v)] = \int_{-\infty}^{\infty} F(v) e^{2\pi i v t}\, dv \tag{1}$$

$$g(t) = \mathscr{F}^{-1}[G(v)] = \int_{-\infty}^{\infty} G(v) e^{2\pi i v t}\, dv, \tag{2}$$

where $\mathscr{F}^{-1}$ denotes the inverse FOURIER TRANSFORM (where the transform pair is defined to have constants $A = 1$ and $B = -2\pi$). Then the CONVOLUTION is

$$f * g \equiv \int_{-\infty}^{\infty} g(t') f(t - t')\, dt'$$

$$= \int_{-\infty}^{\infty} g(t') \left[\int_{-\infty}^{\infty} F(v) e^{2\pi i v(t - t')}\, dv \right] dt'. \tag{3}$$

Interchange the order of integration,

$$f * g = \int_{-\infty}^{\infty} F(v) \left[\int_{-\infty}^{\infty} g(t') e^{-2\pi i v t'}\, dt' \right] e^{2\pi i v t}\, dv$$

$$= \int_{-\infty}^{\infty} F(v) G(v) e^{2\pi i v t}\, dv = \mathscr{F}^{-1}[F(v)G(v)]. \tag{4}$$

So, applying a FOURIER TRANSFORM to each side, we have

$$\mathscr{F}[f * g] = \mathscr{F}[f] \mathscr{F}[g]. \tag{5}$$

The convolution theorem also takes the alternate forms

$$\mathscr{F}[fg] = \mathscr{F}[f] * \mathscr{F}[g] \tag{6}$$

$$\mathscr{F}^{-1}(\mathscr{F}[f] \mathscr{F}[g]) = f * g \tag{7}$$

$$\mathscr{F}^{-1}(\mathscr{F}[f] * \mathscr{F}[g]) = fg. \tag{8}$$

See also AUTOCORRELATION, CONVOLUTION, FOURIER TRANSFORM, WIENER-KHINTCHINE THEOREM

References

Arfken, G. "Convolution Theorem." §15.5 in *Mathematical Methods for Physicists, 3rd ed.* Orlando, FL: Academic Press, pp. 810–14, 1985.

Bracewell, R. "Convolution Theorem." *The Fourier Transform and Its Applications, 3rd ed.* New York: McGraw-Hill, pp. 108–12, 1999.

Conway Groups

The AUTOMORPHISM GROUP Co_1 of the LEECH LATTICE modulo a center of order two is called "the" Conway group. There are 15 exceptional CONJUGACY CLASSES of the Conway group. This group, combined with the GROUPS Co_2 and Co_3 obtained similarly from the LEECH LATTICE by stabilization of the 1-D and 2-D sublattices, are collectively called Conway groups. The Conway groups are SPORADIC GROUPS.

See also LEECH LATTICE, SPORADIC GROUP

References

Wilson, R. A. "ATLAS of Finite Group Representation." http://for.mat.bham.ac.uk/atlas/html/contents.html#spo.

Conway Notation

CONWAY'S KNOT NOTATION, CONWAY POLYHEDRON NOTATION

Conway Polyhedron Notation

A NOTATION for POLYHEDRA which begins by specifying a "seed" polyhedron using a capital letter. The PLATONIC SOLIDS are denoted T (TETRAHEDRON), O (OCTAHEDRON), C (CUBE), I (ICOSAHEDRON), and D (DODECAHEDRON), according to their first letter. Other polyhedra include the PRISMS, Pn, ANTIPRISMS, An, and PYRAMIDS, Yn, where $n \geq 3$ specifies the number of sides of the polyhedron's base.

Operations to be performed on the polyhedron are then specified with lower-case letters preceding the capital letter.

See also POLYHEDRON, SCHLÄFLI SYMBOL, WYTHOFF SYMBOL

References

Hart, G. "Conway Notation for Polyhedra." http://www.geor-gehart.com/virtual-polyhedra/conway_notation.html.

Conway Polynomial

ALEXANDER POLYNOMIAL

Conway Puzzle

Construct a $5 \times 5 \times 5$ cube from thirteen $1 \times 2 \times 4$ blocks, one $2 \times 2 \times 2$ block, one $1 \times 2 \times 2$, and three $1 \times 1 \times 3$ blocks.

See also BOX-PACKING THEOREM, CUBE DISSECTION, DE BRUIJN'S THEOREM, KLARNER'S THEOREM, POLYCUBE, SLOTHOUBER-GRAATSMA PUZZLE

References

Honsberger, R. *Mathematical Gems II.* Washington, DC: Math. Assoc. Amer., pp. 77–0, 1976.

Conway Sequence

The LOOK AND SAY SEQUENCE generated from a starting DIGIT of 3, as given by Vardi (1991).

See also CONWAY'S CONSTANT, COSMOLOGICAL THEOREM, LOOK AND SAY SEQUENCE

References

Vardi, I. *Computational Recreations in Mathematica.* Reading, MA: Addison-Wesley, pp. 13–4, 1991.

Conway Sphere

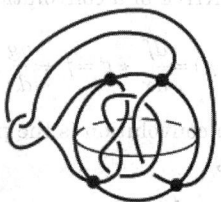

A sphere with four punctures occurring where a KNOT passes through the surface.

References

Adams, C. C. *The Knot Book: An Elementary Introduction to the Mathematical Theory of Knots.* New York: W. H. Freeman, p. 94, 1994.

Conway-Alexander Polynomial

ALEXANDER POLYNOMIAL

Conway's Constant

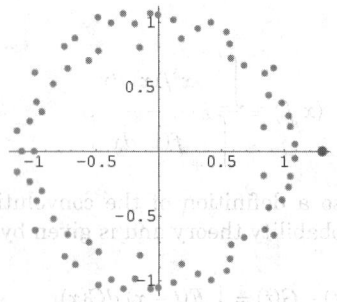

The constant

$$\lambda = 1.303577269034296\ldots$$

(Sloane's A014715) giving the asymptotic rate of growth $C\lambda^n$ of the number of DIGITS in the nth term of the LOOK AND SAY SEQUENCE, given by the unique positive real root of the POLYNOMIAL

$$
\begin{aligned}
0 = {} & x^{71} - x^{69} - 2x^{68} - x^{67} + 2x^{66} + 2x^{65} + x^{64} - x^{63} - x^{62} \\
& - x^{61} - x^{60} - x^{59} + 2x^{58} + 5x^{57} + 3x^{56} - 2x^{55} - 10x^{54} \\
& - 3x^{53} - 2x^{52} + 6x^{51} + 6x^{50} + x^{49} + 9x^{48} - 3x^{47} \\
& - 7x^{46} - 8x^{45} - 8x^{44} + 10x^{43} + 6x^{42} + 8x^{41} - 4x^{40} \\
& - 12x^{39} + 7x^{38} - 7x^{37} + 7x^{36} + x^{35} - 3x^{34} + 10x^{33} \\
& + x^{32} - 6x^{31} - 2x^{30} - 10x^{29} - 3x^{28} + 2x^{27} + 9x^{26} \\
& - 3x^{25} + 14x^{24} - 8x^{23} - 7x^{21} + 9x^{20} - 3x^{19} - 4x^{18} \\
& - 10x^{17} - 7x^{16} + 12x^{15} + 7x^{14} + 2x^{13} - 12x^{12} - 4x^{11} \\
& - 2x^{10} - 5x^9 + x^7 - 7x^6 + 7x^5 - 4x^4 + 12x^3 - 6x^2 \\
& + 3x - 6, \quad\quad\quad\quad\quad\quad\quad\quad\quad\quad\quad\quad\quad (1)
\end{aligned}
$$

illustrated in the figure above. Note that the POLYNOMIAL given in Conway (1987, p. 188) contains a misprint. The CONTINUED FRACTION for λ is 1, 3, 3, 2, 2, 54, 5, 2, 1, 16, 1, 30, 1, 1, 1, 2, 2, 1, 14, 1, ... (Sloane's A014967).

See also CONWAY SEQUENCE, COSMOLOGICAL THEOREM, LOOK AND SAY SEQUENCE

References

Conway, J. H. "The Weird and Wonderful Chemistry of Audioactive Decay." §5.11 in *Open Problems in Communications and Computation* (Ed. T. M. Cover and B. Gopinath). New York: Springer-Verlag, pp. 173–88, 1987.

Conway, J. H. and Guy, R. K. "The Look and Say Sequence." In *The Book of Numbers.* New York: Springer-Verlag, pp. 208–09, 1996.

Finch, S. "Favorite Mathematical Constants." http://www.mathsoft.com/asolve/constant/cnwy/cnwy.html.

Hilgemeier, M. "Die Gleichniszahlen-Reihe." *Bild der Wissensch.*, pp. 194–96, Dec. 1986.

Hilgemeier, M. "'One Metaphor Fits All': A Fractal Voyage with Conway's Audioactive Decay." Ch. 7 in Pickover, C. A. (Ed.). *Fractal Horizons: The Future Use of Fractals.* New York: St. Martin's Press, 1996.

Sloane, N. J. A. Sequences A014715 and A014967 in "An On-Line Version of the Encyclopedia of Integer Sequences." http://www.research.att.com/~njas/sequences/eisonline.html.

Vardi, I. *Computational Recreations in Mathematica.* Reading, MA: Addison-Wesley, pp. 13–4, 1991.

Conway's Game of Life
LIFE

Conway's Knot
The KNOT with BRAID WORD

$$\sigma_2^3 \sigma_1 \sigma_3^{-1} \sigma_2^{-2} \sigma_1 \sigma_2^{-1} \sigma_1 \sigma_3^{-1}.$$

The JONES POLYNOMIAL of Conway's knot is

$$t^{-4}(-1 + 2t - 2t^2 + 2t^3 + t^6 - 2t^7 + 2t^8 - 2t^9 + t^{10}),$$

the *same* as for the KINOSHITA-TERASAKA KNOT.

Conway's Knot Notation
A concise NOTATION based on the concept of the TANGLE used by Conway (1967) to enumerate KNOTS up to 11 crossings. An ALGEBRAIC KNOT containing no NEGATIVE signs in its Conway knot NOTATION is an ALTERNATING KNOT.

References

Conway, J. H. "An Enumeration of Knots and Links, and Some of Their Algebraic Properties." In *Computation Problems in Abstract Algebra* (Ed. J. Leech). Oxford, England: Pergamon Press, pp. 329–58, 1967.

Conway's Life
LIFE

Cookie-Cutter Problem
Maximize the number of cookies you can cut from a given expanse of dough (Hoffman 1998, p. 173).

See also BIN-PACKING PROBLEM, TILING PROBLEM

References

Hoffman, P. *The Man Who Loved Only Numbers: The Story of Paul Erdos and the Search for Mathematical Truth.* New York: Hyperion, 1998.

Coordinate Chart
A coordinate chart is a way of expressing the points of a small NEIGHBORHOOD, usually on a MANIFOLD M, as coordinates in EUCLIDEAN SPACE. An example from geography is the coordinate chart given by the functions of LATITUDE and LONGITUDE. This coordinate chart is not valid on the whole globe, since it doesn't give unique coordinates at the north or south pole (which way is east from the north pole?).

Technically, a coordinate chart is a map

$$\phi : U \to V$$

where U is an open set in M, V is an open set in $\mathbb{R}^n$ and n is the dimension of the manifold. Often, through notational abuse, the open set U is equated with V, and calculations on the manifold are done in the coordinate chart. This technique has the drawback that it must be checked whether a change of coordinates affects the result of a calculation.

The map ϕ must be one-to-one, and in fact must be a HOMEOMORPHISM. On a SMOOTH MANIFOLD, it must be a DIFFEOMORPHISM, although if the chart defines the smooth structure then this is a tautology. Similarly, on a complex manifold, the map ϕ is holomorphic.

If there are two neighborhoods U_1 and U_2 with coordinate charts ϕ_1 and ϕ_2, the TRANSITION FUNCTION $\phi_2 \circ \phi_1^{-1}$ is WELL DEFINED since coordinate charts are one-to-one.

See also ATLAS, CHART, COMPLEX MANIFOLD, EUCLIDEAN SPACE, MANIFOLD, SMOOTH MANIFOLD, TRANSITION FUNCTION

Coordinate Geometry
ANALYTIC GEOMETRY, CARTESIAN GEOMETRY

Coordinate System
A system for specifying points using COORDINATES measured in some specified way. The simplest coordinate system consists of coordinate axes oriented perpendicularly to each other, known as CARTESIAN COORDINATES. Depending on the type of problem under consideration, coordinate systems possessing special properties may allow particularly simple solution.

See also CURVILINEAR COORDINATES, CYCLIDIC COORDINATES, SKEW COORDINATE SYSTEM, ORTHOGONAL COORDINATE SYSTEM

Coordinates

A set of n variables which fix a geometric object. If the coordinates are distances measured along PERPENDICULAR axes, they are known as CARTESIAN COORDINATES. The study of GEOMETRY using one or more coordinate systems is known as ANALYTIC GEOMETRY.

See also AREAL COORDINATES, BARYCENTRIC COORDINATES, BIPOLAR COORDINATES, BIPOLAR CYLINDRICAL COORDINATES, BISPHERICAL COORDINATES, CARTESIAN COORDINATES, CHOW COORDINATES, CIRCULAR CYLINDRICAL COORDINATES, CONFOCAL ELLIPSOIDAL COORDINATES, CONFOCAL PARABOLOIDAL COORDINATES, CONICAL COORDINATES, CURVILINEAR COORDINATES, CYCLIDIC COORDINATES, CYLINDRICAL COORDINATES, ELLIPSOIDAL COORDINATES, ELLIPTIC CYLINDRICAL COORDINATES, GAUSSIAN COORDINATE SYSTEM, GRASSMANN COORDINATES, HARMONIC COORDINATES, HOMOGENEOUS COORDINATES, OBLATE SPHEROIDAL COORDINATES, ORTHOCENTRIC COORDINATES, PARABOLIC COORDINATES, PARABOLIC CYLINDRICAL COORDINATES, PARABOLOIDAL COORDINATES, PEDAL COORDINATES, POLAR COORDINATES, PROLATE SPHEROIDAL COORDINATES, QUADRIPLANAR COORDINATES, RECTANGULAR COORDINATES, SPHERICAL COORDINATES, TOROIDAL COORDINATES, TRILINEAR COORDINATES

References

Arfken, G. "Coordinate Systems." Ch. 2 in *Mathematical Methods for Physicists, 3rd ed.* Orlando, FL: Academic Press, pp. 85–17, 1985.

Woods, F. S. *Higher Geometry: An Introduction to Advanced Methods in Analytic Geometry.* New York: Dover, p. 1, 1961.

Coordination Number

KISSING NUMBER

Copeland-Erdos Constant

The decimal 0.23571113171923... (Sloane's A033308) obtained by concatenating the PRIMES: 2, 23, 235, 2357, 235711, ... (Sloane's A019518; one of the SMARANDACHE SEQUENCES). Copeland and Erdos (1946) showed that it is a NORMAL NUMBER in base 10.

The first few digits of the CONTINUED FRACTION of the Copeland-Erdos constant are 0, 4, 4, 8, 16, 18, 5, 1, ... (Sloane's A030168). The positions of the first occurrence of n in the CONTINUED FRACTION are 8, 16, 20, 2, 7, 15, 12, 4, 17, 254, ... (Sloane's A033309). The incrementally largest terms are 4, 8, 16, 18, 58, 87, 484, ... (Sloane's A033310), which occur at positions 2, 4, 5, 6, 18, 36, 82, 89, ... (Sloane's A033311).

See also CHAMPERNOWNE CONSTANT, PRIME NUMBER

References

Champernowne, D. G. "The Construction of Decimals Normal in the Scale of Ten." *J. London Math. Soc.* **8**, 1933.

Copeland, A. H. and Erdos, P. "Note on Normal Numbers." *Bull. Amer. Math. Soc.* **52**, 857–60, 1946.

Sloane, N. J. A. Sequences A019518, A030168, A033308, A033309, A033310, and A033311 in "An On-Line Version of the Encyclopedia of Integer Sequences." http://www.re-search.att.com/~njas/sequences/eisonline.html.

Coplanar

Three noncollinear points determine a plane and so are trivially coplanar. Four points are coplanar IFF the volume of the TETRAHEDRON defined by them is 0,

$$\begin{vmatrix} x_1 & y_1 & z_1 & 1 \\ x_2 & y_2 & z_2 & 1 \\ x_3 & y_3 & z_3 & 1 \\ x_4 & y_4 & z_4 & 1 \end{vmatrix} = 0.$$

See also PLANE

Copolar Triangles

PERSPECTIVE TRIANGLES

Coprime

RELATIVELY PRIME

Coproduct

Denoted $\coprod$.

Copson-de Bruijn Constant

DE BRUIJN CONSTANT

Copson's Inequality

Let $\{a_n\}$ be a NONNEGATIVE SEQUENCE and $f(x)$ a NONNEGATIVE integrable function. Define

$$A_n = \sum_{k=1}^{n} a_k \tag{1}$$

$$B_n = \sum_{k=n}^{\infty} a_k \tag{2}$$

and

$$F(x) = \int_0^x f(t)\, dt \tag{3}$$

$$G(x) = \int_x^\infty f(t)\, dt, \tag{4}$$

and take $0 < p < 1$. For integrals,

$$\int_0^\infty \left[\frac{G(x)}{x} \right]^p dx > \left(\frac{p}{p-1} \right)^p \int_0^\infty [f(x)]^p\, dx \tag{5}$$

(unless f is identically 0). For sums,

$$\left(1+\frac{1}{p-1}\right)B_1^p + \sum_{n=2}^{\infty}\left(\frac{B_n}{n}\right)^p > \left(\frac{p}{p-1}\right)^p \sum_{n=1}^{\infty} a_n^p \quad (6)$$

(unless all $a_n = 0$).

References

Beesack, P. R. "On Some Integral Inequalities of E. T. Copson." In *General Inequalities 2: Proceedings of the Second International Conference on General Inequalities, held in the Mathematical Research Institut at Oberwolfach, Black Forest, July 30-August 5, 1978* (Ed. E. F. Beckenbach). Basel: Birkhäuser, 1980.

Copson, E. T. "Some Integral Inequalities." *Proc. Royal Soc. Edinburgh* **75A**, 157–64, 1975–976.

Hardy, G. H.; Littlewood, J. E.; and Pólya, G. Theorems 326–27, 337–38, and 345 in *Inequalities.* Cambridge, England: Cambridge University Press, 1934.

Mitrinovic, D. S.; Pecaric, J. E.; and Fink, A. M. *Inequalities Involving Functions and Their Integrals and Derivatives.* Dordrecht, Netherlands: Kluwer, 1991.

Copula

A function that joins univariate distribution functions to form multivariate distribution functions. A 2-D copula is a function $C : I^2 \to I$ such that

$$C(0, t) = C(t, 0) = 0$$

and

$$C(1, t) = C(t, 1) = t$$

for all $t \in I$, and

$$C(u_2, v_2) - C(u_1, v_2) - C(u_2, v_1) + C(u_1, v_1) \geq 0$$

for all $u_1, u_2, v_1, v_2 \in I$ such that $u_1 \leq u_2$ and $v_1 \leq v_2$.

See also SKLAR'S THEOREM

Cordial Graph

A GRAPH is called cordial if it is possible to label its vertices with 0s and 1s so that when the edges are labeled with the difference of the labels at their endpoints, the number of vertices (edges) labeled with ones and zeros differ at most by one. Cordial labelings were introduced by Cahit (1987) as a weakened version of GRACEFUL and HARMONIOUS.

An EULER GRAPH is not cordial if the number of its vertices is multiple of four. For example, all TREES are cordial, CYCLE GRAPHS of length n are cordial if n is not a multiple of four, COMPLETE GRAPHS on n vertices are cordial if $n < 4$, and the WHEEL GRAPH on $n + 1$ vertices is cordial IFF n is not congruent to 3 modulo 4.

See also GRACEFUL GRAPH, HARMONIOUS GRAPH, LABELED GRAPH

References

Cahit, I. "Cordial Graphs: A Weaker Version of Graceful and Harmonious Graphs." *Ars Combin.* **23**, 201–08, 1987.

Cordiform Projection

WERNER PROJECTION

Cork Plug

A 3-D SOLID which can stopper a SQUARE, TRIANGULAR, or CIRCULAR HOLE. There is an infinite family of such shapes. The one with smallest VOLUME has TRIANGULAR CROSS SECTIONS and $V = \pi r^3$; that with the largest VOLUME is made using two cuts from the top diameter to the EDGE and has VOLUME $V = 4\pi r^3/3$.

See also CROSS SECTION, STEREOLOGY, TRIP-LET

Corkscrew Surface

A surface also called the TWISTED SPHERE.

References

Gray, A. "The Corkscrew Surface." *Modern Differential Geometry of Curves and Surfaces with Mathematica, 2nd ed.* Boca Raton, FL: CRC Press, pp. 477–78, 1997.

Cornish-Fisher Asymptotic Expansion

$$y \approx m + \sigma w,$$

where

$$
\begin{aligned}
w = x &+ [\gamma_1 h_1(x)] + [\gamma_2 h_2(x) + \gamma_1^2 h_{11}(x)] \\
&+ [\gamma_3 h_3(x) + \gamma_1 \gamma_2 h_{12}(x) + \gamma_1^3 h_{111}(x)] \\
&+ [\gamma_4 h_4(x) + \gamma_2^2 h_{22}(x) + \gamma_1 \gamma_3 h_{13}(x) + \gamma_1^2 \gamma_2 h_{112}(x) \\
&+ \gamma_1^4 h_{1111}(x)] + \cdots,
\end{aligned}
$$

where

$$h_1(x) = \tfrac{1}{6} \, \mathrm{He}_2(x)$$

$$h_2(x) = \tfrac{1}{24} \, \mathrm{He}_3(x)$$

$$h_{11}(x) = -\tfrac{1}{36}[2\mathrm{He}_3(x) + \mathrm{He}_1(x)]$$

$$h_3(x) = \tfrac{1}{120} \, \mathrm{He}_4(x)$$

$$h_{12}(x) = -\tfrac{1}{24}[\mathrm{He}_4(x) + \mathrm{He}_2(x)]$$

$$h_{111}(x) = \tfrac{1}{324}[12\mathrm{He}_4(x) + 19\mathrm{He}_2(x)]$$

$$h_4(x) = \tfrac{1}{720} \, \mathrm{He}_5(x)$$

$$h_{22}(x) = -\tfrac{1}{384}[3\mathrm{He}_5(x) + 6\mathrm{He}_3(x) + 2\mathrm{He}_1(x)]$$

$$h_{13}(x) = -\tfrac{1}{180}[2\mathrm{He}_5 + 3\mathrm{He}_3(x)]$$

$$h_{112}(x) = \tfrac{1}{288}[14\mathrm{He}_5(x) + 37\mathrm{He}_3(x) + 8\mathrm{He}_1(x)]$$

$$h_{1111}(x) = -\tfrac{1}{7776}[252\mathrm{He}_5(x) + 832\mathrm{He}_3(x) + 227\mathrm{He}_1(x)].$$

See also CHARLIER SERIES, EDGEWORTH SERIES

References

Abramowitz, M. and Stegun, C. A. (Eds.). *Handbook of Mathematical Functions with Formulas, Graphs, and Mathematical Tables, 9th printing.* New York: Dover, p. 935, 1972.

Cornish, E. A. and Fisher, R. A. "Moments and Cumulants in the Specification of Distributions." *Extrait de la Revue de l'Institute International de Statistique* **4**, 1–4, 1937. Reprinted in Fisher, R. A. *Contributions to Mathematical Statistics.* New York: Wiley, 1950.

Wallace, D. L. "Asymptotic Approximations to Distributions." *Ann. Math. Stat.* **29**, 635–54, 1958.

Wasow, W. "On the Asymptotic Transformation of Certain Distributions into the Normal Distribution." *Proceedings of Symposia in Applied Mathematica VI, Numerical Analysis.* New York: McGraw-Hill, pp. 251–59, 1956.

Cornu Spiral

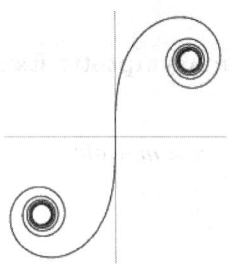

A plot in the COMPLEX PLANE of the points

$$B(t) = S(t) + iC(t), \qquad (1)$$

where $S(t)$ and $C(t)$ are the FRESNEL INTEGRALS (von Seggern 1993, p. 210; Gray 1997, p. 65). The Cornu spiral is also known as the CLOTHOID or EULER'S SPIRAL. It was probably first studied by Johann Bernoulli around 1696 (Bernoulli 1967, pp. 1084–086). A Cornu spiral describes diffraction from the edge of a HALF-PLANE.

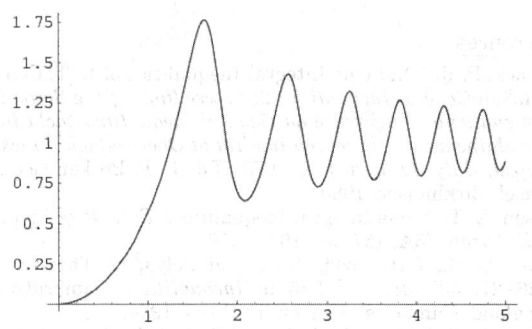

The quantities $C(t)/S(t)$ and $S(t)/C(t)$ are plotted above.

The SLOPE of the curve's TANGENT VECTOR (above right figure) is

$$m_T(t) = \frac{S'(t)}{C'(t)} = \tan\left(\tfrac{1}{2}\pi t^2\right), \qquad (2)$$

plotted below.

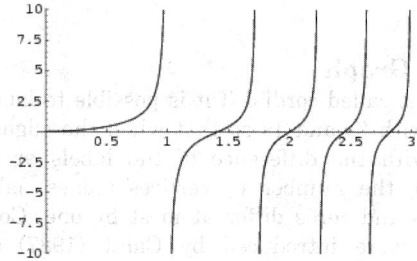

The CESÀRO EQUATION for a Cornu spiral is $\rho = c^2/s$, where ρ is the RADIUS OF CURVATURE and s the ARC LENGTH. The TORSION is $\tau = 0$.

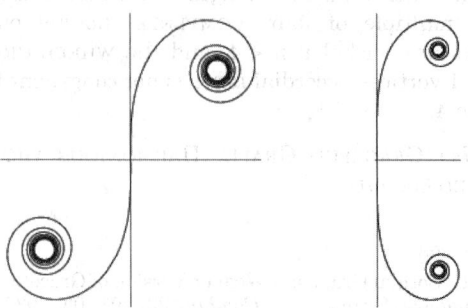

Gray (1997) defines a generalization of the Cornu spiral given by PARAMETRIC EQUATIONS

$$x(t) = a \int_0^t \sin\left(\frac{u^{n+1}}{n+1}\right) du \tag{3}$$

$$= \frac{at^{n+2}}{(n+1)(n+2)}$$

$$\times {}_1F_2\left(\frac{1}{2} + \frac{1}{2(n+1)}; \frac{3}{2}, \frac{3}{2} + \frac{1}{2(n+1)}; -\frac{t^{2(n+1)}}{4(n+1)^2}\right) \tag{4}$$

$$y(t) = a \int_0^t \cos\left(\frac{u^{n+1}}{n+1}\right) du \tag{5}$$

$$= at\, {}_1F_2\left(\frac{1}{2(n+1)}; \frac{1}{2}, 1 + \frac{1}{2(n+1)}; -\frac{t^{2(n+1)}}{4(n+1)^2}\right), \tag{6}$$

where ${}_1F_2(a; b, c; x)$ is a GENERALIZED HYPERGEOMETRIC FUNCTION.

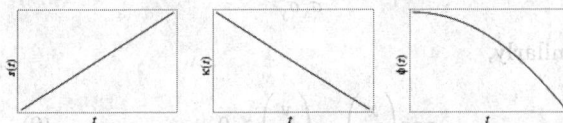

The ARC LENGTH, CURVATURE, and TANGENTIAL ANGLE of this curve are

$$s(t) = at \tag{7}$$

$$\kappa(t) = -\frac{t^n}{a} \tag{8}$$

$$\phi(t) = -\frac{t^{n+1}}{n+1}. \tag{9}$$

The CESÀRO EQUATION is

$$\kappa = -\frac{s^n}{a^{n+1}}. \tag{10}$$

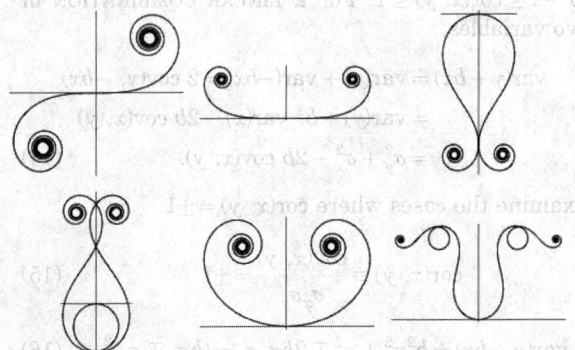

Dillen (1990) describes a class of "polynomial spirals"

for which the CURVATURE is a polynomial function of the ARC LENGTH. These spirals are a further generalization of the Cornu spiral. The curves plotted above correspond to $\kappa = s$, $\kappa = s^2$, $\kappa = s^2 - 2.19$, $\kappa = s^2 - 4$, $\kappa = s^2 + 1$, and $\kappa = 5s^4 - 18s^2 + 5$, respectively.

See also FRESNEL INTEGRALS, NIELSEN'S SPIRAL

References

Bernoulli, J. *Opera, Tomus Secundus.* Brussels, Belgium: Culture er Civilisation, 1967.
Dillen, F. "The Classification of Hypersurfaces of a Euclidean Space with Parallel Higher Fundamental Form." *Math. Z.* **203**, 635–43, 1990.
Gray, A. "Clothoids." §3.7 in *Modern Differential Geometry of Curves and Surfaces with Mathematica,* 2nd ed. Boca Raton, FL: CRC Press, pp. 64–6, 1997.
Lawrence, J. D. *A Catalog of Special Plane Curves.* New York: Dover, pp. 190–91, 1972.
von Seggern, D. *CRC Standard Curves and Surfaces.* Boca Raton, FL: CRC Press, 1993.

Cornucopia

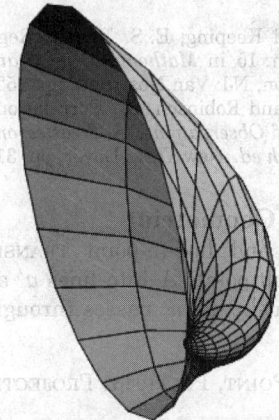

The SURFACE given by the PARAMETRIC EQUATIONS

$$x = e^{bv}\cos v + e^{av}\cos u \cos v$$

$$y = e^{bv}\sin v + e^{av}\cos u \sin v$$

$$z = e^{av}\sin u.$$

References

von Seggern, D. *CRC Standard Curves and Surfaces.* Boca Raton, FL: CRC Press, p. 304, 1993.

Corollary

An immediate consequence of a result already proved. Corollaries usually state more complicated THEOREMS in a language simpler to use and apply.

See also LEMMA, PORISM, THEOREM

Corona (Polyhedron)

AUGMENTED SPHENOCORONA, HEBESPHENOMEGACORONA, SPHENOCORONA, SPHENOMEGACORONA

Corona (Tiling)

The first corona of a TILE is the set of all tiles that have a common boundary point with that tile (including the original tile itself). The second corona is the set of tiles that share a point with something in the first corona, and so on.

References

Eppstein, D. "Heesch's Problem." http://www.ics.uci.edu/~eppstein/junkyard/heesch/.

Correlation

The degree of association between two or more quantities. In a 2-D plot, the degree of correlation between the values on the two axes is quantified by the so-called CORRELATION COEFFICIENT.

See also AUTOCORRELATION, CORRELATION COEFFICIENT, CORRELATION (GEOMETRIC), CORRELATION (STATISTICAL), CROSS-CORRELATION

References

Kenney, J. F. and Keeping, E. S. "Linear Regression and Correlation." Ch. 15 in *Mathematics of Statistics, Pt. 1, 3rd ed.* Princeton, NJ: Van Nostrand, pp. 252–85, 1962.
Whittaker, E. T. and Robinson, G. "Correlation." Ch. 12 in *The Calculus of Observations: A Treatise on Numerical Mathematics, 4th ed.* New York: Dover, pp. 317–42, 1967.

Correlation (Geometric)

A point-to-line and line-to-point TRANSFORMATION which transforms points A into lines a' and lines b into points B' such that a' passes through B' IFF A' lies on b.

See also LINE, POINT, POLARITY, PROJECTIVE CORRELATION

References

Coxeter, H. S. M. "Collineations and Correlations." §14.6 in *Introduction to Geometry, 2nd ed.* New York: Wiley, pp. 247–52, 1969.

Correlation (Statistical)

For two variables x and y, the correlation is defined by

$$\mathrm{cor}(x,\,y) \equiv \frac{\mathrm{cov}(x,\,y)}{\sigma_x \sigma_y}, \tag{1}$$

where σ_x denotes STANDARD DEVIATION and $\mathrm{cov}(x,\,y)$ is the COVARIANCE of these two variables. For the general case of variables x_i and x_j, where $i,\,j = 1, 2, \ldots, n$,

$$\mathrm{cor}(x_i,\,x_j) = \frac{\mathrm{cov}(x_i,\,y_j)}{\sqrt{V_{ii}V_{jj}}}, \tag{2}$$

where V_{ii} are elements of the COVARIANCE MATRIX. In general, a correlation gives the strength of the relationship between variables. For $i = j$,

$$\mathrm{cor}(x_i,\,x_i) = \frac{\mathrm{cov}(x_i,\,x_i)}{\sigma_i} = \frac{\sigma_{ii}}{\sigma_i} = \frac{\sigma_i^2}{\sigma_i} = \sigma_i. \tag{3}$$

The variance of any quantity is always NONNEGATIVE by definition, so

$$\mathrm{var}\left(\frac{x}{\sigma_x} + \frac{y}{\sigma_y}\right) \geq 0. \tag{4}$$

From a property of VARIANCES, the sum can be expanded

$$\mathrm{var}\left(\frac{x}{\sigma_x}\right) + \mathrm{var}\left(\frac{y}{\sigma_y}\right) + 2\mathrm{cov}\left(\frac{x}{\sigma_x},\,\frac{y}{\sigma_y}\right) \geq 0 \tag{5}$$

$$\frac{1}{\sigma_x^2}\,\mathrm{var}(x) + \frac{1}{\sigma_y^2}\,\mathrm{var}(y) + \frac{2}{\sigma_x\sigma_y}\,\mathrm{cov}(x,\,y) \geq 0 \tag{6}$$

$$1 + 1 + \frac{2}{\sigma_x\sigma_y}\,\mathrm{cov}(x,\,y) = 2 + \frac{2}{\sigma_x\sigma_y}\,\mathrm{cov}(x,\,y) \geq 0. \tag{7}$$

Therefore,

$$\mathrm{cor}(x,\,y) = \frac{\mathrm{cov}(x,\,y)}{\sigma_x\sigma_y} \geq -1. \tag{8}$$

Similarly,

$$\mathrm{var}\left(\frac{x}{\sigma_x}\right) - \left(\frac{y}{\sigma_y}\right) \geq 0 \tag{9}$$

$$\mathrm{var}\left(\frac{x}{\sigma_x}\right) + \mathrm{var}\left(-\frac{y}{\sigma_y}\right) + 2\,\mathrm{cov}\left(\frac{x}{\sigma_x},\,-\frac{y}{\sigma_y}\right) \geq 0 \tag{10}$$

$$\frac{1}{\sigma_x^2}\,\mathrm{var}(x) + \frac{1}{\sigma_y^2}\,\mathrm{var}(y) - \frac{2}{\sigma_x\sigma_y}\,\mathrm{cov}(x,\,y) \geq 0 \tag{11}$$

$$1 + 1 - \frac{2}{\sigma_x\sigma_y}\,\mathrm{cov}(x,\,y) = 2 - \frac{2}{\sigma_x\sigma_y}\,\mathrm{cov}(x,\,y) \geq 0. \tag{12}$$

Therefore,

$$\mathrm{cor}(x,\,y) = \frac{\mathrm{cov}(x,\,y)}{\sigma_x\sigma_y} \leq 1, \tag{13}$$

so $-1 \leq \mathrm{cor}(x,\,y) \leq 1$. For a LINEAR COMBINATION of two variables,

$$\begin{aligned}
\mathrm{var}(y - bx) &= \mathrm{var}(y) + \mathrm{var}(-bx) = 2\,\mathrm{cov}(y,\,-bx) \\
&= \mathrm{var}(y) + b^2\,\mathrm{var}(x) - 2b\,\mathrm{cov}(x,\,y) \\
&= \sigma_y^2 + \sigma_x^2 - 2b\,\mathrm{cov}(x,\,y).
\end{aligned} \tag{14}$$

Examine the cases where $\mathrm{cor}(x,\,y) = \pm 1$,

$$\mathrm{cor}(x,\,y) \equiv \frac{\mathrm{cov}(x,\,y)}{\sigma_x\sigma_y} = \pm 1 \tag{15}$$

$$\mathrm{var}(y - bx) = b^2\sigma_x^2 + \sigma_y^2 \mp 2b\sigma_x\sigma_y = (b\sigma_x \mp \sigma_y)^2. \tag{16}$$

The VARIANCE will be zero if $b \equiv \pm \sigma_y/\sigma_x$, which requires that the argument of the VARIANCE is a constant. Therefore, $y - bx = a$, so $y = a + bx$. If $\text{cor}(x, y) = \pm 1$, y is either perfectly correlated ($b > 0$) or perfectly anticorrelated ($b < 0$) with x.

See also COVARIANCE, COVARIANCE MATRIX, VARIANCE

Correlation Coefficient

The correlation coefficient is a quantity which gives the quality of a LEAST SQUARES FITTING to the original data. To define the correlation coefficient, first consider the sum of squared values ss_{xx}, ss_{xy}, and ss_{yy} of a set of n data points (x_i, y_i) about their respective means,

$$ss_{xx} \equiv \sum (x_i - \bar{x})^2 \tag{1}$$

$$= \sum x^2 - 2\bar{x} \sum x + \sum \bar{x}^2$$

$$= \sum x^2 - 2n\bar{x}^2 + n\bar{x}^2 = \sum x^2 - n\bar{x}^2 \tag{2}$$

$$ss_{yy} \equiv \sum (y_i - \bar{y})^2 \tag{3}$$

$$= \sum y^2 - 2\bar{y} \sum y + \sum \bar{y}^2$$

$$= \sum y^2 - 2n\bar{y}^2 + n\bar{y}^2 = \sum y^2 - n\bar{y}^2 \tag{4}$$

$$ss_{xy} \equiv \sum (x_i - \bar{x})(y_i - \bar{y}) \tag{5}$$

$$= \sum (x_i y_i - \bar{x} y_i - x_i \bar{y} + \bar{x}\,\bar{y})$$

$$= \sum xy - n\bar{x}\bar{y} - n\bar{x}\bar{y} + n\bar{x}\bar{y} = \sum xy - n\bar{x}\bar{y}. \tag{6}$$

For linear LEAST SQUARES FITTING, the COEFFICIENT b in

$$y = a + bx \tag{7}$$

is given by

$$b = \frac{n \sum xy - \sum x \sum y}{n \sum x^2 - (\sum x)^2} = \frac{ss_{xy}}{ss_{xx}}, \tag{8}$$

and the COEFFICIENT b' in

$$x = a' + b'y \tag{9}$$

is given by

$$b' = \frac{n \sum xy - \sum x \sum y}{n \sum y^2 - (\sum y)^2}. \tag{10}$$

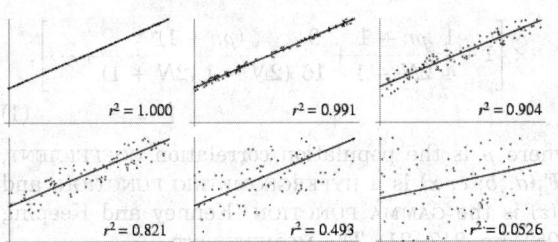

$r^2 = 1.000$ $r^2 = 0.991$ $r^2 = 0.904$

$r^2 = 0.821$ $r^2 = 0.493$ $r^2 = 0.0526$

The correlation coefficient r^2 (sometimes also denoted R^2) is then defined by

$$r \equiv \sqrt{bb'}$$

$$= \frac{n \sum xy - \sum x \sum y}{\sqrt{[n \sum x^2 - (\sum x)^2][n \sum y^2 - (\sum y)^2]}}, \tag{11}$$

which can be written more simply as

$$r^2 = \frac{ss_{xy}^2}{ss_{xx} ss_{yy}}. \tag{12}$$

The correlation coefficient is also known as the PRODUCT-MOMENT COEFFICIENT OF CORRELATION or PEARSON'S CORRELATION. The correlation coefficients for linear fits to increasingly noisy data are shown above.

The correlation coefficient has an important physical interpretation. To see this, define

$$A \equiv \left[\sum x^2 - n\bar{x}^2 \right]^{-1} \tag{13}$$

and denote the "expected" value for y_i as $\hat{y}_i$. Sums of $\hat{y}_i$ are then

$$\hat{y}_i = a + bx_i = \bar{y} - b\bar{x} + bx_i = \bar{x} + b(x_i - \bar{x})$$

$$= A(\bar{y} \sum x^2 - \bar{x} \sum xy + x_i \sum xy - n\bar{x}\bar{y}x_i)$$

$$= A[\bar{y} \sum x^2 + (x_i - \bar{x}) \sum xy - n\bar{x}\bar{y}x_i] \tag{14}$$

$$\sum \hat{y}_i = A(n\bar{y} \sum x^2 - n^2\bar{x}^2\bar{y}) \tag{15}$$

$$\sum \hat{y}_i^2 = A^2[n\bar{y}^2(\sum x^2)^2 - n^2\bar{x}^2\bar{y}^2(\sum x^2)$$
$$- 2n\bar{x}\bar{y}(\sum xy)(\sum x^2) + 2n^2\bar{x}^3\bar{y}(\sum xy)$$
$$+ (\sum x^2)(\sum xy)^2 - n\bar{x}^2(\sum xy)] \tag{16}$$

$$\sum y_i\hat{y}_i = A \sum [y_i\bar{y} \sum x^2 + y_i(x_i - \bar{x})$$
$$\times \sum xy - n\bar{x}\bar{y}x_i y_i]$$
$$= A[n\bar{y}^2 \sum x^2 + (\sum xy)^2 - n\bar{x}\bar{y}$$
$$\times \sum xy - n\bar{x}\bar{y}(\sum xy)]$$
$$= A[n\bar{y}^2 \sum x^2 + (\sum xy)^2 - 2n\bar{x}\bar{y} \sum xy]. \tag{17}$$

The sum of squared residuals is then

$$SSR \equiv \sum (\hat{y}_i - \bar{y})^2 = \sum (\hat{y}_i^2 - 2\bar{y}\hat{y}_i + \bar{y}^2)$$

$$= A^2(\sum xy - n\bar{x}\bar{y})^2(\sum x^2 - n\bar{x}^2) = \frac{(\sum xy - n\bar{x}\bar{y})^2}{\sum x^2 - n\bar{x}^2}$$

$$= b\, ss_{xy} = \frac{ss_{xy}^2}{ss_{xx}} = ss_{yy} r^2 = b^2\, ss_{xx}, \tag{18}$$

and the sum of squared errors is

$$SSE \equiv \sum (y_i - \hat{y}_i)^2 = \sum (y_i - \bar{y} - b\bar{x} - bx_i)^2$$
$$= \sum [y_i - \bar{y} - b(x_i - \bar{x})]^2$$
$$= \sum (y_i - \bar{y})^2 + b^2 \sum (x_i - \bar{x})^2 - 2b$$
$$\times \sum (x_i - \bar{x})(y_i - \bar{y}) = ss_{yy} + b^2 ss_{xx} - 2bss_{xy}. \quad (19)$$

But

$$b = \frac{ss_{xy}}{ss_{xx}} \quad (20)$$

$$r^2 = \frac{ss_{xy}^2}{ss_{xx}ss_{yy}}, \quad (21)$$

so

$$SSE = ss_{yy} + \frac{ss_{xy}^2}{ss_{xx}^2} ss_{xx} - 2 \frac{ss_{xy}}{ss_{xx}} ss_{xy} \quad (22)$$

$$= ss_{yy} - \frac{ss_{xy}^2}{ss_{xx}} \quad (23)$$

$$= ss_{yy}\left(1 - \frac{ss_{xy}^2}{ss_{xx}ss_{yy}}\right) \quad (24)$$

$$= ss_{yy}(1 - r^2), \quad (25)$$

and

$$SSE + SSR = ss_{yy}(1 - r^2) + ss_{yy}r^2 = ss_{yy}. \quad (26)$$

The square of the correlation coefficient r^2 is therefore given by

$$r^2 \equiv \frac{SSR}{ss_{yy}} = \frac{ss_{xy}^2}{ss_{xx}ss_{yy}} = \frac{(\sum xy - n\bar{x}\bar{y})^2}{(\sum x^2 - n\bar{x}^2)(\sum y^2 - n\bar{y}^2)}. \quad (27)$$

In other words, r^2 is the proportion of ss_{yy} which is accounted for by the regression.

If there is complete correlation, then the lines obtained by solving for best-fit (a, b) and (a', b') coincide (since all data points lie on them), so solving (9) for y and equating to (7) gives

$$y = -\frac{a'}{b'} + \frac{x}{b'} = a + bx. \quad (28)$$

Therefore, $a = -a'/b'$ and $b = 1/b'$, giving

$$r^2 = bb' = 1. \quad (29)$$

The correlation coefficient is independent of both origin and scale, so

$$r(u, v) = r(x, y), \quad (30)$$

where

$$u \equiv \frac{x - x_0}{h} \quad (31)$$

$$v \equiv \frac{y - y_0}{h}. \quad (32)$$

See also CORRELATION INDEX, CORRELATION COEFFICIENT–GAUSSIAN BIVARIATE DISTRIBUTION, CORRELATION RATIO, LEAST SQUARES FITTING, REGRESSION COEFFICIENT, SPEARMAN RANK CORRELATION COEFFICIENT

References

Acton, F. S. *Analysis of Straight-Line Data.* New York: Dover, 1966.
Kenney, J. F. and Keeping, E. S. "Linear Regression and Correlation." Ch. 15 in *Mathematics of Statistics, Pt. 1, 3rd ed.* Princeton, NJ: Van Nostrand, pp. 252–85, 1962.
Gonick, L. and Smith, W. "Regression." Ch. 11 in *The Cartoon Guide to Statistics.* New York: Harper Perennial, pp. 187–10, 1993.
Press, W. H.; Flannery, B. P.; Teukolsky, S. A.; and Vetterling, W. T. "Linear Correlation." §14.5 in *Numerical Recipes in FORTRAN: The Art of Scientific Computing, 2nd ed.* Cambridge, England: Cambridge University Press, pp. 630–33, 1992.
Whittaker, E. T. and Robinson, G. "The Coefficient of Correlation for Frequency Distributions which are not Normal." §166 in *The Calculus of Observations: A Treatise on Numerical Mathematics, 4th ed.* New York: Dover, pp. 334–36, 1967.

Correlation Coefficient—Gaussian Bivariate Distribution

For a GAUSSIAN BIVARIATE DISTRIBUTION, the distribution of correlation COEFFICIENTS is given by

$$P(r) = \frac{1}{\pi}(N - 2)(1 - r^2)^{(N-4)/2}$$
$$\times (1 - \rho^2)^{(N-1)/2} \int_0^\infty \frac{d\beta}{(\cosh \beta - \rho r)^{N-1}}$$
$$= \frac{1}{\pi}(N - 2)(1 - r^2)^{(N-4)/2}(1 - \rho^2)^{(N-1)/2} \sqrt{\frac{\pi}{2}} \frac{\Gamma(N - 1)}{\Gamma\left(N - \frac{1}{2}\right)}$$
$$\times (1 - \rho r)^{-(N-3/2)} {}_2F_1\left(\frac{1}{2}, \frac{1}{2}, \frac{2N - 1}{2}; \frac{\rho r + 1}{2}\right)$$
$$= \frac{(N - 2)\Gamma(N - 1)(1 - \rho^2)^{(N-1)/2}(1 - r^2)^{(N-4)/2}}{\sqrt{2\pi}\Gamma\left(N - \frac{1}{2}\right)(1 - \rho r)^{N-3/2}}$$
$$\times \left[1 + \frac{1}{4}\frac{\rho r + 1}{2N - 1} + \frac{9}{16}\frac{(\rho r + 1)^2}{(2N - 1)(2N + 1)} + \cdots\right], \quad (1)$$

where ρ is the population correlation COEFFICIENT, ${}_2F_1(a, b; c; x)$ is a HYPERGEOMETRIC FUNCTION, and $\Gamma(z)$ is the GAMMA FUNCTION (Kenney and Keeping 1951, pp. 217–21). The MOMENTS are

$$\langle r \rangle = \rho - \frac{\rho(1 - \rho^2)}{2n} \qquad (2)$$

$$\mathrm{var}(r) = \frac{(1 - \rho^2)^2}{n}\left(1 + \frac{11\rho^2}{2n} + \cdots\right) \qquad (3)$$

$$\gamma_1 = \frac{6\rho}{\sqrt{n}}\left(1 + \frac{77\rho^2 - 30}{12n} + \cdots\right)$$

$$\gamma_2 = \frac{6}{n}(12\rho^2 - 1) + \ldots, \qquad (4)$$

where $n \equiv n - 1$. If the variates are uncorrelated, then $\rho = 0$ and

$$_2f_1\left(\frac{1}{2}, \frac{1}{2}, \frac{2n - 1}{2}; \frac{\rho r + 1}{2}\right)$$

$$= {}_2F_1\left(\frac{1}{2}, \frac{1}{2}, \frac{2N - 1}{2}; \frac{1}{2}\right)$$

$$= \frac{\Gamma\left(N - \frac{1}{2}\right)2^{3/2-N}\sqrt{\pi}}{\left[\Gamma\left(\dfrac{N}{2}\right)\right]^2}, \qquad (5)$$

so

$$P(r) = \frac{(N - 2)\Gamma(N - 1)}{\sqrt{2\pi}\,\Gamma\left(N - \frac{1}{2}\right)}$$

$$\times(1 - r^2)^{(N-4)/2}\frac{\Gamma\left(N - \frac{1}{2}\right)2^{3/2-N}\sqrt{\pi}}{\left[\Gamma\left(\dfrac{N}{2}\right)\right]^2}$$

$$= \frac{2^{1-N}(N - 2)\Gamma(N - 1)}{\left[\Gamma\left(\dfrac{N}{2}\right)\right]^2}(1 - r^2)^{(N-4/2)}. \qquad (6)$$

But from the LEGENDRE DUPLICATION FORMULA,

$$\sqrt{\pi}\,\Gamma(N - 1) = 2^{N-2}\Gamma\left(\frac{N}{2}\right)\Gamma\left(\frac{N - 1}{2}\right), \qquad (7)$$

so

$$P(r) = \frac{(2^{1-N})(2^{N-2})(N - 2)\Gamma\left(\dfrac{N}{2}\right)\Gamma\left(\dfrac{N - 1}{2}\right)}{\sqrt{\pi}\left[\Gamma\left(\dfrac{N}{2}\right)\right]^2}$$

$$\times(1 - r^2)^{(N-4)/2}$$

$$= \frac{(N - 2)\Gamma\left(\dfrac{N - 1}{2}\right)}{2\sqrt{\pi}\,\Gamma\left(\dfrac{N}{2}\right)}(1 - r^2)^{(N-4)/2}$$

$$= \frac{1}{\sqrt{\pi}}\frac{\dfrac{v}{2}\,\Gamma\left(\dfrac{v + 1}{2}\right)}{\Gamma\left(\dfrac{v}{2} + 1\right)}(1 - r^2)^{(v-2)/2}$$

$$= \frac{1}{\sqrt{\pi}}\frac{\Gamma\left(\dfrac{v + 1}{2}\right)}{\Gamma\left(\dfrac{v}{2}\right)}(1 - r^2)^{(v-2)/2}. \qquad (8)$$

The uncorrelated case can be derived more simply by letting β be the true slope, so that $\eta = \alpha + \beta x$. Then

$$t \equiv (b - \beta)\frac{S_x}{S_y}\sqrt{\frac{N - 2}{1 - r^2}} = \frac{(b - \beta)r}{b}\sqrt{\frac{N - 2}{1 - r^2}} \qquad (9)$$

is distributed as STUDENT'S T with $v \equiv N - 2$ DEGREES OF FREEDOM. Let the population regression COEFFICIENT ρ be 0, then $\beta = 0$, so

$$t = r\sqrt{\frac{v}{1 - r^2}}, \qquad (10)$$

and the distribution is

$$P(t)\,dt = \frac{1}{\sqrt{v\pi}}\frac{\Gamma\left(\dfrac{v + 1}{2}\right)}{\Gamma\left(\dfrac{v}{2}\right)\left(1 + \dfrac{t^2}{v}\right)^{(v+1)/2}}\,dt. \qquad (11)$$

Plugging in for t and using

$$dt = \sqrt{v}\left[\frac{\sqrt{1 - r^2} - r\left(\frac{1}{2}\right)(-2r)(1 - r^2)^{-1/2}}{1 - r^2}\right]dr$$

$$= \sqrt{\frac{v}{1 - r^2}}\left(\frac{1 - r^2 + r^2}{1 - r^2}\right)dr = \sqrt{\frac{v}{(1 - r)^3}}\,dr \qquad (12)$$

gives

$$P(t)\,dt = \frac{1}{\sqrt{v\pi}}\frac{\Gamma\left(\dfrac{v + 1}{2}\right)}{\Gamma\left(\dfrac{v}{2}\right)\left[1 + \dfrac{r^2v}{(1 - r^2)v}\right]^{(v+1)/2}}\sqrt{\frac{v}{(1 - r)^3}}\,dr$$

$$= \frac{(1 - r^2)^{-3/2}}{\sqrt{\pi}}\frac{\Gamma\left(\frac{v+1}{2}\right)}{\Gamma\left(\frac{v}{2}\right)\left(\frac{1}{1-r^2}\right)^{(v+1)/2}}\,dr$$

$$= \frac{1}{\sqrt{\pi}} \frac{\Gamma\left(\frac{v+1}{2}\right)}{\Gamma\left(\frac{v}{2}\right)} (1-r^2)^{-3/2}(1-r^2)^{(v+1)/2} \, dr$$

$$= \frac{1}{\sqrt{\pi}} \frac{\Gamma\left(\frac{v+1}{2}\right)}{\Gamma\left(\frac{v}{2}\right)} (1-r^2)^{(v-2)/2} \, dr, \tag{13}$$

so

$$P(r) = \frac{1}{\sqrt{\pi}} \frac{\Gamma\left(\frac{v+1}{2}\right)}{\Gamma\left(\frac{v}{2}\right)} (1-r^2)^{(v-2)/2} \tag{14}$$

as before. See Bevington (1969, pp. 122–23) or Pugh and Winslow (1966, §12–). If we are interested instead in the probability that a correlation COEFFICIENT would be obtained $\geq |r|$, where r is the observed COEFFICIENT, then 392 Let $I \equiv \frac{1}{2}(v-2)$. For EVEN v, the exponent I is an INTEGER so, by the BINOMIAL THEOREM,

$$(1-r^2)^I = \sum_{k=0}^{I} \binom{I}{k} (-r^2)^k \tag{17}$$

and

$$P_c(r) = 1 - \frac{2}{\sqrt{\pi}} \frac{\Gamma\left(\frac{v+1}{2}\right)}{\Gamma\left(\frac{v}{2}\right)}$$

$$\times (-1)^k \frac{I!}{(I-k)!k!} \int_0^{|r|} \sum_{k=0}^{I} r'^{2k} \, dr'$$

$$= 1 - \frac{2}{\sqrt{\pi}} \frac{\Gamma\left(\frac{v+1}{2}\right)}{\Gamma\left(\frac{v}{2}\right)}$$

$$\times \sum_{k=0}^{I} \left[(-1)^k \frac{I!}{(I-k)!k!} \frac{|r|^{2k+1}}{2k+1} \right]. \tag{18}$$

For ODD v, the integral is

$$P_c(r) = 1 - 2 \int_0^{|r|} P(r') \, dr'$$

$$= 1 - \frac{2}{\sqrt{\pi}} \frac{\Gamma\left(\frac{v+1}{2}\right)}{\Gamma\left(\frac{v}{2}\right)} \int_0^{|r|} (\sqrt{1-r^2})^{v-2} \, dr. \tag{19}$$

Let $r \equiv \sin x$ so $dr = \cos x \, dx$, then

$$P_c(r) = 1 - \frac{2}{\sqrt{\pi}} \frac{\Gamma\left(\frac{v+1}{2}\right)}{\Gamma\left(\frac{v}{2}\right)} \int_0^{\sin^{-1}|r|} \cos^{v-2} x \cos x \, dx$$

$$= 1 - \frac{2}{\sqrt{\pi}} \frac{\Gamma\left(\frac{v+1}{2}\right)}{\Gamma\left(\frac{v}{2}\right)} + \int_0^{\sin^{-1}|r|} \cos^{v-1} x \, dx. \tag{20}$$

But v is ODD, so $v-1 \equiv 2n$ is EVEN. Therefore

$$\frac{2}{\sqrt{\pi}} \frac{\Gamma\left(\frac{v+1}{2}\right)}{\Gamma\left(\frac{v}{2}\right)} = \frac{2}{\sqrt{\pi}} \frac{\Gamma(n+1)}{\Gamma\left(n+\frac{1}{2}\right)} = \frac{2}{\sqrt{\pi}} \frac{n!}{\frac{(2n-1)!!\sqrt{\pi}}{2^n}}$$

$$= \frac{2}{\pi} \frac{2^n n!}{(2n-1)!!} = \frac{2}{\pi} \frac{(2n)!!}{(2n-1)!!}. \tag{21}$$

Combining with the result from the COSINE INTEGRAL gives

$$P_c(r) = 1 - \frac{2}{\pi} \frac{(2n)!!(2n-1)!!}{(2n-1)!!(2n)!!}$$

$$\times \left[\sin x \sum_{k=0}^{n-1} \frac{(2k)!!}{(2k+1)!!} \cos^{2k+1} x + x \right]_0^{\sin^{-1}|r|}. \tag{22}$$

Use

$$\cos^{2k-1} x = (1-r^2)^{(2k-1)/2} = (1-r^2)^{(k-1/2)}, \tag{23}$$

and define $J \equiv n-1 = (v-3)/2$, then

$$P_c(r) = 1 - \frac{2}{\pi}$$

$$\times \left[\sin^{-1}|r| + |r| \sum_{k=0}^{J} \frac{(2k)!!}{(2k+1)!!} (1-r^2)^{k+1/2} \right]. \tag{24}$$

(In Bevington 1969, this is given incorrectly.) Combining the correct solutions

$$P_c(r) = \begin{cases} 1 - \frac{2}{\sqrt{\pi}} \frac{\Gamma[(v+1)/2]}{\Gamma(v/2)} \sum_{k=0}^{I} \left[(-1)^k \frac{I!}{(I-k)!k!} \frac{|r|^{2k+1}}{2k+1} \right] \\ \text{for } v \text{ even} \\ 1 - \frac{2}{\pi} \left[\sin^{-1}|r| + |r| \sum_{k=0}^{J} \frac{(2k)!!}{(2k+1)!!} (1-r^2)^{k+1/2} \right] \\ \text{for } v \text{ odd} \end{cases} \tag{25}$$

If $\rho \neq 0$, a skew distribution is obtained, but the variable z defined by

$$z \equiv \tanh^{-1} r \tag{26}$$

is approximately normal with

$$\mu_z = \tanh^{-1} \rho \qquad (27)$$

$$\sigma_z^2 = \frac{1}{N-3} \qquad (28)$$

(Kenney and Keeping 1962, p. 266).

Let b_j be the slope of a best-fit line, then the multiple correlation COEFFICIENT is

$$R^2 \equiv \sum_{j=1}^{n} \left(b_j \frac{s_{jy}^2}{s_y^2} \right) = \sum_{j=1}^{n} \left(b_j \frac{s_j}{s_y} r_{jy} \right), \qquad (29)$$

where s_{jy} is the sample VARIANCE.

On the surface of a SPHERE,

$$r \equiv \frac{\int fg \, d\Omega}{\int f \, d\Omega \int g \, d\Omega}, \qquad (30)$$

where $d\Omega$ is a differential SOLID ANGLE. This definition guarantees that $-1 < r < 1$. If f and g are expanded in REAL SPHERICAL HARMONICS,

$$f(\theta, \phi) \equiv \sum_{l=0}^{\infty} \sum_{m=0}^{l} [C_l^m Y_l^{mc}(\theta, \phi) \sin(m\phi)$$
$$+ S_l^m Y_l^{ms}(\theta, \phi)] \qquad (31)$$

$$g(\theta, \phi) \equiv \sum_{t=0}^{\infty} \sum_{m=0}^{l} [A_l^m Y_l^{mc}(\theta, \phi) \sin(m\phi)$$
$$+ B_l^m Y_l^{ms}(\theta, \phi)]. \qquad (32)$$

Then

$$r_1 = \frac{\sum_{m=0}^{l} (C_l^m A_l^m + S_l^m B_l^m)}{\sqrt{\sum_{m=0}^{l} (C_l^{m2} + S_l^{m2})} \sqrt{\sum_{m=0}^{l} (A_l^{m2} + B_l^{m2})}}. \qquad (33)$$

The confidence levels are then given by

$$G_1(r) = r$$

$$G_2(r) = r \left(1 + \tfrac{1}{2} s^2 \right) = \tfrac{1}{2} r(3 - r^2)$$

$$G_3(r) = r \left[1 + \tfrac{1}{2} s^2 \left(1 + \tfrac{3}{4} s^2 \right) \right] = \tfrac{1}{8} r(15 - 10r^2 + 3r^4)$$

$$G_4(r) = r \left\{ 1 + \tfrac{1}{2} s^2 \left[1 + \tfrac{3}{4} s^2 \left(1 + \tfrac{5}{6} s^2 \right) \right] \right\}$$
$$= \tfrac{1}{16} r(35 - 35r^2 + 21r^4 - 5r^6),$$

where

$$s \equiv \sqrt{1 - r^2} \qquad (34)$$

(Eckhardt 1984).

See also FISHER'S z'-TRANSFORMATION, SPEARMAN RANK CORRELATION COEFFICIENT, SPHERICAL HARMONIC

References
Bevington, P. R. *Data Reduction and Error Analysis for the Physical Sciences.* New York: McGraw-Hill, 1969.
Eckhardt, D. H. "Correlations Between Global Features of Terrestrial Fields." *Math. Geology* **16**, 155–71, 1984.
Kenney, J. F. and Keeping, E. S. *Mathematics of Statistics, Pt. 1, 3rd ed.* Princeton, NJ: Van Nostrand, 1962.
Kenney, J. F. and Keeping, E. S. *Mathematics of Statistics, Pt. 2, 2nd ed.* Princeton, NJ: Van Nostrand, 1951.
Pugh, E. M. and Winslow, G. H. *The Analysis of Physical Measurements.* Reading, MA: Addison-Wesley, 1966.

Correlation Dimension

Define the correlation integral as

$$C(\epsilon) \equiv \lim_{n \to \infty} \frac{1}{N^2} \sum_{\substack{i,j=1 \\ i \neq j}}^{\infty} H(\epsilon - \|x_i - x_j\|), \qquad (1)$$

where H is the HEAVISIDE STEP FUNCTION. When the below limit exists, the correlation dimension is then defined as

$$D_2 \equiv d_{\text{cor}} \equiv \lim_{\epsilon, \epsilon' \to 0+} \frac{\ln \left[\dfrac{C(\epsilon)}{C(\epsilon')} \right]}{\ln \left(\dfrac{\epsilon}{\epsilon'} \right)}. \qquad (2)$$

If v is the CORRELATION EXPONENT, then

$$\lim_{\epsilon \to 0} v \to D_2. \qquad (3)$$

It satisfies

$$d_{\text{cor}} \leq d_{\text{inf}} \leq d_{\text{cap}} \overset{?}{=} d_{\text{Lya}}. \qquad (4)$$

To estimate the correlation dimension of an M-dimensional system with accuracy $(1 - Q)$ requires $N_{\min}$ data points, where

$$N_{\min} \geq \left[\frac{R(2 - Q)}{2(1 - Q)} \right]^M, \qquad (5)$$

where $R \geq 1$ is the length of the "plateau region." If an ATTRACTOR exists, then an estimate of D_2 saturates above some M given by

$$M \geq 2D + 1, \qquad (6)$$

which is sometimes known as the fractal Whitney embedding prevalence theorem.

See also CORRELATION EXPONENT, Q-DIMENSION

References
Nayfeh, A. H. and Balachandran, B. *Applied Nonlinear Dynamics: Analytical, Computational, and Experimental Methods.* New York: Wiley, pp. 547–48, 1995.

Correlation Exponent

A measure v of a STRANGE ATTRACTOR which allows the presence of CHAOS to be distinguished from random noise. It is related to the CAPACITY DIMENSION D and INFORMATION DIMENSION σ, satisfying

$$v \leq \sigma \leq D. \tag{1}$$

It satisfies

$$v \leq D_{KY}, \tag{2}$$

where D_{KY} is the KAPLAN-YORKE DIMENSION. As the cell size goes to zero,

$$\lim_{\epsilon \to 0} v \to D_2, \tag{3}$$

where D_2 is the CORRELATION DIMENSION.

See also CORRELATION DIMENSION, INFORMATION DIMENSION, KAPLAN-YORKE DIMENSION

References

Grassberger, P. and Procaccia, I. "Measuring the Strangeness of Strange Attractors." *Physica D* **9**, 189–08, 1983.

Correlation Index

Given a *curved* regression, the correlation index is defined by

$$r_c \equiv \frac{s_{y\hat{y}}}{s_y s_{\hat{y}}},$$

where s_y and $s_{\hat{y}}$ are the standard deviations of the data points y and the estimates $\hat{y}$ given by the regression line, and the quantity $s_{y\hat{y}}$ is not defined by Kenney and Keeping 1962. Then

$$r_c^2 = \frac{s_{\hat{y}}^2}{s_y^2} = 1 - \frac{s_{ey}^2}{s_y^2},$$

where s_{ey}^2 is the variance of the observed ys about the best-fitting curved line (Kenney and Keeping 1962, p. 293).

See also CORRELATION COEFFICIENT, REGRESSION

References

Kenney, J. F. and Keeping, E. S. *Mathematics of Statistics, Pt. 1, 3rd ed.* Princeton, NJ: Van Nostrand, 1962.

Correlation Integral

Consider a set of points $\mathbf{X}_i$ on an ATTRACTOR, then the correlation integral is

$$C(l) \equiv \lim_{N \to \infty} \frac{1}{N^2} f,$$

where f is the number of pairs (i, j) whose distance $|\mathbf{X}_i - \mathbf{X}_j| < l$. For small l,

$$C(l) \sim l^v,$$

where v is the CORRELATION EXPONENT.

References

Grassberger, P. and Procaccia, I. "Measuring the Strangeness of Strange Attractors." *Physica D* **9**, 189–08, 1983.

Correlation Ratio

Let there be N_i observations of the ith phenomenon, where $i = 1, ..., p$ and

$$N \equiv \sum N_i \tag{1}$$

$$\bar{y}_i \equiv \frac{1}{N_i} \sum_\alpha y_{i\alpha} \tag{2}$$

$$\bar{y} \equiv \frac{1}{N} \sum_i \sum_\alpha y_{i\alpha}. \tag{3}$$

Then

$$E_{yx}^2 \equiv \frac{\sum_i N_i (\bar{y}_i - \bar{y})^2}{\sum_i \sum_\alpha (y_{i\alpha} - \bar{y})^2}. \tag{4}$$

Let η_{yx} be the population correlation ratio. If $N_i = N_j$ for $i \neq j$, then

$$f(E^2) = \frac{e^{-\lambda}(E^2)^{a-1}(1 - E^2)^{b-1}\,{}_1F_1(a, b; \lambda E^2)}{B(a, b)}, \tag{5}$$

where

$$\lambda \equiv \frac{N\eta^2}{2(1 - \eta^2)} \tag{6}$$

$$a \equiv \frac{n_1}{2} \tag{7}$$

$$b \equiv \frac{n_2}{2} \tag{8}$$

and ${}_1F_1(a, b; z)$ is the CONFLUENT HYPERGEOMETRIC LIMIT FUNCTION. If $\lambda = 0$, then

$$f(E^2) = \beta(a, b) \tag{9}$$

(Kenney and Keeping 1951, pp. 323–24).

See also CORRELATION COEFFICIENT, REGRESSION COEFFICIENT

References

Kenney, J. F. and Keeping, E. S. *Mathematics of Statistics, Pt. 2, 2nd ed.* Princeton, NJ: Van Nostrand, 1951.

Cos

COSINE

Cosecant

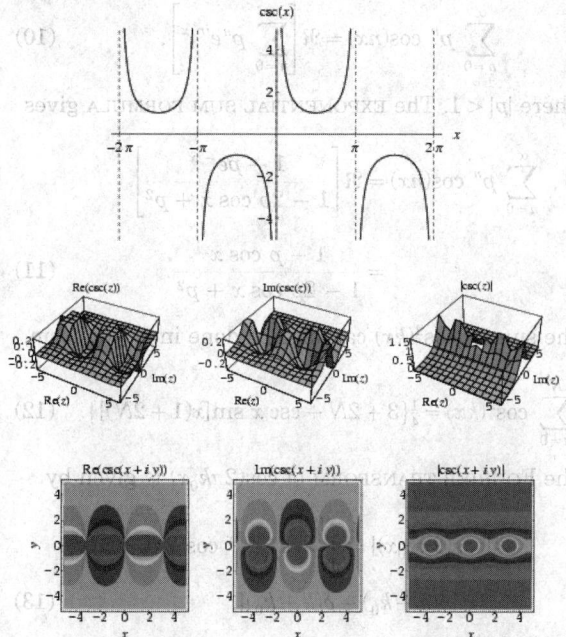

The function defined by $\csc x \equiv 1/\sin x$, where $\sin x$ is the SINE. The MACLAURIN SERIES of the cosecant function is

$$\csc x = \frac{1}{x} + \frac{1}{6}x + \frac{7}{360}x^3 + \frac{31}{15120}x^5 + \dots$$

$$+ \frac{(-1)^{n+1}2(2^{2n-1}-1)B_{2n}}{(2n)!}x^{2n-1} + \dots,$$

where B_{2n} is a BERNOULLI NUMBER.

See also INVERSE COSECANT, SECANT, SINE

References

Abramowitz, M. and Stegun, C. A. (Eds.). "Circular Functions." §4.3 in *Handbook of Mathematical Functions with Formulas, Graphs, and Mathematical Tables, 9th printing.* New York: Dover, pp. 71–9, 1972.
Beyer, W. H. *CRC Standard Mathematical Tables, 28th ed.* Boca Raton, FL: CRC Press, p. 215, 1987.
Spanier, J. and Oldham, K. B. "The Secant sec(x) and Cosecant csc(x) Functions." Ch. 33 in *An Atlas of Functions.* Washington, DC: Hemisphere, pp. 311–18, 1987.

Coset

This entry contributed by NICOLAS BRAY

For a SUBGROUP H of a GROUP G and an element x of G, define xH to be the set $\{xh : h \in H\}$ and Hx to be the set $\{hx : h \in H\}$. A SUBSET of G of the form xH for some $x \in G$ is said to be a LEFT COSET of H and a subset of the form Hx is said to be a RIGHT COSET of H.

For any SUBGROUP H, we can define an EQUIVALENCE RELATION $\sim$ by $x \sim y$ if $x = yh$ for some $h \in H$. The EQUIVALENCE CLASSES of this EQUIVALENCE RELATION are exactly the LEFT COSETS of H, and an element x of G is in the EQUIVALENCE CLASS xH. Thus the LEFT COSETS of H form a partition of G.

It is also true that any two LEFT COSETS of H have the same CARDINALITY, and in particular, every coset of H has the same CARDINALITY as $eH = H$, where e is the IDENTITY ELEMENT. Thus, the CARDINALITY of any LEFT COSET of H has CARDINALITY the order of H.

The same results are true of the RIGHT COSETS of G as well and, in fact, one can prove that the set of LEFT COSETS of H has the same CARDINALITY as the set of RIGHT COSETS of H.

See also EQUIVALENCE CLASS, GROUP, LEFT COSET, QUOTIENT GROUP, RIGHT COSET, SUBGROUP

Cosh
HYPERBOLIC COSINE

CoshIntegral
CHI

Cosine

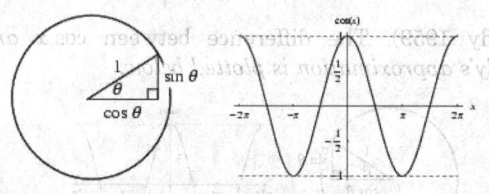

One of the basic TRIGONOMETRIC FUNCTIONS encountered in TRIGONOMETRY. Let θ be an ANGLE measured counterclockwise from the x-AXIS along the arc of the unit CIRCLE. Then $\cos\theta$ is the horizontal coordinate of the arc endpoint. As a result of this definition, the cosine function is periodic with period 2π.

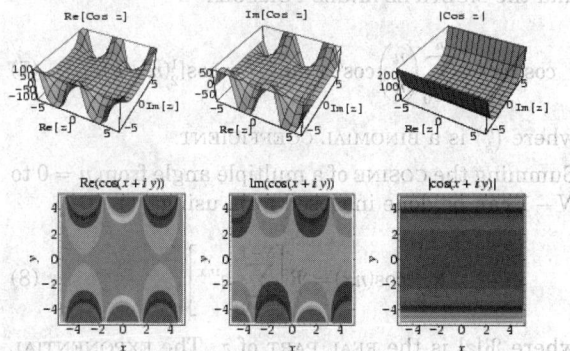

The definition of the cosine function can be extended to complex arguments z using the definition

$$\cos z = \tfrac{1}{2}(e^{iz} + e^{-iz}), \tag{1}$$

where e is the base of the NATURAL LOGARITHM and i is the IMAGINARY NUMBER. A related function known as the HYPERBOLIC COSINE is similarly defined,

$$\cosh z = \tfrac{1}{2}(e^z + e^{-z}). \qquad (2)$$

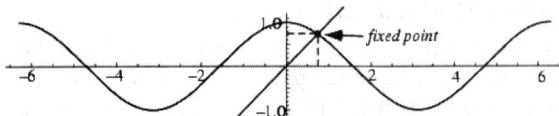

The cosine function has a FIXED POINT at 0.739085.

The cosine function can be defined algebraically using the infinite sum

$$\cos x \equiv \sum_{n=0}^{\infty} \frac{(-1)^n x^{2n}}{(2n)!} = 1 - \frac{x^2}{2!} + \frac{x^4}{4!} - \frac{x^6}{6!} + \ldots, \qquad (3)$$

or the INFINITE PRODUCT

$$\cos x = \prod_{n=1}^{\infty} \left[1 - \frac{4x^2}{\pi^2(2n-1)^2} \right]. \qquad (4)$$

A close approximation to $\cos(x)$ for $x \in [0, \ \pi/2]$ is

$$\cos\left(\frac{\pi}{2} x\right) \approx 1 - \frac{x^2}{x + (1-x)\sqrt{\dfrac{2-x}{3}}} \qquad (5)$$

(Hardy 1959). The difference between $\cos x$ *and Hardy's approximation is plotted below.*

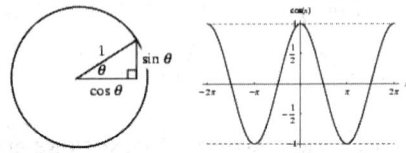

The cosine obeys the identity

$$\cos(n\theta) = 2\cos\theta\cos[(n-1)\theta] - \cos[(n-2)\theta] \qquad (6)$$

and the MULTIPLE-ANGLE FORMULA

$$\cos(nx) = \sum_{k=0}^{n} \binom{n}{k} \cos^k x \, \sin^{n-k} x \, \cos[\tfrac{1}{2}(n-k)\pi], \qquad (7)$$

where $\binom{n}{k}$ is a BINOMIAL COEFFICIENT.

Summing the COSINE of a multiple angle from $n = 0$ to $N - 1$ can be done in closed form using

$$\sum_{n=0}^{N-1} \cos(nx) = \Re \left[\sum_{n=0}^{N-1} e^{inx} \right], \qquad (8)$$

where $\Re[z]$ is the REAL PART of z. The EXPONENTIAL SUM FORMULAS give

$$\sum_{n=1}^{N} \cos(nx) = \Re \left[\frac{\sin(\tfrac{1}{2}Nx)}{\sin(\tfrac{1}{2}x)} e^{i(N+1)x/2} \right]$$

$$= \frac{\sin(\tfrac{1}{2}Nx)}{\sin(\tfrac{1}{2}x)} \cos[\tfrac{1}{2}x(N+1)]. \qquad (9)$$

Similarly,

$$\sum_{n=0}^{\infty} p^n \cos(nx) = \Re \left[\sum_{n=0}^{\infty} p^n e^{in\,x} \right], \qquad (10)$$

where $|p| < 1$. The EXPONENTIAL SUM FORMULA gives

$$\sum_{n=0}^{\infty} p^n \cos(nx) = \Re \left[\frac{1 - pe^{-ix}}{1 - 2p\cos x + p^2} \right]$$

$$= \frac{1 - p\cos x}{1 - 2p\cos x + p^2}. \qquad (11)$$

The sum of $\cos^2(kx)$ can also be done in closed form,

$$\sum_{k=0}^{N} \cos^2(kx) = \tfrac{1}{4}\{3 + 2N + \csc x \, \sin[x(1+2N)]\}. \qquad (12)$$

The FOURIER transform of $\cos(2\pi k_0 x)$ is given by

$$\mathcal{F}[\cos(2\pi k_0 x)] = \int_{-\infty}^{\infty} e^{-2\pi ikx} \cos(2\pi k_0 x) \, dx$$

$$= \tfrac{1}{2}[\delta(k - k_0) + \delta(k + k_0)], \qquad (13)$$

where $\delta(k)$ is the DELTA FUNCTION.

Cvijovic and Klinowski (1995) note that the following series

$$C_\nu(\alpha) = \sum_{k=0}^{\infty} \frac{\cos(2k+1)\alpha}{(2k+1)^\nu} \qquad (14)$$

has closed form for $\nu = 2n$,

$$C_{2n}(\alpha) = \frac{(-1)^n}{4(2n-1)!} \pi^{2n} E_{2n-1}\left(\frac{\alpha}{\pi}\right), \qquad (15)$$

where $E_n(x)$ is an EULER POLYNOMIAL.

See also EULER POLYNOMIAL, EXPONENTIAL SUM FORMULAS, FOURIER TRANSFORM–COSINE, HYPERBOLIC COSINE, SINE, TANGENT, TRIGONOMETRIC FUNCTIONS

References

Abramowitz, M. and Stegun, C. A. (Eds.). "Circular Functions." §4.3 in *Handbook of Mathematical Functions with Formulas, Graphs, and Mathematical Tables, 9th printing.* New York: Dover, pp. 71–9, 1972.

Beyer, W. H. *CRC Standard Mathematical Tables, 28th ed.* Boca Raton, FL: CRC Press, p. 215, 1987.

Hardy, G. H. *Ramanujan: Twelve Lectures on Subjects Suggested by His Life and Work, 3rd ed.* New York: Chelsea, p. 68, 1959.

Cvijovic, D. and Klinowski, J. "Closed-Form Summation of Some Trigonometric Series." *Math. Comput.* **64**, 205–10, 1995.

Hansen, E. R. *A Table of Series and Products.* Englewood Cliffs, NJ: Prentice-Hall, 1975.

Project Mathematics. "Sines and Cosines, Parts I-III." Videotape. http://www.projmath.caltech.edu/sincos1.htm.

Spanier, J. and Oldham, K. B. "The Sine sin(x) and Cosine cos(x) Functions." Ch. 32 in *An Atlas of Functions*. Washington, DC: Hemisphere, pp. 295–10, 1987.

Cosine Apodization Function

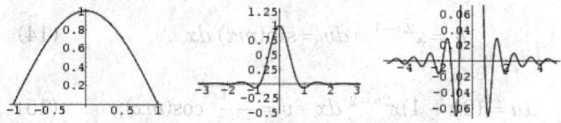

The APODIZATION FUNCTION

$$A(x) = \cos\left(\frac{\pi x}{2a}\right).$$

Its FULL WIDTH AT HALF MAXIMUM is $4a/3$. Its INSTRUMENT FUNCTION is

$$I(k) = \frac{4a\,\cos(2\pi ak)}{\pi(1 - 16a^2k^2)}.$$

See also APODIZATION FUNCTION

Cosine Circle

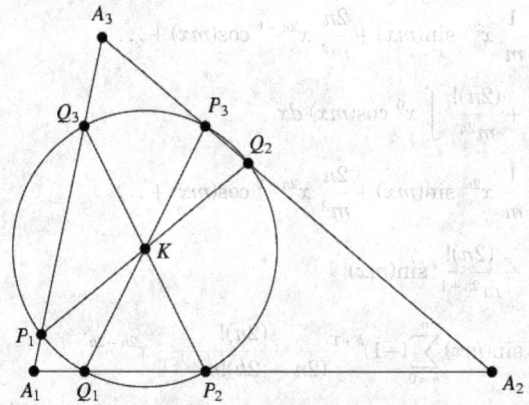

Draw ANTIPARALLELS through the SYMMEDIAN POINT K. The points where these lines intersect the sides then lie on a CIRCLE, known as the cosine circle (or sometimes the second LEMOINE CIRCLE), which has center at K. The CHORDS P_2Q_3, P_3Q_1, and P_1Q_2 are proportional to the COSINES of the ANGLES of $\triangle A_1A_2A_3$, giving the circle its name. The center of the cosine circle is the CIRCUMCENTER O of $\triangle ABC$.

TRIANGLES $P_1P_2P_3$ and $\triangle A_1A_2A_3$ are directly similar, and TRIANGLES $\triangle Q_1Q_2Q_3$ and $A_1A_2A_3$ are similar. The MIQUEL POINT of $\triangle P_1P_2P_3$ is at the BROCARD POINT Ω of $\triangle P_1P_2P_3$.

The cosine circle is a special case of a TUCKER CIRCLE.

See also BROCARD POINTS, EXCOSINE CIRCLE, LEMOINE CIRCLE, MIQUEL POINT, TAYLOR CIRCLE, TUCKER CIRCLES

References

Coolidge, J. L. *A Treatise on the Geometry of the Circle and Sphere*. New York: Chelsea, p. 66, 1971.

Honsberger, R. "The Lemoine Circles." §9.2 in *Episodes in Nineteenth and Twentieth Century Euclidean Geometry*. Washington, DC: Math. Assoc. Amer., pp. 88–9, 1995.

Johnson, R. A. *Modern Geometry: An Elementary Treatise on the Geometry of the Triangle and the Circle*. Boston, MA: Houghton Mifflin, pp. 271–73, 1929.

Lachlan, R. "The Cosine Circle." §129–30 in *An Elementary Treatise on Modern Pure Geometry*. London: Macmillian, p. 75, 1893.

Cosine Hexagon

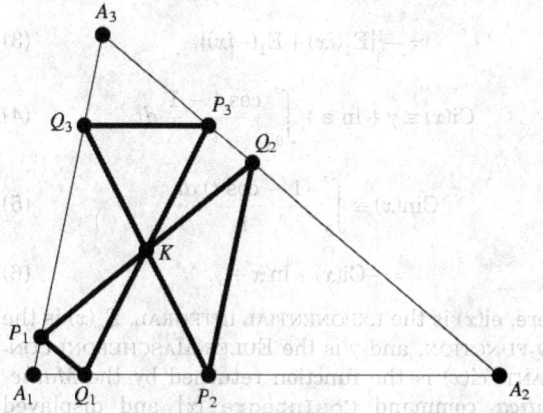

The closed cyclic self-intersecting hexagon formed by joining the adjacent ANTIPARALLELS in the construction of the COSINE CIRCLE. The sides of this hexagon have the property that, in addition to P_1Q_2, P_2Q_3, and P_3Q_1 being ANTIPARALLEL to A_1A_2, A_2A_3, A_1A_3, the remaining sides $P_1Q_1\|A_2A_3$, $P_2Q_2\|A_1A_3$, and $P_3Q_3\|A_1A_2$. The cosine hexagon is a special case of a TUCKER HEXAGON.

See also COSINE CIRCLE, LEMOINE HEXAGON, TUCKER HEXAGON

Cosine Integral

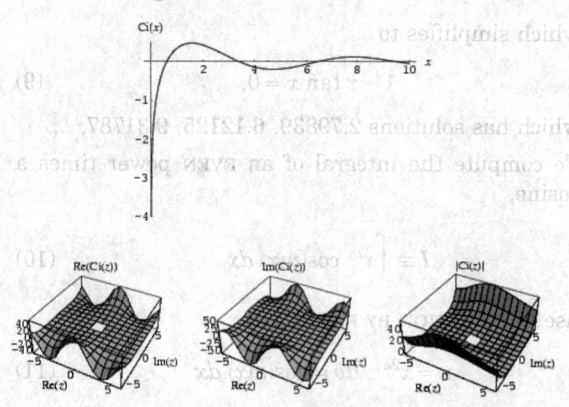

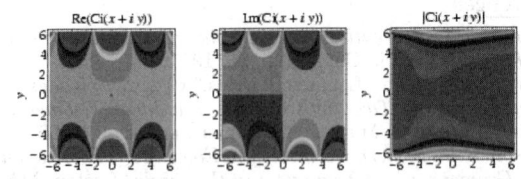

There are (at least) three types of "cosine integrals," denoted ci(x), Ci(x), and Cin(x) :

$$\text{ci}(x) \equiv -\int_x^\infty \frac{\cos t \, dt}{t} \tag{1}$$

$$= \tfrac{1}{2}[\text{ei}(ix) + \text{ei}(-ix)] \tag{2}$$

$$= -\tfrac{1}{2}[\text{E}_1(ix) + \text{E}_1(-ix)], \tag{3}$$

$$\text{Ci}(x) \equiv \gamma + \ln z + \int_0^z \frac{\cos t - 1}{t} \, dt \tag{4}$$

$$\text{Cin}(x) \equiv \int_0^z \frac{(1 - \cos t) \, dt}{t} \tag{5}$$

$$= -\text{Ci}(x) + \ln x + \gamma. \tag{6}$$

Here, ei(x) is the EXPONENTIAL INTEGRAL, $\text{E}_n(x)$ is the *EN*-FUNCTION, and γ is the EULER-MASCHERONI CONSTANT. ci(x) is the function returned by the *Mathematica* command CosIntegral[x] and displayed above.

ci(x) has zeros at 0.616505, 3.38418, 6.42705, ... Extrema occur when

$$\text{ci}'(x) = \frac{\cos x}{x} = 0, \tag{7}$$

or cos x = 0, or $\pi/2$, $3\pi/2$, $5\pi/2$, ..., which are alternately maxima and minima. At these points, ci(x) equals 0.472001, −0.198408, 0.123772, ... Inflection points occur when

$$\text{ci}''(x) = -\frac{\cos x}{x^2} - \frac{\sin x}{x} = 0, \tag{8}$$

which simplifies to

$$1 + x \tan x = 0, \tag{9}$$

which has solutions 2.79839, 6.12125, 9.31787, ...
To compute the integral of an EVEN power times a cosine,

$$I \equiv \int x^{2n} \cos(mx) \, dx, \tag{10}$$

use INTEGRATION BY PARTS. Let

$$u = x^{2n} \quad dv = \cos(mx) \, dx \tag{11}$$

$$du = 2nx^{2n-1} \, dx \quad v = \frac{1}{m} \sin(mx), \tag{12}$$

so

$$I = \frac{1}{m} x^{2n} \sin(mx) - \frac{2n}{m} \int x^{2n-1} \sin(mx) \, dx. \tag{13}$$

Using INTEGRATION BY PARTS again,

$$u = x^{2n-1} \quad dv = \sin(mx) \, dx \tag{14}$$

$$du = (2n-1)x^{2n-2} \, dx \quad v = -\frac{1}{m} \cos(mx), \tag{15}$$

and

$$\int x^{2n} \cos(mx) \, dx$$

$$= \frac{1}{m} x^{2n} \sin(mx) - \frac{2n}{m}$$

$$\times \left[-\frac{1}{m} x^{2n-1} \cos(mx) + \frac{2n-1}{m} \int x^{2n-2} \cos(mx) \, dx \right]$$

$$= \frac{1}{m} x^{2n} \sin(mx) + \frac{2n}{m^2} x^{2n-1} \cos(mx)$$

$$- \frac{(2n)(2n-1)}{m^2} \int x^{2n-2} \cos(mx) \, dx$$

$$= \frac{1}{m} x^{2n} \sin(mx) + \frac{2n}{m^2} x^{2n-1} \cos(mx) + \ldots$$

$$+ \frac{(2n)!}{m^{2n}} \int x^0 \cos(mx) \, dx$$

$$= \frac{1}{m} x^{2n} \sin(mx) + \frac{2n}{m^2} x^{2n-1} \cos(mx) + \ldots$$

$$+ \frac{(2n)!}{m^{2n+1}} \sin(mx)$$

$$= \sin(mx) \sum_{k=0}^n (-1)^{k+1} \frac{(2n)!}{(2n-2k)! m^{2k+1}} x^{2n-2k}$$

$$+ \cos(mx) \sum_{k=1}^n (-1)^{k+1} \frac{(2n)!}{(2k-2n-1)! m^{2k}} x^{2n-2k+1}.$$

$$\tag{16}$$

Letting $k' \equiv n - k$,

$$\int x^{2n} \cos(mx) \, dx$$

$$= \sin(mx) \sum_{k=0}^n (-1)^{n-k+1} \frac{(2n)!}{(2k)! m^{2n-2k+1}} x^{2k}$$

$$+ \cos(mx) \sum_{k=0}^{n-1} (-1)^{n-k+1} \frac{(2n)!}{(2k-1)! m^{2n-2k}} x^{2k+1}$$

$$= (-1)^{n+1}(2n)! \left[\sin(mx) \sum_{k=0}^{n-1} \frac{(-1)^k}{(2k)! m^{2n-2k+1}} x^{2k} \right.$$

$$\left. + \cos(mx) \sum_{k=1}^{n} \frac{(-1)^{k+1}}{(2k-3)! m^{2n-2k+2}} x^{2k+1} \right]. \qquad (17)$$

To find a closed form for an integral power of a cosine function,

$$I \equiv \int \cos^m x \, dx, \qquad (18)$$

perform an INTEGRATION BY PARTS so that

$$u = \cos^{m-1} x \quad dv = \cos x \, dx \qquad (19)$$

$$du = -(m-1) \cos^{m-2} x \sin x \, dx \quad v = \sin x. \qquad (20)$$

Therefore

$$I = \sin x \cos^{m-1} x + (m-1) \int \cos^{m-2} x \sin^2 x \, dx$$

$$= \sin x \cos^{m-1} x + (m-1)$$

$$\times \left[\int \cos^{m-2} x \, dx - \int \cos^m x \, dx \right]$$

$$= \sin x \cos^{m-1} x + (m-1) \left[\int \cos^{m-2} x \, dx - I \right], \qquad (21)$$

so

$$I[1 + (m-1)]$$

$$= \sin x \cos^{m-1} x + (m-1) \int \cos^{m-2} x \, dx \qquad (22)$$

$$I = \int \cos^m x \, dx$$

$$= \frac{\sin x \cos^{m-1} x}{m} + \frac{m-1}{m} \int \cos^{m-2} x \, dx. \qquad (23)$$

Now, if m is EVEN so $m \equiv 2n$, then

$$\int \cos^{2n} x \, dx = \frac{\sin x \cos^{2n-1} x}{2n} + \frac{2n-1}{2n} \int \cos^{2n-2} x \, dx$$

$$= \frac{\sin x \cos^{2n-1} x}{2n}$$

$$+ \frac{2n-1}{2n} \left[\frac{\sin x \cos^{2n-3} x}{2n-2} + \frac{2n-3}{2n-2} \int \cos^{2n-4} x \, dx \right]$$

$$= \sin x \left[\frac{1}{2n} \cos^{2n-1} x + \frac{2n-1}{(2n)(2n-2)} \cos^{2n-3} x \right]$$

$$+ \frac{(2n-1)(2n-3)}{(2n)(2n-2)} \int \cos^{2n-4} x \, dx$$

$$= \sin x \left[\frac{1}{2n} \cos^{2n-1} x + \frac{2n-1}{(2n)(2n-2)} \cos^{2n-3} x + \ldots \right]$$

$$+ \frac{(2n-1)(2n-3) \cdots 1}{(2n)(2n-2) \cdots 2} \int \cos^0 x \, dx$$

$$= \sin x \sum_{k=1}^{n} \frac{(2n-2k)!!}{(2n)!!} \frac{(2n-1)!!}{(2n-2k+1)!!} \cos^{2n-2k+1} x$$

$$+ \frac{(2n-1)!!}{(2n)!!} x. \qquad (24)$$

Now let $k' \equiv n - k + 1$, so $n - k = k' - 1$,

$$\int \cos^{2n} x \, dx$$

$$= \sin x \sum_{k=1}^{n} \frac{(2k-2)!!}{(2n)!!} \frac{(2n-1)!!}{(2k-1)!!} \cos^{2k-1} x$$

$$+ \frac{(2n-1)!!}{(2n)!!} x$$

$$= \frac{(2n-1)!!}{(2n)!!}$$

$$\times \left[\sin x \sum_{k=0}^{n-1} \frac{(2k)!!}{(2k+1)!!} \cos^{2k+1} x + x \right]. \qquad (25)$$

Now if m is ODD so $m \equiv 2n + 1$, then

$$\int \cos^{2n+1} x \, dx = \frac{\sin x \cos^{2n} x}{2n+1} + \frac{2n}{2n+1} \int \cos^{2n-1} x \, dx$$

$$= \frac{\sin x \cos^{2n} x}{2n+1} + \frac{2n}{2n+1}$$

$$\times \left[\frac{\sin x \cos^{2n-2} x}{2n-1} + \frac{2n-2}{2n-1} \int \cos^{2n-3} x \, dx \right]$$

$$= \sin x \left[\frac{1}{2n+1} \cos^{2n} x + \frac{2n}{(2n+1)(2n-1)} \cos^{2n-2} x \right]$$

$$+ \frac{(2n)(2n-2)}{(2n+1)(2n-1)} \int \cos^{2n-3} x \, dx$$

$$= \sin x \left[\frac{1}{2n+1} \cos^{2n} x + \frac{2n}{(2n+1)(2n-1)} \cos^{2n-2} x \right.$$

$$\left. + \ldots \right]$$

$$+ \frac{(2n)(2n-2)\cdots 2}{(2n+1)(2n-1)\cdots 3} \int \cos x \, dx$$

$$= \sin x \sum_{k=0}^{n} \frac{(2n-2k-1)!!}{(2n+1)!!} \frac{(2n)!!}{(2n-2k)!!} \cos^{2n-2k} x. \tag{26}$$

Now let $k' \equiv n - k$,

$$\int \cos^{2n} x \, dx$$

$$= \frac{(2n)!!}{(2n+1)!!} \sin x \sum_{k=0}^{n} \frac{(2k-1)!!}{(2k)!!} \cos^{2k} x. \tag{27}$$

The general result is then

$$\int \cos^m x \, dx$$

$$= \begin{cases} \dfrac{(2n-1)!!}{(2n)!!} \left[\sin x \displaystyle\sum_{k=0}^{n-1} \dfrac{(2k)!!}{(2k+1)!!} \cos^{2k+1} x + x \right] \\ \quad \text{for } m = 2n \\[2mm] \dfrac{(2n)!!}{(2n+1)!!} \sin x \displaystyle\sum_{k=0}^{n} \dfrac{(2k-1)!!}{(2k)!!} \cos^{2k} x \\ \quad \text{for } m = 2n+1. \end{cases} \tag{28}$$

The infinite integral of a cosine times a Gaussian can also be done in closed form,

$$\int_{-\infty}^{\infty} e^{-ax^2} \cos(kx) \, dx = \sqrt{\frac{\pi}{a}} e^{-k^2/4a}. \tag{29}$$

See also CHI, DAMPED EXPONENTIAL COSINE INTEGRAL, NIELSEN'S SPIRAL, SHI, SICI SPIRAL, SINE INTEGRAL

References

Abramowitz, M. and Stegun, C. A. (Eds.). "Sine and Cosine Integrals." §5.2 in *Handbook of Mathematical Functions with Formulas, Graphs, and Mathematical Tables, 9th printing.* New York: Dover, pp. 231–33, 1972.
Arfken, G. *Mathematical Methods for Physicists, 3rd ed.* Orlando, FL: Academic Press, pp. 342–43, 1985.
Press, W. H.; Flannery, B. P.; Teukolsky, S. A.; and Vetterling, W. T. "Fresnel Integrals, Cosine and Sine Integrals." §6.79 in *Numerical Recipes in FORTRAN: The Art of Scientific Computing, 2nd ed.* Cambridge, England: Cambridge University Press, pp. 248–52, 1992.
Spanier, J. and Oldham, K. B. "The Cosine and Sine Integrals." Ch. 38 in *An Atlas of Functions.* Washington, DC: Hemisphere, pp. 361–72, 1987.

Cosines Law

LAW OF COSINES

CosIntegral

COSINE INTEGRAL

Cosmic Figure

PLATONIC SOLID

Cosmological Theorem

There exists an INTEGER N such that every string in the LOOK AND SAY SEQUENCE "decays" in at most N days to a compound of "common" and "transuranic elements."

The table below gives the periodic table of atoms associated with the LOOK AND SAY SEQUENCE as named by Conway (1987). The "abundance" is the average number of occurrences for long strings out of every million atoms. The asymptotic abundances are zero for transuranic elements, and 27.246... for arsenic (As), the next rarest element. The most common element is hydrogen (H), having an abundance of 91,970.383.... The starting element is U, represented by the string "3," and subsequent terms are those giving a description of the current term: one three (13); one one, one three (1113); three ones, one three (3113), etc.

Abundance	n	E_n	E_n is the derivate of E_{n+1}
102.56285249	92	U	3
9883.5986392	91	Pa	13
7581.9047125	90	Th	1113
6926.9352045	89	Ac	3113
5313.7894999	88	Ra	132113
4076.3134078	87	Fr	1113122113
3127.0209328	86	Rn	311311222113
2398.7998311	85	At	Ho.1322113
1840.1669683	84	Po	1113222113
1411.6286100	83	Bi	3113322113
1082.8883285	82	Pb	Pm.123222113
830.70513293	81	Tl	111213322113
637.25039755	80	Hg	31121123222113
488.84742982	79	Au	1321122112133222113
375.00456738	78	Pt	111312212221121123222113
287.67344775	77	Ir	31131122211311222113
220.68001229	76	Os	13211321222113222122211211232221113
169.28801808	75	Re	11131221131211322113322113222112211213322113
315.56655252	74	W	Ge.Ca.3122113222122221121123222113
242.07736666	73	Ta	13112221133211322112211213322113
2669.0970363	72	Hf	11132.Pa.H.Ca.W

2047.5173200	71	Lu	311312
1570.6911808	70	Yb	1321131112
1204.9083841	69	Tm	11131221133112
1098.5955997	68	Er	311311222.Ca.Co
47987.529438	67	Ho	1321132.Pm
36812.186418	66	Dy	111312211312
28239.358949	65	Tb	3113112221131112
21662.972821	64	Gd	Ho.13221133112
20085.668709	63	Eu	1113222.Ca.Co
15408.115182	62	Sm	311332
29820.456167	61	Pm	132.Ca.Zn
22875.863883	60	Nd	111312
17548.529287	59	Pr	31131112
13461.825166	58	Ce	1321133112
10326.833312	57	La	11131.H.Ca.Co
7921.9188284	56	Ba	311311
6077.0611889	55	Cs	13211321
4661.8342720	54	Xe	11131221131211
3576.1856107	53	I	3113112221113111221
2743.3629718	52	Te	Ho.1322113312211
2104.4881933	51	Sb	Eu.Ca.3112221
1614.3946687	50	Sn	Pm.13211
1238.4341972	49	In	11131221
950.02745646	48	Cd	3113112211
728.78492056	47	Ag	132113212221
559.06537946	46	Pd	111312211312113211
428.87015041	45	Rh	311311222113111221131221
328.99480576	44	Ru	Ho.13221133122113112211
386.07704943	43	Tc	Eu.Ca.311322113212221
296.16736852	42	Mo	13211322211312113211
227.19586752	41	Nb	1113122113322113111221131221
174.28645997	40	Zr	Er.12322211331222113112211
133.69860315	39	Y	1112133.H.Ca.Tc
102.56285249	38	Sr	3112112.U
78.678000089	37	Rb	1321122112
60.355455682	36	Kr	11131221222112
46.299868152	35	Br	3113112211322112
35.517547944	34	Se	13211321222113222112
27.246216076	33	As	1113122113121113222112
1887.4372276	32	Ge	3113112221131112211322.Na
1447.8905642	31	Ga	Ho.13221133122211332
23571.391336	30	Zn	Eu.Ca.Ac.H.Ca.312
18082.082203	29	Cu	131112
13871.123200	28	Ni	11133112
45645.877256	27	Co	Zn.32112
35015.858546	26	Fe	13122112
26861.360180	25	Mn	111311222112
20605.882611	24	Cr	31132.Si
15807.181592	23	V	13211312
12126.002783	22	Ti	11131221131112
9302.0974443	21	Sc	3113112221133112
56072.543129	20	Ca	Ho.Pa.H.12.Co
43014.360913	19	K	1112
32997.170122	18	Ar	3112
25312.784218	17	Cl	132112
19417.939250	16	S	1113122112
14895.886658	15	P	311311222112
32032.812960	14	Si	Ho.1322112
24573.006696	13	Al	1113222112
18850.441228	12	Mg	3113322112
14481.448773	11	Na	Pm.123222112
11109.006696	10	Ne	111213322112
8521.9396539	9	F	31121123222112
6537.3490750	8	O	1321112211213322112
5014.9302464	7	N	111311222112111233222112
3847.0525419	6	C	3113112221131122211211213322112
2951.1503716	5	B	132113212221132221222112113222112
2263.8860325	4	Be	111312211312113221133211322211211213322112
4220.0665982	3	Li	Ge.Ca.312211322212221121123222122
3237.2968588	2	He	13112221133211322112211213322112
91790.383216	1	H	Hf.Pa.22.Ca.Li

See also CONWAY'S CONSTANT, LOOK AND SAY SEQUENCE

References

Conway, J. H. "The Weird and Wonderful Chemistry of Audioactive Decay." §5.11 in *Open Problems in Communication and Computation* (Ed. T. M. Cover and B. Gopinath). New York: Springer-Verlag, pp. 173–88, 1987.

Conway, J. H. "The Weird and Wonderful Chemistry of Audioactive Decay." *Eureka,* 5–8, 1985.

Ekhad, S. B. and Zeilberger, D. "Proof of Conway's Lost Cosmological Theorem." *Electronic Research Announcement of the Amer. Math. Soc.* **3**, 78–2, 1997. http://www.math.temple.edu/~zeilberg/mamarim/mamarimhtml/horton.html.

Hilgemeier, M. "Die Gleichniszahlen-Reihe." *Bild der Wissensch.* **12**, 19, 1986.

Hilgemeier, M. "'One Metaphor Fits All': A Fractal Voyage with Conway's Audioactive Decay." Ch. 7 in Pickover, C. A. (Ed.). *Fractal Horizons: The Future Use of Fractals.* New York: St. Martin's Press, 1996.

Costa Minimal Surface

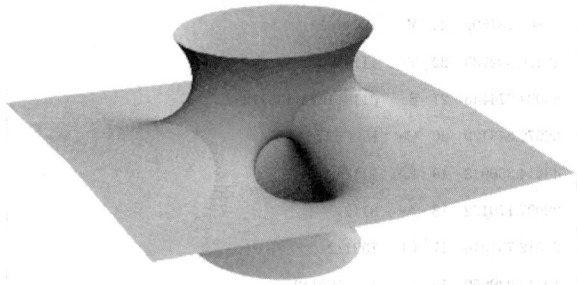

A COMPLETE MINIMAL EMBEDDABLE SURFACE of finite topology (i.e., it has no BOUNDARY and does not intersect itself). Until this surface was discovered by Costa (1984), the only other known complete minimal embeddable surfaces in $\mathbb{R}^3$ with no self-intersections were the PLANE, CATENOID, and HELICOID. The plane is genus 0 and the catenoid and the helicoid are genus 0 with two punctures, but the Costa minimal surface is genus 1 with three punctures (Schwalbe and Wagon 1999). In addition, and rather amazingly, the Costa surface belongs to the D_4 DIHEDRAL GROUP of symmetries. An animation by S. Dickson illustrates the homotopy of the TORUS into a Costa surface (Wolfram Research).

As discovered by Gray (Ferguson *et al.* 1996, Gray 1997), the Costa surface can be represented parametrically explicitly by

$$x = \tfrac{1}{2} \Re \left\{ -\zeta(u+iv) + \pi u + \frac{\pi^2}{4e_1} + \frac{\pi}{2e_1} \left[\zeta(u+iv-\tfrac{1}{2}) - \zeta(u+iv-\tfrac{1}{2}i) \right] \right\}$$

$$y = \tfrac{1}{2} \Re \left\{ -i\zeta(u+iv) + \pi v + \frac{\pi^2}{4e_1} - \frac{\pi}{2e_1} \left[i\zeta(u+iv-\tfrac{1}{2}) - i\zeta(u+iv-\tfrac{1}{2}i) \right] \right\}$$

$$z = \tfrac{1}{4} \sqrt{2\pi} \ln \left| \frac{\wp(u+iv) - e_1}{\wp(u+iv) + e_1} \right|$$

where $\zeta(z)$ is the WEIERSTRASS ZETA FUNCTION, $\wp(g_2, g_3; z)$ is the WEIERSTRASS ELLIPTIC FUNCTION with $(g_2, g_3) = (189.072772\ldots, 0)$ the invariants corresponding to the half-periods $1/2$ and $i/2$, and first root

$$e_1 = \wp(\tfrac{1}{2}; 0, g_3) = \wp(\tfrac{1}{2}|\tfrac{1}{2}, \tfrac{1}{2}i) \approx 6.87519,$$

where $\wp(z; g_2, g_3) = \wp(z|\omega_1, \omega_2)$ is the WEIERSTRASS ELLIPTIC FUNCTION.

See also COMPLETE MINIMAL SURFACE, MINIMAL SURFACE, WEIERSTRASS ELLIPTIC FUNCTION, WEIERSTRASS ZETA FUNCTION

References

Costa, A. "Examples of a Complete Minimal Immersion in R^3 of Genus One and Three Embedded Ends." *Bil. Soc. Bras. Mat.* **15**, 47–4, 1984.
do Carmo, M. P. *Mathematical Models from the Collections of Universities and Museums* (Ed. G. Fischer). Braunschweig, Germany: Vieweg, p. 43, 1986.
Ferguson, H.; Gray, A.; and Markvorsen, S. "Costa's Minimal Surface via *Mathematica*." *Mathematica in Educ. Res.* **5**, 5–0, 1996.
Ferguson, H.; Ferguson, C.; Nemeth, R.; Schwalbe, D.; and Wagon, S. "Invisible Handshake." *Math. Intell.* **21**, 1999. To appear.
Gray, A. "Costa's Minimal Surface." §32.5 in *Modern Differential Geometry of Curves and Surfaces with Mathematica, 2nd ed.* Boca Raton, FL: CRC Press, pp. 747–57, 1997.
Hoffman, D. and Meeks, W. H. III. "A Complete Embedded Minimal Surfaces in $\mathbb{R}^3$ with Genus One and Three Ends." *J. Diff. Geom.* **21**, 109–27, 1985.
Nordstrand, T. "Costa-Hoffman-Meeks Minimal Surface." http://www.uib.no/people/nfytn/costatxt.htm.
Osserman, R. *A Survey of Minimal Surfaces.* New York: Dover, pp. 149–50, 1986.
Peterson, I. "Three Bites in a Doughnut: Computer-Generated Pictures Contribute to the Discovery of a New Minimal Surface." *Sci. News* **127**, 161–76, 1985.
Peterson, I. "The Song in the Stone: Developing the Art of Telecarving a Minimal Surface." *Sci. News* **149**, 110–11, Feb. 17, 1996.
Schwalbe, D. and Wagon, S. "The Costa Surface, in Show and *Mathematica*." *Mathematica in Educ. Res.* **8**, 56–3, 1999.
Wolfram Research, Inc. "3-D Zoetrope at SIGGRAPH 2000." http://www.wolfram.com/news/zoetrope.html.

Costa-Hoffman-Meeks Minimal Surface

COSTA MINIMAL SURFACE

Cosymmedian Triangles

Extend the SYMMEDIANS of a TRIANGLE $\Delta A_1 A_2 A_3$ to meet the CIRCUMCIRCLE at P_1, P_2, P_3. Then the SYMMEDIAN POINT K of $\Delta A_1 A_2 A_3$ is also the SYMMEDIAN POINT of $\Delta P_1 P_2 P_3$. The TRIANGLES $\Delta A_1 A_2 A_3$ and $\Delta P_1 P_2 P_3$ are cosymmedian triangles, and have the same BROCARD CIRCLE, second BROCARD TRIANGLE, BROCARD ANGLE, BROCARD POINTS, and CIRCUMCIRCLE.

See also BROCARD ANGLE, BROCARD CIRCLE, BROCARD POINTS, BROCARD TRIANGLES, CIRCUMCIRCLE, COMEDIAN TRIANGLES, SYMMEDIAN, SYMMEDIAN POINT

References

Lachlan, R. *An Elementary Treatise on Modern Pure Geometry.* London: Macmillian, p. 63, 1893.

Cot

COTANGENT

Cotangent

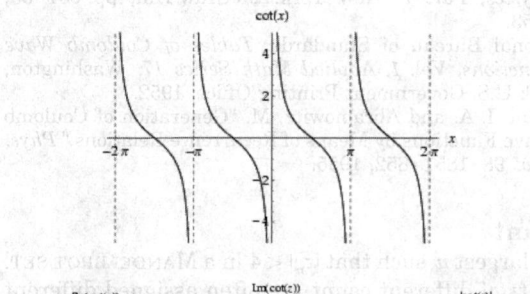

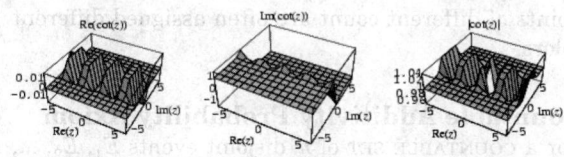

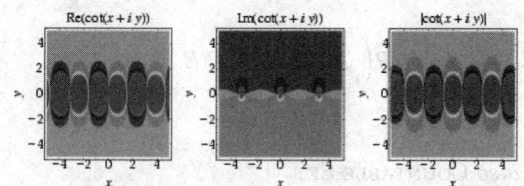

The function defined by $\cot x \equiv 1/\tan x$, where $\tan x$ is the TANGENT. The notations ctn x (Erdélyi *et al.* 1981, p. 7) and ctg x (Gradshteyn and Ryzhik 2000, p. xxix) are sometimes used in place of $\cot x$.

The MACLAURIN SERIES for $\cot x$ is

$$\cot x = \frac{1}{x} - \frac{1}{3}x - \frac{1}{45}x^3 - \frac{2}{945}x^5 - \frac{1}{4725}x^7 - \cdots$$

$$- \frac{(-1)^{n+1}2^{2n}B_{2n}}{(2n)!} - \cdots,$$

where B_n is a BERNOULLI NUMBER.

$$\pi \cot(\pi x) = \frac{1}{x} + 2x \sum_{n=1}^{\infty} \frac{1}{x^2 - n^2}.$$

It is known that, for $n \geq 3$, $\cot(\pi/n)$ is rational only for $n = 4$.

See also HYPERBOLIC COTANGENT, INVERSE COTANGENT, LEHMER'S CONSTANT, TANGENT

References

Abramowitz, M. and Stegun, C. A. (Eds.). "Circular Functions." §4.3 in *Handbook of Mathematical Functions with Formulas, Graphs, and Mathematical Tables, 9th printing.* New York: Dover, pp. 71–9, 1972.

Beyer, W. H. *CRC Standard Mathematical Tables, 28th ed.* Boca Raton, FL: CRC Press, p. 215, 1987.

Erdélyi, A.; Magnus, W.; Oberhettinger, F.; and Tricomi, F. G. *Higher Transcendental Functions, Vol. 1.* New York: Krieger, p. 6, 1981.

Gradshteyn, I. S. and Ryzhik, I. M. *Tables of Integrals, Series, and Products, 6th ed.* San Diego, CA: Academic Press, 2000.

Spanier, J. and Oldham, K. B. "The Tangent tan(x) and Cotangent cot(x) Functions." Ch. 34 in *An Atlas of Functions.* Washington, DC: Hemisphere, pp. 319–30, 1987.

Cotangent Bundle

The cotangent bundle of a MANIFOLD is similar to the TANGENT BUNDLE, except that it is the set (x, f) where $x \in M$ and f is a dual vector in the TANGENT SPACE to $x \in M$. The cotangent bundle is denoted T^*M.

See also TANGENT BUNDLE

Cotes Circle Property

$$x^{2n} + 1 = \left[x^2 - 2x \cos\left(\frac{\pi}{2n}\right) + 1 \right]$$

$$\times \left[x^2 - 2x \cos\left(\frac{3\pi}{2n}\right) + 1 \right] \times \cdots \times$$

$$\times \left[x^2 - 2x \cos\left(\frac{(2n-1)\pi}{2n}\right) + 1 \right].$$

See also COSINE, TRIGONOMETRIC FUNCTIONS

Cotes Number

The numbers λ_{vn} in the GAUSSIAN QUADRATURE formula

$$Q_n(f) = \sum_{v=1}^{n} \lambda_{vn} f(x_{vn}).$$

See also CHRISTOFFEL NUMBER, GAUSSIAN QUADRATURE

References

Cajori, F. *A History of Mathematical Notations, Vols. 1–.* New York: Dover, p. 42, 1993.

Cotes' Spiral

The planar orbit of a particle under a r^{-3} force field. It is an EPISPIRAL.

See also EPISPIRAL

Coth

HYPERBOLIC COTANGENT.

Cotree

The cotree T^* of a spanning tree T in a CONNECTED GRAPH G is the spacing SUBGRAPH of G containing exactly those edges of G which are not in T (Harary 1994, p. 39).

See also TWIG

Coulomb Wave Function

References

Harary, F. *Graph Theory.* Reading, MA: Addison-Wesley, 1994.

Coulomb Wave Function

A special case of the CONFLUENT HYPERGEOMETRIC FUNCTION OF THE FIRST KIND. It gives the solution to the radial Schrödinger equation in the Coulomb potential $(1/r)$ of a point nucleus

$$\frac{d^2W}{d\rho^2} + \left[1 - \frac{2\eta}{\rho} - \frac{L(L+1)}{\rho^2}\right]W = 0 \qquad (1)$$

(Abramowitz and Stegun 1972; Zwillinger 1997, p. 122). The complete solution is

$$W = C_1 F_L(\eta, \; \rho) + C_2 G_L(\eta, \; \rho). \qquad (2)$$

The Coulomb function of the first kind is

$$F_L(\eta, \; \rho) = C_L(\eta)\rho^{L+1}e^{-ip} \, _1F_1(L+1-i\eta; \; 2L \\ +2; \; 2i\rho), \qquad (3)$$

where

$$C_L(\eta) \equiv \frac{2^L e^{-\pi\eta/2}|\Gamma(L+1+i\eta)|}{\Gamma(2L+2)}, \qquad (4)$$

$_1F_1(a; b; z)$ is the CONFLUENT HYPERGEOMETRIC FUNCTION, $\Gamma(z)$ is the GAMMA FUNCTION, and the Coulomb function of the second kind is

$$G_L(\eta, \; \rho) = \frac{2\eta}{C_0^2(\eta)} F_L(\eta, \; \rho)\left[\ln(2\rho) + \frac{q_L(\eta)}{p_L(\eta)}\right] \\ + \frac{1}{(2L+1)C_L(\eta)} \, \rho^{-L} \sum_{K=-L}^{\infty} a_k^L(\eta)\rho^{K+L}, \quad (5)$$

where q_L, p_L, and a_k^L are defined in Abramowitz and Stegun (1972, p. 538).

See also CONFLUENT HYPERGEOMETRIC FUNCTION OF THE FIRST KIND

References

Abramowitz, M. and Antosiewicz, H. A. "Coulomb Wave Functions in the Transition Region." *Phys. Rev.* **96**, 75–7, 1954.

Abramowitz, M. and Rabinowitz, P. "Evaluation of Coulomb Wave Functions along the Transition Line." *Phys. Rev.* **96**, 77–9, 1954.

Abramowitz, M. and Stegun, C. A. (Eds.). "Coulomb Wave Functions." Ch. 14 in *Handbook of Mathematical Functions with Formulas, Graphs, and Mathematical Tables, 9th printing.* New York: Dover, pp. 537–44, 1972.

Biedenharn, L. C.; Gluckstern, R. L.; Hull, M. H. Jr.; and Breit, G. "Coulomb Wave Functions for Large Charges and Small Velocities." *Phys. Rev.* **97**, 542–54, 1955.

Bloch, I.; Hull, M. H. Jr.; Broyles, A. A.; Bouricius, W. G.; Freeman, B. E.; and Breit, G. "Coulomb Functions for Reactions of Protons and Alpha-Particles with the Lighter Nuclei." *Rev. Mod. Phys.* **23**, 147–82, 1951.

Morse, P. M. and Feshbach, H. *Methods of Theoretical Physics, Part I.* New York: McGraw-Hill, pp. 631–33, 1953.

National Bureau of Standards. *Tables of Coulomb Wave Functions, Vol. 1, Applied Math Series 17.* Washington, DC: U.S. Government Printing Office, 1952.

Stegun, I. A. and Abramowitz, M. "Generation of Coulomb Wave Functions by Means of Recurrence Relations." *Phys. Rev.* **98**, 1851–852, 1955.

Count

The largest n such that $|z_n| < 4$ in a MANDELBROT SET. Points of different count are often assigned different colors.

Countable Additivity Probability Axiom

For a COUNTABLE SET of n disjoint events $E_1, E_2, ..., E_n$

$$P\left(\bigcup_{i=1}^{n} E_i\right) = \sum_{i=1}^{n} P(E_i).$$

See also COUNTABLE SET

Countable Set

A SET which is either FINITE or DENUMERABLE. However, some author (Ciesielski 1997, p. 64) use the definition "equipollent to the finite ordinals," commonly used to define a DENUMERABLE SET, to define a countable set.

See also ALEPH-0, ALEPH-1, COUNTABLE INFINITE, DENUMERABLE SET, FINITE, INFINITE, UNCOUNTABLY INFINITE

References

Ciesielski, K. *Set Theory for the Working Mathematician.* Cambridge, England: Cambridge University Press, 1997.

Croft, H. T.; Falconer, K. J.; and Guy, R. K. *Unsolved Problems in Geometry.* New York: Springer-Verlag, p. 2, 1991.

Countable Space

FIRST-COUNTABLE SPACE

Countably Infinite

Any SET which can be put in a ONE-TO-ONE correspondence with the NATURAL NUMBERS (or INTEGERS) so that a prescription can be given for identifying its members one at a time is called a countably infinite (or denumerably infinite) set. Once one countable set S is given, any other set which can be put into a ONE-TO-ONE correspondence with S is also countable. Countably infinite sets have CARDINAL NUMBER ALEPH-0.

Examples of countable sets include the INTEGERS, ALGEBRAIC NUMBERS, and RATIONAL NUMBERS. Georg Cantor showed that the number of REAL NUMBERS is rigorously larger than a countably infinite set, and the postulate that this number, the so-called "CONTINUUM," is equal to ALEPH-1 is called the CONTINUUM HYPOTHESIS. Examples of nondenumerable sets include the REAL, COMPLEX, IRRATIONAL, and TRANSCENDENTAL NUMBERS.

See also ALEPH-0, ALEPH-1, CANTOR DIAGONAL SLASH, CARDINAL NUMBER, CONTINUUM, CONTINUUM HYPOTHESIS, COUNTABLE SET, HILBERT HOTEL, UNCOUNTABLY INFINITE

References

Courant, R. and Robbins, H. "The Denumerability of the Rational Number and the Non-Denumerability of the Continuum." §2.4.2 in *What is Mathematics?: An Elementary Approach to Ideas and Methods, 2nd ed.* Oxford, England: Oxford University Press, pp. 79–3, 1996.
Jeffreys, H. and Jeffreys, B. S. *Methods of Mathematical Physics, 3rd ed.* Cambridge, England: Cambridge University Press, p. 10, 1988.

Counterfeit Coin Problem
WEIGHING

Counting Generalized Principle

If r experiments are performed with n_i possible outcomes for each experiment $i = 1, 2, \ldots, r$, then there are a total of $\prod_{i=1}^{r} n_i$ possible outcomes.

Counting Number

A POSITIVE INTEGER: 1, 2, 3, 4, ... (Sloane's A000027), also called a NATURAL NUMBER. However, zero (0) is sometimes also included in the list of counting numbers. Due to lack of standard terminology, the following terms are recommended in preference to "counting number," "NATURAL NUMBER," and "WHOLE NUMBER."

set	name	symbol
..., -2, -1, 0, 1, 2, ...	INTEGERS	Z
1, 2, 3, 4, ...	POSITIVE INTEGERS	Z+
0, 1, 2, 3, 4, ...	NONNEGATIVE INTEGERS	Z*
0, -1, -2, -3, -4, ...	NONPOSITIVE INTEGERS	
-1, -2, -3, -4, ...	NEGATIVE INTEGERS	Z-

See also NATURAL NUMBER, WHOLE NUMBER, Z, Z-, Z+, Z*

References

Sloane, N. J. A. Sequences A000027/M0472 in "An On-Line Version of the Encyclopedia of Integer Sequences." http://www.research.att.com/~njas/sequences/eisonline.html.

Coupon Collector's Problem

Let n objects be picked repeatedly with probability p_i that object i is picked on a given try, with

$$\sum_i p_i = 1.$$

Find the earliest time at which all n objects have been picked at least once.

References

Hildebrand, M. V. "The Birthday Problem." *Amer. Math. Monthly* **100**, 643, 1993.

Cousin Primes

Pairs of PRIMES OF THE FORM $(p, p+4)$ are called cousin primes. The first few are (3, 7), (7, 11), (13, 17), (19, 23), (37, 41), (43, 47), (67, 71), ... (Sloane's A023200 and A046132). According to the first FIRST HARDY-LITTLEWOOD CONJECTURE, the cousin primes have the same asymptotic density as the TWIN PRIMES,

$$P_x(p, p+4) \sim 2 \prod_{p \geq 3} \frac{p(p-2)}{(p-1)^2} \int_2^x \frac{dx'}{(\ln x')^2}$$

$$= 1.320323632 \int_2^x \frac{dx'}{(\ln x')^2}$$

where $\prod_2 = 1.320323632$ is the TWIN PRIMES CONSTANT.

An analogy to BRUN'S CONSTANT, the constant

$$B_4 \equiv (\tfrac{1}{7} + \tfrac{1}{11}) + (\tfrac{1}{13} + \tfrac{1}{17}) + (\tfrac{1}{19} + \tfrac{1}{23}) + (\tfrac{1}{37} + \tfrac{1}{41}) + \ldots,$$

(omitting the initial term $1/3 + 1/7$) can be defined. Using cousin primes up to 2^{42}, the value of B_4 is estimated as

$$B_4 \approx 1.1970449$$

(Wolf 1996).

See also BRUN'S CONSTANT, PRIME CONSTELLATION, SEXY PRIMES, TWIN PRIMES, TWIN PRIMES CONSTANT

References

Sloane, N. J. A. Sequences A023200 and A046132 in "An On-Line Version of the Encyclopedia of Integer Sequences." http://www.research.att.com/~njas/sequences/eisonline.html.

Covariance

Given n sets of variates denoted $\{x_1\}, \ldots, \{x_n\}$, the covariance $\sigma_{ij} \equiv \mathrm{cov}(x_i, x_j)$ of x_i and x_j is defined by

$$\mathrm{cov}(x_i, x_j) \equiv \langle (x_i - \mu_i)(x_j - \mu_j) \rangle \tag{1}$$

$$= \langle x_i x_j \rangle - \langle x_i \rangle \langle x_j \rangle, \tag{2}$$

where $\mu_i = \langle x_i \rangle$ and $\mu_j = \langle x_j \rangle$ are the MEANS of x_i and x_j, respectively. The matrix (V_{ij}) of the quantities

$V_{ij} = \text{cov}(x_i, x_j)$ is called the COVARIANCE MATRIX. In the special case $i = j$,

$$\text{cov}(x_i, x_i) = \langle x_i^2 \rangle - \langle x_i \rangle^2 = \sigma_i^2, \tag{3}$$

giving the usual VARIANCE $\sigma_{ii} = \sigma_i^2 = \text{var}(x_i)$, .

The covariance of two variates x_i and x_j provides a measure of how strongly correlated these variables are, and the derived quantity

$$\text{cor}(x_i, x_j) \equiv \frac{\text{cov}(x_i, x_j)}{\sigma_i \sigma_j}, \tag{4}$$

where σ_i, σ_j are the STANDARD DEVIATIONS, is called CORRELATION of x_i and x_j. The covariance is symmetric since

$$\text{cov}(x, y) = \text{cov}(y, x). \tag{5}$$

For two variables, the covariance is related to the VARIANCE by

$$\text{var}(x + y) = \text{var}(x) + \text{var}(y) + 2\,\text{cov}(x, y). \tag{6}$$

For two independent variates $x = x_i$ and $y = x_j$,

$$\text{cov}(x, y) = \langle xy \rangle - \mu_x \mu_y = \langle x \rangle \langle y \rangle - \mu_x \mu_y = 0, \tag{7}$$

so the covariance is zero. However, if the variables are correlated in some way, then their covariance will be NONZERO. In fact, if $\text{cov}(x, y) > 0$, then y tends to increase as x increases. If $\text{cov}(x, y) < 0$, then y tends to decrease as x increases.

The covariance obeys the identity

$$\begin{aligned}
\text{cov}(x + z, y) &= \langle (x + z)y - (x + z)(y) \rangle \\
&= \langle xy \rangle + \langle zy \rangle - (\langle x \rangle + \langle z \rangle)\langle y \rangle \\
&= \langle xy \rangle - \langle x \rangle \langle y \rangle + \langle zy \rangle - \langle z \rangle \langle y \rangle \\
&= \text{cov}(x, y) + \text{cov}(z, y). \tag{8}
\end{aligned}$$

By induction, it therefore follows that

$$\text{cov}\left(\sum_{i=1}^{n} x_i, y \right) = \sum_{i=1}^{n} \text{cov}(x_i, y) \tag{9}$$

$$\text{cov}\left(\sum_{i=1}^{n} x_i, \sum_{j=1}^{m} y_j \right) = \sum_{i=1}^{n} \text{cov}\left(x_i, \sum_{j=1}^{m} y_j \right) \tag{10}$$

$$= \sum_{i=1}^{n} \text{cov}\left(\sum_{j=1}^{m} y_j, x_i \right) \tag{11}$$

$$= \sum_{i=1}^{n} \sum_{j=1}^{m} \text{cov}(y_j, x_i) \tag{12}$$

$$= \sum_{i=1}^{n} \sum_{j=1}^{n} \text{cov}(x_i, y_j). \tag{13}$$

See also CORRELATION (STATISTICAL), COVARIANCE MATRIX, VARIANCE

Covariance Matrix

Given n sets of variates denoted $\{x_1\}$, ..., $\{x_n\}$, the first-order covariance matrix is defined by

$$V_{ij} = \text{cov}(x_i, x_j) \equiv \langle (x_i - \mu_i)(x_j - \mu_j) \rangle,$$

where μ_i is the MEAN. Higher order matrices are given by

$$V_{ij}^{mn} = \langle (x_i - \mu_i)^m (x_j - \mu_j)^n \rangle.$$

An individual matrix element $V_{ij} = \text{cov}(x_i, x_j)$ is called the COVARIANCE of x_i and x_j.

See also CORRELATION (STATISTICAL), COVARIANCE, ERROR PROPAGATION, VARIANCE

Covariant Derivative

The covariant derivative of a CONTRAVARIANT TENSOR A^a (also called the "semicolon derivative" since its symbol is a semicolon) is given by

$$\nabla \cdot \mathbf{A} \equiv A^a_{;\,b} = A^a_{\,,\,b} + \Gamma^a_{bk} A^k, \tag{1}$$

where $A^k_{\,k}$ is a COMMA DERIVATIVE and $\nabla \cdot$ is a generalization of the symbol commonly used to denote the DIVERGENCE of a vector function in 3-D, Γ^k_{ij} is a CONNECTION COEFFICIENT, and EINSTEIN SUMMATION has been used in the last term. The covariant derivative of a COVARIANT TENSOR A_a is

$$A_{a;\,b} = \frac{1}{g^{bb}} \frac{\partial A_a}{\partial x_b} - \Gamma^k_{ab} A_k, \tag{2}$$

Schmutzer (1968, p. 72) uses the older notation $A^j_{\,\|k}$ or $A_{j\|k}$.

See also COMMA DERIVATIVE, CONNECTION COEFFICIENT, COVARIANT TENSOR, DIVERGENCE, LEVI-CIVITA CONNECTION

References
Morse, P. M. and Feshbach, H. *Methods of Theoretical Physics, Part I.* New York: McGraw-Hill, pp. 48–0, 1953.
Schmutzer, E. *Relativistische Physik (Klassische Theorie).* Leipzig, Germany: Akademische Verlagsgesellschaft, 1968.

Covariant Tensor

A covariant tensor is a TENSOR having specific transformation properties (cf., a CONTRAVARIANT TENSOR). To examine the transformation properties of a covariant tensor, first consider the GRADIENT

$$\nabla \phi = \frac{\partial \phi}{\partial x_1} \hat{\mathbf{x}}_1 + \frac{\partial \phi}{\partial x_2} \hat{\mathbf{x}}_2 + \frac{\partial \phi}{\partial x_3} \hat{\mathbf{x}}_3, \tag{1}$$

for which

$$\frac{\partial \phi'}{\partial x_i'} = \frac{\partial \phi}{\partial x_j} \frac{\partial x_j}{\partial x_i'}, \tag{2}$$

where $\phi(x_1, x_2, x_3) = \phi'(x_1', x_2', x_3')$. Now let

$$A_i \equiv \frac{\partial \phi}{\partial x_i}, \tag{3}$$

then any set of quantities A_j which transform according to

$$A_i' = \frac{\partial x_j}{\partial x_i'} A_j \tag{4}$$

or, defining

$$a_{ij} \equiv \frac{\partial x_j}{\partial x_i'}, \tag{5}$$

according to

$$A_i' = a_{ij} A_j \tag{6}$$

is a covariant tensor. Covariant tensors are indicated with lowered indices, i.e., a_μ.

CONTRAVARIANT TENSORS are a type of TENSOR with differing transformation properties, denoted a^ν. However, in 3-D CARTESIAN COORDINATES,

$$\frac{\partial x_j}{\partial x_i'} = \frac{\partial x_i'}{\partial x_j} \equiv a_{ij} \tag{7}$$

for $i, j = 1, 2, 3$, meaning that contravariant and covariant tensors are equivalent. The two types of tensors do differ in higher dimensions, however. Covariant FOUR-VECTORS satisfy

$$a_\mu = \Lambda_\mu^\nu a_\nu, \tag{8}$$

where Λ is a LORENTZ TENSOR.

To turn a CONTRAVARIANT TENSOR a^ν into a covariant tensor a_μ (INDEX LOWERING), use the METRIC TENSOR $g_{\mu\nu}$ to write

$$g_{\mu\nu} a^\nu = a_\mu. \tag{9}$$

Covariant and contravariant indices can be used simultaneously in a MIXED TENSOR.

See also CONTRAVARIANT TENSOR, FOUR-VECTOR, INDEX LOWERING, LORENTZ TENSOR, METRIC TENSOR, MIXED TENSOR, TENSOR

References

Arfken, G. "Noncartesian Tensors, Covariant Differentiation." §3.8 in *Mathematical Methods for Physicists, 3rd ed.* Orlando, FL: Academic Press, pp. 158–64, 1985.

Morse, P. M. and Feshbach, H. *Methods of Theoretical Physics, Part I.* New York: McGraw-Hill, pp. 44–6, 1953.

Covariant Vector

A COVARIANT TENSOR of RANK 1, more commonly called a ONE-FORM (or "BRA").

See also BRA, CONTRAVARIANT VECTOR, CONTRAVARIANT TENSOR, KET, ONE-FORM, VECTOR

Cover

A family γ of nonempty SUBSETS of X whose UNION contains the given set X (and which contains no duplicated subsets) is called a cover (or covering) of X. For example, there is only a single cover of $\{1\}$, namely $\{1\}$ itself. However, there are five covers of $\{1, 2\}$, namely $\{\{1\}, \{2\}\}$, $\{\{1, 2\}\}$, $\{\{1\}, \{1, 2\}\}$, $\{\{2\}, \{1, 2\}\}$, and $\{\{1\}, \{2\}, \{1, 2\}\}$.

A MINIMAL COVER is a cover for which removal of one member destroys the covering property. For example, of the five covers of $\{1, 2\}$, only $\{\{1\}, \{2\}\}$ and $\{\{1, 2\}\}$ are minimal covers. There are various other types of specialized covers, including PROPER COVERS, antichain covers, k-covers, and k^*-covers (Macula 1994).

The number of possible covers for a set of N elements are

$$|C(N)| = \frac{1}{2} \sum_{k=0}^{N} (-1)^k \binom{N}{k} 2^{2^{N-k}},$$

the first few of which are 1, 5, 109, 32297, 2147321017, 9223372023970362989, ... (Sloane's A003465).

See also MINIMAL COVER, PROPER COVER

References

Eppstein, D. "Covering and Packing." http://www.ics.uci.edu/~eppstein/junkyard/cover.html.

Macula, A. J. "Covers of a Finite Set." *Math. Mag.* **67**, 141–44, 1994.

Sloane, N. J. A. Sequences A003465/M4024 and A055621 in "An On-Line Version of the Encyclopedia of Integer Sequences." http://www.research.att.com/~njas/sequences/eisonline.html.

Cover Relation

The transitive reflexive reduction of a PARTIAL ORDER. An element z of a POSET $(X, \le)$ covers another element x provided that there exists no third element y in the poset for which $x \le y \le z$. In this case, z is called an "upper cover" of x and x a "lower cover" of z.

See also PARTIAL ORDER

Covering

COVER, COVERING MAP, PACKING

Covering Dimension

LEBESGUE COVERING DIMENSION

Covering Map

A covering map is a SURJECTIVE OPEN MAP $f : X \to Y$ whose preimages $f^{-1}(y)$ are a DISCRETE SET in X. For example, the map $f(z) = z^2$, as a map $f : \mathbb{C} - 0 \to \mathbb{C} -$

0, is a covering. Note that $f^{-1}(w)$ always consists of two points. In general, the cardinality of $f^{-1}(y)$ is independent of $y \in Y$.

Another example is $\pi : \mathbb{C} \to \mathbb{C}/\Gamma \simeq \mathbb{T}$, where $\Gamma = \{(a + bI)|a, b \in \mathbb{Z}\}$. The map π is actually the UNIVERSAL COVER of the torus $\mathbb{T}$. If $f : X \to \mathbb{T}$ is any covering of the torus, then there exists a covering $\tilde{\pi} : \mathbb{C} \to X$ such that π factors through $\tilde{\pi}$, i.e., $\pi = f \circ \tilde{\pi}$.

See also SIMPLY CONNECTED, TOPOLOGICAL SPACE, UNIVERSAL COVER

Covering System
COMPLETE RESIDUE SYSTEM

Coversine

$$\text{covers } A \equiv 1 - \sin A,$$

where $\sin A$ is the SINE.

See also EXSECANT, HAVERSINE, SINE, VERSINE

References
Abramowitz, M. and Stegun, C. A. (Eds.). *Handbook of Mathematical Functions with Formulas, Graphs, and Mathematical Tables, 9th printing.* New York: Dover, p. 78, 1972.

Coxeter Diagram
COXETER-DYNKIN DIAGRAM

Coxeter Graph

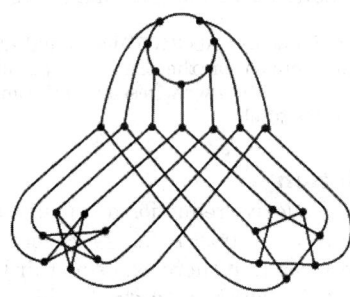

A non-Hamiltonian graph with a high degree of symmetry such that there is a GRAPH AUTOMORPHISM taking any path of length three into any other.

See also COXETER-DYNKIN DIAGRAM, LEVI GRAPH

References
Bondy, J. A. and Murty, U. S. R. *Graph Theory with Applications.* New York: North Holland, p. 241, 1976.
Tutte, W. T. "A Non-Hamiltonian Graph." *Canad. Math. Bull.* **3**, 1–, 1960.

Coxeter Group
A group generated by the elements P_i for $i = 1, ..., n$ subject to

$$(P_i P_j)^{M_{ij}} = 1,$$

where M_{ij} are the elements of a COXETER MATRIX. Coxeter used the NOTATION $[3^{p, q, r}]$ for the Coxeter group generated by the nodes of a Y-shaped COXETER-DYNKIN DIAGRAM whose three arms have p, q, and r EDGES. A Coxeter group of this form is finite IFF

$$\frac{1}{p+1} + \frac{1}{q+1} + \frac{1}{r+1} > 1.$$

See also BIMONSTER, BUILDING, COXETER-DYNKIN DIAGRAM

References
Arnold, V. I. "Snake Calculus and Combinatorics of Bernoulli, Euler, and Springer Numbers for Coxeter Groups." *Russian Math. Surveys* **47**, 3–5, 1992.
Garrett, P. *Buildings and Classical Groups.* Boca Raton, FL: Chapman and Hall, 1997.
Hsiang, W. Y. "Coxeter Groups, Weyl Reduction, and Weyl Formulas." Lec. 4 in *Lectures on Lie Groups.* Singapore: World Scientific, pp. 58–7, 2000.

Coxeter Matrix
An $n \times n$ SQUARE MATRIX M with

$$M_{ii} = 1$$

$$M_{ij} = M_{ji} > 1$$

for all $i, j = 1, ..., n$.

See also COXETER GROUP

Coxeter-Dynkin Diagram
A LABELED GRAPH whose nodes are indexed by the generators of a COXETER GROUP having (P_i, P_j) as an EDGE labeled by M_{ij} whenever $M_{ij} > 2$, where M_{ij} is an element of the COXETER MATRIX. Coxeter-Dynkin diagrams are used to visualize COXETER GROUPS. A Coxeter-Dynkin diagram is associated with each RATIONAL DOUBLE POINT (Fischer 1986), and a Coxeter diagram is sufficient to characterize the algebra of the group.

See also COXETER GROUP, DYNKIN DIAGRAM, RATIONAL DOUBLE POINT

References
Arnold, V. I. "Critical Points of Smooth Functions." *Proc. Int. Congr. Math.* **1**, 19–9, 1974.
Fischer, G. (Ed.). *Mathematical Models from the Collections of Universities and Museums.* Braunschweig, Germany: Vieweg, pp. 12–3, 1986.

Coxeter's Loxodromic Sequence of Tangent Circles
An infinite sequence of CIRCLES such that every four consecutive CIRCLES are mutually tangent, and the CIRCLES' RADII $..., R_{-n}, ..., R_{-1}, R_0, R_1, R_2, R_3, R_4, ...,$

R_n, $R_n + 1$, ..., are in GEOMETRIC PROGRESSION with ratio

$$k \equiv \frac{R_{n+1}}{R_n} = \phi + \sqrt{\phi},$$

where ϕ is the GOLDEN RATIO (Gardner 1979ab). Coxeter (1968) generalized the sequence to SPHERES.

See also ARBELOS, BOWL OF INTEGERS, GOLDEN RATIO, HEXLET, PAPPUS CHAIN, STEINER CHAIN

References

Coxeter, D. "Coxeter on 'Firmament.'" http://www.bangor.-ac.uk/SculMath/image/donald.htm.
Coxeter, H. S. M. "Loxodromic Sequences of Tangent Spheres." *Aequationes Math.* **1**, 112–17, 1968.
Gardner, M. "Mathematical Games: The Diverse Pleasures of Circles that Are Tangent to One Another." *Sci. Amer.* **240**, 18–8, Jan. 1979a.
Gardner, M. "Mathematical Games: How to be a Psychic, Even if You are a Horse or Some Other Animal." *Sci. Amer.* **240**, 18–5, May 1979b.

Coxeter-Todd Lattice

The complex LATTICE Λ_6^ω corresponding to real lattice K_{12} having the densest HYPERSPHERE PACKING (KISSING NUMBER) in 12-D. The associated AUTOMORPHISM GROUP G_0 was discovered by Mitchell (1914). The order of G_0 is given by

$$|\mathrm{Aut}(\Lambda_6^\omega)| = 2^9 \cdot 3^7 \cdot 5 \cdot 7 = 39,191,040.$$

The order of the AUTOMORPHISM GROUP of K_{12} is given by

$$|\mathrm{Aut}(K_{12})| = 2^{10} \cdot 3^7 \cdot 5 \cdot 7$$

(Conway and Sloane 1983).

See also BARNES-WALL LATTICE, LEECH LATTICE

References

Conway, J. H. and Sloane, N. J. A. "The Coxeter-Todd Lattice, the Mitchell Group and Related Sphere Packings." *Math. Proc. Camb. Phil. Soc.* **93**, 421–40, 1983.
Conway, J. H. and Sloane, N. J. A. "The 12-Dimensional Coxeter-Todd Lattice K_{12}." §4.9 in *Sphere Packings, Lattices, and Groups, 2nd ed.* New York: Springer-Verlag, pp. 127–29, 1993.
Coxeter, H. S. M. and Todd, J. A. "As Extreme Duodenary Form." *Canad. J. Math.* **5**, 384–92, 1953.
Mitchell, H. H. "Determination of All Primitive Collineation Groups in More than Four Variables." *Amer. J. Math.* **36**, 1–2, 1914.
Todd, J. A. "The Characters of a Collineation Group in Five Dimensions." *Proc. Roy. Soc. London Ser. A* **200**, 320–36, 1950.

Cox's Theorem

Let σ_1, ..., σ_4 be four PLANES in GENERAL POSITION through a point P and let P_{ij} be a point on the LINE $\sigma_i \cdot \sigma_j$. Let σ_{ijk} denote the PLANE $P_{ij}P_{ik}P_{jk}$. Then the four PLANES σ_{234}, σ_{134}, σ_{124}, σ_{123} all pass through one point P_{1234}. Similarly, let σ_1, ..., σ_5 be five PLANES in

GENERAL POSITION through P. Then the five points P_{2345}, P_{1345}, P_{1245}, P_{1235}, and P_{1234} all lie in one PLANE. And so on.

See also CLIFFORD'S CIRCLE THEOREM, PLANE

Cramér Conjecture

The unproven CONJECTURE that

$$\varlimsup_{n \to \infty} \frac{p_{n+1} - p_n}{(\ln p_n)^2} = 1,$$

where p_n is the nth PRIME.

References

Cramér, H. "On the Order of Magnitude of the Difference Between Consecutive Prime Numbers." *Acta Arith.* **2**, 23–6, 1936.
Guy, R. K. *Unsolved Problems in Number Theory, 2nd ed.* New York: Springer-Verlag, p. 7, 1994.
Riesel, H. "The Cramér Conjecture." *Prime Numbers and Computer Methods for Factorization, 2nd ed.* Boston, MA: Birkhäuser, pp. 79–2, 1994.
Rivera, C. "Problems & Puzzles: Conjecture The Cramer's Conjecture.-007." http://www.primepuzzles.net/conjectures/conj_007.htm.

Cramér-Euler Paradox

A curve of order n is generally determined by $n(n + 3)/2$ points. So a CONIC SECTION is determined by five points and a CUBIC CURVE should require nine. But the MACLAURIN-BÉZOUT THEOREM says that two curves of degree n intersect in n^2 points, so two CUBICS intersect in nine points. This means that $n(n + 3)/2$ points do not always uniquely determine a single curve of order n. The paradox was publicized by Stirling, and explained by Plücker.

See also CUBIC CURVE, MACLAURIN-BÉZOUT THEOREM

Cramer's Rule

Given a set of linear equations

$$\begin{cases} a_1 x + b_1 y + c_1 z = d_1 \\ a_2 x + b_2 y + c_2 z = d_2 \\ a_3 x + b_3 y + c_3 z = d_3, \end{cases} \quad (1)$$

consider the DETERMINANT

$$D \equiv \begin{vmatrix} a_1 & b_1 & c_1 \\ a_2 & b_2 & c_2 \\ a_3 & b_3 & c_3 \end{vmatrix}. \quad (2)$$

Now multiply D by x, and use the property of DETERMINANTS that MULTIPLICATION by a constant is equivalent to MULTIPLICATION of each entry in a given row by that constant

$$x \begin{vmatrix} a_1 & b_1 & c_1 \\ a_2 & b_2 & c_2 \\ a_3 & b_3 & c_3 \end{vmatrix} = \begin{vmatrix} a_1 x & b_1 & c_1 \\ a_2 x & b_2 & c_2 \\ a_3 x & b_3 & c_3 \end{vmatrix}. \quad (3)$$

Another property of DETERMINANTS enables us to add

a constant times any column to any column and obtain the same DETERMINANT, so add y times column 2 and z times column 3 to column 1,

$$xD = \begin{vmatrix} a_1x + b_1y + c_1z & b_1 & c_1 \\ a_2x + b_2y + c_2z & b_2 & c_2 \\ a_3x + b_3x + c_3z & b_3 & c_3 \end{vmatrix} = \begin{vmatrix} d_1 & b_1 & c_1 \\ d_2 & b_2 & c_2 \\ d_3 & b_3 & c_3 \end{vmatrix}. \quad (4)$$

If $\mathbf{d} = \mathbf{0}$, then (4) reduces to $xD = 0$, so the system has nondegenerate solutions (i.e., solutions other than (0, 0, 0)) only if $D = 0$ (in which case there is a family of solutions). If $\mathbf{d} \neq \mathbf{0}$ and $D = 0$, the system has no unique solution. If instead $\mathbf{d} \neq \mathbf{0}$ and $D \neq 0$, then solutions are given by

$$x = \frac{\begin{vmatrix} d_1 & b_1 & c_1 \\ d_2 & b_2 & c_2 \\ d_3 & b_3 & c_3 \end{vmatrix}}{D}, \quad (5)$$

and similarly for

$$y = \frac{\begin{vmatrix} a_1 & d_1 & c_1 \\ a_2 & d_2 & c_2 \\ a_3 & d_3 & c_3 \end{vmatrix}}{D} \quad (6)$$

$$z = \frac{\begin{vmatrix} a_1 & b_1 & d_1 \\ a_2 & b_2 & d_2 \\ a_3 & b_3 & d_3 \end{vmatrix}}{D} \quad (7)$$

This procedure can be generalized to a set of n equations so, given a system of n linear equations

$$\begin{bmatrix} a_{11} & a_{12} & \cdots & a_{1n} \\ \vdots & \vdots & \ddots & \vdots \\ a_{1n1} & a_{n2} & \cdots & a_{nn} \end{bmatrix} \begin{bmatrix} x_1 \\ \vdots \\ x_n \end{bmatrix} = \begin{bmatrix} d_1 \\ \vdots \\ d_n \end{bmatrix}, \quad (8)$$

let

$$D \equiv \begin{vmatrix} a_{11} & a_{12} & \cdots & a_{1n} \\ \vdots & \vdots & \ddots & \vdots \\ a_{1n1} & a_{n2} & \cdots & a_{nn} \end{vmatrix}. \quad (9)$$

If $\mathbf{d} = \mathbf{0}$, then nondegenerate solutions exist only if $D = 0$. If $\mathbf{d} \neq \mathbf{0}$ and $D = 0$, the system has no unique solution. Otherwise, compute

$$D_k \equiv \begin{vmatrix} a_{11} & \cdots & a_{1(k-1)} & d_1 & a_{1(k+1)} & \cdots & a_{1n} \\ \vdots & \ddots & \vdots & \vdots & \vdots & \ddots & \vdots \\ a_{n1} & \cdots & a_{n(k-1)} & d_n & a_{n(k+1)} & \cdots & a_{nn} \end{vmatrix}. \quad (10)$$

Then $x_k = D_k / D$ for $1 \leq k \leq n$. In the 3-D case, the VECTOR analog of Cramer's rule is

$$(\mathbf{A} \times \mathbf{B}) \times (\mathbf{C} \times \mathbf{D}) = (\mathbf{A} \cdot \mathbf{B} \times \mathbf{D})\mathbf{C} - (\mathbf{A} \cdot \mathbf{B} \times \mathbf{C})\mathbf{D}. \quad (11)$$

See also DETERMINANT, LINEAR ALGEBRA, MATRIX, SYSTEM OF EQUATIONS, VECTOR

References

Cramer, G. "Intr. à l'analyse de lignes courbes algébriques." Geneva, 657–59, 1750.

Muir, T. *The Theory of Determinants in the Historical Order of Development, Vol. 1.* New York: Dover, pp. 11–4, 1960.

Cramér's Theorem

If X and Y are INDEPENDENT variates and $X + Y$ is a GAUSSIAN DISTRIBUTION, then both X and Y must have GAUSSIAN DISTRIBUTIONS. This was proved by Cramér in 1936.

Craps

A game played with two DICE. If the total is 7 or 11 (a "natural"), the thrower wins and retains the DICE for another throw. If the total is 2, 3, or 12 ("craps"), the thrower loses but retains the DICE. If the total is any other number (called the thrower's "point"), the thrower must continue throwing and roll the "point" value again before throwing a 7. If he succeeds, he wins and retains the DICE, but if a 7 appears first, the player loses and passes the DICE.

The following table summarizes the probabilities of winning on a roll-by-roll basis, where $P(p = n)$ is the probability of rolling a point n. For rolls that are not naturals (W) or craps (L), the probability that the point $p = n$ will be rolled first is found from

$$P(\text{win}|p = n) = \frac{P(p = n)}{P(p = 7) + P(p = n)}$$

$$= \frac{P(p = n)}{\frac{1}{6}\frac{3}{6} + P(p = n)}.$$

n	$P(p = n)$	W/L	$P(\text{win}|p = n)$
2	$\frac{1}{36}$	L	0
3	$\frac{2}{36}$	L	0
4	$\frac{3}{36}$		$\frac{3}{9}$
5	$\frac{4}{36}$		$\frac{4}{10}$
6	$\frac{5}{36}$		$\frac{5}{11}$
7	$\frac{6}{36}$	W	1
8	$\frac{5}{36}$		$\frac{5}{11}$
9	$\frac{4}{36}$		$\frac{4}{10}$
10	$\frac{3}{36}$		$\frac{3}{9}$
11	$\frac{2}{36}$	W	1
12	$\frac{1}{36}$	L	0

Summing $P(p = n)$ from $n = 1$ to 12 then gives the probability of winning as $244/495 \approx 0.492929$ (Kraitchik 1942), just under 50%.

See also DICE

References

Kenney, J. F. and Keeping, E. S. *Mathematics of Statistics, Pt. 2, 2nd ed.* Princeton, NJ: Van Nostrand, pp. 12–3, 1951.

Kraitchik, M. "Craps." §6.5 in *Mathematical Recreations.* New York: W. W. Norton, pp. 123–26, 1942.

CRC

CYCLIC REDUNDANCY CHECK

Creative Telescoping

TELESCOPING SUM, ZEILBERGER'S ALGORITHM

Cremona Transformation

An entire Cremona transformation is a BIRATIONAL TRANSFORMATION of the PLANE. Cremona transformations are MAPS OF THE FORM

$$x_{i+1} = f(x_i, y_i),$$

$$y_{i+1} = g(x_i, y_i),$$

in which f and g are POLYNOMIALS. A quadratic Cremona transformation is always factorable.

See also NOETHER'S TRANSFORMATION THEOREM

References

Coolidge, J. L. *A Treatise on Algebraic Plane Curves.* New York: Dover, pp. 203–04, 1959.

Cremona-Richmond Configuration

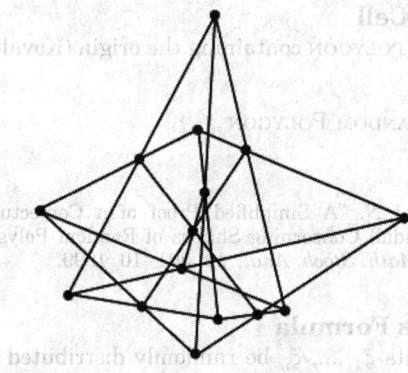

A 15_3 configuration of 15 lines and 15 points, with three lines through three points, three points on every line, and containing no triangles.

See also CONFIGURATION

References

Wells, D. *The Penguin Dictionary of Curious and Interesting Geometry.* London: Penguin, p. 40, 1991.

Cribbage

Cribbage is a game in which each of two players is dealt a hand of six CARDS. Each player then discards two of his six cards to a four-card "crib" which alternates between players. After the discard, the top card in the remaining deck is turned up. Cards are then alternately played out by the two players, with points being scored for pairs, runs, cumulative total of 15 and 31, and playing the last possible card ("go") not giving a total over 31. All face cards are counted as 10 for the purpose of playing out, but the normal values of Jack = 11, Queen = 12, King = 13 are used to determine runs. Aces are always low (ace = 1). After all cards have been played, each player counts the four cards in his hand taken in conjunction with the single top card. Points are awarded for pairs, flushes, runs, and combinations of cards giving 15. A Jack having the same suit as a top card is awarded an additional point for "nobbs." The crib is then also counted and scored. The winner is the first person to "peg" a certain score, as recorded on a "cribbage board."

The best possible score in a hand is 29, corresponding to three 5s and a Jack with a top 5 the same suit as the Jack. Hands with scores of 19, 25, 26, and 27 are not possible. A hand scoring zero points is therefore sometimes humorously referred to as a "19-point" hand.

See also BRIDGE CARD GAME, CARDS, POKER

Criss-Cross Method

A standard form of the LINEAR PROGRAMMING problem of maximizing a linear function over a CONVEX POLYHEDRON is to maximize $\mathbf{c} \cdot \mathbf{x}$ subject to $\mathbf{mx} \le \mathbf{b}$ and $\mathbf{x} \ge \mathbf{0}$, where m is a given $s \times d$ matrix, $\mathbf{c}$ and $\mathbf{b}$ are given d-vector and s-vectors, respectively. The Criss-cross method always finds a VERTEX solution if an optimal solution exists.

See also CONVEX POLYHEDRON, LINEAR PROGRAMMING, VERTEX (POLYHEDRON)

Criterion

A requirement NECESSARY for a given statement or theorem to hold. Also called a CONDITION.

See also BROWN'S CRITERION, CAUCHY CRITERION, EULER'S CRITERION, GAUSS'S CRITERION, KORSELT'S CRITERION, LEIBNIZ CRITERION, POCKLINGTON'S CRITERION, VANDIVER'S CRITERIA, WEYL'S CRITERION

Critical Damping

DAMPED SIMPLE HARMONIC MOTION–CRITICAL DAMPING

Critical Index

Let F be the MACLAURIN SERIES of a MEROMORPHIC FUNCTION f with a finite or infinite number of POLES at points z_k, indexed so that

$$0 < |z_1| \le |z_2| \le |z_3| \le \ldots,$$

then a POLE will occur as many times in the sequence $\{z_k\}$ as indicated by its order. Any index such that

$$|z_m| < |z_{m+1}|$$

holds is then called a critical index of f (Henrici 1988, pp. 641–42).

References

Henrici, P. *Applied and Computational Complex Analysis, Vol. 1: Power Series-Integration-Conformal Mapping-Location of Zeros.* New York: Wiley, pp. 641–42, 1988.

Critical Line

The LINE $\Re(s) = 1/2$ in the COMPLEX PLANE on which the RIEMANN HYPOTHESIS asserts that all nontrivial (COMPLEX) ROOTS of the RIEMANN ZETA FUNCTION $\zeta(s)$ lie. Although it is known that an INFINITE number of zeros lie on the critical line and that these comprise *at least* 40% of all zeros, the RIEMANN HYPOTHESIS is still unproven.

See also CRITICAL STRIP, RIEMANN HYPOTHESIS, RIEMANN ZETA FUNCTION

References

Brent, R. P. "On the Zeros of the Riemann Zeta Function in the Critical Strip." *Math. Comput.* **33**, 1361–372, 1979.
Brent, R. P.; van de Lune, J.; te Riele, H. J. J.; and Winter, D. T. "On the Zeros of the Riemann Zeta Function in the Critical Strip. II." *Math. Comput.* **39**, 681–88, 1982.
Vardi, I. *Computational Recreations in Mathematica.* Reading, MA: Addison-Wesley, p. 142, 1991.

Critical Point

A FUNCTION $y = f(x)$ has critical points at all points x_0 where $f'(x_0) = 0$ or $f(x)$ is not DIFFERENTIABLE. A FUNCTION $z = f(x, y)$ has critical points where the GRADIENT $\nabla f = 0$ or $\partial f / \partial x$ or the PARTIAL DERIVATIVE $\partial f / \partial y$ is not defined.

See also FIXED POINT, INFLECTION POINT, ONLY CRITICAL POINT IN TOWN TEST, STATIONARY POINT

Critical Strip

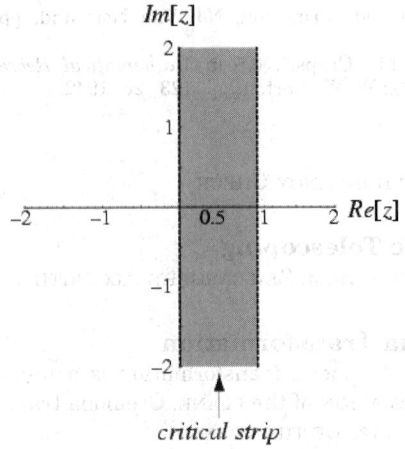

critical strip

The region $0 < \sigma < 1$, where σ is defined as the REAL PART of a COMPLEX NUMBER $s = \sigma + it$. All nontrivial zeros (i.e., those at negative integer) of the RIEMANN ZETA FUNCTION lie inside this strip.

See also CRITICAL LINE, RIEMANN HYPOTHESIS, RIEMANN ZETA FUNCTION

References

Brent, R. P. "On the Zeros of the Riemann Zeta Function in the Critical Strip." *Math. Comput.* **33**, 1361–372, 1979.
Brent, R. P.; van de Lune, J.; te Riele, H. J. J.; and Winter, D. T. "On the Zeros of the Riemann Zeta Function in the Critical Strip. II." *Math. Comput.* **39**, 681–88, 1982.

Crofton Cell

A RANDOM POLYGON containing the origin (Kovalenko 1999).

See also RANDOM POLYGON

References

Kovalenko, I. N. "A Simplified Proof of a Conjecture of D. G. Kendall Concerning Shapes of Random Polygons." *J. Appl. Math. Stoch. Anal.* **12**, 301–10, 1999.

Crofton's Formula

Let n points $\xi_1, \ldots, \xi_n$ be randomly distributed on a domain S, and let H be some event that depends on the positions of the n points. Let S' be a domain slightly smaller than S but contained within it, and let δS be the part of S not in S'. Let $P[H]$ be the probability of event H, s be the measure of S, and δS the measure of δS, then Crofton's formula states that

$$\delta P[H] = n(P[H \xi_1 \in \delta S] - P[H])s^{-1}\delta s$$

(Solomon 1978, p. 99).

See also CROFTON'S INTEGRALS

References

Ruben, H. and Reed, W. J. "A More General Form of the Theory of Crofton." *J. Appl. Prob.* **10**, 479–82, 1973.

Solomon, H. "Crofton's Theorem and Sylvester's Problem in Two and Three Dimensions." Ch. 5 in *Geometric Probability.* Philadelphia, PA: SIAM, pp. 97–25, 1978.

Crofton's Integrals

Consider a convex plane curve K with PERIMETER L, and the set of points P exterior to K. Further, let t_1 and t_2 be the perpendicular distances from P to K (with corresponding tangent points A_1 and A_2 on K), and let $\omega = \angle A_1 P A_2$. Then

$$\int_{P \text{ ext. to } K} \frac{\sin \omega}{t_1 t_2} \, dP = 2\pi^2 \qquad (1)$$

(Crofton 1885; Solomon 1978, p. 28).

If K has a continuous RADIUS OF CURVATURE and the radii of curvature at points A_1 and A_2 are ρ_1 and ρ_2, then

$$\int_{P \text{ ext. to } K} \frac{\sin \omega}{t_1 t_2} \rho_1 \rho_2 \, dP = \tfrac{1}{2} L^2 \qquad (2)$$

(Solomon 1978, p. 28), and furthermore

$$\int_{P \text{ ext. to } K} \frac{\sin \omega}{t_1 t_2} (\rho_1 + \rho_2) \, dP = 2\pi L \qquad (3)$$

(Santaló 1953; Solomon 1978, p. 28).

See also CROFTON'S FORMULA

References

Crofton, M. W. "Probability." *Encyclopaedia Britannica, 9th ed., Vol. 19.* Philadelphia, PA: J. M. Stoddart, pp. 768–88, 1885.
Santaló, L. *Introduction to Integral Geometry.* Paris: Hermann, 1953.
Solomon, H. *Geometric Probability.* Philadelphia, PA: SIAM, 1978.

Crofton's Theorem

CROFTON'S FORMULA

Crook

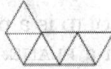

A 6-POLYIAMOND.

See also POLYIAMOND

References

Golomb, S. W. *Polyominoes: Puzzles, Patterns, Problems, and Packings, 2nd ed.* Princeton, NJ: Princeton University Press, p. 92, 1994.

Crookedness

Let a KNOT K be parameterized by a VECTOR FUNCTION $\mathbf{v}(t)$ with $t \in \mathbb{S}^1$, and let $\mathbf{w}$ be a fixed UNIT VECTOR in $\mathbb{R}^3$. Count the number of RELATIVE MINIMA of the projection function $\mathbf{w} \cdot \mathbf{v}(t)$. Then the MINIMUM such

number over all directions $\mathbf{w}$ and all K of the given type is called the crookedness $\mu(K)$. Milnor (1950) showed that $2\pi\mu(K)$ is the INFIMUM of the total curvature of K. For any TAME KNOT K in $\mathbb{R}^3$, $\mu(K) = b(K)$ where $b(K)$ is the BRIDGE INDEX.

See also BRIDGE INDEX

References

Milnor, J. W. "On the Total Curvature of Knots." *Ann. Math.* **52**, 248–57, 1950.
Rolfsen, D. *Knots and Links.* Wilmington, DE: Publish or Perish Press, p. 115, 1976.

Cross

In general, a cross is a figure formed by two intersecting LINE SEGMENTS. IN LINEAR ALGEBRA, a cross is defined as a set of n mutually PERPENDICULAR pairs of VECTORS of equal magnitude from a fixed origin in EUCLIDEAN n-SPACE.

The word "cross" is also used to denote the operation of the CROSS PRODUCT, so $\mathbf{a} \times \mathbf{b}$ would be pronounced "$\mathbf{a}$ cross $\mathbf{b}$."

See also CROSS PRODUCT, DOT, EUTACTIC STAR, GAULLIST CROSS, GREEK CROSS, LATIN CROSS, MALTESE CROSS, PAPAL CROSS, SAINT ANDREW'S CROSS, SAINT ANTHONY'S CROSS, STAR

Cross Curve

CRUCIFORM

Cross Fractal

CANTOR SQUARE FRACTAL

Cross of Lorraine

GAULLIST CROSS

Cross Polytope

A regular POLYTOPE in n-D corresponding to the CONVEX HULL of the points formed by permuting the coordinates ($\pm 1, 0, 0, ..., 0$). A cross-polytope (also called an orthoplex) is denoted ? missing and has $2n$ vertices and SCHLÄFLI SYMBOL

$$\{\underbrace{3, ..., 3}_{n-2}, 4\}.$$

The cross polytope is named because its $2n$ vertices are located equidistant from the origin along the Cartesian axes in n-space, which each such axis perpendicular to all others. A cross polytope is bounded by 2^n $(n-1)$-simplexes, and is a dipyramid erected (in both directions) into the nth dimension, with an $(n-1)$-dimensional cross polytope as its base.

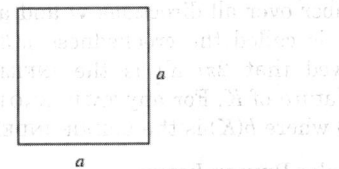

In 1-D, the cross polytope is the LINE SEGMENT $[-1, 1]$. In 2-D, the cross polytope $\{4\}$ is the filled SQUARE with vertices $(-1, 0)$, $(0, -1)$, $(1, 0)$, $(0, 1)$. In 3-D, the cross polytope $(3, 4)$ is the convex hull of the OCTAHEDRON with vertices $(-1, 0, 0)$, $(0, -1, 0)$, $(0, 0, -1)$, $(1, 0, 0)$, $(0, 1, 0)$, $(0, 0, 1)$. In 4-D, the cross polytope $\{3, 3, 4\}$ is the 16-CELL, depicted in the above figure by projecting onto one of the four mutually perpendicular 3-spaces within the 4-space obtained by dropping one of the four vertex components (R. Towle).

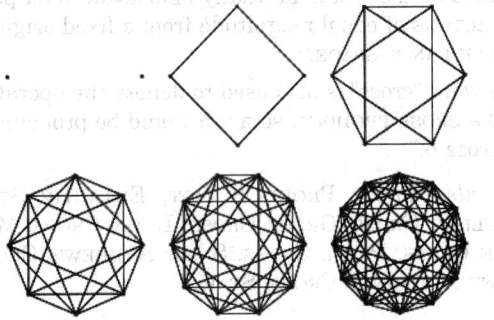

The graph of β_n missing is isomorphic with the CIRCULANT GRAPH $Ci_{1, 2, \ldots, (n-1)}(2n)$.

See also 16-CELL, HYPERCUBE, POLYTOPE, SIMPLEX

Cross Product

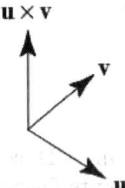

For VECTORS **u** and **v**, the cross product is defined by

$$\mathbf{u} \times \mathbf{v} = \hat{\mathbf{x}}(u_y v_z - u_z v_y) - \hat{\mathbf{y}}(u_x v_z - u_z v_x)$$
$$+ \hat{\mathbf{z}}(u_x v_y - u_y v_x). \qquad (1)$$

This can be written in a shorthand NOTATION which takes the form of a DETERMINANT

$$\mathbf{u} \times \mathbf{v} = \begin{vmatrix} \hat{\mathbf{x}} & \hat{\mathbf{y}} & \hat{\mathbf{z}} \\ u_x & u_y & u_z \\ v_x & v_y & v_z \end{vmatrix}. \qquad (2)$$

Here, $\mathbf{u} \times \mathbf{v}$ is always PERPENDICULAR to both **u** and **v**, with the orientation determinant by the RIGHT-

HAND RULE. It is also true that

$$|\mathbf{u} \times \mathbf{v}| = |\mathbf{u}||\mathbf{v}|\sin\theta, \qquad (3)$$

$$= |\mathbf{u}||\mathbf{v}|\sqrt{1 - (\hat{\mathbf{u}} \cdot \hat{\mathbf{v}})^2}, \qquad (4)$$

where θ is the angle between **u** and **v**, given by the DOT PRODUCT

$$\cos\theta \equiv \hat{\mathbf{u}} \cdot \hat{\mathbf{v}}. \qquad (5)$$

Jeffreys and Jeffreys (1988) use the notation $\mathbf{u} \wedge \mathbf{v}$ to denote the cross product.

The cross product is implemented in *Mathematica* 3.0 and higher as Cross$[a, b]$.

Identities involving the cross product include

$$\frac{d}{dt}[\mathbf{r}_1(t) \times \mathbf{r}_2(t)] = \mathbf{r}_1(t) \times \frac{d\mathbf{r}_2}{dt} + \frac{d\mathbf{r}_1}{dt} \times \mathbf{r}_2(t) \qquad (6)$$

$$\mathbf{A} \times \mathbf{B} = -\mathbf{B} \times \mathbf{A} \qquad (7)$$

$$\mathbf{A} \times (\mathbf{B} + \mathbf{C}) = \mathbf{A} \times \mathbf{B} + \mathbf{A} \times \mathbf{C} \qquad (8)$$

$$(t\mathbf{A}) \times \mathbf{B} = t(\mathbf{A} \times \mathbf{B}). \qquad (9)$$

For a proof that $\mathbf{A} \times \mathbf{B}$ is a PSEUDOVECTOR, see Arfken (1985, pp. 22–3). In TENSOR notation,

$$\mathbf{A} \times \mathbf{B} = \epsilon_{ijk} A^j B^k, \qquad (10)$$

where ϵ_{ijk} is the PERMUTATION SYMBOL.

See also CARTESIAN PRODUCT, DOT PRODUCT, PERMUTATION SYMBOL, RIGHT-HAND RULE, SCALAR TRIPLE PRODUCT, VECTOR, VECTOR DIRECT PRODUCT, VECTOR MULTIPLICATION

References
Arfken, G. "Vector or Cross Product." §1.4 in *Mathematical Methods for Physicists, 3rd ed.* Orlando, FL: Academic Press, pp. 18–6, 1985.
Jeffreys, H. and Jeffreys, B. S. "Vector Product." §2.07 in *Methods of Mathematical Physics, 3rd ed.* Cambridge, England: Cambridge University Press, pp. 67–3, 1988.

Cross Section

The cross section of a SOLID is a plane figure obtained by its intersection with a PLANE. The cross section of an object therefore represents an infinitesimal "slice" of a solid, and may be different depending on the orientation of the slicing plane. While the cross section of a SPHERE is always a DISK, the cross section of a CUBE may be a SQUARE, HEXAGON, or other shape.

See also AXONOMETRY, CAVALIERI'S PRINCIPLE, INNER QUERMASS, LAMINA, PLANE, PROJECTION, RADON TRANSFORM, STEREOLOGY

Cross Sequence

A sequence

$$s_n^{(\lambda)}(x) = [h(t)]^\lambda s_n(x),$$

where $s_n(x)$ is a SHEFFER SEQUENCE, $h(t)$ is invertible, and λ ranges over the real numbers is called a STEFFENSEN SEQUENCE. If $s_n(x)$ is an associated SHEFFER SEQUENCE, then $s_n^{(\lambda)}$ is called a cross sequence.

Examples include the ACTUARIAL POLYNOMIAL and POISSON-CHARLIER POLYNOMIAL.

See also APPELL CROSS SEQUENCE, SHEFFER SEQUENCE, STEFFENSEN SEQUENCE

References

Roman, S. "Cross Sequences and Steffensen Sequences." §5.3 in *The Umbral Calculus*. New York: Academic Press, pp. 140–43, 1984.
Rota, G.-C.; Kahaner, D.; Odlyzko, A. "On the Foundations of Combinatorial Theory. VIII: Finite Operator Calculus." *J. Math. Anal. Appl.* **42**, 684–60, 1973.

Cross Surface

A SPHERE with a single CROSS-CAP. This term is more appropriate in purely topological applications than the more common term REAL PROJECTIVE PLANE, which implies the presence of an affine structure (Francis and Weeks 1999). The double cross surface is the KLEIN BOTTLE and the triple cross surface is called DYCK'S SURFACE (Francis and Collins 1993, Francis and Weeks 1999).

See also CROSS-CAP, REAL PROJECTIVE PLANE

References

Francis, G. and Collins, B. "On Knot-Spanning Surfaces: An Illustrated Essay on Topological Art." Ch. 11 in *The Visual Mind: Art and Mathematics* (Ed. M. Emmer). Cambridge, MA: MIT Press, 1993.
Francis, G. K. and Weeks, J. R. "Conway's ZIP Proof." *Amer. Math. Monthly* **106**, 393–99, 1999.

Cross-Cap

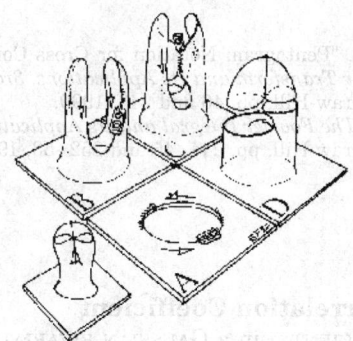

The self-intersection of a one-sided SURFACE. "Cross-cap" is sometimes also written without the hyphen as the single word "crosscap." The cross-cap can be thought of as the object produced by puncturing a surface a single time, attaching two ZIPS around the puncture in the same direction, distorting the hole so that the zips line up, requiring that the surface

intersect itself, and then zipping up. The cross-cap can also be described as a circular HOLE which, when entered, exits from its opposite point (from a topological viewpoint, both singular points on the cross-cap are equivalent).

The cross-cap has a segment of double points which terminates at two "PINCH POINTS" known as WHITNEY SINGULARITIES. A CROSS-HANDLE is homeomorphic to two cross-caps (Francis and Weeks 1999).

A SPHERE with one cross-cap has traditionally been called a REAL PROJECTIVE PLANE. While this is appropriate in the study of PROJECTIVE GEOMETRY when an affine structure is present, J. H. Conway advocates use of the term CROSS SURFACE in a purely topological interpretation (Francis and Weeks 1999). The cross-cap is one of the three possible SURFACES obtained by sewing a MÖBIUS STRIP to the edge of a DISK. The other two are the BOY SURFACE and ROMAN SURFACE.

The cross-cap can be generated using the general method for NONORIENTABLE SURFACES using the polynomial function

$$\mathbf{f}(x,\, y,\, z) = (xz,\; yz,\; \tfrac{1}{2}(z^2 - x^2)) \tag{1}$$

(Pinkall 1986). Transforming to SPHERICAL COORDINATES gives

$$x(u,\, v) = \tfrac{1}{2}\cos u \, \sin(2v) \tag{2}$$

$$y(u,\, v) = \tfrac{1}{2}\sin u \, \sin(2v) \tag{3}$$

$$z(u,\, v) = \tfrac{1}{2}(\cos^2 v - \cos^2 u \, \sin^2 v) \tag{4}$$

for $u \in [0,\, 2\pi)$ and $v \in [0,\, \pi/2]$. To make the equations slightly simpler, all three equations are normally multiplied by a factor of 2 to clear the arbitrary scaling constant. Three views of the cross-cap generated using this equation are shown above. Note that the middle one looks suspiciously like BOUR'S MINIMAL SURFACE.

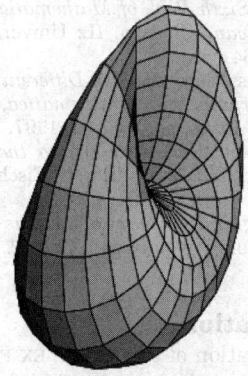

Another representation is

$$\mathbf{f}(x,\, y,\, z) = (yz,\; 2xy,\; x^2 - y^2), \tag{5}$$

(Gray 1997), giving PARAMETRIC EQUATIONS

$$x = \tfrac{1}{2}\sin u \sin(2v) \qquad (6)$$

$$y = \sin(2u)\sin^2 v \qquad (7)$$

$$z = \cos(2u)\sin^2 v, \qquad (8)$$

(Geometry Center) where, for aesthetic reasons, the y- and z-coordinates have been multiplied by 2 to produce a squashed, but topologically equivalent, surface. Nordstrand gives the implicit equation

$$4x^2(x^2+y^2+z^2+z)+y^2(y^2+z^2-1)=0 \qquad (9)$$

which can be solved for z to yield

$$z = \frac{-2x^2 \pm \sqrt{(y^2+2x^2)(1-4x^2-y^2)}}{4x^2+y^2}. \qquad (10)$$

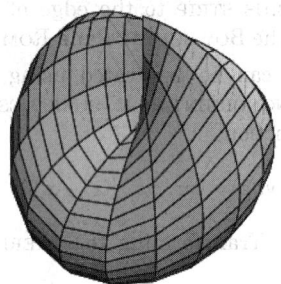

Taking the inversion of a cross-cap such that $(0, 0, -1/2)$ is sent to ∞ gives a CYLINDROID, shown above (Pinkall 1986).

See also BOY SURFACE, CAP, CLASSIFICATION THEOREM OF SURFACES, CROSS-HANDLE, CROSS SURFACE, HANDLE, MÖBIUS STRIP, NONORIENTABLE SURFACE, PROJECTIVE PLANE, ROMAN SURFACE

References

Fischer, G. (Ed.). Plate 107 in *Mathematische Modelle/ Mathematical Models, Bildband/Photograph Volume.* Braunschweig, Germany: Vieweg, p. 108, 1986.
Francis, G. K. and Weeks, J. R. "Conway's ZIP Proof." *Amer. Math. Monthly* **106**, 393–99, 1999.
Gardner, M. *The Sixth Book of Mathematical Games from Scientific American.* Chicago, IL: University of Chicago Press, p. 15, 1984.
Gray, A. "The Cross Cap." *Modern Differential Geometry of Curves and Surfaces with Mathematica, 2nd ed.* Boca Raton, FL: CRC Press, pp. 333–35, 1997.
Pinkall, U. *Mathematical Models from the Collections of Universities and Museums* (Ed. G. Fischer). Braunschweig, Germany: Vieweg, p. 64, 1986.
Wells, D. *The Penguin Dictionary of Curious and Interesting Geometry.* London: Penguin, p. 197, 1991.

Cross-Correlation

The cross-correlation of two COMPLEX FUNCTIONS $f(t)$ and $g(t)$ of a real variable t, denoted $f \star g$ is defined by

$$f \star g \equiv \bar{f}(-t) * g(t), \qquad (1)$$

where $*$ denotes CONVOLUTION and $\bar{f}(t)$ is the COM-

PLEX CONJUGATE of $f(t)$. Since CONVOLUTION is defined by

$$f(t) * g(t) = \int_{-\infty}^{\infty} f(\tau)g(t-\tau)\,d\tau, \qquad (2)$$

it follows that

$$f \star g \equiv \int_{-\infty}^{\infty} \bar{f}(-\tau)g(t-\tau)\,d\tau. \qquad (3)$$

Letting $\tau' = -\tau$, $d\tau' = -d\tau$ so (3) is equivalent to

$$f \star g = \int_{\infty}^{-\infty} \bar{f}(\tau')g(t+\tau')(-d\tau')$$

$$= \int_{-\infty}^{\infty} \bar{f}(\tau)g(t+\tau)\,d\tau. \qquad (4)$$

The cross-correlation satisfies the identity

$$(g \star h) \star (g \star h) = (g \star g) \star (h \star h). \qquad (5)$$

If f or g is EVEN, then

$$f \star g = f * g, \qquad (6)$$

where $*$ again denotes CONVOLUTION.

See also AUTOCORRELATION, CONVOLUTION, CROSS-CORRELATION THEOREM, FOURIER TRANSFORM

References

Bracewell, R. "Pentagram Notation for Cross Correlation." *The Fourier Transform and Its Applications, 3rd ed.* New York: McGraw-Hill, pp. 46 and 243, 1999.
Papoulis, A. *The Fourier Integral and Its Applications.* New York: McGraw-Hill, pp. 244–45 and 252–53, 1962.

Cross-Correlation Coefficient

The COEFFICIENT ρ in a GAUSSIAN BIVARIATE DISTRIBUTION.

Cross-Correlation Theorem

Let $f \star g$ denote the CROSS-CORRELATION of functions $f(t)$ and $g(t)$. Then

$$f \star g = \int_{-\infty}^{\infty} \bar{f}(\tau)g(t+\tau)\,d\tau$$

$$= \int_{-\infty}^{\infty} \left[\int_{-\infty}^{\infty} \bar{F}(v)e^{2\pi i v\tau}\,dv \int_{-\infty}^{\infty} G(v')e^{-2\pi i v'(t+\tau)}\,dv' \right] d\tau$$

$$= \int_{-\infty}^{\infty} \int_{-\infty}^{\infty} \int_{-\infty}^{\infty} \bar{F}(v)G(v')e^{-2\pi i \tau(v'-v)}\,e^{-2\pi i v't}\,d\tau\,dv\,dv'$$

$$= \int_{-\infty}^{\infty} \int_{-\infty}^{\infty} \bar{F}(v)G(v')e^{-2\pi i v't}\left[\int_{-\infty}^{\infty} e^{-2\pi i \tau(v'-v)}\,d\tau\right] dv\,dv'$$

$$= \int_{-\infty}^{\infty} \int_{-\infty}^{\infty} \bar{F}(v)G(v')e^{-2\pi i v't}\delta(v'-v)\,dv'\,dv$$

$$= \int_{-\infty}^{\infty} \bar{F}(v)G(v)e^{-2\pi i vt}\,dv \tag{1}$$

where $\mathscr{F}$ denotes the FOURIER TRANSFORM, $\bar{z}$ is the COMPLEX CONJUGATE, and

$$f(t) \equiv \mathscr{F}[F(v)] = \int_{-\infty}^{\infty} F(v)e^{-2\pi i vt}\,dt \tag{2}$$

$$g(t) \equiv \mathscr{F}[G(v)] = \int_{-\infty}^{\infty} G(v)e^{-2\pi i vt}\,dt. \tag{3}$$

Applying a FOURIER TRANSFORM on each side gives the cross-correlation theorem,

$$f \star g = \mathscr{F}[\bar{F}(v)G(v)]. \tag{4}$$

If $F = G$, then the cross-correlation theorem reduces to the WIENER-KHINTCHINE THEOREM.

See also FOURIER TRANSFORM, WIENER-KHINTCHINE THEOREM

Crosscram
DOMINEERING

Crossed Hyperbolic Rotation
Exchanges branches of the HYPERBOLA $x'y' = xy$.

$$x' = \mu^{-1}x$$

$$y' = -\mu y.$$

See also HYPERBOLIC ROTATION

Crossed Ladders Problem

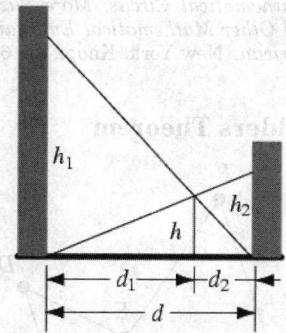

Given two crossed LADDERS resting against two buildings, what is the distance between the buildings? Let the height at which they cross be h and the lengths of the LADDERS l_1 and l_2. The height at which l_2 touches the building h_2 is then obtained by simultaneously solving the equations

$$l_1^2 = h_1^2 + d^2 \tag{1}$$

$$l_2^2 = h_2^2 + d^2 \tag{2}$$

and

$$\frac{1}{h} = \frac{1}{h_1} + \frac{1}{h_2}, \tag{3}$$

the latter of which follows either immediately from the CROSSED LADDERS THEOREM or from similar triangles with $d_1 = dh/h_2$, $d_2 = dh/h_1$, and $d = d_1 + d_2$. Eliminating d gives the equations

$$h_1^4 - 2hh_1^3 + (h-h_1)^2(l_2^2 - l_1^2) = 0. \tag{4}$$

$$h_2^4 - 2hh_2^3 + (h-h_2)^2(l_1^2 - l_2^2) = 0. \tag{5}$$

These quartic equations can be solved for h_1 and h_2 given known values of h, l_1, and l_2.

There are solutions in which not only l_1, l_2, h_1, h_2, and h are all integers, but so are d_1, and d_2. One example is $h_1 = 119$, $h_1 = 70$, $h = 30$, $d_1 = 40$, $d_2 = 16$.

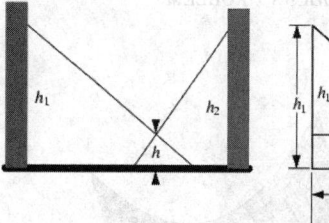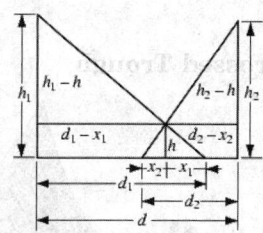

The problem can also be generalized to the situation in which the ends of the ladders are not pinned against the buildings, but propped fixed distances d_1 and d_2 away.

See also CROSSED LADDERS THEOREM, LADDER

References

Gardner, M. *Mathematical Circus: More Puzzles, Games, Paradoxes and Other Mathematical Entertainments from Scientific American.* New York: Knopf, pp. 62–4, 1979.

Crossed Ladders Theorem

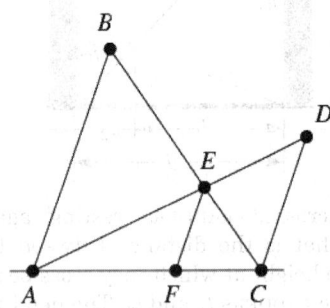

In the above figure, let E be the intersection of AD and BC and specify that $AB \parallel EF \parallel CD$. Then

$$\frac{1}{AB} + \frac{1}{CD} = \frac{1}{EF}.$$

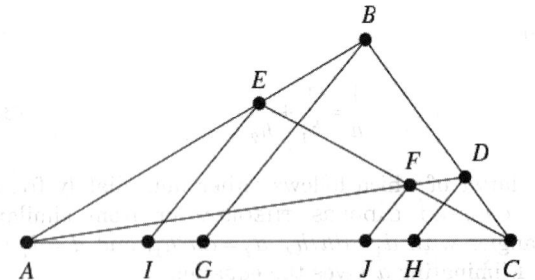

A beautiful related theorem due to H. Stengel can be stated as follows. In the above figure, let E lie on the side AB and D lie on the side BC. Now let EC intersect the line AD at a point F, and construct points H, I, and J so that $EI \parallel DH \parallel FJ \parallel BG$. Then

$$\frac{1}{EI} + \frac{1}{DH} = \frac{1}{FJ} + \frac{1}{BG}.$$

See also CROSSED LADDERS PROBLEM

Crossed Trough

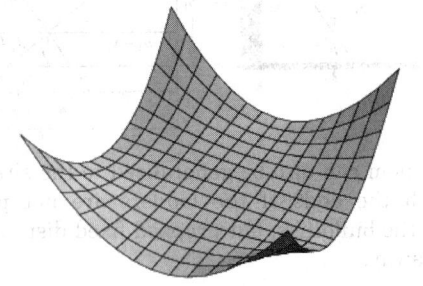

The SURFACE

$$z = cx^2 y^2.$$

See also MONKEY SADDLE

References

von Seggern, D. *CRC Standard Curves and Surfaces.* Boca Raton, FL: CRC Press, p. 286, 1993.

Cross-Handle

A cross-handle is a topological structure which can be thought of as the object produced by puncturing a surface twice, attaching a ZIP around each puncture travelling in the same direction, pulling the edges of the zips together after one tube first passes through itself it order for the direction of the zips to match up, and then zipping up. In 3-space, the cross-handle contains a line of self-intersection.

A cross-handle is homeomorphic to two CROSS-CAPS (Francis and Weeks 1999). DYCK'S THEOREM states that HANDLES and cross-handles are equivalent in the presence of a CROSS-CAP.

See also CAP, CROSS-CAP, DYCK'S THEOREM, HANDLE

References

Francis, G. K. and Weeks, J. R. "Conway's ZIP Proof." *Amer. Math. Monthly* **106**, 393–99, 1999.

Crossing Number (Graph)

Given a "good" GRAPH G (i.e., one for which all intersecting EDGES intersect in a single point and arise from four distinct VERTICES), the crossing number $v(G)$ is the minimum possible number of crossings with which the GRAPH can be drawn. A GRAPH with crossing number 0 is a PLANAR GRAPH. Garey and Johnson (1983) showed that determining the crossing number is an NP-COMPLETE PROBLEM.

GUY'S CONJECTURE suggests that the crossing number for the COMPLETE GRAPH K_n is

$$v(K_n) = \frac{1}{4} \left\lfloor \frac{n}{2} \right\rfloor \left\lfloor \frac{n-1}{2} \right\rfloor \left\lfloor \frac{n-2}{2} \right\rfloor \left\lfloor \frac{n-3}{2} \right\rfloor, \tag{1}$$

which can be rewritten

$$\nu(K_n) = \begin{cases} \frac{1}{64} n(n-2)^2(n-4) & \text{for } n \text{ even} \\ \frac{1}{64}(n-1)^2(n-3)^2 & \text{for } n \text{ odd}. \end{cases} \quad (2)$$

The values of (2) for $n = 1, 2, \ldots$ are then given by 0, 0, 0, 0, 1, 3, 9, 18, 36, 60, 100, 150, 225, 315, 441, 588, ... (Sloane's A000241), although it has not been proven that these agree with the actual crossing numbers for $n \geq 11$.

ZARANKIEWICZ'S CONJECTURE asserts that the crossing number for a COMPLETE BIGRAPH is

$$\nu(K_{m,n}) = \left\lfloor \frac{n}{2} \right\rfloor \left\lfloor \frac{n-1}{2} \right\rfloor \left\lfloor \frac{m}{2} \right\rfloor \left\lfloor \frac{m-1}{2} \right\rfloor. \quad (3)$$

It has been checked up to m, $n = 7$, and Zarankiewicz has shown that, in general, the FORMULA provides an upper bound to the actual number. The table below gives known results. When the number is not known exactly, the prediction of ZARANKIEWICZ'S CONJECTURE is given in parentheses.

	1	2	3	4	5	6		7
1	0	0	0	0	0	0		0
2		0	0	0	0	0		0
3			1	2	4	6		9
4				4	8	12		18
5					16	24		36
6						36		54
7								77, 79, or (81)

Kleitman (1970, 1976) computed the exact crossing numbers $\nu(K_{5,n})$ for all positive n.

See also GUY'S CONJECTURE, RECTILINEAR CROSSING NUMBER, TOROIDAL CROSSING NUMBER, ZARANKIE-WICZ'S CONJECTURE

References

Erdos, P. and Guy, R. K. "Crossing Number Problems." *Amer. Math. Monthly* **80**, 52–7, 1973.

Gardner, M. "Crossing Numbers." Ch. 11 in *Knotted Dough-nuts and Other Mathematical Entertainments.* New York: W. H. Freeman, pp. 133–44, 1986.

Garey, M. R. and Johnson, D. S. "Crossing Number is NP-Complete." *SIAM J. Alg. Discr. Meth.* **4**, 312–16, 1983.

Guy, R. K. "The Crossing Number of the Complete Graph." *Bull. Malayan Math. Soc.* **7**, 68–2, 1960.

Guy, R. K. "Latest Results on Crossing Numbers." In *Recent Trends in Graph Theory, Proc. New York City Graph Theory Conference, 1st, 1970.* (Ed. New York City Graph Theory Conference Staff). New York: Springer-Verlag, 1971.

Guy, R. K. "Crossing Numbers of Graphs." In *Graph Theory and Applications: Proceedings of the Conference at Western Michigan University, Kalamazoo, Mich., May 10–3,* 1972 (Ed. Y. Alavi, D. R. Lick, and A. T. White). New York: Springer-Verlag, pp. 111–24, 1972.

Kleitman, D. J. "The Crossing Number of $K_{5,n}$." *J. Combin. Th.* **9**, 315–23, 1970.

Kleitman, D. J. "A Note on the Parity of the Numbers of Crossings of a Graph." *J. Combin. Th., Ser. B* **21**, 88–9, 1976.

Koman, M. "Extremal Crossing Numbers of Complete k-Chromatic Graphs." *Mat. Casopis Sloven. Akad. Vied.* **20**, 315–25, 1970.

Kovari, T.; Sós, V. T.; and Turán, P. "On a Problem of K. Zarankiewicz." *Colloq. Math.* **3**, 50–7, 1954.

Moon, J. W. "On the Distribution of Crossings in Random Complete Graphs." *SIAM J.* **13**, 506–10, 1965.

Owens, A. "On the Biplanar Crossing Number." *IEEE Trans. Circuit Th.* **18**, 277–80, 1971.

Pach, J. and Tóth, G. "Thirteen Problems on Crossing Numbers." *Geocombin.* **9**, 195–07, 2000.

Richter, R. B. and Thomassen, C. "Relations Between Crossing Numbers of Complete and Complete Bipartite Graphs." *Amer. Math. Monthly* **104**, 131–37, 1997.

Skiena, S. *Implementing Discrete Mathematics: Combinatorics and Graph Theory with Mathematica.* Reading, MA: Addison-Wesley, p. 251, 1990.

Sloane, N. J. A. Sequences A014540 in "An On-Line Version of the Encyclopedia of Integer Sequences." http://www.research.att.com/~njas/sequences/eisonline.html.

Thomassen, C. "Embeddings and Minors." In *Handbook of Combinatorics, 2 vols.* (Ed. R. L. Graham, M. Grötschel, and L. Lovász.) Cambridge, MA: MIT Press, p. 314, 1996.

Tutte, W. T. "Toward a Theory of Crossing Numbers." *J. Comb. Th.* **8**, 45–3, 1970.

Wilf, H. "On Crossing Numbers, and Some Unsolved Problems." In *Combinatorics, Geometry, and Probability: A Tribute to Paul Erdos. Papers from the Conference in Honor of Paul Erdos's 80th Birthday Held at Trinity College, Cambridge, March 1993* (Ed. B. Bollobás and A. Thomason). Cambridge, England: Cambridge University Press, pp. 557–62, 1997.

Crossing Number (Link)

The least number of crossings that occur in any projection of a LINK. In general, it is difficult to find the crossing number of a given LINK. Knots and links are generally tabulated based on their crossing numbers.

See also KNOT, LINK

References

Adams, C. C. *The Knot Book: An Elementary Introduction to the Mathematical Theory of Knots.* New York: W. H. Freeman, pp. 67–9, 1994.

Hoste, J.; Thistlethwaite, M.; and Weeks, J. "The First 1,701,936 Knots." *Math. Intell.* **20**, 33–8, Fall 1998.

Cross-Ratio

$$[a, b, c, d] \equiv \frac{(a-b)(c-d)}{(a-d)(c-b)} \quad (1)$$

For a MÖBIUS TRANSFORMATION f,

$$[a, b, c, d] = [f(a), f(b), f(c), f(d)]. \quad (2)$$

There are six different values which the cross-ratio may take, depending on the order in which the points are chosen. Let $\lambda \equiv [a, b, c, d]$. Possible values of the cross-ratio are then λ, $1 - \lambda$, $1/\lambda$, $(\lambda - 1)/\lambda$, $1/(1 - \lambda)$, and $\lambda/(\lambda - 1)$.

Given lines a, b, c, and d which intersect in a point O, let the lines be cut by a line l, and denote the points of intersection of l with each line by A, B, C, and D. Let the distance between points A and B be denoted AB, etc. Then the cross-ratio

$$[AB, \ CD] \equiv \frac{(AB)(CD)}{(BC)(AD)} \tag{3}$$

is the same for any position of the l (Coxeter and Greitzer 1967). Note that the definitions $(AB/AD)/(BC/CD)$ and $(CA/CB)/(DA/DB)$ are used instead by Kline (1990) and Courant and Robbins (1966), respectively. The identity

$$[AD, \ BC] + [AB, \ DC] = 1 \tag{4}$$

holds IFF $AC//BD$, where $//$ denotes SEPARATION.

The cross-ratio of four points on a radial line of an INVERSION CIRCLE is preserved under INVERSION (Ogilvy 1990, p. 40).

See also BIVALENT RANGE, EQUICROSS, HARMONIC RANGE, HOMOGRAPHIC, MÖBIUS TRANSFORMATION, SEPARATION

References

Anderson, J. W. "The Cross Ratio." §2.3 in *Hyperbolic Geometry.* New York: Springer-Verlag, pp. 30–6, 1999.
Casey, J. "Theory of Anharmonic Section." §6.6 in *A Sequel to the First Six Books of the Elements of Euclid, Containing an Easy Introduction to Modern Geometry with Numerous Examples, 5th ed., rev. enl.* Dublin: Hodges, Figgis, & Co., pp. 126–40, 1888.
Courant, R. and Robbins, H. *What is Mathematics?: An Elementary Approach to Ideas and Methods, 2nd ed.* Oxford, England: Oxford University Press, 1996.
Coxeter, H. S. M. and Greitzer, S. L. *Geometry Revisited.* Washington, DC: Math. Assoc. Amer., pp. 107–08, 1967.
Durell, C. V. *Modern Geometry: The Straight Line and Circle.* London: Macmillan, pp. 73–6, 1928.
Graustein, W. C. "Cross Ratio." Ch. 6 in *Introduction to Higher Geometry.* New York: Macmillan, pp. 72–3, 1930.
Kline, M. *Mathematical Thought from Ancient to Modern Times, Vol. 1.* Oxford, England: Oxford University Press, 1990.
Lachlan, R. "Theory of Cross Ratio." Ch. 16 in *An Elementary Treatise on Modern Pure Geometry.* London: Macmillian, pp. 266–82, 1893.
Möbius, A. F. Ch. 5 in *Der barycentrische Calcul: Ein neues Hülfsmittel zur analytischen Behandlung der Geometrie, dargestellt und insbesondere auf die Bildung neuer Classen von Aufgaben und die Entwickelung mehrerer Eigenschaften der Kegelschnitte angewendet.* Leipzig, Germany: J. A. Barth, 1827.
Ogilvy, C. S. *Excursions in Geometry.* New York: Dover, pp. 39–1, 1990.
Wells, D. *The Penguin Dictionary of Curious and Interesting Geometry.* London: Penguin, p. 41, 1991.

Cross-Stitch Curve

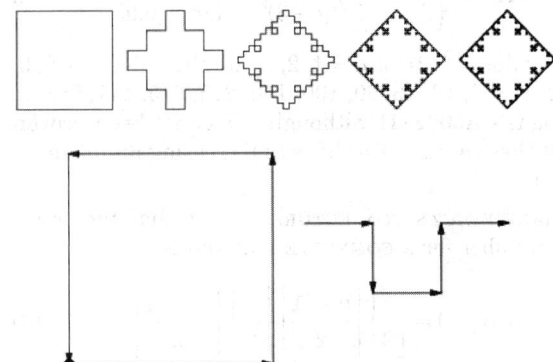

A fractal curve of infinite length which bounds an area twice that of the original square.

See also BOX FRACTAL, CANTOR SQUARE FRACTAL, FRACTAL, SIERPINSKICURVE

References

Gardner, M. *The Sixth Book of Mathematical Games from Scientific American.* Chicago, IL: University of Chicago Press, pp. 228–29, 1984.

Crout's Method

A ROOT finding technique used in LU DECOMPOSITION. It solves the N^2 equations

$$
\begin{aligned}
i < j \quad & l_{i1}u_{1j} + l_{i2}u_{2j} + \cdots + l_{ii}u_{ij} = a_{ij} \\
i = j \quad & l_{i1}u_{1j} + l_{i2}u_{2j} + \cdots + l_{ii}u_{ij} = a_{ij} \\
i > j \quad & l_{i1}u_{1j} + l_{i2}u_{2j} + \cdots + l_{ii}u_{jj} = a_{ij}
\end{aligned}
$$

for the $N^2 + N$ unknowns l_{ij} and u_{ij}.

See also LU DECOMPOSITION

References

Press, W. H.; Flannery, B. P.; Teukolsky, S. A.; and Vetterling, W. T. *Numerical Recipes in FORTRAN: The Art of Scientific Computing, 2nd ed.* Cambridge, England: Cambridge University Press, pp. 36–8, 1992.

Crowd

A group of SOCIABLE NUMBERS of order 3.

Crown

A 6-POLYIAMOND.

See also POLYIAMOND

References

Golomb, S. W. *Polyominoes: Puzzles, Patterns, Problems, and Packings, 2nd ed.* Princeton, NJ: Princeton University Press, p. 92, 1994.

Crucial Point

The HOMOTHETIC CENTER of the ORTHIC TRIANGLE and the triangular hull of the three EXCIRCLES. It has TRIANGLE CENTER FUNCTION

$$\alpha = \tan A = \sin(2B) + \sin(2C) - \sin(2A).$$

References

Kimberling, C. "Central Points and Central Lines in the Plane of a Triangle." *Math. Mag.* **67**, 163–87, 1994.
Lyness, R. and Veldkamp, G. R. Problem 682 and Solution. *Crux Math.* **9**, 23–4, 1983.

Cruciform

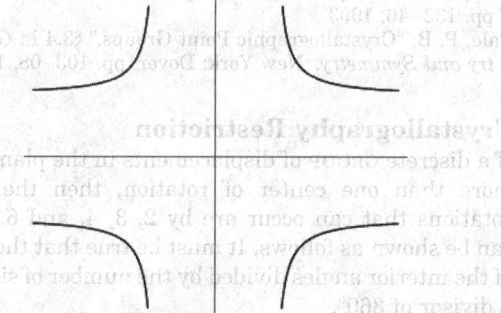

A plane curve also called the CROSS CURVE and POLICEMAN ON POINT DUTY CURVE (Cundy and Rollett 1989). It is given by the equation

$$x^2y^2 - a^2x^2 - b^2y^2 = 0, \tag{1}$$

which is equivalent to

$$1 - \frac{a^2}{x^2} - \frac{b^2}{y^2} = 0 \tag{2}$$

or, rewriting,

$$\frac{a^2}{x^2} + \frac{b^2}{y^2} = 1, \tag{3}$$

$$y^2 = \frac{b^2x^2}{x^2 - a^2}. \tag{4}$$

In parametric form,

$$x = a \sec t \tag{5}$$

$$y = b \csc t. \tag{6}$$

The CURVATURE is

$$\kappa = \frac{3ab \csc^2 t \sec^2 t}{(b^2 \cos^2 t \cos^2 t a^2 \sec^2 t \tan^2 t)^{3/2}}. \tag{7}$$

References

Cundy, H. and Rollett, A. *Mathematical Models, 3rd ed.* Stradbroke, England: Tarquin Pub., p. 71, 1989.
Lawrence, J. D. *A Catalog of Special Plane Curves.* New York: Dover, pp. 127 and 130–31, 1972.

Crunode

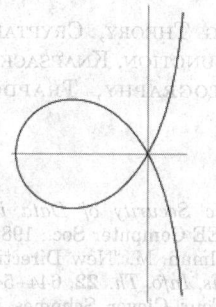

A point where a curve intersects itself so that two branches of the curve have distinct tangent lines. The MACLAURIN TRISECTRIX, shown above, has a crunode at the origin.

See also ACNODE, SPINODE, TACNODE

Cryptarithm

CRYPTARITHMETIC

Cryptarithmetic

A number PUZZLE in which a group of arithmetical operations has some or all of its DIGITS replaced by letters or symbols, and where the original DIGITS must be found. In such a puzzle, each letter represents a unique digit.

See also ALPHAMETIC, DIGIMETIC, SKELETON DIVISION

References

Bogomolny, A. "Cryptarithms." http://www.cut-the-knot.-com/st_crypto.html.
Brooke, M. *One Hundred & Fifty Puzzles in Crypt-Arithmetic.* New York: Dover, 1963.
Kraitchik, M. "Cryptarithmetic." §3.11 in *Mathematical Recreations.* New York: W. W. Norton, pp. 79–3, 1942.
Marks, R. W. *The New Mathematics Dictionary and Handbook.* New York: Bantam Books, 1964.

Cryptographic Hash Function

A cryptographic hash function is most commonly one of the following: a ONE-WAY HASH FUNCTION, a COLLISION-FREE HASH FUNCTION, a TRAPDOOR ONE-WAY HASH FUNCTION, or a function from a class of UNIVERSAL HASH FUNCTIONS.

See also BIRTHDAY ATTACK, COLLISION-FREE HASH FUNCTION, HASH FUNCTION, ONE-WAY HASH FUNCTION, TRAPDOOR ONE-WAY HASH FUNCTION, UNIVERSAL HASH FUNCTION

References

Bakhtiari, S.; Safavi-Naini, R.; and Pieprzyk, J. *Cryptographic Hash Functions: A Survey*. Technical Report 95–9, Department of Computer Science, University of Wollongong, July 1995. ftp://ftp.cs.uow.edu.au/pub/papers/1995/tr-95-9.ps.Z.

Cryptography
The science of adversarial information protection.

See also CODING THEORY, CRYPTARITHM, CRYPTOGRAPHIC HASH FUNCTION, KNAPSACK PROBLEM, PUBLIC-KEY CRYPTOGRAPHY, TRAPDOOR ONE-WAY FUNCTION

References

Davies, D. W. *The Security of Data in Networks*. Los Angeles, CA: IEEE Computer Soc., 1981.
Diffie, W. and Hellman, M. "New Directions in Cryptography." *IEEE Trans. Info. Th.* **22**, 644–54, 1976.
Honsberger, R. "Four Clever Schemes in Cryptography." Ch. 10 in *Mathematical Gems III*. Washington, DC: Math. Assoc. Amer., pp. 151–73, 1985.
Simmons, G. J. "Cryptology, The Mathematics of Secure Communications." *Math. Intel.* **1**, 233–46, 1979.
van Tilborg, H. C. A. *Fundamentals of Cryptography: A Professional Reference and Interactive Tutorial*. Norwell, MA: Kluwer, 1999.

Crystallographic Point Groups
The crystallographic point groups are the POINT GROUPS in which translational periodicity is required (the so-called CRYSTALLOGRAPHY RESTRICTION). There are 32 such groups, summarized in the following table which organized them by SCHÖNFLIES SYMBOL type.

type	point groups
nonaxial	C_i, C_s
cyclic	C_1, C_2, C_3, C_4, C_6
cyclic with horizontal planes	C_{2h}, C_{3h}, C_{4h}, C_{6h}
cyclic with vertical planes	C_{2v}, C_{3v}, C_{4v}, C_{6v}
dihedral	D_2, D_3, D_4, D_6
dihedral with horizontal planes	D_{2h}, D_{3h}, D_{4h}, D_{6h}
dihedral with planes between axes	D_{2d}, D_{3d}
improper rotation	S_4, S_6
cubic groups	T, T_h, T_d, O, O_h

Note that while the TETRAHEDRAL T_d and OCTAHEDRAL O_h POINT GROUPS are also crystallographic point groups, the ICOSAHEDRAL GROUP I_h is not. The orders, classes, and group operations for these groups can be concisely summarized in their CHARACTER TABLES.

See also CHARACTER TABLE, CRYSTALLOGRAPHY RESTRICTION, DIHEDRAL GROUP, GROUP, GROUP THEORY, HERMANN-MAUGUIN SYMBOL, LATTICE GROUPS, OCTAHEDRAL GROUP, POINT GROUPS, SCHÖNFLIES SYMBOL, SPACE GROUPS, TETRAHEDRAL GROUP

References

Arfken, G. "Crystallographic Point and Space Groups." *Mathematical Methods for Physicists, 3rd ed.* Orlando, FL: Academic Press, pp. 248–49, 1985.
Cotton, F. A. *Chemical Applications of Group Theory, 3rd ed.* New York: Wiley, p. 379, 1990.
Hahn, T. (Ed.). *International Tables for Crystallography, vol. A, 4th ed.* Dordrecht, Netherlands: Kluwer, p. 752, 1995.
Lomont, J. S. "Crystallographic Point Groups." §4.4 in *Applications of Finite Groups*. New York: Dover, pp. 132–46, 1993.
Yale, P. B. "Crystallographic Point Groups." §3.4 in *Geometry and Symmetry*. New York: Dover, pp. 103–08, 1988.

Crystallography Restriction
If a discrete GROUP of displacements in the plane has more than one center of rotation, then the only rotations that can occur are by 2, 3, 4, and 6. This can be shown as follows. It must be true that the sum of the interior angles divided by the number of sides is a divisor of $360°$.

$$\frac{180°(n-2)}{n} = \frac{360°}{m},$$

where m is an INTEGER. Therefore, symmetry will be possible only for

$$\frac{2n}{n-2} = m,$$

where m is an INTEGER. This will hold for 1-, 2-, 3-, 4-, and 6-fold symmetry. That it does not hold for $n > 6$ is seen by noting that $n = 6$ corresponds to $m = 3$. The $m = 2$ case requires that $n = n - 2$ (impossible), and the $m = 1$ case requires that $n = -2$ (also impossible).

The POINT GROUPS that satisfy the crystallographic restriction are called CRYSTALLOGRAPHIC POINT GROUPS.

See also CRYSTALLOGRAPHIC POINT GROUPS, POINT GROUPS, SYMMETRY

References

Hilbert, D. and Cohn-Vossen, S. *Geometry and the Imagination*. New York: Chelsea, p. 5, 1999.
Radin, C. *Miles of Tiles*. Providence, RI: Amer. Math. Soc., p. 5, 1999.
Steinhaus, H. *Mathematical Snapshots, 3rd ed.* New York: Dover, p. 304, 1999.

Yale, P. B. *Geometry and Symmetry.* New York: Dover, p. 104, 1988.

Császár Polyhedron

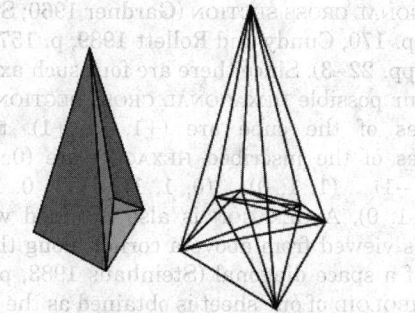

A POLYHEDRON topologically equivalent to a TORUS which was discovered in the late 1940s by Ákos Császár (Gardner 1975). It has 7 VERTICES, 14 faces, and 21 EDGES, and is the DUAL POLYHEDRON of the SZILASSI POLYHEDRON.

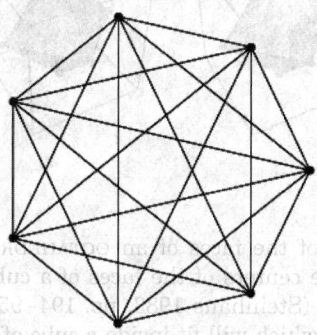

The SKELETON of the Császár polyhedron, illustrated above, is ISOMORPHIC to the COMPLETE GRAPH K_7. Rather surprisingly, the graph of the Császár polyhedron's skeleton and its DUAL GRAPH can be used to find STEINER TRIPLE SYSTEMS (Gardner 1975).

completed base

The figure above shows how to construct the Császár polyhedron.

See also SZILASSI POLYHEDRON, TOROIDAL POLYHEDRON

References

Császár, Á. "A Polyhedron without Diagonals." *Acta Sci. Math.* **13**, 140–42, 1949–1950.

Gardner, M. "Mathematical Games: On the Remarkable Császár Polyhedron and Its Applications in Problem Solving." *Sci. Amer.* **232**, 102–07, May 1975.

Gardner, M. "The Császár Polyhedron." Ch. 11 in *Time Travel and Other Mathematical Bewilderments.* New York: W. H. Freeman, pp. 139–52, 1988.

Gardner, M. *Fractal Music, Hypercards, and More: Mathematical Recreations from Scientific American Magazine.* New York: W. H. Freeman, pp. 118–20, 1992.

Hart, G. "Toroidal Polyhedra." http://www.georgehart.com/virtual-polyhedra/toroidal.html.

Csc

COSECANT

Csch

HYPERBOLIC COSECANT

C-Table

C-DETERMINANT

Ctg

COTANGENT

Cth

HYPERBOLIC COTANGENT

Ctn

COTANGENT

Cubature

Ueberhuber (1997, p. 71) and Krommer and Ueberhuber (1998, pp. 49 and 155–65) use the word "QUADRATURE" to mean numerical computation of a univariate INTEGRAL, and "cubature" to mean numerical computation of a MULTIPLE INTEGRAL. Cubature techniques available in *Mathematica* include MONTE CARLO INTEGRATION, implemented as NIntegrate[f, ..., Method->MonteCarlo] or NIntegrate[f, ..., Method->QuasiMonteCarlo], and the adaptive Genz-Malik algorithm, implemented as NIntegrate[f, ..., Method->MultiDimensional].

See also MONTE CARLO INTEGRATION, NUMERICAL INTEGRATION, QUADRATURE

References

Cools, R. "Monomial Cubature Rules Since "Stroud": A Compilation--Part 2." *J. Comput. Appl. Math.* **112**, 21–7, 1999.

Cools, R. "Encyclopaedia of Cubature Formulas." http://www.cs.kuleuven.ac.be/~nines/research/ecf/ecf.html.

Cools, R. and Rabinowitz, P. "Monomial Cubature Rules Since "Stroud": A Compilation." *J. Comput. Appl. Math.* **48**, 309–26, 1993.

Krommer, A. R. and Ueberhuber, C. W. "Construction of Cubature Formulas." §6.1 in *Computational Integration.* Philadelphia, PA: SIAM, pp. 155–65, 1998.

Ueberhuber, C. W. *Numerical Computation 2: Methods, Software, and Analysis.* Berlin: Springer-Verlag, 1997.

Cube

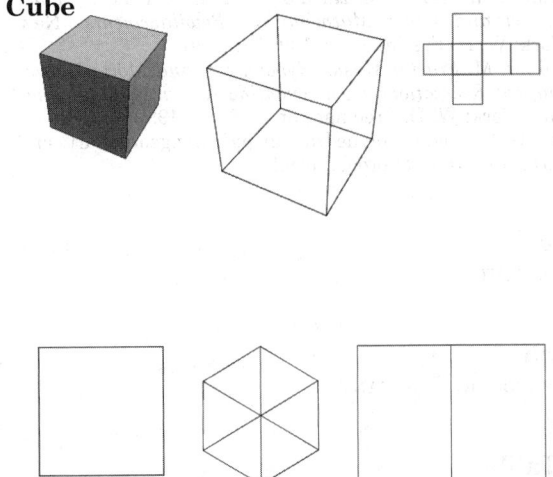

The three-dimensional PLATONIC SOLID P_3 which is also called the HEXAHEDRON. The cube is composed of six SQUARE faces, $6\{4\}$, which meet each other at RIGHT ANGLES, and has eight VERTICES and 12 EDGES. It is also the UNIFORM POLYHEDRON U_6 and Wenninger model W_3. It is described by the SCHLÄFLI SYMBOL $\{4, 3\}$ and WYTHOFF SYMBOL $3|24$.

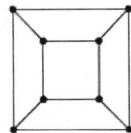

The DUAL POLYHEDRON of the cube is the OCTAHEDRON. It has the O_h OCTAHEDRAL GROUP of symmetries, and is a ZONOHEDRON. The connectivity of the vertices is given by the CUBICAL GRAPH.

Because the VOLUME of a cube of side length n is given by n^3, a number OF THE FORM n^3 is called a CUBIC NUMBER (or sometimes simply "a cube"). Similarly, the operation of taking a number to the third POWER is called CUBING. Sodium chloride (NaCl; common table salt) naturally forms cubic crystals. Using so-called "wallet hinges," a ring of six cubes can be rotated continuously (Wells 1975; Wells 1991, pp. 218–19).

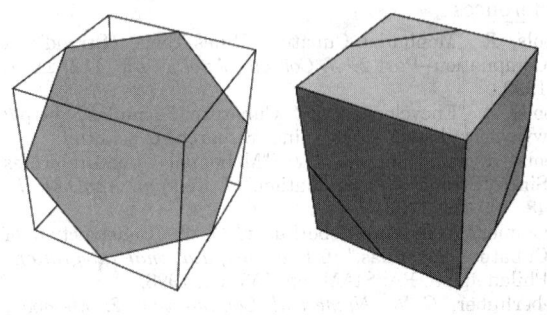

The cube cannot be STELLATED. A PLANE passing through the MIDPOINTS of opposite sides (perpendicular to a C_3 axis) cuts the cube in a regular HEXAGONAL CROSS SECTION (Gardner 1960; Steinhaus 1983, p. 170; Cundy and Rollett 1989, p. 157; Holden 1991, pp. 22–3). Since there are four such axes, there are four possible HEXAGONAL CROSS SECTIONS. If the vertices of the cube are $(\pm1, \pm1, \pm1)$, then the vertices of the inscribed HEXAGON are $(0, -1, -1)$, $(1, 0, -1)$, $(1, 1, 0)$, $(0, 1, 1)$, $(-1, 0, 1)$, and $(-1, -1, 0)$. A HEXAGON is also obtained when the cube is viewed from above a corner along the extension of a space diagonal (Steinhaus 1983, p. 170). A HYPERBOLOID of one sheet is obtained as the envelope of a cube rotated about a space diagonal (Steinhaus 1983, pp. 171–72).

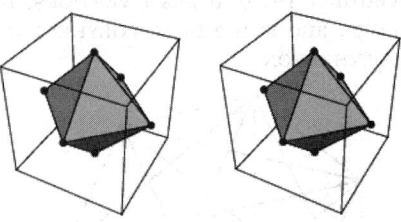

The centers of the faces of an OCTAHEDRON form a cube, and the centers of the faces of a cube form an OCTAHEDRON (Steinhaus 1983, pp. 194–95). The largest SQUARE which will fit inside a cube of side a has each corner a distance $1/4$ from a corner of a cube. The resulting SQUARE has side length $3\sqrt{2}\,a/4$, and the cube containing that side is called PRINCE RUPERT'S CUBE.

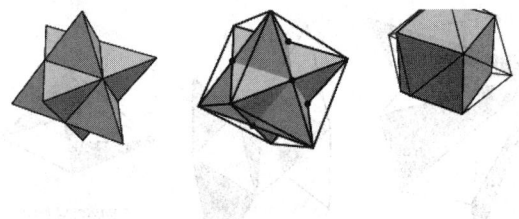

The solid formed by the faces having the sides of the STELLA OCTANGULA (left figure) as DIAGONALS is a cube (right figure; Ball and Coxeter 1987). Affixing a SQUARE PYRAMID of height $1/2$ on each face of a cube having unit edge length results in a RHOMBIC DODECAHEDRON (Brückner 1900, p. 130; Steinhaus 1983, p. 185).

The cube can be constructed by CUMULATION of a unit edge-length TETRAHEDRON by a pyramid with height $\frac{1}{6}\sqrt{6}$. The following table gives polyhedra which can be constructed by CUMULATION of a *cube* by pyramids of given heights h.

h	$(r+h)/h$	Result
$\frac{1}{6}$	$4/3$	TETRAKIS HEXAHEDRON
$\frac{1}{2}$	2	RHOMBIC DODECAHEDRON
$\frac{1}{2}\sqrt{2}$	$1+\sqrt{2}$	24-faced star DELTAHEDRON

The VERTICES of a cube of side length 2 with face-centered axes are given by $(\pm 1,\ \pm 1,\ \pm 1)$. If the cube is oriented with a space diagonal along the z-AXIS, the coordinates are $(0,\ 0,\ \sqrt{3})$, $(0,\ 2\sqrt{2/3},\ 1/\sqrt{3})$, $(\sqrt{2},\ \sqrt{2/3},\ -1/\sqrt{3})$, $(\sqrt{2},\ -\sqrt{2/3},\ 1/\sqrt{3})$, $(0,\ -2\sqrt{2/3},\ -1/\sqrt{3})$, $(-\sqrt{2},\ -\sqrt{2/3},\ 1/\sqrt{3})$, $(-\sqrt{2},\ \sqrt{2/3},\ -1/\sqrt{3})$, and the negatives of these vectors. A FACETED version is the GREAT CUBICUBOCTAHEDRON.

A cube of side length 1 has INRADIUS, MIDRADIUS, and CIRCUMRADIUS of

$$r = \tfrac{1}{2} = 0.5 \tag{1}$$

$$\rho = \tfrac{1}{2}\sqrt{2} \approx 0.70710 \tag{2}$$

$$R = \tfrac{1}{2}\sqrt{3} \approx 0.86602. \tag{3}$$

The cube has a DIHEDRAL ANGLE of

$$\alpha = \tfrac{1}{2}\pi. \tag{4}$$

The SURFACE AREA and VOLUME of the cube are

$$S = 6a^2 \tag{5}$$

$$V = a^3. \tag{6}$$

See also AUGMENTED TRUNCATED CUBE, BIAUGMENTED TRUNCATED CUBE, BIDIAKIS CUBE, BISLIT CUBE, BROWKIN'S THEOREM, CUBE DISSECTION, CUBE DOVETAILING PROBLEM, CUBE DUPLICATION, CUBIC NUMBER, CUBICAL GRAPH, CUBOID, GOURSAT'S SURFACE, HADWIGER PROBLEM, HYPERCUBE, KELLER'S CONJECTURE, PLATONIC SOLID, PRINCE RUPERT'S CUBE, PRISM, RUBIK'S CUBE, SOMA CUBE, STELLA OCTANGULA, TESSERACT, UNIT CUBE

References

Beyer, W. H. (Ed.). *CRC Standard Mathematical Tables, 28th ed.* Boca Raton, FL: CRC Press, pp. 127 and 228, 1987.
Brückner, M. *Vielecke under Vielflache.* Leipzig, Germany: Teubner, 1900.
Cundy, H. and Rollett, A. "Cube. 4^3" and "Hexagonal Section of a Cube." §3.5.2 and 3.15.1 in *Mathematical Models, 3rd ed.* Stradbroke, England: Tarquin Pub., p. 85, 1989.
Davie, T. "The Cube (Hexahedron)." http://www.dcs.st-and.ac.uk/~ad/mathrecs/polyhedra/cube.html.
Eppstein, D. "Rectilinear Geometry." http://www.ics.uci.edu/~eppstein/junkyard/rect.html.

Gardner, M. "Mathematical Games: More About the Shapes that Can Be Made with Complex Dominoes." *Sci. Amer.* **203**, 186–98, Nov. 1960.
Harris, J. W. and Stocker, H. "Cube" and "Cube (Hexahedron)." §4.2.4 and 4.4.3 in *Handbook of Mathematics and Computational Science.* New York: Springer-Verlag, pp. 97–8 and 100, 1998.
Holden, A. *Shapes, Space, and Symmetry.* New York: Dover, 1991.
Kern, W. F. and Bland, J. R. "Cube." §9 in *Solid Mensuration with Proofs, 2nd ed.* New York: Wiley, pp. 19–0, 1948.
Steinhaus, H. *Mathematical Snapshots, 3rd ed.* New York: Dover, pp. 170–72 and 192, 1999.
Wells, D. "Puzzle Page." *Games and Puzzles.* Sep. 1975.
Wells, D. *The Penguin Dictionary of Curious and Interesting Geometry.* London: Penguin, pp. 41–2 and 218–19, 1991.
Wenninger, M. J. "The Hexahedron (Cube)." Model 3 in *Polyhedron Models.* Cambridge, England: Cambridge University Press, p. 16, 1989.

Cube 2-Compound

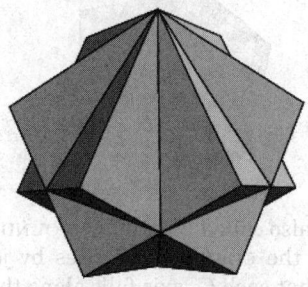

A POLYHEDRON COMPOUND obtained by allowing two CUBES to share opposite VERTICES, then rotating one a sixth of a turn (Holden 1971, p. 34).

See also CUBE, CUBE 3-COMPOUND, CUBE 4-COMPOUND, CUBE 5-COMPOUND, POLYHEDRON COMPOUND

References

Hart, G. "Compound of Two Cubes." http://www.georgehart.com/virtual-polyhedra/vrml/cubes_D6_D3.wrl.
Holden, A. *Shapes, Space, and Symmetry.* New York: Dover, 1991.
Steinhaus, H. *Mathematical Snapshots, 3rd ed.* New York: Dover, p. 213, 1999.
Weisstein, E. W. "Polyhedra." MATHEMATICA NOTEBOOK POLYHEDRA.M.

Cube 3-Compound

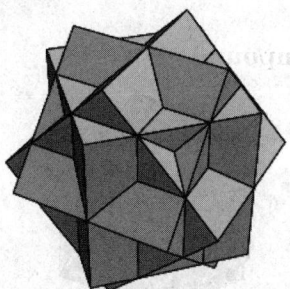

A compound with the symmetry of the CUBE which arises by joining three CUBES such that each shares

two C_2 axes (Holden 1971, p. 35). The solid is depicted atop the left pedestle in M. C. Escher's woodcut *Waterfall*.

See also CUBE, CUBE 2-COMPOUND, CUBE 4-COMPOUND, CUBE 5-COMPOUND, ESCHER'S SOLID, POLYHEDRON COMPOUND

References
Hart, G. "The Compound of Three Cubes." http://www.georgehart.com/virtual-polyhedra/vrml/cubes_S4_D4.wrl.
Holden, A. *Shapes, Space, and Symmetry.* New York: Dover, 1991.
Weisstein, E. W. "Polyhedra." MATHEMATICA NOTEBOOK POLYHEDRA.M.

Cube 4-Compound

A compound also called BAKOS' COMPOUND having the symmetry of the CUBE which arises by joining four CUBES such that each C_3 axis falls along the C_3 axis of one of the other CUBES (Bakos 1959; Holden 1971, p. 35). Let the first cube c_1 consists of a cube in standard position rotated by $\pi/3$ radians around the $(1, 1, 1)$-axis, then the other three cubes are obtained by rotating c_1 around the $(0, 0, 1)$-axis (z-AXIS) by $\pi/2$, $-\pi/2$, and π radians, respectively.

See also CUBE, CUBE 2-COMPOUND, CUBE 3-COMPOUND, CUBE 5-COMPOUND, POLYHEDRON COMPOUND

References
Bakos, T. "Octahedra Inscribed in a Cube." *Math. Gaz.* **43**, 17–0, 1959.
Hart, G. "The Compound of Four Cubes." http://www.georgehart.com/virtual-polyhedra/vrml/cubes_S4_D3.wrl.
Holden, A. *Shapes, Space, and Symmetry.* New York: Dover, 1991.

Cube 5-Compound

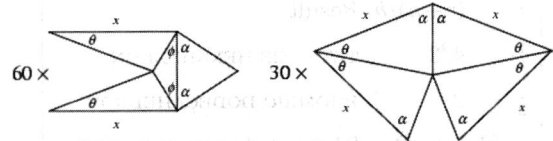

A POLYHEDRON COMPOUND consisting of the arrangement of five CUBES in the VERTICES of a DODECAHEDRON (or the centers of the faces of the ICOSAHEDRON). The cube 5-compound is the dual of the OCTAHEDRON 5-COMPOUND.

In the above figure, let $a = 1$ be the length of a CUBE EDGE. Then

$$x = \tfrac{1}{2}(3 - \sqrt{5})$$

$$\theta = \tan^{-1}\left(\frac{3 - \sqrt{5}}{2}\right) \approx 20°54'$$

$$\phi = \tan^{-1}\left(\frac{\sqrt{5} - 1}{2}\right) \approx 31°43'$$

$$\psi = 90° - \phi \approx 58°17'$$

$$\alpha = 90° - \theta \approx 69°06'.$$

The compound is most easily constructed using pieces like the ones in the above line diagram. The cube 5-compound has the 30 facial planes of the RHOMBIC TRIACONTAHEDRON (Steinhaus 1983, pp. 199 and 209; Ball and Coxeter 1987).

For cubes of unit edge lengths, the resulting compound has edge lengths

$$s_1 = \tfrac{1}{2}\sqrt{\tfrac{1}{2}(65 - 29\sqrt{5})} \tag{1}$$

$$s_2 = \tfrac{1}{2}\sqrt{27 - 12\sqrt{5}} \tag{2}$$

$$s_3 = \tfrac{1}{2}\sqrt{\tfrac{1}{2}(25 - 11\sqrt{5})} \tag{3}$$

$$s_4 = \sqrt{5} - 2 \tag{4}$$

$$s_5 = \tfrac{1}{2}\sqrt{\tfrac{3}{2}(7 - 3\sqrt{5})} \tag{5}$$

$$s_6 = \tfrac{1}{2}\sqrt{5 - 2\sqrt{5}} \tag{6}$$

$$s_7 = \tfrac{1}{2}(3 - \sqrt{5}). \tag{7}$$

The CIRCUMRADIUS is

$$R = \tfrac{1}{2}\sqrt{3}, \tag{8}$$

and the SURFACE AREA and VOLUME are

$$S = 165\sqrt{5} - 360 \tag{9}$$

$$V = \tfrac{1}{2}(55\sqrt{5} - 120). \tag{10}$$

See also CUBE, CUBE 2-COMPOUND, CUBE 3-COMPOUND, CUBE 4-COMPOUND, CUBE 5-COMPOUND–OCTAHEDRON 5-COMPOUND, CUBE 20-COMPOUND, DODECAHEDRON, OCTAHEDRON 5-COMPOUND, POLYHEDRON COMPOUND, RHOMBIC TRIACONTAHEDRON

References

Ball, W. W. R. and Coxeter, H. S. M. *Mathematical Recreations and Essays, 13th ed.* New York: Dover, pp. 135 and 137, 1987.
Cundy, H. and Rollett, A. "Five Cubes in a Dodecahedron." §3.10.6 in *Mathematical Models, 3rd ed.* Stradbroke, England: Tarquin Pub., pp. 135–36, 1989.
Hart, G. "Standard Compound of Five Cubes." http://www.georgehart.com/virtual-polyhedra/vrml/compound_of_5_cubes_(5_colors).wrl.
Weisstein, E. W. "Polyhedra." MATHEMATICA NOTEBOOK POLYHEDRA.M.

Cube 20-Compound

See also CUBE, CUBE 2-COMPOUND, CUBE 3-COMPOUND, CUBE 4-COMPOUND, CUBE 5-COMPOUND, POLYHEDRON COMPOUND

References

Wenninger, M. J. *Dual Models.* Cambridge, England: Cambridge University Press, pp. 139–40, 1983.

Cube 5-Compound–Octahedron 5-Compound

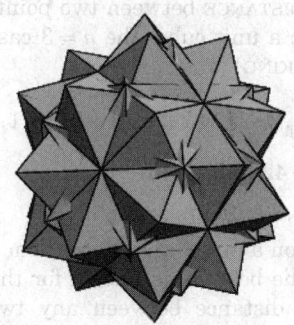

The compound of the CUBE 5-COMPOUND and its dual, the OCTAHEDRON 5-COMPOUND.

See also CUBE 5-COMPOUND, OCTAHEDRON 5-COMPOUND

Cube Dissection

A CUBE can be divided into n subcubes for only $n = 1$, 8, 15, 20, 22, 27, 29, 34, 36, 38, 39, 41, 43, 45, 46, and $n \geq 48$ (Sloane's A014544).

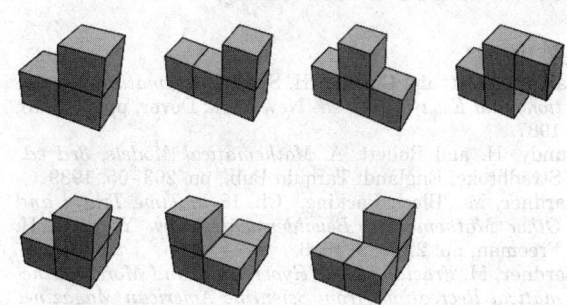

The seven pieces used to construct the $3 \times 3 \times 3$ cube dissection known as the SOMA CUBE are one 3-POLYCUBE and six 4-POLYCUBES ($1 \cdot 3 + 6 \cdot 4 = 27$), illustrated above.

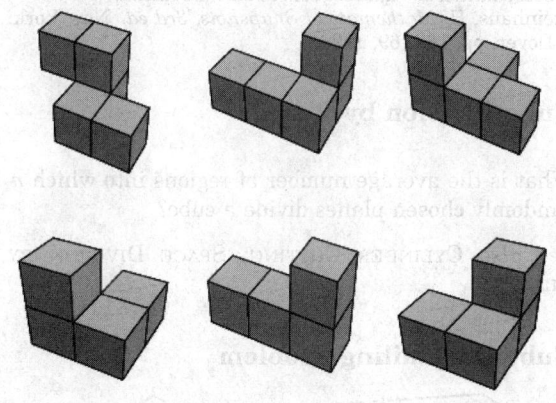

Another $3 \times 3 \times 3$ cube dissection due to Steinhaus (1983) uses three 5-POLYCUBES and three 4-POLYCUBES ($3 \cdot 5 + 3 \cdot 4 = 27$), illustrated above. There are two solutions.

It is possible to cut a 1×3 RECTANGLE into two identical pieces which will form a CUBE (without overlapping) when folded and joined. In fact, an INFINITE number of solutions to this problem were discovered by C. L. Baker (Hunter and Madachy 1975).

Lonke (2000) has considered the number $f(j, k, n)$ of j-dimensional faces of a random k-dimensional central section of the n-cube $B_\infty^n = [-1, 1]^n$, and gives the special result

$$f(0, k, n) = 2^k \binom{n}{k} \sqrt{\frac{2k}{\pi}} \int_0^\infty e^{-kt^2/2} \gamma_{n-k}(tB_\infty^{n-k}) \, dt,$$

where γ_{n-k} is the $(n-k)$-dimensional Gaussian probability measure.

See also CONWAY PUZZLE, DISSECTION, HADWIGER PROBLEM, POLYCUBE, SLOTHOUBER-GRAATSMA PUZZLE, SOMA CUBE

References

Ball, W. W. R. and Coxeter, H. S. M. *Mathematical Recreations and Essays, 13th ed.* New York: Dover, pp. 112–13, 1987.

Cundy, H. and Rollett, A. *Mathematical Models, 3rd ed.* Stradbroke, England: Tarquin Pub., pp. 203–05, 1989.

Gardner, M. "Block Packing." Ch. 18 in *Time Travel and Other Mathematical Bewilderments.* New York: W. H. Freeman, pp. 227–39, 1988.

Gardner, M. *Fractal Music, Hypercards, and More: Mathematical Recreations from Scientific American Magazine.* New York: W. H. Freeman, pp. 297–98, 1992.

Honsberger, R. *Mathematical Gems II.* Washington, DC: Math. Assoc. Amer., pp. 75–0, 1976.

Hunter, J. A. H. and Madachy, J. S. *Mathematical Diversions.* New York: Dover, pp. 69–0, 1975.

Lonke, Y. "On Random Sections of the Cube." *Discr. Comput. Geom.* **23**, 157–69, 2000.

Sloane, N. J. A. Sequences A014544 in "An On-Line Version of the Encyclopedia of Integer Sequences." http://www.research.att.com/~njas/sequences/eisonline.html.

Steinhaus, H. *Mathematical Snapshots, 3rd ed.* New York: Dover, pp. 168–69, 1999.

Cube Division by Planes

What is the average number of regions into which n randomly chosen planes divide a cube?

See also CYLINDER CUTTING, SPACE DIVISION BY PLANES

Cube Dovetailing Problem

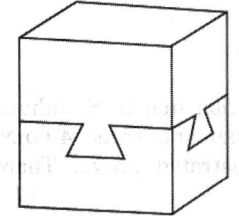 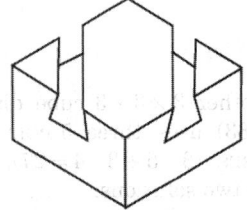

Given the figure on the left (without looking at the solution on the right), determine how to disengage the two slotted CUBE halves without cutting, breaking, or distorting.

References

Dudeney, H. E. *Amusements in Mathematics.* New York: Dover, pp. 145 and 249, 1958.

Ogilvy, C. S. *Excursions in Mathematics.* New York: Dover, pp. 57, 59, and 143, 1994.

Cube Duplication

Also called the DELIAN PROBLEM or DUPLICATION OF THE CUBE. A classical problem of antiquity which, given the EDGE of a CUBE, requires a second CUBE to be constructed having double the VOLUME of the first using only a STRAIGHTEDGE and COMPASS.

Under these restrictions, the problem cannot be solved because the DELIAN CONSTANT $2^{1/3}$ (the required RATIO of sides of the original CUBE and that to

be constructed) is not a EUCLIDEAN NUMBER. The problem can be solved, however, using a NEUSIS CONSTRUCTION.

See also ALHAZEN'S BILLIARD PROBLEM, COMPASS, CUBE, DELIAN CONSTANT, GEOMETRIC PROBLEMS OF ANTIQUITY, NEUSIS CONSTRUCTION, STRAIGHTEDGE

References

Ball, W. W. R. and Coxeter, H. S. M. *Mathematical Recreations and Essays, 13th ed.* New York: Dover, pp. 93–4, 1987.

Bold, B. "The Delian Problem." Ch. 4 in *Famous Problems of Geometry and How to Solve Them.* New York: Dover, pp. 29–1, 1982.

Conway, J. H. and Guy, R. K. *The Book of Numbers.* New York: Springer-Verlag, pp. 190–91, 1996.

Courant, R. and Robbins, H. "Doubling the Cube" and "A Classical Construction for Doubling the Cube." §3.3.1 and 3.5.1 in *What is Mathematics?: An Elementary Approach to Ideas and Methods, 2nd ed.* Oxford, England: Oxford University Press, pp. 134–35 and 146, 1996.

Dörrie, H. "The Delian Cube-Doubling Problem." §35 in *100 Great Problems of Elementary Mathematics: Their History and Solutions.* New York: Dover, pp. 170–72, 1965.

Klein, F. "The Delian Problem and the Trisection of the Angle." Ch. 2 in "Famous Problems of Elementary Geometry: The Duplication of the Cube, the Trisection of the Angle, and the Quadrature of the Circle." In *Famous Problems and Other Monographs.* New York: Chelsea, pp. 13–5, 1980.

Lockwood, E. H. *A Book of Curves.* Cambridge, England: Cambridge University Press, p. 175, 1967.

Wells, D. *The Penguin Dictionary of Curious and Interesting Numbers.* Middlesex, England: Penguin Books, pp. 33–4, 1986.

Wells, D. *The Penguin Dictionary of Curious and Interesting Geometry.* London: Penguin, pp. 49–0, 1991.

Cube Line Picking

The average DISTANCE between two points chosen at random inside a unit cube (the $n = 3$ case of HYPERCUBE LINE PICKING) is

$$\Delta(3) = \tfrac{1}{105}[4 + 17\sqrt{2} - 6\sqrt{3} + 21\ln(1 + \sqrt{2})$$
$$+ 42\ln(2 + \sqrt{3}) - 7\pi]$$

(Robbins 1978, Le Lionnais 1983).

Pick n points on a CUBE, and space them as far apart as possible. The best value known for the minimum straight LINE distance between any two points is given in the following table.

n	$d(n)$
5	1.1180339887498
6	1.0606601482100
7	1
8	1
9	0.86602540378463
10	0.74999998333331

11	0.70961617562351
12	0.70710678118660
13	0.70710678118660
14	0.70710678118660
15	0.625

See also CUBE POINT PICKING, CUBE TRIANGLE PICKING, DISCREPANCY THEOREM, HYPERCUBE LINE PICKING, POINT PICKING, POINT-POINT DISTANCE–1-D

References

Bolis, T. S. Solution to Problem E2629. "Average Distance between Two Points in a Box." *Amer. Math. Monthly* **85**, 277–78, 1978.

Finch, S. "Favorite Mathematical Constants." http://www.mathsoft.com/asolve/constant/geom/geom.html.

Ghosh, B. "Random Distances within a Rectangle and between Two Rectangles." *Bull. Calcutta Math. Soc.* **43**, 17–4, 1951.

Holshouser, A. L.; King, L. R.; and Klein, B. G. Solution to Problem E3217, "Minimum Average Distance between Points in a Rectangle." *Amer. Math. Monthly* **96**, 64–5, 1989.

Le Lionnais, F. *Les nombres remarquables.* Paris: Hermann, p. 30, 1983.

Robbins, D. "Average Distance between Two Points in a Box." *Amer. Math. Monthly* **85**, 278, 1978.

Santaló, L. A. *Integral Geometry and Geometric Probability.* Reading, MA: Addison-Wesley, 1976.

Cube Packing

References

Friedman, E. "Cubes in Cubes." http://www.stetson.edu/~efriedma/cubincub/.

Cube Point Picking

Pick N points $p_1, ..., p_N$ randomly in a unit n-cube. Let C be the CONVEX HULL, so

$$C \equiv \left\{ \sum_{j=1}^{N} \lambda_j p_j : \lambda_j \geq 0 \text{ for all } j \text{ and } \sum_{j=1}^{N} \lambda_j = 1 \right\}. \quad (1)$$

Let $V(n, N)$ be the expected n-D VOLUME (the CONTENT) of C, $S(n, N)$ be the expected $(n-1)$-D SURFACE AREA of C, and $P(n, N)$ the expected number of VERTICES on the POLYGONAL boundary of C. Then

$$\lim_{N \to \infty} \frac{N[1 - V(2, N)]}{\ln N} = \frac{8}{3}$$

$$\lim_{N \to \infty} \sqrt{N}[4 - S(2, N)]$$

$$= \sqrt{2\pi} \left[2 - \int_0^1 (\sqrt{1+t^2} - 1)t^{-3/2} \, dt \right]$$

$$= 4.2472965..., \quad (2)$$

$$\lim_{N \to \infty} P(2, N) - \frac{8}{3} \ln N = \frac{8}{3}(\gamma - \ln 2)$$

$$= -0.309150708... \quad (3)$$

(Rényi and Sulanke 1963, 1964).

See also BALL POINT PICKING, CUBE LINE PICKING, SPHERE POINT PICKING

References

Rényi, A. and Sulanke, R. "Über die konvexe Hülle von n zufällig gewählten Punkten, I." *Z. Wahrscheinlichkeits* **2**, 75–4, 1963.

Rényi, A. and Sulanke, R. "Über die konvexe Hülle von n zufällig gewählten Punkten, II." *Z. Wahrscheinlichkeits* **3**, 138–47, 1964.

Cube Power

A number raised to the third POWER. x^3 is read as "x cubed."

See also CUBIC NUMBER

Cube Root

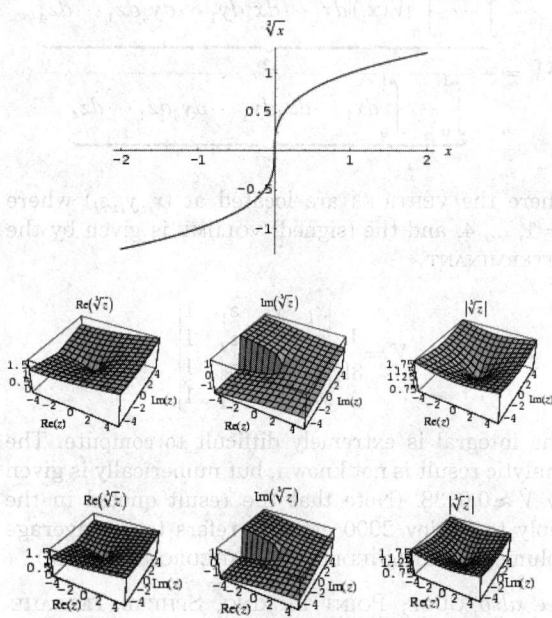

Given a number z, the cube root of z, denoted $\sqrt[3]{z}$ or $z^{1/3}$ (z to the 1/3 POWER), is a number a such that $a^3 = z$. There are three (not necessarily distinct) cube roots for any number.

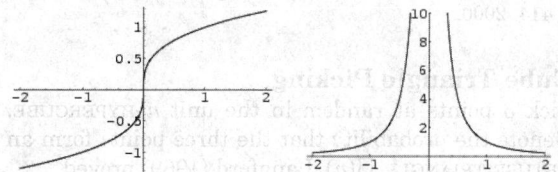

For real arguments, the cube root is an INCREASING FUNCTION, although the usual derivative test cannot be used to establish this fact at the ORIGIN since the

derivative approaches infinity there (as illustrated above).

See also CUBE DUPLICATION, CUBED, DELIAN CONSTANT, GEOMETRIC PROBLEMS OF ANTIQUITY, κ-MATRIX, SQUARE ROOT

Cube Tetrahedron Picking

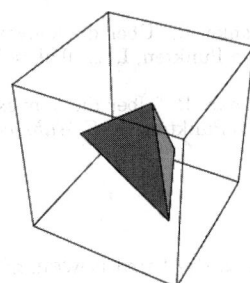

Given four points chosen at random inside a UNIT CUBE, the average VOLUME of the TETRAHEDRON determined by these points is given by

$$\bar{V} = \frac{\underbrace{\int_0^1 \cdots \int_0^1 |V(\mathbf{x}_i)| dx_1 \cdots dx_4 dy_1 \cdots dy_4 dz_1 \cdots dz_4}_{12}}{\underbrace{\int_0^1 \cdots \int_0^1 dx_1 \cdots dx_4 dy_1 \cdots dy_4 dz_1 \cdots dz_4}_{12}}$$

where the VERTICES are located at (x_i, y_i, z_i) where $i = 1, ..., 4$, and the (signed) VOLUME is given by the DETERMINANT

$$V = \frac{1}{3!} \begin{vmatrix} x_1 & y_1 & z_1 & 1 \\ x_2 & y_2 & z_2 & 1 \\ x_3 & y_3 & z_3 & 1 \\ x_4 & y_4 & z_4 & 1 \end{vmatrix}.$$

The integral is extremely difficult to compute. The analytic result is not known, but numerically is given by $\bar{V} \approx 0.0138$. (Note that the result quoted in the reply to Seidov 2000 actually refers to the average volume for TETRAHEDRON TETRAHEDRON PICKING.)

See also CUBE, POINT PICKING, SPHERE TETRAHEDRON PICKING, SQUARE TRIANGLE PICKING, TETRAHEDRON

References
Seidov, Z. F. "Letters: Random Triangle." *Mathematica J.* **7**, 414, 2000.

Cube Triangle Picking

Pick 3 points at random in the unit n-HYPERCUBE. Denote the probability that the three points form an OBTUSE TRIANGLE $\prod(n)$. Langford (1969) proved

$$\Phi(2) = \frac{97}{150} + \frac{1}{40}\pi = 0.725206483\ldots$$

See also BALL TRIANGLE PICKING, CUBE POINT PICKING

References
Finch, S. "Favorite Mathematical Constants." http://www.mathsoft.com/asolve/constant/geom/geom.html.
Langford, E. "The Probability that a Random Triangle is Obtuse." *Biometrika* **56**, 689–90, 1969.
Santaló, L. A. *Integral Geometry and Geometric Probability.* Reading, MA: Addison-Wesley, 1976.

Cubed

A number to the POWER 3 is said to be cubed, so that x^3 is called "x cubed."

See also CUBE ROOT, SQUARED

Cubefree

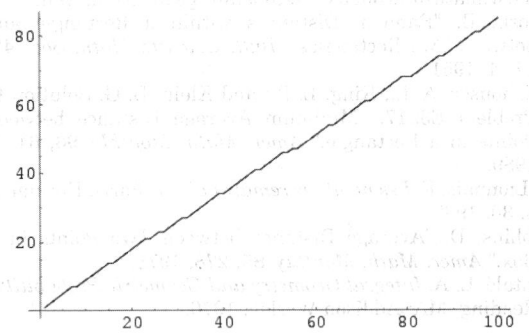

A number is said to be cubefree if its PRIME FACTORIZATION contains no tripled factors. All PRIMES are therefore trivially cubefree. The cubefree numbers are 1, 2, 3, 4, 5, 6, 7, 9, 10, 11, 12, 13, 14, 15, 17, ... (Sloane's A004709). The cubeful numbers (i.e., those that contain at least one cube) are 8, 16, 24, 27, 32, 40, 48, 54, ... (Sloane's A046099). The number of cubefree numbers less than 10, 100, 1000, ... are 9, 85, 833, 8319, 83190, 831910, ..., and their asymptotic density is $1/\zeta(3) \approx 0.831907$, where $\zeta(n)$ is the RIEMANN ZETA FUNCTION.

See also BIQUADRATEFREE, CUBEFREE PART, PRIME NUMBER, RIEMANN ZETA FUNCTION, SQUAREFREE

References
Sloane, N. J. A. Sequences A004709 and A046099 in "An On-Line Version of the Encyclopedia of Integer Sequences." http://www.research.att.com/~njas/sequences/eisonline.html.

Cubefree Part

That part of a POSITIVE INTEGER left after all cubic factors are divided out. For example, the cubefree part of $24 = 2^3 \cdot 3$ is 3. For $n = 1, 2, ...$, the first few are 1, 2, 3, 4, 5, 6, 7, 1, 9, 10, 11, 12, 13, 14, 15, 2, ... (Sloane's A050985). The squarefree part function can be implemented in *Mathematica* as

```
SquarefreePart[n_Integer?Positive] :=
```

```
Times @@ Power @@@ ({#[[1]], Mod[#[[2]], 3]} &
/@ FactorInteger[n])
```

See also CUBEFREE, CUBIC PART, SQUAREFREE PART

References

Sloane, N. J. A. Sequences A050985 in "An On-Line Version of the Encyclopedia of Integer Sequences." http://www.re-search.att.com/~njas/sequences/eisonline.html.

Cubefree Word

A cubefree word contains *no* cubed words as sub-words. The number of binary cubefree words of length $n = 1, 2, \ldots$ are 2, 4, 6, 10, 16, 24, 36, 56, 80, 118, ... (Sloane's A028445). Binary cubefree words satisfy

$$2 \cdot 1.080^n \leq c(n) \leq 2 \cdot 1.522^n. \tag{1}$$

The number of ternary cubefree words of length $n = 1, 2, \ldots$ are 3, 9, 24, 66, 180, 486, 1314, ... (Sloane's A051042). The number of quaternary cubefree words of length $n = 1, 2, \ldots$ are 4, 16, 60, 228, 864, 3264, 12336, ... (Sloane's A051043).

See also OVERLAPFREE WORD, SQUAREFREE WORD, WORD

References

Finch, S. "Favorite Mathematical Constants." http://www.mathsoft.com/asolve/constant/words/words.html.
Sloane, N. J. A. Sequences A028445, A051042, and A051043 in "An On-Line Version of the Encyclopedia of Integer Sequences." http://www.research.att.com/~njas/sequences/eisonline.html.

Cube-Octahedron Compound

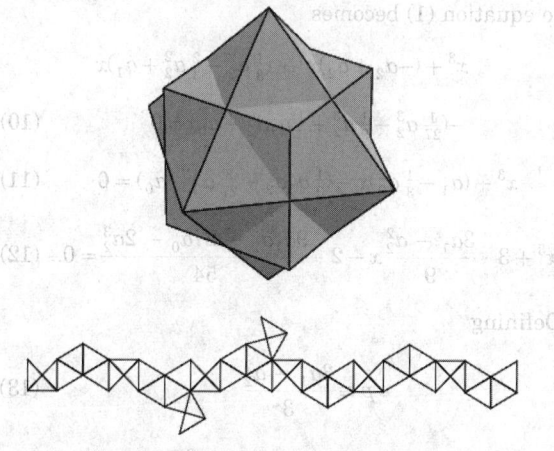

A POLYHEDRON COMPOUND composed of a CUBE and its DUAL POLYHEDRON, the OCTAHEDRON. For a CUBE of edge length 1, the 14 vertices are located at $(\pm 1/2, \pm 1/2, \pm 1/2)$, $(\pm 1, 0, 0)$, $(0, \pm 1, 0)$, $(0, 0, \pm 1)$. Since the edges of the cube and octahedron bisect each other, the resulting solid has side lengths 1/2 and $\sqrt{2}/2$, and SURFACE AREA and VOLUME given by

$$S = 3(1 + \sqrt{3})$$

$$V = \tfrac{3}{2}.$$

The CONVEX HULL of the cube-octahedron compound is a RHOMBIC DODECAHEDRON.

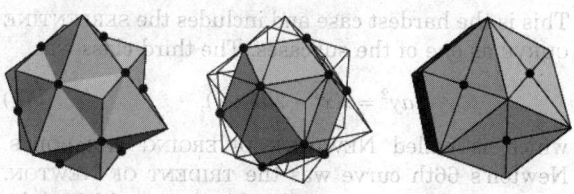

The solid common to both the CUBE and OCTAHEDRON (left figure) in a cube-octahedron compound is a CUBOCTAHEDRON (middle figure). The edges intersecting in the points plotted above are the diagonals of RHOMBUSES, and the 12 RHOMBUSES form a RHOMBIC DODECAHEDRON (right figure; Ball and Coxeter 1987).

See also CUBE, CUBOCTAHEDRON, OCTAHEDRON, POLYHEDRON COMPOUND

References

Ball, W. W. R. and Coxeter, H. S. M. *Mathematical Recreations and Essays, 13th ed.* New York: Dover, p. 137, 1987.
Coxeter, H. S. M. *Introduction to Geometry, 2nd ed.* New York: Wiley, p. 158, 1969.
Cundy, H. and Rollett, A. "Cube Plus Octahedron." §3.10.2 in *Mathematical Models, 3rd ed.* Stradbroke, England: Tarquin Pub., p. 130, 1989.
Weisstein, E. W. "Polyhedra." MATHEMATICA NOTEBOOK POLYHEDRA.M.
Wenninger, M. J. "Compound of a Cube and Octahedron." §43 in *Polyhedron Models.* New York: Cambridge University Press, p. 68, 1989.

Cubic Close Packing

SPHERE PACKING

Cubic Curve

A cubic curve is an ALGEBRAIC CURVE of degree 3. An algebraic curve over a FIELD K is an equation $f(X, Y) = 0$, where $f(X, Y)$ is a POLYNOMIAL in X and Y with COEFFICIENTS in K, and the degree of f is the MAXIMUM degree of each of its terms (MONOMIALS).

Newton showed that all cubics can be generated by the projection of the five divergent cubic parabolas. Newton's classification of cubic curves appeared in the chapter "Curves" in *Lexicon Technicum* by John Harris published in London in 1710. Newton also classified all cubics into 72 types, missing six of them. In addition, he showed that any cubic can be obtained by a suitable projection of the ELLIPTIC CURVE

$$y^2 = ax^3 + bx^2 + cx + d, \tag{1}$$

where the projection is a BIRATIONAL TRANSFORMA-

TION, and the general cubic can also be written as

$$y^2 = x^3 + ax + b. \tag{2}$$

Newton's first class is equations OF THE FORM

$$xy^2 + ey = ax^3 + bx^2 + cx + d. \tag{3}$$

This is the hardest case and includes the SERPENTINE CURVE as one of the subcases. The third class was

$$ay^2 = x(x^2 - 2bx + c), \tag{4}$$

which is called NEWTON'S DIVERGING PARABOLAS. Newton's 66th curve was the TRIDENT OF NEWTON. Newton's classification of cubics was criticized by Euler because it lacked generality. Plücker later gave a more detailed classification with 219 types.

The NINE ASSOCIATED POINTS THEOREM states that Any cubic curve that passes through eight of the nine intersections of two given cubic curves automatically passes through the ninth (Evelyn *et al.* 1974, p. 15).

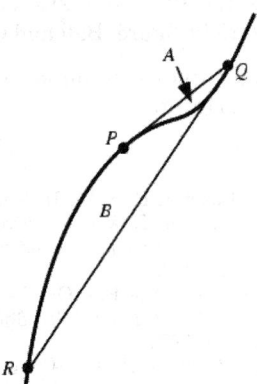

Pick a point P, and draw the tangent to the curve at P. Call the point where this tangent intersects the curve Q. Draw another tangent and call the point of intersection with the curve R. Every curve of third degree has the property that, with the areas in the above labeled figure,

$$B = 16A \tag{5}$$

(Honsberger 1991).

See also CAYLEY-BACHARACH THEOREM, CUBIC EQUATION, ELLIPTIC CURVE, NINE ASSOCIATED POINTS THEOREM, TRIANGLE CUBIC CURVE

References

Evelyn, C. J. A.; Money-Coutts, G. B.; and Tyrrell, J. A. *The Seven Circles Theorem and Other New Theorems*. London: Stacey International, p. 15, 1974.
Honsberger, R. *More Mathematical Morsels*. Washington, DC: Math. Assoc. Amer., pp. 114–18, 1991.
Newton, I. *Mathematical Works, Vol. 2*. New York: Johnson Reprint Corp., pp. 135–61, 1967.
Wall, C. T. C. "Affine Cubic Functions III." *Math. Proc. Cambridge Phil. Soc.* **87**, 1–4, 1980.
Westfall, R. S. *Never at Rest: A Biography of Isaac Newton*. New York: Cambridge University Press, 1988.

Yates, R. C. "Cubic Parabola." *A Handbook on Curves and Their Properties*. Ann Arbor, MI: J. W. Edwards, pp. 56–9, 1952.

Cubic Equation

A cubic equation is a POLYNOMIAL equation of degree three. Given a general cubic equation

$$z^3 + a_2 z^2 + a_1 z + a_0 = 0 \tag{1}$$

(the COEFFICIENT a_3 of z^3 may be taken as 1 without loss of generality by dividing the entire equation through by a_3), first attempt to eliminate the a_2 term by making a substitution OF THE FORM

$$z \equiv x - \lambda. \tag{2}$$

Then

$$(x - \lambda)^3 + a_2(x - \lambda)^2 + a_1(x - \lambda) + a_0 = 0 \tag{3}$$

$$(x^3 - 3\lambda x^2 + 3\lambda^2 x - \lambda^3) + a_2(x^2 - 2\lambda x + \lambda^2)$$
$$+ a_1(x - \lambda) + a_0 = 0 \tag{4}$$

$$x^3 + (a_2 - 3\lambda)x^2 + (a_1 - 2a_2\lambda + 3\lambda^2)x$$
$$+ (a_0 - a_1\lambda + a_2\lambda^2 - \lambda^3) = 0. \tag{5}$$

The x^2 is eliminated by letting $\lambda = a_2/3$, so

$$z \equiv x - \tfrac{1}{3}a_2. \tag{6}$$

Then

$$z^3 = (x - \tfrac{1}{3}a_2)^3 = x^3 - a_2 x^2 + \tfrac{1}{3}a_2^2 x - \tfrac{1}{27}a_2^3. \tag{7}$$

$$a_2 z^2 = a_2(x - \tfrac{1}{3}a_2)^2 = a_2 x^2 - \tfrac{2}{3}a_2^2 x + \tfrac{1}{9}a_2^3 \tag{8}$$

$$a_1 z = a_1(x - \tfrac{1}{3}a_2) = a_1 x - \tfrac{1}{3}a_1 a_2, \tag{9}$$

so equation (1) becomes

$$x^3 + (-a_2 + a_2)x^2 + (\tfrac{1}{3}a_2^2 - \tfrac{2}{3}a_2^2 + a_1)x$$
$$- (\tfrac{1}{27}a_2^3 - \tfrac{1}{9}a_2^3 + \tfrac{1}{3}a_1 a_2 - a_0) = 0 \tag{10}$$

$$x^3 + (a_1 - \tfrac{1}{3}a_2^2)x - (\tfrac{1}{3}a_1 a_2 - \tfrac{2}{27}a_2^3 - a_0) = 0 \tag{11}$$

$$x^3 + 3 \cdot \frac{3a_1 - a_2^2}{9} x - 2 \cdot \frac{9a_1 a_2 - 27a_0 - 2a_2^3}{54} = 0. \tag{12}$$

Defining

$$p \equiv \frac{3a_1 - a_2^2}{3} \tag{13}$$

$$q \equiv \frac{9a_1 a_2 - 27a_0 - 2a_2^3}{27} \tag{14}$$

then allows (12) to be written in the standard form

$$x^3 + px = q. \tag{15}$$

The simplest way to proceed is to make VIETA'S

SUBSTITUTION

$$x = w - \frac{p}{3w}, \tag{16}$$

which reduces the cubic to the equation

$$w^3 - \frac{p^3}{27w^3} - q = 0, \tag{17}$$

which is easily turned into a QUADRATIC EQUATION in w^3 by multiplying through by w^3 to obtain

$$(w^3)^2 - q(w^3) - \tfrac{1}{27}p^3 = 0 \tag{18}$$

(Birkhoff and Mac Lane 1996, p. 106). The result from the QUADRATIC EQUATION is

$$w^3 = \tfrac{1}{2}\left(q \pm \sqrt{q^2 + \tfrac{4}{27}p^3}\right) = \tfrac{1}{2}q \pm \sqrt{\tfrac{1}{4}q^2 + \tfrac{1}{27}p^3}$$
$$= R \pm \sqrt{R^2 + Q^3}, \tag{19}$$

where Q and R are sometimes more useful to deal with than are p and q. There are therefore six solutions for w (two corresponding to each sign for each ROOT of w^3). Plugging w back in to (17) gives three pairs of solutions, but each pair is equal, so there are three solutions to the cubic equation.

Equation (12) may also be explicitly factored by attempting to pull out a term OF THE FORM $(x - B)$ from the cubic equation, leaving behind a quadratic equation which can then be factored using the QUADRATIC FORMULA. This process is equivalent to making VIETA'S SUBSTITUTION, but does a slightly better job of motivating Vieta's "magic" substitution, and also at producing the explicit formulas for the solutions. First, define the intermediate variables

$$Q \equiv \frac{3a_1 - a_2^2}{9} \tag{20}$$

$$R \equiv \frac{9a_2a_1 - 27a_0 - 2a_2^3}{54} \tag{21}$$

(which are identical to p and q up to a constant factor). The general cubic equation (12) then becomes

$$x^3 + 3Qx - 2R = 0. \tag{22}$$

Let B and C be, for the moment, arbitrary constants. An identity satisfied by PERFECT CUBIC POLYNOMIAL equations is that

$$x^3 - B^3 = (x - B)(x^2 + Bx + B^2). \tag{23}$$

The general cubic would therefore be directly factorable if it did not have an x term (i.e., if $Q = 0$). However, since in general $Q \neq 0$, add a multiple of $(x - B)$—say $C(x - B)$—to both sides of (23) to give the slightly messy identity

$$(x^3 - B^3) + C(x - B) = (x - B)(x^2 + Bx + B^2 + C)$$
$$= 0, \tag{24}$$

which, after regrouping terms, is

$$x^3 + Cx - (B^3 + BC) = (x - B)[x^2 + Bx + (B^2 + C)]$$
$$= 0. \tag{25}$$

We would now like to match the COEFFICIENTS C and $-(B^3 + BC)$ with those of equation (22), so we must have

$$C = 3Q \tag{26}$$

$$B^3 + BC = 2R. \tag{27}$$

Plugging the former into the latter then gives

$$B^3 + 3QB = 2R. \tag{28}$$

Therefore, if we can find a value of B satisfying the above identity, we have factored a linear term from the cubic, thus reducing it to a QUADRATIC EQUATION. The trial solution accomplishing this miracle turns out to be the symmetrical expression

$$B = [R + \sqrt{Q^3 + R^2}]^{1/3} + [R - \sqrt{Q^3 + R^2}]^{1/3}. \tag{29}$$

Taking the second and third POWERS of B gives

$$B^2 = [R + \sqrt{Q^3 + R^2}]^{2/3} + 2[R^2 - (Q^3 + R^2)]^{1/3}$$
$$+ [R - \sqrt{Q^3 + R^2}]^{2/3}$$
$$= [R + \sqrt{Q^3 + R^2}]^{2/3} + [R - \sqrt{Q^3 + R^2}]^{2/3} - 2Q \tag{30}$$

$$B^3 = -2QB$$
$$+ \left\{ [R + \sqrt{Q^3 + R^2}]^{1/3} + [R - \sqrt{Q^3 + R^2}]^{1/3} \right\}$$
$$\times \left\{ [R + \sqrt{Q^3 + R^2}]^{2/3} + [R - \sqrt{Q^3 + R^2}]^{2/3} \right\}$$
$$= [R + \sqrt{Q^3 + R^2}] + [R - \sqrt{Q^3 + R^2}]$$
$$+ [R + \sqrt{Q^3 + R^2}]^{1/3}[R - \sqrt{Q^3 + R^2}]^{2/3}$$
$$+ [R + \sqrt{Q^3 + R^2}]^{2/3}[R - \sqrt{Q^3 + R^2}]^{1/3} - 2QB$$
$$= -2QB + 2R + [R^2 - (Q^3 + R^2)]^{1/3}$$
$$\times \left[\left(R + \sqrt{Q^3 + R^2}\right)^{1/3} + \left(R - \sqrt{Q^3 - R^2}\right)^{1/3} \right]$$
$$= -2QB + 2R - QB = -3QB + 2R. \tag{31}$$

Plugging B^3 and B into the left side of (28) gives

$$(-3QB + 2R) + 3QB = 2R, \tag{32}$$

so we have indeed found the factor $(x - B)$ of (22), and we need now only factor the quadratic part. Plugging $C = 3Q$ into the quadratic part of (25) and solving the resulting

$$x^2 + Bx + (B^2 + 3Q) = 0 \tag{33}$$

then gives the solutions

$$x = \frac{1}{2}\left[-B \pm \sqrt{B^2 - 4(B^2 + 3Q)}\right]$$

$$= -\frac{1}{2}B \pm \frac{1}{2}\sqrt{-3B^2 - 12Q}$$

$$= -\frac{1}{2}B \pm \frac{1}{2}\sqrt{3i}\sqrt{B^2 + 4Q}. \tag{34}$$

These can be simplified by defining

$$A \equiv \left[R + \sqrt{Q^3 + R^2}\right]^{1/3} - \left[R - \sqrt{Q^3 + R^2}\right]^{1/3} \tag{35}$$

$$A^2 = \left[R + \sqrt{Q^3 + R^2}\right]^{2/3} - 2\left[R^2 - (Q^3 + R^2)\right]^{1/3}$$

$$+ \left[R - \sqrt{Q^3 + R^2}\right]^{2/3}$$

$$= \left[R + \sqrt{Q^3 + R^2}\right]^{2/3} + \left[R - \sqrt{Q^3 + R^2})\right]^{2/3} + 2Q$$

$$= B^2 + 4Q, \tag{36}$$

so that the solutions to the quadratic part can be written

$$x = -\frac{1}{2}B \pm \frac{1}{2}\sqrt{3}iA. \tag{37}$$

Defining

$$D \equiv Q^3 + R^2 \tag{38}$$

$$S \equiv \sqrt{R + \sqrt{D}} \tag{39}$$

$$T \equiv \sqrt{R - \sqrt{D}}, \tag{40}$$

where D is the DISCRIMINANT (which is defined slightly differently, including the opposite SIGN, by Birkhoff and Mac Lane 1996) then gives very simple expressions for A and B, namely

$$B = S + T \tag{41}$$

$$A = S - T. \tag{42}$$

Therefore, at last, the ROOTS of the original equation in z are then given by

$$z_1 = -\frac{1}{3}a_2 + (S + T) \tag{43}$$

$$z_2 = -\frac{1}{3}a_2 - \frac{1}{2}(S + T) + \frac{1}{2}i\sqrt{3}(S - T) \tag{44}$$

$$z_3 = -\frac{1}{3}a_2 - \frac{1}{2}(S + T) - \frac{1}{2}i\sqrt{3}(S - T), \tag{45}$$

with a_2 the COEFFICIENT of z^2 in the original equation, and S and T as defined above. These three equations giving the three ROOTS of the cubic equation are sometimes known as CARDANO'S FORMULA. Note that if the equation is in the standard form of Vieta

$$x^3 + px = q, \tag{46}$$

in the variable x, then $a_2 = 0$, $a_1 = p$, and $a_0 = -q$, and the intermediate variables have the simple form (cf. Beyer 1987)

$$Q = \frac{1}{3}p \tag{47}$$

$$R = \frac{1}{2}q \tag{48}$$

$$D \equiv Q^3 + R^2 = \left(\frac{p}{3}\right)^2 + \left(\frac{q}{2}\right)^2. \tag{49}$$

The solutions satisfy NEWTON'S RELATIONS

$$z_1 + z_2 + z_3 = -a_2 \tag{50}$$

$$z_1 z_2 + z_2 z_3 + z_1 z_3 = a_1 \tag{51}$$

$$z_1 z_2 z_3 = -a_0. \tag{52}$$

In standard form (46), $a_2 = 0$, $a_1 = p$, and $a_0 = -q$, so eliminating q gives

$$p = -(z_i^2 + z_i z_j + z_j^2) \tag{53}$$

for $i \neq j$, and eliminating p gives

$$q = -z_i z_j (z_i + z_j) \tag{54}$$

for $i \neq j$. In addition, the properties of the SYMMETRIC POLYNOMIALS appearing in NEWTON'S RELATIONS give

$$z_1^2 + z_2^2 + z_3^2 = -2p \tag{55}$$

$$z_1^3 + z_2^3 + z_3^3 = 3q \tag{56}$$

$$z_1^4 + z_2^4 + z_3^4 = 2p^2 \tag{57}$$

$$z_1^5 + z_2^5 + z_3^5 = -5pq. \tag{58}$$

The equation for z_1 in CARDANO'S FORMULA does not have an i appearing in it explicitly while z_2 and z_3 do, but this does not say anything about the number of REAL and COMPLEX ROOTS (since S and T are themselves, in general, COMPLEX). However, determining which ROOTS are REAL and which are COMPLEX can be accomplished by noting that if the DISCRIMINANT $D > 0$, one ROOT is REAL and two are COMPLEX CONJUGATES; if $D = 0$, all ROOTS are REAL and at least two are equal; and if $D < 0$, all ROOTS are REAL and unequal. If $D < 0$, define

$$\theta \equiv \cos^{-1}\left(\frac{R}{\sqrt{-Q^3}}\right). \tag{59}$$

Then the REAL solutions are OF THE FORM

$$z_1 = 2\sqrt{-Q}\cos\left(\frac{\theta}{3}\right) - \frac{1}{3}a_2 \tag{60}$$

$$z_2 = 2\sqrt{-Q}\cos\left(\frac{\theta + 2\pi}{3}\right) - \frac{1}{3}a_2 \tag{61}$$

$$z_3 = 2\sqrt{-Q}\cos\left(\frac{\theta + 4\pi}{3}\right) - \frac{1}{3}a_2. \tag{62}$$

This procedure can be generalized to find the REAL ROOTS for any equation in the standard form (46) by using the identity

$$\sin^3 \theta - \tfrac{3}{4} \sin \theta + \tfrac{1}{4} \sin(3\theta) = 0 \qquad (63)$$

(Dickson 1914) and setting

$$x \equiv \sqrt{\frac{4|p|}{3}}\, y \qquad (64)$$

(Birkhoff and Mac Lane 1996, pp. 90–1), then

$$\left(\frac{4|p|}{3}\right)^{3/2} y^3 + p\sqrt{\frac{4|p|}{3}}\, y = q \qquad (65)$$

$$y^3 + \tfrac{3}{4}\frac{p}{|p|}\, y = \left(\frac{3}{4|p|}\right)^{3/2} q \qquad (66)$$

$$4y^3 + 3\,\mathrm{sgn}(p) y = \tfrac{1}{2} q \left(\frac{3}{|p|}\right)^{3/2} \equiv C. \qquad (67)$$

If $p > 0$, then use

$$\sinh(3\theta) = 4 \sinh^3 \theta + 3 \sinh \theta \qquad (68)$$

to obtain

$$y = \sinh(\tfrac{1}{3} \sinh^{-1} C). \qquad (69)$$

If $p < 0$ and $|C| \geq 1$, use

$$\cosh(3\theta) = 4 \cosh^3 \theta - 3 \cosh \theta, \qquad (70)$$

and if $p < 0$ and $|C| \leq 1$, use

$$\cos(3\theta) = 4 \cos^3 \theta - 3 \cos \theta, \qquad (71)$$

to obtain

$$y = \begin{cases} \cosh(\tfrac{1}{3} \cosh^{-1} C) & \text{for } C \geq 1 \\ -\cosh(\tfrac{1}{3} \cosh^{-1} |C|) & \text{for } C \leq -1 \\ \cos(\tfrac{1}{3} \cos^{-1} C) & \text{for } |C| < 1 \end{cases} \qquad (72)$$

The solutions to the original equation are then

$$x_i = 2\sqrt{\frac{|p|}{3}}\, y_i - \tfrac{1}{3}\, a_2. \qquad (73)$$

An alternate approach to solving the cubic equation is to use LAGRANGE RESOLVENTS (Faucette 1996). Let $\omega \equiv e^{2\pi i/3}$, define

$$(1,\, x_1) = x_1 + x_2 + x_3 \qquad (74)$$

$$(\omega,\, x_1) = x_1 + \omega x_2 + \omega^2 x_3 \qquad (75)$$

$$(\omega^2,\, x_1) = x_1 + \omega^2 x_2 + \omega x_3, \qquad (76)$$

where x_i are the ROOTS of

$$x^3 + px - q = 0, \qquad (77)$$

and consider the equation

$$[x - (u_1 + u_2)][x - (\omega u_1 + \omega^2 u_2)][x - (\omega^2 u_1 + \omega u_2)] = 0, \qquad (78)$$

where u_1 and u_2 are COMPLEX NUMBERS. The ROOTS are then

$$x_j = \omega^j u_1 + \omega^{2j} u_2 \qquad (79)$$

for $j = 0, 1, 2$. Multiplying through gives

$$x^3 - 3u_1 u_2 x - (u_1^3 + u_2^3) = 0, \qquad (80)$$

which can be written in the form (77), where

$$u_1^3 + u_2^3 = q \qquad (81)$$

$$u_1^3 u_2^3 = -\left(\frac{p}{3}\right)^3. \qquad (82)$$

Some curious identities involving the roots of a cubic equation due to Ramanujan are given by Berndt (1994).

See also CASUS IRREDUCIBILIS, DISCRIMINANT (POLYNOMIAL), PERFECT CUBIC POLYNOMIAL, QUADRATIC EQUATION, QUARTIC EQUATION, QUINTIC EQUATION, SEXTIC EQUATION

References

Abramowitz, M. and Stegun, C. A. (Eds.). *Handbook of Mathematical Functions with Formulas, Graphs, and Mathematical Tables, 9th printing.* New York: Dover, p. 17, 1972.

Berger, M. §16.4.1–6.4.11.1 in *Geometry I.* New York: Springer-Verlag, 1994.

Berndt, B. C. *Ramanujan's Notebooks, Part IV.* New York: Springer-Verlag, pp. 22–3, 1994.

Beyer, W. H. *CRC Standard Mathematical Tables, 28th ed.* Boca Raton, FL: CRC Press, pp. 9–1, 1987.

Birkhoff, G. and Mac Lane, S. *A Survey of Modern Algebra, 5th ed.* New York: Macmillan, pp. 90–1, 106–07, and 414–17, 1996.

Borwein, P. and Erdélyi, T. "Cubic Equations." §1.1.E.1b in *Polynomials and Polynomial Inequalities.* New York: Springer-Verlag, p. 4, 1995.

Dickson, L. E. "A New Solution of the Cubic Equation." *Amer. Math. Monthly* **5**, 38–9, 1898.

Dickson, L. E. *Elementary Theory of Equations.* New York: Wiley, pp. 36–7, 1914.

Dunham, W. "Cardano and the Solution of the Cubic." Ch. 6 in *Journey through Genius: The Great Theorems of Mathematics.* New York: Wiley, pp. 133–54, 1990.

Ehrlich, G. §4.16 in *Fundamental Concepts of Abstract Algebra.* Boston, MA: PWS-Kent, 1991.

Faucette, W. M. "A Geometric Interpretation of the Solution of the General Quartic Polynomial." *Amer. Math. Monthly* **103**, 51–7, 1996.

Jones, J. "Omar Khayyám and a Geometric Solution of the Cubic." http://jwilson.coe.uga.edu/emt669/Student.-Folders/Jones.June/omar/omarpaper.html.

Kennedy, E. C. "A Note on the Roots of a Cubic." *Amer. Math. Monthly* **40**, 411–12, 1933.

King, R. B. *Beyond the Quartic Equation.* Boston, MA: Birkhäuser, 1996.

Press, W. H.; Flannery, B. P.; Teukolsky, S. A.; and Vetterling, W. T. "Quadratic and Cubic Equations." §5.6 in *Numerical Recipes in FORTRAN: The Art of Scientific*

Computing, 2nd ed. Cambridge, England: Cambridge University Press, pp. 178–80, 1992.

Spanier, J. and Oldham, K. B. "The Cubic Function $x^3 + ax^2 + bx + c$ and Higher Polynomials." Ch. 17 in *An Atlas of Functions.* Washington, DC: Hemisphere, pp. 131–47, 1987.

van der Waerden, B. L. §64 in *Algebra.* New York: Frederick Ungar, 1970.

Whittaker, E. T. and Robinson, G. "The Solution of the Cubic." §62 in *The Calculus of Observations: A Treatise on Numerical Mathematics, 4th ed.* New York: Dover, pp. 124–26, 1967.

Cubic Graph

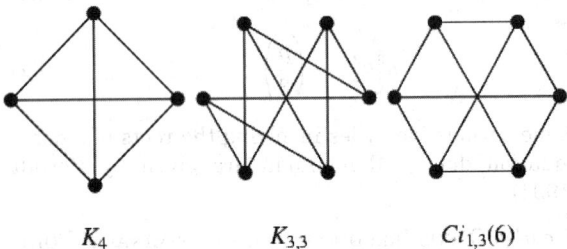

K_4 $K_{3,3}$ $Ci_{1,3}(6)$

Cubic graphs, also called trivalent graphs, are graphs all of whose nodes have degree 3 (i.e., 3-REGULAR GRAPHS). Cubic graphs on n nodes exists only for even n (Harary 1994, p. 15). The numbers of cubic graphs on 2, 4, 6, ... nodes are 0, 1, 2, 6, 21, 94, 540, 4207, ... (Sloane's A005638). The unique 4-node cubic graph is the COMPLETE GRAPH k_4. The two 6-node cubic graphs are the UTILITY GRAPH $K_{3,3}$ and the CIRCULANT GRAPH $Ci_{1,3}(6)$. The connected 3-regular graphs have been determined by Brinkmann (1996) up to 24 nodes.

$(3, g)$-CAGE GRAPHS and UNITRANSITIVE GRAPHS are cubic. In addition, the following tables gives polyhedra whose SKELETONS are cubic.

POLYHEDRON	nodes
TETRAHEDRON	4
CUBE	8
TRUNCATED TETRAHEDRON	12
DODECAHEDRON	20
TRUNCATED CUBE	24
TRUNCATED OCTAHEDRON	24
GREAT RHOMBICUBOCTAHEDRON (ARCHIMEDEAN)	48
TRUNCATED ICOSAHEDRON	60
GREAT RHOMBICOSIDODECAHEDRON (ARCHIMEDEAN)	120

See also BARNETTE'S CONJECTURE, BICUBIC GRAPH, CAGE GRAPH, CUBICAL GRAPH, FRUCHT GRAPH, QUARTIC GRAPH, QUINTIC GRAPH, REGULAR GRAPH, TAIT'S HAMILTONIAN GRAPH CONJECTURE, TUTTE CONJECTURE, UNITRANSITIVE GRAPH

References

Brinkmann, G. "Fast Generation of Cubic Graphs." *J. Graph Th.* **23**, 139–49, 1996.

Harary, F. *Graph Theory.* Reading, MA: Addison-Wesley, 1994.

Read, R. C. and Wilson, R. J. *An Atlas of Graphs.* Oxford, England: Oxford University Press, 1998.

Robinson, R. W.; Wormald, N. C. "Number of Cubic Graphs." *J. Graph. Th.* **7**, 463–67, 1983.

Skiena, S. *Implementing Discrete Mathematics: Combinatorics and Graph Theory with Mathematica.* Reading, MA: Addison-Wesley, p. 177, 1990.

Sloane, N. J. A. Sequences A005638/M1656 in "An On-Line Version of the Encyclopedia of Integer Sequences." http://www.research.att.com/~njas/sequences/eisonline.html.

Tutte, W. T. "A Family of Cubical Graphs." *Proc. Cambridge Philos. Soc.*, 459–74, 1947.

Tutte, W. T. "A Theory of 3-Connected Graphs." *Indag. Math.* **23**, 441–55, 1961.

Cubic Number

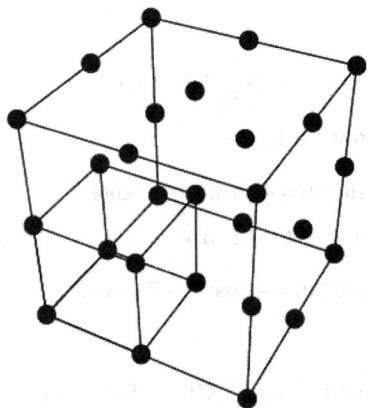

A FIGURATE NUMBER OF THE FORM n^3, for n a POSITIVE INTEGER. The first few are 1, 8, 27, 64, ... (Sloane's A000578). The GENERATING FUNCTION giving the cubic numbers is

$$\frac{x(x^2 + 4x + 1)}{(x - 1)^4} = x + 8x^2 + 27x^3 + \ldots \quad (1)$$

The HEX PYRAMIDAL NUMBERS are equivalent to the cubic numbers (Conway and Guy 1996).

As a part of the study of WARING'S PROBLEM, it is known that every positive integer is a sum of no more than 9 *positive* cubes ($g(3) = 9$, proved by Dickson, Pillai, and Niven in the early twentieth century), that every "sufficiently large" integer is a sum of no more than 7 positive cubes ($G(3) \leq 7$). However, it is not known if 7 can be reduced (Wells 1986, p. 70). The number of *positive* cubes needed to represent the numbers 1, 2, 3, ... are 1, 2, 3, 4, 5, 6, 7, 1, 2, 3, 4, 5, 6, 7, 8, 2, ...(Sloane's A002376), and the number of distinct ways to represent the numbers 1, 2, 3, ... in terms of positive cubes are 1, 1, 1, 1, 1, 1, 1, 2, 2, 2, 2, 2, 2, 2, 2, 3, 3, 3, 3, 3, 3, 3, 3, 4, 4, 4, 5, 5, 5, 5, ... (Sloane's A003108).

In 1939, Dickson proved that the only INTEGERS requiring nine positive cubes are 23 and 239. Wieferich proved that only 15 INTEGERS require eight cubes: 15, 22, 50, 114, 167, 175, 186, 212, 213, 238, 303, 364, 420, 428, and 454 (Sloane's A018889). The quantity $G(3)$ in WARING'S PROBLEM therefore satisfies $G(3) \leq 7$, and the largest number known requiring seven cubes is 8042. Deshouillers *et al.* (1999) conjectured that 7,373,170,279,850 is the largest integer that cannot be expressed as the sum of four nonnegative cubes.

The following table gives the first few numbers which require *at least* $N = 1, 2, 3, ..., 9$ (i.e., N or more) positive cubes to represent them as a sum.

N	Sloane	Numbers
1	Sloane's A000578	1, 8, 27, 64, 125, 216, 343, 512, ...
2	Sloane's A003325	2, 9, 16, 28, 35, 54, 65, 72, 91, ...
3	Sloane's A003072	3, 10, 17, 24, 29, 36, 43, 55, 62, ...
4	Sloane's A003327	4, 11, 18, 25, 30, 32, 37, 44, 51, ...
5	Sloane's A003328	5, 12, 19, 26, 31, 33, 38, 40, 45, ...
6	Sloane's A003329	6, 13, 20, 34, 39, 41, 46, 48, 53, ...
7	Sloane's A018890	7, 14, 21, 42, 47, 49, 61, 77, ...
8	Sloane's A018889	15, 22, 50, 114, 167, 175, 186, ...
9	Sloane's A018888	23, 239

There is a finite set of numbers which cannot be expressed as the sum of *distinct positive* cubes: 2, 3, 4, 5, 6, 7, 10, 11, 12, 13, 14, 15, 16, 17, 18, 19, 20, 21, 22, 23, 24, 25, 26, ...(Sloane's A001476).

It is known that every integer is a sum of at most 5 signed cubes ($eg(3) \leq 5$ in WARING'S PROBLEM). It is *believed* that 5 can be reduced to 4, so that

$$N = A^3 + B^3 + C^3 + D^3 \qquad (2)$$

for any number N, although this has not been proved for numbers OF THE FORM $9n \pm 4$. However, every multiple of 6 can be REPRESENTED AS a sum of four

signed cubes as a result of the algebraic identity

$$6x = (x+1)^3 + (x-1)^3 - x^3 - x^3. \qquad (3)$$

In fact, all numbers $N < 1000$ and *not* OF THE FORM $9n \pm 4$ are known to be expressible as the SUM

$$N = A^3 + B^3 + C^3 \qquad (4)$$

of *three* (positive or negative) cubes with the exception of $N = 30, 33, 42, 52, 74, 110, 114, 156, 165, 195, 290, 318, 366, 390, 420, 444, 452, 478, 501, 530, 534, 564, 579, 588, 600, 606, 609, 618, 627, 633, 732, 735, 758, 767, 786, 789, 795, 830, 834, 861, 894, 903, 906, 912, 921, 933, 948, 964, 969,$ and 975 (Sloane's A046041; Miller and Woollett 1955; Gardiner *et al.* 1964; Guy 1994, p. 151). While it is known that (4) has no solutions for N of the form $9n \pm 4$ (Hardy and Wright 1979, p. 327), there is known reason for excluding the above integers (Gardiner *et al.* 1964). Mahler proved that 1 has infinitely-many representations as 3 signed cubes.

The following table gives the numbers which can be represented in *exactly* W different ways as a sum of N positive cubes. (Combining all Ws for a given N then gives the sequences in the previous table.) For example,

$$157 = 4^3 + 4^3 + 3^3 + 1^3 + 1^3 = 5^3 + 2^3 + 2^3 + 2^3 + 2^3 \quad (5)$$

can be represented in $W = 2$ ways by $N = 5$ cubes. The smallest number representable in $W = 2$ ways as a sum of $N = 2$ cubes,

$$1729 = 1^3 + 12^3 = 9^3 + 10^3, \qquad (6)$$

is called the HARDY-RAMANUJAN NUMBER and has special significance in the history of mathematics as a result of a story told by Hardy about Ramanujan. Note that Sloane's A001235 is defined as the sequence of numbers which are the sum of cubes in *two or more* ways, and so appears identical in the first few terms to the ($N = 2$, $W = 2$) series given below.

N	W	Sloane	numbers
1	0	A007412	2, 3, 4, 5, 6, 7, 9, 10, 11, 12, 13, 14, ...
1	1	A000578	1, 8, 27, 64, 125, 216, 343, 512, ...
2	0	A057903	1, 3, 4, 5, 6, 7, 8, 10, 11, 12, 13, 14, ...
2	1		2, 9, 16, 28, 35, 54, 65, 72, 91, ...
2	2	A018850	1729, 4104, 13832, 20683, 32832, ...
2	3	A003825	87539319, 119824488, 143604279, ...

2	4	A003826	6963472309248, 12625136269928, ...
2	5		48988659276962496, ...
2	6		8230545258248091551205888, ...
3	0	A057904	1, 2, 4, 5, 6, 7, 8, 9, 11, 12, 13, 14, ...
3	1	A025395	3, 10, 17, 24, 29, 36, 43, 55, 62, ...
3	2		251, ...
4	0	A057905	1, 2, 3, 5, 6, 7, 8, 9, 10, 12, 13, 14, ...
4	1	A025403	4, 11, 18, 25, 30, 32, 37, 44, 51, ...
4	2	A025404	219, 252, 259, 278, 315, 376, 467, ...
5	0	A057906	1, 2, 3, 4, 6, 7, 8, 9, 10, 11, 13, 14, 15, ...
5	1	A048926	5, 12, 19, 26, 31, 33, 38, 40, 45, ...
5	2	A048927	157, 220, 227, 246, 253, 260, 267, ...
6	0	A057907	1, 2, 3, 4, 5, 7, 8, 9, 10, 11, 12, 14, 15, ...
6	1	A048929	6, 13, 20, 27, 32, 34, 39, 41, 46, ...
6	2	A048930	158, 165, 184, 221, 228, 235, 247, ...
6	3	A048931	221, 254, 369, 411, 443, 469, 495, ...

10	10	0, 1, 2, 3, 4, 5, 6, 7, 8, 9
11	11	0, 1, 2, 3, 4, 5, 6, 7, 8, 9, 10
12	9	0, 1, 3, 4, 5, 7, 8, 9, 11
13	5	0, 1, 5, 8, 12
14	6	0, 1, 6, 7, 8, 13
15	15	0, 1, 2, 3, 4, 5, 6, 7, 8, 9, 10, 11, 12, 13, 14
16	10	0, 1, 3, 5, 7, 8, 9, 11, 13, 15
17	17	0, 1, 2, 3, 4, 5, 6, 7, 8, 9, 10, 11, 12, 13, 14, 15, 16
18	6	0, 1, 8, 9, 10, 17
19	7	0, 1, 7, 8, 11, 12, 18
20	15	0, 1, 3, 4, 5, 7, 8, 9, 11, 12, 13, 15, 16, 17, 19

Dudeney found two RATIONAL NUMBERS other than 1 and 2 whose cubes sum to 9,

$$\frac{415280564497}{348671682660} \quad \text{and} \quad \frac{676702467503}{348671682660} \tag{7}$$

(Gardner 1958). The problem of finding two RATIONAL NUMBERS whose cubes sum to six was "proved" impossible by Legendre. However, Dudeney found the simple solutions 17/21 and 37/21.

The only three consecutive INTEGERS whose cubes sum to a cube are given by the DIOPHANTINE EQUATION

$$3^3 + 4^3 + 5^3 = 6^3. \tag{8}$$

CATALAN'S CONJECTURE states that 8 and 9 (2^3 and 3^2) are the only consecutive POWERS (excluding 0 and 1), i.e., the only solution to CATALAN'S DIOPHANTINE PROBLEM. This CONJECTURE has not yet been proved or refuted, although R. Tijdeman has proved that there can be only a finite number of exceptions should the CONJECTURE not hold. It is also known that 8 and 9 are the only consecutive cubic and SQUARE NUMBERS (in either order).

There are six POSITIVE INTEGERS equal to the sum of the DIGITS of their cubes: 1, 8, 17, 18, 26, and 27 (Sloane's A046459; Moret Blanc 1879). There are four POSITIVE INTEGERS equal to the sums of the cubes of their digits:

$$153 = 1^3 + 5^3 + 3^3 \tag{9}$$

$$370 = 3^3 + 7^3 + 0^3 \tag{10}$$

$$371 = 3^3 + 7^3 + 1^3 \tag{11}$$

$$407 = 4^3 + 0^3 + 7^3 \tag{12}$$

(Ball and Coxeter 1987). There are two SQUARE

The following table gives the possible residues (mod n) for cubic numbers for $n = 1$ to 20, as well as the number of distinct residues $s(n)$.

n	$s(n)$	$x^3 \pmod{n}$
2	2	0, 1
3	3	0, 1, 2
4	3	0, 1, 3
5	5	0, 1, 2, 3, 4
6	6	0, 1, 2, 3, 4, 5
7	3	0, 1, 6
8	5	0, 1, 3, 5, 7
9	3	0, 1, 8

NUMBERS OF THE FORM $n^3 - 4$: $4 = 2^3 - 4$ and $121 = 5^3 - 4$ (Le Lionnais 1983). A cube cannot be the concatenation of two cubes, since if c^3 is the concatenation of a^3 and b^3, then $c^3 = 10^k a^3 + b^3$, where k is the number of digits in b^3. After shifting any powers of 1000 in 10^k into a^3, the original problem is equivalent to finding a solution to one of the DIOPHANTINE EQUATIONS

$$c^3 - b^3 = a^3 \tag{13}$$

$$c^3 - b^3 = 10a^3 \tag{14}$$

$$c^3 - b^3 = 100a^3. \tag{15}$$

None of these have solutions in integers, as proved independently by Sylvester, Lucas, and Pepin (Dickson 1966, pp. 572–78).

See also BIQUADRATIC NUMBER, CENTERED CUBE NUMBER, CLARK'S TRIANGLE, DIOPHANTINE EQUATION–3RD POWERS, HARDY-RAMANUJAN NUMBER, PARTITION, SQUARE NUMBER

References

Ball, W. W. R. and Coxeter, H. S. M. *Mathematical Recreations and Essays, 13th ed.* New York: Dover, p. 14, 1987.

Bertault, F.; Ramaré, O.; and Zimmermann, P. "On Sums of Seven Cubes." *Math. Comput.* **68**, 1303–310, 1999.

Conway, J. H. and Guy, R. K. *The Book of Numbers.* New York: Springer-Verlag, pp. 42–4, 1996.

Davenport, H. "On Waring's Problem for Cubes." *Acta Math.* **71**, 123–43, 1939.

Deshouillers, J.-M.; Hennecart, F.; and Landreau, B. "7 373 170 279 850." *Math. Comput.* **69**, 421–39, 1999.

Dickson, L. E. *History of the Theory of Numbers, Vol. 2: Diophantine Analysis.* New York: Chelsea, 1966.

Gardiner, V. L.; Lazarus, R. B.; and Stein, P. R. "Solutions of the Diophantine Equation $x^3 + y^3 = z^3 - d$." *Math. Comput.* **18**, 408–13, 1964.

Gardner, M. "Mathematical Games: About Henry Ernest Dudeney, A Brilliant Creator of Puzzles." *Sci. Amer.* **198**, 108–12, Jun. 1958.

Guy, R. K. "Sum of Four Cubes." §D5 in *Unsolved Problems in Number Theory, 2nd ed.* New York: Springer-Verlag, pp. 151–52, 1994.

Hardy, G. H. and Wright, E. M. "Representation by Cubes and Higher Powers." Ch. 21 in *An Introduction to the Theory of Numbers, 5th ed.* Oxford, England: Clarendon Press, pp. 317–39, 1979.

Le Lionnais, F. *Les nombres remarquables.* Paris: Hermann, p. 53, 1983.

Miller, J. C. P. and Woollett, M. F. C. "Solutions of the Diophantine Equation $x^3 + y^3 + z^3 = k$." *J. London Math. Soc.* **30**, 101–10, 1955.

Sloane, N. J. A. Sequences A000578/M4499, A001235, A001476, A002376/M0466, A003108/M0209, A003072, A003325, A003327, A003328, A003825, A003826, A007412/M0493, A011541, A018850, A018888, A018889, A018890, A025395, A046040, A046459, A048926, A048927, A048928, A048929, A048930, A048931, A048932, A057903, A057904, A057905, A057906, and A057907 in "An On-Line Version of the Encyclopedia of Integer Sequences." http://www.research.att.com/~njas/sequences/eisonline.html.

Wells, D. *The Penguin Dictionary of Curious and Interesting Numbers.* Middlesex, England: Penguin Books, p. 70, 1986.

Cubic Part

The largest cube dividing a POSITIVE INTEGER n. For $n = 1, 2, \ldots$, the first few are 1, 1, 1, 1, 1, 1, 1, 8, 1, 1, ... (Sloane's A008834).

See also CUBEFREE PART, CUBIC NUMBER, SQUARE PART

References

Sloane, N. J. A. Sequences A008834 in "An On-Line Version of the Encyclopedia of Integer Sequences." http://www.research.att.com/~njas/sequences/eisonline.html.

Cubic Reciprocity Theorem

A RECIPROCITY THEOREM for the case $n = 3$ solved by Gauss using "INTEGERS" OF THE FORM $a + b\rho$, when ρ is a root of $x^2 + x + 1 = 0$ (i.e., ρ equals $-(-1)^{1/3}$ or $(-1)^{2/3}$) and a, b are INTEGERS.

See also CUBIC RESIDUE, RECIPROCITY THEOREM

References

Ireland, K. and Rosen, M. "Cubic and Biquadratic Reciprocity." Ch. 9 in *A Classical Introduction to Modern Number Theory, 2nd ed.* New York: Springer-Verlag, pp. 108–37, 1990.

Cubic Residue

If there is an INTEGER x such that

$$x^3 \equiv q \pmod{p}, \tag{1}$$

then q is said to be a cubic residue (mod p). If not, q is said to be a cubic nonresidue (mod p).

See also CUBIC RECIPROCITY THEOREM, QUADRATIC RESIDUE

References

Nagell, T. *Introduction to Number Theory.* New York: Wiley, p. 115, 1951.

Cubic Spline

A cubic spline is a SPLINE constructed of piecewise third-order POLYNOMIALS which pass through a set of control points. The second DERIVATIVE of each POLYNOMIAL is commonly set to zero at the endpoints, since this provides a boundary condition that completes the system of $n - 2$ equations, leading to a simple 3-diagonal system which can be solved easily to give the coefficients of the polynomials. However, this choice is not the only one possible, and other boundary conditions can be used instead.

See also SPLINE, THIN PLATE SPLINE

References

Burden, R. L.; Faires, J. D.; and Reynolds, A. C. *Numerical Analysis, 6th ed.* Boston, MA: Brooks/Cole, pp. 120–21, 1997.

Press, W. H.; Flannery, B. P.; Teukolsky, S. A.; and Vetterling, W. T. "Cubic Spline Interpolation." §3.3 in *Numerical*

Recipes in FORTRAN: The Art of Scientific Computing, 2nd ed. Cambridge, England: Cambridge University Press, pp. 107–10, 1992.

Cubic Surface

An ALGEBRAIC SURFACE of ORDER 3. Schläfli and Cayley classified the singular cubic surfaces. On the general cubic, there exists a curious geometrical structure called DOUBLE SIXES, and also a particular arrangement of 27 (possibly complex) lines, as discovered by Schläfli (Salmon 1965, Fischer 1986) and sometimes called SOLOMON'S SEAL LINES. A nonregular cubic surface can contain 3, 7, 15, or 27 real lines (Segre 1942, Le Lionnais 1983). The CLEBSCH DIAGONAL CUBIC contains all possible 27. The maximum number of ORDINARY DOUBLE POINTS on a cubic surface is four, and the unique cubic surface having four ORDINARY DOUBLE POINTS is the CAYLEY CUBIC.

Schoutte (1910) showed that the 27 lines can be put into a ONE-TO-ONE correspondence with the vertices of a particular POLYTOPE in 6-D space in such a manner that all incidence relations between the lines are mirrored in the connectivity of the POLYTOPE and conversely (Du Val 1931). A similar correspondence can be made between the 28 bitangents of the general plane QUARTIC CURVE and a 7-D POLYTOPE (Coxeter 1928) and between the tritangent planes of the canonical curve of genus 4 and an 8-D POLYTOPE (Du Val 1933).

A smooth cubic surface contains 45 TRITANGENTS (Hunt). The Hessian of smooth cubic surface contains at least 10 ORDINARY DOUBLE POINTS, although the Hessian of the CAYLEY CUBIC contains 14 (Hunt).

See also CAYLEY CUBIC, CLEBSCH DIAGONAL CUBIC, DOUBLE SIXES, ECKARDT POINT, ISOLATED SINGULARITY, NORDSTRAND'S WEIRD SURFACE, SOLOMON'S SEAL LINES, TRITANGENT

References

Bruce, J. and Wall, C. T. C. "On the Classification of Cubic Surfaces." *J. London Math. Soc.* **19**, 245–56, 1979.

Cayley, A. "A Memoir on Cubic Surfaces." *Phil. Trans. Roy. Soc.* **159**, 231–26, 1869.

Coxeter, H. S. M. "The Pure Archimedean Polytopes in Six and Seven Dimensions." *Proc. Cambridge Phil. Soc.* **24**, 7–, 1928.

Du Val, P. "On the Directrices of a Set of Points in a Plane." *Proc. London Math. Soc. Ser. 2* **35**, 23–4, 1933.

Fischer, G. (Ed.). *Mathematical Models from the Collections of Universities and Museums.* Braunschweig, Germany: Vieweg, pp. 9–4, 1986.

Fladt, K. and Baur, A. *Analytische Geometrie spezieler Flächen und Raumkurven.* Braunschweig, Germany: Vieweg, pp. 248–55, 1975.

Hunt, B. "Algebraic Surfaces." http://www.mathematik.uni-kl.de/~wwwagag/E/Galerie.html.

Hunt, B. "The 27 Lines on a Cubic Surface" and "Cubic Surfaces." Ch. 4 and Appendix B.4 in *The Geometry of Some Special Arithmetic Quotients.* New York: Springer-Verlag, pp. 108–67 and 302–10, 1996.

Klein, F. "Über Flächen dritter Ordnung." *Gesammelte Abhandlungen, Band II.* Berlin: Springer-Verlag, pp. 11–2, 1973.

Le Lionnais, F. *Les nombres remarquables.* Paris: Hermann, p. 49, 1983.

Rodenberg, C. "Zur Classification der Flächen dritter Ordnung." *Math. Ann.* **14**, 46–10, 1878.

Salmon, G. *Analytic Geometry of Three Dimensions.* New York: Chelsea, 1965.

Schläfli, L. "On the Distribution of Surface of Third Order into Species." *Phil. Trans. Roy. Soc.* **153**, 193–47, 1864.

Schoutte, P. H. "On the Relation Between the Vertices of a Definite Sixdimensional Polytope and the Lines of a Cubic Surface." *Proc. Roy. Acad. Amsterdam* **13**, 375–83, 1910.

Segre, B. *The Nonsingular Cubic Surface.* Oxford, England: Clarendon Press, 1942.

Cubical Conic Section

CUBICAL ELLIPSE, CUBICAL HYPERBOLA, CUBICAL PARABOLA, SKEW CONIC

Cubical Ellipse

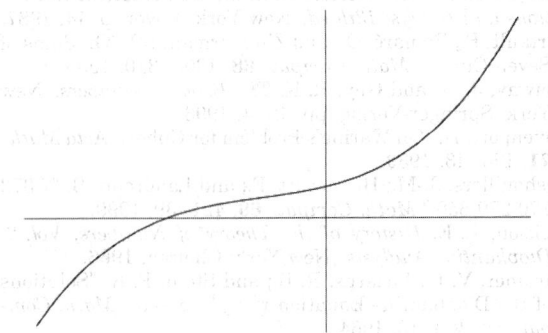

An equation OF THE FORM

$$y = ax^3 + bx^2 + cx + d$$

where only one ROOT is real.

See also CUBICAL CONIC SECTION, CUBICAL HYPERBOLA, CUBICAL PARABOLA, CUBICAL PARABOLIC HYPERBOLA, ELLIPSE, SKEW CONIC

Cubical Graph

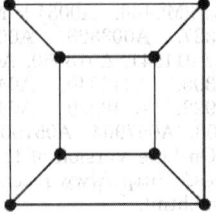

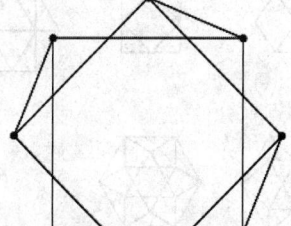

The PLATONIC GRAPH corresponding to the connectivity of the CUBE. Several symmetrical circular embeddings of this graph are illustrated in the second figure above. The cubical graph has 8 nodes, 12 edges, VERTEX CONNECTIVITY 3, and EDGE CONNECTIVITY 3, GRAPH DIAMETER 3, GRAPH RADIUS 3, and GIRTH 4. The cubical graph's CHROMATIC POLYNOMIAL is

$$\pi_G(z) = z^8 - 12z^7 + 66z^6 - 214z^5 + 441z^4 - 572z^3$$
$$+ 423z^2 - 133z,$$

and the CHROMATIC NUMBER is $\chi(G) = 2$.

The maximum number of nodes in a cubical graph which induce a cycle is six (Danzer and Klee 1967; Skiena 1990, p. 149).

See also BIDIAKIS CUBE, BISLIT CUBE, CUBE, DODECAHEDRAL GRAPH, ICOSAHEDRAL GRAPH, OCTAHEDRAL GRAPH, PLATONIC GRAPH, TETRAHEDRAL GRAPH

References

Bondy, J. A. and Murty, U. S. R. *Graph Theory with Applications.* New York: North Holland, p. 234, 1976.
Danzer, L. and Klee, V. "Lengths of Snakes in Boxes." *J. Combin. Th.* **2**, 258–65, 1967.
Skiena, S. *Implementing Discrete Mathematics: Combinatorics and Graph Theory with Mathematica.* Reading, MA: Addison-Wesley, 1990.

Cubical Hyperbola

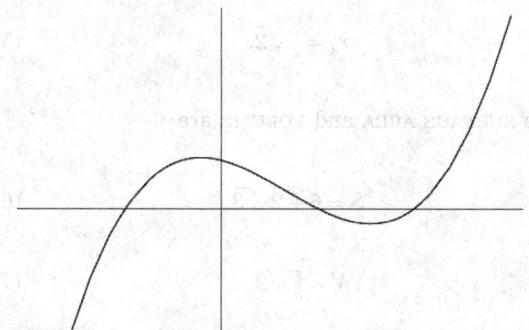

An equation OF THE FORM

$$y = ax^3 + bx^2 + cx + d,$$

where the three ROOTS are REAL and distinct, i.e.,

$$y = a(x - r_1)(x - r_2)(x - r_3)$$
$$= a[x^3 - (r_1 + r_2 + r_3)x^2 + (r_1 r_2 + r_1 r_3 + r_2 r_3)x$$
$$- r_1 r_2 r_3].$$

See also CUBICAL CONIC SECTION, CUBICAL ELLIPSE, CUBICAL HYPERBOLA, CUBICAL PARABOLA, HYPERBOLA

Cubical Parabola

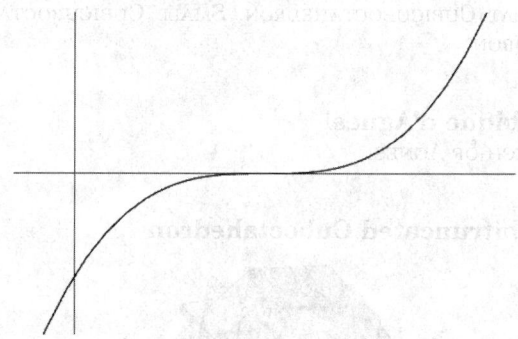

An equation OF THE FORM

$$y = ax^3 + bx^2 + cx + d,$$

where the three ROOTS of the equation coincide (and are therefore real), i.e.,

$$y = a(x - r)^3 = a(x^3 - 3rx^2 - 3r^2x - r^3).$$

See also CUBICAL CONIC SECTION, CUBICAL ELLIPSE, CUBICAL HYPERBOLA, CUBICAL PARABOLIC HYPERBOLA, PARABOLA, SEMICUBICAL PARABOLA

References

Beyer, W. H. *CRC Standard Mathematical Tables, 28th ed.* Boca Raton, FL: CRC Press, pp. 215 and 223, 1987.

Cubical Parabolic Hyperbola

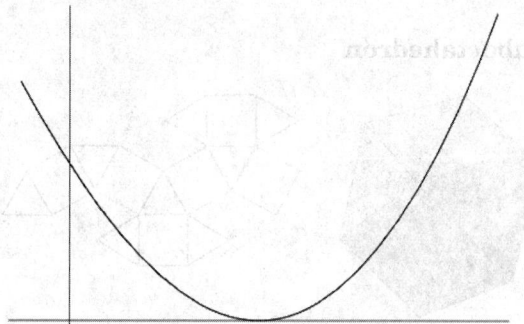

An equation OF THE FORM

$$y = ax^3 + bx^2 + cx + d,$$

where two of the ROOTS of the equation coincide (and all three are therefore real), i.e.,

$$y = a(x - r_1)^2(x - r_2)$$
$$= a[x^3 - (2r_1 + r_2)x^2 + r_1(r_1 + 2r_2)x - r_1^2 r_2].$$

See also CUBICAL CONIC SECTION, CUBICAL ELLIPSE, CUBICAL HYPERBOLA, CUBICAL PARABOLA, HYPERBOLA

Cubicuboctahedron

GREAT CUBICUBOCTAHEDRON, SMALL CUBICUBOCTAHEDRON

Cubique d'Agnesi

WITCH OF AGNESI

Cubitruncated Cuboctahedron

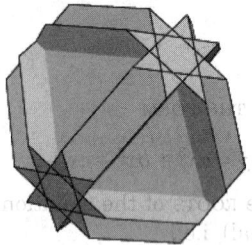

The UNIFORM POLYHEDRON U_{16} whose DUAL is the TETRADYAKIS HEXAHEDRON. It has WYTHOFF SYMBOL $3\frac{4}{3}\,4|$. Its faces are $8\{6\} + 6\{8\} + 6\{\frac{8}{3}\}$. It is a FACETED OCTAHEDRON. the CIRCUMRADIUS for a cubitruncated cuboctahedron of unit edge length is

$$r = \tfrac{1}{2}\sqrt{7}.$$

References

Wenninger, M. J. *Polyhedron Models.* Cambridge, England: Cambridge University Press, pp. 113–14, 1971.

Cuboctahedron

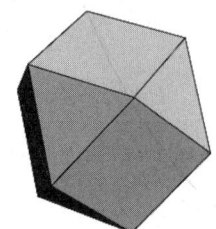
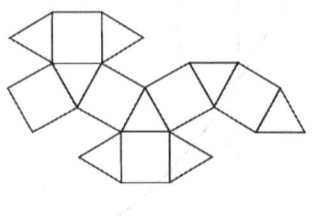

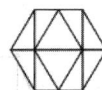

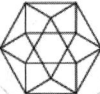

The ARCHIMEDEAN SOLID A_1 (also called the DYMAXION or HEPTAPARALLELOHEDRON) with faces $8\{3\} + 6\{4\}$. It is one of the two convex QUASIREGULAR POLYHEDRA. It is UNIFORM POLYHEDRON U_7 and Wenninger model W_{11}. It has SCHLÄFLI SYMBOL $\{\frac{3}{4}\}$ and WYTHOFF SYMBOL $2\,|\,34$.

The DUAL POLYHEDRON is the RHOMBIC DODECAHEDRON. The cuboctahedron has the O_h OCTAHEDRAL GROUP of symmetries. According to Heron, Archimedes ascribed the cuboctahedron to Plato (Heath 1981; Coxeter 1973, p. 30). The VERTICES of a cuboctahedron with EDGE length of $\sqrt{2}$ are $(0, \pm 1, \pm 1)$, $(\pm 1, 0, \pm 1)$, and $(\pm 1, \pm 1, 0)$.

The INRADIUS r of the dual, MIDRADIUS ρ of the solid and dual, and CIRCUMRADIUS R of the solid for $a = 1$ are

$$r = \tfrac{3}{4} = 0.75 \tag{1}$$

$$\rho = \tfrac{1}{2}\sqrt{3} \approx 0.86602 \tag{2}$$

$$R = 1. \tag{3}$$

The distances from the center of the solid to the centroids of the triangular and square faces are

$$r_3 = \tfrac{1}{3}\sqrt{6} \tag{4}$$

$$r_4 = \tfrac{1}{2}\sqrt{2}. \tag{5}$$

The SURFACE AREA and VOLUME are

$$S = 6 + 2\sqrt{3} \tag{6}$$

$$V = \tfrac{5}{3}\sqrt{2}. \tag{7}$$

FACETED versions of the cuboctahedron include the CUBOHEMIOCTAHEDRON and OCTAHEMIOCTAHEDRON.

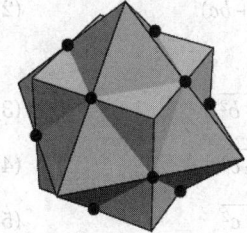 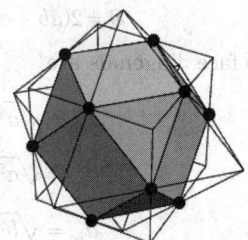

The solid common to both the CUBE and OCTAHEDRON (left figure) in a CUBE-OCTAHEDRON COMPOUND is a CUBOCTAHEDRON (right figure; Ball and Coxeter 1987). The mineral argentite (Ag₂S) forms cuboctahedral crystals (Steinhaus 1983, p. 203). The cuboctahedron can be inscribed in the RHOMBIC DODECAHEDRON (Steinhaus 1983, p. 206).

Wenninger (1989) lists four of the possible STELLATIONS of the cuboctahedron: the CUBE-OCTAHEDRON COMPOUND, a truncated form of the STELLA OCTANGULA, a sort of compound of six intersecting square pyramids, and an attractive concave solid formed of rhombi meeting four at a time.

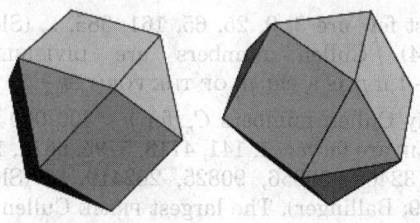

If a cuboctahedron is oriented with triangles on top and bottom, the two halves may be rotated one sixth of a turn with respect to each other to obtain JOHNSON SOLID J_{27}, the TRIANGULAR ORTHOBICUPOLA.

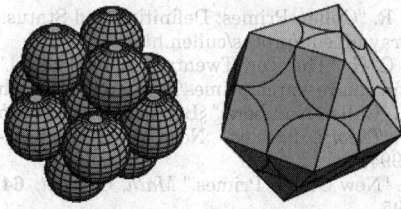

In cubic close packing, each sphere is surrounded by 12 other spheres. Taking a collection of 13 such spheres gives the cluster illustrated above. Connecting the centers of the external 12 spheres gives a cuboctahedron (Steinhaus 1983, pp. 203–07), which is therefore also a SPACE-FILLING POLYHEDRON.

See also ARCHIMEDEAN SOLID, CUBE, CUBE-OCTAHEDRON COMPOUND, CUBOHEMIOCTAHEDRON, OCTAHEDRON, OCTAHEMIOCTAHEDRON, QUASIREGULAR POLYHEDRON, RHOMBIC DODECAHEDRON, RHOMBIC

DODECAHEDRON STELLATIONS, RHOMBUS, SPACE-FILLING POLYHEDRON, SPHERE PACKING, STELLATION, TRIANGULAR ORTHOBICUPOLA

References

Ball, W. W. R. and Coxeter, H. S. M. *Mathematical Recreations and Essays, 13th ed.* New York: Dover, p. 137, 1987.
Coxeter, H. S. M. *Regular Polytopes, 3rd ed.* New York: Dover, 1973.
Cundy, H. and Rollett, A. "Cuboctahedron. (3.4)²." §3.7.2 in *Mathematical Models, 3rd ed.* Stradbroke, England: Tarquin Pub., p. 102, 1989.
Ghyka, M. *The Geometry of Art and Life.* New York: Dover, p. 54, 1977.
Heath, T. L. *A History of Greek Mathematics, Vol. 1: From Thales to Euclid.* New York: Dover, 1981.
Steinhaus, H. *Mathematical Snapshots, 3rd ed.* New York: Dover, pp. 203–05, 1999.
Wenninger, M. J. "The Cuboctahedron." Model 11 in *Polyhedron Models.* Cambridge, England: Cambridge University Press, p. 25, 1989.
Wenninger, M. J. "Commentary on the Stellation of the Archimedean Solids." In *Polyhedron Models.* New York: Cambridge University Press, pp. 66–2, 1989.

Cuboctahedron-Rhombic Dodecahedron Compound

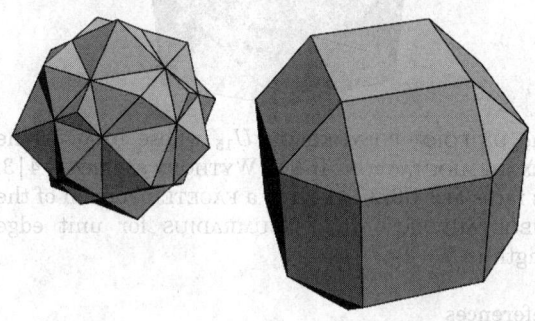

The POLYHEDRON COMPOUND consisting of the CUBOCTAHEDRON and its dual, the RHOMBIC DODECAHEDRON, illustrated in the left figure above. The right figure shows the solid common to the two polyhedra. If the CUBOCTAHEDRON has unit edge length, the compound can be constructed by midpoint CUMULATION with heights

$$h_3 = \tfrac{1}{4}\sqrt{6} \tag{1}$$

$$h_4 = \tfrac{1}{2}\sqrt{2}. \tag{2}$$

The resulting compound has side lengths

$$s_1 = \tfrac{1}{8}\sqrt{6} \tag{3}$$

$$s_2 = \tfrac{1}{2} \tag{4}$$

$$s_3 = \tfrac{1}{4}\sqrt{6} \tag{5}$$

$$s_4 = \tfrac{1}{2}\sqrt{2}, \tag{6}$$

and SURFACE AREA and VOLUME

$$S = \tfrac{3}{4}(4 + 5\sqrt{2} + 2\sqrt{3}) \tag{7}$$

$$V = \tfrac{31}{16}\sqrt{2}. \tag{8}$$

See also CUBOCTAHEDRON, POLYHEDRON COMPOUND, POLYHEDRON DUAL, RHOMBIC DODECAHEDRON

Cuboctatruncated Cuboctahedron
CUBITRUNCATED CUBOCTAHEDRON

Cubocycloid
ASTROID

Cubohemioctahedron

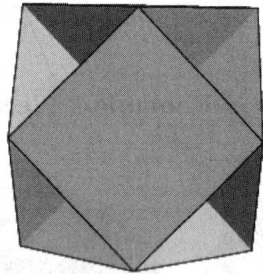

The UNIFORM POLYHEDRON U_{15} whose DUAL is the HEXAHEMIOCTACRON. It has WYTHOFF SYMBOL $\tfrac{4}{3}\,4\,|\,3$. Its faces are $4\{6\} + 6\{4\}$. It is a FACETED version of the CUBOCTAHEDRON. Its CIRCUMRADIUS for unit edge length is $R = 1$.

References
Wenninger, M. J. *Polyhedron Models.* Cambridge, England: Cambridge University Press, pp. 121–22, 1971.

Cuboid

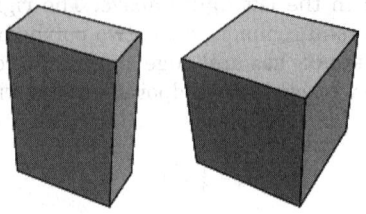

A rectangular PARALLELEPIPED, sometimes also called a brick. A cuboid of side lengths a, b, and c has VOLUME

$$V = abc \tag{1}$$

and SURFACE AREA

$$S = 2(ab + ac + bc). \tag{2}$$

The face diagonals are

$$d_{ab} = \sqrt{a^2 + b^2} \tag{3}$$

$$d_{ac} = \sqrt{a^2 + c^2} \tag{4}$$

$$d_{bc} = \sqrt{b^2 + c^2} \tag{5}$$

and the body diagonal is

$$d_{abc} = \sqrt{a^2 + b^2 + c^2}. \tag{6}$$

A cuboid with all sides equal is called a CUBE.

See also CUBE, EULER BRICK, PARALLELEPIPED, PRISM, SPIDER AND FLY PROBLEM

References
Harris, J. W. and Stocker, H. "Cuboid." §4.2.3 in *Handbook of Mathematics and Computational Science.* New York: Springer-Verlag, p. 97, 1998.

Cullen Number
A number OF THE FORM

$$C_n = 2^n n + 1.$$

The first few are 3, 9, 25, 65, 161, 385, ... (Sloane's A002064). Cullen numbers are DIVISIBLE by $p = 2n - 1$ if p is a PRIME OF THE FORM $8k \pm 3$.

The only Cullen numbers C_n for $n < 300{,}000$ which are PRIME are for $n = 1$, 141, 4713, 5795, 6611, 18496, 32292, 32469, 59656, 90825, 262419, ... (Sloane's A005849; Ballinger). The largest PRIME Cullen number known is for $n = 361275$, but the range 335000–45000 has not yet been fully checked.

See also CUNNINGHAM NUMBER, FERMAT NUMBER, SIERPINSKI NUMBER OF THE FIRST KIND, WOODALL NUMBER

References
Ballinger, R. "Cullen Primes: Definition and Status." http://vamri.xray.ufl.edu/proths/cullen.html.
Caldwell, C. K. "The Top Twenty: Cullen Primes." http://www.utm.edu/research/primes/lists/top20/Cullen.html.
Guy, R. K. "Cullen Numbers." §B20 in *Unsolved Problems in Number Theory, 2nd ed.* New York: Springer-Verlag, p. 77, 1994.
Keller, W. "New Cullen Primes." *Math. Comput.* **64**, 1733–741, 1995.
Leyland, P. ftp://sable.ox.ac.uk/pub/math/factors/cullen/.
Ribenboim, P. *The New Book of Prime Number Records.* New York: Springer-Verlag, pp. 360–61, 1996.
Sloane, N. J. A. Sequences A002064/M2795 and A0058495401 in "An On-Line Version of the Encyclopedia of Integer Sequences." http://www.research.att.com/~njas/sequences/eisonline.html.

Cumulant
Let $\phi(t)$ be the CHARACTERISTIC FUNCTION, defined as the FOURIER TRANSFORM of the PROBABILITY DENSITY

FUNCTION (*using* FOURIER TRANSFORM *parameters* $a = b = 1$),

$$\phi(t) = \mathcal{F}[P(x)] = \int_{-\infty}^{\infty} e^{itx} P(x)\, dx. \qquad (1)$$

Then the cumulants κ_n are then defined by

$$\ln \phi(t) \equiv \sum_{n=0}^{\infty} k_n \frac{(it)^n}{n!} \qquad (2)$$

(Abramowitz and Stegun 1972, p. 928). Taking the MACLAURIN SERIES gives

$$\ln \phi(t) = (it)\mu_1' + \frac{1}{2}(it)^2(\mu_2' - \mu_1'^2) + \frac{1}{3!}(it)^3$$
$$\times (2\mu_1'^3 - 3\mu_1'\mu_2' + \mu_3') + \tfrac{1}{4!}(it)^4$$
$$\times (-6\mu_1'^4 + 12\mu_1'^2\mu_2' - 3\mu_2'^2 - 4\mu_1'\mu_3' + \mu_4') + \frac{1}{5!}$$
$$\times (it)^5$$
$$\times [24\mu_1'^5 - 60\mu_1'^3\mu_2' + 20\mu_1'^2\mu_3' - 10\mu_2'\mu_3'$$
$$+ 5\mu_1'(6\mu_2'^2 - \mu_4') + \mu_5'] + \ldots, \qquad (3)$$

where μ_n' are RAW MOMENTS, so

$$\kappa_1 = \mu_1' \qquad (4)$$

$$\kappa_2 = \mu_2' - \mu_1' \qquad (5)$$

$$\kappa_3 = 2\mu_1'^3 - 3\mu_1'\mu_2' + \mu_3' \qquad (6)$$

$$\kappa_4 = -6\mu_1'^4 + 12\mu_1'^2\mu_2' - 3\mu_2'^2 - 4\mu_1'\mu_3' + \mu_4' \qquad (7)$$

$$\kappa_5 = -24\mu_1'^5 - 60\mu_1'^3\mu_2' + 20\mu_1'^2\mu_3' - 10\mu_2'\mu_3'$$
$$+ 5\mu_1'(6\mu_2'^2 - \mu_4') + \mu_5'. \qquad (8)$$

In terms of the CENTRAL MOMENTS μ_n,

$$\kappa_1 = \mu \qquad (9)$$

$$\kappa_2 = \mu_2 = \sigma^2 \qquad (10)$$

$$\kappa_3 = \mu_3 \qquad (11)$$

$$\kappa_4 = \mu_4 - 3\mu_2^2 \qquad (12)$$

$$\kappa_5 = \mu_5 - 10\mu_2\mu_3, \qquad (13)$$

where μ is the MEAN and $\sigma^2 \equiv \mu_2$ is the VARIANCE.

The K-STATISTIC are UNBIASED ESTIMATORS of the cumulants.

See also CHARACTERISTIC FUNCTION (PROBABILITY), CUMULANT-GENERATING FUNCTION, K-STATISTIC, KURTOSIS, MEAN, MOMENT, SHEPPARD'S CORRECTION, SKEWNESS, UNBIASED ESTIMATOR, VARIANCE

References

Abramowitz, M. and Stegun, C. A. (Eds.). *Handbook of Mathematical Functions with Formulas, Graphs, and Mathematical Tables, 9th printing.* New York: Dover, p. 928, 1972.
Kenney, J. F. and Keeping, E. S. "Cumulants and the Cumulant-Generating Function," "Additive Property of Cumulants," and "Sheppard's Correction." §4.10–.12 in *Mathematics of Statistics, Pt. 2, 2nd ed.* Princeton, NJ: Van Nostrand, pp. 77–2, 1951.

Cumulant-Generating Function

Let $M(h)$ be the MOMENT-GENERATING FUNCTION, then

$$K(h) \equiv \ln M(h) = \kappa_1 h + \frac{1}{2!} h^2 \kappa_2 + \frac{1}{3!} h^3 \kappa_3 + \ldots, \qquad (1)$$

where κ_1, κ_2, ..., are the CUMULANTS.
If

$$L = \sum_{j=1}^{N} c_j x_j \qquad (2)$$

is a function of N independent variables, then the cumulant-generating function for L is given by

$$K(h) = \sum_{j=1}^{N} K_j(c_j h). \qquad (3)$$

See also CUMULANT, MOMENT-GENERATING FUNCTION

References

Abramowitz, M. and Stegun, C. A. (Eds.). *Handbook of Mathematical Functions with Formulas, Graphs, and Mathematical Tables, 9th printing.* New York: Dover, p. 928, 1972.
Kenney, J. F. and Keeping, E. S. "Cumulants and the Cumulant-Generating Function" and "Additive Property of Cumulants." §4.10–.11 in *Mathematics of Statistics, Pt. 2, 2nd ed.* Princeton, NJ: Van Nostrand, pp. 77–0, 1951.

Cumulation

The dual operation of TRUNCATION which replaces the faces of a POLYHEDRON with PYRAMIDS of height h (where h may be positive, zero, or negative) having the face as the base. This operation is implemented in *Mathematica* under the misnomer Stellate[*poly*, *ratio*] in the *Mathematica* add-on package Graphics`Polyhedra` (which can be loaded with the command < <Graphics`). The operation is sometimes also called accretion, or sometimes akisation (since it transforms a regular polygon to an n-akis polyhedron, i.e., quadruples the number of faces).

The following plots show cumulation series for the TETRAHEDRON, CUBE, OCTAHEDRON, DODECAHEDRON, and ICOSAHEDRON.

cumulation allow compounds of Archimedean solids and their duals to be easily constructed.

ARCHIMEDEAN SOLID	dual	face 1	face 2
CUBOCTAHEDRON	RHOMBIC DODECAHEDRON	$3 : \frac{1}{4}\sqrt{6}$	$4 : \frac{1}{2}\sqrt{2}$
ICOSIDODECAHEDRON	RHOMBIC TRIACONTAHEDRON	$3 : \frac{1}{4}\sqrt{\frac{1}{5}(\sqrt{7-3\sqrt{5}})}$	$\frac{1}{4}\sqrt{\frac{1}{5}(5+2\sqrt{5})}$
SMALL RHOMBICUBOCTAHEDRON	DELTOIDAL ICOSITETRAHEDRON	$3 : \frac{1}{42}\sqrt{3}(3-\sqrt{2})$	$4 : \frac{1}{2}(\sqrt{2}-1)$
TRUNCATED CUBE	SMALL TRIAKIS OCTAHEDRON	$3 : \frac{1}{6}\sqrt{3}(3-2\sqrt{2})$	$8 : \frac{1}{2}(1+\sqrt{2})$
TRUNCATED DODECAHEDRON	TRIAKIS ICOSAHEDRON	$3 : \frac{1}{372}\sqrt{3}(1+5\sqrt{5})$	$\frac{1}{2}\sqrt{\frac{1}{2}(6+\sqrt{5})}$
TRUNCATED ICOSAHEDRON	PENTAKIS DODECAHEDRON	$\frac{1}{38}\sqrt{\frac{1}{10}(305+131\sqrt{5})}$	$6 : \frac{1}{4}\sqrt{3}(\sqrt{5}-3)$
TRUNCATED OCTAHEDRON	TETRAKIS HEXAHEDRON	$4 : \frac{1}{8}\sqrt{2}$	$3 : \frac{1}{4}\sqrt{6}$
TRUNCATED TETRAHEDRON	TRIAKIS TETRAHEDRON	$3 : \frac{1}{30}\sqrt{6}$	$6 : \frac{1}{2}\sqrt{6}$

See also ELEVATUM, ESCHER'S SOLID, INVAGINATUM, PYRAMID, STELLATION, TRUNCATION

References

Graziotti, U. *Polyhedra, the Realm of Geometric Beauty.* San Francisco, CA: 1962.

Weisstein, E. W. "Polyhedra." MATHEMATICA NOTEBOOK POLYHEDRA.M.

Cumulative Distribution Function

DISTRIBUTION FUNCTION

Cumulative Frequency

Let the ABSOLUTE FREQUENCIES of occurrence of an event in a number of CLASS INTERVALS be denoted f_1, f_2, ... The cumulative frequency corresponding to the upper boundary of any CLASS INTERVAL c_i in a FREQUENCY DISTRIBUTION is the total absolute frequency of all values less than that boundary, denoted

$$F_< \equiv \sum_{i \le n} f_i.$$

See also ABSOLUTE FREQUENCY, CLASS INTERVAL, CUMULATIVE FREQUENCY POLYGON, FREQUENCY DISTRIBUTION, RELATIVE CUMULATIVE FREQUENCY, RE-

---

Cumulation with $h = 0$ gives a triangulated version of the original solid. The following table gives special solids formed by cumulation of given heights on simple solids. In this table, r is the INRADIUS, and $(r + h)/h$ is the "stellation ratio" as defined in *Mathematica*.

Original	h	$(r+h)/h$	Result
CUBE	$\frac{1}{6}$	$4/3$	TETRAKIS HEXAHEDRON
CUBE	$\frac{1}{2}$	2	RHOMBIC DODECAHEDRON
CUBE	$\frac{1}{2}\sqrt{2}$	$1+\sqrt{2}$	24-faced star DELTAHEDRON
DODECAHEDRON	$\frac{1}{19}\sqrt{\frac{1}{5}(65+22\sqrt{5})}$	$\frac{3}{19}(10-\sqrt{5})$	PENTAKIS DODECAHEDRON
DODECAHEDRON	$\sqrt{\frac{1}{10}(5-\sqrt{5})}$	$2\sqrt{5}-3$	60-faced star DELTAHEDRON
DODECAHEDRON	$\sqrt{\frac{1}{5}(5+2\sqrt{5})}$	$\sqrt{5}$	SMALL STELLATED DODECAHEDRON
ICOSAHEDRON	$\frac{1}{6}\sqrt{3}(\sqrt{5}-3)$	$3(\sqrt{5}-2)$	GREAT DODECAHEDRON
ICOSAHEDRON	$\frac{1}{15}\sqrt{15}$	$\frac{1}{5}(10-3\sqrt{5})$	SMALL TRIAMBIC ICOSAHEDRON
ICOSAHEDRON	$\frac{1}{3}\sqrt{6}$	$1-3\sqrt{2}+\sqrt{10}$	60-faced star DELTAHEDRON
ICOSAHEDRON	$\frac{1}{6}\sqrt{3}(3+\sqrt{5})$	3	GREAT STELLATED DODECAHEDRON
OCTAHEDRON	$\sqrt{3}-\frac{2}{3}\sqrt{6}$	$5-3\sqrt{2}$	SMALL TRIAKIS OCTAHEDRON
OCTAHEDRON	$\frac{1}{3}\sqrt{6}$	3	STELLA OCTANGULA
TETRAHEDRON	$\frac{1}{15}\sqrt{6}$	$\frac{7}{5}$	TRIAKIS TETRAHEDRON
TETRAHEDRON	$\frac{1}{6}\sqrt{6}$	2	CUBE
TETRAHEDRON	$\frac{1}{3}\sqrt{6}$	3	9-faced star DELTAHEDRON

Another type of cumulation (which I call "midpoint cumulation") replaces each facial polygon with triangular polygons joining vertices with the neighboring edge midpoints, and then constructs a pyramid with base determined by the face's midpoints. Midpoint

LATIVE FREQUENCY

References

Kenney, J. F. and Keeping, E. S. "Cumulative Frequencies." §1.11 in *Mathematics of Statistics, Pt. 1, 3rd ed.* Princeton, NJ: Van Nostrand, pp. 17–9, 1962.

Cumulative Frequency Polygon

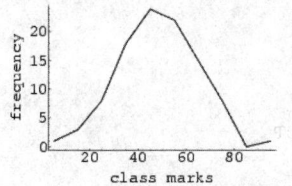 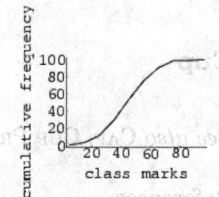

A plot of the cumulative frequency against the upper class boundary with the points joined by line segments. Any continuous cumulative frequency curve, including a cumulative frequency polygon, is called an OGIVE.

See also ABSOLUTE FREQUENCY, CLASS INTERVAL, FREQUENCY DISTRIBUTION, FREQUENCY POLYGON, OGIVE, RELATIVE CUMULATIVE FREQUENCY, RELATIVE FREQUENCY

References

Kenney, J. F. and Keeping, E. S. "Cumulative Frequency Polygons." §2.6 in *Mathematics of Statistics, Pt. 1, 3rd ed.* Princeton, NJ: Van Nostrand, pp. 28–9, 1962.

Cundy and Rollett's Egg

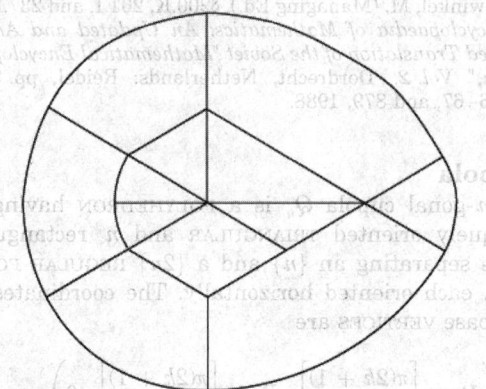

An OVAL dissected into pieces which are to used to create pictures. The resulting figures resemble those constructed out of TANGRAMS.

See also DISSECTION, EGG, OVAL, TANGRAM

References

Cundy, H. and Rollett, A. *Mathematical Models, 3rd ed.* Stradbroke, England: Tarquin Pub., pp. 19–1, 1989.

Dixon, R. *Mathographics.* New York: Dover, p. 11, 1991.

Cunningham Chain

A SEQUENCE of PRIMES $q_1 < q_2 < \ldots < q_k$ is a Cunningham chain of the first kind (second kind) of length k if $q_{1+1} = 2q_i + 1$ ($q_{1+1} = 2q_i - 1$) for $i = 1, \ldots, k-1$. Cunningham PRIMES of the first kind are SOPHIE GERMAIN PRIMES.

The two largest known Cunningham chains (of the first kind) of length three are ($384205437 \cdot 2^{4000} - 1$, $384205437 \cdot 2^{4001} - 1$, $384205437 \cdot 2^{4002} - 1$) and ($651358155 \cdot 2^{3291} - 1$, $651358155 \cdot 2^{3292} - 1$, $651358155 \cdot 2^{3293} - 1$), both discovered by W. Roonguthai in 1998.

See also BITWIN CHAIN, PRIME ARITHMETIC PROGRESSION, PRIME CLUSTER

References

Forbes, T. "Prime Clusters and Cunningham Chains." *Math. Comput.* **68**, 1739–748, 1999.
Guy, R. K. "Cunningham Chains." §A7 in *Unsolved Problems in Number Theory, 2nd ed.* New York: Springer-Verlag, pp. 18–9, 1994.
Ribenboim, P. *The New Book of Prime Number Records.* New York: Springer-Verlag, p. 333, 1996.
Roonguthai, W. "Yves Gallot's Proth and Cunningham Chains." http://ksc9.th.com/warut/cunningham.html.

Cunningham Function

Sometimes also called the PEARSON-CUNNINGHAM FUNCTION. It can be expressed using WHITTAKER FUNCTIONS (Whittaker and Watson 1990, p. 353).

$$\omega_{n,m}(x) \equiv \frac{e^{\pi i (m/2 - n) + x}}{\Gamma(1 + n - \frac{1}{2}m)} U(\tfrac{1}{2}m - n, \; 1 + m, \; x),$$

where $U(a, b, z)$ is a CONFLUENT HYPERGEOMETRIC FUNCTION OF THE SECOND KIND (Abramowitz and Stegun 1972, p. 510).

See also CONFLUENT HYPERGEOMETRIC FUNCTION OF THE SECOND KIND, WHITTAKER FUNCTION

References

Abramowitz, M. and Stegun, C. A. (Eds.). *Handbook of Mathematical Functions with Formulas, Graphs, and Mathematical Tables, 9th printing.* New York: Dover, 1972.
Whittaker, E. T. and Watson, G. N. *A Course in Modern Analysis, 4th ed.* Cambridge, England: Cambridge University Press, 1990.

Cunningham Number

A BINOMIAL NUMBER OF THE FORM $C^{\pm}(b, n) \equiv b^n \pm 1$. Bases b^k which are themselves powers need not be considered since they correspond to $(b^k)^n \pm 1 = b^{kn} \pm 1$. PRIME NUMBERS OF THE FORM $C^{\pm}(b, n)$ are very rare. A NECESSARY (but not SUFFICIENT) condition for $C^{+}(2, n) = 2^n + 1$ to be PRIME is that n be OF THE FORM $n = 2^m$. Numbers OF THE FORM $F_m =$

$C^+(2, 2^m) = 2^{2m} + 1$ are called FERMAT NUMBERS, and the only known PRIMES occur for $C^+(2, 1) = 3$, $C^+(2, 2) = 5$, $C^+(2, 4) = 17$, $C^+(2, 8) = 257$, and $C^+(2, 16) = 65537$ (i.e., $n = 0, 1, 2, 3, 4$). The only other PRIMES $C^+(b, n)$ for nontrivial $b \leq 11$ and $2 \leq n \leq 1000$ are $C^+(6, 2) = 37$, $C^+(6, 4) = 1297$, and $C^+(10, 2) = 101$.

PRIMES OF THE FORM $C^-(b, n)$ are also very rare. The MERSENNE NUMBERS $M_n = C^-(2, n) = 2^n - 1$ are known to be prime only for 37 values, the first few of which are $n = 2, 3, 5, 7, 13, 17, 19, \ldots$ (Sloane's A000043). There are no other PRIMES $C^-(b, n)$ for nontrivial $b \leq 20$ and $2 \leq n \leq 1000$.

In 1925, Cunningham and Woodall (1925) gathered together all that was known about the PRIMALITY and factorization of the numbers $C^\pm(b, n)$ and published a small book of tables. These tables collected from scattered sources the known prime factors for the bases 2 and 10 and also presented the authors' results of 30 years' work with these and other bases.

Since 1925, many people have worked on filling in these tables. D. H. Lehmer, a well-known mathematician who died in 1991, was for many years a leader of these efforts. Lehmer was a mathematician who was at the forefront of computing as modern electronic computers became a reality. He was also known as the inventor of some ingenious pre-electronic computing devices specifically designed for factoring numbers.

Updated factorizations were published in Brillhart *et al.* (1988). The current archive of Cunningham number factorizations for $b = 1, \ldots, \pm 12$ is kept on ftp://sable.ox.ac.uk/pub/math/cunningham/. The tables have been extended by Brent and te Riele (1992) to $b = 13, \ldots, 100$ with $m < 255$ for $b < 30$ and $m < 100$ for $b \geq 30$. All numbers with exponent 58 and smaller, and all composites with ≤ 90 digits have now been factored.

See also BINOMIAL NUMBER, CULLEN NUMBER, FERMAT NUMBER, MERSENNE NUMBER, REPUNIT, RIESEL NUMBER, SIERPINSKI NUMBER OF THE FIRST KIND, WOODALL NUMBER

References

Brent, R. P. and te Riele, H. J. J. "Factorizations of $a^n \pm 1$, $13 \leq a < 100$" Report NM-R9212, *Centrum voor Wiskunde en Informatica.* Amsterdam, June 1992. ftp://sable.ox.ac.uk/pub/math/factors/.

Brillhart, J.; Lehmer, D. H.; Selfridge, J.; Wagstaff, S. S. Jr.; and Tuckerman, B. *Factorizations of $b^n \pm 1$, $b = 2$, 3, 5, 6, 7, 10, 11, 12 Up to High Powers,* rev. ed. Providence, RI: Amer. Math. Soc., 1988. Updates are available electronically from ftp://sable.ox.ac.uk/pub/math/cunningham/.

Cunningham, A. J. C. and Woodall, H. J. *Factorisation of $y^n \mp 1$, $y = 2, 3, 5, 6, 7, 10, 11, 12$ Up to High Powers (n).* London: Hodgson, 1925.

Mudge, M. "Not Numerology but Numeralogy!" *Personal Computer World,* 279–80, 1997.

Ribenboim, P. "Numbers $k \times 2^n \pm 1$." §5.7 in *The New Book of Prime Number Records.* New York: Springer-Verlag, pp. 355–60, 1996.

Sloane, N. J. A. Sequences A000043/M0672 in "An On-Line Version of the Encyclopedia of Integer Sequences." http://www.research.att.com/~njas/sequences/eisonline.html.

Cunningham Project

CUNNINGHAM NUMBER

Cup

See also CAP, CUP PRODUCT

References

Feller, W. *An Introduction to Probability Theory and Its Applications, Vol. 2,* 3rd ed. New York: Wiley, 1971.

Cup Product

The cup product is a product on COHOMOLOGY CLASSES. In the case of DE RHAM COHOMOLOGY, a COHOMOLOGY CLASS can be represented by a CLOSED FORM. The cup product of $[\alpha]$ and $[\beta]$ is represented by the CLOSED FORM $[\alpha \wedge \beta]$, where $\wedge$ is the WEDGE PRODUCT of DIFFERENTIAL K-FORMS. It is the dual operation to intersection in HOMOLOGY.

In general, the cup product is a map

$$\vee: H^p \times H^q \to H^{p+q}$$

which satisfies $a \vee b = (-1)^{pq} b \vee a$.

See also COHOMOLOGY, CUP, DE RHAM COHOMOLOGY, HOMOLOGY

References

Hazewinkel, M. (Managing Ed.). §200.K, 201.I, and 237.D in *Encyclopaedia of Mathematics: An Updated and Annotated Translation of the Soviet "Mathematical Encyclopaedia," Vol. 2.* Dordrecht, Netherlands: Reidel, pp. 756, 766–67, and 879, 1988.

Cupola

An n-gonal cupola Q_n is a POLYHEDRON having n obliquely oriented TRIANGULAR and n rectangular faces separating an $\{n\}$ and a $\{2n\}$ REGULAR POLYGON, each oriented horizontally. The coordinates of the base VERTICES are

$$\left(R \cos\left[\frac{\pi(2k + 1)}{2n}\right], R \sin\left[\frac{\pi(2k + 1)}{2n}\right], 0 \right), \qquad (1)$$

and the coordinates of the top VERTICES are

$$\left(r \cos\left[\frac{2k\pi}{n}\right], r \sin\left[\frac{2k\pi}{n}\right], z \right), \qquad (2)$$

where R and r are the CIRCUMRADII of the base and top

$$R = \tfrac{1}{2}a \csc\left(\frac{\pi}{2n}\right) \qquad (3)$$

$$r = \tfrac{1}{2}a \csc\left(\frac{\pi}{n}\right), \qquad (4)$$

and z is the height.

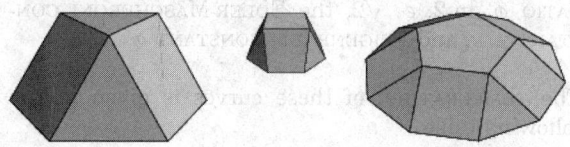

A cupola with all unit edge lengths (in which case the triangles become unit equilateral triangles and the rectangles become unit squares) is possible only for $n = 3, 4, 5$, in which case the height z can be obtained by letting $k = 0$ in the equations (1) and (2) to obtain the coordinates of neighboring bottom and top VERTICES,

$$\mathbf{b} = \begin{bmatrix} R \cos\left(\dfrac{\pi}{2n}\right) \\ R \sin\left(\dfrac{\pi}{2n}\right) \\ 0 \end{bmatrix} \qquad (5)$$

$$\mathbf{t} = \begin{bmatrix} r \\ 0 \\ z \end{bmatrix}. \qquad (6)$$

Since all side lengths are a,

$$|\mathbf{b} - \mathbf{t}|^2 = a^2. \qquad (7)$$

Solving for z then gives

$$\left[R \cos\left(\frac{\pi}{2n}\right) - r \right]^2 + R^2 \sin^2\left(\frac{\pi}{2n}\right) + z^2 = a^2 \qquad (8)$$

$$z^2 + R^2 + r^2 - 2rR \cos\left(\frac{\pi}{2n}\right) = a^2 \qquad (9)$$

$$z = \sqrt{a^2 - 2rR \cos\left(\frac{\pi}{2n}\right) - r^2 - R^2}$$

$$= a\sqrt{1 - \tfrac{1}{4}\csc^2\left(\frac{\pi}{n}\right)} \qquad (10)$$

See also BICUPOLA, ELONGATED CUPOLA, GYROELONGATED CUPOLA, PENTAGONAL CUPOLA, ROTUNDA, SQUARE CUPOLA, TRIANGULAR CUPOLA

References

Johnson, N. W. "Convex Polyhedra with Regular Faces." *Canad. J. Math.* **18**, 169–00, 1966.

Cupolarotunda

A CUPOLA adjoined to a ROTUNDA.

See also GYROCUPOLAROTUNDA, ORTHOCUPOLAROTUNDA

Curl

The curl of a TENSOR field is given by

$$(\nabla \times A)^\alpha = \epsilon^{\alpha\mu\nu} A_{v:\mu}, \qquad (1)$$

where ϵ_{ijk} is the LEVI-CIVITA TENSOR and ";" is the COVARIANT DERIVATIVE. For a VECTOR FIELD, the curl is denoted

$$\mathrm{curl}(\mathbf{F}) \equiv \nabla \times \mathbf{F}, \qquad (2)$$

and $\nabla \times \mathbf{F}$ is normal to the PLANE in which the "circulation" is MAXIMUM. Its magnitude is the limiting value of circulation per unit AREA,

$$(\nabla \times \mathbf{F}) \cdot \hat{\mathbf{n}} \equiv \lim_{A \to 0} \frac{\oint_C \mathbf{F} \cdot ds}{A}. \qquad (3)$$

Let

$$\mathbf{F} \equiv F_1 \hat{\mathbf{u}}_1 + F_2 \hat{\mathbf{u}}_2 + F_3 \hat{\mathbf{u}}_3 \qquad (4)$$

and

$$h_i \equiv \left| \frac{\partial \mathbf{r}}{\partial u_i} \right|, \qquad (5)$$

then

$$\nabla \times \mathbf{F} \equiv \frac{1}{h_1 h_2 h_3} \begin{vmatrix} h_1 \hat{\mathbf{u}}_1 & h_2 \hat{\mathbf{u}}_2 & h_3 \hat{\mathbf{u}}_3 \\ \dfrac{\partial}{\partial u_1} & \dfrac{\partial}{\partial u_2} & \dfrac{\partial}{\partial u_3} \\ h_1 F_1 & h_2 F_2 & h_2 F_2 \end{vmatrix}$$

$$= \frac{1}{h_2 h_3} \left[\frac{\partial}{\partial u_2}(h_3 F_3) - \frac{\partial}{\partial u_3}(h_2 F_2) \right] \hat{\mathbf{u}}_1$$

$$+ \frac{1}{h_1 h_3} \left[\frac{\partial}{\partial u_3}(h_1 F_1) - \frac{\partial}{\partial u_1}(h_3 F_3) \right] \hat{\mathbf{u}}_2$$

$$+ \frac{1}{h_1 h_2} \left[\frac{\partial}{\partial u_1}(h_2 F_2) - \frac{\partial}{\partial u_2}(h_1 F_1) \right] \hat{\mathbf{u}}_3. \qquad (6)$$

Special cases of the curl formulas above can be given for CURVILINEAR COORDINATES.

See also CURL THEOREM, CURVILINEAR COORDINATES, DIVERGENCE, GRADIENT, VECTOR DERIVATIVE

References

Arfken, G. "Curl, $\nabla \times$." §1.8 in *Mathematical Methods for Physicists, 3rd ed.* Orlando, FL: Academic Press, pp. 42–7, 1985.

Curl Theorem

A special case of STOKES' THEOREM in which F is a VECTOR FIELD and M is an oriented, compact embedded 2-MANIFOLD with boundary in $\mathbb{R}^2$, given by

$$\int_S (\nabla \times \mathbf{F}) \cdot d\mathbf{a} = \int_{\partial S} \mathbf{F} \cdot d\mathbf{s}. \qquad (1)$$

There are also alternate forms. If

$$\mathbf{F} \equiv \mathbf{c}F, \qquad (2)$$

then

$$\int_S d\mathbf{a} \times \nabla F = \int_C F \, d\mathbf{s}. \qquad (3)$$

and if

$$\mathbf{F} \equiv \mathbf{c} \times \mathbf{P}, \qquad (4)$$

then

$$\int_S (d\mathbf{a} \times \nabla) \times \mathbf{P} = \int_C d\mathbf{s} \times \mathbf{P}. \qquad (5)$$

See also CHANGE OF VARIABLES THEOREM, CURL, STOKES' THEOREM

References

Arfken, G. "Stokes's Theorem." §1.12 in *Mathematical Methods for Physicists, 3rd ed.* Orlando, FL: Academic Press, pp. 61–4, 1985.

Curlicue Fractal

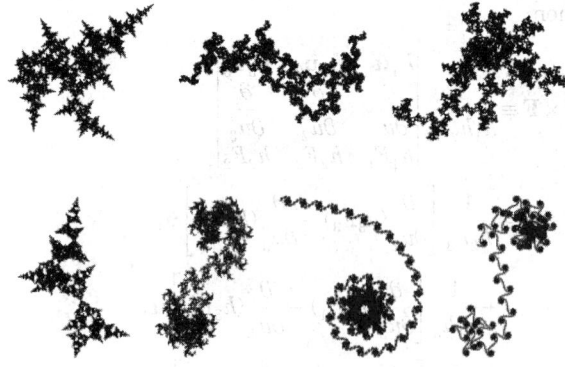

The curlicue fractal is a figure obtained by the following procedure. Let s be an IRRATIONAL NUMBER. Begin with a line segment of unit length, which makes an ANGLE $\phi_0 \equiv 0$ to the horizontal. Then define θ_n iteratively by

$$\theta_{n+1} = (\theta_n + 2\pi s)(\mathrm{mod}\ 2\pi),$$

with $\theta_0 = 0$. To the end of the previous line segment, draw a line segment of unit length which makes an angle

$$\phi_{n+1} = \theta_n + \phi_n (\mathrm{mod}\ 2\pi),$$

to the horizontal (Pickover 1995). The result is a FRACTAL, and the above figures correspond to the curlicue fractals with 10,000 points for the GOLDEN RATIO ϕ, $\ln 2$, e, $\sqrt{2}$, the EULER-MASCHERONI CONSTANT γ, π, and FEIGENBAUM CONSTANT δ.

The TEMPERATURE of these curves is given in the following table.

Constant	Temperature
GOLDEN RATIO ϕ	46
$\ln 2$	51
e	58
$\sqrt{2}$	58
EULER-MASCHERONI CONSTANT γ	63
π	90
FEIGENBAUM CONSTANT α	92

References

Berry, M. and Goldberg, J. "Renormalization of Curlicues." *Nonlinearity* **1**, 1–6, 1988.

Moore, R. and van der Poorten, A. "On the Thermodynamics of Curves and Other Curlicues." *McQuarie Univ. Math. Rep.* 89–031, April 1989.

Pickover, C. A. "The Fractal Golden Curlicue is Cool." Ch. 21 in *Keys to Infinity.* New York: W. H. Freeman, pp. 163–67, 1995.

Pickover, C. A. *Mazes for the Mind: Computers and the Unexpected.* New York: St. Martin's Press, 1993.

Sedgewick, R. *Algorithms in C, 3rd ed.* Reading, MA: Addison-Wesley, 1998.

Stewart, I. *Another Fine Math You've Got Me Into....* New York: W. H. Freeman, 1992.

Stoschek, E. "Module 35: Curlicue Variations: Polygon Patterns in the Gauss Plane of Complex Numbers." http://marvin.sn.schule.de/~inftreff/modul35/task35_e.htm.

Stoschek, E. "Module 36: The Feigenbaum-Constant δ in the Gauss Plane." http://marvin.sn.schule.de/~inftreff/modul36/task36_e.htm.

Curly Brace

BRACE

Current

A linear FUNCTIONAL on a smooth differential form.

See also FLAT NORM, INTEGRAL CURRENT, RECTIFI-

ABLE CURRENT

Curtate Cycloid

The path traced out by a fixed point at a RADIUS $b <$ a, where a is the RADIUS of a rolling CIRCLE, sometimes also called a CONTRACTED CYCLOID.

$$x = a\phi - b \sin \phi \qquad (1)$$

$$y = a - b \cos \phi. \qquad (2)$$

The ARC LENGTH from $\phi = 0$ is

$$s = 2(a+b)E(u), \qquad (3)$$

where

$$\sin(\tfrac{1}{2}\phi) = \text{sn } u \qquad (4)$$

$$k^2 = \frac{4ab}{(a+c)^2}, \qquad (5)$$

and $E(u)$ is a complete ELLIPTIC INTEGRAL OF THE SECOND KIND and sn u is a JACOBI ELLIPTIC FUNCTION.

See also CYCLOID, PROLATE CYCLOID, TROCHOID

References

Beyer, W. H. *CRC Standard Mathematical Tables, 28th ed.* Boca Raton, FL: CRC Press, p. 216, 1987.

Harris, J. W. and Stocker, H. *Handbook of Mathematics and Computational Science.* New York: Springer-Verlag, p. 325, 1998.

Lawrence, J. D. *A Catalog of Special Plane Curves.* New York: Dover, pp. 192 and 194–97, 1972.

Lockwood, E. H. *A Book of Curves.* Cambridge, England: Cambridge University Press, p. 146, 1967.

Steinhaus, H. *Mathematical Snapshots, 3rd ed.* New York: Dover, pp. 147–48, 1999.

Zwillinger, D. (Ed.). *CRC Standard Mathematical Tables and Formulae.* Boca Raton, FL: CRC Press, p. 292, 1995.

Curtate Cycloid Evolute

The EVOLUTE of the CURTATE CYCLOID

$$x = a\phi - b \sin \phi \qquad (1)$$

$$y = a - b \cos \phi. \qquad (2)$$

is given by

$$x = \frac{a[-2b\phi + 2a\phi \cos \phi - 2a \sin \phi + b \sin(2\phi)]}{2(a \cos \phi - b)} \qquad (3)$$

$$y = \frac{a(a - b \cos \phi)^2}{b(a \cos \phi - b)}. \qquad (4)$$

Curvature

In general, there are two important types of curvature: EXTRINSIC CURVATURE and INTRINSIC CURVATURE. The EXTRINSIC CURVATURE of curves in 2- and 3-space was the first type of curvature to be studied historically, culminating in the FRENET FORMULAS, which describe a SPACE CURVE entirely in terms of its "curvature," TORSION, and the initial starting point and direction.

After the curvature of 2- and 3-d curves was studied, attention turned to the curvature of surfaces in 3-space. The main curvatures which emerged from this scrutiny are the MEAN CURVATURE, GAUSSIAN CURVATURE, and the WEINGARTEN MAP. MEAN CURVATURE was the most important for applications at the time and was the most studied, but Gauss was the first to recognize the importance of the GAUSSIAN CURVATURE.

Because GAUSSIAN CURVATURE is "intrinsic," it is detectable to 2-dimensional "inhabitants" of the surface, whereas MEAN CURVATURE and the WEINGARTEN MAP are not detectable to someone who can't study the 3-dimensional space surrounding the surface on which he resides. The importance of GAUSSIAN CURVATURE to an inhabitant is that it controls the surface AREA of SPHERES around the inhabitant.

Riemann and many others generalized the concept of curvature to SECTIONAL CURVATURE, SCALAR CURVATURE, the RIEMANN TENSOR, RICCI CURVATURE, and a host of other INTRINSIC and EXTRINSIC CURVATURES. General curvatures no longer need to be numbers, and can take the form of a MAP, GROUP, GROUPOID, tensor field, etc.

The simplest form of curvature and that usually first encountered in CALCULUS is an EXTRINSIC CURVATURE. In 2-D, let a PLANE CURVE be given by CARTESIAN PARAMETRIC EQUATIONS $x = x(t)$ and $y = y(t)$. Then the curvature κ is defined by

$$\kappa \equiv \frac{d\phi}{ds} = \frac{\dfrac{d\phi}{dt}}{\dfrac{ds}{dt}} = \frac{\dfrac{d\phi}{dt}}{\sqrt{\left(\dfrac{dx}{dt}\right)^2 + \left(\dfrac{dy}{dt}\right)^2}} = \frac{\dfrac{d\phi}{dt}}{\sqrt{x'^2 + y'^2}}, \qquad (1)$$

where ϕ is the TANGENTIAL ANGLE and s is the ARC LENGTH. As can readily be seen from the definition, curvature therefore has units of inverse distance. The $\frac{d\phi}{dt}$ derivative in the above equation can be found using the identity

$$\tan \phi = \frac{dy}{dx} = \frac{dy/dt}{dx/dt} = \frac{y'}{x'}, \qquad (2)$$

so

$$\frac{d}{dt}(\tan \phi) = \sec^2 \phi \, \frac{d\phi}{dt} = \frac{x'y'' - y'x''}{x'^2} \qquad (3)$$

and

$$\frac{d\phi}{dt} = \frac{1}{\sec^2\phi}\frac{d}{dt}(\tan\phi) = \frac{1}{1+\tan^2\phi}\frac{x'y''-y'x''}{x'^2}$$

$$= \frac{1}{1+\frac{y'^2}{x'^2}}\frac{x'y''-y'x''}{x'^2} = \frac{x'y''-y'x''}{x'^2+y'^2}. \qquad (4)$$

Combining (1), (2), and (4) then gives

$$\kappa = \frac{x'y''-y'x''}{(x'^2+y'^2)^{3/2}}. \qquad (5)$$

For a 2-D curve written in the form $y = f(x)$, the equation of curvature becomes

$$\kappa = \frac{\frac{d^2y}{dx^2}}{\left[1+\left(\frac{dy}{dx}\right)^2\right]^{3/2}}. \qquad (6)$$

If the 2-D curve is instead parameterized in POLAR COORDINATES, then

$$\kappa = \frac{r^2 + 2r_\theta^2 - rr_{\theta\theta}}{(r^2+r_\theta^2)^{3/2}}, \qquad (7)$$

where $r_\theta \equiv \partial r/\partial\theta$ (Gray 1997, p. 89). In PEDAL COORDINATES, the curvature is given by

$$\kappa = \frac{1}{r}\frac{dp}{dr}. \qquad (8)$$

The curvature for a 2-D curve given implicitly by $g(x, y) = 0$ is given by

$$\kappa = \frac{g_{xx}g_y^2 - 2g_{xy}g_xg_y + g_{yy}g_x^2}{(g_x^2+g_y^2)^{3/2}} \qquad (9)$$

(Gray 1997).

Now consider a parameterized SPACE CURVE $\mathbf{r}(t)$ in 3-D for which the TANGENT VECTOR $\hat{\mathbf{T}}$ is defined as

$$\hat{\mathbf{T}} \equiv \frac{\frac{d\mathbf{r}}{dt}}{\left|\frac{d\mathbf{r}}{dt}\right|} = \frac{\frac{d\mathbf{r}}{dt}}{\frac{ds}{dt}}. \qquad (10)$$

Therefore,

$$\frac{d\mathbf{r}}{dt} = \frac{ds}{dt}\hat{\mathbf{T}} \qquad (11)$$

$$\frac{d^2\mathbf{r}}{dt^2} = \frac{d^2s}{dt^2}\hat{\mathbf{T}} + \frac{ds}{dt}\frac{d\hat{\mathbf{T}}}{dt} = \frac{d^2s}{dt^2}\hat{\mathbf{T}} + \kappa\hat{\mathbf{N}}\left(\frac{ds}{dt}\right)^2, \qquad (12)$$

where $\hat{\mathbf{N}}$ is the NORMAL VECTOR. But

$$\frac{d\mathbf{r}}{dt} \times \frac{d^2\mathbf{r}}{dt^2} = \frac{ds}{dt}\frac{d^2s}{dt^2}(\hat{\mathbf{T}} \times \hat{\mathbf{T}}) + \kappa\left(\frac{ds}{dt}\right)^3(\hat{\mathbf{T}} \times \hat{\mathbf{N}})$$

$$= \kappa\left(\frac{ds}{dt}\right)^3(\hat{\mathbf{T}} \times \hat{\mathbf{N}}) \qquad (13)$$

$$\left|\frac{d\mathbf{r}}{dt} \times \frac{d^2\mathbf{r}}{dt^2}\right| = \kappa\left(\frac{ds}{dt}\right)^3 = \kappa\left|\frac{d\mathbf{r}}{dt}\right|^3, \qquad (14)$$

so

$$\kappa = \left|\frac{d\hat{\mathbf{T}}}{ds}\right| = \frac{\left|\frac{d\mathbf{r}}{dt} \times \frac{d^2\mathbf{r}}{dt^2}\right|}{\left|\frac{d\mathbf{r}}{dt}\right|^3}. \qquad (15)$$

The curvature of a 2-D curve is related to the RADIUS OF CURVATURE of the curve's OSCULATING CIRCLE. Consider a CIRCLE specified parametrically by

$$x = a\cos t \qquad (16)$$

$$y = a\sin t \qquad (17)$$

which is tangent to the curve at a given point. The curvature is then

$$\kappa = \frac{x'y''-y'x''}{(x'^2+y'^2)^{3/2}} = \frac{a^2}{a^3} = \frac{1}{a}, \qquad (18)$$

or one over the RADIUS OF CURVATURE. The curvature of a CIRCLE can also be repeated in vector notation. For the CIRCLE with $0 \le t < 2\pi$, the ARC LENGTH is

$$s(t) = \int_0^t \sqrt{\left(\frac{dx}{dt}\right)^2 + \left(\frac{dy}{dt}\right)^2}\, dt$$

$$= \int_0^t \sqrt{a^2\cos^2 t + a^2\sin^2 t}\, dt = at, \qquad (19)$$

so $t = s/a$ and the equations of the CIRCLE can be rewritten as

$$x = a\cos\left(\frac{s}{a}\right) \qquad (20)$$

$$y = a\sin\left(\frac{s}{a}\right). \qquad (21)$$

The POSITION VECTOR is then given by

$$\mathbf{r}(s) = a\cos\left(\frac{s}{a}\right)\hat{\mathbf{x}} + a\sin\left(\frac{s}{a}\right)\hat{\mathbf{y}}, \qquad (22)$$

and the TANGENT VECTOR is

$$\hat{\mathbf{T}} = \frac{d\mathbf{r}}{ds} = -\sin\left(\frac{s}{a}\right)\hat{\mathbf{x}} + \cos\left(\frac{s}{a}\right)\hat{\mathbf{y}}, \qquad (23)$$

so the curvature is related to the RADIUS OF CURVATURE a by

$$\kappa = \left|\frac{d\hat{\mathbf{T}}}{ds}\right| = \left|-\frac{1}{a}\cos\left(\frac{s}{a}\right)\hat{\mathbf{x}} - \frac{1}{a}\sin\left(\frac{s}{a}\right)\hat{\mathbf{y}}\right|$$

$$= \sqrt{\frac{\cos^2\left(\frac{s}{a}\right) + \sin^2\left(\frac{s}{a}\right)}{a^2}} = \frac{1}{a}, \tag{24}$$

as expected.

Four very important derivative relations in differential geometry related to the FRENET FORMULAS are

$$\dot{\mathbf{r}} = \mathbf{T} \tag{25}$$

$$\ddot{\mathbf{r}} = \kappa\mathbf{N} \tag{26}$$

$$\dddot{\mathbf{r}} = \dot{\kappa}\mathbf{N} + \kappa(\tau\mathbf{B} - \kappa\mathbf{T}) \tag{27}$$

$$[\dot{\mathbf{r}}, \ddot{\mathbf{r}}, \dddot{\mathbf{r}}] = \kappa^2\tau, \tag{28}$$

where $\mathbf{T}$ is the TANGENT VECTOR, $\mathbf{N}$ is the NORMAL VECTOR, $\mathbf{B}$ is the BINORMAL VECTOR, and τ is the TORSION (Coxeter 1969, p. 322).

The curvature at a point on a surface takes on a variety of values as the PLANE through the normal varies. As κ varies, it achieves a minimum and a maximum (which are in perpendicular directions) known as the PRINCIPAL CURVATURES. As shown in Coxeter (1969, pp. 352–53),

$$\kappa^2 - \sum b_i^i\kappa + \det(b_i^j) = 0 \tag{29}$$

$$\kappa^2 - 2H\kappa + K = 0, \tag{30}$$

where K is the GAUSSIAN CURVATURE, H is the MEAN CURVATURE, and det denotes the DETERMINANT.

The curvature κ is sometimes called the FIRST CURVATURE and the TORSION τ the SECOND CURVATURE. In addition, a THIRD CURVATURE (sometimes called TOTAL CURVATURE)

$$\sqrt{ds_T^2 + ds_B^2} \tag{31}$$

is also defined. A signed version of the curvature of a CIRCLE appearing in the DESCARTES CIRCLE THEOREM for the radius of the fourth of four mutually tangent circles is called the BEND.

See also BEND (CURVATURE), CURVATURE CENTER, CURVATURE SCALAR, EXTRINSIC CURVATURE, FIRST CURVATURE, FOUR-VERTEX THEOREM, GAUSSIAN CURVATURE, INTRINSIC CURVATURE, LANCRET EQUATION, LINE OF CURVATURE, MEAN CURVATURE, NORMAL CURVATURE, PRINCIPAL CURVATURES, RADIUS OF CURVATURE, RICCI CURVATURE, RIEMANN TENSOR, SECOND CURVATURE, SECTIONAL CURVATURE, SODDY CIRCLES, THIRD CURVATURE, TORSION (DIFFERENTIAL GEOMETRY), WEINGARTEN MAP

References

Casey, J. *Exploring Curvature.* Wiesbaden, Germany: Vieweg, 1996.
Coxeter, H. S. M. *Introduction to Geometry, 2nd ed.* New York: Wiley, 1969.

Fischer, G. (Ed.). Plates 79–5 in *Mathematische Modelle/ Mathematical Models, Bildband/Photograph Volume.* Braunschweig, Germany: Vieweg, pp. 74–1, 1986.
Gray, A. "Curvature of Curves in the Plane," "Drawing Plane Curves with Assigned Curvature," and "Drawing Space Curves with Assigned Curvature." §1.5, 6.4, and 10.2 in *Modern Differential Geometry of Curves and Surfaces with Mathematica, 2nd ed.* Boca Raton, FL: CRC Press, pp. 14–7, 140–46, and 222–24, 1997.
Kreyszig, E. "Principal Normal, Curvature, Osculating Circle." §12 in *Differential Geometry.* New York: Dover, pp. 34–6, 1991.
Yates, R. C. "Curvature." *A Handbook on Curves and Their Properties.* Ann Arbor, MI: J. W. Edwards, pp. 60–4, 1952.

Curvature Center

The point on the POSITIVE RAY of the NORMAL VECTOR at a distance $\rho(s)$, where ρ is the RADIUS OF CURVATURE. It is given by

$$\mathbf{z} = \mathbf{x} + \rho\mathbf{N} = \mathbf{x} + \rho^2\frac{\mathbf{T}}{ds}, \tag{1}$$

where $\mathbf{N}$ is the NORMAL VECTOR and $\mathbf{T}$ is the TANGENT VECTOR. It can be written in terms of $\mathbf{x}$ explicitly as

$$\mathbf{z} = \mathbf{x} + \frac{\mathbf{x}''(\mathbf{x}' \cdot \mathbf{x}')^2 - \mathbf{x}'(\mathbf{x}' \cdot \mathbf{x}')(\mathbf{x}' \cdot \mathbf{x}'')}{(\mathbf{x}' \cdot \mathbf{x}')(\mathbf{x}'' \cdot \mathbf{x}'') - (\mathbf{x}' \cdot \mathbf{x}'')^2}. \tag{2}$$

For a CURVE represented parametrically by $(f(t), g(t))$,

$$\alpha = f - \frac{(f'^2 - g'^2)g'}{f'g'' - f''g'} \tag{3}$$

$$\beta = g + \frac{(f'^2 - g'^2)f'}{f'g'' - f''g'} \tag{4}$$

References

Gray, A. *Modern Differential Geometry of Curves and Surfaces with Mathematica, 2nd ed.* Boca Raton, FL: CRC Press, 1997.

Curvature Scalar

SCALAR CURVATURE

Curvature Vector

$$\mathbf{K} \equiv \frac{d\mathbf{T}}{ds},$$

where $\mathbf{T}$ is the TANGENT VECTOR defined by

$$\mathbf{T} \equiv \dfrac{\dfrac{d\mathbf{x}}{ds}}{\left| \dfrac{d\mathbf{x}}{ds} \right|}.$$

Curve

A CONTINUOUS MAP from a 1-D SPACE to an n-D SPACE. Loosely speaking, the word "curve" is often used to mean the GRAPH of a 2- or 3-D curve. The simplest curves can be represented parametrically in n-D SPACE as

$$x_1 = f_1(t)$$
$$x_2 = f_2(t)$$
$$\vdots$$
$$x_n = f_n(t).$$

Other simple curves can be simply defined only implicitly, i.e., in the form

$$f(x_1, x_2, \ldots) = 0.$$

See also PLANE CURVE, SPACE CURVE, SPHERICAL CURVE

References
Cundy, H. and Rollett, A. *Mathematical Models, 3rd ed.* Stradbroke, England: Tarquin Pub., pp. 71–5, 1989.
"Geometry." *The New Encyclopædia Britannica, 15th ed.* **19**, pp. 946–51, 1990.
Gallier, J. H. *Curves and Surfaces for Geometric Design: Theory and Algorithms.* New York: Academic Press, 1999.
Oakley, C. O. *Analytic Geometry.* New York: Barnes and Noble, 1957.
Rutter, J. W. *Geometry of Curves.* Boca Raton, FL: Chapman and Hall/CRC, 2000.
Shikin, E. V. *Handbook and Atlas of Curves.* Boca Raton, FL: CRC Press, 1995.
Seggern, D. von *CRC Standard Curves and Surfaces.* Boca Raton, FL: CRC Press, 1993.
Smith, P. F.; Gale, A. S.; and Neelley, J. H. *New Analytic Geometry, Alternate Edition.* Boston, MA: Ginn and Company, 1938.
Walker, R. J. *Algebraic Curves.* New York: Springer-Verlag, 1978.
Weisstein, E. W. "Books about Curves." http://www.treasure-troves.com/books/Curves.html.
Yates, R. C. *The Trisection Problem.* Reston, VA: National Council of Teachers of Mathematics, 1971.
Zwillinger, D. (Ed.). "Algebraic Curves." §8.1 in *CRC Standard Mathematical Tables and Formulae, 3rd ed.* Boca Raton, FL: CRC Press, 1996.

Curve of Constant Breadth
CURVE OF CONSTANT WIDTH

Curve of Constant Precession

A curve whose CENTRODE revolves about a fixed axis with constant ANGLE and SPEED when the curve is traversed with unit SPEED. The TANGENT INDICATRIX of a curve of constant precession is a SPHERICAL HELIX. An ARC LENGTH parameterization of a curve of constant precession with NATURAL EQUATIONS

$$\kappa(s) = -\omega \sin(\mu s) \tag{1}$$

$$\tau(s) = -\omega \cos(\mu s) \tag{2}$$

is

$$x(s) = \frac{\alpha + \mu}{2\alpha} \frac{\sin[(\alpha - \mu)s]}{\alpha - \mu} - \frac{\alpha - \mu}{2\alpha} \frac{\sin[(\alpha + \mu)s]}{\alpha + \mu} \tag{3}$$

$$y(s) = \frac{\alpha + \mu}{2\alpha} \frac{\sin[(\alpha - \mu)s]}{\alpha - \mu} + \frac{\alpha - \mu}{2\alpha} \frac{\cos[(\alpha + \mu)s]}{\alpha + \mu} \tag{4}$$

$$z(s) = \frac{\omega}{\mu\alpha} \sin(\mu s), \tag{5}$$

where

$$\alpha \equiv \sqrt{\omega^2 + \mu^2} \tag{6}$$

and ω, and μ are constant. This curve lies on a circular one-sheeted HYPERBOLOID

$$x^2 + y^2 - \frac{\mu^2}{\omega^2} z^2 = \frac{4\mu^2}{\omega^4}. \tag{7}$$

The curve is closed IFF μ/α is RATIONAL.

References
Scofield, P. D. "Curves of Constant Precession." *Amer. Math. Monthly* **102**, 531–37, 1995.

Curve of Constant Slope
GENERALIZED HELIX

Curve of Constant Width

Curves which, when rotated in a square, make contact with all four sides. Such curves are sometimes also known as ROLLERS.

The "width" of a closed convex curve is defined as the distance between parallel lines bounding it ("supporting lines"). Every curve of constant width is convex. Curves of constant width have the same "width" regardless of their orientation between the parallel lines. In fact, they also share the same PERIMETER (BARBIER'S THEOREM). Examples include the CIRCLE (with largest AREA), and REULEAUX TRIANGLE (with smallest AREA) but there are an infinite number. A curve of constant width can be used in a special drill chuck to cut square "HOLES."

A generalization gives solids of constant width. These do not have the same surface AREA for a given width, but their shadows are curves of constant width with the *same* width!

See also DELTA CURVE, KAKEYA NEEDLE PROBLEM, REULEAUX TRIANGLE

References

Blaschke, W. "Konvexe Bereiche gegebener konstanter Breite und kleinsten Inhalts." *Math. Ann.* **76**, 504–13, 1915.

Bogomolny, A. "Shapes of Constant Width." http://www.cut-the-knot.com/do_you_know/cwidth.html.

Böhm, J. "Convex Bodies of Constant Width." Ch. 4 in *Mathematical Models from the Collections of Universities and Museums* (Ed. G. Fischer). Braunschweig, Germany: Vieweg, pp. 96–00, 1986.

Croft, H. T.; Falconer, K. J.; and Guy, R. K. *Unsolved Problems in Geometry.* New York: Springer-Verlag, p. 7, 1991.

Fischer, G. (Ed.). Plates 98–02 in *Mathematische Modelle/Mathematical Models, Bildband/Photograph Volume.* Braunschweig, Germany: Vieweg, pp. 89 and 96, 1986.

Gardner, M. "Mathematical Games: Curves of Constant Width, One of which Makes it Possible to Drill Square Holes." *Sci. Amer.* **208**, 148–56, Feb. 1963.

Gardner, M. "Curves of Constant Width." Ch. 18 in *The Unexpected Hanging and Other Mathematical Diversions.* Chicago, IL: Chicago University Press, pp. 212–21, 1991.

Goldberg, M. "Circular-Arc Rotors in Regular Polygons." *Amer. Math. Monthly* **55**, 393–02, 1948.

Kelly, P. *Convex Figures.* New York: Harcourt Brace, 1995.

Rademacher, H. and Toeplitz, O. *The Enjoyment of Mathematics: Selections from Mathematics for the Amateur.* Princeton, NJ: Princeton University Press, 1957.

Steinhaus, H. *Mathematical Snapshots, 3rd ed.* New York: Dover, pp. 150–51, 1999.

Wells, D. *The Penguin Dictionary of Curious and Interesting Geometry.* London: Penguin, pp. 219–20, 1991.

Yaglom, I. M. and Boltyanski, V. G. *Convex Figures.* New York: Holt, Rinehart, and Winston, 1961.

Curvilinear Coordinates

A COORDINATE SYSTEM composed of intersecting surfaces. If the intersections are all at right angles, then the curvilinear coordinates are said to form an ORTHOGONAL COORDINATE SYSTEM. If not, they form a SKEW COORDINATE SYSTEM.

A general METRIC $g_{\mu\nu}$ has a LINE ELEMENT

$$ds^2 = g_{\mu\nu}du^\mu du^\nu, \qquad (1)$$

where EINSTEIN SUMMATION is being used. Curvilinear coordinates are defined as those with a diagonal METRIC so that

$$g_{\mu\nu} \equiv \delta^\mu_\nu h^2_\mu, \qquad (2)$$

where δ^μ_ν is the KRONECKER DELTA. Curvilinear coordinates therefore have a simple LINE ELEMENT

$$ds^2 = \delta^\mu_\nu h^2_\mu du^\mu du^\nu = h^2_\mu du^{\mu 2}, \qquad (3)$$

which is just the PYTHAGOREAN THEOREM, so the differential VECTOR is

$$d\mathbf{r} = h_\mu \, du_\mu \, \hat{\mathbf{u}}_\mu, \qquad (4)$$

or

$$d\mathbf{r} = \frac{\partial \mathbf{r}}{\partial u_1} \, du_1 + \frac{\partial \mathbf{r}}{\partial u_2} \, du_2 + \frac{\partial \mathbf{r}}{\partial u_3} \, du_3, \qquad (5)$$

where the SCALE FACTORS are

$$h_i \equiv \left| \frac{\partial \mathbf{r}}{\partial u_i} \right| \qquad (6)$$

and

$$\hat{\mathbf{u}}_i \equiv \frac{\frac{\partial \mathbf{r}}{\partial u_i}}{\left| \frac{\partial \mathbf{r}}{\partial u_i} \right|} = \frac{1}{h_i} \frac{\partial \mathbf{r}}{\partial u_i}. \qquad (7)$$

Equation (5) may therefore be re-expressed as

$$d\mathbf{r} = h_1 \, du_1 \hat{\mathbf{u}}_1 + h_2 \, du_2 \hat{\mathbf{u}}_2 + h_3 \, du_3 \hat{\mathbf{u}}_3. \qquad (8)$$

The GRADIENT is

$$\mathrm{grad}(\phi) \equiv \nabla\phi$$

$$= \frac{1}{h_1} \frac{\partial \phi}{\partial u_1} \hat{\mathbf{u}}_1 + \frac{1}{h_2} \frac{\partial \phi}{\partial u_2} \hat{\mathbf{u}}_2 + \frac{1}{h_3} \frac{\partial \phi}{\partial u_3} \hat{\mathbf{u}}_3, \qquad (9)$$

the DIVERGENCE is

$$\mathrm{div}(F) \equiv \nabla \cdot \mathbf{F} \equiv \frac{1}{h_1 h_2 h_3}$$

$$\times \left[\frac{\partial}{\partial u_1}(h_2 h_3 F_1) + \frac{\partial}{\partial u_2}(h_3 h_1 F_2) + \frac{\partial}{\partial u_3}(h_1 h_2 F_3) \right], \qquad (10)$$

and the CURL is

$$\nabla \times \mathbf{F} \equiv \frac{1}{h_1 h_2 h_3} \begin{vmatrix} h_1 \hat{\mathbf{u}}_1 & h_2 \hat{\mathbf{u}}_2 & h_3 \hat{\mathbf{u}}_3 \\ \frac{\partial}{\partial u_1} & \frac{\partial}{\partial u_2} & \frac{\partial}{\partial u_3} \\ h_1 F_1 & h_2 F_2 & h_2 F_2 \end{vmatrix}$$

$$= \frac{1}{h_2 h_3} \left[\frac{\partial}{\partial u_2}(h_3 F_3) - \frac{\partial}{\partial u_3}(h_2 F_2) \right] \hat{\mathbf{u}}_1$$

$$+ \frac{1}{h_1 h_3} \left[\frac{\partial}{\partial u_3}(h_1 F_1) - \frac{\partial}{\partial u_1}(h_3 F_3) \right] \hat{\mathbf{u}}_2$$

$$+ \frac{1}{h_1 h_2} \left[\frac{\partial}{\partial u_1}(h_2 F_2) - \frac{\partial}{\partial u_2}(h_1 F_1) \right] \hat{\mathbf{u}}_3. \qquad (11)$$

See also ORTHOGONAL COORDINATE SYSTEM, SKEW COORDINATE SYSTEM

References

Byerly, W. E. "Orthogonal Curvilinear Coördinates." §130 in *An Elementary Treatise on Fourier's Series, and Spherical, Cylindrical, and Ellipsoidal Harmonics, with Applications to Problems in Mathematical Physics.* New York: Dover, pp. 238–39, 1959.

Moon, P. and Spencer, D. E. *Foundations of Electrodynamics.* Princeton, NJ: Van Nostrand, 1960.

Moon, P. and Spencer, D. E. *Field Theory Handbook, Including Coordinate Systems, Differential Equations, and Their Solutions, 2nd ed.* New York: Springer-Verlag, pp. 1–, 1988.

Cushion

The QUARTIC SURFACE resembling a squashed round cushion on a barroom stool and given by the equation

$$z^2 x^2 - z^4 - 2zx^2 + 2z^3 + x^2 - z^2$$
$$-(x^2 - z)^2 - y^4 - 2x^2 y^2 - y^2 z^2 + 2y^2 z + y^2 = 0.$$

See also QUARTIC SURFACE

References
Nordstrand, T. "Surfaces." http://www.uib.no/people/nfytn/surfaces.htm.

Cusp

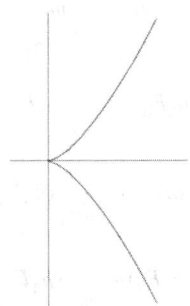

A cusp is a point on a continuous curve where the tangent vector reverses sign as the curve is traversed. A cusp is a type of DOUBLE POINT. The above plot shows the curve $x^3 - y^2 = 0$, which has a cusp at the ORIGIN.

See also CRUNODE, DOUBLE CUSP, DOUBLE POINT, ORDINARY DOUBLE POINT, RAMPHOID CUSP, SALIENT POINT, SPINODE, TACNODE

References
Walker, R. J. *Algebraic Curves.* New York: Springer-Verlag, pp. 57–8, 1978.

Cusp Catastrophe

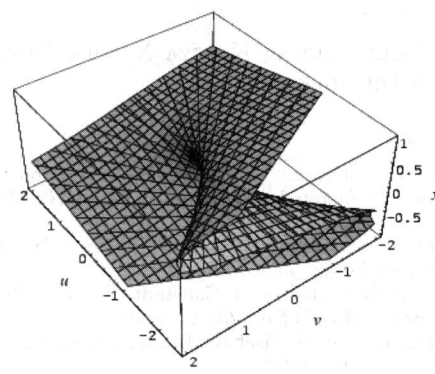

A CATASTROPHE which can occur for two control factors and one behavior axis. The cusp catastrophe is the universal unfolding of the singularity $f(x) = x^4$ and has the equation $F(x, u, v) = x^4 + ux^2 + vx$. The equation $y = x^{2/3}$ also has a cusp catastrophe.

See also CATASTROPHE THEORY

References
Sanns, W. *Catastrophe Theory with Mathematica: A Geometric Approach.* Germany: DAV, 2000.
von Seggern, D. *CRC Standard Curves and Surfaces.* Boca Raton, FL: CRC Press, p. 28, 1993.

Cusp Form

A cusp form is a MODULAR FORM for which the coefficient $c(0) = 0$ in the FOURIER SERIES

$$f(\tau) = \sum_{n=0}^{\infty} c(n) e^{2\pi i n \tau}$$

(Apostol 1997, p. 114). The only entire cusp form of weight $k < 12$ is the zero function (Apostol 1997, p. 116). The set of all cusp forms in M_k (all MODULAR FORMS of weight k) is a linear subspace of M_k which is denoted $M_{k,0}$. The dimension of $M_{k,0}$ is 1 for $k = 12$, 16, 18, 20, 22, and 26 (Apostol 1997, p. 119). For a cusp form $f \in M_{2k,0}$,

$$c(n) = \mathcal{O}(n^k) \tag{1}$$

(Apostol 1997, p. 135) or, more precisely,

$$c(n) = \mathcal{O}(n^{k-1/4+\epsilon}) \tag{2}$$

for every $\epsilon > 0$ (Selberg 1965; Apostol 1997, p. 136). It is conjectured that the $-1/4$ in the exponent can be reduced to $-1/2$ (Apostol 1997, p. 136).

See also MODULAR FORM

References
Apostol, T. M. *Modular Functions and Dirichlet Series in Number Theory, 2nd ed.* New York: Springer-Verlag, pp. 114 and 116, 1997.
Selberg, A. "On the Estimate of Coefficients of Modular Forms." *Proc. Sympos. Pure Math.* **8**, 1–5, 1965.

Cusp Map

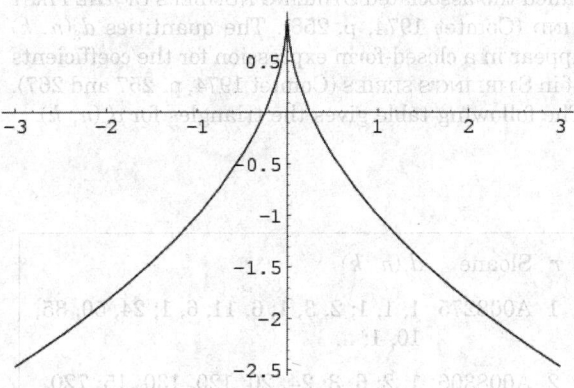

The function

$$f(x) = 1 - 2|x|^{1/2}$$

for $x \in [-1, \ 1]$. The INVARIANT DENSITY is

$$\rho(y) = \tfrac{1}{2}(1 - y).$$

References

Beck, C. and Schlögl, F. *Thermodynamics of Chaotic Systems.* Cambridge, England: Cambridge University Press, p. 195, 1995.

Cusp Point

CUSP

Cut

Given a weighted, UNDIRECTED GRAPH $G = (V, \ E)$ and a GRAPHICAL PARTITION of V into two sets A and B, the cut of G with respect to A and B is defined as

$$\text{cut}(A, \ B) = \sum_{i \in A, \ j \in B} W(i, \ j),$$

where $W(i, \ j)$ denotes the weight for the edge connecting vertices i and j.

See also BRANCH CUT, CUT SET

References

Demmel, J. "CS 267: Lectures 20 and 21, Mar 21, 1996 and Apr 2, 1999. Graph Partitioning, Part 1." http://www.cs.berkeley.edu/~demmel/cs267/lecture18/lecture18.html.

Cut Set

A set of edges of a GRAPH which, if removed (or "cut"), disconnects the graph (i.e., forms a DISCONNECTED GRAPH).

See also ARTICULATION VERTEX, DISCONNECTED GRAPH

References

Skiena, S. "Reconstructing Graphs from Cut-Set Sizes." *Info. Proc. Lett.* **32**, 123–27, 1989.
Skiena, S. *Implementing Discrete Mathematics: Combinatorics and Graph Theory with Mathematica.* Reading, MA: Addison-Wesley, 1990.

Cutpoint

ARTICULATION VERTEX

Cutting

The slicing of a 3-D object by a plane (or more general slice).

See also ARCHIMEDES' HAT-BOX THEOREM, ARRANGEMENT, CAKE CUTTING, CYLINDER CUTTING, DIVISION, HADWIGER PROBLEM, HAM SANDWICH THEOREM, PANCAKE CUTTING, PIE CUTTING, SQUARE DIVISION BY LINES, TORUS CUTTING

Cut-Vertex

ARTICULATION VERTEX

CW-Approximation Theorem

If X is any SPACE, then there is a CW-COMPLEX Y and a MAP $f : Y \to X$ inducing ISOMORPHISMS on all HOMOTOPY, HOMOLOGY, and COHOMOLOGY groups.

CW-Complex

A CW-complex is a homotopy-theoretic generalization of the notion of a SIMPLICIAL COMPLEX. A CW-complex is any SPACE X which can be built by starting off with a discrete collection of points called X^0, then attaching 1-D DISKS D^1 to X^0 along their boundaries $\mathbb{S}^0$, writing X^1 for the object obtained by attaching the $\mathbb{D}^1$s to X^0, then attaching 2-D DISKS $\mathbb{D}^2$ to X^1 along their boundaries $\mathbb{S}^1$, writing X^2 for the new SPACE, and so on, giving spaces X^n for every n. A CW-complex is any SPACE that has this sort of decomposition into SUBSPACES X^n built up in such a hierarchical fashion (so the X^ns must exhaust all of X). In particular, X^n may be built from X^{n-1} by attaching infinitely many n-DISKS, and the attaching MAPS $\mathbb{S}^{n-1} \to X^{n-1}$ may be any continuous MAPS.

The main importance of CW-complexes is that, for the sake of HOMOTOPY, HOMOLOGY, and COHOMOLOGY groups, every SPACE is a CW-complex. This is called the CW-APPROXIMATION THEOREM. Another is WHITEHEAD'S THEOREM, which says that MAPS between CW-complexes that induce ISOMORPHISMS on all HOMOTOPY GROUPS are actually HOMOTOPY equivalences.

See also COHOMOLOGY, CW-APPROXIMATION THEOREM, HOMOLOGY GROUP, HOMOTOPY GROUP, SIMPLICIAL COMPLEX, SPACE, SUBSPACE, WHITEHEAD'S THEOREM

Cycle (Circle)

A CIRCLE with an arrow indicating a direction.

Cycle (Map)

An n-cycle is a finite sequence of points $Y_0, \ldots, Y_{n-1}$ such that, under a MAP G,

$$Y_1 = G(Y_0)$$

$$Y_2 = G(Y_1)$$

$$Y_{n-1} = G(Y_{n-2})$$

$$Y_0 = G(Y_{n-1}).$$

In other words, it is a periodic trajectory which comes back to the same point after n iterations of the cycle. Every point Y_j of the cycle satisfies $Y_j = G^n(Y_j)$ and is therefore a FIXED POINT of the mapping G^n. A fixed point of G is simply a CYCLE of period 1.

Cycle (Permutation)

A SUBSET of a PERMUTATION whose elements trade places with one another. Permutations cycles are called "orbits" by Comtet (1974, p. 256). For example, in the PERMUTATION GROUP $\{4, 2, 1, 3\}$, $\{1, 3, 4\}$ is a 3-cycle ($1 \to 3$, $3 \to 4$, and $4 \to 1$) and $\{2\}$ is a 1-cycle ($2 \to 2$). There is a great deal of freedom in picking the representation of a cyclic decomposition since (1) the cycles are disjoint and can therefore be specified in any order, and (2) any rotation of a given cycle specifies the same cycle (Skiena 1990, p. 20). Therefore, (431)(2), (314)(2), (143)(2), (2)(431), (2)(314), and (2)(143) all describe the same cycle.

The cyclic decomposition of a PERMUTATION can be computed in *Mathematica* with the function ToCycles[p] in the *Mathematica* add-on package DiscreteMath`Permutations` (which can be loaded with the command < <DiscreteMath`) and the PERMUTATION corresponding to a cyclic decomposition can be computed with FromCycles[c1, ..., cn] in the *Mathematica* add-on package DiscreteMath`Permutations` (which can be loaded with the command < <DiscreteMath`). According to Vardi (1991), the *Mathematica* code for ToCycles is one of the most obscure ever written.

Every PERMUTATION GROUP on n symbols can be uniquely expressed as a product of disjoint cycles (Skiena 1990, p. 20). A cycle decomposition of a PERMUTATION can be viewed as a CLASS of a PERMUTATION GROUP.

The number $d_1(n, k)$ of k-cycles in a PERMUTATION GROUP of order n is given by

$$d_1(n, k) = (-1)^{n-k} S_1(n, k) = |S_1(n, k)|, \tag{1}$$

where $S_1(n, m)$ are the STIRLING NUMBERS OF THE FIRST KIND. More generally, let $d_r(n, k)$ be the number of permutations of n having exactly k cycles all of which are of length $\geq r$. $d_2(n, k)$ are sometimes called the associated STIRLING NUMBERS OF THE FIRST KIND (Comtet 1974, p. 256). The quantities $d_3(n, k)$ appear in a closed-form expression for the coefficients of in STIRLING'S SERIES (Comtet 1974, p. 257 and 267). The following table gives the triangles for $d_r(n, k)$.

r	Sloane	$d_r(n, k)$
1	A008275	1; 1, 1; 2, 3, 1; 6, 11, 6, 1; 24, 50, 35, 10, 1; ...
2	A008306	1; 2; 6, 3; 24, 20; 120, 130, 15; 720, 924, 210; ...
3	A050211	2; 6; 24; 120, 40; 720, 420; 5040, 3948; 40320, ...
4	A050212	6; 24; 120; 720; 5040, 1260; 40320, 18144; ...
5	A050213	24; 120; 720; 5040; 40320; 362880, 72576; ...

The functions $d_r(n, k)$ are given by the RECURRENCE RELATION

$$d_r(n, k) = (n-1)d_r(n-1, k) + (n-1)_{r-1}d_r(n-r, k-1), \tag{2}$$

where $(n)_k$ is the FALLING FACTORIAL, combined with the initial conditions

$$d_r(n, k) = 0 \quad \text{for } n \leq kr - 1 \tag{3}$$

$$d_r(n, 1) = (n-1)! \tag{4}$$

(Riordan 1958, p. 85; Comtet 1974, p. 257).

See also GOLOMB-DICKMAN CONSTANT, PERMUTATION, PERMUTATION GROUP, STIRLING NUMBER OF THE FIRST KIND, STIRLING'S SERIES, SUBSET

References

Biggs, N. *Discrete Mathematics, rev. ed.* Oxford, England: Clarendon Press, 1993.

Comtet, L. *Advanced Combinatorics: The Art of Finite and Infinite Expansions, rev. enl. ed.* Dordrecht, Netherlands: Reidel, p. 257, 1974.

Graham, R. L.; Knuth, D. E.; and Patashnik, O. *Concrete Mathematics: A Foundation for Computer Science, 2nd ed.* Reading, MA: Addison-Wesley, 1994.

Knuth, D. E. *The Art of Computer Programming, Vol. 1: Fundamental Algorithms, 3rd ed.* Reading, MA: Addison-Wesley, 1997.

Riordan, J. *Combinatorial Identities.* New York: Wiley, 1958.

Skiena, S. "The Cycle Structure of Permutations." §1.2.4 in *Implementing Discrete Mathematics: Combinatorics and Graph Theory with Mathematica.* Reading, MA: Addison-Wesley, pp. 20–4, 1990.

Sloane, N. J. A. Sequences A008275, A008306, A050211, A050212, A050213 in "An On-Line Version of the Encyclopedia of Integer Sequences." http://www.research.-att.com/~njas/sequences/eisonline.html.

Stanton, D. and White, D. *Constructive Combinatorics.* New York: Springer-Verlag, 1986.

Vardi, I. *Computational Recreations in Mathematica.* Redwood City, CA: Addison-Wesley, p. 223, 1991.

Cycle Decomposition

CYCLE (PERMUTATION)

Cycle Graph

A cycle graph C_n is a graph on n nodes containing a single cycle through all nodes. Cycle graphs can be generated using Cycle[n] in the *Mathematica* add-on package DiscreteMath`Combinatorica` (which can be loaded with the command <<DiscreteMath`). The CHROMATIC NUMBER of C_n is given by

$$\chi(C_n) = \begin{cases} 3 & \text{for } n \text{ odd} \\ 2 & \text{for } n \text{ even}. \end{cases}$$

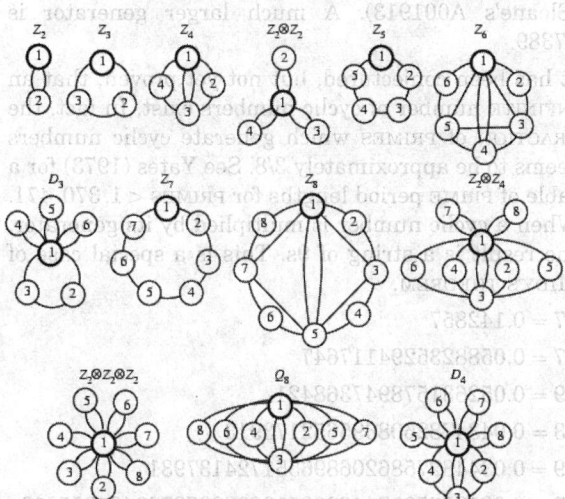

A cycle graph of a GROUP is a GRAPH which shows cycles of a GROUP as well as the connectivity between the cycles. Several examples are shown above. For Z_4, the group elements A_i satisfy $A_i^4 = 1$, where 1 is the IDENTITY ELEMENT, and two elements satisfy $A_1^2 = A_3^2 = 1$.

For a CYCLIC GROUP of COMPOSITE ORDER n (e.g., Z_4, Z_6, Z_8), the degenerate subcycles corresponding to factors dividing n are often not shown explicitly since their presence is implied.

See also CHAIN (GRAPH), CHARACTERISTIC FACTOR, CYCLIC GRAPH, CYCLIC GROUP, GRAPH CYCLE, HAMILTONIAN CYCLE, SQUARE GRAPH, TRIANGLE GRAPH, WALK

References

Shanks, D. *Solved and Unsolved Problems in Number Theory,* 4th ed. New York: Chelsea, pp. 83–8, 1993.

Skiena, S. "Cycles, Stars, and Wheels." §4.2.3 in *Implementing Discrete Mathematics: Combinatorics and Graph Theory with Mathematica.* Reading, MA: Addison-Wesley, pp. 144–47, 1990.

Cyclic Graph

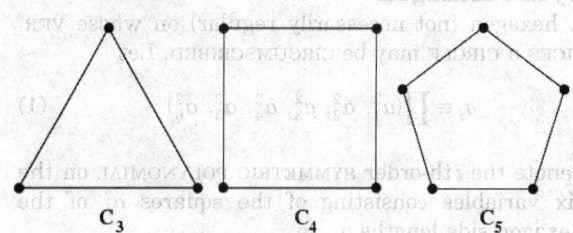

C_3 C_4 C_5

A GRAPH of n nodes and n edges such that node i is connected to the two adjacent nodes $i+1$ and $i-1$ (mod n), where the nodes are numbered $0, 1, ..., n-1$.

See also CYCLE GRAPH, FOREST, GRAPH CYCLE, STAR GRAPH, WHEEL GRAPH

References

Balaban, A. T. "Enumeration of Cyclic Graphs." In *Chemical Applications of Graph Theory* (Ed. A. T. Balaban). London: Academic Press, pp. 63–05, 1976.

Cyclic Group

A cyclic group $\mathbb{Z}_n$ (also commonly denoted Z_n or C_n; Shanks 1993, p. 75) of ORDER n is a GROUP defined by the element X (the GENERATOR) and its n POWERS up to

$$X^n = I,$$

where I is the IDENTITY ELEMENT. Cyclic groups are ABELIAN. There exists a unique cyclic group of every order $n \geq 2$, so cyclic groups of the same order are always isomorphic (Scott 1987, p. 34; Shanks 1993, p. 74). Furthermore, subgroups of cyclic groups are cyclic, and all GROUPS of PRIME ORDER are cyclic. In fact, the only SIMPLE ABELIAN GROUPS are the cyclic groups of order $n = 1$ or a n a prime (Scott 1987, p. 35).

Examples of cyclic groups include $\mathbb{Z}_2$, $\mathbb{Z}_3$, $\mathbb{Z}_4$, and the MODULO MULTIPLICATION GROUPS M_m such that $m = 2, 4, p^n$, or $2p^n$, for p an ODD PRIME and $n \geq 1$ (Shanks 1993, p. 92). By computing the CHARACTERISTIC FACTORS, any ABELIAN GROUP can be expressed as a GROUP DIRECT PRODUCT of cyclic SUBGROUPS, for example, $Z_2 \otimes Z_4$ or $Z_2 \otimes Z_2 \otimes Z_2$.

See also ABELIAN GROUP, CHARACTERISTIC FACTOR, FINITE GROUP Z2, FINITE GROUP Z3, FINITE GROUP Z4, FINITE GROUP Z5, FINITE GROUP Z6, METACYCLIC GROUP, MODULO MULTIPLICATION GROUP, SIMPLE GROUP

References

Lomont, J. S. "Cyclic Groups." §3.10.A in *Applications of Finite Groups*. New York: Dover, p. 78, 1987.

Scott, W. R. "Cyclic Groups." §2.4 in *Group Theory*. New York: Dover, pp. 34–5, 1987.

Shanks, D. *Solved and Unsolved Problems in Number Theory, 4th ed.* New York: Chelsea, 1993.

Cyclic Hexagon

A hexagon (not necessarily regular) on whose VERTICES a CIRCLE may be CIRCUMSCRIBED. Let

$$\sigma_i \equiv \prod_i (a_1^2, \, a_2^2, \, a_3^2, \, a_4^2, \, a_5^2, \, a_6^2) \tag{1}$$

denote the ith-order SYMMETRIC POLYNOMIAL on the six variables consisting of the squares a_i^2 of the hexagon side lengths a_i, so

$$\sigma_1 = a_1^2 + a_2^2 + a_3^2 + a_4^2 + a_5^2 + a_6^2 \tag{2}$$

$$\sigma_2 = a_1^2 a_2^2 + a_1^2 a_3^2 + a_1^2 a_4^2 + a_1^2 a_5^2 + a_1^2 a_6^2$$
$$+ a_2^2 a_3^2 + a_2^2 a_4^2 + a_2^2 a_5^2 + a_2^2 a_6^2$$
$$+ a_3^2 a_4^2 + a_3^2 a_5^2 + a_3^2 a_6^2$$
$$+ a_4^2 a_5^2 + a_4^2 a_6^2 + a_5^2 a_6^2 \tag{3}$$

$$\sigma_3 = a_1^2 a_2^2 a_3^2 + a_1^2 a_2^2 a_4^2 + a_1^2 a_2^2 a_5^2 + a_1^2 a_2^2 a_6^2$$
$$+ a_2^2 a_3^2 a_4^2 + a_2^2 a_3^2 a_5^2 + a_2^2 a_3^2 a_6^2$$
$$+ a_3^2 a_4^2 a_5^2 + a_3^2 a_4^2 a_6^2 + a_4^2 a_5^2 a_6^2 \tag{4}$$

$$\sigma_4 = a_1^2 a_2^2 a_3^2 a_4^2 + a_1^2 a_2^2 a_3^2 a_5^2 + a_1^2 a_2^2 a_3^2 a_6^2$$
$$+ a_1^2 a_3^2 a_4^2 a_5^2 + a_1^2 a_3^2 a_4^2 a_6^2$$
$$+ a_1^2 a_3^2 a_5^2 a_6^2 + a_1^2 a_4^2 a_5^2 a_6^2$$
$$+ a_2^2 a_3^2 a_4^2 a_5^2 + a_2^2 a_3^2 a_4^2 a_6^2 + a_2^2 a_3^2 a_5^2 a_6^2$$
$$+ a_2^2 a_4^2 a_5^2 a_6^2 + a_3^2 a_4^2 a_5^2 a_6^2 \tag{5}$$

$$\sigma_5 = a_1^2 a_2^2 a_3^2 a_4^2 a_5^2 + a_1^2 a_2^2 a_3^2 a_4^2 a_6^2$$
$$+ a_1^2 a_2^2 a_3^2 a_5^2 a_6^2 + a_1^2 a_2^2 a_4^2 a_5^2 a_6^2$$
$$+ a_1^2 a_3^2 a_4^2 a_5^2 a_6^2 + a_2^2 a_3^2 a_4^2 a_5^2 a_6^2 \tag{6}$$

$$\sigma_6 = a_1^2 a_2^2 a_3^2 a_4^2 a_5^2 a_6^2. \tag{7}$$

Then let K be the AREA of the hexagon and define

$$u = 16K^2 \tag{8}$$

$$t_2 = u - 4\sigma_2 + \sigma_1^2 \tag{9}$$

$$t_3 = 8\sigma_3 + \sigma_1 t_2 - 16\sqrt{\sigma_6} \tag{10}$$

$$t_4 = t_2^2 - 64\sigma_4 + 64\sigma_1\sqrt{\sigma_6} \tag{11}$$

$$t_5 = 128\sigma_5 + 32 t_2 \sqrt{\sigma_6}. \tag{12}$$

The AREA of the hexagon then satisfies

$$u t_4^3 + t_3^2 t_4^2 - 16 t_3^3 t_5 - 18 u t_3 t_4 t_5 - 27 u^2 t_5^2 = 0, \tag{13}$$

or this equation with $\sqrt{\sigma_6}$ replaced by $-\sqrt{\sigma_6}$, a seventh order POLYNOMIAL in u. This is $1/(4u^2)$ times the DISCRIMINANT of the CUBIC EQUATION

$$z^3 + 2t_3 z^2 - u t_4 z + 2u^2 t_5. \tag{14}$$

See also CONCYCLIC, CYCLIC PENTAGON, CYCLIC POLYGON, FUHRMANN'S THEOREM

References

Robbins, D. P. "Areas of Polygons Inscribed in a Circle." *Discr. Comput. Geom.* **12**, 223–36, 1994.

Robbins, D. P. "Areas of Polygons Inscribed in a Circle." *Amer. Math. Monthly* **102**, 523–30, 1995.

Cyclic Number

A number having $n - 1$ DIGITS which, when MULTIPLIED by 1, 2, 3, ..., $n - 1$, produces the same digits in a different order. Cyclic numbers are generated by the UNIT FRACTIONS $1/n$ which have maximal period DECIMAL EXPANSIONS (which means n must be PRIME). The first few numbers which generate cyclic numbers are 7, 17, 19, 23, 29, 47, 59, 61, 97, ... (Sloane's A001913). A much larger generator is 17389.

It has been conjectured, but not yet proven, that an INFINITE number of cyclic numbers exist. In fact, the FRACTION of PRIMES which generate cyclic numbers seems to be approximately 3/8. See Yates (1973) for a table of PRIME period lengths for PRIMES $< 1,370,471$. When a cyclic number is multiplied by its generator, the result is a string of 9s. This is a special case of MIDY'S THEOREM.

$07 = 0.142857$

$17 = 0.0588235294117647$

$19 = 0.052631578947368421$

$23 = 0.0434782608695652173913$

$29 = 0.0344827586206896551724137931$

$47 = 0.0212765957446808510638297872340425531 9\text{-} 0212765957446808510638297872340425531914893617$

$59 = 0.0169491525423728813559322033898305084 7\text{-} 0169491525423728813559322033898305084745762711864406779661$

$61 = 0.0163934426229508196721311475409836065 5\text{-} 0163934426229508196721311475409836065573770491803278688524 59$

$97 = 0.0103092783505154639175257731958762886 5\text{-} 01030927835051546391752577319587628865979 38144329896907216494845360824742268041237113402 06185567$

See also DECIMAL EXPANSION, FULL REPTEND PRIME, MIDY'S THEOREM

References

Gardner, M. "Cyclic Numbers." Ch. 10 in *Mathematical Circus: More Puzzles, Games, Paradoxes and Other Mathematical Entertainments from Scientific American.* New York: Knopf, pp. 111–22, 1979.

Guttman, S. "On Cyclic Numbers." *Amer. Math. Monthly* **44**, 159–66, 1934.

Kraitchik, M. "Cyclic Numbers." §3.7 in *Mathematical Recreations.* New York: W. W. Norton, pp. 75–6, 1942.

Rao, K. S. "A Note on the Recurring Period of the Reciprocal of an Odd Number." *Amer. Math. Monthly* **62**, 484–87, 1955.

Rivera, C. "Problems & Puzzles: Puzzle Period Length of $1/p$.-012." http://www.primepuzzles.net/puzzles/puzz_012.htm.

Sloane, N. J. A. Sequences A001913/M4353 in "An On-Line Version of the Encyclopedia of Integer Sequences." http://www.research.att.com/~njas/sequences/eisonline.html.

Yates, S. *Primes with Given Period Length.* Trondheim, Norway: Universitetsforlaget, 1973.

Cyclic Pentagon

A cyclic pentagon is a not necessarily regular PENTAGON on whose VERTICES a CIRCLE may be CIRCUMSCRIBED. Let such a pentagon have edge lengths a_1, ..., a_5, and AREA K, and let

$$\sigma_i \equiv \Pi_i(a_1^2,\ a_2^2,\ a_3^2,\ a_4^2,\ a_5^2) \tag{1}$$

denote the ith-order SYMMETRIC POLYNOMIAL on the five variables consisting of the squares a_i^2 of the pentagon side lengths a_i, so

$$\sigma_1 = a_1^2 + a_2^2 + a_3^2 + a_4^2 + a_5^2 \tag{2}$$

$$\sigma_2 = a_1^2 a_2^2 + a_1^2 a_3^2 + a_1^2 a_4^2 + a_1^2 a_5^2 + a_2^2 a_3^2$$
$$+ a_2^2 a_4^2 + a_2^2 a_5^2 + a_3^2 a_4^2 + a_3^2 a_5^2$$
$$+ a_4^2 a_5^2 \tag{3}$$

$$\sigma_3 = a_1^2 a_2^2 a_3^2 + a_1^2 a_2^2 a_4^2 + a_1^2 a_2^2 a_5^2$$
$$+ a_2^2 a_3^2 a_4^2 + a_2^2 a_3^2 a_5^2 + a_3^2 a_4^2 a_5^2 \tag{4}$$

$$\sigma_4 = a_1^2 a_2^2 a_3^2 a_4^2 + a_1^2 a_2^2 a_3^2 a_5^2 + a_1^2 a_3^2 a_4^2 a_5^2$$
$$+ a_1^2 a_2^2 a_4^2 a_5^2 + a_2^2 a_3^2 a_4^2 a_5^2 \tag{5}$$

$$\sigma_5 = a_1^2 a_2^2 a_3^2 a_4^2 a_5^2. \tag{6}$$

In addition, also define

$$u = 16K^2 \tag{7}$$

$$t_2 = u - 4\sigma_2 + \sigma_1^2 \tag{8}$$

$$t_3 = 8\sigma_3 + \sigma_1 t_2 \tag{9}$$

$$t_4 = -64\sigma_4 + t_2^2 \tag{10}$$

$$t_5 = 128\sigma_5. \tag{11}$$

Then the AREA of the pentagon satisfies

$$ut_4^3 + t_3^2 t_4^2 - 16t_3^3 t_5 - 18ut_3 t_4 t_5 - 27u^2 t_5^2 = 0, \tag{12}$$

a seventh order POLYNOMIAL in u (Robbins 1995). This is also $1/(4u^2)$ times the DISCRIMINANT of the CUBIC EQUATION

$$z^3 + 2t_3 z^2 - ut_4 z + 2u^2 t_5 \tag{13}$$

(Robbins 1995).

See also CONCYCLIC, CYCLIC HEXAGON, CYCLIC POLYGON

References

Robbins, D. P. "Areas of Polygons Inscribed in a Circle." *Discr. Comput. Geom.* **12**, 223–36, 1994.

Robbins, D. P. "Areas of Polygons Inscribed in a Circle." *Amer. Math. Monthly* **102**, 523–30, 1995.

Cyclic Permutation

A PERMUTATION which shifts all elements of a SET by a fixed offset, with the elements shifted off the end inserted back at the beginning. For a SET with elements $a_0, a_1, ..., a_{n-1}$, a cyclic permutation of one place to the left would yield $a_1, ..., a_{n-1}, a_0$, and a cyclic permutation of one place to the right would yield $a_{n-1}, a_0, a_1, ...$.

The mapping can be written as $a_i \to a_{i+k(\bmod\ n)}$ for a shift of k places. A shift of k places to the left is implemented in *Mathematica* as RotateLeft[*list*, k], while a shift of k places to the right is implemented as RotateRight[*list*, k].

See also PERMUTATION

Cyclic Polygon

A cyclic polygon is a POLYGON with VERTICES upon which a CIRCLE can be CIRCUMSCRIBED. Since every TRIANGLE has a CIRCUMCIRCLE, every TRIANGLE is cyclic. It is conjectured that for a cyclic polygon of $2m+1$ sides, $16K^2$ (where K is the AREA) satisfies a MONIC POLYNOMIAL of degree Δ_m, where

$$\Delta_m = \sum_{k=0}^{m-1} (m-k)\binom{2m+1}{k} \tag{1}$$

$$= \frac{1}{2}\left[(2m+1)\binom{2m}{m} - 2^{2m}\right] \tag{2}$$

(Robbins 1995). It is also conjectured that a cyclic polygon with $2m+2$ sides satisfies one of two POLYNOMIALS of degree Δ_m. The first few values of Δ_m are 1, 7, 38, 187, 874, ... (Sloane's A000531).

For TRIANGLES $n = 3 = 2 \cdot 1 + 1$, the POLYNOMIAL is HERON'S FORMULA, which may be written

$$16K^2 = 2a^2 b^2 + 2a^2 c^2 + 2b^2 c^2 - a^4 - b^4 - c^4, \tag{3}$$

and which is of order $\Delta_1 = 1$ in $16K^2$. For a CYCLIC QUADRILATERAL, the POLYNOMIAL is BRAHMAGUPTA'S FORMULA, which may be written

$$16K^2 = -a^4 + 2a^2b^2 - b^4 + 2a^2c^2 + 2b^2c^2 - c^4$$

$$+8abcd + 2a^2d^2 + 2b^2d^2 + 2c^2d^2 - d^4, \qquad (4)$$

which is of order $\Delta_1 = 1$ in $16K^2$. Robbins (1995) gives the corresponding FORMULAS for the CYCLIC PENTAGON and CYCLIC HEXAGON.

See also CONCYCLIC, CYCLIC HEXAGON, CYCLIC PENTAGON, CYCLIC QUADRANGLE, CYCLIC QUADRILATERAL, JAPANESE THEOREM

References

Robbins, D. P. "Areas of Polygons Inscribed in a Circle." *Discr. Comput. Geom.* **12**, 223–36, 1994.
Robbins, D. P. "Areas of Polygons Inscribed in a Circle." *Amer. Math. Monthly* **102**, 523–30, 1995.
Sloane, N. J. A. Sequences A000531 in "An On-Line Version of the Encyclopedia of Integer Sequences." http://www.research.att.com/~njas/sequences/eisonline.html.

Cyclic Quadrangle

Let A_1, A_2, A_3, and A_4 be four POINTS on a CIRCLE, and H_1, H_2, H_3, H_4 the ORTHOCENTERS of TRIANGLES $\Delta A_2 A_3 A_4$, etc. If, from the eight POINTS, four with different subscripts are chosen such that three are from one set and the fourth from the other, these POINTS form an ORTHOCENTRIC SYSTEM. There are eight such systems, which are analogous to the six sets of ORTHOCENTRIC SYSTEMS obtained using the feet of the ANGLE BISECTORS, ORTHOCENTER, and VERTICES of a generic TRIANGLE.

On the other hand, if all the POINTS are chosen from one set, or two from each set, with all different subscripts, the four POINTS lie on a CIRCLE. There are four pairs of such CIRCLES, and eight POINTS lie by fours on eight equal CIRCLES.

The SIMSON LINE of A_4 with regard to TRIANGLE $\Delta A_1 A_2 A_3$ is the same as that of H_4 with regard to the TRIANGLE $\Delta H_1 A_2 A_3$.

See also ANGLE BISECTOR, CONCYCLIC, CYCLIC POLYGON, CYCLIC QUADRILATERAL, ORTHOCENTRIC SYSTEM

References

Coxeter, H. S. M. and Greitzer, S. L. "Cyclic Quadrangles; Brahmagupta's Formula." §3.2 in *Geometry Revisited.* Washington, DC: Math. Assoc. Amer., pp. 56–0, 1967.
Johnson, R. A. *Modern Geometry: An Elementary Treatise on the Geometry of the Triangle and the Circle.* Boston, MA: Houghton Mifflin, pp. 251–53, 1929.

Cyclic Quadrilateral

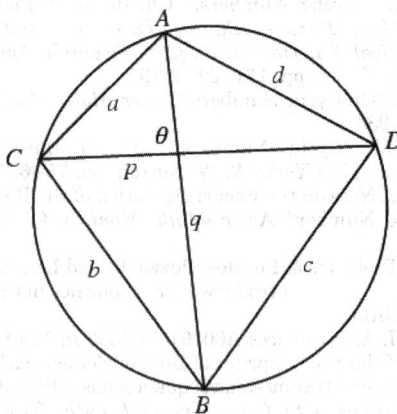

A QUADRILATERAL for which a CIRCLE can be circumscribed so that it touches each VERTEX. The AREA is then given by a special case of BRETSCHNEIDER'S FORMULA. Let the sides have lengths a, b, c, and d, let s be the SEMIPERIMETER

$$s \equiv \tfrac{1}{2}(a + b + c + d), \qquad (1)$$

and let R be the CIRCUMRADIUS. Then

$$A = \sqrt{(s-a)(s-b)(s-c)(s-d)} \qquad (2)$$

$$= \frac{\sqrt{(ac + bd)(ad + bc)(ab + cd)}}{4R}. \qquad (3)$$

Solving for the CIRCUMRADIUS gives

$$R = \tfrac{1}{4}\sqrt{\frac{(ac + bd)(ad + bc)(ab + cd)}{(s - a)(s - b)(s - c)(s - d)}}. \qquad (4)$$

The DIAGONALS of a cyclic quadrilateral have lengths

$$p = \sqrt{\frac{(ab + cd)(ac + bd)}{ad + bc}} \qquad (5)$$

$$q = \sqrt{\frac{(ac + bd)(ad + bc)}{ab + cd}}, \qquad (6)$$

so that $pq = ac + bd$.
In general, there are three essentially distinct cyclic quadrilaterals (modulo ROTATION and REFLECTION) whose edges are permutations of the lengths a, b, c, and d. Of the six corresponding DIAGONAL lengths, three are distinct. In addition to p and q, there is therefore a "third" DIAGONAL which can be denoted r. It is given by the equation

$$r = \sqrt{\frac{(ad + bc)(ab + cd)}{ac + bd}}. \qquad (7)$$

This allows the AREA formula to be written in the particularly beautiful and simple form

$$A = \frac{pqr}{4R}. \qquad (8)$$

The DIAGONALS are sometimes also denoted p, q, and r.

The AREA of a cyclic quadrilateral is the MAXIMUM possible for any QUADRILATERAL with the given side lengths. Also, the opposite ANGLES of a cyclic quadrilateral sum to π RADIANS (Dunham 1990). There exists a closed BILLIARDS path inside a cyclic quadrilateral if its CIRCUMCENTER lies inside the quadrilateral (Wells 1991, p. 11).

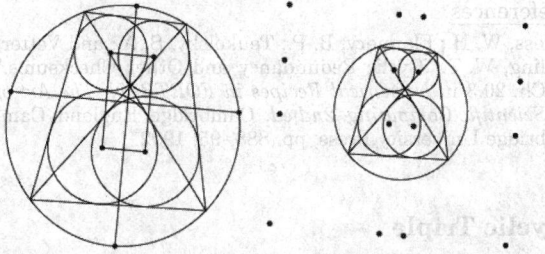

The INCENTERS of the four triangles composing the cyclic quadrilateral form a RECTANGLE. Furthermore, the sides of the RECTANGLE are PARALLEL to the lines connecting the MID-ARC POINTS between each pair of vertices (left figure above; Fuhrmann 1890, p. 50; Johnson 1929, pp. 254–55; Wells 1991). If the EXCENTERS of the triangles constituting the quadrilateral are added to the INCENTERS, a 4×4 rectangular grid is obtained (right figure; Johnson 1929, p. 255; Wells 1991).

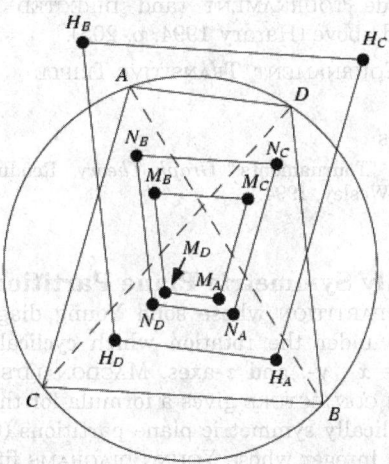

Consider again the four triangles contained in a cyclic quadrilateral. Amazingly, the CENTROIDS M_i, NINE-POINT CENTERS N_i, and ORTHOCENTERS H_i formed by these triangles are similar to the original quadrilateral. In fact, the triangle formed by the ORTHOCENTERS is congruent to it (Wells 1991, p. 44).

A cyclic quadrilateral with RATIONAL sides a, b, c, and d, DIAGONALS p and q, CIRCUMRADIUS r, and AREA a is given by $a = 25$, $b = 33$, $c = 39$, $d = 65$, $p = 60$, $q = 52$, $r = 65/2$, and $a = 1344$.

Let $ahbo$ be a QUADRILATERAL such that the angles $\angle hab$ and $\angle hob$ are RIGHT ANGLES, then $ahbo$ is a cyclic quadrilateral (Dunham 1990). This is a COROLLARY of the theorem that, in a RIGHT TRIANGLE, the MIDPOINT of the HYPOTENUSE is equidistant from the three VERTICES. Since M is the MIDPOINT of both RIGHT TRIANGLES $\triangle AHB$ and $\triangle BOH$, it is equidistant from all four VERTICES, so a CIRCLE centered at M may be drawn through them. This theorem is one of the building blocks of Heron's derivation of HERON'S FORMULA.

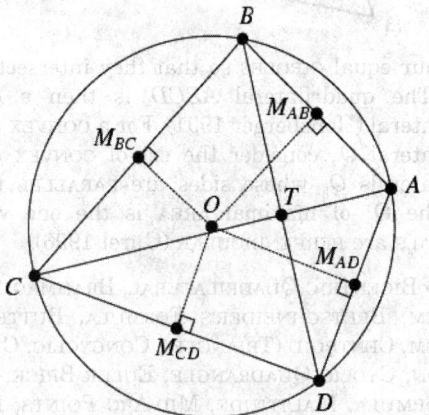

An application of BRAHMAGUPTA'S THEOREM gives the pretty result that, for a cyclic quadrilateral with perpendicular diagonals, the distance from the CIRCUMCENTER O to a side is half the length of the opposite side, so in the above figure,

$$OM_{AB} = \tfrac{1}{2}CD = CM_{CD} = DM_{CD}, \qquad (9)$$

and so on (Honsberger 1995, pp. 37–8).

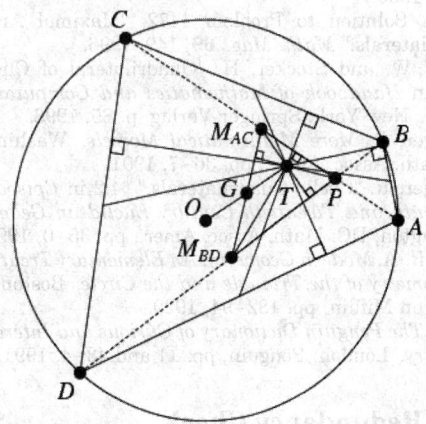

Let M_{AC} and M_{BD} be the MIDPOINTS of the diagonals of a cyclic quadrilateral $ABCD$, and let P be the intersection of the diagonals. Then the ORTHOCENTER of TRIANGLE $\triangle PM_{AC}M_{BD}$ is the ANTICENTER T of $ABCD$ (Honsberger 1995, p. 39).

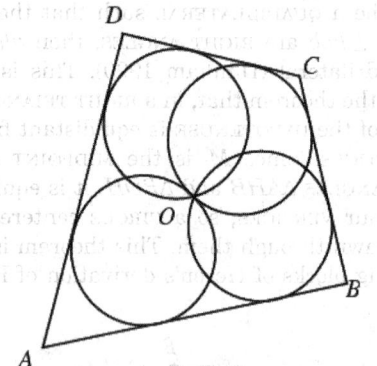

Place four equal CIRCLES so that they intersect in a point. The quadrilateral $ABCD$ is then a cyclic quadrilateral (Honsberger 1991). For a CONVEX cyclic quadrilateral Q, consider the set of CONVEX cyclic quadrilaterals $Q_\parallel$ whose sides are PARALLEL to Q. Then the $Q_\parallel$ of maximal AREA is the one whose DIAGONALS are PERPENDICULAR (Gürel 1996).

See also BICENTRIC QUADRILATERAL, BRAHMAGUPTA'S THEOREM, BRETSCHNEIDER'S FORMULA, BUTTERFLY THEOREM, CENTROID (TRIANGLE), CONCYCLIC, CYCLIC POLYGON, CYCLIC QUADRANGLE, EULER BRICK, HERON'S FORMULA, MALTITUDE, MID-ARC POINTS, NINE-POINT CENTER, ORTHOCENTER, PONCELET TRANSVERSE, PTOLEMY'S THEOREM, QUADRILATERAL, TANGENTIAL QUADRILATERAL

References

Andreescu, T. and Gelca, R. "Cyclic Quadrilaterals." §1.2 in *Mathematical Olympiad Challenges.* Boston, MA: Birkhäuser, pp. 6–, 2000.

Beyer, W. H. *CRC Standard Mathematical Tables, 28th ed.* Boca Raton, FL: CRC Press, p. 123, 1987.

Dunham, W. *Journey through Genius: The Great Theorems of Mathematics.* New York: Wiley, p. 121, 1990.

Fuhrmann, W. *Synthetische Beweise Planimetrischer Sätze.* Berlin, 1890.

Gürel, E. Solution to Problem 1472. "Maximal Area of Quadrilaterals." *Math. Mag.* **69**, 149, 1996.

Harris, J. W. and Stocker, H. "Quadrilateral of Chords." §3.6.7 in *Handbook of Mathematics and Computational Science.* New York: Springer-Verlag, p. 85, 1998.

Honsberger, R. *More Mathematical Morsels.* Washington, DC: Math. Assoc. Amer., pp. 36–7, 1991.

Honsberger, R. "Cyclic Quadrilaterals." §4.2 in *Episodes in Nineteenth and Twentieth Century Euclidean Geometry.* Washington, DC: Math. Assoc. Amer., pp. 35–0, 1995.

Johnson, R. A. *Modern Geometry: An Elementary Treatise on the Geometry of the Triangle and the Circle.* Boston, MA: Houghton Mifflin, pp. 182–94, 1929.

Wells, D. *The Penguin Dictionary of Curious and Interesting Geometry.* London: Penguin, pp. 11 and 43–4, 1991.

Cyclic Redundancy Check

A sophisticated CHECKSUM (often abbreviated CRC), which is based on the algebra of polynomials over the integers (mod 2). It is substantially more reliable in detecting transmission errors, and is one common error-checking protocol used in modems. The CRC is a form of HASH FUNCTION.

To compare large data blocks using the CRC, first precalculate the CRCs for each block. Two blocks can then be rapidly compared by seeing if their CRCs are equal, saving a great deal of calculation time in most cases. The method is not infallible since for an N-bit checksum, $1/2^N$ of random blocks will have the same checksum for inequivalent data blocks. However, if N is large, the probability that two inequivalent blocks have the same CRC can be made very small.

See also CHECKSUM, ERROR-CORRECTING CODE, HASH FUNCTION

References

Press, W. H.; Flannery, B. P.; Teukolsky, S. A.; and Vetterling, W. T. "Cyclic Redundancy and Other Checksums." Ch. 20.3 in *Numerical Recipes in FORTRAN: The Art of Scientific Computing, 2nd ed.* Cambridge, England: Cambridge University Press, pp. 888–95, 1992.

Cyclic Triple

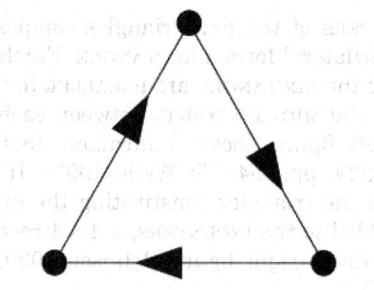

The 3-node TOURNAMENT (and DIRECTED GRAPH) illustrated above (Harary 1994, p. 205).

See also TOURNAMENT, TRANSITIVE TRIPLE

References

Harary, F. "Tournaments." *Graph Theory.* Reading, MA: Addison-Wesley, 1994.

Cyclically Symmetric Plane Partition

A PLANE PARTITION whose solid Young diagram is invariant under the rotation which cyclically permutes the x-, y-, and z-axes. MACDONALD'S PLANE PARTITION CONJECTURE gives a formula for the number of cyclically symmetric plane partitions (CSPPs) of a given integer whose YOUNG DIAGRAMS fit inside an $n \times n \times n$ box. Macdonald gave a product representation for the power series whose coefficients q^n were the number of such partitions of n.

See also MACDONALD'S PLANE PARTITION CONJECTURE, MAGOG TRIANGLE, PLANE PARTITION

References

Bressoud, D. and Propp, J. "How the Alternating Sign Matrix Conjecture was Solved." *Not. Amer. Math. Soc.* **46**, 637–46.

Cyclic-Inscriptable Quadrilateral

BICENTRIC QUADRILATERAL

Cyclid

CYCLIDE

Cyclide

A pair of focal conics which are the envelopes of two one-parameter families of spheres, sometimes also called a CYCLID. The cyclide is a QUARTIC SURFACE, and the lines of curvature on a cyclide are all straight lines or circular arcs (Pinkall 1986). The STANDARD TORI and their INVERSIONS in an INVERSION SPHERE S centered at a point x_0 and of RADIUS r, given by

$$I(x_0, r) = x_0 + \frac{x - x_0 r^2}{|x - x_0|^2},$$

are both cyclides (Pinkall 1986). Illustrated above are RING CYCLIDES, HORN CYCLIDES, and SPINDLE CYCLIDES. The figures on the right correspond to x_0 lying on the torus itself, and are called the PARABOLIC RING CYCLIDE, PARABOLIC HORN CYCLIDE, and PARABOLIC SPINDLE CYCLIDE, respectively.

See also CYCLIDIC COORDINATES, HORN CYCLIDE, INVERSION, INVERSION SPHERE, PARABOLIC HORN CYCLIDE, PARABOLIC RING CYCLIDE, RING CYCLIDE, SPINDLE CYCLIDE, STANDARD TORI

References

Byerly, W. E. *An Elementary Treatise on Fourier's Series, and Spherical, Cylindrical, and Ellipsoidal Harmonics, with Applications to Problems in Mathematical Physics.* New York: Dover, p. 273, 1959.
Eisenhart, L. P. "Cyclides of Dupin." §133 in *A Treatise on the Differential Geometry of Curves and Surfaces.* New York: Dover, pp. 312–14, 1960.
Fischer, G. (Ed.). Plates 71–7 in *Mathematische Modelle / Mathematical Models, Bildband / Photograph Volume.* Braunschweig, Germany: Vieweg, pp. 66–2, 1986.
JavaView. "Classic Surfaces from Differential Geometry: Dupin Cyclide." http://www-sfb288.math.tu-berlin.de/vgp/javaview/demo/surface/common/PaSurface_DupinCycloid.html.
Marsan, A. "Cyclides." http://www.engin.umich.edu/dept/meam/deslab/cadcam/Cyclides/cyclide.html.
Nordstrand, T. "Dupin Cyclide." http://www.uib.no/people/nfytn/dupintxt.htm.
Pinkall, U. "Cyclides of Dupin." §3.3 in *Mathematical Models from the Collections of Universities and Museums* (Ed. G. Fischer). Braunschweig, Germany: Vieweg, pp. 28–0, 1986.

Salmon, G. *Analytic Geometry of Three Dimensions.* New York: Chelsea, p. 527, 1979.
Wells, D. *The Penguin Dictionary of Curious and Interesting Geometry.* London: Penguin, p. 62, 1991.

Cyclidic Coordinates

A general system of fourth-order CURVILINEAR COORDINATES based on the CYCLIDE in which LAPLACE'S EQUATION is SEPARABLE (either simply separable or R-separable). Bôcher (1894) treated all possible systems of this class (Moon and Spencer 1988, p. 49).

See also BICYCLIDE COORDINATES, CAP-CYCLIDE COORDINATES, DISK-CYCLIDE COORDINATES, ORTHOGONAL COORDINATE SYSTEM

References

Bôcher, M. *Über die Reihenentwicklungen der Potentialtheorie.* Leipzig, Germany: Teubner, 1894.
Byerly, W. E. *An Elementary Treatise on Fourier's Series, and Spherical, Cylindrical, and Ellipsoidal Harmonics, with Applications to Problems in Mathematical Physics.* New York: Dover, p. 273, 1959.
Casey, J. "On Cyclides and Sphero-Quartics." *Philos. Trans. Roy. Soc. London* **161**, 585–21, 1871.
Darboux, G. "Remarques sur la théorie des surfaces orthogonales." *Comptes Rendus* **59**, 240–42, 1864.
Darboux, G. "Sur l'application des méthodes de la physique mathématique à l'étude de corps terminés par des cyclides." *Comptes Rendus* **83**, 1037–039, 1864.
Klein, F. *Über lineare Differentialgleichungen der zweiter Ordnung; Vorlesungen gehalten im Sommersemester 1894.* Göttingen, Germany: 1894.
Maxwell, J. C. "On the Cyclide." *Quart. J. Pure Appl. Math.* **9**, 111–26, 1868.
Moon, P. and Spencer, D. E. *Field Theory Handbook, Including Coordinate Systems, Differential Equations, and Their Solutions, 2nd ed.* New York: Springer-Verlag, 1988.
Wangerin. *Preisschriften der Jablanowski'schen Gesellschaft,* No. 18, 1875–876.
Wangerin. *Crelle's J.* **82**, 1875–876.
Wangerin. *Berliner Monatsber.* 1878.

Cycloid

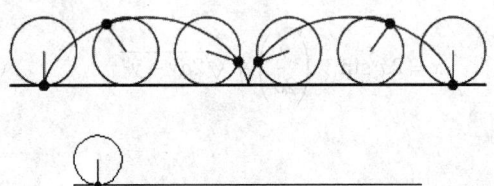

The cycloid is the locus of a point on the rim of a CIRCLE of RADIUS a rolling along a straight LINE. It was studied and named by Galileo in 1599. Galileo attempted to find the AREA by weighing pieces of metal cut into the shape of the cycloid. Torricelli, Fermat, and Descartes all found the AREA. The cycloid was also studied by Roberval in 1634, Wren in 1658, Huygens in 1673, and Johann Bernoulli in 1696. Roberval and Wren found the ARC LENGTH (MacTutor Archive). Gear teeth were also made out of cycloids, as first proposed by Desargues in the 1630s (Cundy and Rollett 1989).

In 1696, Johann Bernoulli challenged other mathematicians to find the curve which solves the BRACHISTOCHRONE PROBLEM, knowing the solution to be a cycloid. Leibniz, Newton, Jakob Bernoulli and L'Hospital all solved Bernoulli's challenge. The cycloid also solves the TAUTOCHRONE PROBLEM, as alluded to in the following passage from *Moby Dick*: "[The try-pot] is also a place for profound mathematical meditation. It was in the left-hand try-pot of the *Pequod*, with the soapstone diligently circling round me, that I was first indirectly struck by the remarkable fact, that in geometry all bodies gliding along a cycloid, my soapstone, for example, will descend from any point in precisely the same time" (Melville 1851). Because of the frequency with which it provoked quarrels among mathematicians in the 17th century, the cycloid became known as the "Helen of Geometers" (Boyer 1968, p. 389).

The cycloid is the CATACAUSTIC of a CIRCLE for a RADIANT POINT on the circumference, as shown by Jakob and Johann Bernoulli in 1692. The CAUSTIC of the cycloid when the rays are parallel to the Y-AXIS is a cycloid with twice as many arches. The RADIAL CURVE of a CYCLOID is a CIRCLE. The EVOLUTE and INVOLUTE of a cycloid are identical cycloids.

If the cycloid has a CUSP at the ORIGIN, its equation in CARTESIAN COORDINATES is

$$x = a \cos^{-1}\left(\frac{a-y}{a}\right) \mp \sqrt{2ay - y^2}. \tag{1}$$

In parametric form, this becomes

$$x = a(t - \sin t) \tag{2}$$

$$y = a(1 - \cos t). \tag{3}$$

If the cycloid is upside-down with a cusp at $(0, a)$, (2) and (3) become

$$x = 2a \sin^{-1}\left(\frac{y}{2a}\right) + \sqrt{2ay - y^2} \tag{4}$$

or

$$x = a(t + \sin t) \tag{5}$$

$$y = a(1 - \cos t) \tag{6}$$

(sign of $\sin t$ flipped for x).

The DERIVATIVES of the parametric representation (2) and (3) are

$$x' = a(1 - \cos t) \tag{7}$$

$$y' = a \sin t \tag{8}$$

$$\frac{dy}{dx} = \frac{y'}{x'} = \frac{a \sin t}{a(1 - \cos t)} = \frac{\sin t}{1 - \cos t} = \frac{2 \sin(\frac{1}{2} t) \cos(\frac{1}{2} t)}{2 \sin^2(\frac{1}{2} t)}$$

$$= \cot(\tfrac{1}{2} t) \tag{9}$$

The squares of the derivatives are

$$x'^2 = a^2(1 - 2 \cos t + \cos^2 t) \tag{10}$$

$$y'^2 = a^2 \sin^2 t, \tag{11}$$

so the ARC LENGTH of a single cycle is

$$L = \int ds = \int_0^{2\pi} \sqrt{x'^2 + y'^2}\, dt$$

$$= a \int_0^{2\pi} \sqrt{(1 - 2 \cos t + \cos^2 t) + \sin^2 t}\, dt$$

$$= a\sqrt{2} \int_0^{2\pi} \sqrt{1 - \cos t}\, dt = 2a \int_0^{2\pi} \sqrt{\frac{1 - \cos t}{2}}\, dt$$

$$= 2a \int_0^{2\pi} \left|\sin(\tfrac{1}{2} t)\right|\, dt. \tag{12}$$

Now let $u \equiv t/2$ so $du = dt/2$. Then

$$L = 4a \int_0^{\pi} \sin u\, du = 4a[-\cos u]_0^{\pi}$$

$$= -4a[(-1) - 1] = 8a. \tag{13}$$

The ARC LENGTH, CURVATURE, and TANGENTIAL ANGLE are

$$s = 8a \sin^2(\tfrac{1}{4} t) \tag{14}$$

$$\kappa = -\tfrac{1}{4} a \csc(\tfrac{1}{2} t) \tag{15}$$

$$\phi = -\tfrac{1}{2} at. \tag{16}$$

The AREA under a single cycle is

$$A = \int_0^{2\pi} y\, dx = a^2 \int_0^{2\pi} (1 - \cos \phi)(1 - \cos \phi)\, d\phi$$

$$= a^2 \int_0^{2\pi} (1 - \cos \phi)^2\, d\phi$$

$$= a^2 \int_0^{2\pi} (1 - 2 \cos \phi + \cos^2 \phi)\, d\phi$$

$$= a^2 \int_0^{2\pi} \{1 - 2 \cos \phi + \tfrac{1}{2}[1 + \cos(2\phi)]\}\, d\phi$$

$$= a^2 \int_0^{2\pi} [\tfrac{3}{2} - 2 \cos \phi + \tfrac{1}{2} \cos(2\phi)]\, d\phi$$

$$= a^2 [\tfrac{3}{2} \phi - 2 \sin \phi + \tfrac{1}{4} \sin(2\phi)]_0^{2\pi}$$

$$= a^2 \tfrac{3}{2} 2\pi = 3\pi a^2. \tag{17}$$

The NORMAL is

$$\hat{\mathbf{T}} = \frac{1}{\sqrt{2 - 2\cos t}} \begin{bmatrix} 1 - \cos t \\ \sin t \end{bmatrix}. \qquad (18)$$

See also BRACHISTOCHRONE PROBLEM, CURTATE CY-
CLOID, CYCLIDE, CYCLOID EVOLUTE, CYCLOID INVO-
LUTE, EPICYCLOID, HYPOCYCLOID, PROLATE CYCLOID,
TAUTOCHRONE PROBLEM, TROCHOID

References

Beyer, W. H. *CRC Standard Mathematical Tables, 28th ed.*
Boca Raton, FL: CRC Press, p. 216, 1987.
Bogomolny, A. "Cycloids." http://www.cut-the-knot.com/
pythagoras/cycloids.html.
Boyer, C. B. *A History of Mathematics.* New York: Wiley,
1968.
Cundy, H. and Rollett, A. "Cycloid." §5.1.6 in *Mathematical
Models, 3rd ed.* Stradbroke, England: Tarquin Pub.,
pp. 215–16, 1989.
Gardner, M. "The Cycloid: Helen of Geometers." Ch. 13 in
*The Sixth Book of Mathematical Games from Scientific
American.* Chicago, IL: University of Chicago Press,
pp. 127–34, 1984.
Gray, A. "Cycloids." §3.1 in *Modern Differential Geometry of
Curves and Surfaces with Mathematica, 2nd ed.* Boca
Raton, FL: CRC Press, pp. 50–2, 1997.
Harris, J. W. and Stocker, H. *Handbook of Mathematics and
Computational Science.* New York: Springer-Verlag,
p. 325, 1998.
Lawrence, J. D. *A Catalog of Special Plane Curves.* New
York: Dover, pp. 192 and 197, 1972.
Lockwood, E. H. "The Cycloid." Ch. 9 in *A Book of Curves.*
Cambridge, England: Cambridge University Press,
pp. 80–9, 1967.
MacTutor History of Mathematics Archive. "Cycloid." http://
www-groups.dcs.st-and.ac.uk/~history/Curves/Cy-
cloid.html.
Melville, H. "The Tryworks." Ch. 96 in *Moby Dick.* New
York: Bantam, 1981. Originally published in 1851.
Muterspaugh, J.; Driver, T.; and Dick, J. E. "The Cycloid
and Tautochronism." http://php.indiana.edu/~jedick/pro-
ject/intro.html.
Pappas, T. "The Cycloid--The Helen of Geometry." *The Joy of
Mathematics.* San Carlos, CA: Wide World Publ./Tetra,
pp. 6–, 1989.
Phillips, J. P. "Brachistochrone, Tautochrone, Cycloid--Ap-
ple of Discord." *Math. Teacher* **60**, 506–08, 1967.
Proctor, R. A. *A Treatise on the Cycloid.* London: Longmans,
Green, 1878.
Steinhaus, H. *Mathematical Snapshots, 3rd ed.* New York:
Dover, p. 147, 1999.
Wagon, S. "Rolling Circles." Ch. 2 in *Mathematica in Action.*
New York: W. H. Freeman, pp. 39–6, 1991.
Wells, D. *The Penguin Dictionary of Curious and Interesting
Geometry.* London: Penguin, pp. 44–7, 1991.
Whitman, E. A. "Some Historical Notes on the Cycloid."
Amer. Math. Monthly **50**, 309–15, 1948.
Yates, R. C. "Cycloid." *A Handbook on Curves and Their
Properties.* Ann Arbor, MI: J. W. Edwards, pp. 65–0,
1952.
Zwillinger, D. (Ed.). *CRC Standard Mathematical Tables
and Formulae.* Boca Raton, FL: CRC Press, pp. 291–92,
1995.

Cycloid Evolute

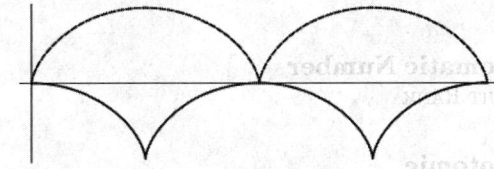

The EVOLUTE of the CYCLOID

$$x(t) = a(t - \sin t)$$
$$y(t) = a(1 - \cos t)$$

is given by

$$x(t) = a(t + \sin t)$$
$$y(t) = a(\cos t - 1).$$

As can be seen in the above figure, the EVOLUTE is
simply a shifted copy of the original CYCLOID, so the
CYCLOID is its own EVOLUTE.

Cycloid Involute

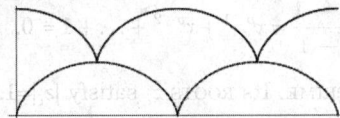

The INVOLUTE of the CYCLOID

$$x(t) = a(t - \sin t)$$
$$y(t) = a(1 - \cos t)$$

is given by

$$x(t) = a(t + \sin t)$$
$$y(t) = a(3 + \cos t).$$

As can be seen in the above figure, the INVOLUTE is
simply a shifted copy of the original CYCLOID, so the
CYCLOID is its own INVOLUTE!

Cycloid Radial Curve

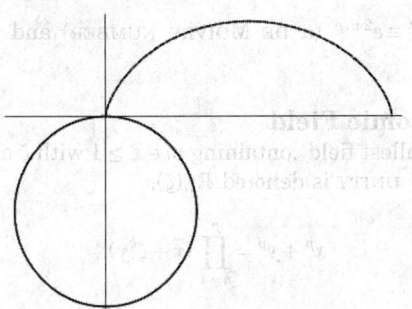

The RADIAL CURVE of the CYCLOID is the CIRCLE

$$x = x_0 + 2a \sin \phi$$

$$y = -2a + y_0 + 2a \cos \phi.$$

Cyclomatic Number
CIRCUIT RANK

Cyclotomic
CYCLOTOMIC POLYNOMIAL

Cyclotomic Equation
The equation

$$x^p = 1,$$

where solutions $\zeta_k = e^{2\pi i k/p}$ are the ROOTS OF UNITY sometimes called DE MOIVRE NUMBERS. Gauss showed that the cyclotomic equation can be reduced to solving a series of QUADRATIC EQUATIONS whenever p is a FERMAT PRIME. Wantzel (1836) subsequently showed that this condition is not only SUFFICIENT, but also NECESSARY. An "irreducible" cyclotomic equation is an expression OF THE FORM

$$\frac{x^p - 1}{x - 1} = x^{p-1} + x^{p-2} + \ldots + 1 = 0,$$

where p is PRIME. Its ROOTS z_i satisfy $|z_i| = 1$.

See also CYCLOTOMIC POLYNOMIAL, DE MOIVRE NUMBER, POLYGON, PRIMITIVE ROOT OF UNITY

References
Courant, R. and Robbins, H. *What is Mathematics?: An Elementary Approach to Ideas and Methods, 2nd ed.* Oxford, England: Oxford University Press, pp. 99–00, 1996.
Scott, C. A. "The Binomial Equation $x^p - 1 = 0$." *Amer. J. Math.* **8**, 261–64, 1886.
Wantzel, M. L. "Recherches sur les moyens de reconnaître si un Problème de Géométrie peut se résoudre avec la règle et le compas." *J. Math. pures appliq.* **1**, 366–72, 1836.

Cyclotomic Factorization

$$z^p - y^p = (z - y)(z - \zeta y) \cdots (z - \zeta^{p-1} y),$$

where $\zeta \equiv e^{2\pi i/p}$ (a DE MOIVRE NUMBER) and p is a PRIME.

Cyclotomic Field
The smallest field containing $m \in \mathbb{Z} \geq 1$ with ζ a PRIME ROOT OF UNITY is denoted $\mathbb{R}_m(\zeta)$,

$$x^p + y^p = \prod_{k=1}^{p} (x + \zeta^k y).$$

Specific cases are

$$\mathbb{R}_3 = \mathbb{Q}(\sqrt{-3})$$

$$\mathbb{R}_4 = \mathbb{Q}(\sqrt{-1})$$

$$\mathbb{R}_6 = \mathbb{Q}(\sqrt{-3}),$$

where $\mathbb{Q}$ denotes a QUADRATIC FIELD.

References
Koch, H. "Cyclotomic Fields." §6.4 in *Number Theory: Algebraic Numbers and Functions.* Providence, RI: Amer. Math. Soc., pp. 180–84, 2000.
Weiss, E. *Algebraic Number Theory.* New York: Dover, 1998.

Cyclotomic Integer
A number OF THE FORM

$$a_0 + a_1 \zeta + \ldots + a_{p-1} \zeta^{p-1},$$

where

$$\zeta \equiv e^{2\pi i/p}$$

is a DE MOIVRE NUMBER and p is a PRIME NUMBER. Unique factorizations of cyclotomic INTEGERS fail for $p > 23$.

Cyclotomic Invariant
Let p be an ODD PRIME and F_n the CYCLOTOMIC FIELD of p^{n+1}th ROOTS of unity over the rational FIELD. Now let $p^{e(n)}$ be the POWER of p which divides the CLASS NUMBER h_n of F_n. Then there exist INTEGERS μ_p, $\lambda_p \geq 0$ and ν_p such that

$$e(n) = \mu_p p^n + \lambda_p n + \nu_p$$

for all sufficiently large n. For REGULAR PRIMES, $\mu_p = \lambda_p = \nu_p = 0$.

References
Johnson, W. "Irregular Primes and Cyclotomic Invariants." *Math. Comput.* **29**, 113–20, 1975.

Cyclotomic Number
DE MOIVRE NUMBER, SYLVESTER CYCLOTOMIC NUMBER

Cyclotomic Polynomial
A polynomial given by

$$\Phi_n(x) = \prod_{k=1}^{n} {}' (x - \zeta_k), \qquad (1)$$

where ζ_k are the ROOTS OF UNITY in $\mathbb{C}$ given by

$$\zeta_k \equiv e^{2\pi i k/n} \qquad (2)$$

and k runs over integers RELATIVELY PRIME to n. The prime may be dropped if the product is instead taken over PRIMITIVE ROOTS OF UNITY, so that

$$\Phi_n(x) = \prod_{\substack{k=1 \\ \text{primitive} \\ \zeta_k}}^{n} (x - \zeta_k). \qquad (3)$$

The notation $F_n(x)$ is also frequently encountered. Dickson *et al.* (1923) and Apostol (1975) give extensive bibliographies for cyclotomic polynomials.

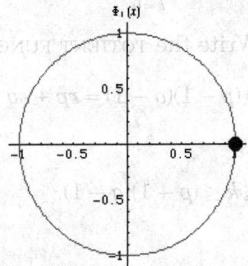

$\Phi_n(x)$ is an INTEGER POLYNOMIAL and an IRREDUCIBLE POLYNOMIAL with DEGREE $\phi(n)$, where $\phi(n)$ is the TOTIENT FUNCTION. Cyclotomic polynomials are returned by the *Mathematica* command Cyclotomic[n, x]. The roots of cyclotomic polynomials lie on the UNIT CIRCLE in the COMPLEX PLANE, as illustrated above for the first few cyclotomic polynomials.

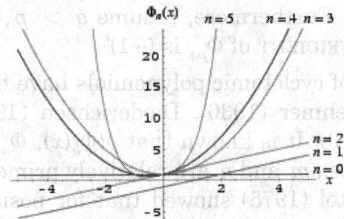

The first few cyclotomic POLYNOMIALS are

$$\Phi_1(x) = x - 1$$

$$\Phi_2(x) = x + 1$$

$$\Phi_3(x) = x^2 + x + 1$$

$$\Phi_4(x) = x^2 + 1$$

$$\Phi_5(x) = x^4 + x^3 + x^2 + x + 1$$

$$\Phi_6(x) = x^2 - x + 1$$

$$\Phi_7(x) = x^6 + x^5 + x^4 + x^3 + x^2 + x + 1$$

$$\Phi_8(x) = x^4 + 1$$

$$\Phi_9(x) = x^6 + x^3 + 1$$

$$\Phi_{10}(x) = x^4 - x^3 + x^2 - x + 1.$$

If p is an ODD PRIME, then

$$\Phi_p(x) = \frac{x^p - 1}{x - 1} = x^{p-1} + x^{p-2} + \ldots + x + 1 \tag{4}$$

$$\Phi_{2p}(x) = \frac{x^{2p} - 1}{x^p - 1} \frac{x - 1}{x^2 - 1} = x^{p-1} - x^{p-2} + \ldots - x + 1 \tag{5}$$

$$\Phi_{4p}(x) = \frac{x^{4p} - 1}{x^{2p} - 1} \frac{x^2 - 1}{x^4 - 1}$$

$$= x^{2p-2} - x^{2p-4} + \ldots - x^2 + 1 \tag{6}$$

(Riesel 1994, p. 306). Similarly, for p again an ODD PRIME,

$$x^p - 1 = \Phi_1(x)\Phi_p(x) \tag{7}$$

$$x^{2p} - 1 = \Phi_1(x)\Phi_2(x)\Phi_p(x)\Phi_{2p}(x) \tag{8}$$

$$x^{4p} - 1 = \Phi_1(x)\Phi_4(x)\Phi_2(x)\Phi_p(x)\Phi_{2p}(x)\Phi_{4p}(x). \tag{9}$$

For the first few remaining values of n,

$$x - 1 = \Phi_1(x) \tag{10}$$

$$x^2 - 1 = \Phi_1(x)\Phi_2(x) \tag{11}$$

$$x^4 - 1 = \Phi_1(x)\Phi_2(x)\Phi_4(x) \tag{12}$$

$$x^8 - 1 = \Phi_1(x)\Phi_2(x)\Phi_4(x)\Phi_8(x) \tag{13}$$

$$x^9 - 1 = \Phi_1(x)\Phi_3(x)\Phi_9(x) \tag{14}$$

$$x^{15} - 1 = \Phi_1(x)\Phi_3(x)\Phi_5(x)\Phi_{15}(x) \tag{15}$$

$$x^{16} - 1 = \Phi_1(x)\Phi_2(x)\Phi_4(x)\Phi_8(x)\Phi_{16}(x) \tag{16}$$

$$x^{18} - 1 = \Phi_1(x)\Phi_2(x)\Phi_3(x)\Phi_6(x)\Phi_9(x)\Phi_{18}(x) \tag{17}$$

(Riesel 1994, p. 307).

For p a PRIME relatively prime to n,

$$F_{np}(x) = \frac{F_n(x^p)}{F_n(x)}, \tag{18}$$

but if $p|n$,

$$F_{np}(x) = F_n(x^p) \tag{19}$$

(Nagell 1951, p. 160).

An explicit equation for $\Phi_n(x)$ for SQUAREFREE n is given by

$$\Phi_n(x) = \sum_{j=0}^{\phi(n)} a_{nj} z^{\phi(n) - j}, \tag{20}$$

where A_{nj} is calculated using the RECURRENCE RELATION

$$a_{nj} =$$

$$-\frac{\mu(n)}{j} \sum_{m=0}^{j-1} a_{nm} \mu(\text{GCD}(n, j-m)) \phi(\text{GCD}(n, j-m)), \tag{21}$$

with $a_{n0} = 1$, where μ_n is the MÖBIUS FUNCTION and GCD(m, n) is the GREATEST COMMON DENOMINATOR of m and n.

The POLYNOMIAL x^{n-1} can be factored as

$$x^n - 1 = \prod_{d|n} \Phi_d(x), \tag{22}$$

where $\Phi_d(x)$ is a CYCLOTOMIC POLYNOMIAL. Furthermore,

$$x^n + 1 = \frac{x^{2n} - 1}{x^n - 1} = \frac{\prod_{d|2n} \Phi_d(x)}{\prod_{d|n} \Phi_d(x)}. \quad (23)$$

The COEFFICIENTS of the inverse of the cyclotomic POLYNOMIAL

$$\frac{1}{1 + x + x^2} = 1 - x + x^3 - x^4 + x^6 - x^7 + x^9 - x^{10} + \dots$$

$$\equiv \sum_{n=0}^{\infty} c_n x^n \quad (24)$$

can also be computed from

$$c_n = 1 - 2\left\lfloor \tfrac{1}{3}(n+2) \right\rfloor + \left\lfloor \tfrac{1}{3}(n+1) \right\rfloor + \left\lfloor \tfrac{1}{3}n \right\rfloor \quad (25)$$

$$= 1 - 3\left\lfloor \tfrac{1}{3}(n+2) \right\rfloor + \lfloor n \rfloor \quad (26)$$

$$= \frac{2}{\sqrt{3}} \sin[\tfrac{2}{3}\pi(n+1)], \quad (27)$$

where $\lfloor x \rfloor$ is the FLOOR FUNCTION.

The LOGARITHM of the cyclotomic polynomial

$$\Phi_n(x) = \prod_{d|n} (1 - x^{n/d})^{\mu(d)} \quad (28)$$

is the MÖBIUS INVERSION FORMULA (Vardi 1991, p. 225).

For p PRIME,

$$\Phi_p(x) = \sum_{k=0}^{p-1} x^k, \quad (29)$$

i.e., the coefficients are all 1. The first cyclotomic polynomial to have a coefficient other than ± 1 and 0 is $\Phi_{105}(x)$, which has coefficients of -2 for x^7 and x^{41}. This is true because 105 is the first number to have three distinct ODD PRIME factors, i.e., T_d (McClellan and Rader 1979, Schroeder 1997). The smallest values of n for which $\Phi_n(x)$ has one or more coefficients ± 1, ± 2, ± 3, ... are 0, 105, 385, 1365, 1785, 2805, 3135, 6545, 6545, 10465, 10465, 10465, 10465, 11305, ... (Sloane's A013594).

It appears to be true that, for m, $n > 1$, if $\Phi_m(x) + \Phi_n(x)$ factors, then the factors contain a cyclotomic polynomial. For example,

$$\Phi_7(x) + \Phi_{22}(x) = (x^2 + 1)(x^8 - x^7 + 2x^4 + 2)$$

$$= \Phi_4(x)(x^8 - x^7 + 2x^4 + 2). \quad (30)$$

This observation has been checked up to m, $n = 150$ (C. Nicol). If m and n are prime, then $C_m + C_n$ is irreducible.

Migotti (1883) showed that COEFFICIENTS of $\Phi_{pq}(x)$ for p and q distinct PRIMES can be only 0, ± 1. Lam and

Leung (1996) considered

$$\Phi_{pq}(x) \equiv \sum_{k=0}^{pq-1} a_k x^k \quad (31)$$

for p, q PRIME. Write the TOTIENT FUNCTION as

$$\phi(pq) = (p-1)(q-1) = rp + sq \quad (32)$$

and let

$$0 \le k \le (p-1)(q-1), \quad (33)$$

then

1. $a_k = 1$ IFF $k = ip + jq$ for some $i \in [0, r]$ and $j \in [0, s]$,
2. $a_k = -1$ IFF $k + pq = ip + jp$ for $i \in [r+1, q-1]$ and $j \in [s+1, p-1]$,
3. otherwise $a_k = 0$.

The number of terms having $a_k = 1$ is $(r+1)(s+1)$, and the number of terms having $a_k = -1$ is $(p-s-1)(q-r-1)$. Furthermore, assume $q > p$, then the middle COEFFICIENT of Φ_{pq} is $(-1)^r$.

Resultants of cyclotomic polynomials have been computed by Lehmer (1930), Diederichsen (1940), and Apostol (1970). It is known that $\rho(\Phi_k(x), \Phi_n(x)) = 1$ if $(m, n) = 1$, i.e., m and n are relatively prime (Apostol 1975). Apostol (1975) showed that for positive integers m and n and arbitrary nonzero complex numbers a and b,

$$\rho(\Phi_m(ax), \Phi_n(bx))$$

$$= b^{\phi(m)\phi(n)} \prod_{d|n} \left[\Phi_{m/\delta}\left(\frac{a^d}{b^d}\right) \right]^{\mu(n/d)\phi(m)/\phi(m/\delta)}, \quad (34)$$

where $\delta = \text{GCD}(m, d)$ is the GREATEST COMMON DIVISOR of m and d, $\phi(n)$ is the TOTIENT FUNCTION, $\mu(n)$ is the MÖBIUS FUNCTION, and the product is over the divisors of n. If m and n are distinct primes p and q, then (34) simplifies to

$$\rho(\Phi_q(ax), \Phi_p(bx))$$

$$= \begin{cases} \dfrac{a^{pq} - b^{pq}}{a^p - b^p} \dfrac{a - b}{a^q - b^q} & \text{for } a \neq b \\ a^{(p-1)(q-1)} & \text{for } a = b. \end{cases} \quad (35)$$

The following table gives the RESULTANTS $\rho(\Phi_k(x), \Phi_n(x))$ (Sloane's A054372).

$k \backslash n$	1	2	3	4	5	6	7
1	0						
2	2	0					
3	3	1	0				
4	2	2	1	0			

```
5 5 1 1 1 0
6 1 3 4 1 1 0
7 7 1 1 1 1 1 0
```

See also AURIFEUILLEAN FACTORIZATION, GAUSS'S CYCLOTOMIC FORMULA, LUCAS'S THEOREM, MÖBIUS INVERSION FORMULA, PRIMITIVE ROOT OF UNITY, ROOT OF UNITY

References

Apostol, T. M. "Resultants of Cyclotomic Polynomials." *Proc. Amer. Math. Soc.* **24**, 457–62, 1970.

Apostol, T. M. "The Resultant of the Cyclotomic Polynomials $F_m(ax)$ and $F_n(bx)$." *Math. Comput.* **29**, 1–, 1975.

Beiter, M. "The Midterm Coefficient of the Cyclotomic Polynomial $F_{pq}(x)$." *Amer. Math. Monthly* **71**, 769–70, 1964.

Beiter, M. "Magnitude of the Coefficients of the Cyclotomic Polynomial F_{pq}." *Amer. Math. Monthly* **75**, 370–72, 1968.

Bloom, D. M. "On the Coefficients of the Cyclotomic Polynomials." *Amer. Math. Monthly* **75**, 372–77, 1968.

Brent, R. P. "On Computing Factors of Cyclotomic Polynomials." *Math. Comput.* **61**, 131–49, 1993.

Carlitz, L. "The Number of Terms in the Cyclotomic Polynomial $F_{pq}(x)$." *Amer. Math. Monthly* **73**, 979–81, 1966.

Conway, J. H. and Guy, R. K. *The Book of Numbers.* New York: Springer-Verlag, 1996.

de Bruijn, N. G. "On the Factorization of Cyclic Groups." *Indag. Math.* **15**, 370–77, 1953.

Dickson, L. E.; Mitchell, H. H.; Vandiver, H. S.; and Wahlin, G. E. *Algebraic Numbers.* Bull Nat. Res. Council, Vol. 5, Part 3, No. 28. Washington, DC: National Acad. Sci., 1923.

Diederichsen, F.-E. "Über die Ausreduktion ganzzahliger Gruppendarstellungen bei arithmetischer Äquivalenz." *Abh. Math. Sem. Hanisches Univ.* **13**, 357–12, 1940.

Lam, T. Y. and Leung, K. H. "On the Cyclotomic Polynomial $\Phi_{pq}(X)$." *Amer. Math. Monthly* **103**, 562–64, 1996.

Lehmer, E. "On the Magnitude of the Coefficients of the Cyclotomic Polynomial." *Bull. Amer. Math. Soc.* **42**, 389–92, 1936.

Lehmer, E. "On the Magnitude of Coefficients of the Cyclotomic Polynomial." *Bull. Amer. Math. Soc.* **42**, 389–92, 1936.

McClellan, J. H. and Rader, C. *Number Theory in Digital Signal Processing.* Englewood Cliffs, NJ: Prentice-Hall, 1979.

Migotti, A. "Zur Theorie der Kreisteilungsgleichung." *Sitzber. Math.-Naturwiss. Classe der Kaiser. Akad. der Wiss., Wien* **87**, 7–4, 1883.

Nagell, T. "The Cyclotomic Polynomials" and "The Prime Divisors of the Cyclotomic Polynomial." §46 and 48 in *Introduction to Number Theory.* New York: Wiley, pp. 158–60 and 164–68, 1951.

Riesel, H. "The Cyclotomic Polynomials" in Appendix 6. *Prime Numbers and Computer Methods for Factorization,* 2nd ed. Boston, MA: Birkhäuser, pp. 305–08, 1994.

Schroeder, M. R. *Number Theory in Science and Communication, with Applications in Cryptography, Physics, Digital Information, Computing, and Self-Similarity,* 3rd ed. New York: Springer-Verlag, p. 245, 1997.

Séroul, R. "Cyclotomic Polynomials." §10.8 in *Programming for Mathematicians.* Berlin: Springer-Verlag, pp. 265–69, 2000.

Sloane, N. J. A. Sequences A013594 and A054372 in "An On-Line Version of the Encyclopedia of Integer Sequences." http://www.research.att.com/~njas/sequences/eisonline.html.

Vardi, I. *Computational Recreations in Mathematica.* Redwood City, CA: Addison-Wesley, pp. 8 and 224–25, 1991.

Cylinder

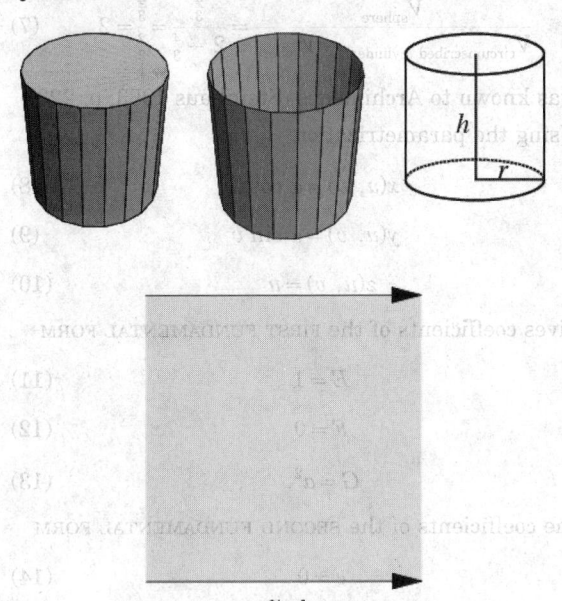

cylinder

In common usage, the term "cylinder" refers to a SOLID of circular CROSS SECTION in which the centers of the CIRCLES all lie on a single LINE (i.e., a right circular cylinder). In mathematical usage, "cylinder" is commonly taken to refer to only the lateral sides of this solid, excluding the top and bottom caps. If a plane inclined with respect to the caps intersects a cylinder, it does so in an ELLIPSE. The cylinder was extensively studied by Archimedes in his two-volume work *On the Sphere and Cylinder* in ca. 225 BC.

A cylinder is called a right cylinder if it is "straight" in the sense that its CROSS SECTIONS lie directly on top of each other; otherwise, the cylinder is called oblique. The lateral surface of a cylinder of height h and RADIUS r can be described parametrically by

$$x = r \cos \theta \qquad (1)$$

$$y = r \sin \theta \qquad (2)$$

$$z = z, \qquad (3)$$

for $z \in [0, h]$ and $\theta \in [0, 2\pi)$. These are the basis for CYLINDRICAL COORDINATES. The SURFACE AREA (of the sides) and VOLUME of the cylinder of height h and RADIUS r are

$$S = 2\pi r h \qquad (4)$$

$$V = \pi r^2 h. \qquad (5)$$

Therefore, if top and bottom caps are added, the volume-to-surface area ratio for a cylindrical solid is

$$\frac{V}{S} = \frac{\pi r^2 h}{2\pi rh + 2\pi r^2} = \frac{1}{2}\left(\frac{1}{r} + \frac{1}{h}\right)^{-1}, \qquad (6)$$

which is related to the HARMONIC MEAN of the radius r and height h. The fact that

$$\frac{V_{\text{sphere}}}{V_{\text{circumscribed cylinder}} - V_{\text{sphere}}} = \frac{\frac{4}{3}}{2 - \frac{4}{3}} = \frac{\frac{4}{3}}{\frac{2}{3}} = 2 \qquad (7)$$

was known to Archimedes (Steinhaus 1983, p. 223).

Using the parametrization

$$x(u, v) = a \cos v \qquad (8)$$

$$y(u, v) = a \sin v \qquad (9)$$

$$z(u, v) = u \qquad (10)$$

gives coefficients of the FIRST FUNDAMENTAL FORM

$$E = 1 \qquad (11)$$

$$F = 0 \qquad (12)$$

$$G = a^2, \qquad (13)$$

the coefficients of the SECOND FUNDAMENTAL FORM

$$e = 0 \qquad (14)$$

$$f = 0 \qquad (15)$$

$$g = a \qquad (16)$$

AREA ELEMENT

$$dS = a \, du \wedge dv, \qquad (17)$$

GAUSSIAN CURVATURE

$$K = 0, \qquad (18)$$

and MEAN CURVATURE

$$H = \frac{1}{2a}. \qquad (19)$$

See also ARCHIMEDES' HAT-BOX THEOREM, BARREL, CONE, CYLINDER DISSECTION, CYLINDER-SPHERE INTERSECTION, CYLINDRICAL SEGMENT, CYLINDRICAL WEDGE, ELLIPTIC CYLINDER, GENERALIZED CYLINDER, SPHERE, STEINMETZ SOLID, VIVIANI'S CURVE

References
Beyer, W. H. (Ed.). *CRC Standard Mathematical Tables, 28th ed.* Boca Raton, FL: CRC Press, p. 129, 1987.
Harris, J. W. and Stocker, H. "Cylinder." §4.6 in *Handbook of Mathematics and Computational Science.* New York: Springer-Verlag, pp. 102–04, 1998.
Hilbert, D. and Cohn-Vossen, S. "The Cylinder, the Cone, the Conic Sections, and Their Surfaces of Revolution." §2 in *Geometry and the Imagination.* New York: Chelsea, pp. 7–1, 1999.

JavaView. "Classic Surfaces from Differential Geometry: Cylinder." http://www-sfb288.math.tu-berlin.de/vgp/javaview/demo/surface/common/PaSurface_Cylinder.html.
Kern, W. F. and Bland, J. R. "Circular Cylinder" and "Right Circular Cylinder." §16–7 in *Solid Mensuration with Proofs, 2nd ed.* New York: Wiley, pp. 36–2, 1948.
Steinhaus, H. *Mathematical Snapshots, 3rd ed.* New York: Dover, 1999.

Cylinder Cutting

The maximum number of pieces into which a cylinder can be divided by n oblique cuts is given by

$$f(n) = \binom{n+1}{3} + n + 1 = \tfrac{1}{6}(n+2)(n+3),$$

where $\binom{a}{b}$ is a BINOMIAL COEFFICIENT. This problem is sometimes also called cake cutting or pie cutting, and has the same solution as SPACE DIVISION BY PLANES. For $n = 1, 2, \ldots$ cuts, the maximum number of pieces is 2, 4, 8, 15, 26, 42, ... (Sloane's A000125).

See also CIRCLE DIVISION BY LINES, CUBE DIVISION BY PLANES, HAM SANDWICH THEOREM, PANCAKE THEOREM, SPACE DIVISION BY PLANES, TORUS CUTTING

References
Bogomolny, A. "Can You Cut a Cake into 8 Pieces with Three Movements." http://www.cut-the-knot.com/do_you_know/cake.html.
Sloane, N. J. A. Sequences A000125/M1100 in "An On-Line Version of the Encyclopedia of Integer Sequences." http://www.research.att.com/~njas/sequences/eisonline.html.

Cylinder Dissection

A cylinder can be dissected into unequal squares, with nine squares required at a minimum. Trivial squarings can be constructed by taking rectangle dissections and matching edges, but there are two nontrivial nine-square tilings (Stewart 1997).

See also MÖBIUS STRIP DISSECTION, PERFECT SQUARE DISSECTION, TORUS DISSECTION

References
Stewart, I. "Squaring the Square." *Sci. Amer.* **277**, 94–6, July 1997.

Cylinder Function

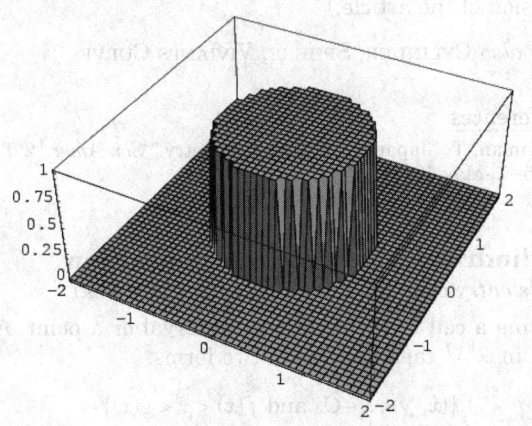

The cylinder function is defined as

$$C(x, y) \equiv \begin{cases} 1 & \text{for } \sqrt{x^2 + y^2} \le a \\ 0 & \text{for } \sqrt{x^2 + y^2} > a. \end{cases} \tag{1}$$

The BESSEL FUNCTIONS are sometimes also called cylinder functions. To find the FOURIER TRANSFORM of the cylinder function, let

$$k_x = k \cos \alpha \tag{2}$$

$$k_y = k \sin \alpha \tag{3}$$

$$x = r \cos \theta \tag{4}$$

$$y = r \sin \theta. \tag{5}$$

Then

$$\begin{aligned} F(k, a) &= \mathscr{F}(C(x, y)) \\ &= \int_0^{2\pi} \int_0^a e^{i(k \cos \alpha \, r \cos \theta + k \sin \alpha \, r \sin \theta)} r \, dr \, d\theta \\ &= \int_0^{2\pi} \int_0^a e^{ikr \cos(\theta - \alpha)} r \, dr \, d\theta. \end{aligned} \tag{6}$$

Let $b = \theta - \alpha$, so $db = d\theta$. Then

$$\begin{aligned} F(k, a) &= \int_{-\alpha}^{2\pi - \alpha} \int_0^a e^{ikr \cos b} r \, dr \, d\theta \\ &= \int_0^{2\pi} \int_0^a e^{ikr \cos b} r \, dr \, d\theta \\ &= 2\pi \int_0^a J_0(kr) r \, dr, \end{aligned} \tag{7}$$

where $J_0(x)$ is a zeroth order BESSEL FUNCTION OF THE FIRST KIND. Let $u \equiv kr$, so $du = k \, dr$, then

$$\begin{aligned} F(k, a) &= \frac{2\pi}{k^2} \int_0^{ka} J_0(u) u \, du = \frac{2\pi}{k^2} [u J_1(u)]_0^{ka} \\ &= \frac{2\pi a}{k} J_1(ka) = 2\pi a^2 \frac{J_1(ka)}{ka}. \end{aligned} \tag{8}$$

As defined by Watson (1966), a "cylinder function" is any function which satisfies the RECURRENCE RELATIONS

$$\mathscr{C}_{\nu-1}(z) + \mathscr{C}_{\nu+1}(z) = \frac{2\nu}{z} \mathscr{C}_\nu(z) \tag{9}$$

$$\mathscr{C}_{\nu-1}(z) - \mathscr{C}_{\nu+1}(z) = 2\mathscr{C}_\nu'(z). \tag{10}$$

This class of functions can be expressed in terms of BESSEL FUNCTIONS.

See also BESSEL FUNCTION OF THE FIRST KIND, CYLINDER FUNCTION, CYLINDRICAL FUNCTION, HEMISPHERICAL FUNCTION

References

Watson, G. N. *A Treatise on the Theory of Bessel Functions,* *2nd ed.* Cambridge, England: Cambridge University Press, 1966.

Cylinder-Cylinder Intersection

STEINMETZ SOLID

Cylinder-Plane Intersection

CYLINDRICAL SECTION

Cylinder-Sphere Intersection

The curve formed by the intersection of a CYLINDER and a SPHERE is known as VIVIANI'S CURVE.

The problem of finding the lateral SURFACE AREA of a CYLINDER of radius r internally tangent to a SPHERE of radius R was given in a SANGAKU PROBLEM from 1825.

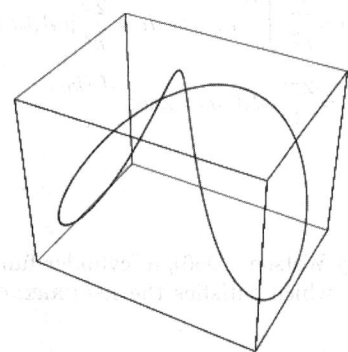

The easiest way to determine the solution is to solve the simultaneous equations

$$x^2 + y^2 + z^2 = R^2 \tag{1}$$

$$y^2 + [z - (R - r)]^2 = r^2 \tag{2}$$

for x and y,

$$x = \pm\sqrt{2(R - r)(R - z)} \tag{3}$$

$$y = \pm\sqrt{(R - z)(2r - R - z)}. \tag{4}$$

These give the PARAMETRIC EQUATIONS for VIVIANI'S CURVE in this case (left figure). The SURFACE AREA can the be found by constructing a series of curved segments (right figure). The arc length element around the surface of the cylinder at a height z is given by

$$ds = \sqrt{1 + \left(\frac{dy}{dz}\right)^2}\, dz = \frac{r}{\sqrt{(R - z)(2r - R + z)}}. \tag{5}$$

The SURFACE AREA of one quarter of the surface is then

$$S_{1/4} = \int x(z)\, ds$$

$$= \int_{R-2r}^{R} \sqrt{2(R - r)(R - z)}\, \frac{r}{\sqrt{(R - z)(2r - R - z)}}\, dz$$

$$= \int_{R-2r}^{R} r\sqrt{\frac{2(R - r)}{2r - R + z}}\, dz, \tag{6}$$

where some care is needed treating the lower limit,

$$S_{1/4} = \lim_{r' \to r^-} 4r[\sqrt{r(R - r)} - \sqrt{(R - r)(r - r')}]$$

$$= 4r^{3/2}\sqrt{R - r}. \tag{7}$$

The total SURFACE AREA is then

$$S = 4S_{1/4} = 16r^{3/2}\sqrt{R - r} \tag{8}$$

a result obtained in a more roundabout geometric arguments by Rothman (1998). (Note that the answer printed in the original Rothman article was incorrect;

the corrected answer has been posted on the Internet version of the article.)

See also CYLINDER, SPHERE, VIVIANI'S CURVE

References

Rothman, T. "Japanese Temple Geometry." *Sci. Amer.* **278**, 85–1, May 1998.

Cylindrical Algebraic Decomposition

This entry contributed by ADAM STRZEBONSKI

Define a cell in $\mathbb{R}^1$ as an open interval or a point. A cell in $\mathbb{R}^{k+1}$ then has one of two forms,

$$\{(x, y) : x \in C, \text{ and } f(x) < y < g(x)\}$$

or

$$\{(x, y) : x \in C, \text{ and } y = f(x)\},$$

where $x = \{x_1, \ldots, x_k\}$, C is a cell in $\mathbb{R}^k$, f and g are either (1) continuous functions on C such that for some polynomials F and G, $F(x, f(x)) = 0$ and $G(x, g(x)) = 0$, or (2) $\pm\infty$, and $f(x) < g(x)$ for all $x \in C$.

A cylindrical algebraic decomposition of $S \subset \mathbb{R}^n$ is a representation of S as a finite union of disjoint cells. Let F be finite set of polynomials in n variables. A cylindrical algebraic decomposition of $S \subset \mathbb{R}^n$ is said to be F-invariant if each of the polynomials from F has a constant sign on each cell of the decomposition.

The cylindrical algebraic decomposition (CAD) algorithm, given a finite set F of polynomials in n variables, computes an F-invariant cylindrical algebraic decomposition of $\mathbb{R}^n$. Given a logical combination of polynomial equations and inequalities in n real unknowns, one can use the CAD algorithm to find a cylindrical algebraic decomposition of its solution set. For example, the decomposition of

$$x^2 + y^2 + z^2 < 1$$

is given by

$$\begin{cases} -1 < x < 1 \\ 1 - \sqrt{1 - x^2} < y < \sqrt{1 - x^2} \\ -\sqrt{1 - x^2 - y^2} < z < \sqrt{1 - x^2 - y^2}. \end{cases}$$

Mathematica 4.0 contains the function Cylindri-calAlgebraicDecomposition which performs cylindrical algebraic decompositions. Although the process is algorithmic, it becomes computationally infeasible for complicated inequalities.

See also CYLINDRICAL PARTS, GENERIC CYLINDRICAL ALGEBRAIC DECOMPOSITION, QUANTIFIER ELIMINATION, TARSKI'S THEOREM

References

Caviness, B. F. and Johnson, J. R. (Eds.). *Quantifier Elimination and Cylindrical Algebraic Decomposition.* New York: Springer-Verlag, 1998.

Collins, G. E. "Quantifier Elimination for the Elementary Theory of Real Closed Fields by Cylindrical Algebraic Decomposition." *Lect. Notes Comput. Sci.* **33**, 134–83, 1975.

Collins, G. E. "Quantifier Elimination by Cylindrical Algebraic Decomposition--Twenty Years of Progress." In *Quantifier Elimination and Cylindrical Algebraic Decomposition* (Ed. B. F. Caviness and J. R. Johnson). New York: Springer-Verlag, pp. 8–3, 1998.

Collins, G. E. and Hong, H. "Partial Cylindrical Algebraic Decomposition for Quantifier Elimination." *J. Symb. Comput.* **12**, 299–28, 1991.

Dolzmann, A. and Sturm, T. "Simplification of Quantifier-Free Formulae Over Ordered Fields." *J. Symb. Comput.* **24**, 209–31, 1997.

Faugere, J. C.; Gianni, P.; Lazard, D.; and Mora, T. "Efficient Computation of Zero-Dimensional Groebner Bases by Change of Ordering." *J. Symb. Comput.* **16**, 329–44, 1993.

Hong, H. "An Improvement of the Projection Operator in Cylindrical Algebraic Decomposition." In *ISSAC '90: Proceedings of the International Symposium on Symbolic and Algebraic Computation, August 20–4, 1990, Tokyo, Japan* (Ed. S. Watanabe and M. Nagata). New York: ACM Press, pp. 261–64, 1990.

Loos, R. and Weispfenning, V. "Applying Lattice Quantifier Elimination." *Comput. J.* **36**, 450–61, 1993.

McCallum, S. "Solving Polynomial Strict Inequalities Using Cylindrical Algebraic Decomposition." *Comput. J.* **36**, 432–38, 1993.

McCallum, S. "An Improved Projection for Cylindrical Algebraic Decomposition of Three Dimensional Space." *J. Symb. Comput.* **5**, 141–61, 1988.

McCallum, S. "An Improved Projection for Cylindrical Algebraic Decomposition." In *Quantifier Elimination and Cylindrical Algebraic Decomposition* (Ed. B. F. Caviness and J. R. Johnson). New York: Springer-Verlag, pp. 242–68, 1998.

Strzebonski, A. "An Algorithm for Systems of Strong Polynomial Inequalities." *Mathematica J.* **4**, 74–7, 1994.

Strzebonski, A. "A Real Polynomial Decision Algorithm Using Arbitrary-Precision Floating Point Arithmetic." *Reliable Comput.* **5**, 337–46, 1999.

Strzebonski, A. "Solving Algebraic Inequalities." *Mathematica J.* **7**, 525–41, 2000.

Cylindrical Coordinates

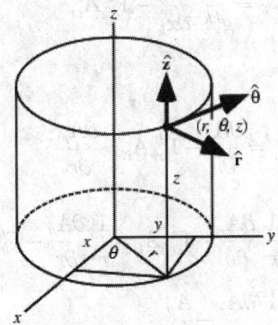

Cylindrical coordinates are a generalization of 2-D POLAR COORDINATES to 3-D by superposing a height (z) axis. Unfortunately, there are a number of different notations used for the other two coordinates. Either r or ρ is used to refer to the radial coordinate and either ϕ or θ to the azimuthal coordinates. Arfken (1985), for instance, uses (ρ, ϕ, z), while

Beyer (1987) uses (r, θ, z). In this work, the NOTATION (r, θ, z) is used.

$$r = \sqrt{x^2 + y^2} \tag{1}$$

$$\theta = \tan^{-1}\left(\frac{y}{x}\right) \tag{2}$$

$$z = z, \tag{3}$$

where $r \in [0, \infty)$, $\theta \in [0, 2\pi)$, and $z \in (-\infty, \infty)$. In terms of x, y, and z

$$x = r \cos\theta \tag{4}$$

$$y = r \sin\theta \tag{5}$$

$$z = z. \tag{6}$$

Morse and Feshbach (1953) define the cylindrical coordinates by

$$x = \xi_1\xi_2 \tag{7}$$

$$y = \xi_1\sqrt{1 - \xi_2^2} \tag{8}$$

$$z = \xi_3, \tag{9}$$

where $\xi_1 = r$ and $\xi_2 = \cos\theta$. The METRIC elements of the cylindrical coordinates are

$$g_{rr} = 1 \tag{10}$$

$$g_{\theta\theta} = r^2 \tag{11}$$

$$g_{zz} = 1, \tag{12}$$

so the SCALE FACTORS are

$$g_r = 1 \tag{13}$$

$$g_\theta = r \tag{14}$$

$$g_z = 1. \tag{15}$$

The LINE ELEMENT is

$$d\mathbf{s} = dr\,\hat{\mathbf{r}} + r\,d\theta\,\hat{\boldsymbol{\theta}} + dz\,\hat{\mathbf{z}}, \tag{16}$$

and the VOLUME ELEMENT is

$$dV = r\,dr\,d\theta\,dz. \tag{17}$$

The JACOBIAN is

$$\left|\frac{\partial(x, y, z)}{\partial(r, \theta, z)}\right| = r. \tag{18}$$

A CARTESIAN VECTOR is given in CYLINDRICAL COORDINATES by

$$\mathbf{r} = \begin{bmatrix} r\cos\theta \\ r\sin\theta \\ z \end{bmatrix}. \tag{19}$$

To find the UNIT VECTORS,

$$\hat{\mathbf{r}} \equiv \frac{\dfrac{d\mathbf{r}}{dr}}{\left|\dfrac{d\mathbf{r}}{dr}\right|} = \begin{bmatrix} \cos\theta \\ \sin\theta \\ 0 \end{bmatrix} \qquad (20)$$

$$\hat{\boldsymbol{\theta}} \equiv \frac{\dfrac{d\mathbf{r}}{d\theta}}{\left|\dfrac{d\mathbf{r}}{d\theta}\right|} = \begin{bmatrix} -\sin\theta \\ \cos\theta \\ 0 \end{bmatrix} \qquad (21)$$

$$\hat{\mathbf{z}} \equiv \frac{\dfrac{d\mathbf{r}}{dz}}{\left|\dfrac{d\mathbf{r}}{dz}\right|} = \begin{bmatrix} 0 \\ 0 \\ 1 \end{bmatrix}. \qquad (22)$$

Derivatives of unit VECTORS with respect to the coordinates are

$$\frac{\partial \hat{\mathbf{r}}}{\partial r} = \mathbf{0} \qquad (23)$$

$$\frac{\partial \hat{\mathbf{r}}}{\partial \theta} = \begin{bmatrix} -\sin\theta \\ \cos\theta \\ 0 \end{bmatrix} = \hat{\boldsymbol{\theta}} \qquad (24)$$

$$\frac{\partial \hat{\mathbf{r}}}{\partial z} = \mathbf{0} \qquad (25)$$

$$\frac{\partial \hat{\boldsymbol{\theta}}}{\partial r} = \mathbf{0} \qquad (26)$$

$$\frac{\partial \hat{\boldsymbol{\theta}}}{\partial \theta} = \begin{bmatrix} -\cos\theta \\ -\sin\theta \\ 0 \end{bmatrix} = -\hat{\mathbf{r}} \qquad (27)$$

$$\frac{\partial \hat{\boldsymbol{\theta}}}{\partial z} = \mathbf{0} \qquad (28)$$

$$\frac{\partial \hat{\mathbf{z}}}{\partial r} = \mathbf{0} \qquad (29)$$

$$\frac{\partial \hat{\mathbf{z}}}{\partial \theta} = \mathbf{0} \qquad (30)$$

$$\frac{\partial \hat{\mathbf{z}}}{\partial z} = \mathbf{0}. \qquad (31)$$

The GRADIENT of a VECTOR FIELD in cylindrical coordinates is given by

$$\nabla \equiv \hat{\mathbf{r}}\,\frac{\partial}{\partial r} + \hat{\boldsymbol{\theta}}\,\frac{1}{r}\frac{\partial}{\partial \theta} + \hat{\mathbf{z}}\,\frac{\partial}{\partial z}, \qquad (32)$$

so the GRADIENT components become

$$\nabla_r \hat{\mathbf{r}} = \mathbf{0} \qquad (33)$$

$$\nabla_\theta \hat{\mathbf{r}} = \frac{1}{r}\,\hat{\boldsymbol{\theta}} \qquad (34)$$

$$\nabla_z \hat{\mathbf{r}} = \mathbf{0} \qquad (35)$$

$$\nabla_r \hat{\boldsymbol{\theta}} = \mathbf{0} \qquad (36)$$

$$\nabla_\theta \hat{\boldsymbol{\theta}} = -\frac{1}{r}\,\hat{\mathbf{r}} \qquad (37)$$

$$\nabla_z \hat{\boldsymbol{\theta}} = \mathbf{0} \qquad (38)$$

$$\nabla_r \hat{\mathbf{z}} = \mathbf{0} \qquad (39)$$

$$\nabla_\theta \hat{\mathbf{z}} = \mathbf{0} \qquad (40)$$

$$\nabla_z \hat{\mathbf{z}} = \mathbf{0}. \qquad (41)$$

Now, since the CONNECTION COEFFICIENTS are defined by

$$\Gamma^i_{jk} = \hat{\mathbf{x}}_i \cdot (\nabla_k \hat{\mathbf{x}}_j), \qquad (42)$$

$$\Gamma^r = \begin{bmatrix} 0 & 0 & 0 \\ 0 & -\dfrac{1}{r} & 0 \\ 0 & 0 & 0 \end{bmatrix} \qquad (43)$$

$$\Gamma^\theta = \begin{bmatrix} 0 & \dfrac{1}{r} & 0 \\ \dfrac{1}{r} & 0 & 0 \\ 0 & 0 & 0 \end{bmatrix} \qquad (44)$$

$$\Gamma^z = \begin{bmatrix} 0 & 0 & 0 \\ 0 & 0 & 0 \\ 0 & 0 & 0 \end{bmatrix}, \qquad (45)$$

the COVARIANT DERIVATIVES, given by

$$A_{j;\,k} = \frac{1}{g^{kk}}\,\frac{\partial A_j}{\partial x_k} - \Gamma^i_{jk} A_i, \qquad (46)$$

are

$$A_{r;\,r} = \frac{\partial A_r}{\partial r} - \Gamma^i_{rr} A_i = \frac{\partial A_r}{\partial r} \qquad (47)$$

$$A_{r;\,\theta} = \frac{1}{r}\frac{\partial A_r}{\partial \theta} - \Gamma^i_{r\theta} A_i = \frac{1}{r}\frac{\partial A_r}{\partial \theta} - \Gamma^\theta_{r\theta} A_\theta$$

$$= \frac{1}{r}\frac{\partial A_r}{\partial \theta} - \frac{A_\theta}{r} \qquad (48)$$

$$A_{r;z} = \frac{\partial A_r}{\partial z} - \Gamma^i_{r\,z} A_i = \frac{\partial A_r}{\partial z} \qquad (49)$$

$$A_{\theta;\,r} = \frac{\partial A_\theta}{\partial r} \; \Gamma^i_{\theta r} A_i = \frac{\partial A_\theta}{\partial r} \qquad (50)$$

$$A_{\theta;\,\theta} = \frac{1}{r}\frac{\partial A_\theta}{\partial \theta} - \Gamma_{\theta\theta}^i A_i = \frac{1}{r}\frac{\partial A_\theta}{\partial_\theta} - \Gamma_{\theta\theta}^r A_r$$

$$= \frac{1}{r}\frac{\partial A_\theta}{\partial \theta} + \frac{A_r}{r} \tag{51}$$

$$A_{\theta;\,z} = \frac{\partial A_\theta}{\partial z} - \Gamma_{\theta z}^i A_i = \frac{\partial A_\theta}{\partial z} \tag{52}$$

$$A_{z;\,r} = \frac{\partial A_z}{\partial r} - \Gamma_{zr}^i A_i = \frac{\partial A_z}{\partial r} \tag{53}$$

$$A_{z;\,\theta} = \frac{\partial A_z}{\partial \theta} - \Gamma_{z\theta}^i A_i = \frac{1}{r}\frac{\partial A_z}{\partial \theta} \tag{54}$$

$$A_{z;\,z} = \frac{\partial A_z}{\partial z} - \Gamma_{zz}^i A_i = \frac{\partial A_z}{\partial z}. \tag{55}$$

CROSS PRODUCTS of the coordinate axes are

$$\hat{\mathbf{r}} \times \hat{\mathbf{z}} = -\hat{\boldsymbol{\theta}} \tag{56}$$

$$\hat{\boldsymbol{\theta}} \times \hat{\mathbf{z}} = \hat{\mathbf{r}} \tag{57}$$

$$\hat{\mathbf{r}} \times \hat{\boldsymbol{\theta}} = \hat{\mathbf{z}}. \tag{58}$$

The COMMUTATION COEFFICIENTS are given by

$$c_{\alpha\beta}^\mu \vec{e}_\mu = [\vec{e}_\alpha,\ \vec{e}_\beta] = \nabla_\alpha \vec{e}_\beta - \nabla_\beta \vec{e}_\alpha, \tag{59}$$

But

$$[\hat{\mathbf{r}},\ \hat{\mathbf{r}}] = [\hat{\boldsymbol{\theta}},\ \hat{\boldsymbol{\theta}}] = [\hat{\boldsymbol{\phi}},\ \hat{\boldsymbol{\phi}}] = 0, \tag{60}$$

so $c_{rr}^\alpha = c_{\theta\theta}^\alpha = c_{\phi\phi}^\alpha = 0$, where $\alpha = r,\ \theta,\ \phi$. Also

$$[\hat{\mathbf{r}},\ \hat{\boldsymbol{\theta}}] = -[\hat{\boldsymbol{\theta}},\ \hat{\mathbf{r}}] = \nabla_r \hat{\boldsymbol{\theta}} - \nabla_\theta \hat{\mathbf{r}} = 0 - \frac{1}{r}\hat{\boldsymbol{\theta}} = -\frac{1}{r}\hat{\boldsymbol{\theta}}, \tag{61}$$

so $c_{r\theta}^\theta = -c_{\theta r}^\theta = -\frac{1}{r}$, $c_{r\theta}^r = c_{r\theta}^\phi = 0$. Finally,

$$[\hat{\mathbf{r}},\ \hat{\boldsymbol{\phi}}] = [\hat{\boldsymbol{\theta}},\ \hat{\boldsymbol{\phi}}] = 0. \tag{62}$$

Summarizing,

$$c^r = \begin{bmatrix} 0 & 0 & 0 \\ 0 & 0 & 0 \\ 0 & 0 & 0 \end{bmatrix} \tag{63}$$

$$c^\theta = \begin{bmatrix} 0 & -\dfrac{1}{r} & 0 \\ \dfrac{1}{r} & 0 & 0 \\ 0 & 0 & 0 \end{bmatrix} \tag{64}$$

$$c^\phi = \begin{bmatrix} 0 & 0 & 0 \\ 0 & 0 & 0 \\ 0 & 0 & 0 \end{bmatrix}. \tag{65}$$

Time DERIVATIVES of the VECTOR are

$$\dot{\mathbf{r}} = \begin{bmatrix} \cos\theta\,\dot{r} - r\sin\theta\,\dot{\theta} \\ \sin\theta\,\dot{r} + r\cos\theta\,\dot{\theta} \\ \dot{z} \end{bmatrix} = \dot{r}\hat{\mathbf{r}} + r\dot{\theta}\hat{\boldsymbol{\theta}} + \dot{z}\hat{\mathbf{z}} \tag{66}$$

$$\ddot{\mathbf{r}} =$$

$$\begin{bmatrix} -\sin\theta\,\dot{r}\dot{\theta} + \cos\theta\,\ddot{r} - \sin\theta\,\dot{r}\dot{\theta} - r\cos\theta\,\dot{\theta}^2 - r\sin\theta\,\ddot{\theta} \\ \cos\theta\,\dot{r}\dot{\theta} + \sin\theta\,\ddot{r} + \cos\theta\,\dot{r}\dot{\theta} - r\sin\theta\,\dot{\theta}^2 + r\cos\theta\,\ddot{\theta} \\ \ddot{z} \end{bmatrix}$$

$$= \begin{bmatrix} -2\sin\theta\,\dot{r}\dot{\theta} + \cos\theta\,\ddot{r} - r\cos\theta\,\dot{\theta}^2 - r\sin\theta\,\ddot{\theta} \\ 2\cos\theta\,\dot{r}\dot{\theta} + \sin\theta\,\ddot{r} - r\sin\theta\,\dot{\theta}^2 + r\cos\theta\,\ddot{\theta} \\ \ddot{z} \end{bmatrix}$$

$$= (\ddot{r} - r\dot{\theta}^2)\hat{\mathbf{r}} + (2\dot{r}\dot{\theta} + r\ddot{\theta})\hat{\boldsymbol{\theta}} + \ddot{z}\hat{\mathbf{z}}. \tag{67}$$

SPEED is given by

$$v \equiv |\dot{\mathbf{r}}| = \sqrt{\dot{r}^2 + r^2\dot{\theta}^2 + \dot{z}^2}. \tag{68}$$

Time derivatives of the UNIT VECTORS are

$$\dot{\hat{\mathbf{r}}} = \begin{bmatrix} -\sin\theta\,\dot{\theta} \\ \cos\theta\,\dot{\theta} \\ 0 \end{bmatrix} = \dot{\theta}\hat{\boldsymbol{\theta}} \tag{69}$$

$$\dot{\hat{\boldsymbol{\theta}}} = \begin{bmatrix} -\cos\theta\,\dot{\theta} \\ -\sin\theta\,\dot{\theta} \\ 0 \end{bmatrix} = -\dot{\theta}\hat{\mathbf{r}} \tag{70}$$

$$\dot{\hat{\mathbf{z}}} = \begin{bmatrix} 0 \\ 0 \\ 0 \end{bmatrix} = \mathbf{0}. \tag{71}$$

CROSS PRODUCTS of the axes are

$$\hat{\mathbf{r}} \times \hat{\mathbf{z}} = -\hat{\boldsymbol{\theta}} \tag{72}$$

$$\hat{\boldsymbol{\theta}} \times \hat{\mathbf{z}} = \hat{\mathbf{r}} \tag{73}$$

$$\hat{\mathbf{r}} \times \hat{\boldsymbol{\theta}} = \hat{\mathbf{z}}. \tag{74}$$

The CONVECTIVE DERIVATIVE is

$$\frac{D\dot{\mathbf{r}}}{Dt} \equiv \left(\frac{\partial}{\partial t} + \dot{\mathbf{r}}\cdot\nabla\right)\dot{\mathbf{r}} = \frac{\partial\dot{\mathbf{r}}}{\partial t} + \dot{\mathbf{r}}\cdot\nabla\dot{\mathbf{r}}. \tag{75}$$

To rewrite this, use the identity

$$\nabla(\mathbf{A}\cdot\mathbf{B}) = \mathbf{A}\times(\nabla\times\mathbf{B}) + \mathbf{B}\times(\nabla\times\mathbf{A}) + (\mathbf{A}\cdot\nabla)\mathbf{B} + (\mathbf{B}\cdot\nabla)\mathbf{A} \tag{76}$$

and set $\mathbf{A} = \mathbf{B}$, to obtain

$$\nabla(\mathbf{A}\cdot\mathbf{A}) = 2\mathbf{A}\times(\nabla\times\mathbf{A}) + 2(\mathbf{A}\cdot\nabla)\mathbf{A}, \tag{77}$$

so

$$(\mathbf{A}\cdot\nabla)\mathbf{A} = (\tfrac{1}{2}\mathbf{A}^2) - \mathbf{A}\times(\nabla\times\mathbf{A}). \tag{78}$$

Then

$$\frac{D\dot{\mathbf{r}}}{Dt} = \ddot{\mathbf{r}} + \nabla(\tfrac{1}{2}\dot{\mathbf{r}}^2) - \dot{\mathbf{r}}\times(\nabla\times\dot{\mathbf{r}})$$

$$= \ddot{\mathbf{r}} + (\nabla\times\dot{\mathbf{r}})\times\dot{\mathbf{r}} + \nabla(\tfrac{1}{2}\dot{\mathbf{r}}^2). \tag{79}$$

The CURL in the above expression gives

$$\nabla\times\dot{\mathbf{r}} = \frac{1}{r}\frac{\partial}{\partial r}(r^2\dot{\theta})\hat{\mathbf{z}} = 2\dot{\theta}\hat{\mathbf{z}}, \tag{80}$$

so

$$-\dot{\mathbf{r}} \times (\nabla \times \dot{\mathbf{r}}) = -2\dot{\theta}(\dot{r}\hat{\mathbf{r}} \times \hat{\mathbf{z}} + r\dot{\theta}\hat{\boldsymbol{\theta}} \times \hat{z}) = -2\dot{\theta}(-\dot{r}\hat{\boldsymbol{\theta}} + r\dot{\theta}\hat{\mathbf{r}})$$
$$= 2\dot{r}\dot{\theta}\hat{\boldsymbol{\theta}} - 2r\dot{\theta}^2\hat{\mathbf{r}}. \tag{81}$$

We expect the gradient term to vanish since SPEED does not depend on position. Check this using the identity $\nabla(f^2) = 2f\nabla f$,

$$\nabla(\tfrac{1}{2}\,\dot{\mathbf{r}}^2) = \tfrac{1}{2}\,\nabla(\dot{r}^2 + r^2\dot{\theta}^2 + \dot{z}^2) = \dot{r}\nabla\dot{r} + r\dot{\theta}\nabla(r\dot{\theta}) + \dot{z}\nabla\dot{z}. \tag{82}$$

Examining this term by term,

$$\dot{r}\nabla\dot{r} = \dot{r}\,\frac{\partial}{\partial t}\,\nabla r = \dot{r}\,\frac{\partial}{\partial t}\,\hat{\mathbf{r}} = \dot{r}\dot{\hat{\mathbf{r}}} = \dot{r}\dot{\theta}\hat{\boldsymbol{\theta}} \tag{83}$$

$$r\dot{\theta}\nabla(r\dot{\theta}) = r\dot{\theta}\left[r\,\frac{\partial}{\partial t}\,\nabla\theta + \dot{\theta}\nabla r\right] = r\dot{\theta}\left[r\,\frac{\partial}{\partial t}\left(\frac{1}{r}\,\hat{\boldsymbol{\theta}}\right) + \dot{\theta}\hat{\mathbf{r}}\right]$$
$$= r\dot{\theta}\left[r\left(-\frac{1}{r^2}\,\dot{r}\hat{\boldsymbol{\theta}} + \frac{1}{r}\,\dot{\hat{\boldsymbol{\theta}}}\right) + \dot{\theta}\hat{\mathbf{r}}\right]$$
$$= -\dot{\theta}\dot{r}\hat{\boldsymbol{\theta}} + r\dot{\theta}(-\dot{\theta}\hat{\mathbf{r}}) + r\dot{\theta}^2\hat{\mathbf{r}} = -\dot{\theta}\dot{r}\hat{\boldsymbol{\theta}} \tag{84}$$

$$\dot{z}\nabla\dot{z} = \dot{z}\,\frac{\partial}{\partial t}\,\nabla z = \dot{z}\,\frac{\partial}{\partial t}\,\hat{\mathbf{z}} = \dot{z}\dot{\hat{\mathbf{z}}} = \mathbf{0}, \tag{85}$$

so, as expected,

$$\nabla(\tfrac{1}{2}\,\dot{\mathbf{r}}^2) = \mathbf{0}. \tag{86}$$

We have already computed $\ddot{\mathbf{r}}$, so combining all three pieces gives

$$\frac{D\dot{\mathbf{r}}}{Dt} = (\ddot{r} - r\dot{\theta}^2 - 2r\dot{\theta}^2)\hat{\mathbf{r}} + (2\dot{r}\dot{\theta} + 2\dot{r}\dot{\theta} + r\ddot{\theta})\hat{\boldsymbol{\theta}} + \ddot{z}\hat{\mathbf{z}}$$
$$= (\ddot{r} - 3r\dot{\theta}^2)\hat{r} + (4\dot{r}\dot{\theta} + r\ddot{\theta})\hat{\boldsymbol{\theta}} + \ddot{z}\hat{\mathbf{z}}. \tag{87}$$

The DIVERGENCE is

$$\nabla \cdot A = A^r_{,r} = A^r_{,r} + (\Gamma^r_{rr}A^t + \Gamma^r_{\theta r}A^\theta + \Gamma^r_{zr}A^z) + A^\theta_{,\theta}$$
$$+ (\Gamma^\theta_{r\theta}A^r + \Gamma^\theta_{\theta\theta}A^\theta + \Gamma^\theta_{z\theta}A^z)$$
$$+ A^z_{,z} + (\Gamma^z_{rz}A^r + \Gamma^z_{\theta z}A^\theta + \Gamma^z_{zz}A^z)$$
$$= A^r_{,r} + A^\theta_{,\theta} + A^z_{,z} + (0 + 0 + 0) + \left(\frac{1}{r} + 0 + 0\right)$$
$$+ (0 + 0 + 0)$$
$$= \frac{1}{g_r}\,\frac{\partial}{\partial r}\,A^r + \frac{1}{g_\theta}\,\frac{\partial}{\partial\theta}\,A^\theta + \frac{1}{g_z}\,\frac{\partial}{\partial z}\,A^z + \frac{1}{r}\,A^r$$
$$= \left(\frac{\partial}{\partial r} + \frac{1}{r}\right)A^r + \frac{1}{r}\,\frac{\partial}{\partial\theta}\,A^\theta + \frac{\partial}{\partial z}\,A^z, \tag{88}$$

or, in VECTOR notation

$$\nabla \cdot \mathbf{F} = \frac{1}{r}\,\frac{\partial}{\partial r}\,(rF_r) + \frac{1}{r}\,\frac{\partial F_\theta}{\partial\theta} + \frac{\partial F_z}{\partial z}. \tag{89}$$

The CROSS PRODUCT is

$$\nabla \times \mathbf{F} = \left(\frac{1}{r}\,\frac{\partial F_z}{\partial\theta} - \frac{\partial F_\theta}{\partial z}\right)\hat{\mathbf{r}} + \left(\frac{\partial F_r}{\partial z} - \frac{\partial F_z}{\partial r}\right)\hat{\boldsymbol{\theta}}$$
$$+ \frac{1}{r}\left[\frac{\partial}{\partial r}\,(rF_\theta) - \frac{\partial F_r}{\partial\theta}\right]\hat{\mathbf{z}}. \tag{90}$$

The scalar LAPLACIAN is

$$\nabla^2 f \equiv \frac{1}{r}\,\frac{\partial}{\partial r}\left(r\,\frac{\partial f}{\partial r}\right) + \frac{1}{r^2}\,\frac{\partial^2 f}{\partial\theta^2} + \frac{\partial^2 f}{\partial z^2}$$
$$= \frac{\partial^2 f}{\partial r^2} + \frac{1}{r}\,\frac{\partial f}{\partial r} + \frac{1}{r^2}\,\frac{\partial^2 f}{\partial\theta^2} + \frac{\partial^2 f}{\partial z^2}. \tag{91}$$

The vector LAPLACIAN is

$$\nabla^2\mathbf{v} = \begin{bmatrix} \dfrac{\partial^2 v_r}{\partial r^2} + \dfrac{1}{r^2}\dfrac{\partial^2 v_r}{\partial\phi^2} + \dfrac{\partial^2 v_r}{z^2} + \dfrac{1}{r}\dfrac{\partial v_r}{\partial r} - \dfrac{2}{r^2}\dfrac{\partial v_\phi}{\partial\phi} - \dfrac{v_r}{r^2} \\[2mm] \dfrac{\partial^2}{\partial r^2} + \dfrac{1}{r^2}\dfrac{\partial^2 v_\phi}{\partial\phi^2} + \dfrac{\partial^2 v_\phi}{\partial z^2} + \dfrac{1}{r}\dfrac{\partial v_\phi}{\partial r} + \dfrac{2}{r^2}\dfrac{\partial v_r}{\partial\phi} - \dfrac{v_\phi}{r^2} \\[2mm] \dfrac{\partial^2 v_z}{\partial r^2} + \dfrac{1}{r^2}\dfrac{\partial^2 v_z}{\partial\phi^2} + \dfrac{\partial^2 v_z}{\partial z^2} + \dfrac{1}{r}\dfrac{\partial v_z}{\partial r} \end{bmatrix}. \tag{92}$$

The HELMHOLTZ DIFFERENTIAL EQUATION is separable in cylindrical coordinates and has STÄCKEL DETERMINANT $S = 1$ (for r, θ, z) or $S = 1/(1 - \xi_2^2)$ (for Morse and Feshbach's ξ_1, ξ_2, ξ_3).

See also ELLIPTIC CYLINDRICAL COORDINATES, HELMHOLTZ DIFFERENTIAL EQUATION–CIRCULAR CYLINDRICAL COORDINATES, POLAR COORDINATES, SPHERICAL COORDINATES

References

Arfken, G. "Circular Cylindrical Coordinates." §2.4 in *Mathematical Methods for Physicists, 3rd ed.* Orlando, FL: Academic Press, pp. 95–01, 1985.
Beyer, W. H. *CRC Standard Mathematical Tables, 28th ed.* Boca Raton, FL: CRC Press, p. 212, 1987.
Moon, P. and Spencer, D. E. "Circular-Cylinder Coordinates (r, ψ, z)." Table 1.02 in *Field Theory Handbook, Including Coordinate Systems, Differential Equations, and Their Solutions, 2nd ed.* New York: Springer-Verlag, pp. 12–7, 1988.
Morse, P. M. and Feshbach, H. *Methods of Theoretical Physics, Part I.* New York: McGraw-Hill, p. 657, 1953.

Cylindrical Equal-Area Projection

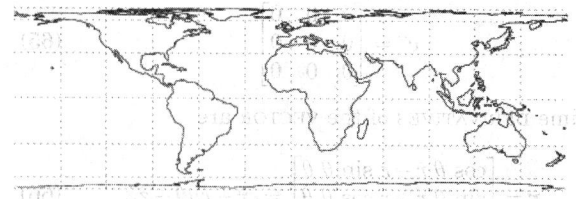

The MAP PROJECTION having transformation equations

$$x = (\lambda - \lambda_0)\cos \phi_s \qquad (1)$$

$$y = \sin \phi \sec \phi_s \qquad (2)$$

for the normal aspect, where λ is the LONGITUDE, λ_0 is the standard LONGITUDE (horizontal center of the projection), ϕ is the LATITUDE, and ϕ_s is the so-called "standard latitude." The inverse transformation equations for the normal aspect are

$$\phi = \sin^{-1}(y \cos \phi_s) \qquad (3)$$

$$\lambda = x \sec \phi_s + \lambda_0. \qquad (4)$$

Special cases of cylindrical equal-area projections are summarized in the following table (Maling 1992).

ψ_s	MAP PROJECTION
0°	LAMBERT CYLINDRICAL EQUAL-AREA PROJECTION
30°	BEHRMANN CYLINDRICAL EQUAL-AREA PROJECTION
37.383°	TRISTAN EDWARDS PROJECTION
44.138°	PETERS PROJECTION
45°	GALL ORTHOGRAPHIC PROJECTION
50°	BALTHASART PROJECTION

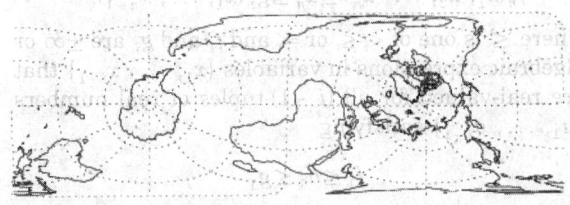

An oblique form of the cylindrical equal-area projection is given by the equations

$$\lambda_p =$$

$$\tan^{-1}\left(\frac{\cos \phi_1 \sin \phi_2 \cos \lambda_1 - \sin \phi_1 \cos \phi_2 \cos \lambda_2}{\sin \phi_1 \cos \phi_2 \sin \lambda_2 - \cos \phi_1 \sin \phi_2 \sin \lambda_1}\right) \qquad (5)$$

$$\phi_p = \tan^{-1}\left[-\frac{\cos(\lambda_p - \lambda_1)}{\tan \phi_1}\right], \qquad (6)$$

and the inverse FORMULAS are

$$\phi = \sin^{-1}(y \sin \phi_p + \sqrt{1 - y^2} \cos \phi_p \sin x) \qquad (7)$$

$$\lambda = \lambda_0 + \tan^{-1}\left(\frac{\sqrt{1 - y^2} \sin \phi_p \sin x - y \cos \phi_p}{\sqrt{1 - y^2} \cos x}\right). \qquad (8)$$

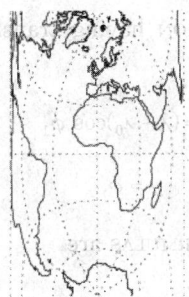

A transverse form of the cylindrical equal-area projection is given by the equations

$$x = \cos \phi \sin(\lambda - \lambda_0) \qquad (9)$$

$$y = \tan^{-1}\left[\frac{\tan \phi}{\cos(\lambda - \lambda_0)}\right] - \phi_0, \qquad (10)$$

and the inverse FORMULAS are

$$\phi = \sin^{-1}[\sqrt{1 - x^2} \sin(y + \phi_0)] \qquad (11)$$

$$\lambda = \lambda_0 + \tan^{-1}\left[\frac{x}{\sqrt{1 - x^2}} \cos(y + \phi_0)\right]. \qquad (12)$$

See also BALTHASART PROJECTION, BEHRMANN CYLINDRICAL EQUAL-AREA PROJECTION, CYLINDRICAL EQUIDISTANT PROJECTION, EQUAL-AREA PROJECTION, GALL ORTHOGRAPHIC PROJECTION, LAMBERT CYLINDRICAL EQUAL-AREA PROJECTION, PETERS PROJECTION TRISTAN EDWARDS PROJECTION

References

Maling, D. H. *Coordinate Systems and Map Projections, 2nd ed, rev.* Woburn, MA: Butterworth-Heinemann, 1993.

Snyder, J. P. *Map Projections--A Working Manual.* U. S. Geological Survey Professional Paper 1395. Washington, DC: U. S. Government Printing Office, pp. 76–5, 1987.

Steinhaus, H. *Mathematical Snapshots, 3rd ed.* New York: Dover, pp. 221–22, 1999.

Cylindrical Equidistant Projection

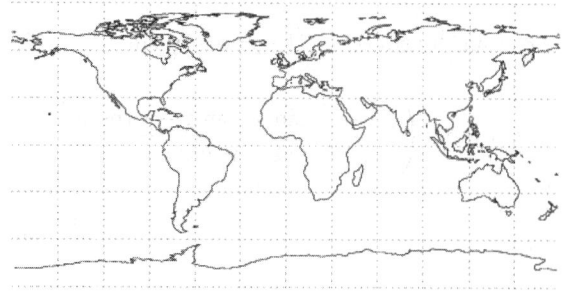

The MAP PROJECTION having transformation equations

$$x = (\lambda - \lambda_0)\cos \phi_1 \tag{1}$$

$$y = \phi, \tag{2}$$

and the inverse FORMULAS are

$$\phi = y \tag{3}$$

$$\lambda = \lambda_0 + x \sec \phi_1, \tag{4}$$

The following table gives special cases of the cylindrical equidistant projection.

ϕ_1	projection name
0°	EQUIRECTANGULAR PROJECTION
37°30′	MILLER EQUIDISTANT PROJECTION
43°	MILLER EQUIDISTANT PROJECTION
45°	GALL ISOGRAPHIC PROJECTION
50°28′	MILLER EQUIDISTANT PROJECTION

See also CYLINDRICAL EQUAL-AREA PROJECTION, EQUIDISTANT PROJECTION, EQUIRECTANGULAR PROJECTION, GALL ISOGRAPHIC PROJECTION, MILLER EQUIDISTANT PROJECTION

References

Snyder, J. P. *Map Projections--A Working Manual.* U. S. Geological Survey Professional Paper 1395. Washington, DC: U. S. Government Printing Office, pp. 90–1, 1987.
Snyder, J. P. *Flattening the Earth: Two Thousand Years of Map Projections.* Chicago, IL: University of Chicago Press, 1993.

Cylindrical Equirectangular Projection
CYLINDRICAL EQUIDISTANT PROJECTION

Cylindrical Function

$$R_m(x, y) \equiv \frac{J'_m(x)Y'_m(y) - J'_m(y)Y'_m(x)}{J_m(x)Y'_m(y) - J'_m(y)Y_m(x)}$$

$$S_m(x, y) \equiv \frac{J'_m(x)Y_m(y) - J_m(y)Y'_m(x)}{J_m(x)Y_m(y) - J_m(y)Y_m(x)}.$$

See also CYLINDER FUNCTION, HEMISPHERICAL FUNCTION

Cylindrical Harmonics
BESSEL FUNCTION OF THE FIRST KIND

Cylindrical Hoof
CYLINDRICAL WEDGE

Cylindrical Parts
The cylindrical parts of a system of real algebraic equations and inequalities in variables $\{x_1, \ldots, x_n\}$ are the terms

$$f_1 \leqq x_1 \leqq g_1$$

$$f_2(x_1) \leqq x_2 \leqq g_2(x_1)$$

$$\vdots$$

$$f_n(x_1, x_2, \ldots, x_n) \leqq x_n \leqq g_n(x_1, \ldots, x_{n-1}),$$

where '$\leqq$' is one of $<$, $\leq$, or $=$, and f_i and g_i are $\pm\infty$ or algebraic expressions in variables $\{x_1, \ldots, x_{i-1}\}$ that are real-valued for all $(i-1)$-tuples of real numbers $\{a_1, \ldots, a_{i-1}\}$ satisfying

$$f_1 \leqq a_1 \leqq g_1$$

$$f_2(a_1) \leqq a_2 \leqq g_2(a_2)$$

$$\vdots$$

$$f_{i-1}(a_1, \ldots, a_{i-2}) \leqq a_{i-1} \leqq g_{i-1}(a_1, \ldots, a_{i-2}).$$

The CONJUNCTION of a finite number of disjoint cylindrical parts is called a CYLINDRICAL ALGEBRAIC DECOMPOSITION.

See also CYLINDRICAL ALGEBRAIC DECOMPOSITION

References

Strzebonski, A. "Solving Algebraic Inequalities." *Mathematica J.* **7**, 525–41, 2000.

Cylindrical Projection

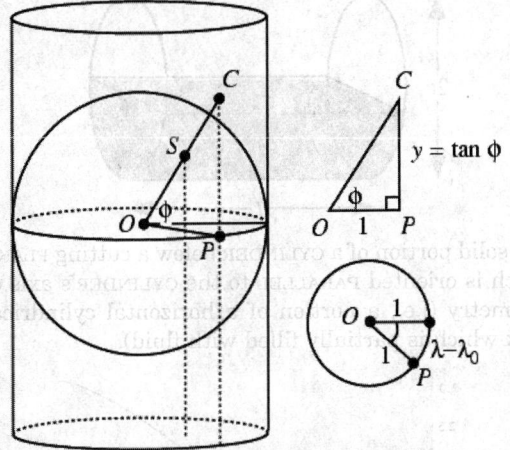

A cylindrical projection of points on a unit sphere centered at O consists of extending the line OS for each point S until it intersects a cylinder tangent to the sphere at its equator at a corresponding point C. If the sphere is tangent to the cylinder at longitude λ_0, then a point on the sphere with latitude ϕ and longitude λ is mapped to a point on the cylinder with height $\tan \phi$.

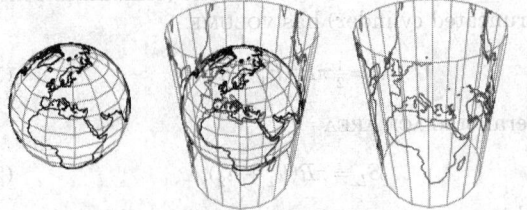

Unwrapping and flattening out the cylinder then gives the Cartesian coordinates

$$x = \lambda - \lambda_0 \qquad (1)$$

$$y = \tan \phi. \qquad (2)$$

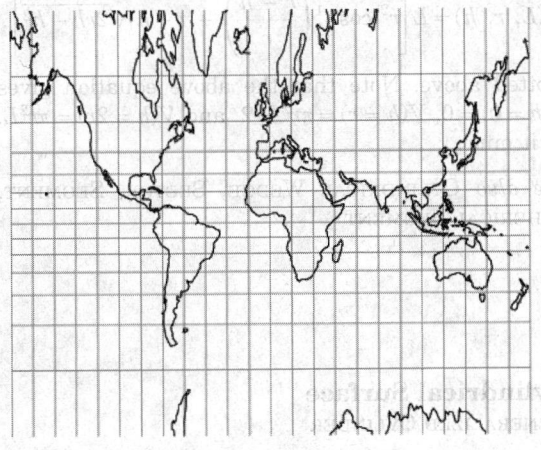

The cylindrical projection of the Earth is illustrated above.

This form of the projection, however, is seldom used in practice, and the term "cylindrical projection" is used instead to refer to *any* projection in which lines of longitude are mapped to equally spaced parallel lines and lines of latitude (parallels) are mapped to parallel lines with arbitrary mathematically spaced separations (Snyder 1987, p. 5). For example, the common MERCATOR PROJECTION uses the complicated transformation

$$y = \ln[\tan(\tfrac{1}{4}\pi + \tfrac{1}{2}\phi)] \qquad (3)$$

instead of $\tan \phi$ in order to achieve certain desirable properties in the projection.

Craig (1882) used the term "cylindric" instead of "cylindrical" (Lee 1944), but this convention did not catch on.

See also BEHRMANN CYLINDRICAL EQUAL-AREA PROJECTION, CYLINDRICAL EQUAL-AREA PROJECTION, CYLINDRICAL EQUIDISTANT PROJECTION, GALL ORTHOGRAPHIC PROJECTION, MERCATOR PROJECTION, MILLER CYLINDRICAL PROJECTION, PETERS PROJECTION, PSEUDOCYLINDRICAL PROJECTION

References

Craig, T. *A Treatise on Projections.* Washington, DC: U. S. Government Printing Office, 1882.

Lee, L. P. "The Nomenclature and Classification of Map Projections." *Empire Survey Rev.* **7**, 190–00, 1944.

Snyder, J. P. *Map Projections--A Working Manual.* U. S. Geological Survey Professional Paper 1395. Washington, DC: U. S. Government Printing Office, 1987.

Cylindrical Section

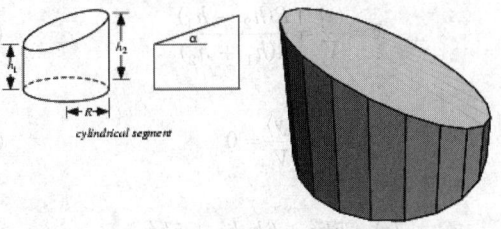

cylindrical segment

The intersection of a PLANE with a right circular CYLINDER is a CIRCLE (if the plane is at a right angle to the axis), an ELLIPSE, or, if the plane is parallel to the axis, a single line (if the plane is tangent to the cylinder), pair of parallel lines bounding an infinite rectangle (if the plane cuts the cylinder), or no intersection at all (if the plane missed the cylinder entirely; Hilbert and Cohn-Vossen 1999, pp. 7–).

The volume of the cylindrical section can be obtained instantly by noting that two such sections can be fitted together to form a cylinder of radius R and height $h_1 + h_2$, so the volume of the original wedge is half that of the cylinder of height $h_1 + h_2$. The volume can be found directly through integration by noting that the height in polar and Cartesian coordinates is given by

$$h(r, \theta) = h_1 + \frac{1}{2}\left(1 + \frac{r}{R} \cos\theta\right)(h_2 - h_1) \qquad (1)$$

$$h(x, y) = h_1 + \frac{1}{2}\left(1 + \frac{x}{R}\right)(h_2 - h_1), \qquad (2)$$

so

$$V = \int_0^R \int_0^{2\pi} \int_0^{h(r,\theta)} r \, dr \, d\theta \, dz \qquad (3)$$

$$= \int_{-R}^R \int_{-\sqrt{R^2 - x^2}}^{\sqrt{R^2 - x^2}} \int_0^{h(x,y)} dx \, dy \, dz, \qquad (4)$$

giving (1). Similarly, the volume-weighted coordinates are given by

$$\langle x \rangle = \tfrac{1}{8}\pi R^3(h_2 - h_1) \qquad (5)$$

$$\langle y \rangle = 0 \qquad (6)$$

$$\langle z \rangle = \tfrac{1}{32}\pi R^2(5h_1^2 + 6h_1 h_2 + 5h_2^2), \qquad (7)$$

so the centroids are given by

$$\bar{x} = \frac{\langle x \rangle}{V} = \frac{R(h_2 - h_1)}{4(h_1 + h_2)} \qquad (8)$$

$$\bar{y} = \frac{\langle y \rangle}{V} = 0 \qquad (9)$$

$$\bar{z} = \frac{\langle z \rangle}{V} = \frac{5h_1^2 + 6h_1 h_2 + 5h_2^2}{16(h_1 + h_2)}, \qquad (10)$$

(cf. the strange parameterization used by Harris and Stocker 1998, p. 103).

See also CONIC SECTION, CYLINDER, CYLINDRICAL SEGMENT, CYLINDRICAL WEDGE, ELLIPSE

References

Hilbert, D. and Cohn-Vossen, S. "The Cylinder, the Cone, the Conic Sections, and Their Surfaces of Revolution." §2 in *Geometry and the Imagination.* New York: Chelsea, pp. 7–1, 1999.

Cylindrical Segment

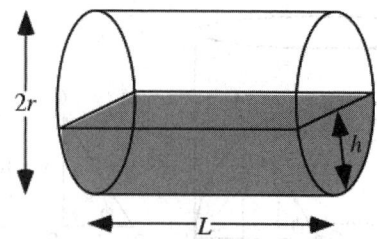

The solid portion of a CYLINDER below a cutting PLANE which is oriented PARALLEL to the CYLINDER's axis of symmetry (i.e., a portion of a horizontal cylindrical tank which is partially filled with fluid).

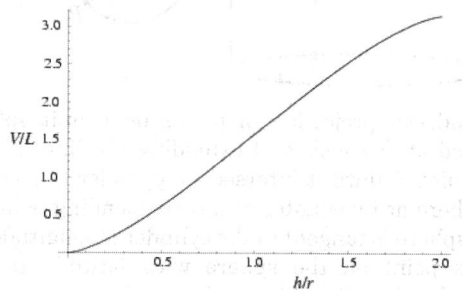

The solid cut from a circular CYLINDER by a tilted PLANE which does not cut the base (sometimes called a truncated cylinder) has VOLUME

$$V = \tfrac{1}{2}\pi R^2(h_1 + h_2), \qquad (1)$$

lateral SURFACE AREA

$$S_L = \pi R(h_1 + h_2), \qquad (2)$$

and top SURFACE AREA

$$S_T = \pi R\sqrt{R^2 + \tfrac{1}{4}(h_2 - h_1)^2} \qquad (3)$$

(Harris and Stocker 1998, p. 103).
For a CYLINDER of RADIUS r and length L, the VOLUME $V(L, r, h)$ of the cylindrical segment is given by multiplying the AREA of a circular SEGMENT of height h by L,

$$V(L, r, h) = L\left[r^2 \cos^{-1}\left(\frac{r - h}{r}\right) - (r - h)\sqrt{2rh - h^2}\right],$$

plotted above. Note that the above equation gives $V(h = 0) = 0$, $V(h = r) = \pi r^2 L/2$, and $V(h = 2r) = \pi r^2 L$, as it must.

See also CYLINDRICAL WEDGE, SECTOR, SEGMENT, SPHERICAL SEGMENT

Cylindrical Surface

GENERALIZED CYLINDER

Cylindrical Wedge

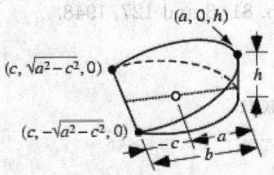

cylindrical wedge

A wedge is cut from a CYLINDER by slicing with a plane that intersects the base of the cylinder. The VOLUME of a cylindrical wedge can be found by noting that the plane cutting the cylinder passes through the three points illustrated above (with $c = a - b$), so the three-point form of the plane gives the equation

$$-hx + bz + (a - b)h = 0. \tag{1}$$

Solving for z gives

$$z = \frac{h(x - a + b)}{b}. \tag{2}$$

The volume is therefore

$$V = 2 \int_0^{\sqrt{a^2 - x^2}} \int_{a-b}^a \frac{h(x+b-a)}{b} \, dx \, dy \tag{3}$$

$$= \frac{h}{6b} \Big[2\sqrt{(2a-b)b}(3a^2 - 2ab + b^2) - 3\pi a^2(a-b)$$

$$+ 6a^2(a-b)\tan^{-1}\Big(\tfrac{a-b}{\sqrt{(2a-b)b}}\Big)\Big], \tag{4}$$

and the lateral SURFACE AREA

$$S_L = \frac{2h}{b}$$

$$\times \left[\sqrt{(2a-b)b} - a(a-b)\cot^{-1}\Big(\frac{a-b}{\sqrt{(2a-b)b}}\Big) \right], \tag{5}$$

(apparently given incorrectly by Harris and Stocker 1998, p. 104).

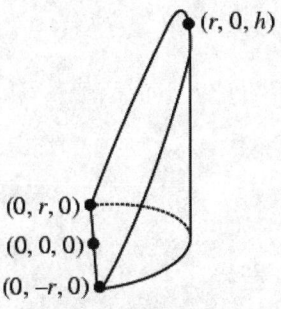

A special case of the cylindrical wedge, also called a cylindrical hoof, is a wedge passing through a DIAMETER of the base (so that $a = b$). Let the height of the wedge be h and the radius of the CYLINDER from which it is cut r. Then plugging the points

$(0, -r, 0)$, $(0, r, 0)$, and $(r, 0, h)$ into the 3-point equation for a PLANE gives the equation for the plane as

$$hx - rz = 0. \tag{6}$$

combining with the equation of the CIRCLE which describes the curved part remaining of the cylinder (and writing $t = x$ then gives the PARAMETRIC EQUATIONS of the "tongue" of the wedge as

$$x = t \tag{7}$$

$$y = \pm\sqrt{r^2 - t^2} \tag{8}$$

$$z = \frac{ht}{r} \tag{9}$$

for $t \in [0, r]$. To examine the form of the tongue, it needs to be rotated into a convenient plane. This can be accomplished by first rotating the plane of the curve by 90° about the X-AXIS using the ROTATION MATRIX $R_x(90°)$ and then by the ANGLE

$$\theta = \tan^{-1}\Big(\frac{h}{r}\Big) \tag{10}$$

above the z-AXIS. The transformed plane now rests in the xz-plane and has PARAMETRIC EQUATIONS

$$x = \frac{t\sqrt{h^2 + r^2}}{r} \tag{11}$$

$$z = \pm\sqrt{r^2 - t^2} \tag{12}$$

and is shown below.

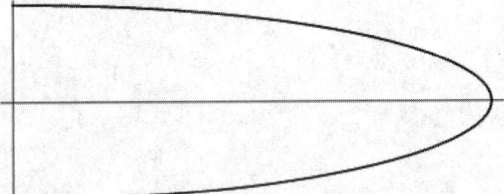

The length of the tongue (measured down its middle) is obtained by plugging $t = r$ into the above equation for x, which becomes

$$L = \sqrt{h^2 + r^2} \tag{13}$$

(and which follows immediately from the PYTHAGOREAN THEOREM). The VOLUME of the wedge is given by

$$V = \tfrac{2}{3} r^2 h \tag{14}$$

and the lateral SURFACE AREA by

$$S_L = 2rh. \tag{15}$$

While the centroid of the general cylindrical wedge is complicated for $a \neq b$, for the cylindrical hoof, the centroid is given by

$$\bar{x} = \int_{a-b}^{a} \int_{-\sqrt{a^2-x^2}}^{\sqrt{a^2-x^2}} \int_{0}^{h(b-a+x)/b} \mathbf{x}\, dz\, dy\, dz, \quad (16)$$

giving

$$\langle x \rangle = \tfrac{3}{16}\,\pi r \quad (17)$$

$$\langle y \rangle = 0 \quad (18)$$

$$\langle z \rangle = \tfrac{3}{32}\,\pi h. \quad (19)$$

See also CONICAL WEDGE, CYLINDRICAL SECTION, CYLINDRICAL SEGMENT, WEDGE

References

Harris, J. W. and Stocker, H. "Obliquely Cut Circular Cylinder" and "Segment of a Cylinder." §4.6.3–.6.4 in *Handbook of Mathematics and Computational Science.* New York: Springer-Verlag, pp. 103–04, 1998.

Kern, W. F. and Bland, J. R. "Truncated Prism (or Cylinder)." §31 in *Solid Mensuration with Proofs, 2nd ed.* New York: Wiley, pp. 81–3 and 127, 1948.

Cylindroid

PLÜCKER'S CONOID

C*

The RIEMANN SPHERE $\mathbb{C}^* = \mathbb{C} \cup \{\infty\}$, also called the EXTENDED COMPLEX PLANE. The notation $\hat{\mathbb{C}}$ is sometimes also used (Krantz 1999, p. 82).

The notation $\mathbb{C}^*$ also stands for $\mathbb{C} - \{0\}$, the punctured plane, which is both a LIE GROUP and an ABELIAN VARIETY.

See also C, COMPLEX NUMBER, COMPLEX PLANE, EXTENDED COMPLEX PLANE, Q, R, RIEMANN SPHERE, Z

References

Krantz, S. G. *Handbook of Complex Analysis.* Boston, MA: Birkhäuser, p. 82, 1999.

D

d'Alembert Ratio Test
RATIO TEST

d'Alembert's Equation
The ORDINARY DIFFERENTIAL EQUATION

$$y = xf(y') + g(y'),$$

where $y \equiv dy/dx$ and f and g are given functions. This equation is sometimes also known as LAGRANGE'S EQUATION (Zwillinger 1997).

See also LAGRANGE'S EQUATION

References
Ince, E. L. *Ordinary Differential Equations.* New York: Dover, pp. 38–9, 1956.
Murphy, G. M. *Ordinary Differential Equations and Their Solution.* Princeton, NJ: Van Nostrand, pp. 65–6, 1960.
Valiron, G. *The Geometric Theory of Ordinary Differential Equations and Algebraic Functions.* Brookline, MA: Math. Sci. Press, pp. 217–18, 1950.
Zwillinger, D. "Lagrange's Equation." §II.A.69 in *Handbook of Differential Equations, 3rd ed.* Boston, MA: Academic Press, pp. 120 and 265–68, 1997.

d'Alembert's Solution
The method of d'Alembert provides a solution to the one-dimensional WAVE EQUATION

$$\frac{\partial^2 y}{\partial x^2} = \frac{1}{c^2} \frac{\partial^2 y}{\partial t^2} \tag{1}$$

that models vibrations of a string.

The general solution can be obtained by introducing new variables $\xi = x - ct$ and $\eta = x + ct$, and applying the CHAIN RULE to obtain

$$\frac{\partial}{\partial x} = \frac{\partial \xi}{\partial x} \frac{\partial}{\partial \xi} + \frac{\partial \eta}{\partial x} \frac{\partial}{\partial \eta} \tag{2}$$

$$= \frac{\partial}{\partial \xi} + \frac{\partial}{\partial \eta} \tag{3}$$

$$\frac{\partial}{\partial t} = \frac{\partial \xi}{\partial t} \frac{\partial}{\partial \xi} + \frac{\partial \eta}{\partial t} \frac{\partial}{\partial \eta} \tag{4}$$

$$= -c \frac{\partial}{\partial \xi} + c \frac{\partial}{\partial \eta}. \tag{5}$$

Using (3) and (5) to compute the left and right sides of (1) then gives

$$\frac{\partial^2 y}{\partial x^2} = \left(\frac{\partial}{\partial \xi} + \frac{\partial}{\partial \eta} \right) \left(\frac{\partial y}{\partial \xi} + \frac{\partial y}{\partial \eta} \right) = \frac{\partial^2 y}{\partial \xi^2} + 2 \frac{\partial^2 y}{\partial \xi \partial \eta} + \frac{\partial^2 y}{\partial \eta^2} \tag{6}$$

$$\frac{\partial^2 y}{\partial t^2} = \left(-c \frac{\partial}{\partial \xi} + c \frac{\partial}{\partial \eta} \right) \left(-c \frac{\partial y}{\partial \xi} + c \frac{\partial y}{\partial \eta} \right)$$

$$= c^2 \frac{\partial^2 y}{\partial \xi^2} - 2c^2 \frac{\partial^2 y}{\partial \xi \partial \eta} + c^2 \frac{\partial^2 y}{\partial \eta^2}. \tag{7}$$

respectively, so plugging in and expanding then gives

$$\frac{\partial^2 y}{\partial \xi \partial \eta} = 0. \tag{8}$$

This partial differential equation has general solution

$$= f(\xi) + g(\eta) \tag{9}$$

$$= f(x - ct) + g(x + ct). \tag{10}$$

where f and g are arbitrary functions, with f representing a right-traveling wave and g a left-traveling wave.

See also WAVE EQUATION

References
Bekefi, G. and Barrett, A. H. *Electromagnetic Vibrations, Waves, and Radiation.* Cambridge, MA: MIT Press, pp. 161–63, 1987.

d'Alembert's Theorem
If three CIRCLES A, B, and C are taken in pairs, the external SIMILARITY POINTS of the three pairs lie on a straight LINE. Similarly, the external SIMILARITY POINT of one pair and the two internal SIMILARITY POINTS of the other two pairs lie upon a straight LINE, forming a SIMILARITY AXIS of the three CIRCLES.

See also SIMILARITY POINT

References
Dörrie, H. *100 Great Problems of Elementary Mathematics: Their History and Solutions.* New York: Dover, p. 155, 1965.

d'Alembertian
Written in the NOTATION of PARTIAL DERIVATIVES, the d'Alembertian $\Box^2$ is defined by

$$\Box^2 \equiv \nabla^2 - \frac{1}{c^2} \frac{\partial^2}{\partial t^2},$$

where c is the speed of light. Writing in TENSOR notation,

$$\Box^2 \phi \equiv \left(g^{\lambda \kappa} \phi_{;\lambda} \right)_{;\kappa} = g^{\lambda \kappa} \frac{\partial^2 \phi}{\partial x^\lambda \partial x^\kappa} - \Gamma^\lambda \frac{\partial \phi}{\partial x^\lambda}.$$

See also GRADIENT FOUR-VECTOR, HARMONIC COORDINATES, LAPLACIAN, WAVE EQUATION

d'Alembertian Operator

Written in the NOTATION of PARTIAL DERIVATIVES,

$$\Box^2$$

where c is the speed of light. Writing in TENSOR notation,

$$\Box^2 \equiv \nabla^2 - \frac{1}{c^2}\frac{\partial^2}{\partial t^2},$$

See also HARMONIC COORDINATES

d'Ocagne's Identity

$$F_m F_{n+1} - F_n F_{m+1} = (-1)^n F_{m-n},$$

where F_n is a FIBONACCI NUMBER.

See also CASSINI'S IDENTITY, CATALAN'S IDENTITY, FIBONACCI NUMBER

© *1999–001 Wolfram Research, Inc.*

d'Octagne's Identity

© *1999–001 Wolfram Research, Inc.*

DAG

ACYCLIC DIGRAPH

Dagger

The symbol † most commonly used in older physics texts to denote the ADJOINT operator. The dagger is also known as the obelisk, obelus, or long cross (Bringhurst 1997, p. 275).

See also ADJOINT, DOUBLE DAGGER

References

Bringhurst, R. *The Elements of Typographic Style, 2nd ed.* Point Roberts, WA: Hartley and Marks, 1997.

Daisy

A figure resembling a daisy or sunflower in which copies of a geometric figure of increasing size are placed at regular intervals along a spiral. The result-ing figure appears to have multiple spirals spreading out from the center.

See also HEXLET, PHYLLOTAXIS, SPIRAL, SWIRL, WHIRL

References

Dixon, R. "On Drawing a Daisy." §5.1 in *Mathographics*. New York: Dover, pp. 122–43, 1991.

Damped Exponential Cosine Integral

$$\int_0^\infty e^{-wT}\cos(\omega t)\,d\omega. \tag{1}$$

Integrate by parts with

$$u \equiv e^{-\omega T} \quad dv = \cos(\omega t)\,d\omega \tag{2}$$

$$du \equiv -Te^{-\omega T}\,d\omega \quad v = \frac{1}{t}\sin(\omega t), \tag{3}$$

so

$$\int e^{-\omega T}\cos(\omega t)\,d\omega$$
$$= \frac{1}{t}e^{-wt}\sin(\omega t) + \frac{T}{t}\int e^{-wT}\sin(\omega t)\,d\omega. \tag{4}$$

Now integrate

$$\int e^{-\omega T}\sin(\omega t)\,d\omega \tag{5}$$

by parts. Let

$$u = e^{-\omega T} \quad dv = \sin(\omega t)\,d\omega \tag{6}$$

$$du = -Te^{-\omega T}\,d\omega \quad v = -\frac{1}{t}\cos(\omega t), \tag{7}$$

so

$$\int e^{-\omega t}\sin(\omega t)\,d\omega$$
$$= -\frac{1}{t}\cos(\omega t) - \frac{T}{t}\int e^{-\omega T}\cos(\omega t)\,d\omega \tag{8}$$

and

$$\int e^{\omega T} \cos(\omega t) d\omega = \frac{1}{t} e^{-\omega t} \sin(\omega t)$$

$$-\frac{T}{t^2} e^{-\omega t} \cos(\omega t) - \frac{T^2}{t^2} \int e^{-\omega T} \cos(\omega t) d\omega \quad (9)$$

$$\left(1 + \frac{T^2}{t^2}\right) \int e^{-\omega T} \cos(\omega t) d\omega$$

$$= e^{-\omega T} \left[\frac{1}{t} \sin(\omega t) - \frac{T}{t^2} \cos(\omega t)\right] \quad (10)$$

$$\frac{t^2 + T^2}{t^2} \int e^{-\omega T} \cos(\omega t) d\omega$$

$$= \frac{e^{-\omega t}}{t^2} [t \sin(\omega T) - T \cos(\omega t)] \quad (11)$$

$$\int e^{-\omega T} \cos(\omega t) d\omega$$

$$= \frac{e^{-\omega T}}{t^2 + T^2} [t \sin(\omega t) - T \cos(\omega T)]. \quad (12)$$

Therefore,

$$\int_0^\infty e^{-\omega T} \cos(\omega t) d\omega = 0 + \frac{T}{t^2 + T^2} = \frac{T}{t^2 + T^2}. \quad (13)$$

See also Cosine Integral, Fourier Transform–
Lorentzian Function, Lorentzian Function

Damped Simple Harmonic Motion

Adding a damping force proportional to $\dot{x}$ to the equation of simple harmonic motion, the first derivative of x with respect to time, the equation of motion for *damped* simple harmonic motion is

$$\ddot{x} + \beta \dot{x} + \omega_0^2 x = 0, \quad (1)$$

where β is the damping constant. This equation arises, for example, in the analysis of the flow of current in an electronic CLR circuit, (which contains a capacitor, an inductor, and a resistor). The curve produced by two damped harmonic oscillators at right angles to each other is called a harmonograph, and simplifies to a Lissajous curve if $\beta_1 = \beta_2 = 0$.

The damped harmonic oscillator can be solved by looking for trial solutions of the form $x = e^{rt}$. Plugging this into (1) gives

$$(r^2 + \beta r + \omega_0^2) e^{rt} = 0 \quad (2)$$

$$r^2 + \beta r + \omega_0^2 = 0. \quad (3)$$

This is a quadratic equation with solutions

$$r = \frac{1}{2} \left(-\beta \pm \sqrt{\beta^2 - 4\omega_0^2}\right). \quad (4)$$

There are therefore three solution regimes depending on the sign of the quantity inside the square root,

$$\alpha \equiv \beta^2 - 4\omega_0^2. \quad (5)$$

The three regimes are summarized in the following table.

α	regime
$\alpha < 0$	UNDERDAMPING
$\alpha = 0$	CRITICAL DAMPING
$\alpha > 0$	OVERDAMPING

If a periodic (sinusoidal) forcing term is added at angular frequency ω, the same three solution regimes are again obtained. Surprisingly, the resulting motion is still periodic (after an initial transient response, corresponding to the solution to the unforced case, has died out), but it has an amplitude different from the forcing amplitude.

The "particular" solution $x_p(t)$ to the forced second-order nonhomogeneous ORDINARY DIFFERENTIAL EQUATION

$$\ddot{x} + p(t)\dot{x} + q(t)x = A \cos(\omega t) \quad (6)$$

due to forcing is given by the equation

$$x_p(t) = -x_1(t) \int \frac{x_2(t)g(t)}{W(t)} dt + x_2(t) \int \frac{x_1(t)g(t)}{W(t)} dt, \quad (7)$$

where x_1 and x_2 are the homogeneous solutions to the unforced equation

$$\ddot{x} + p(t)\dot{x} + q(t)x = 0 \quad (8)$$

and $W(t)$ is the WRONSKIAN of these two functions. Once the sinusoidal case of forcing is solved, it can be generalized to *any* periodic function by expressing the periodic function in a FOURIER SERIES.

See also Damped Simple Harmonic Motion, Damped Simple Harmonic Motion–Critical Damping, Damped Simple Harmonic Motion–Overdamping, Damped Simple Harmonic Motion–Underdamping, Harmonograph, Lissajous Curve, Simple Harmonic Motion

References

Papoulis, A. "Motion of a Harmonically Bound Particle." §15– in *Probability, Random Variables, and Stochastic Processes, 2nd ed.* New York: McGraw-Hill, pp. 524–28, 1984.

Damped Simple Harmonic Motion— Critical Damping

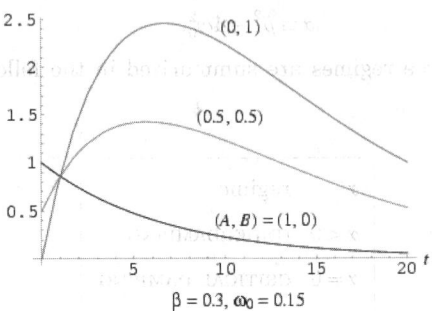

critically damped simple harmonic motion
$(\beta = 2\omega_0)$

$\beta = 0.3, \omega_0 = 0.15$

Critical damping is a special case of damped simple harmonic motion in which

$$\alpha \equiv \beta^2 - 4\omega_0^2 = 0, \qquad (1)$$

so

$$\beta = 2\omega_0. \qquad (2)$$

In this case, $\alpha = 0$ so the solutions OF THE FORM $x = e^{rt}$ satisfy

$$r_{\pm} = \frac{1}{2}(-\beta) = -\frac{1}{2}\beta = -\omega_0. \qquad (3)$$

One of the solutions is therefore

$$x_1 = e^{-\omega_0 t}. \qquad (4)$$

In order to find the other linearly independent solution, we can make use of the identity

$$x_2(t) = x_1(t) \int \frac{e^{-\int p(t)dt}}{[x_1(t)]^2} dt. \qquad (5)$$

Since we have $p(t) = 2\omega_0$, $e^{-\int p(t)dt}$ simplifies to $e^{-2\omega_0 t}$. Equation (5) therefore becomes

$$x_2(t) = e^{-\omega_0 t} \int \frac{e^{-2\omega_0 t}}{[e^{-\omega_0 t}]^2} dt = e^{-\omega_0 t} \int dt = te^{-\omega_0 t}. \qquad (6)$$

The general solution is therefore

$$x = (A + Bt)e^{-\omega_0 t}. \qquad (7)$$

In terms of the constants A and B, the initial values are

$$x(0) = A \qquad (8)$$

$$x'(0) = B - A\omega. \qquad (9)$$

so

$$A = x(0) \qquad (10)$$

$$B = x'(0) + \omega_0 x(0). \qquad (11)$$

The above plot shows a critically damped simple harmonic oscillator with $\omega = 0.3$, $\beta = 0.15$ for a variety of initial conditions (A, B).

For sinusoidally forced simple harmonic motion with critical damping, the equation of motion is

$$\ddot{x} + 2\omega_0 \dot{x} + \omega_0^2 x = A \cos(\omega t), \qquad (12)$$

and the WRONSKIAN is

$$W(t) \equiv x_1 \dot{x}_2 - \dot{x}_1 x_2 = e^{-2\omega_0 t}. \qquad (13)$$

Plugging this into the equation for the particular solution gives

$$x_p(t) = -e^{-\omega_0 t} \int \frac{te^{-\omega_0 t}A \cos(\omega t)}{e^{-2\omega_0 t}} dt$$

$$+ te^{-\omega_0 t} \int \frac{e^{-\omega_0 t}A \cos(\omega t)}{e^{-2\omega_0 t}} dt$$

$$= \frac{A}{(\omega^2 + \omega_0^2)} \left[(\omega_0^2 - \omega^2) \cos(\omega t) + 2\omega\omega_0 \sin(\omega t) \right]. \qquad (14)$$

In order to put this in the desired form, note that we want to equate

$$C \cos\theta + S \sin\theta = Q \cos(\theta + \delta)$$
$$= Q(\cos\theta \cos\delta - \sin\theta \sin\delta). \qquad (15)$$

This means

$$C \equiv Q \cos\delta = \omega_0^2 - \omega^2 \qquad (16)$$

$$S \equiv -Q \sin\delta = 2\omega\omega_0, \qquad (17)$$

so

$$Q = \sqrt{C^2 + S^2} \qquad (18)$$

$$\delta = \tan^{-1}\left(-\frac{S}{C}\right). \qquad (19)$$

Plugging in,

$$Q = \sqrt{\omega_0^4 - 2\omega_0^2\omega^2 + \omega^4 + 4\omega_0^2\omega^2} = \omega_0^2\omega^2. \qquad (20)$$

$$\delta = \tan^{-1}\left(\frac{2\omega\omega_0}{\omega^2 - \omega_0^2}\right). \qquad (21)$$

The solution in the requested form is therefore

$$x_p = \frac{A}{(\omega^2 + \omega_0^2)^2}(\omega_0^2 + \omega^2)\cos(\omega t + \delta)$$

$$\frac{A}{\omega^2 + \omega_0^2}\cos(\omega t + \delta), \qquad (22)$$

where δ is defined by (21).

See also DAMPED SIMPLE HARMONIC MOTION, DAMPED SIMPLE HARMONIC MOTION–OVERDAMPING,

DAMPED SIMPLE HARMONIC MOTION–UNDERDAMPING,
SIMPLE HARMONIC MOTION

References

Papoulis, A. *Probability, Random Variables, and Stochastic Processes, 2nd ed.* New York: McGraw-Hill, p. 528, 1984.

Damped Simple Harmonic Motion— Overdamping

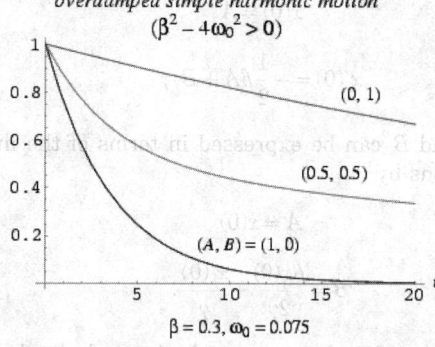

overdamped simple harmonic motion
$(\beta^2 - 4\omega_0^2 > 0)$

$\beta = 0.3, \omega_0 = 0.075$

Overdamped simple harmonic motion occurs when

$$\beta^2 - 4\omega_0^2 > 0, \tag{1}$$

so

$$\alpha \equiv \beta^2 - 4\omega_0^2 > 0. \tag{2}$$

$$x_1 = e^{r-t} \tag{3}$$

$$x_2 = e^{r+t}, \tag{4}$$

where

$$r\pm \equiv \frac{1}{2}\left(-\beta \pm \sqrt{\beta^2 - 4\omega_0^2}\right). \tag{5}$$

The general solution is therefore

$$x = Ae^{r-t} + Be^{r+t}, \tag{6}$$

where A and B are constants. The initial values are

$$x(0) = A + B \tag{7}$$

$$x'(0) = Ar_- + Br_+, \tag{8}$$

so

$$A = x(0) + \frac{r_+ x(0) - x'(0)}{r_- - r_+} \tag{9}$$

$$B = -\frac{r_+ x(0) - x'(0)}{r_- - r_+}. \tag{10}$$

The above plot shows an overdamped simple harmonic oscillator with $\omega = 0.3$, $\beta = 0.075$ and three different initial conditions (A, B).

For a cosinusoidally forced overdamped oscillator with forcing function $g(t) = C\cos(\omega t)$, the particular solutions are

$$y_1(t) = e^{r_1 t} \tag{11}$$

$$y_2(t) = e^{r_2 t}, \tag{12}$$

where

$$r_1 \equiv \frac{1}{2}\left(-\beta + \sqrt{\beta^2 - 4\omega_0^2}\right) \tag{13}$$

$$r_2 \equiv \frac{1}{2}\left(-\beta - \sqrt{\beta^2 - 4\omega_0^2}\right). \tag{14}$$

These give the identities

$$r_1 + r_2 = -\beta \tag{15}$$

$$r_1 - r_2 = \sqrt{\beta^2 - 4\omega_0^2} \tag{16}$$

and

$$\omega_0^2 = \frac{1}{4}\left[\beta - (r_1 - r_2)^2\right] = r_1 r_2. \tag{17}$$

The WRONSKIAN is

$$W(t) = y_1 y_2' - y_1' y_2 = e^{r_1 t} r_2 e^{r_2 t} - r_1 e^{r_1 t} e^{r_2 t}$$

$$= (r_2 - r_1) e^{(r_1 + r_2)t}. \tag{18}$$

The particular solution is

$$y_p = -y_1 v_1 + y_2 v_2, \tag{19}$$

where

$$v_1 \equiv \int \frac{y_2 g(t)}{W(t)} = \frac{C}{r_2 - r_1} \frac{\omega \sin(\omega t) - r_2 \cos(\omega t)}{e^{r_2 t}(r_2^2 + \omega^2)} \tag{20}$$

$$v_2 \equiv \int \frac{y_2 g(t)}{W(t)} = \frac{C}{r_2 - r_1} \frac{\omega \sin(\omega t) - r_1 \cos(\omega t)}{e^{r_1 t}(r_2^2 + \omega^2)}. \tag{21}$$

Therefore,

$$y_p = C\frac{\cos(\omega t)(r_1 r_2 - \omega^2) - \sin(\omega t)\omega(r_1 + r_2)}{(r_1^2 + \omega^2)(r_2^2 + \omega^2)}$$

$$= \frac{C}{\sqrt{\beta^2\omega^2 + (\omega^2 - \omega_0^2)^2}}\cos(\omega t + \delta), \tag{22}$$

where

$$\delta = \tan^{-1}\left(\frac{\beta\omega}{\omega^2 - \omega_0^2}\right). \tag{23}$$

See also DAMPED SIMPLE HARMONIC MOTION, DAMPED SIMPLE HARMONIC MOTION–CRITICAL DAMPING, DAMPED SIMPLE HARMONIC MOTION–UNDERDAMPING, SIMPLE HARMONIC MOTION

References

Papoulis, A. *Probability, Random Variables, and Stochastic Processes, 2nd ed.* New York: McGraw-Hill, pp. 527–28, 1984.

Damped Simple Harmonic Motion— Underdamping

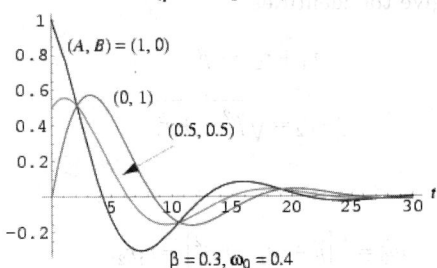

underdamped simple harmonic motion
$(\beta^2 - 4\omega_0^2 < 0)$

$(A, B) = (1, 0)$

$(0, 1)$

$(0.5, 0.5)$

$\beta = 0.3, \omega_0 = 0.4$

Underdamped simple harmonic motion occurs when

$$\beta^2 - 4\omega_0^2 < 0, \tag{1}$$

so

$$\alpha \equiv \beta^2 - 4\omega_0^2 < 0. \tag{2}$$

Define

$$\gamma \equiv \sqrt{-\alpha} = \frac{1}{2}\sqrt{4\omega_0^2 - \beta^2}, \tag{3}$$

then solutions satisfy

$$r\pm = -\frac{1}{2}\beta \pm i\gamma, \tag{4}$$

where

$$r\pm \equiv \frac{1}{2}\left(-\beta \pm \sqrt{\beta^2 - 4\omega_0^2}\right), \tag{5}$$

and are OF THE FORM

$$x = e^{-(\beta/2 \pm i\gamma)t}. \tag{6}$$

Using the EULER FORMULA

$$e^{ix} = \cos x + i\,\sin x, \tag{7}$$

this can be rewritten

$$x = e^{-(\beta/2)t}[\cos(\gamma t) \pm i\,\sin(\gamma t)]. \tag{8}$$

We are interested in the *real* solutions. Since we are dealing here with a *linear homogeneous* ODE, linear sums of LINEARLY INDEPENDENT solutions are also solutions. Since we have a sum of such solutions in (8), it follows that the IMAGINARY and REAL PARTS separately satisfy the ODE and are therefore the solutions we seek. The constant in front of the sine

term is arbitrary, so we can identify the solutions as

$$x_1 = e^{-(\beta/2)t}\cos(\gamma t) \tag{9}$$

$$x_2 = e^{-(\beta/2)t}\sin(\gamma t), \tag{10}$$

so the general solution is

$$x = e^{-(\beta/2)t}[A\cos(\gamma t) + B\sin(\gamma t)]. \tag{11}$$

The initial values are

$$x(0) = A \tag{12}$$

$$x'(0) = -\frac{1}{2}\beta A + B, \gamma \tag{13}$$

so A and B can be expressed in terms of the initial conditions by

$$A = x(0) \tag{14}$$

$$B = \frac{\beta x(0)}{2\gamma} + \frac{x'(0)}{\gamma}. \tag{15}$$

The above plot shows an underdamped simple harmonic oscillator with $\omega = 0.3$, $\beta = 0.4$ for a variety of initial conditions (A, B).

For a cosinusoidally forced underdamped oscillator with forcing function $g(t) = C\cos(\omega t)$, use

$$\gamma \equiv \frac{1}{2}\sqrt{4\omega_0^2 - \beta^2} \tag{16}$$

$$\alpha \equiv \frac{1}{2}\beta \tag{17}$$

to obtain

$$4\omega_0^2 - \beta^2 = 4\gamma^2 \tag{18}$$

$$\omega_0^2 = \gamma^2 + \frac{1}{4}\beta^2 = \gamma^2 + \alpha^2 \tag{19}$$

$$\beta = 2\alpha. \tag{20}$$

The particular solutions are

$$y_1(t) = e^{-\alpha t}\cos(\gamma t) \tag{21}$$

$$y_2(t) = e^{-\alpha t}\sin(\gamma t). \tag{22}$$

The WRONSKIAN is

$$W(t) \equiv y_1 y_2' - y_1' y_2$$

$$= e^{-\alpha t}\cos(\gamma t)[-\alpha e^{-\alpha t}\sin(\gamma t) + e^{-\alpha t}\gamma\cos(\gamma t)]$$

$$- e^{-\alpha t}\sin(\gamma t)[-\alpha e^{-\alpha t}\cos(\gamma t) - e^{-\alpha t}\gamma\sin(\gamma t)]$$

$$= e^{-2\alpha t}\{\alpha[-\sin(\gamma t)\cos(\gamma t) + \sin(\gamma t)\cos(\gamma t)]$$

$$+ \gamma[\cos^2(\gamma t) + \sin^2(\gamma t)]\}$$

$$= \gamma e^{-2\alpha t}. \tag{23}$$

The particular solution is given by

$$y_p = -y_1 v_1 + y_2 v_2, \qquad (24)$$

where

$$v_1 = \int \frac{y_2 g(t)}{W(t)} = \frac{C}{\gamma} \int e^{\alpha t} \cos(\gamma t) \cos(\omega t) dt \qquad (25)$$

$$v_2 = \int \frac{y_2 g(t)}{W(t)} = \frac{C}{\gamma} \int e^{\alpha t} \cos(\gamma t) \cos(\omega t) dt. \qquad (26)$$

Using computer algebra to perform the algebra, the particular solution is

$$y_p(t) = C \frac{(\alpha^2 + \gamma^2 - \omega^2) \cos(\omega t) + 2\alpha\omega \sin(\omega t)}{\left[\alpha^2 + (\gamma - \omega)^2\right]\left[\alpha^2 + (\gamma + \omega)^2\right]}$$

$$= C \frac{\sqrt{(\omega_0^2 - \omega^2)^2 + \beta^2 \omega^2}}{(\omega_0^2 - \omega^2)^2 - \omega^2(4\omega_0^2 - \beta^2)} \cos(\omega t + \delta), \qquad (27)$$

where

$$\delta = \tan^{-1}\left(\frac{\beta\omega}{\omega^2 - \omega_0^2}\right). \qquad (28)$$

If the forcing function is sinusoidal instead of cosinusoidal, then

$$\delta' = \delta - \frac{1}{2}\pi = \tan^{-1} x - \frac{1}{2}\pi = \tan^{-1}\left(-\frac{1}{x}\right), \qquad (29)$$

so

$$\delta' = \tan^{-1}\left(\frac{\omega_0^2 - \omega^2}{\beta\omega}\right). \qquad (30)$$

See also DAMPED SIMPLE HARMONIC MOTION, DAMPED SIMPLE HARMONIC MOTION–CRITICAL DAMPING, DAMPED SIMPLE HARMONIC MOTION–OVERDAMPING, SIMPLE HARMONIC MOTION

References

Papoulis, A. *Probability, Random Variables, and Stochastic Processes*, 2nd ed. New York: McGraw-Hill, pp. 525–27, 1984.

d-Analog

N.B. *A detailed online essay by S. Finch was the starting point for this entry.*

The *d*-analog of a COMPLEX NUMBER *s* is defined as

$$[s]_d = 1 - \frac{2^d}{s^d} \qquad (1)$$

(Flajolet *et al.* 1995). For integer n, $[2]! \equiv 1$ and

$$[n]_d! = [3][4]\cdots[n]$$

$$= \left(1 - \frac{2^d}{3^d}\right)\left(1 - \frac{2^d}{4^d}\right)\cdots\left(1 - \frac{2^d}{n^d}\right). \qquad (2)$$

It can then be extended to complex values via

$$[s]_d! = \prod_{j=1}^{\infty} \frac{[j+2]}{[j+s]} \qquad (3)$$

(Flajolet *et al.* 1995). It satisfies the basic functional identity

$$[s]_d! = [s]_d[s-1]_d!. \qquad (4)$$

The *d*-analog of the POLYGAMMA FUNCTION is

$$[\psi]_d(s+1) = \frac{d}{ds}\ln[s]_d!$$

$$= -d \cdot 2^d \sum_{m=1}^{\infty} \frac{1}{(m+s)\left[(m+s)^d - 2^d\right]}. \qquad (5)$$

The first few values are

$$[\psi]_1(s) = \frac{3 - 2s}{s^2 - 3s + 2} \qquad (6)$$

$$[\psi]_2(s) = \psi_0(s-2) - 2\psi_0(s) + \psi_0(s+2), \qquad (7)$$

where $\psi_0(x)$ is the DIGAMMA FUNCTION.

The *d*-analog of the EULER-MASCHERONI CONSTANT γ is

$$[\gamma]_d = -[\psi]_d(3) = d \cdot 2^d \sum_{m=3}^{\infty} \frac{1}{m(m^d - 2^d)} \qquad (8)$$

(Flajolet *et al.* 1995). The first few values are

$$[\gamma]_1 = \frac{3}{2} \qquad (9)$$

$$[\gamma]_2 = \frac{11}{12} \qquad (10)$$

$$[\gamma]_3 = \frac{9}{2} - H_{3-i\sqrt{3}} - H_{3+i\sqrt{3}} \qquad (11)$$

$$[\gamma]_4 = \frac{47}{12} - H_{2-2i} - H_{2+2i}, \qquad (12)$$

where H_n is a HARMONIC NUMBER.

The *d*-analog of the HARMONIC NUMBERS is $[H_2]_d = 0$ and

$$[H_n]_d = d \cdot 2^d \left(\frac{1}{3^{d+1}[3]} + \frac{1}{4^{d+1}[4]} + \ldots + \frac{1}{n^{d+1}[n]}\right) \qquad (13)$$

$$= [\psi]_d(n+1) + [\gamma]_d \qquad (14)$$

(Flajolet *et al.* 1995).

The d-analog of INFINITY FACTORIAL is given by

$$[\infty!]_d = \prod_{n=3}^{\infty}\left(1-\frac{2^d}{n^d}\right). \tag{15}$$

This INFINITE PRODUCT can be evaluated in closed form in terms of π, the HYPERBOLIC SINE $\sinh x$, and GAMMA FUNCTIONS $\Gamma(x)$ involving roots of unity $\zeta_n^k \equiv (-1)^{k/n}$,

$$d_1 = 0 \tag{16}$$

$$d_2 = \frac{1}{6} \tag{17}$$

$$d_3 = \frac{\sinh\left(\pi\sqrt{3}\right)}{42\pi\sqrt{3}} \tag{18}$$

$$d_4 = \frac{\cosh\pi\,\sinh\pi}{60\pi} \tag{19}$$

$$d_5 = \frac{1}{1240\left|\Gamma\left(2\zeta_5^1\right)\Gamma\left(-2\zeta_5^2\right)\right|^2} \tag{20}$$

$$d_6 = \frac{\sinh^2(\pi\sqrt{3})}{1512\pi^2} \tag{21}$$

$$d_7 = \frac{1}{28448\left|\Gamma\left(2\zeta_7^1\right)\Gamma\left(-2\zeta_7^2\right)\Gamma\left(2\zeta_7^3\right)\right|^2} \tag{22}$$

$$d_8 = \frac{\sinh(2\pi)\left|\sinh\left(2\zeta_4^1\right)\right|^2}{16320\pi^3} \tag{23}$$

$$d_9 = \frac{\sinh\left(\pi\sqrt{3}\right)}{588672\pi\sqrt{3}\left|\Gamma\left(2\zeta_9^1\right)\Gamma\left(-2\zeta_9^2\right)\Gamma\left(-2\zeta_9^4\right)\right|^2}. \tag{24}$$

These are all special cases of a general result for INFINITE PRODUCTS.

See also INFINITE PRODUCT, Q-ANALOG

References

Finch, S. "Favorite Mathematical Constants." http://www.mathsoft.com/asolve/constant/infprd/infprd.html.
Flajolet, P.; Labelle, G.; Laforest, L.; and Salvy, B. "Hypergeometrics and the Cost Structure of Quadtrees." *Random Structure Alg.* **7**, 117–44, 1995. http://pauillac.inria.fr/algo/flajolet/Publications/publist.html.
Kahovec, H. "Basic Infinite Products." http://www.mathsoft.com/asolve/constant/infprd/kahovec/ip.html.
Kahovec, H. "Proof of the Infinite Product Formulas." http://www.mathsoft.com/asolve/constant/infprd/kahovec/proof01.html.

Dandelin Spheres

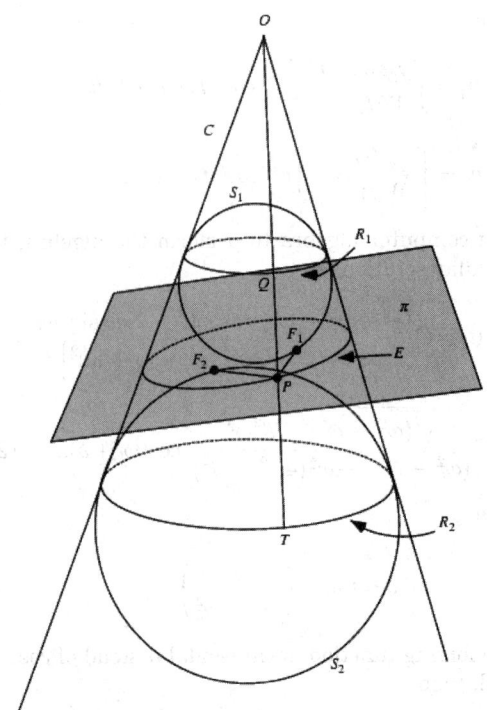

The inner and outer SPHERES TANGENT internally to a CONE and also to a PLANE intersecting the CONE are called Dandelin spheres.

The SPHERES can be used to show that the intersection of the PLANE with the CONE is an ELLIPSE. Let π be a PLANE intersecting a right circular CONE with vertex O in the curve E. Call the SPHERES TANGENT to the CONE and the PLANE S_1 and S_2, and the CIRCLES on which the SPHERES are TANGENT to the CONE R_1 and R_2. Pick a line along the CONE which intersects R_1 at Q, E at P, and R_2 at T. Call the points on the PLANE where the CIRCLES are TANGENT F_1 and F_2. Because intersecting tangents have the same length,

$$F_1 P = QP$$

$$F_2 P = TP.$$

Therefore,

$$PF_1 + PF_2 = QP + PT = QT,$$

which is a constant independent of P, so E is an ELLIPSE with $a = QT/2$.

See also CONE, SPHERE

References

Honsberger, R. "Kepler's Conics." Ch. 9 in *Mathematical Plums* (Ed. R. Honsberger). Washington, DC: Math. Assoc. Amer., p. 170, 1979.
Honsberger, R. *More Mathematical Morsels.* Washington, DC: Math. Assoc. Amer., pp. 40–4, 1991.
Ogilvy, C. S. *Excursions in Geometry.* New York: Dover, pp. 80–1, 1990.

Ogilvy, C. S. *Excursions in Mathematics.* New York: Dover, pp. 68–9, 1994.
Wells, D. *The Penguin Dictionary of Curious and Interesting Geometry.* London: Penguin, p. 48, 1991.

Danielson-Lanczos Lemma

The DISCRETE FOURIER TRANSFORM of length N (where N is EVEN) can be rewritten as the sum of two DISCRETE FOURIER TRANSFORMS, each of length $N/2$. One is formed from the EVEN-numbered points; the other from the ODD-numbered points. Denote the kth point of the DISCRETE FOURIER TRANSFORM by F_n. Then

$$F_n = \sum_{k=0}^{N-1} f_k e^{-2\pi i n k/N}$$

$$= \sum_{k=0}^{N/2-1} e^{-2\pi i k n/(N/2)} f_{2k} + W^n \sum_{k=0}^{N/2-1} e^{-2\pi i k n/(N/2)} f_{2k+1}$$

$$= F_n^e + W_n F_n^o,$$

where $W \equiv e^{-2\pi i/N}$ and $n = 0, \ldots, N$. This procedure can be applied recursively to break up the $N/2$ even and ODD points to their $N/4$ EVEN and ODD points. If N is a POWER of 2, this procedure breaks up the original transform into $1gN$ transforms of length 1. Each transform of an individual point has $F_n^{eeo\cdots} = f_k$ for some k. By reversing the patterns of evens and odds, then letting $e = 0$ and $o = 1$, the value of k in BINARY is produced. This is the basis for the FAST FOURIER TRANSFORM.

See also DISCRETE FOURIER TRANSFORM, FAST FOURIER TRANSFORM, FOURIER TRANSFORM

References

Press, W. H.; Flannery, B. P.; Teukolsky, S. A.; and Vetterling, W. T. *Numerical Recipes in C: The Art of Scientific Computing.* Cambridge, England: Cambridge University Press, pp. 407–11, 1989.

Darboux Integral

A variant of the RIEMANN INTEGRAL defined when the UPPER and LOWER INTEGRALS, taken as limits of the LOWER SUM

$$L(f; \phi; N) = \sum_{r=1}^{n} M(f; \delta_r) - \phi(x_{r-1})$$

and UPPER SUM

$$U(f; \phi; N) = \sum_{r=1}^{n} M(f; \delta_r) - \phi(x_{r-1}),$$

are equal. Here, $f(x)$ is a REAL FUNCTION, $\phi(x)$ is a monotonic increasing function with respect to which the sum is taken, $m(f; S)$ denotes the lower bound of $f(x)$ over the interval S, and $M(f; S)$ denotes the upper bound.

See also LOWER INTEGRAL, LOWER SUM, RIEMANN INTEGRAL, UPPER INTEGRAL, UPPER SUM

References

Kestelman, H. *Modern Theories of Integration, 2nd rev. ed.* New York: Dover, p. 250, 1960.

Darboux Problem

GOURSAT PROBLEM

Darboux Vector

The rotation VECTOR of the TRIHEDRON of a curve with CURVATURE $\kappa \neq 0$ when a point moves along a curve with unit SPEED. It is given by

$$\mathbf{D} = \tau \mathbf{T} + \kappa \mathbf{B}, \tag{1}$$

where τ is the TORSION, $\mathbf{T}$ the TANGENT VECTOR, and $\mathbf{B}$ the BINORMAL VECTOR. The Darboux vector field satisfies

$$\dot{\mathbf{T}} = \mathbf{D} \times \mathbf{T} \tag{2}$$

$$\dot{\mathbf{N}} = \mathbf{D} \times \mathbf{N} \tag{3}$$

$$\dot{\mathbf{B}} = \mathbf{D} \times \mathbf{B}. \tag{4}$$

See also BINORMAL VECTOR, CURVATURE, TANGENT VECTOR, TORSION (DIFFERENTIAL GEOMETRY)

References

Gray, A. *Modern Differential Geometry of Curves and Surfaces with Mathematica, 2nd ed.* Boca Raton, FL: CRC Press, p. 205, 1997.

Darboux's Formula

Darboux's formula is a theorem on the expansion of functions in infinite series. TAYLOR SERIES may be obtained as a special case of the formula, which may be stated as follows.

Let $f(z)$ be analytic at all points of the line joining a to z, and let $\phi(t)$ be any POLYNOMIAL of degree n in t. Then if $0 \leq t \leq 1$, differentiation gives

$$\frac{d}{dt} \sum_{m=1}^{\infty} (-1)^m (z-a)^m B^{(n-m)}(t) f^{(m)}(a + t(a-z))$$

$$= -(z-a)\phi^{(n)}(t) f'(a + t(z-a))$$

$$+ (-1)^n (z-a)^{n+1} \phi(t) f^{(n+1)}(a + t(z-a)). \tag{1}$$

But $\phi^{(n)}(t) = \phi^{(n)}(0)$, so integrating t over the interval 0 to 1 gives

$$\phi^{(n)}(0)[f(z) - f(a)]$$

$$= \sum_{m=1}^{n} (-1)^{m-1} (z-a)^m [\phi^{(n-m)}(1) f^{(m)}(z)$$

$$-\phi^{(n-m)}(0)f^{(m)}(a)]$$

$$+(-1)^n(z-a)^{n+1}\int_0^1 \phi(t)f^{(n+1)}(a+t(z-a))dt. \quad (2)$$

The TAYLOR SERIES follows by letting $\phi(t)=(t-1)^n$ and letting $n\to\infty$ (Whittaker and Watson 1990, p. 125).

See also BÜRMANN'S THEOREM, EULER-MACLAURIN INTEGRATION FORMULAS, MACLAURIN SERIES, TAYLOR SERIES

References

Whittaker, E. T. and Watson, G. N. "A Formula Due to Darboux." §7.1 in *A Course in Modern Analysis, 4th ed.* Cambridge, England: Cambridge University Press, p. 125, 1990.

Darboux-Stieltjes Integral
DARBOUX INTEGRAL

Darling's Products
A generalization of the HYPERGEOMETRIC FUNCTION identity

$$_2F_1(\alpha,\beta;\gamma;z)\,_2F_1(1-\alpha,1-\beta;2-\gamma;z)$$

$$=_2F_1(\alpha+1-\gamma,\beta+1-\gamma;2-\gamma;z)\,_2F_1(\gamma-\alpha,\gamma-\beta;\gamma;z)$$

$$(1)$$

to the GENERALIZED HYPERGEOMETRIC FUNCTION $_3F_2(a,b,c;d,e;x)$. Darling's products are

$$_3F_2\begin{bmatrix}\alpha,\beta,\gamma;z\\\delta,\varepsilon\end{bmatrix}\,_3F_2\begin{bmatrix}1-\alpha,1-\beta,1-\gamma;z\\2-\delta,2-\varepsilon\end{bmatrix}$$

$$=\frac{\varepsilon-1}{\varepsilon-\delta}\,_3F_2\begin{bmatrix}\alpha+1-\delta,\beta+1-\delta,\gamma+1-\delta;z\\2-\delta,\varepsilon+1-\delta\end{bmatrix}$$

$$\times\,_3F_2\begin{bmatrix}\delta-\alpha,\delta-\beta,\delta-\gamma;z\\\delta,\delta+1-\varepsilon\end{bmatrix}$$

$$+\frac{\delta-1}{\delta-\varepsilon}\,_3F_2\begin{bmatrix}\alpha+1-\varepsilon,\beta+1-\varepsilon,\gamma+1-\varepsilon;z\\2-\varepsilon,\delta+1-\varepsilon\end{bmatrix}$$

$$\times\,_3F_2\begin{bmatrix}\varepsilon-\alpha,\varepsilon-\beta,\varepsilon-\gamma;z\\\varepsilon,\varepsilon+1-\delta\end{bmatrix} \quad (2)$$

and

$$(1-z)^{\alpha+\beta+\gamma-\delta-\varepsilon}\,_3F_2\begin{bmatrix}\alpha,\beta,\gamma;z\\\delta,\varepsilon\end{bmatrix}$$

$$=\frac{\varepsilon-1}{\varepsilon-\delta}\,_3F_2\begin{bmatrix}\delta-\alpha,\delta-\beta,\delta-\gamma;z\\\delta,\delta+1-\varepsilon\end{bmatrix}$$

$$\times\,_3F_2\begin{bmatrix}\varepsilon-\alpha,\varepsilon-\beta,\varepsilon-\gamma;z\\\varepsilon-1,\varepsilon+1-\delta\end{bmatrix}$$

$$+\frac{\delta-1}{\delta-\varepsilon}\,_3F_2\begin{bmatrix}\varepsilon-\alpha,\varepsilon-\beta,\varepsilon-\gamma;z\\\varepsilon,\varepsilon+1-\delta\end{bmatrix}$$

$$\times\,_3F_2\begin{bmatrix}\delta-\alpha,\delta-\beta,\delta-\gamma;z\\\delta-1,\delta+1-\varepsilon\end{bmatrix}, \quad (3)$$

which reduce to (1) when $\gamma=\varepsilon\to\infty$.

See also GENERALIZED HYPERGEOMETRIC FUNCTION

References

Bailey, W. N. "Darling's Theorems of Products." §10.3 in *Generalised Hypergeometric Series.* Cambridge, England: Cambridge University Press, pp. 88–2, 1935.

Dart
PENROSE TILES

Darwin's Expansions
Series expansions of the PARABOLIC CYLINDER FUNCTIONS $U(a,x)$ and $W(a,x)$. The formulas can be found in Abramowitz and Stegun (1972).

See also PARABOLIC CYLINDER FUNCTION

References

Abramowitz, M. and Stegun, C. A. (Eds.). *Handbook of Mathematical Functions with Formulas, Graphs, and Mathematical Tables, 9th printing.* New York: Dover, pp. 689–90 and 694–95, 1972.

Darwin-de Sitter Spheroid
A SURFACE OF REVOLUTION OF THE FORM

$$r(\phi)=a\left[1-e\sin^2\phi-\left(\frac{3}{8}e^2+k\right)\sin^2(2\phi)\right],$$

where k is a second-order correction to the figure of a rotating fluid.

See also OBLATE SPHEROID, PROLATE SPHEROID, SPHEROID

References

Zharkov, V. N. and Trubitsyn, V. P. *Physics of Planetary Interiors.* Tucson, AZ: Pachart Publ. House, 1978.

Data Cube
A 3-D data set consisting of stacked 2-D data slices as a function of a third coordinate.

See also GRAPH (FUNCTION)

Data Structure
A formal structure for the organization of information. Examples of data structures include the LIST, QUEUE, STACK, and TREE.

References

Tarjan, R. E. *Data Structures and Network Algorithms.* Philadelphia, PA: SIAM Press, 1983.
Wood, D. *Data Structures, Algorithms, and Performance.* Reading, MA: Addison-Wesley, 1993.

Database

A database can be roughly defined as a structure consisting of

1. A collection of information (the data),
2. A collection of queries that can be submitted, and
3. A collection of algorithms by which the structure responds to queries, searches the data, and returns the results.

References

Petkovsek, M.; Wilf, H. S.; and Zeilberger, D. $A = B$. Wellesley, MA: A. K. Peters, p. 48, 1996.

Daubechies Wavelet Filter

A WAVELET used for filtering signals. Daubechies (1988, p. 980) has tabulated the numerical values up to order $p = 10$.

See also WAVELET

References

Daubechies, I. "Orthonormal Bases of Compactly Supported Wavelets." *Comm. Pure Appl. Math.* **41**, 909–96, 1988.
Press, W. H.; Flannery, B. P.; Teukolsky, S. A.; and Vetterling, W. T. "Interpolation and Extrapolation." Ch. 3 in *Numerical Recipes in FORTRAN: The Art of Scientific Computing, 2nd ed.* Cambridge, England: Cambridge University Press, pp. 584–86, 1992.

Davenport-Schinzel Sequence

Form a sequence from an ALPHABET of letters $[1, n]$ such that there are no consecutive letters and no alternating subsequences of length greater than d. Then the sequence is a Davenport-Schinzel sequence if it has maximal length $N_d(n)$. The value of $N_1(n)$ is the trivial sequence of 1s: 1, 1, 1, ... (Sloane's A000012). The values of $N_2(n)$ are the POSITIVE INTEGERS 1, 2, 3, 4, ... (Sloane's A000027). The values of $N_3(n)$ are the ODD INTEGERS 1, 3, 5, 7, ... (Sloane's A005408). The first nontrivial Davenport-Schinzel sequence $N_4(n)$ is given by 1, 4, 8, 12, 17, 22, 27, 32, ... (Sloane's A002004). Additional sequences are given by Guy (1994, p. 221) and Sloane.

References

Agarwal, P. K. and Sharir, M. "Davenport-Schinzel Sequences and Their Geometric Applications." Ch. 1 in *Handbook of Computational Geometry* (Ed. J.-R. Sack and J. Urrutia). Amsterdam, Netherlands: North-Holland, pp. 1–7, 2000.
Davenport, H. and Schinzel, A. "A Combinatorial Problem Connected with Differential Equations." *Amer. J. Math.* **87**, 684–90, 1965.
Guy, R. K. "Davenport-Schinzel Sequences." §E20 in *Unsolved Problems in Number Theory, 2nd ed.* New York: Springer-Verlag, pp. 220–22, 1994.
Roselle, D. P. and Stanton, R. G. "Results of Davenport-Schinzel Sequences." In *Proc. Louisiana Conference on Combinatorics, Graph Theory, and Computing.* Louisiana State University, Baton Rouge, March 1–, 1970 (Ed. R. C. Mullin, K. B. Reid, and D. P. Roselle). Winnipeg, Manitoba: Utilitas Mathematica, pp. 249–67, 1960.
Sharir, M. and Agarwal, P. *Davenport-Schinzel Sequences and Their Geometric Applications.* New York: Cambridge University Press, 1995.
Sloane, N. J. A. Sequences A000012/M0003, A000027/ M0472, and A002004/M3328 in "An On-Line Version of the Encyclopedia of Integer Sequences." http://www.research.att.com/~njas/sequences/eisonline.html.

Davey-Stewartson Equations

The system of PARTIAL DIFFERENTIAL EQUATIONS

$$iu_t + u_{xx} + \alpha u_{yy} + \beta u |u|^2 - uv = 0$$

$$v_{xx} + \gamma v_{yy} + \delta \left(|u|^2 \right)_{yy} = 0.$$

References

Champagne, B. and Winternitz, P. "On the Infinite-Dimensional Group of the Davey-Stewartson Equations." *J. Math. Phys.* **29**, 1–, 1988.
Zwillinger, D. *Handbook of Differential Equations, 3rd ed.* Boston, MA: Academic Press, p. 137, 1997.

Dawson's Integral

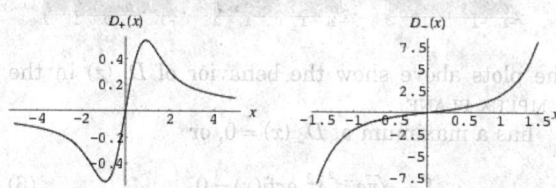

An INTEGRAL which arises in computation of the Voigt lineshape:

$$D(x) \equiv e^{-x^2} \int_0^x e^{y^2} dy. \qquad (1)$$

It is sometimes generalized such that

$$D_{\pm}(x) \equiv e^{\mp x^2} \int_0^x e^{\pm y^2} dy, \qquad (2)$$

giving

$$D_+(x) = \frac{1}{2} \sqrt{\pi} e^{-x^2} \, \mathrm{erfi}(x) \qquad (3)$$

$$D_-(x) = \frac{1}{2} \sqrt{\pi} e^{x^2} \, \mathrm{erf}(x), \qquad (4)$$

where $\mathrm{erf}(z)$ is the ERF function and $\mathrm{erfi}(z)$ is the imaginary error function ERFI. $D_+(x)$ is illustrated in the left figure above, and $D_-(x)$ in the right figure.

$D_+(x)$ has an ASYMPTOTIC SERIES

$$D_+(x) \sim \frac{1}{2x} + \frac{1}{4x^3} + \dots \quad (5)$$

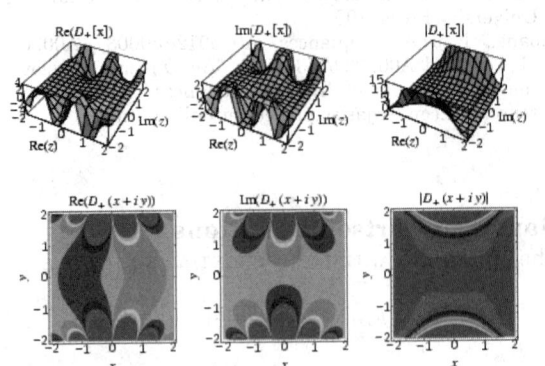

The plots above show the behavior of $D_+(z)$ in the COMPLEX PLANE.

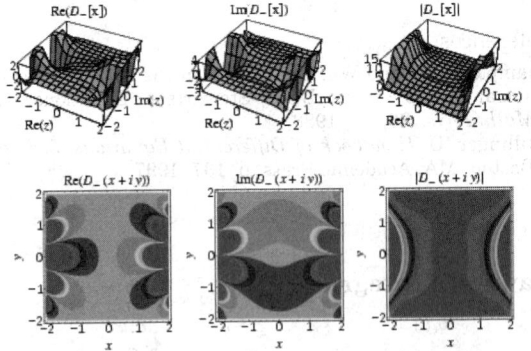

The plots above show the behavior of $D_-(z)$ in the COMPLEX PLANE.

D_+ has a maximum at $D'_+(x) = 0$, or

$$1 - \sqrt{\pi}e^{-x^2}x^2\,\mathrm{erfi}(x) = 0, \quad (6)$$

giving

$$D_+(0.9241388730) = 0.5410442246, \quad (7)$$

and an inflection at $D''_+(x) = 0$, or

$$-2x + \sqrt{\pi}e^{-x^2}(2x^2 - 1)\,\mathrm{erfi}(x) = 0, \quad (8)$$

giving

$$D_+(1.5019752683) = 0.4276866160. \quad (9)$$

See also ERFI, GAUSSIAN FUNCTION

References

Abramowitz, M. and Stegun, C. A. (Eds.). *Handbook of Mathematical Functions with Formulas, Graphs, and Mathematical Tables, 9th printing.* New York: Dover, p. 298, 1972.
Press, W. H.; Flannery, B. P.; Teukolsky, S. A.; and Vetterling, W. T. "Dawson's Integrals." §6.10 in *Numerical Recipes in FORTRAN: The Art of Scientific Computing,* 2nd ed. Cambridge, England: Cambridge University Press, pp. 252–54, 1992.
Spanier, J. and Oldham, K. B. "Dawson's Integral." Ch. 42 in *An Atlas of Functions.* Washington, DC: Hemisphere, pp. 405–10, 1987.

dc

JACOBI ELLIPTIC FUNCTIONS

de Bruijn Constant

Also called the COPSON-DE BRUIJN CONSTANT. It is the minimal constant

$$c = 1.0164957714\dots$$

such that the inequality

$$\sum_{n=1}^{\infty} a_n \le c \sum_{n=1}^{\infty} \sqrt{\frac{a_n^2 + a_{n+1}^2 + a_{n+2}^2 + \dots}{n}}$$

always holds.

References

Copson, E. T. "Note on Series of Positive Terms." *J. London Math. Soc.* **2**, 9–2, 1927.
Copson, E. T. "Note on Series of Positive Terms." *J. London Math. Soc.* **3**, 49–1, 1928.
de Bruijn, N. G. *Asymptotic Methods in Analysis.* New York: Dover, 1981.
Finch, S. "Favorite Mathematical Constants." http://www.mathsoft.com/asolve/constant/copson/copson.html.

de Bruijn Diagram

DE BRUIJN GRAPH

de Bruijn Graph

A graph whose nodes are sequences of symbols from some ALPHABET and whose edges indicate the sequences which might overlap.

References

Golomb, S. W. *Shift Register Sequences.* San Francisco, CA: Holden-Day, 1967.
Ralston, A. "de Bruijn Sequences--A Model Example of the Interaction of Discrete Mathematics and Computer Science." *Math. Mag.* **55**, 131–43, 1982.

de Bruijn Sequence

The shortest circular sequence of length σ^a such that every string of length n on the ALPHABET a of size σ occurs as a contiguous subrange of the sequence described by a. A de Bruijn sequence can be generated using DeBruijnSequence[a, n] in the *Mathematica* add-on package DiscreteMath`Combinatorica` (which can be loaded with the command < <DiscreteMath`). For example, a de Bruijn sequence of order n on the alphabet $\{a, b, c\}$ is given by $\{a, a, c, b, b, c, c, a, b\}$.

Every de Bruijn sequence corresponds to an EULER-IAN CYCLE on a DE BRUIJN GRAPH. Surprisingly, it turns out that the lexicographic sequence of LYNDON WORDS of lengths DIVISIBLE by n gives the lexicographically smallest de Bruijn sequence (Ruskey).

de Bruijn sequences can be generated by feedback shift registers (Golomb 1966; Ronse 1984; Skiena 1990, p. 196).

See also DE BRUIJN GRAPH, LYNDON WORD

References

de Bruijn, N. G. "A Combinatorial Problem." *Koninklijke Nederlandse Akademie v. Wetenschappen* **49**, 758–64, 1946.

Golomb, S. W. *Shift Register Sequences.* San Francisco, CA: Holden-Day, 1967.

Good, I. J. "Normal Recurring Decimals." *J. London Math. Soc.* **21**, 167–72, 1946.

Knuth, D. E. "Oriented Subtrees of an Arc Digraph." *J. Combin. Th.* **3**, 309–14, 1967.

Ronse, C. *Feedback Shift Registers.* Berlin: Springer-Verlag, 1984.

Ruskey, F. "Information on Necklaces, Lyndon Words, de Bruijn Sequences." http://www.theory.csc.uvic.ca/~cos/inf/neck/NecklaceInfo.html.

Skiena, S. *Implementing Discrete Mathematics: Combinatorics and Graph Theory with Mathematica.* Reading, MA: Addison-Wesley, pp. 195–96, 1990.

de Bruijn's Theorem

A box can be packed with a HARMONIC BRICK $a \times ab \times abc$ IFF the box has dimensions $ap \times abq \times abcr$ for some natural numbers p, q, r (i.e., the box is a multiple of the brick).

See also BOX-PACKING THEOREM, CONWAY PUZZLE, KLARNER'S THEOREM

References

Honsberger, R. *Mathematical Gems II.* Washington, DC: Math. Assoc. Amer., pp. 69–2, 1976.

de Bruijn-Newman Constant

N.B. A detailed online essay by S. Finch was the starting point for this entry.

Let Ξ be the XI FUNCTION defined by

$$\Xi(iz) = \frac{1}{2}\left(z^2 - \frac{1}{4}\right)\pi^{-z/2-\frac{1}{4}}\Gamma\left(\frac{1}{2}z + \frac{1}{4}\right)\zeta\left(z + \frac{1}{2}\right). \quad (1)$$

$\Xi(z/2)/8$ can be viewed as the FOURIER TRANSFORM of the signal

$$\Phi(t) = \sum_{n=1}^{\infty}\left(2\pi^2 n^4 e^{9t} - 3\pi n^2 e^{5t}\right)e^{-\pi n^2 e^{4t}} \quad (2)$$

for $t \in \mathbb{R} \geq 0$. Then denote the FOURIER TRANSFORM of $\Phi(t)e^{\lambda t^2}$ as $H(\lambda, z)$,

$$\mathscr{F}\left[\Phi(t)e^{\lambda t^2}\right] = H(\lambda, z). \quad (3)$$

de Bruijn (1950) proved that H has only REAL zeros for $\lambda \geq 1/2$. C. M. Newman (1976) proved that there exists a constant Λ such that H has only REAL zeros IFF $\lambda \geq \Lambda$. The best current lower bound (Csordas *et al.* 1993, 1994) is $\Lambda > -5.895 \times 10^{-9}$. The RIEMANN HYPOTHESIS is equivalent to the conjecture that $\Lambda \leq 0$.

See also XI FUNCTION

References

Csordas, G.; Odlyzko, A.; Smith, W.; and Varga, R. S. "A New Lehmer Pair of Zeros and a New Lower Bound for the de Bruijn-Newman Constant." *Elec. Trans. Numer. Analysis* **1**, 104–11, 1993.

Csordas, G.; Smith, W.; and Varga, R. S. "Lehmer Pairs of Zeros, the de Bruijn-Newman Constant and the Riemann Hypothesis." *Constr. Approx.* **10**, 107–29, 1994.

de Bruijn, N. G. "The Roots of Trigonometric Integrals." *Duke Math. J.* **17**, 197–26, 1950.

Finch, S. "Favorite Mathematical Constants." http://www.mathsoft.com/asolve/constant/dbnwm/dbnwm.html.

Newman, C. M. "Fourier Transforms with only Real Zeros." *Proc. Amer. Math. Soc.* **61**, 245–51, 1976.

de Gua's Theorem

The square of the AREA of the base (i.e., the face opposite the right TRIHEDRAL ANGLE) of a TRIRECTANGULAR TETRAHEDRON is equal to the sum of the squares of the AREAS of its other three faces. This theorem was presented to the Paris Academy of Sciences in 1783 by J. P. de Gua de Malves (1712–785), although it was known to Descartes (1859) and to Faulhaber (Altshiller-Court 1979, p. 300). It is a special case of a general theorem presented by Tinseau to the Paris Academy in 1774 (Osgood and Graustein 1930, p. 517; Altshiller-Court 1979).

See also PYTHAGOREAN THEOREM, TRIRECTANGULAR TETRAHEDRON

References

Altshiller-Court, N. *Modern Pure Solid Geometry.* New York: Chelsea, pp. 92 and 300, 1979.

Descartes, R. *Oeuvres inédites de Descartes.* Paris, 1859.

Osgood, W. F. and Graustein, W. C. *Plane and Solid Analytic Geometry.* New York: Macmillan, Th. 2, p. 517, 1930.

de Jonquières Theorem

For an algebraic curve, the total number of groups of a g_N^r consisting in a point of multiplicity k_1, one of multiplicity k_2, ..., one of multiplicity k_p, where

$$\sum k_i = N \quad (1)$$

$$\sum (k_i - 1) = r, \quad (2)$$

and where α_1 points have one multiplicity, α_2 another, etc., and

$$\prod = k_1 k_2 \ldots k_p \quad (3)$$

is

$$\prod \frac{p(p-1)\ldots(p-\rho)}{\alpha_1!\alpha_2!\cdots}$$

$$\times \left[\frac{\Pi}{p-\rho} - \frac{\sum_i \frac{\partial \Pi}{\partial k_i}}{p-\rho+1} + \frac{\sum_{ij} \frac{\partial^2 \Pi}{\partial k_i \partial k_j}}{p-\rho+2} + \cdots \right]. \quad (4)$$

References

Coolidge, J. L. *A Treatise on Algebraic Plane Curves.* New York: Dover, p. 288, 1959.

de Jonquières Transformation

A transformation of an algebraic curve which is of the same type as its inverse. A de Jonquières transformation is always factorable.

References

Coolidge, J. L. *A Treatise on Algebraic Plane Curves.* New York: Dover, pp. 203–04, 1959.

de la Loubere's Method

A method for constructing MAGIC SQUARES of ODD order, also called the SIAMESE METHOD.

See also MAGIC SQUARE

de Longchamps Point

The reflection of the ORTHOCENTER about the CIRCUMCENTER of a TRIANGLE. This point is also the ORTHOCENTER of the ANTICOMPLEMENTARY TRIANGLE. It has TRIANGLE CENTER FUNCTION

$$\alpha = \cos A - \cos B \cos C.$$

The SODDY LINE intersects the EULER LINE in the de Longchamps point (Oldknow 1996).

See also CIRCUMCENTER, EULER LINE, ORTHOCENTER, SODDY LINE

References

Altshiller-Court, N. "On the de Longchamps Circle of the Triangle." *Amer. Math. Monthly* **33**, 368–75, 1926.
Kimberling, C. "Central Points and Central Lines in the Plane of a Triangle." *Math. Mag.* **67**, 163–87, 1994.
Oldknow, A. "The Euler-Gergonne-Soddy Triangle of a Triangle." *Amer. Math. Monthly* **103**, 319–29, 1996.
Vandeghen, A. "Soddy's Circles and the de Longchamps Point of a Triangle." *Amer. Math. Monthly* **71**, 176–79, 1964.

de Méré's Problem

The probability of getting at least one "6" in four rolls of a single 6-sided DIE is

$$1 - \left(\frac{5}{6}\right)^4 \approx 0.5177, \quad (1)$$

which is slightly higher than the probability of at least one double-six in 24 throws of two dice,

$$1 - \left(\frac{35}{36}\right)^{24} \approx 0.4914. \quad (2)$$

The French nobleman and gambler Chevalier de Méré suspected that (1) was higher than (2), but his mathematical skills were not great enough to demonstrate why this should be so. He posed the question to Pascal, who solved the problem and proved de Méré correct. In fact, de Méré's observation remains true even if two dice are thrown 25 times, since the probability of throwing at least one double-six is then

$$1 - \left(\frac{35}{36}\right) 25 \approx 0.5055. \quad (3)$$

See also BOXCARS, DICE

References

Gonick, L. and Smith, W. *The Cartoon Guide to Statistics.* New York: Harper Perennial, pp. 28–9 and 44–5, 1993.
Kraitchik, M. "A Dice Problem." §6.2 in *Mathematical Recreations.* New York: W. W. Norton, pp. 118–19, 1942.
Uspensky, J. V. *Introduction to Mathematical Probability.* New York: McGraw-Hill, pp. 21–2, 1937.

de Moivre Number

A solution $\zeta_k = e^{2\pi i k/d}$ to the CYCLOTOMIC EQUATION

$$x^d = 1.$$

The de Moivre numbers give the coordinates in the COMPLEX PLANE of the VERTICES of a REGULAR POLYGON with d sides and unit RADIUS.

n	de Moivre Number
2	± 1
3	$1, \frac{1}{2}\left(-1 \pm i\sqrt{3}\right)$
4	$\pm 1, \pm i$

$$5 \quad 1, \frac{1}{4}\left(-1+\sqrt{5}\pm i\sqrt{10+2\sqrt{5}}\right),$$
$$\frac{1}{4}\left(-1-\sqrt{5}\pm i\sqrt{10-2\sqrt{5}}\right)$$
$$6 \quad \pm 1, \pm \frac{1}{2}\left(\pm 1 + i\sqrt{3}\right)$$

See also CYCLOTOMIC EQUATION, CYCLOTOMIC POLYNOMIAL, EUCLIDEAN NUMBER

References

Conway, J. H. and Guy, R. K. *The Book of Numbers.* New York: Springer-Verlag, 1996.

de Moivre's Identity

$$e^{i(n\theta)} = \left(e^{i\theta}\right)^n. \tag{1}$$

From the EULER FORMULA it follows that

$$\cos(n\theta) + i\sin(n\theta) = (\cos\theta + i\sin\theta)^n. \tag{2}$$

A similar identity holds for the HYPERBOLIC FUNCTIONS,

$$(\cosh z + \sinh z)^n = \cosh(nz) + \sinh(nz). \tag{3}$$

See also EULER FORMULA

References

Arfken, G. *Mathematical Methods for Physicists, 3rd ed.* Orlando, FL: Academic Press, pp. 356–57, 1985.
Courant, R. and Robbins, H. *What is Mathematics?: An Elementary Approach to Ideas and Methods, 2nd ed.* Oxford, England: Oxford University Press, pp. 96–00, 1996.
Nagell, T. *Introduction to Number Theory.* New York: Wiley, p. 156, 1951.

de Moivre's Quintic

A QUINTIC EQUATION OF THE FORM

$$x^5 + ax^3 + \frac{1}{5}a^2 x + b = 0.$$

See also QUINTIC EQUATION

de Moivre-Laplace Theorem

The asymptotic form of the n-step BERNOULLI DISTRIBUTION with parameters p and $q = 1 - p$ is given by

$$P_n(k) = \binom{n}{k}p^k q^{n-k} \sim \frac{1}{\sqrt{2\pi npq}}e^{-(k-np)^2/(2npq)} \tag{1}$$

(Papoulis 1984, p. 66).

Uspensky (1937) defines the de Moivre-Laplace theorem as the fact that the sum of those terms of the BINOMIAL SERIES of $(p+q)^n$ for which the number of successes x falls between d_1 and d_2 is approximately

$$Q \approx \frac{1}{\sqrt{2\pi}}\int_{t_1}^{t_2}e^{-t^2/2}dt, \tag{2}$$

where

$$t_1 \equiv \frac{d_1 - \frac{1}{2} - np}{\sigma} \tag{3}$$

$$t_2 \equiv \frac{d_2 + \frac{1}{2} - np}{\sigma} \tag{4}$$

$$\sigma \equiv \sqrt{npq}. \tag{5}$$

More specifically, Uspensky (1937, p. 129) showed that

$$Q = \frac{1}{\sqrt{2\pi}}\int_{t_1}^{t_2}e^{-t^2/2}dt + \frac{q-p}{6\sqrt{2\pi}\sigma}\left[(1-t^2)e^{-t^2/2}\right]_{t_1}^{t_2} + \Omega, \tag{6}$$

where the error term satisfies

$$|\Omega| < \frac{0.13 + 0.18|p\text{-}q|}{\sigma^2} + e^{-3\sigma/2} \tag{7}$$

for $\sigma \geq 5$ (Uspensky 1937, p. 129; Kenney and Keeping 1958, pp. 36–7). Note that Kenney and Keeping (1958, p. 37) give the slightly smaller DENOMINATOR $0.12 + 0.18|p-q|$.

A COROLLARY states that the probability that x successes in n trials will differ from the expected value np by more than d is $P\delta = 1 - Q_\delta$, where

$$Q_\delta = \frac{2}{\sqrt{2\pi}}\int_0^\delta e^{-t^2/2}dt, \tag{8}$$

with

$$\delta \equiv \frac{d + \frac{1}{2}}{\sigma} \tag{9}$$

(Kenney and Keeping 1958, p. 39). Uspensky (1937, p. 130) showed that $Q_{\delta_1} \equiv P(|x - np| \leq d)$ is given by

$$Q_{\delta_1} = \frac{2}{\sqrt{2\pi}}\int_0^{\delta_1}e^{-u^2/2}du + \frac{1-\theta_1-\theta_2}{\sqrt{2\pi}\sigma}e^{-\delta_1^2/2} + \Omega_1, \tag{10}$$

where

$$\delta_1 \equiv \frac{d}{\delta} \tag{11}$$

$$\theta_1 \equiv (nq + d) - \lfloor nq + d \rfloor \qquad (12)$$

$$\theta_2 \equiv (np + d) - \lfloor np + d \rfloor \qquad (13)$$

and the error term satisfies

$$|\Omega_1| < \frac{0.20 + 0.25|p - q|}{\sigma^2} + e^{-3\sigma/2}, \qquad (14)$$

for $\sigma \geq 5$ (Uspensky 1937, p. 130; Kenney and Keeping 1958, pp. 40–1).

See also BERNOULLI DISTRIBUTION, BINOMIAL SERIES, GAUSSIAN DISTRIBUTION, NORMAL DISTRIBUTION, WEAK LAW OF LARGE NUMBERS

References
de la Vallée-Poussin, C. "Demonstration nouvelle du théorème de Bernoulli." *Ann. Soc. Sci. Bruxelles* **31**, 219–36, 1907.
de Moivre, A. *Miscellanea analytica.* Lib. 5, 1730.
de Moivre, A. *The Doctrine of Chances, or, a Method of Calculating the Probabilities of Events in Play, 3rd ed.* New York: Chelsea, 2000. Reprint of 1756 3rd ed. Original ed. published 1716.
Kenney, J. F. and Keeping, E. S. "The DeMoivre-Laplace Theorem" and "Simple Sampling of Attributes." §2.10 and 2.11 in *Mathematics of Statistics, Pt. 2, 2nd ed.* Princeton, NJ: Van Nostrand, pp. 36–1, 1951.
Laplace, P. *Théorie analytiques de probabilités, 3ème éd., revue et augmentée par l'auteur.* Paris: Courcier, 1820. Reprinted in Œuvres complètes de Laplace, tome 7. Paris: Gauthier-Villars, pp. 280–85, 1886.
Mirimanoff, D. "Le jeu de pile ou face et les formules de Laplace et de J. Eggenberger." *Commentarii Mathematici Helvetici* **2**, 133–68, 1930.
Papoulis, A. *Probability, Random Variables, and Stochastic Processes, 2nd ed.* New York: McGraw-Hill, 1984.
Uspensky, J. V. "Approximate Evaluation of Probabilities in Bernoullian Case." Ch. 7 in *Introduction to Mathematical Probability.* New York: McGraw-Hill, pp. 119–38, 1937.

de Morgan's and Bertrand's Test
BERTRAND'S TEST

de Morgan's Duality Law

For every proposition involving logical addition and multiplication ("or" and "and"), there is a corresponding proposition in which the words "addition" and "multiplication" are interchanged.

de Morgan's Laws

Let $\cup$ represent "or", $\cap$ represent "and", and $'$ represent "not." Then, for two logical units E and F,

$$(E \cup F)' = E' \cap F'$$

$$(E \cap F)' = E' \cup F'.$$

These laws also apply in the more general context of BOOLEAN ALGEBRA and, in particular, in the BOOLEAN ALGEBRA of SET THEORY, in which case $\cup$ would denote UNION, $\cap$ INTERSECTION, and $'$ complementation with respect to any superset of E and F.

References
Dugundji, J. *Topology.* Englewood Cliffs, NJ: Prentice-Hall, 1965.
Halmos, P. R. *Naive Set Theory.* New York: Springer-Verlag, 1974.
Kelley, J. L. *General Topology.* New York: Springer-Verlag, 1975.
Papoulis, A. *Probability, Random Variables, and Stochastic Processes, 2nd ed.* New York: McGraw-Hill, p. 23, 1984.
Simpson, R. E. *Introductory Electronics for Scientists and Engineers, 2nd ed.* Boston, MA: Allyn and Bacon, pp. 540–41, 1987.

de Polignac's Conjecture

Every EVEN NUMBER is the difference of two consecutive PRIMES in infinitely many ways (Dickson 1952, p. 424). If true, taking the difference 2, this conjecture implies that there are infinitely many TWIN PRIMES (Ball and Coxeter 1987). The CONJECTURE has never been proven true or refuted.

See also EVEN NUMBER, GOLDBACH CONJECTURE, TWIN PRIMES

References
Ball, W. W. R. and Coxeter, H. S. M. *Mathematical Recreations and Essays, 13th ed.* New York: Dover, p. 64, 1987.
Burton, D. M. *Elementary Number Theory, 4th ed.* Boston, MA: Allyn and Bacon, p. 76, 1989.
de Polignac, A. "Six propositions arithmologiques déduites de crible d'Ératosthène." *Nouv. Ann. Math.* **8**, 423–29, 1849.
de Polignac, A. *Comptes Rendus Paris* **29**, 400 and 738–39, 1849.
Dickson, L. E. *History of the Theory of Numbers, Vol. 1: Divisibility and Primality.* New York: Chelsea, 1952.

de Rham Cohomology

de Rham cohomology is a formal set-up for the analytic problem: If you have a DIFFERENTIAL K-FORM ω on a MANIFOLD M, is it the EXTERIOR DERIVATIVE of another DIFFERENTIAL K-FORM ω'? Formally, if $\omega = d\omega'$ then $d\omega = 0$.. This is more commonly stated as $d \circ d = 0$, meaning that if ω is to be the EXTERIOR DERIVATIVE of a DIFFERENTIAL K-FORM, a NECESSARY condition that ω must satisfy is that its EXTERIOR DERIVATIVE is zero.

de Rham cohomology gives a formalism that aims to answer the question, "Are all differential k-forms on a MANIFOLD with zero EXTERIOR DERIVATIVE the EXTERIOR DERIVATIVES of $(k-1)$-forms?" In particular, the kth de Rham cohomology vector space is defined to be the space of all k-forms with EXTERIOR DERIVATIVE 0, modulo the space of all boundaries of $(k-1)$-forms. This is the trivial VECTOR SPACE IFF the answer to our question is yes.

The fundamental result about de Rham cohomology is that it is a topological invariant of the MANIFOLD, namely: the kth de Rham cohomology VECTOR SPACE of a MANIFOLD M is canonically isomorphic to the ALEXANDER-SPANIER COHOMOLOGY VECTOR SPACE $H^k(M; \mathbb{R})$ (also called cohomology with compact support). In the case that M is compact, ALEXANDER-SPANIER COHOMOLOGY is exactly singular cohomology.

See also ALEXANDER-SPANIER COHOMOLOGY, CHANGE OF VARIABLES THEOREM, COHOMOLOGY, DIFFERENTIAL K-FORM, EXTERIOR DERIVATIVE, VECTOR SPACE

de Sluze Conchoid
CONCHOID OF DE SLUZE

de Sluze Pearls
PEARLS OF SLUZE

Dead Variable
DUMMY VARIABLE

Debye Functions

$$\int_0^x \frac{t^n dt}{e^t - 1} = x^n \left[\frac{1}{n} - \frac{x}{2(n+1)} + \sum_{k=1}^\infty \frac{B_{2k} x^{2k}}{(2k+n)(2k!)} \right], \tag{1}$$

where $|x| < 2\pi$ and B_n are BERNOULLI NUMBERS.

$$\int_x^\infty \frac{t^n dt}{e^t - 1}$$

$$= \sum_{k=1}^\infty e^{-kx} \left[\frac{x^n}{k} + \frac{nx^{n-1}}{k^2} + \frac{n(n-1)x^{n-2}}{k^3} + \ldots + \frac{n!}{k^{n+1}} \right], \tag{2}$$

where $x > 0$. The sum of these two integrals is

$$\int_0^\infty \frac{t^n dt}{e^t - 1} = n! \zeta(n+1), \tag{3}$$

where $\zeta(z)$ is the RIEMANN ZETA FUNCTION.

References
Abramowitz, M. and Stegun, C. A. (Eds.). "Debye Functions." §27.1 in *Handbook of Mathematical Functions with Formulas, Graphs, and Mathematical Tables, 9th printing.* New York: Dover, p. 998, 1972.

Debye's Asymptotic Representation
An asymptotic expansion for a HANKEL FUNCTION OF THE FIRST KIND

$$H_\nu^{(1)}(x) \sim \frac{1}{\sqrt{\pi}} \exp\{ix[\cos\alpha + (\alpha - \pi/2)\sin\alpha]\}$$

$$\times \left[\frac{e^{i\pi/4}}{X} + \left(\frac{1}{8} + \frac{5}{24}\tan^2\alpha \right) \frac{3e^{3\pi i/4}}{2X^3} \right.$$

$$\left. + \left(\frac{3}{128} + \frac{77}{576}\tan^\alpha + \frac{385}{3456}\tan^4\alpha \right) \frac{3 \cdot e^{5\pi i/4}}{2^2 X^5} + \ldots \right],$$

where

$$\frac{\nu}{x} = \sin\alpha,$$

$$1 - \frac{\nu}{x} > \frac{3}{x}\nu^{1/2},$$

and

$$X \equiv \sqrt{-x \cos\left(\frac{1}{2}\alpha \right)}.$$

See also HANKEL FUNCTION OF THE FIRST KIND

References
Iyanaga, S. and Kawada, Y. (Eds.). *Encyclopedic Dictionary of Mathematics.* Cambridge, MA: MIT Press, p. 1475, 1980.

Decade
A power of 10.

See also OCTAVE

Decagon

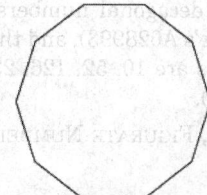

The constructible regular 10-sided POLYGON with SCHLÄFLI SYMBOL $\{10\}$. The INRADIUS r, CIRCUMRADIUS R, and AREA can be computed directly from the formulas for a general REGULAR POLYGON with side length s and $n = 10$ sides,

$$r = \frac{1}{2} s \cot\left(\frac{\pi}{10}\right) = \frac{1}{2}\sqrt{25 - 10\sqrt{5}s} \quad (1)$$

$$R = \frac{1}{2} s \csc\left(\frac{\pi}{10}\right) = \frac{1}{2}\left(1 + \sqrt{5}\right)s = \phi s \quad (2)$$

$$A = \frac{1}{4} n s^2 \cot\left(\frac{\pi}{10}\right) = \frac{5}{2}\sqrt{5 + 2\sqrt{5}s^2}. \quad (3)$$

Here, ϕ is the GOLDEN MEAN.

See also DECAGRAM, DODECAGON, TRIGONOMETRY VALUES PI/10, UNDECAGON

References

Dixon, R. *Mathographics.* New York: Dover, p. 18, 1991.

Decagonal Number

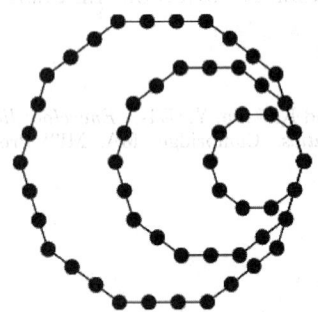

A FIGURATE NUMBER OF THE FORM $4n^2 - 3n$. The first few are 1, 10, 27, 52, 85, ... (Sloane's A001107). The GENERATING FUNCTION giving the decagonal numbers is

$$\frac{x(7x + 1)}{(1 - x)^3} = x + 10x^2 + 27x^3 + 52x^4 + \ldots$$

The first few odd decagonal numbers are 1, 27, 85, 175, 297, ... (Sloane's A028993), and the first few even decagonal numbers are 10, 52, 126, 232, 360, 540, ... (Sloane's A028994).

See also DECAGON, FIGURATE NUMBER

References

Sloane, N. J. A. Sequences A001107/M4690, A028993, and A028994 in "An On-Line Version of the Encyclopedia of Integer Sequences." http://www.research.att.com/~njas/sequences/eisonline.html.

Decagram

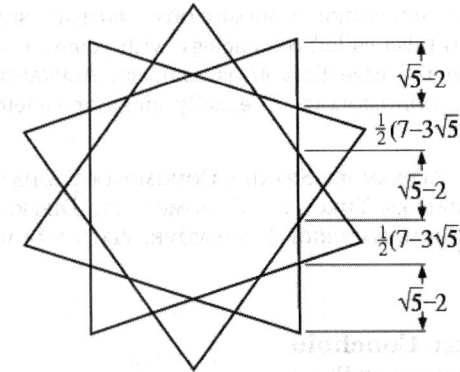

The STAR POLYGON $\{10/3\}$.

See also DECAGON, STAR POLYGON

Decahedral Graph

A POLYHEDRAL GRAPH having 10 vertices. There are 32,300 nonisomorphic nonahedral graphs, as first enumerated by Duijvestijn and Federico (1981).

See also POLYHEDRAL GRAPH

References

Duijvestijn, A. J. W. and Federico, P. J. "The Number of Polyhedral (3-Connected Planar) Graphs." *Math. Comput.* **37**, 523–32, 1981.

Decic Surface

An ALGEBRAIC SURFACE which can be represented implicitly by a POLYNOMIAL of degree 10 in x, y, and z. An example is the BARTH DECIC.

See also ALGEBRAIC SURFACE, BARTH DECIC, CUBIC SURFACE, QUADRATIC SURFACE, QUARTIC SURFACE

Decidable

A THEORY is decidable IFF there is an algorithm which can determine whether or not any SENTENCE r is a member of the THEORY.

See also CHURCH-TURING THESIS, DETERMINISTIC, GÖDEL'S COMPLETENESS THEOREM, GÖDEL'S INCOMPLETENESS THEOREM, KREISEL CONJECTURE, SENTENCE, TARSKI'S THEOREM, THEORY, UNDECIDABLE

References

Enderton, H. B. *Elements of Set Theory.* New York: Academic Press, 1977.

Kemeny, J. G. "Undecidable Problems of Elementary Number Theory." *Math. Ann.* **135**, 160–69, 1958.

Decillion

In the American system, 10^{33}.

See also LARGE NUMBER

Decimal

The BASE-10 notational system for representing REAL NUMBERS. The expression of a number in the decimal system is called its DECIMAL EXPANSION, examples of which are 1, 13, 2028, 12.1, and 3.14159. Each number is called a decimal DIGIT, and the period placed to the right of the units place in a decimal number is called the DECIMAL POINT.

See also 10, BASE (NUMBER), BINARY, DECIMAL POINT, HEXADECIMAL, NEGADECIMAL, OCTAL

References

Pappas, T. "The Evolution of Base Ten." *The Joy of Mathematics.* San Carlos, CA: Wide World Publ./Tetra, pp. 2–, 1989.

Wells, D. *The Penguin Dictionary of Curious and Interesting Numbers.* Middlesex, England: Penguin Books, pp. 78–0, 1986.

Decimal Comma

The symbol used in continental Europe to denote a DECIMAL POINT, point example 3,14159....

See also DECIMAL POINT

Decimal Expansion

The decimal expansion of a number is its representation in base 10. For example, the decimal expansion of 25^2 is 625, of π is 3.14159..., and of 1/9 is 0.1111....

If $r = p/q$ has a finite decimal expansion, then

$$r = \frac{a_1}{10} + \frac{a_2}{10^2} + \ldots + \frac{a_n}{10^n}$$

$$= \frac{a_1 10^{n-1} + a_2 10^{n-2} + \ldots + a_n}{10^n}$$

$$= \frac{a_1 10^{n-1} + a_2 10^{n-2} + \ldots + a_n}{2^n \cdot 5^n}. \tag{1}$$

FACTORING possible common multiples gives

$$r = \frac{p}{2^\alpha 5^\beta}, \tag{2}$$

where $p \not\equiv 0 \pmod{2, 5}$. Therefore, the numbers with finite decimal expansions are fractions of this form. The number of decimals is given by $\max(\alpha, \beta)$ (Wells 1986, p. 60). Numbers which have a finite decimal expansion are called REGULAR NUMBERS.

Any NONREGULAR fraction m/n is periodic, and has a period $\lambda(n)$ independent of m, which is at most $n - 1$ DIGITS long. If n is RELATIVELY PRIME to 10, then the period $\lambda(n)$ of m/n is a divisor of $\phi(n)$ and has at most $\phi(n)$ DIGITS, where ϕ is the TOTIENT FUNCTION. It turns out that $\lambda(n)$ is the HAUPT-EXPONENT of 10 (mod n) (Glaisher 1878, Lehmer 1941). When a rational number m/n with $(m, n) = 1$ is expanded, the period begins after s terms and has length t, where s and t

are the smallest numbers satisfying

$$10^2 \equiv 10^{s+t} \pmod{n}. \tag{3}$$

When $n \not\equiv 0 \pmod{2, 5}$, $s = 0$, and this becomes a purely periodic decimal with

$$10^t \equiv 1 \pmod{n}. \tag{4}$$

As an example, consider $n = 84$.

$$\begin{aligned}
10^0 &\equiv 1 \quad & 10^1 &\equiv 10 \quad & 10^2 &\equiv 16 \quad & 10^3 &\equiv -8 \\
10^4 &\equiv 4 \quad & 10^5 &\equiv 40 \quad & 10^6 &\equiv -20 \quad & 10^7 &\equiv -32, \\
10^8 &\equiv 16
\end{aligned}$$

so $s = 2$, $t = 6$. The decimal representation is $1/84 = 0.01\overline{190476}$. When the DENOMINATOR of a fraction m/n has the form $n = n_0 2^\alpha 5^\beta$ with $(n_0, 10) = 1$, then the period begins after $\max(\alpha, \beta)$ terms and the length of the period is the exponent to which 10 belongs (mod n_0), i.e., the number x such that $10^x \equiv 1 \pmod{n_0}$. If q is PRIME and $\lambda(q)$ is EVEN, then breaking the repeating DIGITS into two equal halves and adding gives all 9s. For example, $1/7 = 0.\overline{142857}$, and $142 + 857 = 999$. For $1/q$ with a PRIME DENOMINATOR other than 2 or 5, all cycles n/q have the same length (Conway and Guy 1996).

If n is a PRIME and 10 is a PRIMITIVE ROOT of n, then the period $\lambda(n)$ of the repeating decimal $1/n$ is given by

$$\lambda(n) = \phi(n), \tag{5}$$

where $\phi(n)$ is the TOTIENT FUNCTION. Furthermore, the decimal expansions for p/n, with $p = 1, 2, \ldots, n-1$ have periods of length $n - 1$ and differ only by a cyclic permutation. Such numbers are called LONG PRIMES by conway and guy (1996). an equivalent definition is that

$$10^i \equiv 1 \pmod{n} \tag{6}$$

for $i = n - 1$ and no i less than this. In other words, a NECESSARY (but not SUFFICIENT) condition is that the number $9R_{n-1}$ (where R_n is a REPUNIT) is DIVISIBLE by n, which means that R_n is DIVISIBLE by n.

The first few numbers with maximal decimal expansions, called FULL REPTEND PRIMES, are 7, 17, 19, 23, 29, 47, 59, 61, 97, 109, 113, 131, 149, 167, ... (Sloane's A001913). The decimals corresponding to these are called CYCLIC NUMBERS. No general method is known for finding FULL REPTEND PRIMES. Artin conjectured that ARTIN'S CONSTANT $C = 0.3739558136\ldots$ is the fraction of PRIMES p for with $1/p$ has decimal maximal period (Conway and Guy 1996). D. Lehmer has generalized this conjecture to other bases, obtaining values which are small rational multiples of C.

To find DENOMINATORS with short periods, note that

$$10^1 - 1 = 3^2$$

$$10^2 - 1 = 3^2 \cdot 11$$

$$10^3 - 1 = 3^3 \cdot 37$$

$$10^4 - 1 = 3^2 \cdot 11 \cdot 101$$

$$10^5 - 1 = 3^2 \cdot 41 \cdot 271$$

$$10^6 - 1 = 3^3 \cdot 7 \cdot 11 \cdot 13 \cdot 37$$

$$10^7 - 1 = 3^2 \cdot 239 \cdot 4649$$

$$10^8 - 1 = 3^2 \cdot 11 \cdot 73 \cdot 101 \cdot 137$$

$$10^9 - 1 = 3^4 \cdot 37 \cdot 333667$$

$$10^{10} - 1 = 3^2 \cdot 11 \cdot 41 \cdot 271 \cdot 9091$$

$$10^{11} - 1 = 3^2 \cdot 21649 \cdot 513239$$

$$10^{12} - 1 = 3^3 \cdot 7 \cdot 11 \cdot 13 \cdot 37 \cdot 101 \cdot 9901.$$

The period of a fraction with DENOMINATOR equal to a PRIME FACTOR above is therefore the POWER of 10 in which the factor first appears. For example, 37 appears in the factorization of $10^3 - 1$ and $10^9 - 1$, so its period is 3. Multiplication of any FACTOR by a $2^\alpha 5^\beta$ still gives the same period as the FACTOR alone. A DENOMINATOR obtained by a multiplication of two FACTORS has a period equal to the first POWER of 10 in which both FACTORS appear. The following table gives the PRIMES having small periods (Sloane's A046106, A046107, and A046108; Ogilvy and Anderson 1988).

period	primes
1	3
2	11
3	37
4	101
5	41, 271
6	7, 13
7	239, 4649
8	73, 137
9	333667
10	9091
11	21649, 513239
12	9901
13	53, 79, 265371653
14	909091
15	31, 2906161
16	17, 5882353
17	2071723, 5363222357
18	19, 52579
19	1111111111111111111
20	3541, 27961

A table of the periods e of small PRIMES other than the special $p = 5$, for which the decimal expansion is not periodic, follows (Sloane's A002371).

p	e	p	e	p	e
3	1	31	15	67	33
7	6	37	3	71	35
11	2	41	5	73	8
13	6	43	21	79	13
17	16	47	46	83	41
19	18	53	13	89	44
23	22	59	58	97	96
29	28	61	60	101	4

Shanks (1873ab) computed the periods for all PRIMES up to 120,000 and published those up to 29,989.

See also DECIMAL, DECIMAL POINT, FRACTION, HAUPT-EXPONENT, MIDY'S THEOREM, REPEATING DECIMAL

References

Conway, J. H. and Guy, R. K. "Fractions Cycle into Decimals." In *The Book of Numbers.* New York: Springer-Verlag, pp. 157–63 and 166–71, 1996.

Das, R. C. "On Bose Numbers." *Amer. Math. Monthly* **56**, 87–9, 1949.

de Polignac, A. "Note sur la divisibilité des nombres." *Nouv. Ann. Math.* **14**, 118–20, 1855.

Dickson, L. E. *History of the Theory of Numbers, Vol. 1: Divisibility and Primality.* New York: Chelsea, pp. 159–79, 1952.

Glaisher, J. W. L. "Periods of Reciprocals of Integers Prime to 10." *Proc. Cambridge Philos. Soc.* **3**, 185–06, 1878.

Lehmer, D. H. "Guide to Tables in the Theory of Numbers." Bulletin No. 105. Washington, DC: National Research Council, pp. 7–2, 1941.

Lehmer, D. H. "A Note on Primitive Roots." *Scripta Math.* **26**, 117–19, 1963.

Ogilvy, C. S. and Anderson, J. T. *Excursions in Number Theory.* New York: Dover, p. 60, 1988.

Rademacher, H. and Toeplitz, O. *The Enjoyment of Mathematics: Selections from Mathematics for the Amateur.* Princeton, NJ: Princeton University Press, pp. 147–63, 1957.

Rao, K. S. "A Note on the Recurring Period of the Reciprocal of an Odd Number." *Amer. Math. Monthly* **62**, 484–87, 1955.

Shanks, W. "On the Number of Figures in the Period of the Reciprocal of Every Prime Number Below 20,000." *Proc. Roy. Soc. London* **22**, 200, 1873a.

Shanks, W. "On the Number of Figures in the Period of the Reciprocal of Every Prime Number Between 20,000 and 30,000." *Proc. Roy. Soc. London* **22**, 384, 1873b.

Shiller, J. K. "A Theorem in the Decimal Representation of Rationals." *Amer. Math. Monthly* **66**, 797–98, 1959.

Sloane, N. J. A. Sequences A001913/M4353, A002329/M4045, A002371/M4050, A046106, A046107, and A046108 in "An On-Line Version of the Encyclopedia of

Integer Sequences." http://www.research.att.com/~njas/sequences/eisonline.html.

Wells, D. *The Penguin Dictionary of Curious and Interesting Numbers.* Middlesex, England: Penguin Books, p. 60, 1986.

Decimal Period

DECIMAL COMMA, DECIMAL EXPANSION, DECIMAL POINT

Decimal Point

The symbol uses to separate the integer part of a decimal number from its fractional part is called the decimal point. In the United States, the decimal point is denoted with a period (e.g., 3.1415), whereas a raised period is used in Britain (e.g., 3.1415), and a DECIMAL COMMA is used in continental Europe (e.g., 3,1415). The number 3.1415 is voiced "three point one four one five," while in continental Europe, 3,1415 would be voiced "three comma one four one five."

See also COMMA, DECIMAL, DECIMAL COMMA, DECIMAL EXPANSION

Decision Problem

Does there exist an ALGORITHM for deciding whether or not a specific mathematical assertion does or does not have a proof? The decision problem is also known as the ENTSCHEIDUNGSPROBLEM (which, not so coincidentally, is German for "decision problem"). Using the concept of the TURING MACHINE, Turing showed the answer to be NEGATIVE for elementary NUMBER THEORY. J. Robinson and Tarski showed the decision problem is undecidable for arbitrary FIELDS.

Decision Theory

A branch of GAME THEORY dealing with strategies to maximize the outcome of a given process in the face of uncertain conditions.

See also NEWCOMB'S PARADOX, OPERATIONS RESEARCH, PRISONER'S DILEMMA

Deck Transformation

The deck transformations of a UNIVERSAL COVER $\tilde{X}$ form a group Γ, which is the FUNDAMENTAL GROUP of the QUOTIENT SPACE

$$X = \tilde{X}/\Gamma.$$

Deck transformations are also called covering transformations, and are defined for any COVER $p : A \to X$. They act on A by homeomorphisms which preserve the projection p.

The UNIVERSAL COVER of X, denoted $\tilde{X}$, is a SIMPLY CONNECTED space and is a COVERING of $\pi : \tilde{X} \to X$. Every loop in X, say a function f on the unit interval with $f(0) = f(1) = p$, lifts to a path $\tilde{f} \in \tilde{X}$, which only depends on the choice of $\tilde{f} \in \pi^{-1}(p)$, i.e., the starting

point in the PREIMAGE of π. Moreover, the endpoint $\tilde{f}(1)$ depends only on the HOMOTOPY CLASS of f and $\tilde{f}(0)$. Given a point $q \in \tilde{X}$, and α, a member of the FUNDAMENTAL GROUP of X, a point $\alpha \cdot q$ is defined to be the endpoint of a LIFT of a path f which represents α.

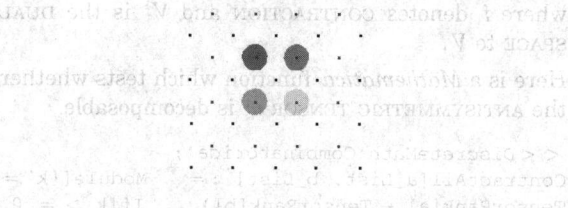

For example, when X is the SQUARE TORUS then $\tilde{X}$ is the plane and the preimage $\pi^{-1}(p)$ is a translation of the integer lattice $\{(n, m)\} \subset \mathbb{R}^2$. Any loop in the torus lifts to a path in the plane, with the endpoints lying in the integer lattice. These translated integer lattices are the ORBITS of the action of $\mathbb{Z} \times \mathbb{Z}$ on $\mathbb{R}^2$ by addition. The above animation shows the action of some deck transformations on some disks in the plane. The spaces are the torus and its UNIVERSAL COVER, the plane. An element of the fundamental group, shown as the path in blue, defines a deck transformation of the universal cover. It moves around the points in the universal cover. The points moved to have the same projection in the torus. The blue path is a loop in the torus, and all of its preimages are shown.

See also COVER, FUNDAMENTAL GROUP, GROUP ACTION, UNIVERSAL COVER

References

Fulton, W. *Algebraic Topology: A First Course.* New York: Springer-Verlag, pp. 163–64, 1995.

Massey, W. S. *A Basic Course in Algebraic Topology.* New York: Springer-Verlag, pp. 130–40, 1991.

Decomposable

A DIFFERENTIAL K-FORM ω of degree p in an EXTERIOR ALGEBRA $\wedge V$ is decomposable if there exist p ONE-FORMS α_i such that

$$\omega = \alpha_1 \wedge \ldots \wedge \alpha_{pi}, \tag{1}$$

where $\alpha \wedge \beta$ denotes a WEDGE PRODUCT. Forms of degree 0, 1, $\dim V - 1$, and $\dim V$ are always decomposable. Hence the first instance of indecomposable forms occurs in $\mathbb{R}^4$, in which case $e_1 \wedge e_2 + e_3 \wedge e_4$ is indecomposable.

If a p-form ω has an ENVELOPE of dimension p then it is decomposable. In fact, the ONE-FORMS in the (dual) basis to the envelope can be used as the α_i above.

The PLÜCKER RELATIONS form a system of quadratic equations on the a_I in

$$\omega = \sum a_I e_{i_1} \wedge \ldots \wedge e_{i_p}, \tag{2}$$

which is equivalent to ω being decomposable. Since a decomposable p-form corresponds to a p-dimensional subspace, these quadratic equations show that the GRASSMANNIAN is a PROJECTIVE VARIETY. In particular, ω is decomposable if for every $\beta \in \wedge^{p+1} V^*$,

$$i(i(\beta)\omega)\omega = 0, \tag{3}$$

where i denotes CONTRACTION and V^* is the DUAL SPACE to V.

Here is a *Mathematica* function which tests whether the ANTISYMMETRIC TENSOR w is decomposable.

```
< <DiscreteMath`Combinatorica`;
ContractAll[a_List, b_List] :=   Module[{k =
TensorRank[a] - TensorRank[b]},    If[k >= 0,
Map[Flatten[#1].Flatten[b]    &,   a,    {k}],
ContractAll[b, a]
    ]
    ] Envelope[a_List?VectorQ] := Select[{a},
#1    !=    Table[0,    {Length[a]}]    &]
Envelope[a_List] := Module[
    {
     z, inds, vects,
     d = Dimensions[a][[1]], r = TensorRank[a]
    },
    z = Table[0, ##1] & @@ Table[{d}, {r - 1}];
    inds = KSubsets[Range[d], r - 1];
    vects = Map[ContractAll[a, ReplacePart[z,
1, #1]] &, inds];
     Select[RowReduce[vects], #1 != Table[0,
{d}] &]
    ]    DecomposableQ[a_?ListQ]    :=
(Length[Envelope[a]]  == TensorRank[a])
```

See also CONTRACTION (TENSOR), EXTERIOR ALGEBRA, GRASSMANNIAN, PLÜCKER RELATIONS, VECTOR SPACE, WEDGE PRODUCT

References

Sternberg, S. *Differential Geometry*. New York: Chelsea, pp. 14–0, 1983.

Decomposition

A rewriting of a given quantity (e.g., a MATRIX) in terms of a combination of "simpler" quantities.

See also CHOLESKY DECOMPOSITION, COMPOSITION, CONNECTED SUM DECOMPOSITION, JACO-SHALEN-JO-HANNSON TORUS DECOMPOSITION, LU DECOMPOSITION, PRIME FACTORIZATION, QR DECOMPOSITION, SINGULAR VALUE DECOMPOSITION

Decomposition Group

References

Koch, H. "Decomposition Group and Ramification Group." §6.1 in *Number Theory: Algebraic Numbers and Functions*. Providence, RI: Amer. Math. Soc., pp. 172–76, 2000.

Deconvolution

The inversion of a CONVOLUTION equation, i.e., the solution for f of an equation OF THE FORM

$$f * g = h + \epsilon,$$

given g and h, where ε is the NOISE and $*$ denotes the CONVOLUTION. Deconvolution is ill-posed and will usually not have a unique solution even in the absence of NOISE.

Linear deconvolution ALGORITHMS include INVERSE FILTERING and WIENER FILTERING. Nonlinear ALGORITHMS include the CLEAN algorithm, MAXIMUM ENTROPY METHOD, and LUCY.

See also CONVOLUTION, LUCY, MAXIMUM ENTROPY METHOD, WIENER FILTER

References

Cornwell, T. and Braun, R. "Deconvolution." Ch. 8 in *Synthesis Imaging in Radio Astronomy: Third NRAO Summer School, 1988* (Ed. R. A. Perley, F. R. Schwab, and A. H. Bridle). San Francisco, CA: Astronomical Society of the Pacific, pp. 167–83, 1989.
Press, W. H.; Flannery, B. P.; Teukolsky, S. A.; and Vetterling, W. T. "Convolution and Deconvolution Using the FFT." §13.1 in *Numerical Recipes in FORTRAN: The Art of Scientific Computing, 2nd ed.* Cambridge, England: Cambridge University Press, pp. 531–37, 1992.

Decreasing Function

A function $f(x)$ decreases on an INTERVAL I if $f(b) < f(a)$ for all $b > a$, where $a, b \in I$. Conversely, a function $f(x)$ increases on an INTERVAL I if $f(b) > f(a)$ for all $b > a$ with $a, b \in I$.

If the DERIVATIVE $f'(x)$ of a CONTINUOUS FUNCTION $f(x)$ satisfies $f'(x) < 0$ on an OPEN INTERVAL (a, b), then $f(x)$ is decreasing on (a, b). However, a function may decrease on an interval without having a derivative defined at all points. For example, the function $-x^{1/3}$ is decreasing everywhere, including the origin $x = 0$, despite the fact that the DERIVATIVE is not defined at that point.

See also DERIVATIVE, INCREASING FUNCTION, NONDECREASING FUNCTION, NONINCREASING FUNCTION

References

Jeffreys, H. and Jeffreys, B. S. "Increasing and Decreasing Functions." §1.065 in *Methods of Mathematical Physics, 3rd ed.* Cambridge, England: Cambridge University Press, p. 22, 1988.

Decreasing Sequence

A SEQUENCE $\{a_1, a_2 ...\}$ for which $a_1 \geq a_2 \geq ....$

See also INCREASING SEQUENCE, SEQUENCE

Decreasing Series

A SERIES $s_1, s_2, ...$ for which $s_1 \geq s_2 \geq ....$

Dedekind Cut

A set partition of the RATIONAL NUMBERS into two nonempty subsets S_1 and S_2 such that all members of S_1 are less than those of S_2 and such that S_1 has no greatest member. REAL NUMBERS can be defined using either Dedekind cuts or CAUCHY SEQUENCES.

See also CANTOR-DEDEKIND AXIOM, CAUCHY SEQUENCE

References

Courant, R. and Robbins, H. "Alternative Methods of Defining Irrational Numbers. Dedekind Cuts." §2.2.6 in *What is Mathematics?: An Elementary Approach to Ideas and Methods, 2nd ed.* Oxford, England: Oxford University Press, pp. 71–2, 1996.

Jeffreys, H. and Jeffreys, B. S. "Nests of Intervals: Dedekind Section." §1.031 in *Methods of Mathematical Physics, 3rd ed.* Cambridge, England: Cambridge University Press, pp. 6–, 1988.

Dedekind Eta

DEDEKIND ETA FUNCTION

Dedekind Eta Function

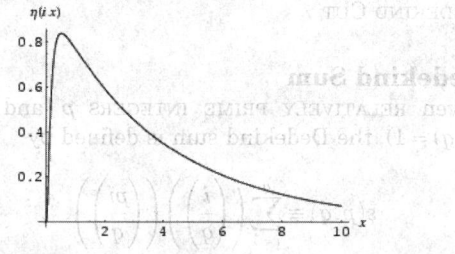

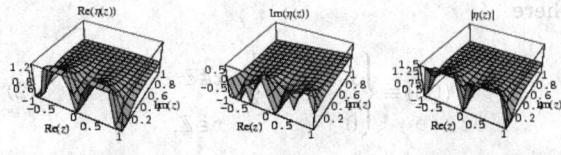

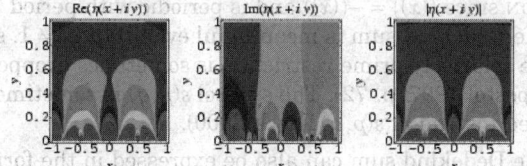

Let

$$q = e^{2\pi i\tau}, \tag{1}$$

then the Dedekind eta function is defined over the UPPER HALF-PLANE $H = \{\tau : \Im[\tau] > 0\}$ by

$$\eta(\tau) \equiv q^{1/24} \prod_{n=1}^{\infty} (1 - q^n) = (q;q)_\infty, \tag{2}$$

which can be written as

$$\eta(\tau) = q^{1/24} \left\{ 1 + \sum_{n=1}^{\infty} (-1)^n \left[q^{n(3n-1)/2} + q^{n(3n+1)/2} \right] \right\} \tag{3}$$

(Weber 1902, pp. 85 and 112; Atkin and Morain 1993). $\eta(\tau)$ is a MODULAR FORM first introduced by Dedekind in 1877, and is related to the MODULAR DISCRIMINANT of the WEIERSTRASS ELLIPTIC FUNCTION by

$$\Delta(\tau) = (2\pi)^{12} [\eta(\pi)]^{24} \tag{4}$$

(Apostol 1997, p. 47).
The derivative of $\eta(t)$ satisfies

$$-4\pi i \frac{d}{d\tau} \ln[\eta(\tau)] = G_2(\tau) \tag{5}$$

$$\frac{d}{d\tau} \ln\left[-\frac{1}{\tau} \right] = \frac{d}{d\tau} \ln[\eta(\tau)] + \frac{1}{2}\frac{d}{dr}\ln(-i\tau), \tag{6}$$

where $G_2(\tau)$ is an EISENSTEIN SERIES.

Letting $\zeta_{24} = e^{2\pi i/24} = e^{\pi i/12}$ be a ROOT OF UNITY, $\eta(\tau)$ satisfies

$$\eta(\tau + 1) = e^{\pi i/12} \eta(\tau) \tag{7}$$

$$\eta(\tau + n) = e^{\pi in/12} \eta(\tau) \tag{8}$$

$$\eta\left(-\frac{1}{\tau}\right) = \sqrt{-i\tau}\,\eta(\tau) \tag{9}$$

where n is an integer (Weber 1902, p. 113; Atkin and Morain 1993; Apostol 1997, p. 47). The Dedekind eta function is related to the JACOBI THETA FUNCTION ϑ_3 by

$$\vartheta_3(0, e^{\pi i\tau}) = \frac{\eta^2\left(\frac{1}{2}(\tau + 1)\right)}{\eta(\tau + 1)} \tag{10}$$

(Apostol 1997, p. 91).

Macdonald (1972) has related most expansions OF THE FORM $(q,q)_\infty^c$ to affine ROOT SYSTEMS. Exceptions not included in Macdonald's treatment include $c = 2$, found by Hecke and Rogers, $c = 4$, found by Ramanujan, and $c = 26$, found by Atkin (Leininger and Milne 1997). Using the Dedekind eta function, the JACOBI TRIPLE PRODUCT identity is written

$$(q,q)_\infty^3 = \sum_{n=0}^{\infty} (-1)^n (2n+1) q^{n(n+1)/2} \tag{11}$$

(Jacobi 1829, Hardy and Wright 1979, Leininger and Milne 1997, Hirschhorn 1999).

Dedekind's functional equation states that if $\begin{bmatrix} ab \\ cd \end{bmatrix} \in \Gamma$, where Γ is the MODULAR GROUP GAMMA, $c > 0$, and $\tau \in H$, then

$$\eta\left(\frac{a\tau + b}{c\tau + d}\right) = \epsilon(a,b,c,d)\left[-i\sqrt{c\tau + d}\right]\eta(\tau), \qquad (12)$$

where

$$\epsilon(a,b,c,d) = \exp\left[\pi i\left(\frac{a+d}{12c} + s(-d,c)\right)\right], \qquad (13)$$

and

$$s(h,k) = \sum_{r=1}^{k-1}\frac{r}{k}\left(\frac{hr}{k} - \left[\frac{hr}{k}\right] - \frac{1}{2}\right) \qquad (14)$$

is a DEDEKIND SUM (Apostol 1997, pp. 52–7), with $\lfloor x \rfloor$ the FLOOR FUNCTION.

See also DIRICHLET ETA FUNCTION, DEDEKIND SUM, ELLIPTIC LAMBDA FUNCTION, INFINITE PRODUCT, INVARIANT (ELLIPTIC FUNCTION), JACOBI THETA FUNCTIONS, KLEIN'S ABSOLUTE INVARIANT, Q-SERIES, TAU FUNCTION, WEBER FUNCTIONS

References

Apostol, T. M. "The Dedekind Eta Function." Ch. 3 in *Modular Functions and Dirichlet Series in Number Theory, 2nd ed.* New York: Springer-Verlag, pp. 47–3, 1997.
Atkin, A. O. L. and Morain, F. "Elliptic Curves and Primality Proving." *Math. Comput.* **61**, 29–8, 1993.
Bhargava, S. and Somashekara, D. "Some Eta-Function Identities Deducible from Ramanujan's $1\psi 1$ Summation." *J. Math. Anal. Appl.* **176**, 554–60, 1993.
Hardy, G. H. and Wright, E. M. *An Introduction to the Theory of Numbers, 5th ed.* Oxford, England: Clarendon Press, 1979.
Hirschhorn, M. D. "Another Short Proof of Ramanujan's Mod 5 Partition Congruences, and More." *Amer. Math. Monthly* **106**, 580–83, 1999.
Jacobi, C. G. J. *Fundamentia Nova Theoriae Functionum Ellipticarum.* Regiomonti, Sumtibus fratrum Borntraeger, p. 90, 1829.
Leininger, V. E. and Milne, S. C. "Some New Infinite Families of Eta Function Identities." Preprint. http://www.math.ohio-state.edu/~milne/preprints.html.
Leininger, V. E. and Milne, S. C. "Expansions for $(q)_\infty^{n^2+n}$ and Basic Hypergeometric Series in $U(n)$." Preprint. http://www.math.ohio-state.edu/~milne/preprints.html.
Köhler, G. "Some Eta-Identities Arising from Theta Series." *Math. Scand.* **66**, 147–54, 1990.
Macdonald, I. G. "Affine Root Systems and Dedekind's η-Function." *Invent. Math.* **15**, 91–43, 1972.
Ramanujan, S. "On Certain Arithmetical Functions." *Trans. Cambridge Philos. Soc.* **22**, 159–84, 1916.
Siegel, C. L. "A Simple Proof of $\eta(-1/\tau) = \eta(\tau)\sqrt{\tau/i}$." *Mathematika* **1**, 4, 1954.
Weber, H. *Lehrbuch der Algebra, Vols. I-II.* New York: Chelsea, 1902.

Dedekind Function

$$\psi(n) = n\prod_{\substack{\text{distinct prime} \\ \text{factors } p \text{ of } n}}(1 + p^{-1})$$

where the PRODUCT is over the distinct PRIME FAC-TORS of n. The first few values are 1, 3, 4, 6, 6, 12, 8, 12, 12, 18, ... (Sloane's A001615).

See also DEDEKIND ETA FUNCTION, EULER PRODUCT, TOTIENT FUNCTION

References

Cox, D. A. *Primes of the Form $x^2 + ny^2$: Fermat, Class Field Theory and Complex Multiplication.* New York: Wiley, p. 228, 1997.
Guy, R. K. *Unsolved Problems in Number Theory, 2nd ed.* New York: Springer-Verlag, p. 96, 1994.
Sloane, N. J. A. Sequences A001615/M2315 in "An On-Line Version of the Encyclopedia of Integer Sequences." http://www.research.att.com/~njas/sequences/eisonline.html.

Dedekind Number

ANTICHAIN

Dedekind Ring

A abstract commutative RING in which every NON-ZERO IDEAL is a unique product of PRIME IDEALS.

References

Noether, E. "Abstract Development of Ideal Theory in Algebraic Number Fields and Function Fields." *Math. Ann.* **96**, 26–1, 1927.

Dedekind Section

DEDEKIND CUT

Dedekind Sum

Given RELATIVELY PRIME INTEGERS p and q (i.e., $(p,q) = 1$), the Dedekind sum is defined by

$$s(p,q) \equiv \sum_{i=1}^{q}\left(\left(\frac{i}{q}\right)\right)\left(\left(\frac{pi}{q}\right)\right), \qquad (1)$$

where

$$((x)) \equiv \begin{cases} x - \lfloor x \rfloor - \dfrac{1}{2} & x \notin \mathbb{Z} \\ 0 & x \in \mathbb{Z}, \end{cases} \qquad (2)$$

with $\lfloor x \rfloor$ the FLOOR FUNCTION. $((x))$ is an ODD FUNCTION since $((x)) = -(x)$ and is periodic with period 1. The Dedekind sum is meaningful even if $(p,q) \neq 1$, so the relatively prime restriction is sometimes dropped (Apostol 1997, p. 72). The symbol $s(p,q)$ is sometimes used instead of $s(p,a)$ (Beck 2000).

The Dedekind sum can also be expressed in the form

$$s(p,q) = \frac{1}{4q}\sum_{r=1}^{q-1}\cot\left(\frac{\pi pr}{k}\right)\cot\left(\frac{\pi r}{q}\right). \qquad (3)$$

If $0 < h < k$, let $r_0, r_1, ..., r_{n+1}$ denote the remainders in the EUCLIDEAN ALGORITHM given by

$$r_0 = k \qquad (4)$$

$$r_1 = h \qquad (5)$$

$$r_{j+1} \equiv r_{j-1} \pmod{r_j} \qquad (6)$$

for $1 \le r_{j+1} < r_j$ and $r_{n+1} = 1$. Then

$$s(h,k) = \frac{1}{12} \sum_{j=1}^{n+1} \left\{ (-1)^{j+1} \frac{r_j^2 + r_{j-1}^2 + 1}{r_j r_{j-1}} \right\}$$
$$- \frac{(-1)^n + 1}{8} \qquad (7)$$

(Apostol 1997, pp. 72–3).

In general, there is no simple formula for closed-form evaluation of $s(p,q)$, but some special cases are

$$s(1,q) = \frac{(q-1)(q-2)}{12q} \qquad (8)$$

$$s(2,q \text{ odd}) = \frac{(q-1)(q-2)}{24q} \qquad (9)$$

(Apostol 1997, p. 62). Apostol (1997, p. 73) gives the additional special cases

$$12hks(h,k) = (k-1)(k-h^2-1) \qquad (10)$$

for $k \equiv 1 \pmod{h}$

$$12hks(h,k) = (k-2)\left[k - \frac{1}{2}(h^2+1)\right] \qquad (11)$$

for $k \equiv 2 \pmod{h}$

$$12hks(h,k) = k^2 + (h^2 - 6h + 2)k + h^2 + 1 \qquad (12)$$

for $k \equiv -1 \pmod{h}$

$$12hks(h,k) = k^2 - \frac{h^2 - t(r-1)(r-2)h + r^2 + 1}{r}k$$
$$+ h^2 + 1 \qquad (13)$$

for $k \equiv r \pmod{h}$ and $h \equiv t \pmod{r}$, where $r \ge 1$ and $t = \pm 1$. Finally,

$$12hks(h,k) = k^2 - \frac{h^2 + 4r(t-2)(t+2)h + 26}{5}k + h^2$$
$$+ 1 \qquad (14)$$

for $k \equiv 5 \pmod{h}$ and $h \equiv t \pmod{5}$, where $t = \pm 1$ or ± 2.

Dedekind sums obey 2-term

$$s(p,q) + s(q,p) = -\frac{1}{4} + \frac{1}{12}\left(\frac{p}{q} + \frac{q}{p} + \frac{1}{pq}\right) \qquad (15)$$

(Dedekind 1953; Rademacher and Grosswald 1972; Pommersheim 1993; Apostol 1997, pp. 62–4) and 3-term

$$s(bc',a) + s(ca',b) + s(ab',c)$$
$$= -\frac{1}{4} + \frac{1}{12}\left(\frac{a}{bc} + \frac{b}{ca} + \frac{c}{ab}\right) \qquad (16)$$

(Rademacher 1954), reciprocity laws, where a, a'; b, b'; and c, c' are pairwise COPRIME, and

$$aa' \equiv 1 \pmod{b} \qquad (17)$$

$$bb' \equiv 1 \pmod{c} \qquad (18)$$

$$cc' \equiv 1 \pmod{a} \qquad (19)$$

(Pommersheim 1993).

$6ps(p,q)$ is an integer, and if $\theta = (3,q)$, then

$$12pqs(p,q) \equiv 0 \pmod{\theta p} \qquad (20)$$

and

$$12pqs(q,p) \equiv q^2 + 1 \pmod{\theta p}. \qquad (21)$$

In addition, $s(p,q)$ satisfies the congruence

$$12qs(p,q) \equiv (q-1)(q+2) - 4p(q-1)$$
$$+ 4 \sum_{r < q/2} \left\lfloor \frac{2pr}{q} \right\rfloor \pmod{8}, \qquad (22)$$

which, if q is odd, becomes

$$12qs(p,q) \equiv q - 1 + 4 \sum_{r < q/2} \left\lfloor \frac{2pr}{q} \right\rfloor \pmod{8} \qquad (23)$$

(Apostol 1997, pp. 65–6). If $q = 3$, 5, 7, or 13, let $r = 24/(q-1)$, let integers a, b, c, d be given with $ad - bc = 1$ such that $c = c_1 q$ and $c_1 > 0$, and let

$$\delta = \left\{ s(a,c) - \frac{a+d}{12c} \right\} - \left\{ s(a_1, c_1) - \frac{a+d}{12c_1} \right\}. \qquad (24)$$

Then $r\delta$ is an even integer (Apostol 1997, pp. 66–9).

Let p, q, u, $v \in \mathbb{N}$ with $(p,q) = (u,v) = 1$ (i.e., are pairwise RELATIVELY PRIME), then the Dedekind sums also satisfy

$$s(p,q) + s(u,v)$$
$$= s(pu' - qv', pv + qu) - \frac{1}{4} + \frac{1}{12}\left(\frac{q}{vt} + \frac{v}{tq} + \frac{t}{qv}\right), \qquad (25)$$

where $t = pv + qu$, and u', v' are any INTEGERS such that $uu' + vv' = 1$ (Pommersheim 1993).

If p is prime, then

$$(p+1)s(h,k) = s(ph,k) + \sum_{m=0}^{p-1} s(h+mk, pk) \qquad (26)$$

(Dedekind 1953; Apostol 1997, p. 73). Moreover, it has been beautifully generalized by Knopp (1980).

See also DEDEKIND ETA FUNCTION, ISEKI'S FORMULA

References

Apostol, T. M. "Properties of Dedekind Sums," "The Reciprocity Law for Dedekind Sums," and "Congruence Properties of Dedekind Sums." §3.7–.9 in *Modular Functions and Dirichlet Series in Number Theory, 2nd ed.* New York: Springer-Verlag, pp. 52 and 61–9, 1997.

Apostol, T. M. Ch. 12 in *Introduction to Analytic Number Theory.* New York: Springer-Verlag, 1976.

Beck, M. "Dedekind Cotangent Sums." Submitted.

Dedekind, R. "Erlauterungen zu den Fragmenten, XXVIII." In *Collected Works of Bernhard Riemann.* New York: Dover, pp. 466–78, 1953.

Iseki, S. "The Transformation Formula for the Dedekind Modular Function and Related Functional Equations." *Duke Math. J.* **24**, 653–62, 1957.

Knopp, M. I. "Hecke Operators and an Identity for Dedekind Sums." *J. Number Th.* **12**, 2–, 1980.

Pommersheim, J. "Toric Varieties, Lattice Points, and Dedekind Sums." *Math. Ann.* **295**, 1–4, 1993.

Rademacher, H. "Generalization of the Reciprocity Formula for Dedekind Sums." *Duke Math. J.* **21**, 391–98, 1954.

Rademacher, H. and Grosswald, E. *Dedekind Sums.* Washington, DC: Math. Assoc. Amer., 1972.

Rademacher, H. and Whiteman, A. L. "Theorems on Dedekind Sums." *Amer. J. Math.* **63**, 377–07, 1941.

Dedekind's Axiom

For every partition of all the points on a line into two nonempty SETS such that no point of either lies between two points of the other, there is a point of one SET which lies between every other point of that SET and every point of the other SET.

Dedekind's Problem

The determination of the number of monotone BOOLEAN FUNCTIONS of n variables (equivalent to the number of ANTICHAINS on the n-set $\{1, 2, ..., n\}$) is called Dedekind's problem.

See also ANTICHAIN, BOOLEAN FUNCTION

References

Dedekind, R. "Über Zerlegungen von Zahlen durch ihre grössten gemeinsammen Teiler." In *Gesammelte Werke, Bd. 1.* pp. 103–48, 1897.

Kleitman, D. "On Dedekind's Problem: The Number of Monotone Boolean Functions." *Proc. Amer. Math. Soc.* **21**, 677–82, 1969 677–82. Kleitman, D. and Markowsky, G. "On Dedekind's Problem: The Number of Isotone Boolean Functions. II." *Trans. Amer. Math. Soc.* **213**, 373–90, 1975.

Deducible

If q is logically deducible from p, this is written $p \vdash q$.

Deep Theorem

Qualitatively, a deep theorem is a theorem whose proof is long, complicated, difficult, or appears to involve branches of mathematics which are not obviously related to the theorem itself (Shanks 1993). Shanks (1993) cites the QUADRATIC RECIPROCITY THEOREM as an example of a deep theorem.

See also THEOREM, TRIVIAL

References

Shanks, D. "Is the Quadratic Reciprocity Law a Deep Theorem?" §2.25 in *Solved and Unsolved Problems in Number Theory, 4th ed.* New York: Chelsea, pp. 64–6, 1993.

Defective Matrix

A MATRIX whose EIGENVECTORS are not COMPLETE.

Defective Number

DEFICIENT NUMBER

Deficiency

Given BINOMIAL COEFFICIENT $\binom{N}{k}$, write

$$N - k + i = a_i b_i,$$

for $1 \leq i \leq k$, where b_i contains only those prime factors $> k$. Then the number of i for which $b_i = 1$ (i.e., for which all the factors of $N - k + i$ are $\leq k$ is called the deficiency of $\binom{N}{k}$ (Erdos *et al.* 1993, Guy 1994). The following table gives the GOOD BINOMIAL COEFFICIENTS (i.e., those with $1 \, \mathrm{pf}\binom{N}{k} > k)$) having deficiency $d \geq 1$ (Erdos *et al.* 1993), and Erdos *et al.* (1993) conjecture that there are no other with $d > 1$.

d	Good Binomial Coefficients
1	$\binom{3}{2}, \binom{7}{3}, \binom{13}{4}, \binom{14}{4}, \binom{23}{5}, \binom{62}{6}, \binom{89}{8}, \cdots$
2	$\binom{7}{4}, \binom{44}{8}, \binom{74}{10}, \binom{174}{12}, \binom{239}{14}, \binom{5179}{27},$ $\binom{8413}{28}, \binom{96622}{42}$
3	$\binom{46}{10}, \binom{47}{10}, \binom{241}{16}, \binom{2105}{25}, \binom{1119}{27},$ $\binom{6459}{33}$
4	$\binom{47}{11}$
9	$\binom{284}{28}$

See also ABUNDANCE, GOOD BINOMIAL COEFFICIENT

References

Erdos, P.; Lacampagne, C. B.; and Selfridge, J. L. "Estimates of the Least Prime Factor of a Binomial Coefficient." *Math. Comput.* **61**, 215–24, 1993.

Guy, R. K. *Unsolved Problems in Number Theory, 2nd ed.* New York: Springer-Verlag, pp. 84–5, 1994.

Deficient Number
Numbers which are not PERFECT and for which

$$s(N) \equiv \sigma(N) - N < N,$$

or equivalently

$$\sigma(n) < 2n,$$

where $\sigma(N)$ is the DIVISOR FUNCTION. Deficient numbers are sometimes called DEFECTIVE NUMBERS (Singh 1997). PRIMES, PRIME POWERS, and any divisors of a PERFECT or deficient number are all deficient. The first few deficient numbers are 1, 2, 3, 4, 5, 7, 8, 9, 10, 11, 13, 14, 15, 16, 17, 19, 21, 22, 23, ... (Sloane's A005100).

See also ABUNDANT NUMBER, LEAST DEFICIENT NUMBER, PERFECT NUMBER

References
Dickson, L. E. *History of the Theory of Numbers, Vol. 1: Divisibility and Primality.* New York: Chelsea, pp. 3–3, 1952.

Guy, R. K. *Unsolved Problems in Number Theory, 2nd ed.* New York: Springer-Verlag, p. 45, 1994.

Singh, S. *Fermat's Enigma: The Epic Quest to Solve the World's Greatest Mathematical Problem.* New York: Walker, p. 11, 1997.

Sloane, N. J. A. Sequences A005100/M0514 in "An On-Line Version of the Encyclopedia of Integer Sequences." http://www.research.att.com/~njas/sequences/eisonline.html.

Souissi, M. *Un Texte Manuscrit d'Ibn Al-Banna' Al-Marrakusi sur les Nombres Parfaits, Abondants, Deficients, et Amiables.* Karachi, Pakistan: Hamdard Nat. Found., 1975.

Definable Set
An ANALYTIC, BOREL, or COANALYTIC SET.

Defined
If A and B are equal by definition (i.e., A is defined as B), then this is written symbolically as $A \equiv B$, $A := B$, or sometimes $\doteq$.

Definite Integral
An INTEGRAL

$$\int_a^b f(x)\,dx$$

with upper and lower limits. The first FUNDAMENTAL THEOREM OF CALCULUS allows definite integrals to be computed in terms of INDEFINITE INTEGRALS, since if F is the INDEFINITE INTEGRAL for $f(x)$, then

$$\int_a^b f(x)\,dx = F(b) - F(a).$$

See also CALCULUS, FUNDAMENTAL THEOREMS OF CALCULUS, INDEFINITE INTEGRAL, INTEGRAL

Degen's Eight-Square Identity

See also EULER FOUR-SQUARE IDENTITY, FIBONACCI IDENTITY

Degeneracy
The property of being DEGENERATE.

See also DEGENERATE

Degenerate
A limiting case in which a class of object changes its nature so as to belong to another, usually simpler, class. For example, the POINT is a degenerate case of the CIRCLE as the RADIUS approaches 0, and the CIRCLE is a degenerate form of an ELLIPSE as the ECCENTRICITY approaches 0. Another example is the two identical ROOTS of the second-order POLYNOMIAL $(x-1)^2$. Since the n ROOTS of an nth degree POLYNOMIAL are usually distinct, ROOTS which coincide are said to be degenerate. Degenerate cases often require special treatment in numerical and analytical solutions. For example, a simple search for both ROOTS of the above equation would find only a single one: 1.

The word degenerate also has several very specific and technical meanings in different branches of mathematics.

See also TRIVIAL

References
Arfken, G. *Mathematical Methods for Physicists, 3rd ed.* Orlando, FL: Academic Press, pp. 513–14, 1985.

Degree
The word "degree" has many meanings in mathematics.

The most common meaning is the unit of ANGLE measure defined such that an entire rotation is 360°. This unit harks back to the Babylonians, who used a base 60 number system. 360° likely arises from the Babylonian year, which was composed of 360 days (12 months of 30 days each). The degree is subdivided into 60 MINUTES per degree, and 60 SECONDS per MINUTE.

The word "degree" is also used in many contexts where it is synonymous with "order," as applied for example to polynomials.

See also ARC MINUTE, ARC SECOND, DEGREE (EXTENSION FIELD), DEGREE OF FREEDOM, DEGREE (MAP), DEGREE (POLYNOMIAL), DEGREE (VERTEX), INDEGREE, LOCAL DEGREE, OUTDEGREE

References
Bringhurst, R. *The Elements of Typographic Style, 2nd ed.* Point Roberts, WA: Hartley and Marks, p. 276, 1997.

Degree (Algebraic Surface)

ORDER (ALGEBRAIC SURFACE)

Degree (Extension Field)

The degree (or relative degree, or index) of an EXTENSION FIELD K/F, denoted $[K:F]$, is the dimension of K as a VECTOR SPACE over F, i.e.,

$$[K:F] = \dim_F K.$$

If $[K:F]$ is finite, then the extension is said to be finite; otherwise, it is said to be infinite.

See also EXTENSION FIELD

References

Dummit, D. S. and Foote, R. M. *Abstract Algebra, 2nd ed.* Englewood Cliffs, NJ: Prentice-Hall, p. 424, 1998.

Degree (Map)

Let $f : M \mapsto N$ be a MAP between two compact, connected, oriented n-D MANIFOLDS without boundary. Then f induces a HOMOMORPHISM f_* from the HOMOLOGY GROUPS $H_n(M)$ to $H_n(N)$, both canonically isomorphic to the INTEGERS, and so f_* can be thought of as a HOMOMORPHISM of the INTEGERS. The INTEGER $d(f)$ to which the number 1 gets sent is called the degree of the MAP f.

There is an easy way to compute $d(f)$ if the MANIFOLDS involved are smooth. Let $x \in \mathbb{N}$, and approximate f by a smooth map HOMOTOPIC to f such that x is a "regular value" of f (which exist and are everywhere by SARD'S THEOREM). By the IMPLICIT FUNCTION THEOREM, each point in $f^{-1}(x)$ has a NEIGHBORHOOD such that f restricted to it is a DIFFEOMORPHISM. If the DIFFEOMORPHISM is orientation preserving, assign it the number $+1$, and if it is orientation reversing, assign it the number -1. Add up all the numbers for all the points in $f^{-1}(x)$, and that is the $d(f)$, the degree of f. One reason why the degree of a map is important is because it is a HOMOTOPY invariant. A sharper result states that two self-maps of the n-sphere are homotopic IFF they have the same degree. This is equivalent to the result that the nth HOMOTOPY GROUP of the n-SPHERE is the set $\mathbb{Z}$ of INTEGERS. The ISOMORPHISM is given by taking the degree of any representation.

One important application of the degree concept is that homotopy classes of maps from n-spheres to n-spheres are classified by their degree (there is exactly one homotopy class of maps for every INTEGER n, and n is the degree of those maps).

Degree (Polynomial)

The highest POWER in a UNIVARIATE POLYNOMIAL is known as its degree, or sometimes "order." For example, the POLYNOMIAL

$$P(x) = a_n x^n + \ldots + a_2 x^2 + a_1 x + a_0$$

is of degree n, denoted $P(x) = n$. The degree of a polynomial is implemented in *Mathematica* as Exponent[$poly, x$].

See also ORDER (POLYNOMIAL)

Degree (Vertex)

VERTEX DEGREE

Degree Matrix

A DIAGONAL MATRIX corresponding to a GRAPH that has the VERTEX DEGREE of v_i in the ith position (Skiena 1990, p. 235).

See also VERTEX DEGREE

References

Skiena, S. *Implementing Discrete Mathematics: Combinatorics and Graph Theory with Mathematica.* Reading, MA: Addison-Wesley, 1990.

Degree of Freedom

The number of degrees of freedom in a problem, distribution, etc., is the number of parameters which may be independently varied.

See also LIKELIHOOD RATIO

Degree Sequence

Given an UNDIRECTED GRAPH, a degree sequence is a monotonic nonincreasing sequence of the VERTEX DEGREES (valencies) of its VERTICES. The number of degree sequences for a graph of a given order is closely related to GRAPHICAL PARTITIONS. The minimum vertex degree in a GRAPH G is denoted $\delta(G)$, and the maximum degree is denoted $\Delta(G)$ (Skiena 1990, p. 157). A GRAPH whose degree sequence contains multiple copies of a single integer is called a REGULAR GRAPH. A graph corresponding to a given degree sequence can be constructed using RealizeDegreeSequence[d] in the *Mathematica* add-on package DiscreteMath`Combinatorica` (which can be loaded with the command < <DiscreteMath`).

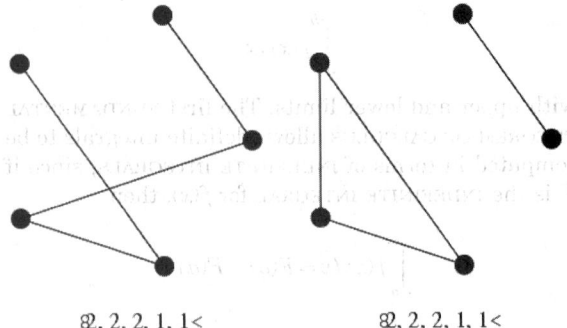

$\{2, 2, 2, 1, 1\}$ $\{2, 2, 2, 1, 1\}$

It is possible for two topologically distinct graphs to have the same DEGREE SEQUENCE.

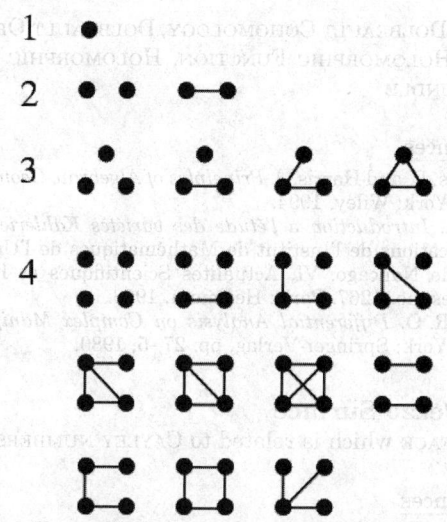

The number of distinct degree sequences for graphs of $n = 1, 2, \ldots$ nodes are given by 1, 2, 4, 11, 31, 102, 342, ... (Sloane's A004251), compared with the total number of nonisomorphic simple undirected graphs with n NODES of 1, 2, 4, 11, 34, 156, 1044, ... (Sloane's A000088). The first order having fewer degree sequences than number of nonisomorphic graphs is therefore $n = 5$. For the graphs illustrated above, the degree sequences are given in the following table.

1	$\{0\}$
2	$\{0,0\}$, $\{1,1\}$
3	$\{0,0,0\}$, $\{1,1,0\}$, $\{2,1,1\}$, $\{2,2,2\}$
4	$\{0,0,0,0\}$, $\{1,1,0,0\}$, $\{2,1,1,0\}$, $\{2,2,2,0\}$, $\{3,2,2,1\}$, $\{3,3,2,2\}$, $\{3,3,3,3\}$, $\{1,1,1,1\}$, $\{2,2,1,1\}$, $\{2,2,2,2\}$, $\{3,1,1,1\}$

The possible sums of elements for a degree sequence of order n are $0, 2, 4, 6, \ldots, n(n-1)$.

A degree sequence is said to be k-connected if there exists some k-CONNECTED GRAPH corresponding to the degree sequence. For example, while the degree sequence $\{1, 2, 1\}$ is 1- but not 2-connected, $\{2, 2, 2\}$ is 2-connected.

See also DEGREE SET, DEGREE (VERTEX), GRAPHIC SEQUENCE, GRAPHICAL PARTITION, K-CONNECTED GRAPH, REGULAR GRAPH

References

Ruskey, F. "Information on Degree Sequences." http://www.theory.csc.uvic.ca/~cos/inf/nump/DegreeSequences.html.
Ruskey, F.; Cohen, R.; Eades, P.; and Scott, A. "Alley CATs in Search of Good Homes." *Congres. Numer.* **102**, 97–10, 1994.

Skiena, S. "Realizing Degree Sequences." §4.4.2 in *Implementing Discrete Mathematics: Combinatorics and Graph Theory with Mathematica.* Reading, MA: Addison-Wesley, pp. 157–60, 1990.
Sloane, N. J. A. Sequences A004251/M1250 in "An On-Line Version of the Encyclopedia of Integer Sequences." http://www.research.att.com/~njas/sequences/eisonline.html.

Degree Set

The set of integers which make up a DEGREE SEQUENCE. Any set of positive integers is the degree set for some graph.

See also DEGREE SEQUENCE

References

Skiena, S. *Implementing Discrete Mathematics: Combinatorics and Graph Theory with Mathematica.* Reading, MA: Addison-Wesley, p. 167, 1990.

Dehn Invariant

An invariant defined using the angles of a 3-D POLYHEDRON. It remains constant under solid DISSECTION and reassembly. Solids with the same VOLUME can have different Dehn invariants.

Two POLYHEDRA can be dissected into each other only if they have the same volume and the same Dehn invariant. In 1902, Dehn showed that two interdissectable polyhedra must have equal Dehn invariants, settling the third of HILBERT'S PROBLEMS, and Sydler (1965) showed that two polyhedra with the same Dehn invariants are interdissectable.

See also DISSECTION, EHRHART POLYNOMIAL, HILBERT'S PROBLEMS

References

Sydler, J.-P. "Conditions nécessaires et suffisantes pour l'équivalence des polyèdres de l'espace euclidean à trois dimensions." *Comment. Math. Helv.* **40**, 43–0, 1965.

Dehn Surgery

The operation of drilling a TUBULAR NEIGHBORHOOD of a KNOT K in $\mathbb{S}^3$ and then gluing in a solid TORUS so that its meridian curve goes to a (p, q)-curve on the TORUS boundary of the KNOT exterior. Every compact connected 3-MANIFOLD comes from Dehn surgery on a LINK in $\mathbb{S}^3$.

See also KIRBY CALCULUS, TUBULAR NEIGHBORHOOD

References

Adams, C. C. "The Poincaré Conjecture, Dehn Surgery, and the Gordon-Luecke Theorem." §9.3 in *The Knot Book: An Elementary Introduction to the Mathematical Theory of Knots.* New York: W. H. Freeman, pp. 257–63, 1994.

Dehn's Lemma

An embedding of a 1-SPHERE in a 3-MANIFOLD which exists continuously over the 2-DISK also extends over the DISK as an embedding. This theorem was proposed by Dehn in 1910, but a correct proof was not obtained until the work of Papakyriakopoulos (1957ab).

References

Hempel, J. *3-Manifolds.* Princeton, NJ: Princeton University Press, 1976.
Papakyriakopoulos, C. D. "On Dehn's Lemma and the Asphericity of Knots." *Proc. Nat. Acad. Sci. USA* **43**, 169–72, 1957a.
Papakyriakopoulos, C. D. "On Dehn's Lemma and the Asphericity of Knots." *Ann. Math.* **66**, 1–6, 1957b.
Rolfsen, D. *Knots and Links.* Wilmington, DE: Publish or Perish Press, pp. 100–01, 1976.

Del

GRADIENT

Del Bar Operator

The operator $\bar{\partial}$ is defined on a COMPLEX MANIFOLD, and is called the 'del bar operator.' The EXTERIOR DERIVATIVE d takes a function and yields a ONE-FORM. It decomposes as

$$d = \partial + \bar{\partial}, \tag{1}$$

as complex ONE-FORMS decompose into TYPE

$$\Lambda^1 = \Lambda^{1,0} \otimes \Lambda^{0,1} \tag{2}$$

where $\otimes$ denotes the DIRECT SUM. More concretely, in coordinates $z_k = x_k + iy_k$,

$$\partial f = \sum \left(\frac{\partial f}{\partial x_k} - i \frac{\partial f}{\partial y_k} \right) dz_k \tag{3}$$

and

$$\bar{\partial} f = \sum \left(\frac{\partial f}{\partial x_k} + i \frac{\partial f}{\partial y_k} \right) d\bar{z}_k. \tag{4}$$

These operators extend naturally to forms of higher degree. In general, if α is a (p, q)-FORM, then $\partial\alpha$ is a $(p+1,q)$-form and $\bar{\partial}\alpha$ is a $(p,q+1)$-form. The equation $\bar{\partial}f = 0$ expresses the condition of f being a HOLOMORPHIC FUNCTION. More generally, a $(p,0)$-FORM α is called HOLOMORPHIC if $\bar{\partial}\alpha = 0$, in which case its coefficients, as written in a COORDINATE CHART, are HOLOMORPHIC FUNCTIONS.

The del bar operator is also well-defined on SECTIONS of a HOLOMORPHIC VECTOR BUNDLE. The reason is because a change in coordinates or trivializations is HOLOMORPHIC.

See also ALMOST COMPLEX STRUCTURE, ANALYTIC FUNCTION, CAUCHY-RIEMANN EQUATIONS, COMPLEX MANIFOLD, COMPLEX FORM (TYPE), DIFFERENTIAL K- FORM, DOLBEAULT COHOMOLOGY, DOLBEAULT OPERATORS, HOLOMORPHIC FUNCTION, HOLOMORPHIC VECTOR BUNDLE

References

Griffiths, P. and Harris, J. *Principles of Algebraic Geometry.* New York: Wiley, 1994.
Weil, A. *Introduction à l'étude des variétès Kähleriennes.* Publications de l'Institut de Mathématiques de l'Université de Nancago, VI, Actualites Scientifiques et Industrielles, no. 1267. Paris: Hermann, 1958.
Wells, R. O. *Differential Analysis on Complex Manifolds.* New York: Springer-Verlag, pp. 27–5, 1980.

Del Pezzo Surface

A SURFACE which is related to CAYLEY NUMBERS.

References

Coxeter, H. S. M. *Regular Polytopes, 3rd ed.* New York: Dover, p. 211, 1973.
Hunt, B. "Del Pezzo Surfaces." §4.1.4 in *The Geometry of Some Special Arithmetic Quotients.* New York: Springer-Verlag, pp. 128–29, 1996.

Delambre's Analogies

GAUSS'S FORMULAS

Delannoy Number

The Delannoy numbers are the number of lattice paths from $(0,0)$ to (b, a) in which only east $(1, 0)$, north $(0, 1)$, and northeast $(1, 1)$ steps are allowed (i.e, $\rightarrow$, $\uparrow$, and $\nearrow$). They are given by the RECURRENCE RELATION

$$D(a,b) = D(a-1,b) + D(a,b-1) + D(a-1,b-1), \tag{1}$$

with $D(0,0) = 1$. They have the GENERATING FUNCTION

$$\sum_{p,q=1}^{\infty} D(p,q)x^p y^q = (1 - x - y - xy)^{-1} \tag{2}$$

(Comtet 1974, p. 81).

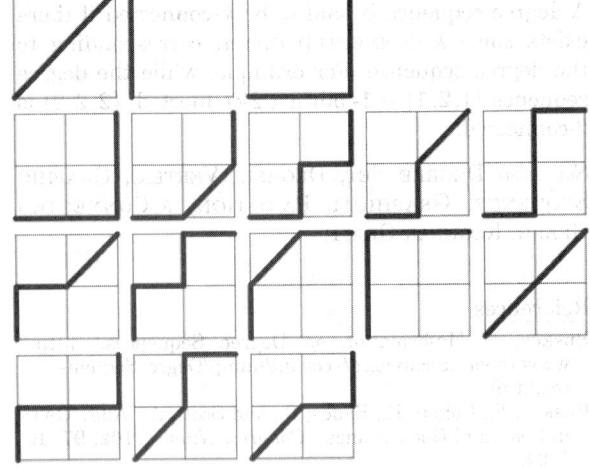

Delaunay Triangulation

For $n \equiv a = b$, the Delannoy numbers are the number of "king walks"

$$D(n,n) = P_n(3),$$

where $P_n(x)$ is a LEGENDRE POLYNOMIAL (Moser 1955; Comtet 1974, p. 81; Vardi 1991). Another expression is

$$D(n,n) = \sum_{k=0}^{n} \binom{n}{k}\binom{n+k}{k} = {}_2F_1(-n, n+1; 1, -1), \quad (3)$$

where $\binom{a}{b}$ is a BINOMIAL COEFFICIENT and $_2F_1(a,b;c;z)$ is a HYPERGEOMETRIC FUNCTION. The values of $D(n,n)$ for $n = 1, 2, \ldots$ are 3, 13, 63, 321, 1683, 8989, 48639, ... (Sloane's A001850).

The SCHRÖDER NUMBERS bear the same relation to the Delannoy numbers as the CATALAN NUMBERS do to the BINOMIAL COEFFICIENTS.

See also BINOMIAL COEFFICIENT, CATALAN NUMBER, MOTZKIN NUMBER, SCHRÖDER NUMBER

References
Comtet, L. *Advanced Combinatorics: The Art of Finite and Infinite Expansions, rev. enl. ed.* Dordrecht, Netherlands: Reidel, pp. 80–1, 1974.
Dickau, R. M. "Delannoy and Motzkin Numbers." http://www.prairienet.org/~pops/delannoy.html.
Goodman and Narayana. "Lattice Paths with Diagonal Steps." *U. Alberta.* No. 39, 1967.
Moser, L. "King Paths on a Chessboard." *Math. Gaz.* **39**, 54, 1955.
Moser, L. and Zayachkowski, H. S. "Lattice Paths with Diagonal Steps." *Scripta Math.* **26**, 223–29, 1963.
Sloane, N. J. A. Sequences A001850/M2942 in "An On-Line Version of the Encyclopedia of Integer Sequences." http://www.research.att.com/~njas/sequences/eisonline.html.
Stocks, D. R. Jr. "Lattice Paths in E^3 with Diagonal Steps." *Canad. Math. Bull.* **10**, 653–58, 1967.
Vardi, I. *Computational Recreations in Mathematica.* Reading, MA: Addison-Wesley, 1991.

Delaunay Triangulation

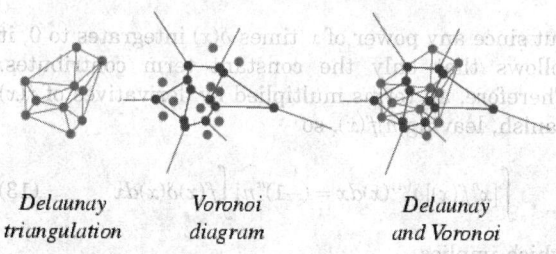

Delaunay triangulation   *Voronoi diagram*   *Delaunay and Voronoi*

The Delaunay triangulation is a TRIANGULATION which is equivalent to the NERVE of the cells in a VORONOI DIAGRAM, i.e., that triangulation of the CONVEX HULL of the points in the diagram in which every CIRCUMCIRCLE of a TRIANGLE is an empty circle (Okabe *et al.* 1992, p. 94). The *Mathematica* command PlanarGraphPlot[*pts*] in the *Mathematica* add-on package DiscreteMath`Computational-Geometry` (which can be loaded with the command

< <DiscreteMath`) plots the Delaunay triangulation of the given list of points.
The Delaunay triangulation and VORONOI DIAGRAM in $\mathbb{R}^2$ are dual to each other.

See also TRIANGULATION, VORONOI DIAGRAM

References
Lee, D. T. and Schachter, B. J. "Two Algorithms for Constructing a Delaunay Triangulation." *Int. J. Computer Information Sci.* **9**, 219–42, 1980.
Okabe, A.; Boots, B.; and Sugihara, K. *Spatial Tessellations: Concepts and Applications of Voronoi Diagrams.* New York: Wiley, 1992.
Preparata, F. R. and Shamos, M. I. *Computational Geometry: An Introduction.* New York: Springer-Verlag, 1985.

Delian Constant

The number $2^{1/3}$ (the CUBE ROOT of 2) which is to be constructed in the CUBE DUPLICATION problem. This number is not a EUCLIDEAN NUMBER although it is an ALGEBRAIC of third degree.

See also CUBE, CUBE DUPLICATION, CUBE ROOT, GEOMETRIC CONSTRUCTION, GEOMETRIC PROBLEMS OF ANTIQUITY

References
Conway, J. H. and Guy, R. K. "Three Greek Problems." In *The Book of Numbers.* New York: Springer-Verlag, pp. 192–94, 1996.
Wells, D. *The Penguin Dictionary of Curious and Interesting Numbers.* Middlesex, England: Penguin Books, pp. 33–4, 1986.

Delian Problem

CUBE DUPLICATION, DELIAN CONSTANT

Delta Amplitude

Given an AMPLITUDE ϕ and a MODULUS m in an ELLIPTIC INTEGRAL,

$$\Delta(\phi) \equiv \sqrt{1 - m \sin^2 \phi}.$$

See also AMPLITUDE, ELLIPTIC INTEGRAL, MODULUS (ELLIPTIC INTEGRAL)

Delta Curve

A curve which can be turned continuously inside an EQUILATERAL TRIANGLE. There are an infinite number of delta curves, but the simplest are the CIRCLE and lens-shaped Δ-biangle. All the Δ curves of height h have the same PERIMETER $2\pi h/3$. Also, at each position of a Δ curve turning in an EQUILATERAL TRIANGLE, the perpendiculars to the sides at the points of contact are CONCURRENT at the instantaneous center of rotation.

See also EQUILATERAL TRIANGLE, LENS, REULEAUX POLYGON, REULEAUX TRIANGLE, ROTOR

References
Honsberger, R. *Mathematical Gems I*. Washington, DC: Math. Assoc. Amer., pp. 56–9, 1973.

Delta Function

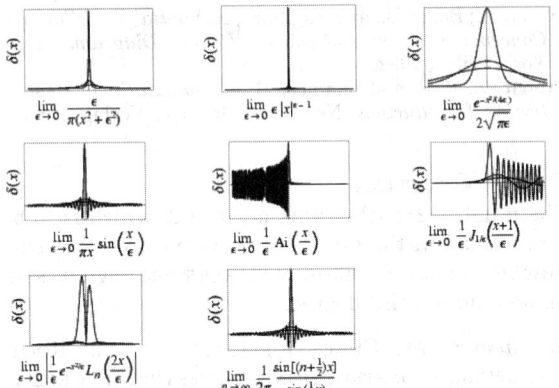

A GENERALIZED FUNCTION which can be defined as the limit of a class of DELTA SEQUENCES. The delta function is sometimes called "Dirac's delta function" or the "impulse symbol" (Bracewell 1999). Formally, δ is a LINEAR FUNCTIONAL from a space (commonly taken as a SCHWARZ SPACE S or the space of all smooth functions of compact support D) of test functions f. The action of δ on f, commonly denoted $\delta[f]$ or $\langle \delta, f \rangle$, then gives the value at 0 of f for any function f.

In engineering contexts, the functional nature of the delta function is often suppressed, and δ is instead viewed as a "special kind" of function, resulting in the useful (but unfortunately deceptive) notation $\delta(x)$. In addition, it is possible to define the delta function as an integral satisfying certain properties at infinity (although this is often not explicitly stated), and commonly used (equivalent) definitions of this type include

$$\delta(x) = \frac{1}{\pi} \lim_{\epsilon \to 0} \frac{\epsilon}{x^2 + \epsilon^2}, \tag{1}$$

$$= \lim_{\epsilon \to 0} \epsilon |x|^{\epsilon - 1} \tag{2}$$

$$= \lim_{\epsilon \to 0^+} \frac{1}{2\sqrt{\pi \epsilon}} e^{-x^2/(4\epsilon)} \tag{3}$$

$$= \lim_{\epsilon \to 0} \frac{1}{\pi x} \sin\left(\frac{x}{\epsilon}\right) \tag{4}$$

$$= \lim_{\epsilon \to 0} \frac{1}{\epsilon} \text{Ai}\left(\frac{x}{\epsilon}\right) \tag{5}$$

$$= \lim_{\epsilon \to 0} \frac{1}{\epsilon} J_{1/\epsilon}\left(\frac{x + 1}{\epsilon}\right) \tag{6}$$

$$= \lim_{\epsilon \to 0} \left| \frac{1}{\epsilon} e^{-x^2/\epsilon} L_n\left(\frac{2x}{\epsilon}\right) \right| \tag{7}$$

$$= \lim_{n \to \infty} \frac{1}{2\pi} \frac{\sin\left[\left(n + \frac{1}{2}\right)x\right]}{\sin\left(\frac{1}{2}x\right)}. \tag{8}$$

Here, Ai(x) is an AIRY FUNCTION, $J_n(x)$ is a BESSEL FUNCTION OF THE FIRST KIND, and $L_n(x)$ is a LAGUERRE POLYNOMIAL of arbitrary positive integer order. (8) is sometimes called the DIRICHLET KERNEL.

The fundamental equation that defines derivatives of the delta function $\delta(x)$ is

$$\int f(x) \delta^{(n)}(x)\, dx \equiv -\int \frac{\partial f}{\partial x} \delta^{(n-1)}(x)\, dx. \tag{9}$$

Letting $f(x) = xg(x)$ in this definition, it follows that

$$\int xg(x)\delta'(x)dx = -\int \delta(x) \frac{\partial}{\partial x}[xg(x)]dx,$$

$$= -\int \delta(x)[g(x) + xg'(x)]dx$$

$$= -\int g(x)\delta(x)dx, \tag{10}$$

where the second term can be dropped since $\int xg'(x)\delta(x)dx = 0$, so (10) implies

$$x\delta'(x) = -\delta(x). \tag{11}$$

In general, the same procedure gives

$$\int [x^n f(x)]\delta^{(n)}(x)dx = (-1)^n \int \frac{\partial^n [x^n f(x)]}{\partial x^n} \delta(x)dx, \tag{12}$$

but since any power of x times $\delta(x)$ integrates to 0, it follows that only the constant term contributes. Therefore, all terms multiplied by derivatives of $f(x)$ vanish, leaving $n!f(x)$, so

$$\int [x^n f(x)]\delta^{(n)}(x)dx = (-1)^n n! \int f(x)\delta(x)dx, \tag{13}$$

which implies

$$x^n \delta^{(n)}(x) = (-1)^n n! \delta(x). \tag{14}$$

Other identities involving the derivative of the delta function include

$$\delta'(-x) = -\delta'(x) \tag{15}$$

$$\int_{-\infty}^{\infty} f(x)\delta'(x - a)dx = -f'(a) \tag{16}$$

$$(\delta' * f)(a) = \int_{-\infty}^{\infty} \delta'(a-x)f(x)dx = f'(a) \qquad (17)$$

where $*$ denotes CONVOLUTION,

$$\int_{-\infty}^{\infty} |\delta'(x)|dx = \infty, \qquad (18)$$

and

$$x^2 \delta'(x) = 0. \qquad (19)$$

The delta function can also be viewed as the DERIVATIVE of the HEAVISIDE STEP FUNCTION,

$$\frac{d}{dx}[H(x)] = \delta(x) \qquad (20)$$

(Bracewell 1999, p. 94).
Additional identities include

$$\delta(x-a) = 0 \qquad (21)$$

for $x \neq a$,

$$\int_{a-\varepsilon}^{a+\varepsilon} \delta(x-a)dx = 1, \qquad (22)$$

where ε is any POSITIVE number, and

$$\int_{-\infty}^{\infty} f(x)\delta(x-a)dx = f(a) \qquad (23)$$

$$\delta(ax) = \frac{1}{|a|}\delta(x) \qquad (24)$$

$$\delta(x^2 - a^2) = \frac{1}{2|a|}[\delta(x+a) + \delta(x-a)] \qquad (25)$$

More generally, the delta function of a function is given by

$$\delta[g(x)] = \sum_i \frac{\delta(x-x_i)}{|g'(x_i)|}, \qquad (26)$$

where the x_is are the ROOTS of g. For example, examine

$$\delta(x^2 + x - 2) = \delta[(x-1)(x+2)]. \qquad (27)$$

Then $g'(x) = 2x + 1$, so $g'(x_1) = g'(1) = 3$ and $g'(x_2) = g'(-2) = -3$, and we have

$$\delta(x^2 + x - 2) = \frac{1}{3}\delta(x-1) + \frac{1}{3}\delta(x+2). \qquad (28)$$

A FOURIER SERIES expansion of $\delta(x-a)$ gives

$$a_n = \frac{1}{\pi}\int_{-\pi}^{\pi} \delta(x-a)\cos(nx)dx = \frac{1}{\pi}\cos(na) \qquad (29)$$

$$b_n = \frac{1}{\pi}\int_{-\pi}^{\pi} \delta(x-a)\sin(nx)dx = \frac{1}{\pi}\sin(na), \qquad (30)$$

so

$$\delta(x-a) = \frac{1}{2\pi} + \frac{1}{\pi}$$

$$\times \sum_{n=1}^{\infty}[\cos(na)\cos(nx) + \sin(na)\sin(nx)]$$

$$= \frac{1}{2\pi} + \frac{1}{\pi}\sum_{n=1}^{\infty}\cos[n(x-a)]. \qquad (31)$$

The delta function is given as a FOURIER TRANSFORM as

$$\delta(x) = \mathcal{F}[1] = \int_{-\infty}^{\infty} e^{-2\pi i k x}dk. \qquad (32)$$

Similarly,

$$\mathcal{F}^{-1}[\delta(x)] = \int_{-\infty}^{\infty} \delta(x)e^{2\pi i k x}dx = 1 \qquad (33)$$

(Bracewell 1999, p. 95). More generally, the FOURIER TRANSFORM of the delta function is

$$\mathcal{F}[\delta(x-x_0)] = \int_{-\infty}^{\infty} e^{-2\pi i k x}\delta(x-x_0)dx = e^{2\pi i k x_0}. \qquad (34)$$

Delta functions can also be defined in 2-D, so that in 2-D CARTESIAN COORDINATES

$$\delta^2(x,y) = \begin{cases} 0 & x^2 + y^2 \neq 0 \\ \infty & x^2 + y^2 = 0, \end{cases} \qquad (35)$$

$$\int_{-\infty}^{\infty}\int_{-\infty}^{\infty} \delta^2(x,y)dxdy = 1 \qquad (36)$$

$$\delta^2(ax,by) = \frac{1}{|ab|}\delta^2(x,y), \qquad (37)$$

and

$$\delta^2(x,y) = \delta(x)\delta(y). \qquad (38)$$

Similarly, in POLAR COORDINATES,

$$\delta^2(x,y) = \frac{\delta(r)}{\pi|r|} \qquad (39)$$

(Bracewell 1999, p. 85).
In 3-D CARTESIAN COORDINATES

$$\delta^3(x,y,z) = \delta^3(\mathbf{x}) = \begin{cases} 0 & x^2 + y^2 + z^2 \neq 0 \\ \infty & x^2 + y^2 + z^2 = 0 \end{cases} \qquad (40)$$

$$\int_{-\infty}^{\infty}\int_{-\infty}^{\infty}\int_{-\infty}^{\infty} \delta^3(x,y,z)dxdydz = 1 \qquad (41)$$

and

$$\delta(x)\delta(y)\delta(z). \qquad (42)$$

in CYLINDRICAL COORDINATES (r, θ, z),

$$\delta^3(r,\theta,z) = \frac{\delta(r)\delta(z)}{\pi r}. \tag{43}$$

In SPHERICAL COORDINATES (r,θ,ϕ),

$$\delta^3(r,\theta,\phi) = \frac{\delta(r)}{2\pi r^2} \tag{44}$$

(Bracewell 1999, p. 85).

A series expansion in CYLINDRICAL COORDINATES gives

$$\delta^3(\mathbf{r}_1 - \mathbf{r}_2) = \frac{1}{r_1}\delta(r_1 - r_2)\delta(\theta_1 - \theta_2)\delta(z_1 - z_2)$$

$$= \frac{1}{r_1}\delta(r_1 - r_2)\frac{1}{2\pi}\sum_{m=-\infty}^{\infty} e^{im(\theta_1-\theta_2)}\frac{1}{2\pi}\int_{-\infty}^{\infty} e^{ik(z_1-z_2)}dk.$$

$$\tag{45}$$

The delta function also obeys the so-called SIFTING PROPERTY

$$\int f(x)\delta(x-x_0)dx = f(x_0) \tag{46}$$

(Bracewell 1999, pp. 74–5).

See also DELTA SEQUENCE, DOUBLET FUNCTION, FOURIER TRANSFORM–DELTA FUNCTION, GENERALIZED FUNCTION, IMPULSE SYMBOL, POINCARÉ-BERTRAND THEOREM, SHAH FUNCTION, SOKHOTSKII'S FORMULA

References

Arfken, G. *Mathematical Methods for Physicists, 3rd ed.* Orlando, FL: Academic Press, pp. 481–85, 1985.
Bracewell, R. "The Impulse Symbol." Ch. 5 in *The Fourier Transform and Its Applications, 3rd ed.* New York: McGraw-Hill, pp. 69–7, 1999.
Dirac, P. A. M. *Quantum Mechanics, 4th ed.* London: Oxford University Press, 1958.
Gasiorowicz, S. *Quantum Physics.* New York: Wiley, pp. 491–94, 1974.
Papoulis, A. *Probability, Random Variables, and Stochastic Processes, 2nd ed.* New York: McGraw-Hill, pp. 97–8, 1984.
Spanier, J. and Oldham, K. B. "The Dirac Delta Function $\delta(x-a)$." Ch. 10 in *An Atlas of Functions.* Washington, DC: Hemisphere, pp. 79–2, 1987.
van der Pol, B. and Bremmer, H. *Operational Calculus Based on the Two-Sided Laplace Integral.* Cambridge, England: Cambridge University Press, 1955.

Delta Operator

A SHIFT-INVARIANT OPERATOR Q for which Qx is a NONZERO constant.

1. $Qa = 0$ for every constant a.
2. If $p(x)$ is a POLYNOMIAL of degree n, $Qp(x)$ is a POLYNOMIAL of degree $n-1$.
3. Every delta sequence has a unique BASIC POLYNOMIAL SEQUENCE.

See also BASIC POLYNOMIAL SEQUENCE, SHIFT-INVARIANT OPERATOR, UMBRAL CALCULUS

References

Roman, S. *The Umbral Calculus.* New York: Academic Press, 1984.
Rota, G.-C.; Kahaner, D.; Odlyzko, A. "On the Foundations of Combinatorial Theory. VIII: Finite Operator Calculus." *J. Math. Anal. Appl.* **42**, 684–60, 1973.

Delta Sequence

A SEQUENCE of strongly peaked functions for which

$$\lim_{n\to\infty}\int_{-\infty}^{\infty}\delta_n(x)f(x)\,dx = f(0) \tag{1}$$

so that in the limit as $n\to\infty$, the sequences become DELTA FUNCTIONS. Examples include

$$\delta^n(x) = \begin{cases} 0 & x < -\frac{1}{2n} \\ n & -\frac{1}{2n} < x < \frac{1}{2n} \\ 0 & x > \frac{1}{2n} \end{cases} \tag{2}$$

$$= \frac{n}{\sqrt{\pi}}e^{-n^2x^2} \tag{3}$$

$$= \frac{n}{\pi}\operatorname{sinc}(ax) \equiv \frac{\sin(nx)}{\pi x} \tag{4}$$

$$= \frac{1}{\pi x}\frac{e^{inx} - e^{-inx}}{2i} \tag{5}$$

$$= \frac{1}{2\pi i x}[e^{ixt}]_{-n}^{n} \tag{6}$$

$$= \frac{1}{2\pi}\int_{-n}^{n} e^{ixt}\,dt \tag{7}$$

$$= \frac{1}{2\pi}\frac{\sin[(n+\frac{1}{2})x]}{\sin(\frac{1}{2}x)}, \tag{8}$$

where (8) is known as the DIRICHLET KERNEL.

See also DELTA FUNCTION

Delta Variation

VARIATION

Deltahedron

A POLYHEDRON whose faces are CONGRUENT EQUILATERAL TRIANGLES (Wells 1986, p. 73). There are an infinite number of deltahedra, but only eight convex ones (Freudenthal and van der Waerden 1947).

Among this list of eight, faces composed of coplanar equilateral triangles sharing an edge (such as the RHOMBIC DODECAHEDRON) are not allowed. The eight convex deltahedra have $n = 4, 6, 8, 10, 12, 14, 16$, and 20 faces. These are summarized in the table below, and illustrated in the following figures.

n	Name
4	TETRAHEDRON
6	TRIANGULAR DIPYRAMID
8	OCTAHEDRON
10	PENTAGONAL DIPYRAMID
12	SNUB DISPHENOID
14	TRIAUGMENTED TRIANGULAR PRISM
16	GYROELONGATED SQUARE DIPYRAMID
20	ICOSAHEDRON

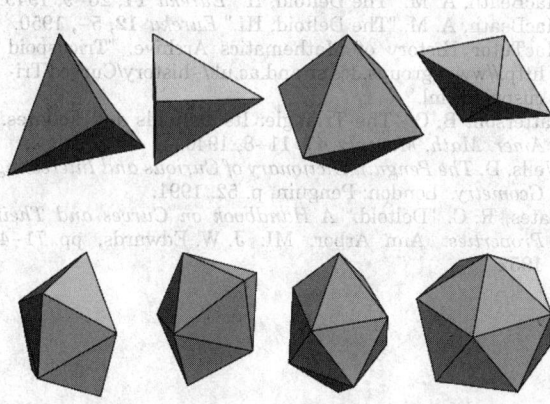

The 24-faced deltahedra formed by (1) CUMULATION of the CUBE and (2) STELLA OCTANGULA are both concave.

The "caved in" CUMULATED DODECAHEDRON is a deltahedron with 60 faces. It is ICOSAHEDRON STELLATION I_{20} (Wells 1991, p. 78).

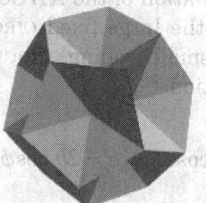

Cundy (1952) identified 17 concave deltahedra with two kinds of VERTICES.

See also CUMULATION, GYROELONGATED SQUARE DIPYRAMID, ICOSAHEDRON, OCTAHEDRON, PENTAGONAL DIPYRAMID, SNUB DISPHENOID TETRAHEDRON, TRIANGULAR DIPYRAMID, TRIAUGMENTED TRIANGULAR PRISM

References
Cundy, H. M. "Deltahedra." *Math. Gaz.* **36**, 263–66, 1952.
Cundy, H. and Rollett, A. "Deltahedra." §3.11 in *Mathematical Models, 3rd ed.* Stradbroke, England: Tarquin Pub., pp. 142–44, 1989.
Freudenthal, H. and van der Waerden, B. L. "On an Assertion of Euclid." *Simon Stevin* **25**, 115–21, 1947.
Gardner, M. *Fractal Music, Hypercards, and More: Mathematical Recreations from Scientific American Magazine.* New York: W. H. Freeman, pp. 40, 53, and 58–0, 1992.
Pugh, A. *Polyhedra: A Visual Approach.* Berkeley, CA: University of California Press, pp. 35–6, 1976.
Wells, D. *The Penguin Dictionary of Curious and Interesting Numbers.* Middlesex, England: Penguin Books, p. 73, 1986.
Wells, D. *The Penguin Dictionary of Curious and Interesting Geometry.* London: Penguin, pp. 51 and 78, 1991.

Deltohedron
TRAPEZOHEDRON

Deltoid

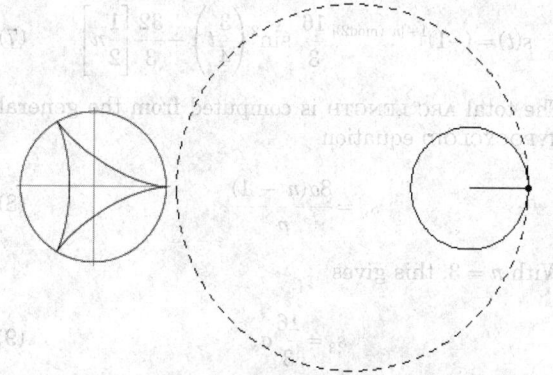

A 3-cusped HYPOCYCLOID, also called a tricuspoid. The deltoid was first considered by Euler in 1745 in connection with an optical problem. It was also investigated by Steiner in 1856 and is sometimes

called Steiner's hypocycloid (Lockwood 1967; Coxeter and Greitzer 1967, p. 44; MacTutor Archive). The equation of the deltoid is obtained by setting $n \equiv a/b = 3$ in the equation of the HYPOCYCLOID, where a is the RADIUS of the large fixed CIRCLE and b is the RADIUS of the small rolling CIRCLE, yielding the parametric equations

$$x = \left[\frac{2}{3}\cos\phi - \frac{1}{3}\cos(2\phi)\right]a = 2b\cos\phi + b\cos(2\phi) \quad (1)$$

$$y = \left[\frac{2}{3}\sin\phi + \frac{1}{3}\sin(2\phi)\right]a = 2b\sin\phi - b\sin(2\phi). \quad (2)$$

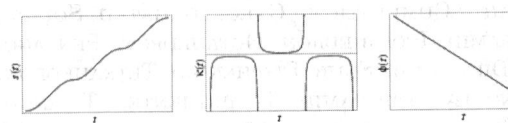

The ARC LENGTH, CURVATURE, and TANGENTIAL ANGLE are

$$s(t) = 4\int_0^t \left|\sin\left(\frac{3}{2}t'\right)\right|dt' = \frac{16}{3}\sin^2\left(\frac{3}{4}t\right) \quad (3)$$

$$\kappa(t) = -\frac{1}{8}\csc\left(\frac{3}{2}t\right) \quad (4)$$

$$\phi(t) = -\frac{1}{2}t. \quad (5)$$

As usual, care must be taken in the evaluation of $s(t)$ for $t > 2\pi/3$. Since the form given above comes from an integral involving the ABSOLUTE VALUE of a function, it must be monotonic increasing. Each branch can be treated correctly by defining

$$n = \left\lfloor\frac{3t}{2\pi}\right\rfloor + 1, \quad (6)$$

where $\lfloor x \rfloor$ is the FLOOR FUNCTION, giving the formula

$$s(t) = (-1)^{1 + [n \ (\text{mod} 2)]}\frac{16}{3}\sin^2\left(\frac{3}{4}t\right) + \frac{32}{3}\left\lfloor\frac{1}{2}n\right\rfloor. \quad (7)$$

The total ARC LENGTH is computed from the general HYPOCYCLOID equation

$$s_n = \frac{8a(n-1)}{n}. \quad (8)$$

With $n = 3$, this gives

$$s_3 = \frac{16}{3}a. \quad (9)$$

The AREA is given by

$$A_n = \frac{(n-a)(n-2)}{n^2}\pi a^2 \quad (10)$$

with $n = 3$

$$A_3 = \frac{2}{9}\pi a^2. \quad (11)$$

The length of the tangent to the tricuspoid, measured between the two points P, Q in which it cuts the curve again, is constant and equal to $4a$. If you draw TANGENTS at P and Q, they are at RIGHT ANGLES.

See also ASTROID, HYPOCYCLOID, SIMSON LINE

References

Beyer, W. H. *CRC Standard Mathematical Tables, 28th ed.* Boca Raton, FL: CRC Press, p. 219, 1987.
Coxeter, H. S. M. and Greitzer, S. L. *Geometry Revisited.* Washington, DC: Math. Assoc. Amer., p. 44, 1967.
Gray, A. *Modern Differential Geometry of Curves and Surfaces with Mathematica, 2nd ed.* Boca Raton, FL: CRC Press, p. 70, 1997.
Lawrence, J. D. *A Catalog of Special Plane Curves.* New York: Dover, pp. 131–35, 1972.
Lockwood, E. H. "The Deltoid." Ch. 8 in *A Book of Curves.* Cambridge, England: Cambridge University Press, pp. 72–9, 1967.
MacBeath, A. M. "The Deltoid." *Eureka* **10**, 20–3, 1948.
MacBeath, A. M. "The Deltoid, II." *Eureka* **11**, 26–9, 1949.
MacBeath, A. M. "The Deltoid, III." *Eureka* **12**, 5–, 1950.
MacTutor History of Mathematics Archive. "Tricuspoid." http://www-groups.dcs.st-and.ac.uk/~history/Curves/Tricuspoid.html.
Patterson, B. C. "The Triangle: Its Deltoids and Foliates." *Amer. Math. Monthly* **47**, 11–8, 1940.
Wells, D. *The Penguin Dictionary of Curious and Interesting Geometry.* London: Penguin, p. 52, 1991.
Yates, R. C. "Deltoid." *A Handbook on Curves and Their Properties.* Ann Arbor, MI: J. W. Edwards, pp. 71–4, 1952.

Deltoid Caustic

The caustic of the DELTOID when the rays are PARALLEL in any direction is an ASTROID.

Deltoid Evolute

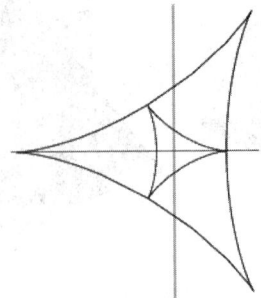

A HYPOCYCLOID EVOLUTE for $n = 3$ is another DEL-

TOID scaled by a factor $n/(n-2) = 3/1 = 3$ and rotated $1/(2\cdot3) = 1/6$ of a turn.

Deltoid Involute

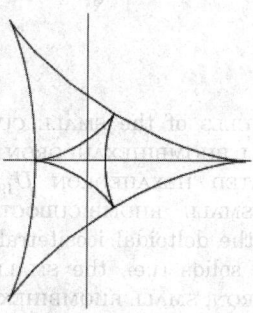

A HYPOCYCLOID INVOLUTE for $n = 3$ is another DEL-TOID scaled by a factor $(n-2)/n = 1/3$ and rotated $1/(2\cdot3) = 1/6$ of a turn.

Deltoid Pedal Curve

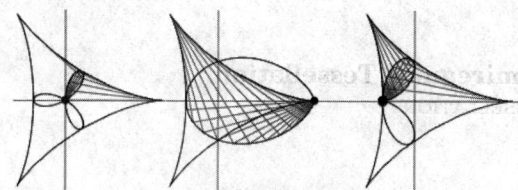

The PEDAL CURVE for a DELTOID with the PEDAL POINT at the CUSP is a FOLIUM. For the PEDAL POINT at the CUSP (NEGATIVE x-intercept), it is a BIFOLIUM. At the center, or anywhere on the inscribed EQUILATERAL TRIANGLE, it is a TRIFOLIUM.

Deltoid Radial Curve

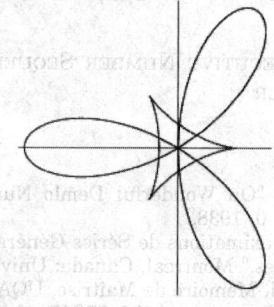

The TRIFOLIUM

$$x = x_0 + 4a\,\cos\phi - 4a\,\cos(2\phi)$$

$$y = y_0 + 4a\,\sin\phi + 4a\,\sin(2\phi).$$

Deltoidal Hexecontahedron

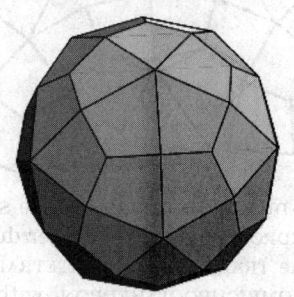

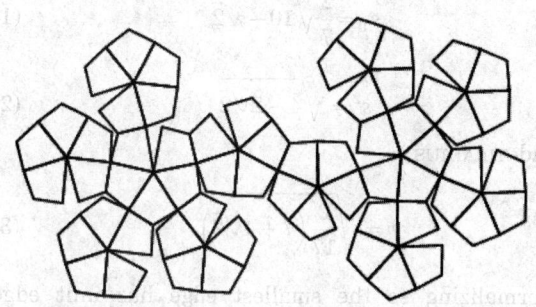

The 60-faced DUAL POLYHEDRON of the SMALL RHOMBICOSIDODECAHEDRON A_5 and Wenninger dual W_{14}. It is sometimes also called the trapezoidal hexecontahedron or strombic hexecontahedron.

See also ARCHIMEDEAN DUAL, ARCHIMEDEAN SOLID, HEXECONTAHEDRON, SMALL RHOMBICOSIDODECAHEDRON

References
Wenninger, M. J. *Dual Models.* Cambridge, England: Cambridge University Press, p. 24, 1983.

Deltoidal Icositetrahedron

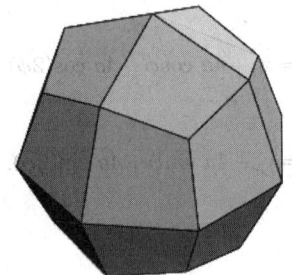

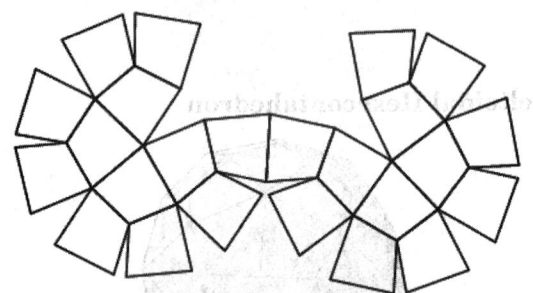

The 24-faced DUAL POLYHEDRON of the SMALL RHOM-BICUBOCTAHEDRON A_6 and Wenninger dual W_{13}. It is also called the TRAPEZOIDAL ICOSITETRAHEDRON. For a SMALL RHOMBICUBOCTAHEDRON with unit edge length, the deltoidal icositetrahedron has edge lengths

$$s_1 = \frac{2}{7}\sqrt{10 - \sqrt{2}} \tag{1}$$

$$s^2 = \sqrt{4 - 2\sqrt{2}} \tag{2}$$

and INRADIUS

$$r = \sqrt{\frac{2}{17}\left(7 + 4\sqrt{2}\right)}. \tag{3}$$

Normalizing so the smallest edge has unit edge length $s_1 = 1$ gives a deltoidal icositetrahedron with SURFACE AREA and VOLUME

$$S = 6\sqrt{29 - 2\sqrt{2}}. \tag{4}$$

$$V = \sqrt{122 + 71\sqrt{2}}. \tag{5}$$

See also ARCHIMEDEAN SOLID, DELTOIDAL ICOSITE-TRAHEDRON STELLATIONS, DELTOIDAL ICOSITETRAHE-DRON STELLATIONS, ICOSITETRAHEDRON, SMALL RHOMBICUBOCTAHEDRON

References
Wenninger, M. J. *Dual Models.* Cambridge, England: Cambridge University Press, p. 23, 1983.

Deltoidal Icositetrahedron Stellations

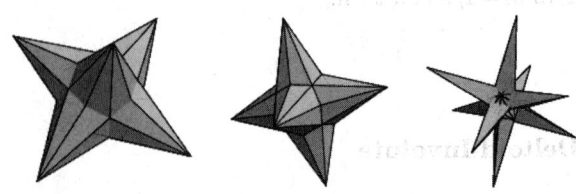

The CONVEX HULLS of the SMALL CUBICUBOCTAHE-DRON U_{13}, SMALL RHOMBIHEXAHEDRON U_{18}, and STEL-LATED TRUNCATED HEXAHEDRON U_{19} are all the Archimedean SMALL RHOMBICUBOCTAHEDRON A_6, whose dual is the deltoidal icositetrahedron, so the duals of these solids (i.e., the SMALL HEXACRONIC ICOSITETRAHEDRON, SMALL RHOMBIHEXAHEDRON, and GREAT TRIAKIS OCTAHEDRON) are all stellations of the deltoidal icositetrahedron (Wenninger 1983, p. 57).

See also ARCHIMEDEAN SOLID, ICOSITETRAHEDRON, SMALL RHOMBICUBOCTAHEDRON

References
Wenninger, M. J. *Dual Models.* Cambridge, England: Cambridge University Press, 1983.

Demiregular Tessellation
TESSELLATION

Demlo Number
The initially PALINDROMIC NUMBERS 1, 121, 12321, 1234321, 123454321, ... (Sloane's A002477). For the first through ninth terms, the sequence is given by the GENERATING FUNCTION

$$-\frac{10x + 1}{(x - 1)(10x - 1)(100x - 1)}$$
$$= 1 + 121x + 12321x^2 + 1234321x^3 + \dots$$

(Plouffe 1992, Sloane and Plouffe 1995). The definition of this sequence is slightly ambiguous from the tenth term on.

See also CONSECUTIVE NUMBER SEQUENCES, PALINDROMIC NUMBER

References
Kaprekar, D. R. "On Wonderful Demlo Numbers." *Math. Student* **6**, 68–0, 1938.

Plouffe, S. "Approximations de Séries Génératrices et quelques conjectures." Montréal, Canada: Université du Québec à Montréal, Mémoire de Maîtrise, UQAM, 1992.

Sloane, N. J. A. Sequences A002477/M5386 in "An On-Line Version of the Encyclopedia of Integer Sequences." http://www.research.att.com/~njas/sequences/eisonline.html.

Dendrite

A system of line segments connecting a given set of points.

See also PLATEAU'S PROBLEM, TRAVELING SALESMAN PROBLEM

References

Steinhaus, H. *Mathematical Snapshots, 3rd ed.* New York: Dover, pp. 120–25, 1999.

Dendrite Fractal

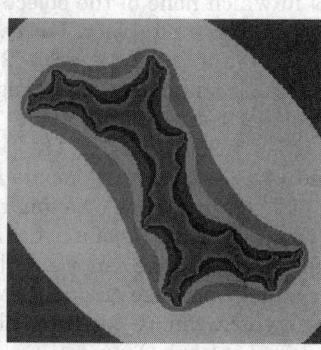

A JULIA SET with constant c chosen at the boundary of the MANDELBROT SET (Branner 1989; Dufner *et al.* 1998, p. 225). The image above was computed using $c = i$.

See also JULIA SET

References

Branner, B. "The Mandelbrot Set." In *Chaos and Fractals: The Mathematics behind the Computer Graphics* (Ed. R. L. Devaney and L. Keen). Providence, RI: Amer. Math. Soc., pp. 75–05, 1989.
Dufner, J.; Roser, A.; and Unseld, F. *Fraktale und Julia-Mengen.* Harri Deutsch, p. 225, 1998.

Denjoy Integral

A type of INTEGRAL which is an extension of both the RIEMANN INTEGRAL and the LEBESGUE INTEGRAL. The original Denjoy integral is now called a Denjoy integral "in the restricted sense," and a more general type is now called a Denjoy integral "in the wider sense." The independently discovered PERRON INTEGRAL turns out to be equivalent to the Denjoy integral "in the restricted sense."

See also INTEGRAL, LEBESGUE INTEGRAL, PERRON INTEGRAL, RIEMANN INTEGRAL

References

Iyanaga, S. and Kawada, Y. (Eds.). "Denjoy Integrals." §103 in *Encyclopedic Dictionary of Mathematics.* Cambridge, MA: MIT Press, pp. 337–40, 1980.
Kestelman, H. "General Denjoy Integral." §9.2 in *Modern Theories of Integration, 2nd rev. ed.* New York: Dover, pp. 217–27, 1960.

Denominator

The number q in a FRACTION p/q.

See also FRACTION, NUMERATOR, RATIO, RATIONAL NUMBER

Dense

A set A in a FIRST-COUNTABLE SPACE is dense in B if $B = A \cup L$, where L is the limit of sequences of elements of A. For example, the rational numbers are dense in the reals. In general, a SUBSET A of X is dense if its CLOSURE $cl(A) = X$.

See also CLOSURE (SET), DENSITY, DERIVED SET, NOWHERE DENSE, PERFECT SET

Density

DENSITY (POLYGON), DENSITY (SEQUENCE), NATURAL DENSITY

Density (Polygon)

The number q in a STAR POLYGON $\{p/q\}$.

See also STAR POLYGON

Density (Sequence)

Let a SEQUENCE $\{a_i\}_{i=1}^{\infty}$ be strictly increasing and composed of NONNEGATIVE INTEGERS. Call $A(x)$ the number of terms not exceeding x. Then the density is given by $\lim_{x \to \infty} A(x)/x$ if the LIMIT exists.

References

Guy, R. K. *Unsolved Problems in Number Theory, 2nd ed.* New York: Springer-Verlag, p. 199, 1994.

Density Function

PROBABILITY FUNCTION

Denumerable Set

A SET is denumerable IFF it is EQUIPOLLENT to the finite ORDINAL NUMBERS. (Moore 1982, p. 6; Rubin 1967, p. 107; Suppes 1972, pp. 151–52). However, Ciesielski (1997, p. 64) calls this property "countable." The set ALEPH-0 is most commonly called "denumerable" to "COUNTABLY INFINITE".

See also COUNTABLE SET, COUNTABLY INFINITE

References

Ciesielski, K. *Set Theory for the Working Mathematician.* Cambridge, England: Cambridge University Press, 1997.
Dauben, J. W. *Georg Cantor: His Mathematics and Philosophy of the Infinite.* Princeton, NJ: Princeton University Press, 1990.
Ferreirós, J. "Non-Denumerability of ℝ." §6.2 in *Labyrinth of Thought: A History of Set Theory and Its Role in Modern Mathematics.* Basel, Switzerland: Birkhäuser, pp. 177–83, 1999.
Moore, G. H. *Zermelo's Axiom of Choice: Its Origin, Development, and Influence.* New York: Springer-Verlag, 1982.

Rubin, J. E. *Set Theory for the Mathematician.* New York: Holden-Day, 1967.
Suppes, P. *Axiomatic Set Theory.* New York: Dover, 1972.

Denumerably Infinite

COUNTABLY INFINITE

Depth (Graph)

GRAPH THICKNESS

Depth (Size)

The depth of a box is the horizontal DISTANCE from front to back (usually not necessarily defined to be smaller than the WIDTH, the horizontal DISTANCE from side to side).

See also HEIGHT, WIDTH (SIZE)

Depth (Statistics)

The smallest RANK (either up or down) of a set of data.

See also RANK (STATISTICS)

References

Tukey, J. W. *Explanatory Data Analysis.* Reading, MA: Addison-Wesley, p. 30, 1977.

Depth (Tree)

The depth of a RESOLVING TREE is the number of levels of links, not including the top. The depth of the link is the minimal depth for any RESOLVING TREE of that link. The only links of length 0 are the trivial links. A KNOT of length 1 is always a trivial KNOT and links of depth one are always HOPF LINKS, possibly with a few additional trivial components (Bleiler and Scharlemann 1988). The LINKS of depth two have also been classified (Scharlemann and Thompson 1991).

References

Adams, C. C. *The Knot Book: An Elementary Introduction to the Mathematical Theory of Knots.* New York: W. H. Freeman, p. 169, 1994.
Bleiler, S. and Scharlemann, M. "A Projective Plane in $\mathbb{R}^4$ with Three Critical Points is Standard. Strongly Invertible Knots have Property P." *Topology* **27**, 519–40, 1988.
Scharlemann, M. and Thompson, A. "Detecting Unknotted Graphs in 3-Space." *J. Diff. Geom.* **34**, 539–60, 1991.

Depth-First Traversal

A search algorithm of a GRAPH which explores the first son of a node before visiting its brothers. Tarjan (1972) and Hopcroft and Tarjan (1973) showed that depth-first search gives linear time algorithms for many problems in graph theory (Skiena 1990).

See also BREADTH-FIRST TRAVERSAL

References

Hopcroft, J. and Tarjan, R. "Algorithm 447: Efficient Algorithms for Graph Manipulation." *Comm. ACM* **16**, 372–78, 1973.
Skiena, S. "Breadth-First and Depth-First Search." §3.2.5 in *Implementing Discrete Mathematics: Combinatorics and Graph Theory with Mathematica.* Reading, MA: Addison-Wesley, pp. 95–7, 1990.
Tarjan, R. E. "Depth-First Search and Linear Graph Algorithms." *SIAM J. Comput.* **1**, 146–60, 1972.

Derangement

A derangement of n ordered objects, denoted $!n$, is a PERMUTATION in which none of the objects appear in their "natural" (i.e., ordered) place. For example, the only derangements of $\{1, 2, 3\}$ are $\{2, 3, 1\}$ and $\{3, 1, 2\}$, so $!3 = 2$. Similarly, the derangements of $\{1, 2, 3, 4\}$ are $\{2, 1, 4, 3\}$, $\{2, 3, 4, 1\}$, $\{2, 4, 1, 3\}$, $\{3, 1, 4, 2\}$, $\{3, 4, 1, 2\}$, $\{3, 4, 2, 1\}$, $\{4, 1, 2, 3\}$, $\{4, 3, 1, 2\}$, and $\{4, 3, 2, 1\}$. Derangements are permutations without fixed points (i.e., having no cycles of length one). The derangements of a list of n elements can be computed using Derangments[n] in the *Mathematica* add-on package DiscreteMath`Combinatorica` (which can be loaded with the command < <DiscreteMath`).

The problem was formulated by P. R. de Montmort in 1708, and solved by him in 1713 (de Montmort 1713–714). Nicholas Bernoulli also solved the problem using the INCLUSION-EXCLUSION PRINCIPLE (de Montmort 1713–714, p. 301; Bhatnagar, p. 8).

The function giving the number of distinct derangements on n elements is called the SUBFACTORIAL $!n$ and is equal to

$$!n \equiv n! \sum_{k=0}^{n} \frac{(-1)^k}{k!} \qquad (1)$$

(Bhatnagar, pp. 8–) or

$$!n \equiv \left[\frac{n!}{e}\right], \qquad (2)$$

where $k!$ is the usual FACTORIAL and $[x]$ is the NEAREST INTEGER FUNCTION. These are also called RENCONTRES NUMBERS (named after rencontres solitaire), or COMPLETE PERMUTATIONS, or derangements. The number of derangements $!n = d(n)$ of length n satisfy the RECURRENCE RELATIONS

$$d(n) = (n-1)[d(n-1) + d(n-2)] \qquad (3)$$

and

$$d(n) = nd(n-1) + (-1)^n, \qquad (4)$$

with $d(1) = 0$ and $d(2) = 1$ (Skiena 1990, p. 33). The first few are 0, 1, 2, 9, 44, 265, 1854, ... (Sloane's A000166). This sequence cannot be expressed as a fixed number of hypergeometric terms (Petkovsek *et al.* 1996, pp. 157–60).

See also MARRIED COUPLES PROBLEM, PERMUTATION, ROOT, SUBFACTORIAL

References

Aitken, A. C. *Determinants and Matrices.* Westport, CT: Greenwood Pub., p. 135, 1983.

Ball, W. W. R. and Coxeter, H. S. M. *Mathematical Recreations and Essays, 13th ed.* New York: Dover, pp. 46–7, 1987.

Bhatnagar, G. *Inverse Relations, Generalized Bibasic Series, and their $U(n)$ Extensions.* Ph.D. thesis. Ohio State University, 1995.

Comtet, L. "The 'Problème des Recontres'." §4.2 in *Advanced Combinatorics: The Art of Finite and Infinite Expansions, rev. enl. ed.* Dordrecht, Netherlands: Reidel, pp. 180–83, 1974.

Coolidge, J. L. *An Introduction to Mathematical Probability.* Oxford, England: Oxford University Press, p. 24, 1925.

Courant, R. and Robbins, H. *What is Mathematics?: An Elementary Approach to Ideas and Methods, 2nd ed.* Oxford, England: Oxford University Press, pp. 115–16, 1996.

de Montmort, P. R. *Essai d'analyse sur les jeux de hasard.* Paris, 1708. Second edition published 1713–714. Third edition reprinted in New York: Chelsea, pp. 131–38, 1980.

Dickau, R. M. "Derangements." http://forum.swarthmore.edu/advanced/robertd/derangements.html.

Durell, C. V. and Robson, A. *Advanced Algebra.* London, p. 459, 1937.

Graham, R. L.; Knuth, D. E.; and Patashnik, O. *Concrete Mathematics: A Foundation for Computer Science, 2nd ed.* Reading, MA: Addison-Wesley, 1994.

Petkovsek, M.; Wilf, H. S.; and Zeilberger, D. *A = B.* Wellesley, MA: A. K. Peters, 1996.

Roberts, F. S. *Applied Combinatorics.* Englewood Cliffs, NJ: Prentice-Hall, 1984.

Ruskey, F. "Information on Derangements." http://www.theory.csc.uvic.ca/~cos/inf/perm/Derangements.html.

Skiena, S. "Derangements." §1.4.2 in *Implementing Discrete Mathematics: Combinatorics and Graph Theory with Mathematica.* Reading, MA: Addison-Wesley, pp. 33–4, 1990.

Sloane, N. J. A. Sequences A000166/M1937 in "An On-Line Version of the Encyclopedia of Integer Sequences." http://www.research.att.com/~njas/sequences/eisonline.html.

Stanley, R. P. *Enumerative Combinatorics, Vol. 1.* New York: Cambridge University Press, p. 67, 1986.

Vardi, I. *Computational Recreations in Mathematica.* Reading, MA: Addison-Wesley, p. 123, 1991.

Derivation

A derivation is a sequence of steps, logical or computational, from one result to another. The word derivation comes from the word "derive."

"Derivation" can also refer to a particular type of operator used to define a DERIVATION ALGEBRA on a ring or algebra.

See also DERIVATION ALGEBRA

Derivation Algebra

Let A be any algebra over a FIELD F, and define a derivation of A as a linear operator D on A satisfying

$$(xy)D = (xD)y + x(yD)$$

for all $x, y \in A$. Then the set $D(A)$ of all derivations of A in a SUBSPACE of the associative algebra of all linear operators on A is a LIE ALGEBRA, called the derivation algebra.

See also LIE ALGEBRA

References

Schafer, R. D. *An Introduction to Nonassociative Algebras.* New York: Dover, pp. 3–, 1996.

Derivative

The derivative of a FUNCTION represents an infinitesimal change in the function with respect to whatever parameters it may have. The "simple" derivative of a function f with respect to x is denoted either $f'(x)$ or

$$\frac{df}{dx} \tag{1}$$

(and often written in-line as df/dx). When derivatives are taken with respect to time, they are often denoted using Newton's OVERDOT notation for FLUXIONS,

$$\frac{dx}{dt} = \dot{x}. \tag{2}$$

When a derivative is taken n times, the notation $x^{(n)}$ or

$$\frac{d^n f}{dx^n} \tag{3}$$

is used, with

$$\dot{x}, \ddot{x}, \dddot{x}, \text{ etc.} \tag{4}$$

the corresponding FLUXION notation. When a function $f(x, y, \ldots)$ depends on more than one variable, a PARTIAL DERIVATIVE

$$\frac{\partial f}{\partial x}, \frac{\partial^2 f}{\partial x \partial y}, \text{ etc.} \tag{5}$$

can be used to specify the derivative with respect to one or more variables.

The derivative of a function $f(x)$ with respect to the variable x is defined as

$$f'(x) \equiv \lim_{h \to 0} \frac{f(x+h) - f(x)}{h}. \tag{6}$$

Note that in order for the limit to exist, both $\lim_{h \to 0^+}$ and $\lim_{h \to 0^-}$ must exist and be equal, so the FUNCTION must be continuous. However, continuity is a NECESSARY but *not* SUFFICIENT condition for differentiability. Since some DISCONTINUOUS functions can be integrated, in a sense there are "more" functions which can be integrated than differentiated. In a letter to Stieltjes, Hermite wrote, "I recoil with dismay and horror at this lamentable plague of functions which do not have derivatives."

A 3-D generalization of the derivative to an arbitrary direction is known as the DIRECTIONAL DERIVATIVE. In general, derivatives are mathematical objects which exist between smooth functions on manifolds. In this formalism, derivatives are usually assembled into "TANGENT MAPS."

Simple derivatives of some simple functions follow.

$$\frac{d}{dx}x^n = nx^{n-1} \tag{7}$$

$$\frac{d}{dx}\ln|x| = \frac{1}{x} \tag{8}$$

$$\frac{d}{dx}\sin x = \cos x \tag{9}$$

$$\frac{d}{dx}\cos x = -\sin x \tag{10}$$

$$\frac{d}{dx}\tan x = \frac{d}{dx}\left(\frac{\sin x}{\cos x}\right) = \frac{\cos x \cos x - \sin x(-\sin x)}{\cos^2 x}$$

$$= \frac{1}{\cos^2 x} = \sec^2 x \tag{11}$$

$$\frac{d}{dx}\csc x = \frac{d}{dx}(\sin x)^{-1} = -(\sin x)^{-2}\cos x = -\frac{\cos x}{\sin^2 x}$$

$$= -\csc x \cot x \tag{12}$$

$$\frac{d}{dx}\sec x = \frac{d}{dx}(\cos x)^{-1} = -(\cos x)^{-2}(-\sin x) = \frac{\sin x}{\cos^2 x}$$

$$= \sec x \tan x \tag{13}$$

$$\frac{d}{dx}\cot x = \frac{d}{dx}\left(\frac{\cos x}{\sin x}\right) = \frac{\sin x(-\sin x) - \cos x \cos x}{\sin^2 x}$$

$$= -\frac{1}{\sin^2 x} = -\csc^2 x \tag{14}$$

$$\frac{d}{dx}e^x = e^x \tag{15}$$

$$\frac{d}{dx}a^x = \frac{d}{dx}e^{\ln a^x} = \frac{d}{dx}e^{x\ln a} = (\ln a)e^{x\ln a} = (\ln a)a^x \tag{16}$$

$$\frac{d}{dx}\sin^{-1}x = \frac{1}{\sqrt{1-x^2}} \tag{17}$$

$$\frac{d}{dx}\cos^{-1}x = -\frac{1}{\sqrt{1-x^2}} \tag{18}$$

$$\frac{d}{dx}\tan^{-1}x = \frac{1}{1+x^2} \tag{19}$$

$$\frac{d}{dx}\cot^{-1}x = -\frac{1}{1+x^2} \tag{20}$$

$$\frac{d}{dx}\sec^{-1}x = \frac{1}{x\sqrt{x^2-1}} \tag{21}$$

$$\frac{d}{dx}\csc^{-1}x = -\frac{1}{x\sqrt{x^2-1}} \tag{22}$$

$$\frac{d}{dx}\sinh x = \cosh x \tag{23}$$

$$\frac{d}{dx}\cosh x = \sinh x \tag{24}$$

$$\frac{d}{dx}\tanh x = \text{sech}^2 x \tag{25}$$

$$\frac{d}{dx}\coth x = -\text{csch}^2 x \tag{26}$$

$$\frac{d}{dx}\text{sech} x = -\text{sech} x \tanh x \tag{27}$$

$$\frac{d}{dx}\text{csch} x = -\text{csch} x \coth x \tag{28}$$

$$\frac{d}{dx}\text{sn} x = \text{cn} x \, \text{dn} x \tag{29}$$

$$\frac{d}{dx}\text{cn} x = -\text{sn} x \, \text{dn} x \tag{30}$$

$$\frac{d}{dx}\text{dn} x = -k^2 \, \text{sn} x \, \text{cn} x. \tag{31}$$

where $\text{sn}(x) \equiv \text{sn}(x, k)$, $\text{cn}(x) \equiv \text{cn}(x, k)$, etc. are JACOBI ELLIPTIC FUNCTIONS, and the PRODUCT RULE and QUOTIENT RULE have been used extensively to expand the derivatives.

There are a number of important rules for computing derivatives of certain combinations of functions. Derivatives of sums are equal to the sum of derivatives so that

$$[f(x) + \cdots + h(x)]' = f'(x) + \cdots + h'(x). \tag{32}$$

In addition, if c is a constant,

$$\frac{d}{dx}[cf(x)] = cf'(x). \tag{33}$$

The PRODUCT RULE for differentiation states

$$\frac{d}{dx}[f(x)g(x)] = f(x)g'(x) + f'(x)g(x), \tag{34}$$

where f' denotes the DERIVATIVE of f with respect to x. This derivative rule can be applied iteratively to yield derivate rules for products of three or more functions, for example,

$$[fgh]' = (fg)h' + (fg)'h = fgh' + (fg' + f'g)h$$

$$= f'gh + fg'h + fgh'. \qquad (35)$$

The QUOTIENT RULE for derivatives states that

$$\frac{d}{dx}\left[\frac{f(x)}{g(x)}\right] = \frac{g(x)f'(x) - f(x)g'(x)}{[g(x)]^2} \qquad (36)$$

while the POWER RULE gives

$$\frac{d}{dx}(x^n) = nx^{n-1} \qquad (37)$$

Other very important rule for computing derivatives is the CHAIN RULE, which states that

$$\frac{dy}{dx} = \frac{dy}{du}\frac{du}{dx} \qquad (38)$$

or more generally,

$$\frac{dz}{dt} = \frac{\partial z}{\partial x}\frac{dx}{dt} + \frac{\partial z}{\partial y}\frac{dy}{dt}, \qquad (39)$$

were $\partial z/\partial x$ denotes a PARTIAL DERIVATIVE. Miscellaneous other derivative identities include

$$\frac{dy}{dx} = \frac{\dfrac{dy}{dt}}{\dfrac{dx}{dt}} \qquad (40)$$

$$\frac{dy}{dx} = \frac{1}{\dfrac{dx}{dy}}. \qquad (41)$$

If $F(x,y) = C$, where C is a constant, then

$$dF = \frac{\partial F}{\partial y}dy + \frac{\partial F}{\partial x}dx = 0, \qquad (42)$$

so

$$\frac{dy}{dx} = -\frac{\dfrac{\partial F}{\partial x}}{\dfrac{\partial F}{\partial y}}. \qquad (43)$$

A vector derivative of a vector function

$$\mathbf{X}(t) \equiv \begin{bmatrix} x_1(t) \\ x_2(t) \\ \vdots \\ x_k(t) \end{bmatrix} \qquad (44)$$

can be defined by

$$\frac{d\mathbf{X}}{dt} = \begin{bmatrix} \dfrac{dx_1}{dt} \\ \dfrac{dx_2}{dt} \\ \vdots \\ \dfrac{dt_k}{dt} \end{bmatrix} \qquad (45)$$

The nth derivatives of $x^n f(x)$ for $n = 1, 2, \ldots$ are

$$\frac{d}{dx}[xf(x)] = f(x) + xf'(x) \qquad (46)$$

$$\frac{d^2}{dx^2}[x^2 f(x)] = 2f(x) + 4xf'(x) + x^2 f''(x) \qquad (47)$$

$$\frac{d^3}{dx^3}[x^3 f(x)] = 6f(x) + 18xf'(x) + 9x^2 f''(x) + x^3 f'''(x). \qquad (48)$$

See also BLANCMANGE FUNCTION, CARATHÉODORY DERIVATIVE, CHAIN RULE, COMMA DERIVATIVE, CONVECTIVE DERIVATIVE, COVARIANT DERIVATIVE, DIRECTIONAL DERIVATIVE, EULER-LAGRANGE DERIVATIVE, FLUXION, FRACTIONAL CALCULUS, FRÉCHET DERIVATIVE, LAGRANGIAN DERIVATIVE, LIE DERIVATIVE, LOGARITHMIC DERIVATIVE, PINCHERLE DERIVATIVE, POWER RULE, PRODUCT RULE, Q-SERIES, QUOTIENT RULE, SCHWARZIAN DERIVATIVE, SEMICOLON DERIVATIVE, WEIERSTRASS FUNCTION

References

Abramowitz, M. and Stegun, C. A. (Eds.). *Handbook of Mathematical Functions with Formulas, Graphs, and Mathematical Tables, 9th printing.* New York: Dover, p. 11, 1972.

Anton, H. *Calculus: A New Horizon, 6th ed.* New York: Wiley, 1999.

Beyer, W. H. "Derivatives." *CRC Standard Mathematical Tables, 28th ed.* Boca Raton, FL: CRC Press, pp. 229–32, 1987.

Griewank, A. *Principles and Techniques of Algorithmic Differentiation.* Philadelphia, PA: SIAM, 2000.

Derivative Test

FIRST DERIVATIVE TEST, SECOND DERIVATIVE TEST

Derived Polygon

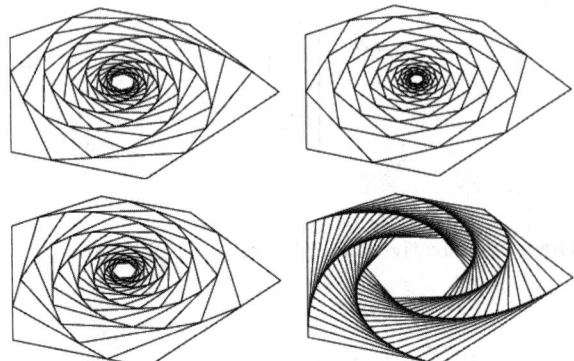

Given a POLYGON with an EVEN NUMBER of sides, the derived polygon is obtained by joining the points which are a fractional distance r along each side. If $r = 1/2$, then the derived polygons are called MID-POINT POLYGONS and tend to a shape with opposite sides parallel and equal in length. Furthermore, alternate polygons have approximately the same length, and the original and all derived polygons have the same centroid.

Amazingly, if $r \neq 1$, the derived polygons still approach a shape with opposite sides parallel and equal in length, and all have the same centroid. The above illustrations show 20 derived polygons for ratios $r = 0.3$, 0.5, 0.7, and 0.9. More amazingly still, if the original polygon is skew, a plane polygonal is approached which has these same properties.

See also MIDPOINT POLYGON, WHIRL

References

Cadwell, J. H. *Topics in Recreational Mathematics.* Cambridge, England: Cambridge University Press, 1966.
Wells, D. *The Penguin Dictionary of Curious and Interesting Geometry.* London: Penguin, pp. 53–4, 1991.

Derived Set

The LIMIT POINTS of a SET P, denoted P'.

See also DENSE, LIMIT POINT, PERFECT SET

References

Ferreirós, J. "Cantor's Derived Sets" and "Derived Sets and Cardinalities." §4.4.3 and 6.6 in *Labyrinth of Thought: A History of Set Theory and Its Role in Modern Mathematics.* Basel, Switzerland: Birkhäuser, pp. 141–44 and 202–08, 1999.

Dervish

A QUINTIC SURFACE having the maximum possible number of ORDINARY DOUBLE POINTS (31), which was constructed by W. Barth in 1994 (Endraß). The implicit equation of the surface is

$$64(x - w)\big[x^4 - 4x^3 w - 10x^2 y^2 - 4x^2 w^2$$

$$+ 16xw^3 - 20xy^2 w + 5y^4 + 16w^4 - 20y^2 w^2\big]$$

$$- 5\sqrt{5 - \sqrt{5}}\left(2z - \sqrt{5 - \sqrt{5}}\,w\right)$$

$$\times \left[4\big(x^2 + y^2 + z^2\big) + (1 + 3\sqrt{5})w^2\right]^2, \qquad (1)$$

where w is a parameter (Endraß). The surface can also be described by the equation

$$aF + q = 0, \qquad (2)$$

where

$$F = h_1 h_2 h_3 h_4 h_5, \qquad (3)$$

$$h_1 = x - z \qquad (4)$$

$$h2 = \cos\left(\frac{2\pi}{5}\right)x - \sin\left(\frac{2\pi}{5}\right)y - z \qquad (5)$$

$$h_3 = \cos\left(\frac{4\pi}{5}\right)x - \sin\left(\frac{4\pi}{5}\right)y - z \qquad (6)$$

$$h_4 = \cos\left(\frac{6\pi}{5}\right)x - \sin\left(\frac{6\pi}{5}\right)y - z \qquad (7)$$

$$h_5 = \cos\left(\frac{8\pi}{5}\right)x - \sin\left(\frac{8\pi}{5}\right)y - z \qquad (8)$$

$$q = (1 - cz)\big(x^2 + y^2 - 1 + rz^2\big)^2, \qquad (9)$$

and

$$r = \frac{1}{4}\left(1 + \sqrt{5}\right) \qquad (10)$$

$$a = -\frac{8}{5}\left(1 + \frac{1}{\sqrt{5}}\right)\sqrt{5 - \sqrt{5}} \qquad (11)$$

$$c = \frac{1}{2}\sqrt{5 - \sqrt{5}} \qquad (12)$$

(Nordstrand).

The dervish is invariant under the GROUP D_5 and contains exactly 15 lines. Five of these are the intersection of the surface with a D_5-invariant cone containing 16 nodes, five are the intersection of the surface with a D_5-invariant plane containing 10 nodes, and the last five are the intersection of the surface with a second D_5-invariant plane containing no nodes (Endraß).

See also ALGEBRAIC SURFACE, QUINTIC SURFACE

References

Endraß, S. "Togliatti Surfaces." http://enriques.mathematik.uni-mainz.de/kon/docs/Etogliatti.shtml.
Endraß, S. "Flächen mit vielen Doppelpunkten." *DMV-Mitteilungen* **4**, 17–0, 4/1995.
Endraß, S. *Symmetrische Fläche mit vielen gewöhnlichen Doppelpunkten.* Ph.D. thesis. Erlangen, Germany, 1996.
Nordstrand, T. "Dervish." http://www.uib.no/people/nfytn/dervtxt.htm.

Desargues' Configuration

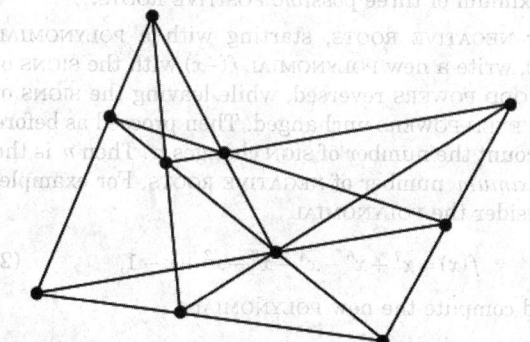

The 10_3 CONFIGURATION of ten lines intersecting three at a time in 10 points which arises in DESARGUES' THEOREM.

See also CONFIGURATION, DESARGUES' THEOREM

Desargues' Theorem

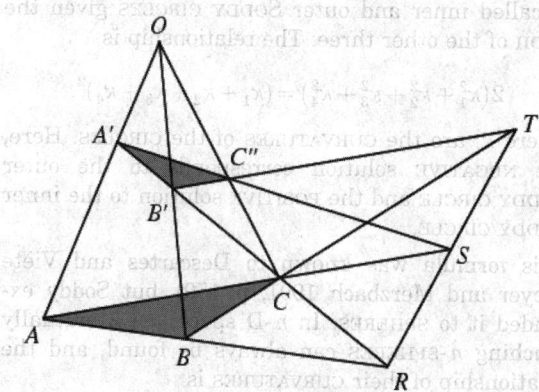

If the three straight LINES joining the corresponding VERTICES of two TRIANGLES ABC and $A'B'C''$ all meet in a point (the PERSPECTIVE CENTER), then the three intersections of pairs of corresponding sides lie on a straight LINE (the PERSPECTIVE AXIS). Equivalently, if two TRIANGLES are PERSPECTIVE from a POINT, they are PERSPECTIVE from a LINE.

The 10 lines and 10 3-line intersections form a 10_3 CONFIGURATION sometimes called DESARGUES' CONFIGURATION.

Desargues' theorem is SELF-DUAL upon application of the DUALITY PRINCIPLE of PROJECTIVE GEOMETRY.

See also DESARGUES' CONFIGURATION, DUALITY PRINCIPLE, PAPPUS'S HEXAGON THEOREM, PASCAL LINES, PASCAL'S THEOREM, PERSPECTIVE AXIS, PERSPECTIVE CENTER, PERSPECTIVE TRIANGLES, SELF-DUAL

References

Coxeter, H. S. M. and Greitzer, S. L. "Perspective Triangles; Desargues's Theorem." §3.6 in *Geometry Revisited.* Washington, DC: Math. Assoc. Amer., pp. 70–2, 1967.
Durell, C. V. *Modern Geometry: The Straight Line and Circle.* London: Macmillan, p. 44, 1928.
Eves, H. "Desargues' Two-Triangle Theorem." §6.2.5 in *A Survey of Geometry, rev. ed.* Boston, MA: Allyn & Bacon, pp. 249–51, 1965.
Graustein, W. C. *Introduction to Higher Geometry.* New York: Macmillan, pp. 23–5, 1930.
Ogilvy, C. S. *Excursions in Geometry.* New York: Dover, pp. 89–2, 1990.
Johnson, R. A. *Modern Geometry: An Elementary Treatise on the Geometry of the Triangle and the Circle.* Boston, MA: Houghton Mifflin, p. 231, 1929.
Wells, D. *The Penguin Dictionary of Curious and Interesting Numbers.* Middlesex, England: Penguin Books, p. 77, 1986.
Wells, D. *The Penguin Dictionary of Curious and Interesting Geometry.* London: Penguin, pp. 54–5, 1991.

Descartes Circle Theorem

A special case of APOLLONIUS' PROBLEM requiring the determination of a CIRCLE touching three mutually TANGENT CIRCLES (also called the KISSING CIRCLES PROBLEM). There are two solutions: a small circle surrounded by the three original CIRCLES, and a large circle surrounding the original three. Frederick

Soddy gave the FORMULA for finding the RADIUS of the so-called inner and outer SODDY CIRCLES given the RADII of the other three. The relationship is

$$2\left(\kappa_1^2 + \kappa_2^2 + \kappa_3^2 + \kappa_4^2\right) = \left(\kappa_1 + \kappa_2 + \kappa_3 + \kappa_4\right)^2,$$

where κ_i are the CURVATURES of the CIRCLES. Here, the NEGATIVE solution corresponds to the outer SODDY CIRCLE and the POSITIVE solution to the inner SODDY CIRCLE.

This formula was known to Descartes and Viète (Boyer and Merzbach 1991, p. 159), but Soddy extended it to SPHERES. In n-D space, $n+2$ mutually touching n-SPHERES can always be found, and the relationship of their CURVATURES is

$$n\left(\sum_{i=1}^{n+2}\kappa_i^2\right) = \left(\sum_{i=1}^{n+2}\kappa_i\right)^2.$$

See also APOLLONIUS' PROBLEM, FOUR COINS PROBLEM, SANGAKU PROBLEM, SODDY CIRCLES, SPHERE PACKING, TANGENT CIRCLES

References
Boyer, C. B. and Merzbach, U. C. *A History of Mathematics, 2nd ed.* New York: Wiley, 1991.
Coxeter, H. S. M. *Introduction to Geometry, 2nd ed.* New York: Wiley, pp. 13–6, 1969.
Fukagawa, H. and Pedoe, D. "The Descartes Circle Theorem." §1.7 in *Japanese Temple Geometry Problems.* Winnipeg, Manitoba, Canada: Charles Babbage Research Foundation, pp. 16–7 and 92, 1989.
Rothman, T. "Japanese Temple Geometry." *Sci. Amer.* **278**, 85–1, May 1998.
Wilker, J. B. "Four Proofs of a Generalization of the Descartes Circle Theorem." *Amer. Math. Monthly* **76**, 278–82, 1969.
Williams, R. *The Geometrical Foundation of Natural Structure: A Source Book of Design.* New York: Dover, pp. 50–1, 1979.

Descartes Folium
FOLIUM OF DESCARTES

Descartes Ovals
CARTESIAN OVALS

Descartes Total Angular Defect
The total angular defect is the sum of the ANGULAR DEFECTS over all VERTICES of a POLYHEDRON, where the ANGULAR DEFECT δ at a given VERTEX is the difference between the sum of face angles and 2π. For any convex POLYHEDRON, the Descartes total angular defect is

$$\Delta = \sum_i \delta i = 4\pi. \tag{1}$$

This is equivalent to the POLYHEDRAL FORMULA for a closed rectilinear surface, which satisfies

$$\Delta = 2\pi(V - E + F). \tag{2}$$

A POLYHEDRON with N_0 equivalent VERTICES is called a PLATONIC SOLID and can be assigned a SCHLÄFLI SYMBOL $\{p, q\}$. It then satisfies

$$N_0 = \frac{4\pi}{\delta} \tag{3}$$

and

$$\delta = 2\pi - q\left(1 - \frac{2}{p}\right)\pi, \tag{4}$$

so

$$N_0 = \frac{4p}{2p + 2q - pq}. \tag{5}$$

See also ANGULAR DEFECT, PLATONIC SOLID, POLYHEDRAL FORMULA, POLYHEDRON

Descartes' Formula
DESCARTES TOTAL ANGULAR DEFECT

Descartes' Sign Rule
A method of determining the maximum number of POSITIVE and NEGATIVE REAL ROOTS of a POLYNOMIAL.

For POSITIVE ROOTS, start with the SIGN of the COEFFICIENT of the lowest (or highest) POWER. Count the number of SIGN changes n as you proceed from the lowest to the highest POWER (ignoring POWERS which do not appear). Then n is the *maximum* number of POSITIVE ROOTS. Furthermore, the number of allowable ROOTS is n, $n-2$, $n-4$, ... For example, consider the POLYNOMIAL

$$f(x) = x^7 + x^6 - x^4 - x^3 - x^2 + x - 1. \tag{1}$$

Since there are three SIGN changes, there are a maximum of three possible POSITIVE ROOTS.

For NEGATIVE ROOTS, starting with a POLYNOMIAL $f(x)$, write a new POLYNOMIAL $f(-x)$ with the SIGNS of all ODD POWERS reversed, while leaving the SIGNS of the EVEN POWERS unchanged. Then proceed as before to count the number of SIGN changes n. Then n is the *maximum* number of NEGATIVE ROOTS. For example, consider the POLYNOMIAL

$$f(x) = x^7 + x^6 - x^4 - x^3 - x^2 + x - 1, \tag{2}$$

and compute the new POLYNOMIAL

$$f(-x) = -x^7 + x^6 - x^4 + x^3 - x^2 - x - 1. \tag{3}$$

In this example, there are four SIGN changes, so there are a maximum of four NEGATIVE ROOTS.

See also BOUND, ROOT, STURM FUNCTION

References

Anderson, B.; Jackson, J.; and Sitharam, M. "Descartes' Rule of Signs Revisited." *Amer. Math. Monthly* **105**, 447–51, 1998.

Grabiner, D. J. "Descartes' Rule of Signs: Another Construction." *Amer. Math. Monthly* **106**, 854–55, 1999.

Hall, H. S. and Knight, S. R. *Higher Algebra: A Sequel to Elementary Algebra for Schools.* London: Macmillan, pp. 459–60, 1950.

Henrici, P. "Sign Changes. The Rule of Descartes." §6.2 in *Applied and Computational Complex Analysis, Vol. 1: Power Series-Integration-Conformal Mapping-Location of Zeros.* New York: Wiley, pp. 439–43, 1988.

Itenberg, U. and Roy, M. F. "Multivariate Descartes' Rule." *Beiträge Algebra Geom.* **37**, 337–46, 1996.

Struik, D. J. (Ed.). *A Source Book in Mathematics 1200–800.* Princeton, NJ: Princeton University Press, pp. 89–3, 1986.

Descartes-Euler Polyhedral Formula

POLYHEDRAL FORMULA

Descending Plane Partition

$$
\begin{array}{cccccc}
7 & 7 & 6 & 6 & 3 & 1 \\
& 6 & 5 & 4 & 2 & \\
& & 3 & 3 & & \\
& & 2 & & &
\end{array}
$$

A descending plane partition of order n is a 2-D array (possibly empty) of positive integers less than or equal to n such that the left-hand edges are successively indented, rows are nonincreasing across, columns are decreasing downwards, and the number of entries in each row is strictly less than the largest entry in that row. Implicit in this definition are the requirements that no "holes" are allowed in the array, all rows are flush against the top, and the diagonal element *must* be filled if *any* element of its row is filled. The above example shows a decreasing plane partition of order seven.

$$
\begin{array}{l}
3\ \ 3 \\
\quad\quad\quad 3\ \ 3\ \ 3\ \ 2\ \ 3\ \ 1\ \ 3\ \ 2\ \ \phi \\
2
\end{array}
$$

The sole descending plane partition of order one is the empty one $\emptyset$, the two of order two are "2" and ϕ, and the seven of order three are illustrated above. In general, the number of descending plane partitions of order n is equal to the number of $+1$-bordered ALTERNATING SIGN MATRICES: 1, 2, 7, 42, 429, ... (Sloane's A005130).

See also ALTERNATING SIGN MATRIX, PLANE PARTITION

References

Andrews, G. E. "Plane Partitions (III): The Weak Macdonald Conjecture." *Invent. Math.* **53**, 193–25, 1979.

Bressoud, D. and Propp, J. "How the Alternating Sign Matrix Conjecture was Solved." *Not. Amer. Math. Soc.* **46**, 637–46.

Sloane, N. J. A. Sequences A005130/M1808 in "An On-Line Version of the Encyclopedia of Integer Sequences." http://www.research.att.com/~njas/sequences/eisonline.html.

Descriptive Geometry

PROJECTIVE GEOMETRY

Descriptive Set Theory

The study of DEFINABLE SETS and functions in POLISH SPACES.

References

Becker, H. and Kechris, A. S. *The Descriptive Set Theory of Polish Group Actions.* New York: Cambridge University Press, 1996.

Design

A formal description of the constraints on the possible configurations of an experiment which is subject to given conditions. A design is sometimes called an EXPERIMENTAL DESIGN.

See also BLOCK DESIGN, COMBINATORICS, DESIGN THEORY, HADAMARD DESIGN, HOWELL DESIGN, SPHERICAL DESIGN, SYMMETRIC BLOCK DESIGN, TRANSVERSAL DESIGN

Design Theory

The study of DESIGNS and, in particular, NECESSARY and SUFFICIENT conditions for the existence of a BLOCK DESIGN.

See also BLOCK DESIGN, BRUCK-RYSER-CHOWLA THEOREM, DESIGN, FISHER'S BLOCK DESIGN INEQUALITY

References

Assmus, E. F. Jr. and Key, J. D. *Designs and Their Codes.* New York: Cambridge University Press, 1993.

Colbourn, C. J. and Dinitz, J. H. *CRC Handbook of Combinatorial Designs.* Boca Raton, FL: CRC Press, 1996.

Dinitz, J. H. and Stinson, D. R. (Eds.). "A Brief Introduction to Design Theory." Ch. 1 in *Contemporary Design Theory: A Collection of Surveys.* New York: Wiley, pp. 1–2, 1992.

Lindner, C. C. and Rodger, C. A. *Design Theory.* Boca Raton, FL: CRC Press, 1997.

Desmic Surface

Let Δ_1, Δ_2, and Δ_3 be tetrahedra in projective 3-space $\mathbb{P}^3$. Then the tetrahedra are said to be desmically related if there exist constants α, β, and γ such that

$$\alpha\Delta_1 + \beta\Delta_2 + \gamma\Delta_3 = 0.$$

A desmic surface is then defined as a QUARTIC SURFACE which can be written as

$$a\Delta_1 + b\Delta_2 + c\Delta_3 = 0$$

for desmically related tetrahedra Δ_1, Δ_2, and Δ_3. Desmic surfaces have 12 ORDINARY DOUBLE POINTS, which are the vertices of three tetrahedra in 3-space (Hunt).

See also QUARTIC SURFACE

References

Hunt, B. "Desmic Surfaces." §B.5.2 in *The Geometry of Some Special Arithmetic Quotients.* New York: Springer-Verlag, pp. 311–15, 1996.

Jessop, C. §13 in *Quartic Surfaces with Singular Points.* Cambridge, England: Cambridge University Press, 1916.

Destructive Dilemma

A formal argument in LOGIC in which it is stated that

1. $P \Rightarrow Q$ and $R \Rightarrow S$ (where $\Rightarrow$ means "IMPLIES"), and
2. Either not-Q or not-S is true, from which two statements it follows that either not-P or not-R is true.

See also CONSTRUCTIVE DILEMMA, DILEMMA

Determinant

Determinants are mathematical objects which are very useful in the analysis and solution of SYSTEMS OF LINEAR EQUATIONS. As shown by CRAMER'S RULE, a nonhomogeneous system of linear equations has a nontrivial solution IFF the determinant of the system's MATRIX is NONZERO (i.e., the MATRIX is nonsingular). For example, eliminating x, y, and z from the equations

$$a_1 x + a_2 y + a_3 z = 0 \tag{1}$$

$$b_1 x + b_2 y + b_3 z = 0 \tag{2}$$

$$c_1 x + c_2 y + c_3 z = 0 \tag{3}$$

gives the expression

$$a_1 b_2 c_3 - a_1 b_3 c_2 + a_2 b_3 c_1 - a_2 b_1 c_3 + a_3 b_1 c_2 - a_3 b_2 c_1$$
$$= 0, \tag{4}$$

which is called the determinant for this system of equation. Determinants are defined only for SQUARE MATRICES. If the determinant of a MATRIX is 0, the MATRIX is said to be a SINGULAR MATRIX.

The determinant of a MATRIX A,

$$\begin{vmatrix} a_1 & a_2 & \cdots & a_n \\ b_1 & b_2 & \cdots & b_n \\ \vdots & \vdots & \ddots & \vdots \\ z_1 & z_2 & \cdots & z_n \end{vmatrix} \tag{5}$$

is commonly denoted det A, $|A|$, or in component notation as $\Sigma(\pm a_1 b_2 c_3 \cdots)$, $D(a_1 b_2 c_3 \cdots)$, or $|a_1 b_2 c_3 \cdots|$ (Muir 1960, p. 17).

A 2×2 determinant is defined to be

$$\det \begin{bmatrix} a & b \\ c & d \end{bmatrix} \equiv \begin{vmatrix} a & b \\ c & d \end{vmatrix} \equiv ad - bc. \tag{6}$$

A $k \times k$ determinant can be expanded "by MINORS" to obtain

$$\begin{vmatrix} a_{11} & a_{12} & a_{13} & \cdots & a_{1k} \\ a_{21} & a_{22} & a_{23} & \cdots & a_{2k} \\ \vdots & \vdots & \vdots & \ddots & \vdots \\ a_{k1} & a_{k2} & a_{k3} & \cdots & a_{kk} \end{vmatrix} = a_{11} \begin{vmatrix} a_{22} & a_{23} & \cdots & a_{2k} \\ \vdots & \vdots & \ddots & \vdots \\ a_{k2} & a_{k3} & \cdots & a_{kk} \end{vmatrix}$$

$$- a_{12} \begin{vmatrix} a_{21} & a_{23} & \cdots & a_{2k} \\ \vdots & \vdots & \ddots & \vdots \\ a_{k1} & a_{k3} & \cdots & a_{kk} \end{vmatrix} + \cdots$$

$$\pm a_{1k} \begin{vmatrix} a_{21} & a_{22} & \cdots & a_{2(k-1)} \\ \vdots & \vdots & \ddots & \vdots \\ a_{k1} & a_{k2} & \cdots & a_{k(k-1)} \end{vmatrix}. \tag{7}$$

A general determinant for a MATRIX A has a value

$$|A| = \sum_i a_{ij} a^{ij}, \tag{8}$$

with no implied summation over j and where a^{ij} is the COFACTOR of a_{ij} defined by

$$a^{ij} \equiv (-1)^{i+j} C_{ij}. \tag{9}$$

Here, C is the $(n-1) \times (n-1)$ MATRIX formed by eliminating row i and column j from A. This process is called DETERMINANT EXPANSION BY MINORS (or "Laplacian expansion by minors," sometimes further shortened to simply "Laplacian expansion").

A determinant can also be computed by writing down all PERMUTATIONS of $\{1, \ldots, n\}$, taking each permutation as the subscripts of the letters a, b, ..., and summing with signs determined by $\epsilon_p = (-1)^{i(p)}$, where $i(p)$ is the number of PERMUTATION INVERSIONS in permutation p (Muir 1960, p. 16), and $\epsilon_{n_1 n_2} \ldots$ is the PERMUTATION SYMBOL. For example, with $n = 3$, the permutations and the number of inversions they contain are 123 (0), 132 (1), 213 (1), 231 (2), 312 (2), and 321 (3), so the determinant is given by

$$\begin{vmatrix} a_1 & a_2 & a_3 \\ b_1 & b_2 & b_3 \\ c_1 & c_2 & c_3 \end{vmatrix}$$

$$= a_1 b_2 c_3 - a_1 b_3 c_2 - a_2 b_1 c_3 + a_2 b_3 c_1 + a_3 b_1 c_2$$
$$- a_3 b_2 c_1. \tag{10}$$

If c is a constant and A an $n \times n$ SQUARE MATRIX, then

$$|aA| = a^n |A|. \tag{11}$$

Given an $n \times n$ determinant, the additive inverse is

$$|-A| = (-1)^n |A|. \tag{12}$$

Determinants are also DISTRIBUTIVE, so

$$|AB| = |A||B|. \tag{13}$$

This means that the determinant of a MATRIX INVERSE can be found as follows:

$$|I| = |AA^{-1}| = |A||A^{-1}| = 1, \tag{14}$$

where I is the IDENTITY MATRIX, so

$$|\mathsf{A}|=\frac{1}{|\mathsf{A}^{-1}|}. \tag{15}$$

Determinants are MULTILINEAR in rows and columns, since

$$\begin{vmatrix} a_1 & a_2 & a_3 \\ a_4 & a_5 & a_6 \\ a_7 & a_8 & a_9 \end{vmatrix} = \begin{vmatrix} a_1 & 0 & 0 \\ a_4 & a_5 & a_6 \\ a_7 & a_8 & a_9 \end{vmatrix} + \begin{vmatrix} 0 & a_2 & 0 \\ a_4 & a_5 & a_6 \\ a_7 & a_8 & a_9 \end{vmatrix} + \begin{vmatrix} 0 & 0 & a_3 \\ a_4 & a_5 & a_6 \\ a_7 & a_8 & a_9 \end{vmatrix} \tag{16}$$

and

$$\begin{vmatrix} a_1 & a_2 & a_3 \\ a_4 & a_5 & a_6 \\ a_7 & a_8 & a_9 \end{vmatrix} = \begin{vmatrix} a_1 & a_2 & a_3 \\ 0 & a_5 & a_6 \\ 0 & a_8 & a_9 \end{vmatrix} + \begin{vmatrix} 0 & a_2 & a_3 \\ a_4 & a_5 & a_6 \\ 0 & a_8 & a_9 \end{vmatrix} + \begin{vmatrix} 0 & a_2 & a_3 \\ 0 & a_5 & a_6 \\ a_7 & a_8 & a_9 \end{vmatrix}. \tag{17}$$

The determinant of the SIMILARITY TRANSFORMATION of a matrix is equal to the determinant of the original MATRIX

$$|\mathsf{B}\mathsf{A}\mathsf{B}^{-1}|=|\mathsf{B}||\mathsf{A}||\mathsf{B}^{-1}|=|\mathsf{B}||\mathsf{A}|\frac{1}{|\mathsf{B}|}=|\mathsf{A}|. \tag{18}$$

The determinant of a similarity transformation minus a multiple of the unit MATRIX is given by

$$|\mathsf{B}^{-1}\mathsf{A}\mathsf{B}-\lambda\mathsf{I}|=|\mathsf{B}^{-1}\mathsf{A}\mathsf{B}-\mathsf{B}^{-1}\lambda\mathsf{I}\mathsf{B}|=|\mathsf{B}^{-1}(\mathsf{A}-\lambda\mathsf{I})\mathsf{B}|$$
$$=|\mathsf{B}^{-1}||\mathsf{A}-\lambda\mathsf{I}||\mathsf{B}|=|\mathsf{A}-\lambda\mathsf{I}|. \tag{19}$$

The determinant of a MATRIX TRANSPOSE equals the determinant of the original MATRIX,

$$|\mathsf{A}|=|\mathsf{A}^{\mathsf{T}}|, \tag{20}$$

and the determinant of a COMPLEX CONJUGATE is equal to the COMPLEX CONJUGATE of the determinant

$$|\bar{\mathsf{A}}|=\overline{|\mathsf{A}|}. \tag{21}$$

Let ε be a small number. Then

$$|\mathsf{I}+\epsilon\mathsf{A}|=1+\epsilon\mathrm{Tr}(\mathsf{A})+\mathcal{O}(\epsilon^2), \tag{22}$$

where $\mathrm{Tr}(\mathsf{A})$ is the TRACE of A. The determinant takes on a particularly simple form for a TRIANGULAR MATRIX

$$\begin{vmatrix} a_{11} & a_{21} & \cdots & a_{k1} \\ 0 & a_{22} & \cdots & a_{k2} \\ \vdots & \vdots & \ddots & \vdots \\ 0 & 0 & \vdots & a_{kk} \end{vmatrix} = \prod_{n=1}^{k} a_{nn}. \tag{23}$$

Important properties of the determinant include the following, which include invariance under ELEMENTARY ROW AND COLUMN OPERATIONS.

1. Switching two rows or columns changes the sign.
2. Scalars can be factored out from rows and columns.
3. Multiples of rows and columns can be added together without changing the determinant's value.
4. Scalar multiplication of a row by a constant c multiplies the determinant by c.
5. A determinant with a row or column of zeros has value 0.
6. Any determinant with two rows or columns equal has value 0.

Property 1 can be established by induction. For a 2×2 MATRIX, the determinant is

$$\begin{vmatrix} a_1 & b_1 \\ a_2 & b_2 \end{vmatrix} = a_1 b_2 - b_1 a_2 = -(b_1 a_2 - a_1 b_2)$$

$$= -\begin{vmatrix} b_1 & a_1 \\ b_2 & a_2 \end{vmatrix} \tag{24}$$

For a 3×3 MATRIX, the determinant is

$$\begin{vmatrix} a_1 & b_1 & c_1 \\ a_2 & b_2 & c_2 \\ a_3 & b_3 & c_3 \end{vmatrix} = a_1 \begin{vmatrix} b_2 & c_2 \\ b_3 & c_3 \end{vmatrix} - b_1 \begin{vmatrix} a_2 & c_2 \\ a_3 & c_3 \end{vmatrix} + c_1 \begin{vmatrix} a_2 & b_2 \\ a_3 & b_3 \end{vmatrix}$$

$$= -\left(a_1 \begin{vmatrix} c_2 & b_2 \\ c_3 & b_3 \end{vmatrix} + b_1 \begin{vmatrix} c_2 & a_2 \\ c_3 & a_3 \end{vmatrix} - c_1 \begin{vmatrix} a_2 & b_2 \\ a_3 & b_3 \end{vmatrix} \right)$$

$$= -\begin{vmatrix} a_1 & c_1 & b_1 \\ a_2 & c_2 & b_2 \\ a_3 & c_3 & b_3 \end{vmatrix}$$

$$= -\left(-a_1 \begin{vmatrix} b_2 & c_2 \\ b_3 & c_3 \end{vmatrix} + b_1 \begin{vmatrix} a_2 & c_2 \\ a_3 & c_3 \end{vmatrix} + c_1 \begin{vmatrix} b_2 & a_2 \\ b_3 & a_3 \end{vmatrix} \right)$$

$$= -\begin{vmatrix} b_1 & a_1 & c_1 \\ b_2 & a_2 & c_2 \\ b_3 & a_3 & c_3 \end{vmatrix}$$

$$= -\left(-a_1 \begin{vmatrix} c_2 & b_2 \\ c_3 & b_3 \end{vmatrix} - b_1 \begin{vmatrix} a_2 & c_2 \\ a_3 & c_3 \end{vmatrix} + c_1 \begin{vmatrix} b_2 & a_2 \\ b_3 & a_3 \end{vmatrix} \right)$$

$$= -\begin{vmatrix} c_1 & b_1 & a_1 \\ c_2 & b_2 & a_2 \\ c_3 & b_3 & a_3 \end{vmatrix}. \tag{25}$$

Property 2 follows likewise. For 2×2 and 3×3 matrices,

$$\begin{vmatrix} ka_1 & b_1 \\ ka_2 & b_2 \end{vmatrix} = k(a_1 b_2) - k(b_1 a_2) = k\begin{vmatrix} a_1 & b_1 \\ a_2 & b_2 \end{vmatrix} \tag{26}$$

and

$$\begin{vmatrix} ka_1 & b_1 & c_1 \\ ka_2 & b_2 & c_2 \\ ka_3 & b_3 & c_3 \end{vmatrix} = ka_1 \begin{vmatrix} b_2 & c_2 \\ b_3 & c_3 \end{vmatrix} - b_1 \begin{vmatrix} ka_2 & c_2 \\ ka_3 & c_3 \end{vmatrix}$$

$$+c_1 \begin{vmatrix} ka_2 & b_2 \\ ka_3 & b_3 \end{vmatrix} = k \begin{vmatrix} a_1 & b_1 & c_1 \\ a_2 & b_2 & c_2 \\ a_3 & b_3 & c_3 \end{vmatrix}. \qquad (27)$$

Property 3 follows from the identity

$$\begin{vmatrix} a_1+kb_1 & b_1 & c_1 \\ a_2+kb_2 & b_2 & c_2 \\ a_3+kb_3 & b_3 & c_3 \end{vmatrix}$$

$$= (a_1+kb_1)$$

$$\times \begin{vmatrix} b_2 & c_2 \\ b_3 & c_3 \end{vmatrix} - b_1 \begin{vmatrix} a+kb_2 & c_2 \\ a_3+kb_3 & c_3 \end{vmatrix} + c_1 \begin{vmatrix} a_2+kb_2 & b_2 \\ a_3+kb_3 & b_3 \end{vmatrix}. \quad (28)$$

If a_{ij} is an $n \times n$ MATRIX with a_{ij} REAL NUMBERS, then $\det[a_{ij}]$ has the interpretation as the oriented n-dimensional CONTENT of the PARALLELEPIPED spanned by the column vectors $[a_{i,1}], ..., [a_{i,n}]$ in $\mathbb{R}^n$.. Here, "oriented" means that, up to a change of $+$ or $-$ SIGN, the number is the n-dimensional CONTENT, but the SIGN depends on the "orientation" of the column vectors involved. If they agree with the standard orientation, there is a $+$ SIGN; if not, there is a $-$ SIGN. The PARALLELEPIPED spanned by the n-D vectors $\mathbf{v}_1$ through $\mathbf{v}_i$ is the collection of points

$$t_1\mathbf{v}_1 + ... + t_i\mathbf{v}_i, \qquad (29)$$

where t_j is a REAL NUMBER in the CLOSED INTERVAL $[0, 1]$..

Several accounts state that Lewis Carroll (Charles Dodgson) sent Queen Victoria a copy of one of his mathematical works, in one account, *An Elementary Treatise on Determinants*. Heath (1974) states, "A well-known story tells how Queen Victoria, charmed by *Alice in Wonderland*, expressed a desire to receive the author's next work, and was presented, in due course, with a loyally inscribed copy of *An Elementary Treatise on Determinants*," while Gattegno (1974) asserts "Queen Victoria, having enjoyed *Alice* so much, made known her wish to receive the author's other books, and was sent one of Dodgson's mathematical works." However, in *Symbolic Logic* (1896), Carroll stated, "I take this opportunity of giving what publicity I can to my contradiction of a silly story, which has been going the round of the papers, about my having presented certain books to Her Majesty the Queen. It is so constantly repeated, and is such absolute fiction, that I think it worth while to state, once for all, that it is utterly false in every particular: nothing even resembling it has occurred" (Mikkelson and Mikkelson).

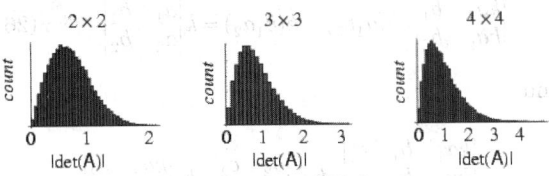

Hadamard (1893) showed that the absolute value of

the determinant of a COMPLEX $n \times n$ matrix with entries in the UNIT DISK satisfies

$$|\det A| \leq n^{n/2} \qquad (30)$$

(Brenner 1972). The plots above show the distribution of determinants for random $n \times n$ complex matrices with entries satisfying $|a_{ij}| < 1$ for $n = 2$, 3, and 4.

There are an infinite number of 3×3 determinants with no 0 or ± 1 entries having unity determinant. One parametric family is

$$\begin{vmatrix} -8n^2-8n & 2n+1 & 4n \\ -4n^2-4n & n+1 & 2n+1 \\ -4n^2-4n-1 & n & 2n-1 \end{vmatrix}. \qquad (31)$$

Specific examples having small entries include

$$\begin{vmatrix} 2 & 3 & 2 \\ 4 & 2 & 3 \\ 9 & 6 & 7 \end{vmatrix}, \begin{vmatrix} 2 & 3 & 5 \\ 3 & 2 & 3 \\ 9 & 5 & 7 \end{vmatrix}, \begin{vmatrix} 2 & 3 & 6 \\ 3 & 2 & 3 \\ 17 & 11 & 16 \end{vmatrix}, \dots \qquad (32)$$

(Guy 1989, 1994).

See also CAYLEY-MENGER DETERMINANT, CIRCULANT DETERMINANT, COFACTOR, CONDENSATION, CRAMER'S RULE, DETERMINANT EXPANSION BY MINORS, DETERMINANT IDENTITIES, ELEMENTARY ROW AND COLUMN OPERATIONS, HADAMARD'S MAXIMUM DETERMINANT PROBLEM, HESSIAN DETERMINANT, HYPERDETERMINANT, IMMANANT, JACOBIAN, KNOT DETERMINANT, MATRIX, MINOR, PERMANENT, PFAFFIAN, SINGULAR MATRIX, SYLVESTER'S DETERMINANT IDENTITY, SYLVESTER MATRIX, SYSTEM OF EQUATIONS, VANDERMONDE DETERMINANT, WRONSKIAN

References

Andrews, G. E. and Burge, W. H. "Determinant Identities." *Pacific J. Math.* **158**, 1–4, 1993.

Arfken, G. "Determinants." §4.1 in *Mathematical Methods for Physicists, 3rd ed.* Orlando, FL: Academic Press, pp. 168–76, 1985.

Brenner, J. and Cummings, L. "The Hadamard Maximum Determinant Problem." *Amer. Math. Monthly* **79**, 626–30, 1972.

Dostor, G. *Eléments de la théorie des déterminants, avec application à l'algèbre, la trigonométrie et la géométrie analytique dans le plan et l'espace, 2ème ed.* Paris: Gauthier-Villars, 1905.

Gattegno, J. *Lewis Carroll: Fragments of a Looking-Glass.* New York: Crowell, 1974.

Guy, R. K. "Unsolved Problems Come of Age." *Amer. Math. Monthly* **96**, 903–09, 1989.

Guy, R. K. "A Determinant of Value One." §F28 in *Unsolved Problems in Number Theory, 2nd ed.* New York: Springer-Verlag, pp. 265–66, 1994.

Hadamard, J. "Résolution d'une question relative aux déterminants." *Bull. Sci. Math.* **17**, 30–1, 1893.

Heath, P. *The Philosopher's Alice: Alice's Adventures in Wonderland and Through the Looking-Glass.* New York: St. Martin's Press, 1974.

Kowalewski, G. *Einführung in die Determinantentheorie.* New York: Chelsea, 1948.

Mikkelson, D. P. and Mikkelson, B. "Fit for a Queen." http://www.snopes.com/errata/carroll.htm.

Muir, T. *A Treatise on the Theory of Determinants.* New York: Dover, 1960.

Whittaker, E. T. and Robinson, G. "Determinants and Linear Equations." Ch. 5 in *The Calculus of Observations: A Treatise on Numerical Mathematics, 4th ed.* New York: Dover, pp. 71–7, 1967.

Yvinec, Y. "Geometric Computing: Exact Sign of a Determinant." http://www-sop.inria.fr/prisme/personnel/yvinec/Determinants/english.html.

Determinant (Binary Quadratic Form)

The determinant of a BINARY QUADRATIC FORM

$$Au^2 + 2Buv + Cv^2$$

is

$$D \equiv B^2 - AC.$$

It is equal to 1/4 of the corresponding DISCRIMINANT.

Determinant (Knot)

KNOT DETERMINANT

Determinant Expansion by Minors

Also known as "Laplacian" determinant expansion by minors, expansion by minors is a technique for computing the DETERMINANT of a given SQUARE MATRIX M. Although efficient for small matrices, techniques such as GAUSSIAN ELIMINATION are much more efficient when the matrix size becomes large.

Let |M| denote the DETERMINANT of a MATRIX M, then

$$|\mathsf{M}| = \sum_{i=1}^{k} (-1)^{i+j} a_{ij} M_{ij}, \tag{1}$$

where M_{ij} is a so-called MINOR of M, obtained by taking the determinant of M with row i and column j "crossed out." For example, for a 3×3 matrix, the above formula gives

$$\begin{vmatrix} a_{11} & a_{12} & a_{13} \\ a_{21} & a_{22} & a_{23} \\ a_{31} & a_{32} & a_{33} \end{vmatrix}$$

$$= a_{11} \begin{vmatrix} a_{22} & a_{23} \\ a_{32} & a_{33} \end{vmatrix} - a_{12} \begin{vmatrix} a_{21} & a_{23} \\ a_{31} & a_{33} \end{vmatrix} + a_{13} \begin{vmatrix} a_{21} & a_{22} \\ a_{31} & a_{32} \end{vmatrix}. \tag{2}$$

The procedure can then be iteratively applied to calculate the minors in terms of subminors, etc. The factor $(-1)^{i+j}$ is sometimes absorbed into the minor as

$$|\mathsf{M}| = \sum_{i=1}^{k} a_{ij} C_{ij}, \tag{3}$$

in which case C_{ij} is called a COFACTOR.

The equation for the determinant can also be formally written as

$$|\mathsf{A}| = \sum_{\pi} (-1)^{I(\pi)} \prod_{i=1}^{n} a_{i,\pi(i)}, \tag{4}$$

where π ranges over all permutations of $\{1, 2, ..., n\}$ and $I(\pi)$ is the INVERSION NUMBER of π (Bressoud and Propp 1999).

See also COFACTOR, CONDENSATION, DETERMINANT, GAUSSIAN ELIMINATION

References

Arfken, G. *Mathematical Methods for Physicists, 3rd ed.* Orlando, FL: Academic Press, pp. 169–70, 1985.

Bressoud, D. and Propp, J. "How the Alternating Sign Matrix Conjecture was Solved." *Not. Amer. Math. Soc.* **46**, 637–46.

Muir, T. "Minors and Expansions." Ch. 4 in *A Treatise on the Theory of Determinants.* New York: Dover, pp. 53–37, 1960.

Determinant Identities

Interesting DETERMINANT identities include

$$\begin{vmatrix} 1 & a & b+c \\ 1 & b & c+a \\ 1 & c & a+b \end{vmatrix} = 0 \tag{1}$$

(Muir 1960, p. 39),

$$\begin{vmatrix} a+b+c+d & b & c & d \\ b+c+d+a & c & d & a \\ c+d+a+b & d & a & c \\ d+a+b+c & a & b & c \end{vmatrix} = \begin{vmatrix} 1 & b & c & d \\ 1 & c & d & a \\ 1 & d & a & b \\ 1 & a & b & c \end{vmatrix}$$

$$\times (a+b+c+d) \tag{2}$$

(Muir 1960, p. 41),

$$\begin{vmatrix} 1 & a & a^2 & a^3 \\ 1 & b & b^2 & b^3 \\ 1 & c & c^2 & c^3 \\ 1 & d & d^2 & d^3 \end{vmatrix} = (b-a)(c-a)(c-b)(d-a)(d-b)$$

$$\times (d-c) \tag{3}$$

(Muir 1960, p. 42),

$$\begin{vmatrix} bcd & a & a^2 & a^3 \\ cda & b & b^2 & b^3 \\ dab & c & c^2 & c^3 \\ abc & d & d^2 & d^3 \end{vmatrix} = \begin{vmatrix} 1 & a^2 & a^3 & a^4 \\ 1 & b^2 & b^3 & b^4 \\ 1 & c^2 & c^3 & c^4 \\ 1 & d^2 & d^3 & d^4 \end{vmatrix} \tag{4}$$

(Muir 1960, p. 47),

$$\begin{vmatrix} 0 & a^2 & b^2 & c^2 \\ a^2 & 0 & \gamma^2 & \beta^2 \\ b^2 & \gamma^2 & 0 & \alpha^2 \\ c^2 & \beta^2 & \alpha^2 & 0 \end{vmatrix} = \begin{vmatrix} 0 & a\alpha & b\beta & c\gamma \\ a\alpha & 0 & c\gamma & a\alpha \\ b\beta & c\gamma & 0 & a\alpha \\ c\gamma & b\beta & a\alpha & 0 \end{vmatrix} \tag{5}$$

(Muir 1960, p. 42),

$$\begin{vmatrix} 1 & 1 & 1 & 1 \\ 1 & 1+x & 1 & 1 \\ 1 & 1 & 1+y & 1 \\ 1 & 1 & 1 & 1+z \end{vmatrix} = xyz \tag{6}$$

(Muir 1960, p. 44), and the CAYLEY-MENGER DETERMINANT

$$\begin{vmatrix} 0 & a & b & c \\ a & 0 & c & b \\ b & c & 0 & a \\ c & b & a & 0 \end{vmatrix} = \begin{vmatrix} 0 & 1 & 1 & 1 \\ 1 & 0 & c^2 & b^2 \\ 1 & c^2 & 0 & a^2 \\ 1 & b^2 & a^2 & 0 \end{vmatrix} \qquad (7)$$

(Muir 1960, p. 46), which is closely related to HERON'S FORMULA.

See also DETERMINANT

References

Muir, T. *A Treatise on the Theory of Determinants.* New York: Dover, 1960.

Determinant Theorem

Given a MATRIX M, the following are equivalent:

1. $|M| \neq 0$.
2. The columns of M are linearly independent.
3. The rows of M are linearly independent.
4. $\text{Range}(M) = \mathbb{R}^n$.
5. $\text{Null}(M) = \{0\}$.
6. M has a MATRIX INVERSE.

See also DETERMINANT, MATRIX INVERSE, NULLSPACE, RANGE (IMAGE)

Deterministic

A TURING MACHINE is called deterministic if there is always at most one instruction associated with a given present internal state/tape state pair (q, s). Otherwise, it is called nondeterministic (Itô 1987, p. 137).

In prediction theory, let $\{X_t\}$ be a weakly stationary process, and let $M_t(X)$ be a subspace spanned by the X_s (with $s \leq t$). If $M_t(X)$ is independent of t so that $M_t(X) = M(X)$ for every t, then $\{X_t\}$ is said to be deterministic (Itô 1987, p. 1463).

See also TURING MACHINE

References

Itô, K. (Ed.). "Turing Machines." §31B in *Encyclopedic Dictionary of Mathematics, 2nd ed., Vol. 1.* Cambridge, MA: MIT Press, pp. 136–37, 1987.
Itô, K. (Ed.). §395D in *Encyclopedic Dictionary of Mathematics, 2nd ed., Vol. 3.* Cambridge, MA: MIT Press, p. 1463, 1987.

Developable Surface

A surface on which the GAUSSIAN CURVATURE K is everywhere 0.

See also BINORMAL DEVELOPABLE, GAUSSIAN CURVATURE, NORMAL DEVELOPABLE, SYNCLASTIC, TANGENT DEVELOPABLE

References

Snyder, J. P. *Map Projections--A Working Manual.* U. S. Geological Survey Professional Paper 1395. Washington, DC: U. S. Government Printing Office, p. 5, 1987.

Deviation

The DIFFERENCE of a quantity from some fixed value, usually the "correct" or "expected" one.

See also ABSOLUTE DEVIATION, AVERAGE ABSOLUTE DEVIATION, DIFFERENCE, DISPERSION (STATISTICS), MEAN DEVIATION, SIGNED DEVIATION, STANDARD DEVIATION

References

Kenney, J. F. and Keeping, E. S. "Deviations." §6.3 in *Mathematics of Statistics, Pt. 1, 3rd ed.* Princeton, NJ: Van Nostrand, p. 76 1962.

Devil on Two Sticks

DEVIL'S CURVE

Devil's Curve

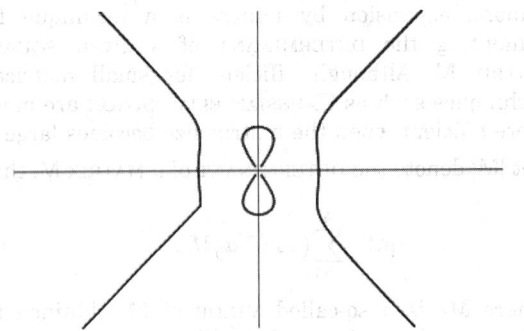

The devil's curve was studied by G. Cramer in 1750 and Lacroix in 1810 (MacTutor Archive). It appeared in *Nouvelles Annales* in 1858. The Cartesian equation is

$$y^4 - a^2 y^2 = x^4 - b^2 x^2, \qquad (1)$$

equivalent to

$$y^2(y^2 - a^2) = x^2(x^2 - b^2), \qquad (2)$$

the polar equation is

$$r^2(\sin^2 \theta - \cos^2 \theta) = a^2 \sin^2 \theta - b^2 \cos^2 \theta, \qquad (3)$$

and the PARAMETRIC EQUATIONS are

$$x = \cos t \sqrt{\frac{a^2 \sin^2 t - b^2 \cos^2 t}{\sin^2 t - \cos^2 t}} \qquad (4)$$

$$y = \sin t \sqrt{\frac{a^2 \sin^2 t - b^2 \cos^2 t}{\sin^2 t - \cos^2 t}}. \qquad (5)$$

The curve illustrated above corresponds to parameters $a^2 = 1$ and $b^2 = 2$.

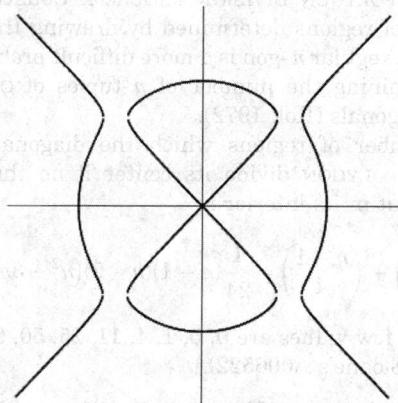

A special case of the Devil's curve is the so-called "electric motor curve":

$$y^2(y^2 - 96) = x^2(x^2 - 100) \qquad (6)$$

(Cundy and Rollett 1989).

References

Cundy, H. and Rollett, A. *Mathematical Models, 3rd ed.* Stradbroke, England: Tarquin Pub., p. 71, 1989.

Gray, A. *Modern Differential Geometry of Curves and Surfaces with Mathematica, 2nd ed.* Boca Raton, FL: CRC Press, pp. 92–3, 1997.

Lawrence, J. D. *A Catalog of Special Plane Curves.* New York: Dover, pp. 151–52, 1972.

MacTutor History of Mathematics Archive. "Devil's Curve." http://www-groups.dcs.st-and.ac.uk/~history/Curves/Devils.html.

Devil's Needle Puzzle

BAGUENAUDIER

Devil's Staircase

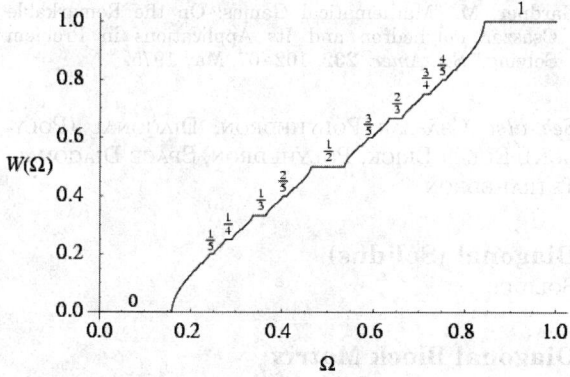

A plot of the WINDING NUMBER W resulting from MODE LOCKING as a function of Ω for the CIRCLE MAP

$$\theta_{n+1} = \theta_n + \Omega - \frac{K}{2\pi}\sin(2\pi\theta_n)$$

with $K = 1$. (Since the CIRCLE MAP becomes MODE-LOCKED, the WINDING NUMBER is independent of the initial starting argument θ_0.) At each value of Ω, the WINDING NUMBER is some RATIONAL NUMBER. The result is a monotonic increasing "staircase" for which the simplest RATIONAL NUMBERS have the largest steps. The Devil's staircase continuously maps the interval $[0, 1]$ onto $[0, 1]$, but is constant almost everywhere (i.e., except on a CANTOR SET).

For $K = 1$, the MEASURE of quasiperiodic states (Ω IRRATIONAL) on the Ω-axis has become zero, and the measure of MODE-LOCKED state has become 1. The DIMENSION of the Devil's staircase $\approx 0.8700 \pm 3.7 \times 10^{-4}$.

See also CANTOR FUNCTION, CIRCLE MAP, MINKOWSKI'S QUESTION MARK FUNCTION, WINDING NUMBER (MAP)

References

Devaney, R. L. *An Introduction to Chaotic Dynamical Systems.* Redwood City, CA: Addison-Wesley, pp. 109–10, 1987.

Mandelbrot, B. B. *The Fractal Geometry of Nature.* New York: W. H. Freeman, 1983.

Ott, E. *Chaos in Dynamical Systems.* New York: Cambridge University Press, 1993.

Rasband, S. N. "The Circle Map and the Devil's Staircase." §6.5 in *Chaotic Dynamics of Nonlinear Systems.* New York: Wiley, pp. 128–32, 1990.

Diabolic Square

The term used by Hunter and Madachy (1975, p. 24) and Madachy (1979, p. 87) to refer to a PANMAGIC SQUARE.

See also PANMAGIC SQUARE

References

Hunter, J. A. H. and Madachy, J. S. "Mystic Arrays." Ch. 3 in *Mathematical Diversions.* New York: Dover, 1975.

Madachy, J. S. *Madachy's Mathematical Recreations.* New York: Dover, 1979.

Diabolical Cube

A 6-piece POLYCUBE DISSECTION of the 3×3 CUBE.

See also CUBE DISSECTION, SOMA CUBE

References

Gardner, M. "Polycubes." Ch. 3 in *Knotted Doughnuts and Other Mathematical Entertainments.* New York: W. H. Freeman, pp. 29–0, 1986.

Diabolical Square

DIABOLIC SQUARE

Diabolo

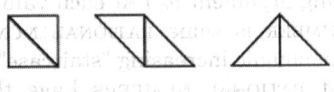

diaboloes

One of the three 2-POLYABOLOES.

See also POLYABOLO

Diacaustic

The ENVELOPE of refracted rays for a given curve.

See also CATACAUSTIC, CAUSTIC

References

Lawrence, J. D. *A Catalog of Special Plane Curves.* New York: Dover, p. 60, 1972.

Diagonal

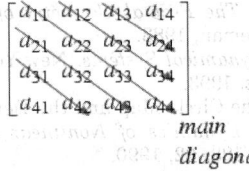

diagonals

$$\begin{bmatrix} a_{11} & a_{12} & a_{13} & a_{14} \\ a_{21} & a_{22} & a_{23} & a_{24} \\ a_{31} & a_{32} & a_{33} & a_{34} \\ a_{41} & a_{42} & a_{43} & a_{44} \end{bmatrix}$$

main diagonal

A diagonal of a SQUARE MATRIX which is traversed in the "southeast" direction. "The" diagonal (or "main diagonal" or "principal diagonal"); of an $n \times n$ square matrix is the diagonal from a_{11} to a_{nn}.

See also DIAGONAL MATRIX, DIAGONAL METRIC, DIAGONAL (POLYGON), DIAGONAL (POLYHEDRON), DIAGONAL RAMSEY NUMBER, DIAGONAL SLASH, DIAGONAL TRIANGLE, DIAGONALIZABLE MATRIX, SHALLOW DIAGONAL, SKEW DIAGONAL, SUBDIAGONAL, SUPERDIAGONAL, TRIDIAGONAL MATRIX

Diagonal (Polygon)

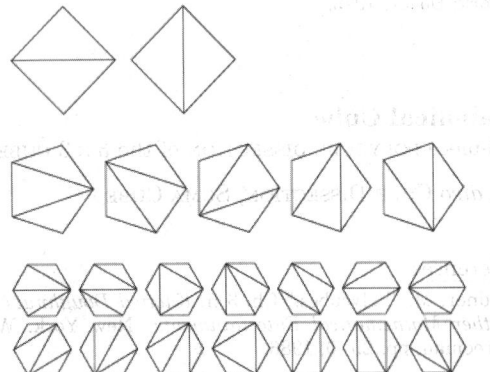

A LINE SEGMENT connecting two nonadjacent VERTICES of a POLYGON. The number of ways a fixed convex n-gon can be divided into TRIANGLES by nonintersecting diagonals is C_{n-2} (with C_{n-3} diag-

onals), where C_n is a CATALAN NUMBER. This is EULER'S POLYGON DIVISION PROBLEM. Counting the number of regions determined by drawing the diagonals of a regular n-gon is a more difficult problem, as is determining the number of n-tuples of CONCURRENT diagonals (Kok 1972).

The number of regions which the diagonals of a CONVEX POLYGON divide its center if no three are concurrent in its interior is

$$N = \binom{n}{4} + \binom{n-1}{4} = \frac{1}{24}(n-1)(n-2)(n^2 - 3n + 12).$$

The first few values are 0, 0, 1, 4, 11, 25, 50, 91, 154, 246, ... (Sloane's A006522).

See also CATALAN NUMBER, DIAGONAL (POLYHEDRON), EULER'S POLYGON DIVISION PROBLEM, POLYGON, VERTEX (POLYGON)

References

Kok, J. Item 2 in Beeler, M.; Gosper, R. W.; and Schroeppel, R. *HAKMEM.* Cambridge, MA: MIT Artificial Intelligence Laboratory, Memo AIM-239, p. 3, Feb. 1972.
Sloane, N. J. A. Sequences A006522/M3413 in "An On-Line Version of the Encyclopedia of Integer Sequences." http://www.research.att.com/~njas/sequences/eisonline.html.

Diagonal (Polyhedron)

A LINE SEGMENT connecting two nonadjacent sides of a POLYHEDRON. Any polyhedron having no diagonals must have a SKELETON which is a COMPLETE GRAPH (Gardner 1975). The only SIMPLE POLYHEDRON with no diagonals is the TETRAHEDRON. The only known TOROIDAL POLYHEDRON with no diagonals is the CSÁSZÁR POLYHEDRON.

See also CSÁSZÁR POLYHEDRON, TETRAHEDRON

References

Gardner, M. "Mathematical Games: On the Remarkable Császár Polyhedron and Its Applications in Problem Solving." *Sci. Amer.* **232**, 102–07, May 1975.

See also CSÁSZÁR POLYHEDRON, DIAGONAL (POLYGON), EULER BRICK, POLYHEDRON, SPACE DIAGONAL, TETRAHEDRON

Diagonal (Solidus)

SOLIDUS

Diagonal Block Matrix

BLOCK DIAGONAL MATRIX

Diagonal Matrix

A diagonal matrix is a SQUARE MATRIX **A** OF THE FORM

$$a_{ij} = c_i \delta_{ij}, \tag{1}$$

where δ_{ij} is the KRONECKER DELTA, c_i are constants, and $i, j = 1, 2, \ldots, n$, with is no implied summation over indices. The general diagonal matrix is therefore OF THE FORM

$$\begin{bmatrix} c_1 & 0 & \cdots & 0 \\ 0 & c_2 & \cdots & 0 \\ \vdots & \vdots & \ddots & \vdots \\ 0 & 0 & \cdots & c_n \end{bmatrix} \tag{2}$$

often denoted $\mathrm{diag}(c_1, c_2, \ldots, c_n)$. The diagonal matrix with elements $l = \{c_1, \ldots, c_n\}$ can be computed in *Mathematica* using `DiagonalMatrix[l]`.

Given a MATRIX EQUATION OF THE FORM

$$\begin{bmatrix} a_{11} & \cdots & a_{1n} \\ \vdots & \ddots & \vdots \\ a_{n1} & \cdots & a_{nn} \end{bmatrix} \begin{bmatrix} \lambda_1 & \cdots & 0 \\ \vdots & \ddots & \vdots \\ 0 & \cdots & \lambda_n \end{bmatrix}$$
$$= \begin{bmatrix} \lambda_1 & \cdots & 0 \\ \vdots & \ddots & \vdots \\ 0 & \cdots & \lambda_n \end{bmatrix} \begin{bmatrix} a_{11} & \cdots & a_{1n} \\ \vdots & \ddots & \vdots \\ a_{n1} & \cdots & a_{nn} \end{bmatrix}, \tag{3}$$

multiply through to obtain

$$\begin{bmatrix} a_{11}\lambda_1 & \cdots & a_{1n}\lambda_n \\ \vdots & \ddots & \vdots \\ a_{n1}\lambda_1 & \cdots & a_{nn}\lambda_n \end{bmatrix} = \begin{bmatrix} a_{11}\lambda_1 & \cdots & a_{1n}\lambda_1 \\ \vdots & \ddots & \vdots \\ a_{n1}\lambda_n & \cdots & a_{nn}\lambda_n \end{bmatrix}. \tag{4}$$

Since in general, $\lambda_i \neq \lambda_j$ for $i \neq j$, this can be true only if off-diagonal components vanish. Therefore, A must be diagonal.

Given a diagonal matrix T, the MATRIX POWER can be computed simply by taking each element to the power in question,

$$T^n = \begin{bmatrix} t_1 & 0 & \cdots & 0 \\ 0 & t_2 & \cdots & 0 \\ \vdots & \vdots & \ddots & \vdots \\ 0 & 0 & \cdots & t_k \end{bmatrix}^n = \begin{bmatrix} t_1^n & 0 & \cdots & 0 \\ 0 & t_2^n & \cdots & 0 \\ \vdots & \vdots & \ddots & \vdots \\ 0 & 0 & \cdots & t_k^n \end{bmatrix}. \tag{5}$$

Similarly, a MATRIX EXPONENTIAL can be performed simply by exponentiating each of the diagonal elements,

$$\exp(A) = \begin{bmatrix} e^{t_1} & 0 & \cdots & 0 \\ 0 & e^{t_2} & \cdots & 0 \\ \vdots & \vdots & \ddots & \vdots \\ 0 & 0 & \cdots & e^{t_k} \end{bmatrix} \tag{6}$$

See also CANONICAL BOX MATRIX DIAGONAL, DIAGONALIZABLE MATRIX, EXPONENTIAL MATRIX, MATRIX, NORMAL MATRIX, PERSYMMETRIC MATRIX, SKEW SYMMETRIC MATRIX, SYMMETRIC MATRIX, TRIANGULAR MATRIX, TRIDIAGONAL MATRIX

References

Arfken, G. *Mathematical Methods for Physicists, 3rd ed.* Orlando, FL: Academic Press, pp. 181–84 and 217–29, 1985.

Diagonal Metric

A METRIC g_{ij} which is zero for $i \neq j$.

See also METRIC

Diagonal Slash

CANTOR DIAGONAL METHOD

Diagonal Triangle

Diagonal Quadratic Form

If $A = (a_{ij})$ is a DIAGONAL MATRIX, a special case of a SYMMETRIC MATRIX, then

$$Q(v) = v^{\mathrm{T}} A v = \sum a_{ii} v_i^2$$

is a diagonal quadratic form, and $Q(v, w) = v^{\mathrm{T}} A w$ is its associated diagonal SYMMETRIC BILINEAR FORM.

For a general SYMMETRIC MATRIX A, a SYMMETRIC BILINEAR FORM Q may be diagonalized by a nondegenerate $n \times n$ matrix C such that $Q(Cv, Cw)$ is a diagonal form. That is, $C^{\mathrm{T}} A C$ is a DIAGONAL MATRIX. Note that C may not be an ORTHOGONAL MATRIX.

Here is a *Mathematica* function to find a matrix C which will diagonalize a symmetric bilinear form, given a SYMMETRIC MATRIX.

```
DiagonalizerMatrix[a_List?MatrixQ] := Module[
  {
    q, ctr, t2,
    v1 = Prepend[Table[0, {Length[a] - 1}], 1]
  },
  q[v_] := v.a.v;
    If[(t2 = q[v1]) != 0, v1 /=
Sqrt[Abs[t2]]];
  ctr = {v1};
  Do[
    v1 = NullSpace[ctr.a][[1]];
      If[(t2 = q[v1]) != 0, v1 /=
Sqrt[Abs[t2]]];
    AppendTo[ctr, v1],
    {Length[a] - 1}
  ];
  Transpose[Sort[ctr, q[#1] > q[#2] &]]
]
```

For example, consider

$$A = \begin{bmatrix} 1 & 2 \\ 2 & 3 \end{bmatrix}.$$

Then taking

$$C = \begin{bmatrix} 1 & -2 \\ 0 & 1 \end{bmatrix}$$

gives

$$C^{\mathrm{T}} A C = \begin{bmatrix} 1 & 0 \\ 0 & -1 \end{bmatrix},$$

so A has SIGNATURE $(1, 1)$.

See also QUADRATIC FORM, SIGNATURE (MATRIX), SYMMETRIC BILINEAR FORM, VECTOR SPACE

Diagonal Ramsey Number

A RAMSEY NUMBER OF THE FORM $R(k, k; 2)$.

See also RAMSEY NUMBER

Diagonal Slash

CANTOR DIAGONAL METHOD

Diagonal Triangle

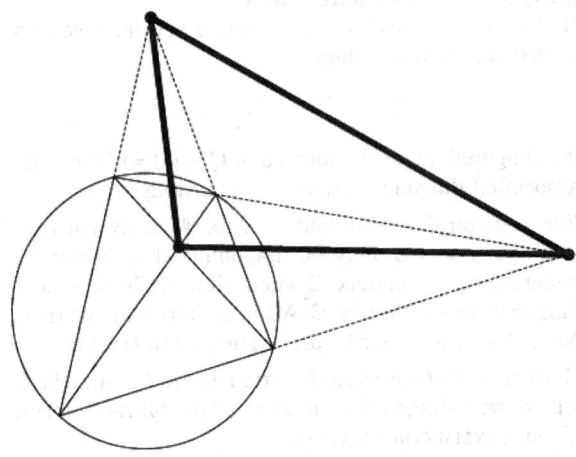

The TRIANGLE determined by the intersections of the sides and diagonals of a CYCLIC QUADRILATERAL. Each vertex is the POLE of the opposite side with respect to the CIRCLE

See also CYCLIC QUADRILATERAL, POLE (INVERSION), TRIANGLE

References

Wells, D. *The Penguin Dictionary of Curious and Interesting Geometry.* London: Penguin, p. 44, 1991.

Diagonalizable Matrix

This entry contributed by VIKTOR BENGTSSON

An $n \times n$-matrix A is said to be diagonalizable if it can be written on the form

$$A = PDP^{-1},$$

where D is a DIAGONAL $n \times n$ matrix with the EIGENVALUES of A as its entries and P is an INVERTIBLE $n \times n$ matrix consisting of the EIGENVECTORS corresponding to the EIGENVALUES in D.

The diagonalization theorem states that a quadratic matrix A is diagonalizable if and only if A has n linearly independent eigenvectors. Diagonalization (and most other forms of matrix factorisation) are particularly useful when studying linear transformations, discrete dynamical systems, continuous systems, and so on.

See also CANTOR DIAGONAL ARGUMENT, DIAGONAL MATRIX, DIAGONAL QUADRATIC FORM, INVERTIBLE MATRIX

Diagonalization

MATRIX DIAGONALIZATION

Diagonals Problem

EULER BRICK

Diagram

A schematic mathematical illustration showing the relationships between or properties of mathematical objects.

See also ALTERNATING KNOT DIAGRAM, ARGAND DIAGRAM, COXETER-DYNKIN DIAGRAM, DE BRUIJN DIAGRAM, DYNKIN DIAGRAM, FERRERS DIAGRAM, HASSE DIAGRAM, HEEGAARD DIAGRAM, KNOT DIAGRAM, LINK DIAGRAM, PLOT, STEM-AND-LEAF DIAGRAM, VENN DIAGRAM, VORONOI DIAGRAM, YOUNG DIAGRAM

Diagrammatic Move

KNOT MOVE

Diameter

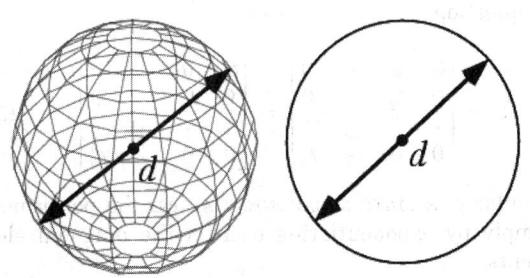

The diameter of a CIRCLE is the DISTANCE from a point on the CIRCLE to a point π RADIANS away, and is the maximum distance from one point on a circle to another. The diameter of a SPHERE is the maximum distance between two ANTIPODAL POINTS on the surface of the sphere.

If r is the RADIUS of a CIRCLE or SPHERE, then $d = 2r$. The ratio of the CIRCUMFERENCE C of a CIRCLE or GREAT CIRCLE of a SPHERE to the diameter d is PI,

$$\pi = \frac{C}{d}.$$

See also BROCARD DIAMETER, CIRCUMFERENCE, GENERALIZED DIAMETER, GRAPH DIAMETER, PI, RADIUS, SPHERE, TRANSFINITE DIAMETER

Diamond

diamond

Another word for a RHOMBUS. The diamond is also the name given to the unique 2-POLYIAMOND.

See also KITE, LOZENGE, PARALLELOGRAM, POLYIA-MOND, QUADRILATERAL, RHOMBUS

Dice

A die (plural "dice") is a SOLID with markings on each of its faces. The faces are usually all the same shape, making PLATONIC SOLIDS and ARCHIMEDEAN SOLID DUALS the obvious choices. The die can be "rolled" by throwing it in the air and allowing it to come to rest on one of its faces. Dice are used in many games of chance as a way of picking RANDOM NUMBERS on which to bet, and are used in board or role-playing games to determine the number of spaces to move, results of a conflict, etc. A COIN can be viewed as a degenerate 2-sided case of a die.

The most common type of die is a six-sided CUBE with the numbers 1– placed on the faces. The value of the roll is indicated by the number of "spots" showing on the top. For the six-sided die, opposite faces are arranged to always sum to seven. This gives two possible MIRROR IMAGE arrangements in which the numbers 1, 2, and 3 may be arranged in a clockwise or counterclockwise order about a corner. Commercial dice may, in fact, have either orientation. The illustrations below show 6-sided dice with counter-clockwise and clockwise arrangements, respectively.

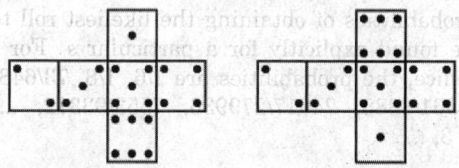

The CUBE has the nice property that there is an upward-pointing face opposite the bottom face from which the value of the "roll" can easily be read. This would not be true, for instance, for a TETRAHEDRAL die, which would have to be picked up and turned over to reveal the number underneath (although it could be determined by noting which number 1– was *not* visible on one of the upper three faces). The arrangement of spots ⁙ corresponding to a roll of 5 on a six-sided die is called the QUINCUNX. There are also special names for certain rolls of two six-sided dice: two 1s are called SNAKE EYES and two 6s are called BOXCARS.

Shapes of dice other than the usual 6-sided CUBE are commercially available from companies such as Dice & Games, Ltd. ® Diaconis and Keller (1989) show that there exist "fair" dice other than the usual PLATONIC SOLIDS and duals of the ARCHIMEDEAN SOLIDS, where

a fair die is one for which its symmetry group acts transitively on its faces (i.e., ISOHEDRA). There are 30 isohedra.

The probability of obtaining p points (a roll of p) on n s-sided dice can be computed as follows. The number of ways in which p can be obtained is the COEFFICIENT of x^p in

$$f(x) = (x + x^2 + \ldots + x^s)^n \qquad (1)$$

since each possible arrangement contributes one term. $f(x)$ can be written as a MULTINOMIAL SERIES

$$f(x) = x^n \left(\sum_{i=0}^{s-1} x^i \right)^n = x^n \left(\frac{1 - x^s}{1 - x} \right)^n, \qquad (2)$$

so the desired number c is the COEFFICIENT of x^p in

$$x^n (1 - x^s)^n (1 - x)^{-n}. \qquad (3)$$

Expanding,

$$x^n \sum_{k=0}^{n} (-1)^k \binom{n}{k} x^{sk} \sum_{l=0}^{\infty} \binom{n + l - 1}{l} x^l, \qquad (4)$$

so in order to get the COEFFICIENT of x^p, include all terms with

$$p = n + sk + l. \qquad (5)$$

c is therefore

$$c = \sum_{k=0}^{n} (-1)^k \binom{n}{k} \binom{p - sk - 1}{p - sk - n}. \qquad (6)$$

But $p - sk - n > 0$ only when $k < (p-n)/s$, so the other terms do not contribute. Furthermore,

$$\binom{p - sk - 1}{p - sk - n} = \binom{p - sk - 1}{n - 1}, \qquad (7)$$

so

$$c = \sum_{k=0}^{\lfloor (p-n)/s \rfloor} (-1)^k \binom{n}{k} \binom{p - sk - 1}{n - 1}, \qquad (8)$$

where $\lfloor x \rfloor$ is the FLOOR FUNCTION, and

$$P(p, n, s) = \frac{1}{s^n} \sum_{k=0}^{\lfloor (p-n)/s \rfloor} (-1)^k \binom{n}{k} \binom{p - sk - 1}{n - 1} \qquad (9)$$

(Uspensky 1937, pp. 23–4).

Consider now $s = 6$. For $n = 2$ six-sided dice,

$$k_{\max} \equiv \left\lfloor \frac{p - 2}{6} \right\rfloor = \begin{cases} 0 & \text{for } 2 \leq p \leq 7 \\ 1 & \text{for } 12 \leq p \leq 8, \end{cases} \qquad (10)$$

and

$$P(p, 2, 6) = \frac{1}{6^2} \sum_{k=0}^{k_{\max}} (-1)^k \binom{2}{k} \binom{p - 6k - 1}{1} \qquad$$

$$= \frac{1}{6^2} \sum_{k=0}^{k_{\max}} (-1)^k \frac{2!}{k!(2-k)!} (p - 6k - 1)$$

$$= \frac{1}{36} \sum_{k=0}^{k_{\max}} (1 - 2k)(k+1)(p - 6k - 1)$$

$$\frac{1}{36} \begin{cases} p - 1 & \text{for } 2 \leq p \leq 7 \\ 13 - p & \text{for } 8 \leq p \leq 12 \end{cases}$$

$$= \frac{6 - |p - 7|}{36} \quad \text{for } 2 \leq p \leq 12. \tag{11}$$

The most common roll is therefore seen to be a 7, with probability $6/36 = 1/6$, and the least common rolls are 2 and 12, both with probability 1/36.

For $n = 3$ six-sided dice,

$$k_{\max} = \left\lfloor \frac{vp - 3}{6} \right\rfloor = \begin{cases} 0 & \text{for } 3 \leq p \leq 8 \\ 1 & \text{for } 9 \leq p \leq 14 \\ 2 & \text{for } 15 \leq p \leq 18 \end{cases}, \tag{12}$$

and

$$P(p, 3, 6)$$

$$= \frac{1}{6^3} \sum_{k=0}^{k_{\max}} (-1)^k \binom{3}{k} \binom{p - 6k - 1}{2}$$

$$= \frac{1}{6^3} \sum_{k=0}^{k_{\max}} (-1)^k \frac{3!}{k!(3-k)!} \frac{(p - 6k - 1)(p - 6k - 2)}{2}$$

$$= \frac{1}{216}$$

$$\times \begin{cases} \dfrac{(p-1)(p-2)}{2} \\ \quad \text{for } 3 \leq p \leq 8 \\ \dfrac{(p-1)(p-2)}{2} - 3 \dfrac{(p-7)(p-8)}{2} \\ \quad \text{for } 9 \leq p \leq 14 \\ \dfrac{(p-1)(p-2)}{2} - 3 \dfrac{(p-7)(p-8)}{2} + 3 \dfrac{(p-13)(p-14)}{2} \\ \quad \text{for } 15 \leq p \leq 18 \end{cases}$$

$$= \frac{1}{216} \begin{cases} \frac{1}{2}(p-1)(p-2) & \text{for } 3 \leq p \leq 8 \\ -p^2 + 21p - 83 & \text{for } 9 \leq p \leq 14 \\ \frac{1}{2}(19-p)(20-p) & \text{for } 15 \leq p \leq 18. \end{cases} \tag{13}$$

For three six-sided dice, the most common rolls are 10 and 11, both with probability 1/8; and the least common rolls are 3 and 18, both with probability 1/216.

For four six-sided dice, the most common roll is 14, with probability 73/648; and the least common rolls are 4 and 24, both with probability 1/1296.

In general, the likeliest roll p_L for n s-sided dice is given by

$$p_L(n, s) = \left\lfloor \frac{1}{2} n(s+1) \right\rfloor, \tag{14}$$

which can be written explicitly as

$$p_L(n, s) = \begin{cases} \frac{1}{2} n(s+1) & \text{for } n \text{ even} \\ \frac{1}{2} [n(s+1) - 1] & \text{for } n \text{ odd, } s \text{ even} \\ \frac{1}{2} n(s+1) & \text{for } n \text{ odd, } s \text{ odd.} \end{cases} \tag{15}$$

For 6-sided dice, the likeliest rolls are given by

$$p_L(n, 6) = \left\lfloor \frac{7}{2} n \right\rfloor = \begin{cases} \frac{7}{2} n & \text{for n even} \\ \frac{1}{2}(7n - 1) & \text{for n odd,} \end{cases} \tag{16}$$

or 7, 10, 14, 17, 21, 24, 28, 31, 35, ... for $n = 2, 3, ...$ (Sloane's A030123) dice. The probabilities corresponding to the most likely rolls can be computed by plugging $p = p_L$ into the general formula together with

$$k_L(n, s) = \begin{cases} \frac{1}{2} n & \text{for } n \text{ even} \\ \left\lfloor \frac{n(s-1) - 1}{2s} \right\rfloor & \text{for } n \text{ odd, } s \text{ even} \\ \left\lfloor \frac{n(s-1)}{2s} \right\rfloor & \text{for } n \text{ odd, } s \text{ odd.} \end{cases} \tag{17}$$

Unfortunately, $P(p_L, n, s)$ does not have a simple closed-form expression in terms of s and n. However, the probabilities of obtaining the likeliest roll totals can be found explicitly for a particular s. For n 6-sided dice, the probabilities are 1/6, 1/8, 73/648, 65/648, 361/3888, 24017/279936, 7553/93312, ... for $n = 2, 3, ...$

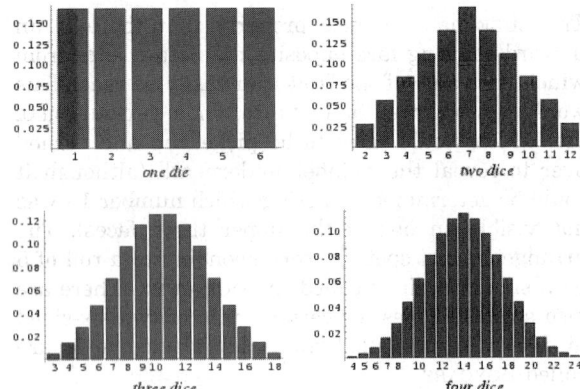

The probabilities for obtaining a given total using n 6-sided dice are shown above for $n = 1, 2, 3,$ and 4 dice. They can be seen to approach a GAUSSIAN DISTRIBUTION as the number of dice is increased.

See also Boxcars, Coin Tossing, Craps, de Méré's Problem, Efron's Dice, Isohedron, Poker, Quincunx, Sicherman Dice, Snake Eyes, Yahtzee

References

Culin, S. "Tjou-sa-a--Dice." §72 in *Games of the Orient: Korea, China, Japan.* Rutland, VT: Charles E. Tuttle, pp. 78–9, 1965.

Diaconis, P. and Keller, J. B. "Fair Dice." *Amer. Math. Monthly* **96**, 337–39, 1989.

Dice & Games, Ltd. "Dice & Games Hobby Games Accessories." http://www.dice.co.uk/hob.htm.

Gardner, M. "Dice." Ch. 18 in *Mathematical Magic Show: More Puzzles, Games, Diversions, Illusions and Other Mathematical Sleight-of-Mind from Scientific American.* New York: Vintage, pp. 251–62, 1978.

Pegg, E. Jr. "Fair Dice." http://www.mathpuzzle.com/Fairdice.htm.

Robertson, L. C.; Shortt, R. M.; Landry, S. G. "Dice with Fair Sums." *Amer. Math. Monthly* **95**, 316–28, 1988.

Sloane, N. J. A. Sequences A030123 in "An On-Line Version of the Encyclopedia of Integer Sequences." http://www.research.att.com/~njas/sequences/eisonline.html.

Uspensky, J. V. *Introduction to Mathematical Probability.* New York: McGraw-Hill, pp. 23–4, 1937.

Dichroic Polynomial

A POLYNOMIAL $Z_G(q,v)$ in two variables for abstract GRAPHS. A GRAPH with one VERTEX has $Z = q$. Adding a VERTEX not attached by any EDGES multiplies the Z by q. Picking a particular EDGE of a GRAPH G, the POLYNOMIAL for G is defined by adding the POLYNOMIAL of the GRAPH with that EDGE deleted to v times the POLYNOMIAL of the graph with that EDGE collapsed to a point. Setting $v = -1$ gives the number of distinct VERTEX colorings of the GRAPH. The dichroic POLYNOMIAL of a PLANAR GRAPH can be expressed as the SQUARE BRACKET POLYNOMIAL of the corresponding ALTERNATING LINK by

$$Z_G(q,v) = q^{N/2} B_{L(G)},$$

where N is the number of VERTICES in G. Dichroic POLYNOMIALS for some simple GRAPHS are

$$Z_{K_1} = q$$

$$Z_{K_2} = q^2 + vq$$

$$Z_{K_3} = q^3 + 3vq^2 + 3v^2q + v^3.$$

References

Adams, C. C. *The Knot Book: An Elementary Introduction to the Mathematical Theory of Knots.* New York: W. H. Freeman, pp. 231–35, 1994.

Dickman Function

The probability that a random integer between 1 and x will have its GREATEST PRIME FACTOR $\leq x^\alpha$ approaches a limiting value $F(\alpha)$ as $x \to \infty$, where $F(\alpha) = 1$ for $\alpha > 1$ and

$$F(\alpha) = \int_0^\alpha F\left(\frac{t}{1-t}\right)\frac{dt}{t}$$

for $0 \leq \alpha \leq 1$ (Dickman 1930, Knuth 1997). Similarly, the second-largest prime factor will be $\leq x^\beta$ with approximate probability $G(\beta)$, where $G(\beta) = 1$ for $\beta \geq 1/2$ and

$$G(\beta) = \int_0^\beta \left[G\left(\frac{t}{1-t}\right) - F\left(\frac{t}{1-t}\right)\right]\frac{dt}{t}$$

for $0 \leq \beta \leq 1/2$.

See also Greatest Prime Factor, Prime Factors

References

Dickman, K. *Arkiv för Mat., Astron. och Fys.* **22A**, 1–4, 1930.

Knuth, D. E. *The Art of Computer Programming, Vol. 2: Seminumerical Algorithms, 3rd ed.* Reading, MA: Addison-Wesley, pp. 382–84, 1998.

Norton, K. K. *Numbers with Small Prime Factors, and the Least kth Power Non-Residue.* Providence, RI: Amer. Math. Soc., 1971.

Ramaswami, V. "On the Number of Positive Integers Less than x and Free of Prime Divisors Greater than x^c." *Bull. Amer. Math. Soc.* **55**, 1122–127, 1949.

Ramaswami, V. "The Number of Positive Integers $\leq X$ and Free of Prime Divisors $> x^G$, and a Problem of S. S. Pillai." *Duke Math. J.* **16**, 99–09, 1949.

Dicone

BICONE

Dictionary Order

LEXICOGRAPHIC ORDER

Dido's Problem

Find the figure bounded by a line which has the maximum AREA for a given PERIMETER. The solution is a SEMICIRCLE. The problem is based on a passage from Virgil's *Aeneid*: "The Kingdom you see is Carthage, the Tyrians, the town of Agenor;

But the country around is Libya, no folk to meet in war.

Dido, who left the city of Tyre to escape her brother,

Rules here–a long a labyrinthine tale of wrong

Is hers, but I will touch on its salient points in order....

Dido, in great disquiet, organised her friends for escape.

They met together, all those who harshly hated the tyrant

Or keenly feared him: they seized some ships which chanced to be ready...

They came to this spot, where to-day you can behold the mighty

Battlements and the rising citadel of New Carthage,

And purchased a site, which was named 'Bull's Hide'
after the bargain
By which they should get as much land as they could
enclose with a bull's hide."

See also ISOPERIMETRIC PROBLEM, ISOVOLUME PROBLEM, PERIMETER, SEMICIRCLE

References

Thomas, I. *Greek Mathematical Works, Vol. 2: From Aristarchus to Pappus.* London: Heinemann, 1980.
Tikhomirov, V. M. *Stories About Maxima and Minima.* Providence, RI: Amer. Math. Soc., pp. 9–8, 1991.
Virgil. Translated by C. D. Lewis. Book I, lines 307–72 in *The Aeneid.* New York: Doubleday, pp. 22–3, 1953.
Wells, D. *The Penguin Dictionary of Curious and Interesting Geometry.* London: Penguin, pp. 122–24, 1991.

Diesis

The symbol ‡, also called the DOUBLE DAGGER (Bringhurst 1997, p. 277).

References

Bringhurst, R. *The Elements of Typographic Style, 2nd ed.* Point Roberts, WA: Hartley and Marks, 1997.

Diffeomorphic

See also DIFFEOMORPHISM

Diffeomorphism

A diffeomorphism is a MAP between MANIFOLDS which
is DIFFERENTIABLE and has a DIFFERENTIABLE inverse.

See also ANOSOV DIFFEOMORPHISM, AXIOM A DIFFEOMORPHISM, DIFFEOMORPHIC, PESIN THEORY, SYMPLECTIC DIFFEOMORPHISM, TANGENT MAP

Difference

The difference of two numbers n_1 and n_2 is $n_1 - n_2$,
where the MINUS sign denotes SUBTRACTION.

See also BACKWARD DIFFERENCE, FINITE DIFFERENCE, FORWARD DIFFERENCE, MINUS, SUBTRACTION, SYMMETRIC DIFFERENCE

Difference Equation

A difference equation is the discrete analog of a
DIFFERENTIAL EQUATION. A difference equation involves a FUNCTION with INTEGER-valued arguments
$f(n)$ in a form like

$$f(n) - f(n-1) = g(n), \tag{1}$$

where g is some FUNCTION. The above equation is the
discrete analog of the first-order ORDINARY DIFFERENTIAL EQUATION

$$f'(x) = g(x) \tag{2}$$

Examples of difference equations often arise in
DYNAMICAL SYSTEMS. Examples include the iteration
involved in the MANDELBROT and JULIA SET definitions,

$$f(n+1) = f(n)^2 + c, \tag{3}$$

with c a constant, as well as the LOGISTIC EQUATION

$$f(n+1) = rf(n)[1 - f(n)], \tag{4}$$

with r a constant.

See also FINITE DIFFERENCE, ORDINARY DIFFERENTIAL EQUATION, RECURRENCE RELATION

References

Agarwal, R. P. *Difference Equations and Inequality: Theory, Methods, and Applications, 2nd ed., rev. exp.* New York: Dekker, 2000.
Batchelder, P. M. *An Introduction to Linear Difference Equations.* New York: Dover, 1967.
Bellman, R. E. and Cooke, K. L. *Differential-Difference Equations.* New York: Academic Press, 1963.
Beyer, W. H. "Finite Differences." *CRC Standard Mathematical Tables, 28th ed.* Boca Raton, FL: CRC Press, pp. 429–60, 1988.
Brand, L. *Differential and Difference Equations.* New York: Wiley, 1966.
Fulford, G.; Forrester, P.; and Jones, A. *Modelling with Differential and Difference Equations.* New York: Cambridge University Press, 1997.
Goldberg, S. *Introduction to Difference Equations, with Illustrative Examples from Economics, Psychology, and Sociology.* New York: Dover, 1986.
Levy, H. and Lessman, F. *Finite Difference Equations.* New York: Dover, 1992.
Richtmyer, R. D. and Morton, K. W. *Difference Methods for Initial-Value Problems, 2nd ed.* New York: Interscience Publishers, 1967.
Weisstein, E. W. "Books about Difference Equations." http://www.treasure-troves.com/books/DifferenceEquations.html.

Difference of Successes

If x_1/n_1 and x_2/n_2 are the observed proportions from
standard NORMALLY DISTRIBUTED samples with proportion of success θ, then the probability that

$$w \equiv \frac{x_1}{n_1} - \frac{x_2}{n_2} \tag{1}$$

will be as great as observed is

$$P_\delta = 1 - 2 \int_0^{|\delta|} \phi(t) dt \tag{2}$$

where

$$\delta \equiv \frac{w}{\sigma_w} \tag{3}$$

$$\sigma_w \equiv \sqrt{\hat{\theta}(1 - \hat{\theta})\left(\frac{1}{n_1} + \frac{1}{n_2}\right)} \tag{4}$$

$$\hat{\theta} \equiv \frac{x_1 + x_2}{n_1 + n_2}. \tag{5}$$

Here, $\hat{\theta}$ is the UNBIASED ESTIMATOR. The SKEWNESS and KURTOSIS of this distribution are

$$\gamma_1^2 = \frac{(n_1 - n_2)^2}{n_1 n_2 (n_1 + n_2)} \frac{1 - 4\hat{\theta}(1 - \hat{\theta})}{\hat{\theta}(1 - \hat{\theta})} \tag{6}$$

$$\gamma_2 = \frac{n_1^2 - n_1 n_2 + n_2^2}{n_1 n_2 (n_1 + n_2)} \frac{1 - 6\hat{\theta}(1 - \hat{\theta})}{\hat{\theta}(1 - \hat{\theta})}. \tag{7}$$

Difference Operator

BACKWARD DIFFERENCE, FORWARD DIFFERENCE

Difference Quotient

$$\Delta_h f(x) \equiv \frac{f(x + h) - f(x)}{h} = \frac{\Delta f}{h}.$$

It gives the slope of the SECANT LINE passing through $f(x)$ and $f(x + h)$. In the limit $n \to 0$, the difference quotient becomes the PARTIAL DERIVATIVE

$$\lim_{h \to 1} \Delta_{x(h)} f(x, y) = \frac{\partial f}{\partial x}.$$

Difference Set

Let G be a GROUP of ORDER h and D be a set of k elements of G. If the set of differences $d_i - d_j$ contains every NONZERO element of G exactly λ times, then D is a (h, k, λ)-difference set in G of ORDER $n = k - \lambda$. If $\lambda = 1$, the difference set is called planar. The quadratic residues in the FINITE FIELD $GF(11)$ form a difference set. If there is a difference set of size k in a group G, then $2\binom{k}{2}$ must be a multiple of $|G| - 1$, where $\binom{k}{2}$ is a BINOMIAL COEFFICIENT.

See also BRUCK-RYSER-CHOWLA THEOREM, FIRST MULTIPLIER THEOREM, PRIME POWER CONJECTURE

References

Gordon, D. M. "The Prime Power Conjecture is True for $n < 2,000,000$." *Electronic J. Combinatorics* **1**, R6 1–, 1994. http://www.combinatorics.org/Volume_1/volume1.html#R6.

Difference Table

A table made by subtracting adjacent entries in a sequence, then repeating the process with those numbers.

See also DIVIDED DIFFERENCE, FINITE DIFFERENCE, INTERPOLATION, QUOTIENT-DIFFERENCE TABLE

References

Sloane, N. J. A. and Plouffe, S. "Analysis of Differences." §2.5 in *The Encyclopedia of Integer Sequences*. San Diego, CA: Academic Press, pp. 10–3, 1995.
Whittaker, E. T. and Robinson, G. "Difference Table." §2 in *The Calculus of Observations: A Treatise on Numerical Mathematics, 4th ed.* New York: Dover, pp. 2–, 1967.

Different

Two quantities are said to be different (or "unequal") if they are not EQUAL.

The term "different" also has a technical usage related to MODULES. Let a MODULE M in an INTEGRAL DOMAIN D_1 for $R\left(\sqrt{D}\right)$ be expressed using a two-element basis as

$$M = [\xi_1, \xi_2],$$

where ξ_1 and ξ_2 are in D_1. Then the different of the MODULE is defined as

$$\Delta = \Delta(M) = \begin{vmatrix} \xi_1 & \xi_2 \\ \xi_1' & \xi_2' \end{vmatrix} = \xi_1 \xi_2' - \xi_1' \xi_2.$$

The different $\Delta \neq 0$ IFF ξ_1 and ξ_2 are linearly independent. The DISCRIMINANT is defined as the square of the different.

See also DISCRIMINANT (MODULE), EQUAL, MODULE

References

Cohn, H. *Advanced Number Theory*. New York: Dover, pp. 72–3, 1980.

Different Prime Factors

DISTINCT PRIME FACTORS

Differentiable

A REAL FUNCTION is said to be differentiable at a point if its DERIVATIVE exists at that point. The notion of differentiability can also be extended to COMPLEX FUNCTIONS (leading to the CAUCHY-RIEMANN EQUATIONS and the theory of HOLOMORPHIC FUNCTIONS), although a few additional subtleties arise in COMPLEX DIFFERENTIABILITY that are not present in the real case.

Amazingly, there exist CONTINUOUS FUNCTIONS which are nowhere differentiable. Two examples are the BLANCMANGE FUNCTION and WEIERSTRASS FUNCTION.

See also ANALYTIC FUNCTION, BLANCMANGE FUNCTION, CAUCHY-RIEMANN EQUATIONS, COMPLEX DIFFERENTIABLE, CONTINUOUS FUNCTION, DERIVATIVE, HOLOMORPHIC FUNCTION, PARTIAL DERIVATIVE, WEAKLY DIFFERENTIABLE, WEIERSTRASS FUNCTION

References

Krantz, S. G. "Alternative Terminology for Holomorphic Functions" and "Differentiable and C^k Curves." §1.3.6 and 2.1.3 in *Handbook of Complex Analysis*. Boston, MA: Birkhäuser, p. 16 and 21, 1999.

Differentiable Manifold
SMOOTH MANIFOLD

Differential
A ONE-FORM.

See also DIFFERENTIAL K-FORM, EXACT DIFFEREN-TIAL, INEXACT DIFFERENTIAL, ONE-FORM

Differential Calculus
That portion of "the" CALCULUS dealing with DERIVA-TIVES.

See also INTEGRAL CALCULUS

Differential Equation
An equation which involves the DERIVATIVES of a function as well as the function itself. If PARTIAL DERIVATIVES are involved, the equation is called a PARTIAL DIFFERENTIAL EQUATION; if only ordinary DERIVATIVES are present, the equation is called an ORDINARY DIFFERENTIAL EQUATION. Differential equations play an extremely important and useful role in applied math, engineering, and physics, and much mathematical and numerical machinery has been developed for the solution of differential equations.

See also ADAMS' METHOD, DIFFERENCE EQUATION, INTEGRAL EQUATION, ORDINARY DIFFERENTIAL EQUA-TION, PARTIAL DIFFERENTIAL EQUATION

References
Arfken, G. "Differential Equations." Ch. 8 in *Mathematical Methods for Physicists, 3rd ed.* Orlando, FL: Academic Press, pp. 437–96, 1985.
Dormand, J. R. *Numerical Methods for Differential Equations: A Computational Approach.* Boca Raton, FL: CRC Press, 1996.

Differential Evolution
A simple EVOLUTION STRATEGY which is fairly fast and reasonably robust.

See also EVOLUTION STRATEGIES, GENETIC ALGO-RITHM, OPTIMIZATION THEORY

References
Price, K. and Storn, R. "Differential Evolution." *Dr. Dobb's J.*, No. 264, 18–8, Apr. 1997.

Differential Form
DIFFERENTIAL K-FORM

Differential Geometry
Differential geometry is the study of RIEMANNIAN MANIFOLDS. Differential geometry deals with metrical notions on MANIFOLDS, while DIFFERENTIAL TOPOLOGY deals with those nonmetrical notions of MANIFOLDS.

See also DIFFERENTIAL TOPOLOGY

References
Dillen, F. J. E. and Verstraelen, L. C.A. (Eds.). *Handbook of Differential Geometry, Vol. 1.* Amsterdam, Netherlands: North-Holland, 2000.
Eisenhart, L. P. *A Treatise on the Differential Geometry of Curves and Surfaces.* New York: Dover, 1960.
Graustein, W. C. *Differential Geometry.* New York: Dover, 1966.
Gray, A. *Modern Differential Geometry of Curves and Surfaces with Mathematica, 2nd ed.* Boca Raton, FL: CRC Press, 1997.
Kreyszig, E. *Differential Geometry.* New York: Dover, 1991.
Lipschutz, M. M. *Theory and Problems of Differential Geometry.* New York: McGraw-Hill, 1969.
Spivak, M. *A Comprehensive Introduction to Differential Geometry, Vol. 1, 2nd ed.* Berkeley, CA: Publish or Perish Press, 1979.
Spivak, M. *A Comprehensive Introduction to Differential Geometry, Vol. 2, 2nd ed.* Berkeley, CA: Publish or Perish Press, 1990.
Spivak, M. *A Comprehensive Introduction to Differential Geometry, Vol. 3, 2nd ed.* Berkeley, CA: Publish or Perish Press, 1990.
Spivak, M. *A Comprehensive Introduction to Differential Geometry, Vol. 4, 2nd ed.* Berkeley, CA: Publish or Perish Press, 1979.
Spivak, M. *A Comprehensive Introduction to Differential Geometry, Vol. 5, 2nd ed.* Berkeley, CA: Publish or Perish Press, 1979.
Struik, D. J. *Lectures on Classical Differential Geometry.* New York: Dover, 1988.
Weatherburn, C. E. *Differential Geometry of Three Dimensions, 2 vols.* Cambridge, England: Cambridge University Press, 1961.
Weisstein, E. W. "Books about Differential Geometry." http://www.treasure-troves.com/books/DifferentialGeometry.html.

Differential Ideal
A differential ideal $\mathfrak{I}$ on a MANIFOLD M is an IDEAL in the EXTERIOR ALGEBRA of DIFFERENTIAL K-FORMS on M which is also CLOSED under the EXTERIOR DERIVA-TIVE d. That is, for any differential form α and any form $\beta \in \mathfrak{I}$, then

1. $\alpha \wedge \beta \in \mathfrak{I}$, and
2. $d\beta \in \mathfrak{I}$

For example, $\mathfrak{I} = \langle xdy, dx \wedge dy \rangle$ is a differential ideal on $M = \mathbb{R}^2$.

A smooth map $f : X \to M$ is called an integral of $\mathfrak{I}$ if the PULLBACK MAP of all forms in $\mathfrak{I}$ vanish on X, i.e., $f^*(\mathfrak{I}) = 0$.

See also DIFFERENTIAL FORM, ENVELOPE (FORM), INTEGRABLE (DIFFERENTIAL IDEAL), MANIFOLD

Differential k-Form
A differential k-form is a TENSOR of RANK k which is antisymmetric under exchange of any pair of indices. The number of ALGEBRAICALLY INDEPENDENT components in n-D is given by the BINOMIAL COEFFICIENT $\binom{n}{k}$. In particular, a ONE-FORM ω^1 (often simply called a "differential") is a quantity

$$\omega^1 = b_1 dx_1 + b_2 dx_2 + \ldots + b_n dx_n, \qquad (1)$$

where $b_1 = b_1(x_1, x_2, \ldots, x_n)$ and $b_2 = b_2(x_1, x_1, \ldots, x_n)$ are the components of a COVARIANT TENSOR. Changing variables from $\mathbf{x}$ to $\mathbf{y}$ gives

$$\omega^1 = \sum_{i=1}^n b_i dx_i = \sum_{i=1}^n \sum_{j=1}^n b_i \frac{\partial x_i}{\partial y_j} dy_j = \sum_{j=1}^n \bar{b}_j dy_j, \qquad (2)$$

where

$$\bar{b}_j \equiv \sum_{i=1}^n b_j \frac{\partial x_i}{\partial y_j}, \qquad (3)$$

which is the covariant transformation law.

A p-ALTERNATING MULTILINEAR FORM on a VECTOR SPACE V corresponds to an element of $\wedge^p V^*$, the pth EXTERIOR POWER of the DUAL SPACE to V. A differential p-form on a MANIFOLD is a SECTION of the VECTOR BUNDLE $\wedge^p T^*M$, the pth EXTERIOR POWER of the COTANGENT BUNDLE. Hence, it is possible to write a p-form in coordinates by

$$\sum_{|I|=p} a_I dx_{i_1} \wedge \ldots \wedge dx_{i_p} \qquad (4)$$

where I ranges over all increasing subsets of p elements from $\{1, \ldots, n\}$, and the a_I are functions.

An important operation on differential forms, the EXTERIOR DERIVATIVE, is used in the celebrated STOKES' THEOREM. The EXTERIOR DERIVATIVE d of a p form is a $(p+1)$-form. In fact, by definition, if x_i is the coordinate function, thought of as a ZERO-FORM, then $d(x_i) = dx_i$.

Another important operation on forms is the WEDGE PRODUCT, or exterior product. If α is a p-form and β is q-form, then $\alpha \wedge \beta$ is a $p+q$ form. Also, a p-form can be CONTRACTED with an r-vector, i.e., a SECTION of $\wedge^r TM$, to give a $(p-r)$-form, or if $r > p$, an $(r-p)$-vector. If the manifold has a METRIC, then there is an operation dual to the exterior product, called the INTERIOR PRODUCT.

In higher dimensions, there are more kinds of differential forms. For instance, on the TANGENT SPACE to $\mathbb{R}^2$ there is the ZERO-FORM 1, two ONE-FORMS dx and dy, and one TWO-FORM $dx \wedge dy$. A ONE-FORM can be written uniquely as $f dx + g dy$. In four dimensions, $dx_1 \wedge dx_2 + dx_3 \wedge dx_4$ is a TWO-FORM which cannot be written as $a \wedge b$.

The minimum number of terms necessary to write a form is sometimes called the rank of the form, usually in the case of a TWO-FORM. When a form has rank one, it is called DECOMPOSABLE. Another meaning for rank of a form is its rank as a TENSOR, in which case a p-form can be described as an ANTISYMMETRIC TENSOR of rank p, in fact of type $(0, p)$. The rank of a form can also mean the dimension of its ENVELOPE, in which case the rank is an integer-valued function. With the

latter definition of rank, a p-form is decomposable IFF it has rank p.

When n is the dimension of a MANIFOLD M, then n is also the dimension of the TANGENT SPACE TM_x. Consequently, an n-form always has rank one, and for $p > n$, a p-form must be zero. Hence, an n-form is called a TOP-DIMENSIONAL FORM. A TOP-DIMENSIONAL FORM can be INTEGRATED without using a METRIC. Consequently, a p-form can be integrated on a p-dimensional SUBMANIFOLD. Differential forms are a VECTOR SPACE (with a C-INFINITY TOPOLOGY) and therefore have a dual space. Submanifolds represent an element of the dual via integration, so it is common to say that they are in the dual space of forms, which is the space of CURRENTS. With a METRIC, the HODGE STAR operator $*$ defines a map from p-forms to $(n-p)$-forms such that $** = (-1)^{p(n-p)}$. When $f : M \to N$ is a SMOOTH MAP, it pushes forward TANGENT VECTORS from TM to TN according to the JACOBIAN f_*. Hence, a differential form on N pulls back to a differential form on M.

$$f^*\alpha(v_1 \wedge \ldots \wedge v_p) = \alpha(f_* v_1 \wedge \ldots \wedge f_* v_p) \qquad (5)$$

The PULLBACK MAP is a linear map which commutes with the EXTERIOR DERIVATIVE,

$$f^*(d\alpha) = df^*(\alpha). \qquad (6)$$

See also ANGLE BRACKET, BRA, COVARIANT TENSOR, EXTERIOR ALGEBRA, EXTERIOR DERIVATIVE, HODGE STAR, INTEGRATION (FORM), JACOBIAN, KET, MANIFOLD, ONE-FORM, STOKES' THEOREM, SYMPLECTIC FORM, TANGENT BUNDLE, TENSOR, TWO-FORM, WEDGE PRODUCT, ZERO-FORM

References

Berger, M. *Differential Geometry.* New York: Springer-Verlag, pp. 146–37, 1988.
Flanders, H. *Differential Forms with Applications to the Physical Sciences.* New York: Academic Press, 1963.
Spivak, M. *A Comprehensive Introduction to Differential Geometry, Vol. 1, 2nd ed.* Houston, TX: Publish or Perish, pp. 273–83, 1999.
Sternberg, S. *Differential Geometry.* New York: Chelsea, pp. 14–0, 1983.
Weintraub, S. H. *Differential Forms: A Complement to Vector Calculus.* San Diego, CA: Academic Press, 1996.

Differential Operator

The OPERATOR representing the computation of a DERIVATIVE,

$$\tilde{D} \equiv \frac{d}{dx}. \qquad (1)$$

The second derivative is then denoted $\tilde{D}^2$, the third $\tilde{D}^3$, etc. The INTEGRAL is denoted $\tilde{D}^{-1}$.

The differential operator satisfies the identity

$$x - \frac{d}{dx} = -e^{x^2/2}\frac{d}{dx}e^{-x^2/2} \tag{2}$$

(Arfken 1985, p. 720). Furthermore,

$$\left(2x - \frac{d}{dx}\right)^n 1 = H_n(x), \tag{3}$$

where $H_n(x)$ is a HERMITE POLYNOMIAL.

The symbol ϑ can be used to denote the operator

$$\vartheta \equiv z\frac{d}{dz} \tag{4}$$

(Bailey 1935, p. 8).

See also CONVECTIVE DERIVATIVE, DERIVATIVE, FRACTIONAL DERIVATIVE, GRADIENT

References

Bailey, W. N. *Generalised Hypergeometric Series.* Cambridge, England: University Press, 1935.
Arfken, G. *Mathematical Methods for Physicists, 3rd ed.* Orlando, FL: Academic Press, 1985.

Differential Structure
EXOTIC R4, EXOTIC SPHERE

Differential Topology
The motivating force of TOPOLOGY, consisting of the study of smooth (differentiable) MANIFOLDS. Differential topology deals with nonmetrical notions of MANIFOLDS, while DIFFERENTIAL GEOMETRY deals with metrical notions of MANIFOLDS.

See also DIFFERENTIAL GEOMETRY

References

Dieudonné, J. *A History of Algebraic and Differential Topology: 1900–960.* Boston, MA: Birkhäuser, 1989.
Munkres, J. R. *Elementary Differential Topology.* Princeton, NJ: Princeton University Press, 1963.

Differentiating Under the Integral Sign
INTEGRATION UNDER THE INTEGRAL SIGN, LEIBNIZ INTEGRAL RULE

Differentiation
The computation of a DERIVATIVE.

See also CALCULUS, DERIVATIVE, INTEGRAL, INTEGRATION

References

Griewank, A. *Principles and Techniques of Algorithmic Differentiation.* Philadelphia, PA: SIAM, 2000.

Digamma Function

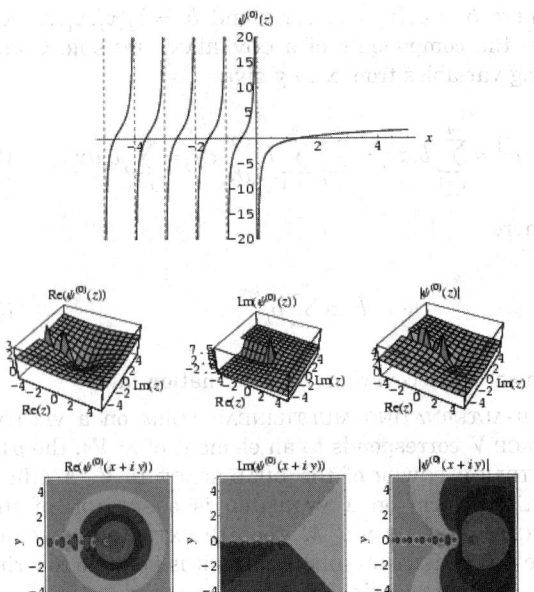

A SPECIAL FUNCTION which is given by the LOGARITHMIC DERIVATIVE of the GAMMA FUNCTION (or, depending on the definition, the LOGARITHMIC DERIVATIVE of the FACTORIAL). Because of this ambiguity, two different notations are sometimes (but not always) used, with

$$\Psi(z) \equiv \frac{d}{dz}\ln\Gamma(z) = \frac{\Gamma'(z)}{\Gamma(z)} \tag{1}$$

defined as the LOGARITHMIC DERIVATIVE of the GAMMA FUNCTION $\Gamma(z)$, and

$$F(z) \equiv \frac{d}{dz}\ln z! \tag{2}$$

defined as the LOGARITHMIC DERIVATIVE of the FACTORIAL function. The two are connected by the relationship

$$F(z) = \Psi(z+1). \tag{3}$$

The nth DERIVATIVE of $\Psi(z)$ is called the POLYGAMMA FUNCTION, denoted $\psi_n(z)$. The notation $\psi_0(z) = \Psi(z)$ is therefore frequently used for the digamma function itself, and Erdélyi *et al.* (1981) use the notation $\psi(z)$ for $\Psi(z)$. The function $\Psi(z) = \psi_0(z)$ is returned by the function `PolyGamma[z]` or `PolyGamma[0, z]` in *Mathematica*.

From a series expansion of the FACTORIAL function,

$$\psi_0(z+1) = \frac{d}{dz}$$
$$\times \lim_{n\to\infty}[\ln n! + z\ln n - \ln(z+1) - \ln(z+2)$$
$$- \ldots - \ln(z+n)] \tag{4}$$

$$= \lim_{n \to \infty} \left(\ln n - \frac{1}{z+1} - \frac{1}{z+2} - \cdots - \frac{1}{z+n} \right) \quad (5)$$

$$= -\gamma - \sum_{n=1}^{\infty} \left(\frac{1}{z+1} - \frac{1}{n} \right) \quad (6)$$

$$= -\gamma + \sum_{n=1}^{\infty} \frac{z}{n(n+z)} \quad (7)$$

$$= \ln z + \frac{1}{2z} - \sum_{n=1}^{\infty} \frac{B_{2n}}{2n z^{2n}}, \quad (8)$$

where γ is the EULER-MASCHERONI CONSTANT and B_{2n} are BERNOULLI NUMBERS.

The digamma function satisfies

$$\psi_0(z) = \int_0^{\infty} \left(\frac{e^{-t}}{t} - \frac{e^{-zt}}{1 - e^{-t}} \right) dt. \quad (9)$$

For integral $z \equiv n$,

$$\psi_0(n) = -\gamma + \sum_{k=1}^{n-1} \frac{1}{k} = -\gamma + H_{n-1}, \quad (10)$$

where γ is the EULER-MASCHERONI CONSTANT and H_n is a HARMONIC NUMBER. Other identities include

$$\frac{d\psi_0}{dz} = \sum_{n=0}^{\infty} \frac{1}{(z+n)^2} \quad (11)$$

$$\psi_0(1-z) - \psi_0(z) = \pi \cot(\pi z) \quad (12)$$

$$\psi_0(z+1) = \psi_0(z) + \frac{1}{z} \quad (13)$$

$$\psi_0(2z) = \frac{1}{2} \psi_0(z) + \frac{1}{2} \psi_0 \left(z + \frac{1}{2} \right) + \ln 2. \quad (14)$$

Special values are

$$\psi_0 \left(\frac{1}{2} \right) = -\gamma - 2 \ln 2 \quad (15)$$

$$\psi_0(1) = -\gamma. \quad (16)$$

At integral values,

$$\psi_0(n+1) = -\gamma + \sum_{k=1}^{n} \frac{1}{k}, \quad (17)$$

and at half-integral values,

$$\psi_0 \left(\frac{1}{2} + n \right) = -\gamma - 2 \ln 2 + 2 \sum_{k=1}^{n} \frac{1}{2k-1}$$

$$= -\gamma + H_{n-1/2}, \quad (18)$$

where H_n is a HARMONIC NUMBER. At rational argu-

ments, $\psi_0(p/q)$ is given by GAUSS'S DIGAMMA THEOREM.

Sums and differences of $\psi_1(r/s)$ for small integral r and s can be expressed in terms of CATALAN'S CONSTANT and π.

See also BARNES' G-FUNCTION, G-FUNCTION, GAMMA FUNCTION, GAUSS'S DIGAMMA THEOREM, HARMONIC NUMBER, HURWITZ ZETA FUNCTION, LOGARITHMIC DERIVATIVE, MELLIN'S FORMULA, POLYGAMMA FUNCTION, RAMANUJAN FUNCTION

References

Abramowitz, M. and Stegun, C. A. (Eds.). "Psi (Digamma) Function." §6.3 in *Handbook of Mathematical Functions with Formulas, Graphs, and Mathematical Tables, 9th printing.* New York: Dover, pp. 258–59, 1972.
Arfken, G. "Digamma and Polygamma Functions." §10.2 in *Mathematical Methods for Physicists, 3rd ed.* Orlando, FL: Academic Press, pp. 549–55, 1985.
Erdélyi, A.; Magnus, W.; Oberhettinger, F.; and Tricomi, F. G. "The ψ Function." §1.7 in *Higher Transcendental Functions, Vol. 1.* New York: Krieger, pp. 15–0, 1981.
Jeffreys, H. and Jeffreys, B. S. "The Digamma ($\mathscr{F}$) and Trigamma ($\mathscr{F}'$) Functions." *Methods of Mathematical Physics, 3rd ed.* Cambridge, England: Cambridge University Press, pp. 465–66, 1988.
Spanier, J. and Oldham, K. B. "The Digamma Function $\psi(x)$." Ch. 44 in *An Atlas of Functions.* Washington, DC: Hemisphere, pp. 423–34, 1987.

Digimetic

A CRYPTARITHM in which DIGITS are used to represent other DIGITS.

See also CRYPTARITHM

Digit

The number of digits D in an INTEGER n is the number of numbers in some base (usually 10) required to represent it. The numbers 1 to 9 are therefore single digits, while the numbers 10 to 99 are double digits. Terms such as "double-digit inflation" are occasionally encountered, although this particular usage has thankfully not been needed in the U.S. for some time. The number of (base 10) digits in a number n can be calculated as

$$D = \lfloor 1 + \log_{10} |n| \rfloor,$$

where $\lfloor x \rfloor$ is the FLOOR FUNCTION.

The number of digits d in the number n represented in base b is given by the *Mathematica* function DigitCount[n, b, d], with DigitCount[n, b] giving a list of the numbers of each digit in n.

Numbers in base-10 which are divisible by their digits are 1, 2, 3, 4, 5, 6, 7, 8, 9, 11, 12, 15, 22, 24, 33, 36, 44, 48, 55, 66, 77, 88, 99, 111, 112, 115, 122, ... (Sloane's A034838). Numbers which are divisible by the sum of their digits are called HARSHAD NUMBERS: 1, 2, 3, 4, 5, 6, 7, 8, 9, 10, 12, 18, 20, 21, 24, ... (Sloane's A005349). Numbers which are divisible by both their digits and the sum of their digits are 1, 2, 3, 4, 5, 6, 7, 8, 9, 12, 24, 36, 48, 111, 112, 126, 132, 135, 144, ... (Sloane's A050104). Numbers which are *equal* to (i.e., not just divisible by) the product of their divisors and the sum of their divisors are called SUM-PRODUCT NUMBERS and are given by 1, 135, 144, ... (Sloane's A038369).

b	order	Sloane	Numbers ($\geq b$)
2	increasing		
2	nondecreasing	A000225	3, 7, 15, 31, 63, 127, 255, 511, 1023, ...
2	nonincreasing	A031997	2, 3, 4, 6, 7, 8, 12, 14, 15, 16, 24, 28, 30, 31, ...
2	decreasing		2
10	increasing	A009993	12, 13, 14, 15, 16, 17, 18, 19, 23, 24, 25, 26, ...
10	nondecreasing	A009994	11, 12, 13, 14, 15, 16, 17, 18, 19, 22, 23, 24, ...
10	nonincreasing	A009996	10, 11, 20, 21, 22, 30, 31, 32, 33, 40, 41, 42, ...
10	decreasing	A009995	10, 20, 21, 30, 31, 32, 40, 41, 42, 43, 50, 51, ...
16	increasing	A023784	18, 19, 20, 21, 22, 23, 24, 25, 26, 27, 28, 29, ...
16	nondecreasing	A023757	17, 18, 19, 20, 21, 22, 23, 24, 25, 26, 27, 28, ...
16	nonincreasing	A023771	17, 32, 33, 34, 48, 49, 50, 51, 64, 65, 66, 67, ...
16	decreasing	A023797	32, 33, 48, 49, 50, 64, 65, 66, 67, 80, 81, 82, ...

In HEXADECIMAL, numbers with increasing digits are called METADROMES, those with nondecreasing digits are called PLAINDRONES, those with nonincreasing digits are called NIALPDROMES, and those with decreasing digits are called KATADROMES.

The count of numbers with strictly increasing digits in base-b is 2^{b-1}, and the number with strictly decreasing digits is 2^{b-1}.

See also 196-ALGORITHM, ADDITIVE PERSISTENCE, DIGIT PRODUCT, DIGIT SERIES, DIGIT-SHIFTING CONSTANTS, DIGITADDITION, DIGITAL ROOT, FACTORION, FIGURES, HARSHAD NUMBER, KATADROME, LENGTH (NUMBER), METADROME, MULTIPLICATIVE PERSISTENCE, NARCISSISTIC NUMBER, NIALPDROME, PLAINDROME, SCIENTIFIC NOTATION, SIGNIFICANT DIGITS, SMITH NUMBER, SUM-PRODUCT NUMBER

References

Bailey, D. H. and Crandall, R. E. "On the Random Character of Fundamental Constant Expansions." Manuscript, Mar. 2000. http://www.nersc.gov/~dhbailey/dhbpapers/dhbpapers.html.

Sloane, N. J. A. Sequences A0053490481, A034838, A038369, and A050104 in "An On-Line Version of the Encyclopedia of Integer Sequences." http://www.research.att.com/~njas/sequences/eisonline.html.

Digit Block

Let $u_B(n)$ be the number of DIGIT BLOCKS of a sequence **B** in the base-b expansion of n, which can be implemented in *Mathematica* as

```
u[n_Integer, b_Integer, block_List] :=
    Count[Partition[IntegerDigits[n,  b],
Length[block],  1],  block]
```

The following table gives the sequence $\{u_{\mathbf{B}}(n)\}$ for a number of blocks **B**.

B	Sloane	sequence
00	A056973	0, 0, 0, 1, 0, 0, 0, 2, 1, 0, 0, 1, 0, 0, 0, 3, ...
01	A037800	0, 0, 0, 0, 1, 0, 0, 0, 1, 1, 1, 0, 1, 0, 0, 0, ...
10	A033264	0, 1, 0, 1, 1, 1, 0, 1, 1, 2, 1, 1, 1, 1, 0, 1, ...
11	A014081	0, 0, 1, 0, 0, 1, 2, 0, 0, 0, 1, 1, 1, 2, 3, 0, ...
000	A056974	0, 0, 0, 0, 0, 0, 0, 1, 0, 0, 0, 0, 0, 0, 0, 2, ...
001	A056975	0, 0, 0, 0, 0, 0, 0, 0, 1, 0, 0, 0, 0, 0, 0, 0, ...
010	A056976	0, 0, 0, 0, 0, 0, 0, 0, 0, 1, 0, 0, 0, 0, 0, 0, ...
011	A056977	0, 0, 0, 0, 0, 0, 0, 0, 0, 0, 1, 0, 0, 0, 0, 0, ...
100	A056978	0, 0, 0, 1, 0, 0, 0, 1, 1, 0, 0, 1, 0, 0, 0, 1, ...
101	A056979	0, 0, 0, 0, 1, 0, 0, 0, 0, 1, 1, 0, 1, 0, 0, 0, ...

| 110 | A056980 | 0, 0, 0, 0, 0, 1, 0, 0, 0, 0, 0, 0, 1, 1, 1, 0, 0, ... |
| 111 | A014082 | 0, 0, 0, 0, 0, 0, 1, 0, 0, 0, 0, 0, 0, 1, 2, 0, ... |

See also DIGIT SERIES, RUDIN-SHAPIRO SEQUENCE

References

Sloane, N. J. A. Sequences A014081, A014082, A033264, A037800, A056973, A056974, A056975, A056976, A056977, A056978, A056979, and A056980 in "An On-Line Version of the Encyclopedia of Integer Sequences." http://www.research.att.com/~njas/sequences/eisonline.html.

Digit Product

Let $s_b(n)$ be the sum of the base-b digits of n, and $\epsilon(n) = (-1)^{S_2(n)}$ the THUE-MORSE SEQUENCE, then

$$\prod_{n=0}^{\infty} \left(\frac{2n+1}{2n+2}\right)^{\epsilon(n)} = \frac{1}{2}\sqrt{2}. \tag{1}$$

See also DIGIT, DIGIT SERIES

References

Allouche, J.-P. "Series and Infinite Products Related to Binary Expansions of Integers." http://algo.inria.fr/seminars/sem92-3/allouche.ps.
Shallit, J. O. "On Infinite Products Associated with Sums of Digits." *J. Number Th.* **21**, 128–34, 1985.

Digit Series

Let $s_b(n)$ be the sum of the base-b digits of n, which can be implemented in *Mathematica* as

```
s[n_, b_] := Plus @@ IntegerDigits[n, b]
```

Then

$$\sum_{n=1}^{\infty} \frac{s_b(n)}{n(n+1)} = \frac{b}{b-1}\ln b, \tag{1}$$

the $b = 2$ case of which was given in the 1981 Putnam competition (Allouche 1992). In addition,

$$\sum_{n=1}^{\infty} s_2 \frac{2n+1}{n^2(n+1)^2} = \frac{\pi^2}{9} \tag{2}$$

$$\sum_{n=2}^{\infty} [s_2(n)]^2 \frac{8n^3+4n^2+n-1}{4n(n^2-1)(4n^2-1)} = \frac{17}{24} + \ln 2 \tag{3}$$

(Allouche 1992, Allouche and Shallit 1992).

Let $u(n)$ be the number of DIGIT BLOCKS of 11 in the binary expansion of n, then

$$\sum_{n=1}^{\infty} \frac{u(n)}{n(n+1)} = \frac{3}{2}\ln 2 - \frac{1}{4}\pi \tag{4}$$

(Allouche 1992).

See also DIGIT, DIGIT BLOCK, DIGIT PRODUCT

References

Allouche, J.-P. "Series and Infinite Products Related to Binary Expansions of Integers." 1992. http://algo.inria.fr/seminars/sem92-3/allouche.ps.
Allouche, J.-P. and Shallit, J. "The Ring of k-Regular Sequences." *Theor. Comput. Sci.* **98**, 163–97, 1992.
Shallit, J. O. "On Infinite Products Associated with Sums of Digits." *J. Number Th.* **21**, 128–34, 1985.

Digitaddition

Start with an INTEGER n, known as the GENERATOR. Add the SUM of the GENERATOR's digits to obtain the digitaddition n'. A number can have more than one GENERATOR. If a number has no GENERATOR, it is called a SELF NUMBER. The sum of all numbers in a digitaddition series is given by the last term minus the first plus the sum of the DIGITS of the last.

If the digitaddition process is performed on n' to yield *its* digitaddition n'', on n'' to yield n''', etc., a single-digit number, known as the DIGITAL ROOT of n, is eventually obtained. The digital roots of the first few integers are 1, 2, 3, 4, 5, 6, 7, 8, 9, 1, 2, 3, 4, 5, 6, 7, 8, 9, 1, ... (Sloane's A010888).

If the process is generalized so that the kth (instead of first) powers of the digits of a number are repeatedly added, a periodic sequence of numbers is eventually obtained for any given starting number n. If the original number n is equal to the sum of the kth powers of its digits, it is called a NARCISSISTIC NUMBER. If the original number is the smallest number in the eventually periodic sequence of numbers in the repeated k-digitadditions, it is called a RECURRING DIGITAL INVARIANT. Both NARCISSISTIC NUMBERS and RECURRING DIGITAL INVARIANTS are relatively rare.

The only possible periods for repeated 2-digitadditions are 1 and 8, and the periods of the first few positive integers are 1, 8, 8, 8, 8, 8, 1, 8, 8, 1, ... The possible periods p for n-digitadditions are summarized in the following table, together with digitadditions for the first few integers and the corresponding sequence numbers. Some periods do not show up for a long time. For example, a period-6 10-digitaddition does not occur until the number 266.

n	Sloane	p	s	n-Digitadditions
2	Sloane's A031176	1, 8		1, 8, 8, 8, 8, 8, 1, 8, 8, 1, ...
3	Sloane's A031178	1, 2, 3		1, 1, 1, 3, 1, 1, 1, 1, 1, 1, 1, 1, 3, ...

n	Sloane		
4	Sloane's A031182	1, 2, 7	1, 7, 7, 7, 7, 7, 7, 7, 7, 1, 7, 1, 7, 7, ...
5	Sloane's A031186	1, 2, 4, 6, 10, 12, 22, 28	1, 12, 22, 4, 10, 22, 28, 10, 22, 1, ...
6	Sloane's A031195	1, 2, 3, 4, 10, 30	1, 10, 30, 30, 30, 10, 10, 10, 3, 1, 10, ...
7	Sloane's A031200	1, 2, 3, 6, 12, 14, 21, 27, 30, 56, 92	1, 92, 14, 30, 92, 56, 6, 92, 56, 1, 92, 27, ...
8	Sloane's A031211	1, 25, 154	1, 25, 154, 154, 154, 154, 25, 154, 154, 1, 25, 154, 154, 1, ...
9	Sloane's A031212	1, 2, 3, 4, 8, 10, 19, 24, 28, 30, 80, 93	1, 30, 93, 1, 19, 80, 4, 30, 80, 1, 30, 93, 4, 10, ...
10	Sloane's A031213	1, 6, 7, 17, 81, 123	1, 17, 123, 17, 17, 123, 123, 123, 123, 1, 17, 123, 17 ...

The numbers having period-1 2-digitadded sequences are also called HAPPY NUMBERS. The first few numbers having period p n-digitadditions are summarized in the following table, together with their sequence numbers.

n	p	Sloane	Members
2	1	Sloane's A007770	1, 7, 10, 13, 19, 23, 28, 31, 32, ...
2	8	Sloane's A031177	2, 3, 4, 5, 6, 8, 9, 11, 12, 14, 15, ...
3	1	Sloane's A031179	1, 2, 3, 5, 6, 7, 8, 9, 10, 11, 12, ...
3	2	Sloane's A031180	49, 94, 136, 163, 199, 244, 316, ...
3	3	Sloane's A031181	4, 13, 16, 22, 25, 28, 31, 40, 46, ...
4	1	Sloane's A031183	1, 10, 12, 17, 21, 46, 64, 71, 100, ...
4	2	Sloane's A031184	66, 127, 172, 217, 228, 271, 282, ...
4	7	Sloane's A031185	2, 3, 4, 5, 6, 7, 8, 9, 11, 13, 14, ...
5	1	Sloane's A031187	1, 10, 100, 145, 154, 247, 274, ...
5	2	Sloane's A031188	133, 139, 193, 199, 226, 262, ...
5	4	Sloane's A031189	4, 37, 40, 55, 73, 124, 142, ...
5	6	Sloane's A031190	16, 61, 106, 160, 601, 610, 778, ...
5	10	Sloane's A031191	5, 8, 17, 26, 35, 44, 47, 50, 53, ...
5	12	Sloane's A031192	2, 11, 14, 20, 23, 29, 32, 38, 41, ...
5	22	Sloane's A031193	3, 6, 9, 12, 15, 18, 21, 24, 27, ...
5	28	Sloane's A031194	7, 13, 19, 22, 25, 28, 31, 34, 43, ...
6	1	Sloane's A011557	1, 10, 100, 1000, 10000, 100000, ...
6	2	Sloane's A031357	3468, 3486, 3648, 3684, 3846, ...
6	3	Sloane's A031196	9, 13, 31, 37, 39, 49, 57, 73, 75, ...
6	4	Sloane's A031197	255, 466, 525, 552, 646, 664, ...
6	10	Sloane's A031198	2, 6, 7, 8, 11, 12, 14, 15, 17, 19, ...
6	30	Sloane's A031199	3, 4, 5, 16, 18, 22, 29, 30, 33, ...
7	1	Sloane's A031201	1, 10, 100, 1000, 1259, 1295, ...
7	2	Sloane's A031202	22, 202, 220, 256, 265, 526, 562, ...
7	3	Sloane's A031203	124, 142, 148, 184, 214, 241, 259, ...
7	6		7, 70, 700, 7000, 70000, 700000, ...
7	12	Sloane's A031204	17, 26, 47, 59, 62, 71, 74, 77, 89, ...
7	14	Sloane's A031205	3, 30, 111, 156, 165, 249, 294, ...
7	21	Sloane's A031206	19, 34, 43, 91, 109, 127, 172, 190, ...
7	27	Sloane's A031207	12, 18, 21, 24, 39, 42, 45, 54, 78, ...
7	30	Sloane's A031208	4, 13, 16, 25, 28, 31, 37, 40, 46, ...
7	56	Sloane's A031209	6, 9, 15, 27, 33, 36, 48, 51, 57, ...

7	92	Sloane's A031210	2, 5, 8, 11, 14, 20, 23, 29, 32, 35, ...
8	1		1, 10, 14, 17, 29, 37, 41, 71, 73, ...
8	25		2, 7, 11, 15, 16, 20, 23, 27, 32, ...
8	154		3, 4, 5, 6, 8, 9, 12, 13, 18, 19, ...
9	1		1, 4, 10, 40, 100, 400, 1000, 1111, ...
9	2		127, 172, 217, 235, 253, 271, 325, ...
9	3		444, 4044, 4404, 4440, 4558, ...
9	4		7, 13, 31, 67, 70, 76, 103, 130, ...
9	8		22, 28, 34, 37, 43, 55, 58, 73, 79, ...
9	10		14, 38, 41, 44, 83, 104, 128, 140, ...
9	19		5, 26, 50, 62, 89, 98, 155, 206, ...
9	24		16, 61, 106, 160, 337, 373, 445, ...
9	28		19, 25, 46, 49, 52, 64, 91, 94, ...
9	30		2, 8, 11, 17, 20, 23, 29, 32, 35, ...
9	80		6, 9, 15, 18, 24, 33, 42, 48, 51, ...
9	93		3, 12, 21, 27, 30, 36, 39, 45, 54, ...
10	1	Sloane's A011557	1, 10, 100, 1000, 10000, 100000, ...
10	6		266, 626, 662, 1159, 1195, 1519, ...
10	7		46, 58, 64, 85, 122, 123, 132, ...
10	17		2, 4, 5, 11, 13, 20, 31, 38, 40, ...
10	81		17, 18, 37, 71, 73, 81, 107, 108, ...
10	123		3, 6, 7, 8, 9, 12, 14, 15, 16, 19, ...

See also 196-ALGORITHM, ADDITIVE PERSISTENCE, DIGIT, DIGITAL ROOT, MULTIPLICATIVE PERSISTENCE, NARCISSISTIC NUMBER, RECURRING DIGITAL INVARIANT

References
Trott, M. "Numerical Computations." §1.2.1 in *The Mathematica Guidebook, Vol. 1: Programming in Mathematica*. New York: Springer-Verlag, 2000.

Digital Root

Consider the process of taking a number, adding its DIGITS, then adding the DIGITS of numbers derived from it, etc., until the remaining number has only one DIGIT. The number of additions required to obtain a single DIGIT from a number n is called the ADDITIVE PERSISTENCE of n, and the DIGIT obtained is called the digital root of n.

For example, the sequence obtained from the starting number 9876 is (9876, 30, 3), so 9876 has an ADDITIVE PERSISTENCE of 2 and a digital root of 3. The digital roots of the first few integers are 1, 2, 3, 4, 5, 6, 7, 8, 9, 1, 2, 3, 4, 5, 6, 7, 9, 1, ... (Sloane's A010888). The digital root of an INTEGER n can therefore be computed without actually performing the iteration using the simple congruence formula

$$\begin{cases} n \pmod 9 & n \not\equiv 0 \pmod 9 \\ 9 & n \equiv 0 \pmod 9. \end{cases}$$

See also ADDITIVE PERSISTENCE, DIGITADDITION, KAPREKAR NUMBER, MULTIPLICATIVE DIGITAL ROOT, MULTIPLICATIVE PERSISTENCE, NARCISSISTIC NUMBER, RECURRING DIGITAL INVARIANT, SELF NUMBER

References
Sloane, N. J. A. Sequences A007612/M1114 and A010888 in "An On-Line Version of the Encyclopedia of Integer Sequences." http://www.research.att.com/~njas/sequences/eisonline.html.
Trott, M. "Numerical Computations." §1.2.1 in *The Mathematica Guidebook, Vol. 1: Programming in Mathematica*. New York: Springer-Verlag, 2000.

Digit-Extraction Algorithm

An algorithm which allows digits of a given number to be calculated without requiring the computation of earlier digits. The BAILEY-BORWEIN-PLOUFFE ALGORITHM for PI is the best-known such algorithm, but an algorithm also exists for E.

See also BAILEY-BORWEIN-PLOUFFE ALGORITHM

Digit-Shifting Constants

Given a REAL NUMBER x, find the powers of a base b that will shift the digits of x a number of places n to the left. This is equivalent to solving

$$b^x = b^n x \qquad (1)$$

or

$$x = n + \log_b x. \tag{2}$$

The solution is given by

$$x = -\frac{W(-b^{-n}\ln b)}{\ln b}, \tag{3}$$

where $W(x)$ is LAMBERT'S W-FUNCTION.

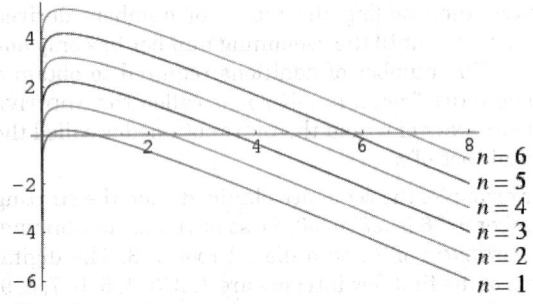

The above plot shows $\log_b x + n - x$ for $b = 10$ and small values of n. As can be seen, there are two distinct solutions, corresponding to two different BRANCHES of $W(x)$ in (3). For $n = 1, 2, ...,$ these solutions are approximately given by 0.137129, 0.0102386, 0.00100231, 0.000100023, 0.0000100002, ..., and 1, 2.37581, 3.55026, 4.66925, 5.76046, ..., respectively. For example,

$$10^{0.0102385...} = 1.02385... \tag{4}$$

and

$$10^{2.37581...} = 237.581... \tag{5}$$

See also BASE (NUMBER), DIGIT, LOGARITHM

Digon

digon

The DEGENERATE POLYGON (corresponding to a LINE SEGMENT) with SCHLÄFLI SYMBOL {2}.

See also LINE SEGMENT, POLYGON, TRIGONOMETRY VALUES PI/2

Digraph

DIRECTED GRAPH

Dihedral Angle

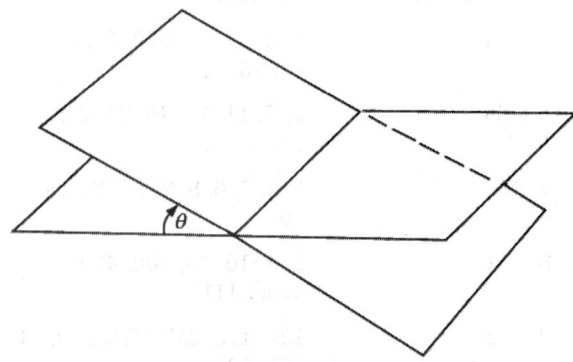

The ANGLE θ between two PLANES. The dihedral angle between the planes

$$A_1 x + B_1 y + C_1 z + D_1 = 0 \tag{1}$$

$$A_2 x + B_2 y + C_2 z + D_2 = 0 \tag{2}$$

which have normal vectors $\mathbf{N}_1 = (A_1, B_1, C_1)$ and $\mathbf{N}_2 = (A_2, B_2, C_2)$ is simply given via the DOT PRODUCT of the normals,

$$\cos\theta = \mathbf{N}_1 \cdot \mathbf{N}_2$$

$$= \frac{A_1 A_2 + B_1 B_2 + C_1 C_2}{\sqrt{A_1^2 + B_1^2 + C_1^2}\sqrt{A_2^2 + B_2^2 + C_2^2}}. \tag{3}$$

The dihedral angle between planes in a general TETRAHEDRON is closely connected with the face areas via a generalization of the LAW OF COSINES.

See also ANGLE, PLANE, TETRAHEDRON, TRIHEDRON, VERTEX ANGLE

References

Kern, W. F. and Bland, J. R. *Solid Mensuration with Proofs, 2nd ed.* New York: Wiley, p. 15, 1948.

Dihedral Group

A GROUP of symmetries for an n-sided REGULAR POLYGON, denoted D_n. The ORDER of D_n is $2n$.

See also FINITE GROUP D3, FINITE GROUP D4

References

Arfken, G. "Dihedral Groups, D_n." *Mathematical Methods for Physicists, 3rd ed.* Orlando, FL: Academic Press, p. 248, 1985.
Lomont, J. S. "Dihedral Groups." §3.10.B in *Applications of Finite Groups.* New York: Dover, pp. 78–0, 1987.

Dihedral Prime

A number n such that the "LED representation" of n (i.e., the arrangement of horizonal and vertical lines seen on a digital clock or pocket calculator), n upside down, n in a mirror, and n upside-down-and-in-a-mirror are all primes. The digits of n are therefore restricted to 0, 1, 2, 5, and 8. The first few dihedral primes are 2, 11, 101, 181, 1181, 1811, 18181, 108881, 110881, 118081, 120121, ... (Sloane's A038136).

References

Rivera, C. "Problems & Puzzles: Puzzle The Mirrorable Numbers (by Mike Keith).-039." http://www.primepuzzles.net/puzzles/puzz_039.htm.

Sloane, N. J. A. Sequences A038136 in "An On-Line Version of the Encyclopedia of Integer Sequences." http://www.research.att.com/~njas/sequences/eisonline.html.

Dijkstra Tree

The shortest path-spanning TREE from a VERTEX of a GRAPH.

Dijkstra's Algorithm

An ALGORITHM for finding a GRAPH GEODESIC, i.e., the shortest path between two VERTICES in a GRAPH. It functions by constructing a shortest-path tree from the initial vertex to every other vertex in the graph. The algorithm is implemented as `Dijkstra[g]` in the *Mathematica* add-on package `DiscreteMath`Combinatorica`` (which can be loaded with the command `< <DiscreteMath``).

See also FLOYD'S ALGORITHM, GRAPH GEODESIC

References

Dijkstra, E. W. "A Note on Two Problems in Connection with Graphs." *Numerische Math.* **1**, 269–71, 1959.

Skiena, S. "Dijkstra's Algorithm." §6.1.1 in *Implementing Discrete Mathematics: Combinatorics and Graph Theory with Mathematica.* Reading, MA: Addison-Wesley, pp. 225–27, 1990.

Whiting, P. D. and Hillier, J. A. "A Method for Finding the Shortest Route through a Road Network." *Operational Res. Quart.* **11**, 37–0, 1960.

Dilation

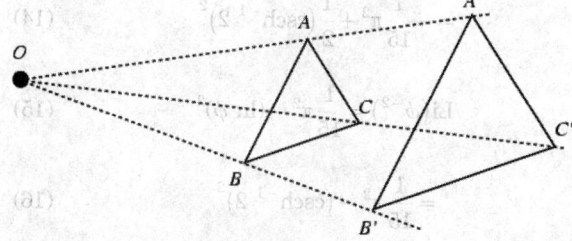

A SIMILARITY TRANSFORMATION which transforms each line to a PARALLEL line whose length is a fixed multiple of the length of the original line. The simplest dilation is therefore a TRANSLATION, and any dilation that is not merely a TRANSLATION is called a CENTRAL DILATION. Two triangles related by a CENTRAL DILATION are said to be PERSPECTIVE TRIANGLES because the lines joining corresponding vertices CONCUR. A dilation corresponds to an EXPANSION plus a TRANSLATION.

See also EXPANSION, PARALLEL, PERSPECTIVE TRIANGLES, TRANSLATION

References

Coxeter, H. S. M. and Greitzer, S. L. "Dilation." §4.7 in *Geometry Revisited.* Washington, DC: Math. Assoc. Amer., pp. 94–5, 1967.

Dilative Rotation

SPIRAL SIMILARITY

Dilcher's Formula

$$\sum_{1 \le k \le n} \binom{n}{k} \frac{(-1)^{k-1}}{k^m}$$

$$= \sum_{1 \le i_1 \le i_2 \le \ldots \le i_m \le n} \frac{1}{i_1 i_2 \cdots i_m}, \quad (1)$$

where $\binom{n}{k}$ is a BINOMIAL COEFFICIENT (Dilcher 1995, Flajolet and Sedgewick 1995, Prodinger 2000). An inverted version is given by

$$\sum_{1 \le k \le n} \binom{n}{k} (-1)^{k-1} \sum_{1 \le i_1 \le i_2 \le \ldots \le i_m \le k} \frac{1}{i_1 i_2 \cdots i_m}$$

$$= \sum_{1 \le k \le n} \frac{1}{k^m} = H_n^{(m)}, \quad (2)$$

where $H_n^{(k)}$ is a HARMONIC NUMBER of order m (Hernández 1999, Prodinger 2000). A q-ANALOG of (1) is given by

$$\sum_{1 \le k \le n} \begin{bmatrix} n \\ k \end{bmatrix}_q (-1)^{k-1} q^{\binom{k+1}{2} + (m-1)k} \over (1-q^k)^m$$

$$= \sum_{1 \le i_1 \le i_2 \le \ldots \le i_m \le n} \frac{q^{i_1}}{1-q^{i_1}} \cdots \frac{q^{i_m}}{1-q^{i_m}}, \quad (3)$$

where

$$\begin{bmatrix} n \\ k \end{bmatrix}_q = \frac{(q;q)_n}{(q;q)_k (q;q)_{n-k}} \quad (4)$$

is a GAUSSIAN POLYNOMIAL (Prodinger 2000).

See also BINOMIAL IDENTITY

References

Dilcher, K. "Some q-Series Identities Related to Divisor Functions." *Disc. Math.* **145**, 83–3, 1995.

Flajolet, P. and Sedgewick, R. "Mellin Transforms and Asymptotics: Finite Differences and Rice's Integrals." *Theor. Comput. Sci.* **144**, 101–24, 1995.

Hernández, V. "Solution IV of Problem 10490: A Reciprocal Summation Identity." *Amer. Math. Monthly* **106**, 589–90, 1999.

Prodinger, H. "A q-Analogue of a Formula of Hernandez Obtained by Inverting a Result of Dilcher." *Austral. J. Combin.* **21**, 271–74, 2000.

Dilemma

Informally, a situation in which a decision must be made from several alternatives, none of which is obviously the optimal one. In formal LOGIC, a dilemma is a specific type of argument using two conditional statements which may take the form of a CONSTRUCTIVE DILEMMA or a DESTRUCTIVE DILEMMA.

See also CONSTRUCTIVE DILEMMA, DESTRUCTIVE DILEMMA, MONTY HALL PROBLEM, PARADOX, PRISONER'S DILEMMA

Dilogarithm

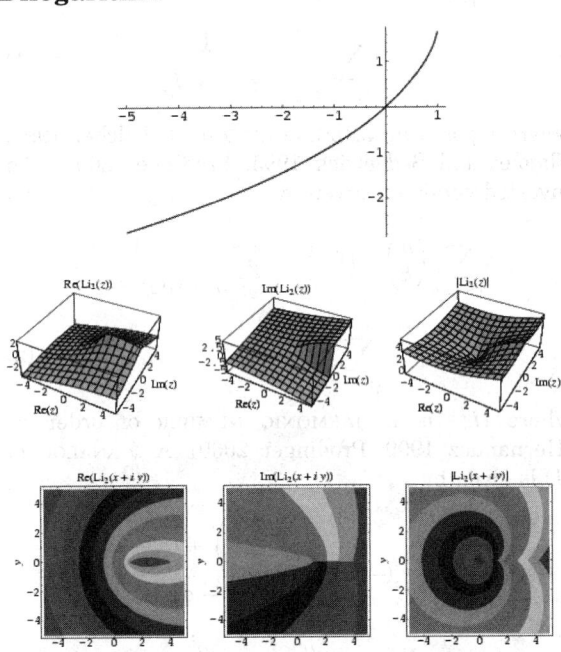

A special case of the POLYLOGARITHM $Li_n(z)$ for $n = 2$. It is denoted $Li_2(z)$, or sometimes $L_2(z)$. The notation $Li_2(x)$ for the dilogarithm is unfortunately similar to that for the LOGARITHMIC INTEGRAL $Li(x)$. The dilogarithm can be defined by the sum

$$Li_2(z) = \sum_{k=1}^{\infty} \frac{z^k}{k^2} \qquad (1)$$

or the integral

$$Li_2(z) \equiv \int_z^0 \frac{\ln(1-t)dt}{t}. \qquad (2)$$

There are also two different commonly encountered normalizations for the $Li_2(z)$ function, both denoted $L(z)$, and one of which is known as the ROGERS L-FUNCTION.

The major functional equations for the dilogarithm are given by

$$Li_2(x) + Li_2(-x) = \frac{1}{2} Li_2(x^2) \qquad (3)$$

$$Li_2(1-x) + Li_2(1-x^{-1}) = -\frac{1}{2}(\ln x)^2 \qquad (4)$$

$$Li_2(x) + Li_2(1-x) = \frac{1}{6}\pi^2 - (\ln x)\ln(1-x) \qquad (5)$$

$$Li_2(-x) - Li_2(1-x) + \frac{1}{2}Li_2(1-x^2)$$
$$= -\frac{1}{12}\pi^2 - (\ln x)\ln(x+1). \qquad (6)$$

A complete list of $Li_2(x)$ which can be evaluated in closed form is given by

$$Li_2(-1) = -\frac{1}{12}\pi^2 \qquad (7)$$

$$Li_2(0) = 0 \qquad (8)$$

$$Li_2\left(\frac{1}{2}\right) = \frac{1}{12}\pi^2 - \frac{1}{2}(\ln 2)^2 \qquad (9)$$

$$Li_2(1) = \frac{1}{6}\pi^2 \qquad (10)$$

$$Li_2(-\phi) = -\frac{1}{10}\pi^2 - (\ln \phi)^2 \qquad (11)$$

$$= -\frac{1}{10}\pi^2 - (csch^{-1} 2)^2 \qquad (12)$$

$$Li_2(-\phi^{-1}) = -\frac{1}{15}\pi^2 + \frac{1}{2}(\ln \phi)^2 \qquad (13)$$

$$= -\frac{1}{15}\pi^2 + \frac{1}{2}(csch^{-1} 2)^2 \qquad (14)$$

$$Li(\phi^{-2}) = \frac{1}{15}\pi^2 - (\ln \phi)^2 \qquad (15)$$

$$= \frac{1}{15}\pi^2 - (csch^{-1} 2)^2 \qquad (16)$$

$$Li(\phi^{-1}) = \frac{1}{10}\pi^2 - (\ln \phi)^2 \qquad (17)$$

$$= \frac{1}{10}\pi^2 - (csch^{-1} 2)^2, \qquad (18)$$

where ϕ is the GOLDEN RATIO (Lewin 1981, Borwein *et al.* 1998).

There are several remarkable identities involving the DILOGARITHM function. Ramanujan gave the identities

$$\text{Li}_2\left(\frac{1}{3}\right) - \frac{1}{6}\text{Li}_2\left(\frac{1}{9}\right) = \frac{1}{18}\pi^2 - \frac{1}{6}(\ln 3)^2 \qquad (19)$$

$$\text{Li}_2\left(-\frac{1}{2}\right) + \frac{1}{5}\text{Li}_2\left(\frac{1}{9}\right)$$
$$= -\frac{1}{18}\pi^2 + \ln 2 \ln 3 - \frac{1}{2}(\ln 2)^2 - \frac{1}{3}(\ln 3)^2 \qquad (20)$$

$$\text{Li}_2\left(\frac{1}{4}\right) + \frac{1}{3}\text{Li}_2\left(\frac{1}{9}\right)$$
$$= \frac{1}{18}\pi^2 + 2\ln 2 \ln 3 - 2(\ln 2)^2 - \frac{2}{3}(\ln 3)^2 \qquad (21)$$

$$\text{Li}_2\left(-\frac{1}{3}\right) - \frac{1}{3}\text{Li}_2\left(\frac{1}{9}\right) = -\frac{1}{18}\pi^2 + \frac{1}{6}(\ln 3)^2 \qquad (22)$$

$$\text{Li}_2\left(-\frac{1}{8}\right) + \text{Li}_2\left(\frac{1}{9}\right) = -\frac{1}{2}\left(\ln\frac{9}{8}\right)^2 \qquad (23)$$

$$\text{Li}_2\left(\frac{1}{2}\left(\sqrt{5}-1\right)\right) = \frac{1}{10}\pi^2 - \left[\ln\left(\frac{1}{2}\left(1+\sqrt{5}\right)\right)\right]^2 \qquad (24)$$

(Berndt 1994, Gordon and McIntosh 1997), and Bailey *et al.* show that

$$\pi^2 = 36\text{Li}_2\left(\frac{1}{2}\right) - 36\text{Li}_2\left(\frac{1}{4}\right) - 12\text{Li}_2\left(\frac{1}{8}\right) + 6\text{Li}_2\left(\frac{1}{64}\right) \qquad (25)$$

$$12\text{Li}_2\left(\frac{1}{2}\right) = \pi^2 - 6(\ln 2)^2 \qquad (26)$$

See also ABEL'S DUPLICATION FORMULA, ABEL'S FUNCTIONAL EQUATION, CLAUSEN FUNCTION, INVERSE TANGENT INTEGRAL, *L*-ALGEBRAIC NUMBER, LEGENDRE'S CHI-FUNCTION, LOGARITHM, POLYLOGARITHM, ROGERS *L*-FUNCTION, SPENCE'S FUNCTION, SPENCE'S INTEGRAL, TRILOGARITHM, WATSON IDENTITIES

References

Abramowitz, M. and Stegun, C. A. (Eds.). "Dilogarithm." §27.7 in *Handbook of Mathematical Functions with Formulas, Graphs, and Mathematical Tables, 9th printing.* New York: Dover, pp. 1004–005, 1972.

Andrews, G. E.; Askey, R.; and Roy, R. *Special Functions.* Cambridge, England: Cambridge University Press, 1999.

Bailey, D.; Borwein, P.; and Plouffe, S. "On the Rapid Computation of Various Polylogarithmic Constants." http://www.cecm.sfu.ca/~pborwein/PAPERS/P123.ps.

Berndt, B. C. *Ramanujan's Notebooks, Part IV.* New York: Springer-Verlag, pp. 323–26, 1994.

Borwein, J. M.; Bradley, D. M.; Broadhurst, D. J.; and Losinek, P. "Special Values of Multidimensional Polylogarithms." CECM-98:106, 14 May 1998. http://www.cecm.sfu.ca/preprints/1998pp.html#98:106.

Bytsko, A. G. *J. Physics A* **32**, 8045, 1999.

Erdélyi, A.; Magnus, W.; Oberhettinger, F.; and Tricomi, F. G. "Euler's Dilogarithm." §1.11.1 in *Higher Transcendental Functions, Vol. 1.* New York: Krieger, pp. 31–2, 1981.

Gordon, B. and McIntosh, R. J. "Algebraic Dilogarithm Identities." *Ramanujan J.* **1**, 431–48, 1997.

Kirillov, A. N. "Dilogarithm Identities." *Progr. Theor. Phys. Suppl.* **118**, 61–42, 1995.

Lewin, L. *Dilogarithms and Associated Functions.* London: Macdonald, 1958.

Lewin, L. *Polylogarithms and Associated Functions.* New York: North-Holland, 1981.

Lewin, L. "The Dilogarithm in Algebraic Fields." *J. Austral. Soc. Ser. A* **33**, 302–30, 1982.

Watson, G. N. *Quart. J. Math. Oxford Ser.* **8**, 39, 1937.

Dilworth's Lemma

The WIDTH of a set P is equal to the minimum number of CHAINS needed to COVER P. Equivalently, if a set P of $ab+1$ elements is PARTIALLY ORDERED, then P contains a CHAIN of size $a+1$ or an ANTICHAIN of size $b+1$. Letting N be the CARDINALITY of P, W the WIDTH, and L the LENGTH, this last statement says $N \le LW$. Dilworth's lemma is a generalization of the ERDOS-SZEKERES THEOREM. RAMSEY'S THEOREM generalizes Dilworth's lemma.

See also ANTICHAIN, CHAIN, COMBINATORICS, ERDOS-SZEKERES THEOREM, RAMSEY'S THEOREM

References

Dilworth, R. P. "A Decomposition Theorem for Partially Ordered Sets." *Ann. Math.* **51**, 161–66, 1950.

Skiena, S. "Dilworth's Lemma." §6.4.2 in *Implementing Discrete Mathematics: Combinatorics and Graph Theory with Mathematica.* Reading, MA: Addison-Wesley, pp. 241–43, 1990.

Dilworth's Theorem

DILWORTH'S LEMMA

Dimension

The dimension of an object is a topological measure of the size of its covering properties. Roughly speaking, it is the number of coordinates needed to specify a point on the object. For example, a RECTANGLE is two-dimensional, while a CUBE is three-dimensional. The dimension of an object is sometimes also called its "dimensionality."

The prefix "hyper-" is usually used to refer to the 4- (and higher-) dimensional analogs of 3-dimensional objects, e.g. HYPERCUBE, HYPERPLANE.

The notion of dimension is important in mathematics because it gives a precise parameterization of the conceptual or visual complexity of any geometric object. In fact, the concept can even be applied to abstract objects which cannot be directly visualized. For example, the notion of time can be considered as one-dimensional, since it can be thought of as consisting of only "now," "before" and "after." Since "before" and "after," regardless of how far back or how far into the future they are, are extensions, time is like a line, a 1-dimensional object.

To see how lower and higher dimensions relate to each other, take any geometric object (like a POINT, LINE, CIRCLE, PLANE, etc.), and "drag" it in an opposing direction (drag a POINT to trace out a LINE, a LINE to trace out a box, a CIRCLE to trace out a CYLINDER, a DISK to a solid CYLINDER, etc.). The result is an object which is qualitatively "larger" than the previous object, "qualitative" in the sense that, regardless of how you drag the original object, you always trace out an object of the same "qualitative size." The POINT could be made into a straight LINE, a CIRCLE, a HELIX, or some other CURVE, but all of these objects are qualitatively of the same dimension. The notion of dimension was invented for the purpose of measuring this "qualitative" topological property.

Finite collections of objects (e.g., points in space) are considered 0-dimensional. Objects that are "dragged" versions of 0-dimensional objects are then called 1-dimensional. Similarly, objects which are dragged 1-dimensional objects are 2-dimensional, and so on. Dimension is formalized in mathematics as the intrinsic dimension of a TOPOLOGICAL SPACE. This dimension is called the LEBESGUE COVERING DIMENSION (also known simply as the TOPOLOGICAL DIMENSION). The archetypal example is EUCLIDEAN n-space $\mathbb{R}^n$, which has topological dimension n. The basic ideas leading up to this result (including the DIMENSION INVARIANCE THEOREM, DOMAIN INVARIANCE THEOREM, and LEBESGUE COVERING DIMENSION) were developed by Poincaré, Brouwer, Lebesgue, Urysohn, and Menger.

There are several branchings and extensions of the notion of topological dimension. Implicit in the notion of the LEBESGUE COVERING DIMENSION is that dimension, in a sense, is a measure of how an object fills space. If it takes up a lot of room, it is higher dimensional, and if it takes up less room, it is lower dimensional. HAUSDORFF DIMENSION (also called FRACTAL DIMENSION) is a fine tuning of this definition that allows notions of objects with dimensions other than INTEGERS. FRACTALS are objects whose HAUSDORFF DIMENSION is different from their TOPOLOGICAL DIMENSION.

The concept of dimension is also used in ALGEBRA, primarily as the dimension of a VECTOR SPACE over a FIELD. This usage stems from the fact that VECTOR SPACES over the reals were the first VECTOR SPACES to be studied, and for them, their topological dimension can be calculated by purely algebraic means as the CARDINALITY of a maximal linearly independent subset. In particular, the dimension of a SUBSPACE of $\mathbb{R}^n$ is equal to the number of LINEARLY INDEPENDENT VECTORS needed to generate it (i.e., the number of VECTORS in its BASIS). Given a transformation A of $\mathbb{R}^n$,

$$\dim[\text{Range}(A)] + \dim[\text{Null}(A)] = \dim(\mathbb{R}^n).$$

See also 4-DIMENSIONAL GEOMETRY, BASIS (VECTOR SPACE), CAPACITY DIMENSION, CODIMENSION, CORRELATION DIMENSION, EXTERIOR DIMENSION, FRACTAL DIMENSION, HAUSDORFF DIMENSION, HAUSDORFF-BESICOVITCH DIMENSION, KAPLAN-YORKE DIMENSION, KRULL DIMENSION, LEBESGUE COVERING DIMENSION, LEBESGUE DIMENSION, LYAPUNOV DIMENSION, POSET DIMENSION, Q-DIMENSION, SIMILARITY DIMENSION, TOPOLOGICAL DIMENSION

References

Abbott, E. A. *Flatland: A Romance of Many Dimensions.* New York: Dover, 1992.

Croft, H. T.; Falconer, K. J.; and Guy, R. K. *Unsolved Problems in Geometry.* New York: Springer-Verlag, p. 8, 1991.

Czyz, J. *Paradoxes of Measures and Dimensions Originating in Felix Hausdorff's Ideas.* Singapore: World Scientific, 1994.

Hinton, C. H. *The Fourth Dimension.* Pomeroy, WA: Health Research, 1993.

Manning, H. *The Fourth Dimension Simply Explained.* Magnolia, MA: Peter Smith, 1990.

Manning, H. *Geometry of Four Dimensions.* New York: Dover, 1956.

Neville, E. H. *The Fourth Dimension.* Cambridge, England: Cambridge University Press, 1921.

Rucker, R. von Bitter. *The Fourth Dimension: A Guided Tour of the Higher Universes.* Boston, MA: Houghton Mifflin, 1984.

Sommerville, D. M. Y. *An Introduction to the Geometry of N Dimensions.* New York: Dover, 1958.

Weisstein, E. W. "Books about Dimensions." http://www.treasure-troves.com/books/Dimensions.html.

Dimension Axiom

One of the EILENBERG-STEENROD AXIOMS. Let X be a single point space. $H_n(X) = 0$ unless $n = 0$, in which case $H_0(X) = 0$ where G are some GROUPS. The H_0 are called the COEFFICIENTS of the HOMOLOGY THEORY $H(\cdot)$.

See also EILENBERG-STEENROD AXIOMS, HOMOLOGY (TOPOLOGY)

Dimension Invariance Theorem

$\mathbb{R}^n$ is HOMEOMORPHIC to $\mathbb{R}^m$ IFF $n = m$. This theorem was first proved by Brouwer.

See also DOMAIN INVARIANCE THEOREM

Dimensionality

DIMENSION

Dimensionality Theorem

For a FINITE GROUP of h elements with an n_ith dimensional ith irreducible representation,

$$\sum_i n_i^2 = h$$

Diminished Polyhedron

A UNIFORM POLYHEDRON with pieces removed.

Diminished Rhombicosidodecahedron

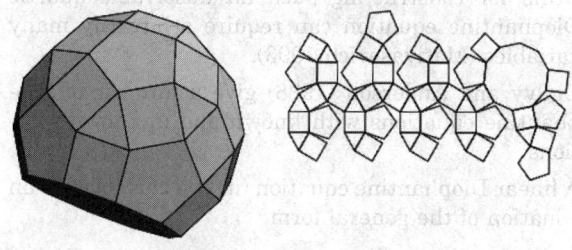

JOHNSON SOLID J_{76}.

References
Weisstein, E. W. "Johnson Solids." MATHEMATICA NOTEBOOK JOHNSONSOLIDS.M.
Weisstein, E. W. "Johnson Solid Netlib Database." MATHEMATICA NOTEBOOK JOHNSONSOLIDS.DAT.

Dini Expansion

An expansion based on the ROOTS of

$$x^{-n}[xJ_n^t(x) + HJ_n(x)] = 0,$$

where $J_n(x)$ is a BESSEL FUNCTION OF THE FIRST KIND, is called a Dini expansion.

See also BESSEL FUNCTION FOURIER EXPANSION

References
Bowman, F. *Introduction to Bessel Functions.* New York: Dover, p. 109, 1958.

Dini's Surface

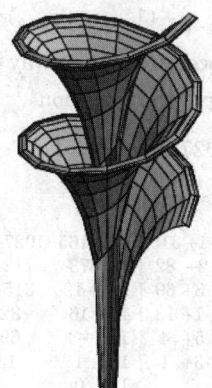

A surface of constant NEGATIVE CURVATURE obtained

by twisting a PSEUDOSPHERE and given by the PARAMETRIC EQUATIONS

$$x = a \cos u \sin v \tag{1}$$

$$y = a \sin u \sin v \tag{2}$$

$$z = a\left\{\cos v + \ln\left[\tan\left(\frac{1}{2}v\right)\right]\right\} + bu. \tag{3}$$

The above figure corresponds to $a = 1$, $b = 0.2$, $u \in [0, 4\pi]$, and $v \in (0, 2]$.

The coefficients of the FIRST FUNDAMENTAL FORM are

$$E = \frac{1}{2}\left[a^2 + 2b^2 - a^2 \cos(2v)\right] \tag{4}$$

$$F = ab \cos v \cot v \tag{5}$$

$$G = a^2 \cot^2 v, \tag{6}$$

the coefficients of the SECOND FUNDAMENTAL FORM are

$$e = -\frac{a^2 \cos v \sin v}{\sqrt{a^2 + b^2}} \tag{7}$$

$$f = \frac{ab \cos v}{\sqrt{a^2 + b^2}} \tag{8}$$

$$g = \frac{a^2 \cot v}{\sqrt{a^2 + b^2}}, \tag{9}$$

and the AREA ELEMENT is

$$dA = a\sqrt{a^2 + b^2} \cos v. \tag{10}$$

The GAUSSIAN and MEAN CURVATURES are given by

$$K = -\frac{1}{a^2 + b^2} \tag{11}$$

$$H = -\frac{\cot(2v)}{\sqrt{a^2 + b^2}}. \tag{12}$$

See also PSEUDOSPHERE

References
Gray, A. "Dini's Surface." §21.5 in *Modern Differential Geometry of Curves and Surfaces with Mathematica, 2nd ed.* Boca Raton, FL: CRC Press, pp. 493–95, 1997.
Nordstrand, T. "Dini's Surface." http://www.uib.no/people/nfytn/dintxt.htm.

Dini's Test

A test for the convergence of FOURIER SERIES. Let

$$\phi_x(t) \equiv f(x + t) + f(x - t) - 2f(x),$$

then if

$$\int_0^\pi \frac{|\phi_x(t)|dt}{t}$$

is FINITE, the FOURIER SERIES converges to $f(x)$ at x.

See also FOURIER SERIES

References

Sansone, G. *Orthogonal Functions, rev. English ed.* New York: Dover, pp. 65–8, 1991.

Dinitz Problem

Given any assignment of n-element sets to the n^2 locations of a square $n \times n$ array, is it always possible to find a PARTIAL LATIN SQUARE? The fact that such a PARTIAL LATIN SQUARE can always be found for a 2×2 array can be proven analytically, and techniques were developed which also proved the existence for 4×4 and 6×6 arrays. However, the general problem eluded solution until it was answered in the affirmative by Galvin in 1993 using results of Janssen (1993ab) and F. Maffray.

See also PARTIAL LATIN SQUARE

References

Chetwynd, A. and Häggkvist, R. "A Note on List-Colorings." *J. Graph Th.* **13**, 87–5, 1989.
Cipra, B. "Quite Easily Done." In *What's Happening in the Mathematical Sciences* **2**, pp. 41–6, 1994.
Erdos, P.; Rubin, A.; and Taylor, H. "Choosability in Graphs." *Congr. Numer.* **26**, 125–57, 1979.
Häggkvist, R. "Towards a Solution of the Dinitz Problem?" *Disc. Math.* **75**, 247–51, 1989.
Janssen, J. C. M. "The Dinitz Problem Solved for Rectangles." *Bull. Amer. Math. Soc.* **29**, 243–49, 1993a.
Janssen, J. C. M. *Even and Odd Latin Squares.* Ph.D. thesis. Lehigh University, 1993b.
Kahn, J. "Recent Results on Some Not-So-Recent Hypergraph Matching and Covering Problems." *Proceedings of the Conference on Extremal Problems for Finite Sets.* Visegràd, Hungary, 1991.
Kahn, J. "Coloring Nearly-Disjoint Hypergraphs with $n + o(n)$ Colors." *J. Combin. Th. Ser. A* **59**, 31–9, 1992.

Diocles's Cissoid

CISSOID OF DIOCLES

Diophantine Equation

An equation in which only INTEGER solutions are allowed. HILBERT'S 10TH PROBLEM asked if a technique for solving a general Diophantine existed. A general method exists for the solution of first degree Diophantine equations. However, the impossibility of obtaining a general solution was proven by Julia Robinson and Martin Davis in 1970, following proof of the result that the relation $n = F_{2m}$ (where F_{2m} is a FIBONACCI NUMBER) is Diophantine by Yuri Matiyasevich (Matiyasevich 1970, Davis 1973, Davis and Hersh 1973, Davis 1982, Matiyasevich 1993). More specifically, Matiyasevich showed that there is a polynomial P in n, m, and a number of other

variables x, y, z, ... having the property that $n = F_{2m}$ IFF there exist integers x, y, z, ... such that $P(n, m, x, y, z, \ldots) = 0$.

Jones and Matiyasevich (1982) proved that no ALGORITHMS can exist to determine if an arbitrary Diophantine equation in nine variables has solutions. As a consequence of this result, it can be proved that there does not exists a general algorithm for solving a QUARTIC DIOPHANTINE EQUATION, although the algorithm for constructing such an unsolvable quartic Diophantine equation can require arbitrarily many variables (Matiyasevich 1993).

Ogilvy and Anderson (1988) give a number of Diophantine equations with known and unknown solutions.

A linear Diophantine equation (in two variables) is an equation of the general form

$$ax + by = c, \tag{1}$$

where solutions are sought with a, b, and c INTEGERS. Such equations can be solved completely, and the first known solution was constructed by Brahmagupta. Consider the equation

$$ax + by = 1. \tag{2}$$

Now use a variation of the EUCLIDEAN ALGORITHM, letting $a = r_1$ and $b = r_2$

$$r_1 = q_1 r_2 + r_3 \tag{3}$$

$$r_2 = q_2 r_3 + r_4 \tag{4}$$

$$r_{n-3} = q_{n-3} r_{n-2} + r_{n-1} \tag{5}$$

$$r_{n-2} = q_{n-2} r_{n-1} + 1. \tag{6}$$

Starting from the bottom gives

$$1 = r_{n-2} - q_{n-2} r_{n-1} \tag{7}$$

$$r_{n-1} = r_{n-3} - q_{n-3} r_{n-2}, \tag{8}$$

so

$$1 = r_{n-2} - q_{n-2}(r_{n-3} - q_{n-3} r_{n-2})$$
$$= -q_{n-2} r_{n-3} + (1 - q_{n-2} q_{n-3}) r_{n-2}. \tag{9}$$

Continue this procedure all the way back to the top. Take as an example the equation

$$1027x + 712y = 1. \tag{10}$$

Proceed as follows.

```
1027 = 712·1+ 315 | 1 = −165·  1027+ 238·712 ↑
 712 = 315·2+  82 | 1 =  73·   712− 165·315 |
 315 =  82·3+  69 | 1 = −19·   315+  73· 82 |
  82 =  69·1+  13 | 1 =  16·    82−  19· 69 |
  69 =  13·5+   4 | 1 =  −3·    69+  16· 13 |
  13 =   4·3+   1 ↓ 1 =   1·    13−   3·  4 |
                    1 =   0·     4+   1·  1 |
```

The solution is therefore $x = -165$, $y = 238$. The

above procedure can be simplified by noting that the two left-most columns are offset by one entry and alternate signs, as they must since

$$1 = -A_{i+1}r_i + A_i r_{i+1} \tag{11}$$

$$r_{i+1} = r_{i-1} - r_i q_{i-1} \tag{12}$$

$$1 = A_i r_{i-1} - (A_i q_{i-1} + A_{i+1}), \tag{13}$$

so the COEFFICIENTS of r_{i-1} and r_{i+1} are the same and

$$A_{i-1} = -(A_i q_{i-1} + A_{i+1}). \tag{14}$$

Repeating the above example using this information therefore gives

$$
\begin{array}{llll}
1027 = & 712 \cdot 1 + 315 & |1 = -165 \cdot & 1027 + 238 \cdot 712\!\uparrow \\
712 = & 315 \cdot 2 + 82 & |1 = 73 \cdot & 712 - 165 \cdot 315| \\
315 = & 82 \cdot 3 + 69 & |1 = -19 \cdot & 315 + 73 \cdot 82 \,| \\
82 = & 69 \cdot 1 + 13 & |1 = 16 \cdot & 82 - 19 \cdot 69 \,| \\
69 = & 13 \cdot 5 + 4 & |1 = -3 \cdot & 69 + 16 \cdot 13 \,| \\
13 = & 4 \cdot 3 + 1 & |1 = 1 \cdot & 13 - 3 \cdot 4 \,| \\
1 = & 0 & & 4 + 1 \cdot 1 \,|
\end{array}
$$

and we recover the above solution.

Call the solutions to

$$ax + by = 1 \tag{15}$$

x_0 and y_0. If the signs in front of ax or by are NEGATIVE, then solve the above equation and take the signs of the solutions from the following table:

equation	x	y
$ax + by = 1$	x_0	y_0
$ax - by = 1$	x_0	$-y_0$
$-ax + by = 1$	$-x_0$	y_0
$-ax - by = 1$	$-x_0$	$-y_0$

In fact, the solution to the equation

$$ax - by = 1 \tag{16}$$

is equivalent to finding the CONTINUED FRACTION for a/b, with a and b RELATIVELY PRIME (Olds 1963). If there are n terms in the fraction, take the $(n-1)$th convergent p_{n-1}/q_{n-1}. But

$$p_n q_{n-1} - p_{n-1} q_n = (-1)^n, \tag{17}$$

so one solution is $x_0 = (-1)^n q_{n-1}$, $y_0 = (-1)^n p_{n-1}$, with a general solution

$$x = x_0 + kb \tag{18}$$

$$y = y_0 + ka \tag{19}$$

with k an arbitrary INTEGER. The solution in terms of smallest POSITIVE INTEGERS is given by choosing an appropriate k.

Now consider the general first-order equation OF THE FORM

$$ax + by = c. \tag{20}$$

The GREATEST COMMON DIVISOR $d \equiv \mathrm{GCD}(a, b)$ can be divided through yielding

$$a'x + b'y = c', \tag{21}$$

where $a' \equiv a/d$, $b' \equiv b/d$, and $c' \equiv c/d$. If $d \nmid c$, then c' is not an INTEGER and the equation cannot have a solution in INTEGERS. A necessary and sufficient condition for the general first-order equation to have solutions in INTEGERS is therefore that $d | c$. If this is the case, then solve

$$a'x + b'y = 1 \tag{22}$$

and multiply the solutions by c', since

$$a'(c'x) + b'(c'y) = c'. \tag{23}$$

D. Wilson has compiled a list of the smallest nth POWERS which are the sums of n distinct smaller nth POWERS. The first few are 3, 5, 6, 15, 12, 25, 40, ...(Sloane's A030052):

$$3^1 = 1^1 + 2^1$$
$$5^2 = 3^2 + 4^2$$
$$6^3 = 3^3 + 4^3 + 5^3$$
$$15^4 = 4^4 + 6^4 + 8^4 + 9^4 + 14^4$$
$$12^5 = 4^5 + 5^5 + 6^5 + 7^5 + 9^5 + 11^5$$
$$25^6 = 1^6 + 2^6 + 3^6 + 5^6 + 6^6 + 7^6 + 8^6 + 9^6 + 10^6$$
$$\qquad + 12^6 + 13^6 + 15^6 + 16^6 + 17^6 + 18^6 + 23^6$$
$$40^7 = 1^7 + 3^7 + 5^7 + 9^7 + 12^7 + 14^7 + 16^7 + 17^7$$
$$\qquad + 18^7 + 20^7 + 21^7 + 22^7 + 25^7 + 28^7 + 39^7$$
$$84^8 = 1^8 + 2^8 + 3^8 + 5^8 + 7^8 + 9^8 + 10^8 + 11^8$$
$$\qquad + 12^8 + 13^8 + 14^8 + 15^8 + 16^8 + 17^8 + 18^8$$
$$\qquad + 19^8 + 21^8 + 23^8 + 24^8 + 25^8 + 26^8 + 27^8$$
$$\qquad + 29^8 + 32^8 + 33^8 + 35^8 + 37^8 + 38^8 + 39^8$$
$$\qquad + 41^8 + 42^8 + 43^8 + 45^8 + 46^8 + 47^8 + 48^8$$
$$\qquad + 49^8 + 51^8 + 52^8 + 53^8 + 57^8 + 58^8 + 59^8$$
$$\qquad + 61^8 + 63^8 + 69^8 + 73^8$$
$$47^9 = 1^9 + 2^9 + 4^9 + 7^9 + 11^9 + 14^9 + 15^9 + 18^9$$
$$\qquad + 26^9 + 27^9 + 30^9 + 31^9 + 32^9 + 33^9$$
$$\qquad + 36^9 + 38^9 + 39^9 + 43^9$$
$$63^{10} = 1^{10} + 2^{10} + 4^{10} + 5^{10} + 6^{10} + 8^{10} + 12^{10}$$
$$\qquad + 15^{10} + 16^{10} + 17^{10} + 20^{10} + 21^{10} + 25^{10}$$
$$\qquad + 26^{10} + 27^{10} + 28^{10} + 30^{10} + 36^{10} + 37^{10}$$
$$\qquad + 38^{10} + 40^{10} + 51^{10} + 62^{10}.$$

See also ABC CONJECTURE, ARCHIMEDES' CATTLE PROBLEM, BACHET EQUATION, BRAHMAGUPTA'S PROBLEM, CANNONBALL PROBLEM, CATALAN'S PROBLEM, DIOPHANTINE EQUATION–2ND POWERS, DIOPHANTINE EQUATION–3RD POWERS, DIOPHANTINE EQUATION–4TH POWERS, DIOPHANTINE EQUATION–5TH POWERS, DIOPHANTINE EQUATION–6TH POWERS, DIOPHANTINE EQUATION–7TH POWERS, DIOPHANTINE EQUATION–8TH POWERS, DIOPHANTINE EQUATION–9TH POWERS,

DIOPHANTINE EQUATION–10TH POWERS, DIOPHANTINE EQUATION *N*TH POWERS, DIOPHANTUS PROPERTY, EULER BRICK, EULER QUARTIC CONJECTURE, FERMAT'S LAST THEOREM, FERMAT ELLIPTIC CURVE THEOREM, GENUS THEOREM, HURWITZ EQUATION, MARKOV NUMBER, MONKEY AND COCONUT PROBLEM, MULTIGRADE EQUATION, *P*-ADIC NUMBER, PELL EQUATION, PYTHAGOREAN QUADRUPLE, PYTHAGOREAN TRIPLE, THUE EQUATION

References

Bashmakova, I. G. *Diophantus and Diophantine Equations.* Washington, DC: Math. Assoc. Amer., 1997.

Beiler, A. H. *Recreations in the Theory of Numbers: The Queen of Mathematics Entertains.* New York: Dover, 1966.

Carmichael, R. D. *The Theory of Numbers, and Diophantine Analysis.* New York: Dover, 1959.

Chen, S. "Equal Sums of Like Powers: On the Integer Solution of the Diophantine System." http://www.nease.net/~chin/eslp/.

Chen, S. "References." http://www.nease.net/~chin/eslp/referenc.htm.

Courant, R. and Robbins, H. "Continued Fractions. Diophantine Equations." §2.4 in Supplement to Ch. 1 in *What is Mathematics?: An Elementary Approach to Ideas and Methods, 2nd ed.* Oxford, England: Oxford University Press, pp. 49–1, 1996.

Davis, M. "Hilbert's Tenth Problem is Unsolvable." *Amer. Math. Monthly* **80**, 233–69, 1973.

Davis, M. and Hersh, R. "Hilbert's 10th Problem." *Sci. Amer.* **229**, 84–1, Nov. 1973.

Davis, M. "Hilbert's Tenth Problem is Unsolvable." Appendix 2 in *Computability and Unsolvability.* New York: Dover, 1999–35, 1982.

Dickson, L. E. "Linear Diophantine Equations and Congruences." Ch. 2 in *History of the Theory of Numbers, Vol. 2: Diophantine Analysis.* New York: Chelsea, pp. 41–9, 1952.

dmoz. "Equal Sums of Like Powers." http://dmoz.org/Science/Math/Number_Theory/Diophantine_Equations/Equal_Sums_of_Like_Powers/.

Dörrie, H. "The Fermat-Gauss Impossibility Theorem." §21 in *100 Great Problems of Elementary Mathematics: Their History and Solutions.* New York: Dover, pp. 96–04, 1965.

Ekl, R. L. "New Results in Equal Sums of Like Powers." *Math. Comput.* **67**, 1309–315, 1998.

Guy, R. K. "Diophantine Equations." Ch. D in *Unsolved Problems in Number Theory, 2nd ed.* New York: Springer-Verlag, pp. 139–98, 1994.

Hardy, G. H. and Wright, E. M. *An Introduction to the Theory of Numbers, 5th ed.* Oxford, England: Clarendon Press, 1979.

Hunter, J. A. H. and Madachy, J. S. "Diophantos and All That." Ch. 6 in *Mathematical Diversions.* New York: Dover, pp. 52–4, 1975.

Ireland, K. and Rosen, M. "Diophantine Equations." Ch. 17 in *A Classical Introduction to Modern Number Theory, 2nd ed.* New York: Springer-Verlag, pp. 269–96, 1990.

Jones, J. P. and Matiyasevich, Yu. V. "Exponential Diophantine Representation of Recursively Enumerable Sets." *Proceedings of the Herbrand Symposium, Marseilles, 1981.* Amsterdam, Netherlands: North-Holland, pp. 159–77, 1982.

Lang, S. *Introduction to Diophantine Approximations, 2nd ed.* New York: Springer-Verlag, 1995.

Matiyasevich, Yu. V. "Solution of the Tenth Problem of Hilbert." *Mat. Lapok* **21**, 83–7, 1970.

Matiyasevich, Yu. V. *Hilbert's Tenth Problem.* Cambridge, MA: MIT Press, 1993. http://www.informatik.uni-stuttgart.de/ifi/ti/personen/Matiyasevich/H10Pbook/.

Meyrignac, J.-C. "Computing Minimal Equal Sums of Like Powers." http://euler.free.fr/.

Mordell, L. J. *Diophantine Equations.* New York: Academic Press, 1969.

Nagell, T. "Diophantine Equations of First Degree." §10 in *Introduction to Number Theory.* New York: Wiley, pp. 29–2, 1951.

Ogilvy, C. S. and Anderson, J. T. "Diophantine Equations." Ch. 6 in *Excursions in Number Theory.* New York: Dover, pp. 65–3, 1988.

Olds, C. D. Ch. 2 in *Continued Fractions.* New York: Random House, 1963.

Sloane, N. J. A. Sequences A030052 in "An On-Line Version of the Encyclopedia of Integer Sequences." http://www.research.att.com/~njas/sequences/eisonline.html.

Weisstein, E. W. "Like Powers." MATHEMATICA NOTEBOOK LIKEPOWERS.M.

Weisstein, E. W. "Books about Diophantine Equations." http://www.treasure-troves.com/books/DiophantineEquations.html.

Diophantine Equation—10th Powers

The 10.1.2 equation

$$A^{10} = B^{10} + C^{10} \tag{1}$$

is a special case of FERMAT'S LAST THEOREM with $n = 10$, and so has no solution. The smallest 10.1.15 solution is

$$100^{10} + 94^{10} + 91^{10} + 2 \cdot 77^{10} + 76^{10} + 63^{10} + 62^{10} + 52^{10}$$

$$+ 45^{10} + 35^{10} + 33^{10} + 16^{10} + 10^{10} + 1^{10} = 108^{10} \tag{2}$$

(J.-C. Meyrignac 1999, PowerSum). The smallest 10.1.22 solution is

$$33^{10} = 2 \cdot 30^{10} + 2 \cdot 26^{10} + 23^{10} + 21^{10} + 19^{10} + 18^{10}$$

$$+ 2 \cdot 13^{10} + 2 \cdot 12^{10} + 5 \cdot 10^{10} + 2 \cdot 9^{10} + 7^{10} + 6^{10} + 3^{10} \tag{3}$$

(Ekl 1998). The smallest 10.1.23 solution is

$$5 \cdot 1^{10} + 2^{10} + 3^{10} + 6^{10} + 6 \cdot 7^{10} + 4 \cdot 9^{10}$$

$$+ 10^{10} + 2 \cdot 12^{10} + 13^{10} + 14^{10} = 15^{10} \tag{4}$$

(Lander *et al.* 1967).

The smallest 10.2.13 solution is

$$51^{10} + 32^{10} = 49^{10} + 43^{10} + 41^{10} + 37^{10} + 28^{10} + 26^{10}$$

$$+ 25^{10} + 15^{10} + 10^{10} + 109^{10} + 5^{10} + 3^{10}. \tag{5}$$

The smallest 10.2.15 solution is

$$35^{10} + 3^{10} = 33^{10} + 32^{10} + 24^{10} + 21^{10} + 2 \cdot 20^{10}$$

$$+ 3 \cdot 13^{10} + 12^{10} + 11^{10} + 9^{10} + 7^{10} + 2 \cdot 1^{10} \tag{6}$$

(Ekl 1998). The smallest 10.2.19 solution is

$$5 \cdot 2^{10} + 5^{10} + 6^{10} + 10^{10} + 6 \cdot 11^{10}$$

$$+2 \cdot 12^{10} + 3 \cdot 15^{10} = 9^{10} + 17^{10} \qquad (7)$$

(Lander *et al.* 1967).

The smallest 10.3.13 solution is

$$46^{10} + 32^{10} + 22^{10}$$
$$= 43^{10} + 43^{10} + 27^{10} + 26^{10} + 17^{10} + 16^{10}$$
$$+ 12^{10} + 9^{10} + 9^{10} + 6^{10} + 4^{10} + 3^{10} + 3^{10}. \qquad (8)$$

The smallest 10.3.14 solution is

$$30^{10} + 28^{10} + 4^{10} = 31^{10} + 23^{10} + 2 \cdot 20^{10} + 2 \cdot 17^{10}$$
$$+ 16^{10} + 10^{10} + 3 \cdot 9^{10} + 5^{10} + 2 \cdot 2^{10} \qquad (9)$$

(Ekl 1998). The smallest 10.3.24 solution is

$$1^{10} + 2^{10} + 3^{10} + 10 \cdot 4^{10} + 7^{10} + 7 \cdot 8^{10}$$
$$+ 10^{10} + 12^{10} + 16^{10} = 11^{10} + 2 \cdot 15^{10} \qquad (10)$$

(Lander *et al.* 1967).

The 10.4.12 equation has solution

$$51^{10} + 49^{10} + 43^{10} + 39^{10} + 29^{10} + 28^{10} + 2 \cdot 17^{10}$$
$$+ 16^{10} + 13^{10} + 7^{10} + 4^{10} = 53^{10} + 244^{10} + 22^{10} \qquad (11)$$

(E. Bainville 1999, PowerSum). The smallest 10.4.15 solution is

$$4 \cdot 23^{10} = 26^{10} + 5 \cdot 18^{10} + 3 \cdot 17^{10} + 15^{10} + 12^{10} + 6^{10}$$
$$+ 3 \cdot 4^{10} \qquad (12)$$

(Ekl 1998). The smallest 10.4.23 solution is

$$5 \cdot 1^{10} + 2 \cdot 2^{10} + 2 \cdot 3^{10} + 4^{10} + 4 \cdot 6^{10} + 3 \cdot 7^{10} + 8^{10}$$
$$+ 2 \cdot 10^{10} + 2 \cdot 14^{10} + 15^{10} = 3 \cdot 11^{10} + 16^{10} \qquad (13)$$

(Lander *et al.* 1967).

The smallest 10.5.16 solutions are

$$4 \cdot 1^{10} + 2^{10} + 2 \cdot 4^{10} + 6^{10} + 2 \cdot 12^{10}$$
$$+ 5 \cdot 13^{10} + 15^{10} = 2 \cdot 3^{10} + 8^{10} + 14^{10} + 16^{10} \qquad (14)$$

$$20^{10} + 11^{10} + 8^{10} + 3^{10} + 1^{10} = 2 \cdot 18^{10} + 17^{10}$$
$$+ 16^{10} + 10^{10} + 2 \cdot 7^{10} + 6 \cdot 4^{10} + 2 \cdot 2^{10} \qquad (15)$$

(Lander *et al.* 1967, Ekl 1998).

The smallest 10.6.6 solution is

$$95^{10} + 71^{10} + 32^{10} + 28^{10} + 25^{10} + 16^{10}$$
$$= 92^{10} + 85^{10} + 34^{10} + 34^{10} + 23^{10} + 5^{10}. \qquad (16)$$

The smallest 10.6.16 solution is

$$18^{10} + 12^{10} + 11^{10} + 10^{10} + 3^{10} + 2^{10}$$
$$= 17^{10} + 16^{10} + 4 \cdot 13^{10} + 4 \cdot 7^{10} + 4 \cdot 6^{10} + 5^{10} + 4^{10} \qquad (17)$$

(Ekl 1998). The smallest 10.6.27 solution is

$$1^{10} + 4 \cdot 3^{10} + 2 \cdot 4^{10} + 2 \cdot 5^{10} + 7 \cdot 6^{10}$$
$$+ 9 \cdot 7^{10} + 10^{10} + 13^{10} = 2 \cdot 2^{10} + 8^{10} + 11^{10} + 2 \cdot 12^{10} \qquad (18)$$

(Lander *et al.* 1967).

The smallest 10.7.7 solutions are

$$38^{10} + 33^{10} + 26^{10} + 26^{10} + 15^{10} + 8^{10} + 1^{10}$$
$$= 36^{10} + 35^{10} + 32^{10} + 29^{10} + 24^{10} + 23^{10} + 22^{10} \qquad (19)$$

$$68^{10} + 61^{10} + 55^{10} + 32^{10} + 31^{10} + 28^{10} + 1^{10}$$
$$= 67^{10} + 64^{10} + 49^{10} + 44^{10} + 23^{10} + 20^{10} + 17^{10} \qquad (20)$$

(Lander *et al.* 1967, Ekl 1998).

References

Ekl, R. L. "New Results in Equal Sums of Like Powers." *Math. Comput.* **67**, 1309–315, 1998.

Lander, L. J.; Parkin, T. R.; and Selfridge, J. L. "A Survey of Equal Sums of Like Powers." *Math. Comput.* **21**, 446–59, 1967.

PowerSum. "Index of Equal Sums of Like Powers." http://www.chez.com/powersum/.

Weisstein, E. W. "Like Powers." MATHEMATICA NOTEBOOK LIKEPOWERS.M.

Diophantine Equation—2nd Powers

A general quadratic Diophantine equation in two variables x and y is given by

$$ax^2 + cy^2 = k, \qquad (1)$$

where a, c, and k are specified (positive or negative) integers and x and y are unknown integers satisfying the equation whose values are sought. The slightly more general second-order equation

$$ax^2 + bxy + cy^2 = k \qquad (2)$$

is one of the principal topics in Gauss's *Disquisitiones arithmeticae*. According to Itô (1987), equation (2) can be solved completely using solutions to the PELL EQUATION. In particular, all solutions of

$$ax^2 + bxy + cy^2 = 1 \qquad (3)$$

are among the CONVERGENTS of the CONTINUED FRACTIONS of the roots of $ax^2 + bx + c$. In *Mathematica* 5.0, solution to the general bivariate quadratic Diophantine equation will be implemented as Reduce[*eqn* && Element[*x* | *y*, Integers], {*x*, *y*}].

For quadratic Diophantine equations in more than two variables, there exist additional deep results due to C. L. Siegel.

An equation OF THE FORM

$$x^2 - Dy^2 = 1, \qquad (4)$$

where D is an INTEGER is a very special type of equation called a PELL EQUATION. Pell equations, as well as the analogous equation with a minus sign on the right, can be solved by finding the CONTINUED FRACTION for $\sqrt{D}$. The more complicated equation

$$x^2 - Dy^2 = c \tag{5}$$

can also be solved for certain values of c and D, but the procedure is more complicated (Chrystal 1961). However, if a single solution to (5) is known, other solutions can be found using the standard technique for the PELL EQUATION.

The following table summarizes possible representation of primes p of given forms, where x and y are positive integers. No odd primes other than those indicated share these properties (Nagell 1951, p. 188).

form	congruence for p
$x^2 + y^2$	$\equiv 1 \pmod{4}$
$x^2 + 2y^2$	$\equiv 1, 3 \pmod{8}$
$x^2 + 3y^2$	$\equiv 1 \pmod{6}$
$x^2 + 7y^2$	$\equiv 1, 9, 11 \pmod{14}$
$2x^2 + 3y^2$	$\equiv 5, 11 \pmod{24}$

As a part of the study of WARING'S PROBLEM, it is known that every positive integer is a sum of no more than 4 positive squares ($g(2) = 4$; LAGRANGE'S FOUR-SQUARE THEOREM), that every "sufficiently large" integer is a sum of no more than 4 positive squares ($G(2) = 4$), and that every integer is a sum of at most 3 signed squares ($eg(2) = 3$). If zero is counted as a square, both POSITIVE and NEGATIVE numbers are included, and the order of the two squares is distinguished, Jacobi showed that the number of ways a number can be written as the sum of two squares (the $r_2(n)$ function) is four times the excess of the number of DIVISORS of the form $4x + 1$ over the number of DIVISORS OF THE FORM $4x - 1$.

In 1769 Euler (1862) noted the identity

$$\alpha b(apr \pm \beta qs)^2 + \alpha\beta(\alpha ps \mp bqr)^2 = (\alpha\alpha p^2 + b\beta q^2)(abr^2 + \alpha\beta s^2), \tag{6}$$

which gives a parametric solution to the equation

$$Ax^2 + By^2 = C \tag{7}$$

for integers A, B, C, x, y with C composite (Dickson 1957, p. 407).

Call a Diophantine equation consisting of finding a sum of m kth POWERS which is equal to a sum of n kth POWERS a "$k.m.n$ equation." The 2.1.2 quadratic Diophantine equation

$$A^2 = B^2 + C^2, \tag{8}$$

corresponds to finding a PYTHAGOREAN TRIPLE (A, B, C) has a well-known general solution (Dickson 1966, pp. 165–70). To solve the equation, note that every PRIME OF THE FORM $4x + 1$ can be expressed as the sum of two RELATIVELY PRIME squares in exactly one way. A set of INTEGERS satisfying the 2.1.3 equation

$$A^2 = B^2 + C^2 + D^2 \tag{9}$$

is called a PYTHAGOREAN QUADRUPLE.

Parametric solutions to the 2.2.2 equation

$$A^2 + B^2 = C^2 + D^2 \tag{10}$$

are known (Dickson 1966; Guy 1994, p. 140). To find in how many ways a general number m can be expressed as a sum of two squares, factor it as follows

$$m = 2^{a_0} p_1^{2a_1} \cdots p_n^{2a_n} q_1^{b_1} \cdots q_r^{b_r}, \tag{11}$$

where the ps are primes OF THE FORM $4x - 1$ and the qs are primes OF THE FORM $x + 1$. If the as are integral, then define

$$B \equiv (2b_1 + 1)(2b_2 + 1) \cdots (2b_r + 1) - 1. \tag{12}$$

Then m is a sum of two *unequal* squares in

$$N(m) = \begin{cases} 0 \\ \quad \text{for any } a_i \text{ half-integral} \\ \frac{1}{2}(b_1 + 1)(b_2 + 1) \cdots (b_r + 1) \\ \quad \text{for all } a_i \text{ integral}, B \text{ odd} \\ \frac{1}{2}(b_1 + 1)(b_2 + 2) \cdots (b_r + 1) - \frac{1}{2} \\ \quad \text{for all } a_i \text{ integral}, B \text{ even.} \end{cases} \tag{13}$$

Solutions to an equation OF THE FORM

$$(A^2 + B^2)(C^2 + D^2) = E^2 + F^2 \tag{14}$$

are given by the FIBONACCI IDENTITY

$$(a^2 + b^2)(c^2 + d^2) = (ac \pm bd)^2 + (bc \mp ad)^2$$
$$\equiv e^2 + f^2. \tag{15}$$

Another similar identity is the EULER FOUR-SQUARE IDENTITY

$$(a_1^2 + a_2^2)(b_1^2 + b_2^2)(c_1^2 + c_2^2)(d_1^2 + d_2^2)$$
$$= e_1^2 + e_2^2 + e_3^2 + e_4^2 \tag{16}$$

$$(a_1^2 + a_2^2 + a_3^2 + a_4^2)(b_1^2 + b_2^2 + b_3^2 + b_4^2)$$
$$= (a_1 b_1 - a_2 b_2 - a_3 b_3 - a_4 b_4)^2$$
$$+ (a_1 b_2 + a_2 b_1 + a_3 b_4 - a_4 b_3)^2$$
$$+ (a_1 b_3 - a_2 b_4 + a_3 b_1 + a_4 b_2)^2$$
$$+ (a_1 b_4 + a_2 b_3 - a_3 b_2 + a_4 b_1)^2. \tag{17}$$

Degen's eight-square identity holds for eight squares, but no other number, as proved by Cayley. The two-square identity underlies much of TRIGONOMETRY, the four-square identity some of QUATERNIONS, and the eight-square identity, the CAYLEY ALGEBRA (a noncommutative nonassociative algebra; Bell 1945).

Chen Shuwen found the 2.6.6 equation

$$87^2 + 233^2 + 264^2 + 396^2 + 496^2 + 540^2$$

$$= 90^2 + 206^2 + 309^2 + 366^2 + 522^2 + 523^2. \quad (18)$$

RAMANUJAN'S SQUARE EQUATION

$$2^n - 7 = x^2 \quad (19)$$

has been proved to have only solutions $n = 3, 4, 5, 7$, and 15 (Schroeppel 1972).

See also ALGEBRA, CANNONBALL PROBLEM, CONTINUED FRACTION, EULER FOUR-SQUARE IDENTITY, FERMAT DIFFERENCE EQUATION, GENUS THEOREM, HILBERT SYMBOL, LAGRANGE NUMBER (DIOPHANTINE EQUATION), LEBESGUE IDENTITY, PELL EQUATION, PYTHAGOREAN QUADRUPLE, PYTHAGOREAN TRIPLE, QUADRATIC RESIDUE, SQUARE NUMBER, SUM OF SQUARES FUNCTION, WARING'S PROBLEM

References

Beiler, A. H. "The Pellian." Ch. 22 in *Recreations in the Theory of Numbers: The Queen of Mathematics Entertains.* New York: Dover, pp. 248–68, 1966.

Bell, E. T. *The Development of Mathematics, 2nd ed.* New York: McGraw-Hill, p. 159, 1945.

Chrystal, G. *Textbook of Algebra, 2 vols.* New York: Chelsea, 1961.

Degan, C. F. *Canon Pellianus.* Copenhagen, Denmark, 1817.

Dickson, L. E. "Number of Representations as a Sum of 5, 6, 7, or 8 Squares." Ch. 13 in *Studies in the Theory of Numbers.* Chicago, IL: University of Chicago Press, 1930.

Dickson, L. E. "Pell Equation; $ax^2 + bx + c$ Made a Square" and "Further Single Equations of the Second Degree." Chs. 12–3 in *History of the Theory of Numbers, Vol. 2: Diophantine Analysis.* New York: Chelsea, pp. 341–34, 1966.

Guy, R. K. *Unsolved Problems in Number Theory, 2nd ed.* New York: Springer-Verlag, 1994.

Itô, K. (Ed.). *Encyclopedic Dictionary of Mathematics, 2nd ed, Vol. 1.* Cambridge, MA: MIT Press, p. 450, 1987.

Lam, T. Y. *The Algebraic Theory of Quadratic Forms.* Reading, MA: W. A. Benjamin, 1973.

Nagell, T. "Diophantine Equations of the Second Degree." Ch. 6 in *Introduction to Number Theory.* New York: Wiley, pp. 188–26, 1951.

Rajwade, A. R. *Squares.* Cambridge, England: Cambridge University Press, 1993.

Scharlau, W. *Quadratic and Hermitian Forms.* Berlin: Springer-Verlag, 1985.

Schroeppel, R. Item 31 in Beeler, M.; Gosper, R. W.; and Schroeppel, R. *HAKMEM.* Cambridge, MA: MIT Artificial Intelligence Laboratory, Memo AIM-239, p. 14, Feb. 1972.

Shapiro, D. B. "Products of Sums and Squares." *Expo. Math.* **2**, 235–61, 1984.

Smarandache, F. "Un metodo de resolucion de la ecuacion diofantica." *Gaz. Math.* **1**, 151–57, 1988.

Smarandache, F. "Method to Solve the Diophantine Equation $ax^2 - by^2 + c = 0$." In *Collected Papers, Vol. 1.* Bucharest, Romania: Tempus, 1996.

Taussky, O. "Sums of Squares." *Amer. Math. Monthly* **77**, 805–30, 1970.

Whitford, E. E. *Pell Equation.* New York: Columbia University Press, 1912.

Diophantine Equation—3rd Powers

As a part of the study of WARING'S PROBLEM, it is known that every positive integer is a sum of no more than 9 positive cubes ($g(3) = 9$), that every "sufficiently large" integer is a sum of no more than 7 positive cubes ($G(3) \leq 7$; although it is not known if 7 can be reduced), and that every integer is a sum of at most 5 signed cubes ($eg(3) \leq 5$; although it is not known if 5 can be reduced to 4).

It is known that every n can be written is the form

$$n = A^2 + B^2 - C^3. \quad (1)$$

The 3.1.2 equation

$$A^3 = B^3 + C^3 \quad (2)$$

is a case of FERMAT'S LAST THEOREM with $n = 3$. In fact, this particular case was known not to have any solutions long before the general validity of FERMAT'S LAST THEOREM was established. Thue showed that a Diophantine equation OF THE FORM

$$AX^3 - BY^3 = 1 \quad (3)$$

for A, B, and l integers, has only finite many solutions (Hardy 1999, pp. 78–9).

Miller and Woollett (1955) and Gardiner *et al.* (1964) investigated integer solutions of

$$A^3 + B^3 + C^3 = D \quad (4)$$

i.e., numbers representable as the sum of three (positive or negative) CUBIC NUMBERS.

The general rational solution to the 3.1.3 equation

$$A^3 = B^3 + C^3 + D^3 \quad (5)$$

was found by Euler and Vieta (Dickson 1966, pp. 550–54; Hardy 1999, pp. 20–1). Hardy and Wright (1979, pp. 199–01) give a solution which can be based on the identities

$$a^3(a^3 + b^3)^3$$

$$= b^3(a^3 + b^3)^3 + a^3(a^3 - 2b^3)^3 + b^3(2a^3 + b^3)^3 \quad (6)$$

$$a^3(a^3 + 2b^3)^3$$

$$= a^3(a^3 - b^3)^3 + b^3(a^3 - b^3)^3 + (2a^3 + b^3)^3. \quad (7)$$

This is equivalent to the general 3.2.2 solution found by Ramanujan (Dickson 1966, pp. 500 and 554; Berndt 1994, pp. 54 and 107; Hardy 1999, p. 11, 68, and 237). The smallest integer solutions are

$$3^3 + 4^3 + 5^3 = 6^3 \tag{8}$$

$$1^3 + 6^3 + 8^3 = 9^3 \tag{9}$$

$$7^3 + 14^3 + 17^3 = 20^3 \tag{10}$$

$$11^3 + 15^3 + 27^3 = 29^3 \tag{11}$$

$$28^3 + 53^3 + 75^3 = 84^3 \tag{12}$$

$$26^3 + 55^3 + 78^3 = 87^3 \tag{13}$$

$$33^3 + 70^3 + 92^3 = 105^3 \tag{14}$$

(Fredkin 1972; Madachy 1979, pp. 124 and 141). Other general solutions have been found by Binet (1841) and Schwering (1902), although Ramanujan's formulation is the simplest. No general solution giving *all* POSITIVE integral solutions is known (Dickson 1966, pp. 550–61). Y. Kohmoto has found a $3.1.3^9$ solution,

$$\begin{aligned} 2100000^3 &= 2046000^3 + 882000^3 + 216000^3 \\ &= 1979600^3 + 1145400^3 + 85000^3 \\ &= 2081100^3 + 628110^3 + 1890^3 \\ &= 2043150^3 + 901200^3 + 30450^3 \\ &= 2002280^3 + 1072480^3 + 30360^3 \\ &= 1960480^3 + 1199520^3 + 15200^3 \\ &= 1948800^3 + 1229760^3 + 30240^3 \\ &= 2078160^3 + 658812^3 + 13188^3 \\ &= 2009112^3 + 1048040^3 + 13888^3. \end{aligned} \tag{15}$$

3.1.4 equations include

$$11^3 + 12^3 + 13^3 + 14^3 = 20^3 \tag{16}$$

$$5^3 + 7^3 + 9^3 + 10^3 = 13^3. \tag{17}$$

3.1.5 equations include

$$1^3 + 3^3 + 4^3 + 5^3 + 8^3 = 9^3 \tag{18}$$

$$3^3 + 4^3 + 5^3 + 8^3 + 10^3 = 12^3, \tag{19}$$

and a 3.1.6 equation is given by

$$1^3 + 5^3 + 6^3 + 7^3 + 8^3 + 10^3 = 13^3. \tag{20}$$

The 3.2.2 equation

$$A^3 + B^3 = C^3 + D^3 \tag{21}$$

has a known parametric solution (Dickson 1966, pp. 550–54; Guy 1994, p. 140), and 10 solutions with sum $< 10^5$,

$$1729 = 1^3 + 12^3 + = 9^3 + 10^3 \tag{22}$$

$$4104 = 2^3 + 16^3 = 9^3 + 15^3 \tag{23}$$

$$13832 = 2^3 + 24^3 = 18^3 + 20^3 \tag{24}$$

$$20683 = 10^3 + 27^3 = 19^3 + 24^3 \tag{25}$$

$$32832 = 4^3 + 32^3 = 18^3 + 30^3 \tag{26}$$

$$39312 = 2^3 + 34^3 = 15^3 + 33^3 \tag{27}$$

$$40033 = 9^3 + 34^3 = 16^3 + 33^3 \tag{28}$$

$$46683 = 3^3 + 36^3 = 27^3 + 30^3 \tag{29}$$

$$64232 = 17^3 + 39^3 = 26^3 + 36^3 \tag{30}$$

$$65728 = 12^3 + 40^3 = 31^3 + 33^3 \tag{31}$$

(Sloane's A001235; Moreau 1898). The first number (Madachy 1979, pp. 124 and 141) in this sequence, the so-called HARDY-RAMANUJAN NUMBER, is associated with a story told about Ramanujan by G. H. Hardy, but was known as early as 1657 (Berndt and Bhargava 1993). The smallest number representable in n ways as a sum of cubes is called the nth TAXICAB NUMBER.

Ramanujan gave a general solution to the 3.2.2 equation as

$$\left(\alpha + \lambda^2 \gamma\right)^3 + \left(\lambda \beta + \gamma\right)^3 = \left(\lambda \alpha + \gamma\right)^3 + \left(\beta + \lambda^2 \gamma\right)^3 \tag{32}$$

where

$$\alpha^2 + \alpha\beta + \beta^2 = 3\lambda\gamma^2 \tag{33}$$

(Berndt 1994, p. 107). Another form due to Ramanujan is

$$\left(A^2 + 7AB - 9B^2\right)^3 + \left(2A^2 - 4AB + 12B^2\right)^3$$
$$= \left(2A^2 + 10B^2\right)^3 + \left(A^2 - 9AB - B^2\right)^3. \tag{34}$$

Hardy and Wright (1979, Theorem 412) prove that there are numbers that are expressible as the sum of two cubes in n ways for any n (Guy 1994, pp. 140–41). The proof is constructive, providing a method for computing such numbers: given RATIONALS NUMBERS r and s, compute

$$t = \frac{r(r^3 + 2s^3)}{r^3 - s^3} \tag{35}$$

$$u = \frac{s(2r^3 + s^3)}{r^3 - s^3} \tag{36}$$

$$v = \frac{t(t^3 - 2u^3)}{t^3 + u^3} \tag{37}$$

$$w = \frac{u(2t^3 - u^3)}{t^3 + u^3}. \tag{38}$$

Then

$$r^3 + s^3 = t^3 - u^3 = v^3 + w^3 \tag{39}$$

The DENOMINATORS can now be cleared to produce an integer solution. If r/s is picked to be large enough, the v and w will be POSITIVE. If r/s is still larger, the v/w will be large enough for v and w to be used as the inputs to produce a third pair, etc. However, the resulting integers may be quite large, even for $n = 2$. E.g., starting with $3^3 + 1^3 = 28$, the algorithm finds

$$28 = \left(\frac{28340511}{21446828}\right)^3 + \left(\frac{63284705}{21446828}\right)^3, \quad (40)$$

giving

$$28 \cdot 21446828^3 = (3 \cdot 21446828)^3 + 21446828^3 \quad (41)$$

$$= 28340511^3 + 63284705^3. \quad (42)$$

The numbers representable in three ways as a sum of two cubes (a 3.2^3 equation) are

$$87539319 = 167^3 + 436^3 = 228^3 + 423^3$$
$$= 255^3 + 414^3 \quad (43)$$

$$119824488 = 11^3 + 493^3 = 90^3 + 492^3$$
$$= 346^3 + 428^3 \quad (44)$$

$$143604279 = 111^3 + 522^3 = 359^3 + 460^3$$
$$= 408^3 + 423^3 \quad (45)$$

$$175959000 = 70^3 + 560^3 = 198^3 + 552^3$$
$$= 315^3 + 525^3 \quad (46)$$

$$327763000 = 300^3 + 670^3 = 339^3 + 661^3$$
$$= 510^3 + 580^3. \quad (47)$$

(Guy 1994, Sloane's A003825). Wilson (1997) found 32 numbers representable in four ways as the sum of two cubes (a 3.2^4 equation). The first is

$$6963472309248 = 2421^3 + 19083^3 = 5436^2 + 18948^3$$
$$= 10200^3 + 18072^3 = 13322^3 + 16630^3. \quad (48)$$

The smallest known numbers so representable are 6963472309248, 12625136269928, 21131226514944, 26059452841000, ... (Sloane's A003826). Wilson also found six five-way sums,

$$48988659276962496 = 38787^3 + 365757^3$$
$$= 107839^3 + 362753^3$$
$$= 205292^3 + 342952^3$$
$$= 221424^3 + 336588^3$$
$$= 231518^3 + 331954^3 \quad (49)$$

$$490593422681271000 = 48369^3 + 788631^3$$
$$= 233775^3 + 781785^3$$
$$= 285120^3 + 776070^3$$
$$= 543145^3 + 691295^3$$

$$= 579240^3 + 666630^3 \quad (50)$$

$$6355491080314102272 = 103113^3 + 1852215^3$$
$$= 580488^3 + 1833120^3$$
$$= 788724^3 + 1803372^3$$
$$= 1150792^3 + 1690544^3$$
$$= 1462050^3 + 1478238^3 \quad (51)$$

$$27365551142421413376 = 167751^3 + 3013305^3$$
$$= 265392^3 + 3012792^3$$
$$= 944376^3 + 2982240^3$$
$$= 1283148^3 + 2933844^3$$
$$= 1872184^3 + 2750288^3 \quad (52)$$

$$119996286021987046963 2 = 591543^3 + 10625865^3$$
$$= 935856^3 + 10624056^3$$
$$3330168^3 + 10516320^3$$
$$= 6601912^3 + 9698384^3$$
$$= 8387550^3 + 8480418^3 \quad (53)$$

$$111549833098123426841016 = 1074073^3 + 48137999^3$$
$$= 8787870^3 + 48040356^3$$
$$= 13950972^3 + 47744382^3$$
$$= 24450192^3 + 45936462^3$$
$$= 33784478^3 + 41791204^3, \quad (54)$$

and a single six-way sum

$$82305452582480915512058 88$$
$$= 11239317^3 + 201891435^3$$
$$= 17781264^3 + 201857064^3$$
$$= 63273192^3 + 199810080^3$$
$$= 85970916^3 + 196567548^3$$
$$= 125436328^3 + 184269296^3$$
$$= 159363450^3 + 161127942^3. \quad (55)$$

A solution to the 3.4.4 equation is

$$2^3 + 3^3 + 10^3 + 11^3 = 1^3 + 5^3 + 8^3 + 12^3 \quad (56)$$

(Madachy 1979, pp. 118 and 133).

3.6.6 equations also exist:

$$1^3 + 2^3 + 4^3 + 8^3 + 9^3 + 12^3$$
$$= 3^3 + 5^3 + 6^3 + 7^3 + 10^3 + 11^3 \quad (57)$$

$$87^3 + 233^3 + 264^3 + 396^3 + 496^3 + 540^3$$

$$= 90^3 + 206^3 + 309^3 + 366^3 + 522^3 + 523^3. \qquad (58)$$

(Madachy 1979, p. 142; Chen Shuwen).

Euler gave the general solution to

$$A^3 + B^3 = C^2 \qquad (59)$$

as

$$A = 3n^2 + 6n^2 - n \qquad (60)$$

$$B = -3n^3 + 6n^2 + n \qquad (61)$$

$$C = 6n^2(3n^2 + 1). \qquad (62)$$

See also CANNONBALL PROBLEM, CUBIC NUMBER, HARDY-RAMANUJAN NUMBER, MULTIGRADE EQUATION, SUPER-*d* NUMBER, TAXICAB NUMBER, TRIMORPHIC NUMBER, WARING'S PROBLEM

References

Berndt, B. C. *Ramanujan's Notebooks, Part IV*. New York: Springer-Verlag, 1994.

Berndt, B. C. and Bhargava, S. "Ramanujan--For Lowbrows." *Amer. Math. Monthly* **100**, 645–56, 1993.

Binet, J. P. M. "Note sur une question relative à la théorie des nombres." *C. R. Acad. Sci. (Paris)* **12**, 248–50, 1841.

Chen, S. "Equal Sums of Like Powers: On the Integer Solution of the Diophantine System." http://www.nease.net/~chin/eslp/

Dickson, L. E. *History of the Theory of Numbers, Vol. 2: Diophantine Analysis*. New York: Chelsea, 1966.

Fredkin, E. Item 58 in Beeler, M.; Gosper, R. W.; and Schroeppel, R. *HAKMEM*. Cambridge, MA: MIT Artificial Intelligence Laboratory, Memo AIM-239, p. 23, Feb. 1972.

Gardiner, V. L.; Lazarus, R. B.; and Stein, P. R. "Solutions of the Diophantine Equation $x^3 + y^3 = z^3 - d$." *Math. Comput.* **18**, 408–13, 1964.

Guy, R. K. "Sums of Like Powers. Euler's Conjecture." §D1 in *Unsolved Problems in Number Theory, 2nd ed.* New York: Springer-Verlag, pp. 139–44, 1994.

Hardy, G. H. *Ramanujan: Twelve Lectures on Subjects Suggested by His Life and Work, 3rd ed.* New York: Chelsea, 1999.

Hardy, G. H. and Wright, E. M. *An Introduction to the Theory of Numbers, 5th ed.* Oxford, England: Clarendon Press, 1979.

Koyama, K.; Tsuruoka, Y.; and Sekigawa, S. "On Searching for Solutions of the Diophantine Equation $x^3 + y^3 + z^3 = n$." *Math. Comput.* **66**, 841–51, 1997.

Kraus, A. "Sur l'équation $a^3 + b^3 = c^p$." *Experim. Math.* **7**, 1–3, 1998.

Madachy, J. S. *Madachy's Mathematical Recreations*. New York: Dover, 1979.

Miller, J. C. P. and Woollett, M. F. C. "Solutions of the Diophantine Equation $x^3 + y^3 + z^3 = k$." *J. London Math. Soc.* **30**, 101–10, 1955.

Moreau, C. "Plus petit nombre égal à la somme de deux cubes de deux façons." *L'Intermediaire Math.* **5**, 66, 1898.

Nagell, T. "The Diophantine Equation $\xi^3 + \eta^3 + \zeta^3$ and Analogous Equations" and "Diophantine Equations of the Third Degree with an Infinity of Solutions." §65 and 66 in *Introduction to Number Theory*. New York: Wiley, pp. 241–48, 1951.

Rivera, C. "Problems & Puzzles: Puzzle $p^3 = a^3 + b^3 + c^3$, pa, b, c Prime.-048." http://www.primepuzzles.net/puzzles/puzz_048.htm.

Schwering, K. "Vereinfachte Lösungen des Eulerschen Aufgabe: $x^3 + y^3 + z^3 + v^3 = 0$." *Arch. Math. Phys.* **2**, 280–84, 1902.

Shanks, D. *Solved and Unsolved Problems in Number Theory, 4th ed.* New York: Chelsea, p. 157, 1993.

Sloane, N. J. A. Sequences A001235 and A003825 in "An On-Line Version of the Encyclopedia of Integer Sequences." http://www.research.att.com/~njas/sequences/eisonline.html.

Weisstein, E. W. "Like Powers." MATHEMATICA NOTEBOOK LIKEPOWERS.M.

Wilson, D. Personal communication, Apr. 17, 1997.

© *1999– 001 Wolfram Research, Inc.*

Diophantine Equation—4th Powers

As a consequence of Matiyasevich's refutation of Hilbert's 10th problem, it can be proved that there does not exists a general algorithm for solving a general quartic Diophantine equation. However, the algorithm for constructing such an unsolvable quartic Diophantine equation can require arbitrarily many variables (Matiyasevich 1993).

As a part of the study of WARING'S PROBLEM, it is known that every positive integer is a sum of no more than 19 positive biquadrates ($g(4) = 19$), that every "sufficiently large" integer is a sum of no more than 16 positive biquadrates ($G(4) = 16$), and that every integer is a sum of at most 10 signed biquadrates ($eg(4) \leq 10$; although it is not known if 10 can be reduced to 9). The first few numbers n which are a sum of four fourth POWERS ($m - 1$ equations) are 353, 651, 2487, 2501, 2829, ... (Sloane's A003294).

The 4.1.2 equation

$$x^4 = y^4 + z^4 \qquad (1)$$

is a case of FERMAT'S LAST THEOREM with $n = 4$ and therefore has no solutions. In fact, the equations

$$x^4 \pm y^4 = z^2 \qquad (2)$$

also have no solutions in INTEGERS (Nagell 1951, pp. 227 and 229). The equation

$$x^4 - y^4 = 2z^2 \qquad (3)$$

has no solutions in integers (Nagell 1951, p. 230). The only number OF THE FORM

$$4x^4 + y^4 \qquad (4)$$

which is PRIME is 5 (Baudran 1885, Le Lionnais 1983).

Let the notation $p.m.n$ stand for the equation consisting of a sum of m pth powers being equal to a sum of n pth powers. In 1772, Euler proposed that the 4.1.3 equation

$$A^4 + B^4 + C^4 = D^4 \qquad (5)$$

had no solutions in INTEGERS (Lander *et al.* 1967). This assertion is known as the EULER QUARTIC

CONJECTURE. Ward (1948) showed there were no solutions for $D \leq 10,000$, which was subsequently improved to $D \leq 220,000$ by Lander *et al.* (1967). However, the EULER QUARTIC CONJECTURE was disproved in 1987 by N. Elkies, who, using a geometric construction, found

$$2,682,440^4 + 15,365,639^4 + 18,796,760^4$$

$$= 20,615,673^4 \tag{6}$$

and showed that infinitely many solutions existed (Guy 1994, p. 140). In 1988, Roger Frye found

$$95,800^4 + 217,519^4 + 414,560^4 = 422,481^4 \tag{7}$$

and proved that there are no solutions in smaller INTEGERS (Guy 1994, p. 140). Another solution was found by Allan MacLeod in 1997,

$$638,523,249^4 = 630,662,624^4$$

$$+ 275,156,240^4 + 219,076,465^4 \tag{8}$$

(Ekl 1998). It is not known if there is a parametric solution. In contrast, there are many solutions to the equation

$$A^4 + B^4 + C^4 = 2D^4 \tag{9}$$

(see below).

The 4.1.4 equation

$$A^4 + B^4 + C^4 + D^4 = E^4 \tag{10}$$

has solutions

$$30^4 + 120^4 + 272^4 + 315^4 = 353^4 \tag{11}$$

$$240^4 + 340^4 + 430^4 + 599^4 = 651^4 \tag{12}$$

$$435^4 + 710^4 + 1384^5 + 2420^4 = 2487^4 \tag{13}$$

$$1130^4 + 1190^4 + 1432^4 + 2365^4 = 2501^4 \tag{14}$$

$$850^4 + 1010^4 + 1546^4 + 2745^4 = 2829^4 \tag{15}$$

$$2270^4 + 2345^4 + 2460^4 + 3152^4 = 3723^4 \tag{16}$$

$$350^4 + 1652^4 + 3230^4 + 3395^4 = 3973^4 \tag{17}$$

$$205^4 + 1060^4 + 2650^4 + 4094^4 = 4267^4 \tag{18}$$

$$1394^4 + 1750^4 + 3545^4 + 3670^4 = 4333^4 \tag{19}$$

$$699^4 + 700^4 + 2840^4 + 4250^4 = 4449^4 \tag{20}$$

$$380^4 + 1660^4 + 1880^4 + 4907^4 = 4949^4 \tag{21}$$

$$1000^4 + 1120^4 + 3233^4 + 5080^4 = 5281^4 \tag{22}$$

$$410^4 + 1412^4 + 3910^4 + 5055^4 = 5463^4 \tag{23}$$

$$955^4 + 1770^4 + 2634^4 + 5400^4 = 5491^4 \tag{24}$$

$$30^4 + 1680^4 + 3043^4 + 5400^4 = 5543^4 \tag{25}$$

$$1354^4 + 1810^4 + 4355^4 + 5150^4 = 5729^4 \tag{26}$$

$$542^4 + 2770^4 + 4280^4 + 5695^4 = 6167^4 \tag{27}$$

$$50^4 + 885^4 + 5000^4 + 5984^4 = 6609^4 \tag{28}$$

$$1490^4 + 3468^4 + 4790^4 + 6185^4 = 6801^4 \tag{29}$$

$$1390^4 + 2850^4 + 5365^4 + 6368^4 = 7101^4 \tag{30}$$

$$160^4 + 1345^4 + 2790^4 + 7166^4 = 7209^4 \tag{31}$$

$$800^4 + 3052^4 + 5440^4 + 6635^4 = 7339^4 \tag{32}$$

$$2230^4 + 3196^4 + 5620^4 + 6995^4 = 7703^4 \tag{33}$$

(Norrie 1911, Patterson 1942, Leech 1958, Brudno 1964, Lander *et al.* 1967), but it is not known if there is a parametric solution (Guy 1994, p. 139).

There are an infinite number of solutions to the 4.1.5 equation

$$A^4 = B^4 + C^4 + D^4 + E^4 + F^4. \tag{34}$$

Some of the smallest are

$$2^4 + 2^4 + 3^4 + 4^4 + 4^4 = 5^4 \tag{35}$$

$$4^4 + 6^4 + 8^4 + 9^4 + 14^4 = 15^4 \tag{36}$$

$$4^4 + 21^4 + 22^4 + 26^4 + 28^4 = 35^4 \tag{37}$$

$$1^4 + 2^4 + 12^4 + 24^4 + 44^4 = 45^4 \tag{38}$$

$$1^4 + 8^4 + 12^4 + 32^4 + 64^4 = 65^4 \tag{39}$$

$$2^4 + 39^4 + 44^4 + 46^4 + 52^4 = 65^4 \tag{40}$$

$$22^4 + 52^4 + 57^4 + 74^4 + 76^4 = 95^4 \tag{41}$$

$$22^4 + 28^4 + 63^4 + 72^4 + 94^4 = 105^4 \tag{42}$$

(Berndt 1994). Berndt and Bhargava (1993) and Berndt (1994, pp. 94–6) give Ramanujan's solutions for arbitrary s, t, m, and n,

$$\left(8s^2 + 40st - 24t^2\right)^4 + \left(6s^2 - 44st - 18t^2\right)^4$$

$$+ \left(14s^2 - 4st - 42t^2\right)^4 + \left(9s^2 + 27t^2\right)^4 + \left(4s^2 + 12t^2\right)^4$$

$$= \left(15s^2 + 45t^2\right)^4, \tag{43}$$

and

$$\left(4m^2 - 12n^2\right)^4 + \left(3m^2 + 9n^2\right)^4 + \left(2m^2 - 12mn - 6n^2\right)^4$$

$$+ \left(4m^2 + 12n^2\right)^4 + \left(2m^2 + 12mn - 6n^2\right)^4$$

$$= \left(5m^2 + 15n^2\right)^4. \tag{44}$$

These are also given by Dickson (1966, p. 649), and two general FORMULAS are given by Beiler (1966, p. 290). Other solutions are given by Fauquembergue (1898), Haldeman (1904), and Martin (1910).

Parametric solutions to the 4.2.2 equation

$$A^4 + B^4 = C^4 + D^4 \tag{45}$$

are known (Euler 1802; Gérardin 1917; Guy 1994, pp. 140–41), but no "general" solution is known (Hardy 1999, p. 21). A few specific solutions are

$$59^4 + 158^4 = 133^4 + 134^4 = 635,318,657 \tag{46}$$

$$7^4 + 239^4 = 157^4 + 227^4 = 3,262,811,042 \tag{47}$$

$$193^4 + 292^4 = 256^4 + 257^4 = 8,657,437,697 \tag{48}$$

$$298^4 + 497^4 = 271^4 + 502^4 = 68,899,596,497 \tag{49}$$

$$514^4 + 359^4 = 103^4 + 542^4 = 86,409,838,577 \tag{50}$$

$$222^4 + 631^4 = 503^4 + 558^4 = 160,961,094,577 \tag{51}$$

$$21^4 + 717^4 = 471^4 + 681^4 = 264,287,694,402 \tag{52}$$

$$76^4 + 1203^4 = 653^4 + 1176^4$$
$$= 2,094,447,251,857 \tag{53}$$

$$997^4 + 1342^4 = 878^4 + 1381^4$$
$$= 4,231,525,221,377 \tag{54}$$

(Sloane's A003824 and A018786; Richmond 1920; Dickson, pp. 60–2; Dickson 1966, pp. 644–47; Leech 1957; Berndt 1994, p. 107; Ekl 1998 [with typo]), the smallest of which is due to Euler (Hardy 1999, p. 21). Lander *et al.* (1967) give a list of 25 primitive 4.2.2 solutions. General (but incomplete) solutions are given by

$$x = a + b \tag{55}$$

$$y = c - d \tag{56}$$

$$u = a - b \tag{57}$$

$$v = c + d, \tag{58}$$

where

$$a = n(m^2 + n^2)(-m^4 + 18m^2n^2 - n^4) \tag{59}$$

$$b = 2m(m^6 + 10m^4n^4 + m^2n^4 + 4n^6) \tag{60}$$

$$c = 2n(4m^6 + m^4n^2 + 10m^2n^4 + n^6) \tag{61}$$

$$d = m(m^2 + n^2)(-m^4 + 18m^2n^2 - n^4) \tag{62}$$

(Hardy and Wright 1979).

Parametric solutions to the 4.2.3 equation

$$A^4 + B^4 = C^4 + D^4 + E^4 \tag{63}$$

are known (Gérardin 1910, Ferrari 1913). The smallest solution is

$$3^4 + 5^4 + 8^4 = 7^4 + 7^4 \tag{64}$$

(Lander *et al.* 1967).

Ramanujan gave the 4.2.4 equation

$$3^4 + 9^4 = 5^4 + 5^4 + 6^4 + 8^4. \tag{65}$$

Ramanujan gave the 4.3.3 equations

$$2^4 + 4^4 + 7^4 = 3^4 + 6^4 + 6^4 \tag{66}$$

$$3^4 + 7^4 + 8^4 = 1^4 + 2^4 + 9^4 \tag{67}$$

$$6^4 + 9^4 + 12^4 = 2^4 + 2^4 + 13^4 \tag{68}$$

(Berndt 1994, p. 101). Similar examples can be found in Martin (1896). Parametric solutions were given by Gérardin (1911).

Ramanujan also gave the general expression

$$3^4 + (2x^4 - 1)^4 + (4x^5 + x)^4$$
$$= (4x^4 + 1)^4 + (6x^4 - 3)^4 + (4x^5 - 5x)^4 \tag{69}$$

(Berndt 1994, p. 106). Dickson (1966, pp. 653–55) cites several FORMULAS giving solutions to the 4.3.3 equation, and Haldeman (1904) gives a general FORMULA.

Ramanujan gave the 4.3.4 identities

$$2^4 + 2^4 + 7^4 = 4^4 + 4^4 + 5^4 + 6^4 \tag{70}$$

$$3^4 + 9^4 + 14^4 = 7^4 + 8^4 + 10^4 + 13^4 \tag{71}$$

$$7^4 + 10^4 + 13^4 = 5^4 + 5^4 + 6^4 + 14^4 \tag{72}$$

(Berndt 1994, p. 101). Haldeman (1904) gives general FORMULAS for 4– and 4– equations.

Ramanujan gave

$$2(ab + ac + bc)^2 = a^4 + b^4 + c^4 \tag{73}$$

$$2(ab + ac + bc)^4 = a^4(b-c)^4 + b^4(c-a)^4 + c^4(a-b)^4 \tag{74}$$

$$2(ab + ac + bc+)^6$$
$$= (a^2b + b^2c + c^2a)^4 + (ab^2 + bc^2 + ca^2)^4 + 3(abc)^4 \tag{75}$$

$$2(ab + ac + bc)^8 = (a^3 + 2abc)^4(b-c)^4$$
$$+ (b^3 + 2abc)^4(c-a)^4 + (c^3 + 2abc)^4(a-b)^4, \tag{76}$$

where

$$a + b + c = 0 \tag{77}$$

(Berndt 1994, pp. 96–7). FORMULA (74) is equivalent to FERRARI'S IDENTITY

$$(a^2 + 2ac - 2bc - b^2)^4 + (b^2 - 2ab - 2ac - c^2)^4$$
$$+ (c^2 + 2ab + 2bc - a^2)^4$$
$$= 2(a^2 + b^2 + c^2 - ab + ac + bc)^4. \tag{78}$$

BHARGAVA'S THEOREM is a general identity which gives the above equations as a special case, and may have been the route by which Ramanujan proceeded. Another identity due to Ramanujan is

$$(a+b+c)^4 + (b+c+d)^4 + (a-d)^4$$

$$= (c+d+a)^4 + (d+a+b)^4 + (b-c)^4, \qquad (79)$$

where $a/b = c/d$, and 4 may also be replaced by 2 (Ramanujan 1957, Hirschhorn 1998).

V. Kyrtatas noticed that $a = 3$, $b = 7$, $c = 20$, $d = 25$, $e = 38$, and $f = 39$ satisfy

$$\frac{a^4 + b^4 + c^4}{d^4 + e^4 + f^4} = \frac{a + b + c}{d + e + f} \qquad (80)$$

and asks if there are any other distinct integer solutions.

See also BHARGAVA'S THEOREM, BIQUADRATIC NUMBER, FORD'S THEOREM, MULTIGRADE EQUATION, WARING'S PROBLEM

References

Barbette, E. *Les sommes de p-iémes puissances distinctes égales à une p-iéme puissance.* Doctoral Dissertation, Liege, Belgium. Paris: Gauthier-Villars, 1910.

Beiler, A. H. *Recreations in the Theory of Numbers: The Queen of Mathematics Entertains.* New York: Dover, 1966.

Berndt, B. C. *Ramanujan's Notebooks, Part IV.* New York: Springer-Verlag, 1994.

Berndt, B. C. and Bhargava, S. "Ramanujan--For Lowbrows." *Am. Math. Monthly* **100**, 645–56, 1993.

Bhargava, S. "On a Family of Ramanujan's Formulas for Sums of Fourth Powers." *Ganita* **43**, 63–7, 1992.

Brudno, S. "A Further Example of $A^4 + B^4 + C^4 + D^4 = E^4$." *Proc. Cambridge Phil. Soc.* **60**, 1027–028, 1964.

Chen, S. "Equal Sums of Like Powers: On the Integer Solution of the Diophantine System." http://www.nease.net/~chin/eslp/

Dickson, L. E. *Introduction to the Theory of Numbers.* New York: Dover.

Dickson, L. E. *History of the Theory of Numbers, Vol. 2: Diophantine Analysis.* New York: Chelsea, 1966.

Ekl, R. L. "New Results in Equal Sums of Like Powers." *Math. Comput.* **67**, 1309–315, 1998.

Euler, L. *Nova Acta Acad. Petrop. as annos 1795–796* **13**, 45, 1802.

Fauquembergue, E. *L'intermédiaire des Math.* **5**, 33, 1898.

Ferrari, F. *L'intermédiaire des Math.* **20**, 105–06, 1913.

Guy, R. K. "Sums of Like Powers. Euler's Conjecture" and "Some Quartic Equations." §D1 and D23 in *Unsolved Problems in Number Theory, 2nd ed.* New York: Springer-Verlag, pp. 139–44 and 192–93, 1994.

Haldeman, C. B. "On Biquadrate Numbers." *Math. Mag.* **2**, 285–96, 1904.

Hardy, G. H. *Ramanujan: Twelve Lectures on Subjects Suggested by His Life and Work, 3rd ed.* New York: Chelsea, 1999.

Hardy, G. H. and Wright, E. M. §13.7 in *An Introduction to the Theory of Numbers, 5th ed.* Oxford, England: Clarendon Press, 1979.

Hirschhorn, M. D. "Two or Three Identities of Ramanujan." *Amer. Math. Monthly* **105**, 52–5, 1998.

Lander, L. J.; Parkin, T. R.; and Selfridge, J. L. "A Survey of Equal Sums of Like Powers." *Math. Comput.* **21**, 446–59, 1967.

Le Lionnais, F. *Les nombres remarquables.* Paris: Hermann, p. 56, 1983.

Leech, J. "Some Solutions of Diophantine Equations." *Proc. Cambridge Phil. Soc.* **53**, 778–80, 1957.

Leech, J. "On $A^4 + B^4 + C^4 + D^4 = E^4$." *Proc. Cambridge Phil. Soc.* **54**, 554–55, 1958.

Martin, A. "About Biquadrate Numbers whose Sum is a Biquadrate." *Math. Mag.* **2**, 173–84, 1896.

Martin, A. "About Biquadrate Numbers whose Sum is a Biquadrate--II." *Math. Mag.* **2**, 325–52, 1904.

Nagell, T. "Some Diophantine Equations of the Fourth Degree with Three Unknowns" and "The Diophantine Equation $2x^4 - y^4 = z^2$." §62 and 63 in *Introduction to Number Theory.* New York: Wiley, pp. 227–35, 1951.

Norrie, R. *University of St. Andrews 500th Anniversary Memorial Volume.* Edinburgh, Scotland: pp. 87–9, 1911.

Patterson, J. O. "A Note on the Diophantine Problem of Finding Four Biquadrates whose Sum is a Biquadrate." *Bull. Amer. Math. Soc.* **48**, 736–37, 1942.

Ramanujan, S. *Notebooks.* New York: Springer-Verlag, pp. 385–86, 1987.

Richmond, H. W. "On Integers Which Satisfy the Equation $t^3 \pm x^3 \pm y^3 \pm z^3 = 0$." *Trans. Cambridge Phil. Soc.* **22**, 389–03, 1920.

Rivera, C. "Problems & Puzzles: Puzzle $p^4 = a^4 + b^4 + c^4 + d^4$, $a, b, c, d > 0$.-047." http://www.primepuzzles.net/puzzles/puzz_047.htm.

Sloane, N. J. A. Sequences A003294/M5446, A003824, and A018786 in "An On-Line Version of the Encyclopedia of Integer Sequences." http://www.research.att.com/~njas/sequences/eisonline.html.

Ward, M. "Euler's Problem on Sums of Three Fourth Powers." *Duke Math. J.* **15**, 827–37, 1948.

Weisstein, E. W. "Like Powers." MATHEMATICA NOTEBOOK LIKEPOWERS.M.

Diophantine Equation—5th Powers

The 5.1.2 fifth-order Diophantine equation

$$A^5 = B^5 + C^5 \qquad (1)$$

is a special case of FERMAT'S LAST THEOREM with $n = 5$, and so has no solution. improving on the results on Lander *et al.* (1967), who checked up to 2.8×10^{14}. (In fact, no solutions are known for POWERS of 6 or 7 either.) No solutions to the 5.1.3 equation

$$A^5 + B^5 + C^5 = D^5 \qquad (2)$$

are known (Lander *et al.* 1967). For 4 fifth POWERS, we have the 5.1.4 equation

$$27^5 + 84^5 + 110^5 + 133^5 = 144^5 \qquad (3)$$

(Lander and Parkin 1967, Lander *et al.* 1967, Ekl 1998), but it is not known if there is a parametric solution (Guy 1994, p. 140). Sastry (1934) found a 2-parameter solution for 5.1.5 equations

$$(75v^5 - u^5)^5 + (u^5 + 25v^5)^5 + (u^5 - 25v^5)^5$$

$$+ (10u^3v^2)^5 + (50uv^4)^5 = (u^5 + 75v^5)^5 \qquad (4)$$

(quoted in Lander and Parkin 1967), and Lander and Parkin (1967) found the smallest numerical solutions. Lander *et al.* (1967) give a list of the smallest solutions, the first few being

$$19^5 + 43^5 + 46^5 + 47^5 + 67^5 = 72^5 \tag{5}$$

$$21^5 + 23^5 + 37^5 + 79^5 + 84^5 = 94^5 \tag{6}$$

$$7^5 + 43^5 + 57^5 + 80^5 + 100^5 = 107^5 \tag{7}$$

$$78^5 + 120^5 + 191^5 + 259^5 + 347^5 = 365^5 \tag{8}$$

$$79^5 + 202^5 + 258^5 + 261^5 + 395^5 = 415^5 \tag{9}$$

$$4^5 + 26^5 + 139^5 + 296^5 + 412^5 = 427^5 \tag{10}$$

$$31^5 + 105^5 + 139^5 + 314^5 + 416^5 = 435^5 \tag{11}$$

$$54^5 + 91^5 + 101^5 + 404^5 + 430^5 = 480^5 \tag{12}$$

$$19^5 + 201^5 + 347^5 + 388^5 + 448^5 = 503^5 \tag{13}$$

$$159^5 + 172^5 + 200^5 + 356^5 + 513^5 = 530^5 \tag{14}$$

$$218^5 + 276^5 + 385^5 + 409^5 + 495^5 = 553^5 \tag{15}$$

$$2^5 + 298^5 + 351^5 + 474^5 + 500^5 = 575^5 \tag{16}$$

(Lander and Parkin 1967, Lander *et al.* 1967). The 5.1.6 equation has solutions

$$4^5 + 5^5 + 6^5 + 7^5 + 9^5 + 11^5 = 12^5 \tag{17}$$

$$5^5 + 10^5 + 11^5 + 16^5 + 19^5 + 29^5 = 30^5 \tag{18}$$

$$15^5 + 16^5 + 17^5 + 22^5 + 24^5 + 28^5 = 32^5 \tag{19}$$

$$13^5 + 18^5 + 23^5 + 31^5 + 36^5 + 66^5 = 67^5 \tag{20}$$

$$7^5 + 20^5 + 29^5 + 31^5 + 34^5 + 66^5 = 67^5 \tag{21}$$

$$22^5 + 35^5 + 48^5 + 58^5 + 61^5 + 64^5 = 78^5 \tag{22}$$

$$4^5 + 13^5 + 19^5 + 20^5 + 67^5 + 96^5 = 99^5 \tag{23}$$

$$6^5 + 17^5 + 60^5 + 64^5 + 73^5 + 89^5 = 99^5 \tag{24}$$

(Martin 1887, 1888, Lander and Parkin 1967, Lander *et al.* 1967). The smallest 5.1.7 solution is

$$1^5 + 7^5 + 8^5 + 14^5 + 15^5 + 18^5 + 20^5 = 23^5 \tag{25}$$

(Lander *et al.* 1967).

No solutions to the 5.2.2 equation

$$A^5 + B^5 = C^5 + D^5 \tag{26}$$

are known, despite the fact that sums up to 1.026×10^{26} have been checked (Guy 1994, p. 140). The smallest 5.2.3 solution is

$$14132^5 + 220^5 = 14068^5 + 6237^5 + 5027^5 \tag{27}$$

(B. Scher and E. Seidl 1996, Ekl 1998). Sastry's (1934) 5.1.5 solution gives some 5.2.4 solutions. The smallest primitive 5.2.4 solutions are

$$4^5 + 10^5 + 20^5 + 28^5 = 3^5 + 29^5 \tag{28}$$

$$5^5 + 13^5 + 25^5 + 37^5 = 12^5 + 38^5 \tag{29}$$

$$26^5 + 29^5 + 35^5 + 50^5 = 28^5 + 52^5 \tag{30}$$

$$5^5 + 25^5 + 62^5 + 63^5 = 61^5 + 64^5 \tag{31}$$

$$6^5 + 50^5 + 53^5 + 82^5 = 16^5 + 85^5 \tag{32}$$

$$56^5 + 63^5 + 72^5 + 86^5 = 31^5 + 96^5 \tag{33}$$

$$44^5 + 58^5 + 67^5 + 94^5 = 14^5 + 99^5 \tag{34}$$

$$11^5 + 13^5 + 37^5 + 99^5 = 63^5 + 97^5 \tag{35}$$

$$48^5 + 57^5 + 76^5 + 100^5 = 25^5 + 106^5 \tag{36}$$

$$58^5 + 76^5 + 79^5 + 102^5 = 54^5 + 111^5 \tag{37}$$

(Rao 1934, Moessner 1948, Lander *et al.* 1967). The smallest primitive 5.2.5 solutions are

$$4^5 + 5^5 + 7^5 + 16^5 + 21^5 = 1^5 + 22^5 \tag{38}$$

$$9^5 + 11^5 + 14^5 + 18^5 + 30^5 = 23^5 + 29^5 \tag{39}$$

$$10^5 + 14^5 + 26^5 + 31^5 + 33^5 = 16^5 + 38^5 \tag{40}$$

$$4^5 + 22^5 + 29^5 + 35^5 + 36^5 = 24^5 + 42^5 \tag{41}$$

$$8^5 + 15^5 + 17^5 + 19^5 + 45^5 = 30^5 + 44^5 \tag{42}$$

$$5^5 + 6^5 + 26^5 + 27^5 + 44^5 = 36^5 + 42^5 \tag{43}$$

(Rao 1934, Lander *et al.* 1967).

Parametric solutions are known for the 5.3.3 (Sastry and Lander 1934; Moessner 1951; Swinnerton-Dyer 1952; Lander 1968; Bremmer 1981; Guy 1994, pp. 140 and 142; Choudhry 1999). Swinnerton-Dyer (1952) gave two parametric solutions to the 5.3.3 equation but, forty years later, W. Gosper discovered that the second scheme has an unfixable bug. Choudhry (1999) gave a parametric solution to the more general equation

$$ax^5 + by^5 + cz^5 = au^5 + bv^5 + cw^5 \tag{44}$$

with $a + b + c = 0$. The smallest primitive solutions to the 5.3.3 equation with unit coefficients are

$$24^5 + 28^5 + 67^5 = 3^5 + 54^5 + 62^5 \tag{45}$$

$$18^5 + 44^5 + 66^5 = 13^5 + 51^5 + 64^5 \tag{46}$$

$$21^5 + 43^5 + 74^5 = 8^5 + 62^5 + 68^5 \tag{47}$$

$$56^5 + 67^5 + 83^5 = 53^5 + 72^5 + 81^5 \tag{48}$$

$$49^5 + 75^5 + 107^5 = 39^5 + 92^5 + 100^5 \tag{49}$$

(Moessner 1939, Moessner 1948, Lander *et al.* 1967, Ekl 1998).

A two-parameter solution to the 5.3.4 equation was given by Xeroudakes and Moessner (1958). Gloden (1949) also gave a parametric solution. The smallest solution is

$$1^5 + 8^5 + 14^5 + 27^5 = 3^5 + 22^5 + 25^5 \qquad (50)$$

(Rao 1934, Lander *et al.* 1967).

Several parametric solutions to the 5.4.4 equation were found by Xeroudakes and Moessner (1958). The smallest 5.4.4 solution is

$$5^5 + 6^5 + 6^5 + 8^5 = 4^5 + 7^5 + 7^5 + 7^5 \qquad (51)$$

(Rao 1934, Lander *et al.* 1967). The first 5.4.4.4 equation is

$$3^5 + 48^5 + 52^5 + 61^5 = 13^5 + 36^5 + 51^5 + 64^5$$
$$= 18^5 + 36^5 + 44^5 + 66^5 \qquad (52)$$

(Lander *et al.* 1967).

Moessner and Gloden (1944) give the 5.5.6 solution

$$22^5 + 17^5 + 16^5 + 6^5 + 5^5$$
$$= 21^5 + 20^5 + 12^5 + 10^5 + 2^5 + 1^5. \qquad (53)$$

Chen Shuwen found the 5.6.6 solution

$$87^5 + 233^5 + 264^5 + 396^5 + 496^5 + 540^5$$

$$= 90^5 + 206^5 + 309^5 + 366^5 + 522^5 + 523^5. \qquad (54)$$

See also MULTIGRADE EQUATION

References

Berndt, B. C. *Ramanujan's Notebooks, Part IV.* New York: Springer-Verlag, p. 95, 1994.

Bremner, A. "A Geometric Approach to Equal Sums of Fifth Powers." *J. Number Th.* **13**, 337–54, 1981.

Chen, S. "Equal Sums of Like Powers: On the Integer Solution of the Diophantine System." http://www.nease.net/~chin/eslp/

Choudhry, A. "The Diophantine Equation $ax^5 + by^5 + cz^5 + = au^5 + bv^5 + cw^5$." *Rocky Mtn. J. Math.* **29**, 459–62, 1999.

Ekl, R. L. "New Results in Equal Sums of Like Powers." *Math. Comput.* **67**, 1309–315, 1998.

Gloden, A. "Uuml;ber mehrgeradige Gleichungen." *Arch. Math.* **1**, 482–83, 1949.

Guy, R. K. "Sums of Like Powers. Euler's Conjecture." §D1 in *Unsolved Problems in Number Theory, 2nd ed.* New York: Springer-Verlag, pp. 139–44, 1994.

Lander, L. J. and Parkin, T. R. "A Counterexample to Euler's Sum of Powers Conjecture." *Math. Comput.* **21**, 101–03, 1967.

Lander, L. J.; Parkin, T. R.; and Selfridge, J. L. "A Survey of Equal Sums of Like Powers." *Math. Comput.* **21**, 446–59, 1967.

Lander, L. J. "Geometric Aspects of Diophantine Equations Involving Equal Sums of Like Power." *Amer. Math. Monthly* **75**, 1061–073, 1968.

Martin, A. "Methods of Finding nth-Power Numbers Whose Sum is an nth Power; With Examples." *Bull. Philos. Soc. Washington* **10**, 107–10, 1887.

Martin, A. *Smithsonian Misc. Coll.* **33**, 1888.

Martin, A. "About Fifth-Power Numbers whose Sum is a Fifth Power." *Math. Mag.* **2**, 201–08, 1896.

Moessner, A. "Einige numerische Identitäten." *Proc. Indian Acad. Sci. Sect. A* **10**, 296–06, 1939.

Moessner, A. "Alcune richerche di teoria dei numeri e problemi diofantei." *Bol. Soc. Mat. Mexicana* **2**, 36–9, 1948.

Moessner, A. "Due Sistemi Diofantei." *Boll. Un. Mat. Ital.* **6**, 117–18, 1951.

Moessner, A. and Gloden, A. "Einige Zahlentheoretische Untersuchungen und Resultate." *Bull. Sci. École Polytech. de Timisoara* **11**, 196–19, 1944.

Rao, K. S. "On Sums of Fifth Powers." *J. London Math. Soc.* **9**, 170–71, 1934.

Sastry, S. and Chowla, S. "On Sums of Powers." *J. London Math. Soc.* **9**, 242–46, 1934.

Swinnerton-Dyer, H. P. F. "A Solution of $A^5 + B^5 + C^5 = D^5 + E^5 + F^5$." *Proc. Cambridge Phil. Soc.* **48**, 516–18, 1952.

Weisstein, E. W. "Like Powers." MATHEMATICA NOTEBOOK LIKEPOWERS.M.

Xeroudakes, G. and Moessner, A. "On Equal Sums of Like Powers." *Proc. Indian Acad. Sci. Sect. A* **48**, 245–55, 1958.

Diophantine Equation—6th Powers

The 6.1.2 equation

$$A^6 = B^6 + C^6 \qquad (1)$$

is a special case of FERMAT'S LAST THEOREM with $n = 6$, and so has no solution. No 6.1.n solutions are known for $n \le 6$ (Lander *et al.* 1967; Guy 1994, p. 140). The smallest 6.1.7 solution is

$$74^6 + 234^6 + 402^6 + 474^6 + 702^6 + 894^6 + 1017^6$$
$$= 1141^6 \qquad (2)$$

(Lander *et al.* 1967; Ekl 1998). The smallest primitive 6.1.8 solutions are

$$8^6 + 12^6 + 30^6 + 78^6 + 102^6 + 138^6 + 165^6 + 246^6$$
$$= 251^6 \qquad (3)$$

$$48^6 + 111^6 + 156^6 + 186^6 + 188^6 + 228^6 + 240^6 + 426^6$$
$$= 431^6 \qquad (4)$$

$$93^6 + 93^6 + 195^6 + 197^6 + 303^6 + 303^6 + 303^6 + 411^6$$
$$= 440^6 \qquad (5)$$

$$219^6 + 255^6 + 261^6 + 267^6 + 289^6 + 351^6 + 351^6 + 351^6$$
$$= 440^6 \qquad (6)$$

$$12^6 + 66^6 + 138^6 + 174^6 + 212^6 + 288^6 + 306^6 + 441^6$$
$$= 455^6 \qquad (7)$$

$$12^6 + 48^6 + 222^6 + 236^6 + 333^6 + 384^6 + 390^6 + 426^6$$
$$= 493^6 \qquad (8)$$

$$66^6 + 78^6 + 144^6 + 228^6 + 256^6 + 288^6 + 435^6 + 444^6$$
$$= 499^6 \qquad (9)$$

$$16^6 + 24^6 + 60^6 + 156^6 + 204^6 + 276^6 + 330^6 + 492^6$$
$$= 502^6 \qquad (10)$$

$$61^6 + 96^6 + 156^6 + 228^6 + 276^6 + 318^6 + 354^6 + 534^6$$
$$= 547^6 \qquad (11)$$

$$170^6 + 177^6 + 276^6 + 312^6 + 312^6 + 408^6 + 450^6 + 498^6$$
$$= 559^6 \tag{12}$$

$$60^6 + 102^6 + 126^6 + 261^6 + 270^6 + 338^6 + 354^6 + 570^6$$
$$= 581^6 \tag{13}$$

$$57^6 + 146^6 + 150^6 + 360^6 + 390^6 + 402^6 + 444^6 + 528^6$$
$$= 583^6 \tag{14}$$

$$33^6 + 72^6 + 122^6 + 192^6 + 204^6 + 390^6 + 534^6 + 534^6$$
$$= 607^6 \tag{15}$$

$$12^6 + 90^6 + 114^6 + 114^6 + 273^6 + 306^6 + 492^6 + 592^6$$
$$= 623^6 \tag{16}$$

(Lander *et al.* 1967). The smallest 6.1.9 solution is

$$1^6 + 17^6 + 19^6 + 22^6 + 31^6 + 37^6 + 37^6 + 41^6 + 49^6$$
$$= 54^6 \tag{17}$$

(Lander *et al.* 1967). The smallest 6.1.10 solution is

$$2^6 + 4^6 + 7^6 + 14^6 + 16^6 + 26^6 + 26^6 + 30^6 + 32^6 + 32^6$$
$$= 39^6 \tag{18}$$

(Lander *et al.* 1967). The smallest 6.1.11 solution is

$$2^6 + 5^6 + 5^6 + 5^6 + 7^6 + 7^6 + 9^6 + 9^6 + 10^6 + 14^6 + 17^6$$
$$= 18^6 \tag{19}$$

(Lander *et al.* 1967). There is also at least one 6.1.16 identity,

$$1^6 + 2^6 + 4^6 + 5^6 + 6^6 + 7^6 + 9^6 + 12^6 + 13^6 + 15^6$$
$$+ 16^6 + 18^6 + 20^6 + 21^6 + 22^6 + 23^6 = 28^6 \tag{20}$$

(Martin 1893). Moessner (1959) gave solutions for 6.1.16, 6.1.18, 6.1.20, and 6.1.23 equations.

Ekl (1996) has searched and found no solutions to the 6.2.2

$$A^6 + B^6 = C^6 + D^6 \tag{21}$$

with sums less than 7.25×10^{26}. No solutions are known to the 6.2.3 or 6.2.4 equations. The smallest primitive 6.2.5 equations are

$$1092^6 + 861^6 + 602^6 + 212^6 + 84^6 = 1117^6 + 770^6 \tag{22}$$

$$1893^6 + 1468^6 + 1407^6 + 1302^6 + 1246^6$$
$$= 2041^6 + 691^6 \tag{23}$$

$$2184^6 + 2096^6 + 1484^6 + 1266^6 + 1239^6$$
$$= 2441^6 + 752^6 \tag{24}$$

$$2653^6 + 2962^6 + 1488^6 + 1281^6 + 390^6$$
$$= 2827^6 + 151^6 \tag{25}$$

$$2954^6 + 2481^6 + 850^6 + 798^6 + 420^6$$
$$= 2959^6 + 2470^6 \tag{26}$$

(E. Brisse 1999 Resta 1999, PowerSum). The smallest 6.2.6 equation is

$$241^6 + 17^6 = 218^6 + 210^6 + 118^6 + 2.63^6 + 42^6 \tag{27}$$

(Ekl 1998). The smallest 6.2.7 solution is

$$18^6 + 22^6 + 36^6 + 58^6 + 69^6 + 78^6 + 78^6$$
$$= 56^6 + 91^6 \tag{28}$$

(Lander *et al.* 1967). The smallest 6.2.8 solution is

$$8^6 + 10^6 + 12^6 + 15^6 + 24^6 + 30^6 + 33^6 + 36^6$$
$$= 35^6 + 37^6 \tag{29}$$

(Lander *et al.* 1967). The smallest 6.2.9 solution is

$$1^6 + 5^6 + 5^6 + 7^6 + 13^6 + 13^6 + 13^6 + 17^6 + 19^6$$
$$= 6^6 + 21^6 \tag{30}$$

(Lander *et al.* 1967). The smallest 6.2.10 solution is

$$1^6 + 1^6 + 1^6 + 4^6 + 4^6 + 7^6 + 9^6 + 11^6 + 11^6 + 11^6$$
$$= 12^6 + 12^6 \tag{31}$$

(Lander *et al.* 1967).

Parametric solutions are known for the 6.3.3 equation

$$A^6 + B^6 + C^6 = D^6 + E^6 + F^6 \tag{32}$$

(Guy 1994, pp. 140 and 142). Known solutions are

$$3^6 + 19^6 + 22^6 = 10^6 + 15^6 + 23^6 \tag{33}$$

$$36^6 + 37^6 + 67^6 = 15^6 + 52^6 + 65^6 \tag{34}$$

$$33^6 + 47^6 + 74^6 = 23^6 + 54^6 + 73^6 \tag{35}$$

$$32^6 + 43^6 + 81^6 = 3^6 + 55^6 + 80^6 \tag{36}$$

$$37^6 + 50^6 + 81^6 = 11^6 + 65^6 + 78^6 \tag{37}$$

$$25^6 + 62^6 + 138^6 = 82^6 + 92^6 + 135^6 \tag{38}$$

$$51^6 + 113^6 + 136^6 = 40^6 + 125^6 + 129^6 \tag{39}$$

$$71^6 + 92^6 + 147^6 = 1^6 + 132^6 + 133^6 \tag{40}$$

$$111^6 + 121^6 + 230^6 = 26^6 + 169^6 + 225^6 \tag{41}$$

$$75^6 + 142^6 + 245^6 = 14^6 + 163^6 + 243^6 \tag{42}$$

(Rao 1934, Lander *et al.* 1967, Ekl 1998). Ekl (1998) mentions but does not list the 87 smallest solutions to the 6.2.6 equation. The smallest primitive 6.3.4 solutions are

$$73^6 + 58^6 + 41^6 = 70^6 + 65^6 + 32^6 + 15^6 \tag{43}$$

$$85^6 + 62^6 + 61^6 = 83^6 + 69^6 + 56^6 + 52^6 \tag{44}$$

$$85^6 + 74^6 + 61^6 = 87^6 + 71^6 + 56^6 + 26^6 \tag{45}$$

$$90^6 + 88^6 + 11^6 = 92^6 + 78^6 + 74^6 + 21^6 \tag{46}$$

$$95^6 + 83^6 + 26^6 = 101^6 + 28^6 + 24^6 + 23^6 \tag{47}$$

$$130^6 + 44^6 + 23^6 = 119^6 + 108^6 + 86^6 + 38^6 \tag{48}$$

$$125^6 + 114^6 + 38^6 = 126^6 + 104^6 + 93^6 + 68^6 \tag{49}$$

$$205^6 + 113^6 + 18^6 = 198^6 + 148^6 + 133^6 + 39^6 \quad (50)$$

$$211^6 + 123^6 + 34^6 = 210^6 + 134^6 + 73^6 + 39^6 \quad (51)$$

$$212^6 + 164^6 + 103^6 = 217^6 + 130^6 + 114^6 + 8^6 \quad (52)$$

$$222^6 + 34^6 + 25^6 = 217^6 + 156^6 + 96^6 + 68^6 \quad (53)$$

$$218^6 + 167^6 + 29^6 = 224^6 + 107^6 + 102^6 + 65^6 \quad (54)$$

$$226^6 + 110^6 + 17^6 = 224^6 + 143^6 + 72^6 + 34^6 \quad (55)$$

$$244^6 + 123^6 + 112^6 = 238^6 + 180^6 + 91^6 + 72^6 \quad (56)$$

$$241^6 + 172^6 + 156^6 = 246^6 + 145^6 + 132^6 + 56^6 \quad (57)$$

$$257^6 + 155^6 + 6^6 = 252^6 + 181^6 + 143^6 + 114^6 \quad (58)$$

$$265^6 + 147^6 + 12^6 = 231^6 + 221^6 + 210^6 + 114^6 \quad (59)$$

$$260^6 + 218^6 + 185^6 = 276^6 + 152^6 + 112^6 + 25^6 \quad (60)$$

$$305^6 + 85^6 + 66^6 = 273^6 + 267^6 + 172^6 + 122^6 \quad (61)$$

$$312^6 + 241^6 + 33^6 = 315^6 + 228^6 + 99^6 + 2^6 \quad (62)$$

$$331^6 + 234^6 + 59^6 = 306^6 + 294^6 + 151^6 + 95^6 \quad (63)$$

$$332^6 + 243^6 + 43^6 = 338^6 + 177^6 + 168^6 + 95^6 \quad (64)$$

$$351^6 + 265^6 + 221^6 = 336^6 + 309^6 + 169^6 + 73^6 \quad (65)$$

$$365^6 + 137^6 + 126^6 = 360^6 + 234^6 + 175^6 + 133^6 \quad (66)$$

$$360^6 + 265^6 + 200^6 = 336^6 + 318^6 + 212^6 + 169^6 \quad (67)$$

$$348^6 + 325^6 + 36^6 = 357^6 + 276^6 + 276^6 + 82^6 \quad (68)$$

$$373^6 + 288^6 + 104^6 = 363^6 + 292^6 + 266^6 + 120^6 \quad (69)$$

$$386^6 + 113^6 + 62^6 = 378^6 + 260^6 + 209^6 + 88^6 \quad (70)$$

(Lander *et al.* 1967, Ekl 1998).

Moessner (1947) gave three parametric solutions to the 6.4.4 equation. The smallest 6.4.4 solution is

$$2^6 + 2^6 + 9^6 + 9^6 = 3^6 + 5^6 + 6^6 + 10^6 \quad (71)$$

(Rao 1934, Lander *et al.* 1967). The smallest 6.4.4.4 solution is

$$1^6 + 34^6 + 49^6 + 111^6 = 7^6 + 43^6 + 69^6 + 110^6$$
$$= 18^6 + 25^6 + 77^6 + 109^6 \quad (72)$$

(Lander *et al.* 1967).

Moessner and Gloden (1944) give the 6.7.8 solution

$$32^6 + 31^6 + 23^6 + 22^6 + 13^6 + 6^6 + 5^6$$
$$= 33^6 + 28^6 + 27^6 + 20^6 + 11^6 + 10^6 + 2^6 + 1^6. \quad (73)$$

References

Ekl, R. L. "Equal Sums of Four Seventh Powers." *Math. Comput.* **65**, 1755–756, 1996.

Ekl, R. L. "New Results in Equal Sums of Like Powers." *Math. Comput.* **67**, 1309–315, 1998.

Guy, R. K. "Sums of Like Powers. Euler's Conjecture." §D1 in *Unsolved Problems in Number Theory, 2nd ed.* New York: Springer-Verlag, pp. 139–44, 1994.

Lander, L. J.; Parkin, T. R.; and Selfridge, J. L. "A Survey of Equal Sums of Like Powers." *Math. Comput.* **21**, 446–59, 1967.

Martin, A. "On Powers of Numbers Whose Sum is the Same Power of Some Number." *Quart. J. Math.* **26**, 225–27, 1893.

Moessner, A. "On Equal Sums of Like Powers." *Math. Student* **15**, 83–8, 1947.

Moessner, A. "Einige zahlentheoretische Untersuchungen und diophantische Probleme." *Glasnik Mat.-Fiz. Astron. Drustvo Mat. Fiz. Hrvatske Ser. 2* **14**, 177–82, 1959.

Moessner, A. and Gloden, A. "Einige Zahlentheoretische Untersuchungen und Resultate." *Bull. Sci. École Polytech. de Timisoara* **11**, 196–19, 1944.

PowerSum. "Index of Equal Sums of Like Powers." http://www.chez.com/powersum/.

Rao, S. K. "On Sums of Sixth Powers." *J. London Math. Soc.* **9**, 172–73, 1934.

Resta, G. "New Results on Equal Sums of Sixth Powers." Instituto di Matematica Computazionale, Pisa, Italy. April 1999. http://www.chez.com/powersum/Tr-b4-8.zip.

Weisstein, E. W. "Like Powers." MATHEMATICA NOTEBOOK LIKEPOWERS.M.

Diophantine Equation—7th Powers

The 7.1.2 equation

$$A^7 + B^7 = C^7 \quad (1)$$

is a special case of FERMAT'S LAST THEOREM with $n = 7$, and so has no solution. No solutions to the 7.1.3, 7.1.4, 7.1.5, 7.1.6 equations are known. There is now a known solutions to the 7.1.7 equation,

$$568^7 = 525^7 + 439^7 + 430^7 + 413^7 + 266^7 + 258^7 + 127^7 \quad (2)$$

(M. Dodrill 1999, PowerSum), requiring an update by Guy (1994, p. 140). The smallest 7.1.8 solution is

$$12^7 + 35^7 + 53^7 + 58^7 + 64^7 + 83^7 + 85^7 + 90^7$$
$$= 102^7 \quad (3)$$

(Lander *et al.* 1967, Ekl 1998). The smallest 7.1.9 solution is

$$6^7 + 14^7 + 20^7 + 22^7 + 27^7 + 33^7 + 41^7 + 50^7 + 59^7$$
$$= 62^7 \quad (4)$$

(Lander *et al.* 1967).

No solutions to the 7.2.2, 7.2.3, 7.2.4, or 7.2.5 equations are known. The smallest 7.2.6 equation is

$$125^7 + 24^7 = 121^7 + 94^7 + 83^7 + 61^7 + 57^7 + 27^7 \quad (5)$$

(Meyrignac). The smallest 7.2.8 solution is

$$5^7 + 6^7 + 7^7 + 15^7 + 15^7 + 20^7 + 28^7 + 31^7$$
$$= 10^7 + 33^7 \quad (6)$$

(Lander *et al.* 1967, Ekl 1998). A 7.2.10.10 solution is

$$2^7 + 27^7 = 4^7 + 8^7 + 13^7 + 14^7 + 14^7 + 16^7 + 18^7 + 22^7$$
$$+ 23^7 + 23^7$$
$$= 7^7 + 7^7 + 9^7 + 13^7 + 14^7 + 18^7 + 20^7 + 22^7$$
$$+ 22^7 + 23^7 \qquad (7)$$

(Lander *et al.* 1967).

No solutions to the 7.3.3 equation are known (Ekl 1996), nor are any to 7.3.4. The smallest 7.3.5 equations are

$$96^7 + 41^7 + 17^7 = 87^7 + 2\cdot 77^7 + 68^7 + 56^7 \qquad (8)$$

$$153^7 + 43^7 + 14^7 = 140^7 + 137^7 + 59^7 + 42^7 + 42^7. \qquad (9)$$

No solutions are known to the 7.3.6 equation. The smallest 7.3.7 solution is

$$7^7 + 7^7 + 12^7 + 16^7 + 27^7 + 28^7 + 31^7$$
$$= 26^7 + 30^7 + 30^7 \qquad (10)$$

(Lander *et al.* 1967).

Guy (1994, p. 140) asked if a 7.4.4 equation exists. The following solution provide an affirmative answer

$$149^7 + 123^7 + 14^7 + 10^7 = 146^7 + 129^7 + 90^7 + 15^7 \quad (11)$$

$$194^7 + 150^7 + 105^7 + 23^7$$
$$= 192^7 + 152^7 + 132^7 + 38^7 \qquad (12)$$

$$354^7 + 112^7 + 52^7 + 19^7 = 343^7 + 281^7 + 46^7 + 35^7 \quad (13)$$

(Ekl 1996, Elk 1998, M. Lau 1999, PowerSum). Numerical solutions to the 7.4.5 equation are given by Gloden (1948). The smallest primitive 7.4.5 solutions are

$$50^7 + 43^7 + 16^7 + 12^7 = 52^7 + 29^7 + 26^7 + 11^7 + 3^7 \quad (14)$$

$$81^7 + 58^7 + 19^7 + 9^7 = 77^7 + 68^7 + 56^7 + 48^7 + 2^7 \quad (15)$$

$$87^7 + 74^7 + 69^7 + 40^7$$
$$= 82^7 + 79^7 + 75^7 + 25^7 + 9^7 \qquad (16)$$

$$99^7 + 76^7 + 32^7 + 29^7$$
$$= 93^7 + 88^7 + 66^7 + 36^7 + 35^7 \qquad (17)$$

$$98^7 + 82^7 + 58^7 + 34^7$$
$$= 99^7 + 75^7 + 69^7 + 16^7 + 13^7 \qquad (18)$$

$$104^7 + 96^7 + 60^7 + 14^7$$
$$= 102^7 + 95^7 + 81^7 + 57^7 + 23^7 \qquad (19)$$

$$111^7 + 102^7 + 40^7 + 29^7$$
$$= 112^7 + 96^7 + 82^7 + 55^7 + 21^7 \qquad (20)$$

$$113^7 + 102^7 + 86^7 + 23^7$$
$$= 120^7 + 81^7 + 58^7 + 55^7 + 10^7 \qquad (21)$$

(Lander *et al.* 1967, Ekl 1998).

Gloden (1949) gives parametric solutions to the 7.5.5 equation. The first few 7.5.5 solutions are

$$8^7 + 8^7 + 13^7 + 16^7 + 19^7$$
$$= 2^7 + 12^7 + 15^7 + 17^7 + 18^7 \qquad (22)$$

$$4^7 + 8^7 + 14^7 + 16^7 + 23^7$$
$$= 7^7 + 7^7 + 9^7 + 20^7 + 22^7 \qquad (23)$$

$$11^7 + 12^7 + 18^7 + 21^7 + 26^7$$
$$= 9^7 + 10^7 + 22^7 + 23^7 + 24^7 \qquad (24)$$

$$6^7 + 12^7 + 20^7 + 22^7 + 27^7$$
$$= 10^7 + 13^7 + 13^7 + 25^7 + 26^7 \qquad (25)$$

$$3^7 + 13^7 + 17^7 + 24^7 + 38^7$$
$$= 14^7 + 26^7 + 32^7 + 32^7 + 33^7 \qquad (26)$$

(Lander *et al.* 1967). Ekl (1998) mentions but does not list 107 primitive solutions to 7.5.5.

A parametric solution to the 7.6.6 equation was given by Sastry and Rai (1948). The smallest is

$$2^7 + 3^7 + 6^7 + 6^7 + 10^7 + 13^7$$
$$= 1^7 + 1^7 + 7^7 + 7^7 + 12^7 + 12^7 \qquad (27)$$

(Lander *et al.* 1967). Another found by Chen Shuwen is

$$87^7 + 233^7 + 264^7 + 396^7 + 496^7 + 540^7$$
$$= 90^7 + 206^7 + 309^7 + 366^7 + 522^7 + 523^7. \qquad (28)$$

Moessner and Gloden (1944) gave the 7.9.10 solution

$$42^7 + 37^7 + 36^7 + 29^7 + 23^7 + 19^7 + 13^7 + 6^7 + 5^7$$
$$= 41^7 + 40^7 + 33^7 + 28^7 + 27^7 + 15^7 + 14^7 + 9^7 + 2^7$$
$$+ 1^7. \qquad (29)$$

References

Ekl, R. L. "Equal Sums of Four Seventh Powers." *Math. Comput.* **65**, 1755–756, 1996.

Ekl, R. L. "New Results in Equal Sums of Like Powers." *Math. Comput.* **67**, 1309–315, 1998.

Gloden, A. "Zwei Parameterlösungen einer mehrgeradigen Gleichung." *Arch. Math.* **1**, 480–82, 1949.

Guy, R. K. "Sums of Like Powers. Euler's Conjecture." §D1 in *Unsolved Problems in Number Theory, 2nd ed.* New York: Springer-Verlag, pp. 139–44, 1994.

Lander, L. J.; Parkin, T. R.; and Selfridge, J. L. "A Survey of Equal Sums of Like Powers." *Math. Comput.* **21**, 446–59, 1967.

Moessner, A. and Gloden, A. "Einige Zahlentheoretische Untersuchungen und Resultate." *Bull. Sci. École Polytech. de Timisoara* **11**, 196–19, 1944.

Nagell, T. "The Diophantine Equation $x^7 + y^7 + z^7 = 0$." §67 in *Introduction to Number Theory.* New York: Wiley, pp. 248–51, 1951.

PowerSum. "Index of Equal Sums of Like Powers." http://www.chez.com/powersum/.

Sastry, S. and Rai, T. "On Equal Sums of Like Powers." *Math. Student* **16**, 18–9, 1948.

Weisstein, E. W. "Like Powers." MATHEMATICA NOTEBOOK LIKEPOWERS.M.

Diophantine Equation—8th Powers
The 8.1.2 equation

$$A^8 + B^8 = C^8 \tag{1}$$

is a special case of FERMAT'S LAST THEOREM with $n = 8$, and so has no solution. No 8.1.3, 8.1.4, 8.1.5, 8.1.6, 8.1.7, or 8.1.8 solutions are known. The smallest 8.1.9 is

$$1167^8 = 10948 + 1040^8 + 560^8 + 558^8$$

$$+366^8 + 348^8 + 284^8 + 271^8 + 190^8 \tag{2}$$

(N. Kuosa). The smallest 8.1.10 is

$$235^8 = 226^8 + 184^8 + 171^8 + 152^8 + 142^8$$

$$+66^8 + 58^8 + 34^8 + 16^8 + 6^8 \tag{3}$$

(N. Kuosa, PowerSum). The smallest 8.1.11 solution is

$$14^8 + 18^8 + 44^8 + 44^8 + 66^8 + 70^8 + 92^8 + 93^8$$

$$+96^8 + 106^8 + 112^8 = 125^8 \tag{4}$$

(Lander *et al.* 1967, Ekl 1998). The smallest 8.1.12 solution is

$$8^8 + 8^8 + 10^8 + 24^8 + 24^8 + 248 + 26^8 + 30^8 + 34^8$$

$$+44^8 + 52^8 + 63^8 = 65^8 \tag{5}$$

(Lander *et al.* 1967). The general identity

$$\left(2^{8k+4} + 1\right)^8 = \left(2^{8k+4} - 1\right)^8 + \left(2^{7k+4}\right)^8 + \left(2^{k+1}\right)^8$$

$$+7\left[\left(2^{5k+3}\right)^8 + \left(2^{3k+2}\right)^8\right] \tag{6}$$

gives a solution to the 8.1.17 equation (Lander *et al.* 1967).

No 8.2.2, 8.2.3, 8.2.4, 8.2.5, 8.2.6, or 8.2.7 solutions are known. The smallest 8.2.8 solution is

$$129^8 + 95^8 = 128^8 + 92^8 + 86^8 + 82^8 + 74^8 + 57^8 + 55^8$$

$$+ 20^8. \tag{7}$$

The smallest 8.2.9 solution is

$$2^8 + 7^8 + 8^8 + 16^8 + 17^8 + 20^8 + 20^8 + 24^8 + 24^8$$

$$= 11^8 + 27^8 \tag{8}$$

(Lander *et al.* 1967, Ekl 1998).

No 8.3.3, 8.3.4, 8.3.5, or 8.3.6 solutions are known. The smallest 8.3.7 solution is

$$108^8 + 68^8 + 5^8$$

$$= 102^8 + 88^8 + 88^8 + 52^8 + 37^8 + 26^8 + 6^8. \tag{9}$$

The smallest 8.3.8 solution is

$$6^8 + 12^8 + 16^8 + 16^8 + 38^8 + 38^8 + 40^8 + 47^8$$

$$= 8^8 + 17^8 + 50^8 \tag{10}$$

(Lander *et al.* 1967, Ekl 1998).

No 8.4.4 solutions is known. The smallest 8.4.5 solution is

$$221^8 + 108^8 + 94^8 + 94^8$$

$$= 195^8 + 194^8 + 188^8 + 126^8 + 38^8. \tag{11}$$

The smallest 8.4.6 solution is

$$47^8 + 29^8 + 12^8 + 5^8$$

$$= 45^8 + 40^8 + 30^8 + 268 + 23^8 + 3^8 \tag{12}$$

(Ekl 1998). The smallest 8.4.7 solution is

$$7^8 + 9^8 + 16^8 + 22^8 + 22^8 + 28^8 + 34^8$$

$$= 6^8 + 11^8 + 20^8 + 35^8 \tag{13}$$

(Lander *et al.* 1967).

The smallest 8.5.5 solutions are

$$43^8 + 20^8 + 11^8 + 10^8 + 1^8$$

$$= 41^8 + 35^8 + 32^8 + 28^8 + 5^8 \tag{14}$$

$$42^8 + 41^8 + 35^8 + 9^8 + 6^8$$

$$= 45^8 + 36^8 + 27^8 + 13^8 + 8^8 \tag{15}$$

$$63^8 + 63^8 + 31^8 + 15^8 + 6^8$$

$$= 65^8 + 59^8 + 48^8 + 37^8 + 7^8 \tag{16}$$

$$75^8 + 47^8 + 39^8 + 26^8 + 6^8$$

$$= 67^8 + 67^8 + 62^8 + 20^8 + 11^8 \tag{17}$$

$$77^8 + 76^8 + 71^8 + 42^8 + 28^8$$

$$= 86^8 + 41^8 + 36^8 + 32^8 + 29^8 \tag{18}$$

$$90^8 + 81^8 + 10^8 + 4^8 + 3^8$$

$$= 92^8 + 74^8 + 55^8 + 50^8 + 37^8 \tag{19}$$

$$93^8 + 65^8 + 65^8 + 41^8 + 13^8$$

$$= 81^8 + 81^8 + 79^8 + 75^8 + 45^8 \tag{20}$$

$$89^8 + 87^8 + 28^8 + 14^8 + 14^8$$

$$= 96^8 + 36^8 + 33^8 + 31^8 + 24^8 \tag{21}$$

$$93^8 + 90^8 + 32^8 + 18^8 + 9^8$$

$$= 94^8 + 86^8 + 71^8 + 60^8 + 19^8 \tag{22}$$

$$104^8 + 73^8 + 36^8 + 17^8 + 3^8$$

$$= 103^8 + 78^8 + 68^8 + 11^8 + 9^8 \tag{23}$$

$$103^8 + 86^8 + 58^8 + 11^8 + 8^8$$

$$= 104^8 + 78^8 + 69^8 + 62^8 + 9^8 \tag{24}$$

$$108^8 + 101^8 + 88^8 + 45^8 + 1^8$$

$$= 116^8 + 59^8 + 46^8 + 15^8 + 3^8 \tag{25}$$

$$116^8 + 92^8 + 79^8 + 33^8 + 25^8$$

$$= 113^8 + 103^8 + 60^8 + 44^8 + 31^8 \tag{26}$$

$$123^8 + 97^8 + 71^8 + 10^8 + 2^8$$
$$= 125^8 + 77^8 + 48^8 + 37^8 + 26^8 \qquad (27)$$

$$121^8 + 109^8 + 71^8 + 70^8 + 40^8$$
$$= 120^8 + 104^8 + 99^8 + 75^8 + 61^8 \qquad (28)$$

$$127^8 + 43^8 + 26^8 + 10^8 + 3^8$$
$$= 123^8 + 105^8 + 69^8 + 42^8 + 14^8 \qquad (29)$$

(Letac 1942, Lander *et al.* 1967, Ekl 1998). The smallest 8.5.6 solutions are

$$36^8 + 36^8 + 33^8 + 25^8 + 21^8$$
$$= 38^8 + 34^8 + 32^8 + 15^8 + 15^8 + 13^8 \qquad (30)$$

$$39^8 + 33^8 + 32^8 + 25^8 + 19^8$$
$$= 37^8 + 35^8 + 35^8 + 17^8 + 16^8 + 2^8 \qquad (31)$$

$$41^8 + 21^8 + 20^8 + 19^8 + 16^8$$
$$= 40^8 + 31^8 + 30^8 + 17^8 + 9^8 + 8^8 \qquad (32)$$

$$43^8 + 34^8 + 24^8 + 8^8 + 1^8$$
$$= 42^8 + 37^8 + 28^8 + 16^8 + 16^8 + 15^8 \qquad (33)$$

$$44^8 + 42^8 + 24^8 + 17^8 + 4^8$$
$$= 47^8 + 20^8 + 18^8 + 8^8 + 6^8 + 6^8 \qquad (34)$$

$$49^8 + 29^8 + 22^8 + 1^8 + 1^8$$
$$= 47^8 + 42^8 + 26^8 + 23^8 + 17^8 + 5^8 \qquad (35)$$

$$46^8 + 46^8 + 33^8 + 30^8 + 9^8$$
$$= 45^8 + 45^8 + 36^8 + 36^8 + 34^8 + 32^8 \qquad (36)$$

$$51^8 + 48^8 + 39^8 + 21^8 + 10^8$$
$$= 53^8 + 45^8 + 25^8 + 22^8 + 22^8 + 6^8 \qquad (37)$$

$$55^8 + 37^8 + 19^8 + 19^8 + 18^8$$
$$= 51^8 + 50^8 + 35^8 + 26^8 + 11^8 + 9^8 \qquad (38)$$

$$58^8 + 17^8 + 13^8 + 10^8 + 7^8$$
$$= 56^8 + 45^8 + 41^8 + 40^8 + 8^8 + 1^8 \qquad (39)$$

$$55^8 + 53^8 + 24^8 + 21^8 + 2^8$$
$$= 52^8 + 52^8 + 50^8 + 25^8 + 17^8 + 7^8 \qquad (40)$$

$$58^8 + 51^8 + 17^8 + 11^8 + 11^8$$
$$= 60^8 + 37^8 + 34^8 + 29^8 + 23^8 + 3^8 \qquad (41)$$

$$54^8 + 51^8 + 51^8 + 43^8 + 4^8$$
$$= 59^8 + 46^8 + 41^8 + 30^8 + 17^8 + 2^8 \qquad (42)$$

$$58^8 + 53^8 + 35^8 + 19^8 + 17^8$$
$$= 61^8 + 30^8 + 25^8 + 23^8 + 16^8 + 1^8 \qquad (43)$$

$$61^8 + 29^8 + 28^8 + 27^8 + 26^8$$
$$= 57^8 + 52^8 + 48^8 + 17^8 + 14^8 + 5^8 \qquad (44)$$

$$58^8 + 51^8 + 49^8 + 8^8 + 6^8$$
$$= 61^8 + 44^8 + 32^8 + 26^8 + 10^8 + 1^8 \qquad (45)$$

$$62^8 + 53^8 + 38^8 + 32^8 + 23^8$$
$$= 61^8 + 52^8 + 50^8 + 34^8 + 24^8 + 1^8 \qquad (46)$$

$$59^8 + 57^8 + 47^8 + 40^8 + 8^8$$
$$= 62^8 + 52^8 + 45^8 + 17^8 + 15^8 + 2^8 \qquad (47)$$

$$63^8 + 62^8 + 55^8 + 43^8 + 27^8$$
$$= 65^8 + 59^8 + 56^8 + 17^8 + 13^8 + 10^8 \qquad (48)$$

(Ekl 1998).

Moessner and Gloden (1944) found solutions to the 8.6.6 equation. The smallest 8.6.6 solution is

$$3^8 + 6^8 + 8^8 + 10^8 + 15^8 + 23^8$$
$$= 5^8 + 9^8 + 9^8 + 12^8 + 20^8 + 22^8 \qquad (49)$$

(Lander *et al.* 1967). Ekl (1998) mentions but does not list 204 primitive solutions to the 8.6.6 equation. Moessner and Gloden (1944) found solutions to the 8.6.7 equation.

Parametric solutions to the 8.7.7 equation were given by Moessner (1947) and Gloden (1948). The smallest 8.7.7 solution is

$$1^8 + 3^8 + 5^8 + 6^8 + 6^8 + 8^8 + 13^8$$
$$= 4^8 + 7^8 + 9^8 + 9^8 + 10^8 + 11^8 + 12^8 \qquad (50)$$

(Lander *et al.* 1967).

Sastry (1934) used the smallest 17– solution to give a parametric 8.8.8 solution. The smallest 8.8.8 solution is

$$1^8 + 3^8 + 7^8 + 7^8 + 7^8 + 10^8 + 10^8 + 12^8$$
$$= 4^8 + 5^8 + 5^8 + 6^8 + 6^8 + 11^8 + 11^8 + 11^8 \qquad (51)$$

(Lander *et al.* 1967).

Letac (1942) found solutions to the 8.9.9 equation.

Moessner and Gloden (1944) found the 8.9.10 solution

$$54^8 + 53^8 + 46^8 + 37^8 + 29^8 + 23^8 + 22^8 + 6^8 + 5^8$$
$$= 55^8 + +50^8 + 49^8 + 33^8 + 32^8 + 26^8 + 18^8 + 9^8 + 2^8$$
$$+ 1^8. \qquad (52)$$

References

Ekl, R. L. "New Results in Equal Sums of Like Powers." *Math. Comput.* **67**, 1309–315, 1998.

Gloden, A. "Parametric Solutions of Two Multi-Degreed Equalities." *Amer. Math. Monthly* **55**, 86–8, 1948.

Lander, L. J.; Parkin, T. R.; and Selfridge, J. L. "A Survey of Equal Sums of Like Powers." *Math. Comput.* **21**, 446–59, 1967.

Letac, A. *Gazetta Mathematica* **48**, 68–9, 1942.

Moessner, A. "On Equal Sums of Like Powers." *Math. Student* **15**, 83–8, 1947.

Moessner, A. and Gloden, A. "Einige Zahlentheoretische Untersuchungen und Resultate." *Bull. Sci. École Polytech. de Timisoara* **11**, 196–19, 1944.

Sastry, S. "On Sums of Powers." *J. London Math. Soc.* **9**, 242–46, 1934.

Weisstein, E. W. "Like Powers." Mathematica notebook LikePowers.m.

Diophantine Equation—9th Powers

The 9.1.2 equation

$$A^9 = B^9 + C^9 \tag{1}$$

is a special case of Fermat's last theorem with $n = 9$, and so has no solution. No 9.1.3, 9.1.4, 9.1.5, 9.1.6, 9.1.7, 9.1.8, 9.1.9, 9.1.10, or 9.1.11 solutions are known. The smallest 9.1.12 solution is

$$103^9 = 91^9 + 91^9 + 89^9 + 71^9 + 68^9 + 65^9$$
$$+ 43^9 + 42^9 + 19^9 + 16^9 + 13^9 + 5^9. \tag{2}$$

To 9.1.13 solution is known. The smallest 9.1.14 solution is

$$66^9 = 63^9 + 54^9 + 51^9 + 49^9 + 38^9 + 35^9 + 29^9$$
$$+ 24^9 + 21^9 + 12^9 + 10^9 + 7^9 + 2^9 + 1^9 \tag{3}$$

(Ekl 1998).

No 9.2.2, 9.2.3, 9.2.4,. 9.2.5, 9.2.6, 9.2.7, 9.2.8, or 9.2.9 solutions are known. A 9.2.10 solution is given by

$$121^9 + 2 \cdot 116^9 + 115^9 + 89^9 + 52^9 + 28^9$$
$$+ 26^9 + 14^9 + 9^9 = 137^9 + 69^9 \tag{4}$$

(L. Morelli 1999, PowerSum). No 9.2.11 solutions are known. The smallest 9.2.12 solution is

$$4 \cdot 2^9 + 2 \cdot 3^9 + 4^9 + 7^9 + 16^9 + 17^9 + 2 \cdot 19^9$$
$$= 15^9 + 21^9 \tag{5}$$

(Lander *et al.* 1967, Ekl 1998). There are no known 9.1.13 or 9.1.14 solutions. The smallest 9.1.15 solution is

$$2^9 + 2^9 + 4^9 + 6^9 + 6^9 + 7^9 + 9^9 + 9^9 + 10^9 + 15^9$$
$$+ 18^9 + 21^9 + 21^9 + 23^9 + 23^9 = 26^9 \tag{6}$$

(Lander *et al.* 1967).

There are no known 9.3.3, 9.3.4, 9.3.5, 9.3.6, 9.3.7, or 9.3.8 solutions. The smallest 9.3.9 solution is

$$2 \cdot 38^9 + 3^9 = 41^9 + 23^9 + 2 \cdot 20^9 + 18^9 + 2 \cdot 13^9 + 12^9 + 9^9 \tag{7}$$

(Ekl 1998). There is no known 9.3.10 solution. The smallest 9.3.11 solution is

$$2^9 + 3^9 + 6^9 + 7^9 + 9^9 + 9^9 + 19^9 + 19^9 + 21^9 + 25^9$$
$$+ 29^9 = 13^9 + 16^9 + 30^9 \tag{8}$$

(Lander *et al.* 1967).

There are no known 9.4.4 or 9.4.5 solutions are known. The smallest 9.4.6 solution is

$$90^9 + 64^9 + 35^9 + 35^9$$
$$= 86^9 + 80^9 + 62^9 + 43^9 + 27^9 + 16^9. \tag{9}$$

There are no known 9.4.7 or 9.4.8 solutions. The smallest 9.4.9 solution is

$$38^9 + 31^9 + 12^9 + 2^9$$
$$= 36^9 + 2 \cdot 32^9 + 30^9 + 15^9 + 13^9 + 8^9 + 4^9 + 3^9 \tag{10}$$

(Ekl 1998). The smallest 9.4.10 solutions are

$$2^9 + 6^9 + 6^9 + 9^9 + 10^9 + 11^9 + 14^9 + 18^9 + 19^9 + 19^9$$
$$= 5^9 + 12^9 + 16^9 + 21^9 \tag{11}$$

(Lander *et al.* 1967).

The smallest 9.5.5 solution is

$$192^9 + 101^9 + 91^9 + 30^9 + 26^9$$
$$= 180^9 + 175^9 + 116^9 + 17^9 + 12^9. \tag{12}$$

There is no known 9.5.6 solution. The smallest 9.5.7 solution is

$$35^9 + 26^9 + 2 \cdot 15^9 + 12^9$$
$$= 33^9 + 32^9 + 24^9 + 16^9 + 14^9 + 8^9 + 6^9 \tag{13}$$

(Ekl 1998). There are no known 9.5.8, 9.5.9, or 9.5.10 solutions. The smallest 9.5.11 solution is

$$3^9 + 5^9 + 5^9 + 9^9 + 9^9 + 12^9 + 15^9 + 15^9 + 16^9 + 21^9$$
$$+ 21^9 = 7^9 + 8^9 + 14^9 + 20^9 + 22^9 \tag{14}$$

(Lander *et al.* 1967).

The smallest 9.6.6 solutions are

$$23^9 + 18^9 + 14^9 + 13^9 + 13^9 + 1^9$$
$$= 22^9 + 21^9 + 15^9 + 10^9 + 9^9 + 5^9 \tag{15}$$

$$31^9 + 23^9 + 21^9 + 14^9 + 9^9 + 2^9$$
$$= 29^9 + 29^9 + 15^9 + 11^9 + 10^9 + 6^9 \tag{16}$$

$$46^9 + 44^9 + 27^9 + 27^9 + 27^9 + 9^9$$
$$= 48^9 + 39^9 + 23^9 + 15^9 + 13^9 + 12^9 \tag{17}$$

$$47^9 + 47^9 + 22^9 + 22^9 + 12^9 + 4^9$$
$$= 50^9 + 39^9 + 35^9 + 13^9 + 10^9 + 7^9 \tag{18}$$

$$54^9 + 52^9 + 48^9 + 47^9 + 46^9 + 14^9$$
$$= 60^9 + 18^9 + 17^9 + 16^9 + 15^9 + 15^9 \tag{19}$$

$$70^9 + 44^9 + 36^9 + 33^9 + 19^9 + 4^9$$
$$= 64^9 + 63^9 + 57^9 + 47^9 + 22^9 + 13^9 \tag{20}$$

$$68^9 + 58^9 + 50^9 + 46^9 + 41^9 + 7^9$$
$$= 70^9 + 48^9 + 26^9 + 25^9 + 23^9 + 18^9 \tag{21}$$

(Lander *et al.* 1967, Ekl 1998).

Ekl (1998) mentions but does not list nine primitive solutions to the 9.7.7 equation.

Moessner (1947) gives a parametric solution to the 9.10.10 equation.

Palamá (1953) gave a solution to the 9.11.11 equation.

Moessner and Gloden (1944) give the 9.11.12 solution

$$72^9 + 67^9 + 66^9 + 53^9 + 43^9 + 37^9 + 35^9 + 29^9 + 19^9$$
$$+ 6^9 + 5^9$$
$$= 71^9 + 70^9 + 63^9 + 55^9 + 40^9 + 39^9 + 33^9 + 32^9$$
$$+ 17^9 + 9^9 + 2^9 + 1^9. \tag{22}$$

References

Ekl, R. L. "New Results in Equal Sums of Like Powers." *Math. Comput.* **67**, 1309–315, 1998.

Lander, L. J.; Parkin, T. R.; and Selfridge, J. L. "A Survey of Equal Sums of Like Powers." *Math. Comput.* **21**, 446–59, 1967.

Moessner, A. "On Equal Sums of Like Powers." *Math. Student* **15**, 83–8, 1947.

Moessner, A. and Gloden, A. "Einige Zahlentheoretische Untersuchungen und Resultate." *Bull. Sci. École Polytech. de Timisoara* **11**, 196–19, 1944.

Palamá, G. "Diophantine Systems of the Type $\Sigma_{i=1}^p a_i^k = \Sigma_{i=1}^p b_i^k$ $(k = 1, 2, ..., n, n+2, n+4, ..., n+2r)$." *Scripta Math.* **19**, 132–34, 1953.

PowerSum. "Index of Equal Sums of Like Powers." http://www.chez.com/powersum/.

Weisstein, E. W. "Like Powers." MATHEMATICA NOTEBOOK LIKEPOWERS.M.

Diophantine Equation—nth Powers

The 2– equation

$$A^n + B^n + = C^n \tag{1}$$

is a special case of FERMAT'S LAST THEOREM and so has no solutions for $n \geq 3$. Lander *et al.* (1967) give a table showing the smallest n for which a solution to

$$x_1^k + x_2^k + \ldots + x_m^k = y_1^k + y_2^k + \ldots + y_n^k, \tag{2}$$

with $1 \leq m \leq n$ is known. An updated table is given below; a more extensive table may be found at the PowerSum web site.

	k								
m	2	3	4	5	6	7	8	9	10
1	2	3	3	4	7	8	11	15	23
2	2	2	2	4	7	8	9	12	19
3			3	3	7		8	11	24
4				4		7	10	23	
5				5		5	11	16	
6						6	27		
7						7			

Take the results from the RAMANUJAN 6–0– IDENTITY that for $ad = bc$, with

$$F_{2m}(a, b, c, d)$$
$$= (a + b + c)^{2m} + (b + c + d)^{2m} - (c + d + a)^{2m}$$
$$- (d + a + b)^{2m} + (a - d)^{2m} - (b - c)^{2m} \tag{3}$$

and

$$f_{2m}(x, y) = (1 + x + y)^{2m} + (x + y + xy)^{2m} - (y + xy + 1)^{2m}$$
$$- (xy + 1 + x)^{2m} + (1 - xy)^{2m} - (x - y)^{2m}, \tag{4}$$

then

$$F_{2m}(a, b, c, d) = a^{2m} f_{2m}(x, y). \tag{5}$$

Using

$$f_2(x, y) = 0 \tag{6}$$

$$f_4(x, y) = 0 \tag{7}$$

now gives

$$(a + b + c)^n + (b + c + d)^n + (a - d)^n$$
$$= (c + d + a)^n + (d + a + b)^n + (b - c)^n \tag{8}$$

for $n = 2$ or 4.

See also DIOPHANTINE EQUATION, RAMANUJAN 6–0– IDENTITY

References

Berndt, B. C. *Ramanujan's Notebooks, Part IV.* New York: Springer-Verlag, p. 101, 1994.

Berndt, B. C. and Bhargava, S. "Ramanujan--For Lowbrows." *Amer. Math. Monthly* **100**, 644–56, 1993.

Dickson, L. E. *History of the Theory of Numbers, Vol. 2: Diophantine Analysis.* New York: Chelsea, pp. 653–57, 1966.

Gloden, A. *Mehrgradige Gleichungen.* Groningen, Netherlands: P. Noordhoff, 1944.

Guy, R. K. *Unsolved Problems in Number Theory, 2nd ed.* New York: Springer-Verlag, 1994.

Lander, L. J.; Parkin, T. R.; and Selfridge, J. L. "A Survey of Equal Sums of Like Powers." *Math. Comput.* **21**, 446–59, 1967.

PowerSum. "Index of Equal Sums of Like Powers." http://www.chez.com/powersum/.

Reznick, B. *Sums of Even Powers of Real Linear Forms.* Providence, RI: Amer. Math. Soc., 1992.

Sekigawa, H. and Koyama, K. "Nonexistence Conditions of a Solution for the Congruence $x_1^k + \ldots + x_s^k \equiv N \pmod{p^n}$." *Math. Comput.* **68**, 1283–297, 1999.

Diophantine Quadruple

DIOPHANTINE SET

Diophantine Set

A set S of POSITIVE INTEGERS is said to be Diophantine IFF there exists a POLYNOMIAL Q with integral coefficients in $m \geq 1$ indeterminates such that

$$S = \{Q(x_1, ..., x_m) \geq 1 : x_1 \geq 1, ..., x_m \geq 1\}.$$

It has been proved that the set of PRIME NUMBERS is a Diophantine set.

References

Ribenboim, P. *The New Book of Prime Number Records.* New York: Springer-Verlag, pp. 189–92, 1995.

Diophantus Property

A set of m distinct POSITIVE INTEGERS $S = \{a_1, ..., a_m\}$ satisfies the Diophantus property $D(n)$ of order n (a positive integer) if, for all $i, j = 1, ..., m$ with $i \neq j$,

$$a_i a_j + n = b_{ij}^2, \tag{1}$$

the b_{ij}s are INTEGERS. The set S is called a Diophantine n-tuple.

Diophantine 1-doubles are abundant: $(1, 3)$, $(2, 4)$, $(3, 5)$, $(4, 6)$, $(5, 7)$, $(1, 8)$, $(3, 8)$, $(6, 8)$, $(7, 9)$, $(8, 10)$, $(9, 11)$, ... (Sloane's A050269 and A050270). Diophantine 1-triples are less abundant: $(1, 3, 8)$, $(2, 4, 12)$, $(1, 8, 15)$, $(3, 5, 16)$, $(4, 6, 20)$, ... (Sloane's A050273, A050274, and A050275).

Fermat found the smallest Diophantine 1-quadruple: $\{1, 3, 8, 120\}$ (Davenport and Baker 1969, Jones 1976). There are no others with largest term ≤ 200, and Davenport and Baker (1969) showed that if $c + 1$, $3c + 1$, and $8c + 1$ are all squares, then $c = 120$. Jones (1976) derived an infinite sequence of polynomials $S = \{x, x + 2, c_1(x), c_2(x), ...\}$ such that the product of any two, increased by 1, is the square of a polynomial. Letting $c_{-1}(x) = c_0(x) = 0$, then the general $c_k(x)$ is given by the RECURRENCE RELATION

$$c_k = (4x^2 + 8x + 2)c_{k-1} - c_{k-2} + 4(x + 1). \tag{2}$$

The first few c_k are

$$c_1 = 4(x + 1)$$

$$c_2 = 4(3 + 11x + 12x^2 + 4x^3)$$

$$c_3 = 8(3 + 23x + 62x^2 + 74x^3 + 40x^4 + 8x^5).$$

Letting $x = 1$ gives the sequence $s_n = 1$, 3, 8, 120, 1680, 23408, 326040, ... (Sloane's A051047), for which $\sqrt{s_n s_{n+1} + 1}$ is 2, 5, 31, 449, 6271, 87361, ... (Sloane's A051048).

General $D(1)$ quadruples are

$$\{F_{2n}, F_{2n+2}, F_{2n+4}, 4F_{2n+1}F_{2n+2}F_{2n+3},\} \tag{3}$$

where F_n are FIBONACCI NUMBERS, and

$$\{n, n + 2, 4n + 4, 4(n + 1), (2n + 1)(2n + 3)\}. \tag{4}$$

The quadruplet

$$\{2F_{n-1}, 2F_{n+1}, 2F_n^3 F_{n+1}F_{n+2},$$

$$2F_{n+1}F_{n+2}F_{n+3}(2F_{n+1}^2 - F_n^2)\} \tag{5}$$

is $D(F_n^2)$ (Dujella 1996). Dujella (1993) showed there exist no Diophantine quadruples $D(4k + 2)$.

References

Brown, E. "Sets in Which $xy + k$ is Always a Square." *Math. Comput.* **45**, 613–20, 1985.
Davenport, H. and Baker, A. "The Equations $3x^2 - 2 = y^2$ and $8x^2 - 7 = z^2$." *Quart. J. Math. (Oxford) Ser. 2* **20**, 129–37, 1969.
Diofant Aleksandriĭskiĭ. *Arifmetika i kniga o mnogougol'nyh chislakh* [Russian]. Moscow: Nauka, 1974.
Dujella, A. "Generalization of a Problem of Diophantus." *Acta Arith.* **65**, 15–7, 1993.
Dujella, A. "Diophantine Quadruples for Squares of Fibonacci and Lucas Numbers." *Portugaliae Math.* **52**, 305–18, 1995.
Dujella, A. "Generalized Fibonacci Numbers and the Problem of Diophantus." *Fib. Quart.* **34**, 164–75, 1996.
Hoggatt, V. E. Jr. and Bergum, G. E. "A Problem of Fermat and the Fibonacci Sequence." *Fib. Quart.* **15**, 323–30, 1977.
Jones, B. W. "A Variation of a Problem of Davenport and Diophantus." *Quart. J. Math. (Oxford) Ser. (2)* **27**, 349–53, 1976.
Morgado, J. "Generalization of a Result of Hoggatt and Bergum on Fibonacci Numbers." *Portugaliae Math.* **42**, 441–45, 1983–984.
Sloane, N. J. A. Sequences A050269, A050269, A050273, A050274, A050275, A051047, and A051048 in "An On-Line Version of the Encyclopedia of Integer Sequences." http://www.research.att.com/~njas/sequences/eisonline.html.

Diophantus's Riddle

"Diophantus's youth lasts 1/6 of his life. He grew a beard after 1/12 more of his life. After 1/7 more of his life, Diophantus married. Five years later, he had a son. The son lived exactly half as long as his father, and Diophantus died just four years after his son's death. All of this totals the years Diophantus lived."

Let D be the number of years Diophantus lived, and let S be the number of years his son lived. Then the above word problem gives the two equations

$$D = \left(\frac{1}{6} + \frac{1}{12} + \frac{1}{7}\right)D + 5 + S + 4$$

$$S = \frac{1}{2}D.$$

Solving this simultaneously gives $S = 42$ as the age of the son and $D = 84$ as the age of Diophantus.

References

Hoffman, P. *The Man Who Loved Only Numbers: The Story of Paul Erdos and the Search for Mathematical Truth.* New York: Hyperion, pp. 186–87, 1998.
Pappas, T. "Diophantus' Riddle." *The Joy of Mathematics.* San Carlos, CA: Wide World Publ./Tetra, pp. 123 and 232, 1989.

Dipyramid

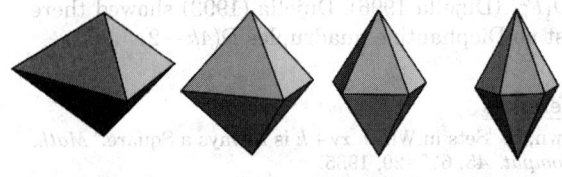

 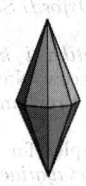 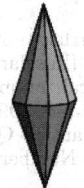 

Two PYRAMIDS symmetrically placed base-to-base, also called a BIPYRAMID. The dipyramids are DUALS of the regular PRISMS.

Consider the dipyramids generated by taking the duals of the n-PRISMS. The edge lengths of the base S_n^b and slant edges S_n^s, half-height (half the distance from peak to peak) h_n, surface areas S_n and volumes V_n (after scaling so that the *smallest* edge length is 1) are given by

$$S_3^b, S_3^s = 2, \frac{4}{3} \tag{1}$$

$$h_3 = \frac{2}{3} \tag{2}$$

$$S_3 = \frac{9}{8}\sqrt{7} \tag{3}$$

$$V_3 = \frac{3}{16}\sqrt{3} \tag{4}$$

$$s_4^b, s_4^s = \sqrt{2}, \sqrt{2} \tag{5}$$

$$h_4 = 1 \tag{6}$$

$$S_4 = 2\sqrt{3} \tag{7}$$

$$V_4 = \frac{1}{3}\sqrt{2} \tag{8}$$

$$S_4^b, S_4^s = \sqrt{5} - 1, \frac{4}{5}\sqrt{5} \tag{9}$$

$$h_4 = \frac{1}{2}\sqrt{2} \tag{10}$$

$$S_4 = 2\sqrt{3} \tag{11}$$

$$V_4 = \frac{1}{3}\sqrt{2} \tag{12}$$

$$S_5^b, S_5^s = \sqrt{5} - 1, \frac{4}{5}\sqrt{5} \tag{13}$$

$$h_5 = \frac{1}{5}(5 + \sqrt{5}) \tag{14}$$

$$S_5 = \frac{1}{2}\sqrt{95 + 40\sqrt{5}} \tag{15}$$

$$V_5 = \frac{1}{6}\sqrt{\frac{1}{2}\left(65 + 29\sqrt{5}\right)} \tag{16}$$

$$S_6^b, S_6^s = \frac{2}{3}\sqrt{3}, \frac{4}{3}\sqrt{3} \tag{17}$$

$$h_6 = 2 \tag{18}$$

$$S_6 = 3\sqrt{15} \tag{19}$$

$$V_6 = 3 \tag{20}$$

$$S_8^b, S_8^s = \sqrt{2\left(2 - \sqrt{2}\right)}, 2\sqrt{2 + \sqrt{2}} \tag{21}$$

$$h_8 = 2 + \sqrt{2} \tag{22}$$

$$S_8 = 4\sqrt{23 + 16\sqrt{2}} \tag{23}$$

$$V_8 = \frac{2}{3}\sqrt{2\left(58 + 41\sqrt{2}\right)} \tag{24}$$

$$S_{10}^b, S_{10}^s = \sqrt{\frac{2}{5}\left(5 - \sqrt{5}\right)}, 4\sqrt{\frac{1}{5}\left(5 + 2\sqrt{5}\right)} \tag{25}$$

$$h_{10} = 3 + \sqrt{5} \tag{26}$$

$$S_{10} = 5\sqrt{55 + 24\sqrt{5}} \tag{27}$$

$$V_{10} = \frac{5}{6} + \left(15 + 7\sqrt{5}\right). \tag{28}$$

JOHNSON SOLID J_{12} is a triangular dipyramid, the OCTAHEDRON is a square dipyramid, and JOHNSON SOLID J_{13} is a pentagonal dipyramid.

See also DELTAHEDRON, ELONGATED DIPYRAMID, JOHNSON SOLID, OCTAHEDRON, PENTAGONAL DIPYRAMID, PRISM, PYRAMID, TRAPEZOHEDRON, TRIANGULAR DIPYRAMID, TRIGONAL DIPYRAMID

References

Cundy, H. and Rollett, A. *Mathematical Models, 3rd ed.* Stradbroke, England: Tarquin Pub., p. 117, 1989.
Pedagoguery Software. `Poly`. http://www.peda.com/poly/.

Dirac Delta Function
DELTA FUNCTION

Dirac Distribution
DELTA FUNCTION

© *1999–001 Wolfram Research, Inc.*

Dirac Equation
The quantum electrodynamical law which applies to spin-1/2 particles and is the relativistic generalization of the SCHRÖDINGER EQUATION. In $3+1$ dimensions (three space dimensions and one time dimension), it is given by

$$\frac{ih}{c}\frac{\partial \psi}{\partial t} = \left[\alpha_x p_x + \alpha_y p_y + \alpha_z p_z + \alpha_4(mc)\right]\psi, \qquad (1)$$

where h is h-bar, c is the speed of light, ψ is the wavefunction , m is the mass of the particle, α_i are the DIRAC MATRICES, σ_i are PAULI SPIN MATRICES, and

$$p_i = \begin{bmatrix} p_i & 0 & 0 & 0 \\ 0 & p_i & 0 & 0 \\ 0 & 0 & p_i & 0 \\ 0 & 0 & 0 & p_i \end{bmatrix}. \qquad (2)$$

In $1+1$ dimensions, the Dirac equation is the system of PARTIAL DIFFERENTIAL EQUATIONS

$$u_t + v_x + imu + 2i\lambda\left(|u|^2 - |v|^2\right)u = 0 \qquad (3)$$

$$v_t + u_x + imv + 2i\lambda\left(|v|^2 - |u|^2\right)v = 0 \qquad (4)$$

(Alvarez *et al.* 1982; Zwillinger 1997, p. 137);

See also SCHRÖDINGER EQUATION

References
Alvarez, A.; Pen-Yu, K.; and Vazquez, L. "The Numerical Study of a Nonlinear One-Dimensional Dirac Equation." *Appl. Math. Comput.* **18**, 1–5, 1983.
Zwillinger, D. *Handbook of Differential Equations, 3rd ed.* Boston, MA: Academic Press, p. 137, 1997.

Dirac Gamma Matrices
DIRAC MATRICES

Dirac Matrices
The Dirac matrices are a class of 4×4 matrices which arise in quantum electrodynamics. There are a variety of different symbols used, and Dirac matrices are also known as gamma matrices or Dirac gamma matrices.

The Dirac matrices are defined as the 4×4 matrices

$$\sigma_i = I_2 \otimes \sigma_{i,\text{Pauli}} \qquad (1)$$

$$\rho_i = \sigma_{i,\text{Pauli}} \otimes I_2, \qquad (2)$$

where σ_i, Pauli are the (2×2) PAULI MATRICES, I_2 is the (2×2) IDENTITY MATRIX, $i = 1, 2, 3$, and $A \otimes B$ is the MATRIX DIRECT PRODUCT. Explicitly, this set of Dirac matrices is then given by

$$I = \begin{bmatrix} 1 & 0 & 0 & 0 \\ 0 & 1 & 0 & 0 \\ 0 & 0 & 1 & 0 \\ 0 & 0 & 0 & 1 \end{bmatrix} \qquad (3)$$

$$\sigma_1 = \begin{bmatrix} 0 & 1 & 0 & 0 \\ 1 & 0 & 0 & 0 \\ 0 & 0 & 0 & 1 \\ 0 & 0 & 1 & 0 \end{bmatrix} \qquad (4)$$

$$\sigma_2 = \begin{bmatrix} 0 & -i & 0 & 0 \\ i & 0 & 0 & 0 \\ 0 & 0 & 0 & -i \\ 0 & 0 & i & 0 \end{bmatrix} \qquad (5)$$

$$\sigma_3 = \begin{bmatrix} 1 & 0 & 0 & 0 \\ 0 & -1 & 0 & 0 \\ 0 & 0 & 1 & 0 \\ 0 & 0 & 0 & -1 \end{bmatrix} \qquad (6)$$

$$\rho_1 = \begin{bmatrix} 0 & 0 & 1 & 0 \\ 0 & 0 & 0 & 1 \\ 1 & 0 & 0 & 0 \\ 0 & 1 & 0 & 0 \end{bmatrix} \qquad (7)$$

$$\rho_2 = \begin{bmatrix} 0 & 0 & -i & 0 \\ 0 & 0 & 0 & -i \\ i & 0 & 0 & 0 \\ 0 & i & 0 & 0 \end{bmatrix} \qquad (8)$$

$$\rho_3 = \begin{bmatrix} 1 & 0 & 0 & 0 \\ 0 & 1 & 0 & 0 \\ 0 & 0 & -1 & 0 \\ 0 & 0 & 0 & -1 \end{bmatrix} \qquad (9)$$

These matrices satisfy the anticommutation identities

$$\sigma_i \sigma_j + \sigma_j \sigma_i = 2\delta_{ij} I \qquad (10)$$

$$\rho_i \rho_j + \rho_j \rho_i = 2\delta_{ij} I, \qquad (11)$$

where δ_{ij} is the KRONECKER DELTA, the commutation identity

$$[\sigma_i, \rho_j] = \sigma_i \rho_j + \sigma_j \rho_i = 0, \qquad (12)$$

and are cyclic under permutations of indices

$$\sigma_i \sigma_i = i\sigma_k \qquad (13)$$

$$\rho_i \rho_i = i\rho_k. \qquad (14)$$

A total of 16 Dirac matrices can be defined via

$$E_{ij} = \sigma_i \rho_j \qquad (15)$$

for $i,j = 0, 1, 2, 3$ and where $\sigma_0 = \rho_0 \equiv I$. These matrices satisfy

1. $|E_{ij}|=1$, where $|A|$ is the DETERMINANT,
2. $E_{ij}^2 = I$,
3. $E_{ij}^* = E_{ij}$, where A^* denotes the ADJOINT MATRIX, making them Hermitian, and therefore unitary,
4. $Tr(E_{ij}) = 0$, except $Tr(E_{00}) = 4$,
5. Any two E_{ij} multiplied together yield a Dirac matrix to within a multiplicative factor of -1 or $\pm i$,
6. The E_{ij} are linearly independent,
7. The E_{ij} form a complete set, i.e., any 4×4 constant matrix may be written as

$$A = \sum_{i,j=0}^{3} c_{ij} E_{ij}, \tag{16}$$

where the c_{ij} are real or complex and are given by

$$c_{mn} = \frac{1}{4} Tr(A E_{mn}) \tag{17}$$

(Arfken 1985).

Dirac's original matrices were written α_i and were defined by

$$\alpha_i = E_{1i} = \rho_1 \sigma_i \tag{18}$$

$$\alpha_4 = E_{30} = \rho_3, \tag{19}$$

for $i = 1, 2, 3$, giving

$$\alpha_1 = E_{11} = \begin{bmatrix} 0 & 0 & 0 & 1 \\ 0 & 0 & 1 & 0 \\ 0 & 1 & 0 & 0 \\ 1 & 0 & 0 & 0 \end{bmatrix} \tag{20}$$

$$\alpha_2 = E_{12} = \begin{bmatrix} 0 & 0 & 0 & -i \\ 0 & 0 & i & 0 \\ 0 & -i & 0 & 0 \\ i & 0 & 0 & 0 \end{bmatrix} \tag{21}$$

$$\alpha_3 = E_{13} = \begin{bmatrix} 0 & 0 & 1 & 0 \\ 0 & 0 & 0 & -1 \\ 1 & 0 & 0 & 0 \\ 0 & -1 & 0 & 0 \end{bmatrix} \tag{22}$$

$$\alpha_4 = E_{30} = \begin{bmatrix} 1 & 0 & 0 & 1 \\ 0 & 1 & 0 & 0 \\ 0 & 0 & -1 & 0 \\ 0 & 0 & 0 & -1 \end{bmatrix} \tag{23}$$

The additional matrix

$$\alpha_5 = E_{20} = \rho_2 = \begin{bmatrix} 0 & 0 & -i & 0 \\ 0 & 0 & 0 & -i \\ i & 0 & 0 & 0 \\ 0 & i & 0 & 0 \end{bmatrix} \tag{24}$$

is sometimes defined.

A closely related set of Dirac matrices is defined by

$$\gamma_i = \begin{bmatrix} 0 & \sigma_i \\ -\sigma_i & 0 \end{bmatrix} \tag{25}$$

$$\gamma_4 = \begin{bmatrix} I & 0 \\ 2I & -I \end{bmatrix} \tag{26}$$

for $i = 1, 2, 3$ (Goldstein 1980). Instead of γ_4, γ_0, is commonly used. Unfortunately, there are two different conventions for its definition, the "chiral basis"

$$\gamma_0 = \begin{bmatrix} 0 & I \\ I & 0 \end{bmatrix} \tag{27}$$

and the "Dirac basis"

$$\gamma_0 = \begin{bmatrix} I & 0 \\ 0 & -I \end{bmatrix}. \tag{28}$$

Other sets of Dirac matrices are sometimes defined as

$$y_i = E_{2i} \tag{29}$$

$$y_4 = E_{30} \tag{30}$$

$$y_5 = -E_{10} \tag{31}$$

and

$$\delta_i = E_{3i} \tag{32}$$

for $i = 1, 2, 3$ (Arfken 1985).

Any of the 15 Dirac matrices (excluding the identity matrix) commute with eight Dirac matrices and anticommute with the other eight. Let $M \equiv \frac{1}{2}(1 + E_{ij})$, then

$$M^2 = M \tag{33}$$

(Arfken 1985, p. 216). In addition

$$\begin{bmatrix} \alpha_1 \\ \alpha_2 \\ \alpha_3 \end{bmatrix} \times \begin{bmatrix} \alpha_1 \\ \alpha_2 \\ \alpha_3 \end{bmatrix} = 2i\sigma. \tag{34}$$

The products of α_i and y_i satisfy

$$\alpha_1 \alpha_2 \alpha_3 \alpha_4 \alpha_5 = 1 \tag{35}$$

$$y_1 y_2 y_3 y_4 y_5 = 1. \tag{36}$$

The 16 Dirac matrices form six anticommuting sets of five matrices each:

1. $\alpha_1, \alpha_2, \alpha_3, \alpha_4, \alpha_5$,
2. y_1, y_2, y_3, y_4, y_5,
3. $\delta_1, \delta_2, \delta_3, \rho_1, \rho_2$,
4. $\alpha_1, y_1, \delta_1, \sigma_2, \sigma_3$,
5. $\alpha_2, y_2, \delta_2, \sigma_1, \sigma_3$,
6. $\alpha_3, y_3, \delta_3, \sigma_1, \sigma_2$.

See also PAULI MATRICES

Dirac Notation

References

Arfken, G. *Mathematical Methods for Physicists, 3rd ed.* Orlando, FL: Academic Press, pp. 211–17, 1985.

Dirac, P. A. M. *Principles of Quantum Mechanics, 4th ed.* Oxford, England: Oxford University Press, 1982.

Goldstein, H. *Classical Mechanics, 2nd ed.* Reading, MA: Addison-Wesley, p. 580, 1980.

Dirac Notation

A notation invented by Dirac which is very useful in quantum mechanics. The notation defines the "KET" vector, denoted $|\psi\rangle$, and its transpose, called the "BRA" vector and denoted $\langle\psi|$.. The "bracket" is then defined by $\langle\phi|\psi\rangle$.. Dirac notation satisfies the identities

$$\langle\phi|\hat{O}|\psi\rangle \equiv \langle\phi|\hat{O}\psi\rangle$$

$$\langle\phi|\psi\rangle \equiv \int_{-\infty}^{\infty} \bar{\phi}\psi\,dx,$$

where $\bar{\psi}$ is the COMPLEX CONJUGATE.

See also ANGLE BRACKET, BRA, DIFFERENTIAL K-FORM, KET, L2-SPACE, ONE-FORM

Dirac Operator

The operator $D = -i(d + d^*)$, where d^* is the ADJOINT.

Dirac's Theorem

A GRAPH with $n \geq 3$ VERTICES in which each VERTEX has VERTEX DEGREE $\geq n/2$ has a HAMILTONIAN CIRCUIT.

See also HAMILTONIAN CIRCUIT

Direct Analytic Continuation

If (f, U) and (g, V) are FUNCTIONS ELEMENTS, then (g, V) is a direct analytic continuation of (f, U) if $U \cap V \neq 0\varnothing$ and f and G are equal on $U \cap V$..

See also ANALYTIC CONTINUATION, GLOBAL ANALYTIC CONTINUATION

References

Krantz, S. G. "Direct Analytic Continuation." §10.1.4 in *Handbook of Complex Analysis.* Boston, MA: Birkhäuser, p. 128, 1999.

Direct Product

The direct product is defined for a number of classes of algebraic objects, including GROUPS, RINGS, and MODULES. In each case, the direct product of an algebraic object is given by the CARTESIAN PRODUCT of its elements, considered as sets, and its algebraic operations are defined componentwise. For instance, the direct product of two VECTOR SPACES of DIMENSIONS n and m is a VECTOR SPACE of DIMENSION $n + m$.

Direct products satisfy the property that, given maps $\alpha : S \to A$ and $\beta : S \to B$, there exists a unique map $S \to A \times B$ given by $(\alpha(s), \beta(s))$.. The notion of map is determined by the CATEGORY, and this definition extends to other CATEGORIES such as TOPOLOGICAL SPACES. Note that no notion of commutativity is necessary, in contrast to the case for the COPRODUCT. In fact, when A and B are ABELIAN, as in the cases of MODULES (e.g., VECTOR SPACES) or ABELIAN GROUPS) (which are MODULES over the integers), then the DIRECT SUM $A \oplus B$ is well-defined and is the same as the direct product. Although the terminology is slightly confusing because of the distinction between the elementary operations of addition and multiplication, the term "direct sum" is used in these cases instead of "direct product" because of the implicit connotation that addition is always commutative.

Note that direct products and DIRECT SUMS differ for infinite indices. An element of the DIRECT SUM is zero for all but a finite number of entries, while an element of the direct product can have all nonzero entries.

Some other unrelated objects are sometimes also called a direct product. For example, the TENSOR DIRECT PRODUCT is the same as the TENSOR PRODUCT, in which case the dimensions multiply instead of add. Here, "direct" may be used to distinguish it from the EXTERNAL TENSOR PRODUCT.

See also CARTESIAN PRODUCT, CATEGORY THEORY, COPRODUCT, DIRECT SUM, GROUP DIRECT PRODUCT, MATRIX DIRECT PRODUCT, PRODUCT (CATEGORY THEORY), RING DIRECT PRODUCT, SET DIRECT PRODUCT, TENSOR DIRECT PRODUCT, TENSOR PRODUCT (VECTOR SPACE)

Direct Proportion

DIRECTLY PROPORTIONAL

Direct Search Factorization

Direct search factorization is the simplest (and most simple-minded) PRIME FACTORIZATION ALGORITHM. It consists of searching for factors of a number by systematically performing TRIAL DIVISIONS, usually using a sequence of increasing numbers. Multiples of small PRIMES are commonly excluded to reduce the number of trial DIVISORS, but just including them is sometimes faster than the time required to exclude them. Direct search factorization is very inefficient, and can be used only with fairly small numbers.

When using this method on a number n, only DIVISORS up to $\lfloor\sqrt{n}\rfloor$ (where $\lfloor x \rfloor$ is the FLOOR FUNCTION) need to be tested. This is true since if all INTEGERS less than this had been tried, then

$$\frac{n}{\lfloor\sqrt{n}\rfloor + 1} < \sqrt{n}. \tag{1}$$

In other words, all possible FACTORS have had their COFACTORS already tested. It is also true that, when the smallest PRIME FACTOR p of n is $> \sqrt[3]{n}$, then its COFACTOR m (such that $n = pm$) must be PRIME. To prove this, suppose that the smallest p is $> \sqrt[3]{n}$,. If $m = ab$, then the smallest value a and b could assume is p. But then

$$n = pm = pab = p^3 > n, \tag{2}$$

which cannot be true. Therefore, m must be PRIME, so

$$n = p_1 p_2 \tag{3}$$

See also PRIME FACTORIZATION ALGORITHMS, TRIAL DIVISION

Direct Sum

The direct sum $A \oplus B$ of two sets of integers A and B consists of the set $\{a + b : a \in A, b \in B\}$, and can be generalized to an arbitrary number of sets $A \oplus B \oplus \cdots$ in the obvious way. For example, the direct sum of $A = \{1, 2\}$, $B = \{1, 2\}$, and $C = \{2, 3\}$ is $A \oplus B \oplus C = \{4, 5, 5, 6, 5, 6, 6, 7\}$,. The direct sum of a sequence of sets l can be implemented in *Mathematica* as follows.

```
DirectSum[l__] := Flatten[Outer[Plus, l]]
```

The significant property of the direct sum is that it is the COPRODUCT in the CATEGORY of MODULES (i.e., a MODULE DIRECT SUM). This general definition gives as a consequence the definition of the direct sum $A \oplus B$ of ABELIAN GROUPS A and B (since they are $\mathbb{Z}$-modules, i.e., MODULES over the INTEGERS) and the direct sum of VECTOR SPACES (since they are MODULES over a FIELD). Note that the direct sum of Abelian groups is the same as the GROUP DIRECT PRODUCT, but that the term direct sum is not used for groups which are NON-ABELIAN.

Note that DIRECT PRODUCTS and direct sums differ for infinite indices. An element of the direct sum is zero for all but a finite number of entries, while an element of the DIRECT PRODUCT can have all nonzero entries.

See also ABELIAN GROUP, DIRECT PRODUCT, GROUP DIRECT PRODUCT, MATRIX DIRECT SUM, MODULE, MODULE DIRECT SUM

Direct Variation

DIRECTLY PROPORTIONAL

Directed Acyclic Graph

ACYCLIC DIGRAPH

Directed Angle

The symbol $\angle ABC$ denotes the directed angle from AB to BC, which is the signed angle through which AB must be rotated about B to coincide with BC. Four points $ABCD$ lie on a CIRCLE (i.e., are CONCYCLIC) IFF $\angle ABC = \angle ADC$.. It is also true that

$$\angle l_1 l_2 + \angle l_2 l_1 = 0° \text{ or } 360°.$$

Three points A, B, and C are COLLINEAR IFF $\angle ABC = 0 \text{ or } 180.$ or 180°. For any four points, A, B, C, and D,

$$\angle ABC + \angle CDA = \angle BAD + \angle DCB.$$

See also ANGLE, COLLINEAR, CONCYCLIC, MIQUEL EQUATION

References

Johnson, R. A. *Modern Geometry: An Elementary Treatise on the Geometry of the Triangle and the Circle.* Boston, MA: Houghton Mifflin, pp. 11–5, 1929.

Directed Convex Polyomino

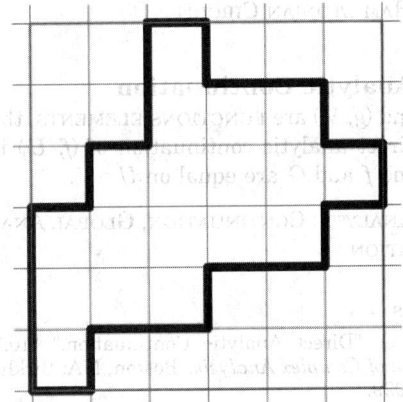

A CONVEX POLYOMINO containing at least one edge of its minimal bounding rectangle. The perimeter and area generating function for directed polygons of width m, height n, and area q is given by

$$G(x, y, q) = \sum_{x \geq 1} \sum_{y \geq 1} \sum_{q \geq 1} C(m, n, a) x^m y^n q^n$$

$$= y \frac{R(x) - \hat{N}(x)}{N(x)} \tag{1}$$

where

$$N(x) = \sum_{n \geq 0} \frac{(-1)^n x^n q^{\binom{n+1}{2}}}{(q)_n (yq)_n} \tag{2}$$

$$\hat{N}(x) = \sum_{n \geq 1} \frac{(-1)^n x^n q^{\binom{n+1}{2}}}{(q)_{n-1} (yq)_n} \tag{3}$$

$$R(x) = y \sum_{n \geq 2} \left[\frac{x^n q^n}{(yq)_n} \left(\frac{\sum_{m=0}^{n-2} (-1)^m q \binom{m+2}{2}}{(q)_m (yq^{m+1})_{n-m-1}} \right) \right] \tag{4}$$

(Bousquet-Mélou 1992).
The anisotropic perimeter generating function for directed convex polygons of width x and height y is given by

$$G(x,y) = \sum_{x \geq 1} \sum_{y \geq 1} C(m,n) x^m y^n = \frac{xy}{\sqrt{\Delta(x,y)}}, \tag{5}$$

where

$$\Delta(x,y) = 1 - 2x - 2y - 2xy + x^2 + y^2$$

$$= (1-y)^2 \left[1 - \frac{x(2+2y-x)}{(1-y)^2} \right] \tag{6}$$

(Lin and Chang 1988, Bousquet 1992, Bousquet-Mélou et al. 1999). This can be solved to explicitly give

$$C(m,n) = \binom{m+n-2}{m-1} \binom{m+n-2}{n-1} \tag{7}$$

(Bousquet-Mélou 1992). Expanding the generating function gives

$$G(x,y) = \sum_{m \geq 1} H_m(y) x^m \tag{8}$$

$$= \frac{y}{1-y} x + \frac{y(1+y)}{(1-y)^3} x^2 + \frac{y(1+4y+y^2)}{(1-y)^5} x^3 + \ldots \tag{9}$$

$$= (y + y^2 + y^3 + y^4 + y^5 + \ldots)x$$

$$+ (y + 4y^2 + 9y^3 + 16y^4 + 25y^5 + \ldots)x^2$$

$$+ (y + 9y^2 + 36y^3 + 100y^4 + 225y^5 + \ldots)x^3$$

$$+ (y + 16y^2 + 100y^3 + 400y^4 + 1225y^5 + \ldots)x^4 + \ldots \tag{10}$$

An explicit formula of $H_m(y)$ is given by Bousquet-Mélou (1992). These functions satisfy the reciprocity relations

$$H_m(1/y) = -y^{m-2} H_m(y) \tag{11}$$

$$G(x,y) + y^2 G(x/y, 1/y) = 0 \tag{12}$$

(Bousquet-Mélou et al. 1999).

The anisotropic area and horizontal perimeter generating function $G(x,q)$ and partial generating functions $H_m(q)$, connected by

$$G(x,q) = \sum_{m \geq 1} H_m(q) x^m,$$

satisfy the self-reciprocity and inversion relations

$$H_m(1/q) = -\frac{1}{q} H_m(q)$$

and

$$G(x,q) + qG(x, 1/q) = 0$$

(Bousquet-Mélou et al. 1999).

See also CONVEX POLYOMINO, LATTICE POLYGON

References

Bousquet-Mélou, M. "Convex Polyominoes and Heaps of Segments." *J. Phys. A: Math. Gen.* **25**, 1925–934, 1992.

Bousquet-Mélou, M. "Convex Polyominoes and Algebraic Languages." *J. Phys. A: Math. Gen.* **25**, 1935–944, 1992.

Bousquet-Mélou, M.; Guttmann, A. J.; Orrick, W. P.; and Rechnitzer, A. Inversion Relations, Reciprocity and Polyominoes. 23 Aug 1999. http://xxx.lanl.gov/abs/math.CO/9908123/.

Lin, K. Y. and Chang, S. J. "Rigorous Results for the Number of Convex Polygons on the Square and Honeycomb Lattices." *J. Phys. A: Math. Gen.* **21**, 2635–642, 1988.

Directed Graph

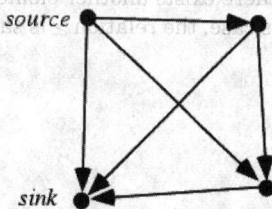

A GRAPH in which each EDGE is replaced by a directed EDGE, also called a digraph or reflexive graph. A COMPLETE directed graph is called a TOURNAMENT. A directed graph having no symmetric pair of directed edges is called an ORIENTED GRAPH.

If G is an undirected connected GRAPH, then one can always direct the circuit EDGES of G and leave the SEPARATING EDGES undirected so that there is a directed path from any node to another. Such a GRAPH is said to be transitive if the adjacency relation is transitive.

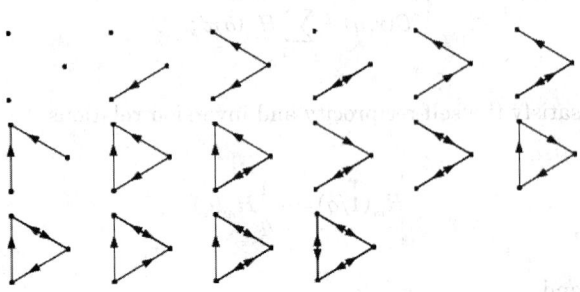

The number of directed graphs of n nodes for $n = 1, 2,$ ... are 1, 3, 16, 218, 9608, ... (Sloane's A000273).

See also ACYCLIC DIGRAPH, ARBORESCENCE, CAYLEY GRAPH, GRAPH, INDEGREE, NETWORK, ORIENTED GRAPH, OUTDEGREE, SINK (DIRECTED GRAPH), SOURCE, STRONGLY CONNECTED DIGRAPH, TOPOLOGY (DIGRAPH) TOURNAMENT, WEAKLY CONNECTED DIGRAPH

References

Chartrand, G. "Directed Graphs as Mathematical Models." §1.5 in *Introductory Graph Theory*. New York: Dover, pp. 16–9, 1985.
Harary, F. "Digraphs." Ch. 16 in *Graph Theory*. Reading, MA: Addison-Wesley, pp. 10 and 198–11, 1994.
Saaty, T. L. and Kainen, P. C. *The Four-Color Problem: Assaults and Conquest.* New York: Dover, p. 122, 1986.
Sloane, N. J. A. Sequences A000273/M3032 in "An On-Line Version of the Encyclopedia of Integer Sequences." http://www.research.att.com/~njas/sequences/eisonline.html.

Directed Set

A set S together with a RELATION $\geq$ which is both transitive and reflexive such that for any two elements $a, b \in S$, there exists another element $c \in S$ with $a \geq c \geq b$. In this case, the relation $\geq$ is said to "direct" the set.

See also NET

Direction Cosine

Let a be the ANGLE between $\mathbf{v}$ and $\mathbf{x}$, b the ANGLE between $\mathbf{v}$ and $\mathbf{y}$, and c the ANGLE between $\mathbf{v}$ and $\mathbf{z}$. Then the direction cosines are equivalent to the (x, y, z) coordinates of a UNIT VECTOR $\hat{\mathbf{v}}$,

$$\alpha \equiv \cos a \equiv \frac{\mathbf{v} \cdot \hat{\mathbf{x}}}{|\mathbf{v}|} \tag{1}$$

$$\beta \equiv \cos b \equiv \frac{\mathbf{v} \cdot \hat{\mathbf{y}}}{|\mathbf{v}|} \tag{2}$$

$$\gamma \equiv \cos c \equiv \frac{\mathbf{v} \cdot \hat{\mathbf{z}}}{|\mathbf{v}|}. \tag{3}$$

From these definitions, it follows that

$$\alpha^2 + \beta^2 + \gamma^2 = 1. \tag{4}$$

To find the JACOBIAN when performing integrals over direction cosines, use

$$\theta = \sin^{-1}\left(\sqrt{\alpha^2 + \beta^2}\right) \tag{5}$$

$$\phi = \tan^{-1}\left(\frac{\beta}{\alpha}\right) \tag{6}$$

$$\gamma = \sqrt{1 - \alpha^2 - \beta^2}. \tag{7}$$

The JACOBIAN is

$$\left|\frac{\partial(\theta, \phi)}{\partial(\alpha, \beta)}\right| = \begin{vmatrix} \dfrac{\partial \theta}{\partial \alpha} & \dfrac{\partial \theta}{\partial \beta} \\ \dfrac{\partial \phi}{\partial \alpha} & \dfrac{\partial \phi}{\partial \beta} \end{vmatrix}. \tag{8}$$

Using

$$\frac{d}{dx}\left(\sin^{-1} x\right) = \frac{1}{\sqrt{1 - x^2}} \tag{9}$$

$$\frac{d}{dx}\left(\tan^{-1} x\right) = \frac{1}{1 + x^2}, \tag{10}$$

$$\left|\frac{\partial(\theta, \phi)}{\partial(\alpha, \beta)}\right| = \begin{vmatrix} \dfrac{\frac{1}{2}\left(\alpha^2 + \beta^2\right)^{-1/2} 2\alpha}{\sqrt{1 - \alpha^2 - \beta^2}} & \dfrac{\frac{1}{2}\left(\alpha^2 + \beta^2\right)^{-1/2} 2\beta}{\sqrt{1 - \alpha^2 - \beta^2}} \\ \dfrac{-\alpha^{-2}\beta}{1 + \dfrac{\beta^2}{\alpha^2}} & \dfrac{\alpha^{-1}}{1 + \dfrac{\beta^2}{\alpha^2}} \end{vmatrix}$$

$$= \frac{1}{\sqrt{1 - \alpha^2 - \beta^2}} \frac{\left(\alpha^2 + \beta^2\right)^{-1/2}}{1 + \dfrac{\beta^2}{\alpha^2}}\left(1 + \frac{\beta^2}{\alpha^2}\right)$$

$$= \frac{1}{\sqrt{\left(\alpha^2 + \beta^2\right)\left(1 - \alpha^2 - \beta^2\right)}}, \tag{11}$$

so

$$d\Omega = \sin\theta \, d\phi \, d\theta = \sqrt{\alpha^2 + \beta^2} \left|\frac{\partial(\theta, \phi)}{\partial(\alpha, \beta)}\right| d\alpha \, d\beta$$

$$= \frac{d\alpha \, d\beta}{\sqrt{1 - \alpha^2 - \beta^2}} = \frac{d\alpha \, d\beta}{\gamma}. \tag{12}$$

Direction cosines can also be defined between two sets of CARTESIAN COORDINATES,

$$\alpha_1 \equiv \hat{\mathbf{x}}' \cdot \hat{\mathbf{x}} \tag{13}$$

$$\alpha_2 \equiv \hat{\mathbf{x}}' \cdot \hat{\mathbf{y}} \tag{14}$$

$$\alpha_3 \equiv \hat{\mathbf{x}}' \cdot \hat{\mathbf{z}} \tag{15}$$

$$\beta_1 \equiv \hat{\mathbf{y}}' \cdot \hat{\mathbf{x}} \tag{16}$$

$$\beta_2 \equiv \hat{\mathbf{y}}' \cdot \hat{\mathbf{y}} \tag{17}$$

$$\beta_3 \equiv \hat{\mathbf{y}}' \cdot \hat{\mathbf{z}} \tag{18}$$

$$\gamma_1 \equiv \hat{\mathbf{z}}' \cdot \hat{\mathbf{x}} \tag{19}$$

$$\gamma_2 \equiv \hat{\mathbf{z}}' \cdot \hat{\mathbf{y}} \tag{20}$$

$$\gamma_3 \equiv \hat{\mathbf{z}}' \cdot \hat{\mathbf{z}}. \tag{21}$$

Projections of the unprimed coordinates onto the primed coordinates yield

$$\hat{\mathbf{x}}' = (\hat{\mathbf{x}}' \cdot \hat{\mathbf{x}})\hat{\mathbf{x}} + (\hat{\mathbf{x}}' \cdot \hat{\mathbf{y}})\hat{\mathbf{y}} + (\hat{\mathbf{x}}' \cdot \hat{\mathbf{z}})\hat{\mathbf{z}} = \alpha_1\hat{\mathbf{x}} + \alpha_2\hat{\mathbf{y}} + \alpha_3\hat{\mathbf{z}} \tag{22}$$

$$\hat{\mathbf{y}}' = (\hat{\mathbf{y}}' \cdot \hat{\mathbf{x}})\hat{\mathbf{x}} + (\hat{\mathbf{y}}' \cdot \hat{\mathbf{y}})\hat{\mathbf{y}} + (\hat{\mathbf{y}}' \cdot \hat{\mathbf{z}})\hat{\mathbf{z}} = \beta_1\hat{\mathbf{x}} + \beta_2\hat{\mathbf{y}} + \beta_3\hat{\mathbf{z}} \tag{23}$$

$$\hat{\mathbf{z}}' = (\hat{\mathbf{z}}' \cdot \hat{\mathbf{x}})\hat{\mathbf{x}} + (\hat{\mathbf{z}}' \cdot \hat{\mathbf{y}})\hat{\mathbf{y}} + (\hat{\mathbf{z}}' \cdot \hat{\mathbf{z}})\hat{\mathbf{z}} = \gamma_1\hat{\mathbf{x}} + \gamma_2\hat{\mathbf{y}} + \gamma_3\hat{\mathbf{z}}, \tag{24}$$

and

$$x' = \mathbf{r} \cdot \hat{\mathbf{x}}' = \alpha_1 x + \alpha_2 y + \alpha_3 z \tag{25}$$

$$y' = \mathbf{r} \cdot \hat{\mathbf{y}}' = \beta_1 x + \beta_2 y + \beta_3 z \tag{26}$$

$$z' = \mathbf{r} \cdot \hat{\mathbf{z}}' = \gamma_1 x + \gamma_2 y + \gamma_3 z. \tag{27}$$

Projections of the primed coordinates onto the unprimed coordinates yield

$$\hat{\mathbf{x}} = (\hat{\mathbf{x}} \cdot \hat{\mathbf{x}}')\hat{\mathbf{x}}' + (\hat{\mathbf{x}} \cdot \hat{\mathbf{y}}')\hat{\mathbf{y}}' + (\hat{\mathbf{x}} \cdot \hat{\mathbf{z}}')\hat{\mathbf{z}}'$$
$$= \alpha_1\hat{\mathbf{x}}' + \beta_1\hat{\mathbf{y}}' + \gamma_1\hat{\mathbf{z}}' \tag{28}$$

$$\hat{\mathbf{y}} = (\hat{\mathbf{y}} \cdot \hat{\mathbf{x}}')\hat{\mathbf{x}}' + (\hat{\mathbf{y}} \cdot \hat{\mathbf{y}}')\hat{\mathbf{y}}' + (\hat{\mathbf{y}} \cdot \hat{\mathbf{z}}')\hat{\mathbf{z}}'$$
$$= \alpha_2\hat{\mathbf{x}}' + \beta_2\hat{\mathbf{y}}' + \gamma_2\hat{\mathbf{z}}' \tag{29}$$

$$\hat{\mathbf{z}} = (\hat{\mathbf{z}} \cdot \hat{\mathbf{x}}')\hat{\mathbf{x}}' + (\hat{\mathbf{z}} \cdot \hat{\mathbf{x}}')\hat{\mathbf{y}}' + (\hat{\mathbf{z}} \cdot \hat{\mathbf{z}}')\hat{\mathbf{z}}'$$
$$= \alpha_3\hat{\mathbf{x}}' + \beta_3\hat{\mathbf{y}}' + \gamma_3\hat{\mathbf{z}}', \tag{30}$$

and

$$x = \mathbf{r} \cdot \hat{\mathbf{x}} = \alpha_1 x + \beta_1 y + \gamma_1 z \tag{31}$$

$$y = \mathbf{r} \cdot \hat{\mathbf{y}} = \alpha_2 x + \beta_2 y + \gamma_2 z \tag{32}$$

$$z = \mathbf{r} \cdot \hat{\mathbf{z}} = \alpha_3 x + \beta_3 y + \gamma_3 z. \tag{33}$$

Using the orthogonality of the coordinate system, it must be true that

$$\hat{\mathbf{x}} \cdot \hat{\mathbf{y}} = \hat{\mathbf{y}} \cdot \hat{\mathbf{z}} = \hat{\mathbf{z}} \cdot \hat{\mathbf{x}} = 0 \tag{34}$$

$$\hat{\mathbf{x}} \cdot \hat{\mathbf{x}} = \hat{\mathbf{y}} \cdot \hat{\mathbf{y}} = \hat{\mathbf{z}} \cdot \hat{\mathbf{z}} = 1, \tag{35}$$

giving the identities

$$\alpha_l \alpha_m + \beta_l \beta_m + \gamma_l \gamma_m = 0 \tag{36}$$

for $l, m = 1, 2, 3$ and $l \neq m$, and

$$\alpha_l^2 + \beta_l^2 + \gamma_l^2 = 1 \tag{37}$$

for $l = 1, 2, 3$. These two identities may be combined into the single identity

$$\alpha_l \alpha_m + \beta_l \beta_m + \gamma_l \gamma_m = \delta_{lm}, \tag{38}$$

where δ_{lm} is the KRONECKER DELTA.

Direction Vector
UNIT VECTOR

© 1999–001 Wolfram Research, Inc.

Directional Derivative

$$\nabla_{\mathbf{u}} f \equiv \nabla f \cdot \frac{\mathbf{u}}{|\mathbf{u}|} \propto \lim_{h \to 0} \frac{f(\mathbf{x} + h\mathbf{u}) - f(\mathbf{x})}{h}. \tag{1}$$

$\nabla_{\mathbf{u}} f(x_0, y_0, z_0)$ is the rate at which the function $w = f(x, y, z)$ changes at (x_0, y_0, z_0) in the direction $\mathbf{u}$. Let $\mathbf{u}$ be a UNIT VECTOR in CARTESIAN COORDINATES, so

$$|\mathbf{u}| = \sqrt{u_x^2 + u_y^2 + u_z^2} = 1, \tag{2}$$

then

$$\nabla_{\mathbf{u}} f = \frac{\partial f}{\partial x} u_x + \frac{\partial f}{\partial y} u_y + \frac{\partial f}{\partial z} u_z. \tag{3}$$

The directional derivative is often written in the notation

$$\frac{d}{ds} \equiv \hat{s} \cdot \mathbf{V} = s_x \frac{\partial}{\partial x} + s_y \frac{\partial}{\partial y} + s_z \frac{\partial}{\partial z}. \tag{4}$$

Directly Proportional

Two quantities y and x are said to be directly proportional, proportional, or "in direct proportion" if y is given by a constant multiple of x, i.e., $y = cx$ for c a constant. This relationship is commonly written $y \propto x$.

See also INVERSELY PROPORTIONAL, PROPORTIONAL

© 1999–001 Wolfram Research, Inc.

Directly Similar

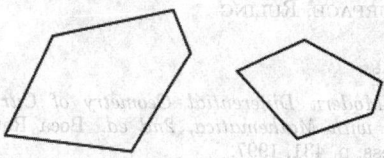

directly similar

Two figures are said to be SIMILAR when all corresponding ANGLES are equal, and are directly similar when all corresponding ANGLES are equal and described in the same rotational sense.

Any two directly similar figures are related either by a TRANSLATION or by a SPIRAL SIMILARITY (Coxeter and Greitzer 1967, p. 97).

See also DOUGLAS-NEUMANN THEOREM, FUNDAMENTAL THEOREM OF DIRECTLY SIMILAR FIGURES, HOMOTHETIC, INVERSELY SIMILAR, SIMILAR, SPIRAL SIMILARITY

References

Casey, J. "Two Figures Directly Similar." Supp. Ch. §2 in *A Sequel to the First Six Books of the Elements of Euclid, Containing an Easy Introduction to Modern Geometry with Numerous Examples, 5th ed., rev. enl.* Dublin: Hodges, Figgis, & Co., pp. 173–79, 1888.
Coxeter, H. S. M. and Greitzer, S. L. *Geometry Revisited.* Washington, DC: Math. Assoc. Amer., p. 95, 1967.
Lachlan, R. "Properties of Two Figures Directly Similar" and "Properties of Three Figures Directly Similar." §213–19 and 223–43 in *An Elementary Treatise on Modern Pure Geometry.* London: Macmillian, pp. 135–38 and 140–43, 1893.
Wells, D. *The Penguin Dictionary of Curious and Interesting Geometry.* London: Penguin, p. 12, 1991.

Director

A PLANE parallel to two (or more) SKEW LINES, also called a director plane. The orientation of a director is fixed, but it is specified uniquely only if a point lying on it is also specified.

A director of two SKEW LINES is perpendicular to the line of shortest distance of these two lines (Altshiller-Court 1979, p. 1).

See also SKEW LINES

References

Altshiller-Court, N. *Modern Pure Solid Geometry.* New York: Chelsea, p. 1, 1979.

Director Curve

The curve $\mathbf{d}(u)$ in the RULED SURFACE parameterization

$$\mathbf{x}(u,v) = \mathbf{b}(u) + v\mathbf{d}(u).$$

See also DIRECTOR, DIRECTRIX (RULED SURFACE), RULED SURFACE, RULING

References

Gray, A. *Modern Differential Geometry of Curves and Surfaces with Mathematica, 2nd ed.* Boca Raton, FL: CRC Press, p. 431, 1997.

Director Plane

DIRECTOR

Directrix

DIRECTRIX (CONIC SECTION), DIRECTRIX (GRAPH), DIRECTRIX (RULED SURFACE)

Directrix (Conic Section)

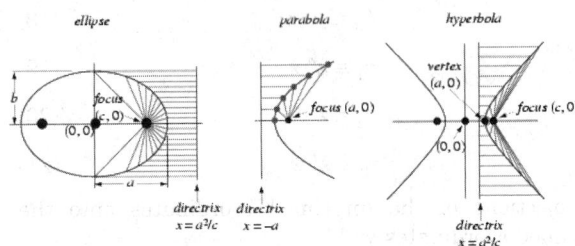

The LINE which, together with the point known as the FOCUS, serves to define a CONIC SECTION as the LOCUS of points whose distance from the FOCUS is proportional to the horizontal distance from the directrix. If the ratio $r = 1$, the conic is a PARABOLA, if $r < 1$, it is an ELLIPSE, and if $r > 1$, it is a HYPERBOLA (Hilbert and Cohn-Vossen 1999, p. 27).

HYPERBOLAS and noncircular ELLIPSES have two distinct FOCI and two associated DIRECTRICES, each DIRECTRIX being PERPENDICULAR to the line joining the two foci (Eves 1965, p. 275).

See also CONIC SECTION, ELLIPSE, FOCUS, HYPERBOLA, PARABOLA

References

Coxeter, H. S. M. *Introduction to Geometry, 2nd ed.* New York: Wiley, pp. 115–16, 1969.
Coxeter, H. S. M. and Greitzer, S. L. *Geometry Revisited.* Washington, DC: Math. Assoc. Amer., pp. 141–44, 1967.
Eves, H. "The Focus-Directrix Property." §6.8 in *A Survey of Geometry, rev. ed.* Boston, MA: Allyn & Bacon, pp. 272–75, 1965.
Hilbert, D. and Cohn-Vossen, S. "The Directrices of the Conics." Ch. 1, Appendix 2 in *Geometry and the Imagination.* New York: Chelsea, pp. 27–9, 1999.

Directrix (Graph)

A GRAPH CYCLE.

See also GRAPH CYCLE

Directrix (Ruled Surface)

The curve $\mathbf{b}(u)$ in the RULED SURFACE parameterization

$$\mathbf{x}(u,v) = \mathbf{b}(u) + v\mathbf{d}(u)$$

is called the directrix (or BASE CURVE).

See also DIRECTOR CURVE, RULED SURFACE

Dirichlet Beta Function

References

Gray, A. *Modern Differential Geometry of Curves and Surfaces with Mathematica, 2nd ed.* Boca Raton, FL: CRC Press, p. 431, 1997.

Dirichlet Beta Function

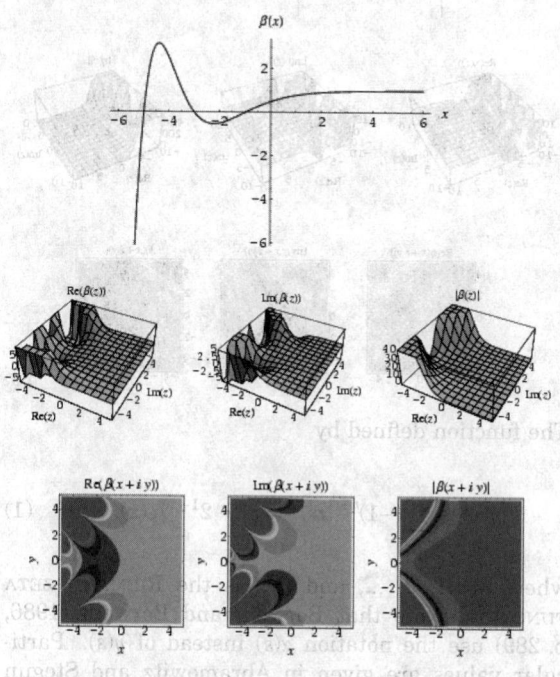

$$\beta(x) \equiv \sum_{n=0}^{\infty} (-1)^n (2n+1)^{-x} \quad (1)$$

$$\beta(x) = 2^{-x} \Phi\left(-1, x, \frac{1}{2}\right), \quad (2)$$

where $\Phi(z, s, a)$ is the LERCH TRANSCENDENT. The beta function can be written in terms of the HURWITZ ZETA FUNCTION $\zeta(x, a)$ by

$$\beta(x) = \frac{1}{4^x}\left[\zeta\left(x, \frac{1}{4}\right) - \zeta\left(x, \frac{3}{4}\right)\right]. \quad (3)$$

The beta function can be evaluated directly for POSITIVE ODD x as

$$\beta(2k+1) = \frac{(-1)^k E_{2k}}{2(2k)!}\left(\frac{1}{2}\pi\right)^{2k+1}, \quad (4)$$

where E_n is an EULER NUMBER. The beta function can be defined over the whole COMPLEX PLANE using ANALYTIC CONTINUATION,

$$\beta(1-z) = \left(\frac{2}{\pi}\right)^z \sin\left(\frac{1}{2}\pi z\right)\Gamma(z)\beta(z), \quad (5)$$

where $\Gamma(z)$ is the GAMMA FUNCTION.
Particular values for β are

$$\beta(1) = \frac{1}{4}\pi \quad (6)$$

$$\beta(2) \equiv K \quad (7)$$

$$\beta(3) = \frac{1}{32}\pi^3, \quad (8)$$

where K is CATALAN'S CONSTANT.

See also CATALAN'S CONSTANT, DIRICHLET ETA FUNCTION, DIRICHLET LAMBDA FUNCTION, HURWITZ ZETA FUNCTION, LEGENDRE'S CHI-FUNCTION, LERCH TRANSCENDENT, RIEMANN ZETA FUNCTION, ZETA FUNCTION

References

Abramowitz, M. and Stegun, C. A. (Eds.). *Handbook of Mathematical Functions with Formulas, Graphs, and Mathematical Tables, 9th printing.* New York: Dover, pp. 807–08, 1972.
Spanier, J. and Oldham, K. B. "The Zeta Numbers and Related Functions." Ch. 3 in *An Atlas of Functions.* Washington, DC: Hemisphere, pp. 25–3, 1987.

Dirichlet Boundary Conditions

PARTIAL DIFFERENTIAL EQUATION BOUNDARY CONDITIONS which give the value of the function on a surface, e.g., $T = f(\mathbf{r}, t)$.

See also BOUNDARY CONDITIONS, CAUCHY BOUNDARY CONDITIONS

References

Morse, P. M. and Feshbach, H. *Methods of Theoretical Physics, Part I.* New York: McGraw-Hill, p. 679, 1953.

Dirichlet Conditions

DIRICHLET BOUNDARY CONDITIONS, DIRICHLET FOURIER SERIES CONDITIONS

Dirichlet Divisor Problem

Let the DIVISOR FUNCTION $d(n) = \nu(n) = \sigma_0(n)$ be the number of DIVISORS of n (including n itself). For a PRIME p, $\nu(p) = 2$. In general,

$$\sum_{k=1}^{n} v(k) = n \ln n + (2\gamma - 1)n + \mathcal{O}(n^{\theta}),$$

where γ is the EULER-MASCHERONI CONSTANT. Dirichlet originally gave $\theta \approx 1/2$ (Hardy 1999, pp. 67–8), and Landau (1916) showed than $\theta \geq 1/4$ (Hardy 1999, p. 81). The following table summarizes incremental progress on the upper limit (Hardy 1999, p. 81).

θ	approx.	citation
7/22	0.31818	1988
27/82	0.32927	van der Corput 1928
33/100	0.33000	van der Corput 1922
1/3	0.33333	Voronoi 1903
1/2	0.50000	Dirichlet

See also DIVISOR FUNCTION, GAUSS'S CIRCLE PROBLEMGauss's Circle Problem

References

Bohr, H. and Cramér. *Enzykl. d. Math. Wiss.* **II C 8**, 815–22, 1922.

Hardy, G. H. *Ramanujan: Twelve Lectures on Subjects Suggested by His Life and Work, 3rd ed.* New York: Chelsea, 1999.

Hardy, G. H. and Wright, E. M. *An Introduction to the Theory of Numbers, 5th ed.* Oxford, England: Clarendon Press, pp. 262–63, 1979.

van der Corput. *Math. Ann.* **98**, 697–17, 1928.

Dirichlet Energy

Let h be a real-valued HARMONIC FUNCTION on a bounded DOMAIN Ω, then the Dirichlet energy is defined as $\int_{\Omega} |\nabla h|^2 dx$, where ∇ is the GRADIENT.

See also ENERGY

Dirichlet Eta Function

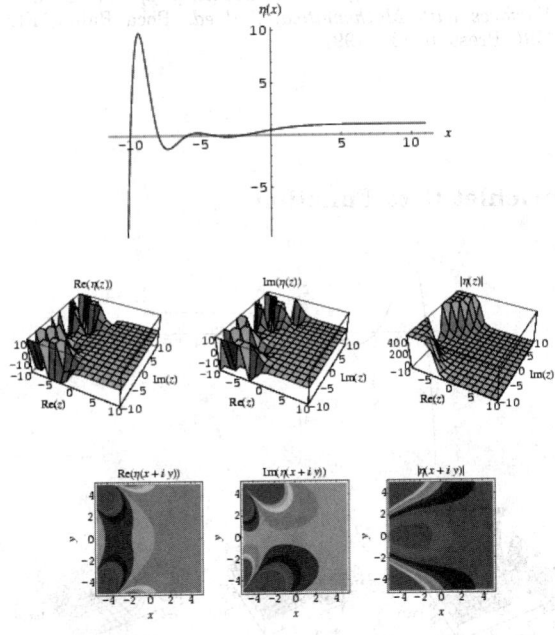

The function defined by

$$\eta(x) \equiv \sum_{n=1}^{\infty} (-1)^{n-1} n^{-x} = (1 - 2^{1-x}) \zeta(x), \qquad (1)$$

where $n = 1, 2, \ldots$, and $\zeta(x)$ is the RIEMANN ZETA FUNCTION. Note that Borwein and Borwein (1986, p. 289) use the notation $\alpha(s)$ instead of $\eta(s)$.. Particular values are given in Abramowitz and Stegun (1972, p. 811).

The eta function is related to the RIEMANN ZETA FUNCTION and DIRICHLET LAMBDA FUNCTION by

$$\frac{\zeta(v)}{2^v} = \frac{\lambda(v)}{2^v - 1} = \frac{\eta(v)}{2^v - 2} \qquad (2)$$

and

$$\zeta(v) + \eta(v) = 2\lambda(v) \qquad (3)$$

(Spanier and Oldham 1987). The eta function is also a special case of the POLYLOGARITHM function,

$$\eta(x) = -Li_x(-1). \qquad (4)$$

The value $\eta(1)$ may be computed by noting that the MACLAURIN SERIES for $\ln(1 + x)$ for $-1 \leq x \leq 1$ is

$$\ln(1 + x) = x - \frac{1}{2}x^2 + \frac{1}{3}x^3 - \frac{1}{4}x^4 + \cdots \qquad (5)$$

Therefore,

$$\ln 2 = \ln(1 + 1) = 1 - \frac{1}{2} + \frac{1}{3} - \frac{1}{4} + \cdots$$

$$= \sum_{n=1}^{\infty} \frac{(-1)^{n-1}}{n} = \eta(1). \tag{6}$$

The derivative of the eta function is given by

$$\eta'(x) = -2^{1-x} \ln 2 \zeta(x) + (1 - 2^{1-x}) \zeta'(x), \tag{7}$$

or in the special case $x = 0$, by

$$\lim_{x \to 0} \left[\frac{d}{dx} \eta(x) \right] = -\ln 2 - \zeta'(0) = -\ln 2 + \frac{1}{2} \ln(2\pi)$$

$$= -\ln \left(\sqrt{\frac{2}{\pi}} \right) = \frac{1}{2} \ln \left(\frac{1}{2} \pi \right). \tag{8}$$

This latter fact provides a remarkable proof of the WALLIS FORMULA.

Values for EVEN INTEGERS are related to the analytical values of the RIEMANN ZETA FUNCTION. $\eta(0)$ is defined to be $\frac{1}{2}$.

$$\eta(0) = \frac{1}{2}$$

$$\eta(1) = \ln 2$$

$$\eta(2) = \frac{\pi^2}{12}$$

$$\eta(3) = 0.90154\ldots$$

$$\eta(4) = \frac{7\pi^4}{720}.$$

See also DEDEKIND ETA FUNCTION, DIRICHLET BETA FUNCTION, DIRICHLET *L*-SERIES, DIRICHLET LAMBDA FUNCTION, RIEMANN ZETA FUNCTION, ZETA FUNCTION

References

Abramowitz, M. and Stegun, C. A. (Eds.). *Handbook of Mathematical Functions with Formulas, Graphs, and Mathematical Tables, 9th printing.* New York: Dover, pp. 807–08, 1972.

Borwein, J. M. and Borwein, P. B. *Pi & the AGM: A Study in Analytic Number Theory and Computational Complexity.* New York: Wiley, 1987.

Spanier, J. and Oldham, K. B. "The Zeta Numbers and Related Functions." Ch. 3 in *An Atlas of Functions.* Washington, DC: Hemisphere, pp. 25–3, 1987.

Dirichlet Fourier Series Conditions

A piecewise regular function which

1. Has a finite number of finite discontinuities and
2. Has a finite number of extrema

can be expanded in a FOURIER SERIES which converges to the function at continuous points and the mean of the POSITIVE and NEGATIVE limits at points of discontinuity.

See also FOURIER SERIES

Dirichlet Function

Let c and $d \neq c$ be REAL NUMBERS (usually taken as $c = 1$ and $d = 0$). The Dirichlet function is defined by

$$D(x) = \begin{cases} c & \text{for} \quad x \quad \text{rational} \\ d & \text{for} \quad x \quad \text{irrational} \end{cases} \tag{1}$$

and is discontinuous everywhere. The Dirichlet function can be written analytically as

$$D(x) = \lim_{m \to \infty} \lim_{n \to \infty} \cos^{2n}(m! \pi x). \tag{2}$$

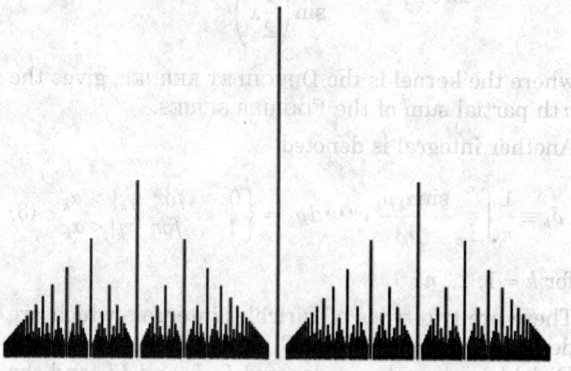

Because the Dirichlet function cannot be plotted without producing a solid blend of lines, a modified version can be defined as

$$D_M(x) = \begin{cases} 0 & \text{for} \quad x \quad \text{irrational} \\ 1/b & \text{for} \quad x = a/b \text{ a reduced fraction} \end{cases} \tag{3}$$

(Dixon 1991), illustrated above. This function is continuous at irrational x and discontinuous at rational x (although a small interval around an irrational point x contains infinitely many ration points, these rationals will have very large denominators). When viewed from a corner along the line $y = x$ in normal perspective, a QUADRANT of EUCLID'S ORCHARD turns into the modified Dirichlet function (Gosper).

See also CONTINUOUS FUNCTION, EUCLID'S ORCHARD, IRRATIONAL NUMBER, RATIONAL NUMBER

References

Dixon, R. *Mathographics.* New York: Dover, pp. 177 and 184–86, 1991.

Tall, D. "The Gradient of a Graph." *Math. Teaching* **111**, 48–2, 1985.

Trott, M. "Numerical Computations." §1.2.1 in *The Mathematica Guidebook, Vol. 1: Programming in Mathematica.* New York: Springer-Verlag, 2000.

See also Fourier series.

Dirichlet Integrals

There are several types of integrals which go under the name of a "Dirichlet integral." The integral

$$D[u] = \int_\Omega |\nabla u|^2 dV \tag{1}$$

appears in DIRICHLET'S PRINCIPLE.

The integral

$$\frac{1}{2\pi} \int_{-\pi}^{\pi} f(x) \frac{\sin\left[\left(n + \frac{1}{2}\right)x\right]}{\sin\left(\frac{1}{2}x\right)} dx, \tag{2}$$

where the kernel is the DIRICHLET KERNEL, gives the nth partial sum of the FOURIER SERIES.

Another integral is denoted

$$\delta_k \equiv \frac{1}{\pi} \int_{-\infty}^{\infty} \frac{\sin \alpha_k \rho_k}{\rho_k} e^{i\rho_k \gamma_k} d\rho_k = \begin{cases} 0 & \text{for } |\gamma_k| > \alpha_k \\ 1 & \text{for } |\gamma_k| < \alpha_k \end{cases} \tag{3}$$

for $k = 1, ..., n$.

There are two types of Dirichlet integrals which are denoted using the letters C, D, I, and J. The type 1 Dirichlet integrals are denoted I, J, and IJ, and the type 2 Dirichlet integrals are denoted C, D, and CD.

The type 1 integrals are given by

$$I \equiv \int \int \cdots \int f(t_1 + t_2 + ... + t_n) t_1^{a_1-1} t_2^{a_2-1} ... t_n^{a_n-1} dt_1 dt_2 dt_n$$

$$= \frac{\Gamma(\alpha_1)\Gamma(\alpha_2)...\Gamma(\alpha_n)}{\Gamma(\sum_n \alpha_n)} \int_0^1 f(r) r \left(\sum_n \alpha\right)^{-1} dr, \tag{4}$$

where $\Gamma(z)$ is the GAMMA FUNCTION. In the case $n = 2$,

$$I = \int \int_T x^p y^q dx dy = \frac{p!q!}{(p+q+2)!} = \frac{B(p+1, q+1)}{p+q+2}, \tag{5}$$

where the integration is over the TRIANGLE T bounded by the x-AXIS, y-AXIS, and line $x + y = 1$ and $B(x, y)$ is the BETA FUNCTION.

The type 2 integrals are given for b-D vectors $\mathbf{a}$ and $\mathbf{r}$, and $0 \le c \le b$,

$$C_\mathbf{a}^{(b)}(\mathbf{r}, m) = \frac{\Gamma(m+R)}{\Gamma(m) \prod_{i=1}^b \Gamma(r_i)} \int_0^{a_1} \cdots \int_0^{a_b}$$

$$\times \frac{\prod_{i=1}^b x_i^{r_i-1} dx_i}{\left(1 + \sum_{i=1}^b x_i\right)^{m+R}} \tag{6}$$

$$D_\mathbf{a}^{(b)}(\mathbf{r}, m) = \frac{\Gamma(m+R)}{\Gamma(m) \prod_{i=1}^b \Gamma(r_i)} \int_{a1}^{\infty} \cdots \int_{ak}^{\infty}$$

$$\times \frac{\prod_{i=1}^b x_i^{r_i-1} dx_i}{\left(1 + \sum_{i=1}^b xi\right)^{m+R}} \tag{7}$$

$$CD_\mathbf{a}^{(c,d-c)}(\mathbf{r}, m)$$

$$= \frac{\Gamma(m+R)}{\Gamma(m) \prod_{i=1}^b \Gamma(r_i)} \int_0^{a_c} \int_{a_{c+1}}^{\infty} \int_{ab}^{\infty} \frac{\prod_{i=1}^b x_i^{r_i-1} dx_i}{\left(1 + \sum_{i=1}^b x_i\right)^{m+R}}, \tag{8}$$

where

$$R \equiv \sum_{i=1}^k r_i \tag{9}$$

$$a_i \equiv \frac{p_i}{1 - \sum_{i=1}^k p_i}, \tag{10}$$

and p_i are the cell probabilities. For equal probabilities, $a_i = 1$. The Dirichlet D integral can be expanded as a MULTINOMIAL SERIES as

$$D_\mathbf{a}^{(b)}(\mathbf{r}, m) = \frac{1}{\left(1 + \sum_{i=1}^b \right)^m}$$

$$\times \sum_{x_1 < r_1} \cdots \sum_{x_b < r_b} \binom{m - 1 + \sum_{a=1}^b x_i}{m - 1, x_1 ..., x_b}$$

$$\prod_{i=1}^b \left(\frac{a_i}{1 + \sum_{k=1}^b ak}\right)^{x_i}. \tag{11}$$

For small b, C and D can be expressed analytically either partially or fully for general arguments and $a_i = 1$.

$$C_1^{(1)}(r_2; r_1) = \frac{\Gamma(r_1 + r_2) \, {}_2F_i(r_2, r_1 + r_2; 1 + r_2; -1)}{r_2 \Gamma(r_1) \Gamma(r_2)} \tag{12}$$

$$C_1^{(2)}(r_2, r_3; r_1) = \frac{\Gamma(r_1 + r_2 + r_3)}{r_2 \Gamma(r_1) \Gamma(r_2) \Gamma(r_3)}$$

$$\times \int_0^1 {}_2F_1 y^{r_a-1} (1+y)^{-(r_1+r_2+r_3)} dy, \tag{13}$$

where

$$_2F_1 \equiv {}_2F_1\left(r_2, r_1 + r_2 + r_3; 1 + r_2, -(1+y)^{-1}\right) \tag{14}$$

is a HYPERGEOMETRIC FUNCTION.

$$D_1^{(1)}(r_2; r_1) = \frac{\Gamma(r_1 + r_2) \, {}_2F_1(r_1, r_1 + r_2; 1 + r_1; -1)}{r_1 \Gamma(r_1) \Gamma(r_2)} \tag{15}$$

$$D_1^{(2)}(r_2, r_3; r_1)$$

$$= \frac{\Gamma(r_1 + r_2 + r_3)}{(r_1 + r_3)\Gamma(r_1)\Gamma(r_2)\Gamma(r_3)} \int_1^\infty {}_2F_1 y^{r_3 - 1} dy, \quad (16)$$

where

$${}_2F_1 \equiv {}_2F_1(r_1 + r_3, r_1 + r_2 + r_3; 1 + r_1 + r_3; -1 - y). \quad (17)$$

References

Jeffreys, H. and Jeffreys, B. S. "Dirichlet Integrals." §15.08 in *Methods of Mathematical Physics, 3rd ed.* Cambridge, England: Cambridge University Press, pp. 468–70, 1988.

Sobel, M.; Uppuluri, R. R.; and Frankowski, K. *Selected Tables in Mathematical Statistics, Vol. 4: Dirichlet Distribution--Type 1.* Providence, RI: Amer. Math. Soc., 1977.

Sobel, M.; Uppuluri, R. R.; and Frankowski, K. *Selected Tables in Mathematical Statistics, Vol. 9: Dirichlet Integrals of Type 2 and Their Applications.* Providence, RI: Amer. Math. Soc., 1985.

Weisstein, E. W. "Dirichlet Integrals." MATHEMATICA NOTEBOOK DIRICHLETINTEGRALS.M.

Dirichlet Kernel

The Dirichlet kernel D_n^M is obtained by integrating the CHARACTER $e^{i(\xi, x)}$ over the BALL $|\xi| \leq M$,

$$D_n^M = -\frac{1}{2\pi r} \frac{d}{dr} D_{n-2}^M.$$

The Dirichlet kernel of a DELTA SEQUENCE is given by

$$\delta_n(x) \equiv \frac{1}{2\pi} \frac{\sin\left[\left(n + \frac{1}{2}\right)x\right]}{\sin\left(\frac{1}{2}x\right)}.$$

The integral of this kernel is called the DIRICHLET INTEGRAL $D[u]$.

See also DELTA SEQUENCE, DIRICHLET INTEGRALS, DIRICHLET'S LEMMA

Dirichlet Lambda Function

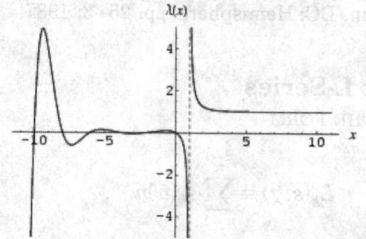

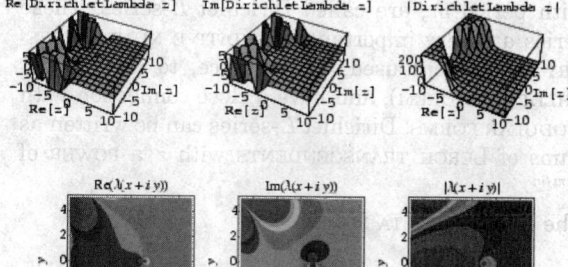

$$\lambda(x) \equiv \sum_{n=0}^\infty (2n = 1)^{-x} = (1 - 2^{-x})\zeta(x) \quad (1)$$

for $x = 2, 3, \ldots$, where $\zeta(x)$ is the RIEMANN ZETA FUNCTION. The function is undefined at $x = 1$. It can be computed in closed form where $\zeta(x)$ can, that is for EVEN POSITIVE n. It is related to the RIEMANN ZETA FUNCTION and DIRICHLET ETA FUNCTION by

$$\frac{\zeta(v)}{2^v} = \frac{\lambda(v)}{2^v - 1} = \frac{\eta(v)}{2^v - 2} \quad (2)$$

and

$$\zeta(v) + \eta(v) = 2\lambda(v) \quad (3)$$

(Spanier and Oldham 1987). Special values of $\lambda(n)$ include

$$\lambda(2) = \frac{\pi^2}{8} \quad (4)$$

$$\lambda(4) = \frac{\pi^4}{96} \quad (5)$$

See also DIRICHLET BETA FUNCTION, DIRICHLET ETA FUNCTION, LEGENDRE'S CHI-FUNCTION, RIEMANN ZETA FUNCTION, ZETA FUNCTION

References

Abramowitz, M. and Stegun, C. A. (Eds.). *Handbook of Mathematical Functions with Formulas, Graphs, and Mathematical Tables, 9th printing.* New York: Dover, pp. 807–08, 1972.

Spanier, J. and Oldham, K. B. "The Zeta Numbers and Related Functions." Ch. 3 in *An Atlas of Functions*. Washington, DC: Hemisphere, pp. 25–3, 1987.

Dirichlet L-Series

Series OF THE FORM

$$L_k(s, \chi) \equiv \sum_{n=1}^{\infty} \chi_k(n) n^{-s}, \tag{1}$$

where the CHARACTER $\chi_k(n)$ is an INTEGER FUNCTION with period m, are called Dirichlet L-series. These series are very important in ADDITIVE NUMBER THEORY (they were used, for instance, to prove DIRICHLET'S THEOREM), and have a close connection with MODULAR FORMS. Dirichlet L-series can be written as sums of LERCH TRANSCENDENTS with z a POWER of $e^{2\pi i/m}$.

The DIRICHLET ETA FUNCTION

$$\eta(s) \equiv \sum_{n=1}^{\infty} \frac{(-1)^{n+1}}{n^s} = (1 - 2^{1-s})\zeta(s) \tag{2}$$

(for $s \neq 1$), DIRICHLET BETA FUNCTION

$$L_{-4}(s) = \beta(s) \equiv \sum_{n=0}^{\infty} \frac{(-1)^n}{(2n+1)^s}, \tag{3}$$

and RIEMANN ZETA FUNCTION

$$L_{+1}(s) = \zeta(s) \equiv \sum_{n=0}^{\infty} \frac{1}{n^s} \tag{4}$$

are all Dirichlet L-series (Borwein and Borwein 1987, p. 289).

Hecke found a remarkable connection between each MODULAR FORM with FOURIER SERIES

$$f(r) = c(0) + \sum_{n=1}^{\infty} c(n) e^{2\pi i \tau} \tag{5}$$

and the Dirichlet L-series

$$\phi(s) = \sum_{m=1}^{\infty} \frac{c(n)}{n^s} \tag{6}$$

This Dirichlet series converges absolutely for $\sigma = \Re[s] > k+1$ (if f is a CUSP FORM) and $\sigma > 2k$ if f is not a CUSP FORM. In particular, if the coefficients $c(n)$ satisfy the multiplicative property

$$c(m)c(n) = \sum_{d|(m,n)} d^{2k-1} c\left(\frac{mn}{d^2}\right), \tag{7}$$

then the Dirichlet L-series will have a representation OF THE FORM

$$\phi(s) = \prod_p \frac{1}{1 - c(p)p^{-s} + p^{2k-1}p^{-2s}}, \tag{8}$$

which is absolutely convergent with the Dirichlet series (Apostol 1997, pp. 136–37). In addition, let $k \geq 4$ be an EVEN integer, then $\phi(s)$ can be ANALYTICALLY CONTINUED beyond the line $\sigma = k$ such that

1. If $c(0) = 0$, then $\phi(s)$ is an ENTIRE FUNCTION of s,
2. If $c(0) \neq 0$, $\phi(s)$ is analytic for all s except a single SIMPLE POLE at $s = k$ with RESIDUE

$$\frac{(-1)^{k/2} c(0)(2\pi)^k}{\Gamma(k)}, \tag{9}$$

where $\Gamma(k)$ is the GAMMA FUNCTION, and
3. $\phi(s)$ satisfies

$$(2\pi)^{-s}\Gamma(s)\phi(s) = (-1)^{k \cdot 2}(2\pi)^{s-k}\Gamma(k-s)\phi(k-s) \tag{10}$$

(Apostol 1997, p. 137).

The CHARACTER χ_k is called primitive if the CONDUCTOR $f(x) = k$. Otherwise, χ_k is imprimitive. A primitive L-series modulo k is then defined as one for which $\chi_k(n)$ is primitive. All imprimitive L-series can be expressed in terms of primitive L-series.

Let $P = 1$ or $P = \prod_{i=1}^{t} p_i$, where p_i are distinct ODD PRIMES. Then there are three possible types of primitive L-series with REAL COEFFICIENTS. The requirement of REAL COEFFICIENTS restricts the CHARACTER to $\chi_k(n) = \pm 1$ for all k and n. The three type are then

1. If $k = P$ (e.g., $k = 1, 3, 5, ...$) or $k = 4P$ (e.g., $k = 4, 12, 20, ...$), there is exactly one primitive L-series.
2. If $k = 8P$ (e.g., $k = 8, 24, ...$), there are two primitive L-series.
3. If $k = 2P, Pp_i$, or $2^x P$ where $\alpha > 3$ (e.g., $k = 2, 6, 9, ...$), there are no primitive L-series

(Zucker and Robertson 1976). All primitive L-series are ALGEBRAICALLY INDEPENDENT and divide into two types according to

$$\chi_k(k-1) = \pm 1. \tag{11}$$

Primitive L-series of these types are denoted $L \pm$. For a primitive L-series with REAL CHARACTER (NUMBER THEORY), if $k = P$, then

$$L_k = \begin{cases} L_{-k} & if \quad P \equiv 3 \pmod 4 \\ L_k & if \quad P \equiv 1 \pmod 4 \end{cases}. \tag{12}$$

If $k = 4P$, then

$$L_k = \begin{cases} L_{-k} & if \quad P \equiv 1 \pmod 4 \\ L_k & if \quad P \equiv 3 \pmod 4 \end{cases}, \tag{13}$$

and if $k = 8P$, then there is a primitive function of each type (Zucker and Robertson 1976).

The first few primitive NEGATIVE L-series are L_{-3}, L_{-4}, L_{-7}, L_{-8}, L_{11}, L_{-15}, L_{-19}, L_{-20}, L_{-23}, L_{-24}, L_{-31},

L_{-35}, L_{-39}, L_{-40}, L_{-43}, L_{-47}, L_{-51}, L_{-52}, L_{-55}, L_{-56}, L_{-59}, L_{-67}, L_{-68}, L_{-71}, L_{-79}, L_{-83}, L_{-84}, L_{-87}, L_{-88}, L_{-91}, L_{-95}, ... (Sloane's A003657), corresponding to the negated discriminants of IMAGINARY QUADRATIC FIELDS. The first few primitive POSITIVE L-series are L_{+1}, L_{+5}, L_{+8}, L_{+12}, L_{+13}, L_{+17}, L_{+21}, L_{+24}, L_{+28}, L_{+29}, L_{+33}, L_{+37}, L_{+40}, L_{+41}, L_{+44}, L_{+53}, L_{+56}, L_{+57}, L_{+60}, L_{+61}, L_{+65}, L_{+69}, L_{+73}, L_{+76}, L_{+77}, L_{+85}, L_{+88}, L_{+89}, L_{+92}, L_{+93}, L_{+97}, ... (Sloane's A046113).

The KRONECKER SYMBOL is a REAL CHARACTER modulo k, and is in fact essentially the only type of REAL primitive CHARACTER (Ayoub 1963). Therefore,

$$L_{+d}(s) = \sum_{n=1}^{\infty} (d|n) n^{-s} \qquad (14)$$

$$L_{-d}(s) = \sum_{n=1}^{\infty} (-d|n) n^{-s}, \qquad (15)$$

where $(d|n)$ is the KRONECKER SYMBOL (Borwein and Borwein 1986, p. 293). The functional equations for $L_{\pm}$ are

$$L_{-k}(s) = 2^s \pi^{s-1} k^{-s+1/2} \Gamma(1-s) \cos\left(\frac{1}{2}s\pi\right) L - k(1-s) \qquad (16)$$

$$L_{+k}(s) = 2^s \pi^{s-1} k^{-s+1/2} \Gamma(1-s) \sin\left(\frac{1}{2}s\pi\right) L_{+k}(1-s) \qquad (17)$$

For m a POSITIVE INTEGER

$$L_{+k}(-2m) = 0 \qquad (18)$$

$$L_{-k}(1-2m) = 0 \qquad (19)$$

$$L_{+k}(2m) = R k^{-1/2} \pi^{2m} \qquad (20)$$

$$L_{-k}(2m-1) = R' k^{-1/2} \pi^{2m-1} \qquad (21)$$

$$L_{+k}(1-2m) = \frac{(-1)^m (2m-1)! R}{(2k)^{2m-1}} \qquad (22)$$

$$L_{-k}(-2k) = \frac{(-1)^m R'(2m)!}{(2k)^{2m}} \qquad (23)$$

where R and R' are RATIONAL NUMBERS. Nothing general appears to be known about $L_{-k}(2m)$ or $L_{+k}(2m-1)$, although it is possible to express all $L_{\pm}(1)$ in terms of known transcendentals (Zucker and Robertson 1976).

$L_{+k}(1)$ can be expressed in terms of transcendentals by

$$L_d(1) = h(d)\kappa(d), \qquad (24)$$

where $h(d)$ is the CLASS NUMBER and $\kappa(d)$ is the DIRICHLET STRUCTURE CONSTANT. Some specific va-

lues of primitive L-series are

$$L_{-15}(1) = \frac{2\pi}{\sqrt{15}}$$

$$L_{-11}(1) = \frac{\pi}{\sqrt{11}}$$

$$L_{-8}(1) = \frac{\pi}{2\sqrt{2}}$$

$$L_{-7}(1) = \frac{\pi}{\sqrt{7}}$$

$$L_{-4}(1) = \frac{1}{4}\pi$$

$$L_{-3}(1) = \frac{\pi}{3\sqrt{3}}$$

$$L_{+5}(1) = \frac{2}{\sqrt{5}} \ln\left(\frac{1+\sqrt{5}}{2}\right)$$

$$L_{+s}(1) = \frac{\ln(1+\sqrt{2})}{\sqrt{2}}$$

$$L_{+12}(1) = \frac{\ln(2+\sqrt{3})}{\sqrt{3}}$$

$$L_{+13}(1) = \frac{2}{\sqrt{13}} \ln\left(\frac{3+\sqrt{13}}{2}\right)$$

$$L_{+17}(1) = \frac{2}{\sqrt{17}} \ln(4+\sqrt{17})$$

$$L_{+21}(1) = \frac{2}{\sqrt{21}} \ln\left(\frac{5+\sqrt{21}}{2}\right)$$

$$L_{+24}(1) = \frac{\ln(5+2\sqrt{6})}{\sqrt{6}}.$$

In particular,

$$L_{-3}(1) = L(1,\chi) = \sum_{n=0}^{\infty} \frac{1}{(3n+1)(3n+2)} \qquad (25)$$

for χ a nontrivial Dirichlet character modulo 3 (Ireland and Rosen 1990, p. 266).

No general forms are known for $L_{-k}(2m)$ and $L_{+k}(2m-1)$ in terms of known transcendentals. For example,

$$L_{-4}(2) = \beta(2) \equiv K, \qquad (26)$$

where K is defined as CATALAN'S CONSTANT.

See also DIRICHLET BETA FUNCTION, DIRICHLET ETA FUNCTION, DIRICHLET SERIES, DOUBLE SUM, HECKE L-SERIES, MODULAR FORM, PETERSSON CONJECTURE

References

Apostol, T. M. *Introduction to Analytic Number Theory.* New York: Springer-Verlag, 1976.

Apostol, T. M. "Modular Forms and Dirichlet Series" and "Equivalence of Ordinary Dirichlet Series." §6.16 and §8.8 in *Modular Functions and Dirichlet Series in Number Theory, 2nd ed.* New York: Springer-Verlag, pp. 136–37 and 174–76, 1997.

Ayoub, R. G. *An Introduction to the Analytic Theory of Numbers.* Providence, RI: Amer. Math. Soc., 1963.

Borwein, J. M. and Borwein, P. B. *Pi & the AGM: A Study in Analytic Number Theory and Computational Complexity.* New York: Wiley, 1987.

Buell, D. A. "Small Class Numbers and Extreme Values of *L*-Functions of Quadratic Fields." *Math. Comput.* **139**, 786–96, 1977.

Hecke, E. "Über die Bestimmung Dirichletscher Reihen durch ihre Funktionalgleichung." *Math. Ann.* **112**, 664–99, 1936.

Ireland, K. and Rosen, M. "Dirichlet *L*-Functions." Ch. 16 in *A Classical Introduction to Modern Number Theory, 2nd ed.* New York: Springer-Verlag, pp. 249–68, 1990.

Koch, H. "*L*-Series." Ch. 7 in *Number Theory: Algebraic Numbers and Functions.* Providence, RI: Amer. Math. Soc., pp. 203–58, 2000.

Sloane, N. J. A. Sequences A003657/M2332 and A046113 in "An On-Line Version of the Encyclopedia of Integer Sequences." http://www.research.att.com/~njas/sequences/eisonline.html.

Weisstein, E. W. "Class Numbers." MATHEMATICA NOTEBOOK CLASSNUMBERS.M.

Zucker, I. J. and Robertson, M. M. "Some Properties of Dirichlet *L*-Series." *J. Phys. A: Math. Gen.* **9**, 1207–214, 1976.

Dirichlet Problem

The problem of finding the connection between a continuous function f on the boundary ∂R of a region R with a HARMONIC FUNCTION taking on the value f on ∂R. In general, the problem asks if such a solution exists and, if so, if it is unique. The Dirichlet problem is extremely important in mathematical physics (Courant and Hilbert 1989, pp. 179–80 and 240; Logan 1997; Krantz 1999b).

If f is a CONTINUOUS FUNCTION on the boundary of the open unit disk $\partial D(0,1)$, then define

$$u(z) = \begin{cases} \dfrac{1}{2\pi} \displaystyle\int_0^{2\pi} f(e^{i\psi}) \dfrac{1-|z|^2}{|z-e^{i\psi}|^2} \, d\psi & \text{if } z \in D(0,\, 1) \\ f(z) & \text{if } z \in \partial D(0,\, 1) \end{cases},$$

where $\partial D(0,1)$, is the boundary of $D(0,1)$. Then u is continuous on the closed unit disk $\overline{D}(0,1)$ and harmonic on $D(0,1)$ (Krantz 1999a, p. 93).

See also POISSON INTEGRAL, POISSON KERNEL

References

Courant, R. and Hilbert, D. *Methods of Mathematical Physics, Vol. 1.* New York: Wiley, pp. 179–80 and 240, 1989.

Krantz, S. G. "The Dirichlet Problem" and "Application of Conformal Mapping to the Dirichlet Problem." §7.3.3, 7.7.1, and 14.2 in *Handbook of Complex Analysis.* Boston, MA: Birkhäuser, pp. 93, 97–8, and 164–68, 1999a.

Krantz, S. G. *A Panorama of Harmonic Analysis.* Washington, DC: Math. Assoc. Amer., 1999b.

Logan, J. D. *Applied Mathematics, 2nd ed.* New York: Wiley, 1997.

Dirichlet Region

VORONOI POLYGON

Dirichlet Series

A series

$$\sum a(n) e^{-\lambda(n)z},$$

where $a(n)$ and z are COMPLEX and $\{\lambda(n)\}$ is a MONOTONIC increasing sequence of REAL NUMBERS is called a general Dirichlet series. The numbers $\lambda(n)$ are called the exponents, and $a(n)$ are called the coefficients. When $\lambda(n) = \ln n$, then $e^{-\lambda(n)z} = n^{-z}$, the series is a normal DIRICHLET L-SERIES. The Dirichlet series is a special case of the LAPLACE-STIELTJES TRANSFORM.

See also DIRICHLET L-SERIES, LAPLACE-STIELTJES TRANSFORM, MODULAR FORM, MODULAR FUNCTION

References

Apostol, T. M. "General Dirichlet Series and Bohr's Equivalence Theorem." Ch. 8 in *Modular Functions and Dirichlet Series in Number Theory, 2nd ed.* New York: Springer-Verlag, pp. 161–89, 1997.

Bohr, H. "Zur Theorie der allgemeinen Dirichletschen Reihen." *Math. Ann.* **79**, 136–56, 1919.

Dirichlet Structure Constant

$$\kappa(d) = \begin{cases} \dfrac{2 \ln \eta(d)}{\sqrt{d}} & \text{for } d > 0 \\ \dfrac{2\pi}{w(d)\sqrt{|d|}} & \text{for } d > 0 \end{cases}$$

where $\eta(d)$ is the FUNDAMENTAL UNIT and $w(d)$ is the number of substitutions which leave the BINARY QUADRATIC FORM unchanged

$$w(d) = \begin{cases} 6 & \text{for } d = -3 \\ 4 & \text{for } d = -4 \\ 2 & \text{otherwise.} \end{cases}$$

See also CLASS NUMBER, DIRICHLET L-SERIES

References

Weisstein, E. W. "Class Numbers." MATHEMATICA NOTEBOOK CLASSNUMBERS.M.

Dirichlet Tessellation

VORONOI DIAGRAM

Dirichlet's Approximation Theorem

Given any REAL NUMBER θ and any POSITIVE INTEGER N, there exist integers h and k with $0 \le k \le N$ such that

$$|k\theta - h| < \frac{1}{N}.$$

A slightly weaker form of the theorem states that for every real θ, there exist integers h and k with $k > 0$ and $(h,k)1 = 1$ such that

$$\left|\theta - \frac{h}{k}\right| < \frac{1}{k^2}.$$

See also HURWITZ'S IRRATIONAL NUMBER THEOREM, IRRATIONALITY MEASURE, LIOUVILLE'S APPROXIMATION THEOREM, RATIONAL APPROXIMATION, ROTH'S THEOREM, THUE-SIEGEL-ROTH THEOREM

References
Apostol, T. M. "Dirichlet's Approximation Theorem." §7.2 in *Modular Functions and Dirichlet Series in Number Theory*, 2nd ed. New York: Springer-Verlag, pp. 143–45, 1997.

Dirichlet's Box Principle

A.k.a. the PIGEONHOLE PRINCIPLE. Given n boxes and $m > n$ objects, at least one box must contain more than one object. This statement has important applications in NUMBER THEORY and was first stated by Dirichlet in 1834.

See also FUBINI PRINCIPLE

References
Chartrand, G. *Introductory Graph Theory*. New York: Dover, p. 38, 1985.
Nagell, T. *Introduction to Number Theory*. New York: Wiley, p. 38, 1951.
Shanks, D. *Solved and Unsolved Problems in Number Theory*, 4th ed. New York: Chelsea, p. 161, 1993.

Dirichlet's Boxing-In Principle

DIRICHLET'S BOX PRINCIPLE

Dirichlet's Formula

If g is continuous and $\mu, \nu > 0$, then

$$\int_0^t (t-\xi)^{\mu-1} d\xi \int_0^\xi (\xi-x)^{\nu+1} g(\xi,x) dx$$
$$= \int_0^t dx \int_x^t (t-\xi)^{\mu-1} (\xi-x)^{\nu-1} g(\xi,x) d\xi.$$

Dirichlet's Lemma

$$\int_0^\pi \frac{\sin\left[\left(n + \frac{1}{2}\right)x\right]}{2 \sin\left(\frac{1}{2}x\right)} dx = \frac{1}{2}\pi,$$

where the KERNEL is the DIRICHLET KERNEL.

See also DIRICHLET KERNEL

References
Cohn, H. *Advanced Number Theory*. New York: Dover, p. 37, 1980.
Gradshteyn, I. S. and Ryzhik, I. M. *Tables of Integrals, Series, and Products*, 6th ed. San Diego, CA: Academic Press, p. 1101, 2000.

Dirichlet's Principle

Dirichlet's principle, also known as Thomson's principle, states that here exists a function u that minimizes the functional

$$D[u] = \int_\Omega |\nabla u|^2 dV$$

(called the DIRICHLET INTEGRAL) for $\Omega \subset \mathbb{R}^2$ or $\mathbb{R}^3$ among all the functions $u \in C^{(1)}(\Omega) \cap C^{(0)}(\bar{\Omega})$ which take on given values f on the boundary $\partial\Omega$ of Ω, and that function u satisfies $\nabla^2 = 0$ in Ω, $u|\partial\Omega = f$, $u \in C^{(2)}(\Omega) \cap C^{(0)}(\bar{\Omega})$. Weierstrass showed that Dirichlet's argument contained a subtle fallacy. As a result, it can be claimed only that there exists a lower bound to which $D[u]$ comes arbitrarily close without being forced to actually reach it. Kneser, however, obtained a valid proof of Dirichlet's principle.

See also DIRICHLET'S BOX PRINCIPLE, DIRICHLET INTEGRALS

References
Monna, A. F. *Dirichlet's Principle: A Mathematical Comedy of Errors and Its Influence on the Development of Analysis*. Utrecht, Netherlands: Osothoek, Scheltema, and Holkema, 1975.

Dirichlet's Test

Let

$$\left|\sum_{n=1}^p a_n\right| < K,$$

where K is independent of p. Then if $f_n \ge f_{n+1} > 0$ and

$$\lim_{n\to\infty} f_n = 0,$$

it follows that

$$\sum_{n=1}^{\infty} a_n f_n$$

CONVERGES.

See also CONVERGENCE TESTS

Dirichlet's Theorem

Given an ARITHMETIC SERIES of terms $an + b$, for $n = 1, 2, ...$, the series contains an infinite number of PRIMES if a and b are RELATIVELY PRIME, i.e., $(a, b) = 1$. Dirichlet proved this theorem using DIRICHLET L-SERIES, but the proof is challenging enough that, in their classic text on NUMBER THEORY, the usually explicit Hardy and Wright (1979) report "this theorem is too difficult for insertion in this book."

See also PRIME ARITHMETIC PROGRESSION, PRIME PATTERNS CONJECTURE, RELATIVELY PRIME, SIERPINSKI'S PRIME SEQUENCE THEOREM

References

Courant, R. and Robbins, H. "Primes in Arithmetical Progressions." §1.2b in Supplement to Ch. 1 in *What is Mathematics?: An Elementary Approach to Ideas and Methods, 2nd ed.* Oxford, England: Oxford University Press, pp. 26–7, 1996.
Hardy, G. H. and Wright, E. M. *An Introduction to the Theory of Numbers, 5th ed.* Oxford, England: Clarendon Press, pp. 13–4, 1979.
Landau, E. *Vorlesungen über Zahlentheorie, Vol. 1.* New York: Chelsea, pp. 79–6, 1970.
Landau, E. *Handbuch der Lehre von der Verteilung der Primzahlen, 3rd ed.* New York: Chelsea, pp. 422–46, 1974.
Shanks, D. *Solved and Unsolved Problems in Number Theory, 4th ed.* New York: Chelsea, pp. 22–3, 1993.

Dirichlet-Hardy Test

If, in an interval of x, $\sum_{r=1}^{n}$ is uniformly bounded with respect to n and x, and $\{v_r\}$ is a sequence of positive non-increasing quantities tending to zero, then $\sum a_r(x) v_r$ is uniformly convergent in the interval.

References

Jeffreys, H. and Jeffreys, B. S. "Dirichlet-Hardy Test." §1.1155 in *Methods of Mathematical Physics, 3rd ed.* Cambridge, England: Cambridge University Press, pp. 42–3, 1988.

Disc

DISK

Disconnected Form

A FORM which is the sum of two FORMS involving separate sets of variables.

Disconnected Graph

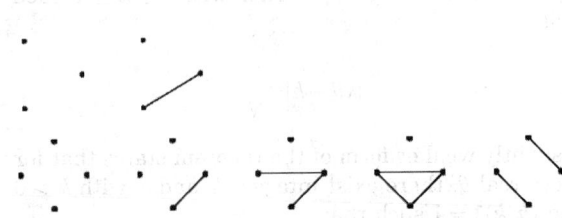

A graph is said to be disconnected if it is not CONNECTED, i.e., if there exist two nodes is G such that no edge in G having those nodes as endpoints. The numbers of disconnected simple unlabeled graphs on $n = 1, 2, ...$ nodes are 0, 1, 2, 5, 13, 44, 191, ... (Sloane's A000719).
If G is disconnected, then its complement $\bar{G}$ is connected (Skiena 1990, p. 171; Bollobás 1998). However, the converse is not true, as can be seen using the example of the CYCLE GRAPH C_5 which is connected and isomorphic to its complement.

See also CONNECTED GRAPH, CUT SET, K-CONNECTED GRAPH

References

Bollobás, B. *Modern Graph Theory.* New York: Springer-Verlag, 1998.
Harary, F. "The Number of Linear, Directed, Rooted, and Connected Graphs." *Trans. Amer. Math. Soc.* **78**, 445–63, 1955.
Read, R. C. and Wilson, R. J. *An Atlas of Graphs.* Oxford, England: Oxford University Press, 1998.
Skiena, S. *Implementing Discrete Mathematics: Combinatorics and Graph Theory with Mathematica.* Reading, MA: Addison-Wesley, 1990.
Sloane, N. J. A. Sequences A000719/M1452 in "An On-Line Version of the Encyclopedia of Integer Sequences." http://www.research.att.com/~njas/sequences/eisonline.html.
Stein, M. L. and Stein, P. R. "Enumeration of Linear Graphs and Connected Linear Graphs Up to $p = 18$ Points." Report LA-3775. Los Alamos, NM: Los Alamos National Laboratory, Oct. 1967.

Disconnectivity

Disconnectivities are mathematical entities which stand in the way of a SPACE being contractible (i.e., shrunk to a point, where the shrinking takes place inside the SPACE itself). When dealing with TOPOLOGICAL SPACES, a disconnectivity is interpreted as a "HOLE" in the space. Disconnectivities in SPACE are studied through the EXTENSION PROBLEM or the LIFTING PROBLEM.

See also EXTENSION PROBLEM, HOLE, LIFTING PROBLEM

Discontinuity

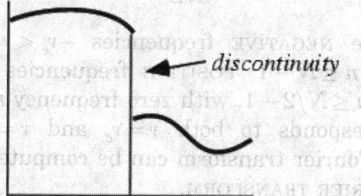

discontinuity

A point at which a mathematical object is DISCONTINUOUS.

Discontinuous

Not CONTINUOUS. A point at which a function is discontinuous is called a DISCONTINUITY, or sometimes a JUMP.

See also CONTINUOUS, DISCONTINUITY

References
Yates, R. C. "Functions with Discontinuous Properties." *A Handbook on Curves and Their Properties.* Ann Arbor, MI: J. W. Edwards, pp. 100–07, 1952.

Discordant Permutation
MARRIED COUPLES PROBLEM

Discrepancy Theorem

Let $s_1, s_2, \ldots$ be an infinite series of real numbers lying between 0 and 1. Then corresponding to any arbitrarily large K, there exists a positive integer n and two subintervals of equal length such that the number of s_v with $v = 1, 2, \ldots, n$ which lie in one of the subintervals differs from the number of such s_v that lie in the other subinterval by more than K (van der Corput 1935ab, van Aardenne-Ehrenfest 1945, 1949, Roth 1954).

This statement can be refined as follows. Let N be a large integer and $s_1, s_2, \ldots, s_N$ be a sequence of N real numbers lying between 0 and 1. Then for any integer $1 \le n \le N$ and any real number α satisfying $0 < \alpha < 1$, let $D_n(\alpha)$ denote the number of s_v with $v = 1, 2, \ldots, n$ that satisfy $0 \le s_v < \alpha$. Then there exist n and α such that

$$|D_n(\alpha) - n\alpha| > c_1 \frac{\ln \ln N}{\ln \ln \ln N}$$

where c_1 is a positive constant.

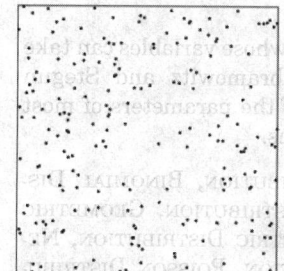 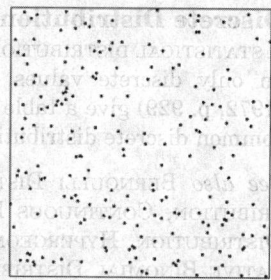

This result can be further strengthened, which is most easily done by reformulating the problem. Let $N > 1$ be an integer and $P_1, P_2, \ldots, P_N$ be N (not necessarily distinct) points in the square $0 \le x \le 1$, $0 \le y \le 1$. Then

$$\int_0^1 \int_0^1 [S(x,y) - Nxy]^2 \, dx \, dy > c_2 \ln N,$$

where c_2 is a positive constant and $S(u,v)$ is the number of points in the rectangle $0 \le x < u$, $0 \le y < v$ (Roth 1954). Therefore,

$$|S(x,y) - Nxy| > c_3 \sqrt{\ln N},$$

and the original result can be stated as the fact that there exist n and α such that

$$|D_n(\alpha) - n\alpha| > c_4 \sqrt{\ln N}.$$

The randomly distributed points shown in the above squares have $|S(x,y) - Nxy|^2 = 6.40$ and 9.11, respectively.

Similarly, the discrepancy of a set of N points in a unit d-HYPERCUBE satisfies

$$|S(x,y) - Nxy| > c(\ln N)^{(d-1)/2}$$

(Roth 1954, 1976, 1979, 1980).

See also 18-POINT PROBLEM, CUBE POINT PICKING

References
Berlekamp, E. R. and Graham, R. L. "Irregularities in the Distributions of Finite Sequences." *J. Number Th.* **2**, 152–61, 1970.
Roth, K. F. "On Irregularities of Distribution." *Mathematika* **1**, 73–9, 1954.
Roth, K. F. "On Irregularities of Distribution. II." *Comm. Pure Appl. Math.* **29**, 739–44, 1976.
Roth, K. F. "On Irregularities of Distribution. III." *Acta Arith.* **35**, 373–84, 1979.
Roth, K. F. "On Irregularities of Distribution. IV." *Acta Arith.* **37**, 67–5, 1980.
van Aardenne-Ehrenfest, T. "Proof of the Impossibility of a Just Distribution of an Infinite Sequence Over an Interval." *Proc. Kon. Ned. Akad. Wetensch.* **48**, 3–, 1945.
van Aardenne-Ehrenfest, T. *Proc. Kon. Ned. Akad. Wetensch.* **52**, 734–39, 1949.
van der Corput, J. G. *Proc. Kon. Ned. Akad. Wetensch.* **38**, 813–21, 1935a.
van der Corput, J. G. *Proc. Kon. Ned. Akad. Wetensch.* **38**, 1058–066, 1935b.

Discrete Distribution

A STATISTICAL DISTRIBUTION whose variables can take on only discrete values. Abramowitz and Stegun (1972, p. 929) give a table of the parameters of most common discrete distributions.

See also BERNOULLI DISTRIBUTION, BINOMIAL DISTRIBUTION, CONTINUOUS DISTRIBUTION, GEOMETRIC DISTRIBUTION, HYPERGEOMETRIC DISTRIBUTION, NEGATIVE BINOMIAL DISTRIBUTION, POISSON DISTRIBUTION, PROBABILITY, STATISTICAL DISTRIBUTION, STATISTICS, UNIFORM DISTRIBUTION

References
Abramowitz, M. and Stegun, C. A. (Eds.). *Handbook of Mathematical Functions with Formulas, Graphs, and Mathematical Tables, 9th printing.* New York: Dover, pp. 927 and 929, 1972.
Evans, M.; Hastings, N.; and Peacock, B. *Statistical Distributions, 3rd ed.* New York: Wiley, 2000.
McLaughlin, M. "Common Probability Distributions." http://www.geocities.com/~mikemclaughlin/math_stat/Dists/Compendium.html.
Wilmmer, G. and Altmann, G. *Thesaurus of Univariate Discrete Probability Distributions.* Essen, Germany: STAMM, 1999.

Discrete Fourier Transform

The FOURIER TRANSFORM is defined as

$$f(v) = \mathcal{F}[f(t)] = \int_{-\infty}^{\infty} f(t) e^{-2\pi i v t} dt. \tag{1}$$

Now consider generalization to the case of a discrete function, $f(t) \to f(t_k)$ by letting $f_k \equiv f(t_k)$, where $t_k \equiv k\Delta$, with $k = 0, \ldots, N-1$. Choose the frequency step such that

$$v_n = \frac{n}{N\Delta}, \tag{2}$$

with $n = -N/2, \ldots, 0, \ldots, N/2$. There are $N+1$ values of n, so there is one relationship between the frequency components. Writing this out as per Press *et al.* (1989)

$$\mathcal{F}[f(t)] = \sum_{k=0}^{N-1} f_k e^{-2\pi i (n/N\Delta) k \Delta} \Delta = \Delta \sum_{k=0}^{N-1} f_k e^{-2\pi i n k/N}, \tag{3}$$

and

$$F_n \equiv \sum_{k=0}^{N-1} f_k e^{-2\pi i n k/N}. \tag{4}$$

The inverse transform is

$$f_k = \frac{1}{N} \sum_{n=0}^{N-1} F_n e^{2\pi i n k/N}. \tag{5}$$

Note that $F_{-n} = F_{N-n}$, $n = 1, 2, \ldots$, so an alternate formulation is

$$v_n = \frac{n}{N\Delta}, \tag{6}$$

where the NEGATIVE frequencies $-v_c < v < 0$ have $N/2 + 1 \leq n \leq N-1$, POSITIVE frequencies $0 < v < v_c$ have $1 \leq n \leq N/2 - 1$, with zero frequency $n = 0$. $n = N/2$ corresponds to both $v = v_c$ and $v = -v_c$. The discrete Fourier transform can be computed using a FAST FOURIER TRANSFORM.

The discrete Fourier transform is a special case of the Z-TRANSFORM. It can be computed for a list l of COMPLEX NUMBERS using the *Mathematica* command `Fourier[l]`.

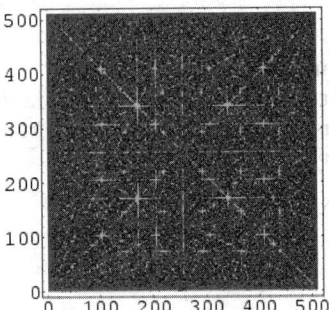

The above plot shows the 2-D discrete Fourier transform of the reciprocals of the greatest common divisor GCD (i, j) for $i, j \in [1, 512]$ (Trott 2000).

See also FAST FOURIER TRANSFORM, FOURIER TRANSFORM, HARTLEY TRANSFORM, WINOGRAD TRANSFORM, Z-TRANSFORM

References
Arfken, G. "Discrete Orthogonality--Discrete Fourier Transform." §14.6 in *Mathematical Methods for Physicists, 3rd ed.* Orlando, FL: Academic Press, pp. 787–92, 1985.
Press, W. H.; Flannery, B. P.; Teukolsky, S. A.; and Vetterling, W. T. "Fourier Transform of Discretely Sampled Data." §12.1 in *Numerical Recipes in C: The Art of Scientific Computing.* Cambridge, England: Cambridge University Press, pp. 494–98, 1989.
Trott, M. "Numerical Computations." §1.2.1 in *The Mathematica Guidebook, Vol. 1: Programming in Mathematica.* New York: Springer-Verlag, 2000.

Discrete Geometry

See also COMPUTATIONAL GEOMETRY

References
Goodman, J. E. and O'Rourke, J. *Handbook of Discrete and Computational Geometry.* Boca Raton, FL: CRC Press, 1997.

Discrete Group

See also CONTINUOUS GROUP, FINITE GROUP

Discrete Logarithm

MULTIPLICATIVE ORDER

Discrete Mathematics

The branch of mathematics dealing with objects which can assume only certain "discrete" values. Discrete objects can be characterized by INTEGERS, whereas continuous objects require REAL NUMBERS. The study of how discrete objects combine with one another and the probabilities of various outcomes is known as COMBINATORICS.

See also COMBINATORICS, DISCRETE DISTRIBUTION, DISCRETE FOURIER TRANSFORM, DISCRETE GEOMETRY, DISCRETE LOGARITHM

References
Balakrishnan, V. K. *Introductory Discrete Mathematics.* New York: Dover, 1997.

Bobrow, L. S. and Arbib, M. A. *Discrete Mathematics: Applied Algebra for Computer and Information Science.* Philadelphia, PA: Saunders, 1974.

Dossey, J. A.; Otto, A. D.; Spence, L.; and Eynden, C. V. *Discrete Mathematics, 3rd ed.* Reading, MA: Addison-Wesley, 1997.

Graham, R. L.; Knuth, D. E.; and Patashnik, O. *Concrete Mathematics: A Foundation for Computer Science, 2nd ed.* Reading, MA: Addison-Wesley, 1994.

Hall, C. and O'Donnell, J. *Discrete Mathematics Using a Computer.* London: Springer-Verlag, 2000.

Lipschutz, S. and Lipson, M. L. *2000 Solved Problems in Discrete Mathematics.* New York: McGraw-Hill, 1991.

Lipschutz, S. and Lipson, M. L. *Schaum's Outline of Discrete Mathematics, 2nd ed.* New York: McGraw-Hill, 1997.

Rosenstein, J. G.; Franzblau, D. S.; and Roberts, F. S. *Discrete Mathematics in the Schools.* Providence, RI: Amer. Math. Soc., 1997.

Skiena, S. *Implementing Discrete Mathematics.* Reading, MA: Addison-Wesley, 1990.

Weisstein, E. W. "Books about Discrete Mathematics." http://www.treasure-troves.com/books/DiscreteMathematics.html.

Discrete Set

A set S is discrete in a larger TOPOLOGICAL SPACE X if every point $x \in S$ has a NEIGHBORHOOD U such that $S \cap U = \{x\}$.. The points of S are then said to be ISOLATED (Krantz 1999, p. 63). Typically, a discrete set is either finite or COUNTABLY INFINITE. For example, the set of integers is discrete on the REAL LINE. Another example of an infinite discrete set is the set $\{1/n$ for all integers $n > 1\}$. On any reasonable space, a finite set is discrete. A set is discrete if it has the DISCRETE TOPOLOGY, that is, if every subset is open.

In the case of a subset S, as in the examples above, one uses the RELATIVE TOPOLOGY on S. Sometimes a discrete set is also closed. Then there cannot be any ACCUMULATION POINTS of a discrete set. On a COMPACT SET such as the SPHERE, a closed discrete set must be finite because of this.

See also ACCUMULATION POINT, COMPACT SPACE, DISCRETE TOPOLOGY, ISOLATED POINT, NEIGHBORHOOD, TOPOLOGICAL SPACE

References
Krantz, S. G. "Discrete Sets and Isolated Points." §4.6.2 in *Handbook of Complex Analysis.* Boston, MA: Birkhäuser, pp. 63-4, 1999.

Discrete Topology

A topology is given by a collection of subsets of a TOPOLOGICAL SPACE X. The smallest topology has two OPEN SETS, ϕ and X. The largest topology contains all subsets as open sets, and is called the discrete topology. In particular, every point in X is an OPEN SET in the discrete topology.

See also DISCRETE MATHEMATICS, DISCRETE SET, TOPOLOGICAL SPACE

Discrete Uniform Distribution

EQUALLY LIKELY OUTCOMES DISTRIBUTION

DiscreteDelta

KRONECKER DELTA

Discriminant

A discriminant is a quantity (usually invariant under certain classes of transformations) which characterizes certain properties of a quantity's ROOTS. The concept of the discriminant is used for BINARY QUADRATIC FORMS, ELLIPTIC CURVES, METRICS, MODULES, POLYNOMIALS, QUADRATIC CURVES, QUADRATIC FIELDS, QUADRATIC FORMS, and in the SECOND DERIVATIVE TEST.

See also DISCRIMINANT (BINARY QUADRATIC FORM), DISCRIMINANT (CIRCLE), DISCRIMINANT (CONIC SECTION), DISCRIMINANT (ELLIPTIC CURVE), DISCRIMINANT (METRIC), MODULAR DISCRIMINANT, DISCRIMINANT (MODULE), DISCRIMINANT (POLYNOMIAL), DISCRIMINANT (QUADRATIC CURVE), DISCRIMINANT (SECOND DERIVATIVE TEST)

Discriminant (Binary Quadratic Form)

The discriminant of a BINARY QUADRATIC FORM

$$au^2 + buv + cv^2$$

is defined by

$$d \equiv b^2 - 4ac.$$

It is equal to four times the corresponding DETERMINANT.

See also CLASS NUMBER

Discriminant (Circle)

In HOMOGENEOUS COORDINATES (x_1, x_2, x_3), the equation of a CIRCLE C is

$$a(x_1^2 + x_2^2) + 2fx_2x_3 + 2gx_1x_3 + cx_3^2 = 0.$$

The discriminant of this circle is defined as

$$\Delta = \begin{vmatrix} a & 0 & g \\ 0 & a & f \\ g & f & c \end{vmatrix} = a(ac - f^2 - g^2),$$

and the quadratic form $q(C) = ac - f^2 - g^2$ is the basic invariant.

See also DISCRIMINANT (CONIC SECTION)

References
Barth, W. and Bauer, T. "Poncelet Theorems." *Expos. Math.* **14**, 125–44, 1996.

Discriminant (Conic Section)
The discriminant of the general CONIC SECTION

$$ax_1^2 + bx_2^2 + cx_3^2 + 2fx_2x_3 + 2gx_1x_3 + 2hx_1x_2 = 0$$

is defined as

$$\Delta = \begin{vmatrix} a & h & g \\ h & b & f \\ g & f & c \end{vmatrix} = abc + 2fgh - af^2 - bg^2 - ch^2.$$

If $b = a$ and $g = h = 0$, then simplifies to the DISCRIMINANT of a CIRCLE.

See also DISCRIMINANT (CIRCLE)

References
Salmon, G. *Conic Sections, 6th ed.* New York: Chelsea, p. 266, 1960.

Discriminant (Elliptic Curve)
An ELLIPTIC CURVE is the set of solutions to an equation of the form

$$y^2 + a_1xy + a_3y = x^3 + a_2x^2 + a_4x + a_6. \tag{1}$$

By changing variables, $y \to 2y + a_1x + a_3$, assuming the CHARACTERISTIC is not 2, the equation becomes

$$y^2 = 4x^3 + b_2x^2 + 2b_4x + b_6 \tag{2}$$

where

$$b_2 = a_1^2 + 4a_2 \tag{3}$$

$$b_4 = 2a_4 + a_1a_3 \tag{4}$$

$$b_6 = a_3^2 + 4a_6. \tag{5}$$

Define also the quantity

$$b_8 = a_1^2a_6 + 4a_2a_6 - a_1a_3a_4 + a_2a_3^2 - a_4^2, \tag{6}$$

then the discriminant is given by

$$\Delta = -b_2^2b_8 - 8b_4^3 - 27b_6^2 + 9b_2b_4b_6. \tag{7}$$

The discriminant depends on the choice of equations, and can change after a change of variables, unlike the J-INVARIANT.

If the CHARACTERISTIC of the FIELD is neither 2 or 3, then its equation can be written as

$$y^2 = x^3 + Ax + B, \tag{8}$$

in which case, the discriminant is given by

$$\Delta = -16(4A^3 + 27B^2). \tag{9}$$

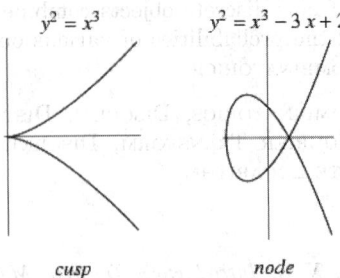

cusp   *node*

Algebraically, the discriminant is nonzero when the right-hand side has three distinct roots. In the classical case of an ELLIPTIC CURVE over the COMPLEX NUMBERS, the discriminant has a geometric interpretation. If $\Delta \neq 0$, then the elliptic curve is nonsingular and has GENUS 1, i.e., it is a TORUS. If $\Delta = 0$ and $A = 0$, then it has a CUSP singularity, in which case there is one tangent direction at the singularity. If $\Delta = 0$ and $A \neq 0$, then its singularity is called an ORDINARY DOUBLE POINT (or node), in which case the singularity has two distinct tangent directions.

Note that the discriminant of an ELLIPTIC CURVE is not the same as the DISCRIMINANT of the corresponding polynomial, but the two kinds of discriminants vanish for the same values of A and B.

See also ALGEBRAIC GEOMETRY, ELLIPTIC CURVE, FREY CURVE, ISOGENY, J-INVARIANT, LEGENDRE FORM, MINIMAL DISCRIMINANT, WEIERSTRASS FORM

References
Silverman, J. *The Arithmetic of Elliptic Curves.* New York: Springer-Verlag, 1986.

Discriminant (Elliptic Function)
If $y^2 = 4x^3 + b_2x^2 + 2b_4x + b_6$ and b_2 are the INVARIANTS of a WEIERSTRASS ELLIPTIC FUNCTION $a_1^2 + 4a_2$ with periods b_4 and $2a_4 + a_{1a_3}$, then the discriminant is defined by

$$b_6 \tag{1}$$

Letting $a_3^2 + 4a_6.$, then

$$b_8 = a_1^2a_6 + 4a_2a_6 - a_1a_3a_4 + a_2a_3^2 - a_4^{2, r > 1}\Delta$$
$$= -b_2^2b_8 - 8b_4^3 - 27b_6^2 + 9b_2b_4b_6.$$

$$= y^2 = x^3 + Ax + B, \tag{2}$$

$$= \Delta = -16(4A^3 + 27B^2) \tag{3}$$

The FOURIER SERIES of for $\Delta \neq 0$, where H is the UPPER HALF-PLANE, is

$$A = 0 \qquad (4)$$

where $A \neq 0$, is the TAU FUNCTION, and $A \neq 0$, are integers (Apostol 1997, p. 20). The discriminant can also be expressed in terms of the DEDEKIND ETA FUNCTION $g\alpha\beta$ by

$$g \equiv \det(g_{\alpha,\beta}) = \begin{vmatrix} g_{11} & g_{12} \\ g_{21} & g_{22} \end{vmatrix} = g_{11}g_{22} - (g_{12})^2. \qquad (5)$$

(Apostol 1997, p. 51).

See also DEDEKIND ETA FUNCTION, INVARIANT (ELLIPTIC FUNCTION), KLEIN'S ABSOLUTE INVARIANT, TAU FUNCTION, WEIERSTRASS ELLIPTIC FUNCTION

References

Apostol, T. M. "The Discriminant $\bar{g}$" and "The Fourier Expansions of and $\bar{g} = D^2g$." §1.11 and 1.15 in *Modular Functions and Dirichlet Series in Number Theory, 2nd ed.* New York: Springer-Verlag, pp. 14 and 20–2, 1997.

Discriminant (Metric)

Given a METRIC $g_{\alpha\beta}$, the discriminant is defined by

$$g \equiv \det(g_{\alpha,\beta}) = \begin{vmatrix} g_{11} & g_{12} \\ g_{21} & g_{22} \end{vmatrix} = g_{11}g_{22} - (g_{12})^2. \qquad (1)$$

Let g be the discriminant and $\bar{g}$ the transformed discriminant, then

$$\bar{g} = D^2 g \qquad (2)$$

$$g = \bar{D}^2 \bar{g} \qquad (3)$$

where

$$D \equiv \frac{\partial(u^1, u^2)}{\partial(\bar{u}^1, \bar{u}^2)} = \begin{vmatrix} \dfrac{\partial u^1}{\partial \bar{u}^1} & \dfrac{\partial u^1}{\partial \bar{u}^2} \\ \dfrac{\partial u^2}{\partial \bar{u}^1} & \dfrac{\partial u^2}{\partial \bar{u}^2} \end{vmatrix}. \qquad (4)$$

$$\bar{D} \equiv \frac{\partial(\bar{u}^1, \bar{u}^2)}{\partial(u^1, u^2)} = \begin{vmatrix} \dfrac{\partial \bar{u}^1}{\partial u^1} & \dfrac{\partial \bar{u}^1}{\partial u^2} \\ \dfrac{\partial \bar{u}^2}{\partial u^1} & \dfrac{\partial \bar{u}^2}{\partial u^2} \end{vmatrix}. \qquad (5)$$

Discriminant (Module)

Let a MODULE M in an INTEGRAL DOMAIN D_1 for $R(\sqrt{D})$ be expressed using a two-element basis as

$$M = [\xi_1, \xi_2],$$

where ξ_1 and ξ_2 are in D_1. Then the DIFFERENT of the MODULE is defined as

$$\Delta = \Delta(M) = \begin{vmatrix} \xi_1 & \xi_2 \\ \xi_2' & \xi_2' \end{vmatrix} = \xi_1 \xi_2' - \xi_1' \xi_2$$

and the discriminant is defined as the square of the DIFFERENT (Cohn 1980).

For IMAGINARY QUADRATIC FIELDS $\mathbb{Q}(\sqrt{n})$ (with $n < 0$), the discriminants are given in the following table.

−1	-2^2	−33	$-2^2 \cdot 3 \cdot 11$	−67	−67
−2	-2^3	−34	$-2^3 \cdot 17$	−69	$-2^2 \cdot 3 \cdot 23$
−3	−3	−35	$-5 \cdot 7$	−70	$-2^3 \cdot 5 \cdot 7$
−5	$-2^2 \cdot 5$	−37	$-2^2 \cdot 37$	−71	−71
−6	$-2^3 \cdot 3$	−39	$-3 \cdot 13$	−73	$-2^2 \cdot 73$
−7	−7	−41	$-2^2 \cdot 41$	−74	$-2^3 \cdot 37$
−10	$-2^3 \cdot 5$	−42	$-2^3 \cdot 3 \cdot 7$	−77	$-2^2 \cdot 7 \cdot 11$
−11	−11	−43	−43	−78	$-2^3 \cdot 3 \cdot 13$
−13	$-2^2 \cdot 13$	−46	$-2^3 \cdot 23$	−79	−79
−14	$-2^3 \cdot 7$	−47	−47	−82	$-2^3 \cdot 41$
−15	$-3 \cdot 5$	−51	$-3 \cdot 17$	−83	−83
−17	$-2^2 \cdot 17$	−53	$-2^2 \cdot 53$	−85	$-2^2 \cdot 5 \cdot 17$
−19	−19	−55	$-5 \cdot 11$	−86	$-2^3 \cdot 43$
−21	$-2^2 \cdot 3 \cdot 7$	−57	$-2^2 \cdot 3 \cdot 19$	−87	$-3 \cdot 29$
−22	$-2^3 \cdot 11$	−58	$-2^3 \cdot 29$	−89	$-2^2 \cdot 89$
−23	−23	−59	−59	−91	$-7 \cdot 13$
−26	$-2^3 \cdot 13$	−61	$-2^2 \cdot 61$	−93	$-2^2 \cdot 3 \cdot 31$
−29	$-2^2 \cdot 29$	−62	$-2^3 \cdot 31$	−94	$-2^3 \cdot 47$
−30	$-2^3 \cdot 3 \cdot 5$	−65	$-2^2 \cdot 5 \cdot 13$	−95	$-5 \cdot 19$
−31	−31	−66	$-2^3 \cdot 3 \cdot 11$	−97	$-2^2 \cdot 97$

The discriminants of REAL QUADRATIC FIELDS $\mathbb{Q}(\sqrt{n})$ ($n > 0$) are given in the following table.

2	2^3	34	$2^3 \cdot 17$	67	$67 \cdot 2^2$
3	$3 \cdot 2^2$	35	$7 \cdot 2^2 \cdot 5$	69	$3 \cdot 23$
5	5	37	37	70	$7 \cdot 2^3 \cdot 5$
6	$3 \cdot 2^3$	38	$19 \cdot 2^3$	71	$71 \cdot 2^2$
7	$7 \cdot 2^2$	39	$3 \cdot 2^2 \cdot 13$	73	73
10	$2^3 \cdot 5$	41	41	74	$2^3 \cdot 37$
11	$11 \cdot 2^2$	42	$3 \cdot 2^3 \cdot 7$	77	$7 \cdot 11$
13	13	43	$43 \cdot 2^2$	78	$3 \cdot 2^3 \cdot 13$
14	$7 \cdot 2^3$	46	$23 \cdot 2^3$	79	$79 \cdot 2^2$
15	$3 \cdot 2^2 \cdot 5$	47	$47 \cdot 2^2$	82	$2^3 \cdot 41$
17	17	51	$3 \cdot 2^2 \cdot 17$	83	$83 \cdot 2^2$
19	$19 \cdot 2^2$	53	53	85	$5 \cdot 17$

21	$3\cdot7$	55	$11\cdot2^2\cdot5$	86	$43\cdot2^3$
22	$11\cdot2^3$	57	$3\cdot19$	87	$3\cdot2^2\cdot13$
23	$23\cdot2^2$	58	$2^3\cdot29$	89	89
26	$2^3\cdot13$	59	$59\cdot2^2$	91	$7\cdot2^2\cdot13$
29	29	61	61	93	$3\cdot31$
30	$3\cdot2^3\cdot5$	62	$31\cdot2^3$	94	$47\cdot2^3$
31	$31\cdot2^2$	65	$5\cdot13$	95	$19\cdot2^2\cdot5$
33	$3\cdot11$	66	$3\cdot2^3\cdot11$	97	97

See also Different, Fundamental Discriminant, Module

References
Cohn, H. *Advanced Number Theory.* New York: Dover, pp. 72–3 and 261–74, 1980.

Discriminant (Polynomial)
The product of the squares of the differences of the polynomial roots r_i. The discriminant of a polynomial is only defined up to sign. For a polynomial

$$a_n z^n + a_{n-1} z^{n-1} + \cdots + a_1 z + a_0 = 0 \qquad (1)$$

of degree n,

$$D_n = \prod_{\substack{i,j \\ i<j}}^{n} (r_i - r_j)^2. \qquad (2)$$

It is also common to consider discriminants D'_n for $a_n \equiv 1$ or discriminants D''_n obtained from D_n by multiplying by $a_n^{2(n-1)}$. If desired, powers a_n can be inserted mentally so that each term is of degree $2(n-1)$ and the whole expression is divided by $a_n^{2(n-1)}$.

The discriminant is closely related to resultants and can be implemented in *Mathematica* as

```
Discriminant[p_?PolynomialQ,x_] :=
  With[{n = Exponent[p,x]},    Cancel[
    ((-1)^(n(n-1)/2)Resultant[p,D[p,x],x])/
    Coefficient[p,x,n]^(2n-1)
  ]
]
```

The discriminant of the quadratic equation

$$a_2 z^2 + a_1 z + a_0 = 0 \qquad (3)$$

is given by

$$D_2 = \frac{a_1^2 - 4a_0 a_2}{a_2^2}. \qquad (4)$$

The discriminant of the cubic equation

$$a_3 z^3 + a_2 z^2 + a_1 z + a_0 = 0 \qquad (5)$$

is given by

$$D_3 = \frac{a_1^2 a_2^2 - 4a_0 a_2^3 - 4a_1^3 a_3 + 18a_0 a_1 a_2 a_3 - 27a_0^2 a_3^2}{a_3^4} \qquad (6)$$

The discriminant of a quartic equation

$$z^4 + a_3 z^3 + a_2 z^2 + a_1 z + a_0 = 0 \qquad (7)$$

is

$$\begin{aligned}
D_4 = \frac{1}{a_4^6} [& (a_1^2 a_2^2 a_3^2 - 4a_1^3 a_3^3 - 4a_1^2 a_2^3 a_4 \\
& + 18a_1^3 a_2 a_3 a_4 - 27a_1^4 a_4^2 + 256a_0^3 a_4^3) \\
& + a_0(-4a_2^3 a_3^3 + 18a_1 a_2 a_3^3 + 16a_2^4 a_4 \\
& -80a_1 a_2^2 a_3 a_4 - 6a_1^2 a_3^2 a_4 + 144a_1^2 a_2 a_4^2) \\
& + a_0^2(-27a_3^4 + 144a_2 a_3^2 a_4 - 128a_2^2 a_4^2 - 192a_1 a_3 a_4^2)]
\end{aligned}$$

(Beeler *et al.* 1972, Item 4).

See also Cubic Equation, Newton's Relations, Polynomial, Quadratic Equation, Quartic Equation, Resultant, Subresultant

References
Schroeppel, R. Item 4 in Beeler, M.; Gosper, R. W.; and Schroeppel, R. *HAKMEM.* Cambridge, MA: MIT Artificial Intelligence Laboratory, Memo AIM-239, p. 4, Feb. 1972.

Discriminant (Quadratic Curve)
Given a general quadratic curve

$$Ax^2 + Bxy + Cy^2 + Dx + Ey + F = 0, \qquad (1)$$

the quantity X is known as the discriminant, where

$$X \equiv B^2 - 4AC, \qquad (2)$$

and is invariant under rotation. Using the coefficients from quadratic equations for a rotation by an angle θ,

$$\begin{aligned}
A' &= \frac{1}{2}A[1 + \cos(2\theta)] + \frac{1}{2}B\sin(2\theta) + \frac{1}{2}C[1 - \cos(2\theta)] \\
&= \frac{A+C}{2} + \frac{B}{2}\sin(2\theta) + \frac{A-C}{2}\cos(2\theta) \qquad (3)
\end{aligned}$$

$$B' = G\cos\left(2\theta + \delta - \frac{\pi}{2}\right) = G\sin(2\theta + \delta) \qquad (4)$$

$$\begin{aligned}
C' &= \frac{1}{2}A[1 - \cos(2\theta)] - \frac{1}{2}B\sin\left(2\theta + \frac{1}{2}\right)C[1 + \cos(2\theta)] \\
&= \frac{A+C}{2} - \frac{B}{2}\sin(2\theta) + \frac{C-A}{2}\cos(2\theta). \qquad (5)
\end{aligned}$$

Now let

$$G \equiv \sqrt{B^2 + (A-C)^2} \qquad (6)$$

$$\delta \equiv \tan^{-1}\left(\frac{B}{C-A}\right) \qquad (7)$$

$$\delta_2 \equiv \tan^{-1}\left(\frac{A-C}{B}\right) = -\cot^{-1}\left(\frac{B}{C-A}\right), \qquad (8)$$

and use

$$\cot^{-1}(x) \equiv \frac{1}{2}\pi - \tan^{-1}(x) \qquad (9)$$

$$\delta_2 = \delta - \frac{1}{2}\pi \qquad (10)$$

to rewrite the primed variables

$$A' = \frac{A+C}{2} + \frac{1}{2}G\cos(2\theta+\delta) \qquad (11)$$

$$B' = B\cos(2\theta) + (C-A)\sin(2\theta) = G(2\theta + \delta_2) \qquad (12)$$

$$C' = \frac{A+C}{2} - \frac{1}{2}G\cos(2\theta+\delta). \qquad (13)$$

From (11) and (13), it follows that

$$4A'C' = (A+C)^2 - G^2\cos(2\theta+\delta). \qquad (14)$$

Combining with (12) yields, for an arbitrary θ

$$X \equiv B'^2 - 4A'C'$$

$$= G^2\sin^2(2\theta+\delta) + G^2\cos^2(2\theta+\delta) - (A+C)^2$$

$$= G^2 - (A+C)^2 = B^2 + (A-C)^2 - (A+C)^2$$

$$= B^2 - 4AC, \qquad (15)$$

which is therefore invariant under rotation. This invariant therefore provides a useful shortcut to determining the shape represented by a QUADRATIC CURVE. Choosing θ to make $B' = 0$ (see QUADRATIC EQUATION), the curve takes on the form

$$A'x^2 + C'y^2 + D'x + E'y + F = 0. \qquad (16)$$

COMPLETING THE SQUARE and defining new variables gives

$$A'x'^2 + C'y'^2 = H. \qquad (17)$$

Without loss of generality, take the sign of H to be positive. The discriminant is

$$X = B'^2 + 4A'C' = -4A'C'. \qquad (18)$$

Now, if $-4A'C' < 0$, then A' and C' both have the same sign, and the equation has the general form of an ELLIPSE (if A' and B' are positive). If $-4A'C' > 0$, then A' and C' have opposite signs, and the equation has the general form of a HYPERBOLA. If $-4A'C' = 0$, then

either A' or C' is zero, and the equation has the general form of a PARABOLA (if the NONZERO A' or C' is positive). Since the discriminant is invariant, these conclusions will also hold for an arbitrary choice of θ, so they also hold when $-4A'C'$ is replaced by the original $B^2 - 4AC$. The general result is

1. If $B^2 - 4AC < 0$, the equation represents an ELLIPSE, a CIRCLE (degenerate ELLIPSE), a POINT (degenerate CIRCLE), or has no graph.
2. If $B^2 - 4AC > 0$, the equation represents a HYPERBOLA or pair of intersecting lines (degenerate HYPERBOLA).
3. If $B^2 - 4AC = 0$, the equation represents a PARABOLA, a LINE (degenerate PARABOLA), a pair of PARALLEL lines (degenerate PARABOLA), or has no graph.

Discriminant (Quadratic Form)
DISCRIMINANT (BINARY QUADRATIC FORM)

Discriminant (Second Derivative Test)

$$D \equiv f_{xx}f_{yy} - f_{xy}f_{yx} = f_{xx}f_{yy} - f_{xy}^2,$$

where f_{ij} are PARTIAL DERIVATIVES.
See also SECOND DERIVATIVE TEST

Disdyakis Dodecahedron

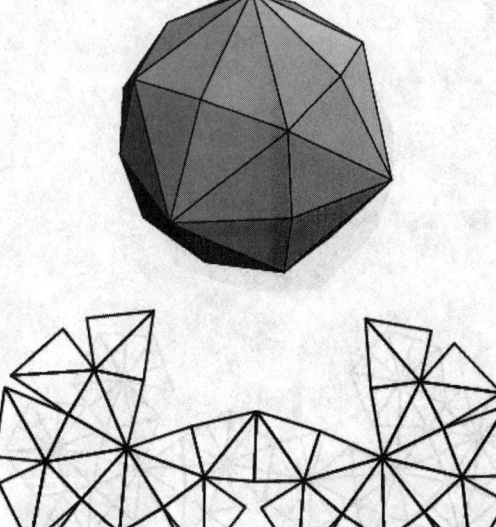

The DUAL POLYHEDRON of the Archimedean GREAT RHOMBICUBOCTAHEDRON A_3 and Wenninger dual W_{15}, also called the HEXAKIS OCTAHEDRON. If the original GREAT RHOMBICUBOCTAHEDRON has unit side lengths, then the resulting dual has edge lengths

$$s_1 = \frac{2}{7}\sqrt{30 - 3\sqrt{2}} \qquad (1)$$

$$s_2 = \frac{3}{7}\sqrt{6\left(2 + \sqrt{2}\right)} \qquad (2)$$

$$s_3 = \frac{2}{7}\sqrt{6\left(10 + \sqrt{2}\right)}. \qquad (3)$$

The INRADIUS is

$$r = 3\sqrt{\frac{2}{97}\left(15 + 8\sqrt{2}\right)}. \qquad (4)$$

Scaling the disdyakis dodecahedron so that $s_1 = 1$ gives a solid with SURFACE AREA and VOLUME

$$S = \frac{6}{7}\sqrt{783 + 436\sqrt{2}} \qquad (5)$$

$$V = \frac{1}{7}\sqrt{3\left(2194 + 1513\sqrt{2}\right)}. \qquad (6)$$

See also ARCHIMEDEAN DUAL, ARCHIMEDEAN SOLID, GREAT DISDYAKIS DODECAHEDRON, OCTATETRAHEDRON

References

Wenninger, M. J. *Dual Models.* Cambridge, England: Cambridge University Press, p. 25–6, 1983.

Disdyakis Triacontahedron

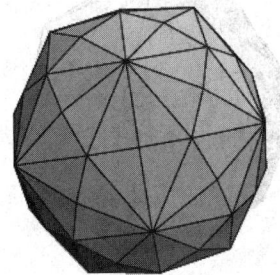

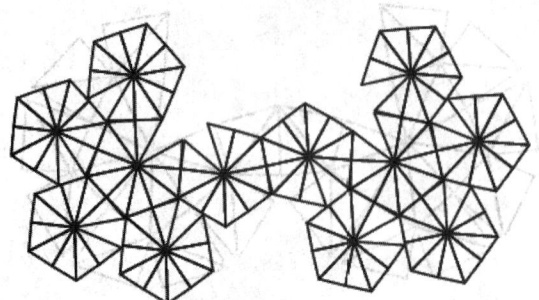

The DUAL POLYHEDRON of the Archimedean GREAT RHOMBICOSIDODECAHEDRON A_2 and Wenninger dual W_{16}. It is also called the HEXAKIS ICOSAHEDRON.

See also ARCHIMEDEAN DUAL, ARCHIMEDEAN SOLID

References

Wenninger, M. J. *Dual Models.* Cambridge, England: Cambridge University Press, pp. 25 and 27, 1983.

Disjoint Sets

Two SETS A_1 and A_2 are disjoint if their INTERSECTION $A_1 \cap A_2 \equiv \emptyset$, where $\emptyset$ is the EMPTY SET. n sets $A_1, A_2, ..., A_n$ are disjoint if $A_i \cap A_j \equiv \emptyset$ for $i \neq j$. For example, $\{A, B, C\}$ and $\{D, E\}$ are disjoint, but $\{A, B, C\}$ and $\{C, D, E\}$ are not. Disjoint sets are also said to be mutually exclusive or independent.

See also EMPTY SET, INDEPENDENT SET, INTERSECTION, SET

Disjoint Union

The disjoint union of two SETS A and B is a BINARY OPERATOR that combines all distinct elements of a pair of given sets, while retaining the original set membership as a distinguishing characteristic of the union set. The disjoint union is denoted

$$A \cup^* B = (A \times \{0\}) \cup (B \times \{1\}) \equiv A^* \cup B^*,$$

where $A \times S$ is a SET DIRECT PRODUCT. For example, the disjoint union of sets $A = \{1, 2, 3, 4, 5\}$ and $B = \{1, 2, 3, 4, 5\}$ can be computed by finding

$$A^* = \{(1, 0), (2, 0), (3, 0), (4, 0), (5, 0)\}$$

$$B^* = \{(1, 1), (2, 1), (3, 1), (4, 1)\},$$

so

$$A \cup^* B = A^* \cup B^*$$

$$= \{(1, 0), (2, 0), (3, 0), (4, 0), (5, 0),$$

$$(1, 1), (2, 1), (3, 1), (4, 1)\}$$

See also UNION

References

Armstrong, M. A. *Basic Topology, rev. ed.* New York: Springer-Verlag, 1997.

Disjunction

The term in logic used to describe the operation commonly known as OR.

See also CONJUNCTION, DISJUNCTIVE NORMAL FORM, DISJUNCTIVE SYLLOGISM, OR

Disjunctive Game

NIM-HEAP

Disjunctive Normal Form

A statement is in disjunctive normal form if it is a DISJUNCTION (sequence of ORs) consisting of one or more disjuncts, each of which is a CONJUNCTION (AND) of one or more statement letters and negations of statement letters. Examples of disjunctive normal forms include

$$A \tag{1}$$

$$(A \wedge B) \vee (!A \wedge C) \tag{2}$$

$$(A \wedge B \wedge !A) \vee (C \wedge !B) \vee (A \wedge !C) \tag{3}$$

$$(A \wedge B) \tag{4}$$

$$A \vee (B \wedge C), \tag{5}$$

where $\vee$ denotes OR, $\wedge$ denotes AND, and ! denotes NOT. Every statement in logic consisting of a combination of multiple $\wedge$, $\vee$, and !s can be written in conjunctive normal form.

See also CONJUNCTIVE NORMAL FORM

References

Mendelson, E. *Introduction to Mathematical Logic, 4th ed.* London: Chapman & Hall, pp. 27, 1997.

Disk

An n-D disk (or DISC) of RADIUS r is the collection of points of distance $\leq r$ (CLOSED DISK) or $< r$ (OPEN DISK) from a fixed point in EUCLIDEAN n-space. A disk is the SHADOW of a BALL on a PLANE PERPENDICULAR to the BALL-RADIANT POINT line.

The n-disk for $n \geq 3$ is called a BALL, and the boundary of the n-disk is a $(n-1)$-HYPERSPHERE. The standard n-disk, denoted $\mathbb{D}^n$ (or $\mathbb{B}^n$), has its center at the ORIGIN and has RADIUS $r = 1$.

See also BALL, CLOSED DISK, DISK COVERING PROBLEM, FIVE DISKS PROBLEM, HYPERSPHERE, LOWER HALF-DISK, MERGELYAN-WESLER THEOREM, OPEN DISK, POLYDISK, SPHERE, UNIT DISK, UPPER HALF-DISK

Disk Algebra

This entry contributed by RONALD M. AARTS

A disk algebra is an ALGEBRA of functions which are analytic on the OPEN UNIT DISK in C and continuous up to the boundary. A representative measure for a point x in the CLOSED DISK is a nonnegative MEASURE m such that $\mathrm{Int}(f\,dm) = f(x)$ for all f in A. These measures form a COMPACT, CONVEX SET M_x in the linear space of all measures.

See also ALGEBRA

Disk Covering Problem

N.B. A detailed online essay by S. Finch was the starting point for this entry.

Given a UNIT DISK, find the smallest RADIUS $r(n)$ required for n equal disks to completely cover the UNIT DISK. For a symmetrical arrangement with $n = 5$ (the FIVE DISKS PROBLEM), $r(5) = \phi - 1 = 1/\phi = 0.6180340\ldots$, where ϕ is the GOLDEN RATIO. However, the radius can be reduced in the general disk covering problem where symmetry is not required. The first few such values are

$$r(1) = 1$$

$$r(2) = 1$$

$$r(3) = \frac{1}{2}\sqrt{3}$$

$$r(4) = \frac{1}{2}\sqrt{2}$$

$$r(5) = 0.609382864\ldots$$

$$r(6) = 0.555$$

$$r(7) = \frac{1}{2}$$

$$r(8) = 0.437$$

$$r(9) = 0.422$$

$$r(10) = 0.398.$$

Here, values for $n = 6, 8, 9, 10$ were obtained using computer experimentation by Zahn (1962). The value $r(5)$ is equal to $\cos(\theta + \phi/2)$, where θ and ϕ are solutions to

$$2\sin\theta - \sin\left(\theta + \frac{1}{2}\phi + \psi\right) - \sin\left(\psi - \theta - \frac{1}{2}\phi\right) = 0 \tag{1}$$

$$2\sin\phi - \sin\left(\theta + \frac{1}{2}\phi + \chi\right) - \sin\left(\chi - \theta - \frac{1}{2}\phi\right) = 0 \tag{2}$$

$$2\sin\theta + \sin(\chi + \theta) - \sin(\chi - \theta) - \sin(\psi + \phi)$$
$$-\sin(\psi - \phi) - 2\sin(\psi - 2\theta) = 0 \tag{3}$$

$$\cos(2\psi - \chi + \phi) - \cos(2\psi + \chi - \phi) - 2\cos\chi$$
$$+\cos(2\psi + \chi - 2\theta) + \cos(2\psi - \chi - 2\theta) = 0 \tag{4}$$

(Neville 1915). It is also given by $1/x$, where x is the largest real root of

$$a(y)x^6 - b(y)x^5 + c(y)x^4 - d(y)x^3 + e(y)x^2 - f(y)x + g(y)$$
$$= 0 \tag{5}$$

maximized over all y, subject to the constraints

$$\sqrt{2} < x < 2y + 1 \tag{6}$$

$$-1 < y < 1, \tag{7}$$

and with

$$a(y) = 80y^2 + 64y \tag{8}$$

$$b(y) = 416y^3 + 384y^2 + 64y \tag{9}$$

$$c(y) = 848y^4 + 928y^3 + 352y^2 + 32y \tag{10}$$

$$d(y) = 768y^5 + 992y^4 + 736y^3 + 288y^2 + 96y$$

$$e(y) = 256y^6 + 384y^5 + 592y^4 + 480y^3 + 336y^2 + 96y + 16 \tag{11}$$

$$f(y) = 128y^5 + 192y^4 + 256y^3 + 160y^2 + 96y + 32 \tag{12}$$

$$g(y) = 64y^2 + 64y + 16 \tag{13}$$

(Bezdek 1983, 1984).

Letting $N(\varepsilon)$ be the smallest number of DISKS of RADIUS ε needed to cover a disk D, the limit of the ratio of the AREA of D to the AREA of the disks is given by

$$\lim_{\varepsilon \to 0^+} \frac{1}{\varepsilon^2 N(\varepsilon)} = \frac{3\sqrt{3}}{2\pi} \tag{14}$$

(Kershner 1939, Verblunsky 1949).

See also CIRCLE COVERING, FIVE DISKS PROBLEM

References

Ball, W. W. R. and Coxeter, H. S. M. "The Five-Disc Problem." In *Mathematical Recreations and Essays, 13th ed.* New York: Dover, pp. 97–9, 1987.

Bezdek, K. "Uuml;ber einige Kreisüberdeckungen." *Beiträge Algebra Geom.* **14**, 7–3, 1983.

Bezdek, K. "Über einige optimale Konfigurationen von Kreisen." *Ann. Univ. Sci. Budapest Eotvos Sect. Math.* **27**, 141–51, 1984.

Finch, S. "Favorite Mathematical Constants." http://www.mathsoft.com/asolve/constant/circle/circle.html.

Kershner, R. "The Number of Circles Covering a Set." *Amer. J. Math.* **61**, 665–71, 1939.

Neville, E. H. "On the Solution of Numerical Functional Equations, Illustrated by an Account of a Popular Puzzle and of its Solution." *Proc. London Math. Soc.* **14**, 308–26, 1915.

Verblunsky, S. "On the Least Number of Unit Circles which Can Cover a Square." *J. London Math. Soc.* **24**, 164–70, 1949.

Zahn, C. T. "Black Box Maximization of Circular Coverage." *J. Res. Nat. Bur. Stand. B* **66**, 181–16, 1962.

Disk Lattice Points

GAUSS'S CIRCLE PROBLEM

Disk Line Picking

Using DISK POINT PICKING,

$$x = \sqrt{r} \cos\theta \tag{1}$$

$$y = \sqrt{r} \sin\theta \tag{2}$$

for $r \in [0, 1]$, $\theta \in [0, 2\pi)$, choose two points at random in a UNIT DISK and find the distribution of distances s between the two points. Without loss of generality, take the first point as $(r, \theta) = (r_1, 0)$ and the second point as (r_2, θ). Then>

$$\bar{s} = \frac{v \int_0^1 \int_0^1 \int_0^{2\pi} \sqrt{r_1 + r_2 - 2\sqrt{r_1 r_2 \cos\theta}} \, dr_1 dr_2 d\theta}{\int_0^1 \int_0^1 \int_0^{2\pi} dr_1 dr_2 d\theta} \tag{3}$$

$$= \frac{128}{45\pi} \tag{4}$$

(Uspensky 1937, p. 258).

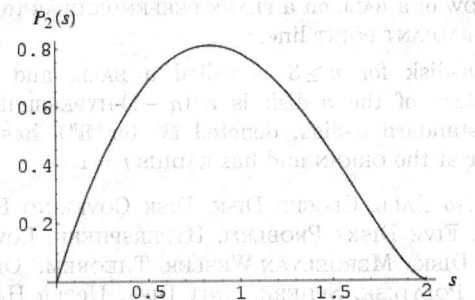

This is a special case of BALL LINE PICKING with $n = 2$, so the full probability function for a disk of radius R is

$$P_2(s) = \frac{4s}{\pi R^2} \cos^{-1}\left(\frac{s}{2R}\right) - \frac{2s^2}{\pi R^3} \sqrt{1 - \frac{s^2}{4R^2}} \tag{5}$$

(Solomon 1978, p. 129).

See also BALL LINE PICKING, CIRCLE LINE PICKING

References

Solomon, H. *Geometric Probability*. Philadelphia, PA: SIAM, 1978.

Uspensky, J. V. Ch. 12, Problem 5 in *Introduction to Mathematical Probability*. New York: McGraw-Hill, pp. 257–58, 1937.

Disk Packing
CIRCLE PACKING

Disk Point Picking

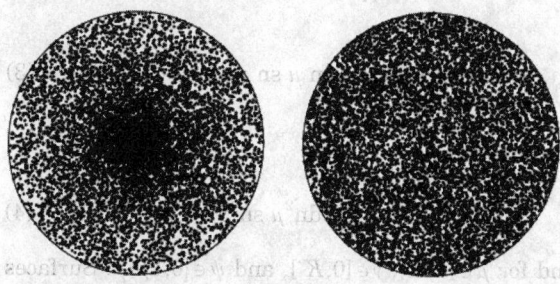

To generate random points over the UNIT DISK, it is *incorrect* to use two uniformly distributed variables $r \in [0,1]$, and $\theta \in [0, 2\pi)$, and then take

$$x = r \cos\theta \qquad (1)$$

$$y = r \sin\theta. \qquad (2)$$

Because the area element is given by

$$dA = 2\pi r \, dr, \qquad (3)$$

this gives a concentration of points in the center (left figure above).

The correct transformation is instead given by

$$x = \sqrt{r} \cos\theta \qquad (4)$$

$$y = \sqrt{r} \sin\theta \qquad (5)$$

(right figure above).

See also CIRCLE POINT PICKING, DISK LINE PICKING, POINT PICKING, SPHERE POINT PICKING

Disk Triangle Picking

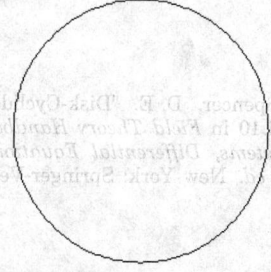

Pick three points $P = (x_1, y_1)$, $Q = (x_2, y_2)$, and $R = (x_3, y_3)$ distributed independently and uniformly in a UNIT DISK K. Then the average area of the TRIANGLE determined by these points is

$$\bar{A} = \frac{\displaystyle\iint_{P \in K} \iint_{Q \in K} \iint_{R \in K} \frac{1}{2} \begin{vmatrix} x_1 & y_1 & 1 \\ x_2 & y_2 & 1 \\ x_3 & y_3 & 1 \end{vmatrix} dy_3 dy_3 dy_1 dx_3 dx_2 dx_1}{\displaystyle\iint_{P \in K} \iint_{Q \in K} \iint_{R \in K} dy_3 dy_3 dy_1 dx_3 dx_2 dx_1}$$

$$(1)$$

which can be evaluated using CROFTON'S FORMULA and polar coordinates to yield $\bar{A} = 35/(48\pi^2)$ (Woolhouse 1967; Solomon 1987; Pfiefer 1989). This problem is very closely related to SYLVESTER'S FOUR-POINT PROBLEM, and can be derived as the limit as $n \to \infty$ of the general POLYGON TRIANGLE PICKING problem.

The probability P_2 that three random points in a disk form an ACUTE TRIANGLE is

$$P_2 = \frac{4}{\pi^2} - \frac{1}{8} \qquad (2)$$

(Woolhouse 1886). The problem was generalized by Hall (1982) to n-D BALL TRIANGLE PICKING, and Buchta (1986) gave closed form evaluations for Hall's integrals.

Let the VERTICES of a triangle in n-D be NORMAL (GAUSSIAN) variates. The probability that a Gaussian triangle in n-D is OBTUSE is

$$P_n = \frac{3\Gamma(n)}{\Gamma^2\left(\frac{1}{2}n\right)} \int_0^{1/3} \frac{x^{(n-2)/2}}{(1+x)^n} \, dx$$

$$= \frac{3\Gamma(n)}{\Gamma^2\left(\frac{1}{2}n\right)2^{n-1}} \int_0^{\pi/3} \sin^{n-1}\theta \, d\theta$$

$$= \frac{6\Gamma(n)\,_2F_1\left(\frac{1}{2}n, n; 1 + \frac{1}{2}n; -\frac{1}{3}\right)}{3^{n/2} n \Gamma^2\left(\frac{1}{2}n\right)}, \qquad (3)$$

where $\Gamma(n)$ is the GAMMA FUNCTION and $_2F_1(a,b;c;x)$ is the HYPERGEOMETRIC FUNCTION. For EVEN $n \equiv 2k$,

$$P_{2k} = 3 \sum_{j=k}^{2k-1} \binom{2k-1}{j} \left(\frac{1}{4}\right)^j \left(\frac{3}{4}\right)^{2k-1-j} \qquad (4)$$

(Eisenberg and Sullivan 1996). The first few cases are explicitly

$$P_2 = \frac{3}{4} = 0.75 \qquad (5)$$

$$P_3 = 1 - \frac{3\sqrt{3}}{4\pi} = 0.586503\ldots \qquad (6)$$

$$P_4 = \frac{15}{32} = 0.46875 \qquad (7)$$

$$P_5 = 1 - \frac{9\sqrt{3}}{8\pi} = 0.37975499\ldots \qquad (8)$$

See also BALL TRIANGLE PICKING, HEXAGON TRIANGLE PICKING, OBTUSE TRIANGLE, SQUARE TRIANGLE PICKING, SYLVESTER'S FOUR-POINT PROBLEM, TRIANGLE TRIANGLE PICKING

References

Buchta, C. "Zufallspolygone in konvexen Vielecken." *J. reine angew. Math.* **347**, 212–20, 1984.

Buchta, C. "A Note on the Volume of a Random Polytope in a Tetrahedron." *Ill. J. Math.* **30**, 653–59, 1986.

Eisenberg, B. and Sullivan, R. "Random Triangles *n* Dimensions." *Amer. Math. Monthly* **103**, 308–18, 1996.

Guy, R. K. "There are Three Times as Many Obtuse-Angled Triangles as There are Acute-Angled Ones." *Math. Mag.* **66**, 175–78, 1993.

Hall, G. R. "Acute Triangles in the *n*-Ball." *J. Appl. Prob.* **19**, 712–15, 1982.

Pfiefer, R. E. "The Historical Development of J. J. Sylvester's Four Point Problem." *Math. Mag.* **62**, 309–17, 1989.

Solomon, H. *Geometric Probability*. Philadelphia, PA: SIAM, 1978.

Woolhouse, W. S. B. Solution to Problem 1350. *Mathematical Questions, with Their Solutions, from the Educational Times, Vol. 1.* London: F. Hodgson and Son, pp. 49–1, 1886.

Woolhouse, W. S. B. "Some Additional Observations on the Four-Point Problem." *Mathematical Questions, with Their Solutions, from the Educational Times, Vol. 7.* London: F. Hodgson and Son, p. 81, 1867.

Disk-Cyclide Coordinates

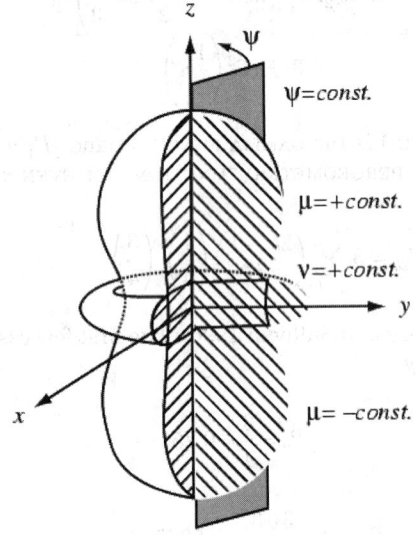

A coordinate system defined by the transformation

equations

$$x = \frac{a}{\Lambda} \operatorname{cn} \mu \operatorname{cn} v \cos\psi \qquad (1)$$

$$y = \frac{a}{\Lambda} \operatorname{cn} \mu \operatorname{cn} v \sin\psi \qquad (2)$$

$$z = \frac{a}{\Lambda} \operatorname{sn} \mu \operatorname{dn} \mu \operatorname{sn} v \operatorname{dn} v, \qquad (3)$$

where

$$\Lambda \equiv 1 - \operatorname{dn}^2 \mu \operatorname{sn}^2 v \qquad (4)$$

and for $\mu \in [0, K]$, $v \in [0, K']$, and $\psi \in [0, 2pi)$.. Surfaces of constant μ are given by the cyclides of rotation

$$\left(\frac{x^2 + y^2}{a^2 \operatorname{cn}^2 \mu} + \frac{k^2 \operatorname{sn}^2 \mu}{a^2 \operatorname{dn}^2 \mu} z^2 \right)^2$$

$$- \frac{2(x^2 + y^2)}{a^2 \operatorname{cn}^2 \mu} - \frac{2k^2 \operatorname{sn}^2 \mu}{a^2 \operatorname{dn}^2 \mu} z^2 + 1 = 0' \qquad (5)$$

surfaces of constant v by the disk cyclides

$$\left[\frac{\operatorname{cn}^2 v}{a^2} \left(x^2 + y^2 \right) + \frac{k'^2 \operatorname{sn}^2 v}{a^2 \operatorname{dn}^2 v} z^2 \right]^2$$

$$- \frac{2 \operatorname{cn}^2 v}{a^2} \left(x^2 + y^2 \right) - \frac{2k'^2 \operatorname{sn}^2 v}{a^2 \operatorname{dn}^2 v} z^2 + 1 = 0, \qquad (6)$$

and surfaces of constant ψ by the half-planes

$$\tan \psi = \frac{y}{x}. \qquad (7)$$

See also CAP-CYCLIDE COORDINATES, CYCLIDIC COORDINATES, FLAT-RING CYCLIDE COORDINATES

References

Moon, P. and Spencer, D. E. "Disk-Cyclide Coordinates (μ, v, ψ)." Fig. 4.10 in *Field Theory Handbook, Including Coordinate Systems, Differential Equations, and Their Solutions, 2nd ed.* New York: Springer-Verlag, pp. 129–32, 1988.

Dispersion (Sequence)

An array $B = b_{ij}$, $i, j \geq 1$ of POSITIVE INTEGERS is called a dispersion if

1. The first column of B is a strictly increasing sequence, and there exists a strictly increasing sequence $\{s_k\}$ such that
2. $b_{12} = s_1 \geq 2$,
3. The complement of the SET $\{b_{i1} : i \geq 1\}$ is the SET $\{s_k\}$,
4. $b_{ij} = s_{b_{i,j-1}}$ for all $j \geq 3$ for $i = 1$ and for all $g \geq 2$ for all $i \geq 2$..

If an array $B = b_{ij}$, is a dispersion, then it is an INTERSPERSION.

See also INTERSPERSION

References

Kimberling, C. "Interspersions and Dispersions." *Proc. Amer. Math. Soc.* **117**, 313–21, 1993.

Dispersion (Statistics)

$$(\Delta u)_i^2 \equiv (u_i - \bar{u})^2,$$

where $\bar{u}$ is the average of $\{u_i\}$..

See also ABSOLUTE DEVIATION, SIGNED DEVIATION, VARIANCE

Dispersion Numbers

MAGIC GEOMETRIC CONSTANTS

Dispersion Relation

Any pair of equations giving the REAL PART of a function as an integral of its IMAGINARY PART and the IMAGINARY PART as an integral of its REAL PART. Dispersion relationships imply causality in physics. Let

$$f(x_0) \equiv u(x_0) + iv(x_0), \tag{1}$$

then

$$u(x_0) = \frac{1}{\pi} PV \int_{-\infty}^{\infty} \frac{v(x)dx}{x - x_0} \tag{2}$$

$$v(x_0) = -\frac{1}{\pi} PV \int_{-\infty}^{\infty} \frac{u(x)dx}{x - x_0}, \tag{3}$$

where PV denotes the CAUCHY PRINCIPAL VALUE and $u(x_0)$ and $v(x_0)$ are HILBERT TRANSFORMS of each other. If the COMPLEX function is symmetric such that $f(-x) = f^*(x)$, then

$$u(x_0) = \frac{2}{\pi} PV \int_0^{\infty} \frac{xv(x)dx}{x^2 - x_0^2} \tag{4}$$

$$v(x_0) = -\frac{2}{\pi} PV \int_0^{\infty} \frac{xu(x)dx}{x^2 - x_0^2}. \tag{5}$$

See also HILBERT TRANSFORM

Dispersive Long-Wave Equation

The system of PARTIAL DIFFERENTIAL EQUATIONS

$$u_t = (u^2 - v_x + 2v)_x$$

$$v_t = (2uv + v_x)_x.$$

References

Boiti, M.; Leon, J. J.-P.; and Pempinelli, F. "Integrable Two-Dimensional Generalisation of the Sine- and Sinh-Gordon Equations." *Inverse Prob.* **3**, 37–9, 1987.
Zwillinger, D. *Handbook of Differential Equations*, 3rd ed. Boston, MA: Academic Press, p. 137, 1997.

Disphenocingulum

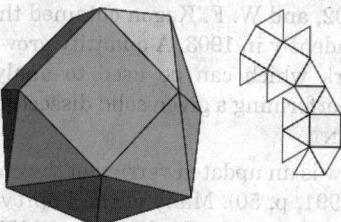

JOHNSON SOLID J_{90}..

References

Weisstein, E. W. "Johnson Solids." MATHEMATICA NOTEBOOK JOHNSONSOLIDS.M.
Weisstein, E. W. "Johnson Solid Netlib Database." MATHEMATICA NOTEBOOK JOHNSONSOLIDS.DAT.

Disphenoid

A TETRAHEDRON with identical ISOSCELES or SCALENE faces.

See also SNUB DISPHENOID

Dissection

Any two rectilinear figures with equal AREA can be dissected into a finite number of pieces to form each other. This is the WALLACE-BOLYAI-GERWEIN THEOREM. For minimal dissections of a TRIANGLE, PENTAGON, and OCTAGON into a SQUARE, see Stewart (1987, pp. 169–70) and Ball and Coxeter (1987, pp. 89–1). The TRIANGLE to SQUARE dissection (HABERDASHER'S PROBLEM) is particularly interesting because it can be built from hinged pieces which can be folded and unfolded to yield the two shapes (Gardner 1961; Stewart 1987, p. 169; Pappas 1989; Steinhaus 1983, pp. 3–; Wells 1991, pp. 61–2).

	{3}	{4}	{5}	{6}	{7}	{8}	{9}	{10}	{12}	GR	GC	LC	MC	SW	{5/2}	{6/2}
{4}	4															
{5}	6	6														
{6}	5	5	7													
{7}	8	7	9	8												
{8}	7	5	9	8	11											
{9}	8	9	12	11	14	13										
{10}	7	7	10	9	11	10	13									
{12}	8	6	10	6	11	10	14	12								
GR	4	3	6	5	7	6	9	6	7							
GC	5	4	7	7	9	9	12	10	6	5						
LC	5	5	8	6	8	8	11	10	7	5	7					
MC		7		14											8	
SW		6		12											8	9
{5/2}	7	7	9	9	11	10	14	6	12	7	10	10				
{6/2}	5	5	8	6	9	8	11	9	9	5	8	8			11	
{8/3}	8	8	9	9	12	6	13	12	12	7	10	11			13	10

Laczkovich (1988) proved that the CIRCLE can be squared in a finite number of dissections (($\sim 10^{50}$).). Furthermore, any shape whose boundary is composed of smoothly curving pieces can be dissected into a SQUARE.

The situation becomes considerably more difficult moving from 2-D to 3-D. In general, a POLYHEDRON cannot be dissected into other POLYHEDRA of a specified type. A CUBE *can* be dissected into n^3 CUBES, where n is any INTEGER. In 1900, Dehn proved that not every PRISM can be dissected into a TETRAHEDRON (Lenhard 1962, Ball and Coxeter 1987) The third of HILBERT'S PROBLEMS asks for the determination of two TETRAHEDRA which cannot be decomposed into congruent TETRAHEDRA directly or by adjoining congruent TETRAHEDRA. Max Dehn showed this could not be done in 1902, and W. F. Kagon obtained the same result independently in 1903. A quantity growing out of Dehn's work which can be used to analyze the possibility of performing a given solid dissection is the DEHN INVARIANT.

The table below is an updated version of the one given in Gardner (1991, p. 50). Many of the improvements are due to G. Theobald (Frederickson 1997). The minimum number of pieces known to dissect a regular n-gon (where n is a number in the first column) into a k-gon (where k is a number is the bottom row) is read off by the intersection of the corresponding row and column. In the table, $\{n\}$ denotes a regular n-gon, GR a GOLDEN RECTANGLE, GC a GREEK CROSS, LC a LATIN CROSS, MC a MALTESE CROSS, SW a SWASTIKA, $\{5/2\}$ a five-point star (solid PENTAGRAM), $\{6/2\}$ a six-point star (i.e., HEXAGRAM or solid STAR OF DAVID), and $\{8/3\}$ the solid OCTAGRAM.

Wells (1991) gives several attractive dissections of the regular DODECAGON. The best-known dissections of one regular convex n-gon into another are shown for $n = 3$, 4, 5, 6, 7, 8, 9, 10, and 12 in the following illustrations due to Theobald.

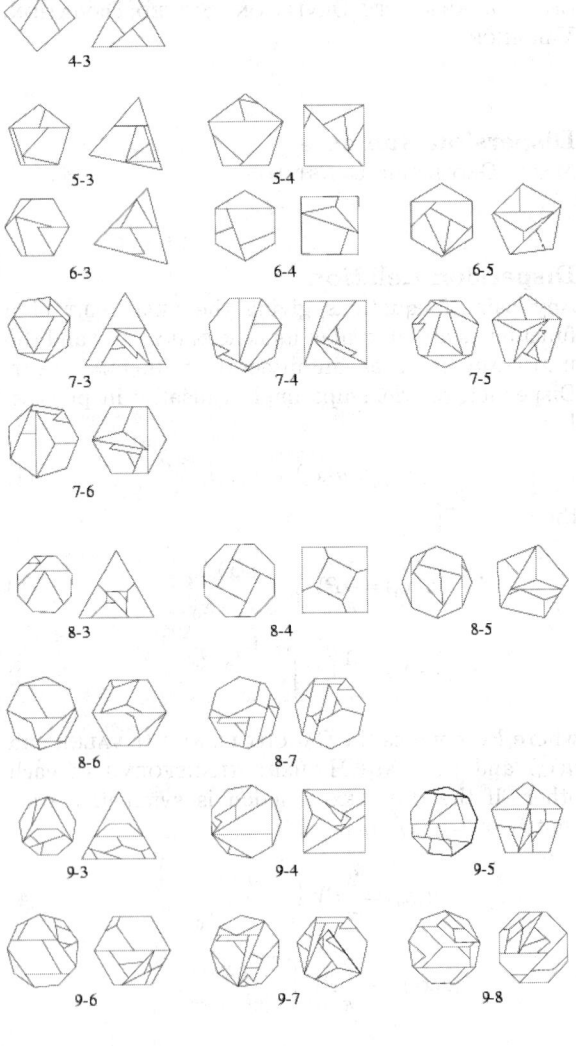

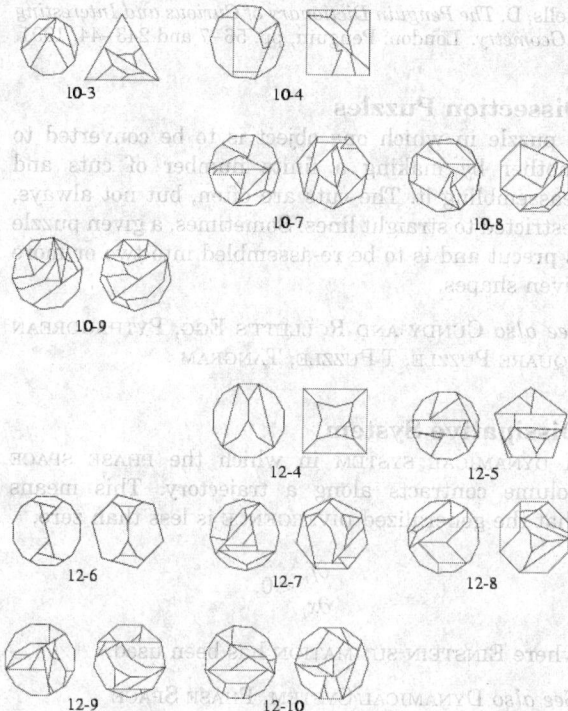

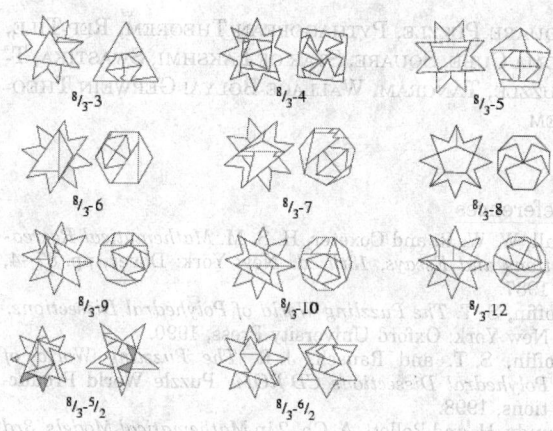

The best-known dissections of regular concave polygons are illustrated below for {5/2}, {6/2}, and {8/3} (Theobald).

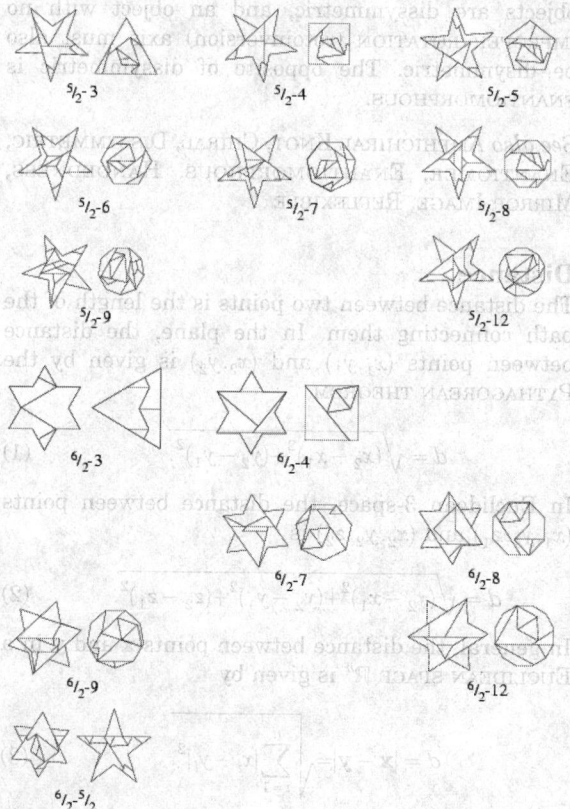

The best-known dissections of various crosses are illustrated below (Theobald).

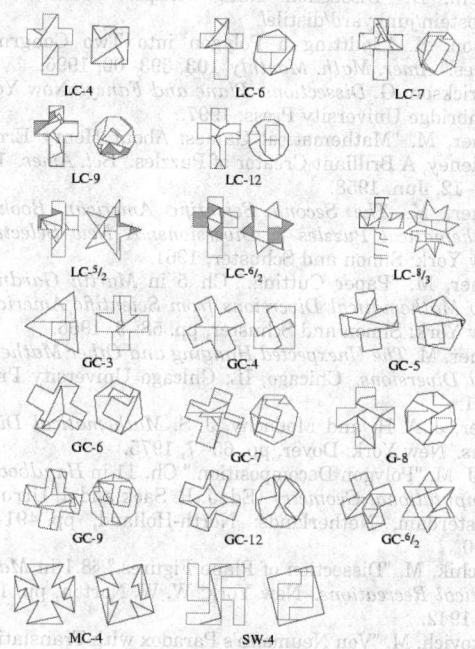

The best-known dissections of the GOLDEN RECTANGLE are illustrated below (Theobald).

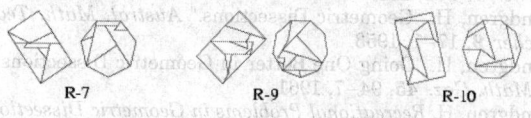

See also BANACH-TARSKI PARADOX, BLANCHE'S DISSECTION, CUNDY AND ROLLETT'S EGG, DECAGON, DEHN INVARIANT, DIABOLICAL CUBE, DISSECTION PUZZLES, DODECAGON, EHRHART POLYNOMIAL, EQUIDECOMPOSABLE, EQUILATERAL TRIANGLE, GOLDEN RECTANGLE, HEPTAGON, HEXAGON, HEXAGRAM, HILBERT'S PROBLEMS, LATIN CROSS, MALTESE CROSS, NONAGON, OCTAGON, OCTAGRAM, PENTAGON, PENTAGRAM, POLYHEDRON DISSECTION, PYTHAGOREAN

Square Puzzle, Pythagorean Theorem, Rep-Tile, Soma Cube, Square, Star of Lakshmi, Swastika, T-Puzzle, Tangram, Wallace-Bolyai-Gerwein Theorem

References

Ball, W. W. R. and Coxeter, H. S. M. *Mathematical Recreations and Essays, 13th ed.* New York: Dover, pp. 87–4, 1987.

Coffin, S. T. *The Puzzling World of Polyhedral Dissections.* New York: Oxford University Press, 1990.

Coffin, S. T. and Rausch, J. R. *The Puzzling World of Polyhedral Dissections CD-ROM.* Puzzle World Productions, 1998.

Cundy, H. and Rollett, A. Ch. 2 in *Mathematical Models, 3rd ed.* Stradbroke, England: Tarquin Pub., 1989.

Eppstein, D. "Dissection." http://www.ics.uci.edu/~eppstein/junkyard/dissect.html.

Eppstein, D. "Dissection Tiling." http://www.ics.uci.edu/~eppstein/junkyard/distile/.

Eriksson, K. "Splitting a Polygon into Two Congruent Pieces." *Amer. Math. Monthly* **103**, 393–00, 1996.

Frederickson, G. *Dissections: Plane and Fancy.* New York: Cambridge University Press, 1997.

Gardner, M. "Mathematical Games: About Henry Ernest Dudeney, A Brilliant Creator of Puzzles." *Sci. Amer.* **198**, 108–12, Jun. 1958.

Gardner, M. *The Second Scientific American Book of Mathematical Puzzles & Diversions: A New Selection.* New York: Simon and Schuster, 1961.

Gardner, M. "Paper Cutting." Ch. 5 in *Martin Gardner's New Mathematical Diversions from Scientific American.* New York: Simon and Schuster, pp. 58–9, 1966.

Gardner, M. *The Unexpected Hanging and Other Mathematical Diversions.* Chicago, IL: Chicago University Press, 1991.

Hunter, J. A. H. and Madachy, J. S. *Mathematical Diversions.* New York: Dover, pp. 65–7, 1975.

Keil, J. M. "Polygon Decomposition." Ch. 11 in *Handbook of Computational Geometry* (Ed. J.-R. Sack and J. Urrutia). Amsterdam, Netherlands: North-Holland, pp. 491–18, 2000.

Kraitchik, M. "Dissection of Plane Figures." §8.1 in *Mathematical Recreations.* New York: W. W. Norton, pp. 193–98, 1942.

Laczkovich, M. "Von Neumann's Paradox with Translation." *Fund. Math.* **131**, 1–2, 1988.

Lenhard, H.-C. "Über fünf neue Tetraeder, die einem Würfel äquivalent sind." *Elemente Math.* **17**, 108–09, 1962.

Lindgren, H. "Geometric Dissections." *Austral. Math. Teacher* **7**, 7–0, 1951.

Lindgren, H. "Geometric Dissections." *Austral. Math. Teacher* **9**, 17–1, 1953.

Lindgren, H. "Going One Better in Geometric Dissections." *Math. Gaz.* **45**, 94–7, 1961.

Lindgren, H. *Recreational Problems in Geometric Dissection and How to Solve Them.* New York: Dover, 1972.

Madachy, J. S. "Geometric Dissection." Ch. 1 in *Madachy's Mathematical Recreations.* New York: Dover, pp. 15–3, 1979.

Pappas, T. "A Triangle to a Square." *The Joy of Mathematics.* San Carlos, CA: Wide World Publ./Tetra, pp. 9 and 230, 1989.

Steinhaus, H. *Mathematical Snapshots, 3rd ed.* New York: Dover, 1999.

Stewart, I. *The Problems of Mathematics, 2nd ed.* Oxford, England: Oxford University Press, 1987.

Weisstein, E. W. "Books about Dissections." http://www.treasure-troves.com/books/Dissections.html.

Wells, D. *The Penguin Dictionary of Curious and Interesting Geometry.* London: Penguin, pp. 56–7 and 243–44, 1991.

Dissection Puzzles

A puzzle in which one object is to be converted to another by making a finite number of cuts and reassembling it. The cuts are often, but not always, restricted to straight lines. Sometimes, a given puzzle is precut and is to be re-assembled into two or more given shapes.

See also Cundy and Rollett's Egg, Pythagorean Square Puzzle, T-Puzzle, Tangram

Dissipative System

A dynamical system in which the phase space volume contracts along a trajectory. This means that the generalized divergence is less than zero,

$$\frac{\partial f_i}{\partial x_i} < 0,$$

where Einstein summation has been used.

See also Dynamical System, Phase Space

Dissymmetric

An object that is not superimposable on its mirror image is said to be disymmetric. All asymmetric objects are dissymmetric, and an object with no improper rotation (rotoinversion) axis must also be dissymmetric. The opposite of dissymmetric is enantiomorphous.

See also Amphichiral Knot, Chiral, Dissymmetric, Enantiomer, Enantiomorphous, Handedness, Mirror Image, Reflexible

Distance

The distance between two points is the length of the path connecting them. In the plane, the distance between points (x_1, y_1) and (x_2, y_2) is given by the Pythagorean Theorem,

$$d = \sqrt{(x_2 - x_1)^2 + (y_2 - y_1)^2}. \tag{1}$$

In Euclidean 3-space, the distance between points (x_1, y_1, z_1) and (x_2, y_2, z_2) is

$$d = \sqrt{(x_2 - x_1)^2 + (y_2 - y_1)^2 + (z_2 - z_1)^2}. \tag{2}$$

In general, the distance between points **x** and **y** in a Euclidean space $\mathbb{R}^n$ is given by

$$d = |\mathbf{x} - \mathbf{y}| = \sqrt{\sum_{i=1}^{n} |x_i - y_i|^2}. \tag{3}$$

For curved or more complicated surfaces, the so-called metric can be used to compute the distance between two points by integration. When unqualified, "the" distance generally means the *shortest* distance between two points. For example, there are an infinite number of paths between two points on a SPHERE but, in general, only a single shortest path. The *shortest* distance between two points is the length of a so-called GEODESIC between the points. In the case of the sphere, the geodesic is a segment of a GREAT CIRCLE containing the two points.

Let $\gamma(t)$ be a smooth curve in a MANIFOLD M from x to y with $\gamma(0) = x.$ and $\gamma(1) = y.$. Then $\gamma'(t) \in T_{\gamma(t)}$, where T_x is the TANGENT SPACE of M at x. The LENGTH of γ with respect to the Riemannian structure is given by

$$\int_0^1 \|\gamma'(t)\|_{\gamma(t)} dt, \tag{4}$$

and the distance $d(x, y)$ between x and y is the shortest distance between x and y given by

$$d(x, y) = \inf_{\gamma^{ix\ to\ y}} \int \|\gamma'(t)\|_{\gamma(t)} dt. \tag{5}$$

In order to specify the relative distances of $n > 1$ points in the plane, $1 + 2(n - 2) = 2n - 3$ coordinates are needed, since the first can always be taken as $(0, 0)$ and the second as $(x, 0)$, which defines the x-AXIS. The remaining $n - 2$ points need two coordinates each. However, the total number of distances is

$$\binom{n}{2} = \frac{n!}{2!(n-2)!} = \frac{1}{2}n(n-1), \tag{6}$$

where $\binom{n}{k}$ is a BINOMIAL COEFFICIENT. The distances between $n > 1$ points are therefore subject to m relationships, where

$$m \equiv \frac{1}{2}n(n-1) - (2n-3) = \frac{1}{2}(n-2)(n-3). \tag{7}$$

For $n = 1, 2, \ldots$, this gives 0, 0, 0, 1, 3, 6, 10, 15, 21, 28, ... (Sloane's A000217) relationships, and the number of relationships between n points is the TRIANGULAR NUMBER T_{n-3}.

Although there are no relationships for $n = 2$ and $n = 3$ points, for $n = 4$ (a QUADRILATERAL), there is one (Weinberg 1972):

$$0 = d_{12}^4 d_{34}^2 + d_{13}^4 d_{24}^2 + d_{14}^4 d_{23}^2 + d_{23}^4 d_{14}^2 + d_{24}^4 d_{13}^2 + d_{34}^4 d_{12}^2$$

$$+ d_{12}^2 d_{23}^2 d_{31}^2 + d_{12}^2 d_{24}^2 d_{41}^2 + d_{13}^2 d_{34}^2 d_{41}^2$$

$$+ d_{23}^2 d_{34}^2 d_{42}^2 - d_{12}^2 d_{23}^2 d_{34}^2 - d_{13}^2 d_{32}^2 d_{24}^2$$

$$- d_{12}^2 d_{24}^2 d_{43}^2 - d_{14}^2 d_{42}^2 d_{23}^2 - d_{13}^2 d_{34}^2 d_{42}^2$$

$$- d_{14}^2 d_{43}^2 d_{32}^2 - d_{23}^2 d_{31}^2 d_{14}^2 - d_{21}^2 d_{13}^2 d_{34}^2$$

$$- d_{24}^2 d_{41}^2 d_{13}^2 - d_{21}^2 d_{14}^2 d_{43}^2 - d_{31}^2 d_{12}^2 d_{24}^2$$

$$- d_{32}^2 d_{21}^2 d_{14}^2. \tag{8}$$

This equation can be derived by writing

$$d_{ij} \equiv \sqrt{(x_i - x_j)^2 + (y_i - y_j)^2} \tag{9}$$

and eliminating x_i and y_j from the equations for d_{12}, d_{13}, d_{14}, d_{23}, d_{24}, and $d_{34}.$. This results in a CAYLEY-MENGER DETERMINANT

$$0 = \begin{vmatrix} 0 & 1 & 1 & 1 & 1 \\ 1 & 0 & d_{12}^2 & d_{13}^2 & d_{14}^2 \\ 1 & d_{21}^2 & 0 & d_{23}^2 & d_{24}^2 \\ 1 & d_{31}^2 & d_{32}^2 & 0 & d_{34}^2 \\ 1 & d_{41}^2 & d_{42}^2 & d_{43}^2 & 0 \end{vmatrix}, \tag{10}$$

as observed by Uspensky (1948, p. 256).

See also ARC LENGTH, CUBE POINT PICKING, EXPANSIVE, GEODESIC, LENGTH (CURVE), METRIC, PLANAR DISTANCE, POINT DISTANCES, POINT-LINE DISTANCE-2-D, POINT-LINE DISTANCE-3-D, POINT-PLANE DISTANCE, POINT-POINT DISTANCE-1-D, POINT-POINT DISTANCE-2-D, POINT-POINT DISTANCE-3-D, SPHERE

References

Gray, A. "The Intuitive Idea of Distance on a Surface." §15.1 in *Modern Differential Geometry of Curves and Surfaces with Mathematica, 2nd ed.* Boca Raton, FL: CRC Press, pp. 341–45, 1997.

Sloane, N. J. A. Sequences A000217/M2535 in "An On-Line Version of the Encyclopedia of Integer Sequences." http://www.research.att.com/~njas/sequences/eisonline.html.

Uspensky, J. V. *Theory of Equations.* New York: McGraw-Hill, p. 256, 1948.

Weinberg, S. *Gravitation and Cosmology: Principles and Applications of the General Theory of Relativity.* New York: Wiley, p. 7, 1972.

Distance Graph

Let D be a set of positive numbers containing 1, then the D-distance graph $X(D)$ on a nonempty subset X of Euclidean space is the GRAPH with vertex set X and edge set $\{(x, y) : d(x, y) \in D\}$, where $d(x, y)$ is the Euclidean distance between vertices x and y.

See also PRIME-DISTANCE GRAPH, UNIT-DISTANCE GRAPH, UNIT NEIGHBORHOOD GRAPH

References

Maehara, H. "Distance Graphs in Euclidean Space." *Ryukyu Math. J.* **5**, 33–1, 1992.

Distance-Regular Graph

A CONNECTED GRAPH G is called distance-regular if there are integers $d(x, y)$ such that for any two vertices $x, y \in G$ ar distance $i = d(x, y)$, there are exactly c_i neighbors of $y \in G_{i-1}(x)$ and b_i neighbors of $y \in G_{i+1}(x).$.

See also INTERSECTION ARRAY, MOORE GRAPH, REGULAR GRAPH

References

Bendito, E.; Carmona, A.; and Encinas, A. M. "Shortest Paths in Distance-Regular Graphs." *Europ. J. Combin.* **21**, 153–66, 2000.

Brouwer, A. E.; Cohen, A. M.; and Neumaier, A. *Distance Regular Graphs.* New York: Springer-Verlag, 1989.

Distinct Prime Factors

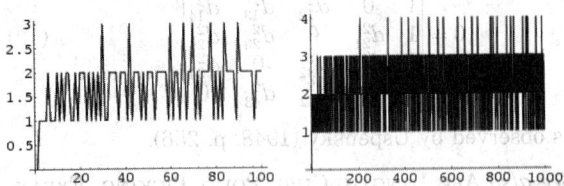

The number of distinct prime factors of a number n is denoted (n). The first few values for $n = 1, 2, \ldots$ are 0, 1, 1, 1, 1, 2, 1, 1, 1, 2, 1, 2, 1, 2, 2, 1, 1, 2, 1, 2, ... (Sloane's A001221; Abramowitz and Stegun 1972, Kac 1959). This sequence is given by the inverse MÖBIUS TRANSFORM of $b_n = 1$ for n prime and $b_n = 0$ for n (Sloane and Plouffe 1995, p. 22).

The first few values of the SUMMATORY FUNCTION

$$\sum_{k=2}^{n} \omega(k)$$

are 1, 2, 3, 4, 6, 7, 8, 9, 11, 12, 14, 15, 17, 19, 20, 21, ... (Sloane's A013939), and the asymptotic value is

$$\sum_{k=2}^{n} \omega(k) = n \ln \ln n + B_1 n + o(n),$$

where B_1 is MERTENS CONSTANT. In addition,

$$\sum_{k=2}^{n} [\omega(k)]^2 = n(\ln \ln n)^2 + O(\ln \ln n).$$

The numbers consisting only of distinct prime factors are precisely the SQUAREFREE numbers.

See also DIVISOR FUNCTION, ERDOS-KAC THEOREM, GREATEST PRIME FACTOR, HARDY-RAMANUJAN THEOREM, HETEROGENEOUS NUMBERS, LEAST PRIME FACTOR, MERTENS CONSTANT, PRIME FACTORS, SQUAREFREE

References

Abramowitz, M. and Stegun, C. A. (Eds.). *Handbook of Mathematical Functions with Formulas, Graphs, and Mathematical Tables, 9th printing.* New York: Dover, p. 844, 1972.

Hardy, G. H. and Wright, E. M. "The Number of Prime Factors of n" and "The Normal Order of $\sigma(n)$ and $\Omega(n)$.." §22.10 and 22.11 in *An Introduction to the Theory of Numbers, 5th ed.* Oxford, England: Clarendon Press, pp. 354–58, 1979.

Kac, M. *Statistical Independence in Probability, Analysis and Number Theory.* Washington, DC: Math. Assoc. Amer., p. 64, 1959.

Sloane, N. J. A. Sequences A001221/M0056 and A013939 in "An On-Line Version of the Encyclopedia of Integer Sequences." http://www.research.att.com/~njas/sequences/eisonline.html.

Sloane, N. J. A. and Plouffe, S. *The Encyclopedia of Integer Sequences.* San Diego, CA: Academic Press, 1995.

Distribution (Generalized Function)

The class of all regular sequences of PARTICULARLY WELL-BEHAVED FUNCTIONS equivalent to a given regular sequence. A distribution is sometimes also called a "generalized function" or "ideal function." As its name implies, a generalized function is a generalization of the concept of a FUNCTION. For example, in physics, a baseball being hit by a bat encounters a force from the bat, as a function of time. Since the transfer of momentum from the bat is modeled as taking place at an instant, the force is not actually a function. Instead, it is a multiple of the DELTA FUNCTION. The set of distributions contains functions (LOCALLY INTEGRABLE) and RADON MEASURES. Note that the term "distribution" is closely related to STATISTICAL DISTRIBUTIONS.

Generalized functions are defined as continuous linear FUNCTIONALS over a SPACE of infinitely differentiable functions such that all continuous functions have derivatives which are themselves generalized functions. The most commonly encountered generalized function is the DELTA FUNCTION. Vladimirov (1984) contains a nice treatment of distributions from a physicist's point of view, while the multivolume work by Gel'fand and Shilov (1977) is a classic and rigorous treatment of the field.

While it is possible to add distributions, it is not possible to multiply distributions when they have coinciding singular support. Despite this, it is possible to take the DERIVATIVE of a distribution, to get another distribution. Consequently, they may satisfy a linear PARTIAL DIFFERENTIAL EQUATION, in which case the distribution is called a weak solution. For example, given any locally integrable function f it makes sense to ask for solutions u of POISSON'S EQUATION

$$\nabla^2 u = f \qquad (1)$$

by only requiring the equation to hold in the sense of distributions, that is, both sides are the same distribution. The definitions of the derivatives of a distribution $p(x)$ are given by

$$\int_{-\infty}^{\infty} p'(x) f(x) dx = -\int_{-\infty}^{\infty} p(x) f'(x) dx \qquad (2)$$

$$\int_{-\infty}^{\infty} p^{(n)}(x) f(x) dx = (-1)^n \int_{-\infty}^{\infty} p(x) f^{(n)}(x) dx. \qquad (3)$$

Distributions also differ from functions because they are COVARIANT, that is, they push forward. Given a SMOOTH FUNCTION $\alpha : \Omega_1 \to \Omega_2$, a distribution T on Ω_1

pushes forward to a distribution on Ω_2. In contrast, a REAL FUNCTION f on Ω_2. pulls back to a function on Ω_1, namely $f(\alpha(x))$.

Distributions are, by definition, the dual to the SMOOTH FUNCTIONS of COMPACT SUPPORT, with a particular TOPOLOGY. For example, the DELTA FUNCTION δ is the LINEAR FUNCTIONAL $\delta(f) = f(0)$. The distribution corresponding to a function g is

$$T_g(f) = \int_\Omega fg, \tag{4}$$

and the distribution corresponding to a MEASURE μ is

$$T_\mu(f) = \int_\Omega fd\mu. \tag{5}$$

The PUSHFORWARD MAP of a distribution T along α is defined by

$$\alpha_* T(f) = T(f \circ \alpha), \tag{6}$$

and the derivative of T is defined by $DT(f) = T(D^*f)$ where D^* is the FORMAL ADJOINT of D. For example, the first derivative of the DELTA FUNCTION is given by

$$\frac{d}{dx}[\delta(f)] = -\frac{df}{dx}\bigg|_{x=0}. \tag{7}$$

As is the case for any function space, the topology determines which LINEAR FUNCTIONALS are continuous, that is, are in the DUAL SPACE. The topology is defined by the family of SEMINORMS,

$$N_{K,\alpha}(f) = \sup_k \|D^{\alpha}f\|, \tag{8}$$

where sup denotes the SUPREMUM. It agrees with the C-INFINITY TOPOLOGY on compact subsets. In this topology, a sequence converges, $f_n \to f$, IFF there is a compact set K such that all f_n are supported in K and every derivative $D^x f_n$ converges uniformly to $D^x f$ in K. Therefore, the constant function 1 is a distribution, because if $f_n \to f$, then

$$T_1(f_n) = \int_K f_n \to \int_K f = T_1(f). \tag{9}$$

See also CONVOLUTION, DELTA FUNCTION, DELTA SEQUENCE, FOURIER SERIES, FUNCTIONAL, LINEAR FUNCTIONAL, MICROLOCAL ANALYSIS, STATISTICAL ANALYSIS, TEMPERED DISTRIBUTION, ULTRADISTRIBUTION

References

Friedlander, F. G. *Introduction to the Theory of Distributions, 2nd ed.* Cambridge, England: Cambridge University Press, 1999.

Gel'fand, I. M.; Graev, M. I.; and Vilenkin, N. Ya. *Generalized Functions, Vol. 5: Integral Geometry and Representation Theory.* New York: Harcourt Brace, 1977.

Gel'fand, I. M. and Shilov, G. E. *Generalized Functions, Vol. 1: Properties and Operations.* New York: Harcourt Brace, 1977.

Gel'fand, I. M. and Shilov, G. E. *Generalized Functions, Vol. 2: Spaces of Fundamental and Generalized Functions.* New York: Harcourt Brace, 1977.

Gel'fand, I. M. and Shilov, G. E. *Generalized Functions, Vol. 3: Theory of Differential Equations.* New York: Harcourt Brace, 1977.

Gel'fand, I. M. and Vilenkin, N. Ya. *Generalized Functions, Vol. 4: Applications of Harmonic Analysis.* New York: Harcourt Brace, 1977.

Griffel, D. H. *Applied Functional Analysis.* Englewood Cliffs, NJ: Prentice-Hall, 1984.

Halperin, I. and Schwartz, L. *Introduction to the Theory of Distributions, Based on the Lectures Given by Laurent Schwarz.* Toronto, Canada: University of Toronto Press, 1952.

Lighthill, M. J. *Introduction to Fourier Analysis and Generalised Functions.* Cambridge, England: Cambridge University Press, 1958.

Richards, I. and Young, H. *The Theory of Distributions: A Nontechnical Introduction.* Cambridge, England: Cambridge University Press, 1990.

Rudin, W. *Functional Analysis, 2nd ed.* New York: McGraw-Hill, 1991.

Strichartz, R. *Fourier Transforms and Distribution Theory.* Boca Raton, FL: CRC Press, 1993.

Vladimirov, V. S. *Equations of Mathematical Physics.* Moscow: Mir, 1984.

Weisstein, E. W. "Books about Generalized Functions." http://www.treasure-troves.com/books/GeneralizedFunctions.html.

Yoshida, K. *Functional Analysis.* Berlin: Springer-Verlag, pp. 28–9 and 46–2, 1974.

Zemanian, A. H. *Distribution Theory and Transform Analysis: An Introduction to Generalized Functions, with Applications.* New York: Dover, 1987.

Distribution (Statistical)

STATISTICAL DISTRIBUTION

Distribution Function

The distribution function $D(x)$, sometimes also called the PROBABILITY DISTRIBUTION FUNCTION, describes the probability that a trial X takes on a value less than or equal to a number x. The distribution function is therefore related to a continuous PROBABILITY DENSITY FUNCTION $P(x)$ by

$$D(x) = P(X \le x) \equiv \int_{-\infty}^x P(x')dx', \tag{1}$$

so $P(x)$ (when it exists) is simply the derivative of the distribution function

$$P(x) = D'(x) = [P(x')]_{-\infty}^x = P(x) - P(-\infty). \tag{2}$$

Similarly, the distribution function is related to a discrete probability $P(x)$ by

$$D(x) = P(X \le x) = \sum_{X \le x} P(x). \tag{3}$$

In general, there exist distributions which are neither continuous nor discrete.

A JOINT DISTRIBUTION FUNCTION can be defined if outcomes are dependent on two parameters:

$$D(x,y) \equiv P(X \leq x, Y \leq y) \qquad (4)$$

$$D_x(x) \equiv D(x, \infty) \qquad (5)$$

$$D_y(y) \equiv D(\infty, y). \qquad (6)$$

Similarly, a multiple distribution function can be defined if outcomes depend on n parameters:

$$D(a_1, ..., a_n) \equiv P(x_1 \leq a_1, ..., x_n \leq a_n). \qquad (7)$$

Given a continuous $P(x)$, assume you wish to generate numbers distributed as $P(x)$ using a random number generator. If the random number generator yields a uniformly distributed value y_i in $[0,1]$ for each trial i, then compute

$$D(x) \equiv \int^x P(x')dx'. \qquad (8)$$

The FORMULA connecting y_i with a variable distributed as $P(x)$ is then

$$x_i = D^{-1}(y_i), \qquad (9)$$

where $D^{-i}(x)$ is the inverse function of $D(x)$,. For example, if $P(x)$ were a GAUSSIAN DISTRIBUTION so that

$$D(x) = \frac{1}{2}\left[1 + \mathrm{erf}\left(\frac{x-\mu}{\sigma\sqrt{2}}\right)\right], \qquad (10)$$

then

$$x_i = \sigma\sqrt{2}\mathrm{erf}^{-1}(2y_i - 1) + \mu. \qquad (11)$$

A distribution with constant VARIANCE of y for all values of x is known as a HOMOSCEDASTIC distribution. The method of finding the value at which the distribution is a maximum is known as the MAXIMUM LIKELIHOOD method.

See also BERNOULLI DISTRIBUTION, BETA DISTRIBUTION, BINOMIAL DISTRIBUTION, BIVARIATE DISTRIBUTION, CAUCHY DISTRIBUTION, CHI DISTRIBUTION, CHI-SQUARED DISTRIBUTION, CORNISH-FISHER ASYMPTOTIC EXPANSION, CORRELATION COEFFICIENT, DOUBLE EXPONENTIAL DISTRIBUTION, EQUALLY LIKELY OUTCOMES DISTRIBUTION, EXPONENTIAL DISTRIBUTION, EXTREME VALUE DISTRIBUTION, F-DISTRIBUTION, FERMI-DIRAC DISTRIBUTION, FISHER'S z-DISTRIBUTION, FISHER-TIPPETT DISTRIBUTION, GAMMA DISTRIBUTION, GAUSSIAN DISTRIBUTION, GEOMETRIC DISTRIBUTION, HALF-NORMAL DISTRIBUTION, HYPERGEOMETRIC DISTRIBUTION, JOINT DISTRIBUTION FUNCTION, LAPLACE DISTRIBUTION, LATTICE DISTRIBUTION, LÉVY DISTRIBUTION, LOGARITHMIC DISTRIBUTION, LOG-SERIES DISTRIBUTION, LOGISTIC DISTRIBUTION, LORENTZIAN DISTRIBUTION, MAXWELL DISTRIBUTION, NEGATIVE BINOMIAL DISTRIBUTION, NORMAL DISTRIBUTION, PARETO DISTRIBUTION, PASCAL DISTRIBUTION, PEARSON TYPE III DISTRIBUTION, POISSON DISTRIBUTION, PÓLYA DISTRIBUTION, RANDOM NUMBER, RATIO DISTRIBUTION, RAYLEIGH DISTRIBUTION, RICE DISTRIBUTION, SNEDECOR'S F-DISTRIBUTION, STATISTICAL DISTRIBUTION, STUDENT'S t-DISTRIBUTION, STUDENT'S z-DISTRIBUTION, UNIFORM DISTRIBUTION, WEIBULL DISTRIBUTION

References

Abramowitz, M. and Stegun, C. A. (Eds.). "Probability Functions." Ch. 26 in *Handbook of Mathematical Functions with Formulas, Graphs, and Mathematical Tables, 9th printing.* New York: Dover, pp. 925–64, 1972.
Iyanaga, S. and Kawada, Y. (Eds.). "Distribution of Typical Random Variables." Appendix A, Table 22 in *Encyclopedic Dictionary of Mathematics.* Cambridge, MA: MIT Press, pp. 1483–486, 1980.
Papoulis, A. *Probability, Random Variables, and Stochastic Processes, 2nd ed.* New York: McGraw-Hill, pp. 92–4, 1984.

Distribution Parameter

The distribution parameter of a NONCYLINDRICAL RULED SURFACE parameterized by

$$\mathbf{x}(u,v) = \sigma(u) + v\delta(u), \qquad (1)$$

where σ is the STRICTION CURVE and δ the DIRECTOR CURVE, is the function p defined by

$$p = \frac{\det(\sigma'\delta\delta')}{\delta'.\delta'}. \qquad (2)$$

The GAUSSIAN CURVATURE of a RULED SURFACE is given in terms of its distribution parameter by

$$K = -\frac{[p(u)]^2}{\left\{[p(u)]^2 + v^2\right\}^2}. \qquad (3)$$

See also NONCYLINDRICAL RULED SURFACE, RULED SURFACE, STRICTION CURVE

References

Gray, A. *Modern Differential Geometry of Curves and Surfaces with Mathematica, 2nd ed.* Boca Raton, FL: CRC Press, p. 447, 1997.

Distributive

Elements of an ALGEBRA which obey the identity

$$A(B + C) = AB = AC$$

are said to be distributive over the operation +.

See also ASSOCIATIVE, COMMUTATIVE, TRANSITIVE

Distributive Lattice

A LATTICE which satisfies the identities

$$(x \wedge y) \vee (x \wedge y) = x \wedge (y \vee z)$$

$$(x \vee y) \wedge (x \vee z) = x \vee (y \wedge z)$$

is said to be distributive.

See also LATTICE, MODULAR LATTICE

References
Grätzer, G. *Lattice Theory: First Concepts and Distributive Lattices.* San Francisco, CA: W. H. Freeman, pp. 35–6, 1971.

Ditrigonal Dodecadodecahedron

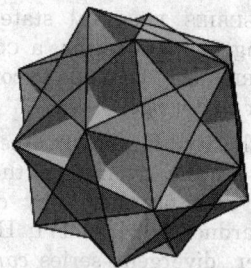

The UNIFORM POLYHEDRON U_{41}, also called the DITRIGONAL DODECAHEDRON, whose DUAL POLYHEDRON is the MEDIAL TRIAMBIC ICOSAHEDRON. It has WYTHOFF SYMBOL $3|\frac{5}{3}5$. Its faces are $12\{\frac{5}{2}\} + 12\{5\}$. It is a FACETED version of the SMALL DITRIGONAL ICOSIDODECAHEDRON. The CIRCUMRADIUS for unit edge length is

$$R = \frac{1}{2}\sqrt{3}.$$

References
Wenninger, M. J. *Polyhedron Models.* Cambridge, England: Cambridge University Press, pp. 123–24, 1989.

Ditrigonal Dodecahedron

DITRIGONAL DODECADODECAHEDRON

Divergence

The divergence of a VECTOR FIELD $\mathbf{F}$ is given by

$$div(\mathbf{F}) \equiv \nabla \cdot \mathbf{F} \equiv \lim_{V \to 0} \frac{\oint_S \mathbf{F} \cdot d\mathbf{a}}{V}. \tag{1}$$

Define

$$\mathbf{F} \equiv F_1 \hat{\mathbf{u}}_1 + F_2 \hat{\mathbf{u}}_2 + F_3 \hat{\mathbf{u}}_3. \tag{2}$$

Then in arbitrary orthogonal CURVILINEAR COORDINATES,

$$div(F) \equiv \nabla \cdot \mathbf{F} \equiv \frac{1}{h_1 h_2 h_3} \left[\frac{\partial}{\partial u_1}(h_2 h_3 F_1) + \frac{\partial}{\partial u_2}(h_3 h_1 F_2) \right.$$
$$\left. + \frac{\partial}{\partial u_3}(h_1 h_2 F_3) \right]. \tag{3}$$

If $\nabla \cdot \mathbf{F} = 0$, then the field is said to be a DIVERGENCELESS FIELD. For divergence in individual coordinate systems, see CURVILINEAR COORDINATES.

$$\nabla \cdot \frac{\mathbf{A}\mathbf{x}}{|\mathbf{x}|} = \frac{\mathrm{Tr}(\mathbf{A})}{|\mathbf{x}|} - \frac{\mathbf{x}^{\mathrm{T}}(\mathbf{A}\mathbf{x})}{|\mathbf{x}|^3}. \tag{4}$$

The divergence of a TENSOR A is

$$\nabla \cdot A \equiv A_{i\alpha}^{\alpha} \tag{5}$$

$$= A_{,k}^k + \Gamma_{jk}^k A^j, \tag{6}$$

$$= \frac{1}{g^{1/2}} (g^{1/2} A^k)_{,k} \tag{7}$$

where $A_{i\alpha}^{\alpha}$ is the COVARIANT DERIVATIVE, $A_{,k}^k$ is the COMMA DERIVATIVE, g_{ij} is the METRIC TENSOR, and $g = \det(g_{ij})$, (Arfken 1985, p. 165). Expanding the terms gives

$$A_{;\alpha}^{\alpha} = A_{,\alpha}^{\alpha} + \left(\Gamma_{\alpha\alpha}^{\alpha} A^{\alpha} + \Gamma_{\beta\alpha}^{\alpha} A^{\beta} + \Gamma_{\gamma\alpha}^{\alpha} A^{\gamma} \right)$$
$$+ A_{,\beta}^{\beta} + \left(\Gamma_{\alpha\beta}^{\beta} A^{\alpha} + \Gamma_{\beta\beta}^{\beta} A^{\beta} + \Gamma_{\gamma\beta}^{\beta} A^{\gamma} \right)$$
$$+ A_{,\gamma}^{\gamma} + \left(\Gamma_{\alpha\gamma}^{\gamma} A^{\alpha} + \Gamma_{\beta\gamma}^{\gamma} A^{\beta} + \Gamma_{\gamma\gamma}^{\gamma} A^{\gamma} \right). \tag{8}$$

See also COMMA DERIVATIVE, COVARIANT DERIVATIVE, CURL, CURL THEOREM, DIVERGENCE THEOREM, GRADIENT, GREEN'S THEOREM, VECTOR DERIVATIVE

References
Arfken, G. "Divergence, $\nabla \cdot$." §1.7 in *Mathematical Methods for Physicists, 3rd ed.* Orlando, FL: Academic Press, pp. 37–2, 1985.

Divergence Tests
If

$$\lim_{k \to \infty} u_k \neq 0,$$

then the series $\{u_n\}$ diverges.

See also CONVERGENCE TESTS, CONVERGENT SERIES, DINI'S TEST, SERIES

Divergence Theorem
A.k.a. GAUSS'S THEOREM. Let V be a region in space with boundary ∂V. Then

$$\int_V (\nabla \cdot \mathbf{F}) dV = \int_{\partial V} \mathbf{F} \cdot d\mathbf{a}. \tag{1}$$

Let S be a region in the plane with boundary ∂S.

$$\int_S \nabla \cdot \mathbf{F} dA = \int_{\partial S} \mathbf{F} \cdot \mathbf{n} ds. \tag{2}$$

If the VECTOR FIELD $\mathbf{F}$ satisfies certain constraints, simplified forms can be used. If $\mathbf{F}(x,y,z) = v(x,y,z)\mathbf{c}$ where $\mathbf{c}$ is a constant vector $\neq 0$, then

$$\int_S \mathbf{F}.d\mathbf{a} = \mathbf{c} \cdot \int_S v d\mathbf{a}. \tag{3}$$

But

$$\nabla \cdot (f\mathbf{v}) = (\nabla f) \cdot \mathbf{v} + f(\nabla \cdot \mathbf{v}), \tag{4}$$

so

$$\int_V \nabla \cdot (\mathbf{cv}) dV = \mathbf{c} \cdot \int_V (\nabla v + v\nabla \cdot \mathbf{c}) dV = \mathbf{c} \cdot \int_V \nabla v dV \tag{5}$$

$$\mathbf{c} \cdot \left(\int_S v d\mathbf{a} - \int_V \nabla v dV \right) = 0. \tag{6}$$

But $\mathbf{c} \neq 0$, and $\mathbf{c}.\mathbf{f}(v)$ must vary with v so that $\mathbf{c}.\mathbf{f}(v)$ cannot always equal zero. Therefore,

$$\int_S v d\mathbf{a} = \int_V \nabla v dV. \tag{7}$$

If $\mathbf{F}(x,y,z) = \mathbf{c} \times P(x,y,z)$, where $\mathbf{c}$ is a constant vector $\neq 0$, then

$$\int_S d\mathbf{a} \times \mathbf{P} \int_V \nabla \times \mathbf{P} dV. \tag{8}$$

See also CURL THEOREM, GRADIENT, GREEN'S THEOREM

References

Arfken, G. "Gauss's Theorem." §1.11 in *Mathematical Methods for Physicists, 3rd ed.* Orlando, FL: Academic Press, pp. 57–1, 1985.

Divergenceless Field

A divergenceless field, also called a SOLENOIDAL FIELD, is a FIELD for which $\nabla \cdot \mathbf{F} \equiv 0$. Therefore, there exists a $\mathbf{G}$ such that $\mathbf{F} = \nabla \times \mathbf{G}$. Furthermore, $\mathbf{F}$ can be written as

$$\mathbf{F} = \nabla \times (T\mathbf{r}) + \nabla^2(S\mathbf{r}) \equiv \mathbf{T} + \mathbf{S}, \tag{1}$$

where

$$\mathbf{T} \equiv \nabla \times (T\mathbf{r}) = -\mathbf{r} \times (\nabla T) \tag{2}$$

$$\mathbf{S} \equiv \nabla^2(S\mathbf{r}) = \nabla \left[\frac{\partial}{\partial r}(rS) \right] - \mathbf{r}\nabla^2 S. \tag{3}$$

Following Lamb, $\mathbf{T}$ and $\mathbf{S}$ are called TOROIDAL FIELD and POLOIDAL FIELD.

See also BELTRAMI FIELD, IRROTATIONAL FIELD, POLOIDAL FIELD, SOLENOIDAL FIELD, TOROIDAL FIELD

Divergent Sequence

A divergent sequence is a SEQUENCE for which the LIMIT exists but is not CONVERGENT.

See also CONVERGENT SEQUENCE, DIVERGENT SERIES

Divergent Series

A SERIES which is not CONVERGENT. Series may diverge by marching off to infinity or by oscillating. Divergent series have some curious properties. For example, rearranging the terms of $1 - 1 + 1 - 1 + 1 - \cdots$ gives both $(1-1) + (1-1) + (1-1) + \cdots = 0$ and $1 - (1-1) - (1-1) + \cdots = 1$.

The RIEMANN SERIES THEOREM states that, by a suitable rearrangement of terms, a CONDITIONALLY CONVERGENT SERIES may be made to converge to any desired value, or to diverge.

No less an authority than N. H. Abel wrote "The divergent series are the invention of the devil, and it is a shame to base on them any demonstration whatsoever" (Gardner 1984, p. 171; Hoffman 1998, p. 218). However, divergent series *can* actually be "summed" rigorously by using extensions to the usual summation rules (e.g., so-called Abel and Cesàro sums). For example, the divergent series $1 - 1 + 1 - 1 + 1 - \cdots$ has both Abel and Cesàro sums of 1/2.

See also ABSOLUTE CONVERGENCE, CONDITIONAL CONVERGENCE, CONVERGENT SERIES, DIVERGENT SEQUENCE

References

Bromwich, T. J. I'a and MacRobert, T. M. *An Introduction to the Theory of Infinite Series, 3rd ed.* New York: Chelsea, 1991.
Gardner, M. *The Sixth Book of Mathematical Games from Scientific American.* Chicago, IL: University of Chicago Press, pp. 170–71, 1984.
Hardy, G. H. *Divergent Series.* New York: Oxford University Press, 1949.
Hoffman, P. *The Man Who Loved Only Numbers: The Story of Paul Erdos and the Search for Mathematical Truth.* New York: Hyperion, 1998.

Diversity Condition

For any group of k men out of N, there must be at least k jobs for which they are collectively qualified.

Divide

To divide is to perform the operation of DIVISION, i.e., to see how many times a DIVISOR d goes into another number n. n divided by d is written n/d or $n \div d$. The result need not be an INTEGER, but if it is, some additional terminology is used. $d|n$ is read "d divides n" and means that d is a DIVISOR of n. In this case, n is said to be DIVISIBLE by d. Clearly, $1|n$ and $n|n$. By convention, $n|0$ for every n except 0 (Hardy and Wright 1979). The "divisibility" relation satisfies

$$b|a \quad for \quad c|b \Rightarrow c|a$$

$$b|a \Rightarrow bc|ac$$

$$c|a \quad \text{and} \quad c|b \quad \Rightarrow c|(ma+nb),$$

where the symbol $\Rightarrow$ means IMPLIES.

$d' \nmid n$ is read "d' does not divide n" and means that d' is not a DIVISOR of n. $a^k \| b$ means a^k divides b exactly. If n and d are RELATIVELY PRIME, the notation $(n, d) = 1$ or sometimes $n \perp d$ is used.

See also CONGRUENCE, DIVISIBLE, DIVISIBILITY TESTS, DIVISION, DIVISOR, GREATEST DIVIDING EXPONENT, RELATIVELY PRIME

References
Hardy, G. H. and Wright, E. M. *An Introduction to the Theory of Numbers, 5th ed.* Oxford, England: Clarendon Press, p. 1, 1979.

Divided Difference

The divided difference $f[x_1, x_2, ..., x_n]$ on n points $x_1, x_2, ..., x_n$ of a function $f(x)$ is defined by $f[x_1] \equiv f(x_1)$ and

$$f[x_1, x_2, ..., x_n] = \frac{f[x_1, ..., x_n] - f[x_2, ..., x_n]}{x_1 - x_n} \tag{1}$$

for $n \geq 2$. The first few differences are

$$[x_0, x_1] = \frac{f_0 - f_1}{x_0 - x_1} \tag{2}$$

$$[x_0, x_1, x_2] = \frac{[x_0, x_1] - [x_1, x_2]}{x_0 - x_2} \tag{3}$$

$$[x_0, x_1, ..., x_n] = \frac{[x_0, ..., x_{n-1}] - [x_1, ..., x_n]}{x_0 - x_n}. \tag{4}$$

Defining

$$\pi_n(x) \equiv (x - x_0)(x - x_1) \cdots (x - x_n) \tag{5}$$

and taking the DERIVATIVE

$$\pi'_n(x_k) = (x_k - x_0)...(x_k - x_{k-1})...(x_k - x_n) \tag{6}$$

gives the identity

$$[x_0, x_1, ..., x_n] = \sum_{k=0}^{n} \frac{f_k}{\pi'_n(x_k)}. \tag{7}$$

Consider the following question: does the property

$$f[x_1, x_2, ..., x_n] = h(x_1 + x_2 + ... + x_n) \tag{8}$$

for $n \geq 2$ and $h(x)$ a given function guarantee that $f(x)$ is a POLYNOMIAL of degree $\leq n$? Aczél (1985) showed that the answer is "yes" for $n = 2$, and Bailey (1992) showed it to be true for $n = 3$ with differentiable $f(x)$. Schwaiger (1994) and Andersen (1996) subsequently showed the answer to be "yes" for all $n \geq 3$ with restrictions on $f(x)$ or $h(x)$.

See also HORNER'S METHOD, INTERPOLATION, NEWTON'S DIVIDED DIFFERENCE INTERPOLATION FORMULA, RECIPROCAL DIFFERENCE

References
Abramowitz, M. and Stegun, C. A. (Eds.). *Handbook of Mathematical Functions with Formulas, Graphs, and Mathematical Tables, 9th printing.* New York: Dover, pp. 877–78, 1972.
Aczél, J. "A Mean Value Property of the Derivative of Quadratic Polynomials--Without Mean Values and Derivatives." *Math. Mag.* **58**, 42–5, 1985.
Andersen, K. M. "A Characterization of Polynomials." *Math. Mag.* **69**, 137–42, 1996.
Bailey, D. F. "A Mean-Value Property of Cubic Polynomials--Without Mean Values." *Math. Mag.* **65**, 123–24, 1992.
Beyer, W. H. (Ed.). *CRC Standard Mathematical Tables, 28th ed.* Boca Raton, FL: CRC Press, pp. 439–40, 1987.
Jeffreys, H. and Jeffreys, B. S. "Divided Differences." §9.012 in *Methods of Mathematical Physics, 3rd ed.* Cambridge, England: Cambridge University Press, pp. 260–64, 1988.
Schwaiger, J. "On a Characterization of Polynomials by Divided Differences." *Aequationes Math.* **48**, 317–23, 1994.
Whittaker, E. T. and Robinson, G. "Divided Differences" and "Theorems on Divided Differences." §11–2 in *The Calculus of Observations: A Treatise on Numerical Mathematics, 4th ed.* New York: Dover, pp. 20–4, 1967.

Dividend

A quantity that is divided by another quantity.

See also DIVISION, DIVISOR

Divine Proportion

GOLDEN RATIO

Divisibility Tests

Write a positive decimal integer a out digit by digit in the form $a_n \cdots a_3 a_2 a_1 a_0$. The following rules then determine if a is DIVISIBLE by another number by examining the CONGRUENCE properties of its digits. In CONGRUENCE notation, $n \equiv k \pmod{m}$ means that the remainder when n is divided by a modulus m is k. (Note that it is always true that $10^0 = 1 \equiv 1$ for any base.)

1. All integers are DIVISIBLE by 1.

2. $10^1 \equiv 0 \pmod 2$, so $10^n \equiv 0 \pmod 2$ for $n \geq 1$. Therefore, if the last digit a_0 is DIVISIBLE by 2 (i.e., is EVEN), then so is a.

3. $10^0 \equiv 1$, $10^1 \equiv 1$, $10^2 \equiv 1$, ..., $10^n \equiv 1 \pmod 3$. Therefore, if $\Sigma_{i=0}^{n} a_i$ is DIVISIBLE by 3, so is a (Wells 1986, p. 48).

4. $10^1 \equiv 2$, $10^2 \equiv 0$, ...$10^n \equiv 0 \pmod 4$. So if the last two digits are DIVISIBLE by 4, more specifically if $r \equiv a_0 + 2a_1$ is, then so is a.

5. $10^1 \equiv 0 \pmod 5$, so $10^n \equiv 0 \pmod 5$ for $n \geq 1$. Therefore, if the last digit a_0 is DIVISIBLE by 5 (i.e., is 5 or 0), then so is a.

6. $10^1 \equiv -2$, $10^2 \equiv -2$, ..., $10^n \equiv -2 \pmod 6$. Therefore, if $r \equiv a_0 - 2\Sigma_{i=1}^{n} a_i$ is DIVISIBLE by 6, so is a. A

simpler rule states that if a is DIVISIBLE by 3 and is EVEN, then a is also DIVISIBLE by 6.

7a. $10^1 \equiv 3$, $10^2 \equiv 2$, $10^3 \equiv -1$, $10^4 \equiv -3$, $10^5 \equiv -2$, $10^6 \equiv 1$ (mod 7), and the sequence then repeats. Therefore, if $r \equiv (a_0 + 3a_1 + 2a_2 - a_3 - 3a_4 - 2a_5) + (a_6 + 3a_7 + \cdots) + \cdots$ is DIVISIBLE by 7, so is a.

7b. An alternate test proceeds by multiplying a_n by 3 and adding to a_{n-1}, then repeating the procedure up through a_0. The final number can then, of course, be further reduced using the same procedure. If the result is divisible by 7, then so is the original number (Wells 1986, p. 70).

7c. A third test multiplies a_0 by 5 and adds it to a_1, proceeding up through a_n. The final number can then, of course, be further reduced using the same procedure. If the result is divisible by 7, then so is the original number (Wells 1986, p. 70).

8. $10^1 \equiv 2$, $10^2 \equiv 4$, $10^3 \equiv 0$, ..., $10^n \equiv 0$ (mod 8). Therefore, if the last three digits are DIVISIBLE by 8, more specifically if $r \equiv a_0 + 2a_1 + 4a_2$ is, then so is a (Wells 1986, p. 72).

9. $10^0 \equiv 1$, $10^1 \equiv 1$, $10^2 \equiv 1$, ..., $10^n \equiv 1$ (mod 9). Therefore, if $\Sigma_{i=0}^n a_i$ is DIVISIBLE by 9, so is a (Wells 1986, p. 74).

10. $10^1 \equiv 0$ (mod 10), so if the last digit is 0, then a is DIVISIBLE by 10.

11. $10^1 \equiv -1$, $10^2 \equiv 1$, $10^3 \equiv -1$, $10^4 \equiv 1$, ... (mod 11). Therefore, if $r \equiv a_0 - a_1 + a_2 - a_3 + \cdots$ is DIVISIBLE by 11, then so is a.

12. $10^1 \equiv -2$, $10^2 \equiv 4$, $10^3 \equiv 4$, ... (mod 12). Therefore, if $r \equiv a_0 - 2a_1 + 4(a_2 + a_3 + \cdots)$ is DIVISIBLE by 12, then so is a. Divisibility by 12 can also be checked by seeing if a is DIVISIBLE by 3 and 4.

13. $10^1 \equiv -3$, $10^2 \equiv -4$, $10^3 \equiv -1$, $10^4 \equiv 3$, $10^5 \equiv 4$, $10^6 \equiv 1$ (mod 13), and the pattern repeats. Therefore, if $r \equiv (a_0 - 3a_1 - 4a_2 - a_3 + 3a_4 + 4a_5) + (a_6 - 3a_7 + \ldots) + \cdots$ is DIVISIBLE by 13, so is a.

For additional tests for 13, see Gardner (1991).

See also CONGRUENCE, DIVISIBLE, DIVISOR, MODULUS (CONGRUENCE)

References

Burton, D. M. "Special Divisibility Tests." §4.3 in *Elementary Number Theory, 4th ed.* Boston, MA: Allyn and Bacon, pp. 89–6, 1989.

Dickson, L. E. *History of the Theory of Numbers, Vol. 1: Divisibility and Primality.* New York: Chelsea, pp. 337–46, 1952.

Gardner, M. "Tests of Divisibility." Ch. 14 in *The Unexpected Hanging and Other Mathematical Diversions.* Chicago, IL: Chicago University Press, pp. 160–69, 1991.

Wells, D. *The Penguin Dictionary of Curious and Interesting Numbers.* Middlesex, England: Penguin Books, p. 48, 1986.

Divisible

A number n is said to be divisible by d if d is a DIVISOR of n.

The product of any n consecutive integers is divisible by $n!$. The sum of any n consecutive integers is divisible by n if n is ODD, and by $n/2$ if n is EVEN.

See also DIVIDE, DIVISIBILITY TESTS, DIVISOR, DIVISOR FUNCTION

References

Guy, R. K. "Divisibility." Ch. B in *Unsolved Problems in Number Theory, 2nd ed.* New York: Springer-Verlag, pp. 44–04, 1994.

Nagell, T. "Divisibility." Ch. 1 in *Introduction to Number Theory.* New York: Wiley, pp. 11–6, 1951.

Division

Taking the RATIO x/y of two numbers x and y, also written $x \div y$. Here, x is called the DIVIDEND, y is called the DIVISOR, and x/y is called a QUOTIENT. The symbol "/" is called a SOLIDUS (or DIAGONAL), and the symbol "$\div$" is called the OBELUS. If left unevaluated, x/y is called a FRACTION, with x known as the NUMERATOR and y known as the DENOMINATOR.

Division in which the fractional (remainder) is discarded is called INTEGER DIVISION, and is sometimes denoted using a backslash, \.

Division is the inverse operation of MULTIPLICATION, so that if

$$a \times b = c,$$

then a can be recovered as

$$a = c \div b$$

as long as $b \neq 0$. In general, DIVISION BY ZERO is not defined since the ability to "invert" $a \times b = c$ to recover a breaks down if $b = 0$ (in which case c is always 0, independent of a).

Cutting or separating an object into two or more parts is also called division.

See also ADDITION, COMPLEX DIVISION, CUTTING, DENOMINATOR, DIVIDE, DIVIDEND, DIVISION BY ZERO, DIVISOR, INTEGER DIVISION, LONG DIVISION, MULTIPLICATION, NUMERATOR, OBELUS, ODDS, PLANE DIVISION BY LINES, QUOTIENT, RATIO, SKELETON DIVISION, SOLIDUS, SPACE DIVISION BY SPHERES, SUBTRACTION, TRIAL DIVISION, VECTOR DIVISION

Division Algebra

A division algebra, also called a "division ring" or "skew field," is a RING in which every NONZERO element has a multiplicative inverse, but multiplication is not COMMUTATIVE. In French, the term "corps non commutatif" is used to mean division algebra, while "corps" alone means FIELD.

Explicitly, a division algebra is a set together with two BINARY OPERATORS $S(+, *)$ satisfying the following conditions:

1. Additive associativity: For all $a, b, c \in S$, $(a + b) + c = a + (b + c)$,

2. Additive commutativity: For all $a, b \in S$, $a + b = b + a$,

3. Additive identity: There exists an element $0 \in S$ such that for all $a \in S$, $0 + a = a + 0 = a$,

4. Additive inverse: For every $a \in S$ there exists an element $-a \in S$ such that $a = (-a) = (-a) + a = 0$,

5. Multiplicative associativity: For all $a, b, c \in S$, $(a * b) * c = a * (b * c)$,

6. Multiplicative identity: There exists an element $1 \in S$ not equal to 0 such that for all $a \in S$, $1 * a = a * 1 = a$,

7. Multiplicative inverse: For every $a \in S$ not equal to 0, there exists $a^{-1} \in S$ such that $a * a^{-1} = a^{-1} * a = 1$,

8. Left and right distributivity: For all $a, b, c \in S$, $a * (b + c) = (a * b) + (a * c)$ and $(b + c) * a = (b * a) + (c * a)$.

Thus a division algebra $(S, +, *)$ is a UNIT RING for which $(S - \{0\}, *)$ is a GROUP. A division algebra must contain at least two elements. A COMMUTATIVE division algebra is called a FIELD.

In 1878 and 1880, Frobenius and Peirce proved that the only associative REAL division algebras are REAL NUMBERS, COMPLEX NUMBERS, and QUATERNIONS (Mishchenko and Solovyov 2000). The CAYLEY ALGEBRA is the only NONASSOCIATIVE DIVISION ALGEBRA. Hurwitz (1898) proved that the ALGEBRAS of REAL NUMBERS, COMPLEX NUMBERS, QUATERNIONS, and CAYLEY NUMBERS are the only ones where multiplication by unit "vectors" is distance-preserving.

Adams (1956) proved that n-dimensional vectors form an ALGEBRA in which division (except by 0) is always possible only for $n = 1, 2, 4$, and 8. Bott and Milnor (1958) proved that the only finite dimensional real division algebras occur for dimensions $n = 1, 2, 4$, and 8. Each gives rise to an ALGEBRA with particularly useful physical applications (which, however, is not itself necessarily nonassociative), and these four cases correspond to REAL NUMBERS, COMPLEX NUMBERS, QUATERNIONS, and CAYLEY NUMBERS, respectively.

See also ALTERNATIVE ALGEBRA, CAYLEY NUMBER, FIELD, GROUP, JORDAN ALGEBRA, LIE ALGEBRA, NONASSOCIATIVE ALGEBRA, POWER ASSOCIATIVE ALGEBRA, QUATERNION, SCHUR'S LEMMA, UNIT RING

References

Albert, A. A. (Ed.). *Studies in Modern Algebra.* Washington, DC: Math. Assoc. Amer., 1963.
Bott, R. and Milnor, J. "On the Parallelizability of the Spheres." *Bull. Amer. Math. Soc.* **64**, 87–9, 1958.
Dickson, L. E. *Algebras and Their Arithmetics.* Chicago, IL: University of Chicago Press, 1923.
Dixon, G. M. *Division Algebras: Octonions, Quaternions, Complex Numbers and the Algebraic Design of Physics.* Dordrecht, Netherlands: Kluwer, 1994.
Herstein, I. N. *Topics in Algebra, 2nd ed.* New York: Wiley, pp. 326–29, 1975.
Hurwitz, A. "Ueber die Composition der quadratischen Formen von beliebig vielen Variabeln." *Nachr. Königl. Gesell. Wiss. Göttingen. Math.-phys. Klasse,* 309–16, 1898.
Joye, M. "Introduction élémentaire à la théorie des courbes elliptiques." http://www.dice.ucl.ac.be/crypto/introductory/courbes_elliptiques.html.
Kurosh, A. G. *General Algebra.* New York: Chelsea, pp. 221–43, 1963.
Mishchenko, A. and Solovyov, Y. "Quaternions." *Quantum* **11**, 4– and 18, 2000.
Petro, J. "Real Division Algebras of Dimension > 1 contain $\mathbb{C}$." *Amer. Math. Monthly* **94**, 445–49, 1987.
Saltman, D. D. *Lectures on Division Algebras.* Providence, RI: Amer. Math. Soc., 1999.

Division by Zero

Division by zero is the operation of taking the QUOTIENT of any number x and 0, i.e., $x/0$. The uniqueness of DIVISION breaks down when dividing by zero, since the product $0 \cdot y = 0$ is the same for any y, so y cannot be recovered by inverting the process of MULTIPLICATION. 0 is the only number with this property and, as a result, division by zero is UNDEFINED for REAL NUMBERS and can produce a fatal condition called a "division by zero error" in computer programs.

There are, however, contexts in which division by zero can be considered as defined. For example, division by zero $z/0$ for $z \in \mathbb{C}^* \neq 0$ in the EXTENDED COMPLEX PLANE $\mathbb{C}^*$ is defined to be a quantity known as COMPLEX INFINITY. This definition expresses the fact that, for $z \neq 0$, $\lim_{w \to 0} z/w = \infty$ (i.e., COMPLEX INFINITY). However, even though the formal statement $1/0 = \infty$ is permitted in $\mathbb{C}^*$, note that this does *not* mean that $1 = 0 \cdot \infty$. Zero does not have a multiplicative inverse under any circumstances.

Although division by zero is not defined for reals, LIMITS involving division by a real quantity x which *approaches* zero may be in fact be WELL DEFINED. For example,

$$\lim_{x \to 0} \frac{\sin x}{x} = 1.$$

Of course, such limits may also approach INFINITY,

$$\lim_{x \to 0^+} \frac{1}{x} = \infty.$$

See also $\mathbb{C}^*$, COMPLEX INFINITY, COMPLEX NUMBER, DIVISION, EXTENDED COMPLEX PLANE, FALLACY, FIELD, LIMIT REAL NUMBER, RING, ZERO

Division Lemma

When ac is DIVISIBLE by a number b that is RELATIVELY PRIME to a, then c must be DIVISIBLE by b.

Division Ring

DIVISION ALGEBRA

Divisor

A divisor of a number N is a number d which DIVIDES N, also called a FACTOR. The total number of divisors for a given number N can be found as follows. Write a number in terms of its PRIME FACTORIZATION

$$N = p_1^{\alpha_1} p_2^{\alpha_2} \cdots p_r^{\alpha_r}. \tag{1}$$

For any divisor d of N, $N = dd'$ where

$$d = p_1^{\delta_1} p_2^{\delta_2} \cdots p_r^{\delta_r}, \tag{2}$$

so

$$d' = p_1^{\alpha_1 - \delta_1} p_2^{\alpha_2 - \delta_2} \cdots p_r^{\alpha_r - \delta_r}. \tag{3}$$

Now, $\delta_1 = 0, 1, \ldots, \alpha_1$, so there are $\alpha_1 + 1$ possible values. Similarly, for δ_n, there are $\alpha_n + 1$ possible values, so the total number of divisors $v(N)$ of N is given by

$$v(N) = \prod_{n=1}^{r} (\alpha_n + 1). \tag{4}$$

The function $v(N)$ is also sometimes denoted $d(N)$ or $\sigma_0(N)$. The product of divisors can be found by writing the number N in terms of all possible products

$$N = \begin{cases} d^{(1)} d'^{(1)} \\ \vdots \\ d^{(v)} d'^{(v)} \end{cases}, \tag{5}$$

so

$$N^{v(N)} = \left[d^{(1)} \cdots d^{(v)} \right] \left[d'^{(1)} d'^{(v)} \right]$$

$$= \prod_{i=1}^{v} d_i \prod_{i=1}^{v} d_i' = \left(\prod d \right)^2, \tag{6}$$

and

$$\prod d = N^{v(N)/2}. \tag{7}$$

The GEOMETRIC MEAN of divisors is

$$G \equiv \left(\prod d \right)^{1/v(N)} = \left[N^{v(n)/2} \right]^{1/v(N)} = \sqrt{N}. \tag{8}$$

The ARITHMETIC MEAN is

$$A(N) \equiv \frac{\sigma(N)}{v(N)}. \tag{9}$$

The HARMONIC MEAN is

$$\frac{1}{H} \equiv \frac{1}{N} \left(\sum \frac{1}{d} \right). \tag{10}$$

But $N = dd'$, so $1/d = d'/N$ and

$$\sum \frac{1}{d} = \frac{1}{N} \sum d' = \frac{1}{N} \sum d = \frac{\sigma(N)}{N}, \tag{11}$$

and we have

$$\frac{1}{H(N)} = \frac{1}{v(N)} \frac{\sigma(N)}{N} = \frac{A(N)}{N} \tag{12}$$

$$N = A(N)H(N). \tag{13}$$

Given three INTEGERS chosen at random, the probability that no common factor will divide them all is

$$[\zeta(3)]^{-1} \approx 1.20206^{-1} \approx 0.831907, \tag{14}$$

where $\zeta(3)$ is APÉRY'S CONSTANT.

The smallest numbers having exactly 0, 1, 2, ... divisors (other than 1) are 1, 2, 4, 6, 16, 12, 64, 24, 36, ... (Sloane's A005179).

Let $f(n)$ be the number of elements in the greatest subset of $[1, n]$ such that none of its elements are divisible by two others. For n sufficiently large,

$$0.6725 \ldots \leq \frac{f(n)}{n} \leq 0.673 \ldots \tag{15}$$

(Le Lionnais 1983, Lebensold 1976/1977).

See also ALIQUANT DIVISOR, ALIQUOT DIVISOR, ALIQUOT SEQUENCE, DIRICHLET DIVISOR PROBLEM, DIVIDEND, DIVISION, DIVISOR (CURVE), DIVISOR FUNCTION, DIVISOR THEORY, E-DIVISOR, EXPONENTIAL DIVISOR, GREATEST COMMON DIVISOR, IMPROPER DIVISOR, INFINARY DIVISOR, K-ARY DIVISOR, PERFECT NUMBER, PROPER DIVISOR, UNITARY DIVISOR

References

Guy, R. K. "Solutions of $d(n) = d(n+1)$." §B18 in *Unsolved Problems in Number Theory, 2nd ed.* New York: Springer-Verlag, pp. 73–5, 1994.

Le Lionnais, F. *Les nombres remarquables.* Paris: Hermann, p. 43, 1983.

Lebensold, K. "A Divisibility Problem." *Studies Appl. Math.* **56**, 291–94, 1976/1977.

Nagell, T. "Divisors." §1 in *Introduction to Number Theory.* New York: Wiley, pp. 11–2, 1951.

Sloane, N. J. A. Sequences A005179/M1026 in "An On-Line Version of the Encyclopedia of Integer Sequences." http://www.research.att.com/~njas/sequences/eisonline.html.

Divisor Function

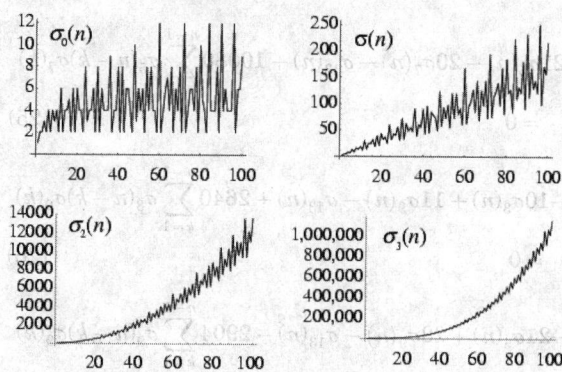

$\sigma_k(n)$ for n an integer is defined as the sum of the kth POWERS of the DIVISORS of n. As an illustrative example, consider the number 140, which has DIVISORS $d_i = 1, 2, 4, 5, 7, 10, 14, 20, 28, 35, 70,$ and 140 (for a total of $N = 12$ of them). Therefore,

$$d(140) = \sigma_0(140) = N = 12 \qquad (1)$$

$$\sigma(140) = \sigma_1(140) = \sum_{i=1}^{N} d_i = 336 \qquad (2)$$

$$\sigma_2(140) = \sum_{i=1}^{N} d_i^2 = 27,300 \qquad (3)$$

$$\sigma_3(140) = \sum_{i=1}^{N} d_i^3 = 3,164,112. \qquad (4)$$

The divisor function can also be generalized to GAUSSIAN INTEGERS.

The function $\sigma_0(n)$ gives the total number of DIVISORS of n and is often denoted $d(n)$, $\nu(n)$, $\tau(n)$, or $\Omega(n)$. (Hardy and Wright 1979, pp. 354–55). The first few values of $\sigma_0(n)$ are 1, 2, 2, 3, 2, 4, 2, 4, 3, 4, 2, 6, ... (Sloane's A000005). These values can be found as the inverse MÖBIUS TRANSFORM of 1, 1, 1, ... (Sloane and Plouffe 1995, p. 22). Heath-Brown (1984) proved that $\sigma_0(n) = \sigma_0(n + 1)$ infinitely often.

The function $\sigma_1(n)$ is equal to the sum of DIVISORS of n and is often denoted $\sigma(n)$. The first few values of $\sigma(n)$ are 1, 3, 4, 7, 6, 12, 8, 15, 13, 18, ... (Sloane's A000203). The first few values of $\sigma_2(n)$ are 1, 5, 10, 21, 26, 50, 50, 85, 91, 130, ... (Sloane's A001157). The first few values of $\sigma_3(n)$ are 1, 9, 28, 73, 126, 252, 344, 585, 757, 1134, ... (Sloane's A001158).

The sum of the DIVISORS of n excluding n itself (i.e., the PROPER DIVISORS of n) is called the RESTRICTED DIVISOR FUNCTION and is denoted $s(n)$. The first few values are 0, 1, 1, 3, 1, 6, 1, 7, 4, 8, 1, 16, ... (Sloane's A001065).

The sum of divisors $\sigma(N)$ can be found as follows. Let $N \equiv ab$ with $a \neq b$ and $(a, b) = 1$. For any divisor d of N, $d = a_i b_i$, where a_i is a divisor of a and b_i is a divisor

of b. The divisors of a are 1, a_1, a_2, ..., and a. The divisors of b are 1, b_1, b_2, ..., b. The sums of the divisors are then

$$\sigma(a) = 1 + a_1 + a_2 + ... + a \qquad (5)$$

$$\sigma(b) = 1 + b_1 + b_2 + ... + b. \qquad (6)$$

For a given a_i,

$$a_i(1 + b_1 + b_2 + ... + b) = a_i \sigma(b). \qquad (7)$$

Summing over all a_i,

$$(1 + a_1 + a_2 + ... + a)\sigma(b) = \sigma(a)\sigma(b), \qquad (8)$$

so $\sigma(N) = \sigma(ab) = \sigma(a)\sigma(b)$. Splitting a and b into prime factors,

$$\sigma(N) = \sigma(p_1^{\alpha_1})\sigma(p_2^{\alpha_2})\dots\sigma(p_r^{\alpha_r}). \qquad (9)$$

For a prime POWER $p_i^{\alpha_i}$, the divisors are $1, p_i, p_i^2, ..., p_i^{\alpha_i}$, so

$$\sigma(p_i^{\alpha_i}) = 1 + p_i + p_i^2 + ... + p_i^{\alpha_i} = \frac{p_i^{\alpha_i+1} - 1}{p_i - 1}. \qquad (10)$$

For N, therefore,

$$\sigma(N) = \prod_{i=1}^{r} \frac{p_i^{\alpha_i+1} - 1}{p_i - 1}. \qquad (11)$$

For the special case of N a PRIME, (11) simplifies to

$$\sigma(p) = \frac{p^2 - 1}{p - 1} = p + 1. \qquad (12)$$

For N a POWER of two, (11) simplifies to

$$\sigma(2^\alpha) = \frac{2^{\alpha+1} - 1}{2 - 1} = 2^{\alpha+1} - 1. \qquad (13)$$

The identity (9) can be generalized to

$$\sigma_k(N) = \sigma_k(p_1^{\alpha_1})\sigma_k(p_2^{\alpha_2})\dots\sigma_k(p_r^{\alpha_r}). \qquad (14)$$

In general,

$$\sigma_k(n) \equiv \sum_{d|n} d^k. \qquad (15)$$

The $\sigma(n)$ function has the series expansion

$$\sigma(n) = \frac{1}{6}\pi^2 n \left[1 + \frac{(-1)^n}{2^2} + \frac{2\cos\left(\frac{2}{3}n\pi\right)}{3^2} \right.$$

$$\left. + \frac{2\cos\left(\frac{1}{2}n\pi\right)}{4^2} + 2\left[\frac{\cos\left(\frac{2}{5}n\pi\right) + \cos\left(\frac{4}{5}n\pi\right)}{5^2 + ...} \right] \right] \qquad (16)$$

(Hardy 1999). Ramanujan gave the beautiful formula

$$\sum_{n=1}^{\infty} \frac{\sigma_a(n)\sigma_b(n)}{n^s}$$

$$= \frac{\zeta(s)\zeta(s-a)\zeta(s-b)\zeta(s-a-b)}{\zeta(2s-a-b)}, \qquad (17)$$

where $\zeta(n)$ is the ZETA FUNCTION and $\Re[s]$, $\Re[s-a]$, $\Re[s-b]$, $\Re[s-a-b] > 1$ (Wilson 1923), which was used by Ingham in a proof of the PRIME NUMBER THEOREM (Hardy 1999, pp. 59–0). This gives the special case

$$\sum_{n=1}^{\infty} \frac{[d(n)]^2}{n^s} = \frac{[\zeta(s)]^4}{\zeta(2s)} \qquad (18)$$

(Hardy 1999, p. 59).

The divisor function also satisfies the INEQUALITY

$$\frac{\sigma(n)}{n \ln \ln n} \le e^{\gamma} + \frac{2(1-\sqrt{2})+\gamma-\ln(4\pi)}{\sqrt{\ln n}\,\ln \ln n}$$

$$+ o\left(\frac{1}{\sqrt{\ln n}(\ln \ln n)^2}\right), \qquad (19)$$

where γ is the EULER-MASCHERONI CONSTANT (Robin 1984, Erdos 1989).

Let a number n have prime factorization

$$n = \prod_{j=1}^{r} p_j^{\alpha_j}, \qquad (20)$$

then

$$\sigma(n) = \prod_{j=1}^{r} \frac{p_j^{\alpha_j+1}-1}{p_j-1} \qquad (21)$$

(Berndt 1985). GRONWALL'S THEOREM states that

$$\varlimsup_{n \to \infty} \frac{\sigma(n)}{n \ln \ln n} = e^{\gamma}, \qquad (22)$$

where γ is the EULER-MASCHERONI CONSTANT. $\sigma(n)$ is a power of 2 IFF $n=1$ or n is a product of distinct MERSENNE PRIMES (Sierpinski 1958/59, Sivaramakrishnan 1989, Kaplansky 1999). The first few such n are 1, 3, 7, 21, 31, 93, 127, 217, 381, 651, 889, 2667, ... (Sloane's A046528), and the powers of 2 these correspond to are 0, 2, 3, 5, 5, 7, 7, 8, 9, 10, 10, 12, 12, 13, 14, ... (Sloane's A048947).

Curious identities derived using MODULAR FORM theory are given by

$$\sigma_3(n) - \sigma_7(n) + 120 \sum_{k=1}^{n-1} \sigma_3(k)\sigma_3(n-k) = 0 \qquad (23)$$

$$-10\sigma_3(n) + 21\sigma_5(n) - 11\sigma_9(n) + 5040 \sum_{k=1}^{n-1} \sigma_3(k)\sigma_5(n-k)$$

$$= 0 \qquad (24)$$

(Apostol 1997, p. 140), together with

$$21\sigma_5(n) - 20\sigma_7(n) - \sigma_{13}(n) + 10080 \sum_{k=1}^{n-1} \sigma_5(n-k)\sigma_7(k)$$

$$= 0 \qquad (25)$$

$$-10\sigma_3(n) + 11\sigma_9(n) - \sigma_{13}(n) + 2640 \sum_{k=1}^{n-1} \sigma_3(n-k)\sigma_9(k)$$

$$= 0 \qquad (26)$$

$$-21\sigma_5(n) + 22\sigma_9(n) - \sigma_{13}(n) - 2904 \sum_{k=1}^{n-1} \sigma_9(n-k)\sigma_9(k)$$

$$+ 504 \sum_{k=1}^{n-1} \sigma_5(n-k)\sigma_{13}(k) = 0 \qquad (27)$$

(M. Trott).

The divisor function is ODD IFF n is a SQUARE NUMBER or twice a SQUARE NUMBER. The divisor function satisfies the CONGRUENCE

$$n\sigma(n) \equiv 2 \pmod{\phi(n)}, \qquad (28)$$

for all PRIMES and no COMPOSITE NUMBERS with the exception of 4, 6, and 22 (Subbarao 1974). $r(n)$ is PRIME whenever $\sigma(n)$ is (Honsberger 1991). Factorizations of $\sigma(p^a)$ for PRIME p are given by Sorli.

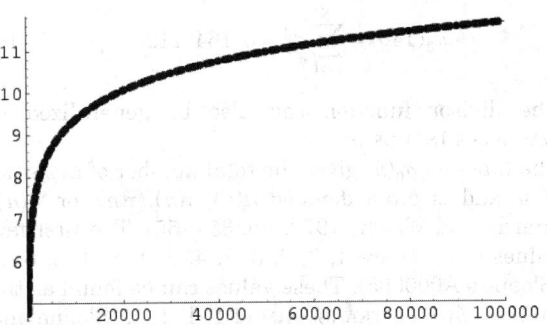

In 1838, Dirichlet showed that the average number of DIVISORS of all numbers from 1 to n is asymptotic to

$$\frac{\sum_{i=1}^{n} \sigma_0(i)}{n} \sim \ln n + 2\gamma - 1 \qquad (29)$$

(Conway and Guy 1996; Hardy 1999, p. 55), as illustrated above, where the thin solid curve plots the actual values and the thick dashed curve plots the asymptotic function. This is related to the DIRICHLET DIVISOR PROBLEM, which seeks to find the "best" coefficient θ in

$$\sum_{k=1}^{n} \nu(k) = n \ln n + (2\gamma-1)n + \mathcal{O}(n^{\theta}). \qquad (30)$$

A more precise formula is given by

$$\sum_{k=2}^{n} \sigma_0(k) = n \ln \ln n + B_2 n + o(n), \qquad (31)$$

where

$$B_2 = \gamma + \sum_{p \text{ prime}} \left[\ln\left(1 - p^{-1}\right) + \frac{1}{p-1} \right] \approx 1.034653 \quad (32)$$

(Hardy and Wright 1979, p. 355). The SUMMATORY FUNCTIONS for σ_a with $a > 1$ are

$$\sum_{k=1}^{n} \sigma_a(k) = \frac{\zeta(a+b)}{a+1} n^{a+1} + \mathcal{O}(n^a). \qquad (33)$$

For $a = 1$,

$$\sum_{k=1}^{n} \sigma_1(k) = \frac{\pi^2}{12} n^2 + \mathcal{O}(n \ln n). \qquad (34)$$

See also DIRICHLET DIVISOR PROBLEM, DIVISOR, DIVISOR PRODUCT, EVEN DIVISOR FUNCTION, FACTOR, GREATEST PRIME FACTOR, GRONWALL'S THEOREM, LEAST PRIME FACTOR, MULTIPLY PERFECT NUMBER, ODD DIVISOR FUNCTION, ORE'S CONJECTURE, PERFECT NUMBER, RESTRICTED DIVISOR FUNCTION, SILVERMAN CONSTANT, SUM OF SQUARES FUNCTION, TAU FUNCTION, TOTIENT FUNCTION, TOTIENT VALENCE FUNCTION, TWIN PEAKS, UNITARY DIVISOR FUNCTION

References

Abramowitz, M. and Stegun, C. A. (Eds.). "Divisor Functions." §24.3.3 in *Handbook of Mathematical Functions with Formulas, Graphs, and Mathematical Tables, 9th printing.* New York: Dover, p. 827, 1972.

Apostol, T. M. *Modular Functions and Dirichlet Series in Number Theory, 2nd ed.* New York: Springer-Verlag, p. 140, 1997.

Berndt, B. C. *Ramanujan's Notebooks: Part I.* New York: Springer-Verlag, p. 94, 1985.

Conway, J. H. and Guy, R. K. *The Book of Numbers.* New York: Springer-Verlag, pp. 260–61, 1996.

Dickson, L. E. *History of the Theory of Numbers, Vol. 1: Divisibility and Primality.* New York: Chelsea, pp. 279–25, 1952.

Dirichlet, G. L. "Sur l'usage des séries infinies dans la théorie des nombres." *J. reine angew. Math.* **18**, 259–74, 1838.

Erdos, P. "Ramanujan and I." In *Proceedings of the International Ramanujan Centenary Conference held at Anna University, Madras, Dec. 21, 1987.* (Ed. K. Alladi). New York: Springer-Verlag, pp. 1–0, 1989.

Guy, R. K. "Solutions of $m\sigma(m) = n\sigma(n)$," "Analogs with $d(n)$, $\sigma_k(n)$," "Solutions of $\sigma(n) = \sigma(n+1)$," and "Solutions of $\sigma(q) + \sigma(r) = \sigma(q+r)$." §B11, B12, B13 and B15 in *Unsolved Problems in Number Theory, 2nd ed.* New York: Springer-Verlag, pp. 67–0, 1994.

Hardy, G. H. *Ramanujan: Twelve Lectures on Subjects Suggested by His Life and Work, 3rd ed.* New York: Chelsea, pp. 55 and 141, 1999.

Hardy, G. H. and Weight, E. M. *An Introduction to the Theory of Numbers, 5th ed.* Oxford, England: Oxford University Press, pp. 354–55, 1979.

Heath-Brown, D. R. "A Parity Problem from Sieve Theory." *Mathematika* **29**, 1–, 1982.

Heath-Brown, D. R. "The Divisor Function at Consecutive Integers." *Mathematika* **31**, 141–49, 1984.

Honsberger, R. *More Mathematical Morsels.* Washington, DC: Math. Assoc. Amer., pp. 250–51, 1991.

Kaplansky, I. "The First Two Chapters of Dickson's History." Unpublished manuscript, Apr. 1999.

Nagell, T. *Introduction to Number Theory.* New York: Wiley, pp. 26–7, 1951.

Robin, G. "Grandes valeurs de la fonction somme des diviseurs et hypothese de Riemann." *J. Math. Pures Appl.* **63**, 187–13, 1984.

Sierpinski, W. "Sur les nombres dont la somme de diviseurs est une puissance du nombre 2." *Calcutta Math. Soc. Golden Jubilee Commemoration 1958/59, Part I.* Calcutta: Calcutta Math. Soc., pp. 7–, 1963.

Sloane, N. J. A. Sequences A000005, A000203, A001065, A001157, A001158, A046528, and A048947 in "An On-Line Version of the Encyclopedia of Integer Sequences." http://www.research.att.com/~njas/sequences/eisonline.html.

Sloane, N. J. A. and Plouffe, S. *The Encyclopedia of Integer Sequences.* San Diego, CA: Academic Press, 1995.

Sivaramakrishnan, R. *Classical Theory of Arithmetic Functions.* New York: Dekker, 1989.

Subbarao, M. V. "On Two Congruences for Primality." *Pacific J. Math.* **52**, 261–68, 1974.

Wilson, B. M. "Proofs of Some Formulae Enunciated by Ramanujan." *Proc. London Math. Soc.* **21**, 235–55, 1923.

Divisor Product

Let $\pi(n)$ denote the product of the divisors of n including n itself. For $n = 1, 2, \ldots$, the first few values are 1, 2, 3, 8, 5, 36, 7, 64, 27, 100, 11, 1728, 13, 196, ... (Sloane's A007955). The following table gives values of n for which $\pi(n)$ is a Pth power. Lionnet (1879) considered the case $P = 1$.

P	Sloane	n
1	Sloane's A048943	1, 6, 8, 10, 14, 15, 16, 21, 22, 24, 26, ...
2	Sloane's A048944	1, 4, 8, 9, 12, 18, 20, 25, 27, 28, 32, ...
3	Sloane's A048945	1, 24, 30, 40, 42, 54, 56, 66, 70, 78, ...
4	Sloane's A048946	1, 16, 32, 48, 80, 81, 112, 144, 162, ...

Write the prime factorization of a number n,

$$n = p_1^{\alpha_1} p_2^{\alpha_2} \cdots p_r^{\alpha_r}.$$

Then the power of p_i occurring in $\pi(n)$ is

$$\frac{1}{2} a_i (a_1 + 1)(a_2 + 1) \cdots (a_r + 1)$$

(Kaplansky 1999). This allows rules for determining when $\pi(n)$ is a power of n to be determined, as considered by Halcke (1719) and Lionnet (1879). Let p, q, and r be distinct primes, then the following table gives the conditions and first few n for which $\pi(n)$ is a given power P of n (Dickson 1952, Ireland and Rosen 1990, Kaplansky 1999). The case of third powers corresponds to numbers having exactly six divisors, the case of forth powers to numbers having eight divisors, and so on.

P	Forms	Sloane	n
2	p^3, pq	A007422	6, 8, 10, 14, 15, 21, 22, ...
3	p^5, p^2q	A030515	12, 18, 20, 28, 32, 44, ...
4	p^7, p^3q, pqr	A030626	24, 30, 40, 42, 54, 56, ...
5	p^9, p^4q	A030628	48, 80, 112, 162, 176, ...

References

Dickson, L. E. *History of the Theory of Numbers, Vol. 1: Divisibility and Primality.* New York: Chelsea, p. 58, 1952.
Halcke, P. Exs. 150–52 in *Deliciae Mathematicae; oder, Mathematisches sinnen-confect.* Hamburg, Germany: N. Sauer, p. 197, 1719.
Ireland, K. and Rosen, M. *A Classical Introduction to Modern Number Theory, 2nd ed.* New York: Springer-Verlag, p. 19, 1990.
Kaplansky, I. "The First Two Chapters of Dickson's History." Unpublished manuscript, Apr. 1999.
Lionnet, E. "Note sur les nombres parfaits." *Nouv. Ann. Math.* **18**, 306–08, 1879.
Lucas, E. Ex. 6 in *Théorie des nombres.* Paris: Gauthier-Villars, p. 373, 1891.
Sloane, N. J. A. Sequences A007422/M4068, A007955, A030515, A030626, A030628, A048943, A048944, A048945, and A048946 in "An On-Line Version of the Encyclopedia of Integer Sequences." http://www.research.att.com/~njas/sequences/eisonline.html.
Smarandache, F. *Only Problems, Not Solutions!, 4th ed.* Phoenix, AZ: Xiquan, 1993.

Divisor Theory

A generalization by Kronecker of Kummer's theory of PRIME IDEAL factors. A divisor on a full subcategory C of mod(A) is an additive mapping χ on C with values in a SEMIGROUP of IDEALS on A.

See also IDEAL, IDEAL NUMBER, PRIME IDEAL, SEMI-GROUP

References

Edwards, H. M. *Divisor Theory.* Boston, MA: Birkhäuser, 1989.
Vasconcelos, W. V. *Divisor Theory in Module Categories.* Amsterdam, Netherlands: North-Holland, pp. 63–4, 1974.

Divorce Digraph

A binary relation associated with an instance of the STABLE MARRIAGE PROBLEM. Stable marriages correspond to vertices with outdegree 0 in the divorce digraph (Skiena 1990, p. 252).

See also STABLE MARRIAGE PROBLEM

References

Gusfield, D. and Irving, R. W. *The Stable Marriage Problem: Structure and Algorithms.* Cambridge, MA: MIT Press, 1989.
Skiena, S. *Implementing Discrete Mathematics: Combinatorics and Graph Theory with Mathematica.* Reading, MA: Addison-Wesley, 1990.

Dixon's Factorization Method

In order to find INTEGERS x and y such that

$$x^2 \equiv y^2 \pmod{n} \tag{1}$$

(a modified form of FERMAT'S FACTORIZATION METHOD), in which case there is a 50% chance that $GCD(n, x - y)$ is a FACTOR of n, choose a RANDOM INTEGER r_i, compute

$$g(r_i) \equiv r_i^2 \pmod{n}, \tag{2}$$

and try to factor $g(r_i)$. If $g(r_i)$ is not easily factorable (up to some small trial divisor d), try another r_i. In practice, the trial rs are usually taken to be $\lfloor \sqrt{n} \rfloor + k$, with $k = 1, 2, \ldots$, which allows the QUADRATIC SIEVE factorization method to be used. Continue finding and factoring $g(r_i)$s until $N \equiv \pi d$ are found, where π is the PRIME COUNTING FUNCTION. Now for each $g(r_i)$, write

$$g(r_i) = p_{1i}^{a_{1i}} p_{2i}^{a_{2i}} \ldots p_{Ni}^{a_{Ni}}, \tag{3}$$

and form the EXPONENT VECTOR

$$\mathbf{v}(r_i) = \begin{bmatrix} a_{1i} \\ a_{2i} \\ \vdots \\ a_{Ni} \end{bmatrix}. \tag{4}$$

Now, if a_{ki} are even for any k, then $g(r_i)$ is a SQUARE NUMBER and we have found a solution to (1). If not, look for a LINEAR COMBINATION $\Sigma_i c_i \mathbf{v}(r_i)$ such that the elements are all even, i.e.,

$$c_1 \begin{bmatrix} a_{11} \\ a_{21} \\ \vdots \\ a_{N1} \end{bmatrix} + c_2 \begin{bmatrix} a_{12} \\ a_{22} \\ \vdots \\ a_{N2} \end{bmatrix} + \cdots + c_N \begin{bmatrix} a_{1N} \\ a_{2N} \\ \vdots \\ a_{NN} \end{bmatrix} = \begin{bmatrix} 0 \\ 0 \\ \vdots \\ 0 \end{bmatrix} \tag{5}$$

$$(\text{mod} 2)$$

$$\begin{bmatrix} a_{11} & a_{12} & \cdots & a_{1N} \\ a_{21} & a_{22} & \cdots & a_{2N} \\ \vdots & \vdots & \ddots & \vdots \\ a_{N1} & a_{N2} & \cdots & a_{NN} \end{bmatrix} \begin{bmatrix} c_1 \\ c_2 \\ \vdots \\ c_N \end{bmatrix} = \begin{bmatrix} 0 \\ 0 \\ \vdots \\ 0 \end{bmatrix} \pmod{2}. \tag{6}$$

Since this must be solved only mod 2, the problem can be simplified by replacing the a_{ij}s with

$$b_{ij} = \begin{cases} 0 & \text{for } a_{ij} \text{ even} \\ 1 & \text{for } a_{ij} \text{ odd.} \end{cases} \quad (7)$$

GAUSSIAN ELIMINATION can then be used to solve

$$\mathbf{bc} = \mathbf{z} \quad (8)$$

for **c**, where **z** is a VECTOR equal to 0 (mod2) . Once **c** is known, then we have

$$\prod_k g(r_k) \equiv \prod_k r_k^2 \ (\text{mod} n), \quad (9)$$

where the products are taken over all k for which $c_k = 1$. Both sides are PERFECT SQUARES, so we have a 50% chance that this yields a nontrivial factor of n. If it does not, then we proceed to a different **z** and repeat the procedure. There is no guarantee that this method will yield a factor, but in practice it produces factors faster than any method using trial divisors. It is especially amenable to parallel processing, since each processor can work on a different value of r.

References

Bressoud, D. M. *Factorization and Prime Testing.* New York: Springer-Verlag, pp. 102–04, 1989.

Dixon, J. D. "Asymptotically Fast Factorization of Integers." *Math. Comput.* **36**, 255–60, 1981.

Lenstra, A. K. and Lenstra, H. W. Jr. "Algorithms in Number Theory." In *Handbook of Theoretical Computer Science, Volume A: Algorithms and Complexity* (Ed. J. van Leeuwen). New York: Elsevier, pp. 673–15, 1990.

Pomerance, C. "A Tale of Two Sieves." *Not. Amer. Math. Soc.* **43**, 1473–485, 1996.

Dixon's Identity

$$\sum_{k=-n}^{n} (-1)^k \binom{n+b}{n+k}\binom{n+c}{c+k}\binom{b+c}{b+k}$$
$$= \frac{\Gamma(b+c+n+1)}{n!\,\Gamma(b+1)\Gamma(c+1)}, \quad (1)$$

where $\binom{n}{k}$ is a BINOMIAL COEFFICIENT and $\Gamma(x)$ is a GAMMA FUNCTION.

See also DIXON'S THEOREM

References

Koepf, W. *Hypergeometric Summation: An Algorithmic Approach to Summation and Special Function Identities.* Braunschweig, Germany: Vieweg, pp. 11 and 18–9, 1998.

Dixon's Random Squares Factorization Method

DIXON'S FACTORIZATION METHOD

Dixon's Theorem

$${}_3F_2\left[\begin{matrix} n,-x,-y \\ x+n+1, y+n+1 \end{matrix}\right]$$
$$= \Gamma(x+n+1)\Gamma(y+n+1)\Gamma$$

$$\times \left(\frac{1}{2}n+1\right)\Gamma\left(x+y+\frac{1}{2}n+1\right)$$
$$\times \Gamma(n+1)\Gamma(x+y+n+1)\Gamma$$
$$\times \left(x+\frac{1}{2}n+1\right)\Gamma\left(y+\frac{1}{2}n+1\right), \quad (1)$$

where ${}_3F_2(a,b,c;d,e;z)$ is a GENERALIZED HYPERGEOMETRIC FUNCTION and $\Gamma(z)$ is the GAMMA FUNCTION. It can be derived from the DOUGALL-RAMANUJAN IDENTITY. It can be written more symmetrically as

$${}_3F_2(a,b,c;d,e;1)$$
$$= \frac{\left(\frac{1}{2}a\right)!(a-b)!(a-c)!\left(\frac{1}{2}a-b-c\right)!}{a!\left(\frac{1}{2}a-b\right)!\left(\frac{1}{2}a-c\right)!(a-b-c)!}, \quad (2)$$

where $1+a/2-b-c$ has a positive REAL PART, $d = a-b+1$, and $e=a-c+1$ (Bailey 1935, p. 13; Petkovsek 1996; Koepf 1998, p. 32). The identity can also be written as the beautiful symmetric sum

$$\sum_k (-1)^k \binom{a+b}{a+k}\binom{a+c}{c+k}\binom{b+c}{b+k} = \frac{(a+b+c)!}{a!b!c!} \quad (3)$$

(Petkovsek 1996). In this form, it closely resembles DIXON'S IDENTITY.

See also DIXON'S IDENTITY, DOUGALL-RAMANUJAN IDENTITY, GENERALIZED HYPERGEOMETRIC FUNCTION, ZEILBERGER-BRESSOUD THEOREM

References

Bailey, W. N. "Dixon's Theorem." §3.1 in *Generalised Hypergeometric Series.* Cambridge, England: Cambridge University Press, pp. 13–4, 1935.

Cartier, P. and Foata, D. *Problèmes combinatoires de commutation et réarrangements.* New York: Springer-Verlag, 1969.

Dixon, A. C. "On the Sum of the Cubes of the Coefficients in Certain Expansion by the Binomial Theorem." *Messenger Math.* **20**, 79–0, 1891.

Dixon, A. C. "Summation of Certain Series." *Proc. London Math. Soc.* **35**, 285–89, 1903.

Hardy, G. H. *Ramanujan: Twelve Lectures on Subjects Suggested by His Life and Work,* 3rd ed. New York: Chelsea, pp. 104 and 111, 1999.

Knuth, D. E. *The Art of Computer Programming, Vol. 1: Fundamental Algorithms,* 3rd ed. Reading, MA: Addison-Wesley, 1997.

Koepf, W. *Hypergeometric Summation: An Algorithmic Approach to Summation and Special Function Identities.* Braunschweig, Germany: Vieweg, pp. 18–9, 1998.

MacMahon P. A. "The Sums of the Powers of the Binomial Coefficients." *Quart. J. Math.* **33**, 274–88, 1902.

Morley, F. "On the Series $1+\left(\frac{p}{1}\right)^3+\left\{\frac{p(p+1)}{1\cdot2}\right\}^2+\dots$." *Proc. London Math. Soc.* **34**, 397–02, 1902.

Petkovsek, M.; Wilf, H. S.; and Zeilberger, D. *A=B.* Wellesley, MA: A. K. Peters, p. 43, 1996.

Richmond, H. W. "The Sum of the Cubes of the Coefficients in $(1-x)^{2n}$." *Messenger Math.* **21**, 77–8, 1892.

Watson, G. N. "Dixon's Theorem on Generalized Hypergeometric Functions." *Proc. London Math. Soc.* **22**, xxxii-xxxiii (Records for 17 May, 1923), 1924.

Zeilberger, D. and Bressoud, D. "A Proof of Andrews' q-Dyson Conjecture." *Disc. Math.* **54**, 201–24, 1985.

Dixon-Ferrar Formula

Let $J_v(z)$ be a BESSEL FUNCTION OF THE FIRST KIND, $Y_v(z)$ a BESSEL FUNCTION OF THE SECOND KIND, and $K_v(z)$ a MODIFIED BESSEL FUNCTION OF THE FIRST KIND. Also let $\Re[z] > 0$ and $|\Re[z]| < 1/2$. Then

$$J_v^2(z) + Y_v^2(z) = \frac{8\cos(v\pi)}{\pi^2} \int_0^\infty K_{2v}(2z \sinh dt).$$

See also NICHOLSON'S FORMULA, WATSON'S FORMULA

References

Gradshteyn, I. S. and Ryzhik, I. M. Eqn. 6.518 in *Tables of Integrals, Series, and Products, 6th ed.* San Diego, CA: Academic Press, p. 671, 2000.

Iyanaga, S. and Kawada, Y. (Eds.). *Encyclopedic Dictionary of Mathematics.* Cambridge, MA: MIT Press, p. 1476, 1980.

dn

JACOBI ELLIPTIC FUNCTIONS

D-Number

A NATURAL NUMBER $n > 3$ such that

$$n \mid (a^{n-2} - a)$$

whenever $(a, n) = 1$ (a and n are RELATIVELY PRIME) and $a \le n$. (Here, $n \mid m$ means that n DIVIDES m.) There are an infinite number of such numbers, the first few being 9, 15, 21, 33, 39, 51, ... (Sloane's A033553).

See also DIVIDE, KNÖDEL NUMBERS

References

Makowski, A. "Generalization of Morrow's D-Numbers." *Simon Stevin* **36**, 71, 1962/1963.

Sloane, N. J. A. Sequences A033553 in "An On-Line Version of the Encyclopedia of Integer Sequences." http://www.research.att.com/~njas/sequences/eisonline.html.

Dobinski's Formula

The general formula states that

$$\phi_n(x) = e^{-x} \sum_{k=0}^\infty \frac{k^n}{k!} x^k, \tag{1}$$

where $\phi_n(x)$ is an EXPONENTIAL POLYNOMIAL (Roman 1984, p. 66). Setting $x = 1$ gives the special case of the nth BELL NUMBER,

$$B_n = \frac{1}{e} \sum_{k=0}^\infty \frac{k^n}{k!}. \tag{2}$$

It can be derived by dividing the formula for a STIRLING NUMBER OF THE SECOND KIND by $m!$, yielding

$$\frac{m^n}{m!} = \sum_{k=1}^m \left\{ {n \atop k} \right\} \frac{1}{(m-k)!}. \tag{3}$$

Then

$$\sum_{k=1}^\infty \frac{m^n}{m!} \lambda^m = \left(\sum_{k=1}^n \left\{ {n \atop k} \right\} \lambda^k \right) \left(\sum_{k=0}^\infty \frac{\lambda^j}{j!} \right), \tag{4}$$

and

$$\sum_{k=1}^n \left\{ {n \atop k} \right\} \lambda^k = e^{-\lambda} \sum_{m=1}^\infty \frac{m^n}{m!} \lambda^m. \tag{5}$$

Now setting $\lambda = 1$ gives the identity (Dobinski 1877; Rota 1964; Berge 1971, p. 44; Comtet 1974, p. 211; Roman 1984, p. 66; Lupas 1988; Wilf 1990, p. 106; Chen and Yeh 1994; Pitman 1997).

References

Berge, C. *Principles of Combinatorics.* New York: Academic Press, 1971.

Chen, B. and Yeh, Y.-N. "Some Explanations of Dobinski's Formula." *Studies Appl. Math.* **92**, 191–99, 1994.

Comtet, L. *Advanced Combinatorics: The Art of Finite and Infinite Expansions, rev. enl. ed.* Dordrecht, Netherlands: Reidel, 1974.

Dobinski, G. "Summierung der Reihe $\Sigma m^m/n!$ für $m = 1, 2, 3, 4, 5, ...$" *Grunert Archiv (Arch. Math. Phys.)* **61**, 333–36, 1877.

Foata, D. *La série génératrice exponentielle dans les problèmes d'énumération.* Vol. 54 of *Séminaire de Mathématiques supérieures.* Montréal, Canada: Presses de l'Université de Montréal, 1974.

Lupas, A. "Dobinski-Type Formula for Binomial Polynomials." *Stud. Univ. Babes-Bolyai Math.* **33**, 30–4, 1988.

Pitman, J. "Some Probabilistic Aspects of Set Partitions." *Amer. Math. Monthly* **104**, 201–09, 1997.

Roman, S. *The Umbral Calculus.* New York: Academic Press, p. 66, 1984.

Rota, G.-C. "The Number of Partitions of a Set." *Amer. Math. Monthly* **71**, 498–04, 1964.

Wilf, H. *Generatingfunctionology, 2nd ed.* San Diego, CA: Academic Press, 1990.

Dodecadodecahedron

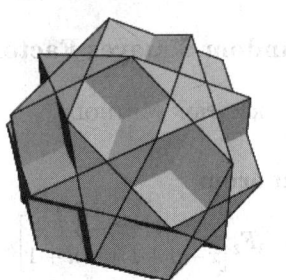

The UNIFORM POLYHEDRON U_{36} whose DUAL POLYHEDRON is the MEDIAL RHOMBIC TRIACONTAHEDRON. The solid is also called the GREAT DODECADODECAHEDRON, and its DUAL POLYHEDRON is also called the SMALL STELLATED TRIACONTAHEDRON. The dodecadodecahedron has SCHLÄFLI SYMBOL $\left\{\frac{5}{2}, 5\right\}$ and WYTHOFF SYMBOL $2\left|\frac{5}{2}5\right.$. Its faces are $12\left\{\frac{5}{2}\right\} + 12\{5\}$, and its CIRCUMRADIUS for unit edge length is

$$R = 1.$$

It can be obtained by TRUNCATING a GREAT DODECAHEDRON or FACETING a ICOSIDODECAHEDRON with PENTAGONS and covering remaining open spaces with PENTAGRAMS (Holden 1991, p. 103).

A FACETED version is the GREAT DODECAHEMICOSAHEDRON. The CONVEX HULL of the dodecadodecahedron is an ICOSIDODECAHEDRON and the dual of the ICOSIDODECAHEDRON is the RHOMBIC TRIACONTAHEDRON, so the dual of the dodecadodecahedron is one of the RHOMBIC TRIACONTAHEDRON STELLATIONS (Wenninger 1983, p. 41).

References

Cundy, H. and Rollett, A. "Great Dodecadodecahedron. $(5 \cdot \frac{5}{2})^2$." §3.9.1 in *Mathematical Models, 3rd ed.* Stradbroke, England: Tarquin Pub., p. 123, 1989.

Holden, A. *Shapes, Space, and Symmetry.* New York: Dover, 1991.

Wenninger, M. J. *Dual Models.* Cambridge, England: Cambridge University Press, p. 41, 1983.

Wenninger, M. J. *Polyhedron Models.* Cambridge, England: Cambridge University Press, p. 112, 1989.

Dodecagon

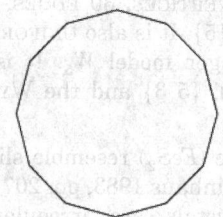

A 12-sided polygon. The regular dodecagon is CONSTRUCTIBLE denoted using the SCHLÄFLI SYMBOL $\{12\}$. The INRADIUS r, CIRCUMRADIUS R, and AREA A can be computed directly from the formulas for a general REGULAR POLYGON with side length s and $n = 12$ sides,

$$r = \frac{1}{2}s \cot\left(\frac{\pi}{12}\right) = \frac{1}{2}\left(2 + \sqrt{3}\right)s \tag{1}$$

$$R = \frac{1}{2}s \csc\left(\frac{\pi}{12}\right) = \frac{1}{2}\left(\sqrt{2} + \sqrt{6}\right)s \tag{2}$$

$$A = \frac{1}{4}ns^2 \cot\left(\frac{\pi}{12}\right) = 3\left(2 + \sqrt{3}\right)s^2. \tag{3}$$

KURSCHÁK'S THEOREM gives the AREA of the dodecagon inscribed in a UNIT CIRCLE with $R = 1$,

$$A = \frac{1}{2}nR^2 \sin\left(\frac{2\pi}{n}\right) = 3 \tag{4}$$

(Wells 1991, p. 137).

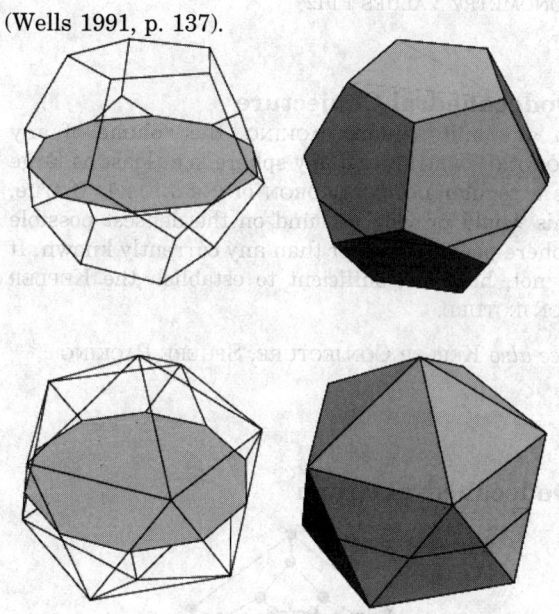

A PLANE PERPENDICULAR to a C_5 axis of a DODECAHEDRON or ICOSAHEDRON cuts the solid in a regular DECAGONAL CROSS SECTION (Holden 1991, pp. 24–5).

The GREEK, LATIN, and MALTESE CROSSES are all irregular dodecagons.

See also DECAGON, DODECAGRAM, DODECAHEDRON, GREEK CROSS, KURSCHÁK'S THEOREM, KURSCHÁK'S TILE, LATIN CROSS, MALTESE CROSS, TRIGONOMETRY VALUES PI/12, UNDECAGON

References

Holden, A. *Shapes, Space, and Symmetry.* New York: Dover, 1991.

Wells, D. *The Penguin Dictionary of Curious and Interesting Geometry.* London: Penguin, pp. 56–7 and 137, 1991.

Dodecagram

The STAR POLYGON $\{12/5\}$.

See also POLYGON, POLYGRAM, STAR POLYGON, TRIGONOMETRY VALUES PI/12

Dodecahedral Conjecture

In any unit SPHERE PACKING, the volume of any VORONOI CELL around any sphere is at least as large as a regular DODECAHEDRON of INRADIUS 1. If true, this would provide a bound on the densest possible sphere packing greater than any currently known. It is not, however, sufficient to establish the KEPLER CONJECTURE.

See also KEPLER CONJECTURE, SPHERE PACKING

Dodecahedral Graph

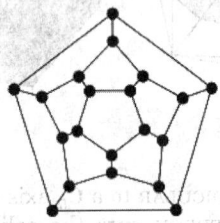

The PLATONIC GRAPH corresponding to the connectivity of the vertices of a DODECAHEDRON. Finding a HAMILTONIAN CIRCUIT on this graph is known as the ICOSIAN GAME. The dodecahedral graph has 20 nodes, 30 edges, VERTEX CONNECTIVITY 3, EDGE CONNECTIVITY 3, GRAPH DIAMETER 5, GRAPH RADIUS 5, and GIRTH 5.

See also CUBICAL GRAPH, ICOSAHEDRAL GRAPH, ICOSIAN GAME, OCTAHEDRAL GRAPH, PLATONIC GRAPH, TETRAHEDRAL GRAPH

References

Ball, W. W. R. and Coxeter, H. S. M. *Mathematical Recreations and Essays, 13th ed.* New York: Dover, 1987.
Bondy, J. A. and Murty, U. S. R. *Graph Theory with Applications.* New York: North Holland, p. 234, 1976.
Chartrand, G. *Introductory Graph Theory.* New York: Dover, 1985.
Skiena, S. *Implementing Discrete Mathematics: Combinatorics and Graph Theory with Mathematica.* Reading, MA: Addison-Wesley, p. 198, 1990.

Dodecahedral Space

POINCARÉ MANIFOLD

Dodecahedron

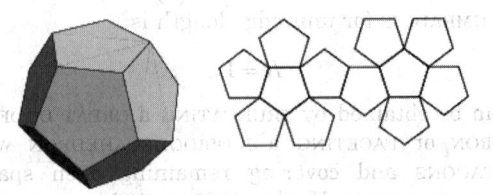

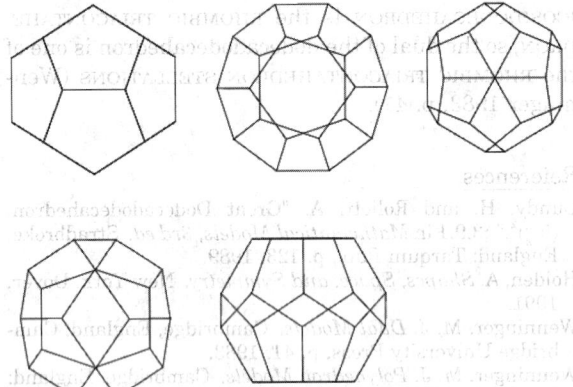

The regular dodecahedron is the PLATONIC SOLID P_4 composed of 20 VERTICES, 30 EDGES, and 12 PENTAGONAL FACES, $12\{5\}$. It is also UNIFORM POLYHEDRON U_{23} and Wenninger model W_5. It is given by the SCHLÄFLI SYMBOL $\{5,3\}$ and the WYTHOFF SYMBOL $3|25$.

Crystals of pyrite (FeS_2) resemble slightly distorted dodecahedra (Steinhaus 1983, pp. 207–08), and sphalerite (ZnS) crystals are irregular dodecahedra bounded by congruent deltoids (Steinhaus 1983, pp. 207 and 209). The HEXAGONAL SCALENOHEDRON is another irregular dodecahedron. The DELTOIDAL HEXECONTAHEDRON and TRIAKIS TETRAHEDRON are irregular dodecahedra composed of a single type of face, and the CUBOCTAHEDRON and TRUNCATED TETRAHEDRON are dodecahedral ARCHIMEDEAN SOLIDS consisting of multiple types of faces.

Dodecahedra were known to the Greeks, and 90 models of dodecahedra with knobbed vertices have been found in a number of archaeological excavations in Europe dating from the Gallo-Roman period in locations ranging from military camps to public bath houses to treasure chests (Schuur).

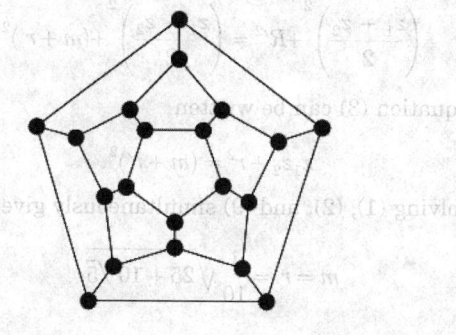

The dodecahedron has the ICOSAHEDRAL GROUP I_h of symmetries. The connectivity of the vertices is given by the DODECAHEDRAL GRAPH. There are three DODE-CAHEDRON STELLATIONS.

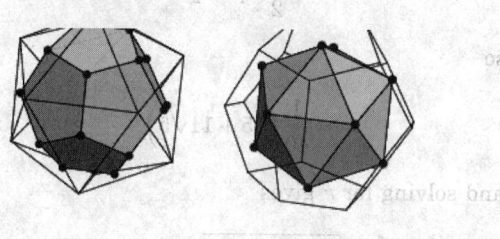

The DUAL POLYHEDRON of the dodecahedron is the ICOSAHEDRON, so the centers of the faces of an ICOSAHEDRON form a dodecahedron, and vice versa (Steinhaus 1983, pp. 199–01).

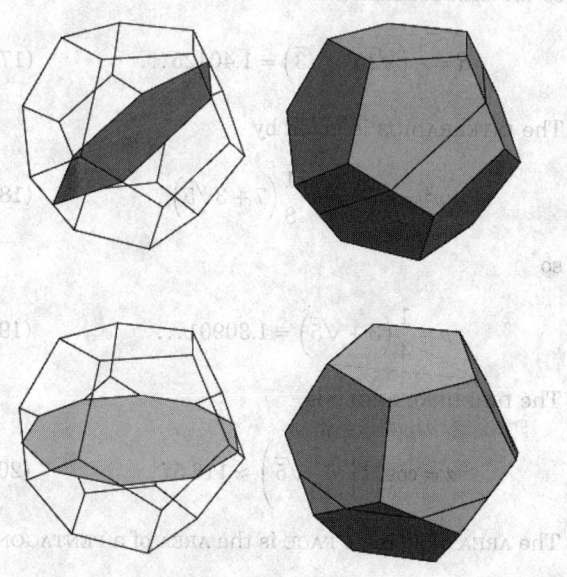

A PLANE PERPENDICULAR to a C_3 axis of a dodecahedron cuts the solid in a regular HEXAGONAL CROSS SECTION (Holden 1991, p. 27). A PLANE PERPENDICULAR to a C_5 axis of a dodecahedron cuts the solid in a regular DECAGONAL CROSS SECTION (Holden 1991, p. 24).

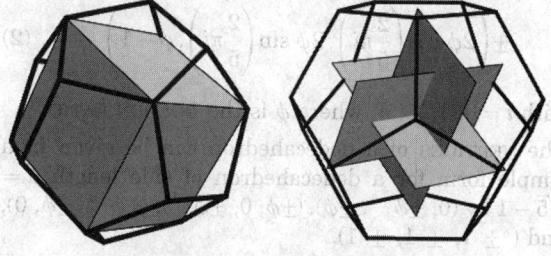

A CUBE can be constructed from the dodecahedron's vertices taken eight at a time (above left figure; Steinhaus 1983, pp. 198–99; Wells 1991). Five such cubes can be constructed, forming the CUBE 5-COM-POUND. In addition, joining the centers of the faces gives three mutually PERPENDICULAR GOLDEN REC-TANGLES (right figure; Wells 1991).

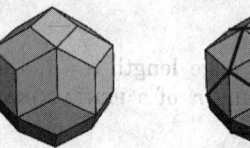  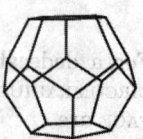

The short diagonals of the faces of the RHOMBIC TRIACONTAHEDRON give the edges of a dodecahedron (Steinhaus 1983, pp. 209–10).

The following table gives polyhedra which can be constructed by CUMULATION of a dodecahedron by pyramids of given heights h.

h	$(r+h)/h$	Result
$-\sqrt{\frac{1}{10}(5-\sqrt{5})}$	$2\sqrt{5}-3$	60-faced dimpled DELTA-HEDRON
$\frac{1}{19}\sqrt{\frac{1}{5}(65+22\sqrt{5})}$	$\frac{3}{19}(10-\sqrt{5})$	PENTAKIS DODE-CAHEDRON
$\sqrt{\frac{1}{10}(5-\sqrt{5})}$	$2\sqrt{5}-3$	60-faced star DELTAHEDRON
$\sqrt{\frac{1}{5}(5+2\sqrt{5})}$	$\sqrt{5}$	SMALL STEL-LATED DODECA-HEDRON

When the dodecahedron with edge length $\sqrt{10-2\sqrt{5}}$ is oriented with two opposite faces parallel to the xy-PLANE, the vertices of the top and bottom faces lie at $z = \pm(\phi+1)$ and the other VERTICES lie at $z = \pm(\phi-1)$, where ϕ is the GOLDEN RATIO. The explicit coordinates are

$$\pm\left(2\cos\left(\frac{2}{5}\pi i\right), 2\sin\left(\frac{2}{5}\pi i\right), \phi+1\right) \quad (1)$$

$$\pm\left(2\phi\cos\left(\frac{2}{5}\pi i\right), 2\phi\sin\left(\frac{2}{5}\pi i\right), \phi-1\right) \quad (2)$$

with $i = 0, 1, \ldots, 4$, where ϕ is the GOLDEN RATIO.

The VERTICES of a dodecahedron can be given in a simple form for a dodecahedron of side length $a = \sqrt{5}-1$ by $(0, \pm\phi^{-1}, \pm\phi)$, $(\pm\phi, 0, \pm\phi^{-1})$, $(\pm\phi^{-1}, \pm\phi, 0)$, and $(\pm 1, \pm 1, \pm 1)$.

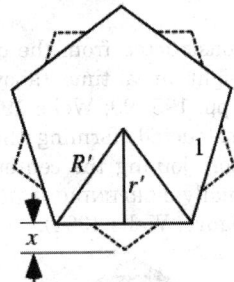

For a dodecahedron of unit edge length $a = 1$, the CIRCUMRADIUS R' and INRADIUS r' of a PENTAGONAL FACE are

$$R' = \frac{1}{10}\sqrt{50+10\sqrt{5}} \quad (3)$$

$$r' = \frac{1}{10}\sqrt{25+10\sqrt{5}}. \quad (4)$$

The SAGITTA x is then given by

$$x \equiv R' - r' = \frac{1}{10}\sqrt{125-10\sqrt{5}}. \quad (5)$$

Now consider the following figure.

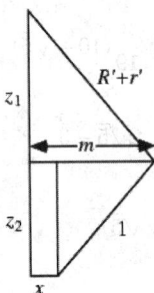

Using the PYTHAGOREAN THEOREM on the figure then gives

$$z_1^2 + m^2 = (R'+r)^2 \quad (6)$$

$$z_2^2 + (m-x)^2 = 1 \quad (7)$$

$$\left(\frac{z_1+z_2}{2}\right)^2 + R'^2 = \left(\frac{z_1-z_2}{2}\right)^2 + (m+r')^2. \quad (8)$$

Equation (3) can be written

$$z_1 z_2 + r^2 = (m+r')^2. \quad (9)$$

Solving (1), (2), and (9) simultaneously gives

$$m = r' = \frac{1}{10}\sqrt{25+10\sqrt{5}} \quad (10)$$

$$z_1 = 2r' = \frac{1}{5}\sqrt{25+10\sqrt{5}} \quad (11)$$

$$z_2 = R' = \frac{1}{10}\sqrt{50+10\sqrt{5}}. \quad (12)$$

The INRADIUS of the dodecahedron is then given by

$$r = \frac{1}{2}(z_1+z_2), \quad (13)$$

so

$$r^2 = \frac{1}{40}\left(25+11\sqrt{5}\right), \quad (14)$$

and solving for r gives

$$r = \frac{1}{20}\sqrt{250+110\sqrt{5}} = 1.11351\ldots \quad (15)$$

Now,

$$R^2 = R'^2 + r^2 = \frac{3}{8}\left(3+\sqrt{5}\right), \quad (16)$$

so the CIRCUMRADIUS is

$$R = \frac{1}{4}\left(\sqrt{15}+\sqrt{3}\right) = 1.40125\ldots \quad (17)$$

The INTERRADIUS is given by

$$\rho^2 = r'^2 + r^2 = \frac{1}{8}\left(7+3\sqrt{5}\right), \quad (18)$$

so

$$\rho = \frac{1}{4}\left(3+\sqrt{5}\right) = 1.30901\ldots \quad (19)$$

The DIHEDRAL ANGLE is

$$\alpha = \cos^{-1}\left(-\frac{1}{5}\sqrt{5}\right) \approx 116.57°. \quad (20)$$

The AREA of a single FACE is the AREA of a PENTAGON,

$$A = \frac{1}{4}\sqrt{25+10\sqrt{5}}. \quad (21)$$

The VOLUME of the dodecahedron can be computed by summing the volume of the 12 constituent PENTAGONAL PYRAMIDS,

$$V = 12\left(\frac{1}{3}Ar\right) = \frac{1}{4}\left(15 + 7\sqrt{5}\right). \tag{22}$$

Apollonius showed that the VOLUME V and SURFACE AREA A of the dodecahedron and its DUAL the ICOSAHEDRON are related by

$$\frac{V_{\text{icosahedron}}}{V_{\text{dodecahedron}}} = \frac{A_{\text{icosahedron}}}{A_{\text{dodecahedron}}} \tag{23}$$

See also AUGMENTED DODECAHEDRON, AUGMENTED TRUNCATED DODECAHEDRON, CAIRO TESSELLATION, CUBOCTAHEDRON, DELTOIDAL HEXECONTAHEDRON, DODECAGON, DODECAHEDRON 2-COMPOUND, DODECAHEDRON 3-COMPOUND, DODECAHEDRON 5-COMPOUND, DODECAHEDRON-ICOSAHEDRON COMPOUND, DODECAHEDRON-SMALL TRIAMBIC ICOSAHEDRON COMPOUND, DODECAHEDRON STELLATIONS, ELONGATED DODECAHEDRON, GREAT DODECAHEDRON, GREAT STELLATED DODECAHEDRON, HYPERBOLIC DODECAHEDRON, ICOSAHEDRON, METABIAUGMENTED DODECAHEDRON, METABIAUGMENTED TRUNCATED DODECAHEDRON, PARABIAUGMENTED DODECAHEDRON, PARABIAUGMENTED TRUNCATED DODECAHEDRON, PYRITOHEDRON, RHOMBIC DODECAHEDRON, RHOMBIC TRIACONTAHEDRON, SMALL STELLATED DODECAHEDRON, STELLATION, TRIAKIS TETRAHEDRON, TRIAUGMENTED DODECAHEDRON, TRIAUGMENTED TRUNCATED DODECAHEDRON, TRIGONAL DODECAHEDRON, TRIGONOMETRY VALUES PI/5, TRUNCATED DODECAHEDRON, TRUNCATED TETRAHEDRON

References

Beyer, W. H. *CRC Standard Mathematical Tables, 28th ed.* Boca Raton, FL: CRC Press, p. 228, 1987.

Cundy, H. and Rollett, A. "Dodecahedron. 5^3." §3.5.4 in *Mathematical Models, 3rd ed.* Stradbroke, England: Tarquin Pub., p. 87, 1989.

Davie, T. "The Dodecahedron." http://www.dcs.st-and.ac.uk/~ad/mathrecs/polyhedra/dodecahedron.html.

Harris, J. W. and Stocker, H. "Dodecahedron." §4.4.5 in *Handbook of Mathematics and Computational Science.* New York: Springer-Verlag, p. 101, 1998.

Holden, A. *Shapes, Space, and Symmetry.* New York: Dover, 1991.

Schuur, W. A. "Pentagonale Dodecaeder." http://home.wxs.nl/~wschuur/dcaeder.htm.

Steinhaus, H. *Mathematical Snapshots, 3rd ed.* New York: Dover, pp. 195–99, 1999.

Weisstein, E. W. "Polyhedra." MATHEMATICA NOTEBOOK POLYHEDRA.M.

Wells, D. *The Penguin Dictionary of Curious and Interesting Geometry.* London: Penguin, pp. 57–8, 1991.

Wenninger, M. J. "The Dodecahedron." Model 5 in *Polyhedron Models.* Cambridge, England: Cambridge University Press, p. 19, 1989.

Dodecahedron 2-Compound

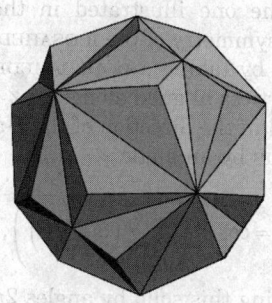

A compound of two dodecahedra having the symmetry of the CUBE arises by combining two dodecahedra rotated $90°$ with respect to each other about a common C_2 axis (Holden 1991, p. 37).

See also DODECAHEDRON, DODECAHEDRON 3-COMPOUND, DODECAHEDRON 5-COMPOUND, POLYHEDRON COMPOUND

References

Holden, A. *Shapes, Space, and Symmetry.* New York: Dover, p. 37, 1991.

Dodecahedron 3-Compound

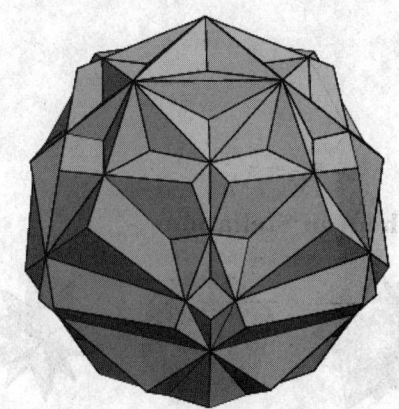

See also DODECAHEDRON, DODECAHEDRON 2-COMPOUND, DODECAHEDRON 5-COMPOUND

Dodecahedron 5-Compound

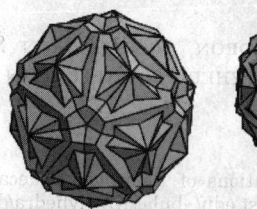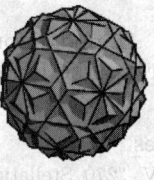

There are at least two attractive 5-dodecahedra compounds. The one illustrated in the left figure above has the symmetry of the ICOSAHEDRON and can be constructed by taking a DODECAHEDRON with top and bottom vertices aligned along the z-AXIS and one vertex oriented in the direction of the x-axis, rotating about the y-AXIS by an angle

$$\alpha = \cos^{-1}\left(\sqrt{\frac{2}{15}\left(5+\sqrt{5}\right)}\right),$$

and then rotating this solid by angles $2\pi i/5$ for $i = 0$, 1, ..., 4.

The compound shown at right can be obtained by combining five dodecahedra, each rotated by 1/10 of a turn about the line joining the centroids of opposite faces.

See also DODECAHEDRON, DODECAHEDRON 2-COMPOUND, DODECAHEDRON 3-COMPOUND, POLYHEDRON COMPOUND

References

Wenninger, M. J. *Dual Models.* Cambridge, England: Cambridge University Press, pp. 145–47, 1983.

Dodecahedron Stellations

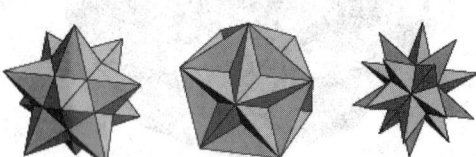

The dodecahedron has three STELLATIONS: the SMALL STELLATED DODECAHEDRON, GREAT DODECAHEDRON, and GREAT STELLATED DODECAHEDRON (Wenninger 1989, pp. 35 and 38–0). Bulatov has produced 270 stellations of a deformed dodecahedron.

See also DODECAHEDRON, ICOSAHEDRON STELLATIONS, STELLATED POLYHEDRON, STELLATION

References

Bulatov, V. "270 Stellations of Deformed Dodecahedron." http://www.physics.orst.edu/~bulatov/polyhedra/dodeca270/.
Wenninger, M. J. *Polyhedron Models.* New York: Cambridge University Press, pp. 35 and 38–0, 1989.

Dodecahedron-Icosahedron Compound

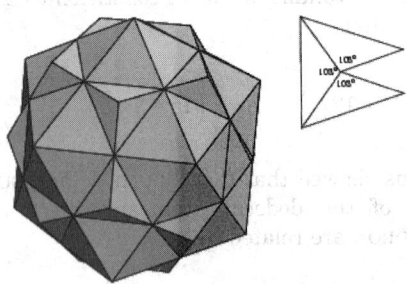

A POLYHEDRON COMPOUND consisting of a DODECAHEDRON and its dual the ICOSAHEDRON. It is most easily constructed by adding 20 triangular PYRAMIDS, constructed as above, to an ICOSAHEDRON. In the compound, the DODECAHEDRON and ICOSAHEDRON are rotated $\pi/5$ radians with respect to each other, and the ratio of the ICOSAHEDRON to DODECAHEDRON edges lengths are the GOLDEN RATIO ϕ.

If the DODECAHEDRON is chosen to have unit edge length, the resulting compound has side lengths

$$s_1 = \frac{1}{2} \tag{1}$$

$$s_2 = \frac{1}{4}\left(1+\sqrt{5}\right). \tag{2}$$

Normalizing so that $s_1 = 1$ gives SURFACE AREA and VOLUME

$$S = 15\sqrt{13 + 5\sqrt{5}} + \sqrt{6(25 + 11\sqrt{5})} \tag{3}$$

$$V = \frac{5}{2}\left(15 + 7\sqrt{5}\right). \tag{4}$$

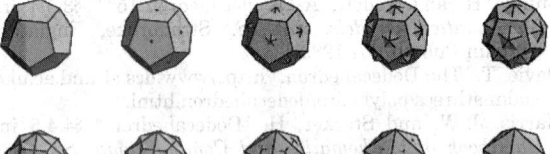

The above figure shows compounds composed of a DODECAHEDRON of unit edge length and ICOSAHEDRA having edge lengths varying from $\sqrt{5}/2$ (inscribed in the dodecahedron) to 2 (circumscribed about the dodecahedron).

The intersecting edges of the compound form the DIAGONALS of the 30 RHOMBUSES constituting the TRIACONTAHEDRON, which is the DUAL POLYHEDRON of the ICOSIDODECAHEDRON (Ball and Coxeter 1987). The dodecahedron-icosahedron compound is also the first STELLATION of the ICOSIDODECAHEDRON.

See also DUAL POLYHEDRON, DODECAHEDRON, ICOSA-HEDRON, ICOSIDODECAHEDRON, PLATONIC SOLID, POLYHEDRON COMPOUND, RHOMBIC TRIACONTAHE-DRON

References

Cundy, H. and Rollett, A. "Dodecahedron Plus Icosahedron." §3.10.3 in *Mathematical Models, 2nd ed.* Stradbroke, England: Tarquin Pub., p. 131, 1989.
Weisstein, E. W. "Polyhedra." MATHEMATICA NOTEBOOK POLYHEDRA.M.
Wenninger, M. J. "First Stellation of the Icosidodecahe-dron." §47 in *Polyhedron Models.* Cambridge, England: Cambridge University Press, p. 76, 1989.

Dodecahedron-Small Triambic Icosahedron Compound

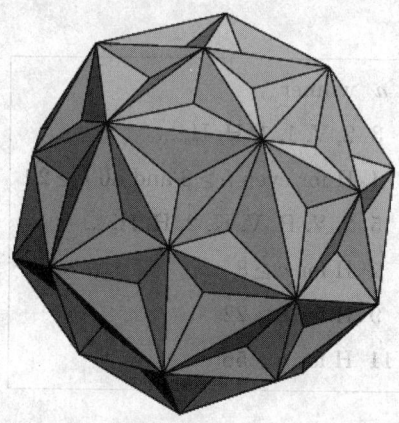

A stellated form of a truncated icosahedron, but a *different* truncation than in the TRUNCATED ICOSAHE-DRON ARCHIMEDEAN SOLID. It contains curious but attractive patterns of raised regular pentagrams and irregular hexagrams. For the solid constructed from a DODECAHEDRON with unit edge lengths, the SURFACE AREA is given by the root of a 10 order polynomial with large integer coefficients, and the VOLUME is given by

$$V = \frac{1}{20}\left(35 + 15\sqrt{15} - 4\sqrt{650 - 290\sqrt{5}}\right).$$

See also DODECAHEDRON, SMALL TRIAMBIC ICOSAHE-DRON

References

Wenninger, M. J. *Dual Models.* Cambridge, England: Cambridge University Press, pp. 51–2 1983.

Dodecic Surface

An ALGEBRAIC SURFACE of degree 12.

See also ALGEBRAIC SURFACE, SARTI DODECIC

Dolbeault Cohomology

See also CALABI-YAU SPACE, DOLBEAULT OPERATORS

Dolbeault Operators

See also DEL BAR OPERATOR, DOLBEAULT COHOMOL-OGY

Domain

A CONNECTED OPEN SET. The term domain is also used to describe the set of values D for which a FUNCTION is defined. The set of values to which D is sent by the function (MAP) is then called the RANGE.

See also CODOMAIN, CONNECTED SET, MAP, ONE-TO-ONE, ONTO, RANGE (IMAGE), REINHARDT DOMAIN

References

Krantz, S. G. *Handbook of Complex Analysis.* Boston, MA: Birkhäuser, p. 76, 1999.

Domain Invariance Theorem

The Invariance of domain theorem states that if $f : A \to \mathbb{R}^n$ is a ONE-TO-ONE continuous MAP from A, then a compact subset of $\mathbb{R}^n$, then the interior of A is mapped to the interior of $f(A)$.

See also DIMENSION INVARIANCE THEOREM

Dome

BOHEMIAN DOME, GEODESIC DOME, HEMISPHERE, SPHERICAL CAP, TORISPHERICAL DOME, VAULT

Dominance

The dominance RELATION on a SET of points in EUCLIDEAN n-space is the INTERSECTION of the n coordinate-wise orderings. A point p dominates a point q provided that every coordinate of p is at least as large as the corresponding coordinate of q.

A PARTITION p_a dominates a PARTITION p_b if, for all k, the sum of the k largest parts of p_a is $\geq$ the sum of the k largest parts of p_b. For example, for $n = 7$, $\{7\}$ dominates all other PARTITIONS, while $\{1,1,1,1,1,1,1\}$ is dominated by all others. In contrast, $\{3,1,1,1,1\}$ and $\{2,2,2,1\}$ do not dominate each other (Skiena 1990, p. 52).

The dominance orders in $\mathbb{R}^n$ are precisely the POSETS of DIMENSION at most n.

See also DOMINATING SET, DOMINATION NUMBER, PARTIALLY ORDERED SET, REALIZER

References

Skiena, S. *Implementing Discrete Mathematics: Combinatorics and Graph Theory with Mathematica.* Reading, MA: Addison-Wesley, 1990.

Stanton, D. and White, D. *Constructive Combinatorics.* New York: Springer-Verlag, 1986.

Dominant Set

DOMINANCE, DOMINATING SET

Dominating Set

This entry contributed by NICOLAS BRAY

For a GRAPH G and a subset S of the VERTEX SET $V(G)$, denote by $N_G[S]$ the set of vertices in G which are in S or adjacent to a vertex in S. If $N_G[S] = V(G)$, then S is said to be a dominating set (of vertices in G).

See also DOMINANCE, DOMINATION NUMBER

Domination Number

This entry contributed by NICOLAS BRAY

The domination number of a graph G, denoted $\gamma(G)$, is the minimum size of a DOMINATING SET of vertices in G.

See also DOMINANCE, DOMINATING SET, VIZING CONJECTURE

References

Clark, W. E. and Suen, S. "An Inequality Related to Vizing's Conjecture." *Electronic J. Combinatorics* **7**, No. 1, N4, 1–, 2000. http://www.combinatorics.org/Volume_7/ v7i1toc.html#N4.
Haynes, T. W.; Hedetniemi, S. T.; and Slater, P. J. *Domination in Graphs--Advanced Topics.* New York: Dekker, 1998.
Haynes, T. W.; Hedetniemi, S. T.; and Slater, P. J. *Fundamentals of Domination in Graphs.* New York: Dekker, 1998.

Domineering

A two-player game, also called crosscram, in which player H has horizontal DOMINOES and player V has vertical DOMINOES. The two players alternately place a domino on a BOARD until the other cannot move, in which case the player having made the last move wins (Gardner 1974, Lachmann *et al.* 2000). Depending on the dimension of the board, the winner will be H, V, 1 (the player making the first move), or 2 (the player making the second move). For example, the (2×2) board is a win for the first player.

Berlekamp (1988) solved the general problem for $2 \times n$ board for odd n. Solutions for the $2 \times n$ board are summarized in the following table, with $2 \times n$ a win for H for $n \geq 28$.

n	win	n	win	n	win
0	2	10	1	20	H

1	V	11	1	21	H
2	1	12	H	22	H
3	1	13	2	23	1
4	H	14	1	24	H
5	V	15	1	25	H
6	1	16	H	26	H
7	1	17	H	27	1
8	H	18	1	28	H
9	V	19	1	29	H

Lachmann *et al.* (2000) have solved the game $k \times n$ for widths of $n = 2, 3, 4, 5, 7, 9,$ and 11, obtaining the results summarized in the following table for $k = 0, 1,$ ....

n	winner
3	2, V, 1, 1, H, H, ...
4	H for even $k \geq 8$ and all $k \geq 22$
5	2, V, H, V, H, 2, H, H, ...
7	H for $n \geq 8$
9	H for $n \geq 22$
11	H for $n \geq 56$

See also DOMINO

References

Berlekamp, E. R. "Blockbuster and Domineering." *J. Combin. Th. Ser. A* **49**, 67–16, 1988.
Berlekamp, E. R.; Conway, J. H.; and Guy, R. K. *Winning Ways for Your Mathematical Plays, Vol. 2: Games in Particular.* London: Academic Press, 1982.
Breuker, D. M.; Uiterwijk, J. W. H. M.; van den Herik, H. J. "Solving 8×8 Domineering." *Theor. Comput. Sci.* **122**, 43–8, 2000.
Conway, J. H. *On Numbers and Games.* New York: Academic Press, 1976.
Gardner, M. "Mathematical Games: Cram, Crosscram and Quadraphage: New Games having Elusive Winning Strategies." *Sci. Amer.* **230**, 106–08, Feb. 1974.
Lachmann, M.; Moore, C.; and Rapaport, I. Who Wins Domineering on Rectangular Boards? 8 Jun 2000. http://xxx.lanl.gov/abs/math.CO/0006066/.
Uiterwijk, J. W. H. M. and van den Herik, H. J. "The Advantage of the Initiative." *Info. Sci.* **122**, 43–8, 2000.
Wolfe, D. "The Gamesman's Toolkit." In *Games of No Chance.* (Ed. R. J. Nowakowski). Cambridge, England: Cambridge University Press, 1998.

Domino

The unique 2-POLYOMINO consisting of two equal squares connected along a complete EDGE.

The FIBONACCI NUMBER F_{n+1} gives the number of ways for 2×1 dominoes to cover a $2 \times n$ CHECKERBOARD, as illustrated in the following diagrams (Dickau).

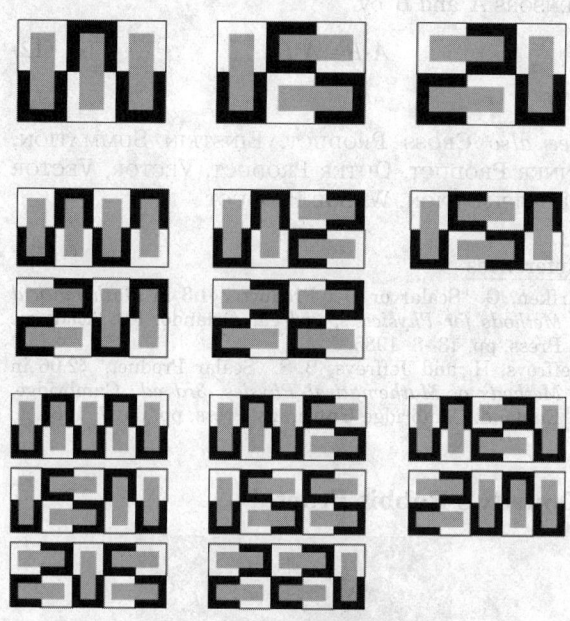

See also DOMINEERING, FIBONACCI NUMBER, GOMORY'S THEOREM, HEXOMINO, PENTOMINO, POLYOMINO, POLYOMINO TILING, TETROMINO, TRIOMINO

References

Culin, S. "Kol-hpai, Bone Tablets--Dominoes." §81 in *Games of the Orient: Korea, China, Japan.* Rutland, VT: Charles E. Tuttle, pp. 102–03, 1965.

Cohn, H. "2-adic Behavior of Numbers of Domino Tilings." *Electronic J. Combinatorics* **6**, No. 1, R14, 1–, 1999. http://www.combinatorics.org/Volume_6/v6i1toc.html#R14.

Dickau, R. M. "Fibonacci Numbers." http://www.prairienet.org/~pops/fibboard.html.

Gardner, M. "Polyominoes." Ch. 13 in *The Scientific American Book of Mathematical Puzzles & Diversions.* New York: Simon and Schuster, pp. 124–40, 1959.

Kraitchik, M. "Dominoes." §12.1.22 in *Mathematical Recreations.* New York: W. W. Norton, pp. 298–02, 1942.

Lei, A. "Domino." http://www.cs.ust.hk/~philipl/omino/domino.html.

Madachy, J. S. "Domino Recreations." *Madachy's Mathematical Recreations.* New York: Dover, pp. 209–19, 1979.

Schroeppel, R. Item 111 in Beeler, M.; Gosper, R. W.; and Schroeppel, R. *HAKMEM.* Cambridge, MA: MIT Artificial Intelligence Laboratory, Memo AIM-239, p. 48, Feb. 1972.

Domino Problem

WANG'S CONJECTURE

Donaldson Invariants

Distinguish between smooth MANIFOLDS in 4-D.

See also DONALDSON THEORY

Donaldson Theory

See also DONALDSON INVARIANTS

Donkin's Theorem

The product of three translations along the directed sides of a TRIANGLE through twice the lengths of these sides is the IDENTITY MAP.

Donut

TORUS

Doob's Theorem

A theorem proved by Doob (1942) which states that any random process which is both GAUSSIAN and MARKOV has the following forms for its correlation function $C_y(\tau)$, spectral density $G_y(f)$, and probability densities $p_1(y)$ and $p_2(y_1|y_2, \tau)$ ::

$$C_y \tau = \sigma_y^2 e^{-\tau/\tau_r}$$

$$G_y(f) = \frac{4\tau_\tau^{-1}\sigma_y^2}{(2\pi f)^2 + \tau_\tau^{-2}}$$

$$p_1(y) = \frac{1}{\sqrt{2\pi\sigma_y^2}} e^{-(y-\bar{y})^2/2\sigma_y^2}$$

$$p_2(y_1/y_2, \tau) = \frac{1}{\sqrt{2\pi(1 - e^{-\tau/\tau_\tau})\sigma_y^2}} \exp$$

$$\times \left\{ -\frac{\left[(y_2 - \bar{y}) - e^{-\tau/\tau_\tau}(y_1 - \bar{y})\right]^2}{2(1 - e^{-2\tau/\tau_\tau})\sigma_y^2} \right\},$$

where $\bar{y}$ is the MEAN, σ_y the STANDARD DEVIATION, and τ_r the relaxation time.

References

Doob, J. L. "Topics in the Theory of Markov Chains." *Trans. Amer. Math. Soc.* **52**, 37–4, 1942.

Dorman-Luke Construction

DUAL POLYHEDRON

Dot

The "dot" · has several meanings in mathematics, including MULTIPLICATION ($a.b$ is pronounced "a times b"), computation of a DOT PRODUCT (**ab** is pronounced "**a** dot **b**").

See also DERIVATIVE, DOT PRODUCT, OVERDOT, TIMES

Dot Product

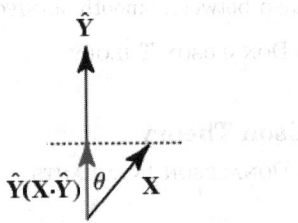

The dot product can be defined for two VECTORS $\mathbf{X}$ and $\mathbf{Y}$ by

$$\mathbf{X}\cdot\mathbf{Y} = |\mathbf{X}||\mathbf{Y}|\cos\theta, \qquad (1)$$

where θ is the ANGLE between the VECTORS. It follows immediately that $\mathbf{X}\cdot\mathbf{Y} = 0$ if $\mathbf{X}$ is PERPENDICULAR to $\mathbf{Y}$. The dot product therefore has the geometric interpretation as the length of the PROJECTION of $\mathbf{X}$ onto the UNIT VECTOR $\overline{\mathbf{Y}}$ when the two vectors are placed so that their tails coincide.

By writing

$$A_x = A\cos\theta_A \quad B_x = B\cos\theta_B \qquad (2)$$

$$A_y = A\sin\theta_A \quad B_y = B\sin\theta_B, \qquad (3)$$

it follows that (1) yields

$$\mathbf{A}\cdot\mathbf{B} = AB\cos(\theta_A - \theta_B)$$

$$= AB(\cos\theta_A\cos\theta_B + \sin\theta_A\sin\theta_B)$$

$$= A\cos\theta_A B\cos\theta_B + A\sin\theta_A B\sin\theta_B$$

$$= A_x B_x + A_y B_y. \qquad (4)$$

So, in general,

$$\mathbf{X}\cdot\mathbf{Y} = \sum_{i=1}^{n} x_i y_i = x_1 y_1 + \cdots + x_n y_n. \qquad (5)$$

This can be written very succinctly using EINSTEIN SUMMATION notation as

$$\mathbf{X}\cdot\mathbf{Y} = x_i y_i. \qquad (6)$$

The dot product is implemented in *Mathematica* as Dot[a, b], or simply by using a period, a . b.

The dot product is COMMUTATIVE

$$\mathbf{X}\cdot\mathbf{Y} = \mathbf{Y}\cdot\mathbf{X}, \qquad (7)$$

ASSOCIATIVE

$$(r\mathbf{X})\cdot\mathbf{Y} = r(\mathbf{X}\cdot\mathbf{Y}), \qquad (8)$$

and DISTRIBUTIVE

$$\mathbf{X}\cdot(\mathbf{Y} + \mathbf{Z}) = \mathbf{X}\cdot\mathbf{Y} + \mathbf{X}\cdot\mathbf{Z}. \qquad (9)$$

The DERIVATIVE of a dot product of VECTORS is

$$\frac{d}{dt}[\mathbf{r}_1(t)\cdot\mathbf{r}_2(t)] = \mathbf{r}_1(t)\cdot\frac{d\mathbf{r}_2}{dt} + \frac{d\mathbf{r}_1}{dt}\cdot\mathbf{r}_2(t). \qquad (10)$$

The dot product is invariant under rotations

$$\mathbf{A}'\cdot\mathbf{B}' = A_i' B_i' = a_{ij}A_j a_{ik}B_k = (a_{ij}a_{ik})A_j B_k$$

$$= \delta_{jk}A_j B_k = A_j B_j = \mathbf{A}\cdot\mathbf{B}, \qquad (11)$$

where EINSTEIN SUMMATION has been used.

The dot product is also called the scalar product and INNER PRODUCT. In the latter context, it is usually written $\langle a, b \rangle$. The dot product is also defined for TENSORS A and B by

$$A\cdot B \equiv A^{\alpha}B_{\alpha}. \qquad (12)$$

See also CROSS PRODUCT, EINSTEIN SUMMATION, INNER PRODUCT, OUTER PRODUCT, VECTOR, VECTOR MULTIPLICATION, WEDGE PRODUCT

References

Arfken, G. "Scalar or Dot Product." §1.3 in *Mathematical Methods for Physicists, 3rd ed.* Orlando, FL: Academic Press, pp. 13–8, 1985.
Jeffreys, H. and Jeffreys, B. S. "Scalar Product." §2.06 in *Methods of Mathematical Physics, 3rd ed.* Cambridge, England: Cambridge University Press, pp. 65–7, 1988.

Douady's Rabbit Fractal

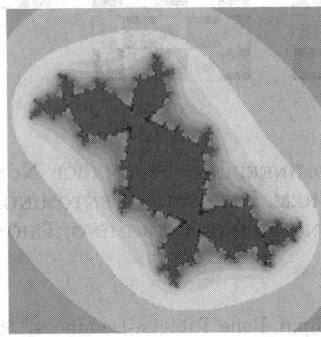

A JULIA SET with $c = -0.123 + 0.745i$, also known as the dragon fractal.

See also DENDRITE FRACTAL, JULIA SET, SAN MARCO FRACTAL, SIEGEL DISK FRACTAL

References

Wagon, S. *Mathematica in Action.* New York: W. H. Freeman, p. 176, 1991.

Double Bar

The symbol $\|$ used to denote certain kinds of NORMS in mathematics $((\|x\|).)$.

See also BAR

References

Bringhurst, R. *The Elements of Typographic Style, 2nd ed.* Point Roberts, WA: Hartley and Marks, p. 277, 1997.

Double Bubble

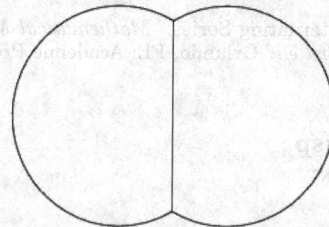

A double bubble is pair of BUBBLES which intersect and are separated by a membrane bounded by the intersection. The usual double bubble is illustrated in the left figure above. A more exotic configuration in which one bubble is torus-shaped and the other is shaped like a dumbbell is illustrated at right (illustrations courtesy of J. M. Sullivan).

In the plane, the analog of the double bubble consists of three circular arcs meeting in two points. It has been proved that the configuration of arcs meeting at equal 120° ANGLES) has the minimum PERIMETER for enclosing two equal areas (Alfaro *et al.* 1993, Morgan 1995).

It had been conjectured that two *equal* partial SPHERES sharing a boundary of a flat disk separate two volumes of air using a total SURFACE AREA that is less than any other boundary. This equal-volume case was proved by Hass *et al.* (1995), who reduced the problem to a set of 200,260 integrals which they carried out on an ordinary PC. Frank Morgan, Michael Hutchings, Manuel Ritoré, and Antonio Ros finally proved the conjecture for arbitrary double bubbles in early 2000. In this case of two unequal partial spheres, Morgan *et al.* showed that the separating boundary which minimizes total surface area is a portion of a SPHERE which meets the outer spherical surfaces at DIHEDRAL ANGLES of 120°. Furthermore, the CURVATURE of the partition is simply the difference of the CURVATURES of the two bubbles.

Amazingly, a group of undergraduates has extended the theorem to 4-dimensional double bubbles, as well as certain cases in 5-space and higher dimensions. The corresponding triple bubble conjecture remains open (Cipra 2000).

See also APPLE, BUBBLE, CIRCLE-CIRCLE INTERSECTION, ISOVOLUME PROBLEM, SPHERE-SPHERE INTERSECTION

References

Alfaro, M.; Brock, J.; Foisy, J.; Hodges, N.; and Zimba, J. "The Standard Double Bubble in ℝ² Uniquely Minimized Perimeter." *Pacific J. Math.* **159**, 47–9, 1993.
Almgren, F. J. and Taylor, J. "The Geometry of Soap Films and Soap Bubbles." *Sci. Amer.* **235**, 82–3, 1976.
Campbell, P. J. (Ed.). Reviews. *Math. Mag.* **68**, 321, 1995.
Cipra, B. "Rounding Out Solutions to Three Conjectures." *Science* **287**, 1910–911, 2000.
Haas, J.; Hutchings, M.; and Schlafy, R. "The Double Bubble Conjecture." *Electron. Res. Announc. Amer. Math. Soc.* **1**, 98–02, 1995.
Haas, J. "General Double Bubble Conjecture in ℝ³ Solved." *Focus: The Newsletter of the Math. Assoc. Amer.*, No. 5, pp. 4–, May/June 2000.
Hutchings, M.; Morgan, F.; Ritoré, M.; and Ros, A. "Proof of the Double Bubble Conjecture." http://www.williams.edu/Mathematics/fmorgan/ann.html.
Morgan, F. "The Double Bubble Conjecture." *FOCUS* **15**, 6–, 1995.
Morgan, F. "Double Bubble Conjecture Proved." http://www.maa.org/features/mathchat/math-chat_3_18_00.html.
Peterson, I. "Toil and Trouble over Double Bubbles." *Sci. News* **148**, 101, Aug. 12, 1995.
Ritoré, M. "Proof of the Double Bubble Conjecture Preprint." http://www.ugr.es/~ritore/bubble/bubble.htm.
Sullivan, J. M. "Double Bubble Images." http://www.math.uiuc.edu/~jms/Images/dubble.html.

Double Bubble Conjecture

DOUBLE BUBBLE

Double Cone

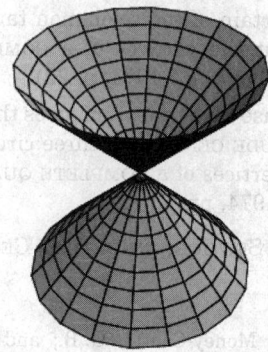

Two CONES placed apex to apex. The double cone is given by algebraic equation

$$\frac{x^2}{c^2} = \frac{x^2 + y^2}{a^2}.$$

See also BICONE, CONE, NAPPE

Double Contact Theorem

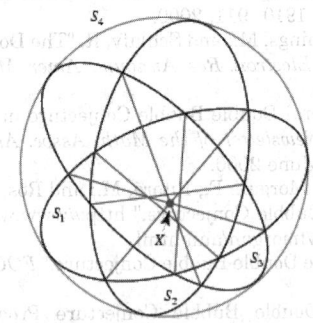

If S_1, S_2, and S_3 are three conics having the property that there is a point X, not on any of the conics, lying on a common chord of each pair of the three conics (with the chords in question being distinct), then there exists a conic S_4 that has a double contact with each of S_1, S_2, and S_3 (Evelyn *et al.* 1974, p. 18).

The converse of the theorem states that if three conics S_1, S_2, and S_3 all have double contact with another S_4 then each two of S_1, S_2, and S_3 have a "distinguished" pair of opposite common chords, the three such pairs of common chords being the pairs of opposite sides of a COMPLETE QUADRANGLE (Evelyn *et al.* 1974, p. 19).

The dual theorems are stated as follows. If three conics are such that, taken by pairs, they have couples of common tangents intersecting at three distinct points on a line (that is not itself a tangent to any of the conics), then (a) the conics have this property in four different ways, and (b) the conics all have double contact with a fourth. And, conversely, if three conics each have double contact with a fourth, then certain of their common tangents intersect by pairs at the vertices of a COMPLETE QUADRILATERAL (Evelyn *et al.* 1974, p. 22).

A degenerate case of the theorem gives the result that the six SIMILITUDE CENTERS of three circles taken by pairs are the vertices of a COMPLETE QUADRILATERAL (Evelyn *et al.* 1974, pp. 21–2).

See also CONIC SECTION, SIMILITUDE CENTER

References

Evelyn, C. J. A.; Money-Coutts, G. B.; and Tyrrell, J. A. "The Double-Contact Theorem." §2.3 in *The Seven Circles Theorem and Other New Theorems.* London: Stacey International, pp. 18–2, 1974.

Double Contraction Relation

A TENSOR t is said to satisfy the double contraction relation when

$$\bar{t}_{ij}^m t_{ij}^n = \delta_{mn}.$$

This equation is satisfied by

$$\hat{t}^0 = \frac{2\hat{z}\hat{z} - \hat{x}\hat{x} - \hat{y}\hat{y}}{\sqrt{6}}$$

$$\hat{t}^{\pm 1} = \mp\frac{1}{2}(\hat{x}\hat{z} + \hat{z}\hat{x}) - \frac{1}{2}i(\hat{y}\hat{z}\text{-}\hat{z}\hat{y})$$

$$\hat{t}^{\pm 2} = \mp\frac{1}{2}(\hat{x}\hat{x} + \hat{y}\hat{y}) - \frac{1}{2}i(\hat{x}\hat{y}\text{-}\hat{y}\hat{x}),$$

where the hat denotes zero trace, symmetric unit TENSORS. These TENSORS are used to define the SPHERICAL HARMONIC TENSOR.

See also SPHERICAL HARMONIC TENSOR, TENSOR

References

Arfken, G. "Alternating Series." *Mathematical Methods for Physicists, 3rd ed.* Orlando, FL: Academic Press, p. 140, 1985.

Double Cusp

DOUBLE POINT

Double Dagger

The symbol ‡ which is not used very commonly in mathematics. The double dagger is also known as the double obelisk or diesis (Bringhurst 1997, p. 275).

See also DAGGER

References

Bringhurst, R. *The Elements of Typographic Style, 2nd ed.* Point Roberts, WA: Hartley and Marks, p. 277, 1997.

Double Dot

A pair of OVERDOTS placed over a symbol, as in $\ddot{x}$, most commonly used to denote a second derivative with respect to time, i.e., $\ddot{x} = d^2x/dt^2$.

See also OVERDOT

Double Exponential Distribution

FISHER-TIPPETT DISTRIBUTION, LAPLACE DISTRIBUTION

Double Exponential Integration

An fairly good NUMERICAL INTEGRATION technique used by *Maple V R4* ® (Waterloo Maple Inc.) for numerical computation of integrals. The method is also available in *Mathematica* using the option Method- > DoubleExponential to NIntegrate.

See also INTEGRAL, INTEGRATION, NUMERICAL INTEGRATION, QUADRATURE

References

Davis, P. J. and Rabinowitz, P. *Methods of Numerical Integration, 2nd ed.* New York: Academic Press, p. 214, 1984.

Di Marco, G.; Favati, P.; Lotti, G.; and Romani, F. "Asymptotic Behaviour of Automatic Quadrature." *J. Complexity* **10**, 296–40, 1994.

Mori, M. *Developments in the Double Exponential Formula for Numerical Integration. Proceedings of the International Congress of Mathematicians, Kyoto 1990.* New York: Springer-Verlag, pp. 1585–594, 1991.

Mori, M. and Ooura, T. "Double Exponential Formulas for Fourier Type Integrals with a Divergent Integrand." In *Contributions in Numerical Mathematics* (Ed. R. P. Agarwal). New York: World Scientific, pp. 301–08, 1993.

Ooura, T. and Mori, M. "The Double Exponential Formula for Oscillatory Functions over the Half Infinite Interval." *J. Comput. Appl. Math.* **38**, 353–60, 1991.

Takahasi, H. and Mori, M. "Double Exponential Formulas for Numerical Integration." *Pub. RIMS Kyoto Univ.* **9**, 721–41, 1974.

Toda, H. and Ono, H. "Some Remarks for Efficient Usage of the Double Exponential Formulas." *Kokyuroku RIMS Kyoto Univ.* **339**, 74–09, 1978.

Double Factorial

The double factorial is a generalization of the usual FACTORIAL $n!$ defined by

$$n!! \equiv \begin{cases} n \cdot (n-2) \ldots 5.3.1 & n \text{ odd} \\ n \cdot (n-2) \ldots 6.4.2 & n \text{ even} \\ 1 & n = -1, 0 \end{cases} \quad (1)$$

Note that $-1!! = 0!! = 1$, by definition (Arfken 1985, p. 547). For $n = 0, 1, 2, \ldots$, the first few values are 1, 1, 2, 3, 8, 15, 48, 105, 384, ... (Sloane's A006882). The double factorial is implemented in *Mathematica* as $n!!$ or Factorial2[n]. The double factorial is a special case of the MULTIFACTORIAL.

The double factorial can be expressed in terms of the GAMMA FUNCTION by

$$\Gamma\left(n + \frac{1}{2}\right) = \frac{(2n-1)!!}{2^n}\sqrt{\pi} \quad (2)$$

(Arfken 1985, p. 548).

There are many identities relating double factorials to FACTORIALS. Since

$$(2n+1)!!2^n n!$$

$$= [(2n+1)(2n-1)\ldots 1][2n][2(n-1)][2(n-2)]\ldots 2(1)$$

$$= [(2n+1)(2n-1)\cdots 1][2n(2n-2)(2n-4)\cdots 2]$$

$$= (2n+1)(2n)(2n-1)(2n-2)(2n-3)(2n-4)\cdots 2(1)$$

$$= (2n+1)!, \quad (3)$$

it follows that $(2n+1)!! = \frac{(2n+1)!}{2^n n!}$. For $n = 0, 1, \ldots$, the first few values are 1, 3, 15, 105, 945, 10395, ... (Sloane's A001147).

Also, since

$$(2n+1)!! = (2n)(2n-2)(2n-4)\cdots 2$$

$$= [(2n)][2(n-1)][2(n-2)]\cdots 2 = 2^n n!, \quad (4)$$

it follows that $(2n)!! = 2^n n!$. For $n = 0, 1, \ldots$, the first

few values are 1, 2, 8, 48, 384, 3840, 46080, ... (Sloane's A000165).

Finally, since

$$(2n-1)!!2^n n!$$

$$= [(2n-1)(2n-3)\cdots 1][(2n)][2(n-1)]$$
$$\times [2(n-2)]\cdots 2(1)$$

$$= (2n-1)(2n-3)\cdots 1[2n(2n-2)(2n-4)\cdots 2]$$

$$= 2n(2n-1)(2n-2)(2n-3)(2n-4)\cdots 2(1)$$

$$= (2n)!, \quad (5)$$

it follows that

$$(2n-1)!! = \frac{(2n)!}{2^n n!} \quad (6)$$

The double factorial can also be extended to negative odd integers using the definition

$$(-2n-1)!! = \frac{(-1)^n}{(2n-1)!!} = \frac{(-1)^n 2^n n!}{(2n)!} \quad (7)$$

for $n = 0, 1, \ldots$ (Arfken 1985, p. 547). Similarly, the double factorial can be extended to complex arguments as

$$z!! = 2^{[1+2x-\cos(\pi x)]/4}\pi^{[\cos(\pi x)-1]/4}\Gamma\left(1 + \frac{1}{2}x\right). \quad (8)$$

For n ODD,

$$\frac{n!}{n!!} = \frac{n(n-1)(n-2)\cdots(1)}{n(n-2)(n-4)\cdots(1)}$$

$$= (n-1)(n-3)\cdots(1) = (n-1)!!. \quad (9)$$

For n EVEN,

$$\frac{n!}{n!!} = \frac{n(n-1)(n-2)\cdots(2)}{n(n-2)(n-4)\cdots(2)}$$

$$(n-1)(n-3)\cdots(2) = (n-1)!!. \quad (10)$$

Therefore, for any n,

$$\frac{n!}{n!!} = (n-1)!! \quad (11)$$

$$n! = n!!(n-1)!!. \quad (12)$$

A closed-form sum due to Ramanujan is given by

$$\sum_{n=0}^{\infty}(-1)^n\left[\frac{(2n-1)!!}{(2n)!!}\right]^3 = \left[\frac{\Gamma\left(\frac{9}{8}\right)}{\Gamma\left(\frac{5}{4}\right)\Gamma\left(\frac{7}{8}\right)}\right]^2 \quad (13)$$

(Hardy 1999, p. 106). Whipple (1926) gives a generalization of this sum (Hardy 1999, pp. 111–12).

See also FACTORIAL, GAMMA FUNCTION, MULTIFACTORIAL

References

Arfken, G. *Mathematical Methods for Physicists, 3rd ed.* Orlando, FL: Academic Press, pp. 544–45 and 547–48, 1985.

Sloane, N. J. A. Sequences A000165/M1878, A001147/M3002, and A006882/M0876 in "An On-Line Version of the Encyclopedia of Integer Sequences." http://www.research.att.com/~njas/sequences/eisonline.html.

Whipple, F. J. W. "On Well-Poised Series, Generalised Hypergeometric Series Having Parameters in Pairs, Each Pair with the Same Sum." *Proc. London Math. Soc.* **24**, 247–63, 1926.

Double Folium

Bifolium

Double Gamma Function

Barnes G-Function, Digamma Function

Double Integral

Multiple Integral

Double Mersenne Number

A number of the form

$$M_{M_n} = 2(2^n - 1) - 1,$$

where M_n is a Mersenne number (T. Forbes). The following table gives known factors of these numbers.

n	factors	reference
2	prime	
3	prime	
5	prime	
7	prime	
13	338193759479	Wilfrid Keller (1976)
17	231733529	Raphael Robinson (1957)
19	62914441	Raphael Robinson (1957)
31	295257526626031	Guy Haworth (1983)

See also Mersenne Number, Mersenne Prime

Double Normal

A chord which is a normal at each end. A centrosymmetric set $K \subset \mathbb{R}^d$ has d double normals through the center (Croft *et al.* 1991). For a curve of constant width, all normals are double normals.

See also Centrosymmetric Set

References

Croft, H. T.; Falconer, K. J.; and Guy, R. K. *Unsolved Problems in Geometry.* New York: Springer-Verlag, p. 15, 1991.

Kuiper, N. H. "Double Normals of Convex Bodies." *Israel J. Math.* **2**, 71–0, 1964.

Double Obelisk

Double Dagger

Double Overdot

Double Dot

Double Point

A point traced out twice as a closed curve is traversed. The maximum number of double points for a non-degenerate quartic curve is three. An ordinary double point is called a node.

Arnold (1994) gives pictures of spherical and plane curves with up to five double points, as well as other curves.

See also Biplanar Double Point, Conic Double Point, Crunode, Cusp, Elliptic Cone Point, Gauss's Double Point Theorem, Node (Algebraic Curve), Ordinary Double Point, Quadruple Point, Rational Double Point, Spinode, Tacnode, Triple Point, Uniplanar Double Point

References

Aicardi, F. Appendix to "Plane Curves, Their Invariants, Perestroikas, and Classifications." In *Singularities & Bifurcations* (Ed. V. I. Arnold). Providence, RI: Amer. Math. Soc., pp. 80–1, 1994.

Fischer, G. (Ed.). *Mathematical Models from the Collections of Universities and Museums.* Braunschweig, Germany: Vieweg, pp. 12–3, 1986.

Double Prime

A symbol used to distinguish a third quantity x'' ("x double prime") from two other related quantities x and x' ("x prime"). Double primes are most commonly used to denote transformed coordinates, conjugate points, and derivatives. A double prime is also used to denote the number of arc seconds in an angle measure, or the number of inches in a length.

See also Prime

References

Bringhurst, R. *The Elements of Typographic Style, 2nd ed.* Point Roberts, WA: Hartley and Marks, p. 277, 1997.

Double Series

A SERIES having terms depending on two indices,

$$\sum_{ij} a_{ij}.$$

Identities involving double sums include the following:

$$\sum_{p=0}^{\infty}\sum_{q=0}^{p} a_{q,p-q} = \sum_{m=0}^{\infty}\sum_{n=0}^{\infty} a_{n,m} = \sum_{r=0}^{\infty}\sum_{s=0}^{\lfloor r/2\rfloor} a_{s,r-2s}, \qquad (1)$$

where

$$\lfloor r/2 \rfloor = \begin{cases} \dfrac{1}{2}r & r \text{ even} \\[2mm] \dfrac{1}{2}(r-1) & r \text{ odd} \end{cases} \qquad (2)$$

is the FLOOR FUNCTION, and

$$\sum_{i=1}^{\infty}\sum_{j=1}^{p} x_i x_j = n^2 \langle x^2 \rangle. \qquad (3)$$

Consider the series

$$S(a,b,c;s) = \sum_{(m,n)\neq(0,0)} \left(am^2 + bmn + cn^2\right)^{-s} \qquad (4)$$

over binary QUADRATIC FORMS. If S can be decomposed into a linear sum of products of DIRICHLET L-SERIES, it is said to be solvable. The related sums

$$S_1(a,b,c;s) = \sum_{(m,n)\neq(0,0)} (-1)^m \left(am^2 + bmn + cn^2\right)^{-s} \qquad (5)$$

$$S_2(a,b,c;s) = \sum_{(m,n)\neq(0,0)} (-1)^n \left(am^2 + bmn + cn^2\right)^{-s} \qquad (6)$$

$$S_{1,2}(a,b,c;s)$$
$$= \sum_{(m,n)\neq(0,0)} (-1)^{m+n} \left(am^2 + bmn + cn^2\right)^{-s} \qquad (7)$$

can also be defined, which gives rise to such impressive FORMULAS as

$$S_1(1,0,58;1) = -\frac{\pi \ln\left(27 + 5\sqrt{29}\right)}{\sqrt{58}} \qquad (8)$$

(Glasser and Zucker 1976b). A complete table of the principal solutions of all solvable $S(a,b,c;s)$ is given in Glasser and Zucker (1980, pp. 126–31).

The LATTICE SUM $b_2(2s)$ can be separated into two pieces,

$$b_2(2s) = \sum_{i,j=-\infty}^{\infty} \frac{(-1)^{i+j}}{(i^2+j^2)^s}$$

$$= \sum_{i=1}^{\infty}\sum_{j=1}^{\infty} \frac{(-1)^{i+j}}{(i^2+j^2)^2} + \sum_{i=1}^{\infty}\sum_{j=-1}^{-\infty} \frac{(-1)^{i+j}}{(i^2+j^2)^2}$$

$$+ \sum_{i=-1}^{-\infty}\sum_{j=1}^{\infty} \frac{(-1)^{i+j}}{(i^2+j^2)^2} + \sum_{i=-1}^{-\infty}\sum_{j=-1}^{-\infty} \frac{(-1)^{i+j}}{(i^2+j^2)^2}$$

$$+ \sum_{j=-\infty}^{-1} \frac{(-1)^j}{j^{2s}} + \sum_{j=1}^{\infty} \frac{(-1)^j}{j^{2s}} + \sum_{j=-\infty}^{-1} \frac{(-1)^i}{i^{2s}} + \sum_{i=1}^{\infty} \frac{(-1)^i}{i^{2s}}$$

$$= 4\left[\sum_{i,j=1}^{\infty} \frac{(-1)^{i+j}}{(i^2+j^2)^s} + \sum_{i=1}^{\infty} \frac{(-1)^i}{i^{2s}}\right]$$

$$= 4\left[\sum_{i,j=1}^{\infty} \frac{(-1)^{i+j}}{(i^2+j^2)^s} + \eta(2s)\right] \qquad (9)$$

where $\eta(n)$ is the DIRICHLET ETA FUNCTION. Using the analytic form of the LATTICE SUM

$$b_2(s) = -4\beta(s)\eta(s) = 4\left[S_{1,2}(1,0,1;s) - \eta(2s)\right], \qquad (10)$$

where $\beta(s)$ is the DIRICHLET BETA FUNCTION gives the sum

$$S_{1,2}(1,0,1;s) = \sum_{i,j=1}^{\infty} \frac{(-1)^{i+j}}{(i^2+j^2)^s} = \eta(2s) - \eta(s)\beta(s). \qquad (11)$$

Borwein and Borwein (1986, p. 291) show that for $\Re[s] > 1$,

$$\sum_{i,j=-\infty}^{\infty} \frac{1}{(i^2+j^2)^s} = 4\beta(s)\zeta(s) \qquad (12)$$

$$\sum_{i,j=-\infty}^{\infty} \frac{(-1)^j}{(i^2+j^2)^s} = 2^{-s} b_2(2s), \qquad (13)$$

where $\zeta(s)$ is the RIEMANN ZETA FUNCTION, and for appropriate s,

$$\sum_{i,j=-1}^{\infty} \frac{(-1)^{i+j}}{(i+j)^s} = \eta(s) - \eta(s-1) \qquad (14)$$

$$\sum_{i,j=1}^{\infty} \frac{(-1)^{i+j}}{(i+j)^s} = 2^{-s}\zeta(s) \qquad (15)$$

$$\sum_{i,j=1}^{\infty} \frac{1}{(i+j)^s} = \zeta(s-1) - \zeta(s) \qquad (16)$$

$$\sum_{i,j=-\infty}^{\infty} \frac{(-1)^{i+j+1}}{(|i|+|j|)^s} = 4\eta(s-1) \qquad (17)$$

$$\sum_{i,j=-\infty}^{\infty} \frac{1}{(i+j)^s} = 4\zeta(s-1) \qquad (18)$$

$$\sum_{i,j=-\infty}^{\infty} \frac{(-1)^{i+j}}{(2i+j+1)^s} = \frac{1}{2}(1-2^{-s})\eta(s)+\frac{1}{2}\beta(s) \quad (19)$$

(Borwein and Borwein 1986, p. 305).

Another double series reduction is given by

$$\sum_{m,n=-\infty}^{\infty} \frac{F(|2m+2n+1|)}{\cosh[(2n+1)u]\cosh(2nu)}$$

$$= 2\sum_{n=0}^{\infty} \frac{(2n+1)F(2n+1)}{\sinh[(2n+1)u]}, \quad (20)$$

where F denotes any function (Glasser 1974).

See also EULER SUM, LATTICE SUM, MADELUNG CONSTANTS, SERIES, WEIERSTRASS'S DOUBLE SERIES THEOREM

References

Borwein, J. M. and Borwein, P. B. *Pi & the AGM: A Study in Analytic Number Theory and Computational Complexity.* New York: Wiley, 1987.
Glasser, M. L. "Reduction Formulas for Multiple Series." *Math. Comp.* **28**, 265–66, 1974.
Glasser, M. L. and Zucker, I. J. "Lattice Sums." In *Perspectives in Theoretical Chemistry: Advances and Perspectives, Vol. 5* (Ed. H. Eyring). New York: Academic Press, pp. 67–39, 1980.
Hardy, G. H. "On the Convergence of Certain Multiple Series." *Proc. London Math. Soc.* **2**, 24–8, 1904.
Hardy, G. H. "On the Convergence of Certain Multiple Series." *Proc. Cambridge Math. Soc.* **19**, 86–5, 1917.
Jeffreys, H. and Jeffreys, B. S. "Double Series." §1.053 in *Methods of Mathematical Physics, 3rd ed.* Cambridge, England: Cambridge University Press, pp. 16–7, 1988.
Meyer, B. "On the Convergence of Alternating Double Series." *Amer. Math. Monthly* **60**, 402–04, 1953.
Móricz, F. "Some remarks on the notion of regular convergence of multiple series." *Acta Math. Hungar.* **41**, 161–68, 1983.
Wilansky, A. "On the Convergence of Double Series." *Bull. Amer. Math. Soc.* **53**, 793–99, 1947.
Zucker, I. J. and Robertson, M. M. "Some Properties of Dirichlet L-Series." *J. Phys. A: Math. Gen.* **9**, 1207–214, 1976a.
Zucker, I. J. and Robertson, M. M. "A Systematic Approach to the Evaluation of $\Sigma_{(m,n\neq 0,0)}(am^2+bmn+cn^2)^{-s}$." *J. Phys. A: Math. Gen.* **9**, 1215–225, 1976b.

Double Sixes

Two sextuples of SKEW LINES on the general CUBIC SURFACE such that each line of one is SKEW to one LINE in the other set. In all, there are 30 points, with two lines through each point, and 12 lines with five points on each line. Two lines can be placed in the plane of each of the faces of a cube. The double sixes were discovered by Schläfli.

See also BOXCARS, CONFIGURATION, CUBIC SURFACE, SKEW LINES, SOLOMON'S SEAL LINES

References

Fischer, G. (Ed.). *Mathematical Models from the Collections of Universities and Museums.* Braunschweig, Germany: Vieweg, p. 11, 1986.
Wells, D. *The Penguin Dictionary of Curious and Interesting Geometry.* London: Penguin, p. 224, 1991.

Double Sum

DOUBLE SERIES

Double Torus

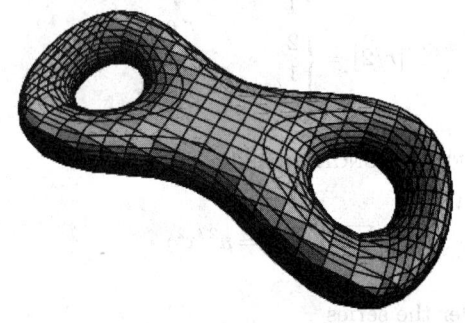

A SPHERE with two HANDLES, i.e., a genus-2 TORUS.

See also HANDLE, TORUS, TRIPLE TORUS

Double-Angle Formulas

Formulas expressing trigonometric functions of an angle $2x$ in terms of functions of an angle x,

$$\sin(2x) = 2\sin x \cos x \quad (1)$$

$$\cos(2x) = \cos^2 x - \sin^2 x \quad (2)$$

$$= 2\cos^2 x - 1 \quad (3)$$

$$= 1 - 2\sin^2 x \quad (4)$$

$$\tan(2x) = \frac{2\tan x}{1-\tan^2 x}. \quad (5)$$

The corresponding hyperbolic function double-angle formulas are

$$\sinh(2x) = 2\sinh x \cosh x \quad (6)$$

$$\cosh(2x) - 2\cosh^2 x - 1 \quad (7)$$

$$\tanh(2x) = \frac{2\tanh x}{1+\tanh^2 x}. \quad (8)$$

See also HALF-ANGLE FORMULAS, HYPERBOLIC FUNCTIONS, MULTIPLE-ANGLE FORMULAS, PROSTHAPHAERESIS FORMULAS, TRIGONOMETRIC ADDITION FORMULAS, TRIGONOMETRIC FUNCTIONS, TRIGONOMETRY

Double-Free Set

A SET of POSITIVE INTEGERS is double-free if, for any integer x, the SET $\{x, 2x\} \nsubseteq S$ (or equivalently, $x \in S$ IMPLIES $2x \notin S$). For example, of the subsets of $\{1, 2, 3\}$, the sets $\varnothing$, $\{1\}$, $\{2\}$, $\{2, 3\}$, $\{1, 3\}$, and $\{3\}$ are double-free, while $\{1, 2\}$ and $\{1, 2, 3\}$ are not.

The number $a(n)$ of double-free subsets of $\{1, 2, \ldots, n\}$ can be computed using $a(1) = 2$ and the RECURRENCE RELATION

$$a(n) = a(n-1) \frac{F_{b(n)+3}}{F_{b(n)+2}}, \tag{1}$$

where F_n is a FIBONACCI NUMBER, 1, 1, 2, 3, 5, 8, ... (Sloane's A000045), and $b(n)$ is the BINARY CARRY SEQUENCE giving the number of trailing 0s is the BINARY representation of n, 0, 1, 0, 2, 0, 1, 3, 0, 1, ... (Sloane's A007814) (C. Bower). For $n = 1, 2, ..., a(n)$ are given by are 2, 3, 6, 10, 20, 30, 60, 96, 192, ... (Sloane's A050291).

Define

$$r(n) = \max\{|s| : S \subset \{1, 2, \ldots, n\} \text{ is double-free}\}, \tag{2}$$

where $|S|$ is the CARDINAL NUMBER of (number of members in) S. Then for $n = 1, 2, ..., r(n)$ is given by 1, 1, 2, 3, 4, 4, 5, 5, 6, 6, 7, 8, 9, 9, 10, ... (Sloane's A050292). An explicit formula for $r(n)$ is given by

$$r(n) = \sum_{i=1}^{n} p(i), \tag{3}$$

where

$$p(i) = \begin{cases} 1 & \text{if } b(i) \text{ is even} \\ 0 & \text{if } b(i) \text{ is odd} \end{cases} \tag{4}$$

where $b(n)$ is defined above and the first few values of $p(i)$ are 1, 0, 1, 1, 1, 0, 1, 0, 1, 0, 1, 1, 1, ... (Sloane's A035263; C. Bower). A simple RECURRENCE RELATION for $r(n)$ is given by

$$f(n) = \left\lceil \frac{1}{2} n \right\rceil + f\left(\left\lfloor \frac{1}{4} n \right\rfloor \right) \tag{5}$$

with $f(0) = 0$ (Wang 1989), where $\lfloor x \rfloor$ is the FLOOR FUNCTION and $\lceil x \rceil$ is the CEILING FUNCTION. An asymptotic formula for $r(n)$ is given by

$$r(n) \sim \frac{2}{3} n + \mathcal{O}(\log_4 n) \tag{6}$$

(Wang 1989).

See also A-SEQUENCE, KLARNER-RADO SEQUENCE, SUM-FREE SET, TRIPLE-FREE SET

References

Finch, S. "Favorite Mathematical Constants." http://www.mathsoft.com/asolve/constant/triple/triple.html.

Sloane, N. J. A. Sequences A000045/M0692, A007814, A035263, A050291 and A050292 in "An On-Line Version of the Encyclopedia of Integer Sequences." http://www.research.att.com/~njas/sequences/eisonline.html.

Wang, E. T. H. "On Double-Free Sets of Integers." *Ars Combin.* **28**, 97–00, 1989.

Doublestruck

A letter of the alphabet drawn with doubled vertical strokes is called doublestruck, or sometimes blackboard bold (because doublestruck characters provide a means of indicating bold font weight when writing on a blackboard). For example, $\mathbb{A}$, $\mathbb{B}$, $\mathbb{C}$, $\mathbb{D}$, $\mathbb{E}$, ... Important SETS in mathematics are commonly denoted using doublestruck characters, e.g., $\mathbb{C}$ for the set of complex numbers and $\mathbb{R}$ for the real numbers.

Doublestruck characters can be encoded using the AMSFonts extended fonts for L^AT_EX using the syntax \mathbb{C}, and typed in *Mathematica* using the syntax \[DoubleStruckC] or \[DoundStruckCapitalC], where C denotes any letter.

Doublet Function

$$y = \delta'(x - a),$$

where $\delta(x)$ is the DELTA FUNCTION.

See also DELTA FUNCTION

References

von Seggern, D. *CRC Standard Curves and Surfaces.* Boca Raton, FL: CRC Press, p. 324, 1993.

Doubly Even Number

An even number N for which $N \equiv 0 \pmod 4$. The first few POSITIVE doubly even numbers are 4, 8, 12, 16, ... (Sloane's A008586).

See also EVEN FUNCTION, ODD NUMBER, SINGLY EVEN NUMBER

References

Sloane, N. J. A. Sequences A008586 in "An On-Line Version of the Encyclopedia of Integer Sequences." http://www.research.att.com/~njas/sequences/eisonline.html.

Doubly Magic Square

BIMAGIC SQUARE

Doubly Periodic Function

A function $f(z)$ is said to be doubly periodic if it has two periods ω_1 and ω_2 whose ratio ω_2/ω_1 is not real.

See also ELLIPTIC FUNCTION, PERIODIC FUNCTION

References

Apostol, T. M. "Doubly Periodic Functions." §1.2 in *Modular Functions and Dirichlet Series in Number Theory, 2nd ed.* New York: Springer-Verlag, pp. 1–, 1997.

Knopp, K. "Doubly-Periodic Functions; in Particular, Elliptic Functions." §9 in *Theory of Functions Parts I and II, Two Volumes Bound as One, Part II.* New York: Dover, pp. 73–2, 1996.

Doubly Ruled Surface

A surface that contains two families of rulings. The only three doubly ruled surfaces are the PLANE, HYPERBOLIC PARABOLOID, and single-sheeted HYPERBOLOID.

See also HYPERBOLIC PARABOLOID, HYPERBOLOID, PLANE, RULED SURFACE

References

Hilbert, D. and Cohn-Vossen, S. *Geometry and the Imagination.* New York: Chelsea, p. 15, 1999.

Doubly Stochastic Matrix

A doubly stochastic matrix is a matrix $\mathbf{A} = (a_{ij})$ such that $a_{ij} \geq 0$ and

$$\sum_i a_{ij} = \sum_j a_{ij} = 1$$

is some field for all i and j. In other words, both the matrix itself and its transpose are STOCHASTIC.

The following tables give the number of distinct doubly stochastic matrices (and distinct nonsingular doubly stochastic matrices) over $\mathbb{Z}_m$ for small m.

m	doubly stochastic $n \times n$ matrices over $\mathbb{Z}_m$
2	1, 2, 16, 512, ...
3	1, 3, 81, ...
4	1, 4, 256, ...

m	doubly stochastic nonsingular $n \times n$ matrices over $\mathbb{Z}_m$
2	1, 2, 6, 192, ...
3	1, 2, 54, ...
4	1, 4, 192, ...

Horn (1954) proved that if $\mathbf{y} = \mathbf{A}\mathbf{x}$, where $\mathbf{x}$ and $\mathbf{y}$ are complex n-vectors, $\mathbf{A}$ is doubly stochastic, and c_1, c_2, ..., C_n are any complex numbers, then $\Sigma_{i=1}^{n} c_i y_i$ lies in the CONVEX HULL of all the points $\Sigma_{i=1}^{n} c_i x_{ai}$, $a \in R^n$,

where R^n is the set of all permutations of $\{1, ..., n\}$. Sherman (1955) also proved the converse.

Birkhoff (1946) proved that any doubly stochastic $n \times n$ matrix is in the CONVEX HULL of m PERMUTATION MATRICES for $m \leq (n-1)^2 + 1$. There are several proofs and extensions of this result (Dulmage and Halperin 1955, Mendelsohn and Dulmage 1958, Mirsky 1958, Marcus 1960).

See also STOCHASTIC MATRIX

References

Birkhoff, G. "Three Observations on Linear Algebra." *Univ. Nac. Tucumán. Rev. Ser. A* **5**, 147–51, 1946.

Dulmage, L. and Halperin, I. "On a Theorem of Frobenius-König and J. von Neumann's Game of Hide and Seek." *Trans. Roy. Soc. Canada Sect. III* **49**, 23–9, 1955.

Horn, A. "Doubly Stochastic Matrices and the Diagonal of a Rotation Matrix." *Amer. J. Math.* **76**, 620–30, 1954.

Marcus, M. "Some Properties and Applications of Doubly Stochastic Matrices." *Amer. Math. Monthly* **67**, 215–21, 1960.

Mendelsohn, N. S. and Dulmage, A. L. "The Convex Hull of Subpermutation Matrices." *Proc. Amer. Math. Soc.* **9**, 253–54, 1958.

Mirsky, L. "Proofs of Two Theorems on Doubly Stochastic Matrices." *Proc. Amer. Math. Soc.* **9**, 371–74, 1958.

Schreiber, S. "On a Result of S. Sherman Concerning Doubly Stochastic Matrices." *Proc. Amer. Math. Soc.* **9**, 350–53, 1958.

Sherman, S. "A Correction to 'On a Conjecture Concerning Doubly Stochastic Matrices.'" *Proc. Amer. Math. Soc.* **5**, 998–99, 1954.

Sherman, S. "Doubly Stochastic Matrices and Complex Vector Spaces." *Amer. J. Math.* **77**, 245–46, 1955.

Dougall's Formula

For $\Re[a + b - c - d] < -1$ and a and b not integers,

$$\sum_{n=-\infty}^{\infty} \frac{\Gamma(a+n)\Gamma(b+n)}{\Gamma(c+n)\Gamma(d+n)}$$
$$= \frac{\pi^2 \csc(\pi a)\csc(\pi b)\Gamma(c+d-a-b-1)}{\Gamma(c-a)\Gamma(d-a)\Gamma(c-b)\Gamma(d-b)}.$$

See also GAMMA FUNCTION

References

Erdélyi, A.; Magnus, W.; Oberhettinger, F.; and Tricomi, F. G. *Higher Transcendental Functions, Vol. 1.* New York: Krieger, p. 7, 1981.

Dougall's Theorem

$${}_5F_4 \left[\begin{array}{c} \frac{1}{2}n+1, n, -x, -y, -z \\ \frac{1}{2}n, x+n+1, y+n+1, z+n+1 \end{array} \right]$$

$$= \frac{\Gamma(x+n+1)\Gamma(y+n+1)\Gamma(z+n+1)\Gamma(x+y+z+n+1)}{\Gamma(n+1)\Gamma(x+y+n+1)\Gamma(y+z+n+1)\Gamma(x+z+n+1)},$$

where $_5F_4(a,b,c,d,e;f,g,h,i;z)$ is a GENERALIZED HYPERGEOMETRIC FUNCTION and $\Gamma(z)$ is the GAMMA FUNCTION.

Bailey (1935, pp. 25–6) called the DOUGALL-RAMANUJAN IDENTITY "Dougall's theorem."

See also DOUGALL-RAMANUJAN IDENTITY, GENERALIZED HYPERGEOMETRIC FUNCTION

References

Bailey, W. N. *Generalised Hypergeometric Series.* Cambridge, England: Cambridge University Press, pp. 25–7, 1935.

Dougall, J. "On Vandermonde's Theorem and Some More General Expansions." *Proc. Edinburgh Math. Soc.* **25**, 114–32, 1907.

Hardy, G. H. "A Chapter from Ramanujan's Note-Book." *Proc. Cambridge Philos. Soc.* **21**, 492–03, 1923.

Koepf, W. *Hypergeometric Summation: An Algorithmic Approach to Summation and Special Function Identities.* Braunschweig, Germany: Vieweg, p. 84, 1998.

Whipple, F. J. W. "On Well-Poised Series, Generalized Hypergeometric Series Having Parameters in Pairs, Each Pair with the Same Sum." *Proc. London Math. Soc.* **24**, 247–63, 1926.

Dougall-Ramanujan Identity

A hypergeometric identity discovered by Ramanujan around 1910. From Hardy (1999, pp. 13 and 102–03),

$$\sum_{n=0}^{\infty} (-1)^n (s+2n) \frac{s^{(n)} (x+y+z+u+2s+1)^{(n)}}{(x+y+z+u-s)_{(n)}} \prod_{x,y,x,u}$$
$$\times \frac{x_{(n)}}{(x+s+1)^{(n)}}$$
$$= \frac{s}{\Gamma(s+1)\Gamma(x+y+z+u+s+1)} \prod_{x,y,z,u}$$
$$\times \frac{\Gamma(x+s+1)\Gamma(y+z+u+s+1)}{\Gamma(z+u+s+1)}. \tag{1}$$

where

$$a^{(n)} \equiv a(a+1)\cdots(a+n-1) \tag{2}$$

is the RISING FACTORIAL (a.k.a. POCHHAMMER SYMBOL,

$$a_{(n)} \equiv a(a-1)\cdots(a-n+1) \tag{3}$$

is the FALLING FACTORIAL (Hardy 1999, p. 101), $\Gamma(z)$ is a GAMMA FUNCTION, and one of

$$x,y,z,u,-x-y-z-u-2s-1 \tag{4}$$

is a POSITIVE INTEGER.

Equation (1) can also be rewritten as

$$_7F_6 \left[\begin{array}{c} s, 1+\frac{1}{2}s, -x, -y, -z, -u, x-y-z+u+2s+1 \\ \frac{1}{2}s, x+s+1, y+s+1, z+s+1, u+s+1, \\ -x-y-z-u-s \end{array} ; 1 \right]$$
$$= \frac{1}{\Gamma(s+1)\Gamma(x+y+z+u+s+1)} \prod_{x,y,z,u}$$
$$\times \frac{\Gamma(x+s+1)\Gamma(y+z+u+s+1)}{\Gamma(z+u+s+1)}. \tag{5}$$

(Hardy 1999, p. 102). In a more symmetric form, if $n = 2a_1 + 1 = a_2 + a_3 + a_4 + a_5$, $a_6 = 1+a_1/2$, $a_7 = -n$, and $b_i = 1 + a_1 - a_{i+1}$ for $i = 1, 2, ..., 6$, then

$$_7F_6 \left[\begin{array}{c} a_1, a_2, a_3, a_4, a_5, a_6, a_7 \\ b_1, b_2, b_3, b_4, b_5, b_6 \end{array} ; 1 \right]$$
$$= \frac{(a_1+1)_n (a_1-a_2-a_3+1)_n}{(a_1-a_2+1)_n (a_1-a_3+1)_n}$$
$$\times \frac{(a_1-a_2-a_4+1)_n (a_1-a_3-a_4+1)_n}{(a_1-a_4+1)_n (a_1-a_2-a_3-a_4+1)_n}, \tag{6}$$

where $(a)_n$ is the POCHHAMMER SYMBOL (Petkovsek *et al.* 1996).

The identity is a special case of JACKSON'S IDENTITY, and gives DIXON'S THEOREM, SAALSCHÜTZ'S THEOREM, and MORLEY'S FORMULA as special cases.

See also BAILEY'S TRANSFORMATION, DIXON'S THEOREM, DOUGALL'S THEOREM, GENERALIZED HYPERGEOMETRIC FUNCTION, HYPERGEOMETRIC FUNCTION, JACKSON'S IDENTITY, MORLEY'S FORMULA, ROGERS-RAMANUJAN IDENTITIES, SAALSCHÜTZ'S THEOREM

References

Bailey, W. N. "An Elementary Proof of Dougall's Theorem." §5.1 in *Generalised Hypergeometric Series.* Cambridge, England: Cambridge University Press, pp. 25–6 and 34, 1935.

Dixon, A. C. "Summation of a Certain Series." *Proc. London Math. Soc.* **35**, 285–89, 1903.

Dougall, J. "On Vandermonde's Theorem and Some More General Expansions." *Proc. Edinburgh Math. Soc.* **25**, 114–32, 1907.

Hardy, G. H. "A Chapter from Ramanujan's Note-Book." *Proc. Cambridge Philos. Soc.* **21**, 492–03, 1923.

Hardy, G. H. *Ramanujan: Twelve Lectures on Subjects Suggested by His Life and Work,* 3rd ed. New York: Chelsea, 1999.

Petkovsek, M.; Wilf, H. S.; and Zeilberger, D. *A = B.* Wellesley, MA: A. K. Peters, pp. 43, 126–27, and 183–84, 1996.

Doughnut

TORUS

Douglas-Neumann Theorem

If the lines joining corresponding points of two DIRECTLY SIMILAR figures are divided proportionally, then the LOCUS of the points of the division will be a figure DIRECTLY SIMILAR to the given figures.

See also DIRECTLY SIMILAR

References

Eves, H. "Solution to Problem E521." *Amer. Math. Monthly* **50**, 64, 1943.
Musselman, J. R. "Problem E521." *Amer. Math. Monthly* **49**, 335, 1942.

Dovetailing Problem

CUBE DOVETAILING PROBLEM

Dowker Notation

A simple way to describe a knot projection. The advantage of this notation is that it enables a KNOT DIAGRAM to be drawn quickly.

For an oriented ALTERNATING KNOT with n crossings, begin at an arbitrary crossing and label it 1. Now follow the undergoing strand to the next crossing, and denote it 2. Continue around the knot following the same strand until each crossing has been numbered twice. Each crossing will have one even number and one odd number, with the numbers running from 1 to $2n$.

Now write out the ODD NUMBERS 1, 3, ..., $2n-1$ in a row, and underneath write the even crossing number corresponding to each number. The Dowker NOTATION is this bottom row of numbers. When the sequence of even numbers can be broken into two permutations of consecutive sequences (such as $\{4, 6, 2\}\{10, 12, 8\}$), the knot is composite and is not uniquely determined by the Dowker notation. Otherwise, the knot is prime and the NOTATION uniquely defines a single knot (for amphichiral knots) or corresponds to a single knot or its MIRROR IMAGE (for chiral knots).

For general nonalternating knots, the procedure is modified slightly by making the sign of the even numbers POSITIVE if the crossing is on the top strand, and NEGATIVE if it is on the bottom strand.

These data are available for knots, but not for links, from Berkeley's gopher site.

References

Adams, C. C. *The Knot Book: An Elementary Introduction to the Mathematical Theory of Knots.* New York: W. H. Freeman, pp. 35–0, 1994.
Dowker, C. H. and Thistlethwaite, M. B. "Classification of Knot Projections." *Topol. Appl.* **16**, 19–1, 1983.
Hoste, J.; Thistlethwaite, M.; and Weeks, J. "The First 1,701,936 Knots." *Math. Intell.* **20**, 33–8, Fall 1998.
Thistlethwaite, M. B. "Knot Tabulations and Related Topics." In *Aspects of Topology in Memory of Hugh Dowker 1912–982* (Ed. I. M. James and E. H. Kronheimer). Cambridge, England: Cambridge University Press, pp. 2–6, 1985.

Down Arrow Notation

An inverse of the up ARROW NOTATION defined by

$$e \downarrow n = \ln n$$

$$e \downarrow\downarrow n = \ln^* n$$

$$e \downarrow\downarrow\downarrow n = \ln^{**} n,$$

where $\ln^* n$ is the number of times the NATURAL LOGARITHM must be iterated to obtain a value $\leq e$.

See also ARROW NOTATION

References

Vardi, I. *Computational Recreations in Mathematica.* Redwood City, CA: Addison-Wesley, pp. 12 and 231–32, 1991.

Dozen

12.

See also BAKER'S DOZEN, DUODECIMAL, GROSS

Dragon Curve

Nonintersecting curves which can be iterated to yield more and more sinuosity. They can be constructed by taking a path around a set of dots, representing a left turn by 1 and a right turn by 0. The first-order curve is then denoted 1. For higher order curves, add a 1 to the end, then copy the string of digits preceding it to the end but switching its center digit. For example, the second-order curve is generated as follows: (1)1 → (1)1(0) → 110, and the third as: (110)1 → (110)1(100) → 1101100. Continuing gives 110110011100100... (Sloane's A014577). The OCTAL representation sequence is 1, 6, 154, 66344, ...(Sloane's A003460). The dragon curves of orders 1 to 9 are illustrated below.

This procedure is equivalent to drawing a RIGHT ANGLE and subsequently replacing each RIGHT ANGLE with another smaller RIGHT ANGLE (Gardner 1978). In fact, the dragon curve can be written as a LINDEN-MAYER SYSTEM with initial string "FX", STRING REWRITING rules "X" → "X+YF+", "Y" → "−FX−Y", and angle 90°.

See also LINDENMAYER SYSTEM, PEANO CURVE

References

Bulaevsky, J. "The Dragon Curve or Jurassic Park Fractal." http://www.best.com/~ejad/java/fractals/jurasic.shtml.
Dickau, R. M. "Two-Dimensional L-Systems." http://forum.s-warthmore.edu/advanced/robertd/lsys2d.html.
Dixon, R. *Mathographics.* New York: Dover, pp. 180–81, 1991.
Dubrovsky, V. "Nesting Puzzles, Part I: Moving Oriental Towers." *Quantum* **6**, 53–7 (Jan.) and 49–1 (Feb.), 1996.
Dubrovsky, V. "Nesting Puzzles, Part II: Chinese Rings Produce a Chinese Monster." *Quantum* **6**, 61–5 (Mar.) and 58–9 (Apr.), 1996.
Gardner, M. *Mathematical Magic Show: More Puzzles, Games, Diversions, Illusions and Other Mathematical Sleight-of-Mind from Scientific American.* New York: Vintage, pp. 207–09 and 215–20, 1978.
Lauwerier, H. *Fractals: Endlessly Repeated Geometric Figures.* Princeton, NJ: Princeton University Press, pp. 48–3, 1991.
Peitgen, H.-O. and Saupe, D. (Eds.). *The Science of Fractal Images.* New York: Springer-Verlag, p. 284, 1988.
Sloane, N. J. A. Sequences A003460/M4300 and A014577 in "An On-Line Version of the Encyclopedia of Integer Sequences." http://www.research.att.com/~njas/sequences/eisonline.html.
Vasilyev, N. and Gutenmacher, V. "Dragon Curves." *Quantum* **6**, 5–0, 1995.
Weisstein, E. W. "Fractals." MATHEMATICA NOTEBOOK FRACTAL.M.
Wells, D. *The Penguin Dictionary of Curious and Interesting Geometry.* London: Penguin, p. 59, 1991.

Dragon Fractal
DOUADY'S RABBIT FRACTAL

Draughts
CHECKERS

Draw
The ending of a GAME in which neither of two players wins, sometimes also called a "tie." A GAME in which no draw is possible is called a CATEGORICAL GAME.

See also CATEGORICAL GAME, GAME, UNFAIR GAME

Drinfel'd-Sokolov-Wilson Equation
The system of PARTIAL DIFFERENTIAL EQUATIONS

$$u_t = 3ww_x$$

$$w_t = 2w_{xxx} + 2uw_x + u_xw.$$

References

Hirota, R.; Grammaticos, B.; and Ramani, A. "Soliton Structure of the Drinfel'd-Sokolov-Wilson Equation." *J. Math. Phys.* **27**, 1499–505, 1986.
Zwillinger, D. *Handbook of Differential Equations*, 3rd ed. Boston, MA: Academic Press, p. 138, 1997.

Drinfeld Module
See also MODULE

References

Gekeler, E.-U.; van der Put, M.; Reversat, M.; and van Geel, J. (Eds.). *Proceedings of the Workshop on Drinfeld Modules, Modular Schemes and Applications: Alden-Biesen, 9–4 September 1996.* Singapore: World Scientific, 1997.

Drinfeld's Symmetric Space
A set of points which do not lie on any of a certain class of HYPERPLANES.

References

Teitelbaum, J. "The Geometry of *p*-adic Symmetric Spaces." *Not. Amer. Math. Soc.* **42**, 1120–126, 1995.

Droz-Farny Circles

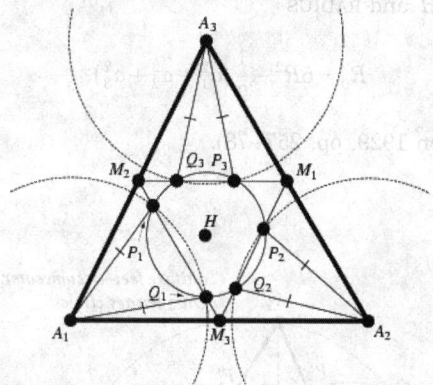

The following amazing property of a triangle, first given by Steiner and then proved by Droz-Farny (1901), is related to the so-called Droz-Farny circles. Draw a CIRCLE with center at the ORTHOCENTER H which cuts the lines M_2M_3, M_3M_1, and M_1M_2 (where M_i are the MIDPOINTS of their respective sides) at P_1, Q_1; P_2, Q_2; and P_3, Q_3 respectively, then the line segments $A_iP_i = A_iQ_i$ are all equal:

$$A_1P_1 = A_2P_2 = A_3P_3 = A_1Q_1 = A_2Q_2 = A_3Q_3.$$

Conversely, if equal CIRCLES are drawn about the VERTICES of a TRIANGLE (dashed circles in the above figure), they cut the lines joining the MIDPOINTS of the corresponding sides in six points P_1, Q_1, P_2, Q_2, P_3, and Q_3, which lie on a CIRCLE whose center is the ORTHOCENTER. If r is the RADIUS of the equal CIRCLES centered on the vertices A_1, A_2, and A_3, and R_0 is the

RADIUS of the CIRCLE about H, then

$$R_0^2 = 4R^2 + r^2 - \frac{1}{2}\left(a_1^2 + a_2^2 + a_3^2\right)$$

(Johnson 1929, p. 257).

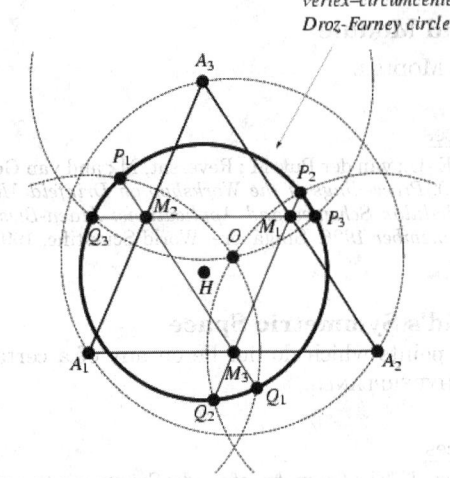

vertex–circumcenter Droz-Farney circle

In the special case that r is taken as the CIRCUMRADIUS of the original triangle, then a circle D_1, known as the Droz-Farny circle (in particular, the "vertex-circumcenter Droz-Farny circle"), is obtained, having center H and RADIUS

$$R_0^2 = 5R^2 - \frac{1}{2}\left(a_1^2 + a_2^2 + a_3^2\right)$$

(Johnson 1929, pp. 257–78).

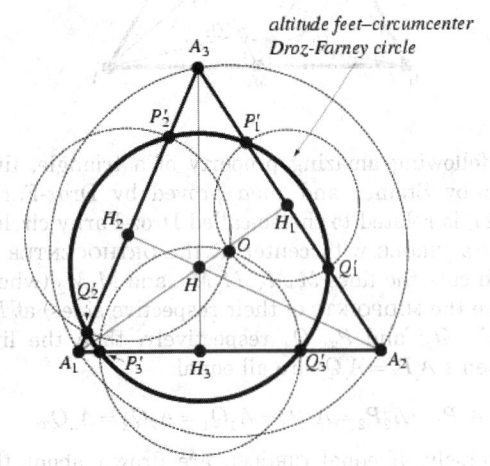

altitude feet–circumcenter Droz-Farney circle

The "altitude feet-circumcenter" Droz-Farny circle D_1' is obtained by drawing circles with centers at the feet of the altitudes and passing through the CIRCUMCEN-

TER. These circles cut the corresponding sides in six concyclic points, having *the same center H* and *the same radius R_0* as the vertex-circumcenter Droz-Farny circle. This is the first Droz-Farny circle.

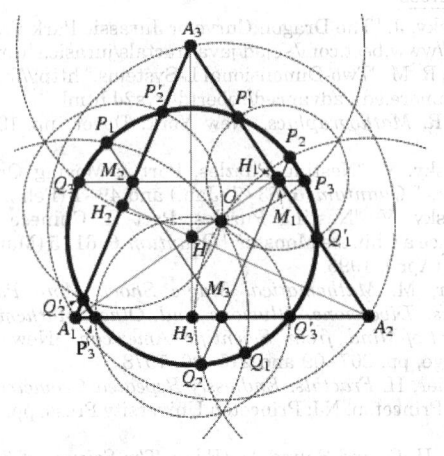

The first Droz-Farny circle D_1 therefore passes through 12 notable points, two on each of the sides and two on each of the lines joining midpoints of the sides, as illustrated in the rather busy figure above.

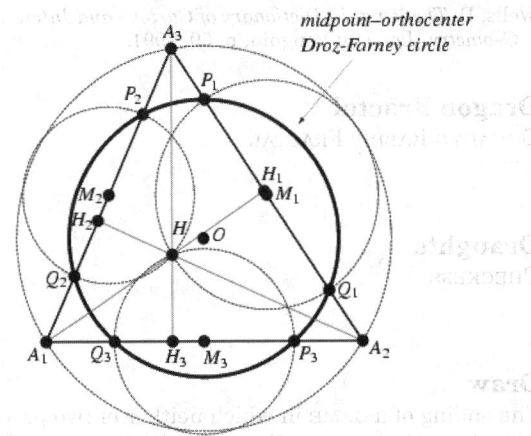

midpoint–orthocenter Droz-Farney circle

The circles about the midpoints of the sides and passing though H cut the sides in six points lying on another circle D_2. This is the second Droz-Farny circle, which has RADIUS equal to that of D_1, but whose center is the CIRCUMCENTER O instead of the ORTHOCENTER H.

There is a beautiful generalization of the Droz-Farny circles motivated by the observation that the ORTHOCENTER and CIRCUMCENTER are ISOGONAL CONJU-

GATES. Let P and Q be any pair of ISOGONAL CONJUGATES of a triangle ΔABC, and let D, E, and F be the feet of the perpendiculars to the sides from one of the points (say, P), and let circles with centers D, E, and F be drawn to pass through Q. Then the three pairs of points on the sides of ΔABC which are determined by these circles always lie on a circle with center P, and the two circles constructed in this way are congruent (Honsberger 1995).

See also CIRCUMCENTER, ORTHOCENTER

References

Droz-Farny. "Notes sur un théorème de Steiner." *Mathesis* **21**, 22–4, 1901.

Goormaghtigh, R. "Droz-Farny's Theorem." *Scripta Math.* **16**, 268–71, 1950.

Honsberger, R. "The Droz-Farny Circles." §7.4 (ix) in *Episodes in Nineteenth and Twentieth Century Euclidean Geometry.* Washington, DC: Math. Assoc. Amer., pp. 69–2, 1995.

Johnson, R. A. *Modern Geometry: An Elementary Treatise on the Geometry of the Triangle and the Circle.* Boston, MA: Houghton Mifflin, pp. 256–58, 1929.

Droz-Farny Theorem

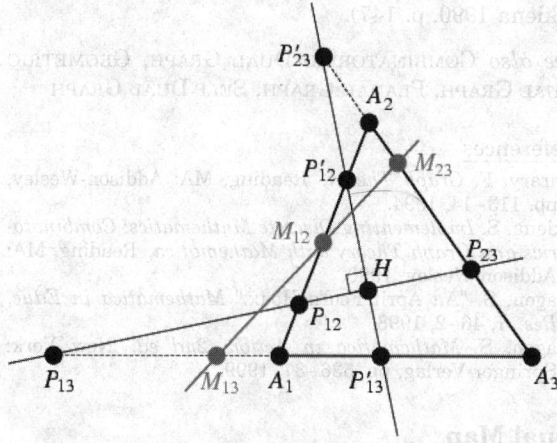

If two perpendicular lines are drawn through the ORTHOCENTER H of any triangle, these lines intercept each side (or its extension) in two points (labeled P_{12}, P'_{12}, P_{13}, P'_{13}, P_{23}, P'_{23}). Then the MIDPOINTS M_{12}, M_{12}, and M_{23} of these three segments are COLLINEAR.

See also COLLINEAR, MIDPOINT

References

Honsberger, R. *Episodes in Nineteenth and Twentieth Century Euclidean Geometry.* Washington, DC: Math. Assoc. Amer., p. 73, 1995.

Drum

ISOSPECTRAL MANIFOLDS

JACOBI ELLIPTIC FUNCTIONS

D-Statistic

KOLMOGOROV-SMIRNOV TEST

D-Triangle

Let the CIRCLES c_2 and c'_3 used in the construction of the BROCARD POINTS which are tangent to A_2A_3 at A_2 and A_3, respectively, meet again at D_1. The points $D_1D_2D_3$ then define the D-triangle. The VERTICES of the D-triangle lie on the respective APOLLONIUS CIRCLES.

See also APOLLONIUS CIRCLES, BROCARD POINTS

References

Johnson, R. A. *Modern Geometry: An Elementary Treatise on the Geometry of the Triangle and the Circle.* Boston, MA: Houghton Mifflin, pp. 284–85, 296 and 307, 1929.

Du Bois Reymond Constants

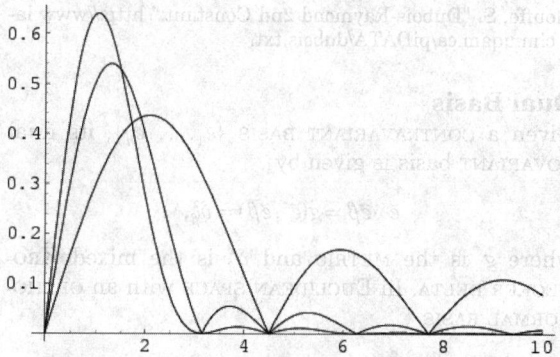

The constants C_n defined by

$$C_n \equiv \int_0^\infty \left| \frac{d}{dt}\left(\frac{\sin t}{t}\right)^n \right| dt - 1. \qquad (1)$$

These constants can also be written as

$$C_n = 2\sum_{k=1}^\infty \left(1 + x_k^2\right)^{-n/2}, \qquad (2)$$

where x_k is the kth root of

$$t = \tan t. \qquad (3)$$

C_1 diverges, and the first few constant are numerically given by

$$C_2 \approx 0.1945280494 \qquad (4)$$

$$C_3 \approx 0.028254 \qquad (5)$$

$$C_4 \approx 0.005240704678. \qquad (6)$$

Rather surprisingly, the even-ordered du Bois Reymond constants (and, in particular, C_2; Le Lionnais 1983) can be computed analytically as polynomials in e^2,

$$C_2 = \frac{1}{2}\left(e^2 - 7\right) \tag{7}$$

$$C_4 = \frac{1}{8}\left(e^4 - 4e^2 - 25\right) \tag{8}$$

$$C_6 = \frac{1}{32}\left(e^6 - 6e^4 + 3e^2 - 98\right) \tag{9}$$

These have the explicit formula

$$C_n = -3 - 2\operatorname*{Res}_{x=i}\left(\frac{x^2}{(1+x^2)^n(\tan x - x)}\right), \tag{10}$$

where n is even and Res denotes a RESIDUE (V. Adamchik).

See also INFINITE SERIES

References

Le Lionnais, F. *Les nombres remarquables.* Paris: Hermann, p. 23, 1983.
Plouffe, S. "Dubois-Raymond 2nd Constant." http://www.lacim.uqam.ca/piDATA/dubois.txt.

Dual Basis

Given a CONTRAVARIANT BASIS $\{\vec{e}_1, \ldots, \vec{e}_n\}$, its dual COVARIANT basis is given by

$$\vec{e}^\alpha \cdot \vec{e}\beta = g(\vec{e}^\alpha, \vec{e}\beta) = \delta^\alpha_\beta,$$

where g is the METRIC and δ^α_β is the mixed KRONECKER DELTA. In EUCLIDEAN SPACE with an ORTHONORMAL BASIS,

$$\vec{e}^j = \vec{e}_j,$$

so the BASIS and its dual are the same.

See also DUAL SPACE

Dual Bivector

A dual BIVECTOR is defined by

$$\tilde{X}_{ab} \equiv \frac{1}{2}\epsilon_{abcd}X^{cd},$$

and a self-dual BIVECTOR by

$$X^*_{ab} \equiv X_{ab} + i\tilde{X}_{ab}.$$

See also BIVECTOR

Dual Bundle

Given a VECTOR BUNDLE $\pi: E \to M$, its dual bundle is a VECTOR BUNDLE $\pi^*: E^* \to M$. The FIBER BUNDLE of

E^* over a point $p \in M$ is the DUAL VECTOR SPACE to the fiber of E.

See also DUAL SPACE, VECTOR BUNDLE

Dual Graph

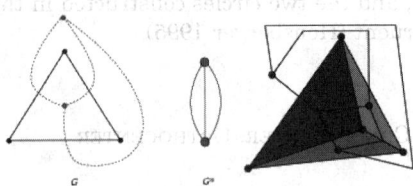

Given a PLANAR GRAPH G, a GEOMETRIC DUAL GRAPH and COMBINATORIAL DUAL GRAPH can be defined. Whitney showed that these are equivalent (Harary 1994), so that one make speak of "the" dual graph G^*. The illustration above shows the process of constructing a GEOMETRIC DUAL GRAPH.

The dual graph G^* of a POLYHEDRAL GRAPH G has VERTICES each of which corresponds to a face of G and each of whose faces corresponds to a VERTEX of G. Two nodes in G^* are connected by an EDGE if the corresponding faces in G have a boundary EDGE in common.

The dual graph of a WHEEL GRAPH is itself a wheel (Skiena 1990, p. 147).

See also COMBINATORIAL DUAL GRAPH, GEOMETRIC DUAL GRAPH, PLANAR GRAPH, SELF-DUAL GRAPH

References

Harary, F. *Graph Theory.* Reading, MA: Addison-Wesley, pp. 113–14, 1994.
Skiena, S. *Implementing Discrete Mathematics: Combinatorics and Graph Theory with Mathematica.* Reading, MA: Addison-Wesley, 1990.
Wagon, S. "An April Fool's Hoax." *Mathematica in Educ. Res.* **7**, 46–2, 1998.
Wagon, S. *Mathematica in Action, 2nd ed.* New York: Springer-Verlag, pp. 536–37, 1999.

Dual Map

PULLBACK MAP

Dual Number

A number $x + \epsilon y$, where $x, y \in \mathbb{R}$ and ε is a UNIT with the property that $\epsilon^2 = 0$.

References

Brand, L. *Vector and Tensor Analysis.* New York: Wiley, 1947.

Dual Polyhedron

By the DUALITY PRINCIPLE, for every POLYHEDRON, there exists another POLYHEDRON in which faces and VERTICES occupy complementary locations. This POLY-

HEDRON is known as the dual, or RECIPROCAL. The process of taking the dual is also called RECIPROCATION, or polar reciprocation. Brückner (1900) was among the first to give a precise definition of duality (Wenninger 1983, p. 1).

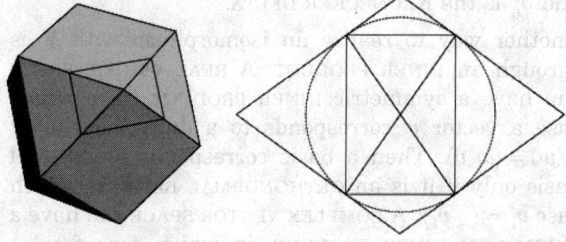

The dual of a PLATONIC SOLID or ARCHIMEDEAN SOLID can be computed by connecting the midpoints of the sides surrounding each VERTEX (the VERTEX FIGURE; left figure), and constructing the corresponding TANGENTIAL POLYGON (tangent to the CIRCUMCIRCLE of the VERTEX FIGURE; right figure.) This is sometimes called the Dorman-Luke construction (Wenninger 1983, p. 30).

The dual polyhedron of a PLATONIC SOLID or ARCHIMEDEAN SOLID can be also drawn by constructing EDGES tangent to the MIDSPHERE (sometimes also known as the reciprocating sphere or intersphere) which are PERPENDICULAR to the original EDGES. Furthermore, let r be the INRADIUS of the dual polyhedron (corresponding to the INSPHERE, which touches the faces of the dual solid), ρ be the MIDRADIUS of both the polyhedron and its dual (corresponding to the MIDSPHERE, which touches the edges of both the polyhedron and its duals), and R the CIRCUMRADIUS (corresponding to the CIRCUMSPHERE of the solid which touches the vertices of the solid). Since the CIRCUMSPHERE and INSPHERE are dual to each other, r, R, and ρ obey the polar relationship

$$Rr = \rho^2$$

(Cundy and Rollett 1989, Table II following p. 144).

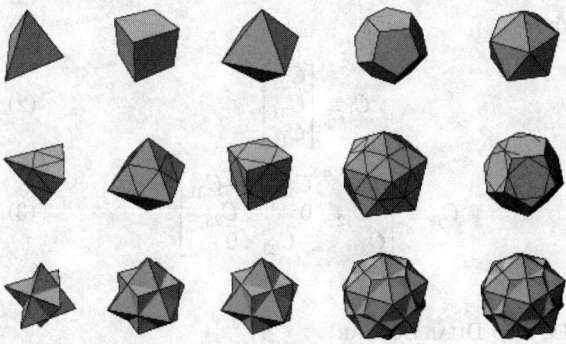

The process of forming duals is illustrated above for the PLATONIC SOLIDS. The top row shows the original solid, the middle row shows the vertex figures of the original solid as lines superposed on the tangential

polygons forming the dual faces. The POLYHEDRON COMPOUNDS consisting of a POLYHEDRON and its dual are generally very attractive, and are illustrated in the bottom row.

For an ARCHIMEDEAN SOLID with v vertices, f faces, and e edges, the dual polyhedron has f vertices, v faces, and e edges. The dual of an isogonal solid (i.e., all vertices are alike) is isohedral (i.e., all faces are alike) (Wenninger 1983, p. 5).

The dual of any non-convex UNIFORM POLYHEDRON is a stellated form of the CONVEX HULL of the given polyhedron (Wenninger 1983, pp. 3– and 40).

The following table gives a list of the duals of the PLATONIC SOLIDS and KEPLER-POINSOT SOLIDS, together with the names of the POLYHEDRON-dual COMPOUNDS. (Note that the duals of the PLATONIC SOLIDS are themselves PLATONIC SOLIDS, so no new solids are formed by taking the duals of the Platonic solids.)

Duals can also be taken of other polyhedrons, including the Archimedean solids and Uniform solids. The names of some solids and their duals are given in the table below.

POLYHEDRON	Dual	POLYHEDRON COMPOUND
CSÁSZÁR POLYHEDRON	SZILASSI POLYHEDRON	
CUBE	OCTAHEDRON	CUBE-OCTAHEDRON COMPOUND
CUBOCTAHEDRON	RHOMBIC DODECAHEDRON	
DODECAHEDRON	ICOSAHEDRON	DODECAHEDRON-ICOSAHEDRON COMPOUND
GREAT DODECAHEDRON	SMALL STELLATED DODECAHEDRON	GREAT DODECAHEDRON-SMALL STELLATED DODECAHEDRON COMPOUND
GREAT ICOSAHEDRON	GREAT STELLATED DODECAHEDRON	GREAT ICOSAHEDRON-GREAT STELLATED DODECAHEDRON COMPOUND
GREAT STELLATED DODECAHEDRON	GREAT ICOSAHEDRON	GREAT ICOSAHEDRON-GREAT STELLATED DODECAHEDRON COMPOUND
ICOSAHEDRON	DODECAHEDRON	DODECAHEDRON-ICOSAHEDRON COMPOUND
OCTAHEDRON	CUBE	CUBE-OCTAHEDRON COMPOUND

SMALL STEL-LATED DODECA-HEDRON	GREAT DODECA-HEDRON	GREAT DODECA-HEDRON-SMALL STELLATED DODE-CAHEDRON COM-POUND
SZILASSI POLYHE-DRON	CSÁSZÁR POLY-HEDRON	
TETRAHEDRON	TETRAHEDRON	STELLA OCTANGU-LA

When a POLYCHORON with SCHLÄFLI SYMBOL $\{p,q,r\}$ and its dual are in reciprocal positions, the vertices of $\{p,q,r\}$'s bounding polyhedra can be found by selecting those vertices of $\{p,q,r\}$ closest to each vertex of $\{r,q,p\}$.

See also ARCHIMEDEAN SOLID, DUALITY PRINCIPLE, PLATONIC SOLID, POLYHEDRON, POLYHEDRON COMPOUND, RECIPROCATING SPHERE, RECIPROCATION, SELF-DUAL POLYHEDRON, UNIFORM POLYHEDRON, ZONOHEDRON

References

Brückner, M. *Vielecke under Vielflache.* Leipzig, Germany: Teubner, 1900.
Cundy, H. and Rollett, A. *Mathematical Models, 3rd ed.* Stradbroke, England: Tarquin Pub., 1989.
Hart, G. "Duality." http://www.georgehart.com/virtual-polyhedra/duality.html.
Weisstein, E. W. "Polyhedron Duals." MATHEMATICA NOTEBOOK DUALS.M.
Wells, D. *The Penguin Dictionary of Curious and Interesting Geometry.* London: Penguin, p. 60, 1991.
Wenninger, M. J. *Dual Models.* Cambridge, England: Cambridge University Press, 1983.

Dual Scalar

Given a third RANK TENSOR,

$$V_{ijk} \equiv \det[\mathbf{A} \quad \mathbf{B} \quad \mathbf{C}],$$

where det is the DETERMINANT, the dual scalar is defined as

$$V \equiv \frac{1}{3!} \epsilon_{ijk} V_{ijk},$$

where ϵ_{ijk} is the LEVI-CIVITA TENSOR.

See also DUAL TENSOR, LEVI-CIVITA TENSOR

Dual Solid

DUAL POLYHEDRON

Dual Space

The dual space to a real VECTOR SPACE V is the VECTOR SPACE of LINEAR FUNCTIONS $f : V \to \mathbb{R}$, and is denoted V^*. In the dual to a COMPLEX VECTOR SPACE, the linear functions take complex values.

In either case, the dual space has the same DIMENSION as V. Given a BASIS $v_1, \ldots, v_n$ for V there exists a DUAL BASIS for V^*, written $v_1^*, \ldots, v_n^*$, where $v_i^*(v_j) = \delta_{ij}$ and δ_{ij} is the KRONECKER DELTA.

Another way to realize an isomorphism with V is through an INNER PRODUCT. A REAL VECTOR SPACE can have a symmetric INNER PRODUCT $\langle,\rangle$ in which case a vector v corresponds to a dual element by $f_v(w) = \langle w, v \rangle$. Then a basis corresponds to its dual basis only if it is an ORTHONORMAL BASIS, in which case $v_i^* = \langle \_, v_i \rangle$. A COMPLEX VECTOR SPACE can have a HERMITIAN INNER PRODUCT, in which case $f_v(w) = \langle w, v \rangle$ is a conjugate-linear isomorphism of V with V^*, i.e., $f_{xv} = \bar{x} f_v$.

Dual spaces can describe many objects in linear algebra. When V and W are finite dimensional vector spaces, an element of the tensor product $V^* \otimes W$, say $\Sigma a_{ij} v_j^* \otimes w_i$, corresponds to the linear transformation $T(v) = \Sigma a_{ij} v_j^*(w) w_i$. That is, $V^* \otimes W \simeq \mathrm{Hom}(V, W)$. For example, the identity transformation is $v_1 \otimes v_1^* + \ldots + v_n \otimes v_n^*$. A BILINEAR FORM on V, such as an inner product, is an element of $V^* \otimes V^*$.

When V is infinite dimensional, care has to be taken of the topology. The dual space of V is the VECTOR SPACE of CONTINUOUS LINEAR FUNCTIONALS on V.

See also BASIS (VECTOR SPACE), BILINEAR FORM, DISTRIBUTION (GENERALIZED FUNCTION), DUAL VECTOR SPACE, LINEAR FUNCTIONAL, MATRIX, SELF-DUAL, VECTOR SPACE

Dual Tensor

Given an antisymmetric second RANK TENSOR C_{ij}, a dual pseudotensor C_i is defined by

$$C_i \equiv \frac{1}{2} \epsilon_{ijk} C_{jk}, \tag{1}$$

where

$$C_i \equiv \begin{bmatrix} C_{23} \\ C_{31} \\ C_{12} \end{bmatrix} \tag{2}$$

$$C_{jk} \equiv \begin{bmatrix} 0 & C_{12} & -C_{31} \\ -C_{12} & 0 & C_{23} \\ C_{31} & -C_{23} & 0 \end{bmatrix}. \tag{3}$$

See also DUAL SCALAR

References

Arfken, G. "Pseudotensors, Dual Tensors." §3.4 in *Mathematical Methods for Physicists, 3rd ed.* Orlando, FL: Academic Press, pp. 128–37, 1985.

Dual Tessellation

The dual of a regular TESSELLATION is formed by taking the center of each polygon as a vertex and joining the centers of adjacent polygons.

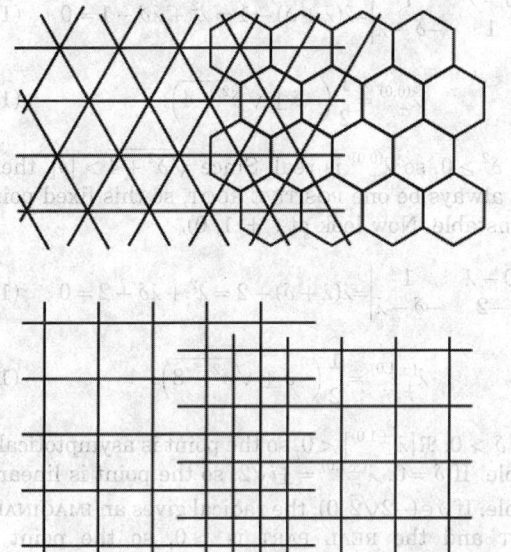

The triangular and hexagonal tessellations are duals of each other, while the square tessellation it its own dual.

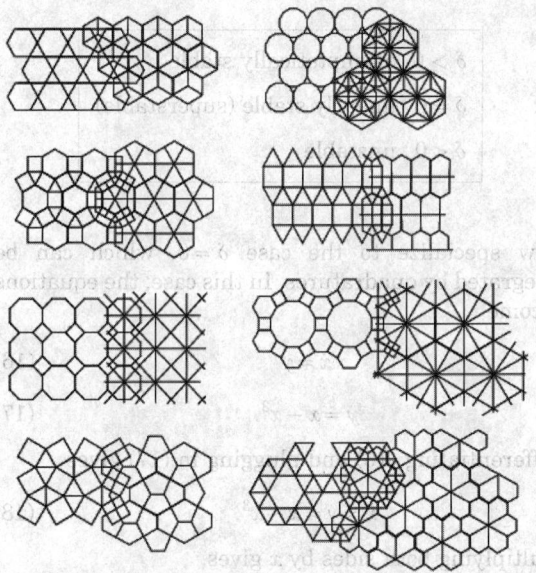

Williams (1979, pp. 37–1) illustrates the dual tessellations of the semiregular tessellations.

See also CAIRO TESSELLATION, TESSELLATION

References

Wells, D. *The Penguin Dictionary of Curious and Interesting Geometry.* London: Penguin, pp. 60–1, 1991.

Williams, R. *The Geometrical Foundation of Natural Structure: A Source Book of Design.* New York: Dover, p. 37, 1979.

Dual Vector Space

Given a VECTOR SPACE X, the dual vector space $X*$ is the set of all bounded LINEAR FUNCTIONALS on X.

See also DUAL SPACE, LINEAR FUNCTIONAL, VECTOR SPACE

Dual Voting

A term in SOCIAL CHOICE THEORY meaning each alternative receives equal weight for a single vote.

See also ANONYMOUS, MONOTONIC VOTING

Duality Principle

All the propositions in PROJECTIVE GEOMETRY occur in dual pairs which have the property that, starting from either proposition of a pair, the other can be immediately inferred by interchanging the parts played by the words "point" and "line." The principle was enunciated by Gergonne (1826; Cremona 1960, p. x). A similar duality exists for RECIPROCATION as first enunciated by Poncelet (1818; Casey 1893; Lachlan 1893; Cremona 1960, p. x).

Example of dual geometric objects include BRIANCHON'S THEOREM and PASCAL'S THEOREM, the 15 PLÜCKER LINES and 15 SALMON POINTS, the 20 CAYLEY LINES and 20 STEINER POINTS, the 60 PASCAL LINES and 60 KIRKMAN POINTS, DUAL POLYHEDRA, and DUAL TESSELLATIONS.

Propositions which are equivalent to their duals are said to be SELF-DUAL.

See also BRIANCHON'S THEOREM, CONSERVATION OF NUMBER PRINCIPLE, DESARGUES' THEOREM, DUAL POLYHEDRON, PAPPUS'S HEXAGON THEOREM, PASCAL'S THEOREM, PERMANENCE OF MATHEMATICAL RELATIONS PRINCIPLE, PROJECTIVE GEOMETRY, RECIPROCAL, RECIPROCATION, SELF-DUAL

References

Casey, J. "Theory of Duality and Reciprocal Polars." Ch. 13 in *A Treatise on the Analytical Geometry of the Point, Line, Circle, and Conic Sections, Containing an Account of Its Most Recent Extensions, with Numerous Examples,* 2nd ed., rev. enl. Dublin: Hodges, Figgis, & Co., pp. 382–92, 1893.
Cremona, L. *Elements of Projective Geometry, 3rd ed.* New York: Dover, 1960.
Durell, C. V. *Modern Geometry: The Straight Line and Circle.* London: Macmillan, p. 78, 1928.
Gergonne, J. D. *Ann. Math.* **16**, 209, 1826.
Graustein, W. C. *Introduction to Higher Geometry.* New York: Macmillan, pp. 26–7 and 41–3, 1930.

Lachlan, R. "The Principle of Duality." §7 and 284–99 in *An Elementary Treatise on Modern Pure Geometry*. London: Macmillian, pp. 3– and 174–82, 1893.

Ogilvy, C. S. *Excursions in Geometry*. New York: Dover, pp. 107–10, 1990.

Poncelet, J.-V. *Ann. Math.* **8**, 201, 1818.

Duality Theorem

Dual pairs of LINEAR PROGRAMS are in "strong duality" if both are possible. The theorem was first conceived by John von Neumann. The first written proof was an Air Force report by George Dantzig, but credit is usually given to Tucker, Kuhn, and Gale.

See also LINEAR PROGRAMMING

Duffing Differential Equation

The most general forced form of the Duffing equation is

$$\ddot{x} + \delta\dot{x} + \left(\beta x^3 \pm \omega_0^2 x\right) = A\,\sin(\omega t + \phi). \tag{1}$$

If there is no forcing, the right side vanishes, leaving

$$\ddot{x} + \delta\dot{x} + \left(\beta x^3 \pm \omega_0^2 x\right) = 0. \tag{2}$$

If $\delta = 0$ and we take the plus sign,

$$\ddot{x} + \omega_0^2 x + \beta x^3 = 0 \tag{3}$$

(Bender and Orszag 1978, p. 547; Zwillinger 1997, p. 122).

This equation can display chaotic behavior. For $\beta > 0$, the equation represents a "hard spring," and for $\beta < 0$, it represents a "soft spring." If $\beta < 0$, the phase portrait curves are closed. Returning to (1), take $\beta = 1$, $\omega_0 = 1$, $A = 0$, and use the minus sign. Then the equation is

$$\ddot{x} + \delta\dot{x} + \left(x^3 - x\right) = 0 \tag{4}$$

(Ott 1993, p. 3). This can be written as a system of first-order ordinary differential equations by writing

$$\dot{x} = y, \tag{5}$$

$$\dot{y} = x - x^3 - \delta y. \tag{6}$$

The fixed points of these differential equations

$$\dot{x} = y = 0, \tag{7}$$

so $y = 0$, and

$$\dot{y} = x - x^3 - \delta y = x\left(1 - x^2\right) - 0 \tag{8}$$

giving $x = 0, \pm 1$. Differentiating,

$$\ddot{x} = \dot{y} = x - x^3 - \delta y \tag{9}$$

$$\ddot{y} = \left(1 - 3x^2\right)\dot{x} - \delta\dot{y} \tag{10}$$

$$\begin{bmatrix} \ddot{x} \\ \ddot{y} \end{bmatrix} = \begin{bmatrix} 0 & 1 \\ 1 - 3x^2 & -\delta \end{bmatrix} \begin{bmatrix} \dot{x} \\ \dot{y} \end{bmatrix}. \tag{11}$$

Examine the stability of the point $(0,0)$:

$$\begin{vmatrix} 0 - \lambda & 1 \\ 1 & -\delta - \lambda \end{vmatrix} = \lambda(\lambda + \delta) - 1 = \lambda^2 + \lambda\delta - 1 = 0 \tag{12}$$

$$\lambda_\pm^{(0,0)} = \frac{1}{2}\left(-\delta \pm \sqrt{\delta^2 + 4}\right). \tag{13}$$

But $\delta^2 \geq 0$, so $\lambda_\pm^{(0,0)}$ is real. Since $\sqrt{\delta^2 + 4} > |\delta|$, there will always be one POSITIVE ROOT, so this fixed point is unstable. Now look at $(\pm 1, 0)$.

$$\begin{vmatrix} 0 - \lambda & 1 \\ -2 & -\delta - \lambda \end{vmatrix} = \lambda(\lambda + \delta) + 2 = \lambda^2 + \lambda\delta + 2 = 0 \tag{14}$$

$$\lambda_\pm^{(\pm 1,0)} = \frac{1}{2}\left(-\delta \pm \sqrt{\delta^2 - 8}\right). \tag{15}$$

For $\delta > 0$, $\Re\left[\lambda_\pm^{(\pm 1,0)}\right] < 0$, so the point is asymptotically stable. If $\delta = 0$, $\lambda_+^{(\pm 1,0)} = \pm i\sqrt{2}$, so the point is linearly stable. If $\delta \in (-2\sqrt{2}, 0)$, the radical gives an IMAGINARY PART and the REAL PART is > 0, so the point is unstable. If $\delta = -2\sqrt{2}$, $\lambda_\pm^{(\pm 1,0)} = \sqrt{2}$, which has a POSITIVE REAL ROOT, so the point is unstable. If $\delta < -2\sqrt{2}$, then $|\delta| < \sqrt{\delta^2 - 8}$, so both ROOTS are POSITIVE and the point is unstable. The following table summarizes these results.

$\delta > 0$	asymptotically stable
$\delta = 0$	linearly stable (superstable)
$\delta < 0$	unstable

Now specialize to the case $\delta = 0$, which can be integrated by quadratures. In this case, the equations become

$$\dot{x} = y \tag{16}$$

$$\dot{y} = x - x^3. \tag{17}$$

Differentiating (16) and plugging in (17) gives

$$\ddot{x} = \dot{y} = x - x^3. \tag{18}$$

Multiplying both sides by $\dot{x}$ gives

$$\ddot{x}\dot{x} - \dot{x}x + \dot{x}x^3 = 0 \tag{19}$$

$$\frac{d}{dt}\left(\frac{1}{2}\dot{x}^2 - \frac{1}{2}x^2 - \frac{1}{4}x^4\right) = 0, \tag{20}$$

so we have an invariant of motion h,

$$h \equiv \frac{1}{2}\dot{x}^2 - \frac{1}{2}x + \frac{1}{4}x^4. \tag{21}$$

Solving for $\dot{x}^2$ gives

$$\dot{x}^2 = \left(\frac{dx}{dt}\right)^2 = 2h + x^2 - \frac{1}{2}x^4, \tag{22}$$

$$\frac{dx}{dt} = \sqrt{2h + x^2 + \frac{1}{2}x^2}, \tag{23}$$

so

$$t = \int dt = \int \frac{dx}{\sqrt{2h + x^2 + \frac{1}{2}x^2}}. \tag{24}$$

Note that the invariant of motion h satisfies

$$\dot{x} = \frac{\partial h}{\partial \dot{x}} = \frac{\partial h}{\partial y} \tag{25}$$

$$\frac{\partial h}{\partial x} = -x + x^3 = -\dot{y}, \tag{26}$$

so the equations of the Duffing oscillator are given by the HAMILTONIAN SYSTEM

$$\begin{cases} \dot{x} = \dfrac{\partial h}{\partial y} \\[2mm] \dot{y} = -\dfrac{\partial h}{\partial x}. \end{cases} \tag{27}$$

References

Bender, C. M. and Orszag, S. A. *Advanced Mathematical Methods for Scientists and Engineers.* New York: McGraw-Hill, p. 547, 1978.

Ott, E. *Chaos in Dynamical Systems.* New York: Cambridge University Press, 1993.

Zwillinger, D. (Ed.). *CRC Standard Mathematical Tables and Formulae.* Boca Raton, FL: CRC Press, p. 413, 1995.

Zwillinger, D. *Handbook of Differential Equations, 3rd ed.* Boston, MA: Academic Press, p. 122, 1997.

Duhamel's Convolution Principle

Can be used to invert a LAPLACE TRANSFORM.

Dumbbell Curve

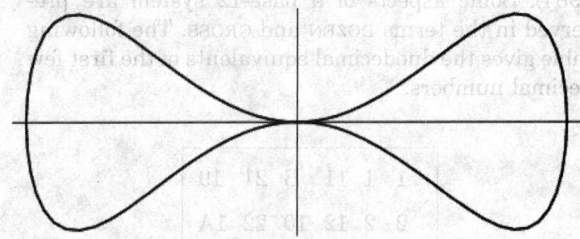

$$y^2 = a^2\left(x^4 - x^6\right).$$

See also BUTTERFLY CURVE, EIGHT CURVE, PIRIFORM

References

Cundy, H. and Rollett, A. *Mathematical Models, 3rd ed.* Stradbroke, England: Tarquin Pub., p. 72, 1989.

Dummy Variable

A variable that appears in a calculation only as a placeholder and which disappears completely in the final result. For example, in the integral

$$\int_0^x f(x')\,dx',$$

x' is a dummy variable since it is "integrated out" in the final answer. Any variable name other than x could therefore be used in the above expression, e.g. $\int_0^x f(\lambda)\,d\lambda$, $\int_0^x f(q)\,dq$, etc.

Dummy variables are also called BOUND VARIABLES or dead variables. Comtet (1974) adopts a notation in which dummy variable appearing as indices in sums are denoted by placing a dot underneath them (i.e., indicating them with an UNDERDOT), e.g.,

$$\sum_{c_1 + c_2 = n} c_1 c_2 = \frac{1}{6}n\left(n^2 - 1\right)$$

(Comtet 1974, p. 33).

See also BOUND VARIABLE, UNDERDOT

References

Comtet, L. *Advanced Combinatorics: The Art of Finite and Infinite Expansions, rev. enl. ed.* Dordrecht, Netherlands: Reidel, pp. 32–3, 1974.

Duodecillion

In the American system, 10^{39}.

See also LARGE NUMBER

Duodecimal

The base-12 number system composed of the digits 1, 2, 3, 4, 5, 6, 7, 8, 9, A, B. Such a system has been advocated by no less than Herbert Spencer, John

Quincy Adams, and George Bernard Shaw (Gardner 1984). Some aspects of a base-12 system are preserved in the terms DOZEN and GROSS. The following table gives the duodecimal equivalents of the first few decimal numbers.

1	1	11	B	21	19
2	2	12	10	22	1A
3	3	13	11	23	1B
4	4	14	12	24	20
5	5	15	13	25	21
6	6	16	14	26	22
7	7	17	15	27	23
8	8	18	16	28	24
9	9	19	17	29	25
10	A	20	18	30	26

See also BASE (NUMBER), DOZEN, GROSS

References

Gardner, M. *The Sixth Book of Mathematical Games from Scientific American.* Chicago, IL: University of Chicago Press, pp. 104–05, 1984.

Dupin's Cyclide
CYCLIDE

Dupin's Indicatrix
A pair of conics obtained by expanding an equation in MONGE'S FORM $z = F(x, y)$ in a MACLAURIN SERIES

$$z = z(0,0) + z_1 x + z_2 y + \frac{1}{2}\left(z_{11}x^2 + 2z_{12}xy + z_{22}y^2\right) + \ldots$$

$$= \frac{1}{2}\left(b_{11}x^2 + 2b_{12}xy + b_{22}y^2\right).$$

This gives the equation

$$b_{11}x^2 + 2b_{12}xy + b_{22}y^2 = \pm 1.$$

Amazingly, the radius of the indicatrix in any direction is equal to the SQUARE ROOT of the RADIUS OF CURVATURE in that direction (Coxeter 1969).

References

Coxeter, H. S. M. "Dupin's Indicatrix" §19.8 in *Introduction to Geometry, 2nd ed.* New York: Wiley, pp. 363–65, 1969.

Dupin's Theorem
In three mutually orthogonal systems of surfaces, the LINES OF CURVATURE on any surface in one of the systems are its intersections with the surfaces of the other two systems.

Duplication Formula
ABEL'S DUPLICATION FORMULA, DOUBLE-ANGLE FORMULAS, LEGENDRE DUPLICATION FORMULA

Duplication of the Cube
CUBE DUPLICATION

Durand's Rule
Let the values of a function $f(x)$ be tabulated at points x_i equally spaced by $h = x_{i+1} - x_i$, so $f_1 = f(x_1)$, $f_2 = f(x_2)$, ..., $f_n = f(x_n)$. Then Durand's rule approximating the integral of $f(x)$ is given by the NEWTON-COTES-like formula

$$\int_{x_i}^{x_1} f(x)dx = h\left(\frac{2}{5}f_1 + \frac{11}{10}f_2 + f_3 + \ldots + f_{n-2} + \frac{11}{10}f_{n-1} + \frac{2}{5}f_n\right).$$

See also BODE'S RULE, HARDY'S RULE, NEWTON-COTES FORMULAS, SIMPSON'S 3/8 RULE, SIMPSON'S RULE, TRAPEZOIDAL RULE, WEDDLE'S RULE

References

Beyer, W. H. (Ed.). *CRC Standard Mathematical Tables, 28th ed.* Boca Raton, FL: CRC Press, p. 127, 1987.

Dürer's Conchoid

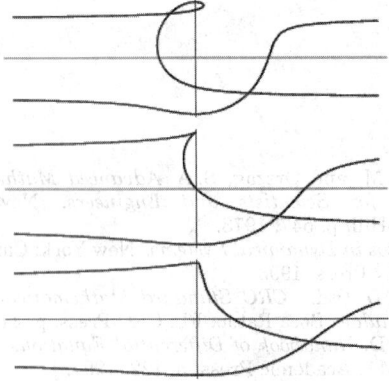

These curves appear in Dürer's work *Instruction in Measurement with Compasses and Straight Edge* (1525) and arose in investigations of perspective. Dürer constructed the curve by drawing lines QRP and $P'QR$ of length 16 units through $Q(q,0)$ and $R(r,0)$, where $q + r = 13$. The locus of P and P' is the curve, although Dürer found only one of the two branches of the curve.

The ENVELOPE of the lines QRP and $P'QR$ is a PARABOLA, and the curve is therefore a GLISSETTE of a point on a line segment sliding between a PARABOLA and one of its TANGENTS.

Dürer called the curve "muschellini," which means CONCHOID. However, it is not a true CONCHOID and so is sometimes called DÜRER'S SHELL CURVE. The Cartesian equation is

$$2y^2(x^2+y^2) - 2by^2(x+y) + (b^2-3a^2)y^2 - a^2x^2$$

$$+2a^2b(x+y) + a^2(a^2-b^2) = 0.$$

The above curves are for $(a,b) = (3,1)$, $(3,3)$, $(3,5)$. There are a number of interesting special cases. If $b = 0$, the curve becomes two coincident straight lines $x = 0$. For $a = 0$, the curve becomes the line pair $x = b/2$, $x = -b/2$, together with the CIRCLE $x+y=b$. If $a = b/2$, the curve has a CUSP at $(-2a,a)$.

References

Lawrence, J. D. *A Catalog of Special Plane Curves*. New York: Dover, pp. 157–59, 1972.
Lockwood, E. H. *A Book of Curves*. Cambridge, England: Cambridge University Press, p. 163, 1967.
MacTutor History of Mathematics Archive. "Dürer's Shell Curves." http://www-groups.dcs.st-and.ac.uk/~history/ Curves/Durers.html.

Dürer's Magic Square

16	3	2	13
5	10	11	8
9	6	7	12
4	15	14	1

Dürer's magic square is a MAGIC SQUARE with MAGIC CONSTANT 34 used in an engraving entitled *Melencolia I* by Albrecht Dürer (The British Museum, Burton 1989, Gellert *et al.* 1989). The engraving shows a disorganized jumble of scientific equipment lying unused while an intellectual sits absorbed in thought. Dürer's magic square is located in the upper right-hand corner of the engraving. The numbers 15 and 14 appear in the middle of the bottom row, indicating the date of the engraving, 1514.

16	3	2	13		16	3	2	13
5	10	11	8		5	10	11	8
9	6	7	12		9	6	7	12
4	15	14	1		4	15	14	1

Dürer's magic square has the additional property that the sums in any of the four quadrants, as well as

the sum of the middle four numbers, are all 34 (Hunter and Madachy 1975, p. 24).

See also DÜRER'S SOLID, MAGIC SQUARE

References

Boyer, C. D. and Merzbach, U. C. *A History of Mathematics*. New York: Wiley, pp. 296–97, 1991.
Burton, D. M. Cover illustration of *Elementary Number Theory, 4th ed.* Boston, MA: Allyn and Bacon, 1989.
Gellert, W.; Gottwald, S.; Hellwich, M.; Kästner, H.; and Künstner, H. (Eds.). Appendix, Plate 19. *VNR Concise Encyclopedia of Mathematics, 2nd ed.* New York: Van Nostrand Reinhold, 1989.
Hunter, J. A. H. and Madachy, J. S. *Mathematical Diversions*. New York: Dover, p. 24, 1975.
Rivera, C. "Melancholia." http://www.primepuzzles.net/melancholia.htm.

Dürer's Shell Curve

DÜRER'S CONCHOID

Dürer's Solid

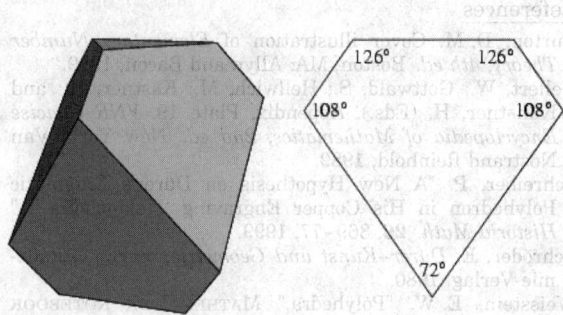

The 8-faced solid depicted in an engraving entitled *Melencolia I* by Albrecht Dürer (The British Museum, Burton 1989, Gellert *et al.* 1989), the same engraving in which DÜRER'S MAGIC SQUARE appears, which depicts a disorganized jumble of scientific equipment lying unused while an intellectual sits absorbed in thought. Although Dürer does not specify how his solid is constructed, Schreiber (1999) has noted that it appears to consist of a distorted CUBE which is first stretched to give rhombic faces with angles of 72°, and then truncated on top and bottom to yield bounding triangular faces whose vertices lie on the CIRCUMSPHERE of the azimuthal cube vertices.

Starting with a unit cube oriented parallel to the axes of the coordinate system, rotate it by EULER ANGLES $\psi = \pi/4$ and $\theta = \sec^{-1}\sqrt{3}$ to align a threefold symmetry axis along the z-axis. The stretch factor needed to produce rhombic angles of 72° is then

$$s = \sqrt{1 + \frac{3}{\sqrt{5}}}. \qquad (1)$$

The azimuthal points are a distance $d = s/2$ away from the origin, and in order for the vertices of the triangles obtained by truncation to lie at this same distance, the TRUNCATION must be done a distance $(3 - \sqrt{5})/2$ along the edge from one of the azimuthal points, which corresponds to a height

$$h = \sqrt{\frac{23}{\sqrt{5}} - \frac{1}{4}}. \qquad (2)$$

The resulting solid has six 126–08–2–08–26° pentagonal faces and two equilateral triangular faces, and the lengths of the sides are in the ratio

$$1 : \tfrac{1}{2}(3 + \sqrt{5}) : \sqrt{\tfrac{1}{2}(5 + \sqrt{5})}. \qquad (3)$$

Examination of this solid shows it to be identical to the dimensions of the solid reconstructed from its perspective picture (Schröder 1980, p. 70; Schreiber 1999).

See also DÜRER'S MAGIC SQUARE

References

Burton, D. M. Cover illustration of *Elementary Number Theory, 4th ed.* Boston, MA: Allyn and Bacon, 1989.

Gellert, W.; Gottwald, S.; Hellwich, M.; Kästner, H.; and Künstner, H. (Eds.). Appendix, Plate 19. *VNR Concise Encyclopedia of Mathematics, 2nd ed.* New York: Van Nostrand Reinhold, 1989.

Schreiber, P. "A New Hypothesis on Dürer's Enigmatic Polyhedron in His Copper Engraving 'Melancholia I'." *Historia Math.* **26**, 369–77, 1999.

Schröder, E. *Dürer--Kunst und Geometrie.* Berlin: Akademie-Verlag, 1980.

Weisstein, E. W. "Polyhedra." MATHEMATICA NOTEBOOK POLYHEDRA.M.

Durfee Polynomial

Let $F(n)$ be a family of PARTITIONS of n and let $F(n, d)$ denote the set of PARTITIONS in $F(n)$ with DURFEE SQUARE of size d. The Durfee polynomial of $F(n)$ is then defined as the polynomial

$$P_{F,n} = \sum |F(n, d)| y^d,$$

where $0 \le d \le \sqrt{n}$.

See also DURFEE SQUARE, PARTITION

References

Canfield, E. R.; Corteel, S.; and Savage, C. D. "Durfee Polynomials." *Electronic J. Combinatorics* **5**, No. 1, R32, 1–1, 1998. http://www.combinatorics.org/Volume_5/v5i1toc.html#R32.

Durfee Square

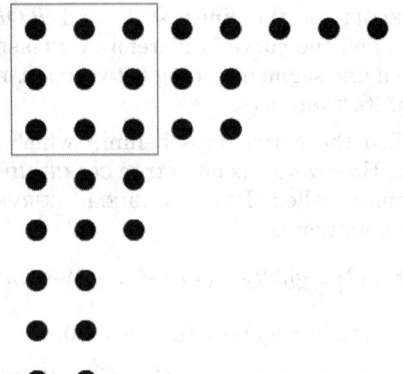

The length of the largest-sized SQUARE contained within the FERRERS DIAGRAM of a PARTITION. Its size can be determined using DurfeeSquare[f] in the *Mathematica* add-on package DiscreteMath`Combinatorica` (which can be loaded with the command <<DiscreteMath`). The size of the Durfee square remains unchanged between a partition and its CONJUGATE PARTITION (Skiena 1990, p. 57). In the plot above, the Durfee square has size 3.

See also CONJUGATE PARTITION, DURFEE POLYNOMIAL, FERRERS DIAGRAM, PARTITION

References

Skiena, S. *Implementing Discrete Mathematics: Combinatorics and Graph Theory with Mathematica.* Reading, MA: Addison-Wesley, 1990.

Dust

CANTOR DUST, FATOU DUST

Dvoretzky's Theorem

Each centered convex body of sufficiently high dimension has an "almost spherical" k-dimensional central section.

Dyad

Dyads extend VECTORS to provide an alternative description to second RANK TENSORS. A dyad $D(\mathbf{A}, \mathbf{B})$ of a pair of VECTORS $\mathbf{A}$ and $\mathbf{B}$ is defined by $D(\mathbf{A}, \mathbf{B}) \equiv \mathbf{AB}$. The DOT PRODUCT is defined by

$$\mathbf{A} \cdot \mathbf{BC} \equiv (\mathbf{A} \cdot \mathbf{B})\mathbf{C}$$

$$\mathbf{AB} \cdot \mathbf{C} \equiv \mathbf{A}(\mathbf{B} \cdot \mathbf{C}),$$

and the COLON PRODUCT by

$$\mathbf{AB} : \mathbf{CD} \equiv \mathbf{C} \cdot \mathbf{AB} \cdot \mathbf{D} = (\mathbf{A} \cdot \mathbf{C})(\mathbf{B} \cdot \mathbf{D})$$

See also DYADIC, TENSOR

References

Morse, P. M. and Feshbach, H. "Dyadics and Other Vector Operators." §1.6 in *Methods of Theoretical Physics, Part I.* New York: McGraw-Hill, pp. 54–2, 1953.

Dyadic

A linear POLYNOMIAL of DYADS **AB** + **CD** + ... consisting of nine components A_{ij} which transform as

$$\left(A_{ij}\right)' = \sum_{m,n} \frac{h_m h_n}{h_i' h_j'} \frac{\partial x_m}{\partial x_i'} \frac{\partial x_n}{\partial x_j'} A_{mn} \qquad (1)$$

$$= \sum_{m,n} \frac{h_i' h_j'}{h_m h_n} \frac{\partial x_i'}{\partial x_m} \frac{\partial x_j}{\partial x_n} A_{mn} \qquad (2)$$

$$= \sum_{m,n} \frac{h_i' h_n}{h_m h_j'} \frac{\partial x_i'}{\partial x_m} \frac{\partial x_m}{\partial x_j'} A_{mn}. \qquad (3)$$

Dyadics are often represented by Gothic capital letters. The use of dyadics is nearly archaic since TENSORS perform the same function but are notationally simpler.

A unit dyadic is also called the IDEMFACTOR and is defined such that

$$\mathbf{I} \cdot \mathbf{A} \equiv \mathbf{A}. \qquad (4)$$

In CARTESIAN COORDINATES,

$$\mathbf{I} = \hat{\mathbf{x}}\hat{\mathbf{x}} + \hat{\mathbf{y}}\hat{\mathbf{y}} + \hat{\mathbf{z}}\hat{\mathbf{z}}, \qquad (5)$$

and in SPHERICAL COORDINATES

$$\mathbf{I} = \nabla \mathbf{r}. \qquad (6)$$

See also DYAD, TENSOR, TETRADIC

References

Arfken, G. "Dyadics." §3.5 in *Mathematical Methods for Physicists, 3rd ed.* Orlando, FL: Academic Press, pp. 137–40, 1985.

Jeffreys, H. and Jeffreys, B. S. "Dyadic Notation." §3.04 in *Methods of Mathematical Physics, 3rd ed.* Cambridge, England: Cambridge University Press, p. 89, 1988.

Morse, P. M. and Feshbach, H. "Dyadics and Other Vector Operators." §1.6 in *Methods of Theoretical Physics, Part I.* New York: McGraw-Hill, pp. 54–2, 1953.

Dyck Language

The simplest ALGEBRAIC LANGUAGE, denoted $\mathfrak{D}$. If X is the alphabet $\{x, \bar{x}\}$, then $\mathfrak{D}$ is the set of words u of X which satisfy

1. $|u|_x = |u|_{\bar{x}}$, where $|u|_x$ is the numbers of letters x in the word u, and
2. if u is factored as vw, where v and w are words of X^*, then $|v|_x \geq |v|_{\bar{x}}$.

See also ALGEBRAIC LANGUAGE

References

Bousquet-Mélou, M. "Convex Polyominoes and Heaps of Segments." *J. Phys. A: Math. Gen.* **25**, 1925–934, 1992.

Dyck Path

A LATTICE PATH from $(0,0)$ to (n, n) which never crosses (but may touch) the line $y = x$. There are

$$C_n = \frac{1}{n+1}\binom{2n}{n}$$

Dyck paths, where C_n is a CATALAN NUMBER.

See also LATTICE PATH

References

Degenhardt, S. L. and Milne, S. C. "Weighted Inversion Statistics and Their Symmetry Groups." Preprint.

Dyck's Surface

The surface with three CROSS-CAPS (Francis and Collins 1993, Francis and Weeks 1999).

See also CROSS-CAP

References

Francis, G. and Collins, B. "On Knot-Spanning Surfaces: An Illustrated Essay on Topological Art." Ch. 11 in *The Visual Mind: Art and Mathematics* (Ed. M. Emmer). Cambridge, MA: MIT Press, 1993.

Francis, G. K. and Weeks, J. R. "Conway's ZIP Proof." *Amer. Math. Monthly* **106**, 393–99, 1999.

Dyck's Theorem

HANDLES and CROSS-HANDLES are equivalent in the presence of a CROSS-CAP.

See also CROSS-CAP, CROSS-HANDLE, HANDLE, VON DYCK'S THEOREM

References

Dyck, W. "Beiträge zur Analysis situs I." *Math. Ann.* **32**, 459–12, 1888.

Francis, G. K. and Weeks, J. R. "Conway's ZIP Proof." *Amer. Math. Monthly* **106**, 393–99, 1999.

Dye's Theorem

For any two ergodic measure-preserving transformations on nonatomic PROBABILITY SPACES, there is an ISOMORPHISM between the two PROBABILITY SPACES carrying orbits onto orbits.

See also ERGODIC THEORY

Dyet

INEXACT DIFFERENTIAL

Dymaxion

Buckminster Fuller's term for the CUBOCTAHEDRON.

See also CUBOCTAHEDRON, MECON

Dynamical System

A means of describing how one state develops into another state over the course of time. Technically, a dynamical system is a smooth action of the reals or the INTEGERS on another object (usually a MANIFOLD). When the reals are acting, the system is called a continuous dynamical system, and when the INTEGERS are acting, the system is called a discrete dynamical system. If f is any CONTINUOUS FUNCTION, then the evolution of a variable x can be given by the formula

$$x_{n+1} = f(x_n). \qquad (1)$$

This equation can also be viewed as a difference equation

$$x_{n+1} - x_n = f(x_n) - x_n, \qquad (2)$$

so defining

$$g(x) \equiv f(x) - x \qquad (3)$$

gives

$$x_{n+1} - x_n = g(x_n) * 1, \qquad (4)$$

which can be read "as n changes by 1 unit, x changes by $g(x)$." This is the discrete analog of the DIFFERENTIAL EQUATION

$$x'(n) = g(x(n)). \qquad (5)$$

See also ANOSOV DIFFEOMORPHISM, ANOSOV FLOW, AXIOM A DIFFEOMORPHISM, AXIOM A FLOW, BIFURCATION THEORY, CHAOS, ERGODIC THEORY, GEODESIC FLOW

References

Aoki, N. and Hiraide, K. *Topological Theory of Dynamical Systems.* Amsterdam, Netherlands: North-Holland, 1994.
Golubitsky, M. *Introduction to Applied Nonlinear Dynamical Systems and Chaos.* New York: Springer-Verlag, 1997.
Guckenheimer, J. and Holmes, P. *Nonlinear Oscillations, Dynamical Systems, and Bifurcations of Vector Fields, 3rd ed.* New York: Springer-Verlag, 1997.
Jordan, D. W. and Smith, P. *Nonlinear Ordinary Differential Equations: An Introduction to Dynamical Systems, 3rd ed.* Oxford, England: Oxford University Press, 1999.
Lichtenberg, A. and Lieberman, M. *Regular and Stochastic Motion, 2nd ed.* New York: Springer-Verlag, 1994.
Ott, E. *Chaos in Dynamical Systems.* New York: Cambridge University Press, 1993.
Rasband, S. N. *Chaotic Dynamics of Nonlinear Systems.* New York: Wiley, 1990.
Strogatz, S. H. *Nonlinear Dynamics and Chaos, with Applications to Physics, Biology, Chemistry, and Engineering.* Reading, MA: Addison-Wesley, 1994.
Tabor, M. *Chaos and Integrability in Nonlinear Dynamics: An Introduction.* New York: Wiley, 1989.

Dynkin Diagram

Every SEMISIMPLE LIE ALGEBRA g is classified by its Dynkin diagram. A Dynkin diagram is a GRAPH with a few different kinds of possible edges. The CONNECTED COMPONENTS of the graph correspond to the irreducible subalgebras of g. So a SIMPLE LIE ALGEBRA's Dynkin diagram has only one component. The rules are restrictive. In fact, there are only certain possibilities for each component, corresponding to the classification of SEMI-SIMPLE LIE ALGEBRAS.

The roots of a complex LIE ALGEBRA form a LATTICE of rank k in a CARTAN SUBALGEBRA h ⊂ g, where k is the RANK of g. Hence, the ROOT LATTICE can be considered a lattice in $\mathbb{R}^k$. A vertex, or node, in the Dynkin diagram is drawn for each SIMPLE ROOT, which corresponds to a generator of the ROOT LATTICE. Between two nodes α and β, an edge is drawn if the simple roots are not perpendicular. One line is drawn if the angle between them is $2\pi/3$, two lines if the angle is $3\pi/3$, and three lines are drawn if the angle is $5\pi/6$. There are no other possible angles between SIMPLE ROOTS. Alternatively, the number of lines N between the simple roots α and β is given by

$$N = A_{\alpha\beta}A_{\beta\alpha} = \frac{2\langle\alpha,\beta\rangle}{|\alpha|^2}\frac{2\langle\beta,\alpha\rangle}{|\beta|^2} = 4\cos^2\theta,$$

where $A_{\alpha\beta}$ is an entry in the CARTAN MATRIX. In a Dynkin diagram, an arrow is drawn from the longer root to the shorter root (when the angle is $3\pi/3$ or $5\pi/6$).

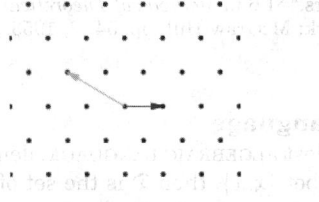

The picture above shows the two simple roots for G_2, at an angle of $5\pi/6$, in the ROOT LATTICE. Therefore, the Dynkin diagram for G_2 has two nodes, with three lines between them.

Here are some properties of admissible Dynkin diagrams.

1. A diagram obtained by removing a node from an admissible node is admissible.
2. An admissible diagram has no loops.
3. No node has more than three lines attached to it.
4. A sequence of nodes with only two single lines can be collapsed to give an admissible diagram.
5. The only connected diagram with a triple line has two nodes.

A COXETER-DYNKIN DIAGRAM, also called a Coxeter graph, is the same as a Dynkin diagram, without the arrows, although sometimes these are also called Dynkin diagrams. The Coxeter diagram is sufficient to characterize the algebra, as can be seen by enumerating connected diagrams.

The simplest way to recover a SIMPLE LIE ALGEBRA from its Dynkin diagram is to first reconstruct its CARTAN MATRIX (A_{ij}). The ith node and jth node are connected by $A_{ij}A_{ji}$ lines. Since $A_{ij} = 0$ IFF $A_{ji} = 0$, and otherwise $A_{ji} \in \{-3, -2, -1\}$, it is easy to find A_{ij} and A_{ji}, up to order, from their product. The arrow in the diagram indicates which is larger. For example, if node 1 and node 2 have two lines between them, from node 1 to node 2, then $A_{12} = -1$ and $A_{21} = -2$.

However, it is worth pointing out that each SIMPLE LIE ALGEBRA can be constructed concretely. For instance, the infinite families A_n, B_n, C_n, and D_n correspond to $\mathfrak{sl}_{n+1}\mathbb{C}$ the SPECIAL LINEAR LIE ALGEBRA, $\mathfrak{so}_{2n+1}\mathbb{C}$ the odd ORTHOGONAL LIE ALGEBRA, $\mathfrak{sp}_{2n}\mathbb{C}$ the SYMPLECTIC LIE ALGEBRA, and $\mathfrak{so}_{2n}\mathbb{C}$ the even ORTHOGONAL LIE ALGEBRA. The other simple Lie algebras are called EXCEPTIONAL LIE ALGEBRAS, and have constructions related to the OCTONIONS.

See also CARTAN MATRIX, COXETER-DYNKIN DIAGRAM, KILLING FORM, LIE ALGEBRA, LIE GROUP, ROOT LATTICE, ROOT (LIE ALGEBRA), SIMPLE LIE ALGEBRA, WEYL GROUP

References

Fulton, W. and Harris, J. *Representation Theory.* New York: Springer-Verlag, 1991.
Hsiang, W. Y. *Lectures on Lie Groups.* Singapore: World Scientific, pp. 98–02, 2000.
Huang, J.-S. "Dynkin Diagrams." §4.6 in *Lectures on Representation Theory.* Singapore: World Scientific, pp. 39–4, 1999.
Jacobson, N. "The Determination of the Cartan Matrices." §4.5 in *Lie Algebras.* New York: Dover, pp. 128–35, 1979.
Knapp, A. *Lie Groups Beyond an Introduction.* Boston, MA: Birkhäuser, 1996.

Dyson's Conjecture

Based on a problem in particle physics, Dyson (1962abc) conjectured that the constant term in the LAURENT SERIES

$$\prod_{1 \le i \ne j \le n} \left(1 - \frac{x_i}{x_j}\right)^{a_i}$$

is the MULTINOMIAL COEFFICIENT

$$\frac{(a_1 + a_2 + \ldots + a_n)}{a_1! a_2! \ldots a_n!}$$

The theorem was proved by Wilson (1962) and independently by Gunson (1962). A definitive proof was subsequently published by Good (1970).

See also MACDONALD'S CONSTANT-TERM CONJECTURE, ZEILBERGER-BRESSOUD THEOREM

References

Andrews, G. E. "The Zeilberger-Bressoud Theorem." §4.3 in *q*-Series: Their Development and Application in Analysis, Number Theory, Combinatorics, Physics, and Computer Algebra. Providence, RI: Amer. Math. Soc., pp. 36–8, 1986.
Dyson, F. "Statistical Theory of the Energy Levels of Complex Systems. I." *J. Math. Phys.* **3**, 140–56, 1962a.
Dyson, F. "Statistical Theory of the Energy Levels of Complex Systems. II." *J. Math. Phys.* **3**, 157–65, 1962b.
Dyson, F. "Statistical Theory of the Energy Levels of Complex Systems. III." *J. Math. Phys.* **3**, 166–75, 1962c.
Good, I. J. "Short Proof of a Conjecture by Dyson." *J. Math. Phys.* **11**, 1884, 1970.
Gunson, J. "Proof of a Conjecture of Dyson in the Statistical Theory of Energy Levels." *J. Math. Phys.* **3**, 752–53, 1962.
Wilson, K. G. "Proof of a Conjecture by Dyson." *J. Math. Phys.* **3**, 1040–043, 1962.

E

Ear

A PRINCIPAL VERTEX x_i of a SIMPLE POLYGON P is called an ear if the diagonal $[x_{i-1}, x_{i+1}]$ that bridges x_i lies entirely in P. Two ears x_i and x_j are said to overlap if

$$\text{int}[x_{i-1}, x_i, x_{i+1}] \cap \text{int}[x_{j-1}, x_j, x_{j+1}] \neq \varnothing.$$

The TWO-EARS THEOREM states that, except for TRIANGLES, every SIMPLE POLYGON has at least two nonoverlapping ears.

See also ANTHROPOMORPHIC POLYGON, MOUTH, TWO-EARS THEOREM

References

Meisters, G. H. "Polygons Have Ears." *Amer. Math. Monthly* **82**, 648–51, 1975.
Meisters, G. H. "Principal Vertices, Exposed Points, and Ears." *Amer. Math. Monthly* **87**, 284–85, 1980.
Toussaint, G. "Anthropomorphic Polygons." *Amer. Math. Monthly* **122**, 31–5, 1991.

Early Election Results

Let Jones and Smith be the only two contestants in an election that will end in a deadlock when all votes for Jones (J) and Smith (S) are counted. What is the EXPECTATION VALUE of $X_k \equiv |S - J|$ after k votes are counted? The solution is

$$\langle X_k \rangle = \frac{2N \binom{N-1}{\lfloor k/2 \rfloor} \binom{N-1}{\lfloor k/2 \rfloor - 1}}{\binom{2N}{k}}$$

$$= \begin{cases} \dfrac{k(2N-k)}{2N} \binom{N}{k/2}^2 \binom{2N}{k}^{-1} \\ \quad \text{for } k \text{ even} \\ \dfrac{k(2N-k+1)}{2N} \binom{N}{(k-1)/2}^2 \binom{2N}{k-1}^{-1} \\ \quad \text{for } k \text{ odd.} \end{cases}$$

References

Handelsman, M. B. Solution to Problem 10248. "Early Returns in a Tied Election." *Amer. Math. Monthly* **102**, 554–56, 1995.

Eban Number

The sequence of numbers whose names (in English) do not contain the letter "e" (i.e., "e" is "banned"). The first few eban numbers are 2, 4, 6, 30, 32, 34, 36, 40, 42, 44, 46, 50, 52, 54, 56, 60, 62, 64, 66, 2000, 2002, 2004, ... (Sloane's A006933); i.e., two, four, six, thirty, etc.

References

Sloane, N. J. A. Sequences A006933/M1030 in "An On-Line Version of the Encyclopedia of Integer Sequences." http://www.research.att.com/~njas/sequences/eisonline.html.

Eberhart's Conjecture

If q_n is the nth prime such that M_{q_n} is a MERSENNE PRIME, then

$$q_n \sim (3/2)^n.$$

It was modified by Wagstaff (1983) to yield WAGSTAFF'S CONJECTURE,

$$q_n \sim (2^{e^{-\gamma}})^n,$$

where γ is the EULER-MASCHERONI CONSTANT.

See also WAGSTAFF'S CONJECTURE

References

Ribenboim, P. *The New Book of Prime Number Records.* New York: Springer-Verlag, p. 412, 1996.
Wagstaff, S. S. "Divisors of Mersenne Numbers." *Math. Comput.* **40**, 385–97, 1983.

Eccentric

Not CONCENTRIC.

See also CONCENTRIC, CONCYCLIC

Eccentric Angle

The angle θ measured from the CENTER of an ELLIPSE to a point on the ELLIPSE.

See also ECCENTRICITY, ELLIPSE

Eccentric Anomaly

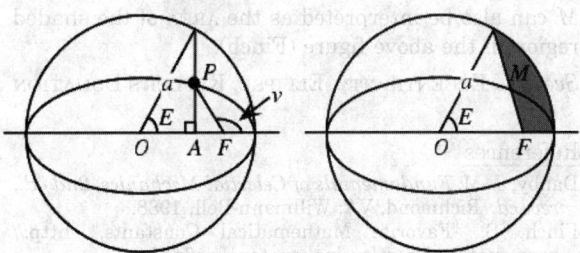

The ANGLE obtained by drawing the AUXILIARY CIRCLE of an ELLIPSE with center O and FOCUS F, and drawing a LINE PERPENDICULAR to the SEMIMAJOR AXIS and intersecting it at A. The ANGLE E is then defined as illustrated above. Then for an ELLIPSE with ECCENTRICITY e,

$$AF = OF - AO = ae - a\cos E \qquad (1)$$

But the distance AF is also given in terms of the distance from the FOCUS $r = FP$ and the SUPPLEMENT of the ANGLE from the SEMIMAJOR AXIS v by

$$AF = r\cos(\pi - v) = -r\cos v. \qquad (2)$$

Equating these two expressions gives

$$r = \frac{a(\cos E - e)}{\cos v}, \tag{3}$$

which can be solved for $\cos v$ to obtain

$$\cos v = \frac{a(\cos E - e)}{r}. \tag{4}$$

To get E in terms of r, plug (4) into the equation of the ELLIPSE

$$r = \frac{a(1 - e^2)}{1 + \cos v}. \tag{5}$$

Rearranging,

$$r(1 + e \cos v) = a(1 - e^2) \tag{6}$$

and plugging in (4) then gives

$$r\left(1 + \frac{ae \cos E}{r} - \frac{e^2}{r}\right) = r + ae \cos E - e^2 a$$

$$= a(1 - e^2). \tag{7}$$

Solving for r gives

$$r = a(1 - e^2) - ea \cos E + e^2 a = a(1 - e \cos E), \tag{8}$$

so differentiating yields the result

$$\dot{r} = ae\dot{E} \sin E. \tag{9}$$

The eccentric anomaly is a very useful concept in orbital mechanics, where it is related to the so-called mean anomaly M by KEPLER'S EQUATION

$$M = E - e \sin E. \tag{10}$$

M can also be interpreted as the AREA of the shaded region in the above figure (Finch).

See also ECCENTRICITY, ELLIPSE, KEPLER'S EQUATION

References

Danby, J. M. *Fundamentals of Celestial Mechanics, 2nd ed., rev. ed.* Richmond, VA: Willmann-Bell, 1988.

Finch, S. "Favorite Mathematical Constants." http://www.mathsoft.com/asolve/constant/lpc/lpc.html.

Montenbruck, O. and Pfleger, T. *Astronomy on the Personal Computer, 4th ed.* Berlin: Springer-Verlag, p. 62, 2000.

Eccentricity

A quantity defined for a CONIC SECTION which can be given in terms of SEMIMAJOR a and SEMIMINOR AXES b.

interval	curve	e
$e = 0$	CIRCLE	0
$0 < e < 1$	ELLIPSE	$\sqrt{1 - \dfrac{b^2}{a^2}}$
$e = 1$	PARABOLA	1
$e > 1$	HYPERBOLA	$\sqrt{1 + \dfrac{b^2}{a^2}}$

The eccentricity can also be interpreted as the fraction of the distance to the semimajor axis at which the FOCUS lies,

$$e = \frac{c}{a},$$

where c is the distance from the center of the CONIC SECTION to the FOCUS.

See also CIRCLE, CONIC SECTION, ECCENTRIC ANOMALY, ELLIPSE, FLATTENING, FOCAL PARAMETER, HYPERBOLA, PARABOLA, SEMIMAJOR AXIS, SEMIMINOR AXIS

Echidnahedron

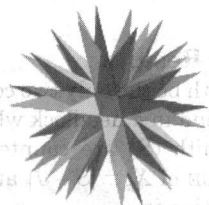

ICOSAHEDRON STELLATION #4.

References

Wenninger, M. J. *Polyhedron Models.* Cambridge, England: Cambridge University Press, p. 65, 1971.

Eckardt Point

On the CLEBSCH DIAGONAL CUBIC, all 27 of the complex lines present on a general smooth CUBIC SURFACE are real. In addition, there are 10 points on the surface where three of the 27 lines meet. These points are called Eckardt points (Fischer 1986).

See also CLEBSCH DIAGONAL CUBIC, CUBIC SURFACE

References

Fischer, G. (Ed.). *Mathematical Models from the Collections of Universities and Museums.* Braunschweig, Germany: Vieweg, p. 11, 1986.

Eckart Differential Equation

The second-order ORDINARY DIFFERENTIAL EQUATION

$$y'' + \left[\frac{\alpha\eta}{1 + \eta} + \frac{\beta\eta}{(1 + \eta)^2} + \gamma\right]y = 0,$$

where $\eta = e^{\delta x}$.

Eckert IV Projection

References

Barut, A. O.; Inomata, A.; and Wilson, R. "Algebraic Treatment of Second Pöschl-Teller, Morse-Rosen, and Eckart Equations." *J. Phys. A: Math. Gen.* **20**, 4083–096, 1987.
Zwillinger, D. *Handbook of Differential Equations, 3rd ed.* Boston, MA: Academic Press, p. 122, 1997.

Eckert IV Projection

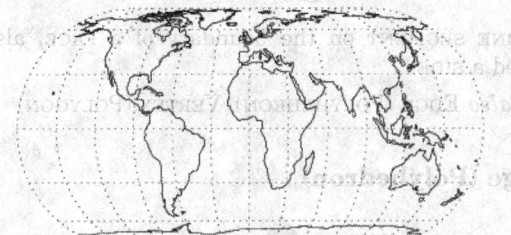

The equations are

$$x = \frac{2}{\sqrt{\pi(4+\pi)}}\,(\lambda - \lambda_0)(1 + \cos\theta) \qquad (1)$$

$$y = 2\sqrt{\frac{\pi}{4+\pi}}\,\sin\theta, \qquad (2)$$

where θ is the solution to

$$\theta + \sin\theta\cos\theta + 2\sin\theta = (2 + \tfrac{1}{2}\pi)\sin\phi. \qquad (3)$$

This can be solved iteratively using NEWTON'S METHOD with $\theta_0 = \phi/2$ to obtain

$$\Delta\theta = -\frac{\theta + \sin\theta\cos\theta + 2\sin\theta - (2 - \tfrac{1}{2}\pi)\sin\phi}{2\cos\theta(1 + \cos\theta)}. \qquad (4)$$

The inverse FORMULAS are

$$\phi = \sin^{-1}\left(\frac{\theta + \sin\theta\cos\theta + 2\sin\theta}{2 + \tfrac{1}{2}\pi}\right) \qquad (5)$$

$$\lambda = \lambda_0 + \frac{\pi\sqrt{4+\pi}\,x}{1 + \cos\theta}, \qquad (6)$$

where

$$\theta = \sin^{-1}\left(\frac{y}{2}\sqrt{\frac{4+\pi}{\pi}}\right). \qquad (7)$$

References

Snyder, J. P. *Map Projections--A Working Manual.* U. S. Geological Survey Professional Paper 1395. Washington, DC: U. S. Government Printing Office, pp. 253–58, 1987.

Eckert VI Projection

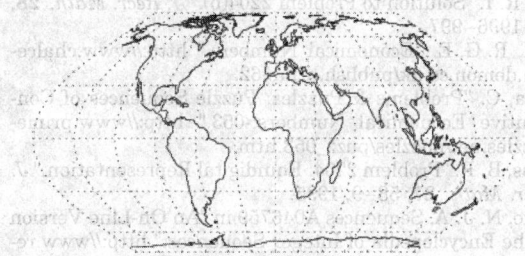

The equations are

$$x = \frac{(\lambda - \lambda_0)(1 + \cos\theta)}{\sqrt{2 + \pi}} \qquad (1)$$

$$y = \frac{2\theta}{\sqrt{2 + \pi}}, \qquad (2)$$

where θ is the solution to

$$\theta + \sin\theta = (1 + \tfrac{1}{2}\pi)\sin\phi. \qquad (3)$$

This can be solved iteratively using NEWTON'S METHOD with $\theta_0 = \phi$ to obtain

$$\Delta\theta = -\frac{\theta + \sin\theta - (1 + \tfrac{1}{2}\pi)\sin\phi}{1 + \cos\theta}. \qquad (4)$$

The inverse FORMULAS are

$$\phi = \sin^{-1}\left(\frac{\theta + \sin\theta}{1 + \tfrac{1}{2}\pi}\right) \qquad (5)$$

$$\lambda = \lambda_0 + \frac{\sqrt{2 + \pi}\,x}{1 + \cos\theta}, \qquad (6)$$

where

$$\theta = \tfrac{1}{2}\sqrt{2 + \pi}\,y. \qquad (7)$$

References

Snyder, J. P. *Map Projections--A Working Manual.* U. S. Geological Survey Professional Paper 1395. Washington, DC: U. S. Government Printing Office, pp. 253–58, 1987.

Economical Number

A number n is called an economical number if the number of digits in the prime factorization of n (including powers) uses fewer digits than the number of digits in n. The first few economical numbers are 125, 128, 243, 256, 343, 512, 625, 729, ... (Sloane's A046759). Pinch shows that, under a plausible hypothesis related to the TWIN PRIME CONJECTURE, there are arbitrarily long sequences of consecutive economical numbers, and exhibits such a sequence of length nine starting at 1034429177995381247.

See also EQUIDIGITAL NUMBER, WASTEFUL NUMBER

References

Hess, R. I. "Solution to Problem 2204(b)." *J. Recr. Math.* **28**, 67, 1996–997.

Pinch, R. G. E. "Economical Numbers." http://www.chalcedon.demon.co.uk/publish.html#62.

Rivera, C. "Problems & Puzzles: Puzzle Sequences of Consecutive Economical Numbers.-053." http://www.prime-puzzles.net/puzzles/puzz_053.htm.

Santos, B. R. "Problem 2204. Equidigital Representation." *J. Recr. Math.* **27**, 58–9, 1995.

Sloane, N. J. A. Sequences A046759 in "An On-Line Version of the Encyclopedia of Integer Sequences." http://www.research.att.com/~njas/sequences/eisonline.html.

Weisstein, E. W. "Integer Sequences." MATHEMATICA NOTEBOOK INTEGERSEQUENCES.M.

Economized Rational Approximation

A PADÉ APPROXIMANT perturbed with a CHEBYSHEV POLYNOMIAL OF THE FIRST KIND to reduce the leading COEFFICIENT in the ERROR.

See also PADÉ APPROXIMANT

Eddington Number

$$136 \cdot 2^{256} \approx 1.575 \times 10^{79}.$$

According to Eddington, the exact number of protons in the universe, where 136 was the RECIPROCAL of the fine structure constant as best as it could be measured in his time.

See also LARGE NUMBER

References

Hardy, G. H. *Ramanujan: Twelve Lectures on Subjects Suggested by His Life and Work*, 3rd ed. New York: Chelsea, pp. 15 and 49, 1999.

Edge (Graph)

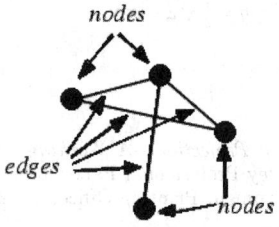

For an UNDIRECTED GRAPH, an unordered pair of nodes which specify the line connecting them are said to form an edge. For a DIRECTED GRAPH, the edge is an ordered pair of nodes. The terms "line," "arc," "branch," and "1-simplex" are sometimes used instead of edge (Skiena 1990, p. 80; Harary 1994). Harary (1994) calls an edge of a graph a "line."

See also EDGE NUMBER, HYPEREDGE, NULL GRAPH, TAIT COLORING, TAIT CYCLE, VERTEX (GRAPH)

References

Harary, F. *Graph Theory*. Reading, MA: Addison-Wesley, 1994.

Skiena, S. *Implementing Discrete Mathematics: Combinatorics and Graph Theory with Mathematica*. Reading, MA: Addison-Wesley, 1990.

Edge (Polygon)

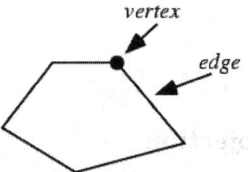

A LINE SEGMENT on the boundary of a FACE, also called a SIDE.

See also EDGE (POLYHEDRON), VERTEX (POLYGON)

Edge (Polyhedron)

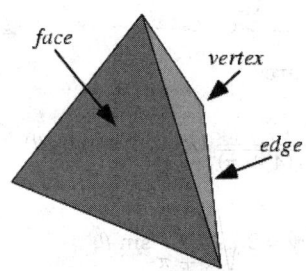

A LINE SEGMENT where two FACES of a POLYHEDRON meet, also called a SIDE.

See also EDGE (POLYGON), VERTEX (POLYHEDRON)

Edge (Polytope)

A 1-D LINE SEGMENT where two 2-D FACES of an n-D POLYTOPE meet, also called a SIDE.

See also EDGE (POLYGON), EDGE (POLYHEDRON)

Edge Chromatic Number

The fewest number of colors necessary to color each EDGE of a GRAPH so that no two EDGES incident on the same VERTEX have the same color. The edge chromatic number of a graph must be at least Δ, the largest VERTEX DEGREE of the graph (Skiena 1990, p. 216). However, Vizing (1964) and Gupta (1966) showed that any graph can be edge-colored with at most $\Delta + 1$ colors.

The edge chromatic number of a COMPLETE BIPARTITE GRAPH is Δ.

Determining the edge chromatic number of a graph is an NP-COMPLETE PROBLEM (Holyer 1981; Skiena 1990, p. 216). The edge chromatic number of a graph can be computed using EdgeChromaticNumber[g] in the *Mathematica* add-on package DiscreteMath`Combinatorica` (which can be loaded with the command < < DiscreteMath`).

See also CHROMATIC NUMBER, EDGE COLORING

References

Gupta, R. P. "The Chromatic Index and the Degree of a Graph." *Not. Amer. Math. Soc.* **13**, 719, 1966.
Holyer, I. "The NP-Completeness of Edge Colorings." *SIAM J. Comput.* **10**, 718–20, 1981.
Skiena, S. "Edge Colorings." §5.5.4 in *Implementing Discrete Mathematics: Combinatorics and Graph Theory with Mathematica.* Reading, MA: Addison-Wesley, p. 216, 1990.
Vizing, V. G. "On an Estimate of the Chromatic Class of a *p*-Graph" [Russian]. *Diskret. Analiz* **3**, 23–0, 1964.
© 1999–001 Wolfram Research, Inc.

Edge Coloring

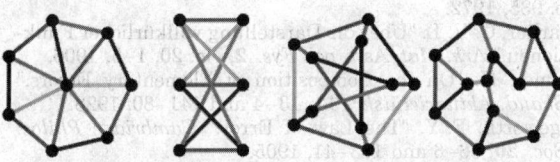

An edge coloring of a GRAPH G is a coloring of the edges of G such that adjacent edges (or the edges bounding different regions) receive different colors. BRELAZ'S HEURISTIC ALGORITHM can be used to find a good, but not necessarily minimal, edge coloring. Finding the minimum vertex coloring is equivalent to finding the minimum VERTEX COLORING of its LINE GRAPH (Skiena 1990, p. 216). The EDGE CHROMATIC NUMBER gives the minimum number of colors with which a graph can be colored.

An edge coloring of a graph can be computed using EdgeColoring[g] in the *Mathematica* add-on package DiscreteMath`Combinatorica` (which can be loaded with the command < <DiscreteMath`).

See also BRELAZ'S HEURISTIC ALGORITHM, CHROMATIC NUMBER, EDGE CHROMATIC NUMBER, K-COLORING

References

Saaty, T. L. and Kainen, P. C. *The Four-Color Problem: Assaults and Conquest.* New York: Dover, p. 13, 1986.
Skiena, S. "Edge Colorings." §5.5.4 in *Implementing Discrete Mathematics: Combinatorics and Graph Theory with Mathematica.* Reading, MA: Addison-Wesley, p. 216, 1990.
© 1999–001 Wolfram Research, Inc.

Edge Connectivity

The minimum number of edges $\lambda(G)$ whose deletion from a GRAPH G disconnects G, also called the line connectivity. The edge connectivity of a DISCONNECTED GRAPH is 0, while that of a CONNECTED GRAPH with a BRIDGE is 1.

Let $\kappa(G)$ be the VERTEX CONNECTIVITY of a graph G and $\delta(G)$ its minimum degree, then for any graph,

$$\kappa(G) \leq \lambda(G) \leq \delta(G)$$

(Whitney 1932, Harary 1994, p. 43).

The edge-connectivity of a graph can be determined with the command EdgeConnectivity[g] in the *Mathematica* add-on package DiscreteMath`Combinatorica` (which can be loaded with the command < <DiscreteMath`).

See also DISCONNECTED GRAPH, K-CONNECTED GRAPH, VERTEX CONNECTIVITY

References

Harary, F. *Graph Theory.* Reading, MA: Addison-Wesley, p. 43, 1994.
Skiena, S. *Implementing Discrete Mathematics: Combinatorics and Graph Theory with Mathematica.* Reading, MA: Addison-Wesley, pp. 177–78, 1990.
Whitney, H. "Congruent Graphs and the Connectivity of Graphs." *Amer. J. Math.* **54**, 150–68, 1932.

Edge Cover

A subset of edges defined similarly to the VERTEX COVER (Skiena 1990, p. 219). Gallai (1959) showed that the size of the minimum edge cover plus the side of the maximum number of independent edges equals the number of vertices of a graph.

See also VERTEX COVER

References

Gallai, T. "Über extreme Punkt- und Kantenmengen." *Ann. Univ. Sci. Budapest, Eotvos Sect. Math.* **2**, 133–38, 1959.
Skiena, S. *Implementing Discrete Mathematics: Combinatorics and Graph Theory with Mathematica.* Reading, MA: Addison-Wesley, p. 178, 1990.
© 1999–001 Wolfram Research, Inc.

Edge Number

The number of EDGES in a GRAPH, denoted $|E|$.

See also EDGE (GRAPH)

Edge Set

The edge set of a GRAPH is simply a set of all edges of the graph.

See also VERTEX SET

© 1999–001 Wolfram Research, Inc.

Edge-Graceful Graph

A generalization of the GRACEFUL GRAPH.

See also GRACEFUL GRAPH, SKOLEM-GRACEFUL GRAPH, SUPER-EDGE-GRACEFUL GRAPH

References

Sheng-Ping, L. "One Edge-Graceful Labeling of Graphs." *Congressus Numer.* **50**, 31–41, 1985.

Edge-Transitive Graph

A GRAPH such that any two edges are equivalent under some element of its automorphism group. Every nontrivial graph that is edge-transitive but not VERTEX-TRANSITIVE contains at least 20 vertices (Skiena 1990, p. 186). The smallest known CUBIC GRAPH that is edge- but not VERTEX-TRANSITIVE is the GRAY GRAPH.

See also GRAY GRAPH, FOLKMAN GRAPH, VERTEX-TRANSITIVE GRAPH

References

Skiena, S. *Implementing Discrete Mathematics: Combinatorics and Graph Theory with Mathematica.* Reading, MA: Addison-Wesley, 1990.
© *1999–001 Wolfram Research, Inc.*

Edgeworth Series

Let a distribution to be approximated be the distribution F_n of standardized sums

$$Y_n = \frac{\sum_{i=1}^n (X_i - \bar{X})}{\sqrt{\sum_{i=1}^n \sigma_X^2}}. \tag{1}$$

In the CHARLIER SERIES, take the component random variables identically distributed with mean μ, variance σ^2, and higher cumulants $\sigma^r \lambda_r$ for $r \geq 3$. Also, take the developing function $\Psi(t)$ as the standard NORMAL DISTRIBUTION FUNCTION $\Phi(t)$, so we have

$$\kappa_1 - \gamma_1 = 0 \tag{2}$$

$$\kappa_2 - \gamma_2 = 0 \tag{3}$$

$$\kappa_3 - \gamma_3 = \frac{\lambda^r}{n^{r/2-1}}. \tag{4}$$

Then the Edgeworth series is obtained by collecting terms to obtain the asymptotic expansion of the CHARACTERISTIC FUNCTION (PROBABILITY) OF THE FORM

$$f_n(t) = \left[1 + \sum_{r=1}^\infty \frac{P_r(it)}{n^{r/2}}\right] e^{-t^2/2}, \tag{5}$$

where P_r is a polynomial of degree $3r$ with coefficients depending on the cumulants of orders 3 to $r + 2$. If the powers of Ψ are interpreted as derivatives, then the distribution function expansion is given by

$$F_n(x) = \Psi(x) + \sum_{r=1}^\infty \frac{P_r(-\Phi(x))}{n^{r/2}} \tag{6}$$

(Wallace 1958). The first few terms of this expansion are then given by

$$f(t) = \Psi(t) - \frac{\lambda_3 \Psi^{(3)}(t)}{6\sqrt{n}} + \frac{1}{n}\left[\frac{\lambda_4 \Psi^{(4)}(t)}{24} + \frac{\lambda_3^2 \Psi^{(6)}(t)}{72}\right] + \cdots \tag{7}$$

Cramér (1928) proved that this series is uniformly valid in t.

See also CHARLIER SERIES, CORNISH-FISHER ASYMPTOTIC EXPANSION

References

Abramowitz, M. and Stegun, C. A. (Eds.). *Handbook of Mathematical Functions with Formulas, Graphs, and Mathematical Tables, 9th printing.* New York: Dover, p. 935, 1972.
Charlier, C. V. L. "Über dir Darstellung willkürlicher Funktionen." *Ark. Mat. Astr. och Fys.* **2**, No. 20, 1–5, 1906.
Cramér, H. "On the Composition of Elementary Errors." *Skand. Aktuarietidskr.* **11**, 13–4 and 141–80, 1928.
Edgeworth, F. Y. "The Law of Error." *Cambridge Philos. Soc.* **20**, 36–6 and 113–41, 1905.
Esseen, C. G. "Fourier Analysis of Distribution Functions." *Acta Math.* **77**, 1–25, 1945.
Hsu, P. L. "The Approximate Distribution of the Mean and Variance of a Sample of Independent Variables." *Ann. Math. Stat.* **16**, 1–9, 1945.
Kenney, J. F. and Keeping, E. S. *Mathematics of Statistics, Pt. 2, 2nd ed.* Princeton, NJ: Van Nostrand, pp. 107–08, 1951.
Wallace, D. L. "Asymptotic Approximations to Distributions." *Ann. Math. Stat.* **29**, 635–54, 1958.

e-Divisor

d is called an e-divisor (or exponential divisor) of a number n with PRIME FACTORIZATION

$$n = p_1^{a_1} p_2^{a_2} \cdots p_r^{a_r}$$

if $d \mid n$ and

$$d = p_1^{b_1} p_2^{b_2} \cdots p_r^{b_r},$$

where $b_j \mid a_j$ for $1 \leq j \leq r$. For example, the e-divisors of 36 are $2 \cdot 3$, $4 \cdot 3$, $2 \cdot 9$, and $4 \cdot 9$.

See also E-PERFECT NUMBER

References

Guy, R. K. "Exponential-Perfect Numbers." §B17 in *Unsolved Problems in Number Theory, 2nd ed.* New York: Springer-Verlag, p. 73, 1994.
Straus, E. G. and Subbarao, M. V. "On Exponential Divisors." *Duke Math. J.* **41**, 465–71, 1974.

Edmonds' Map

A nonreflexible regular map of GENUS 7 with eight VERTICES, 28 EDGES, and eight HEPTAGONAL faces.

Effective Action

A GROUP ACTION $G \times X \to X$ is effective if there are no nontrivial actions. In particular, this means that there is no element of the GROUP (besides the

IDENTITY ELEMENT) which does nothing, leaving every point where it is. This can be expressed as $\cap_{x \in X} G_x = \{e\}$, where G_x is the ISOTROPY GROUP at x and e is the identity of G.

It is possible for a LIE GROUP G to have an effective action on a smaller dimensional space M. However,

$$N(M) = \max\{\dim G | G \text{ is a compact Lie group,}$$
$$\text{acting effectively on } M\}$$

is finite, and is called the degree of symmetry of M.

See also FREE ACTION, GROUP, ISOTROPY GROUP, MATRIX GROUP, ORBIT (GROUP), QUOTIENT SPACE (LIE GROUP), REPRESENTATION, TOPOLOGICAL GROUP, TRANSITIVE

References

Kawakubo, K. *The Theory of Transformation Groups.* Oxford, England: Oxford University Press, pp. 4– and 221–24, 1987.

Efron's Dice

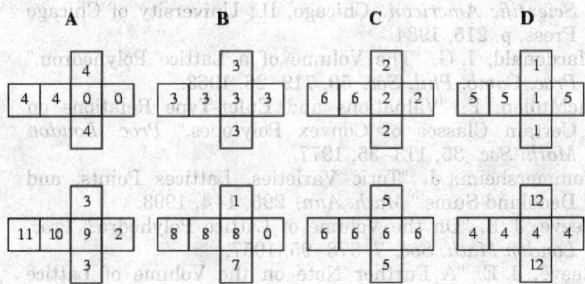

A set of four nontransitive DICE such that the probabilities of A winning against B, B against C, C against D, and D against A are all 2:1. A set in which ties may occur, in which case the DICE are rolled again, which gives ODDS of 11:6 is

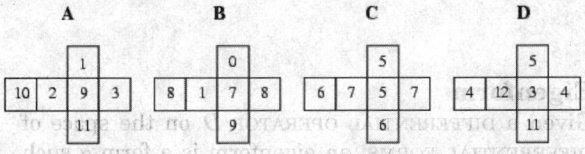

See also DICE, SICHERMAN DICE

References

Gardner, M. "Mathematical Games: The Paradox of the Nontransitive Dice and the Elusive Principle of Indifference." *Sci. Amer.* **223**, 110–14, Dec. 1970.
Honsberger, R. "Some Surprises in Probability." Ch. 5 in *Mathematical Plums* (Ed. R. Honsberger). Washington, DC: Math. Assoc. Amer., pp. 94–7, 1979.

E-Function

For any $\alpha \in \mathbb{A}$ (where $\mathbb{A}$ denotes the set of ALGEBRAIC NUMBERS), let $\overline{|\alpha|}$ denote the maximum of moduli of all conjugates of α. Then a function

$$f(z) = \sum_{n=0}^{\infty} c_n \frac{z^n}{n!}$$

is said to be an E-function if the following conditions hold (Nesterenko 1999).

1. All coefficients c_n belong to the same ALGEBRAIC NUMBER FIELD K of finite degree over $\mathbb{Q}$.
2. If $\epsilon > 0$ is any positive number, then $\overline{|c_n|} = \mathcal{O}(n^{\epsilon n})$ as $n \to \infty$.
3. For any $\epsilon > 0$, there exists a sequence of natural numbers $\{q_n\}_{n \geq 1}$ such that $q_n c_k \in \mathbb{Z}_K$ for $k = 0, ..., n$ and that $q_n = \mathcal{O}(n^{\epsilon n})$.

Every E-function is an ENTIRE FUNCTION, and the set of E-functions is a RING under the operations of ADDITION and MULTIPLICATION. Furthermore, if $f(z)$ is an E-function, then $f'(z)$ and $\int_0^z f(t) \, dt$ are E-functions, and for any ALGEBRAIC NUMBER α, the function $f(\alpha z)$ is also an E-function (Nesterenko 1999).

See also SHIDLOVSKII THEOREM

References

Nesterenko, Yu. V. §1.2 in *A Course on Algebraic Independence: Lectures at IHP 1999.* http://www.math.jussieu.fr/~nesteren/.
Siegel, C. L. *Transcendental Numbers.* New York: Chelsea, 1965.

Egg

An OVAL with one end more pointed than the other.

See also ELLIPSE, MOSS'S EGG, OVAL, OVOID, THOM'S EGGS

Egyptian Fraction

EGYPTIAN NUMBER, UNIT FRACTION

Egyptian Number

A number n is called an Egyptian number if it is the sum of the DENOMINATORS in some UNIT FRACTION representation of a positive whole number not consisting entirely of 1s. For example,

$$1 = \frac{1}{2} + \frac{1}{3} + \frac{1}{6},$$

so $2 + 3 + 6 = 11$ is an Egyptian number. The numbers which are *not* Egyptian are 2, 3, 5, 6, 7, 8, 12, 13, 14, 15, 19, 21, and 23 (Sloane's A028229; Konhauser *et al.* 1996, p. 147).

If n is the sum of denominators of a unit fraction representation composed of *distinct* denominators which are not all 1s, then it is called a strictly Egyptian number. For example, by virtue of

$$1 = \frac{1}{2} + \frac{1}{2},$$

$2 + 2 = 4$ is Egyptian, but it is *not* strictly Egyptian. Graham (1963) proved that every number ≥ 78 is strictly Egyptian. Numbers which are strictly Egyptian are 11, 24, 30, 31, 32, 37, 38, 43, ... (Sloane's A052428), and those which are not are 2, 3, 4, 5, 6, 7, 8, 9, 10, 12, ... (Sloane's A051882).

See also UNIT FRACTION

References

Graham, R. L. "A Theorem on Partitions." *J. Austral. Math. Soc.* **3**, 435–41, 1963.

Konhauser, J. D. E.; Vellman, D.; and Wagon, S. *Which Way Did the Bicycle Go and Other Intriguing Mathematical Mysteries.* Washington, DC: Amer. Math. Soc., 1996.

Sloane, N. J. A. Sequences A028229, A051882, and A052428 in "An On-Line Version of the Encyclopedia of Integer Sequences." http://www.research.att.com/~njas/sequences/eisonline.html.

Ehrhart Polynomial

Let Δ denote an integral convex POLYTOPE of DIMENSION n in a lattice M, and let $l_\Delta(k)$ denote the number of LATTICE POINTS in Δ dilated by a factor of the integer k,

$$l_\Delta(k) = \#(k\Delta \cap M) \tag{1}$$

for $k \in \mathbb{Z}^+$. Then l_Δ is a polynomial function in k of degree n with rational coefficients

$$l_\Delta(k) = a_n k^n + a_{n-1} k^{n-1} + \ldots + a_0 \tag{2}$$

called the Ehrhart polynomial (Ehrhart 1967, Pommersheim 1993). Specific coefficients have important geometric interpretations.

1. a_n is the CONTENT of Δ.
2. a_{n-1} is half the sum of the CONTENTS of the $(n-1)$-D faces of Δ.
3. $a_0 = 1$.

Let $S_2(\Delta)$ denote the sum of the lattice lengths of the edges of Δ, then the case $n = 2$ corresponds to PICK'S THEOREM,

$$l_\Delta(k) = \mathrm{Vol}(\Delta)k^2 + \tfrac{1}{2} S_2(\Delta) + 1. \tag{3}$$

Let $S_3(\Delta)$ denote the sum of the lattice volumes of the 2-D faces of Δ, then the case $n = 3$ gives

$$l_\Delta(k) = \mathrm{Vol}(\Delta)k^3 + \tfrac{1}{2} S_3(\Delta)k^2 + a_1 k + 1, \tag{4}$$

where a rather complicated expression is given by Pommersheim (1993), since a_1 can unfortunately *not* be interpreted in terms of the edges of Δ. The Ehrhart polynomial of the tetrahedron with vertices at $(0, 0, 0)$, $(a, 0, 0)$, $(0, b, 0)$, $(0, 0, c)$ is

$$l_\Delta(k) = \tfrac{1}{6} abck^3 + \tfrac{1}{4}(ab + ac + bc + d)k^2$$
$$+ \left[\frac{1}{12}\left(\frac{ac}{b} + \frac{bc}{a} + \frac{ab}{c} + \frac{d^2}{abc} \right) + \tfrac{1}{4}(a + b + c + A + B + C) \right.$$
$$- As\left(\frac{bc}{d}, \frac{aA}{d} \right) - Bs\left(\frac{ac}{d}, \frac{bB}{d} \right)$$
$$\left. - Cs\left(\frac{ab}{d}, \frac{cC}{d} \right) \right] k + 1, \tag{5}$$

where $s(x, y)$ is a DEDEKIND SUM, $A = \mathrm{GCD}(b, c)$, $B = \mathrm{GCD}(a, c)$, $C = \mathrm{GCD}(a, b)$ (here, GCD is the GREATEST COMMON DIVISOR), and $d = ABC$ (Pommersheim 1993).

See also DEHN INVARIANT, PICK'S THEOREM

References

Ehrhart, E. "Sur une problème de géométrie diophantine linéaire." *J. reine angew. Math.* **227**, 1–9, 1967.

Gardner, M. *The Sixth Book of Mathematical Games from Scientific American.* Chicago, IL: University of Chicago Press, p. 215, 1984.

Macdonald, I. G. "The Volume of a Lattice Polyhedron." *Proc. Camb. Phil. Soc.* **59**, 719–26, 1963.

McMullen, P. "Valuations and Euler-Type Relations on Certain Classes of Convex Polytopes." *Proc. London Math. Soc.* **35**, 113–35, 1977.

Pommersheim, J. "Toric Varieties, Lattices Points, and Dedekind Sums." *Math. Ann.* **295**, 1–4, 1993.

Reeve, J. E. "On the Volume of Lattice Polyhedra." *Proc. London Math. Soc.* **7**, 378–95, 1957.

Reeve, J. E. "A Further Note on the Volume of Lattice Polyhedra." *Proc. London Math. Soc.* **34**, 57–2, 1959.

Ei

EXPONENTIAL INTEGRAL, E_N-FUNCTION

Eigenform

Given a DIFFERENTIAL OPERATOR D on the space of DIFFERENTIAL FORMS, an eigenform is a form α such that

$$D\alpha = \lambda\alpha$$

for some constant λ. For example, on the TORUS, the DIRAC OPERATOR $D = -i(d + d^*)$ acts on the form

$$\beta = 3e^{i(3x+4y)} + 5e^{i(3x+4y)}\, dx - 4e^{i(3x+4y)}\, dx \wedge dy,$$

giving

$$D\beta = 15e^{i(3x+4y)} + 25e^{i(3x+4y)}\, dx - 20e^{i(3x+4y)}\, dx \wedge dy,$$

i.e., $D\beta = 5\beta$.

See also DIRAC OPERATOR, LAPLACIAN, SPECTRUM (OPERATOR)

Eigenfunction

If $\check{L}$ is a linear OPERATOR on a FUNCTION SPACE, then f is an eigenfunction for $\check{L}$ and λ is the associated EIGENVALUE whenever $\check{L}f = \lambda f$.

See also EIGENVALUE, EIGENVECTOR, FUNCTIONAL

Eigenspace

If A is an $n \times n$ matrix, and λ is an EIGENVALUE of A, then the union of the ZERO VECTOR and the set of all EIGENVECTORS corresponding to λ is a SUBSPACE of $\mathbb{R}^n$ known as the EIGENSPACE of λ.

Eigenvalue

Let A be a linear transformation represented by a MATRIX A. If there is a VECTOR $\mathbf{X} \in \mathbb{R}^n \neq 0$ such that

$$\mathbf{AX} = \lambda \mathbf{X} \qquad (1)$$

for some SCALAR λ, then λ is called the eigenvalue of A with corresponding (right) EIGENVECTOR $\mathbf{X}$. Eigenvalues are also known as characteristic roots, proper values, or latent roots (Marcus and Minc 1988, p. 144).

Letting A be a $k \times k$ MATRIX,

$$\begin{bmatrix} a_{11} & a_{12} & \cdots & a_{1k} \\ a_{21} & a_{22} & \cdots & a_{2k} \\ \vdots & \vdots & \ddots & \vdots \\ a_{k1} & a_{k2} & \cdots & a_{kk} \end{bmatrix} \qquad (2)$$

with eigenvalue λ, then the corresponding EIGENVECTORS satisfy

$$\begin{bmatrix} a_{11} & a_{12} & \cdots & a_{1k} \\ a_{21} & a_{22} & \cdots & a_{2k} \\ \vdots & \vdots & \ddots & \vdots \\ a_{k1} & a_{k2} & \cdots & a_{kk} \end{bmatrix} \begin{bmatrix} x_1 \\ x_2 \\ \vdots \\ x_k \end{bmatrix} = \lambda \begin{bmatrix} x_1 \\ x_2 \\ \vdots \\ x_k \end{bmatrix}, \qquad (3)$$

which is equivalent to the homogeneous system

$$\begin{bmatrix} a_{11} - \lambda & a_{12} & \cdots & a_{1k} \\ a_{21} & a_{22} - \lambda & \cdots & a_{2k} \\ \vdots & \vdots & \ddots & \vdots \\ a_{k1} & a_{k2} & \cdots & a_{kk} - \lambda \end{bmatrix} \begin{bmatrix} x_1 \\ x_2 \\ \vdots \\ x_k \end{bmatrix} = \begin{bmatrix} 0 \\ 0 \\ \vdots \\ 0 \end{bmatrix}. \qquad (4)$$

Equation (4) can be written compactly as

$$(\mathbf{A} - \lambda \mathbf{I})\mathbf{X} = 0, \qquad (5)$$

where I is the IDENTITY MATRIX. This MATRIX EQUATION can then be solved for λ.

Eigenvalues are given by the solutions of the CHARACTERISTIC EQUATION of a given matrix. For example, for a 2×2 matrix, the eigenvalues are

$$\lambda_\pm = \tfrac{1}{2}\left[(a_{11} + a_{22}) \pm \sqrt{4a_{12}a_{21} + (a_{11} - a_{22})^2}\right], \qquad (6)$$

which arises as the solutions of the CHARACTERISTIC EQUATION

$$x^2 - x(a_{11} + a_{22}) + (a_{11}a_{22} - a_{12}a_{21}) = 0, \qquad (7)$$

which can be written

$$x^2 - x\mathrm{Tr}(\mathsf{A}) + \det(\mathsf{A}) = 0, \qquad (8)$$

where $\mathrm{Tr}(\mathsf{A})$ is the TRACE of A and $\det(\mathsf{A})$ is its DETERMINANT. The CHARACTERISTIC EQUATION for the 3×3 case is

$$x^3 - \mathrm{Tr}(\mathsf{A})x^2 - \tfrac{1}{2}(a_{ij}a_{ji} - a_{ii}a_{jj})(1 - \delta_{ij})x - \det(\mathsf{A}) = 0, \quad (9)$$

where δ_{ij} is the KRONECKER DELTA and EINSTEIN SUMMATION has been used. The corresponding analytic eigenvalue expressions for 4×4 and larger matrices are very complicated.

As shown in CRAMER'S RULE, a system of linear equations has nontrivial solutions only if the DETERMINANT vanishes, so we obtain the CHARACTERISTIC EQUATION

$$|\mathsf{A} - \lambda \mathsf{I}| = 0. \qquad (10)$$

If all k λs are different, then plugging these back in gives $k - 1$ independent equations for the k components of each corresponding EIGENVECTOR. The EIGENVECTORS will then be orthogonal and the system is said to be nondegenerate. If the eigenvalues are n-fold DEGENERATE, then the system is said to be degenerate and the EIGENVECTORS are not linearly independent. In such cases, the additional constraint that the EIGENVECTORS be ORTHOGONAL,

$$\mathbf{X}_i \cdot \mathbf{X}_j = |\mathbf{X}_i||\mathbf{X}_j|\delta_{ij}, \qquad (11)$$

where δ_{ij} is the KRONECKER DELTA, can be applied to yield n additional constraints, thus allowing solution for the EIGENVECTORS.

Assume A has nondegenerate eigenvalues $\lambda_1, \lambda_2, \ldots, \lambda_k$ and corresponding linearly independent EIGENVECTORS $\mathbf{X}_1, \mathbf{X}_2, \ldots, \mathbf{X}_k$ which can be denoted

$$\begin{bmatrix} x_{11} \\ x_{12} \\ \vdots \\ x_{1k} \end{bmatrix}, \begin{bmatrix} x_{21} \\ x_{22} \\ \vdots \\ x_{2k} \end{bmatrix}, \ldots \begin{bmatrix} x_{k1} \\ x_{k2} \\ \vdots \\ x_{kk} \end{bmatrix}. \qquad (12)$$

Define the matrices composed of eigenvectors

$$\mathsf{P} \equiv [\mathbf{X}_1 \quad \mathbf{X}_2 \quad \cdots \quad \mathbf{X}_k] = \begin{bmatrix} x_{11} & x_{21} & \cdots & x_{k1} \\ x_{12} & x_{22} & \cdots & x_{k2} \\ \vdots & \vdots & \ddots & \vdots \\ x_{1k} & x_{2k} & \cdots & x_{kk} \end{bmatrix} \qquad (13)$$

and eigenvalues

$$\mathsf{D} \equiv \begin{bmatrix} \lambda_1 & 0 & \cdots & 0 \\ 0 & \lambda_2 & \cdots & 0 \\ \vdots & \vdots & \ddots & \vdots \\ 0 & 0 & \cdots & \lambda_k \end{bmatrix}, \qquad (14)$$

where D is a DIAGONAL MATRIX. Then

$$\mathbf{AP} = \mathbf{A}[\mathbf{X}_1 \quad \mathbf{X}_2 \quad \cdots \quad \mathbf{X}_k]$$

$$= [\mathbf{AX}_1 \quad \mathbf{AX}_2 \quad \cdots \quad \mathbf{AX}_k]$$

$$= [\lambda_1 \mathbf{X}_1 \quad \lambda_2 \mathbf{X}_2 \quad \cdots \quad \lambda_k \mathbf{X}_k]$$

$$= \begin{bmatrix} \lambda_1 x_{11} & \lambda_2 x_{21} & \cdots & \lambda_k x_{k1} \\ \lambda_1 x_{12} & \lambda_2 x_{22} & \cdots & \lambda_k x_{k2} \\ \vdots & \vdots & \ddots & \vdots \\ \lambda_1 x_{1k} & \lambda_2 x_{2k} & \cdots & \lambda_k x_{kk} \end{bmatrix}$$

$$= \begin{bmatrix} x_{11} & x_{21} & \cdots & x_{k1} \\ x_{12} & x_{22} & \cdots & x_{k2} \\ \vdots & \vdots & \ddots & \vdots \\ x_{1k} & x_{2k} & \cdots & x_{kk} \end{bmatrix} \begin{bmatrix} \lambda_1 & 0 & \cdots & 0 \\ 0 & \lambda_2 & \cdots & 0 \\ \vdots & \vdots & \ddots & \vdots \\ 0 & 0 & \cdots & \lambda_k \end{bmatrix}$$

$$= \mathbf{PD}, \tag{15}$$

so

$$\mathbf{A} = \mathbf{PDP}^{-1}. \tag{16}$$

Furthermore,

$$\mathbf{A}^2 = (\mathbf{PDP}^{-1})(\mathbf{PDP}^{-1}) = \mathbf{PD}(\mathbf{P}^{-1}\mathbf{P})\mathbf{DP}^{-1}$$

$$= \mathbf{PD}^2\mathbf{P}^{-1}. \tag{17}$$

By induction, it follows that for $n > 0$,

$$\mathbf{A}^n = \mathbf{PD}^n\mathbf{P}^{-1}. \tag{18}$$

The inverse of $\mathbf{A}$ is

$$\mathbf{A}^{-1} = (\mathbf{PDP}^{-1})^{-1} = \mathbf{PD}^{-1}\mathbf{P}^{-1}, \tag{19}$$

where the inverse of the DIAGONAL MATRIX $\mathbf{D}$ is trivially given by

$$\mathbf{D}^{-1} = \begin{bmatrix} \lambda_1^{-1} & 0 & \cdots & 0 \\ 0 & \lambda_2^{-1} & \cdots & 0 \\ \vdots & \vdots & \ddots & \vdots \\ 0 & 0 & \cdots & \lambda_k^{-1} \end{bmatrix}. \tag{20}$$

Equation (18) therefore holds for both POSITIVE and NEGATIVE n.

A further remarkable result involving the matrices $\mathbf{P}$ and $\mathbf{D}$ follows from the definition

$$e^{\mathbf{A}} \equiv \sum_{n=0}^{\infty} \frac{\mathbf{A}^n}{n!} = \sum_{n=0}^{\infty} \frac{\mathbf{PD}^n\mathbf{P}^{-1}}{n!}$$

$$= \mathbf{P}\left(\frac{\sum_{n=0}^{\infty} \mathbf{D}^n}{n!} \right) \mathbf{P}^{-1} = \mathbf{P}e^{\mathbf{D}}\mathbf{P}^{-1}. \tag{21}$$

Since $\mathbf{D}$ is a DIAGONAL MATRIX,

$$e^{\mathbf{D}} = \sum_{n=0}^{\infty} \frac{\mathbf{D}^n}{n!} = \sum_{n=0}^{\infty} \frac{1}{n!} \begin{bmatrix} \lambda_1^n & 0 & \cdots & 0 \\ 0 & \lambda_2^n & \cdots & 0 \\ \vdots & \vdots & \ddots & \vdots \\ 0 & 0 & \cdots & \lambda_k^n \end{bmatrix}$$

$$= \begin{bmatrix} \sum_{n=0}^{\infty} \dfrac{\lambda_1^n}{n!} & 0 & \cdots & 0 \\ 0 & \sum_{n=0}^{\infty} \dfrac{\lambda_2^n}{n!} & \cdots & 0 \\ \vdots & \vdots & \ddots & \vdots \\ 0 & 0 & \cdots & \sum_{n=0}^{\infty} \dfrac{\lambda_k^n}{n!} \end{bmatrix}$$

$$= \begin{bmatrix} e^{\lambda_1} & 0 & \cdots & 0 \\ 0 & e^{\lambda_2} & \cdots & 0 \\ \vdots & \vdots & \ddots & \vdots \\ 0 & 0 & \cdots & e^{\lambda_k} \end{bmatrix}, \tag{22}$$

$e^{\mathbf{D}}$ can be found using

$$\mathbf{D}^n = \begin{bmatrix} \lambda_1^n & 0 & \cdots & 0 \\ 0 & \lambda_2^n & \cdots & 0 \\ \vdots & \vdots & \ddots & \vdots \\ 0 & 0 & \cdots & \lambda_k^n \end{bmatrix} \tag{23}$$

Assume we know the eigenvalue for

$$\mathbf{AX} = \lambda \mathbf{X}. \tag{24}$$

Adding a constant times the IDENTITY MATRIX to $\mathbf{A}$,

$$(\mathbf{A} + c\mathbf{I})\mathbf{X} = (\lambda + c)\mathbf{X} \equiv \lambda'\pi\mathbf{X}, \tag{25}$$

so the new eigenvalues equal the old plus c. Multiplying $\mathbf{A}$ by a constant c

$$(c\mathbf{A})\mathbf{X} = c(\lambda\mathbf{X}) \equiv \lambda'\mathbf{X}, \tag{26}$$

so the new eigenvalues are the old multiplied by c.

Now consider a SIMILARITY TRANSFORMATION of $\mathbf{A}$. Let $|\mathbf{A}|$ be the DETERMINANT of $\mathbf{A}$, then

$$|\mathbf{Z}^{-1}\mathbf{AZ} - \lambda\mathbf{I}| = |\mathbf{Z}^{-1}(\mathbf{A} - \lambda\mathbf{I})\mathbf{Z}|$$

$$= |\mathbf{Z}||\mathbf{A} - \lambda\mathbf{I}||\mathbf{Z}^{-1}| = |\mathbf{A} - \lambda\mathbf{I}|, \tag{27}$$

so the eigenvalues are the same as for $\mathbf{A}$.

See also BRAUER'S THEOREM, COMPLEX MATRIX, CONDITION NUMBER, EIGENFUNCTION, EIGENVECTOR, FROBENIUS THEOREM, GERSGORIN CIRCLE THEOREM, LYAPUNOV'S FIRST THEOREM, LYAPUNOV'S SECOND THEOREM, OSTROWSKI'S THEOREM, PERRON'S THEOREM, PERRON-FROBENIUS THEOREM, POINCARÉ SEPARATION THEOREM, RANDOM MATRIX, REAL MATRIX, SCHUR'S INEQUALITIES, STURMIAN SEPARATION THEOREM, SYLVESTER'S INERTIA LAW, WIELANDT'S THEOREM

References

Arfken, G. "Eigenvectors, Eigenvalues." §4.7 in *Mathematical Methods for Physicists, 3rd ed.* Orlando, FL: Academic Press, pp. 229–37, 1985.

Marcus, M. and Minc, H. *Introduction to Linear Algebra.* New York: Dover, p. 145, 1988.

Nash, J. C. "The Algebraic Eigenvalue Problem." Ch. 9 in *Compact Numerical Methods for Computers: Linear Algebra and Function Minimisation, 2nd ed.* Bristol, England: Adam Hilger, pp. 102–18, 1990.

Press, W. H.; Flannery, B. P.; Teukolsky, S. A.; and Vetterling, W. T. "Eigensystems." Ch. 11 in *Numerical Recipes in FORTRAN: The Art of Scientific Computing, 2nd ed.* Cambridge, England: Cambridge University Press, pp. 449–89, 1992.

Eigenvector

A right eigenvector satisfies

$$A\mathbf{X} = \lambda\mathbf{X}, \tag{1}$$

where $\mathbf{X}$ is a column VECTOR. The right EIGENVALUES therefore satisfy

$$|A - \lambda I| = 0. \tag{2}$$

A left eigenvector satisfies

$$\mathbf{X}A = \lambda\mathbf{X}, \tag{3}$$

where $\mathbf{X}$ is a row VECTOR, so

$$(\mathbf{X}A)^{T} = \lambda_L\mathbf{X}^{T}, \tag{4}$$

$$A^{T}\mathbf{X}^{T} = \lambda_L\mathbf{X}_T, \tag{5}$$

where $\mathbf{X}^{T}$ is the transpose of $\mathbf{X}$.
The left EIGENVALUES satisfy

$$\left|A^{T} - \lambda_L I\right| = \left|A^{T} - \lambda_L I^{T}\right| = \left|(A - \lambda_L I)^{T}\right| = |(A - \lambda_L I)|, \tag{6}$$

(since $|A| = |A^{T}|$) where $|A|$ is the DETERMINANT of A. But this is the same equation satisfied by the right EIGENVALUES, so the left and right EIGENVALUES are the same. Let $\mathbf{X}_R$ be a MATRIX formed by the columns of the right eigenvectors and $\mathbf{X}_L$ be a MATRIX formed by the rows of the left eigenvectors. Let

$$D \equiv \begin{bmatrix} \lambda_1 & \cdots & 0 \\ \vdots & \ddots & \vdots \\ 0 & \cdots & \lambda_n \end{bmatrix}. \tag{7}$$

Then

$$A\mathbf{X}_R = \mathbf{X}_R D \quad \mathbf{X}_L A = D\mathbf{X}_L \tag{8}$$

$$\mathbf{X}_L A\mathbf{X}_R = \mathbf{X}_L \mathbf{X}_R D \quad \mathbf{X}_L A\mathbf{X}_R = D\mathbf{X}_L \mathbf{X}_R, \tag{9}$$

so

$$\mathbf{X}_L \mathbf{X}_R D = D\mathbf{X}_L \mathbf{X}_R. \tag{10}$$

But this equation is OF THE FORM CD = DC where D is a DIAGONAL MATRIX, so it must be true that $C \equiv \mathbf{X}_L\mathbf{X}_R$ is also diagonal. In particular, if A is a SYMMETRIC MATRIX, then the left and right eigenvectors are transposes of each other. If A is a SELF-ADJOINT MATRIX, then the left and right eigenvectors are conjugate HERMITIAN MATRICES.

Eigenvectors are sometimes known as characteristic vectors, proper vectors, or latent vectors (Marcus and Minc 1988, p. 144).

Given a 3×3 MATRIX A with eigenvectors $\mathbf{x}_1$, $\mathbf{x}_2$, and $\mathbf{x}_3$ and corresponding EIGENVALUES λ_1, λ_2, and λ_3,

then an arbitrary VECTOR $\mathbf{y}$ can be written

$$y = b_1\mathbf{x}_1 + b_2\mathbf{x}_2 + b_3\mathbf{x}_3. \tag{11}$$

Applying the MATRIX A,

$$A\mathbf{y} = b_1 A\mathbf{x}_1 + b_2 A\mathbf{x}_2 + b_3 A\mathbf{x}_3$$

$$= \lambda_1\left(b_1\mathbf{x}_1 + \frac{\lambda_2}{\lambda_1}b_2\mathbf{x}_2 + \frac{\lambda_3}{\lambda_1}b_3\mathbf{x}_3\right), \tag{12}$$

so

$$A^{n}\mathbf{y} = \lambda_1^{n}\left[b_1\mathbf{x}_1 + \left(\frac{\lambda_2}{\lambda_1}\right)^{n}b_2\mathbf{x}_2 + \left(\frac{\lambda_3}{\lambda_1}\right)^{n}b_3\mathbf{x}_3\right]. \tag{13}$$

If $\lambda_1 > \lambda_2$, λ_3, it therefore follows that

$$\lim_{n\to\infty} A^{n}\mathbf{y} = \lambda_1^{n}b_1\mathbf{x}_I, \tag{14}$$

so repeated application of the matrix to an arbitrary vector results in a vector proportional to the EIGENVECTOR having the largest EIGENVALUE.

See also EIGENFUNCTION, EIGENVALUE

References

Arfken, G. "Eigenvectors, Eigenvalues." §4.7 in *Mathematical Methods for Physicists, 3rd ed.* Orlando, FL: Academic Press, pp. 229–37, 1985.
Marcus, M. and Minc, H. *Introduction to Linear Algebra.* New York: Dover, p. 145, 1988.
Press, W. H.; Flannery, B. P.; Teukolsky, S. A.; and Vetterling, W. T. "Eigensystems." Ch. 11 in *Numerical Recipes in FORTRAN: The Art of Scientific Computing, 2nd ed.* Cambridge, England: Cambridge University Press, pp. 449–89, 1992.

Eight Curve

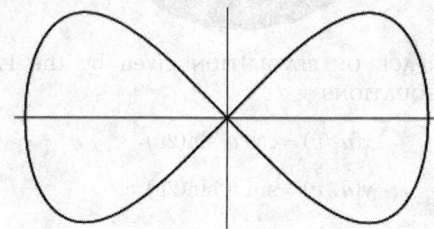

A curve also known as the GERONO LEMNISCATE. It is given by CARTESIAN COORDINATES

$$x^4 = a^2(x^2 - y^2), \tag{1}$$

POLAR COORDINATES,

$$r^2 = a^2 \sec^4 \theta \cos(2\theta), \tag{2}$$

and PARAMETRIC EQUATIONS

$$x = a \sin t \qquad (3)$$

$$y = a \sin t \cos t. \qquad (4)$$

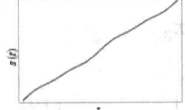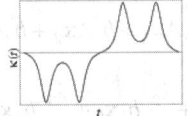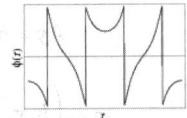

The CURVATURE and TANGENTIAL ANGLE are

$$\kappa(t) = -\frac{3 \sin t + \sin(3t)}{2[\cos^2 t + \cos^2(2t)]^{3/2}} \qquad (5)$$

$$\phi(t) = -\tan^{-1}[\cos t \sec(2t)]. \qquad (6)$$

See also BUTTERFLY CURVE, DUMBBELL CURVE, EIGHT SURFACE, PIRIFORM

References

Cundy, H. and Rollett, A. *Mathematical Models, 3rd ed.* Stradbroke, England: Tarquin Pub., p. 71, 1989.
Lawrence, J. D. *A Catalog of Special Plane Curves.* New York: Dover, pp. 124–26, 1972.
MacTutor History of Mathematics Archive. "Eight Curve." http://www-groups.dcs.st-and.ac.uk/~history/Curves/Eight.html.

Eight Surface

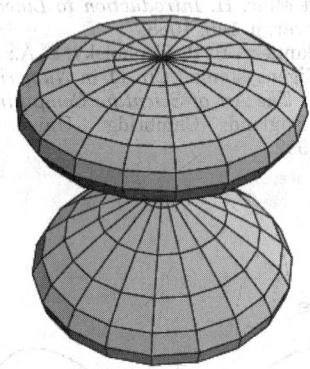

The SURFACE OF REVOLUTION given by the PARAMETRIC EQUATIONS

$$x(u, v) = \cos u \sin(2v) \qquad (1)$$

$$y(u, v) = \sin u \sin(2v) \qquad (2)$$

$$z(u, v) = \sin v \qquad (3)$$

for $u \in [0, 2\pi)$ and $v \in [-\pi/2, \pi/2]$.

See also EIGHT CURVE

References

Gray, A. *Modern Differential Geometry of Curves and Surfaces with Mathematica, 2nd ed.* Boca Raton, FL: CRC Press, p. 310, 1997.

Eight-Point Circle Theorem

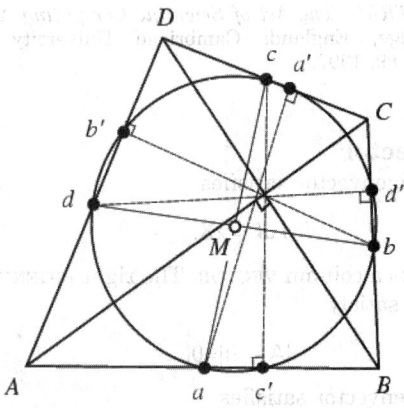

Let *ABCD* be a QUADRILATERAL with PERPENDICULAR DIAGONALS. The MIDPOINTS of the sides (a, b, c, and d) determine a PARALLELOGRAM (the VARIGNON PARALLELOGRAM) with sides PARALLEL to the DIAGONALS. The eight-point circle passes through the four MIDPOINTS and the four feet of the PERPENDICULARS from the opposite sides a', b', c', and d'.

See also FEUERBACH'S THEOREM

References

Brand, L. "The Eight-Point Circle and the Nine-Point Circle." *Amer. Math. Monthly* **51**, 84–5, 1944.
Honsberger, R. *Mathematical Gems II.* Washington, DC: Math. Assoc. Amer., pp. 11–3, 1976.

Eikonal Equation

$$\sum_{i=1}^{n} \left(\frac{\partial u}{\partial x_i} \right)^2 = 1.$$

Eilenberg-Mac Lane Space

For any ABELIAN GROUP G and any NATURAL NUMBER n, there is a unique SPACE (up to HOMOTOPY type) such that all HOMOTOPY GROUPS except for the nth are trivial (including the 0th HOMOTOPY GROUPS, meaning the SPACE is path-connected), and the nth HOMOTOPY GROUP is ISOMORPHIC to the GROUP G. In the case where $n = 1$, the GROUP G can be non-ABELIAN as well.

Eilenberg-Mac Lane spaces have many important applications. One of them is that every TOPOLOGICAL SPACE has the HOMOTOPY type of an iterated FIBRATION of Eilenberg-Mac Lane spaces (called a POSTNIKOV SYSTEM). In addition, there is a spectral sequence relating the COHOMOLOGY of Eilenberg-Mac Lane spaces to the HOMOTOPY GROUPS of SPHERES.

Eilenberg-Mac Lane-Steenrod-Milnor Axioms

EILENBERG-STEENROD AXIOMS

Eilenberg-Steenrod Axioms

A family of FUNCTORS $H_n(\cdot)$ from the CATEGORY of pairs of TOPOLOGICAL SPACES and continuous maps, to the CATEGORY of ABELIAN GROUPS and group homomorphisms satisfies the Eilenberg-Steenrod axioms if the following conditions hold.

1. LONG EXACT SEQUENCE OF A PAIR AXIOM. For every pair (X, A), there is a natural long exact sequence

$$\ldots \to H_n(A) \to H_n(X) \to H_n(X, A) \to H_{n-1}(A)$$
$$\to \ldots, \tag{1}$$

where the MAP $H_n(A) \to H_n(X)$ is induced by the INCLUSION MAP $A \to X$ and $H_n(X) \to H_n(X, A)$ is induced by the INCLUSION MAP $(X, \phi) \to (X, A)$. The MAP $H_n(X, A) \to H_{n-1}(A)$ is called the BOUNDARY MAP.

2. HOMOTOPY AXIOM. If $f : (X, A) \to (Y, B)$ is homotopic to $g : (X, A) \to (Y, B)$, then their INDUCED MAPS $f_* : H_n(X, A) \to H_n(Y, B)$ and $g_* : H_n(X, A) \to H_n(Y, B)$ are the same.

3. EXCISION AXIOM. If X is a SPACE with SUBSPACES A and U such that the CLOSURE of A is contained in the interior of U, then the INCLUSION MAP $(X \setminus U, A \setminus U) \to (X, A)$ induces an isomorphism $H_n(X \setminus U, A \setminus U) \to H_n(X, A)$.

4. DIMENSION AXIOM. Let X be a single point space. $H_n(X) = 0$ unless $n = 0$, in which case $H_0(X) = G$ where G are some GROUPS. The H_0 are called the COEFFICIENTS of the HOMOLOGY theory $H(\cdot)$.

These are the axioms for a generalized homology theory. For a cohomology theory, instead of requiring that $H(\cdot)$ be a FUNCTOR, it is required to be a cofunctor (meaning the INDUCED MAP points in the opposite direction). With that modification, the axioms are essentially the same (except that all the induced maps point backwards).

See also ALEKSANDROV-CECH COHOMOLOGY

Ein Function

$$\mathrm{Ein}(z) \equiv \int_0^z \frac{(1 - e^{-t})\, dt}{t} = E_1(z) + \ln z + \gamma,$$

where γ is the EULER-MASCHERONI CONSTANT and E^1 is the E_N-FUNCTION with $n = 1$.

See also E_N-FUNCTION

Einstein Field Equations

The 16 coupled hyperbolic-elliptic nonlinear PARTIAL DIFFERENTIAL EQUATIONS that describe the gravitational effects produced by a given mass in general relativity. The equations state that

$$G_{\mu\nu} = 8\pi T_{\mu\nu},$$

where $T_{\mu\nu}$ is the stress-energy tensor, and

$$G_{\mu\nu} = R_{\mu\nu} - \tfrac{1}{2} g_{\mu\nu} R$$

is the EINSTEIN TENSOR, with $R_{\mu\nu}$ the RICCI TENSOR and R the SCALAR CURVATURE.

© *1999–001 Wolfram Research, Inc.*

Einstein Functions

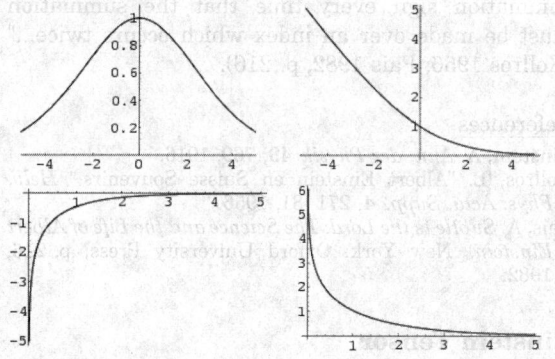

The functions

$$E_1(x) = \frac{x^2 e^x}{(e^x - 1)^2} \tag{1}$$

$$E_2(x) = \frac{x}{e^x - 1} \tag{2}$$

$$E_3(x) = \ln(1 - e^{-x}) \tag{3}$$

$$E_4(x) = \frac{x}{e^x - 1} - \ln(1 - e^{-x}). \tag{4}$$

$E_1(x)$ has an inflection point at

$$E_1''(x) = \tfrac{1}{8} \operatorname{csch}^4(\tfrac{1}{2} x)[(x^2 + 2)\cosh x$$
$$+ 2(x^2 - 2x \sinh x - 1)] = 0 \tag{5}$$

which can be solved numerically to give $x \approx 2.34693$. $E_1(x)$ has an inflection point at

$$E_2''(x) = \frac{e^x[x + 2 + e^x(x - 2)]}{(e^x - 1)^3} = 0, \tag{6}$$

which can be solved numerically to give $x \approx 17.5221$.

References

Abramowitz, M. and Stegun, C. A. (Eds.). "Debye Functions." §27.1 in *Handbook of Mathematical Functions with*

Formulas, Graphs, and Mathematical Tables, 9th printing. New York: Dover, pp. 999–000, 1972.

Einstein Summation

The convention that repeated indices are implicitly summed over. This can greatly simplify and shorten equations involving TENSORS. For example, using Einstein summation,

$$a_i a_i \equiv \sum_i a_i a_i$$

and

$$a_{ik} a_{ij} = \sum_i a_{ik} a_{ij}.$$

The convention was introduced by Einstein (1916), who later jested to a friend,"I have made a great discovery in mathematics; I have suppressed the summation sign every time that the summation must be made over an index which occurs twice..." (Kollros 1956; Pais 1982, p. 216).

References

Einstein, A. *Ann. der Physik* **49**, 769, 1916.
Kollros, L. "Albert Einstein en Suisse Souvenirs." *Helv. Phys. Acta. Supp.* **4**, 271–81, 1956.
Pais, A. *Subtle is the Lord: The Science and the Life of Albert Einstein.* New York: Oxford University Press, p. 216, 1982.

Einstein Tensor

$$G_{ab} = R_{ab} - \tfrac{1}{2} R g_{ab},$$

where R_{ab} is the RICCI TENSOR, R is the SCALAR CURVATURE, and g_{ab} is the METRIC TENSOR. (Wald 1984, pp. 40–1). It satisfies

$$G^{\mu\nu}{}_{;\nu} = 0$$

(Misner *et al.* 1973, p. 222).

See also METRIC TENSOR, RICCI TENSOR, SCALAR CURVATURE

References

Misner, C. W.; Thorne, K. S.; and Wheeler, J. A. *Gravitation.* San Francisco: W. H. Freeman, 1973.
Wald, R. M. *General Relativity.* Chicago, IL: University of Chicago Press, 1984.
© 1999–001 Wolfram Research, Inc.

Eisenstein Integer

The numbers $a + b\omega$, where

$$\omega \equiv \tfrac{1}{2}(-1 + i\sqrt{3})$$

is one of the ROOTS of $z^3 = 1$, the others being 1 and

$$\omega^2 \equiv \tfrac{1}{2}(-1 - i\sqrt{3}).$$

Eisenstein integers are members of the IMAGINARY QUADRATIC FIELD $\mathbb{Q}(\sqrt{-3})$, and the COMPLEX NUMBERS $\mathbb{Z}[\omega]$. Every Eisenstein integer has a unique factorization. Specifically, any NONZERO Eisenstein integer is uniquely the product of POWERS of -1, ω, and the "positive" EISENSTEIN PRIMES (Conway and Guy 1996). Every Eisenstein integer is within a distance $|n|/\sqrt{3}$ of some multiple of a given Eisenstein integer n.

Dörrie (1965) uses the alternative notation

$$J \equiv \tfrac{1}{2}(1 + i\sqrt{3}) \tag{1}$$

$$O \equiv \tfrac{1}{2}(1 - i\sqrt{3}). \tag{2}$$

for $-\omega^2$ and $-\omega$, and calls numbers OF THE FORM $aJ + bO$ G-NUMBERS. O and J satisfy

$$J + O = 1 \tag{3}$$

$$JO = 1 \tag{4}$$

$$J^2 + O = 0 \tag{5}$$

$$O^2 + J = 0 \tag{6}$$

$$J^3 = -1 \tag{7}$$

$$O^3 = -1. \tag{8}$$

The sum, difference, and products of G numbers are also G numbers. The norm of a G number is

$$N(aJ + bO) = a^2 + b^2 - ab. \tag{9}$$

The analog of FERMAT'S THEOREM for Eisenstein integers is that a PRIME NUMBER p can be written in the form

$$a^2 - ab + b^2 = (a + b\omega)(a + b\omega^2)$$

IFF $3 \nmid p + 1$. These are precisely the PRIMES OF THE FORM $3m^2 + n^2$ (Conway and Guy 1996).

See also EISENSTEIN PRIME, EISENSTEIN UNIT, GAUSSIAN INTEGER, INTEGER

References

Conway, J. H. and Guy, R. K. *The Book of Numbers.* New York: Springer-Verlag, pp. 220–23, 1996.
Cox, D. A. §4A in *Primes of the Form $x^2 + ny^2$: Fermat, Class Field Theory and Complex Multiplication.* New York: Wiley, 1989.
Dörrie, H. "The Fermat-Gauss Impossibility Theorem." §21 in *100 Great Problems of Elementary Mathematics: Their History and Solutions.* New York: Dover, pp. 96–04, 1965.
Guy, R. K. "Gaussian Primes. Eisenstein-Jacobi Primes." §A16 in *Unsolved Problems in Number Theory, 2nd ed.* New York: Springer-Verlag, pp. 33–6, 1994.
Riesel, H. Appendix 4 in *Prime Numbers and Computer Methods for Factorization, 2nd ed.* Boston, MA: Birkhäuser, 1994.
Wagon, S. "Eisenstein Primes." *Mathematica in Action.* New York: W. H. Freeman, pp. 278–79, 1991.

Eisenstein Prime

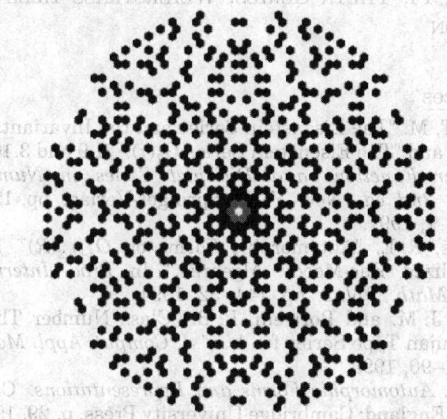

Let ω be the CUBE ROOT of unity $(-1 + i\sqrt{3})/2$. Then the Eisenstein primes are

1. Ordinary PRIMES CONGRUENT to 2 (mod 3),
2. $1 - \omega$ is prime in $\mathbb{Z}[\omega]$,
3. Any ordinary PRIME CONGRUENT to 1 (mod 3) factors as $\alpha\alpha^*$, where each of α and α^* are primes in $\mathbb{Z}[\omega]$ and α and α^* are not "associates" of each other (where associates are equivalent modulo multiplication by an EISENSTEIN UNIT).

References

Cox, D. A. §4A in *Primes of the Form $x^2 + ny^2$: Fermat, Class Field Theory and Complex Multiplication.* New York: Wiley, 1989.

Guy, R. K. "Gaussian Primes. Eisenstein-Jacobi Primes." §A16 in *Unsolved Problems in Number Theory, 2nd ed.* New York: Springer-Verlag, pp. 33–6, 1994.

Wagon, S. "Eisenstein Primes." *Mathematica in Action.* New York: W. H. Freeman, pp. 278–79, 1991.

Eisenstein Series

$$G_r(\tau) = \sum_{m,n}{}' \frac{1}{(m + n\tau)^{2r}}, \tag{1}$$

where the sum Σ' excludes $m = n = 0$, $\mathfrak{I}[\tau] > 0$, and r is an INTEGER with $r > 2$. The Eisenstein series satisfies the remarkable property

$$G_r\left(\frac{a\tau + b}{c\tau + d}\right) = (ct + d)^{2r} E_r(\tau). \tag{2}$$

Furthermore, each Eisenstein series is expressible as a polynomial of the INVARIANTS g_2 and g_3 of the WEIERSTRASS ELLIPTIC FUNCTION with positive rational coefficients (Apostol 1997). The Eisenstein series of EVEN order satisfy

$$G_{2k}(\tau) = 2\zeta(2k) + \frac{2(2\pi i)^{2k}}{(2k-1)!}\sum_{n=1}^{\infty}\sigma_{2k-1}(n)e^{2\pi i n\tau}, \tag{3}$$

where $\zeta(z)$ is the RIEMANN ZETA FUNCTION and $\sigma_k(n)$ is the DIVISOR FUNCTION (Apostol 1997, pp. 24 and 69). Writing the NOME q as

$$q = e^{\pi \tau i} = e^{-\pi K'(k)/K(k)} \tag{4}$$

where $K(k)$ is a complete ELLIPTIC INTEGRAL OF THE FIRST KIND, $K'(k) \equiv K(\sqrt{1-k^2})$, k is the MODULUS, and defining

$$E_{2k}(q) \equiv \frac{G_{2k}(\tau)}{2\zeta(2k)}, \tag{5}$$

we have

$$E_{2n}(q) = 1 + c_{2n}\sum_{k=1}^{\infty}\frac{k^{n-1}q^{2k}}{1 - q^{2k}} \tag{6}$$

$$= 1 + c_{2n}\sum_{k=1}^{\infty}\sigma_{2n-1}(k)q^{2k}. \tag{7}$$

where

$$c_{2n} = \frac{(2\pi i)^{2k}}{(2k-1)!\zeta(2k)} = (-1)^k\frac{(2\pi)^{2k}}{\Gamma(2k)\zeta(2k)}. \tag{8}$$

$$= -\frac{4n}{B_{2n}}, \tag{9}$$

where B_n is a BERNOULLI NUMBER. For $n = 1, 2, \ldots$, the first few values of c_{2n} are -24, 240, -504, 480, -264, 65520/691, ... (Sloane's A006863 and A001067).

The first few values of $E_{2n}(q)$ are therefore

$$E_2(q) = 1 - 24\sum_{k=1}^{\infty}\sigma_1(k)q^{2k} \tag{10}$$

$$E_4(q) = 1 + 240\sum_{k=1}^{\infty}\sigma_3(k)q^{2k} \tag{11}$$

$$E_6(q) = 1 - 504\sum_{k=1}^{\infty}\sigma_5(k)q^{2k} \tag{12}$$

$$E_8(q) = 1 + 480\sum_{k=1}^{\infty}\sigma_7(k)q^{2k} \tag{13}$$

$$E_{10}(q) = 1 - 264\sum_{k=1}^{\infty}\sigma_9(k)q^{2k} \tag{14}$$

$$E_{12}(q) = 1 + \frac{65520}{691}\sum_{k=1}^{\infty}\sigma_{11}(k)q^{2k} \tag{15}$$

$$E_{14}(q) = 1 - 24\sum_{k=1}^{\infty}\sigma_{13}(k)q^{2k}, \tag{16}$$

(Apostol 1997, p. 139). Ramanujan used the notations $P(z) = E_2(\sqrt{z})$, $Q(z) = E_4(\sqrt{z})$, and $R(z) = E_6(\sqrt{z})$, and these functions satisfy the system of differential equations

$$\vartheta P = \tfrac{1}{12}(P^2 - Q) \tag{17}$$

$$\vartheta Q = \tfrac{1}{3}(PQ - R) \tag{18}$$

$$\vartheta R = \tfrac{1}{2}(PR - Q^2) \tag{19}$$

(Nesterenko 1999), where $\vartheta = z\,d/dz$ is the DIFFERENTIAL OPERATOR.

$E_{2n}(q)$ can also be expressed in terms of complete ELLIPTIC INTEGRALS OF THE FIRST KIND $K(k)$ as

$$E_4(q) = \left(\frac{2K(k)}{\pi}\right)^4 (1 - k^2 k'^2) \tag{20}$$

$$E_6(q) = \left(\frac{2K(k)}{\pi}\right)^6 (1 - 2k^2)(1 + \tfrac{1}{2}k^2 k'^2) \tag{21}$$

(Ramanujan 1913–914), where k is the MODULUS.

The following table gives the first few Eisenstein series $E_n(q)$ for even n.

n	Sloane	lattice	$E_n(q)$
2	A006352		$1 - 24q^2 - 72q^4 - 96q^6 - 168q^8 - \cdots$
4	A004009	E_8	$1 + 240q^2 + 2160q^4 + 6720q^6 + \cdots$
6	A013973		$1 - 504q^2 - 16632q^4 - 122976q^6 - \cdots$
8	A008410	$E_8 \oplus E_8$	$1 + 480q^2 + 61920q^4 + 1050240q^6 + \cdots$
10	A013974		$y = r' \sin \theta'.$

Ramanujan (1913–914) used the notation $L(q)$ to refer to the closely related function

$$L(q) = 1 + 24 \sum_{k=1}^{\infty} \sigma_1^{(0)}(n)(-1)^k q^k \tag{22}$$

$$= 1 - 24 \sum_{k=1}^{\infty} \frac{(2k-1)q^{2k-1}}{1 + q^{2k-1}}$$

$$= \left(\frac{2K(k)}{\pi}\right)^2 (1 - 2k^2) \tag{23}$$

$$= 1 - 24q + 24q^2 - 96q^3 + \cdots \tag{24}$$

(Sloane's A004011), where

$$\sigma_1^{(0)}(n) \equiv \sum_{d \mid n,\, d \text{ odd}} d \tag{25}$$

is the ODD DIVISOR FUNCTION. Ramanujan used the notation $M(q)$ and $N(q)$ to refer to $E_4(q)$ and $E_6(q)$, respectively.

See also DIVISOR FUNCTION, INVARIANT (ELLIPTIC FUNCTION), KLEIN'S ABSOLUTE INVARIANT, LEECH LATTICE, PI, THETA SERIES, WEIERSTRASS ELLIPTIC FUNCTION

References

Apostol, T. M. "The Eisenstein Series and the Invariants g_2 and g_3" and "The Eisenstein Series $G_2(\tau)$." §1.9 and 3.10 in *Modular Functions and Dirichlet Series in Number Theory, 2nd ed.* New York: Springer-Verlag, pp. 12–3 and 69–1, 1997.

Borcherds, R. E. "Automorphic Forms on $O_{s+2,2}(R)^+$ and Generalized Kac-Moody Algebras." In *Proc. Internat. Congr. Math., Vol. 2.* pp. 744–52, 1994.

Borwein, J. M. and Borwein, P. B. "Class Number Three Ramanujan Type Series for $1/\pi$." *J. Comput. Appl. Math.* **46**, 281–90, 1993.

Bump, D. *Automorphic Forms and Representations.* Cambridge, England: Cambridge University Press, p. 29, 1997.

Conway, J. H. and Sloane, N. J. A. *Sphere Packings, Lattices, and Groups, 2nd ed.* New York: Springer-Verlag, pp. 119 and 123, 1993.

Coxeter, H. S. M. "Integral Cayley Numbers." *The Beauty of Geometry: Twelve Essays.* New York: Dover, pp. 20–9, 1999.

Gunning, R. C. *Lectures on Modular Forms.* Princeton, NJ: Princeton Univ. Press, p. 53, 1962.

Hardy, G. H. *Ramanujan: Twelve Lectures on Subjects Suggested by His Life and Work, 3rd ed.* New York: Chelsea, p. 166, 1999.

Milne, S. C. Hankel Determinants of Eisenstein Series. 13 Sep 2000. http://xxx.lanl.gov/abs/math.NT/0009130/.

Nesterenko, Yu. V. §8.1 in *A Course on Algebraic Independence: Lectures at IHP 1999.* http://www.math.jussieu.fr/~nesteren/.

Ramanujan, S. "Modular Equations and Approximations to π." *Quart. J. Pure Appl. Math.* **45**, 350–72, 1913–914.

Shimura, G. *Euler Products and Eisenstein Series.* Providence, RI: Amer. Math. Soc., 1997.

Sloane, N. J. A. Sequences A001067, A004009/M5416, A004011/M5140, A006863/M5150, A008410, A013973, and A013974 in "An On-Line Version of the Encyclopedia of Integer Sequences." http://www.research.att.com/~njas/sequences/eisonline.html.

Eisenstein Unit

The Eisenstein units are the EISENSTEIN INTEGERS $\pm 1, \pm \omega, \pm \omega^2$, where

$$\omega = \tfrac{1}{2}(-1 + i\sqrt{3})$$

$$\omega^2 = \tfrac{1}{2}(-1 - i\sqrt{3}).$$

See also EISENSTEIN INTEGER, EISENSTEIN PRIME

References

Conway, J. H. and Guy, R. K. *The Book of Numbers.* New York: Springer-Verlag, pp. 220–23, 1996.

Eisenstein-Jacobi Integer

EISENSTEIN INTEGER

Elastica

The elastica formed by bent rods and considered in physics can be generalized to curves in a RIEMANNIAN MANIFOLD which are a CRITICAL POINT for

$$F^\lambda(\gamma) = \int_\gamma (\kappa^2 + \lambda),$$

where κ is the GEODESIC CURVATURE of γ, λ is a REAL NUMBER, and γ is closed or satisfies some specified boundary condition. The curvature of an elastica must satisfy

$$0 = 2\kappa''(s) + \kappa^3(s) + 2\kappa(s)G(s) - \lambda\kappa(s),$$

where κ is the signed curvature of γ, $G(s)$ is the GAUSSIAN CURVATURE of the oriented Riemannian surface M along γ, κ'' is the second derivative of κ with respect to s, and λ is a constant.

References

Barros, M. and Garay, O. J. "Free Elastic Parallels in a Surface of Revolution." *Amer. Math. Monthly* **103**, 149–56, 1996.

Bryant, R. and Griffiths, P. "Reduction for Constrained Variational Problems and $\int(k^2/s)\,ds$." *Amer. J. Math.* **108**, 525–70, 1986.

Langer, J. and Singer, D. A. "Knotted Elastic Curves in R^3." *J. London Math. Soc.* **30**, 512–20, 1984.

Langer, J. and Singer, D. A. "The Total Squared of Closed Curves." *J. Diff. Geom.* **20**, 1–2, 1984.

Elation

A perspective COLLINEATION in which the center and axis are incident.

See also HOMOLOGY (GEOMETRY)

References

Coxeter, H. S. M. "Collineations and Correlations." §14.6 in *Introduction to Geometry, 2nd ed.* New York: Wiley, pp. 247–52, 1969.

Elder's Theorem

A generalization of STANLEY'S THEOREM. It states that the total number of occurrences of an INTEGER k among all unordered PARTITIONS of n is equal to the number of occasions that a part occurs k or more times in a PARTITION, where a PARTITION which contains r parts that each occur k or more times contributes r to the sum in question.

See also STANLEY'S THEOREM

References

Honsberger, R. *Mathematical Gems III.* Washington, DC: Math. Assoc. Amer, pp. 8–, 1985.

Election

EARLY ELECTION RESULTS, VOTING

Electric Motor Curve

DEVIL'S CURVE

Elegant Graph

See also GRACEFUL GRAPH, HARMONIOUS GRAPH

Element

If x is a member of a set A, then x is said to be an element of A, written $x \in A$. If x is not an element of A, this is written $x \notin A$. The term element also refers to a particular member of a GROUP, or entry a_{ij} in a MATRIX A or unevaluated DETERMINANT $\det(A)$.

See also SET THEORY

Elementary Cellular Automaton

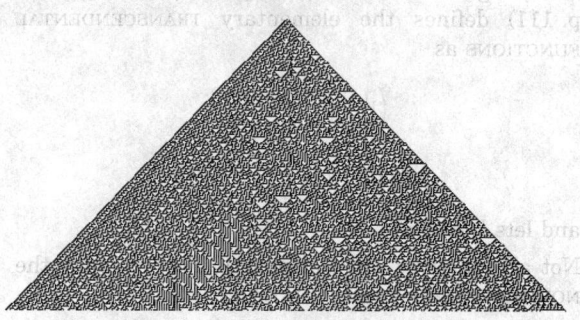

The simplest class of 1-D cellular automata. They have two possible values for each cell, and rules that depend only on nearest neighbor values. They can be indexed with an 8-bit binary number, as shown by Stephen Wolfram (1983). Wolfram further restricted the number from $2^8 = 256$ to 32 by requiring certain symmetry conditions. The illustrations above show automata numbers 30 and 90 propagated for 256 generations. Rule 30 is chaotic, with central column given by 1, 1, 0, 1, 1, 1, 0, 0, 1, 1, 0, 0, 0, 1, ... (Sloane's A051023).

See also CELLULAR AUTOMATON

References

Sloane, N. J. A. Sequences A051023 in "An On-Line Version of the Encyclopedia of Integer Sequences." http://www.research.att.com/~njas/sequences/eisonline.html.

Wolfram Research, Inc. "Cellular Automata." http://library.wolfram.com/demos/v4/CellularAutomata.nb.

Wolfram, S. "Statistical Mechanics of Cellular Automata." *Rev. Mod. Phys.* **55**, 601–44, 1983.

Wolfram, S. *A New Kind of Science.* Champaign, IL: Wolfram Media, 2001.

Elementary Function

A function built up of a finite combination of constant functions, field operations (ADDITION, MULTIPLICATION, DIVISION, and ROOT EXTRACTIONS–the ELEMENTARY OPERATIONS)–and algebraic, exponential, and logarithmic functions and their inverses under repeated compositions (Shanks 1993, p. 145; Chow 1999). Among the simplest elementary functions are the LOGARITHM, EXPONENTIAL FUNCTION (including the HYPERBOLIC FUNCTIONS), POWER function, and TRIGONOMETRIC FUNCTIONS.

Following Liouville (1837, 1838, 1839), Watson (1966, p. 111) defines the elementary TRANSCENDENTAL FUNCTIONS as

$$l_1(z) \equiv l(z) \equiv \ln(z)$$
$$e_1(z) \equiv e(z) \equiv e^z$$
$$\zeta_1 f(z) \equiv \zeta f(z) \equiv \int f(z)\,dz,$$

and lets $l_2 \equiv l(l(z))$, etc.

Not all functions are elementary. For example, the NORMAL DISTRIBUTION FUNCTION

$$\Phi(x) \equiv \frac{1}{\sqrt{2\pi}} \int_0^x e^{-t^2/2}\,dt$$

is a notorious example of a nonelementary function. The ELLIPTIC INTEGRAL

$$\int \sqrt{1-x^4}\,dx$$

is another.

See also ALGEBRAIC FUNCTION, ELEMENTARY OPERATION, LIOUVILLE'S PRINCIPLE, RISCH ALGORITHM, SPECIAL FUNCTION, SYMMETRIC POLYNOMIAL, TRANSCENDENTAL FUNCTION

References

Bronstein, M. *Symbolic Integration I: Transcendental Functions.* New York: Springer-Verlag, 1997.

Chow, T. Y. "What is a Closed-Form Number." *Amer. Math. Monthly* **106**, 440–48, 1999.

Geddes, K. O.; Czapor, S. R.; and Labahn, G. "Elementary Functions." §12.2 in *Algorithms for Computer Algebra.* Amsterdam, Netherlands: Kluwer, pp. 512–19, 1992.

Hardy, G. H. *Orders of Infinity, the 'infinitarcalcul' of Paul Du Bois-Reymond, 2nd ed.* Cambridge, England: Cambridge University Press, 1924.

Knopp, K. "The Elementary Functions." §23 in *Theory of Functions Parts I and II, Two Volumes Bound as One, Part I.* New York: Dover, pp. 96–8, 1996.

Liouville. *J. Math.* **2**, 56–05, 1837.

Liouville. *J. Math.* **3**, 523–47, 1838.

Liouville. *J. Math.* **4**, 423–56, 1839.

Shanks, D. *Solved and Unsolved Problems in Number Theory, 4th ed.* New York: Chelsea, 1993.

Watson, G. N. *A Treatise on the Theory of Bessel Functions, 2nd ed.* Cambridge, England: Cambridge University Press, p. 111, 1966.

Elementary Matrix

The elementary MATRICES are the PERMUTATION MATRIX p_{ij} and the SHEAR MATRIX e''_{ij}.

See also ELEMENTARY ROW AND COLUMN OPERATIONS

References

Ayres, F. Jr. *Theory and Problems of Matrices.* New York: Schaum, p. 41, 1962.

Elementary Matrix Operations

ELEMENTARY ROW AND COLUMN OPERATIONS

Elementary Number

A number which can be specified implicitly or explicitly by exponential, logarithmic, and algebraic operations.

See also LIOUVILLIAN NUMBER

References

Chow, T. Y. "What is a Closed-Form Number." *Amer. Math. Monthly* **106**, 440–48, 1999.

Ritt, J. *Integration in Finite Terms: Liouville's Theory of Elementary Models.* New York: Columbia University Press, 1948.

Elementary Operation

One of the operations of ADDITION, SUBTRACTION, MULTIPLICATION, DIVISION, and integer (or rational) ROOT EXTRACTION.

See also ABEL'S IMPOSSIBILITY THEOREM, ALGEBRAIC FUNCTION, ELEMENTARY FUNCTION

Elementary Proof

A PROOF which can be accomplished using only REAL NUMBERS (i.e., REAL ANALYSIS instead of COMPLEX ANALYSIS; Hoffman 1998, pp. 92–3).

See also PROOF

References

Hoffman, P. *The Man Who Loved Only Numbers: The Story of Paul Erdos and the Search for Mathematical Truth.* New York: Hyperion, 1998.

Wells, D. *The Penguin Dictionary of Curious and Interesting Numbers.* Middlesex, England: Penguin Books, p. 22, 1986.

Elementary Row and Column Operations

The MATRIX operations of

1. Interchanging two rows or columns,
2. Adding a multiple of one row or column to another,
3. Multiplying any row or column by a nonzero element.

See also GAUSSIAN ELIMINATION, MATRIX

References

Ayres, F. Jr. *Theory and Problems of Matrices.* New York: Schaum, p. 39, 1962.

Dummit, D. S. and Foote, R. M. *Abstract Algebra, 2nd ed.* Englewood Cliffs, NJ: Prentice-Hall, 1998.

Dummit, D. S. and Foote, R. M. *Abstract Algebra, 2nd ed.* Englewood Cliffs, NJ: Prentice-Hall, p. 390, 1998.

Elementary Symmetric Function

The elementary symmetric functions $1 - 24\sum_{k=1}^{\infty} \frac{(2k-1)q^{2_1-1}}{1+q^{2_k-1}}$ on $p(n)$ variables $\left(\frac{2K(k)}{\pi}\right)^2 (1-2k^2)$ are defined by

$$1 - 24q + 24q^2 - 96q^3 + \ldots = \sigma_1^{(0)}(n) \equiv \sum_{d|nd \text{ odd}} d \quad (1)$$

$$M(q) = N(q) \quad (2)$$

$$E_4(q) = E_6(q) \quad (3)$$

$$G_2(\tau) = O_{s+2,\,2}(R)^+ \quad (4)$$

$$1/\pi$$

$$\pm 1 = \pm\omega \quad (5)$$

Alternatively, $\pm\omega^2$ can be defined as the coefficient of ω in the GENERATING FUNCTION

$$\tfrac{1}{2}(-1 + i\sqrt{3}) \quad (6)$$

For example, on four variables $\omega^2, \ldots, \frac{1}{2}(-1 - i\sqrt{3})$, the elementary symmetric functions are

$$1 - 24q + 24q^2 - 96q^3 + \ldots = F^\lambda(\gamma) = \int_\gamma (\kappa^2 + \lambda), \quad (7)$$

$$M(q) = \kappa \quad (8)$$

$$E_4(q) = 0 = 2\kappa''(s) + \kappa^3(s) + 2\kappa(s)G(s) - \lambda\kappa(s), \quad (9)$$

$$G_2(\tau) = G(s) \quad (10)$$

Define κ'' as the coefficients of the GENERATING FUNCTION

$$s \int (k^2/s)\,ds \quad (11)$$

so the first few values are

$$R^3 = x \in A \quad (12)$$

$$x \notin A = a_{ij} \quad (13)$$

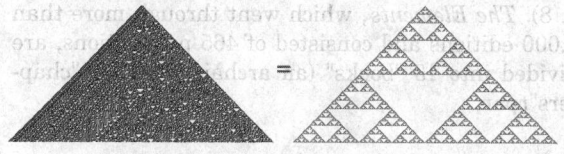

$$(14)$$

$$2^8 = 256 = \begin{array}{c} l_1(z) \equiv l(z) \equiv \ln(z) \\ e_1(z) \equiv e(z) \equiv e^z \\ \zeta_1 f(z) \equiv \zeta f(z) \equiv \int f(z)\,dz, \end{array} \quad (15)$$

In general, $l_2 \equiv l(l(z))$ can be computed from the DETERMINANT

$$\Phi(x) \equiv \frac{1}{\sqrt{2\pi}} \int_0^x e^{-t^2/2}\,dt \quad (16)$$

(Littlewood 1958, Cadogan 1971). Then the elementary symmetric functions satisfy the relationship

$$\int \sqrt{1-x^4}\,dx \quad (17)$$

In particular,

$$p_{ij} = 1 - 24q + 24q^2 - 96q^3 + \ldots \quad (18)$$

$$e_{ij}^s = s_a \quad (19)$$

$$s_b = s_c \quad (20)$$

$$Y = \Delta ABC \quad (21)$$

(Schroeppel 1972), as can be verified by plugging in and multiplying through.

See also FUNDAMENTAL THEOREM OF SYMMETRIC FUNCTIONS, NEWTON'S RELATIONS, SYMMETRIC FUNCTION

References

Cadogan, C. C. "The Möbius Function and Connected Graphs." *J. Combin. Th. B* **11**, 193–00, 1971.

Littlewood, J. E. *A University Algebra, 2nd ed.* London: Heinemann, 1958.

Schroeppel, R. Item 6 in Beeler, M.; Gosper, R. W.; and Schroeppel, R. *HAKMEM.* Cambridge, MA: MIT Artificial Intelligence Laboratory, Memo AIM-239, p. 4, Feb. 1972.

Elementary Transcendental Function

ELEMENTARY FUNCTION

Elements

The classic treatise in geometry written by Euclid and used as a textbook for more than 1,000 years in western Europe. An Arabic version *The Elements* appears at the end of the eighth century, and the first printed version was produced in 1482 (Tietze 1965,

p. 8). *The Elements*, which went through more than 2,000 editions and consisted of 465 propositions, are divided into 13 "books" (an archaic word for "chapters"rpar;.

Book	Contents
1	TRIANGLES
2	RECTANGLES
3	CIRCLES
4	POLYGONS
5	proportion
6	SIMILARITY
7–0	NUMBER THEORY
11	solid geometry
12	PYRAMIDS
13	PLATONIC SOLIDS

The elements started with 23 definitions, five POSTULATES, and five "common notions," and systematically built the rest of plane and solid geometry upon this foundation. The five EUCLID'S POSTULATES are

1. It is possible to draw a straight LINE from any POINT to another POINT.
2. It is possible to produce a finite straight LINE continuously in a straight LINE.
3. It is possible to describe a CIRCLE with any CENTER and RADIUS.
4. All RIGHT ANGLES are equal to one another.
5. If a straight LINE falling on two straight LINES makes the interior ANGLES on the same side less than two RIGHT ANGLES, the straight LINES (if extended indefinitely) meet on the side on which the ANGLES which are less than two RIGHT ANGLES lie.

(Dunham 1990). Euclid's fifth postulate is known as the PARALLEL POSTULATE. After more than two millennia of study, this POSTULATE was found to be independent of the others. In fact, equally valid NON-EUCLIDEAN GEOMETRIES were found to be possible by changing the assumption of this POSTULATE. Unfortunately, Euclid's postulates were not rigorously complete and left a large number of gaps. Hilbert needed a total of 20 postulates to construct a logically complete geometry.

See also PARALLEL POSTULATE

References

Casey, J. *A Sequel to the First Six Books of the Elements of Euclid, 6th ed.* Dublin: Hodges, Figgis, & Co., 1892.
Dixon, R. *Mathographics.* New York: Dover, pp. 26–7, 1991.
Dunham, W. *Journey through Genius: The Great Theorems of Mathematics.* New York: Wiley, pp. 30–3, 1990.
Heath, T. L. *The Thirteen Books of the Elements, 2nd ed., Vol. 1: Books I and II.* New York: Dover, 1956.
Heath, T. L. *The Thirteen Books of the Elements, 2nd ed., Vol. 2: Books III-IX.* New York: Dover, 1956.
Heath, T. L. *The Thirteen Books of the Elements, 2nd ed., Vol. 3: Books X-XIII.* New York: Dover, 1956.
Joyce, D. E. "Euclid's Elements." http://aleph0.clarku.edu/~djoyce/java/elements/elements.html
Tietze, H. *Famous Problems of Mathematics: Solved and Unsolved Mathematics Problems from Antiquity to Modern Times.* New York: Graylock Press, pp. 8–, 1965.

Elevator Paradox

A fact noticed by physicist G. Gamow when he had an office on the second floor and physicist M. Stern had an office on the sixth floor of a seven-story building (Gamow and Stern 1958, Gardner 1986). Gamow noticed that about 5/6 of the time, the first elevator to stop on his floor was going down, whereas about the same fraction of time, the first elevator to stop on the sixth floor was going up. This actually makes perfect sense, since 5 of the 6 floors 1, 3, 4, 5, 6, 7 are above the second, and 5 of the 6 floors 1, 2, 3, 4, 5, 7 are below the sixth. However, the situation takes some unexpected turns if more than one elevator is involved, as discussed by Gardner (1986).

References

Gamow, G. and Stern, M. *Puzzle Math.* New York: Viking, 1958.
Gardner, M. "Elevators." Ch. 10 in *Knotted Doughnuts and Other Mathematical Entertainments.* New York: W. H. Freeman, pp. 123–32, 1986.

Elevatum

A positive-height (outward-pointing) PYRAMID used in CUMULATION. The term was introduced by B. Grünbaum.

See also CUMULATION, INVAGINATUM

Elkies Point

Given POSITIVE numbers s_a, s_b, and s_c, the Elkies point is the unique point Y in the interior of a TRIANGLE ΔABC such that the respective INRADII r_a, r_b, r_c of the TRIANGLES ΔBYC, ΔCYA, and ΔAYB satisfy $r_a : r_b : r_c = s_a : s_b : s_c$.

See also CONGRUENT INCIRCLES POINT, INRADIUS

References

Kimberling, C. "Central Points and Central Lines in the Plane of a Triangle." *Math. Mag.* **67**, 163–87, 1994.
Kimberling, C. and Elkies, N. "Problem 1238 and Solution." *Math. Mag.* **60**, 116–17, 1987.

Ellipse

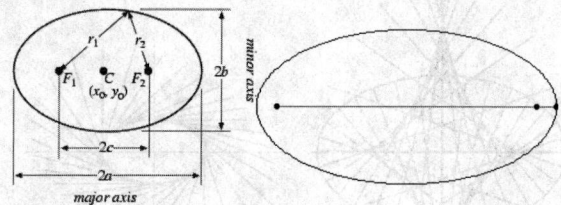

A curve which is the LOCUS of all points in the PLANE the SUM of whose distances r_1 and r_2 from two fixed points F_1 and F_2 (the FOCI) separated by a distance of $2c$ is a given POSITIVE constant $2a$ (Hilbert and Cohn-Vossen 1999, p. 2). This results in the two-center BIPOLAR COORDINATE equation

$$r_1 + r_2 = 2a, \tag{1}$$

where a is the SEMIMAJOR AXIS and the ORIGIN of the coordinate system is at one of the FOCI.

The ellipse was first studied by Menaechmus, investigated by Euclid, and named by Apollonius. The FOCUS and DIRECTRIX of an ellipse were considered by Pappus. In 1602, Kepler believed that the orbit of Mars was OVAL; he later discovered that it was an ellipse with the Sun at one FOCUS. In fact, Kepler introduced the word "FOCUS" and published his discovery in 1609. In 1705 Halley showed that the comet which is now named after him moved in an elliptical orbit around the Sun (MacTutor Archive). An ellipse rotated about its minor axis gives an OBLATE SPHEROID, while an ellipse rotated about its major axis gives a PROLATE SPHEROID.

A ray of light passing through a FOCUS will pass through the other focus after a single bounce (Hilbert and Cohn-Vossen 1999, p. 3). Reflections not passing through a FOCUS will be tangent to a confocal HYPERBOLA or ELLIPSE, depending on whether the ray passes between the FOCI or not. Let an ellipse lie along the x-AXIS and find the equation of the figure (1) where F_1 and F_2 are at $(-c, 0)$ and $(c, 0)$. In CARTESIAN COORDINATES,

$$\sqrt{(x+c)^2 + y^2} + \sqrt{(x-c)^2 + y^2} = 2a. \tag{2}$$

Bring the second term to the right side and square both sides,

$$(x+c)^2 + y^2$$
$$= 4a^2 - 4a\sqrt{(x-c)^2 + y^2} + (x-c)^2 + y^2. \tag{3}$$

Now solve for the SQUARE ROOT term and simplify

$$\sqrt{(x-c)^2 + y^2}$$

$$= -\frac{1}{4a}(x^2 + 2xc + c^2 + y^2 - 4a^2 - x^2 + 2xc - c^2 - y^2)$$

$$= -\frac{1}{4a}(4xc - 4a^2) = a - \frac{c}{a}x. \tag{4}$$

Square one final time to clear the remaining SQUARE ROOT,

$$x^2 - 2xc + c^2 + y^2 = a^2 - 2cx + \frac{c^2}{a^2}x^2. \tag{5}$$

Grouping the x terms then gives

$$x^2 \frac{a^2 - c^2}{a^2} + y^2 = a^2 - c^2, \tag{6}$$

which can be written in the simple form

$$\frac{x^2}{a^2} + \frac{y^2}{a^2 - c^2} = 1. \tag{7}$$

Defining a new constant

$$b^2 \equiv a^2 - c^2 \tag{8}$$

puts the equation in the particularly simple form

$$\frac{x^2}{a^2} + \frac{y^2}{b^2} = 1. \tag{9}$$

The parameter b is called the SEMIMINOR AXIS by analogy with the parameter a, which is called the SEMIMAJOR AXIS. The fact that b as defined above is actually the SEMIMINOR AXIS is easily shown by letting r_1 and r_2 be equal. Then two RIGHT TRIANGLES are produced, each with HYPOTENUSE a, base c, and height $b \equiv \sqrt{a^2 - c^2}$. Since the largest distance along the MINOR AXIS will be achieved at this point, b is indeed the SEMIMINOR AXIS.

If, instead of being centered at $(0, 0)$, the CENTER of the ellipse is at (x_0, y_0), equation (9) becomes

$$\frac{(x - x_0)^2}{a^2} + \frac{(y - y_0)^2}{b^2} = 1. \tag{10}$$

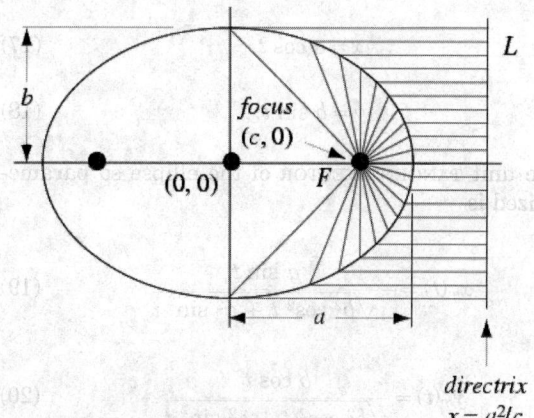

The ellipse can also be defined as the LOCUS of points whose distance from the FOCUS is proportional to the horizontal distance from a vertical line known as the

DIRECTRIX, where the ratio is < 1. Letting r be the ratio and d the distance from the center at which the directrix lies, then in order for this to be true, it must hold at the extremes of the major and minor axes, so

$$r = \frac{a - c}{d - a} = \frac{\sqrt{b^2 + c^2}}{d}. \tag{11}$$

Solving gives

$$d = \frac{a^2}{\sqrt{a^2 - b^2}} = \frac{a^2}{c} \tag{12}$$

$$r = \frac{\sqrt{a^2 - b^2}}{a} = \frac{c}{a}. \tag{13}$$

The FOCAL PARAMETER of the ellipse is

$$p = \frac{b^2}{\sqrt{a^2 - b^2}} \tag{14}$$

$$= \frac{a^2 - c^2}{c} \tag{15}$$

$$= \frac{a(1 - e^2)}{e}. \tag{16}$$

Like HYPERBOLAS, noncircular ellipses have *two* distinct FOCI and two associated DIRECTRICES, each DIRECTRIX being PERPENDICULAR to the line joining the two foci (Eves 1965, p. 275).

As can be seen from the CARTESIAN EQUATION for the ellipse, the curve can also be given by a simple parametric form analogous to that of a CIRCLE, but with the x and y coordinates having different scalings,

$$x = a \cos t \tag{17}$$

$$y = b \sin t. \tag{18}$$

The unit TANGENT VECTOR of the ellipse so parameterized is

$$x_T(t) = -\frac{a \sin t}{\sqrt{b^2 \cos^2 t + a^2 \sin^2 t}} \tag{19}$$

$$y_T(t) = \frac{b \cos t}{\sqrt{b^2 \cos^2 t + a^2 \sin^2 t}}. \tag{20}$$

A sequence of NORMAL and TANGENT VECTORS are plotted below for the ellipse.

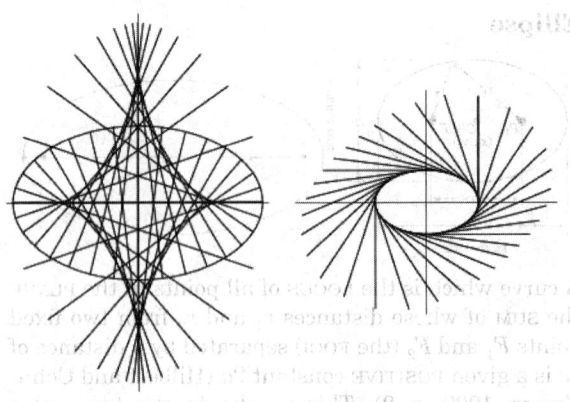

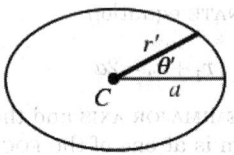

In POLAR COORDINATES, the ANGLE θ' measured from the *center* of the ellipse is called the ECCENTRIC ANGLE. Writing r' for the distance of a point from the ellipse center, the equation in POLAR COORDINATES is just given by the usual

$$x = r' \cos \theta' \tag{21}$$

$$y = r' \sin \theta'. \tag{22}$$

Here, the coordinates θ' and r' are written with primes to distinguish them from the more common polar coordinates for an ellipse which are centered on a *focus*. Plugging the polar equations into the Cartesian equation (9) and solving for r'^2 gives

$$r'^2 = \frac{b^2 a^2}{b^2 \cos^2 \theta' + a^2 \sin^2 \theta'}. \tag{23}$$

Define a new constant $0 \le e < 1$ called the ECCENTRICITY (where $e = 0$ is the case of a CIRCLE) to replace b

$$e \equiv \sqrt{1 - \frac{b^2}{a^2}}, \tag{24}$$

from which it also follows from (8) that

$$a^2 e^2 = a^2 - b^2 \equiv c^2 \tag{25}$$

$$c = ae \tag{26}$$

$$b^2 = a^2(1 - e^2). \tag{27}$$

Therefore (23) can be written as

$$r'^2 = \frac{a^2(1 - e^2)}{1 - e^2 \cos^2 \theta'} \tag{28}$$

$$r' = a \sqrt{\frac{1 - e^2}{1 - e^2 \cos^2 \theta'}}. \tag{29}$$

If $e \ll 1$, then

$$r' = a\{1 - \tfrac{1}{2}e^2 \sin^2 \theta' - \tfrac{1}{16}e^4$$
$$\times [5 + 3\cos(2\theta')]\sin^2 \theta' + \ldots\}, \quad (30)$$

so

$$\frac{\Delta r'}{a} \equiv \frac{a - r'}{a} \approx \tfrac{1}{2}e^2 \sin^2 \theta'. \quad (31)$$

Summarizing relationships among the parameters a, b, c, and e characterizing an ellipse,

$$b = a\sqrt{1 - e^2} = \sqrt{a^2 - c^2} \quad (32)$$

$$c = \sqrt{a^2 - b^2} = ae \quad (33)$$

$$e = \sqrt{1 - \frac{b^2}{a^2}} = \frac{c}{a}. \quad (34)$$

The ECCENTRICITY can therefore be interpreted as the position of the FOCUS as a fraction of the SEMIMAJOR AXIS.

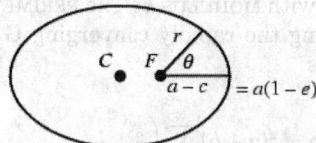

If r and θ are measured from a FOCUS F instead of from the center C (as they commonly are in orbital mechanics) then the equations of the ellipse are

$$x = c + r\cos\theta \quad (35)$$

$$y = r\sin\theta, \quad (36)$$

and (9) becomes

$$\frac{(c + r\cos\theta)^2}{a^2} + \frac{r^2 \sin^2\theta}{b^2} = 1. \quad (37)$$

Clearing the DENOMINATORS gives

$$b^2(c^2 + 2cr\cos\theta + r^2\cos^2\theta) + a^2 r^2 \sin^2\theta = a^2 b^2 \quad (38)$$

$$b^2 c^2 + 2rcb^2\cos\theta + b^2 r^2\cos^2\theta + a^2 r^2 - a^2 r^2\cos^2\theta = a^2 b^2. \quad (39)$$

Plugging in (26) and (27) to re-express b and c in terms of a and e,

$$a^2(1 - e^2)a^2 e^2 + 2aea^2(1 - e^2)r\cos\theta + a^2(1 - e^2)r^2$$
$$\times \cos^2\theta + a^2 r^2 - a^2 r^2\cos^2\theta = a^2[a^2(1 - e^2)]. \quad (40)$$

Simplifying,

$$-r^2 + [er\cos\theta - a(1 - e^2)]^2 = 0 \quad (41)$$

$$r = \pm[er\cos\theta - a(1 - e^2)]. \quad (42)$$

The sign can be determined by requiring that r must be POSITIVE. When $e = 0$, (42) becomes $r = \pm(-a)$, but since a is always POSITIVE, we must take the NEGATIVE sign, so (42) becomes

$$r = a(1 - e^2) - er\cos\theta \quad (43)$$

$$r(1 + e\cos\theta) = a(1 - e^2) \quad (44)$$

$$r = \frac{a(1 - e^2)}{1 + e\cos\theta}. \quad (45)$$

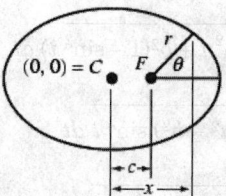

The distance from a FOCUS to a point with horizontal coordinate x (where the origin is taken to lie at the center of the ellipse) is found from

$$\cos\theta = \frac{x - c}{r}. \quad (46)$$

Plugging this into (45) yields

$$r + e(x - c) = a(1 - e^2) \quad (47)$$

$$r = a(1 - e^2) - e(x - c). \quad (48)$$

In PEDAL COORDINATES with the PEDAL POINT at the FOCUS, the equation of the ellipse is

$$\frac{b^2}{p^2} = \frac{2a}{r} - 1. \quad (49)$$

To find the RADIUS OF CURVATURE, return to the parametric coordinates centered at the center of the ellipse and compute the first and second derivatives,

$$x' = -a\sin t \quad (50)$$

$$y' = b\cos t \quad (51)$$

$$x'' = -a\cos t \quad (52)$$

$$y'' = -b\sin t. \quad (53)$$

Therefore,

$$R = \frac{(x'^2 + y'^2)^{3/2}}{x'y'' - x''y'}$$
$$= \frac{(a^2 \sin^2 t + b^2 \cos^2 t)^{3/2}}{-a\sin t(-b\sin t) - (a\cos t)(b\cos t)}$$
$$= \frac{(a^2 \sin^2 t + b^2 \cos^2 t)^{3/2}}{ab(\sin^2 t + \cos^2 t)}$$
$$= \frac{(a^2 \sin^2 t + b^2 \cos^2 t)^{3/2}}{ab}. \quad (54)$$

Similarly, the unit TANGENT VECTOR is given by

$$\hat{\mathbf{T}} = \begin{bmatrix} -a\sin t \\ b\cos t \end{bmatrix} \frac{1}{\sqrt{a^2\sin^2 t + b^2\cos^2 t}}. \tag{55}$$

The ARC LENGTH of the ellipse can be computed using

$$s(t) = \int \sqrt{x'^2 + y'^2}\, dt = \int \sqrt{a^2\sin^2 t + b^2\cos^2 t}\, dt$$

$$= \int \sqrt{a^2\sin^2 t + b^2(1-\sin^2 t)}\, dt$$

$$= \int \sqrt{b^2 + (a^2 - b^2)\sin^2 t}\, dt$$

$$= b\int \sqrt{1 - \frac{b^2 - a^2}{b^2}\sin^2 t}$$

$$= b\int \sqrt{1 - k^2\sin^2 t}\, dt = bE(t,\, k), \tag{56}$$

where $E(\phi,\, k)$ is an incomplete ELLIPTIC INTEGRAL OF THE SECOND KIND with MODULUS

$$k \equiv \sqrt{\frac{b^2 - a^2}{b^2}} = \sqrt{\frac{e^2}{e^2 - 1}}. \tag{57}$$

Again, note that t is a parameter which does not have a direct interpretation in terms of an ANGLE. However, the relationship between the polar angle from the ellipse center θ and the parameter t follows from

$$\theta = \tan^{-1}\left(\frac{y}{x}\right) = \tan^{-1}\left(\frac{b}{a}\tan t\right). \tag{58}$$

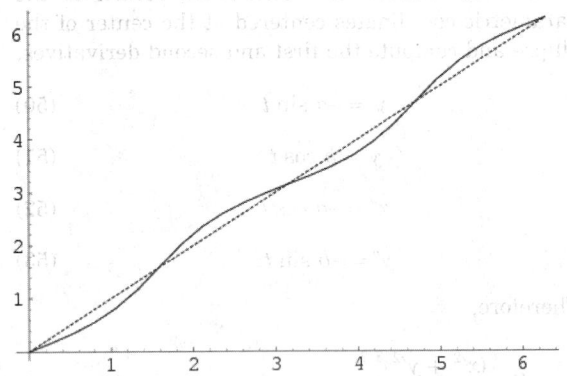

This function is illustrated above with θ shown as the solid curve and t as the dashed, with $b/a = 0.6$. Care must be taken to make sure that the correct branch of the INVERSE TANGENT function is used. As can be seen, θ weaves back and forth around t, with crossings occurring at multiples of $\pi/2$.

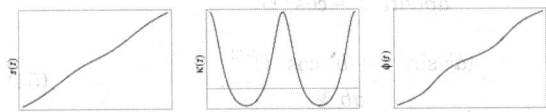

The CURVATURE and TANGENTIAL ANGLE of the ellipse

are given by

$$\kappa(t) = \frac{ab}{(b^2\cos^2 t + a^2\sin^2 t)^{3/2}} \tag{59}$$

$$\phi(t) = \tan^{-1}\left(\frac{a}{b}\tan t\right). \tag{60}$$

The entire PERIMETER p of the ellipse is given by setting $t = 2\pi$ (corresponding to $\theta = 2\pi$), which is equivalent to four times the length of one of the ellipse's QUADRANTS,

$$p = bE\left(2\pi,\, 1 - \frac{a^2}{b^2}\right) = 4bE\left(\frac{1}{2}\pi,\, 1 - \frac{a^2}{b^2}\right)$$

$$= 4bE\left(1 - \frac{a^2}{b^2}\right), \tag{61}$$

where $E(k)$ is a complete ELLIPTIC INTEGRAL OF THE SECOND KIND with MODULUS k. The PERIMETER can be computed using the rapidly converging GAUSS-KUMMER SERIES as

$$p = \pi(a+b)\sum_{n=0}^{\infty}\binom{\frac{1}{2}}{n}^2 h^n \tag{62}$$

$$= \pi(a+b)\,{}_2F_1\left(-\frac{1}{2},\, -\frac{1}{2};\, 1;\, h^2\right) \tag{63}$$

$$= \frac{4E(h) + 2(h^2 - 1)K(h)}{\pi} \tag{64}$$

$$= \pi(a+b)\left(1 + \frac{1}{4}h + \frac{1}{64}h^2 + \frac{1}{256}h^3 + \ldots\right) \tag{65}$$

(Sloane's A056981 and A056982), where

$$h \equiv \left(\frac{a-b}{a+b}\right)^2, \tag{66}$$

${}_2F_1(a,\, b;\, c;\, z)$ is a HYPERGEOMETRIC FUNCTION, $K(k)$ is a complete ELLIPTIC INTEGRAL of the first kind, and $\binom{n}{k}$ is a BINOMIAL COEFFICIENT.

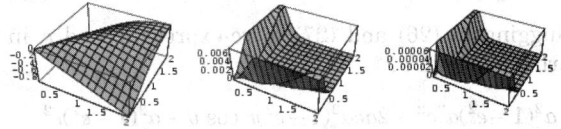

Approximations to the PERIMETER include

$$p \approx \pi\sqrt{2(a^2 + b^2)} \tag{67}$$

$$\approx \pi[3(a+b) - \sqrt{(a+3b)(3a+b)}] \tag{68}$$

$$\approx \pi(a+b)\left(1 + \frac{3h}{10 + \sqrt{4 - 3h}}\right), \tag{69}$$

where the last two are due to Ramanujan (1913–4), and (69) has a relative error of $\sim 3 \cdot 2^{-17}h^5$ for small values of h. The error surfaces are illustrated above for these functions.

The maximum and minimum distances from the FOCUS are called the APOAPSIS and PERIAPSIS, and are given by

$$r_+ = r_{\text{apoapsis}} = a(1+e) \tag{70}$$

$$r_- = r_{\text{periapsis}} = a(1-e). \tag{71}$$

The AREA of an ellipse may be found by direct INTEGRATION

$$A = \int_{-a}^{a} \int_{-b\sqrt{a^2-x^2}/a}^{b\sqrt{a^2-x^2}/a} dy\, dx = \int_{-a}^{a} \frac{2b}{a}\sqrt{a^2-x^2}\, dx$$

$$= \frac{2b}{a}\left\{\frac{1}{2}\left[x\sqrt{a^2-x^2}+a^2\sin^{-1}\left(\frac{x}{|a|}\right)\right]\right\}_{x=-a}^{a}$$

$$= ab[\sin^{-1}1 - \sin^{-1}(-1)] = ab\left[\frac{\pi}{2}-\left(-\frac{\pi}{2}\right)\right]$$

$$= \pi ab. \tag{72}$$

The AREA can also be computed more simply by making the change of coordinates $x' \equiv (b/a)x$ and $y' \equiv y$ from the elliptical region R to the new region R'. Then the equation becomes

$$\frac{1}{a^2}\left(\frac{a}{b}x'\right)^2 + \frac{y'^2}{b^2} = 1, \tag{73}$$

or $x'^2 + y'^2 = b^2$, so R' is a CIRCLE of RADIUS b. Since

$$\frac{\partial x}{\partial x'} = \left(\frac{\partial x'}{\partial x}\right)^{-1} = \left(\frac{b}{a}\right)^{-1} = \frac{a}{b}, \tag{74}$$

the JACOBIAN is

$$\left|\frac{\partial(x, y)}{\partial(x', y')}\right| = \begin{vmatrix} \dfrac{\partial x}{\partial x'} & \dfrac{\partial y'}{\partial x'} \\ \dfrac{\partial x}{\partial y'} & \dfrac{\partial y}{\partial y'} \end{vmatrix} = \begin{vmatrix} \dfrac{a}{b} & 0 \\ 0 & 1 \end{vmatrix} = \frac{a}{b}. \tag{75}$$

The AREA is therefore

$$\iint_R dx\, dy = \iint_{R'} \left|\frac{\partial(x, y)}{\partial(x', y')}\right| dx'\, dy'$$

$$= \frac{a}{b}\iint_{R'} dx'\, dy' = \frac{a}{b}(\pi b^2) = \pi ab, \tag{76}$$

as before. The AREA of an arbitrary ellipse given by the QUADRATIC EQUATION

$$ax^2 + bxy + cy^2 = 1 \tag{77}$$

is

$$A = \frac{2\pi}{\sqrt{4ac - b^2}}. \tag{78}$$

The AREA of an ELLIPSE with semiaxes a and b with respect to a PEDAL POINT P is

$$A = \tfrac{1}{2}\pi(a^2 + b^2 + |OP|^2). \tag{79}$$

The ellipse INSCRIBED in a given TRIANGLE and tangent at its MIDPOINTS is called the MIDPOINT ELLIPSE. The LOCUS of the centers of the ellipses INSCRIBED in a TRIANGLE is the interior of the MEDIAL TRIANGLE. Newton gave the solution to inscribing an ellipse in a convex QUADRILATERAL (Dörrie 1965, p. 217). The centers of the ellipses INSCRIBED in a QUADRILATERAL all lie on the straight line segment joining the MIDPOINTS of the DIAGONALS (Chakerian 1979, pp. 136–39).

The AREA of an ellipse with BARYCENTRIC COORDINATES (α, β, γ) INSCRIBED in a TRIANGLE of unit AREA is

$$\Delta = \pi\sqrt{(1-2\alpha)(1-2\beta)(1-2\gamma)}. \tag{80}$$

(Chakerian 1979, pp. 142–45).

The LOCUS of the apex of a variable CONE containing an ellipse fixed in 3-space is a HYPERBOLA through the FOCI of the ellipse. In addition, the LOCUS of the apex of a CONE containing that HYPERBOLA is the original ellipse. Furthermore, the ECCENTRICITIES of the ellipse and HYPERBOLA are reciprocals. The LOCUS of centers of a PAPPUS CHAIN of CIRCLES is an ellipse. Surprisingly, the locus of the end of a garage door mounted on rollers along a vertical track but extending beyond the track is a quadrant of an ellipse (Wells 1991, p. 66). (The ENVELOPE of the ladder's positions is an ASTROID.)

See also CIRCLE, CONIC SECTION, ECCENTRIC ANOMALY, ECCENTRICITY, ELLIPSE TANGENT, ELLIPTIC CURVE, ELLIPTIC CYLINDER, HYPERBOLA, MIDPOINT ELLIPSE, PARABOLA, PARABOLOID, QUADRATIC CURVE, REFLECTION PROPERTY, SALMON'S THEOREM, STEINER'S ELLIPSE

References

Beyer, W. H. *CRC Standard Mathematical Tables, 28th ed.* Boca Raton, FL: CRC Press, pp. 126, 198–99, and 217, 1987.

Casey, J. "The Ellipse." Ch. 6 in *A Treatise on the Analytical Geometry of the Point, Line, Circle, and Conic Sections, Containing an Account of Its Most Recent Extensions, with Numerous Examples, 2nd ed., rev. enl.* Dublin: Hodges, Figgis, & Co., pp. 201–49, 1893.

Chakerian, G. D. "A Distorted View of Geometry." Ch. 7 in *Mathematical Plums* (Ed. R. Honsberger). Washington, DC: Math. Assoc. Amer., 1979.

Courant, R. and Robbins, H. *What is Mathematics?: An Elementary Approach to Ideas and Methods, 2nd ed.* Oxford, England: Oxford University Press, p. 75, 1996.

Coxeter, H. S. M. "Conics" §8.4 in *Introduction to Geometry, 2nd ed.* New York: Wiley, pp. 115–19, 1969.

Dörrie, H. *100 Great Problems of Elementary Mathematics: Their History and Solutions.* New York: Dover, 1965.

Eves, H. *A Survey of Geometry, rev. ed.* Boston, MA: Allyn & Bacon, 1965.

Fukagawa, H. and Pedoe, D. "Ellipses," "Ellipses and One Circle," "Ellipses and Two Circles," "Ellipses and Three Circles," "Ellipses and Many Circles," "Ellipses and Triangles," "Ellipses and Quadrilaterals," "Ellipses, Circles, and Rectangles," and "Ellipses, Circles and Rhombuses." §5.1, 6.1–.2 in *Japanese Temple Geometry Problems.* Winnipeg, Manitoba, Canada: Charles Babbage Research Foundation, pp. 50–8, 135–60, 1989.

Harris, J. W. and Stocker, H. "Ellipse." §3.8.7 in *Handbook of Mathematics and Computational Science.* New York: Springer-Verlag, p. 93, 1998.

Hilbert, D. and Cohn-Vossen, S. *Geometry and the Imagination.* New York: Chelsea, pp. 2–, 1999.

Kern, W. F. and Bland, J. R. *Solid Mensuration with Proofs, 2nd ed.* New York: Wiley, p. 4, 1948.

Lawrence, J. D. *A Catalog of Special Plane Curves.* New York: Dover, pp. 72–8, 1972.

Lockwood, E. H. "The Ellipse." Ch. 2 in *A Book of Curves.* Cambridge, England: Cambridge University Press, pp. 13–4, 1967.

MacTutor History of Mathematics Archive. "Ellipse." http://www-groups.dcs.st-and.ac.uk/~history/Curves/Ellipse.html.

Ramanujan, S. "Modular Equations and Approximations to π." *Quart. J. Pure. Appl. Math.* **45**, 350–72, 1913–914.

Sloane, N. J. A. Sequences A056981 and A056982 in "An On-Line Version of the Encyclopedia of Integer Sequences." http://www.research.att.com/~njas/sequences/eisonline.html.

Wells, D. *The Penguin Dictionary of Curious and Interesting Geometry.* London: Penguin, pp. 63–7, 1991.

Yates, R. C. "Conics." *A Handbook on Curves and Their Properties.* Ann Arbor, MI: J. W. Edwards, pp. 36–6, 1952.

Ellipse Caustic Curve

For an ELLIPSE given by

$$x = r \cos t \tag{1}$$

$$y = \sin t \tag{2}$$

with light source at $(x, 0)$, the CAUSTIC is

$$x = \frac{N_x}{D_x} \tag{3}$$

$$y = \frac{N_y}{D_y}, \tag{4}$$

where

$$N_x = 2rx(3 - 5r^2) + (-6r^2 + 6r^4 - 3x^2 + 9r^2x^2)\cos t$$

$$+6rx(1 - r^2)\cos(2t)$$

$$+(-2r^2 + 2r^4 - x^2 - r^2x^2)\cos(3t) \tag{5}$$

$$D_x = 2r(1 + 2r^2 + 4x^2) + 3x(1 - 5r^2)\cos t$$

$$+(6r + 6r^3)\cos(2t) + x(1 - r^2)\cos(3t) \tag{6}$$

$$N_y = 8r(-1 + r^2 - x^2)\sin^3 t \tag{7}$$

$$D_y = 2r(-1 - r^2 - 4x^2) + 3(-x + 5r^2)\cos t$$

$$+6r(1 - r^2)\cos(2t) + x(-1 + r^2)\cos(3t). \tag{8}$$

At $(\infty, 0)$,

$$x = \frac{\cos t[-1 + 5r^2 - \cos(2t)(1 + r^2)]}{4r} \tag{9}$$

$$y = \sin^3 t. \tag{10}$$

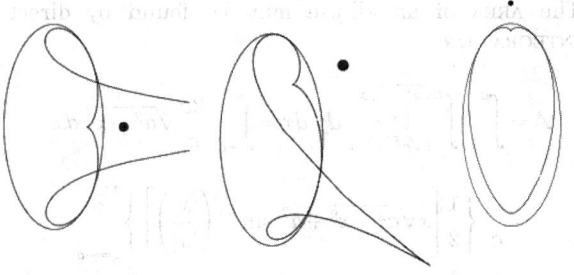

Ellipse Envelope

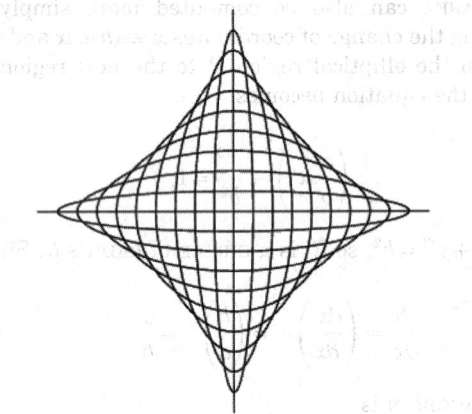

Consider the family of ELLIPSES

$$\frac{x^2}{c^2} + \frac{y^2}{(1 - c)^2} - 1 = 0 \tag{1}$$

for $c \in [0, 1]$. The PARTIAL DERIVATIVE with respect to c is

$$-\frac{2x^2}{c^3} + \frac{2y^2}{(1 - c)^3} = 0 \tag{2}$$

$$\frac{x^2}{c^3} - \frac{y^2}{(1 - c)^3} = 0. \tag{3}$$

Combining (1) and (3) gives the set of equations

$$\begin{bmatrix} \dfrac{1}{c^2} & \dfrac{1}{(1 - c)^2} \\[2ex] \dfrac{1}{c^3} & -\dfrac{1}{(1 - c)^3} \end{bmatrix} \begin{bmatrix} x^2 \\ y^2 \end{bmatrix} = \begin{bmatrix} 1 \\ 0 \end{bmatrix} \tag{4}$$

$$\begin{bmatrix} x^2 \\ y^2 \end{bmatrix} = \frac{1}{\Delta} \begin{bmatrix} -\dfrac{1}{(1-c)^3} & \dfrac{1}{(1-c)^2} \\ -\dfrac{1}{c^3} & \dfrac{1}{c^2} \end{bmatrix} \begin{bmatrix} 1 \\ 0 \end{bmatrix}$$

$$= \frac{1}{\Delta} \begin{bmatrix} -\dfrac{1}{(1-c)^3} \\ -\dfrac{1}{c^3} \end{bmatrix}, \tag{5}$$

where the DISCRIMINANT is

$$\Delta = -\frac{1}{c^2(1-c)^3} - \frac{1}{c^3(1-c)^2} = -\frac{1}{c^3(1-c)^3}, \tag{6}$$

so (5) becomes

$$\begin{bmatrix} x^2 \\ y^2 \end{bmatrix} = \begin{bmatrix} c^3 \\ (1-c)^3 \end{bmatrix}. \tag{7}$$

Eliminating c then gives

$$x^{2/3} + y^{2/3} = 1, \tag{8}$$

which is the equation of the ASTROID. If the curve is instead represented parametrically, then

$$x = c \cos t \tag{9}$$

$$y = (1-c) \sin t. \tag{10}$$

Solving

$$\frac{\partial x}{\partial t} \frac{\partial y}{\partial c} - \frac{\partial x}{\partial c} \frac{\partial y}{\partial t}$$

$$= (-c \sin t)(-\sin t) - (\cos t)[(1-c) \cos t]$$

$$= c(\sin^2 t + \cos^2 t) - \cos^2 t = c - \cos^2 t = 0 \tag{11}$$

for c gives

$$c = \cos^2 t, \tag{12}$$

so substituting this back into (9) and (10) gives

$$x = (\cos^2 t) \cos t = \cos^3 t \tag{13}$$

$$y = (1 - \cos^2 t) \sin t = \sin^3 t, \tag{14}$$

the PARAMETRIC EQUATIONS of the ASTROID.

See also ASTROID, ELLIPSE, ENVELOPE

Ellipse Evolute

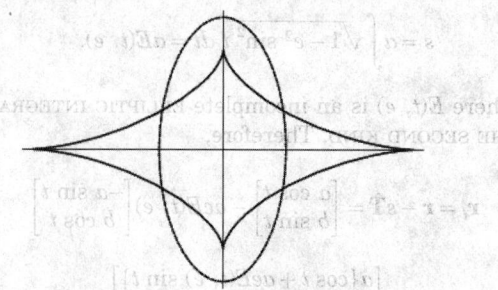

The EVOLUTE of an ELLIPSE is given by the PARAMETRIC EQUATIONS

$$x = \frac{a^2 - b^2}{a} \cos^3 t \tag{1}$$

$$y = \frac{b^2 - a^2}{b} \sin^3 t, \tag{2}$$

which can be combined and written

$$(ax)^{2/3} + (by)^{2/3}$$

$$= [(a^2 - b^2) \cos^3 t]^{2/3} + [(b^2 - a^2)] \sin^3 t]^{2/3}$$

$$= (a^2 - b^2)^{2/3}(\sin^2 t + \cos^2 t) = (a^2 - b^2)^{2/3} = c^{4/3}, \tag{3}$$

which is a stretched ASTROID sometimes called the LAMÉ CURVE. From a point inside the EVOLUTE, four NORMALS can be drawn to the ellipse, but from a point outside, only two NORMALS can be drawn.

See also ASTROID, ELLIPSE, EVOLUTE, LAMÉ CURVE

References

Beyer, W. H. *CRC Standard Mathematical Tables, 28th ed.* Boca Raton, FL: CRC Press, p. 217, 1987.

Gray, A. *Modern Differential Geometry of Curves and Surfaces with Mathematica, 2nd ed.* Boca Raton, FL: CRC Press, pp. 99–01, 1997.

Ellipse Involute

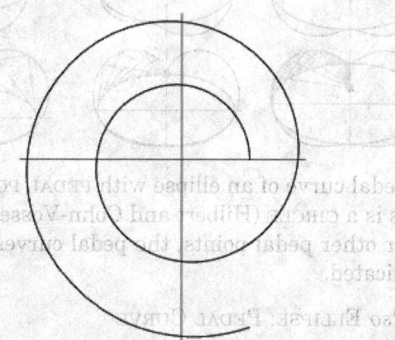

From ELLIPSE, the TANGENT VECTOR is

$$\mathbf{T} = \begin{bmatrix} -a \sin t \\ b \cos t \end{bmatrix}, \tag{1}$$

and the ARC LENGTH is

$$s = a \int \sqrt{1 - e^2 \sin^2 t} \, dt = aE(t, e), \qquad (2)$$

where $E(t, e)$ is an incomplete ELLIPTIC INTEGRAL OF THE SECOND KIND. Therefore,

$$\mathbf{r}_i = \mathbf{r} - s\hat{\mathbf{T}} = \begin{bmatrix} a \cos t \\ b \sin t \end{bmatrix} - aeE(t, e) \begin{bmatrix} -a \sin t \\ b \cos t \end{bmatrix} \qquad (3)$$

$$= \begin{bmatrix} a\{\cos t + aeE(t, e) \sin t\} \\ b\{\sin t - aeE(t, e) \cos t\} \end{bmatrix}. \qquad (4)$$

Ellipse Pedal Curve

The pedal curve of an ellipse with semimajor axis a, semiminor axis b, and PEDAL POINT (x_0, y_0) is given by

$$f = \frac{a[ax_0 \sin^2 t + b \cos t(b - y_0 \sin t)]}{b^2 \cos^2 t + a^2 \sin^2 t}$$

$$g = \frac{b[a^2 \sin^2 t - ax_0 \cos t \sin t + by_0 \cos^2 t]}{b^2 \cos^2 t + a^2 \sin^2 t}.$$

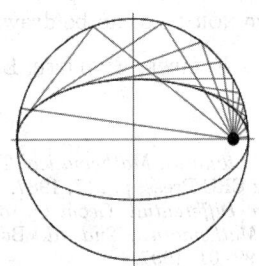

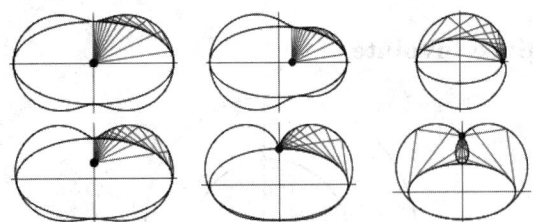

The pedal curve of an ellipse with PEDAL POINT at the FOCUS is a CIRCLE (Hilbert and Cohn-Vossen, pp. 25–6). For other pedal points, the pedal curves are more complicated.

See also ELLIPSE, PEDAL CURVE

References
Hilbert, D. and Cohn-Vossen, S. *Geometry and the Imagination.* New York: Chelsea, 1999.

Ellipse Point Picking

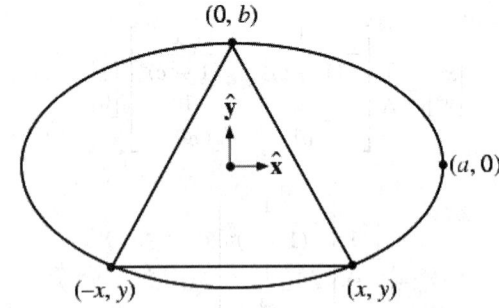

To inscribe an EQUILATERAL TRIANGLE in an ELLIPSE, place the top VERTEX at $(0, b)$, then solve to find the (x, y) coordinate of the other two VERTICES.

$$\sqrt{x^2 + (b - y)^2} = 2x \qquad (1)$$

$$x^2 + (b - y)^2 = 4x^2 \qquad (2)$$

$$3x^2 = (b - y)^2. \qquad (3)$$

Now plugging in the equation of the ELLIPSE

$$\frac{x^2}{a^2} + \frac{y^2}{b^2} = 1, \qquad (4)$$

gives

$$3a^2 \left(1 - \frac{y^2}{b^2}\right) = b^2 - 2by + y^2 \qquad (5)$$

$$y^2 \left(1 + 3\frac{a^2}{b^2}\right) - 2by + (b^2 - 3a^2) = 0 \qquad (6)$$

$$y = \frac{2b - \sqrt{4b^2 - 4(b^2 - 3a^2)\left(1 + 3\frac{a^2}{b^2}\right)}}{2\left(1 + 3\frac{a^2}{b^2}\right)}$$

$$= \frac{1 - \sqrt{1 - \left(1 - 3\frac{a^2}{b^2}\right)\left(1 + 3\frac{a^2}{b^2}\right)}}{1 + 3\frac{a^2}{b^2}} b, \qquad (7)$$

and

$$x = \pm a\sqrt{1 - \frac{y^2}{b^2}}. \qquad (8)$$

See also ELLIPSE, EQUILATERAL TRIANGLE

Ellipse Tangent

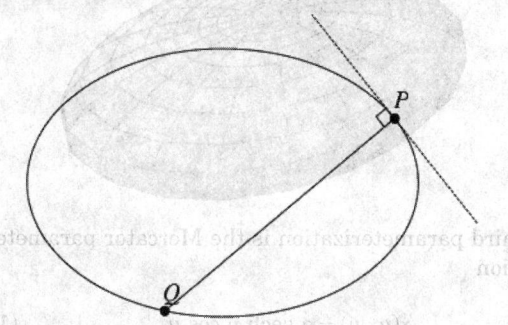

The normal to an ellipse at a point P intersects the ellipse at another point Q. The angle corresponding to Q can be found by solving the equation

$$(\mathbf{P} - \mathbf{Q}) \cdot \frac{d\mathbf{P}}{dt} = 0 \tag{1}$$

for t', where $\mathbf{P}(t) = (a \cos t,\ b \sin t)$ and $\mathbf{Q}(t) = (a \cos t',\ b \sin t')$. This gives solutions

$$t' = \pm \cos^{-1}\left[\pm \frac{N(t)}{a^4 \sin^2 t + b^4 \cos^2 t}\right], \tag{2}$$

where

$$N(t) \equiv b^2 \cos t[a^2 + b^2(b^2 - a)^2 \cos(2t)]$$

$$+ a^2(a - b)(a + b) \cos t \sin^2 t, \tag{3}$$

of which $(+, -)$ gives the valid solution. Plugging this in to obtain Q then gives

$$d(t) = |\mathbf{P} - \mathbf{Q}|$$

$$= \frac{\sqrt{2}\, ab[a^2 + b^2 + (b^2 - a^2) \cos(2t)]^{3/2}}{a^4 + b^4 + (b^4 - a^4) \cos(2t)} \tag{4}$$

$$= \frac{2ab(b^2 \cos^2 t + a^2 \sin^2 t)^{3/2}}{b^4 \cos^2 t + a^4 \sin^2 t}. \tag{5}$$

To find the maximum distance, take the derivative and set equal to zero,

$$d'(t)$$

$$= \frac{2ab(a - b)(a + b) \cos t \sin t \sqrt{b^2 \cos^2 t + a^2 \sin^2 t}}{(b^4 \cos^2 t + a^4 \sin^2 t)^2}$$

$$\times (a^4 \sin^2 t + b^4 \cos^2 t - 2a^2 b^2) = 0, \tag{6}$$

which simplifies to

$$a^4 \sin^2 t + b^4 \cos^2 t - 2a^2 b^2 = 0. \tag{7}$$

Substituting for $\sin^2 t$ and solving gives

$$\cos^2 t = \frac{a^4 - 2a^2 b^2}{a^4 - b^4} \tag{8}$$

$$\sin^2 t = \frac{2a^2 b^2 - b^4}{a^4 - b^4}. \tag{9}$$

Plugging these into $d(t)$ then gives

$$d_{\min} = \frac{3\sqrt{3}\, a^2 b^2}{(a^2 + b^2)^{3/2}}. \tag{10}$$

This problem was given as a SANGAKU PROBLEM on a tablet from Miyagi Prefecture in 1912 (Rothman 1998). There is probably a clever solution to this problem which does not require calculus, but it is unknown if calculus was used in the solution by the original authors (Rothman 1998).

See also ELLIPSE

References
Rothman, T. "Japanese Temple Geometry." *Sci. Amer.* **278**, 85–1, May 1998.

Ellipsoid

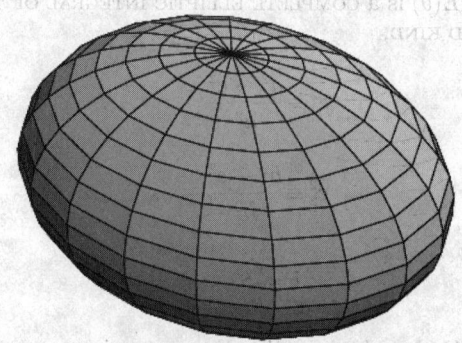

A QUADRATIC SURFACE which is given in CARTESIAN COORDINATES by

$$\frac{x^2}{a^2} + \frac{y^2}{b^2} + \frac{z^2}{c^2} = 1, \tag{1}$$

where the semi-axes are of lengths a, b, and c. In SPHERICAL COORDINATES, this becomes

$$\frac{r^2 \cos^2 \theta \sin^2 \phi}{a^2} + \frac{r^2 \sin^2 \theta \sin^2 \phi}{b^2} + \frac{r^2 \cos^2 \phi}{c^2} = 1. \tag{2}$$

The PARAMETRIC EQUATIONS are

$$x = a \cos \theta \sin \phi \tag{3}$$

$$y = b \sin \theta \sin \phi \tag{4}$$

$$z = c \cos \phi. \tag{5}$$

for $\theta \in [0,\ 2\pi)$ and $\phi \in [0,\ \pi]$.

If the lengths of two axes of an ellipsoid are the same, the figure is called a SPHEROID (depending on whether $c < a$ or $c > a$, an OBLATE SPHEROID or PROLATE SPHEROID, respectively), and if all three are the same, it is a SPHERE. Tietze (1965, p. 28) calls the general ellipsoid a "triaxial ellipsoid."

There are two families of parallel CIRCULAR CROSS SECTIONS in every ellipsoid. However, the two coincide for SPHEROIDS (Hilbert and Cohn-Vossen 1999, pp. 17–9). If the two sets of circles are fastened together by suitably chosen slits so that are free to rotate without sliding, the model is movable. Furthermore, the disks can always be moved into the shape of a SPHERE (Hilbert and Cohn-Vossen 1999, p. 18).

In 1882, Staude discovered a "thread" construction for an ellipsoid analogous to the taught pencil and string construction of the ELLIPSE (Hilbert and Cohn-Vossen 1999, pp. 19–2). This construction makes use of a fixed framework consisting of an ELLIPSE and a HYPERBOLA.

The SURFACE AREA of an ellipsoid (Bowman 1961, pp. 31–2) is given by

$$S = 2\pi c^2 + \frac{2\pi b}{\sqrt{a^2 - c^2}}[(a^2 - c^2)E(\theta) + c^2\theta], \qquad (6)$$

where $E(\theta)$ is a COMPLETE ELLIPTIC INTEGRAL OF THE SECOND KIND,

$$e_1^2 \equiv \frac{a^2 - c^2}{a^2} \qquad (7)$$

$$e_2^2 \equiv \frac{b^2 - c^2}{b^2} \qquad (8)$$

$$k \equiv \frac{e_2}{e_1}, \qquad (9)$$

and θ is given by inverting the expression

$$e_1 = \operatorname{sn}(\theta, k), \qquad (10)$$

where $\operatorname{sn}(\theta, k)$ is a JACOBI ELLIPTIC FUNCTION. The VOLUME of an ellipsoid is

$$V = \tfrac{4}{3}\pi abc. \qquad (11)$$

A different parameterization of the ellipsoid is the so-called stereographic ellipsoid, given by the PARA-METRIC EQUATIONS

$$x(u, v) = \frac{a(1 - u^2 - v^2)}{1 + u^2 + v^2} \qquad (12)$$

$$y(u, v) = \frac{2bu}{1 + u^2 + v^2} \qquad (13)$$

$$z(u, v) = \frac{2cv}{1 + u^2 + v^2}. \qquad (14)$$

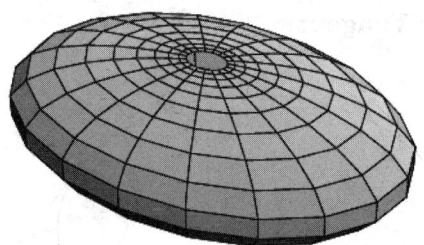

A third parameterization is the Mercator parameterization

$$x(u, v) = a \operatorname{sech} v \cos u \qquad (15)$$

$$y(u, v) = b \operatorname{sech} v \sin u \qquad (16)$$

$$z(u, v) = c \tanh v \qquad (17)$$

(Gray 1997).

The SUPPORT FUNCTION of the ellipsoid is

$$h = \left(\frac{x^2}{a^4} + \frac{y^2}{b^4} + \frac{z^2}{c^4}\right)^{-1/2}, \qquad (18)$$

and the GAUSSIAN CURVATURE is

$$K = \frac{h^4}{a^2 b^2 c^2} \qquad (19)$$

(Gray 1997, p. 296).

See also CONFOCAL ELLIPSOIDAL COORDINATES, CONFOCAL QUADRICS, CONVEX OPTIMIZATION THEORY, ELLIPSOID PACKING, GOURSAT'S SURFACE, OBLATE SPHEROID, PROLATE SPHEROID, SPHERE, SPHEROID, SUPERELLIPSOID

References

Beyer, W. H. *CRC Standard Mathematical Tables, 28th ed.* Boca Raton, FL: CRC Press, pp. 131 and 226, 1987.

Bowman, F. *Introduction to Elliptic Functions, with Applications.* New York: Dover, 1961.

Fischer, G. (Ed.). Plate 65 in *Mathematische Modelle/Mathematical Models, Bildband/Photograph Volume.* Braunschweig, Germany: Vieweg, p. 60, 1986.

Gray, A. "The Ellipsoid" and "The Stereographic Ellipsoid." §13.2 and 13.3 in *Modern Differential Geometry of Curves and Surfaces with Mathematica, 2nd ed.* Boca Raton, FL: CRC Press, pp. 301–03, 1997.

Harris, J. W. and Stocker, H. "Ellipsoid." §4.10.1 in *Handbook of Mathematics and Computational Science.* New York: Springer-Verlag, p. 111, 1998.

Hilbert, D. and Cohn-Vossen, S. "The Thread Construction of the Ellipsoid, and Confocal Quadrics." §4 in *Geometry and the Imagination.* New York: Chelsea, pp. 19–5, 1999.

JavaView. "Classic Surfaces from Differential Geometry: Ellipsoid." http://www-sfb288.math.tu-berlin.de/vgp/javaview/demo/surface/common/PaSurface_Ellipsoid.html.

Tietze, H. *Famous Problems of Mathematics: Solved and Unsolved Mathematics Problems from Antiquity to Modern Times.* New York: Graylock Press, pp. 28 and 40–1, 1965.

Ellipsoid Geodesic

An ELLIPSOID can be specified parametrically by

$$x = a \cos u \sin v \tag{1}$$

$$y = b \sin u \sin v \tag{2}$$

$$z = c \cos v. \tag{3}$$

The GEODESIC parameters are then

$$P = \sin^2 v(b^2 \cos^2 u + a^2 \sin^2 u) \tag{4}$$

$$Q = \tfrac{1}{4}(b^2 - a^2) \sin(2u) \sin(2v) \tag{5}$$

$$R = \cos^2 v(a^2 \cos^2 u + b^2 \sin^2 u) + c^2 \sin^2 v. \tag{6}$$

When the coordinates of a point are on the QUADRIC

$$\frac{x^2}{a} + \frac{y^2}{b} + \frac{z^2}{c} = 1 \tag{7}$$

and expressed in terms of the parameters p and q of the confocal quadrics passing through that point (in other words, having $a + p$, $b + p$, $c + p$, and $a + q$, $b + q$, $c + q$ for the squares of their semimajor axes), then the equation of a GEODESIC can be expressed in the form

$$\frac{q \, dq}{\sqrt{q(a+q)(b+q)(c+q)(\theta+q)}}$$

$$\pm \frac{p \, dp}{\sqrt{p(a+p)(b+p)(c+p)(\theta+p)}} = 0, \tag{8}$$

with θ an arbitrary constant, and the ARC LENGTH element ds is given by

$$-2\frac{ds}{pq} = \frac{dq}{\sqrt{q(a+q)(b+q)(c+q)(\theta+q)}}$$

$$\pm \frac{dp}{\sqrt{p(a+p)(b+p)(c+p)(\theta+p)}}, \tag{9}$$

where upper and lower signs are taken together.

See also OBLATE SPHEROID GEODESIC, SPHERE GEODESIC

References

Eisenhart, L. P. *A Treatise on the Differential Geometry of Curves and Surfaces.* New York: Dover, pp. 236–41, 1960.
Forsyth, A. R. *Calculus of Variations.* New York: Dover, p. 447, 1960.
Tietze, H. *Famous Problems of Mathematics: Solved and Unsolved Mathematics Problems from Antiquity to Modern Times.* New York: Graylock Press, pp. 28–9 and 40–1, 1965.

Ellipsoid of Revolution

OBLATE SPHEROID, PROLATE SPHEROID, SPHEROID

Ellipsoid Packing

Bezdek and Kuperberg (1991) have constructed packings of identical ellipsoids of densities > 0.75, greater than the maximum density possible for identical spheres (Sloane 1998).

See also SPHERE PACKING

References

Bezdek, A. and Kuperberg, W. In *Applied Geometry and Discrete Mathematics: The Victor Klee Festschrift* (Ed. P. Gritzmann and B. Sturmfels). Providence, RI: Amer. Math. Soc., pp. 71–0, 1991.
Sloane, N. J. A. "Kepler's Conjecture Confirmed." *Nature* **395**, 435–36, 1998.

Ellipsoidal Calculus

Ellipsoidal calculus is a method for solving problems in control and estimation theory having unknown but bounded errors in terms of sets of approximating ellipsoidal-value functions. Ellipsoidal calculus has been especially useful in the study of LINEAR PROGRAMMING.

References

Kurzhanski, A. B. and Vályi, I. *Ellipsoidal Calculus for Estimation and Control.* Boston, MA: Birkhäuser, 1996.
Papadimitriou, C. H. and Steiglitz, K. *Combinatorial Optimization: Algorithms and Complexity.* New York: Dover, 1998.

Ellipsoidal Coordinates

CONFOCAL ELLIPSOIDAL COORDINATES

Ellipsoidal Harmonic

ELLIPSOIDAL HARMONIC OF THE FIRST KIND, ELLIPSOIDAL HARMONIC OF THE SECOND KIND

Ellipsoidal Harmonic of the First Kind

The first solution to LAMÉ'S DIFFERENTIAL EQUATION, denoted $E_n^m(x)$ for $m = 1, ..., 2n + 1$. They are also called LAMÉ FUNCTIONS. The product of two ellipsoidal harmonics of the first kind is a SPHERICAL HARMONIC. Whittaker and Watson (1990, pp. 536–37) write

$$\Theta_p = \frac{x^2}{a^2 + \theta_p} + \frac{y^2}{b^2 + \theta_p} + \frac{z^2}{c^2 + \theta_p} - 1 \tag{1}$$

$$\Pi(\Theta) \equiv \Theta_1 \Theta_2 \cdots \Theta_m, \tag{2}$$

and give various types of ellipsoidal harmonics and their highest degree terms as

1. $\Pi(\Theta) : 2m$
2. $x\Pi(\Theta),\ y\Pi(\Theta),\ z\Pi(\Theta) : 2m+1$
3. $yz\Pi(\Theta),\ zx\Pi(\Theta),\ xy\Pi(\Theta) : 2m+2$
4. $xyz\Pi(\Theta) : 2m+3$.

A Lamé function of degree n may be expressed as

$$(\theta+a^2)^{\kappa_1}(\theta+b^2)^{\kappa_2}(\theta+c^2)^{\kappa_3}\prod_{p=1}^{m}(\theta-\theta_p), \qquad (3)$$

where $\kappa_i = 0$ or $1/2$, θ_i are REAL and unequal to each other and to $-a^2$, $-b^2$, and $-c^2$, and

$$\tfrac{1}{2}n = m+\kappa_1+\kappa_2+\kappa_3. \qquad (4)$$

Byerly (1959) uses the RECURRENCE RELATIONS to explicitly compute some ellipsoidal harmonics, which he denotes by $K(x)$, $L(x)$, $M(x)$, and $N(x)$,

$$K_0(x) = 1$$

$$L_0(x) = 0$$

$$M_0(x) = 0$$

$$N_0(x) = 0$$

$$K_1(x) = x$$

$$L_1(x) = \sqrt{x^2-b^2}$$

$$M_1(x) = \sqrt{x^2-c^2}$$

$$N_1(x) = 0$$

$$K_2^{p_1}(x) = x^2 - \tfrac{1}{3}[b^2+c^2 - \sqrt{(b^2+c^2)^2-3b^2c^2}]$$

$$K_2^{p_2}(x) = x^2 - \tfrac{1}{3}[b^2+c^2 + \sqrt{(b^2+c^2)^2-3b^2c^2}]$$

$$L_2(x) = x\sqrt{x^2-b^2}$$

$$M_2(x) = x\sqrt{x^2-c^2}$$

$$N_2(x) = \sqrt{(x^2-b^2)(x^2-c^2)}$$

$$K_3^{p_1}(x) = x^3 - \tfrac{1}{5}x[2(b^2+c^2) - \sqrt{4(b^2+c^2)^2-15b^2c^2}]$$

$$K_3^{p_2}(x) = x^3 - \tfrac{1}{5}x[2(b^2+c^2) + \sqrt{4(b^2+c^2)^2-15b^2c^2}]$$

$$L_3^{q_1}(x) = \sqrt{x^2-b^2}[x^2$$
$$-\tfrac{1}{5}(b^2+2c^2 - \sqrt{(b^2+2c^2)^2-5b^2c^2})]$$

$$L_3^{q_2}(x) = \sqrt{x^2-b^2}[x^2 - \tfrac{1}{5}(b^2+2c^2$$
$$+\sqrt{(b^2+2c^2)^2-5b^2c^2})]$$

$$M_3^{q_1}(x) = \sqrt{x^2-c^2}[x^2 - \tfrac{1}{5}(2b^2+c^2$$
$$-\sqrt{(2b^2+c^2)^2-5b^2c^2})]$$

$$M_3^{q_2}(x) = \sqrt{x^2-c^2}[x^2 - \tfrac{1}{5}(2b^2+c^2$$
$$+\sqrt{(2b^2+c^2)^2-5b^2c^2})]$$

$$M_3^{q_3}(x) = x\sqrt{(x^2-b^2)(x^2-c^2)}$$

See also ELLIPSOIDAL HARMONIC OF THE SECOND KIND, STIELTJES' THEOREM

References

Byerly, W. E. "Laplace's Equation in Curvilinear Coördinates. Ellipsoidal Harmonics." Ch. 8 in *An Elementary Treatise on Fourier's Series, and Spherical, Cylindrical, and Ellipsoidal Harmonics, with Applications to Problems in Mathematical Physics.* New York: Dover, pp. 238–66, 1959.
Humbert, P. *Fonctions de Lamé et Fonctions de Mathieu.* Paris: Gauthier-Villars, 1926.
Whittaker, E. T. and Watson, G. N. *A Course in Modern Analysis, 4th ed.* Cambridge, England: Cambridge University Press, 1990.

Ellipsoidal Harmonic of the Second Kind

Given by

$$F_m^p(x) = (2m+1)E_m^p(x)\int_x^\infty \frac{dx}{(x^2-b^2)(x^2-c^2)[E_m^p(x)]^2}.$$

Ellipsoidal Wave Equation

The ORDINARY DIFFERENTIAL EQUATION

$$y'' - (a+bk^2\,\mathrm{sn}^2\,x + qk^4\,\mathrm{sn}^4\,x)y = 0,$$

where $\mathrm{sn}\,x = \mathrm{sn}(x,k)$ is a JACOBI ELLIPTIC FUNCTION (Arscott 1981).

See also LAMÉ'S DIFFERENTIAL EQUATION

References

Arscott, F. M. "The Land beyond Bessel: A Survey of Higher Special Functions." In *Ordinary and Partial Differential Equations: Proceeding of the Sixth Conference held at the University of Dundee, March 31–April 4, 1980* (Ed. W. N. Everitt and B. D. Sleeman). New York: Springer-Verlag, pp. 26–5, 1981.
Zwillinger, D. *Handbook of Differential Equations, 3rd ed.* Boston, MA: Academic Press, p. 122, 1997.

Elliptic Alpha Function

Elliptic alpha functions relate the complete ELLIPTIC INTEGRALS OF THE FIRST $K(k_r)$ and SECOND KINDS $E(k_r)$ at ELLIPTIC INTEGRAL SINGULAR VALUES k_r

according to

$$\alpha(r) = \frac{e'(k_r)}{k(k_r)} - \frac{\pi}{4[k(k_r)]^2} \tag{1}$$

$$= \frac{\pi}{4[k(k_r)]^2} + \sqrt{r} - \frac{e(k_r)\sqrt{r}}{k(k_r)} \tag{2}$$

$$= \frac{\pi^{-1} - 4\sqrt{r}\, q \dfrac{d\vartheta_4(q)}{dq} \dfrac{1}{\vartheta_4(q)}}{\vartheta_3^4(q)}, \tag{3}$$

where $\vartheta_3(q)$ is a JACOBI THETA FUNCTION and

$$k_r = \lambda^*(r) \tag{4}$$

$$q = e^{-\pi\sqrt{r}}, \tag{5}$$

and $\lambda^*(r)$ is the ELLIPTIC LAMBDA FUNCTION. The elliptic alpha function is related to the ELLIPTIC DELTA FUNCTION by

$$\alpha(r) = \tfrac{1}{2}[\sqrt{r} - \delta(r)]. \tag{6}$$

It satisfies

$$\alpha(4r) = (1 + k_r)^2 \alpha(r) - 2\sqrt{r}\, k_r, \tag{7}$$

and has the limit

$$\lim_{r \to \infty}\left[\alpha(r) - \frac{1}{\pi}\right] \approx 8\left(\sqrt{r} - \frac{1}{\pi}\right)e^{-\pi\sqrt{r}} \tag{8}$$

(Borwein *et al.* 1989). A few specific values (Borwein and Borwein 1987, p. 172) are

$$\alpha(1) = \tfrac{1}{2}$$

$$\alpha(2) = \sqrt{2} - 1$$

$$\alpha(3) = \tfrac{1}{2}(\sqrt{3} - 1)$$

$$\alpha(4) = 2(\sqrt{2} - 1)^2$$

$$\alpha(5) = \tfrac{1}{2}(\sqrt{5} - \sqrt{2\sqrt{5} - 2})$$

$$\alpha(6) = 5\sqrt{6} + 6\sqrt{3} - 8\sqrt{2} - 11$$

$$\alpha(7) = \tfrac{1}{2}(\sqrt{7} - 2)$$

$$\alpha(8) = 2(10 + 7\sqrt{2})(1 - \sqrt{\sqrt{8} - 2})^2$$

$$\alpha(9) = \tfrac{1}{2}[3 - 3^{3/4}\sqrt{2}(\sqrt{3} - 1)]$$

$$\alpha(10) = -103 + 72\sqrt{2} - 46\sqrt{5} + 33\sqrt{10}$$

$$\alpha(12) = 264 + 154\sqrt{3} - 188\sqrt{2} - 108\sqrt{6}$$

$$\alpha(13) = \tfrac{1}{2}(\sqrt{13} - \sqrt{74\sqrt{13} - 258})$$

$$\alpha(15) = \tfrac{1}{2}(\sqrt{15} - \sqrt{5} - 1)$$

$$\alpha(16) = \frac{4(\sqrt{8} - 1)}{(2^{1/4} + 1)^4}$$

$$\alpha(18) = -3057 + 2163\sqrt{2} + 1764\sqrt{3} - 1248\sqrt{6}$$

$$\alpha(22) = -12479 - 8824\sqrt{2} + 3762\sqrt{11} + 2661\sqrt{22}$$

$$\alpha(25) = \tfrac{5}{2}[1 - 25^{1/4}(7 - 3\sqrt{5})]$$

$$\alpha(27) = 3[\tfrac{1}{2}(\sqrt{3} + 1) - 2^{1/3}]$$

$$\alpha(30) = \tfrac{1}{2}\sqrt{30} - (2 + \sqrt{5})^2(3 + \sqrt{10})^2$$
$$\times(-6 - 5\sqrt{2} - 3\sqrt{5} - 2\sqrt{10} + \sqrt{6}\sqrt{57 + 40\sqrt{2}}$$
$$\times[56 + 38\sqrt{2} + \sqrt{30}(2 + \sqrt{5})(3 + \sqrt{10})]\}$$

$$\alpha(37) = \tfrac{1}{2}\left[\sqrt{37} - (171 - 25\sqrt{37})\sqrt{\sqrt{37} - 6}\right]$$

$$\alpha(46) = \tfrac{1}{2}[\sqrt{46} + (18 + 13\sqrt{2} + \sqrt{661 + 468\sqrt{2}})^2$$
$$\times(18 + 13\sqrt{2} - 3\sqrt{2}\sqrt{147 + 104\sqrt{2}} + \sqrt{661 + 468\sqrt{2}})$$
$$\times(200 + 14\sqrt{2} + 26\sqrt{23} + 18\sqrt{46} + \sqrt{46}\sqrt{661 + 468\sqrt{2}})]$$

$$\alpha(49) = \tfrac{7}{2} - \sqrt{7[\sqrt{27^{3/4}(33011 + 12477\sqrt{7})} - 21(9567 + 3616\sqrt{7})]}$$

$$\alpha(58) = [\tfrac{1}{2}(\sqrt{29} + 5)]^6(99\sqrt{29} - 444)(99\sqrt{2} - 70 - 13\sqrt{29})$$
$$= 3(-40768961 + 2882008\sqrt{2} - 7570606\sqrt{29} + 5353227$$
$$\times\sqrt{58})$$

$$\alpha(64) = \frac{8[2(\sqrt{8} - 1) - (2^{1/4} - 1)^4]}{(\sqrt{\sqrt{2} + 1} + 2^{5/8})^4}.$$

J. Borwein has written an ALGORITHM which uses lattice basis reduction to provide algebraic values for $\alpha(n)$.

See also ELLIPTIC INTEGRAL OF THE FIRST KIND, ELLIPTIC INTEGRAL OF THE SECOND KIND, ELLIPTIC INTEGRAL SINGULAR VALUE, ELLIPTIC LAMBDA FUNCTION

References

Borwein, J. M. and Borwein, P. B. *Pi & the AGM: A Study in Analytic Number Theory and Computational Complexity.* New York: Wiley, 1987.

Borwein, J. M.; Borwein, P. B.; and Bailey, D. H. "Ramanujan, Modular Equations, and Approximations to Pi, or How to Compute One Billion Digits of Pi." *Amer. Math. Monthly* **96**, 201–19, 1989.

Weisstein, E. W. "Elliptic Singular Values." MATHEMATICA NOTEBOOK ELLIPTICSINGULAR.M.

Elliptic Cone

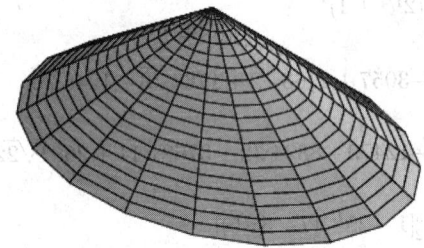

A CONE with ELLIPTICAL CROSS SECTION. The PARAMETRIC EQUATIONS for an elliptic cone of height h, SEMIMAJOR AXIS a, and SEMIMINOR AXIS b are

$$x = (h-z)a \cos \theta$$

$$y = (h-z)b \sin \theta$$

$$z = z,$$

where $\theta \in [0, 2\pi)$ and $z \in [0, h]$. The elliptic cone is a QUADRATIC RULED SURFACE, and has VOLUME

$$V = \tfrac{1}{3}\pi ab.$$

See also CONE, ELLIPTIC CYLINDER, ELLIPTIC PARABOLOID, HYPERBOLIC PARABOLOID, QUADRATIC SURFACE, RULED SURFACE

References

Beyer, W. H. *CRC Standard Mathematical Tables, 28th ed.* Boca Raton, FL: CRC Press, p. 226, 1987.

Fischer, G. (Ed.). Plate 68 in *Mathematische Modelle/ Mathematical Models, Bildband/Photograph Volume.* Braunschweig, Germany: Vieweg, p. 63, 1986.

Elliptic Cone Point

ISOLATED SINGULARITY

Elliptic Coordinates

CONFOCAL ELLIPSOIDAL COORDINATES

Elliptic Curve

Informally, an elliptic curve is a type of CUBIC CURVE whose solutions are confined to a region of space which is topologically equivalent to a TORUS. The WEIERSTRASS ELLIPTIC FUNCTION $\wp(z; g_2, g_3)$ describes how to get from this TORUS to the algebraic form of an elliptic curve.

Formally, an elliptic curve over a FIELD K is a nonsingular CUBIC CURVE in two variables, $f(X, Y) = 0$, with a K-rational point (which may be a POINT AT INFINITY). The FIELD K is usually taken to be the COMPLEX NUMBERS $\mathbb{C}$, REALS $\mathbb{R}$, RATIONALS $\mathbb{Q}$, algebraic extensions of $\mathbb{Q}$, P-ADIC NUMBERS $\mathbb{Q}_p$, or a FINITE FIELD.

By an appropriate change of variables, a general elliptic curve over a FIELD of CHARACTERISTIC $\neq 2, 3$

$$Ax^3 + Bx^2y + Cxy^2 + Dy^3 + Ex^2 + Fxy + Gy^2 + Hx$$
$$+ Iy + J = 0, \tag{1}$$

where A, B, ..., are elements of K, can be written in the form

$$y^2 = x^3 + ax + b, \tag{2}$$

where the right side of (2) has no repeated factors. If K has CHARACTERISTIC three, then the best that can be done is to transform the curve into

$$y^2 = x^3 + ax^2 + bx + c \tag{3}$$

(the x^2 term cannot be eliminated). If K has CHARACTERISTIC two, then the situation is even worse. A general form into which an elliptic curve over any K can be transformed is called the WEIERSTRASS FORM, and is given by

$$y^2 + ay = x^3 + bx^2 + cxy + dx + e, \tag{4}$$

where a, b, c, d, and e are elements of K. Luckily, $\mathbb{Q}$, $\mathbb{R}$, and $\mathbb{C}$ all have CHARACTERISTIC zero.

Whereas CONIC SECTIONS can be parameterized by the rational functions, elliptic curves cannot. The simplest parameterization functions are ELLIPTIC FUNCTIONS. ABELIAN VARIETIES can be viewed as generalizations of elliptic curves.

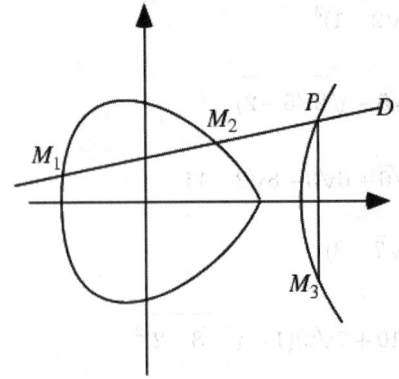

If the underlying FIELD of an elliptic curve is algebraically closed, then a straight line cuts an elliptic curve at three points (counting multiple roots at points of tangency). If two are known, it is possible to compute the third. If two of the intersection points are K-RATIONAL, then so is the third. Mazur and Tate (1973/74) proved that there is no elliptic curve over $\mathbb{Q}$ having a RATIONAL POINT of order 13.

Let (x_1, y_1) and (x_2, y_2) be two points on an elliptic curve E with DISCRIMINANT

$$\Delta_E = -16(4a^3 + 27b^2) \tag{5}$$

satisfying

$$\Delta_E \neq 0. \tag{6}$$

A related quantity known as the J-INVARIANT of E is defined as

$$j(E) \equiv \frac{2^8 3^3 a^3}{4a^3 + 27b^2}. \tag{7}$$

Now define

$$\lambda = \begin{cases} \dfrac{y_1 - y_2}{x_1 - x_2} & \text{for } x_1 \neq x_2 \\[2mm] \dfrac{3x_1^2 + \alpha}{2y_1} & \text{for } x_1 = x_2. \end{cases} \tag{8}$$

Then the coordinates of the third point are

$$x_3 = \lambda^2 - x_1 - x_2 \tag{9}$$

$$y_3 = \lambda(x_3 - x_1) + y_1. \tag{10}$$

For elliptic curves over $\mathbb{Q}$, Mordell proved that there are a finite number of integral solutions. The MOR-DELL-WEIL THEOREM says that the GROUP of RATIONAL POINTS of an elliptic curve over $\mathbb{Q}$ is finitely generated. Let the ROOTS of y^2 be r_1, r_2, and r_3. The discriminant is then

$$\Delta = k(r_1 - r_2)^2(r_1 - r_3)^2(r_2 - r_3)^2. \tag{11}$$

The amazing TANIYAMA-SHIMURA CONJECTURE states that all rational elliptic curves are also modular. This fact is far from obvious, and despite the fact that the conjecture was proposed in 1955, it was not even partially proved until 1995. Even so, Wiles' proof for the semistable case surprised most mathematicians, who had believed the conjecture unassailable. As a side benefit, Wiles' proof of the TANIYAMA-SHIMURA CONJECTURE also laid to rest the famous and thorny problem which had baffled mathematicians for hundreds of years, FERMAT'S LAST THEOREM.

Curves with small CONDUCTORS are listed in Swin-nerton-Dyer (1975) and Cremona (1997). Methods for computing integral points (points with integral co-ordinates) are given in Gebel *et al.* and Stroeker and Tzanakis (1994). The SCHOOF-ELKIES-ATKIN ALGO-RITHM can be used to determine the order of an elliptic curve E/F_p over the FINITE FIELD F_p.

See also CUBIC CURVE, ELLIPTIC CURVE GROUP LAW, FERMAT'S LAST THEOREM, FREY CURVE, J-INVARIANT,

MINIMAL DISCRIMINANT, MORDELL-WEIL THEOREM, OCHOA CURVE, RIBET'S THEOREM, SCHOOF-ELKIES-ATKIN ALGORITHM, SIEGEL'S THEOREM, SWINNERTON-DYER CONJECTURE, TANIYAMA-SHIMURA CONJEC-TURE, WEIERSTRASS ELLIPTIC FUNCTION, WEIER-STRASS FORM

References
Atkin, A. O. L. and Morain, F. "Elliptic Curves and Prim-ality Proving." *Math. Comput.* **61**, 29–8, 1993.
Cassels, J. W. S. *Lectures on Elliptic Curves.* New York: Cambridge University Press, 1991.
Cremona, J. E. *Algorithms for Modular Elliptic Curves, 2nd ed.* Cambridge, England: Cambridge University Press, 1997.
Du Val, P. *Elliptic Functions and Elliptic Curves.* Cam-bridge, England: Cambridge University Press, 1973.
Gebel, J.; Petho, A.; and Zimmer, H. G. "Computing Integral Points on Elliptic Curves." *Acta Arith.* **68**, 171–92, 1994.
Ireland, K. and Rosen, M. "Elliptic Curves." Ch. 18 in *A Classical Introduction to Modern Number Theory, 2nd ed.* New York: Springer-Verlag, pp. 297–18, 1990.
Joye, M. "Some Interesting References on Elliptic Curves." http://www.dice.ucl.ac.be/crypto/joye/biblio_ell.html.
Katz, N. M. and Mazur, B. *Arithmetic Moduli of Elliptic Curves.* Princeton, NJ: Princeton University Press, 1985.
Knapp, A. W. *Elliptic Curves.* Princeton, NJ: Princeton University Press, 1992.
Koblitz, N. *Introduction to Elliptic Curves and Modular Forms.* New York: Springer-Verlag, 1993.
Lang, S. *Elliptic Curves: Diophantine Analysis.* Berlin: Springer-Verlag, 1978.
Mazur, B. and Tate, J. "Points of Order 13 on Elliptic Curves." *Invent. Math.* **22**, 41–9, 1973/74.
Riesel, H. "Elliptic Curves." Appendix 7 in *Prime Numbers and Computer Methods for Factorization, 2nd ed.* Boston, MA: Birkhäuser, pp. 317–26, 1994.
Silverman, J. H. *The Arithmetic of Elliptic Curves.* New York: Springer-Verlag, 1986.
Silverman, J. H. *The Arithmetic of Elliptic Curves II.* New York: Springer-Verlag, 1994.
Silverman, J. H. and Tate, J. T. *Rational Points on Elliptic Curves.* New York: Springer-Verlag, 1992.
Stillwell, J. "Elliptic Curves." *Amer. Math. Monthly* **102**, 831–37, 1995.
Stroeker, R. J. and Tzanakis, N. "Solving Elliptic Diophan-tine Equations by Estimating Linear Forms in Elliptic Logarithms." *Acta Arith.* **67**, 177–96, 1994.
Swinnerton-Dyer, H. P. F. "Correction to: 'On 1-adic Repre-sentations and Congruences for Coefficients of Modular Forms.'" In *Modular Functions of One Variable, Vol. 4, Proc. Internat. Summer School for Theoret. Phys., Univ. Antwerp, Antwerp, RUCA, July-Aug. 1972.* Berlin: Springer-Verlag, 1975.
Weisstein, E. W. "Books about Elliptic Curves." http://www.treasure-troves.com/books/EllipticCurves.html.

Elliptic Curve Factorization Method

A factorization method, abbreviated ECM, which computes a large multiple of a point on a random ELLIPTIC CURVE modulo the number to be factored N. It tends to be faster than the POLLARD RHO FACTOR-IZATION and POLLARD P-1 FACTORIZATION METHODS.

Zimmermann maintains a table of the largest factors found using the ECM. The largest factor found using this algorithm is a prime factor of 54 digits of the 127-

digit cofactor C of

$$n = b^4 - b^2 + 1 = 13 \cdot 733 \cdot 7177 \cdot C,$$

where $b = 63^{43} - 1$, found by N. Lygeros and M. Mizony in Dec. 1999.

See also ATKIN-GOLDWASSER-KILIAN-MORAIN CERTIFICATE, ELLIPTIC CURVE PRIMALITY PROVING, ELLIPTIC PSEUDOPRIME

References

Atkin, A. O. L. and Morain, F. "Finding Suitable Curves for the Elliptic Curve Method of Factorization." *Math. Comput.* **60**, 399–05, 1993.

Brent, R. P. "Some Integer Factorization Algorithms Using Elliptic Curves." *Austral. Comp. Sci. Comm.* **8**, 149–63, 1986.

Brent, R. P. "Parallel Algorithms for Integer Factorisation." In *Number Theory and Cryptography* (Ed. J. H. Loxton). New York: Cambridge University Press, pp. 26–7, 1990.

Brillhart, J.; Lehmer, D. H.; Selfridge, J.; Wagstaff, S. S. Jr.; and Tuckerman, B. *Factorizations of* $b^n \pm 1$, $b = 2,3,5,6,7,10,11,12$ Up to High Powers, rev. ed. Providence, RI: Amer. Math. Soc., p. lxxxiii, 1988.

Eldershaw, C. and Brent, R. P. "Factorization of Large Integers on Some Vector and Parallel Computers."

Lenstra, A. K. and Lenstra, H. W. Jr. "Algorithms in Number Theory." In *Handbook of Theoretical Computer Science, Volume A: Algorithms and Complexity* (Ed. J. van Leeuwen). Amsterdam: Netherlands, Elsevier, pp. 673–15, 1990.

Lenstra, H. W. Jr. "Factoring Integers with Elliptic Curves." *Ann. Math.* **126**, 649–73, 1987.

Montgomery, P. L. "Speeding the Pollard and Elliptic Curve Methods of Factorization." *Math. Comput.* **48**, 243–64, 1987.

Zimmermann, P. "The ECMNET Project." http://www.loria.fr/~zimmerma/records/ecmnet.html.

Zimmermann, P. "ECM Top 100 Table." http://www.loria.fr/~zimmerma/records/top100.html.

Elliptic Curve Group Law

The GROUP of an ELLIPTIC CURVE which has been transformed to the form

$$y^2 = x^3 + ax + b$$

is the set of K-RATIONAL POINTS, including the single POINT AT INFINITY. The group law (addition) is defined as follows: Take 2 K-RATIONAL POINTS P and Q. Now 'draw' a straight line through them and compute the third point of intersection R (also a K-RATIONAL POINT). Then

$$P + Q + R = 0$$

gives the identity POINT AT INFINITY. Now find the inverse of R, which can be done by setting $R = (a, b)$ giving $-R = (a, -b)$.

This remarkable result is only a special case of a more general procedure. Essentially, the reason is that this type of ELLIPTIC CURVE has a single POINT AT INFINITY which is an inflection point (the line at infinity meets the curve at a single POINT AT INFINITY, so it must be an intersection of multiplicity three).

Elliptic Curve Primality Proving

A class of algorithm, abbreviated ECPP, which provides certificates of primality using sophisticated results from the theory of ELLIPTIC CURVES. A detailed description and list of references are given by Atkin and Morain (1990, 1993).

Adleman and Huang (1987) designed an independent algorithm using elliptic curves of genus two.

See also ATKIN-GOLDWASSER-KILIAN-MORAIN CERTIFICATE, ELLIPTIC CURVE FACTORIZATION METHOD, ELLIPTIC PSEUDOPRIME

References

Adleman, L. M. and Huang, M. A. "Recognizing Primes in Random Polynomial Time." In *Proc. 19th STOC, New York City, May 25–7, 1986.* New York: ACM Press, pp. 462–69, 1987.

Atkin, A. O. L. Lecture notes of a conference, Boulder, CO, Aug. 1986.

Atkin, A. O. L. and Morain, F. "Elliptic Curves and Primality Proving." Res. Rep. 1256, INRIA, June 1990.

Atkin, A. O. L. and Morain, F. "Elliptic Curves and Primality Proving." *Math. Comput.* **61**, 29–8, 1993.

Bosma, W. "Primality Testing Using Elliptic Curves." Techn. Rep. 85–2, Math. Inst., Univ. Amsterdam, 1985.

Chudnovsky, D. V. and Chudnovsky, G. V. "Sequences of Numbers Generated by Addition in Formal Groups and New Primality and Factorization Tests." Res. Rep. RC 11262, IBM, Yorktown Heights, NY, 1985.

Cohen, H. *Cryptographie, factorisation et primalité: l'utilisation des courbes elliptiques.* Paris: C. R. J. Soc. Math. France, Jan. 1987.

Kaltofen, E.; Valente, R.; and Yui, N. "An Improved Las Vegas Primality Test." Res. Rep. 89–2, Rensselaer Polytechnic Inst., Troy, NY, May 1989.

Elliptic Cylinder

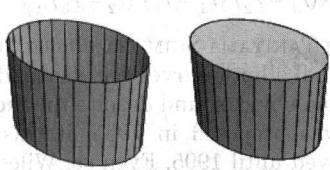

A CYLINDER with ELLIPTICAL CROSS SECTION. The PARAMETRIC EQUATIONS for the laterals sides of an elliptic cylinder of height h, SEMIMAJOR AXIS a, and SEMIMINOR AXIS b are

$$x = a \cos \theta$$
$$y = b \sin \theta$$
$$z = z,$$

where $\theta \in [0, 2\pi)$ and $z \in [0, h]$.

The elliptic cylinder is a QUADRATIC RULED SURFACE.

See also CONE, CYLINDER, ELLIPTIC CONE, ELLIPTIC PARABOLOID, QUADRATIC SURFACE, RULED SURFACE

References

Beyer, W. H. *CRC Standard Mathematical Tables, 28th ed.* Boca Raton, FL: CRC Press, p. 227, 1987.

Hilbert, D. and Cohn-Vossen, S. *Geometry and the Imagination.* New York: Chelsea, p. 12, 1999.

Elliptic Cylindrical Coordinates

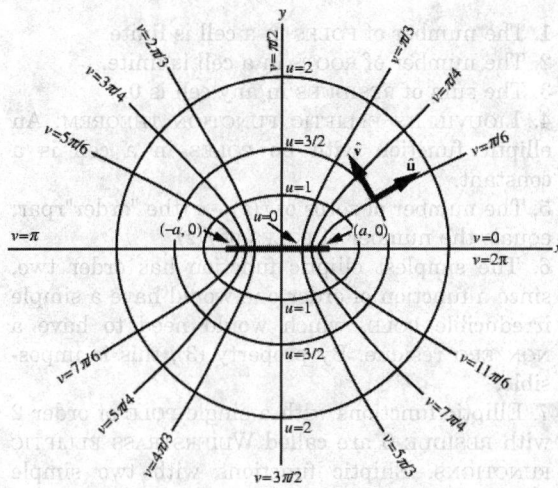

The v coordinates are the asymptotic angle of confocal HYPERBOLIC CYLINDERS symmetrical about the x-AXIS. The u coordinates are confocal ELLIPTIC CYLINDERS centered on the origin.

$$x = a \cosh u \cos v \qquad (1)$$

$$y = a \sinh u \sin v \qquad (2)$$

$$z = z, \qquad (3)$$

where $u \in [0, \infty)$, $v \in [0, 2\pi)$, and $z \in (-\infty, \infty)$. They are related to CARTESIAN coordinates by

$$\frac{x^2}{a^2 \cosh^2 u} + \frac{y^2}{a^2 \sinh^2 u} = 1 \qquad (4)$$

$$\frac{x^2}{a^2 \cos^2 v} - \frac{y^2}{a^2 \sin^2 v} = 1. \qquad (5)$$

The SCALE FACTORS are

$$h_1 = a\sqrt{\cosh^2 u \sin^2 v + \sinh^2 u \cos^2 v} \qquad (6)$$

$$= a\sqrt{\frac{\cosh(2u) - \cos(2v)}{2}} \qquad (7)$$

$$= a\sqrt{\sinh^2 u + \sin^2 v} \qquad (8)$$

$$h_2 = a\sqrt{\sinh^2 u \sin^2 v + \sinh^2 u \cos^2 v} \qquad (9)$$

$$= a\sqrt{\frac{\cosh(2u) - \cos(2v)}{2}} \qquad (10)$$

$$= a\sqrt{\sinh^2 u + \sin^2 v} \qquad (11)$$

$$h_3 = 1. \qquad (12)$$

The LAPLACIAN is

$$\nabla^2 = \frac{1}{a^2(\sinh^2 u + \sin^2 v)}\left(\frac{\partial^2}{\partial u^2} + \frac{\partial^2}{\partial v^2}\right) + \frac{\partial^2}{\partial z^2}. \qquad (13)$$

Let

$$q_1 = \cosh u \qquad (14)$$

$$q_2 = \cos v \qquad (15)$$

$$q_3 = z. \qquad (16)$$

Then the new SCALE FACTORS are

$$h_{q_1} = a\sqrt{\frac{q_1^2 - q_2^2}{q_1^2 - 1}} \qquad (17)$$

$$h_{q_2} = a\sqrt{\frac{q_1^2 - q_2^2}{1 - q_1^2}} \qquad (18)$$

$$h_{q_3} = 1. \qquad (19)$$

The HELMHOLTZ DIFFERENTIAL EQUATION IS SEPARABLE.

See also CYLINDRICAL COORDINATES, HELMHOLTZ DIFFERENTIAL EQUATION–ELLIPTIC CYLINDRICAL COORDINATES

References

Arfken, G. "Elliptic Cylindrical Coordinates (u, v, z)." §2.7 in *Mathematical Methods for Physicists, 2nd ed.* Orlando, FL: Academic Press, pp. 95–7, 1970.

Moon, P. and Spencer, D. E. "Elliptic-Cylinder Coordinates (η, ϕ, z)." Table 1.03 in *Field Theory Handbook, Including Coordinate Systems, Differential Equations, and Their Solutions, 2nd ed.* New York: Springer-Verlag, pp. 17–0, 1988.

Morse, P. M. and Feshbach, H. *Methods of Theoretical Physics, Part I.* New York: McGraw-Hill, p. 657, 1953.

Elliptic Delta Function

$$\delta(r) = \sqrt{r} - 2\alpha(r),$$

where $\alpha(r)$ is the ELLIPTIC ALPHA FUNCTION.

See also ELLIPTIC ALPHA FUNCTION, ELLIPTIC INTEGRAL SINGULAR VALUE

References

Borwein, J. M. and Borwein, P. B. *Pi & the AGM: A Study in Analytic Number Theory and Computational Complexity.* New York: Wiley, 1987.
Weisstein, E. W. "Elliptic Singular Values." MATHEMATICA NOTEBOOK ELLIPTICSINGULAR.M.

Elliptic Exponential Function

The inverse of the ELLIPTIC LOGARITHM

$$\text{eln}(x) \equiv \int_x^\infty \frac{dt}{\sqrt{t^3 + at^2 + bt}}.$$

It is doubly periodic in the COMPLEX PLANE.

Elliptic Fixed Point (Differential Equations)

A FIXED POINT for which the STABILITY MATRIX is purely IMAGINARY, $\lambda_\pm = \pm i\omega$ (for $\omega > 0$).

See also DIFFERENTIAL EQUATION, FIXED POINT, HYPERBOLIC FIXED POINT (DIFFERENTIAL EQUATIONS), PARABOLIC FIXED POINT, STABLE IMPROPER NODE, STABLE NODE, STABLE SPIRAL POINT, STABLE STAR, UNSTABLE IMPROPER NODE, UNSTABLE NODE, UNSTABLE SPIRAL POINT, UNSTABLE STAR

References

Tabor, M. "Classification of Fixed Points." §1.4.b in *Chaos and Integrability in Nonlinear Dynamics: An Introduction.* New York: Wiley, pp. 22–5, 1989.

Elliptic Fixed Point (Map)

A FIXED POINT of a LINEAR TRANSFORMATION (MAP) for which the rescaled variables satisfy

$$(\delta - \alpha)^2 + 4\beta\gamma < 0.$$

See also HYPERBOLIC FIXED POINT (MAP), LINEAR TRANSFORMATION, PARABOLIC FIXED POINT

Elliptic Function

A DOUBLY PERIODIC FUNCTION with periods $2\omega_1$ and $2\omega_2$ such that

$$f(z + 2\omega_1) = f(z + 2\omega_2) = f(z), \tag{1}$$

which is ANALYTIC and has no singularities except for POLES in the finite part of the COMPLEX PLANE. The

HALF-PERIOD RATIO $\tau \equiv \omega_2/\omega_1$ must not be purely real, because if it is, the function reduces to a singly periodic function if τ is rational, and a constant if τ is irrational (Jacobi 1835). ω_1 and ω_2 are labeled such that $\Im[\tau] \equiv \Im[\omega_2/\omega_1] > 0$, where $\Im[z]$ is the IMAGINARY PART.

A "cell" of an elliptic function is defined as a parallelogram region in the COMPLEX PLANE in which the function is not multi-valued. Properties obeyed by elliptic functions include

1. The number of POLES in a cell is finite.
2. The number of ROOTS in a cell is finite.
3. The sum of RESIDUES in any cell is 0.
4. LIOUVILLE'S ELLIPTIC FUNCTION THEOREM: An elliptic function with no POLES in a cell is a constant.
5. The number of zeros of $f(z) - c$ (the "order") equals the number of POLES of $f(z)$.
6. The simplest elliptic function has order two, since a function of order one would have a simple irreducible POLE, which would need to have a NONZERO residue. By property (3), this is impossible.
7. Elliptic functions with a single POLE of order 2 with RESIDUE 0 are called WEIERSTRASS ELLIPTIC FUNCTIONS. Elliptic functions with two simple POLES having residues a_0 and $-a_0$ are called JACOBI ELLIPTIC FUNCTIONS.
8. Any elliptic function is expressible in terms of either WEIERSTRASS ELLIPTIC FUNCTION or JACOBI ELLIPTIC FUNCTIONS.
9. The sum of the AFFIXES of ROOTS equals the sum of the AFFIXES of the POLES.
10. An algebraic relationship exists between any two elliptic functions with the same periods.

The elliptic functions are inversions of the ELLIPTIC INTEGRALS. The two standard forms of these functions are known as JACOBI ELLIPTIC FUNCTIONS and WEIERSTRASS ELLIPTIC FUNCTIONS. JACOBI ELLIPTIC FUNCTIONS arise as solutions to differential equations OF THE FORM

$$\frac{d^2x}{dt^2} = A + Bx + Cx^2 + Dx^3, \tag{2}$$

and WEIERSTRASS ELLIPTIC FUNCTIONS arise as solutions to differential equations OF THE FORM

$$\frac{d^2x}{dt^2} = A + Bx + Cx^2. \tag{3}$$

See also DOUBLY PERIODIC FUNCTION, ELLIPTIC CURVE, ELLIPTIC INTEGRAL, HALF-PERIOD RATIO, JACOBI ELLIPTIC FUNCTIONS, JACOBI THETA FUNCTIONS, LIOUVILLE'S ELLIPTIC FUNCTION THEOREM, MODULAR FORM, MODULAR FUNCTION, NEVILLE THE-

TA FUNCTIONS, THETA FUNCTIONS, WEIERSTRASS ELLIPTIC FUNCTIONS

References

Akhiezer, N. I. *Elements of the Theory of Elliptic Functions.* Providence, RI: Amer. Math. Soc., 1990.

Apostol, T. M. "Elliptic Functions." §1.4 in *Modular Functions and Dirichlet Series in Number Theory, 2nd ed.* New York: Springer-Verlag, pp. 4-, 1997.

Bellman, R. E. *A Brief Introduction to Theta Functions.* New York: Holt, Rinehart and Winston, 1961.

Borwein, J. M. and Borwein, P. B. *Pi & the AGM: A Study in Analytic Number Theory and Computational Complexity.* New York: Wiley, 1987.

Bowman, F. *Introduction to Elliptic Functions, with Applications.* New York: Dover, 1961.

Byrd, P. F. and Friedman, M. D. *Handbook of Elliptic Integrals for Engineers and Scientists, 2nd ed., rev.* Berlin: Springer-Verlag, 1971.

Cayley, A. *An Elementary Treatise on Elliptic Functions, 2nd ed.* London: G. Bell, 1895.

Chandrasekharan, K. *Elliptic Functions.* Berlin: Springer-Verlag, 1985.

Du Val, P. *Elliptic Functions and Elliptic Curves.* Cambridge, England: Cambridge University Press, 1973.

Dutta, M. and Debnath, L. *Elements of the Theory of Elliptic and Associated Functions with Applications.* Calcutta, India: World Press, 1965.

Eagle, A. *The Elliptic Functions as They Should Be: An Account, with Applications, of the Functions in a New Canonical Form.* Cambridge, England: Galloway and Porter, 1958.

Greenhill, A. G. *The Applications of Elliptic Functions.* London: Macmillan, 1892.

Hancock, H. *Lectures on the Theory of Elliptic Functions.* New York: Wiley, 1910.

Jacobi, C. G. J. *Fundamenta Nova Theoriae Functionum Ellipticarum.* Regiomonti, Sumtibus fratrum Borntraeger, 1829.

King, L. V. *On the Direct Numerical Calculation of Elliptic Functions and Integrals.* Cambridge, England: Cambridge University Press, 1924.

Knopp, K. "Doubly-Periodic Functions; in Particular, Elliptic Functions." §9 in *Theory of Functions Parts I and II, Two Volumes Bound as One, Part II.* New York: Dover, pp. 73-2, 1996.

Lang, S. *Elliptic Functions, 2nd ed.* New York: Springer-Verlag, 1987.

Lawden, D. F. *Elliptic Functions and Applications.* New York: Springer Verlag, 1989.

Morse, P. M. and Feshbach, H. *Methods of Theoretical Physics, Part I.* New York: McGraw-Hill, pp. 427 and 433-34, 1953.

Murty, M. R. (Ed.). *Theta Functions.* Providence, RI: Amer. Math. Soc., 1993.

Neville, E. H. *Jacobian Elliptic Functions, 2nd ed.* Oxford, England: Clarendon Press, 1951.

Oberhettinger, F. and Magnus, W. *Anwendung der Elliptischen Funktionen in Physik und Technik.* Berlin: Springer-Verlag, 1949.

Petkovsek, M.; Wilf, H. S.; and Zeilberger, D. "Elliptic Function Identities." §1.8 in *A = B.* Wellesley, MA: A. K. Peters, pp. 13-5, 1996.

Prasolov, V. and Solovyev, Y. *Elliptic Functions and Elliptic Integrals.* Providence, RI: Amer. Math. Soc., 1997.

Siegel, C. L. *Topics in Complex Function Theory, Vol. 1: Elliptic Functions and Uniformization Theory.* New York: Wiley, 1988.

Walker, P. L. *Elliptic Functions: A Constructive Approach.* New York: Wiley, 1996.

Weisstein, E. W. "Books about Elliptic Functions." http://www.treasure-troves.com/books/EllipticFunctions.html.

Whittaker, E. T. and Watson, G. N. Chs. 20-2 in *A Course of Modern Analysis, 4th ed.* Cambridge, England: University Press, 1943.

Elliptic Functional

COERCIVE FUNCTIONAL

Elliptic Geometry

A constant curvature NON-EUCLIDEAN GEOMETRY which replaces the PARALLEL POSTULATE with the statement "through any point in the plane, there exist no lines PARALLEL to a given line." Elliptic geometry is sometimes also called RIEMANNIAN GEOMETRY. It can be visualized as the surface of a SPHERE on which "lines" are taken as GREAT CIRCLES. In elliptic geometry, the sum of angles of a TRIANGLE is > 180°.

See also EUCLIDEAN GEOMETRY, HYPERBOLIC GEOMETRY, NON-EUCLIDEAN GEOMETRY

References

Gray, A. *Modern Differential Geometry of Curves and Surfaces with Mathematica, 2nd ed.* Boca Raton, FL: CRC Press, p. 422, 1997.

Elliptic Group Modulo p

$E(a, b)/p$ denotes the elliptic GROUP modulo p whose elements are 1 and ∞ together with the pairs of INTEGERS (x, y) with $0 \leq x, y < p$ satisfying

$$y^2 \equiv x^3 + ax + b \pmod{p} \tag{1}$$

with a and b INTEGERS such that

$$4a^3 + 27b^2 \not\equiv 0 \pmod{p}. \tag{2}$$

Given (x_1, y_1), define

$$(x_i, y_i) \equiv (x_1, y_1)^i \pmod{p}. \tag{3}$$

The ORDER h of $E(a, b)/p$ is given by

$$h = 1 + \sum_{x=1}^{p} \left[\left(\frac{x^3 + ax + b}{p} \right) + 1 \right], \tag{4}$$

where $x^3 + ax + b/p$ is the LEGENDRE SYMBOL, although this FORMULA quickly becomes impractical. However, it has been proven that

$$p + 1 - 2\sqrt{p} \leq h(E(a, b)/p) \leq p + 1 + 2\sqrt{p}. \tag{5}$$

Furthermore, for p a PRIME > 3 and INTEGER n in the above interval, there exists a and b such that

$$h(E(a, b)/p) = n, \tag{6}$$

and the orders of elliptic GROUPS mod p are nearly uniformly distributed in the interval.

Elliptic Helicoid

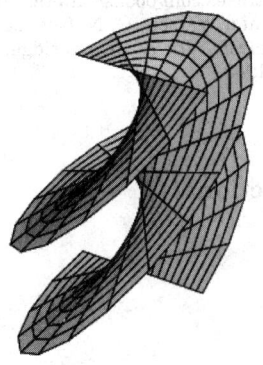

A generalization of the HELICOID to the PARAMETRIC EQUATIONS

$$x(u, v) = av \cos u$$
$$y(u, v) = bv \sin u$$
$$z(u, v) = cu.$$

See also HELICOID

References

Gray, A. *Modern Differential Geometry of Curves and Surfaces with Mathematica, 2nd ed.* Boca Raton, FL: CRC Press, p. 422, 1997.

Elliptic Hyperboloid

The elliptic hyperboloid is the generalization of the HYPERBOLOID to three distinct semimajor axes. The elliptic hyperboloid of one sheet is a RULED SURFACE and has Cartesian equation

$$\frac{x^2}{a^2} + \frac{y^2}{b^2} - \frac{z^2}{c^2} = 1, \tag{1}$$

and PARAMETRIC EQUATIONS

$$x(u, v) = a\sqrt{1 + u^2} \cos v \tag{2}$$

$$y(u, v) = b\sqrt{1 + u^2} \sin v \tag{3}$$

$$z(u, v) = cu \tag{4}$$

for $v \in [0, 2\pi)$, or

$$x(u, v) = a(\cos u \mp v \sin u) \tag{5}$$

$$y(u, v) = b(\sin u \pm v \cos u) \tag{6}$$

$$z(u, v) = \pm cv, \tag{7}$$

or

$$x(u, v) = a \cosh v \cos u \tag{8}$$

$$y(u, v) = b \cosh v \sin u \tag{9}$$

$$z(u, v) = c \sinh v. \tag{10}$$

The two-sheeted elliptic hyperboloid oriented along the z-AXIS has Cartesian equation

$$\frac{x^2}{a^2} + \frac{y^2}{a^2} - \frac{z^2}{c^2} = -1, \tag{11}$$

and PARAMETRIC EQUATIONS

$$x = a \sinh u \cos v \tag{12}$$

$$y = b \sinh u \sin v \tag{13}$$

$$z = c \pm \cosh u. \tag{14}$$

The two-sheeted elliptic hyperboloid oriented along the x-AXIS has Cartesian equation

$$\frac{x^2}{a^2} - \frac{y^2}{a^2} - \frac{z^2}{c^2} = 1 \tag{15}$$

and PARAMETRIC EQUATIONS

$$x = a \cosh u \cosh v \tag{16}$$

$$y = b \sinh u \cosh v \tag{17}$$

$$z = c \sinh v. \tag{18}$$

See also HYPERBOLOID, RULED SURFACE

References

Gray, A. *Modern Differential Geometry of Curves and Surfaces with Mathematica, 2nd ed.* Boca Raton, FL: CRC Press, pp. 404–06 and 470, 1997.

Elliptic Integral

An elliptic integral is an INTEGRAL OF THE FORM

$$\int \frac{A(x) + B(x)\sqrt{S(x)}}{A(x) + D(x)\sqrt{S(x)}} \, dx, \tag{1}$$

or

$$\int \frac{A(x) \, dx}{B(x)\sqrt{S(x)}}, \tag{2}$$

where $A(x)$, $B(x)$, $C(x)$, and $D(x)$ are POLYNOMIALS in x, and $S(x)$ is a POLYNOMIAL of degree 3 or 4. Stated more simply, an elliptic integral is an integral OF THE FORM

$$\int R(w,\, x)\, dx, \tag{3}$$

where $R(w,\, x)$ is a RATIONAL FUNCTION of x and w, w^2 is a function of x that is CUBIC or QUARTIC in x, $R(w,\, x)$ contains at least one ODD POWER of w, and w^2 has no repeated factors (Abramowitz and Stegun 1972, p. 589).

Elliptic integrals can be viewed as generalizations of the inverse TRIGONOMETRIC FUNCTIONS and provide solutions to a wider class of problems. For instance, while the ARC LENGTH of a CIRCLE is given as a simple function of the parameter, computing the ARC LENGTH of an ELLIPSE requires an elliptic integral. Similarly, the position of a pendulum is given by a TRIGONO-METRIC FUNCTION as a function of time for small angle oscillations, but the full solution for arbitrarily large displacements requires the use of elliptic integrals. Many other problems in electromagnetism and grav-itation are solved by elliptic integrals.

A very useful class of functions known as ELLIPTIC FUNCTIONS is obtained by inverting elliptic integrals to obtain generalizations of the trigonometric func-tions. ELLIPTIC FUNCTIONS (among which the JACOBI ELLIPTIC FUNCTIONS and WEIERSTRASS ELLIPTIC FUNCTION are the two most common forms) provide a powerful tool for analyzing many deep problems in NUMBER THEORY, as well as other areas of mathe-matics.

All elliptic integrals can be written in terms of three "standard" types. To see this, write

$$R(w,\, x) \equiv \frac{P(w,\, x)}{Q(w,\, x)} = \frac{wP(w,\, x)Q(-w,\, x)}{wQ(w,\, x)Q(-w,\, x)}. \tag{4}$$

But since $w^2 = f(x)$,

$$Q(w,\, x)Q(-w,\, x) \equiv Q_1(w,\, x) = Q_1(-w,\, x), \tag{5}$$

then

$$wP(w,\, x)Q(-w,\, x) = A + Bx + Cw + Dx^2 + Ewx$$

$$+ Fw^2 + Gw^2x + Hw^3x$$

$$= (A + Bx + Dx^2 + Fw^2 + Gw^2x)$$

$$+ w(c + Ex + Hw^2x + \ldots)$$

$$= P_1(x) + wP_2(x), \tag{6}$$

so

$$R(w,\, x) = \frac{P_1(x) + wP_2(x)}{wQ_1(w)} = \frac{R_1(x)}{w} + R_2(x). \tag{7}$$

But any function $\int R_2(x)\, dx$ can be evaluated in terms

of elementary functions, so the only portion that need be considered is

$$\int \frac{R_1(x)}{w}\, dx. \tag{8}$$

Now, any quartic can be expressed as $S_1 S_2$ where

$$S_1 \equiv a_1 x^2 + 2b_1 x + c_1 \tag{9}$$

$$S_2 \equiv a_2 x^2 + 2b_2 x + c_2. \tag{10}$$

The COEFFICIENTS here are real, since pairs of COMPLEX ROOTS are COMPLEX CONJUGATES

$$[x - (R + Ii)][x - (R - Ii)]$$

$$= x^2 + x(-R + Ii - R - Ii) + (R^2 - I^2 i)$$

$$= x^2 - 2Rx + (R^2 + I^2). \tag{11}$$

If all four ROOTS are real, they must be arranged so as not to interleave (Whittaker and Watson 1990, p. 514). Now define a quantity λ such that $S_1 + \lambda S_2$

$$(a_1 - \lambda a_2)x^2 - (2b_1 - 2b_2\lambda)x + (c_1 - \lambda c_2) \tag{12}$$

is a SQUARE NUMBER and

$$2\sqrt{(a_1 - \lambda a_2)(c_1 - \lambda_2)} = 2(b_1 - b_2\lambda) \tag{13}$$

$$(a_1 - \lambda a_2)(c_1 - \lambda c_2) - (b_1 - \lambda b_2)^2 = 0. \tag{14}$$

Call the ROOTS of this equation λ_1 and λ_2, then

$$S_1 - \lambda_1 S_2 = \left[\sqrt{(a_1 - \lambda_1 a_2)}x + \sqrt{c_1 - \lambda c_2} \right]^2$$

$$= (a_1 - \lambda_1 a_2)\left(x + \sqrt{\frac{c_1 - \lambda_1 c_2}{a_1 - \lambda_1 a_2}} \right)$$

$$\equiv (a_1 - \lambda_1 a_2)(x - \alpha)^2 \tag{15}$$

$$S_1 - \lambda_2 S_2 = \left[\sqrt{(a_1 - \lambda_1 a_2)}x + \sqrt{c_1 - \lambda c_2} \right]^2$$

$$= (a_1 - \lambda_1 a_2)\left(x + \sqrt{\frac{c_1 - \lambda_2 c_2}{a_1 - \lambda_2 a_2}} \right)$$

$$\equiv (a_1 - \lambda_2 a_2)(x - \beta)^2. \tag{16}$$

Taking (15)-(16) and $\lambda_2(1) - \lambda_1(2)$ gives

$$S_2(\lambda_2 - \lambda_1) = (a_1 - \lambda_1 a_2)(x - \alpha)^2 - (a_1 - \lambda_2 a_2)$$

$$\times (x - \beta)^2 \tag{17}$$

$$S_1(\lambda_2 - \lambda_1) = \lambda_2(a_1 - \lambda_1 a_2)(x - \alpha)^2 - \lambda_1(a_1 - \lambda_2 a_2)$$

$$\times (x - \beta^2). \tag{18}$$

Solving gives

$$S_1 = \frac{a_1 - \lambda_1 a_2}{\lambda_2 - \lambda_1}(x - \alpha)^2 - \frac{a_1 - \lambda_2 a_2}{\lambda_2 - \lambda_1}(x - \beta)^2$$

$$\equiv A_1(x - \alpha)^2 + B_1(x - \beta)^2 \tag{19}$$

$$S_2 = \frac{\lambda_2(a_1 - \lambda_1 a_2)}{\lambda_2 - \lambda_1}(x - \alpha)^2 - \frac{\lambda_1(a_1 - \lambda_2 a_2)}{\lambda_2 - \lambda_1}(x - \beta)^2$$

$$\equiv A_2(x - \alpha)^2 + B_2(x - \beta)^2, \tag{20}$$

so we have

$$w^2 = S_1 S_2 = [A_1(x - \alpha)^2 + B_1(x - \beta)^2]$$
$$\times [A^2(x - \alpha)^2 + B^2(x - \beta)^2]. \tag{21}$$

Now let

$$t \equiv \frac{x - \alpha}{x - \beta} \tag{22}$$

$$dy = [(x - \beta)^{-1} - (x - \alpha)(x - \beta)^{-2}]\,dx$$
$$= \frac{(x - \beta) - (x - \alpha)}{(x - \beta)^2}\,dx$$
$$= \frac{\alpha - \beta}{(x - \beta)^2}\,dx, \tag{23}$$

so

$$w^2 = (x - \beta)^4\left[A_1\left(\frac{x - \alpha}{x - \beta}\right)^2 + B_1\right]\left[A_2\left(\frac{x - \alpha}{x - \beta}\right)^2 + B_2\right]$$
$$= (x - \beta)^4(A_1 t^2 + B_1)(A_2 t^2 + B_2), \tag{24}$$

and

$$w = (x - \beta)^2\sqrt{(A_1 t^2 + B_1)(A_2 t^2 + B_2)} \tag{25}$$

$$\frac{dx}{w} = \left[\frac{(x - \beta)^2}{\alpha - \beta}\,dt\right]\frac{1}{(x - \beta)^2\sqrt{(A_1 t^2 + B_1)(A_2 t^2 + B_2)}}$$

$$= \frac{dt}{(\alpha - \beta)\sqrt{(A_1 t^2 + B_1)(A_2 t^2 + B_2)}}. \tag{26}$$

Now let

$$R_3(t) \equiv \frac{R_1(x)}{\alpha - \beta}, \tag{27}$$

so

$$\int \frac{R_1(x)\,dx}{w} = \int \frac{R_3(t)\,dt}{\sqrt{(A_1 t^2 + B_1)(A_2 t^2 + B_2)}}. \tag{28}$$

Rewriting the EVEN and ODD parts

$$R_3(t) + R_3(-t) \equiv 2R_4(t^2) \tag{29}$$

$$R_3(t) - R_3(-t) \equiv 2t R_5(t^2), \tag{30}$$

gives

$$R_3(t) \equiv \tfrac{1}{2}(R_{\text{even}} - R_{\text{odd}}) = R_4(t^2) + t R_5(t^2), \tag{31}$$

so we have

$$\int \frac{R_1(x)\,dx}{w} = \int \frac{R_4(t^2)\,dt}{\sqrt{(A_1 t^2 + B_1)(A_2 t^2 + B_2)}}$$
$$+ \int \frac{R_5(t^2)t\,dt}{\sqrt{(A_1 t^2 + B_1)(A_2 t^2 + B_2)}}. \tag{32}$$

Letting

$$u \equiv t^2 \tag{33}$$

$$du = 2t\,dt \tag{34}$$

reduces the second integral to

$$\frac{1}{2}\int \frac{R_5(u)\,du}{\sqrt{(A_1 u + B_1)(A_2 u + B_2)}}, \tag{35}$$

which can be evaluated using elementary functions. The first integral can then be reduced by INTEGRATION BY PARTS to one of the three Legendre elliptic integrals (also called Legendre-Jacobi ELLIPTIC INTEGRALS), known as incomplete elliptic integrals of the first, second, and third kind, denoted $F(\phi, k)$, $E(\phi, k)$, and $\prod(n; \phi, k)$, respectively (von Kármán and Biot 1940, Whittaker and Watson 1990, p. 515). If $\phi = \pi/2$, then the integrals are called complete elliptic integrals and are denoted $K(k)$, $E(k)$, $\prod(n; k)$.

Incomplete elliptic integrals are denoted using a MODULUS k, PARAMETER $m \equiv k^2$, or MODULAR ANGLE $\alpha \equiv \sin^{-1} k$. An elliptic integral is written $I(\phi|m)$ when the PARAMETER is used, $I(\phi, k)$ when the MODULUS is used, and $I(\phi \backslash \alpha)$ when the MODULAR ANGLE is used. Complete elliptic integrals are defined when $\phi = \pi/2$ and can be expressed using the expansion

$$(1 - k^2 \sin^2 \theta)^{-1/2} = \sum_{n=0}^{\infty} \frac{(2n - 1)!!}{(2n)!!} k^{2n} \sin^{2n} \theta. \tag{36}$$

An elliptic integral in standard form

$$\int_a^x \frac{dx}{\sqrt{f(x)}}, \tag{37}$$

where

$$f(x) = a_4 x^4 + a_3 x^3 + a_2 x^2 + a_1 x + a_0, \tag{38}$$

can be computed analytically (Whittaker and Watson 1990, p. 453) in terms of the WEIERSTRASS ELLIPTIC FUNCTION with invariants

$$g_2 = a_0 a_4 - 4a_1 a_3 + 3a_2^2 \tag{39}$$

$$g_3 = a_0 a_2 a_4 - 2a_1 a_2 a_3 - a_4 a_1^2 - a_3^2 a_0. \tag{40}$$

If $a \equiv x_0$ is a root of $f(x) = 0$, then the solution is

$$x = x_0 + \tfrac{1}{4}f'(x_0)[\wp(z; g_2, g_3) - \tfrac{1}{24}f''(x_0)]^{-1}. \tag{41}$$

For an arbitrary lower bound,

$$x = a$$

$$+ \frac{\sqrt{f(a)}\wp'(z)\frac{1}{2}f'(a)[\wp(z) - \frac{1}{24}f''(a)] + \frac{1}{24}f(a)f'''(a)}{2[\wp(z) - \frac{1}{24}f''(a)]^2 - \frac{1}{48}f(a)f^{(iv)}(a)}, \tag{42}$$

where $\wp(z) \equiv \wp(z; g_2, g_3)$ is a WEIERSTRASS ELLIPTIC FUNCTION (Whittaker and Watson 1990, p. 454).

A generalized elliptic integral can be defined by the function

$$T(a, b) \equiv \frac{2}{\pi} \int_0^{\pi/2} \frac{d\theta}{\sqrt{a^2 \cos^2 \theta + b^2 \sin^2 \theta}} \tag{43}$$

$$= \frac{2}{\pi} \int_0^{\pi/2} \frac{d\theta}{\cos\theta \sqrt{a^2 + b^2 \tan^2 \theta}} \tag{44}$$

(Borwein and Borwein 1987). Now let

$$t \equiv b \tan\theta \tag{45}$$

$$dt = b \sec^2 \theta \, d\theta. \tag{46}$$

But

$$\sec\theta = \sqrt{1 + \tan^2 \theta}, \tag{47}$$

so

$$dt = \frac{b}{\cos\theta} \sec\theta \, d\theta = \frac{b}{\cos\theta} \sqrt{1 + \tan^2 \theta} \, d\theta$$

$$= \frac{b}{\cos\theta} \sqrt{1 + \left(\frac{t}{b}\right)^2} \, d\theta$$

$$= \frac{d\theta}{\cos\theta} \sqrt{b^2 + t^2}, \tag{48}$$

and

$$\frac{d\theta}{\cos\theta} = \frac{dt}{\sqrt{b^2 + t^2}}, \tag{49}$$

and the equation becomes

$$T(a, b) = \frac{2}{\pi} \int_0^\infty \frac{dt}{\sqrt{(a^2 + t^2)(b^2 + t^2)}}$$

$$= \frac{1}{\pi} \int_{-\infty}^\infty \frac{dt}{\sqrt{(a^2 + t^2)(b^2 + t^2)}}. \tag{50}$$

Now we make the further substitution $u \equiv \frac{1}{2}(t - ab/t)$. The differential becomes

$$du = \frac{1}{2}(1 + ab/t^2) \, dt, \tag{51}$$

but $2u = t - ab/t$, so

$$2u/t = 1 - ab/t^2 \tag{52}$$

$$ab/t^2 = 1 - 2u/t \tag{53}$$

and

$$1 + ab/t^2 = 2 - 2u/t = 2(1 - u/t). \tag{54}$$

However, the left side is always positive, so

$$1 + ab/t^2 = 2 - 2u/t = 2|1 - u/t| \tag{55}$$

and the differential is

$$dt = \frac{du}{\left|1 - \dfrac{u}{t}\right|}. \tag{56}$$

We need to take some care with the limits of integration. Write (50) as

$$\int_{-\infty}^\infty f(t) \, dt = \int_{-\infty}^{0-} f(t) \, dt + \int_{0+}^\infty f(t) \, dt. \tag{57}$$

Now change the limits to those appropriate for the u integration

$$\int_{-\infty}^\infty g(u) \, du + \int_{-\infty}^\infty g(u) \, du = 2 \int_{-\infty}^\infty g(u) \, du, \tag{58}$$

so we have picked up a factor of 2 which must be included. Using this fact and plugging (56) in (50) therefore gives

$$T(a, b) = \frac{2}{\pi} \int_{-\infty}^\infty \frac{du}{\left|1 - \dfrac{u}{t}\right| \sqrt{a^2 b^2 + (a^2 + b^2)t^2 + t^4}}. \tag{59}$$

Now note that

$$u^2 = \frac{t^4 - 2abt^2 + a^2 b^2}{4t^2} \tag{60}$$

$$4u^2 t^2 = t^4 - 2abt^2 + a^2 b^2 \tag{61}$$

$$a^2 b^2 + t^4 = 4u^2 t^2 + 2abt^2. \tag{62}$$

Plug (62) into (59) to obtain

$$T(a, b) = \frac{2}{\pi} \int_{-\infty}^\infty \frac{du}{\left|1 - \dfrac{u}{t}\right| \sqrt{4u^2 t^2 + 2abt^2 + (a^2 + b^2)t^2}}$$

$$= \frac{2}{\pi} \int_{-\infty}^\infty \frac{du}{|t - u| \sqrt{4u^2 + (a + b)^2}}. \tag{63}$$

But

$$2ut = t^2 - ab \tag{64}$$

$$t^2 - 2ut - ab = 0 \tag{65}$$

$$t = \frac{1}{2}(2u \pm \sqrt{4u^2 + 4ab}) = u \pm \sqrt{u^2 + ab}, \tag{66}$$

so

$$t - u = \pm\sqrt{u^2 + ab}, \tag{67}$$

and (63) becomes

$$T(a,\ b) = \frac{2}{\pi} \int_{-\infty}^{\infty} \frac{du}{\sqrt{[4u^2 + (a+b)^2] + (u^2 + ab)}}$$

$$= \frac{1}{\pi} \int_{-\infty}^{\infty} \frac{du}{\sqrt{\left[u^2 + \left(\frac{a+b}{2}\right)^2\right](u^2 + ab)}}. \tag{68}$$

We have therefore demonstrated that

$$T(a,\ b) = T(\tfrac{1}{2}(a+b),\ \sqrt{ab}). \tag{69}$$

We can thus iterate

$$a_{i+1} = \tfrac{1}{2}(a_i + b_i) \tag{70}$$

$$b_{i+1} = \sqrt{a_i b_i}, \tag{71}$$

as many times as we wish, without changing the value of the integral. But this iteration is the same as and therefore converges to the ARITHMETIC-GEO-METRIC MEAN, so the iteration terminates at $a_i = b_i = M(a_0,\ b_0)$, and we have

$$T(a_0,\ b_0) = T(M(a_0,\ b_0),\ M(a_0,\ b_0))$$

$$= \frac{1}{\pi} \int_{-\infty}^{\infty} \frac{dt}{M^2(a_0,\ b_0) + t^2}$$

$$= \frac{1}{\pi M(a_0,\ b_0)} \left[\tan^{-1}\left(\frac{t}{M(a_0,\ b_0)}\right)\right]_{-\infty}^{\infty}$$

$$= \frac{1}{\pi M(a_0,\ b_0)} \left[\frac{\pi}{2} - \left(-\frac{\pi}{2}\right)\right]$$

$$= \frac{1}{M(a_0,\ b_0)}. \tag{72}$$

Complete elliptic integrals arise in finding the arc length of an ELLIPSE and the period of a pendulum. They also arise in a natural way from the theory of THETA FUNCTIONS. Complete elliptic integrals can be computed using a procedure involving the ARITH-METIC-GEOMETRIC MEAN. Note that

$$T(a,\ b) \equiv \frac{2}{\pi} \int_0^{\pi/2} \frac{d\theta}{\sqrt{a^2 \cos^2\theta + b^2 \sin^2\theta}}$$

$$= \frac{2}{\pi} \int_0^{\pi/2} \frac{d\theta}{a\sqrt{\cos^2\theta + \left(\frac{b}{a}\right)^2 \sin^2\theta}}$$

$$= \frac{2}{a\pi} \int_0^{\pi/2} \frac{d\theta}{\sqrt{1 - \left(1 - \frac{b^2}{a^2}\right)^2 \sin^2\theta}}. \tag{73}$$

So we have

$$T(a,\ b) = \frac{2}{a\pi} K\left(1 - \frac{b^2}{a^2}\right) = \frac{1}{M(a,\ b)}, \tag{74}$$

where $K(k)$ is the complete ELLIPTIC INTEGRAL OF THE FIRST KIND. We are free to let $a \equiv a_0 \equiv 1$ and $b \equiv b_0 \equiv k'$, so

$$\frac{2}{\pi} K(\sqrt{1 - k'^2}) = \frac{2}{\pi} K(k) = \frac{1}{M(1,\ k')}, \tag{75}$$

since $k \equiv \sqrt{1 - k'^2}$, so

$$K(k) = \frac{\pi}{2M(1,\ k')}. \tag{76}$$

But the ARITHMETIC-GEOMETRIC MEAN is defined by

$$a_i = \tfrac{1}{2}(a_{i-1} + b_{i-1}) \tag{77}$$

$$b_i = \sqrt{a_{i-1} + b_{i-1}} \tag{78}$$

$$c_i = \begin{cases} \tfrac{1}{2}(a_{i-1} - b_{i-1}) & i > 0 \\ \sqrt{a_0^2 - b_0^2} & i = 0 \end{cases}, \tag{79}$$

where

$$c_{n-1} = \tfrac{1}{2} a_n - b_n = \frac{c_n^2}{4a_{n+1}} \leq \frac{c_n^2}{4M(a_0,\ b_0)}, \tag{80}$$

so we have

$$K(k) = \frac{\pi}{2a_N}, \tag{81}$$

where a_N is the value to which a_n converges. Similarly, taking instead $a_0' = 1$ and $b_0' = k$ gives

$$K'(k) = \frac{\pi}{2a_N'}. \tag{82}$$

Borwein and Borwein (1987) also show that defining

$$U(a,\ b) \equiv \frac{\pi}{2} \int_0^{\pi/2} \sqrt{a^2 \cos^2\theta + b^2 \sin^2\theta}\ d\theta$$

$$= aE'\left(\frac{b}{a}\right) \tag{83}$$

leads to

$$2U(a_{n+1},\ b_{n+1}) - U(a_n,\ b_n) = a_n b_n T(a_n,\ b_n), \tag{84}$$

so

$$\frac{K(k) - E(k)}{K(k)} = \tfrac{1}{2}(c_0^2 + 2c_1^2 + 2^2 c_2^2 + \ldots + 2^n c_n^2) \tag{85}$$

for $a_0 \equiv 1$ and $b_0 \equiv k'$, and

$$\frac{K'(k) - E'(k)}{K'(k)} = \tfrac{1}{2}(c_0'^2 + 2c_1'^2 + 2^2 c_2'^2 + \ldots + 2^n c_n'^2). \quad (86)$$

The elliptic integrals satisfy a large number of identities. The complementary functions and moduli are defined by

$$K'(k) \equiv K(\sqrt{1 - k^2}) = K(k'). \quad (87)$$

Use the identity of generalized elliptic integrals

$$T(a, b) = T(\tfrac{1}{2}(a + b), \sqrt{ab}) \quad (88)$$

to write

$$\frac{1}{a} K\!\left(\sqrt{1 - \frac{b^2}{a^2}}\right) = \frac{2}{a + b} K\!\left(\sqrt{1 - \frac{4ab}{(a + b)^2}}\right)$$

$$= \frac{2}{a + b} K\!\left(\sqrt{\frac{a^2 + b^2 - 2ab}{(a + b)^2}}\right)$$

$$= \frac{2}{a + b} K\!\left(\frac{a - b}{a + b}\right) \quad (89)$$

$$K\!\left(\sqrt{1 - \frac{b^2}{a^2}}\right) = \frac{2}{1 + \frac{b}{a}} K\!\left(\frac{1 - \frac{b}{a}}{1 + \frac{b}{a}}\right). \quad (90)$$

Define

$$k' \equiv \frac{b}{a}, \quad (91)$$

and use

$$k \equiv \sqrt{1 - k'^2}, \quad (92)$$

so

$$K(k) = \frac{2}{1 + k'} K\!\left(\frac{1 - k'}{1 + k'}\right). \quad (93)$$

Now letting $l \equiv (1 - k')/(1 + k')$ gives

$$l(1 + k') = 1 - k' \Rightarrow k'(l + 1) = 1 - l \quad (94)$$

$$k' = \frac{1 - l}{1 + l} \quad (95)$$

$$k = \sqrt{1 - k'^2} = \sqrt{1 - \left(\frac{1 - l}{1 + l}\right)^2}$$

$$= \sqrt{\frac{(1 + l)^2 - (1 - l)^2}{(1 + l)^2}} = \frac{2\sqrt{l}}{1 + l}, \quad (96)$$

and

$$\tfrac{1}{2}(1 + k') = \frac{1}{2}\left(1 + \frac{1 - l}{1 + l}\right) = \frac{1}{2}\left[\frac{(1 + l) + (1 - l)}{1 + l}\right]$$

$$= \frac{1}{1 + l}. \quad (97)$$

Writing k instead of l,

$$k(k) = \frac{1}{k + 1} K\!\left(\frac{2\sqrt{k}}{1 + k}\right). \quad (98)$$

Similarly, from Borwein and Borwein (1987),

$$E(k) = \frac{1 + k}{2} E\!\left(\frac{2\sqrt{k}}{1 + k}\right) + \frac{k'^2}{2} K(k) \quad (99)$$

$$E(k) = (1 + k')E\!\left(\frac{1 - k'}{1 + k'}\right) - k'K(k). \quad (100)$$

Expressions in terms of the complementary function can be derived from interchanging the moduli and their complements in (93), (98), (99), and (100).

$$K'(k) = K(k') = \frac{2}{1 + k} K\!\left(\frac{1 - k}{1 + k}\right)$$

$$= \frac{2}{1 + k} K'\!\left(\sqrt{1 - \left(\frac{1 - k}{1 + k}\right)^2}\right)$$

$$= \frac{2}{1 + k} K'\!\left(\frac{2\sqrt{k}}{1 + k}\right) \quad (101)$$

$$K'(k) = \frac{1}{1 + k'} K\!\left(\frac{2\sqrt{k'}}{1 + k'}\right)$$

$$= \frac{1}{1 + k'} K'\!\left(\frac{1 - k'}{1 + k'}\right), \quad (102)$$

and

$$E'(k) = (1 + k)E'\!\left(\frac{2\sqrt{k}}{1 + k}\right) - kK'(k) \quad (103)$$

$$E'(k) = \left(\frac{1 + k'}{2}\right)E'\!\left(\frac{1 - k'}{1 + k'}\right) + \frac{k^2}{2} K'(k). \quad (104)$$

Taking the ratios

$$\frac{K'(k)}{K(k)} = 2\frac{K'\!\left(\frac{2\sqrt{k}}{1 + k}\right)}{K\!\left(\frac{2\sqrt{k}}{1 + k}\right)} = \frac{1}{2}\frac{K'\!\left(\frac{1 - k'}{1 + k'}\right)}{K\!\left(\frac{1 - k'}{1 + k'}\right)} \quad (105)$$

gives the MODULAR EQUATION of degree 2. It is also true that

$$K(x) = \frac{4}{(1 + \sqrt{x'})^2} \, K\left(\left[\frac{1 - \sqrt{1 - x^4}}{1 + \sqrt{1 - x^4}}\right]^2\right). \qquad (106)$$

See also ABELIAN INTEGRAL, AMPLITUDE, ARGUMENT (ELLIPTIC INTEGRAL), CHARACTERISTIC (ELLIPTIC INTEGRAL), DELTA AMPLITUDE, ELLIPTIC FUNCTION, ELLIPTIC INTEGRAL OF THE FIRST KIND, ELLIPTIC INTEGRAL OF THE SECOND KIND, ELLIPTIC INTEGRAL OF THE THIRD KIND, ELLIPTIC INTEGRAL SINGULAR VALUE, HEUMAN LAMBDA FUNCTION, JACOBI ZETA FUNCTION, MODULAR ANGLE, MODULUS (ELLIPTIC INTEGRAL), NOME, PARAMETER

References

Abramowitz, M. and Stegun, C. A. (Eds.). "Elliptic Integrals." Ch. 17 in *Handbook of Mathematical Functions with Formulas, Graphs, and Mathematical Tables, 9th printing.* New York: Dover, pp. 587–07, 1972.
Arfken, G. "Elliptic Integrals." §5.8 in *Mathematical Methods for Physicists, 3rd ed.* Orlando, FL: Academic Press, pp. 321–27, 1985.
Borwein, J. M. and Borwein, P. B. *Pi & the AGM: A Study in Analytic Number Theory and Computational Complexity.* New York: Wiley, 1987.
Hancock, H. *Elliptic Integrals.* New York: Wiley, 1917.
Kármán, T. von and Biot, M. A. *Mathematical Methods in Engineering: An Introduction to the Mathematical Treatment of Engineering Problems.* New York: McGraw-Hill, p. 121, 1940.
King, L. V. *The Direct Numerical Calculation of Elliptic Functions and Integrals.* London: Cambridge University Press, 1924.
Prasolov, V. and Solovyev, Y. *Elliptic Functions and Elliptic Integrals.* Providence, RI: Amer. Math. Soc., 1997.
Press, W. H.; Flannery, B. P.; Teukolsky, S. A.; and Vetterling, W. T. "Elliptic Integrals and Jacobi Elliptic Functions." §6.11 in *Numerical Recipes in FORTRAN: The Art of Scientific Computing, 2nd ed.* Cambridge, England: Cambridge University Press, pp. 254–63, 1992.
Prudnikov, A. P.; Brychkov, Yu. A.; and Marichev, O. I. *Integrals and Series, Vol. 1: Elementary Functions.* New York: Gordon & Breach, 1986.
Timofeev, A. F. *Integration of Functions.* Moscow and Leningrad: GTTI, 1948.
Weisstein, E. W. "Books about Elliptic Integrals." http://www.treasure-troves.com/books/EllipticIntegrals.html.
Whittaker, E. T. and Watson, G. N. *A Course in Modern Analysis, 4th ed.* Cambridge, England: Cambridge University Press, 1990.
Woods, F. S. "Elliptic Integrals." Ch. 16 in *Advanced Calculus: A Course Arranged with Special Reference to the Needs of Students of Applied Mathematics.* Boston, MA: Ginn, pp. 365–86, 1926.

Elliptic Integral of the First Kind

Let the MODULUS k satisfy $0 < k^2 < 1$, and the AMPLITUDE be given by $\phi = \text{am } u$. The incomplete elliptic integral of the first kind is then defined as

$$u = F(\phi, k) = \int_0^\phi \frac{d\theta}{\sqrt{1 - k^2 \sin^2 \theta}}. \qquad (1)$$

Let

$$t \equiv \sin \theta \qquad (2)$$

$$dt = \cos \theta \, d\theta = \sqrt{1 - t^2} \, d\theta, \qquad (3)$$

then (1) can be written as

$$F(\phi, k) = \int_0^{\sin \phi} \frac{1}{\sqrt{1 - k^2 t^2}} \frac{dt}{\sqrt{1 - t^2}}$$

$$= \int_0^{\sin \phi} \frac{dt}{\sqrt{1 - k^2 t^2} \sqrt{1 - t^2}}. \qquad (4)$$

Let

$$v \equiv \tan \theta \qquad (5)$$

$$dv \equiv \sec^2 \theta \, d\theta = (1 + v^2) \, d\theta, \qquad (6)$$

then the integral can also be written as

$$F(\phi, k) = \int_0^{\tan \phi} \frac{1}{\sqrt{1 - k^2 \dfrac{v^2}{1 + u^2}}} \frac{du}{1 + v^2}$$

$$= \int_0^{\tan \phi} \frac{dv}{\sqrt{1 + v^2}\sqrt{(1 + v^2) - k^2 v^2}} \qquad (7)$$

$$= \int_0^{\tan \phi} \frac{dv}{\sqrt{(1 + v^2)(1 + k' v^2)}}, \qquad (8)$$

where $k'^2 \equiv 1 - k^2$ is the complementary MODULUS. The elliptic integral of the first kind is implemented in *Mathematica* as EllipticK[*phi*, *m*] (*note the use of the parameter* $m = k^2$ *instead of the modulus* k).

The inverse function of $F(\phi, k)$ is given by the AMPLITUDE

$$F^{-1}(u, k) = \phi = \text{am}(u, k) = \text{am } u. \qquad (9)$$

The integral

$$I = \frac{1}{\sqrt{2}} \int_0^{\theta_0} \frac{d\theta}{\sqrt{\cos \theta - \cos \theta_0}}, \qquad (10)$$

which arises in computing the period of a pendulum, is also an elliptic integral of the first kind. Use

$$\cos \theta = 1 - 2 \sin^2(\tfrac{1}{2} \theta) \qquad (11)$$

$$\sin(\tfrac{1}{2} \theta) = \sqrt{\frac{1 - \cos \theta}{2}} \qquad (12)$$

to write

$$\sqrt{\cos \theta - \cos \theta_0} = \sqrt{1 - 2 \sin^2(\tfrac{1}{2} \theta) - \cos \theta_0}$$

$$= \sqrt{1 - \cos \theta_0}\sqrt{1 - \frac{2}{1 - \cos \theta_0} \sin^2(\tfrac{1}{2} \theta)}$$

$$= \sqrt{2} \sin(\tfrac{1}{2} \theta_0)\sqrt{1 - \csc^2(\tfrac{1}{2} \theta_0) \sin^2(\tfrac{1}{2} \theta)}, \qquad (13)$$

so

$$I = \frac{1}{2} \int_0^{\theta_0} \frac{d\theta}{\sin(\frac{1}{2}\theta_0)\sqrt{1 - \csc^2(\frac{1}{2}\theta_0)\sin^2(\frac{1}{2}\theta)}}. \quad (14)$$

Now let

$$\sin(\tfrac{1}{2}\theta) = \sin(\tfrac{1}{2}\theta_0)\sin\phi, \quad (15)$$

so the angle θ is transformed to

$$\phi = \sin^{-1}\left[\frac{\sin(\frac{1}{2}\theta)}{\sin(\frac{1}{2}\theta_0)}\right], \quad (16)$$

which ranges from 0 to $\pi/2$ as θ varies from 0 to θ_0. Taking the differential gives

$$\tfrac{1}{2}\cos(\tfrac{1}{2}\theta)\,d\theta = \sin(\tfrac{1}{2}\theta_0)\cos\phi\,d\phi, \quad (17)$$

or

$$\tfrac{1}{2}\sqrt{1 - \sin^2(\tfrac{1}{2}\theta_0)\sin^2\phi}\,d\theta = \sin(\tfrac{1}{2}\theta_0)\cos\phi\,d\phi. \quad (18)$$

Plugging this in gives

$$I = \int_0^{\pi/2} \frac{1}{\sqrt{1 - \sin^2(\frac{1}{2}\theta_0)\sin^2\phi}} \frac{\sin(\frac{1}{2}\theta_0)\cos\phi\,d\phi}{\sin(\frac{1}{2}\theta_0)\sqrt{1 - \sin^2\phi}}$$

$$= \int_0^{\pi/2} \frac{d\phi}{\sqrt{1 - \sin^2(\frac{1}{2}\theta_0)\sin^2\phi}} = K(\sin(\tfrac{1}{2}\theta_0)), \quad (19)$$

so

$$I = \frac{1}{\sqrt{2}} \int_0^{\theta_0} \frac{d\theta}{\sqrt{\cos\theta - \cos\theta_0}} = K(\sin(\tfrac{1}{2}\theta_0)). \quad (20)$$

Making the slightly different substitution $\phi = \theta/2$, so $d\theta = 2\,d\phi$ leads to an equivalent, but more complicated expression involving an *incomplete* elliptic integral of the first kind,

$$I = 2\frac{1}{\sqrt{2}}\frac{1}{\sqrt{2}}\csc(\tfrac{1}{2}\theta_0)\int_0^{\theta_0} \frac{d\theta}{\sqrt{1 - \csc^2(\frac{1}{2}\theta_0)\sin^2\phi}}$$

$$= \csc(\tfrac{1}{2}\theta_0)F(\tfrac{1}{2}\theta_0,\ \csc(\tfrac{1}{2}\theta_0)). \quad (21)$$

Therefore, we have proven the identity

$$\csc x F(x,\ \csc x) = K(\sin x). \quad (22)$$

The elliptic integral of the first kind satisfies

$$F(-\phi,\ k) = -F(\phi,\ k). \quad (23)$$

Special values of $F(\phi,\ k)$ include

$$F(0,\ k) = 0 \quad (24)$$

$$F(\tfrac{1}{2}\pi,\ k) = K(k), \quad (25)$$

where $K(k)$ is known as the complete elliptic integral of the first kind.

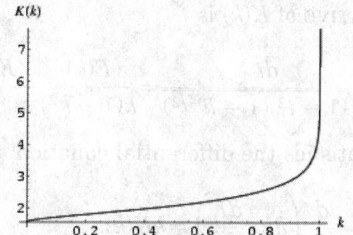

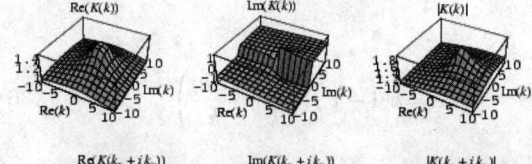

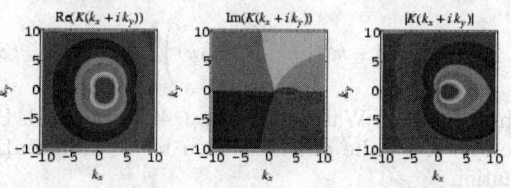

The complete elliptic integral of the first kind, illustrated above as a function of $m = k^2$, is defined by

$$K(k) \equiv F(\tfrac{1}{2}\pi,\ k) \quad (26)$$

$$= \sum_{n=0}^{\infty} \frac{(2n-1)!!}{(2n)!!}k^{2n}\int_0^{2\pi}\sin^{2n}\theta\,d\theta \quad (27)$$

$$= \tfrac{1}{2}\pi\vartheta_3^2(q) \quad (28)$$

$$= \sum_{n=0}^{\infty} \frac{(2n-1)!!}{(2n)!!}k^{2n}\frac{\pi}{2}\frac{(2n-1)!!}{(2n)!!}$$

$$= \frac{\pi}{2}\sum_{n=0}^{\infty}\left[\frac{(2n-1)!!}{(2n)!!}\right]^2 k^{2n} \quad (29)$$

$$= \tfrac{1}{2}\pi\,{}_2F_1(\tfrac{1}{2},\ \tfrac{1}{2};\ 1;\ k^2) \quad (30)$$

$$= \frac{\pi}{2\sqrt{1 - k^2}}P_{-1/2}\left(\frac{1 + k^2}{1 - k^2}\right), \quad (31)$$

where

$$q = e^{-\lambda K'(k)/K(k)} \quad (32)$$

is the NOME (for $|q| < 1$), ${}_2F_1(a,\ b;\ c;\ x)$ is the HYPERGEOMETRIC FUNCTION, and $P_n(x)$ is a LEGENDRE POLYNOMIAL. $K(k)$ satisfies the LEGENDRE RELATION

$$E(k)K'(k) + E'(k)K(k) - K(k)K'(k) = \tfrac{1}{2}\pi, \quad (33)$$

where $K(k)$ and $E(k)$ are complete elliptic integrals of the first and SECOND KINDS, respectively, and $K'(k)$ and $E'(k)$ are the complementary integrals. The modulus k is often suppressed for conciseness, so that $K(k)$ and $E(k)$ are often simply written K and E, respectively.

The DERIVATIVE of $K(k)$ is

$$\frac{dK}{dk} \equiv \int_0^1 \frac{dt}{\sqrt{(1-t^2)(1-k'^2 t^2)}} = \frac{E(k)}{k(1-k^2)} - \frac{K(k)}{k} \quad (34)$$

and $K(k)$ satisfies the differential equation

$$\frac{d}{dk}\left(kk'^2 \frac{dK}{dk}\right) = kK(k), \quad (35)$$

so

$$E = k(1-k^2)\left(\frac{dK}{dk} + \frac{K(k)}{k}\right) \quad (36)$$

$$= (1-k^2)\left(k\frac{dK}{dk} + K(k)\right) \quad (37)$$

(Whittaker and Watson 1990, pp. 499 and 521). Besides $y = K(k)$, the other solution to the differential equation

$$\frac{d}{dk}\left[k(1-k^2)\frac{dy}{dk}\right] - ky = 0 \quad (38)$$

(Zwillinger 1997, p. 122; Gradshteyn and Ryzhik 2000, p. 907) is MEIJER'S G-FUNCTION

$$y = G_{2,2}^{2,0}\left(k^2 \left| \begin{matrix} \frac{1}{2}, \frac{1}{2} \\ 0, 0 \end{matrix} \right.\right). \quad (39)$$

See also AMPLITUDE, CHARACTERISTIC (ELLIPTIC INTEGRAL), ELLIPTIC INTEGRAL OF THE SECOND KIND, ELLIPTIC INTEGRAL OF THE THIRD KIND, ELLIPTIC INTEGRAL SINGULAR VALUE, GAUSS'S TRANSFORMATION, LANDEN'S TRANSFORMATION, LEGENDRE RELATION, MODULAR ANGLE, MODULUS (ELLIPTIC INTEGRAL), PARAMETER

References

Abramowitz, M. and Stegun, C. A. (Eds.). "Elliptic Integrals." Ch. 17 in *Handbook of Mathematical Functions with Formulas, Graphs, and Mathematical Tables, 9th printing.* New York: Dover, pp. 587–07, 1972.

Gradshteyn, I. S. and Ryzhik, I. M. *Tables of Integrals, Series, and Products, 6th ed.* San Diego, CA: Academic Press, 2000.

Spanier, J. and Oldham, K. B. "The Complete Elliptic Integrals $K(p)$ and $E(p)$" and "The Incomplete Elliptic Integrals $F(p; \phi)$ and $E(p; \phi)$." Chs. 61–2 in *An Atlas of Functions.* Washington, DC: Hemisphere, pp. 609–33, 1987.

Tölke, F. "Parameterfunktionen." Ch. 3 in *Praktische Funktionenlehre, zweiter Band: Theta-Funktionen und spezielle Weierstraßsche Funktionen.* Berlin: Springer-Verlag, pp. 83–15, 1966.

Tölke, F. "Umkehrfunktionen der Jacobischen elliptischen Funktionen und elliptische Normalintegrale erster Gattung. Elliptische Amplitudenfunktionen sowie Legendresche F- und E-Funktion. Elliptische Normalintegrale zweiter Gattung. Jacobische Zeta- und Heumansche Lambda-Funktionen," and "Normalintegrale dritter Gattung. Legendresche ∏-Funktion. Zurückführung des allgemeinen elliptischen Integrals auf Normalintegrale erster, zweiter, und dritter Gattung." Chs. 6– in *Praktische Funktionenlehre, dritter Band: Jacobische elliptische Funktionen, Legendresche elliptische Normalintegrale und spezielle Weierstraßsche Zeta- und Sigma Funktionen.* Berlin: Springer-Verlag, pp. 58–44, 1967.

Whittaker, E. T. and Watson, G. N. *A Course in Modern Analysis, 4th ed.* Cambridge, England: Cambridge University Press, 1990.

Zwillinger, D. *Handbook of Differential Equations, 3rd ed.* Boston, MA: Academic Press, p. 122, 1997.

Elliptic Integral of the Second Kind

Let the MODULUS k satisfy $0 < k^2 < 1$. (This may also be written in terms of the PARAMETER $m \equiv k^2$ or MODULAR ANGLE $\alpha \equiv \sin^{-1} k$.) The incomplete elliptic integral of the second kind is then defined as

$$E(\phi, k) \equiv \int_0^\phi \sqrt{1 - k^2 \sin^2 \theta}\, d\theta. \quad (1)$$

The elliptic integral of the second kind is implemented in *Mathematica* as EllipticE[phi, m] (*note the use of the parameter $m = k^2$ instead of the modulus k*).

To place the elliptic integral of the second kind in a slightly different form, let

$$t \equiv \sin \theta \quad (2)$$

$$dt = \cos \theta\, d\theta = \sqrt{1 - t^2}\, d\theta, \quad (3)$$

so the elliptic integral can also be written as

$$E(\phi, k) = \int_0^{\sin \phi} \sqrt{1 - k^2 t^2}\, \frac{dt}{\sqrt{1-t^2}}$$

$$= \int_0^{\sin \phi} \sqrt{\frac{1 - k^2 t^2}{1 - t^2}}\, dt. \quad (4)$$

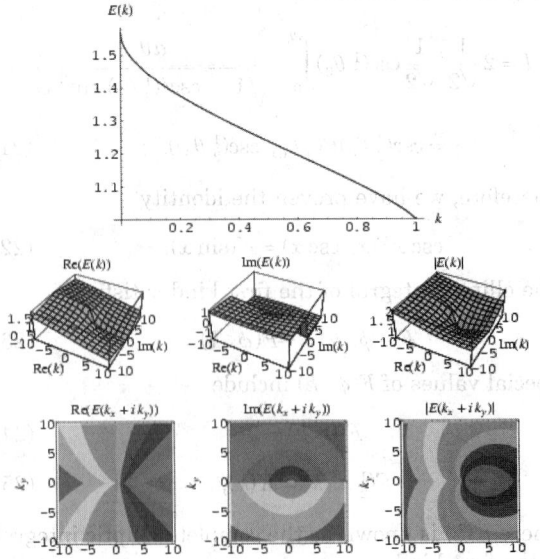

The complete elliptic integral of the second kind, illustrated above as a function of the PARAMETER m, is defined by

$$E(k) \equiv E(\tfrac{1}{2}\pi, \ k) \tag{5}$$

$$= \frac{\pi}{2}\left\{1 - \sum_{n=1}^{\infty}\left[\frac{(2n-1)!!}{(2n)!!}\right]^2 \frac{k^{2n}}{2n-1}\right\} \tag{6}$$

$$= \tfrac{1}{2}\pi \ {}_2F_1(-\tfrac{1}{2}, \ \tfrac{1}{2}; \ 1; \ k^2) \tag{7}$$

$$= \int_0^K \mathrm{dn}^2 \ u \ du, \tag{8}$$

where ${}_2F_1(a, \ b; \ c; \ x)$ is the HYPERGEOMETRIC FUNCTION and $\mathrm{dn}\ u$ is a JACOBI ELLIPTIC FUNCTION. The complete elliptic integral of the second kind satisfies the LEGENDRE RELATION

$$E(k)K'(k) + E'(k)K(k) - K(k)K'(k) = \tfrac{1}{2}\pi, \tag{9}$$

where $K(k)$ and $E(k)$ are complete ELLIPTIC INTEGRALS OF THE FIRST and second kinds, respectively, and $K'(k)$ and $E'(k)$ are the complementary integrals. The DERIVATIVE is

$$\frac{dE}{dk} = \frac{E(k) - K(k)}{k} \tag{10}$$

(Whittaker and Watson 1990, p. 521). Besides $y = E(k)$, the other solution to the differential equation

$$k'^2 \frac{d}{dk}\left(k\frac{dy}{dk}\right) + ky = 0 \tag{11}$$

(Zwillinger 1997, p. 122; Gradshteyn and Ryzhik 2000, p. 907) is MEIJER'S G-FUNCTION

$$y = G_{2,\,2}^{2,\,0}\left(k^2 \left|\begin{matrix}\tfrac{1}{2}, \ \tfrac{3}{2} \\ 0, \ 0\end{matrix}\right.\right). \tag{12}$$

If k_r is a singular value (i.e.,

$$k_r = \lambda^*(r), \tag{13}$$

where λ^* is the ELLIPTIC LAMBDA FUNCTION), and $K(k_r)$ and the ELLIPTIC ALPHA FUNCTION $\alpha(r)$ are also known, then

$$E(k) = \frac{K(k)}{\sqrt{r}}\left[\frac{\pi}{3[K(k)]^2} - \alpha(r)\right] + K(k). \tag{14}$$

A generalization replacing $\sin\theta$ with $\sinh\theta$ in (1) gives

$$-iE(i\phi, \ -k) = \int_0^{\phi}\sqrt{1 - k^2\sinh^2\theta}\ d\theta. \tag{15}$$

See also ELLIPTIC INTEGRAL OF THE FIRST KIND, ELLIPTIC INTEGRAL OF THE THIRD KIND, ELLIPTIC

INTEGRAL SINGULAR VALUE

References

Abramowitz, M. and Stegun, C. A. (Eds.). "Elliptic Integrals." Ch. 17 in *Handbook of Mathematical Functions with Formulas, Graphs, and Mathematical Tables, 9th printing.* New York: Dover, pp. 587–07, 1972.

Spanier, J. and Oldham, K. B. "The Complete Elliptic Integrals $K(p)$ and $E(p)$" and "The Incomplete Elliptic Integrals $F(p; \ \phi)$ and $E(p; \ \phi)$." Chs. 61 and 62 in *An Atlas of Functions.* Washington, DC: Hemisphere, pp. 609–33, 1987.

Tölke, F. "Parameterfunktionen." Ch. 3 in *Praktische Funktionenlehre, zweiter Band: Theta-Funktionen und spezielle Weierstraßsche Funktionen.* Berlin: Springer-Verlag, pp. 83–15, 1966.

Tölke, F. "Umkehrfunktionen der Jacobischen elliptischen Funktionen und elliptische Normalintegrale erster Gattung. Elliptische Amplitudenfunktionen sowie Legendresche F- und E-Funktion. Elliptische Normalintegrale zweiter Gattung. Jacobische Zeta- und Heumansche Lambda-Funktionen," and "Normalintegrale dritter Gattung. Legendresche $\prod$-Funktion. Zurückführung des allgemeinen elliptischen Integrals auf Normalintegrale erster, zweiter, und dritter Gattung." Chs. 6– in *Praktische Funktionenlehre, dritter Band: Jacobische elliptische Funktionen, Legendresche elliptische Normalintegrale und spezielle Weierstraßsche Zeta- und Sigma Funktionen.* Berlin: Springer-Verlag, pp. 58–44, 1967.

Whittaker, E. T. and Watson, G. N. *A Course in Modern Analysis, 4th ed.* Cambridge, England: Cambridge University Press, 1990.

Elliptic Integral of the Third Kind

Let $0 < k^2 < 1$. The incomplete elliptic integral of the third kind is then defined as

$$\Pi(n; \ \phi, \ k) = \int_0^{\phi}\frac{d\theta}{(1 - n\sin^2\theta)\sqrt{1 - k^2\sin^2\theta}} \tag{1}$$

$$= \int_0^{\sin\phi}\frac{dt}{(1 - nt^2)\sqrt{(1 - t^2)(1 - k^2t^2)}}, \tag{2}$$

where n is a constant known as the CHARACTERISTIC.

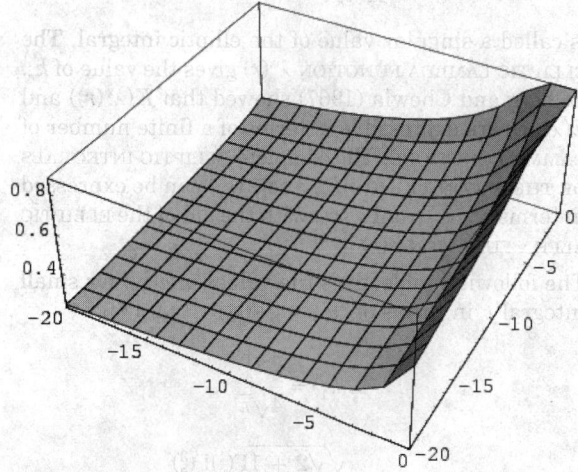

The complete elliptic integral of the third kind

$$\Pi(n|m) = \Pi(n; \tfrac{1}{2}\pi|m) \qquad (3)$$

is illustrated above.

See also ELLIPTIC INTEGRAL OF THE FIRST KIND, ELLIPTIC INTEGRAL OF THE SECOND KIND, ELLIPTIC INTEGRAL SINGULAR VALUE

References

Abramowitz, M. and Stegun, C. A. (Eds.). "Elliptic Integrals" and "Elliptic Integrals of the Third Kind." Ch. 17 and §17.7 in *Handbook of Mathematical Functions with Formulas, Graphs, and Mathematical Tables, 9th printing.* New York: Dover, pp. 587–07, 1972.

Tölke, F. "Normalintegrale dritter Gattung. Legendresche $\prod$-Funktion. Zurückführung des allgemeinen elliptischen Integrals auf Normalintegrale erster, zweiter, und dritter Gattung." Ch. 7 in *Praktische Funktionenlehre, dritter Band: Jacobische elliptische Funktionen, Legendresche elliptische Normalintegrale und spezielle Weierstraßsche Zeta- und Sigma Funktionen.* Berlin: Springer-Verlag, pp. 100–44, 1967.

Elliptic Integral Singular Value

When the MODULUS k has a singular value, the complete elliptic integrals may be computed in analytic form in terms of GAMMA FUNCTIONS. Abel (quoted in Whittaker and Watson 1990, p. 525) proved that whenever

$$\frac{K'(k)}{K(k)} = \frac{a + b\sqrt{n}}{c + d\sqrt{n}}, \qquad (1)$$

where a, b, c, d, and n are INTEGERS, $K(k)$ is a complete ELLIPTIC INTEGRAL OF THE FIRST KIND, and $K'(k) \equiv K(\sqrt{1-k^2})$ is the complementary complete ELLIPTIC INTEGRAL OF THE FIRST KIND, then the MODULUS k is the ROOT of an algebraic equation with INTEGER COEFFICIENTS.

A MODULUS k_r such that

$$\frac{K'(k_r)}{K(k_r)} = \sqrt{r}, \qquad (2)$$

is called a singular value of the elliptic integral. The ELLIPTIC LAMBDA FUNCTION $\lambda^*(r)$ gives the value of k_r. Selberg and Chowla (1967) showed that $K(\lambda^*(r))$ and $E(\lambda^*(r))$ are expressible in terms of a finite number of GAMMA FUNCTIONS. The complete ELLIPTIC INTEGRALS OF THE SECOND KIND $e(k_r)$ and $e'(k_r)$ can be expressed in terms of $k(k_r)$ and $k'(k_r)$ with the aid of the ELLIPTIC ALPHA FUNCTION $\alpha(r)$.

The following table gives the values of $k(k_r)$ for small integral r in terms of GAMMA FUNCTIONS $\Gamma(z)$.

$$K(k_1) = \frac{\Gamma^2(\frac{1}{4})}{4\sqrt{\pi}}$$

$$K(k_2) = \frac{\sqrt{\sqrt{2}+1}\,\Gamma(\frac{1}{8})\Gamma(\frac{3}{8})}{2^{13/4}\sqrt{\pi}}$$

$$K(k_3) = \frac{3^{1/4}\Gamma^3(\frac{1}{3})}{2^{7/3}\pi}$$

$$K(k_4) = \frac{(\sqrt{2}+1)\Gamma^2(\frac{1}{4})}{2^{7/2}\sqrt{\pi}}$$

$$K(k_5) = (\sqrt{5}+2)^{1/4}\sqrt{\frac{\Gamma(\frac{1}{20})\Gamma(\frac{3}{20})\Gamma(\frac{7}{20})\Gamma(\frac{9}{20})}{160\pi}}$$

$$K(k_6) = \sqrt{(\sqrt{2}-1)(\sqrt{3}+\sqrt{2})(2+\sqrt{3})}$$
$$\times \sqrt{\frac{\Gamma(\frac{1}{24})\Gamma(\frac{5}{24})\Gamma(\frac{7}{24})\Gamma(\frac{11}{24})}{384\pi}}$$

$$K(k_7) = \frac{\Gamma(\frac{1}{7})\Gamma(\frac{2}{7})\Gamma(\frac{4}{7})}{7^{1/4}4\pi}$$

$$K(k_8) = \sqrt{\frac{2\sqrt{2}+\sqrt{1+5\sqrt{2}}}{4\sqrt{2}}}\,\frac{(\sqrt{2}+1)^{1/4}\Gamma(\frac{1}{8})\Gamma(\frac{3}{8})}{8\sqrt{\pi}}$$

$$K(k_9) = \frac{3^{1/4}\sqrt{2+\sqrt{3}}}{12\sqrt{\pi}\Gamma^2(\frac{1}{4})}$$

$$K(k_{10}) = \sqrt{(2+3\sqrt{2}+\sqrt{5})}$$
$$\times \sqrt{\frac{\Gamma(\frac{1}{40})\Gamma(\frac{7}{40})\Gamma(\frac{9}{40})\Gamma(\frac{11}{40})\Gamma(\frac{13}{40})\Gamma(\frac{19}{40})\Gamma(\frac{23}{40})\Gamma(\frac{37}{40})}{256\pi^3}}$$

$$K(k_{11}) = [2+(17+3\sqrt{33})^{1/3}-(3\sqrt{33}-17)^{1/3}]^2$$
$$\times \frac{\Gamma(\frac{1}{11})\Gamma(\frac{3}{11})\Gamma(\frac{4}{11})\Gamma(\frac{5}{11})\Gamma(\frac{9}{11})}{11^{1/4}144\pi^2}$$

$$K(k_{12}) = \frac{3^{1/4}(\sqrt{2}+1)(\sqrt{3}+\sqrt{2})\sqrt{2-\sqrt{3}}\,\Gamma^3(\frac{1}{3})}{2^{13/3}\pi}$$

$$K(k_{13}) = \frac{(18+5\sqrt{13})^{1/4}}{\sqrt{6656\pi^5}}$$
$$\times \sqrt{\Gamma(\frac{1}{52})\Gamma(\frac{7}{52})\Gamma(\frac{9}{52})\Gamma(\frac{11}{52})\Gamma(\frac{15}{52})\Gamma(\frac{17}{52})\Gamma(\frac{19}{52})\Gamma(\frac{25}{52})\Gamma(\frac{29}{52})\Gamma(\frac{31}{52})\Gamma}$$

$$K(k_{15}) = \sqrt{\frac{(\sqrt{5}+1)\Gamma(\frac{1}{15})\Gamma(\frac{2}{15})\Gamma(\frac{4}{15})\Gamma(\frac{8}{15})}{240\pi}}$$

$$K(k_{16}) = \frac{(2^{1/4}+1)^2\Gamma^2(\frac{1}{4})}{2^{9/2}\sqrt{\pi}}$$

$$K(k_{17}) = C_1\left[\frac{\Gamma(\frac{1}{68})\Gamma(\frac{3}{68})\Gamma(\frac{7}{68})\Gamma(\frac{11}{68})\Gamma(\frac{13}{68})}{\Gamma(\frac{5}{68})\Gamma(\frac{15}{68})\Gamma(\frac{19}{68})\Gamma(\frac{29}{68})}\right]^{1/4}$$
$$\times [\Gamma(\frac{21}{68})\Gamma(\frac{25}{68})\Gamma(\frac{27}{68})\Gamma(\frac{31}{68})\Gamma(\frac{33}{68})]^{1/4}$$

$$K(k_{25}) = \frac{\sqrt{5}+2}{20}\frac{\Gamma^2(\frac{1}{4})}{\sqrt{\pi}},$$

where $\Gamma(z)$ is the GAMMA FUNCTION and C_1 is an algebraic number (Borwein and Borwein 1987, p. 298).

Borwein and Zucker (1992) give amazing expressions for singular values of complete elliptic integrals in terms of CENTRAL BETA FUNCTIONS

$$\beta(p) \equiv B(p, p). \tag{3}$$

Furthermore, they show that $K(k_n)$ is *always* expressible in terms of these functions for $n \equiv 1, 2 \pmod 4$. In such cases, the $\Gamma(z)$ functions appearing in the expression are OF THE FORM $\Gamma(t/4n)$ where $1 \leq t \leq (2n-1)$ and $(t, 4n) = 1$. The terms in the numerator depend on the sign of the KRONECKER SYMBOL $\{t/4n\}$. Values for the first few n are

$$K(k_1) = 2^{-2}\beta(\tfrac{1}{4})$$

$$K(k_2) = 2^{-13/4}\beta(\tfrac{1}{8})$$

$$K(k_3) = 2^{-4/3}3^{-1/4}\beta(\tfrac{1}{3}) = 2^{-5/3}3^{-3/4}\beta(\tfrac{1}{6})$$

$$K(k_5) = 2^{-33/20}5^{-5/8}(11+5\sqrt{5})^{1/4}\sin(\tfrac{1}{20}\pi)\beta(\tfrac{1}{2})$$

$$= 2^{-29/20}5^{-3/8}(1+\sqrt{5})^{1/4}\sin(\tfrac{3}{20}\pi)\beta(\tfrac{3}{20})$$

$$K(k_6) = 2^{-47/12}3^{-3/4}(\sqrt{2}-1)(\sqrt{3}+1)\beta(\tfrac{1}{24})$$

$$= 2^{-43/12}3^{-1/4}(\sqrt{3}-1)\beta(\tfrac{5}{24})$$

$$K(k_7) = 2 \cdot 7^{-3/4}\sin(\tfrac{1}{7}\pi)\sin(\tfrac{2}{7}\pi)B(\tfrac{1}{7}, \tfrac{2}{7})$$

$$= 2^{-2/7}7^{-1/4}\frac{\beta(\tfrac{1}{7})\beta(\tfrac{2}{7})}{\beta(\tfrac{1}{14})}$$

$$K(k_{10}) = 2^{-61/20}5^{-1/4}(\sqrt{5}-2)^{1/2}(\sqrt{10}+3)\frac{\beta(\tfrac{1}{8})\beta(\tfrac{7}{40})}{\beta(\tfrac{3}{40})}$$

$$= 2^{-15/4}5^{-3/4}(\sqrt{5}-2)^{1/2}\frac{\beta(\tfrac{1}{40})\beta(\tfrac{9}{40})}{\beta(\tfrac{3}{8})}$$

$$K(k_{11}) = R \cdot 2^{-7/11}\sin(\tfrac{1}{11}\pi)\sin(\tfrac{3}{11}\pi)B(\tfrac{1}{22}, \tfrac{3}{22})$$

$$K(k_{13}) = 2^{-3}13^{-5/8}(5\sqrt{13}+18)^{1/4}$$

$$\times [\tan(\tfrac{1}{52}\pi)\tan(\tfrac{3}{52}\pi)\tan(\tfrac{9}{52}\pi)]^{1/2}\frac{\beta\left(\tfrac{1}{52}\right)\beta\left(\tfrac{9}{52}\right)}{\beta\left(\tfrac{23}{52}\right)}$$

$$K(k_{14}) = \sqrt{\sqrt{4\sqrt{2}+2}+\sqrt{2}+\sqrt{2\sqrt{2}-1}}$$

$$\cdot 2^{-13/4}7^{-3/8}\left[\frac{\tan(\tfrac{5}{56}\pi)\tan(\tfrac{13}{56}\pi)}{\tan(\tfrac{11}{56}\pi)}\right]^{1/4}$$

$$\times \sqrt{\frac{\beta(\tfrac{5}{56})\beta(\tfrac{13}{56})\beta(\tfrac{1}{8})}{\beta(\tfrac{11}{56})}}$$

$$K(k_{15}) = 2^{-1}3^{-3/4}5^{-7/12}B(\tfrac{1}{15}, \tfrac{4}{15})$$

$$= \frac{2^{-2}3^{-3/4}5^{-3/4}(\sqrt{5}-1)\beta(\tfrac{1}{15})\beta(\tfrac{4}{15})}{\beta(\tfrac{1}{3})}$$

$$K(k_{17}) = C_2\left[\frac{\beta(\tfrac{1}{68})\beta(\tfrac{3}{68})\beta(\tfrac{7}{68})\beta(\tfrac{9}{68})\beta(\tfrac{11}{68})\beta(\tfrac{13}{68})}{\beta(\tfrac{5}{68})\beta(\tfrac{15}{68})}\right]^{1/4},$$

where R is the REAL ROOT of

$$x^3 - 4x = 4 = 0 \tag{4}$$

and C_2 is an algebraic number (Borwein and Zucker 1992). Note that $K(k_{11})$ is the only value in the above list which cannot be expressed in terms of CENTRAL BETA FUNCTIONS.

Using the ELLIPTIC ALPHA FUNCTION, the ELLIPTIC INTEGRALS OF THE SECOND KIND can also be found from

$$E = \frac{\pi}{4\sqrt{r}K} + \left[1 - \frac{\alpha(r)}{\sqrt{r}}\right]K \tag{5}$$

$$E' = \frac{\pi}{4k} + \alpha(r)K, \tag{6}$$

and by definition,

$$K' = K\sqrt{n}. \tag{7}$$

See also CENTRAL BETA FUNCTION, ELLIPTIC ALPHA FUNCTION, ELLIPTIC DELTA FUNCTION, ELLIPTIC INTEGRAL OF THE FIRST KIND, ELLIPTIC INTEGRAL OF THE SECOND KIND, ELLIPTIC LAMBDA FUNCTION, GAMMA FUNCTION, MODULUS (ELLIPTIC INTEGRAL)

References

Abel, N. H. "Recherches sur les fonctions elliptiques." *J. reine angew. Math.* **3**, 160–90, 1828. Reprinted in Abel, N. H. *Oeuvres Completes* (Ed. L. Sylow and S. Lie). New York: Johnson Reprint Corp., p. 377, 1988.

Borwein, J. M. and Borwein, P. B. *Pi & the AGM: A Study in Analytic Number Theory and Computational Complexity.* New York: Wiley, pp. 139 and 298, 1987.

Borwein, J. M. and Zucker, I. J. "Elliptic Integral Evaluation of the Gamma Function at Rational Values of Small Denominator." *IMA J. Numerical Analysis* **12**, 519–26, 1992.

Bowman, F. *Introduction to Elliptic Functions, with Applications.* New York: Dover, pp. 75, 95, and 98, 1961.

Glasser, M. L. and Wood, V. E. "A Closed Form Evaluation of the Elliptic Integral." *Math. Comput.* **22**, 535–36, 1971.

Selberg, A. and Chowla, S. "On Epstein's Zeta-Function." *J. reine angew. Math.* **227**, 86–10, 1967.

Weisstein, E. W. "Elliptic Singular Values." MATHEMATICA NOTEBOOK ELLIPTICSINGULAR.M.

Whittaker, E. T. and Watson, G. N. *A Course in Modern Analysis, 4th ed.* Cambridge, England: Cambridge University Press, pp. 524–28, 1990.

Wrigge, S. "An Elliptic Integral Identity." *Math. Comput.* **27**, 837–40, 1973.

Zucker, I. J. "The Evaluation in Terms of Γ-Functions of the Periods of Elliptic Curves Admitting Complex Multiplication." *Math. Proc. Cambridge Phil. Soc.* **82**, 111–18, 1977.

Elliptic Integral Singular Value k1

The first singular value k_1 of the ELLIPTIC INTEGRAL OF THE FIRST KIND $K(k)$, corresponding to

$$K'(k_1) = K(k_1), \tag{1}$$

is given by

$$k_1 = \frac{1}{\sqrt{2}} \tag{2}$$

$$k_1' = \frac{1}{\sqrt{2}}. \tag{3}$$

The value $K(k_1)$ is given by

$$K\left(\frac{1}{\sqrt{2}}\right) \equiv \int_0^1 \frac{dt}{\sqrt{(1-t^2)(1-\frac{1}{2}t^2)}}, \tag{4}$$

which can be transformed to

$$K\left(\frac{1}{\sqrt{2}}\right) = \sqrt{2} \int_0^1 \frac{dt}{\sqrt{1-t^4}}. \tag{5}$$

Let

$$u \equiv t^4 \tag{6}$$

$$du = 4t^3\, dt = 4u^{3/4}\, dt \tag{7}$$

$$dt = \tfrac{1}{4} u^{-3/4}\, du, \tag{8}$$

then

$$k\left(\frac{1}{\sqrt{2}}\right) = \frac{\sqrt{2}}{4} \int_0^1 u^{-3/4}(1-u)^{-1/2}\, du$$

$$= \frac{\sqrt{2}}{4} B(\tfrac{1}{4}, \tfrac{1}{2}) = \frac{\Gamma(\tfrac{1}{4})\Gamma(\tfrac{1}{2})}{\Gamma(\tfrac{3}{4})} \frac{\sqrt{2}}{4}. \tag{9}$$

where $B(a, b)$ is the BETA FUNCTION and $\Gamma(z)$ is the GAMMA FUNCTION. Now use

$$\Gamma(\tfrac{1}{2}) = \sqrt{\pi} \tag{10}$$

and

$$\frac{1}{\Gamma(1-x)} = \frac{\sin(\pi x)}{\pi}\, \Gamma(x), \tag{11}$$

so

$$\frac{1}{\Gamma(\tfrac{3}{4})} = \frac{1}{\Gamma(1-\tfrac{1}{4})} = \frac{\sin\left(\dfrac{\pi}{4}\right)}{\pi}\, \Gamma(\tfrac{1}{4}) = \frac{1}{\pi\sqrt{2}}\, \Gamma(\tfrac{1}{4}). \tag{12}$$

Therefore,

$$K\left(\frac{1}{\sqrt{2}}\right) = \frac{\Gamma^2(\tfrac{1}{4})\sqrt{\pi}\sqrt{2}}{4\pi\sqrt{2}} = \frac{\Gamma^2(\tfrac{1}{4})}{4\sqrt{\pi}}. \tag{13}$$

Now consider

$$E\left(\frac{1}{\sqrt{2}}\right) \equiv \int_0^1 \sqrt{\frac{1-\tfrac{1}{2}t^2}{1-t^2}}\, dt. \tag{14}$$

Let

$$t^2 \equiv 1 - u^2 \tag{15}$$

$$2t\, dt = -2u\, du \tag{16}$$

$$dt = -\frac{1}{t} u\, du = u(1-u^2)^{-1/2}\, du, \tag{17}$$

so

$$E\left(\frac{1}{\sqrt{2}}\right) = \int_0^1 \sqrt{\frac{1-\tfrac{1}{2}(1-u^2)}{1-(1-u^2)}}\, u(1-u^2)^{-1/2}\, du$$

$$= \int_0^1 \sqrt{\frac{\tfrac{1}{2}(1+u^2)}{u}}\, u(1-u^2)^{-1/2}\, du$$

$$= \frac{1}{\sqrt{2}} \int_0^1 \sqrt{\frac{(1+u^2)}{(1-u^2)}}\, du. \tag{18}$$

Now note that

$$\left(\frac{1}{\sqrt{1-u^4}} + \frac{u^2}{\sqrt{1-u^4}}\right)^2 = \frac{(1+u^2)^2}{1-u^4} = \frac{(1+u^2)^2}{(1+u^2)(1-u^2)}$$

$$= \frac{1+u^2}{1-u^2}, \tag{19}$$

so

$$E\left(\frac{1}{\sqrt{2}}\right) = \frac{1}{\sqrt{2}} \int_0^1 \sqrt{\frac{1+u^2}{1-u^2}}\, du$$

$$= \frac{1}{\sqrt{2}} \int_0^1 \left(\frac{1}{\sqrt{1-u^4}} + \frac{u^2}{\sqrt{1-u^4}}\right) du$$

$$= \frac{1}{2} K\left(\frac{1}{\sqrt{2}}\right) + \frac{1}{\sqrt{2}} \int_0^1 \frac{u^2\, du}{\sqrt{1-u^4}}. \tag{20}$$

Now let

$$t \equiv u^4 \tag{21}$$

$$dt = 4u^3 \, du, \tag{22}$$

so

$$\int_0^1 \frac{u^2 \, du}{\sqrt{1-u^4}} = \frac{1}{4} \int_0^1 t^{1/2} t^{-3/4} (1-t)^{-1/2} \, dt$$

$$= \frac{1}{4} \int_0^1 t^{-1/4} (1-t)^{-1/2} \, dt$$

$$= \frac{1}{4} B(\tfrac{3}{4}, \tfrac{1}{2}) = \frac{\Gamma(\tfrac{3}{4}) \Gamma(\tfrac{1}{2})}{4 \Gamma(\tfrac{5}{4})}. \tag{23}$$

But

$$[\Gamma(\tfrac{5}{4})]^{-1} = [\tfrac{1}{4} \Gamma(\tfrac{1}{4})]^{-1} \tag{24}$$

$$\Gamma(\tfrac{3}{4}) = \pi \sqrt{2} [\Gamma(\tfrac{1}{4})]^{-1} \tag{25}$$

$$\Gamma(\tfrac{1}{2}) = \sqrt{\pi}, \tag{26}$$

so

$$\int_0^1 \frac{u^2 \, du}{\sqrt{1-u^4}} = \frac{1}{4} \frac{\pi \sqrt{2} \cdot 4 \sqrt{\pi}}{\Gamma^2(\tfrac{1}{4})} = \frac{\sqrt{2} \, \pi^{3/2}}{\Gamma^2(\tfrac{1}{4})} \tag{27}$$

$$E\left(\frac{1}{\sqrt{2}}\right) = \frac{1}{2} K + \frac{\pi^{3/2}}{\Gamma^2(\tfrac{1}{4})} = \frac{\Gamma^2(\tfrac{1}{4})}{8\sqrt{\pi}} + \frac{\pi^{3/2}}{\Gamma^2(\tfrac{1}{4})}$$

$$= \frac{1}{4} \sqrt{\frac{\pi}{2}} \left[\frac{\Gamma(\tfrac{1}{4})}{\Gamma(\tfrac{3}{4})} + \frac{\Gamma(\tfrac{3}{4})}{\Gamma(\tfrac{5}{4})} \right]. \tag{28}$$

Summarizing (13) and (28) gives

$$K\left(\frac{1}{\sqrt{2}}\right) = \frac{\Gamma^2(\tfrac{1}{4})}{4\sqrt{\pi}}$$

$$K'\left(\frac{1}{\sqrt{2}}\right) = \frac{\Gamma^2(\tfrac{1}{4})}{4\sqrt{\pi}}$$

$$E\left(\frac{1}{\sqrt{2}}\right) = \frac{\Gamma^2(\tfrac{1}{4})}{8\sqrt{\pi}} + \frac{\pi^{3/2}}{\Gamma^2(\tfrac{1}{4})}$$

$$E'\left(\frac{1}{\sqrt{2}}\right) = \frac{\Gamma^2(\tfrac{1}{4})}{8\sqrt{\pi}} + \frac{\pi^{3/2}}{\Gamma^2(\tfrac{1}{4})}.$$

Elliptic Integral Singular Value k2

The second SINGULAR VALUE k_2, corresponding to

$$K'(k_2) = \sqrt{2} K(k_2), \tag{1}$$

is given by

$$k_2 = \tan\left(\frac{\pi}{8}\right) = \sqrt{2} - 1 \tag{2}$$

$$k_2' = \sqrt{2}(\sqrt{2} - 1). \tag{3}$$

For this modulus,

$$E(\sqrt{2} - 1) = \frac{1}{4} \sqrt{\frac{\pi}{4}} \left[\frac{\Gamma(\tfrac{1}{8})}{\Gamma(\tfrac{5}{8})} + \frac{\Gamma(\tfrac{5}{8})}{\Gamma(\tfrac{9}{8})} \right]. \tag{4}$$

Elliptic Integral Singular Value k3

The third SINGULAR VALUE k_3, corresponding to

$$K'(k_3) = \sqrt{3} K(k_3), \tag{1}$$

is given by

$$k_3 = \sin\left(\frac{\pi}{12}\right) = \tfrac{1}{4}(\sqrt{6} - \sqrt{2}). \tag{2}$$

As shown by Legendre,

$$K(k_3) = \frac{\sqrt{\pi}}{2 \cdot 3^{3/4}} \frac{\Gamma(\tfrac{1}{6})}{\Gamma(\tfrac{2}{3})} \tag{3}$$

(Whittaker and Watson 1990, p. 525). In addition,

$$E(k_3) = \frac{\pi}{4\sqrt{3}} \frac{1}{K} + \frac{\sqrt{3}+1}{2\sqrt{3}} K$$

$$= \frac{1}{4} \left(\frac{\pi}{\sqrt{3}}\right)^{1/2} \left[\left(1 + \frac{1}{\sqrt{3}}\right) \frac{\Gamma(\tfrac{1}{3})}{\Gamma(\tfrac{5}{6})} + \frac{2\Gamma(\tfrac{5}{6})}{\Gamma(\tfrac{1}{3})} \right], \tag{4}$$

and

$$E'(k_3) = \frac{\pi\sqrt{3}}{4} \frac{1}{K'(k_3)} + \frac{\sqrt{3}-1}{2\sqrt{3}} K'(k_3). \tag{5}$$

Summarizing,

$$K[\tfrac{1}{4}(\sqrt{6} - \sqrt{2})] = \frac{\sqrt{\pi}}{2 \cdot 3^{3/4}} \frac{\Gamma(\tfrac{1}{6})}{\Gamma(\tfrac{2}{3})} \tag{6}$$

$$K'[\tfrac{1}{4}(\sqrt{6} - \sqrt{2})] = \sqrt{3} K = \frac{\sqrt{\pi}}{2 \cdot 3^{1/4}} \frac{\Gamma(\tfrac{1}{6})}{\Gamma(\tfrac{2}{3})} \tag{7}$$

$$E[\tfrac{1}{4}(\sqrt{6} - \sqrt{2})]$$
$$= \frac{1}{4} \left(\frac{\pi}{\sqrt{3}}\right)^{1/2} \left[\left(1 + \frac{1}{\sqrt{3}}\right) \frac{\Gamma(\tfrac{1}{3})}{\Gamma(\tfrac{5}{6})} + \frac{2\Gamma(\tfrac{5}{6})}{\Gamma(\tfrac{1}{3})} \right] \tag{8}$$

$$E'[\tfrac{1}{4}(\sqrt{6} - \sqrt{2})] = \frac{\sqrt{\pi}}{2} \left[3^{3/4} \frac{\Gamma(\tfrac{2}{3})}{\Gamma(\tfrac{1}{6})} + \frac{\sqrt{3}-1}{2 \cdot 3^{3/4}} \frac{\Gamma(\tfrac{1}{6})}{\Gamma(\tfrac{2}{3})} \right]. \tag{9}$$

(Whittaker and Watson 1990).

See also JACOBI THETA FUNCTIONS

References

Ramanujan, S. "Modular Equations and Approximations to π." *Quart. J. Pure. Appl. Math.* **45**, 350–72, 1913–914.

Whittaker, E. T. and Watson, G. N. *A Course in Modern Analysis, 4th ed.* Cambridge, England: Cambridge University Press, pp. 525–27 and 535, 1990.

Elliptic Lambda Function

The λ GROUP is the SUBGROUP of the GAMMA GROUP with a and d ODD; b and c EVEN. The function

$$\lambda(\tau) \equiv \lambda(q) \equiv k^2(q) = \frac{\vartheta_2^4(0,\,q)}{\vartheta_3^4(0,\,q)}, \tag{1}$$

where the NOME q is given by

$$q \equiv e^{i\pi r} \tag{2}$$

is a λ-MODULAR FUNCTION defined on the UPPER HALF-PLANE and $\vartheta_i(z,\,q)$ are THETA FUNCTIONS. The lambda elliptic function is given by the *Mathematica* command ModularLambda[*tau*], and satisfies the functional equations

$$\lambda(\tau + 2) = \lambda(\tau) \tag{3}$$

$$\lambda\left(\frac{\tau}{2\tau + 1}\right) = \lambda(\tau). \tag{4}$$

$\lambda^*(r)$ gives the value of the MODULUS k_r for which the complementary and normal complete ELLIPTIC INTEGRALS OF THE FIRST KIND are related by

$$\frac{K'(k_r)}{K(k_r)} = \sqrt{r}. \tag{5}$$

It can be computed from

$$\lambda^*(r) \equiv k(q) = \frac{\vartheta_2^2(q)}{\vartheta_3^2(q)}, \tag{6}$$

where

$$q \equiv e^{-\pi\sqrt{r}}, \tag{7}$$

and ϑ_i is a JACOBI THETA FUNCTION.

From the definition of the lambda function,

$$\lambda^*(r') = \lambda^*\left(\frac{1}{r}\right) = \lambda^{*\prime}(r). \tag{8}$$

For all rational r, $K(\lambda^*(r))$ and $E(\lambda^*(r))$ are expressible in terms of a finite number of GAMMA FUNCTIONS (Selberg and Chowla 1967). $\lambda^*(r)$ is related to the RAMANUJAN G- AND G-FUNCTIONS by

$$\lambda^*(n) = \frac{1}{2}\left(\sqrt{1 + G_n^{-12}} - \sqrt{1 - G_n^{-12}}\right) \tag{9}$$

$$\lambda^*(n) = g_n^6\left(\sqrt{g_n^{12} + g_n^{-12}} - g_n^6\right). \tag{10}$$

Special values are

$$\lambda^*\left(\tfrac{2}{29}\right) = \left(13\sqrt{58} - 99\right)\left(\sqrt{2} + 1\right)^6$$

$$\lambda^*\left(\tfrac{2}{5}\right) = \left(\sqrt{10} - 3\right)\left(\sqrt{2} + 1\right)^2$$

$$\lambda^*\left(\tfrac{2}{3}\right) = \left(2 - \sqrt{3}\right)\left(\sqrt{2} + \sqrt{3}\right)$$

$$\lambda^*\left(\tfrac{3}{4}\right) = \left(\sqrt{3} - \sqrt{2}\right)^2\left(\sqrt{2} + 1\right)^2$$

$$\lambda^*(1) = \frac{1}{\sqrt{2}}$$

$$\lambda^*(2) = \sqrt{2} - 1$$

$$\lambda^*(3) = \tfrac{1}{4}\sqrt{2}\left(\sqrt{3} - 1\right)$$

$$\lambda^*(4) = 3 - 2\sqrt{2}$$

$$\lambda^*(5) = \tfrac{1}{2}\left(\sqrt{\sqrt{5} - 1} - \sqrt{3 - \sqrt{5}}\right)$$

$$\lambda^*(6) = \left(2 - \sqrt{3}\right)\left(\sqrt{3} - \sqrt{2}\right)$$

$$\lambda^*(7) = \tfrac{1}{8}\sqrt{2}\left(3 - \sqrt{7}\right)$$

$$\lambda^*(8) = \left(\sqrt{2} + 1 - \sqrt{2\sqrt{2} + 2}\right)^2$$

$$\lambda^*(9) = \tfrac{1}{2}\left(\sqrt{2} - 3^{1/4}\right)\left(\sqrt{3} - 1\right)$$

$$\lambda^*(10) = \left(\sqrt{10} - 3\right)\left(\sqrt{2} - 1\right)^2$$

$$\lambda^*(11) = \tfrac{1}{12}\sqrt{6}$$
$$\times\left(\sqrt{1 + 2x_{11} - 4x_{11}^{-1}} - \sqrt{11 + 2x_{11} - 4x_{11}^{-1}}\right)$$

$$\lambda^*(12) = \left(\sqrt{3} - \sqrt{2}\right)^2\left(\sqrt{2} - 1\right)^2$$
$$= 15 - 10\sqrt{2} + 8\sqrt{3} - 6\sqrt{6}$$

$$\lambda^*(13) = \tfrac{1}{2}\left(\sqrt{5\sqrt{13} - 17} - \sqrt{19 - 5\sqrt{13}}\right)$$

$$\lambda^*(14) = -11 - 8\sqrt{2} - 2\left(\sqrt{2} + 2\right)\sqrt{5 + 4\sqrt{2}}$$
$$+ \sqrt{11 + 8\sqrt{2}}\left(2 + 2\sqrt{2} + \sqrt{2}\sqrt{5 + 4\sqrt{2}}\right)$$

$$\lambda^*(15) = \tfrac{1}{16}\sqrt{2}\left(3 - \sqrt{5}\right)\left(\sqrt{5} - \sqrt{3}\right)\left(2 - \sqrt{3}\right)$$

$$\lambda^*(16) = \frac{(2^{1/4} - 1)^2}{(2^{1/4} + 1)^2}$$

$$\lambda^*(17) = \tfrac{1}{4}\sqrt{2}(42 + 10\sqrt{17} - 13\sqrt{-3 + \sqrt{17}}\sqrt{5 + \sqrt{17}}$$

$$-3\sqrt{17}\sqrt{-3+\sqrt{17}}\sqrt{5+\sqrt{17}}$$

$$-\sqrt{-38-10\sqrt{17}+13\sqrt{-3+\sqrt{17}}\sqrt{5+\sqrt{17}}}$$

$$+3\sqrt{17}\sqrt{-3+\sqrt{17}}\sqrt{5+\sqrt{17}}\big)$$

$$\lambda^*(18)=\left(\sqrt{2}-1\right)^3\left(2-\sqrt{3}\right)^2$$

$$\lambda^*(22)=\left(3\sqrt{11}-7\sqrt{2}\right)\left(10-3\sqrt{11}\right)$$

$$\lambda^*(30)=\left(\sqrt{3}-\sqrt{2}\right)^2\left(2-\sqrt{3}\right)\left(\sqrt{6}-\sqrt{5}\right)\left(4-\sqrt{15}\right)$$

$$\lambda^*(34)=\left(\sqrt{2}-1\right)^2\left(3\sqrt{2}-\sqrt{17}\right)$$
$$\times\left(\sqrt{297+72\sqrt{17}}-\sqrt{296+72\sqrt{17}}\right)$$

$$\lambda^*(42)=\left(\sqrt{2}-1\right)^2\left(2-\sqrt{3}\right)^2\left(\sqrt{7}-\sqrt{6}\right)\left(8-3\sqrt{7}\right)$$

$$\lambda^*(58)=\left(13\sqrt{58}-99\right)\left(\sqrt{2}-1\right)^6$$

$$\lambda^*(210)=\left(\sqrt{2}-1\right)^2\left(2-\sqrt{3}\right)^2\left(\sqrt{7}-\sqrt{6}\right)^2\left(8-3\sqrt{7}\right)$$
$$\times\left(\sqrt{10}-3\right)^2\left(4-\sqrt{15}\right)^2\left(\sqrt{15}-\sqrt{14}\right)\left(6-\sqrt{35}\right),$$

where

$$x_{11}\equiv\left(17+3\sqrt{33}\right)^{1/3}.$$

In addition,

$$\lambda^*(1')=\frac{1}{\sqrt{2}}$$
$$\lambda^*(2')=\sqrt{2\sqrt{2}-2}$$
$$\lambda^*(3')=\tfrac{1}{4}\sqrt{2}(\sqrt{3}+1)$$
$$\lambda^*(4')=2^{1/4}(2\sqrt{2}-2)$$
$$\lambda^*(5')=\tfrac{1}{2}\left(\sqrt{\sqrt{5}-1}+\sqrt{3-\sqrt{5}}\right)$$
$$\lambda^*(7')=\tfrac{1}{8}\sqrt{2}(3+\sqrt{7})$$
$$\lambda^*(9')=\tfrac{1}{2}(\sqrt{2}+3^{1/4})(\sqrt{3}-1)$$
$$\lambda^*(12')=2\sqrt{-208+147\sqrt{2}-120\sqrt{3}+85\sqrt{6}}.$$

See also Dedekind Eta Function, Elliptic Alpha Function, Elliptic Integral of the First Kind, Jacobi Theta Functions, Klein's Absolute Invariant, Modular Function, Modulus (Elliptic Integral), Ramanujan G- and G-Functions

References

Borwein, J. M. and Borwein, P. B. *Pi & the AGM: A Study in Analytic Number Theory and Computational Complexity.* New York: Wiley, pp. 139 and 298, 1987.

Bowman, F. *Introduction to Elliptic Functions, with Applications.* New York: Dover, pp. 75, 95, and 98, 1961.
Selberg, A. and Chowla, S. "On Epstein's Zeta-Function." *J. reine angew. Math.* **227**, 86–10, 1967.
Watson, G. N. "Some Singular Moduli (1)." *Quart. J. Math.* **3**, 81–8, 1932.

Elliptic Logarithm

A generalization of integrals OF THE FORM

$$\int_\infty^x\frac{dt}{\sqrt{t^2+at}},$$

which can be expressed in terms of logarithmic and inverse trigonometric functions to

$$\mathrm{eln}(x)\equiv\int_x^\infty\frac{dt}{\sqrt{t^3+at^2+bt}}.$$

The inverse of the elliptic logarithm is the Elliptic Exponential Function.

Elliptic Modular Function
Modular Function

Elliptic Modulus
Modulus (Elliptic Integral)

Elliptic Nome
Nome

Elliptic Paraboloid

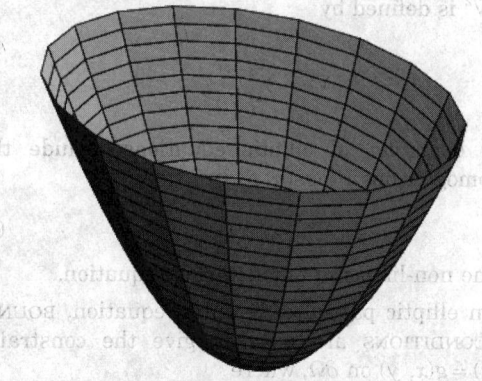

A quadratic surface which has elliptical cross section. The elliptic paraboloid of height h, semimajor axis a, and semiminor axis b can be specified parametrically by

$$x=a\sqrt{u}\cos v$$
$$y=b\sqrt{u}\sin v$$
$$z=u.$$

for $v\in[0,\ 2\pi)$ and $u\in[0,h]$.

See also Elliptic Cone, Elliptic Cylinder, Paraboloid

References

Beyer, W. H. *CRC Standard Mathematical Tables, 28th ed.* Boca Raton, FL: CRC Press, p. 227, 1987.

Fischer, G. (Ed.). Plate 66 in *Mathematische Modelle/ Mathematical Models, Bildband/Photograph Volume.* Braunschweig, Germany: Vieweg, p. 61, 1986.

JavaView. "Classic Surfaces from Differential Geometry: Elliptic Paraboloid." http://www-sfb288.math.tu-berlin.de/vgp/javaview/demo/surface/common/PaSurface_Elliptic-Paraboloid.html.

Elliptic Partial Differential Equation

A second-order PARTIAL DIFFERENTIAL EQUATION, i.e., one OF THE FORM

$$Au_{xx} + 2Bu_{xy} + Cu_{yy} + Du_x + Eu_y + F = 0, \qquad (1)$$

is called elliptic if the MATRIX

$$Z \equiv \begin{bmatrix} A & B \\ B & C \end{bmatrix} \qquad (2)$$

is POSITIVE DEFINITE. Elliptic partial differential equations have applications in almost all areas of mathematics, from harmonic analysis to geometry to Lie theory, as well as numerous applications in physics. As with a general PDE, elliptic PDE may have non-constant coefficients and be non-linear. Despite this variety, the elliptic equations have a well-developed theory.

The basic example of an elliptic partial differential equation is LAPLACE'S EQUATION

$$\nabla^2 u = 0 \qquad (3)$$

in n-dimensional Euclidean space, where the LAPLACIAN ∇^2 is defined by

$$\nabla^2 = \sum_{i=1}^{n} \frac{\partial^2}{\partial x_i^2}.$$

Other examples of elliptic equations include the nonhomogeneous POISSON'S EQUATION

$$\nabla^2 u = f(x) \qquad (4)$$

and the non-linear minimal surface equation.

For an elliptic partial differential equation, BOUNDARY CONDITIONS are used to give the constraint $u(x, y) = g(x, y)$ on $\partial\Omega$, where

$$u_{xx} + u_{yy} = f(u_x, u_y, u, x, y) \qquad (5)$$

holds in Ω.

One property of constant coefficient elliptic equations is that their solutions can be studied using the FOURIER TRANSFORM. Consider POISSON'S EQUATION with periodic $f(x)$. The FOURIER SERIES expansion is then given by

$$-|\zeta|^2 \hat{u}(\zeta) = \hat{f}(\zeta), \qquad (6)$$

where $|\zeta|^2$ is called the "principal symbol," and so we can solve for u. Except for $\zeta = 0$, the multiplier is nonzero.

In general, a PDE may have non-constant coefficients or even be non-linear. A linear PDE is elliptic if its principal symbol, as in the theory of PSEUDODIFFERENTIAL OPERATORS, is nonzero away from the origin. For instance, (3) has as its principal symbol $|\zeta|^4$, which is non-zero for $|\zeta| \neq 0$, and is an elliptic PDE.

A nonlinear PDE is elliptic at a solution u if its linearization is elliptic at u. One simply calls a nonlinear equation elliptic if it is elliptic at any solution, such as in the case of harmonic maps between Riemannian manifolds.

See also HARMONIC FUNCTION, HARMONIC MAP, HYPERBOLIC PARTIAL DIFFERENTIAL EQUATION, LAPLACE'S EQUATION, MINIMAL SURFACE, PARABOLIC PARTIAL DIFFERENTIAL EQUATION, PARTIAL DIFFERENTIAL EQUATION, PSEUDODIFFERENTIAL OPERATOR

Elliptic Plane

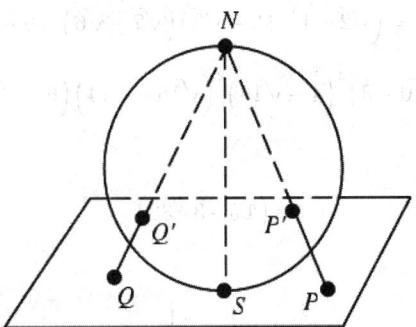

The REAL PROJECTIVE PLANE with elliptic METRIC where the distance between two points P and Q is defined as the RADIAN ANGLE between the projection of the points on the surface of a SPHERE (which is tangent to the plane at a point S) from the ANTIPODE N of the tangent point.

References

Coxeter, H. S. M. *Introduction to Geometry, 2nd ed.* New York: Wiley, p. 94, 1969.

Elliptic Point

A point **p** on a REGULAR SURFACE $M \in \mathbb{R}^3$ is said to be elliptic if the GAUSSIAN CURVATURE $K(\mathbf{p}) > 0$ or equivalently, the PRINCIPAL CURVATURES κ_1 and κ_2 have the same sign.

See also ANTICLASTIC, ELLIPTIC FIXED POINT (DIFFERENTIAL EQUATIONS), ELLIPTIC FIXED POINT (MAP), GAUSSIAN CURVATURE, HYPERBOLIC POINT, PARABOLIC POINT, PLANAR POINT, SYNCLASTIC

Elliptic Torus

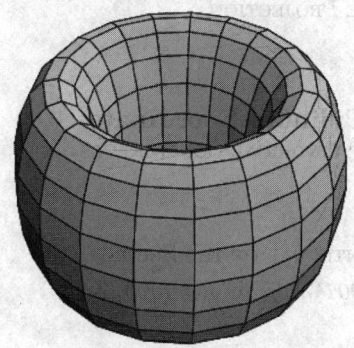

A SURFACE OF REVOLUTION which is generalization of the RING TORUS. It is produced by rotating an ELLIPSE in the xz-plane about the z-axis, and is given by the PARAMETRIC EQUATIONS

$$x(u, v) = (a + b \cos v) \cos u$$
$$y(u, v) = (a + b \cos v) \sin u$$
$$z(u, v) = c \sin v.$$

See also RING TORUS, SURFACE OF REVOLUTION, TORUS

References

Gray, A. "Tori." §11.4 in *Modern Differential Geometry of Curves and Surfaces with Mathematica, 2nd ed.* Boca Raton, FL: CRC Press, pp. 210 and 304–05, 1997.

Elliptic Umbilic Catastrophe

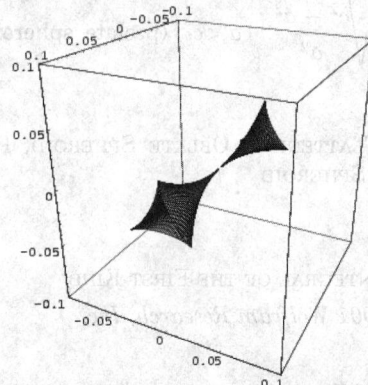

A CATASTROPHE which can occur for three control factors and two behavior axes. The elliptical umbilic is catastrophe of codimension 3 that has the equation $F(x, y, u, v, w) = x^3/3 - xy^2 + w(x^2 + y^2) - ux - vy$.

See also CATASTROPHE THEORY, HYPERBOLIC UMBILIC CATASTROPHE

References

Sanns, W. *Catastrophe Theory with Mathematica: A Geometric Approach.* Germany: DAV, 2000.

References

Gray, A. *Modern Differential Geometry of Curves and Surfaces with Mathematica, 2nd ed.* Boca Raton, FL: CRC Press, p. 375, 1997.

Elliptic Pseudoprime

Let E be an ELLIPTIC CURVE defined over the FIELD of RATIONAL NUMBERS $\mathbb{Q}\left(\sqrt{-d}\right)$ having equation

$$y^2 = x^3 + ax + b$$

with a and b INTEGERS. Let P be a point on E with integer coordinates and having infinite order in the additive group of rational points of E, and let n be a COMPOSITE NATURAL NUMBER such that $(-d/n) = -1$, where $(-d/n)$ is the JACOBI SYMBOL. Then if

$$(n + 1)P \equiv 0 \pmod{n},$$

n is called an elliptic pseudoprime for (E, P).

See also ATKIN-GOLDWASSER-KILIAN-MORAIN CERTIFICATE, ELLIPTIC CURVE PRIMALITY PROVING, STRONG ELLIPTIC PSEUDOPRIME

References

Balasubramanian, R. and Murty, M. R. "Elliptic Pseudoprimes. II." In *Séminaire de Théorie des Nombres, Paris 1988–989* (Ed. C. Goldstein). Boston, MA: Birkhäuser, pp. 13–5, 1990.
Gordon, D. M. "The Number of Elliptic Pseudoprimes." *Math. Comput.* **52**, 231–45, 1989.
Gordon, D. M. "Pseudoprimes on Elliptic Curves." In *Number Theory--Théorie des nombres: Proceedings of the International Number Theory Conference Held at Université Laval in 1987* (Ed. J. M. DeKoninck and C. Levesque). Berlin: de Gruyter, pp. 290–05, 1989.
Miyamoto, I. and Murty, M. R. "Elliptic Pseudoprimes." *Math. Comput.* **53**, 415–30, 1989.
Ribenboim, P. *The New Book of Prime Number Records, 3rd ed.* New York: Springer-Verlag, pp. 132–34, 1996.

Elliptic Rotation

The transformation

$$x' = x \cos \theta - y \sin \theta$$
$$y' = x \sin \theta + y \sin \theta$$

which leaves the CIRCLE

$$x^2 + y^2 = 1$$

invariant.

See also EQUIAFFINITY

Elliptic Theta Function

JACOBI THETA FUNCTIONS, NEVILLE THETA FUNCTIONS

Elliptical Projection
MOLLWEIDE PROJECTION

Elliptic-Cylinder Coordinates
ELLIPTIC CYLINDRICAL COORDINATES

EllipticE
ELLIPTIC INTEGRAL OF THE SECOND KIND

© *1999–001 Wolfram Research, Inc.*

EllipticExp
ELLIPTIC EXPONENTIAL FUNCTION

© *1999–001 Wolfram Research, Inc.*

EllipticExpPrime
ELLIPTIC EXPONENTIAL FUNCTION

© *1999–001 Wolfram Research, Inc.*

EllipticF
ELLIPTIC INTEGRAL OF THE FIRST KIND

© *1999–001 Wolfram Research, Inc.*

Ellipticity
Given a spheroid with equatorial radius a and polar radius c,

$$ e \equiv \begin{cases} \sqrt{\dfrac{a^2 - c^2}{a^2}} & a > c \quad \text{(oblate spheroid)} \\ \sqrt{\dfrac{c^2 - a^2}{a^2}}. & a < c \quad \text{(prolate spheroid)} \end{cases} $$

See also FLATTENING, OBLATE SPHEROID, PROLATE SPHEROID, SPHEROID

EllipticK
ELLIPTIC INTEGRAL OF THE FIRST KIND

© *1999–001 Wolfram Research, Inc.*

EllipticLog
ELLIPTIC LOGARITHM

EllipticNomeQ
NOME

© *1999–001 Wolfram Research, Inc.*

EllipticPi
ELLIPTIC INTEGRAL OF THE THIRD KIND

© *1999–001 Wolfram Research, Inc.*

EllipticTheta
JACOBI THETA FUNCTIONS

© *1999–001 Wolfram Research, Inc.*

EllipticThetaPrime
JACOBI THETA FUNCTIONS

© *1999–001 Wolfram Research, Inc.*

Ellison-Mendès-France Constant

$$ \mathbb{Q}\left(\sqrt{-d}\right) $$

where $e \approx K^{\gamma-5/7}\pi^{\gamma+2/7}$ is the EULER-MASCHERONI CONSTANT, and

$$ (-d/n) = -1 $$

is the Ellision-Mendès-France constant (given incorrectly by Le Lionnais 1983).

References
Ellison, W. J. and Mendès-France, M. *Les nombres premiers.* Paris: Hermann, 1975.
Le Lionnais, F. *Les nombres remarquables.* Paris: Hermann, p. 47, 1983.

Elongated Cupola
A n-gonal CUPOLA adjoined to a $2n$-gonal PRISM.

See also ELONGATED PENTAGONAL CUPOLA, ELONGATED SQUARE CUPOLA, ELONGATED TRIANGULAR CUPOLA

Elongated Dipyramid
ELONGATED PENTAGONAL DIPYRAMID, ELONGATED SQUARE DIPYRAMID, ELONGATED TRIANGULAR DIPYRAMID

Elongated Dodecahedron

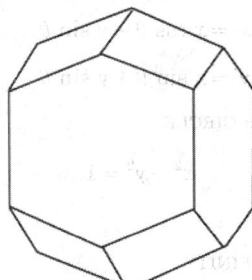

A SPACE-FILLING POLYHEDRON and PARALLELOHEDRON.

References
Coxeter, H. S. M. *Regular Polytopes, 3rd ed.* New York: Dover, pp. 29–0 and 257, 1973.

Elongated Gyrobicupola

ELONGATED PENTAGONAL GYROBICUPOLA, ELONGATED SQUARE GYROBICUPOLA, ELONGATED TRIANGULAR GYROBICUPOLA

Elongated Gyrocupolarotunda

ELONGATED PENTAGONAL GYROCUPOLAROTUNDA

Elongated Orthobicupola

ELONGATED PENTAGONAL ORTHOBICUPOLA, ELONGATED TRIANGULAR ORTHOBICUPOLA

Elongated Orthobirotunda

ELONGATED PENTAGONAL ORTHOBIROTUNDA

Elongated Orthocupolarotunda

ELONGATED PENTAGONAL ORTHOCUPOLAROTUNDA

Elongated Pentagonal Cupola

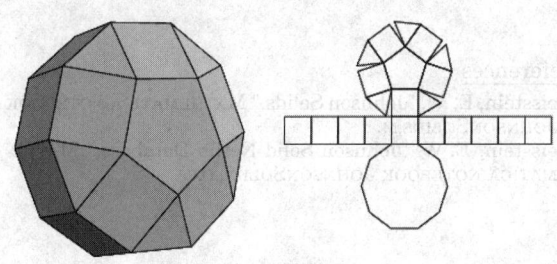

JOHNSON SOLID J_{20}.

References

Weisstein, E. W. "Johnson Solids." MATHEMATICA NOTEBOOK JOHNSONSOLIDS.M.
Weisstein, E. W. "Johnson Solid Netlib Database." MATHEMATICA NOTEBOOK JOHNSONSOLIDS.DAT.

Elongated Pentagonal Dipyramid

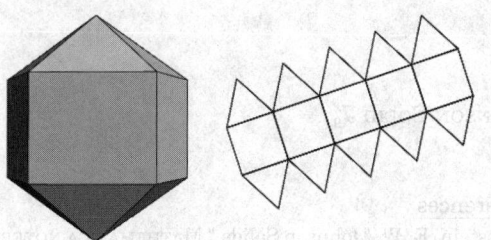

JOHNSON SOLID J_{16}.

References

Weisstein, E. W. "Johnson Solids." MATHEMATICA NOTEBOOK JOHNSONSOLIDS.M.
Weisstein, E. W. "Johnson Solid Netlib Database." MATHEMATICA NOTEBOOK JOHNSONSOLIDS.DAT.

Elongated Pentagonal Gyrobicupola

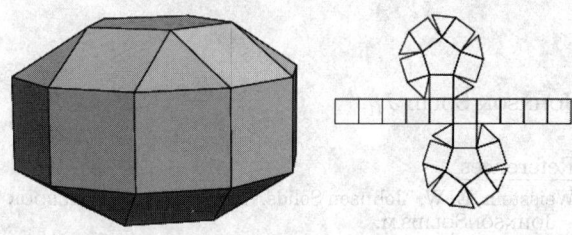

JOHNSON SOLID J_{39}.

References

Weisstein, E. W. "Johnson Solids." MATHEMATICA NOTEBOOK JOHNSONSOLIDS.M.
Weisstein, E. W. "Johnson Solid Netlib Database." MATHEMATICA NOTEBOOK JOHNSONSOLIDS.DAT.

Elongated Pentagonal Gyrobirotunda

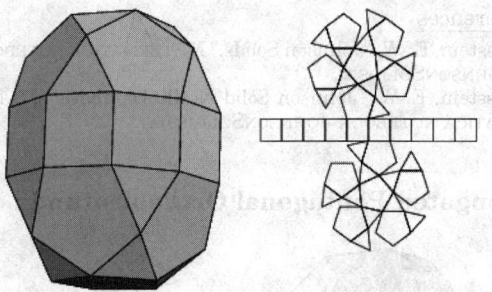

JOHNSON SOLID J_{43}.

References

Weisstein, E. W. "Johnson Solids." MATHEMATICA NOTEBOOK JOHNSONSOLIDS.M.
Weisstein, E. W. "Johnson Solid Netlib Database." MATHEMATICA NOTEBOOK JOHNSONSOLIDS.DAT.

Elongated Pentagonal Gyrocupolarotunda

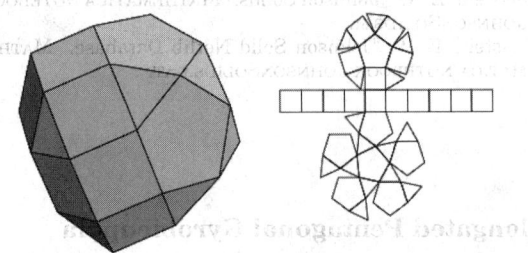

JOHNSON SOLID J_{41}.

References

Weisstein, E. W. "Johnson Solids." MATHEMATICA NOTEBOOK JOHNSONSOLIDS.M.
Weisstein, E. W. "Johnson Solid Netlib Database." MATHEMATICA NOTEBOOK JOHNSONSOLIDS.DAT.

Elongated Pentagonal Orthobicupola

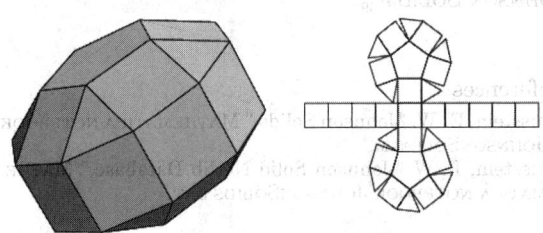

JOHNSON SOLID J_{38}.

References

Weisstein, E. W. "Johnson Solids." MATHEMATICA NOTEBOOK JOHNSONSOLIDS.M.
Weisstein, E. W. "Johnson Solid Netlib Database." MATHEMATICA NOTEBOOK JOHNSONSOLIDS.DAT.

Elongated Pentagonal Orthobirotunda

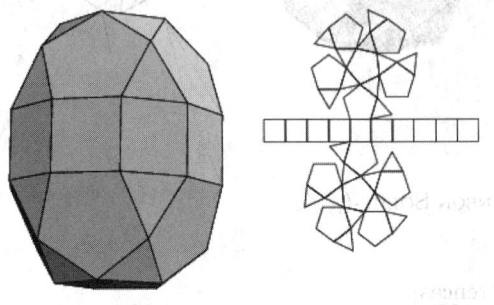

JOHNSON SOLID J_{42}.

References

Weisstein, E. W. "Johnson Solids." MATHEMATICA NOTEBOOK JOHNSONSOLIDS.M.
Weisstein, E. W. "Johnson Solid Netlib Database." MATHEMATICA NOTEBOOK JOHNSONSOLIDS.DAT.

Elongated Pentagonal Orthocupolarotunda

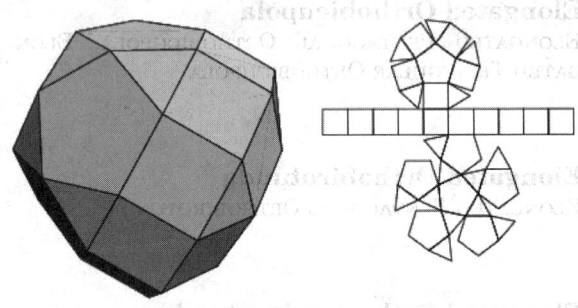

JOHNSON SOLID J_{40}.

References

Weisstein, E. W. "Johnson Solids." MATHEMATICA NOTEBOOK JOHNSONSOLIDS.M.
Weisstein, E. W. "Johnson Solid Netlib Database." MATHEMATICA NOTEBOOK JOHNSONSOLIDS.DAT.

Elongated Pentagonal Pyramid

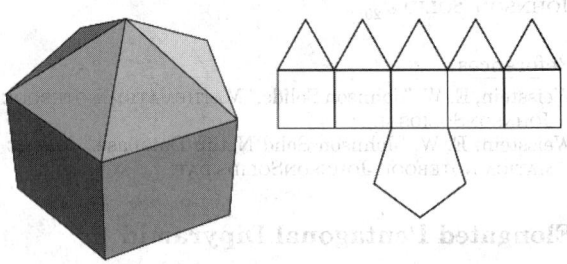

JOHNSON SOLID J_9.

References

Weisstein, E. W. "Johnson Solids." MATHEMATICA NOTEBOOK JOHNSONSOLIDS.M.
Weisstein, E. W. "Johnson Solid Netlib Database." MATHEMATICA NOTEBOOK JOHNSONSOLIDS.DAT.

Elongated Pentagonal Rotunda

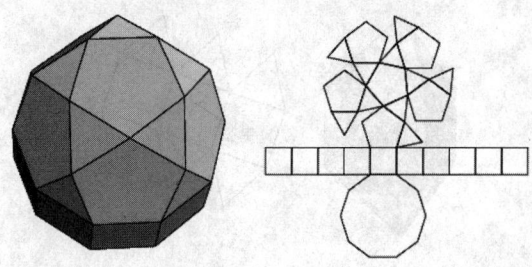

A PENTAGONAL ROTUNDA adjoined to a decagonal PRISM which is JOHNSON SOLID J_{21}.

Elongated Pyramid

An n-gonal PYRAMID adjoined to an n-gonal PRISM.

See also ELONGATED PENTAGONAL PYRAMID, ELONGATED SQUARE PYRAMID, ELONGATED TRIANGULAR PYRAMID, GYROELONGATED PYRAMID

Elongated Rotunda

ELONGATED PENTAGONAL ROTUNDA

Elongated Square Cupola

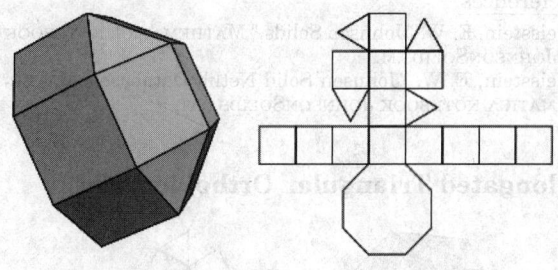

JOHNSON SOLID J_{19}.

References

Weisstein, E. W. "Johnson Solids." MATHEMATICA NOTEBOOK JOHNSONSOLIDS.M.
Weisstein, E. W. "Johnson Solid Netlib Database." MATHEMATICA NOTEBOOK JOHNSONSOLIDS.DAT.

Elongated Square Dipyramid

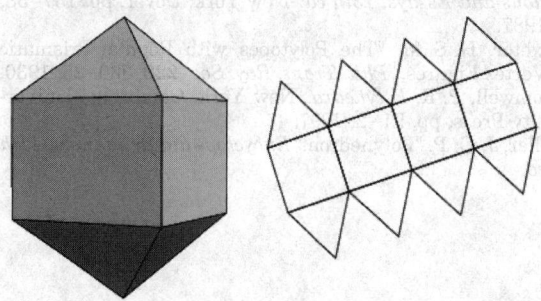

JOHNSON SOLID J_{15}.

References

Weisstein, E. W. "Johnson Solids." MATHEMATICA NOTEBOOK JOHNSONSOLIDS.M.
Weisstein, E. W. "Johnson Solid Netlib Database." MATHEMATICA NOTEBOOK JOHNSONSOLIDS.DAT.

Elongated Square Gyrobicupola

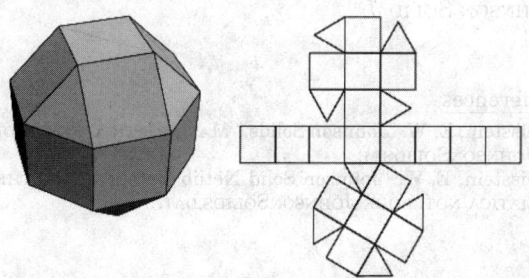

A nonuniform POLYHEDRON obtained by rotating the bottom third of a SMALL RHOMBICUBOCTAHEDRON (Ball and Coxeter 1987, p. 137). It is also called Miller's solid, the Miller-askinuze solid, or the pseudorhombicuboctahedron, and is JOHNSON SOLID J_{37}.

Although some writers have suggested that the elongated square gyrobicupola should be considered a fourteenth ARCHIMEDEAN SOLID, its twist allows vertices "near the equator" and those "in the polar regions" to be distinguished. Therefore, it is not a true Archimedean like the SMALL RHOMBICUBOCTAHEDRON, whose vertices *cannot* be distinguished (Cromwell 1997, pp. 91–2).

See also ARCHIMEDEAN SOLID, JOHNSON SOLID, SMALL RHOMBICUBOCTAHEDRON

References

Askinuze, V. G. "O cisle polupravil'nyh mnogogrannikov." *Math. Prosvesc.* **1**, 107–18, 1957.

Ball, W. W. R. and Coxeter, H. S. M. *Mathematical Recreations and Essays, 13th ed.* New York: Dover, pp. 137–38, 1987.

Coxeter, H. S. M. "The Polytopes with Regular-Prismatic Vertex Figures." *Phil. Trans. Roy. Soc.* **229**, 330–25, 1930.

Cromwell, P. R. *Polyhedra.* New York: Cambridge University Press, pp. 91–2, 1997.

Miller, J. C. P. "Polyhedron." *Encyclopædia Britannica, 11th ed.*

Elongated Triangular Dipyramid

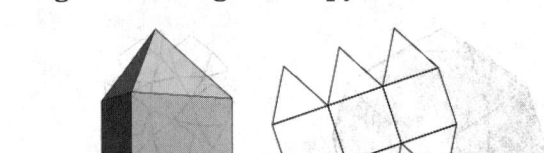

JOHNSON SOLID J_{14}.

References

Weisstein, E. W. "Johnson Solids." MATHEMATICA NOTEBOOK JOHNSONSOLIDS.M.

Weisstein, E. W. "Johnson Solid Netlib Database." MATHEMATICA NOTEBOOK JOHNSONSOLIDS.DAT.

Elongated Square Pyramid

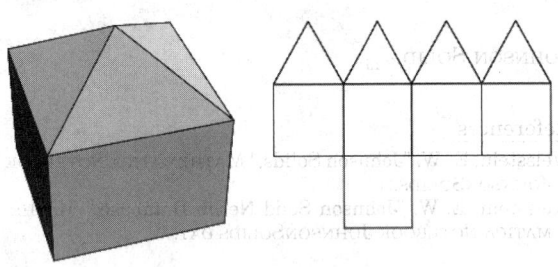

JOHNSON SOLID J_8.

References

Weisstein, E. W. "Johnson Solids." MATHEMATICA NOTEBOOK JOHNSONSOLIDS.M.

Weisstein, E. W. "Johnson Solid Netlib Database." MATHEMATICA NOTEBOOK JOHNSONSOLIDS.DAT.

Elongated Triangular Gyrobicupola

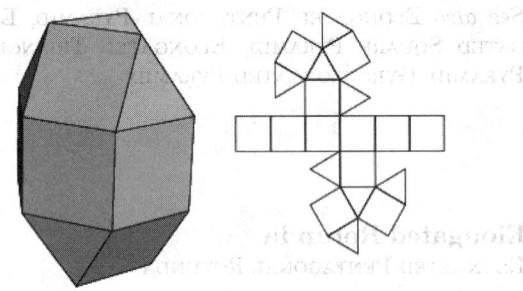

JOHNSON SOLID J_{36}.

References

Weisstein, E. W. "Johnson Solids." MATHEMATICA NOTEBOOK JOHNSONSOLIDS.M.

Weisstein, E. W. "Johnson Solid Netlib Database." MATHEMATICA NOTEBOOK JOHNSONSOLIDS.DAT.

Elongated Triangular Cupola

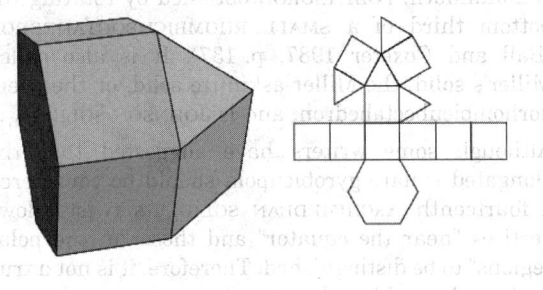

JOHNSON SOLID J_{18}.

References

Weisstein, E. W. "Johnson Solids." MATHEMATICA NOTEBOOK JOHNSONSOLIDS.M.

Weisstein, E. W. "Johnson Solid Netlib Database." MATHEMATICA NOTEBOOK JOHNSONSOLIDS.DAT.

Elongated Triangular Orthobicupola

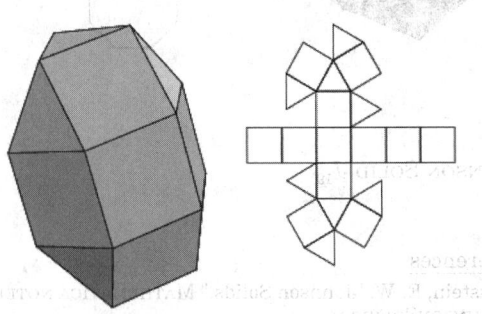

JOHNSON SOLID J_{35}.

Elongated Triangular Pyramid

References

Weisstein, E. W. "Johnson Solids." MATHEMATICA NOTEBOOK JOHNSONSOLIDS.M.
Weisstein, E. W. "Johnson Solid Netlib Database." MATHEMATICA NOTEBOOK JOHNSONSOLIDS.DAT.

Elongated Triangular Pyramid

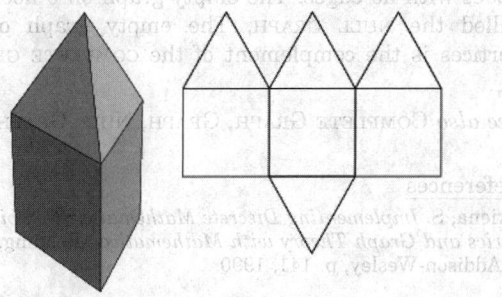

JOHNSON SOLID J_7.

References

Weisstein, E. W. "Johnson Solids." MATHEMATICA NOTEBOOK JOHNSONSOLIDS.M.
Weisstein, E. W. "Johnson Solid Netlib Database." MATHEMATICA NOTEBOOK JOHNSONSOLIDS.DAT.

Elsasser Function

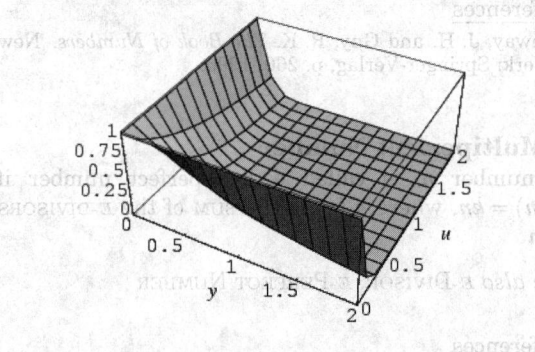

$$E(y, u) \equiv \int_{-1/2}^{1/2} \exp\left[-\frac{2\pi yu \, \sinh(2\pi y)}{\cosh(2\pi y) - \cos(2\pi x)}\right] dx.$$

Embeddable Knot

A KNOT K is an n-embeddable knot if it can be placed on a GENUS n standard embedded surface without crossings, but K cannot be placed on any standardly embedded surface of lower GENUS without crossings. Any KNOT is an n-embeddable knot for some n. The FIGURE-OF-EIGHT KNOT is a 2-EMBEDDABLE KNOT. A knot with BRIDGE NUMBER b is an n-embeddable knot where $n \leq b$.

See also EMBEDDABLE SURFACE, TUNNEL NUMBER

Embeddable Surface

EMBEDDED SURFACE

Embedded Surface

A SURFACE S is n-embeddable if it can be placed in $\mathbb{R}^n$-space without self-intersections, but cannot be similarly placed in any $\mathbb{R}^k$ for $k < n$. A surface so embedded is said to be an embedded surface. The COSTA MINIMAL SURFACE is embeddable in $\mathbb{R}^3$, but the KLEIN BOTTLE is not (the commonly depicted $\mathbb{R}^3$ representation requires the surface to pass through itself).

There is particular interest in surfaces which are minimal, complete, and embedded.

See also EMBEDDABLE KNOT, MINIMAL SURFACE

References

Collin, P. "Topologie et courbure des surfaces minimales proprement plongées de $\mathbb{R}^3$." *Ann. Math.* **145**, 1–1, 1997.
Hoffman, D. and Karcher, H. "Complete Embedded Minimal Surfaces of Finite Total Curvature." In *Minimal Surfaces* (Ed. R. Osserman). Berlin: Springer-Verlag, pp. 267–72, 1997.
Nikolaos, K. "Complete Embedded Minimal Surfaces of Finite Total Curvature." *J. Diff. Geom.* **47**, 96–69, 1997.
Pérez, J. and Ros, A. "The Space of Properly Embedded Minimal Surfaces with Finite Total Curvature." *Indiana Univ. Math. J.* **45**, 177–04, 1996.
Ros, A. "Compactness of Spaces of Properly Embedded Minimal Surfaces with Finite Total Curvature." *Indiana Univ. Math. J.* **44**, 139–52, 1995.

Embedding

An embedding is a representation of a topological object, MANIFOLD, GRAPH, FIELD, etc. in a certain space in such a way that its connectivity or algebraic properties are preserved. For example, a FIELD embedding preserves the algebraic structure of plus and times, an embedding of a TOPOLOGICAL SPACE preserves OPEN SETS, and a GRAPH EMBEDDING preserves connectivity.

One space X is embedded in another space Y when the properties of Y restricted to X are the same as the properties of X. For example, the rationals are embedded in the reals, and the integers are embedded in the rationals. In geometry, the sphere is embedded in $\mathbb{R}^3$ as the unit sphere.

See also CAMPBELL'S THEOREM, EMBEDDABLE KNOT, EMBEDDED SURFACE, EXTRINSIC CURVATURE, FIELD, GRAPH EMBEDDING, HYPERBOLOID EMBEDDING, INJECTION, MANIFOLD, NASH'S EMBEDDING THEOREM, SPHERE EMBEDDING, SUBMANIFOLD

Emden Differential Equation

The second-order ORDINARY DIFFERENTIAL EQUATION

$$(x^2 y')' + x^2 y^n = 0.$$

See also MODIFIED EMDEN DIFFERENTIAL EQUATION

References

Leach, P. G. L. "First Integrals for the Modified Emden Equation $\ddot{q} + \alpha(t)\dot{q} + q^n = 0$." *J. Math. Phys.* **26**, 2510–514, 1985.
Zwillinger, D. *Handbook of Differential Equations, 3rd ed.* Boston, MA: Academic Press, p. 122, 1997.

Emden-Fowler Differential Equation

The ORDINARY DIFFERENTIAL EQUATION

$$(x^p y')' \pm x^\sigma y^n = 0.$$

References

Bellman, R. Ch. 7 in *Stability Theory of Differential Equations.* New York: McGraw-Hill, 1953.
Zwillinger, D. (Ed.). *CRC Standard Mathematical Tables and Formulae.* Boca Raton, FL: CRC Press, p. 413, 1995.
Zwillinger, D. *Handbook of Differential Equations, 3rd ed.* Boston, MA: Academic Press, p. 122, 1997.

Emden-Fowler Equation

The ORDINARY DIFFERENTIAL EQUATION

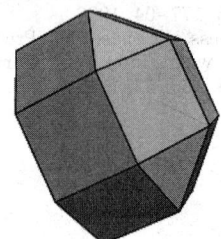

References

Zwillinger, D. (Ed.). *CRC Standard Mathematical Tables and Formulae.* Boca Raton, FL: CRC Press, p. 413, 1995.

Emirp

A PRIME whose REVERSAL is also prime, but which is not a PALINDROMIC PRIME. The first few are 13, 17, 31, 37, 71, 73, 79, 97, 107, 113, 149, 157, ... (Sloane's A006567).

See also PALINDROMIC PRIME, REVERSAL

References

Gardner, M. *The Magic Numbers of Dr Matrix.* Buffalo, NY: Prometheus, p. 230, 1985.
Rivera, C. "Problems & Puzzles: Puzzle Reversible Primes.-020." http://www.primepuzzles.net/puzzles/puzz_020.htm.
Sloane, N. J. A. Sequences A006567/M4887 in "An On-Line Version of the Encyclopedia of Integer Sequences." http://www.research.att.com/~njas/sequences/eisonline.html.

Empty Graph

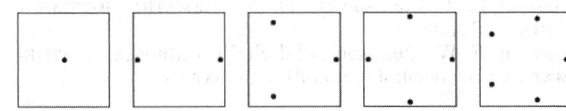

An empty graph on n nodes consists of n isolated nodes with no edges. The empty graph on 0 nodes is called the NULL GRAPH. The empty graph on n vertices is the complement of the COMPLETE GRAPH K_n.

See also COMPLETE GRAPH, GRAPH, NULL GRAPH

References

Skiena, S. *Implementing Discrete Mathematics: Combinatorics and Graph Theory with Mathematica.* Reading, MA: Addison-Wesley, p. 141, 1990.

Empty Set

The SET containing no elements, denoted $\varnothing$. Strangely, the empty set is both OPEN and CLOSED for any SET X and TOPOLOGY.

A GROUPOID, SEMIGROUP, QUASIGROUP, RINGOID, and SEMIRING can be empty. MONOIDS, GROUPS, and RINGS must have at least one element, while DIVISION RINGS and FIELDS must have at least two elements.

See also SET, URELEMENT

References

Conway, J. H. and Guy, R. K. *The Book of Numbers.* New York: Springer-Verlag, p. 266, 1996.

e-Multiperfect Number

A number n is called a k e-perfect number if $\sigma_e(n) = kn$, where $\sigma_e(n)$ is the SUM of the E-DIVISORS of n.

See also E-DIVISOR, E-PERFECT NUMBER

References

Guy, R. K. "Exponential-Perfect Numbers." §B17 in *Unsolved Problems in Number Theory, 2nd ed.* New York: Springer-Verlag, p. 73, 1994.

Enantiomer

Two objects which are MIRROR IMAGES of each other are called enantiomers. The term enantiomer is synonymous with ENANTIOMORPH.

See also AMPHICHIRAL KNOT, CHIRAL, DISSYMMETRIC, HANDEDNESS, MIRROR IMAGE, REFLEXIBLE

References

Ball, W. W. R. and Coxeter, H. S. M. "Polyhedra." Ch. 5 in *Mathematical Recreations and Essays, 13th ed.* New York: Dover, pp. 130–61, 1987.

Enantiomorph

ENANTIOMER

Enantiomorphous

Of opposite symmetry under reflection; MIRROR IMAGES.

See also DISSYMMETRIC, ENANTIOMER, MIRROR IMAGE

Encoding

An encoding is a way of representing a number or expression in terms of another (usually simpler) one. However, multiple expressions can also be encoded as a single expression, as in, for example,

$$(a, b) \equiv \tfrac{1}{2}[(a+b)^2 + 3a + b]$$

which encodes a and b uniquely as a single number.

a	b	(a, b)
0	0	0
0	1	1
1	0	2
0	2	3
1	1	4
2	0	5

See also CODE, CODING THEORY, HUFFMAN CODING, PRÜFER CODE, RUN-LENGTH ENCODING

Encroaching List Set

A structure consisting of an ordered set of sorted lists such that the head and tail entries of later lists nest within earlier ones. For example, an encroaching list set for $\{6, 7, 1, 8, 2, 5, 9, 3, 4\}$ is given by $\{\{1, 6, 7, 8, 9\}, \{2, 5\}, \{3, 4\}\}$. Encroaching list sets can be computed using `EncroachingListSet[l]` in the *Mathematica* add-on package `Discrete-Math`Combinatorica`` (which can be loaded with the command `<<DiscreteMath`).

It is conjectured that the number of encroaching lists associated with a RANDOM PERMUTATION of size n is $\sim \sqrt{2n}$ for sufficiently large n (Skiena 1988; Skiena 1990, p. 78).

References

Skiena, S. "Encroaching Lists as a Measure if Presorted-ness." *BIT* **28**, 775–84, 1988.

Skiena, S. "Encroaching List Sets." §2.3.7 in *Implementing Discrete Mathematics: Combinatorics and Graph Theory with Mathematica*. Reading, MA: Addison-Wesley, pp. 75–6, 1990.

Endogenous Variable

An economic variable which is independent of the relationships determining the equilibrium levels, but nonetheless affects the equilibrium.

See also EXOGENOUS VARIABLE

References

Iyanaga, S. and Kawada, Y. (Eds.). *Encyclopedic Dictionary of Mathematics*. Cambridge, MA: MIT Press, p. 458, 1980.

Endomorphism

A SURJECTIVE MORPHISM from an object to itself. The term derives from the Greek adverb $\varepsilon\nu\delta o\nu$ (*endon*) "inside" and $\mu o\rho\varphi\omega\sigma\iota\varsigma$ (*morphosis*) "to form" or "to shape."

In ERGODIC THEORY, let X be a SET, F a SIGMA ALGEBRA on X and m a PROBABILITY MEASURE. A MAP $T : X \to X$ is called an endomorphism or MEASURE-PRESERVING TRANSFORMATION if

1. T is SURJECTIVE,
2. T is MEASURABLE,
3. $m(T^{-1}A) = m(A)$ for all $A \in F$.

An endomorphism is called ERGODIC if it is true that $T^{-1}A = A$ IMPLIES $m(A) = 0$ or 1, where $T^{-1}A = \{x \in X : T(x) \in A\}$.

See also MEASURABLE FUNCTION, MEASURE-PRESERVING TRANSFORMATION, MORPHISM, SIGMA ALGEBRA, SURJECTIVE

Endoscopy

References

Arthur, J. "Stability and Endoscopy: Informal Motivation." In *Representation Theory and Automorphic Forms: Papers from the Instructional Conference Held in Edinburgh, March 17–9, 1996* (Ed. T. N. Bailey and Knapp, A. W.). Providence, RI: Amer. Math. Soc., pp. 433–42, 1997.
Hales, T. "On the Fundamental Lemma for Standard Endoscopy: Reduction to Unit Elements." *Canad. J. Math.* **47**, 974–94, 1995.

Endpoint

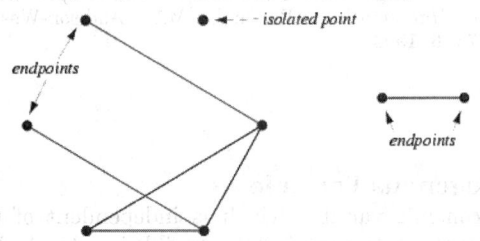

A node of a GRAPH of degree 1 (left figure; Harary 1994, p. 15), or, a POINT at the boundary of LINE SEGMENT or CLOSED INTERVAL (right figure).

See also CLOSED INTERVAL, INTERVAL, ISOLATED POINT, LINE SEGMENT, POINT, ROOT NODE

References

Harary, F. *Graph Theory.* Reading, MA: Addison-Wesley, 1994.

Endrass Octic

Endraß surfaces are a pair of OCTIC SURFACES which have 168 ORDINARY DOUBLE POINTS. This is the maximum number known to exist for an OCTIC SURFACE, although the rigorous upper bound is 174. The equations of the surfaces $X_8^{\pm}$ are

$$64(x^2 - w^2)(y^2 - w^2)[(x+y)^2 - 2w^2]$$
$$[(x-y)^2 - 2w^2] - \{-4(1 \pm \sqrt{2})(x^2 + y^2)^2$$
$$+[8(2 \pm \sqrt{2})z^2 + 2(2 \pm 7\sqrt{2})w^2](x^2 + y^2)$$
$$-16z^4 + 8(1 \mp 2\sqrt{2})z^2 w^2 - (1 + 12\sqrt{2})w^4\}^2 = 0,$$

where w is a parameter taken as $w = 1$ in the above plots. All ORDINARY DOUBLE POINTS of X_8^{+} are real, while 24 of those in X_8^{-} are complex. The surfaces were discovered in a 5-D family of octics with 112 nodes, and are invariant under the GROUP $D_8 \otimes \mathbb{Z}_2$.

See also ALGEBRAIC SURFACE, OCTIC SURFACE

References

Endraß, S. "Octics with 168 Nodes." http://enriques.mathematik.uni-mainz.de/kon/docs/Eendrassoctic.shtml.
Endraß, S. "Flächen mit vielen Doppelpunkten." *DMV-Mitteilungen* **4**, 17–0, 4/1995.
Endraß, S. "A Proctive Surface of Degree Eight with 168 Nodes." *J. Algebraic Geom.* **6**, 325–34, 1997.

Energy

The term energy has an important physical meaning in physics and is an extremely useful concept. A much more abstract mathematical generalization is defined as follows. Let Ω be a SPACE with MEASURE $\mu \geq 0$ and let $\Phi(P, Q)$ be a real function on the PRODUCT SPACE $\Omega \times \Omega$. When

$$(\mu, \nu) = \int \int \Phi(P, Q) \, d\mu(Q) \, d\nu(P)$$

$$= \int \Phi(P, \mu) \, d\nu(P)$$

exists for measures $\mu, \nu \geq 0$, (μ, ν) is called the MUTUAL ENERGY and (μ, μ) is called the ENERGY.

See also DIRICHLET ENERGY, MUTUAL ENERGY

References

Iyanaga, S. and Kawada, Y. (Eds.). "General Potential." §335.B in *Encyclopedic Dictionary of Mathematics.* Cambridge, MA: MIT Press, p. 1038, 1980.

En-Function

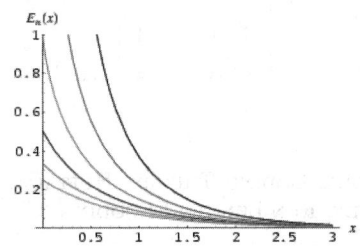

The $\mathrm{E}_n(x)$ function is defined by the integral

$$\mathrm{E}_n(x) \equiv \int_1^{\infty} \frac{e^{-xt} \, dt}{t^n} \tag{1}$$

and is given by the *Mathematica* function ExpIntegralE[n, x]. Defining $t \equiv \eta^{-1}$ so that $dt = -\eta^{-2} \, d\eta$,

$$\mathrm{E}_n(x) = \int_0^1 e^{-x/\eta} \, \eta^{\eta-2} \, d\eta \tag{2}$$

$$\mathrm{E}_n(0) = \frac{1}{n-1}. \tag{3}$$

The function satisfies the RECURRENCE RELATIONS

$$\mathrm{E}'_n(x) = -\mathrm{E}_{n-1}(x) \tag{4}$$

$$n\mathrm{E}_{n+1}(x) = e^{-x} - x\mathrm{E}_n(x). \tag{5}$$

Equation (4) can be derived from

$$E_n(x) = \int_1^\infty \frac{e^{-tx}}{t^n}\,dt \qquad (6)$$

$$E_n'(x) = \frac{d}{dx}\int_1^\infty \frac{e^{-tx}}{t^n}\,dt = \int_1^\infty \frac{d}{dx}\left(\frac{e^{-tx}}{t^n}\right)dt$$

$$= -\int_1^\infty t\,\frac{e^{-tx}}{t^n}\,dt$$

$$= -\int_1^\infty \frac{e^{-tx}}{t^{n-1}}\,dt = -E_{n-1}(x), \qquad (7)$$

and (5) using INTEGRATION BY PARTS, letting

$$u = \frac{1}{t^n} \qquad dv = e^{-tx}\,dt \qquad (8)$$

$$du = -\frac{n}{t^{n+1}}\,dt \qquad v = -\frac{e^{-tx}}{x} \qquad (9)$$

gives

$$E_n(x) = \int_1^\infty u\,dv = [uv]_1^\infty - \int_1^\infty v\,du$$

$$= \left[-\frac{e^{-tx}}{xt^n}\right]_{t=1}^\infty - \frac{n}{x}\int_1^\infty \frac{e^{-tx}}{t^{n+1}}\,dt$$

$$= \left[0 - \left(-\frac{e^{-x}}{x}\right)\right] - \frac{n}{x}\int_1^\infty \frac{e^{-tx}}{t^{n+1}}\,dt$$

$$= \frac{e^{-x}}{x} - \frac{n}{x}\,E_{n+1}(x). \qquad (10)$$

Solving (10) for $nE_{n+1}(x)$ then gives (5).
An ASYMPTOTIC SERIES is given by

$$(n-1)!\,E_n(x)$$

$$= (-x)^{n-1}E_1(x) + e^{-x}\sum_{s=0}^n -2(n-s-2)!(-x)^s, \quad (11)$$

so

$$E_n(x) = \frac{e^{-x}}{x}\left[1 - \frac{n}{x} + \frac{n(n+1)}{x^2} + \cdots\right]. \qquad (12)$$

The special case $n = 1$ gives

$$E_1(x) \equiv -\text{ei}(-x) = \int_1^\infty \frac{e^{-tx}\,dt}{t} = \int_x^\infty \frac{e^{-u}\,du}{u}, \qquad (13)$$

where $\text{ei}(x)$ is the EXPONENTIAL INTEGRAL, which is also equal to

$$E_1(x) = -\gamma - \ln x - \sum_{n=1}^\infty \frac{(-1)^n x^n}{n!\,n}, \qquad (14)$$

where γ is the EULER-MASCHERONI CONSTANT.

$$E_1(0) = \infty \qquad (15)$$

$$E_1(ix) = -\text{ci}(x) + i\ \text{si}(x), \qquad (16)$$

where $\text{ci}(x)$ and $\text{si}(x)$ are the COSINE INTEGRAL and SINE INTEGRAL.

See also COSINE INTEGRAL, *Et*-FUNCTION, EXPONENTIAL INTEGRAL, GOMPERTZ CONSTANT, SINE INTEGRAL

References
Abramowitz, M. and Stegun, C. A. (Eds.). "Exponential Integral and Related Functions." Ch. 5 in *Handbook of Mathematical Functions with Formulas, Graphs, and Mathematical Tables, 9th printing.* New York: Dover, pp. 227–33, 1972.

Press, W. H.; Flannery, B. P.; Teukolsky, S. A.; and Vetterling, W. T. "Exponential Integrals." §6.3 in *Numerical Recipes in FORTRAN: The Art of Scientific Computing, 2nd ed.* Cambridge, England: Cambridge University Press, pp. 215–19, 1992.

Spanier, J. and Oldham, K. B. "The Exponential Integral Ei(x) and Related Functions." Ch. 37 in *An Atlas of Functions.* Washington, DC: Hemisphere, pp. 351–60, 1987.

Engel's Theorem

A finite-dimensional LIE ALGEBRA all of whose elements are ad-NILPOTENT is itself a NILPOTENT LIE ALGEBRA.

Enlargement

See also EXPANSION

Enneacontagon

A 90-sided POLYGON. The regular enneacontagon is CONSTRUCTIBLE.

Enneacontahedron

A ZONOHEDRON constructed from the 10 diameters of the DODECAHEDRON which has 90 faces, 30 of which are RHOMBS of one type and the other 60 of which are RHOMBS of another. The enneacontahedron somewhat resembles a figure of Sharp.

See also DODECAHEDRON, RHOMB, ZONOHEDRON

References
Ball, W. W. R. and Coxeter, H. S. M. *Mathematical Recreations and Essays, 13th ed.* New York: Dover, pp. 142–43, 1987.

Sharp, A. *Geometry Improv'd: 1. By a Large and Accurate Table of Segments of Circles, with Compendious Tables for Finding a True Proportional Part, Exemplify'd in Making out Logarithms from them, there Being a Table of them for all Primes to 1100, True to 61 Figures. 2. A Concise Treatise of Polyhedra, or Solid Bodies, of Many Bases.* London: R. Mount, p. 87, 1717.

Enneadecagon

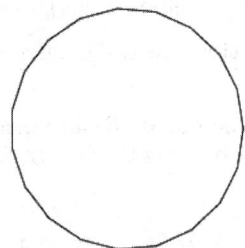

A 19-sided POLYGON, sometimes also called the ENNEAKAIDECAGON.

Enneagon
NONAGON

Enneagonal Number
NONAGONAL NUMBER

Enneakaidecagon
ENNEADECAGON

Enneper's Minimal Surface

A self-intersecting MINIMAL SURFACE which can be generated using the ENNEPER-WEIERSTRASS PARAMETERIZATION with

$$f(z) = 1 \tag{1}$$

$$g(z) = \zeta. \tag{2}$$

Letting $z = re^{i\phi}$ and taking the REAL PART give

$$x = \Re\left[re^{i\phi} - \tfrac{1}{3}r^3 e^{3i\phi}\right] \tag{3}$$

$$= r\cos\phi - \tfrac{1}{3}r^3\cos(3\phi) \tag{4}$$

$$y = \Re[ire^{i\phi} + \tfrac{1}{3}ir^3 e^{3i\phi}] \tag{5}$$

$$= -\tfrac{1}{3}r[3\sin\phi + r^2\sin(3\phi)] \tag{6}$$

$$z = \Re[r^2 e^{2i\phi}] \tag{7}$$

$$= r^2\cos(2\phi), \tag{8}$$

where $r \in [0, 1]$ and $\phi \in [-\pi, \pi)$. The coefficients of the FIRST FUNDAMENTAL FORM are

$$E = -2\cos(2\phi) \tag{9}$$

$$F = 4r\cos\phi\sin\phi \tag{10}$$

$$G = 2r^2\cos(2\phi), \tag{11}$$

the SECOND FUNDAMENTAL FORM coefficients are

$$e = (1 + r^2)^2 \tag{12}$$

$$f = 0 \tag{13}$$

$$g = r^2(1 + r^2)^2, \tag{14}$$

and the GAUSSIAN and MEAN CURVATURES are

$$K = -\frac{4}{(1 + r^2)^4} \tag{15}$$

$$H = 0. \tag{16}$$

Letting $z = u + iv$ gives the figure above, with parametrization

$$x = u - \tfrac{1}{3}u^3 + uv^2 \tag{17}$$

$$y = -v - u^2 v + \tfrac{1}{3}v^3 \tag{18}$$

$$z = u^2 - v^2 \tag{19}$$

(do Carmo 1986, Gray 1997, Nordstrand). In this parameterization, the coefficients of the FIRST FUNDAMENTAL FORM are

$$E = (1 + u^2 + v^2)^2 \tag{20}$$

$$F = 0 \tag{21}$$

$$G = (1 + u^2 + v^2)^2, \tag{22}$$

the SECOND FUNDAMENTAL FORM coefficients are

$$e = -2 \tag{23}$$

$$f = 0 \tag{24}$$

$$g = 2, \tag{25}$$

the AREA ELEMENT is

$$dA = (1 + u^2 + v^2)\,du \wedge dv, \tag{26}$$

and the GAUSSIAN and MEAN CURVATURES are

$$K = -\frac{4}{(1 + u^2 + v^2)^4} \tag{27}$$

$$H = 0. \tag{28}$$

Nordstrand gives the implicit form

$$\left(\frac{y^2 - x^2}{2z} + \tfrac{2}{9}z^2 + \tfrac{2}{3}\right)^3$$

$$-6\left[\frac{(y^2-x^2)}{4z}-\tfrac{1}{4}(x^2+y^2+\tfrac{8}{9}z^2)+\tfrac{2}{9}\right]^2=0. \qquad (29)$$

See also ENNEPER-WEIERSTRASS PARAMETERIZATION

References

Dickson, S. "Minimal Surfaces." *Mathematica J.* **1**, 38–0, 1990.

do Carmo, M. P. "Enneper's Surface." §3.5C in *Mathematical Models from the Collections of Universities and Museums* (Ed. G. Fischer). Braunschweig, Germany: Vieweg, p. 43, 1986.

Enneper, A. "Analytisch-geometrische Untersuchungen." *Z. Math. Phys.* **9**, 96–25, 1864.

Gray, A. "Examples of Minimal Surfaces," "The Associated Family of Enneper's Surface," and "Enneper's Surface of Degree *n*." §30.2 and 31.7 in *Modern Differential Geometry of Curves and Surfaces with Mathematica, 2nd ed.* Boca Raton, FL: CRC Press, pp. 358, 684–85, and 726–32, 1997.

JavaView. "Classic Surfaces from Differential Geometry: Enneper." http://www-sfb288.math.tu-berlin.de/vgp/javaview/demo/surface/common/PaSurface_Enneper.html.

Maeder, R. *The Mathematica Programmer.* San Diego, CA: Academic Press, pp. 150–51, 1994.

Nordstrand, T. "Enneper's Minimal Surface." http://www.uib.no/people/nfytn/enntxt.htm.

Osserman, R. *A Survey of Minimal Surfaces.* New York: Dover, p. 65, 87, and 143, 1986.

Wolfram Research "Mathematica Version 2.0 Graphics Gallery." http://www.mathsource.com/cgi-bin/msitem22?0207-55.

Enneper's Negative Curvature Surfaces

The Enneper surfaces are a three-parameter family of surfaces with constant negative curvature (and nonconstant MEAN CURVATURE). In general, they are described by ELLIPTIC FUNCTIONS. However, a special case which can be specified parametrically using ELEMENTARY FUNCTIONS is the KUEN SURFACE.

See also KUEN SURFACE

References

Enneper, A. "Analytisch-geometrische Untersuchungen." *Nachr. Königl. Gesell. Wissensch. Georg-Augustus-Univ. Göttingen* **12**, 258–77, 1868.

Fischer, G. (Ed.). Plate 92 in *Mathematische Modelle/ Mathematical Models, Bildband/Photograph Volume.* Braunschweig, Germany: Vieweg, p. 88, 1986.

Reckziegel, H. "Enneper's Surfaces." §3.4.4 in *Mathematical Models from the Collections of Universities and Museums* (Ed. G. Fischer). Braunschweig, Germany: Vieweg, pp. 37–9, 1986.

Enneper-Weierstrass Parameterization

A parameterization of a MINIMAL SURFACE in terms of two functions $f(z)$ and $g(z)$ as

$$\begin{bmatrix} x(r,\ \phi) \\ y(r,\ \phi) \\ z(r,\ \phi) \end{bmatrix} = \Re \int \begin{bmatrix} f(1-g^2) \\ if(1+g^2) \\ 2fg \end{bmatrix} dz,$$

where $z = re^{i\phi}$ and $\Re$ is the REAL PART. Examples are given in the following table.

Surface	$f(z)$	$g(z)$
ENNEPER'S MINIMAL SURFACE	1	z
HENNEBERG'S MINIMAL SURFACE	$2(1-z^{-4})$	z
BOUR'S MINIMAL SURFACE	1	$\sqrt{z}$
TRINOID	$(z^3-1)^{-2}$	z^2

See also BOUR'S MINIMAL SURFACE, ENNEPER'S MINIMAL SURFACE, HENNEBERG'S MINIMAL SURFACE, MINIMAL SURFACE, TRINOID

References

Dickson, S. "Minimal Surfaces." *Mathematica J.* **1**, 38–0, 1990.

do Carmo, M. P. *Mathematical Models from the Collections of Universities and Museums* (Ed. G. Fischer). Braunschweig, Germany: Vieweg, p. 41, 1986.

Gray, A. "Minimal Surfaces via the Weierstrass Representation." Ch. 32 in *Modern Differential Geometry of Curves and Surfaces with Mathematica, 2nd ed.* Boca Raton, FL: CRC Press, pp. 735–60, 1997.

Weierstrass, K. "Über die Flächen deren mittlere Krümmung überall gleich null ist." *Monatsber. Berliner Akad.*, 612–25, 1866.

Wolfram Research, Inc. "Minimal Surfaces à la Weierstrass." http://library.wolfram.com/demos/WeierstrassSurfaces.nb.

Enormous Theorem

CLASSIFICATION THEOREM

Enriques Surfaces

An Enriques surface X is a smooth compact complex surface having irregularity $q(X) = 0$ and nontrivial canonical sheaf K_X such that $K_X^2 = O_X$ (Endraß). Such surfaces cannot be embedded in projective 3-space, but there nonetheless exist transformations onto singular surfaces in projective 3-space. There exists a family of such transformed surfaces of degree six which passes through each edge of a TETRAHEDRON twice. A subfamily with tetrahedral symmetry is given by the two-parameter $(r,\ c)$ family of surfaces

$$f_r x_0 x_1 x_2 x_3 + c(x_0^2 x_1^2 x_2^2 + x_0^2 x_1^2 x_3^2 + x_0^2 x_2^2 x_3^2 + x_1^2 x_2^2 x_3^2) = 0$$

and the polynomial f_r is a sphere with radius r,

$$f_r = (3-r)(x_0^2 + x_1^2 + x_2^2 + x_3^2)$$

$$-2(1+r)(x_0 x_1 + x_0 x_2 + x_0 x_3 + x_1 x_2 + x_1 x_3 + x_2 x_3)$$

(Endraß).

References

Angermüller, G. and Barth, W. "Elliptic Fibres on Enriques Surfaces." *Compos. Math.* **47**, 317–32, 1982.

Barth, W. and Peters, C. "Automorphisms of Enriques Surfaces." *Invent. Math.* **73**, 383–11, 1983.

Barth, W. P.; Peters, C. A.; and van de Ven, A. A. *Compact Complex Surfaces.* New York: Springer-Verlag, 1984.

Barth, W. "Lectures on K3- and Enriques Surfaces." In *Algebraic Geometry, Sitges (Barcelona) 1983, Proceedings of a Conference Held in Sitges (Barcelona), Spain, October 5–2, 1983* (Ed. E. Casas-Alvero, G. E. Welters, and S. Xambó-Descamps). New York: Springer-Verlag, pp. 21–7, 1983.

Endraß, S. "Enriques Surfaces." http://enriques.mathematik.uni-mainz.de/kon/docs/enriques.shtml.

Enriques, F. *Le superficie algebriche.* Bologna, Italy: Zanichelli, 1949.

Enriques, F. "Sulla classificazione." *Atti Accad. Naz. Lincei* **5**, 1914.

Hunt, B. *The Geometry of Some Special Arithmetic Quotients.* New York: Springer-Verlag, p. 317, 1996.

Kim, Y. "Normal Quintic Enriques Surfaces." *J. Korean Math. Soc.* **36**, 545–66, 1999.

Entire Function

If a COMPLEX FUNCTION is ANALYTIC at all finite points of the COMPLEX PLANE $\mathbb{C}$, then it is said to be entire, sometimes also called "integral" (Knopp 1996, p. 112).

See also ANALYTIC FUNCTION, FINITE ORDER, HADAMARD FACTORIZATION THEOREM, HOLOMORPHIC FUNCTION, LIOUVILLE'S BOUNDEDNESS THEOREM, MEROMORPHIC FUNCTION, WEIERSTRASS FACTOR THEOREM

References

Knopp, K. "Entire Transcendental Functions." Ch. 9 in *Theory of Functions Parts I and II, Two Volumes Bound as One, Part I.* New York: Dover, pp. 112–16, 1996.

Krantz, S. G. "Entire Functions and Liouville's Theorem." §3.1.3 in *Handbook of Complex Analysis.* Boston, MA: Birkhäuser, pp. 31–2, 1999.

Entire Modular Form

A MODULAR FORM which is not allowed to have poles in the UPPER HALF-PLANE H or at $i\infty$.

See also MODULAR FORM

Entringer Number

The Entringer numbers $E(n, k)$ are the number of PERMUTATIONS of $\{1, 2, \ldots, n+1\}$, starting with $k+1$, which, after initially falling, alternately fall then rise. The Entringer numbers are given by

$$E(0, 0) = 1$$
$$E(n, 0) = 0$$

together with the RECURRENCE RELATION

$$E(n, k) = E(n, k+1) + E(n-1, n-k).$$

The numbers $E(n) = E(n, n)$ are the SECANT and TANGENT NUMBERS given by the MACLAURIN SERIES

$$\sec x + \tan x$$
$$= A_0 + A_1 x + A_2 \frac{x^2}{2!} + A_3 \frac{x^3}{3!} A_4 \frac{x^4}{4!} + A_5 \frac{x^5}{5!} + \ldots.$$

See also ALTERNATING PERMUTATION, BOUSTROPHEDON TRANSFORM, EULER ZIGZAG NUMBER, PERMUTATION, SECANT NUMBER, SEIDEL-ENTRINGER-ARNOLD TRIANGLE, TANGENT NUMBER, ZAG NUMBER, ZIG NUMBER

References

Bauslaugh, B. and Ruskey, F. "Generating Alternating Permutations Lexographically." *BIT* **80**, 17–6, 1990.

Entringer, R. C. "A Combinatorial Interpretation of the Euler and Bernoulli Numbers." *Nieuw. Arch. Wisk.* **14**, 241–46, 1966.

Millar, J.; Sloane, N. J. A.; and Young, N. E. "A New Operation on Sequences: The Boustrophedon Transform." *J. Combin. Th. Ser. A* **76**, 44–4, 1996.

Poupard, C. "De nouvelles significations enumeratives des nombres d'Entringer." *Disc. Math.* **38**, 265–71, 1982.

Ruskey, F. "Information of Alternating Permutations." http://www.theory.csc.uvic.ca/~cos/inf/perm/Alternating.html.

Sloane, N. J. A. Sequences A000111/M1492 in "An On-Line Version of the Encyclopedia of Integer Sequences." http://www.research.att.com/~njas/sequences/eisonline.html.

Entropy

In physics, the word entropy has important physical implications as the amount of "disorder" of a system. In mathematics, a more abstract definition is used. The (Shannon) entropy of a variable X is defined as

$$H(X) \equiv -\sum_x p(x) \ln[p(x)],$$

where $p(x)$ is the probability that X is in the state x, and $p \ln p$ is defined as 0 if $p = 0$. The joint entropy of variables $X_1, \ldots, X_n$ is then defined by

$$H(X_1, \ldots, X_n)$$
$$\equiv -\sum_{x_1} \cdots \sum_{x_n} p(x_1, \ldots, x_n) \ln[p(x_1, \ldots, x_n)].$$

See also INFORMATION THEORY, KOLMOGOROV ENTROPY, KOLMOGOROV-SINAI ENTROPY, MAXIMUM ENTROPY METHOD, METRIC ENTROPY, ORNSTEIN'S THEOREM, REDUNDANCY, RELATIVE ENTROPY, SHANNON ENTROPY, TOPOLOGICAL ENTROPY

References

Ellis, R. S. *Entropy, Large Deviations, and Statistical Mechanics.* New York: Springer-Verlag, 1985.

Khinchin, A. I. *Mathematical Foundations of Information Theory.* New York: Dover, 1957.

Lasota, A. and Mackey, M. C. *Chaos, Fractals, and Noise: Stochastic Aspects of Dynamics, 2nd ed.* New York: Springer-Verlag, 1994.

Ott, E. "Entropies." §4.5 in *Chaos in Dynamical Systems.* New York: Cambridge University Press, pp. 138–44, 1993.

Rothstein, J. *Science* **114**, 171, 1951.

Schnakenberg, J. "Network Theory of Microscopic and Macroscopic Behavior of Master Equation Systems." *Rev. Mod. Phys.* **48**, 571–85, 1976.

Shannon, C. E. "A Mathematical Theory of Communication." *The Bell System Technical J.* **27**, 379–23 and 623–56, July and Oct. 1948. http://cm.bell-labs.com/cm/ms/what/shannonday/shannon1948.pdf.

Shannon, C. E. and Weaver, W. *Mathematical Theory of Communication.* Urbana, IL: University of Illinois Press, 1963.

Entscheidungsproblem

DECISION PROBLEM

Enumerable

DENUMERABLE SET

Enumerate

A GENERATING FUNCTION

$$F(x) = \sum_n a_n x^n$$

is said to enumerate a_n (Hardy 1999, p. 85).

See also GENERATING FUNCTION

References

Hardy, G. H. *Ramanujan: Twelve Lectures on Subjects Suggested by His Life and Work, 3rd ed.* New York: Chelsea, 1999.

Enumeration Problem

The problem of determining (or counting) the set of all solutions to a given problem.

See also CLASSIFICATION, COMBINATORICS, EXISTENCE PROBLEM

References

Gardner, M. *The Sixth Book of Mathematical Games from Scientific American.* Chicago, IL: University of Chicago Press, p. 22, 1984.

Enumerative Geometry

Schubert's application of the CONSERVATION OF NUMBER PRINCIPLE.

See also CONSERVATION OF NUMBER PRINCIPLE, DUALITY PRINCIPLE, HILBERT'S PROBLEMS, PERMANENCE OF MATHEMATICAL RELATIONS PRINCIPLE

References

Bell, E. T. *The Development of Mathematics, 2nd ed.* New York: McGraw-Hill, p. 340, 1945.

Envelope

The envelope of a one-parameter family of curves given implicitly by

$$U(x, y, c) = 0, \tag{1}$$

or in parametric form by $(f(t, c), g(t, c))$, is a curve which touches every member of the family. For a curve represented by $(f(t, c), g(t, c))$, the envelope is found by solving

$$0 = \frac{\partial f}{\partial t} \frac{\partial g}{\partial c} - \frac{\partial f}{\partial c} \frac{\partial g}{\partial t}. \tag{2}$$

For a curve represented implicitly, the envelope is given by simultaneously solving

$$\frac{\partial U}{\partial c} = 0 \tag{3}$$

$$U(x, y, c) = 0. \tag{4}$$

See also ASTROID, CARDIOID, CATACAUSTIC, CAUSTIC, CAYLEYIAN CURVE, DÜRER'S CONCHOID, ELLIPSE ENVELOPE, ENVELOPE THEOREM, EVOLUTE, GLISSETTE, HEDGEHOG, KIEPERT'S PARABOLA, LINDELOF'S THEOREM, NEGATIVE PEDAL CURVE

References

Lawrence, J. D. *A Catalog of Special Plane Curves.* New York: Dover, pp. 33–4, 1972.

Yates, R. C. "Envelopes." *A Handbook on Curves and Their Properties.* Ann Arbor, MI: J. W. Edwards, pp. 75–0, 1952.

Envelope (Form)

Given a DIFFERENTIAL P-FORM q in the EXTERIOR ALGEBRA $\wedge^p V^*$, its envelope is the smallest SUBSPACE W such that q is in the subspace $\wedge^p W^* \subset \wedge^p V^*$. Alternatively, W is spanned by the vectors that can be written as the CONTRACTION of q with an element of $\wedge^{p-1} V$.

For example, the envelope of dx in $V = \mathbb{R}^2$ is $W = \langle \partial/\partial x \rangle$, and the envelope of $dx_1 \wedge dx_2 + dx_3 \wedge dx_4$ in $V = \mathbb{R}^4$ is all of V.

Here is a *Mathematica* function which will compute the envelope of an ANTISYMMETRIC TENSOR.

```
< <DiscreteMath`Combinatorica`;
ContractAll[a_List, b_List] :=   Module[{k =
TensorRank[a] - TensorRank[b]},    If[k >= 0,
Map[Flatten[#1].Flatten[b]   &,   a,    {k}],
ContractAll[b, a]
   ]
  ]  Envelope[a_List?VectorQ] := Select[{a},
#1   !=    Table[0,    {Length[a]}]   &]
Envelope[a_List] := Module[
  {
   z, inds, vects,
   d = Dimensions[a][[1]], r = TensorRank[a]
```

```
    },
   z = Table[0, ##1] & @@ Table[{d}, {r - 1}];
   inds = KSubsets[Range[d], r - 1];
   vects = Map[ContractAll[a, ReplacePart[z, 1,
#1]] &, inds];
   Select[RowReduce[vects], #1 != Table[0, {d}]
&]
```

See also DECOMPOSABLE, DIFFERENTIAL FORM, DIF-
FERENTIAL IDEAL, EXTERIOR ALGEBRA, VECTOR
SPACE, WEDGE PRODUCT

Envelope Theorem

Relates EVOLUTES to single paths in the CALCULUS OF
VARIATIONS. Proved in the general case by Darboux
and Zermelo (1894) and Kneser (1898). It states:
"When a single parameter family of external paths
from a fixed point O has an ENVELOPE, the integral
from the fixed point to any point A on the ENVELOPE
equals the integral from the fixed point to any second
point B on the ENVELOPE plus the integral along the
envelope to the first point on the ENVELOPE,
$J_{OA} = J_{OB} + J_{BA}$."

References

Kimball, W. S. *Calculus of Variations by Parallel Displacement.* London: Butterworth, p. 292, 1952.

Envyfree

An agreement in which all parties feel as if they have
received the best deal.

See also CAKE CUTTING

References

Robertson, J. and Webb, W. *Cake Cutting Algorithms: Be
Fair If You Can.* Natick, MA: Peters, 1998.
Stewart, I. "Mathematical Recreations." *Sci. Amer.*, p. 86,
Jan. 1999.

E-Operator
SUMMATION BY PARTS

e-Perfect Number

A number n is called an e-perfect number if $\sigma_e(n) =
2n$, where $\sigma_e(n)$ is the SUM of the E-DIVISORS of n. If m
is SQUAREFREE, then $\sigma_e(m) = m$. As a result, if n is e-
perfect and m is SQUAREFREE with $m \perp b$, then mn is
e-perfect.

The first few e-perfect numbers are 36, 180, 252, 396,
468, ... (Sloane's A054979). There are no ODD e-
perfect numbers. The first few primitive e-perfect
numbers are 36, 1800, 2700, 17424, ... (Sloane's
A054980).

See also E-DIVISOR

References

Guy, R. K. "Exponential-Perfect Numbers." §B17 in *Unsolved Problems in Number Theory, 2nd ed.* New York:
Springer-Verlag, p. 73, 1994.
Sloane, N. J. A. Sequences A054979 and A054980 in "An
On-Line Version of the Encyclopedia of Integer Sequences." http://www.research.att.com/~njas/sequences/
eisonline.html.
Subbarao, M. V. and Suryanarayan, D. "Exponential Perfect
and Unitary Perfect Numbers." *Not. Amer. Math. Soc.* **18**,
798, 1971.

Epicycloid

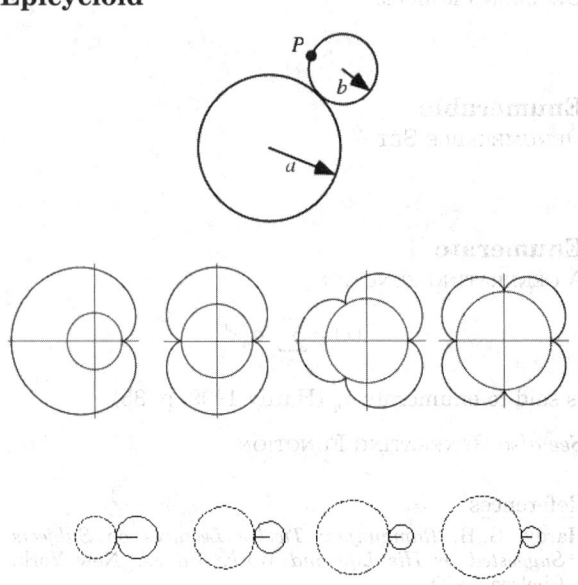

The path traced out by a point P on the edge of a
CIRCLE of RADIUS b rolling on the outside of a CIRCLE
of RADIUS a. An epicycloid is therefore an EPITRO-
CHOID with $h = b$. Epicycloids are given by the
PARAMETRIC EQUATIONS

$$x = (a+b)\cos\phi - b\cos\left(\frac{a+b}{b}\phi\right) \quad (1)$$

$$y = (a+b)\sin\phi - b\sin\left(\frac{a+b}{b}\phi\right). \quad (2)$$

A polar equation can be derived by computing

$$x^2 = (a+b)^2\cos^2\phi - 2b(a+b)\cos\phi\cos\left(\frac{a+b}{b}\phi\right)$$

$$+ b^2\cos^2\left(\frac{a+b}{b}\phi\right) \quad (3)$$

$$y^2 = (a+b)^2 \sin^2 \phi - 2b(a+b) \sin \phi \sin\left(\frac{a+b}{b}\phi\right)$$

$$+ b^2 \sin^2\left(\frac{a+b}{b}\phi\right), \tag{4}$$

so

$$r^2 = x^2 + y^2 = (a+b)^2 + b^2 - 2b(a+b)$$

$$\times \left\{ \cos\left[\left(\frac{a}{b}+1\right)\phi\right]\cos\phi + \sin\left[\left(\frac{a}{b}+1\right)\phi\right]\sin\phi \right\}. \tag{5}$$

But

$$\cos\alpha\cos\beta + \sin\alpha\sin\beta = \cos(\alpha - \beta), \tag{6}$$

so

$$r^2 = (a+b)^2 + b^2 - 2b(a+b)\cos\left[\left(\frac{a}{b}+1\right)\phi - \phi\right]$$

$$= (a+b)^2 + b^2 - 2b(a+b)\cos\left(\frac{a}{b}\phi\right). \tag{7}$$

Note that ϕ is the parameter here, *not* the polar angle. The polar angle from the center is

$$\tan\theta = \frac{y}{x} = \frac{(a+b)\sin\phi - b\sin\left(\dfrac{a+b}{b}\phi\right)}{(a+b)\cos\phi - b\cos\left(\dfrac{a+b}{b}\phi\right)}. \tag{8}$$

To get n CUSPS in the epicycloid, $b = a/n$, because then n rotations of b bring the point on the edge back to its starting position.

$$r^2 = a^2\left[\left(1+\frac{1}{n}\right)^2 + \left(\frac{1}{n}\right)^2 - 2\left(\frac{1}{n}\right)\left(1+\frac{1}{n}\right)\cos(n\phi)\right]$$

$$= a^2\left[1 + \frac{2}{n} + \frac{1}{n^2} + \frac{1}{n^2} - \left(\frac{2}{n}\right)\left(\frac{n+1}{n}\right)\cos(n\phi)\right]$$

$$= a^2\left[\frac{n^2 + 2n + 2}{n^2} - \frac{2(n+1)}{n^2}\cos(n\phi)\right]$$

$$= \frac{a^2}{n^2}\left[(n^2 + 2n + 2) - 2(n+1)\cos(n\phi)\right], \tag{9}$$

so

$$\tan\theta = \frac{a\left(\dfrac{n+1}{n}\right)\sin\phi - \dfrac{a}{n}\sin[(n+1)\phi]}{a\left(\dfrac{n+1}{n}\right)\cos\phi - \dfrac{a}{n}\cos[(n+1)\phi]}$$

$$= \frac{(n+1)\sin\phi - \sin[(n+1)\phi]}{(n+1)\cos\phi - \cos[(n+1)\phi]}. \tag{10}$$

An epicycloid with one cusp is called a CARDIOID, one with two cusps is called a NEPHROID, and one with five cusps is called a RANUNCULOID.

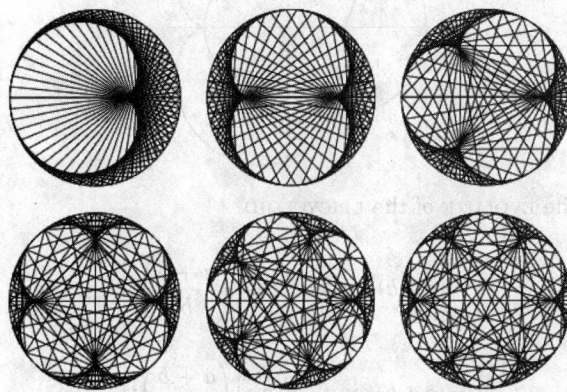

n-epicycloids can also be constructed by beginning with the DIAMETER of a CIRCLE, offsetting one end by a series of steps while at the same time offsetting the other end by steps n times as large. After traveling around the CIRCLE once, an n-cusped epicycloid is produced, as illustrated above (Madachy 1979).

Epicycloids have TORSION

$$\tau = 0 \tag{11}$$

and satisfy

$$\frac{s^2}{a^2} + \frac{\rho^2}{b^2} = 1, \tag{12}$$

where ρ is the RADIUS OF CURVATURE ($1/\kappa$).

See also CARDIOID, CYCLIDE, CYCLOID, EPICYCLOID–1-CUSPED, EPICYCLOID EVOLUTE, EPICYCLOID INVOLUTE, EPICYCLOID PEDAL CURVE, EPITROCHOID, HYPOCYCLOID, NEPHROID, RANUNCULOID

References

Beyer, W. H. *CRC Standard Mathematical Tables*, 28th ed. Boca Raton, FL: CRC Press, p. 217, 1987.
Bogomolny, A. "Cycloids." http://www.cut-the-knot.com/pythagoras/cycloids.html.
Lawrence, J. D. *A Catalog of Special Plane Curves*. New York: Dover, pp. 160–64 and 169, 1972.
Lemaire, J. *Hypocycloïdes et epicycloïdes*. Paris: Albert Blanchard, 1967.
MacTutor History of Mathematics Archive. "Epicycloid." http://www-groups.dcs.st-and.ac.uk/~history/Curves/Epicycloid.html.
Madachy, J. S. *Madachy's Mathematical Recreations*. New York: Dover, pp. 219–25, 1979.
Wagon, S. *Mathematica in Action*. New York: W. H. Freeman, pp. 50–2, 1991.
Yates, R. C. "Epi- and Hypo-Cycloids." *A Handbook on Curves and Their Properties*. Ann Arbor, MI: J. W. Edwards, pp. 81–5, 1952.

Epicycloid Evolute

The EVOLUTE of the EPICYCLOID

$$x = (a+b)\cos t - b\cos\left[\left(\frac{a+b}{b}\right)t\right]$$

$$y = (a+b)\sin t - b\sin\left[\left(\frac{a+b}{b}\right)t\right]$$

is another EPICYCLOID given by

$$x = \frac{a}{a+2b}\left\{(a+b)\cos t + b\cos\left[\left(\frac{a+b}{b}\right)t\right]\right\}$$

$$y = \frac{a}{a+2b}\left\{(a+b)\sin t + b\cos\left[\left(\frac{a+b}{b}\right)t\right]\right\}.$$

Epicycloid Involute

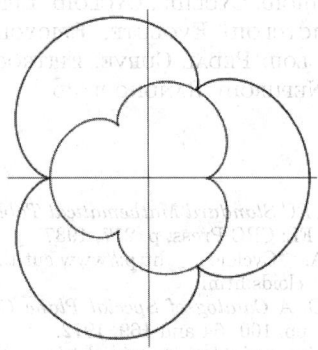

The INVOLUTE of the EPICYCLOID

$$x = (a+b)\cos t - b\cos\left[\left(\frac{a+b}{b}\right)t\right]$$

$$y = (a+b)\sin t - b\sin\left[\left(\frac{a+b}{b}\right)t\right]$$

is another EPICYCLOID given by

$$x = \frac{a+2b}{a}\left\{(a+b)\cos t + b\cos\left[\left(\frac{a+b}{b}\right)t\right]\right\}$$

$$y = \frac{a+2b}{a}\left\{(a+b)\sin t + b\cos\left[\left(\frac{a+b}{b}\right)t\right]\right\}.$$

Epicycloid Pedal Curve

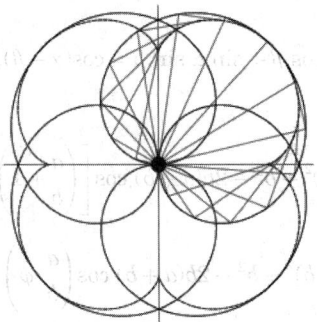

The PEDAL CURVE of an EPICYCLOID with PEDAL POINT at the center, shown for an epicycloid with four cusps, is not a ROSE as claimed by Lawrence (1972).

References

Lawrence, J. D. *A Catalog of Special Plane Curves.* New York: Dover, p. 204, 1972.

Epicycloid Radial Curve

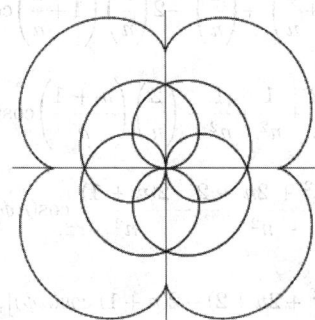

The RADIAL CURVE of an EPICYCLOID is shown above for an epicycloid with four cusps. It is not a ROSE, as claimed by Lawrence (1972).

References

Lawrence, J. D. *A Catalog of Special Plane Curves.* New York: Dover, p. 202, 1972.

Epicycloid1-Cusped

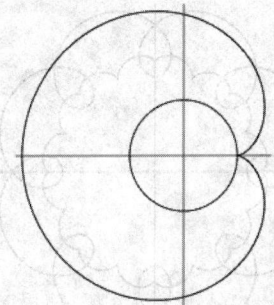

A 1-cusped epicycloid has $b = a$, so $n = 1$. The radius measured from the center of the large circle for a 1-cusped epicycloid is given by EPICYCLOID equation (9) with $n = 1$ so

$$r^2 = \frac{a^2}{n^2}[(n^2 + 2n + 2) - 2(n + 1)\cos(n\phi)]$$

$$= a^2[(1^2 + 2 \cdot 1 + 2) - 2(1 + 1)\cos(1 \cdot \phi)]$$

$$= a^2(5 - 4\cos\phi) \qquad (1)$$

$$r = a\sqrt{5 - 4\cos\phi}, \qquad (2)$$

and

$$\tan\theta = \frac{2\sin\phi - \sin(2\phi)}{2\cos\phi - \cos(2\phi)}. \qquad (3)$$

The 1-cusped epicycloid is just an offset CARDIOID.

Epicycloid–2-Cusped

NEPHROID

Epimenides Paradox

A version of the LIAR'S PARADOX, attributed to the philosopher Epimenides in the sixth century BC. "All Cretans are liars...One of their own poets has said so." This is not a true paradox since the poet may have knowledge that at least one Cretan is, in fact, honest, and so be lying when he says that *all* Cretans are liars. There therefore need be no self-contradiction in what could simply be a false statement by a person who is himself a liar.

A sharper version of the paradox (which has no such loophole) is the EUBULIDES PARADOX, "This statement is false."

See also EUBULIDES PARADOX, LIAR'S PARADOX, SOCRATES' PARADOX

References
Curry, H. B. *Foundations of Mathematical Logic.* New York: Dover, pp. 5–, 1977.

Erickson, G. W. and Fossa, J. A. *Dictionary of Paradox.* Lanham, MD: University Press of America, pp. 58–0, 1998.
Hoffman, P. *The Man Who Loved Only Numbers: The Story of Paul Erdos and the Search for Mathematical Truth.* New York: Hyperion, p. 115, 1998.
Hofstadter, D. R. *Gödel, Escher, Bach: An Eternal Golden Braid.* New York: Vintage Books, p. 17, 1989.
Prior, A. N. "Epimenides the Cretan." *J. Symb. Logic* **23**, 261–66, 1958.

Epimorphism

A MORPHISM $f : Y \to X$ in a CATEGORY is an epimorphism if, for any two morphisms u, $v : X \to Z$, $uf = vf$ implies $u = v$.

See also CATEGORY, MORPHISM

Epispiral

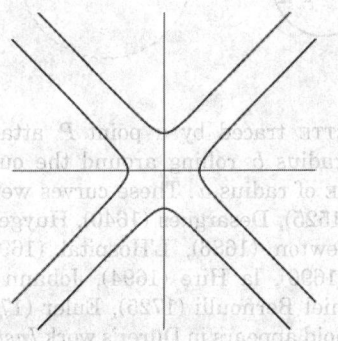

A plane curve with polar equation

$$r = a\sec(n\theta).$$

There are n sections if n is ODD and $2n$ if n is EVEN.

References
Lawrence, J. D. *A Catalog of Special Plane Curves.* New York: Dover, pp. 192–93, 1972.

Epispiral Inverse Curve

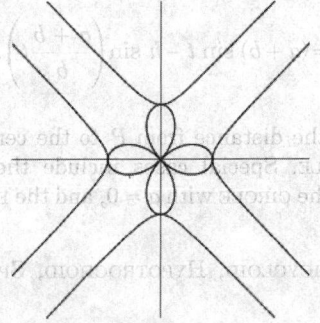

The INVERSE CURVE of the EPISPIRAL

$$r = a\sec(n\theta)$$

with INVERSION CENTER at the origin and inversion

radius k is the ROSE

$$r = \frac{k \cos(n\theta)}{a}.$$

See also EPISPIRAL, INVERSE CURVE, ROSE

Epitrochoid

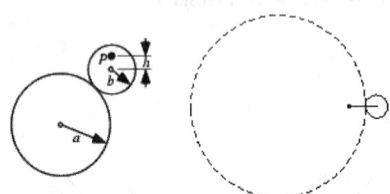

The ROULETTE traced by a point P attached to a CIRCLE of radius b rolling around the outside of a fixed CIRCLE of radius a. These curves were studied by Dürer (1525), Desargues (1640), Huygens (1679), Leibniz, Newton (1686), L'Hospital (1690), Jakob Bernoulli (1690), la Hire (1694), Johann Bernoulli (1695), Daniel Bernoulli (1725), Euler (1745, 1781). An epitrochoid appears in Dürer's work *Instruction in Measurement with Compasses and Straight Edge* (1525). He called epitrochoids SPIDER LINES because the lines he used to construct the curves looked like a spider.

The PARAMETRIC EQUATIONS for an epitrochoid are

$$x = (a+b) \cos t - h \cos\left(\frac{a+b}{b} t\right)$$

$$y = (a+b) \sin t - h \sin\left(\frac{a+b}{b} t\right),$$

where h is the distance from P to the center of the rolling CIRCLE. Special cases include the LIMAÇON with $a = b$, the CIRCLE with $a = 0$, and the EPICYCLOID with $h = b$.

See also EPICYCLOID, HYPOTROCHOID, SPIROGRAPH, TROCHOID

References

Lawrence, J. D. *A Catalog of Special Plane Curves.* New York: Dover, pp. 168–70, 1972.

Epitrochoid Evolute

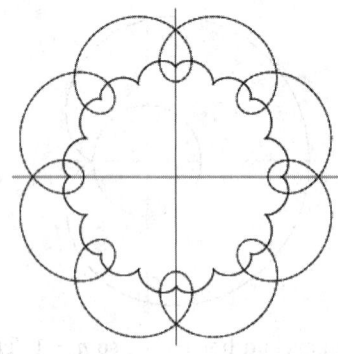

The PARAMETRIC EQUATIONS of the EVOLUTE of an EPITROCHOID specified by circle radii a and b with offset h are

$$x = \frac{ah(a+b)c_1(t) \cos t + bc_2(t) \cos\left[\dfrac{(a+b)t}{b}\right]}{b^3 + (a+b)h^2 - b(a+2b)h \cos\left(\dfrac{at}{b}\right)} \quad (1)$$

$$y = \frac{ah(a+b)c_1(t) \sin t + bc_2(t) \sin\left[\dfrac{(a+b)t}{b}\right]}{b^3 + (a+b)h^2 - b(a+2b)h \cos\left(\dfrac{at}{b}\right)}, \quad (2)$$

where

$$c_1(t) \equiv h - b \cos\left(\frac{at}{b}\right) \quad (3)$$

$$c_2(t) \equiv b - h \cos\left(\frac{at}{b}\right). \quad (4)$$

See also EPITROCHOID, EVOLUTE

Epsilon

In mathematics, a small POSITIVE INFINITESIMAL quantity, usually denoted ϵ or ε, whose LIMIT is usually taken as $\epsilon \to 0$.

The late mathematician P. Erdos also used the term "epsilons" to refer to children (Hoffman 1998, p. 4).

See also EPSILON CONJECTURE, WYNN'S EPSILON METHOD

References

Hoffman, P. *The Man Who Loved Only Numbers: The Story of Paul Erdos and the Search for Mathematical Truth.* New York: Hyperion, 1998.

Epsilon Conjecture

The conjecture that Frey's ELLIPTIC CURVE was *not* modular. The conjecture was quickly proved by Ribet (RIBET'S THEOREM) in 1986, and was an important step in the proof of FERMAT'S LAST THEOREM and the TANIYAMA-SHIMURA CONJECTURE.

See also FERMAT'S LAST THEOREM, RIBET'S THEOREM, TANIYAMA-SHIMURA CONJECTURE

Epsilon-Delta Definition

CONTINUOUS FUNCTION, LIMIT

Epsilon-Neighborhood

NEIGHBORHOOD

Epstein Zeta Function

$$Z\begin{vmatrix}\mathbf{g}\\\mathbf{h}\end{vmatrix}(q;s) = \sum_1 \frac{e^{-2\pi i \mathbf{h} \cdot \mathbf{1}}}{[q(\mathbf{1}+\mathbf{g})]^{s/2}},$$

where $\mathbf{g}$ and $\mathbf{h}$ are arbitrary VECTORS, the SUM runs over a d-dimensional LATTICE, and $\mathbf{1} = -\mathbf{g}$ is omitted if $\mathbf{g}$ is a lattice VECTOR.

See also ZETA FUNCTION

References

Glasser, M. L. and Zucker, I. J. "Lattice Sums in Theoretical Chemistry." In *Theoretical Chemistry: Advances and Perspectives, Vol. 5* (Ed. H. Eyring). New York: Academic Press, pp. 69–0, 1980.
Shanks, D. "Calculation and Applications of Epstein Zeta Functions." *Math. Comput.* **29**, 271–87, 1975.

Equal

Two quantities are said to be equal if they are, in some WELL DEFINED sense, equivalent. Equality of quantities a and b is written $a = b$. Equal is implemented in *Mathematica* as Equal[A, B, ...], or $A == B == ....$

A symbol with three horizontal line segments ($\equiv$) resembling the equals sign is used to denote both equality by definition (e.g., $A \equiv B$ means A is DEFINED to be equal to B) and CONGRUENCE (e.g., $13 \equiv 1 \pmod{12}$ means 13 divided by 12 leaves a REMAINDER of 1–a fact known to all readers of analog clocks).

See also CONGRUENCE, DEFINED, DIFFERENT, EQUAL BY DEFINITION, EQUALITY, EQUIVALENT, ISOMORPHISM, UNEQUAL

Equal by Definition

DEFINED

Equal Detour Point

The center of an outer SODDY CIRCLE. It has TRIANGLE CENTER FUNCTION

$$\alpha = 1 + \frac{2\Delta}{a(b+c-a)} = \sec(\tfrac{1}{2}A)\cos(\tfrac{1}{2}B)\cos(\tfrac{1}{2}C) + 1.$$

Given a point Y not between A and B, a detour of length

$$|AY| + |YB| - |AB|$$

is made walking from A to B via Y, the point is of equal detour if the three detours from one side to another via Y are equal. If ABC has no ANGLE $> 2\sin^{-1}(4/5)$, then the point given by the above TRILINEAR COORDINATES is the unique equal detour point. Otherwise, the ISOPERIMETRIC POINT is also equal detour.

References

Kimberling, C. "Central Points and Central Lines in the Plane of a Triangle." *Math. Mag.* **67**, 163–87, 1994.
Kimberling, C. "Isoperimetric Point and Equal Detour Point." http://cedar.evansville.edu/~ck6/tcenters/recent/isoper.html.
Veldkamp, G. R. "The Isoperimetric Point and the Point(s) of Equal Detour." *Amer. Math. Monthly* **92**, 546–58, 1985.

Equal Incircles Theorem

INCIRCLE

Equal Parallelians Point

The point of intersection of the three LINE SEGMENTS, each parallel to one side of a TRIANGLE and touching the other two, such that all three segments are of the same length. The TRILINEAR COORDINATES are

$$bc(ca + ab - bc) : ca(ab + bc - ca) : ab(bc + ca - ab).$$

References

Kimberling, C. "Equal Parallelians Point." http://cedar.evansville.edu/~ck6/tcenters/recent/eqparal.html.

Equal-Area Projection

A MAP PROJECTION in which areas on a sphere, and the areas of any features contained on it, are mapped to the plane in such a way that two are related by a *constant* scaling factor. No projection can be both equal-area and CONFORMAL, and projections which are neither equal-area nor CONFORMAL are sometimes called APHYLACTIC (Snyder 1987, p. 4). Equal-area projections are also called EQUIVALENT, HOMOLOGRAPHIC, HOMALOGRAPHIC, AUTHALIC, or EQUIAREAL (Lee 1944; Snyder 1987, p. 4).

See also ALBERS EQUAL-AREA CONIC PROJECTION, APHYLACTIC PROJECTION, BEHRMANN CYLINDRICAL

EQUAL-AREA PROJECTION, CONFORMAL PROJECTION, CYLINDRICAL EQUAL-AREA PROJECTION, EQUIDISTANT PROJECTION, HAMMER-AITOFF EQUAL-AREA PROJECTION, LAMBERT AZIMUTHAL EQUAL-AREA PROJECTION, MAP PROJECTION

References

Lee, L. P. "The Nomenclature and Classification of Map Projections." *Empire Survey Rev.* **7**, 190–00, 1944.
Snyder, J. P. *Map Projections--A Working Manual.* U. S. Geological Survey Professional Paper 1395. Washington, DC: U. S. Government Printing Office, 1987.

Equality

A mathematical statement of the equivalence of two quantities. The equality "A is equal to B" is written $A = B$.

See also EQUAL, FORMULA, INEQUALITY

Equally Likely Outcomes Distribution

Let there be a set S with N elements, each of them having the same probability. Then

$$P(S) = P\left(\bigcup_{i=1}^{N} E_i\right) = \sum_{i=1}^{N} P(E_i)$$

$$= P(E_i) \sum_{i=1}^{N} 1 = NP(E_i).$$

Using $P(S) \equiv 1$ gives

$$P(E_i) = \frac{1}{N}.$$

See also UNIFORM DISTRIBUTION

Equation

A mathematical expression stating that two or more quantities are the same as one another, also called an EQUALITY, FORMULA, or IDENTITY.

See also EQUALITY, FORMULA, IDENTITY, INEQUATION

Equiaffinity

An AREA-preserving AFFINITY. Equiaffinities include the CROSSED HYPERBOLIC ROTATION, ELLIPTIC ROTATION, HYPERBOLIC ROTATION, and PARABOLIC ROTATION.

Equiangular Polygon

A POLYGON whose vertex angles are equal (Williams 1979, p. 32).

See also EQUILATERAL POLYGON, POLYGON, REGULAR POLYGON

References

Williams, R. *The Geometrical Foundation of Natural Structure: A Source Book of Design.* New York: Dover, 1979.

Equiangular Spiral

LOGARITHMIC SPIRAL

Equianharmonic Case

The case of the WEIERSTRASS ELLIPTIC FUNCTION with invariants $g_2 = 0$ and $g_3 = 1$.

See also LEMNISCATE CASE, PSEUDOLEMNISCATE CASE

References

Abramowitz, M. and Stegun, C. A. (Eds.). "Equianharmonic Case ($g_2 = 0$, $g_3 = 1$)." §18.13 in *Handbook of Mathematical Functions with Formulas, Graphs, and Mathematical Tables, 9th printing.* New York: Dover, p. 652, 1972.

Equiareal Projection

EQUAL-AREA PROJECTION

Equi-Brocard Center

The point Y for which the TRIANGLES BYC, CYA, and AYB have equal BROCARD ANGLES.

References

Kimberling, C. "Central Points and Central Lines in the Plane of a Triangle." *Math. Mag.* **67**, 163–87, 1994.

Equichordal Point

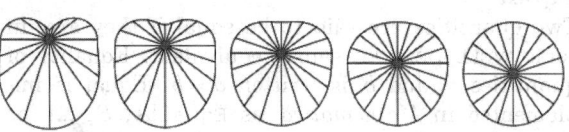

A point $\mathbf{p}$ for which all the CHORDS of a curve C passing through $\mathbf{p}$ are of the same length. In other words, $\mathbf{p}$ is an equichordal point if, for every chord $[\mathbf{x}, \mathbf{y}]$ of length p of the curve C, $\mathbf{p}$ satisfies

$$|\mathbf{x} - \mathbf{p}| + |\mathbf{y} - \mathbf{p}| = p.$$

A function $r(\theta)$ satisfying

$$r(0) = p - r(\pi)$$

corresponds to a curve with equichordal point $(0, 0)$ and chord length p defined by letting $r(\theta)$ be the polar equation of the half-curve for $0 \leq \theta \leq \pi$ and then superimposing the polar equation $r(\theta) - p$ over the same range. The curves illustrated above correspond to polar equations OF THE FORM

$$r(\theta) = x + (\tfrac{1}{2} - x) \cos(2\theta)$$

for various values of x.

Although it long remained an outstanding problem (the EQUICHORDAL POINT PROBLEM), it is now known that a plane convex region can have two equichordal points.

See also CHORD, EQUICHORDAL POINT PROBLEM, EQUIPRODUCT POINT, EQUIRECIPROCAL POINT

References

Croft, H. T.; Falconer, K. J.; and Guy, R. K. *Unsolved Problems in Geometry*. New York: Springer-Verlag, p. 9, 1991.

Dirac, G. A. "Ovals with Equichordal Points." *J. London Math. Soc.* **27**, 429–37, 1952.

Dirac, G. A. *J. London Math. Soc.* **28**, 245, 1953.

Hallstrom, A. P. "Equichordal and Equireciprocal Points." *Bogasici Univ. J. Sci.* **2**, 83–8, 1974.

Steinhaus, H. *Mathematical Snapshots, 3rd ed.* New York: Dover, p. 152, 1999.

Zindler, K. "Uuml;ber konvexe Gebilde, II." *Monatshefte f. Math. u. Phys.* **3**, 25–9, 1921.

Equichordal Point Problem

Is there a plane CONVEX SET having two distinct EQUICHORDAL POINTS? The problem was first proposed by Fujiwara (1916) and Blaschke *et al.* (1917), but long defied solution. Rogers went so far as to remark, "If you are interested in studying the problem, my first advice is: 'Don't'" (Croft *et al.* 1991, p. 9). This advice to the contrary, the problem was recently solved by Rychlik (1997).

See also EQUICHORDAL POINT

References

Blaschke, W.; Rothe, W.; and Weitzenböck, R. "Aufgabe 552." *Arch. Math. Phys.* **27**, 82, 1917.

Croft, H. T.; Falconer, K. J.; and Guy, R. K. "The Equichordal Point Problem." §A1 in *Unsolved Problems in Geometry*. New York: Springer-Verlag, pp. 9–1, 1991.

Fujiwara, M. "Über die Mittelkurve zweier geschlossenen konvexen Kurven in Bezug auf einen Punkt." *Tôhoku Math. J.* **10**, 99–03, 1916.

Rychlik, M. "The Equichordal Point Problem." *Elec. Res. Announcements Amer. Math. Soc.* **2**, 108–23, 1996.

Rychlik, M. "A Complete Solution to the Equichordal Problem of Fujiwara, Blaschke, Rothe, and Weitzenböck." *Invent. Math.* **129**, 141–12, 1997.

Wirsing, E. "Zur Analytisität von Doppelspeichkurven." *Arch. Math.* **9**, 300–07, 1958.

Equicross

RANGES and PENCILS which have equal CROSS-RATIOS are said to be equicross.

See also CROSS-RATIO, PENCIL, RANGE (LINE SEGMENT)

References

Durell, C. V. *Modern Geometry: The Straight Line and Circle.* London: Macmillan, pp. 74–6, 1928.

Lachlan, R. §422–28 in *An Elementary Treatise on Modern Pure Geometry.* London: Macmillian, pp. 269–74, 1893.

Equidecomposable

The ability of two plane or space regions to be DISSECTED into each other.

Equidigital Number

A number n is called equidigital if the number of digits in the prime factorization of n (including powers) uses the same number of digits as the number of digits in n. The first few equidigital numbers are 1, 2, 3, 5, 7, 10, 11, 13, 14, 15, 16, 17, 19, 21, 23, ... (Sloane's A046758).

See also ECONOMICAL NUMBER, WASTEFUL NUMBER

References

Pinch, R. G. E. "Economical Numbers." http://www.chalcedon.demon.co.uk/publish.html#62.

Santos, B. R. "Problem 2204. Equidigital Representation." *J. Recr. Math.* **27**, 58–9, 1995.

Sloane, N. J. A. Sequences A046758 in "An On-Line Version of the Encyclopedia of Integer Sequences." http://www.research.att.com/~njas/sequences/eisonline.html.

Weisstein, E. W. "Integer Sequences." MATHEMATICA NOTEBOOK INTEGERSEQUENCES.M.

Equidistance Postulate

PARALLEL lines are everywhere equidistant. This POSTULATE is equivalent to the PARALLEL AXIOM.

References

Dunham, W. "Hippocrates' Quadrature of the Lune." Ch. 1 in *Journey through Genius: The Great Theorems of Mathematics.* New York: Wiley, p. 54, 1990.

Equidistant Projection

A MAP PROJECTION in which the distances between one or two points and every other point on the map differ from the corresponding distances on the sphere by only a constant scaling factor (Snyder 1987, p. 4).

See also AZIMUTHAL EQUIDISTANT PROJECTION, CONFORMAL PROJECTION, CONIC EQUIDISTANT PROJECTION, CYLINDRICAL EQUIDISTANT PROJECTION, EQUAL-AREA PROJECTION, EQUIDISTANT PROJECTION, MILLER EQUIDISTANT PROJECTION

References

Snyder, J. P. *Map Projections--A Working Manual.* U. S. Geological Survey Professional Paper 1395. Washington, DC: U. S. Government Printing Office, 1987.

Equidistributed Sequence

A sequence of REAL NUMBERS $\{x_n\}$ is equidistributed if the probability of finding x_n in any subinterval is proportional to the subinterval length.

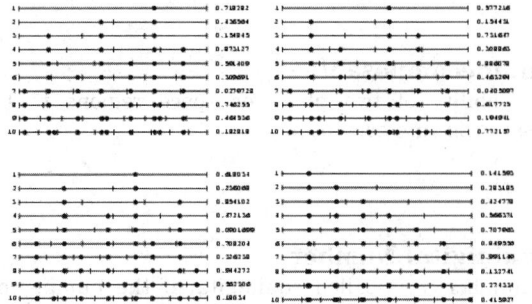

Consider the distribution of the FRACTIONAL PARTS of nr in the intervals bounded by 0, $1/n$, $2/n$, ..., $(n-1)/n$, 1. In particular, the number of empty intervals for $n = 1, 2, ...,$ are given below for E, the EULER-MASCHERONI CONSTANT γ, the GOLDEN RATIO ϕ, and PI.

r	Sloane	# Empty Intervals for $n = 1$, $2, ...,$
e	Sloane's A036412	0, 0, 0, 0, 1, 0, 0, 1, 1, 3, 1, 4, 4, 7, 5, ...
γ	Sloane's A046157	0, 0, 0, 1, 0, 0, 0, 1, 2, 2, 3, 0, 3, 5, 3, ...
ϕ	Sloane's A036414	0, 0, 0, 0, 0, 0, 1, 0, 2, 0, 1, 1, 0, 2, 2, ...
π	Sloane's A036416	0, 1, 1, 1, 1, 0, 0, 1, 2, 3, 4, 4, 5, 7, 7, ...

The values of n for which *no* bins are left blank are given in the following table.

r	Sloane	n with no empty intervals
e	Sloane's A036413	1, 2, 3, 4, 6, 7, 32, 35, 39, 71, 465, 536, 1001, ...
γ	Sloane's A046158	1, 2, 3, 5, 6, 7, 12, 19, 26, 97, 123, 149, 272, 395, ...
ϕ	Sloane's A036415	1, 2, 3, 4, 5, 6, 8, 10, 13, 16, 21, 34, 55, 89, 144, ...
π	Sloane's A036417	1, 6, 7, 106, 112, 113, 33102, 33215, ...

Steinhaus (1983) remarks that the highly uniform distribution of $\mathrm{frac}(n\phi)$ has its roots in the form of the CONTINUED FRACTION for ϕ.

See also PISOT-VIJAYARAGHAVAN CONSTANT, UNIFORM DISTRIBUTION, WEYL'S CRITERION

References

Kuipers, L. and Niederreiter, H. *Uniform Distribution of Sequences.* New York: Wiley, 1974.
Pólya, G. and Szego, G. *Problems and Theorems in Analysis I.* New York: Springer-Verlag, p. 88, 1972.
Sloane, N. J. A. Sequences A036412, A036413, A036414, A036415, A036416, A036417, A046157, and A046158 in "An On-Line Version of the Encyclopedia of Integer Sequences." http://www.research.att.com/~njas/sequences/eisonline.html.
Vardi, I. *Computational Recreations in Mathematica.* Reading, MA: Addison-Wesley, pp. 155–56, 1991.

Equilateral Hyperbola

RECTANGULAR HYPERBOLA

Equilateral Polygon

A POLYGON whose side are equal (Williams 1979, pp. 31–2).

See also EQUIANGULAR POLYGON, EQUILATERAL TRIANGLE, POLYGON, REGULAR POLYGON

References

Williams, R. *The Geometrical Foundation of Natural Structure: A Source Book of Design.* New York: Dover, 1979.

Equilateral Triangle

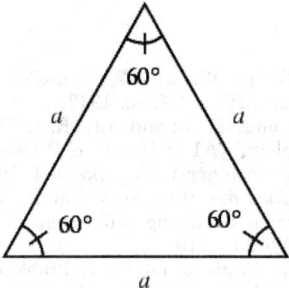

An equilateral triangle is a TRIANGLE with all three sides of equal length a. An equilateral triangle also has three equal $60°$ ANGLES.

The ALTITUDE h of an equilateral triangle is

$$h = \tfrac{1}{2}\sqrt{3}\,a, \tag{1}$$

where a is the side length, so the AREA is

$$A = \tfrac{1}{2}ah = \tfrac{1}{4}\sqrt{3}\,a^2. \tag{2}$$

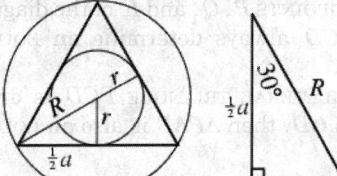

The INRADIUS r, CIRCUMRADIUS R, and AREA A can be computed directly from the formulas for a general REGULAR POLYGON with side length a and $n = 3$ sides,

$$r = \tfrac{1}{2} a \cot\left(\frac{\pi}{3}\right) = \tfrac{1}{2} a \tan\left(\frac{\pi}{6}\right) = \tfrac{1}{6}\sqrt{3}a \qquad (3)$$

$$R = \tfrac{1}{2} a \csc\left(\frac{\pi}{3}\right) = \tfrac{1}{2} a \sec\left(\frac{\pi}{6}\right) = \tfrac{1}{3}\sqrt{3}a \qquad (4)$$

$$A = \tfrac{1}{4} na^2 \cot\left(\frac{\pi}{3}\right) = \tfrac{1}{4}\sqrt{3}a^2. \qquad (5)$$

The AREAS of the INCIRCLE and CIRCUMCIRCLE are

$$A_r = \pi r^2 = \tfrac{1}{12}\pi a^2 \qquad (6)$$

$$A_R = \pi R^2 = \tfrac{1}{3}\pi a^2. \qquad (7)$$

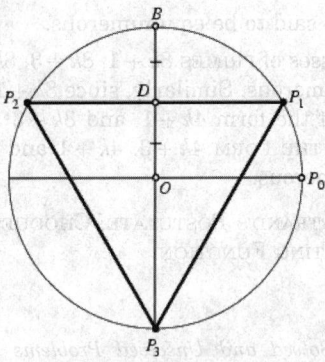

equilateral triangle construction

GEOMETRIC CONSTRUCTION of an equilateral consists of drawing a diameter of a circle OP_0 and then constructing its perpendicular bisector P_3OB. Bisect OB in point D, and extend the line P_1P_2 through D. The resulting figure $P_1P_2P_3$ is then an equilateral triangle. An equilateral triangle may also be constructed (although not using the usual Greek rules, which do *not* permit angle trisection) by TRISECTING all three ANGLES of any TRIANGLE (MORLEY'S THEOREM).

NAPOLEON'S THEOREM states that if three equilateral triangles are drawn on the LEGS of any TRIANGLE (either all drawn inwards or outwards) and the centers of these triangles are connected, the result is another equilateral triangle.

Given the distances of a point from the three corners of an equilateral triangle, a, b, and c, the length of a side s is given by

$$3(a^4 + b^4 + c^4 + s^4) = (a^2 + b^2 + c^2 + s^2)^2 \qquad (8)$$

(Gardner 1977, pp. 56–7 and 63). There are infinitely many solutions for which a, b, and c are INTEGERS. In these cases, one of a, b, c, and s is DIVISIBLE by 3, one by 5, one by 7, and one by 8 (Guy 1994, p. 183).

Begin with an arbitrary TRIANGLE and find the EXCENTRAL TRIANGLE. Then find the EXCENTRAL TRIANGLE of that triangle, and so on. Then the resulting triangle approaches an equilateral triangle. The only RATIONAL TRIANGLE is the equilateral triangle (Conway and Guy 1996). A POLYHEDRON composed of only equilateral triangles is known as a DELTAHEDRON.

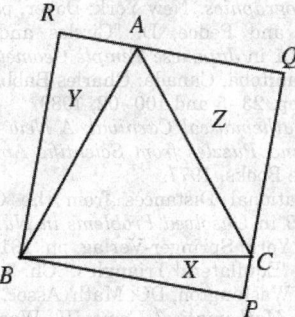

Let any RECTANGLE be circumscribed about an EQUILATERAL TRIANGLE. Then

$$X + Y = Z, \qquad (9)$$

where X, Y, and Z are the AREAS of the triangles in the figure (Honsberger 1985).

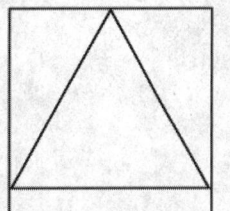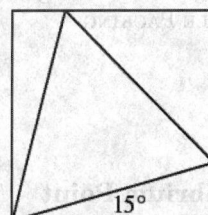

The smallest equilateral triangle which can be inscribed in a UNIT SQUARE (left figure) has side length and area

$$s = 1 \qquad (10)$$

$$A = \tfrac{1}{4}\sqrt{3} \approx 0.4330. \qquad (11)$$

The largest equilateral triangle which can be inscribed (right figure) is oriented at an angle of 15° and has side length and area

$$s = \sec(15°) = \sqrt{6} - \sqrt{2} \qquad (12)$$

$$A = 2\sqrt{3} - 3 \approx 0.4641 \qquad (13)$$

(Madachy 1979).

See also ACUTE TRIANGLE, DELTAHEDRON, EQUILIC QUADRILATERAL, FERMAT POINTS, GYROELONGATED

SQUARE DIPYRAMID, ICOSAHEDRON, ISOSCELES TRIAN-
GLE, MORLEY'S THEOREM, OCTAHEDRON, PENTAGONAL
DIPYRAMID, REULEAUX TRIANGLE, RIGHT TRIANGLE,
SCALENE TRIANGLE, SNUB DISPHENOID, TETRAHE-
DRON, TRIANGLE, TRIANGLE PACKING, TRIANGULAR
DIPYRAMID, TRIAUGMENTED TRIANGULAR PRISM, VI-
VIANI'S THEOREM

References

Beyer, W. H. (Ed.). *CRC Standard Mathematical Tables,
28th ed.* Boca Raton, FL: CRC Press, p. 121, 1987.
Conway, J. H. and Guy, R. K. "The Only Rational Triangle."
In *The Book of Numbers.* New York: Springer-Verlag,
pp. 201 and 228–39, 1996.
Dixon, R. *Mathographics.* New York: Dover, p. 33, 1991.
Fukagawa, H. and Pedoe, D. "Circles and Equilateral
Triangles." §2.1 in *Japanese Temple Geometry Problems.*
Winnipeg, Manitoba, Canada: Charles Babbage Research
Foundation, pp. 23–5 and 100–02, 1989.
Gardner, M. *Mathematical Carnival: A New Round-Up of
Tantalizers and Puzzles from Scientific American.* New
York: Vintage Books, 1977.
Guy, R. K. "Rational Distances from the Corners of a
Square." §D19 in *Unsolved Problems in Number Theory,
2nd ed.* New York: Springer-Verlag, pp. 181–85, 1994.
Honsberger, R. "Equilateral Triangles." Ch. 3 in *Mathema-
tical Gems I.* Washington, DC: Math. Assoc. Amer., 1973.
Honsberger, R. *Mathematical Gems III.* Washington, DC:
Math. Assoc. Amer., pp. 19–1, 1985.
Madachy, J. S. *Madachy's Mathematical Recreations.* New
York: Dover, pp. 115 and 129–31, 1979.

Equilateral Triangle Packing

TRIANGLE PACKING

Equilibrium Point

An equilibrium point in GAME THEORY is a set of
strategies $\{\hat{x}_1, \ldots, \hat{x}_n\}$ such that the ith payoff
function $K_i(\mathbf{x})$ is larger or equal for any other ith
strategy, i.e.,

$$K_i(\hat{x}_1, \ldots, \hat{x}_n) \geq K_i(\hat{x}_1, \ldots, \hat{x}_{i-1}, x_i, \hat{x}_{i+1}, \ldots, \hat{x}_n).$$

NASH EQUILIBRIUM

Equilic Quadrilateral

A QUADRILATERAL in which a pair of opposite sides
have the same length and are inclined at 60° to each
other (or equivalently, satisfy $\langle A \rangle + \langle B \rangle = 120°$).
Some interesting theorems hold for such quadrilat-
erals. Let *ABCD* be an equilic quadrilateral with
$AD = BC$ and $\langle A \rangle + \langle B \rangle = 120°$. Then

1. The MIDPOINTS P, Q, and R of the diagonals and
the side CD always determine an EQUILATERAL
TRIANGLE.
2. If EQUILATERAL TRIANGLE PCD is drawn outwardly on CD, then ΔPAB is also an EQUILATERAL
TRIANGLE.
3. If EQUILATERAL TRIANGLES are drawn on AC,
DC, and DB away from AB, then the three new
VERTICES P, Q, and R are COLLINEAR.

See Honsberger (1985) for additional theorems.

References

Garfunkel, J. "The Equilic Quadrilateral." *Pi Mu Epsilon J.*
7, 317–29, 1981.
Honsberger, R. *Mathematical Gems III.* Washington, DC:
Math. Assoc. Amer., pp. 32–5, 1985.

Equinumerous

Let A and B be two classes of POSITIVE INTEGERS. Let
$A(n)$ be the number of integers in A which are less
than or equal to n, and let $B(n)$ be the number of
integers in B which are less than or equal to n. Then
if

$$A(n) \sim B(n),$$

A and B are said to be equinumerous.

The four classes of PRIMES $8k + 1$, $8k + 3$, $8k + 5$, $8k +
7$ are equinumerous. Similarly, since $8k + 1$ and $8k +
5$ are both of the form $4k + 1$, and $8k + 3$ and $8k + 7$
are both OF THE FORM $4k + 3$, $4k + 1$ and $4k + 3$ are
also equinumerous.

See also BERTRAND'S POSTULATE, CHOQUET THEORY,
PRIME COUNTING FUNCTION

References

Shanks, D. *Solved and Unsolved Problems in Number
Theory, 4th ed.* New York: Chelsea, pp. 21–2 and 31–2,
1993.

Equipollent

Two statements in LOGIC are said to be equipollent if
they are deducible from each other.

Two sets A and B are said to be equipollent IFF there
is a one-to-one function (i.e., a BIJECTION) from A onto
B (Moore 1982, p. 10; Rubin 1967, p. 67; Suppes
1972, p. 91).

The term equipotent is sometimes used instead of
equipollent.

References

Moore, G. H. *Zermelo's Axiom of Choice: Its Origin, Devel-
opment, and Influence.* New York: Springer-Verlag, 1982.
Rubin, J. E. *Set Theory for the Mathematician.* New York:
Holden-Day, 1967.
Suppes, P. *Axiomatic Set Theory.* New York: Dover, 1972.

Equipotent
EQUIPOLLENT

Equipotential Curve
A curve in 2-D on which the value of a function $f(x, y)$ is a constant. Other synonymous terms are ISARITHM and ISOPLETH. A plot of several equipotential curves is called a CONTOUR PLOT.

See also CONTOUR PLOT, LEMNISCATE

Equiproduct Point
A point, such as interior points of a disk, such that

$$(px)(py) = [\text{const}],$$

where p is the CHORD length.

See also EQUICHORDAL POINT, EQUIRECIPROCAL POINT

Equireciprocal Point
p is an equireciprocal point if, for every chord [**x**, **y**] of a curve C, **p** satisfies

$$|\mathbf{x} - \mathbf{p}|^{-1} + |\mathbf{y} - \mathbf{p}|^{-1} = c$$

for some constant c. The FOCI of an ELLIPSE are equichordal points.

See also EQUICHORDAL POINT, EQUIPRODUCT POINT

References
Croft, H. T.; Falconer, K. J.; and Guy, R. K. *Unsolved Problems in Geometry.* New York: Springer-Verlag, p. 10, 1991.
Falconer, K. J. "On the Equireciprocal Point Problem." *Geom. Dedicata* **14**, 113–26, 1983.
Hallstrom, A. P. "Equichordal and Equireciprocal Points." *Bogasici Univ. J. Sci.* **2**, 83–8, 1974.
Klee, V. "Can a Plane Convex Body have Two Equireciprocal Points?" *Amer. Math. Monthly* **76**, 54–5, 1969.
Klee, V. "Correction to 'Can a Plane Convex Body have Two Equireciprocal Points?'" *Amer. Math. Monthly* **78**, 114, 1971.

Equirectangular Projection

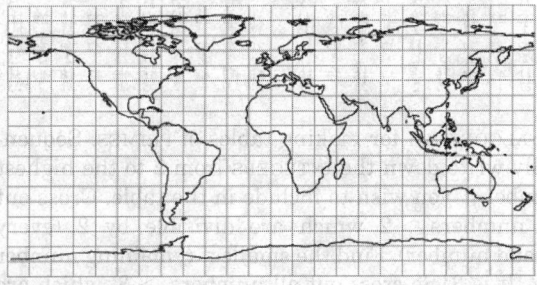

A CYLINDRICAL EQUIDISTANT PROJECTION, also called a RECTANGULAR PROJECTION, PLANE CHART, PLATE CARRE, or UNPROJECTED MAP, in which the horizontal coordinate is the longitude and the vertical coordinate is the latitude, so the standard parallel is taken as $\phi_1 = 0$.

See also CYLINDRICAL EQUIDISTANT PROJECTION

Equiripple
A distribution of ERROR such that the ERROR remaining is always given approximately by the last term dropped.

Equitangential Curve
TRACTRIX

Equivalence
BICONDITIONAL, EQUIVALENT

Equivalence Class
An equivalence class is defined as a SUBSET OF THE FORM $\{x \in X : xRa\}$, where a is an element of X and the NOTATION "xRy" is used to mean that there is an EQUIVALENCE RELATION between x and y. It can be shown that any two equivalence classes are either equal or disjoint, hence the collection of equivalence classes forms a partition of X. For all $a, b \in X$, we have aRb IFF a and b belong to the same equivalence class.

A set of CLASS REPRESENTATIVES is a SUBSET of X which contains EXACTLY ONE element from each equivalence class.

For n a POSITIVE INTEGER, and a, b INTEGERS, consider the CONGRUENCE $a \equiv b \pmod{n}$, then the equivalence classes are the sets $\{\ldots, -2n, -n, 0, n, 2n, \ldots\}$, $\{\ldots, 1-2n, 1-n, 1, 1+n, 1+2n, \ldots\}$ etc. The standard CLASS REPRESENTATIVES are taken to be $0, 1, 2, \ldots, n-1$.

See also CONGRUENCE, COSET

References
Shanks, D. *Solved and Unsolved Problems in Number Theory, 4th ed.* New York: Chelsea, pp. 56–7, 1993.

Equivalence Moves
REIDEMEISTER MOVES

Equivalence Problem
METRIC EQUIVALENCE PROBLEM

Equivalence Relation
An equivalence relation on a set X is a SUBSET of $X \times X$, i.e., a collection R of ordered pairs of elements of X, satisfying certain properties. Write "xRy" to mean (x, y) is an element of R, and we say "x is related to y," then the properties are

1. Reflexive: aRa for all $a \in X$,
2. Symmetric: aRb IMPLIES bRa for all a, $b \in X$
3. Transitive: aRb and bRc imply aRc for all a, b, $c \in X$,

where these three properties are completely independent. Other notations are often used to indicate a relation, e.g., $a \equiv b$ or $a \sim b$.

See also EQUIVALENCE CLASS, TEICHMÜLLER SPACE

References

Skiena, S. *Implementing Discrete Mathematics: Combinatorics and Graph Theory with Mathematica.* Reading, MA: Addison-Wesley, p. 18, 1990.
Stewart, I. and Tall, D. *The Foundations of Mathematics.* Oxford, England: Oxford University Press, 1977.

Equivalent

If $A \Rightarrow B$ and $B \Rightarrow A$ (i.e, $A \Rightarrow B \wedge B \Rightarrow A$, where $\Rightarrow$ denotes IMPLIES), then A and B are said to be equivalent, a relationship which is written symbolically as $A \equiv B$ (Carnap 1958, p. 8), $A \Leftrightarrow B$, or $A \rightleftharpoons B$. Equivalence is implemented in *Mathematica* as `Equal[A, B, ...]`. Binary equivalence has the following TRUTH TABLE (Carnap 1958, p. 10).

A	B	$A \equiv B$
T	T	T
T	F	F
F	T	F
F	F	T

Similarly, ternary equivalence has the following TRUTH TABLE.

A	B	C	$A \equiv B \equiv C$
T	T	T	T
T	T	F	F
T	F	T	F
T	F	F	F
F	T	T	F
F	T	F	F
F	F	T	F
F	F	F	T

The opposite of being equivalent is being NONEQUIVALENT.

Note that the symbol $\equiv$ is confusingly used in at least two other different contexts. If A and B are "equivalent by definition" (i.e., A is DEFINED to be B), this is written $A \equiv B$, and "a is CONGRUENT to b modulo m" is written $a \equiv b \pmod{m}$.

See also BICONDITIONAL, CONNECTIVE, DEFINED, IFF, IMPLIES, NONEQUIVALENT

References

Carnap, R. *Introduction to Symbolic Logic and Its Applications.* New York: Dover, p. 8, 1958.

Equivalent Matrix

Two matrices A and B are equal to each other, written $A = B$, if they have the same dimensions $m \times n$ and the same elements $a_{ij} = b_{ij}$ for $i = 1, ..., n$ and $j = 1, ..., m$.

Gradshteyn and Ryzhik (2000) call an $m \times n$ MATRIX A "equivalent" to another $m \times n$ MATRIX B IFF

$$B = PAQ$$

for P and Q any suitable nonsingular $m \times n$ and $n \times n$ MATRICES, respectively.

See also MATRIX

References

Gradshteyn, I. S. and Ryzhik, I. M. *Tables of Integrals, Series, and Products, 6th ed.* San Diego, CA: Academic Press, p. 1103, 2000.

Equivalent Projection

EQUAL-AREA PROJECTION

Eratosthenes Sieve

An ALGORITHM for making tables of PRIMES. Sequentially write down the INTEGERS from 2 to the highest number n you wish to include in the table. Cross out all numbers > 2 which are divisible by 2 (every second number). Find the smallest remaining number > 2. It is 3. So cross out all numbers > 3 which are divisible by 3 (every third number). Find the smallest remaining number > 3. It is 5. So cross out all numbers > 5 which are divisible by 5 (every fifth number).

Continue until you have crossed out all numbers divisible by $\lfloor \sqrt{n} \rfloor$, where $\lfloor x \rfloor$ is the FLOOR FUNCTION. The numbers remaining are PRIME. This procedure is illustrated in the above diagram which sieves up to 50, and therefore crosses out PRIMES up to $\lfloor \sqrt{50} \rfloor = 7$. If the procedure is then continued up to n, then the number of cross-outs gives the number of distinct PRIME FACTORS of each number.

References

Conway, J. H. and Guy, R. K. *The Book of Numbers.* New York: Springer-Verlag, pp. 127–30, 1996.
Ribenboim, P. *The New Book of Prime Number Records.* New York: Springer-Verlag, pp. 20–1, 1996.

Erdos Number

The number of "hops" needed to connect the author of a paper with the prolific late mathematician Paul Erdos. An author's Erdos number is 1 if he has co-authored a paper with Erdos, 2 if he has co-authored a paper with someone who has co-authored a paper with Erdos, etc. (Hoffman 1998, p. 13).

References

de Castro, R. and Grossman, J. W. "Famous Trails to Paul Erdos." *Math. Intell.* **21**, 51–3, 1999.
Grossman, J. and Ion, P. "The Erdos Number Project." http://www.acs.oakland.edu/~grossman/erdoshp.html.
Hoffman, P. *The Man Who Loved Only Numbers: The Story of Paul Erdos and the Search for Mathematical Truth.* New York: Hyperion, 1998.
Lewandowski, J.; Nurowski, P.; and Abramowicz, M. A. "Erdos Number Updates." *Math. Intell.* **22**, 3, 2000.

Erdos Reciprocal Sum Constants

A-SEQUENCE, B_2-SEQUENCE, NONAVERAGING SEQUENCE

Erdos Squarefree Conjecture

The CENTRAL BINOMIAL COEFFICIENT $\binom{2n}{n}$ is never SQUAREFREE for $n > 4$. This was proved true for all sufficiently large n by SÁRKOZY'S THEOREM. Goetgheluck (1988) proved the CONJECTURE true for $4 < n \le 2^{42205184}$ and Vardi (1991) for $4 < n < 2^{774840978}$. The conjecture was proved true in its entirety by Granville and Ramare (1996).

See also CENTRAL BINOMIAL COEFFICIENT

References

Erdos, P. and Graham, R. L. *Old and New Problems and Results in Combinatorial Number Theory.* Geneva, Switzerland: L'Enseignement Mathématique Université de Genève, Vol. 28, p. 71, 1980.
Goetgheluck, P. "Prime Divisors of Binomial Coefficients." *Math. Comput.* **51**, 325–29, 1988.
Granville, A. and Ramare, O. "Explicit Bounds on Exponential Sums and the Scarcity of Squarefree Binomial Coefficients." *Mathematika* **43**, 73–07, 1996.

Sander, J. W. "On Prime Divisors of Binomial Coefficients." *Bull. London Math. Soc.* **24**, 140–42, 1992.
Sander, J. W. "A Story of Binomial Coefficients and Primes." *Amer. Math. Monthly* **102**, 802–07, 1995.
Sárkozy, A. "On Divisors of Binomial Coefficients. I." *J. Number Th.* **20**, 70–0, 1985.
Vardi, I. "Applications to Binomial Coefficients." *Computational Recreations in Mathematica.* Reading, MA: Addison-Wesley, pp. 25–8, 1991.

Erdos-Anning Theorem

If an infinite number of points in the PLANE are all separated by INTEGER distances, then all the points lie on a straight LINE.

Erdos-Heilbronn Conjecture

Erdos and Heilbronn (Erdos and Graham 1980) posed the problem of estimating from below the number of sums $a + b$ where $a \in A$ and $b \in B$ range over given sets $A, B \subseteq \mathbb{Z}/p\mathbb{Z}$ of residues modulo a prime p, so that $a \ne b$. Dias da Silva and Hamidoune (1994) gave a solution, and Alon *et al.* (1995) developed a polynomial method that allows one to handle restrictions of the type $f(a, b) \ne 0$, where f is a polynomial in two variables over $\mathbb{Z}/p\mathbb{Z}$.

References

Alon, N.; Nathanson, M. B.; and Ruzsa, I. Z. "Adding Distinct Congruence Classes Modulo a Prime." *Amer. Math. Monthly* **102**, 250–55, 1995.
Dias da Silva, J. A. and Hamidoune, Y. O. "Cyclic Spaces for Grassmann Derivatives and Additive Theory." *Bull. London Math. Soc.* **26**, 140–46, 1994.
Erdos, P. and Graham, R. L. *Old and New Problems and Results in Combinatorial Number Theory.* Geneva, Switzerland: L'Enseignement Mathématique Université de Genève, Vol. 28, 1980.
Lev, V. F. "Restricted Set Addition in Groups, II. A Generalization of the Erdos-Heilbronn Conjecture.." *Electronic J. Combinatorics* **7**, No. 1, R4, 1–0, 2000. http://www.combinatorics.org/Volume_7/v7i1toc.html.

Erdos-Ivic Conjecture

There are infinitely many primes m which divide some value of the PARTITION FUNCTION P.

See also NEWMAN'S CONJECTURE, PARTITION FUNCTION P

References

Erdos, P. and Ivic, A. "The Distribution of Certain Arithmetical Functions at Consecutive Integers." In *Proc. Budapest Conf. Number Th., Coll. Math. Soc. J. Bolyai* **51**, 45–1, 1989.
Ono, K. "Distribution of the Partition Functions Modulo m." *Ann. Math.* **151**, 293–07, 2000.

Erdos-Kac Theorem

A deeper result than the HARDY-RAMANUJAN THEOREM. Let $N(x, a, b)$ be the number of INTEGERS in $[3, x]$ such that inequality

$$a \leq \frac{\omega(n) - \ln \ln n}{\sqrt{\ln \ln n}} \leq b$$

holds, where $\omega(n)$ is the number of DISTINCT PRIME FACTORS of n. Then

$$\lim_{x \to \infty} N(x, a, b) = \frac{(x + o(x))}{\sqrt{2\pi}} \int_a^b e^{-t^2/2} \, dt.$$

The theorem is discussed in Kac (1959).

See also DISTINCT PRIME FACTORS

References

Kac, M. *Statistical Independence in Probability, Analysis and Number Theory.* New York: Wiley, 1959.
Riesel, H. "The Erdos-Kac Theorem." *Prime Numbers and Computer Methods for Factorization, 2nd ed.* Boston, MA: Birkhäuser, pp. 158–59, 1994.

Erdos-Mordell Theorem

If O is any point inside a TRIANGLE ΔABC, and P, Q, and R are the feet of the perpendiculars from O upon the respective sides BC, CA, and AB, then

$$OA + OB + OC \geq 2(OP + OQ + OR).$$

Oppenheim (1961) and Mordell (1962) also showed that

$$OA \times OB \times OC \geq (OQ + OR)(OR + OP)(OP + OQ).$$

References

Bankoff, L. "An Elementary Proof of the Erdos-Mordell Theorem." *Amer. Math. Monthly* **65**, 521, 1958.
Brabant, H. "The Erdos-Mordell Inequality Again." *Nieuw Tijdschr. Wisk.* **46**, 87, 1958/1959.
Casey, J. *A Sequel to the First Six Books of the Elements of Euclid, 6th ed.* Dublin: Hodges, Figgis, & Co., p. 253, 1892.
Coxeter, H. S. M. *Introduction to Geometry, 2nd ed.* New York: Wiley, p. 9, 1969.
Erdos, P. "Problem 3740." *Amer. Math. Monthly* **42**, 396, 1935.
Fejes-Tóth, L. *Lagerungen in der Ebene auf der Kugel und im Raum.* Berlin: Springer, 1953.
Mordell, L. J. "On Geometric Problems of Erdos and Oppenheim." *Math. Gaz.* **46**, 213–15, 1962.
Mordell, L. J. and Barrow, D. F. "Solution to Problem 3740." *Amer. Math. Monthly* **44**, 252–54, 1937.
Oppenheim, A. "The Erdos Inequality and Other Inequalities for a Triangle." *Amer. Math. Monthly* **68**, 226–30 and 349, 1961.
Veldkamp, G. R. "The Erdos-Mordell Inequality." *Nieuw Tijdschr. Wisk.* **45**, 193–96, 1957/1958.

Erdos-Moser Equation

The DIOPHANTINE EQUATION

$$\sum_{j=1}^{m-1} j^n = m^n.$$

Erdos conjectured that there is no solution to this

equation other than the trivial solution $1^1 + 2^1 = 3^1$, although this remains unproved (Guy 1994, pp. 153–54). Moser (1953) proved that there is no solution for $m < 10^{10^6}$, and Butske *et al.* (1999) extended this to $m < 10^{9.3 \times 10^6}$, or more specifically, $m < 1.485 \times 10^{9321155}$.

References

Butske, W.; Jaje, L. M.; and Mayernik, D. R. "The Equation $\Sigma_{p|N} 1/p + 1/N = 1$, Pseudoperfect Numbers, and Partially Weighted Graphs." *Math. Comput.* **69**, 407–20, 1999.
Guy, R. K. *Unsolved Problems in Number Theory, 2nd ed.* New York: Springer-Verlag, 1994.
Moree, P. "Diophantine Equations of Erdos-Moser Type." *Bull. Austral. Math. Soc.* **53**, 281–92, 1996.
Moser, L. "On the Diophantine Equation $1^n + 2^n + 3^n + \ldots + (m-1)^n = m^n$." *Scripta Math.* **19**, 84–8, 1953.

Erdos-Selfridge Function

The Erdos-Selfridge function $g(k)$ is defined as the least integer bigger than $k + 1$ such that the LEAST PRIME FACTOR of $\binom{g(k)}{k}$ exceeds k (Ecklund *et al.* 1974, Erdos *et al.* 1993). The best lower bound known is

$$g(k) \geq \exp\left(c \sqrt{\frac{[\ln k]^3}{\ln \ln k}}\right)$$

(Granville and Ramare 1996). Scheidler and Williams (1992) tabulated $g(k)$ up to $k = 140$, and Lukes *et al.* (1997) tabulated $g(k)$ for $135 \leq k \leq 200$. The values for $n = 2, 3, \ldots$ are 4, 7, 7, 23, 62, 143, 44, 159, 46, 47, 174, 2239, ... (Sloane's A046105).

See also BINOMIAL COEFFICIENT, GOOD BINOMIAL COEFFICIENT, LEAST PRIME FACTOR

References

Ecklund, E. F. Jr.; Erdos, P.; and Selfridge, J. L. "A New Function Associated with the prime factors of $\binom{n}{k}$." *Math. Comput.* **28**, 647–49, 1974.
Erdos, P.; Lacampagne, C. B.; and Selfridge, J. L. "Estimates of the Least Prime Factor of a Binomial Coefficient." *Math. Comput.* **61**, 215–24, 1993.
Granville, A. and Ramare, O. "Explicit Bounds on Exponential Sums and the Scarcity of Squarefree Binomial Coefficients." *Mathematika* **43**, 73–07, 1996.
Lukes, R. F.; Scheidler, R.; and Williams, H. C. "Further Tabulation of the Erdos-Selfridge Function." *Math. Comput.* **66**, 1709–717, 1997.
Scheidler, R. and Williams, H. C. "A Method of Tabulating the Number-Theoretic Function $g(k)$." *Math. Comput.* **59**, 251–57, 1992.
Sloane, N. J. A. Sequences A046105 in "An On-Line Version of the Encyclopedia of Integer Sequences." http://www.research.att.com/~njas/sequences/eisonline.html.

Erdos-Stone Theorem

A generalization of TURÁN'S THEOREM to non-COMPLETE GRAPHS.

See also CLIQUE, EXTREMAL GRAPH THEORY, TURÁN'S THEOREM

References

Chvátal, V. and Szemerédi, E. "On the Erdos-Stone Theorem." *J. London Math. Soc.* **23**, 207–14, 1981.

Pach, J. and Agarwal, P. K. *Combinatorial Geometry.* New York: Wiley, 1995.

Erdos-Szekeres Theorem

Suppose a, $b \in \mathbb{N}$, $n = ab + 1$, and x_1, ..., x_n is a sequence of n REAL NUMBERS. Then this sequence contains a MONOTONIC increasing (decreasing) subsequence of $a + 1$ terms or a MONOTONIC decreasing (increasing) subsequence of $b + 1$ terms. DILWORTH'S LEMMA is a generalization of this theorem.

See also COMBINATORICS, DILWORTH'S LEMMA

References

Hoffman, P. *The Man Who Loved Only Numbers: The Story of Paul Erdos and the Search for Mathematical Truth.* New York: Hyperion, pp. 54–5, 1998.

Erdos-Turán Theorem

For any integers a_i with

$$1 \le a_1 < a_2 < \cdots < a_k \le n,$$

the proportion of PERMUTATIONS in the SYMMETRIC GROUP S_n whose cyclic decompositions contain no cycles of lengths $a_1, a_2, \ldots, a_k$ is at most

$$\left(\sum_{i=1}^{k} \frac{1}{a_i} \right)^{-1}$$

(Erdos and Turán 1967, Dixon 1969).

See also CYCLE (PERMUTATION), SYMMETRIC GROUP

References

Dixon, J. D. "The Probability of Generating the Symmetric Group." *Math. Z.* **110**, 199–05, 1969.

Erdos, P. and Turán, P. "On Some Problems in Statistical Group Theory. II." *Acta Math. Acad. Sci. Hung.* **18**, 151–63, 1867.

Erf

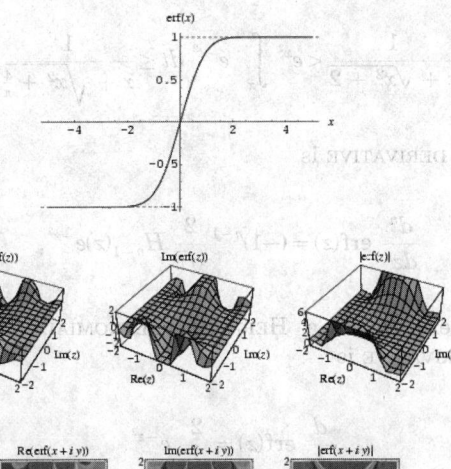

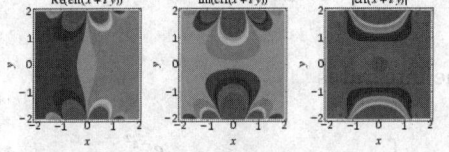

The "error function" encountered in integrating the GAUSSIAN DISTRIBUTION (which is a normalized form of the GAUSSIAN FUNCTION),

$$\operatorname{erf}(z) \equiv \frac{2}{\sqrt{\pi}} \int_0^z e^{-t^2}\, dt \tag{1}$$

$$= 1 - \operatorname{erfc}(z) \tag{2}$$

$$= \pi^{-1/2} \gamma(\tfrac{1}{2},\, z^2), \tag{3}$$

where ERFC is the complementary error function and $\gamma(x,\, a)$ is the incomplete GAMMA FUNCTION. It can also be defined as a MACLAURIN SERIES

$$\operatorname{erf}(z) = \frac{2}{\sqrt{\pi}} \sum_{n=0}^{\infty} \frac{(-1)^n z^{2n+1}}{n!(2n+1)}. \tag{4}$$

Erf has the values

$$\operatorname{erf}(0) = 0 \tag{5}$$

$$\operatorname{erf}(\infty) = 1. \tag{6}$$

It is an ODD FUNCTION

$$\operatorname{erf}(-z) = -\operatorname{erf}(z), \tag{7}$$

and satisfies

$$\operatorname{erf}(z) + \operatorname{erfc}(z) = 1. \tag{8}$$

Erf may be expressed in terms of a CONFLUENT HYPERGEOMETRIC FUNCTION OF THE FIRST KIND M as

$$\operatorname{erf}(z) = \frac{2z}{\sqrt{\pi}} M(\tfrac{1}{2}, \tfrac{3}{2}, -z^2) = \frac{2z}{\sqrt{\pi}} e^{-z^2} M(1, \tfrac{3}{2}, z^2). \tag{9}$$

Erf is bounded by

$$\frac{1}{x + \sqrt{x^2 + 2}} < e^{x^2} \int_x^\infty e^{-t^2}\, dt \le \frac{1}{x + \sqrt{x^2 + \frac{4}{\pi}}}. \quad (10)$$

Its DERIVATIVE is

$$\frac{d^n}{dz^n}\, \mathrm{erf}(z) = (-1)^{n-1} \frac{2}{\sqrt{\pi}}\, H_{n-1}(z)e^{-z^2}, \quad (11)$$

where H_n is a HERMITE POLYNOMIAL. The first DERIVATIVE is

$$\frac{d}{dz}\, \mathrm{erf}(z) = \frac{2}{\sqrt{\pi}}\, e^{-z^2}, \quad (12)$$

and the integral is

$$\int \mathrm{erf}(z)\, dz = z\, \mathrm{erf}(z) + \frac{e^{-z^2}}{\sqrt{\pi}}. \quad (13)$$

For $x \ll 1$, erf may be computed from

$$\mathrm{erf}(x) = \frac{2}{\sqrt{\pi}} \int_0^x e^{-t^2}\, dt \quad (14)$$

$$= \frac{2}{\sqrt{\pi}} \int_0^x \sum_{k=0}^\infty \frac{(-t^2)^k}{k!}\, dt$$

$$= \frac{2}{\sqrt{\pi}} \int_0^x \sum_{k=0}^\infty \frac{(-1)^k t^{2k}}{k!}\, dt$$

$$= \frac{2}{\sqrt{\pi}} \sum_{k=0}^\infty \frac{x^{2k+1}(-1)^k}{k!(2k+1)} \quad (15)$$

$$= \frac{2}{\sqrt{\pi}}(x - \tfrac{1}{3}x^3 + \tfrac{1}{10}x^5 - \tfrac{1}{42}x^7 + \tfrac{1}{216}x^9 - \tfrac{1}{1320}x^{11} + \ldots) \quad (16)$$

$$= \frac{2}{\sqrt{\pi}}\, e^{-x^2} x \left[1 + \frac{2x^2}{1 \cdot 3} + \frac{(2x^2)^2}{1 \cdot 3 \cdot 5} + \ldots\right] \quad (17)$$

(Acton 1990). For $x \gg 1$,

$$\mathrm{erf}(x) = \frac{2}{\sqrt{\pi}} \left(\int_0^\infty e^{-t^2}\, dt - \int_x^\infty e^{-t^2}\, dt\right)$$

$$= 1 - \frac{2}{\sqrt{\pi}} \int_x^\infty e^{-t^2}\, dt. \quad (18)$$

Using INTEGRATION BY PARTS gives

$$\int_x^\infty e^{-t^2}\, dt = -\frac{1}{2} \int_x^\infty \frac{1}{t}\, d(e^{-t^2})$$

$$= -\frac{1}{2}\left[\frac{e^{-t^2}}{t}\right]_x^\infty - \frac{1}{2} \int_x^\infty \frac{e^{-t^2}\, dt}{t^2}$$

$$= \frac{e^{-x^2}}{2x} + \frac{1}{4} \int_x^\infty \frac{1}{t^3}\, d(e^{-t^2})$$

$$= \frac{e^{-x^2}}{2x} - \frac{e^{-x^2}}{4x^3} - \ldots, \quad (19)$$

so

$$\mathrm{erf}(x) = 1 - \frac{e^{-x^2}}{\sqrt{\pi} x}\left(1 - \frac{1}{2x^2} - \ldots\right) \quad (20)$$

and continuing the procedure gives the ASYMPTOTIC SERIES

$$\mathrm{erf}(x) = 1 - \frac{e^{-x^2}}{\sqrt{\pi}}$$
$$\times (x^{-1} - \tfrac{1}{2}x^{-3} + \tfrac{3}{4}x^{-5} - \tfrac{15}{8}x^{-7} + \tfrac{105}{16}x^{-9} + \ldots). \quad (21)$$

Ramanujan rediscovered the CONTINUED FRACTION formula

$$\int_0^a e^{-t^2}\, dt = \tfrac{1}{2}\sqrt{\pi}\, \mathrm{erf}\, a$$

$$= \tfrac{1}{2}\sqrt{\pi} - \frac{e^{-a^2}}{2a+}\, \frac{1}{a+}\, \frac{2}{2a+}\, \frac{3}{a+}\, \frac{4}{2a+\ldots}, \quad (22)$$

first stated by Laplace and proved by Jacobi (Watson 1928; Hardy 1999, pp. 8–).

A COMPLEX generalization of erf x is defined as

$$w(z) = e^{-z^2}\, \mathrm{erfc}(-iz) \quad (23)$$

$$= e^{-z^2}\left(1 + \frac{2i}{\sqrt{\pi}} + \frac{2i}{\sqrt{\pi}} \int_0^z e^{t^2}\, dt\right) \quad (24)$$

$$= \frac{i}{\pi} \int_{-\infty}^\infty \frac{e^{-t^2}\, dt}{z - t} = \frac{2iz}{\pi} \int_0^\infty \frac{e^{-t^2}\, dt}{z - t}. \quad (25)$$

See also DAWSON'S INTEGRAL, ERFC, ERFI, FRESNEL INTEGRALS, GAUSSIAN FUNCTION < GAUSSIAN INTEGRAL, NORMAL DISTRIBUTION FUNCTION, PROBABILITY INTEGRAL

References

Abramowitz, M. and Stegun, C. A. (Eds.). "Error Function and Fresnel Integrals." Ch. 7 in *Handbook of Mathema-*

tical Functions with Formulas, Graphs, and Mathematical Tables, 9th printing. New York: Dover, pp. 297–09, 1972.

Acton, F. S. *Numerical Methods That Work, 2nd printing.* Washington, DC: Math. Assoc. Amer., p. 16, 1990.

Arfken, G. *Mathematical Methods for Physicists, 3rd ed.* Orlando, FL: Academic Press, pp. 568–69, 1985.

Hardy, G. H. *Ramanujan: Twelve Lectures on Subjects Suggested by His Life and Work, 3rd ed.* New York: Chelsea, 1999.

Spanier, J. and Oldham, K. B. "The Error Function erf(*x*) and Its Complement erfc(*x*)." Ch. 40 in *An Atlas of Functions.* Washington, DC: Hemisphere, pp. 385–93, 1987.

Watson, G. N. "Theorems Stated by Ramanujan (IV): Theorems on Approximate Integration and Summation of Series." *J. London Math. Soc.* **3**, 282–89, 1928.

Whittaker, E. T. and Robinson, G. "The Error Function." §92 in *The Calculus of Observations: A Treatise on Numerical Mathematics, 4th ed.* New York: Dover, pp. 179–82, 1967.

Erfc

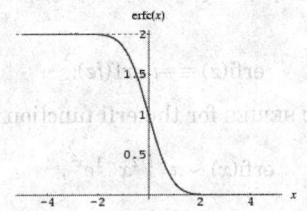

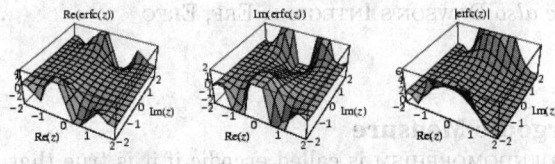

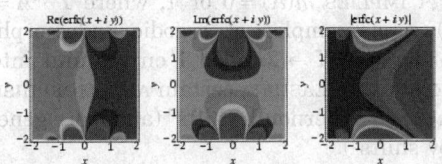

The "complementary error function" defined by

$$\operatorname{erfc}(x) \equiv 1 - \operatorname{erf}(x) \tag{1}$$

$$= \frac{2}{\sqrt{\pi}} \int_x^\infty e^{-t^2}\, dt \tag{2}$$

$$= \sqrt{\pi}\,\gamma(\tfrac{1}{2},\, z^2), \tag{3}$$

where γ is the incomplete GAMMA FUNCTION. It has the values

$$\operatorname{erfc}(0) = 1 \tag{4}$$

$$\lim_{x \to \infty} \operatorname{erfc}(x) = 0 \tag{5}$$

$$\operatorname{erfc}(-x) = 2 - \operatorname{erfc}(x) \tag{6}$$

$$\int_0^\infty \operatorname{erfc}(x)\, dx = \frac{1}{\sqrt{\pi}} \tag{7}$$

$$\int_0^\infty \operatorname{erfc}^2(x)\, dx = \frac{2 - \sqrt{2}}{\sqrt{\pi}}. \tag{8}$$

See also Erf, SERF, DIFFERENTIAL EQUATION, ERFI

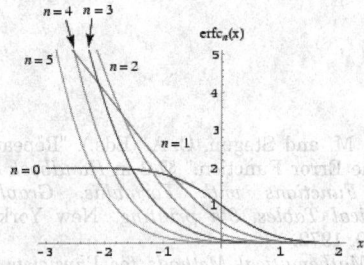

A generalization is obtained from the ERFC DIFFERENTIAL EQUATION

$$\frac{d^2 y}{dz^2} + 2z\,\frac{dy}{dz} - 2ny = 0 \tag{9}$$

(Abramowitz and Stegun 1972, p. 299; Zwillinger 1997, p. 122). The general solution is then

$$y = A\,\operatorname{erfc}_n(z) + B\,\operatorname{erfc}_n(-z), \tag{10}$$

where $\operatorname{erfc}_n(z)$ is the repeated erfc integral. For integral $n \geq 1$,

$$\operatorname{erfc}_n(z) = \underbrace{\int \cdots \int}_{n} \operatorname{erfc}(z)\, dz \tag{11}$$

$$= \frac{2}{\sqrt{2}} \int_z^\infty \frac{(t - z)^n}{n!}\, e^{-t^2}\, dt \tag{12}$$

$$= 2^{-n} e^{-z^2} \left[\frac{{}_1F_1(\tfrac{1}{2}(n+1);\ \tfrac{1}{2};\ z^2)}{\Gamma(1 + \tfrac{1}{2}n)} - \frac{2z\,{}_1F_1(1 + \tfrac{1}{2}n;\ \tfrac{3}{2};\ z^2)}{\Gamma(\tfrac{1}{2}(n+1))} \right] \tag{13}$$

(Abramowitz and Stegun 1972), where ${}_1F_1(a;\ b;\ z)$ is a CONFLUENT HYPERGEOMETRIC FUNCTION OF THE FIRST KIND and $\Gamma(z)$ is a GAMMA FUNCTION. The first few values, extended by the definition for $n = -1$ and 0, are given by

$$\operatorname{erfc}_{-1}(z) = \frac{2}{\sqrt{\pi}}\, e^{-z^2} \tag{14}$$

$$\operatorname{erfc}_0(z) = \operatorname{erfc}(z) \tag{15}$$

$$\operatorname{erfc}_1(z) = \frac{e^{-z^2}}{\sqrt{\pi}} - z\,\operatorname{erfc}(z) \tag{16}$$

$$\text{erfc}_2(z) = \frac{1}{4}\left[(1+2z^2)\,\text{erfc}(z) - \frac{2ze^{-z^2}}{\sqrt{\pi}}\right]. \qquad (17)$$

See also ERF, ERFC DIFFERENTIAL EQUATION, ERFI

References

Abramowitz, M. and Stegun, C. A. (Eds.). "Repeated Integrals of the Error Function." §7.2 in *Handbook of Mathematical Functions with Formulas, Graphs, and Mathematical Tables, 9th printing.* New York: Dover, pp. 299–00, 1972.

Arfken, G. *Mathematical Methods for Physicists, 3rd ed.* Orlando, FL: Academic Press, pp. 568–69, 1985.

Press, W. H.; Flannery, B. P.; Teukolsky, S. A.; and Vetterling, W. T. "Incomplete Gamma Function, Error Function, Chi-Square Probability Function, Cumulative Poisson Function." §6.2 in *Numerical Recipes in FORTRAN: The Art of Scientific Computing, 2nd ed.* Cambridge, England: Cambridge University Press, pp. 209–14, 1992.

Spanier, J. and Oldham, K. B. "The Error Function erf(x) and Its Complement erfc(x)" and "The exp(x) and erfc($\sqrt{x}$) and Related Functions." Chs. 40 and 41 in *An Atlas of Functions.* Washington, DC: Hemisphere, pp. 385–93 and 395–03, 1987.

Zwillinger, D. *Handbook of Differential Equations, 3rd ed.* Boston, MA: Academic Press, p. 122, 1997.

Erfc Differential Equation

The second-order ORDINARY DIFFERENTIAL EQUATION

$$y'' + 2xy' - 2ny = 0, \qquad (1)$$

whose solutions may be written either

$$y = A\,\text{erfc}_n(x) + B\,\text{erfc}_n(-x), \qquad (2)$$

where $\text{erfc}_n(x)$ is the repeated integral of the ERFC function (Abramowitz and Stegun 1972, p. 299), or

$$y = C_1 e^{-x^2} H_{-n-1}(x) + C_2\,{}_1F_1(\tfrac{1}{2}(n+1);\ \tfrac{1}{2};\ x^2), \qquad (3)$$

where $H_n(x)$ is a HERMITE POLYNOMIAL and ${}_1F_1(a;\ b;\ z)$ is a CONFLUENT HYPERGEOMETRIC FUNCTION OF THE FIRST KIND.

See also ERFC

References

Abramowitz, M. and Stegun, C. A. (Eds.). *Handbook of Mathematical Functions with Formulas, Graphs, and Mathematical Tables, 9th printing.* New York: Dover, p. 299, 1972.

Zwillinger, D. *Handbook of Differential Equations, 3rd ed.* Boston, MA: Academic Press, p. 122, 1997.

Erfi

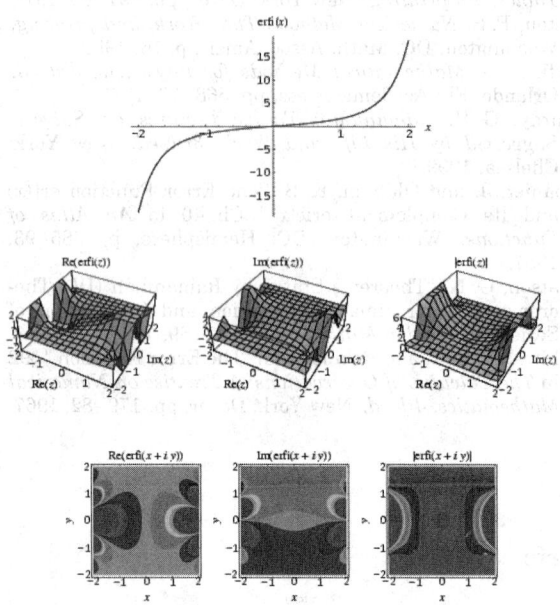

$$\text{erfi}(z) \equiv -i\,\text{erf}(iz).$$

A ASYMPTOTIC SERIES for the erfi function is given by

$$\text{erfi}(x) \sim \pi^{-1/2} x^{-1} e^{x^2}.$$

See also DAWSON'S INTEGRAL, ERF, ERFC

Ergodic Measure

An ENDOMORPHISM is called ergodic if it is true that $T^{-1}A = A$ IMPLIES $m(A) = 0$ or 1, where $T^{-1}A = \{x \in X : T(x) \in A\}$. Examples of ergodic endomorphisms include the MAP $X \to 2x$ mod 1 on the unit interval with LEBESGUE MEASURE, certain AUTOMORPHISMS of the TORUS, and "Bernoulli shifts" (and more generally "Markov shifts").

Given a MAP T and a SIGMA ALGEBRA, there may be many ergodic measures. If there is only one ergodic measure, then T is called uniquely ergodic. An example of a uniquely ergodic transformation is the MAP $x \mapsto x + a$ mod 1 on the unit interval when a is irrational. Here, the unique ergodic measure is LEBESGUE MEASURE.

Ergodic Theory

Ergodic theory can be described as the statistical and qualitative behavior of measurable group and semigroup actions on MEASURE SPACES. The GROUP is most commonly N, R, R+, and Z.

Ergodic theory had its origins in the work of Boltzmann in statistical mechanics problems where time-

and space-distribution averages are equal. Steinhaus (1983, pp. 237–39) gives a practical application to ergodic theory to keeping one's feet dry (when walking along a shoreline without having to constantly turn one's head to anticipate incoming waves. The mathematical origins of ergodic theory are due to von Neumann, Birkhoff, and Koopman in the 1930s. It has since grown to be a huge subject and has applications not only to statistical mechanics, but also to NUMBER THEORY, DIFFERENTIAL GEOMETRY, FUNCTIONAL ANALYSIS, etc. There are also many internal problems (e.g., ergodic theory being applied to ergodic theory) which are interesting.

See also AMBROSE-KAKUTANI THEOREM, BIRKHOFF'S ERGODIC THEOREM, DYE'S THEOREM, DYNAMICAL SYSTEM, HOPF'S THEOREM, ORNSTEIN'S THEOREM

References

Billingsley, P. *Ergodic Theory and Information.* New York: Wiley, 1965.

Cornfeld, I.; Fomin, S.; and Sinai, Ya. G. *Ergodic Theory.* New York: Springer-Verlag, 1982.

Katok, A. and Hasselblatt, B. *An Introduction to the Modern Theory of Dynamical Systems.* Cambridge, England: Cambridge University Press, 1996.

Nadkarni, M. G. *Basic Ergodic Theory.* India: Hindustan Book Agency, 1995.

Parry, W. *Topics in Ergodic Theory.* Cambridge, England: Cambridge University Press, 1982.

Petersen, K. *Ergodic Theory.* Cambridge, England: Cambridge University Press, 1983.

Radin, C. "Ergodic Theory." Ch. 1 in *Miles of Tiles.* Providence, RI: Amer. Math. Soc., pp. 17–4, 1999.

Sinai, Ya. G. *Topics in Ergodic Theory.* Princeton, NJ: Princeton University Press, 1993.

Smorodinsky, M. *Ergodic Theory, Entropy.* Berlin: Springer-Verlag, 1971.

Steinhaus, H. *Mathematical Snapshots, 3rd ed.* New York: Dover, pp. 237–39, 1999.

Walters, P. *Ergodic Theory: Introductory Lectures.* New York: Springer-Verlag, 1975.

Walters, P. *Introduction to Ergodic Theory.* New York: Springer-Verlag, 2000.

Ergodic Transformation

A transformation which has only trivial invariant SUBSETS is said to be ergodic.

Erlang Distribution

Given a POISSON DISTRIBUTION with a rate of change λ, the DISTRIBUTION FUNCTION $D(x)$ giving the waiting times until the hth Poisson event is

$$D(x) = 1 - \frac{\Gamma(h, x\lambda)}{\Gamma(h)} \qquad (1)$$

for $x \in [0, \infty)$, where $\Gamma(x)$ is a complete GAMMA FUNCTION, and $\Gamma(a, x)$ an INCOMPLETE GAMMA FUNCTION. With h explicitly an integer, this distribution is known as the Erlang distribution, and has probability function

$$P(x) = \frac{\lambda(\lambda x)^{h-1}}{(h-1)!} e^{-\lambda x}. \qquad (2)$$

It is closely related to the GAMMA DISTRIBUTION, which is obtained by letting $\alpha \equiv h$ (not necessarily an integer) and defining $\theta \equiv 1/\lambda$. When $h = 1$, it simplifies to the EXPONENTIAL DISTRIBUTION.

See also EXPONENTIAL DISTRIBUTION, GAMMA DISTRIBUTION

Erlanger Program

A program initiated by F. Klein in an 1872 lecture to describe geometric structures in terms of their AUTOMORPHISM GROUPS.

References

Klein, F. "Vergleichende Betrachtungen über neuere geometrische Forschungen." 1872.

Yaglom, I. M. *Felix Klein and Sophus Lie: Evolution of the Idea of Symmetry in the Nineteenth Century.* Boston, MA: Birkhäuser, 1988.

Ermakoff's Test

The series $\Sigma f(n)$ for a monotonic nonincreasing $f(x)$ is convergent if

$$\overline{\lim_{x \to \infty}} \frac{e^x f(e^x)}{f(x)} < 1$$

and divergent if

$$\lim_{x \to \infty} \frac{e^x f(e^x)}{f(x)} > 1.$$

References

Bromwich, T. J. I'a and MacRobert, T. M. *An Introduction to the Theory of Infinite Series, 3rd ed.* New York: Chelsea, p. 43, 1991.

Ernst Equation

The PARTIAL DIFFERENTIAL EQUATION

$$\Re[u]\left(u_{rr} + \frac{u_r}{r} + u_{zz}\right) = u_r^2 + u_z^2,$$

where $\Re[u]$ is the REAL PART of u (Calogero and Degasperis 1982, p. 62; Zwillinger 1997, p. 131).

References

Calogero, F. and Degasperis, A. *Spectral Transform and Solitons: Tools to Solve and Investigate Nonlinear Evolution Equations.* New York: North-Holland, 1982.

Zwillinger, D. *Handbook of Differential Equations, 3rd ed.* Boston, MA: Academic Press, p. 131, 1997.

Errera Graph

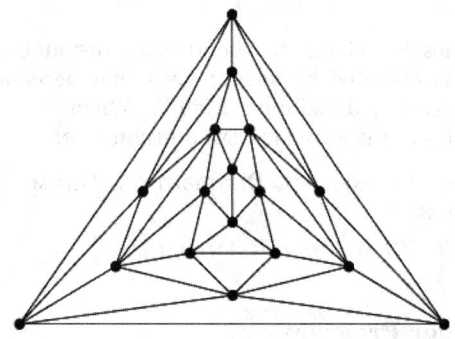

The 17-node PLANAR GRAPH illustrated above which tangles the Kempe chains in Kempe's algorithm and thus provides an example of how Kempe's supposed proof of the FOUR-COLOR THEOREM fails.

See also FOUR-COLOR THEOREM, KITTELL GRAPH

References

Wagon, S. *Mathematica in Action, 2nd ed.* New York: Springer-Verlag, pp. 522–24, 1999.
© 1999–001 Wolfram Research, Inc.

Error

The difference between a quantity and its estimated or measured quantity.

See also ABSOLUTE ERROR, PERCENTAGE ERROR, RELATIVE ERROR

Error Curve

GAUSSIAN FUNCTION

Error Function

ERF, ERFC

Error Function Distribution

A NORMAL DISTRIBUTION with MEAN 0,

$$P(x) = \frac{h}{\sqrt{\pi}} e^{-h^2 x^2}. \tag{1}$$

The CHARACTERISTIC FUNCTION is

$$\phi(t) = e^{-t^2/(4h^2)}. \tag{2}$$

The MEAN, VARIANCE, SKEWNESS, and KURTOSIS are

$$\mu = 0 \tag{3}$$

$$\sigma^2 = \frac{1}{2h^2} \tag{4}$$

$$\gamma_1 = 0 \tag{5}$$

$$\gamma_2 = 0. \tag{6}$$

The CUMULANTS are

$$\kappa_1 = 0 \tag{7}$$

$$\kappa_2 = \frac{1}{2h^2} \tag{8}$$

$$\kappa_n = 0 \tag{9}$$

for $n \geq 3$.

Error Propagation

Given a FORMULA $y = f(x)$ with an ABSOLUTE ERROR in x of dx, the ABSOLUTE ERROR is dy. The RELATIVE ERROR is dy/y. If $x = f(u, v)$, then

$$x_i - \bar{x} = (u_i - \bar{u})\frac{\partial x}{\partial u} + (v_i - \bar{v})\frac{\partial x}{\partial v} + \dots, \tag{1}$$

where $\bar{x}$ denotes the MEAN, so

$$\sigma_x^2 \equiv \frac{1}{N-1} \sum_{i=1}^{N} (x_i - \bar{x})^2$$

$$= \frac{1}{N-1} \sum_{i=1}^{N} \left[(u_i - \bar{u})^2 \left(\frac{\partial x}{\partial u}\right)^2 + (v_i - \bar{v})^2 \left(\frac{\partial x}{\partial v}\right)^2 \right.$$

$$\left. + 2(u_i - \bar{u})(v_i - \bar{v}) \left(\frac{\partial x}{\partial u}\right) \left(\frac{\partial x}{\partial v}\right) + \dots \right]. \tag{2}$$

The definitions of VARIANCE and COVARIANCE then give

$$\sigma_u^2 \equiv \frac{1}{N-1} \sum_{i=1}^{N} (u_i - \bar{u})^2 \tag{3}$$

$$\sigma_v^2 \equiv \frac{1}{N-1} \sum_{i=1}^{N} (v_i - \bar{v})^2 \tag{4}$$

$$\sigma_{uv} \equiv \frac{1}{N-1} \sum_{i=1}^{N} (u_i - \bar{u})(v_i - \bar{v}) \tag{5}$$

(where $\sigma_{ii} \equiv \sigma_i^2$), so

$$\sigma_x^2 = \sigma_u^2 \left(\frac{\partial x}{\partial u}\right)^2 + \sigma_v^2 \left(\frac{\partial x}{\partial v}\right)^2 + 2\sigma_{uv} \left(\frac{\partial x}{\partial u}\right) \left(\frac{\partial x}{\partial v}\right) + \dots. \tag{6}$$

If u and v are uncorrelated, then $\sigma_{uv} = 0$ so

$$\sigma_x^2 = \sigma_u^2 \left(\frac{\partial x}{\partial u}\right)^2 + \sigma_v^2 \left(\frac{\partial x}{\partial v}\right)^2. \tag{7}$$

Now consider addition of quantities with errors. For $x = au \pm bv$, $\partial x/\partial u = a$ and $\partial x/\partial v = \pm b$, so

$$\sigma_x^2 = a^2 \sigma_u^2 + b^2 \sigma_v^2 \pm 2ab\sigma_{uv}. \tag{8}$$

For division of quantities with $x = \pm au/v$, $\partial x/\partial u = \pm a/v$ and $\partial x/\partial v = \mp au/v^2$, so

$$\sigma_x^2 = \frac{a^2}{v^2}\sigma_u^2 + \frac{a^2 u^2}{v^4}\sigma_v^2 - 2\frac{a}{v}\frac{au}{v^2}\sigma_{uv}. \tag{9}$$

$$\left(\frac{\sigma_x}{x}\right)^2 = \frac{a^2}{v^2}\frac{v^2}{a^2 u^2}\sigma_u^2 + \frac{a^2 u^2}{v^4}\frac{v^2}{a^2 u^2} - 2\left(\frac{a}{v}\right)\left(\frac{au}{v^2}\right)\sigma_{uv}$$

$$= \left(\frac{\sigma_u}{u}\right)^2 + \left(\frac{\sigma_v}{v}\right)^2 - 2\left(\frac{\sigma_{uv}}{u}\right)\left(\frac{\sigma_{uv}}{v}\right). \tag{10}$$

For exponentiation of quantities with

$$x = a^{\pm bu} = (e^{\ln a})^{\pm bu} = e^{\pm b(\ln a)u}, \tag{11}$$

$$\frac{\partial x}{\partial u} = \pm b(\ln a)e^{\pm b \ln au} = \pm b(\ln a)x, \tag{12}$$

so

$$\sigma_x = \sigma_u b(\ln a)x \tag{13}$$

$$\frac{\sigma_x}{x} = b \ln a\sigma_u. \tag{14}$$

If $a = e$, then

$$\frac{\sigma_x}{x} = b\sigma_u. \tag{15}$$

For LOGARITHMS of quantities with $x = a\ln(\pm bu)$, $\partial x/\partial u = a(\pm b)/(\pm bu) = a/u$, so

$$\sigma_x^2 = \sigma_u^2\left(\frac{a^2}{u^2}\right) \tag{16}$$

$$\sigma_x = a\frac{\sigma_u}{u}. \tag{17}$$

For multiplication with $x = \pm auv$, $\partial x/\partial u = \pm av$ and $\partial x/\partial v = \pm au$, so

$$\sigma_x^2 = a^2 v^2 \sigma_u^2 + a^2 u^2 \sigma_v^2 + 2a^2 uv\sigma_{uv} \tag{18}$$

$$\left(\frac{\sigma_x}{x}\right)^2 = \frac{a^2 v^2}{a^2 u^2 v^2}\sigma_u^2 + \frac{a^2 u^2}{a^2 u^2 v^2}\sigma_v^2 + \frac{2a^2 uv}{a^2 u^2 v^2}\sigma_{uv}$$

$$= \left(\frac{\sigma_u}{u}\right)^2 + \left(\frac{\sigma_v}{v}\right)^2 + 2\left(\frac{\sigma_{uv}}{u}\right)\left(\frac{\sigma_{uv}}{v}\right). \tag{19}$$

For POWERS, with $x = au^{\pm b}$, $\partial x/\partial u = \pm abu^{\pm b-1} = \pm bx/u$, so

$$\sigma_x^2 = \sigma_u^2 \frac{b^2 x^2}{u^2} \tag{20}$$

$$\frac{\sigma_x}{x} = b\frac{\sigma_u}{u}. \tag{21}$$

See also ABSOLUTE ERROR, COVARIANCE, PERCENTAGE ERROR, RELATIVE ERROR, VARIANCE

References

Abramowitz, M. and Stegun, C. A. (Eds.). *Handbook of Mathematical Functions with Formulas, Graphs, and Mathematical Tables, 9th printing.* New York: Dover, p. 14, 1972.

Bevington, P. R. *Data Reduction and Error Analysis for the Physical Sciences.* New York: McGraw-Hill, pp. 58–4, 1969.

Error-Correcting Code

An error-correcting code is an algorithm for expressing a sequence of numbers such that any errors which are introduced can be detected and corrected (within certain limitations) based on the remaining numbers. The study of error-correcting codes and the associated mathematics is known as CODING THEORY.

Error detection is much simpler than error correction, and one or more "check" digits are commonly embedded in credit card numbers in order to detect mistakes. Early space probes like Mariner used a type of error-correcting code called a block code, and more recent space probes use convolution codes. Error-correcting codes are also used in CD players, high speed modems, and cellular phones. Modems use error detection when they compute CHECKSUMS, which are sums of the digits in a given transmission modulo some number. The ISBN used to identify books also incorporates a check DIGIT.

A powerful check for 13 DIGIT numbers consists of the following. Write the number as a string of DIGITS $a_1, a_2, a_3 \ldots a_{13}$. Take $a_1 + a_3 + \cdots + a_{13}$ and double. Now add the number of DIGITS in ODD positions which are > 4 to this number. Now add $a_2 + a_4 + \cdots + a_{12}$. The check number is then the number required to bring the last DIGIT to 0. This scheme detects all single DIGIT errors and all TRANSPOSITIONS of adjacent DIGITS except 0 and 9.

Let $A(n, d)$ denote the maximal number of n (0,1)-vectors having the property that any two of the set differ in at least d places. The corresponding vectors can correct $\lfloor (d-1)/2 \rfloor$ errors. $A(n, d, w)$ is the number of $A(n, d)$s with precisely w 1s (Sloane and Plouffe 1995). Since it is not possible for n-vectors to differ in $d > n$ places and since n-vectors which differ in all n places partition into disparate sets of two,

$$A(n,\,d) = \begin{cases} 1 & n < d \\ 2 & n = d. \end{cases}$$

Values of $A(n,\,d)$ can be found by labeling the 2^n (0,1)-n-vectors, finding all unordered pairs $(a_i,\,a_j)$ of n-vectors which differ from each other in at least d places, forming a GRAPH from these unordered pairs, and then finding the CLIQUE NUMBER of this graph. Unfortunately, finding the size of a clique for a given GRAPH is an NP-COMPLETE PROBLEM.

d	Sloane	$A(n,\,d)$
1	A000079	2, 4, 8, 16, 32, 64, 128, ...
2		1, 2, 4, 8, ...
3		1, 1, 2, 2, ...
4	A005864	1, 1, 1, 2, 4, 8, 16, 20, 40, ...
5		1, 1, 1, 1, 2, ...
6	A005865	1, 1, 1, 1, 1, 2, 2, 2, 4, 6, 12, ...
7		1, 1, 1, 1, 1, 1, 2, ...
8	A005866	1, 1, 1, 1, 1, 1, 1, 2, 2, 2, 2, 4, ...

See also CHECKSUM, CLIQUE, CLIQUE NUMBER, CODING THEORY, FINITE FIELD, HADAMARD MATRIX, HAMMING CODE, ISBN, UPC

References

Baylis, J. *Error Correcting Codes: A Mathematical Introduction.* Boca Raton, FL: CRC Press, 1998.
Berlekamp, E. R. *Algebraic Coding Theory, rev. ed.* New York: McGraw-Hill, 1968.
Brouwer, A. E.; Shearer, J. B.; Sloane, N. J. A.; and Smith, W. D. "A New Table of Constant Weight Codes." *IEEE Trans. Inform. Th.* **36**, 1334–380, 1990.
Calderbank, A. R.; Hammons, A. R. Jr.; Kumar, P. V.; Sloane, N. J. A.; and Solé, P. "A Linear Construction for Certain Kerdock and Preparata Codes." *Bull. Amer. Math. Soc.* **29**, 218–22, 1993.
Conway, J. H. and Sloane, N. J. A. "Quaternary Constructions for the Binary Single-Error-Correcting Codes of Julin, Best and Others." *Des. Codes Cryptogr.* **4**, 31–2, 1994.
Conway, J. H. and Sloane, N. J. A. "Error-Correcting Codes." §3.2 in *Sphere Packings, Lattices, and Groups, 2nd ed.* New York: Springer-Verlag, pp. 75–8, 1993.
Gallian, J. "How Computers Can Read and Correct ID Numbers." *Math Horizons*, pp. 14–5, Winter 1993.
Guy, R. K. *Unsolved Problems in Number Theory, 2nd ed.* New York: Springer-Verlag, pp. 119–21, 1994.
MacWilliams, F. J. and Sloane, N. J. A. *The Theory of Error-Correcting Codes.* Amsterdam, Netherlands: North-Holland, 1977.
Sloane, N. J. A. Sequences A000079/M1129, A005864/M1111, A005865/M0240, and A005866/M0226 in "An On-Line Version of the Encyclopedia of Integer Sequences." http://www.research.att.com/~njas/sequences/eisonline.html.
Sloane, N. J. A. and Plouffe, S. Figure M0240 in *The Encyclopedia of Integer Sequences.* San Diego: Academic Press, 1995.

Escher's Map

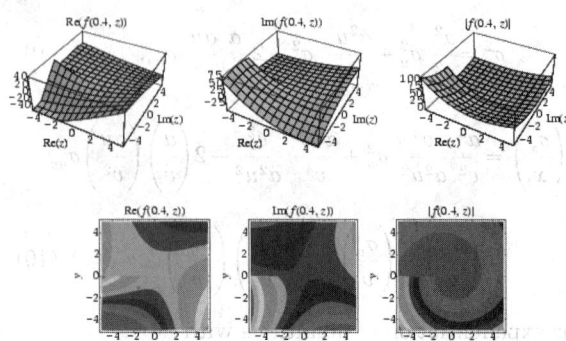

The function

$$f(\beta, z) \mapsto z^{(1+\cos\beta + i\,\sin\beta)/2},$$

illustrated above for $\beta = 0.4$.

Escher's Solid

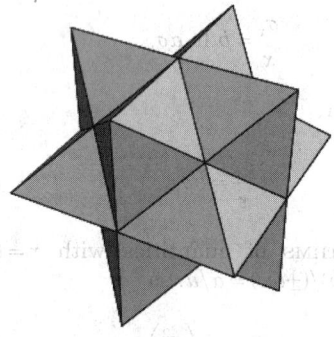

The solid illustrated on the right pedestal in M. C. Escher's *Waterfall* woodcut. It can be constructed by CUMULATION of the RHOMBIC DODECAHEDRON with cumulation height 5/2.

See also CUBE 3-COMPOUND, CUMULATION, RHOMBIC DODECAHEDRON

Escribed Circle

EXCIRCLE

Essential Singularity

A SINGULAR POINT a for which $f(z)(z-a)^n$ is not DIFFERENTIABLE for any INTEGER $n > 0$.

See also PICARD'S THEOREM, POLE, REMOVABLE SINGULARITY, SINGULAR POINT (FUNCTION), WEIERSTRASS-CASORATI THEOREM

Essential Supremum

References

Knopp, K. "Essential and Non-Essential Singularities or Poles." §31 in *Theory of Functions Parts I and II, Two Volumes Bound as One, Part I.* New York: Dover, pp. 123–26, 1996.

Krantz, S. G. "Removable Singularities, Poles, and Essential Singularities." §4.1.4 in *Handbook of Complex Analysis.* Boston, MA: Birkhäuser, p. 42, 1999.

Essential Supremum

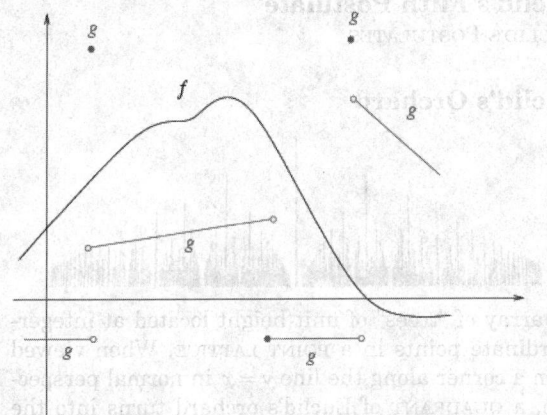

The functions f and g have the same essential supremum

The essential supremum is the proper generalization to MEASURABLE FUNCTIONS of the MAXIMUM. The technical difference is that the values of a function on a set of MEASURE ZERO don't affect the essential supremum.

Given a MEASURABLE FUNCTION $f : X \to \mathbb{R}$, where X is a MEASURE SPACE with measure μ, the essential supremum is the smallest number α such that

$$\mu(\{x \text{ such that } f(x) > \alpha\}$$

has MEASURE ZERO. If no such number exists, as in the case of $f(x) = 1/x$ on $(0, 1)$, then the essential supremum is ∞.

The essential supremum of the absolute value of a function $|f|$ is usually denoted $\|f\|_\infty$, and this serves as the norm for L-INFINITY-SPACE.

See also L-INFINITY-SPACE, L_p-SPACE, L_2-SPACE, MEASURE, MEASURABLE FUNCTION, MEASURE SPACE

Estimate

An estimate is an educated guess for an unknown quantity or outcome based on known information. The making of estimates is an important part of statistics, since care is needed to provide as accurate an estimate as possible using as little input data as possible. Often, an estimate for the uncertainty ΔE of an estimate E can also be determined statistically. A rule that tells how to calculate an estimate based on the measurements contained in a sample is called an ESTIMATOR.

See also BIAS (ESTIMATOR), ERROR, ESTIMATOR

References

Iyanaga, S. and Kawada, Y. (Eds.). "Statistical Estimation and Statistical Hypothesis Testing." Appendix A, Table 23 in *Encyclopedic Dictionary of Mathematics.* Cambridge, MA: MIT Press, pp. 1486–489, 1980.

Estimator

An estimator is a rule that tells how to calculate an ESTIMATE based on the measurements contained in a sample. For example, the "sample MEAN" AVERAGE $\bar{x}$ is an estimator for the population MEAN μ.

The mean square error of an estimator $\tilde{\theta}$ is defined by

$$\text{MSE} \equiv \left\langle (\tilde{\theta} - \theta)^2 \right\rangle.$$

Let B be the BIAS, then

$$\text{MSE} = \left\langle [(\tilde{\theta} - \langle \tilde{\theta} \rangle) + B(\tilde{\theta})]^2 \right\rangle$$

$$= \left\langle (\tilde{\theta} - \langle \tilde{\theta} \rangle)^2 \right\rangle + B^2(\tilde{\theta}) \equiv V(\tilde{\theta}) + B^2(\tilde{\theta}),$$

where V is the estimator VARIANCE.

See also BIAS (ESTIMATOR), ERROR, ESTIMATE, K-STATISTIC, UNBIASED ESTIMATOR

Eta Function

DEDEKIND ETA FUNCTION, DIRICHLET ETA FUNCTION, JACOBI THETA FUNCTIONS

Et-Function

A function which arises in FRACTIONAL CALCULUS.

$$E_t(v, a) = \frac{1}{\Gamma(v)} e^{at} \int_0^t x^{v-1} e^{-ax} \, dx = t^v e^{at} \gamma(v, at), \quad (1)$$

where $\gamma(a, \xi)$ is the incomplete GAMMA FUNCTION and $\Gamma(z)$ the complete GAMMA FUNCTION. The E_t function satisfies the RECURRENCE RELATION

$$E_t(v, a) = aE_t(v+1, a) + \frac{t^v}{\Gamma(v+1)}. \quad (2)$$

A special value is

$$E_t(0, a) = e^{at}. \quad (3)$$

See also E_N-FUNCTION, FRACTIONAL CALCULUS

References

Abramowitz, M. and Stegun, C. A. (Eds.). "Exponential Integral and Related Functions." Ch. 5 in *Handbook of Mathematical Functions with Formulas, Graphs, and Mathematical Tables, 9th printing.* New York: Dover, pp. 227–33, 1972.

Ethiopian Multiplication
RUSSIAN MULTIPLICATION

Etruscan Venus Surface
A 3-D shadow of a 4-D KLEIN BOTTLE.

See also IDA SURFACE, KLEIN BOTTLE

References
Peterson, I. *Islands of Truth: A Mathematical Mystery Cruise.* New York: W. H. Freeman, pp. 42–4, 1990.

Eubulides Paradox
The PARADOX "This statement is false," stated in the fourth century BC. It is a sharper version of the EPIMENIDES PARADOX, "All Cretans are liars...One of their own poets has said so."

See also EPIMENIDES PARADOX, SOCRATES' PARADOX

References
Erickson, G. W. and Fossa, J. A. *Dictionary of Paradox.* Lanham, MD: University Press of America, pp. 63–4, 1998.
Hofstadter, D. R. *Gödel, Escher, Bach: An Eternal Golden Braid.* New York: Vintage Books, p. 17, 1989.

Euclid Number
The nth Euclid number is defined by

$$E_n \equiv 1 + \prod_{i=1}^{n} p_i = 1 + p_n\#,$$

where p_i is the ith PRIME and $p_n\#$ is the PRIMORIAL. The first few E_n are 3, 7, 31, 211, 2311, 30031, 510511, 9699691, 223092871, 6469693231, ... (Sloane's A006862; Tietze 1965, p. 19).

The largest factors of E_n for $n = 1, 2, ...$ are 3, 7, 31, 211, 2311, 509, 277, 27953, ... (Sloane's A002585). The n of the first few PRIME Euclid numbers E_n are 1, 2, 3, 4, 5, 11, 75, 171, 172, 384, 457, 616, 643, ... (Sloane's A014545), and the largest known Euclid number is E_{4413}. It is not known if there are an INFINITE number of PRIME Euclid numbers (Guy 1994, Ribenboim 1996).

See also EUCLID-MULLIN SEQUENCE, PRIMORIAL, SMARANDACHE SEQUENCES

References
Guy, R. K. *Unsolved Problems in Number Theory, 2nd ed.* New York: Springer-Verlag, 1994.
Ribenboim, P. *The New Book of Prime Number Records.* New York: Springer-Verlag, 1996.
Sloane, N. J. A. Sequences A006862/M2698, A002585/M2697, and A014545 in "An On-Line Version of the Encyclopedia of Integer Sequences." http://www.research.att.com/~njas/sequences/eisonline.html.
Tietze, H. *Famous Problems of Mathematics: Solved and Unsolved Mathematics Problems from Antiquity to Modern Times.* New York: Graylock Press, 1965.

Wagon, S. *Mathematica in Action.* New York: W. H. Freeman, pp. 35–7, 1991.

Euclid's Axioms
EUCLID'S POSTULATES

Euclid's Elements
ELEMENTS

Euclid's Fifth Postulate
EUCLID'S POSTULATES

Euclid's Orchard

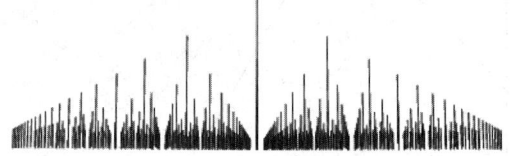

An array of "trees" of unit height located at integer-coordinate points in a POINT LATTICE. When viewed from a corner along the line $y = x$ in normal perspective, a QUADRANT of Euclid's orchard turns into the modified DIRICHLET FUNCTION (Gosper).

See also DIRICHLET FUNCTION, GREATEST COMMON DIVISOR, ORCHARD-PLANTING PROBLEM

Euclid's Postulates

1. A straight LINE SEGMENT can be drawn joining any two points.
2. Any straight LINE SEGMENT can be extended indefinitely in a straight LINE.
3. Given any straight LINE SEGMENT, a CIRCLE can be drawn having the segment as RADIUS and one endpoint as center.
4. All RIGHT ANGLES are congruent.
5. If two lines are drawn which intersect a third in such a way that the sum of the inner angles on one side is less than two RIGHT ANGLES, then the two lines inevitably must intersect each other on that side if extended far enough. This postulate is equivalent to what is known as the PARALLEL POSTULATE.

Euclid's fifth postulate cannot be proven as a theorem, although this was attempted by many people. Euclid himself used only the first four postulates (for the first 28 propositions of the ELEMENTS, but was forced to invoke the PARALLEL POSTULATE on the 29th. In 1823, Janos Bolyai and Nicolai Lobachevsky independently realized that entirely self-consistent "NON-EUCLIDEAN GEOMETRIES" could be created in which the parallel postulate *did not hold*. (Gauss had also discovered but suppressed the existence of non-Euclidean geometries.)

See also ABSOLUTE GEOMETRY, CIRCLE, ELEMENTS, LINE SEGMENT, NON-EUCLIDEAN GEOMETRY, PARALLEL POSTULATE, PASCH'S THEOREM, RIGHT ANGLE

References

Hofstadter, D. R. *Gödel, Escher, Bach: An Eternal Golden Braid.* New York: Vintage Books, pp. 88–2, 1989.

Euclid's Principle
EUCLID'S THEOREMS

Euclid's Theorems

A theorem sometimes called "Euclid's First Theorem" or EUCLID'S PRINCIPLE states that if p is a PRIME and $p|ab$, then $p|a$ or $p|b$ (where | means DIVIDES). A COROLLARY is that $p|a^n \Rightarrow p|a$ (Conway and Guy 1996). The FUNDAMENTAL THEOREM OF ARITHMETIC is another COROLLARY (Hardy and Wright 1979).

Euclid's Second Theorem states that the number of PRIMES is INFINITE. This theorem, also called the INFINITUDE OF PRIMES theorem, was proved by Euclid in Proposition IX.20 of the ELEMENTS (Tietze 1965, pp. 7–). Ribenboim (1989) gives nine (and a half) proofs of this theorem. Euclid's elegant proof proceeds as follows. Given a finite sequence of consecutive PRIMES 2, 3, 5, ..., p, the number

$$N = 2 \cdot 3 \cdot 5 \cdots p + 1, \qquad (1)$$

known as the ith EUCLID NUMBER when $p = p_i$ is the ith PRIME, is either a new PRIME or the product of PRIMES. If N is a PRIME, then it must be greater than the previous PRIMES, since one plus the product of PRIMES must be greater than each PRIME composing the product. Now, if N is a product of PRIMES, then at least one of the PRIMES must be greater than p. This can be shown as follows.

If N is COMPOSITE and has no prime factors greater than p, then one of its factors (say F) must be one of the PRIMES in the sequence, 2, 3, 5, ..., p. It therefore DIVIDES the product $2 \cdot 3 \cdot 5 \cdots p$. However, since it is a factor of N, it also DIVIDES N. But a number which DIVIDES two numbers a and $b < a$ also DIVIDES their difference $a - b$, so F must also divide

$$N - (2 \cdot 3 \cdot 5 \cdots p) = (2 \cdot 3 \cdot 5 \cdots p + 1) - (2 \cdot 3 \cdot 5 \cdots p) = 1. \qquad (2)$$

However, in order to divide 1, F must be 1, which is contrary to the assumption that it is a PRIME in the sequence 2, 3, 5, ... It therefore follows that if N is composite, it has at least one factor greater than p. Since N is either a PRIME greater than p or contains a prime factor greater than p, a PRIME larger than the largest in the finite sequence can always be found, so there are an infinite number of PRIMES. Hardy (1967) remarks that this proof is "as fresh and significant as when it was discovered" so that "two thousand years have not written a wrinkle" on it.

A similar argument shows that $p! \pm 1$ and

$$1 \cdot 3 \cdot 5 \cdot 7 \cdots p + 1 \qquad (3)$$

must be either PRIME or be divisible by a PRIME $> p$. Kummer used a variation of this proof, which is also a proof by contradiction. It assumes that there exist only a finite number of PRIMES $N = p_1, p_2, ..., p_r$. Now consider $N - 1$. It must be a product of PRIMES, so it has a PRIME divisor p_i in common with N. Therefore, $p_i | N - (N - 1) = 1$ which is nonsense, so we have proved the initial assumption is wrong by contradiction.

It is also true that there are runs of COMPOSITE NUMBERS which are arbitrarily long. This can be seen by defining

$$n \equiv j! = \prod_{i=1}^{j} i, \qquad (4)$$

where $j!$ is a FACTORIAL. Then the $j - 1$ consecutive numbers $n + 2, n + 3, ..., n + j$ are COMPOSITE, since

$$n + 2 = (1 \cdot 2 \cdots j) + 2 = 2(1 \cdot 3 \cdot 4 \cdots n + 1) \qquad (5)$$

$$n + 3 = (1 \cdot 2 \cdots j) + 3 = 3(1 \cdot 2 \cdot 4 \cdot 5 \cdots n + 1) \qquad (6)$$

$$n + j = (1 \cdot 2 \cdots j) + j = j[1 \cdot 2 \cdots (j - 1) + 1]. \qquad (7)$$

Guy (1981, 1988) points out that while $p_1 p_2 \cdots p_n + 1$ is not necessarily PRIME, letting q be the next PRIME after $p_1 p_2 \cdots p_n + 1$, the number $q - p_1 p_2 \cdots p_n + 1$ is almost always a PRIME, although it has not been proven that this must *always* be the case.

See also DIVIDE, EUCLID NUMBER, PRIME NUMBER

References

Ball, W. W. R. and Coxeter, H. S. M. *Mathematical Recreations and Essays, 13th ed.* New York: Dover, p. 60, 1987.
Conway, J. H. and Guy, R. K. "There are Always New Primes!" In *The Book of Numbers.* New York: Springer-Verlag, pp. 133–34, 1996.
Cosgrave, J. B. "A Remark on Euclid's Proof of the Infinitude of Primes." *Amer. Math. Monthly* **96**, 339–41, 1989.
Courant, R. and Robbins, H. *What is Mathematics?: An Elementary Approach to Ideas and Methods, 2nd ed.* Oxford, England: Oxford University Press, p. 22, 1996.
Dunham, W. "Great Theorem: The Infinitude of Primes." *Journey through Genius: The Great Theorems of Mathematics.* New York: Wiley, pp. 73–5, 1990.
Guy, R. K. §A12 in *Unsolved Problems in Number Theory.* New York: Springer-Verlag, 1981.
Guy, R. K. "The Strong Law of Small Numbers." *Amer. Math. Monthly* **95**, 697–12, 1988.
Hardy, G. H. *A Mathematician's Apology.* Cambridge, England: Cambridge University Press, 1992.
Ribenboim, P. *The Book of Prime Number Records, 2nd ed.* New York: Springer-Verlag, pp. 3–2, 1989.
Tietze, H. *Famous Problems of Mathematics: Solved and Unsolved Mathematics Problems from Antiquity to Modern Times.* New York: Graylock Press, pp. 7–, 1965.

Euclidean Algorithm

An ALGORITHM for finding the GREATEST COMMON DIVISOR of two numbers a and b, also called Euclid's algorithm. The algorithm can also be defined for more general RINGS than just the integers $\mathbb{Z}$. There are even PRINCIPAL RINGS which are not EUCLIDEAN but where one can define the equivalent of the Euclidean algorithm. The algorithm for rational numbers was given in Book VII of Euclid's *Elements*, and the algorithm for reals appeared in Book X, and is the earliest example of an INTEGER RELATION algorithm (Ferguson *et al.* 1999).

The Euclidean algorithm is an example of a P-PROBLEM whose time complexity is bounded by a quadratic function of the length of the input values (Banach and Shallit). Let $a = bq + r$, then find a number u which DIVIDES both a and b (so that $a = su$ and $b = tu$), then u also DIVIDES r since

$$r = a - bq = su - qtu = (s - qt)u. \tag{1}$$

Similarly, find a number v which DIVIDES b and r (so that $b = s'v$ and $r = t'v$), then v DIVIDES a since

$$a = bq + r = s'vq + t'v = (s'q + t')v. \tag{2}$$

Therefore, every common DIVISOR of a and b is a common DIVISOR of b and r, so the procedure can be iterated as follows.

$$q_1 = \left\lfloor \frac{a}{b} \right\rfloor \quad a = bq_1 + r_1 \quad r_1 = a - bq_1 \tag{3}$$

$$q_2 = \left\lfloor \frac{b}{r_1} \right\rfloor \quad b = q_2 r_1 + r_2 \quad r_2 = b - q_2 r_1 \tag{4}$$

$$q_3 = \left\lfloor \frac{r_1}{r_2} \right\rfloor \quad r_1 = q_3 r_2 + r_3 \quad r_3 = r_1 - q_3 r_2 \tag{5}$$

$$q_4 = \left\lfloor \frac{r_2}{r_3} \right\rfloor \quad r_2 = q_4 r_3 + r_4 \quad r_4 = r_2 - q_4 r_3 \tag{6}$$

$$q_n = \left\lfloor \frac{r_{n-2}}{r_{n-1}} \right\rfloor \quad r_{n-2} = q_n r_{n-1} + r_n \quad r_n = r_{n-2} - q_n r_{n-1} \tag{7}$$

$$q_{n+1} = \left\lfloor \frac{r_{n-1}}{r_n} \right\rfloor \quad r_{n-1} = q_{n+1} r_n + 0 \quad r_n = r_{n-1}/q_{n+1}. \tag{8}$$

For integers, the algorithm terminates when q_{n+1} divides r_{n-1} exactly, at which point r_n corresponds to the GREATEST COMMON DIVISOR of a and b, $\mathrm{GCD}(a, b) = r_n$. For real numbers, the algorithm yields either an exact relation or an infinite sequence of approximate relations (Ferguson *et al.* 1999).

Lamé showed that the number of steps needed to arrive at the GREATEST COMMON DIVISOR for two numbers less than n is

$$\text{steps} \leq \frac{\log_{10} n}{\log_{10} \phi} + \frac{\log_{10} \sqrt{5}}{\log_{10} \phi} \tag{9}$$

where ϕ is the GOLDEN MEAN, or ≤ 5 times the number of digits in the smaller number (Wells 1986, p. 59). Numerically, Lamé's expression evaluates to

$$\text{steps} \leq 4.785 \log_{10} n + 1.6723. \tag{10}$$

As shown by LAMÉ'S THEOREM, the worst case occurs when the ALGORITHM is applied to two consecutive FIBONACCI NUMBERS. Heilbronn showed that the average number of steps is $12 \ln 2/\pi^2 \log_{10} n = 0.843 \log_{10} n$ for all pairs (n, b) with $b < n$. Kronecker showed that the shortest application of the ALGORITHM uses least absolute remainders. The QUOTIENTS obtained are distributed as shown in the following table (Wagon 1991).

Quotient	%
1	41.5
2	17.0
3	9.3

For details, see Uspensky and Heaslet (1939) or Knuth (1973). Let $T(m, n)$ be the number of divisions required to compute $\mathrm{GCD}(m, n)$ using the Euclidean algorithm, and define $T(m, 0) = 0$ if $m \geq 0$. Then the function $T(m, n)$ is given by the RECURRENCE RELATION

$$T(m, n) = \begin{cases} 1 + T(n, m \bmod n) & \text{for } m \geq n \\ 1 + T(n, m) & \text{for } m < n. \end{cases} \tag{11}$$

Tabulating this function for $0 \leq m < n$ gives

$$
\begin{array}{ccccccc}
0 & & & & & & \\
0 & 1 & & & & & \\
0 & 1 & 2 & & & & \\
0 & 1 & 1 & 2 & & & \\
0 & 1 & 2 & 3 & 2 & & \\
0 & 1 & 1 & 1 & 2 & 2 &
\end{array}
$$

(Sloane's A051010). The maximum numbers of steps for a given $n = 1, 2, 3, \ldots$ are $1, 2, 2, 3, 2, 3, 4, 3, 3, 4, 4, 5, \ldots$ (Sloane's A034883).

Define the functions

$$T(n) = \frac{1}{n} \sum_{0 \leq m < n} T(m, n) \tag{12}$$

$$\tau(n) = \frac{1}{\phi(n)} \sum_{\substack{0 < m < n \\ \mathrm{GCD}(m, n)=1}} T(m, n) \tag{13}$$

$$A(N) = \frac{1}{N^2} \sum_{\substack{1 \leq m < N \\ 1 \leq n \leq N}} T(m, n), \tag{14}$$

where $\phi(n)$ is the TOTIENT FUNCTION, $T(n)$ is the average number of divisions when n is fixed and m chosen at random, $\tau(n)$ is the average number of divisions when n is fixed and m is a random number coprime to n, and $A(N)$ is the average number of divisions when m and n are both chosen at random in $[1, N]$. The first few values of $T(n)$ are 0, 1/2, 1, 1, 8/5, 7/6, 13/7, 7/4, ... (Sloane's A051011 and A051012).

Norton (1990) showed that

$$T(n) = \frac{12 \ln 2}{\pi^2} \left[\ln n - \sum_{d|n} \frac{\Lambda(d)}{d} \right] + C$$
$$+ \frac{1}{n} \sum_{d|n} \phi(d) \mathcal{O}(d^{-1/6+\epsilon}), \qquad (15)$$

where $\Lambda(d)$ is the VON MANGOLDT FUNCTION and C is PORTER'S CONSTANT. Porter (1975) showed that

$$\tau(n) = \frac{12 \ln 2}{\pi^2} \ln n + C + \mathcal{O}(n^{-1/6} + \epsilon), \qquad (16)$$

and Norton (1990) proved that

$$A(N) = \frac{12 \ln 2}{\pi^2} \left[\ln N - \frac{1}{2} + \frac{6}{\pi^2} \zeta'(2) \right] + C - \frac{1}{2}$$
$$+ \mathcal{O}(N^{-1/6+\epsilon}), \qquad (17)$$

where $\zeta'(z)$ is the derivative of the RIEMANN ZETA FUNCTION.

There exist 21 QUADRATIC FIELDS in which there is a Euclidean algorithm (Inkeri 1947, Barnes and Swinnerton-Dyer 1952).

Although various attempts were made to generalize the algorithm to find INTEGER RELATIONS between $n \geq 3$ variables, none were successful until the discovery of the FERGUSON-FORCADE ALGORITHM (Ferguson *et al.* 1999). Several other INTEGER RELATION algorithms have now been discovered.

See also BLANKINSHIP ALGORITHM, EUCLIDEAN RING, FERGUSON-FORCADE ALGORITHM, INTEGER RELATION, QUADRATIC FIELD

References

Bach, E. and Shallit, J. *Algorithmic Number Theory, Vol. 1: Efficient Algorithms.* Cambridge, MA: MIT Press, 1996.

Barnes, E. S. and Swinnerton-Dyer, H. P. F. "The Inhomogeneous Minima of Binary Quadratic Forms. I." *Acta Math* **87**, 259–23, 1952.

Chabert, J.-L. (Ed.). "Euclid's Algorithm." Ch. 4 in *A History of Algorithms: From the Pebble to the Microchip.* New York: Springer-Verlag, pp. 113–38, 1999.

Cohen, H. *A Course in Computational Algebraic Number Theory.* New York: Springer-Verlag, 1993.

Courant, R. and Robbins, H. "The Euclidean Algorithm." §2.4 in Supplement to Ch. 1 in *What is Mathematics?: An Elementary Approach to Ideas and Methods, 2nd ed.* Oxford, England: Oxford University Press, pp. 42–1, 1996.

Dunham, W. *Journey through Genius: The Great Theorems of Mathematics.* New York: Wiley, pp. 69–0, 1990.

Ferguson, H. R. P.; Bailey, D. H.; and Arno, S. "Analysis of PSLQ, An Integer Relation Finding Algorithm." *Math. Comput.* **68**, 351–69, 1999.

Finch, S. "Favorite Mathematical Constants." http://www.mathsoft.com/asolve/constant/porter/porter.html.

Inkeri, K. "Über den Euklidischen Algorithmus in quadratischen Zahlkörpern." *Ann. Acad. Sci. Fennicae. Ser. A. I. Math.-Phys.* **1947**, 1–5, 1947.

Knuth, D. E. *The Art of Computer Programming, Vol. 1: Fundamental Algorithms, 3rd ed.* Reading, MA: Addison-Wesley, 1997.

Knuth, D. E. *The Art of Computer Programming, Vol. 2: Seminumerical Algorithms, 3rd ed.* Reading, MA: Addison-Wesley, 1998.

Motzkin, T. "The Euclidean Algorithm." *Bull. Amer. Math. Soc.* **55**, 1142–146, 1949.

Nagell, T. "Euclid's Algorithm." §7 in *Introduction to Number Theory.* New York: Wiley, pp. 21–3, 1951.

Norton, G. H. "On the Asymptotic Analysis of the Euclidean Algorithm." *J. Symb. Comput.* **10**, 53–8, 1990.

Porter, J. W. "On a Theorem of Heilbronn." *Mathematika* **22**, 20–8, 1975.

Séroul, R. "Euclidean Division" and "The Euclidean Algorithm." §2.1 and 8.1 in *Programming for Mathematicians.* Berlin: Springer-Verlag, pp. 5 and 169–61, 2000.

Sloane, N. J. A. Sequences A034883, A051010, A051011, and A051012 in "An On-Line Version of the Encyclopedia of Integer Sequences." http://www.research.att.com/~njas/sequences/eisonline.html.

Uspensky, J. V. and Heaslet, M. A. *Elementary Number Theory.* New York: McGraw-Hill, 1939.

Wagon, S. "The Ancient and Modern Euclidean Algorithm" and "The Extended Euclidean Algorithm." §8.1 and 8.2 in *Mathematica in Action.* New York: W. H. Freeman, pp. 247–52 and 252–56, 1991.

Wells, D. *The Penguin Dictionary of Curious and Interesting Numbers.* Middlesex, England: Penguin Books, p. 59, 1986.

Euclidean Construction

GEOMETRIC CONSTRUCTION

Euclidean Domain

A more common way to describe a EUCLIDEAN RING.

See also ALGEBRAIC NUMBER THEORY, EUCLIDEAN RING

Euclidean Geometry

A GEOMETRY in which EUCLID'S FIFTH POSTULATE holds, sometimes also called PARABOLIC GEOMETRY. 2-D Euclidean geometry is called PLANE GEOMETRY, and 3-D Euclidean geometry is called SOLID GEOMETRY. Hilbert proved the CONSISTENCY of Euclidean geometry.

See also ELLIPTIC GEOMETRY, GEOMETRIC CONSTRUCTION, GEOMETRY, HYPERBOLIC GEOMETRY, NON-EUCLIDEAN GEOMETRY, PLANE GEOMETRY

References

Altshiller-Court, N. *College Geometry: A Second Course in Plane Geometry for Colleges and Normal Schools, 2nd ed., rev. enl.* New York: Barnes and Noble, 1952.

Casey, J. *A Treatise on the Analytical Geometry of the Point, Line, Circle, and Conic Sections, Containing an Account of Its Most Recent Extensions with Numerous Examples, 2nd rev. enl. ed.* Dublin: Hodges, Figgis, & Co., 1893.

Coxeter, H. S. M. and Greitzer, S. L. *Geometry Revisited.* Washington, DC: Math. Assoc. Amer., 1967

Coxeter, H. S. M. *Introduction to Geometry, 2nd ed.* New York: Wiley, 1969.

Gallatly, W. *The Modern Geometry of the Triangle, 2nd ed.* London: Hodgson, 1913.

Greenberg, M. J. *Euclidean and Non-Euclidean Geometries: Development and History, 3rd ed.* San Francisco, CA: W. H. Freeman, 1994.

Heath, T. L. *The Thirteen Books of the Elements, 2nd ed., Vol. 1: Books I and II.* New York: Dover, 1956.

Heath, T. L. *The Thirteen Books of the Elements, 2nd ed., Vol. 2: Books III-IX.* New York: Dover, 1956.

Heath, T. L. *The Thirteen Books of the Elements, 2nd ed., Vol. 3: Books X-XIII.* New York: Dover, 1956.

Honsberger, R. *Episodes in Nineteenth and Twentieth Century Euclidean Geometry.* Washington, DC: Math. Assoc. Amer., 1995.

Johnson, R. A. *Modern Geometry: An Elementary Treatise on the Geometry of the Triangle and the Circle.* Boston, MA: Houghton Mifflin, 1929.

Johnson, R. A. *Modern Geometry: An Elementary Treatise on the Geometry of the Triangle and the Circle.* Boston, MA: Houghton Mifflin, 1929.

Klee, V. "Some Unsolved Problems in Plane Geometry." *Math. Mag.* **52**, 131–45, 1979.

Klee, V. and Wagon, S. *Old and New Unsolved Problems in Plane Geometry and Number Theory, rev. ed.* Washington, DC: Math. Assoc. Amer., 1991.

Weisstein, E. W. "Books about Plane Geometry." http://www.treasure-troves.com/books/PlaneGeometry.html.

Euclidean Graph

A WEIGHTED GRAPH in which the weights are equal to the Euclidean lengths of the edges in a specified embedding (Skiena 1990, pp. 201 and 252).

References

Skiena, S. *Implementing Discrete Mathematics: Combinatorics and Graph Theory with Mathematica.* Reading, MA: Addison-Wesley, 1990.

Euclidean Group

The GROUP of ROTATIONS and TRANSLATIONS.

See also ROTATION, TRANSLATION

References

Lomont, J. S. *Applications of Finite Groups.* New York: Dover, 1987.

Euclidean Metric

The FUNCTION $f : \mathbb{R}^n \times \mathbb{R}^n \to \mathbb{R}$ that assigns to any two VECTORS $(x_1, ..., x_n)$ and $(y_1, ..., y_n)$ the number

$$\sqrt{(x_1 - y_1)^2 + \ldots + (x_n - y_n)^2},$$

and so gives the "standard" distance between any two VECTORS in $\mathbb{R}^n$.

Euclidean Motion

A Euclidean motion of $\mathbb{R}^n$ is an AFFINE TRANSFORMATION whose linear part is an ORTHOGONAL TRANSFORMATION.

See also RIGID MOTION

References

Gray, A. "Euclidean Motions." §6.1 in *Modern Differential Geometry of Curves and Surfaces with Mathematica, 2nd ed.* Boca Raton, FL: CRC Press, pp. 128–34, 1997.

Euclidean Norm

L2-NORM

Euclidean Number

A Euclidean number is a number which can be obtained by repeatedly solving the QUADRATIC EQUATION. Euclidean numbers, together with the RATIONAL NUMBERS, can be constructed using classical GEOMETRIC CONSTRUCTIONS. However, the cases for which the values of the TRIGONOMETRIC FUNCTIONS SINE, COSINE, TANGENT, etc., can be written in closed form involving square roots of REAL NUMBERS are much more restricted.

See also ALGEBRAIC INTEGER, ALGEBRAIC NUMBER, CONSTRUCTIBLE NUMBER, RADICAL INTEGER

References

Conway, J. H. and Guy, R. K. "Three Greek Problems." In *The Book of Numbers.* New York: Springer-Verlag, pp. 192–94, 1996.

Klein, F. "Algebraic Equations Solvable by Square Roots." Part I, Ch. 1 in "Famous Problems of Elementary Geometry: The Duplication of the Cube, the Trisection of the Angle, and the Quadrature of the Circle." In *Famous Problems and Other Monographs.* New York: Chelsea, pp. 5–2, 1980.

Euclidean Plane

The 2-D EUCLIDEAN SPACE denoted $\mathbb{R}^2$.

See also COMPLEX PLANE, EUCLIDEAN SPACE

Euclidean Ring

A RING without zero divisors in which an integer norm and an associated division algorithm (i.e., a EUCLIDEAN ALGORITHM) can be defined. For signed integers, the usual norm is the ABSOLUTE VALUE and the division algorithm gives the ordinary QUOTIENT and REMAINDER. For polynomials, the norm is the degree.

Important examples of Euclidean rings (besides $\mathbb{Z}$) are the GAUSSIAN INTEGERS and $\mathbb{C}[x]$, the RING of polynomials with complex coefficients. All Euclidean rings are also PRINCIPAL RINGS.

See also EUCLIDEAN ALGORITHM, PRINCIPAL RING, RING

Euclidean Space

References
Wilson, J. C. "A Principle Ring that is Not a Euclidean Ring." *Math. Mag.* 34–8, 1973.

Euclidean Space

Euclidean n-space is the SPACE of all n-tuples of REAL NUMBERS, $(x_1, x_2, ..., x_n)$ and is denoted $\mathbb{R}^n$. It is sometimes also called Cartesian space. $\mathbb{R}^n$ is a VECTOR SPACE and has LEBESGUE COVERING DIMENSION n. Elements of $\mathbb{R}^n$ are called n-VECTORS. $\mathbb{R}^1 = \mathbb{R}$ is the set of REAL NUMBERS (i.e., the REAL LINE), and $\mathbb{R}^2$ is called the EUCLIDEAN PLANE. In Euclidean space, COVARIANT and CONTRAVARIANT quantities are equivalent so $\vec{e}^j = \vec{e}_j$.

See also EUCLIDEAN PLANE, PSEUDO-EUCLIDEAN SPACE, REAL LINE, VECTOR

References
Gray, A. "Euclidean Spaces." §1.1 in *Modern Differential Geometry of Curves and Surfaces with Mathematica, 2nd ed.* Boca Raton, FL: CRC Press, pp. 2–, 1997.

Euclid-Mullin Sequence

The sequence of numbers obtained by letting $a_i = 2$, and defining

$$a_n = \mathrm{1pf}\left(1 + \prod_{k=1}^{n-1} a_k\right)$$

where $\mathrm{lpf}(n)$ is the LEAST PRIME FACTOR. The first few terms are 2, 3, 7, 43, 13, 53, 5, 6221671, 38709183810571, 139, ... (Sloane's A000945). Only 43 terms of the sequence are known; the 44th requires factoring a composite 180-digit number.

See also EUCLID NUMBER, LEAST PRIME FACTOR

References
Guy, R. K. and Nowakowski, R. "Discovering Primes with Euclid." *Delta (Waukesha)* **5**, 49–3, 1975.
Mullin, A. A. "Recursive Function Theory." *Bull. Amer. Math. Soc.* **69**, 737, 1963.
Naur, T. "Mullin's Sequence of Primes Is Not Monotonic." *Proc. Amer. Math. Soc.* **90**, 43–4, 1984.
Sloane, N. J. A. Sequences A000945/M0863 in "An On-Line Version of the Encyclopedia of Integer Sequences." http://www.research.att.com/~njas/sequences/eisonline.html.
Wagstaff, S. S. "Computing Euclid's Primes." *Bull. Institute Combin. Applications* **8**, 23–2, 1993.

Eudoxus's Kampyle

KAMPYLE OF EUDOXUS

Euler Angles

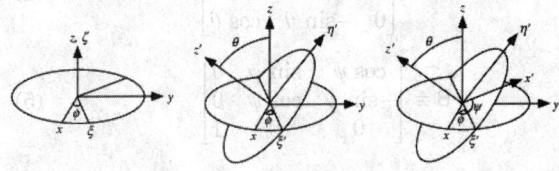

According to EULER'S ROTATION THEOREM, any ROTATION may be described using three ANGLES. If the ROTATIONS are written in terms of ROTATION MATRICES B, C, and D, then a general ROTATION A can be written as

$$A = BCD. \tag{1}$$

The three angles giving the three rotation matrices are called Euler angles. There are several conventions for Euler angles, depending on the axes about which the rotations are carried out. Write the MATRIX A as

$$A \equiv \begin{bmatrix} a_{11} & a_{12} & a_{13} \\ a_{21} & a_{22} & a_{23} \\ a_{31} & a_{32} & a_{33} \end{bmatrix}. \tag{2}$$

The so-called "x-convention," illustrated above, is the most common definition. In this convention, the rotation given by Euler angles (ϕ, θ, ψ), where the first rotation is by an angle ϕ about the z-AXIS, the second is by an angle $\theta \in [0, \pi]$ about the x-AXIS, and the third is by an angle ψ about the z-AXIS (again). Note, however, that several notational conventions for the angles are in common use. Goldstein (1960, pp. 145–48) and Landau and Lifschitz (1976) use (ϕ, θ, ψ), Tuma (1974) says (ψ, θ, ϕ) is used in aeronautical engineering in the analysis of space vehicles (but claims that (ϕ, θ, ψ) is used in the analysis of gyroscopic motion), while Bate *et al.* (1971) use (Ω, i, ω). Goldstein remarks that continental authors usually use (ψ, θ, ϕ), and warns that left-handed coordinate systems are also in occasional use (Osgood 1937, Margenau and Murphy 1956–4). Here, the notation (ϕ, θ, ψ) is used, a convention also followed by *Mathematica*'s RotateMatrix3D[*phi*, *theta*, *psi*] in the *Mathematica* add-on package Geometry`Rotations` (which can be loaded with the command < <Geometry`) and RotateShape[*g, phi, theta, psi*] in the *Mathematica* add-on package Graphics`Shapes` (which can be loaded with the command < <Graphics`) commands. In the x-convention, the component rotations are then given by

$$D \equiv \begin{bmatrix} \cos\phi & \sin\phi & 0 \\ -\sin\phi & \cos\phi & 0 \\ 0 & 0 & 1 \end{bmatrix} \tag{3}$$

$$C \equiv \begin{bmatrix} 1 & 0 & 0 \\ 0 & \cos\theta & \sin\theta \\ 0 & -\sin\theta & \cos\theta \end{bmatrix} \qquad (4)$$

$$B \equiv \begin{bmatrix} \cos\psi & \sin\psi & 0 \\ -\sin\psi & \cos\psi & 0 \\ 0 & 0 & 1 \end{bmatrix}, \qquad (5)$$

so

$$a_{11} = \cos\psi\cos\phi - \cos\theta\sin\phi\sin\psi$$

$$a_{12} = \cos\psi\sin\phi + \cos\theta\cos\phi\sin\psi$$

$$a_{13} = \sin\psi\sin\theta$$

$$a_{21} = -\sin\psi\cos\phi - \cos\theta\sin\phi\cos\psi$$

$$a_{22} = -\sin\psi\sin\phi + \cos\theta\cos\phi\cos\psi$$

$$a_{23} = \cos\psi\sin\theta$$

$$a_{31} = \sin\theta\sin\phi$$

$$a_{32} = -\sin\theta\cos\phi$$

$$a_{33} = \cos\theta$$

To obtain the components of the ANGULAR VELOCITY ω in the body axes, note that for a MATRIX

$$A \equiv [\mathbf{A}_1 \quad \mathbf{A}_2 \quad \mathbf{A}_3], \qquad (6)$$

it is true that

$$\begin{bmatrix} a_{11} & a_{12} & a_{13} \\ a_{21} & a_{22} & a_{23} \\ a_{31} & a_{32} & a_{33} \end{bmatrix} \begin{bmatrix} \omega_x \\ \omega_y \\ \omega_z \end{bmatrix} = \begin{bmatrix} a_{11}\omega_x + a_{12}\omega_y + a_{13}\omega_z \\ a_{21}\omega_x + a_{22}\omega_y + a_{23}\omega_z \\ a_{31}\omega_x + a_{32}\omega_y + a_{33}\omega_z \end{bmatrix} \qquad (7)$$

$$= \mathbf{A}_1\omega_x + \mathbf{A}_2\omega_y + \mathbf{A}_3\omega_z. \qquad (8)$$

Now, ω_z corresponds to rotation about the ϕ axis, so look at the ω_z component of $A\omega$,

$$\omega_\phi = \mathbf{A}_1\omega_z = \begin{bmatrix} \sin\psi\sin\theta \\ \cos\psi\sin\theta \\ \cos\theta \end{bmatrix}\dot\phi. \qquad (9)$$

The line of nodes corresponds to a rotation by θ about the ξ-axis, so look at the ω_ξ component of $B\omega$,

$$\omega_\theta = \mathbf{B}_1\omega_\xi = \mathbf{B}_1\dot\theta = \begin{bmatrix} \cos\psi \\ -\sin\psi \\ 0 \end{bmatrix}\dot\theta. \qquad (10)$$

Similarly, to find rotation by ψ about the remaining axis, look at the ω_ψ component of $B\omega$,

$$\omega_\psi = \mathbf{B}_3\omega_\psi = \mathbf{B}_3\dot\psi = \begin{bmatrix} 0 \\ 0 \\ 1 \end{bmatrix}\dot\psi. \qquad (11)$$

Combining the pieces gives

$$\omega = \begin{bmatrix} \sin\psi\sin\theta\,\dot\phi + \cos\psi\dot\theta \\ \cos\psi\sin\theta\,\dot\phi - \sin\psi\dot\theta \\ \cos\theta\dot\phi + \dot\psi. \end{bmatrix} \qquad (12)$$

For more details, see Goldstein (1980, p. 176) and Landau and Lifschitz (1976, p. 111).

The x-convention Euler angles are given in terms of the CAYLEY-KLEIN PARAMETERS by

$$\phi = -2i\ln\left[\pm\frac{\alpha^{1/2}\gamma^{1/4}}{\beta^{1/4}(1+\beta\gamma)^{1/4}}\right], \; -2i\ln\left[\pm\frac{i\alpha^{1/2}\gamma^{1/4}}{\beta^{1/4}(1+\beta\gamma)^{1/4}}\right] \qquad (13)$$

$$\psi = -2i\ln\left[\pm\frac{\alpha^{1/2}\beta^{1/4}}{\gamma^{1/4}(1+\beta\gamma)^{1/4}}\right], \; -2i\ln\left[\pm\frac{i\alpha^{1/2}\beta^{1/4}}{\gamma^{1/4}(1+\beta\gamma)^{1/4}}\right] \qquad (14)$$

$$\theta = \pm 2\cos^{-1}\left(\pm\sqrt{1+\beta\gamma}\right). \qquad (15)$$

In the "y-convention,"

$$\phi_x \equiv \phi_y + \tfrac{1}{2}\pi \qquad (16)$$

$$\psi_x \equiv \psi_y - \tfrac{1}{2}\pi. \qquad (17)$$

Therefore,

$$\sin\phi_x = \cos\phi_y \qquad (18)$$

$$\cos\phi_x = -\sin\phi_y \qquad (19)$$

$$\sin\psi_x = -\cos\psi_y \qquad (20)$$

$$\cos\psi_x = \sin\psi_y, \qquad (21)$$

giving rotation matrices

$$D \equiv \begin{bmatrix} -\sin\phi & \cos\phi & 0 \\ -\cos\phi & -\sin\phi & 0 \\ 0 & 0 & 1 \end{bmatrix} \qquad (22)$$

$$C \equiv \begin{bmatrix} 1 & 0 & 0 \\ 0 & \cos\theta & \sin\theta \\ 0 & -\sin\theta & \cos\theta \end{bmatrix} \qquad (23)$$

$$B \equiv \begin{bmatrix} \sin\psi & -\cos\psi & 0 \\ \cos\psi & \sin\psi & 0 \\ 0 & 0 & 1 \end{bmatrix} \qquad (24)$$

and A is given by

$$a_{11} = -\sin\psi\sin\phi + \cos\theta\cos\phi\cos\psi$$

$$a_{12} = \sin\psi\cos\phi + \cos\theta\sin\phi\cos\psi$$

$$a_{13} = -\cos\psi\sin\theta$$

$$a_{21} = -\cos\psi\sin\phi - \cos\theta\cos\phi\sin\psi$$

$$a_{22} = \cos\psi\cos\phi - \cos\theta\sin\phi\sin\psi$$

$$a_{23} = \sin\psi\sin\theta$$

$$a_{31} = \sin\theta\cos\phi$$

$$a_{32} = \sin\theta\sin\phi$$

$$a_{33} = \cos\theta.$$

In the "xyz" (pitch-roll-yaw) convention, θ is pitch, ψ is roll, and ϕ is yaw.

$$D \equiv \begin{bmatrix} \cos\phi & \sin\phi & 0 \\ -\sin\phi & \cos\phi & 0 \\ 0 & 0 & 1 \end{bmatrix} \tag{25}$$

$$C \equiv \begin{bmatrix} \cos\theta & 0 & -\sin\theta \\ 0 & 1 & 0 \\ \sin\theta & 0 & \cos\theta \end{bmatrix} \tag{26}$$

$$B \equiv \begin{bmatrix} 1 & 0 & 0 \\ 0 & \cos\psi & \sin\psi \\ 0 & -\sin\psi & \cos\psi \end{bmatrix} \tag{27}$$

and A is given by

$$\begin{aligned}
a_{11} &= \cos\theta\cos\phi \\
a_{12} &= \cos\theta\sin\phi \\
a_{13} &= -\sin\theta \\
a_{21} &= \sin\psi\sin\theta\cos\phi - \cos\psi\sin\phi \\
a_{22} &= \sin\psi\sin\theta\sin\phi + \cos\psi\cos\phi \\
a_{23} &= \cos\theta\sin\psi \\
a_{31} &= \cos\psi\sin\theta\cos\phi + \sin\psi\sin\phi \\
a_{32} &= \cos\psi\sin\theta\sin\phi - \sin\psi\cos\phi \\
a_{33} &= \cos\theta\cos\psi.
\end{aligned}$$

Varshalovich (1988, pp. 21–3) use the notation (α, β, γ) or $(\alpha', \beta', \gamma')$ to denote the Euler angles, and give three different angle conventions, none of which corresponds to the x-convention.

A set of parameters sometimes used instead of angles are the EULER PARAMETERS e_0, e_1, e_2 and e_3, defined by

$$e_0 \equiv \cos\left(\frac{\phi}{2}\right) \tag{28}$$

$$\mathbf{e} \equiv \begin{bmatrix} e_1 \\ e_2 \\ e_3 \end{bmatrix} = \hat{\mathbf{n}}\sin\left(\frac{\phi}{2}\right). \tag{29}$$

Using EULER PARAMETERS (which are QUATERNIONS), an arbitrary ROTATION MATRIX can be described by

$$\begin{aligned}
a_{11} &= e_0^2 + e_1^2 - e_2^2 - e_3^2 \\
a_{12} &= 2(e_1 e_2 + e_0 e_3) \\
a_{13} &= 2(e_1 e_3 - e_0 e_2) \\
a_{21} &= 2(e_1 e_2 - e_0 e_3) \\
a_{22} &= e_0^2 - e_1^2 + e_2^2 - e_3^2 \\
a_{23} &= 2(e_2 e_3 + e_0 e_1) \\
a_{31} &= 2(e_1 e_3 + e_0 e_2) \\
a_{32} &= 2(e_2 e_3 - e_0 e_1) \\
a_{33} &= e_0^2 - e_1^2 - e_2^2 + e_3^2
\end{aligned}$$

(Goldstein 1960, p. 153).

If the coordinates of two pairs of n points $\mathbf{x}_i$ and $\mathbf{x}'_i$ are known, one rotated with respect to the other, then the Euler rotation matrix can be obtained in a straightforward manner using LEAST SQUARES FITTING. Write the points as arrays of vectors, so

$$[\mathbf{x}'_i \cdots \mathbf{x}'_n] = A[\mathbf{x}_1 \cdots \mathbf{x}_n]. \tag{30}$$

Writing the arrays of vectors as matrices gives

$$X' = AX \tag{31}$$

$$X'X^T = AXX^T, \tag{32}$$

and solving for A gives

$$A = X'X^T(XX^T)^{-1}. \tag{33}$$

However, we want the angles θ, ϕ, and ψ, not their combinations contained in the MATRIX A. Therefore, write the 3×3 MATRIX

$$A = \begin{bmatrix} f_1(\theta, \phi, \psi) & f_2(\theta, \phi, \psi) & f_3(\theta, \phi, \psi) \\ f_4(\theta, \phi, \psi) & f_5(\theta, \phi, \psi) & f_6(\theta, \phi, \psi) \\ f_7(\theta, \phi, \psi) & f_7(\theta, \phi, \psi) & f_9(\theta, \phi, \psi) \end{bmatrix} \tag{34}$$

as a 1×9 VECTOR

$$\mathbf{f} = \begin{bmatrix} f_1(\theta, \phi, \psi) \\ \vdots \\ f_9(\theta, \phi, \psi) \end{bmatrix}. \tag{35}$$

Now set up the matrices

$$\begin{bmatrix} \left.\frac{\partial f_1}{\partial\theta}\right|_{\theta_i,\,\phi_i,\,\psi_i} & \left.\frac{\partial f_1}{\partial\phi}\right|_{\theta_i,\,\phi_i,\,\psi_i} & \left.\frac{\partial f_1}{\partial\psi}\right|_{\theta_i,\,\phi_i,\,\psi_i} \\ \vdots & \vdots & \vdots \\ \left.\frac{\partial f_9}{\partial\theta}\right|_{\theta_i,\,\phi_i,\,\psi_i} & \left.\frac{\partial f_9}{\partial\phi}\right|_{\theta_i,\,\phi_i,\,\psi_i} & \left.\frac{\partial f_9}{\partial\psi}\right|_{\theta_i,\,\phi_i,\,\psi_i} \end{bmatrix} \begin{bmatrix} d\theta \\ d\phi \\ d\psi \end{bmatrix} = d\mathbf{f}. \tag{36}$$

Using NONLINEAR LEAST SQUARES FITTING then gives solutions which converge to (θ, ϕ, ψ).

See also CAYLEY-KLEIN PARAMETERS, EULER PARAMETERS, EULER'S ROTATION THEOREM, INFINITESIMAL ROTATION, QUATERNION, ROTATION, ROTATION FORMULA, ROTATION MATRIX

References

Arfken, G. *Mathematical Methods for Physicists, 3rd ed.* Orlando, FL: Academic Press, pp. 198–00, 1985.

Bate, R. R.; Mueller, D. D.; and White, J. E. *Fundamentals of Astrodynamics.* New York: Dover, 1971.

Goldstein, H. "The Euler Angles" and "Euler Angles in Alternate Conventions." §4– and Appendix B in *Classical Mechanics, 2nd ed.* Reading, MA: Addison-Wesley, pp. 143–48 and 606–10, 1980.

Kraus, M. "LiveGraphics3D Example: Euler Angles." http:// wwwvis.informatik.uni-stuttgart.de/~kraus/LiveGraphics3D/examples/Euler.html.

Landau, L. D. and Lifschitz, E. M. *Mechanics, 3rd ed.* Oxford, England: Pergamon Press, 1976.

Margenau, H. and Murphy, G. M. *The Mathematics of Physics and Chemistry, 2 vols.* Princeton, NJ: Van Nostrand, 1956–4.

Osgood, W. F. *Mechanics.* New York: Macmillan, 1937.

Tuma, J. J. *Dynamics.* New York: Quantum Publishers, 1974.

Varshalovich, D. A.; Moskalev, A. N.; and Khersonskii, V. K. "Description of Rotation in Terms of the Euler Angles." §1.4.1 in *Quantum Theory of Angular Momentum.* Singapore: World Scientific, pp. 21–3, 1988.

Euler Brick

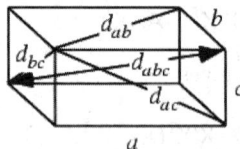

A RECTANGULAR PARALLELEPIPED ("BRICK") with integer edges $a > b > c$ and face diagonals d_{ij} given by

$$d_{ab} = \sqrt{a^2 + b^2} \qquad (1)$$

$$d_{ac} = \sqrt{a^2 + c^2} \qquad (2)$$

$$d_{bc} = \sqrt{b^2 + c^2}. \qquad (3)$$

The problem is also called the brick problem, diagonals problem, perfect box problem, perfect cuboid problem, or rational cuboid problem.

The smallest solution with integer edges and face diagonals has sides $(a, b, c) = (240, 117, 44)$ and face DIAGONALS $d_{ab} = 267$, $d_{ac} = 244$, and $d_{bc} = 125$, and was discovered by Halcke (1719; Dickson 1952, pp. 497–00). Interest in this problem was high during the 18th century, and Saunderson (1740) found a parametric solution, while Euler (1770, 1772) found at least two parametric solutions. Kraitchik gave 257 cuboids with the ODD edge less than 1 million (Guy 1994, p. 174). F. Helenius has compiled a list of the 5003 smallest (measured by the longest edge) Euler bricks. The first few are (240, 117, 44), (275, 252, 240), (693, 480, 140), (720, 132, 85), (792, 231, 160), ... (Sloane's A031173, A031174, and A031175). Parametric solutions for Euler bricks are also known.

No solution is known to the more general problem in which the oblique SPACE DIAGONAL

$$d_{abc} = \sqrt{a^2 + b^2 + c^2} \qquad (4)$$

is also an INTEGER. If such a brick exists, the smallest side must be at least 1,281,000,000 (R. Rathbun 1996). Such a solution is equivalent to solving the DIOPHANTINE EQUATIONS

$$A^2 + B^2 = C^2 \qquad (5)$$

$$A^2 + D^2 = E^2 \qquad (6)$$

$$B^2 + D^2 = F^2 \qquad (7)$$

$$B^2 + E^2 = G^2. \qquad (8)$$

A solution with integral SPACE DIAGONAL and two out of three face diagonals is $a = 672$, $b = 153$, and $c = 104$, giving $d_{ab} = 3\sqrt{52777}$, $d_{ac} = 680$, $d_{bc} = 185$, and $d_{abc} = 697$, which was known to Euler. A solution giving integral space and face diagonals with only a single nonintegral EDGE is $a = 18720$, $b = \sqrt{211773121}$, and $c = 7800$, giving $d_{ab} = 23711$, $d_{ac} = 20280$, $d_{bc} = 16511$, and $d_{abc} = 24961$.

See also CUBOID, CYCLIC QUADRILATERAL, DIAGONAL (POLYHEDRON), PARALLELEPIPED, PYTHAGOREAN QUADRUPLE

References

Dickson, L. E. *History of the Theory of Numbers, Vol. 2: Diophantine Analysis.* New York: Chelsea, 1952.

Guy, R. K. "Is There a Perfect Cuboid? Four Squares whose Sums in Pairs are Square. Four Squares whose Differences are Square." §D18 in *Unsolved Problems in Number Theory, 2nd ed.* New York: Springer-Verlag, pp. 173–81, 1994.

Halcke, P. *Deliciae Mathematicae; oder, Mathematisches sinnen-confect.* Hamburg, Germany: N. Sauer, p. 265, 1719.

Helenius, F. First 1000 Primitive Euler Bricks. NOTEBOOKS/EULERBRICKS.DAT.

Leech, J. "The Rational Cuboid Revisited." *Amer. Math. Monthly* **84**, 518–33, 1977. Erratum in *Amer. Math. Monthly* **85**, 472, 1978.

Sloane, N. J. A. Sequences A031173, A031174, and A031175 in "An On-Line Version of the Encyclopedia of Integer Sequences." http://www.research.att.com/~njas/sequences/eisonline.html.

Rathbun, R. L. Personal communication, 1996.

Saunderson, N. *The Elements of Algebra in 10 Books, Vol. 2.* Cambridge, England: University Press, pp. 429–31, 1740.

Spohn, W. G. "On the Integral Cuboid." *Amer. Math. Monthly* **79**, 57–9, 1972.

Spohn, W. G. "On the Derived Cuboid." *Canad. Math. Bull.* **17**, 575–77, 1974.

Wells, D. G. *The Penguin Dictionary of Curious and Interesting Numbers.* London: Penguin, p. 127, 1986.

Euler Chain

A CHAIN whose EDGES consist of all graph EDGES.

Euler Characteristic

Let a closed surface have GENUS g. Then the POLYHEDRAL FORMULA generalizes to the POINCARÉ FORMULA

$$\chi \equiv V - E + F = \chi(g), \qquad (1)$$

where

$$\chi(g) = 2 - 2g \qquad (2)$$

is the Euler characteristic, sometimes also known as the EULER-POINCARÉ CHARACTERISTIC. The POLYHEDRAL FORMULA corresponds to the special case $g = 0$. The only compact closed surfaces with Euler characteristic 0 are the KLEIN BOTTLE and TORUS (Dodson and Parker 1997, p. 125).

In terms of the INTEGRAL CURVATURE of the surface K,

$$\iint K\, da = 2\pi\chi. \qquad (3)$$

The Euler characteristic is sometimes also called the EULER NUMBER. It can also be expressed as

$$\chi = p_0 - p_1 + p_2, \qquad (4)$$

where p_i is the ith BETTI NUMBER of the space.

See also CHROMATIC NUMBER, EULER NUMBER (FINITE COMPLEX), MAP COLORING, POINCARÉ FORMULA, POLYHEDRAL FORMULA

References

Coxeter, H. S. M. "Poincaré's Proof of Euler's Formula." Ch. 9 in *Regular Polytopes, 3rd ed.* New York: Dover, pp. 165–72, 1973.

Dodson, C. T. J. and Parker, P. E. *A User's Guide to Algebraic Topology.* Dordrecht, Netherlands: Kluwer, 1997.

Gray, A. *Modern Differential Geometry of Curves and Surfaces with Mathematica, 2nd ed.* Boca Raton, FL: CRC Press, p. 635, 1997.

Euler Constant

e, EULER-MASCHERONI CONSTANT, MACLAURIN-CAUCHY THEOREM

Euler Curvature Formula

The curvature of a surface satisfies

$$\kappa = \kappa_1 \cos^2 \theta + \kappa_2 \sin^2 \theta,$$

where κ is the normal CURVATURE in a direction making an ANGLE θ with the first principal direction and κ_1 and κ_2 are the PRINCIPAL CURVATURES.

See also PRINCIPAL CURVATURES

Euler Differential Equation

The general nonhomogeneous differential equation is given by

$$x^2 \frac{d^2y}{dx^2} + \alpha x \frac{dy}{dx} + \beta y = S(x), \qquad (1)$$

and the homogeneous equation is

$$x^2 y'' + \alpha x y' + \beta y = 0 \qquad (2)$$

$$y'' + \frac{\alpha}{x} y' + \frac{\beta}{x^2} y = 0. \qquad (3)$$

Now attempt to convert the equation from

$$y'' + p(x)y' + q(x)y = 0 \qquad (4)$$

to one with constant COEFFICIENTS

$$\frac{d^2y}{dz^2} + A \frac{dy}{dz} + By = 0 \qquad (5)$$

by using the standard transformation for linear SECOND-ORDER ORDINARY DIFFERENTIAL EQUATIONS. Comparing (3) and (5), the functions $p(x)$ and $q(x)$ are

$$p(x) \equiv \frac{\alpha}{x} = \alpha x^{-1} \qquad (6)$$

$$q(x) \equiv \frac{\beta}{x^2} = \beta x^{-2}. \qquad (7)$$

Let $B \equiv \beta$ and define

$$z \equiv B^{-1/2} \int \sqrt{q(x)}\, dx = \beta^{-1/2} \int \sqrt{\beta x^{-2}}\, dx$$

$$= \int x^{-1}\, dx = \ln x. \qquad (8)$$

Then A is given by

$$A \equiv \frac{q'(x) + 2p(x)q(x)}{2[q(x)]^{3/2}} B^{1/2}$$

$$= \frac{-2\beta x^{-3} + 2(\alpha x^{-1})(\beta x^{-2})}{2(\beta x^{-2})^{3/2}} \beta^{1/2}$$

$$= \alpha - 1, \qquad (9)$$

which is a constant. Therefore, the equation becomes a second-order ODE with constant COEFFICIENTS

$$\frac{d^2y}{dz^2} + (\alpha - 1)\frac{dy}{dz} + \beta y = 0. \qquad (10)$$

Define

$$r_1 \equiv \tfrac{1}{2}\left(-A + \sqrt{A^2 - 4B}\right)$$

$$= \tfrac{1}{2}\left[1 - \alpha + \sqrt{(\alpha - 1)^2 - 4\beta}\right] \qquad (11)$$

$$r_2 \equiv \tfrac{1}{2}\left(-A - \sqrt{A^2 - 4B}\right)$$

$$= \tfrac{1}{2}\left[1 - \alpha - \sqrt{(\alpha - 1)^2 - 4\beta}\right] \qquad (12)$$

and

$$a \equiv \tfrac{1}{2}(1 - \alpha) \qquad (13)$$

$$b \equiv \tfrac{1}{2}\sqrt{4\beta - (\alpha - 1)^2}. \qquad (14)$$

The solutions are

$$y = \begin{cases} c_1 e^{r_1 z} + c_2 e^{r_2 z} & (\alpha - 1)^2 > 4\beta \\ c_1 + c_2 z)e^{az} & (\alpha - 1)^2 = 4 \\ c^{az}[c_1 \cos(bz) + c_2 \sin(bz)] & (\alpha - 1)^2 < 4\beta. \end{cases} \qquad (15)$$

In terms of the original variable x,

$$y = \begin{cases} c_1 |x|^{r_1} + c_2 |x|^{r_2} & (\alpha - 1)^2 > 4\beta \\ (c_1 + c_2 \ln|x|)|x|^a & (\alpha - 1)^2 = 4\beta \\ |x|^a[c_1 \cos(b \ln|x|) + c_2 \sin(b \ln|x|)] & (\alpha - 1)^2 < 4\beta. \end{cases}$$

$$(16)$$

Zwillinger (1997, p. 120) gives two other types of equations known as Euler differential equations,

$$y' = \pm \sqrt{\frac{ay^4 + by^3 + cy^2 + dy + e}{ax^3 + bx^3 + cx^2 + dx + e}} \qquad (17)$$

(Valiron 1950, p. 201) and

$$y' + y^2 = \alpha x^m \qquad (18)$$

(Valiron 1950, p. 212), the latter of which can be solved in terms of Bessel functions.

See also EULER'S EQUATIONS OF INVISCID MOTION

References

Valiron, G. *The Geometric Theory of Ordinary Differential Equations and Algebraic Functions.* Brookline, MA: Math. Sci. Press, 1950.

Zwillinger, D. *Handbook of Differential Equations, 3rd ed.* Boston, MA: Academic Press, p. 120, 1997.

Euler Equation

EULER DIFFERENTIAL EQUATION, EULER'S EQUATIONS OF INVISCID MOTION, EULER FORMULA, EULER-LAGRANGE DIFFERENTIAL EQUATION

Euler Formula

The Euler formula states

$$e^{ix} = \cos x + i \sin x, \qquad (1)$$

where i is the IMAGINARY NUMBER. Note that Euler's POLYHEDRAL FORMULA is sometimes also called the Euler formula, as is the EULER CURVATURE FORMULA. The equivalent expression

$$ix = \ln(\cos x + i \sin x) \qquad (2)$$

had previously been published by Cotes (1714). The special case of the formula with $x = \pi$ gives the beautiful identity

$$e^{i\pi} + 1 = 0, \qquad (3)$$

an equation connecting the fundamental numbers i, PI, e, 1, and 0 (ZERO).

The Euler formula can be demonstrated using a series expansion

$$e^{ix} = \sum_{n=0}^{\infty} \frac{(ix)^n}{n!}$$

$$= \sum_{n=0}^{\infty} \frac{(-1)^n x^{2n}}{(2n)!} + i \sum_{n=1}^{\infty} \frac{(-1)^{n-1} x^{2n-1}}{(2n-1)!}$$

$$= \cos x + i \sin x. \qquad (4)$$

It can also be proven using a COMPLEX integral. Let

$$z \equiv \cos \theta + i \sin \theta \qquad (5)$$

$$dz = (-\sin \theta + i \cos \theta)\, d\theta = i(\cos \theta + i \sin \theta)\, d\theta$$

$$= iz\, d\theta \qquad (6)$$

$$\int \frac{dz}{z} = \int i\, d\theta \qquad (7)$$

$$\ln z = i\theta, \qquad (8)$$

so

$$z = e^{i\theta} \equiv \cos \theta + i \sin \theta. \qquad (9)$$

See also DE MOIVRE'S IDENTITY, POLYHEDRAL FORMULA

References

Castellanos, D. "The Ubiquitous Pi. Part I." *Math. Mag.* **61**, 67–8, 1988.

Conway, J. H. and Guy, R. K. "Euler's Wonderful Relation." *The Book of Numbers.* New York: Springer-Verlag, pp. 254–56, 1996.

Cotes, R. *Philosophical Transactions* **29**, 32, 1714.

Euler, L. *Miscellanea Berolinensia* **7**, 179, 1743.

Euler, L. *Introductio in Analysin Infinitorum, Vol. 1.* Lausanne, p. 104, 1748.

Hoffman, P. *The Man Who Loved Only Numbers: The Story of Paul Erdos and the Search for Mathematical Truth.* New York: Hyperion, p. 212, 1998.

Euler Four-Square Identity

The amazing polynomial identity

$$(a_1^2 + a_2^2 + a_3^2 + a_4^2)(b_1^2 + b_2^2 + b_3^2 + b_4^2)$$

$$= (a_1 b_1 - a_2 b_2 - a_3 b_3 - a_4 b_4)^2$$

$$+ (a_1 b_2 + a_2 b_1 + a_3 b_4 - a_4 b_3)^2$$

$$+ (a_1 b_3 - a_2 b_4 + a_3 b_1 + a_4 b_2)^2$$

$$+ (a_1 b_4 + a_2 b_3 - a_3 b_2 + a_4 b_1)^2,$$

communicated by Euler in a letter to Goldbach on April 15, 1750 (incorrectly given as April 15, 1705–before Euler was born–in Conway and Guy 1996, p. 232). The identity also follows from the fact that the norm of the product of two QUATERNIONS is the product of the norms (Conway and Guy 1996).

See also FIBONACCI IDENTITY, LAGRANGE'S FOUR-SQUARE THEOREM, LEBESGUE IDENTITY

References

Conway, J. H. and Guy, R. K. *The Book of Numbers.* New York: Springer-Verlag, p. 232, 1996.

Nagell, T. *Introduction to Number Theory.* New York: Wiley, pp. 191–92, 1951.

Petkovsek, M.; Wilf, H. S.; and Zeilberger, D. *A = B.* Wellesley, MA: A. K. Peters, p. 8, 1996.

Euler Graph

EULERIAN GRAPH

Euler Identity

For $|z| < 1$,

$$\prod_{k=1}^{\infty}(1 + z^k) = \prod_{k=1}^{\infty}(1 - z^{2k-1})^{-1}.$$

Euler Integral

Expanding and taking a series expansion about zero for either side gives

$$1 + z + z^2 + 2z^3 + 2z^4 + 3z^5 + 4z^6 + 5z^7 + \ldots,$$

giving 1, 1, 1, 2, 2, 3, 4, 5, 6, 8, 10, 12, 15, 18, 22, 27, ... (Sloane's A000009), the number of partitions of n into distinct parts.

See also JACOBI TRIPLE PRODUCT, PARTITION FUNCTION P, q-SERIES

References

Bailey, W. N. *Generalised Hypergeometric Series.* Cambridge, England: Cambridge University Press, p. 72, 1935.
Franklin. *Comptes Rendus* **92**, 448–50, 1881.
Hardy, G. H. §6.2 in *Ramanujan: Twelve Lectures on Subjects Suggested by His Life and Work, 3rd ed.* New York: Chelsea, pp. 83–5, 1999.
Hardy, G. H. and Wright, E. M. §19.11 in *An Introduction to the Theory of Numbers, 5th ed.* Oxford, England: Clarendon Press, 1979.
MacMahon, P. A. *Combinatory Analysis, Vol. 2.* New York: Chelsea, pp. 21–3, 1960.
Nagell, T. *Introduction to Number Theory.* New York: Wiley, p. 55, 1951.
Sloane, N. J. A. Sequences A000009/M0281 in "An On-Line Version of the Encyclopedia of Integer Sequences." http://www.research.att.com/~njas/sequences/eisonline.html.

Euler Integral

Euler integration was defined by Schanuel and subsequently explored by Rota, Chen, and Klain. The Euler integral of a FUNCTION $f : \mathbb{R} \to \mathbb{R}$ (assumed to be piecewise-constant with finitely many discontinuities) is the sum of

$$f(x) - \tfrac{1}{2}[f(x_+) + f(x_-)]$$

over the finitely many discontinuities of f. The n-D Euler integral can be defined for classes of functions $\mathbb{R}^n \to \mathbb{R}$. Euler integration is additive, so the Euler integral of $f + g$ equals the sum of the Euler integrals of f and g.

See also EULER MEASURE

Euler Law

POLYHEDRAL FORMULA

Euler L-Function

A special case of the ARTIN L-FUNCTION for the POLYNOMIAL $x^2 + 1$. It is given by

$$L(s) = \prod_{p \text{ odd prime}} \frac{1}{1 - \chi^-(p) p^{-s}},$$

where

$$\chi^-(p) \equiv \begin{cases} 1 & \text{for } p \equiv 1 \pmod 4 \\ -1 & \text{for } p \equiv 3 \pmod 4 \end{cases} = \left(\frac{-1}{p} \right),$$

where $(-1/p)$ is a LEGENDRE SYMBOL.

References

Knapp, A. W. "Group Representations and Harmonic Analysis, Part II." *Not. Amer. Math. Soc.* **43**, 537–49, 1996.

Euler Line

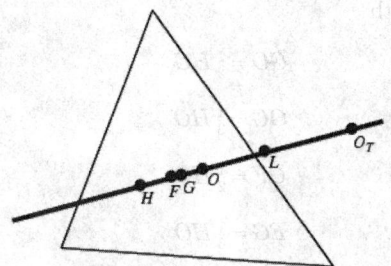

The line on which the ORTHOCENTER H, CENTROID G, CIRCUMCENTER O, DE LONGCHAMPS POINT L, NINE-POINT CENTER F, and the TANGENTIAL TRIANGLE CIRCUMCIRCLE O_T of a TRIANGLE lie. The INCENTER lies on the Euler line only if the TRIANGLE is an ISOSCELES TRIANGLE. The Euler line consists of all points with TRILINEAR COORDINATES $\alpha : \beta : \gamma$ which satisfy

$$\begin{vmatrix} \alpha & \beta & \gamma \\ \cos A & \cos B & \cos C \\ \cos B \cos C & \cos C \cos A & \cos A \cos B \end{vmatrix} = 0, \qquad (1)$$

which simplifies to

$$\alpha \cos A(\cos^2 B - \cos^2 C) + \beta \cos B(\cos^2 C - \cos^2 A)$$
$$+ \gamma \cos C(\cos^2 A - \cos^2 B) = 0. \qquad (2)$$

This can also be written

$$\alpha \sin(2A) \sin(B - C) + \beta \sin(2B) \sin(C - A)$$
$$+ \gamma \sin(2C) \sin(A - B) = 0. \qquad (3)$$

The Euler line may also be given parametrically in EXACT TRILINEAR COORDINATES by

$$P(\lambda) = O + \lambda H \qquad (4)$$

where the following table summarized important TRIANGLES CENTERS corresponding to various values of λ (including the factor of 1/2 omitted by Oldknow 1996).

λ	TRIANGLE CENTER
-1	POINT AT INFINITY
$-\tfrac{1}{2}$	DE LONGCHAMPS POINT L
0	CIRCUMCENTER O
$\tfrac{1}{2}$	CENTROID G
1	NINE-POINT CENTER F
∞	ORTHOCENTER H

The CIRCUMCENTER O, NINE-POINT CENTER F, CENTROID G, and ORTHOCENTER H form a HARMONIC RANGE with

$$GO = \tfrac{1}{2}HG \qquad (5)$$

$$OG = \tfrac{1}{3}HO \qquad (6)$$

$$OF = \tfrac{1}{2}HO \qquad (7)$$

$$FG = \tfrac{1}{6}HO \qquad (8)$$

(Honsberger 1995, p. 7).

The Euler line intersects the SODDY LINE in the DE LONGCHAMPS POINT, and the GERGONNE LINE in the EVANS POINT. The ISOTOMIC CONJUGATE of the Euler line is called JERABEK'S HYPERBOLA (Casey 1893, Vandeghen 1965).

See also CENTROID (TRIANGLE), CIRCUMCENTER, EVANS POINT, GERGONNE LINE, JERABEK'S HYPERBOLA, DE LONGCHAMPS POINT, NINE-POINT CENTER, ORTHOCENTER, SODDY LINE, TANGENTIAL TRIANGLE

References

Casey, J. *A Treatise on the Analytical Geometry of the Point, Line, Circle, and Conic Sections, Containing an Account of Its Most Recent Extensions with Numerous Examples, 2nd rev. enl. ed.* Dublin: Hodges, Figgis, & Co., 1893.

Coxeter, H. S. M. and Greitzer, S. L. "The Medial Triangle and Euler Line." §1.7 in *Geometry Revisited*. Washington, DC: Math. Assoc. Amer., pp. 18–0, 1967.

Dörrie, H. "Euler's Straight Line." §27 in *100 Great Problems of Elementary Mathematics: Their History and Solutions*. New York: Dover, pp. 141–42, 1965.

Durell, C. V. *Modern Geometry: The Straight Line and Circle*. London: Macmillan, p. 28, 1928.

Honsberger, R. *Episodes in Nineteenth and Twentieth Century Euclidean Geometry*. Washington, DC: Math. Assoc. Amer., p. 7, 1995.

Ogilvy, C. S. *Excursions in Geometry*. New York: Dover, pp. 117–19, 1990.

Oldknow, A. "The Euler-Gergonne-Soddy Triangle of a Triangle." *Amer. Math. Monthly* **103**, 319–29, 1996.

Vandeghen, A. "Some Remarks on the Isogonal and Cevian Transforms. Alignments of Remarkable Points of a Triangle." *Amer. Math. Monthly* **72**, 1091–094, 1965.

Wells, D. *The Penguin Dictionary of Curious and Interesting Geometry*. London: Penguin, p. 69, 1991.

Euler Measure

Define the Euler measure of a polyhedral set as the EULER INTEGRAL of its indicator function. It is easy to show by induction that the Euler measure of a closed bounded convex POLYHEDRON is always 1 (independent of dimension), while the Euler measure of a d-D relative-open bounded convex POLYHEDRON is $(-1)^d$.

Euler Number

The Euler numbers, also called the SECANT NUMBERS or ZIG NUMBERS, are defined for $|x| < \pi/2$ by

$$\operatorname{sech} x - 1 \equiv -\frac{E_1^* x^2}{2!} + \frac{E_2^* x^4}{4!} - \frac{E_3^* x^6}{6!} + \ldots \qquad (1)$$

$$\sec x - 1 \equiv \frac{E_1^* x^2}{2!} + \frac{E_2^* x^4}{4!} - \frac{E_3^* x^6}{6!} + \ldots, \qquad (2)$$

where sech is the HYPERBOLIC SECANT and sec is the SECANT. Euler numbers give the number of ODD ALTERNATING PERMUTATIONS and are related to GENOCCHI NUMBERS. The base e of the NATURAL LOGARITHM is sometimes known as Euler's number.

Some values of the Euler numbers are

$$E_1^* = 1$$

$$E_2^* = 5$$

$$E_3^* = 61$$

$$E_4^* = 1,385$$

$$E_5^* = 50,521$$

$$E_6^* = 2,702,765$$

$$E_7^* = 199,360,981$$

$$E_8^* = 19,391,512,145$$

$$E_9^* = 2,404,879,675,441$$

$$E_{10}^* = 370,371,188,237,525$$

$$E_{11}^* = 69,348,874,393,137,901$$

$$E_{12}^* = 15,514,534,163,557,086,905$$

(Sloane's A000364). The first few PRIME Euler numbers E_n^* occur for $n = 2, 3, 19, 227, 255, \ldots$ (Sloane's A014547) up to a search limit of $n = 1415$.

The slightly different convention defined by

$$E_{2n} = (-1)^n E_n^* \qquad (3)$$

$$E_{2n+1} = 0 \qquad (4)$$

is frequently used. These are, for example, the Euler numbers computed by the *Mathematica* function EulerE[n]. This definition has the particularly simple series definition

$$\operatorname{sech} x \equiv \sum_{k=0}^{\infty} \frac{E_k x^k}{k!} \qquad (5)$$

and is equivalent to

$$E_n = 2^n E_n(\tfrac{1}{2}), \tag{6}$$

where $E_n(x)$ is an EULER POLYNOMIAL. The Euler numbers have the ASYMPTOTIC SERIES

$$E_{2n} \sim (-1)^n 8 \sqrt{\frac{n}{\pi}} \left(\frac{4n}{\pi e}\right)^{2n}. \tag{7}$$

To confuse matters further, the EULER CHARACTERISTIC is sometimes also called the "Euler number."

See also BERNOULLI NUMBER, EULER NUMBER (FINITE COMPLEX), EULERIAN NUMBER, EULER POLYNOMIAL, EULER ZIGZAG NUMBER, GENOCCHI NUMBER

References

Abramowitz, M. and Stegun, C. A. (Eds.). "Bernoulli and Euler Polynomials and the Euler-Maclaurin Formula." §23.1 in *Handbook of Mathematical Functions with Formulas, Graphs, and Mathematical Tables, 9th printing.* New York: Dover, pp. 804–06, 1972.
Conway, J. H. and Guy, R. K. In *The Book of Numbers.* New York: Springer-Verlag, pp. 110–11, 1996.
Guy, R. K. "Euler Numbers." §B45 in *Unsolved Problems in Number Theory, 2nd ed.* New York: Springer-Verlag, p. 101, 1994.
Hauss, M. *Verallgemeinerte Stirling, Bernoulli und Euler Zahlen, deren Anwendungen und schnell konvergente Reihen für Zeta Funktionen.* Aachen, Germany: Verlag Shaker, 1995.
Knuth, D. E. and Buckholtz, T. J. "Computation of Tangent, Euler, and Bernoulli Numbers." *Math. Comput.* **21,** 663–88, 1967.
Sloane, N. J. A. Sequences A0003644019 and A014547 in "An On-Line Version of the Encyclopedia of Integer Sequences." http://www.research.att.com/~njas/sequences/eisonline.html.
Spanier, J. and Oldham, K. B. "The Euler Numbers, E_n." Ch. 5 in *An Atlas of Functions.* Washington, DC: Hemisphere, pp. 39–2, 1987.
Young, P. T. "Congruences for Bernoulli, Euler, and Stirling Numbers." *J. Number Th.* **78,** 204–27, 1999.

Euler Number (Finite Complex)

The Euler number of a finite complex K is defined by

$$\chi(K) = \sum (-1)^p \operatorname{rank}(C_p(K)).$$

The Euler number is a topological invariant.

See also EULER CHARACTERISTIC, LEFSCHETZ NUMBER

References

Munkres, J. R. *Elements of Algebraic Topology.* Perseus Press, p. 124, 1993.

Euler Parameters

The four parameters e_0, e_1, e_2, and e_3 describing a finite rotation about an arbitrary axis. The Euler parameters are defined by

$$e_0 \equiv \cos\left(\frac{\phi}{2}\right) \tag{1}$$

$$\mathbf{e} \equiv \begin{bmatrix} e_1 \\ e_2 \\ e_3 \end{bmatrix} = \hat{\mathbf{n}} \sin\left(\frac{\phi}{2}\right), \tag{2}$$

and are a QUATERNION in scalar-vector representation

$$(e_0, \mathbf{e}) = e_0 + e_1 i + e_2 j + e_3 k. \tag{3}$$

Because EULER'S ROTATION THEOREM states that an arbitrary rotation may be described by only three parameters, a relationship must exist between these four quantities

$$e_0^2 + \mathbf{e} \cdot \mathbf{e} = e_0^2 + e_1^2 + e_2^2 + e_3^2 = 1 \tag{4}$$

(Goldstein 1980, p. 153). The rotation angle is then related to the Euler parameters by

$$\cos\phi = 2e_0^2 - 1 = e_0^2 - \mathbf{e} \cdot \mathbf{e} = e_0^2 - e_1^2 - e_2^2 - e_3^2 \tag{5}$$

$$\hat{\mathbf{n}} \sin\phi = 2\mathbf{e} e_0. \tag{6}$$

The Euler parameters may be given in terms of the EULER ANGLES by

$$e_0 = \cos[\tfrac{1}{2}(\phi + \psi)] \cos(\tfrac{1}{2}\theta) \tag{7}$$

$$e_1 = \sin[\tfrac{1}{2}(\phi - \psi)] \sin(\tfrac{1}{2}\theta) \tag{8}$$

$$e_2 = \cos[\tfrac{1}{2}(\phi - \psi)] \sin(\tfrac{1}{2}\theta) \tag{9}$$

$$e_3 = \sin[\tfrac{1}{2}(\phi + \psi)] \cos(\tfrac{1}{2}\theta) \tag{10}$$

(Goldstein 1980, p. 155).

Using the Euler parameters, the ROTATION FORMULA becomes

$$\mathbf{r}' = \mathbf{r}(e_0^2 - e_1^2 - e_2^2 - e_3^2) + 2\mathbf{e}(\mathbf{e} \cdot \mathbf{r}) + (\mathbf{r} \times \hat{\mathbf{n}})\sin\phi, \tag{11}$$

and the ROTATION MATRIX becomes

$$\begin{bmatrix} x' \\ y' \\ z' \end{bmatrix} = \mathsf{A} \begin{bmatrix} x \\ y \\ z \end{bmatrix}, \tag{12}$$

where the elements of the matrix are

$$a_{ij} = \delta_{ij}(e_0^2 - e_k e_k) + 2e_i e_j + 2\epsilon_{ijk} e_0 e_k. \tag{13}$$

Here, EINSTEIN SUMMATION has been used, δ_{ij} is the KRONECKER DELTA, and ϵ_{ijk} is the PERMUTATION SYMBOL. Written out explicitly, the matrix elements are

$$a_{11} = e_0^2 + e_1^2 - e_2^2 - e_3^2 \tag{14}$$

$$a_{12} = 2(e_1 e_2 + e_0 e_3) \tag{15}$$

$$a_{13} = 2(e_1 e_3 - e_0 e_2) \tag{16}$$

$$a_{21} = 2(e_1 e_2 - e_0 e_3) \tag{17}$$

$$a_{22} = e_0^2 - e_1^2 + e_2^2 - e_3^2 \qquad (18)$$

$$a_{23} = 2(e_2 e_3 + e_0 e_1) \qquad (19)$$

$$a_{31} = 2(e_1 e_3 + e_0 e_2) \qquad (20)$$

$$a_{32} = 2(e_2 e_3 - e_0 e_1) \qquad (21)$$

$$a_{33} = e_0^2 - e_1^2 - e_2^2 + e_3^2. \qquad (22)$$

See also EULER ANGLES, QUATERNION, ROTATION FORMULA, ROTATION MATRIX

References

Arfken, G. *Mathematical Methods for Physicists, 3rd ed.* Orlando, FL: Academic Press, pp. 198–00, 1985.
Goldstein, H. *Classical Mechanics, 2nd ed.* Reading, MA: Addison-Wesley, 1980.
Landau, L. D. and Lifschitz, E. M. *Mechanics, 3rd ed.* Oxford, England: Pergamon Press, 1976.

Euler Point

The MIDPOINTS M_{HA}, M_{HB}, M_{HC} of the segments which join the VERTICES of a triangle and the ORTHOCENTER H are called Euler points. They are three of the nine prominent points of a triangle through which the NINE-POINT CIRCLE passes.

See also FEUERBACH'S THEOREM, NINE-POINT CIRCLE

References

Honsberger, R. *Episodes in Nineteenth and Twentieth Century Euclidean Geometry.* Washington, DC: Math. Assoc. Amer., p. 6, 1995.

Euler Polyhedral Formula

POLYHEDRAL FORMULA

Euler Polynomial

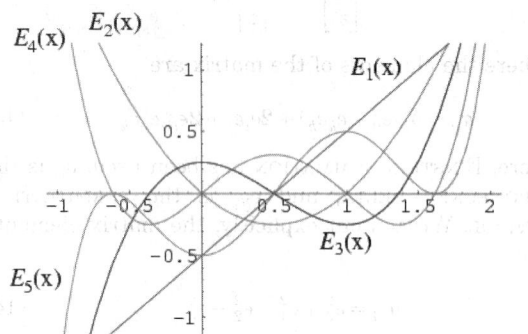

The Euler polynomial $E_n(x)$ is given by the APPELL SEQUENCE with

$$g(t) = \tfrac{1}{2}(e^t + 1), \qquad (1)$$

giving the GENERATING FUNCTION

$$\frac{2e^{xt}}{e^t + 1} \equiv \sum_{n=0}^{\infty} E_n(x) \frac{t^n}{n!}. \qquad (2)$$

Roman (1984, p. 100) defines a generalization $E_n^{(x)}(x)$ for which $E_n(x) = E_n^{(1)}(x)$. Euler polynomials are related to the BERNOULLI NUMBERS by

$$E_{n-1}(x) = \frac{2^n}{n}\left[B_n\left(\frac{x+1}{2}\right) - B_n\left(\frac{x}{2}\right) \right] \qquad (3)$$

$$= \frac{2}{n}\left[B_n(x) - 2^n B_n\left(\frac{x}{2}\right) \right] \qquad (4)$$

$$E_{n-2}(x) = 2\binom{n}{2}^{-1} \sum_{k=0}^{n-2} \binom{n}{2}[(2^{n-k}-1)B_{n-k}B_k(x)], \quad (5)$$

where $\binom{n}{k}$ is a BINOMIAL COEFFICIENT. Setting $x = 1/2$ and normalizing by 2^n gives the EULER NUMBER

$$E_n = 2^n E_n(\tfrac{1}{2}). \qquad (6)$$

Call $E_n' = E_n(0)$, then the first few terms are $-1/2$, 0, $1/4$, $-1/2$, 0, $17/8$, 0, $31/2$, 0, ... The terms are the same but with the SIGNS reversed if $x = 1$. These values can be computed using the double sum

$$E_n(0) = 2^{-n} \sum_{j=1}^{n}\left[(-1)^{j+n+1} j^k \sum_{k=0}^{n-j} \binom{n+1}{k} \right]. \qquad (7)$$

The BERNOULLI NUMBERS B_n for $n > 1$ can be expressed in terms of the E_n' by

$$B_n = -\frac{n E_{n-1}'}{2(2^n - 1)}. \qquad (8)$$

The Newton expansion of the Euler polynomials is given by

$$E_n(x) = \sum_{j=0}^{n} \sum_{k=j}^{n} \binom{-1}{j} \frac{1}{2^j} (k)_j S(n, k)(x)_{k-j}, \qquad (9)$$

where $\binom{n}{k}$ is a BINOMIAL COEFFICIENT, $(k)_j$ is a FALLING FACTORIAL, and $S(n, k)$ is a STIRLING NUMBER OF THE SECOND KIND (Roman 1984, p. 101).

The Euler polynomials satisfy the identity

$$\sum_{k=0}^{n} \binom{n}{2} E_k(z) E_{n-k}(w)$$

$$= 2(1 - w - z)E_n(z + w) + 2E_{n+1}(z + w) \qquad (10)$$

for n a NONNEGATIVE INTEGER.

See also APPELL SEQUENCE, BERNOULLI POLYNOMIAL, EULER NUMBER, GENOCCHI NUMBER

References

Abramowitz, M. and Stegun, C. A. (Eds.). "Bernoulli and Euler Polynomials and the Euler-Maclaurin Formula."

§23.1 in *Handbook of Mathematical Functions with Formulas, Graphs, and Mathematical Tables, 9th printing.* New York: Dover, pp. 804–06, 1972.

Gradshteyn, I. S. and Ryzhik, I. M. *Tables of Integrals, Series, and Products, 6th ed.* San Diego, CA: Academic Press, 2000.

Prudnikov, A. P.; Marichev, O. I.; and Brychkov, Yu. A. "The Generalized Zeta Function $\zeta(s, x)$, Bernoulli Polynomials $B_n(x)$, Euler Polynomials $E_n(x)$, and Polylogarithms $\text{Li}_\nu(x)$." §1.2 in *Integrals and Series, Vol. 3: More Special Functions.* Newark, NJ: Gordon and Breach, pp. 23–4, 1990.

Roman, S. "The Euler Polynomials." §4.2.3 in *The Umbral Calculus.* New York: Academic Press, pp. 100–06, 1984.

Spanier, J. and Oldham, K. B. "The Euler Polynomials $E_n(x)$." Ch. 20 in *An Atlas of Functions.* Washington, DC: Hemisphere, pp. 175–81, 1987.

Euler Polynomial Identity

EULER FOUR-SQUARE IDENTITY

Euler Power Conjecture

EULER'S SUM OF POWERS CONJECTURE

Euler Product

For $s > 1$, the RIEMANN ZETA FUNCTION is given by

$$\zeta(s) \equiv \sum_{n=1}^{\infty} \frac{1}{n^s} = \prod_{n=1}^{\infty} \frac{1}{1 - \dfrac{1}{p_n^s}},$$

where p_i is the ith PRIME. This is Euler's product (Whittaker and Watson 1990).

Let $s \to 1$, then the terms in the product for upper limits $n = 1, 2, \ldots$, are given by 2, 4, 6, 15/2, 35/4, 77/8, 1001/96, 17017/1536, ... (Sloane's A050298 and A050299). The limiting case as $n \to \infty$ gives MERTENS THEOREM,

$$e^{\gamma} = \lim_{n \to \infty} \frac{1}{\ln n} \prod_{i=1}^{n} \frac{1}{1 - \dfrac{1}{pi}},$$

where γ is the EULER-MASCHERONI CONSTANT.

See also DEDEKIND FUNCTION, EULER-MASCHERONI CONSTANT, MERTENS THEOREM, RIEMANN ZETA FUNCTION, STIELTJES CONSTANTS

References

Hardy, G. H. and Wright, E. M. "The Zeta Function." §17.2 in *An Introduction to the Theory of Numbers, 5th ed.* Oxford, England: Clarendon Press, pp. 245–47, 1979.

Ribenboim, P. *The New Book of Prime Number Records, 3rd ed.* New York: Springer-Verlag, p. 216, 1996.

Shimura, G. *Euler Products and Eisenstein Series.* Providence, RI: Amer. Math. Soc., 1997.

Sloane, N. J. A. Sequences A050298 and A050299 in "An On-Line Version of the Encyclopedia of Integer Sequences." http://www.research.att.com/~njas/sequences/eisonline.html.

Whittaker, E. T. and Watson, G. N. "Euler's Product for $\zeta(s)$." §13.3 in *A Course in Modern Analysis, 4th ed.*

Cambridge, England: Cambridge University Press, pp. 271–72, 1990.

Euler Pseudoprime

An Euler pseudoprime is a composite number n which satisfies

$$2^{(n-1)/2} \equiv \pm 1 \pmod{n}.$$

The first few base-2 Euler pseudoprimes are 341, 561, 1105, 1729, 1905, 2047, ... (Sloane's A006970).

See also EULER-JACOBI PSEUDOPRIME, PSEUDOPRIME, STRONG PSEUDOPRIME

References

Sloane, N. J. A. Sequences A006970/M5442 in "An On-Line Version of the Encyclopedia of Integer Sequences." http://www.research.att.com/~njas/sequences/eisonline.html.

Euler Quartic Conjecture

Euler conjectured that there are no POSITIVE INTEGER solutions to the quartic DIOPHANTINE EQUATION

$$A^4 = B^4 + C^4 + D^4.$$

This conjecture was disproved by Elkies (1988), who found an infinite class of solutions.

See also DIOPHANTINE EQUATION–4TH POWERS, EULER'S SUM OF POWERS CONJECTURE

References

Berndt, B. C. and Bhargava, S. "Ramanujan--For Lowbrows." *Amer. Math. Monthly* **100**, 644–56, 1993.

Elkies, N. "On $A^4 + B^4 + C^4 = D^4$." *Math. Comput.* **51**, 825–35, 1988.

Guy, R. K. *Unsolved Problems in Number Theory, 2nd ed.* New York: Springer-Verlag, pp. 139–40, 1994.

Hoffman, P. *The Man Who Loved Only Numbers: The Story of Paul Erdos and the Search for Mathematical Truth.* New York: Hyperion, p. 201, 1998.

Lander, L. J.; Parkin, T. R.; and Selfridge, J. L. "A Survey of Equal Sums of Like Powers." *Math. Comput.* **21**, 446–59, 1967.

Ward, M. "Euler's Problem on Sums of Three Fourth Powers." *Duke Math. J.* **15**, 827–37, 1948.

Wiles, A. "The Birch and Swinnerton-Dyer Conjecture." http://www.claymath.org/prize_problems/birchsd.pdf.

Euler Square

A square ARRAY made by combining n objects of two types such that the first and second elements form LATIN SQUARES. Euler squares are also known as GRAECO-LATIN SQUARES, GRAECO-ROMAN SQUARES, or LATIN-GRAECO SQUARES. For many years, Euler squares were known to exist for $n = 3$, 4, and for every ODD n except $n = 3k$. EULER'S GRAECO-ROMAN SQUARES CONJECTURE maintained that there do not exist Euler squares of order $n = 4k + 2$ for $k = 1, 2, \ldots$. However, such squares were found to exist in 1959, refuting the CONJECTURE.

See also LATIN RECTANGLE, LATIN SQUARE, ROOM SQUARE

References

Beezer, R. "Graeco-Latin Squares." http://buzzard.ups.edu/squares.html.

Fisher, R. A. *The Design of Experiments, 8th ed.* New York: Hafner, 1971.

Kraitchik, M. "Euler (Graeco-Latin) Squares." §7.12 in *Mathematical Recreations.* New York: W. W. Norton, pp. 179–82, 1942.

Steinhaus, H. *Mathematical Snapshots, 3rd ed.* New York: Dover, pp. 31–3, 1999.

Euler Sum

In response to a letter from Goldbach, Euler considered DOUBLE SUMS OF THE FORM

$$s_h(m, n) = \sum_{k=1}^{\infty} \left(1 + \frac{1}{2} + \ldots + \frac{1}{k}\right)^m (k+1)^{-n} \quad (1)$$

$$= \sum_{k=1}^{\infty} [\gamma + \psi_0(k+1)]^m (k+1)^{-n} \quad (2)$$

with $m \geq 1$ and $n \geq 2$ and where γ is the EULER-MASCHERONI CONSTANT and $\Psi(x) = \psi_0(x)$ is the DI-GAMMA FUNCTION. Euler found explicit formulas in terms of the RIEMANN ZETA FUNCTION for $s(1, n)$ with $n \geq 2$, and E. Au-Yeung numerically discovered

$$\sum_{k=1}^{\infty} \left(1 + \frac{1}{2} + \ldots + \frac{1}{k}\right)^2 k^{-2} = \frac{17}{4}\zeta(4), \quad (3)$$

where $\zeta(z)$ is the RIEMANN ZETA FUNCTION, which was subsequently rigorously proven true (Borwein and Borwein 1995). Sums involving k^{-n} can be re-expressed in terms of sums the form $(k+1)^{-n}$ via

$$\sum_{k=1}^{\infty} \left(1 + \frac{1}{2^m} + \ldots + \frac{1}{k^m}\right) k^{-n}$$

$$= \sum_{k=0}^{\infty} \left[1 + \frac{2}{2^m} + \ldots + \frac{1}{(k+1)^m}\right](k+1)^{-n}$$

$$= \sum_{k=1}^{\infty} \left(1 + \frac{1}{2^m} + \ldots + \frac{1}{k^m}\right)(k+1)^{-n} + \sum_{k=1}^{\infty} k^{-(m+n)}$$

$$\equiv \sigma_h(m, n) + \zeta(m+n) \quad (4)$$

$$\sum_{k=1}^{\infty} \left(1 + \frac{1}{2} + \ldots + \frac{1}{k}\right)^2 k^{-n}$$

$$= s_h(2, n) + 2s_h(1, n+1) + \zeta(n+2), \quad (5)$$

where σ_h is defined below.

Bailey *et al.* (1994) subsequently considered sums OF THE FORMS

$$s_h(m, n) = \sum_{k=1}^{\infty} \left(1 + \frac{1}{2} + \ldots + \frac{1}{k}\right)^m (k+1)^{-n} \quad (6)$$

$$s_a(m, n) = \sum_{k=1}^{\infty} \left[1 - \frac{1}{2} + \ldots + \frac{(-1)^{k+1}}{k}\right]^m (k+1)^{-n} \quad (7)$$

$$a_h(m, n) = \sum_{k=1}^{\infty} \left(1 + \frac{1}{2} + \ldots + \frac{1}{k}\right)^m (-1)^{k+1}(k+1)^{-n} \quad (8)$$

$$a_a(m, n) = \sum_{k=1}^{\infty} \left(1 - \frac{1}{2} + \ldots + \frac{(-1)^{k+1}}{k}\right)^m (-1)^{k+1}$$
$$\times (k+1)^{-n} \quad (9)$$

$$\sigma_h(m, n) = \sum_{k=1}^{\infty} \left(1 + \frac{1}{2^m} + \ldots + \frac{1}{k^m}\right)(k+1)^{-n} \quad (10)$$

$$\sigma_a(m, n) = \sum_{k=1}^{\infty} \left(1 - \frac{1}{2^m} + \ldots + \frac{(-1)^{k+1}}{k^m}\right)(k+1)^{-n} \quad (11)$$

$$\alpha_h(m, n) = \sum_{k=1}^{\infty} \left(1 + \frac{1}{2^m} + \ldots + \frac{1}{k^m}\right)(-1)^{k+1}$$
$$\times (k+1)^{-n} \quad (12)$$

$$\alpha_a(m, n) = \sum_{k=1}^{\infty} \left(1 - \frac{1}{2^m} + \ldots + \frac{(-1)^{k+1}}{k^m}\right)(-1)^{k+1}$$
$$\times (k+1)^{-n}, \quad (13)$$

where s_h and s_a have the special forms

$$s_h = \sum_{k=1}^{\infty} [\gamma + \psi_0(n+1)]^m (k+1)^{-n} \quad (14)$$

$$a_a = \sum_{k=1}^{\infty} \{\ln 2 + \tfrac{1}{2}(-1)^n [\psi_0(\tfrac{1}{2}n + \tfrac{1}{2}) - \psi_0(\tfrac{1}{2}n + 1)]\}^m$$
$$\times (k+1)^{-m}. \quad (15)$$

Analytic single or double sums over $\zeta(z)$ can be constructed for

$$s_h(2, n) = \tfrac{1}{3}n(n+1)\zeta(n+2) + \zeta(2)\zeta(n)$$
$$- \tfrac{1}{2}n \sum_{k=0}^{n-2} \zeta(n-k)\zeta(k+2) \quad (16)$$

$$s_h(2, 2n-1) = \tfrac{1}{6}(2n^2 - 7n - 3)\zeta(2n+1) + \zeta(2)\zeta(2n-1)$$
$$- \tfrac{1}{2}\sum_{k=1}^{n-2}(2k-1)\zeta(2n-1-2k)\zeta(2k+2) \quad (17)$$

$$\sigma_h(2, 2n-1)$$
$$= -\tfrac{1}{2}(2n^2 + n + 1)\zeta(2n+1) + \zeta(2)\zeta(2n-1) \quad (18)$$

$\sigma_h(m \text{ even}, n \text{ odd})$

$$= \frac{1}{2}\left[\binom{m+n}{m} - 1\right]\zeta(m+n) + \zeta(m)\zeta(n)$$

$$- \sum_{j=1}^{m+n}\left[\binom{2j-2}{m-1} + \binom{2j-2}{n-1}\right] \tag{19}$$

$\sigma_h(m \text{ odd}, n \text{ even})$

$$= -\frac{1}{2}\left[\binom{m+n}{m} + 1\right]\zeta(m+n)$$

$$+ \sum_{k=1}^{m+n}\left[\binom{2j-2}{m-1} + \binom{2j-2}{n-1}\right] \tag{20}$$

where $\binom{n}{m}$ is a BINOMIAL COEFFICIENT. Explicit formulas inferred using the PSLQ ALGORITHM include

$$s_h(2, 2) = \frac{3}{2}\zeta(4) + \frac{1}{2}[\zeta(2)]^2 \tag{21}$$

$$= \frac{11}{360}\pi^4 \tag{22}$$

$$s_h(2, 4) = \frac{2}{3}\zeta(6) - \frac{1}{3}\zeta(2)\zeta(4) + \frac{1}{3}[\zeta(2)]^3 - [\zeta(3)]^2 \tag{23}$$

$$= \frac{37}{22680}\pi^6 - [\zeta(3)]^2 \tag{24}$$

$$s_h(3, 2) = \frac{15}{2}\zeta(5) + \zeta(2)\zeta(3) \tag{25}$$

$$s_h(3, 3) = -\frac{33}{16}\zeta(6) + 2[\zeta(3)]^2 \tag{26}$$

$$s_h(3, 4) = \frac{119}{16}\zeta(7) - \frac{33}{4}\zeta(3)\zeta(4) + 2\zeta(2)\zeta(5) \tag{27}$$

$$s_h(3, 6) = \frac{197}{24}\zeta(9) - \frac{33}{4}\zeta(4)\zeta(5) - \frac{37}{8}\zeta(3)\zeta(6) + [\zeta(3)]^3$$

$$+ 3\zeta(2)\zeta(7) \tag{28}$$

$$s_h(4, 2) = \frac{859}{24}\zeta(6) + 3[\zeta(3)]^2 \tag{29}$$

$$s_h(4, 3) = -\frac{109}{8}\zeta(7) + \frac{37}{2}\zeta(3)\zeta(4) - 5\zeta(2)\zeta(5) \tag{30}$$

$$s_h(4, 5) = -\frac{29}{2}\zeta(9) + \frac{37}{2}\zeta(4)\zeta(5) + \frac{33}{4}\zeta(3)\zeta(6) - \frac{8}{3}[\zeta(3)]^3$$

$$- 7\zeta(2)\zeta(7) \tag{31}$$

$$s_h(5, 2) = \frac{1855}{16}\zeta(7) + 33\zeta(3)\zeta(4) + \frac{57}{2}\zeta(2)\zeta(5) \tag{32}$$

$$s_h(5, 4) = \frac{890}{9}\zeta(9) + 66\zeta(4)\zeta(5) - \frac{4295}{24}\zeta(3)\zeta(6) - 5[\zeta(3)]^3$$

$$+ \frac{265}{8}\zeta(2)\zeta(7) \tag{33}$$

$$s_h(6, 3) = -\frac{3073}{12}\zeta(9) - 243\zeta(4)\zeta(5) + \frac{2097}{4}\zeta(3)\zeta(6)$$

$$+ \frac{67}{3}[\zeta(3)]^3 - \frac{651}{8}\zeta(2)\zeta(7) \tag{34}$$

$$s_h(7, 2) = \frac{134701}{36}\zeta(9) + \frac{15697}{8}\zeta(4)\zeta(5) + \frac{29555}{24}\zeta(3)\zeta(6)$$

$$+ 56[\zeta(3)]^3 + \frac{3287}{4}\zeta(2)\zeta(7), \tag{35}$$

$$a_h(2, 2) = -2\mathrm{Li}_4(\tfrac{1}{2}) - \frac{1}{12}(\ln 2)^4 + \frac{99}{48}\zeta(4) - \frac{7}{4}\zeta(3)\ln 2$$

$$+ \frac{1}{2}\zeta(2)(\ln 2)^2 \tag{36}$$

$$a_h(2, 3) = -4\mathrm{Li}_5(\tfrac{1}{2}) - 4(\ln 2)\mathrm{Li}_4(\tfrac{1}{2}) - \frac{2}{15}(\ln 2)^5 + \frac{107}{32}\zeta(5)$$

$$- \frac{7}{4}\zeta(3)(\ln 2)^2 + \frac{2}{3}\zeta(2)(\ln 2)^3 + \frac{3}{8}\zeta(2)\zeta(3) \tag{37}$$

$$a_h(3, 2) = 6\mathrm{Li}_5(\tfrac{1}{2}) + 6(\ln 2)\mathrm{Li}_4(\tfrac{1}{2}) + \frac{1}{5}(\ln 2)^5 - \frac{33}{8}\zeta(5)$$

$$+ \frac{21}{8}\zeta(3)(\ln 2)^2 - \zeta(2)(\ln 2)^3 - \frac{15}{16}\zeta(2)\zeta(3), \tag{38}$$

and

$$a_a(2, 2) = -4\mathrm{Li}_4(\tfrac{1}{2}) - \frac{1}{6}(\ln 2)^4 + \frac{37}{16}\zeta(4) + \frac{7}{4}\zeta(3)(\ln 2)$$

$$- 2\zeta(\ln 2)^2 \tag{39}$$

$$a_a(2, 3) = 4(\ln 2)\mathrm{Li}_4(\tfrac{1}{2}) + \frac{1}{6}(\ln 2)^5 - \frac{79}{32}\zeta(5) + \frac{11}{8}\zeta(4)(\ln 2)$$

$$- \zeta(2)(\ln 2)^3 \tag{40}$$

$$a_a(3, 2) = 30\mathrm{Li}_5(\tfrac{1}{2}) - \frac{1}{4}(\ln 2)^5 - \frac{1813}{64}\zeta(5) + \frac{285}{16}\zeta(4)(\ln 2)$$

$$+ \frac{21}{8}\zeta(3)(\ln 2)^2 - \frac{7}{2}\zeta(2)(\ln 2)^3 + \frac{3}{4}\zeta(2)\zeta(3), \tag{41}$$

where Li_n is a POLYLOGARITHM, and $\zeta(z)$ is the RIEMANN ZETA FUNCTION (Bailey and Plouffe). Of these, only $s_h(3, 2)$, $s_h(3, 3)$ and the identities for $s_a(m, n)$, $a_h(m, n)$ and $a_a(m, n)$ have been rigorously established.

References

Adamchik, V. "On Stirling Numbers and Euler Sums." *J. Comput. Appl. Math.* **79**, 119–30, 1197. http://members.wri.com/victor/articles/stirling.html.

Bailey, D. and Plouffe, S. "Recognizing Numerical Constants." http://www.cecm.sfu.ca/organics/papers/bailey/.

Bailey, D. H.; Borwein, J. M.; and Girgensohn, R. "Experimental Evaluation of Euler Sums." *Exper. Math.* **3**, 17–0, 1994.

Berndt, B. C. *Ramanujan's Notebooks: Part I.* New York: Springer-Verlag, 1985.

Borwein, D. and Borwein, J. M. "On an Intriguing Integral and Some Series Related to $\zeta(4)$." *Proc. Amer. Math. Soc.* **123**, 1191–198, 1995.

Borwein, D.; Borwein, J. M.; and Girgensohn, R. "Explicit Evaluation of Euler Sums." *Proc. Edinburgh Math. Soc.* **38**, 277–94, 1995.

de Doelder, P. J. "On Some Series Containing $\Psi(x) - \Psi(y)$ and $(\Psi(x) - \Psi(y))^2$ for Certain Values of x and y." *J. Comp. Appl. Math.* **37**, 125–41, 1991.

Ferguson, H. R. P.; Bailey, D. H.; and Arno, S. "Analysis of PSLQ, An Integer Relation Finding Algorithm." *Math. Comput.* **68**, 351–69, 1999.

Flajolet, P. and Salvy, B. "Euler Sums and Contour Integral Representation." *Experim. Math.* **7**, 15–5, 1998.

Euler System

A mathematical structure first introduced by Kolyvagin (1990) and defined as follows. Let T be a finite-dimensional p-adic representation of the GALOIS GROUP of a NUMBER FIELD K. Then an Euler system for T is a collection of COHOMOLOGY CLASSES $\mathbf{c}_F \in H^1(F, T)$ for a family of Abelian extensions F of K, with a relation between $\mathbf{c}_{F'}$ and $\mathbf{c}_F$ whenever $F \subset F'$ (Rubin 2000, p. 4).

Wiles' proof of FERMAT'S LAST THEOREM via the TANIYAMA-SHIMURA CONJECTURE made use of Euler systems.

References

Kolyvagin, V. A. "Euler Systems." In *The Grothendieck Festschrift, Vol. 2* (Ed. P. Cartier *et al.*). Boston, MA: Birkhäuser, pp. 435–83, 1990.

Rubin, K. *Euler Systems.* Princeton, NJ: Princeton University Press, 2000.

Euler Totient Function
TOTIENT FUNCTION

Euler Transform

There are (at least) three types of Euler transforms (or transformations). The first is a set of transformations of HYPERGEOMETRIC FUNCTIONS, called EULER'S HYPERGEOMETRIC TRANSFORMATIONS.

The second type of Euler transform is a technique for SERIES CONVERGENCE IMPROVEMENT which takes a convergent alternating series

$$\sum_{k=0}^{\infty} (-1)^k a_k = a_0 - a_1 + a_2 - \dots \qquad (1)$$

into a series with more rapid convergence to the same value to

$$s = \sum_{k=0}^{\infty} \frac{(-1)^k \Delta^k a_0}{2^{k+1}}, \qquad (2)$$

where the FORWARD DIFFERENCE is defined by

$$\Delta^k a_0 = \sum_{m=0}^{k} \equiv (-1)^m \binom{k}{m} a_{k-m} \qquad (3)$$

(Abramowitz and Stegun 1972; Beeler *et al.* 1972).

The third type of Euler transform is a relationship between certain types of INTEGER SEQUENCES (Sloane and Plouffe 1995, pp. 20–1). If $a_1, a_2, \dots$ and $b_1, b_2, \dots$ are related by

$$1 + \sum_{n=1}^{\infty} b_n x^n = \prod_{i=1}^{\infty} \frac{1}{(1 - x^i)^{a_1}} \qquad (4)$$

or, in terms of GENERATING FUNCTIONS $A(x)$ and $B(x)$,

$$1 + B(x) = \exp\left[\sum_{k=1}^{\infty} \frac{A(x^k)}{k}\right], \qquad (5)$$

then $\{b_n\}$ is said to be the Euler transform of $\{a_n\}$ (Sloane and Plouffe 1995, p. 20). The Euler transform can be effected by introducing the intermediate series $c_1, c_2, \dots$ given by

$$c_n = \sum_{d \mid n} d a_d, \qquad (6)$$

then

$$b_n = \frac{1}{n}\left[c_n + \sum_{k=1}^{n-1} c_k b_{n-k}\right], \qquad (7)$$

with $b_1 = c_1$. Similarly, the inverse transform can be effected by computing the intermediate series as

$$c_n = n b_n - \sum_{k=1}^{n-1} c_k b_{n-1}, \qquad (8)$$

then

$$a_n = \frac{1}{n} \sum_{d \mid n} \mu\left(\frac{n}{d}\right) c_d, \qquad (9)$$

where $\mu(n)$ is the MÖBIUS FUNCTION.

In GRAPH THEORY, if a_n is the number of UNLABELED CONNECTED GRAPHS on n nodes satisfying some property, then b_n is the *total* number of UNLABELED GRAPHS (connected or not) with the same property. This application of the Euler transform is called RIDDELL'S FORMULA for unlabeled graph (Sloane and Plouffe 1995, p. 20).

There are also important number theoretic applications of the Euler transform. For example, if there are a_1 kinds of parts of size 1, a_2 kinds of parts of size 2, etc., in a given type of partition, then the Euler transform b_n of a_n is the number of partitions of n into these integer parts. For example, if $a_n = 1$ for all n, then b_n is the number of partitions of n into integer parts. Similarly, if $a_n = 1$ for n PRIME and $a_n = 0$ for n composite, then b_n is the number of partitions of n into prime parts (Sloane and Plouffe 1995, p. 21). Other applications are given by Andrews (1986), Andrews and Baxter (1989), and Cameron (1989).

See also BINOMIAL TRANSFORM, EULER'S HYPERGEOMETRIC TRANSFORMATIONS, FORWARD DIFFERENCE, INTEGER SEQUENCE, MÖBIUS TRANSFORM, RIDDELL'S FORMULA, STIRLING TRANSFORM

References

Abramowitz, M. and Stegun, C. A. (Eds.). *Handbook of Mathematical Functions with Formulas, Graphs, and Mathematical Tables, 9th printing.* New York: Dover, p. 16, 1972.

Andrews, G. E. *q*-Series: Their Development and Application in Analysis, Number Theory, Combinatorics, Physics, and Computer Algebra. Providence, RI: Amer. Math. Soc., 1986.

Andrews, G. E. and Baxter, R. J. "A Motivated Proof of the Rogers-Ramanujan Identities." *Amer. Math. Monthly* **96**, 401–09, 1989.

Beeler, M. *et al.* Item 120 in Beeler, M.; Gosper, R. W.; and Schroeppel, R. *HAKMEM.* Cambridge, MA: MIT Artificial Intelligence Laboratory, Memo AIM-239, p. 55, Feb. 1972.

Bernstein, M. and Sloane, N. J. A. "Some Canonical Sequences of Integers." *Linear Algebra Appl.* **226/228**, 57–2, 1995.

Cameron, P. J. "Some Sequences of Integers." *Disc. Math.* **75**, 89–02, 1989.

Iyanaga, S. and Kawada, Y. (Eds.). *Encyclopedic Dictionary of Mathematics.* Cambridge, MA: MIT Press, p. 1163, 1980.

Sloane, N. J. A. and Plouffe, S. *The Encyclopedia of Integer Sequences.* San Diego, CA: Academic Press, pp. 20–1, 1995.

Euler Triangle Formula

Let O and I be the CIRCUMCENTER and INCENTER of a TRIANGLE with CIRCUMRADIUS R and INRADIUS r. Let d be the distance between O and I. Then

$$d^2 = R^2 - 2rR.$$

This is the simplest case of PONCELET'S PORISM.

See also PONCELET'S PORISM

Euler Walk

EULERIAN TRAIL

Euler Zigzag Number

The number of ALTERNATING PERMUTATIONS for n elements is sometimes called an Euler zigzag number. Denote the number of ALTERNATING PERMUTATIONS on n elements for which the first element is k by $E(n, k)$. Then $E(1, 1) = 1$ and

$$E(n, k) =$$
$$\begin{cases} 0 & \text{for } k \geq n \text{ or } k < 1 \\ E(n, k+1) + E(n-1, n-k) & \text{otherwise.} \end{cases}$$

where $E(n, k)$ is an ENTRINGER NUMBER.

See also ALTERNATING PERMUTATION, ENTRINGER NUMBER, SECANT NUMBER, TANGENT NUMBER

References

Ruskey, F. "Information of Alternating Permutations." http://www.theory.csc.uvic.ca/~cos/inf/perm/Alternating.html.

Sloane, N. J. A. Sequences A000111/M1492 in "An On-Line Version of the Encyclopedia of Integer Sequences." http://www.research.att.com/~njas/sequences/eisonline.html.

Euler's 6n + 1 Theorem

Every PRIME OF THE FORM $6n + 1$ can be written in the form $x^2 + 3y^2$.

Euler's Addition Theorem

Let $g(x) \equiv (1 - x^2)(1 - k^2 x^2)$. Then

$$\int_0^a \frac{dx}{\sqrt{g(x)}} + \int_0^b \frac{dx}{\sqrt{g(x)}} = \int_0^c \frac{dx}{\sqrt{g(x)}},$$

where

$$c \equiv \frac{b\sqrt{g(a)} + a\sqrt{g(b)}}{\sqrt{1 - k^2 a^2 b^2}}.$$

Euler's Circle

NINE-POINT CIRCLE

Euler's Conjecture

Define $g(k)$ as the quantity appearing in WARING'S PROBLEM, then Euler conjectured that

$$g(k) = 2^k + \left\lfloor \left(\frac{3}{2} \right)^k \right\rfloor - 2,$$

where $\lfloor x \rfloor$ is the FLOOR FUNCTION.

See also WARING'S PROBLEM

Euler's Criterion

For p an ODD PRIME and a POSITIVE INTEGER a which is not a multiple of p,

$$a^{(p-1)/2} \equiv \left(\frac{a}{p} \right) \pmod{p},$$

where $(a|p)$ is the LEGENDRE SYMBOL.

See also LEGENDRE SYMBOL, QUADRATIC RESIDUE

References

Nagell, T. "Euler's Criterion and Legendre's Symbol." §38 in *Introduction to Number Theory.* New York: Wiley, pp. 133–36, 1951.

Rosen, K. H. Ch. 9 in *Elementary Number Theory and Its Applications, 3rd ed.* Reading, MA: Addison-Wesley, 1993.

Shanks, D. *Solved and Unsolved Problems in Number Theory, 4th ed.* New York: Chelsea, pp. 33–7, 1993.

Wagon, S. *Mathematica in Action.* New York: W. H. Freeman, p. 293, 1991.

Euler's Dilogarithm

DILOGARITHM

Euler's Displacement Theorem

The general displacement of a rigid body (or coordinate frame) with one point fixed is a ROTATION about some axis. Furthermore, a ROTATION may be described in any basis using three ANGLES.

See also EUCLIDEAN MOTION, EULER ANGLES, RIGID MOTION, ROTATION, TRANSLATION

Euler's Distribution Theorem

For signed distances on a LINE SEGMENT,

$$\overline{AB} \cdot \overline{CD} + \overline{AC} \cdot \overline{DB} + \overline{AD} \cdot \overline{BC} = 0,$$

since

$$(b-a)(d-c)+(c-a)(b-d)+(d-a)(c-b)=0.$$

References

Johnson, R. A. *Modern Geometry: An Elementary Treatise on the Geometry of the Triangle and the Circle.* Boston, MA: Houghton Mifflin, p. 3, 1929.

Euler's Equations of Inviscid Motion

The system of PARTIAL DIFFERENTIAL EQUATIONS describing fluid flow in the absence of viscosity, given by

$$\frac{\partial \mathbf{u}}{\partial t}+(\mathbf{u}\cdot\nabla)\mathbf{u}=-\frac{\nabla P}{\rho},$$

where $\mathbf{u}$ is the fluid velocity, P is the pressure, and ρ is the fluid density.

See also EULER DIFFERENTIAL EQUATION

References

Landau, L. D. and Lifschitz, E. M. *Fluid Mechanics, 2nd ed.* Oxford, England: Pergamon Press, p. 3, 1982.
Zwillinger, D. *Handbook of Differential Equations, 3rd ed.* Boston, MA: Academic Press, p. 138, 1997.

Euler's Factorization Method

A factorization algorithm which works by expressing N as a QUADRATIC FORM in two different ways. Then

$$N=a^2+b^2=c^2+d^2, \tag{1}$$

so

$$a^2-c^2=d^2-b^2 \tag{2}$$

$$(a-c)(a+c)=(d-b)(d+b). \tag{3}$$

Let k be the GREATEST COMMON DIVISOR of $a-c$ and $d-b$ so

$$a-c=kl \tag{4}$$

$$d-b=km \tag{5}$$

$$(l,m)=1, \tag{6}$$

(where (l,m) denotes the GREATEST COMMON DIVISOR of l and m), and

$$l(a+c)=m(d+b). \tag{7}$$

But since $(l,m)=1$, $m|a+c$ and

$$a+c=mn, \tag{8}$$

which gives

$$b+d=ln, \tag{9}$$

so we have

$$[(\tfrac{1}{2}k)^2+(\tfrac{1}{2}n)^2](l^2+m^2)=\tfrac{1}{4}(k^2+n^2)(l^2+m^2)$$

Euler's Homogeneous Function

$$=\tfrac{1}{4}[(kn)^2+(kl)^2+(nm)^2+(nl)^2]$$

$$=\tfrac{1}{4}[(d-b)^2+(a-c)^2+(a+c)^2+(d+b)^2]$$

$$=\tfrac{1}{4}(2a^2+2b^2+2c^2+2d^2)$$

$$=\tfrac{1}{4}(2N+2N)=N. \tag{10}$$

See also PRIME FACTORIZATION ALGORITHMS

Euler's Graeco-Roman Squares Conjecture

46	57	68	70	81	02	13	24	35	99
71	94	37	65	12	40	29	06	88	53
93	26	54	01	38	19	85	77	60	42
15	43	80	27	09	74	66	58	92	31
32	78	16	89	63	55	47	91	04	20
67	05	79	52	44	36	90	83	21	18
84	69	41	33	25	98	72	10	56	07
59	30	22	14	97	61	08	45	73	86
28	11	03	96	50	87	34	62	49	75
00	82	95	48	76	23	51	39	17	64

Euler conjectured that there do not exist GRAECO-ROMAN SQUARES (now known as EULER SQUARES) of order $n=4k+2$ for $k=1$, 2, ... In fact, MacNeish (1921–922) published a purported proof of this conjecture (Bruck and Ryser 1949). While it is true that no such square of order six exists, such squares were found to exist for *all* other orders of the form $4k+2$ by Bose, Shrikhande, and Parker in 1959 (Wells 198, p. 77), refuting the CONJECTURE (and establishing unequivocally the invalidity of MacNeish's "proof").

See also 36 OFFICER PROBLEM, EULER SQUARE, LATIN SQUARE

References

Bose, R. C. "On the Application of the Properties of Galois Fields to the Problem of Construction of Hyper-Graeco-Latin Squares." *Indian J. Statistics* **3**, 323–38, 1938.
Bose, R. C.; Shrikhande, S. S.; and Parker, E. T. "Further Results on the Construction of Mutually Orthogonal Latin Squares and the Falsity of Euler's Conjecture." *Canad. J. Math.* **12**, 189, 1960.
Bruck, R. H. and Ryser, H. J. "The Nonexistence of Certain Finite Projective Planes." *Canad. J. Math.* **1**, 88–3, 1949.
Levi, F. W. Second lecture in *Finite Geometrical Systems.* Calcutta, India: University of Calcutta, 1942.
MacNeish, H. F. "Euler Squares." *Ann. Math.* **23**, 221–27, 1921–922.
Mann, H. B. "On Orthogonal Latin Squares." *Bull. Amer. Math. Soc.* **51**, 185–97, 1945.
Wells, D. *The Penguin Dictionary of Curious and Interesting Numbers.* Middlesex, England: Penguin Books, p. 77, 1986.

Euler's Homogeneous Function Theorem

Let $f(x,y)$ be a HOMOGENEOUS FUNCTION of order n so that

$$f(tx, \; ty) = t^n f(x, \; y). \qquad (1)$$

Then define $x' \equiv xt$ and $y' \equiv yt$. Then

$$nt^{n-1} f(x, \; y) = \frac{\partial f}{\partial x'} \frac{\partial x'}{\partial t} + \frac{\partial f}{\partial y'} \frac{\partial y'}{\partial t}$$

$$= x \frac{\partial f}{\partial x'} + y \frac{\partial f}{\partial y'} = x \frac{\partial f}{\partial (xt)} + y \frac{\partial f}{\partial (yt)}. \qquad (2)$$

Let $t = 1$, then

$$x \frac{\partial f}{\partial x} + y \frac{\partial f}{\partial y} = n f(x, \; y). \qquad (3)$$

This can be generalized to an arbitrary number of variables

$$x_i \frac{\partial f}{\partial x_i} = n f(\mathbf{x}), \qquad (4)$$

where EINSTEIN SUMMATION has been used.

Euler's Hypergeometric Transformations

$$_2F_1(a, \; b; \; c; \; z) = \int_0^1 \frac{t^{b-1} (1-t)^{c-b-1}}{(1-tz)^a} \, dt, \qquad (1)$$

where $_2F_1(a, \; b; \; c; \; z)$ is a HYPERGEOMETRIC FUNCTION. The solution can be written using the Euler's transformations

$$t \to t \qquad (2)$$

$$t \to 1 - t \qquad (3)$$

$$t \to (1 - z - tz)^{-1} \qquad (4)$$

$$t \to \frac{1-t}{1-tz} \qquad (5)$$

in the equivalent forms

$$_2F_1(a, \; b; \; c; \; z)$$

$$= (1-z)^{-a} \, _2F_1(a, \; c-b; \; c; \; z/(z-1)) \qquad (6)$$

$$= (1-z)^{-b} \, _2F_1(c-a, \; b; \; c; \; z/(z-1)) \qquad (7)$$

$$= (1-z)^{c-a-b} \, _2F_1(c-a, \; c-b; \; c; \; z). \qquad (8)$$

See also HYPERGEOMETRIC FUNCTION

References
Euler, L. *Nova Acta Acad. Petropol.* **7**, p. 58, 1778.
Morse, P. M. and Feshbach, H. *Methods of Theoretical Physics, Part I.* New York: McGraw-Hill, pp. 585–91, 1953.

Euler's Ideoneal Number
IDONEAL NUMBER

Euler's Machin-Like Formula
The MACHIN-LIKE FORMULA

$$\tfrac{1}{4} \pi = \tan^{-1}(\tfrac{1}{2}) + \tan^{-1}(\tfrac{1}{3}).$$

The other 2-term MACHIN-LIKE FORMULAS are HERMANN'S FORMULA, HUTTON'S FORMULA, and MACHIN'S FORMULA.

See also INVERSE TANGENT

Euler's Pentagonal Number Theorem
PENTAGONAL NUMBER THEOREM

Euler's Phi Function
TOTIENT FUNCTION

Euler's Polygon Division Problem
The problem of finding in how many ways E_n a PLANE convex POLYGON of n sides can be divided into TRIANGLES by diagonals. Euler first proposed it to Christian Goldbach in 1751, and the solution is the CATALAN NUMBER $E_n = C_{n-2}$.

See also CATALAN NUMBER, CATALAN'S PROBLEM

References
Forder, H. G. "Some Problems in Combinatorics." *Math. Gaz.* **41**, 199–01, 1961.
Guy, R. K. "Dissecting a Polygon Into Triangles." *Bull. Malayan Math. Soc.* **5**, 57–0, 1958.

Euler's Quadratic Residue Theorem
A number D that possesses no common divisor with a prime number p is either a QUADRATIC RESIDUE or nonresidue of p, depending whether $D^{(p-1)/2}$ is congruent mod p to ± 1.

Euler's Rotation Theorem
An arbitrary ROTATION may be described by only three parameters.

See also EULER ANGLES, EULER PARAMETERS, ROTATION MATRIX

Euler's Rule
The numbers $2^n pq$ and $2^n r$ are an AMICABLE PAIR if the three INTEGERS

$$p \equiv 2^m (2^{n-m} + 1) - 1 \qquad (1)$$

$$q \equiv 2^n (2^{n-m} + 1) - 1 \qquad (2)$$

$$r \equiv 2^{n+m} (2^{n-m} + 1)^2 - 1 \qquad (3)$$

are all PRIME NUMBERS for some POSITIVE INTEGER m satisfying $1 \le m \le n - 1$ (Dickson 1952, p. 42). However, there are many AMICABLE PAIRS which do not satisfy Euler's rule, so it is a SUFFICIENT but not NECESSARY condition for amicability. Euler's rule is a generalization of THÂBIT IBN KURRAH RULE.

For example, Euler's rule is satisfied for $(n, m) = (2, 1), (4, 4), (6, 7), (8, 1), (40, 29), \ldots$, corresponding to the triples $(p, q, r) = (5, 11, 71), (23, 47, 1151), (191, 383, 73727), \ldots$, giving the AMICABLE PAIRS (220, 284), (17296, 18416), (9363584, 9437056), ...

See also AMICABLE PAIR, THÂBIT IBN KURRAH RULE

References
Borho, W. "On Thabit ibn Kurrah's Formula for Amicable Numbers." *Math. Comput.* **26**, 571–78, 1972.
Dickson, L. E. *History of the Theory of Numbers, Vol. 1: Divisibility and Primality.* New York: Chelsea, 1952.
Euler, L. "De Numeris Amicabilibus." In *Leonhardi Euleri Opera Omnia, Ser. 1, Vol. 2.* Leipzig, Germany: Teubner, pp. 63–62, 1915.
te Riele, H. J. J. "Four Large Amicable Pairs." *Math. Comput.* **28**, 309–12, 1974.

Euler's Series Transformation

Accelerates the rate of CONVERGENCE for an ALTERNATING SERIES

$$S = \sum_{s=0}^{\infty} (-1)^s u_s$$

$$= u_0 - u_1 + u_2 - \ldots - u_{n-1} + \sum_{s=0}^{\infty} \frac{(-1)^2}{2^{s+1}} [\Delta^s u_n] \quad (1)$$

for n EVEN and Δ the FORWARD DIFFERENCE operator

$$\Delta^k u_n \equiv \sum_{m=0}^{k} (-1)^m \binom{k}{m} u_{n+k-m}, \quad (2)$$

where $\binom{k}{m}$ are BINOMIAL COEFFICIENTS. The POSITIVE terms in the series can be converted to an ALTERNATING SERIES using

$$\sum_{r=1}^{\infty} v_r = \sum_{r=1}^{\infty} (-1)^{r-1} w_r, \quad (3)$$

where

$$w_r \equiv v_r + 2v_{2r} + 4v_{4r} + 8v_{8r} + \ldots. \quad (4)$$

See also ALTERNATING SERIES

References
Abramowitz, M. and Stegun, C. A. (Eds.). *Handbook of Mathematical Functions with Formulas, Graphs, and Mathematical Tables, 9th printing.* New York: Dover, p. 16, 1972.

Euler's Spiral
CORNU SPIRAL

Euler's Sum of Powers Conjecture

Euler conjectured that at least n nth POWERS are required for $n > 2$ to provide a sum that is itself an nth POWER. The conjecture was disproved by Lander and Parkin (1967) with the counterexample

$$27^5 + 84^5 + 110^5 + 133^5 = 144^5.$$

Ekl (1998) defined Euler's extended conjecture as the assertion that there are no solutions to the $k.m.n$ DIOPHANTINE EQUATION

$$a_1^k + a_2^k + \ldots + a_m^k = b_1^k + b_2^k + \ldots + b_n^k,$$

with a_i and b_i not necessarily distinct, such that $m + n < k$. There are no known counterexamples to this conjecture (Ekl 1998). Ekl (1998) defines the Euler conjecture number as the minimum known value of $\Delta \equiv m + n - k$. The following table gives the smallest known values.

k	Soln.	Δ	Reference
4	4.1.3	0	Elkies 1988
5	5.1.4	0	Lander *et al.* 1967
6	6.3.3	0	Subba Rao 1934
7	7.4.4	1	Ekl 1996
8	8.5.5	2	Letac 1942
9	9.6.6	3	Lander *et al.* 1967
10	10.7.7	4	Moessner 1939

See also DIOPHANTINE EQUATION–5TH POWERS, EULER QUARTIC CONJECTURE

References
Ekl, R. L. "Equal Sums of Four Seventh Powers." *Math. Comput.* **65**, 1755–756, 1996.
Ekl, R. L. "New Results in Equal Sums of Like Powers." *Math. Comput.* **67**, 1309–315, 1998.
Elkies, N. "On $A^4 + B^4 + C^4 = D^4$." *Math. Comput.* **51**, 828–38, 1988.
Hoffman, P. *The Man Who Loved Only Numbers: The Story of Paul Erdos and the Search for Mathematical Truth.* New York: Hyperion, p. 195, 1998.
Lander, L. J. and Parkin, T. R. "A Counterexample to Euler's Sum of Powers Conjecture." *Math. Comput.* **21**, 101–03, 1967.
Lander, L. J.; Parkin, T. R.; and Selfridge, J. L. "A Survey of Equal Sums of Like Powers." *Math. Comput.* **21**, 446–59, 1967.
Letac, A. *Gazetta Mathematica* **48**, 68–9, 1942.
Moessner, A. "Einige Numerische Identitaten." *Proc. Indian Acad. Sci. Sect. A* **10**, 296–06, 1939.
Subba Rao, K. "On Sums of Sixth Powers." *J. London Math. Soc.* **9**, 172–73, 1934.

Euler's Theorem

A generalization of FERMAT'S LITTLE THEOREM. Euler published a proof of the following more general theorem in 1736. Let $\phi(n)$ denote the TOTIENT FUNCTION. Then

$$a^{\phi(n)} \equiv 1 \pmod{n}$$

for all a RELATIVELY PRIME to n.

See also CHINESE HYPOTHESIS, EULER'S DISPLACEMENT THEOREM, EULER'S DISTRIBUTION THEOREM, FERMAT'S LITTLE THEOREM, TOTIENT FUNCTION

References

Séroul, R. "The Theorems of Fermat and Euler." §2.8 in *Programming for Mathematicians.* Berlin: Springer-Verlag, p. 15, 2000.
Shanks, D. *Solved and Unsolved Problems in Number Theory, 4th ed.* New York: Chelsea, p. 21 and 23–5, 1993.

Euler's Totient Rule

The number of bases in which $1/p$ is a REPEATING DECIMAL (actually, repeating b-ary) of length l is the same as the number of FRACTIONS $0/(p-1)$, $1/(p-1)$, ..., $(p-2)/(p-1)$ which have reduced DENOMINATOR l. For example, in bases 2, 3, ..., 6, 1/7 is given by

$$\frac{1}{7} = 0.001001001001\ldots_2$$

$$= 0.010212010212\ldots_3$$

$$= 0.021021021020\ldots_4$$

$$= 0.032412032412\ldots_5$$

$$= 0.050505050505\ldots_6,$$

which have periods 3, 6, 3, 6, and 2, respectively, corresponding to the DENOMINATORS 6, 3, 2, 3, and 6 of

$$\frac{1}{6}, \frac{1}{3}, \frac{1}{2}, \frac{2}{3}, \text{ and } \frac{5}{6}.$$

See also CYCLIC NUMBER, REPEATING DECIMAL, TOTIENT FUNCTION

References

Conway, J. H. and Guy, R. K. *The Book of Numbers.* New York: Springer-Verlag, pp. 167–68, 1996.

Euler's Triangle

The triangle of numbers $A_{n,k}$ given by

$$A_{n,1} = A_{n,n} = 1$$

and the RECURRENCE RELATION

$$A_{n+1,k} = kA_{n,k} + (n+2-k)A_{n,k-1}$$

for $k \in [2, n]$, where $A_{n,k}$ are EULERIAN NUMBERS.

$$1$$
$$1 \quad 1$$
$$1 \quad 4 \quad 1$$
$$1 \quad 11 \quad 11 \quad 1$$
$$1 \quad 26 \quad 66 \quad 26 \quad 1$$
$$1 \quad 57 \quad 302 \quad 302 \quad 57 \quad 1$$

The numbers 1, 1, 1, 1, 4, 1, 1, 11, 11, 1, ... are Sloane's A008292. Amazingly, the Z-TRANSFORMS of t^n

$$\frac{(z-1)^n}{T^n z} Z[t^n] = \frac{(1-z)^n}{T^n z} \lim_{x \to 0} \frac{\partial^n}{\partial x^n}\left(\frac{z}{z - e^{-xT}}\right)$$

are generators for Euler's triangle.

A SPHERICAL TRIANGLE is sometimes also called Euler's triangle.

See also CLARK'S TRIANGLE, EULERIAN NUMBER, LEIBNIZ HARMONIC TRIANGLE, LOSSNITSCH'S TRIANGLE, NUMBER TRIANGLE, PASCAL'S TRIANGLE, SEIDEL-ENTRINGER-ARNOLD TRIANGLE, SPHERICAL TRIANGLE, Z-TRANSFORM

References

Sloane, N. J. A. Sequences A008292 in "An On-Line Version of the Encyclopedia of Integer Sequences." http://www.research.att.com/~njas/sequences/eisonline.html.

Euler-Bernoulli Triangle

SEIDEL-ENTRINGER-ARNOLD TRIANGLE

Euler-Darboux Equation

The PARTIAL DIFFERENTIAL EQUATION

$$u_{xy} + \frac{\alpha u_x - \beta u_y}{x - y} = 0.$$

See also EULER-POISSON-DARBOUX EQUATION

References

Miller, W. Jr. *Symmetry and Separation of Variables.* Reading, MA: Addison-Wesley, 1977.
Zwillinger, D. *Handbook of Differential Equations, 3rd ed.* Boston, MA: Academic Press, p. 129, 1997.

EulerE

EULER NUMBER, EULER POLYNOMIAL

EulerGamma

EULER-MASCHERONI CONSTANT

Eulerian Circuit

An EULERIAN TRAIL which starts and ends at the same VERTEX. In other words, it is a GRAPH CYCLE which uses each EDGE exactly once. The term EULERIAN CYCLE is also used synonymously with Eulerian circuit. For technical reasons, Eulerian circuits are easier to study mathematically than are HAMILTONIAN CIRCUITS. As a generalization of the KÖNIGSBERG BRIDGE PROBLEM, Euler showed (without proof) that a CONNECTED GRAPH has an Eulerian circuit IFF it has no VERTICES of ODD DEGREE.

FLEURY'S ALGORITHM is an elegant, but inefficient, method of generating Eulerian circuit. An Eulerian cycle of a graph may be found using `EulerianCycle[g]` in the *Mathematica* add-on package `DiscreteMath`Combinatorica`` (which can be loaded with the command `<<DiscreteMath``).

See also CHINESE POSTMAN PROBLEM, EULER GRAPH, HAMILTONIAN CIRCUIT, UNICURSAL CIRCUIT

References

Bollobás, B. *Graph Theory: An Introductory Course.* New York: Springer-Verlag, p. 12, 1979.
Gardner, M. *The Sixth Book of Mathematical Games from Scientific American.* Chicago, IL: University of Chicago Press, pp. 94–6, 1984.
Hierholzer, C. "Über die Möglichkeit, einen Linienzug ohne Wiederholung und ohne Unterbrechung zu umfahren." *Math. Ann.* **6**, 30–2, 1873.
Lucas, E. *Récréations Mathématiques.* Paris: Gauthier-Villars, 1891.
Skiena, S. "Eulerian Cycles." §5.3.3 in *Implementing Discrete Mathematics: Combinatorics and Graph Theory with Mathematica.* Reading, MA: Addison-Wesley, pp. 192–96, 1990.

Eulerian Cycle

EULERIAN CIRCUIT

Eulerian Graph

A GRAPH containing an EULERIAN CIRCUIT. Finding the largest SUBGRAPH of graph having an odd number of vertices which is Eulerian is an NP-COMPLETE PROBLEM (Skiena 1990, p. 194).

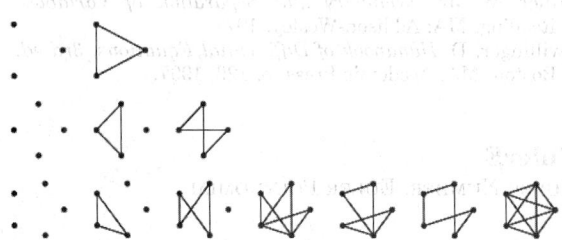

An UNDIRECTED GRAPH is Eulerian IFF every VERTEX has EVEN DEGREE. The numbers of Eulerian graphs with $n = 1, 2, \ldots$ nodes are 1, 1, 2, 3, 7, 16, 54, 243, ...

(Sloane's A002854; Robinson 1969; Mallows and Sloane 1975; Buekenhout 1995, p. 881; Colbourn and Dinitz 1996, p. 687). There is an explicit formula giving these numbers.

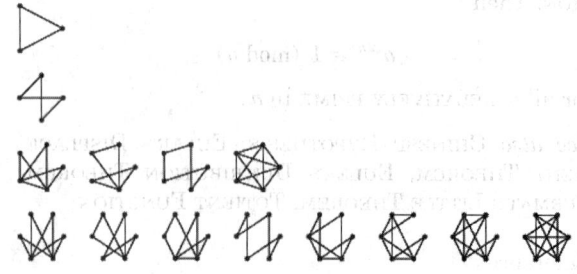

Euler showed (without proof) that a CONNECTED GRAPH is Eulerian IFF it has no VERTICES of ODD DEGREE. The numbers of connected Eulerian graphs with $n = 1, 2, \ldots$ nodes are 1, 0, 1, 1, 4, 8, 37, 184, ... (Sloane's A003049; Robinson 1969; Liskovec 1972; Harary and Palmer 1973, p. 117).

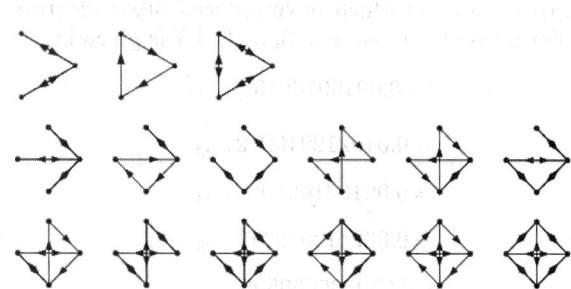

A DIRECTED GRAPH is Eulerian IFF every VERTEX has equal INDEGREE and OUTDEGREE. A planar BIPARTITE GRAPH is DUAL to a PLANAR Eulerian graph and vice versa. The numbers of Eulerian digraphs on $n = 1, 2, \ldots$ nodes are 1, 1, 3, 12, ...

See also HAMILTONIAN GRAPH, TWO-GRAPH

References

Bollobás, B. *Graph Theory: An Introductory Course.* New York: Springer-Verlag, p. 12, 1979.
Buekenhout, F. (Ed.). *Handbook of Incidence Geometry: Building and Foundations.* Amsterdam, Netherlands: North-Holland, 1995.
Colbourn, C. J. and Dinitz, J. H. (Eds.). *CRC Handbook of Combinatorial Designs.* Boca Raton, FL: CRC Press, 1996.
Gardner, M. *The Sixth Book of Mathematical Games from Scientific American.* Chicago, IL: University of Chicago Press, p. 94, 1984.
Harary, F. and Palmer, E. M. *Graphical Enumeration.* New York: Academic Press, p. 117, 1973.
Liskovec, V. A. "Enumeration of Euler Graphs" [Russian]. Review MR#6557 in *Math. Rev.* **44**, 1195, 1972.
Mallows, C. L. and Sloane, N. J. A. "Two-Graphs, Switching Classes, and Euler Graphs are Equal in Number." *SIAM J. Appl. Math.* **28**, 876–80, 1975.

Robinson, R. W. "Enumeration of Euler Graphs." In *Proof Techniques in Graph Theory* (Ed. F. Harary). New York: Academic Press, pp. 147–53, 1969.

Skiena, S. "Eulerian Cycles." §5.3.3 in *Implementing Discrete Mathematics: Combinatorics and Graph Theory with Mathematica.* Reading, MA: Addison-Wesley, pp. 192–96, 1990.

Sloane, N. J. A. Sequences A002854/M0846 and A003049/M3344 in "An On-Line Version of the Encyclopedia of Integer Sequences." http://www.research.att.com/~njas/sequences/eisonline.html.

Eulerian Integral of the First Kind

Legendre and Whittaker and Watson's (1990) term for the BETA INTEGRAL

$$\int_0^1 x^p (1-x)^q \, dx,$$

whose solution is the BETA FUNCTION $B(p+1, q+1)$.

See also BETA FUNCTION, BETA INTEGRAL, EULERIAN INTEGRAL OF THE SECOND KIND

References

Whittaker, E. T. and Watson, G. N. *A Course in Modern Analysis, 4th ed.* Cambridge, England: Cambridge University Press, 1990.

Eulerian Integral of the Second Kind

For $\Re[n] > -1$ and $\Re[z] > 0$,

$$\prod(z, n) = n^z \int_0^1 (1-x)^n x^{z-1} \, dx \tag{1}$$

$$= \frac{n!}{(z)_{n+1}} n^z \tag{2}$$

$$= B(z, n+1), \tag{3}$$

where $(z)_n$ is the POCHHAMMER SYMBOL and $B(p, q)$ is the BETA FUNCTION.

See also BETA FUNCTION, BETA INTEGRAL, EULERIAN INTEGRAL OF THE FIRST KIND

Eulerian Number

The number of PERMUTATION RUNS of length n with $k \leq n$, denoted $\left\langle {n \atop k} \right\rangle$, $A_{n,k}$, or $A(n, k)$. The Eulerian numbers are given explicitly by the sum

$$\left\langle {n \atop k} \right\rangle = \sum_{j=0}^k (-1)^j \binom{n+1}{j} (k-j)^n. \tag{1}$$

Making the definition

$$b_{n,1} = 1 \tag{2}$$

$$b_{1,n} = 1 \tag{3}$$

together with the RECURRENCE RELATION

$$b_{n,k} = n b_{n,k-1} + k b_{n-1,k} \tag{4}$$

for $n > k$ then gives

$$\left\langle {n \atop k} \right\rangle = b_{k,\,n-k+1}. \tag{5}$$

The arrangement of the numbers into a triangle gives EULER'S TRIANGLE, whose entries are 1, 1, 1, 1, 4, 1, 1, 11, 11, 1, ... (Sloane's A008292). Therefore, they represent a sort of generalization of the BINOMIAL COEFFICIENTS where the defining RECURRENCE RELATION weights the sum of neighbors by their row and column numbers, respectively.

The Eulerian numbers satisfy

$$\sum_{k=1}^n \left\langle {n \atop k} \right\rangle = n!. \tag{6}$$

Eulerian numbers also arise in the surprising context of integrating the SINC FUNCTION, and also in sums of the form

$$\sum_{k=1}^\infty k^n r^k = \mathrm{Li}_{-n}(r) = \frac{r}{(1-r)^{n+1}} \sum_{i=1}^n \left\langle {n \atop k} \right\rangle r^{n-i}, \tag{7}$$

where $\mathrm{Li}_m(z)$ is the POLYLOGARITHM function.

See also COMBINATION LOCK, EULER NUMBER, EULER'S TRIANGLE, EULER ZIGZAG NUMBER, PERMUTATION RUN, POLYLOGARITHM, SIMON NEWCOMB'S PROBLEM, SINC FUNCTION, WORPITZKY'S IDENTITY, Z-TRANSFORM

References

Abramson, M. and Moser, W. O. J. "Permutations without Rising or Falling ω-Sequences." *Ann. Math. Statist.* **38**, 1245–254, 1967.

André, D. "Mémoir sur les couples actifs de permutations." *Mem. della Pontificia Acad. Romana dei Nuovo Lincei* **23**, 189–23, 1906.

Carlitz, L. "Note on a Paper of Shanks." *Amer. Math. Monthly* **59**, 239–41, 1952.

Carlitz, L. "Eulerian Numbers and Polynomials." *Math. Mag.* **32**, 247–60, 1959.

Carlitz, L. "Eulerian Numbers and Polynomials of Higher Order." *Duke Math. J.* **27**, 401–23, 1960.

Carlitz, L. "A Note on the Eulerian Numbers." *Arch. Math.* **14**, 383–90, 1963.

Carlitz, L. and Riordan, J. "Congruences for Eulerian Numbers." *Duke Math. J.* **20**, 339–43, 1953.

Carlitz, L.; Roselle, D. P.; and Scoville, R. "Permutations and Sequences with Repetitions by Number of Increase." *J. Combin. Th.* **1**, 350–74, 1966.

Cesàro, E. "Dérivées des fonctions de fonctions." *Nouv. Ann.* **5**, 305–12, 1886.

Comtet, L. "Permutations by Number of Rises; Eulerian Numbers." §6.5 in *Advanced Combinatorics: The Art of Finite and Infinite Expansions, rev. enl. ed.* Dordrecht, Netherlands: Reidel, pp. 240–46, 1974.

David, F. N.; Kendall, M. G.; and Barton, D. E. *Symmetric Function and Allied Tables.* Cambridge, England: Cambridge University Press, p. 260, 1966.

Dillon, J. F.; Roselle, D. P. "Eulerian Numbers of Higher Order." *Duke Math. J.* **35**, 247–56, 1968.

Foata, D. and Schützenberger, M.-P. *Théorie Géométrique des Polynômes Eulériens.* Berlin: Springer-Verlag, 1970.

Frobenius, F. G. "Ueber die Bernoullischen Zahlen und die Eulerischen Polynome." *Sitzungsber. Preuss. Akad. Wiss.*, pp. 808–47, 1910.

Graham, R. L.; Knuth, D. E.; and Patashnik, O. "Eulerian Numbers." §6.2 in *Concrete Mathematics: A Foundation for Computer Science, 2nd ed.* Reading, MA: Addison-Wesley, pp. 267–72, 1994.

Kimber, A. C. "Eulerian Numbers." Supplement to *Encyclopedia of Statistical Sciences.* (Eds. S. Kotz, N. L. Johnson, and C. B. Read). New York: Wiley, pp. 59–0, 1989.

Poussin, F. "Sur une propriété arithmétique de certains polynomes associés aux nombres d'Euler." *C. R. Acad. Sci. Paris Sér. A-B* **266**, A392-A393, 1968.

Salama, I. A. and Kupper, L. L. "A Geometric Interpretation for the Eulerian Numbers." *Amer. Math. Monthly* **93**, 51–2, 1986.

Schrutka, L. "Eine neue Einleitung der Permutationen." *Math. Ann.* **118**, 246–50, 1941.

Shanks, E. B. "Iterated Sums of Powers of the Binomial Coefficients." *Amer. Math. Monthly* **58**, 404–07, 1951.

Sloane, N. J. A. Sequences A008292 in "An On-Line Version of the Encyclopedia of Integer Sequences." http://www.research.att.com/~njas/sequences/eisonline.html.

Tomic, M. "Sur une nouvelle classe de polynômes de la théorie des fonctions spéciales." *Publ. Fac. Elect. U. Belgrade*, No. 38, 1960.

Toscano, L. "Su due sviluppi della potenza di un binomio, q-coefficienti di Eulero." *Bull. S. M. Calabrese* **16**, 1–, 1965.

Eulerian Tour

EULERIAN TRAIL

Eulerian Trail

A WALK on the EDGES of a GRAPH which uses each EDGE exactly once. A CONNECTED GRAPH has an Eulerian trail IFF it has at most two VERTICES of ODD DEGREE.

See also EULERIAN CIRCUIT, EULERIAN GRAPH KÖNIGSBERG BRIDGE PROBLEM

References

Edmonds, J. and Johnson, E. L. "Matching, Euler Tours, and the Chinese Postman." *Math. Programm.* **5**, 88–24, 1973.

Wilson, R. J. "An Eulerian Trail through Königsberg." *J. Graph Th.* **10**, 265–75, 1986.

Euler-Jacobi Pseudoprime

An Euler-Jacobi pseudoprime to a base a is an ODD COMPOSITE numbers such that $(a, n) = 1$ and the JACOBI SYMBOL (a/n) satisfies

$$\left(\frac{a}{n}\right) \equiv a^{(n-1)/2} \pmod{n}.$$

(Guy 1994; but note that Guy calls these simply "Euler pseudoprimes"). No ODD COMPOSITE number is an Euler-Jacobi pseudoprime for all bases a RELATIVELY PRIME to it. This class includes some CARMICHAEL NUMBERS, all STRONG PSEUDOPRIMES to base a, and all EULER PSEUDOPRIMES to base a. An Euler

pseudoprime is pseudoprime to at most 1/2 of all possible bases less than itself.

The first few base-2 Euler-Jacobi pseudoprimes are 561, 1105, 1729, 1905, 2047, 2465, ... (Sloane's A047713), and the first few base-3 Euler-Jacobi pseudoprimes are 121, 703, 1729, 1891, 2821, 3281, 7381, ... (Sloane's A048950). The number of base-2 Euler-Jacobi primes less than 10^2, 10^3, ... are 0, 1, 12, 36, 114, ... (Sloane's A055551).

See also EULER PSEUDOPRIME, PSEUDOPRIME

References

Guy, R. K. "Pseudoprimes. Euler Pseudoprimes. Strong Pseudoprimes." §A12 in *Unsolved Problems in Number Theory, 2nd ed.* New York: Springer-Verlag, pp. 27–0, 1994.

Pinch, R. G. E. "The Pseudoprimes Up to 10^{13}." ftp://ftp.dpmms.cam.ac.uk/pub/PSP/.

Riesel, H. *Prime Numbers and Computer Methods for Factorization, 2nd ed.* Boston, MA: Birkhäuser, 1994.

Sloane, N. J. A. Sequences A047713/M5461, A048950, and A055551 in "An On-Line Version of the Encyclopedia of Integer Sequences." http://www.research.att.com/~njas/sequences/eisonline.html.

Euler-Lagrange Derivative

The derivative

$$\frac{\delta L}{\delta q} \equiv \frac{\partial L}{\partial q} - \frac{d}{dt}\left(\frac{\partial L}{\partial \dot{q}}\right)$$

appearing in the EULER-LAGRANGE DIFFERENTIAL EQUATION.

Euler-Lagrange Differential Equation

A fundamental equation of CALCULUS OF VARIATIONS which states that if J is defined by an INTEGRAL OF THE FORM

$$J = \int f(x, y, \dot{y}) \, dx, \tag{1}$$

where

$$\dot{y} \equiv \frac{dy}{dt}, \tag{2}$$

then J has a STATIONARY VALUE if the Euler-Lagrange differential equation

$$\frac{\partial f}{\partial y} - \frac{d}{dt}\left(\frac{\partial f}{\partial \dot{y}}\right) = 0 \tag{3}$$

is satisfied. If time DERIVATIVE NOTATION is replaced instead by space variable notation, the equation becomes

$$\frac{\partial f}{\partial y} - \frac{d}{dx}\frac{\partial f}{\partial y_x} = 0. \tag{4}$$

In many physical problems, f_x (the PARTIAL DERIVA-

TIVE of f with respect to x) turns out to be 0, in which case a manipulation of the Euler-Lagrange differential equation reduces to the greatly simplified and partially integrated form known as the BELTRAMI IDENTITY,

$$f - y_x \frac{\partial f}{\partial y_x} = C. \tag{5}$$

For three independent variables (Arfken 1985, pp. 924–44), the equation generalizes to

$$\frac{\partial f}{\partial u} - \frac{\partial}{\partial x}\frac{\partial f}{\partial u_x} - \frac{\partial}{\partial y}\frac{\partial f}{\partial u_y} - \frac{\partial}{\partial z}\frac{\partial f}{\partial u_z} = 0. \tag{6}$$

Problems in the CALCULUS OF VARIATIONS often can be solved by solution of the appropriate Euler-Lagrange equation.

To derive the Euler-Lagrange differential equation, examine

$$\delta J \equiv \delta \int L(q, \dot{q}, t)\, dt = \int \left(\frac{\partial L}{\partial q}\delta q + \frac{\partial L}{\partial \dot{q}}\delta \dot{q} \right) dt$$

$$= \int \left[\frac{\partial L}{\partial q}\delta q + \frac{\partial L}{\partial \dot{q}}\frac{d(\delta q)}{dt} \right] dt, \tag{7}$$

since $\delta \dot{q} = d(\delta q)/dt$. Now, integrate the second term by PARTS using

$$u = \frac{\partial L}{\partial \dot{q}} \qquad dv = d(\delta q) \tag{8}$$

$$du = \frac{d}{dt}\left(\frac{\partial L}{\partial \dot{q}} \right) dt \qquad v = \delta q, \tag{9}$$

so

$$\int \frac{\partial L}{\partial \dot{q}}\frac{d(\delta q)}{dt}\, dt = \int \frac{\partial L}{\partial \dot{q}}\, d(\delta q)$$

$$= \left[\frac{\partial L}{\partial \dot{q}}\delta q \right]_{t_1}^{t_2} - \int_{t_1}^{t_2}\left(\frac{d}{dt}\frac{\partial L}{\partial \dot{q}}\, dt \right)\delta q. \tag{10}$$

Combining (7) and (10) then gives

$$\delta J = \left[\frac{\partial L}{\partial \dot{q}}\delta q \right]_{t_1}^{t_2} + \int_{t_1}^{t_2}\left(\frac{\partial L}{\partial q} - \frac{d}{dt}\frac{\partial L}{\partial \dot{q}} \right)\delta q\, dt. \tag{11}$$

But we are varying the path only, not the endpoints, so $\delta q(t_1) = \delta q(t_2) = 0$ and (11) becomes

$$\delta J = \int_{t_1}^{t_2}\left(\frac{\partial L}{\partial q} - \frac{d}{dt}\frac{\partial L}{\partial \dot{q}} \right)\delta q\, dt. \tag{12}$$

We are finding the STATIONARY VALUES such that $\delta J = 0$. These must vanish for any small change δq, which gives from (12),

$$\frac{\partial L}{\partial q} - \frac{d}{dt}\left(\frac{\partial L}{\partial \dot{q}} \right) = 0. \tag{13}$$

This is the Euler-Lagrange differential equation.

The variation in J can also be written in terms of the parameter κ as

$$\delta J = \int [f(x,\ y + \kappa v,\ \dot{y} + \kappa \dot{v}) - f(x,\ y,\ \dot{y})]\, dt$$

$$= \kappa I_1 + \tfrac{1}{2}\kappa^2 I_2 + \tfrac{1}{6}\kappa^3 I_3 + \tfrac{1}{24}\kappa^4 I_4 + \ldots, \tag{14}$$

where

$$v = \delta y \tag{15}$$

$$\dot{v} = \delta \dot{y} \tag{16}$$

and the first, second, etc., variations are

$$I_1 = \int (vf_y + \dot{v}f_{\dot{y}})\, dt \tag{17}$$

$$I_2 = \int (v^2 f_{yy} + 2v\dot{v}f_{y\dot{y}} + \dot{v}^2 f_{\dot{y}\dot{y}})\, dt \tag{18}$$

$$I_3 = \int (v^3 f_{yyy} + 3v^2\dot{v}f_{yy\dot{y}} + 3v\dot{v}^2 f_{y\dot{y}\dot{y}} + \dot{v}^3 f_{\dot{y}\dot{y}\dot{y}})\, dt \tag{19}$$

$$I_4 = \int (v^4 f_{yyyy} + 4v^3\dot{v}f_{yyy\dot{y}} + 6v^2\dot{v}^2 f_{yy\dot{y}\dot{y}} + 4v\dot{v}^3 f_{y\dot{y}\dot{y}\dot{y}}$$
$$+ \dot{v}^4 f_{\dot{y}\dot{y}\dot{y}\dot{y}})\, dt. \tag{20}$$

The second variation can be re-expressed using

$$\frac{d}{dt}(v^2\lambda) = v^2\dot{\lambda} + 2v\dot{v}\lambda, \tag{21}$$

so

$$I_2 + [v^2\lambda]_{\frac{1}{2}}^{\frac{1}{2}} = \int_1^2 [v^2(f_{yy} + \dot{\lambda}) + 2v\dot{v}(f_{y\dot{y}} + \lambda) + \dot{v}^2 f_{\dot{y}\dot{y}}]\, dt. \tag{22}$$

But

$$[v^2\lambda]_{\frac{1}{2}}^{\frac{1}{2}} = 0. \tag{23}$$

Now choose λ such that

$$f_{\dot{y}\dot{y}}(f_{yy} + \dot{\lambda}) = (f_{y\dot{y}} + \lambda)^2 \tag{24}$$

and z such that

$$f_{y\dot{y}} + \lambda = -\frac{f_{\dot{y}\dot{y}}}{z}\frac{dz}{dt} \tag{25}$$

so that z satisfies

$$f_{\dot{y}\dot{y}}\ddot{z} + \dot{f}_{\dot{y}\dot{y}}\dot{z} - (f_{yy} - \dot{f}_{y\dot{y}})z = 0. \tag{26}$$

It then follows that

$$I_2 = \int f_{\dot{y}\dot{y}} \left(\dot{v} + \frac{f_{y\dot{y}} + \lambda}{f_{\dot{y}\dot{y}} v} \right)^2 dt = \int f_{\dot{y}\dot{y}} \left(\dot{v} - \frac{v}{z} \frac{dz}{dt} \right)^2. \quad (27)$$

See also BELTRAMI IDENTITY, BRACHISTOCHRONE PROBLEM, CALCULUS OF VARIATIONS, EULER-LAGRANGE DERIVATIVE

References

Arfken, G. *Mathematical Methods for Physicists, 3rd ed.* Orlando, FL: Academic Press, 1985.

Forsyth, A. R. *Calculus of Variations.* New York: Dover, pp. 17–0 and 29, 1960.

Morse, P. M. and Feshbach, H. "The Variational Integral and the Euler Equations." §3.1 in *Methods of Theoretical Physics, Part I.* New York: McGraw-Hill, pp. 276–80, 1953.

Euler-Lucas Pseudoprime

Let $U(P, Q)$ and $V(P, Q)$ be LUCAS SEQUENCES generated by P and Q, and define

$$D \equiv P^2 - 4Q.$$

Then

$$\begin{cases} U_{(n-(D/n))/2} \equiv 0 \pmod{n} & \text{when } (Q/n) = 1 \\ V_{(n-(D/n))/2} \equiv D \pmod{n} & \text{when } (Q/n) = -1, \end{cases}$$

where (Q/n) is the LEGENDRE SYMBOL. An ODD COMPOSITE NUMBER n such that $(n, QD) = 1$ (i.e., n and QD are RELATIVELY PRIME) is called an Euler-Lucas pseudoprime with parameters (P, Q).

See also PSEUDOPRIME, STRONG LUCAS PSEUDOPRIME

References

Ribenboim, P. "Euler-Lucas Pseudoprimes (elpsp(P, Q)) and Strong Lucas Pseudoprimes (slpsp(P, Q))." §2.X.C in *The New Book of Prime Number Records.* New York: Springer-Verlag, pp. 130–31, 1996.

Euler-Maclaurin Integration Formulas

The Euler-Maclaurin integration and sums formulas can be derived from DARBOUX'S FORMULA by substituting the BERNOULLI POLYNOMIAL $B_n(t)$ in for the function $\phi(t)$. Differentiating the identity

$$B_n(t + 1) - B_n(t) = nt^{n-1} \quad (1)$$

$n - k$ times gives

$$B_n^{(n-k)}(t + 1) - \phi_n^{(n-k)}(t) = n(n-1)\cdots kt^{k-1}. \quad (2)$$

Plugging in $t = 0$ gives $B_n^{(n-k)}(1) = B_n^{(n-k)}(0)$. From the Maclaurin series of $B_n(z)$ with $k > 0$, we have

$$B_n^{(n-2k-1)}(0) = 0 \quad (3)$$

$$B_n^{(n-2k)}(0) = \frac{n!}{(2k)!} B_{2k} \quad (4)$$

$$B_n^{(n-1)}(0) = \tfrac{1}{2} n! \quad (5)$$

$$B_n^{(n)}(0) = n!, \quad (6)$$

where B_n is a BERNOULLI NUMBER, and substituting these values of $B_n^{(n-k)}(1)$ and $B_n^{(n-k)}(0)$ into DARBOUX'S FORMULA gives

$$(z - a)f'(a) = f(z) - f(a) - \frac{z-a}{2}[f'(z) - f'(a)]$$

$$+ \sum_{m=1}^{n-1} \frac{B_{2m}(z-a)^{2m}}{(2m)!}[f^{(2m)}(z) - f^{(2m)}(a)]$$

$$- \frac{(z-a)^{2n+1}}{(2n)!} \int_0^1 B_{2n}(t)f^{(2n+1)}[a - (z-a)t]\, dt, \quad (7)$$

which is the Euler-Maclaurin integration formula (Whittaker and Watson 1990, p. 128).

In certain cases, the last term tends to 0 as $n \to \infty$, and an infinite series can then be obtained for $f(z) - f(a)$. In such cases, SUMS may be converted to INTEGRALS by inverting the formula to obtain the Euler-Maclaurin sum formula

$$\sum_{k=1}^{n-1} f_k = \int_0^n f(k)dk - \tfrac{1}{2}[f(0) + f(n)]$$

$$+ \sum_{k=1}^{\infty} \frac{B_{2n}}{(2n)!}[f^{(2n-1)}(n) - f^{(2n-1)}(0)], \quad (8)$$

which, when expanded, gives

$$\sum_{k=1}^{n-1} f_k = \int_0^n f(k)\, dk - \tfrac{1}{2}[f(0) + f(n)] + \tfrac{1}{12}[f'(n) - f'(0)]$$

$$- \tfrac{1}{720}[f'''(n) - f'''(0)] + \tfrac{1}{30240}[f^{(5)}(n) - f^{(5)}(0)]$$

$$- \tfrac{1}{1209600}[f^{(7)} - f^{(7)}(0)] + \dots \quad (9)$$

(Abramowitz and Stegun 1972, p. 16). The Euler-Maclaurin sum formula is implemented in *Mathematica* as the function NSum with option Method-> Integrate.

The second Euler-Maclaurin integration formula is used when $f(x)$ is tabulated at n values $f_{3/2}, f_{5/2}, \dots, f_{n-1/2}$:

$$\int_{x_1}^{x_n} f(x)\, dx = h[f_{3/2} + f_{5/2} + f_{7/2} + \dots + f_{n-3/2} + f_{n-1/2}]$$

$$- \sum_{k=1}^{\infty} \frac{B_{2k}h^{2k}}{(2k)!}(1 - 2^{-2k+1})[f_n^{(2k-1)} - f_1^{(2k-1)}]. \quad (10)$$

See also DARBOUX'S FORMULA, SUM, WYNN'S EPSILON METHOD

References

Abramowitz, M. and Stegun, C. A. (Eds.). *Handbook of Mathematical Functions with Formulas, Graphs, and Mathematical Tables, 9th printing.* New York: Dover, pp. 16 and 806, 1972.

Apostol, T. M. "An Elementary View of Euler's Summation Formula." *Amer. Math. Monthly* **106**, 409–18, 1999.

Arfken, G. "Bernoulli Numbers, Euler-Maclaurin Formula." §5.9 in *Mathematical Methods for Physicists, 3rd ed.* Orlando, FL: Academic Press, pp. 327–38, 1985.

Borwein, J. M.; Borwein, P. B.; and Dilcher, K. "Pi, Euler Numbers, and Asymptotic Expansions." *Amer. Math. Monthly* **96**, 681–87, 1989.

Euler, L. *Comm. Acad. Sci. Imp. Petrop.* **6**, 68, 1738.

Knopp, K. *Theory and Application of Infinite Series.* New York: Hafner, 1951.

Maclaurin, C. *Treatise of Fluxions.* Edinburgh, p. 672, 1742.

Vardi, I. "The Euler-Maclaurin Formula." §8.3 in *Computational Recreations in Mathematica.* Reading, MA: Addison-Wesley, pp. 159–63, 1991.

Whittaker, E. T. and Robinson, G. "The Euler-Maclaurin Formula." §67 in *The Calculus of Observations: A Treatise on Numerical Mathematics, 4th ed.* New York: Dover, pp. 134–36, 1967.

Whittaker, E. T. and Watson, G. N. "The Euler-Maclaurin Expansion." §7.21 in *A Course in Modern Analysis, 4th ed.* Cambridge, England: Cambridge University Press, pp. 127–28, 1990.

Euler-Maclaurin Sum Formula

EULER-MACLAURIN INTEGRATION FORMULAS

Euler-Mascheroni Constant

The Euler-Mascheroni constant is denoted γ (or sometimes C) and has the numerical value

$$\gamma \approx 0.5772156649015328606065120900824024310 42\ldots \tag{1}$$

(Sloane's A001620). The Euler-Mascheroni constant was denoted γ and calculated to 16 digits by Euler in 1781. It is therefore sometimes known as Euler's constant. No quadratically converging algorithm for computing γ is known (Bailey 1988). X. Gourdon and P. Demichel computed a record 108 million digits of γ in October 1999 (Gourdon and Sebah).

The Euler-Mascheroni constant is implemented in *Mathematica* as EulerGamma. It is not known if this constant is IRRATIONAL, let alone TRANSCENDENTAL (Wells 1986, p. 28). If γ is a simple fraction a/b, then it is known that $b > 10^{10,000}$ (Brent 1977; Wells 1986, p. 28). Conway and Guy (1996) are "prepared to bet that it is transcendental," although they do not expect a proof to be achieved within their lifetimes.

The CONTINUED FRACTION of the Euler-Mascheroni constant is [0, 1, 1, 2, 1, 2, 1, 4, 3, 13, 5, 1, 1, 8, 1, 2, 4, 1, 1, 40, ...] (Sloane's A002852). The first few CONVERGENTS are 1, 1/2, 3/5, 4/7, 11/19, 15/26, 71/123, 228/395, 3035/5258, 15403/26685, ... (Sloane's A046114 and A046115). The positions at which the digits 1, 2, ... first occur in the CONTINUED FRACTION are 2, 4, 9, 8, 11, 69, 24, 14, 139, 52, 22, ... (Sloane's A033149). The sequence of largest terms in the CONTINUED FRACTION is 1, 2, 4, 13, 40, 49, 65, 399, 2076, ... (Sloane's A033091), which occur at positions 2, 4, 8, 10, 20, 31, 34, 40, 529, ... (Sloane's A033092).

The Euler-Mascheroni constant arises in many integrals

$$\gamma \equiv -\int_0^\infty e^{-x} \ln x \, dx \tag{2}$$

$$= \int_0^\infty \left(\frac{1}{1 - e^{-x}} - \frac{1}{x} \right) e^{-x} \, dx \tag{3}$$

$$= \int_0^\infty \frac{1}{x} \left(\frac{1}{1+x} - e^{-x} \right) dx \tag{4}$$

(Whittaker and Watson 1990, p. 246), and sums

$$\gamma \equiv 1 + \sum_{k=2}^\infty \left[\frac{1}{k} + \ln\left(\frac{k-1}{k} \right) \right] \tag{5}$$

$$= \lim_{n \to \infty} (H_n - \ln n) \tag{6}$$

$$= \sum_{n=2}^\infty (-1)^n \frac{\zeta(n)}{n} \tag{7}$$

$$= \ln\left(\frac{4}{\pi} \right) - \sum_{n=1}^\infty \frac{(-1)^n \zeta(n+1)}{2^n(n+1)}, \tag{8}$$

where H_n is a HARMONIC NUMBER (Graham *et al.* 1994, p. 278) and $\zeta(z)$ is the RIEMANN ZETA FUNCTION. γ is also given by the EULER PRODUCT

$$e^\gamma = \lim_{n \to \infty} \frac{1}{\ln n} \prod_{i=1}^n \frac{1}{1 - \frac{1}{pi}}, \tag{9}$$

where the product is over PRIMES p. Another connection with the PRIMES was provided by Dirichlet's 1838 proof that the average number of DIVISORS of all numbers from 1 to n is asymptotic to

$$\frac{\sum_{i=1}^n \sigma_0(i)}{n} \sim \ln n + 2\gamma - 1 \tag{10}$$

(Conway and Guy 1996). de la Vallée Poussin (1898) proved that, if a large number n is divided by all PRIMES $\leq n$, then the average amount by which the QUOTIENT is less than the next whole number is γ.

INFINITE PRODUCTS involving γ also arise from the BARNES' G-FUNCTION with POSITIVE INTEGER n. The cases $G(2)$ and $G(3)$ give

$$\prod_{n=1}^\infty e^{-1+1/2(n)} \left(1 + \frac{1}{n} \right)^n = \frac{e^{1+\gamma/2}}{\sqrt{2\pi}} \tag{11}$$

$$\prod_{n=1}^\infty e^{-2+2/n} \left(1 + \frac{2}{n} \right)^n = \frac{e^{3+2\gamma}}{\sqrt{2\pi}}. \tag{12}$$

The Euler-Mascheroni constant is also given by the limits

$$\gamma = -\Gamma'(1) \tag{13}$$

(Whittaker and Watson 1990, p. 236),

$$\gamma = \lim_{s \to 1} \zeta(s) - \frac{1}{s-1} \tag{14}$$

(Whittaker and Watson 1990, p. 271), and

$$\gamma = \lim_{x \to \infty} \left[x - \Gamma\left(\frac{1}{x}\right) \right] \tag{15}$$

(Le Lionnais 1983).

The difference between the nth convergent in (6) and γ is given by

$$\sum_{k=1}^{n} \frac{1}{k} - \ln n - \gamma = \int_{n}^{\infty} \frac{x - \lfloor x \rfloor}{x^2} \, dx, \tag{16}$$

where $\lfloor x \rfloor$ is the FLOOR FUNCTION, and satisfies the INEQUALITY

$$\frac{1}{2(n+1)} < \sum_{k=1}^{n} \frac{1}{k} - \ln n - \gamma < \frac{1}{2n} \tag{17}$$

(Young 1991). A series with accelerated convergence is

$$\gamma = \frac{3}{2} - \ln 2 - \sum_{m=2}^{\infty} (-1)^m \frac{m-1}{m} [\zeta(m) - 1] \tag{18}$$

(Flajolet and Vardi 1996). Another series is

$$\gamma = \sum_{n=1}^{\infty} (-1)^n \frac{\lfloor \lg n \rfloor}{n} \tag{19}$$

(Vacca 1910, Gerst 1969), where LG is the LOGARITHM to base 2. The convergence of this series can be greatly improved using Euler's CONVERGENCE IMPROVEMENT transformation to

$$\gamma = \sum_{k=1}^{\infty} 2^{-(k+1)} \sum_{j=0}^{k-1} \frac{1}{\binom{2^{k-j}+j}{j}}, \tag{20}$$

where $\binom{a}{b}$ is a BINOMIAL COEFFICIENT (Beeler *et al.* 1972, with $k-j$ replacing the undefined i). Bailey (1988) gives

$$\gamma = \frac{2^n}{e^{2n}} \sum_{m=0}^{\infty} \frac{2^{mn}}{(m+1)!} \sum_{t=0}^{m} \frac{1}{t+1} - n \ln 2 + \mathcal{O}\left(\frac{1}{2^n e^{2n}}\right), \tag{21}$$

which is an improvement over Sweeney (1963).

The symbol γ is sometimes also used for

$$\gamma' \equiv e^\gamma \approx 1.781072 \tag{22}$$

(Gradshteyn and Ryzhik 2000, p. xxvii).

Odena (1982–983) gave the strange approximation

$$(0.11111111)^{1/4} = 0.577350\ldots, \tag{23}$$

and Castellanos (1988) gave

$$\left(\tfrac{7}{83}\right)^{2/9} = 0.57721521\ldots \tag{24}$$

$$\left(\frac{520^2 + 22}{52^4}\right)^{1/6} = 0.5772156634\ldots \tag{25}$$

$$\left(\frac{80^3 + 92}{61^4}\right)^{1/6} = 0.57721566457\ldots \tag{26}$$

$$\frac{990^3 - 55^3 - 79^2 - 4^2}{70^5} = 0.5772156649015295\ldots. \tag{27}$$

See also EULER PRODUCT, MERTENS THEOREM, STIELTJES CONSTANTS

References

Anastassow, T. *Die Mascheroni'sche Konstante: Eine historisch-analytisch zusammenfassende Studie.* Thesis. Bonn, Germany: Universität Bonn. Wetzikon: J. Wirz, 1914.

Bailey, D. H. "Numerical Results on the Transcendence of Constants Involving π, e, and Euler's Constant." *Math. Comput.* **50**, 275–81, 1988.

Beeler, M. *et al.* Item 120 in Beeler, M.; Gosper, R. W.; and Schroeppel, R. *HAKMEM.* Cambridge, MA: MIT Artificial Intelligence Laboratory, Memo AIM-239, p. 55, Feb. 1972.

Brent, R. P. "Computation of the Regular Continued Fraction for Euler's Constant." *Math. Comput.* **31**, 771–77, 1977.

Brent, R. P. and McMillan, E. M. "Some New Algorithms for High-Precision Computation of Euler's Constant." *Math. Comput.* **34**, 305–12, 1980.

Castellanos, D. "The Ubiquitous Pi. Part I." *Math. Mag.* **61**, 67–8, 1988.

Conway, J. H. and Guy, R. K. "The Euler-Mascheroni Number." In *The Book of Numbers.* New York: Springer-Verlag, pp. 260–61, 1996.

de la Vallée Poussin, C.-J. Untitled communication. *Annales de la Soc. Sci. Bruxelles* **22**, 84–0, 1898.

DeTemple, D. W. "A Quicker Convergence to Euler's Constant." *Amer. Math. Monthly* **100**, 468–70, 1993.

Dirichlet, G. L. "Sur l'usage des séries infinies dans la théorie des nombres." *J. reine angew. Math.* **18**, 259–74, 1838.

Erdélyi, A.; Magnus, W.; Oberhettinger, F.; and Tricomi, F. G. *Higher Transcendental Functions, Vol. 1.* New York: Krieger, p. 1, 1981.

Finch, S. "Favorite Mathematical Constants." http://www.mathsoft.com/asolve/constant/euler/euler.html.

Flajolet, P. and Vardi, I. "Zeta Function Expansions of Classical Constants." Unpublished manuscript, 1996. http://pauillac.inria.fr/algo/flajolet/Publications/landau.ps.

Gerst, I. "Some Series for Euler's Constant." *Amer. Math. Monthly* **76**, 273–75, 1969.

Glaisher, J. W. L. "On the History of Euler's Constant." *Messenger of Math.* **1**, 25–0, 1872.

Gourdon, X. and Sebah, P. "The Euler Constant: γ." http://xavier.gourdon.free.fr/Constants/Gamma/gamma.html.

Gradshteyn, I. S. and Ryzhik, I. M. *Tables of Integrals, Series, and Products, 6th ed.* San Diego, CA: Academic Press, 2000.

Graham, R. L.; Knuth, D. E.; and Patashnik, O. *Concrete Mathematics: A Foundation for Computer Science, 2nd ed.* Reading, MA: Addison-Wesley, 1994.

Knuth, D. E. "Euler's Constant to 1271 Places." *Math. Comput.* **16**, 275–81, 1962.

Krantz, S. G. "The Euler-Mascheroni Constant." §13.1.7 in *Handbook of Complex Analysis.* Boston, MA: Birkhäuser, pp. 156–57, 1999.

Le Lionnais, F. *Les nombres remarquables.* Paris: Hermann, p. 28, 1983.

Plouffe, S. "Plouffe's Inverter: Table of Current Records for the Computation of Constants." http://www.lacim.u-qam.ca/pi/records.html.

Sloane, N. J. A. Sequences A001620/M3755, A002852/M0097, A033091, A033092, A033149, A046114, and A046115 in "An On-Line Version of the Encyclopedia of Integer Sequences." http://www.research.att.com/~njas/sequences/eisonline.html.

Sweeney, D. W. "On the Computation of Euler's Constant." *Math. Comput.* **17**, 170–78, 1963.

Vacca, G. "A New Series for the Eulerian Constant." *Quart. J. Pure Appl. Math.* **41**, 363–68, 1910.

Wells, D. *The Penguin Dictionary of Curious and Interesting Numbers.* Middlesex, England: Penguin Books, p. 28, 1986.

Whittaker, E. T. and Watson, G. N. *A Course in Modern Analysis, 4th ed.* Cambridge, England: Cambridge University Press, pp. 235–36 and 271, 1990.

Young, R. M. "Euler's Constant." *Math. Gaz.* **75**, 187–90, 1991.

Euler-Mascheroni Integrals

Define

$$I_n \equiv (-1)^n \int_0^\infty (\ln z)^n e^{-z}\, dz, \tag{1}$$

then

$$I_0 = \int_0^\infty e^{-z}\, dz = [-e^{-z}]_0^\infty = (0+1) = 1 \tag{2}$$

$$I_1 = -\int_0^\infty (\ln z) e^{-z}\, dz = \gamma \tag{3}$$

$$I_2 = \gamma^2 + \tfrac{1}{6}\pi^2 \tag{4}$$

$$I_3 = \gamma^3 + \tfrac{1}{2}\gamma\pi^2 + 2\zeta(3) \tag{5}$$

$$I_4 = \gamma^4 + \gamma^2\pi^2 - \tfrac{3}{20}\pi^4 + 8\gamma\zeta(3), \tag{6}$$

where γ is the EULER-MASCHERONI CONSTANT and $\zeta(3)$ is APÉRY'S CONSTANT.

EulerPhi

TOTIENT FUNCTION

Euler-Poincaré Characteristic

EULER CHARACTERISTIC

Euler-Poisson-Darboux Equation

The PARTIAL DIFFERENTIAL EQUATION

$$u_{xy} + \frac{N(u_x + u_y)}{x + y} = 0.$$

See also EULER-DARBOUX EQUATION

References

Ames, W. F. "Ad Hoc Exact Techniques for Nonlinear Partial Differential Equations." §3.3 in *Nonlinear Partial Differential Equations in Engineering* (Ed. W. F. Ames). New York: Academic Press, 1967.

Zwillinger, D. *Handbook of Differential Equations, 3rd ed.* Boston, MA: Academic Press, p. 129, 1997.

Eutactic Star

An orthogonal projection of a CROSS onto a 3-D SUBSPACE. It is said to be normalized if the CROSS vectors are all of unit length.

See also HADWIGER'S PRINCIPAL THEOREM

Evans Point

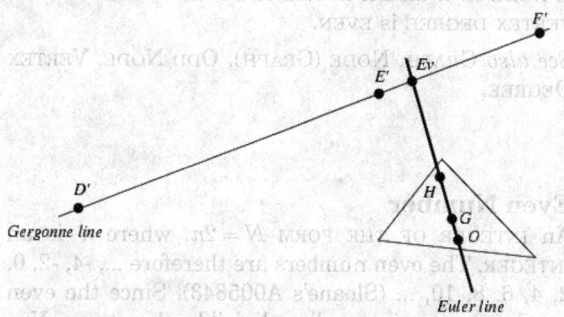

Gergonne line

Euler line

The intersection of the GERGONNE LINE and the EULER LINE. It does not appear to have a simple parametric representation.

See also EULER LINE, GERGONNE LINE

References

Oldknow, A. "The Euler-Gergonne-Soddy Triangle of a Triangle." *Amer. Math. Monthly* **103**, 319–29, 1996.

Eve

APPLE, ROOT, SNAKE, SNAKE EYES, SNAKE OIL METHOD, SNAKE POLYIAMOND

Even Divisor Function

The sum of powers of EVEN DIVISORS of a number. It is the analog of the DIVISOR FUNCTION for even divisors only and is written $\sigma_k^{(e)}(n)$. It is given simply in terms of the usual DIVISOR FUNCTION by

$$\sigma_k^{(e)}(n) = \begin{cases} 0 & \text{for } n \text{ odd} \\ 2^k \sigma_k(n/2) & \text{for } n \text{ even.} \end{cases}$$

See also DIVISOR FUNCTION, ODD DIVISOR FUNCTION

Even Function

A function $f(x)$ such that $f(x) = f(-x)$. An even function times an ODD FUNCTION is odd.

Even Node

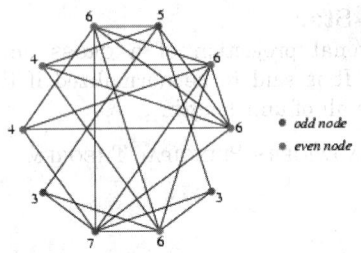

A NODE in a GRAPH is said to be an even node if its VERTEX DEGREE is EVEN.

See also GRAPH, NODE (GRAPH), ODD NODE, VERTEX DEGREE

Even Number

An INTEGER OF THE FORM $N = 2n$, where n is an INTEGER. The even numbers are therefore ..., -4, -2, 0, 2, 4, 6, 8, 10, ... (Sloane's A005843). Since the even numbers are integrally divisible by two, $N \equiv 0 \pmod 2$ for even N. An even number N for which $N \equiv 2 \pmod 4$ is called a SINGLY EVEN NUMBER, and an even number N for which $N \equiv 0 \pmod 4$ is called a DOUBLY EVEN NUMBER. An integer which is not even is called an ODD NUMBER. The GENERATING FUNCTION of the even numbers is

$$\frac{2x}{(x-1)^2} = 2x + 4x^2 + 6x^3 + 8x^4 + \dots.$$

See also DOUBLY EVEN NUMBER, EVEN FUNCTION, ODD NUMBER, SINGLY EVEN NUMBER

References

Commission on Mathematics of the College Entrance Examination Board. *Informal Deduction in Algebra: Properties of Odd and Even Numbers.* Princeton, NJ, 1959.
Sloane, N. J. A. Sequences A005843/M0985 in "An On-Line Version of the Encyclopedia of Integer Sequences." http://www.research.att.com/~njas/sequences/eisonline.html.

Even Part

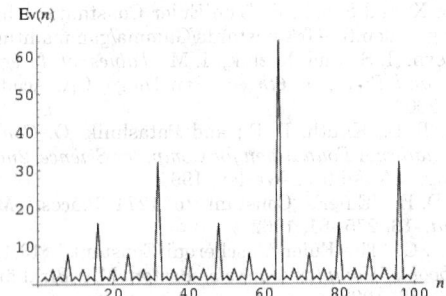

The even part $\text{Ev}(n)$ of a positive integer n is defined by

$$\text{Ev}(n) = 2^{b(n)},$$

where $b(n)$ is the EXPONENT of the exact power of 2 dividing n. The values for $n = 1, 2, \dots,$ are 1, 2, 1, 4, 1, 2, 1, 8, 1, 2, 1, ... (Sloane's A006519). The even part function can be implemented in *Mathematica* as

```
EvenPart[0] := 1
EvenPart[n_Integer] := 2^IntegerExponent[n,2]
```

See also GREATEST DIVIDING EXPONENT, ODD PART

References

Sloane, N. J. A. Sequences A006519/M0162 in "An On-Line Version of the Encyclopedia of Integer Sequences." http://www.research.att.com/~njas/sequences/eisonline.html.

Even Prime

The unique EVEN PRIME NUMBER 2. All other PRIMES are ODD PRIMES.

The sequence 2, 4, 6, 10, 14, 22, 26, 34, 38, ... (Sloane's A001747) consisting of the number 2 together with the PRIMES multiplied by 2 is sometimes also called the even primes, since these are the even numbers $n = 2k$ that are divisible by just 1, 2, k, and $2k$.

See also EVEN NUMBER, ODD PRIME, PRIME NUMBER

References

Sloane, N. J. A. Sequences A001747 in "An On-Line Version of the Encyclopedia of Integer Sequences." http://www.research.att.com/~njas/sequences/eisonline.html.
Wells, D. *The Penguin Dictionary of Curious and Interesting Numbers.* Middlesex, England: Penguin Books, p. 44, 1986.

Event

An event is a certain subset of a PROBABILITY SPACE. Events are therefore collections of OUTCOMES on which probabilities have been assigned. Events are sometimes assumed to form a BOREL FIELD (Papoulis 1984, p. 29).

See also EXPERIMENT, INDEPENDENT EVENTS, MUTUALLY EXCLUSIVE EVENTS, OUTCOME, TRIAL

References

Papoulis, A. *Probability, Random Variables, and Stochastic Processes, 2nd ed.* New York: McGraw-Hill, pp. 24 and 29–0, 1984.

Eventually Periodic

A PERIODIC SEQUENCE such as $\{1, 1, 1, 2, 1, 2, 1, 2, 1, 2, 1, 1, 2, 1, \ldots\}$ which is periodic from some point onwards.

See also PERIODIC SEQUENCE

Everett Interpolation

EVERETT'S FORMULA

Everett's Formula

$$f_p = (1-p)f_0 + pf_1 + E_2\delta_0^2 + F_2\delta_1^2 + E_4\delta_0^4 + F_4\delta_1^4$$
$$+ E_6\delta_0^6 + F_6\delta_1^6 + \ldots, \tag{1}$$

for $p \in [0, 1]$, where δ is the CENTRAL DIFFERENCE and

$$E_{2n} \equiv G_{2n} - G_{2n+1} \equiv B_{2n} - B_{2n+1} \tag{2}$$

$$F_{2n} \equiv G_{2n+1} \equiv B_{2n} + B_{2n+1}, \tag{3}$$

where G_k are the COEFFICIENTS from GAUSS'S BACKWARD FORMULA and GAUSS'S FORWARD FORMULA and B_k are the COEFFICIENTS from BESSEL'S FINITE DIFFERENCE FORMULA. The E_ks and F_ks also satisfy

$$E_{2n}(p) = F_{2n}(q) \tag{4}$$

$$F_{2n}(p) = E_{2n}(q), \tag{5}$$

for

$$q \equiv 1 - p. \tag{6}$$

See also BESSEL'S FINITE DIFFERENCE FORMULA

References

Abramowitz, M. and Stegun, C. A. (Eds.). *Handbook of Mathematical Functions with Formulas, Graphs, and Mathematical Tables, 9th printing.* New York: Dover, pp. 880–81, 1972.

Acton, F. S. *Numerical Methods That Work, 2nd printing.* Washington, DC: Math. Assoc. Amer., pp. 92–3, 1990.

Beyer, W. H. *CRC Standard Mathematical Tables, 28th ed.* Boca Raton, FL: CRC Press, p. 433, 1987.

Whittaker, E. T. and Robinson, G. "The Laplace-Everett Formula." §25 in *The Calculus of Observations: A Treatise on Numerical Mathematics, 4th ed.* New York: Dover, pp. 40–1, 1967.

Eversion

A curve on the unit sphere $\mathbb{S}^2$ is an eversion if it has no corners or cusps (but it may be self-intersecting). These properties are guaranteed by requiring that the curve's velocity never vanishes. A mapping $\sigma : \mathbb{S}^1 \to \mathbb{S}^2$ forms an immersion of the CIRCLE into the SPHERE IFF, for all $\theta \in \mathbb{R}$,

$$\left| \frac{d}{d\theta}[\sigma(e^{i\theta})] \right| > 0.$$

Smale (1958) showed it is possible to turn a SPHERE inside out (SPHERE EVERSION) using eversion.

See also SPHERE EVERSION

References

Smale, S. "A Classification of Immersions of the Two-Sphere." *Trans. Amer. Math. Soc.* **90**, 281–90, 1958.

Evolute

An evolute is the locus of centers of curvature (the envelope) of a plane curve's normals. The original curve is then said to be the INVOLUTE of its evolute. Given a plane curve represented parametrically by $(f(t), g(t))$, the equation of the evolute is given by

$$x = f - R \sin \tau \tag{1}$$

$$y = g + R \cos \tau, \tag{2}$$

where (x, y) are the coordinates of the running point, R is the RADIUS OF CURVATURE

$$R = \frac{(f'^2 + g'^2)^{3/2}}{f'g'' - f''g'}, \tag{3}$$

and τ is the angle between the unit TANGENT VECTOR

$$\hat{\mathbf{T}} = \frac{\mathbf{x}'}{|\mathbf{x}'|} = \frac{1}{\sqrt{f'^2 + g'^2}} \begin{bmatrix} f' \\ g' \end{bmatrix} \tag{4}$$

and the X-AXIS,

$$\cos \tau = \hat{\mathbf{T}} \cdot \hat{\mathbf{x}} \tag{5}$$

$$\sin \tau = \hat{\mathbf{T}} \cdot \hat{\mathbf{y}}. \tag{6}$$

Combining gives

$$x = f - \frac{(f'^2 + g'^2)g'}{f'g'' - f''g'} \tag{7}$$

$$y = g + \frac{(f'^2 + g'^2)f'}{f'g'' - f''g'}. \tag{8}$$

The definition of the evolute of a curve is independent of parameterization for any differentiable function (Gray 1997). If E is the evolute of a curve I, then I is said to be the INVOLUTE of E. The centers of the OSCULATING CIRCLES to a curve form the evolute to that curve (Gray 1997, p. 111).

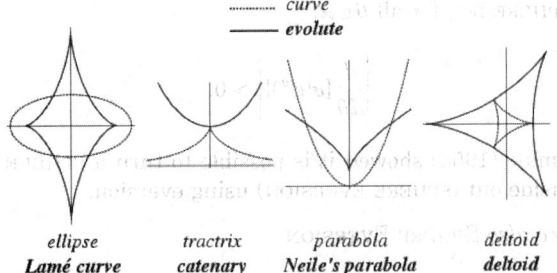

| curve | evolute |

| ellipse | tractrix | parabola | deltoid |
| **Lamé curve** | **catenary** | **Neile's parabola** | **deltoid** |

The following table lists the evolutes of some common curves, some of which are illustrated above.

Curve	Evolute
ASTROID	ASTROID 2 times as large
CARDIOID	CARDIOID 1/3 as large
CAYLEY'S SEXTIC	NEPHROID
CIRCLE	point $(0, 0)$
CYCLOID	equal CYCLOID
DELTOID	DELTOID 3 times as large
ELLIPSE	ELLIPSE EVOLUTE
EPICYCLOID	enlarged EPICYCLOID
HYPOCYCLOID	similar HYPOCYCLOID
LIMAÇON	CIRCLE CATACAUSTIC for a point source
LOGARITHMIC SPIRAL	equal LOGARITHMIC SPIRAL
NEPHROID	NEPHROID 1/2 as large
PARABOLA	NEILE'S PARABOLA
TRACTRIX	CATENARY

See also ENVELOPE, INVOLUTE, OSCULATING CIRCLE, ROULETTE

References

Cayley, A. "On Evolutes of Parallel Curves." *Quart. J. Pure Appl. Math.* **11**, 183–99, 1871.
Dixon, R. "String Drawings." Ch. 2 in *Mathographics.* New York: Dover, pp. 75–8, 1991.
Gray, A. "Evolutes." §5.1 in *Modern Differential Geometry of Curves and Surfaces with Mathematica, 2nd ed.* Boca Raton, FL: CRC Press, pp. 98–03, 1997.
Jeffrey, H. M. "On the Evolutes of Cubic Curves." *Quart. J. Pure Appl. Math.* **11**, 78–1 and 145–55, 1871.
Lawrence, J. D. *A Catalog of Special Plane Curves.* New York: Dover, pp. 40 and 202, 1972.
Lockwood, E. H. "Evolutes and Involutes." Ch. 21 in *A Book of Curves.* Cambridge, England: Cambridge University Press, pp. 166–71, 1967.

Yates, R. C. "Evolutes." *A Handbook on Curves and Their Properties.* Ann Arbor, MI: J. W. Edwards, pp. 86–2, 1952.

Evolution Strategies

A DIFFERENTIAL EVOLUTION method used to minimize functions of real variables. Evolution strategies are significantly faster at numerical optimization than traditional GENETIC ALGORITHMS and also more likely to find a function's true GLOBAL EXTREMUM.

See also DIFFERENTIAL EVOLUTION, GENETIC ALGORITHM, OPTIMIZATION THEORY

References

Price, K. and Storn, R. "Differential Evolution." *Dr. Dobb's J.*, 18–8, Apr. 1997.

Exact Covering System

A system of congruences a_i mod n_i with $1 \leq i \leq k$ is called a COVERING SYSTEM if every INTEGER y satisfies $y \equiv a_i \pmod{n}$ for at least one value of i. A covering system in which each integer is covered by just one congruence is called an exact covering system.

See also COVERING SYSTEM

References

Guy, R. K. "Exact Covering Systems." §F14 in *Unsolved Problems in Number Theory, 2nd ed.* New York: Springer-Verlag, pp. 253–56, 1994.

Exact Differential

A differential OF THE FORM

$$df = P(x, y)\, dx + Q(x, y)\, dy \tag{1}$$

is exact (also called a TOTAL DIFFERENTIAL) if $\int df$ is path-independent. This will be true if

$$df = \frac{\partial f}{\partial x}\, dx + \frac{\partial f}{\partial y}\, dy, \tag{2}$$

so P and Q must be OF THE FORM

$$P(x, y) = \frac{\partial f}{\partial x} \qquad Q(x, y) = \frac{\partial f}{\partial y}. \tag{3}$$

But

$$\frac{\partial P}{\partial y} = \frac{\partial^2 f}{\partial y \partial x} \tag{4}$$

$$\frac{\partial Q}{\partial x} = \frac{\partial^2 f}{\partial x \partial y}, \tag{5}$$

so

$$\frac{\partial P}{\partial y} = \frac{\partial Q}{\partial x}. \tag{6}$$

See also Pfaffian Form, Inexact Differential

Exact Period

Least Period

Exact Sequence

An exact sequence is a sequence of maps

$$\alpha_i : A_i \to A_{i+1} \tag{1}$$

between a sequence of spaces A_i, which satisfies

$$\text{im } \alpha_i = \ker \alpha_{i+1}, \tag{2}$$

where "im" denotes the IMAGE and "ker" the KERNEL. That is, for $a \in A_i$, $\alpha_i(a) = 0$ IFF $a = \alpha_{i-1}(b)$ for some $b \in A_{i-1}$. It follows that $\alpha_{i+1} \circ \alpha_i = 0$. The notion of exact sequence makes sense when the spaces are GROUPS, MODULES, CHAIN COMPLEXES, or SHEAVES. The notation for the maps may be suppressed and the sequence written on a single line as

$$\ldots \to A_{i-1} \to A_i \to A_{i+1} \to \ldots. \tag{3}$$

An exact sequence may be of either finite or infinite length. The special case of length five,

$$0 \to A \to B \to C \to 0, \tag{4}$$

beginning and ending with zero, meaning the zero module $\{0\}$, is called a SHORT EXACT SEQUENCE. An infinite exact sequence is called a LONG EXACT SEQUENCE. For example, the sequence where $A_i = \mathbb{Z}/4\mathbb{Z}$ and α_i is given by multiplying by 2,

$$\ldots \overset{\times 2}{\to} \mathbb{Z}/4\mathbb{Z} \overset{\times 2}{\to} \mathbb{Z}/4\mathbb{Z} \overset{\times 2}{\to} \ldots, \tag{5}$$

is a long exact sequence because at each stage the kernel and image are equal to the SUBGROUP $\{0, 2\}$.

Special information is conveyed when one of the spaces A_i is the ZERO MODULE. For instance, the sequence

$$0 \to A \to B \tag{6}$$

is exact IFF the map $A \to B$ is INJECTIVE. Similarly,

$$A \to B \to 0 \tag{7}$$

is exact IFF the map $A \to B$ is SURJECTIVE.

See also Chain Complex, Homology, Long Exact Sequence, Short Exact Sequence

References

Atiyah, M. F. and MacDonald, I. G. *Introduction to Commutative Algebra.* Reading, MA: Addison-Wesley, pp. 22–4, 1969.

Fulton, W. *Algebraic Topology: A First Course.* New York: Springer-Verlag, p. 144, 1995.

Hilton, P. and Stammbach, U. *A Course in Homological Algebra.* New York: Springer-Verlag, 1997.

Munkres, J. *Elements of Algebraic Topology.* Reading, MA: Addison-Wesley, pp. 130–33, 1984.

Exact Trilinear Coordinates

The TRILINEAR COORDINATES $\alpha : \beta : \gamma$ of a point P relative to a TRIANGLE are PROPORTIONAL to the directed distances $a' : b' : c'$ from P to the side lines (i.e, $a' = k\alpha$, $b' = k\beta$, $c' = k\gamma$). Letting k be the constant of proportionality,

$$k \equiv \frac{2\Delta}{a\alpha + b\beta + c\gamma},$$

where Δ is the AREA of ΔABC and a, b, and c are the lengths of its sides. When the trilinears are chosen so that $k = 1$, the coordinates are known as exact trilinear coordinates.

See also Trilinear Coordinates

Exactly One

"Exactly one" means "one and only one," sometimes also referred to as "JUST ONE." J. H. Conway has also humorously suggested "onee" (one and only one) by analogy with IFF (if and only if), "twoo" (two and only two), and "threee" (three and only three). This refinement is sometimes needed in formal mathematical discourse because, for example, if you have two apples, you also have one apple, but you do not have *exactly one* apple.

In 2-valued LOGIC, exactly one is equivalent to the exclusive or operator XOR,

$$P(E) \text{ XOR } P(F) = P(E) + P(F) - 2P(E \cap F).$$

See also Iff, Precisely Unless, Xnor, Xor

Exactly When

Iff

Excenter

The center J_i of an EXCIRCLE. There are three excenters for a given TRIANGLE, denoted J_1, J_2, J_3. The INCENTER I and excenters J_i of a TRIANGLE are an ORTHOCENTRIC SYSTEM.

$$\overline{OI}^2 + \overline{OJ_1}^2 + \overline{OJ_2}^2 + \overline{OJ_3}^2 = 12R^2,$$

where O is the CIRCUMCENTER, J_i are the excenters, and R is the CIRCUMRADIUS (Johnson 1929, p. 190). Denote the MIDPOINTS of the original TRIANGLE M_1, M_2, and M_3. Then the lines J_1M_1, J_2M_2, and J_3M_3 intersect in a point known as the MITTENPUNKT.

See also Centroid (Orthocentric System), Excen-

TER-EXCENTER CIRCLE, EXCENTRAL TRIANGLE, EXCIRCLE, INCENTER, MITTENPUNKT

References

Coxeter, H. S. M. and Greitzer, S. L. *Geometry Revisited.* Washington, DC: Math. Assoc. Amer., p. 13, 1967.
Dixon, R. *Mathographics.* New York: Dover, pp. 58–9, 1991.
Johnson, R. A. *Modern Geometry: An Elementary Treatise on the Geometry of the Triangle and the Circle.* Boston, MA: Houghton Mifflin, 1929.
Wells, D. *The Penguin Dictionary of Curious and Interesting Geometry.* London: Penguin, pp. 115–16, 1991.

Excenter-Excenter Circle

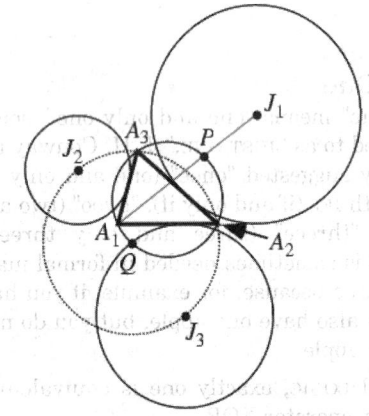

Given a TRIANGLE $\Delta A_1 A_2 A_3$, the points A_1, I, and J_1 lie on a line, where I is the INCENTER and J_1 is the EXCENTER corresponding to A_1. Furthermore, the circle with $J_2 J_3$ as the diameter has Q as its center, where P is the intersection of $A_1 J_1$ with the CIRCUMCIRCLE of $A_1 A_2 A_3$ and Q is the point opposite P on the CIRCUMCIRCLE. The circle with diameter $J_2 J_3$ also passes through A_2 and A_3 and has radius

$$r = \tfrac{1}{2} a_1 \csc\left(\tfrac{1}{2} \alpha_1\right) = 2R \cos\left(\tfrac{1}{2} \alpha_1\right).$$

It arises because the points I, J_1, J_2, and J_3 form an ORTHOCENTRIC SYSTEM.

See also EXCENTER, INCENTER-EXCENTER CIRCLE, ORTHOCENTRIC SYSTEM

References

Johnson, R. A. *Modern Geometry: An Elementary Treatise on the Geometry of the Triangle and the Circle.* Boston, MA: Houghton Mifflin, pp. 185–86, 1929.

Excentral Triangle

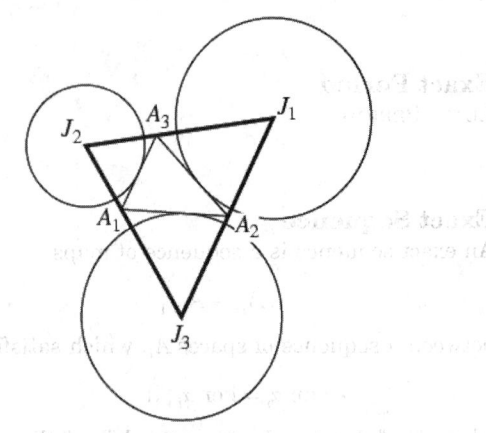

The TRIANGLE $J = \Delta J_1 J_2 J_3$ with VERTICES corresponding to the EXCENTERS of a given TRIANGLE A, also called the TRITANGENT TRIANGLE.

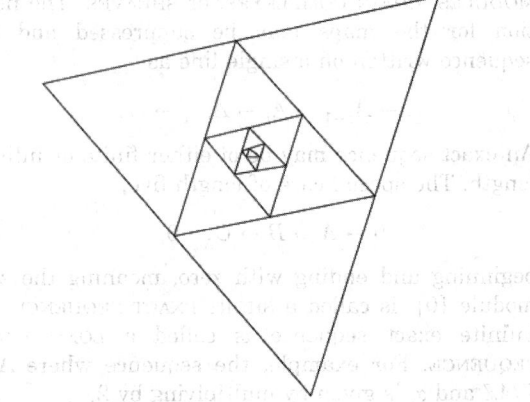

Beginning with an arbitrary TRIANGLE A, find the excentral triangle J. Then find the excentral triangle J' of that TRIANGLE, and so on. Then the resulting TRIANGLE $J^{(\infty)}$ approaches an EQUILATERAL TRIANGLE.

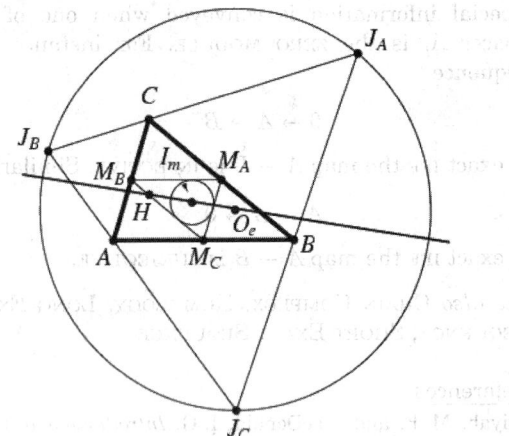

Given a triangle ΔABC, draw the excentral triangle $\Delta J_A J_B J_C$ and MEDIAL TRIANGLE $\Delta M_A M_B M_C$. Then the ORTHOCENTER H of ΔABC, INCENTER I_m of $\Delta M_A M_B M_C$, and CIRCUMCENTER O_e of $\Delta J_A J_B J_C$ are

COLLINEAR with I_m the MIDPOINT of HO_e (Honsberger 1995).

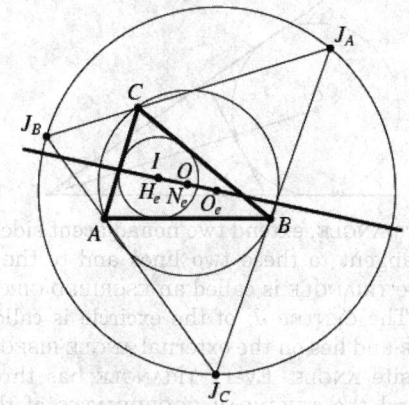

The INCENTER I of ΔABC coincides with the ORTHO-CENTER H_e of $\Delta J_A J_B J_C$, and the CIRCUMCENTER O of ΔABC coincides with the NINE-POINT CENTER N_e of $\Delta J_A J_B J_C$. Furthermore, $N_e = O$ is the MIDPOINT of the line segment joining the ORTHOCENTER H_e and CIRCUMCENTER O_e of $\Delta J_A J_B J_C$ (Honsberger 1995).

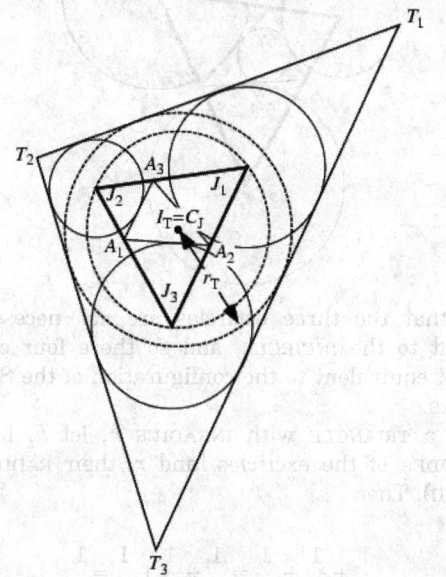

Call T the TRIANGLE tangent externally to the EXCIRCLES of A. Then the INCENTER I_T of K coincides with the CIRCUMCENTER C_J of TRIANGLE $\Delta J_1 J_2 J_3$, where J_i are the EXCENTERS of A. The INRADIUS r_T of the INCIRCLE of T is

$$r_T = 2R + r = \tfrac{1}{2}(r + r_1 + r_2 + r_3),$$

where R is the CIRCUMRADIUS of A, r is the INRADIUS, and r_i are the EXRADII (Johnson 1929, p. 192).

See also EXCENTER, EXCENTER-EXCENTER CIRCLE, EXCIRCLE, GERGONNE POINT, MITTENPUNKT, SODDY CIRCLES

References

Honsberger, R. "A Trio of Nested Triangles." §3.2 in *Episodes in Nineteenth and Twentieth Century Euclidean Geometry.* Washington, DC: Math. Assoc. Amer., pp. 27–0, 1995.

Johnson, R. A. *Modern Geometry: An Elementary Treatise on the Geometry of the Triangle and the Circle.* Boston, MA: Houghton Mifflin, 1929.

Exceptional Binomial Coefficient

A BINOMIAL COEFFICIENT $\binom{N}{k}$ is said to be exceptional if $\mathrm{lpf}\binom{N}{k} > N/k$. The following tables gives the exception binomial coefficients which are also GOOD BINOMIAL COEFFICIENTS, are not OF THE FORM $\binom{N}{N-1}$, and have specified least prime factors $p > 5$.

p	Exceptional Binomial Coefficients
13	$\binom{3574}{406}$
17	$\binom{241}{16}$, $\binom{439}{33}$, $\binom{317}{56}$, $\binom{482}{130}$, $\binom{998}{256}$, $\binom{998}{260}$, $\binom{14273}{896}$, $\binom{13277}{900}$
19	$\binom{62}{6}$, $\binom{959}{56}$
23	$\binom{474}{66}$
29	$\binom{284}{28}$

See also GOOD BINOMIAL COEFFICIENT, LEAST PRIME FACTOR

References

Erdos, P.; Lacampagne, C. B.; and Selfridge, J. L. "Estimates of the Least Prime Factor of a Binomial Coefficient." *Math. Comput.* **61**, 215–24, 1993.

Exceptional Jordan Algebra

A JORDAN ALGEBRA which is not isomorphic to a subalgebra.

See also JORDAN ALGEBRA, SPECIAL JORDAN ALGEBRA

References

Albert, A. A. "A Construction of Exceptional Jordan Division Algebras." *Ann. Math.* **67**, 1–8, 1958.

Albert, A. A. and Jacobson, N. "On Reduced Exceptional Simple Jordan Algebra." *Ann. Math.* **66**, 400–17, 1957.

Exceptional Set of Goldbach Numbers

GOLDBACH NUMBER

Excess

The KURTOSIS of a distribution is sometimes called the excess, or excess coefficient. The term is also used to refer to the quantity

$$e \equiv n - f_0(n, g)$$

for a GRAPH G with n vertices and GIRTH g, where

$$f_0(v, g) = \begin{cases} \dfrac{v(v-1)^r - 2}{v-2} & \text{for } g = 2r+1 \\[3mm] \dfrac{2(v-1)^r - 2}{v-2} & \text{for } g = 2r \end{cases}$$

(Biggs and Ito 1980, Wong 1982). A (v, g)-CAGE GRAPH having $f(v, g) = f_0(v, g)$ vertices (i.e., the minimal number, so that the excess is $e = 0$) is called a MOORE GRAPH.

See also CAGE GRAPH, KURTOSIS, MOORE GRAPH

References

Biggs, N. L. and Ito, T. "Graphs with Even Girth and Small Excess." *Math. Proc. Cambridge Philos. Soc.* **88**, 1–0, 1980.
Wong, P. K. "Cages--A Survey." *J. Graph Th.* **6**, 1–2, 1982.

Excess Coefficient
KURTOSIS

Excessive Number
ABUNDANT NUMBER

Exchange Shuffle

A SHUFFLE of a deck of cards obtained by successively exchanging the cards in position 1, 2, ..., n with cards in randomly chosen positions. For $4 \leq n \leq 17$, the most frequent permutation is $(n, \ldots, m+1)(m, \ldots, 1)$, where $m = n/2$ if n is even and either $(n-1)/2$ or $(n+1)/2$ if n is odd (Goldstine and Moews 2000). Amazingly, for $n \geq 18$ cards, the identity permutation (i.e., the original state before the cards were shuffled) is the most likely (Goldstine and Moews 2000).

See also SHUFFLE

References

Goldstein, D. ad Moews, D. The Identity Is the Most Likely Exchange Shuffle for Large n. 6 Oct 2000. http://xxx.lanl.-gov/abs/math.CO/0010066/.
Robbins, D. P. and Bolker, E. D. "The Bias of Three Pseudo-Random Shuffles." *Aeq. Math* **22**, 268–92, 1981.
Schmidt, F. and Simion, R. "Card Shuffling and a Transformation on S_n." *Aeq. Math* **44**, 11–4, 1992.

Excircle

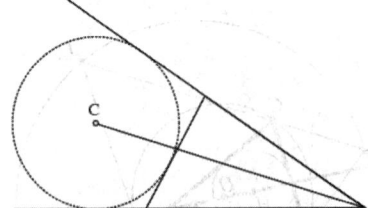

Given a TRIANGLE, extend two nonadjacent sides. The CIRCLE tangent to these two lines and to the other side of the TRIANGLE is called an ESCRIBED CIRCLE, or excircle. The CENTER J_i of the excircle is called the EXCENTER and lies on the external ANGLE BISECTOR of the opposite ANGLE. Every TRIANGLE has three excircles, and the TRILINEAR COORDINATES of the EXCENTERS are $-1:1:1$, $1:-1:1$, and $1:1:-1$. The RADIUS r_i of the excircle i is called its EXRADIUS.

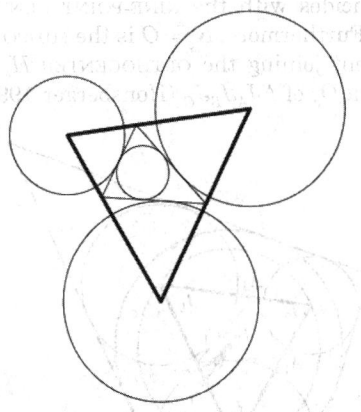

Note that the three excircles are *not* necessarily tangent to the INCIRCLE, and so these four circles are not equivalent to the configuration of the SODDY CIRCLES.

Given a TRIANGLE with INRADIUS r, let h_i be the ALTITUDES of the excircles, and r_i their RADII (the EXRADII). Then

$$\frac{1}{h_1} + \frac{1}{h_2} + \frac{1}{h_3} = \frac{1}{r_1} + \frac{1}{r_2} + \frac{1}{r_3} = \frac{1}{r}$$

(Johnson 1929, p. 189).

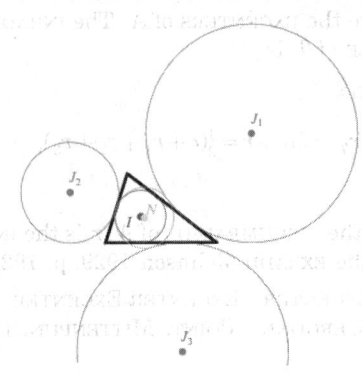

There are four CIRCLES that are tangent all three sides (or their extensions) of a given TRIANGLE: the INCIRCLE I and three excircles J_1, J_2, and J_3. These four circles are, in turn, all touched by the NINE-POINT CIRCLE N.

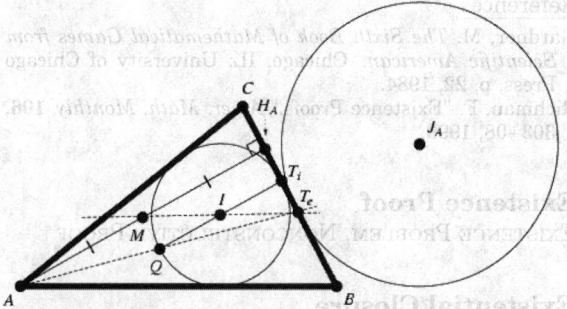

Given a TRIANGLE ΔABC, construct the INCIRCLE with INCENTER I and EXCIRCLE with EXCENTER J_A. Let T_i be the tangent point of ΔABC with its incircle, T_e be the tangent point of ΔABC with its EXCIRCLE J_A, H_A the foot of the ALTITUDE to vertex A, M the MIDPOINT of AH_A, and construct Q such that QT_i is a DIAMETER of the INCIRCLE. Then M, I, and T_e are COLLINEAR, as are A, Q, and T_e (Honsberger 1995).

See also EXCENTER, EXCENTER-EXCENTER CIRCLE, EXCENTRAL TRIANGLE, FEUERBACH'S THEOREM, NAGEL POINT, TRIANGLE TRANSFORMATION PRINCIPLE

References

Coxeter, H. S. M. and Greitzer, S. L. "The Incircle and Excircles." §1.4 in *Geometry Revisited.* Washington, DC: Math. Assoc. Amer., pp. 10–3, 1967.

Honsberger, R. "An Unlikely Collinearity." §3.3 in *Episodes in Nineteenth and Twentieth Century Euclidean Geometry.* Washington, DC: Math. Assoc. Amer., pp. 30–1, 1995.

Johnson, R. A. *Modern Geometry: An Elementary Treatise on the Geometry of the Triangle and the Circle.* Boston, MA: Houghton Mifflin, pp. 176–77 and 182–94, 1929.

Lachlan, R. "The Inscribed and the Escribed Circles." §126–28 in *An Elementary Treatise on Modern Pure Geometry.* London: Macmillian, pp. 72–4, 1893.

Excision Axiom

One of the EILENBERG-STEENROD AXIOMS which states that, if X is a SPACE with SUBSPACES A and U such that the CLOSURE of A is contained in the interior of U, then the INCLUSION MAP $(X\ U, A\ U) \to (X, A)$ induces an isomorphism $H_n(X\ U, A\ U) \to H_n(X, A)$.

Excluded Middle Law

A law in (2-valued) LOGIC which states there is no third alternative to TRUTH or FALSEHOOD. In other words, for any statement A, either A or not-A must be true and the other must be false. This law no longer holds in THREE-VALUED LOGIC or FUZZY LOGIC.

See also BIVALENT, FUZZY LOGIC, THREE-VALUED LOGIC

References

Erickson, G. W. and Fossa, J. A. *Dictionary of Paradox.* Lanham, MD: University Press of America, pp. 64–5, 1998.

Excludent

A method which can be used to solve any QUADRATIC CONGRUENCE EQUATION. This technique relies on the fact that solving

$$x^2 \equiv b \pmod{p}$$

is equivalent to finding a value y such that

$$b + py = x^2.$$

Pick a few small moduli m. If $y \bmod m$ does not make $b + py$ a quadratic residue of m, then this value of y may be excluded. Furthermore, values of $y > p/4$ are never necessary.

See also QUADRATIC CONGRUENCE EQUATION

Excludent Factorization Method

Also known as the difference of squares method. It was first used by Fermat and improved by Gauss. Gauss looked for INTEGERS x and y satisfying

$$y^2 \equiv x^2 - N \pmod{E}$$

for various moduli E. This allowed the exclusion of many potential factors. This method works best when factors are of approximately the same size, so it is sometimes better to attempt mN for some suitably chosen value of m.

See also PRIME FACTORIZATION ALGORITHMS

Exclusion

METHOD OF EXCLUSIONS

Exclusive Disjunction

A DISJUNCTION that is true if only one, but not both, of its arguments are true, and is false if neither or both are true, which is equivalent to the XOR connective.

By contrast, the INCLUSIVE DISJUNCTION is true if either *or both* of its arguments are true. This is equivalent to the OR CONNECTIVE.

See also DISJUNCTION, INCLUSIVE DISJUNCTION, OR, XOR

Exclusive Nor

XNOR

Exclusive Or
XOR

Excosine Circle
If the tangents at B and C to the CIRCUMCIRCLE of a TRIANGLE $\triangle ABC$ intersect in a point K_1, then the CIRCLE with center K_1 and which passes through B and C is called the excosine circle, and cuts AB and AC in two points which are extremities of a DIAMETER.

See also COSINE CIRCLE

References
Lachlan, R. *An Elementary Treatise on Modern Pure Geometry.* London: Macmillian, p. 75, 1893.

Exeter Point
Define A' to be the point (other than the VERTEX A) where the MEDIAN through A meets the CIRCUMCIRCLE of ABC, and define B' and C' similarly. Then the Exeter point is the PERSPECTIVE CENTER of the TRIANGLE $A'B'C'$ and the TANGENTIAL TRIANGLE. It has TRIANGLE CENTER FUNCTION

$$\alpha = a(b^4 + c^4 - a^4).$$

References
Kimberling, C. "Central Points and Central Lines in the Plane of a Triangle." *Math. Mag.* **67**, 163–87, 1994.
Kimberling, C. "Exeter Point." http://cedar.evansville.edu/~ck6/tcenters/recent/exeter.html.
Kimberling, C. and Lossers, O. P. "Problem 6557 and Solution." *Amer. Math. Monthly* **97**, 535–37, 1990.

Exhaustion Method
The method of exhaustion was a INTEGRAL-like limiting process used by Archimedes to compute the AREA and VOLUME of 2-D LAMINA and 3-D SOLIDS.

See also INTEGRAL, LIMIT

Existence
If at least one solution can be determined for a given problem, a solution to that problem is said to exist. Frequently, mathematicians seek to prove the existence of solutions (the EXISTENCE PROBLEM) and then investigate their UNIQUENESS.

See also EXISTENCE PROBLEM, EXISTS, PICARD'S EXISTENCE THEOREM, UNIQUE

Existence Problem
The question of whether a solution to a given problem exists. The existence problem can be solved in the affirmative without actually finding a solution to the original problem. Such a demonstration is said to be nonconstructive, and is called a NONCONSTRUCTIVE PROOF or an existence proof.

See also ENUMERATION PROBLEM, EXISTENCE, NONCONSTRUCTIVE PROOF, PICARD'S EXISTENCE THEOREM

References
Gardner, M. *The Sixth Book of Mathematical Games from Scientific American.* Chicago, IL: University of Chicago Press, p. 22, 1984.
Richman, F. "Existence Proofs." *Amer. Math. Monthly* **106**, 303–08, 1999.

Existence Proof
EXISTENCE PROBLEM, NONCONSTRUCTIVE PROOF

Existential Closure
A class of processes which attempt to round off a domain and simplify its theory by adjoining elements.

See also MODEL COMPLETION

References
Manders, K. L. "Domain Extension and the Philosophy of Mathematics." *J. Philos.* **86**, 553–62, 1989.

Existential Formula
UNIVERSAL FORMULA

Existential Quantifier
The EXISTS QUANTIFIER $\exists$.

See also EXISTS, FOR ALL, GENERAL QUANTIFIER, QUANTIFIER

Existential Sentence

See also UNIVERSAL SENTENCE

References
Carnap, R. *Introduction to Symbolic Logic and Its Applications.* New York: Dover, p. 34, 1958.

Exists
If there exists an A, this is written $\exists A$. Similarly, "A does not exist" is written $\nexists A$. $\exists$ is one of the two mathematical objects known as QUANTIFIERS.

In *Mathematica* 4.0, the command ExistsRealQ[*ineqs*, *vars*] can be used to determine if there exist real values of the variables *vars* satisfying the system of real equations and inequalities *ineqs*.

See also EXISTENCE, FOR ALL, IMPLIES, QUANTIFIER

Exmedian
The line through the VERTEX of a TRIANGLE which is PARALLEL to the opposite side.

References

Johnson, R. A. *Modern Geometry: An Elementary Treatise on the Geometry of the Triangle and the Circle.* Boston, MA: Houghton Mifflin, p. 176, 1929.

Exmedian Point

The point of intersection of two EXMEDIANS.

References

Johnson, R. A. *Modern Geometry: An Elementary Treatise on the Geometry of the Triangle and the Circle.* Boston, MA: Houghton Mifflin, p. 176, 1929.

Exogenous Variable

An economic variable that is related to other economic variables and determines their equilibrium levels.

See also ENDOGENOUS VARIABLE

References

Iyanaga, S. and Kawada, Y. (Eds.). *Encyclopedic Dictionary of Mathematics.* Cambridge, MA: MIT Press, p. 458, 1980.

Exotic R4

Donaldson (1983) showed there exists an exotic smooth DIFFERENTIAL STRUCTURE on $\mathbb{R}^4$. Donaldson's result has been extended to there being precisely a CONTINUUM of nondiffeomorphic DIFFERENTIAL STRUCTURES on $\mathbb{R}^4$.

See also EXOTIC SPHERE, SMOOTH STRUCTURE

References

Donaldson, S. K. "Self-Dual Connections and the Topology of Smooth 4-Manifold." *Bull. Amer. Math. Soc.* **8**, 81–3, 1983.
Monastyrsky, M. *Modern Mathematics in the Light of the Fields Medals.* Wellesley, MA: A. K. Peters, 1997.

Exotic Sphere

Milnor (1963) found more than one smooth structure on the 7-D HYPERSPHERE. Generalizations have subsequently been found in other dimensions. Using SURGERY theory, it is possible to relate the number of DIFFEOMORPHISM classes of exotic spheres to higher homotopy groups of spheres (Kosinski 1992).

Kervaire and Milnor (1963) computed a list of the number $N(d)$ of distinct (up to DIFFEOMORPHISM) DIFFERENTIAL STRUCTURES on spheres indexed by the DIMENSION d of the sphere. For $d = 1$, 2, ..., assuming the POINCARÉ CONJECTURE, they are 1, 1, 1, ≥ 2, 1, 1, 28, 2, 8, 6, 992, 1, 3, 2, 16256, 2, 16, 16, ... (Sloane's A001676). The status of $d = 4$ is still unresolved: at least one exotic structure exists, but it is not known if others do as well.

The only exotic Euclidean spaces are a CONTINUUM of EXOTIC R4 structures.

See also EXOTIC R4, HYPERSPHERE, SMOOTH STRUC-

TURE

References

Kervaire, M. A. and Milnor, J. W. "Groups of Homotopy Spheres: I." *Ann. Math.* **77**, 504–37, 1963.
Kosinski, A. A. §X.6 in *Differential Manifolds.* Boston, MA: Academic Press, 1992.
Milnor, J. "Topological Manifolds and Smooth Manifolds." In *Proc. Internat. Congr. Mathematicians (Stockholm, 1962).* Djursholm: Inst. Mittag-Leffler, pp. 132–38, 1963.
Milnor, J. W. and Stasheff, J. D. *Characteristic Classes.* Princeton, NJ: Princeton University Press, 1973.
Monastyrsky, M. *Modern Mathematics in the Light of the Fields Medals.* Wellesley, MA: A. K. Peters, 1997.
Novikov, S. P. (Ed.). *Topology I.* New York: Springer-Verlag, 1996.
Sloane, N. J. A. Sequences A001676/M5197 in "An On-Line Version of the Encyclopedia of Integer Sequences." http://www.research.att.com/~njas/sequences/eisonline.html.

Exp

EXPONENTIAL FUNCTION

Expansion

An AFFINE TRANSFORMATION (sometimes called an enlargement or dilation) in which the scale is increased. It is the opposite of a CONTRACTION, and is also sometimes called an enlargement. A CENTRAL DILATION corresponds to an expansion plus a TRANSLATION.

See also AFFINE TRANSFORMATION, CENTRAL DILATION, CONTRACTION (GEOMETRY), DILATION, HOMOTHETIC, TRANSFORMATION

References

Coxeter, H. S. M. and Greitzer, S. L. "Dilation." §4.7 in *Geometry Revisited.* Washington, DC: Math. Assoc. Amer., pp. 94–5, 1967.
Hilbert, D. and Cohn-Vossen, S. *Geometry and the Imagination.* New York: Chelsea, p. 13, 1999.

Expansive

Let ϕ be a MAP. Then ϕ is expansive if the statement that the DISTANCE $d(\phi^n x,\ \phi^n y) < \delta$ for all $n \in \mathbb{Z}$ implies that $x = y$. Equivalently, ϕ is expansive if the orbits of two points x and y are never very close.

Expectation Value

The expectation value of a function $f(x)$ in a variable x is denoted $\langle f(x) \rangle$ or $E\{f(x)\}$. For a single discrete variable, it is defined by

$$\langle f(x) \rangle = \sum_x f(x) P(x). \qquad (1)$$

For a single continuous variable it is defined by,

$$\langle f(x) \rangle = \int f(x) P(x)\, dx. \qquad (2)$$

The expectation value satisfies

$$\langle ax + by \rangle = a\langle x \rangle + b\langle y \rangle \tag{3}$$

$$\langle a \rangle = a \tag{4}$$

$$\left\langle \sum x \right\rangle = \sum \langle x \rangle. \tag{5}$$

For multiple discrete variables

$$\langle f(x_1, \ldots, x_n) \rangle$$
$$= \sum_{x_1, \ldots, x_n} f(x_1, \ldots, x_n) P(x_1, \ldots, x_n). \tag{6}$$

For multiple continuous variables

$$\langle f(x_1, \ldots, x_n) \rangle$$
$$= \int f(x_1, \ldots, x_n) P(x_1, \ldots, x_n) \, dx_1 \cdots dx_n. \tag{7}$$

The (multiple) expectation value satisfies

$$\langle (x - \mu_x)(y - \mu_y) \rangle = \langle xy - \mu_x y - \mu_y x + \mu_x \mu_y \rangle$$
$$= \langle xy \rangle - \mu_x \mu_y - \mu_y \mu_x + \mu_x \mu_y$$
$$= \langle xy \rangle - \langle x \rangle \langle y \rangle, \tag{8}$$

where μ_i is the MEAN for the variable i.

See also CENTRAL MOMENT, ESTIMATOR, MAXIMUM LIKELIHOOD, MEAN, MOMENT, RAW MOMENT, WALD'S EQUATION

References

Papoulis, A. "Expected Value; Dispersion; Moments." §5– in *Probability, Random Variables, and Stochastic Processes, 2nd ed.* New York: McGraw-Hill, pp. 139–52, 1984.

Expected Value

EXPECTATION VALUE

Experiment

An experiment $E(S, F, P)$ is defined (Papoulis 1984, p. 30) as a mathematical object consisting of the following elements.

1. A set S (the PROBABILITY SPACE) of elements.
2. A BOREL FIELD F consisting of certain subsets of S called EVENTS.
3. A number $P(X)$ satisfying the PROBABILITY AXIOMS, called the probability, that is assigned to every event A.

See also EVENT, OUTCOME, PROBABILITY AXIOMS, PROBABILITY SPACE, TRIAL

References

Papoulis, A. *Probability, Random Variables, and Stochastic Processes, 2nd ed.* New York: McGraw-Hill, 1984.

Experimental Design

DESIGN

ExpIntegralE

EN-FUNCTION

ExpIntegralEi

EXPONENTIAL INTEGRAL

Exploration Problem

JEEP PROBLEM

Exponent

The POWER p in an expression a^p.

See also BASE (NUMBER), POWER, EXPONENT LAWS, EXPONENT VECTOR, HAUPT-EXPONENT

Exponent Laws

The laws governing the combination of EXPONENTS (POWERS), sometimes called the laws of indices (Higgens 1998). The laws are given by

$$x^m \cdot x^n = x^{m+n} \tag{1}$$

$$\frac{x^m}{x^n} = x^{m-n} \tag{2}$$

$$(x^m)^n = x^{mn} \tag{3}$$

$$(xy)^m = x^m y^m \tag{4}$$

$$\left(\frac{x}{y}\right)^n = \frac{x^n}{y^n} \tag{5}$$

$$x^{-n} = \frac{1}{x^n} \tag{6}$$

$$\left(\frac{x}{y}\right)^{-n} = \left(\frac{y}{x}\right)^n, \tag{7}$$

where quantities in the DENOMINATOR are taken to be nonzero. Special cases include

$$x^1 = x \tag{8}$$

and

$$x^0 = 1 \tag{9}$$

for $x \neq 0$. The definition $0^0 = 1$ is sometimes used to simplify formulas, but it should be kept in mind that this equality is a definition and not a fundamental mathematical truth.

See also EXPONENT, EXPONENTIAL FUNCTION, POWER

References

Higgins, P. M. *Mathematics for the Curious.* Oxford, England: Oxford University Press, 1998.
Krantz, S. G. "Laws of Exponentiation." §1.2.3 in *Handbook of Complex Analysis.* Boston, MA: Birkhäuser, p. 8, 1999.

Exponent Vector

Let p_i denote the ith PRIME, and write

$$m = \prod_i p_i^{v_i}.$$

Then the exponent vector is $\mathbf{v}(m) = (v_1, v_2, \ldots)$.

See also DIXON'S FACTORIZATION METHOD

References

Pomerance, C. "A Tale of Two Sieves." *Not. Amer. Math. Soc.* **43**, 1473–485, 1996.

Exponential

EXPONENTIAL FUNCTION

Exponential Digital Invariant

NARCISSISTIC NUMBER

Exponential Distribution

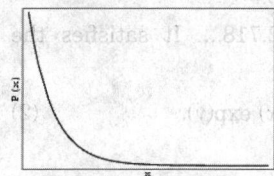 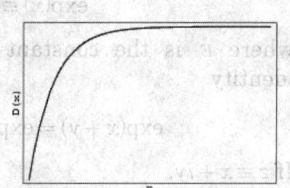

Given a POISSON DISTRIBUTION with rate of change λ, the distribution of waiting times between successive changes (with $k = 0$) is

$$D(x) \equiv P(X \leq x) = 1 - P(X > x)$$

$$= 1 - \frac{(\lambda x)^0 e^{-\lambda x}}{0!} = 1 - e^{-\lambda x} \tag{1}$$

$$P(x) = D'(x) = \lambda e^{-\lambda x}, \tag{2}$$

which is normalized since

$$\int_0^\infty P(x)\, dx = \lambda \int_0^\infty e^{-\lambda x}\, dx$$

$$= -[e^{-\lambda x}]_0^\infty = -(0 - 1) = 1. \tag{3}$$

This is the only MEMORYLESS RANDOM DISTRIBUTION. Define the MEAN waiting time between successive changes as $\theta \equiv \lambda^{-1}$. Then

$$P(x) = \begin{cases} \frac{1}{\theta} e^{-x/\theta} & x \geq 0 \\ 0 & x < 0. \end{cases} \tag{4}$$

The MOMENT-GENERATING FUNCTION is

$$M(t) = \int_0^\infty e^{tx}\left(\frac{1}{\theta}\right) e^{-x/\theta}\, dx = \frac{1}{\theta} \int_0^\infty e^{-(1-\theta t)x/\theta}\, dx$$

$$= \left[\frac{e^{-(1-\theta t)x/\theta}}{1 - \theta t}\right]_0^\infty = \frac{1}{1 - \theta t} \tag{5}$$

$$M'(t) = \frac{\theta}{(1 - \theta t)^2} \tag{6}$$

$$M''(t) = \frac{2\theta^2}{(1 - \theta t)^3}, \tag{7}$$

so

$$R(t) \equiv \ln M(t) = -\ln(1 - \theta t) \tag{8}$$

$$R'(t) = \frac{\theta}{1 - \theta t} \tag{9}$$

$$R''(t) = \frac{\theta^2}{(1 - \theta t)^2} \tag{10}$$

$$\mu = R'(0) = \theta \tag{11}$$

$$\sigma^2 = R''(0) = \theta^2. \tag{12}$$

The CHARACTERISTIC FUNCTION is

$$\phi(t) = \mathscr{F}\{\lambda e^{-\lambda x}[\tfrac{1}{2}(1 + \operatorname{sgn} x)]\} \tag{13}$$

$$= \frac{i\lambda}{t + i\lambda}, \tag{14}$$

where $\mathscr{F}[f]$ is the FOURIER TRANSFORM with parameters $a = b = 1$.

The SKEWNESS and KURTOSIS are given by

$$\gamma_1 = 2 \tag{15}$$

$$\gamma_2 = 6. \tag{16}$$

The MEAN and VARIANCE can also be computed directly

$$\langle x \rangle \equiv \int_0^\infty P(x)\, dx = \frac{1}{s} \int_0^\infty x e^{-x/s}\, dx. \tag{17}$$

Use the integral

$$\int x e^{ax}\, dx = \frac{e^{ax}}{a^2}(ax - 1) \tag{18}$$

to obtain

$$\langle x \rangle = \frac{1}{s}\left[\frac{e^{-x/s}}{\left(-\frac{1}{s}\right)^2}\left\{\left(-\frac{1}{s}\right)x - 1\right\}\right]_0^\infty$$

$$= -s\left[e^{-x/s}\left(1+\frac{x}{s}\right)\right]_0^\infty$$

$$= -s(0-1) = s. \tag{19}$$

Now, to find

$$\langle x^2\rangle = \frac{1}{s}\int_0^\infty x^2 e^{-x/s}\,dx, \tag{20}$$

use the integral

$$\int x^2 e^{-x/s}\,dx = \frac{e^{ax}}{a^3}(2-2ax+a^2x^2) \tag{21}$$

$$\langle x^2\rangle = \frac{1}{s}\left[\frac{e^{-x/s}}{\left(-\frac{1}{s}\right)^3}\left(2+\frac{2}{s}x+\frac{1}{s^2}x^2\right)\right]_0^\infty$$

$$= -s^2(0-2) = 2s^2, \tag{22}$$

giving

$$\sigma^2 \equiv \langle x^2\rangle - \langle x\rangle^2$$

$$= 2s^2 - s^2 = s^2 \tag{23}$$

$$\sigma \equiv \sqrt{\mathrm{var}(x)} = s. \tag{24}$$

If a generalized exponential probability function is defined by

$$P_{(\alpha,\,\beta)}(x) = \frac{1}{\beta}\,e^{-(x-\alpha)/\beta}, \tag{25}$$

for $x \geq \alpha$, then the CHARACTERISTIC FUNCTION is

$$\phi(t) = \frac{e^{i\alpha t}}{1-i\beta t}, \tag{26}$$

and the MEAN, VARIANCE, SKEWNESS, and KURTOSIS are

$$\mu = \alpha + \beta \tag{27}$$

$$\sigma^2 = \beta^2 \tag{28}$$

$$\gamma_1 = 2 \tag{29}$$

$$\gamma_2 = 6. \tag{30}$$

See also DOUBLE EXPONENTIAL DISTRIBUTION

References

Balakrishnan, N. and Basu, A. P. *The Exponential Distribution: Theory, Methods, and Applications.* New York: Gordon and Breach, 1996.

Beyer, W. H. *CRC Standard Mathematical Tables, 28th ed.* Boca Raton, FL: CRC Press, pp. 534–35, 1987.

Spiegel, M. R. *Theory and Problems of Probability and Statistics.* New York: McGraw-Hill, p. 119, 1992.

Exponential Divisor

E-DIVISOR

Exponential Function

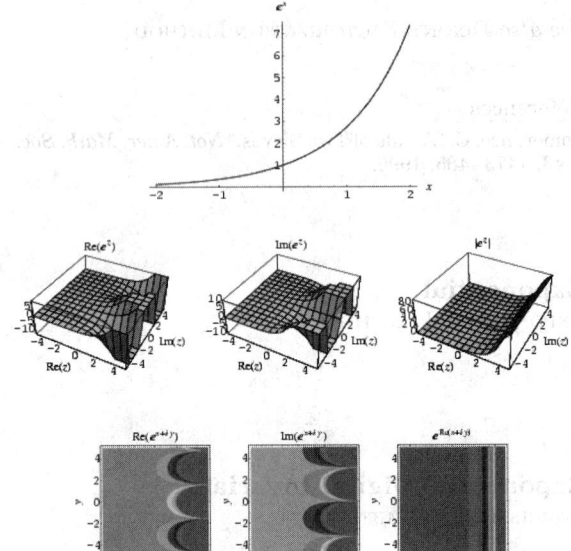

The exponential function is defined by

$$\exp(x) \equiv e^x, \tag{1}$$

where e is the constant 2.718.... It satisfies the identity

$$\exp(x+y) = \exp(x)\exp(y). \tag{2}$$

If $z \equiv x + iy$,

$$e^z = e^{x+iy} = e^x e^{iy} = e^x(\cos y + i\sin y). \tag{3}$$

The exponential function satisfies the identities

$$e^x = \cosh x + \sinh x \tag{4}$$

$$= \sec(\mathrm{gd}\,x) + \tan(\mathrm{gd}\,x) \tag{5}$$

$$= \tan\left(\tfrac{1}{4}\pi + \tfrac{1}{2}\,\mathrm{gd}\,x\right) \tag{6}$$

$$= \frac{1+\sin(\mathrm{gd}\,x)}{\cos(\mathrm{gd}\,x)}, \tag{7}$$

where $\mathrm{gd}\,x$ is the GUDERMANNIAN FUNCTION (Beyer 1987, p. 164; Zwillinger 1995, p. 485).

The exponential function has MACLAURIN SERIES

$$\exp(x) = \sum_{n=0}^\infty \frac{x^n}{n!}, \tag{8}$$

and satisfies the LIMIT

$$\exp(x) = \lim_{n\to\infty}\left(1+\frac{x}{n}\right)^n. \tag{9}$$

If

$$a + bi = e^{x+iy}, \qquad (10)$$

then

$$y = \tan^{-1}\left(\frac{b}{a}\right) \qquad (11)$$

$$x = \ln\left\{ b \csc\left[\tan^{-1}\left(\frac{b}{a}\right)\right]\right\}$$

$$= \ln\left\{ a \sec\left[\tan^{-1}\left(\frac{b}{a}\right)\right]\right\}. \qquad (12)$$

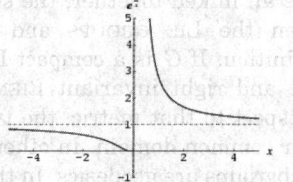

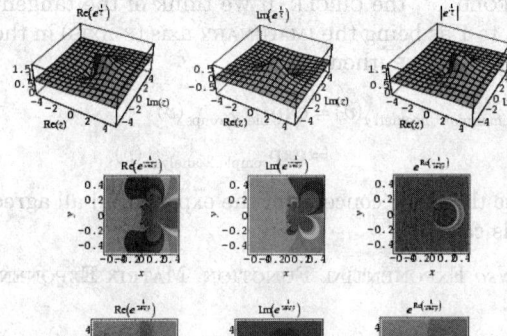

The above plot shows the function $e^{1/z}$.

See also CIS, *E*, EULER FORMULA, EXPONENT LAWS, EXPONENTIAL RAMP, FOURIER TRANSFORM–EXPONENTIAL FUNCTION, GUDERMANNIAN FUNCTION, PHASOR, POWER, SIGMOID FUNCTION

References

Abramowitz, M. and Stegun, C. A. (Eds.). "Exponential Function." §4.2 in *Handbook of Mathematical Functions with Formulas, Graphs, and Mathematical Tables, 9th printing*. New York: Dover, pp. 69–1, 1972.

Beyer, W. H. *CRC Standard Mathematical Tables, 28th ed.* Boca Raton, FL: CRC Press, p. 217, 1987.

Finch, S. "Unsolved Mathematics Problems: Linear Independence of Exponential Functions." http://www.math-soft.com/asolve/sstein/sstein.html.

Fischer, G. (Ed.). Plates 127–28 in *Mathematische Modelle / Mathematical Models, Bildband / Photograph Volume*. Braunschweig, Germany: Vieweg, pp. 124–25, 1986.

Krantz, S. G. "The Exponential and Applications." §1.2 in *Handbook of Complex Analysis*. Boston, MA: Birkhäuser, pp. 7–2, 1999.

Spanier, J. and Oldham, K. B. "The Exponential Function $\exp(bx+c)$" and "Exponentials of Powers $\exp(-ax^\nu)$."

Chs. 26–7 in *An Atlas of Functions*. Washington, DC: Hemisphere, pp. 233–61, 1987.

Yates, R. C. "Exponential Curves." *A Handbook on Curves and Their Properties*. Ann Arbor, MI: J. W. Edwards, pp. 86–7, 1952.

Zwillinger, D. (Ed.). *CRC Standard Mathematical Tables and Formulae*. Boca Raton, FL: CRC Press, 1995.

Exponential Generating Function

An exponential generating function for the integer sequence $a_0, a_1, \ldots$ is a function $E(x)$ such that

$$E(x) = \sum_{k=0}^{\infty} a_k \frac{x^k}{k!} = a_0 + a_1 \frac{x}{1!} + a_2 \frac{x^2}{2!} + \ldots .$$

See also GENERATING FUNCTION

References

Sloane, N. J. A. and Plouffe, S. *The Encyclopedia of Integer Sequences*. San Diego, CA: Academic Press, p. 9, 1995.

Exponential Inequality

For $c < 1$,

$$x^c < 1 + c(x-1).$$

For $c > 1$,

$$x^c > 1 + c(x-1).$$

Exponential Integral

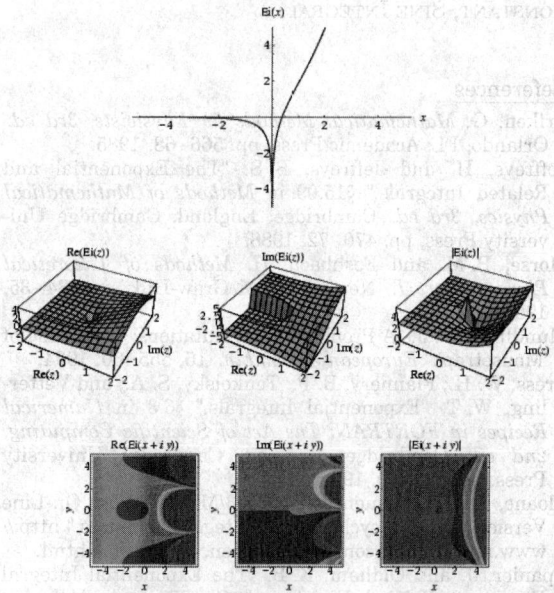

Let $\mathrm{E}_1(x)$ be the E_N-FUNCTION with $n = 1$,

$$\mathrm{E}_1(x) \equiv \int_1^\infty \frac{e^{-tx}\, dt}{t} = \int_x^\infty \frac{e^{-u}\, du}{u}. \qquad (1)$$

Then define the exponential integral ei(x) by

$$E_1(x) = -\text{ei}(-x), \qquad (2)$$

where the retention of the $-\text{ei}(-x)$ NOTATION is a historical artifact. Then ei(x) is given by the integral

$$\text{ei}(x) = -\int_{-x}^{\infty} \frac{e^{-t}\, dt}{t}. \qquad (3)$$

This function is given by the *Mathematica* function ExpIntegralEi[x]. The exponential integral can also be written

$$\text{ei}(ix) = \text{ci}(x) + i\,\text{si}(x), \qquad (4)$$

where ci(x) and si(x) are COSINE and SINE INTEGRAL. The real ROOT of the exponential integral occurs at 0.37250741078..., which is not known to be expressible in terms of other standard constants. The quantity $-e\,\text{ei}(-1) = 0.596347362\ldots$ is known as the GOMPERTZ CONSTANT.

$$\lim_{x \to 0+} \frac{e^{2\text{ei}(-x)}}{x^2} = e^{2\gamma}, \qquad (5)$$

where γ is the EULER-MASCHERONI CONSTANT. The TAYLOR SERIES of ei($-x$) is given by

$$\text{ei}(-x) = \gamma + i\pi + \ln x - x + \tfrac{1}{4}x^2 - \tfrac{1}{18}x^3 + \tfrac{1}{96}x^4 - \tfrac{1}{600}x^5$$
$$+ \ldots, \qquad (6)$$

where the denominators of the coefficients are given by $n \cdot n!$ (Sloane's A001563; van Heemert 1957, Mundfrom 1994).

See also COSINE INTEGRAL, E_n-FUNCTION, GOMPERTZ CONSTANT, SINE INTEGRAL

References

Arfken, G. *Mathematical Methods for Physicists, 3rd ed.* Orlando, FL: Academic Press, pp. 566–68, 1985.

Jeffreys, H. and Jeffreys, B. S. "The Exponential and Related Integrals." §15.09 in *Methods of Mathematical Physics, 3rd ed.* Cambridge, England: Cambridge University Press, pp. 470–72, 1988.

Morse, P. M. and Feshbach, H. *Methods of Theoretical Physics, Part I.* New York: McGraw-Hill, pp. 434–35, 1953.

Mundfrom, D. J. "A Problem in Permutations: The Game of 'Mousetrap'." *European J. Combin.* **15**, 555–60, 1994.

Press, W. H.; Flannery, B. P.; Teukolsky, S. A.; and Vetterling, W. T. "Exponential Integrals." §6.3 in *Numerical Recipes in FORTRAN: The Art of Scientific Computing, 2nd ed.* Cambridge, England: Cambridge University Press, pp. 215–19, 1992.

Sloane, N. J. A. Sequences A001563/M3545 in "An On-Line Version of the Encyclopedia of Integer Sequences." http://www.research.att.com/~njas/sequences/eisonline.html.

Spanier, J. and Oldham, K. B. "The Exponential Integral Ei(x) and Related Functions." Ch. 37 in *An Atlas of Functions.* Washington, DC: Hemisphere, pp. 351–60, 1987.

van Heemert, A. "Cyclic Permutations with Sequences and Related Problems." *J. reine angew. Math.* **198**, 56–2, 1957.

Exponential Map

On a LIE GROUP, exp is a MAP from the LIE ALGEBRA to its LIE GROUP. If you think of the LIE ALGEBRA as the TANGENT SPACE to the identity of the LIE GROUP, exp(v) is defined to be $h(1)$, where h is the unique LIE GROUP HOMEOMORPHISM from the REAL NUMBERS to the LIE GROUP such that its velocity at time 0 is v.

On a RIEMANNIAN MANIFOLD, exp is a MAP from the TANGENT BUNDLE of the MANIFOLD to the MANIFOLD, and exp(v) is defined to be $h(1)$, where h is the unique GEODESIC traveling through the base-point of v such that its velocity at time 0 is v.

The three notions of exp (exp from COMPLEX ANALYSIS, exp from LIE GROUPS, and exp from Riemannian geometry) are all linked together, the strongest link being between the LIE GROUPS and Riemannian geometry definition. If G is a compact LIE GROUP, it admits a left and right invariant RIEMANNIAN METRIC. With respect to that metric, the two exp maps agree on their common domain. In other words, one-parameter subgroups are geodesics. In the case of the MANIFOLD $\mathbb{S}^1$, the CIRCLE, if we think of the tangent space to 1 as being the IMAGINARY axis (Y-AXIS) in the COMPLEX PLANE, then

$$\exp_{\text{Riemannian geometry}}(v) = \exp_{\text{Lie Groups}}(v)$$
$$= \exp_{\text{complex analysis}}(v),$$

and so the three concepts of the exponential all agree in this case.

See also EXPONENTIAL FUNCTION, MATRIX EXPONENTIAL

References

Huang, J.-S. "The Exponential Map." §7.3 in *Lectures on Representation Theory.* Singapore: World Scientific, pp. v, 1999.

Exponential Map Matrix

MATRIX EXPONENTIAL

Exponential Matrix

MATRIX EXPONENTIAL

Exponential Polynomial

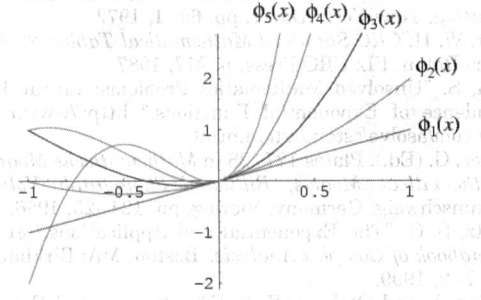

Polynomials $\phi_n(x)$ (sometimes called the BELL POLY-

NOMIALS) which form the associated SHEFFER SE-
QUENCE for

$$f(t) = \ln(1+t), \tag{1}$$

and therefore have GENERATING FUNCTION

$$\sum_{k=0}^{n} \frac{\phi_k(x)}{k!} t^k = e^{(e^t-1)x}. \tag{2}$$

Additional GENERATING FUNCTIONS are given by

$$\phi_n(x) \equiv e^{-x} \sum_{k=0}^{\infty} \frac{k^n x^k}{k!} \tag{3}$$

or

$$\phi_n(x) = x \sum_{k=1}^{n} \binom{n-1}{k-1} \phi_{k-1}(x), \tag{4}$$

with $\phi_0(x) = 1$, where $\binom{n}{k}$ is a BINOMIAL COEFFICIENT. The exponential polynomials have the explicit formula

$$\phi_n(x) = \sum_{k=0}^{n} S(n, k) x^k, \tag{5}$$

where $S(n, k)$ is a STIRLING NUMBER OF THE SECOND KIND. The binomial identity

$$\phi_n(x+y) = \sum_{k=0}^{n} \binom{n}{k} \phi_k(x) \phi_{n-k}(y), \tag{6}$$

where $\binom{n}{k}$ is a BINOMIAL COEFFICIENT, and the recurrence formula is

$$\phi_{n+1}(x) = x[\phi_n(x) + \phi_n'(x)]. \tag{7}$$

The Bell polynomials are defined such that $\phi_n(1) = B_n$, where B_n is a BELL NUMBER. The first few Bell polynomials are

$$\phi_0(x) = 1$$

$$\phi_1(x) = x$$

$$\phi_2(x) = x + x^2$$

$$\phi_3(x) = x + 3x^2 + x^3$$

$$\phi_4(x) = x + 7x^2 + 6x^3 + x^4$$

$$\phi_5(x) = x + 15x^2 + 25x^3 + 10x^4 + x^5$$

$$\phi_6(x) = x + 31x^2 + 90x^3 + 65x^4 + 15x^5 + x^6.$$

See also ACTUARIAL POLYNOMIAL, BELL NUMBER, DOBINSKI'S FORMULA, LAH NUMBER, SHEFFER SE-

QUENCE, STIRLING NUMBER OF THE SECOND KIND

References
Bell, E. T. "Exponential Polynomials." *Ann. Math.* **35**, 258–77, 1934.
Roman, S. "The Exponential Polynomials." §4.1.3. in *The Umbral Calculus.* New York: Academic Press, pp. 63–7, 1984.

Exponential Ramp

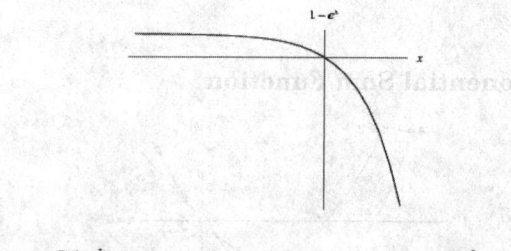

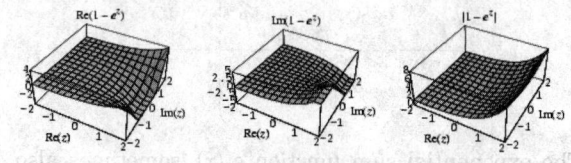

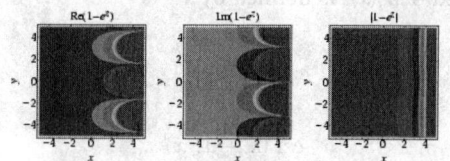

The curve

$$y = 1 - e^{ax}$$

illustrated above.

See also EXPONENTIAL FUNCTION, SIGMOID FUNCTION

References
von Seggern, D. *CRC Standard Curves and Surfaces.* Boca Raton, FL: CRC Press, p. 158, 1993.

Exponential Sum Formulas

$$\sum_{n=0}^{N-1} e^{inx} = \frac{1 - e^{iNx}}{1 - e^{ix}} = \frac{-e^{iNx/2}(e^{-iNx/2} - e^{iNx/2})}{-e^{ix/2}(e^{-ix/2} - e^{ix/2})}$$

$$= \frac{\sin\left(\frac{1}{2}Nx\right)}{\sin\left(\frac{1}{2}x\right)} e^{ix(N-1)/2}, \tag{1}$$

where

$$\sum_{n=0}^{N-1} r^n = \frac{1 - r^N}{1 - r} \tag{2}$$

has been used. Similarly,

$$\sum_{n=0}^{N-1} p^n e^{inx} = \frac{1 - p^N e^{iNx}}{1 - p e^{ix}} \qquad (3)$$

$$\sum_{n=0}^{\infty} p^n e^{inx} = \frac{1}{e^{ipx} - 1} = \frac{1 - p e^{-ix}}{1 - 2p \cos x + p^2}. \qquad (4)$$

By looking at the REAL and IMAGINARY PARTS of these FORMULAS, sums involving sines and cosines can be obtained.

Exponential Sum Function

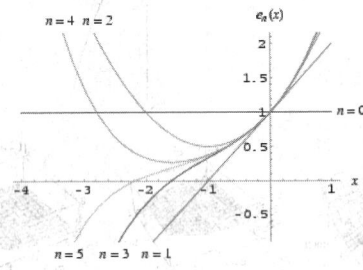

The exponential sum function $e_n(x)$, sometimes also denoted $\exp_n(x)$, is defined by

$$e_n(x) \equiv \sum_{k=0}^{n} \frac{x^k}{k!}$$

$$= \frac{e^x \Gamma(n+1, x)}{\Gamma(n+1)},$$

where $\Gamma(a, x)$ is the upper INCOMPLETE GAMMA FUNCTION and $\Gamma(x)$ is the (complete) GAMMA FUNCTION.

See also GAMMA FUNCTION, INCOMPLETE GAMMA FUNCTION

Exponential Transform

The exponential transform is the transformation of a sequence $a_1, a_2, \dots$ into a sequence $b_1, b_2, \dots$ according to the equation

$$1 + \sum_{n=1}^{\infty} \frac{b_n x^n}{n!} = \exp\left(\sum_{n=1}^{\infty} \frac{a_n x^n}{n!}\right).$$

The inverse ("logarithmic"rpar; transform is then given by

$$\sum_{n=1}^{\infty} \frac{a_n x^n}{n!} = \ln\left(1 + \sum_{n=1}^{\infty} \frac{b_n x^n}{n!}\right).$$

The exponential transform relates the number a_n of labeled CONNECTED GRAPHS on n nodes satisfying some property with the corresponding total number b_n (not necessarily connected) of labeled GRAPHS on n nodes. In this application, the transform is called RIDDELL'S FORMULA for labeled graphs.

See also BINOMIAL TRANSFORM, EULER TRANSFORM, LOGARITHMIC TRANSFORM, MÖBIUS TRANSFORM, RIDDELL'S FORMULA, STIRLING TRANSFORM

References
Sloane, N. J. A. and Plouffe, S. *The Encyclopedia of Integer Sequences.* San Diego, CA: Academic Press, pp. 19–0, 1995.

Expression

See also QUANTITY

Exradius

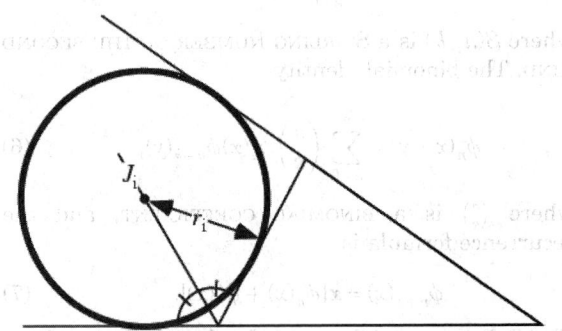

The RADIUS of an EXCIRCLE. Let a TRIANGLE have exradius r_1 (sometimes denoted ρ_1), opposite side of length a_1 and angle α_1, AREA Δ, and SEMIPERIMETER s. Then

$$r_1^2 = \left(\frac{\Delta}{s - a_1}\right)^2 \qquad (1)$$

$$= \frac{s(s - a_2)(s - a_3)}{s - a_1} \qquad (2)$$

$$= 4R \sin\left(\tfrac{1}{2}\alpha_1\right) \cos\left(\tfrac{1}{2}\alpha_2\right) \cos\left(\tfrac{1}{2}\alpha_3\right) \qquad (3)$$

(Johnson 1929, p. 189), where R is the CIRCUMRADIUS. Let r be the INRADIUS, then

$$4R = r_1 + r_2 + r_3 - r \qquad (4)$$

$$\frac{1}{r_1} + \frac{1}{r_2} + \frac{1}{r_3} = \frac{1}{r} \qquad (5)$$

$$r r_1 r_2 r_3 = \Delta^2. \qquad (6)$$

Some fascinating FORMULAS due to Feuerbach are

$$r(r_2 r_3 + r_3 r_1 + r_1 r_2) = s\Delta = r_1 r_2 r_3 \tag{7}$$

$$r(r_1 + r_2 + r_3) = a_2 a_3 + a_3 a_1 + a_1 a_2 - s^2 \tag{8}$$

$$rr_1 + rr_2 + rr_3 + r_1 r_2 + r_2 r_3 + r_3 r_1$$
$$= a_2 a_3 + a_3 a_1 + a_1 a_2 \tag{9}$$

$$r_2 r_3 + r_3 r_1 + r_1 r_2 - rr_1 - rr_2 - rr_3 = \tfrac{1}{2}(a_1^2 + a_2^2 + a_3^2) \tag{10}$$

(Johnson 1929, pp. 190–91).

See also CIRCLE, CIRCUMRADIUS, EXCIRCLE, INRADIUS, RADIUS

References

Johnson, R. A. *Modern Geometry: An Elementary Treatise on the Geometry of the Triangle and the Circle.* Boston, MA: Houghton Mifflin, 1929.
Mackay, J. S. "Formulas Connected with the Radii of the Incircle and Excircles of a Triangle." *Proc. Edinburgh Math. Soc.* **12**, 86–05.
Mackay, J. S. "Formulas Connected with the Radii of the Incircle and Excircles of a Triangle." *Proc. Edinburgh Math. Soc.* **13**, 103–04.

Exsecant

$$\operatorname{exsec} x \equiv \sec x - 1,$$

where $\sec x$ is the SECANT.

See also COVERSINE, HAVERSINE, SECANT, VERSINE

References

Abramowitz, M. and Stegun, C. A. (Eds.). *Handbook of Mathematical Functions with Formulas, Graphs, and Mathematical Tables, 9th printing.* New York: Dover, p. 78, 1972.

Extended Binary Tree

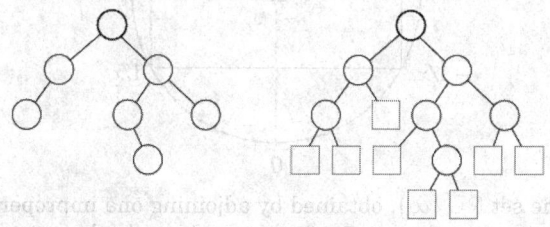

A BINARY TREE in which special nodes are added wherever a null subtree was present in the original tree so that each node in the original tree (except the root node) has degree three (Knuth 1997, p. 399).

See also BINARY TREE

References

Knuth, D. E. *The Art of Computer Programming, Vol. 1: Fundamental Algorithms, 3rd ed.* Reading, MA: Addison-Wesley, 1997.

Extended Complex Plane

The COMPLEX PLANE with a POINT AT INFINITY attached: $\mathbb{C} \cup \{\infty\}$, where ∞ denotes COMPLEX INFINITY. The extended complex plane is denoted $\mathbb{C}*$.

See also $\mathbb{C}*$, COMPLEX INFINITY, COMPLEX PLANE, RIEMANN SPHERE

References

Krantz, S. G. "The Topology of the Extended Complex Plane." §6.3.2 in *Handbook of Complex Analysis.* Boston, MA: Birkhäuser, p. 83, 1999.

Extended Cycloid

PROLATE CYCLOID

Extended Goldbach Conjecture

GOLDBACH CONJECTURE

Extended Greatest Common Divisor

GREATEST COMMON DIVISOR

Extended Mean-Value Theorem

Let the functions f and g be DIFFERENTIABLE on the OPEN INTERVAL (a, b) and CONTINUOUS on the CLOSED INTERVAL $[a, b]$. If $g'(x) \neq 0$ for any $x \in (a, b)$, then there is at least one point $c \in (a, b)$ such that

$$\frac{f'(c)}{g'(c)} = \frac{f(b) - f(a)}{g(b) - g(a)}.$$

See also MEAN-VALUE THEOREM

Extended Real Number (Affine)

This entry contributed by DAVID W. CANTRELL

The set $\mathbb{R} \cup \{+\infty, -\infty\}$ obtained by adjoining two improper elements to the set $\mathbb{R}$ of real numbers is normally called the set of (affinely) extended real numbers. Although the notation for this set is not completely standardized, $\bar{\mathbb{R}}$ is commonly used. The set may also be written in interval notation as $[-\infty, +\infty]$. With an appropriate topology, $\bar{\mathbb{R}}$ is the two-point COMPACTIFICATION (or affine closure) of $\mathbb{R}$. The improper elements, the affine infinities $+\infty$ and $-\infty$, correspond to ideal points of the number line. Note that these improper elements are *not* real numbers, and that this system of extended real numbers is not a FIELD.

Instead of writing $+\infty$, many authors write simply ∞. However, the compound symbol $+\infty$ will be used here to represent the positive improper element of $\bar{\mathbb{R}}$, allowing the individual symbol ∞ to be used unambiguously to represent the unsigned improper element of $\mathbb{R}^*$, the one-point COMPACTIFICATION (or projective closure) of $\mathbb{R}$.

A very important property of $\bar{\mathbb{R}}$, which $\mathbb{R}$ lacks, is that every subset S of $\bar{\mathbb{R}}$ has an INFIMUM (greatest lower bound) and a SUPREMUM (least upper bound). In particular, sup $\varnothing = -\infty$ and, if S is unbounded above, then sup $S = +\infty$. Similarly, inf $\varnothing = +\infty$ and, if S is unbounded below, then inf $S = -\infty$.

Order relations can be extended from $\mathbb{R}$ to $\bar{\mathbb{R}}$, and arithmetic operations can be partially extended. For $x \in \bar{\mathbb{R}}$,

$$-\infty < x < +\infty \text{ if } x \neq \pm\infty, \ -\infty < +\infty \quad (1)$$

$$-(+\infty) = -\infty, \ -(-\infty) = +\infty \quad (2)$$

$$x + (+\infty) = +\infty + x = +\infty \text{ if } x \neq -\infty \quad (3)$$

$$x + (-\infty) = -\infty + x = -\infty \text{ if } x \neq +\infty \quad (4)$$

$$x \cdot (\pm\infty) = \pm\infty \cdot x = \pm\infty \text{ if } x > 0 \quad (5)$$

$$x \cdot (\pm\infty) = \pm\infty \cdot x = \mp\infty \text{ if } x < 0 \quad (6)$$

$$\frac{x}{\pm\infty} = 0 \text{ if } x \neq \pm\infty \quad (7)$$

$$\left|\frac{x}{0}\right| = +\infty \text{ if } x \neq 0 \quad (8)$$

However, the expressions $+\infty + (-\infty)$, $-\infty + (+\infty)$, and $x/0$ are UNDEFINED.

The above statements which define results of arithmetic operations on $\bar{\mathbb{R}}$ may be considered as abbreviations of statements about determinate LIMIT forms. For example, $-(+\infty) = -\infty$ may be considered as an abbreviation for "If x increases without bound, then $-x$ decreases without bound." Most descriptions of $\bar{\mathbb{R}}$ also make a statement concerning the products of the improper elements and 0, but there is no consensus as to what that statement should be. Some authors (e.g., Kolmogorov 1995, p. 193) state that, like $+\infty + (-\infty)$ and $-\infty + (+\infty)$, $0 \cdot (\pm\infty)$ and $\pm\infty \cdot 0$ should be UNDEFINED, presumably because of the INDETERMINATE status of the corresponding LIMIT forms. Other authors (such as McShane 1983, p. 2) accept $0 \cdot (\pm\infty) = \pm\infty \cdot 0 = 0$, at least as a convention which is useful in certain contexts.

Many results for other operations and functions can be obtained by considering determinate LIMIT forms. For example, a partial extension of the function $f(x, y) = x^y$ can be obtained for $x, y \in \bar{\mathbb{R}}$ as

$$(+\infty)^y = \begin{cases} 0 & \text{if } y < 0 \\ +\infty & \text{if } y > 0 \end{cases} \quad (9)$$

$$x^{+\infty} = \begin{cases} 0 & \text{if } 0 < x < 1 \\ +\infty & \text{if } x > 1 \end{cases} \quad (10)$$

$$x^{-\infty} = \begin{cases} +\infty & \text{if } 0 < x < 1 \\ 0 & \text{if } x > 1. \end{cases} \quad (11)$$

The functions e^x and $\ln|x|$ can be fully extended to $\bar{\mathbb{R}}$, with

$$e^{-\infty} = 0 \quad (12)$$

$$e^{+\infty} = +\infty \quad (13)$$

$$\ln|0| = -\infty \quad (14)$$

$$\ln|\pm\infty| = +\infty. \quad (15)$$

Some other important functions (e.g., $\tanh(\pm\infty) = \pm 1$ and $\tan^{-1}(\pm\infty) = \pm\pi/2$) can be extended to $\bar{\mathbb{R}}$, while others (e.g., $\sin x$, $\cos x$) cannot. Evaluations of expressions involving $+\infty$ and $-\infty$, derived by considering determinate LIMIT forms, are routinely used by computer algebra systems such as *Mathematica* when performing simplifications.

See also CLOSURE (SET), COMPACTIFICATION, EXTENDED REAL NUMBER (PROJECTIVE), INDETERMINATE, LIMIT, R, R-, R+, REAL NUMBER

References

Kolmogorov, N. A. "Infinity." *Encyclopaedia of Mathematics: An Updated and Annotated Translation of the Soviet "Mathematical Encyclopaedia,"* 2nd ed., Vol. 3. (Managing Ed. M. Hazewinkel). Dordrecht, Netherlands: Reidel, 1995.

McShane, E. J. *Unified Integration.* Orlando, FL: Academic Press, p. 2, 1983.

Extended Real Number (Projective)

This entry contributed by DAVID W. CANTRELL

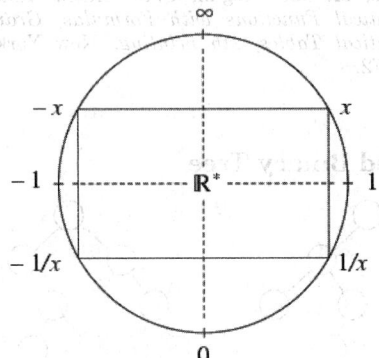

The set $\mathbb{R} \cup \{\infty\}$, obtained by adjoining one improper element to the set $\mathbb{R}$ of real numbers, is the set of projectively extended real numbers. Although notation is not completely standardized, $\mathbb{R}^*$ is used here to denote this set of extended real numbers. With an appropriate topology, $\mathbb{R}^*$ is the one-point COMPACTIFICATION (or projective closure) of $\mathbb{R}$. As shown above, the cross section of the RIEMANN SPHERE consisting of its "real axis" and "north pole" can be used to visualize $\mathbb{R}^*$. The improper element, projective infinity (∞), then corresponds with the ideal point, the "north pole."

In contrast to the signed affine infinities ($+\infty$ and $-\infty$) of the affinely EXTENDED REAL NUMBERS $\bar{\mathbb{R}}$, projective infinity, ∞, is unsigned, like 0. Regrettably, ∞ is also unordered, i.e., for $x \in \mathbb{R}^*$ it can be said neither that $x < \infty$ nor that $x > \infty$. For this reason, $\mathbb{R}^*$ is used much less often in real analysis than is $\bar{\mathbb{R}}$. Thus, if context is not specified, "the extended real numbers" normally refers to $\bar{\mathbb{R}}$, not $\mathbb{R}^*$.

Arithmetic operations can be partially extended from $\mathbb{R}$ to $\mathbb{R}^*$,

$$-(\infty) = \infty, \ x + \infty = \infty + x = \infty \quad \text{if } x \neq \infty,$$

$$x \cdot \infty = \infty \cdot x = \infty \quad \text{if } x \neq 0,$$

$$x/\infty = 0 \quad \text{if } x \neq \infty,$$

and

$$x/0 = \infty \quad \text{if } x \neq 0$$

(by contrast, $x/0$ is UNDEFINED in $\bar{\mathbb{R}}$). The expressions K_n and $0 \cdot \infty$ are most often left UNDEFINED in $\mathbb{R}^*$.

The exponential function e^x cannot be extended to $\mathbb{R}^*$. On the other hand, $\mathbb{R}^*$ is useful when dealing with rational functions and certain other functions. For example, if $\mathbb{R}^*$ is used as the range of $\tan x$, then by taking $\tan((2n + 1)\pi/2) = \infty$ for integer n, the domain of the function can be extended to all of $\mathbb{R}$. Extended real numbers are sometimes used in the implementation of FLOATING-POINT ARITHMETIC (Hauser 1996, pp. 158–59).

See also COMPACTIFICATION, CLOSURE (SET), EXTENDED REAL NUMBER (AFFINE), REAL NUMBER, RIEMANN SPHERE

References

Hauser, J. R. "Handling Floating-Point Exceptions in Numeric Programs." *ACM Trans. Program. Lang. Sys.* **18**, 139–74, 1996. http://www.cs.berkeley.edu/~jhauser/exceptions/HandlingFloatingPointExceptions.html.

Hazewinkel, M. (Managing Ed.). *Encyclopaedia of Mathematics: An Updated and Annotated Translation of the Soviet "Mathematical Encyclopaedia,"* Vol. 3. Dordrecht, Netherlands: Reidel, p. 193, 1988.

Extended Riemann Hypothesis

The first quadratic nonresidue mod p of a number is always less than $2(\ln p)^2$.

See also RIEMANN HYPOTHESIS

References

Bach, E. *Analytic Methods in the Analysis and Design of Number-Theoretic Algorithms.* Cambridge, MA: MIT Press, 1985.

Wagon, S. *Mathematica in Action.* New York: W. H. Freeman, p. 295, 1991.

ExtendedGCD

GREATEST COMMON DIVISOR

Extension (Ideal)

The extension of $\mathfrak{a}$, an IDEAL in COMMUTATIVE RING A, in a RING B, is the IDEAL generated by its image $f(\mathfrak{a})$ under a RING HOMOMORPHISM f. Explicitly, it is any finite sum OF THE FORM $\Sigma\, y_i f(x_i)$ where y_i is in B and x_i is in $\mathfrak{a}$. Sometimes the extension of $\mathfrak{a}$ is denoted $\mathfrak{a}^e$.

The image $f(\mathfrak{a})$ may not be an ideal if f is not SURJECTIVE. For instance, $f : \mathbb{Z} \to \mathbb{Z}[x]$ is a ring homomorphism and the image of the even integers is not an ideal since it does not contain any non-constant polynomials. The extension of the even integers in this case is the set of polynomials with even coefficients.

The extension of a PRIME IDEAL may not be prime. For example, consider $f : \mathbb{Z} \to \mathbb{Z}[\sqrt{2}]$. Then the extension of the even integers is not a prime ideal since $2 = \sqrt{2} \cdot \sqrt{2}$.

See also ALGEBRAIC NUMBER THEORY, CONTRACTION (IDEAL), IDEAL, PRIME IDEAL, RING

References

Atiyah, M. F. and MacDonald, I. G. *Introduction to Commutative Algebra.* Reading, MA: Addison-Wesley, pp. 9–0, 1969.

Extension (Set)

The definition of a SET by enumerating its members. An extensional definition can always be reduced to an INTENTIONAL one.

An EXTENSION FIELD is sometimes also called simply an extension.

See also EXTENSION FIELD, INTENSION

References

Russell, B. "Definition of Number." *Introduction to Mathematical Philosophy.* New York: Simon and Schuster, 1971.

Extension Field

A FIELD K is said to be an extension field (or field extension, or extension), denoted K/F, of a field F if F is a SUBFIELD of K. The COMPLEX NUMBERS are an extension field of the REAL NUMBERS, and the REAL NUMBERS are an extension field of the RATIONAL NUMBERS.

The DEGREE) (or relative degree, or index) of an extension field K/F, denoted $[K : F]$, is the dimension of K as a VECTOR SPACE over F, i.e.,

$$[K : F] = \dim_F K.$$

See also DEGREE (EXTENSION FIELD), FIELD, PYTHAGOREAN EXTENSION, SPLITTING FIELD, SUBFIELD

References
Dummit, D. S. and Foote, R. M. "Basic Theory of Field Extensions." §13.1 in *Abstract Algebra, 2nd ed.* Englewood Cliffs, NJ: Prentice-Hall, pp. 422–32, 1998.

Extension Problem

Given a SUBSPACE A of a SPACE X and a MAP from A to a SPACE Y, is it possible to extend that MAP to a MAP from X to Y?

See also LIFTING PROBLEM

Extensions Calculus

EXTERIOR ALGEBRA

Extent

The RADIUS of the smallest CIRCLE centered at one of the points of an N-CLUSTER, which contains all the points in the N-CLUSTER.

See also N-CLUSTER

Exterior

That portion of a region lying "outside" a specified boundary.

See also INTERIOR

Exterior Algebra

The ALGEBRA of the EXTERIOR PRODUCT, also called an alternating algebra or Grassmann algebra. The study of exterior algebra is also called Ausdehnungslehre and extensions calculus. Exterior algebras are GRADED ALGEBRAS.

In particular, the exterior algebra of a VECTOR SPACE is the DIRECT SUM over k in the natural numbers of the VECTOR SPACES of alternating k-forms on that VECTOR SPACE. The product on this algebra is then the wedge product of forms. The exterior algebra for a VECTOR SPACE V is constructed by forming monomials u, $v \wedge w$, $x \wedge y \wedge z$, etc., where u, v, w, x, y, and z are vectors in V and $\wedge$ is asymmetric multiplication. The sums formed from LINEAR COMBINATIONS of the MONOMIALS are the elements of an exterior algebra.

The exterior algebra of a VECTOR SPACE can also be described as a QUOTIENT VECTOR SPACE,

$$\Lambda^p V = \otimes^p V / W_p, \qquad (1)$$

where W_p is the subspace of p-tensors generated by transpositions such as $W_2 = \langle x \otimes y + y \otimes x \rangle$ and $\otimes$ denotes the TENSOR PRODUCT. The EQUIVALENCE CLASS $[x_1 \otimes \ldots \otimes x_p]$ is denoted $x_1 \wedge \ldots \wedge x_p$. For instance,

$$x \wedge y + y \wedge x = 0, \qquad (2)$$

since the representatives add to an element of W_2. Consequently, $x \wedge y = -y \wedge x$. Sometimes $\Lambda^p V$ is called the pth exterior power of V, and may also be denoted by $\mathrm{Alt}^p V$.

The alternating products are a SUBSPACE of the tensor products. Define the linear map

$$\mathrm{Alt} : \otimes^p V \to \otimes^p V \qquad (3)$$

by

$$\mathrm{Alt}(v_{i_1} \otimes \ldots \otimes v_{i_p}) = \frac{1}{p!} \sum_\sigma \pi(\sigma) v_{i_{\sigma(1)}} \otimes \ldots \otimes v_{i_{\sigma(p)}}, \qquad (4)$$

where σ ranges over all PERMUTATIONS of $\{1, \ldots, p\}$, and $\pi(\sigma)$ is the signature of the PERMUTATION, given by the PERMUTATION SYMBOL. Then $\Lambda^p V$ is the image of Alt, as W_p is its NULLSPACE. The constant factor $1/p!$, which is sometimes not used, makes Alt into a PROJECTION OPERATOR.

For example, if V has the BASIS $\{e_1, e_2, e_3, e_4\}$, then

$$\Lambda^0 V = \langle 1 \rangle \qquad (5)$$

$$\Lambda^1 V = \langle e_1, e_2, e_3, e_4 \rangle \qquad (6)$$

$$\Lambda^2 V = \langle e_1 \wedge e_2, e_1 \wedge e_3, e_1 \wedge e_4, e_2 \wedge e_3, e_2 \wedge e_4, e_3 \wedge e_4 \rangle \qquad (7)$$

$$\Lambda^3 V = \langle e_1 \wedge e_2 \wedge e_3, e_1 \wedge e_2 \wedge e_4, e_1 \wedge e_3 \wedge e_3, \times e_2 \wedge e_3 \wedge e_4 \rangle \qquad (8)$$

$$\Lambda^4 V = \langle e_1 \wedge e_2 \wedge e_3 \wedge e_4 \rangle, \qquad (9)$$

and $\Lambda^k V = \{0\}$ where $k > \dim V$. For a general VECTOR SPACE V of dimension n, the space $\Lambda^p V$ has dimension $\binom{n}{p}$.

Here is a *Mathematica* function that implements the Alt operator, whose image is the alternating subspace of the p-tensors.

```
Alt[x_] := Module[
   {p = TensorRank[x], perms},
   perms = Permutations[Range[p]];
   Sum[
      Signature[perms[[i]]]   Transpose[x,
perms[[i]]],
     {i, p!}
   ]/p!
  ]
```

Here is a *Mathematica* function which tests whether a p-tensor is alternating by testing transpositions.

```
Transpositions[n_] := Module[{i},
  Table[Range[n] /. {i -> i + 1, i + 1 -> i},
{i, n - 1}]
 ] AlternatingQ[a_] := (And[##1] &) @@ ((a
 ==         -Transpose[a,      #1]     &)     /@
Transpositions[TensorRank[a]])
```

The space $\Lambda^* = \otimes_p \Lambda^p V$ becomes an ALGEBRA with the WEDGE PRODUCT, defined using the function Alt. Also, if $T : V \to W$ is a LINEAR TRANSFORMATION, then the map $T^*,_p : \Lambda^p V \to \Lambda^p W$ sends $v_1 \wedge \ldots \wedge v_p$ to $T(v_1) \wedge \ldots \wedge T(v_p)$. If $n = \dim V$ and $T(v) = Av$ where A is a SQUARE MATRIX, then $T^*,_n (e_1 \wedge \ldots \wedge e_n) = (\det A) e_1 \wedge \ldots \wedge e_n$.

The alternating algebra, also called the exterior algebra, $\Lambda^* V$ is a 2^n dimensional ALGEBRA. In *Mathematica*, an element of the alternating algebra can be represented by an n-nested binary list. For example, $\{\{\{1, 2\}, \{0, 0\}\}, \{\{3, 0\}, \{4, 5\}\}\}$ represents $e_1 \wedge e_2 \wedge e_3 + 2e_1 \wedge e_3 + 3e_2 \wedge e_3 + 4e_3 + 5$. The WEDGE PRODUCT can defined by the following *Mathematica* function

```
sgntmp[a_, b_] := (-1)^(Mod[Sum[b[[i]], {i,
Length[b]}], 2]) a sgn[a_] := Module[{d =
TensorRank[a]},
  MapIndexed[sgntmp, a, {d}]
] wedge[{a_, b_}, {c_, d_}] := Module[{rnk
= TensorRank[a]},
  If[rnk == 0,
  {a d + b c, b d},
  {wedge[a, d] + wedge[ sgn[b], c], wedge[b,
d]}}
  ]
]
```

The following *Mathematica* function gives the p powers of an element a in the exterior algebra as a tensor.

```
ExtToTensor[a_, p_] := Module[{d =
TensorRank[a], tmp, ind, indices},
  tmp = Table[2, {d}];
  If[p == 0, (a[[##1]] &) @@ tmp,
  Array[
  (Block[{b},
    b = {##1};
        ind = ReplacePart[tmp, 1,
Transpose[{b}]];
     Signature[b]/p! (a[[##1]] &) @@ ind] &),
   Table[d, {p}]]]
]
```

The rank of an alternating form has a couple different definitions. The rank of a form, used in studying integral manifolds of differential ideals, is the dimension of its ENVELOPE. Another definition is its rank as a TENSOR.

The DIFFERENTIAL K-FORMS in modern geometry are an exterior algebra, and play a role in multivariable calculus. In general, it is only necessary for V to have the structure of a MODULE. So exterior algebras come up in REPRESENTATION THEORY. For example, if V is a REPRESENTATION of a group G, then $\text{Sym}_2 V \otimes \Lambda^2 V$ is a decomposition of $V \otimes V$ into two representations.

See also DIFFERENTIAL FORM, ENVELOPE (FORM), REPRESENTATION, SYMMETRIC GROUP, TENSOR PRODUCT, VECTOR SPACE, WEDGE PRODUCT

References

Flanders, H. *Differential Forms with Applications to the Physical Sciences.* New York: Academic Press, 1963.
Forder, H. G. *The Calculus of Extension.* Cambridge, England: Cambridge University Press, 1941.
Fulton, W. and Harris, J. *Representation Theory.* New York: Springer-Verlag, pp. 472–75, 1991.
Lounesto, P. "Counterexamples to Theorems Published and Proved in Recent Literature on Clifford Algebras, Spinors, Spin Groups, and the Exterior Algebra." http://www.hit.fi/~lounesto/counterexamples.htm.
Peano, G. *Geometric Calculus According to the Ausdehnungslehre of H. Grassmann.* Boston: Birkhäuser, 2000.
Sternberg, S. *Differential Geometry.* New York: Chelsea, pp. 14–0, 1983.

Exterior Angle

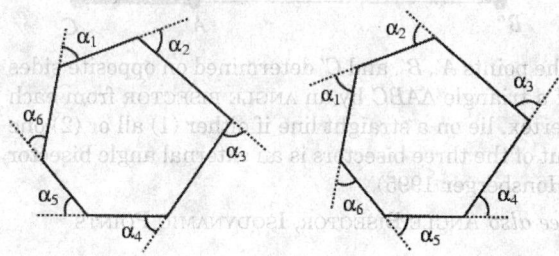

The angle α_i formed between a side of a polygon and the extension of an adjacent side. Since there are two directions in which a side can be extended, there are two exterior angles at each vertex. However, since corresponding angles are opposite, they are also equal. The sum of exterior angles in a convex polygon is equal to 2π RADIANS ($360°$), since this corresponds to one complete rotation of the polygon.

See also ANGLE, EXTERIOR ANGLE BISECTOR

Exterior Angle Bisector

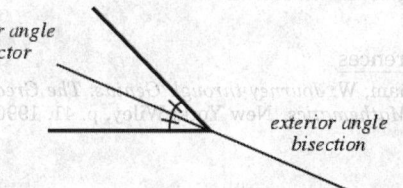

The exterior bisector of an ANGLE is the LINE or LINE SEGMENT which cuts it into two equal ANGLES on the opposite "side" as the ANGLE.

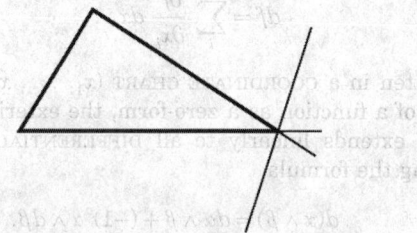

For a TRIANGLE, the exterior angle bisector bisects the SUPPLEMENTARY ANGLE at a given VERTEX. It also

divides the opposite side externally in the ratio of adjacent sides.

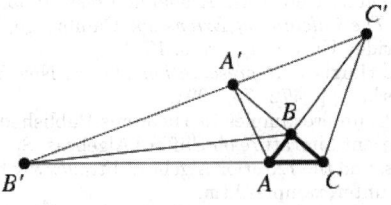

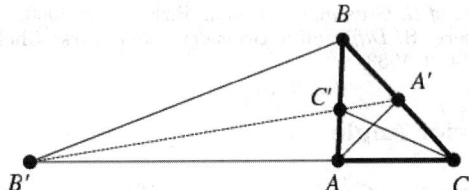

The points A', B', and C' determined on opposite sides of a triangle $\triangle ABC$ by an ANGLE BISECTOR from each vertex, lie on a straight line if either (1) all or (2) one out of the three bisectors is an external angle bisector (Honsberger 1995).

See also ANGLE BISECTOR, ISODYNAMIC POINTS

References
Coxeter, H. S. M. and Greitzer, S. L. *Geometry Revisited.* Washington, DC: Math. Assoc. Amer., p. 12, 1967.
Honsberger, R. *Episodes in Nineteenth and Twentieth Century Euclidean Geometry.* Washington, DC: Math. Assoc. Amer., pp. 149–50, 1995.

Exterior Angle Theorem
In any TRIANGLE, if one of the sides is extended, the exterior angle is greater than both the interior and opposite angles.

See also EXTERIOR ANGLE

References
Dunham, W. *Journey through Genius: The Great Theorems of Mathematics.* New York: Wiley, p. 41, 1990.

Exterior Derivative
The exterior derivative of a function f is the ONE-FORM

$$df = \sum_i \frac{\partial f}{\partial x_i}\, dx_i \tag{1}$$

written in a COORDINATE CHART $(x_1, \ldots, x_n)$. Thinking of a function as a zero-form, the exterior derivative extends linearly to all DIFFERENTIAL K-FORMS using the formula

$$d(\alpha \wedge \beta) = d\alpha \wedge \beta + (-1)^p \alpha \wedge d\beta,$$

when α is a k-form and where $\wedge$ is the WEDGE PRODUCT.

The exterior derivative of a k-form is a $(k+1)$-form. For example, for a DIFFERENTIAL K-FORM

$$\omega^1 = b_1\, dx_1 + b_2\, dx_2, \tag{2}$$

the exterior derivative is

$$d\omega^1 = db_1 \wedge dx_1 + db_2 \wedge dx_2. \tag{3}$$

Similarly, consider

$$\omega^1 = b_1(x_1,\, x_2)\, dx_1 + b_2(x_1,\, x_2)\, dx_2. \tag{4}$$

Then

$$\begin{aligned}
d\omega^1 &= db_1 \wedge dx_1 + db_2 \wedge dx_2 \\
&= \left(\frac{\partial b_1}{\partial x_1}\, dx_1 + \frac{\partial b_1}{\partial x_2}\, dx_2 \right) \wedge dx_1 \\
&\quad + \left(\frac{\partial b_2}{\partial x_1}\, dx_1 + \frac{\partial b_2}{\partial x_2}\, dx_2 \right) \wedge dx_2.
\end{aligned} \tag{5}$$

Denote the exterior derivative by

$$Dt \equiv \frac{\partial}{\partial x} \wedge t. \tag{6}$$

Then for a 0-form t,

$$(Dt)_\mu \equiv \frac{\partial t}{\partial x^\mu}, \tag{7}$$

for a 1-form t,

$$(Dt)_{\mu\nu} \equiv \frac{1}{2}\left(\frac{\partial t_\nu}{\partial x^\mu} - \frac{\partial t_\mu}{\partial x^\nu} \right), \tag{8}$$

and for a 2-form t,

$$(Dt)_{ijk} \equiv \tfrac{1}{3}\, \epsilon_{ijk} \left(\frac{\partial t_{23}}{\partial x^1} + \frac{\partial t_{31}}{\partial x^2} + \frac{\partial t_{12}}{\partial x^3} \right), \tag{9}$$

where ϵ_{ijk} is the PERMUTATION TENSOR.

It is always the case that $d(d\alpha) = 0$. When $d\alpha = 0$, then α is called a CLOSED FORM. A TOP-DIMENSIONAL FORM is always a CLOSED FORM. When $\alpha = d\eta$ then α is called an EXACT FORM, so any EXACT FORM is also CLOSED. An example of a CLOSED FORM which is not EXACT is $d\theta$ on the circle. Since θ is a function defined up to a constant multiple of 2π, $d\theta$ is a WELL DEFINED ONE-FORM, but there is no function for which it is the EXTERIOR DERIVATIVE.

The exterior derivative is linear and commutes with the PULLBACK ω^* of DIFFERENTIAL K-FORMS ω. That is,

$$df^*(\alpha) = f^*(d\alpha). \tag{10}$$

Hence the PULLBACK of a CLOSED FORM is closed and the PULLBACK of an EXACT FORM is exact. Moreover, a DE RHAM COHOMOLOGY class $[\alpha]$ has a WELL DEFINED PULLBACK MAP $[f^*(\alpha)]$.

In *Mathematica*, a k-form can be written as an ANTISYMMETRIC k-tensor. Using this format, the following *Mathematica* function computes the exterior derivative of the form a in the (ordered) variables vars.

```
Alt[x_List] := Module[
  {
    p = TensorRank[x],      perms
  },
  perms = Permutations[Range[p]];
    Sum[Signature[perms[[i]]]  Transpose[x,
perms[[i]]],{i, p!}]/p!
  ] ExtD1[a_List, vars_?List] :=
  Alt[Outer[D[#2, #1] &, vars , a]]
```

It is also possible to use an n-nested binary tree to represent the algebra of differential forms. Using this format, the following *Mathematica* function computes the exterior derivative recursively.

```
ExtD2[{a_List, b_List}, vars_List] :=
  {D[b, First[vars]] - ExtD2[a, Rest[vars]],
ExtD2[b, Rest[vars]]}  ExtD2[{a_?(! ListQ[#1]
&), b_?(! ListQ[#1] &)}, var_?ListQ] :=
  {D[b, First[var]], 0}
```

See also DIFFERENTIAL κ-FORM, EXTERIOR ALGEBRA, HODGE STAR, JACOBIAN, MANIFOLD, POINCARÉ'S LEMMA, STOKES' THEOREM, TANGENT BUNDLE, TENSOR, WEDGE PRODUCT

References

Berger, M. *Differential Geometry.* New York: Springer-Verlag, p. 152, 1988.
Spivak, M. *A Comprehensive Introduction to Differential Geometry, Vol. 1, 2nd ed.* Houston, TX: Publish or Perish Press, pp. 286–05, 1999.
Sternberg, S. *Differential Geometry.* New York: Chelsea, pp. 99–04, 1983.

Exterior Dimension

A type of DIMENSION which can be used to characterize FAT FRACTALS.

See also FAT FRACTAL

References

Grebogi, C.; McDonald, S. W.; Ott, E.; and Yorke, J. A. "Exterior Dimension of Fat Fractals." *Phys. Let. A* **110**, 1–, 1985.
Grebogi, C.; McDonald, S. W.; Ott, E.; and Yorke, J. A. Erratum to "Exterior Dimension of Fat Fractals." *Phys. Let. A* **113**, 495, 1986.
Ott, E. *Chaos in Dynamical Systems.* New York: Cambridge University Press, p. 98, 1993.

Exterior Power

The kth exterior power of an element α in an EXTERIOR ALGEBRA ΛV is given by the WEDGE PRODUCT of α with itself k times. Note that if α has odd degree, then any higher power of α must be zero. The

situation for even degree forms is different. For example, if

$$\alpha = e_1 \wedge e_2 + e_3 \wedge e_4 + e_5 \wedge e_6, \qquad (1)$$

then

$$\alpha^2 = 2e_1 \wedge e_2 \wedge e_3 \wedge e_4 + 2e_1 \wedge e_2 \wedge e_5 \wedge e_6 + 2e_3 \wedge e_4 \\ \wedge e_5 \wedge e_6 \quad (2)$$

$$\alpha^3 = 6e_1 \wedge e_2 \wedge e_3 \wedge e_4 \wedge e_5 \wedge e_6, \qquad (3)$$

$$\alpha^4 = 0. \qquad (4)$$

See also EXTERIOR ALGEBRA, WEDGE PRODUCT

Exterior Product

WEDGE PRODUCT

Exterior Snowflake

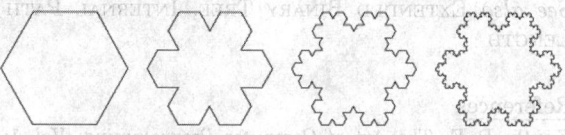

The FRACTAL illustrated above.

See also FLOWSNAKE FRACTAL, KOCH ANTISNOWFLAKE, KOCH SNOWFLAKE, PENTAFLAKE

References

Wagon, S. *Mathematica in Action.* New York: W. H. Freeman, pp. 193–95, 1991.
Weisstein, E. W. "Fractals." MATHEMATICA NOTEBOOK FRACTAL.M.

External Contact

TANGENT EXTERNALLY

External Direct Product

The term external direct product is used to refer to either the EXTERNAL DIRECT SUM of groups under the group operation of multiplication, or over infinitely many spaces in which the sum is not required to be finite. In the latter case, the operation is also called the CARTESIAN PRODUCT.

See also CARTESIAN PRODUCT, EXTERNAL DIRECT SUM

External Direct Sum

The CARTESIAN PRODUCT of a finite or infinite set of modules over a ring with only finitely many nonzero entries in each sequence.

See also CARTESIAN PRODUCT, EXTERNAL DIRECT PRODUCT

External Path Length

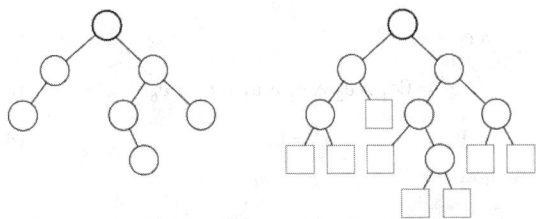

The sum over all external (square) nodes of the paths from the root of an EXTENDED BINARY TREE to each node. For example, in the tree above, the external path length is 25 (Knuth 1997, p. 399–00). The INTERNAL and external path lengths are related by

$$E = I + 2n,$$

where n is the number of internal nodes.

See also EXTENDED BINARY TREE, INTERNAL PATH LENGTH

References
Knuth, D. E. *The Art of Computer Programming, Vol. 1: Fundamental Algorithms, 3rd ed.* Reading, MA: Addison-Wesley, 1997.

External Tensor Product

Suppose that V is a REPRESENTATION of G, and W is a REPRESENTATION of H. Then the TENSOR PRODUCT $V \otimes W$ is a REPRESENTATION of the GROUP DIRECT PRODUCT $G \times H$. An element (g, h) of $G \times H$ acts on a basis element $v \otimes w$ by

$$(g, h)(v \otimes w) = gv \otimes hw.$$

To distinguish from the TENSOR PRODUCT of representations, the external tensor product is denoted $V \bar{\otimes} W$, although the only possible confusion would occur when $G = H$.

When V and W are IRREDUCIBLE REPRESENTATIONS of G and H respectively, then so is the external tensor product. In fact, all IRREDUCIBLE REPRESENTATIONS of $G \times H$ arise as external direct products of IRREDUCIBLE REPRESENTATIONS.

See also GROUP, IRREDUCIBLE REPRESENTATION, REPRESENTATION, TENSOR PRODUCT (REPRESENTATION), TENSOR PRODUCT (VECTOR SPACE), VECTOR SPACE

Externally Tangent

TANGENT EXTERNALLY

Extra Strong Lucas Pseudoprime

Given the LUCAS SEQUENCE $U_n(b, -1)$ and $V_n(b, -1)$, define $\Delta = b^2 - 4$. Then an extra strong Lucas pseudoprime to the base b is a COMPOSITE NUMBER $n = 2^r s + (\Delta/n)$, where s is ODD and $(n, 2\Delta) = 1$ such that either $U_s \equiv 0 \pmod{n}$ and $V_s \equiv \pm 2 \pmod{n}$, or $V_{2^t s} \equiv 0 \pmod{n}$ for some t with $0 \leq t < r - 1$. An extra strong Lucas pseudoprime is a STRONG LUCAS PSEUDOPRIME with parameters $(b, -1)$. COMPOSITE n are extra strong pseudoprimes for at most 1/8 of possible bases (Grantham 1997).

See also LUCAS PSEUDOPRIME, STRONG LUCAS PSEUDOPRIME

References
Grantham, J. "Frobenius Pseudoprimes." http://www.clark.net/pub/grantham/pseudo/pseudo1.ps
Grantham, J. "A Frobenius Probable Prime Test with High Confidence." 1997. http://www.clark.net/pub/grantham/pseudo2.ps
Jones, J. P. and Mo, Z. "A New Primality Test Using Lucas Sequences." Preprint.

Extrapolation

RICHARDSON EXTRAPOLATION

Extremal Coloring

EXTREMAL GRAPH

Extremal Graph

In general, an extremal graph is the largest graph of order n which does not contain a given graph G as a SUBGRAPH (Skiena 1990, p. 143). Turán studied extremal graphs that do not contain a COMPLETE GRAPH K_p as a SUBGRAPH.

One much-studied type of extremal graph is a two-coloring of a COMPLETE GRAPH K_n of n nodes which contains exactly the number $N \equiv (R + B)_{\min}$ of MONOCHROMATIC FORCED TRIANGLES and no more (i.e., a minimum of $R + B$ where R and B are the numbers of red and blue TRIANGLES). Goodman (1959) showed that for an extremal graph of this type,

$$N(n) = \begin{cases} \frac{1}{3} m(m-1)(m-2) & \text{for } n = 2m \\ \frac{1}{3} 2m(m-1)(4m+1) & \text{for } n = 4m+1 \\ \frac{1}{3} 2m(m+1)(4m-1) & \text{for } n = 4m+3. \end{cases}$$

This is sometimes known as GOODMAN'S FORMULA. Schwenk (1972) rewrote it in the form

$$N(n) = \binom{n}{3} - \left\lfloor \frac{1}{2} n \left\lfloor \frac{1}{4}(n-1)^2 \right\rfloor \right\rfloor,$$

sometimes known as SCHWENK'S FORMULA, where $\lfloor x \rfloor$ is the FLOOR FUNCTION. The first few values of $N(n)$

for $n = 1, 2, \ldots$ are 0, 0, 0, 0, 0, 2, 4, 8, 12, 20, 28, 40, 52, 70, 88, ... (Sloane's A014557).

See also Bichromatic Graph, Blue-Empty Graph, Extremal Graph Theory, Goodman's Formula, Monochromatic Forced Triangle, Schwenk's Formula, Turán Graph

References

Goodman, A. W. "On Sets of Acquaintances and Strangers at Any Party." *Amer. Math. Monthly* **66**, 778–83, 1959.

Schwenk, A. J. "Acquaintance Party Problem." *Amer. Math. Monthly* **79**, 1113–117, 1972.

Skiena, S. *Implementing Discrete Mathematics: Combinatorics and Graph Theory with Mathematica.* Reading, MA: Addison-Wesley, p. 143, 1990.

Sloane, N. J. A. Sequences A014557 in "An On-Line Version of the Encyclopedia of Integer Sequences." http://www.research.att.com/~njas/sequences/eisonline.html.

Extremal Graph Theory

The study of how the intrinsic structure of graphs ensures certain types of properties (e.g., clique-formation and graph colorings) under appropriate conditions.

See also Erdos-Stone Theorem, Extremal Graph, Ramsey Theory, Structural Ramsey Theory, Szemerédi's Regularity Lemma, Turán Graph, Turán's Theorem

References

Bollobás, B. *Extremal Graph Theory.* New York: Academic Press, 1978.

Bollobás, B. *Extremal Graph Theory with Emphasis on Probabilistic Methods.* Providence, RI: Amer. Math. Soc., 1986.

Skiena, S. *Implementing Discrete Mathematics: Combinatorics and Graph Theory with Mathematica.* Reading, MA: Addison-Wesley, p. 143, 1990.

Extremals

A field of extremals is a plane region which is simply connected by a one-parameter family of extremals. The concept was invented by Weierstrass.

Extreme and Mean Ratio

Golden Mean

Extreme Value Distribution

N.B. A detailed online essay by S. Finch was the starting point for this entry.

Let M_n denote the "extreme" (i.e., largest) order statistic $X^{\langle n \rangle}$ for a distribution of n elements X_i taken from a continuous uniform distribution. Then the distribution of the M_n is

$$P(M_n < x) = \begin{cases} 0 & \text{if } x < 0 \\ x^n & \text{if } 0 \le x \le 1 \\ 1 & \text{if } x > \end{cases} \qquad (1)$$

and the mean and variance are

$$\mu = \frac{n}{n+1} \qquad (2)$$

$$\sigma^2 = \frac{n}{(n+1)^2 (n+2)}. \qquad (3)$$

If X_i are taken from a standard normal distribution, then its cumulative distribution is

$$F(x) = \frac{1}{\sqrt{2x}} \int_{-\infty}^{x} e^{-t^2/2}\, dt = \tfrac{1}{2} + \Phi(x), \qquad (4)$$

where $\Phi(x)$ is the normal distribution function. The probability distribution of M_n is then

$$P(M_n < x) = [F(x)]^n = \frac{n}{\sqrt{2n}} \int_{-\infty}^{x} [F(t)]^{n-1} e^{-t^2/2}\, dt. \qquad (5)$$

The mean $\mu(n)$ and variance $\sigma^2(n)$ are expressible in closed form for small n,

$$\mu(1) = 0 \qquad (6)$$

$$\mu(2) = \frac{1}{\sqrt{\pi}} \qquad (7)$$

$$\mu(3) = \frac{3}{2\sqrt{\pi}} \qquad (8)$$

$$\mu(4) = \frac{3}{2\sqrt{\pi}} \left[1 + \frac{2}{\pi} \sin^{-1}\left(\tfrac{1}{3}\right) \right] \qquad (9)$$

$$\mu(5) = \frac{5}{4\sqrt{\pi}} \left[1 + \frac{6}{\pi} \sin^{-1}\left(\tfrac{1}{3}\right) \right] \qquad (10)$$

and

$$\sigma^2(1) = 1 \qquad (11)$$

$$\sigma^2(2) = 1 - \frac{1}{\pi} \qquad (12)$$

$$\sigma^2(3) = \frac{4\pi - 9 + 2\sqrt{3}}{4\pi} \qquad (13)$$

$$\sigma^2(4) = 1 + \frac{\sqrt{3}}{\pi} - [\mu(4)]^2 \qquad (14)$$

$$\sigma^2(5) = 1 + \frac{5\sqrt{3}}{4\pi} + \frac{5\sqrt{3}}{2\pi^2} \sin^{-1}\left(\tfrac{1}{4}\right) - [\mu(5)]^2. \qquad (15)$$

No exact expression is known for $\mu(6)$ or $\sigma^2(6)$, but there is an equation connecting them

$$[\mu(6)]^2 + \sigma^2(6) = 1 + \frac{5\sqrt{3}}{4\pi} + \frac{15\sqrt{3}}{2\pi^2} \sin^{-1}\left(\tfrac{1}{4}\right). \qquad (16)$$

An analog to the central limit theorem states that the asymptotic normalized distribution of M_n satisfies one of the three distributions

$$P(y) = \exp(-e^{-y}) \qquad (17)$$

$$P(y) = \begin{cases} 0 & \text{if } y \le 0 \\ \exp[-(-y^{-a})] & \text{if } y > 0 \end{cases} \qquad (18)$$

$$P(y) = \begin{cases} \exp[-(-y)^{a}] & \text{if } y \le 0 \\ 1 & \text{if } y > 0, \end{cases} \qquad (19)$$

also known as GUMBEL, Fréchet, and WEIBULL DISTRIBUTIONS, respectively.

See also FISHER-TIPPETT DISTRIBUTION, ORDER STATISTIC

References

Balakrishnan, N. and Cohen, A. C. *Order Statistics and Inference.* New York: Academic Press, 1991.
David, H. A. *Order Statistics, 2nd ed.* New York: Wiley, 1981.
Finch, S. "Favorite Mathematical Constants." http://www.mathsoft.com/asolve/constant/extval/extval.html.
Gibbons, J. D. and Chakraborti, S. *Nonparametric Statistical Inference, 3rd rev. ext. ed.* New York: Dekker, 1992.

Extreme Value Theorem

If a function $f(x)$ is continuous on a closed interval $[a, b]$, then $f(x)$ has both a MAXIMUM and a MINIMUM on $[a, b]$. If $f(x)$ has an extreme value on an open interval (a, b), then the extreme value occurs at a CRITICAL POINT. This theorem is sometimes also called the WEIERSTRASS EXTREME VALUE THEOREM.

Extremum

A MAXIMUM or MINIMUM. An extremum may be LOCAL (a.k.a. a RELATIVE EXTREMUM; an extremum in a given region which is not the overall MAXIMUM or MINIMUM) or GLOBAL. Functions with many extrema can be very difficult to GRAPH. Notorious examples include the functions $\cos(1/x)$ and $\sin(1/x)$ near $x = 0$

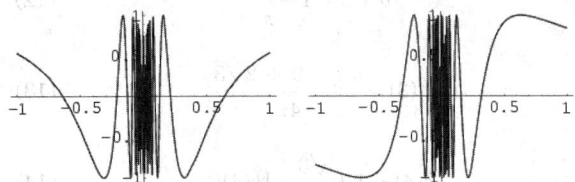

and $\sin(e^{2x+9})$ near 0 and 1.

The latter has

$$\left\lfloor \frac{e^{11}}{\pi} - \frac{1}{2} \right\rfloor - \left\lfloor \frac{e^{9}}{\pi} - \frac{1}{2} \right\rfloor + 1 = 19085 - 2579 + 1 = 16480$$

extrema in the CLOSED INTERVAL $[0,1]$ (Mulcahy 1996).

See also GLOBAL EXTREMUM, GLOBAL MAXIMUM, GLOBAL MINIMUM, KUHN-TUCKER THEOREM, LAGRANGE MULTIPLIER, LOCAL EXTREMUM, LOCAL MAXIMUM, LOCAL MINIMUM, MAXIMUM, MINIMUM

References

Abramowitz, M. and Stegun, C. A. (Eds.). *Handbook of Mathematical Functions with Formulas, Graphs, and Mathematical Tables, 9th printing.* New York: Dover, p. 14, 1972.
Mulcahy, C. "Plotting and Scheming with Wavelets." *Math. Mag.* **69**, 323–43, 1996.
Tikhomirov, V. M. *Stories About Maxima and Minima.* Providence, RI: Amer. Math. Soc., 1991.

Extremum Test

Consider a function $f(x)$ in 1-D. If $f(x)$ has a relative extremum at (x_0), then either $f'(x_0) = 0$ or f is not DIFFERENTIABLE at (x_0). Either the first or second DERIVATIVE tests may be used to locate relative extrema of the first kind.

A NECESSARY condition for $f(x)$ to have a MINIMUM (MAXIMUM) at (x_0) is

$$f'(x_0) = 0,$$

and

$$f''(x_0) \ge 0 \quad (f''(x_0) \le 0).$$

A SUFFICIENT condition is $f'(x_0) = 0$ and $f''(x_0) > 0$ $(f''(x_0) < 0)$. Let $f'(x_0) = 0$, $f''(x_0) = 0$, ..., $f^{(n)}(x_0) = 0$, but $f^{(n+1)}(x_0) \ne 0$. Then $f(x)$ has a RELATIVE MAXIMUM at (x_0) if n is ODD and $f^{(n+1)}(x_0) > 0$, and $f(x)$ has a RELATIVE MINIMUM at (x_0) if n is ODD and $f^{(n+1)}(x_0) > 0$. There is a SADDLE POINT at (x_0) if n is EVEN.

See also EXTREMUM, FIRST DERIVATIVE TEST, RELATIVE MAXIMUM, RELATIVE MINIMUM, SADDLE POINT (FUNCTION), SECOND DERIVATIVE TEST

Extrinsic Curvature

A curvature of a SUBMANIFOLD of a MANIFOLD which depends on its particular EMBEDDING. Examples of extrinsic curvature include the CURVATURE and TORSION of curves in 3-space, or the mean curvature of surfaces in 3-space.

See also CURVATURE, INTRINSIC CURVATURE, MEAN CURVATURE

Eyeball Theorem

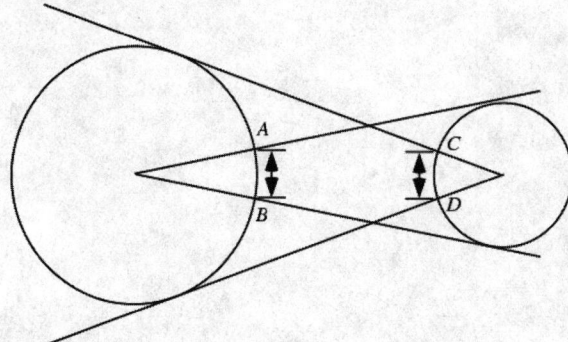

Given two circles, draw the tangents from the center of each circle to the sides of the other. Then the line segments AB and CD are of equal length.

See also CIRCLE

References

Wells, D. *The Penguin Dictionary of Curious and Interesting Geometry.* London: Penguin, p. 70, 1991.

F

Faá di Bruno's Formula

If $f(t)$ and $g(t)$ are functions for which all necessary derivatives are defined, then

$$D^n f(g(t)) =$$

$$\sum \frac{n!}{k_1! \cdots k_n!} (D^k f)(g(t)) \left(\frac{D_g(t)}{1!}\right)^{k_1} \cdots \left(\frac{D^n g(t)}{n!}\right)^{k_n},$$

where $k = k_1 + \ldots + k_n$ and the sum of over all $k_1, \ldots, k_n$ for which

$$k_1 + 2k_2 + \ldots + nk_n = n$$

(Roman 1980).

See also LEIBNIZ IDENTITY, UMBRAL CALCULUS

References

Bertrand, J. *Cours de calcul différentiel er intégral, tome I.* Paris: Gauthier-Villars, p. 138, 1864.

Cesàro. "Dérivées des fonctions de fonctions." *Nouvelles Ann.* **4**, 41–5, 1885.

Comtet, L. *Advanced Combinatorics: The Art of Finite and Infinite Expansions, rev. enl. ed.* Dordrecht, Netherlands: Reidel, pp. 137–39, 1974.

Dederick. "Successive Derivatives of a Function of Several Functions." *Ann. Math.* **27**, 385–94, 1926.

Faá di Bruno. "Sullo sviluppo delle funzione." *Ann. di Scienze Matem. et Fisiche di Tortoloni* **6**, 479–80, 1855.

Faá di Bruno. "Note sur un nouvelle formule de calcul différentiel." *Quart. J. Math.* **1**, 359–60, 1857.

Français. "Du calcul des dérivations rameré à ses véritables principes...." *Ann. Gergonne* **6**, 61–11, 1815.

Joni, S. A. and Rota, C.-G. "The Faá di Bruno Bialgebra." §IX in "Coalgebras and Bialgebras in Combinatorics." *Umbral Calculus and Hopf Algebras. Contemp. Math.* **6**, 18–1, 1982.

Jordan, C. *Calculus of Finite Differences, 3rd ed.* New York: Chelsea, p. 33, 1965.

Knuth, D. E. *The Art of Computer Programming, Vol. 1: Fundamental Algorithms, 3rd ed.* Reading, MA: Addison-Wesley, p. 50, 1997.

Marchand. "Sur le changement de variables." *Ann. École Normale Sup.* **3**, 137–88 and 343–88, 1886.

Riordan, J. *An Introduction to Combinatorial Analysis.* New York: Wiley, pp. 35–7, 1958.

Roman, S. "The Formula of Faa di Bruno." *Amer. Math. Monthly* **87**, 805–09, 1980.

Teixeira. "Sur les dérivées d'ordre quelconque." *Giornale di Matem. di Battaglini* **18**, 306–16, 1880.

Wall. "On the n-th Derivative of $f(x)$." *Bull. Amer. Math. Soc.* **44**, 395–98, 1938.

Faber Polynomial

Let

$$f(x) = z + a_1 + a_2 z^{-1} + a_3 z^{-2} + \ldots = z \sum_{n=0}^{\infty} a_n z^{-n}$$

$$\equiv zg(1/z) \qquad (1)$$

be a LAURENT POLYNOMIAL with $a_0 = 1$. Then the Faber polynomial $P_m(f)$ in $f(z)$ of degree m is defined such that

$$P_m(f) = z^m + c_{m1} z^{-1} + c_{m2} z^{-2} + \ldots = z^m + G_m(1/z), \quad (2)$$

where

$$G_m(x) = \sum_{n=1}^{\infty} c_{mn} x^n \qquad (3)$$

(Schur 1945). Writing

$$[g(x)]^m = \sum_{k=0}^{\infty} a_{mk} x^l \qquad (4)$$

for $m = 1, 2, \ldots$ gives the relationship

$$a_{m,m+n} = c_{mn} + a_{m1} c_{m-1,n} + a_{m2} c_{m-2,n}$$

$$+ \ldots + a_{m,m-1} c_{1n}. \qquad (5)$$

connecting a_{mn} and c_{mn}.

This polynomial can be used to calculate the number of LATTICE PATHS from a point $(r, 0)$ to a point (a, b) that remain below the line $y = cx$.

See also LATTICE PATH

References

Gessel, I. M. Ree, S. "Lattice Paths and Faber Polynomials." In *Advances in Combinatorial Methods and Applications to Probability and Statistics* (Ed. N. Balakrishnan). Boston, MA: Birkhäuser, 1997.

Pommerenke, C. "Über die Faberschen Polynome schlichter Funktionen." *Math. Z.* **85**, 197–08, 1964.

Schiffer, M. "Faber Polynomials in the Theory of Univalent Functions." *Bull. Amer. Math. Soc.* **54**, 503–17, 1948.

Schur, I. "On Faber Polynomials." *Amer. J. Math.* **67**, 33–1, 1945.

Fabry Imbedding

A representation of a PLANAR GRAPH as a planar straight line graph such that no two EDGES cross.

See also PLANAR GRAPH

Face

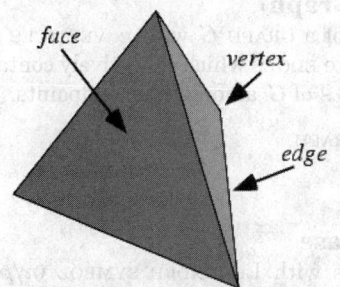

The intersection of an n-D POLYTOPE with a tangent HYPERPLANE. 0-D faces are known as VERTICES (nodes), 1-D faces as EDGES, $(n-2)$-D faces as RIDGES, and $(n-1)$-D faces as FACETS.

See also EDGE (POLYHEDRON), FACET, POLYTOPE, RIDGE, VERTEX (POLYHEDRON)

Face-Regular Polyhedron

JOHNSON SOLID

Facet

An $(n-1)$-D FACE of an n-D POLYTOPE. A procedure for generating facets is known as FACETING.

Faceting

Using a set of corners of a SOLID that lie in a plane to form the VERTICES of a new POLYGON is called faceting. Such POLYGONS may outline new FACES that join to enclose a new SOLID, even if the sides of the POLYGONS do not fall along EDGES of the original SOLID.

References

Holden, A. *Shapes, Space, and Symmetry.* New York: Columbia University Press, p. 94, 1971.

Factor

A factor is a portion of a quantity, usually an INTEGER or POLYNOMIAL that, when MULTIPLIED by all other factors, give the entire quantity. The determination of factors is called FACTORIZATION (or sometimes "FACTORING"). It is usually desired to break factors down into the smallest possible pieces so that no factor is itself factorable. For INTEGERS, the determination of factors is called PRIME FACTORIZATION. For large quantities, the determination of all factors is usually very difficult except in exceptional circumstances.

See also DIVISOR, FACTORIZATION, GREATEST PRIME FACTOR, LEAST PRIME FACTOR, MULTIPLICATION, POLYNOMIAL FACTORIZATION, PRIME FACTORIZATION, PRIME FACTORIZATION ALGORITHMS

Factor (Graph)

A 1-factor of a GRAPH G with n VERTICES is a set of $n/2$ separate EDGES which collectively contain all n of the VERTICES of G among their endpoints.

See also GRAPH

Factor Base

The primes with LEGENDRE SYMBOL $(n/p)=1$ (less than $N=\pi(d)$ for trial divisor d) which need be considered when using the QUADRATIC SIEVE factorization method.

See also DIXON'S FACTORIZATION METHOD

References

Morrison, M. A. and Brillhart, J. "A Method of Factoring and the Factorization of F_7." *Math. Comput.* **29**, 183–05, 1975.

Factor Group

QUOTIENT GROUP

Factor Level

A grouping of statistics.

Factor Ring

QUOTIENT RING

Factor Space

QUOTIENT SPACE

Factorial

The factorial $n!$ is defined for a POSITIVE INTEGER n as

$$n! \equiv \begin{cases} n \cdot (n-1) \cdots 2 \cdot 1 & n = 1, 2, \ldots \\ 1 & n = 0. \end{cases} \quad (1)$$

The factorial $n!$ gives the number of ways in which n objects can be permuted. For example, $3! = 6$, since the six possible permutations of $\{1,2,3\}$ are $\{1,2,3\}$, $\{1,3,2\}$, $\{2,1,3\}$, $\{2,3,1\}$, $\{3,1,2\}$, $\{3,2,1\}$. Since there is a single permutation of zero elements (the EMPTY SET $\varnothing$), $0! = 1$. The first few factorials for $\lfloor n = 0, 1, 2, \ldots$ are 1, 1, 2, 6, 24, 120, ... (Sloane's A000142). An older NOTATION for the factorial is n (Mellin 1909; Lewin 1958, p. 19; Dudeney 1970; Gardner 1978; Conway and Guy 1996).

As n grows large, factorials begin acquiring tails of trailing ZEROS. To calculate the number Z of trailing ZEROS for $n!$, use

$$Z = \sum_{k=1}^{k_{\max}} \left\lfloor \frac{n}{5^k} \right\rfloor, \quad (2)$$

where

$$k_{\max} \equiv \left\lfloor \frac{\ln n}{\ln 5} \right\rfloor \quad (3)$$

and $\lfloor x \rfloor$ is the FLOOR FUNCTION (Gardner 1978, p. 63; Ogilvy and Anderson 1988, pp. 112–14). For $n = 1, 2, \ldots$, the number of trailing zeros are 0, 0, 0, 0, 1, 1, 1, 1, 1, 2, 2, 2, 2, 2, 3, 3, ... (Sloane's A027868). This is a special application of the general result that the POWER of a PRIME p dividing $n!$ is

$$\in_p (n) = \sum_{k \geq 0} \left\lfloor \frac{n}{p^k} \right\rfloor \qquad (4)$$

(Landau 1974, pp. 75–6; Hardy and Wright 1979, pp. 342; Ingham 1990, p. 20; Graham *et al.* 1994; Vardi 1991; Hardy 1999, pp. 18 and 21). Stated another way, the exact POWER of a PRIME p which divides $n!$ is

$$\frac{n - \text{sum of digits of the base} - p \text{ representation of } n}{p - 1} \qquad (5)$$

Let $a(n)$ be the *last* nonzero digit in $n!$, then the first few values are 2, 6, 4, 2, 2, 4, 2, 8, 8, 8, 6, 8, ... (Sloane's A008904). This sequence was studied by Kakutani (1967), who showed that this sequence is "5-automatic," meaning roughly that there exists a finite automaton which, when given the digits of n in base-5, will wind up in a state for which an output mapping specifies $a(n)$. The exact distribution of digits follows from this result.

By noting that

$$n! \equiv \Gamma(n + 1), \qquad (6)$$

where $\Gamma(n)$ is the GAMMA FUNCTION for INTEGERS n, the definition can be generalized to COMPLEX values

$$z! \equiv \Gamma(z + 1) \equiv \int_0^\infty e^{-t} t^z dt. \qquad (7)$$

This defines $z!$ for all COMPLEX values of z, except when z is a NEGATIVE INTEGER, in which case $z! = \infty$. Using the identities for GAMMA FUNCTIONS, the values of $(\frac{1}{2}n)!$ (half integral values) can be written explicitly

$$\left(-\frac{1}{2}\right)! = \sqrt{\pi} \qquad (8)$$

$$\left(\frac{1}{2}\right)! = \frac{1}{2}\sqrt{\pi} \qquad (9)$$

$$\left(n - \frac{1}{2}\right)! = \frac{\sqrt{\pi}}{2^n}(2n - 1)!! \qquad (10)$$

$$\left(n + \frac{1}{2}\right)! = \frac{\sqrt{\pi}}{2^{n+1}}(2n - 1)!!, \qquad (11)$$

where $n!!$ is a DOUBLE FACTORIAL.

For INTEGERS s and n with $s < n$,

$$\frac{(s - n)!}{(2s - 2n)!} = \frac{(-1)^{n-s}(2n - 2s)!}{(n - s)!}. \qquad (12)$$

The LOGARITHM of $z!$ is frequently encountered

$$\ln(z!) = \frac{1}{2}\ln\left[\frac{\pi z}{\sin(\pi z)}\right] - \gamma - \sum_{n=1}^\infty \frac{\zeta(2n + 1)}{2n + 1}z^{2n+1} \qquad (13)$$

$$= \frac{1}{2}\ln\left[\frac{\pi z}{\sin(\pi z)}\right] - \frac{1}{2}\ln\left(\frac{1 + z}{1 - z}\right) + (1 - \gamma)z$$

$$- \sum_{n=1}^\infty [\zeta(2n + 1) - 1]\frac{z^{2n+1}}{2n + 1} \qquad (14)$$

$$= \ln\left[\lim_{n \to \infty} \frac{n!}{(z + 1)(z + 2)\cdots(z + n)}n^z\right] \qquad (15)$$

$$= \lim_{n \to \infty}[\ln(n!) + z \ln n - \ln(z + 1) - \ln(z + 2) - \ldots$$

$$- \ln(z + n)] \qquad (16)$$

$$= \sum_{n=1}^\infty \frac{z^n}{n!}F_{n-1}(0) \qquad (17)$$

$$= -\gamma z + \sum_{n=2}^\infty (-1)^n \frac{z^n}{n}\zeta(n) \qquad (18)$$

$$= -\ln(1 + z) + z(1 - \gamma) + \sum_{n=2}^\infty (-1)^n [\zeta(n) - 1]\frac{z^n}{n}, \qquad (19)$$

where γ is the EULER-MASCHERONI CONSTANT, $\zeta(z)$ is the RIEMANN ZETA FUNCTION, and $F_n(z)$ is the POLYGAMMA FUNCTION. The factorial can be expanded in a series

$$z! =$$

$$\sqrt{2\pi}z^{z+1/2}e^{-z}\left(1 + \frac{1}{2}z^{-1} + \frac{1}{288}z^{-2} - \frac{139}{51840}z^{-3} + \ldots\right) \qquad (20)$$

(Sloane's A001163 and A001164). STIRLING'S SERIES gives the series expansion for $\ln(z!)$,

$$\ln(z!) = \frac{1}{2}\ln(2\pi) + \left(z + \frac{1}{2}\right)\ln z - z + \frac{B_2}{2z} + \ldots$$

$$+ \frac{B_{2n}}{2n(2n - 1)z^{2n-1}} + \ldots$$

$$= \frac{1}{2}\ln(2\pi) + \left(z + \frac{1}{2}\right)\ln z - z + \frac{1}{12}z^{-1} - \frac{1}{360}z^{-3}$$

$$+ \frac{1}{1260}z^{-5} - \ldots \qquad (21)$$

(Sloane's A046968 and A046969), where B_n is a BERNOULLI NUMBER.

Let h be the exponent of the greatest POWER of a PRIME p dividing $n!$. Then

$$h = \sum_{\substack{i=1 \\ p^i \le n}} \left\lfloor \frac{n}{p^i} \right\rfloor. \tag{22}$$

Let g be the number of 1s in the BINARY representation of n. Then

$$g + h = n \tag{23}$$

(Honsberger 1976). In general, as discovered by Legendre in 1808, the POWER m of the PRIME p dividing $n!$ is given by

$$m = \sum_{k=0}^{\infty} \left\lfloor \frac{n}{p^k} \right\rfloor = \frac{n - (n_0 + n_1 + \ldots + n_N)}{p - 1}, \tag{24}$$

where the INTEGERS $n_1, \ldots, n_N$ are the digits of n in base p (Ribenboim 1989).

The numbers $n! + 1$ are prime for $n = 1, 2, 3, 11, 27, 37, 41, 73, 77, 116, 154, \ldots$ (Sloane's A002981; Wells 1986, p. 70), and the numbers $n! - 1$ are prime for $n = 3, 4, 6, 7, 12, 14, 30, 32, 33, 38, 94, 166, \ldots$ (Sloane's A002982). In general, the power-product sequences (Mudge 1997) are given by $S_k^{\pm}(n) = (n!)^k \pm 1$. The first few terms of $S_2^+(n)$ are 2, 5, 37, 577, 14401, 518401, ... (Sloane's A020549), and $S_2^+(n)$ is PRIME for $n = 1, 2, 3, 4, 5, 9, 10, 11, 13, 24, 65, 76, \ldots$ (Sloane's A046029). The first few terms of $S_2^-(n)$ are 0, 3, 35, 575, 14399, 518399, ... (Sloane's A046032), but $S_2^-(n)$ is PRIME for only $n = 2$ since $S_2^-(n) = (n!)^2 - 1 = (n!+1)(n!-1)$ for $n > 2$. The first few terms of $S_3^-(n)$ are 0, 7, 215, 13823, 1727999, ... (Sloane's A046033), and the first few terms of $S_3^+(n)$ are 2, 9, 217, 13825, 1728001, ... (Sloane's A019514).

The first few numbers n such that the sum of the factorials of their digits is equal to the PRIME COUNTING FUNCTION $\pi(n)$ are 6500, 6501, 6510, 6511, 6521, 12066, 50372, ... (Sloane's A049529). This sequence is finite, with the largest term being $a_{23} = 11,071,599$.

There are three numbers less than 200,000 for which

$$(n-1)! + 1 \equiv 0 \pmod{n^2}, \tag{25}$$

namely 5, 13, and 563 (Le Lionnais 1983). BROWN NUMBERS are pairs (m, n) of INTEGERS satisfying the condition of BROCARD'S PROBLEM, i.e., such that

$$n! + 1 = m^2, \tag{26}$$

Only three such numbers are known: (5, 4), (11, 5), (71, 7). Erdos conjectured that these are the only three such pairs (Guy 1994, p. 193).

See also ALLADI-GRINSTEAD CONSTANT, BROCARD'S PROBLEM, BROWN NUMBERS, CENTRAL FACTORIAL, DOUBLE FACTORIAL, FACTORIAL PRIME, FACTORIAL PRODUCTS, FACTORIAL SUMS, FACTORION, FALLING FACTORIAL, GAMMA FUNCTION, HYPERFACTORIAL, MULTIFACTORIAL, POCHHAMMER SYMBOL, PRIMORIAL, RISING FACTORIAL, ROMAN FACTORIAL, STIRLING'S SERIES, SUBFACTORIAL, SUPERFACTORIAL, WILSON PRIME

References

Caldwell, C. K. "The Top Twenty: Primorial and Factorial Primes." http://www.utm.edu/research/primes/lists/top20/PrimorialFactorial.html.

Conway, J. H. and Guy, R. K. "Factorial Numbers." In *The Book of Numbers.* New York: Springer-Verlag, pp. 65–6, 1996.

Dudeney, H. E. *Amusements in Mathematics.* New York: Dover, p. 96, 1970.

Gardner, M. "Factorial Oddities." Ch. 4 in *Mathematical Magic Show: More Puzzles, Games, Diversions, Illusions and Other Mathematical Sleight-of-Mind from Scientific American.* New York: Vintage, pp. 50–5, 1978.

Graham, R. L.; Knuth, D. E.; and Patashnik, O. "Factorial Factors." §4.4 in *Concrete Mathematics: A Foundation for Computer Science, 2nd ed.* Reading, MA: Addison-Wesley, pp. 111--115, 1994.

Guy, R. K. "Equal Products of Factorials," "Alternating Sums of Factorials," and "Equations Involving Factorial n." §B23, B43, and D25 in *Unsolved Problems in Number Theory, 2nd ed.* New York: Springer-Verlag, pp. 80, 100, and 193–94, 1994.

Hardy, G. H. and Wright, E. M. *An Introduction to the Theory of Numbers, 5th ed.* Oxford, England: Clarendon Press, 1979.

Honsberger, R. *Mathematical Gems II.* Washington, DC: Math. Assoc. Amer., p. 2, 1976.

Ingham, A. E. *The Distribution of Prime Numbers.* Cambridge, England: Cambridge University Press, 1990.

Jeffreys, H. and Jeffreys, B. S. *Methods of Mathematical Physics, 3rd ed.* Cambridge, England: Cambridge University Press, pp. 462–63, 1988.

Kakutani, S. "Ergodic Theory of Shift Transformations." In *Proc. 5th Berkeley Symposium on Mathematical Statistics and Probability, Vol. 2.* Berkeley, CA: University of California Press, pp. 405–14, 1967.

Landau, E. *Handbuch der Lehre von der Verteilung der Primzahlen, 3rd ed.* New York: Chelsea, 1974.

Le Lionnais, F. *Les nombres remarquables.* Paris: Hermann, p. 56, 1983.

Lewin, L. *Dilogarithms and Associated Functions.* London: Macdonald, 1958.

Leyland, P. ftp://sable.ox.ac.uk/pub/math/factors/factorial-.Z and ftp://sable.ox.ac.uk/pub/math/factors/factorial+.Z.

Madachy, J. S. *Madachy's Mathematical Recreations.* New York: Dover, p. 174, 1979.

Mellin, H. "Abrißeiner einheitlichen Theorie der Gamma- und der hypergeometrischen Funktionen." *Math. Ann.* **68**, 305–37, 1909.

Mudge, M. "Not Numerology but Numeralogy!" *Personal Computer World,* 279–80, 1997.

Ogilvy, C. S. and Anderson, J. T. *Excursions in Number Theory.* New York: Dover, 1988.

Petkovsek, M.; Wilf, H. S.; and Zeilberger, D. *A = B.* Wellesley, MA: A. K. Peters, p. 86, 1996.

Press, W. H.; Flannery, B. P.; Teukolsky, S. A.; and Vetterling, W. T. "Gamma Function, Beta Function, Factorials, Binomial Coefficients." §6.1 in *Numerical Recipes in FORTRAN: The Art of Scientific Computing, 2nd ed.* Cambridge, England: Cambridge University Press, pp. 206–09, 1992.

Ribenboim, P. *The Book of Prime Number Records, 2nd ed.* New York: Springer-Verlag, pp. 22–4, 1989.

Sloane, N. J. A. Sequences A000142/M1675, A001163/M5400, A001164/M4878, A002981/M0908, A002982/M2321, A008904, A019514, A020549, A027868, A046029, A046032, A046033, A046968, A046969, and A049529 in

"An On-Line Version of the Encyclopedia of Integer Sequences." http://www.research.att.com/~njas/se-quences/eisonline.html.

Spanier, J. and Oldham, K. B. "The Factorial Function $n!$ and Its Reciprocal." Ch. 2 in *An Atlas of Functions.* Washington, DC: Hemisphere, pp. 19–3, 1987.

Vardi, I. *Computational Recreations in Mathematica.* Reading, MA: Addison-Wesley, p. 67, 1991.

Wells, D. *The Penguin Dictionary of Curious and Interesting Numbers.* Middlesex, England: Penguin Books, p. 70, 1986.

Factorial Moment

$$v_{(r)} \equiv \sum_x x^{(r)} f(x),$$

where

$$x^{(r)} \equiv x(x-1) \cdots (x-r+1).$$

See also MOMENT

Factorial Number

FACTORIAL

Factorial Prime

A PRIME OF THE FORM $n! \pm 1$. $n! + 1$ is PRIME for 1, 2, 3, 11, 27, 37, 41, 73, 77, 116, 154, 320, 340, 399, 427, 872, 1477, 6380, ... (Sloane's A002981). No others are known, but N. Kuosa is coordinating a search in the range $23,000 < n < 30,000$.

$n! - 1$ is PRIME for 3, 4, 6, 7, 12, 14, 30, 32, 33, 38, 94, 166, 324, 379, 469, 546, 974, 1963, 3507, 3610, 6917, ... (Sloane's A002982).

See also FACTORIAL, PRIME NUMBER, PRIMORIAL

References

Borning, A. "Some Results for $k! + 1$ and $2 \cdot 3 \cdot 5 \cdot p + 1$." *Math. Comput.* **26**, 567–70, 1972.

Buhler, J. P.; Crandall, R. E.; and Penk, M. A. "Primes of the Form $M! + 1$ and $2 \cdot 3 \cdot 5 \cdots p + 1$." *Math. Comput.* **38**, 639–43, 1982.

Caldwell, C. K. "Prime Links++: Resources in theory: special_forms: near_products: factorial." http://primes.utm.edu/links/theory/special_forms/near_products/factorial/.

Caldwell, C. K. "On the Primality of $N! + 1$ and $2 \cdot 3 \cdot 5 \cdots p \pm 1$." *Math. Comput.* **64**, 889–90, 1995.

Dubner, H. "Factorial and Primorial Primes." *J. Rec. Math.* **19**, 197–03, 1987.

Guy, R. K. *Unsolved Problems in Number Theory, 2nd ed.* New York: Springer-Verlag, p. 7, 1994.

Kuosa, N. "Search of [sic] the Next Prime of the Form $n! + 1$." http://www.hut.fi/~nkuosa/primeform/.

Sloane, N. J. A. Sequences A002981/M0908 and A0029822321 in "An On-Line Version of the Encyclopedia of Integer Sequences." http://www.research.att.com/~njas/sequences/eisonline.html.

Temper, M. "On the Primality of $k! + 1$ and $\cdot 3 \cdot 5 \cdots p + 1$." *Math. Comput.* **34**, 303–04, 1980.

Factorial Products

The only known factorials which are products of factorials in an ARITHMETIC SEQUENCE are

$$0!1! = 1!$$
$$1!2! = 2!$$
$$0!1!2! = 2!$$
$$6!7! = 10!$$
$$1!3!5! = 6!$$
$$1!3!5!7! = 10!$$

(Madachy 1979).

There are no identities OF THE FORM

$$n! = a_1! a_2! \cdots a_r! \tag{1}$$

for $r \geq 2$ with $a_i \geq a_j \geq 2$ for $i < j$ for $n \leq 18160$ except

$$9! = 7!3!3!2! \tag{2}$$

$$10! = 7!6! = 7!5!3! \tag{3}$$

$$16! = 14!5!2! \tag{4}$$

(Guy 1994, p. 80).

See also FACTORIAL, FACTORIAL SUMS

References

Guy, R. K. "Equal Products of Factorials," "Alternating Sums of Factorials," and "Equations Involving Factorial n." §B23, B43, and D25 in *Unsolved Problems in Number Theory, 2nd ed.* New York: Springer-Verlag, pp. 80, 100, and 193–94, 1994.

Madachy, J. S. *Madachy's Mathematical Recreations.* New York: Dover, p. 174, 1979.

Factorial Sums

The sum-of-factorials function is defined by

$$\sum(n) \equiv \sum_{k=1}^n k!$$

$$= \frac{-e + \operatorname{ei}(1) + \pi i + E_{2n+1}(-1)]\Gamma(n+2)}{e} \tag{1}$$

$$= \frac{-e + \operatorname{ei}(1) + \Re[E_{2n+1}(-1)]\Gamma(n+2)}{e}, \tag{2}$$

where $\operatorname{ei}(1) \approx 1.89512$ is the EXPONENTIAL INTEGRAL, E_n is the E_N-FUNCTION, $\Re[z]$ is the REAL PART of z, and I is the IMAGINARY NUMBER. The first few values are 1, 3, 9, 33, 153, 873, 5913, 46233, 409113, ... (Sloane's A007489). $\Sigma(n)$ cannot be written as a hypergeometric term plus a constant (Petkovsek *et al.* 1996). However the sum

$$\sum{}'(n) \equiv \sum_{k=1}^{n} kk! = (n+1)! - 1 \qquad (3)$$

has a simple form, with the first few values being 1, 5, 23, 119, 719, 5039, ... (Sloane's A033312).

There are only four INTEGERS equal to the sum of the factorials of their digits. Such numbers are called FACTORIONS. While no factorial greater than 1! is a SQUARE NUMBER, D. Hoey listed sums $< 10^{12}$ of distinct factorials which give SQUARE NUMBERS, and J. McCranie gave the one additional sum less than $21! = 5.1 \times 10^{19}$:

$$0! + 1! + 2! = 2^2$$
$$1! + 2! + 3! = 3^2$$
$$1! + 4! = 5^2$$
$$1! + 5! = 11^2$$
$$4! + 5! = 12^2$$
$$1! + 2! + 3! + 6! = 27^2$$
$$1! + 5! + 6! = 29^2$$
$$1! + 7! = 71^2$$
$$4! + 5! + 7! = 72^2$$
$$1! + 2! + 3! + 7! + 8! = 213^2$$
$$1! + 4! + 5! + 6! + 7! + 8! = 215^2$$
$$1! + 2! + 3! + 6! + 9! = 603^2$$
$$1! + 4! + 8! + 9! = 635^2$$
$$1! + 2! + 3! + 6! + 7! + 8! + 10! = 1917^2$$

$$1! + 2! + 3! + 7! + 8! + 9! + 10! + 11! + 12! + 13! + 14! + 15!$$
$$= 1183893^2$$

(Sloane's A014597).

The first few values of the alternating SUM

$$a(n) \equiv \sum_{i=1}^{n} (-1)^{n-i} i! \qquad (4)$$

$$= (-1)^n \left[-1 - e\,\mathrm{ei}(-1) + (-1)^n \mathrm{E}_{n+2}(1)\Gamma(n+2) \right], \qquad (5)$$

where $\mathrm{ei}(x)$ is the EXPONENTIAL INTEGRAL, $E_n(x)$ is the E_N-FUNCTION, and $\Gamma(x)$ is the GAMMA FUNCTION, are 1, 1, 5, 19, 101, 619, 4421, 35899, ... (Sloane's A005165), and the first few values n for which $a(n)$ are prime are $n = 3, 4, 5, 6, 7, 8, 10, 15, 19, 41, 59, 61, 105, 160, 661, 2653, 3069, 3943, 4053, 4998, ...$ (Sloane's A001272, Guy 1994, p. 100). Zivkovic (1999) has shown that the number of such primes is finite.

Sums with powers of an index in the NUMERATOR and products of FACTORIALS in the DENOMINATOR can often be done analytically. For example, for numerator 1,

$$\sum_{i=1}^{n} \frac{1}{(k_{1+i})!(k_{2+i})!} = {}_1\tilde{F}_2(1; 2+k_1, 2+k_2; 1)$$
$$- {}_1\tilde{F}_2(1; n+k_1+2, n+k_2+2; 1) \qquad (6)$$

$$\sum_{i=1}^{n} \frac{1}{(k_1-i)!(k_2+i)!} = {}_2\tilde{F}\frac{(1, 1-k_1; k_2+2; -1)}{\Gamma(k_1)}$$

$$-{}_2\tilde{F}\frac{(1, n-k_1+1; n+k_2+2; -1)}{\Gamma(k_1-n)} \qquad (7)$$

where ${}_p\tilde{F}_q$ is a REGULARIZED HYPERGEOMETRIC FUNCTION. For numerator i,

$$\sum_{i=1}^{n} \frac{i}{(k_1+i)!(k_2+i)!}$$
$$= -(n+1)\,{}_1\tilde{F}_2(1; n+k_1+2, n+k_2+2; 1)$$
$$+ {}_1\tilde{F}_2(2; k_1+2, k_2+2; 1)$$
$$- {}_1\tilde{F}_2(2; n+k_1+3, n+k_2+3; 1) \qquad (8)$$

$$\sum_{i=1}^{n} \frac{i}{(k_1-i)!(k_2+i)!}$$
$$- \frac{(n+1)\,{}_2\tilde{F}_1(1, n-k_1+1; n+k_2+2; -1)}{\Gamma(k_1-n)}$$
$$+ \frac{{}_2\tilde{F}_1(2, 1-k_1; k_2+2; -1)}{\Gamma(k_1)}$$
$$- \frac{{}_2\tilde{F}_1(2, n-k_1+2; n+k_2+3; -1)}{\Gamma(k_1-n-1)}. \qquad (9)$$

These sums simplify substantially for special values of k_1 and k_2. For example, with $k_1 = k_2 = n$,

$$\sum_{i=1}^{n} \frac{1}{(n-i)!(n+i)!} = \frac{2^{2n-1}}{\Gamma(2n+1)} - \frac{1}{2[\Gamma(n)]^2} \qquad (10)$$

$$\sum_{i=1}^{n} \frac{i}{(n-i)!(n+i)!} = \frac{1}{2\Gamma(n)\Gamma(n+1)} \qquad (11)$$

$$\sum_{i=1}^{n} \frac{i^2}{(n-i)!(n+i)!}$$
$$= \frac{1}{2\Gamma(n)\Gamma(n+1)} + \frac{2\,{}_2\tilde{F}_1(3, 2-n; n+3; -1)}{\Gamma(n-1)}. \qquad (12)$$

With $k_1 = n$ and $k_2 = n-1$,

$$\sum_{i=1}^{n} \frac{1}{(n-i)!(n-1+i)!} = \frac{4^{n-1}}{\Gamma(2n)} \qquad (13)$$

$$\sum_{i=1}^{n} \frac{i}{(n-i)!(n-1+i)!} = \frac{1}{2[\Gamma(n)]^2} + \frac{2^{2n-3}}{\Gamma(2n)}. \qquad (14)$$

With $k_1 = n$ and $k_2 = n+1$,

$$\sum_{i=1}^{n} \frac{1}{(n-i)!(n+1+i)!}$$
$$= \frac{4^n}{\Gamma(2n+2)} - \frac{1}{\Gamma(n+1)\Gamma(n+2)} \qquad (15)$$

$$\sum_{i=1}^{n} \frac{1}{(n-i)!(n+1+i)!}$$

$$= \frac{\Gamma(n) + \Gamma(n+1)}{2\Gamma(n)\Gamma(n+1)\Gamma(n+2)} - \frac{2^{2n-1}}{\Gamma(2n+2)} \qquad (16)$$

Sums of factorial POWERS include

$$\sum_{n=0}^{\infty} \frac{(n!)^2}{(2n)!} = \frac{2}{27}\left(18 + \sqrt{3}\pi\right) \qquad (17)$$

$$\sum_{n=0}^{\infty} \frac{(n!)^3}{(3n)!} = {}_3F_2\left(1,1,1;\frac{1}{3},\frac{2}{3};\frac{1}{27}\right) \qquad (18)$$

$$= \int_0^1 \left[P(t) + Q(t)\cos^{-1}R(t)\right]dt, \qquad (19)$$

where

$$P(t) = \frac{2(8 + 7t^2 - 7t^3)}{(4 - t^2 + t^3)^2} \qquad (20)$$

$$Q(t) = \frac{4t(1-t)(5 + t^2 - t^3)}{(4 - t^2 + t^3)^2\sqrt{(1-t)(4 - t^2 + t^3)}} \qquad (21)$$

$$R(t) = 1 - \frac{1}{2}(t^2 - t^3) \qquad (22)$$

(Schroeppel and Gosper 1972). In general,

$$\sum_{n=0}^{\infty} \frac{(n!)^k}{(kn)!} = {}_kF_{k-1}\left(\underbrace{1,\ldots,1}_{k}; \frac{1}{k},\frac{2}{k},\ldots,\frac{k-1}{k};\frac{1}{k^k}\right). \qquad (23)$$

Identities satisfied by sums of factorials include

$$\sum_{k=0}^{\infty} \frac{1}{k!} = e = 2.718281828\ldots \qquad (24)$$

$$\sum_{k=0}^{\infty} \frac{(-1)^k}{k!} = e^{-1} = 0.3678794411\ldots \qquad (25)$$

$$\sum_{k=0}^{\infty} \frac{1}{(k!)^2} = I_0(2) = 2.279585302\ldots \qquad (26)$$

$$\sum_{k=0}^{\infty} \frac{(-1)^k}{(k!)^2} = J_0(2) = 0.2238907791\ldots \qquad (27)$$

$$\sum_{k=0}^{\infty} \frac{1}{(2k)!} = \cosh 1 = 1.543080634\ldots \qquad (28)$$

$$\sum_{k=0}^{\infty} \frac{(-1)^k}{(2k)!} = \cos 1 = 0.5403023058\ldots \qquad (29)$$

$$\sum_{k=0}^{\infty} \frac{1}{(2k+1)!} = \sinh 1 = 1.175201193\ldots \qquad (30)$$

$$\sum_{k=0}^{\infty} \frac{(-1)^k}{(2k+1)!} = \sin 1 = 0.8414709848\ldots \qquad (31)$$

(Spanier and Oldham 1987), where $I_0(x)$ is a MODIFIED BESSEL FUNCTION OF THE FIRST KIND, $J_0(x)$ is a BESSEL FUNCTION OF THE FIRST KIND, $\cosh x$ is the HYPERBOLIC COSINE, $\cos x$ is the COSINE, $\sinh x$ is the HYPERBOLIC SINE, and $\sin x$ is the SINE.

See also BINOMIAL SUMS, FACTORIAL, FACTORIAL PRODUCTS

References

Guy, R. K. "Equal Products of Factorials," "Alternating Sums of Factorials," and "Equations Involving Factorial n." §B23, B43, and D25 in *Unsolved Problems in Number Theory, 2nd ed.* New York: Springer-Verlag, pp. 80, 100, and 193–94, 1994.

Schroeppel, R. and Gosper, R. W. Item 116 in Beeler, M.; Gosper, R. W.; and Schroeppel, R. *HAKMEM.* Cambridge, MA: MIT Artificial Intelligence Laboratory, Memo AIM-239, p. 54, Feb. 1972.

Sloane, N. J. A. Sequences A001272, A005165/M3892, A007489/M2818, A014597, and A033312 in "An On-Line Version of the Encyclopedia of Integer Sequences." http://www.research.att.com/~njas/sequences/eisonline.html.

Spanier, J. and Oldham, K. B. "The Factorial Function $n!$ and Its Reciprocal." Ch. 2 in *An Atlas of Functions.* Washington, DC: Hemisphere, pp. 19–3, 1987.

Zivkovic, M. "The Number of Primes $\Sigma_{i=1}^{n}(-1)^{n-i}i!$ is Finite." *Math. Comput.* **68**, 403–09, 1999.

Factorial2

DOUBLE FACTORIAL

Factoring

FACTORIZATION

Factorion

A factorion is an INTEGER which is equal to the sum of FACTORIALS of its digits. There are exactly four such numbers:

$$1 = 1! \qquad (1)$$

$$2 = 2! \qquad (2)$$

$$145 = 1! + 4! + 5! \qquad (3)$$

$$40,585 = 4! + 0! + 5! + 8! + 5! \qquad (4)$$

(Sloane's A014080; Gardner 1978, Madachy 1979, Pickover 1995). Obviously, the factorion of an n-digit number cannot exceed $n \cdot 9!$.

See also FACTORIAL, FACTORIAL SUMS

References

Gardner, M. "Factorial Oddities." Ch. 4 in *Mathematical Magic Show: More Puzzles, Games, Diversions, Illusions and Other Mathematical Sleight-of-Mind from Scientific American.* New York: Vintage, pp. 61 and 64, 1978.

Madachy, J. S. *Madachy's Mathematical Recreations.* New York: Dover, p. 167, 1979.

Pickover, C. A. "The Loneliness of the Factorions." Ch. 22 in *Keys to Infinity.* New York: W. H. Freeman, pp. 169–71 and 319–20, 1995.

Sloane, N. J. A. Sequences A014080 in "An On-Line Version of the Encyclopedia of Integer Sequences." http://www.research.att.com/~njas/sequences/eisonline.html.

Factorization

The determination of FACTORS (DIVISORS) of a given INTEGER("PRIME FACTORIZATION"), POLYNOMIAL ("POLYNOMIAL FACTORIZATION"), etc. In many cases of interest (particularly PRIME FACTORIZATION, factorization is unique, and so gives the "simplest" representation of a given quantity in terms of smaller parts.

The terms "factorization" and "factoring" are used synonymously.

See also FACTOR, POLYNOMIAL FACTORIZATION, PRIME FACTORIZATION, PRIME FACTORIZATION ALGORITHMS

Fagnano's Point

The point of coincidence of P and p' in FAGNANO'S THEOREM.

See also FAGNANO'S THEOREM

Fagnano's Problem

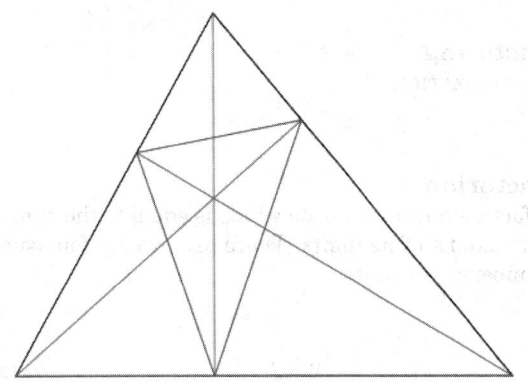

In a given ACUTE TRIANGLE ΔABC, find the INSCRIBED TRIANGLE whose PERIMETER is as small as possible. The answer is the ORTHIC TRIANGLE of ΔABC. The problem was proposed and solved using calculus by Fagnano in 1775 (Coxeter and Greitzer 1967, p. 88).

See also ACUTE TRIANGLE, ORTHIC TRIANGLE, PERIMETER

References

Coxeter, H. S. M. *Introduction to Geometry, 2nd ed.* New York: Wiley, p. 21, 1969.

Coxeter, H. S. M. and Greitzer, S. L. "Fagnano's Problem." §4.5 in *Geometry Revisited.* Washington, DC: Math. Assoc. Amer., pp. 88–9, 1967.

Courant, R. and Robbins, H. *What is Mathematics?: An Elementary Approach to Ideas and Methods, 2nd ed.* Oxford, England: Oxford University Press, p. 347, 1996.

Kazarinoff, N. D. *Geometric Inequalities.* New York: Random House, pp. 76–7, 1961.

Morley, F. and Morley, F. V. *Inversive Geometry.* Boston, MA: Ginn, p. 37, 1933.

Fagnano's Theorem

If $P(x,y)$ and $P(x',y')$ are two points on an ELLIPSE

$$\frac{x^2}{a^2}+\frac{y^2}{b^2}=1, \tag{1}$$

with ECCENTRIC ANGLES ϕ and ϕ' such that

$$\tan\phi\tan\phi'=\frac{b}{a} \tag{2}$$

and $A=P(a,0)$ and $B=P(0,b)$. Then

$$\text{arc}BP+\text{arc}BP'=\frac{e^2xx'}{a}. \tag{3}$$

This follows from the identity

$$E(u,k)+E(v,k)-E(k)=k^2\,\text{sn}(u,k)\,\text{sn}(v,k), \tag{4}$$

where $E(u,k)$ is an incomplete ELLIPTIC INTEGRAL OF THE SECOND KIND, $E(k)$ is a complete ELLIPTIC INTEGRAL OF THE SECOND KIND, and $\text{sn}(v,k)$ is a JACOBI ELLIPTIC FUNCTION. If P and p' coincide, the point where they coincide is called FAGNANO'S POINT.

See also ELLIPSE, FAGNANO'S POINT

Fair Dice

DICE, ICOSAHEDRON

Fair Division

CAKE CUTTING

Fair Game

A GAME which is not biased toward any player.

See also FUTILE GAME, GAME, MARTINGALE

Fairy Chess

A variation of CHESS involving a change in the form of the board, the rules of play, or the pieces used. For example, the normal rules of chess can be used but with a cylindrical or MÖBIUS STRIP connection of the edges.

See also CHESS

References

Kraitchik, M. "Fairy Chess." §12.2 in *Mathematical Recreations.* New York: W. W. Norton, pp. 276–79, 1942.

Faithful Group Action

A GROUP ACTION $\phi : G \times X \to X$ is called faithful if there are no group elements g such that $gx = x$ for all $x \in X$. Equivalently, the map ϕ induces an INJECTION of G into the SYMMETRIC GROUP Sx. So G can be identified with a PERMUTATION SUBGROUP.

Most actions that arise naturally are faithful. An example of an action which is not faithful is the action $e^{i(x+y)}$ of $G = \mathbb{R}^2 = \{(x,y)\}$ on $X = \mathbb{S}^1 = \{e^{i\theta}\}$, i.e., $\phi(x, y, e^{i\theta}) = e^{i(\theta + x + y)}$.

See also ADO'S THEOREM, EFFECTIVE ACTION, FREE ACTION, GROUP, IWASAWA'S THEOREM, ORBIT (GROUP), QUOTIENT SPACE (LIE GROUP), TRANSITIVE

References

Huang, J.-S. "Faithful Irreducible Representations." §9.3 in *Lectures on Representation Theory.* Singapore: World Scientific, pp. 124–28, 1999.
Rotman, J. *Theory of Groups.* New York: Allyn and Bacon, p. 180, 1984.

Falkner-Skan Differential Equation

The third-order ORDINARY DIFFERENTIAL EQUATION

$$y''' + \alpha y y'' + \beta \left(1 - y'^2\right) = 0.$$

References

Cebeci, T. and Keller, H. B. "Shooting and Parallel Shooting Methods for Solving Falkner-Shan Boundary Layer Equation." *J. Comput. Phys.* **71**, 289–00, 1971.
Zwillinger, D. *Handbook of Differential Equations, 3rd ed.* Boston, MA: Academic Press, p. 128, 1997.

Fallacy

A fallacy is an incorrect result arrived at by apparently correct, though actually specious reasoning. The great Greek geometer Euclid wrote an entire book on geometric fallacies which, unfortunately, has not survived (Gardner 1984, p. ix).

The most common example of a mathematical fallacy is the "proof" that $1 = 2$ as follows. Let $a = b$, then

$$ab = a^2 \tag{1}$$

$$ab - b^2 = a^2 - b^2 \tag{2}$$

$$b(a - b) = (a + b)(a - b) \tag{3}$$

$$b = a + b \tag{4}$$

$$b = 2b \tag{5}$$

$$1 = 2. \tag{6}$$

The incorrect step is (4), in which DIVISION BY ZERO

$(a - b = 0)$ is performed, which is not an allowed algebraic operation. Similarly flawed reasoning can be used to show that $0 = 1$, or any number equals any other number.

Ball and Coxeter (1987) give other such examples in the areas of both arithmetic and geometry.

See also DIVISION BY ZERO

References

Ball, W. W. R. and Coxeter, H. S. M. *Mathematical Recreations and Essays, 13th ed.* New York: Dover, pp. 41–5 and 76–4, 1987.
Barbeau, E. J. *Mathematical Fallacies, Flaws, and Flimflam.* Washington, DC: Math. Assoc. Amer., 1999.
Gardner, M. *The Sixth Book of Mathematical Games from Scientific American.* Chicago, IL: University of Chicago Press, 1984.
Pappas, T. "Geometric Fallacy & the Fibonacci Sequence." *The Joy of Mathematics.* San Carlos, CA: Wide World Publ./Tetra, p. 191, 1989.

Falling Factorial

For $n \geq 0$, the falling factorial is defined by

$$(x)_n = x(x - 1) \cdots (x - n + 1), \tag{1}$$

and is related to the RISING FACTORIAL $x^{(n)}$ (a.k.a. POCHHAMMER SYMBOL) by

$$(x)_n = (-1)^n (-x)^{(n)}. \tag{2}$$

The falling factorial can be implemented in *Mathematica* as

```
FallingFactorial[x_, n_] := (-1) Pochhammer[-x, n]
```

The falling factorial is also called a binomial polynomial or lower factorial.

Unfortunately, there are two notations used for the falling and rising factorials, $(x)_n$ and $x^{(n)}$, which are unfortunately polar opposites of one another. In combinatorial usage, the falling factorial is denoted $(x)_n$ and the RISING FACTORIAL is denoted $(x)^{(n)}$ (Comtet 1974, p. 6; Roman 1984, p. 5; Hardy 1999, p. 101), whereas in the calculus of FINITE DIFFERENCES and the theory of special functions, the falling factorial is denoted $x^{(n)}$ and the RISING FACTORIAL is denoted $(x)_n$ (Roman 1984, p. 5; Abramowitz and Stegun 1972, p. 256; Spanier 1987). Extreme caution is therefore needed in interpreting the meanings of the notations $(x)_n$ and $x^{(n)}$. *In this work, the notation $(x)_n$ is used for the falling factorial*, potentially causing confusion with the POCHHAMMER SYMBOL (another name for the RISING FACTORIAL, which is universally denoted $(x)_n$).

The first few falling factorials are

$$(x)_0 = 1$$

$$(x)_1 = x$$

$$(x)_2 = x(x-1) = x^2 - x$$

$$(x)_3 = x(x-1)(x-2) = x^3 - 3x^2 + 2x$$

$$(x)_4 = x(-1)(x-2)(x-3) = x^4 - 6x^3 + 11x^2 - 6x.$$

A sum formula connecting the falling factorial $(x)_n$ and rising factorial $x^{(n)}$,

$$(x)_n = \sum_{k=0}^{n} c_{nk} x^{(k)}, \tag{3}$$

is given using the Sheffer formalism with

$$g(t) = 1 \tag{4}$$

$$f(t) = e^t - 1 \tag{5}$$

$$h(t) = 1 \tag{6}$$

$$l(t) = 1 - e^{-t}, \tag{7}$$

which gives the GENERATING FUNCTION

$$\sum_{n=0}^{\infty} \frac{t_n(x)}{n!} t^n = \sum_{n=0}^{\infty} \frac{1}{n!} \sum_{k=0}^{n} c_{nk} x^k t^k = e^{tx/(1+t)}, \tag{8}$$

$$= 1 + xt + \frac{1}{2}(x^2 - 2x)t^2 + \frac{1}{6}(x^3 - 6x^2 + 6x)t^3$$

$$+ \frac{1}{24}(x^4 - 12x^3 + 36x^2 - 24x)t^4 + \cdots, \tag{9}$$

where

$$t_n(x) = \sum_{k=0}^{n} c_{nk} x^k. \tag{10}$$

Reading the coefficients off gives

$$c_{00} = 1$$

$$c_{11} = 1 \quad c_{10} = 0$$

$$c_{22} = 1 \quad c_{21} = -2 \quad c_{20} = 0$$

$$c_{33} = 1 \quad c_{32} = -6 \quad c_{31} = 6 \quad c_{30} = 0,$$

so,

$$(x)_0 = x^{(0)} \tag{11}$$

$$(x)_1 = x^{(1)} \tag{12}$$

$$(x)_2 = x^{(2)} - 2x^{(1)} \tag{13}$$

$$(x)_3 = x^{(3)} - 6x^{(2)} + 6x^{(1)}, \tag{14}$$

etc. (and the formula given by Roman 1984, p. 133, is incorrect).

The falling factorial is an associated SHEFFER SEQUENCE with

$$f(t) = e^t - 1 \tag{15}$$

(Roman 1984, p. 29), and has GENERATING FUNCTION

$$\sum_{k=0}^{\infty} \frac{(x)_k}{k!} t^k = e^{x \ln(1+t)} = (1+t)^x, \tag{16}$$

which is equivalent to the BINOMIAL THEOREM

$$\sum_{k=0}^{\infty} \binom{x}{k} t^k = (1+t)^x. \tag{17}$$

The binomial identity of the SHEFFER SEQUENCE is

$$(x+y)_n = \sum_{k=0}^{n} \binom{n}{k} (x)_k (y)_{n-k}, \tag{18}$$

where $\binom{n}{k}$ is a BINOMIAL COEFFICIENT, which can be rewritten as

$$\binom{x+y}{n} = \sum_{k=0}^{\infty} \binom{x}{k} \binom{y}{n-k}, \tag{19}$$

known as the CHU-VANDERMONDE IDENTITY. The falling factorials obey the RECURRENCE RELATION

$$x(x)_n = (x)_{n+1} + n(x)_n \tag{20}$$

(Roman 1984, p. 61).

See also BINOMIAL THEOREM, CENTRAL FACTORIAL, CHU-VANDERMONDE IDENTITY, RISING FACTORIAL, SHEFFER SEQUENCE

References

Abramowitz, M. and Stegun, C. A. (Eds.). *Handbook of Mathematical Functions with Formulas, Graphs, and Mathematical Tables, 9th printing.* New York: Dover, 1972.

Comtet, L. *Advanced Combinatorics: The Art of Finite and Infinite Expansions, rev. enl. ed.* Dordrecht, Netherlands: Reidel, 1974.

Hardy, G. H. *Ramanujan: Twelve Lectures on Subjects Suggested by His Life and Work, 3rd ed.* New York: Chelsea, p. 101, 1999.

Roman, S. "The Lower Factorial Polynomial." §1.2 in *The Umbral Calculus.* New York: Academic Press, pp. 5, 28–9, and 56–3, 1984.

Spanier, J. and Oldham, K. B. "The Pochhammer Polynomials $(x)_n$." Ch. 18 in *An Atlas of Functions.* Washington, DC: Hemisphere, pp. 149–65, 1987.

False

A statement which is rigorously not TRUE. Regular two-valued LOGIC allows statements to be only TRUE or false, but FUZZY LOGIC treats "truth" as a continuum which can have a value between 0 and 1. The symbol λ is sometimes used to denote "false," although "F" is more commonly used in TRUTH TABLES.

See also ALETHIC, BOOLEANS, FUZZY LOGIC, LOGIC, TRUE, TRUTH TABLE, UNDECIDABLE

False Position Method
METHOD OF FALSE POSITION

False Spiral

References
Fraser, J. *Brit. J. Psychol.* Jan. 1908.
Pappas, T. "The False Spiral Optical Illusion." *The Joy of Mathematics.* San Carlos, CA: Wide World Publ./Tetra, p. 114, 1989.

Faltung (Form)
Let A and B be bilinear forms

$$A = A(x,y) = \sum \sum a_{ij} x_i y_i$$

$$B = B(x,y) = \sum \sum b_{ij} x_i y_i$$

and suppose that A and B are bounded in $[p, p']$ with bounds M and N. Then

$$F = F(A,B) = \sum \sum f_{ij} x_i y_j,$$

where the series

$$f_{ij} = \sum_k a_{ik} b_{kj}$$

is absolutely convergent, is called the faltung of A and B. F is bounded in $[p, p']$, and its bound does not exceed MN.

References
Hardy, G. H.; Littlewood, J. E.; and Pólya, G. *Inequalities, 2nd ed.* Cambridge, England: Cambridge University Press, pp. 210–11, 1988.

Faltung (Function)
CONVOLUTION

Family Number
HOME PRIME

Fan
A SPREAD in which each node has a FINITE number of children.

See also SPREAD (TREE)

Fano Configuration
FANO PLANE

Fano Plane

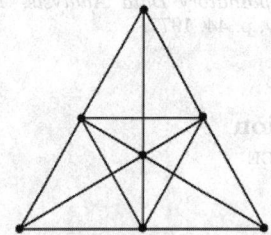

The 2-D finite PROJECTIVE PLANE over $GF(2)$ ("of order two"), illustrated above. It is a BLOCK DESIGN with $v = 7$, $k = 3$, $\lambda = 1$, $r = 3$, and $b = 7$, the STEINER TRIPLE SYSTEM $S(7)$, and the unique 7_3 CONFIGURATION.

The Fano plane also solves the TRANSYLVANIA LOTTERY, which picks three numbers from the INTEGERS 1–4. Using two Fano planes we can guarantee matching two by playing just 14 times as follows. Label the VERTICES of one Fano plane by the INTEGERS 1–, the other plane by the INTEGERS 8–4. The 14 tickets to play are the 14 lines of the two planes. Then if (a, b, c) is the winning ticket, at least two of a, b, c are either in the interval $[1, 7]$ or $[8, 14]$. These two numbers are on exactly one line of the corresponding plane, so one of our tickets matches them.

The Lehmers (1974) found an application of the Fano plane for factoring INTEGERS via QUADRATIC FORMS. Here, the triples of forms used form the lines of the PROJECTIVE GEOMETRY on seven points, whose planes are Fano configurations corresponding to pairs of residue classes mod 24 (Lehmer and Lehmer 1974, Guy 1975, Shanks 1985). The group of AUTOMORPHISMS (incidence-preserving BIJECTIONS) of the Fano plane is the SIMPLE GROUP of ORDER 168 (Klein 1870).

See also CONFIGURATION, DESIGN, PROJECTIVE PLANE, STEINER TRIPLE SYSTEM, TRANSYLVANIA LOTTERY

References
Guy, R. "How to Factor a Number." *Proc. Fifth Manitoba Conf. on Numerical Math.*, 49–9, 1975.
Lehmer, D. H. and Lehmer, E. "A New Factorization Technique Using Quadratic Forms." *Math. Comput.* **28**, 625–35, 1974.
Shanks, D. *Solved and Unsolved Problems in Number Theory, 3rd ed.* New York: Chelsea, pp. 202 and 238, 1985.
Wells, D. *The Penguin Dictionary of Curious and Interesting Geometry.* London: Penguin, p. 72, 1991.

Fano's Axiom
The three diagonal points of a COMPLETE QUADRILATERAL are never COLLINEAR.

Far Out
A phrase used by Tukey to describe data points which are outside the outer FENCES.

See also FENCE

References
Tukey, J. W. *Explanatory Data Analysis.* Reading, MA: Addison-Wesley, p. 44, 1977.

Farey Fraction
FAREY SEQUENCE

Farey Sequence

The Farey sequence F_n for any POSITIVE INTEGER n is the set of irreducible RATIONAL NUMBERS a/b with $0 \le a \le b \le n$ and $(a, b) = 1$ arranged in increasing order. The first few are

$$F_1 = \left\{ \frac{0}{1}, \frac{1}{1} \right\} \tag{1}$$

$$F_2 = \left\{ \frac{0}{1}, \frac{1}{2}, \frac{1}{1} \right\} \tag{2}$$

$$F_3 = \left\{ \frac{0}{1}, \frac{1}{3}, \frac{1}{2}, \frac{2}{3}, \frac{1}{1} \right\} \tag{3}$$

$$F_4 = \left\{ \frac{0}{1}, \frac{1}{4}, \frac{1}{3}, \frac{1}{2}, \frac{2}{3}, \frac{3}{4}, \frac{1}{1} \right\} \tag{4}$$

$$F_5 = \left\{ \frac{0}{1}, \frac{1}{5}, \frac{1}{4}, \frac{1}{3}, \frac{2}{5}, \frac{1}{2}, \frac{3}{5}, \frac{2}{3}, \frac{3}{4}, \frac{4}{5}, \frac{1}{1} \right\} \tag{5}$$

(Sloane's A006842 and A006843). Except for F_1, each F_n has an ODD number of terms and the middle term is always 1/2.

Let p/q, p'/q', and p''/q'' be three successive terms in a Farey series. Then

$$qp' - pq' = 1 \tag{6}$$

$$\frac{p'}{q'} = \frac{p + p''}{q + q''}. \tag{7}$$

These two statements are actually equivalent (Hardy and Wright 1979, p. 24). For a method of computing a successive sequence from an existing one of n terms, insert the MEDIANT fraction $(a + b)/(c + d)$ between terms a/c and b/d when $c + d \le n$ (Hardy and Wright 1979, pp. 25–6; Conway and Guy 1996; Apostol 1997). Given $0 \le a/b < c/d \le 1$ with $bc - ad = 1$, let h/k be the MEDIANT of a/b and c/d. Then $a/b < h/k < c/d$, and these fractions satisfy the unimodular relations

$$bh - ak = 1 \tag{8}$$

$$ck - dh = 1 \tag{9}$$

(Apostol 1997, p. 99).

The number of terms $N(n)$ in the Farey sequence for the INTEGER n is

$$N(n) = 1 + \sum_{k=1}^{n} \phi(k) = 1 + \Phi(n), \tag{10}$$

where $\phi(k)$ is the TOTIENT FUNCTION and $\Phi(n)$ is the SUMMATORY FUNCTION of $\phi(k)$, giving 2, 3, 5, 7, 11, 13, 19, ... (Sloane's A005728). The asymptotic limit for the function $N(n)$ is

$$N(n) \sim \frac{3n^2}{\pi^2} = 0.3039635509 n^2 \tag{11}$$

(Vardi 1991, p. 155).

FORD CIRCLES provide a method of visualizing the Farey sequence. The Farey sequence F_n defines a subtree of the STERN-BROCOT TREE obtained by pruning unwanted branches (Graham *et al.* 1994).

See also FORD CIRCLE, MEDIANT, MINKOWSKI'S QUESTION MARK FUNCTION, RANK (SEQUENCE), STERN-BROCOT TREE

References

Apostol, T. M. "Farey Fractions." §5.4 in *Modular Functions and Dirichlet Series in Number Theory, 2nd ed.* New York: Springer-Verlag, pp. 97–9, 1997.

Beiler, A. H. "Farey Tails." Ch. 16 in *Recreations in the Theory of Numbers: The Queen of Mathematics Entertains.* New York: Dover, 1966.

Bogomolny, A. "Farey Series, A Story." http://www.cut-the-knot.com/blue/FareyHistory.html.

Conway, J. H. and Guy, R. K. "Farey Fractions and Ford Circles." *The Book of Numbers.* New York: Springer-Verlag, pp. 152–54 and 156, 1996.

Devaney, R. "The Mandelbrot Set and the Farey Tree, and the Fibonacci Sequence." *Amer. Math. Monthly* **106**, 289–02, 1999.

Dickson, L. E. *History of the Theory of Numbers, Vol. 1: Divisibility and Primality.* New York: Chelsea, pp. 155–58, 1952.

Farey, J. "On a Curious Property of Vulgar Fractions." *London, Edinburgh and Dublin Phil. Mag.* **47**, 385, 1816.

Graham, R. L.; Knuth, D. E.; and Patashnik, O. *Concrete Mathematics: A Foundation for Computer Science, 2nd ed.* Reading, MA: Addison-Wesley, pp. 118–19, 1994.

Guy, R. K. "Mahler's Generalization of Farey Series." §F27 in *Unsolved Problems in Number Theory, 2nd ed.* New York: Springer-Verlag, pp. 263–65, 1994.

Hardy, G. H. and Wright, E. M. "Farey Series and a Theorem of Minkowski." Ch. 3 in *An Introduction to the Theory of Numbers, 5th ed.* Oxford, England: Clarendon Press, pp. 23–7, 1979.

Sloane, N. J. A. Sequences A005728/M0661, A006842/M0041, and A006843/M0081 in "An On-Line Version of the Encyclopedia of Integer Sequences." http://www.research.att.com/~njas/sequences/eisonline.html.

Sylvester, J. J. "On the Number of Fractions Contained in Any Farey Series of Which the Limiting Number is Given." *London, Edinburgh and Dublin Phil. Mag. (5th Series)* **15**, 251, 1883.

Vardi, I. *Computational Recreations in Mathematica.* Reading, MA: Addison-Wesley, p. 155, 1991.

Weisstein, E. W. "Plane Geometry." MATHEMATICA NOTEBOOK PLANEGEOMETRY.M.

Farey Series
FAREY SEQUENCE

Farkas's Lemma

The system

$$Ax = x, \quad x \geq 0$$

has no solution IFF the system

$$A^T w \leq 0, \quad b^T > 0$$

has a solution (Fang and Puthenpura 1993, p. 60). This LEMMA is used in the proof of the KUHN-TUCKER THEOREM.

See also KUHN-TUCKER THEOREM, LAGRANGE MULTIPLIER

References

Fang, S.-C. and Puthenpura, S. *Linear Optimization and Extensions: Theory and Algorithms.* Englewood Cliffs, NJ: Prentice-Hall, p. 60, 1993.

Faro Shuffle

RIFFLE SHUFFLE

Far-Out Point

For a TRIANGLE with side lengths a, b, and c, the far-out point has TRIANGLE CENTER FUNCTION

$$\alpha = a(b^4 + c^4 - a^4 - b^2 c^2).$$

As $a : b : c$ approaches $1 : 1 : 1$, this point moves out along the EULER LINE to infinity.

References

Kimberling, C. "Central Points and Central Lines in the Plane of a Triangle." *Math. Mag.* **67**, 163–87, 1994.
Kimberling, C.; Lyness, R. C.; and Veldkamp, G. R. "Problem 1195 and Solution." *Crux Math.* **14**, 177–79, 1988.

Fast Fibonacci Transform

For a general second-order RECURRENCE RELATION

$$f_{n+1} = x f_n + y f_{n-1}, \tag{1}$$

define a multiplication rule on ordered pairs by

$$(A, B)(C, D) = (AD + BC + xAC, BD + yAC). \tag{2}$$

The inverse is then given by

$$(A, B)^{-1} = \frac{(-A, xA + B)}{B^2 + xAB - yA^2}, \tag{3}$$

and we have the identity

$$(f_1, y f_0)(1, 0)^n = (f_{n+1}, y f_n) \tag{4}$$

(Beeler *et al.* 1972, Item 12).

References

Gosper, R. W. and Salamin, G. Item 12 in Beeler, M.; Gosper, R. W.; and Schroeppel, R. *HAKMEM.* Cambridge,

MA: MIT Artificial Intelligence Laboratory, Memo AIM-239, p. 6, Feb. 1972.

Fast Fourier Transform

The fast Fourier transform (FFT) is a DISCRETE FOURIER TRANSFORM ALGORITHM which reduces the number of computations needed for N points from $2N^2$ to $2N \lg N$, where LG is the base-2 LOGARITHM. If the function to be transformed is not harmonically related to the sampling frequency, the response of an FFT looks like a SINC FUNCTION (although the integrated POWER is still correct). ALIASING (LEAKAGE) can be reduced by APODIZATION using a TAPERING FUNCTION. However, ALIASING reduction is at the expense of broadening the spectral response.

FFTs were first discussed by Cooley and Tukey (1965), although Gauss had actually described the critical factorization step as early as 1805 (Gergkand 1969, Strang 1993). A DISCRETE FOURIER TRANSFORM can be computed using an FFT by means of the DANIELSON-LANCZOS LEMMA if the number of points N is a POWER of two. If the number of points N is not a POWER of two, a transform can be performed on sets of points corresponding to the prime factors of N which is slightly degraded in speed. An efficient real Fourier transform algorithm or a fast HARTLEY TRANSFORM (Bracewell 1999) gives a further increase in speed by approximately a factor of two. Base-4 and base-8 fast Fourier transforms use optimized code, and can be 20–0% faster than base-2 fast Fourier transforms. PRIME factorization is slow when the factors are large, but discrete Fourier transforms can be made fast for $N = 2, 3, 4, 5, 7, 8, 11, 13$, and 16 using the WINOGRAD TRANSFORM ALGORITHM (Press *et al.* 1992, pp. 412–13, Arndt).

Fast Fourier transform algorithms generally fall into two classes: decimation in time, and decimation in frequency. The Cooley-Tukey FFT ALGORITHM first rearranges the input elements in bit-reversed order, then builds the output transform (decimation in time). The basic idea is to break up a transform of length N into two transforms of length $N/2$ using the identity

$$\sum_{n=0}^{N-1} a_n e^{-2\pi i n k / N}$$

$$= \sum_{n=0}^{N/2-1} a_{2n} e^{-2\pi i (2n) k / N} + \sum_{n=0}^{N/2-1} a_{2n+1} e^{-2\pi i (2n+1) k / N}$$

$$= \sum_{n=0}^{N/2-1} a_n^{\text{even}} e^{-2\pi i n k / (N/2)} + e^{-2\pi i k / N}$$

$$\times \sum_{n=0}^{N/2-1} a_n^{\text{odd}} e^{-2\pi i n k / (N/2)},$$

sometimes called the DANIELSON-LANCZOS LEMMA.

The easiest way to visualize this procedure is perhaps via the FOURIER MATRIX.

The Sande-Tukey ALGORITHM (Stoer and Bulirsch 1980) first transforms, then rearranges the output values (decimation in frequency).

See also DANIELSON-LANCZOS LEMMA, DISCRETE FOURIER TRANSFORM, FOURIER MATRIX, FOURIER TRANSFORM, HARTLEY TRANSFORM, NUMBER THEORETIC TRANSFORM, WINOGRAD TRANSFORM

References

Arndt, J. "FFT Code and Related Stuff." http://www.jjj.de/fxt/.
Bell Laboratories. "Netlib FFTPack." http://netlib.bell-labs.com/netlib/fftpack/.
Blahut, R. E. *Fast Algorithms for Digital Signal Processing.* New York: Addison-Wesley, 1984.
Bracewell, R. *The Fourier Transform and Its Applications, 3rd ed.* New York: McGraw-Hill, 1999.
Brigham, E. O. *The Fast Fourier Transform and Applications.* Englewood Cliffs, NJ: Prentice Hall, 1988.
Chu, E. and George, A. *Inside the FFT Black Box: Serial and Parallel Fast Fourier Transform Algorithms.* Boca Raton, FL: CRC Press, 2000.
Cooley, J. W. and Tukey, O. W. "An Algorithm for the Machine Calculation of Complex Fourier Series." *Math. Comput.* **19**, 297–01, 1965.
Duhamel, P. and Vetterli, M. "Fast Fourier Transforms: A Tutorial Review." *Signal Processing* **19**, 259–99, 1990.
Gergkand, G. D. "A Guided Tour of the Fast Fourier Transform." *IEEE Spectrum* **6**, 41–2, July 1969.
Lipson, J. D. *Elements of Algebra and Algebraic Computing.* Reading, MA: Addison-Wesley, 1981.
Nussbaumer, H. J. *Fast Fourier Transform and Convolution Algorithms, 2nd ed.* New York: Springer-Verlag, 1982.
Papoulis, A. *The Fourier Integral and its Applications.* New York: McGraw-Hill, 1962.
Press, W. H.; Flannery, B. P.; Teukolsky, S. A.; and Vetterling, W. T. "Fast Fourier Transform." Ch. 12 in *Numerical Recipes in FORTRAN: The Art of Scientific Computing, 2nd ed.* Cambridge, England: Cambridge University Press, pp. 490–29, 1992.
Ramirez, R. W. *The FFT: Fundamentals and Concepts.* Englewood Cliffs, NJ: Prentice-Hall, 1985.
Stoer, J. and Bulirsch, R. *Introduction to Numerical Analysis.* New York: Springer-Verlag, 1980.
Strang, G. "Wavelet Transforms Versus Fourier Transforms." *Bull. Amer. Math. Soc.* **28**, 288–05, 1993.
Van Loan, C. *Computational Frameworks for the Fast Fourier Transform.* Philadelphia, PA: SIAM, 1992.
Walker, J. S. *Fast Fourier Transform, 2nd ed.* Boca Raton, FL: CRC Press, 1996.

Fast Gossiping

GOSSIPING

Fat Fractal

A CANTOR SET with LEBESGUE MEASURE greater than 0.

See also CANTOR SET, EXTERIOR DERIVATIVE, FRACTAL, LEBESGUE MEASURE

References

Ott, E. "Fat Fractals." §3.9 in *Chaos in Dynamical Systems.* New York: Cambridge University Press, pp. 97–00, 1993.

Fatou Dust

FATOU SET

Fatou Set

A JULIA SET J consisting of a set of isolated points which is formed by taking a point outside an underlying set M (e.g., the MANDELBROT SET). If the point is outside but near the boundary of M, the Fatou set resembles the JULIA SET for nearby points within M. As the point moves further away, however, the set becomes thinner and is called FATOU DUST.

See also JULIA SET

References

Schroeder, M. *Fractals, Chaos, Power Laws.* New York: W. H. Freeman, p. 39, 1991.
Wells, D. *The Penguin Dictionary of Curious and Interesting Geometry.* London: Penguin, pp. 72–3, 1991.

Fatou's Lemma

If $\{f_n\}$ is a SEQUENCE of NONNEGATIVE measurable functions, then

$$\int \liminf_{n \to \infty} f_n \, d\mu \le \liminf_{n \to \infty} \int f_n \, d\mu.$$

See also ALMOST EVERYWHERE CONVERGENCE, MEASURE THEORY, POINTWISE CONVERGENCE

References

Browder, A. *Mathematical Analysis: An Introduction.* New York: Springer-Verlag, 1996.
Zeidler, E. *Applied Functional Analysis: Applications to Mathematical Physics.* New York: Springer-Verlag, 1995.

Fatou's Theorems

Let $f(\theta)$ be LEBESGUE INTEGRABLE and let

$$f(r, \theta) = \frac{1}{2\pi} \int_{-\pi}^{\pi} f(t) \frac{1 - r^2}{1 - 2r \cos(t - \theta) + r^2} dt \quad (1)$$

be the corresponding POISSON INTEGRAL. Then ALMOST EVERYWHERE in $-\pi \le \theta \le \pi$

$$\lim_{r \to 0^-} f(r, \theta) = f(\theta). \quad (2)$$

Let

$$F(z) = c_0 + c_1 z + c_2 z^2 + \ldots + c_n z^n + \ldots \quad (3)$$

be regular for $|z| < 1$, and let the integral

$$\frac{1}{2\pi}\int_{-\pi}^{\pi}|F(re^{i\theta})|^2 \, d\theta \qquad (4)$$

be bounded for $r < 1$. This condition is equivalent to the convergence of

$$|C_0|^2 + |C_1|^2 + \ldots + |C_n|^2 + \ldots \qquad (5)$$

Then almost everywhere in $-\pi \le \theta \le \pi$,

$$\lim_{r \to 0^-} F(re^{i\theta}) = F(e^{i\theta}). \qquad (6)$$

Furthermore, $F(e^{i\theta})$ is measurable, $|F(e^{i\theta})|^2$ is LEBESGUE INTEGRABLE, and the FOURIER SERIES of $F(e^{i\theta})$ is given by writing $z = e^{i\theta}$.

References

Szego, G. *Orthogonal Polynomials, 4th ed.* Providence, RI: Amer. Math. Soc., p. 274, 1975.

Faulhaber's Formula

In a 1631 edition of *Academiae Algebrae*, J. Faulhaber published the general formula for the POWER SUM of the first n POSITIVE INTEGERS,

$$\sum_{k=1}^{n} k^p = \frac{1}{p+1} \sum_{i=1}^{p+1} (-1)^{\delta_{ip}} \binom{p+1}{i} B_{p+1-i} n^i, \qquad (1)$$

where δ_{ip} is the KRONECKER DELTA, $\binom{n}{i}$ is a BINOMIAL COEFFICIENT, and B_i is the ith BERNOULLI NUMBER. Computing the sums for $p = 1, \ldots, 10$ gives

$$\sum_{k=1}^{n} k = \frac{1}{2}(n^2 + n) \qquad (2)$$

$$\sum_{k=1}^{n} k^2 = \frac{1}{6}(2n^3 + 3n^2 + n) \qquad (3)$$

$$\sum_{k=1}^{n} k^3 = \frac{1}{4}(n^4 + 2n^3 + n^2) \qquad (4)$$

$$\sum_{k=1}^{n} k^4 = \frac{1}{30}(6n^5 + 15n^4 + 10n^3 - n) \qquad (5)$$

$$\sum_{k=1}^{n} k^5 = \frac{1}{12}(2n^6 + 6n^5 + 5n^4 - n^2) \qquad (6)$$

$$\sum_{k=1}^{n} k^6 = \frac{1}{42}(6n^7 + 21n^6 + 21n^5 - 7n^3 + n) \qquad (7)$$

$$\sum_{k=1}^{n} k^7 = \frac{1}{24}(3n^8 + 12n^7 + 14n^6 - 7n^4 + 2n^2) \qquad (8)$$

$$\sum_{k=1}^{n} k^8 = \frac{1}{90}(10n^9 + 45n^8 + 60n^7 - 42n^5 + 20n^3 - 3n) \qquad (9)$$

$$\sum_{k=1}^{n} k^9 = \frac{1}{20}(2n^{10} + 10n^9 + 15n^8 - 14n^6 + 10n^4 - 3n^2) \qquad (10)$$

$$\sum_{k=1}^{n} k^{10} = \frac{1}{66}(6n^{11} + 33n^{10} + 55n^9 - 66n^5 - 33n^3 + 5n). \qquad (11)$$

See also POWER, POWER SUM, SUM

References

Conway, J. H. and Guy, R. K. *The Book of Numbers.* New York: Springer-Verlag, p. 106, 1996.

Fault-Free Rectangle

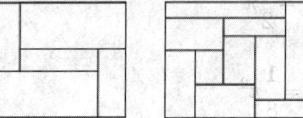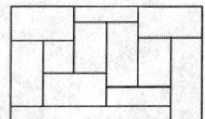

A DISSECTION of a RECTANGLE into smaller RECTANGLES such that the original rectangle is not divided into two subrectangles. Rectangle dissections into 3, 4, or 6 pieces cannot be fault-free but, as illustrated above, a dissection into five or more pieces may be fault-free.

See also BLANCHE'S DISSECTION, MRS. PERKINS' QUILT, PERFECT SQUARE DISSECTION, RECTANGLE

References

Steinhaus, H. *Mathematical Snapshots, 3rd ed.* New York: Dover, p. 85, 1999.
Wells, D. *The Penguin Dictionary of Curious and Interesting Geometry.* London: Penguin, p. 73, 1991.

Favard Constants

N.B. A detailed online essay by S. Finch was the starting point for this entry.

Let $T_n(x)$ be an arbitrary trigonometric POLYNOMIAL

$$T_n(x) = \frac{1}{2}a_0 + \left\{ \sum_{k=1}^{n} [a_k \cos(kx) + b_k \sin(kx)] \right\}, \qquad (1)$$

where the COEFFICIENTS are real. Let the rth derivative of $T_n(x)$ be bounded in $[-1, 1]$, then there exists a POLYNOMIAL $T_n(x)$ for which

$$|f(x) - T_n(x)| \le \frac{K_r}{(n+1)^r}, \qquad (2)$$

for all x, where K_r is the rth Favard constant, which is the smallest constant possible,

$$K_r = \frac{4}{\pi} \sum_{k=0}^{\infty} \left[\frac{(-1)^k}{2k+1} \right]^{r+1}, \qquad (3)$$

which can be written in terms of the LERCH TRANSCENDENT as

$$K_r = 2^{-(r+1)} \Phi\left((-1)^{r+1}, r+1, \frac{1}{2}\right). \quad (4)$$

These can be expressed by

$$K_r = \begin{cases} \dfrac{4}{\pi} \lambda(r+1) & \text{for r odd} \\ \dfrac{4}{\pi} \beta(r+1) & \text{for r even,} \end{cases} \quad (5)$$

where $\lambda(x)$ is the DIRICHLET LAMBDA FUNCTION and $\beta(x)$ is the DIRICHLET BETA FUNCTION. Explicitly,

$$K_0 = 1$$

$$K_1 = \frac{1}{2}\pi$$

$$K_2 = \frac{1}{8}\pi^2$$

$$K_3 = \frac{1}{24}\pi^3$$

$$K_4 = \frac{5}{384}\pi^4$$

$$K_5 = \frac{1}{240}\pi^5$$

(Sloane's A050970 and A050971).

See also DIRICHLET BETA FUNCTION, DIRICHLET LAMBDA FUNCTION

References

Finch, S. "Favorite Mathematical Constants." http://www.mathsoft.com/asolve/constant/favard/favard.html.
Kolmogorov, A. N. "Zur Grössenordnung des Restgliedes Fourierscher reihen differenzierbarer Funktionen." *Ann. Math.* **36**, 521–26, 1935.
Sloane, N. J. A. Sequences A050970 and A050970 in "An On-Line Version of the Encyclopedia of Integer Sequences." http://www.research.att.com/~njas/sequences/eisonline.html.
Zygmund, A. G. *Trigonometric Series, Vols. 1–, 2nd ed.* New York: Cambridge University Press, 1959.

F-Distribution

A continuous statistical distribution which arises in the testing of whether two observed samples have the same VARIANCE. Let χ_m^2 and χ_n^2 be independent variates distributed as CHI-SQUARED with m and n DEGREES OF FREEDOM. Define a statistic $F_{n,m}$ as the ratio of the dispersions of the two distributions

$$F_{n,m} \equiv \frac{\chi_n^2/n}{\chi_m^2/m}. \quad (1)$$

This statistic then has an F-distribution with probability function $f_{n,m}(x)$ and cumulative distribution function $F_{n,m}(x)$ given by

$$f_{n,m}(x) = \frac{\Gamma\left(\dfrac{n+m}{2}\right) n^{n/2} m^{m/2}}{\Gamma\left(\dfrac{n}{2}\right)\Gamma\left(\dfrac{m}{2}\right)} \frac{x^{n/2-1}}{(m+nx)^{(n+m)/2}} \quad (2)$$

$$= \frac{m^{m/2} n^{n/2} x^{n/2-1}}{(m+nx)^{(n+m)/2} B\left(\dfrac{1}{2}n, \dfrac{1}{2}m\right)} \quad (3)$$

$$F_{n,m}(x) = I\left(1; \frac{1}{2}m, \frac{1}{2}n\right) - I\left(\frac{m}{m+nx}; \frac{1}{2}m, \frac{1}{2}n\right), \quad (4)$$

where $\Gamma(z)$ is the GAMMA FUNCTION, $B(a,b)$ is the BETA FUNCTION, and $I(x; a, b)$ is the REGULARIZED BETA FUNCTION. The MEAN, VARIANCE, SKEWNESS and KURTOSIS are

$$\mu = \frac{m}{m-2} \quad (5)$$

$$\sigma^2 = \frac{2m^2(m+n-2)}{n(m-2)^2(m-4)} \quad (6)$$

$$\gamma_1 = \frac{2(m+2n-2)}{m-6}\sqrt{\frac{2(m-4)}{n(m+n-2)}} \quad (7)$$

$$\gamma_2 = \frac{12(-16+20m-8m^2+m^3+44n)}{n(m-6)(m-8)(n+m-2)}$$
$$+ \frac{12(-32mn+5m^2n-22n^2+5mn^2)}{n(m-6)(m-8)(n+m-2)}. \quad (8)$$

The probability that F would be as large as it is if the first distribution has a smaller variance than the second is denoted $Q(F_{n,m})$.

The noncentral F-distribution is given by

$$P(x) = e^{-\lambda/2 + (\lambda n_1 x)/[2(n_2+n_1 x)]} n_1^{n_1/2} n_2^{n_2/2} x^{n_1/2-1}$$

$$\times (n_2 + n_1 x)^{-(n_1+n_2)/2}$$

$$\times \frac{\Gamma\left(\dfrac{1}{2}n_1\right)\Gamma\left(1+\dfrac{1}{2}n_2\right) L_{n_2/2}^{n_1/2-1}\left(-\dfrac{\lambda n_1 x}{2(n_2+n_1 x)}\right)}{B\left(\dfrac{1}{2}n_1, \dfrac{1}{2}n_2\right)\Gamma\left[\dfrac{1}{2}(n_1+n_2)\right]}, \quad (9)$$

where $\Gamma(z)$ is the GAMMA FUNCTION, $B(\alpha, \beta)$ is the BETA FUNCTION, and $L_m^n(z)$ is an associated LAGUERRE POLYNOMIAL.

See also BETA FUNCTION, GAMMA FUNCTION, HOTELLING T-SQUARED DISTRIBUTION, REGULARIZED BETA FUNCTION, SNEDECOR'S F-DISTRIBUTION

References

Abramowitz, M. and Stegun, C. A. (Eds.). *Handbook of Mathematical Functions with Formulas, Graphs, and Mathematical Tables, 9th printing.* New York: Dover, pp. 946–49, 1972.

David, F. N. "The Moments of the z and F Distributions." *Biometrika* **36**, 394–03, 1949.

Press, W. H.; Flannery, B. P.; Teukolsky, S. A.; and Vetterling, W. T. "Incomplete Beta Function, Student's Distribution, F-Distribution, Cumulative Binomial Distribution." §6.2 in *Numerical Recipes in FORTRAN: The Art of Scientific Computing, 2nd ed.* Cambridge, England: Cambridge University Press, pp. 219–23, 1992.

Spiegel, M. R. *Theory and Problems of Probability and Statistics.* New York: McGraw-Hill, pp. 117–18, 1992.

Feigenbaum Constant

A universal constant for functions approaching CHAOS via period doubling. It was discovered by Feigenbaum in 1975 and demonstrated rigorously by Lanford (1982) and Collet and Eckmann (1979, 1980). The Feigenbaum constant δ characterizes the geometric approach of the bifurcation parameter to its limiting value. Let μ_k be the point at which a period 2^k cycle becomes unstable. Denote the converged value by μ_∞. Assuming geometric convergence, the difference between this value and μ_k is denoted

$$\lim_{k \to \infty} \mu_\infty - \mu_k = \frac{\Gamma}{\delta^k}, \qquad (1)$$

where Γ is a constant and δ is a constant > 1. Solving for δ gives

$$\delta = \lim_{n \to \infty} \frac{\mu_{n+1} - \mu_n}{\mu_{n+2} - \mu_{n+1}} \qquad (2)$$

(Rasband 1990, p. 23). For the LOGISTIC EQUATION,

$$\delta = 4.669201609102990\ldots \qquad (3)$$

$$\Gamma = 2.637\ldots \qquad (4)$$

$$\mu_\infty = 3.5699456\ldots \qquad (5)$$

Stoschek gives the approximation

$$\delta = 4 \; \cfrac{1 + \cfrac{12^2}{163} + \cfrac{4.12^2 + 31}{4.163^2} + \cdots}{1 + \cfrac{10^2}{163} + \cfrac{10^2 + 30}{163^2} + \cdots} \qquad (6)$$

$$\approx 4.66920160933975.$$

Amazingly, the Feigenbaum constant $\delta \approx 4.669$ is "universal" (i.e., the same) for all 1-D MAPS $f(x)$ if $f(x)$ has a single locally quadratic MAXIMUM. More specifically, the Feigenbaum constant is universal for 1-D MAPS if the SCHWARZIAN DERIVATIVE

$$D_{\text{Schwarzian}} = \frac{f'''(x)}{f'(x)} - \frac{3}{2}\left[\frac{f''(x)}{f'(x)}\right]^2 \qquad (7)$$

is NEGATIVE in the bounded interval (Tabor 1989, p. 220). Examples of maps which are universal include the HÉNON MAP, LOGISTIC MAP, LORENZ SYSTEM, Navier-Stokes truncations, and sine map $x_{n+1} = a \sin(\pi x_n)$. The value of the Feigenbaum constant can be computed explicitly using functional group renormalization theory. The universal constant also occurs in phase transitions in physics and, curiously, is very nearly equal to

$$\pi + \tan^{-1}(e^\pi) = 4.669201932\ldots \qquad (8)$$

For an AREA-PRESERVING 2-D MAP with

$$x_{n+1} = f(x_n, y_n) \qquad (9)$$

$$y_{n+1} = g(x_n, y_n), \qquad (10)$$

the Feigenbaum constant is $\delta = 8.7210978\ldots$ (Tabor 1989, p. 225). For a function OF THE FORM

$$f(x) = 1 - a|x|^n \qquad (11)$$

with a and n constant and n an INTEGER, the Feigenbaum constant for various n is given in the following table (Briggs 1991, Briggs *et al.* 1991, Finch), which updates the values in Tabor (1989, p. 225).

n	δ	α
3	5.9679687038...	1.9276909638...
4	7.2846862171...	1.6903029714...
5	8.3494991320...	1.5557712501...
6	9.2962468327...	1.4677424503...

An additional constant α, defined as the separation of adjacent elements of PERIOD DOUBLED ATTRACTORS from one double to the next, has a value

$$\lim_{n \to \infty} \frac{d_n}{d_{n+1}} \equiv -\alpha = -2.502907875\ldots \qquad (12)$$

for "universal" maps (Rasband 1990, p. 37). This value may be approximated from functional group renormalization theory to the zeroth order by

$$1 - \alpha^{-1} = \frac{1 - \alpha^{-2}}{[1 - \alpha^{-2}(1 - \alpha^{-1})]^2}, \qquad (13)$$

which, when the QUINTIC EQUATION is numerically solved, gives $\alpha = -2.48634\ldots$, only 0.7% off from the actual value (Feigenbaum 1988).

See also ATTRACTOR, BIFURCATION, FEIGENBAUM FUNCTION, LINEAR STABILITY, LOGISTIC EQUATION, PERIOD DOUBLING

References

Briggs, K. "A Precise Calculation of the Feigenbaum Constants." *Math. Comput.* **57**, 435–39, 1991.

Briggs, K.; Quispel, G.; and Thompson, C. "Feigenvalues for Mandelsets." *J. Phys. A: Math. Gen.* **24** 3363–368, 1991.

Collet, P. and Eckmann, J.-P. "Properties of Continuous Maps of the Interval to Itself." *Mathematical Problems in Theoretical Physics* (Ed. K. Osterwalder). New York: Springer-Verlag, 1979.

Collet, P. and Eckmann, J.-P. *Iterated Maps on the Interval as Dynamical Systems.* Boston, MA: Birkhäuser, 1980.

Eckmann, J.-P. and Wittwer, P. *Computer Methods and Borel Summability Applied to Feigenbaum's Equations.* New York: Springer-Verlag, 1985.

Feigenbaum, M. J. "Presentation Functions, Fixed Points, and a Theory of Scaling Function Dynamics." *J. Stat. Phys.* **52**, 527–69, 1988.

Finch, S. "Favorite Mathematical Constants." http://www.mathsoft.com/asolve/constant/fgnbaum/fgnbaum.html.

Finch, S. "Generalized Feigenbaum Constants." http://www.mathsoft.com/asolve/constant/fgnbaum/general.html.

Lanford, O. E. "A Computer-Assisted Proof of the Feigenbaum Conjectures." *Bull. Amer. Math. Soc.* **6**, 427–34, 1982.

Rasband, S. N. *Chaotic Dynamics of Nonlinear Systems.* New York: Wiley, 1990.

Stephenson, J. W. and Wang, Y. "Numerical Solution of Feigenbaum's Equation." *Appl. Math. Notes* **15**, 68–8, 1990.

Stephenson, J. W. and Wang, Y. "Relationships Between the Solutions of Feigenbaum's Equations." *Appl. Math. Let.* **4**, 37–9, 1991.

Stoschek, E. "Modul 33: Algames with Numbers." http://marvin.sn.schule.de/~inftreff/modul33/task33.htm.

Tabor, M. *Chaos and Integrability in Nonlinear Dynamics: An Introduction.* New York: Wiley, 1989.

Feigenbaum Function

Consider an arbitrary 1-D MAP

$$x_{n+1} = F(x_n) \tag{1}$$

at the onset of CHAOS. After a suitable rescaling, the Feigenbaum function

$$g(x) = \lim_{n \to \infty} \frac{1}{F^{(2^n)}(0)} F^{(2^n)}\left(x F^{(2^n)}(0)\right) \tag{2}$$

is obtained. This function satisfies

$$g(g(x)) = -\frac{1}{\alpha} g(\alpha x), \tag{3}$$

with $\alpha = 2.50290\ldots$, a quantity related to the FEIGENBAUM CONSTANT.

See also BIFURCATION, CHAOS, FEIGENBAUM CONSTANT

References

Grassberger, P. and Procaccia, I. "Measuring the Strangeness of Strange Attractors." *Physica D* **9**, 189–08, 1983.

Feit-Thompson Conjecture

The conjecture that there are no PRIMES p and q for which $(p^q - 1)/(p - 1)$ and $(q^p - 1)/(q - 1)$ have a common factor. Parker noticed that if this were true, it would greatly simplify the lengthy proof of the FEIT-THOMPSON THEOREM (Guy 1994, p. 81). However, the counterexample $(p = 17, q = 3313)$ with a common factor 112,643 was subsequently found by Stephens (1971). There are no other such pairs with both values less than 400,000.

See also FEIT-THOMPSON THEOREM

References

Apostol, T. M. "The Resultant of the Cyclotomic Polynomials $F_m(ax)$ and $F_n(bx)$." *Math. Comput.* **29**, 1–, 1975.

Feit, W. and Thompson, J. G. "A Solvability Criterion for Finite Groups and Some Consequences." *Proc. Nat. Acad. Sci. USA* **48**, 968–70, 1962.

Feit, W. and Thompson, J. G. "Solvability of Groups of Odd Order." *Pacific J. Math.* **13**, 775–029, 1963.

Guy, R. K. *Unsolved Problems in Number Theory, 2nd ed.* New York: Springer-Verlag, p. 81, 1994.

Stephens, N. M. "On the Feit-Thompson Conjecture." *Math. Comput.* **25**, 625, 1971.

Wells, D. G. *The Penguin Dictionary of Curious and Interesting Numbers.* London: Penguin, p. 17, 1986.

Feit-Thompson Theorem

Every FINITE SIMPLE GROUP (which is not CYCLIC) has EVEN ORDER, and the ORDER of every FINITE SIMPLE noncommutative group is DOUBLY EVEN, i.e., divisible by 4 (Feit and Thompson 1963).

See also BURNSIDE PROBLEM, FEIT-THOMPSON CONJECTURE, FINITE GROUP, ORDER (GROUP), SIMPLE GROUP

References

Guy, R. K. *Unsolved Problems in Number Theory, 2nd ed.* New York: Springer-Verlag, p. 81, 1994.

Feit, W. and Thompson, J. G. "A Solvability Criterion for Finite Groups and Some Consequences." *Proc. Nat. Acad. Sci. USA* **48**, 968–70, 1962.

Feit, W. and Thompson, J. G. "Solvability of Groups of Odd Order." *Pacific J. Math.* **13**, 775–029, 1963.

Fejes Tóth's Integral

$$\frac{1}{2\pi(n+1)} \int_{-\pi}^{\pi} f(x) \left\{ \frac{\sin\left[\frac{1}{2}(n+1)x\right]}{\sin\left(\frac{1}{2}x\right)} \right\}^2 dx$$

gives the nth CESÀRO MEAN of the FOURIER SERIES of $f(x)$.

References

Szego, G. *Orthogonal Polynomials, 4th ed.* Providence, RI: Amer. Math. Soc., p. 12, 1975.

Fejes Tóth's Problem

SPHERICAL CODE

Feldman's Theorem

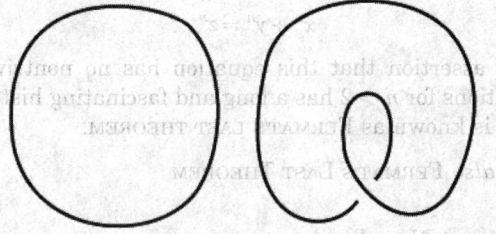

Any nondegenerate closed SPACE CURVE may be nondegenerately deformed into either of the two curves illustrated above. Neither of these can be nondegenerately transformed into the other.

References

Feldman, E. A. "Deformations of Closed Space Curves." *J. Diff. Geom.* **2**, 67–5, 1968.

Pohl, W. F. "The Self-Linking Number of a Closed Space Curve." *J. Math. Mech.* **17**, 975–85, 1968.

Feller's Coin-Tossing Constants

COIN TOSSING

Feller-Lévy Condition

Given a sequence of independent random variates X_1, X_2, ..., if $\sigma_k^2 = \mathrm{var}(X_k)$ and

$$\rho_n^2 \equiv \max_{k \le n} \left(\frac{\sigma_k^2}{s_n^2} \right),$$

then

$$\lim_{n \to \infty} \rho_n^2 = 0.$$

This means that if the LINDEBERG CONDITION holds for the sequence of variates $X_1, ...,$ then the VARIANCE of an individual term in the sum S_n of X_k is asymptotically negligible. For such sequences, the LINDEBERG CONDITION is NECESSARY as well as SUFFICIENT for the LINDEBERG-FELLER CENTRAL LIMIT THEOREM to hold.

See also BERRY-ESSÉEN THEOREM, CENTRAL LIMIT THEOREM, LINDEBERG CONDITION

References

Lindeberg, J. W. "Eine neue Herleitung des Exponentialgesetzes in der Wahrschienlichkeitsrechnung." *Math. Z.* **15**, 211–25, 1922.

Zabell, S. L. "Alan Turing and the Central Limit Theorem." *Amer. Math. Monthly* **102**, 483–94, 1995.

Fence

Values one STEP outside the HINGES are called inner fences, and values two steps outside the HINGES are called outer fences. Tukey calls values outside the outer fences FAR OUT.

See also ADJACENT VALUE

References

Tukey, J. W. *Explanatory Data Analysis.* Reading, MA: Addison-Wesley, p. 44, 1977.

Fence Poset

A PARTIAL ORDER defined by $(i-1)$, i), $(i+1)$, i) for ODD i.

See also PARTIAL ORDER

References

Ruskey, F. "Information on Ideals of Partially Ordered Sets." http://www.theory.csc.uvic.ca/~cos/inf/pose/Ideals.html.

Ferguson-Forcade Algorithm

The first practical algorithm for determining if there exist integers a_i for given real numbers x_i such that

$$a_1 x_1 + a_2 x_2 + \ldots + a_n x_n = 0,$$

or else establish bounds within which no such INTEGER RELATION can exist (Ferguson and Forcade 1979). The algorithm therefore became the first viable generalization of the EUCLIDEAN ALGORITHM to $n \ge 3$ variables.

A nonrecursive variant of the original algorithm was subsequently devised by Ferguson (1987). The Ferguson-Forcade algorithm has been shown to be polynomial-time in the logarithm in the size of a smallest relation, but has not been shown to be polynomial in dimension (Ferguson *et al.* 1999).

See also CONSTANT PROBLEM, EUCLIDEAN ALGORITHM, INTEGER RELATION, PSLQ ALGORITHM

References

Bailey, D. H. "Numerical Results on the Transcendence of Constants Involving π, e, and Euler's Constant." *Math. Comput.* **50**, 275–81, 1988.

Bergman, G. "Notes on Ferguson and Forcade's Generalized Euclidean Algorithm." Unpublished notes. Berkeley, CA: University of California at Berkeley, Nov. 1980.

Ferguson, H. R. P. "A Short Proof of the Existence of Vector Euclidean Algorithms." *Proc. Amer. Math. Soc.* **97**, 8–0, 1986.

Ferguson, H. R. P. "A Non-Inductive GL(n, Z) Algorithm that Constructs Linear Relations for n Z-Linearly Dependent Real Numbers." *J. Algorithms* **8**, 131–45, 1987.

Ferguson, H. R. P.; Bailey, D. H.; and Arno, S. "Analysis of PSLQ, An Integer Relation Finding Algorithm." *Math. Comput.* **68**, 351–69, 1999.

Ferguson, H. R. P. and Forcade, R. W. "Generalization of the Euclidean Algorithm for Real Numbers to All Dimensions Higher than Two." *Bull. Amer. Math. Soc.* **1**, 912–14, 1979.

Ferguson, H. R. P. and Forcade, R. W. "Multidimensional Euclidean Algorithms." *J. reine angew. Math.* **334**, 171–81, 1982.

Fermat 4n + 1 Theorem

Every PRIME p OF THE FORM $p = 4n + 1$ is a sum of two SQUARE NUMBERS in one unique way (up to the order of SUMMANDS). The theorem was stated by Fermat, but the first published proof was by Euler.

The first few primes p which are 1 or 2 (mod 4) are 2, 5, 13, 17, 29, 37, 41, 53, 61, ... (Sloane's A002313) (with the only prime congruent to 2 mod 4 being 2). The numbers (x, y) such that $x^2 + y^2$ equal these primes are (1, 1), (1, 2), (2, 3), (1, 4), (2, 5), (1, 6), ... (Sloane's A002331 and A002330).

See also SIERPINSKI'S PRIME SEQUENCE THEOREM, SQUARE NUMBER

References

Conway, J. H. and Guy, R. K. *The Book of Numbers.* New York: Springer-Verlag, pp. 146–47, 1996.
Hardy, G. H. and Wright, E. M. *An Introduction to the Theory of Numbers, 5th ed.* Oxford, England: Clarendon Press, pp. 13 and 219, 1979.
Séroul, R. "Prime Number and Sum of Two Squares." §2.11 in *Programming for Mathematicians.* Berlin: Springer-Verlag, pp. 18–9, 2000.
Sloane, N. J. A. Sequences A002313/M1430, A002330/M000462, and A002331/M0096 in "An On-Line Version of the Encyclopedia of Integer Sequences." http://www.research.att.com/~njas/sequences/eisonline.html.

Fermat Compositeness Test

The COMPOSITENESS TEST consisting of the application of FERMAT'S LITTLE THEOREM

Fermat Conic

A PLANE CURVE OF THE FORM $y = x^n$. For $n > 0$, the curve is a generalized PARABOLA; for $n < 0$ it is a generalized HYPERBOLA.

See also CONIC SECTION, HYPERBOLA, PARABOLA

Fermat Difference Equation

PELL EQUATION

Fermat Diophantine Equation

PELL EQUATION

Fermat Elliptic Curve Theorem

The only whole number solution to the DIOPHANTINE EQUATION

$$y^3 = x^2 + 2$$

is $y = 3$, $x = \pm 5$. This theorem was offered as a problem by Fermat, who suppressed his own proof.

Fermat Equation

The DIOPHANTINE EQUATION

$$x^n + y^n = z^n.$$

The assertion that this equation has no nontrivial solutions for $n > 2$ has a long and fascinating history and is known as FERMAT'S LAST THEOREM.

See also FERMAT'S LAST THEOREM

Fermat Number

A BINOMIAL NUMBER OF THE FORM $F_n = 2^{2^n} + 1$. The first few for $n = 0, 1, 2, ...$ are 3, 5, 17, 257, 65537, 4294967297, ... (Sloane's A000215). The number of DIGITS for a Fermat number is

$$D(n) = \left\lfloor \left[\log(2^{2^n} + 1) \right] + 1 \right\rfloor \approx \left\lfloor \log(2^{2^n}) + 1 \right\rfloor$$
$$= \lfloor 2^n \log 2 + 1 \rfloor. \tag{1}$$

Being a Fermat number is the NECESSARY (but not SUFFICIENT) form a number

$$N_n \equiv 2^n + 1 \tag{2}$$

must have in order to be PRIME. This can be seen by noting that if $N_n = 2^n + 1$ is to be PRIME, then n cannot have any ODD factors b or else N_n would be a factorable number OF THE FORM

$$2^n + 1 = (2^a)^b + 1 = (2^a + 1)$$
$$\times \left[2^{a(b-1)} - 2^{a(b-2)} + 2^{a(b-3)} - \dots + 1 \right]. \tag{3}$$

Therefore, for a PRIME N_n, n must be a POWER of 2. No two Fermat numbers have a common divisor greater than 1 (Hardy and Wright 1979, p. 14).

Fermat conjectured in 1650 that every Fermat number is PRIME and Eisenstein (1844) proposed as a problem the proof that there are an infinite number of Fermat primes (Ribenboim 1996, p. 88). At present, however, only COMPOSITE Fermat numbers F_n are known for $n \geq 5$. An anonymous writer proposed that numbers OF THE FORM $2^2 + 1$, $2^{2^2} + 1$, $2^{2^{2^2}} + 1$ were PRIME. However, this conjecture was refuted when Selfridge (1953) showed that

$$F_{16} = 2^{2^{16}} + 1 = 2^{2^{2^{2^2}}} + 1 \tag{4}$$

is COMPOSITE (Ribenboim 1996, p. 88). Numbers OF THE FORM $a^{2^n} + b^{2^n}$ are called generalized Fermat numbers (Ribenboim 1996, pp. 359–60).

Fermat numbers satisfy the RECURRENCE RELATION

$$F_m = F_0 F_1 \dots F_{m-1} + 2. \tag{5}$$

F_n can be shown to be PRIME IFF it satisfies PÉPIN'S TEST

$$3^{(F_n-1)/2} \equiv -1 \pmod{F_n}. \qquad (6)$$

PÉPIN'S THEOREM

$$3^{2^{2^{n-1}}} \equiv -1 \pmod{F_n} \qquad (7)$$

is also NECESSARY and SUFFICIENT.

In 1770, Euler showed that any FACTOR of F_n must have the form

$$2^{n+1}K + 1, \qquad (8)$$

where K is a POSITIVE INTEGER. In 1878, Lucas increased the exponent of 2 by one, showing that FACTORS of Fermat numbers must be OF THE FORM

$$2^{n+2}L + 1. \qquad (9)$$

If

$$F = p_1 p_2 \cdots p_r \qquad (10)$$

is the factored part of $F_n = FC$ (where C is the cofactor to be tested for primality), compute

$$A \equiv 3^{F_n - 1} \pmod{F_n} \qquad (11)$$

$$B \equiv 3^{F - 1} \pmod{F_n} \qquad (12)$$

$$R \equiv A - B \pmod{C}. \qquad (13)$$

Then if $R \equiv 0$, the cofactor is a PROBABLE PRIME to the base 3^F;; otherwise C is COMPOSITE.

In order for a POLYGON to be circumscribed about a CIRCLE (i.e., a CONSTRUCTIBLE POLYGON), it must have a number of sides N given by

$$N = 2^k F_0 \ldots F_n, \qquad (14)$$

where the F_n are *distinct* Fermat primes (as stated by Gauss and first published by Wantzel 1836). This is equivalent to the statement that the trigonometric functions $\sin(k\pi/N)$, $\cos(k\pi/N)$, etc., can be computed in terms of finite numbers of additions, multiplications, and square root extractions IFF N is of the above form. The only known Fermat PRIMES are

$$F_0 = 3$$

$$F_1 = 5$$

$$F_2 = 17$$

$$F_3 = 257$$

$$F_4 = 65537$$

and it seems unlikely that any more exist.

Factoring Fermat numbers is extremely difficult as a result of their large size. In fact, only F_5 to F_{11} have been complete factored, as summarized in the following table. Written out explicitly, the complete factorizations are

$$F_5 = 641 \cdot 6700417$$

$$F_6 = 274177 \cdot 67280421310721$$

$$F_7 = 59649589127497217 \cdot 5704689200685129054721$$

$$F_8 = 1238926361552897$$
$$\cdot 93461639715357977769163 \cdots$$
$$\cdots 5581996068965840512375416381885802 80321$$

$$F_9 = 2424833$$
$$\cdot 7455602825647884208337395736200 4 \cdots$$
$$\cdots 54918783366342657 \cdot P99$$

$$F_{10} = 45592577 \cdot 6487031809 \cdot 46597757852200185 \cdots$$
$$\cdots 4326456074307677819 2897 \cdot P252$$

$$F_{11} = 319489 \cdot 974849 \cdot 167988556341760475137$$
$$\cdot 3560841906445833920513 \cdot P564.$$

Here, the final large PRIME is not explicitly given since it can be computed by dividing F_n by the other given factors.

The following table summarizes the properties of completely factored Fermat numbers.

F_n	Digits	Factors	Digits	Reference
5	10	2	3, 7	Euler 1732
6	20	2	6, 14	Landry 1880
7	39	2	7, 22	Morrison and Brillhart 1975
8	78	2	16, 62	Brent and Pollard 1981
9	155	3	7, 49, 99	Manasse and Lenstra (In Cipra 1993)
10	309	4	8, 10, 40, 252	Brent 1995
11	617	5	6, 6, 21, 22, 564	Brent 1988

Tables of known factors of Fermat numbers are given by Keller (1983), Brillhart *et al.* (1988), Young and Buell (1988), Riesel (1994), and Pomerance (1996). Young and Buell (1988) discovered that F_{20} is COMPOSITE, and Crandall *et al.* (1995) that F_{22} is COMPOSITE. In 1999, Crandall *et al.* showed that F_{24} is COMPOSITE. A current list of the known factors of Fermat numbers is maintained by Keller, and reproduced in the form of a *Mathematica* notebook by Weisstein. In these tables, since all factors are OF THE FORM $k2^n + 1$, the known factors are expressed in the concise form (k, n). The number of factors for Fermat numbers F_n for $n = 0, 1, 2, \ldots$ are 1, 1, 1, 1, 1, 2, 2, 2, 2, 3, 4, 5, ...

See also CULLEN NUMBER, PÉPIN'S TEST, PÉPIN'S THEOREM, POCKLINGTON'S THEOREM, POLYGON, PROTH'S THEOREM, SELFRIDGE-HURWITZ RESIDUE, WOODALL NUMBER

References

Ball, W. W. R. and Coxeter, H. S. M. *Mathematical Recreations and Essays, 13th ed.* New York: Dover, pp. 68–9 and 94–5, 1987.

Brent, R. P. "Factorization of the Eighth Fermat Number." *Amer. Math. Soc. Abstracts* **1**, 565, 1980.

Brent, R. P. "Factorisation of F10." http://cslab.anu.edu.au/~rpb/F10.html.

Brent, R. P. "Factorization of the Tenth Fermat Number." *Math. Comput.* **68**, 429–51, 1999.

Brent, R. P. and Pollard, J. M. "Factorization of the Eighth Fermat Number." *Math. Comput.* **36**, 627–30, 1981.

Brillhart, J.; Lehmer, D. H.; Selfridge, J.; Wagstaff, S. S. Jr.; and Tuckerman, B. *Factorizations of $b^n \pm 1$, $b = 2, 3, 5, 6, 7, 10, 11, 12$ Up to High Powers, rev. ed.* Providence, RI: Amer. Math. Soc., pp. lxxxvii and 2– of Update 2.2, 1988.

Caldwell, C. K. "The Top Twenty: Fermat Divisors." http://www.utm.edu/research/primes/lists/top20/FermatDivisor.html.

Cipra, B. "Big Number Breakdown." *Science* **248**, 1608, 1990.

Conway, J. H. and Guy, R. K. "Fermat's Numbers." In *The Book of Numbers.* New York: Springer-Verlag, pp. 137–41, 1996.

Cormack, G. V. and Williams, H. C. "Some Very Large Primes of the Form $k \cdot 2^m + 1$." *Math. Comput.* **35**, 1419–421, 1980.

Courant, R. and Robbins, H. *What is Mathematics?: An Elementary Approach to Ideas and Methods, 2nd ed.* Oxford, England: Oxford University Press, pp. 25–6 and 119, 1996.

Crandall, R.; Doenias, J.; Norrie, C.; and Young, J. "The Twenty-Second Fermat Number is Composite." *Math. Comput.* **64**, 863–68, 1995.

Crandall, R. "F24 Resolved--Official Announcement." *NMBRTHRY@listserv.nodak.edu* posting, 29 Sep 1999.

Dickson, L. E. "Fermat Numbers $F_n = 2^{2^n} + 1$." Ch. 15 in *History of the Theory of Numbers, Vol. 1: Divisibility and Primality.* New York: Chelsea, pp. 375–80, 1952.

Dixon, R. *Mathographics.* New York: Dover, p. 53, 1991.

Euler, L. "Observationes de theoremate quodam Fermatiano aliisque ad numeros primos spectantibus." *Acad. Sci. Petropol.* **6**, 103–07, ad annos 1732–3 (1738). In *Leonhardi Euleri Opera Omnia*, Ser. I, Vol. II. Leipzig: Teubner, pp. 1–, 1915.

Gardner, M. "Patterns in Primes are a Clue to the Strong Law of Small Numbers." *Sci. Amer.* **243**, 18–8, Dec. 1980.

Gostin, G. B. "A Factor of F_{17}." *Math. Comput.* **35**, 975–76, 1980.

Gostin, G. B. "New Factors of Fermat Numbers." *Math. Comput.* **64**, 393–95, 1995.

Gostin, G. B. and McLaughlin, P. B. Jr. "Six New Factors of Fermat Numbers." *Math. Comput.* **38**, 645–49, 1982.

Guy, R. K. "Mersenne Primes. Repunits. Fermat Numbers. Primes of Shape $k \cdot 2^n + 2$." §A3 in *Unsolved Problems in Number Theory, 2nd ed.* New York: Springer-Verlag, pp. 8–3, 1994.

Hallyburton, J. C. Jr. and Brillhart, J. "Two New Factors of Fermat Numbers." *Math. Comput.* **29**, 109–12, 1975.

Hardy, G. H. and Wright, E. M. *An Introduction to the Theory of Numbers, 5th ed.* Oxford, England: Clarendon Press, pp. 14–5 and 19, 1979.

Hoffman, P. *The Man Who Loved Only Numbers: The Story of Paul Erdos and the Search for Mathematical Truth.* New York: Hyperion, p. 200, 1998.

Keller, W. "Factor of Fermat Numbers and Large Primes of the Form $k \cdot 2^n + 1$." *Math. Comput.* **41**, 661–73, 1983.

Keller, W. "Factors of Fermat Numbers and Large Primes of the Form $k \cdot 2^n + 1$, II." In prep.

Keller, W. "Prime Factors $k \cdot 2^n + 1$ of Fermat Numbers F_m and Complete Factoring Status." http://vamri.xray.ufl.edu/proths/fermat.html.

Kraitchik, M. "Fermat Numbers." §3.6 in *Mathematical Recreations.* New York: W. W. Norton, pp. 73–5, 1942.

Landry, F. "Note sur la décomposition du nombre $2^{64} + 1$ (Extrait)." *C. R. Acad. Sci. Paris*, **91**, 138, 1880.

Lenstra, A. K.; Lenstra, H. W. Jr.; Manasse, M. S.; and Pollard, J. M. "The Factorization of the Ninth Fermat Number." *Math. Comput.* **61**, 319–49, 1993.

Morrison, M. A. and Brillhart, J. "A Method of Factoring and the Factorization of F_7." *Math. Comput.* **29**, 183–05, 1975.

Pólya, G. and Szego, G. Problem 94, Part 8 in *Problems and Theorems in Analysis.* Berlin: Springer-Verlag, 1976.

Pomerance, C. "A Tale of Two Sieves." *Not. Amer. Math. Soc.* **43**, 1473–485, 1996.

Ribenboim, P. "Fermat Numbers" and "Numbers $k \times 2^n \pm 1$." §2.6 and 5.7 in *The New Book of Prime Number Records.* New York: Springer-Verlag, pp. 83–0 and 355–60, 1996.

Riesel, H. *Prime Numbers and Computer Methods for Factorization, 2nd ed.* Basel: Birkhäuser, pp. 384–88, 1994.

Robinson, R. M. "A Report on Primes of the Form $k \cdot 2^n + 1$ and on Factors of Fermat Numbers." *Proc. Amer. Math. Soc.* **9**, 673–81, 1958.

Selfridge, J. L. "Factors of Fermat Numbers." *Math. Comput.* **7**, 274–75, 1953.

Shanks, D. *Solved and Unsolved Problems in Number Theory, 4th ed.* New York: Chelsea, pp. 13 and 78–0, 1993.

Shorey, T. N. and Stewart, C. L. "On Divisors of Fermat, Fibonacci, Lucas and Lehmer Numbers, 2." *J. London Math. Soc.* **23**, 17–3, 1981.

Stewart, C. L. "On Divisors of Fermat, Fibonacci, Lucas and Lehmer Numbers." *Proc. London Math. Soc.* **35**, 425–47, 1977.

Sloane, N. J. A. Sequences A000215/M2503 in "An On-Line Version of the Encyclopedia of Integer Sequences." http://www.research.att.com/~njas/sequences/eisonline.html.

Wantzel, M. L. "Recherches sur les moyens de reconnaître si un problème de géométrie peut se résoudre avec la règle et le compas." *J. Math. pures appliq.* **1**, 366–72, 1836.

Weisstein, E. W. "Fermat Numbers." MATHEMATICA NOTEBOOK FERMAT.M.

Wrathall, C. P. "New Factors of Fermat Numbers." *Math. Comput.* **18**, 324–25, 1964.

Young, J. and Buell, D. A. "The Twentieth Fermat Number is Composite." *Math. Comput.* **50**, 261–63, 1988.

Fermat Number (Lucas)

A number OF THE FORM $2^n - 1$ obtained by setting $x = 1$ in a FERMAT POLYNOMIAL is called a MERSENNE NUMBER.

See also FERMAT-LUCAS NUMBER, MERSENNE NUMBER

Fermat Points

In a given ACUTE TRIANGLE ΔABC, the Fermat point X (or "first Fermat point" F_1, also called the Torricelli point) is the point which minimizes the sum of distances from A, B, and C,

$$|AX|+|BX|+|CX|. \tag{1}$$

This problem is called FERMAT'S PROBLEM or STEINER'S PROBLEM (Courant and Robbins 1941) and was proposed by Fermat to Torricelli. Torricelli's solution was published by his pupil Viviani in 1659 (Johnson 1929).

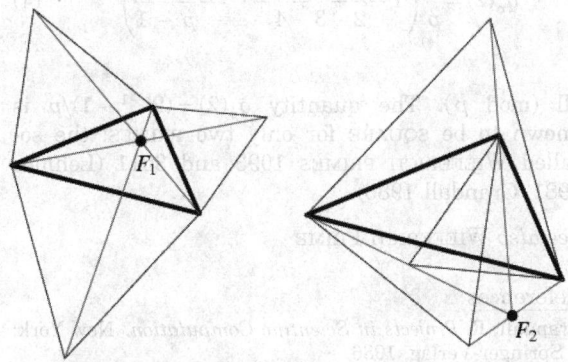

If all ANGLES of the TRIANGLE are less than $120°$ $(2\pi/3)$, then the Fermat point is the interior point X from which each side subtends an ANGLE of $120°$, i.e.,

$$\angle BXC = \angle CXA = \angle AXB = 120°. \tag{2}$$

The Fermat point can be constructed by drawing EQUILATERAL TRIANGLES on the outside of the given TRIANGLE and connecting opposite VERTICES. The three diagonals in the figure then intersect in the Fermat point. Similarly, the second Fermat point F_2 is constructed using equilateral triangles pointing inwards. The Fermat points are also known as the isogonic centers, since they are ISOGONAL CONJUGATES of the ISODYNAMIC POINTS.

The TRIANGLE CENTER FUNCTIONS of the Fermat points are

$$\alpha_1 = \csc\left(A + \frac{1}{3}\pi\right) \tag{3}$$

$$bc\left[c^2a^2 + (c^2 + a^2 - b^2)^2\right]\left[a^2b^2 - (a^2 + b^2 - c^2)^2\right]$$
$$\times \left[4\Delta - \sqrt{3}(b^2 + c^2 - d^2)\right] \tag{4}$$

$$\alpha_2 = \csc\left(A - \frac{1}{3}\pi\right) \tag{5}$$

The ANTIPEDAL TRIANGLE of F_1 is EQUILATERAL and has AREA

$$\Delta' = 2\Delta\left[1 + \cot\omega\cot\left(\frac{\pi}{3}\right)\right], \tag{6}$$

where ω is the BROCARD ANGLE. The ANTIPEDAL TRIANGLE of F_2 is also an EQUILATERAL and has AREA

$$2\Delta = \left[-1 + \cot\omega\cot\left(\frac{1}{3}\pi\right)\right]. \tag{7}$$

Given three POSITIVE REAL NUMBERS l, m, n, the "generalized" Fermat point is the point P of a given ACUTE TRIANGLE ΔABC such that

$$l \cdot PA + m \cdot PB + n \cdot PC \tag{8}$$

is a minimum (Greenberg and Robertello 1965, van de Lindt 1966, Tong and Chua 1995)

See also BROCARD ANGLE, EQUILATERAL TRIANGLE, FERMAT POINTS, ISODYNAMIC POINTS, ISOGONAL CONJUGATE, LESTER CIRCLE

References

Courant, R. and Robbins, H. *What is Mathematics?*, 2nd ed. Oxford, England: Oxford University Press, 1941.
Gallatly, W. *The Modern Geometry of the Triangle, 2nd ed.* London: Hodgson, p. 107, 1913.
Greenberg, I. and Robertello, R. A. "The Three Factory Problem." *Math. Mag.* **38**, 67–2, 1965.
Honsberger, R. *Mathematical Gems I.* Washington, DC: Math. Assoc. Amer., pp. 24–4, 1973.
Johnson, R. A. *Modern Geometry: An Elementary Treatise on the Geometry of the Triangle and the Circle.* Boston, MA: Houghton Mifflin, pp. 221–22, 1929.
Kimberling, C. "Central Points and Central Lines in the Plane of a Triangle." *Math. Mag.* **67**, 163–87, 1994.
Kimberling, C. "Fermat Point." http://cedar.evansville.edu/~ck6/tcenters/class/fermat.html.
Mowaffaq, H. "An Advanced Calculus Approach to Finding the Fermat Point." *Math. Mag.* **67**, 29–4, 1994.
Nelson, D. "Napoleon Revisited." *Math. Gaz.* No. 404, 1974.
Pottage, J. *Geometrical Investigations.* Reading, MA: Addison-Wesley, 1983.
Spain, P. G. "The Fermat Point of a Triangle." *Math. Mag.* **69**, 131–33, 1996.
Tong, J. and Chua, Y. S. "The Generalized Fermat's Point." *Math. Mag.* **68**, 214–15, 1995.
van de Lindt, W. J. "A Geometrical Solution of the Three Factory Problem." *Math. Mag.* **39**, 162–65, 1966.
Wells, D. *The Penguin Dictionary of Curious and Interesting Geometry.* Middlesex, England: Penguin Books, pp. 75–6, 1991.

Fermat Polynomial

The POLYNOMIALS obtained by setting $p(x) = 3x$ and $q(x) = -2$ in the LUCAS POLYNOMIAL SEQUENCES. The first few Fermat polynomials are

$$\mathcal{F}(x) = 1$$

$$\mathcal{F}_2(x) = 3x$$

$$\mathcal{F}_3(x) = 9x^2 - 2$$

$$\mathcal{F}_4(x) = 27x^3 - 12x$$

$$\mathscr{F}_5(x) = 81x^4 - 54x^2 + 4,$$

and the first few Fermat-Lucas polynomials are

$$f_1(x) = 3x$$
$$f_2(x) = 9x^2 - 4$$
$$f_3 = 27x^3 - 18x$$
$$f_4(x) = 81x^4 - 72x^2 + 8$$
$$f_5(x) = 243x^5 - 270x^3 + 60x.$$

Fermat and Fermat-Lucas POLYNOMIALS satisfy

$$\mathscr{F}_n(1) = \mathscr{F}_n$$
$$f_n(1) = f_n$$

where $\mathscr{F}_n$ are FERMAT NUMBERS and f_n are FERMAT-LUCAS NUMBERS.

Fermat Prime

A FERMAT NUMBER $F_n = 2^{2^n} + 1$ which is PRIME.

See also CONSTRUCTIBLE POLYGON, FERMAT NUMBER

Fermat Pseudoprime

A Fermat pseudoprime to a base a, written psp(a), is a COMPOSITE NUMBER n such that $a^{n-1} \equiv 1 \pmod{n}$ (i.e., it satisfies FERMAT'S LITTLE THEOREM, sometimes with the requirement that n must be ODD; Pomerance *et al.* 1980). psp(2)s are called POULET NUMBERS or, less commonly, SARRUS NUMBERS or FERMATIANS (Shanks 1993). The first few EVEN psp(2)s (including the PRIME 2 as a pseudoprime) are 2, 161038, 215326, ... (Sloane's A006935).

If base 3 is used in addition to base 2 to weed out potential COMPOSITE NUMBERS, only 4709 COMPOSITE NUMBERS remain $< 25 \times 10^9$. Adding base 5 leaves 2552, and base 7 leaves only 1770 COMPOSITE NUMBERS.

See also CARMICHAEL NUMBER, FERMAT'S LITTLE THEOREM, POULET NUMBER, PSEUDOPRIME

References
Hoffman, P. *The Man Who Loved Only Numbers: The Story of Paul Erdos and the Search for Mathematical Truth.* New York: Hyperion, p. 182, 1998.
Pomerance, C.; Selfridge, J. L.; and Wagstaff, S. S. "The Pseudoprimes to 25·10⁹." *Math. Comput.* **35**, 1003–026, 1980. Available electronically from ftp://sable.ox.ac.uk/pub/math/primes/ps2.Z.
Shanks, D. *Solved and Unsolved Problems in Number Theory, 4th ed.* New York: Chelsea, p. 115, 1993.
Sloane, N. J. A. Sequences A006935/M2190 in "An On-Line Version of the Encyclopedia of Integer Sequences." http://www.research.att.com/~njas/sequences/eisonline.html.

Fermat Quotient

The Fermat quotient for a number a and a PRIME base p is defined as

$$q_p(a) \equiv \frac{a^{p-1} - 1}{p}. \tag{1}$$

If $p \nmid ab$, then

$$q_p(ab) = q_p(a) + q_p(b) \tag{2}$$

$$q_p(p \pm 1) = \mp 1 \tag{3}$$

$$q_p(2) = \frac{1}{p}\left(1 - \frac{1}{2} + \frac{1}{3} - \frac{1}{4} + \cdots - \frac{1}{p-1}\right) \tag{4}$$

all (mod p). The quantity $q_p(2) = (2^{p-1} - 1)/p$ is known to be SQUARE for only two PRIMES: the so-called WIEFERICH PRIMES 1093 and 3511 (Lehmer 1981, Crandall 1986).

See also WIEFERICH PRIME

References
Crandall, R. *Projects in Scientific Computation.* New York: Springer-Verlag, 1986.
Lehmer, D. H. "On Fermat's Quotient, Base Two." *Math. Comput.* **36**, 289–90, 1981.
Wells, D. *The Penguin Dictionary of Curious and Interesting Numbers.* Middlesex, England: Penguin Books, p. 70, 1986.

Fermat's Algorithm
FERMAT'S FACTORIZATION METHOD

Fermat's Congruence
FERMAT'S LITTLE THEOREM

Fermat's Conjecture
FERMAT'S LAST THEOREM

Fermat's Divisor Problem

In 1657, Fermat posed the problem of finding solutions to

$$\sigma(x^3) = y^2 \tag{1}$$

and

$$\sigma(x^2) = y^3, \tag{2}$$

where $\sigma(n)$ is the DIVISOR FUNCTION (Dickson 1952).

The first few solutions to $\sigma(x^3) = y^2$ are $(x, y) = (1, 1)$, (7, 20), (751530, 1292054400) (Sloane's A008849 and A048948) .... Lucas stated that there are an infinite number of solutions (Dickson 1952, p. 56), but only solutions up to the fourth are known to be complete.

The first few solutions to $\sigma(x^2) = y^3$ are $(x, y) = (1, 1)$, (43098, 1729), ... (Sloane's A008850 and A048949), with only solutions up to the second known to be complete.

See also DIVISOR FUNCTION, WALLIS'S PROBLEM

References

Beiler, A. H. *Recreations in the Theory of Numbers: The Queen of Mathematics Entertains.* New York: Dover, p. 9, 1966.

Dickson, L. E. *History of the Theory of Numbers, Vol. 1: Divisibility and Primality.* New York: Chelsea, pp. 54–8, 1952.

Sloane, N. J. A. Sequences A008849, A008850, A048948, and A048949 in "An On-Line Version of the Encyclopedia of Integer Sequences." http://www.research.att.com/~njas/sequences/eisonline.html.

Fermat's Factorization Method

Given a number n, look for INTEGERS x and y such that $n = x^2 - y^2$. Then

$$n = (x - y)(x + y) \tag{1}$$

and n is factored. Any ODD NUMBER can be represented in this form since then $n = ab$, a and b are ODD, and

$$a = x + y \tag{2}$$

$$b = x - y. \tag{3}$$

Adding and subtracting,

$$a + b = 2x \tag{4}$$

$$a - b = 2y, \tag{5}$$

so solving for x and y gives

$$x = \frac{1}{2}(a + b) \tag{6}$$

$$y = \frac{1}{2}(a - b). \tag{7}$$

Therefore,

$$x^2 - y^2 = \frac{1}{4}\left[(a + b)^2 - (a - b)^2\right] = ab. \tag{8}$$

As the first trial for x, try $x_1 \lceil \sqrt{n} \rceil$, where $\lceil x \rceil$ is the CEILING FUNCTION. Then check if

$$\Delta x_1 = x_1^2 - n \tag{9}$$

is a SQUARE NUMBER. There are only 22 combinations of the last two digits which a SQUARE NUMBER can assume, so most combinations can be eliminated. If Δx_1 is not a SQUARE NUMBER, then try

$$x_2 = x_1 + 1, \tag{10}$$

so

$$\begin{aligned}
\Delta x_2 &= x_2^2 - n \\
&= (x_1 + 1)^2 - n = x_1^2 + 2x_1 + 1 - n \\
&= \Delta x_1 + 2x_1 + 1. \tag{11}
\end{aligned}$$

Continue with

$$\begin{aligned}
\Delta x_3 &= x_3^2 - n \\
&= (x_2 + 1)^2 - n = x_2^2 + 2x_2 + 1 - n = \Delta x_2 + 2x_2 + 1 \\
&= \Delta x_2 + 2x_1 + 3, \tag{12}
\end{aligned}$$

so subsequent differences are obtained simply by adding two.

Maurice Kraitchik sped up the ALGORITHM by looking for x and y satisfying

$$x^2 \equiv y^2 \pmod{n}, \tag{13}$$

i.e., $n | (x^2 - y^2)$. This congruence has uninteresting solutions $x \equiv \pm y \pmod{n}$ and interesting solutions $x \not\equiv \pm y \pmod{n}$. It turns out that if n is ODD and DIVISIBLE by at least two different PRIMES, then at least half of the solutions to $x^2 \equiv y^2 \pmod{n}$ with xy COPRIME to n are interesting. For such solutions, $(n, x - y)$ is neither n nor 1 and is therefore a nontrivial factor of n (Pomerance 1996). This ALGORITHM can be used to prove primality, but is not practical. In 1931, Lehmer and Powers discovered how to search for such pairs using CONTINUED FRACTIONS. This method was improved by Morrison and Brillhart (1975) into the CONTINUED FRACTION FACTORIZATION ALGORITHM, which was the fastest ALGORITHM in use before the QUADRATIC SIEVE factorization method was developed.

See also PRIME FACTORIZATION ALGORITHMS, SMOOTH NUMBER

References

Lehmer, D. H. and Powers, R. E. "On Factoring Large Numbers." *Bull. Amer. Math. Soc.* **37**, 770–76, 1931.

McKee, J. "Speeding Fermat's Factoring Method." *Math. Comput.* **68**, 1729–738, 1999.

Morrison, M. A. and Brillhart, J. "A Method of Factoring and the Factorization of F_7." *Math. Comput.* **29**, 183–05, 1975.

Pomerance, C. "A Tale of Two Sieves." *Not. Amer. Math. Soc.* **43**, 1473–485, 1996.

Fermat's Last Theorem

A theorem first proposed by Fermat in the form of a note scribbled in the margin of his copy of the ancient Greek text *Arithmetica* by Diophantus. The scribbled note was discovered posthumously, and the original is now lost. However, a copy was preserved in a book published by Fermat's son. In the note, Fermat claimed to have discovered a proof that the DIOPHANTINE EQUATION $x^n + y^n = z^n$ has no INTEGER solutions for $n > 2$.

The full text of Fermat's statement, written in Latin, reads "Cubum autem in duos cubos, aut quadrato-quadratum in duos quadrato-quadratos, et generaliter nullam in infinitum ultra quadratum potestatem in duos eiusdem nominis fas est dividere cuius rei demonstrationem mirabilem sane detexi. Hanc marginis exiguitas non caperet" (Nagell 1951, p. 252). In translation, "It is impossible for a cube to be the sum of two cubes, a fourth power to be the sum of two fourth powers, or in general for any number that is a power greater than the second to be the sum of two like powers. I have discovered a truly marvelous demonstration of this proposition that this margin is too narrow to contain."

As a result of Fermat's marginal note, the proposition that the DIOPHANTINE EQUATION

$$x^n + y^n = z^n, \tag{1}$$

where x, y, z, and n are INTEGERS, has no NONZERO solutions for $n > 2$ has come to be known as Fermat's Last Theorem. It was called a "THEOREM" on the strength of Fermat's statement, despite the fact that no other mathematician was able to prove it for hundreds of years.

Note that the restriction $n > 2$ is obviously necessary since there are a number of elementary formulas for generating an infinite number of PYTHAGOREAN TRIPLES (x, y, z) satisfying the equation for $n = 2$,

$$x^2 + y^2 = z^2. \tag{2}$$

A first attempt to solve the equation can be made by attempting to factor the equation, giving

$$\left(z^{n/2} + y^{n/2}\right)\left(z^{n/2} - y^{n/2}\right) = x^n. \tag{3}$$

Since the product is an exact POWER,

$$\begin{cases} z^{n/2} + y^{n/2} = 2^{n-1}p^n \\ z^{n/2} - y^{n/2} = 2q^n \end{cases} \text{ or } \begin{cases} z^{n/2} + y^{n/2} = 2p^n \\ z^{n/2} - y^{n/2} = 2^{n-1}q^n \end{cases}. \tag{4}$$

Solving for y and z gives

$$\begin{cases} z^{n/2} = 2^{n-2}p^n + q^n \\ y^{n/2} = 2^{n-2}p^n - q^n \end{cases} \text{ or } \begin{cases} z^{n/2} = p^n + 2^{n-2}q^n \\ y^{n/2} = p^n - 2^{n-2}q^n \end{cases}, \tag{5}$$

which give

$$\begin{cases} z = \left(2^{n-2}p^n + q^n\right)^{2/n} \\ y = \left(2^{n-2}p^n - q^n\right)^{2/n} \end{cases} \text{ or } \begin{cases} z = \left(p^n + 2^{n-2}q^n\right)^{2/n} \\ y = \left(p^n - 2^{n-2}q^n\right)^{2/n} \end{cases}. \tag{6}$$

However, since solutions to these equations in RATIONAL NUMBERS are no easier to find than solutions to the original equation, this approach unfortunately does not provide any additional insight.

It is sufficient to prove Fermat's Last Theorem by considering PRIME POWERS only, since the arguments can otherwise be written

$$(x^m)^p + (y^m)^p = (z^m)^p, \tag{7}$$

so redefining the arguments gives

$$z^p + y^p = z^p. \tag{8}$$

The so-called "first case" of the theorem is for exponents which are RELATIVELY PRIME to x, y, and z ($p \nmid x, y, z$) and was considered by Wieferich. Sophie Germain proved the first case of Fermat's Last Theorem for any ODD PRIME p when $2p + 1$ is also a PRIME. Legendre subsequently proved that if p is a PRIME such that $4p + 1$, $8p + 1$, $10p + 1$, $14p + 1$, or $16p + 1$ is also a PRIME, then the first case of Fermat's Last Theorem holds for p. This established Fermat's Last Theorem for $p < 100$. In 1849, Kummer proved it for all REGULAR PRIMES and COMPOSITE NUMBERS of which they are factors (Vandiver 1929, Ball and Coxeter 1987).

Kummer's attack led to the theory of IDEALS, and Vandiver developed VANDIVER'S CRITERIA for deciding if a given IRREGULAR PRIME satisfies the theorem. Genocchi (1852) proved that the first case is true for p if $(p, p - 3)$ is not an IRREGULAR PAIR. In 1858, Kummer showed that the first case is true if either $(p, p - 3)$ or $(p, p - 5)$ is an IRREGULAR PAIR, which was subsequently extended to include $(p, p - 7)$ and $(p, p - 9)$ by Mirimanoff (1905). Vandiver (1920ab) pointed out gaps and errors in Kummer's memoir which, in his view, invalidate Kummer's proof of Fermat's Last Theorem for the irregular primes 37, 59, and 67, although he claims Mirimanoff's proof of FLT for exponent 37 is still valid.

Wieferich (1909) proved that if the equation is solved in integers RELATIVELY PRIME to an ODD PRIME p, then

$$2^{p-1} \equiv 1 \pmod{p^2}. \tag{9}$$

(Ball and Coxeter 1987). Such numbers are called WIEFERICH PRIMES. Mirimanoff (1909) subsequently showed that

$$3^{p-1} \equiv 1 \pmod{p^2} \tag{10}$$

must also hold for solutions RELATIVELY PRIME to an ODD PRIME p, which excludes the first two WIEFERICH PRIMES 1093 and 3511. Vandiver (1914) showed

$$5^{p-1} \equiv 1 \pmod{p^2}, \tag{11}$$

and Frobenius extended this to

$$11^{p-1}, 17^{p-1} \equiv 1 \pmod{p^2}. \tag{12}$$

It has also been shown that if p were a PRIME OF THE FORM $6x - 1$, then

$$7^{p-1}, 13^{p-1}, 19^{p-1} \equiv 1 \pmod{p^2}, \tag{13}$$

which raised the smallest possible p in the "first case" to 253,747,889 by 1941 (Rosser 1941). Granville and Monagan (1988) showed if there exists a PRIME p satisfying Fermat's Last Theorem, then

$$q^{p-1} \equiv 1 \pmod{p^2} \qquad (14)$$

for $q = 5, 7, 11, ..., 71$. This establishes that the first case is true for all PRIME exponents up to 714,591,416,091,398 (Vardi 1991).

The "second case" of Fermat's Last Theorem (for $p|x,y,z$) proved harder than the first case.

Euler proved the general case of the theorem for $n = 3$, Fermat $n = 4$, Dirichlet and Lagrange $n = 5$. In 1832, Dirichlet established the case $n = 14$. The $n = 7$ case was proved by Lamé (1839; Wells 1986, p. 70), using the identity

$$(X+Y+Z)^7 - (X^7+Y^7+Z^7) = 7(X+Y)(X+Z)(Y+Z)$$
$$\times \left[\left(X^2+Y^2+Z^2+XY+XZ+YZ\right)^2 + XYZ(X+Y+Z) \right].$$
$$(15)$$

Although some errors were present in this proof, these were subsequently fixed by Lebesgue (1840). Much additional progress was made over the next 150 years, but no completely general result had been obtained. Buoyed by false confidence after his proof that PI IS TRANSCENDENTAL, the mathematician Lindemann proceeded to publish several proofs of Fermat's Last Theorem, all of them invalid (Bell 1937, pp. 464–65). A prize of 100,000 German marks, known as the Wolfskehl Prize, was also offered for the first valid proof (Ball and Coxeter 1987, p. 72; Barner 1997; Hoffman 1998, pp. 193–94 and 199).

A recent false alarm for a general proof was raised by Y. Miyaoka (Cipra 1988) whose proof, however, turned out to be flawed. Other attempted proofs among both professional and amateur mathematicians are discussed by vos Savant (1993), although vos Savant erroneously claims that work on the problem by Wiles (discussed below) is invalid. By the time 1993 rolled around, the general case of Fermat's Last Theorem had been shown to be true for all exponents up to 4×10^6 (Cipra 1993). However, given that a proof of Fermat's Last Theorem requires truth for *all* exponents, proof for any finite number of exponents does not constitute any significant progress towards a proof of the general theorem (although the fact that no counterexamples were found for this many cases is highly suggestive).

In 1993, a bombshell was dropped. In that year, the general theorem was partially proven by Andrew Wiles (Cipra 1993, Stewart 1993) by proving the SEMISTABLE case of the TANIYAMA-SHIMURA CONJEC-TURE. Unfortunately, several holes were discovered in the proof shortly thereafter when Wiles' approach via the TANIYAMA-SHIMURA CONJECTURE became hung up on properties of the SELMER GROUP using a tool called an EULER SYSTEM. However, the difficulty was circumvented by Wiles and R. Taylor in late 1994 (Cipra 1994, 1995ab) and published in Taylor and Wiles (1995) and Wiles (1995). Wiles' proof succeeds by (1) replacing ELLIPTIC CURVES with Galois representa-tions, (2) reducing the problem to a CLASS NUMBER FORMULA, (3) proving that FORMULA, and (4) tying up loose ends that arise because the formalisms fail in the simplest degenerate cases (Cipra 1995a).

The proof of Fermat's Last Theorem marks the end of a mathematical era. Since virtually all of the tools which were eventually brought to bear on the problem had yet to be invented in the time of Fermat, it is interesting to speculate about whether he actually was in possession of an elementary proof of the theorem. Judging by the temerity with which the problem resisted attack for so long, Fermat's alleged proof seems likely to have been illusionary. This conclusion is further supported by the fact that Fermat searched for proofs for the cases $n = 4$ and $n = 5$, which would have been superfluous had he actually been in possession of a general proof.

See also ABC CONJECTURE, BEAL'S CONJECTURE, BOGOMOLOV-MIYAOKA-YAU INEQUALITY, EULER SYS-TEM, FERMAT-CATALAN CONJECTURE, GENERALIZED FERMAT EQUATION, MORDELL CONJECTURE, PYTHA-GOREAN TRIPLE, RIBET'S THEOREM, SELMER GROUP, SOPHIE GERMAIN PRIME, SZPIRO'S CONJECTURE, TA-NIYAMA-SHIMURA CONJECTURE, VOJTA'S CONJECTURE, WARING FORMULA

References

Ball, W. W. R. and Coxeter, H. S. M. *Mathematical Recrea-tions and Essays, 13th ed.* New York: Dover, pp. 69–3, 1987.

Barner, K. "Paul Wolfskehl and the Wolfskehl Prize." *Not. Amer. Math. Soc.* **44**, 1294–303, 1997.

Beiler, A. H. "The Stone Wall." Ch. 24 in *Recreations in the Theory of Numbers: The Queen of Mathematics Enter-tains.* New York: Dover, 1966.

Bell, E. T. *Men of Mathematics.* New York: Simon and Schuster, 1937.

Bell, E. T. *The Last Problem.* New York: Simon and Schuster, 1961.

Cipra, B. A. "Fermat Theorem Proved." *Science* **239**, 1373, 1988.

Cipra, B. A. "Mathematics--Fermat's Last Theorem Finally Yields." *Science* **261**, 32–3, 1993.

Cipra, B. A. "Is the Fix in on Fermat's Last Theorem?" *Science* **266**, 725, 1994.

Cipra, B. A. "Fermat's Theorem--At Last." *What's Happen-ing in the Mathematical Sciences, 1995–996, Vol. 3.* Providence, RI: Amer. Math. Soc., pp. 2–4, 1996.

Cipra, B. A. "Princeton Mathematician Looks Back on Fermat Proof." *Science* **268**, 1133–134, 1995b.

Courant, R. and Robbins, H. "Pythagorean Numbers and Fermat's Last Theorem." §2.3 in Supplement to Ch. 1 in *What is Mathematics?: An Elementary Approach to Ideas and Methods, 2nd ed.* Oxford, England: Oxford University Press, pp. 40–2, 1996.

Cox, D. A. "Introduction to Fermat's Last Theorem." *Amer. Math. Monthly* **101**, 3–4, 1994.

Darmon, H. and Merel, L. "Winding Quotients and Some Variants of Fermat's Last Theorem." *J. reine angew. Math.* **490**, 81–00, 1997.

Dickson, L. E. "Fermat's Last Theorem, $ax^r + by^s = cz^t$, and the Congruence $x^n + y^n \equiv z^n \pmod{p}$." Ch. 26 in *History of the Theory of Numbers, Vol. 2: Diophantine Analysis.* New York: Chelsea, pp. 731–76, 1952.

Edwards, H. M. *Fermat's Last Theorem: A Genetic Introduction to Algebraic Number Theory.* New York: Springer-Verlag, 1977.

Edwards, H. M. "Fermat's Last Theorem." *Sci. Amer.* **239**, 104–22, Oct. 1978.

Granville, A. "Review of BBC's Horizon Program, 'Fermat's Last Theorem'." *Not. Amer. Math. Soc.* **44**, 26–8, 1997.

Granville, A. and Monagan, M. B. "The First Case of Fermat's Last Theorem is True for All Prime Exponents up to 714,591,416,091,389." *Trans. Amer. Math. Soc.* **306**, 329–59, 1988.

Guy, R. K. "The Fermat Problem." §D2 in *Unsolved Problems in Number Theory, 2nd ed.* New York: Springer-Verlag, pp. 144–46, 1994.

Hanson, A. "Fermat Project." http://www.cica.indiana.edu/ projects/Fermat/.

Hoffman, P. *The Man Who Loved Only Numbers: The Story of Paul Erdos and the Search for Mathematical Truth.* New York: Hyperion, pp. 183–99, 1998.

Kolata, G. "Andrew Wiles: A Math Whiz Battles 350-Year-Old Puzzle." *New York Times*, June 29, 1993.

Lynch, J. "Fermat's Last Theorem." BBC Horizon television documentary. http://www.bbc.co.uk/horizon/fermat.shtml.

Lynch, J. (Producer and Writer). "The Proof." NOVA television episode. 52 mins. Broadcast by the U. S. Public Broadcasting System on Oct. 28, 1997.

Mirimanoff, D. "Sur le dernier théorème de Fermat et le critérium de Wiefer." *Enseignement Math.* **11**, 455–59, 1909.

Mordell, L. J. *Fermat's Last Theorem.* New York: Chelsea, 1956.

Murty, V. K. (Ed.). *Fermat's Last Theorem: Proceedings of the Fields Institute for Research in Mathematical Sciences on Fermat's Last Theorem, Held 1993–994 Toronto, Ontario, Canada.* Providence, RI: Amer. Math. Soc., 1995.

Nagell, T. "Fermat's Last Theorem." §68 in *Introduction to Number Theory.* New York: Wiley, pp. 251–53, 1951.

Osserman, R. (Ed.). *Fermat's Last Theorem. The Theorem and Its Proof: An Exploration of Issues and Ideas.* 98 min. videotape and 56 pp. book. 1994.

Ribenboim, P. *13 Lectures on Fermat's Last Theorem.* New York: Springer-Verlag, 1979.

Ribenboim, P. *Fermat's Last Theorem for Amateurs.* New York: Springer-Verlag, 1999.

Ribet, K. A. and Hayes, B. "Fermat's Last Theorem and Modern Arithmetic." *Amer. Sci.* **82**, 144–56, March/April 1994.

Ribet, K. A. and Hayes, B. Correction to "Fermat's Last Theorem and Modern Arithmetic." *Amer. Sci.* **82**, 205, May/June 1994.

Rosser, B. "On the First Case of Fermat's Last Theorem." *Bull. Amer. Math. Soc.* **45**, 636–40, 1939.

Rosser, B. "A New Lower Bound for the Exponent in the First Case of Fermat's Last Theorem." *Bull. Amer. Math. Soc.* **46**, 299–04, 1940.

Rosser, B. "An Additional Criterion for the First Case of Fermat's Last Theorem." *Bull. Amer. Math. Soc.* **47**, 109–10, 1941.

Shanks, D. *Solved and Unsolved Problems in Number Theory, 4th ed.* New York: Chelsea, pp. 144–49, 1993.

Singh, S. *Fermat's Enigma: The Quest to Solve the World's Greatest Mathematical Problem.* New York: Walker & Co., 1997.

Stewart, I. "Fermat's Last Time-Trip." *Sci. Amer.* **269**, 112–15, 1993.

Swinnerton-Dwyer, P. *Nature* **364**, 13–4, 1993.

Taylor, R. and Wiles, A. "Ring-Theoretic Properties of Certain Hecke Algebras." *Ann. Math.* **141**, 553–72, 1995.

van der Poorten, A. *Notes on Fermat's Last Theorem.* New York: Wiley, 1996.

Vandiver, H. S. "On Kummer's Memoir of 1857 Concerning Fermat's Last Theorem." *Proc. Nat. Acad. Sci.* **6**, 266–69, 1920a.

Vandiver, H. S. "On the Class Number of the Field $\Omega(e^{2i\pi/p^n})$ and the Second Case of Fermat's Last Theorem." *Proc. Nat. Acad. Sci.* **6**, 416–21, 1920b.

Vandiver, H. S. "On Fermat's Last Theorem." *Trans. Amer. Math. Soc.* **31**, 613–42, 1929.

Vandiver, H. S. *Fermat's Last Theorem and Related Topics in Number Theory.* Ann Arbor, MI: 1935.

Vandiver, H. S. "Fermat's Last Theorem: Its History and the Nature of the Known Results Concerning It." *Amer. Math. Monthly,* **53**, 555–78, 1946.

Vandiver, H. S. "A Supplementary Note to a 1946 Article on Fermat's Last Theorem." *Amer. Math. Monthly* **60**, 164–67, 1953.

Vandiver, H. S. "Examination of Methods of Attack on the Second Case of Fermat's Last Theorem." *Proc. Nat. Acad. Sci.* **40**, 732–35, 1954.

Vardi, I. *Computational Recreations in Mathematica.* Reading, MA: Addison-Wesley, pp. 59–1, 1991.

vos Savant, M. *The World's Most Famous Math Problem.* New York: St. Martin's Press, 1993.

Weisstein, E. W. "Books about Fermat's Last Theorem." http://www.treasure-troves.com/books/FermatsLastTheorem.html.

Wieferich, A. "Zum letzten Fermat'schen Theorem." *J. reine angew. Math.* **136**, 293–02, 1909.

Wiles, A. "Modular Elliptic-Curves and Fermat's Last Theorem." *Ann. Math.* **141**, 443–51, 1995.

Fermat's Lesser Theorem

FERMAT'S LITTLE THEOREM

Fermat's Little Theorem

If p is a PRIME NUMBER and a a NATURAL NUMBER, then

$$a^p \equiv a \pmod{p}. \tag{1}$$

Furthermore, if $p \nmid a$ (p does not divide a), then there exists some smallest exponent d such that

$$a^d - 1 \equiv 0 \pmod{p} \tag{2}$$

and d divides $p - 1$. Hence,

$$a^{p-1} - 1 \equiv 0 \pmod{p}. \tag{3}$$

This is a generalization of the CHINESE HYPOTHESIS and a special case of EULER'S THEOREM. It is sometimes called FERMAT'S PRIMALITY TEST and is a NECESSARY but not SUFFICIENT test for primality. Although it was presumably proved (but suppressed) by Fermat, the first proof was published by Euler in 1749.

The theorem is easily proved using mathematical INDUCTION. Suppose $p \mid a^p - a$. Then examine

$$(a + 1)^p - (a + 1). \tag{4}$$

From the BINOMIAL THEOREM,

$$(a+1)^p$$
$$= a^p + \binom{p}{1}a^{p-1} + \binom{p}{2}a^{p-2} + \cdots + \binom{p}{p-1}a + 1.$$
$$(5)$$

Rewriting,

$$(a+1)^p - a^p - 1$$
$$= \binom{p}{1}a^{p-1} + \binom{p}{2}a^{p-2} + \cdots + \binom{p}{p-1}a. \quad (6)$$

But p divides the right side, so it also divides the left side. Combining with the induction hypothesis gives that p divides the sum

$$[(a+1)^p - a^p - 1] + (a^p - a) = (a+1)^p - (a+1), \quad (7)$$

as assumed, so the hypothesis is true for any a. The theorem is sometimes called FERMAT'S SIMPLE THEOREM. WILSON'S THEOREM follows as a COROLLARY of Fermat's little theorem.

Fermat's little theorem shows that, if p is PRIME, there does not exist a base $a < p$ with $(a, p) = 1$ such that $a^{p-1} - 1$ possesses a nonzero residue modulo p. If such base a exists, p is therefore guaranteed to be composite. However, the lack of a nonzero residue in Fermat's little theorem does *not* guarantee that p is PRIME. The property of unambiguously certifying composite numbers while passing some PRIMES make Fermat's little theorem a COMPOSITENESS TEST which is sometimes called the FERMAT COMPOSITENESS TEST. A number satisfying Fermat's little theorem for some nontrivial base and which is not known to be composite is called a PROBABLE PRIME.

COMPOSITE NUMBERS known as FERMAT PSEUDOPRIMES (or sometimes simply "PSEUDOPRIMES") have zero residue for some as and so are not identified as composite. Worse still, there exist numbers known as CARMICHAEL NUMBERS (the smallest of which is 561) which give zero residue for *any* choice of the base a RELATIVELY PRIME to p. However, FERMAT'S LITTLE THEOREM CONVERSE provides a criterion for certifying the primality of a number. A table of the smallest PSEUDOPRIMES P for the first 100 bases a follows (Sloane's A007535; Beiler 1966, p. 42 with typos corrected).

a	P	a	P	a	P	a	P	a	P
2	341	22	69	42	205	62	63	82	91
3	91	23	33	43	77	63	341	83	105
4	15	24	25	44	45	64	65	84	85
5	124	25	28	45	76	65	112	85	129
6	35	26	27	46	133	66	91	86	87
7	25	27	65	47	65	67	85	87	91
8	9	28	45	48	49	68	69	88	91
9	28	29	35	49	66	69	85	89	99
10	33	30	49	50	51	70	169	90	91
11	15	31	49	51	65	71	105	91	115
12	65	32	33	52	85	72	85	92	93
13	21	33	85	53	65	73	111	93	301
14	15	34	35	54	55	74	75	94	95
15	341	35	51	55	63	75	91	95	141
16	51	36	91	56	57	76	77	96	133
17	45	37	45	57	65	77	247	97	105
18	25	38	39	58	133	78	341	98	99
19	45	39	95	59	87	79	91	99	145
20	21	40	91	60	341	80	81	100	153
21	55	41	105	61	91	81	85		

See also BINOMIAL THEOREM, CARMICHAEL NUMBER, CHINESE HYPOTHESIS, COMPOSITE NUMBER, COMPOSITENESS TEST, EULER'S THEOREM, FERMAT'S LITTLE THEOREM CONVERSE, FERMAT PSEUDOPRIME, MODULO MULTIPLICATION GROUP, PRATT CERTIFICATE, PRIMALITY TEST, PRIME NUMBER, PSEUDOPRIME, RELATIVELY PRIME, TOTIENT FUNCTION, WIEFERICH PRIME, WILSON'S THEOREM, WITNESS

References

Ball, W. W. R. and Coxeter, H. S. M. *Mathematical Recreations and Essays, 13th ed.* New York: Dover, p. 61, 1987.

Beiler, A. H. *Recreations in the Theory of Numbers: The Queen of Mathematics Entertains.* New York: Dover, 1966.

Conway, J. H. and Guy, R. K. *The Book of Numbers.* New York: Springer-Verlag, pp. 141–42, 1996.

Courant, R. and Robbins, H. "Fermat's Theorem." §2.2 in Supplement to Ch. 1 in *What is Mathematics?: An Elementary Approach to Ideas and Methods, 2nd ed.* Oxford, England: Oxford University Press, pp. 37–8, 1996.

Nagell, T. "Fermat's Theorem and Its Generalization by Euler." §21 in *Introduction to Number Theory.* New York: Wiley, pp. 71–3, 1951.

Séroul, R. "The Theorems of Fermat and Euler." §2.8 in *Programming for Mathematicians.* Berlin: Springer-Verlag, p. 15, 2000.

Shanks, D. *Solved and Unsolved Problems in Number Theory, 4th ed.* New York: Chelsea, p. 20, 1993.

Sloane, N. J. A. Sequences A007535/M5440 in "An On-Line Version of the Encyclopedia of Integer Sequences." http://www.research.att.com/~njas/sequences/eisonline.html.

Fermat's Little Theorem Converse

The converse of FERMAT'S LITTLE THEOREM is also known as LEHMER'S THEOREM. It states that, if an INTEGER x is PRIME to m and $x^{m-1} \equiv 1 \pmod{m}$ and there is no INTEGER $e < m - 1$ for which $x^e \equiv 1 \pmod{m}$, then m is PRIME. Here, x is called a WITNESS to the primality of m. This theorem is the basis for the PRATT PRIMALITY CERTIFICATE.

See also FERMAT'S LITTLE THEOREM, PRATT CERTIFICATE, PRIMALITY CERTIFICATE, WITNESS

References

Riesel, H. *Prime Numbers and Computer Methods for Factorization, 2nd ed.* Boston, MA: Birkhäuser, p. 96, 1994.

Wagon, S. *Mathematica in Action.* New York: W. H. Freeman, pp. 278–79, 1991.

Fermat's Polygonal Number Theorem

In 1638, Fermat proposed that every POSITIVE INTEGER is a sum of *at most* three TRIANGULAR NUMBERS, four SQUARE NUMBERS, five PENTAGONAL NUMBERS, and n n-POLYGONAL NUMBERS. Fermat claimed to have a proof of this result, although Fermat's proof has never been found. Gauss proved the triangular case, and noted the event in his diary on July 10, 1796, with the notation

$$* * E Y R H K A \quad num = \Delta + \Delta + \Delta.$$

This case is equivalent to the statement that every number OF THE FORM $8m + 3$ is a sum of three ODD SQUARES (Duke 1997). More specifically, a number is a sum of three SQUARES IFF it is not OF THE FORM $4^b(8m + 7)$ for $b \geq 0$, as first proved by Legendre in 1798.

Euler was unable to prove the square case of Fermat's theorem, but he left partial results which were subsequently used by Lagrange. The square case was finally proved by Jacobi and independently by Lagrange in 1772. It is therefore sometimes known as LAGRANGE'S FOUR-SQUARE THEOREM. In 1813, Cauchy proved the proposition in its entirety.

See also FIFTEEN THEOREM, LAGRANGE'S FOUR-SQUARE THEOREM, SUM OF SQUARES FUNCTION, VINOGRADOV'S THEOREM, WARING'S PROBLEM

References

Cassels, J. W. S. *Rational Quadratic Forms.* New York: Academic Press, 1978.

Cauchy, A. "Démonstration du théorème général de Fermat sur les nombres polygones." In *Oeuvres complètes d'Augustin Cauchy, Vol. VI (II Série).* Paris: Gauthier-Villars, pp. 320–53, 1905.

Conway, J. H.; Guy, R. K.; Schneeberger, W. A.; and Sloane, N. J. A. "The Primary Pretenders." *Acta Arith.* **78**, 307–13, 1997.

Duke, W. "Some Old Problems and New Results about Quadratic Forms." *Not. Amer. Math. Soc.* **44**, 190–96, 1997.

Nathanson, M. B. "A Short Proof of Cauchy's Polygonal Number Theorem." *Proc. Amer. Math. Soc.* **9**, 22–4, 1987.

Savin, A. "Shape Numbers." *Quantum* **11**, 14–8, 2000.

Shanks, D. *Solved and Unsolved Problems in Number Theory, 4th ed.* New York: Chelsea, pp. 143–44, 1993.

Smith, D. E. *A Source Book in Mathematics.* New York: Dover, p. 91, 1984.

Fermat's Primality Test

FERMAT'S LITTLE THEOREM

Fermat's Principle of Conjunctive Probability

The probability that two events will both happen is hk, where h is the probability that the first event will happen, and k is the probability that the second event will happen when the first even is known to have happened.

See also CONDITIONAL PROBABILITY

References

Whittaker, E. T. and Robinson, G. *The Calculus of Observations: A Treatise on Numerical Mathematics, 4th ed.* New York: Dover, p. 317, 1967.

Fermat's Problem

In a given ACUTE TRIANGLE $\triangle ABC$, locate a point whose distances from A, B, and C have the smallest possible sum. The solution is the point from which each side subtends an angle of $120°$, known as the first FERMAT POINT.

See also ACUTE TRIANGLE, FERMAT POINTS

Fermat's Right Triangle Theorem

The AREA of a RATIONAL RIGHT TRIANGLE cannot be a SQUARE NUMBER. This statement is equivalent to "a CONGRUUM cannot be a SQUARE NUMBER."

See also CONGRUUM, RATIONAL TRIANGLE, RIGHT TRIANGLE, SQUARE NUMBER

Fermat's Simple Theorem

FERMAT'S LITTLE THEOREM

Fermat's Spiral

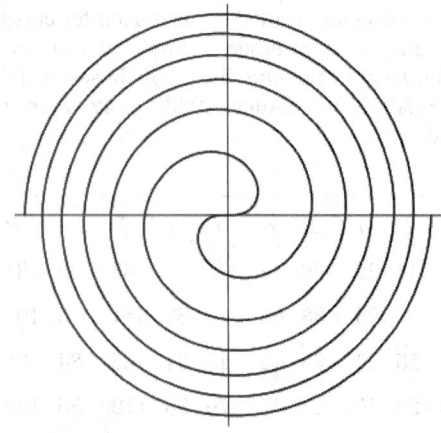

An ARCHIMEDEAN SPIRAL with $m = 2$ having polar

equation

$$r = a\theta^{1/2},$$

discussed by Fermat in 1636 (MacTutor Archive). It is also known as the PARABOLIC SPIRAL. For any given POSITIVE value of θ, there are two corresponding values of r of opposite signs. The resulting spiral is therefore symmetrical about the origin. The CURVATURE is

$$\kappa(\theta) = \frac{\dfrac{3a^2}{4\theta} + a^2\theta}{\left(\dfrac{a^2}{4\theta} + a^2\theta\right)^{3/2}}.$$

See also ARCHIMEDEAN SPIRAL, FERMAT'S SPIRAL INVERSE CURVE

References

Beyer, W. H. *CRC Standard Mathematical Tables, 28th ed.* Boca Raton, FL: CRC Press, p. 225, 1987.

Dixon, R. "The Mathematics and Computer Graphics of Spirals in Plants." *Leonardo* **16**, 86–0, 1983.

Dixon, R. *Mathographics.* New York: Dover, p. 121, 1991.

Gray, A. *Modern Differential Geometry of Curves and Surfaces with Mathematica, 2nd ed.* Boca Raton, FL: CRC Press, pp. 90 and 96, 1997.

Lockwood, E. H. *A Book of Curves.* Cambridge, England: Cambridge University Press, p. 175, 1967.

MacTutor History of Mathematics Archive. "Fermat's Spiral." http://www-groups.dcs.st-and.ac.uk/~history/Curves/Fermats.html.

Wells, D. *The Penguin Dictionary of Curious and Interesting Geometry.* Middlesex, England: Penguin Books, pp. 74–5, 1991.

Fermat's Spiral Inverse Curve

The INVERSE CURVE of FERMAT'S SPIRAL with the origin taken as the INVERSION CENTER is the LITUUS.

References

Lawrence, J. D. *A Catalog of Special Plane Curves.* New York: Dover, pp. 186–87, 1972.

Fermat's Theorem

A PRIME p can be represented in an essentially unique manner in the form $x^2 + y^2$ for integral x and y IFF $p \equiv 1 \pmod{4}$ or $p = 2$. It can be restated by letting

$$Q(x,y) \equiv x^2 + y^2,$$

then all RELATIVELY PRIME solutions (x, y) to the problem of representing $Q(x,y) = m$ for m any INTEGER are achieved by means of successive applications of the GENUS THEOREM and COMPOSITION THEOREM. There is an analog of this theorem for EISENSTEIN INTEGERS.

See also EISENSTEIN INTEGER, SQUARE NUMBER

References

Shanks, D. *Solved and Unsolved Problems in Number Theory, 4th ed.* New York: Chelsea, pp. 142–43, 1993.

Fermat's Two-Square Theorem

FERMAT'S THEOREM

Fermat-Catalan Conjecture

The conjecture that there are only finitely many triples of RELATIVELY PRIME integer powers x^p, y^q, z^r for which

$$x^p + y^q = z^r$$

with

$$\frac{1}{p} + \frac{1}{q} + \frac{1}{r} < 1.$$

Darmon and Merel (1997) have shown that there are no relatively prime solutions $(x, x, 3)$ with $x \geq 3$. Ten solutions are known,

$$1 + 2^3 = 3^2$$

$$2^5 + 7^2 = 3^4$$

$$7^3 + 13^2 = 2^9$$

$$2^7 + 17^3 = 71^2$$

$$3^5 + 11^4 = 122^2$$

$$17^7 + 76271^3 = 21063928^2$$

$$1414^3 + 2213459^2 = 65^7$$

$$9262^3 + 15312283^2 = 113^7$$

$$43^8 + 96222^3 = 30042907^2$$

$$33^8 + 1549034^2 = 15613^3$$

(Mauldin 1997).

See also FERMAT'S LAST THEOREM

References

Darmon, H. and Granville, A. "On the Equations $z^m = F(x,y)$ and $Ax^p + By^q = Cz^r$." *Bull. London Math. Soc.* **27**, 513–43, 1995.

Darmon, H. and Merel, L. "Winding Quotients and Some Variants of Fermat's Last Theorem." *J. reine angew. Math.* **490**, 81–00, 1997.

Mauldin, R. D. "A Generalization of Fermat's Last Theorem: The Beal Conjecture and Prize Problem." *Not. Amer. Math. Soc.* **44**, 1436–437, 1997.

Fermat-Euler Theorem

FERMAT'S LITTLE THEOREM

Fermatian
POULET NUMBER

Fermat-Lucas Number
A number OF THE FORM $2^n + 1$ obtained by setting $x = 1$ in a FERMAT-LUCAS POLYNOMIAL. The first few are 3, 5, 9, 17, 33, ... (Sloane's A000051).

See also FERMAT NUMBER (LUCAS)

References
Shorey, T. N. and Stewart, C. L. "On Divisors of Fermat, Fibonacci, Lucas and Lehmer Numbers, 2." *J. London Math. Soc.* **23**, 17–3, 1981.

Stewart, C. L. "On Divisors of Fermat, Fibonacci, Lucas and Lehmer Numbers." *Proc. London Math. Soc.* **35**, 425–47, 1977.

Sloane, N. J. A. Sequences A000051/M0717 in "An On-Line Version of the Encyclopedia of Integer Sequences." http://www.research.att.com/~njas/sequences/eisonline.html.

Fermi-Dirac Distribution
A distribution which arises in the study of half-integral spin particles in physics,

$$R(k) = \frac{k^a}{e^{k-\mu} + 1}.$$

Its integral is

$$\int_0^\infty \frac{k^a dk}{e^{k-\mu} + 1} = e^\mu \Gamma(s+1) \Phi(-e^\mu, s+1, 1),$$

where $\Phi(z, s, a)$ is the LERCH TRANSCENDENT.

Fern
BARNSLEY'S FERN

Ferrari's Identity

$$\left(a^2 + 2ac - 2bc - b^2\right)^4 + \left(b^2 - 2ab - 2ac - c^2\right)^4$$
$$+ \left(c^2 + 2ab + 2bc - a^2\right)^4$$
$$= 2\left(a^2 + b^2 + c^2 - ab + ac + bc\right)^4.$$

See also DIOPHANTINE EQUATION–4TH POWERS

References
Berndt, B. C. *Ramanujan's Notebooks, Part IV.* New York: Springer-Verlag, pp. 96–7, 1994.

Ferrars Diagram
FERRERS DIAGRAM

Ferrers Diagram

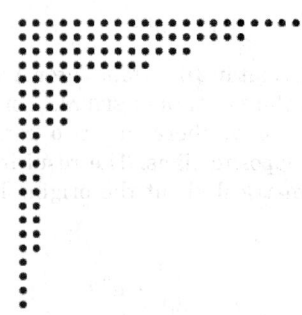

A Ferrers diagram represents PARTITIONS as patterns of dots, with the nth row having the same number of dots as the nth term in the PARTITION. The spelling "Ferrars" (Skiena 1990, pp. 53 and 78) is sometimes also used, and the diagram is sometimes called a graphical representation or **Ferrers graph** (Andrews 1998, p. 6). A Ferrers diagram of the PARTITION

$$n = a + b + \ldots + c,$$

for a list a, b, ..., c of k POSITIVE INTEGERS with $a \geq b \geq \ldots \geq c$ is therefore the arrangement of n dots or square boxes in k rows, such that the dots or boxes are left-justified, the first row is of length a, the second row is of length b, and so on, with the kth row of length c. The above diagram corresponds to one of the possible partitions of 100.

See also CONJUGATE PARTITION, DURFEE SQUARE, SELF-CONJUGATE PARTITION, YOUNG DIAGRAM

References
Andrews, G. E. *The Theory of Partitions.* Cambridge, England: Cambridge University Press, pp. 6–, 1998.

Comtet, L. "Ferrers Diagrams." §2.4 in *Advanced Combinatorics: The Art of Finite and Infinite Expansions, rev. enl. ed.* Dordrecht, Netherlands: Reidel, pp. 98–02, 1974.

Liu, C. L. *Introduction to Combinatorial Mathematics.* New York: McGraw-Hill, 1968.

MacMahon, P. A. *Combinatory Analysis, Vol. 2.* New York: Chelsea, pp. 3–, 1960.

Propp, J. "Some Variants of Ferrers Diagrams." *J. Combin. Th. A* **52**, 98–28, 1989.

Riordan, J. *An Introduction to Combinatorial Analysis.* New York: Wiley, pp. 108–09, 1980.

Skiena, S. "Ferrers Diagrams." §2.1.2 in *Implementing Discrete Mathematics: Combinatorics and Graph Theory with Mathematica.* Reading, MA: Addison-Wesley, pp. 53–5, 1990.

Stanley, R. P. *Enumerative Combinatorics, Vol. 1.* Cambridge, England: Cambridge University Press, 1999.

Stanton, D. and White, D. *Constructive Combinatorics.* New York: Springer-Verlag, 1986.

Ferrers Graph
FERRERS DIAGRAM

Ferrers Graph Polygon

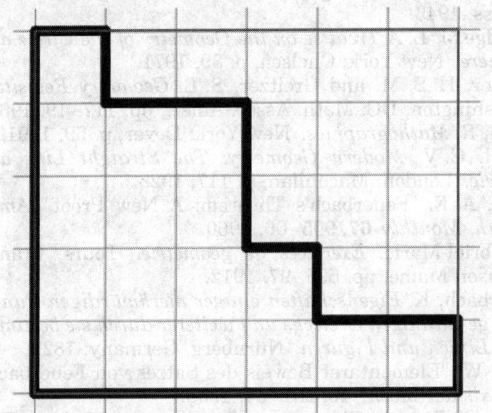

A SELF-AVOIDING POLYGON containing three corners of its minimal bounding rectangle. The anisotropic area and perimeter generating function $G(x,y)$ and partial generating functions $H_m(y)$, connected by

$$G(x,y,q) = \sum_{m \geq 1} H_m(y,q)x^m,$$

satisfy the self-reciprocity and inversion relations

$$H_m(1/y, 1/q) = (-1)^m y^{m-2} q^{(m^3-3m)/2} H_m(y,q)$$

and

$$G(x,y) - y^2 G(-x/y, 1/y) = 0$$

(Bousquet-Mélou *et al.* 1999).

See also LATTICE POLYGON, SELF-AVOIDING POLYGON

References

Bousquet-Mélou, M.; Guttmann, A. J.; Orrick, W. P.; and Rechnitzer, A. Inversion Relations, Reciprocity and Polyominoes. 23 Aug 1999. http://xxx.lanl.gov/abs/math.CO/9908123/.

© 1999–001 Wolfram Research, Inc.

Ferrers' Function

An alternative name for an associated LEGENDRE POLYNOMIAL.

See also LEGENDRE POLYNOMIAL

References

Sansone, G. *Orthogonal Functions, rev. English ed.* New York: Dover, p. 246, 1991.

Ferrier's Prime

According to Hardy and Wright (1979), the largest PRIME found before the days of electronic computers is the 44-digit number

$$F \equiv \frac{1}{17}(2^{148} + 1)$$

$$= 20988936657440586486151264256610222593863921,$$

which was found using only a mechanical calculator. *Mathematica* can verify primality of this number in a (small) fraction of a second, showing how far the art of numerical computation has advanced in the intervening years,

```
In[1]:=
PrimeQ[(2^148 + 1)/17] // Timing
Out[1]=
{0.0333333 Second, True}
```

See also PRIME NUMBER

References

Hardy, G. H. and Wright, E. M. *An Introduction to the Theory of Numbers, 5th ed.* Oxford, England: Clarendon Press, pp. 16–2, 1979.

Feuerbach Circle

NINE-POINT CIRCLE

Feuerbach Point

The point F at which the INCIRCLE and NINE-POINT CIRCLE are tangent. It has TRIANGLE CENTER FUNCTION

$$\alpha = 1 - \cos(B-C).$$

See also FEUERBACH'S THEOREM

References

Johnson, R. A. *Modern Geometry: An Elementary Treatise on the Geometry of the Triangle and the Circle.* Boston, MA: Houghton Mifflin, p. 200, 1929.
Kimberling, C. "Central Points and Central Lines in the Plane of a Triangle." *Math. Mag.* **67**, 163–87, 1994.
Salmon, G. *Conic Sections, 6th ed.* New York: Chelsea, p. 127, 1960.

Feuerbach's Conic Theorem

The LOCUS of the centers of all CONICS through the VERTICES and ORTHOCENTER of a TRIANGLE (which are RECTANGULAR HYPERBOLAS when not degenerate), is a CIRCLE through the MIDPOINTS of the sides, the points half way from the ORTHOCENTER to the VERTICES, and the feet of the ALTITUDE.

See also ALTITUDE, CONIC SECTION, FEUERBACH'S THEOREM, KIEPERT'S HYPERBOLA, MIDPOINT, ORTHOCENTER, RECTANGULAR HYPERBOLA

References

Coolidge, J. L. *A Treatise on Algebraic Plane Curves.* New York: Dover, p. 198, 1959.

Feuerbach's Theorem

There are two theorems commonly known as Feuerbach's theorem. The first states that CIRCLE which passes through the feet of the PERPENDICULARS dropped from the VERTICES of any TRIANGLE on the sides opposite them passes also through the MIDPOINTS of these sides as well as through the MIDPOINT of the segments which join the VERTICES to the point of intersection of the PERPENDICULAR. Such a circle is called a NINE-POINT CIRCLE.

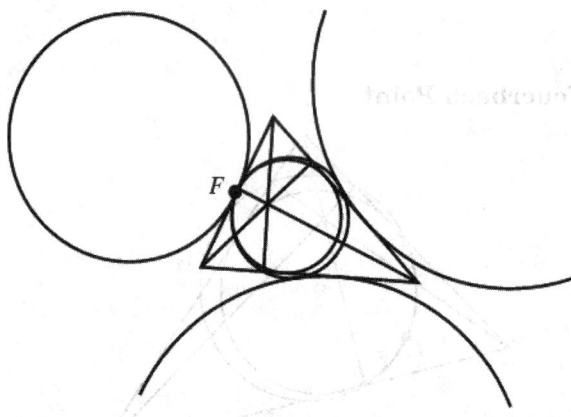

The proposition most frequently called Feuerbach's theorem states that the NINE-POINT CIRCLE of any TRIANGLE is TANGENT internally to the INCIRCLE and TANGENT externally to the three EXCIRCLES. This theorem was first published by Feuerbach (1822). Many proofs have been given (Elder 1960), with the simplest being the one presented by McClelland (1891, p. 225) and Lachlan (1893, p. 74).

See also EXCIRCLE, FEUERBACH POINT, HART CIRCLE, INCIRCLE, MIDPOINT, NINE-POINT CIRCLE, PERPENDICULAR, TANGENT

References

Altshiller-Court, N. *College Geometry: A Second Course in Plane Geometry for Colleges and Normal Schools*, 2nd ed., rev. enl. New York: Barnes and Noble, pp. 107, 273, and 290, 1952.

Baker, H. F. Appendix to Ch. 12 in *An Introduction to Plane Geometry.* Cambridge, England: Cambridge University Press, 1943.
Coolidge, J. L. *A Treatise on the Geometry of the Circle and Sphere.* New York: Chelsea, p. 39, 1971.
Coxeter, H. S. M. and Greitzer, S. L. *Geometry Revisited.* Washington, DC: Math. Assoc. Amer., pp. 117–19, 1967.
Dixon, R. *Mathographics.* New York: Dover, p. 59, 1991.
Durell, C. V. *Modern Geometry: The Straight Line and Circle.* London: Macmillan, p. 117, 1928.
Elder, A. E. "Feuerbach's Theorem: A New Proof." *Amer. Math. Monthly* **67**, 905–06, 1960.
F. Gabriel-Marie. *Exercices de géométrie.* Tours, France: Maison Mame, pp. 595–97, 1912.
Feuerbach, K. *Eigenschaften einiger merkwürdigen Punkte des geradlinigen Dreiecks und weiterer durch sie bestimmten Linien und Figuren.* Nürnberg, Germany: 1822.
Kroll, W. "Elementarer Beweis des Satzes von Feuerbach." *Praxis der Math.* **40**, 251–54, 1998.
Lachlan, R. *An Elementary Treatise on Modern Pure Geometry.* London: Macmillan, 1893.
McClelland, W. J. *Geometry of the Circle.* London, 1891.
Rouché, E. and de Comberousse, C. *Traité de géométrie plane.* Paris: Gauthier-Villars, pp. 307–09, 1900.
Sawayama, Y. "Démonstration élémentaire du théorème de Feuerbach." *L'enseign. math.* **7**, 479–82, 1905.
Sawayama, Y. "8 nouvelles démonstrations d'un théorème relatif au cercle des 9 points." *L'enseign. math.* **13**, 31–9, 1911.
Wells, D. *The Penguin Dictionary of Curious and Interesting Geometry.* Middlesex, England: Penguin Books, pp. 76–7, 1991.

Feynman Point

The sequence of six 9s which begins at the 762nd decimal place of PI,

$$\pi = 3.14159\ldots 134\,\underbrace{999999}_{\text{six 9s}}\,837\ldots$$

(Wells 1986, p. 51). The positions of the first occurrences of strings of 1, 2, ... consecutive 9s are 5, 44, 762, 762, 762, 762, 1722776, ... (Sloane's A048940). There is no string of seven 9s in the first million digits of PI.

See also PI DIGITS

References

Sloane, N. J. A. Sequences A048940 in "An On-Line Version of the Encyclopedia of Integer Sequences." http://www.research.att.com/~njas/sequences/eisonline.html.
Wells, D. *The Penguin Dictionary of Curious and Interesting Numbers.* Middlesex, England: Penguin Books, p. 51, 1986.

FFT

FAST FOURIER TRANSFORM

Fiber

A fiber of a map $f : X \to Y$ is the PREIMAGE of an element $y \in Y$. That is,

$$f^{-1}(y) = \{x \in X \text{ such that } f(x) = y\}.$$

For instance, let X and Y be the COMPLEX NUMBERS $\mathbb{C}$. When $f(z) = z^2$, every fiber consists of two points $\{z, -z\}$, except for the fiber over 0, which has one point. Note that a fiber may be the EMPTY SET.

In special cases, the fiber may be independent, in some sense, of the choice of $y \in Y$. For instance, if f is a COVERING MAP, then the fibers are all DISCRETE and have the same CARDINALITY. The example $f(z) = z^2$ is a covering map away from zero, i.e., $f(z) = z^2$ from the punctured plane $\mathbb{C} - \{0\}$ to itself has a fiber consisting of two points.

When $\pi : E \to M$ is a FIBER BUNDLE, then every fiber is ISOMORPHIC, in whatever CATEGORY is being used. For instance, when E is a REAL VECTOR BUNDLE of RANK k, every fiber is isomorphic to $\mathbb{R}^k$.

See also COMPLEX NUMBER, COVERING MAP, FIBER BUNDLE, MAP, RANK (BUNDLE), WHITNEY SUM

Fiber Bundle

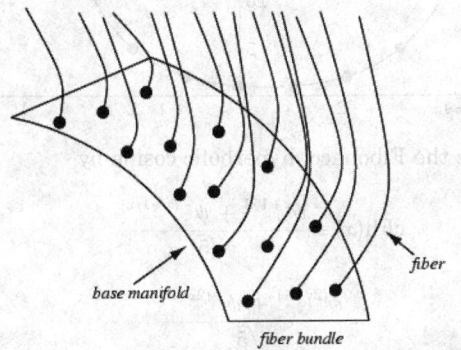

base manifold

fiber

fiber bundle

A fiber bundle (also called simply a BUNDLE) with FIBER F is a MAP $f : E \to B$ where E is called the TOTAL SPACE of the fiber bundle and B the BASE SPACE of the fiber bundle. The main condition for the MAP to be a fiber bundle is that every point in the BASE SPACE $b \in B$ has a NEIGHBORHOOD U such that $f^{-1}(U)$ is HOMEOMORPHIC to $U \times F$ in a special way. Namely, if

$$h : f^{-1}(U) \to U \times F$$

is the HOMEOMORPHISM, then

$$\text{proj}_U \circ h = f_{|f^{-1}(U)|},$$

where the MAP proj_U means projection onto the U component. The homeomorphisms h which "commute with projection" are called local TRIVIALIZATIONS for the fiber bundle f. In other words, E looks like the product $B \times F$ (at least locally), except that the fibers $f^{-1}(x)$ for $x \in B$ may be a bit "twisted."

A fiber bundle is the most general kind of BUNDLE. Special cases are often described by replacing the word "fiber" with a word that describes the fiber being used, e.g., VECTOR BUNDLES and PRINCIPAL BUNDLES.

Examples of fiber bundles include any product $B \times F \to B$ (which is a bundle over B with FIBER F), the MÖBIUS STRIP (which is a fiber bundle over the CIRCLE with FIBER given by the unit interval [0,1]; i.e, the BASE SPACE is the CIRCLE), and $\mathbb{S}^3$ (which is a bundle over $\mathbb{S}^2$ with fiber $\mathbb{S}^1$). A special class of fiber bundle is the VECTOR BUNDLE, in which the FIBER is a VECTOR SPACE. A basic example of a nontrivial bundle is the MÖBIUS STRIP, which is a fiber bundle with the circle as its base, $B = \mathbb{S}^{-1}$, and the interval $F = (-1, 1)$ as its fiber.

Some of the properties of graphs of functions $f : B \to F$ carry over to fiber bundles. A GRAPH of such a function sits in $B \times F$ as $(b, f(b))$. A graph always projects ONTO the base B and is ONE-TO-ONE.

A fiber bundle E is a TOTAL SPACE and, like $B \times F$, it has a projection $\pi : E \to B$. The PREIMAGE, $\pi^{-1}(b)$, of any point b is isomorphic to F. Unlike $B \times F$, there is no canonical projection from E to F. Instead, maps to F only make sense locally on B. Near any point b in the base B, there is a TRIVIALIZATION of E in which there are actual functions from a neighborhood to F.

These local functions can sometimes be patched together to give a (GLOBAL) SECTION $s : B \to E$ such that the projection of s is the identity. This is analogous to the map from a domain X of a function $f : X \to Y$ to its graph in $X \times Y$ by $\tilde{f}(x) = (x, f(x))$.

A fiber bundle also comes with a GROUP ACTION on the fiber. This group action represents the different ways the fiber can be viewed as equivalent. For instance, in topology, the GROUP might be the group of HOMEOMORPHISMS of the fiber. The group on a vector bundle is the group of INVERTIBLE LINEAR MAPS, which reflects the equivalent descriptions of a VECTOR SPACE using different BASES.

Fiber bundles are not always used to generalize functions. Sometimes they are convenient descriptions of interesting manifolds. A common example in GEOMETRIC TOPOLOGY is a torus bundle on the circle.

See also BUNDLE, FIBER SPACE, FIBRATION, GEOMETRIC TOPOLOGY, PRINCIPAL BUNDLE, SHEAF, TANGENT BUNDLE, VECTOR BUNDLE

Fiber Direct Sum

See also DIRECT SUM

© *1999–001 Wolfram Research, Inc.*

Fiber Space

A fiber space, depending on context, means either a FIBER BUNDLE or a FIBRATION.

See also FIBER BUNDLE, FIBRATION

Fibonacci
FIBONACCI NUMBER, FIBONACCI POLYNOMIAL

Fibonacci Coefficient

The coefficient defined by

$$\begin{bmatrix} m \\ k \end{bmatrix}_F = \frac{F_m F_{m-1} \cdots F_{m-k+1}}{F_1 F_2 \cdots F_k},$$

where $\begin{bmatrix} m \\ 0 \end{bmatrix}_F = 1$ and F_n is a FIBONACCI NUMBER. This coefficient satisfies

$$2\begin{bmatrix} n \\ m \end{bmatrix}_F = L_n \begin{bmatrix} m-1 \\ n \end{bmatrix} + L_{m-n} \begin{bmatrix} m-1 \\ n-1 \end{bmatrix}_F,$$

where L_n is a LUCAS NUMBER.

See also FIBONACCI NUMBER, LUCAS NUMBER

Fibonacci Dual Theorem

Let F_n be the nth FIBONACCI NUMBER. Then the sequence $\{F_n\}_{n=2}^{\infty} = \{1, 2, 3, 5, 8, \ldots\}$ is COMPLETE, even if one is restricted to subsequences in which no two consecutive terms are both passed over (until the desired total is reached; Brown 1965, Honsberger 1985).

See also COMPLETE SEQUENCE, FIBONACCI NUMBER.

References
Brown, J. L. Jr. "A New Characterization of the Fibonacci Numbers." *Fib. Quart.* **3**, 1–, 1965.
Honsberger, R. *Mathematical Gems III.* Washington, DC: Math. Assoc. Amer., p. 130, 1985.

Fibonacci Hyperbolic Functions

Let

$$\psi \equiv 1 + \phi = \frac{1}{2}(3 + \sqrt{5}) \approx 2.618034 \quad (1)$$

where ϕ is the GOLDEN RATIO, and

$$\alpha = \ln \phi \approx 0.4812118. \quad (2)$$

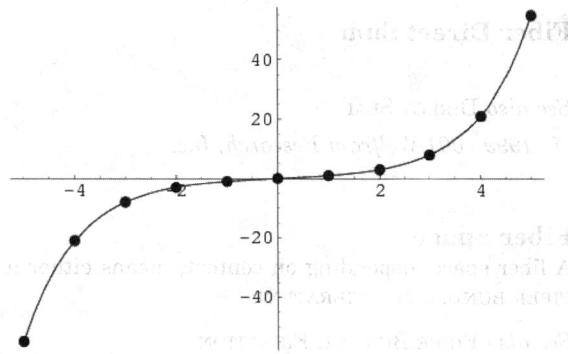

Define the Fibonacci hyperbolic sine by

$$sFh(x) \equiv \frac{\psi^x - \psi^{-x}}{\sqrt{5}} \quad (3)$$

$$= \frac{\phi^{2x} - \phi^{-2x}}{\sqrt{5}} \quad (4)$$

$$= \frac{2}{\sqrt{5}} \sinh[2x\alpha]. \quad (5)$$

The function satisfies

$$sFh(-x) = -sFh(x), \quad (6)$$

and for $n \in \mathbb{Z}$, $sFh(n) = F_{2n}$ where F_n is a FIBONACCI NUMBER.

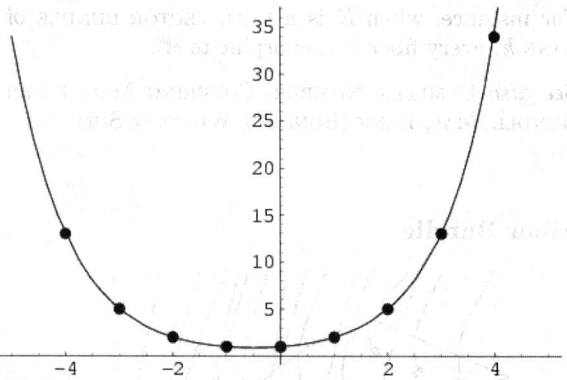

Define the Fibonacci hyperbolic cosine by

$$cFh(x) \equiv \frac{\psi^{x+1/2} + \psi^{-(x+1/2)}}{\sqrt{5}} \quad (7)$$

$$= \frac{\phi^{(2x+1)} + \phi^{-(2x+1)}}{\sqrt{5}} \quad (8)$$

$$= \frac{2}{\sqrt{5}} \cosh[(2x+1)\alpha]. \quad (9)$$

This function satisfies

$$cFh(-x) = cFh(x-1), \quad (10)$$

and for $n \in \mathbb{Z}$, $cFh(n) = F_{2n+1}$ where F_n is a FIBONACCI NUMBER.

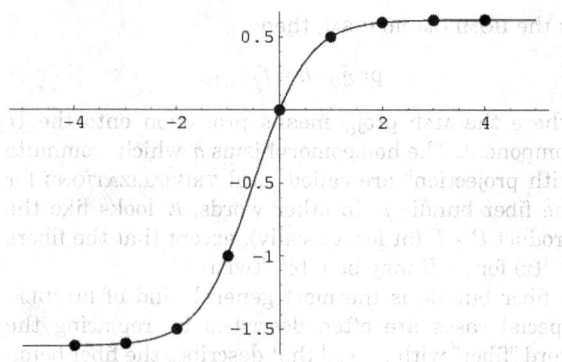

Similarly, the Fibonacci hyperbolic tangent is defined

by

$$sFh(x) \equiv \frac{cFh(x)}{cFh(x)},$$

and for $x \in \mathbb{Z}$, $cFh(n) = F_{2n}/F_{2n+1}$.

References

Trzaska, Z. W. "On Fibonacci Hyperbolic Trigonometry and Modified Numerical Triangles." *Fib. Quart.* **34**, 129–38, 1996.
© 1999–001 Wolfram Research, Inc.

Fibonacci Identity

Since

$$|(a+ib)(c+id)| = |a+ib||c+di| \qquad (1)$$

$$|(ac-bd)+i(bc+ad)| = \sqrt{a^2+b^2}\sqrt{c^2+d^2}, \qquad (2)$$

it follows that

$$(a^2+b^2)(c^2+d^2) = (ac-bd)^2 + (bc+ad)^2 \equiv e^2 + f^2. \quad (3)$$

This identity implies the 2-dimensional CAUCHY'S INEQUALITY.

See also CAUCHY'S INEQUALITY, EULER FOUR-SQUARE IDENTITY, LEBESGUE IDENTITY

References

Petkovsek, M.; Wilf, H. S.; and Zeilberger, D. *A = B*. Wellesley, MA: A. K. Peters, p. 9, 1996.

Fibonacci Matrix

A SQUARE MATRIX related to the FIBONACCI NUMBERS. The simplest is the FIBONACCI Q-MATRIX.

Fibonacci n-Step Number

An n-step Fibonacci sequence is given by defining $F_k = 0$ for $k \leq 0$, $F_1 = F_2 = 1$, $F_3 = 2$, and

$$F_k = \sum_{i=1}^{k} F_{n-i} \qquad (1)$$

for $k > 3$. The case $n = 1$ corresponds to the degenerate 1, 1, 2, 2, 2, 2 ..., $n = 2$ to the usual FIBONACCI NUMBERS 1, 1, 2, 3, 5, 8, ... (Sloane's A000045), $n = 3$ to the TRIBONACCI NUMBERS 1, 1, 2, 4, 7, 13, 24, 44, 81, ... (Sloane's A000073), $n = 4$ to the TETRANACCI NUMBERS 1, 1, 2, 4, 8, 15, 29, 56, 108, ... (Sloane's A000078), etc.

The limit $\lim_{k \to \infty} F_k/F_{k-1}$ is given by solving

$$x^n(2-x) = 1, \qquad (2)$$

or equivalently

$$x^n - x^{n-1} - x^{n-2} - \cdots - x - 1 = 0, \qquad (3)$$

for x and then taking the REAL ROOT $x > 1$. For EVEN n, there are exactly two real roots, one greater than 1 and one less than 1, and for ODD n, there is exactly one real root, which is always ≥ 1.

If $n = 2$, equation (2) reduces to

$$x^2(2-x) = 1 \qquad (4)$$

$$x^3 - 2x^2 + 1 = (x-1)(x^2 - x - 1) = 0, \qquad (5)$$

giving solutions

$$x = 1, \frac{1}{2}\left(1 \pm \sqrt{5}\right). \qquad (6)$$

The ratio is therefore

$$x = \frac{1}{2}\left(1 + \sqrt{5}\right) = \phi = 1.618..., \qquad (7)$$

which is the GOLDEN RATIO, as expected.

The analytic solutions for $n = 1, 2, ...$ are given by

$$x_1 = 1$$

$$x_2 = \frac{1}{2}\left(1 + \sqrt{5}\right)$$

$$x_3 = \frac{1}{3}\left[1 + \left(19 - 3\sqrt{33}\right)^{1/3} + \left(19 + 3\sqrt{33}\right)^{1/3}\right]$$

and numerically by 1, 1.61803, 1.83929, 1.92756, 1.96595, ..., approaching 2 as $n \to \infty$.

See also FIBONACCI NUMBER, TRIBONACCI NUMBER

References

Sloane, N. J. A. Sequences A000045/M0692, A000073/M1074, and A000078/M1108 in "An On-Line Version of the Encyclopedia of Integer Sequences." http://www.research.att.com/~njas/sequences/eisonline.html.

Fibonacci Number

The sequence of numbers F_n defined by the U_n in the LUCAS SEQUENCE, which can be viewed as a particular case of the FIBONACCI POLYNOMIALS $F_n(x)$ with $F_n = F_n(1)$. They are companions to the LUCAS NUMBERS and satisfy the same RECURRENCE RELATION,

$$F_n \equiv F_{n-2} + F_{n-1} \qquad (1)$$

for $n = 3$, 4, ..., with $F_1 = F_2 = 1$. The first few Fibonacci numbers are 1, 1, 2, 3, 5, 8, 13, 21, ... (Sloane's A000045). The Fibonacci numbers give the number of pairs of rabbits n months after a single pair begins breeding (and newly born bunnies are assumed to begin breeding when they are two months old), as first described by Leonardo of Pisa in his book *Liber Abaci*. Kepler also described the Fibonacci numbers (Kepler 1966; Wells 1986, pp. 61–2 and 65).

The ratios of successive Fibonacci numbers F_n/F_{n-1} approaches the GOLDEN RATIO ϕ as n approaches infinity, as first proved by Scottish mathematician Robert Simson in 1753 (Wells 1986, p. 62). The ratios of alternate Fibonacci numbers are given by the CONVERGENTS to ϕ^{-2}, where ϕ is the GOLDEN RATIO, and are said to measure the fraction of a turn between successive leaves on the stalk of a plant (PHYLLOTAXIS): 1/2 for elm and linden, 1/3 for beech and hazel, 2/5 for oak and apple, 3/8 for poplar and rose, 5/13 for willow and almond, etc. (Coxeter 1969, Ball and Coxeter 1987). The Fibonacci numbers are sometimes called PINE CONE NUMBERS (Pappas 1989, p. 224). The role of the Fibonacci numbers in botany is sometimes called LUDWIG'S LAW (Szymkiewicz 1928; Wells 1986, p. 66; Steinhaus 1983, p. 299).

Another RECURRENCE RELATION for the Fibonacci numbers is

$$F_{n+1} = \left\lfloor \frac{F_n\left(1 + \sqrt{5}\right) + 1}{2} \right\rfloor = \left\lfloor \phi F_n + \frac{1}{2} \right\rfloor, \quad (2)$$

where $\lfloor x \rfloor$ is the FLOOR FUNCTION and ϕ is the GOLDEN RATIO. This expression follows from the more general RECURRENCE RELATION that

$$\begin{vmatrix} F_{n+1} & F_{n+2} & \cdots & F_{n+k} \\ F_{n+k+1} & F_{n+k+2} & \cdots & F_{n+2k} \\ \vdots & \vdots & \ddots & \vdots \\ F_{n+k(k-1)+1} & F_{n+k(k-1)+2} & \cdots & F_{n+k^2} \end{vmatrix} = 0. \quad (3)$$

The GENERATING FUNCTION for the Fibonacci numbers is

$$g(x) = \sum_{n=0}^{\infty} F_n x^n = \frac{x}{1 - x - x^2}$$

$$= x + x^2 + 2x^3 + 3x^4 + 5x^5 + \ldots. \quad (4)$$

By plugging in $x = 1/10$, this gives the curious addition tree illustrated below,

$$\sum_{n=0}^{\infty} \frac{F_n}{10^n} = \frac{10}{89}, \quad (5)$$

so

$$\sum_{n=0}^{\infty} \frac{F_n}{10^{n+1}} = \frac{1}{89} \quad (6)$$

```
        0
        1
        1
        2
        3
        5
        8
       13
       21
       34
       55
       89
  0112359550561...
```

Yuri Matiyasevich (1970) showed that there is a

polynomial P in n, m, and a number of other variables x, y, z, ... having the property that $n = F_{2m}$ IFF there exist integers x, y, z, ... such that $p(n, m, x, y, z, \ldots) = 0$. This led to the proof of the impossibility of the tenth of HILBERT'S PROBLEMS (does there exist a general method for solving DIOPHANTINE EQUATIONS?) by Julia Robinson and Martin Davis in 1970 (Reid 1997, p. 107).

The Fibonacci number F_{n+1} gives the number of ways for 2×1 DOMINOES to cover a $2 \times n$ CHECKERBOARD, as illustrated in the following diagrams (Dickau).

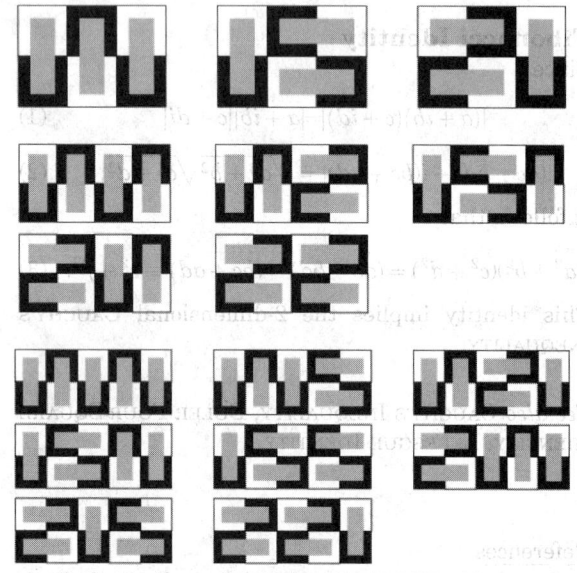

The number of ways of picking a SET (including the EMPTY SET) from the numbers 1, 2, ..., n without picking two consecutive numbers is F_{n+2}. The number of ways of picking a set (including the EMPTY SET) from the numbers 1, 2, ..., n without picking two consecutive numbers (where 1 and n are now consecutive) is $L_n = F_{n+1} + F_{n-1}$, where L_n is a LUCAS NUMBER. The probability of not getting two heads in a row in n tosses of a COIN is $F_{n+2}/2^n$ (Honsberger 1985, pp. 120–22). Fibonacci numbers are also related to the number of ways in which n COIN TOSSES can be made such that there are not three consecutive heads or tails. The number of ideals of an n-element FENCE POSET is the Fibonacci number F_n.

Given a RESISTOR NETWORK of n 1-Ω resistors, each incrementally connected in series or parallel to the preceding resistors, then the net resistance is a RATIONAL NUMBER having maximum possible denominator of F_{n+1}.

The Fibonacci numbers are given in terms of the CHEBYSHEV POLYNOMIAL OF THE SECOND KIND by

$$F_n = i^{n-1} U_{n-1}\left(-\frac{1}{2}i\right). \quad (7)$$

Sum identities include

$$\sum_{k=1}^{n} F_k = F_{n+2} - 1. \tag{8}$$

$$F_1 + F_3 + F_5 + \ldots + F_{2k+1} = F_{2k+2} \tag{9}$$

$$1 + F_2 + F_4 + F_6 + \ldots + F_{2k} = F_{2k+1} \tag{10}$$

$$\sum_{k=1}^{n} F_k^2 = F_n F_{n+1} \tag{11}$$

$$F_{2n} = F_{n+1}^2 - F_{n-1}^2 \tag{12}$$

$$F_{3n} = F_{n+1}^3 + F_n^3 + F_{n-1}^3. \tag{13}$$

There are a number of particular pretty algebraic identities involving the Fibonacci numbers, including

$$F_{n+1}^2 = 4F_n F_{n-1} + F_{n-2}^2 \tag{14}$$

(Brousseau 1972), CATALAN'S IDENTITY

$$F_n^2 - F_{n+r}F_{n-r} = (-1)^{n-r} F_r^2, \tag{15}$$

D'OCAGNE'S IDENTITY

$$F_m F_{n+1} - F_n F_{m+1} = (-1)^n F_{m-n}, \tag{16}$$

and the GELIN-CESÀRO IDENTITY

$$F_n^4 - F_{n-2}F_{n-1}F_{n+1}F_{n+2} = 1. \tag{17}$$

Letting $r = 1$ in (15) gives CASSINI'S IDENTITY

$$F_{n-1}F_{n+1} - F_n^2 = (-1)^n, \tag{18}$$

sometimes also called Simson's formula since it was also discovered by Simson (Coxeter and Greitzer 1967, p. 41; Coxeter 1969, pp. 165–68; Petkovsek *et al.* 1996, p. 12).

The Fibonacci numbers obey the negation formula

$$F_{-n} = (-1)^{n+1} F_n, \tag{19}$$

the addition formula

$$F_{m+n} = \frac{1}{2}(F_m L_n + L_m F_n), \tag{20}$$

where L_n is a LUCAS NUMBER, the subtraction formula

$$F_{m-n} = \frac{1}{2}(-1)(F_m L_n - L_m F_n), \tag{21}$$

the fundamental identity

$$L_n^2 - 5F_n^2 = 4(-1)^n \tag{22}$$

conjugation relation

$$F_n - \frac{1}{5}(L_{n-1} + L_{n+1}), \tag{23}$$

successor relation

$$F_{n+1} = \frac{1}{2}(F_n + L_n), \tag{24}$$

double-angle formula

$$F_{2n} = F_n L_n, \tag{25}$$

multiple-angle recurrence

$$F_{kn} = L_k F_{k(n-1)} - (-1)^k F_{k(n-2)}, \tag{26}$$

multiple-angle formulas

$$F_{kn} = \frac{1}{2^{k-1}} \sum_{i=0}^{\lfloor (k-1)/2 \rfloor} \binom{k}{2i+1} 5^i F_n^{2i+1} L_n^{k-1-2i} \tag{27}$$

$$= F_n \sum_{i=0}^{\lfloor (k-1)/2 \rfloor} \binom{k-1-i}{i} (-1)^{i(n+1)} L_n^{k-1-2i} \tag{28}$$

$$= \begin{cases} L_n \sum_{i=0}^{(k-2)/2} \binom{k-1-i}{i} (-1)^{in} 5^{k/2-1-i} F_n^{k-1-2i} & \text{for } k \text{ even} \\ \sum_{i=0}^{\lfloor k/2 \rfloor} \dfrac{k}{k-i} \binom{k-i}{i} (-1)^{in} 5^{\lfloor k/2 \rfloor - i} F_n^{k-2i} & \text{for } k \text{ odd} \end{cases} \tag{29}$$

$$= \sum_{i=0}^{k} \binom{k}{i} F_i F_n^i F_{n-1}^{k-i}, \tag{30}$$

product expansions

$$F_m F_n = \frac{1}{5}\left[L_{m+n} - (-1)^n L_{m-n}\right] \tag{31}$$

and

$$F_m L_n = F_{m+n} + (-1)^n F_{m-n}, \tag{32}$$

square expansion,

$$F_n^2 = \frac{1}{5}[L_{2n} - 2(-1)^n], \tag{33}$$

and power expansion

$$F_n^k = \frac{1}{2 \cdot 5^{\lfloor k/2 \rfloor}} \sum_{i=0}^{k} \binom{k}{i} (-1)^{i(n+1)}$$

$$\times \begin{cases} F_{(k-2i)n} & \text{for } k \text{ odd} \\ L_{(k-2i)n} & \text{for } k \text{ even}. \end{cases} \tag{34}$$

Honsberger (1985, p. 107) gives the general relations

$$F_{n+m} = F_{n-1}F_m + F_n F_{m+1} \tag{35}$$

$$F_{(k+1)n} = F_{n-1}F_{kn} + F_n F_{kn+1} \tag{36}$$

$$F_n = F_l F_{n-l+1} + F_{l-1}F_{n-l}. \tag{37}$$

In the case $l = n - l + 1$, then $l = (n+1)/2$ and for n ODD,

$$F_n = F_{(n+1)/2}^2 + F_{(n-1)/2}^2. \tag{38}$$

Similarly, for n EVEN,

$$F_n = F_{n/2+1}^2 - F_{n/2-1}^2. \tag{39}$$

Letting $k \equiv (n-1)/2$ gives the identities

$$F_{2k+1} = F_{k+1}^2 + F_k^2 \tag{40}$$

$$F_{n+2}^2 - F_{n+1}^2 = F_n F_{n+3} \tag{41}$$

$$F_n^2 = F_{n-1}^2 + 3F_{n-2}^2 + 2F_{n-2}F_{n-3}. \tag{42}$$

Sum FORMULAS for F_n include

$$F_n = \frac{1}{2^{n-1}}\left[\binom{n}{1} + 5\binom{n}{3} + 5^2\binom{n}{5} + \ldots\right] \tag{43}$$

$$F_{n+1} = \binom{n}{0} + \binom{n-1}{1} + \binom{n-2}{2} + \ldots \tag{44}$$

(Wells 1986, p. 63). Additional identities can be found throughout the *Fibonacci Quarterly* journal. A list of 47 generalized identities are given by Halton (1965). In terms of the LUCAS NUMBER L_n,

$$F_{2n} = F_n L_n \tag{45}$$

$$F_{2n}(L_{2n}^2 - 1) = F_{6n} \tag{46}$$

$$F_{m+p} + (-1)^{p+1}F_{m-p} = F_p L_m \tag{47}$$

$$\sum_{k=a+1}^{a+4n} F_k = F_{a+4n+2} - F_{a+2} = F_{2n}L_{a+2n+2} \tag{48}$$

(Honsberger 1985, pp. 111–13). A remarkable identity is

$$\exp\left(L_{1x} + \frac{1}{2}L_2 x^2 + \frac{1}{3}L_3 x^3 + \ldots\right)$$
$$= F_1 + F_2 x + F_3 x^3 + \ldots \tag{49}$$

(Honsberger 1985, pp. 118–19). It is also true that

$$\frac{L_n^2 - (-1)^a L_{n+a}^2}{F_n^2 - (-1)^a F_{n+a}^2} = 5 \tag{50}$$

for a ODD, and

$$\frac{L_n^2 + L_{n+a}^2 - 8(-1)^n}{F_n^2 + F_{n+a}^2} = 5 \tag{51}$$

for a EVEN (Freitag 1996).

The equation (1) is a LINEAR RECURRENCE SEQUENCE

$$x_n = A x_{x-1} + B x_{n-2} \quad n \ge 3, \tag{52}$$

so the closed form for F_n is given by

$$F_n = \frac{\alpha^n - \beta^n}{\alpha - \beta}, \tag{53}$$

where α and β are the roots of $x^2 = Ax + B$. Here, $A = B = 1$, so the equation becomes

$$x^2 - x - 1 = 0, \tag{54}$$

which has ROOTS

$$x = \frac{1}{2}\left(1 \pm \sqrt{5}\right). \tag{55}$$

The closed form is therefore given by

$$F_n = \frac{(1 + \sqrt{5})^n - (1 - \sqrt{5})^n}{2^n \sqrt{5}}, \tag{56}$$

This is known as BINET'S FIBONACCI NUMBER FORMULA (Wells 1986, p. 62). Another closed form is

$$F_n = \left[\frac{1}{\sqrt{5}}\left(\frac{1 + \sqrt{5}}{2}\right)^n\right] = \left[\frac{\phi^n}{\sqrt{5}}\right], \tag{57}$$

where $[x]$ is the NINT function (Wells 1986, p. 62). From (1), the RATIO of consecutive terms is

$$\frac{F_n}{F_{n-1}} = 1 + \frac{F_{n-2}}{F_{n-1}} = 1 + \frac{1}{\dfrac{F_{n-1}}{F_{n-2}}}$$

$$= 1 + \frac{1}{1 + \dfrac{1}{\dfrac{F_{n-3}}{F_{n-2}}}} = \left[1, 1, \ldots, \frac{F_2}{F_1}\right]$$

$$= \underbrace{[1, 1, \ldots, 1]}_{n-1}, \tag{58}$$

which is just the first few terms of the CONTINUED FRACTION for the GOLDEN RATIO ϕ. Therefore,

$$\lim_{n \to \infty} \frac{F_n}{F_{n-1}} = \phi. \tag{59}$$

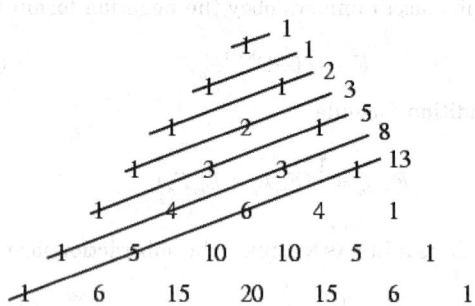

The "SHALLOW DIAGONALS" of PASCAL'S TRIANGLE sum to Fibonacci numbers (Pappas 1989),

$$\sum_{k=1}^{n} \binom{k}{n-k}$$

$$= \frac{(-1)^n \, {}_3F_2\left(1, 2, 1-n; \frac{1}{2}(3-n), 2 - \frac{1}{2}n; -\frac{1}{4}\right)}{\pi(2 - 3n + n^2)}$$

$$= F_{n+1}, \tag{60}$$

where $_3F_2(a,b,c;d,e;z)$ is a GENERALIZED HYPERGEO-METRIC FUNCTION.

Guy (1990) notes the curious fact that $\lceil e^{(n-1)/2} \rceil$ for $n = 0, 1, \ldots$ gives 1, 1, 2, 5, 8, 13, 21, 34, 55, ..., but then continues 91, 149, ... (Sloane's A005181). Taking the product of the first n Fibonacci numbers and adding 1 for $n = 1, 2, \ldots$ gives the sequence 2, 2, 3, 77, 31, 241, ... (Sloane's A052449). If these, 2, 2, 3, 7, 31, 241, 3121, ... (Sloane's A053413) are prime, i.e., the terms 1, 2, 3, 4, 5, 6, 7, 8, 22, 28, ... (Sloane's A053408).

The sequence of final digits in Fibonacci numbers repeats in cycles of 60. The last two digits repeat in 300, the last three in 1500, the last four in 15,000, etc. The number of Fibonacci numbers between n and $2n$ is either 1 or 2 (Wells 1986, p. 65).

Cesàro derived the finite sums

$$\sum_{k=0}^{n} \binom{n}{k} F_k = F_{2n} \tag{61}$$

$$\sum_{k=0}^{n} \binom{n}{k} 2^k F_k = F_{3n} \tag{62}$$

(Honsberger 1985, pp. 109–10). The Fibonacci numbers satisfy the power recurrence

$$\sum_{j=0}^{t+1} (-1)^{j(j+1)/2} \begin{bmatrix} t+1 \\ j \end{bmatrix}_F F_{n-j}^t = 0, \tag{63}$$

where $\begin{bmatrix} a \\ b \end{bmatrix}_F$ is a FIBONACCI COEFFICIENT, the reciprocal sum

$$\sum_{k=1}^{n} \frac{(-1)^k}{F_k F_{k+a}} = \frac{F_n}{F_a} \sum_{k=1}^{a} \frac{(-1)^k}{F_k F_{k+n}}, \tag{64}$$

the convolution

$$\sum_{k=0}^{n} F_k F_{n-k} = \frac{1}{5}(nL_n - F_n), \tag{65}$$

the partial fraction decomposition

$$\frac{1}{F_{n+a} F_{n+b} F_{n+c}} = \frac{A}{F_{n+a}} + \frac{B}{F_{n+b}} + \frac{C}{F_{n+c}}, \tag{66}$$

where

$$A = \frac{(-1)^{n-a}}{F_{b-a} F_{c-a}} \tag{67}$$

$$B = \frac{(-1)^{n-b}}{F_{c-b} F_{a-b}} \tag{68}$$

$$C = \frac{(-1)^{n-c}}{F_{a-c} F_{b-c}}, \tag{69}$$

and the summation formula

$$\sum_{k=0}^{n} x^k F_{ak+b} = \frac{g(n+1) - g(0)}{1 - L_a x + (-1)^a x^2}, \tag{70}$$

where

$$g(n) = (-1)^a F_{a(n-1)+b} x^{n+1} - F_{an+b} x^n. \tag{71}$$

Infinite sums include

$$\sum_{n=1}^{\infty} \frac{(-1)^n}{F_n F_{n+2}} = 2 - \sqrt{5} \tag{72}$$

(Clark 1995) and

$$\sum_{n=1}^{\infty} \frac{(-1)^{n+1}}{F_{n+1} F_{n+2}} = \phi^{-2} \tag{73}$$

$$\sum_{n=1}^{\infty} \frac{1}{F_{2n} F_{2n+2}} = \phi^{-2} \tag{74}$$

where ϕ is the GOLDEN RATIO (Wells 1986, p. 65).

For $n \geq 3$, $F_n | F_m$ IFF $n | m$ (Wells 1986, p. 65). $L_n | L_m$ IFF n divides into m an EVEN number of times. $(F_m, F_n) = F_{(m,n)}$ (Michael 1964; Honsberger 1985, pp. 131–32). No ODD Fibonacci number is divisible by 17 (Honsberger 1985, pp. 132 and 242). No Fibonacci number > 8 is ever OF THE FORM $p - 1$ or $p + 1$ where p is a PRIME NUMBER (Honsberger 1985, p. 133).

Consider the sum

$$s_k = \sum_{n=2}^{k} \frac{1}{F_{n-1} F_{n+1}} = \sum_{n=2}^{k} \left(\frac{1}{F_{n-1} F_n} - \frac{1}{F_n F_{n+1}} \right). \tag{75}$$

This is a TELESCOPING SUM, so

$$s_k = 1 - \frac{1}{F_{k+1} F_{k+2}}, \tag{76}$$

thus

$$S \equiv \lim_{k \to \infty} s_k = 1 \tag{77}$$

(Honsberger 1985, pp. 134–35). Using BINET'S FIBONACCI NUMBER FORMULA, it also follows that

$$\frac{F_{n+r}}{F_n} = \frac{\alpha^{n+r} - \beta^{n+r}}{\alpha^n - \beta^n} = \frac{\alpha^{n+r}}{\alpha^n} \frac{1 - \left(\frac{\beta}{\alpha}\right)^{n+r}}{1 - \left(\frac{\beta}{\alpha}\right)^n}, \tag{78}$$

where

$$\alpha = \frac{1}{2}\left(1 + \sqrt{5}\right) \tag{79}$$

$$\beta = \frac{1}{2}\left(1 - \sqrt{5}\right) \tag{80}$$

so

$$\lim_{n \to \infty} \frac{F_{n+r}}{F_n} = \alpha^r. \tag{81}$$

$$S' = \sum_{n=1}^{\infty} \frac{F_n}{F_{n+1}F_{n+2}} = 1 \tag{82}$$

(Honsberger 1985, pp. 138 and 242–43). The MILLIN SERIES has sum

$$S'' \equiv \sum_{n=0}^{\infty} \frac{1}{F_{2^n}} = \frac{1}{2}\left(7 - \sqrt{5}\right) \tag{83}$$

(Honsberger 1985, pp. 135–37).

The Fibonacci numbers are COMPLETE. In fact, dropping one number still leaves a COMPLETE SEQUENCE, although dropping two numbers does not (Honsberger 1985, pp. 123 and 126). Dropping two terms from the Fibonacci numbers produces a sequence which is not even WEAKLY COMPLETE (Honsberger 1985, p. 128). However, the sequence

$$F'_n \equiv F_n - (-1)^n \tag{84}$$

is WEAKLY COMPLETE, even with any finite subsequence deleted (Graham 1964). $\{F_n^2\}$ is not COMPLETE, but $\{F_n^2\} + \{F_n^2\}$ are. 2^{N-1} copies of $\{F_n^N\}$ are COMPLETE.

For a discussion of SQUARE Fibonacci numbers, see Cohn (1964), who proved that the only SQUARE NUMBER Fibonacci numbers are 1 and $F_{12} = 144$ (Cohn 1964, Guy 1994). Ming (1989) proved that the only TRIANGULAR Fibonacci numbers are 1, 3, 21, and 55. The Fibonacci and LUCAS NUMBERS have no common terms except 1 and 3. The only CUBIC Fibonacci numbers are 1 and 8.

$$\left(F_n F_{n+3}, 2F_{n+1}F_{n+2}, F_{2n+3} = F_{n+1}^2 + F_{n+2}^2\right) \tag{85}$$

is a PYTHAGOREAN TRIPLE.

$$F_{4n}^2 + 8F_{2n}(F_{2n} + F_{6n}) = (3F_{4n})^2 \tag{86}$$

is always a SQUARE NUMBER (Honsberger 1985, p. 243).

In 1975, James P. Jones showed that the Fibonacci numbers are the POSITIVE INTEGER values of the POLYNOMIAL

$$P(x,y) = -y^5 + 2y^4x + y^3x^2 - 2y^2x^3 - y(x^4 - 2) \tag{87}$$

for GAUSSIAN INTEGERS x and y (Le Lionnais 1983). If n and k are two POSITIVE INTEGERS, then between n^k and n^{k+1}, there can never occur more than n Fibonacci numbers (Honsberger 1985, pp. 104–05).

Every F_n that is PRIME has a PRIME index n, with the exception of $F_4 = 3$. However, the converse is not true (i.e., not every prime index p gives a PRIME F_p). The first few PRIME Fibonacci numbers F_n are 2, 3, 5, 13, 89, 233, 1597, 28657, 514229, ... (Sloane's A005478), which occur for $n = 3, 4, 5, 7, 11, 13, 17, 23, 29, 43,$ 47, 83, 131, 137, 359, 431, 433, 449, 509, 569, 571, ... (Sloane's A001605; Dubner and Keller 1999). Gardner's statement that F_{531} is prime is incorrect, especially since 531 is not even PRIME (Gardner 1979, p. 161). It is not known if there are an INFINITE number of Fibonacci primes.

The Fibonacci numbers F_n, are SQUAREFUL for $n = 6,$ 12, 18, 24, 25, 30, 36, 42, 48, 50, 54, 56, 60, 66, ..., 372, 375, 378, 384, ... (Sloane's A037917) and SQUAREFREE for $n = 1, 2, 3, 4, 5, 7, 8, 9, 10, 11, 13, ...$ (Sloane's A037918). $4|F_{6n}$ and $25|F_{25n}$ for all n, and there is at least one $n \le 2m$ such that $m|F_n$. No SQUAREFUL Fibonacci numbers F_p are known with p PRIME.

See also CASSINI'S IDENTITY, CATALAN'S IDENTITY, D'OCAGNE'S IDENTITY, FAST FIBONACCI TRANSFORM, FIBONACCI COEFFICIENT, FIBONACCI DUAL THEOREM, FIBONACCI N-STEP NUMBER, FIBONACCI POLYNOMIAL, FIBONACCI Q-MATRIX, GELIN-CESÀRO IDENTITY, GENERALIZED FIBONACCI NUMBER, INVERSE TANGENT, LINEAR RECURRENCE SEQUENCE, LUCAS SEQUENCE, NEAR NOBLE NUMBER, PELL SEQUENCE, RABBIT CONSTANT, RANDOM FIBONACCI SEQUENCE, STOLARSKY ARRAY, TETRANACCI NUMBER, TRIBONACCI NUMBER, WYTHOFF ARRAY, ZECKENDORF REPRESENTATION, ZECKENDORF'S THEOREM

References

Ball, W. W. R. and Coxeter, H. S. M. *Mathematical Recreations and Essays, 13th ed.* New York: Dover, pp. 56–7, 1987.

Basin, S. L. and Hoggatt, V. E. Jr. "A Primer on the Fibonacci Sequence." *Fib. Quart.* **1**, 1963.

Basin, S. L. and Hoggatt, V. E. Jr. "A Primer on the Fibonacci Sequence--Part II." *Fib. Quart.* **1**, 61–8, 1963.

Borwein, J. M. and Borwein, P. B. *Pi & the AGM: A Study in Analytic Number Theory and Computational Complexity.* New York: Wiley, pp. 94–01, 1987.

Brillhart, J.; Montgomery, P. L.; and Silverman, R. D. "Tables of Fibonacci and Lucas Factorizations." *Math. Comput.* **50**, 251–60 and S1-S15, 1988.

Brook, M. "Fibonacci Formulas." *Fib. Quart.* **1**, 60, 1963.

Brousseau, A. "Fibonacci Numbers and Geometry." *Fib. Quart.* **10**, 303–18, 1972.

Clark, D. Solution to Problem 10262. *Amer. Math. Monthly* **102**, 467, 1995.

Cohn, J. H. E. "On Square Fibonacci Numbers." *J. London Math. Soc.* **39**, 537–41, 1964.

Conway, J. H. and Guy, R. K. "Fibonacci Numbers." In *The Book of Numbers.* New York: Springer-Verlag, pp. 111–13, 1996.

Coxeter, H. S. M. "The Golden Section and Phyllotaxis." Ch. 11 in *Introduction to Geometry, 2nd ed.* New York: Wiley, 1969.

Coxeter, H. S. M. and Greitzer, S. L. *Geometry Revisited.* Washington, DC: Math. Assoc. Amer., p. 41, 1967.

Devaney, R. "The Mandelbrot Set and the Farey Tree, and the Fibonacci Sequence." *Amer. Math. Monthly* **106**, 289–02, 1999.

Dickau, R. M. "Fibonacci Numbers." http://www.prairienet.org/~pops/fibboard.html.

Dubner, H. and Keller, W. "New Fibonacci and Lucas Primes." *Math. Comput.* **68**, 417–27 and S1-S12, 1999.

Freitag, H. Solution to Problem B-772. "An Integral Ratio." *Fib. Quart.* **34**, 82, 1996.

Gardner, M. *Mathematical Circus: More Puzzles, Games, Paradoxes and Other Mathematical Entertainments from Scientific American.* New York: Knopf, 1979.

Graham, R. "A Property of Fibonacci Numbers." *Fib. Quart.* **2**, 1–0, 1964.

Graham, R. L.; Knuth, D. E.; and Patashnik, O. "Fibonacci Numbers." §6.6 in *Concrete Mathematics: A Foundation for Computer Science, 2nd ed.* Reading, MA: Addison-Wesley, pp. 290–01, 1994.

Guy, R. K. "The Second Strong Law of Small Numbers." *Math. Mag.* **63**, 3–0, 1990.

Guy, R. K. "Fibonacci Numbers of Various Shapes." §D26 in *Unsolved Problems in Number Theory, 2nd ed.* New York: Springer-Verlag, pp. 194–95, 1994.

Halton, J. H. "On a General Fibonacci Identity." *Fib. Quart.* **3**, 31–3, 1965.

Hilton, P.; Holton, D.; and Pedersen, J. "Fibonacci and Lucas Numbers." Ch. 3 in *Mathematical Reflections in a Room with Many Mirrors.* New York: Springer-Verlag, pp. 61–5, 1997.

Hilton, P. and Pedersen, J. "Fibonacci and Lucas Numbers in Teaching and Research." *J. Math. Informatique* **3**, 36–7, 1991–992.

Hilton, P. and Pedersen, J. "A Note on a Geometrical Property of Fibonacci Numbers." *Fib. Quart.* **32**, 386–88, 1994.

Hoffman, P. *The Man Who Loved Only Numbers: The Story of Paul Erdos and the Search for Mathematical Truth.* New York: Hyperion, p. 208, 1998.

Hoggatt, V. E. Jr. *The Fibonacci and Lucas Numbers.* Boston, MA: Houghton Mifflin, 1969.

Hoggatt, V. E. Jr. and Ruggles, I. D. "A Primer on the Fibonacci Sequence--Part III." *Fib. Quart.* **1**, 61–5, 1963.

Hoggatt, V. E. Jr. and Ruggles, I. D. "A Primer on the Fibonacci Sequence--Part IV." *Fib. Quart.* **1**, 65–1, 1963.

Hoggatt, V. E. Jr. and Ruggles, I. D. "A Primer on the Fibonacci Sequence--Part V." *Fib. Quart.* **2**, 59–6, 1964.

Hoggatt, V. E. Jr.; Cox, N.; and Bicknell, M. "A Primer for the Fibonacci Numbers: Part XII." *Fib. Quart.* **11**, 317–31, 1973.

Honsberger, R. "A Second Look at the Fibonacci and Lucas Numbers." Ch. 8 in *Mathematical Gems III.* Washington, DC: Math. Assoc. Amer., 1985.

Kepler, J. *The Six-Cornered Snowflake.* Oxford, England: Oxford University Press, 1966.

Knott, R. "Fibonacci Numbers and the Golden Section." http://www.mcs.surrey.ac.uk/Personal/R.Knott/Fibonacci/fib.html.

Le Lionnais, F. *Les nombres remarquables.* Paris: Hermann, p. 146, 1983.

Leyland, P. ftp://sable.ox.ac.uk/pub/math/factors/fibonacci.Z.

Matiyasevich, Yu. V. "Solution to of the Tenth Problem of Hilbert." *Mat. Lapok* **21**, 83–7, 1970.

Matijasevich, Yu. V. *Hilbert's Tenth Problem.* Cambridge, MA: MIT Press, 1993. http://www.informatik.uni-stuttgart.de/ifi/ti/personen/Matiyasevich/H10Pbook/.

Michael, G. "A New Proof for an Old Property." *Fib. Quart.* **2**, 57–8, 1964.

Ming, L. "On Triangular Fibonacci Numbers." *Fib. Quart.* **27**, 98–08, 1989.

Ogilvy, C. S. and Anderson, J. T. "Fibonacci Numbers." Ch. 11 in *Excursions in Number Theory.* New York: Dover, pp. 133–44, 1988.

Pappas, T. "Fibonacci Sequence," "Pascal's Triangle, the Fibonacci Sequence & Binomial Formula," "The Fibonacci Trick," and "The Fibonacci Sequence & Nature." *The Joy of Mathematics.* San Carlos, CA: Wide World Publ./Tetra, pp. 28–9, 40–1, 51, 106, and 222–25, 1989.

Petkovsek, M.; Wilf, H. S.; and Zeilberger, D. *A = B.* Wellesley, MA: A. K. Peters, p. 12, 1996.

Ram, R. "Fibonacci Formulae." http://users.tellurian.net/hsejar/maths/fibonacci/.

Reid, C. *Julia: A Life in Mathematics.* Washington, DC: Math. Assoc. Amer., 1997.

Reiter, C. "Fast Fibonacci Numbers." *Mathematica J.* **2**, 58–0, 1992.

Schroeder, M. *Fractals, Chaos, Power Laws: Minutes from an Infinite Paradise.* New York: W. H. Freeman, pp. 49–7, 1991.

Séroul, R. "The Fibonacci Numbers." §2.13 in *Programming for Mathematicians.* Berlin: Springer-Verlag, pp. 21–2, 2000.

Shorey, T. N. and Stewart, C. L. "On Divisors of Fermat, Fibonacci, Lucas and Lehmer Numbers, 2." *J. London Math. Soc.* **23**, 17–3, 1981.

Sloane, N. J. A. Sequences A000045/M0692, A001605/M2309, A005181/M0693, A005478/M0741, A037917, A037918, A053408, A052449, and A053413 in "An On-Line Version of the Encyclopedia of Integer Sequences." http://www.research.att.com/~njas/sequences/eisonline.html.

Smith, H. J. "Fibonacci Numbers." http://pweb.netcom.com/~hjsmith/Fibonacc.html.

Steinhaus, H. *Mathematical Snapshots, 3rd ed.* New York: Dover, pp. 46–7 and 299, 1999.

Stewart, C. L. "On Divisors of Fermat, Fibonacci, Lucas and Lehmer Numbers." *Proc. London Math. Soc.* **35**, 425–47, 1977.

Szymkiewicz, D. "Sur la portée de la loi de Ludwig." *Acta Soc. Botanicorum Poloniae* **5**, 390–95, 1928.

Vogler, P. "Das ‚Ludwig'sche Gipfelgesetz' und seine Tragweite." *Flora* **104**, 123–28, 1912.

Vorob'ev, N. N. *Fibonacci Numbers.* New York: Blaisdell, 1961.

Weisstein, E. W. "Books about Fibonacci Numbers." http://www.treasure-troves.com/books/FibonacciNumbers.html.

Wells, D. *The Penguin Dictionary of Curious and Interesting Numbers.* Middlesex, England: Penguin Books, pp. 61–7, 1986.

Zylinski, E. "Numbers of Fibonacci in Biological Statistics." *Atti del Congr. internaz. matematici* **4**, 153–56, 1928.

Fibonacci Polynomial

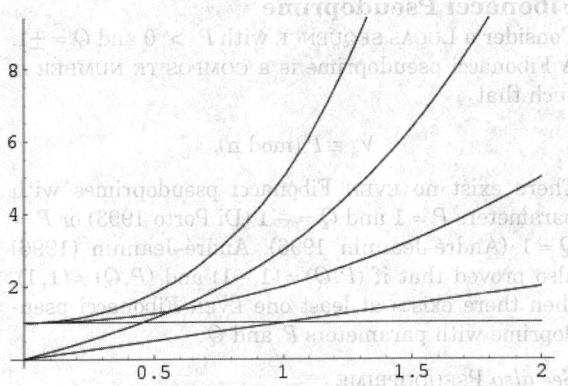

The W POLYNOMIALS obtained by setting $p(x) = x$ and $q(x) = 1$ in the LUCAS POLYNOMIAL SEQUENCE. (The corresponding w POLYNOMIALS are called LUCAS POLYNOMIALS.) The Fibonacci polynomials are defined by the RECURRENCE RELATION

$$F_{n+1}(x) = xF_n(x) + F_{n-1}(x), \qquad (1)$$

with $F_1(x) = 1$ and $F_2(x) = x$. They are also given by

the explicit sum formula

$$F_n(x) = \sum_{j=0}^{\lfloor (n-1)/2 \rfloor} \binom{n-j-1}{j} x^{n-2j-1}, \qquad (2)$$

where $\lfloor x \rfloor$ is the FLOOR FUNCTION and $\binom{n}{m}$ is a BINOMIAL COEFFICIENT. The first few Fibonacci polynomials are

$$F_1(x) = 1$$

$$F_2(x) = x$$

$$F_3(x) = x^2 + 1$$

$$F_4(x) = x^3 + 2x$$

$$F_5(x) = x^4 + 3x^2 + 1.$$

The Fibonacci polynomials are normalized so that

$$F_n(1) = F_n, \qquad (3)$$

where the F_ns are FIBONACCI NUMBERS.
The Fibonacci polynomials are related to the MORGAN-VOYCE POLYNOMIALS by

$$F_{2n+1}(x) = b_n(x^2) \qquad (4)$$

$$F_{2n+n2}(x) = x B_n(x^2) \qquad (5)$$

(Swamy 1968).

See also BRAHMAGUPTA POLYNOMIAL, FIBONACCI NUMBER, MORGAN-VOYCE POLYNOMIAL

References

Swamy, M. N. S. "Further Properties of Morgan-Voyce Polynomials." *Fib. Quart.* **6**, 167–75, 1968.

Fibonacci Pseudoprime

Consider a LUCAS SEQUENCE with $P > 0$ and $Q = \pm 1$. A Fibonacci pseudoprime is a COMPOSITE NUMBER n such that

$$V_n \equiv P \pmod{n}.$$

There exist no EVEN Fibonacci pseudoprimes with parameters $P = 1$ and $Q = -1$ (Di Porto 1993) or $P = Q = 1$ (André-Jeannin 1996). André-Jeannin (1996) also proved that if $(P, Q) \neq (1, -1)$ and $(P, Q) \neq (1, 1)$, then there exists at least one EVEN Fibonacci pseudoprime with parameters P and Q.

See also PSEUDOPRIME

References

André-Jeannin, R. "On the Existence of Even Fibonacci Pseudoprimes with Parameters P and Q." *Fib. Quart.* **34**, 75–8, 1996.
Di Porto, A. "Nonexistence of Even Fibonacci Pseudoprimes of the First Kind." *Fib. Quart.* **31**, 173–77, 1993.
Ribenboim, P. "Fibonacci Pseudoprimes." §2.X.A in *The New Book of Prime Number Records, 3rd ed.* New York: Springer-Verlag, pp. 127–29, 1996.

Fibonacci Q-Matrix

A FIBONACCI MATRIX OF THE FORM

$$M = \begin{bmatrix} m & 1 \\ 1 & 0 \end{bmatrix}. \qquad (1)$$

If U and V are defined as BINET FORMS

$$U_n = m U_{n-1} + U_{n-2} \; (U_0 = 0, U_1 = 1) \qquad (2)$$

$$V_n = m V_{n-1} + V_{n-2} \; (V_0 = 2, V_1 = m), \qquad (3)$$

then

$$M = \begin{bmatrix} U_{n+1} & U_n \\ U_n & U_{n-1} \end{bmatrix} \qquad (4)$$

$$M^{-1} = M - ml = \begin{bmatrix} 0 & 1 \\ 1 & -m \end{bmatrix}. \qquad (5)$$

Defining

$$Q \equiv \begin{bmatrix} F_2 & F_1 \\ F_1 & F_0 \end{bmatrix} = \begin{bmatrix} 1 & 1 \\ 1 & 0 \end{bmatrix}, \qquad (6)$$

then

$$Q^n = \begin{bmatrix} F_{n+1} & F_n \\ F_n & F_{n-1} \end{bmatrix} \qquad (7)$$

(Honsberger 1985, pp. 106–07).

See also BINET FORMS, FIBONACCI NUMBER

References

Honsberger, R. "A Second Look at the Fibonacci and Lucas Numbers." Ch. 8 in *Mathematical Gems III*. Washington, DC: Math. Assoc. Amer., 1985.

Fibonacci Sequence

FIBONACCI NUMBER

Fibration

If $f : E \to B$ is a FIBER BUNDLE with B a PARACOMPACT TOPOLOGICAL SPACE, then f satisfies the HOMOTOPY LIFTING PROPERTY with respect to all TOPOLOGICAL SPACES. In other words, if $g : [0, 1] \times X \to B$ is a HOMOTOPY from g_0 to g_1, and if g_0' is a LIFT of the MAP g_0 with respect to f, then g has a LIFT to a MAP g' with respect to f. Therefore, if you have a HOMOTOPY of a MAP into B, and if the beginning of it has a LIFT, then that LIFT can be extended to a LIFT of the HOMOTOPY itself.

A fibration is a MAP between TOPOLOGICAL SPACES $f : E \to B$ such that it satisfies the HOMOTOPY LIFTING PROPERTY.

See also FIBER BUNDLE, FIBER SPACE

Fiedler Vector

The EIGENVECTOR corresponding to the second smallest EIGENVALUE (i.e., the ALGEBRAIC CONNECTIVITY) of the LAPLACIAN MATRIX of a graph G. The Fiedler vector is used in SPECTRAL GRAPH PARTITIONING.

See also ALGEBRAIC CONNECTIVITY, CONNECTED GRAPH, LAPLACIAN MATRIX, SPECTRAL GRAPH PARTITIONING

References

Chung, F. R. K. *Spectral Graph Theory*. Providence, RI: Amer. Math. Soc., 1997.

Demmel, J. "CS 267: Notes for Lecture 23, April 9, 1999. Graph Partitioning, Part 2." http://www.cs.berkeley.edu/~demmel/cs267/lecture20/lecture20.html.

© 1999–001 Wolfram Research, Inc.

Field

A field is any set of elements which satisfies the FIELD AXIOMS for both addition and multiplication and is a commutative DIVISION ALGEBRA. An archaic name for a field is RATIONAL DOMAIN. The French term for a field is *corps* and the German word is *Körper*, both meaning "body." A field with a finite number of members is known as a FINITE FIELD or Galois field.

Because the identity condition must be different for addition and multiplication, every field must have at least two elements. Examples include the COMPLEX NUMBERS ($\mathbb{C}$), RATIONAL NUMBERS ($\mathbb{Q}$), and REAL NUMBERS ($\mathbb{R}$), but *not* the INTEGERS (F), which form only a RING. It has been proven by Hilbert and Weierstrass that all generalizations of the field concept to triplets of elements are equivalent to the field of COMPLEX NUMBERS.

See also ADJUNCTION, CHARACTERISTIC (FIELD), COEFFICIENT FIELD, CYCLOTOMIC FIELD, DIVISION ALGEBRA, EXTENSION FIELD, FIELD AXIOMS, FINITE FIELD, FUNCTION FIELD, LOCAL FIELD, MAC LANE'S THEOREM, MODULE, NUMBER FIELD, PYTHAGOREAN FIELD, QUADRATIC FIELD, RING, SKEW FIELD, SPLITTING FIELD, SUBFIELD, VECTOR FIELD

References

Allenby, R. B. *Rings, Fields, and Groups: An Introduction to Abstract Algebra, 2nd ed.* Oxford, England: Oxford University Press, 1991.

Dummit, D. S. and Foote, R. M. "Field Theory." Ch. 13 in *Abstract Algebra, 2nd ed.* Englewood Cliffs, NJ: Prentice-Hall, pp. 422–70, 1998.

Ellis, G. *Rings and Fields*. Oxford, England: Oxford University Press, 1993.

Ferreirós, J. "A New Fundamental Notion for Algebra: Fields." §3.2 in *Labyrinth of Thought: A History of Set Theory and Its Role in Modern Mathematics*. Basel, Switzerland: Birkhäuser, pp. 90–4, 1999.

Joye, M. "Introduction élémentaire à la théorie des courbes elliptiques." http://www.dice.ucl.ac.be/crypto/introductory/courbes_elliptiques.html.

Nagell, T. "Moduls, Rings, and Fields." §6 in *Introduction to Number Theory*. New York: Wiley, pp. 19–1, 1951.

Field Axioms

The field axioms are generally written in additive and multiplicative pairs.

Name	Addition	Multiplication
Commutativity	$a+b=b+a$	$ab=ba$
Associativity	$(a+b)+c=a+(b+c)$	$(ab)c=a(bc)$
Distributivity	$a(b+c)=ab+ac$	$(a+b)c=ac+bc$
Identity	$a+0=a=0+a$	$a\cdot1=a=1\cdot a$
Inverses	$a+(-a)=0=(-a)+a$	$aa^{-1}=1=a^{-1}a$ if $a\neq0$

See also ALGEBRA, FIELD

References

Apostol, T. M. "The Field Axioms." §I 3.2 in *Calculus, 2nd ed., Vol. 1: One-Variable Calculus, with an Introduction to Linear Algebra*. Waltham, MA: Blaisdell, pp. 17–9, 1967.

Field Extension

EXTENSION FIELD

Fields Medal

Portions of this entry contributed by MICHEL BARRAN

The mathematical equivalent of the Nobel Prize (there is no Nobel Prize in mathematics) which is awarded by the International Mathematical Union every four years to one or more outstanding researchers. "Fields Medals" are more properly known by their official name, "International medals for outstanding discoveries in mathematics."

The Field medals were first proposed at the 1924 International Congress of Mathematicians in Toronto, where a resolution was adopted stating that at each subsequent conference, two gold medals should be awarded to recognize outstanding mathematical achievement. Professor J. C. Fields, a Canadian mathematician who was secretary of the 1924 Congress, later donated funds establishing the medals which were named in his honor. Consistent with Fields' wish that the awards recognize both existing work and the promise of future achievement, it was agreed to restrict the medals to mathematicians not over forty at the year of the Congress. In 1966 it was agreed that, in light of the great expansion of mathematical research, up to four medals could be awarded at each Congress.

The Fields Medal is the highest scientific award for mathematicians, and is presented every four years at the International Congress of Mathematicians, together with a prize of 15,000 Canadian dollars. The first Fields Medal was awarded in 1936 at the World Congress in Oslo. The Fields Medal is made of gold, and shows the head of Archimedes (287–12 BC) together with a quotation attributed to him: "Transire suum pectus mundoque potiri" ("Rise above oneself and grasp the world"). The reverse side bears the inscription: "Congregati ex toto orbe mathematici ob scripta insignia tribuere" ("the mathematicians assembled here from all over the world pay tribute for outstanding work").

Nobel prizes were created in the will of the Swedish chemist and inventor of dynamite Alfred Nobel, but Nobel, who was an inventor and industrialist, did not create a prize in mathematics because he was not particularly interested in mathematics or theoretical science. In fact, his will speaks of prizes for those "inventions or discoveries" of greatest practical benefit to mankind. While it is commonly stated that Nobel decided against a Nobel prize in math because of anger over the romantic attentions of a famous mathematician (often claimed to be Gosta Mittag-Leffler) to a women in his life, there is no historical evidence to support the story. Furthermore, Nobel was a lifelong batchelor, although he did has a Viennese woman named Sophie Hess as his mistress (Lopez-Ortiz).

The following table summarizes Fields Medals winners together with their institutions.

year	winners
1936	Lars Valerian Ahlfors (Harvard University)
	Jesse Douglas (Massachusetts Institute of Technology)
1950	Laurent Schwartz (University of Nancy)
	Alte Selberg (Institute for Advanced Study, Princeton)
1954	Kunihiko Kodaira (Princeton University)
	Jean-Pierre Serre (University of Paris)
1958	Klaus Friedrich Roth (University of London)
	René Thom (University of Strasbourg)
1962	Lars V. Hörmander (University of Stockholm)
	John Willard Milnor (Princeton University)
1966	Michael Francis Atiyah (Oxford University)
	Paul Joseph Cohen (Stanford University)
	Alexander Grothendieck (University of Paris)
	Stephen Smale (University of California, Berkeley)
1970	Alan Baker (Cambridge University)
	Heisuke Hironaka (Harvard University)
	Serge P. Novikov (Moscow University)
	John Griggs Thompson (Cambridge University)
1974	Enrico Bombieri (University of Pisa)
	David Bryant Mumford (Harvard University)
1978	Pierre René Deligne (Institut des Hautes Études Scientifiques)
	Charles Louis Fefferman (Princeton University)
	Gregori Alexandrovitch Margulis (Moscow University)
	Daniel G. Quillen (Massachusetts Institute of Technology)
1982	Alain Connes (Institut des Hautes Études Scientifiques)
	William P. Thurston (Princeton University)
	Shing-Tung Yau (Institute for Advanced Study, Princeton)
1986	Simon Donaldson (Oxford University)
	Gerd Faltings (Princeton University)
	Michael Freedman (University of California, San Diego)
1990	Vladimir Drinfeld (Phys. Inst. Kharkov)
	Vaughan Jones (University of California, Berkeley)
	Shigefumi Mori (University of Kyoto?)
	Edward Witten (Institute for Advanced Study, Princeton)
1994	Pierre-Louis Lions (Université de Paris-Dauphine)
	Jean-Christophe Yoccoz (Université de Paris-Sud)
	Jean Bourgain (Institute for Advanced Study, Princeton)
	Efim Zelmanov (University of Wisconsin)
1998	Richard E. Borcherds (Cambridge University)
	W. Timothy Gowers (Cambridge University)
	Maxim Kontsevich (IHES Bures-sur-Yvette)
	Curtis T. McMullen (Harvard University)

See also BURNSIDE PROBLEM, MATHEMATICS PRIZES, POINCARÉ CONJECTURE, ROTH'S THEOREM, TAU CONJECTURE

References

Albers, D. J.; Alexanderson, G. L.; and Reid, C. *International Mathematical Congresses, An Illustrated History 1893–986, rev. ed., incl. 1986.* New York: Springer Verlag, 1987.

Fields Institute. "Fields Medal Winners." http://www.fields.toronto.edu/medal.html.

International Mathematical Union. "Fields Medals and Rolf Nevanlinna Prize." http://elib.zib.de/IMU/medals/.

Joyce, D. "History of Mathematics: Fields Medals." http://aleph0.clarku.edu/~djoyce/mathhist/fieldsmedal.html.

Lopez-Ortiz, A. "Fields Medal: Historical Introduction." http://www.cs.unb.ca/~alopez-o/math-faq/mathtext/node19.html.

Lopez-Ortiz, A. "Why Is There No Nobel In Mathematics?" http://www.cs.unb.ca/~alopez-o/math-faq/mathtext/node21.html.

MacTutor History of Mathematics Archives. "The Fields Medal." http://www-groups.dcs.st-and.ac.uk/~history/Societies/FieldsMedal.html.

Monastyrsky, M. *Modern Mathematics in the Light of the Fields Medals.* Wellesley, MA: A. K. Peters, 1997.

Technische Universität Berlin. "The Four Fields Medallists and the Nevanlinna Prize Winner of The International Congress of Mathematicians, Berlin 1998." http://www.tu-berlin.de/presse/pi/1998/pi182e.htm.

Tropp, H. S. "The Origins and History of the Fields Medal." *Historia Math.* **3**, 167–81, 1976.

Fifteen Theorem

A theorem due to Conway *et al.* (1997) which states that, if a positive definite QUADRATIC FORM with INTEGER MATRIX entries represents all natural numbers up to 15, then it represents all natural numbers. This theorem contains LAGRANGE'S FOUR-SQUARE THEOREM, since every number up to 15 is the sum of at most four SQUARES.

See also INTEGER MATRIX, INTEGER-MATRIX FORM, LAGRANGE'S FOUR-SQUARE THEOREM, QUADRATIC FORM

References

Conway, J. H.; Guy, R. K.; Schneeberger, W. A.; and Sloane, N. J. A. "The Primary Pretenders." *Acta Arith.* **78**, 307–13, 1997.

Duke, W. "Some Old Problems and New Results about Quadratic Forms." *Not. Amer. Math. Soc.* **44**, 190–96, 1997.

Figurate Number

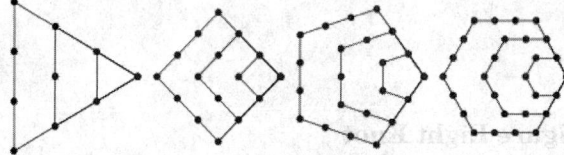

A number which can be represented by a regular geometrical arrangement of equally spaced points. If the arrangement forms a REGULAR POLYGON, the number is called a POLYGONAL NUMBER. The polygonal numbers illustrated above are called triangular, square, pentagonal, and hexagon numbers,

respectively. Figurate numbers can also form other shapes such as centered polygons, L-shapes, 3-dimensional solids, etc.

The nth regular r-polytopic number is given by

$$P_r(n) = \binom{n+r-1}{n} = \frac{1}{r!} n^{(r)},$$

where $\binom{n}{k}$ is a BINOMIAL COEFFICIENT and $n^{(k)}$ is a RISING FACTORIAL, so

$$P_2(n) = \frac{1}{2} n(n+1)$$

are the TRIANGULAR NUMBERS,

$$P_3(n) = \frac{1}{6} n(n+1)(n+2)$$

the TETRAHEDRAL NUMBERS,

$$P_4(n) = \frac{1}{24} n(n+1)(n+1)(n+3)$$

the PENTATOPE NUMBERS, and so on (Dickson 1952, p. 7).

The following table lists the most common types of figurate numbers.

Name	FORMULA
BIQUADRATIC NUMBER	n^4
CENTERED CUBE NUMBER	$(2n-1)(n^2-n+1)$
CENTERED PENTAGONAL NUMBER	$\frac{1}{2}(5n^2+5n+2)$
CENTERED SQUARE NUMBER	$n^2+(n-1)^2$
CENTERED TRIANGULAR NUMBER	$\frac{1}{2}(3n^2-3n+2)$
CUBIC NUMBER	n^3
DECAGONAL NUMBER	$4n^2-3n$
GNOMONIC NUMBER	$2n-1$
Hauy OCTAHEDRAL NUMBER	$\frac{1}{3}(2n-1)(2n^2-2n+3)$
Hauy RHOMBIC DODECAHEDRAL NUMBER	$(2n-1)(8n^2-14n+7)$
HEPTAGONAL NUMBER	$\frac{1}{2}n(5n-3)$
HEX NUMBER	$3n^2-3n+1$
HEPTAGONAL PYRAMIDAL NUMBER	$\frac{1}{6}n(n+1)(5n-2)$
HEXAGONAL NUMBER	$n(2n-1)$
HEXAGONAL PYRAMIDAL NUMBER	$\frac{1}{6}n(n+1)(4n-1)$

OCTAGONAL NUMBER	$n(3n-2)$
OCTAHEDRAL NUMBER	$\frac{1}{3}n(2n^2+1)$
PENTAGONAL NUMBER	$\frac{1}{2}n(3n-1)$
PENTAGONAL PYRAMIDAL NUMBER	$\frac{1}{2}n^2(n+1)$
PENTATOPE NUMBER	$\frac{1}{24}n(n+1)(n+2)(n+3)$
PRONIC NUMBER	$n(n+1)$
RHOMBIC DODECAHEDRAL NUMBER	$(2n-1)(2n^2-2n+1)$
SQUARE NUMBER	n^2
SQUARE PYRAMIDAL NUMBER	$\frac{1}{6}n(n+1)(2n+1)$
STELLA OCTANGULA NUMBER	$n(2n^2-1)$
TETRAHEDRAL NUMBER	$\frac{1}{6}n(n+1)(n+2)$
TRIANGULAR NUMBER	$\frac{1}{2}n(n+1)$
TRUNCATED OCTAHEDRAL NUMBER	$16n^3-33n^2+24n-6$
TRUNCATED TETRAHEDRAL NUMBER	$\frac{1}{6}n(23n^2-27n+10)$

See also BIQUADRATIC NUMBER, CENTERED CUBE NUMBER, CENTERED PENTAGONAL NUMBER, CENTERED POLYGONAL NUMBER, CENTERED SQUARE NUMBER, CENTERED TRIANGULAR NUMBER, CUBIC NUMBER, DECAGONAL NUMBER, FIGURATE NUMBER TRIANGLE, GNOMONIC NUMBER, HEPTAGONAL NUMBER, HEPTAGONAL PYRAMIDAL NUMBER, HEX NUMBER, HEX PYRAMIDAL NUMBER, HEXAGONAL NUMBER, HEXAGONAL PYRAMIDAL NUMBER, NEXUS NUMBER, OCTAGONAL NUMBER, OCTAHEDRAL NUMBER, PENTAGONAL NUMBER, PENTAGONAL PYRAMIDAL NUMBER, PENTATOPE NUMBER, POLYGONAL NUMBER, PRONIC NUMBER, PYRAMIDAL NUMBER, RHOMBIC DODECAHEDRAL NUMBER, SQUARE NUMBER, SQUARE PYRAMIDAL NUMBER, STELLA OCTANGULA NUMBER, TETRAHEDRAL NUMBER, TRIANGULAR NUMBER, TRUNCATED OCTAHEDRAL NUMBER, TRUNCATED TETRAHEDRAL NUMBER

References

Conway, J. H. and Guy, R. K. *The Book of Numbers.* New York: Springer-Verlag, pp. 30–2, 1996.
Dickson, L. E. "Polygonal, Pyramidal, and Figurate Numbers." Ch. 1 in *History of the Theory of Numbers, Vol. 2: Diophantine Analysis.* New York: Chelsea, pp. 1–9, 1952.
Goodwin, P. "A Polyhedral Sequence of Two." *Math. Gaz.* **69**, 191–97, 1985.
Guy, R. K. "Figurate Numbers." §D3 in *Unsolved Problems in Number Theory, 2nd ed.* New York: Springer-Verlag, pp. 147–50, 1994.
Kraitchik, M. "Figurate Numbers." §3.4 in *Mathematical Recreations.* New York: W. W. Norton, pp. 66–9, 1942.

Savin, A. "Shape Numbers." *Quantum* **11**, 14–8, 2000.

Figurate Number Triangle

A PASCAL'S TRIANGLE written in a square grid and padded with zeroes, as written by Jakob Bernoulli (Smith 1984). The figurate number triangle therefore has entries

$$a_{ij} = \binom{i}{j}.$$

where i is the row number, j the column number, and $\binom{i}{j}$ a BINOMIAL COEFFICIENT. Written out explicitly (beginning each row with $j = 0$),

$$\begin{bmatrix} 1 & 0 & 0 & 0 & 0 & 0 & 0 & \cdots \\ 1 & 1 & 0 & 0 & 0 & 0 & 0 & \cdots \\ 1 & 2 & 1 & 0 & 0 & 0 & 0 & \cdots \\ 1 & 3 & 3 & 1 & 0 & 0 & 0 & \cdots \\ 1 & 4 & 6 & 4 & 1 & 0 & 0 & \cdots \\ 1 & 5 & 10 & 10 & 5 & 1 & 0 & \cdots \\ 1 & 6 & 15 & 20 & 15 & 6 & 1 & \cdots \\ 1 & 7 & 21 & 35 & 35 & 21 & 7 & \ddots \\ \vdots & \vdots & \vdots & \vdots & \vdots & \vdots & \vdots & \ddots \end{bmatrix}$$

Then we have the sum identities

$$\sum_{j=0}^{i} a_{ij} = 2^i$$

$$\sum_{j=1}^{i} a_{ij} = 2^i - 1$$

$$\sum_{i=0}^{n} a_{ij} = a_{(n+1),(j+1)} = \frac{n+1}{j+1}a_{nj}.$$

See also BINOMIAL COEFFICIENT, FIGURATE NUMBER, PASCAL'S TRIANGLE

References

Smith, D. E. *A Source Book in Mathematics.* New York: Dover, p. 86, 1984.

Figure Eight Knot

FIGURE-OF-EIGHT KNOT

Figure Eight Surface

EIGHT SURFACE

Figure-of-Eight Knot

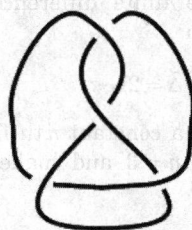

The knot 04–01, which is the unique PRIME KNOT of four crossings, and which is a 2-EMBEDDABLE KNOT. It is AMPHICHIRAL. It is also known as the FLEMISH KNOT and SAVOY KNOT, and it has BRAID WORD $\sigma_1 \sigma_2^{-1} \sigma_1 \sigma_2^{-1}$.

References

Francis, G. K. *A Topological Picture Book.* New York: Springer-Verlag, 1987.

Owen, P. *Knots.* Philadelphia, PA: Courage, p. 16, 1993.

Wells, D. *The Penguin Dictionary of Curious and Interesting Geometry.* Middlesex, England: Penguin Books, pp. 78–9, 1991.

Figures

A number x is said to have "n figures" if it takes n DIGITS to express it. The number of figures is therefore equal to one more than the POWER of 10 in the SCIENTIFIC NOTATION representation of the number. The word is most frequently used in reference to monetary amounts, e.g., a "six-figure salary" would fall in the range of \$100,000 to \$999,999.

See also DIGIT, SCIENTIFIC NOTATION, SIGNIFICANT FIGURES

Filon's Integration Formula

A formula for NUMERICAL INTEGRATION,

$$\int_{x_0}^{x_n} f(x)\cos(tx)dx$$

$$= h\{\alpha(th)[f_{2n}\sin(tx_{2n}) - f_0\sin(tx_0)] + \beta(th)C_{2n}$$

$$+ \gamma(th)C_{2n-1} + \frac{2}{45}th^4 S'_{2n-1}\} - R_n, \tag{1}$$

where

$$C_{2n} = \sum_{i=0}^{n} f_{2i}\cos(tx_{2i}) - \frac{1}{2}[f_{2n}\cos(tx_{2n})$$

$$+ f_0\cos(tx_0)] \tag{2}$$

$$C_{2n-1} = \sum_{i=1}^{n} f_{2i-1}\cos(tx_{2i-1}) \tag{3}$$

$$S'_{2n-1} = \sum_{i=1}^{n} f_{2i-1}^{(3)}\sin(tx_{2i} - 1) \tag{4}$$

$$\alpha(\theta) = \frac{1}{\theta} + \frac{\sin(2\theta)}{2\theta^2} - \frac{2\sin^2\theta}{\theta^3} \tag{5}$$

$$\beta(\theta) = 2\left[\frac{1 + \cos^2\theta}{\theta^2} - \frac{\sin(2\theta)}{\theta^3}\right] \tag{6}$$

$$\gamma(\theta) = 4\left(\frac{\sin\theta}{\theta^3} - \frac{\cos\theta}{\theta^2}\right), \tag{7}$$

and the remainder term is

$$R_n = \frac{1}{90}nh^5 f^{(4)}(\xi) + \mathcal{O}(th^7). \tag{8}$$

See also NUMERICAL INTEGRATION

References

Abramowitz, M. and Stegun, C. A. (Eds.). *Handbook of Mathematical Functions with Formulas, Graphs, and Mathematical Tables, 9th printing.* New York: Dover, pp. 890–91, 1972.

Tukey, J. W. In *On Numerical Approximation: Proceedings of a Symposium Conducted by the Mathematics Research Center, United States Army, at the University of Wisconsin, Madison, April 21–3, 1958* (Ed. R. E. Langer). Madison, WI: University of Wisconsin Press, p. 400, 1959.

Filter

Let S be a nonempty set, then a filter on S is a nonempty collection F of subsets of S having the following properties:

1. $\emptyset \notin F$,
2. If $A, B \in F$, then $A \cap B \in F$,
3. If $A \in F$ and $A \subseteq B \subseteq S$ then $B \in F$

If S is an infinite set, then the collection $F_S = \{A \subseteq S : S - A \text{ is finite}\}$ is a filter called the COFINITE (or Fréchet) filter on S.

In signal processing, a filter is a function or procedure which removes unwanted parts of a signal. The concept of filtering and filter functions is particularly useful in engineering. One particularly elegant method of filtering FOURIER transforms a signal into frequency space, performs the filtering operation there, then transforms back into the original space (Press *et al.* 1992).

See also COFINITE FILTER, REMEZ ALGORITHM, SAVITZKY-GOLAY FILTER, ULTRAFILTER, WIENER FILTER

References

Hamming, R. W. *Digital Filters.* New York: Dover, 1998.

Press, W. H.; Flannery, B. P.; Teukolsky, S. A.; and Vetterling, W. T. "Digital Filtering in the Time Domain." §13.5 in *Numerical Recipes in FORTRAN: The Art of Scientific Computing, 2nd ed.* Cambridge, England: Cambridge University Press, pp. 551–56, 1992.

Filtration

Fine's Equation
The q-SERIES identity

$$\prod_{n=1} \frac{(1-q^{2n})(1-q^{3n})(1-q^{8n})(1-q^{12n})}{(1-q^n)(1-q^{24n})}$$
$$= 1 + \sum_{N=1} E_{1,5,7,11}(N;24)q^N,$$

where $E_{1,5,7,11}(N;24)$ is the sum of the DIVISORS of N CONGRUENT to 1, 5, 7, and 11 (mod 24) minus the sum of DIVISORS of N CONGRUENT to -1, -5, -7, and -11 (mod 24).

See also Q-SERIES

Finite
A SET which contains a NONNEGATIVE integral number of elements is said to be finite. A SET which is not finite is said to be INFINITE. A finite or COUNTABLY INFINITE set is said to be COUNTABLE. While the meaning of the term "finite" is fairly clear in common usage, precise definitions of FINITE and INFINITE are needed in technical mathematics and especially in SET THEORY.

See also COUNTABLE SET, COUNTABLY INFINITE, INFINITE, SET THEORY, UNCOUNTABLY INFINITE

Finite Difference
The finite difference is the discrete analog of the DERIVATIVE. The finite FORWARD DIFFERENCE of a function f_p is defined as

$$\Delta f_p \equiv f_{p+1} - f_p, \tag{1}$$

and the finite BACKWARD DIFFERENCE as

$$\nabla f_p \equiv f_p - f_{p-1}. \tag{2}$$

If the values are tabulated at spacings h, then the notation

$$f_p \equiv f(x_0 + ph) \equiv f(x) \tag{3}$$

is used. The kth FORWARD DIFFERENCE would then be written as $\Delta^k f_p$, and similarly, the kth BACKWARD DIFFERENCE as $\nabla^k f_p$.

However, when f_p is viewed as a discretization of the continuous function $f(x)$, then the finite difference is sometimes written

$$\Delta f(x) \equiv f\left(x + \frac{1}{2}\right) - f\left(x - \frac{1}{2}\right) = 2I_I(x) * f(x), \tag{4}$$

where $*$ denotes CONVOLUTION and $I_I(x)$ is the odd IMPULSE PAIR. The finite difference operator can therefore be written

$$\tilde{\Delta} = 2I_I *. \tag{5}$$

An nth POWER has a constant nth finite difference. For example, take $n = 3$ and make a DIFFERENCE TABLE,

$$
\begin{array}{cccccc}
x & x^3 & \Delta & \Delta^2 & \Delta^3 & \Delta^4 \\
1 & 1 & & & & \\
 & & 7 & 12 & & \\
2 & 8 & & & 6 & \\
 & & 19 & 18 & & 0 \\
3 & 27 & & & 6 & \\
 & & 37 & 24 & & \\
4 & 64 & & & & \\
 & & 61 & & & \\
5 & 125 & & & & \\
\end{array}
\tag{6}
$$

The Δ^3 column is the constant 6.

Finite difference formulas can be very useful for extrapolating a finite amount of data in an attempt to find the general term. Specifically, if a function $f(n)$ is known at only a few discrete values $n = 0, 1, 2, \ldots$ and it is desired to determine the analytical form of f, the following procedure can be used if f is assumed to be a POLYNOMIAL function. Denote the nth value in the SEQUENCE of interest by a_n. Then define b_n as the FORWARD DIFFERENCE $\Delta_n \equiv a_{n+1} - a_n$, c_n as the second FORWARD DIFFERENCE $\Delta_n^2 \equiv b_{n+1} - b_n$, etc., constructing a table as follows

$$a_0 \equiv f(0) \quad a_1 \equiv f(1) \quad a_2 \equiv f(2) \quad \ldots \quad a_p \equiv f(p)$$
$$b_0 \equiv a_1 - a_0 \quad b_1 \equiv a_2 - a_1 \ldots \quad b_{p-1} \equiv a_p - a_{p-1}$$
$$c_0 \equiv b_1 - b_0 \ldots \ldots$$
$$\ddots \tag{7}$$

Continue computing d_0, e_0, etc., until a 0 value is obtained. Then the POLYNOMIAL function giving the values a_n is given by

$$f(n) = \sum_{k=0}^{p} a_k \binom{n}{k}$$
$$= a_0 + b_0 n + \frac{c_0 n(n-1)}{2} + \frac{d_0 n(n-1)(n-2)}{2 \cdot 3}$$
$$+ \cdots \tag{8}$$

When the notation $\Delta_0 \equiv a_0$, $\Delta_0^2 \equiv b_0$, etc., is used, this beautiful equation is called NEWTON'S FORWARD DIFFERENCE FORMULA. To see a particular example, consider a SEQUENCE with first few values of 1, 19, 143, 607, 1789, 4211, and 8539. The difference table is then given by

$$
\begin{array}{ccccccc}
1 & 19 & 143 & 607 & 1789 & 4211 & 8539 \\
 & 18 & 124 & 464 & 1182 & 2422 & 4328 \\
 & & 106 & 340 & 718 & 1240 & 1906 \\
 & & & 234 & 378 & 522 & 666 \\
\end{array}
$$

$$144 \quad 144 \quad 144$$

$$0 \quad 0$$

Reading off the first number in each row gives $a_0 = 1$, $b_0 = 18$, $c_0 = 106$, $d_0 = 234$, $e_0 = 144$. Plugging these in gives the equation

$$f(n) = 1 + 18n + 53n(n-1) + 39n(n-1)(n-2)$$
$$+ 6n(n-1)(n-2)(n-3), \tag{9}$$

which simplifies to $f(n) = 6n^4 + 3n^3 + 2n^2 + 7n + 1$, and indeed fits the original data exactly!

Beyer (1987) gives formulas for the derivatives

$$h^n \frac{d^n f(x_0 + ph)}{dx^n} \equiv h^n \frac{d^n f_p}{dx^n} \equiv \frac{d^n f_p}{dp^n} \tag{10}$$

(Beyer 1987, pp. 449–51) and integrals

$$\int_{x_0}^{x_\infty} f(x)dx = h \int_0^n f_p \, dp \tag{11}$$

(Beyer 1987, pp. 455–56) of finite differences.

Finite differences lead to DIFFERENCE EQUATIONS, finite analogs of DIFFERENTIAL EQUATIONS. In fact, UMBRAL CALCULUS displays many elegant analogs of well-known identities for continuous functions. Common finite difference schemes for PARTIAL DIFFERENTIAL EQUATIONS include the so-called Crank-Nicholson, Du Fort-Frankel, and Laasonen methods.

See also BACKWARD DIFFERENCE, BESSEL'S FINITE DIFFERENCE FORMULA, DIFFERENCE EQUATION, DIFFERENCE TABLE, EVERETT'S FORMULA, FINITE ELEMENT METHOD, FORWARD DIFFERENCE, GAUSS'S BACKWARD FORMULA, GAUSS'S FORWARD FORMULA, INTERPOLATION, JACKSON'S DIFFERENCE FAN, NEWTON'S BACKWARD DIFFERENCE FORMULA, NEWTON-COTES FORMULAS, NEWTON'S DIVIDED DIFFERENCE INTERPOLATION FORMULA, NEWTON'S FORWARD DIFFERENCE FORMULA, QUOTIENT-DIFFERENCE TABLE, STEFFENSON'S FORMULA, STIRLING'S FINITE DIFFERENCE FORMULA, UMBRAL CALCULUS

References

Abramowitz, M. and Stegun, C. A. (Eds.). "Differences." §25.1 in *Handbook of Mathematical Functions with Formulas, Graphs, and Mathematical Tables, 9th printing.* New York: Dover, pp. 877–78, 1972.

Beyer, W. H. *CRC Standard Mathematical Tables, 28th ed.* Boca Raton, FL: CRC Press, pp. 429–15, 1987.

Boole, G. and Moulton, J. F. *A Treatise on the Calculus of Finite Differences, 2nd rev. ed.* New York: Dover, 1960.

Conway, J. H. and Guy, R. K. "Newton's Useful Little Formula." In *The Book of Numbers.* New York: Springer-Verlag, pp. 81–3, 1996.

Iyanaga, S. and Kawada, Y. (Eds.). "Interpolation." Appendix A, Table 21 in *Encyclopedic Dictionary of Mathematics.* Cambridge, MA: MIT Press, pp. 1482–483, 1980.

Jordan, C. *Calculus of Finite Differences, 3rd ed.* New York: Chelsea, 1965.

Levy, H. and Lessman, F. *Finite Difference Equations.* New York: Dover, 1992.

Milne-Thomson, L. M. *The Calculus of Finite Differences.* London: Macmillan, 1951.

Richardson, C. H. *An Introduction to the Calculus of Finite Differences.* New York: Van Nostrand, 1954.

Spiegel, M. *Calculus of Finite Differences and Differential Equations.* New York: McGraw-Hill, 1971.

Stirling, J. *Methodus differentialis, sive tractatus de summation et interpolation serierum infinitarium.* London, 1730. English translation by Holliday, J. *The Differential Method: A Treatise of the Summation and Interpolation of Infinite Series.* 1749.

Tweedie, C. *James Stirling: A Sketch of his Life and Works Along with his Scientific Correspondence.* Oxford, England: Oxford University Press, pp. 30–5, 1922.

Weisstein, E. W. "Books about Finite Difference Equations." http://www.treasure-troves.com/books/FiniteDifferenceEquations.html.

Zwillinger, D. (Ed.). "Difference Equations." §3.9 in *CRC Standard Mathematical Tables and Formulae.* Boca Raton, FL: CRC Press, pp. 228–35, 1995.

Finite Element Method

A method for solving an equation by approximating continuous quantities as a set of quantities at discrete points, often regularly spaced into a so-called GRID or MESH. Because finite element methods can be adapted to problems of great complexity and unusual geometry, they are an extremely powerful tool in the solution of important problems in heat transfer, fluid mechanics, and mechanical systems. Furthermore, the availability of fast and inexpensive computers allows problems which are intractable using analytic or mechanical methods to be solved in a straightforward manner using finite element methods.

See also FINITE DIFFERENCE, LATTICE POINT

References

Akin, J. E. *Finite Elements for Analysis and Design.* San Diego: Academic Press, 1994.

Brenner, S. C. and Scott, L. R. *The Mathematical Theory of Finite Element Methods.* New York: Springer-Verlag, 1994.

Gallagher, R. H. *Finite Element Analysis: Fundamentals.* Englewood Cliffs, NJ: Prentice-Hall, 1975.

Kwon, Y. W. and Bang, H. *The Finite Element Method Using MATLAB.* Boca Raton, FL: CRC Press, 1996.

Özisik, M. N. *Finite Difference Methods in Heat Transfer.* Boca Raton, FL: CRC Press, 1994.

Reddy, J. N. and Gartling, D. K. *The Finite Element Method in Heat Transfer and Fluid Dynamics.* Boca Raton, FL: CRC Press, 1994.

White, R. E. *An Introduction to the Finite Element Method with Applications to Nonlinear Problems.* New York: Wiley, 1985.

Finite Field

A finite field is a FIELD with a finite ORDER (number of elements), also called a Galois field. The order of a finite field is always a PRIME or a POWER of a PRIME (Birkhoff and Mac Lane 1996). For each PRIME POWER, there exists exactly one (with the usual caveat that "exactly one" means "exactly one up to an ISOMORPHISM") finite field GF(p^n), often written as $\mathbb{F}_{p^n}$ in current usage.

GF(p) is called the PRIME FIELD of order p, and is the FIELD of RESIDUE CLASSES modulo p, where the p elements are denoted 0, 1, ..., $p-1$. $a = b$ in GF(p) means the same as $a \equiv b \pmod{p}$. Note, however, that $2 \times 2 \equiv 0 \pmod{4}$ in the RING of residues modulo 4, so 2 has no reciprocal, and the RING of residues modulo 4 is distinct from the finite field with four elements. Finite fields are therefore denoted GF(p^n), instead of GF(k), where $k = p^n$, for clarity.

The finite field GF(2) consists of elements 0 and 1 which satisfy the following addition and multiplication tables.

+	0	1
0	0	1
1	1	0

×	0	1
0	0	0
1	0	1

If a subset S of the elements of a finite field F satisfies the axioms above with the same operators of F, then S is called a SUBFIELD. Finite fields are used extensively in the study of ERROR-CORRECTING CODES.

When $n > 1$, GF(p^n) can be REPRESENTED AS the FIELD of EQUIVALENCE CLASSES of POLYNOMIALS whose COEFFICIENTS belong to GF(p). Any IRREDUCIBLE POLYNOMIAL of degree n yields the same FIELD up to an ISOMORPHISM. For example, for GF(2^3), the modulus can be taken as $x^3 + x^2 + 1$, $x^3 + x + 1$, or any other IRREDUCIBLE POLYNOMIAL of degree 3. Using the modulus $x^3 + x + 1$, the elements of GF(2^3)–written 0, x^0, x^1, ...–can be REPRESENTED AS POLYNOMIALS with degree less than 3. For instance,

$$x^3 \equiv -x - 1 \equiv x + 1$$

$$x^4 \equiv x(x^3) \equiv x(x+1) \equiv x^3 + x$$

$$x^5 \equiv x\left(x^2 + x\right) \equiv x^3 + x^2 \equiv x^2 - x - 1 \equiv x^2 + x + 1$$

$$x^6 \equiv x(x^2 + x + 1) \equiv x^3 + x^2 + x \equiv x^2 - 1 \equiv x^2 + 1$$

$$x^7 \equiv x(x^2 + 1) \equiv x^3 + x \equiv -1 \equiv 1 \equiv x_0.$$

Now consider the following table which contains several different representations of the elements of a finite field. The columns are the power, polynomial representation, triples of polynomial representation COEFFICIENTS (the vector representation), and the binary INTEGER corresponding to the vector representation (the regular representation).

Power	Polynomial	Vector	Regular
0	0	(000)	0
x^0	1	(001)	1
x^1	x	(010)	2
x^2	x^2	(100)	4
x^3	$x + 1$	(011)	3
x^4	$x^2 + x$	(110)	6
x^5	$x^2 + x + 1$	(111)	7
x^6	$x^2 + 1$	(101)	5

The set of POLYNOMIALS in the second column is CLOSED under ADDITION and MULTIPLICATION modulo $x^3 + x + 1$, and these operations on the set satisfy the AXIOMS of finite field. This particular finite field is said to be an extension field of degree 3 of GF(2), written GF(2^3), and the field GF(2) is called the base field of GF(2^3). If an IRREDUCIBLE POLYNOMIAL generates all elements in this way, it is called a PRIMITIVE POLYNOMIAL. For any PRIME or PRIME POWER q and any POSITIVE INTEGER n, there exists a primitive irreducible polynomial of degree n over GF(q).

For any element c of GF(q), $c^q = c$, and for any NONZERO element d of GF(q), $d^{q-1} = 1$. There is a smallest POSITIVE INTEGER n satisfying the sum condition $\underbrace{e + e + \ldots + e = 0}_{n \text{ times}}$ for some element e in GF(q). This number is called the CHARACTERISTIC of the finite field GF(q). The CHARACTERISTIC is a PRIME NUMBER for every finite field, and it is true that

$$(x + y)^p = x^p + y^p$$

over a finite field with characteristic p.

See also CHARACTERISTIC (FIELD), FIELD, HADAMARD MATRIX, IRREDUCIBLE POLYNOMIAL, PRIMITIVE POLYNOMIAL, RING, SUBFIELD

References

Ball, W. W. R. and Coxeter, H. S. M. *Mathematical Recreations and Essays, 13th ed.* New York: Dover, pp. 73–5, 1987.
Birkhoff, G. and Mac Lane, S. *A Survey of Modern Algebra, 5th ed.* New York: Macmillan, p. 413, 1996.
Dickson, L. E. *History of the Theory of Numbers, Vol. 1: Divisibility and Primality.* New York: Chelsea, p. viii, 1952.
Dummit, D. S. and Foote, R. M. "Finite Fields." §14.3 in *Abstract Algebra, 2nd ed.* Englewood Cliffs, NJ: Prentice-Hall, pp. 499–05, 1998.
Lidl, R. and Niederreiter, H. *Introduction to Finite Fields and Their Applications, rev. ed.* Cambridge, England: Cambridge University Press, 1994.
Lidl, R. and Niederreiter, H. (Eds.). *Finite Fields, 2nd ed.* Cambridge, England: Cambridge University Press, 1997.

Finite Game

A GAME in which each player has a finite number of moves and a finite number of choices at each move.

See also GAME, HYPERGAME, ZERO-SUM GAME

References

Dresher, M. *The Mathematics of Games of Strategy: Theory and Applications.* New York: Dover, p. 2, 1981.

Finite Group

A GROUP of finite ORDER. Examples of finite groups are the MODULO MULTIPLICATION GROUPS and the POINT GROUPS. The CLASSIFICATION THEOREM of finite SIMPLE GROUPS states that the finite SIMPLE GROUPS can be classified completely into one of five types.

The following table gives the numbers and names of the first few groups of ORDER h. In the table, *NA* denotes the number of non-Abelian groups, *A* denotes the number of ABELIAN GROUPS, and N the total number of groups. In addition, $\mathbb{Z}_n$ denotes a CYCLIC GROUP of ORDER n, A_n an ALTERNATING GROUP, D_n a DIHEDRAL GROUP, Q_8 the group of the QUATERNIONS, T the cubic group, and $\times$ denotes GROUP DIRECT PRODUCT.

h	Name	A	NA	N
1	FINITE GROUP e	1	0	1
2	FINITE GROUP Z_2	1	0	1
3	FINITE GROUP Z_3	1	0	1
4	FINITE GROUP $Z_2 Z_2$, FINITE GROUP Z_4	2	0	2
5	FINITE GROUP Z_5	1	0	1
6	FINITE GROUP Z_6, FINITE GROUP D_3	1	1	2
7	FINITE GROUP Z_7	1	0	1
8	FINITE GROUP $Z_2 Z_2 Z_2$, FINITE GROUP $Z_2 Z_4$, FINITE GROUP Z_8, FINITE GROUP Q_8, FINITE GROUP D_4	3	2	5
9	$\mathbb{Z}_3 \times \mathbb{Z}_3, \mathbb{Z}_9$	2	0	2
10	$\mathbb{Z}_{10}, D_5$	1	1	2
11	$\mathbb{Z}_{11}$	1	0	1
12	$\mathbb{Z}_2 \times \mathbb{Z}_6, \mathbb{Z}_{12}, A_4, D_6, T$	2	3	5
13	$\mathbb{Z}_{13}$	1	0	1
14	$\mathbb{Z}_{14}, D_7$	1	1	2
15	$\mathbb{Z}_{15}$	1	0	1

The problem of determining the nonisomorphic finite groups of order h was first considered by Cayley (1854). There is no known FORMULA to give the number of possible finite groups $g(h)$ as a function of the ORDER h. However, there are simple formulas for special forms of h.

$$g(1) = 1 \tag{1}$$

$$g(p) = 1 \tag{2}$$

$$g(pq) = \begin{cases} 1 & \text{if } p \nmid (q-1) \\ 2 & \text{if } p | (q-1) \end{cases} \tag{3}$$

$$g(p^2) = 2 \tag{4}$$

$$g(p^3) = 5, \tag{5}$$

where p and $q > p$ are distinct primes. In addition, there is a beautiful algorithm due to Hölder (Hölder 1895, Alonso 1976) for determining $g(h)$ for square-free h, namely

$$g(h) = \sum_{d|n} \prod_{\substack{p|d \\ p \neq 1}} \frac{p^{o_p(n/d)} - 1}{p - 1}, \tag{6}$$

where $o_p(m)$ is the number of primes p such that $q|m$ and $p|(q-1)$ (Dennis).

Miller (1930) gave the number of groups for orders 1–00, including an erroneous 297 as the number of groups of ORDER 64. Senior and Lunn (1934, 1935) subsequently completed the list up to 215, but omitted 128 and 192. The number of groups of ORDER 64 was corrected in Hall and Senior (1964). James *et al.* (1990) found 2328 groups in 115 ISOCLINISM families of ORDER 128, correcting previous work, and O'Brien (1991) found the number of groups of ORDER 256. Currently, the number of groups is known for orders up to 2000, excluding 1024 (Besche and Eick 1999a), with the difficult cases of orders 512 ($g(512) = 10,494,213$; Eick and O'Brien 1999b) and 768 (Besche and Eick 2000) now put to rest. The numbers of nonisomorphic finite groups N of each ORDER h for the first few hundred orders are given in the table below (Sloane's A000001–the very first sequence). The number of nonisomorphic groups of orders 2^n for $n = 0, 1, \ldots$ are 1, 1, 2, 5, 14, 51, 267, 2328, 56092, ... (Sloane's A000679).

The smallest orders h for which there exist $n = 1, 2, \ldots$ nonisomorphic groups are 1, 4, 75, 28, 8, 42, ... (Sloane's A046057). The incrementally largest number of nonisomorphic finite groups are 1, 2, 5, 14, 15, 51, 52, 267, 2328, ... (Sloane's A046058), which occur for orders 1, 4, 8, 16, 24, 32, 48, 64, 128, ... (Sloane's A046059). Dennis has conjectured that the number of groups $g(h)$ of order h assumes every positive integer as a value an infinite number of times.

It is simple to determine the number of ABELIAN GROUPS using the KRONECKER DECOMPOSITION THEO-

REM, and there is at least one ABELIAN GROUP for every finite order h. The number A of ABELIAN GROUPS of ORDER $h = 1, 2, \ldots$ are given by 1, 1, 1, 2, 1, 1, 1, 3, ... (Sloane's A000688). The following table summarizes the total number of finite groups N and the number of Abelian finite groups A for orders h from 1 to 400. A table of orders up to 1000 is given by Royle; the GAP software package includes a table of the number of finite groups up to order 2000, excluding 1024.

h	N	A	h	N	A	h	N	A	h	N	A
1	1	1	51	1	1	101	1	1	151	1	1
2	1	1	52	5	2	102	4	1	152	12	3
3	1	1	53	1	1	103	1	1	153	2	2
4	2	2	54	15	3	104	14	3	154	4	1
5	1	1	55	2	1	105	2	1	155	2	1
6	2	1	56	13	3	106	2	1	156	18	2
7	1	1	57	2	1	107	1	1	157	1	1
8	5	3	58	2	1	108	45	6	158	2	1
9	2	2	59	1	1	109	1	1	159	1	1
10	2	1	60	13	2	110	6	1	160	238	7
11	1	1	61	1	1	111	2	1	161	1	1
12	5	2	62	2	1	112	43	5	162	55	5
13	1	1	63	4	2	113	1	1	163	1	1
14	2	1	64	267	11	114	6	1	164	5	2
15	1	1	65	1	1	115	1	1	165	2	1
16	14	5	66	4	1	116	5	2	166	2	1
17	1	1	67	1	1	117	4	2	167	1	1
18	5	2	68	5	2	118	2	1	168	57	3
19	1	1	69	1	1	119	1	1	169	2	2
20	5	2	70	4	1	120	47	3	170	4	1
21	2	1	71	1	1	121	2	2	171	5	2
22	2	1	72	50	6	122	2	1	172	4	2
23	1	1	73	1	1	123	1	1	173	1	1
24	15	3	74	2	1	124	4	2	174	4	1
25	2	2	75	3	2	125	5	3	175	2	2
26	2	1	76	4	2	126	16	2	176	42	5
27	5	3	77	1	1	127	1	1	177	1	1
28	4	2	78	6	1	128	2328	15	178	2	1
29	1	1	79	1	1	129	2	1	179	1	1
30	4	1	80	52	5	130	4	1	180	37	4
31	1	1	81	15	5	131	1	1	181	1	1
32	51	7	82	2	1	132	10	2	182	4	1
33	1	1	83	1	1	133	1	1	183	2	1
34	2	1	84	15	2	134	2	1	184	12	3
35	1	1	85	1	1	135	5	3	185	1	1
36	14	4	86	2	1	136	15	3	186	6	1
37	1	1	87	1	1	137	1	1	187	1	1
38	2	1	88	12	3	138	4	1	188	4	2
39	2	1	89	1	1	139	1	1	189	13	3
40	14	3	90	10	2	140	11	2	190	4	1
41	1	1	91	1	1	141	1	1	191	1	1
42	6	1	92	4	2	142	2	1	192	1543	11
43	1	1	93	2	1	143	1	1	193	1	1
44	4	2	94	2	1	144	197	1	194	2	1
45	2	2	95	1	1	145	1	1	195	2	1
46	2	1	96	230	7	146	2	1	196	17	4
47	1	1	97	1	1	147	6	2	197	1	1
48	52	5	98	5	2	148	5	2	198	10	2
49	2	2	99	2	2	149	1	1	199	1	1
50	2	2	100	16	4	150	13	2	200	52	6

h	N	A	h	N	A	h	N	A	h	N	A
201	2	1	251	1	1	301	2	1	351	14	3
202	2	1	252	46	4	302	2	1	352	195	7
203	2	1	253	2	1	303	1	1	353	1	1
204	12	2	254	2	1	304	42	5	354	4	1
205	2	1	255	1	1	305	2	1	355	2	1
206	2	1	256	56092	22	306	10	2	356	5	2
207	2	2	257	1	1	307	1	1	357	2	1
208	51	5	258	6	1	308	9	2	358	2	1
209	1	1	259	1	1	309	2	1	359	1	1
210	12	1	260	15	2	310	6	1	360	162	6
211	1	1	261	2	2	311	1	1	361	2	2
212	5	2	262	2	1	312	61	3	362	2	1
213	1	1	263	1	1	313	1	1	363	3	2
214	2	1	264	39	3	314	2	1	364	11	2

215	1	1	265	1	1	315	4	2	365	1	1
216	177	9	266	4	1	316	4	2	366	6	1
217	1	1	267	1	1	317	1	1	367	1	1
218	2	1	268	4	2	318	4	1	368	42	5
219	2	1	269	1	1	319	1	1	369	2	2
220	15	2	270	30	3	320	1640	11	370	4	1
221	1	1	271	1	1	321	1	1	371	1	1
222	6	1	272	54	5	322	4	1	372	15	2
223	1	1	273	5	1	323	1	1	373	1	1
224	197	7	274	2	1	324	176	10	374	4	1
225	6	4	275	4	2	325	2	2	375	7	3
226	2	1	276	10	2	326	2	1	376	12	3
227	1	1	277	1	1	327	2	1	377	1	1
228	15	2	278	2	1	328	15	3	378	60	3
229	1	1	279	4	2	329	1	1	379	1	1
230	4	1	280	40	3	330	12	1	380	11	2
231	2	1	281	1	1	331	1	1	381	2	1
232	14	3	282	4	1	332	4	2	382	2	1
233	1	1	283	1	1	333	5	2	383	1	1
234	16	2	284	4	2	334	2	1	384	20169	15
235	1	1	285	2	1	335	1	1	385	2	1
236	4	2	286	4	1	336	228	5	386	2	1
237	2	1	287	1	1	337	1	1	387	4	2
238	4	1	288	1045	14	338	5	2	388	5	2
239	1	1	289	2	2	339	1	1	389	1	1
240	208	5	290	4	1	340	15	2	390	12	1
241	1	1	291	2	1	341	1	1	391	1	1
242	5	2	292	5	2	342	18	2	392	44	6
243	67	7	293	1	1	343	5	3	393	1	1
244	5	2	294	23	2	344	12	3	394	2	1
245	2	2	295	1	1	345	1	1	395	1	1
246	4	1	296	14	3	346	2	1	396	30	4
247	1	1	297	5	3	347	1	1	397	1	1
248	12	3	298	2	1	348	12	2	398	2	1
249	1	1	299	1	1	349	1	1	399	5	1
250	15	3	300	49	4	350	10	2	400	221	10

See also ABELIAN GROUP, ABHYANKAR'S CONJECTURE,

ALTERNATING GROUP, BURNSIDE'S LEMMA, BURNSIDE PROBLEM, CHEVALLEY GROUPS, CLASSIFICATION THEOREM, COMPOSITION SERIES, CONTINUOUS GROUP, DIHEDRAL GROUP, DISCRETE GROUP, FEIT-THOMPSON THEOREM, GROUP, INFINITE GROUP, JORDAN-HÖLDER THEOREM, KRONECKER DECOMPOSITION THEOREM, LIE GROUP, LIE-TYPE GROUP, LINEAR GROUP, MODULO MULTIPLICATION GROUP, ORDER (GROUP), ORTHOGONAL GROUP, P-GROUP, POINT GROUPS, SIMPLE GROUP, SPORADIC GROUP, SYMMETRIC GROUP, SYMPLECTIC GROUP, TWISTED CHEVALLEY GROUPS, UNITARY GROUP

References

Alonso, J. "Groups of Square-Free Order, an Algorithm." *Math. Comput.* **30**, 632–37, 1976.

Arfken, G. "Discrete Groups." §4.9 in *Mathematical Methods for Physicists, 3rd ed.* Orlando, FL: Academic Press, pp. 243–51, 1985.

Artin, E. "The Order of the Classical Simple Groups." *Comm. Pure Appl. Math.* **8**, 455–72, 1955.

Aschbacher, M. *Finite Group Theory, 2nd ed.* Cambridge, England: Cambridge University Press, 2000.

Aschbacher, M. *The Finite Simple Groups and Their Classification.* New Haven, CT: Yale University Press, 1980.

Ball, W. W. R. and Coxeter, H. S. M. *Mathematical Recreations and Essays, 13th ed.* New York: Dover, pp. 73–5, 1987.

Besche, H.-U. and Eick, B. "Construction of Finite Groups." *J. Symb. Comput.* **27**, 387–04, 1999.

Besche, H.-U. and Eick, B. "The Groups of Order at Most 1000 Except 512 and 768." *J. Symb. Comput.* **27**, 405–13, 1999.

Besche, H.-U. and Eick, B. "The Groups of Order $q^n \cdot p$." In preparation, 2000.

Cayley, A. "On the Theory of Groups as Depending on the Symbolic Equation $\theta^n = 1$." *Philos. Mag.* **7**, 33–9, 1854.

Cayley, A. "On the Theory of Groups as Depending on the Symbolic Equation $\theta^n = 1$.--Part II." *Philos. Mag.* **7**, 408–09, 1854.

Cayley, A. "On the Theory of Groups as Depending on the Symbolic Equation $\theta^n = 1$.--Part III." *Philos. Mag.* **18**, 34–7, 1859.

Conway, J. H.; Curtis, R. T.; Norton, S. P.; Parker, R. A.; and Wilson, R. A. *Atlas of Finite Groups: Maximal Subgroups and Ordinary Characters for Simple Groups.* Oxford, England: Clarendon Press, 1985.

Dennis, K. "The Number of Groups of Order n." Preprint.

Eick, B. and O'Brien, E. A. "Enumerating p-Groups." *J. Austral. Math. Soc. Ser. A* **67**, 191–05, 1999a.

Eick, B. and O'Brien, E. A. "The Groups of Order 512." In *Algorithmic Algebra and Number Theory: Selected Papers from the Conference held at the University of Heidelberg, Heidelberg, October 1997* (Ed. B. H. Matzat, G.-M. Greuel, and G. Hiss). Berlin: Springer-Verlag, pp. 379–80, 1999b.

GAP Group. "GAP--Groups, Algorithms, and Programming." http://www-history.mcs.st-and.ac.uk/~gap/.

Hall, M. Jr. and Senior, J. K. *The Groups of Order 2^n ($n \le 6$).* New York: Macmillan, 1964.

Hölder, O. "Die Gruppen der Ordnung p^3, pq^2, pqr, p^4." *Math. Ann.* **43**, 300–12, 1893.

Hölder, O. "Die Gruppen mit quadratfreier Ordnungszahl." *Nachr. Königl. Gesell. Wissenschaft. Göttingen, Math.-Phys. Kl.*, 211–29, 1895.

Huang, J.-S. "Finite Groups." Part I in *Lectures on Representation Theory.* Singapore: World Scientific, pp. 1–5, 1999.

James, R. "The Groups of Order p^6 (p an Odd Prime)." *Math. Comput.* **34**, 613–37, 1980.

James, R.; Newman, M. F.; and O'Brien, E. A. "The Groups of Order 128." *J. Algebra* **129**, 136–58, 1990.

Laue, R. "Zur Konstruktion und Klassifikation endlicher auflösbarer Gruppen." *Bayreuther Mathemat. Schriften* **9**, 1982.

Miller, G. A. "Determination of All the Groups of Order 64." *Amer. J. Math.* **52**, 617–34, 1930.

Miller, G. A. "Orders for which a Given Number of Groups Exist." *Proc. Nat. Acad. Sci.* **18**, 472–75, 1932.

Miller, G. A. "Orders for which There Exist Exactly Four or Five Groups." *Proc. Nat. Acad. Sci.* **18**, 511–14, 1932.

Miller, G. A. "Groups whose Orders Involve a Small Number of Unity Congruences." *Amer. J. Math.* **55**, 22–8, 1933.

Miller, G. A. "Historical Note on the Determination of Abstract Groups of Given Orders." *J. Indian Math. Soc.* **19**, 205–10, 1932.

Miller, G. A. "Enumeration of Finite Groups." *Math. Student* **8**, 109–11, 1940.

Murty, M. R. and Murty, V. K. "On the Number of Groups of a Given Order." *J. Number Th.* **18**, 178–91, 1984.

Neubüser, J. *Die Untergruppenverbände der Gruppen der Ordnung $\leq$ 100 mit Ausnahme der Ordnungen 64 und 96.* Habilitationsschrift. Kiel, Germany: Universität Kiel, 1967.

O'Brien, E. A. "The Groups of Order 256." *J. Algebra* **143**, 219–35, 1991.

O'Brien, E. A. and Short, M. W. "Bibliography on Classification of Finite Groups." Manuscript, Australian National University, 1988.

Royle, G. "Numbers of Small Groups." http://www.cs.uwa.edu.au/~gordon/remote/group1000.html.

Senior, J. K. and Lunn, A. C. "Determination of the Groups of Orders 101–61, Omitting Order 128." *Amer. J. Math.* **56**, 328–38, 1934.

Senior, J. K. and Lunn, A. C. "Determination of the Groups of Orders 162–15, Omitting Order 192." *Amer. J. Math.* **57**, 254–60, 1935.

Simon, B. *Representations of Finite and Compact Groups.* Providence, RI: Amer. Math. Soc., 1996.

Sloane, N. J. A. Sequences A000001/M0098, A000679/M1470, A000688/M0064, A046057, A046058, and A046059 in "An On-Line Version of the Encyclopedia of Integer Sequences." http://www.research.att.com/~njas/sequences/eisonline.html.

Spiro, C. A. "Local Distribution Results for the Group-Counting Function at Positive Integers." *Congr. Numer.* **50**, 107–10, 1985.

University of Sydney Computational Algebra Group. "The Magma Computational Algebra for Algebra, Number Theory and Geometry." http://www.maths.usyd.edu.au:8000/u/magma/.

Weisstein, E. W. "Groups." Mathematica notebook Groups.m.

Wilson, R. A. "ATLAS of Finite Group Representation." http://for.mat.bham.ac.uk/atlas/.

Finite Group D3

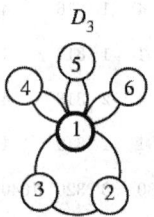

The DIHEDRAL GROUP D_3 is one of the two groups of ORDER 6. It is the non-Abelian group of smallest ORDER. Examples of D_3 include the POINT GROUPS known as C_{3h}, C_{3v}, S_3, D_3, the symmetry group of the EQUILATERAL TRIANGLE, and the group of permutation of three objects. Its elements A_i satisfy $A_i^3 = 1$, and four of its elements satisfy $A_i^2 = 1$, where 1 is the IDENTITY ELEMENT. The CYCLE GRAPH is shown above, and the MULTIPLICATION TABLE is given below (Cotton 1990, p. 12).

D_3	1	A	B	C	D	E
1	1	A	B	C	D	E
A	A	1	D	E	B	C
B	B	E	1	D	C	A
C	C	D	E	1	A	B
D	D	C	A	B	E	1
E	E	B	C	A	1	D

The CONJUGACY CLASSES are $\{1\}$ (which is always in a class by itself), $\{A, B, C\}$,

$$A^{-1}AA = A \tag{1}$$

$$B^{-1}AB = C \tag{2}$$

$$C^{-1}AC = B \tag{3}$$

$$D^{-1}AD = C \tag{4}$$

$$E^{-1}AE = B, \tag{5}$$

and $\{D, E\}$,

$$A^{-1}DA = E \tag{6}$$

$$B^{-1}DB = D. \tag{7}$$

A reducible 2-D representation using REAL MATRICES can be found by performing the spatial rotations corresponding to the symmetry elements of C_{3v}. Take the z-AXIS along the C_3 axis.

$$I = R_z(0) = \begin{bmatrix} 1 & 0 \\ 0 & 1 \end{bmatrix} \tag{8}$$

$$A = R_z\left(\frac{2}{3}\Pi\right) = \begin{bmatrix} \cos\left(\frac{2}{3}\Pi\right) & \sin\left(\frac{2}{3}\Pi\right) \\ -\sin\left(\frac{2}{3}\Pi\right) & \cos\left(\frac{2}{3}\Pi\right) \end{bmatrix}$$

$$= \begin{bmatrix} -\frac{1}{2} & -\frac{1}{2}\sqrt{3} \\ \frac{1}{2}\sqrt{3} & -\frac{1}{2} \end{bmatrix} \tag{9}$$

$$B = R_z\left(\frac{4}{3}\Pi\right) = \begin{bmatrix} -\frac{1}{2} & \frac{1}{2}\sqrt{3} \\ -\frac{1}{2}\sqrt{3} & -\frac{1}{2} \end{bmatrix} \tag{10}$$

$$C = R_c(\Pi) = \begin{bmatrix} -1 & 0 \\ 0 & 1 \end{bmatrix} \tag{11}$$

$$D = R_D(\Pi) = CB = \begin{bmatrix} \frac{1}{2} & -\frac{1}{2}\sqrt{3} \\ -\frac{1}{2}\sqrt{3} & -\frac{1}{2} \end{bmatrix} \tag{12}$$

$$E = R_E(\Pi) = CA = \begin{bmatrix} \frac{1}{2} & \frac{1}{2}\sqrt{3} \\ \frac{1}{2}\sqrt{3} & -\frac{1}{2} \end{bmatrix} \tag{13}$$

To find the irreducible representation, note that there are three CONJUGACY CLASSES. GROUP rule 5 requires that there be three irreducible representations satisfying

$$h = l_1^2 + l_2^2 + l_3^2 = 6, \tag{14}$$

so it must be true that

$$l_1 = l_2 = 1, l_3 = 2. \tag{15}$$

By GROUP rule 6, we can let the first representation have all 1s.

D_3	1	A	B	C	D	E
Γ_1	1	1	1	1	1	1

To find a representation orthogonal to the totally symmetric representation, we must have three +1 and three −1 CHARACTERS. We can also add the constraint that the components of the IDENTITY ELEMENT 1 be positive. The three CONJUGACY CLASSES have 1, 2, and 3 elements. Since we need a total of three +1s and we have required that a +1 occur for the CONJUGACY CLASS of ORDER 1, the remaining +1s must be used for the elements of the CONJUGACY CLASS of ORDER 2, i.e., D and E.

D_3	1	A	B	C	D	E
Γ_1	1	1	1	1	1	1
Γ_2	1	−1	−1	−1	1	1

Using GROUP rule 1, we see that

$$1^2 + 1^2 + \chi_3^2(1) = 6 \tag{16}$$

so the final representation for 1 has CHARACTER 2. Orthogonality with the first two representations (GROUP rule 3) then yields the following constraints:

$$1 \cdot 1 \cdot 2 + 1 \cdot 2 \cdot \chi_2 + 1 \cdot 3 \cdot \chi_3 = 2 + 2\chi_2 + 3\chi_3 = 0 \tag{17}$$

$$1 \cdot 1 \cdot 2 + 1 \cdot 2 \cdot \chi_2 + (-1) \cdot 3 \cdot \chi_3 = 2 + 2\chi_2 - 3\chi_3 = 0. \tag{18}$$

Solving these simultaneous equations by adding and subtracting (18) from (17), we obtain $\chi_2 = -1$, $\chi_3 = 0$. The full CHARACTER TABLE is then

D_3	1	A	B	C	D	E
Γ_1	1	1	1	1	1	1
Γ_2	1	−1	−1	−1	1	1
Γ_3	2	0	0	0	−1	−1

Since there are only three CONJUGACY CLASSES, this table is conventionally written simply as

D_3	1	A = B = C	D = E
Γ_1	1	1	1
Γ_2	1	−1	1
Γ_3	2	0	−1

Writing the irreducible representations in matrix form then yields

$$1 = \begin{bmatrix} 1 & 0 & 0 & 0 \\ 0 & 1 & 0 & 0 \\ 0 & 0 & 1 & 0 \\ 0 & 0 & 0 & 1 \end{bmatrix} \tag{19}$$

$$A = \begin{bmatrix} 1 & 0 & 0 & 0 \\ 0 & 1 & 0 & 0 \\ 0 & 0 & -\frac{1}{2} & -\frac{1}{2}\sqrt{3} \\ 0 & 0 & \frac{1}{2}\sqrt{3} & -\frac{1}{2} \end{bmatrix} \tag{20}$$

$$B = \begin{bmatrix} 1 & 0 & 0 & 0 \\ 0 & 1 & 0 & 0 \\ 0 & 0 & -\dfrac{1}{2} & \dfrac{1}{2}\sqrt{3} \\ 0 & 0 & -\dfrac{1}{2}\sqrt{3} & -\dfrac{1}{2} \end{bmatrix} \qquad (21)$$

$$C = \begin{bmatrix} 1 & 0 & 0 & 0 \\ 0 & -1 & 0 & 0 \\ 0 & 0 & -1 & 0 \\ 0 & 0 & 0 & 1 \end{bmatrix} \qquad (22)$$

$$D = \begin{bmatrix} 1 & 0 & 0 & 0 \\ 0 & 1 & 0 & 0 \\ 0 & 0 & -\dfrac{1}{2} & \dfrac{1}{2}\sqrt{3} \\ 0 & 0 & -\dfrac{1}{2}\sqrt{3} & -\dfrac{1}{2} \end{bmatrix} \qquad (23)$$

$$E = \begin{bmatrix} 1 & 0 & 0 & 0 \\ 0 & -1 & 0 & 0 \\ 0 & 0 & \dfrac{1}{2} & \dfrac{1}{2}\sqrt{3} \\ 0 & 0 & \dfrac{1}{2}\sqrt{3} & -\dfrac{1}{2} \end{bmatrix} \qquad (24)$$

See also DIHEDRAL GROUP, FINITE GROUP D4, FINITE GROUP Z6

Finite Group D4

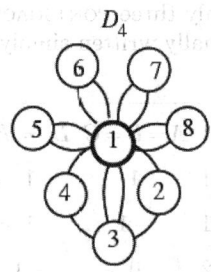

The DIHEDRAL GROUP D_4 is one of the two non-Abelian groups of the five groups total of ORDER 8. It is sometimes called the octic group. Examples of D_4 include the symmetry group of the SQUARE. The CYCLE GRAPH is shown above.

See also DIHEDRAL GROUP, FINITE GROUP D3, FINITE GROUP Z8, FINITE GROUP Z2Z2Z2, FINITE GROUP Z2Z4, FINITE GROUP Z8

References
Cotton, F. A. *Chemical Applications of Group Theory, 3rd ed.* New York: Wiley, 1990.

Finite Group e

The unique (and trivial) group of ORDER 1 is denoted $\langle e \rangle$. It is (trivially) ABELIAN and CYCLIC. Examples include the POINT GROUP C_1 and the integers modulo 1 under addition.

$\langle e \rangle$	1
1	1

Its only conjugacy class is $\{1\}$.

Finite Group Q8

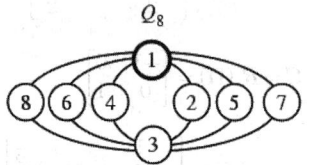

One of the two non-Abelian groups of the five groups total of ORDER 8. The group Q_8 has the MULTIPLICATION TABLE of $\pm 1, i, j, k$, where 1, i, j, and k are the QUATERNIONS. The CYCLE GRAPH is shown above.

See also FINITE GROUP D4, FINITE GROUP Z2Z2Z2, FINITE GROUP Z2Z4, FINITE GROUP Z8, QUATERNION

Finite Group Z2

The unique group of ORDER 2. $\mathbb{Z}_2$ is both ABELIAN and CYCLIC. Examples include the POINT GROUPS C_s, C_i, and C_2, the integers modulo 2 under addition, and the MODULO MULTIPLICATION GROUPS M_3, M_4, and M_6. The elements A_i satisfy $A_i^2 = 1$, where 1 is the IDENTITY ELEMENT. The CYCLE GRAPH is shown above, and the MULTIPLICATION TABLE is given below.

$\mathbb{Z}_2$	1	A
1	1	A
A	A	1

The CONJUGACY CLASSES are $\{1\}$ and $\{A\}$. The irreducible representation for the C_2 group is $\{1, -1\}$.

Finite Group Z2Z2

$$Z_2 \otimes Z_2$$

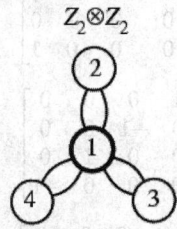

One of the two groups of ORDER 4. The name of this group derives from the fact that it is a GROUP DIRECT PRODUCT of two $\mathbb{Z}_2$ SUBGROUPS. Like the group $\mathbb{Z}_4$, $\mathbb{Z}_2 \times \mathbb{Z}_2$ is an ABELIAN GROUP. Unlike $\mathbb{Z}_4$, however, it is not CYCLIC. In addition to satisfying $A_i^4 = 1$ for each element A_i, it also satisfies $A_i^2 = 1$, where 1 is the IDENTITY ELEMENT. Examples of the $\mathbb{Z}_2 \times \mathbb{Z}_2$ group include the VIERGRUPPE, POINT GROUPS D_2, C_{2h}, and C_{2v}, and the MODULO MULTIPLICATION GROUPS M_8 and M_{12}. That M_8, the RESIDUE CLASSES prime to 8 given by $\{1, 3, 5, 7\}$, are a group of type $\mathbb{Z}_2 \times \mathbb{Z}_2$ can be shown by verifying that

$$1^2 = 1 \quad 3^2 = 9 \equiv 1 \quad 5^2 = 25 \equiv 1$$
$$7^2 = 49 \equiv 1 \pmod 8 \tag{1}$$

and

$$3 \cdot 5 = 15 \equiv 7 \quad 3 \cdot 7 = 21 \equiv 5 \quad 5 \cdot 7 = 35 \equiv 3 \pmod 8. \tag{2}$$

$\mathbb{Z}_2 \times \mathbb{Z}_2$ is therefore a MODULO MULTIPLICATION GROUP.

The CYCLE GRAPH is shown above, and the multiplication table for the $\mathbb{Z}_2 \times \mathbb{Z}_2$ group is given below (Cotton 1990, p. 11).

$\mathbb{Z}_2 \times \mathbb{Z}_2$	1	A	B	C
1	1	A	B	C
A	A	1	C	B
B	B	C	1	A
C	C	B	A	1

The CONJUGACY CLASSES are $\{1\}$, $\{A\}$,

$$A^{-1}AA = A \tag{3}$$

$$B^{-1}AB = A \tag{4}$$

$$C^{-1}AC = A, \tag{5}$$

$\{B\}$,

$$A^{-1}BA = B \tag{6}$$

$$C^{-1}BC = B, \tag{7}$$

and $\{C\}$.

Now explicitly consider the elements of the C_{2v} POINT GROUP.

C_{2v}	E	C_2	σ_v	σ_v
E	E	C_2	σ_v	σ_v'
C_2	C_2	E	σ_v'	σ_v
σ_v	σ_v	σ_v'	E	C_2
σ_v'	σ_v'	σ_v	C_2	E

In terms of the VIERGRUPPE elements

V	I	V_1	V_2	V_3
I	V_1	V_2	V_3	V_4
V_1	V_1	I	V_3	V_2
V_2	V_2	V_3	I	V_1
V_3	V_3	V_2	V_1	I

A reducible representation using 2-D REAL MATRICES is

$$1 = \begin{bmatrix} 1 & 0 \\ 0 & 1 \end{bmatrix} \tag{8}$$

$$A = \begin{bmatrix} -1 & 0 \\ 0 & -1 \end{bmatrix} \tag{9}$$

$$B = \begin{bmatrix} 0 & 1 \\ 1 & 0 \end{bmatrix} \tag{10}$$

$$C = \begin{bmatrix} 0 & -1 \\ -1 & 0 \end{bmatrix}. \tag{11}$$

Another reducible representation using 3-D REAL MATRICES can be obtained from the symmetry elements of the D_2 group (1, $C_2(z)$, $C_2(y)$, and $C_2(x)$) or C_{2v} group (1, C_2, σ_v, and σ_v'). Place the C_2 axis along the z-AXIS, σ_v in the x-y plane, and σ_v' in the y-z plane.

$$1 = E = E = \begin{bmatrix} 1 & 0 & 0 \\ 0 & 1 & 0 \\ 0 & 0 & 1 \end{bmatrix} \tag{12}$$

$$A = R_x(\Pi) = \sigma_v = \begin{bmatrix} 1 & 0 & 0 \\ 0 & -1 & 0 \\ 0 & 0 & 1 \end{bmatrix} \tag{13}$$

$$C = R_z(\Pi) = C_2 = \begin{bmatrix} -1 & 0 & 0 \\ 0 & -1 & 0 \\ 0 & 0 & 1 \end{bmatrix} \tag{14}$$

$$B = R_y(\Pi) = \sigma_v' = \begin{bmatrix} -1 & 0 & 0 \\ 0 & 1 & 0 \\ 0 & 0 & 1 \end{bmatrix}. \qquad (15)$$

In order to find the irreducible representations, note that the traces are given by $\chi(1) = 3$, $\chi(C_2) = -1$ and $\chi(\sigma_v) = \chi(\sigma_v') = 1$ Therefore, there are at least three distinct CONJUGACY CLASSES. However, we see from the MULTIPLICATION TABLE that there are actually four CONJUGACY CLASSES, so GROUP rule 5 requires that there must be four irreducible representations. By GROUP rule 1, we are looking for POSITIVE INTEGERS which satisfy

$$l_1^2 + l_2^2 + l_3^2 + l_4^2 = 4. \qquad (16)$$

The only combination which will work is

$$l_1 = l_2 = l_3 = l_4 = 1, \qquad (17)$$

so there are four one-dimensional representations. GROUP rule 2 requires that the sum of the squares equal the ORDER $h = 4$, so each 1-D representation must have CHARACTER ± 1. GROUP rule 6 requires that a totally symmetric representation always exists, so we are free to start off with the first representation having all 1s. We then use orthogonality (GROUP rule 3) to build up the other representations. The simplest solution is then given by

C_{2v}	1	C_2	σ_v	σ_v'
Γ_1	1	1	1	1
Γ_2	1	-1	-1	1
Γ_3	1	-1	1	-1
Γ_4	1	1	-1	-1

These can be put into a more familiar form by switching Γ_1 and Γ_3, giving the CHARACTER TABLE

C_{2v}	1	C_2	σ_v	σ_v'
Γ_3	1	-1	1	-1
Γ_2	1	-1	-1	1
Γ_1	1	1	1	1
Γ_4	1	1	-1	-1

The matrices corresponding to this representation are now

$$1 = \begin{bmatrix} 1 & 0 & 0 & 0 \\ 0 & 1 & 0 & 0 \\ 0 & 0 & 1 & 0 \\ 0 & 0 & 0 & 1 \end{bmatrix} \qquad (18)$$

$$C_2 = \begin{bmatrix} -1 & 0 & 0 & 0 \\ 0 & -1 & 0 & 0 \\ 0 & 0 & 1 & 0 \\ 0 & 0 & 0 & 1 \end{bmatrix} \qquad (19)$$

$$\sigma_v = \begin{bmatrix} 1 & 0 & 0 & 0 \\ 0 & -1 & 0 & 0 \\ 0 & 0 & 1 & 0 \\ 0 & 0 & 0 & -1 \end{bmatrix} \qquad (20)$$

$$\sigma_v' = \begin{bmatrix} -1 & 0 & 0 & 0 \\ 0 & 1 & 0 & 0 \\ 0 & 0 & 1 & 0 \\ 0 & 0 & 0 & -1 \end{bmatrix} \qquad (21)$$

which consist of the previous representation with an additional component. These matrices are now orthogonal, and the order equals the matrix dimension. As before, $\chi(\sigma_v) = \chi(\sigma_v^1)$.

See also CYCLIC GROUP, FINITE GROUP Z_4

References
Cotton, F. A. *Chemical Applications of Group Theory*, 3rd ed. New York: Wiley, 1990.

Finite Group Z2Z2Z2

$$Z_2 \otimes Z_2 \otimes Z_2$$

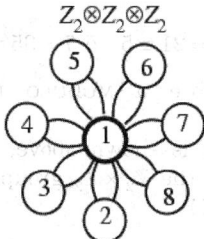

One of the three Abelian groups of the five groups total of ORDER 8. Examples include the MODULO MULTIPLICATION GROUP M_{24}. The elements A_i of this group satisfy $A_i^2 = 1$, where 1 is the IDENTITY ELEMENT. The CYCLE GRAPH is shown above.

See also FINITE GROUP D_4, FINITE GROUP Q_8, FINITE GROUP Z_2Z_4, FINITE GROUP Z_8

Finite Group Z2Z4

$$Z_2 \otimes Z_4$$

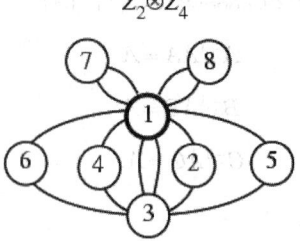

One of the three Abelian groups of the five groups total of ORDER 8. Examples include the MODULO MULTIPLICATION GROUPS M_{15}, M_{16}, M_{20}, and M_{30}. The elements A_i of this group satisfy $A_i^4 = 1$, where

1 is the IDENTITY ELEMENT, and four of the elements satisfy $A_i^2 = 1$. The CYCLE GRAPH is shown above.

See also FINITE GROUP D_4, FINITE GROUP Q_8, FINITE GROUP $Z_2Z_2Z_2$, FINITE GROUP Z_8

Finite Group Z3

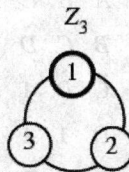

The unique group of ORDER 3. It is both ABELIAN and CYCLIC. Examples include the POINT GROUPS C_3 and D_3 and the integers under addition modulo 3. The elements A_i of the group satisfy $A_i^3 = 1$ where 1 is the IDENTITY ELEMENT. The CYCLE GRAPH is shown above, and the MULTIPLICATION TABLE is given below (Cotton 1990, p. 10).

$\mathbb{Z}_3$	1	A	B
1	1	A	B
A	A	B	1
B	B	1	A

The CONJUGACY CLASSES are $\{1\}$, $\{A\}$,

$$A^{-1}AA = A$$

$$B^{-1}AB = A,$$

and $\{B\}$,

$$A^{-1}BA = B$$

$$B^{-1}BB = B.$$

The irreducible representation (CHARACTER TABLE) is therefore

Γ	1	A	B
Γ_1	1	1	1
Γ_2	1	1	-1
Γ_3	1	-1	1

See also CYCLIC GROUP

References

Cotton, F. A. *Chemical Applications of Group Theory, 3rd ed.* New York: Wiley, 1990.

Finite Group Z4

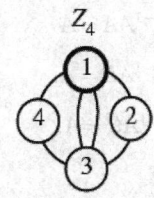

One of the two groups of ORDER 4. Like $\mathbb{Z}_2 \times \mathbb{Z}_2$, it is ABELIAN, but unlike $\mathbb{Z}_2 \times \mathbb{Z}_2$, it is a CYCLIC. Examples include the POINT GROUPS C_4 and S_4 and the MODULO MULTIPLICATION GROUPS M_5 and M_{10}. Elements A_i of the group satisfy $A_i^4 = 1$, where 1 is the IDENTITY ELEMENT, and two of the elements satisfy $A_i^2 = 1$. The CYCLE GRAPH is shown above. The MULTIPLICATION TABLE for this group may be written in three equivalent ways—denoted here by $\mathbb{Z}_4^{(1)}$, $\mathbb{Z}_4^{(2)}$, and $\mathbb{Z}_4^{(3)}$—by permuting the symbols used for the group elements. (Cotton 1990, p. 11).

$\mathbb{Z}_4^{(1)}$	1	A	B	C
1	1	A	B	C
A	A	B	C	1
B	B	C	1	A
C	C	1	A	B

The MULTIPLICATION TABLE for $\mathbb{Z}_4^{(2)}$ is obtained from $\mathbb{Z}_4^{(1)}$ by interchanging A and B.

$\mathbb{Z}_4^{(2)}$	1	A	B	C
1	1	A	B	C
A	A	1	C	B
B	B	C	A	1
C	C	B	1	A

The MULTIPLICATION TABLE for $\mathbb{Z}_4^{(3)}$ is obtained from $\mathbb{Z}_4^{(1)}$ by interchanging A and C.

$\mathbb{Z}_4^{(3)}$	1	A	B	C
1	1	A	B	C
A	A	C	1	B
B	B	1	C	A
C	C	B	A	1

The CONJUGACY CLASSES of $\mathbb{Z}_4$ are $\{1\}$, $\{A\}$,

$$A^{-1}AA = A \qquad (1)$$

$$B^{-1}AB = A \qquad (2)$$

$$C^{-1}AC = A, \qquad (3)$$

$\{B\}$,

$$A^{-1}BA = B \qquad (4)$$

$$B^{-1}BB = B \qquad (5)$$

$$C^{-1}BC = B, \qquad (6)$$

and $\{C\}$.

The group may be given a reducible representation using COMPLEX NUMBERS

$$1 = 1 \qquad (7)$$

$$A = i \qquad (8)$$

$$B = -1 \qquad (9)$$

$$C = -i, \qquad (10)$$

or REAL MATRICES

$$1 = \begin{bmatrix} 1 & 0 \\ 0 & 1 \end{bmatrix} \qquad (11)$$

$$A = \begin{bmatrix} 0 & -1 \\ 1 & 0 \end{bmatrix} \qquad (12)$$

$$B = \begin{bmatrix} -1 & 0 \\ 0 & -1 \end{bmatrix} \qquad (13)$$

$$C = \begin{bmatrix} 0 & 1 \\ -1 & 0 \end{bmatrix}. \qquad (14)$$

See also CYCLIC GROUP, FINITE GROUP Z_2Z_2

References

Cotton, F. A. *Chemical Applications of Group Theory, 3rd ed.* New York: Wiley, 1990.

Finite Group Z5

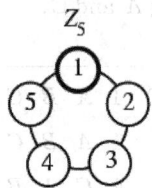

Z_5

The unique GROUP of ORDER 5, which is ABELIAN. Examples include the POINT GROUP C_5 and the integers mod 5 under addition. The elements A_i

satisfy $A_i^5 = 1$, where 1 is the IDENTITY ELEMENT. The CYCLE GRAPH is shown above, and the MULTIPLICATION TABLE is illustrated below.

$\mathbb{Z}_5$	1	A	B	C	D
1	1	A	B	C	D
A	A	B	C	D	1
B	B	C	D	1	A
C	C	D	1	A	B
D	D	1	A	B	C

The CONJUGACY CLASSES are $\{1\}$, $\{A\}$, $\{B\}$, $\{C\}$, and $\{D\}$.

See also CYCLIC GROUP

Finite Group Z6

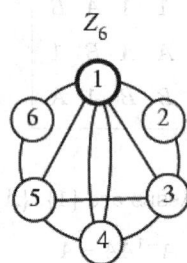

Z_6

One of the two groups of ORDER 6 which, unlike D_3, is ABELIAN. It is also a CYCLIC. It is isomorphic to $\mathbb{Z}_2 \times \mathbb{Z}_3$. Examples include the POINT GROUPS C_6 and S_6, the integers modulo 6 under addition, and the MODULO MULTIPLICATION GROUPS M_7, M_9, and M_{14}. The elements A_i of the group satisfy $A_i^6 = 1$, where 1 is the IDENTITY ELEMENT, three elements satisfy $A_i^3 = 1$, and two elements satisfy $A_i^2 = 1$. The CYCLE GRAPH is shown above, and the MULTIPLICATION TABLE is given below.

$\mathbb{Z}_6$	1	A	B	C	D	E
1	1	A	B	C	D	E
A	A	B	C	D	E	1
B	B	C	D	E	1	A
C	C	D	E	1	A	B
D	D	E	1	A	B	C
E	E	1	A	B	C	D

The CONJUGACY CLASSES are {1}, {A}, {B}, {C}, {D}, and {E}.

See also CYCLIC GROUP, FINITE GROUP D_3

Finite Group Z7

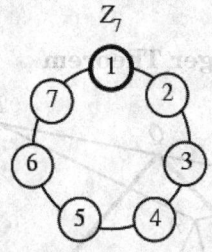

Z_7

The unique GROUP of ORDER 7. It is ABELIAN and CYCLIC. Examples include the POINT GROUP C_7 and the integers modulo 7 under addition. The elements A_i of the group satisfy $A_i^7 = 1$, where 1 is the IDENTITY ELEMENT. The CYCLE GRAPH is shown above.

$\mathbb{Z}_7$	1	A	B	C	D	E	F
1	1	A	B	C	D	E	F
A	A	B	C	D	E	F	1
B	B	C	D	E	F	1	A
C	C	D	E	F	1	A	B
D	D	E	F	1	A	B	C
E	E	F	1	A	B	C	D
F	F	1	A	B	C	D	E

The CONJUGACY CLASSES are {1}, {A}, {B}, {C}, {D}, {E}, and {F}.

See also CYCLIC GROUP

Finite Group Z8

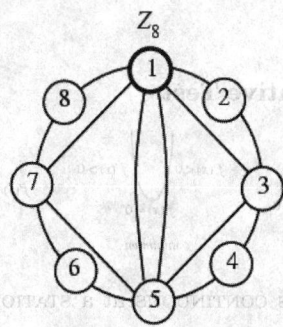

Z_8

One of the three Abelian groups of the five groups total of ORDER 8. An example is the residue classes modulo 17 which QUADRATIC RESIDUES, i.e., {1, 2, 4, 8, 9, 13, 15, 16} under multiplication modulo

17. The elements A_i satisfy $A_i^8 = 1$, four of them satisfy $A_i^4 = 1$, and two satisfy $A_i^2 = 1$. The CYCLE GRAPH is shown above.

See also CYCLIC GROUP, FINITE GROUP D_4, FINITE GROUP Q_8, FINITE GROUP Z_2Z_4, FINITE GROUP $Z_2Z_2Z_2$

Finite Mathematics
The branch of mathematics which does not involve infinite sets, limits, or continuity.

See also COMBINATORICS, DISCRETE MATHEMATICS

References
Hildebrand, F. H. and Johnson, C. G. *Finite Mathematics.* Boston, MA: Prindle, Weber, and Schmidt, 1970.
Kemeny, J. G.; Snell, J. L.; and Thompson, G. L. *Introduction to Finite Mathematics, 3rd ed.* Englewood Cliffs, NJ: Prentice-Hall, 1974.
Marcus, M. *A Survey of Finite Mathematics.* New York: Dover, 1993.
Weisstein, E. W. "Books about Finite Mathematics." http://www.treasure-troves.com/books/FiniteMathematics.html.

Finite Order
An ENTIRE FUNCTION f is said to be of finite order if there exist numbers $a, r > 0$ such that

$$|f(z)| \leq \exp(|z|^a)$$

for all $|z| > r$. The INFIMUM of all numbers a for which this inequality holds is called the ORDER of f, denoted $\lambda = \lambda(f)$.

See also ENTIRE FUNCTION, ORDER (FUNCTION)

References
Krantz, S. G. "Finite Order." §9.3.2 in *Handbook of Complex Analysis.* Boston, MA: Birkhäuser, p. 121, 1999.

Finite Projective Plane
PROJECTIVE PLANE

Finite Simple Group
SIMPLE GROUP

Finite Simple Group Classification Theorem
CLASSIFICATION THEOREM

Finitely Generated
A GROUP G is said to be finitely generated if there exists a finite set of GENERATORS for G.

See also GENERATOR (GROUP)

Finite-to-One Factor
A MAP $\psi : M \to M$, where M is a MANIFOLD, is a finite-to-one factor of a MAP $\Psi : X \to X$ if there exists a

continuous ONTO MAP $\Pi : X \to M$ such that $\psi \circ \Pi = \Pi \circ \Psi$ and $\Pi^{-1}(x) \subset X$ is finite for each $x \in M$.

Finsler Geometry

The geometry of FINSLER SPACE.

Finsler Manifold

FINSLER SPACE

Finsler Metric

A continuous real function $L(x,y)$ defined on the TANGENT BUNDLE $T(M)$ of an n-D DIFFERENTIABLE MANIFOLD M is said to be a Finsler metric if

1. $L(x,y)$ is DIFFERENTIABLE at $x \neq y$,
2. $L(x, \lambda y) = |\lambda| L(x,y)$ for any element $(x,y) \in T(M)$ and any REAL NUMBER λ,
3. Denoting the METRIC

$$g_{ij}(x,y) = \frac{1}{2} \frac{\partial^2 [L(x,y)]^2}{\partial y^i \partial y^j},$$

then g_{ij} is a POSITIVE DEFINITE MATRIX.

A DIFFERENTIABLE MANIFOLD M with a Finsler metric is called a FINSLER SPACE.

See also DIFFERENTIABLE MANIFOLD, FINSLER SPACE, TANGENT BUNDLE

References

Iyanaga, S. and Kawada, Y. (Eds.). "Finsler Spaces." §161 in *Encyclopedic Dictionary of Mathematics*. Cambridge, MA: MIT Press, pp. 540–42, 1980.

Finsler Space

A general space based on the LINE ELEMENT

$$ds = F\left(x^1, \ldots, x^n; dx^1, \ldots, dx^n\right),$$

with $F(x,y) > 0$ for $y \neq 0$ a function on the TANGENT BUNDLE $T(M)$, and homogeneous of degree 1 in y. Formally, a Finsler space is a DIFFERENTIABLE MANIFOLD possessing a FINSLER METRIC. Finsler geometry is RIEMANNIAN GEOMETRY *without* the restriction that the LINE ELEMENT be quadratic and OF THE FORM

$$F^2 = g_{ij}(x) dx^i dx^j.$$

A compact boundaryless Finsler space is locally Minkowskian IFF it has 0 "flag curvature."

See also FINSLER METRIC, HODGE'S THEOREM, RIEMANNIAN GEOMETRY, TANGENT BUNDLE

References

Akbar-Zadeh, H. "Sur les espaces de Finsler à courbures sectionnelles constantes." *Acad. Roy. Belg. Bull. Cl. Sci.* **74**, 281–22, 1988.

Bao, D.; Chern, S.-S.; and Shen, Z. (Eds.). *Finsler Geometry*. Providence, RI: Amer. Math. Soc., 1996.
Chern, S.-S. "Finsler Geometry is Just Riemannian Geometry without the Quadratic Restriction." *Not. Amer. Math. Soc.* **43**, 959–63, 1996.
Iyanaga, S. and Kawada, Y. (Eds.). "Finsler Spaces." §161 in *Encyclopedic Dictionary of Mathematics*. Cambridge, MA: MIT Press, pp. 540–42, 1980.

Finsler-Hadwiger Theorem

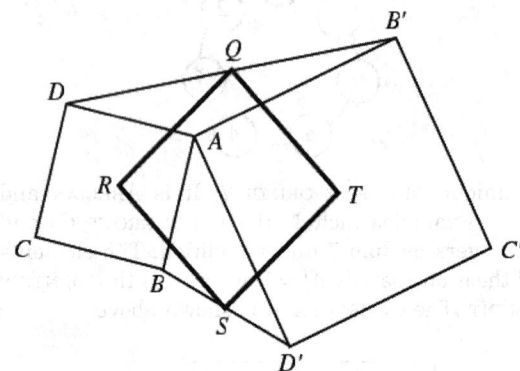

Let the SQUARES $\square ABCD$ and $\square AB'C'D'$ share a common VERTEX A. The midpoints Q and S of the segments $B'D$ and BD' together with the centers of the original squares R and T then form another square $\square QRST$. This theorem is a special case of the FUNDAMENTAL THEOREM OF DIRECTLY SIMILAR FIGURES (Detemple and Harold 1996).

See also DIRECTLY SIMILAR, FUNDAMENTAL THEOREM OF DIRECTLY SIMILAR FIGURES, SQUARE

References

Detemple, D. and Harold, S. "A Round-Up of Square Problems." *Math. Mag.* **69**, 15–7, 1996.
Finsler, P. and Hadwiger, H. "Einige Relationen im Dreieck." *Comment. Helv.* **10**, 316–26, 1937.
Fisher, J. C.; Ruoff, D.; and Shileto, J. "Polygons and Polynomials." In *The Geometric Vein: The Coxeter Festschrift*. New York: Springer-Verlag, 321–33, 1981.

First Curvature

CURVATURE

First Derivative Test

Suppose $f(x)$ is CONTINUOUS at a STATIONARY POINT x_0.

1. If $f'(x) > 0$ on an OPEN INTERVAL extending left from x_0 and $f'(x) < 0$ on an OPEN INTERVAL extend-

ing right from x_0, then $f(x)$ has a RELATIVE MAXIMUM (possibly a GLOBAL MAXIMUM) at x_0.

2. If $f'(x) < 0$ on an OPEN INTERVAL extending left from x_0 and $f'(x) > 0$ on an OPEN INTERVAL extending right from x_0, then $f(x)$ has a RELATIVE MINIMUM (possibly a GLOBAL MINIMUM) at x_0.

3. If $f'(x)$ has the same sign on an OPEN INTERVAL extending left from x_0 and on an OPEN INTERVAL extending right from x_0, then $f(x)$ does not have a RELATIVE EXTREMUM at x_0.

See also EXTREMUM, GLOBAL MAXIMUM, GLOBAL MINIMUM, INFLECTION POINT, MAXIMUM, MINIMUM, RELATIVE EXTREMUM, RELATIVE MAXIMUM, RELATIVE MINIMUM, SECOND DERIVATIVE TEST, STATIONARY POINT

References

Abramowitz, M. and Stegun, C. A. (Eds.). *Handbook of Mathematical Functions with Formulas, Graphs, and Mathematical Tables, 9th printing.* New York: Dover, p. 14, 1972.

First Digit Law

BENFORD'S LAW

First Digit Phenomenon

BENFORD'S LAW

First Fundamental Form

Let M be a REGULAR SURFACE with $\mathbf{v_p}, \mathbf{w_p}$ points in the TANGENT SPACE $M_\mathbf{p}$ of M. Then the first fundamental form is the INNER PRODUCT of tangent vectors,

$$\mathbf{I}(\mathbf{v_p}, \mathbf{w_p}) = \mathbf{v_p} \cdot \mathbf{w_p}. \qquad (1)$$

The first fundamental form satisfies

$$\mathbf{I}(a\mathbf{x}_u + b\mathbf{x}_v, a\mathbf{x}_u + b\mathbf{x}_v) = Ea^2 + 2Fab + Gb^2. \qquad (2)$$

The first fundamental form (or LINE ELEMENT) is given explicitly by the RIEMANNIAN METRIC

$$ds^2 = Edu^2 + 2Fdudv + Gdv^2. \qquad (3)$$

It determines the ARC LENGTH of a curve on a surface. The coefficients are given by

$$E = \mathbf{x}_{uu} = \left|\frac{\partial \mathbf{x}}{\partial u}\right|^2 \qquad (4)$$

$$F = \mathbf{x}_{uv} = \frac{\partial \mathbf{x}}{\partial u} \cdot \frac{\partial \mathbf{x}}{\partial v} \qquad (5)$$

$$G = \mathbf{x}_{vv} = \left|\frac{\partial \mathbf{x}}{\partial v}\right|^2. \qquad (6)$$

The coefficients are also denoted $g_{uu} = E$, $g_{uv} = F$, and $g_{vv} = G$. In CURVILINEAR COORDINATES (where $F = 0$), the quantities

$$h_u \equiv \sqrt{g_{uu}} = \sqrt{E} \qquad (7)$$

$$h_v \equiv \sqrt{g_{vv}} = \sqrt{G} \qquad (8)$$

are called SCALE FACTORS.

See also FUNDAMENTAL FORMS, SECOND FUNDAMENTAL FORM, THIRD FUNDAMENTAL FORM

References

Gray, A. "The Three Fundamental Forms." §16.6 in *Modern Differential Geometry of Curves and Surfaces with Mathematica, 2nd ed.* Boca Raton, FL: CRC Press, pp. 380–82, 1997.

First Kind

Special functions which arise as solutions to second order ordinary differential equations are commonly said to be "of the first kind" if they are *nonsingular* at the origin, while the corresponding linearly independent solutions which are *singular* are said to be "of the second kind." Common examples of functions of the first kind defined in this way include the BESSEL FUNCTION OF THE FIRST KIND, CHEBYSHEV POLYNOMIAL OF THE FIRST KIND, CONFLUENT HYPERGEOMETRIC FUNCTION OF THE FIRST KIND, HANKEL FUNCTION OF THE FIRST KIND, and so on.

The term "first kind" is also used in a more general context to distinguish between two or more types of mathematical objects which, however, all satisfy some common overall property. Examples of objects of this kind include the CHRISTOFFEL SYMBOL OF THE FIRST KIND, ELLIPTIC INTEGRAL OF THE FIRST KIND, FREDHOLM INTEGRAL EQUATION OF THE FIRST KIND, STIRLING NUMBER OF THE FIRST KIND, VOLTERRA INTEGRAL EQUATION OF THE FIRST KIND, and so on.

See also BESSEL FUNCTION OF THE FIRST KIND, CHEBYSHEV POLYNOMIAL OF THE FIRST KIND, CONFLUENT HYPERGEOMETRIC FUNCTION OF THE FIRST KIND, ELLIPTIC INTEGRAL OF THE FIRST KIND, FREDHOLM INTEGRAL EQUATION OF THE FIRST KIND, HANKEL FUNCTION OF THE FIRST KIND, SECOND KIND, SPECIAL FUNCTION, STIRLING NUMBER OF THE FIRST KIND, THIRD KIND, VOLTERRA INTEGRAL EQUATION OF THE FIRST KIND

First Multiplier Theorem

Let D be a planar Abelian DIFFERENCE SET and t be any DIVISOR of n. Then t is a numerical multiplier of D, where a multiplier is defined as an automorphism α of a GROUP G which takes D to a translation $g + D$ of itself for some $g \in G$. If α is OF THE FORM $\alpha : x \to tx$ for $t \in \mathbb{Z}$ relatively prime to the order of G, then α is called a numerical multiplier.

References

Gordon, D. M. "The Prime Power Conjecture is True for $n < 2,000,000$." *Electronic J. Combinatorics* **1**, R6 1–,

1994.   http://www.combinatorics.org/Volume_1/volume1.html#R6.

First-Countable Space
A TOPOLOGICAL SPACE in which every point has a countable BASE for its neighborhood system.

Fischer Groups
The SPORADIC GROUPS Fi_{22}, Fi_{23}, and Fi'_{24}. These groups were discovered during the investigation of 3-TRANSPOSITION GROUPS.

See also SPORADIC GROUP

References
Wilson, R. A. "ATLAS of Finite Group Representation." http://for.mat.bham.ac.uk/atlas/html/contents.html#spo.

Fischer's Baby Monster Group
BABY MONSTER GROUP

Fish Bladder
LENS

Fisher Index
The statistical INDEX

$$P_B \equiv \sqrt{P_L P_P},$$

where P_L is LASPEYRES' INDEX and P_P is PAASCHE'S INDEX.

See also INDEX

References
Kenney, J. F. and Keeping, E. S. *Mathematics of Statistics, Pt. 1, 3rd ed.* Princeton, NJ: Van Nostrand, p. 66, 1962.

Fisher Kurtosis

$$\gamma_2 \equiv b_2 \equiv \frac{\mu_4}{\mu_2^2} - 3 = \frac{\mu_4}{\sigma^4} - 3,$$

where μ_i is the ith MOMENT about the MEAN and $\sigma = \sqrt{\mu_2}$ is the STANDARD DEVIATION.

See also FISHER SKEWNESS, KURTOSIS, PEARSON KURTOSIS

Fisher Sign Test
A robust nonparametric test which is an alternative to the PAIRED T-TEST. This test makes the basic assumption that there is information only in the signs of the differences between paired observations, not in their sizes. Take the paired observations, calculate the differences, and count the number of $+$s n_+ and $-$s n_-, where

$$N \equiv n_+ + n_-$$

is the sample size. Calculate the BINOMIAL COEFFICIENT

$$B \equiv \binom{N}{n_+}.$$

Then $B/2^N$ gives the probability of getting exactly this many $+$s and $-$s if POSITIVE and NEGATIVE values are equally likely. Finally, to obtain the P-VALUE for the test, sum all the COEFFICIENTS that are $\leq B$ and divide by 2^N.

See also HYPOTHESIS TESTING

Fisher Skewness

$$\gamma_1 = \frac{\mu_3}{\mu_2^{3/2}} = \frac{\mu_3}{\sigma^3},$$

where μ_i is the i MOMENT about the MEAN, and $\sigma = \sqrt{\mu_2}$ is the STANDARD DEVIATION.

See also FISHER KURTOSIS, MOMENT, SKEWNESS, STANDARD DEVIATION

Fisher's Block Design Inequality
A balanced incomplete BLOCK DESIGN (v, k, λ, r, b) exists only for $b \geq v$ (or, equivalently, $r \geq k$).

See also BRUCK-RYSER-CHOWLA THEOREM

References
Dinitz, J. H. and Stinson, D. R. "A Brief Introduction to Design Theory." Ch. 1 in *Contemporary Design Theory: A Collection of Surveys* (Ed. J. H. Dinitz and D. R. Stinson). New York: Wiley, pp. 1–2, 1992.

Fisher's Equation
The PARTIAL DIFFERENTIAL EQUATION

$$u_t = Du_{xx} + u - u^2.$$

References
Kaliappan, P. "An Exact Solution for Travelling Waves of $u_t = Du_{xx} + u - u^k$." *Physica D* **11**, 368–74, 1984.
Zwillinger, D. *Handbook of Differential Equations, 3rd ed.* Boston, MA: Academic Press, p. 131, 1997.

Fisher's Estimator Inequality
Given T an UNBIASED ESTIMATOR of θ so that $\langle T \rangle = \theta$. Then

$$\text{var}(T) \geq \frac{1}{N \int_{-\infty}^{\infty} \left[\frac{\partial(\ln f)}{\partial \theta}\right]^2 f \, dx},$$

where var is the VARIANCE.

Fisher's Exact Test

A STATISTICAL TEST used to determine if there are nonrandom associations between two CATEGORICAL VARIABLES.

Let there exist two such variables X and Y, with m and n observed states, respectively. Now form an $n \times m$ MATRIX in which the entries a_{ij} represent the number of observations in which $x = i$ and $y = j$. Calculate the row and column sums R_i and C_j, respectively, and the total sum

$$N = \sum_i R_i = \sum_j C_j \tag{1}$$

of the MATRIX. Then calculate the CONDITIONAL PROBABILITY of getting the actual matrix given the particular row and column sums, given by

$$P_{\text{cutoff}} = \frac{(R_1! R_2! \ldots R_m!)(C_1! C_2! \ldots C_n!)}{N! \prod_{i,j} a_{ij}!}, \tag{2}$$

which is a multivariate generalization of the HYPER-GEOMETRIC probability function. Now find all possible MATRICES of NONNEGATIVE INTEGERS consistent with the row and column sums R_i and C_j. For each one, calculate the associated CONDITIONAL PROBABILITY using (2), where the sum of these probabilities must be 1.

To compute the P-VALUE of the test, the tables must then be ordered by some criterion that measures dependence, and those tables that represent equal or greater deviation from independence than the observed table are the ones whose probabilities are added together. There are a variety of criteria that can be used to measure dependence. In the 2×2 case, which is the one Fisher looked at when he developed the exact test, either the Pearson chi-square or the difference in proportions (which are equivalent) is typically used. Other measures of association, such as the likelihood-ratio-test, G-squared, or any of the other measures typically used for association in contingency tables, can also be used.

The test is most commonly applied to 2×2 MATRICES, and is computationally unwieldy for large m or n. For tables larger than 2×2, the difference in proportion can no longer be used, but the other measures mentioned above remain applicable (and in practice, the Pearson statistic is most often used to order the tables). In the case of the 2×2 matrix, the P-VALUE of the test can be simply computed by the sum of all P-values which are $\leq P_{\text{cutoff}}$.

For an example application of the 2×2 test, let X be a journal, say either *Mathematics Magazine* or *Science*, and let Y be the number of articles on the topics of mathematics and biology appearing in a given issue of one of these journals. If *Mathematics Magazine* has five articles on math and one on biology, and *Science* has none on math and four on biology, then the relevant matrix would be

	Math. Mag.	Science	
math	5	0	$R_1 = 5$
biology	1	4	$R_2 = 5$
	$C_1 = 6$	$C_2 = 4$	$N = 10$.

Computing P_{cutoff} gives

$$P_{\text{cutoff}} = \frac{5!^2 6! 4!}{10!(5! 0! 1! 4!)} = 0.0238,$$

and the other possible matrices and their Ps are

$$\begin{bmatrix} 4 & 1 \\ 2 & 3 \end{bmatrix} \quad P = 0.2381$$

$$\begin{bmatrix} 3 & 2 \\ 3 & 2 \end{bmatrix} \quad P = 0.4762$$

$$\begin{bmatrix} 2 & 3 \\ 4 & 1 \end{bmatrix} \quad P = 0.2381$$

$$\begin{bmatrix} 1 & 4 \\ 5 & 0 \end{bmatrix} \quad P = 0.0238,$$

which indeed sum to 1, as required. The sum of P-values less than or equal to $P_{\text{cutoff}} = 0.0238$ is then 0.0476 which, because it is less than 0.05, is SIGNIFICANT. Therefore, in this case, there would be a statistically significant association between the journal and type of article appearing.

Fisher's Theorem

Let A be a sum of squares of n independent normal standardized variates x_i, and suppose $A = B + C$ where B is a quadratic form in the x_i, distributed as CHI-SQUARED with h DEGREES OF FREEDOM. Then C is distributed as χ^2 with $n - h$ DEGREES OF FREEDOM and is independent of B. The converse of this theorem is known as COCHRAN'S THEOREM.

See also CHI-SQUARED DISTRIBUTION, COCHRAN'S THEOREM

Fisher's z'-Transformation

Let r be the CORRELATION COEFFICIENT. Then defining

$$z' \equiv \tanh^{-1} r \tag{1}$$

$$\zeta \equiv \tanh^{-1} p, \tag{2}$$

gives

$$\sigma_{z'} = (N-3)^{-1/2} \tag{3}$$

$$\text{var}(z') = \frac{1}{n} + \frac{4-\rho^2}{2n^2} + \cdots \tag{4}$$

$$\gamma_1 = \frac{\rho \left| \rho^2 - \dfrac{9}{16} \right|}{n^{3/2}} \tag{5}$$

$$\gamma_2 = \frac{32 - 3\rho^4}{16N}, \tag{6}$$

where $n \equiv N - 1$.

See also CORRELATION COEFFICIENT

References

David, F. N. "The Moments of the z and F Distributions." *Biometrika* **36**, 394–03, 1949.

Fisher's z-Distribution

$$g(z) = \frac{2 n_1^{n_1/2} n_2^{n_2/2}}{B\left(\dfrac{n_1}{2}, \dfrac{n_2}{2}\right)} \frac{e^{n_1 z}}{(n_1 e^{2z} + n_2)^{(n_1 + n_1)/2}} \tag{1}$$

(Kenney and Keeping 1951). This general distribution includes the CHI-SQUARED DISTRIBUTION and STUDENT'S T-DISTRIBUTION as special cases. Let u^2 and v^2 be INDEPENDENT UNBIASED ESTIMATORS of the VARIANCE of a NORMALLY DISTRIBUTED variate. Define

$$z \equiv \ln\left(\frac{u}{v}\right) = \frac{1}{2} \ln\left(\frac{u^2}{v^2}\right). \tag{2}$$

Then let

$$F \equiv \frac{u^2}{v^2} = \frac{\dfrac{N s_1^2}{n_1}}{\dfrac{N s_2^2}{n_2}} \tag{3}$$

so that $n_1 F / n_2$ is a ratio of CHI-SQUARED variates

$$\frac{n_1 F}{n_2} = \frac{\chi^2(n_1)}{\chi^2(n_2)}, \tag{4}$$

which makes it a ratio of GAMMA DISTRIBUTION variates, which is itself a BETA PRIME DISTRIBUTION variate,

$$\frac{\gamma\left(\dfrac{n_1}{2}\right)}{\gamma\left(\dfrac{n_2}{2}\right)} = \beta'\left(\frac{n_1}{2}, \frac{n_2}{2}\right) \tag{5}$$

giving

$$f(F) = \frac{\left(\dfrac{n_1 F}{n_2}\right)^{n_1/2 - 1} \left(1 + \dfrac{n_1 F}{n_2}\right)^{-(n_1 + n_2)/2}}{B\left(\dfrac{n_1}{2}, \dfrac{n_2}{2}\right)} \frac{n_1}{n_2}. \tag{6}$$

The MEAN is

$$\langle F \rangle = \frac{n2}{n_2 - 2}, \tag{7}$$

and the MODE is

$$\frac{n_2}{n_2 + 2} \frac{n_1 - 2}{n_1}. \tag{8}$$

See also BETA DISTRIBUTION, BETA PRIME DISTRIBUTION, CHI-SQUARED DISTRIBUTION, GAMMA DISTRIBUTION, NORMAL DISTRIBUTION, STUDENT'S T-DISTRIBUTION

References

Kenney, J. F. and Keeping, E. S. *Mathematics of Statistics, Pt. 2, 2nd ed.* Princeton, NJ: Van Nostrand, pp. 180–81, 1951.

Fisher-Behrens Problem

The determination of a test for the equality of MEANS for two NORMAL DISTRIBUTIONS with different VARIANCES given samples from each. There exists an exact test which, however, does not give a unique answer because it does not use all the data. There also exist approximate tests which do not use all the data.

See also NORMAL DISTRIBUTION

References

Aspin, A. A. "An Examination and Further Development of a Formula Arising in the Problem of Comparing Two Mean Values." *Biometrika* **35**, 88–6, 1948.
Chernoff, H. "Asymptotic Studentization in Testing of Hypothesis." *Ann. Math. Stat.* **20**, 268–78, 1949.
Fisher, R. A. "The Fiducial Argument in Statistical Inference." *Ann. Eugenics* **6**, 391–98, 1935.
Kenney, J. F. and Keeping, E. S. "The Behrens-Fisher Test." §9.8 in *Mathematics of Statistics, Pt. 2, 2nd ed.* Princeton, NJ: Van Nostrand, pp. 257–60 and 261–64, 1951.
Sukhatme, P. V. "On Fisher and Behrens' Test of Significance of the Difference in Means of Two Normal Samples." *Sankhya* **4**, 39, 1938.
Trickett, W. H. and Welch, B. L. "On the Comparison of Two Means: Further Discussion of Iterative Methods for Calculating Tables." *Biometrika* **41**, 361–74, 1954.
Trickett, W. H.; Welch, B. L.; and James, G. S. "Further Critical Values for the Two-Means Problems." *Biometrika* **43**, 203–05, 1956.
Wallace, D. L. "Asymptotic Approximations to Distributions." *Ann. Math. Stat.* **29**, 635–54, 1958.
Wald, A. "Testing the Difference Between the Means of Two Normal Populations with Unknown Standard Deviations."

In *Selected Papers in Statistics and Probability by Abraham Wald.* New York: McGraw-Hill, pp. 669–95, 1955.
Welch, B. L. "The Generalization of 'Student's' Problem when Several Different Populations are Involved." *Biometrika* **34**, 28–5, 1947.

Fisher-Tippett Distribution

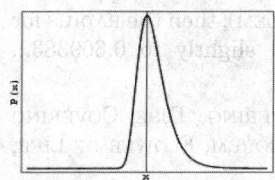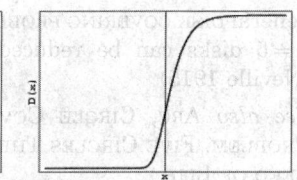

Also called the EXTREME VALUE DISTRIBUTION and LOG-WEIBULL DISTRIBUTION. It is the limiting distribution for the smallest or largest values in a large sample drawn from a variety of distributions.

$$P(x) = \frac{e^{(a-x)/b} - e^{(a-x)/b}}{b} \tag{1}$$

$$D(x) = e^{-e^{(a-x)/b}}. \tag{2}$$

These can be computed directly be defining

$$z \equiv \exp\left(\frac{a-x}{b}\right) \tag{3}$$

$$x = a - b \ln z \tag{4}$$

$$dz = -\frac{1}{b}\exp\left(\frac{a-x}{b}\right)dx. \tag{5}$$

Then the MOMENTS about the origin are

$$\mu'_n \equiv \int_{-\infty}^{\infty} x^n P(x)\,dx$$

$$= \frac{1}{b}\int_{-\infty}^{\infty} x^n \exp\left(\frac{a-x}{b}\right)\exp\left[-e^{(a-x)/b}\right]dx$$

$$= -\int_{\infty}^{0}(a - b \ln z)^n e^{-z}\,dz$$

$$= \int_{0}^{\infty}(a - b \ln z)^n e^{-z}\,dz$$

$$= \sum_{k=0}^{n}\binom{n}{k}(-1)^k a^{n-k} b^k \int_{0}^{\infty}(\ln z)^k e^{-z}\,dz$$

$$= \sum_{k=0}^{n}\binom{n}{k}a^{n-k}b^k I(k), \tag{6}$$

where $I(k)$ are EULER-MASCHERONI INTEGRALS. Plug-

ging in the EULER-MASCHERONI INTEGRALS $I(k)$ gives

$$\mu'_0 = 1 \tag{7}$$

$$\mu'_1 = a + b\gamma \tag{8}$$

$$\mu'_2 = a^2 + 2ab\gamma + b^2\left(\gamma^2 + \frac{1}{6}\pi^2\right) \tag{9}$$

$$\mu'_3 = a^3 + 3a^2 b\gamma + 3ab^2\left(\gamma^2 + \frac{1}{6}\pi^2\right)$$
$$+ b^3\left[\gamma^3 + \frac{1}{2}\gamma\pi^2 + 2\zeta(3)\right] \tag{10}$$

$$\mu'_4 = a^4 + 4a^3 b\gamma + 6a^2 b^2\left(\gamma^2 + \frac{1}{6}\pi^2\right)$$
$$+ 4ab^3\left[\gamma^3 + \frac{1}{2}\gamma\pi^2 + 2\zeta(3)\right]$$
$$+ b^4\left[\gamma^4 + \gamma^2\pi^2 + \frac{3}{20}\pi^4 + 8\gamma\zeta(3)\right], \tag{11}$$

where γ is the EULER-MASCHERONI CONSTANT and $\zeta(3)$ is APÉRY'S CONSTANT. The corresponding moments about the mean $\mu = \mu'_1$ are therefore

$$\mu_2 = \frac{1}{6}b^2\pi^2 \tag{12}$$

$$\mu_3 = 2\zeta(3)b^3 \tag{13}$$

$$\mu_4 = \frac{3}{20}b^4\pi^2, \tag{14}$$

giving MEAN, VARIANCE, SKEWNESS, and KURTOSIS of

$$\mu = a + b\gamma \tag{15}$$

$$\sigma^2 = \mu_2 - \mu_1^2 = \frac{1}{6}\pi^2 b^2 \tag{16}$$

$$\gamma_1 = \frac{\mu_3}{\sigma^3} = \frac{12\sqrt{6}\zeta(3)}{\pi^3} \tag{17}$$

$$\gamma_2 = \frac{\mu_4}{\sigma^4} - 3 = \frac{12}{5}. \tag{18}$$

The CHARACTERISTIC FUNCTION is

$$\phi(t) = \Gamma(1 - i\beta t)e^{i\alpha t}, \tag{19}$$

where $\Gamma(z)$ is the GAMMA FUNCTION (Abramowitz and Stegun 1972, p. 930).
The special case of the Fisher-Tippett distribution with $a = 0$, $b = 1$ is called GUMBEL'S DISTRIBUTION.

See also EULER-MASCHERONI INTEGRALS, GUMBEL'S DISTRIBUTION

Fitting Subgroup

The unique smallest NORMAL NILPOTENT SUBGROUP of H, denoted $F(H)$. The generalized fitting subgroup is defined by $F^*(H) = F(H)E(H)$, where $E(H)$ is the commuting product of all components of H, and F is the fitting subgroup of H.

Fitzhugh-Nagumo Equations

The system of PARTIAL DIFFERENTIAL EQUATIONS

$$u_t = u_{xx} + u(u-a)(1-u) + w$$

$$w_t = eu.$$

References

Sherman, A. S. and Peskin, C. S. "A Monte Carlo Method for Scalar Reaction Diffusion Equations." *SIAM J. Sci. Stat. Comput.* **7**, 1360–372, 1986.

Zwillinger, D. *Handbook of Differential Equations, 3rd ed.* Boston, MA: Academic Press, p. 138, 1997.

Five Circles Theorem

MIQUEL FIVE CIRCLES THEOREM

Five Cubes

CUBE 5-COMPOUND

Five Disks Problem

Given five *equal* DISKS placed *symmetrically* about a given center, what is the smallest RADIUS r for which the RADIUS of the circular AREA covered by the five disks is 1? The answer is $r = \phi - 1 = 1/\phi = 0.6180339\ldots$, where ϕ is the GOLDEN RATIO, and

the centers c_i of the disks $i = 1, \ldots, 5$ are located at

$$c_i = \begin{bmatrix} \dfrac{1}{\phi}\cos\left(\dfrac{2\pi i}{5}\right) \\ \dfrac{1}{\phi}\sin\left(\dfrac{2\pi i}{5}\right) \end{bmatrix}.$$

The GOLDEN RATIO enters here through its connection with the regular PENTAGON. If the requirement that the disks be symmetrically placed is dropped (the general DISK COVERING PROBLEM), then the RADIUS for $n = 5$ disks can be reduced slightly to 0.609383... (Neville 1915).

See also ARC, CIRCLE COVERING, DISK COVERING PROBLEM, FIVE CIRCLES THEOREM, FLOWER OF LIFE, SEED OF LIFE

References

Ball, W. W. R. and Coxeter, H. S. M. "The Five-Disc Problem." In *Mathematical Recreations and Essays, 13th ed.* New York: Dover, pp. 97–9, 1987.

Neville, E. H. "On the Solution of Numerical Functional Equations, Illustrated by an Account of a Popular Puzzle and of its Solution." *Proc. London Math. Soc.* **14**, 308–26, 1915.

Five Tetrahedra Compound

TETRAHEDRON 5-COMPOUND

Fixed

When referring to a planar object, "fixed" means that the object is regarded as fixed in the plane so that it may not be picked up and flipped. As a result, MIRROR IMAGES are not necessarily equivalent for fixed objects.

See also FREE, MIRROR IMAGE

Fixed Element

FIXED POINT (MAP)

Fixed Point

A point which does not change upon application of a MAP, system of DIFFERENTIAL EQUATIONS, etc.

See also FIXED POINT (DIFFERENTIAL EQUATIONS), FIXED POINT (GROUP), FIXED POINT (MAP), FIXED POINT THEOREM

References

Shashkin, Yu. A. *Fixed Points.* Providence, RI: Amer. Math. Soc., 1991.

Fixed Point (Differential Equations)

Points of an AUTONOMOUS system of ordinary differential equations at which

$$\begin{cases} \dfrac{dx_1}{dt} = f_1(x_1, \ldots, x_n) = 0 \\ \vdots \\ \dfrac{dx_n}{dt} = f_n(x_1, \ldots, x_n) = 0 \end{cases}$$

If a variable is slightly displaced from a FIXED POINT, it may (1) move back to the fixed point ("asymptotically stable" or "superstable"), (2) move away ("unstable"), or (3) move in a neighborhood of the fixed point but not approach it ("stable" but not "asymptotically stable"). Fixed points are also called CRITICAL POINTS or EQUILIBRIUM POINTS. If a variable starts at a point that is not a CRITICAL POINT, it cannot reach a critical point in a finite amount of time. Also, a trajectory passing through at least one point that is not a CRITICAL POINT cannot cross itself unless it is a CLOSED CURVE, in which case it corresponds to a periodic solution.

A fixed point can be classified into one of several classes using LINEAR STABILITY analysis and the resulting STABILITY MATRIX.

See also ELLIPTIC FIXED POINT (DIFFERENTIAL EQUATIONS), HYPERBOLIC FIXED POINT (DIFFERENTIAL EQUATIONS), STABLE IMPROPER NODE, STABLE NODE, STABLE SPIRAL POINT, STABLE STAR, UNSTABLE IMPROPER NODE, UNSTABLE NODE, UNSTABLE SPIRAL POINT, UNSTABLE STAR

Fixed Point (Group)

The set of points of X fixed by a GROUP ACTION are called the group's set of fixed points, defined by

$$\{x : gx = x \text{ for all } g \in G\}.$$

In some cases, there may not be a group action, but a single operator T. Then $\{x : x \in X, Tx = x\}$ still makes sense even when T is not invertible (as is the case in a GROUP ACTION).

See also FIXED POINT, GROUP, GROUP ACTION

References

Kawakubo, K. *The Theory of Transformation Groups.* Oxford, England: Oxford University Press, pp. 4– and 31–5, 1987.

Fixed Point (Map)

A point x^* which is mapped to itself under a MAP G, so that $x^* = G(x^*)$. Such points are sometimes also called INVARIANT POINTS, or FIXED ELEMENTS (Woods 1961). Stable fixed points are called elliptical. Unstable fixed points, corresponding to an intersection of a stable and unstable invariant MANIFOLD, are called HYPERBOLIC (or SADDLE). Points may also be called asymptotically stable (a.k.a. superstable).

See also CRITICAL POINT, INVOLUTORY

References

Shashkin, Yu. A. *Fixed Points.* Providence, RI: Amer. Math. Soc., 1991.
Woods, F. S. *Higher Geometry: An Introduction to Advanced Methods in Analytic Geometry.* New York: Dover, p. 14, 1961.

Fixed Point (Transformation)

FIXED POINT (MAP)

Fixed Point Theorem

If g is a continuous function $g(x) \in [a, b]$ FOR ALL $x \in [a, b]$, then g has a FIXED POINT in $[a, b]$. This can be proven by noting that

$$g(a) \geq a \qquad g(b) \leq b$$
$$g(a) - a \geq 0 \qquad g(b) - b \leq 0.$$

Since g is continuous, the INTERMEDIATE VALUE THEOREM guarantees that there exists a $c \in [a, b]$ such that

$$g(c) - c = 0,$$

so there must exist a c such that

$$g(c) = c,$$

so there must exist a FIXED POINT $\in [a, b]$.

See also BANACH FIXED POINT THEOREM, BROUWER FIXED POINT THEOREM, HAIRY BALL THEOREM, KAKUTANI'S FIXED POINT THEOREM, LEFSHETZ FIXED POINT FORMULA, LEFSHETZ TRACE FORMULA, POINCARÉ-BIRKHOFF FIXED POINT THEOREM, SCHAUDER FIXED POINT THEOREM

References

Wells, D. *The Penguin Dictionary of Curious and Interesting Geometry.* Middlesex, England: Penguin Books, p. 80, 1991.

Flag

A collection of FACES of an n-D POLYTOPE or SIMPLICIAL COMPLEX, one of each DIMENSION 0, 1, ..., $n-1$, which all have a common nonempty INTERSECTION. In normal 3-D, the flag consists of a half-plane, its bounding RAY, and the RAY's endpoint.

Flag Manifold

For any SEQUENCE of INTEGERS $0 < n_1 < \ldots < n_k$, there is a flag manifold of type $(n_1, \ldots, n_k)$ which is the collection of ordered pairs of vector SUBSPACES of $\mathbb{R}^{n_k}$ $(V_1, \ldots, V_k)$ with $\dim(V_i) = n_i$ and V_i a SUBSPACE of V_{i+1}. There are also COMPLEX flag manifolds with COMPLEX subspaces of $\mathbb{C}^{n_k}$ instead of REAL SUBSPACES of a REAL n_k-space.

These flag manifolds admit the structure of MANI-FOLDS in a natural way and are used in the theory of LIE GROUPS.

See also GRASSMANN MANIFOLD

References

Lu, J.-H. and Weinstein, A. "Poisson Lie Groups, Dressing Transformations, and the Bruhat Decomposition." *J. Diff. Geom.* **31**, 501–26, 1990.

Flat

A set in $\mathbb{R}^d$ formed by translating an affine subspace or by the intersection of a set of HYPERPLANES.

See also FLAT (MANIFOLD)

Flat (Manifold)

See also FLAT

Flat Norm

The flat norm on a CURRENT is defined by

$$\mathscr{F}(S) = \int \{\text{Area } T + \text{vol } R : S - T = \partial R\},$$

where ∂R is the boundary of R.

See also COMPACTNESS THEOREM, CURRENT

References

Morgan, F. "What Is a Surface?" *Amer. Math. Monthly* **103**, 369–76, 1996.

Flat Space Theorem

If it is possible to transform a coordinate system to a form where the metric elements $g_{\mu\nu}$ are constants independent of x^μ, then the space is flat.

Flat Surface

A REGULAR SURFACE and special class of MINIMAL SURFACE for which the GAUSSIAN CURVATURE vanishes everywhere. A TANGENT DEVELOPABLE, GENERALIZED CONE, and GENERALIZED CYLINDER are all flat surfaces.

See also GAUSSIAN CURVATURE, MINIMAL SURFACE, PLANE

References

Gray, A. *Modern Differential Geometry of Curves and Surfaces with Mathematica*, 2nd ed. Boca Raton, FL: CRC Press, p. 374, 1997.

Flat-Ring Cyclide Coordinates

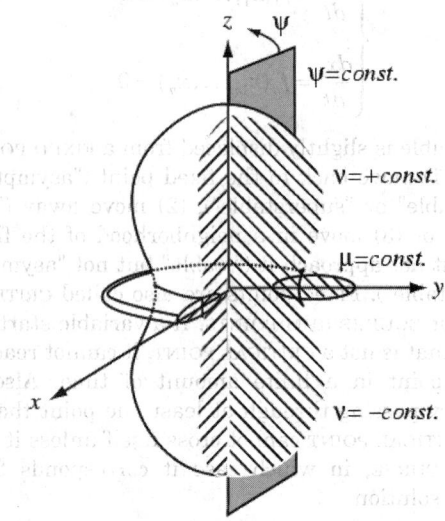

A coordinate system similar to TOROIDAL COORDINATES but with fourth-degree instead of second-degree surfaces for constant μ so that the toroids of circular CROSS SECTION are replaced by flattened rings, and the spherical bowls are replaced by cyclides of rotation for constant ν. The transformation equations are

$$x = \frac{a}{\Lambda} \operatorname{sn} \mu \operatorname{dn} \nu \cos\psi \qquad (1)$$

$$y = \frac{a}{\Lambda} \operatorname{sn} \mu \operatorname{dn} \nu \sin\psi \qquad (2)$$

$$z = \frac{a}{\Lambda} \operatorname{cn} \mu \operatorname{dn} \mu \operatorname{sn} \nu \operatorname{cn} \nu, \qquad (3)$$

where

$$\Lambda \equiv 1 - \operatorname{dn}^2 \mu \operatorname{sn}^2 \nu \qquad (4)$$

and with $\mu \in [0, K]$, $\nu \in [0, K']$, and $\psi \in [0, 2\Pi]$. Surfaces of constant μ are given by the flat-ring cyclides

$$\left(x^2 + y^2 + z^2\right)^2 + \frac{a^2}{k^4}$$

$$\times \frac{(1 - k^2)^2 - 2(1 - k^2)\operatorname{dn}^2 \mu + (1 + k^2)\operatorname{dn}^4 \mu}{\operatorname{dn}^2 \mu \operatorname{cn}^2 \mu} z^2$$

$$- a^2 \left(\operatorname{sn}^2 \mu + \frac{1}{\operatorname{sn}^2 \mu}\right)\left(x^2 + y^2\right) + \frac{a^4}{k^2}$$

$$= 0, \qquad (5)$$

surfaces of constant v by the cyclides of rotation

$$\left[\frac{\mathrm{dn}^2 v}{a^2}(x^2+y^2)+\frac{\mathrm{cn}^2 v}{a^2\,\mathrm{sn}^2\,v}z^2\right]^2-\frac{2\mathrm{cn}^2 v}{a^2\mathrm{sn}^2 v}z^2-\frac{2\mathrm{dn}^2 v}{a^2}$$

$$\times\,(x^2+y^2)+1$$

$$=0,\tag{6}$$

and surfaces of constant ψ by the half-planes

$$\tan\psi=\frac{x}{y}.\tag{7}$$

See also CYCLIDIC COORDINATES, TOROIDAL COORDINATES

References

Moon, P. and Spencer, D. E. "Flat-Ring Cyclide Coordinates (μ, v, ψ)." Fig. 4.09 in *Field Theory Handbook, Including Coordinate Systems, Differential Equations, and Their Solutions, 2nd ed.* New York: Springer-Verlag, pp. 126–29, 1988.

Flattening

The flattening of a SPHEROID (also called OBLATENESS) is denoted $\in$ or f. It is defined as

$$\in\;\equiv\;\begin{cases}\dfrac{a-c}{a}\;=\;1-\dfrac{c}{a}&\text{oblate}\\[2ex]\dfrac{c-a}{a}\;=\;\dfrac{c}{a}-1&\text{prolate,}\end{cases}$$

where c is the polar RADIUS and a is the equatorial RADIUS.

See also ECCENTRICITY, ELLIPSOID, OBLATE SPHEROID, PROLATE SPHEROID, SPHEROID

Flemish Knot

FIGURE-OF-EIGHT KNOT

Fletcher Point

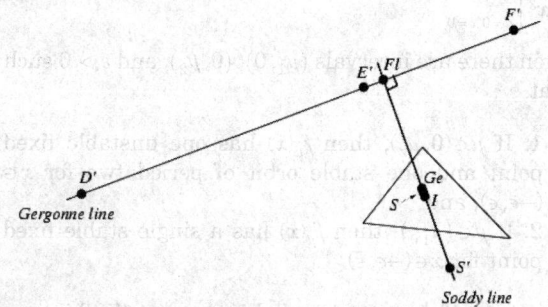

Gergonne line

Soddy line

The intersection Fl of the GERGONNE LINE and the SODDY LINE. In the above figure, D', E', and F' are the

NOBBS POINTS, I is the INCENTER, Ge is the GERGONNE POINT, and S and S' are the SODDY POINTS.

See also GERGONNE LINE, SODDY LINE, SODDY POINTS

References

Oldknow, A. "The Euler-Gergonne-Soddy Triangle of a Triangle." *Amer. Math. Monthly* **103**, 319–29, 1996.

Fleury's Algorithm

An elegant algorithm for constructing an EULERIAN CIRCUIT (Skiena 1990, p. 193).

See also EULERIAN CIRCUIT

References

Lucas, E. *Récréations Mathématiques.* Paris: Gauthier-Villars, 1891.
Skiena, S. *Implementing Discrete Mathematics: Combinatorics and Graph Theory with Mathematica.* Reading, MA: Addison-Wesley, 1990.

Flexagon

An object created by FOLDING a piece of paper along certain lines to form loops. The number of states possible in an n-FLEXAGON is a CATALAN NUMBER. By manipulating the folds, it is possible to hide and reveal different faces.

See also FLEXATUBE, FOLDING, HEXAFLEXAGON, TETRAFLEXAGON

References

Crampin, J. "On Note 2449." *Math. Gazette* **41**, 55–6, 1957.
Cundy, H. and Rollett, A. *Mathematical Models, 3rd ed.* Stradbroke, England: Tarquin Pub., pp. 205–07, 1989.
Madachy, J. S. *Madachy's Mathematical Recreations.* New York: Dover, pp. 62–4, 1979.
Gardner, M. "Hexaflexagons." Ch. 1 in *The Scientific American Book of Mathematical Puzzles & Diversions.* New York: Simon and Schuster, pp. 1–4, 1959.
Gardner, M. "Tetraflexagons." Ch. 2 in *The Second Scientific American Book of Mathematical Puzzles & Diversions: A New Selection.* New York: Simon and Schuster, pp. 24–1, 1961.
Maunsell, F. G. "The Flexagon and the Hexaflexagon." *Math. Gazette* **38**, 213–14, 1954.
Oakley, C. O. and Wisner, R. J. "Flexagons." *Amer. Math. Monthly* **64**, 143–54, 1957.
Wheeler, R. F. "The Flexagon Family." *Math. Gaz.* **42**, 1–, 1958.

Flexatube

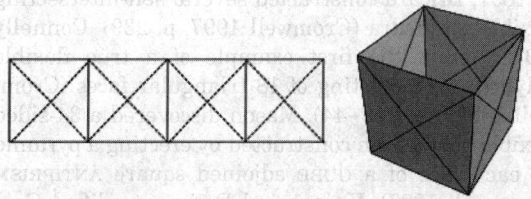

A FLEXAGON-like structure created by connecting the ends of a strip of four squares after folding along 45°

diagonals. Using a number of folding movements, it is possible to flip the flexatube inside out so that the faces originally facing inward face outward. Gardner (1961) illustrated one possible solution, and Steinhaus (1983) gives a second.

See also FLEXAGON, HEXAFLEXAGON, TETRAFLEXAGON

References

Cundy, H. and Rollett, A. *Mathematical Models, 3rd ed.* Stradbroke, England: Tarquin Pub., p. 205, 1989.
Gardner, M. *The Second Scientific American Book of Mathematical Puzzles & Diversions: A New Selection.* New York: Simon and Schuster, pp. 29–1, 1961.
Steinhaus, H. *Mathematical Snapshots, 3rd ed.* New York: Dover, pp. 177–81 and 190, 1999.

Flexible Graph

A GRAPH G is said to be flexible if the vertices of G can be moved continuously so that (1) the distances between adjacent vertices are unchanged, and (2) at least two nonadjacent vertices change their mutual distances. A graph which is not flexible is said to be RIGID.

See also RIGID GRAPH

References

Maehara, H. "Distance Graphs in Euclidean Space." *Ryukyu Math. J.* **5**, 33–1, 1992.

Flexible Polyhedron

Although the RIGIDITY THEOREM states that if the faces of a *convex* POLYHEDRON are made of metal plates and the EDGES are replaced by hinges, the POLYHEDRON would be RIGID, concave polyhedra need not be RIGID. A nonrigid polyhedron may be "SHAKY" (infinitesimally movable) or flexible (continuously movable; Wells 1991).

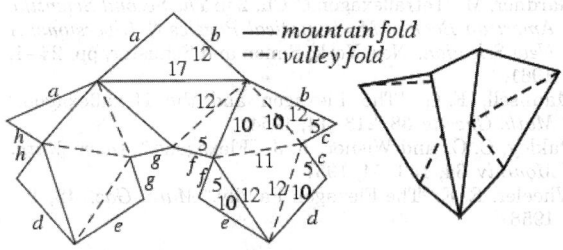

In 1897, Bricard constructed several self-intersecting flexible octahedra (Cromwell 1997, p. 239). Connelly (1978) found the first example of a true flexible polyhedron, consisting of 18 triangular faces (Cromwell 1997, pp. 242–44). Mason discovered a 34-sided flexible polyhedron constructed by erecting a pyramid on each face of a CUBE adjoined square ANTIPRISM (Cromwell 1997). Kuiper and Deligne modified Connelly's polyhedron to create a flexible polyhedron having 18 faces and 11 vertices (Cromwell 1997,

p. 245), and Steffen found a flexible polyhedron with only 14 triangular faces and 9 vertices (shown above; Cromwell 1997, pp. 244–47; Mackenzie 1998). Maksimov (1995) proved that Steffen's is the simplest possible flexible polyhedron composed of only triangles (Cromwell 1997, p. 245).

Connelly *et al.* (1997) proved that a flexible polyhedron must keep its VOLUME constant, confirming the so-called BELLOWS CONJECTURE (Mackenzie 1998).

See also BELLOWS CONJECTURE, POLYHEDRON, QUADRICORN, RIGID POLYHEDRON, RIGIDITY THEOREM, SHAKY POLYHEDRON

References

Cauchy, A. L. "Sur les polygones et les polyèdres." *XVIe Cahier* **IX**, 87–9, 1813.
Connelly, R. "A Flexible Sphere." *Math. Intel.* **1**, 130–31, 1978.
Connelly, R.; Sabitov, I.; and Walz, A. "The Bellows Conjecture." *Contrib. Algebra Geom.* **38**, 1–0, 1997.
Cromwell, P. R. *Polyhedra.* New York: Cambridge University Press, pp. 222, 224, and 239–47, 1997.
Mackenzie, D. "Polyhedra Can Bend But Not Breathe." *Science* **279**, 1637, 1998.
Maksimov, I. G. "Polyhedra with Bendings and Riemann Surfaces." *Uspekhi Matemat. Nauk* **50**, 821–23, 1995.
Wells, D. *The Penguin Dictionary of Curious and Interesting Geometry.* London: Penguin, pp. 161–62, 1991.

Flip Bifurcation

Let $f : \mathbb{R} \times \mathbb{R} \to \mathbb{R}$ be a one-parameter family of C^3 maps satisfying

$$f(0,0) = 0$$

$$\left[\frac{\partial f}{\partial x}\right]_{\mu=0, x=0} = -1$$

$$\left[\frac{\partial^2 f}{\partial x^2}\right]_{\mu=o, x=0} < 0$$

$$\left[\frac{\partial^3 f}{\partial x^3}\right]_{\mu=0, x=0} < 0.$$

Then there are intervals $(\mu_1, 0)$, $(0, \mu_2)$, and $\varepsilon > 0$ such that

1. If $\mu \in (0, \mu_2)$, then $f_\mu(x)$ has one unstable fixed point and one stable orbit of period two for $x \in (-\epsilon, \epsilon)$, and

2. If $\mu \in (\mu_{1,0})$, then $f_\mu(x)$ has a single stable fixed point for $x \in (-\epsilon, \epsilon)$.

This type of BIFURCATION is known as a flip bifurcation. An example of an equation displaying a flip bifurcation is

$$f(x) = \mu - x - x^2.$$

See also BIFURCATION

References

Rasband, S. N. *Chaotic Dynamics of Nonlinear Systems.* New York: Wiley, pp. 27–0, 1990.

Floating-Point Arithmetic

ARITHMETIC performed on real numbers by computers or other automated devices using a fixed number of bits.

ARITHMETIC

References

Hauser, J. R. "Handling Floating-Point Exceptions in Numeric Programs." *ACM Trans. Program. Lang. Sys.* **18**, 139–74, 1996. http://www.cs.berkeley.edu/~jhauser/exceptions/HandlingFloatingPointExceptions.html.

Severance, C. (Ed.). "IEEE 754: An Interview with William Kahan." *Computer*, 114–15, Mar. 1998.

Stevenson, D. "A Proposed Standard for Binary Floating-Point Arithmetic: Draft 8.0 of IEEE Task P754." *IEEE Comput.* **14** 51–2, 1981.

Floor

FLOOR FUNCTION

Floor Function

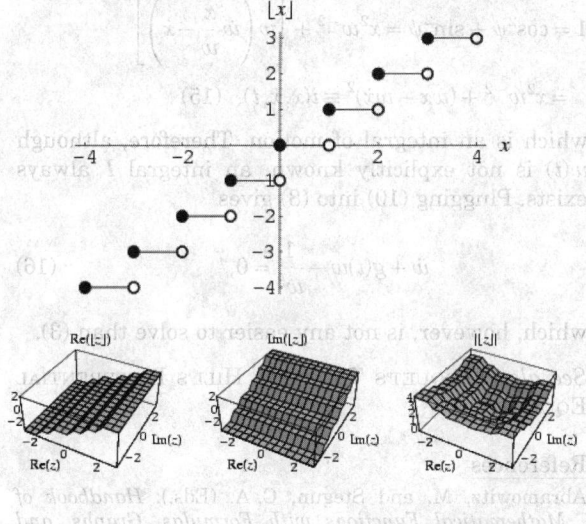

The function floor function $\lfloor x \rfloor$, also called the greatest integer function, gives the largest INTEGER less than or equal to x. In many computer languages, the floor function is called the INTEGER PART function and is denoted int(x). The name and symbol for the floor function were coined by K. E. Iverson (Graham *et al.* 1990).

Unfortunately, in many older and current works (e.g., Steinhaus 1983, p. 300; Shanks 1993; Ribenboim

1996; Hilbert and Cohn-Vossen 1999, p. 38; Hardy 1999, p. 18), the symbol $[x]$ is used instead of $\lfloor x \rfloor$ (Graham *et al.* 1990, p. 67). Because of the elegant symmetry of the floor function and CEILING FUNCTION symbols $\lfloor x \rfloor$ and $\lceil x \rceil$, and because $[x]$ is such a useful symbol when interpreted as an IVERSON BRACKET, the use of $[x]$ to denote the floor function should be deprecated. In this work, the symbol $[x]$ is used to denote the NEAREST INTEGER FUNCTION since it naturally falls between the $\lfloor x \rfloor$ and $\lceil x \rceil$ symbols.

Since usage concerning fractional part/value and integer part/value can be confusing, the following table gives a summary of names and notations used (D. W. Cantrell). Here, S&O indicates Spanier and Oldham (1987).

notation	name	S&O	Graham et al.	*Mathematica*
$\lfloor x \rfloor$	integer-value	$\mathrm{Int}(x)$	floor or integer part	$\mathrm{Floor}[x]$
$\mathrm{sgn}(x)\lfloor\lvert x\rvert\rfloor$	integer-part	$\mathrm{Ip}(x)$	no name	Integer-Part [x]
$x - \lfloor x \rfloor$	fractional-value	$\mathrm{frac}(x)$	fractional part or $\{x\}$	no name
$\mathrm{sgn}(x)(\lvert x\rvert - \lfloor\lvert x\rvert\rfloor)$	fractional-part	$\mathrm{F_p}(x)$	no name	Fractional-Part [x]

There are infinitely many integers OF THE FORM $\lfloor (3/2)^n \rfloor$ and $\lfloor (4/3)^n \rfloor$ which are composite, where $\lfloor x \rfloor$ is the FLOOR FUNCTION (Forman and Shapiro, 1967; Guy 1994, p. 220). The first few composite $\lfloor (3/2)^n \rfloor$ occur for $n = 8, 9, 10, 11, 12, 13, 14, 15, 16, 17, 18, 19, 20, 23, \ldots$ (Sloane's A046037), and the few composite $\lfloor (4/3)^n \rfloor$ occur for $n = 5, 8, 13, 14, 15, 16, 17, 18, 19, 20, 21, 22, \ldots$ (Sloane's A046038). Numbers OF THE FORM $\mathrm{frac}((3/2)^n)$, where $\mathrm{frac}(x)$ is the FRACTIONAL PART also appear in WARING'S PROBLEM.

See also CEILING FUNCTION, FRACTIONAL PART, INT, IVERSON BRACKET, NEAREST INTEGER FUNCTION, QUOTIENT, SHIFT TRANSFORMATION, STAIRCASE FUNCTION

References

Croft, H. T.; Falconer, K. J.; and Guy, R. K. *Unsolved Problems in Geometry.* New York: Springer-Verlag, p. 2, 1991.

Forman, W. and Shapiro, H. N. "An Arithmetic Property of Certain Rational Powers." *Comm. Pure Appl. Math.* **20**, 561–73, 1967.

Graham, R. L.; Knuth, D. E.; and Patashnik, O. "Integer Functions." Ch. 3 in *Concrete Mathematics: A Foundation for Computer Science, 2nd ed.* Reading, MA: Addison-Wesley, pp. 67–01, 1994.

Guy, R. K. *Unsolved Problems in Number Theory, 2nd ed.* New York: Springer-Verlag, 1994.

Hardy, G. H. *Ramanujan: Twelve Lectures on Subjects Suggested by His Life and Work, 3rd ed.* New York: Chelsea, 1999.

Hilbert, D. and Cohn-Vossen, S. *Geometry and the Imagination.* New York: Chelsea, 1999.

Iverson, K. E. *A Programming Language.* New York: Wiley, p. 12, 1962.

Ribenboim, P. *The New Book of Prime Number Records.* New York: Springer-Verlag, pp. 180–82, 1996.

Shanks, D. *Solved and Unsolved Problems in Number Theory, 4th ed.* New York: Chelsea, p. 14, 1993.

Sloane, N. J. A. Sequences A046037 and A046038 in "An On-Line Version of the Encyclopedia of Integer Sequences." http://www.research.att.com/~njas/sequences/eisonline.html.

Spanier, J. and Oldham, K. B. "The Integer-Value Int(*x*) and Fractional-Value frac(*x*) Functions." Ch. 9 in *An Atlas of Functions.* Washington, DC: Hemisphere, pp. 71–8, 1987.

Floquet Analysis

Given a system of periodic ORDINARY DIFFERENTIAL EQUATIONS OF THE FORM

$$\frac{d}{dt}\begin{bmatrix} x \\ y \\ v_x \\ v_y \end{bmatrix} = -\begin{bmatrix} 0 & 0 & -1 & 0 \\ 0 & 0 & 0 & -1 \\ \Phi_{xx} & \Phi_{yy} & 0 & 0 \\ \Phi_{xy} & \Phi_{yy} & 0 & 0 \end{bmatrix}\begin{bmatrix} x \\ y \\ v_x \\ v_y \end{bmatrix}, \tag{1}$$

the solution can be written as a LINEAR COMBINATION of functions OF THE FORM

$$\begin{bmatrix} x(t) \\ y(t) \\ v_x \\ v_y \end{bmatrix} = \begin{bmatrix} x_0 \\ y_0 \\ v_{x0} \\ v_{y0} \end{bmatrix} e^{\mu t} P_\mu(t), \tag{2}$$

where $P_\mu(t)$ is a function periodic with the same period T as the equations themselves. Given an ORDINARY DIFFERENTIAL EQUATION OF THE FORM

$$\ddot{x} + g(t)x = 0, \tag{3}$$

where $g(t)$ is periodic with period T, the ODE has a pair of independent solutions given by the REAL and IMAGINARY PARTS of

$$x + w(t)e^{i\psi(t)} \tag{4}$$

$$\dot{x} = (\dot{w} + iw\dot{\psi})e^{i\psi} \tag{5}$$

$$\ddot{x} = \left[\ddot{w} + i\dot{w}\dot{\psi} + i(\dot{w}\dot{\psi} + w\ddot{\psi} + iw\dot{\psi}^2)\right]e^{i\psi}$$

$$= \left[(\ddot{w} - w\dot{\psi}^2) + i(2\dot{w}\dot{\psi} + w\ddot{\psi})\right]e^{i\psi}. \tag{6}$$

Plugging these into (3) gives

$$\ddot{w} + 2i\dot{w}\dot{\psi} + w(g + i\ddot{\psi} - \dot{\psi}^2) = 0, \tag{7}$$

so the REAL and IMAGINARY PARTS are

$$\ddot{w} + w(g - \dot{\psi}^2) = 0 \tag{8}$$

$$2\dot{w}\dot{\psi} + w\ddot{\psi} = 0. \tag{9}$$

From (9),

$$\frac{2\dot{w}}{w} + \frac{\ddot{\psi}}{\dot{\psi}} = 2\frac{d}{dt}(\ln w) + \frac{d}{dt}[\ln(\dot{\psi})]$$

$$= \frac{d}{dt}\ln(\dot{\psi}w^2) = 0. \tag{10}$$

Integrating gives

$$\dot{\psi} = \frac{c}{w^2}, \tag{11}$$

where C is a constant which must equal 1, so ψ is given by

$$\psi = \int_{t_0}^t \frac{dt}{w^2}. \tag{12}$$

The REAL solution is then

$$x(t) = w(t)\cos[\psi(t)], \tag{13}$$

so

$$\dot{x} = \dot{w}\cos\psi - w\dot{\psi}\sin\psi = \dot{w}\frac{x}{w} - w\dot{\psi}\sin\psi$$

$$= \dot{w}\frac{x}{w} - w\frac{1}{w^2}\sin\psi = \dot{w}\frac{x}{w} - \frac{1}{w}\sin\psi \tag{14}$$

and

$$1 = \cos^2\psi + \sin^2\psi = x^2 w^{-2} + \left[w\left(\dot{w}\frac{x}{w} - \dot{x}\right)\right]^2$$

$$= x^2 w^{-2} + (\dot{w}x - w\dot{x})^2 \equiv i(x,\dot{x},t), \tag{15}$$

which is an integral of motion. Therefore, although $w(t)$ is not explicitly known, an integral I always exists. Plugging (10) into (8) gives

$$\ddot{w} + g(t)w - \frac{1}{w^3} = 0, \tag{16}$$

which, however, is not any easier to solve than (3).

See also FLOQUET'S THEOREM, HILL'S DIFFERENTIAL EQUATION

References

Abramowitz, M. and Stegun, C. A. (Eds.). *Handbook of Mathematical Functions with Formulas, Graphs, and Mathematical Tables, 9th printing.* New York: Dover, p. 727, 1972.

Binney, J. and Tremaine, S. *Galactic Dynamics.* Princeton, NJ: Princeton University Press, p. 175, 1987.

Lichtenberg, A. and Lieberman, M. *Regular and Stochastic Motion.* New York: Springer-Verlag, p. 32, 1983.

Margenau, H. and Murphy, G. M. *The Mathematics of Physics and Chemistry, 2 vols.* Princeton, NJ: Van Nostrand, 1956–4.

Morse, P. M. and Feshbach, H. *Methods of Theoretical Physics, Part I.* New York: McGraw-Hill, pp. 556–57, 1953.

Floquet's Theorem

Let $Q(x)$ be a real or complex piecewise-continuous function of the real variable x defined for all values of x that is periodic with minimum period π so that

$$Q(x + \pi) = Q(x). \tag{1}$$

Then the differential equation

$$y^n + Q(x)y = 0 \tag{2}$$

has two continuously differentiable solutions $y_1(x)$ and $y_2(x)$, and the characteristic equation is

$$\rho^2 - [y_1(\pi) + y_2'(\pi)]\rho + 1 = 0, \tag{3}$$

with eigenvalues $\rho_1 = e^{i\alpha\pi}$ and $\rho_2 = e^{-i\alpha\pi}$. The Floquet's theorem states that if the roots ρ_1 and ρ_2 are different from each other, then (2) has two linearly independent solutions

$$f_1(x) = e^{i\alpha x}p_1(x) \tag{4}$$

$$f_2(x) = e^{-i\alpha x}p_2(x), \tag{5}$$

where $p_1(x)$ and $p_2(x)$ are period with period π (Magnus and Winkler 1979, p. 4).

See also FLOQUET ANALYSIS, HILL'S DIFFERENTIAL EQUATION

References

Magnus, W. and Winkler, S. "Floquet's Theorem." §1.2 in *Hill's Equation*. New York: Dover, pp. 3–, 1979.

Flow

An ACTION with $G = \mathbb{R}$. Flows are generated by VECTOR FIELDS and vice versa.

See also ACTION, AMBROSE-KAKUTANI THEOREM, ANOSOV FLOW, AXIOM A FLOW, CASCADE, GEODESIC FLOW, SEMIFLOW

Flow Line

A flow line for a map on a VECTOR FIELD $\mathbf{F}$ is a path $\sigma(t)$ such that $\sigma'(t) = \mathbf{F}(\sigma(t))$.

Flower

DAISY, FLOWER OF LIFE, ROSE

Flower of Life

One of the beautiful arrangements of CIRCLES found at the Temple of Osiris at Abydos, Egypt (Rawles 1997). The CIRCLES are placed with six-fold symmetry, forming a mesmerizing pattern of CIRCLES and LENSES.

See also CIRCLE COVERING, FIVE DISKS PROBLEM, REULEAUX TRIANGLE, SEED OF LIFE, VENN DIAGRAM

References

Rawles, B. *Sacred Geometry Design Sourcebook: Universal Dimensional Patterns*. Nevada City, CA: Elysian Pub., p. 15, 1997.
Wein, J. "La Fleur de Vie." http://www2.cruzio.com/~flower/fleur.htm.
Weisstein, E. W. "Flower of Life." MATHEMATICA NOTEBOOK FLOWEROFLIFE.M.

Flowsnake

PEANO-GOSPER CURVE

Flowsnake Fractal

GOSPER ISLAND

Floyd's Algorithm

An algorithm for finding the shortest path between two VERTICES.

See also DIJKSTRA'S ALGORITHM

Fluent

Newton's term for a variable in his method of FLUXIONS (differential calculus).

See also CALCULUS, FLUXION

References

Newton, I. *Methodus fluxionum et serierum infinitarum.* 1664–671.

Fluxion

The term for DERIVATIVE in Newton's CALCULUS.

See also CALCULUS, DERIVATIVE, FLUENT

References

Newton, I. *Methodus fluxionum et serierum infinitarum.* 1664–671.

Flype

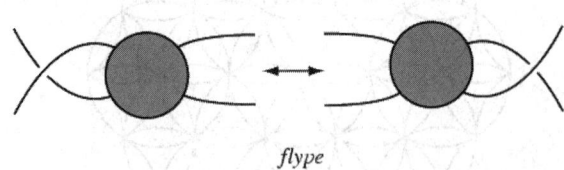

flype

A 180° rotation of a TANGLE. The word "flype" is derived from the old Scottish verb meaning "to turn or fold back." Tait (1898) used this word to indicate a different knot transformation than the one understood in the modern definition, illustrated above (Hoste *et al.* 1998).

See also FLYPING CONJECTURE, TANGLE

References

Hoste, J.; Thistlethwaite, M.; and Weeks, J. "The First 1,701,936 Knots." *Math. Intell.* **20**, 33–8, Fall 1998.
Tait, P. G. "On Knots I, II, and III." *Scientific Papers, Vol. 1.* Cambridge, England: University Press, pp. 273–47, 1898.

Flyping Conjecture

Also called the TAIT FLYPING CONJECTURE. Given two reduced alternating projections of the same KNOT, they are equivalent on the SPHERE IFF they are related by a series of FLYPES. The conjecture was proved by Menasco and Thistlethwaite (1991, 1993) using properties of the JONES POLYNOMIAL. It allows all possible REDUCED alternating projections of a given ALTERNATING KNOT to be drawn.

See also ALTERNATING KNOT, FLYPE, REDUCIBLE CROSSING, TAIT'S KNOT CONJECTURES

References

Adams, C. C. *The Knot Book: An Elementary Introduction to the Mathematical Theory of Knots.* New York: W. H. Freeman, pp. 164–65, 1994.
Hoste, J.; Thistlethwaite, M.; and Weeks, J. "The First 1,701,936 Knots." *Math. Intell.* **20**, 33–8, Fall 1998.
Menasco, W. and Thistlethwaite, M. "The Tait Flyping Conjecture." *Bull. Amer. Math. Soc.* **25**, 403–12, 1991.
Menasco, W. and Thistlethwaite, M. "The Classification of Alternating Links." *Ann. Math.* **138**, 113–71, 1993.
Stewart, I. *The Problems of Mathematics, 2nd ed.* Oxford, England: Oxford University Press, pp. 284–85, 1987.
The following table gives properties of different types of conic sections, where k is the

Focal Parameter

The distance p (sometimes also denoted k) from the FOCUS to the DIRECTRIX of a CONIC SECTION. The following table gives the focal parameter for the different types of conics, where a is the SEMIMAJOR AXIS, c is the distances from the origin to the FOCUS, and e is the ECCENTRICITY.

conic	e	$p(a,b)$	$p(a,c)$	$p(a,e)$
ELLIPSE	$0 < e < 1$	$\dfrac{b^2}{\sqrt{a^2 - b^2}}$	$\dfrac{a^2 - c^2}{c}$	$\dfrac{a(1 - e^2)}{e}$
PARABOLA	$e = 1$	$2a$	$2a$	$2a$
HYPERBOLA	$e > 1$	$\dfrac{b^2}{\sqrt{a^2 + b^2}}$	$\dfrac{c^2 - a^2}{c}$	$\dfrac{a(e^2 - 1)}{e}$

See also CONIC SECTION, DIRECTRIX (CONIC SECTION), ECCENTRICITY, FOCUS

Focus

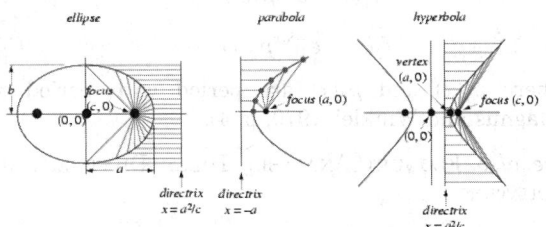

A point related to the construction and properties of CONIC SECTIONS. HYPERBOLAS and noncircular ELLIPSES have two distinct foci and two associated DIRECTRICES, each DIRECTRIX being PERPENDICULAR to the line joining the two foci (Eves 1965, p. 275).

See also DIRECTRIX (CONIC SECTION), ELLIPSE, ELLIPSOID, FOCAL PARAMETER, HYPERBOLA, HYPERBOLOID, PARABOLA, PARABOLOID, REFLECTION PROPERTY

References

Coxeter, H. S. M. and Greitzer, S. L. *Geometry Revisited.* Washington, DC: Math. Assoc. Amer., pp. 141–44, 1967.
Eves, H. "The Focus-Directrix Property." §6.8 in *A Survey of Geometry, rev. ed.* Boston, MA: Allyn & Bacon, pp. 272–75, 1965.

Foias Constant

A problem listed in a fall issue of *Gazeta Matematica* in the mid-1970s posed the question if $x_1 > 0$ and

$$x_{n+1} = \left(1 + \frac{1}{x_n}\right)^n \tag{1}$$

for $n = 1, 2, \ldots$, then are there any values for which $x_n \to \infty$? The problem, listed as one given on an entrance exam to prospective freshman in the mathematics department at the University of Bucharest, was solved by C. Foias.

It turns out that there exists exactly one real number

$$\alpha \approx 1.187452351126501 \tag{2}$$

such that if $x_1 = \alpha$, then $x_n \to \infty$. However, no analytic form is known for this constant, either as the root of a function or as a combination of other constants. Moreover, in this case,

$$\lim_{n \to \infty} x_n \frac{\ln n}{n} = 1, \tag{3}$$

which can be rewritten as

$$\lim_{n \to \infty} \frac{x_n}{\pi(n)} = 1, \tag{4}$$

where $\pi(n)$ is the PRIME COUNTING FUNCTION. However, Ewing and Foias (2000) believe that this connection with the PRIME NUMBER THEOREM is fortuitous.

Foias also discovered that the problem stated in the journal was a misprint of the actual exam problem, which used the recurrence $x_{n+1} = (1 + 1/x_n)^{x_n}$ (Ewing and Foias 2000). In this form, the recurrence converges to

$$x_\infty \approx 2.29316628741186103150802829912508 \tag{5}$$

for all starting values of x_1, which is simply the root of

$$x = \left(1 + \frac{1}{x}\right)^x. \tag{6}$$

See also GROSSMAN'S CONSTANT

References

Ewing, J. and Foias, C. "An Interesting Serendipitous Real Number." In *Finite versus Infinite: Contributions to an Eternal Dilemma* (Ed. C. Caluse and G. Paun). London: Springer-Verlag, pp. 119–26, 2000.

Fold Bifurcation

Let $f : \mathbb{R} \times \mathbb{R} \to \mathbb{R}$ be a one-parameter family of C^2 MAP satisfying

$$f(0,0) = 0$$

$$\left[\frac{\partial f}{\partial x}\right]_{\mu=0, x=0} = 0$$

$$\left[\frac{\partial^2 f}{\partial x^2}\right]_{\mu=0, x=0} > 0$$

$$\left[\frac{\partial f}{\partial \mu}\right]_{\mu=0, x=0} > 0,$$

then there exist intervals $(\mu_1, 0)$, $(0, \mu_2)$ and $\varepsilon > 0$ such that

1. If $\mu \in (\mu_1, 0)$, then $f_\mu(x)$ has two fixed points in $(-\epsilon, \epsilon)$ with the positive one being unstable and the negative one stable, and
2. If $\mu \in (0, \mu_2)$, then $f_\mu(x)$ has no fixed points in $(-\epsilon, \epsilon)$.

This type of BIFURCATION is known as a fold bifurcation, sometimes also called a SADDLE-NODE BIFURCATION or TANGENT BIFURCATION. An example of an equation displaying a fold bifurcation is

$$\dot{x} = \mu - x^2$$

(Guckenheimer and Holmes 1997, p. 145).

See also BIFURCATION

References

Guckenheimer, J. and Holmes, P. *Nonlinear Oscillations, Dynamical Systems, and Bifurcations of Vector Fields*, 3rd ed. New York: Springer-Verlag, pp. 145–49, 1997.
Rasband, S. N. *Chaotic Dynamics of Nonlinear Systems.* New York: Wiley, pp. 27–8, 1990.

Fold Catastrophe

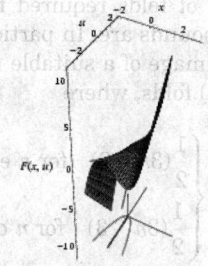

A catastrophe which can occur for one control factor and one behavior axis. It is the universal unfolding of the singularity $f(x) = x^3$ and has the equation $F(x, u) = x^3 + ux$.

See also CATASTROPHE THEORY

References

Sanns, W. *Catastrophe Theory with Mathematica: A Geometric Approach.* Germany: DAV, 2000.

Folding

The points accessible from c by a single fold which leaves $a_1, ..., a_n$ fixed are exactly those points interior to or on the boundary of the intersection of the CIRCLES through c with centers at a_i, for $i = 1, ..., n$. Given any three points in the plane a, b, and c, there is an EQUILATERAL TRIANGLE with VERTICES x, y, and z for which a, b, and c are the images of x, y, and z under a single fold.

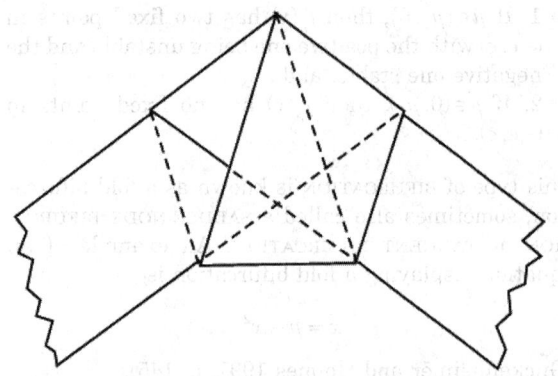

Given any four points in the plane a, b, c, and d, there is some SQUARE with VERTICES x, y, z, and w for which a, b, c, and d are the images of x, y, z, and w under a sequence of at most three folds. In addition, any four collinear points are the images of the VERTICES of a suitable SQUARE under at most two folds. Every five (six) points are the images of the VERTICES of suitable regular PENTAGON (HEXAGON) under at most five (six) folds. Wells (1991) illustrates a PENTAGON, HEXAGON, HEPTAGON, and OCTAGON constructed using paper folding.

The least number of folds required for $n \geq 4$ is not known, but some bounds are. In particular, every set of n points is the image of a suitable REGULAR n-gon under at most $F(n)$ folds, where

$$F(n) \leq \begin{cases} \frac{1}{2}(3n-2) & \text{for } n \text{ even} \\ \frac{1}{2}(3n-3) & \text{for } n \text{ odd.} \end{cases}$$

The first few values are 0, 2, 3, 5, 6, 8, 9, 11, 12, 14, 15, 17, 18, 20, 21, ... (Sloane's A007494).

See also FLEXAGON, MAP FOLDING, ORIGAMI, RUDIN-SHAPIRO SEQUENCE, STAMP FOLDING

References

Cundy, H. and Rollett, A. *Mathematical Models, 3rd ed.* Stradbroke, England: Tarquin Pub., 1989.
Hilton, P.; Holton, D.; and Pedersen, J. "Paper-Folding and Number Theory." Ch. 4 in *Mathematical Reflections in a Room with Many Mirrors.* New York: Springer-Verlag, pp. 87–42, 1997.
Klein, F. "Famous Problems of Elementary Geometry: The Duplication of the Cube, the Trisection of the Angle, and the Quadrature of the Circle." In *Famous Problems and Other Monographs.* New York: Chelsea, p. 42, 1980.
Sabinin, P. and Stone, M. G. "Transforming n-gons by Folding the Plane." *Amer. Math. Monthly* **102**, 620–27, 1995.
Sloane, N. J. A. Sequences A007494 in "An On-Line Version of the Encyclopedia of Integer Sequences." http://www.research.att.com/~njas/sequences/eisonline.html.
Wells, D. *The Penguin Dictionary of Curious and Interesting Geometry.* London: Penguin, pp. 191–92, 1991.

Foliation

Let M^n be an n-MANIFOLD and let $\mathsf{F} = \{F_\alpha\}$ denote a PARTITION of M^n into DISJOINT path-connected SUB-SETS. Then F is called a foliation of M^n of codimension c (with $0 < c < n$) if there exists a COVER of M^n by OPEN SETS U, each equipped with a HOMEOMORPHISM $h : U \to \mathbb{R}^n$ or $h : U \to \mathbb{R}^n_+$ which throws each nonempty component of $F_\alpha \cap U$ onto a parallel translation of the standard HYPERPLANE $\mathbb{R}^{n-c}$ in $\mathbb{R}^n$. Each F_α is then called a LEAF and is not necessarily closed or compact.

See also CONFOLIATION, COVER, HOMEOMORPHISM, LEAF (FOLIATION), MANIFOLD, REEB FOLIATION

References

Candel, A. and Conlon, L. *Foliations I.* Providence, RI: Amer. Math. Soc., 1999.
Rolfsen, D. *Knots and Links.* Wilmington, DE: Publish or Perish Press, p. 284, 1976.

Folium

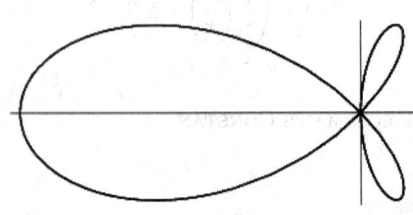

The word "folium" means leaf-shaped. The polar equation is

$$r = \cos\theta(4a\sin^2\theta - b).$$

If $b \geq 4a$, it is a single folium. If $b = 0$, it is a BIFOLIUM. If $0 < b < 4a$, it is a TRIFOLIUM. The simple folium is the PEDAL CURVE of the DELTOID where the PEDAL POINT is one of the CUSPS.

See also BIFOLIUM, FOLIUM OF DESCARTES, KEPLER'S FOLIUM, QUADRIFOLIUM, ROSE, TRIFOLIUM

References

Lawrence, J. D. *A Catalog of Special Plane Curves.* New York: Dover, pp. 152–53, 1972.
MacTutor History of Mathematics Archive. "Folium." http://www-groups.dcs.st-and.ac.uk/~history/Curves/Folium.html.

Folium of Descartes

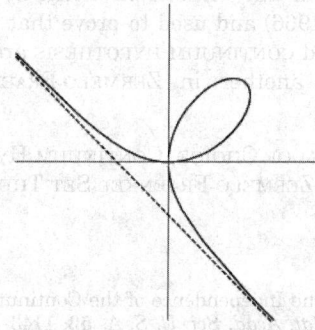

A plane curve proposed by Descartes to challenge Fermat's extremum-finding techniques. In parametric form,

$$x = \frac{3at}{1 + t^3} \qquad (1)$$

$$y = \frac{3at^2}{1 + t^3}. \qquad (2)$$

The curve has a discontinuity at $t = -1$. The left wing is generated as t runs from -1 to 0, the loop as t runs from 0 to ∞, and the right wing as t runs from $-\infty$ to -1.

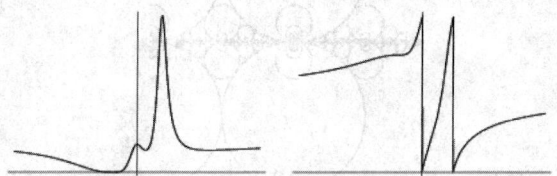

The CURVATURE and TANGENTIAL ANGLE of the folium of Descartes, illustrated above, are

$$\kappa(t) = \frac{2(1 + t^3)^4}{3(1 + 4t^2 - 4t^3 - 4t^5 + 4t^6 + t^8)^{3/2}} \qquad (3)$$

$$\phi(t) = \frac{1}{2} \left[\pi + \tan^{-1} \left(\frac{1 - 2t^3}{t^4 - 2t} \right) - \tan^{-1} \left(\frac{2t^3 - 1}{t^4 - 2t} \right) \right].$$

$$17 + 4\sqrt{18} \qquad (4)$$

Converting the PARAMETRIC EQUATIONS to POLAR COORDINATES gives

$$r^2 = \frac{(3at)^2(1 + t^2)}{(1 + t^3)^2} \qquad (5)$$

$$\theta = \tan^{-1} \left(\frac{y}{x} \right) = \tan^{-1} t, \qquad (6)$$

so

$$d\theta = \frac{dt}{1 + t^2}. \qquad (7)$$

The AREA enclosed by the curve is

$$A = \frac{1}{2} \int r^2 \, d\theta = \frac{1}{2} \int_0^\infty \frac{(3at)^2(1 + t^2)}{(1 + t^3)^2} \frac{dt}{1 + t^2}$$

$$= \frac{3}{2} a^2 \int_0^\infty \frac{3t^2 dt}{(1 + t^3)^2}. \qquad (8)$$

Now let $u \equiv 1 + t^3$ so $du = 3t^2 dt$

$$A = \frac{3}{2} a^2 \int_1^\infty \frac{du}{u^2} = \frac{3}{2} a^2 \left[-\frac{1}{u} \right]_1^\infty = \frac{3}{2} a^2 (-0 + 1) = \frac{3}{2} a^2 \qquad (9)$$

In CARTESIAN COORDINATES,

$$x^3 + y^3 = \frac{(3at)^3(1 + t^3)}{(1 + t^3)^3} = \frac{(3at)^3}{(1 + t^3)^2} = 3axy \qquad (10)$$

(MacTutor Archive). The equation of the ASYMPTOTE is

$$y = -a - x. \qquad (11)$$

References

Beyer, W. H. *CRC Standard Mathematical Tables, 28th ed.* Boca Raton, FL: CRC Press, p. 218, 1987.

Gray, A. *Modern Differential Geometry of Curves and Surfaces with Mathematica, 2nd ed.* Boca Raton, FL: CRC Press, pp. 77–2, 1997.

Lawrence, J. D. *A Catalog of Special Plane Curves.* New York: Dover, pp. 106–09, 1972.

MacTutor History of Mathematics Archive. "Folium of Descartes." http://www-groups.dcs.st-and.ac.uk/~history/Curves/Foliumd.html.

Stroeker, R. J. "Brocard Points, Circulant Matrices, and Descartes' Folium." *Math. Mag.* **61**, 172–87, 1988.

Yates, R. C. "Folium of Descartes." In *A Handbook on Curves and Their Properties.* Ann Arbor, MI: J. W. Edwards, pp. 98–9, 1952.

Folkman Graph

A graph which is EDGE-TRANSITIVE but not VERTEX-TRANSITIVE, and has the minimum possible number of nodes (20) for a nontrivial graph satisfying these properties (Skiena 1990, p. 186).

See also EDGE-TRANSITIVE GRAPH, VERTEX-TRANSITIVE GRAPH

References

Bondy, J. A. and Murty, U. S. R. *Graph Theory with Applications.* New York: North Holland, p. 235, 1976.

Folkman, J. "Regular Line-Symmetric Graphs." *J. Combin. Th.* **3**, 215–32, 1967.

Skiena, S. *Implementing Discrete Mathematics: Combinatorics and Graph Theory with Mathematica.* Reading, MA: Addison-Wesley, pp. 186–87, 1990.

Follows

SUCCEEDS

Fontené Theorems

1. If the sides of the PEDAL TRIANGLE of a point P meet the corresponding sides of a TRIANGLE $\Delta O_1 O_2 O_3$ at X_1, X_2, and X_3, respectively, then $P_1 X_1$, $P_2 X_2$, $P_3 X_3$ meet at a point L common to the CIRCLES $O_1 O_2 O_3$ and $P_1 P_2 P_3$. In other words, L is one of the intersections of the NINE-POINT CIRCLE of $A_1 A_2 A_3$ and the PEDAL CIRCLE of P.

2. If a point moves on a fixed line through the CIRCUMCENTER, then its PEDAL CIRCLE passes through a fixed point on the NINE-POINT CIRCLE.

3. The PEDAL CIRCLE of a point is tangent to the NINE-POINT CIRCLE IFF the point and its ISOGONAL CONJUGATE lie on a LINE through the ORTHOCENTER. FEUERBACH'S THEOREM is a special case of this theorem.

See also CIRCUMCENTER, FEUERBACH'S THEOREM, ISOGONAL CONJUGATE, NINE-POINT CIRCLE, ORTHOCENTER, PEDAL CIRCLE

References

Bricard, R. "Note au sujet de l'article précédent." *Nouv. Ann. Math.* **6**, 59–1, 1906.
Coolidge, J. L. *A Treatise on the Geometry of the Circle and Sphere.* New York: Chelsea, p. 52, 1971.
Fontené, G. "Extension du théorème de Feuerbach." *Nouv. Ann. Math.* **5**, 504–06, 1905.
Fontené, G. "Sur les points de contact du cercle des neuf point d'un triangle avec les cercles tangents aux trois côtés." *Nouv. Ann. Math.* **5**, 529–38, 1905.
Fontené, G. "Sur le cercle pédal." *Nouv. Ann. Math.* **65**, 55–8, 1906.
Johnson, R. A. *Modern Geometry: An Elementary Treatise on the Geometry of the Triangle and the Circle.* Boston, MA: Houghton Mifflin, pp. 245–47, 1929.

Foot

PERPENDICULAR FOOT

Football

LEMON

For All

If a proposition P is true for all B, this is written $P \forall B$. $\forall$ is one of the two so-called QUANTIFIERS.

In *Mathematica* 4.0, the command `ForAllRealQ[ineqs, vars]` can be used to determine if the system of real equations and inequalities *ineqs* is satisfied for all real values of the variables *vars*.

See also ALMOST ALL, EXISTS, IMPLIES, QUANTIFIER, UNIVERSAL QUANTIFIER

Forced Polygon

HAPPY END PROBLEM

Forcing

A technique in SET THEORY invented by P. Cohen (1963, 1964, 1966) and used to prove that the AXIOM OF CHOICE and CONTINUUM HYPOTHESIS are independent of one another in ZERMELO-FRAENKEL SET THEORY.

See also AXIOM OF CHOICE, CONTINUUM HYPOTHESIS, SET THEORY, ZERMELO-FRAENKEL SET THEORY

References

Cohen, P. J. "The Independence of the Continuum Hypothesis." *Proc. Nat. Acad. Sci. U. S. A.* **50**, 1143–148, 1963.
Cohen, P. J. "The Independence of the Continuum Hypothesis. II." *Proc. Nat. Acad. Sci. U. S. A.* **51**, 105–10, 1964.
Cohen, P. J. *Set Theory and the Continuum Hypothesis.* New York: W. A. Benjamin, 1966.
Todorchevich, S. and Farah, I. *Some Applications of the Method of Forcing.* Moscow: Yenisei, 1995.

Ford Circle

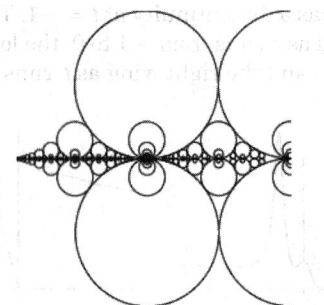

Pick any two INTEGERS h and k, then the CIRCLE $C(h, k)$ of RADIUS $1/(2k^2)$ centered at $(h/k, \pm 1/(2k^2))$ is known as a Ford circle. No matter what and how many hs and ks are picked, none of the Ford circles intersect (and all are tangent to the x-AXIS). This can be seen by examining the squared distance between the centers of the circles with (h, k) and (h', k'),

$$d^2 = \left(\frac{h'}{k'} - \frac{h}{k} \right)^2 + \left(\frac{1}{2k'^2} - \frac{1}{2k^2} \right)^2. \tag{1}$$

Let s be the sum of the radii

$$s = r_1 + r_2 = \frac{1}{2k^2} + \frac{1}{2k'^2}, \tag{2}$$

then

$$d^2 - s^2 = \frac{\left(h'k - hk' \right)^2 - 1}{k^2 k'^2}. \tag{3}$$

But $\left(h'k - k'h \right)^2 \geq 1$, so $d^2 - s^2 \geq 0$ and the distance between circle centers is $\geq$ the sum of the CIRCLE RADII, with equality (and therefore tangency) IFF $|h'k - k'h| = 1$. Ford circles are related to the FAREY

SEQUENCE (Conway and Guy 1996).

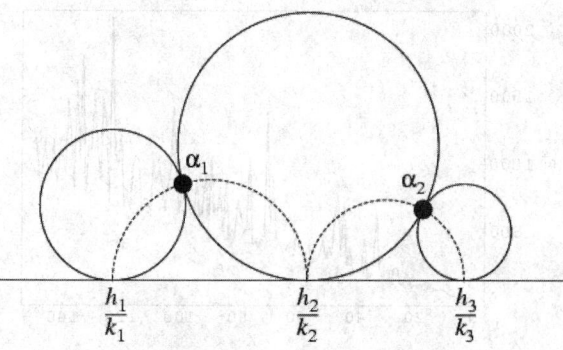

If h_1/k_1, h_2/k_2, and h_3/k_3 are three consecutive terms in a FAREY SEQUENCE, then the circles $c(h_1, k_1)$ and $c(h_2, k_2)$ are tangent at

$$\alpha_1 = \left(\frac{h_2}{k_2} - \frac{k_1}{k_2(k_2^2 + k_1^2)}, \frac{1}{k_2^2 + k_1^2} \right) \quad (4)$$

and the circles $c(h_2, k_2)$ and $C(h_3, k_3)$ intersect in

$$\alpha_2 = \left(\frac{h_2}{k_2} - \frac{k_3}{k_2(k_2^2 + k_3^2)}, \frac{1}{k_2^2 + k_3^2} \right). \quad (5)$$

Moreover, α_1 lies on the circumference of the SEMI-CIRCLE with diameter $(h_1/k_1, 0) - (h_2/k_2, 0)$ and α_2 lies on the circumference of the SEMICIRCLE with diameter $(h_2/k_2, 0) - (h_3/k_3, 0)$ (Apostol 1997, p. 101).

See also ADJACENT FRACTION, APOLLONIAN GASKET, FAREY SEQUENCE, STERN-BROCOT TREE

References

Apostol, T. M. "Ford Circles." §5.5 in *Modular Functions and Dirichlet Series in Number Theory, 2nd ed.* New York: Springer-Verlag, pp. 99–02, 1997.
Conway, J. H. and Guy, R. K. "Farey Fractions and Ford Circles." *The Book of Numbers.* New York: Springer-Verlag, pp. 152–54, 1996.
Ford, L. R. "Fractions." *Amer. Math. Monthly* **45**, 586–01, 1938.
Pickover, C. A. "Fractal Milkshakes and Infinite Archery." Ch. 14 in *Keys to Infinity.* New York: W. H. Freeman, pp. 117–25, 1995.
Rademacher, H. *Higher Mathematics from an Elementary Point of View.* Boston, MA: Birkhäuser, 1983.

Ford's Theorem

Let a, b, and k be INTEGERS with $k \geq 1$. For $j = 0, 1, 2$, let

$$S_j \equiv \sum_{i \equiv j \pmod{3}} (-1)^j \binom{k}{i} a^{k-i} b^i.$$

Then

$$2(a^2 + ab + b^2)^{2k}$$
$$= (S_0 - S_1)^4 + (S_1 - S_2)^4 + (S_2 - S_0)^4$$

See also BHARGAVA'S THEOREM, DIOPHANTINE EQUA-TION–4TH POWERS

References

Berndt, B. C. *Ramanujan's Notebooks, Part IV.* New York: Springer-Verlag, pp. 100–01, 1994.

Forest

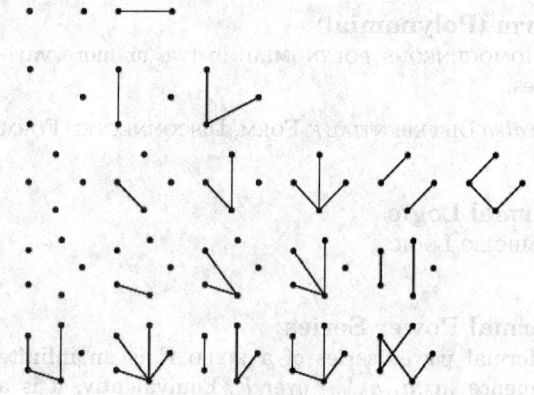

An acyclic graph (i.e., a GRAPH without any CIRCUITS). Forests therefore consist only of (possibly discon-nected) TREES, hence the name "forest." A forest with k components and n nodes has $n - k$ EDGES. The numbers of forests on $n = 1, 2, \ldots$ nodes are 1, 2, 3, 6, 10, 20, 37, ... (Sloane's A005195). A graph can be tested to determine if it is acyclic using AcylicQ[g] in the *Mathematica* add-on package Discrete-Math`Combinatorica` (which can be loaded with the command < <DiscreteMath`).
CONNECTED forests are TREES.

See also ACYCLIC DIGRAPH, CONNECTED GRAPH, GRAPH CYCLE, TREE

References

Harary, F. *Graph Theory.* Reading, MA: Addison-Wesley, p. 32, 1994.
Palmer, E. M. and Schwenk, A. J. "On the Number of Trees in a Random Forest." *J. Combin. Th. B* **27**, 109–21, 1979.
Skiena, S. "Acyclic Graphs." §5.3.1 in *Implementing Discrete Mathematics: Combinatorics and Graph Theory with Mathematica.* Reading, MA: Addison-Wesley, pp. 188–90, 1990.
Sloane, N. J. A. Sequences A005195/M0776 in "An On-Line Version of the Encyclopedia of Integer Sequences." http://www.research.att.com/~njas/sequences/eisonline.html.

Fork

A fork of a TREE T is a node of T which is the endpoint of two or more BRANCHES.

See also BRANCH, TREE

Form

CANONICAL FORM, CUSP FORM, DIFFERENTIAL K-FORM, FORM (GEOMETRIC), FORM (POLYNOMIAL),

MODULAR FORM, NORMAL FORM, PFAFFIAN FORM, QUADRATIC FORM

Form (Geometric)
A 1-D geometric object such as a PENCIL or RANGE.

Form (Polynomial)
A HOMOGENEOUS POLYNOMIAL in two or more variables.

See also DIFFERENTIAL K-FORM, DISCONNECTED FORM

Formal Logic
SYMBOLIC LOGIC

Formal Power Series
A formal power series of a FIELD F is an infinite sequence $\{a_0, a_1, a_2, ...\}$ over F. Equivalently, it is a function from the set of nonnegative integers to F, $\{0, 1, 2, ...\} \to F$. A formal power series is often written

$$a_0 + a_1 x + a_2 x^2 + ... + a_n x^n + ...,$$

but with the understanding that no value is assigned to the symbol x.

See also POWER SERIES

References
Henrici, P. "Definition and Algebraic Properties of Formal Series." §1.2 in *Applied and Computational Complex Analysis, Vol. 1: Power Series-Integration-Conformal Mapping-Location of Zeros.* New York: Wiley, pp. 9–3, 1988.

Formosa Theorem
CHINESE REMAINDER THEOREM

Formula
A mathematical equation or a formal logical expression. The correct Latin plural form of formula is "formulae," although the less pretentious-sounding "formulas" is more commonly used.

See also EQUALITY, EQUATION, IDENTITY

References
Carr, G. S. *Formulas and Theorems in Pure Mathematics.* New York: Chelsea, 1970.
Spiegel, M. R. *Mathematical Handbook of Formulas and Tables.* New York: McGraw-Hill, 1968.
Tallarida, R. J. *Pocket Book of Integrals and Mathematical Formulas, 3rd ed.* Boca Raton, FL: CRC Press, 1992.
Weisstein, E. W. "Books about Handbooks of Mathematics." http://www.treasure-troves.com/books/Handbooksof-Mathematics.html.

Fortunate Prime

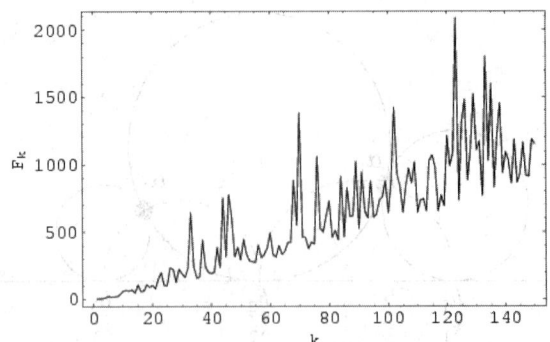

Let

$$X_k \equiv 1 + p_k\#,$$

where p_k is the kth PRIME and p is the PRIMORIAL, and let q_k be the NEXT PRIME (i.e., the smallest PRIME greater than X_k),

$$q_k = p_{1+\pi}(x_k) = p_{1+\pi(1+p_k\#)}$$

where $\pi(n)$ is the PRIME COUNTING FUNCTION. Then R. F. Fortune conjectured that $F_k \equiv q_k - X_k + 1$ is PRIME for all k. The first values of F_k are 3, 5, 7, 13, 23, 17, 19, 23, ... (Sloane's A005235), and all known values of F_k are indeed PRIME (Guy 1994). The indices of these primes are 2, 3, 4, 6, 9, 7, 8, 9, 12, 18, ... In numerical order with duplicates removed, the Fortunate primes are 3, 5, 7, 13, 17, 19, 23, 37, 47, 59, 61, 67, 71, 79, 89, ... (Sloane's A046066).

See also ANDRICA'S CONJECTURE, PRIMORIAL

References
Gardner, M. "Patterns in Primes are a Clue to the Strong Law of Small Numbers." *Sci. Amer.* **243**, 18–8, Dec. 1980.
Guy, R. K. *Unsolved Problems in Number Theory, 2nd ed.* New York: Springer-Verlag, p. 7, 1994.
Sloane, N. J. A. Sequences A005235/M2418 and A046066 in "An On-Line Version of the Encyclopedia of Integer Sequences." http://www.research.att.com/~njas/sequences/eisonline.html.

Forward Difference
The forward difference is a FINITE DIFFERENCE defined by

$$\Delta a_n \equiv a_{n+1} - a_n. \tag{1}$$

Higher order differences are obtained by repeated operations of the forward difference operator,

$$\Delta^k a_n = \Delta^{k-1} a_{n+1} - \Delta^{k-1} a_n, \tag{2}$$

so

$$\Delta^2 a_n = \Delta_n^2 = \Delta(\Delta_n) = \Delta(a_{n+1} - a_n)$$

$$= \Delta_{n+1} - \Delta_n = a_{n+2} - 2a_{n+1} + a_n. \tag{3}$$

In general,

$$\Delta_n^k \equiv \Delta^k a_n \equiv \sum_{i=0}^{k} (-1)^i \binom{k}{i} a_{n+k-i}, \qquad (4)$$

where $\binom{k}{m}$ is a BINOMIAL COEFFICIENT (Sloane and Plouffe 1985, p. 10).

NEWTON'S FORWARD DIFFERENCE FORMULA expresses a_n as the sum of the nth forward differences

$$a_n = a_0 + n\Delta_0 + \frac{1}{2!}n(n+1)\Delta_0^2 + \frac{1}{3!}n(n+1)(n+2)\Delta_0^3$$
$$+ \dots \qquad (5)$$

where Δ_0^n is the first nth difference computed from the difference table. Furthermore, if the differences a_m, Δa_m, $\Delta^2 a_m$, ..., are known for some fixed value of m, then a formula for the nth term is given by

$$a_{n+m} = \sum_{k=0}^{n} \binom{n}{k} \Delta^k a_m \qquad (6)$$

(Sloane and Plouffe 1985, p. 10).

See also BACKWARD DIFFERENCE, CENTRAL DIFFERENCE, DIFFERENCE EQUATION, DIVIDED DIFFERENCE, RECIPROCAL DIFFERENCE

References

Abramowitz, M. and Stegun, C. A. (Eds.). *Handbook of Mathematical Functions with Formulas, Graphs, and Mathematical Tables, 9th printing.* New York: Dover, p. 877, 1972.
Sloane, N. J. A. and Plouffe, S. *The Encyclopedia of Integer Sequences.* San Diego, CA: Academic Press, p. 10, 1995.

Fountain

An (n, k) fountain is an arrangement of n coins in rows such that exactly k coins are in the bottom row and each coin in the $(i+1)$st row touches exactly two in the ith row. A generalized Rogers-Ramanujan-type continued fraction is closely related to the enumeration of coins in a fountain (Berndt 1991, 1985).

References

Berndt, B. C. *Ramanujan's Notebooks, Part III.* New York: Springer-Verlag, p. 79, 1985.
Berndt, B. C.; Huang, S.-S.; Sohn, J.; and Son, S. H. "Some Theorems on the Rogers-Ramanujan Continued Fraction in Ramanujan's Lost Notebook." To appears in *Trans. Amer. Math. Soc.*

Four Coins Problem

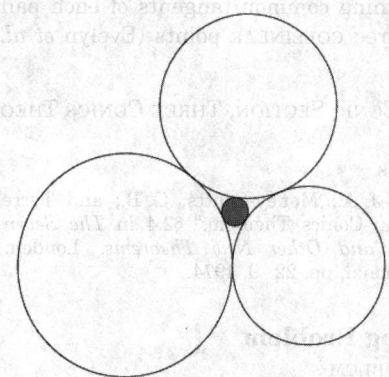

Given three coins of possibly different sizes which are arranged so that each is tangent to the other two, find the coin which is tangent to the other three coins. The solution is the inner SODDY CIRCLE, illustrated above.

See also APOLLONIUS CIRCLES, APOLLONIUS' PROBLEM, ARBELOS, BEND (CURVATURE), CIRCUMCIRCLE, COIN, DESCARTES CIRCLE THEOREM, HART'S THEOREM, PAPPUS CHAIN, SODDY CIRCLES, SPHERE PACKING, STEINER CHAIN, TANGENT CIRCLES

References

Oldknow, A. "The Euler-Gergonne-Soddy Triangle of a Triangle." *Amer. Math. Monthly* **103**, 319–29, 1996.

Four Conics Theorem

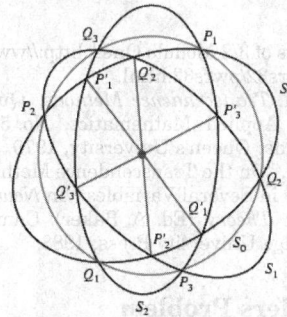

If two intersections of each pair of three conics S_1, S_2, and S_3 lie on a conic S_0, then the lines joining the other two intersections of each pair are CONCURRENT (Evelyn *et al.* 1974, pp. 23 and 25).

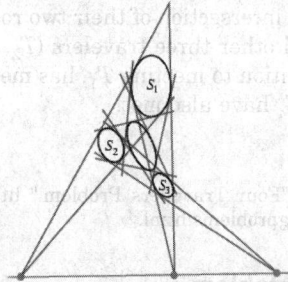

The dual theorem states that if two common tangents

of each pair of three conics touch a fourth conic, then the remaining common tangents of each pair intersect in three COLLINEAR points (Evelyn *et al.* 1974, pp. 24–5).

See also CONIC SECTION, THREE CONICS THEOREM

References

Evelyn, C. J. A.; Money-Coutts, G. B.; and Tyrrell, J. A. "The Four-Conics Theorem." §2.4 in *The Seven Circles Theorem and Other New Theorems.* London: Stacey International, pp. 22–9, 1974.

Four Dog Problem

MICE PROBLEM

Four Exponentials Conjecture

Let x_1 and x_2 be two linearly independent complex numbers, and let y_1 and y_2 be two linearly independent complex numbers. Then the four exponential conjecture posits that at least one of

$$e^{x_1 y_1}, e^{x_1 y_2}, e^{x_2 y_1}, e^{x_2 y_2}$$

is TRANSCENDENTAL (Waldschmidt 1979, p. 3.5). The corresponding statement obtained by replacing y_1, y_2 with y_1, y_2, y_3 has been proven and is known as the SIX EXPONENTIALS THEOREM.

See also HERMITE-LINDEMANN THEOREM, SIX EXPONENTIALS THEOREM, TRANSCENDENTAL NUMBER

References

Finch, S. "Powers of 3/2 Modulo One." http://www.mathsoft.com/asolve/pwrs32/pwrs32.html.
Waldschmidt, M. *Transcendence Methods.* Queen's Papers in Pure and Applied Mathematics, No. 52. Kingston, Ontario, Canada: Queen's University, 1979.
Waldschmidt, M. "On the Transcendence Method of Gelfond and Schneider in Several Variables." In *New Advances in Transcendence Theory* (Ed. A. Baker). Cambridge, England: Cambridge University Press, 1988.

Four Travelers Problem

Let four LINES in a PLANE represent four roads in GENERAL POSITION, and let one traveler T_i be walking along each road at a constant (but not necessarily equal to any other traveler's) speed. Say that two travelers T_i and T_j have "met" if they were simultaneously at the intersection of their two roads. Then if T_1 has met all other three travelers (T_2, T_3, and T_4) and T_2, in addition to meeting T_1, has met T_3 and T_4, then T_3 and T_4 have also met!

References

Bogomolny, A. "Four Travellers Problem." http://www.cut-the-knot.com/gproblems.html.

Four-Bug Problem

MICE PROBLEM

Four-Color Problem

FOUR-COLOR THEOREM

Four-Color Theorem

The four-color theorem states that any map in a PLANE can be colored using four-colors in such a way that regions sharing a common boundary (other than a single point) do not share the same color. This problem is sometimes also called GUTHRIE'S PROBLEM after F. Guthrie, who first conjectured the theorem in 1853. The CONJECTURE was then communicated to de Morgan and thence into the general community. In 1878, Cayley wrote the first paper on the conjecture.

Fallacious proofs were given independently by Kempe (1879) and Tait (1880). Kempe's proof was accepted for a decade until Heawood showed an error using a map with 18 faces (although a map with nine faces suffices to show the fallacy). The HEAWOOD CONJECTURE provided a very general assertion for map coloring, showing that in a GENUS 0 SPACE (i.e., either the SPHERE or PLANE), six colors suffice. This number can easily be reduced to five, but reducing the number of colors all the way to four proved very difficult. (The KLEIN BOTTLE is the sole exception to the HEAWOOD CONJECTURE, requiring five colors instead of the six expected for a surface of genus 0.)

Finally, Appel and Haken (1977) announced a computer-assisted proof that four colors were SUFFICIENT. However, because part of the proof consisted of an exhaustive analysis of many discrete cases by a computer, some mathematicians do not accept it. However, no flaws have yet been found, so the proof appears valid. A potentially independent proof has recently been constructed by N. Robertson, D. P. Sanders, P. D. Seymour, and R. Thomas.

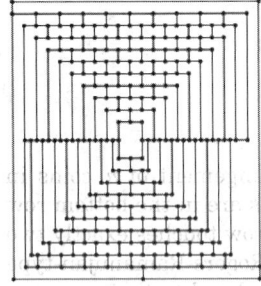 

Martin Gardner (1975) played an April Fool's joke by (incorrectly) claiming that the map of 110 regions illustrated above requires five colors and constitutes a counterexample to the four-color theorem. However, the coloring of Wagon (1998; 1999, pp. 535–36) clearly shows that this map is, in fact, four-colorable.

See also CHROMATIC NUMBER, ERRERA GRAPH, GRAPH

COLORING, HEAWOOD CONJECTURE, KITTELL GRAPH, MAP COLORING, SIX-COLOR THEOREM, TORUS COLORING

References

Appel, K. and Haken, W. "Every Planar Map is Four-Colorable, II: Reducibility." *Illinois J. Math.* **21**, 491–67, 1977.
Appel, K. and Haken, W. "The Solution of the Four-Color Map Problem." *Sci. Amer.* **237**, 108–21, 1977.
Appel, K. and Haken, W. "The Four Color Proof Suffices." *Math. Intell.* **8**, 10–0 and 58, 1986.
Appel, K. and Haken, W. *Every Planar Map is Four-Colorable.* Providence, RI: Amer. Math. Soc., 1989.
Appel, K.; Haken, W.; and Koch, J. "Every Planar Map is Four Colorable. I: Discharging." *Illinois J. Math.* **21**, 429–90, 1977.
Barnette, D. *Map Coloring, Polyhedra, and the Four-Color Problem.* Providence, RI: Math. Assoc. Amer., 1983.
Birkhoff, G. D. "The Reducibility of Maps." *Amer. Math. J.* **35**, 114–28, 1913.
Chartrand, G. "The Four Color Problem." §9.3 in *Introductory Graph Theory.* New York: Dover, pp. 209–15, 1985.
Coxeter, H. S. M. "The Four-Color Map Problem, 1840–890." *Math. Teach.* **52**, 283–89, 1959.
Franklin, P. "Note on the Four Color Problem." *J. Math. Phys.* **16**, 172–84, 1937–1938.
Franklin, P. *The Four-Color Problem.* New York: Scripta Mathematica, Yeshiva College, 1941.
Gardner, M. "Mathematical Games: The Celebrated Four-Color Map Problem of Topology." *Sci. Amer.* **203**, 218–22, Sep. 1960.
Gardner, M. "The Four-Color Map Theorem." Ch. 10 in *Martin Gardner's New Mathematical Diversions from Scientific American.* New York: Simon and Schuster, pp. 113–23, 1966.
Gardner, M. "Mathematical Games: Six Sensational Discoveries that Somehow or Another have Escaped Public Attention." *Sci. Amer.* **232**, 127–31, Apr. 1975.
Gardner, M. "Mathematical Games: On Tessellating the Plane with Convex Polygons." *Sci. Amer.* **232**, 112–17, Jul. 1975.
Harary, F. "The Four Color Conjecture." *Graph Theory.* Reading, MA: Addison-Wesley, p. 5, 1994.
Heawood, P. J. "Map Colour Theorems." *Quart. J. Math.* **24**, 332–38, 1890.
Kempe, A. B. "On the Geographical Problem of Four-Colors." *Amer. J. Math.* **2**, 193–00, 1879.
Kraitchik, M. §8.4.2 in *Mathematical Recreations.* New York: W. W. Norton, p. 211, 1942.
May, K. O. "The Origin of the Four-Color Conjecture." *Isis* **56**, 346–48, 1965.
Morgenstern, C. and Shapiro, H. "Heuristics for Rapidly 4-Coloring Large Planar Graphs." *Algorithmica* **6**, 869–91, 1991.
Ore, Ø. *The Four-Color Problem.* New York: Academic Press, 1967.
Ore, Ø. and Stemple, G. J. "Numerical Methods in the Four Color Problem." *Recent Progress in Combinatorics* (Ed. W. T. Tutte). New York: Academic Press, 1969.
Pappas, T. "The Four-Color Map Problem: Topology Turns the Tables on Map Coloring." *The Joy of Mathematics.* San Carlos, CA: Wide World Publ./Tetra, pp. 152–53, 1989.
Robertson, N.; Sanders, D. P.; and Thomas, R. "The Four-Color Theorem." http://www.math.gatech.edu/~thomas/FC/fourcolor.html.
Saaty, T. L. and Kainen, P. C. *The Four-Color Problem: Assaults and Conquest.* New York: Dover, 1986.
Skiena, S. *Implementing Discrete Mathematics: Combinatorics and Graph Theory with Mathematica.* Reading, MA: Addison-Wesley, p. 210, 1990.
Steinhaus, H. *Mathematical Snapshots, 3rd ed.* New York: Dover, pp. 274–75, 1999.
Tait, P. G. "Note on a Theorem in Geometry of Position." *Trans. Roy. Soc. Edinburgh* **29**, 657–60, 1880.
Wagon, S. "An April Fool's Hoax." *Mathematica in Educ. Res.* **7**, 46–2, 1998.
Wagon, S. *Mathematica in Action, 2nd ed.* New York: Springer-Verlag, pp. 535–36, 1999.
Weisstein, E. W. "Books about Four-Color Problem." http://www.treasure-troves.com/books/Four-ColorProblem.html.
Wells, D. *The Penguin Dictionary of Curious and Interesting Numbers.* Middlesex, England: Penguin Books, p. 57, 1986.
Wells, D. *The Penguin Dictionary of Curious and Interesting Geometry.* London: Penguin, pp. 81–2, 1991.

Four-Dimensional Geometry

4-DIMENSIONAL GEOMETRY

Fourier Analysis

FOURIER SERIES

Fourier Cosine Series

If $f(x)$ is an EVEN FUNCTION, then $b_n = 0$ and the FOURIER SERIES collapses to

$$f(x) = \frac{1}{2}a_0 + \sum_{n=1}^{\infty} a_n \cos(nx), \tag{1}$$

where

$$a_0 = \frac{1}{\pi}\int_{-\pi}^{\pi} f(x)\,dx = \frac{2}{\pi}\int_0^{\pi} f(x)\,dx \tag{2}$$

$$a_n = \frac{1}{\pi}\int_{-\pi}^{\pi} f(x)\cos(nx)\,dx$$

$$= \frac{2}{\pi}\int_0^{\pi} f(x)\cos(nx)\,dx \tag{3}$$

where the last equality is true because

$$f(x)\cos(nx) = f(-x)\cos(-nx) \tag{4}$$

Letting the range go to L,

$$a_0 = \frac{2}{L}\int_0^L f(x)\,dx \tag{5}$$

$$a_n = \frac{2}{L}\int_0^L f(x)\cos\left(\frac{n\pi x}{L}\right)dx. \tag{6}$$

See also EVEN FUNCTION, FOURIER COSINE TRANSFORM, FOURIER SERIES, FOURIER SINE SERIES

Fourier Cosine Transform

The Fourier cosine transform is the REAL PART of the full complex FOURIER TRANSFORM,

$$\mathscr{F}_c[f(x)] = \Re[\mathscr{F}[f(x)]].$$

In *Mathematica* 4.0, the Fourier cosine transform $F_c(k)$ of a function $f(x)$ is implemented as `FourierCosTransform[f, x, k]`, and different choices of a and b can be used by passing the optional `FourierParameters->{a, b}` option. In this work, $a = 0$ and $b = -2\pi$.

In version 4.1, the discrete Fourier cosine transform of a list l of real numbers can be computed using `FourierCos[l]` in the *Mathematica* add-on package `LinearAlgebra'FourierTrig'` (which can be loaded with the command $<<$ `LinearAlgebra'`).

See also FOURIER SINE TRANSFORM, FOURIER TRANSFORM

References

Press, W. H.; Flannery, B. P.; Teukolsky, S. A.; and Vetterling, W. T. "FFT of Real Functions, Sine and Cosine Transforms." §12.3 in *Numerical Recipes in FORTRAN: The Art of Scientific Computing, 2nd ed.* Cambridge, England: Cambridge University Press, pp. 504–15, 1992.

Fourier Integral

FOURIER TRANSFORM

Fourier Matrix

The $n \times n$ SQUARE MATRIX F_n with entries given by

$$F_{jk} = e^{2\pi ijk/n} \equiv \omega^{jk} \tag{1}$$

for $j, k = 0, 1, 2, \ldots, n-1$, where i is the IMAGINARY NUMBER $i = \sqrt{-1}$, and normalized by $1\sqrt{n}$ to make it a UNITARY. The Fourier matrix F_2 is given by

$$\mathsf{F}_2 = \frac{1}{\sqrt{2}} \begin{bmatrix} 1 & 1 \\ 1 & i^2 \end{bmatrix}, \tag{2}$$

and the F_4 matrix by

$$\mathsf{F}_4 = \frac{1}{\sqrt{4}} \begin{bmatrix} 1 & 1 & 1 & 1 \\ 1 & i & i^2 & i^3 \\ 1 & i^2 & i^4 & i^6 \\ 1 & i^3 & i^6 & i^9 \end{bmatrix}$$

$$= \frac{1}{2} \begin{bmatrix} 1 & & 1 & \\ & 1 & & i \\ 1 & & -1 & \\ & 1 & & -i \end{bmatrix} \begin{bmatrix} 1 & 1 & & \\ 1 & i^2 & & \\ & & 1 & 1 \\ & & 1 & i^2 \end{bmatrix} \begin{bmatrix} 1 & & & \\ & & 1 & \\ & 1 & & \\ & & & 1 \end{bmatrix}. \tag{3}$$

In general,

$$\mathsf{F}_{2n} = \begin{bmatrix} \mathsf{I}_n & \mathsf{D}_n \\ \mathsf{I}_n & -\mathsf{D}_n \end{bmatrix} \begin{bmatrix} \mathsf{F}_n & \\ & \mathsf{F}_n \end{bmatrix} \begin{bmatrix} \text{even-odd} \\ \text{shuffle} \end{bmatrix}, \tag{4}$$

with

$$\begin{bmatrix} \mathsf{F}_n & \\ & \mathsf{F}_n \end{bmatrix} = \begin{bmatrix} \mathsf{I}_{n/2} & \mathsf{D}_{n/2} \\ \mathsf{I}_{n/2} & -\mathsf{D}_{n/2} \end{bmatrix}$$

$$\times \begin{bmatrix} \mathsf{F}_{n/2} & & & \\ & \mathsf{F}_{n/2} & & \\ & & \mathsf{F}_{n/2} & \\ & & & \mathsf{F}_{n/2} \end{bmatrix} \begin{bmatrix} \mathsf{I}_{n/2} & \mathsf{D}_{n/2} \\ \mathsf{I}_{n/2} & -\mathsf{D}_{n/2} \end{bmatrix} \begin{bmatrix} \text{even-odd} \\ 0, 2 (\text{mod}4) \\ \text{even-odd} \\ 1, 3 (\text{mod}4) \end{bmatrix}, \tag{5}$$

where I_n is the $n \times n$ IDENTITY MATRIX and D_n is the DIAGONAL MATRIX with entries $1, \omega, \ldots, \omega^{n-1}$. Note that the factorization (which is the basis of the FAST FOURIER TRANSFORM) has two copies of F_2 in the center factor MATRIX.

See also FAST FOURIER TRANSFORM, FOURIER TRANSFORM

References

Strang, G. "Wavelet Transforms Versus Fourier Transforms." *Bull. Amer. Math. Soc.* **28**, 288–05, 1993.

Fourier Series

Fourier series are expansions of PERIODIC FUNCTIONS $f(x)$ in terms of an infinite sum of SINES and COSINES OF THE FORM

$$f(x) = \sum_{n=0}^{\infty} a_n' \cos(nx) + \sum_{n=0}^{\infty} b_n' \sin(nx). \tag{1}$$

Fourier series make use of the ORTHOGONALITY relationships of the SINE and COSINE functions, which can be used to calculate the coefficients a_n and b_n in the sum. The computation and study of Fourier series is known as HARMONIC ANALYSIS.

To compute a Fourier series, use the integral identities

$$\int_{-\pi}^{\pi} \sin(mx) \sin(nx) dx = \pi \delta_{mn} \quad \text{for } n, m \neq 0 \tag{2}$$

$$\int_{-\pi}^{\pi} \cos(mx) \cos(nx) dx = \pi \delta_{mn} \quad \text{for } n, m \neq 0 \tag{3}$$

$$\int_{-\pi}^{\pi} \sin(mx) \cos(nx) dx = 0 \tag{4}$$

$$\int_{-\pi}^{\pi} \sin(mx) dx = 0 \tag{5}$$

$$\int_{-\pi}^{\pi} \cos(mx) dx = 0, \tag{6}$$

where δ_{mn} is the KRONECKER DELTA. Now, expand your function $f(x)$ as an infinite series OF THE FORM

$$f(x) = \sum_{n=0}^{\infty} a'_n \cos(nx) + \sum_{n=0}^{\infty} b'_n \sin(nx)$$

$$= \frac{1}{2} a_0 + \sum_{n=1}^{\infty} a_n \cos(nx) + \sum_{n=1}^{\infty} b_n \sin(nx) \quad (7)$$

where we have relabeled the $a_0 = 2a'_0$ term for future convenience but set $b_n = b'_n$ and left $a_n = a'_n$ for $n \geq 1$. Assume the function is periodic in the interval $[-\pi, \pi]$. Now use the orthogonality conditions to obtain

$$\int_{-\pi}^{\pi} f(x) dx$$

$$= \int_{-\pi}^{\pi} \left[\sum_{n=1}^{\infty} a_n \cos(nx) + \sum_{n=1}^{\infty} b_n \sin(nx) + \frac{1}{2} a_0 \right] dx$$

$$= \sum_{n=1}^{\infty} \int_{-\pi}^{\pi} [a_n \cos(nx) + b_n \sin(nx)] dx + \frac{1}{2} a_0 \int_{-\pi}^{\pi} dx$$

$$= \sum_{n=1}^{\infty} (0 + 0) + \pi a_0 = \pi a_0 \quad (8)$$

and

$$\int_{-\pi}^{\pi} f(x) \sin(mx) dx$$

$$= \int_{-\pi}^{\pi} \left[\sum_{n=1}^{\infty} a_n \cos(nx) + \sum_{n=1}^{\infty} b_n \sin(nx) + \frac{1}{2} a_0 \right] \times \sin(mx) dx$$

$$= \sum_{n=1}^{\infty} \int_{-\pi}^{\pi} [a_n \cos(nx) \sin(mx) + b_n \sin(nx) \sin(mx)] dx$$

$$+ \frac{1}{2} a_0 \int_{-\pi}^{\pi} \sin(mx) dx$$

$$= \sum_{n=1}^{\infty} (0 + b_n \pi \delta_{mn}) + 0 = \pi b_n, \quad (9)$$

so

$$\int_{-\pi}^{\pi} f(x) \cos(mx) dx = \int_{-\pi}^{\pi} \left[\sum_{n=1}^{\infty} a_n \cos(nx) \right.$$

$$+ \sum_{n=1}^{\infty} b_n \sin(nx) + \frac{1}{2} a_0 \right] \cos(mx) dx$$

$$= \sum_{n=1}^{\infty} \int_{-\pi}^{\pi} [a_n \cos(nx) \cos(mx)$$

$$+ b_n \sin(nx) \cos(mx)] dx + \frac{1}{2} a_0 \int_{-\pi}^{\pi} \cos(mx) dx$$

$$= \sum_{n=1}^{\infty} (a_n \pi \delta_{mn} + 0) + 0 = \pi a_n. \quad (10)$$

Plugging back into the original series then gives

$$a_0 = \frac{1}{\pi} \int_{-\pi}^{\pi} f(x) dx \quad (11)$$

$$a_n = \frac{1}{\pi} \int_{-\pi}^{\pi} f(x) \cos(nx) dx \quad (12)$$

$$b_n = \frac{1}{\pi} \int_{-\pi}^{\pi} f(x) \sin(nx) dx \quad (13)$$

for $n = 1, 2, 3, \ldots$. The series expansion converges to the function $\bar{f}$ (equal to the original function at points of continuity or to the average of the two limits at points of discontinuity)

$$\bar{f} \equiv \begin{cases} \frac{1}{2} \left[\lim_{x \to x_0^-} f(x) + \lim_{x \to x_0^+} f(x) \right] \\ \qquad \text{for } -\pi < x_0 < \pi \\ \frac{1}{2} \left[\lim_{x \to \pi^+} f(x) + \lim_{x \to \pi^-} f(x) \right] \\ \qquad \text{for } x_0 = -\pi, \pi \end{cases} \quad (14)$$

if the function satisfies the DIRICHLET CONDITIONS.

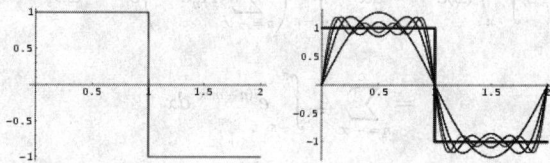

Near points of discontinuity, a "ringing" known as the GIBBS PHENOMENON, illustrated above, occurs. For a function $f(x)$ periodic on an interval $[-L, L]$, use a change of variables to transform the interval of integration to $[-1, 1]$. Let

$$x \equiv \frac{\pi x'}{L} \quad (15)$$

$$dx = \frac{\pi \, dx'}{L}. \quad (16)$$

Solving for x', $x' = Lx/\pi$. Plugging this in gives

$$f(x') = \frac{1}{2} a_0 + \sum_{n=1}^{\infty} a_n \cos\left(\frac{n\pi x'}{L}\right) + \sum_{n=1}^{\infty} b_n \sin\left(\frac{n\pi x'}{L}\right) \quad (17)$$

$$\begin{cases} a_0 = \frac{1}{L} \int_{-L}^{L} f(x') \, dx' \\ a_n = \frac{1}{L} \int_{-L}^{L} f(x') \cos\left(\frac{n\pi x'}{L}\right) dx' \\ b_n = \frac{1}{L} \int_{-L}^{L} f(x') \sin\left(\frac{n\pi x'}{L}\right) dx' \end{cases} \quad (18)$$

If a function is EVEN so that $f(x) = f(-x)$, then $f(x) \sin(nx)$ is ODD. (This follows since $\sin(nx)$ is ODD and an EVEN FUNCTION times an ODD FUNCTION is an ODD FUNCTION.) Therefore, $b_n = 0$ for all n. Similarly, if a function is ODD so that $f(x) = -f(-x)$, then $f(x) \cos(nx)$ is ODD. (This follows since $\cos(nx)$ is EVEN and an EVEN FUNCTION times an ODD FUNCTION is an ODD FUNCTION.) Therefore, $a_n = 0$ for all n.

Because the SINES and COSINES form a COMPLETE ORTHOGONAL BASIS, the SUPERPOSITION PRINCIPLE holds, and the Fourier series of a LINEAR COMBINATION of two functions is the same as the LINEAR COMBINATION of the corresponding two series. The COEFFICIENTS for Fourier series expansions for a few common functions are given in Beyer (1987, pp. 411–12) and Byerly (1959, p. 51).

The notion of a Fourier series can also be extended to COMPLEX COEFFICIENTS. Consider a real-valued function $f(x)$. Write

$$f(x) = \sum_{n=-\infty}^{\infty} A_n e^{inx}. \qquad (19)$$

Now examine

$$\int_{-\pi}^{\pi} f(x) e^{-imx}\, dx = \int_{-\pi}^{\pi} \left(\sum_{n=-\infty}^{\infty} A_n e^{inx} \right) e^{-imx}\, dx$$

$$= \sum_{n=-\infty}^{\infty} A_n \int_{-\pi}^{\pi} e^{i(n-m)x}\, dx$$

$$= \sum_{n=-\infty}^{\infty} A_n \int_{-\pi}^{\pi} \{\cos[(n-m)x] + i \sin[(n-m)x]\}\, dx$$

$$= \sum_{m=-\infty}^{\infty} A_n 2\pi \delta_{mn} = 2\pi A_m, \qquad (20)$$

so

$$A_n = \frac{1}{2\pi} \int_{-\pi}^{\pi} f(x) e^{-inx}\, dx. \qquad (21)$$

The COEFFICIENTS can be expressed in terms of those in the FOURIER SERIES

$$A_n = \frac{1}{2\pi} \int_{-\pi}^{\pi} f(x)[\cos(nx) - i \sin(nx)]\, dx$$

$$= \begin{cases} \dfrac{1}{2\pi} \displaystyle\int_{-\pi}^{\pi} f(x)[\cos(nx) + i \sin(nx)]\, dx & n < 0 \\[2mm] \dfrac{1}{2\pi} \displaystyle\int_{-\pi}^{\pi} f(x)\, dx & n = 0 \\[2mm] \dfrac{1}{2\pi} \displaystyle\int_{-\pi}^{\pi} f(x)[\cos(nx) - i \sin(nx)]\, dx & n > 0 \end{cases}$$

$$= \begin{cases} \tfrac{1}{2}(a_n + ib_n) & \text{for } n < 0 \\ \tfrac{1}{2}a_0 & \text{for } n = 0 \\ \tfrac{1}{2}(a_n - ib_n) & \text{for } n > 0 \end{cases} \qquad (22)$$

For a function periodic in $[-L/2,\ L/2]$, these become

$$f(x) = \sum_{n=-\infty}^{\infty} A_n e^{i(2\pi nx/L)} \qquad (23)$$

$$A_n = \frac{1}{L} \int_{-L/2}^{L/2} f(x) e^{-i(2\pi nx/L)}\, dx. \qquad (24)$$

These equations are the basis for the extremely important FOURIER TRANSFORM, which is obtained by transforming A_n from a discrete variable to a continuous one as the length $L \to \infty$.

See also DIRICHLET FOURIER SERIES CONDITIONS, FOURIER COSINE SERIES, FOURIER SINE SERIES, FOURIER TRANSFORM, GIBBS PHENOMENON, LEBESGUE CONSTANTS (FOURIER SERIES), LEGENDRE SERIES, RIESZ-FISCHER THEOREM, SCHLÖMILCH'S SERIES

References

Arfken, G. "Fourier Series." Ch. 14 in *Mathematical Methods for Physicists, 3rd ed.* Orlando, FL: Academic Press, pp. 760–93, 1985.

Beyer, W. H. (Ed.). *CRC Standard Mathematical Tables, 28th ed.* Boca Raton, FL: CRC Press, 1987.

Brown, J. W. and Churchill, R. V. *Fourier Series and Boundary Value Problems, 5th ed.* New York: McGraw-Hill, 1993.

Byerly, W. E. *An Elementary Treatise on Fourier's Series, and Spherical, Cylindrical, and Ellipsoidal Harmonics, with Applications to Problems in Mathematical Physics.* New York: Dover, 1959.

Carslaw, H. S. *Introduction to the Theory of Fourier's Series and Integrals, 3rd ed., rev. and enl.* New York: Dover, 1950.

Davis, H. F. *Fourier Series and Orthogonal Functions.* New York: Dover, 1963.

Dym, H. and McKean, H. P. *Fourier Series and Integrals.* New York: Academic Press, 1972.

Folland, G. B. *Fourier Analysis and Its Applications.* Pacific Grove, CA: Brooks/Cole, 1992.

Groemer, H. *Geometric Applications of Fourier Series and Spherical Harmonics.* New York: Cambridge University Press, 1996.

Körner, T. W. *Fourier Analysis.* Cambridge, England: Cambridge University Press, 1988.

Körner, T. W. *Exercises for Fourier Analysis.* New York: Cambridge University Press, 1993.

Krantz, S. G. "Fourier Series." §15.1 in *Handbook of Complex Analysis.* Boston, MA: Birkhäuser, pp. 195–02, 1999.

Lighthill, M. J. *Introduction to Fourier Analysis and Generalised Functions.* Cambridge, England: Cambridge University Press, 1958.

Morrison, N. *Introduction to Fourier Analysis.* New York: Wiley, 1994.

Sansone, G. "Expansions in Fourier Series." Ch. 2 in *Orthogonal Functions, rev. English ed.* New York: Dover, pp. 39–68, 1991.

Weisstein, E. W. "Books about Fourier Transforms." http://www.treasure-troves.com/books/FourierTransforms.html.

Whittaker, E. T. and Robinson, G. "Practical Fourier Analysis." Ch. 10 in *The Calculus of Observations: A Treatise on Numerical Mathematics, 4th ed.* New York: Dover, pp. 260–84, 1967.

Fourier Series—Power Series

For $f(x) = x^k$ on the INTERVAL $[-L, L]$ and periodic with period $2L$, the FOURIER SERIES is given by

$$a_n = \frac{1}{L} \int_{-L}^{L} x^k \cos\left(\frac{n\pi x}{L}\right) dx$$

$$= \frac{2L^k}{1+k} {}_1F_2\left(\begin{array}{c} 1+\frac{1}{2}k \\ \frac{1}{2} \quad \frac{1}{2}(3+k) \end{array}; -\frac{1}{4}\pi^2 n^2\right)$$

$$b_n = \frac{1}{L} \int_{-L}^{L} x^k \sin\left(\frac{n\pi x}{L}\right) dx$$

$$= \frac{2n\pi L^k}{2+k} {}_1F_2\left(\begin{array}{c} 1+\frac{1}{2}k \\ \frac{3}{2} \quad 2+\frac{1}{2}k \end{array}; -\frac{1}{4}\pi^2 n^2\right),$$

where ${}_1F_2(a; b, c; x)$ is a generalized HYPERGEOMETRIC FUNCTION.

Fourier Series—Sawtooth Wave

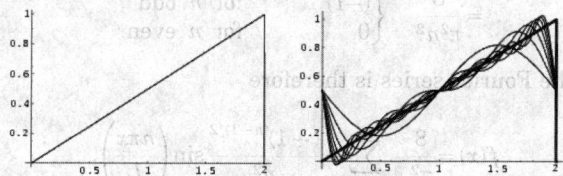

Consider a string of length $2L$ plucked at the right end, then

$$a_0 = \frac{1}{L} \int_0^{2L} \frac{x}{2L} dx = \frac{1}{2L^2}\left[\frac{1}{2}x^2\right]_0^L = \frac{1}{4L^2}(2L)^2 = 1$$

$$a_n = \frac{1}{L} \int_0^{2L} \frac{x}{2L} \cos\left(\frac{n\pi x}{L}\right) dx$$

$$= \frac{[2n\pi \cos(n\pi) - \sin(n\pi)] \sin(n\pi)}{n^2 \pi^2} = 0$$

$$b_n = \frac{1}{L} \int_0^{2L} \frac{x}{2L} \sin\left(\frac{n\pi x}{L}\right) dx$$

$$= \frac{-2n\pi \cos(2n\pi) + \sin(2n\pi)}{2n^2 \pi^2} = -\frac{1}{n\pi}.$$

The Fourier series is therefore

$$f(x) = \frac{1}{2} - \frac{1}{\pi} \sum_{n=1}^{\infty} \frac{1}{n} \sin\left(\frac{n\pi x}{L}\right).$$

See also FOURIER SERIES, FOURIER SERIES–SQUARE WAVE, SAWTOOTH WAVE

Fourier Series—Square Wave

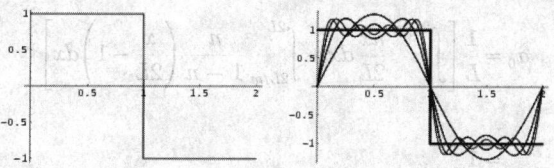

Consider a square wave of length $2L$. Since the function is ODD, $a_0 = a_n = 0$, and

$$b_n = \frac{2}{L} \int_0^L \sin\left(\frac{n\pi x}{L}\right) dx$$

$$= \frac{4}{n\pi} \sin^2(\tfrac{1}{2}n\pi) = \frac{4}{n\pi} \begin{cases} 0 & n \text{ even} \\ 1 & n \text{ odd} \end{cases}.$$

The Fourier series is therefore

$$f(x) = \frac{4}{\pi} \sum_{n=1,3,5,\dots}^{\infty} \frac{1}{n} \sin\left(\frac{n\pi x}{L}\right).$$

See also FOURIER SERIES, FOURIER SERIES–SAWTOOTH WAVE, SQUARE WAVE

Fourier Series—Triangle

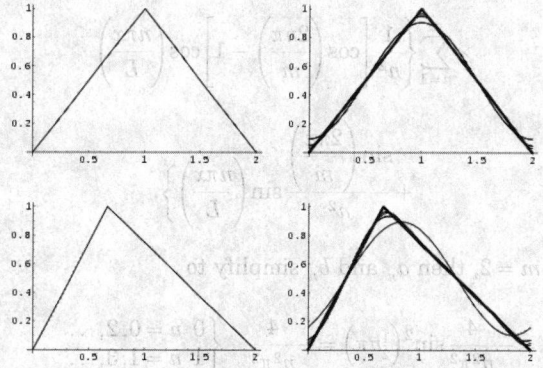

Let a string of length $2L$ have a y-displacement of unity when it is pinned an x-distance which is $((1/m))$ th of the way along the string. The displacement as a function of x is then

$$f_m(x) = \begin{cases} \dfrac{mx}{2L} & 0 \leq x \leq \dfrac{2L}{m} \\ \dfrac{m}{1-m}\left(\dfrac{x}{2L} - 1\right) & \dfrac{2L}{m} \leq x \leq 2L. \end{cases}$$

The COEFFICIENTS are therefore

$$a_0 = \frac{1}{L}\left[\int_0^{2L/m} \frac{nx}{2L}dx + \int_{2L/m}^{2L} \frac{n}{1-n}\left(\frac{x}{2L}-1\right)dx\right]$$

$$= 1$$

$$a_n = \frac{m\left[1 - m - \cos(2\pi n) + m\cos\left(\dfrac{2n\pi}{m}\right)\right]}{2(m-1)n^2\pi^2}$$

$$= \frac{m^2\left[\cos\left(\dfrac{2n\pi}{m}\right) - 1\right]}{2(m-1)m^2\pi^2}$$

$$b_n = \frac{m\left[m\sin\left(\dfrac{2\pi n}{m}\right) - \sin(2\pi n)\right]}{2(m-1)n^2\pi^2}$$

$$= \frac{m^2 \sin\left(\dfrac{2\pi n}{m}\right)}{2(m-1)n^2\pi^2}.$$

The Fourier series is therefore

$$f_m(x) = \tfrac{1}{2} + \frac{m^2}{2(m-1)\pi^2}$$

$$\times \sum_{n=1}^{\infty}\left\{\frac{1}{n^2}\left[\cos\left(\frac{2n\pi}{m}\right) - 1\right]\cos\left(\frac{n\pi x}{L}\right)\right.$$

$$\left. + \frac{\sin\left(\dfrac{2\pi n}{m}\right)}{n^2}\sin\left(\frac{n\pi x}{L}\right)\right\}.$$

If $m = 2$, then a_n and b_n simplify to

$$a_n = -\frac{4}{n^2\pi^2}\sin^2\left(\tfrac{1}{2}n\pi\right) = -\frac{4}{n^2\pi^2}\begin{cases} 0 & n = 0, 2, \dots \\ 1 & n = 1, 3, \dots \end{cases}$$

$$b_n = 0,$$

giving

$$f_2(x) = \tfrac{1}{2} - \frac{4}{\pi^2}\sum_{n=1,3,\dots}^{\infty}\frac{1}{n^2}\cos\left(\frac{n\pi x}{L}\right).$$

See also FOURIER SERIES

Fourier Series—Triangle Wave

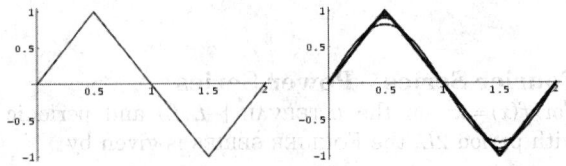

Consider a triangle wave of length $2L$. Since the function is ODD, $a_0 = a_n = 0$, and

$$b_n = \frac{2}{L}\left\{\int_0^{L/2}\frac{x}{L/2}\sin\left(\frac{n\pi x}{L}\right)dx\right.$$

$$\left. + \int_{L/2}^0\left[1 - \frac{2}{L}\left(x - \frac{1}{2}L\right)\right]\sin\left(\frac{n\pi x}{L}\right)dx\right\}dx$$

$$= \frac{32}{\pi^2 n^2}\cos\left(\frac{1}{4}n\pi\right)\sin^3\left(\frac{1}{4}n\pi\right)$$

$$= \frac{32}{\pi^2 n^2}\begin{cases} 0 & n = 0, 4, \dots \\ \frac{1}{4} & n = 1, 5, \dots \\ 0 & n = 2, 6, \dots \\ -\frac{1}{4} & n = 3, 7, \dots \end{cases}$$

$$= \frac{8}{\pi^2 n^2}\begin{cases} (-1)^{(n-1)/2} & \text{for } n \text{ odd} \\ 0 & \text{for } n \text{ even.} \end{cases}$$

The Fourier series is therefore

$$f(x) = \frac{8}{\pi^2}\sum_{n=1,3,5,\dots}^{\infty}\frac{(-1)^{(n-1)/2}}{n^2}\sin\left(\frac{n\pi x}{L}\right).$$

See also FOURIER SERIES

Fourier Sine Series

If $f(x)$ is an ODD FUNCTION, then $a_n = 0$ and the FOURIER SERIES collapses to

$$f(x) = \sum_{n=1}^{\infty} b_n \sin(nx), \tag{1}$$

where

$$b_n = \frac{1}{\pi}\int_{-\pi}^{\pi} f(x)\sin(nx)\,dx = \frac{2}{\pi}\int_0^{\pi} f(x)\sin(nx)\,dx \tag{2}$$

for $n = 1, 2, 3, \dots$. The last EQUALITY is true because

$$f(x)\sin(nx) = [-f(-x)][-\sin(-nx)]$$

$$= f(-x)\sin(-nx). \tag{3}$$

Letting the range go to L,

$$b_n = \frac{2}{L} \int_0^L f(x) \sin\left(\frac{n\pi x}{L}\right) dx. \tag{4}$$

See also FOURIER COSINE SERIES, FOURIER SERIES, FOURIER SINE TRANSFORM

Fourier Sine Transform

The Fourier sine transform is the IMAGINARY PART of the full complex FOURIER TRANSFORM,

$$\mathscr{F}_s[f(x)] = \Im[\mathscr{F}[f(x)]].$$

In *Mathematica* 4.0, the Fourier sine transform $F_s(k)$ of a function $f(x)$ is implemented as FourierSin-Transform[f, x, k], and different choices of a and b can be used by passing the optional FourierPara-meters->{a, b} option. In this work, $a = 0$ and $b = -2\pi$.

In version 4.1, the discrete Fourier sine transform of a list l of real numbers can be computed using FourierSin[l] in the *Mathematica* add-on package LinearAlgebra`FourierTrig` (which can be loaded with the command < <LinearAlgebra`).

See also FOURIER COSINE TRANSFORM, FOURIER TRANSFORM

References
Press, W. H.; Flannery, B. P.; Teukolsky, S. A.; and Vetterling, W. T. "FFT of Real Functions, Sine and Cosine Transforms." §12.3 in *Numerical Recipes in FORTRAN: The Art of Scientific Computing, 2nd ed.* Cambridge, England: Cambridge University Press, pp. 504–15, 1992.

Fourier Transform

The Fourier transform is a generalization of the COMPLEX FOURIER SERIES in the limit as $L \to \infty$. Replace the discrete A_n with the continuous $F(k)dk$ while letting $n/L \to k$. Then change the sum to an INTEGRAL, and the equations become

$$f(x) = \int_{-\infty}^{\infty} F(k)e^{2\pi i k x} dk \tag{1}$$

$$F(k) = \int_{-\infty}^{\infty} f(x)e^{-2\pi i k x} dx \tag{2}$$

Here,

$$F(k) = \mathscr{F}[f(x)] = \int_{-\infty}^{\infty} f(x)e^{-2\pi i k x} dx \tag{3}$$

is called the *forward* $(-i)$ Fourier transform, and

$$f(x) = \mathscr{F}^{-1}[F(k)] = \int_{-\infty}^{\infty} F(k)e^{2\pi i k x} dk \tag{4}$$

is called the *inverse* $(+i)$ Fourier transform. The notation $f^\wedge(k)$ and $f^\vee(x)$ are sometimes used for the

Fourier transform and inverse Fourier transform, respectively (Krantz 1999, p. 202).

Note that some authors (especially physicists) prefer to write the transform in terms of angular frequency $\omega \equiv 2\pi\nu$ instead of the oscillation frequency ν. However, this destroys the symmetry, resulting in the transform pair

$$H(\omega) = \mathscr{F}[h(t)] = \int_{-\infty}^{\infty} h(t)e^{-i\omega t} dt \tag{5}$$

$$h(t) = \mathscr{F}^{-1}[H(\omega)] = \frac{1}{2\pi} \int_{-\infty}^{\infty} H(\omega)e^{i\omega t} d\omega. \tag{6}$$

To restore the symmetry of the transforms, the convention

$$g(y) = \mathscr{F}[f(t)] = \frac{1}{\sqrt{2\pi}} \int_{-\infty}^{\infty} f(t)e^{-iyt} dt \tag{7}$$

$$f(t) = \mathscr{F}^{-1}[g(y)] = \frac{1}{\sqrt{2\pi}} \int_{-\infty}^{\infty} g(y)e^{iyt} dy \tag{8}$$

is sometimes used (Mathews and Walker 1970, p. 102). In general, the Fourier transform pair may be defined using two arbitrary constants a and b as

$$F(\omega) = \sqrt{\frac{|b|}{(2\pi)^{1-a}}} \int_{-\infty}^{\infty} f(t)e^{ib\omega t} dt \tag{9}$$

$$f(t) = \sqrt{\frac{|b|}{(2\pi)^{1+a}}} \int_{-\infty}^{\infty} F(\omega)e^{-ib\omega t} dw. \tag{10}$$

In *Mathematica* 4.0, the Fourier transform $F(k)$ of a function $f(x)$ is implemented as FourierTrans-form[f, x, k], and different choices of a and b can be used by passing the optional FourierPara-meters->{a, b} option. By default, *Mathematica* takes FourierParameters as $(0, 1)$. Unfortunately, a number of other conventions are in widespread use. For example, $(0, 1)$ is used in modern physics, $(1, -1)$ is used in pure mathematics and systems engineering, $(1, 1)$ is used in probability theory for the computation of the CHARACTERISTIC FUNCTION, $(-1, 1)$ is used in classical physics, and $(0, -2\pi)$ is used in signal processing. In this work, following Bracewell (1999, pp. 6–), *it is always assumed that* $a = 0$ and $b = -2\pi$ unless otherwise stated. This choice often results in greatly simplified transforms of common functions such as 1, $\cos(2\pi k_0 x)$, etc.

Since any function can be split up into EVEN and ODD portions $E(x)$ and $O(x)$,

$$f(x) = \tfrac{1}{2}[f(x) + f(-x)] + \tfrac{1}{2}[f(x) - f(-x)] = E(x) + O(x), \tag{11}$$

a Fourier transform can always be expressed in terms of the FOURIER COSINE TRANSFORM and FOURIER SINE TRANSFORM as

$$\mathscr{F}[f(x)] = \int_{-\infty}^{\infty} E(x)\cos(2\pi kx)dx$$

$$-i\int_{-\infty}^{\infty} O(x)\sin(2\pi kx)dx. \qquad (12)$$

A function $f(x)$ has a forward and inverse Fourier transform such that

$$f(x) = \begin{cases} \int_{-\infty}^{\infty} e^{2\pi ikx}\left[\int_{-\infty}^{\infty} f(x)e^{-2\pi ikx}dx\right]dk \\ \quad \text{for } f(x) \text{ continuous at } x \\ \frac{1}{2}[f(x_+) + f(x_-)] \\ \quad \text{for } f(x) \text{ discontinous at } x, \end{cases} \qquad (13)$$

provided that

1. $\int_{-\infty}^{\infty} |f(x)|dx$ exists.
2. There are a finite number of discontinuities.
3. The function has bounded variation. A SUFFICIENT weaker condition is fulfillment of the LIPSCHITZ CONDITION

(Ramirez 1985, p. 29). The smoother a function (i.e., the larger the number of continuous DERIVATIVES), the more compact its Fourier transform.

The Fourier transform is linear, since if $f(x)$ and $g(x)$ have Fourier transforms $F(k)$ and $G(k)$, then

$$\int_{-\infty}^{\infty} [af(x) + bg(x)]e^{-2\pi ikx}dx$$

$$= a\int_{-\infty}^{\infty} f(x)e^{-2\pi ikx}dx + b\int_{-\infty}^{\infty} g(x)e^{-2\pi ikx}dx$$

$$aF(k) + bG(k). \qquad (14)$$

Therefore,

$$\mathscr{F}[af(x) + bg(x)] = a\mathscr{F}[f(x)] + b\mathscr{F}[g(x)]$$
$$= aF(k) + bG(k). \qquad (15)$$

The Fourier transform is also symmetric since $F(k) = \mathscr{F}[f(x)]$ implies $F(-k) = \mathscr{F}[f(-x)]$.

Let $f * g$ denote the CONVOLUTION, then the transforms of convolutions of functions have particularly nice transforms,

$$\mathscr{F}(f * g) = \mathscr{F}[f]\mathscr{F}[g] \qquad (16)$$

$$\mathscr{F}[fg] = \mathscr{F}[f] * \mathscr{F}[g] \qquad (17)$$

$$\mathscr{F}^{-1}[\mathscr{F}(f)\mathscr{F}(g)] = f * g \qquad (18)$$

$$\mathscr{F}^{-1}[\mathscr{F}(f) * \mathscr{F}(g)] = fg. \qquad (19)$$

The first of these is derived as follows:

$$\mathscr{F}[f * g] = \int_{-\infty}^{\infty}\int_{-\infty}^{\infty} e^{-2\pi ikx}f(x')g(x$$
$$-x')dx'dx$$

$$= \int_{-\infty}^{\infty}\int_{-\infty}^{\infty} [e^{-2\pi ikx'}f(x')dx']$$
$$\times [e^{-2\pi ik(x-x')}g(x-x')dx]$$

$$= \left[\int_{-\infty}^{\infty} e^{-2\pi ikx'}f(x')dx'\right]\left[\int_{-\infty}^{\infty} e^{-2\pi ikx''}g(x'')dx''\right]$$

$$= \mathscr{F}[f]\mathscr{F}[g], \qquad (20)$$

where $x'' \equiv x - x'$.

There is also a somewhat surprising and extremely important relationship between the AUTOCORRELATION and the Fourier transform known as the WIENER-KHINTCHINE THEOREM. Let $\mathscr{F}[f(x)] = F(k)$, and $\bar{f}$ denote the COMPLEX CONJUGATE of f, then the Fourier transform of the ABSOLUTE SQUARE of $F(k)$ is given by

$$\mathscr{F}[|F(k)|^2] = \int_{-\infty}^{\infty} \bar{f}(\tau)f(\tau+x)d\tau. \qquad (21)$$

The Fourier transform of a DERIVATIVE $f'(x)$ of a function $f(x)$ is simply related to the transform of the function $f(x)$ itself. Consider

$$\mathscr{F}[f'(x)] = \int_{-\infty}^{\infty} f'(x)e^{-2\pi ikx}dx. \qquad (22)$$

Now use INTEGRATION BY PARTS

$$\int vdu = [uv] - \int udv \qquad (23)$$

with

$$du = f'(x)dx \quad v = e^{-2\pi ikx} \qquad (24)$$
$$u = f(x) \quad dv = -2\pi ike^{-2\pi ikx}dx, \qquad (25)$$

then

$$\mathscr{F}[f'(x)] = [f(x)e^{-2\pi ikx}]_{-\infty}^{\infty} - \int_{-\infty}^{\infty} f(x)(-2\pi ike^{-2\pi ikx}dx). \qquad (26)$$

The first term consists of an oscillating function times $f(x)$. But if the function is bounded so that

$$\lim_{x \to \pm\infty} f(x) = 0 \qquad (27)$$

(as any physically significant signal must be), then the term vanishes, leaving

$$\mathscr{F}[f'(x)] = 2\pi ik\int_{-\infty}^{\infty} f(x)e^{-2\pi ikx}dx = 2\pi ikF[f(x)]. \qquad (28)$$

This process can be iterated for the nth DERIVATIVE to yield

$$F[f^{(n)}(x)] = (2\pi ik)^n F[f(x)]. \qquad (29)$$

The important MODULATION THEOREM of Fourier transforms allows $F[\cos(2\pi k_0 x)f(x)]$ to be expressed in terms of $\mathscr{F}[f(x)] = F(k)$ as follows,

$$\mathscr{F}[\cos(2\pi k_0 x)f(x)] \equiv \int_{-\infty}^{\infty} f(x)\cos(2\pi k_0 x)e^{-2\pi ikx}dx$$

$$= \frac{1}{2}\int_{-\infty}^{\infty} f(x)e^{2\pi ik_0 x}e^{-2\pi ikx}dx + \frac{1}{2}\int_{-\infty}^{\infty} f(x)e^{-2\pi ik_0 x}e^{-2\pi ikx}dx$$

$$= \frac{1}{2}\int_{-\infty}^{\infty} f(x)e^{-2\pi i(k-k_0)x}dx + \frac{1}{2}\int_{-\infty}^{\infty} f(x)e^{-2\pi i(k+k_0)x}dx$$

$$= \frac{1}{2}[F(k-k_0) + F(k+k_0)]. \tag{30}$$

Since the DERIVATIVE of the Fourier transform is given by

$$F'(k) \equiv \frac{d}{dk}\mathscr{F}[f(x)] = \int_{-\infty}^{\infty} (-2\pi ix)f(x)e^{-2\pi ikx}dx, \tag{31}$$

it follows that

$$F'(0) = -2\pi i\int_{-\infty}^{\infty} xf(x)dx. \tag{32}$$

Iterating gives the general FORMULA

$$\mu_n \equiv \int_{-\infty}^{\infty} x^n f(x)dx = \frac{F^{(n)}(0)}{(-2\pi i)^n}. \tag{33}$$

The VARIANCE of a FOURIER TRANSFORM is

$$\sigma_f^2 = \langle(xf - \langle xf\rangle)^2\rangle, \tag{34}$$

and it is true that

$$\sigma_{f+g} = \sigma_f + \sigma_g. \tag{35}$$

If $f(x)$ has the Fourier transform $F(k)$, then the Fourier transform has the shift property

$$\int_{-\infty}^{\infty} f(x-x_0)e^{-2\pi ikx}dx$$

$$= \int_{-\infty}^{\infty} f(x-x_0)e^{-2\pi i(x-x_0)k}e^{-2\pi i(kx_0)}d(x-x_0)$$

$$= e^{-2\pi ikx_0}F(k), \tag{36}$$

so $f(x-x_0)$ has the Fourier transform

$$F[f(x-x_0)] = e^{-2\pi ikx_0}F(k). \tag{37}$$

If $f(x)$ has a Fourier transform $F(k)$, then the Fourier transform obeys a similarity theorem.

$$\int_{-\infty}^{\infty} f(ax)e^{-2\pi ikx}dx = \frac{1}{|a|}\int_{-\infty}^{\infty} f(ax)e^{-2\pi i(ax)(k/a)}d(ax)$$

$$= \frac{1}{|a|}F\left(\frac{k}{a}\right), \tag{38}$$

so $f(ax)$ has the Fourier transform $|a|^{-1}F(k/a)$.

The "equivalent width" of a Fourier transform is

$$w_\varepsilon \equiv \frac{\int_{-\infty}^{\infty} f(x)dx}{f(0)} = \frac{F(0)}{\int_{-\infty}^{\infty} F(k)dk}. \tag{39}$$

The "autocorrelation width" is

$$w_a \equiv \frac{\int_{-\infty}^{\infty} f*\bar{f}dx}{[f*\bar{f}]_0} = \frac{\int_{-\infty}^{\infty} fdx\int_{-\infty}^{\infty} \bar{f}dx}{\int_{-\infty}^{\infty} f\bar{f}dx}, \tag{40}$$

where $f * g$ denotes the CROSS-CORRELATION of f and g and $\bar{f}$ is the COMPLEX CONJUGATE.

Any operation on $f(x)$ which leaves its AREA unchanged leaves $F(0)$ unchanged, since

$$\int_{-\infty}^{\infty} f(x)dx = F[f(0)] = f(0). \tag{41}$$

In 2-D, the Fourier transform becomes

$$F(x,y) = \int_{-\infty}^{\infty}\int_{-\infty}^{\infty} f(k_x,k_y)e^{-2\pi i(k_x x+k_y y)}dk_x dk_y \tag{42}$$

$$F(k_x,k_y) = \int_{-\infty}^{\infty}\int_{-\infty}^{\infty} f(x,y)e^{2\pi i(k_x x+k_y y)}dxdy. \tag{43}$$

Similarly, the n-D Fourier transform can be defined for $\mathbf{k}, \mathbf{x} \in \mathbb{R}^n$ by

$$F(\mathbf{x}) = \underbrace{\int_{-\infty}^{\infty}\cdots\int_{-\infty}^{\infty}}_{n} f(\mathbf{k})e^{-2\pi i\mathbf{k}\cdot\mathbf{x}}d^n\mathbf{k} \tag{44}$$

$$f(\mathbf{k}) = \underbrace{\int_{-\infty}^{\infty}\cdots\int_{-\infty}^{\infty}}_{n} F(\mathbf{x})e^{-2\pi i\mathbf{k}\cdot\mathbf{x}}d^n\mathbf{x}. \tag{45}$$

See also AUTOCORRELATION, CONVOLUTION, DISCRETE FOURIER TRANSFORM, FAST FOURIER TRANSFORM, FOURIER SERIES, FOURIER-STIELTJES TRANSFORM, FOURIER TRANSFORM–1, FOURIER TRANSFORM–COSINE, FOURIER TRANSFORM–DELTA FUNCTION, FOURIER TRANSFORM–EXPONENTIAL FUNCTION, FOURIER TRANSFORM–GAUSSIAN, FOURIER TRANSFORM–HEAVISIDE STEP FUNCTION, FOURIER TRANSFORM–INVERSE FUNCTION, FOURIER TRANSFORM–LORENTZIAN FUNCTION, FOURIER TRANSFORM–RAMP FUNCTION, FOURIER TRANSFORM–RECTANGLE FUNCTION, HANKEL TRANSFORM, HARTLEY TRANSFORM, INTEGRAL TRANSFORM, LAPLACE TRANSFORM, STRUCTURE FACTOR, WINOGRAD TRANSFORM

References

Arfken, G. "Development of the Fourier Integral," "Fourier Transforms--Inversion Theorem," and "Fourier Transform of Derivatives." §15.2–5.4 in *Mathematical Methods for*

Physicists, 3rd ed. Orlando, FL: Academic Press, pp. 794–10, 1985.

Blackman, R. B. and Tukey, J. W. *The Measurement of Power Spectra, From the Point of View of Communications Engineering.* New York: Dover, 1959.

Bracewell, R. *The Fourier Transform and Its Applications, 3rd ed.* New York: McGraw-Hill, 1999.

Brigham, E. O. *The Fast Fourier Transform and Applications.* Englewood Cliffs, NJ: Prentice Hall, 1988.

Folland, G. B. *Real Analysis: Modern Techniques and their Applications, 2nd ed.* New York: Wiley, 1999.

James, J. F. *A Student's Guide to Fourier Transforms with Applications in Physics and Engineering.* New York: Cambridge University Press, 1995.

Körner, T. W. *Fourier Analysis.* Cambridge, England: Cambridge University Press, 1988.

Krantz, S. G. "The Fourier Transform." §15.2 in *Handbook of Complex Analysis.* Boston, MA: Birkhäuser, pp. 202–12, 1999.

Mathews, J. and Walker, R. L. *Mathematical Methods of Physics, 2nd ed.* Reading, MA: W. A. Benjamin/Addison-Wesley, 1970.

Morrison, N. *Introduction to Fourier Analysis.* New York: Wiley, 1994.

Morse, P. M. and Feshbach, H. "Fourier Transforms." §4.8 in *Methods of Theoretical Physics, Part I.* New York: McGraw-Hill, pp. 453–71, 1953.

Oberhettinger, F. *Fourier Transforms of Distributions and Their Inverses: A Collection of Tables.* New York: Academic Press, 1973.

Papoulis, A. *The Fourier Integral and Its Applications.* New York: McGraw-Hill, 1962.

Press, W. H.; Flannery, B. P.; Teukolsky, S. A.; and Vetterling, W. T. *Numerical Recipes in C: The Art of Scientific Computing.* Cambridge, England: Cambridge University Press, 1989.

Ramirez, R. W. *The FFT: Fundamentals and Concepts.* Englewood Cliffs, NJ: Prentice-Hall, 1985.

Sansone, G. "The Fourier Transform." §2.13 in *Orthogonal Functions, rev. English ed.* New York: Dover, pp. 158–68, 1991.

Sneddon, I. N. *Fourier Transforms.* New York: Dover, 1995.

Sogge, C. D. *Fourier Integrals in Classical Analysis.* New York: Cambridge University Press, 1993.

Spiegel, M. R. *Theory and Problems of Fourier Analysis with Applications to Boundary Value Problems.* New York: McGraw-Hill, 1974.

Stein, E. M. and Weiss, G. L. *Introduction to Fourier Analysis on Euclidean Spaces.* Princeton, NJ: Princeton University Press, 1971.

Strichartz, R. *Fourier Transforms and Distribution Theory.* Boca Raton, FL: CRC Press, 1993.

Titchmarsh, E. C. *Introduction to the Theory of Fourier Integrals, 3rd ed.* Oxford, England: Clarendon Press, 1948.

Tolstov, G. P. *Fourier Series.* New York: Dover, 1976.

Walker, J. S. *Fast Fourier Transforms, 2nd ed.* Boca Raton, FL: CRC Press, 1996.

Weisstein, E. W. "Books about Fourier Transforms." http://www.treasure-troves.com/books/FourierTransforms.html.

Fourier Transform—1

The FOURIER TRANSFORM of the CONSTANT FUNCTION $f(x) = 1$ is given by

$$\mathscr{F}[1] = \int_{-\infty}^{\infty} e^{-2\pi i k x} dx = \delta(k),$$

according to the definition of the DELTA FUNCTION.

See also DELTA FUNCTION, FOURIER TRANSFORM

Fourier Transform—Cosine

$$\mathscr{F}[\cos(2\pi k_0 x)] = \int_{-\infty}^{\infty} e^{2\pi i k x} \left(\frac{e^{2\pi i k_0 x} + e^{-2\pi i k_0 x}}{2} \right) dx$$

$$= \frac{1}{2} \int_{-\infty}^{\infty} \left[e^{-2\pi i (k - k_0) x} + e^{-2\pi i (k + k_0) x} \right] dx$$

$$= \frac{1}{2} [\delta(k - k_0) + \delta(k + k_0)],$$

where $\delta(x)$ is the DELTA FUNCTION.

See also COSINE, FOURIER TRANSFORM, FOURIER TRANSFORM–SINE

Fourier Transform—Delta Function

The FOURIER TRANSFORM of the DELTA FUNCTION is given by

$$\mathscr{F}[\delta(x - x_0)] = \int_{-\infty}^{\infty} \delta(x - x_0) e^{-2\pi i k x} dx = e^{-2\pi i k x_0}.$$

See also DELTA FUNCTION, FOURIER TRANSFORM

Fourier Transform—Exponential Function

The FOURIER TRANSFORM of $e^{-k_0 |x|}$ is given by

$$\mathscr{F}\left[e^{-k_0 |x|}\right] = \int_{-\infty}^{\infty} e^{-k_0 |x|} e^{-2\pi i k x} dx$$

$$= \int_{-\infty}^{0} e^{-2\pi i k x} e^{2\pi x k_0} dx + \int_{0}^{\infty} e^{-2\pi i k x} e^{-2\pi k_0 x} dx$$

$$= \int_{-\infty}^{0} [\cos(2\pi k x) - i \sin(2kx)] e^{2\pi k_0 x} dx.$$

$$+ \int_{0}^{\infty} [\cos(2\pi k x) - i \sin(2\pi k x)] e^{-2\pi k_0 x} dx. \tag{1}$$

Now let $u \equiv -x$ so $du = -dx$, then

$$\mathscr{F}\left[e^{-k_0 |x|}\right] = \int_{0}^{\infty} [\cos(2\pi k u) + i \sin(2\pi k u)] e^{-2\pi k_0 u} du$$

$$+ \int_{0}^{\infty} [\cos(2\pi k u) - i \sin(2\pi k u)] e^{-2\pi k_0 u} du$$

$$= 2 \int_{0}^{\infty} \cos(2\pi k u) e^{-2\pi k_0 u} du, \tag{2}$$

which, from the DAMPED EXPONENTIAL COSINE INTEGRAL, gives

$$\mathcal{F}\left[e^{-2\pi k_0 |x|}\right] = \frac{1}{\pi}\frac{k_0}{k^2 + k_0^2}, \tag{3}$$

which is a LORENTZIAN FUNCTION.

See also DAMPED EXPONENTIAL COSINE INTEGRAL, EXPONENTIAL FUNCTION, FOURIER TRANSFORM, LORENTZIAN FUNCTION

Fourier Transform—Gaussian

The FOURIER TRANSFORM of a GAUSSIAN FUNCTION $f(x) \equiv e^{-ax^2}$ is given by

$$F(k) = \int_{-\infty}^{\infty} e^{-ax^2} e^{-2\pi ikx} dx$$

$$= \int_{-\infty}^{\infty} e^{-ax^2} [\cos(2\pi kx) - i\sin(2\pi ix)] dx$$

$$= \int_{-\infty}^{\infty} e^{-ax^2}\cos(2\pi kx)dx - i\int_{-\infty}^{\infty} e^{-ax^2}\sin(2\pi kx)dx.$$

The second integrand is ODD, so integration over a symmetrical range gives 0. The value of the first integral is given by Abramowitz and Stegun (1972, p. 302, equation 7.4.6), so

$$F(k) = \sqrt{\frac{\pi}{a}} e^{-\pi^2 k^2 / a},$$

and a GAUSSIAN transforms to a GAUSSIAN.

See also GAUSSIAN FUNCTION, FOURIER TRANSFORM

References

Abramowitz, M. and Stegun, C. A. (Eds.). *Handbook of Mathematical Functions with Formulas, Graphs, and Mathematical Tables, 9th printing.* New York: Dover, p. 302, 1972.

Fourier Transform—Heaviside Step Function

The FOURIER TRANSFORM of the HEAVISIDE STEP FUNCTION $H(x)$ is given by

$$\mathcal{F}[H(x)] = \int_{-\infty}^{\infty} e^{-2\pi ikx} H(x)dx = \frac{1}{2}\left[\delta(k) - \frac{i}{\pi k}\right],$$

where $\delta(k)$ is the DELTA FUNCTION.

See also FOURIER TRANSFORM, HEAVISIDE STEP FUNCTION

Fourier Transform—Inverse Function

The FOURIER TRANSFORM of the GENERALIZED FUNCTION $1/x$ is given by

$$\mathcal{F}\left(-PV\frac{1}{\pi x}\right) = -\frac{1}{\pi}PV\int_{-\infty}^{\infty}\frac{e^{-2\pi ikx}}{x}dx \tag{1}$$

$$= PV\int_{-\infty}^{\infty}\frac{\cos(2\pi kx) - i\sin(2\pi kx)}{x}dx \tag{2}$$

$$= \begin{cases} -\dfrac{2i}{\pi}\displaystyle\int_0^{\infty}\dfrac{\sin(2\pi kx)}{x}dx & \text{for } k < 0 \\[2ex] \dfrac{2i}{\pi}\displaystyle\int_0^{\infty}\dfrac{\sin(2\pi kx)}{x}dx & \text{for } k > 0 \end{cases} \tag{3}$$

$$= \begin{cases} -i & \text{for } k < 0 \\ i & \text{for } k > 0, \end{cases} \tag{4}$$

where PV denotes the CAUCHY PRINCIPAL VALUE. Equation (4) can also be written as the single equation

$$\mathcal{F}\left(-PV\frac{i}{\pi x}\right) = i[1 - 2H(-k)], \tag{5}$$

where $H(x)$ is the HEAVISIDE STEP FUNCTION. The integrals follow from the identity

$$\int_0^{\infty}\frac{\sin(2\pi kx)}{x}dx = \int_0^{\infty}\frac{\sin(2\pi kx)}{2\pi kx}d(2\pi kx)$$

$$= \int_0^{\infty}\text{sinc}\, z\, dz = \frac{1}{2}\pi. \tag{6}$$

See also FOURIER TRANSFORM

Fourier Transform—Lorentzian Function

$$\mathcal{F}\left[\frac{1}{\pi}\frac{\frac{1}{2}\Gamma}{(x - x_0)^2 + \left(\frac{1}{2}\Gamma\right)^2}\right] = e^{-2\pi ikx_0 - \Gamma\pi|k|}.$$

This transform arises in the computation of the CHARACTERISTIC FUNCTION of the CAUCHY DISTRIBUTION.

See also FOURIER TRANSFORM, LORENTZIAN FUNCTION

Fourier Transform—Ramp Function

Let $R(x)$ be the RAMP FUNCTION, then the FOURIER TRANSFORM of $R(x)$ is given by

$$\mathcal{F}[R(x)] = \int_{-\infty}^{\infty} e^{-2\pi ikx} R(x)dx = \pi i\delta'(2\pi k) - \frac{1}{4\pi^2 k^2},$$

where $\delta'(x)$ is the DERIVATIVE of the DELTA FUNCTION.

See also RAMP FUNCTION

Fourier Transform—Rectangle Function

Let $\Pi(x)$ be the RECTANGLE FUNCTION, then the FOURIER TRANSFORM is

$$\mathscr{F}[\Pi(x)] = \mathrm{sinc}(\pi k),$$

where $\mathrm{sinc}(x)$ is the SINC FUNCTION.

See also FOURIER TRANSFORM, RECTANGLE FUNCTION, SINC FUNCTION

Fourier Transform—Sine

$$\mathscr{F}[\sin(2\pi k_0 x)] = \int_{-\infty}^{\infty} e^{-2\pi i k x}\left(\frac{e^{2\pi i k_0 x} - e^{-2\pi i k_0 x}}{2i}\right)dx$$

$$= \tfrac{1}{2}i\int_{-\infty}^{\infty}\left[-e^{-2\pi i(k-k_0)x} + e^{-2\pi i(k+k_0)x}\right]dt$$

$$= \tfrac{1}{2}i[\delta(k+k_0) - \delta(k-k_0)],$$

where $\delta(x)$ is the DELTA FUNCTION.

See also FOURIER TRANSFORM, FOURIER TRANSFORM–COSINE, SINE

Fourier-Bessel Series

BESSEL FUNCTION FOURIER EXPANSION, SCHLÖMILCH'S SERIES

Fourier-Bessel Transform

HANKEL TRANSFORM

Fourier-Budan Theorem

For any real α and β such that $\beta > \alpha$, let $p(\alpha) \neq 0$ and $p(\beta) \neq 0$ be real polynomials of degree n, and $v(x)$ denote the number of sign changes in the sequence $\{p(x), p'(x), ..., p^{(n)}(x)\}$. Then the number of zeros in the interval $[\alpha, \beta]$ (each zero counted with proper multiplicity) equals $v(\alpha) - v(\beta)$ minus an even non-negative integer.

References
Henrici, P. *Applied and Computational Complex Analysis, Vol. 1: Power Series-Integration-Conformal Mapping-Location of Zeros.* New York: Wiley, p. 443, 1988.

Fourier-Mellin Integral

The inverse of the LAPLACE TRANSFORM

$$\mathrm{F}(t) = \mathscr{L}^{-1}[f(s)] = \frac{1}{2\pi i}\int_{\gamma-i\infty}^{\gamma+i\infty} e^{st}f(s)ds$$

$$f(s) = \mathscr{L}[\mathrm{F}(t)] = \int_{0}^{\infty}\mathrm{F}(t)e^{-st}dt.$$

which is a LORENTZIAN FUNCTION.

See also BROMWICH INTEGRAL, LAPLACE TRANSFORM

Fourier-Stieltjes Transform

Let $f(x)$ be a positive definite, measurable function on the INTERVAL $(-\infty, \infty)$. Then there exists a monotone increasing, real-valued bounded function $\alpha(t)$ such that

$$f(x) = \int_{-\infty}^{\infty} e^{itx}d\alpha(t)$$

for "ALMOST ALL" x. If $\alpha(t)$ is nondecreasing and bounded and $f(x)$ is defined as above, then $f(x)$ is called the Fourier-Stieltjes transform of $\alpha(t)$, and is both continuous and positive definite.

See also FOURIER TRANSFORM, LAPLACE TRANSFORM

References
Iyanaga, S. and Kawada, Y. (Eds.). *Encyclopedic Dictionary of Mathematics.* Cambridge, MA: MIT Press, p. 618, 1980.

Four-Knot

FIGURE-OF-EIGHT KNOT

Four-Square Theorem

LAGRANGE'S FOUR-SQUARE THEOREM

Four-Vector

A four-element vector

$$a^\mu = \begin{bmatrix} a^0 \\ a^1 \\ a^2 \\ a^3 \end{bmatrix}, \tag{1}$$

which transforms under a LORENTZ TRANSFORMATION like the POSITION FOUR-VECTOR. This means it obeys

$$a'^\mu = \Lambda^\mu_v a^v \tag{2}$$

$$a_\mu \cdot b_\mu \equiv a_\mu b^\mu \tag{3}$$

$$a_\mu \cdot b^\mu = a'_\mu b'_\mu \tag{4}$$

where Λ^μ_μ is the LORENTZ TENSOR. Multiplication of two four-vectors with the METRIC $g_{\mu v}$ gives products OF THE FORM

$$g_{\mu v}x^\mu x^v = (x^0)^2 - (x^1)^2 - (x^2)^2 - (x^3)^2. \tag{5}$$

In the case of the POSITION FOUR-VECTOR, $x^0 = ct$ (where c is the speed of light) and this product is an invariant known as the spacetime interval.

See also GRADIENT FOUR-VECTOR, LORENTZ TRANSFORMATION, POSITION FOUR-VECTOR, QUATERNION, TENSOR, VECTOR

References

Morse, P. M. and Feshbach, H. "The Lorentz Transformation, Four-Vectors, Spinors." §1.7 in *Methods of Theoretical Physics, Part I.* New York: McGraw-Hill, pp. 93–07, 1953.

Four-Vertex Theorem

A closed embedded smooth PLANE CURVE has at least four vertices, where a vertex is defined as an extremum of CURVATURE.

See also CURVATURE

References

Tabachnikov, S. "The Four-Vertex Theorem Revisited--Two Variations on the Old Theme." *Amer. Math. Monthly* **102**, 912–16, 1995.

Fox's H-Function

A very general function defined by

$$H(z) = H_{p,q}^{m,n}\left[z\left|\begin{matrix}(a_1,\alpha_1),\ldots,(a_p,\alpha_p)\\(b_1,\beta_1),\ldots,(b_p,\beta_p)\end{matrix}\right.\right]$$

$$= \frac{1}{2\pi i}\int_C \frac{\Pi_{j=1}^m \Gamma(b_j - \beta_i s)\Pi_{j=1}^n \Gamma(1 - a_j + a_j s)}{\Pi_{j=m+1}^q \Gamma(1 - b_j + \beta_j s)\Pi_{j=n+1}^{qp} \Gamma(a_j - a_j s)}$$

$$\times z^s \, ds,$$

where $0 \le m \le q$, $0 \le n \le p$, $\alpha_j, \beta_j > 0$, and a_j, b_j are COMPLEX NUMBERS such that the pole of $\Gamma(b_j - \beta_j s)$ for $j = 1, 2, \ldots, m$ coincides with any POLE of $\Gamma(1 - a_j + \alpha_j s)$ for $j = 1, 2, \ldots, n$. In addition C, is a CONTOUR in the complex s-plane from $\omega - i\infty$ to $\omega + i\infty$ such that $(b_j + k)/\beta_j$ and $(a_j - 1 - k)/\alpha_j$ lie to the right and left of C, respectively.

A. Kilbas has derived a complete description for the asymptotic expansion of the H-function.

See also KAMPE DE FERIET FUNCTION, MACROBERT'S *E*-FUNCTION, MEIJER'S *G*-FUNCTION

References

Carter, B. D. and Springer, M. D. "The Distribution of Products, Quotients, and Powers of Independent H-Functions." *SIAM J. Appl. Math.* **33**, 542–58, 1977.
Fox, C. "The G and H-Functions as Symmetrical Fourier Kernels." *Trans. Amer. Math. Soc.* **98**, 395–29, 1961.
Prudnikov, A. P.; Brychkov, Yu. A.; and Marichev, O. I. "Evaluation of Integrals and the Mellin Transform." *Itogi Nauki i Tekhniki, Seriya Matemat. Analiz* **27**, 3–46, 1989.
Yakubovich, S. B. and Luchko, Y. F. *The Hypergeometric Approach to Integral Transforms and Convolutions.* Amsterdam, Netherlands: Kluwer, 1994.

F-Polynomial

KAUFFMAN POLYNOMIAL F

Frac

FRACTIONAL PART

Fractal

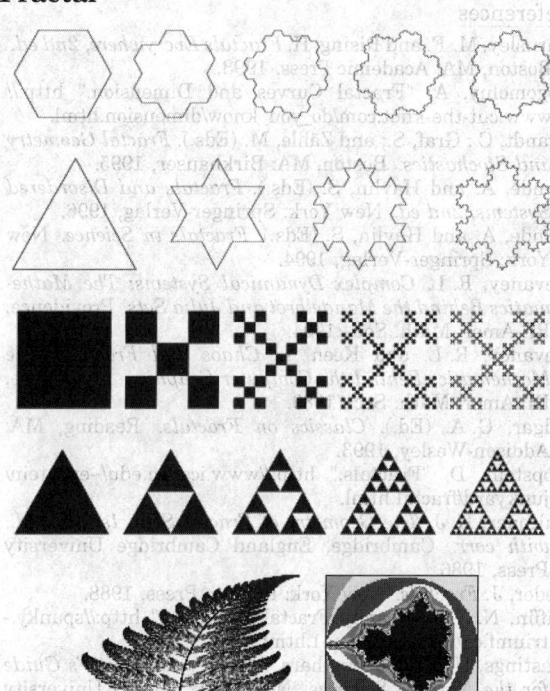

An object or quantity which displays SELF-SIMILARITY, in a somewhat technical sense, on all scales. The object need not exhibit *exactly* the same structure at all scales, but the same "type" of structures must appear on all scales. A plot of the quantity on a log-log graph versus scale then gives a straight line, whose slope is said to be the FRACTAL DIMENSION. The prototypical example for a fractal is the length of a coastline measured with different length RULERS. The shorter the RULER, the longer the length measured, a PARADOX known as the COASTLINE PARADOX.

Illustrated above are the fractals known as the GOSPER ISLAND, KOCH SNOWFLAKE, BOX FRACTAL, SIERPINSKI SIEVE, BARNSLEY'S FERN, and MANDELBROT SET.

See also BACKTRACKING, BARNSLEY'S FERN, BOX FRACTAL, BUTTERFLY FRACTAL, CACTUS FRACTAL, CANTOR SET, CANTOR SQUARE FRACTAL, CAROTID-KUNDALINI FRACTAL, CESÀRO FRACTAL, CHAOS GAME, CIRCLES-AND-SQUARES FRACTAL, COASTLINE PARADOX, DRAGON CURVE, FAT FRACTAL, FATOU SET, FRACTAL DIMENSION, GOSPER ISLAND, H-FRACTAL, HÉNON MAP, ITERATED FUNCTION SYSTEM, JULIA FRACTAL, KAPLAN-YORKE MAP, KOCH ANTISNOWFLAKE, KOCH SNOWFLAKE, LÉVY FRACTAL, LÉVY TAPESTRY, LINDENMAYER SYSTEM, MANDELBROT SET, MANDELBROT TREE, MENGER SPONGE, MINKOWS-

KI SAUSAGE, MIRA FRACTAL, NESTED SQUARE, NEW-TON'S METHOD, PENTAFLAKE, PYTHAGORAS TREE, RABINOVICH-FABRIKANT EQUATION, SAN MARCO FRACTAL, SIERPINSKI CARPET, SIERPINSKI CURVE, SIERPINSKI SIEVE, STAR FRACTAL, ZASLAVSKII MAP

References

Barnsley, M. F. and Rising, H. *Fractals Everywhere, 2nd ed.* Boston, MA: Academic Press, 1993.

Bogomolny, A. "Fractal Curves and Dimension." http://www.cut-the-knot.com/do_you_know/dimension.html.

Brandt, C.; Graf, S.; and Zähle, M. (Eds.). *Fractal Geometry and Stochastics.* Boston, MA: Birkhäuser, 1995.

Bunde, A. and Havlin, S. (Eds.). *Fractals and Disordered Systems, 2nd ed.* New York: Springer-Verlag, 1996.

Bunde, A. and Havlin, S. (Eds.). *Fractals in Science.* New York: Springer-Verlag, 1994.

Devaney, R. L. *Complex Dynamical Systems: The Mathematics Behind the Mandelbrot and Julia Sets.* Providence, RI: Amer. Math. Soc., 1994.

Devaney, R. L. and Keen, L. *Chaos and Fractals: The Mathematics Behind the Computer Graphics.* Providence, RI: Amer. Math. Soc., 1989.

Edgar, G. A. (Ed.). *Classics on Fractals.* Reading, MA: Addison-Wesley, 1993.

Eppstein, D. "Fractals." http://www.ics.uci.edu/~eppstein/junkyard/fractal.html.

Falconer, K. J. *The Geometry of Fractal Sets, 1st pbk. ed., with corr.* Cambridge, England Cambridge University Press, 1986.

Feder, J. *Fractals.* New York: Plenum Press, 1988.

Giffin, N. "The Spanky Fractal Database." http://spanky.-triumf.ca/www/welcome1.html.

Hastings, H. M. and Sugihara, G. *Fractals: A User's Guide for the Natural Sciences.* New York: Oxford University Press, 1994.

Kaye, B. H. *A Random Walk Through Fractal Dimensions, 2nd ed.* New York: Wiley, 1994.

Lauwerier, H. A. *Fractals: Endlessly Repeated Geometrical Figures.* Princeton, NJ: Princeton University Press, 1991.

le Méhaute, A. *Fractal Geometries: Theory and Applications.* Boca Raton, FL: CRC Press, 1992.

Mandelbrot, B. B. *Fractals: Form, Chance, & Dimension.* San Francisco, CA: W. H. Freeman, 1977.

Mandelbrot, B. B. *The Fractal Geometry of Nature.* New York: W. H. Freeman, 1983.

Massopust, P. R. *Fractal Functions, Fractal Surfaces, and Wavelets.* San Diego, CA: Academic Press, 1994.

Pappas, T. "Fractals--Real or Imaginary." *The Joy of Mathematics.* San Carlos, CA: Wide World Publ./Tetra, pp. 78-9, 1989.

Peitgen, H.-O.; Jürgens, H.; and Saupe, D. *Chaos and Fractals: New Frontiers of Science.* New York: Springer-Verlag, 1992.

Peitgen, H.-O.; Jürgens, H.; and Saupe, D. *Fractals for the Classroom, Part 1: Introduction to Fractals and Chaos.* New York: Springer-Verlag, 1992.

Peitgen, H.-O. and Richter, D. H. *The Beauty of Fractals: Images of Complex Dynamical Systems.* New York: Springer-Verlag, 1986.

Peitgen, H.-O. and Saupe, D. (Eds.). *The Science of Fractal Images.* New York: Springer-Verlag, 1988.

Pickover, C. A. (Ed.). *The Pattern Book: Fractals, Art, and Nature.* World Scientific, 1995.

Pickover, C. A. (Ed.). *Fractal Horizons: The Future Use of Fractals.* New York: St. Martin's Press, 1996.

Rietman, E. *Exploring the Geometry of Nature: Computer Modeling of Chaos, Fractals, Cellular Automata, and Neural Networks.* New York: McGraw-Hill, 1989.

Russ, J. C. *Fractal Surfaces.* New York: Plenum, 1994.

Schroeder, M. *Fractals, Chaos, Power Law: Minutes from an Infinite Paradise.* New York: W. H. Freeman, 1991.

Sprott, J. C. "Sprott's Fractal Gallery." http://sprott.physics.wisc.edu/fractals.htm.

Stauffer, D. and Stanley, H. E. *From Newton to Mandelbrot, 2nd ed.* New York: Springer-Verlag, 1995.

Stevens, R. T. *Fractal Programming in C.* New York: Henry Holt, 1989.

Takayasu, H. *Fractals in the Physical Sciences.* Manchester, England: Manchester University Press, 1990.

Tricot, C. *Curves and Fractal Dimension.* New York: Springer-Verlag, 1995.

Triumf Mac Fractal Programs. http://spanky.triumf.ca/pub/fractals/programs/MAC/.

Vicsek, T. *Fractal Growth Phenomena, 2nd ed.* Singapore: World Scientific, 1992.

Weisstein, E. W. "Fractals." MATHEMATICA NOTEBOOK FRAC-TAL.M.

Weisstein, E. W. "Books about Fractals." http://www.treasure-troves.com/books/Fractals.html.

Yamaguti, M.; Hata, M.; and Kigami, J. *Mathematics of Fractals.* Providence, RI: Amer. Math. Soc., 1997.

Fractal Dimension

The term "fractal dimension" is sometimes used to refer to what is more commonly called the CAPACITY DIMENSION (which is, roughly speaking, the exponent D in the expression $n(\epsilon) = \epsilon^{-D}$, where $n(\epsilon)$ is the minimum number of OPEN SETS of diameter ϵ needed to cover the set). However, it can more generally refer to any of the dimensions commonly used to characterize fractals (e.g., CAPACITY DIMENSION, CORRELATION DIMENSION, INFORMATION DIMENSION, LYAPUNOV DIMENSION, MINKOWSKI-BOULIGAND DIMENSION).

See also BOX-COUNTING DIMENSION, CAPACITY DIMENSION, CORRELATION DIMENSION, FRACTAL DIMENSION, HAUSDORFF DIMENSION, INFORMATION DIMENSION, LYAPUNOV DIMENSION, MINKOWSKI-BOULIGAND DIMENSION, POINTWISE DIMENSION, Q-DIMENSION

References

Rasband, S. N. "Fractal Dimension." Ch. 4 in *Chaotic Dynamics of Nonlinear Systems.* New York: Wiley, pp. 71-3, 1990.

Fractal Land

CAROTID-KUNDALINI FRACTAL

Fractal Process

A 1-D MAP whose increments are distributed according to a NORMAL DISTRIBUTION. Let $y(t - \Delta t)$ and $y(t + \Delta t)$ be values, then their correlation is given by the BROWN FUNCTION

$$r = 2^{2H-1} - 1.$$

When $H = 1/2$, $r = 0$ and the fractal process corresponds to 1-D Brownian motion. If $H > 1/2$, then $r > 0$ and the process is called a PERSISTENT PROCESS.

If $H < 1/2$, then $r < 0$ and the process is called an ANTIPERSISTENT PROCESS.

See also ANTIPERSISTENT PROCESS, PERSISTENT PROCESS

References
von Seggern, D. *CRC Standard Curves and Surfaces*. Boca Raton, FL: CRC Press, 1993.

Fractal Sequence
Given an INFINITIVE SEQUENCE $\{x_n\}$ with associated array $a(i,j)$, then $\{x_n\}$ is said to be a fractal sequence

1. If $i + 1 = x_n$, then there exists $m < n$ such that $i = x_m$,
2. If $h < i$, then, for every j, there is exactly one k such that $a(i,j) < a(h,k) < a(i,j+1)$.

(As i and j range through N, the array $A = a(i,j)$, called the associative array of x, ranges through all of N.) An example of a fractal sequence is 1, 1, 1, 1, 2, 1, 2, 1, 3, 2, 1, 3, 2, 1, 3, ...

If $\{x_n\}$ is a fractal sequence, then the associated array is an INTERSPERSION. If x is a fractal sequence, then the UPPER-TRIMMED SUBSEQUENCE is given by $\lambda(x) = x$, and the LOWER-TRIMMED SUBSEQUENCE $V(x)$ is another fractal sequence. The SIGNATURE of an IRRATIONAL NUMBER is a fractal sequence.

See also INFINITIVE SEQUENCE

References
Kimberling, C. "Fractal Sequences and Interspersions." *Ars Combin.* **45**, 157–68, 1997.

Fractal Valley
CAROTID-KUNDALINI FUNCTION

Fractile
QUANTILE

Fraction
A RATIONAL NUMBER expressed in the form a/b (in-line notation) or $\frac{a}{b}$ (traditional "display" notation), where a is called the NUMERATOR and b is called the DENOMINATOR. When written in-line, the slash "/" between NUMERATOR and DENOMINATOR is called a SOLIDUS.

A PROPER FRACTION is a fraction such that $a/b < 1$, and a LOWEST TERMS FRACTION is a fraction with common terms canceled out of the NUMERATOR and DENOMINATOR.

The Egyptians expressed their fractions as sums (and differences) of UNIT FRACTIONS. Conway and Guy (1999) give a table of Roman NOTATION for fractions, in which multiples of 1/12 (the UNCIA) were given separate names.

See also ADJACENT FRACTION, ANOMALOUS CANCELLATION, COMMON FRACTION, COMPLEX FRACTION, CONTINUED FRACTION, DENOMINATOR, EGYPTIAN FRACTION, FAREY SEQUENCE, GOLDEN RULE, HALF, LOWEST TERMS FRACTION, MATRIX FRACTION, MEDIANT, MIXED FRACTION, NUMERATOR, PANDIGITAL FRACTION, PROPER FRACTION, PYTHAGOREAN FRACTION, QUARTER, RATIONAL NUMBER, SOLIDUS, UNIT FRACTION

References
Conway, J. H. and Guy, R. K. *The Book of Numbers.* New York: Springer-Verlag, pp. 22–3, 1996.
Courant, R. and Robbins, H. "Decimal Fractions. Infinite Decimals." §2.2.2 in *What is Mathematics?: An Elementary Approach to Ideas and Methods, 2nd ed.* Oxford, England: Oxford University Press, pp. 61–3, 1996.

Fractional Calculus
The study of an extension of derivatives and integrals to noninteger orders. Fractional calculus is based on the definition of the FRACTIONAL INTEGRAL as

$$D^{-v}f(t) = \frac{1}{\Gamma(v)} \int_0^t (t - \xi)^{v-1} f(\xi)d\xi,$$

where $\Gamma(v)$ is the GAMMA FUNCTION. From this equation, FRACTIONAL DERIVATIVES can also be defined.

See also DERIVATIVE, FRACTIONAL DERIVATIVE, FRACTIONAL DIFFERENTIAL EQUATION, FRACTIONAL INTEGRAL, INTEGRAL, MULTIPLE INTEGRAL

References
Butzer, P. L. and Westphal, U. "An Introduction to Fractional Calculus." Ch. 1 in *Applications of Fractional Calculus in Physics* (Ed. R. Hilfer). Singapore: World Scientific, pp. 1–5, 2000.
McBride, A. C. *Fractional Calculus.* New York: Halsted Press, 1986.
Nishimoto, K. *Fractional Calculus.* New Haven, CT: University of New Haven Press, 1989.
Samko, S. G.; Kilbas, A. A.; and Marichev, O. I. *Fractional Integrals and Derivatives.* Yverdon, Switzerland: Gordon and Breach, 1993.
Spanier, J. and Oldham, K. B. *The Fractional Calculus: Integrations and Differentiations of Arbitrary Order.* New York: Academic Press, 1974.

Fractional Derivative
The fractional derivative of $f(t)$ of order $\mu > 0$ (if it exists) can be defined in terms of the FRACTIONAL INTEGRAL $D^{-v}f(t)$ as

$$D^\mu f(t) = D^m \left[D^{-(m-\mu)} f(t) \right], \qquad (1)$$

where m is an integer $\geq \lceil \mu \rceil$, where $\lceil x \rceil$ is the CEILING FUNCTION. The SEMIDERIVATIVE corresponds to $\mu = 1/2$.

The fractional derivative of the function t^λ is given by

$$D^\mu t^\lambda = D^m \left[D^{-(m-\mu)} t^\lambda \right]$$

$$= D^n \left[\frac{\Gamma(\lambda+1)}{\Gamma(\lambda+m-\mu+1)} t^{\lambda+m-\mu} \right]$$

$$= \frac{\Gamma(\lambda+1)(\lambda-\mu+m)(\lambda-\mu+m-1)\cdots(\lambda-\mu+1)}{\Gamma(1+m+\lambda-\mu)} t^{\lambda-\mu}$$

$$= \frac{\Gamma(\lambda+1)(1+\lambda-\mu)_m}{\Gamma(1+m+\lambda-\mu)} t^{\lambda-\mu}$$

$$= \frac{\Gamma(\lambda+1)}{\Gamma(\lambda-\mu+1)} t^{\lambda-\mu} \tag{2}$$

for $\lambda > -1, \mu > 0$. The fractional derivative of the CONSTANT FUNCTION $f(t) = c$ is then given by

$$D^\mu c = c \lim_{\lambda \to 0} \frac{\Gamma(\lambda+1)}{\Gamma(\lambda-\mu+1)} t^{\lambda-\mu} = \frac{ct^{-\mu}}{\Gamma(1-\mu)}. \tag{3}$$

The fractional derivate of the E_T-FUNCTION is given by

$$D^\rho E_t(v, a) = E_t(v - \rho, a) \tag{4}$$

for $v > 0, \rho \neq 0$.

It is always true that, for $\mu, v > 0$,

$$D^{-\mu} D^{-v} f(t) = D^{-(\mu+v)} \tag{5}$$

but *not* always true that

$$D^\mu D^v = D^{\mu+v} \tag{6}$$

A FRACTIONAL INTEGRAL can also be similarly defined. The study of fractional derivatives and integrals is called FRACTIONAL CALCULUS.

See also FRACTIONAL CALCULUS, SEMIDERIVATIVE

References

Love, E. R. "Fractional Derivatives of Imaginary Order." *J. London Math. Soc.* **3**, 241–59, 1971.
Miller, K. S. "Derivatives of Noninteger Order." *Math. Mag.* **68**, 183–92, 1995.
Samko, S. G.; Kilbas, A. A.; and Marichev, O. I. *Fractional Integrals and Derivatives.* Yverdon, Switzerland: Gordon and Breach, 1993.
Spanier, J. and Oldham, K. B. *The Fractional Calculus: Integrations and Differentiations of Arbitrary Order.* New York: Academic Press, 1974.

Fractional Differential Equation

The solution to the differential equation

$$[D^{2v} + aD^v + bD^0] y(t) = 0$$

is

$$y(t) = \begin{cases} e_\alpha(t) - e_\beta(t) \\ \quad \text{for } \alpha \neq \beta \\ te^{\alpha t}, \sum_{k=-(q-1)}^{q-1} \alpha^k (q-|k|) D^{1-(k+1)v}(te^{\alpha^q t}) \\ \quad \text{for } \alpha = \beta \neq 0 \\ \dfrac{t^{2v-1}}{\Gamma(2v)} \\ \quad \text{for } \alpha = \beta = 0, \end{cases}$$

where

$$q = \frac{1}{v}$$

$$e_\beta(t) = \sum_{k=0}^{q-1} \beta^{q-k-1} E_t(-kv, \beta^q),$$

$E_t(a, x)$ is the E_T-FUNCTION, and $\Gamma(n)$ is the GAMMA FUNCTION.

See also FRACTIONAL CALCULUS

References

Miller, K. S. "Derivatives of Noninteger Order." *Math. Mag.* **68**, 183–92, 1995.

Fractional Fourier Transform

The fractional Fourier transform is generally understood to correspond to a rotation in time-frequency phase space, where the usual FOURIER TRANSFORM corresponds to a rotation of 90° ($\pi/2$ radians). A fractional Fourier transform can be used to detect frequencies which are not INTEGER multiples of the lowest DISCRETE FOURIER TRANSFORM frequency.

See also DISCRETE FOURIER TRANSFORM, FOURIER TRANSFORM

References

Namias, V. "The Fractional Fourier Transform and Its Application to Quantum Mechanics." *J. Inst. Math. Appl.* **25**, 241–65, 1980.
Ozaktas, H. M. "Fractional Fourier Transform and Its Applications in Optics and Signal Processing--A Bibliography." http://www.ee.bilkent.edu.tr/~haldun/ffbiblio.ps.
Ozaktas, H. M. "Publications Related to Fractional Fourier Transforms." http://www.ee.bilkent.edu.tr/~haldun/fracfourpub.ps.

Fractional Integral

Denote the nth DERIVATIVE D^n and the n-fold INTEGRAL D^{-n}. Then

$$D^{-1} f(t) = \int_0^t f(\xi) d\xi. \tag{1}$$

Now, if the equation

$$D^{-n} f(t) = \frac{1}{(n-1)!} \int_0^t (t-\xi)^{n-1} f(\xi) \, d\xi \tag{2}$$

for the MULTIPLE INTEGRAL is true for n, then

$$D^{-(n+1)}f(t) = D^{-1}\left[\frac{1}{(n-1)!}\int_0^t (t-\xi)^{n-1}f(\xi)\,d\xi\right]$$

$$= \int_0^t\left[\frac{1}{(n-1)!}\int_0^x (x-\xi)^{n-1}f(\xi)\,d\xi\right]dx. \quad (3)$$

Interchanging the order of integration gives

$$D^{-(n+1)}f(t) = \frac{1}{n!}\int_0^t (t-\xi)^n f(\xi)\,d\xi. \quad (4)$$

But (2) is true for $n = 1$, so it is also true for all n by INDUCTION. The fractional integral of $f(t)$ of order $v > 0$ can then be defined by

$$D^{-v}f(t) = \frac{1}{\Gamma(v)}\int_0^t (t-\xi)^{v-1}f(\xi)\,d\xi, \quad (5)$$

where $\Gamma(v)$ is the GAMMA FUNCTION.

The fractional integral of order 1/2 is called a SEMI-INTEGRAL.

The fractional integral can only be given in terms of elementary functions for a small number of functions. For example,

$$D^{-v}t^\lambda = \frac{\Gamma(\lambda+1)}{\Gamma(\lambda+v+1)}t^{\lambda+v} \quad \text{for } \lambda > -1, v > 0 \quad (6)$$

$$D^{-v}e^{at} = \frac{1}{\Gamma(v)}e^{at}\int_0^t x^{v-1}e^{-ax}dx$$

$$= \frac{a^{-v}e^{at}\gamma(v,at)}{\Gamma(v)} \equiv E_t(v,a), \quad (7)$$

where $\gamma(a,x)$ is a lower incomplete GAMMA FUNCTION and $E_t(v,a)$ is the E_T-FUNCTION. From (6), the fractional integral of the CONSTANT FUNCTION $f(t) = c$ is given by

$$D^{-v}c = c\lim_{\lambda\to 0}\frac{\Gamma(\lambda+1)}{\Gamma(\lambda+v+1)}t^{\lambda+v} = \frac{t^\mu}{\Gamma(v+1)}. \quad (8)$$

A FRACTIONAL DERIVATIVE can also be similarly defined. The study of fractional derivatives and integrals is called FRACTIONAL CALCULUS.

See also FRACTIONAL CALCULUS, SEMI-INTEGRAL

References

Samko, S. G.; Kilbas, A. A.; and Marichev, O. I. *Fractional Integrals and Derivatives.* Yverdon, Switzerland: Gordon and Breach, 1993.

Spanier, J. and Oldham, K. B. *The Fractional Calculus: Integrations and Differentiations of Arbitrary Order.* New York: Academic Press, 1974.

Fractional Part

The function frac x giving the fractional (noninteger) part of a REAL NUMBER x. The symbol $\{x\}$ is sometimes used instead of frac x (Graham *et al.* 1994, p. 70), but this notation is not used in this work due to possible confusion with the SET containing the element x.

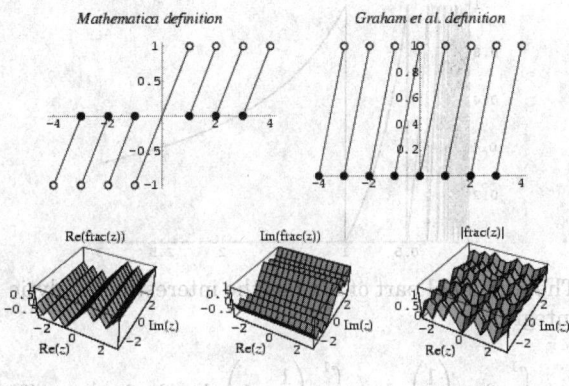

Unfortunately, there is no universal agreement on the meaning of frac x for $x < 0$ and there are two common definitions. Let $\lfloor x\rfloor$ be the FLOOR FUNCTION, then the *Mathematica* command FractionalPart[x] is defined as

$$\text{frac } x \equiv \begin{cases} x - \lfloor x\rfloor & x \geq 0 \\ x - \lfloor x\rfloor - 1 & x < 0 \end{cases} \quad (1)$$

(left figure). This definition has the benefit that frac x + int $x = x$, where int x is the INTEGER PART of x. Although Spanier and Oldham (1987) use the same definition as *Mathematica*, they mention the formula only very briefly and then say it will not be used further. Graham *et al.* (1994, p. 70), and perhaps most other mathematicians, use the different definition

$$\text{frac } x = x - \lfloor x\rfloor, \quad (2)$$

(right figure).

Since usage concerning fractional part/value and integer part/value can be confusing, the following table gives a summary of names and notations used (D. W. Cantrell). Here, S&O indicates Spanier and Oldham (1987).

notation	name	S&O	Graham *et al.*	*Mathematica*
$\lfloor x\rfloor$	integer-value	Int(x)	floor or integer part	Floor[x]

sgn(x)⌊|x|⌋	integer-part	Ip(x)	no name	Integer-Part[x]
x − ⌊x⌋	fractional-value	frac(x)	fractional part or {x}	no name
sgn(x)(|x|−⌊|x|⌋)	fractional-part	F_P(x)	no name	Fractional Part[x]

The (possibly scaled) periodic waveform corresponding to the latter definition is known as the SAWTOOTH WAVE.

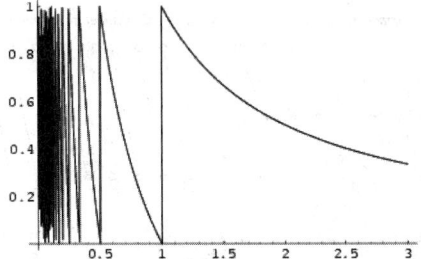

The fractional part of $1/x$ has the interesting analytic integrals

$$\int_{1/2}^{1} \text{frac}\left(\frac{1}{x}\right) dx = \int_{1/2}^{1} \left(\frac{1}{x} - 1\right) dx = \ln 2 - \frac{1}{2} \quad (3)$$

$$\int_{1/3}^{1/2} \text{frac}\left(\frac{1}{x}\right) dx = \int_{1/3}^{1/2} \left(\frac{1}{x} - 2\right) dx = \ln 3 - \ln 2 - \frac{1}{3} \quad (4)$$

$$\int_{1/4}^{1/3} \text{frac}\left(\frac{1}{x}\right) dx = \int_{1/4}^{1/3} \left(\frac{1}{x} - 3\right) dx$$
$$= \ln 4 - \ln 3 - \frac{1}{4}. \quad (5)$$

The integral

$$I = \int_{0}^{1} \text{frac}\left(\frac{1}{x}\right) dx \quad (6)$$

is therefore a TELESCOPING SUM given by

$$I = \int_{0}^{1} \text{frac}\left(\frac{1}{x}\right) dx = \lim_{n \to \infty}\left[\ln n - \sum_{k=2}^{n} \frac{1}{k}\right]$$

$$= 1 - \gamma + \lim_{n \to \infty}(\ln n - \Psi_0(1+n)), \quad (7)$$

where γ is the EULER-MASCHERONI CONSTANT and $\Psi_k(x)$ is the POLYGAMMA FUNCTION. The quantity on the right is 0, so

$$I = 1 - \gamma. \quad (8)$$

A consequence of WEYL'S CRITERION is that the sequence $\{\text{frac}(nx)\}$ is dense and EQUIDISTRIBUTED in the interval [0, 1] for irrational x, where $n = 1, 2, ...$ (finch).

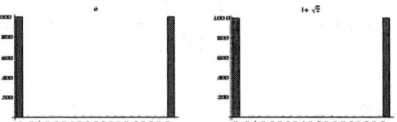

Hardy and Littlewood (1914) proved that the sequence $\{\text{frac}(x^n)\}$ is EQUIDISTRIBUTED for *almost all* real numbers $x > 1$ (i.e., the exceptional set has LEBESGUE MEASURE ZERO). Exceptional numbers include the positive integers, $1 + \sqrt{2}$ (Finch), and the GOLDEN RATIO ϕ. The plots above illustrate the distribution of $\text{frac}(x^n)$ for $x = e$, ϕ, and $1 + \sqrt{2}$. Candidate members of the *measure one* set are easy to find, but difficult to proven. However, Levin has explicitly constructed such an example (Drmota and Tichy 1997).

The properties of $\{\text{frac}((3/2)^n)\}$, the simplest such sequence for a rational number $x > 1$ have been extensively studied (Finch). For example, $\{\text{frac}((3/2)^n)\}$ has infinitely many ACCUMULATION POINTS in both $[0, 1/2]$ and $[1/2, 1]$ (Pisot 1938, Vijayaraghavan 1941). Furthermore, Flatto *et al.* (1995) proved that any subinterval of [0, 1] containing all but at most finitely many ACCUMULATION POINTS of $\text{frac}((3/2)^n)$ must have length at least 1/3. Surprisingly, the sequence $\{\text{frac}((3/2)^n)\}$ is also connected with the COLLATZ PROBLEM and with WARING'S PROBLEM.

In particular, WARING'S PROBLEM can be solved completely if the inequality

$$\text{frac}\left[\left(\frac{3}{2}\right)^n\right] \leq 1 - \left(\frac{3}{4}\right)^n \quad (9)$$

holds. No counterexample to this inequality is known, and it is even believed that can be extended to

$$\left(\frac{3}{4}\right)^n < \text{frac}\left[\left(\frac{3}{2}\right)^n\right] < 1 - \left(\frac{3}{4}\right)^n \quad (10)$$

for $n > 7$ (Finch; Bennett 1993, 1994). Furthermore, the constant 3/4 can be decreased to 0.5769 (Beukers 1981 and Dubitskas 1990). Unfortunately, these inequalities have not been proved.

See also BEATTY SEQUENCE, CEILING FUNCTION, EQUIDISTRIBUTED SEQUENCE, FLOOR FUNCTION, INTEGER PART, NEAREST INTEGER FUNCTION, ROUND, SAWTOOTH WAVE, SHIFT TRANSFORMATION, TRUNCATE, WHOLE NUMBER

References

Bennett, M. A. "Fractional Parts of Powers of Rational Numbers." *Math. Proc. Cambridge Philos. Soc.* **114**, 191–01, 1993.

Bennett, M. A. "An Ideal Waring Problem with Restricted Summands." *Acta Arith.* **66**, 125–32, 1994.

Beukers, F. "Fractional Parts of Powers of Rational Numbers." *Math. Proc. Cambridge Philos. Soc.* **90**, 13–0, 1981.

Drmota, M. and Tichy, R. F. *Sequences, Discrepancies and Applications.* New York: Springer-Verlag, 1997.

Dubitskas, A. K. "A Lower Bound for the Quantity $\{(3/2)^n\}$." *Russian Math. Survey* **45**, 163–64, 1990.

Finch, S. "Powers of 3/2 Modulo One." http://www.mathsoft.com/asolve/pwrs32/pwrs32.html.

Flatto, L.; Lagarias, J. C.; Pollington, A. D. "On the Range of Fractional Parts $\{\xi(p/q)^n\}$." *Acta Arith.* **70**, 125–47, 1995.

Graham, R. L.; Knuth, D. E.; and Patashnik, O. *Concrete Mathematics: A Foundation for Computer Science, 2nd ed.* Reading, MA: Addison-Wesley, 1994.

Miklavc, A. "Elementary Proofs of Two Theorems on the Distribution of Numbers $\{nx\}$ (mod 1)." *Proc. Amer. Math. Soc.* **39**, 279–80, 1973.

Spanier, J. and Oldham, K. B. "The Integer-Value Int(x) and Fractional-Value frac(x) Functions." Ch. 9 in *An Atlas of Functions.* Washington, DC: Hemisphere, pp. 71–8, 1987.

Vijayaraghavan, T. "On the Fractional Parts of the Powers of a Number (I)." *J. London Math. Soc.* **15**, 159–60, 1940.

Vijayaraghavan, T. "On the Fractional Parts of the Powers of a Number (II)." *Proc. Cambridge Phil. Soc.* **37**, 349–57, 1941.

Vijayaraghavan, T. "On the Fractional Parts of the Powers of a Number (III)." *J. London Math. Soc.* **17**, 137–38, 1942.

Fractran

Fractran is an algorithm applied to a given list $f_1, f_2, ..., f_k$ of FRACTIONS. Given a starting INTEGER N, the Fractran algorithm proceeds by repeatedly multiplying the integer at a given stage by the first element f_t given an integer PRODUCT. The algorithm terminates when there is no such f_t.

The list

$$\frac{17}{91}, \frac{78}{85}, \frac{19}{51}, \frac{23}{38}, \frac{29}{33}, \frac{77}{29}, \frac{95}{23}, \frac{77}{19}, \frac{1}{17}, \frac{11}{13}, \frac{13}{11}, \frac{15}{2}, \frac{1}{7}, \frac{55}{1}$$

with starting integer $N = 2$ generates a sequence 2, 15, 825, 725, 1925, 2275, 425, 390, 330, 290, 770, ... (Sloane's A007542). Conway (1987) showed that the only other powers of 2 which occur are those with PRIME exponent: $2^2, 2^3, 2^5, 2^7, \ldots$.

References

Conway, J. H. "Unpredictable Iterations." In *Proceedings of the 1972 Number Theory Conference Held at the University of Colorado, Boulder, Colo., Aug. 14–8, 1972.* Boulder, CO: University of Colorado, pp. 49–2, 1972.

Conway, J. H. "Fractran: A Simple Universal Programming Language for Arithmetic." Ch. 2 in *Open Problems in Communication and Computation* (Ed. T. M. Cover and B. Gopinath). New York: Springer-Verlag, pp. 4–6, 1987.

Sloane, N. J. A. Sequences A007542/M2084 in "An On-Line Version of the Encyclopedia of Integer Sequences." http://www.research.att.com/~njas/sequences/eisonline.html.

Frame

A closed curve associated with a knot which is displaced along the normal by a small amount. For K is parameterized by $x^\mu(s)$ for $0 \le s \le L$ along the length of the knot by parameter s, the frame K_f associated with K is

$$y^\mu = x^\mu(s) + \epsilon n^\mu(s),$$

where ϵ is a small parameter, $n^\mu(s)$ is a unit VECTOR FIELD normal to the curve at s.

See also FRAMEWORK

References

Kaul, R. K. Topological Quantum Field Theories--A Meeting Ground for Physicists and Mathematicians. 15 Jul 1999. http://xxx.lanl.gov/abs/hep-th/9907119/.

Framework

Consider a finite collection of points $p = (p_1, ..., p_n)$, $p_i \in \mathbb{R}^d$ EUCLIDEAN SPACE (known as a CONFIGURATION) and a graph G whose VERTICES correspond to pairs of points that are constrained to stay the same distance apart. Then the graph G together with the configuration p, denoted $G(p)$, is called a framework.

See also BAR (EDGE), CONFIGURATION, RIGID GRAPH, TENSEGRITY

References

Coxeter, H. S. M. and Greitzer, S. L. *Geometry Revisited.* Washington, DC: Math. Assoc. Amer., p. 56, 1967.

Franel Number

One of the numbers $\sum_{k=0}^{n} \binom{n}{k}^3$, where $\binom{n}{k}$ is a BINOMIAL COEFFICIENT. The first few values for $n = 0, 1, \ldots$ are 1, 2, 10, 56, 346, ... (Sloane's A000172).

See also BINOMIAL SUMS

References

Franel, J. "On a Question of Laisant." *L'intermédiaire des mathématiciens* **1**, 45–7, 1894.

Franel, J. "On a Question of J. Franel." *L'intermédiaire des mathématiciens* **2**, 33–5, 1895.

Sloane, N. J. A. Sequences A000172/M1971 in "An On-Line Version of the Encyclopedia of Integer Sequences." http://www.research.att.com/~njas/sequences/eisonline.html.

Franklin Graph

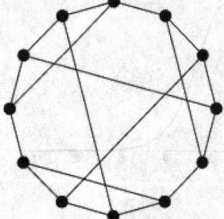 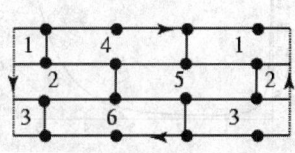

Franklin graph　　　　*coloring of the Klein bottle*

The 12-vertex graph illustrated above which provides the minimal coloring of the KLEIN BOTTLE using six colors, providing the sole counterexample to the HEAWOOD CONJECTURE.

See also HEAWOOD CONJECTURE, KLEIN BOTTLE

Franklin Magic Square

References
Bondy, J. A. and Murty, U. S. R. *Graph Theory with Applications.* New York: North Holland, p. 244, 1976.
Franklin, P. "A Six Color Problem." *J. Math. Phys.* **13**, 363–79, 1934.

52	61	4	13	20	29	36	45
14	3	62	51	46	35	30	19
53	60	5	12	21	28	37	44
11	6	59	54	43	38	27	22
55	58	7	10	23	26	39	42
9	8	57	56	41	40	25	24
50	63	2	15	18	31	34	47
16	1	64	49	48	33	32	17

Benjamin Franklin constructed the above 8×8 PAN-MAGIC SQUARE having MAGIC CONSTANT 260. Any half-row or half-column in this square totals 130, and the four corners plus the middle total 260. In addition, bent diagonals (such as 52–-5-4-0-7-3-6) also total 260 (Madachy 1979, p. 87).

See also MAGIC SQUARE, PANMAGIC SQUARE

References
Madachy, J. S. "Magic and Antimagic Squares." Ch. 4 in *Madachy's Mathematical Recreations.* New York: Dover, pp. 103–13, 1979.
Pappas, T. "The Magic Square of Benjamin Franklin." *The Joy of Mathematics.* San Carlos, CA: Wide World Publ./Tetra, p. 97, 1989.

Fransén-Robinson Constant

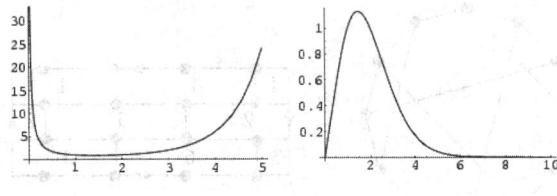

$$F \equiv \int_0^\infty \frac{dx}{\Gamma(x)} = 2.8077702420...,$$

where $\Gamma(x)$ is the GAMMA FUNCTION. The above plots show the functions $\Gamma(x)$ and $1/\Gamma(x)$. No closed-form expression in terms of other constants in known for F.

See also GAMMA FUNCTION

References
Finch, S. "Favorite Mathematical Constants." http://www.mathsoft.com/asolve/constant/fran/fran.html.
Fransén, A. "Accurate Determination of the Inverse Gamma Integral." *BIT* **19**, 137–38, 1979.
Fransén, A. "Addendum and Corrigendum to 'High-Precision Values of the Gamma Function and of Some Related Coefficients.'" *Math. Comput.* **37**, 233–35, 1981.
Fransén, A. and Wrigge, S. "High-Precision Values of the Gamma Function and of Some Related Coefficients." *Math. Comput.* **34**, 553–66, 1980.
Plouffe, S. "Fransen-Robinson Constant." http://www.lacim.uqam.ca/piDATA/fransen.txt.

F-Ratio

The RATIO of two independent estimates of the VARIANCE of a NORMAL DISTRIBUTION.

See also F-DISTRIBUTION, NORMAL DISTRIBUTION, VARIANCE

F-Ratio Distribution

F-DISTRIBUTION

Frattini Extension

If F is a group, then the extensions G of F of order o with $G/\phi(G) \cong F$, where $\phi(G)$ is the FRATTINI SUBGROUP, are called Frattini extensions.

See also FRATTINI FACTOR, FRATTINI SUBGROUP

References
Besche, H.-U. and Eick, B. "Construction of Finite Groups." *J. Symb. Comput.* **27**, 387–04, 1999.
Gaschütz, W. "Über Φ-Untergruppen endlicher Gruppen." *Math. Z.* **58**, 160–70, 1953.

Frattini Factor

A group given by $G/\phi(G)$, where $\phi(G)$ is the FRATTINI SUBGROUP of a given group G.

See also FRATTINI EXTENSION, FRATTINI SUBGROUP

References
Besche, H.-U. and Eick, B. "Construction of Finite Groups." *J. Symb. Comput.* **27**, 387–04, 1999.
Gaschütz, W. "Über Φ-Untergruppen endlicher Gruppen." *Math. Z.* **58**, 160–70, 1953.

Frattini Subgroup

The intersection $\phi(G)$ of all maximal subgroups of a given group G.

See also FRATTINI EXTENSION, FRATTINI FACTOR

References
Besche, H.-U. and Eick, B. "Construction of Finite Groups." *J. Symb. Comput.* **27**, 387–04, 1999.
Gaschütz, W. "Über Φ-Untergruppen endlicher Gruppen." *Math. Z.* **58**, 160–70, 1953.

Fréchet Bounds

Any bivariate distribution function with marginal distribution functions F and G satisfies

$$\max\{F(x) + G(y) - 1, 0\} \le H(x, y) \le \min\{F(x), G(y)\}.$$

Fréchet Derivative

A function f is Fréchet differentiable at a if

$$\lim_{x \to a} \frac{f(x) - f(a)}{x - a}$$

exists. This is equivalent to the statement that ϕ has a removable DISCONTINUITY at a, where

$$\phi(x) \equiv \frac{f(x) - f(a)}{x - a}.$$

Every function which is Fréchet differentiable is also Carathéodory differentiable.

See also CARATHÉODORY DERIVATIVE, DERIVATIVE

Fréchet Filter

COFINITE FILTER

Fréchet Space

A complete metrizable space, sometimes also with the restriction that the space be locally convex. A Fréchet space is a TOPOLOGICAL VECTOR SPACE which is COMPLETE. Its topology is also defined by a COUNTABLE family of SEMINORMS.

For example, the space of SMOOTH FUNCTIONS on $[0, 1]$ is a Fréchet space. Its topology is the C-INFINITY TOPOLOGY, which is given by the countable family of SEMINORMS,

$$\|f\|_\alpha = \sup |D^\alpha f|.$$

Because $f_n \to f$ in this topology implies that f is smooth, i.e.,

$$D^\alpha f_n \to D^\alpha f,$$

any CAUCHY SEQUENCE has a limit in the space of SMOOTH FUNCTIONS, i.e., it is COMPLETE.

See also BANACH SPACE, HILBERT SPACE, TOPOLOGICAL VECTOR SPACE

Fredholm Alternative

See also SPECTRAL THEORY

Fredholm Integral Equation of the First Kind

An INTEGRAL EQUATION OF THE FORM

$$f(x) = \int_{-\infty}^{\infty} K(x, t)\phi(t)\, dt$$

$$\phi(x) = \frac{1}{2\pi} \int_{-\infty}^{\infty} \frac{F(\omega)}{K(\omega)} e^{-i\omega x}\, d\omega.$$

See also FREDHOLM INTEGRAL EQUATION OF THE SECOND KIND, INTEGRAL EQUATION, VOLTERRA INTEGRAL EQUATION OF THE FIRST KIND, VOLTERRA INTEGRAL EQUATION OF THE SECOND KIND

References

Arfken, G. *Mathematical Methods for Physicists, 3rd ed.* Orlando, FL: Academic Press, p. 865, 1985.

Fredholm Integral Equation of the Second Kind

An INTEGRAL EQUATION OF THE FORM

$$\phi(x) = f(x) + \lambda \int_{-\infty}^{\infty} K(x, t)\phi(t)\, dt$$

$$\phi(x) = \frac{1}{\sqrt{2\pi}} \int_{-\infty}^{\infty} \frac{F(t) e^{-ixt}}{1 - \sqrt{2\pi}\lambda K(t)}\, dt.$$

See also FREDHOLM INTEGRAL EQUATION OF THE FIRST KIND, INTEGRAL EQUATION, NEUMANN SERIES (INTEGRAL EQUATION), VOLTERRA INTEGRAL EQUATION OF THE FIRST KIND, VOLTERRA INTEGRAL EQUATION OF THE SECOND KIND

References

Arfken, G. *Mathematical Methods for Physicists, 3rd ed.* Orlando, FL: Academic Press, p. 865, 1985.

Press, W. H.; Flannery, B. P.; Teukolsky, S. A.; and Vetterling, W. T. "Fredholm Equations of the Second Kind." §18.1 in *Numerical Recipes in FORTRAN: The Art of Scientific Computing, 2nd ed.* Cambridge, England: Cambridge University Press, pp. 782–85, 1992.

Fredholm's Theorem

This entry contributed by VIKTOR BENGTSSON

Fredholm's theorem states that, if A is an $m \times n$ matrix, then the ORTHOGONAL COMPLEMENT of the ROW SPACE of A is the NULLSPACE of A, and the ORTHOGONAL COMPLEMENT of the COLUMN SPACE of A is the NULLSPACE of $A^\perp$,

$$(\text{Row } A)^\perp = \text{Null } A$$

$$(\text{Col } A)^\perp = \text{Null } A^\perp.$$

See also COLUMN SPACE, NULLSPACE, ORTHOGONAL DECOMPOSITION, ROW SPACE

Free

When referring to a planar object, "free" means that the object is regarded as capable of being picked up out of the plane and flipped over. As a result, MIRROR IMAGES are equivalent for free objects.

The word "free" is also used in technical senses to refer to a FREE GROUP, FREE SEMIGROUP, FREE TREE, FREE VARIABLE, etc.

In ALGEBRAIC TOPOLOGY, a free abstract mathematical object is generated by n elements in a "free manner" ("FREELY"), i.e., such that the n elements satisfy no nontrivial relations among themselves. To make this more formal, an algebraic GADGET X is freely generated by a SUBSET G if, for any function $f : G \to Y$ where Y is any other algebraic GADGET, there exists a unique HOMOMORPHISM (which has different meanings depending on what kind of GADGETS you're dealing with) $g : X \to Y$ such that g restricted to G is f.

If the algebraic GADGETS are VECTOR SPACES, then G freely generates X IFF G is a BASIS for X. If the algebraic GADGETS are ABELIAN GROUPS, then G freely generates X IFF X is a DIRECT SUM of the INTEGERS, with G consisting of the standard BASIS.

See also FIXED, FREE GROUP, FREE VARIABLE, FREELY, GADGET, MIRROR IMAGE, RANK

Free Abelian Group

A free Abelian group is a group G with a subset which generates the group G with the only relation being $ab = ba$. That is, it has no TORSION. All such groups are a DIRECT PRODUCT of the INTEGERS $\mathbb{Z}$, and have rank given by the number of copies of $\mathbb{Z}$. For example, $\mathbb{Z} \times \mathbb{Z} = \{(n, m)\}$ is a free Abelian group of rank 2. A minimal subset $b_1, ..., b_n$ that generates a free Abelian group is called a basis, and gives G as

$$G = \mathbb{Z}b_1 + \cdots + \mathbb{Z}b_n.$$

A free Abelian group is an ABELIAN GROUP, but is not a FREE GROUP (except when it has rank one, i.e., $\mathbb{Z}$). Free Abelian groups are the FREE MODULES in the case when the RING is the ring of integers $\mathbb{Z}$.

See also ABELIAN GROUP, FREE GROUP, FREE MODULE, GROUP, TORSION (GROUP)

Free Action

A group action $G \times X \to X$ is called free when there are no FIXED POINTS. That is, for any point x there is at least one transformation which does not fix x. The group is said to act freely.

The basic example of a free group action is the action of a group on itself by left multiplication $L : G \times G \to G$. As long as the group has more than the IDENTITY ELEMENT, there is no element h which satisfies $gh = h$ for all g. An example of a free action which is not TRANSITIVE is the action of $\mathbb{S}^1$ on $\mathbb{S}^3 \subset \mathbb{C}^2$ by $e^{i\theta} \cdot (\mathbb{Z}_1, \mathbb{Z}_2) = (e^{i\theta}\mathbb{Z}_1, e^{i\theta}\mathbb{Z}_2)$, which defines the HOPF FIBRATION.

See also EFFECTIVE ACTION, FREE ACTION, GROUP, ISOTROPY GROUP, MATRIX GROUP, ORBIT (GROUP), QUOTIENT SPACE (LIE GROUP), REPRESENTATION, TOPOLOGICAL GROUP, TRANSITIVE GROUP ACTION

Free Group

The generators of a group G are defined to be the smallest subset of group elements such that all other elements of G can be obtained from them and their inverses. A GROUP is a free group if no relation exists between its generators (other than the relationship between an element and its inverse required as one of the defining properties of a group). For example, the additive group of whole numbers is free with a single generator, 1.

See also FREE ABELIAN GROUP, FREE SEMIGROUP

Free Semigroup

A SEMIGROUP with a noncommutative product in which no PRODUCT can ever be expressed more simply in terms of other ELEMENTS.

See also FREE GROUP, SEMIGROUP

Free Tree

A TREE which is not ROOTED, i.e., a normal TREE with no node singled out for special treatment (Skiena 1990, p. 107).

See also ROOTED TREE, TREE

References
Skiena, S. *Implementing Discrete Mathematics: Combinatorics and Graph Theory with Mathematica.* Reading, MA: Addison-Wesley, 1990.

Free Variable

An occurrence of a variable in a LOGIC FORMULA which is not inside the scope of a QUANTIFIER.

See also BOUND, QUANTIFIER, SENTENCE

References
Curry, H. B. *Foundations of Mathematical Logic.* New York: Dover, p. 112, 1977.

Freely

A group acts freely if there are no FIXED POINTS. A point which is fixed by every group element would not be free to move.

See also EFFECTIVE ACTION, FIXED POINT (GROUP), FREE ACTION, GROUP, GROUP ACTION, ISOTROPY GROUP, MATRIX GROUP, ORBIT (GROUP), QUOTIENT SPACE (LIE GROUP), REPRESENTATION, TOPOLOGICAL GROUP, TRANSITIVE

Freemish Crate

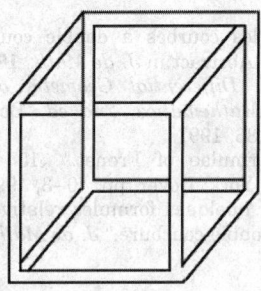

An IMPOSSIBLE FIGURE box which can be drawn but not built.

References

Fineman, M. *The Nature of Visual Illusion.* New York: Dover, pp. 120–22, 1996.
Jablan, S. "Are Impossible Figures Possible?" http://members.tripod.com/~modularity/kulpa.htm.
Pappas, T. "The Impossible Tribar." *The Joy of Mathematics.* San Carlos, CA: Wide World Publ./Tetra, p. 13, 1989.

Freeth's Nephroid

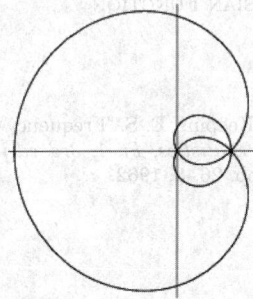

A STROPHOID of a CIRCLE with the POLE O at the CENTER of the CIRCLE and the fixed point P on the CIRCUMFERENCE of the CIRCLE. In a paper published by the London Mathematical Society in 1879, T. J. Freeth described it and various other STROPHOIDS (MacTutor Archive). If the line through P PARALLEL to the y-AXIS cuts the NEPHROID at A, then ANGLE AOP is $3\pi/7$, so this curve can be used to construct a regular HEPTAGON. The POLAR equation is

$$r = a\left[1 + 2\sin\left(\tfrac{1}{2}\theta\right)\right].$$

See also STROPHOID

References

Lawrence, J. D. *A Catalog of Special Plane Curves.* New York: Dover, pp. 175 and 177–78, 1972.
MacTutor History of Mathematics Archive. "Freeth's Nephroid." http://www-groups.dcs.st-and.ac.uk/~history/Curves/Freeths.html.

Frégier's Theorem

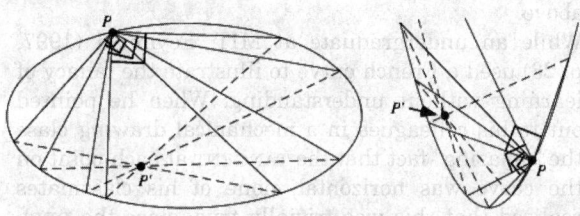

Pick any point P on a CONIC SECTION, and draw a series of RIGHT ANGLES having this point as their vertices. Then the line segments connecting the rays of the RIGHT ANGLES where they intersect the conic section concur in a point p', as illustrated above.

See also CONIC SECTION, RIGHT ANGLE

References

Weisstein, E. W. "Plane Geometry." MATHEMATICA NOTEBOOK PlaneGeometry.m.
Wells, D. *The Penguin Dictionary of Curious and Interesting Geometry.* Middlesex, England: Penguin Books, p. 83, 1991.

Freiman's Constant

The end of the last gap in the LAGRANGE SPECTRUM, given by

$$F \equiv \frac{2221564096 + 283748\sqrt{462}}{491993569} = 4.5278295661\ldots.$$

REAL NUMBERS greater than F are members of the MARKOV SPECTRUM.

See also LAGRANGE SPECTRUM, MARKOV SPECTRUM

References

Conway, J. H. and Guy, R. K. *The Book of Numbers.* New York: Springer-Verlag, pp. 188–89, 1996.

French Curve

French curves are plastic (or wooden) templates having an edge composed of several different curves. French curves are used in drafting (or were before computer-aided design) to draw smooth curves of almost any desired curvature in mechanical drawings. Several typical French curves are illustrated above.

While an undergraduate at MIT, Feynman (1997, p. 23) used a French curve to illustrate the fallacy of learning without understanding. When he pointed out to his colleagues in a mechanical drawing class the "amazing" fact that the TANGENT at each point on the curve was horizontal, none of his classmates realized that this was trivially true, since the DERIVATIVE (tangent) at an extremum (lowest or highest point) of *any* curve is zero (horizontal), as they had already learned in CALCULUS class.

See also CORNU SPIRAL

References

Feynman, R. P. and Leighton, R. "Who Stole the Door?" In *'Surely You're Joking, Mr. Feynman!': Adventures of a Curious Character*. New York: W. W. Norton, 1997.

French Metro Metric

The French metro metric is an example for disproving apparently intuitive but false properties of METRIC SPACES. The metric consists of a distance function on the plane such that for all $a, b \in \mathbb{R}^2$,

$$d(a, b) = \begin{cases} |a - b| & \text{if } a = cb \quad \text{for some} \quad c \in \mathbb{R} \\ |a| + |b| & \text{otherwise,} \end{cases}$$

where $|a|$ is the normal distance function on the plane. This metric has the property that for $r < |a|$, the OPEN BALL of radius r around a is an open line segment along vector a, while for $r > |a|$, the OPEN BALL is the union of a line segment and an OPEN DISK around the origin.

Frenet Formulas

Also known as the Serret-Frenet formulas, these vector differential equations relate inherent properties of a parametrized curve. In matrix form, they can be written

$$\begin{bmatrix} \dot{\mathbf{T}} \\ \dot{\mathbf{N}} \\ \dot{\mathbf{B}} \end{bmatrix} = \begin{bmatrix} 0 & \kappa & 0 \\ -\kappa & 0 & \tau \\ 0 & -\tau & 0 \end{bmatrix} \begin{bmatrix} \mathbf{T} \\ \mathbf{N} \\ \mathbf{B} \end{bmatrix},$$

where $\mathbf{T}$ is the unit TANGENT VECTOR, $\mathbf{N}$ is the unit NORMAL VECTOR, $\mathbf{B}$ is the unit BINORMAL VECTOR, τ is the TORSION, κ is the CURVATURE, and $\dot{\mathbf{x}}$ denotes $d\mathbf{x}/ds$.

See also CENTRODE, FUNDAMENTAL THEOREM OF SPACE CURVES, NATURAL EQUATION

References

Frenet, F. "Sur les courbes à double courbure." Thèse. Toulouse, 1847. Abstract in *J. de Math.* **17**, 1852.
Gray, A. *Modern Differential Geometry of Curves and Surfaces with Mathematica, 2nd ed.* Boca Raton, FL: CRC Press, p. 186, 1997.
Kreyszig, E. "Formulae of Frenet." §15 in *Differential Geometry*. New York: Dover, pp. 40–3, 1991.
Serret, J. A. "Sur quelques formules relatives à la théorie des courbes à double courbure." *J. de Math.* **16**, 1851.

Frequency Curve

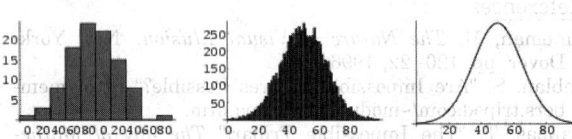

A smooth curve which corresponds to the limiting case of a HISTOGRAM computed for a frequency distribution of a continuous distribution as the number of data points becomes very large.

See also FREQUENCY DISTRIBUTION, FREQUENCY POLYGON, GAUSSIAN FUNCTION

References

Kenney, J. F. and Keeping, E. S. "Frequency Curves." §2.5 in *Mathematics of Statistics, Pt. 1, 3rd ed.* Princeton, NJ: Van Nostrand, pp. 26–8, 1962.

Frequency Distribution

The tabulation of raw data obtained by dividing it into CLASSES of some size and computing the number of data elements (or their fraction out of the total) falling within each pair of CLASS BOUNDARIES. The following table shows the frequency distribution of the data set illustrated by the histogram below.

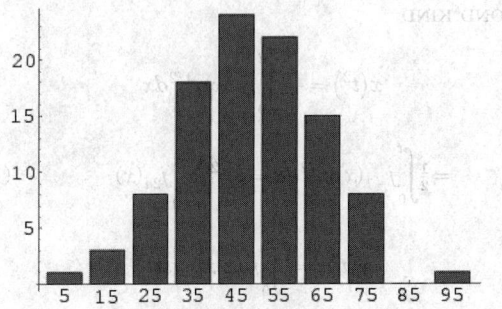

class interval	class mark	absolute frequency	relative frequency	cumulative absolute frequency	relative cumulative frequency
0.00–9.99	5	1	0.01	1	0.01
10.00–9.99	15	3	0.03	4	0.04
20.00–9.99	25	8	0.08	12	0.12
30.00–9.99	35	18	0.18	30	0.30
40.00–9.99	45	24	0.24	54	0.54
50.00–9.99	55	22	0.22	76	0.76
60.00–9.99	65	15	0.15	91	0.91
70.00–9.99	75	8	0.08	99	0.99
80.00–9.99	85	0	0.00	99	0.99
90.00–9.99	95	1	0.01	100	1.00

See also ABSOLUTE FREQUENCY, CLASS, CUMULATIVE FREQUENCY, CLASS BOUNDARIES, HISTOGRAM, RELATIVE FREQUENCY, RELATIVE CUMULATIVE FREQUENCY

References

Kenney, J. F. and Keeping, E. S. "Frequency Distributions." §1.8 in *Mathematics of Statistics, Pt. 1, 3rd ed.* Princeton, NJ: Van Nostrand, pp. 12–9, 1962.

Frequency Polygon

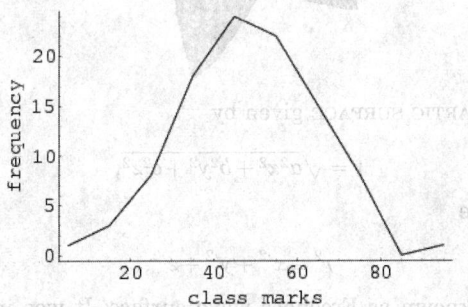

A distribution of values of a discrete variate represented graphically by plotting points (x_1, f_1), (x_2, f_2), ..., (x_k, f_k), and drawing a set of straight line segments connecting adjacent points. It is usually preferable to use a HISTOGRAM for grouped distributions.

See also FREQUENCY CURVE, FREQUENCY DISTRIBUTION, HISTOGRAM, OGIVE

References

Kenney, J. F. and Keeping, E. S. "Frequency Polygons" and "Cumulative Frequency Polygons." §2.3 and 2.6 in *Mathematics of Statistics, Pt. 1, 3rd ed.* Princeton, NJ: Van Nostrand, pp. 24–5 and 28–9, 1962.

Fresnel Integrals

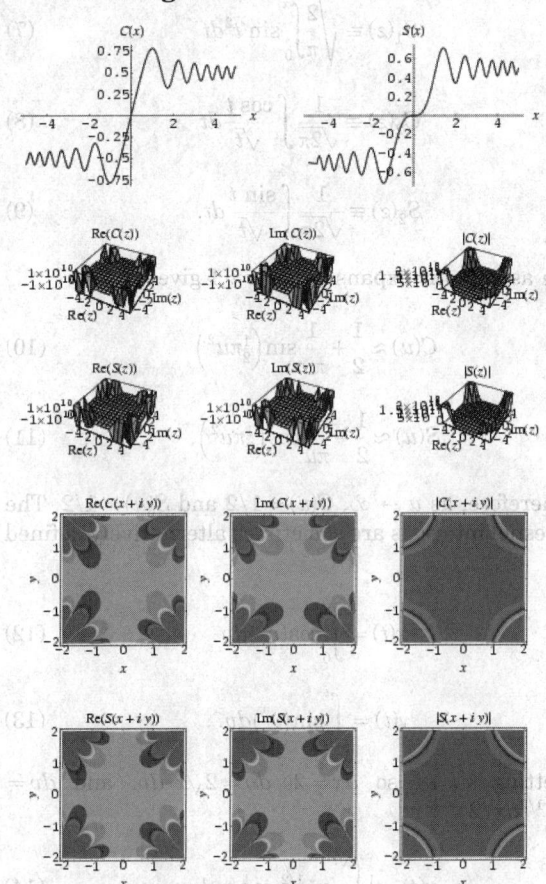

In physics, the Fresnel integrals are most often defined by

$$C(u) + iS(u) \equiv \int_0^u e^{i\pi x^2/2} dx$$

$$= \int_0^u \cos\left(\tfrac{1}{2}\pi x^2\right)dx + i\int_0^u \sin\left(\tfrac{1}{2}\pi x^2\right)dx, \quad (1)$$

so

$$C(u) \equiv \int_0^u \cos\left(\tfrac{1}{2}\pi x^2\right)dx \quad (2)$$

$$S(u) \equiv \int_0^u \sin\left(\tfrac{1}{2}\pi x^2\right)dx. \quad (3)$$

The Fresnel integrals are implemented in *Mathematica* as `FresnelC[z]` and `FresnelC[z]`
They satisfy

$$C(\pm\infty) = -\tfrac{1}{2} \tag{4}$$

$$S(\pm\infty) = \tfrac{1}{2}. \tag{5}$$

Related functions are defined as

$$C_1(z) \equiv \sqrt{\frac{2}{\pi}} \int_0^x \cos t^2 dt \tag{6}$$

$$S_1(z) \equiv \sqrt{\frac{2}{\pi}} \int_0^x \sin t^2 dt \tag{7}$$

$$C_2(z) \equiv \frac{1}{\sqrt{2\pi}} \int \frac{\cos t}{\sqrt{t}} dt \tag{8}$$

$$S_2(z) \equiv \frac{1}{\sqrt{2\pi}} \int \frac{\sin t}{\sqrt{t}} dt. \tag{9}$$

An asymptotic expansion for $x \gg 1$ gives

$$C(u) \approx \frac{1}{2} + \frac{1}{\pi u} \sin\left(\tfrac{1}{2}\pi u^2\right) \tag{10}$$

$$S(u) \approx \frac{1}{2} - \frac{1}{\pi u} \cos\left(\tfrac{1}{2}\pi u^2\right). \tag{11}$$

Therefore, as $u \to \infty$, $C(u) = 1/2$ and $S(u) = 1/2$. The Fresnel integrals are sometimes alternatively defined as

$$x(t) = \int_0^t \cos(v^2) dv \tag{12}$$

$$y(t) = \int_0^t \sin(v^2) dv. \tag{13}$$

Letting $x = v^2$ so $dx = 2v\ dv = 2\sqrt{x}\ dv$, and $dv = x^{-1/2} dx/2$

$$x(t) = \tfrac{1}{2} \int_0^{\sqrt{t}} x^{-1/2} \cos x\ dx \tag{14}$$

$$y(t) = \tfrac{1}{2} \int_0^{\sqrt{t}} x^{-1/2} \sin x\ dx. \tag{15}$$

In this form, they have a particularly simple expansion in terms of SPHERICAL BESSEL FUNCTIONS OF THE FIRST KIND. Using

$$j_0(x) = \frac{\sin x}{x} \tag{16}$$

$$n_1(x) = -j_{-1}(x) = -\frac{\cos x}{x}, \tag{17}$$

where $n_1(x)$ is a SPHERICAL BESSEL FUNCTION OF THE

SECOND KIND

$$x(t^2) = -\tfrac{1}{2} \int_0^t n_1(x) x^{1/2} dx$$

$$= \tfrac{1}{2} \int_0^t j_{-1}(x) x^{1/2} dx = x^{1/2} \sum_{n=0}^{\infty} j_{2n}(x) \tag{18}$$

$$y(t^2) = \tfrac{1}{2} \int_0^t j_0(x) x^{1/2} dx$$

$$= x^{1/2} \sum_{n=0}^{\infty} j_{2n+1}(x). \tag{19}$$

See also CORNU SPIRAL

References

Abramowitz, M. and Stegun, C. A. (Eds.). "Fresnel Integrals." §7.3 in *Handbook of Mathematical Functions with Formulas, Graphs, and Mathematical Tables, 9th printing.* New York: Dover, pp. 300–02, 1972.

Leonard, I. E. "More on Fresnel Integrals." *Amer. Math. Monthly* **95**, 431–33, 1988.

Press, W. H.; Flannery, B. P.; Teukolsky, S. A.; and Vetterling, W. T. "Fresnel Integrals, Cosine and Sine Integrals." §6.79 in *Numerical Recipes in FORTRAN: The Art of Scientific Computing, 2nd ed.* Cambridge, England: Cambridge University Press, pp. 248–52, 1992.

Prudnikov, A. P.; Marichev, O. I.; and Brychkov, Yu. A. "The Generalized Fresnel Integrals $S(x, v)$ and $C(x, v)$." §1.3 in *Integrals and Series, Vol. 3: More Special Functions.* Newark, NJ: Gordon and Breach, p. 24, 1990.

Spanier, J. and Oldham, K. B. "The Fresnel Integrals $S(x)$ and $C(x)$." Ch. 39 in *An Atlas of Functions.* Washington, DC: Hemisphere, pp. 373–83, 1987.

Fresnel's Elasticity Surface

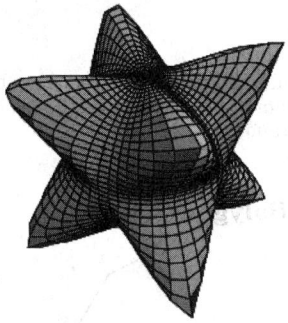

A QUARTIC SURFACE given by

$$r = \sqrt{a^2 x^2 + b^2 y^2 + c^2 z^2},$$

where

$$r^2 \equiv x'^2 + y'^2 + z'^2,$$

also known as Fresnel's wave surface. It was introduced by Fresnel in his studies of crystal optics. The image above shows one particular case of the Fresnel surface (JavaView).

See also QUARTIC SURFACE

References

Fischer, G. (Ed.). *Mathematical Models from the Collections of Universities and Museums.* Braunschweig, Germany: Vieweg, p. 16, 1986.

Fischer, G. (Ed.). Plates 38–9 in *Mathematische Modelle/Mathematical Models, Bildband/Photograph Volume.* Braunschweig, Germany: Vieweg, pp. 38–9, 1986.

JavaView. "Classic Surfaces from Differential Geometry: Fresnel (Single Eigenvalue)." http://www-sfb288.math.tu-berlin.de/vgp/javaview/demo/surface/common/PaSurface_-Fresnel.html.

von Seggern, D. *CRC Standard Curves and Surfaces.* Boca Raton, FL: CRC Press, p. 304, 1993.

Fresnel's Wave Surface

FRESNEL'S ELASTICITY SURFACE

FresnelC

FRESNEL INTEGRALS

FresnelS

FRESNEL INTEGRALS

Frey Curve

Let $a^p + b^p = c^p$ be a solution to FERMAT'S LAST THEOREM. Then the corresponding Frey curve is

$$y^2 = x(x - a^p)(x + b^p). \tag{1}$$

Frey showed that such curves cannot be MODULAR, so if the TANIYAMA-SHIMURA CONJECTURE were true, Frey curves couldn't exist and FERMAT'S LAST THEOREM would follow with b EVEN and $a \equiv -1 \pmod 4$. Frey curves are SEMISTABLE. Invariants include the DISCRIMINANT

$$(a^p - 0)^2 (-b^p - 0)[a^p - (-b)^p]^2 = a^{2p} b^{2p} c^{2p}. \tag{2}$$

The MINIMAL DISCRIMINANT is

$$\Delta = 2^{-8} a^{2p} b^{2p} c^{2p}, \tag{3}$$

the CONDUCTOR is

$$N = \prod_{l|abc} l, \tag{4}$$

and the J-INVARIANT is

$$j = \frac{2^8 (a^{2p} + b^{2p} + a^p b^p)^3}{a^{2p} b^{2p} c^{2p}} = \frac{2^8 (c^{2p} - b^p c^p)^3}{(abc)^{2p}}. \tag{5}$$

See also ELLIPTIC CURVE, FERMAT'S LAST THEOREM, TANIYAMA-SHIMURA CONJECTURE

References

Cox, D. A. "Introduction to Fermat's Last Theorem." *Amer. Math. Monthly* **101**, 3–4, 1994.

Gouvêa, F. Q. "A Marvelous Proof." *Amer. Math. Monthly* **101**, 203–22, 1994.

Frey Elliptic Curve

FREY CURVE

Friend

A friend of a number n is another number m such that (m, n) is a FRIENDLY PAIR.

See also FRIENDLY PAIR, SOLITARY NUMBER

References

Anderson, C. W. and Hickerson, D. Problem 6020. "Friendly Integers." *Amer. Math. Monthly* **84**, 65–6, 1977.

Friendly Giant Group

MONSTER GROUP

Friendly Number

AMICABLE PAIR, FRIENDLY NUMBER

Friendly Pair

Define

$$\sum(n) \equiv \frac{\sigma(n)}{n},$$

where $\sigma(n)$ is the DIVISOR FUNCTION. Then a PAIR of distinct numbers (k, m) is a friendly pair (and k is said to be a FRIEND of m) if

$$\sum(k) = \sum(m).$$

For example, $(4320, 4680)$ are a friendly pair, since $\sigma(4320) = 15120$, $\sigma(4680) = 16380$, and

$$\sum(4320) \equiv \frac{15120}{4320} = \frac{7}{2}$$

$$\sum(4680) \equiv \frac{16380}{4680} = \frac{7}{2}.$$

The first few friendly pairs, ordered by smallest maximum element are $(6, 28)$, $(30, 140)$, $(80, 200)$, $(40, 224)$, $(12, 234)$, $(84, 270)$, $(66, 308)$, ... (Sloane's A050972 and A050973).

Numbers which do not have FRIENDS are called SOLITARY NUMBERS. A sufficient (but not necessary) condition for n to be a SOLITARY NUMBER is that $(\sigma(n), n) = 1$, where (a, b) is the GREATEST COMMON DIVISOR of a and b.

Hoffman (1998, p. 45) uses the term "friendly numbers" to describe AMICABLE PAIRS.

See also ALIQUOT SEQUENCE, AMICABLE PAIR, FRIEND, SOLITARY NUMBER

References

Anderson, C. W. and Hickerson, D. Problem 6020. "Friendly Integers." *Amer. Math. Monthly* **84**, 65–6, 1977.
Hoffman, P. *The Man Who Loved Only Numbers: The Story of Paul Erdos and the Search for Mathematical Truth.* New York: Hyperion, 1998.
Sloane, N. J. A. Sequences A050972 and A050973 in "An On-Line Version of the Encyclopedia of Integer Sequences." http://www.research.att.com/~njas/sequences/eisonline.html.

Frieze Pattern

In general, a frieze consists of repeated copies of a single motif.

$$
\begin{array}{ccc}
 & b & \\
a & & d \\
 & c &
\end{array}
$$

Conway and Guy (1996) define a frieze pattern as an arrangement of numbers at the intersection of two sets of perpendicular diagonals such that $a + d = b + c + 1$ (for an additive frieze pattern) or $ad = bc + 1$ (for a multiplicative frieze pattern) in each diamond.

See also TESSELLATION, TILING

References

Conway, J. H. and Coxeter, H. S. M. "Triangulated Polygons and Frieze Patterns." *Math. Gaz.* **57**, 87–4, 1973.
Conway, J. H. and Guy, R. K. In *The Book of Numbers.* New York: Springer-Verlag, pp. 74–6 and 96–7, 1996.
Wells, D. *The Penguin Dictionary of Curious and Interesting Geometry.* London: Penguin, pp. 83–4, 1991.

Frivolous Theorem of Arithmetic

Almost all natural numbers are very, very, very large.

See also LARGE NUMBER

References

Steinbach, P. *Field Guide to Simple Graphs.* Albuquerque, NM: Design Lab, 1990.

Frobenius Map

A map $x \mapsto x^p$ where p is a PRIME.

Frobenius Method

If x_0 is an ordinary point of the ORDINARY DIFFERENTIAL EQUATION, expand y in a TAYLOR SERIES about x_0, letting

$$y = \sum_{n=0}^{\infty} a_n x^n. \tag{1}$$

Plug y back into the ODE and group the COEFFICIENTS by POWER. Now, obtain a RECURRENCE RELATION for the nth term, and write the TAYLOR SERIES in terms of the a_ns. Expansions for the first few derivatives are

$$y = \sum_{n=0}^{\infty} a_n x^n \tag{2}$$

$$y' = \sum_{n=1}^{\infty} n a_n x^{n-1} = \sum_{n=0}^{\infty} (n+1) a_{n+1} x^n \tag{3}$$

$$y'' = \sum_{n=2}^{\infty} n(n-1) a_n x^{n-2} = \sum_{n=0}^{\infty} (n+2)(n+1) a_{n+2} x^n. \tag{4}$$

If x_0 is a regular singular point of the ORDINARY DIFFERENTIAL EQUATION,

$$P(x) y'' + Q(x) y' + R(x) y = 0, \tag{5}$$

solutions may be found by the Frobenius method or by expansion in a LAURENT SERIES. In the Frobenius method, assume a solution OF THE FORM

$$y = x^k \sum_{n=0}^{\infty} a_n x^n, \tag{6}$$

so that

$$y = x^k \sum_{n=0}^{\infty} a_n x^n = \sum_{n=0}^{\infty} a_n x^{n+k} \tag{7}$$

$$y' = \sum_{n=0}^{\infty} a_n (n+k) x^{k+n-1} \tag{8}$$

$$y'' = \sum_{n=0}^{\infty} a_n (n+k)(n+k-1) x^{k+n-2}. \tag{9}$$

Now, plug y back into the ODE and group the COEFFICIENTS by POWER to obtain a recursion FORMULA for the a_nth term, and then write the TAYLOR SERIES in terms of the a_ns. Equating the a_0 term to 0 will produce the so-called INDICIAL EQUATION, which will give the allowed values of k in the TAYLOR SERIES.

FUCHS'S THEOREM guarantees that at least one POWER SERIES solution will be obtained when applying the Frobenius method if the expansion point is an ordinary, or regular, SINGULAR POINT. For a regular SINGULAR POINT, a LAURENT SERIES expansion can also be used. Expand y in a LAURENT SERIES, letting

$$y = c_{-n} x^{-n} + \cdots + c_0 + c_1 x + \cdots + c_n x^n + \cdots \tag{10}$$

Plug y back into the ODE and group the COEFFICIENTS by POWER. Now, obtain a recurrence FORMULA for the c_nth term, and write the TAYLOR EXPANSION in terms of the c_ns.

Frobenius Pseudoprime

See also FUCHS'S THEOREM, ORDINARY DIFFERENTIAL EQUATION

References

Arfken, G. "Series Solutions--Frobenius' Method." §8.5 in *Mathematical Methods for Physicists, 3rd ed.* Orlando, FL: Academic Press, pp. 454–67, 1985.

Frobenius Pseudoprime

Let $f(x)$ be a MONIC POLYNOMIAL of degree d with discriminant Δ. Then an ODD INTEGER n with $(n, f(0)\Delta) = 1$ is called a Frobenius pseudoprime with respect to $f(x)$ if it passes a certain algorithm given by Grantham (1996). A Frobenius pseudoprime with respect to a POLYNOMIAL $f(x) \in \mathbb{Z}[x]$ is then a composite Frobenius probably prime with respect to the POLYNOMIAL $x - a$.

While 323 is the first LUCAS PSEUDOPRIME with respect to the Fibonacci polynomial $x^2 - x - 1$, the first Frobenius pseudoprime is 5777. If $f(x) = x^3 - rx^2 + sx - 1$, then any Frobenius pseudoprime n with respect to $f(x)$ is also a PERRIN PSEUDOPRIME. Grantham (1997) gives a test based on Frobenius pseudoprimes which is passed by COMPOSITE NUMBERS with probability at most 1/7710.

See also PERRIN PSEUDOPRIME, PSEUDOPRIME, STRONG FROBENIUS PSEUDOPRIME

References

Grantham, J. "Frobenius Pseudoprimes." 1996. http://www.clark.net/pub/grantham/pseudo/pseudo1.ps
Grantham, J. "A Frobenius Probable Prime Test with High Confidence." 1997. http://www.clark.net/pub/grantham/pseudo/pseudo2.ps
Grantham, J. "Pseudoprimes/Probable Primes." http://www.clark.net/pub/grantham/pseudo/.

Frobenius Theorem

Let $A = a_{ij}$ be a MATRIX with POSITIVE COEFFICIENTS so that $a_{ij} > 0$ for all $i, j = 1, 2, \ldots, n$, then A has a POSITIVE EIGENVALUE λ_0, and all its EIGENVALUES lie on the CLOSED DISK

$$|z| \le \lambda_0.$$

See also CLOSED DISK, OSTROWSKI'S THEOREM

References

Gradshteyn, I. S. and Ryzhik, I. M. *Tables of Integrals, Series, and Products, 6th ed.* San Diego, CA: Academic Press, p. 1121, 2000.

Frobenius Triangle Identities

Let C_{LM} be a PADÉ APPROXIMANT. Then

$$C_{(L+1)/M}S_{(L-1)/M} - C_{L/(M+1)}S_{L/(M+1)} = C_{L/M}S_{L/M} \quad (1)$$

$$C_{L/(M+1)}S_{(L+1)/M} - C_{(L+1)/M}S_{L/(M+1)} = C_{(L+1)/(M+1)}XS_{L/M} \quad (2)$$

$$C_{(L+1)/M}S_{L/M} - C_{L/M}S_{(L+1)/M} = C_{(L+1)/(M+1)}xS_{L/(M-1)} \quad (3)$$

$$C_{L/(M+1)}S_{L/M} - C_{L/M}S_{L/(M+1)} = C_{(L+1)/(M+1)}xS_{(L-1)/M}, \quad (4)$$

where

$$S_{L/M} = G(x)P_L(x) + H(x)Q_M(x) \quad (5)$$

and C is the C-DETERMINANT.

See also C-DETERMINANT, PADÉ APPROXIMANT

References

Baker, G. A. Jr. *Essentials of Padé Approximants in Theoretical Physics.* New York: Academic Press, p. 31, 1975.

Frobenius-König Theorem

The PERMANENT of an $n \times n$ INTEGER MATRIX with all entries either 0 or 1 is 0 IFF the MATRIX contains an $r \times s$ submatrix of 0s with $r + s = n + 1$. This result follows from the KÖNIG-EGEVÁRY THEOREM.

See also INTEGER MATRIX, KÖNIG-EGEVÁRY THEOREM, PERMANENT

Frobenius-Perron Equation

$$\rho_{n+1}(x) = \int \rho_n(y)\delta[x - M(y)]dy,$$

where $\delta(x)$ is a DELTA FUNCTION, $M(x)$ is a map, and ρ is the NATURAL INVARIANT.

See also NATURAL INVARIANT, PERRON-FROBENIUS OPERATOR

References

Ott, E. *Chaos in Dynamical Systems.* New York: Cambridge University Press, p. 51, 1993.

Frontier

BOUNDARY

Frucht Graph

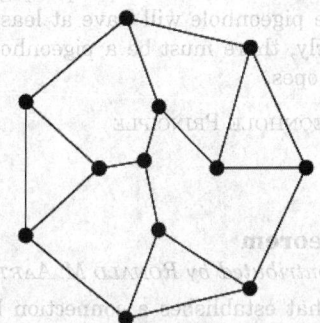

The smallest CUBIC GRAPH whose automorphism

group consists only of the IDENTITY ELEMENT (Skiena 1990, p. 185).

See also CUBIC GRAPH, GRAPH AUTOMORPHISM

References

Bondy, J. A. and Murty, U. S. R. *Graph Theory with Applications.* New York: North Holland, p. 235, 1976.
Frucht, R. "Herstellung von Graphen mit vorgegebener abstrakter Gruppe." *Compos. Math.* **6**, 239–50, 1939.
Skiena, S. *Implementing Discrete Mathematics: Combinatorics and Graph Theory with Mathematica.* Reading, MA: Addison-Wesley, 1990.

Frugal Number

WASTEFUL NUMBER

Frullani's Integral

If S' is continuous and the integral converges,

$$\int_0^\infty \frac{f(ax) - f(bx)}{x}\,dx = [f(0) - f(\infty)]\ln\left(\frac{b}{a}\right).$$

References

Jeffreys, H. and Jeffreys, B. S. "Frullani's Integrals." §12.16 in *Methods of Mathematical Physics, 3rd ed.* Cambridge, England: Cambridge University Press, pp. 406–07, 1988.
Spiegel, M. R. *Mathematical Handbook of Formulas and Tables.* New York: McGraw-Hill, 1968.

Frustum

The portion of a solid which lies between two PARALLEL PLANES cutting the solid. Degenerate cases are obtained for finite solids by cutting with a single PLANE only.

See also CONICAL FRUSTUM, PYRAMIDAL FRUSTUM, SPHERICAL SEGMENT

Fubini Principle

If the average number of envelopes per pigeonhole is a, then some pigeonhole will have at least a envelopes. Similarly, there must be a pigeonhole with at most a envelopes.

See also PIGEONHOLE PRINCIPLE

Fubini Theorem

This entry contributed by RONALD M. AARTS

A theorem that establishes a connection between a MULTIPLE INTEGRAL and a REPEATED one. Under certain assumptions the following equality holds:

$$\iint_{R^{m+n}} f(x,y)d(x,y) = \int_{R^n} dy \int_{R^m} f(x,y)dx.$$

See also MULTIPLE INTEGRAL, REPEATED INTEGRAL

References

Fubine, G. "Sugli integrali multipli." *Opere scelte, Vol. 2.* Cremonese, pp. 243–49, 1958.
Samko, S. G.; Kilbas, A. A.; and Marichev, O. I. *Fractional Integrals and Derivatives.* Yverdon, Switzerland: Gordon and Breach, p. 9, 1993.

Fuchs's Theorem

At least one POWER SERIES solution will be obtained when applying the FROBENIUS METHOD if the expansion point is an ordinary, or regular, SINGULAR POINT. The number of ROOTS is given by the ROOTS of the INDICIAL EQUATION.

References

Arfken, G. *Mathematical Methods for Physicists, 3rd ed.* Orlando, FL: Academic Press, pp. 462–63, 1985.

Fuchsian System

A system of linear differential equations

$$\frac{dy}{dz} = A(z)y,$$

with $A(z)$ an ANALYTIC $n \times n$ MATRIX, for which the MATRIX $A(z)$ is ANALYTIC in $\mathbb{C}\backslash\{a_1,\ldots,a_N\}$ and has a POLE of order 1 at a_j for $j = 1$, ..., N. A system is Fuchsian IFF there exist $n \times n$ matrices B_1, ..., B_N with entries in $\mathbb{Z}$ such that

$$A(z) = \sum_{j=1}^N \frac{B_j}{z - a_j}$$

$$\sum_{j=1}^N B_j = v.$$

Fuglede's Conjecture

Fuglede (1974) conjectured that a domain Ω admits a SPECTRUM IFF it is possible to tile $\mathbb{R}^d$ by a family of translates of Ω. Fuglede proved the conjecture in the special case that the tiling set or the spectrum are lattice subsets of $\mathbb{R}^d$ and Iosevich *et al.* (1999) proved that no smooth symmetric convex body Ω with at least one point of nonvanishing GAUSSIAN CURVATURE can admit an orthogonal basis of exponentials. However, the general conjecture is still far from being proved (Iosevich *et al.* 1999).

See also SPECTRUM (OPERATOR)

References

Fuglede, B. "Commuting Self-Adjoint Partial Differential Operators and a Group Theoretic Problem." *J. Func. Anal.* **16**, 101–21, 1974.

Iosevich, A.; Katz, N. H.; and Tao, T. Convex Bodies with a Point of Curvature Do Not Have Fourier Bases. 23 Nov 1999. http://xxx.lanl.gov/abs/math.CA/9911167/.

Jorgensen, P. E. T. and Pedersen, S. "Orthogonal Harmonic Analysis of Fractal Measures." *Elec. Res. Announc. Amer. Math. Soc.* **4**, 35–2, 1998.

Lagarias, J. and Wang, Y. "Spectral Sets and Factorizations of Finite Abelian Groups." *J. Func. Anal.* **145**, 73–8, 1997.

Fuhrmann Center

The center of the FUHRMANN CIRCLE, given by the MIDPOINT of the line joining the NAGEL POINT and ORTHOCENTER (which forms a DIAMETER of the FUHRMANN CIRCLE).

See also FUHRMANN CIRCLE, NAGEL POINT, ORTHOCENTER

Fuhrmann Circle

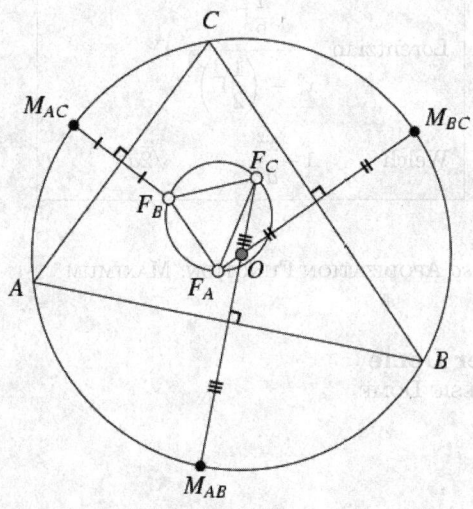

The CIRCUMCIRCLE of the FUHRMANN TRIANGLE. The ORTHOCENTER H, NAGEL POINT Na, and at least six other noteworthy points lie on the Fuhrmann circle (Honsberger 1995, p. 49). In particular, HNa is a DIAMETER of the Fuhrmann circle. It also passes through the points T, U, and V which are a distance $2r$ along the ALTITUDES from the vertices, where r is the INRADIUS of $\triangle ABC$ (Honsberger 1995, p. 52).

See also ALTITUDE, FUHRMANN TRIANGLE, INRADIUS, MID-ARC POINTS, NAGEL POINT, ORTHOCENTER

References

Coolidge, J. L. *A Treatise on the Geometry of the Circle and Sphere.* New York: Chelsea, p. 58, 1971.

Fuhrmann, W. *Synthetische Beweise Planimetrischer Sätze.* Berlin, p. 107, 1890.

Honsberger, R. "The Fuhrmann Circle." Ch. 6 in *Episodes in Nineteenth and Twentieth Century Euclidean Geometry.* Washington, DC: Math. Assoc. Amer., pp. 49–2, 1995.

Johnson, R. A. *Modern Geometry: An Elementary Treatise on the Geometry of the Triangle and the Circle.* Boston, MA: Houghton Mifflin, pp. 228–29, 1929.

Fuhrmann Triangle

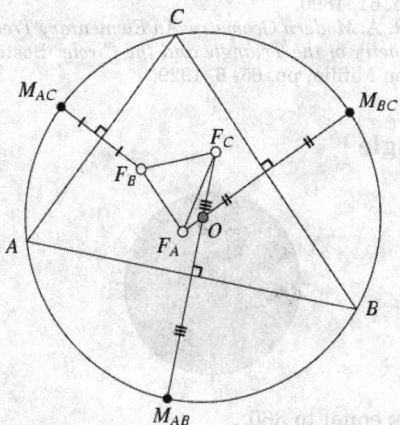

The Fuhrmann triangle of a TRIANGLE $\triangle ABC$ is the TRIANGLE $\triangle F_C F_B F_A$ formed by reflecting the MID-ARC POINTS M_{AB}, M_{AC}, M_{BC} about the lines AB, AC, and BC. The CIRCUMCIRCLE of the Fuhrmann triangle is called the FUHRMANN CIRCLE, and the lines $F_A M_{BC}$, $F_B M_{AC}$, and $F_C M_{AB}$ concur at the CIRCUMCENTER O.

See also FUHRMANN CENTER, FUHRMANN CIRCLE, MID-ARC POINTS

References

Fuhrmann, W. *Synthetische Beweise Planimetrischer Sätze.* Berlin, p. 107, 1890.

Johnson, R. A. *Modern Geometry: An Elementary Treatise on the Geometry of the Triangle and the Circle.* Boston, MA: Houghton Mifflin, pp. 228–29, 1929.

Fuhrmann's Theorem

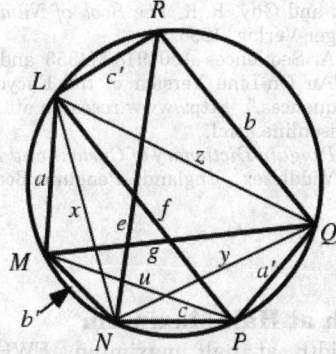

Let the opposite sides of a convex CYCLIC HEXAGON be a, a', b, b', c, and c', and let the DIAGONALS e, f, and g be so chosen that a, a', and e have no common VERTEX (and likewise for b, b', and f), then

$$efg = aa'e + bb'f + cc'g + abc + a'b'c'.$$

This is an extension of PTOLEMY'S THEOREM to the HEXAGON.

See also CYCLIC HEXAGON, HEXAGON, PTOLEMY'S THEOREM

References
Fuhrmann, W. *Synthetische Beweise Planimetrischer Sätze.* Berlin, p. 61, 1890.
Johnson, R. A. *Modern Geometry: An Elementary Treatise on the Geometry of the Triangle and the Circle.* Boston, MA: Houghton Mifflin, pp. 65–6, 1929.

Full Angle

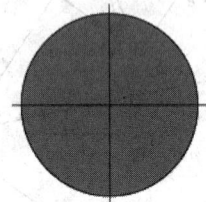

An ANGLE equal to 360°.

See also ACUTE ANGLE, ANGLE, OBTUSE ANGLE, REFLEX ANGLE, RIGHT ANGLE, STRAIGHT ANGLE

Full Reptend Prime
A PRIME p for which $1/p$ has a maximal period DECIMAL EXPANSION of $p - 1$ DIGITS, sometimes called a long prime (Conway and Guy 1996, pp. 157–63 and 166–71). A prime is full reptend IFF 10 is a PRIMITIVE ROOT modulo p. No general method is known for finding full reptend primes. The first few numbers with maximal decimal expansions are 7, 17, 19, 23, 29, 47, 59, 61, 97, ... (Sloane's A001913).

See also DECIMAL EXPANSION, PRIMITIVE ROOT

References
Conway, J. H. and Guy, R. K. *The Book of Numbers.* New York: Springer-Verlag, 1996.
Sloane, N. J. A. Sequences A001913/M4353 and A006883/M1745 in "An On-Line Version of the Encyclopedia of Integer Sequences." http://www.research.att.com/~njas/sequences/eisonline.html.
Wells, D. *The Penguin Dictionary of Curious and Interesting Numbers.* Middlesex, England: Penguin Books, p. 71, 1986.

Full Width at Half Maximum
The full width at half maximum (FWHM) is a parameter commonly used to describe the width of a "bump" on a curve or function. It is given by the distance between points on the curve at which the function reaches half its maximum value. The following table gives the analytic and numerical full widths for several common curves.

Function	Formula	FWHM
Bartlett	$1 - \dfrac{|x|}{a}$	a
Blackman		$0.810957a$
Connes	$\left(1 - \dfrac{x^2}{a^2}\right)$	$\sqrt{4 - 2\sqrt{2}a}$
Cosine	$\cos\left(\dfrac{\pi x}{2a}\right)$	$\dfrac{4}{3}a$
Gaussian	$e^{-x^2/(2\sigma^2)}$	$2\sqrt{2\ln 2}\,\sigma$
Hamming		$1.05543a$
Hanning		a
Lorentzian	$\dfrac{\frac{1}{2}\Gamma}{x^2 + \left(\frac{1}{2}\Gamma\right)^2}$	Γ
Welch	$1 - \dfrac{x^2}{a^2}$	$\sqrt{2}a$

See also APODIZATION FUNCTION, MAXIMUM

Fuller Dome
GEODESIC DOME

Function
A relation which uniquely associates members of one SET with members of another SET. More formally, a function from A to B is an object f such that every $a \in A$ is uniquely associated with an object $f(a) \in B$. A function is therefore a MANY-TO-ONE (or sometimes ONE-TO-ONE) relation. Examples of functions include $\sin x$ (MANY-TO-ONE), x (ONE-TO-ONE), x^2 (two-to-one except for the single point $x = 0$), etc. The term "MAP" is synonymous with function.

Several notations are commonly used to represent functions. The most rigorous notation is $f : x \rightarrow f(x)$, which specifies that f is function acting upon a single number x (i.e., f is a univariate, or one-variable, function) and returning a value $f(x)$. To be even more precise, a notation like "$f : \mathbb{R} \rightarrow \mathbb{R}$, where $f(x) = x^2$" is sometimes used to explicitly specify the domain and range of the function. The slightly different "maps to" notation $f : x \mapsto f(x)$ is sometimes also used when the function is explicitly considered as a "map."

Generally speaking, the symbol f refers to the function itself, while $f(x)$ refers to the *value* taken by the function when evaluated at a point x. However, especially in more introductory texts, the notation $f(x)$ is commonly used to refer to the *function* f itself (as opposed to the value of the function evaluated at x). In this context, the argument x is considered to be a DUMMY VARIABLE whose presence indicates that the function f takes a single argument (as opposed to $f(x,y)$, etc.). While this notation is deprecated by professional mathematicians, it is the more familiar one for most nonprofessionals. Therefore, unless indicated otherwise by context, the notation $f(x)$ is taken in this work to be a shorthand for the more rigorous $f : x \to f(x)$.

Poincaré remarked with regard to the proliferation of pathological functions, "Formerly, when one invented a new function, it was to further some practical purpose; today one invents them in order to make incorrect the reasoning of our fathers, and nothing more will ever be accomplished by these inventions."

References

Abramowitz, M. and Stegun, C. A. (Eds.). "Miscellaneous Functions." Ch. 27 in *Handbook of Mathematical Functions with Formulas, Graphs, and Mathematical Tables, 9th printing.* New York: Dover, pp. 997–010, 1972.

Arfken, G. "Special Functions." Ch. 13 in *Mathematical Methods for Physicists, 3rd ed.* Orlando, FL: Academic Press, pp. 712–59, 1985.

Press, W. H.; Flannery, B. P.; Teukolsky, S. A.; and Vetterling, W. T. "Special Functions." Ch. 6 in *Numerical Recipes in FORTRAN: The Art of Scientific Computing, 2nd ed.* Cambridge, England: Cambridge University Press, pp. 205–65, 1992.

Weisstein, E. W. "Books about Special Functions." http://www.treasure-troves.com/books/SpecialFunctions.html.

Function Element

A function element is an ORDERED PAIR (f, U) where U is a disk $D(Z_0, r)$ and f is an ANALYTIC FUNCTION defined on U. If W is an OPEN SET, then a function element in W is a pair (f, U) such that $U \subseteq W$.

References

Krantz, S. G. "Function Elements." §10.1.3 in *Handbook of Complex Analysis.* Boston, MA: Birkhäuser, p. 128, 1999.

Function Field

A finite extension $K = Z(z)(w)$ of the FIELD $\mathbb{C}(z)$ of RATIONAL FUNCTIONS in the indeterminate z, i.e., w is a ROOT of a POLYNOMIAL $a_0 + a_1\alpha + a_2\alpha^2 + \ldots + a_n\alpha^n$, where $a_i \in \mathbb{C}(z)$. Function fields are sometimes called algebraic function fields.

See also LOCAL FIELD, NUMBER FIELD, RIEMANN SURFACE

Function of the First Kind

FIRST KIND

Function of the Second Kind

SECOND KIND

Function of the Third Kind

THIRD KIND

Function Space

$f(I)$ is the collection of all real-valued continuous functions defined on some interval I. $f^{(n)}(I)$ is the collection of all functions $\in f(I)$ with continuous nth DERIVATIVES. A function space is a TOPOLOGICAL VECTOR SPACE whose "points" are functions.

See also FUNCTIONAL, FUNCTIONAL ANALYSIS, OPERATOR

Functional

A functional is a real-valued function on a VECTOR SPACE V, usually of functions. For example, the ENERGY functional on the UNIT DISK D assigns a number to any differentiable function $f : D \to \mathbb{R}$,

$$E(f) : \int D \|\nabla f\|^2 dA.$$

For the functional to be continuous, it is necessary for the VECTOR SPACE V of functions to have an appropriate TOPOLOGY. The widespread use of functionals in applications, such as the CALCULUS OF VARIATIONS, gave rise to FUNCTIONAL ANALYSIS.

The reason the term "functional" is used is because V can be a space of functions, e.g.,

$$V = \{f : [0, 1] \to \mathbb{R} \text{ such that } f \text{ is continuous}\}$$

in which case $T(f) = f(0)$ is a LINEAR FUNCTIONAL on V.

See also CALCULUS OF VARIATIONS, COERCIVE FUNCTIONAL, CURRENT, ELLIPTIC FUNCTIONAL, EULER-LAGRANGE DIFFERENTIAL EQUATION, FUNCTIONAL ANALYSIS, FUNCTIONAL EQUATION, GENERALIZED FUNCTION, LAPLACIAN, LAX-MILGRAM THEOREM, LINEAR FUNCTIONAL, OPERATOR, RIESZ REPRESENTATION THEOREM, VECTOR SPACE

Functional Analysis

A branch of mathematics concerned with infinite dimensional spaces (mainly FUNCTION SPACES) and mappings between them. The SPACES may be of different, and possibly INFINITE, DIMENSIONS. These mappings are called OPERATORS or, if the range is on the REAL line or in the COMPLEX PLANE, FUNCTIONALS.

See also FUNCTIONAL, FUNCTIONAL EQUATION, GENERALIZED FUNCTION, OPERATOR

References
Balakrishnan, A. V. *Applied Functional Analysis, 2nd ed.* New York: Springer-Verlag, 1981.
Berezansky, Y. M.; Us, G. F.; and Sheftel, Z. G. *Functional Analysis, Vol. 1.* Boston, MA: Birkhäuser, 1996.
Berezansky, Y. M.; Us, G. F.; and Sheftel, Z. G. *Functional Analysis, Vol. 2.* Boston, MA: Birkhäuser, 1996.
Birkhoff, G. and Kreyszig, E. "The Establishment of Functional Analysis." *Historia Math.* **11**, 258–21, 1984.
Hutson, V. and Pym, J. S. *Applications of Functional Analysis and Operator Theory.* New York: Academic Press, 1980.
Kreyszig, E. *Introductory Functional Analysis with Applications.* New York: Wiley, 1989.
Yoshida, K. *Functional Analysis and Its Applications.* New York: Springer-Verlag, 1971.
Zeidler, E. *Nonlinear Functional Analysis and Its Applications.* New York: Springer-Verlag, 1989.
Zeidler, E. *Applied Functional Analysis: Applications to Mathematical Physics.* New York: Springer-Verlag, 1995.

Functional Calculus
An early name for CALCULUS OF VARIATIONS. The term is also sometimes used in place of PREDICATE CALCULUS.

Functional Congruence
A CONGRUENCE OF THE FORM

$$f(x) \equiv g(x) \pmod{n}$$

where $f(x)$ and $g(x)$ are both INTEGER POLYNOMIALS. Functional congruences are sometimes also called "identical congruences" (Nagell 1951, p. 74).

See also CONGRUENCE

References
Nagell, T. "Algebraic Congruences and Functional Congruences." §22 in *Introduction to Number Theory.* New York: Wiley, pp. 73–6, 1951.

Functional Derivative
A generalization of the concept of the DERIVATIVE to GENERALIZED FUNCTIONS.

Functional Distribution
GENERALIZED FUNCTION

Functional Equation
An equation OF THE FORM $f(x, y, ...) = 0$, where f contains a finite number of independent variables, known functions, and unknown functions which are

to be solved for. Many properties of functions can be determined by studying the types of functional equations they satisfy. For example, the GAMMA FUNCTION $\Gamma(z)$ satisfies the functional equations

$$\Gamma(1 + z) = z\Gamma(z)$$
$$\Gamma(1 - z) = -z\Gamma(-z).$$

See also ABEL'S DUPLICATION FORMULA, ABEL'S FUNCTIONAL EQUATION, FUNCTIONAL ANALYSIS

References
Kuczma, M. *Functional Equations in a Single Variable.* Warsaw, Poland: Polska Akademia Nauk, 1968.
Kuczma, M. *An Introduction to the Theory of Functional Equations and Inequalities: Cauchy's Equation and Jensen's Inequality.* Warsaw, Poland: Uniwersitet Slaski, 1985.
Kuczma, M.; Choczewski, B.; and Ger, R. *Iterative Functional Equations.* Cambridge, England: Cambridge University Press, 1990.

Functional Graph
A functional graph is a DIGRAPH in which each vertex has outdegree one, and can therefore be specified by a function mapping $\{1, ..., n\}$ onto itself. Functional graphs are implemented as FunctionalGraph[*f*, *n*] in the *Mathematica* add-on package Discrete-Math`Combinatorica` (which can be loaded with the command < <DiscreteMath`).

References
Skiena, S. "Functional Graphs." §4.5.2 in *Implementing Discrete Mathematics: Combinatorics and Graph Theory with Mathematica.* Reading, MA: Addison-Wesley, pp. 164–65, 1990.

Functor
A function between CATEGORIES which maps objects to objects and MORPHISMS to MORPHISMS. Functors exist in both covariant and contravariant types.

See also CATEGORY, EILENBERG-STEENROD AXIOMS, MORPHISM, SCHUR FUNCTOR

Fundamental Class
The canonical generator of the nonvanishing HOMOLOGY GROUP on a TOPOLOGICAL MANIFOLD.

See also CHERN NUMBER, PONTRYAGIN NUMBER, STIEFEL-WHITNEY NUMBER

Fundamental Continuity Theorem
Given two UNIVARIATE POLYNOMIALS of the same order whose first p COEFFICIENTS (but *not* the first $p - 1$) are 0 where the COEFFICIENTS of the second approach the corresponding COEFFICIENTS of the first as limits, the second POLYNOMIAL will have exactly p

roots that increase indefinitely. Furthermore, exactly k ROOTS of the second will approach each ROOT of multiplicity k of the first as a limit.

References

Coolidge, J. L. *A Treatise on Algebraic Plane Curves.* New York: Dover, p. 4, 1959.

Fundamental Discriminant

$-D$ is a fundamental discriminant if D is a POSITIVE INTEGER which is not DIVISIBLE by any square of an ODD PRIME and which satisfies $D \equiv 3 \pmod 4$ or $D \equiv 4, 8 \pmod{16}$.

See also DISCRIMINANT

References

Atkin, A. O. L. and Morain, F. "Elliptic Curves and Primality Proving." *Math. Comput.* **61**, 29–8, 1993.

Borwein, J. M. and Borwein, P. B. *Pi & the AGM: A Study in Analytic Number Theory and Computational Complexity.* New York: Wiley, p. 294, 1987.

Cohn, H. *Advanced Number Theory.* New York: Dover, 1980.

Dickson, L. E. *History of the Theory of Numbers, Vols. 1–.* New York: Chelsea, 1952.

Fundamental Forms

There are three types of so-called fundamental forms. The most important are the first and second (since the third can be expressed in terms of these). The fundamental forms are extremely important and useful in determining the metric properties of a surface, such as LINE ELEMENT, AREA ELEMENT, NORMAL CURVATURE, GAUSSIAN CURVATURE, and MEAN CURVATURE. Let M be a REGULAR SURFACE with $\mathbf{v}_P, \mathbf{w}_P$ points in the TANGENT SPACE M_P of M. Then the FIRST FUNDAMENTAL FORM is the INNER PRODUCT of tangent vectors,

$$\mathbf{I}(\mathbf{v}_P, \mathbf{w}_P) = \mathbf{v}_P \cdot \mathbf{w}_P. \qquad (1)$$

For $M \in \mathbb{R}^3$, the SECOND FUNDAMENTAL FORM is the symmetric bilinear form on the TANGENT SPACE M_P,

$$\mathbf{II}(\mathbf{v}_p, \mathbf{w}_p) = S(\mathbf{v}_p) \cdot \mathbf{w}_p, \qquad (2)$$

where S is the SHAPE OPERATOR. The THIRD FUNDAMENTAL FORM is given by

$$\mathbf{III}(\mathbf{v}_p, \mathbf{w}_p) = S(\mathbf{v}_p) \cdot S(\mathbf{w}_p). \qquad (3)$$

The FIRST and SECOND FUNDAMENTAL FORMS satisfy

$$\mathbf{I}(a\mathbf{X}_u + b\mathbf{X}_v, a\mathbf{X}_u + b\mathbf{X}_v) = Ea^2 + 2Fab + Gb^2 \qquad (4)$$

$$\mathbf{II}(a\mathbf{X}_u + b\mathbf{X}_v, a\mathbf{X}_u + b\mathbf{X}_v) = ea^2 + 2fab + gb^2 \qquad (5)$$

where $x : U \to \mathbb{R}^3$ is a REGULAR PATCH and $\mathbf{x}_u$ and $\mathbf{x}_v$ are the partial derivatives of $\mathbf{x}$ with respect to parameters u and v, respectively. Their ratio is simply the NORMAL CURVATURE

$$k(\mathbf{v_p}) = \frac{\mathbf{II}(\mathbf{v_p})}{\mathbf{I}(\mathbf{v_p})} \qquad (6)$$

for any nonzero TANGENT VECTOR. The third fundamental form is given in terms of the first and second forms by

$$\mathbf{III} - 2H\mathbf{II} + K\mathbf{I} = 0, \qquad (7)$$

where H is the MEAN CURVATURE and K is the GAUSSIAN CURVATURE.

The first fundamental form (or LINE ELEMENT) is given explicitly by the RIEMANNIAN METRIC

$$ds^2 = E\,du^2 + 2F\,du\,dv + G\,dv^2. \qquad (8)$$

It determines the ARC LENGTH of a curve on a surface. The coefficients are given by

$$E = \mathbf{x}_{uu} = \left| \frac{\partial \mathbf{x}}{\partial u} \right|^2 \qquad (9)$$

$$F = \mathbf{x}_{uv} = \frac{\partial \mathbf{x}}{\partial u} \cdot \frac{\partial \mathbf{x}}{\partial v} \qquad (10)$$

$$G = \mathbf{x}_{vv} = \left| \frac{\partial \mathbf{x}}{\partial v} \right|^2. \qquad (11)$$

The coefficients are also denoted $g_{uu} = E$, $g_{uv} = F$, and $g_{vv} = G$. In CURVILINEAR COORDINATES (where $F = 0$), the quantities

$$h_u \equiv \sqrt{g_{uu}} = \sqrt{E} \qquad (12)$$

$$h_v \equiv \sqrt{g_{vv}} = \sqrt{G} \qquad (13)$$

are called SCALE FACTORS.

The second fundamental form is given explicitly by

$$e\,du^2 + 2f\,du\,dv + g\,dv^2 \qquad (14)$$

where

$$e = \sum_i X_i \frac{\partial^2 x_i}{\partial u^2} \qquad (15)$$

$$f = \sum_i X_i \frac{\partial^2 x_i}{\partial u \partial v} \qquad (16)$$

$$g = \sum_i X_i \frac{\partial^2 x_i}{\partial v^2}, \qquad (17)$$

and X_i are the DIRECTION COSINES of the surface normal. The second fundamental form can also be written

$$e = -\mathbf{N}_u \cdot \mathbf{x}_u = \mathbf{N} \cdot \mathbf{x}_{uu} \qquad (18)$$

$$f = -\mathbf{N}_v \cdot \mathbf{x}_u = \mathbf{N} \cdot \mathbf{x}_{uv} = \mathbf{N}_{vu} \cdot \mathbf{x}_{vu}$$

$$= \mathbf{N}_u \cdot \mathbf{x}_v \qquad (19)$$

$$g = -\mathbf{N}_v \cdot \mathbf{x}_v = \mathbf{N} \cdot \mathbf{x}_{vv}, \qquad (20)$$

where $\mathbf{N}$ is the NORMAL VECTOR, or

$$e = \frac{\det(\mathbf{x}_{uu}\mathbf{x}_u\mathbf{x}_v)}{\sqrt{EG - F^2}} \qquad (21)$$

$$f = \frac{\det(\mathbf{x}_{uv}\mathbf{x}_u\mathbf{x}_v)}{\sqrt{EG - F^2}} \qquad (22)$$

$$g = \frac{\det(\mathbf{x}_{vv}\mathbf{x}_u\mathbf{x}_v)}{\sqrt{EG - F^2}}. \qquad (23)$$

See also ARC LENGTH, AREA ELEMENT, FIRST FUNDA-MENTAL FORM, GAUSSIAN CURVATURE, GEODESIC, KÄHLER MANIFOLD, LINE OF CURVATURE, LINE ELE-MENT, MEAN CURVATURE, NORMAL CURVATURE, RIE-MANNIAN METRIC, SCALE FACTOR, SECOND FUNDAMENTAL FORM, SURFACE AREA, THIRD FUNDA-MENTAL FORM, WEINGARTEN EQUATIONS

References

Gray, A. "The Three Fundamental Forms." §16.6 in *Modern Differential Geometry of Curves and Surfaces with Mathematica, 2nd ed.* Boca Raton, FL: CRC Press, pp. 380–82, 1997.

Fundamental Group

The fundamental group of an ARCWISE-CONNECTED set X is the GROUP formed by the sets of EQUIVALENCE CLASSES of the set of all LOOPS, i.e., paths with initial and final points at a given BASEPOINT p, under the EQUIVALENCE RELATION of HOMOTOPY. The IDENTITY ELEMENT of this group is the set of all paths HOMO-TOPIC to the degenerate path consisting of the point p. The fundamental groups of HOMEOMORPHIC spaces are ISOMORPHIC. In fact, the fundamental group only depends on the HOMOTOPY TYPE of X. The fundamental group of a TOPOLOGICAL SPACE was introduced by Poincaré (Munkres 1993, p. 1).

The following is a table of the fundamental group for some common spaces, where π_1 denotes the fundamental group, H_1 is the first integral HOMOLOGY, $\times$ denotes the GROUP DIRECT PRODUCT, $\mathbb{Z}$ denotes the RING of integers, and $\mathbb{Z}_n$ is the CYCLIC GROUP of order n.

space	symbol	π_1	H_1
CIRCLE	$\mathbb{S}^1$	$\mathbb{Z}$	$\mathbb{Z}$
figure eight		$\mathbb{Z} \coprod \mathbb{Z}$	$\mathbb{Z} \times \mathbb{Z}$
SPHERE	$\mathbb{S}^2$	0	0
TORUS	$\mathbb{T}$	$\mathbb{Z} \times \mathbb{Z}$	$\mathbb{Z} \times \mathbb{Z}$
TORUS of genus g	Σ_g	F_g	$\mathbb{Z}^{2g}$
REAL PROJECTIVE PLANE	$\mathbb{RP}^2$	$\mathbb{Z}_2$	$\mathbb{Z}_2$
KLEIN BOTTLE		$\dfrac{\mathbb{Z} \coprod \mathbb{Z}}{(aba^{-1}b)}$	$\mathbb{Z} \times \mathbb{Z}_2$
COMPLEX PROJECTIVE SPACE	$\mathbb{CP}^n$	0	0
n-torus	$\mathbb{T}^n$	$\mathbb{Z}^n$	$\mathbb{Z}^n$

The group product $a * b$ of LOOP a and LOOP b is given by the path of a followed by the path of b. The identity element is represented by the constant path, and the inverse of a is given by traversing a in the opposite direction. The fundamental group is inde-pendent of the choice of basepoint because any loop through p is HOMOTOPIC to a loop through any other point q. So it makes sense to say the "fundamental group of X."

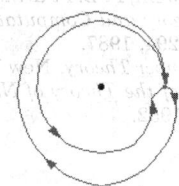

The diagram above shows that a loop followed by the opposite loop is homotopic to the constant loop, i.e., the identity. That is, it starts by traversing the path a, and then turns around and goes the other way, a^{-1}. The composition is deformed, or homotoped, to the constant path, along the original path a.

A space with a trivial fundamental group (i.e., every loop is homotopic to the constant loop), is called SIMPLY CONNECTED. For instance, any CONTRACTIBLE space, like EUCLIDEAN SPACE, is simply connected. The SPHERE is SIMPLY CONNECTED, but not CONTRAC-TIBLE. By definition, the UNIVERSAL COVER $\tilde{X}$ is simply connected, and loops in X lift to paths in $\tilde{X}$. The lifted paths in the universal cover define the DECK TRANSFORMATIONS, which form a GROUP iso-morphic to the fundamental group.

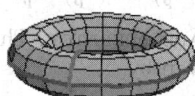

The underlying set of the fundamental group of X is the set of based HOMOTOPY CLASSES from the circle to X, denoted $[\mathbb{S}^1, X]$. For general spaces X and Y, there is no natural group structure on $[X, Y]$, but when there is, X is called a H-SPACE. Besides the circle, every SPHERE $\mathbb{S}^n$ is a H-SPACE, defining the HOMO-TOPY GROUPS. In general, the fundamental group is NON-ABELIAN. However, the higher HOMOTOPY GROUPS are Abelian. In some special cases, the fundamental group is Abelian. For example, the animation above shows that $a*b=b*a$ in the TORUS. The red path goes before the green path. The animation is a homotopy between the loop that goes around the inside first and the loop that goes around the outside first.

Since the first integral HOMOLOGY $H_1(X,\mathbb{Z})$ of X is also represented by loops, which are the only 1-dimensional objects with no boundary, there is a GROUP HOMOMORPHISM

$$\alpha : \pi_1(X) \to H_1(X,\mathbb{Z}),$$

which is SURJECTIVE. In fact, the KERNEL of α is the COMMUTATOR SUBGROUP and α is called ABELIANIZATION.

The fundamental group of X can be computed using VAN KAMPEN'S THEOREM, when X can be written as a union $X = \cup_i X_i$ of spaces whose fundamental groups are known.

When $f:X \to Y$ is a continuous map, then the fundamental group pushes forward. That is, there is a map $f_* : \pi_1(X) \to \pi_1(Y)$ defined by taking the image of loops from X. The pushforward is natural, i.e., $(f \circ g)_* = f_* \circ g_*$ whenever the composition of two maps is defined.

See also ALGEBRAIC FUNDAMENTAL GROUP, CAYLEY GRAPH, CONNECTED SET, DECK TRANSFORMATION, HOMOLOGY, HOMOTOPY GROUP, GROUP, MILNOR'S THEOREM, UNIVERSAL COVER, VAN KAMPEN'S THEOREM

References

Dodson, C. T. J. and Parker, P. E. "The Fundamental Group." §2.5 in *A User's Guide to Algebraic Topology.* Dordrecht, Netherlands: Kluwer, pp. 45–7, 1997.
Fulton, W. *Algebraic Topology: A First Course.* New York: Springer-Verlag, pp. 165–03, 1995.
Massey, W. S. *A Basic Course in Algebraic Topology.* New York: Springer-Verlag, pp. 35–8, 1991.
Munkres, J. R. *Elements of Algebraic Topology.* Perseus Press, 1993.

Fundamental Homology Class

FUNDAMENTAL CLASS

Fundamental Lemma of Calculus of Variations

If

$$\int_a^b M(x)h(x)\,dx = 0$$

$\forall h(x)$ with CONTINUOUS second PARTIAL DERIVATIVES, then

$$M(x) = 0$$

on the OPEN INTERVAL (a, b).

Fundamental Polytope

PRIMITIVE POLYTOPE

Fundamental Region

Let G be a SUBGROUP of the MODULAR GROUP GAMMA. Then an open subset R_G of the UPPER HALF-PLANE H is called a fundamental region of G if

1. No two distinct points of R_G are equivalent under G,
2. If $\tau \in H$, then there is a point τ' in the closure of R_G such that τ' is equivalent to τ under G.

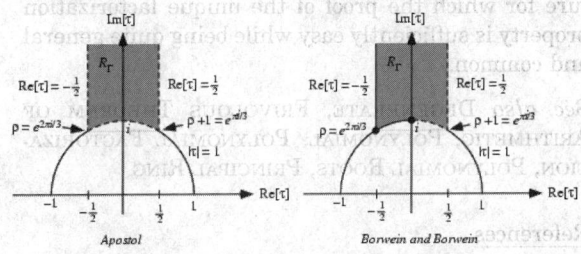

Apostol   *Borwein and Borwein*

A fundamental region R_Γ of the MODULAR GROUP GAMMA is given by $\tau \in H$ such that $|\tau| > 1$ and $|\tau + \bar{\tau}| < 1$, illustrated above, where $\bar{\tau}$ is the COMPLEX CONJUGATE of τ (Apostol 1997, p. 31). Borwein and Borwein (1987, p. 113) define the boundaries of the region slightly differently by including the boundary points with $\Re[\tau] \le 0$.

See also MODULAR GROUP GAMMA, MODULAR GROUP LAMBDA, UPPER HALF-PLANE, VALENCE

References

Apostol, T. M. "Fundamental Region." §2.3 in *Modular Functions and Dirichlet Series in Number Theory,* 2nd ed. New York: Springer-Verlag, pp. 30–4, 1997.
Borwein, J. M. and Borwein, P. B. *Pi & the AGM: A Study in Analytic Number Theory and Computational Complexity.* New York: Wiley, pp. 112–13, 1987.

Fundamental System

A set of ALGEBRAIC INVARIANTS for a QUANTIC such that any invariant of the QUANTIC is expressible as a POLYNOMIAL in members of the set. In 1868, Gordan proved the existence of finite fundamental systems of algebraic invariants and covariants for any binary QUANTIC. In 1890, Hilbert (1890) proved the HILBERT BASIS THEOREM, which is a finiteness theorem for the related concept of SYZYGIES.

See also HILBERT BASIS THEOREM, SYZYGY

References

Hilbert, D. "Über die Theorie der algebraischen Formen." *Math. Ann.* **36**, 473–34, 1890.

Fundamental Theorem of Algebra

Every POLYNOMIAL EQUATION having COMPLEX COEFFICIENTS and degree ≥ 1 has at least one COMPLEX ROOT. This theorem was first proven by Gauss. It is equivalent to the statement that a POLYNOMIAL $P(z)$ of degree n has n values z_i (some of them possibly degenerate) for which $P(z_i) = 0$. Such values are called POLYNOMIAL ROOTS. An example of a POLYNOMIAL with a single ROOT of multiplicity > 1 is $z^2 - 2z + 1 = (z - 1)(z - 1)$, which has $z = 1$ as a ROOT of multiplicity 2.

For RINGS more general than the complex polynomials $\mathbb{C}[x]$, there does not necessarily exist a unique factorization. However, a PRINCIPAL RING is a structure for which the proof of the unique factorization property is sufficiently easy while being quite general and common.

See also DEGENERATE, FRIVOLOUS THEOREM OF ARITHMETIC, POLYNOMIAL, POLYNOMIAL FACTORIZATION, POLYNOMIAL ROOTS, PRINCIPAL RING

References

Courant, R. and Robbins, H. "The Fundamental Theorem of Algebra." §2.5.4 in *What is Mathematics?: An Elementary Approach to Ideas and Methods, 2nd ed.* Oxford, England: Oxford University Press, pp. 101–03, 1996.
Krantz, S. G. "The Fundamental Theorem of Algebra." §1.1.7 and 3.1.4 in *Handbook of Complex Analysis*. Boston, MA: Birkhäuser, pp. 7 and 32–3, 1999.

Fundamental Theorem of Arithmetic

Any POSITIVE INTEGER can be represented in exactly one way as a PRODUCT of PRIMES. The theorem is also called the UNIQUE FACTORIZATION THEOREM. The fundamental theorem of arithmetic is a COROLLARY of the first of EUCLID'S THEOREMS (Hardy and Wright 1979).

See also ABNORMAL NUMBER, EUCLID'S THEOREMS, INTEGER, PRIME NUMBER

References

Courant, R. and Robbins, H. *What is Mathematics?: An Elementary Approach to Ideas and Methods, 2nd ed.* Oxford, England: Oxford University Press, p. 23, 1996.
Davenport, H. *The Higher Arithmetic: An Introduction to the Theory of Numbers, 6th ed.* Cambridge, England: Cambridge University Press, p. 20, 1992.
Hardy, G. H. and Wright, E. M. "Statement of the Fundamental Theorem of Arithmetic," "Proof of the Fundamental Theorem of Arithmetic," and "Another Proof of the Fundamental Theorem of Arithmetic." §1.3, 2.10 and 2.11 in *An Introduction to the Theory of Numbers, 5th ed.* Oxford, England: Clarendon Press, pp. 3 and 21, 1979.
Hasse, H. "Über eindeutige Zerlegung in Primelemente oder in Primhauptideale in Integritätsbereichen." *J. reine angew. Math.* **159**, 3–2, 1928.
Lindemann, F. A. "The Unique Factorization of a Positive Integer." *Quart. J. Math.* **4**, 319–20, 1933.
Nagell, T. "The Fundamental Theorem." §4 in *Introduction to Number Theory*. New York: Wiley, pp. 14–6, 1951.

Zermelo, E. "Elementare Betrachtungen zur Theorie der Primzahlen." *Nachr. Gesellsch. Wissensch. Göttingen* **1**, 43–6, 1934.

Fundamental Theorem of Curves

The CURVATURE and TORSION functions along a SPACE CURVE determine it up to an orientation-preserving ISOMETRY.

Fundamental Theorem of Directly Similar Figures

Let F_0 and F_1 denote two DIRECTLY SIMILAR figures in the plane, where $P_1 \in F_1$ corresponds to $P_1 \in F_0$ under the given similarity. Let $r \in (0, 1)$, and define $F_r = \{(1 - r)P_0 + rP_1 : P_0 \in F_0,\ P_1 \in F_1\}$. Then F_r is also directly similar to F_0.

See also DIRECTLY SIMILAR, FINSLER-HADWIGER THEOREM

References

Detemple, D. and Harold, S. "A Round-Up of Square Problems." *Math. Mag.* **69**, 15–7, 1996.
Eves, H. Solution to Problem E521. *Amer. Math. Monthly* **50**, 64, 1943.

Fundamental Theorem of Gaussian Quadrature

The ABSCISSAS of the N-point GAUSSIAN QUADRATURE FORMULA are precisely the ROOTS of the ORTHOGONAL POLYNOMIAL for the same INTERVAL and WEIGHTING FUNCTION.

See also GAUSSIAN QUADRATURE

Fundamental Theorem of Genera

Consider $h_+(d)$ proper equivalence classes of forms with discriminant d equal to the field discriminant, then they can be subdivided equally into 2^{r-1} genera of $h_+(d)/2^{r-1}$ forms which form a SUBGROUP of the proper equivalence class group under composition (Cohn 1980, p. 224), where r is the number of distinct prime divisors of d. This theorem was proved by Gauss in 1801.

See also GENUS (FORM), GENUS THEOREM

References

Arno, S.; Robinson, M. L.; and Wheeler, F. S. "Imaginary Quadratic Fields with Small Odd Class Number." http://www.math.uiuc.edu/Algebraic-Number-Theory/0009/.
Cohn, H. *Advanced Number Theory*. New York: Dover, 1980.
Gauss, C. F. *Disquisitiones Arithmeticae*. New Haven, CT: Yale University Press, 1966.

Fundamental Theorem of Number Theory

FUNDAMENTAL THEOREM OF ARITHMETIC

Fundamental Theorem of Plane Curves

Two unit-speed plane curves which have the same CURVATURE differ only by a EUCLIDEAN MOTION.

See also FUNDAMENTAL THEOREM OF SPACE CURVES

References

Gray, A. *Modern Differential Geometry of Curves and Surfaces with Mathematica, 2nd ed.* Boca Raton, FL: CRC Press, pp. 136–38, 1997.

Fundamental Theorem of Projective Geometry

A PROJECTIVITY is determined when three points of one RANGE and the corresponding three points of the other are given.

See also PROJECTIVE GEOMETRY

Fundamental Theorem of Riemannian Geometry

On a RIEMANNIAN MANIFOLD, there is a unique CONNECTION which is TORSION-free and compatible with the METRIC. This CONNECTION is called the LEVI-CIVITA CONNECTION.

See also COVARIANT DERIVATIVE, LEVI-CIVITA CONNECTION, RIEMANNIAN MANIFOLD, RIEMANNIAN METRIC

Fundamental Theorem of Space Curves

If two single-valued continuous functions $\kappa(s)$ (CURVATURE) and $\tau(s)$ (TORSION) are given for $s > 0$, then there exists EXACTLY ONE SPACE CURVE, determined except for orientation and position in space (i.e., up to a EUCLIDEAN MOTION), where s is the ARC LENGTH, κ is the CURVATURE, and τ is the TORSION.

See also ARC LENGTH, CURVATURE, EUCLIDEAN MOTION, FUNDAMENTAL THEOREM OF PLANE CURVES, TORSION (DIFFERENTIAL GEOMETRY)

References

Gray, A. "The Fundamental Theorem of Space Curves." §7.7 in *Modern Differential Geometry of Curves and Surfaces with Mathematica, 2nd ed.* Boca Raton, FL: CRC Press, pp. 219–22, 1997.
Struik, D. J. *Lectures on Classical Differential Geometry.* New York: Dover, p. 29, 1988.

Fundamental Theorems of Calculus

The first fundamental theorem of calculus states that, if f is CONTINUOUS on the CLOSED INTERVAL $[a, b]$ and F is the ANTIDERIVATIVE (INDEFINITE INTEGRAL) of f on $[a, b]$, then

$$\int_a^b f(x)\,dx = F(b) - F(a). \tag{1}$$

The second fundamental theorem of calculus lets f be CONTINUOUS on an OPEN INTERVAL I and lets a be any point in I. If F is defined by

$$F(x) = \int_a^x f(t)\,dt, \tag{2}$$

then

$$F'(x) = f(x) \tag{3}$$

at each point in I.

The fundamental theorem of calculus along curves states that if $f(z)$ has a CONTINUOUS ANTIDERIVATIVE $F(z)$ in a region R containing a parameterized curve $\gamma : z = z(t)$ for $\alpha \le t \le \beta$, then

$$\int_\gamma f(z)\,dz = F(z(\beta)) - F(z(\alpha)). \tag{4}$$

See also CALCULUS, DEFINITE INTEGRAL, INDEFINITE INTEGRAL, INTEGRAL

References

Krantz, S. G. "The Fundamental Theorem of Calculus along Curves." §2.1.5 in *Handbook of Complex Analysis.* Boston, MA: Birkhäuser, p. 22, 1999.

Fundamental Unit

In a REAL QUADRATIC FIELD, there exists a special UNIT η known as the fundamental unit such that all units ρ are given by $\rho = \pm \eta^m$, for $m = 0, \pm 1, \pm 2, \dots$. The notation ε_0 is sometimes used instead of η (Zucker and Robertson 1976). The fundamental units for REAL QUADRATIC FIELDS $\mathbb{Q}(\sqrt{D})$ may be computed from the fundamental solution of the PELL EQUATION

$$T^2 - DU^2 = \pm 4,$$

where the sign is taken such that the solution (T, U) has smallest possible positive T (LeVeque 1977; Cohn 1980, p. 101; Hua 1982; Borwein and Borwein 1986, p. 294). If the positive sign is taken, then one solution is simply given by $(T, U) = (2x, 2y)$, where (x, y) is the solution to the PELL EQUATION

$$x^2 - Dy^2 = 1$$

However, this need not be the *minimal* solution. For example, the solution to Pell equation

$$x^2 - 21y^2 = 1$$

is $(x, y) = (55, 12)$, so $(T, U) = (2x, 2y) = (110, 24)$, but $(T, U) = (5, 1)$ is the minimal solution. Given a minimal (T, U) (Sloane's A048941 and A048942), the fundamental unit is given by

$$\eta = \frac{1}{2}(T + U\sqrt{D})$$

(Cohn 1980, p. 101).

The following table gives fundamental units for small D.

D	$\eta(D)$	D	$\eta(D)$
2	$1 + \sqrt{2}$	54	$485 + 66\sqrt{54}$
3	$2 + \sqrt{3}$	55	$89 + 12\sqrt{55}$
5	$\frac{1}{2}(1 + \sqrt{5})$	56	$15 + 2\sqrt{56}$
6	$5 + 2\sqrt{6}$	57	$151 + 20\sqrt{57}$
7	$8 + 3\sqrt{7}$	58	$99 + 13\sqrt{58}$
8	$\frac{1}{2}(1 + 2\sqrt{8})$	59	$530 + 69\sqrt{59}$
10	$3 + \sqrt{10}$	60	$\frac{1}{2}(8 + \sqrt{60})$
11	$10 + 3\sqrt{11}$	61	$\frac{1}{2}(39 + 5\sqrt{61})$
12	$7 + 2\sqrt{12}$	62	$63 + 8\sqrt{62}$
13	$\frac{1}{2}(3 + \sqrt{13})$	63	$8 + \sqrt{63}$
14	$15 + 4\sqrt{14}$	65	$8 + \sqrt{65}$
15	$4 + \sqrt{15}$	66	$65 + 8\sqrt{66}$
17	$4 + \sqrt{17}$	67	$48842 + 5967\sqrt{67}$
18	$17 + 4\sqrt{18}$	68	$\frac{1}{2}(8 + \sqrt{68})$
19	$170 + 39\sqrt{19}$	69	$\frac{1}{2}(25 + 3\sqrt{69})$
20	$\frac{1}{2}(4 + \sqrt{20})$	70	$251 + 30\sqrt{70}$
21	$\frac{1}{2} + (5 + \sqrt{21})$	71	$3480 + 413\sqrt{71}$
22	$197 + 42\sqrt{22}$	72	$17 + 2\sqrt{72}$
23	$24 + 5\sqrt{23}$	73	$1068 + 125\sqrt{73}$
24	$5 + \sqrt{24}$	74	$43 + 5\sqrt{74}$
26	$5 + \sqrt{26}$	75	$26 + 3\sqrt{75}$
27	$26 + 5\sqrt{27}$	76	$170 + 39\sqrt{19}$
28	$\frac{1}{2}(16 + 3\sqrt{28})$	77	$\frac{1}{2}(9 + \sqrt{77})$
29	$\frac{1}{2}(5 + \sqrt{29})$	78	$53 + 6\sqrt{78}$
30	$11 + 2\sqrt{30}$	79	$80 + 9\sqrt{79}$
31	$1520 + 273\sqrt{31}$	80	$9 + \sqrt{80}$
32	$\frac{1}{2}(6 + \sqrt{32})$	82	$9 + \sqrt{82}$
33	$23 + 4\sqrt{33}$	83	$82 + 9\sqrt{83}$
34	$35 + 6\sqrt{34}$	84	$55 + 6\sqrt{84}$
35	$6 + \sqrt{35}$	85	$\frac{1}{2}(9 + \sqrt{85})$
37	$6 + \sqrt{37}$	86	$10405 + 1122\sqrt{86}$
38	$37 + 6\sqrt{38}$	87	$28 + 3\sqrt{87}$
39	$25 + 4\sqrt{39}$	88	$197 + 21\sqrt{88}$
40	$\frac{1}{2}(6 + \sqrt{40})$	89	$500 + 53\sqrt{89}$
41	$32 + 5\sqrt{41}$	90	$19 + 2\sqrt{90}$
42	$13 + 2\sqrt{42}$	91	$1574 + 165\sqrt{91}$
43	$3482 + 531\sqrt{43}$	92	$\frac{1}{2}(48 + 5\sqrt{92})$
44	$\frac{1}{2}(20 + 3\sqrt{44})$	93	$\frac{1}{2}(29 + 3\sqrt{93})$
45	$\frac{1}{2}(7 + \sqrt{45})$	94	$2143295 + 221064\sqrt{94}$
46	$24335 + 3588\sqrt{46}$	95	$39 + 4\sqrt{95}$
47	$48 + 7\sqrt{47}$	96	$\frac{1}{2}(10 + \sqrt{96})$
48	$7 + \sqrt{48}$	97	$5604 + 569\sqrt{97}$
50	$7 + \sqrt{50}$	98	$99 + 10\sqrt{98}$
51	$50 + 7\sqrt{51}$	99	$10 + \sqrt{99}$
52	$18 + 5\sqrt{13}$	101	$10 + \sqrt{101}$
53	$\frac{1}{2}(7 + \sqrt{53})$	102	$101 + 10\sqrt{102}$

See also PELL EQUATION, REAL QUADRATIC FIELD, UNIT

References

Borwein, J. M. and Borwein, P. B. *Pi & the AGM: A Study in Analytic Number Theory and Computational Complexity.* New York: Wiley, 1987.

Cohn, H. "Fundamental Units" and "Construction of Fundamental Units." §6.4 and 6.5 in *Advanced Number Theory.* New York: Dover, pp. 98–02, and 261–74, 1980.

Hua, L. K. *Introduction to Number Theory.* Berlin: Springer-Verlag, 1982.

Ireland, K. and Rosen, M. *A Classical Introduction to Modern Number Theory*, 2nd ed. New York: Springer-Verlag, p. 192, 1990.

LeVeque, W. J. *Fundamentals of Number Theory.* Reading, MA: Addison-Wesley, 1977.

Narkiewicz, W. *Elementary and Analytic Number Theory of Algebraic Numbers.* Warsaw: Polish Scientific Publishers, 1974.

Stark, H. M. *An Introduction to Number Theory.* Chicago, IL: Markham, 1970.

Weisstein, E. W. "Class Numbers." MATHEMATICA NOTEBOOK CLASSNUMBERS.M.

Zucker, I. J. and Robertson, M. M. "Some Properties of Dirichlet *L*-Series." *J. Phys. A: Math. Gen.* **9**, 1207–214, 1976.

Funnel

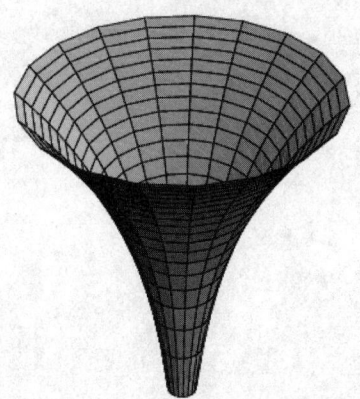

The funnel surface is a REGULAR SURFACE and SURFACE OF REVOLUTION defined by the Cartesian equation

$$Z = \frac{1}{2}\ln\left(x^2 + y^2\right) \tag{1}$$

and the PARAMETRIC EQUATIONS

$$x(u,v) = u\cos v \tag{2}$$

$$y(u,v) = u\sin v \tag{3}$$

$$z(u,v) = \ln u \tag{4}$$

for $u > 0$ and $v \in [0, 2\pi)$. The coefficients of the FIRST FUNDAMENTAL FORM are

$$E = 1 + \frac{1}{u^2} \tag{5}$$

$$F = 0 \tag{6}$$

$$G = u^2, \tag{7}$$

the coefficients of the SECOND FUNDAMENTAL FORM

are

$$e = -\frac{1}{u\sqrt{1 + u^2}} \tag{8}$$

$$f = 0 \tag{9}$$

$$g = \frac{u}{\sqrt{1 + u^2}}, \tag{10}$$

the AREA ELEMENT is

$$dA = \sqrt{1 + u^2}\,du \wedge dv, \tag{11}$$

and the Gaussian and mean curvatures are

$$K = -\frac{1}{\left(1 + u^2\right)^2} \tag{12}$$

$$H = \frac{1}{2u(1 + u^2)^{3/2}}. \tag{13}$$

Both the surface area and volume of the solid are infinite.

See also GABRIEL'S HORN, PSEUDOSPHERE, SINCLAIR'S SOAP FILM PROBLEM

References

Gray, A. "The Funnel Surface." *Modern Differential Geometry of Curves and Surfaces with Mathematica*, 2nd ed. Boca Raton, FL: CRC Press, pp. 423–26, 1997.

Fuss's Problem

BICENTRIC POLYGON

Futile Game

A GAME which permits a draw ("tie") when played properly by both players.

See also CATEGORICAL GAME, FAIR GAME, GAME

References

Steinhaus, H. *Mathematical Snapshots*, 3rd ed. New York: Dover, p. 16, 1999.

Fuzzy Logic

An extension of two-valued LOGIC such that statements need not be TRUE or FALSE, but may have a degree of truth between 0 and 1. Such a system can be extremely useful in designing control logic for real-world systems such as elevators.

See also ALETHIC, FALSE, LOGIC, TRUE

References

McNeill, D. *Fuzzy Logic: A Practical Approach.* New York: Academic Press, 1994.

McNeill, D. and Freiberger, P. *Fuzzy Logic: The Discovery of a Revolutionary Computer Technology and How It is*

Changing Our World. New York: Simon and Schuster, 1993.

Nguyen, H. T. and Walker, E. A. *A First Course in Fuzzy Logic.* Boca Raton, FL: CRC Press, 1996.

Weisstein, E. W. "Books about Fuzzy Logic." http://www.treasure-troves.com/books/FuzzyLogic.html.

Yager, R. R. and Zadeh, L. A. (Eds.). *An Introduction to Fuzzy Logic Applications in Intelligent Systems.* Boston, MA: Kluwer, 1992.

Zadeh, L. and Kacprzyk, J. (Eds.). *Fuzzy Logic for the Management of Uncertainty.* New York: Wiley, 1992.

FWHM

Full Width at Half Maximum

G

Gabor Function

The computer animation format MPEG-7 uses Gabor functions to specify texture descriptors.

References

Gabor, D. "Theory of Communication." *J. Inst. Electr. Engineering, London* **93**, 429–57, 1946.
Hubbard, B. B. *The World According to Wavelets: The Story of a Mathematical Technique in the Making,* 2nd rev. upd. ed. New York: A. K. Peters, pp. 26, 28, and 187–88, 1998.
International Organisation for Standardisation. "MPEG-7 Frequently Asked Questions." http://www.cselt.it/mpeg/faq/faq_mpeg-7.htm.

Gabriel's Horn

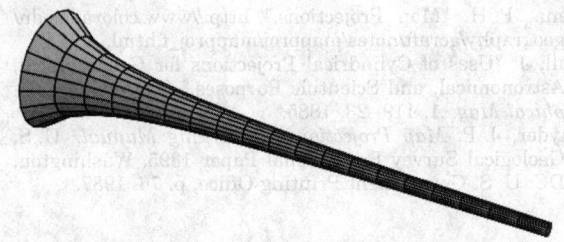

The SURFACE OF REVOLUTION of the function $y = 1/x$ about the x-AXIS for $x \geq 1$. It has FINITE VOLUME

$$V = \int_1^\infty \pi y^2 \, dx = \pi \int_1^\infty \frac{dx}{x^2}$$

$$= \pi \left[-\frac{1}{x} \right]_1^\infty = \pi[0 - (-1)] = \pi,$$

but INFINITE SURFACE AREA, since

$$S = \int_1^\infty 2\pi y \sqrt{1 + y'^2} \, dx$$

$$> 2\pi \int_1^\infty y \, dx = 2\pi \int_1^\infty \frac{dx}{x} = 2\pi[\ln x]_1^\infty$$

$$= 2\pi[\ln \infty - 0] = \infty.$$

This leads to the paradoxical consequence that while Gabriel's horn can be filled up with π cubic units of paint, an INFINITE number of square units of paint are needed to cover its surface!

See also FUNNEL, PSEUDOSPHERE

Gabriel's Staircase

The SUM

$$\sum_{k=1}^\infty k r^k = \frac{r}{(1 - r)^2},$$

valid for $0 < r < 1$.

Gadget

A term of endearment used by ALGEBRAIC TOPOLOGISTS when talking about their favorite power tools such as ABELIAN GROUPS, BUNDLES, HOMOLOGY GROUPS, HOMOTOPY GROUPS, K-THEORY, MORSE THEORY, OBSTRUCTIONS, stable homotopy theory, VECTOR SPACES, etc.

See also ABELIAN GROUP, ALGEBRAIC TOPOLOGY, BUNDLE, FREE, HOMOLOGY GROUP, HOMOTOPY GROUP, K-THEORY, OBSTRUCTION, MORSE THEORY, VECTOR SPACE

References

Page, W. *Topological Uniform Structures.* New York: Dover, 1994.

Galerkin Method

A method of determining coefficients α_k in a power series solution

$$y(x) = y_0(x) + \sum_{k=1}^n \alpha_k \, y_k(x)$$

of the ORDINARY DIFFERENTIAL EQUATION $L[y(x)] = 0$ so that the DIFFERENTIAL OPERATOR $L[y(x)]$ is orthogonal to every $y_k(x)$ for $k = 1, ..., n$.

References

Itô, K. (Ed.). "Methods Other than Difference Methods." §303I in *Encyclopedic Dictionary of Mathematics,* 2nd ed., Vol. 2. Cambridge, MA: MIT Press, p. 1139, 1980.

Gale-Ryser Theorem

Let p and q be PARTITIONS of a POSITIVE INTEGER, then there exists a (0,1)-matrix (i.e., a BINARY MATRIX) such that $c() = p$, $r() = q$ IFF q is dominated by p^*.

See also BINARY MATRIX, PARTITION

References

Brualdi, R. and Ryser, H. J. §6.2.4 in *Combinatorial Matrix Theory.* New York: Cambridge University Press, 1991.
Krause, M. "A Simple Proof of the Gale-Ryser Theorem." *Amer. Math. Monthly* **103**, 335–37, 1996.
Robinson, G. §1.4 in *Representation Theory of the Symmetric Group.* Toronto, Canada: University of Toronto Press, 1961.
Ryser, H. J. "The Class $\mathscr{A}(\mathbf{R}, \mathbf{S})$." *Combinatorial Mathematics.* Buffalo, NY: Math. Assoc. Amer., pp. 61–5, 1963.

Galilean Transformation

A transformation from one reference frame to another moving with a constant VELOCITY v with respect to the first for classical motion. However, special relativity shows that the transformation must be modified to the LORENTZ TRANSFORMATION for relativistic motion. The forward Galilean transformation is

$$\begin{bmatrix} t' \\ x' \\ y' \\ z' \end{bmatrix} = \begin{bmatrix} 1 & 0 & 0 & 0 \\ -v & 1 & 0 & 0 \\ 0 & 0 & 1 & 0 \\ 0 & 0 & 0 & 1 \end{bmatrix} \begin{bmatrix} t \\ x \\ y \\ z \end{bmatrix},$$

and the inverse transformation is

$$\begin{bmatrix} t \\ x \\ y \\ z \end{bmatrix} = \begin{bmatrix} 1 & 0 & 0 & 0 \\ v & 1 & 0 & 0 \\ 0 & 0 & 1 & 0 \\ 0 & 0 & 0 & 1 \end{bmatrix} \begin{bmatrix} t' \\ x' \\ y' \\ z' \end{bmatrix}.$$

See also LORENTZ TRANSFORMATION

Gall Isographic Projection

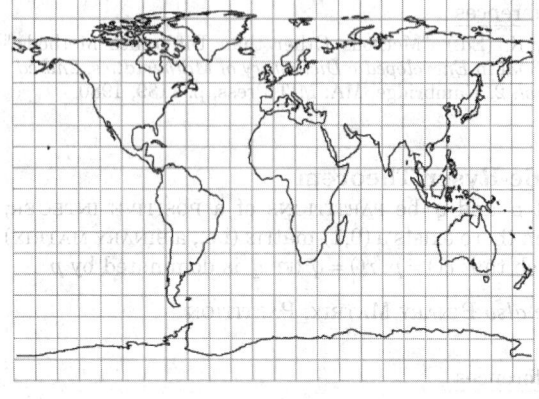

A CYLINDRICAL EQUIDISTANT PROJECTION with standard parallel $\phi_1 = 45°$.

See also CYLINDRICAL EQUIDISTANT PROJECTION

Gall Orthographic Projection

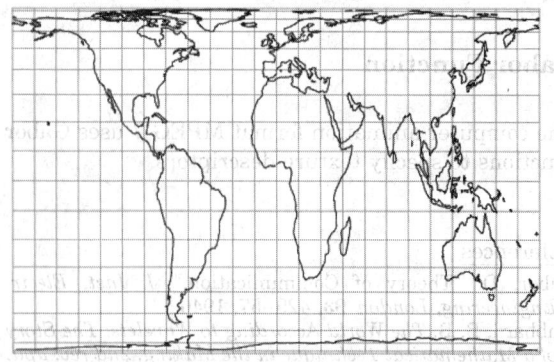

A CYLINDRICAL EQUAL-AREA PROJECTION with standard parallel of 45°.

See also BALTHASART PROJECTION, BEHRMANN CYLINDRICAL EQUAL-AREA PROJECTION, CYLINDRICAL EQUAL-AREA PROJECTION, EQUAL-AREA PROJECTION, GALL ISOGRAPHIC PROJECTION, LAMBERT AZIMUTHAL EQUAL-AREA PROJECTION, PETERS PROJECTION, STEREOGRAPHIC PROJECTION, TRISTAN EDWARDS PROJECTION

References

Dana, P. H. "Map Projections." http://www.colorado.edu/geography/gcraft/notes/mapproj/mapproj_f.html.
Gall, J. "Uses of Cylindrical Projections for Geographical, Astronomical, and Scientific Purposes." *Scottish Geographical Mag.* **1**, 119–23, 1885.
Snyder, J. P. *Map Projections--A Working Manual.* U. S. Geological Survey Professional Paper 1395. Washington, DC: U. S. Government Printing Office, p. 76, 1987.

Gall Stereographic Projection
GALL ORTHOGRAPHIC PROJECTION

Gallows

Schroeder (1991) calls the CEILING FUNCTION symbols ⌈ and ⌉ the "gallows" because of their similarity in appearance to the structure used for hangings.

See also CEILING FUNCTION

References

Schroeder, M. *Fractals, Chaos, Power Laws: Minutes from an Infinite Paradise.* New York: W. H. Freeman, p. 57, 1991.

Gallucci's Theorem

If three SKEW LINES all meet three other SKEW LINES, any TRANSVERSAL to the first set of three meets any TRANSVERSAL to the second set of three.

See also SKEW LINES, TRANSVERSAL LINE

Galois Extension

This entry contributed by NICOLAS BRAY

An extension F of a field K is said to be a Galois extension of K, if for every $x \in F - K$, there is an element of the GALOIS GROUP of the extension which does not fix x (i.e., there exits $\sigma \in \text{Aut}_K F$ such that $\sigma(x) \neq x$)).

See also GALOIS EXTENSION FIELD

Galois Extension Field

If K is the SPLITTING FIELD over a FIELD F of a separable POLYNOMIAL $f(x)$, then the EXTENSION FIELD K/F is a Galois extension field.

See also EXTENSION FIELD, GALOIS EXTENSION, SPLITTING FIELD

References
Dummit, D. S. and Foote, R. M. *Abstract Algebra, 2nd ed.* Englewood Cliffs, NJ: Prentice-Hall, pp. 475–76, 1998.

Galois Field

FINITE FIELD

Galois Group

Let L be a FIELD EXTENSION of K, denoted L/K, and let G be the set of AUTOMORPHISMS of L/K, that is, the set of AUTOMORPHISMS σ of L such that $\sigma(x) = x$ for every $x \in K$, so that K is fixed. Then G is a GROUP of transformations of L, called the Galois group of L/K.

The Galois group of $(\mathbb{C}/\mathbb{R})$ consists of the IDENTITY ELEMENT and COMPLEX CONJUGATION. These functions both take a given REAL to the same real.

See also ABHYANKAR'S CONJECTURE, FINITE GROUP, GROUP

References
Birkhoff, G. and Mac Lane, S. "The Galois Group." §15.2 in *A Survey of Modern Algebra, 5th ed.* New York: Macmillan, pp. 397–01, 1996.
Jacobson, N. *Basic Algebra I, 2nd ed.* New York: W. H. Freeman, p. 234, 1985.

Galois Imaginary

A mathematical object invented to solve irreducible CONGRUENCES OF THE FORM

$$F(x) \equiv 0 \pmod{p},$$

where p is PRIME.

Galois Theory

If there exists a ONE-TO-ONE correspondence between two SUBGROUPS and SUBFIELDS such that

$$G(E(G')) = G'$$

$$E(G(E')) = E',$$

then E is said to have a Galois theory.

See also ABEL'S IMPOSSIBILITY THEOREM, SUBFIELD

References
Artin, E. *Galois Theory, 2nd ed.* Notre Dame, IN: Edwards Brothers, 1944.
Birkhoff, G. and Mac Lane, S. "Galois Theory." Ch. 15 in *A Survey of Modern Algebra, 5th ed.* New York: Macmillan, pp. 395–21, 1996.
Dummit, D. S. and Foote, R. M. "Galois Theory." Ch. 14 in *Abstract Algebra, 2nd ed.* Englewood Cliffs, NJ: Prentice-Hall, pp. 471–70, 1998.

Galois's Theorem

An algebraic equation is algebraically solvable IFF its GROUP is SOLVABLE. In order that an irreducible equation of PRIME degree be solvable by radicals, it is NECESSARY and SUFFICIENT that all its ROOTS be rational functions of two ROOTS.

See also ABEL'S IMPOSSIBILITY THEOREM, SOLVABLE GROUP

Galoisian

An algebraic extension E of F for which every IRREDUCIBLE POLYNOMIAL in F which has a single ROOT in E has *all* its ROOTS in E is said to be Galoisian. Galoisian extensions are also called algebraically normal.

Gambler's Ruin

Let two players each have a finite number of pennies (say, n_1 for player one and n_2 for player two). Now, flip one of the pennies (from either player), with each player having 50% probability of winning, and give the penny to the winner. Now repeat the process until one player has all the pennies.

If the process is repeated indefinitely, the probability that *one* of the two player will *eventually* lose all his pennies must be 100%. In fact, the chances P_1 and P_2 that players one and two, respectively, will be rendered penniless are

$$P_1 = \frac{n_2}{n_1 + n_2}$$

$$P_2 = \frac{n_1}{n_1 + n_2},$$

i.e., your chances of going bankrupt are equal to the ratio of pennies your opponent starts out to the total number of pennies.

Therefore, the player starting out with the smallest number of pennies has the greatest chance of going bankrupt. Even with equal odds, the longer you gamble, the greater the chance that the player starting out with the most pennies wins. Since casinos have more pennies than their individual patrons, this principle allows casinos to always come out ahead in the long run. And the common practice of playing games with odds skewed in favor of the house makes this outcome just that much quicker.

See also COIN TOSSING, MARTINGALE, SAINT PETERS-BURG PARADOX

References

Cover, T. M. "Gambler's Ruin: A Random Walk on the Simplex." §5.4 in *Open Problems in Communications and Computation.* (Ed. T. M. Cover and B. Gopinath). New York: Springer-Verlag, p. 155, 1987.

Hajek, B. "Gambler's Ruin: A Random Walk on the Simplex." §6.3 in *Open Problems in Communications and Computation.* (Ed. T. M. Cover and B. Gopinath). New York: Springer-Verlag, pp. 204–07, 1987.

Kraitchik, M. "The Gambler's Ruin." §6.20 in *Mathematical Recreations.* New York: W. W. Norton, p. 140, 1942.

Game

A game is defined as a conflict involving gains and losses between two or more opponents who follow formal rules. The study of games belongs to a branch of mathematics known as GAME THEORY.

See also BOARD, CARDS, CATEGORICAL GAME, DRAW, FAIR GAME, FINITE GAME, FUTILE GAME, GAME THEORY, HYPERGAME, UNFAIR GAME

References

Falkener, E. *Games Ancient and Oriental and How to Play Them.* New York: Dover, 1961.

Sackson, S. *A Gamut of Games.* New York: Random House, 1969.

University of Waterloo. "Museum and Archive of Games." http://www.ahs.uwaterloo.ca/~museum/.

Game Expectation

Let the elements in a PAYOFF MATRIX be denoted a_{ij}, where the is are player A's STRATEGIES and the js are player B's STRATEGIES. Player A can get at least

$$\min_{j \leq n} a_{ij} \qquad (1)$$

for STRATEGY i. Player B can force player A to get no more than $\max_{j \leq m} a_{ij}$ for a STRATEGY j. The best STRATEGY for player A is therefore

$$\max_{i \leq m} \min_{j \leq n} a_{ij}, \qquad (2)$$

and the best STRATEGY for player B is

$$\min_{j \leq n} \max_{i \leq m} a_{ij}. \qquad (3)$$

In general,

$$\max_{i \leq m} \min_{j \leq n} a_{ij} \leq \min_{j \leq n} \max_{i \leq m} a_{ij}. \qquad (4)$$

Equality holds only if a SADDLE POINT is present, in which case the quantity is called the VALUE of the game.

See also GAME, PAYOFF MATRIX, SADDLE POINT (GAME), STRATEGY, VALUE

Game Matrix

PAYOFF MATRIX

Game of Life

LIFE

Game Theory

A branch of MATHEMATICS and LOGIC which deals with the analysis of GAMES (i.e., situations involving parties with conflicting interests). In addition to the mathematical elegance and complete "solution" which is possible for simple games, the principles of game theory also find applications to complicated games such as cards, checkers, and chess, as well as real-world problems as diverse as economics, property division, politics, and warfare.

See also BOREL DETERMINACY THEOREM, CATEGORICAL GAME, CHECKERS, CHESS, DECISION THEORY, EQUILIBRIUM POINT, FINITE GAME, FUTILE GAME, GAME EXPECTATION, GO, HI-Q, IMPARTIAL GAME, MEX, MINIMAX THEOREM, MIXED STRATEGY, NASH EQUILIBRIUM, NASH'S THEOREM, NIM, NIM-VALUE, PARTISAN GAME, PAYOFF MATRIX, PEG SOLITAIRE, PERFECT INFORMATION, SADDLE POINT (GAME), SAFE, SPRAGUE-GRUNDY FUNCTION, STRATEGY, TACTIX, TIT-FOR-TAT, UNSAFE, VALUE, WYTHOFF'S GAME, ZERO-SUM GAME

References

Ahrens, W. *Mathematische Unterhaltungen und Spiele.* Leipzig, Germany: Teubner, 1910.

Berlekamp, E. R.; Conway, J. H; and Guy, R. K. *Winning Ways for Your Mathematical Plays, Vol. 1: Games in General.* London: Academic Press, 1982.

Berlekamp, E. R.; Conway, J. H; and Guy, R. K. *Winning Ways for Your Mathematical Plays, Vol. 2: Games in Particular.* London: Academic Press, 1982.

Conway, J. H. *On Numbers and Games.* New York: Academic Press, 1976.

Dresher, M. *The Mathematics of Games of Strategy: Theory and Applications.* New York: Dover, 1981.

Eppstein, D. "Combinatorial Game Theory." http://www.ics.uci.edu/~eppstein/cgt/.

Gardner, M. "Game Theory, Guess It, Foxholes." Ch. 3 in *Mathematical Magic Show: More Puzzles, Games, Diversions, Illusions and Other Mathematical Sleight-of-Mind from Scientific American.* New York: Vintage, pp. 35–9, 1978.

Gardner, R. *Games for Business and Economics.* New York: Wiley, 1994.

Isaacs, R. *Differential Games: A Mathematical Theory with Applications to Warfare and Pursuit, Control and Optimization.* New York: Dover, 1999.

Karlin, S. *Mathematical Methods and Theory in Games, Programming, and Economics, 2 Vols. Vol. 1: Matrix Games, Programming, and Mathematical Economics. Vol. 2: The Theory of Infinite Games.* New York: Dover, 1992.

Kuhn, H. W. (Ed.). *Classics in Game Theory.* Princeton, NJ: Princeton University Press, 1997.

McKinsey, J. C. C. *Introduction to the Theory of Games.* New York: McGraw-Hill, 1952.

Mérö, L. *Moral Calculations: Game Theory, Logic and Human Frailty.* New York: Springer-Verlag, 1998.

Neumann, J. von and Morgenstern, O. *Theory of Games and Economic Behavior, 3rd ed.* New York: Wiley, 1964.

Packel, E. *The Mathematics of Games and Gambling.* Washington, DC: Math. Assoc. Amer., 1981.

Stahl, S. *A Gentle Introduction to Game Theory.* Providence, RI: Amer. Math. Soc., 1999.

Straffin, P. D. Jr. *Game Theory and Strategy.* Washington, DC: Math. Assoc. Amer., 1993.

Vajda, S. *Mathematical Games and How to Play Them.* New York: Routledge, 1992.

Walker, P. "An Outline of the History of Game Theory." http://william-king.www.drexel.edu/top/class/histf.html.

Weisstein, E. W. "Books about Game Theory." http://www.treasure-troves.com/books/GameTheory.html.

Williams, J. D. *The Compleat Strategyst, Being a Primer on the Theory of Games of Strategy.* New York: Dover, 1986.

Gamma

GAMMA FUNCTION, INCOMPLETE GAMMA FUNCTION

Gamma Distribution

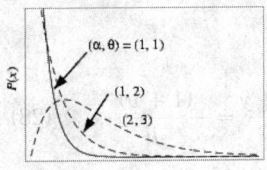
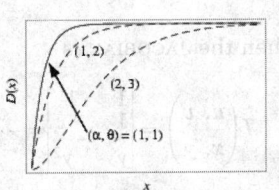

A general type of STATISTICAL DISTRIBUTION which is related to the BETA DISTRIBUTION and arises naturally in processes for which the waiting times between POISSON DISTRIBUTED events are relevant. Gamma distributions have two free parameters, labeled α and θ, a few of which are illustrated above.

Given a POISSON DISTRIBUTION with a rate of change λ, the DISTRIBUTION FUNCTION $D(x)$ giving the waiting times until the hth Poisson event is

$$D(x) = P(X \leq x) = 1 - P(x > x) = 1 - \sum_{k=0}^{h-1} \frac{(\lambda x)^k e^{-\lambda x}}{k!}$$

$$= 1 - e^{-\lambda x} \sum_{k=0}^{h-1} \frac{(\lambda x)^k}{k!} = 1 - \frac{\Gamma(h, x\lambda)}{\Gamma(h)} \quad (1)$$

for $x \in [0, \infty)$, where $\Gamma(x)$ is a complete GAMMA FUNCTION, and $\Gamma(a, x)$ an INCOMPLETE GAMMA FUNCTION. With h an integer, this distribution is a DISCRETE DISTRIBUTION known as the ERLANG DISTRIBUTION. The probability function $P(x)$ is then obtained by differentiating $D(x)$,

$$P(x) = D'(x) = \lambda e^{-\lambda x} \sum_{k=0}^{h-1} \frac{(\lambda x)^k}{k!} - e^{-\lambda x} \sum_{k=0}^{h-1} \frac{k(\lambda x)^{k-1}\lambda}{k!}$$

$$= \lambda e^{-\lambda x} + \lambda e^{-\lambda x} \sum_{k=1}^{h-1} \frac{(\lambda x)^k}{k!} - e^{-\lambda x} \sum_{k=1}^{h-1} \frac{k(\lambda x)^{k-1}\lambda}{k!}$$

$$= \lambda e^{-\lambda x} - \lambda e^{-\lambda x} \sum_{k=1}^{h-1} \left[\frac{k(\lambda x)^{k-1}}{k!} - \frac{(\lambda x)^k}{k!} \right]$$

$$= \lambda e^{-\lambda x} \left\{ 1 - \sum_{k=1}^{h-1} \left[\frac{(\lambda x)^{k-1}}{(k-1)!} - \frac{(\lambda x)^k}{k!} \right] \right\}$$

$$= \lambda e^{-\lambda x} \left\{ 1 - \left[1 - \frac{(\lambda x)^{h-1}}{(h-1)!} \right] \right\} = \frac{\lambda(\lambda x)^{h-1}}{(h-1)!} e^{-\lambda x}. \quad (2)$$

Now let $\alpha \equiv h$ (not necessarily an integer) and define $\theta \equiv 1/\lambda$ to be the time between changes. Then the above equation can be written

$$P(x) \frac{x^{\alpha-1} e^{-x/\theta}}{\Gamma(\alpha)\theta^\alpha} \quad (3)$$

for $x \in [0, \infty)$. The CHARACTERISTIC FUNCTION describing this distribution is

$$\phi(t) = \mathscr{F}\left\{ \frac{x^{-x/\theta} x^{\alpha-1}}{\Gamma(\alpha)\theta^\alpha} [\tfrac{1}{2}(1 + \operatorname{sgn} x)] \right\} = (1 - it\theta)^{-\alpha}, \quad (4)$$

where $\mathscr{F}[f]$ is the FOURIER TRANSFORM with parameters $a = b = 1$, and the MOMENT-GENERATING FUNCTION is

$$M(t) = \int_0^\infty \frac{e^{tx} x^{\alpha-1} e^{-x/\theta}}{\Gamma(\alpha)\theta^\alpha} \, dx = \int_0^\infty \frac{x^{\alpha-1} e^{-(1-\theta t)x/\theta}}{\Gamma(\alpha)\theta^\alpha} \, dx. \quad (5)$$

giving moments about 0 of

$$\mu_r' = \frac{\theta^r \Gamma(\alpha + r)}{\Gamma(\alpha)} \quad (6)$$

(Papoulis 1984, p. 147).

In order to explicitly find the MOMENTS of the distribution using the MOMENT-GENERATING FUNCTION, let

$$y \equiv \frac{(1 - \theta t)x}{\theta} \quad (7)$$

$$dy = \frac{1 - \theta t}{\theta} \, dx, \quad (8)$$

so

$$M(t) = \int_0^\infty \left(\frac{\theta y}{1 - \theta t} \right)^{\alpha-1} \frac{e^{-y}}{\Gamma(\alpha)\theta^\alpha} \frac{\theta \, dy}{1 - \theta t}$$

$$= \frac{1}{(1 - \theta t)^\alpha \Gamma(\alpha)} \int_0^\infty y^{\alpha-1} e^{-y} \, dy$$

$$= \frac{1}{(1 - \theta t)^\alpha}, \qquad (9)$$

giving the logarithmic MOMENT-GENERATING FUNCTION as

$$R(t) \equiv \ln M(t) = -\alpha \ln(1 - \theta t) \qquad (10)$$

$$R'(t) = \frac{\alpha \theta}{1 - \theta t} \qquad (11)$$

$$R''(t) = \frac{\alpha \theta^2}{(1 - \theta t)^2}. \qquad (12)$$

The MEAN, VARIANCE, SKEWNESS, and KURTOSIS are then

$$\mu = R'(0) = \alpha \theta \qquad (13)$$

$$\sigma^2 = R''(0)\alpha \theta^2 \qquad (14)$$

$$\gamma_1 = \frac{2}{\sqrt{\alpha}} \qquad (15)$$

$$\gamma_2 = \frac{6}{\alpha}. \qquad (16)$$

The gamma distribution is closely related to other statistical distributions. If $X_1, X_2, ..., X_n$ are independent random variates with a gamma distribution having parameters $(\alpha_1, \theta), (\alpha_2, \theta), ..., (\alpha_n, \theta)$, then $\Sigma_{i=1}^n X_i$ is distributed as gamma with parameters

$$\alpha = \sum_{i=1}^n \alpha_i \qquad (17)$$

$$\theta = \theta. \qquad (18)$$

Also, if X_1 and X_2 are independent random variates with a gamma distribution having parameters (α_1, θ) and (α_2, θ), then $X_1/(X_1 + X_2)$ is a BETA DISTRIBUTION variate with parameters (α_1, α_2). Both can be derived as follows.

$$P(x, y) = \frac{1}{\Gamma(\alpha_1)\Gamma(\alpha_2)} e^{x_1 + x_2} x_1^{\alpha_1 - 1} x_2^{\alpha_2 - 1}. \qquad (19)$$

Let

$$u = x_1 + x_2 \qquad x_1 = uv \qquad (20)$$

$$v = \frac{x_1}{x_1 + x_2} \qquad x_2 = u(1 - v), \qquad (21)$$

then the JACOBIAN is

$$J\left(\frac{x_1, x_2}{u, v}\right) = \begin{vmatrix} v & u \\ 1 - v & -u \end{vmatrix} = -u, \qquad (22)$$

so

$$g(u, v) \, du \, dv = f(x, y) \, dx \, dy = f(x, y)u \, du \, dv. \qquad (23)$$

$$g(u, v) = \frac{u}{\Gamma(\alpha_1)\Gamma(\alpha_2)} e^{-u}(uv)^{\alpha_1 - 1} u^{\alpha_2 - 1}(1 - v)^{\alpha_2 - 1}$$

$$= \frac{1}{\Gamma(\alpha_1)\Gamma(\alpha_2)} e^{-u} u^{\alpha_1 + \alpha_2 - 1} v^{\alpha_1 - 1}(1 - v)^{\alpha_2 - 1}. \qquad (24)$$

The sum $X_1 + X_2$ therefore has the distribution

$$f(u) = f(x_1 + x_2) = \int_0^1 g(u, v) \, dv = \frac{e^{-u} u^{\alpha_1 + \alpha_2 - 1}}{\Gamma(\alpha_1 + \alpha_2)}, \qquad (25)$$

which is a gamma distribution, and the ratio $X_1/(X_1 + X_2)$ has the distribution

$$h(v) = h\left(\frac{x_1}{x_1 + x_2}\right) = \int_0^\infty g(u, v) \, du$$

$$= \frac{v^{\alpha_1 - 1}(1 - v)^{\alpha_2 - 1}}{B(\alpha_1, \alpha_2)}, \qquad (26)$$

where B is the BETA FUNCTION, which is a BETA DISTRIBUTION.

If X and Y are gamma variates with parameters α_1 and α_2, the X/Y is a variate with a BETA PRIME DISTRIBUTION with parameters α_1 and α_2. Let

$$u = x + y \qquad v = \frac{x}{y}, \qquad (27)$$

then the JACOBIAN is

$$J\left(\frac{u, v}{x, y}\right) = \begin{vmatrix} 1 & 1 \\ \frac{1}{y} & -\frac{x}{y^2} \end{vmatrix} = -\frac{x + y}{y^2} = -\frac{(1 + v)^2}{u}, \qquad (28)$$

so

$$dx \, dy = \frac{u}{(1 + v)^2} \, du \, dv \qquad (29)$$

$$g(u, v) = \frac{1}{\Gamma(\alpha_1)\Gamma(\alpha_2)} e^{-u} \left(\frac{uv}{1 + v}\right)^{\alpha_1 - 1} \left(\frac{u}{1 + v}\right)^{\alpha_2 - 1}$$

$$\times \frac{u}{(1 + v)^2}$$

$$= \frac{1}{\Gamma(\alpha_1)\Gamma(\alpha_2)} e^{-u} u^{\alpha_1 + \alpha_2 - 1} v^{\alpha_1 - 1}(1 + v)^{-\alpha_1 - \alpha_2}. \qquad (30)$$

The ratio X/Y therefore has the distribution

$$h(v) = \int_0^\infty (g(u, v)) \, du = \frac{v^{\alpha_1 - 1}(1 + v)^{-\alpha_1 - \alpha_2}}{B(\alpha_1, \alpha_2)}, \qquad (31)$$

which is a BETA PRIME DISTRIBUTION with parameters (α_1, α_2).

The "standard form" of the gamma distribution is given by letting $y \equiv x/\theta$, so $dy = dx/\theta$ and

$$P(y) \, dy = \frac{x^{\alpha - 1} e^{-x/\theta}}{\Gamma(\alpha)\theta^\alpha} \, dx = \frac{(\theta y)\alpha^{-1} e^{-y}}{\Gamma(\alpha)\theta^\alpha} (\theta \, dy)$$

$$= \frac{y^{\alpha-1}e^{-y}}{\Gamma(\alpha)}\, dy, \qquad (32)$$

so the MOMENTS about 0 are

$$v_r = \frac{1}{\Gamma(\alpha)} \int_0^\infty e^{-x} x^{\alpha-1+r}\, dx = \frac{\Gamma(\alpha+r)}{\Gamma(\alpha)} = (\alpha)_r, \qquad (33)$$

where $(\alpha)_r$ is the POCHHAMMER SYMBOL. The MOMENTS about $\mu = \mu_1$ are then

$$\mu_1 = \alpha \qquad (34)$$

$$\mu_2 = \alpha \qquad (35)$$

$$\mu_3 = 2\alpha \qquad (36)$$

$$\mu_4 = 3\alpha^2 + 6\alpha. \qquad (37)$$

The MOMENT-GENERATING FUNCTION is

$$M(t) = \frac{1}{(1-t)^\alpha}, \qquad (38)$$

and the CUMULANT-GENERATING FUNCTION is

$$K(t) = \alpha \ln(1-t) = \alpha\left(t + \tfrac{1}{2}t^2 + \tfrac{1}{3}t^3 + \ldots\right), \qquad (39)$$

so the CUMULANTS are

$$\kappa_r = \alpha\Gamma(r). \qquad (40)$$

If x is a NORMAL variate with MEAN μ and STANDARD DEVIATION σ, then

$$y \equiv \frac{(x-\mu)^2}{2\sigma^2} \qquad (41)$$

is a standard gamma variate with parameter $\alpha = 1/2$.

See also BETA DISTRIBUTION, CHI-SQUARED DISTRIBUTION, ERLANG DISTRIBUTION

References

Beyer, W. H. *CRC Standard Mathematical Tables, 28th ed.* Boca Raton, FL: CRC Press, p. 534, 1987.

Jambunathan, M. V. "Some Properties of Beta and Gamma Distributions." *Ann. Math. Stat.* **25**, 401–05, 1954.

Papoulis, A. *Probability, Random Variables, and Stochastic Processes, 2nd ed.* New York: McGraw-Hill, pp. 103–04, 1984.

Gamma Function

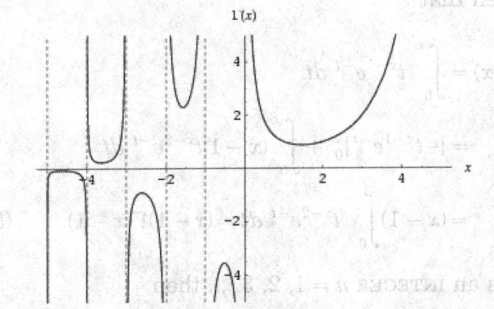

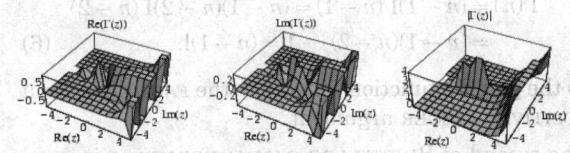

The complete gamma function $\Gamma(n)$ is defined to be an extension of the FACTORIAL to COMPLEX and REAL NUMBER arguments. It is related to the FACTORIAL by $\Gamma(n) = (n-1)!$. It is ANALYTIC everywhere except at $z = 0, -1, -2, \ldots$, and the residue at $z = -k$ is

$$\operatorname*{Res}_{z=-k} \Gamma(z) = \frac{(-1)^k}{k!}. \qquad (1)$$

There are no points z at which $\Gamma(z) = 0$. The gamma function is implemented in *Mathematica* as Gamma[z].

The gamma function can be defined as a DEFINITE INTEGRAL for $\Re[z] > 0$ (Euler's integral form)

$$\Gamma(z) \equiv \int_0^\infty t^{z-1} e^{-t}\, dt \qquad (2)$$

$$= 2\int_0^\infty e^{-t^2} t^{2z-1}\, dt, \qquad (3)$$

or

$$\Gamma(z) \equiv \int_0^1 \left[\ln\left(\frac{1}{t}\right)\right]^{z-1}\, dt. \qquad (4)$$

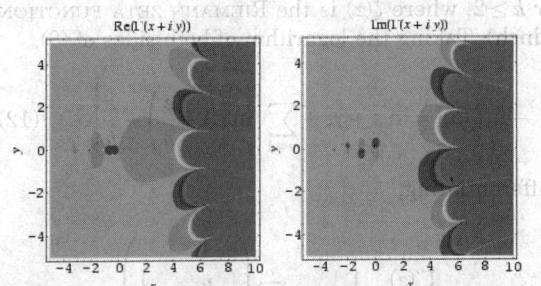

Plots of the real and imaginary parts of $\Gamma(z)$ in the complex plane are illustrated above.

INTEGRATING (2) by parts for a REAL argument, it can be seen that

$$\Gamma(x) = \int_0^\infty t^{x-1} e^{-t}\, dt$$

$$= [-t^{x-1} e^{-t}]_0^\infty + \int_0^\infty (x-1) t^{x-2} e^{-t}\, dt$$

$$= (x-1)\int_0^\infty t^{x-2} e^{-t}\, dt = (x-1)\Gamma(x-1). \quad (5)$$

If x is an INTEGER $n = 1, 2, 3, \ldots$ then

$$\Gamma(n) = (n-1)\Gamma(n-1) = (n-1)(n-2)\Gamma(n-2)$$
$$= (n-1)(n-2)\cdots 1 = (n-1)!, \quad (6)$$

so the gamma function reduces to the FACTORIAL for a POSITIVE INTEGER argument.

The second of BINET'S LOG GAMMA FORMULAS is

$$\ln \Gamma(a) = (a - \tfrac{1}{2})\ln a - a + \tfrac{1}{2}\ln(2\pi)$$

$$+ 2\int_0^\infty \frac{\tan(\frac{z}{a})}{e^{2\pi z} - 1}\, dz \quad (7)$$

for $\Re[a] > 0$ (Whittaker and Watson 1990, p. 251). Another formula for $\ln \Gamma(z)$ is given by MALMSTÉN'S FORMULA, and $\ln \Gamma(z)$ is implemented in *Mathematica* as LogGamma[z]. The gamma function can also be defined by an INFINITE PRODUCT form (Weierstrass Form)

$$\Gamma(z) \equiv \left[z e^{\gamma z} \prod_{r=1}^\infty \left(1 + \frac{z}{r} \right) e^{-z/r} \right]^{-1}, \quad (8)$$

where γ is the EULER-MASCHERONI CONSTANT (Krantz 1999, p. 157). This can be written

$$\Gamma(z) = \frac{1}{z}\exp\left[\sum_{k=1}^\infty \frac{(-1)^k s_k}{k} z^k \right], \quad (9)$$

where

$$s_1 \equiv \gamma \quad (10)$$

$$s_k \equiv \zeta(k) \quad (11)$$

for $k \geq 2$, where $\zeta(z)$ is the RIEMANN ZETA FUNCTION (Finch). Taking the logarithm of both sides of (8),

$$-\ln[\Gamma(z)] = \ln z + \gamma z + \sum_{n=1}^\infty \left[\ln\left(1 + \frac{z}{n} \right) - \frac{z}{n} \right]. \quad (12)$$

Differentiating,

$$-\frac{\Gamma'(z)}{\Gamma(z)} = \frac{1}{z} + \gamma + \sum_{n=1}^\infty \left(\frac{\frac{1}{n}}{1 + \frac{z}{n}} - \frac{1}{n} \right)$$

$$= \frac{1}{z} + \gamma + \sum_{n=1}^\infty \left(\frac{1}{n+z} - \frac{1}{n} \right) \quad (13)$$

$$\Gamma'(z) = -\Gamma(z)\left[\frac{1}{z} + \gamma + \sum_{n=1}^\infty \left(\frac{1}{n+z} - \frac{1}{n} \right) \right] \quad (14)$$

$$\equiv \Gamma(z)\Psi(z) = \Gamma(z)\psi_0(z) \quad (15)$$

$$\Gamma'(1) = -\Gamma(1)$$
$$-\left\{ 1 + \gamma + \left[(\tfrac{1}{2} - 1) + (\tfrac{1}{3} - \tfrac{1}{2}) + \ldots + \left(\frac{1}{n+1} - \frac{1}{n} \right) + \ldots \right] \right\}$$
$$= -(1 + \gamma - 1) = -\gamma \quad (16)$$

$$\Gamma'(n) = -\Gamma(n)$$
$$\times \left\{ \frac{1}{n} + \gamma + \left[\left(\frac{1}{1+n} - 1 \right) + \left(\frac{1}{2+n} - \frac{1}{2} \right) \right.\right.$$
$$\left.\left. + \left(\frac{1}{3+n} - \frac{1}{3} \right) + \ldots \right] \right\}$$
$$= -(n-1)!\left(\frac{1}{n} + \gamma - \sum_{k=1}^n \frac{1}{k} \right), \quad (17)$$

where $\Psi(z)$ is the DIGAMMA FUNCTION and $\psi_0(z)$ is the POLYGAMMA FUNCTION. nth derivatives are given in terms of the POLYGAMMA FUNCTIONS ψ_n, ψ_{n-1}, ..., ψ_0. The minimum value x_0 of $\Gamma(x)$ for REAL POSITIVE $x = x_0$ is achieved when

$$\Gamma'(x_0) = \Gamma(x_0)\psi_0(x_0) = 0 \quad (18)$$

$$\psi_0(x_0) = 0, \quad (19)$$

This can be solved numerically to give $x_0 = 1.46163\ldots$ (Sloane's A030169; Wrench 1968), which has CONTINUED FRACTION [1, 2, 6, 63, 135, 1, 1, 1, 1, 4, 1, 38, ...] (Sloane's A030170). At x_0, $\Gamma(x_0)$ achieves the value $0.8856031944\ldots$ (Sloane's A030171), which has CONTINUED FRACTION [0, 1, 7, 1, 2, 1, 6, 1, 1, ...] (Sloane's A030172).

The Euler limit form is

$$\frac{1}{\Gamma(z)} = z \left[\lim_{m \to \infty} e^{(1 + 1/2 + \ldots + 1/m - \ln m)z} \right]$$

$$\times \left[\lim_{m \to \infty} \prod_{n=1}^m \left\{ \left(1 + \frac{z}{n} \right) e^{-z/n} \right\} \right]$$

$$= \frac{1}{z} \prod_{n=1}^\infty \left[\left(1 + \frac{1}{n} \right)^z \left(1 + \frac{z}{n} \right)^{-1} \right], \quad (20)$$

so

$$\Gamma(z) \equiv \lim_{n \to \infty} \frac{1 \cdot 2 \cdot 3 \cdots n}{z(z+1)(z+2)\cdots(z+n)}\, n^z \quad (21)$$

(Krantz 1999, p. 156). One over the gamma function is also given by

$$\frac{1}{\Gamma(z)} = z \exp\left[\gamma z - \sum_{k=2}^{\infty} \frac{(-1)^k \zeta(k) z^k}{k}\right], \qquad (22)$$

where γ is the EULER-MASCHERONI CONSTANT and $\zeta(z)$ is the RIEMANN ZETA FUNCTION (Wrench 1968). An ASYMPTOTIC SERIES for $1/\Gamma(z)$ is given by

$$\frac{1}{\Gamma(z)} \sim z + \gamma z^2 + \tfrac{1}{12}(6\gamma^2 - \pi^2)z^3 + \tfrac{1}{12}[2\gamma^3 - \gamma\pi^2 + 4\zeta(3)]z^4$$

$$+ \dots \qquad (23)$$

Writing

$$\frac{1}{\Gamma(z)} = \sum_{k=1}^{\infty} a_k z^k, \qquad (24)$$

the a_k satisfy

$$a_n = na_1a_n - a_2a_{n-1} + \sum_{k=2}^{n}(-1)^k\zeta(k)a_{n-k} \qquad (25)$$

(Bourget 1883, Isaacson and Salzer 1942, Wrench 1968). Wrench (1968) numerically computed the coefficients for the series expansion about 0 of

$$\frac{1}{z(1+z)\Gamma(z)}$$

$$= 1 + (\gamma - 1)z + \left[1 + \tfrac{1}{2}(\gamma - 2)\gamma - \tfrac{1}{12}\pi^2\right]z^2 + \dots. \quad (26)$$

The LANCZOS APPROXIMATION for $z > 0$ is

$$\Gamma(z + 1) = (z + \gamma + \tfrac{1}{2})^{z+1/2}e^{z+\gamma+1/2}\sqrt{2\pi}$$

$$\times \left[c_0 + \frac{c_1}{z+1} + \frac{c_2}{z+2} + \dots + \frac{c_n}{z+n} + \varepsilon\right], \qquad (27)$$

where γ is the EULER-MASCHERONI CONSTANT. The gamma function satisfies the FUNCTIONAL EQUATIONS

$$\Gamma(1 + z) = z\Gamma(z) \qquad (28)$$

$$\Gamma(1 - z) = -z\Gamma(-z). \qquad (29)$$

Additional identities are

$$\Gamma(x)\Gamma(-x) = -\frac{\pi}{x\sin(\pi x)} \qquad (30)$$

$$\Gamma(x)\Gamma(1-x) = \frac{\pi}{\sin(\pi x)} \qquad (31)$$

$$\ln[\Gamma(x + iy + 1)]$$

$$= \ln(x^2 + y^2) + i\tan^{-1}\left(\frac{y}{x}\right) + \ln[\Gamma(x + iy)] \qquad (32)$$

$$|(ix)!|^2 = \frac{\pi x}{\sinh(\pi x)} \qquad (33)$$

$$|(n + ix)!| = \sqrt{\frac{\pi x}{\sinh(\pi x)}} \prod_{s=1}^{n} \sqrt{s^2 + x^2}. \qquad (34)$$

For integer $n = 1, 2, \dots$, the first few values of $\Gamma(n)$ are 1, 1, 2, 6, 24, 120, 720, 5040, 40320, 362880, ... (Sloane's A000142). For half integer arguments, $\Gamma(n/2)$ has the special form

$$\Gamma\left(\tfrac{1}{2}n\right) = \frac{(n-2)!!\sqrt{\pi}}{2^{(n-1)/2}}, \qquad (35)$$

where $n!!$ is a DOUBLE FACTORIAL. The first few values for $n = 1, 3, 5, \dots$, are therefore

$$\Gamma(\tfrac{1}{2}) = \sqrt{\pi} \qquad (36)$$

$$\Gamma(\tfrac{3}{2}) = \tfrac{1}{2}\sqrt{\pi} \qquad (37)$$

$$\Gamma(\tfrac{5}{2}) = \tfrac{3}{4}\sqrt{\pi}, \qquad (38)$$

$15\sqrt{\pi}/8$, $105\sqrt{\pi}/16$, ... (Sloane's A001147 and A000079; Wells 1986, p. 40). In general, for n a POSITIVE INTEGER $n = 1, 2, \dots$

$$\Gamma\left(\tfrac{1}{2} + n\right) = \frac{1 \cdot 3 \cdot 5 \cdots (2n - 1)}{2^n}\sqrt{\pi}$$

$$= \frac{(2n - 1)!!}{2^n}\sqrt{\pi} \qquad (39)$$

$$\Gamma(\tfrac{1}{2} - n) = \frac{(-1)^n 2^n}{1 \cdot 3 \cdot 5 \cdots (2n - 1)}\sqrt{\pi}$$

$$= \frac{(-1)^n 2^n}{(2n - 1)!!}\sqrt{\pi}. \qquad (40)$$

For $\Re[x] = -\tfrac{1}{2}$,

$$|(-\tfrac{1}{2} + iy)!|^2 = \frac{\pi}{\cosh(\pi y)}. \qquad (41)$$

Gamma functions of argument $2z$ can be expressed using the LEGENDRE DUPLICATION FORMULA

$$\Gamma(2z) = (2\pi)^{-1/2} 2^{2z-1/2}\Gamma(z)\Gamma(z + \tfrac{1}{2}). \qquad (42)$$

Gamma functions of argument $3z$ can be expressed using a triplication FORMULA

$$\Gamma(3z) = (2\pi)^{-1} 3^{3z-1/2}\Gamma(z)\Gamma(z + \tfrac{1}{3})\Gamma(z + \tfrac{2}{3}). \qquad (43)$$

The general result is the GAUSS MULTIPLICATION FORMULA

$$\Gamma(z)\Gamma(z + \tfrac{1}{n})\cdots\Gamma(z + \tfrac{n-1}{n}) = (2\pi)^{(n-1)/2}n^{1/2-nz}\Gamma(nz). \quad (44)$$

The gamma function is also related to the RIEMANN ZETA FUNCTION $\zeta(z)$ by

$$\Gamma\left(\frac{s}{2}\right)\pi^{-s/2}\zeta(s) = \Gamma\left(\frac{1-s}{2}\right)\pi^{-(1-s)/2}\zeta(1-s). \qquad (45)$$

Borwein and Zucker (1992) give a variety of identities relating gamma functions to square roots and ELLIPTIC INTEGRAL SINGULAR VALUES k_n, i.e., MODULI k_n such that

$$\frac{K'(k_n)}{K(k_n)} = \sqrt{n}, \tag{46}$$

where $K(k)$ is a complete ELLIPTIC INTEGRAL OF THE FIRST KIND and $K'(k) = K(k)' = K(\sqrt{1-k^2})$ is the complementary integral. M. Trott has developed an algorithm for automatically generating hundreds of such identities.

$$\Gamma(\tfrac{1}{3}) = 2^{7/9} 3^{-1/12} \pi^{1/3} [K(k_3)]^{1/3} \tag{47}$$

$$\Gamma(\tfrac{1}{4}) = 2\pi^{1/4} [K(k_1)]^{1/2} \tag{48}$$

$$\Gamma(\tfrac{1}{6}) = 2^{-1/3} 3^{1/2} \pi^{-1/2} [\Gamma(\tfrac{1}{3})]^2 \tag{49}$$

$$\Gamma(\tfrac{1}{8})\Gamma(\tfrac{3}{8}) = (\sqrt{2}-1)^{1/2} 2^{13/4} \pi^{1/2} K(k_2) \tag{50}$$

$$\frac{\Gamma(\tfrac{1}{8})}{\Gamma(\tfrac{3}{8})} = 2(\sqrt{2}+1)^{1/2} \pi^{-1/4} [K(k_1)]^{1/2} \tag{51}$$

$$\Gamma(\tfrac{1}{12}) = 2^{-1/4} 3^{3/8} (\sqrt{3}+1)^{1/2} \pi^{-1/2} \Gamma(\tfrac{1}{4})\Gamma(\tfrac{1}{3}) \tag{52}$$

$$\Gamma(\tfrac{5}{12}) = 2^{1/4} 3^{-1/8} (\sqrt{3}-1)^{1/2} \pi^{1/2} \frac{\Gamma(\tfrac{1}{4})}{\Gamma(\tfrac{1}{3})} \tag{53}$$

$$\frac{\Gamma(\tfrac{1}{24})\Gamma(\tfrac{11}{24})}{\Gamma(\tfrac{5}{24})\Gamma(\tfrac{7}{24})} = \sqrt{3}\sqrt{2+\sqrt{3}} \tag{54}$$

$$\frac{\Gamma(\tfrac{1}{24})\Gamma(\tfrac{5}{24})}{\Gamma(\tfrac{7}{24})\Gamma(\tfrac{11}{24})} = 4 \cdot 3^{1/4} (\sqrt{3}+\sqrt{2}) \pi^{-1/2} K(k_1) \tag{55}$$

$$\frac{\Gamma(\tfrac{1}{24})\Gamma(\tfrac{7}{24})}{\Gamma(\tfrac{5}{24})\Gamma(\tfrac{11}{24})} = 2^{25/18} 3^{1/3} (\sqrt{2}+1) \pi^{-1/3} [K(k_3)]^{2/3} \tag{56}$$

$$\Gamma(\tfrac{1}{24})\Gamma(\tfrac{5}{24})\Gamma(\tfrac{7}{24})\Gamma(\tfrac{11}{24})$$
$$= 384(\sqrt{2}+1)(\sqrt{3}-\sqrt{2})(2-\sqrt{3})\pi[K(k_6)]^2 \tag{57}$$

$$\Gamma(\tfrac{1}{10}) = 2^{-7/10} 5^{1/4} (\sqrt{5}+1)^{1/2} \pi^{-1/2} \Gamma(\tfrac{1}{5})\Gamma(\tfrac{2}{5}) \tag{58}$$

$$\Gamma(\tfrac{3}{10}) = 2^{-3/5} (\sqrt{5}-1) \pi^{1/2} \frac{\Gamma(\tfrac{1}{5})}{\Gamma(\tfrac{2}{5})} \tag{59}$$

$$\frac{\Gamma(\tfrac{1}{15})\Gamma(\tfrac{4}{15})\Gamma(\tfrac{7}{15})}{\Gamma(\tfrac{2}{15})} = 2 \cdot 3^{1/2} 5^{1/6} \sin(\tfrac{2}{15}\pi)[\Gamma(\tfrac{1}{3})]^2 \tag{60}$$

$$\frac{\Gamma(\tfrac{1}{15})\Gamma(\tfrac{2}{15})\Gamma(\tfrac{7}{15})}{\Gamma(\tfrac{4}{15})} = 2^2 \cdot 3^{2/5} \sin(\tfrac{1}{5}\pi) \sin(\tfrac{4}{15}\pi)[\Gamma(\tfrac{1}{5})]^2 \tag{61}$$

$$\frac{\Gamma(\tfrac{2}{15})\Gamma(\tfrac{4}{15})\Gamma(\tfrac{7}{15})}{\Gamma(\tfrac{1}{15})} = \frac{2^{-3/2} 3^{-1/5} 5^{1/4} (\sqrt{5}-1)^{1/2} [\Gamma(\tfrac{2}{5})]^2}{\sin(\tfrac{4}{15}\pi)} \tag{62}$$

$$\frac{\Gamma(\tfrac{1}{15})\Gamma(\tfrac{2}{15})\Gamma(\tfrac{4}{15})}{\Gamma(\tfrac{7}{15})} = 60(\sqrt{5}-1)\sin(\tfrac{7}{15}\pi)[K(k_{15})]^2 \tag{63}$$

$$\frac{\Gamma(\tfrac{1}{20})\Gamma(\tfrac{9}{20})}{\Gamma(\tfrac{3}{20})\Gamma(\tfrac{7}{20})} = 2^{-1} 5^{1/4} (\sqrt{5}+1) \tag{64}$$

$$\frac{\Gamma(\tfrac{1}{20})\Gamma(\tfrac{3}{20})}{\Gamma(\tfrac{7}{20})\Gamma(\tfrac{9}{20})} = 2^{4/5} (10-2\sqrt{5})^{1/2} \pi^{-1} \sin(\tfrac{7}{20}\pi) \sin(\tfrac{9}{20}\pi)$$
$$\times [\Gamma(\tfrac{1}{5})]^2 \tag{65}$$

$$\frac{\Gamma(\tfrac{1}{20})\Gamma(\tfrac{7}{20})}{\Gamma(\tfrac{3}{20})\Gamma(\tfrac{9}{20})} = 2^{3/5} (10+2\sqrt{5})^{1/2} \pi^{-1} \sin(\tfrac{3}{20}\pi) \sin(\tfrac{9}{20}\pi)$$
$$\times [\Gamma(\tfrac{2}{5})]^2 \tag{66}$$

$$\Gamma(\tfrac{1}{20})\Gamma(\tfrac{3}{20})\Gamma(\tfrac{7}{20})\Gamma(\tfrac{9}{20}) = 160(\sqrt{5}-2)^{1/2} \pi[K(k_5)]^2. \tag{67}$$

Several of these are also given in Campbell (1966, p. 31).

A few curious identities include

$$\prod_{n=1}^{8} \Gamma\left(\tfrac{1}{3}n\right) = \frac{640}{3^6} \left(\frac{\pi}{\sqrt{3}}\right)^3 \tag{68}$$

$$\frac{[\Gamma(\tfrac{1}{4})]^4}{16\pi^2} = \frac{3^2}{3^2-1} \frac{5^2-1}{5^2} \frac{7^2}{7^2-1} \cdots \tag{69}$$

$$\frac{\Gamma'(1)}{\Gamma(1)} - \frac{\Gamma'(\tfrac{1}{2})}{\Gamma(\tfrac{1}{2})} = 2\ln 2 \tag{70}$$

(Magnus and Oberhettinger 1949, p. 1). Ramanujan also gave a number of fascinating identities:

$$\frac{\Gamma^2(n+1)}{\Gamma(n+xi+1)\Gamma(n-xi+1)} = \prod_{k=1}^{\infty} \left[1 + \frac{x^2}{(n+k)^2}\right] \tag{71}$$

$$\phi(m,n)\phi(n,m) = \frac{\Gamma^3(m+1)\Gamma^3(n+1)}{\Gamma(2m+n+1)\Gamma(2n+m+1)}$$
$$\times \frac{\cosh[\pi(m+n)\sqrt{3}] - \cos[\pi(m-n)]}{2\pi^2(m^2+mn+n^2)}, \tag{72}$$

where

$$\phi(m,n) \equiv \prod_{k=1}^{\infty} \left[1 + \left(\frac{m+n}{k+m}\right)^3\right], \tag{73}$$

$$\prod_{k=1}^{\infty} \left[1 + \left(\frac{n}{k}\right)^3\right] \prod_{k=1}^{\infty} \left[1 + 3\left(\frac{n}{n+2k}\right)^2\right]$$

$$= \frac{\Gamma\left(\frac{1}{2}n\right)}{\Gamma\left[\frac{1}{2}(n+1)\right]} \frac{\cosh\left(\pi n\sqrt{3}\right) - \cos(\pi n)}{2^{n+2}\pi^{3/2}n} \qquad (74)$$

(Berndt 1994).

Ramanujan gave the infinite sums

$$1 + 9\left(\frac{1}{4}\right)^4 + 17\left(\frac{1 \cdot 5}{4 \cdot 8}\right)^4 + 25\left(\frac{1 \cdot 5 \cdot 9}{4 \cdot 8 \cdot 12}\right)^4 + \cdots$$

$$= \sum_{k=0}^{\infty} (8k+1)\left[\frac{\Gamma\left(k+\frac{1}{4}\right)}{k!\,\Gamma\left(\frac{1}{4}\right)}\right]^4 = \frac{2^{3/2}}{\sqrt{\pi}\left[\Gamma\left(\frac{3}{4}\right)\right]^2} \qquad (75)$$

and

$$1 - 5\left(\frac{1}{2}\right)^5 + 9\left(\frac{1 \cdot 3}{2 \cdot 4}\right)^5 - 13\left(\frac{1 \cdot 3 \cdot 5}{2 \cdot 4 \cdot 6}\right)^5 + \cdots$$

$$= \sum_{k=0}^{\infty} (-1)^k (4k+1)\left[\frac{(2k-1)!!}{(2k)!!}\right]^5 = \frac{2}{\left[\Gamma\left(\frac{3}{4}\right)\right]^4}. \qquad (76)$$

(Hardy 1923; Hardy 1924; Whipple 1926; Watson 1931; Bailey 1935; Hardy 1999, p. 7).

The following ASYMPTOTIC SERIES is occasionally useful in probability theory (e.g., the 1-D RANDOM WALK):

$$\frac{\Gamma\left(J+\frac{1}{2}\right)}{\Gamma(J)}$$

$$= \sqrt{J}\left(1 - \frac{1}{8J} + \frac{1}{128J^2} + \frac{5}{1024J^3} - \frac{21}{32768J^4} + \cdots\right) \qquad (77)$$

(Graham *et al.* 1994). This series also gives a nice asymptotic generalization of STIRLING NUMBERS OF THE FIRST KIND to fractional values.

It has long been known that $\Gamma(\frac{1}{4})\pi^{-1/4}$ is TRANSCENDENTAL (Davis 1959), as is $\Gamma(\frac{1}{3})$ (Le Lionnais 1983), and Chudnovsky has apparently recently proved that $\Gamma(\frac{1}{4})$ is itself TRANSCENDENTAL.

The complete gamma function $\Gamma(x)$ can be generalized to the upper INCOMPLETE GAMMA FUNCTION $\Gamma(a, x)$ and lower INCOMPLETE GAMMA FUNCTION $\gamma(a, x)$.

See also BAILEY'S THEOREM, BARNES' G-FUNCTION, BINET'S FIBONACCI NUMBER FORMULA, BOHR-MOLLERUP THEOREM, DIGAMMA FUNCTION, DOUBLE GAMMA FUNCTION, FRANSÉN-ROBINSON CONSTANT GAUSS MULTIPLICATION FORMULA, INCOMPLETE GAMMA FUNCTION, KNAR'S FORMULA, LAMBDA FUNCTION, LANCZOS APPROXIMATION, LEGENDRE DUPLICATION FORMULA, MALMSTÉN'S FORMULA, MELLIN'S FORMULA, MU FUNCTION, NU FUNCTION, PEARSON'S FUNCTION, POLYGAMMA FUNCTION, REGULARIZED GAMMA FUNCTION, STIRLING'S SERIES, SUPERFACTORIAL

References

Abramowitz, M. and Stegun, C. A. (Eds.). "Gamma (Factorial) Function" and "Incomplete Gamma Function." §6.1 and 6.5 in *Handbook of Mathematical Functions with Formulas, Graphs, and Mathematical Tables, 9th printing.* New York: Dover, pp. 255–58 and 260–63, 1972.

Arfken, G. "The Gamma Function (Factorial Function)." Ch. 10 in *Mathematical Methods for Physicists, 3rd ed.* Orlando, FL: Academic Press, pp. 339–41 and 539–72, 1985.

Artin, E. *The Gamma Function.* New York: Holt, Rinehart, and Winston, 1964.

Bailey, W. N. *Generalised Hypergeometric Series.* Cambridge, England: Cambridge University Press, 1935.

Berndt, B. C. *Ramanujan's Notebooks, Part IV.* New York: Springer-Verlag, pp. 334–42, 1994.

Beyer, W. H. *CRC Standard Mathematical Tables, 28th ed.* Boca Raton, FL: CRC Press, p. 218, 1987.

Borwein, J. M. and Zucker, I. J. "Elliptic Integral Evaluation of the Gamma Function at Rational Values of Small Denominator." *IMA J. Numerical Analysis* **12**, 519–26, 1992.

Bourguet, L. "Sur les intégrales Eulériennes et quelques autres fonctions uniformes." *Acta Math.* **2**, 261–95, 1883.

Campbell, R. *Les intégrales eulériennes et leurs applications.* Paris: Dunod, 1966.

Davis, H. T. *Tables of the Higher Mathematical Functions.* Bloomington, IN: Principia Press, 1933.

Davis, P. J. "Leonhard Euler's Integral: A Historical Profile of the Gamma Function." *Amer. Math. Monthly* **66**, 849–69, 1959.

Erdélyi, A.; Magnus, W.; Oberhettinger, F.; and Tricomi, F. G. "The Gamma Function." Ch. 1 in *Higher Transcendental Functions, Vol. 1.* New York: Krieger, pp. 1–5, 1981.

Finch, S. "Favorite Mathematical Constants." http://www.mathsoft.com/asolve/constant/fran/fran.html.

Graham, R. L.; Knuth, D. E.; and Patashnik, O. Answer to problem 9.60 in *Concrete Mathematics: A Foundation for Computer Science, 2nd ed.* Reading, MA: Addison-Wesley, 1994.

Hardy, G. H. "Some Formulae of Ramanujan." *Proc. London Math. Soc.* (Records of Proceedings at Meetings) **22**, xii-xiii, 1924.

Hardy, G. H. "A Chapter from Ramanujan's Note-Book." *Proc. Cambridge Philos. Soc.* **21**, 492–03, 1923.

Hardy, G. H. *Ramanujan: Twelve Lectures on Subjects Suggested by His Life and Work, 3rd ed.* New York: Chelsea, 1999.

Isaacson and Salzer. *Math. Tab. Aids Comput.* **1**, 124, 1943.

Koepf, W. "The Gamma Function." Ch. 1 in *Hypergeometric Summation: An Algorithmic Approach to Summation and Special Function Identities.* Braunschweig, Germany: Vieweg, pp. 4–0, 1998.

Krantz, S. G. "The Gamma and Beta Functions." §13.1 in *Handbook of Complex Analysis.* Boston, MA: Birkhäuser, pp. 155–58, 1999.

Le Lionnais, F. *Les nombres remarquables.* Paris: Hermann, p. 46, 1983.

Magnus, W. and Oberhettinger, F. *Formulas and Theorems for the Special Functions of Mathematical Physics.* New York: Chelsea, 1949.

Nielsen, N. "Handbuch der Theorie der Gammafunktion." Part I in *Die Gammafunktion.* New York: Chelsea, 1965.

Press, W. H.; Flannery, B. P.; Teukolsky, S. A.; and Vetterling, W. T. "Gamma Function, Beta Function, Factorials, Binomial Coefficients" and "Incomplete Gamma Function, Error Function, Chi-Square Probability Function, Cumulative Poisson Function." §6.1 and 6.2 in *Numerical Recipes in FORTRAN: The Art of Scientific Computing,*

2nd ed. Cambridge, England: Cambridge University Press, pp. 206–09 and 209–14, 1992.

Sloane, N. J. A. Sequences A000079/M1129, A000142/M1675, A001147/M3002, A030169/M030170, and A030171/M030172 in "An On-Line Version of the Encyclopedia of Integer Sequences." http://www.research.att.com/~njas/sequences/eisonline.html.

Spanier, J. and Oldham, K. B. "The Gamma Function $\Gamma(x)$" and "The Incomplete Gamma $\gamma(v; x)$ and Related Functions." Chs. 43 and 45 in *An Atlas of Functions.* Washington, DC: Hemisphere, pp. 411–21 and 435–43, 1987.

Watson, G. N. "Theorems Stated by Ramanujan (XI)." *J. London Math. Soc.* **6**, 59–5, 1931.

Wells, D. *The Penguin Dictionary of Curious and Interesting Numbers.* Middlesex, England: Penguin Books, p. 40, 1986.

Whipple, F. J. W. "A Fundamental Relation Between Generalised Hypergeometric Series." *J. London Math. Soc.* **1**, 138–45, 1926.

Whittaker, E. T. and Watson, G. N. *A Course in Modern Analysis, 4th ed.* Cambridge, England: Cambridge University Press, 1990.

Wrench, J. W. Jr. "Concerning Two Series for the Gamma Function." *Math. Comput.* **22**, 617–26, 1968.

Gamma Group

MODULAR GROUP

Gamma Matrices

DIRAC MATRICES

Gamma Statistic

$$\gamma_r \equiv \frac{\kappa_r}{\sigma^{r+2}},$$

where κ_r are CUMULANTS and σ is the STANDARD DEVIATION.

See also KURTOSIS, SKEWNESS

Gamma-Modular Function

The GAMMA GROUP Γ is the set of all transformations w OF THE FORM

$$w(t) = \frac{at + b}{ct + d},$$

where a, b, c, and d are INTEGERS and $ad - bc = 1$. Γ-modular functions are then defined as in Borwein and Borwein (1987, p. 114).

See also JACOBI THETA FUNCTIONS, KLEIN'S ABSOLUTE INVARIANT, LAMBDA GROUP

References

Borwein, J. M. and Borwein, P. B. *Pi & the AGM: A Study in Analytic Number Theory and Computational Complexity.* New York: Wiley, pp. 127–32, 1987.

GammaRegularized

REGULARIZED GAMMA FUNCTION

Garage Door

ASTROID

Gårding's Inequality

Gives a lower bound for the inner product (Lu, u), where L is a linear elliptic real differential operator of order m, and u has compact support.

References

Knapp, A. W. "Group Representations and Harmonic Analysis, Part II." *Not. Amer. Math. Soc.* **43**, 537–49, 1996.

Garman-Kohlhagen Formula

$$V_t = e^{-y\tau} S_t N(d_1) - e^{-r\tau} K N(d_2),$$

where N is the cumulative NORMAL DISTRIBUTION and

$$d_1, d_2 = \frac{\log\left(\frac{S_t}{K}\right) + \left(r - y \pm \frac{1}{2}\sigma^2\right)\tau}{\sigma\sqrt{\tau}}.$$

If $y = 0$, this is the standard form of the Black-Scholes formula.

See also BLACK-SCHOLES THEORY

References

Garman, M. B. and Kohlhagen, S. W. "Foreign Currency Option Values." *J. International Money and Finance* **2**, 231–37, 1983.

Price, J. F. "Optional Mathematics is Not Optional." *Not. Amer. Math. Soc.* **43**, 964–71, 1996.

Garsia-Haiman Conjecture

$N!$ THEOREM

Garsia-Milne Involution Principle

Let $C = C^+ \cup C^-$ (where $C^+ \cap C^- = \phi$) be the DISJOINT UNION of two finite components C^+ and C^-. Let α and β be two involutions on C, each of whose fixed points lie in C^+. Let F_α (respectively, F_β) denote the fixed point set of α (respectively, β). Stipulate that $\alpha(C^+ - F_\alpha) \subset C^-$ and $\alpha(C^-) \subset C^+$, and similarly $\beta(C^+ - F_\beta) \subset C^-$ and $\beta(C^-) \subset C^+$ (i.e., outside the fixed point sets), both α and β map each component into the other. Then either a cycle of the PERMUTATION $\Delta = \alpha\beta$ contains no fixed points of either α or β, or it contains exactly one element of F_α and one of F_β.

References

Andrews, G. E. "q-Series and Schur's Theorem" and "Bressoud's Proof of Schur's Theorem." §6.2–.3 in *q*-Series: Their Development and Application in Analysis, Number Theory, Combinatorics, Physics, and Computer Algebra.* Providence, RI: Amer. Math. Soc., pp. 53–8, 1986.

Gasket

APOLLONIAN GASKET, SIERPINSKI GASKET

Gasser-Müller Technique

References

Gasser, T. and Müller, H. "Kernel Estimation of Regression Functions." In *Smoothing Techniques for Curve Estimation: Proceedings of a Workshop Held in Heidelberg, April 2-, 1979* (Ed. T. Gasser and M. Rosenblatt). Berlin: Springer-Verlag, pp. 23–8, 1979.

Gate Function

Bracewell's term for the RECTANGLE FUNCTION.

References

Bracewell, R. *The Fourier Transform and Its Applications,* 3rd ed. New York: McGraw-Hill, 1999.

Gauche Conic

SKEW CONIC

Gauge Theory

References

Friedman, R. and Morgan, J. W. (Eds.). *Gauge Theory and the Topology of Four-Manifolds.* Providence, RI: Amer. Math. Soc., 1998.

Gaullist Cross

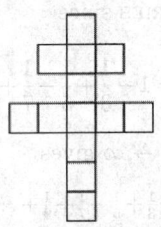

A CROSS also called the CROSS OF LORRAINE or PATRIARCHAL CROSS.

See also CROSS, DISSECTION

Gauss Equations

If **x** is a regular patch on a REGULAR SURFACE in $\mathbb{R}^3$ with normal $\hat{\mathbf{N}}$, then

$$\mathbf{x}_{uu} = \Gamma_{11}^1 \mathbf{x}_u + \Gamma_{11}^2 \mathbf{x}_v + e\hat{\mathbf{N}} \qquad (1)$$

$$\mathbf{x}_{uv} = \Gamma_{12}^1 \mathbf{x}_u + \Gamma_{12}^2 \mathbf{x}_v + f\hat{\mathbf{N}} \qquad (2)$$

$$\mathbf{x}_{vv} = \Gamma_{22}^1 \mathbf{x}_u + \Gamma_{22}^2 \mathbf{x}_v + g\hat{\mathbf{N}}, \qquad (3)$$

where e, f, and g are coefficients of the second FUNDAMENTAL FORM and Γ_{ij}^k are CHRISTOFFEL SYMBOLS OF THE SECOND KIND.

See also CHRISTOFFEL SYMBOL OF THE SECOND KIND, FUNDAMENTAL FORMS, MAINARDI-CODAZZI EQUATIONS

References

Gray, A. *Modern Differential Geometry of Curves and Surfaces with Mathematica,* 2nd ed. Boca Raton, FL: CRC Press, pp. 511–12, 1997.

Gauss Integral

Consider two closed oriented SPACE CURVES $f_1 : C_1 \to \mathbb{R}^3$ and $f_2 : C_2 \to \mathbb{R}^3$, where C_1 and C_2 are distinct CIRCLES, f_1 and f_2 are differentiable C^1 functions, and $f_1(C_1)$ and $f_2(C_3)$ are disjoint loci. Let $\mathrm{Lk}(f_1, f_2)$ be the LINKING NUMBER of the two curves, then the Gauss integral is

$$\mathrm{Lk}(f_1, f_2) = \frac{1}{4\pi} \int_{C_1 \times C_2} dS.$$

See also CALUGAREANU THEOREM, LINKING NUMBER

References

Pohl, W. F. "The Self-Linking Number of a Closed Space Curve." *J. Math. Mech.* **17**, 975–85, 1968.

Gauss Map

The Gauss map is a function from an ORIENTABLE SURFACE M in EUCLIDEAN SPACE to a SPHERE. It associates to every point on the surface its oriented NORMAL VECTOR. For a COMPACT SURFACE M in 3-space, the Gauss map of M has DEGREE given by half the EULER CHARACTERISTIC of the surface

$$\iint_M K \, dA = 2\pi\chi(M) - \sum \alpha_i - \int_{\partial T} \kappa_g \, ds,$$

where this formula holds only for ORIENTABLE SURFACES.

See also CURVATURE, NIRENBERG'S CONJECTURE, PATCH

References

Gray, A. "The Local Gauss Map" and "The Gauss Map via Mathematica." §12.3 and §17.4 in *Modern Differential Geometry of Curves and Surfaces with Mathematica,* 2nd ed. Boca Raton, FL: CRC Press, pp. 279–80 and 403–08, 1997.

Gauss Measure

The standard Gauss measure of a finite dimensional REAL HILBERT SPACE H with norm $\|\cdot\|_H$ has the BOREL MEASURE

$$\mu_H(dh) = (\sqrt{2\pi})^{-\dim(H)} \exp(\tfrac{1}{2}\|h\|_H^2)\lambda_H(dh),$$

where λ_H is the LEBESGUE MEASURE on H.

Gauss Multiplication Formula

$$(2n\pi)^{(n-1)/2}n^{1/2-nz}\Gamma(nz)$$

$$=\Gamma(z)\Gamma\left(z+\frac{1}{n}\right)\Gamma\left(z+\frac{2}{n}\right)\cdots\Gamma\left(z+\frac{n-1}{n}\right)$$

$$=\prod_{k=0}^{n-1}\Gamma\left(z+\frac{k}{n}\right),$$

where $\Gamma(z)$ is the GAMMA FUNCTION.

See also GAMMA FUNCTION, LEGENDRE DUPLICATION FORMULA, POLYGAMMA FUNCTION

References

Abramowitz, M. and Stegun, C. A. (Eds.). *Handbook of Mathematical Functions with Formulas, Graphs, and Mathematical Tables, 9th printing.* New York: Dover, p. 256, 1972.
Erdélyi, A.; Magnus, W.; Oberhettinger, F.; and Tricomi, F. G. *Higher Transcendental Functions, Vol. 1.* New York: Krieger, pp. 4–, 1981.

Gauss Plane

COMPLEX PLANE

Gauss's Backward Formula

$$f_p = f_0 + p\delta_{-1/2} + G_2^*\delta_0^2 + G_3\delta_{-1/2}^3 + G_4^*\delta_0^4$$
$$+ G_5\delta_{-1/2}^5 + \cdots,$$

for $p \in [0, 1]$, where δ is the CENTRAL DIFFERENCE and

$$G_{2n}^* = \binom{p+n}{2n}$$

$$G_{2n+1} = \binom{p+n}{2n+1},$$

where $\binom{n}{k}$ is a BINOMIAL COEFFICIENT.

See also CENTRAL DIFFERENCE, GAUSS'S FORWARD FORMULA

References

Beyer, W. H. *CRC Standard Mathematical Tables, 28th ed.* Boca Raton, FL: CRC Press, p. 433, 1987.
Whittaker, E. T. and Robinson, G. "The Newton-Gauss Backward Formula." §22 in *The Calculus of Observations: A Treatise on Numerical Mathematics, 4th ed.* New York: Dover, pp. 37–8, 1967.

Gauss's Circle Problem

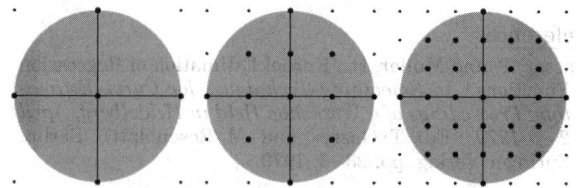

Count the number of LATTICE POINTS $N(r)$ inside the boundary of a CIRCLE of RADIUS r with center at the origin. The exact solution is given by the SUM

$$N(r) = 1 + 4\lfloor r \rfloor + 4\sum_{i=1}^{\lfloor r \rfloor}\left\lfloor\sqrt{r^2-i^2}\right\rfloor \tag{1}$$

$$= 1 + 4\sum_{i=1}^{r^2}(-1)^{i-1}\left\lfloor\frac{r^2}{2i-1}\right\rfloor \tag{2}$$

(Hilbert and Cohn-Vossen 1999, p. 39). The first few values for $r = 0, 1, \ldots$ are 1, 5, 13, 29, 49, 81, 113, 149, ... (Sloane's A000328).
The series for $N(r)$ is intimately connected with $r(n)$, the number of representations of n by two squares, since

$$N(r) = \sum_{n=0}^{r^2} r(n) \tag{3}$$

(Hardy 1999, p. 67). $N(r)$ is also closely connected with the LEIBNIZ SERIES since

$$\frac{1}{4}\left[\frac{N(r)}{r^2}-\frac{1}{r^2}\right] = 1 - \frac{1}{3} + \frac{1}{5} - \frac{1}{7} + \ldots \pm \frac{1}{r}, \tag{4}$$

so taking the limit $r \to \infty$ gives

$$\tfrac{1}{4}\pi = 1 - \tfrac{1}{3} + \tfrac{1}{5} - \tfrac{1}{7} + \tfrac{1}{9} + \ldots \tag{5}$$

(Hilbert and Cohn-Vossen 19991, p. 39).

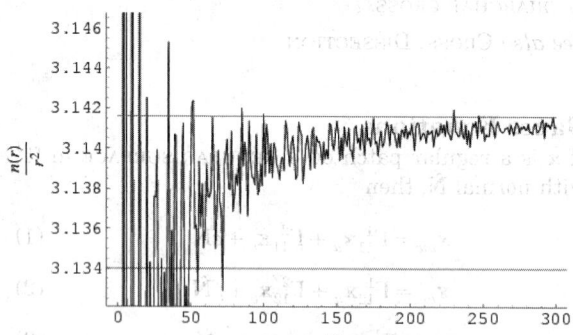

Gauss showed that

$$N(r) = \pi r^2 + E(r), \tag{6}$$

where

$$|E(r)| \le 2\sqrt{2}\pi r \tag{7}$$

(Hardy 1999, p. 67). Writing $|E(r)| \leq Cr^{\theta}$, the best bounds on θ are

$$1/2 < \theta \leq 46/73 \approx 0.630137$$

(Huxley 1990). The lower limit 1/2 was obtained independently by Hardy and Landau in 1915. The following table summarizes incremental improvements in the upper limit (Hardy 1999, p. 81).

θ	approx.	citation
46/73	0.63014	Huxley 1990
7/11	0.63636	
24/37	0.64864	Cheng 1963
34/53	0.64150	Vinogradov
37/56	0.66071	Littlewood and Walfisz 1924
2/3	0.66667	Sierpinski1906, van der Corput 1923

The problem has also been extended to CONICS, ellipsoids (Hardy 1915), and higher dimensions.

See also CIRCLE LATTICE POINTS, DIRICHLET DIVISOR PROBLEM, LEIBNIZ SERIES, SUM OF SQUARES FUNCTION

References

Bohr, H. and Cramér. *Enzykl. d. Math. Wiss.* **II C 8**, 823–24, 1922.
Cheng, J. R. "The Lattice Points in a Circle." *Sci. Sinica* **12**, 633–49, 1963.
Cilleruello, J. "The Distribution of Lattice Points on Circles." *J. Number Th.* **43**, 198–02, 1993.
Guy, R. K. "Gauß's Lattice Point Problem." §F1 in *Unsolved Problems in Number Theory, 2nd ed.* New York: Springer-Verlag, pp. 240–417, 1994.
Hardy, G. H. *Quart. J. Math.* **46**, 283, 1915.
Hardy, G. H. *Ramanujan: Twelve Lectures on Subjects Suggested by His Life and Work, 3rd ed.* New York: Chelsea, 1999.
Hardy, G. H. and Wright, E. M. *An Introduction to the Theory of Numbers, 5th ed.* Oxford, England: Clarendon Press, pp. 268–69, 1979.
Hilbert, D. and Cohn-Vossen, S. *Geometry and the Imagination.* New York: Chelsea, pp. 33–5, 1999.
Huxley, M. N. "Exponential Sums and Lattice Points." *Proc. London Math. Soc.* **60**, 471–02, 1990.
Huxley, M. N. "Corrigenda: 'Exponential Sums and Lattice Points'." *Proc. London Math. Soc.* **66**, 70, 1993.
Landau, E. *Vorlesungen über Zahlentheorie, Vol. 2.* New York: Chelsea, pp. 183–08 1970.
Le Lionnais, F. *Les nombres remarquables.* Paris: Hermann, p. 24, 1983.
Littlewood, J. E. and Walfisz. *Proc. Roy. Soc. (A)* **106**, 478–88, 1924.
Sloane, N. J. A. Sequences A000328/M3829 in "An On-Line Version of the Encyclopedia of Integer Sequences." http://www.research.att.com/~njas/sequences/eisonline.html.
Titchmarsh. *Quart. J. Math. (Oxford)* **2**, 161–73, 1931.
Titchmarsh. *Proc. London Math. Soc.* **38**, 96–15 and 555, 1935.
Weisstein, E. W. "Circle Lattice Points." MATHEMATICA NOTEBOOK CIRCLELATTICEPOINTS.M.

Gauss's Class Number Conjecture

In his monumental treatise *Disquisitiones Arithmeticae,* Gauss conjectured that the CLASS NUMBER $h(-d)$ of an IMAGINARY QUADRATIC FIELD with DISCRIMINANT $-d$ tends to infinity with d. A proof was finally given by Heilbronn (1934), and Siegel (1936) showed that for any $\epsilon > 0$, there exists a constant $c_{\epsilon} > 0$ such that

$$h(-d) > c_{\epsilon} d)^{1/2 - \epsilon}$$

as $d \to \infty$. However, these results were not effective in actually determining the values for a given m of a complete list of fundamental discriminants $-d$ such that $h(-d) = m$, a problem known as GAUSS'S CLASS NUMBER PROBLEM.

Goldfeld (1976) showed that if there exists a "Weil curve" whose associated DIRICHLET L-SERIES has a zero of at least third order at $s = 1$, then for any $\epsilon > 0$, there exists an effectively computable constant c_{ϵ} such that

$$h(-d) > c_{\epsilon} (\ln d)^{1 - \epsilon}.$$

Gross and Zaiger (1983) showed that certain curves must satisfy the condition of Goldfeld, and Goldfeld's proof was simplified by Oesterlé (1985).

See also CLASS NUMBER, GAUSS'S CLASS NUMBER PROBLEM, HEEGNER NUMBER

References

Arno, S.; Robinson, M. L.; and Wheeler, F. S. "Imaginary Quadratic Fields with Small Odd Class Number." http://www.math.uiuc.edu/Algebraic-Number-Theory/0009/.
Böcherer, S. "Das Gauß'sche Klassenzahlproblem." *Mitt. Math. Ges. Hamburg* **11**, 565–89, 1988.
Gauss, C. F. *Disquisitiones Arithmeticae.* New Haven, CT: Yale University Press, 1966.
Goldfeld, D. M. "The Class Number of Quadratic Fields and the Conjectures of Birch and Swinnerton-Dyer." *Ann. Scuola Norm. Sup. Pisa* **3**, 623–63, 1976.
Gross, B. and Zaiger, D. "Points de Heegner et derivées de fonctions L." *C. R. Acad. Sci. Paris* **297**, 85–7, 1983.
Heilbronn, H. "On the Class Number in Imaginary Quadratic Fields." *Quart. J. Math. Oxford Ser.* **25**, 150–60, 1934.
Oesterlé, J. "Nombres de classes des corps quadratiques imaginaires." *Astérique* **121–22**, 309–23, 1985.
Siegel, C. L. "Uuml;ber die Klassenzahl quadratischer Zahlkörper." *Acta. Arith.* **1**, 83–6, 1936.

Gauss's Class Number Problem

For a given m, determine a complete list of fundamental DISCRIMINANTS $-d$ such that the CLASS NUMBER is given by $h(-d) = m$. Heegner (1952) gave a solution for $m = 1$, but it was not completely accepted due to a number of apparent gaps. However, subsequent examination of Heegner's proof showed it to be "essentially" correct (Conway and Guy 1996). Conway and Guy (1996) therefore call the nine values of $n(-d)$ having $h(-d) = 1$ where $-d$ is the DISCRIMINANT corresponding to an QUADRATIC FIELD $a + b\sqrt{-n}$ ($n = -1$, -2, -3, -7, -11, -19, -43, -67, and -163; Sloane's A003173) the HEEGNER NUMBERS. The HEEGNER NUMBERS have a number of fascinating properties.

Stark (1967) and Baker (1966) gave independent proofs of the fact that only nine such numbers exist; both proofs were accepted. Baker (1971) and Stark (1975) subsequently and independently solved the generalized class number problem completely for $m = 2$. Oesterlé (1985) solved the case $m = 3$, and Arno (1992) solved the case $m = 4$. Wagner (1996) solve the cases $n = 5$, 6, and 7. Arno *et al.* (1993) solved the problem for ODD m satisfying $5 \le m \le 23$. In his thesis, M. Watkins has solved the problem for all $m \le 16$.

See also CLASS NUMBER, GAUSS'S CLASS NUMBER CONJECTURE, HEEGNER NUMBER

References

Arno, S. "The Imaginary Quadratic Fields of Class Number 4." *Acta Arith.* **40**, 321–34, 1992.
Arno, S.; Robinson, M. L.; and Wheeler, F. S. "Imaginary Quadratic Fields with Small Odd Class Number." Dec. 1993. http://www.math.uiuc.edu/Algebraic-Number-Theory/0009/.
Baker, A. "Linear Forms in the Logarithms of Algebraic Numbers. I." *Mathematika* **13**, 204–16, 1966.
Baker, A. "Imaginary Quadratic Fields with Class Number 2." *Ann. Math.* **94**, 139–52, 1971.
Conway, J. H. and Guy, R. K. "The Nine Magic Discriminants." In *The Book of Numbers.* New York: Springer-Verlag, pp. 224–26, 1996.
Goldfeld, D. M. "Gauss' Class Number Problem for Imaginary Quadratic Fields." *Bull. Amer. Math. Soc.* **13**, 23–7, 1985.
Heegner, K. "Diophantische Analysis und Modulfunktionen." *Math. Z.* **56**, 227–53, 1952.
Heilbronn, H. A. and Linfoot, E. H. "On the Imaginary Quadratic Corpora of Class-Number One." *Quart. J. Math. (Oxford)* **5**, 293–01, 1934.
Ireland, K. and Rosen, M. *A Classical Introduction to Modern Number Theory,* 2nd ed. New York: Springer-Verlag, p. 192, 1990.
Lehmer, D. H. "On Imaginary Quadratic Fields whose Class Number is Unity." *Bull. Amer. Math. Soc.* **39**, 360, 1933.
Montgomery, H. and Weinberger, P. "Notes on Small Class Numbers." *Acta. Arith.* **24**, 529–42, 1974.
Oesterlé, J. "Nombres de classes des corps quadratiques imaginaires." *Astérique* **121–22**, 309–23, 1985.
Oesterlé, J. "Le problème de Gauss sur le nombre de classes." *Enseign Math.* **34**, 43–7, 1988.
Serre, J.-P. $\Delta = b^2 - 4ac$." *Math. Medley* **13**, 1–0, 1985.
Shanks, D. "On Gauss's Class Number Problems." *Math. Comput.* **23**, 151–63, 1969.
Sloane, N. J. A. Sequences A003173/M0827 in "An On-Line Version of the Encyclopedia of Integer Sequences." http://www.research.att.com/~njas/sequences/eisonline.html.
Stark, H. M. "A Complete Determination of the Complex Quadratic Fields of Class Number One." *Michigan Math. J.* **14**, 1–7, 1967.

Stark, H. M. "On Complex Quadratic Fields with Class Number Two." *Math. Comput.* **29**, 289–02, 1975.
Wagner, C. "Class Number 5, 6, and 7." *Math. Comput.* **65**, 785–00, 1996.

Gauss's Constant

The RECIPROCAL of the ARITHMETIC-GEOMETRIC MEAN of 1 and $\sqrt{2}$,

$$G \equiv \frac{1}{M(1,\ \sqrt{2})} \tag{1}$$

$$= \frac{2}{\pi} \int_0^1 \frac{1}{\sqrt{1 - x^4}}\, dx \tag{2}$$

$$= \frac{2}{\pi} \int_0^{\pi/2} \frac{d\theta}{\sqrt{1 + \sin^2 \theta}} \tag{3}$$

$$= \frac{\sqrt{2}}{\pi} K\left(\frac{1}{\sqrt{2}}\right) \tag{4}$$

$$= \frac{1}{(2\pi)^{3/2}} [\Gamma(\tfrac{1}{4})]^2 \tag{5}$$

$$= 0.83462684167\ldots \tag{6}$$

(Sloane's A014549), where $K(k)$ is the complete ELLIPTIC INTEGRAL OF THE FIRST KIND and $\Gamma(z)$ is the GAMMA FUNCTION. Gauss's constant has CONTINUED FRACTION [0, 1, 5, 21, 3, 4, 14, 1, 1, 1, 1, 1, 3, 1, 15, ...] (Sloane's A053002).

The inverse of Gauss's constant is given by

$$\frac{1}{G} = 1.1981402347355922074399\ldots \tag{7}$$

(Sloane's A053004), and has [1, 5, 21, 3, 4, 14, 1, 1, 1, 1, 1, 3, 1, 15, 1, ...] (Sloane's A053003).

See also ARITHMETIC-GEOMETRIC MEAN, GAUSS-KUZMIN-WIRSING CONSTANT, PYTHAGORAS'S CONSTANT

References

Borwein, J. M. and Borwein, P. B. *Pi & the AGM: A Study in Analytic Number Theory and Computational Complexity.* New York: Wiley, p. 5, 1987.
Goldman, J. R. *The Queen of Mathematics: An Historically Motivated Guide to Number Theory.* Natick, MA: A. K. Peters, p. 92, 1997.
Finch, S. "Favorite Mathematical Constants." http://www.mathsoft.com/asolve/constant/gauss/gauss.html.
Sloane, N. J. A. Sequences A014549, A053002, A053003, and A053004 in "An On-Line Version of the Encyclopedia of Integer Sequences." http://www.research.att.com/~njas/sequences/eisonline.html.

Gauss's Criterion

Let p be an ODD PRIME and b a POSITIVE INTEGER not divisible by p. Then for each POSITIVE ODD INTEGER $2k - 1 < p$, let r_k be

$$r_k \equiv (2k-1)b \pmod{p}$$

with $0 < r_k < p$, and let t be the number of EVEN r_ks. Then

$$(b/p) = (-1)^t,$$

where (b/p) is the LEGENDRE SYMBOL.

References

Shanks, D. "Gauss's Criterion." §1.17 in *Solved and Unsolved Problems in Number Theory, 4th ed.* New York: Chelsea, pp. 38–0, 1993.

Gauss's Cyclotomic Formula

Let $p > 3$ be a PRIME NUMBER, then

$$4\,\frac{x^p - y^p}{x - y} = R^2(x,\,y) - (-1)^{(p-1)/2} p S^2(x,\,y),$$

where $R(x,\,y)$ and $S(x,\,y)$ are HOMOGENEOUS POLYNOMIALS in x and y with integer COEFFICIENTS. Gauss (1965, p. 467) gives the coefficients of R and S up to $p = 23$.

Kraitchik (1924) generalized Gauss's formula to odd SQUAREFREE integers $n > 3$. Then Gauss's formula can be written in the slightly simpler form

$$4\Phi_n(z) = A_n^2(z) - (-1)^{(n-1)/2} n z^2 B_n^2(z),$$

where $A_n(z)$ and $B_n(z)$ have integer coefficients and are of degree $\phi(n)/2$ and $\phi(n)/2 - 2$, respectively, with $\phi(n)$ the TOTIENT FUNCTION and $\Phi_n(z)$ a CYCLOTOMIC POLYNOMIAL. In addition, $A_n(z)$ is symmetric if n is EVEN; otherwise it is antisymmetric. $B_n(z)$ is symmetric in most cases, but it antisymmetric if n is OF THE FORM $4k + 3$ (Riesel 1994, p. 436). The following table gives the first few $A_n(z)$ and $B_n(z)$s (Riesel 1994, pp. 436–42).

n	$A_n(z)$	$B_n(z)$
5	$2z^2 + z + 2$	1
7	$2z^3 + z^2 - z - 2$	$z + 1$
11	$2z^5 + z^4 - 2z^3 + 2z^2 - z - 2$	$z^3 + 1$

See also AURIFEUILLEAN FACTORIZATION, CYCLOTOMIC POLYNOMIAL, LUCAS'S THEOREM

References

Gauss, C. F. §356–57 in *Untersuchungen über höhere Arithmetik.* New York: Chelsea, pp. 425–28 and 467, 1965.
Kraitchik, M. *Recherches sue la théorie des nombres, tome I.* Paris: Gauthier-Villars, pp. 93–29, 1924.
Kraitchik, M. *Recherches sue la théorie des nombres, tome II.* Paris: Gauthier-Villars, pp. 1–, 1929.
Riesel, H. "Gauss's Formula for Cyclotomic Polynomials." In tables at end of *Prime Numbers and Computer Methods*

for Factorization, 2nd ed. Boston, MA: Birkhäuser, pp. 436–42, 1994.

Gauss's Digamma Theorem

At rational arguments p/q, the DIGAMMA FUNCTION $\psi_0(p/q)$ is given by

$$\psi_0\left(\frac{p}{q}\right) = -\gamma - \ln(2q) - \frac{1}{2}\pi \cot\left(\frac{p}{q}\pi\right)$$
$$+ 2 \sum_{k=1}^{\lceil q/2 \rceil - 1} \cos\left(\frac{2\pi pk}{q}\right) \ln\left[\sin\left(\frac{\pi k}{q}\right)\right] \quad (1)$$

for $0 < p < q$ (Knuth 1997, p. 94). These give the special values

$$\psi_o(\tfrac{1}{2}) = -\gamma - 2\ln 2 \quad (2)$$

$$\psi_0(\tfrac{1}{3}) = \tfrac{1}{6}(-6\gamma - \pi\sqrt{3} - 9\ln 3) \quad (3)$$

$$\psi_0(\tfrac{2}{3}) = \tfrac{1}{6}(-6\gamma + \pi\sqrt{3} - 9\ln 3) \quad (4)$$

$$\psi_0(\tfrac{1}{4}) = \tfrac{1}{2}(-2\gamma - \pi - 6\ln 2) \quad (5)$$

$$\psi_0(\tfrac{3}{4}) = \tfrac{1}{2}(-2\gamma + \pi - 6\ln 2) \quad (6)$$

$$\psi_0(\tfrac{1}{6}) = -\gamma - \tfrac{1}{2}\sqrt{3}\pi - 2\ln 2 - \tfrac{3}{2}\ln 3) \quad (7)$$

$$\psi_0(\tfrac{5}{6}) = -\gamma + \tfrac{1}{2}\sqrt{3}\pi - 2\ln 2 - \tfrac{3}{2}\ln 3) \quad (8)$$

$$\psi_0(1) = -\gamma, \quad (9)$$

where γ is the EULER-MASCHERONI CONSTANT.

See also DIGAMMA FUNCTION

References

Böhmer, E. *Differenzengleichungen und bestimmte Integrale.* Leipzig, Germany: Teubner, p. 77, 1939.
Erdélyi, A.; Magnus, W.; Oberhettinger, F.; and Tricomi, F. G. "The ψ Function." §1.7 in *Higher Transcendental Functions, Vol. 1.* New York: Krieger, pp. 15–0, 1981.
Knuth, D. E. *The Art of Computer Programming, Vol. 1: Fundamental Algorithms, 3rd ed.* Reading, MA: Addison-Wesley, 1997.

Gauss's Double Point Theorem

If a sequence of DOUBLE POINTS is passed as a CLOSED CURVE is traversed, each DOUBLE POINT appears once in an EVEN place and once in an ODD place.

References

Rademacher, H. and Toeplitz, O. *The Enjoyment of Mathematics: Selections from Mathematics for the Amateur.* Princeton, NJ: Princeton University Press, pp. 61–6, 1957.

Gauss's Equation (Radius Derivatives)

Expresses the second derivatives of the RADIUS VECTOR **r** in terms of the CHRISTOFFEL SYMBOL OF THE SECOND KIND.

$$\mathbf{r}_{ij} = \Gamma_{ij}^{k}\mathbf{r}_{k} + (\mathbf{r}_{ij} \cdot \mathbf{n})\mathbf{n}.$$

Gauss's Formulas

Let a SPHERICAL TRIANGLE have sides a, b, and c with A, B, and C the corresponding opposite angles. Then

$$\frac{\sin[\frac{1}{2}(a-b)]}{\sin(\frac{1}{2}c)} = \frac{\sin[\frac{1}{2}(A-B)]}{\cos(\frac{1}{2}C)} \tag{1}$$

$$\frac{\sin[\frac{1}{2}(a+b)]}{\sin(\frac{1}{2}c)} = \frac{\cos[\frac{1}{2}(A-B)]}{\sin(\frac{1}{2}C)} \tag{2}$$

$$\frac{\cos[\frac{1}{2}(a-b)]}{\cos(\frac{1}{2}c)} = \frac{\sin[\frac{1}{2}(A+B)]}{\cos(\frac{1}{2}C)} \tag{3}$$

$$\frac{\cos[\frac{1}{2}(a+b)]}{\cos(\frac{1}{2}c)} = \frac{\cos[\frac{1}{2}(A+B)]}{\sin(\frac{1}{2}C)}. \tag{4}$$

These formulas are also known as Delambre's analogies (Smart 1960, p. 22).

See also SPHERICAL TRIGONOMETRY

References
Beyer, W. H. *CRC Standard Mathematical Tables, 28th ed.* Boca Raton, FL: CRC Press, pp. 131 and 147–50, 1987.
Smart, W. M. *Text-Book on Spherical Astronomy, 6th ed.* Cambridge, England: Cambridge University Press, 1960.
Zwillinger, D. (Ed.). "Spherical Geometry and Trigonometry." §6.4 in *CRC Standard Mathematical Tables and Formulae.* Boca Raton, FL: CRC Press, pp. 468–71, 1995.

Gauss's Forward Formula

$$f_p = f_0 + p\delta_{1/2} + G_2\delta_0^2 + G_3\delta_{1/2}^3 + G_4\delta_0^4 + G_5\delta_{1/2}^5$$
$$+ \ldots,$$

for $p \in [0, 1]$, where δ is the CENTRAL DIFFERENCE and

$$G_{2n} = \binom{p+n-1}{2n}$$

$$G_{2n+1} = \binom{p+n}{2n+1},$$

where $\binom{n}{k}$ is a BINOMIAL COEFFICIENT.

See also CENTRAL DIFFERENCE, GAUSS'S BACKWARD FORMULA

References
Beyer, W. H. *CRC Standard Mathematical Tables, 28th ed.* Boca Raton, FL: CRC Press, p. 433, 1987.
Whittaker, E. T. and Robinson, G. "The Newton-Gauss Formula for Interpolation." §21 in *The Calculus of Observations: A Treatise on Numerical Mathematics, 4th ed.* New York: Dover, pp. 36–7, 1967.

Gauss's Harmonic Function Theorem

If a function ϕ is HARMONIC in a SPHERE, then the value of ϕ at the center of the SPHERE is the ARITHMETIC MEAN of its value on the surface.

Gauss's Hypergeometric Theorem

$${}_2F_1(a, b; c; 1) = \frac{(c-b)_{-a}}{(c)_{-a}} = \frac{\Gamma(c)\Gamma(c-a-b)}{\Gamma(c-a)\Gamma(c-b)}$$

for $\Re[c-a-b] > 0$, where ${}_2F_1(a, b; c; x)$ is a (Gauss) HYPERGEOMETRIC FUNCTION. If a is a NEGATIVE INTEGER $-n$, this becomes

$${}_2F_1(-n, b; c; 1) = \frac{(c-b)_n}{(c)_n},$$

which is known as the VANDERMONDE THEOREM.

See also DOUGALL'S FORMULA, GENERALIZED HYPERGEOMETRIC FUNCTION, HYPERGEOMETRIC FUNCTION, THOMAE'S THEOREM, VANDERMONDE THEOREM

References
Bailey, W. N. "Gauss's Theorem." §1.3 in *Generalised Hypergeometric Series.* Cambridge, England: Cambridge University Press, pp. 2–, 1935.
Hardy, G. H. *Ramanujan: Twelve Lectures on Subjects Suggested by His Life and Work, 3rd ed.* New York: Chelsea, p. 104, 1999.
Koepf, W. *Hypergeometric Summation: An Algorithmic Approach to Summation and Special Function Identities.* Braunschweig, Germany: Vieweg, p. 31, 1998.
Petkovšek, M.; Wilf, H. S.; and Zeilberger, D. *A = B.* Wellesley, MA: A. K. Peters, pp. 42 and 126, 1996.

Gauss's Inequality

If a distribution has a single MODE at μ_0, then

$$P(|x - \mu_0| \geq \lambda\tau) \leq \frac{4}{9\lambda^2},$$

where

$$\tau^2 \equiv \sigma^2 + (\mu - \mu_0)^2.$$

See also ...

Gauss's Interpolation Formula

$$f(x) \approx t_n(x) = \sum_{k=0}^{2n} f_k \zeta_k(x),$$

where $t_n(x)$ is a trigonometric POLYNOMIAL of degree n

such that $t_n(x_k) = f_k$ for $k = 0, ..., 2n$, and

$$\zeta_k(x) = \frac{\sin\left[\frac{1}{2}(x - x_0)\right] \cdots \sin\left[\frac{1}{2}(x - x_{k-1})\right]}{\sin\left[\frac{1}{2}(x_k - x_0)\right] \cdots \sin\left[\frac{1}{2}(x_k - x_{k-1})\right]}$$

$$\times \frac{\sin\left[\frac{1}{2}(x - x_{k+1})\right] \cdots \sin\left[\frac{1}{2}(x - x_{2n})\right]}{\sin\left[\frac{1}{2}(x_k - x_{k+1})\right] \cdots \sin\left[\frac{1}{2}(x_k - x_{2n})\right]}.$$

References

Abramowitz, M. and Stegun, C. A. (Eds.). *Handbook of Mathematical Functions with Formulas, Graphs, and Mathematical Tables, 9th printing.* New York: Dover, p. 881, 1972.

Beyer, W. H. (Ed.). *CRC Standard Mathematical Tables, 28th ed.* Boca Raton, FL: CRC Press, pp. 442–43, 1987.

Gauss's Lemma

Let the multiples m, $2m$, ..., $[(p-1)/2]m$ of an INTEGER such that $p \nmid m$ be taken. If there are an EVEN NUMBER r of least POSITIVE RESIDUES mod p of these numbers $> p/2$, then m is a QUADRATIC RESIDUE of p. If r is ODD, m is a QUADRATIC NONRESIDUE. Gauss's lemma can therefore be stated as $(m|p) = (-1)^r$, where $(m|p)$ is the LEGENDRE SYMBOL. It was proved by Gauss as a step along the way to the QUADRATIC RECIPROCITY THEOREM (Nagell 1951).

Another result known as Gauss's lemma states that for any two integer a and b, suppose $d|ab$. Then if d is RELATIVELY PRIME to a, then d divides b (Séroul 2000, p. 10).

See also LEGENDRE SYMBOL, QUADRATIC RECIPROCITY THEOREM

References

Nagell, T. "Gauss's Lemma." §40 in *Introduction to Number Theory.* New York: Wiley, pp. 139–41, 1951.

Séroul, R. "Gauss's Lemma." §2.4.2 in *Programming for Mathematicians.* Berlin: Springer-Verlag, pp. 10–1, 2000.

Gauss's Machin-Like Formula

The MACHIN-LIKE FORMULA

$$\tfrac{1}{4}\pi = 12 \cot^{-1} 18 + 8 \cot^{-1} 57 - 5 \cot^{-1} 239.$$

Gauss's Mean-Value Theorem

Let $f(z)$ be an ANALYTIC FUNCTION in $|z - a| < R$. Then

$$f(z) = \frac{1}{2\pi} \int_0^{2\pi} f(z + re^{i\theta}) \, d\theta$$

for $0 < r < R$.

Gauss's Polynomial Identity

For even h,

$$1 - \frac{1 - x^h}{1 - x} + \frac{(1 - x^h)(1 - x^{h-1})}{(1 - x)(1 - x^2)}$$

$$- \frac{(1 - x^h)(1 - x^{h-1})(1 - x^{h-2})}{(1 - x)(1 - x^2)(1 - x^3)} + \cdots$$

$$= (1 - x)(1 - x^3)(1 - x^5) \cdots (1 - x^{h-1}) \qquad (1)$$

(Nagell 1951, p. 176). Writing out explicitly,

$$\sum_{n=0}^{h} \frac{(-1)^n \Pi_{k=0}^{n-1}(1 - x^{h-k})}{\Pi_{k=1}^{n}} = \prod_{k=0}^{(h-1)/2} 1 - x^{2k+1}. \qquad (2)$$

For example, for $h = 2$,

$$1 - \frac{1 - x^2}{1 - x} + \frac{(1 - x)(1 - x^2)}{(1 - x)(1 - x^2)} = 2 - \frac{1 - x^2}{1 - x} = 1 - x, \qquad (3)$$

and for $h = 4$,

$$1 - \frac{1 - x^4}{1 - x} + \frac{(1 - x^4)(1 - x^3)}{(1 - x)(1 - x^2)}$$

$$- \frac{(1 - x^4)(1 - x^3)(1 - x^2)}{(1 - x)(1 - x^2)(1 - x^3)}$$

$$+ \frac{(1 - x)(1 - x^2)(1 - x^3)(1 - x^4)}{(1 - x)(1 - x^2)(1 - x^3)(1 - x^4)}$$

$$= 2 - \frac{2(1 - x^4)}{1 - x} + \frac{(1 - x^3)(1 - x^4)}{(1 - x)(1 - x^2)}$$

$$= (1 - x)(1 - x^3). \qquad (4)$$

See also Q-SERIES

References

Nagell, T. "A Polynomial Identity of Gauss." §52 in *Introduction to Number Theory.* New York: Wiley, pp. 174–76, 1951.

Gauss's Polynomial Theorem

If an INTEGER POLYNOMIAL

$$f(x) = x^N + C_1 x^{N-1} + C_2 x^{N-2} + \ldots + C_N$$

is divisible into a product of two POLYNOMIALS $f = \psi\phi$

$$\psi = x^m + \alpha_1 x^{m-1} + \ldots + \alpha_m$$

$$\phi = x^n + \beta_1 x^{n-1} + \ldots + \beta_n,$$

then the COEFFICIENTS of these POLYNOMIALS are INTEGERS.

See also ABEL'S IRREDUCIBILITY THEOREM, ABEL'S LEMMA, KRONECKER'S POLYNOMIAL THEOREM, POLYNOMIAL, SCHÖNEMANN'S THEOREM

References

Dörrie, H. *100 Great Problems of Elementary Mathematics: Their History and Solutions.* New York: Dover, p. 119, 1965.

Gauss's Reciprocity Theorem

QUADRATIC RECIPROCITY THEOREM

Gauss's Test

If $u_n > 0$ and given $B(n)$ a bounded function of n as $n \to \infty$, express the ratio of successive terms as

$$\left| \frac{u_n}{u_{n+1}} \right| = 1 + \frac{h}{n} + \frac{B(n)}{n^r}$$

for $r > 1$. The SERIES converges for $h > 1$ and diverges for $h \leq 1$ (Courant and John 1999, p. 567).

See also CONVERGENCE TESTS

References

Arfken, G. *Mathematical Methods for Physicists, 3rd ed.* Orlando, FL: Academic Press, pp. 287–88, 1985.
Courant, R. and John, F. *Introduction to Calculus and Analysis, Vol. 1.* New York: Springer-Verlag, 1999.

Gauss's Theorem

DIVERGENCE THEOREM, GAUSS'S DIGAMMA THEOREM, GAUSS'S DOUBLE POINT THEOREM, GAUSS'S HYPERGEOMETRIC THEOREM, GAUSS'S THEOREMA EGREGIUM

Gauss's Theorema Egregium

Gauss's theorema egregium states that the GAUSSIAN CURVATURE of a surface embedded in 3-space may be understood intrinsically to that surface. "Residents" of the surface may observe the GAUSSIANCURVATURE of the surface without ever venturing into full 3-dimensional space; they can observe the curvature of the surface they live in without even knowing about the 3-dimensional space in which they are embedded.

In particular, GAUSSIAN CURVATURE can be measured by checking how closely the ARC LENGTH of small RADIUS CIRCLES correspond to what they should be in EUCLIDEAN SPACE, $2\pi r$. If the ARC LENGTH of CIRCLES tends to be smaller than what is expected in EUCLIDEAN SPACE, then the space is positively curved; if larger, negatively; if the same, 0 GAUSSIAN CURVATURE.

Gauss (effectively) expressed the theorema egregium by saying that the GAUSSIAN curvature at a point is given by $-R(v, w)v, w$ where R is the RIEMANN TENSOR, and v and w are an orthonormal basis for the TANGENT SPACE.

See also CHRISTOFFEL SYMBOL OF THE SECOND KIND, GAUSS EQUATIONS, GAUSSIAN CURVATURE

References

Gray, A. "Gauss's Theorema Egregium." §22.2 in *Modern Differential Geometry of Curves and Surfaces with Mathematica, 2nd ed.* Boca Raton, FL: CRC Press, pp. 507–09, 1997.
Reckziegel, H. In *Mathematical Models from the Collections of Universities and Museums* (Ed. G. Fischer). Braunschweig, Germany: Vieweg, pp. 31–2, 1986.

Gauss's Transformation

If

$$(1 + x \sin^2 \alpha)\sin \beta = (1 + x)\sin \alpha,$$

then

$$(1 + x) \int_0^\alpha \frac{d\phi}{\sqrt{1 - x^2 \sin^2 \phi}} = \int_0^\beta \frac{d\phi}{\sqrt{1 - \frac{4x}{(1 + x)^2} \sin^2 \phi}}.$$

See also ELLIPTIC INTEGRAL OF THE FIRST KIND, LANDEN'S TRANSFORMATION

Gauss-Bodenmiller Theorem

The CIRCLES on the DIAGONALS of a COMPLETE QUADRILATERAL as DIAMETERS are COAXAL. Furthermore, the ORTHOCENTERS of the four TRIANGLES of a COMPLETE QUADRILATERAL are COLLINEAR on the RADICAL AXIS of the COAXAL CIRCLES.

See also COAXAL CIRCLES, COLLINEAR, COMPLETE QUADRILATERAL, DIAGONAL (POLYGON), ORTHOCENTER, RADICAL AXIS

References

Johnson, R. A. *Modern Geometry: An Elementary Treatise on the Geometry of the Triangle and the Circle.* Boston, MA: Houghton Mifflin, p. 172, 1929.

Gauss-Bolyai-Lobachevsky Space

A non-Euclidean space with constant NEGATIVE GAUSSIAN CURVATURE.

See also LOBACHEVSKY-BOLYAI-GAUSS GEOMETRY, NON-EUCLIDEAN GEOMETRY

Gauss-Bonnet Formula

The Gauss-Bonnet formula has several formulations. The simplest one expresses the total GAUSSIAN CURVATURE of an embedded triangle in terms of the total GEODESIC CURVATURE of the boundary and the JUMP ANGLES at the corners.

More specifically, if M is any 2-D RIEMANNIAN MANIFOLD (like a surface in 3-space) and if T is an embedded triangle, then the Gauss-Bonnet formula states that the integral over the whole triangle of the GAUSSIAN CURVATURE with respect to AREA is given by 2π minus the sum of the JUMP ANGLES minus the integral of the GEODESIC CURVATURE over the whole of the boundary of the triangle (with respect to ARC LENGTH),

$$\iint_T K \, dA = 2\pi - \sum \alpha_i - \int_{\partial T} k_g \, ds, \qquad (1)$$

where K is the GAUSSIAN CURVATURE, dA is the AREA measure, the α_is are the JUMP ANGLES of ∂T, and k_g is the GEODESIC CURVATURE of ∂T, with ds the ARC LENGTH measure.

The next most common formulation of the Gauss-Bonnet formula is that for any compact, boundaryless 2-D RIEMANNIAN MANIFOLD, the integral of the GAUSSIAN CURVATURE over the entire MANIFOLD with respect to AREA is 2π times the EULER CHARACTERISTIC of the MANIFOLD,

$$\iint_M K \, dA = 2\pi\chi(M). \qquad (2)$$

This is somewhat surprising because the total GAUSSIAN CURVATURE is differential-geometric in character, but the EULER CHARACTERISTIC is topological in character and does not depend on differential geometry at all. So if you distort the surface and change the curvature at any location, regardless of how you do it, the same total curvature is maintained.

Another way of looking at the Gauss-Bonnet theorem for surfaces in 3-space is that the GAUSS MAP of the surface has DEGREE given by half the EULER CHARACTERISTIC of the surface

$$\iint_M K \, dA = 2\pi\chi(M) - \sum \alpha_i - \int_{\partial M} k_g \, ds, \qquad (3)$$

which works only for ORIENTABLE SURFACES where M is COMPACT. This makes the Gauss-Bonnet theorem a simple consequence of the POINCARE-HOPF INDEX THEOREM, which is a nice way of looking at things if you're a topologist, but not so nice for a differential geometer. This proof can be found in Guillemin and Pollack (1974). Millman and Parker (1977) give a standard differential-geometric proof of the Gauss-Bonnet theorem, and Singer and Thorpe (1996) give a GAUSS'S THEOREMA EGREGIUM-inspired proof which is entirely intrinsic, without any reference to the ambient EUCLIDEAN SPACE.

A general Gauss-Bonnet formula that takes into account both formulas can also be given. For any compact 2-D RIEMANNIAN MANIFOLD with corners, the integral of the GAUSSIAN CURVATURE over the 2-MANIFOLD with respect to AREA is 2π times the EULER CHARACTERISTIC of the MANIFOLD minus the sum of the JUMP ANGLES and the total GEODESIC CURVATURE of the boundary.

References

Chavel, I. *Riemannian Geometry: A Modern Introduction.* New York: Cambridge University Press, 1994.

Guillemin, V. and Pollack, A. *Differential Topology.* Englewood Cliffs, NJ: Prentice-Hall, 1974.

Millman, R. S. and Parker, G. D. *Elements of Differential Geometry.* Prentice-Hall, 1977.

Reckziegel, H. In *Mathematical Models from the Collections of Universities and Museums* (Ed. G. Fischer). Braunschweig, Germany: Vieweg, p. 31, 1986.

Singer, I. M. and Thorpe, J. A. *Lecture Notes on Elementary Topology and Geometry.* New York: Springer-Verlag, 1996.

Gauss-Bonnet Theorem

GAUSS-BONNET FORMULA

Gaussian Approximation Algorithm

ARITHMETIC-GEOMETRIC MEAN

Gaussian Bivariate Distribution

The Gaussian bivariate distribution is given by

$$P(x_1, x_2) = \frac{1}{2\pi\sigma_1\sigma_2\sqrt{1-\rho^2}} \exp\left[-\frac{z}{2(1-\rho^2)}\right], \qquad (1)$$

where

$$z \equiv \frac{(x_1-\mu_1)^2}{\sigma_1^2} - \frac{2\rho(x_1-\mu_1)(x_2-\mu_2)}{\sigma_1\sigma_2} + \frac{(x_2-\mu_2)^2}{\sigma_2^2}, \qquad (2)$$

and

$$\rho \equiv \mathrm{cor}(x_1, x_2) = \frac{\sigma_{12}}{\sigma_1\sigma_2} \qquad (3)$$

is the CORRELATION of x_1 and x_2 (Kenney and Keeping 1951, pp. 92 and 202–05; Whittaker and Robinson 1967, p. 329). The Gaussian bivariate distribution is implemented in *Mathematica* as MultinormalDistribution[{*mu1*, *mu2*}, {{*sigma11*, *sigma12*}, {*sigma12*, *sigma22*}}, {*x1*, *x2*}] in the *Mathematica* add-on package Statistics`MultinormalDistribution` (which can be loaded with the command <<Statistics`).

The MARGINAL PROBABILITIES are then

$$P(x_1) = \int_{-\infty}^{\infty} P(x_1, x_2) \, dx_2 = \frac{1}{\sigma_1\sqrt{2\pi}} e^{-(x_1-\mu_1)^2/(2\sigma_1^2)} \qquad (4)$$

and

$$P(x_2) = \int_{-\infty}^{\infty} P(x_1, x_2) \, dx_1$$

$$= \frac{1}{\sigma_2\sqrt{2\pi}} \exp\left[-\frac{(x_2-\mu_2)^2}{(2\sigma_2^2)}\right] \qquad (5)$$

(Kenney and Keeping 1951, p. 202).

Let z_1 and z_2 be two independent Gaussian variables with MEANS $\mu_i = 0$ and $\sigma_i^2 = 1$ for $i = 1, 2$. Then the variables a_1 and a_2 defined below are Gaussian bivariates with unit VARIANCE and CROSS-CORRELATION COEFFICIENT ρ :

$$a_1 = \sqrt{\frac{1+\rho}{2}}\, z_1 + \sqrt{\frac{1-\rho}{2}}\, z_2 \qquad (6)$$

$$a_2 = \sqrt{\frac{1+\rho}{2}}\, z_1 - \sqrt{\frac{1-\rho}{2}}\, z_2. \qquad (7)$$

To derive the Gaussian bivariate probability function, let X_1 and X_2 be normally and independently distributed variates with MEAN 0 and VARIANCE 1, then define

$$Y_1 \equiv \mu_1 + \sigma_{11} X_1 + \sigma_{12} X_2 \qquad (8)$$

$$Y_2 \equiv \mu_2 + \sigma_{21} X_1 + \sigma_{22} X_2 \qquad (9)$$

(Kenney and Keeping 1951, p. 92). The variates Y_1 and Y_2 are then themselves normally distributed with MEANS μ_1 and μ_2, VARIANCES

$$\sigma_1^2 \equiv \sigma_{11}^2 + \sigma_{12}^2 \qquad (10)$$

$$\sigma_2^2 \equiv \sigma_{21}^2 + \sigma_{22}^2, \qquad (11)$$

and COVARIANCE

$$V_{12} \equiv \sigma_{11}\sigma_{21} + \sigma_{12}\sigma_{22}. \qquad (12)$$

The COVARIANCE matrix is defined by

$$V_{ij} = \begin{bmatrix} \sigma_1^2 & \rho\sigma_1\sigma_2 \\ \rho\sigma_1\sigma_2 & \sigma_2^2 \end{bmatrix}, \qquad (13)$$

where

$$\rho \equiv \frac{V_{12}}{\sigma_1\sigma_2} = \frac{\sigma_{11}\sigma_{21} + \sigma_{12}\sigma_{22}}{\sigma_1\sigma_2}. \qquad (14)$$

Now, the joint probability density function for x_1 and x_2 is

$$f(x_1, x_2)\, dx_1\, dx_2 = \frac{1}{2\pi}\, e^{-(x_1^2 + x_2^2)/2}\, dx_1 dx_2, \qquad (15)$$

but from (8) and (9), we have

$$\begin{bmatrix} y_1 - \mu_1 \\ y_2 - \mu_2 \end{bmatrix} = \begin{bmatrix} \sigma_{11} & \sigma_{12} \\ \sigma_{21} & \sigma_{22} \end{bmatrix} \begin{bmatrix} x_1 \\ x_2 \end{bmatrix}. \qquad (16)$$

As long as

$$\begin{bmatrix} \sigma_{11} & \sigma_{12} \\ \sigma_{21} & \sigma_{22} \end{bmatrix} \neq 0, \qquad (17)$$

this can be inverted to give

$$\begin{bmatrix} x_1 \\ x_2 \end{bmatrix} = \begin{bmatrix} \sigma_{11} & \sigma_{12} \\ \sigma_{21} & \sigma_{22} \end{bmatrix}^{-1} \begin{bmatrix} y_1 - \mu_1 \\ y_2 - \mu_2 \end{bmatrix}$$

$$= \frac{1}{\sigma_{11}\sigma_{22} - \sigma_{12}\sigma_{21}} \begin{bmatrix} \sigma_{22} & -\sigma_{12} \\ -\sigma_{21} & \sigma_{11} \end{bmatrix} \begin{bmatrix} y_1 - \mu_1 \\ y_2 - \mu_2 \end{bmatrix}. \qquad (18)$$

Therefore,

$$x_1^2 + x_2^2 = \frac{[\sigma_{22}(y_1 - \mu_1) - \sigma_{12}(y_2 - \mu_2)]^2}{(\sigma_{11}\sigma_{22} - \sigma_{12}\sigma_{21})^2}$$

$$+ \frac{[-\sigma_{21}(y_1 - \mu_1) - \sigma_{11}(y_2 - \mu_2)]^2}{(\sigma_{11}\sigma_{22} - \sigma_{12}\sigma_{21})^2}, \qquad (19)$$

and expanding the NUMERATOR of (19) gives

$$\sigma_{22}^2(y_1 - \mu_1)^2 - 2\sigma_{12}\sigma_{22}(y_1 - \mu_1)(y_2 - \mu_2) + \sigma_{12}^2(y_2 - \mu_2)^2$$
$$+ \sigma_{22}^2(y_1 - \mu_1)^2 - 2\sigma_{11}\sigma_{21}(y_1 - \mu_1)(y_2 - \mu_2)$$
$$+ \sigma_{11}^2(y_2 - \mu_2)^2, \qquad (20)$$

so

$$(x_1^2 + x_2^2)(\sigma_{11}\sigma_{22} - \sigma_{12}\sigma_{21})^2$$
$$= (y_1 - \mu_1)^2(\sigma_{21}^2 + \sigma_{22}^2) - 2(y_1 - \mu_1)(y_2 - \mu_2)$$
$$\times (\sigma_{11}\sigma_{21} + \sigma_{12}\sigma_{22}) + (y_2 - \mu_2)^2(\sigma_{21}^2 + \sigma_{12}^2)$$
$$= \sigma_2^2(y_1 - \mu_1)^2 - 2(y_1 - \mu_1)(y_2 - \mu_2)(\rho\sigma_1\sigma_2) + \sigma_1^2(y_2 - \mu_2)^2$$
$$= \sigma_1^2\sigma_2^2 \left[\frac{(y_1 - \mu_1)^2}{\sigma_1^2} - \frac{2\rho(y_1 - \mu_1)(y_2 - \mu_2)}{\sigma_1\sigma_2} + \frac{(y_2 - \mu_2)^2}{\sigma_2^2} \right] \qquad (21)$$

Now, the DENOMINATOR of (19) is

$$\sigma_{11}^2\sigma_{21}^2 + \sigma_{11}^2\sigma_{22}^2 + \sigma_{12}^2\sigma_{21}^2 + \sigma_{12}^2\sigma_{22}^2 - \sigma_{11}^2\sigma_{21}^2$$
$$- 2\sigma_{11}\sigma_{12}\sigma_{21}\sigma_{22} - \sigma_{12}^2\sigma_{22}^2$$
$$= (\sigma_{11}\sigma_{22} - \sigma_{12}\sigma_{21})^2, \qquad (22)$$

so

$$\frac{1}{1 - \rho^2} = \frac{1}{1 - \dfrac{V_{12}^2}{\sigma_1^2\sigma_2^2}} = \frac{\sigma_1^2\sigma_2^2}{\sigma_1^2\sigma_2^2 - V_{12}^2}$$

$$= \frac{\sigma_1^2\sigma_2^2}{(\sigma_{11}^2 + \sigma_{12}^2)(\sigma_{21}^2 + \sigma_{22}^2) - (\sigma_{11}\sigma_{21} + \sigma_{12}\sigma_{22})^2} \qquad (23)$$

can be written simply as

$$\frac{1}{1 - \rho^2} = \frac{\sigma_1^2\sigma_2^2}{(\sigma_{11}\sigma_{22} - \sigma_{12}\sigma_{21})^2}, \qquad (24)$$

and

$$x_1^2 + x_2^2 = \frac{1}{1 - \rho^2}$$
$$\times \left[\frac{(y_1 - \mu_1)^2}{\sigma_1^2} - \frac{2\rho(y_1 - \mu_1)(y_2 - \mu_2)}{\sigma_1\sigma_2} + \frac{(y_2 - \mu_2)^2}{\sigma_2^2} \right]. \qquad (25)$$

Solving for x_1 and x_2 and defining

$$\rho' \equiv \frac{\sigma_1\sigma_2\sqrt{1 - \rho^2}}{\sigma_{11}\sigma_{22} - \sigma_{12}\sigma_{21}} \qquad (26)$$

gives

$$x_1 = \frac{\sigma_{22}(y_1 - \mu_1) - \sigma_{12}(y_2 - \mu_2)}{\rho'} \qquad (27)$$

$$x_2 = \frac{-\sigma_{21}(y_1 - \mu_1) + \sigma_{11}(y_2 - \mu_2)}{\rho'}. \qquad (28)$$

But the JACOBIAN is

$$J\left(\frac{x_1, x_2}{y_1, y_2}\right) = \begin{vmatrix} \dfrac{\partial x_1}{\partial y_1} & \dfrac{\partial x_1}{\partial y_2} \\ \dfrac{\partial x_2}{\partial y_1} & \dfrac{\partial x_2}{\partial y_2} \end{vmatrix} = \begin{vmatrix} \dfrac{\sigma_{22}}{\rho'} & -\dfrac{\sigma_{12}}{\rho'} \\ -\dfrac{\sigma_{21}}{\rho'} & \dfrac{\sigma_{11}}{\rho'} \end{vmatrix}$$

$$= \frac{1}{\rho'^2}(\sigma_{11}\sigma_{22} - \sigma_{12}\sigma_{21}) = \frac{1}{\rho'} = \frac{1}{\sigma_1 \sigma_2 \sqrt{1 - \rho^2}}, \qquad (29)$$

so

$$dx_1\,dx_2 = \frac{dy_1\,dy_2}{\sigma_1\sigma_2\sqrt{1-\rho^2}} \qquad (30)$$

and

$$\frac{1}{2\pi}e^{-(x_1^2+x_2^2)/2}\,dx_1\,dx_2$$

$$= \frac{1}{2\pi\sigma_1\sigma_2\sqrt{1-\rho^2}}\exp\left[-\frac{z}{2(1-\rho^2)}\right]dy_1\,dy_2, \qquad (31)$$

where

$$z \equiv \frac{(y_1-\mu_1)^2}{\sigma_1^2} - \frac{2\rho(y_1-\mu_1)(y_2-\mu_2)}{\sigma_1\sigma_2} + \frac{(y_2-\mu_2)^2}{\sigma_2^2}. \qquad (32)$$

Q.E.D.

In the singular case that

$$\begin{vmatrix} \sigma_{11} & \sigma_{12} \\ \sigma_{21} & \sigma_{22} \end{vmatrix} = 0 \qquad (33)$$

(Kenney and Keeping 1951, p. 94), it follows that

$$\sigma_{11}\sigma_{12} = \sigma_{12}\sigma_{21} \qquad (34)$$

$$y_1 = mu_1 + \sigma_{11}x_1 + \sigma_{12}x_2 \qquad (35)$$

$$y_2 = \mu_2 + \frac{\sigma_{12}\sigma_{21}}{\sigma_{11}}\,x_2 = \mu_2 + \frac{\sigma_{11}\sigma_{21}x_1 + \sigma_{12}\sigma_{21}x_2}{\sigma_{11}}$$

$$= \mu_2 + \frac{\sigma_{21}}{\sigma_{11}}(\sigma_{11}x_1 + \sigma_{12}x_2), \qquad (36)$$

so

$$y_1 = \mu_1 + x_3 \qquad (37)$$

$$y_2 = \mu_2 + \frac{\sigma_{21}}{\sigma_{11}}\,x_3, \qquad (38)$$

where

$$x_3 = y_1 - \mu_1 = \frac{\sigma_{11}}{\sigma_{21}}(y_2 - \mu_2). \qquad (39)$$

The CHARACTERISTIC FUNCTION of the Gaussian bivariate distribution is given by

$$\phi(t_1,\,t_2) \equiv \int_{-\infty}^{\infty}\int_{-\infty}^{\infty} e^{i(t_1x_1+t_2x_2)}P(x_1,\,x_2)\,dx_1\,dx_2$$

$$= N\int_{-\infty}^{\infty}\int_{-\infty}^{\infty} e^{i(t_1x_1+t_2x_2)}\exp\left[-\frac{z}{2(1-\rho^2)}\right]dx_1\,dx_2, \qquad (40)$$

where

$$z \equiv \left[\frac{(x_1-\mu_1)^2}{\sigma_1^2} - \frac{2\rho(x_1-\mu_1)(x_2-\mu_2)}{\sigma_1\sigma_2} + \frac{(x_2-\mu_2)^2}{\sigma_2^2}\right] \qquad (41)$$

and

$$N \equiv \frac{1}{2\pi\sigma_1\sigma_2\sqrt{1-\rho^2}}. \qquad (42)$$

Now let

$$u \equiv x_1 - \mu_1 \qquad (43)$$

$$w \equiv x_2 - \mu_2. \qquad (44)$$

Then

$$\phi(t_1,\,t_2)$$

$$= N'\int_{-\infty}^{\infty}\left(e^{it_2w}\exp\left[-\frac{1}{2(1-\rho^2)}\frac{w^2}{\sigma_2^2}\right]\right)\int_{-\infty}^{\infty} e^v e^{t_1 u}\,du\,dw, \qquad (45)$$

where

$$v \equiv -\frac{1}{2(1-\rho^2)}\frac{1}{\sigma_1^2}\left[u^2 - \frac{2\rho\sigma_1 w}{\sigma_2}\,u\right]$$

$$N' \equiv \frac{e^{i(t_1\mu_1+t_2\mu_2)}}{2\pi\sigma_1\sigma_2\sqrt{1-\rho^2}}. \qquad (46)$$

COMPLETE THE SQUARE in the inner integral

$$\int_{-\infty}^{\infty}\exp\left\{-\frac{1}{2(1-\rho^2)}\frac{1}{\sigma_1^2}\left[u^2 - \frac{2\rho\sigma_1 w}{\sigma_2}\,u\right]\right\}e^{t_1 u}\,du$$

$$= \int_{-\infty}^{\infty}\exp\left\{-\frac{1}{2\sigma_1^2(1-\rho^2)}\left[u - \frac{\rho_1\sigma_1 w}{\sigma^2}\right]^2\right\}$$

$$\times \left\{\frac{1}{2\sigma_1^2(1-\rho^2)}\left(\frac{\rho_1\sigma_1 w}{\sigma_2}\right)^2\right\}e^{it_1 u}\,du. \qquad (47)$$

Rearranging to bring the exponential depending on w outside the inner integral, letting

$$v \equiv u - \rho\,\frac{\sigma_1 w}{\sigma_2}, \qquad (48)$$

and writing

$$e^{it_1 u} = \cos(t_1 u) + i \sin(t_1 u) \qquad (49)$$

gives

$$\phi(t_1, t_2) = N' \int_{-\infty}^{\infty} e^{it_2 w} \exp\left[-\frac{1}{2\sigma_2^2(1-\rho^2)} w^2\right]$$

$$\times \exp\left[\frac{\rho^2}{2\sigma_2^2(1-\rho^2)} w^2\right] \int_{-\infty}^{\infty} \exp\left[-\frac{1}{2\sigma_2^2(1-\rho^2)} v^2\right]$$

$$\times \left\{ \cos\left[t_1\left(v + \frac{\rho\sigma_1 w}{\sigma_2}\right)\right] + i \sin\left[t_1\left(v + \frac{\rho\sigma_1 w}{\sigma_2}\right)\right]\right\} dv\, dw.$$

$$(50)$$

Expanding the term in braces gives

$$\left[\cos(t_1 v)\cos\left(\frac{\rho\sigma_1 w t_1}{\sigma_2}\right) - \sin(t_1 v)\sin\left(\frac{\rho\sigma_1 w}{\sigma_2 t_1}\right)\right]$$

$$+ i\left[\sin(t_1 v)\cos\left(\frac{\rho\sigma_1 w}{\sigma_2 t_1}\right) + \cos(t_1 v)\sin\left(\frac{\rho\sigma_1 w t_1}{\sigma_2}\right)\right]$$

$$= \left[\cos\left(\frac{\rho\sigma_1 w t_1}{\sigma_2}\right) + i \sin\left(\frac{\rho\sigma_1 w t_1}{\sigma_2}\right)\right]$$

$$\times [\cos(t_1 v) + i \sin(t_1 v)]$$

$$= \exp\left(\frac{i\rho\sigma_1 w}{\sigma_2} t_1\right)[\cos(t_1 v) + i \sin(t_1 v)]. \qquad (51)$$

But $e^{-ax^2}\sin(bx)$ is ODD, so the integral over the sine term vanishes, and we are left with

$$\phi(t_1, t_2) = N' \int_{-\infty}^{\infty} e^{it_2 w} \exp\left[-\frac{w^2}{2\sigma_2^2}\right]\exp\left[\frac{\rho^2 w^2}{2\sigma_2^2(1-\rho^2)}\right]$$

$$\times \exp\left[\frac{i\rho\sigma_1 w t_1}{\sigma_2}\right] dw \int_{-\infty}^{\infty} \exp\left[-\frac{v^2}{2\sigma_1^2(1-\rho^2)}\right]\cos(t_1 v)\, dv$$

$$= N' \int_{-\infty}^{\infty} \exp\left[iw\left(t_2 + t_1\left(\rho\,\frac{\sigma_1}{\sigma_2}\right)\right)\right]\exp\left[-\frac{w^2}{2\sigma_2^2}\right] dw$$

$$\int_{-\infty}^{\infty} \exp\left[-\frac{v^2}{2\sigma_1^2(1-\rho^2)}\right]\cos(t_1 v)\, dv. \qquad (52)$$

Now evaluate the GAUSSIAN INTEGRAL

$$\int_{-\infty}^{\infty} e^{ikx} e^{-ax^2}\, dx = \int_{-\infty}^{\infty} e^{-ax^2}\cos(kx)\, dx$$

$$= \sqrt{\frac{\pi}{a}}\, e^{-k^2/4a} \qquad (53)$$

to obtain the explicit form of the CHARACTERISTIC FUNCTION,

$$\phi(t_1, t_2) = \frac{e^{i(t_1\mu_1 + t_2 + \mu_2)}}{2\pi\sigma_1\sigma_2\sqrt{1-\rho^2}}$$

$$\times \left\{\sigma_2\sqrt{2\pi}\exp\left[-\frac{1}{4}\left(t_2 + \rho\,\frac{\sigma_1}{\sigma_2}\, t_1\right)^2 2\sigma_2^2\right]\right\}$$

$$\times \left\{\sigma_1\sqrt{2\pi(1-p^2)}\exp[-\tfrac{2}{1}2\sigma_1^2(1-\rho^2)]\right\}$$

$$= e^{i(t_1\mu_1 + t_2\mu_2)}\exp\{-\tfrac{1}{2}[t_2^2\sigma_2^2 + 2\rho\sigma_1\sigma_2 t_1 t_2 + \rho^2\sigma_1^2 t_1^2$$

$$+ (1-\rho^2)\sigma_1^2 t_1^2]\}$$

$$= \exp[i(t_1\mu_1 + t_2\mu_2) - \tfrac{1}{2}(\sigma_1^2 t_1^2 + 2\rho\sigma_1\sigma_2 t_1 t_2 + \sigma_1^2 t_1^2)]. \qquad (54)$$

See also BOX-MULLER TRANSFORMATION, GAUSSIAN DISTRIBUTION, GAUSSIAN MULTIVARIATE DISTRIBUTION, NORMAL DISTRIBUTION, PRICE'S THEOREM

References

Abramowitz, M. and Stegun, C. A. (Eds.). *Handbook of Mathematical Functions with Formulas, Graphs, and Mathematical Tables, 9th printing.* New York: Dover, pp. 936–37, 1972.

Kenney, J. F. and Keeping, E. S. *Mathematics of Statistics, Pt. 2, 2nd ed.* Princeton, NJ: Van Nostrand, 1951.

Kotz, S.; Balakrishnan, N.; and Johnson, N. L. "Bivariate and Trivariate Normal Distributions." Ch. 46 in *Continuous Multivariate Distributions, Vol. 1: Models and Applications, 2nd ed.* New York: Wiley, pp. 251–48, 2000.

Spiegel, M. R. *Theory and Problems of Probability and Statistics.* New York: McGraw-Hill, p. 118, 1992.

Whittaker, E. T. and Robinson, G. "Determination of the Constants in a Normal Frequency Distribution with Two Variables" and "The Frequencies of the Variables Taken Singly." §161–62 in *The Calculus of Observations: A Treatise on Numerical Mathematics, 4th ed.* New York: Dover, pp. 324–28, 1967.

Gaussian Brackets

A notation published by Gauss in *Disquisitiones Arithmeticae* and defined by

$$[\ ] = 1 \qquad (1)$$

$$[a_1] = a_1 \qquad (2)$$

$$[a_1, a_2] = [a_1]a_2 + [\ ] \qquad (3)$$

$$[a_1, a_2, \ldots, a_n]$$

$$= [a_1, a_2, \ldots, a_{n-1}]a_n + [a_1, a_2, \ldots, a_{n-2}]. \qquad (4)$$

Gaussian brackets are useful for treating CONTINUED FRACTIONS because

$$\cfrac{1}{a_1 + \cfrac{1}{a_2 + \cfrac{1}{a_3 + \ldots + \cfrac{1}{a_n}}}} = \frac{[a_2, an]}{[a_1, an]}. \qquad (5)$$

The NOTATION $[x]$ conflicts with that of GAUSSIAN POLYNOMIALS and the NINT function.

References

Herzberger, M. *Modern Geometrical Optics.* New York: Interscience Publishers, pp. 457–62, 1958.

Gaussian Coefficient

Q-BINOMIAL COEFFICIENT

Gaussian Coordinate System

A coordinate system which has a METRIC satisfying $g_{ii} = -1$ and $\partial g_{ij}/\partial x_j = 0$.

Gaussian Curvature

An intrinsic property of a space independent of the coordinate system used to describe it. The Gaussian curvature of a REGULAR SURFACE in $\mathbb{R}^3$ at a point $\mathbf{p}$ is formally defined as

$$K(\mathbf{p}) = \det(S(\mathbf{p})), \tag{1}$$

where S is the SHAPE OPERATOR and det denotes the DETERMINANT.

If $\mathbf{x} : U \to \mathbb{R}^3$ is a REGULAR PATCH, then the Gaussian curvature is given by

$$K = \frac{eg - f^2}{EG - F^2}, \tag{2}$$

where E, F, and G are coefficients of the first FUNDAMENTAL FORM and e, f, and g are coefficients of the second FUNDAMENTAL FORM (Gray 1997, p. 377). The Gaussian curvature can be given entirely in terms of the first FUNDAMENTAL FORM

$$ds^2 = E\, du^2 + 2F\, du\, dv + G\, dv^2 \tag{3}$$

and the DISCRIMINANT

$$g \equiv EG - F^2 \tag{4}$$

by

$$K = \frac{1}{\sqrt{g}} \left[\frac{\partial}{\partial v} \left(\frac{\sqrt{g}}{E} \Gamma_{11}^2 \right) - \frac{\partial}{\partial u} \left(\frac{\sqrt{g}}{E} \Gamma_{12}^2 \right) \right], \tag{5}$$

where Γ_{ij}^k are the CONNECTION COEFFICIENTS. Equivalently,

$$K = \frac{1}{g^2} \begin{vmatrix} E & F & \frac{\partial F}{\partial v} - \frac{1}{2}\frac{\partial G}{\partial u} \\ F & G & \frac{1}{2}\frac{\partial G}{\partial v} \\ \frac{1}{2}\frac{\partial E}{\partial u} & k_{23} & k_{33} \end{vmatrix} - \frac{1}{g^2} \begin{vmatrix} E & F & \frac{1}{2}\frac{\partial E}{\partial v} \\ F & G & \frac{1}{2}\frac{\partial G}{\partial u} \\ \frac{1}{2}\frac{\partial E}{\partial v} & \frac{1}{2}\frac{\partial G}{\partial v} & 0 \end{vmatrix}, \tag{6}$$

where

$$k_{23} \equiv \frac{\partial F}{\partial u} - \frac{1}{2}\frac{\partial E}{\partial v} \tag{7}$$

$$k_{33} \equiv -\frac{1}{2}\frac{\partial^2 E}{\partial v^2} + \frac{\partial^2 F}{\partial u \partial v} - \frac{1}{2}\frac{\partial^2 G}{\partial u^2}. \tag{8}$$

Writing this out,

$$
\begin{aligned}
K = \frac{1}{2g} & \left[2\,\frac{\partial^2 F}{\partial u \partial v} - \frac{\partial^2 E}{\partial v^2} - \frac{\partial^2 G}{\partial u^2} \right] \\
& - \frac{G}{4g^2} \left[\frac{\partial E}{\partial u} \left(2\,\frac{\partial F}{\partial v} - \frac{\partial G}{\partial u} \right) - \left(\frac{\partial E}{\partial v} \right)^2 \right] + \frac{F}{4g_2} \\
& \times \left[\frac{\partial E}{\partial u}\frac{\partial G}{\partial v} - 2\frac{\partial E}{\partial v}\frac{\partial G}{\partial u} + \left(2\frac{\partial F}{\partial u} - \frac{\partial E}{\partial v} \right)\left(2\frac{\partial F}{\partial v} - \frac{\partial G}{\partial u} \right) \right] \\
& - \frac{E}{4g^2} \left[\frac{\partial G}{\partial v} \left(2\,\frac{\partial F}{\partial u} - \frac{\partial E}{\partial v} \right) - \left(\frac{\partial G}{\partial u} \right)^2 \right].
\end{aligned}
\tag{9}
$$

The Gaussian curvature is also given by

$$K = \frac{\det(\mathbf{x}_{uu}\mathbf{x}_u\mathbf{x}_v)\det(\mathbf{x}_{vv}\mathbf{x}_u\mathbf{x}_v) - [\det(\mathbf{x}_{uv}\mathbf{x}_u\mathbf{x}_v)]^2}{[|\mathbf{x}_u|^2|\mathbf{x}_v|^2 - (\mathbf{x}_u \cdot \mathbf{x}_v)^2]^2} \tag{10}$$

(Gray 1997, p. 380), as well as

$$K = \frac{[\hat{\mathbf{N}}\hat{\mathbf{N}}_1\hat{\mathbf{N}}_2]}{\sqrt{g}} = \frac{\epsilon^{ij}[\hat{\mathbf{N}}\hat{\mathbf{T}}\hat{\mathbf{T}}_i]_j}{\sqrt{g}}, \tag{11}$$

where ϵ^{ij} is the LEVI-CIVITA SYMBOL, $\hat{\mathbf{N}}$ is the unit NORMAL VECTOR and $\hat{\mathbf{T}}$ is the unit TANGENT VECTOR. The Gaussian curvature is also given by

$$K = -\frac{R}{2} = \kappa_1\kappa_2 = \frac{1}{R_1 R_2}, \tag{12}$$

where R is the CURVATURE SCALAR, κ_1 and κ_2 the PRINCIPAL CURVATURES, and R_1 and R_2 the PRINCIPAL RADII OF CURVATURE. For a MONGE PATCH with $z = h(u, v)$,

$$K = \frac{h_{uu}h_{vv} - h_{uv}^2}{(1 + h_u^2 + h_v^2)^2}. \tag{13}$$

The Gaussian curvature K and MEAN CURVATURE H satisfy

$$H^2 \geq K, \tag{14}$$

with equality only at UMBILIC POINTS, since

$$H^2 - K = \tfrac{1}{4}(\kappa_1 - \kappa_2)^2. \tag{15}$$

If $\mathbf{p}$ is a point on a REGULAR SURFACE $M \subset \mathbb{R}^3$ and $\mathbf{v}_\mathbf{p}$ and $\mathbf{w}_\mathbf{p}$ are tangent vectors to M at $\mathbf{p}$, then the Gaussian curvature of M at $\mathbf{p}$ is related to the SHAPE OPERATOR S by

$$S(\mathbf{v}_\mathbf{P}) \times S(\mathbf{w}_\mathbf{P}) = K(\mathbf{p})\mathbf{v}_\mathbf{P} \times \mathbf{w}_\mathbf{P}. \tag{16}$$

Let $\mathbf{Z}$ be a nonvanishing VECTOR FIELD on M which is everywhere PERPENDICULAR to M, and let V and W be VECTOR FIELDS tangent to M such that $V \times W = \mathbf{Z}$, then

$$K = \frac{\mathbf{Z} \cdot (D_V \mathbf{Z} \times D_W \mathbf{Z})}{2|\mathbf{Z}|^4} \qquad (17)$$

(Gray 1997, p. 410).

For a SPHERE, the Gaussian curvature is $K = 1/a^2$. For EUCLIDEAN SPACE, the Gaussian curvature is $K = 0$. For GAUSS-BOLYAI-LOBACHEVSKY SPACE, the Gaussian curvature is $K = -1/a^2$. A FLAT SURFACE is a REGULAR SURFACE and special class of MINIMAL SURFACE on which Gaussian curvature vanishes everywhere.

A point $\mathbf{p}$ on a REGULAR SURFACE $M \in \mathbb{R}^3$ is classified based on the sign of $K(\mathbf{p})$ as given in the following table (Gray 1997, p. 375), where S is the SHAPE OPERATOR.

Sign	Point
$K(\mathbf{p}) > 0$	ELLIPTIC POINT
$K(\mathbf{p}) < 0$	HYPERBOLIC POINT
$K(\mathbf{p}) = 0$ but $S(\mathbf{p}) \neq 0$	PARABOLIC POINT
$K(\mathbf{p}) = 0$ and $S(\mathbf{p}) = 0$	PLANAR POINT

A surface on which the Gaussian curvature K is everywhere POSITIVE is called SYNCLASTIC, while a surface on which K is everywhere NEGATIVE is called ANTICLASTIC. Surfaces with constant Gaussian curvature include the CONE, CYLINDER, KUEN SURFACE, PLANE, PSEUDOSPHERE, and SPHERE. Of these, the CONE and CYLINDER are the only FLAT SURFACES OF REVOLUTION.

See also ANTICLASTIC, BRIOSCHI FORMULA, DEVELOP-ABLE SURFACE, ELLIPTIC POINT, FLAT SURFACE, HYPERBOLIC POINT, INTEGRAL CURVATURE, MEAN CURVATURE, METRIC TENSOR, MINIMAL SURFACE, PARABOLIC POINT, PLANAR POINT, SYNCLASTIC, UM-BILIC POINT

References

Gray, A. "The Gaussian and Mean Curvatures" and "Surfaces of Constant Gaussian Curvature." §16.5 and Ch. 21 in *Modern Differential Geometry of Curves and Surfaces with Mathematica, 2nd ed.* Boca Raton, FL: CRC Press, pp. 373–80 and 481–00, 1997.

Gaussian Curve

GAUSSIAN DISTRIBUTION

Gaussian Differential Equation

HYPERGEOMETRIC DIFFERENTIAL EQUATION

Gaussian Distribution

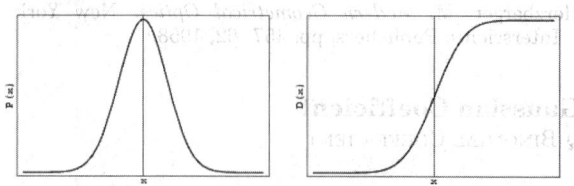

The Gaussian probability distribution with MEAN μ and STANDARD DEVIATION σ is a normalized GAUSSIAN FUNCTION OF THE FORM

$$P(x) = \frac{1}{\sigma\sqrt{2\pi}} e^{-(x-\mu)^2/(2\sigma^2)}, \qquad (1)$$

where $P(x)\,dx$ gives the probability that a variate with a Gaussian distribution takes on a value in the range $[x, x+dx]$. Statisticians commonly call this distribution the NORMAL DISTRIBUTION and, because of its curved flaring shape, social scientists refer to it as the "bell curve." The distribution $P(x)$ is properly normalized for $x \in (-\infty\ \infty)$ since

$$\int_{-\infty}^{\infty} P(x)\,dx = 1. \qquad (2)$$

The cumulative DISTRIBUTION FUNCTION, which gives the probability that a variate will assume a value $\leq x$, is then the integral of the GAUSSIAN FUNCTION,

$$D(x) = \int_{-\infty}^{x} P(x)\,dx = \frac{1}{\sigma\sqrt{2\pi}} \int_{-\infty}^{x} e^{-(x-\mu)^2/(2\sigma^2)}\,dx. \qquad (3)$$

Gaussian distributions have many convenient properties, so random variates with unknown distributions are often assumed to be Gaussian, especially in physics and astronomy. Although this can be a dangerous assumption, it is often a good approximation due to a surprising result known as the CENTRAL LIMIT THEOREM. This theorem states that the MEAN of any set of variates with any distribution having a finite MEAN and VARIANCE tends to the Gaussian distribution. Many common attributes such as test scores, height, etc., follow roughly Gaussian distributions, with few members at the high and low ends and many in the middle. Gaussian distributions are frequently invoked in situations where they may not be applicable. As Lippmann stated, "Everybody believes in the exponential law of errors: the experimenters, because they think it can be proved by mathematics; and the mathematicians, because they believe it has been established by observation" (Whittaker and Robinson 1967, p. 179).

Making the transformation

$$z \equiv \frac{x - \mu}{\sigma}, \qquad (4)$$

so that $dz = dx/\sigma$, gives a variate with VARIANCE $\sigma^2 = 1$ and MEAN $\mu = 0$, transforming $P(x)\,dx$ into

$$P(z)\,dz = \frac{1}{\sqrt{2\pi}}\,e^{-z^2/2}\,dz. \qquad (5)$$

The distribution having this probability function is known as a standard NORMAL DISTRIBUTION, and z defined in this way is known as a z-SCORE.

The NORMAL DISTRIBUTION FUNCTION $\Phi(z)$ gives the probability that a standard normal variate assumes a value in the interval $[0,\,z]$,

$$\Phi(z) \equiv \frac{1}{\sqrt{2\pi}} \int_0^z e^{-x^2/2}\,dx = \tfrac{1}{2}\,\mathrm{erf}\left(\frac{z}{\sqrt{2}}\right), \qquad (6)$$

where ERF is a function sometimes called the error function. Neither $\Phi(z)$ nor ERF can be expressed in terms of finite additions, subtractions, multiplications, and ROOT EXTRACTIONS, and so both must be either computed numerically or otherwise approximated. The value of a for which $P(x)$ falls within the interval $[-a,\,a]$ with a given probability P is called the P CONFIDENCE INTERVAL.

The Gaussian distribution is also a special case of the CHI-SQUARED DISTRIBUTION, since making the substitution

$$\tfrac{1}{2}z \equiv \frac{(x-\mu)^2}{2\sigma^2} \qquad (7)$$

gives

$$d(\tfrac{1}{2}z) = \frac{(x-\mu)}{\sigma^2}\,dx = \frac{\sqrt{z}}{\sigma}\,dx. \qquad (8)$$

Now, the real line $x \in (-\infty\ \infty)$ is mapped onto the half-infinite interval $z \in [0,\ \infty)$ by this transformation, so an extra factor of 2 must be added to $d(z/2)$, transforming $P(x)\,dx$ into

$$P(z)\,dz = \frac{1}{\sigma\sqrt{2\pi}}\,e^{-z/2}\,\frac{\sigma}{\sqrt{z}}\,2(\tfrac{1}{2}\,dz) = \frac{e^{-z/2}z^{-1/2}}{2^{1/2}\Gamma\left(\tfrac{1}{2}\right)}\,dz \qquad (9)$$

(Kenney and Keeping 1951, p. 98), where use has been made of the identity $\Gamma(1/2) = \sqrt{\pi}$. As promised, (9) is a CHI-SQUARED DISTRIBUTION in z with $r = 1$ (and also a GAMMA DISTRIBUTION with $\alpha = 1/2$ and ($\theta = 2$)).

The ratio X/Y of independent Gaussian-distributed variates with zero MEAN is distributed with a CAUCHY DISTRIBUTION. This can be seen as follows. Let X and Y both have MEAN 0 and standard deviations of σ_x and σ_y, respectively, then the joint probability density function is the GAUSSIAN BIVARIATE DISTRIBUTION with $\rho = 0$,

$$f(x,\,y) = \frac{1}{2\pi\sigma_x\sigma_y}\,e^{-[x^2/(2\sigma_x^2)+y^2/(2\sigma_y^2)]}. \qquad (10)$$

From RATIO DISTRIBUTION, the distribution of $U = Y/X$ is

$$P(u) = \int_{-\infty}^{\infty} |x| f(x,\,ux)\,dx$$

$$= \frac{1}{2\pi\sigma_x\sigma_y} \int_{-\infty}^{\infty} |x| e^{-[x^2/(2\sigma_x^2)+u^2x^2/(2\sigma_y^2)]}\,dx$$

$$= \frac{1}{\pi\sigma_x\sigma_y} \int_0^{\infty} x \exp\left[-x^2\left(\frac{1}{2\sigma_x^2}+\frac{u^2}{2\sigma_y^2}\right)\right]\,dx. \qquad (11)$$

But

$$\int_0^{\infty} xe^{-ax^2}\,dx$$

$$= \left[-\frac{1}{2a}\,e^{-ax^2}\right]_0^{\infty} = \frac{1}{2a}\,[0-(-1)] = \frac{1}{2a}, \qquad (12)$$

so

$$P(u) = \frac{1}{\pi\sigma_x\sigma_y}\,\frac{1}{2\left(\dfrac{1}{2\sigma_x^2}+\dfrac{u^2}{2\sigma_y^2}\right)} = \frac{1}{\pi}\,\frac{\sigma_x\sigma_y}{u^2\sigma_x^2+\sigma_y^2}$$

$$= \frac{1}{\pi}\,\frac{\dfrac{\sigma_y}{\sigma_x}}{u^2+\left(\dfrac{\sigma_y}{\sigma_x}\right)^2}, \qquad (13)$$

which is a CAUCHY DISTRIBUTION with MEAN $\mu = 0$ and full width

$$\Gamma = \frac{2\sigma_y}{\sigma_x}. \qquad (14)$$

The CHARACTERISTIC FUNCTION for the Gaussian distribution is

$$\phi(t) = e^{imt-\sigma^2t^2/2}, \qquad (15)$$

and the MOMENT-GENERATING FUNCTION is

$$M(t) = \langle e^{tx}\rangle = \int_{-\infty}^{\infty} \frac{e^{tx}}{\sigma\sqrt{2\pi}}\,e^{-(x-\mu)^2/2\sigma^2}\,dx.$$

$$= \frac{1}{\sigma\sqrt{2\pi}} \int_{-\infty}^{\infty} \exp\left\{-\frac{1}{2\sigma^2}\,[x^2-2(\mu+\sigma^2t)x+\mu^2]\right\}\,dx. \qquad (16)$$

COMPLETING THE SQUARE in the exponent,

$$\frac{1}{2\sigma^2}\,[x^2-2(\mu+\sigma^2t)x+\mu^2]$$

$$= \frac{1}{2\sigma^2}\,\{[x-(\mu+\sigma^2t)]^2+[\mu^2-(\mu+\sigma^2t)^2]\} \qquad (17)$$

Let

$$y \equiv x-(\mu+\sigma^2t) \qquad (18)$$

$$dy = dx \qquad (19)$$

$$a \equiv \frac{1}{2\sigma^2}. \tag{20}$$

The integral then becomes

$$M(t) = \frac{1}{\sigma\sqrt{2\pi}} \int_{-\infty}^{\infty} \exp\left[-ay^2 + \frac{2\mu\sigma^2 t + \sigma^4 t^2}{2\sigma^2}\right] dy$$

$$= \frac{1}{\sigma\sqrt{2\pi}} \int_{-\infty}^{\infty} \exp[-ay^2 + \mu t + \tfrac{1}{2}\sigma^2 t^2]\, dy$$

$$= \frac{1}{\sigma\sqrt{2\pi}}\, e^{\mu t + \sigma^2 t^2/2} \int_{-\infty}^{\infty} e^{-ay^2}\, dy$$

$$= \frac{1}{\sigma\sqrt{2\pi}} \sqrt{\frac{\pi}{a}}\, e^{\mu t + \sigma^2 t^2/2} = \frac{\sqrt{2\sigma^2\pi}}{\sigma\sqrt{2\pi}}\, e^{\mu t + \sigma^2 t^2/2}$$

$$= e^{\mu t + \sigma^2 t^2/2}, \tag{21}$$

so

$$M'(t) = (\mu + \sigma^2 t)e^{\mu t + \sigma^2 t^2/2} \tag{22}$$

$$M'(t) = \sigma^2 e^{\mu t + \sigma^2 t^2/2} + e^{\mu t + \sigma^2 t^2/2}(\mu + t\sigma^2)^2, \tag{23}$$

and

$$\mu = M'(0) = \mu \tag{24}$$

$$\sigma^2 = M''(0) - [M'(0)]^2 = (\sigma^2 + \mu^2) - \mu^2 = \sigma^2. \tag{25}$$

These can also be computed using

$$R(t) = \ln[M(t)] = \mu t + \tfrac{1}{2}\sigma^2 t^2 \tag{26}$$

$$R'(t) = \mu + \sigma^2 t \tag{27}$$

$$R''(t) = \sigma^2, \tag{28}$$

yielding, as before,

$$\mu = R'(0) = \mu \tag{29}$$

$$\sigma^2 = R''(0) = \sigma^2. \tag{30}$$

The raw moments can also be computed directly by computing the MOMENTS about the origin $\mu'_n \equiv \langle x^n \rangle$,

$$\mu'_n = \frac{1}{\sigma\sqrt{2\pi}} \int_{-\infty}^{\infty} x^n e^{-(x-\mu)^2/2\sigma^2}\, dx. \tag{31}$$

(Papoulis 1984, pp. 147–48). Now let

$$u \equiv \frac{x - \mu}{\sqrt{2}\sigma} \tag{32}$$

$$du = \frac{dx}{\sqrt{2}\sigma} \tag{33}$$

$$x = \sigma u \sqrt{2} + \mu, \tag{34}$$

giving the raw moments in terms of GAUSSIAN INTEGRALS,

$$\mu'_n = \frac{\sqrt{2}\sigma}{\sigma\sqrt{2\pi}} \int_{-\infty}^{\infty} x^n e^{-u^2}\, du = \frac{1}{\sqrt{\pi}} \int_{-\infty}^{\infty} x^n e^{-u^2}\, du. \tag{35}$$

Evaluating these integrals gives

$$\mu'_0 = 1 \tag{36}$$

$$\mu'_1 = \mu \tag{37}$$

$$\mu'_2 = \mu^2 + \sigma^2 \tag{38}$$

$$\mu'_3 = \mu(\mu^2 + 3\sigma^2) \tag{39}$$

$$\mu'_4 = \mu^4 + 6\mu^2\sigma^2 + 3\sigma^4. \tag{40}$$

Now find the MOMENTS about the MEAN,

$$\mu_1 = 0 \tag{41}$$

$$\mu_2 = \sigma^2 \tag{42}$$

$$\mu_3 = 0 \tag{43}$$

$$\mu_4 = 3\sigma^4, \tag{44}$$

so the VARIANCE, SKEWNESS, and KURTOSIS are given by

$$\mathrm{var}(x) = \sigma^2 \tag{45}$$

$$\gamma_1 = \frac{\mu_3}{\sigma^3} = 0 \tag{46}$$

$$\gamma_2 = \frac{\mu_4}{\sigma^4} - 3 = \frac{3\sigma^4}{\sigma^4} - 3 = 0 \tag{47}$$

Cramer showed in 1936 that if X and Y are INDEPENDENT variates and $X + Y$ has a Gaussian distribution, then both X and Y must be Gaussian (CRAMER'S THEOREM). An easier result states that the sum of n variates each with is Gaussian distribution also has a Gaussian distribution. This follows from the result

$$P_n(x) = \mathscr{F}^{-1}\{[\phi(t)]^n\} = \frac{e^{-(x-n\mu)^2/(2n\sigma^2)}}{\sqrt{2\pi n\sigma^2}}, \tag{48}$$

where $\phi(t)$ is the CHARACTERISTIC FUNCTION and $\mathscr{F}^{-1}[f]$ is the inverse FOURIER TRANSFORM, taken with parameters $a = b = 1$.

The VARIANCE of the SAMPLE VARIANCE s^2 for a general distribution is given by

$$\mathrm{var}(s^2) = \frac{(N-1)[(N-1)\mu_4 - (N-3)\mu_2'^2]}{N^3}, \tag{49}$$

which simplifies in the case of a Gaussian distribution to

$$\mathrm{var}(s^2) = \frac{2(N-1)(\mu^4 + 2N\mu^2\sigma^2 + N\sigma^4)}{N^3} \tag{50}$$

which, if $\mu = 0$, further simplifies to

$$\text{var}(s^2) = \frac{2\sigma^4(N-1)}{N^2} \qquad (51)$$

(Kenney and Keeping 1951, p. 164).

The CUMULANT-GENERATING FUNCTION for a Gaussian distribution is

$$K(h) = \ln(e^{\nu 1 h} e^{\sigma^2 h^2/2}) = \nu_1 h + \tfrac{1}{2}\,\sigma^2 h^2, \qquad (52)$$

so

$$\kappa_1 = \nu_1 \qquad (53)$$

$$\kappa_2 = \sigma^2 \qquad (54)$$

$$\kappa_r = 0 \quad \text{for } r > 2. \qquad (55)$$

For Gaussian variates, $\kappa_r = 0$ for $r > 2$, so the variance of K-STATISTIC k_3 is

$$\text{var}(k_3) = \frac{\kappa_6}{N} + \frac{9\kappa_2\kappa_4}{N-1} + \frac{9\kappa_3^2}{N-1} + \frac{6\kappa_2^3}{N(N-1)(N-2)}$$

$$= \frac{6\kappa_2^3}{N(N-1)(N-2)}. \qquad (56)$$

Also,

$$\text{var}(k_4) = \frac{24 k_2^4 N(N-1)^2}{(N-3)(N-2)(N+3)(N+5)} \qquad (57)$$

$$\text{var}(g_1) = \frac{6N(N-1)}{(N-2)(N+1)(N+3)} \qquad (58)$$

$$\text{var}(g_2) = \frac{24N(N-1)^2}{(N-3)(N-2)(N+3)(N+5)}, \qquad (59)$$

where

$$g_1 \equiv \frac{k_3}{k_2^{3/2}} \qquad (60)$$

$$g_2 \equiv \frac{k_4}{k_2^2}. \qquad (61)$$

If $P(x)$ is a Gaussian distribution, then

$$D(x) = \frac{1}{2}\left[1 + \text{erf}\left(\frac{x-\mu}{\sigma\sqrt{2}}\right)\right], \qquad (62)$$

so variates x_i with a Gaussian distribution can be generated from variates y_i having a UNIFORM DISTRIBUTION in $(0,1)$ via

$$x_i = \sigma\sqrt{2}\,\text{erf}^{-1}(2y_i - 1) + \mu. \qquad (63)$$

However, a simpler way to obtain numbers with a Gaussian distribution is to use the BOX-MULLER TRANSFORMATION.

The Gaussian distribution is an approximation to the BINOMIAL DISTRIBUTION in the limit of large numbers,

$$P(n_1) = \frac{1}{\sqrt{2\pi Npq}}\exp\left[-\frac{(n_1 - Np)^2}{2Npq}\right], \qquad (64)$$

where n_1 is the number of steps in the POSITIVE direction, N is the number of trials ($(N \equiv n_1 + n_2)$), and p and q are the probabilities of a step in the POSITIVE direction and NEGATIVE direction ($(q \equiv 1 - p)$).

The differential equation having a Gaussian distribution as its solution is

$$\frac{dy}{dx} = \frac{y(\mu - x)}{\sigma^2}, \qquad (65)$$

since

$$\frac{dy}{y} = \frac{\mu - x}{\sigma^2}\,dx \qquad (66)$$

$$\ln\left(\frac{y}{y_0}\right) = -\frac{1}{2\sigma^2}(\mu - x)^2 \qquad (67)$$

$$y = y_0 e^{-(x-\mu)^2/2\sigma^2}. \qquad (68)$$

This equation has been generalized to yield more complicated distributions which are named using the so-called PEARSON SYSTEM.

See also BINOMIAL DISTRIBUTION, BOX-MULLER TRANSFORMATION, CENTRAL LIMIT THEOREM, ERF, GAUSSIAN BIVARIATE DISTRIBUTION, GAUSSIAN DISTRIBUTION–LINEAR COMBINATION OF VARIATES, GAUSSIAN FUNCTION, LOGIT TRANSFORMATION, NORMAL DEVIATES, NORMAL DISTRIBUTION, NORMAL DISTRIBUTION FUNCTION, PEARSON SYSTEM, RATIO DISTRIBUTION, z-SCORE

References

Beyer, W. H. *CRC Standard Mathematical Tables, 28th ed.* Boca Raton, FL: CRC Press, pp. 533–34, 1987.

Kenney, J. F. and Keeping, E. S. *Mathematics of Statistics, Pt. 2, 2nd ed.* Princeton, NJ: Van Nostrand, 1951.

Kraitchik, M. "The Error Curve." §6.4 in *Mathematical Recreations.* New York: W. W. Norton, pp. 121–23, 1942.

Spiegel, M. R. *Theory and Problems of Probability and Statistics.* New York: McGraw-Hill, pp. 109–11, 1992.

Steinhaus, H. *Mathematical Snapshots, 3rd ed.* New York: Dover, pp. 285–90, 1999.

Whittaker, E. T. and Robinson, G. "Normal Frequency Distribution." Ch. 8 in *The Calculus of Observations: A Treatise on Numerical Mathematics, 4th ed.* New York: Dover, pp. 164–08, 1967.

Gaussian Distribution Linear Combination of Variates

If x is NORMALLY DISTRIBUTED with MEAN μ and VARIANCE σ^2, then a linear function of x,

$$y = ax + b, \qquad (1)$$

is also NORMALLY DISTRIBUTED. The new distribution has MEAN $a\mu + b$ and VARIANCE $a^2\sigma^2$, as can be derived using the MOMENT-GENERATING FUNCTION

$$M(t) = \left\langle e^{t(ax+b)} \right\rangle = e^{tb} \left\langle e^{atx} \right\rangle = e^{tb} e^{\mu at + \sigma^2(at)^2/2}$$

$$e^{tb + \mu at + \sigma^2 a^2 t^2/2} = e^{(b+a\mu)t + a^2\sigma^2 t^2/2}, \qquad (2)$$

which is of the standard form with

$$\mu' = b + a \qquad (3)$$

$$\sigma'^2 = a^2 \sigma^2. \qquad (4)$$

For a weighted sum of independent variables

$$y \equiv \sum_{i=1}^{n} a_i x_i, \qquad (5)$$

the expectation is given by

$$M(t) = \left\langle e^{yt} \right\rangle = \left\langle \exp\left(t \sum_{i=1}^{n} a_i x_i \right) \right\rangle$$

$$= \left\langle e^{a_1 t x_1} e^{a_2 t x_2} \cdots e^{a_n t x_n} \right\rangle = \prod_{i=1}^{n} \left\langle e^{a_i t x_i} \right\rangle$$

$$= \prod_{i=1}^{n} \exp(a_i \mu_i t + \tfrac{1}{2} a_i^2 \sigma_i^2 t^2). \qquad (6)$$

Setting this equal to

$$\exp(\mu t + \tfrac{1}{2}\sigma^2 t^2) \qquad (7)$$

gives

$$\mu \equiv \sum_{i=1}^{n} a_i \mu_i \qquad (8)$$

$$\sigma^2 \equiv \sum_{i=1}^{n} a_i^2 \sigma_i^2. \qquad (9)$$

Therefore, the MEAN and VARIANCE of the weighted sums of n RANDOM VARIABLES are their weighted sums.

If x_i are INDEPENDENT and NORMALLY DISTRIBUTED with MEAN 0 and VARIANCE σ^2, define

$$y_i \equiv \sum_j c_{ij} x_j, \qquad (10)$$

where c obeys the ORTHOGONALITY CONDITION

$$c_{ik} c_{jk} = \delta_{ij}, \qquad (11)$$

with δ_{ij} the KRONECKER DELTA. Then y_i are also independent and normally distributed with MEAN 0 and VARIANCE σ^2.

See also GAUSSIAN DISTRIBUTION

Gaussian Elimination

A method for solving MATRIX EQUATIONS OF THE FORM

$$A\mathbf{x} = \mathbf{b}. \qquad (1)$$

To perform Gaussian elimination starting with the system of equations

$$\begin{bmatrix} a_{11} & a_{12} & \cdots & a_{1k} \\ a_{21} & a_{22} & \cdots & a_{2k} \\ \vdots & \vdots & \ddots & \vdots \\ a_{k1} & a_{k2} & \cdots & a_{kk} \end{bmatrix} \begin{bmatrix} x_1 \\ x_2 \\ \vdots \\ x_k \end{bmatrix} = \begin{bmatrix} b_1 \\ b_2 \\ \vdots \\ b_k \end{bmatrix}, \qquad (2)$$

compose the "augmented matrix equation"

$$\begin{bmatrix} a_{11} & a_{12} & \cdots & a_{1k} & \big| & b_1 \\ a_{21} & a_{22} & \cdots & a_{2k} & \big| & b_2 \\ \vdots & \vdots & \ddots & \vdots & \big| & \vdots \\ a_{k1} & a_{k2} & \cdots & a_{kk} & \big| & b_k \end{bmatrix} \begin{bmatrix} x_1 \\ x_2 \\ \vdots \\ x_k \end{bmatrix}. \qquad (3)$$

Here, the COLUMN VECTOR in the variables $\mathbf{x}$ is carried along for labeling the matrix rows. Now, perform ELEMENTARY ROW AND COLUMN OPERATIONS to put the augmented matrix into the UPPER TRIANGULAR form

$$\begin{bmatrix} a'_{11} & a'_{12} & \cdots & a'_{1k} & \big| & b'_1 \\ 0 & a'_{22} & \cdots & a'_{2k} & \big| & b'_2 \\ \vdots & \vdots & \ddots & \vdots & \big| & \vdots \\ 0 & 0 & \cdots & a'_{kk} & \big| & b'_k \end{bmatrix}. \qquad (4)$$

Solve the equation of the kth row for x_k, then substitute back into the equation of the $(k-1)$st row to obtain a solution for x_{k-1}, etc., according to the formula

$$x_i = \frac{1}{a'_{ii}} \left(b'_i - \sum_{j=i+1}^{k} a'_{ij} x_j \right). \qquad (5)$$

For example, consider the MATRIX EQUATION

$$\begin{bmatrix} 9 & 3 & 4 \\ 4 & 3 & 4 \\ 1 & 1 & 1 \end{bmatrix} \begin{bmatrix} x_1 \\ x_2 \\ x_3 \end{bmatrix} = \begin{bmatrix} 7 \\ 8 \\ 3 \end{bmatrix}. \qquad (6)$$

In augmented form, this becomes

$$\begin{bmatrix} 9 & 3 & 4 & \big| & 7 \\ 4 & 3 & 4 & \big| & 8 \\ 1 & 1 & 1 & \big| & 3 \end{bmatrix} \begin{bmatrix} x_1 \\ x_2 \\ x_3 \end{bmatrix}. \qquad (7)$$

Switching the first and third rows gives

$$\begin{bmatrix} 1 & 1 & 1 & \big| & 3 \\ 4 & 3 & 4 & \big| & 8 \\ 9 & 3 & 4 & \big| & 7 \end{bmatrix} \begin{bmatrix} x_1 \\ x_2 \\ x_3 \end{bmatrix}. \qquad (8)$$

Subtracting 9 times the first row from the third row gives

$$\begin{bmatrix} 1 & 1 & 1 & 3 \\ 4 & 3 & 4 & 8 \\ 0 & -6 & -5 & -20 \end{bmatrix} \begin{bmatrix} x_1 \\ x_2 \\ x_3 \end{bmatrix}. \qquad (9)$$

Subtracting 4 times the first row from the second row gives

$$\begin{bmatrix} 1 & 1 & 1 & 3 \\ 0 & -1 & 0 & -4 \\ 0 & -6 & -5 & -20 \end{bmatrix} \begin{bmatrix} x_1 \\ x_2 \\ x_3 \end{bmatrix}. \qquad (10)$$

Finally, adding -6 times the second column to the third one gives

$$\begin{bmatrix} 1 & 1 & 1 & 3 \\ 0 & -1 & 0 & -4 \\ 0 & 0 & -5 & 4 \end{bmatrix} \begin{bmatrix} x_1 \\ x_2 \\ x_3 \end{bmatrix}. \qquad (11)$$

Restoring the transformed matrix equation gives

$$\begin{bmatrix} 1 & 1 & 1 \\ 0 & -1 & 0 \\ 0 & 0 & -5 \end{bmatrix} \begin{bmatrix} x_1 \\ x_2 \\ x_3 \end{bmatrix} = \begin{bmatrix} 3 \\ -4 \\ 4 \end{bmatrix}, \qquad (12)$$

which can be solved immediately to give $x_3 = -4/5$, back-substituting to obtain $x_2 = 4$ (which actually follows trivially in this example), and then again back-substituting to find $x_1 = -1/5$.

See also CONDENSATION, ELEMENTARY ROW AND COLUMN OPERATIONS, GAUSS-JORDAN ELIMINATION, LU DECOMPOSITION, MATRIX EQUATION, SQUARE ROOT METHOD

References

Bareiss, E. H. "Multistep Integer-Preserving Gaussian Elimination." Argonne National Laboratory Report ANL-7213, May 1966.

Bareiss, E. H. "Sylvester's Identity and Multistep Integer-Preserving Gaussian Elimination." *Math. Comput.* **22**, 565–78, 1968.

Garbow, B. S. "Integer-Preserving Gaussian Elimination." Program P-158 (3600F), Applied Mathematics Division, Argonne National Laboratory, Nov. 21, 1966.

Gentle, J. E. "Gaussian Elimination." §3.1 in *Numerical Linear Algebra for Applications in Statistics.* Berlin: Springer-Verlag, pp. 87–1, 1998.

Gaussian Function

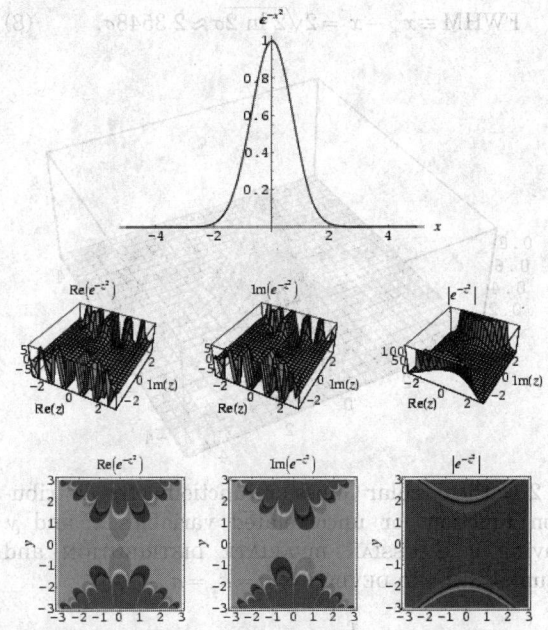

In 1-D, the Gaussian function is the function from the GAUSSIAN DISTRIBUTION,

$$f(x) = \frac{1}{\sigma\sqrt{2\pi}} e^{-(x-\mu)^2/2\sigma^2}, \qquad (1)$$

sometimes also called the FREQUENCY CURVE. The FULL WIDTH AT HALF MAXIMUM (FWHM) for a Gaussian is found by finding the half-maximum points x_0. The constant scaling factor can be ignored, so we must solve

$$e^{-(x_0-\mu)^2/2\sigma^2} = \tfrac{1}{2} f(x_{\max}) \qquad (2)$$

But $f(x_{\max})$ occurs at $x_{\max} = \mu$, so

$$e^{-(x_0-\mu)^2/2\sigma^2} = \tfrac{1}{2} f(\mu) = \tfrac{1}{2}. \qquad (3)$$

Solving,

$$e^{-(x_0-\mu)^2/2\sigma^2} = 2^{-1} \qquad (4)$$

$$-\frac{(x_0-\mu)^2}{2\sigma^2} = -\ln 2 \qquad (5)$$

$$(x_0-\mu)^2 = 2\sigma^2 \ln 2 \qquad (6)$$

$$x_0 \pm \sigma\sqrt{2\ln 2} + \mu. \qquad (7)$$

The FULL WIDTH AT HALF MAXIMUM is therefore given

by

$$\text{FWHM} \equiv x_+ - x_- = 2\sqrt{2 \ln 2}\,\sigma \approx 2.3548\sigma. \qquad (8)$$

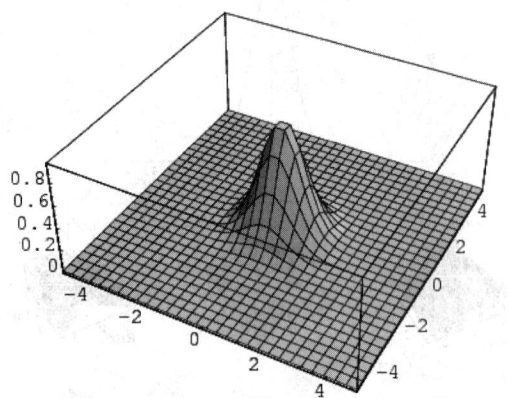

In 2-D, the circular Gaussian function is the distribution function for uncorrelated variables x and y having a GAUSSIAN BIVARIATE DISTRIBUTION and equal STANDARD DEVIATION $\sigma = \sigma_x = \sigma_y$,

$$f(x,\,y) = \frac{1}{2\pi\sigma^2} e^{-[(x-\mu_x)^2 + (y-\mu_y)^2]/2\sigma^2}. \qquad (9)$$

The corresponding elliptical Gaussian function corresponding to $\sigma_x \neq \sigma_y$ is given by

$$f(x,\,y) = \frac{1}{2\pi\sigma_x\sigma_y} e^{-[(x-\mu_x)^2/2\sigma_x^2 + (y-\mu_y)^2/2\sigma_y^2]}. \qquad (10)$$

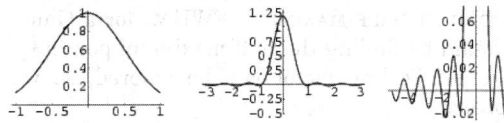

The Gaussian function can also be used as an APODIZATION FUNCTION, shown above with the corresponding INSTRUMENT FUNCTION.

The HYPERGEOMETRIC FUNCTION is also sometimes known as the Gaussian function.

See also ERF, ERFC, FOURIER TRANSFORM–GAUSSIAN, GAUSSIAN BIVARIATE DISTRIBUTION, GAUSSIAN DISTRIBUTION, NORMAL DISTRIBUTION

References

MacTutor History of Mathematics Archive. "Frequency Curve." http://www-groups.dcs.st-and.ac.uk/~history/Curves/Frequency.html.

Gaussian Hypergeometric Series

HYPERGEOMETRIC FUNCTION

Gaussian Integer

A COMPLEX NUMBER $a + bi$ where a and b are INTEGERS. The Gaussian integers are members of the IMAGINARY QUADRATIC FIELD $\mathbb{Q}(\sqrt{-1})$ and form a RING often denoted $\mathbb{Z}[i]$. The sum, difference, and product of two Gaussian integers are Gaussian integers, but $(a + bi)|(c + di)$ only if there is an $e + fi$ such that

$$(a + bi)(e + fi) = (ae - bf) + (af + be)i = c + di.$$

Gaussian integers can be uniquely factored in terms of other Gaussian integers (known as GAUSSIAN PRIMES) up to POWERS of i and rearrangements. The units of $\mathbb{Z}[i]$ are ± 1 and $\pm i$, and the norm of a Gaussian integer is defined by

$$n(x + iy) = x^2 + y^2.$$

Every Gaussian integer is within $|n|/\sqrt{2}$ of a multiple of a Gaussian integer n.

See also COMPLEX NUMBER, EISENSTEIN INTEGER, GAUSSIAN PRIME, INTEGER, OCTONION

References

Conway, J. H. and Guy, R. K. "Gauss's Whole Numbers." In *The Book of Numbers.* New York: Springer-Verlag, pp. 217–23, 1996.
Séroul, R. "The Gaussian Integers." §9.1 in *Programming for Mathematicians.* Berlin: Springer-Verlag, pp. 225–34, 2000.
Shanks, D. "Gaussian Integers and Two Applications." §50 in *Solved and Unsolved Problems in Number Theory, 4th ed.* New York: Chelsea, pp. 149–51, 1993.

Gaussian Integral

The Gaussian integral, also called the PROBABILITY INTEGRAL and closely related to the ERF function, is the integral of the 1-D GAUSSIAN FUNCTION over $(-\infty, \infty)$. It can be computed using the trick of combining two 1-D Gaussians

$$\int_{-\infty}^{\infty} e^{-x^2}\,dx = \sqrt{\left(\int_{-\infty}^{\infty} e^{-y^2}\,dy\right)\left(\int_{-\infty}^{\infty} e^{-x^2}\,dx\right)}$$

$$= \sqrt{\int_{-\infty}^{\infty}\int_{-\infty}^{\infty} e^{-(x^2+y^2)}dy\,dx} \qquad (1)$$

and switching to POLAR COORDINATES,

$$\int_{-\infty}^{\infty} e^{-x^2}\,dx = \sqrt{\int_0^{2\pi}\int_0^{\infty} e^{-r^2} r\,dr\,d\theta} = \sqrt{2\pi\left[-\tfrac{1}{2}e^{-r^2}\right]_0^{\infty}}$$

$$= \sqrt{\pi}. \qquad (2)$$

However, a simple proof can also be given which does not require transformation to POLAR COORDINATES (Nicholas and Yates 1950).

The integral from 0 to a finite upper limit a can be given by the CONTINUED FRACTION

$$\int_0^a e^{-t^2} dt = \tfrac{1}{2}\sqrt{\pi}\,\mathrm{erf}\,a$$

$$\tfrac{1}{2}\sqrt{\pi}-\frac{e^{-a^2}}{2a+}\ \frac{1}{a+}\ \frac{2}{2a}\ \frac{3}{a+}\ \frac{4}{2a+\dots},\qquad (3)$$

first stated by Laplace, proved by Jacobi, and rediscovered by Ramanujan (Watson 1928; Hardy 1999, pp. 8–).

The general class of integrals OF THE FORM

$$I_n(a)\equiv\int_0^\infty e^{-ax^2}x^n\,dx \qquad (4)$$

can be solved analytically by setting

$$x\equiv a^{-1/2}y \qquad (5)$$

$$dx=a^{-1/2}dy \qquad (6)$$

$$y^2=ax^2. \qquad (7)$$

Then

$$I_n(a)=a^{-1/2}\int_0^\infty e^{-y^2}(a^{-1/2}y)^n\,dy$$

$$=a^{-(n+1)/2}\int_0^\infty e^{-y^2}y^n\,dy. \qquad (8)$$

For $n=0$, this is just the usual Gaussian integral, so

$$I_0(a)=\frac{\sqrt{\pi}}{2}a^{-1/2}=\frac{1}{2}\sqrt{\frac{\pi}{a}} \qquad (9)$$

For $n=1$, the integrand is integrable by quadrature,

$$I_1(a)=a^{-1}\int_0^\infty e^{-y^2}y\,dy=a^{-1}\left[-\tfrac{1}{2}e^{-y^2}\right]_0^\infty=\tfrac{1}{2}a^{-1}. \quad (10)$$

To compute $I_n(a)$ for $n>1$, use the identity

$$-\frac{\partial}{\partial a}I_{n-2}(a)=-\frac{\partial}{\partial a}\int_0^\infty e^{-ax^2}x^{n-2}\,dx$$

$$=-\int_0^\infty -x^2e^{-ax^2}x^{n-2}\,dx$$

$$=\int_0^\infty e^{-ax^2}x^n\,dx=I_n(a). \qquad (11)$$

For $n=2s$ EVEN,

$$I_n(a)=\left(-\frac{\partial}{\partial a}\right)I_{n-2}(a)=\left(-\frac{\partial}{\partial a}\right)^2 I_{n-4}$$

$$=\dots=\left(-\frac{\partial}{\partial a}\right)^{n/2}I_0(a)$$

$$=\frac{\partial^{n/2}}{\partial a^{n/2}}I_0(a)=\frac{\sqrt{\pi}}{2}\frac{\partial^{n/2}}{\partial a^{n/2}}a^{-1/2}, \qquad (12)$$

so

$$\int_0^\infty x^{2s}e^{-ax^2}\,dx=\frac{(s-\tfrac{1}{2})!}{2a^{s+1/2}}=\frac{(2s-1)!!}{2^{s+1}a^s}\sqrt{\frac{\pi}{a}}. \qquad (13)$$

If $n=2s+1$ is ODD, then

$$I_n(a)=\left(-\frac{\partial}{\partial a}\right)I_{n-2}(a)=\left(-\frac{\partial}{\partial a}\right)^2 I_{n-4}(a)$$

$$=\dots=\left(-\frac{\partial}{\partial a}\right)^{(n-1)/2}I_1(a)$$

$$=\frac{\partial^{(n-1)/2}}{\partial a^{(n-1)/2}}I_1(a)=\frac{1}{2}\frac{\partial^{(n-1)/2}}{\partial a^{(n-1)/2}}a^{-1}, \qquad (14)$$

so

$$\int_0^\infty x^{2s+1}e^{-ax^2}\,dx=\frac{s!}{2a^{s+1}}. \qquad (15)$$

The solution is therefore

$$\int_0^\infty e^{-ax^2}x^n\,dx$$

$$=\begin{cases}\dfrac{(n-1)!!}{2^{n/2+1}a^{n/2}}\sqrt{\dfrac{\pi}{a}} & \text{for }n\text{ even}\\[3mm]\dfrac{\left[\tfrac{1}{2}(n-1)\right]!}{2a^{(n+1/2)}} & \text{for }n\text{ odd.}\end{cases} \qquad (16)$$

The first few values are therefore

$$I_0(a)=\frac{1}{2}\sqrt{\frac{\pi}{a}} \qquad (17)$$

$$I_1(a)=\frac{1}{2a} \qquad (18)$$

$$I_2(a)=\frac{1}{4a}\sqrt{\frac{\pi}{a}} \qquad (19)$$

$$I_3(a)=\frac{1}{2a^2} \qquad (20)$$

$$I_4(a)=\frac{3}{8a^2}\sqrt{\frac{\pi}{a}} \qquad (21)$$

$$I_5(a)=\frac{1}{a^3} \qquad (22)$$

$$I_6(a)=\frac{15}{16a^3}\sqrt{\frac{\pi}{a}}. \qquad (23)$$

A related, often useful integral is

$$H_n(a)\equiv\frac{1}{\sqrt{\pi}}\int_{-\infty}^\infty e^{-ax^2}x^n\,dx, \qquad (24)$$

which is simply given by

$$H_n(a) = \begin{cases} \dfrac{2I_n(a)}{\sqrt{\pi}} & \text{for } n \text{ even} \\ 0 & \text{for } n \text{ odd.} \end{cases} \quad (25)$$

The more general integral of $x^n e^{-ax^2+bx}$ has the following closed forms

$$\int_{-\infty}^{\infty} x^n e^{-ax^2+bx}\, dx$$

$$= i^{-n} a^{-(n+1)/2} \sqrt{\pi}\, e^{b^2/(4a)} U(-\tfrac{1}{2}n;\ \tfrac{1}{2};\ -b^2/4a) \quad (26)$$

$$= \sqrt{\frac{\pi}{a}} e^{b^2/(4a)} \sum_{k=0}^{n-1} \frac{n!}{k!(n-2k)!} \frac{(2b)^{n-2k}}{(4a)^{n-k}} \quad (27)$$

$$= \sqrt{\frac{\pi}{a}} e^{b^2/(4a)} \sum_{k=0}^{n-1} \binom{n}{2k}(2k-1)!!(2a)^{k-n}b^{n-2k} \quad (28)$$

for integer $n > 0$ (F. Pilolli), where $U(a;\ b;\ x)$ is a CONFLUENT HYPERGEOMETRIC FUNCTION OF THE SECOND KIND and $\binom{n}{k}$ is a BINOMIAL COEFFICIENT.

See also DIFFERENTIATING UNDER THE INTEGRAL SIGN, ERF, GAUSSIAN DISTRIBUTION, GAUSSIAN FUNCTION

References

Guitton, E. "Démonstration de la formule." *Nouv. Ann. Math.* **65**, 237–39, 1906.
Hardy, G. H. *Ramanujan: Twelve Lectures on Subjects Suggested by His Life and Work, 3rd ed.* New York: Chelsea, 1999.
Nicholas, C. B. and Yates, R. C. "The Probability Integral." *Amer. Math. Monthly* **57**, 412–13, 1950.
Papoulis, A. *Probability, Random Variables, and Stochastic Processes, 2nd ed.* New York: McGraw-Hill, pp. 147–48, 1984.
Watson, G. N. "Theorems Stated by Ramanujan (IV): Theorems on Approximate Integration and Summation of Series." *J. London Math. Soc.* **3**, 282–89, 1928.

Gaussian Joint Variable Theorem

Also called the MULTIVARIATE THEOREM. Given an EVEN number of variates from a NORMAL DISTRIBUTION with MEANS all 0,

$$\langle x_1 x_2 \rangle = \langle x_1 \rangle \langle x_2 \rangle, \quad (1)$$

$$\langle x_1 x_2 x_3 x_4 \rangle = \langle x_1 x_2 \rangle \langle x_3 x_4 \rangle + \langle x_1 x_3 \rangle \langle x_2 x_4 \rangle + \langle x_1 x_4 \rangle \langle x_2 x_3 \rangle, \quad (2)$$

etc. Given an ODD number of variates,

$$\langle x_1 \rangle = 0, \quad (3)$$

$$\langle x_1 x_2 x_3 \rangle = 0, \quad (4)$$

etc.

Gaussian Mountain Range

CAROTID-KUNDALINI FUNCTION

Gaussian Multinormal Distribution

GAUSSIAN MULTIVARIATE DISTRIBUTION

Gaussian Multivariate Distribution

A Gaussian p-variate multinormal (or multivariate) distribution is a generalization of the GAUSSIAN BIVARIATE DISTRIBUTION. The p-multivariate distribution with mean vector μ and COVARIANCE MATRIX Σ is denoted $N_p(\mu,\ \text{Sigma})$. The Gaussian multivariate distribution is implemented in *Mathematica* as `MultinormalDistribution[{mu1, mu2, ...}, {{sigma11, sigma12, ...}, {sigma12, sigma22, ...}...}, {x1, x2, ...}]` in the *Mathematica* add-on package `Statistics`MultinormalDistribution`` (which can be loaded with the command `<<Statistics``) (where the matrix Σ is symmetrical since $\sigma_{ij} = \sigma{ji}$).

See also GAUSSIAN BIVARIATE DISTRIBUTION, GAUSSIAN DISTRIBUTION, JOINT THEOREM, MULTIVARIATE THEOREM

Gaussian Polynomial

Q-BINOMIAL COEFFICIENT, Q-BRACKET

Gaussian Prime

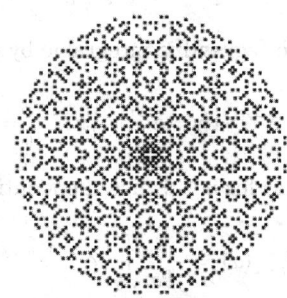

Gaussian primes are GAUSSIAN INTEGERS $z = a + bi$ satisfying one of the following properties.

1. If both a and b are nonzero then, $a + bi$ is a Gaussian prime IFF $a^2 + b^2$ is an ordinary PRIME.
2. If $a = 0$, then bi is a Gaussian prime IFF $|b|$ is an ordinary PRIME and $b \equiv 3$.
3. If $b = 0$, then a is a Gaussian prime IFF $|a|$ is an ordinary PRIME and $a \equiv 3$.

The above plot of the COMPLEX PLANE shows the Gaussian primes as filled squares.

The primes which are also Gaussian primes are 3, 7, 11, 19, 23, 31, 43, ... (Sloane's A002145). The Gaussian primes with $|a|, |b| \leq 5$ are given by $-5 - 4i, -5 - 2i, -5 + 2i, -5 + 4i, -4 - 5i, -4 - i, -4 + i, -4 + 5i, -3 - 2i, -3, -3 + 2i, -2 - 5i, -2 - 3i, -2 - i, -2 + i, -2 + 3i, -2 + 5i, -1 - 4i, -1 - 2i, -1 - i, -1 + i, -1 + 2i, -1 + 4i, -3i, 3i, 1 - 4i, 1 - 2i, 1 - i, 1 + i, 1 + 2i, 1 +$

$4i, 2 - 5i, 2 - 3i, 2 - i, 2 + i, 2 + 3i, 2 + 5i, 3 - 2i, 3, 3 + 2i, 4 - 5i, 4 - i, 4 + i, 4 + 5i, 5 - 4i, 5 - 2i, 5 + 2i, 5 + 4i.$

See also EISENSTEIN INTEGER, GAUSSIAN INTEGER, MOAT-CROSSING PROBLEM

References

Gethner, E.; Wagon, S.; and Wick, B. "A Stroll Through the Gaussian Primes." *Amer. Math. Monthly* **105**, 327–37, 1998.

Guy, R. K. "Gaussian Primes. Eisenstein-Jacobi Primes." §A16 in *Unsolved Problems in Number Theory, 2nd ed.* New York: Springer-Verlag, pp. 33–6, 1994.

Hardy, G. H. and Wright, E. M. *An Introduction to the Theory of Numbers, 5th ed.* Oxford, England: Clarendon Press, 1979.

Rademacher, H. *Topics in Analytic Number Theory.* New York: Springer-Verlag, 1973.

Sloane, N. J. A. Sequences A002145/M2624 in "An On-Line Version of the Encyclopedia of Integer Sequences." http://www.research.att.com/~njas/sequences/eisonline.html.

Smith, H. J. "Gaussian Primes." http://pweb.netcom.com/~hjsmith/GPrimes.html.

Wagon, S. "Gaussian Primes." §9.4 in *Mathematica in Action.* New York: W. H. Freeman, pp. 298–03, 1991.

Wells, D. *The Penguin Dictionary of Curious and Interesting Geometry.* London: Penguin, p. 85, 1991.

Zariski, O. and Samuel, P. *Commutative Algebra I.* New York: Springer-Verlag, 1958.

Gaussian Quadrature

Seeks to obtain the best numerical estimate of an integral by picking optimal ABSCISSAS x_i at which to evaluate the function $f(x)$. The FUNDAMENTAL THEOREM OF GAUSSIAN QUADRATURE states that the optimal ABSCISSAS of the m-point GAUSSIAN QUADRATURE FORMULAS are precisely the roots of the orthogonal POLYNOMIAL for the same interval and WEIGHTING FUNCTION. Gaussian quadrature is optimal because it fits all POLYNOMIALS up to degree $2m$ exactly. Slightly less optimal fits are obtained from RADAU QUADRATURE and LAGUERRE QUADRATURE.

$W(x)$	interval	x_i are roots of
1	$(-1, 1)$	$P_n(x)$
e^{-t}	$(0, \infty)$	$L_n(x)$
e^{-t^2}	$(-\infty, \infty)$	$H_n(x)$
$(1 - t^2)^{-1/2}$	$(-1, 1)$	$T_n(x)$
$(1 - t^2)^{1/2}$	$(-1, 1)$	$U_n(x)$
$x^{1/2}$	$(0, 1)$	$x^{-1/2}P_{2n+1}(\sqrt{x})$
$x^{-1/2}$	$(0, 1)$	$P_n(\sqrt{x})$

To determine the weights corresponding to the Gaussian ABSCISSAS x_i, compute a LAGRANGE INTERPOLATING POLYNOMIAL for $f(x)$ by letting

$$\pi(x) = \prod_{j=1}^{m} (x - x_j) \qquad (1)$$

(where Chandrasekhar 1967 uses F instead of π), so

$$\pi'(x_j) = \left[\frac{d\pi}{dx}\right]_{x=x_j} = \prod_{\substack{i=1 \\ i \neq j}}^{m} (x_j - x_i). \qquad (2)$$

Then fitting a LAGRANGE INTERPOLATING POLYNOMIAL through the m points gives

$$\phi(x) = \sum_{j=1}^{m} \frac{\pi(x)}{(x - x_j)\pi'(x_j)} f(x_j) \qquad (3)$$

for arbitrary points x. We are therefore looking for a set of points x_j and weights w_j such that for a WEIGHTING FUNCTION $W(x)$,

$$\int_a^b \phi(x)W(x)\, dx = \int_a^b \sum_{j=1}^{m} \frac{\pi(x)W(x)}{(x - x_j)\pi'(x_j)}\, dx\, f(x_j)$$
$$\equiv \sum_{j=1}^{m} w_j f(x_j), \qquad (4)$$

with WEIGHT

$$w_j = \frac{1}{\pi'(x_j)} \int_a^b \frac{\pi(x)W(x)}{x - x_j}\, dx. \qquad (5)$$

The weights w_j are sometimes also called the CHRISTOFFEL NUMBER (Chandrasekhar 1967). For orthogonal POLYNOMIALS $\phi_j(x)$ with $j = 1, ..., n$,

$$\phi_j(x) = A_j \pi(x) \qquad (6)$$

(Hildebrand 1956, p. 322), where A_n is the COEFFICIENT of x^n in $\phi_n(x)$, then

$$w_j = \frac{1}{\phi_n'(x_j)} \int_a^b W(x) \frac{\phi(x)}{x - x_j}\, dx$$
$$= -\frac{A_{n+1}\gamma_n}{A_n \phi_n'(x_j)\phi_{n+1}(x_j)}, \qquad (7)$$

where

$$\gamma_m = \int [\phi_m(x)]^2 W(x)\, dx. \qquad (8)$$

Using the relationship

$$\phi_{n+1}(x_i) = -\frac{A_{n+1}A_{n-1}}{A_n^2} \frac{\gamma_n}{\gamma_{n-1}} \phi_{n-1}(x_i) \qquad (9)$$

(Hildebrand 1956, p. 323) gives

$$w_j = \frac{A_n}{A_{n-1}} \frac{\gamma_{n-1}}{\phi_n'(x_j)\phi_{n-1}(x_j)}. \qquad (10)$$

(Note that Press *et al.* 1992 omit the factor A_n/A_{n-1}.)

In Gaussian quadrature, the weights are all POSITIVE. The error is given by

$$E_n = \frac{f^{(2n)}(\xi)}{(2n)!} \int_a^b W(x)[\pi(x)]^2 \, dx = \frac{\gamma_n}{A_n^2} \frac{f^{(2n)}(\xi)}{(2n)!}, \quad (11)$$

where $a < \xi < b$ (Hildebrand 1956, pp. 320–21).

Other curious identities are

$$\sum_{k=0}^m \frac{[\phi_k(x)]^2}{\gamma^k}$$

$$= \frac{A_m}{A_{m+1}\gamma_m} [\phi'_{m+1}(x)\phi_m(x) - \phi'_m(x)\phi_{m+1}(x)] \quad (12)$$

and

$$\sum_{k=0}^m \frac{[\phi_k(x)]^2}{\gamma^k} = -\frac{A_m \phi'_m(x_i)\phi_{m+1}(x_i)}{A_{m+1}\gamma_m} = \frac{1}{w_i} \quad (13)$$

(Hildebrand 1956, p. 323).

In the NOTATION of Szego (1975), let $x_{1n} < \ldots < x_{nn}$ be an ordered set of points in $[a, b]$, and let $\lambda_{1n}, \ldots, \lambda_{nn}$ be a set of REAL NUMBERS. If $f(x)$ is an arbitrary function on the CLOSED INTERVAL $[a, b]$, write the MECHANICAL QUADRATURE as

$$Q_n(f) = \sum_{\nu=1}^n \lambda_{\nu n} f(x_{\nu n}). \quad (14)$$

Here $x_{\nu n}$ are the ABSCISSAS and $\lambda_{\nu n}$ are the COTES NUMBERS.

See also CHEBYSHEV QUADRATURE, CHEBYSHEV-GAUSS QUADRATURE, CHEBYSHEV-RADAU QUADRATURE, FUNDAMENTAL THEOREM OF GAUSSIAN QUADRATURE, HERMITE-GAUSS QUADRATURE, JACOBI-GAUSS QUADRATURE, LAGUERRE-GAUSS QUADRATURE, LEGENDRE-GAUSS QUADRATURE, LOBATTO QUADRATURE, MEHLER QUADRATURE, RADAU QUADRATURE

References

Abramowitz, M. and Stegun, C. A. (Eds.). *Handbook of Mathematical Functions with Formulas, Graphs, and Mathematical Tables, 9th printing.* New York: Dover, pp. 887–88, 1972.
Acton, F. S. *Numerical Methods That Work, 2nd printing.* Washington, DC: Math. Assoc. Amer., p. 103, 1990.
Arfken, G. "Appendix 2: Gaussian Quadrature." *Mathematical Methods for Physicists, 3rd ed.* Orlando, FL: Academic Press, pp. 968–74, 1985.
Beyer, W. H. *CRC Standard Mathematical Tables, 28th ed.* Boca Raton, FL: CRC Press, p. 461, 1987.
Chandrasekhar, S. *An Introduction to the Study of Stellar Structure.* New York: Dover, 1967.
Gauss, C. F. "Methodus nova integralium valores per approx. inveniendi." *Werke, Vol. 3.* p. 163.
Hildebrand, F. B. *Introduction to Numerical Analysis.* New York: McGraw-Hill, pp. 319–23, 1956.
Press, W. H.; Flannery, B. P.; Teukolsky, S. A.; and Vetterling, W. T. "Gaussian Quadratures and Orthogonal Polynomials." §4.5 in *Numerical Recipes in FORTRAN: The Art*

of Scientific Computing, 2nd ed. Cambridge, England: Cambridge University Press, pp. 140–55, 1992.
Szego, G. *Orthogonal Polynomials, 4th ed.* Providence, RI: Amer. Math. Soc., pp. 37–8 and 340–49, 1975.
Whittaker, E. T. and Robinson, G. "Gauss's Formula of Numerical Integration." §80 in *The Calculus of Observations: A Treatise on Numerical Mathematics, 4th ed.* New York: Dover, pp. 152–63, 1967.

Gaussian Sum

A sum OF THE FORM

$$S(p, q) \equiv \sum_{r=0}^{q-1} e^{-\pi i r^2 p/q}, \quad (1)$$

where p and q are RELATIVELY PRIME INTEGERS. The symbol φ is sometimes used instead of S. Although the restriction to RELATIVELY PRIME INTEGERS is often useful, it is not necessary, and Gaussian sums can be written so as to be valid for all integer q (Borwein and Borwein 1987, pp. 83 and 86).

If $(n, n') = 1$, then

$$S(m, nn') = S(mn', n)S(mn, n') \quad (2)$$

(Nagell 1951, p. 178). Gauss showed that

$$S(1, q) = \frac{1 - i^q}{1 - i} \sqrt{q} \quad (3)$$

for ODD q. Written explicitly

$$S(1, q) = \begin{cases} (i+1)\sqrt{q} & \text{for } q \equiv 0 \pmod 4 \\ \sqrt{q} & \text{for } q \equiv 1 \pmod 4 \\ 0 & \text{for } q \equiv 2 \pmod 4 \\ i\sqrt{q} & \text{for } q \equiv 3 \pmod 4 \end{cases} \quad (4)$$

(Nagell 1951, p. 177).

For p and q of opposite PARITY (i.e., one is EVEN and the other is ODD), SCHAAR'S IDENTITY states

$$\frac{1}{\sqrt{q}} \sum_{r=0}^{q-1} e^{-\pi i r^2/q} = \frac{e^{-\pi i/4}}{\sqrt{p}} \sum_{r=0}^{p-1} e^{\pi i r^2 q/p}. \quad (5)$$

Such sums are important in the theory of QUADRATIC RESIDUES.

See also KLOOSTERMAN'S SUM, QUADRATIC RESIDUE, SCHAAR'S IDENTITY, SINGULAR SERIES

References

Borwein, J. M. and Borwein, P. B. *Pi & the AGM: A Study in Analytic Number Theory and Computational Complexity.* New York: Wiley, 1987.
Evans, R. and Berndt, B. "The Determination of Gauss Sums." *Bull. Amer. Math. Soc.* **5**, 107–29, 1981.
Katz, N. M. *Gauss Sums, Kloosterman Sums, and Monodromy Groups.* Princeton, NJ: Princeton University Press, 1987.
Nagell, T. "The Gaussian Sums." §53 in *Introduction to Number Theory.* New York: Wiley, pp. 177–80, 1951.
Riesel, H. *Prime Numbers and Computer Methods for Factorization, 2nd ed.* Boston, MA: Birkhäuser, pp. 132–34, 1994.

Gauss-Jackson Method

A method for numerical solution of a second-order ordinary differential equation

$$y'' = f(x, y)$$

first expounded by Gauss. It proceeds by introducing a function $\delta^{-2}f$ whose second differences are f. The advantage of this method is that summation to get δ^{-2} can be done exactly and that each rounding-off error in the correction term arises only a single time (Jeffreys and Jeffreys 1988, p. 300).

References

Cowell. Appendix to *Greenwich Observations.* 1909.
Jackson, J. *Monthly Not. Roy. Astron. Soc.* **84**, 602–06, 1924.
Jeffreys, H. and Jeffreys, B. S. "The Gauss-Jackson Method." §9.14 in *Methods of Mathematical Physics, 3rd ed.* Cambridge, England: Cambridge University Press, pp. 300–01, 1988.

Gauss-Jacobi Mechanical Quadrature

If $x_1 < x_2 < \ldots < x_n$ denote the zeros of $p_n(x)$, there exist REAL NUMBERS $\lambda_1, \lambda_2, \ldots, \lambda_n$ such that

$$\int_a^b \rho(x)\, d\alpha(x) = \lambda_1 \rho(x_1) + \lambda_2 \rho(x_2) + \ldots + \lambda_n \rho(x_n),$$

for an arbitrary POLYNOMIAL of order $2n - 1$ and the $\lambda_n s$ are called CHRISTOFFEL NUMBERS. The distribution $d\alpha(x)$ and the INTEGER n uniquely determine these numbers λ_v.

References

Szego, G. *Orthogonal Polynomials, 4th ed.* Providence, RI: Amer. Math. Soc., p. 47, 1975.

Gauss-Jordan Elimination

A method for finding a MATRIX INVERSE. To apply Gauss-Jordan elimination, operate on a MATRIX

$$[A \;\; I] \equiv \begin{bmatrix} a_{11} & \cdots & a_{1n} & 1 & 0 & \cdots & 0 \\ a_{21} & \cdots & a_{2n} & 0 & 1 & \cdots & 0 \\ \vdots & \ddots & \vdots & \vdots & \vdots & \ddots & \vdots \\ a_{n1} & \cdots & a_{nn} & 0 & 0 & \cdots & 1 \end{bmatrix},$$

where I is the IDENTITY MATRIX, to obtain a MATRIX OF THE FORM

$$\begin{bmatrix} 1 & 0 & \cdots & 0 & b_{11} & \cdots & b_{1n} \\ 0 & 1 & \cdots & 0 & b_{21} & \cdots & b_{2n} \\ \vdots & \vdots & \ddots & \vdots & \vdots & \ddots & \vdots \\ 0 & 0 & \cdots & 1 & b_{n1} & \cdots & b_{nn} \end{bmatrix}.$$

The MATRIX

$$B \equiv \begin{bmatrix} b_{11} & \cdots & b_{1n} \\ b_{21} & \cdots & b_{2n} \\ \vdots & \ddots & \vdots \\ b_{n1} & \cdots & b_{nn} \end{bmatrix}$$

is then the MATRIX INVERSE of A. The procedure is numerically unstable unless PIVOTING (exchanging rows and columns as appropriate) is used. Picking the largest available element as the pivot is usually a good choice.

See also CONDENSATION, GAUSSIAN ELIMINATION, LU DECOMPOSITION, MATRIX EQUATION

References

Press, W. H.; Flannery, B. P.; Teukolsky, S. A.; and Vetterling, W. T. "Gauss-Jordan Elimination" and "Gaussian Elimination with Backsubstitution." §2.1 and 2.2 in *Numerical Recipes in FORTRAN: The Art of Scientific Computing, 2nd ed.* Cambridge, England: Cambridge University Press, pp. 27–2 and 33–4, 1992.

Gauss-Kronrod Quadrature

An adaptive GAUSSIAN QUADRATURE method for numerical integration in which error is estimation based on evaluation at special points known as "Kronrod points." By suitably picking these points, abscissas from previous iterations can be reused as part of the new set of points, whereas usual GAUSSIAN QUADRATURE would require recomputation of all abscissas at each iteration. This is particularly important when some specified degree of accuracy is needed but the number of points needed to achieve this accuracy is not known ahead of time. Kronrod (1964) showed how to pick Kronrod points optimally from Gauss-Legendre quadrature, and Patterson (1968, 1969) showed how to compute continued extensions of this kind (Press *et al.* 1992, p. 154).

With `Method->Automatic`, the *Mathematica* `NIntegrate` command uses Gauss-Kronrod quadrature for 1-D integrals.

See also GAUSSIAN QUADRATURE, NUMERICAL INTEGRATION, QUADRATURE

References

Calvetti, D.; Golub, G. H.; Gragg, W. B. and Reichel, L. "Computation of Gauss-Kronrod Quadrature Rules." *Math. Comput.* **69**, 1035–052, 2000.
Calvetti, D.; Golub, G. H.; Gragg, W. B. and Reichel, L. "Computation of Gauss-Kronrod Quadrature Rules." Stanford University Scientific Computing/Computational Mathematics Report SCCM-98–9. http://www-sccm.stanford.edu/nflash/nf-publications-tech.html#start-1998.
Kronrod, A. S. [Russian]. *Doklady Akad. Nauk SSSR* **154**, 283–86, 1964.
Patterson, T. N. L. *Math. Comput.* **22**, 847–56 and C1-C11, 1968.
Patterson, T. N. L. *Math. Comput.* **23**, 892, 1969.
Pessens, R.; de Doncker, E.; Uberhuber, C. W.; and Kahaner, D. K. *QUADPACK: A Subroutine Package for Automatic Integration.* New York: Springer-Verlag, 1983.
Press, W. H.; Flannery, B. P.; Teukolsky, S. A.; and Vetterling, W. T. *Numerical Recipes in FORTRAN: The Art of Scientific Computing, 2nd ed.* Cambridge, England: Cambridge University Press, p. 154, 1992.

Ueberhuber, C. W. *Numerical Computation 2: Methods, Software, and Analysis.* Berlin: Springer-Verlag, pp. 105–06, 1997.

Gauss-Kummer Series

$$_2F_1\left(-\tfrac{1}{2}, -\tfrac{1}{2}; 1; h^2\right)$$

$$= \sum_{n=0}^{\infty} \binom{\tfrac{1}{2}}{n}^2 h^{2n} = 1 + \tfrac{1}{4} h^2 + \tfrac{1}{64} h^4 + \tfrac{1}{256} h^6 + \cdots$$

(Sloane's A056981 and A056982), where $_2F_1(a, b; c; x)$ is a HYPERGEOMETRIC FUNCTION. This can be derived using KUMMER'S QUADRATIC TRANSFORMATION. The Gauss-Kummer series is closely related to the PERIMETER of an ellipse.

See also ELLIPSE

References

Sloane, N. J. A. Sequences A056981 and A056982 in "An On-Line Version of the Encyclopedia of Integer Sequences." http://www.research.att.com/~njas/sequences/eisonline.html.

Gauss-Kuzmin-Wirsing Constant

N.B. A detailed online essay by S. Finch was the starting point for this entry.

Let x_0 be a random number from $[0, 1]$ written as a simple CONTINUED FRACTION

$$x_0 = 0 + \cfrac{1}{a_1 + \cfrac{1}{a_2 + \cfrac{1}{a_3 + \cdots}}}. \qquad (1)$$

Define the SHIFT TRANSFORMATION by

$$x_n = 0 + \cfrac{1}{a_{n+1} + \cfrac{1}{a_{n+2} + \cfrac{1}{a_{n+3} + \cdots}}}. \qquad (2)$$

$$= \frac{1}{x_{n-1}} - \left\lfloor \frac{1}{x_{n-1}} \right\rfloor, \qquad (3)$$

where $\lfloor x \rfloor$ is the FLOOR FUNCTION. In a letter to Laplace dated January 30, 1812, Gauss said that he could prove by a simple argument that if $F(n, x)$ is the probability that $x_n < x$, then

$$\lim_{n \to \infty} F(n, x) = \frac{\ln(1 + x)}{\ln 2} \qquad (4)$$

(Rockett and Szüsz 1992, pp. 151–52).

However, Gauss was unable to describe the behavior of the correction term in

$$F(n, x) = \frac{\ln(1 + x)}{\ln 2} + \epsilon(n). \qquad (5)$$

Kuzmin (1928) published the first analysis of the asymptotic behavior of $F(n, x)$, obtaining

$$F(n, x) = \frac{\ln(1 + x)}{\ln 2} + \mathcal{O}(q^{\sqrt{n}}) \qquad (6)$$

with $0 < q < 1$. Using a different method, Lévy (1929) obtained

$$F(n, x) = \frac{\ln(1 + x)}{\ln 2} + \mathcal{O}(q^n) \qquad (7)$$

with $q = 0.7$. Wirsing (1974) subsequently showed, among other results, that

$$\lim_{n \to \infty} \frac{F(n, x) - \dfrac{\ln(1 + x)}{\ln 2}}{(-\lambda)^n} = \Psi(x), \qquad (8)$$

where $\lambda = 0.3036630029\ldots$ and $\Psi(x)$ is an analytic function with $\Psi(0) = \Psi(1) = 0$. This constant is connected to the efficiency of the EUCLIDEAN ALGORITHM (Knuth 1981).

See also CONTINUED FRACTION, EUCLIDEAN ALGORITHM, SHIFT TRANSFORMATION

References

Babenko, K. I. "On a Problem of Gauss." *Soviet Math. Dokl.* **19**, 136–40, 1978.
Daudé, H.; Flajolet, P.; and Vallée, B. "An Average-Case Analysis of the Gaussian Algorithm for Lattice Reduction." Submitted.
Durner, A. "On a Theorem of Gauss-Kuzmin-Lévy." *Arch. Math.* **58**, 251–56, 1992.
Finch, S. "Favorite Mathematical Constants." http://www.mathsoft.com/asolve/constant/kuzmin/kuzmin.html.
Flajolet, P. and Vallée, B. "On the Gauss-Kuzmin-Wirsing Constant." Unpublished memo. 1995. http://pauillac.inria.fr/algo/flajolet/Publications/gauss-kuzmin.ps.
Knuth, D. E. *The Art of Computer Programming, Vol. 2: Seminumerical Algorithms, 3rd ed.* Reading, MA: Addison-Wesley, 1998.
Kuzmin, R. O. "Sur un problème de Gauss." *Anni Congr. Intern. Bologne* **6**, 83–9, 1928.
MacLeod, A. J. "High-Accuracy Numerical Values of the Gauss-Kuzmin Continued Fraction Problem." *Computers Math. Appl.* **26**, 37–4, 1993.
Rockett, A. M. and Szüsz, P. "The Gauss-Kuzmin Theorem." §5.5 in *Continued Fractions.* New York: World Scientific, pp. 151–55, 1992.
Wirsing, E. "On the Theorem of Gauss-Kuzmin-Lévy and a Frobenius-Type Theorem for Function Spaces." *Acta Arith.* **24**, 507–28, 1974.

Gauss-Laguerre Quadrature

LAGUERRE-GAUSS QUADRATURE

Gauss-Manin Connection

A connection defined on a smooth ALGEBRAIC VARIETY defined over the COMPLEX NUMBERS.

References

Iyanaga, S. and Kawada, Y. (Eds.). *Encyclopedic Dictionary of Mathematics.* Cambridge, MA: MIT Press, p. 81, 1980.

Gauss-Salamin Formula

BRENT-SALAMIN FORMULA

GCD

GREATEST COMMON DIVISOR

GCD-Closed Set

A set S is said to be GCD-closed if $GCD(x_i, x_j) \in S$ for $1 \leq i, j \leq n$.

See also BOURQUE-LIGH CONJECTURE

References

Hong, S. "On the Bourque-Ligh Conjecture of Least Common Multiple Matrices." *J. Algebra* **218**, 216–28, 1999.

Gear Curve

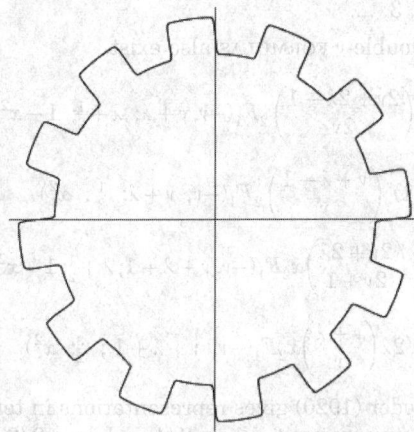

A curve resembling a gear with n teeth given by the PARAMETRIC EQUATIONS

$$x = r \cos t$$

$$y = r \sin t,$$

where

$$r = a + \frac{1}{b} \tanh[b \sin(nt)].$$

The above curve has $n = 12$, $a = 1$, and $b = 10$.

Gear Graph

A WHEEL GRAPH with a VERTEX added between each pair of adjacent VERTICES.

Gegenbauer Differential Equation

The second-order ORDINARY DIFFERENTIAL EQUATION

$$(1-x^2)y'' - 2(\mu+1)xy' + (v-\mu)(v+\mu+1)y = 0 \quad (1)$$

sometimes called the hyperspherical differential equation (Iyanaga and Kawada 1980, p. 1480; Zwillinger 1997, p. 123). The solution to this equation is

$$y = (x^2 - 1)^{-\mu/2}[C_1 P_v^\mu(x) + C_2 Q_v^\mu(x)], \quad (2)$$

where $P_v^\mu(x)$ is an associated LEGENDRE FUNCTION OF THE FIRST KIND and $Q_v^\mu(x)$ is an associated LEGENDRE FUNCTION OF THE SECOND KIND.

A number of other forms of this equation are sometimes also known as the ultraspherical or Gegenbauer differential equation, including

$$(1-x^2)y'' - (2\mu+1)xy' + v(v+2\mu)y = 0. \quad (3)$$

The general solutions to this equation are

$$y = (x^2-1)^{(1-2\mu)/4}$$
$$\times [C_1 P_{-1/2+\mu+v}^{1/2-\mu}(x) + C_2 Q_{-1/2+\mu+v}^{1/2-\mu}(x)]. \quad (4)$$

However, if μ is an integer, then the second part of this equation no longer provides a solution, and the solutions are known as the GEGENBAUER POLYNOMIALS $C_v^{(\mu)}(x)$, also known as ultraspherical polynomials (possibly depending on normalization).

The form

$$(1-x^2)y'' - (2m+3)xy' + \lambda y = 0 \quad (5)$$

is also given by Infeld and Hull (1951, pp. 21–8) and Zwillinger (1997, p. 122). It has the solution

$$y = (x^2-1)^{-(2m+1)/4}$$
$$\times \left[C_1 P_{-1/2+\sqrt{(1+m)^2+\lambda}}^{1/2+m}(x) + C_2 Q_{-1/2+\sqrt{(1+m)^2+\lambda}}^{1/2+m}(x) \right]. \quad (6)$$

See also GEGENBAUER POLYNOMIAL

References

Abramowitz, M. and Stegun, C. A. (Eds.). *Handbook of Mathematical Functions with Formulas, Graphs, and Mathematical Tables, 9th printing.* New York: Dover, 1972.
Morse, P. M. and Feshbach, H. *Methods of Theoretical Physics, Part I.* New York: McGraw-Hill, pp. 547–49, 1953.
Zwillinger, D. *Handbook of Differential Equations, 3rd ed.* Boston, MA: Academic Press, p. 127, 1997.

Gegenbauer Function

GEGENBAUER POLYNOMIAL

Gegenbauer Polynomial

The Gegenbauer polynomials $C_n^{(\lambda)}(x)$ are solutions to the GEGENBAUER DIFFERENTIAL EQUATION for INTEGER n and $\lambda < 1/2$. They are generalizations of the associated LEGENDRE POLYNOMIALS to $(n+2)$-D space, and are proportional to (or, depending on the normalization, equal to) the ultraspherical polynomials $P_n^{(\lambda)}(x)$.

Following Szego, in this work, Gegenbauer polynomials are given in terms of the JACOBI POLYNOMIALS $P_n^{(\alpha, \beta)}(x)$ with $\alpha = \beta = \lambda - 1/2$ by

$$C_n^{(\lambda)}(x) = \frac{\Gamma(\lambda + \frac{1}{2})}{\Gamma(2\lambda)} \frac{\Gamma(n + 2\lambda)}{\Gamma(n + \lambda + \frac{1}{2})} P_n^{(\lambda - 1/2, \lambda - 1/2)}(x) \quad (1)$$

(Szego 1975, p. 80), thus making them equivalent to the Gegenbauer polynomials implemented in *Mathematica* as GegenbauerC[n, $lambda$, x]. These polynomials are also given by the GENERATING FUNCTION

$$\frac{1}{(1 - 2xt + t^2)^{\lambda}} = \sum_{n=0}^{\infty} C_n^{(\lambda)}(x) t^n. \quad (2)$$

The first few Gegenbauer polynomials are

$$C_0^{(\lambda)}(x) = 1 \quad (3)$$

$$C_1^{(\lambda)}(x) = 2\lambda x \quad (4)$$

$$C_2^{(\lambda)}(x) = -\lambda + 2\lambda(1 + \lambda)x^2 \quad (5)$$

$$C_3^{(\lambda)}(x) = -2\lambda(1 + \lambda)x + \tfrac{4}{3}\lambda(1 + \lambda)(2 + \lambda)x^3. \quad (6)$$

In terms of the HYPERGEOMETRIC FUNCTIONS,

$$C_n^{(\lambda)}(x) = \binom{n + 2\lambda - 1}{n}$$

$$\times {}_2F_1(-n, n + 2\lambda; \lambda + \tfrac{1}{2}; \tfrac{1}{2}(1 - x)) \quad (7)$$

$$= 2^n \binom{n + \lambda - 1}{n}(x - 1)^n {}_2F_1$$

$$\times \left(-n, -n - \lambda + \tfrac{1}{2}; -2n - 2\lambda + 1; \frac{2}{1 - x}\right) \quad (8)$$

$$= \binom{n + 2\lambda + 1}{n}\left(\frac{x + 1}{2}\right)^n {}_2F_1$$

$$\times \left(-n, -n - \lambda + \tfrac{1}{2}; \lambda + \tfrac{1}{2}; \frac{x - 1}{x + 1}\right). \quad (9)$$

They are normalized by

$$\int_{-1}^{1}(1 - x^2)^{\lambda - 1/2}[C_n^{(\lambda)}]^2 \, dx$$

$$= 2^{1 - 2\lambda}\pi \frac{\Gamma(n + 2\lambda)}{(n + \lambda)\Gamma^2(\lambda)\Gamma(n + 1)}. \quad (10)$$

Derivative identities include

$$\frac{d}{dx}C_n^{(\lambda)}(x) = 2\lambda C_{n-1}^{(\lambda+1)}(x) \quad (11)$$

$$(1 - x^2)\frac{d}{dx}[C_n^{(\lambda)}] = [2(n + \lambda)]^{-1}[(n + 2\lambda - 1)$$

$$\times(n + 2\lambda)C_{n-1}^{(\lambda)}(x) - n(n + 1)C_{n+1}^{(\lambda)}(x)] \quad (12)$$

$$= -nxC_n^{(\lambda)}(x) + (n + 2\lambda - 1)C_{n-1}^{(\lambda)}(x) \quad (13)$$

$$= (n + 2\lambda)xC_n^{(\lambda)}(x) - (n + 1)C_{n+1}^{(\lambda)}(x) \quad (14)$$

$$nC_n^{(\lambda)}(x) = x\frac{d}{dx}[C_n^{(\lambda)}(x)] - \frac{d}{dx}[C_{n-1}^{(\lambda)}(x)] \quad (15)$$

$$(n + 2\lambda)C_n^{(\lambda)}(x) = \frac{d}{dx}[C_{n+1}^{(\lambda)}(x)] - x\frac{d}{dx}[C_n^{(\lambda)}(x)] \quad (16)$$

$$\frac{d}{dx}[C_{n+1}^{(\lambda)}(x) - C_{n-1}^{(\lambda)}(x)] = 2(n + \lambda)C_n^{(\lambda)}C_n^{(\lambda)}(x) \quad (17)$$

$$= 2\lambda[C_n^{(\lambda+1)}(x) - C_{n-2}^{(\lambda+1)}(x)] \quad (18)$$

(Szego 1975, pp. 80–3).

A RECURRENCE RELATION is

$$nC_n^{(\lambda)}(x) = 2(n + \lambda - 1)xC_{n-1}^{(\lambda)}(x)$$

$$- (n + 2\lambda - 2)C_{n-2}^{(\lambda)}(x) \quad (19)$$

for $n = 2, 3, \ldots$.

Special double-ν FORMULAS also exist

$$C_{2\nu}^{(\lambda)}(x) = \binom{2\nu + 2\lambda - 1}{2\nu} {}_2F_1(-\nu, \nu + \lambda; \lambda + \tfrac{1}{2}; 1 - x^2) \quad (20)$$

$$= (-1)^{\nu}\binom{\nu + \lambda - 1}{\nu} {}_2F_1(-\nu, \nu + \lambda; \tfrac{1}{2}; x^2) \quad (21)$$

$$C_{2\nu+1}^{(\lambda)}(x) = \binom{2\nu + 2\lambda}{2\nu + 1}x\,{}_2F_1(-\nu, \nu + \lambda + 1; \lambda + \tfrac{1}{2}; 1 - x^2) \quad (22)$$

$$= (-1)^{\nu}2\lambda\binom{\nu + \lambda}{\nu}x\,{}_2F_1(-\nu, \nu + \lambda + 1; \tfrac{3}{2}; x^2). \quad (23)$$

Koschmieder (1920) gives representations in terms of ELLIPTIC FUNCTIONS for $\lambda = -3/4$ and $\lambda = -2/3$.

See also BIRTHDAY PROBLEM, CHEBYSHEV POLYNOMIAL OF THE SECOND KIND, ELLIPTIC FUNCTION, GEGENBAUER DIFFERENTIAL EQUATION, HYPERGEOMETRIC FUNCTION, JACOBI POLYNOMIAL

References

Abramowitz, M. and Stegun, C. A. (Eds.). "Orthogonal Polynomials." Ch. 22 in *Handbook of Mathematical Functions with Formulas, Graphs, and Mathematical Tables, 9th printing.* New York: Dover, pp. 771–02, 1972.

Arfken, G. *Mathematical Methods for Physicists, 3rd ed.* Orlando, FL: Academic Press, p. 643, 1985.

Erdélyi, A.; Magnus, W.; Oberhettinger, F.; and Tricomi, F. G. *Higher Transcendental Functions, Vol. 2.* New York: Krieger, p. 175, 1981.

Infeld, L. and Hull, T. E. "The Factorization Method." *Rev. Mod. Phys.* **23**, 21–8, 1951.

Iyanaga, S. and Kawada, Y. (Eds.). "Gegenbauer Polynomials (Gegenbauer Functions)." Appendix A, Table 20.I in *Encyclopedic Dictionary of Mathematics.* Cambridge, MA: MIT Press, pp. 1477–478, 1980.

Koekoek, R. and Swarttouw, R. F. "Gegenbauer / Ultraspherical." §1.8.1 in *The Askey-Scheme of Hypergeometric*

Orthogonal Polynomials and its q -Analogue. Delft, Netherlands: Technische Universiteit Delft, Faculty of Technical Mathematics and Informatics Report 98–7, pp. 40–1, 1998. ftp://www.twi.tudelft.nl/publications/tech-reports/1998/DUT-TWI-98–7.ps.gz.

Koschmieder, L. "Uuml;ber besondere Jacobische Polynome." *Math. Zeitschrift* **8**, 123–37, 1920.

Morse, P. M. and Feshbach, H. *Methods of Theoretical Physics, Part I.* New York: McGraw-Hill, pp. 547–49 and 600–04, 1953.

Roman, S. "A Particular Delta Series and the Gegenbauer Polynomials." §6.3 in *The Umbral Calculus.* New York: Academic Press, pp. 166–74, 1984.

Szego, G. *Orthogonal Polynomials, 4th ed.* Providence, RI: Amer. Math. Soc., 1975.

Zwillinger, D. *Handbook of Differential Equations, 3rd ed.* Boston, MA: Academic Press, pp. 122–23, 1997.

Gegenbauer C

ULTRASPHERICAL POLYNOMIAL

Gelfand Space

References

Stengers, I. and Prigogine, I. *The End of Certainty: Time, Chaos, and the New Laws of Nature.* Free Press, p. 96, 1997.

Gelfand Transform

The Gelfand transform $x \mapsto \hat{x}$ is defined as follows. If $\phi : B \to \mathbb{C}$ is linear and multiplicative in the senses

$$\phi(ax + by) = a\phi(x) + b\phi(y)$$

and

$$\phi(xy) = \phi(x)\phi(y),$$

where B is a commutative BANACH ALGEBRA, then write $\hat{x}(\phi) = \phi(x)$. The Gelfand transform is automatically bounded.

For example, if $B = L^1(\mathbb{R})$ with the usual norm, then B is a BANACH ALGEBRA under convolution and the Gelfand transform is the FOURIER TRANSFORM. (In fact, $\mathbb{R}$ may be replaced by any locally compact Abelian group, and then B has a unit if and only if the group is discrete.)

See also BANACH ALGEBRA

References

Katznelson, Y. *An Introduction to Harmonic Analysis.* New York: Dover, 1976.

Rudin, W. *Real and Complex Analysis, 3rd ed.* New York: McGraw-Hill, 1987.

Gelfond's Theorem

Also called the Gelfond-Schneider theorem, Gelfond's theorem states that a^b is TRANSCENDENTAL if

1. a is ALGEBRAIC $\neq 0$, 1 and
2. b is ALGEBRAIC and IRRATIONAL.

This provides a partial solution to the seventh of HILBERT'S PROBLEMS. Gelfond's theorem is implied by SCHANUEL'S CONJECTURE (Chow 1999).

See also ALGEBRAIC NUMBER, HILBERT'S PROBLEMS, IRRATIONAL NUMBER, SCHANUEL'S CONJECTURE, TRANSCENDENTAL NUMBER

References

Baker, A. *Transcendental Number Theory.* London: Cambridge University Press, 1990.

Chow, T. Y. "What is a Closed-Form Number?" *Amer. Math. Monthly* **106**, 440–48, 1999.

Courant, R. and Robbins, H. *What is Mathematics?: An Elementary Approach to Ideas and Methods, 2nd ed.* Oxford, England: Oxford University Press, p. 107, 1996.

Gelfond-Schneider Constant

The number $2^{\sqrt{2}} = 2.66514414\ldots$ which is known to be TRANSCENDENTAL by GELFOND'S THEOREM.

References

Courant, R. and Robbins, H. *What is Mathematics?: An Elementary Approach to Ideas and Methods, 2nd ed.* Oxford, England: Oxford University Press, p. 107, 1996.

Wells, D. *The Penguin Dictionary of Curious and Interesting Numbers.* Middlesex, England: Penguin Books, p. 45, 1986.

Gelfond-Schneider Theorem

GELFOND'S THEOREM

Gelin-Cesàro Identity

The identity

$$F_n^4 - F_{n-2}F_{n-1}F_{n+1}F_{n+2} = 1,$$

where F_n is a FIBONACCI NUMBER.

See also FIBONACCI NUMBER

Genaille Rods

Numbered rods which can be used to perform multiplication.

See also NAPIER'S BONES

References

Gardner, M. "Napier's Bones." Ch. 7 in *Knotted Doughnuts and Other Mathematical Entertainments.* New York: W. H. Freeman, pp. 85–3, 1986.

Genera

FUNDAMENTAL THEOREM OF GENERA

General Confluent Hypergeometric Differential Equation

$$y'' + \left(\frac{2a}{x} + 2f' + \frac{bh'}{h} - h' - \frac{h''}{h}\right)y'$$

$$+ \left[\left(\frac{bh'}{h} - h' - \frac{h''}{h'}\right)\left(\frac{a}{x} + f'\right) + \frac{a(a-1)}{x^2} + \frac{2af'}{x}\right.$$

$$\left. + f'' + f'^2 - \frac{ah'^2}{h}\right] = 0.$$

See also CONFLUENT HYPERGEOMETRIC DIFFERENTIAL EQUATION

References

Abramowitz, M. and Stegun, C. A. (Eds.). *Handbook of Mathematical Functions with Formulas, Graphs, and Mathematical Tables, 9th printing.* New York: Dover, p. 505, 1972.

Zwillinger, D. *Handbook of Differential Equations, 3rd ed.* Boston, MA: Academic Press, p. 123, 1997.

General Linear Group

The general linear group $GL_n(q)$ is the set of $n \times n$ MATRICES with entries in the FIELD $\mathbb{F}_q$ which have NONZERO DETERMINANT.

See also LANGLANDS RECIPROCITY, PROJECTIVE GENERAL LINEAR GROUP, PROJECTIVE SPECIAL LINEAR GROUP, SPECIAL LINEAR GROUP

References

Conway, J. H.; Curtis, R. T.; Norton, S. P.; Parker, R. A.; and Wilson, R. A. "The Groups $GL_n(q)$, $SL_n(q)$, $PGL_n(q)$, and $PSL_n(q) = L_n(q)$." §2.1 in *Atlas of Finite Groups: Maximal Subgroups and Ordinary Characters for Simple Groups.* Oxford, England: Clarendon Press, p. x, 1985.

General Orthogonal Group

The general orthogonal group $GO_n(q, F)$ is the SUBGROUP of all elements of the PROJECTIVE GENERAL LINEAR GROUP that fix the particular nonsingular QUADRATIC FORM F. The determinant of such an element is ± 1.

See also PROJECTIVE GENERAL LINEAR GROUP

References

Conway, J. H.; Curtis, R. T.; Norton, S. P.; Parker, R. A.; and Wilson, R. A. "The Groups $GO_n(q)$, $SO_n(q)$, $PGO_n(q)$, and $PSO_n(q)$, and $O_n(q)$." §2.4 in *Atlas of Finite Groups: Maximal Subgroups and Ordinary Characters for Simple Groups.* Oxford, England: Clarendon Press, pp. xi-xii, 1985.

General Position

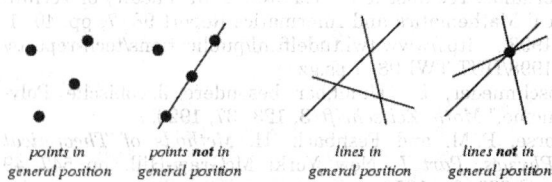

| points in general position | points *not* in general position | lines in general position | lines *not* in general position |

An arrangement of points with no three COLLINEAR, or of lines with no three CONCURRENT.

See also CONCURRENT, ORDINARY LINE, NEAR-PENCIL

References

Guy, R. K. "Unsolved Problems Come of Age." *Amer. Math. Monthly* **96**, 903–09, 1989.

General Prismatoid

A solid such that the AREA A_y of any section parallel to and a distance y from a fixed PLANE can be expressed as

$$A_y = ay^3 + by^2 + cy + d.$$

The volume of such a solid is the same as for a PRISMATOID,

$$V = \tfrac{1}{6}h(A_1 + 4M + A_2).$$

Examples include the CONE, CONICAL FRUSTUM, CYLINDER, PRISMATOID, PYRAMIDAL FRUSTUM, SPHERE, SPHERICAL SEGMENT, and SPHEROID.

See also PRISMATOID, PRISMOID

References

Beyer, W. H. *CRC Standard Mathematical Tables, 28th ed.* Boca Raton, FL: CRC Press, p. 132, 1987.

Kern, W. F. and Bland, J. R. "The General Prismatoid." Ch. 8 in *Solid Mensuration with Proofs, 2nd ed.* New York: Wiley, pp. 120–30, 1948.

General Quantifier

The FOR ALL QUANTIFIER ∀.

See also EXISTENTIAL QUANTIFIER, EXISTS, FOR ALL, QUANTIFIER

General Unitary Group

The general unitary group $GU_n(q)$ is the SUBGROUP of all elements of the GENERAL LINEAR GROUP $GL(q^2)$ that fix a given nonsingular Hermitian form. This is equivalent, in the canonical case, to the definition of GU_n as the group of UNITARY MATRICES.

References

Conway, J. H.; Curtis, R. T.; Norton, S. P.; Parker, R. A.; and Wilson, R. A. "The Groups $GU_n(q)$, $SU_n(q)$, $PGU_n(q)$, and $PSU_n(q) = U_n(q)$." §2.2 in *Atlas of Finite Groups: Maximal Subgroups and Ordinary Characters for Simple Groups.* Oxford, England: Clarendon Press, p. x, 1985.

Generalized Completeness Theorem

The proposition that every CONSISTENT generalized theory has a MODEL. The theorem is true if the AXIOM OF CHOICE is assumed.

See also AXIOM OF CHOICE

References

Mendelson, E. *Introduction to Mathematical Logic, 4th ed.* London: Chapman & Hall, p. 121, 1997.

Generalized Cone

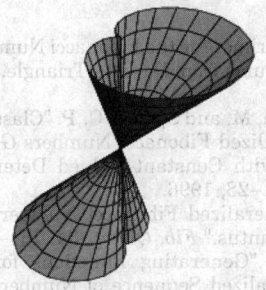

A RULED SURFACE is called a generalized cone if it can be parameterized by $\mathbf{x}(u, v) = \mathbf{p} + v\mathbf{y}(u)$, where $\mathbf{p}$ is a fixed point which can be regarded as the vertex of the cone. A generalized cone is a REGULAR SURFACE wherever $v\mathbf{y} \times \mathbf{y}' \neq \mathbf{0}$. The above surface is a generalized cone over a CARDIOID. A generalized cone is a FLAT SURFACE, and is sometimes called "conical surface."

See also CONE

References

Gray, A. *Modern Differential Geometry of Curves and Surfaces with Mathematica, 2nd ed.* Boca Raton, FL: CRC Press, pp. 439–41, 1997.
Kern, W. F. and Bland, J. R. "Conical Surfaces." §23 in *Solid Mensuration with Proofs, 2nd ed.* New York: Wiley, p. 57, 1948.

Generalized Cylinder

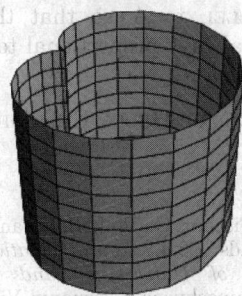

A RULED SURFACE is called a generalized cylinder if it can be parameterized by $\mathbf{x}(u, v) = v\mathbf{p} + \mathbf{y}(u)$, where $\mathbf{p}$ is a fixed point. A generalized cylinder is a REGULAR SURFACE wherever $\mathbf{y}' \times \mathbf{p} \neq \mathbf{0}$. The above surface is a generalized cylinder over a CARDIOID. A generalized

cylinder is a FLAT SURFACE, and is sometimes called a "cylindrical surface."

See also CYLINDER

References

Gray, A. *Modern Differential Geometry of Curves and Surfaces with Mathematica, 2nd ed.* Boca Raton, FL: CRC Press, pp. 439–41, 1997.
Harris, J. W. and Stocker, H. "General Cylinder." §4.6.1 in *Handbook of Mathematics and Computational Science.* New York: Springer-Verlag, p. 103, 1998.
Kern, W. F. and Bland, J. R. "Cylindrical Surface." §14 in *Solid Mensuration with Proofs, 2nd ed.* New York: Wiley, pp. 32–6, 1948.

Generalized Diameter

The farthest DISTANCE between two points on the boundary of a closed figure. The diameter of a SUBSET E of a EUCLIDEAN SPACE $\mathbb{R}^n$ is therefore given by

$$\text{diam } E = \sup\{|\mathbf{x} - \mathbf{y}| : \mathbf{x}, \mathbf{y} \in E\},$$

where sup denotes the SUPREMUM (Croft *et al.* 1991).

See also BLASCHKE'S THEOREM, BORSUK'S CONJECTURE, DIAMETER

References

Croft, H. T.; Falconer, K. J.; and Guy, R. K. *Unsolved Problems in Geometry.* New York: Springer-Verlag, p. 2, 1991.
Eppstein, D. "Width, Diameter, and Geometric Inequalities." http://www.ics.uci.edu/~eppstein/junkyard/diam.html.

Generalized Euclidean Algorithm

INTEGER RELATION

Generalized Fermat Equation

A generalization of the equation whose solution is desired in FERMAT'S LAST THEOREM

$$x^n + y^n = z^n$$

to

$$x^n + y^n = cz^n$$

for x, y, z, and c positive constants, with trivial solutions having $x = 0$, $y = 0$, or $z = 0$ being excluded.

$n = 1$ is trivial to solve by taking $x = y = c$ and $z = 2$. $n = 2$ is more difficult, but can be solved by noting that solutions exist for values of c which can be written as a sum of two SQUARES, the first few of which are 1, 2, 4, 5, 8, 9, 10, 13, 16, 17, 18, 20, 25, 26, ... (Sloane's A001481).

See also FERMAT'S LAST THEOREM, SQUARE NUMBER

References

Finch, S. "Unsolved Mathematics Problems: On a Generalized Fermat-Wiles Equation." http://www.mathsoft.com/asolve/fermat/fermat.html.

Sloane, N. J. A. Sequences A001481/M0968 in "An On-Line Version of the Encyclopedia of Integer Sequences." http://www.research.att.com/~njas/sequences/eisonline.html.

Generalized Fibonacci Number

A generalization of the FIBONACCI NUMBERS defined by $1 = G_1 = G_2 = \ldots = G_{c-1}$ and the RECURRENCE RELATION

$$G_n = G_{n-1} + G_{n-c}. \tag{1}$$

These are the sums of elements on successive diagonals of a left-justified PASCAL'S TRIANGLE beginning in the left-most column and moving in steps of $c - 1$ up and 1 right. The case $c = 2$ equals the usual FIBONACCI NUMBER. These numbers satisfy the identities

$$G_1 + G_2 + G_3 + \ldots + G_n = G_{n+3} - 1 \tag{2}$$

$$G_3 + G_6 + G_9 + \ldots + G_{3k} = G_{3k+1} - 1 \tag{3}$$

$$G_1 + G_4 + G_7 + \ldots + G_{3k+1} = G_{3k+2} \tag{4}$$

$$G_2 + G_5 + G_8 + \ldots + G_{3k+2} = G_{3k+3} \tag{5}$$

(Bicknell-Johnson and Spears 1996). For the special case $c = 3$,

$$G_{n+w} = G_{w-2}G_n + G_{w-3}G_{n+1} + G_{w-1}G_{n+2}. \tag{6}$$

Bicknell-Johnson and Spears (1996) give many further identities.

Horadam (1965) defined the generalized Fibonacci numbers $\{w_n\}$ as $w_n = w_n(a, b; \, p, \, q)$, where a, b, p, and q are INTEGERS, $w_0 = a$, $w_1 = b$, and $w_n = pw_{n-1} - qw_{n-2}$ for $n \geq 2$. They satisfy the identities

$$w_n w_{n+2r} - eq^n U_r = w_{n+r}^2 \tag{7}$$

$$4w_n w_{n+1}^2 w_{n+2} + (wq^n)^2 = (w_n w_{n+2} + w_{n+1}^2)^2 \tag{8}$$

$$w_n w_{n+1} w_{n+3} w_{n+4}$$
$$= w_{n+2}^4 + eq^n(p^2 + q)w_{n+2}^2 + e^2 q^{2n+1}p^2 \tag{9}$$

$$4w_n w_{n+1} w_{n+2} w_{n+4} w_{n+5} w_{n+6}$$
$$+ e^2 q^{2n}(w_n U_4 U_5 - w_{n+1} U_2 U_6 - w_n U_1 U_8)^2$$
$$= (w_{n+1} w_{n+2} w_{n+6} + w_n w_{n+4} w_{n+5})^2, \tag{10}$$

where

$$e \equiv pab - qa^2 - b^2 \tag{11}$$

$$U_n \equiv w_n(0, \, 1; \, p, \, q) \tag{12}$$

(Dujella 1996). The final above result is due to Morgado (1987) and is called the MORGADO IDENTITY.

Another generalization of the Fibonacci numbers is denoted x_n. Given x_1 and x_2, define the generalized Fibonacci number by $x_n \equiv x_{n-2} + x_{n-1}$ for $n \geq 3$,

$$\sum_{i=1}^{n} x_n = x_{n+2} - x_2 \tag{13}$$

$$\sum_{i=1}^{10} x_n = 11x_7 \tag{14}$$

$$x_n^2 - x_{n-1}x_{n+2} = (-1)^n(x_2^2 - x_1^2 - x_1 x_2), \tag{15}$$

where the plus and minus signs alternate.

See also FIBONACCI N-STEP NUMBER, FIBONACCI NUMBER

References

Bicknell, M. "A Primer for the Fibonacci Numbers, Part VIII: Sequences of Sums from Pascal's Triangle." *Fib. Quart.* **9**, 74–1, 1971.
Bicknell-Johnson, M. and Spears, C. P. "Classes of Identities for the Generalized Fibonacci Numbers $G_n = G_{n-1} + G_{n-c}$ for Matrices with Constant Valued Determinants." *Fib. Quart.* **34**, 121–28, 1996.
Dujella, A. "Generalized Fibonacci Numbers and the Problem of Diophantus." *Fib. Quart.* **34**, 164–75, 1996.
Horadam, A. F. "Generating Functions for Powers of a Certain Generalized Sequence of Numbers." *Duke Math. J.* **32**, 437–46, 1965.
Horadam, A. F. "Generalization of a Result of Morgado." *Portugaliae Math.* **44**, 131–36, 1987.
Horadam, A. F. and Shannon, A. G. "Generalization of Identities of Catalan and Others." *Portugaliae Math.* **44**, 137–48, 1987.
Morgado, J. "Note on Some Results of A. F. Horadam and A. G. Shannon Concerning a Catalan's Identity on Fibonacci Numbers." *Portugaliae Math.* **44**, 243–52, 1987.

Generalized Function

DISTRIBUTION (GENERALIZED FUNCTION)

Generalized Helicoid

The SURFACE generated by a twisted curve C when rotated about a fixed axis A and, at the same time, displaced PARALLEL to A so that the velocity of displacement is always proportional to the ANGULAR VELOCITY of ROTATION.

See also GENERALIZED HELIX, HELICOID, HELIX

References

do Carmo, M. P.; Fischer, G.; Pinkall, U.; and Reckziegel, H. "General Helicoids." §3.4.3 in *Mathematical Models from the Collections of Universities and Museums* (Ed. G. Fischer). Braunschweig, Germany: Vieweg, pp. 36–7, 1986.
Fischer, G. (Ed.). Plate 89 in *Mathematische Modelle/Mathematical Models, Bildband/Photograph Volume.* Braunschweig, Germany: Vieweg, p. 85, 1986.
Kreyszig, E. *Differential Geometry.* New York: Dover, p. 88, 1991.

Generalized Helix

The GEODESICS on a general cylinder generated by lines PARALLEL to a line l with which the TANGENT makes a constant ANGLE.

See also HELIX

Generalized Hyperbolic Functions

In 1757, V. Riccati first recorded the generalizations of the HYPERBOLIC FUNCTIONS defined by

$$F_{n,r}^{\alpha}(x) \equiv \sum_{k=0}^{\infty} \frac{\alpha^{k}}{(nk+r)!} x^{nk+r}, \tag{1}$$

for $r = 0, ..., n-1$, where α is COMPLEX, with the value at $x = 0$ defined by

$$F_{n,0}^{\alpha}(0) = 1. \tag{2}$$

This is called the α-hyperbolic function of order n of the rth kind. The functions $F_{n,r}^{\alpha}$ satisfy

$$f^{(k)}(x) = \alpha f(x), \tag{3}$$

where

$$f^{(k)}(0) = \begin{cases} 0 & k \neq r, \ 0 \leq k \leq n-1 \\ 1 & k = r. \end{cases} \tag{4}$$

In addition,

$$\frac{d}{dx} F_{n,\,r}^{\alpha}(x) = \begin{cases} F_{n,r-1}^{\alpha}(x) & \text{for } 0 < r \leq n-1 \\ \alpha F_{n,n-1}^{\alpha}(x) & \text{for } r = 0. \end{cases} \tag{5}$$

The functions give a generalized EULER FORMULA

$$e^{\sqrt{\alpha}} = \sum_{r=0}^{n-1} (\sqrt{\alpha})^{r} F_{n,\,r}^{\alpha}(x). \tag{6}$$

Since there are n nth roots of α, this gives a system of n linear equations. Solving for $F_{n,\,r}^{\alpha}$ gives

$$F_{n,\,r}^{\alpha}(x) = \frac{1}{n} (\sqrt{\alpha})^{-r} \sum_{k=0}^{n-1} \omega_{n}^{-rk} \exp(\omega_{n}^{k} \sqrt{\alpha} x), \tag{7}$$

where

$$\omega_{n} = \exp\left(\frac{2\pi i}{n}\right) \tag{8}$$

is a PRIMITIVE ROOT OF UNITY.

The LAPLACE TRANSFORM is

$$\int_{0}^{\infty} e^{-st} F_{n,r}^{\alpha}(at)\, dt = \frac{s^{n-r-1} a^{r}}{s^{n} + \alpha a_{n}}. \tag{9}$$

The generalized hyperbolic function is also related to the MITTAG-LEFFLER FUNCTION $E_{\gamma}(x)$ by

$$F_{n,0}^{1}(x) = E_{n}(x^{n}). \tag{10}$$

The values $n = 1$ and $n = 2$ give the exponential and circular/hyperbolic functions (depending on the sign of α), respectively.

$$F_{1,0}^{\alpha}(x) = e^{\alpha x} \tag{11}$$

$$F_{2,0}^{\alpha}(x) = \cosh(\sqrt{\alpha} x) \tag{12}$$

$$F_{2,1}^{\alpha}(x) = \frac{\sinh(\sqrt{\alpha} x)}{\sqrt{\alpha}}. \tag{13}$$

For $\alpha = 1$, the first few functions are

$$F_{1,0}^{1}(x) = e^{x}$$

$$F_{2,0}^{1}(x) = \cosh x$$

$$F_{2,1}^{1}(x) = \sinh x$$

$$F_{3,0}^{1}(x) = \tfrac{1}{3}[e^{x} + 2e^{-x/2} \cos(\tfrac{1}{2}\sqrt{3}x)]$$

$$F_{3,1}^{1}(x) = \tfrac{1}{3}\left[e^{x} + 2e^{-x/2} \cos\left(\tfrac{1}{2}\sqrt{3}x + \tfrac{1}{3}\pi\right)\right]$$

$$F_{3,2}^{1}(x) = \tfrac{1}{3}\left[e^{x} + 2e^{-x/2} \cos\left(\tfrac{1}{2}\sqrt{3}x - \tfrac{1}{3}\pi\right)\right]$$

$$F_{4,0}^{1}(x) = \tfrac{1}{2}(\cosh x + \cos x)$$

$$F_{4,1}^{1}(x) = \tfrac{1}{2}(\sinh x + \sin x)$$

$$F_{4,2}^{1}(x) = \tfrac{1}{2}(\cosh x - \cos x)$$

$$F_{4,3}^{1}(x) = \tfrac{1}{2}(\sinh x + \sin x).$$

See also HYPERBOLIC FUNCTIONS, MITTAG-LEFFLER FUNCTION

References
Kaufman, H. "A Biographical Note on the Higher Sine Functions." *Scripta Math.* **28**, 29–6, 1967.
Muldoon, M. E. and Ungar, A. A. "Beyond Sin and Cos." *Math. Mag.* **69**, 3–4, 1996.
Petkovsek, M.; Wilf, H. S.; and Zeilberger, D. *A = B.* Wellesley, MA: A. K. Peters, 1996.
Ungar, A. "Generalized Hyperbolic Functions." *Amer. Math. Monthly* **89**, 688–91, 1982.
Ungar, A. "Higher Order Alpha-Hyperbolic Functions." *Indian J. Pure. Appl. Math.* **15**, 301–04, 1984.

Generalized Hypergeometric Differential Equation

The GENERALIZED HYPERGEOMETRIC FUNCTION

$$F(x) = {}_{p}F_{q}\begin{bmatrix} \alpha_{1}, \alpha_{2}, \ldots, \alpha_{p}; \ x \\ \beta_{1}, \beta_{2}, \ldots, \beta_{q}; \end{bmatrix}$$

satisfies the equation

$$\tilde{D}(\tilde{D} + \beta_{1} - 1) \cdots (\tilde{D} + \beta_{q} - 1) F(x)$$

$$= x(\check{D} + \alpha_1)(\check{D} + \alpha_2) \cdots (\check{D} + \alpha_p) F(x),$$

where $\check{D}$ is the DIFFERENTIAL OPERATOR.

See also GENERALIZED HYPERGEOMETRIC FUNCTION

References

Koepf, W. *Hypergeometric Summation: An Algorithmic Approach to Summation and Special Function Identities.* Braunschweig, Germany: Vieweg, p. 26, 1998.

Miller, W. Jr. *Symmetry and Separation of Variables.* Reading, MA: Addison-Wesley, p. 271, 1977.

Rainville, E. D. *Special Functions.* New York: Chelsea, 1971.

Zwillinger, D. *Handbook of Differential Equations, 3rd ed.* Boston, MA: Academic Press, p. 128, 1997.

Generalized Hypergeometric Function

The generalized hypergeometric function is given by a HYPERGEOMETRIC SERIES, i.e., a series for which the ratio of successive terms can be written

$$\frac{a_{k+1}}{a_k} = \frac{P(k)}{Q(k)}$$

$$= \frac{(k + a_1)(k + a_2) \cdots (k + a_p)}{(k + b_1)(k + b_2) \cdots (k + b_q)(k + 1)} x. \quad (1)$$

(The factor of $k + 1$ in the DENOMINATOR is present for historical reasons of notation.) The resulting generalized hypergeometric function is written

$$\sum_{k+0} a_k x^k = {}_p F_q \begin{bmatrix} a_1, a_2, \ldots, a_p; \\ b_1, b_2, \ldots, b_q; \end{bmatrix} \quad (2)$$

$$= \sum_{k=0}^{\infty} \frac{(a_1)_k (a_2)_k \cdots (a_p)_k}{(b_1)_k b(b_2)_k \cdots (b_q)_k} \frac{x^k}{k!}, \quad (3)$$

where $(a)_k$ is the POCHHAMMER SYMBOL or RISING FACTORIAL

$$(a)_k \equiv \frac{\Gamma(a + k)}{\Gamma(a)} = a(a+1) \cdots (a+k-1). \quad (4)$$

This notation was introduced by Barnes (1907) (Hardy 1999, p. 111). If the argument $x = 1$, then the function is abbreviated

$${}_p F_q \begin{bmatrix} a_1, a_2, \ldots, a_p \\ b_1, b_2, \ldots, b_q \end{bmatrix} \equiv {}_p F_q \begin{bmatrix} a_1, a_2, \ldots, a_p; \\ b_1, b_2, \ldots, b_q; \end{bmatrix}. \quad (5)$$

The KAMPE DE FERIET FUNCTION is a generalization of the generalized hypergeometric function to two variables.

The generalized hypergeometric function $F_n(x) = {}_p F_q \begin{bmatrix} a_1, a_2, \ldots, a_p; \\ b_1, b_2, \ldots, b_q; \end{bmatrix}$ satisfies

$$\vartheta F_n(x) = n[F_{n+1}(x) - F_n(x)] \quad (6)$$

for any of its numerator parameters $n = \alpha_k$, and

$$\vartheta F_n(x) = (n-1)[F_{n-1}(x) - F_n(x)] \quad (7)$$

for any of its denominator parameters $n = \beta_k$, where

$$\vartheta = z \frac{d}{dz} \quad (8)$$

(Rainville 1971, Koepf 1998, p. 27).

${}_2 F_1(a, b; c; z)$ is "the" HYPERGEOMETRIC FUNCTION, and ${}_1 F_1(a; b; z) \equiv M(z)$ is the CONFLUENT HYPERGEOMETRIC FUNCTION. A function OF THE FORM ${}_0 F_1(; b; z)$ is called a CONFLUENT HYPERGEOMETRIC LIMIT FUNCTION.

The generalized hypergeometric function

$${}_{p+1} F_p \begin{bmatrix} a_1, a_2, \ldots, a_{p+1}; \\ b_1, b_2, \ldots, b_p; \end{bmatrix} \quad (9)$$

is a solution to the DIFFERENTIAL EQUATION

$$[\vartheta(\vartheta + b - 1) \cdots (\vartheta + b_p - 1) - z(\vartheta + a_1)$$
$$\times (\vartheta + a_2) \cdots (\vartheta + a_{p+1})] y$$
$$= 0. \quad (10)$$

The other linearly independent solution is

$$z^{1-b_1} {}_{p+1} F_p$$
$$\times \begin{bmatrix} 1 + a_1 - b_1, 1 - a_2 - b_2, \ldots, 1 + a_{p+1} - b_1; \\ 2 - b_1, 1 - b_2 - b_1, \ldots, 1 - b_p - b_1; \end{bmatrix}. \quad (11)$$

A generalized hypergeometric function ${}_{q+1} F_p$ converges absolutely on the unit circle if

$$\Re \left(\sum_{j=1}^{q} \beta_j - \sum_{j=1}^{q+1} \alpha_j \right) > 0 \quad (12)$$

(Rainville 1971, Koepf 1998).

Many sums can be written as generalized hypergeometric functions by inspection of the ratios of consecutive terms in the generating HYPERGEOMETRIC SERIES. For example, for

$$f(n) \equiv \sum_k (-1)^k \binom{2n}{k}^2, \quad (13)$$

the ratio of successive terms is

$$\frac{a_{k+1}}{a_k} = \frac{(-1)^{k+1} \binom{2n}{k+1}^2}{(-1)^k \binom{2n}{k}^2} = -\frac{(k - 2n)^2}{(k+1)^2}, \quad (14)$$

yielding

$$f(n) = {}_2 F_1 \begin{bmatrix} -2n, -2n; \\ 1 \end{bmatrix}$$
$$= {}_2 F_1(-2n, -2n; 1; -1) \quad (15)$$

(Petkovsek 1996, pp. 44–5).

Gosper (1978) discovered a slew of unusual hypergeometric function identities, many of which were subsequently proven by Gessel and Stanton (1982). An important generalization of Gosper's technique, called ZEILBERGER'S ALGORITHM, in turn led to the powerful machinery of the WILF-ZEILBERGER PAIR (Zeilberger 1990).

Special hypergeometric identities include GAUSS'S HYPERGEOMETRIC THEOREM

$$_2F_1(a,\ b,\ c;\ 1) = \frac{\Gamma(c)\Gamma(c-a-b)}{\Gamma(c-a)\Gamma(c-b)} \qquad (16)$$

for $\Re[c-a-b] > 0$, KUMMER'S FORMULA

$$_2F_1(a,\ b,\ c;\ -1) = \frac{\Gamma(\frac{1}{2}b+1)\Gamma(b-a+1)}{\Gamma(b+1)\Gamma(\frac{1}{2}b-a+1)}, \qquad (17)$$

where $a-b+c=1$ and b is a positive integer, SAALSCHÜTZ'S THEOREM

$$_3F_2(a,\ b,\ c;\ d,\ e;\ 1) = \frac{(d-a)_{|c|}(d-b)_{|c|}}{(d)_{|c|}(d-a-b)_{|c|}} \qquad (18)$$

for $d+e=a+b+c+1$ with c a negative integer and $(a)_n$ the POCHHAMMER SYMBOL, DIXON'S THEOREM

$$_3F_2(a,\ b,\ c;\ d,\ e;\ 1)$$
$$= \frac{(\frac{1}{2}a)!(a-b)!(a-c)!(\frac{1}{2}a-b-c)!}{a!(\frac{1}{2}a-b)!(\frac{1}{2}a-c)!(a-b-c)!}, \qquad (19)$$

where $1+a/2-b-c$ has a positive REAL PART, $d = a-b+1$, and $e=a-c+1$, the CLAUSEN FORMULA

$$_4F_3\begin{bmatrix} a,\ b,\ c,\ d \\ e,\ f,\ g \end{bmatrix}; 1 \end{bmatrix} = \frac{(2a)_{|d|}(a+b)_{|d|}(2b)_{|d|}}{(2a+2b)_{|d|}a_{|d|}b_{|d|}}, \qquad (20)$$

for $a+b+c-d=1/2$, $e=a+b+1/2$, $a+f=d+1=b+g$, d a nonpositive integer, and the DOUGALL-RAMANUJAN IDENTITY

$$_7F_6\begin{bmatrix} a_1,\ a_2,\ a_3,\ a_4,\ a_5,\ a_6,\ a_7, \\ b_1,\ b_2,\ b_3,\ b_4,\ b_5,\ b_6 \end{bmatrix}; 1 \end{bmatrix}$$
$$= \frac{(a_1+1)_n(a_1-a_2-a_3+1)_n}{(a_1-a_2+1)_n(a_1-a_3+1)_n}$$
$$\times \frac{(a_1-a_2-a_4+1)_n(a_1-a_3-a_4+1)_n}{(a_1-a_4+1)_n(a_1-a_2-a_3-a_4+1)_n}, \qquad (21)$$

where $n=2a_1+1=a_2+a_3+a_4+a_5$, $a_6=1+a_1/2$, $a_7=-n$, and $b_i=1+a_1-a_{i+1}$ for $i=1, 2, ..., 6$. For all these identities, $(a)_n$ is the POCHHAMMER SYMBOL.

Gessel (1994) found a slew of new identities using WILF-ZEILBERGER PAIRS, including the following:

$$_5F_4\begin{bmatrix} -a-b,\ n+1,\ n+c+1,\ 2n-a-b+1,\ n+\frac{1}{2}(3-a-b) \\ n-a-b-c+1,\ n-a-b+1,\ 2n+2,\ n+\frac{1}{2}(1-a-b) \end{bmatrix}; 1 \end{bmatrix}$$
$$= 0 \qquad (22)$$

$$_3F_2\begin{bmatrix} -3n,\ \frac{2}{3}-c,\ 3n+2 \\ \frac{3}{2},\ 1-3c \end{bmatrix}; \frac{3}{4} \end{bmatrix} = \frac{(c+\frac{2}{3})_n(\frac{1}{3})_n}{(1-c)_n(\frac{4}{3})_n} \qquad (23)$$

$$_3F_2\begin{bmatrix} -3b,\ -\frac{3}{2}n,\ \frac{1}{2}(1-3n) \\ -3n,\ \frac{2}{3}-b-n \end{bmatrix}; \frac{4}{3} \end{bmatrix} = \frac{(\frac{1}{3}-b)_n}{(\frac{1}{3}+b)_n} \qquad (24)$$

$$_4F_3\begin{bmatrix} \frac{3}{2}+\frac{1}{5}n,\ \frac{2}{3},\ -n,\ 2n+2 \\ n+\frac{11}{6},\ \frac{4}{3},\ \frac{1}{5}n+\frac{1}{2} \end{bmatrix}; \frac{2}{27} \end{bmatrix} = \frac{(\frac{5}{2})_n(\frac{11}{6})_n}{(\frac{3}{2})_n(\frac{7}{6})_n} \qquad (25)$$

(Petkovsek *et al.* 1996, pp. 135–37).

The following table gives various named identities ordered by the orders (p, q) of the $_pF_q$s they involve. Bailey (1935) gives a large number of such identities.

$_2F_1$	GAUSS'S HYPERGEOMETRIC THEOREM, KUMMER'S THEOREM, ORR'S THEOREM, RAMANUJAN'S HYPERGEOMETRIC IDENTITY
$_3F_2$	DARLING'S PRODUCTS, DIXON'S THEOREM, RAMANUJAN'S HYPERGEOMETRIC IDENTITY, SAALSCHÜTZ'S THEOREM, THOMAE'S THEOREM, WATSON'S THEOREM, WHIPPLE'S IDENTITY
$_4F_3$	CLAUSEN FORMULA, WHIPPLE'S TRANSFORMATION
$_5F_4$	DOUGALL'S THEOREM
$_6F_5$	WHIPPLE'S IDENTITY
$_7F_6$	DOUGALL-RAMANUJAN IDENTITY, WHIPPLE'S TRANSFORMATION
$_9F_8$	BAILEY'S TRANSFORMATION

Nørlund (1955) gave the general transformation

$$_nF_{n-1}\begin{bmatrix} a_1,\ a_2,\ ...,\ a_n \\ b_1,\ b_2,\ ...,\ b_{n-1} \end{bmatrix}; xz \end{bmatrix}$$
$$= (1-z)^{-a_1} \sum_{v=0}^{\infty} \frac{(a_1)_v}{v!}\ _nF_n\begin{bmatrix} -v,\ a_2,\ a_3,\ ...,\ a_n; \\ b_1,\ b_2,\ ...,\ b_{n-1} \end{bmatrix} x \end{bmatrix}$$
$$\times \left(\frac{z}{z-1}\right)^v \qquad (26)$$

where $(a)_n$ is the POCHHAMMER SYMBOL. This identity is based on the transformation due to Euler

$$\sum_{n=0}^{\infty} \frac{(a)_n}{n!}\ a_n z^n = (1-z)^{-a} \sum_{n=0}^{\infty} \frac{(a)_n}{n!}\ \Delta^n a_0 \left(\frac{z}{1-z}\right)^n, \qquad (27)$$

where Δ is the FORWARD DIFFERENCE and

$$\Delta^k a_0 = \sum_{m=0}^{k} (-1)^m \binom{k}{m} a_{k-m} \qquad (28)$$

(Nørlund 1955).

See also CARLSON'S THEOREM, CLAUSEN FORMULA, CONFLUENT HYPERGEOMETRIC FUNCTION, CONFLUENT HYPERGEOMETRIC LIMIT FUNCTION, DIXON'S

THEOREM, DOUGALL-RAMANUJAN IDENTITY, DOUGALL'S THEOREM, GOSPER'S ALGORITHM, HEINE HYPERGEOMETRIC SERIES, HYPERGEOMETRIC FUNCTION, HYPERGEOMETRIC IDENTITY, HYPERGEOMETRIC SERIES, JACKSON'S IDENTITY, K-BALANCED, KAMPE DE FERIET FUNCTION, KUMMER'S THEOREM, LAURICELLA FUNCTIONS, NEARLY-POISED, RAMANUJAN'S HYPERGEOMETRIC IDENTITY, SAALSCHÜTZ'S THEOREM, SAALSCHÜTZIAN, SISTER CELINE'S METHOD, THOMAE'S THEOREM, WATSON'S THEOREM, WELL-POISED, WHIPPLE'S IDENTITY, WHIPPLE'S TRANSFORMATION, WILF-ZEILBERGER PAIR, ZEILBERGER'S ALGORITHM

References

Bailey, W. N. "Some Identities Involving Generalized Hypergeometric Series." *Proc. London Math. Soc. Ser. 2* **29**, 503–16, 1929.

Bailey, W. N. *Generalised Hypergeometric Series.* Cambridge, England: Cambridge University Press, 1935.

Barnes. *Proc. London Math. Soc.* **5**, 59–16 1907.

Dwork, B. *Generalized Hypergeometric Functions.* Oxford, England: Clarendon Press, 1990.

Exton, H. *Multiple Hypergeometric Functions and Applications.* New York: Wiley, 1976.

Exton, H. *Handbook of Hypergeometric Integrals: Theory, Applications, Tables, Computer Programs.* Chichester, England: Ellis Horwood, 1978.

Gessel, I. "Finding Identities with the WZ Method." *Theoret. Comput. Sci.* To appear.

Gessel, I. M. "Finding Identities with the WZ Method. Symbolic Computation in Combinatorics Δ_1 (Ithaca, NY, 1993)." *J. Symbolic Comput.* **20**, 537–66, 1995.

Gessel, I. and Stanton, D. "Strange Evaluations of Hypergeometric Series." *SIAM J. Math. Anal.* **13**, 295–08, 1982.

Gosper, R. W. "Decision Procedures for Indefinite Hypergeometric Summation." *Proc. Nat. Acad. Sci. USA* **75**, 40–2, 1978.

Hardy, G. H. "Hypergeometric Series." Ch. 7 in *Ramanujan: Twelve Lectures on Subjects Suggested by His Life and Work, 3rd ed.* New York: Chelsea, pp. 101–12, 1999.

Klein, F. *Vorlesungen über die hypergeometrische Funktion.* Berlin: J. Springer, 1933.

Koekoek, R. and Swarttouw, R. F. *The Askey-Scheme of Hypergeometric Orthogonal Polynomials and its q-Analogue.* Delft, Netherlands: Technische Universiteit Delft, Faculty of Technical Mathematics and Informatics Report 98–7, 1–68, 1998. ftp://www.twi.tudelft.nl/publications/tech-reports/1998/DUT-TWI-98-7.ps.gz.

Koepf, W. "Hypergeometric Database." Ch. 3 in *Hypergeometric Summation: An Algorithmic Approach to Summation and Special Function Identities.* Braunschweig, Germany: Vieweg, pp. 12 and 31–3, 1998.

Nørlund, N. E. "Hypergeometric Functions." *Acta Math.* **94**, 289–49, 1955.

Petkovsek, M.; Wilf, H. S.; and Zeilberger, D. *A = B.* Wellesley, MA: A. K. Peters, 1996.

Rainville, E. D. *Special Functions.* New York: Chelsea, 1971.

Saxena, R. K. and Mathai, A. M. *Generalized Hypergeometric Functions with Applications in Statistics and Physical Sciences.* New York: Springer-Verlag, 1973.

Slater, L. J. *Generalized Hypergeometric Functions.* Cambridge, England: Cambridge University Press, 1966.

Zeilberger, D. "A Fast Algorithm for Proving Terminating Hypergeometric Series Identities." *Discrete Math.* **80**, 207–11, 1990.

Generalized Matrix Inverse

MOORE-PENROSE GENERALIZED MATRIX INVERSE

Generalized Mean

A generalized version of the MEAN

$$m(t) \equiv \left(\frac{1}{n} \sum_{k=1}^{n} a_k^t \right)^{1/t} \tag{1}$$

with parameter t which gives the GEOMETRIC MEAN, ARITHMETIC MEAN, and HARMONIC MEAN as special cases:

$$\lim_{t \to 0} m(t) = G \tag{2}$$

$$m(1) = A \tag{3}$$

$$m(-1) = H. \tag{4}$$

See also MEAN

Generalized Polygon

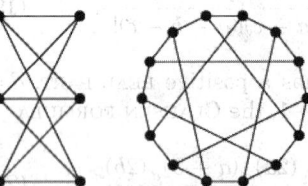 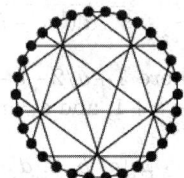

Let O be an incidence geometry, i.e., a set with a symmetric, reflexive binary relation I. Let e and f be elements of O. Let an incidence plane be an incidence geometry whose object set is the disjoint union of two sets P and L such that for e, $f \in P$ or e, $f \in L$, $(e, f) \in I$ only if $e = f$. Then a generalized polygon is an incidence plane such that for all e, $f \in O$,

1. There exists a CHAIN of length at most n from e to f, and.
2. There exists at most one irreducible CHAIN of length less than n from e to f.

(Feit and Higman 1964).

The only CUBIC generalized polygons are the generalized 2-gon $K_{3,3}$ (UTILITY GRAPH), generalized triangle $PG_{2,2}$ (HEAWOOD GRAPH), generalized quadrangle W_2 (the LEVI GRAPH), and generalized hexagon $GH_{2,2}$ (Feit and Higman 1964, Royle).

See also CAGE GRAPH, MOORE GRAPH

References

Feit, W. and Higman, G. "The Non-Existence of Certain Generalized Polygons." *J. Algebra* **1**, 114–31, 1964.

Royle, G. "Cubic Cages." http://www.cs.uwa.edu.au/~gordon/cages/.

Tits, J. "Sur la trialité et certains groupes qui s'en déduisent." *Publ. Math. I.H.E.S. Paris* **2**, 14–0, 1959.

Tits, J. "Théorème de Bruhat er sous-groupes paraboliques." *C. R. Acad. Sci. Paris* **254**, 2910–912, 1962.

Generalized Remainder Method

An algorithm for computing a UNIT FRACTION.

See also UNIT FRACTION

References
Eppstein, D. Egypt.ma Mathematica notebook. http://www.ics.uci.edu/~eppstein/numth/egypt/egypt.ma.

Generating Function

A POWER SERIES

$$f(x) = \sum_{n=0}^{\infty} a_n x^n \qquad (1)$$

whose COEFFICIENTS give the SEQUENCE $\{a_0, a_1, \ldots\}$. The *Mathematica* function PowerSum in the *Mathematica* add-on package DiscreteMath`RSolve` (which can be loaded with the command << DiscreteMath`) gives the generating function of a given expression, and ExponentialPowerSum in the *Mathematica* add-on package DiscreteMath`RSolve` (which can be loaded with the command << DiscreteMath`) gives the so-called EXPONENTIAL GENERATING FUNCTION. The generating function $f(x)$ is sometimes said to "ENUMERATE" a_n (Hardy 1999, p. 85).

Generating functions for the first few powers $a_n^{(p)}$ are given in the following table.

n^p	$f(x)$	series
1	$\frac{x}{1-x}$	$x + x^2 + x^3 + \ldots$
n	$\frac{x}{(1-x)^2}$	$x + 2x^2 + 3x^3 + 4x^4 + \ldots$
n^2	$\frac{x(x+1)}{(1-x)^3}$	$x + 4x^2 + 9x^3 + 16x^4 + \ldots$
n^3	$\frac{x(x^2+4x+1)}{(1-x)^4}$	$x + 8x^2 + 27x^3 + \ldots$
n^4	$\frac{x(x+1)(x^2+10x+1)}{(1-x)^5}$	$x + 16x^2 + 81x^3 + \ldots$

There are many beautiful generating functions for special functions in number theory. A few particularly nice examples are

$$f(x) = \frac{1}{\prod_{k=1}^{\infty} 1 - x^k} = 1 + x + 2x^2 + 3x^3 + \ldots \qquad (2)$$

for the PARTITION FUNCTION P, and

$$f(x) = \sum_{n=0}^{\infty} F_n x^n = \frac{x}{1 - x - x^2}$$
$$= x + x^2 + 2x^3 + 3x^4 + \ldots \qquad (3)$$

for the FIBONACCI NUMBERS F_n.

The generating function of $G(t)$ of a sequence of numbers $f(n)$ given by the Z-TRANSFORM of $f(n)$ in the variable $1/t$ (Germundsson 2000).

See also CUMULANT-GENERATING FUNCTION, ENUMERATE, EXPONENTIAL GENERATING FUNCTION, MOMENT-GENERATING FUNCTION, RECURRENCE RELATION, Z-TRANSFORM

References
Bender, E. A. and Goldman, J. R. "Enumerative Uses of Generating Functions." *Indiana U. Math. J.* **20**, 753–65, 1970/1971.
Bergeron, F.; Labelle, G.; and Leroux, P. "Théorie des espèces er Combinatoire des Structures Arborescentes." Publications du LACIM. Québec, Montréal, Canada: Univ. Québec Montréal, 1994.
Cameron, P. J. "Some Sequences of Integers." *Disc. Math.* **75**, 89–02, 1989.
Doubilet, P.; Rota, G.-C.; and Stanley, R. P. "The Idea of Generating Function." Ch. 3 in *Finite Operator Calculus* (Ed. G.-C. Rota). New York: Academic Press, pp. 83–34, 1975.
Germundsson, R. "*Mathematica* Version 4." *Mathematica J.* **7**, 497–24, 2000.
Graham, R. L.; Knuth, D. E.; and Patashnik, O. *Concrete Mathematics: A Foundation for Computer Science, 2nd ed.* Reading, MA: Addison-Wesley, 1994.
Harary, F. and Palmer, E. M. *Graphical Enumeration.* New York: Academic Press, 1973.
Hardy, G. H. *Ramanujan: Twelve Lectures on Subjects Suggested by His Life and Work, 3rd ed.* New York: Chelsea, p. 85, 1999.
Leroux, P. and Miloudi, B. "Généralisations de la formule d'Otter." *Ann. Sci. Math. Québec* **16**, 53–0, 1992.
Riordan, J. *Combinatorial Identities.* New York: Wiley, 1979.
Riordan, J. *An Introduction to Combinatorial Analysis.* New York: Wiley, 1980.
Sloane, N. J. A. and Plouffe, S. "Recurrences and Generating Functions." §2.4 in *The Encyclopedia of Integer Sequences.* San Diego, CA: Academic Press, pp. 9–0, 1995.
Stanley, R. P. *Enumerative Combinatorics, Vol. 1.* Cambridge, England: Cambridge University Press, p. 63, 1996.
Viennot, G. "Une Théorie Combinatoire des Polynômes Orthogonaux Généraux." Publications du LACIM. Québec, Montréal, Canada: Univ. Québec Montréal, 1983.
Wilf, H. S. *Generatingfunctionology, 2nd ed.* New York: Academic Press, 1990.

Generation

In population studies, the direct offspring of a reference population (roughly) constitutes a single generation. For a CELLULAR AUTOMATON, the fundamental unit of time during which the rules of reproduction are applied once is called a generation.

Generator (Digitaddition)

An INTEGER used to generate a DIGITADDITION. A number can have more than one generator. If a number has no generator, it is called a SELF NUMBER.

Generator (Group)

A member of a CYCLIC GROUP, the POWERS of which generate the entire GROUP.

See also FINITELY GENERATED

References

Arfken, G. "Generators." §4.11 in *Mathematical Methods for Physicists, 3rd ed.* Orlando, FL: Academic Press, pp. 261–67, 1985.

Generic Character

For a form Q, the generic character $\chi_i(Q)$ OF THE FORM is defined as the values of $\chi_i(m)$ where $(m, 2d) = 1$ and Q represents m: $\chi_1(Q)$, $\chi_2(Q)$, ..., $\chi_r(Q)$ (Cohn 1980, p. 223). The characters apply to the class of properly equivalent forms as they represent the same numbers.

See also GENUS (FORM)

References

Cohn, H. "Compositions, Order, and Genera." Ch. 8 in *Advanced Number Theory.* New York: Dover, 1980.

Generic Cylindrical Algebraic Decomposition

A CYLINDRICAL ALGEBRAIC DECOMPOSITION that omits sets of measure zero. Generic cylindrical algebraic decompositions are generally much quicker to compute than are normal decompositions. Generic cylindrical algebraic decomposition is implemented in *Mathematica* as `GenericCylindricalAlgebraicDecomposition[ineqs, vars]`.

See also CYLINDRICAL ALGEBRAIC DECOMPOSITION

References

Strzebonski, A. "Solving Algebraic Inequalities." *Mathematica J.* **7**, 525–41, 2000.

Genetic Algorithm

An adaptive STOCHASTIC OPTIMIZATION ALGORITHM involving search and optimization that was first used by John Holland. Holland created an electronic organism as a binary string ("chromosome"), and then used genetic and evolutionary principles of fitness-proportionate selection for reproduction (including random crossover and mutation) to search enormous solution spaces efficiently. So-called genetic programming languages apply the same principles, using an expression tree instead of a bit string as the "chromosome."

See also CELLULAR AUTOMATON, DIFFERENTIAL EVOLUTION, EVOLUTION STRATEGIES, OPTIMIZATION THEORY, STOCHASTIC OPTIMIZATION

References

Bengtsson, M. "Genetic Algorithms Notebook." http://www.mathsource.com/cgi-bin/msitem?0204–47.

Genocchi Number

A number given by the GENERATING FUNCTION

$$\frac{2t}{e^t + 1} = \sum_{n=1}^{\infty} G_n \frac{t^n}{n!}$$

It satisfies $G_1 = 1$, $G_3 = G_5 = G_7 = \ldots = 0$, and even coefficients are given by

$$G_{2n} = 2(1 - 2^{2n})B_{2n} = 2nE_{2n-1}(0),$$

where B_n is a BERNOULLI NUMBER and $E_n(x)$ is an EULER POLYNOMIAL. The first few Genocchi numbers for n EVEN are -1, 1, -3, 17, -155, 2073, ... (Sloane's A001469).

See also BERNOULLI NUMBER, EULER POLYNOMIAL

References

Comtet, L. *Advanced Combinatorics: The Art of Finite and Infinite Expansions, rev. enl. ed.* Dordrecht, Netherlands: Reidel, p. 49, 1974.
Kreweras, G. "An Additive Generation for the Genocchi Numbers and Two of its Enumerative Meanings." *Bull. Inst. Combin. Appl.* **20**, 99–03, 1997.
Kreweras, G. "Sur les permutations comptées par les nombres de Genocchi de 1-ière et 2-ième espèce." *Europ. J. Comb.* **18**, 49–8, 1997.
Rota, G.-C.; Kahaner, D.; Odlyzko, A. "On the Foundations of Combinatorial Theory, VIII: Finite Operator Calculus." *J. Math. Anal. Appl.* **42**, 684–60, 1973.
Sloane, N. J. A. Sequences A001469/M3041 in "An On-Line Version of the Encyclopedia of Integer Sequences." http://www.research.att.com/~njas/sequences/eisonline.html.

Gentle Diagonal

PASCAL'S TRIANGLE

Gentle Giant Group

MONSTER GROUP

Genus (Curve)

One of the PLÜCKER CHARACTERISTICS, defined by

$$p \equiv \tfrac{1}{2}(n-1)(n-2) - (\delta + \kappa) = \tfrac{1}{2}(m-1)(m-2) - (\tau + \iota),$$

where m is the class, n the order, δ the number of nodes, κ the number of CUSPS, ι the number of stationary tangents (INFLECTION POINTS), and τ the number of BITANGENTS.

See also RIEMANN CURVE THEOREM

References

Coolidge, J. L. *A Treatise on Algebraic Plane Curves.* New York: Dover, p. 100, 1959.

Genus (Form)

Consider the forms Q for which the GENERIC CHARACTERS $\chi_i(Q)$ are equal to some preassigned array of signs $e_i = 1$ or -1,

$$e_1, e_2, \ldots, e_r,$$

subject to $\prod_{i=1}^{r} e_i = 1$. There are 2^{r-1} possible arrays, where r is the number of distinct prime divisors of a field discriminant d, and the set of forms corresponding to each array is called a genus of forms. The forms for which all $e_i = 1$ are called the principal genus of forms, and each genus is also a collection of proper EQUIVALENCE CLASSES (Cohn 1980, pp. 223–24).

See also EQUIVALENCE CLASS, FUNDAMENTAL THEOREM OF GENERA, GENERIC CHARACTER

References

Cohn, H. "Compositions, Order, and Genera." Ch. 8 in *Advanced Number Theory.* New York: Dover, pp. 212–30, 1980.

Genus (Knot)

The least genus of any SEIFERT SURFACE for a given KNOT. The UNKNOT is the only KNOT with genus 0.

Genus (Surface)

A topologically invariant property of a surface defined as the largest number of nonintersecting simple closed curves that can be drawn on the surface without separating it. Roughly speaking, it is the number of HOLES in a surface. The genus of a surface, also called the geometric genus, is related to the EULER CHARACTERISTIC χ by

$$\chi = 2 - 2g.$$

See also EULER CHARACTERISTIC

References

Gray, A. *Modern Differential Geometry of Curves and Surfaces with Mathematica, 2nd ed.* Boca Raton, FL: CRC Press, p. 635, 1997.

Genus Theorem

The DIOPHANTINE EQUATION

$$x^2 + y^2 = p$$

can be solved for p a PRIME IFF $p \equiv 1 \pmod 4$ or $p = 2$. The representation is unique except for changes of sign or rearrangements of x and y. This theorem is intimately connected with the QUADRATIC RECIPROCITY THEOREM, and generalizes to the QUARTIC RECIPROCITY THEOREM.

See also COMPOSITION THEOREM, DIOPHANTINE EQUATION–4TH POWERS, FERMAT'S THEOREM, FUNDAMENTAL THEOREM OF GENERA, GENUS (FORM), QUADRATIC RECIPROCITY THEOREM

Geocentric Latitude

An AUXILIARY LATITUDE given by

$$\phi_g = \tan^{-1}\left[(1 - e^2)\right]\tan \phi].$$

The series expansion is

$$\phi_g = \phi - e_2 \sin(2\phi) + \tfrac{1}{2} e_2^2 \sin(4\phi) + \tfrac{1}{3} e_2^3 \sin(6\phi) + \ldots,$$

where

$$e_2 \equiv \frac{e^2}{2 - e^2}.$$

See also LATITUDE

References

Adams, O. S. "Latitude Developments Connected with Geodesy and Cartography with Tables, Including a Table for Lambert Equal-Area Meridional Projections." Spec. Pub. No. 67. U. S. Coast and Geodetic Survey, 1921.
Snyder, J. P. *Map Projections--A Working Manual.* U. S. Geological Survey Professional Paper 1395. Washington, DC: U. S. Government Printing Office, pp. 17–8, 1987.

Geodesic

Given two points on a surface, the geodesic is defined as the shortest path on the surface connecting them. Geodesics also preserve a direction on a surface (Tietze 1965, pp. 26–7) and have many other interesting properties. The NORMAL VECTOR to any point of a GEODESIC arc lies along the normal to a surface at that point (Weinstock 1974, p. 65).

Furthermore, no matter how badly a SPHERE is distorted, there exist an infinite number of closed geodesics on it. This general result, demonstrated in the early 1990s, extended earlier work by Birkhoff, who proved in 1917 that there exists at least one closed geodesic on a distorted sphere, and Lyusternik and Schnirelmann, who proved in 1923 that there exist at least three closed geodesics on such a sphere (Cipra 1993, p. 28).

For a surface given parametrically by $x = x(u, v)$, $y = y(u, v)$, and $z = z(u, v)$, the geodesic can be found by minimizing the ARC LENGTH

$$L \equiv \int ds = \int \sqrt{dx^2 + dy^2 + dz^2}. \qquad (1)$$

But

$$dx = \frac{\partial x}{\partial u}\, du + \frac{\partial x}{\partial v}\, dv \qquad (2)$$

$$dx^2 = \left(\frac{\partial x}{\partial u}\right)^2 du^2 + 2\,\frac{\partial x}{\partial u}\,\frac{\partial x}{\partial v}\, du\, dv + \left(\frac{\partial x}{\partial v}\right)^2 dv^2, \qquad (3)$$

and similarly for dy^2 and dz^2. Plugging in,

$$L = \int \left\{ \left[\left(\frac{\partial x}{\partial u} \right)^2 + \left(\frac{\partial y}{\partial u} \right)^2 + \left(\frac{\partial z}{\partial u} \right)^2 \right] du^2 \right.$$

$$+ 2 \left[\frac{\partial x}{\partial u} \frac{\partial x}{\partial v} + \frac{\partial y}{\partial u} \frac{\partial y}{\partial v} + \frac{\partial z}{\partial u} \frac{\partial z}{\partial v} \right] du \, dv$$

$$+ \left. \left[\left(\frac{\partial x}{\partial v} \right)^2 + \left(\frac{\partial y}{\partial v} \right)^2 + \left(\frac{\partial z}{\partial v} \right)^2 \right] dv^2 \right\}^{1/2}. \quad (4)$$

This can be rewritten as

$$L = \int \sqrt{P + 2Qv' + Rv'^2} \, du \quad (5)$$

$$= \int \sqrt{Pu'^2 + 2Qu' + R} \, dv, \quad (6)$$

where

$$v' \equiv \frac{dv}{du} \quad (7)$$

$$u' \equiv \frac{du}{dv} \quad (8)$$

and

$$P \equiv \left(\frac{\partial x}{\partial u} \right)^2 + \left(\frac{\partial y}{\partial u} \right)^2 + \left(\frac{\partial z}{\partial u} \right)^2 \quad (9)$$

$$Q \equiv \frac{\partial x}{\partial u} \frac{\partial x}{\partial v} + \frac{\partial y}{\partial u} \frac{\partial y}{\partial v} + \frac{\partial z}{\partial u} \frac{\partial z}{\partial v} \quad (10)$$

$$R \equiv \left(\frac{\partial x}{\partial v} \right)^2 + \left(\frac{\partial y}{\partial v} \right)^2 + \left(\frac{\partial z}{\partial v} \right)^2. \quad (11)$$

Taking derivatives,

$$\frac{\partial L}{\partial v} = \frac{1}{2} (P + 2Qv' + Rv'^2)^{-1/2} \left(\frac{\partial P}{\partial v} + 2 \frac{\partial Q}{\partial v} v' + \frac{\partial R}{\partial v} v'^2 \right)$$

$$(12)$$

$$\frac{\partial L}{\partial v'} = \frac{1}{2} (P + 2Qv' + Rv'^2)^{-1/2} (2Q + 2Rv'), \quad (13)$$

so the EULER-LAGRANGE DIFFERENTIAL EQUATION then gives

$$\frac{\frac{\partial P}{\partial v} + 2v' \frac{\partial Q}{\partial v} + v'^2 \frac{\partial R}{\partial v}}{2\sqrt{P + 2Qv' + Rv'^2}} - \frac{d}{du} \left(\frac{Q + Rv'}{\sqrt{P + 2Qv' + Rv'^2}} \right)$$

$$= 0. \quad (14)$$

In the special case when P, Q, and R are explicit functions of u only,

$$\frac{Q + Rv'}{\sqrt{P + 2Qv' + Rv'^2}} = c_1 \quad (15)$$

$$\frac{Q^2 + 2QRv' + R^2v'^2}{P + 2Qv' + Rv'^2} = c_1^2 \quad (16)$$

$$v'^2 R(R - c_1^2) + 2v'Q(R - c_1^2) + (Q^2 - Pc_1^2) = 0 \quad (17)$$

$$v' = \frac{1}{2R(R - c_1^2)}$$

$$\times \left[2Q(c_1^2 - R) \pm \sqrt{4Q^2(R - c_1^2) - 4R(R - c_1^2)(Q^2 - Pc_1^2)} \right].$$

$$(18)$$

Now, if P and R are explicit functions of u only *and* $Q = 0$,

$$v' = \frac{\sqrt{4R(R - c_1^2)Pc_1^2}}{2R(R - c_1^2)} = c_1 \sqrt{\frac{P}{R(R - c_1^2)}}, \quad (19)$$

so

$$v = c_1 \int \sqrt{\frac{P}{R(R - c_1^2)}} \, du. \quad (20)$$

In the case $Q = 0$ where P and R are explicit functions of v only, then

$$\frac{\frac{\partial P}{\partial v} + v'^2 \frac{\partial R}{\partial v}}{2\sqrt{P + Rv'^2}} - \frac{d}{du} \left(\frac{Rv'}{\sqrt{P + Rv'^2}} \right) = 0, \quad (21)$$

so

$$\frac{\partial P}{\partial v} + v'^2 \frac{\partial R}{\partial v}$$

$$-2\sqrt{P + Rv'^2} R \left[\frac{v''}{\sqrt{P + Rv'^2}} + \left(-\frac{1}{2} \right) \frac{v'(2Rv'v'')}{(P + Rv'^2)^{3/2}} \right]$$

$$= 0 \quad (22)$$

$$\frac{\partial P}{\partial v} + v'^2 \frac{\partial R}{\partial v} - 2Rv'' + \frac{2R^2v'^2v''}{P + Rv'^2} = 0 \quad (23)$$

$$\frac{Rv'^2}{\sqrt{P + Rv'^2}} - \sqrt{P + Rv'^2} = c_1 \quad (24)$$

$$Rv'^2 - (P + Rv'^2) = c_1 \sqrt{P + Rv'^2} \quad (25)$$

$$\left(\frac{p}{c_1} \right)^2 = P + Rv'^2 \quad (26)$$

$$\frac{P^2 - c_1^2 P}{Rc_1^2} = v'^2, \quad (27)$$

and

$$u = c_1 \int \sqrt{\frac{R}{P^2 - c_1^2 P}} \, dv. \tag{28}$$

For a SURFACE OF REVOLUTION in which $y = g(x)$ is rotated about the x-AXIS so that the equation of the surface is

$$y^2 + z^2 = g^2(x), \tag{29}$$

the surface can be parameterized by

$$x = u \tag{30}$$

$$y = g(u) \cos v \tag{31}$$

$$z = g(u) \sin v. \tag{32}$$

The equation of the geodesics is then

$$v = c_1 \int \frac{\sqrt{1 + [g'(u)]^2} \, du}{g(u)\sqrt{[g(u)]^2 - c_1^2}}. \tag{33}$$

See also ELLIPSOID GEODESIC, GEODESIC CURVATURE, GEODESIC DOME, GEODESIC EQUATION, GEODESIC MAPPING, GEODESIC TRIANGLE, GRAPH GEODESIC, GREAT CIRCLE, HARMONIC MAP, OBLATE SPHEROID GEODESIC, PARABOLOID GEODESIC

References

2- Cipra, B. *What's Happening in the Mathematical Sciences, Vol. 1.* Providence, RI: Amer. Math. Soc., p. 28, 1993.
Tietze, H. *Famous Problems of Mathematics: Solved and Unsolved Mathematics Problems from Antiquity to Modern Times.* New York: Graylock Press, pp. 27 and 40, 1965.
Tietze, H. *Mathematische Analyse des Raumproblems.* Berlin, 1923.
Weinstock, R. *Calculus of Variations, with Applications to Physics and Engineering.* New York: Dover, pp. 26-8 and 45-6, 1974.
Weyl, H. §17 in *Space--Time--Matter.* New York: Dover, 1952.

Geodesic Curvature

For a unit speed curve on a surface, the length of the surface-tangential component of acceleration is the geodesic curvature κ_g. Curves with $\kappa_g = 0$ are called GEODESICS. For a curve parameterized as $\alpha(t) = \mathbf{x}(u(t), v(t))$, the geodesic curvature is given by

$$\kappa_g = \sqrt{EG - F^2} \big[-\Gamma_{11}^2 u'^3 + \Gamma_{22}^1 v'^3 - (2\Gamma_{12}^2 - \Gamma_{11}^1)u'^2 v'$$
$$+ (2\Gamma_{12}^1 - \Gamma_{22}^2)u'v'^2 + u''v' - v''u' \big],$$

where E, F, and G are coefficients of the first FUNDAMENTAL FORM and Γ_{ij}^k are CHRISTOFFEL SYMBOLS OF THE SECOND KIND.

See also GEODESIC

References

Gray, A. "Geodesic Curvature and Torsion." §22.4 in *Modern Differential Geometry of Curves and Surfaces with Mathematica, 2nd ed.* Boca Raton, FL: CRC Press, pp. 513–18, 1997.

Geodesic Dome

A TRIANGULATION of a PLATONIC SOLID or other POLYHEDRON to produce a close approximation to a SPHERE (or HEMISPHERE). The nth order geodesation operation replaces each polygon of the polyhedron by the projection onto the CIRCUMSPHERE of the order-n regular tessellation of that polygon. The above figure shows geodesations of orders 1 to 3 (from top to bottom) of the TETRAHEDRON, CUBE, OCTAHEDRON, DODECAHEDRON, and ICOSAHEDRON (from left to right), computed using Geodesate[*poly*, *n*] in the *Mathematica* add-on package Graphics`Polyhedra` (which can be loaded with the command << Graphics`).

R. Buckminster Fuller designed the first geodesic dome (i.e., geodesation of a HEMISPHERE). Fuller's dome was constructed from an ICOSAHEDRON by adding ISOSCELES TRIANGLES about each VERTEX and slightly repositioning the VERTICES. In such domes, neither the VERTICES nor the centers of faces necessarily lie at exactly the same distances from the center. However, these conditions are approximately satisfied.

In the geodesic domes discussed by Kniffen (1994), the sum of VERTEX angles is chosen to be a constant. Given a PLATONIC SOLID, let $e' \equiv 2e/v$ be the number of EDGES meeting at a VERTEX and n be the number of EDGES of the constituent POLYGON. Call the angle of the old VERTEX point A and the angle of the new VERTEX point F. Then

$$A = B \tag{1}$$

$$2e'A = nF \tag{2}$$

$$2A + F = 180°. \tag{3}$$

Solving for A gives

$$2A + \frac{2e'}{n} A = 2A\left(1 + \frac{e'}{n}\right) = 180° \tag{4}$$

$$A = 90° \frac{n}{e' + n}, \tag{5}$$

and

$$F = \frac{2e'}{n} A = 180° \frac{e'}{e' + n}. \tag{6}$$

The VERTEX sum is

$$\Sigma = nF = 180° \frac{e'n}{e' + n}. \tag{7}$$

Solid	f	v	e'	n	A	F	Σ
TETRAHEDRON			3	3	$45°$	$90°$	$270°$
CUBE	24	14	3	4	$51\frac{3}{7}°$	$81\frac{3}{7}°$	$308\frac{4}{7}°$
OCTAHEDRON			4	3	$38\frac{4}{7}°$	$108\frac{4}{7}°$	$308\frac{4}{7}°$
DODECAHEDRON	60	32	3	5	$56\frac{1}{4}°$	$71\frac{1}{4}°$	$337\frac{1}{2}°$
ICOSAHEDRON			5	3	$33\frac{3}{4}°$	$118\frac{3}{4}°$	$337\frac{1}{2}°$

Wenninger and Messer (1996) give general formulas for solving any geodesic chord factor and dihedral angle in a geodesic dome.

See also SPHERE, SPHERICAL TRIANGLE, TRIANGULAR SYMMETRY GROUP

References

Kenner, H. *Geodesic Math and How to Use It.* Berkeley, CA: University of California Press, 1976.
Kniffen, D. "Geodesic Domes for Amateur Astronomers." *Sky & Telescope* **88**, 90–4, Oct. 1994.
Messer, P. W. "Mathematical Formulas for Geodesic Domes." Appendix to Wenninger, M. *Spherical Models.* New York: Dover, pp. 145–49, 1999.
Pappas, T. "Geodesic Dome of Leonardo da Vinci." *The Joy of Mathematics.* San Carlos, CA: Wide World Publ./Tetra, p. 81, 1989.
Wells, D. *The Penguin Dictionary of Curious and Interesting Geometry.* London: Penguin, pp. 85–6, 1991.
Wenninger, M. J. and Messer, P. W. "Patterns on the Spherical Surface." *Internat. J. Space Structures* **11**, 183–92, 1996.
Wenninger, M. "Geodesic Domes." Ch. 4 in *Spherical Models.* New York: Dover, pp. 80–24, 1999.

Geodesic Equation

$$d\tau^2 = -\eta_{\alpha\beta}\, d\xi^\alpha\, d\xi^\beta,$$

or

$$\frac{d^2\xi^\alpha}{dr^2} = 0.$$

See also GEODESIC

Geodesic Flow

A type of FLOW technically defined in terms of the TANGENT BUNDLE of a MANIFOLD.

See also DYNAMICAL SYSTEM

Geodesic Mapping

A geodesic mapping $f : M \to N$ between two RIEMANNIAN MANIFOLDS is a DIFFEOMORPHISM sending GEODESICS of M into GEODESICS of N, whose inverse also sends GEODESICS to GEODESICS (Ambartzumian 1982, p. 26).

See also BELTRAMI'S THEOREM, GEODESIC

References

Ambartzumian, R. V. *Combinatorial Integral Geometry.* Chichester, England: Wiley, 1982.
Kreyszig, E. *Differential Geometry.* New York: Dover, 1991.

Geodesic Triangle

A TRIANGLE formed by the arcs of three GEODESICS on a smooth surface.

See also INTEGRAL CURVATURE, SPHERICAL TRIANGLE

Geodetic Latitude

LATITUDE

Geodetic Number

Let $I(x, y)$ denote the set of all vertices lying on an (x, y)-GRAPH GEODESIC in G, then a set S with $I(S) = V(G)$ is called a geodetic set in G and is denoted $g(G)$.

See also HULL NUMBER

References

Chartrand, G.; Harary, F.; and Zhang, P. "The Forcing Hull Number of a Graph." To appear in *J. Comb. Math. Comb. Combin.*
Chartrand, G. and Zhang, P. "The Geodetic Number of a Graph." To appear in *Networks.*
Chartrand, G. and Zhang, P. "The Forcing Geodetic Number of a Graph." *Discuss. Math. Graph Th.* **19**, 45–8, 1999.
Chartrand, G. and Zhang, P. "Realizable Ratios in Graph Theory: Geodesic Parameters." *Bull. Inst. Comb. Appl.* **27**, 69–0, 1999.
Chartrand, G. and Zhang, P. "The Geodetic Number of an Oriented Graph." *Europ. J. Combin.* **21**, 181–89, 2000.

Geographic Latitude

LATITUDE

Geometric Construction

In antiquity, geometric constructions of figures and lengths were restricted to the use of only a STRAIGHT-EDGE and COMPASS (or in Plato's case, a COMPASS only; a so-called MASCHERONI CONSTRUCTION). Although the term "RULER" is sometimes used instead of "STRAIGHTEDGE," no markings which could be used to make measurements were allowed according to the Greek prescription. Furthermore, the "COMPASS" could not even be used to mark off distances by setting it and then "walking" it along, so the COMPASS had to be considered to automatically collapse when not in the process of drawing a CIRCLE.

Because of the prominent place Greek geometric constructions held in Euclid's *ELEMENTS*, these constructions are sometimes also known as EUCLIDEAN CONSTRUCTIONS. Such constructions lay at the heart of the GEOMETRIC PROBLEMS OF ANTIQUITY of CIRCLE SQUARING, CUBE DUPLICATION, and TRISECTION of an ANGLE. The Greeks were unable to solve these problems, but it was not until hundreds of years later that the problems were proved to be actually impossible under the limitations imposed.

Simple algebraic operations such as $a+b$, $a-b$, ra (for r a RATIONAL NUMBER), a/b, ab, and $\sqrt{x}$ can be performed using geometric constructions (bold 1982, Courant and Robbins 1996). Other more complicated constructions, such as the solution of APOLLONIUS' PROBLEM and the construction of INVERSE POINTS can also accomplished.

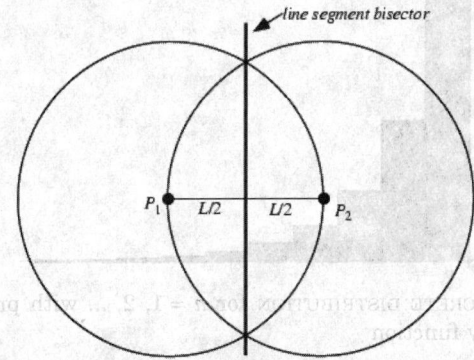

One of the simplest geometric constructions is the construction of a BISECTOR of a LINE SEGMENT, illustrated above.

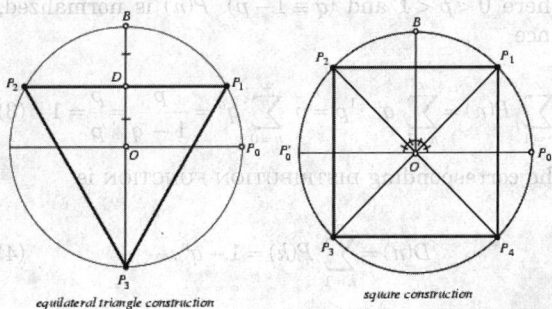

equilateral triangle construction   *square construction*

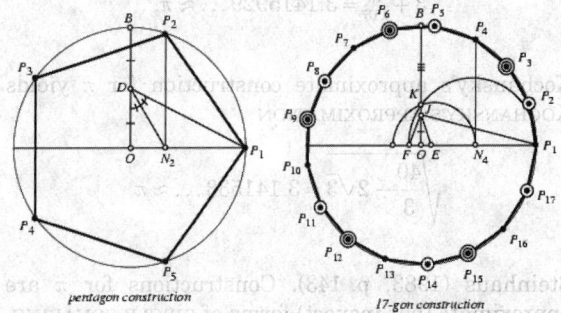

pentagon construction   *17-gon construction*

The Greeks were very adept at constructing POLYGONS, but it took the genius of Gauss to mathematically determine which constructions were possible and which were not. As a result, Gauss determined that a series of POLYGONS (the smallest of which has 17 sides; the HEPTADECAGON) had constructions unknown to the Greeks. Gauss showed that the CONSTRUCTIBLE POLYGONS (several of which are illustrated above) were closely related to numbers called the FERMAT PRIMES.

Wernick (1982) gave a list of 139 sets of three located points from which a TRIANGLE was to be constructed. Of Wernick's original list of 139 problems, 20 had not yet been solved as of 1996 (Meyers 1996).

It is possible to construct RATIONAL NUMBERS and EUCLIDEAN NUMBERS using a STRAIGHTEDGE and COMPASS construction. In general, the term for a number which can be constructed using a COMPASS and STRAIGHTEDGE is a CONSTRUCTIBLE NUMBER. Some IRRATIONAL NUMBERS, but *no* TRANSCENDENTAL NUMBERS, can be constructed.

It turns out that all constructions possible with a COMPASS and STRAIGHTEDGE can be done with a COMPASS alone, as long as a line is considered constructed when its two endpoints are located. The reverse is also true, since Jacob Steiner showed that all constructions possible with STRAIGHTEDGE and COMPASS can be done using only a straightedge, as long as a fixed CIRCLE and its center (or two intersecting CIRCLES without their centers, or three nonintersecting CIRCLES) have been drawn beforehand. Such a construction is known as a STEINER CONSTRUCTION.

GEOMETROGRAPHY is a quantitative measure of the simplicity of a geometric construction. It reduces geometric constructions to five types of operations, and seeks to reduce the total number of operations (called the "SIMPLICITY"rpar; needed to effect a geometric construction.

Dixon (1991, pp. 34–1) gives approximate constructions for some figures (the HEPTAGON and NONAGON) and lengths (PI) which cannot be rigorously constructed. Ramanujan (1913–4) and Olds (1963) give geometric constructions for $355/113 \approx \pi$. Gardner (1966, pp. 92–3) gives a geometric construction for

$$3 + \frac{16}{113} = 3.1415929\ldots \approx \pi.$$

Kochansky's approximate construction for π yields KOCHANSKY'S APPROXIMATION

$$\sqrt{\frac{40}{3} - 2\sqrt{3}} = 3.141533\ldots \approx \pi$$

Steinhaus (1983, p. 143). Constructions for π are approximate (but inexact) forms of CIRCLE SQUARING.

See also CIRCLE SQUARING, COMPASS, CONSTRUCTIBLE NUMBER, CONSTRUCTIBLE POLYGON, CUBE DUPLICATION, ELEMENTS, FERMAT PRIME, GEOMETRIC PROBLEMS OF ANTIQUITY, GEOMETROGRAPHY, KOCHANSKY'S APPROXIMATION, MASCHERONI CONSTRUCTION, MATCHSTICK CONSTRUCTION, NAPOLEON'S PROBLEM, NEUSIS CONSTRUCTION, PLANE GEOMETRY, POLYGON, PONCELET-STEINER THEOREM, RECTIFICATION, SIMPLICITY, STEINER CONSTRUCTION, STRAIGHTEDGE, TRISECTION

References

Ball, W. W. R. and Coxeter, H. S. M. *Mathematical Recreations and Essays, 13th ed.* New York: Dover, pp. 96–7, 1987.

Bold, B. "Achievement of the Ancient Greeks" and "An Analytic Criterion for Constructibility." Chs. 1– in *Famous Problems of Geometry and How to Solve Them.* New York: Dover, pp. 1–7, 1982.

Conway, J. H. and Guy, R. K. *The Book of Numbers.* New York: Springer-Verlag, pp. 191–02, 1996.

Coolidge, J. L. "Famous Problems in Construction." Ch. 3 in *A Treatise on the Geometry of the Circle and Sphere.* New York: Chelsea, pp. 166–88, 1971.

Courant, R. and Robbins, H. "Geometric Constructions. The Algebra of Number Fields." Ch. 3 in *What is Mathematics?: An Elementary Approach to Ideas and Methods, 2nd ed.* Oxford, England: Oxford University Press, pp. 117–64, 1996.

Dantzig, T. *Number, The Language of Science.* New York: Macmillan, p. 316, 1954.

Dickson, L. E. "Constructions with Ruler and Compasses; Regular Polygons." Ch. 8 in *Monographs on Topics of Modern Mathematics Relevant to the Elementary Field* (Ed. J. W. A. Young). New York: Dover, pp. 352–86, 1955.

Dixon, R. *Mathographics.* New York: Dover, 1991.

Dummit, D. S. and Foote, R. M. "Classical Straightedge and Compass Constructions." §13.3 in *Abstract Algebra, 2nd ed.* Englewood Cliffs, NJ: Prentice-Hall, pp. 443–48, 1998.

Eppstein, D. "Geometric Models." http://www.ics.uci.edu/~eppstein/junkyard/model.html.

Gardner, M. "The Transcendental Number Pi." Ch. 8 in *Martin Gardner's New Mathematical Diversions from Scientific American.* New York: Simon and Schuster, pp. 91–02, 1966.

Gardner, M. "Mascheroni Constructions." Ch. 17 in *Mathematical Circus: More Puzzles, Games, Paradoxes and Other Mathematical Entertainments from Scientific American.* New York: Knopf, pp. 216–31, 1979.

Harris, J. W. and Stocker, H. "Basic Constructions." §3.2 in *Handbook of Mathematics and Computational Science.* New York: Springer-Verlag, pp. 60–2, 1998.

Herterich, K. *Die Konstruktion von Dreiecken.* Stuttgart: Ernst Klett Verlag, 1986.

Krötenheerdt, O. "Zur Theorie der Dreieckskonstruktionen." *Wissenschaftliche Zeitschrift der Martin-Luther-Univ. Halle-Wittenberg, Math. Naturw. Reihe* **15**, 677–00, 1966.

Meyers, L. F. "Update on William Wernick's 'Triangle Constructions with Three Located Points.'" *Math. Mag.* **69**, 46–9, 1996.

Olds, C. D. *Continued Fractions.* New York: Random House, pp. 59–0, 1963.

Petersen, J. *Methods and Theories for the Solution of Problems of Geometrical Constructions Applied to 410 Problems.* New York: Stechert, 1923. Reprinted in *String Figures and Other Monographs.* New York: Chelsea, 1960.

Plouffe, S.. "The Computation of Certain Numbers Using a Ruler and Compass." *J. Integer Sequences* **1**, No. 98.1.3, 1998. http://www.research.att.com/~njas/sequences/JIS/compass.html.

Posamentier, A. S. and Wernick, W. *Advanced Geometric Constructions.* Palo Alto, CA: Dale Seymour, 1988.

Ramanujan, S. "Modular Equations and Approximations to π." *Quart. J. Pure. Appl. Math.* **45**, 350–72, 1913–914.

Smogorzhevskii, A. S. *The Ruler in Geometrical Constructions.* New York: Blaisdell, 1961.

Steinhaus, H. *Mathematical Snapshots, 3rd ed.* New York: Dover, 1999.

Sykes, M. *Source Book of Problems for Geometry.* Palo Alto, CA: Dale Seymour, 1997.

Weisstein, E. W. "Books about Geometric Construction." http://www.treasure-troves.com/books/GeometricConstruction.html.

Wernick, W. "Triangle Constructions with Three Located Points." *Math. Mag.* **55**, 227–30, 1982.

Geometric Distribution

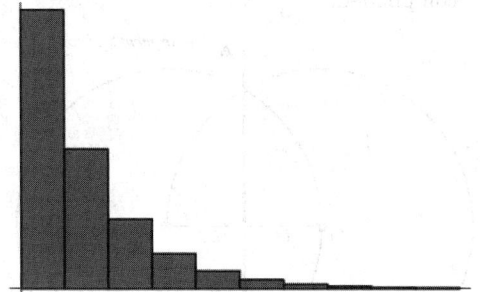

A DISCRETE DISTRIBUTION for $n = 1, 2, \ldots$ with probability function

$$P(n) = q^{n-1}P \tag{1}$$

$$= p(1-p)^{n-1}, \tag{2}$$

where $0 < p < 1$ and $(q \equiv 1 - p)$. $P(n)$ is normalized, since

$$\sum_{n=1}^{\infty} P(n) = \sum_{n=1}^{\infty} q^{n-1}p = p \sum_{n=0}^{\infty} q^n = \frac{p}{1-q} = \frac{p}{p} = 1 \tag{3}$$

The corresponding DISTRIBUTION FUNCTION is

$$D(n) = \sum_{k=1}^{n} P(k) = 1 - q^n. \tag{4}$$

The MOMENT-GENERATING FUNCTION is given by

$$\phi(t) = p\left[1 - (1-p)e^{it}\right]^{-1}, \quad (5)$$

or

$$M(t) = \langle e^{tn} \rangle = \sum_{n=1}^{\infty} e^{tn} pq^{n-1} = p\sum_{n=0}^{\infty} e^{t(n+1)} q^n$$

$$= pe^t \sum_{n=0}^{\infty} (e^t)^n = \frac{pe^t}{1 - e^t q} \quad (6)$$

$$M'(t) = \frac{pe^t}{(1 - e^t q)^2} \quad (7)$$

$$M''(t) = \frac{pe^t(1 + qe^t)}{(1 - e^t q)^3} \quad (8)$$

$$M'''(t) = \frac{pe^t\left[1 + 4e^t(1-p) + e^{2t}(1-p)^2\right]}{(1 - e^t + e^t p)^4}. \quad (9)$$

Therefore, the RAW MOMENTS are

$$M'(0) = \mu_1' = \mu = \frac{p}{(1-q)^2} = \frac{p}{p^2} = \frac{1}{p} \quad (10)$$

$$M''(0) = \mu_2' = \frac{p(1+q)}{(1-q)^3} = \frac{p(2-p)}{p^3} = \frac{2-p}{p^2} \quad (11)$$

$$M'''(0) = \mu_3' = \frac{(6 - 6p + p^2)}{p^3} \quad (12)$$

$$M^4(0) = \mu_4' = \frac{(p-2)(-p^2 + 12p - 12)}{p^4}, \quad (13)$$

giving CENTRAL MOMENTS

$$\mu_2 = \frac{q}{p^2} \quad (14)$$

$$\mu_3 = \frac{(p-1)(p-2)}{p^3} \quad (15)$$

$$\mu_4 = \frac{(p-1)(-p^2 + 9p - 9)}{p^4}, \quad (16)$$

so the MEAN, VARIANCE, SKEWNESS, and KURTOSIS are

given by

$$\mu \equiv \mu_1' = \frac{1}{p} \quad (17)$$

$$\sigma^2 = \mu_2 = \frac{q}{p^2} \quad (18)$$

$$\gamma_1 = \frac{\mu_3}{\mu_2^{3/2}} = \frac{2-p}{\sqrt{q}} \quad (19)$$

$$\gamma_2 = \frac{\mu_4}{\mu_2^2} - 3 = \frac{p^2 - 6p + 6}{1-p}. \quad (20)$$

In fact, the moments of the distribution are given analytically in terms of the POLYLOGARITHM function,

$$\mu_k' \equiv \sum_{n=1}^{\infty} p(n)n^k = \sum_{n=1}^{\infty} p(1-p)^{n-1}n^k$$

$$= \frac{p\mathrm{Li}_{-k}(1-p)}{1-p}. \quad (21)$$

For the case $p = 1/2$ (corresponding to the distribution of the number of COIN TOSSES needed to win in the SAINT PETERSBURG PARADOX) the formula (21) gives

$$\mu_k'|_{p=1/2} = \mathrm{Li}_{-k}\left(\tfrac{1}{2}\right). \quad (22)$$

The first few raw moments are therefore 2, 6, 26, 150, 1082, ... (Sloane's A000629), which have EXPONENTIAL GENERATING FUNCTIONS $f(x) = -\ln(2 - e^x)$ and $g(x) = e^x/(2 - e^x)$. From (22), the MEAN, VARIANCE, SKEWNESS, and KURTOSIS are

$$\mu = 2 \quad (23)$$

$$\sigma^2 = 2 \quad (24)$$

$$\gamma_1 = \tfrac{3}{2}\sqrt{2} \quad (25)$$

$$\gamma_2 = \tfrac{13}{2}. \quad (26)$$

The first CUMULANT of the geometric distribution is

$$\kappa_1 = \frac{1-p}{p}, \quad (27)$$

and subsequent CUMULANTS are given by the RECURRENCE RELATION

$$\kappa_{r+1} = (1-p)\frac{d\kappa_r}{dp} \quad (28)$$

See also SAINT PETERSBURG PARADOX

References
Beyer, W. H. *CRC Standard Mathematical Tables, 28th ed.* Boca Raton, FL: CRC Press, pp. 531–32, 1987.

Sloane, N. J. A. Sequences A000629 in "An On-Line Version of the Encyclopedia of Integer Sequences." http://www.research.att.com/~njas/sequences/eisonline.html.
Spiegel, M. R. *Theory and Problems of Probability and Statistics.* New York: McGraw-Hill, p. 118, 1992.

Geometric Dual Graph

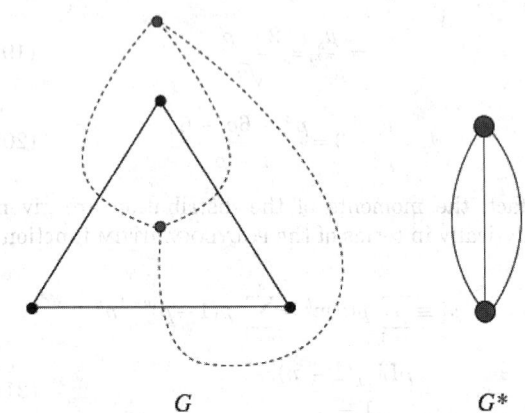

$$G \qquad\qquad G^*$$

Given a PLANAR GRAPH G, its geometric dual G^* is constructed by placing a vertex in each region of G (including the exterior region) and, if two regions have an edge x in common, joining the corresponding vertices by an edge X^* crossing only x. The result is always a planar PSEUDOGRAPH. However, an abstract graph with more than one embedding on the sphere can give rise to more than one dual.

Whitney showed that the geometric dual graph and COMBINATORIAL DUAL GRAPH are equivalent (Harary 1994, p. 115), and so may simply be called "the" DUAL GRAPH.

See also COMBINATORIAL DUAL GRAPH, DUAL GRAPH

References

Harary, F. *Graph Theory.* Reading, MA: Addison-Wesley, pp. 113–15, 1994.

Geometric Genus

GENUS (SURFACE)

Geometric Invariant Theory

INVARIANT

Geometric Mean

The geometric mean of a sequence $\{a_i\}_{i=1}^n$ is defined by

$$G(a_1, \ldots, a_n) \equiv \left(\prod_{i=1}^n a_i\right)^{1/n}. \qquad (1)$$

Thus,

$$G(a_1, a_2) = \sqrt{a_1 a_2} \qquad (2)$$

$$G(a_1, a_2, a_3) = (a_1 a_2 a_3)^{1/3}, \qquad (3)$$

and so on.

Hoehn and Niven (1985) show that

$$\begin{aligned} G(a_1 + c, &\ a_2 + c,\ \ldots,\ a_n + c) \\ &> c + G(a_1, a_2, \ldots, a_n) \end{aligned} \qquad (4)$$

for any POSITIVE constant c.

See also ARITHMETIC MEAN, ARITHMETIC-GEOMETRIC MEAN, CARLEMAN'S INEQUALITY, HARMONIC MEAN, MEAN, ROOT-MEAN-SQUARE

References

Abramowitz, M. and Stegun, C. A. (Eds.). *Handbook of Mathematical Functions with Formulas, Graphs, and Mathematical Tables, 9th printing.* New York: Dover, p. 10, 1972.
Hoehn, L. and Niven, I. "Averages on the Move." *Math. Mag.* **58**, 151–56, 1985.
Kenney, J. F. and Keeping, E. S. "Geometric Mean." §4.10 in *Mathematics of Statistics, Pt. 1, 3rd ed.* Princeton, NJ: Van Nostrand, pp. 54–5, 1962.
Zwillinger, D. (Ed.). *CRC Standard Mathematical Tables and Formulae.* Boca Raton, FL: CRC Press, p. 602, 1995.

Geometric Mean Index

The statistical INDEX

$$P_G \equiv \left[\prod \left(\frac{p_n}{p_0}\right)^{v_0}\right]^{1/\Sigma\, v_0},$$

where p_n is the price per unit in period n, q_n is the quantity produced in period n, and $v_n \equiv p_n q_n$ the value of the n units.

See also INDEX

References

Kenney, J. F. and Keeping, E. S. *Mathematics of Statistics, Pt. 1, 3rd ed.* Princeton, NJ: Van Nostrand, p. 69, 1962.

Geometric Modeling

References

Strasser, W.; Klein, R.; and Rau, R. (Eds.). *Geometric Modeling: Theory and Practice, the State of the Art.* Berlin: Springer-Verlag, 1997.

Geometric Probability

The study of the probabilities involved in geometric problems, e.g., the distributions of length, area, volume, etc. for geometric objects under stated conditions.

See also BERTRAND'S PROBLEM, BUFFON-LAPLACE NEEDLE PROBLEM, BUFFON'S NEEDLE PROBLEM, CIRCLE INSCRIBING, COMPUTATIONAL GEOMETRY, INTEGRAL GEOMETRY, POINT PICKING, STOCHASTIC GEOMETRY, SYLVESTER'S FOUR-POINT PROBLEM

Geometric Problems of Antiquity

References

Ambartzumian, R. V. (Ed.). *Stochastic and Integral Geometry.* Dordrecht, Netherlands: Reidel, 1987.

Isaac, R. *The Pleasures of Probability.* New York: Springer-Verlag, 1995.

Kendall, M. G. and Moran, P. A. P. *Geometric Probability.* New York: Hafner, 1963.

Kendall, W. S.; Barndorff-Nielson, O.; and van Lieshout, M. C. *Current Trends in Stochastic Geometry: Likelihood and Computation.* Boca Raton, FL: CRC Press, 1998.

Klain, D. A. and Rota, G.-C. *Introduction to Geometric Probability.* New York: Cambridge University Press, 1997.

Santaló, L. A. *Introduction to Integral Geometry.* Paris: Hermann, 1953.

Santaló, L. A. *Integral Geometry and Geometric Probability.* Reading, MA: Addison-Wesley, 1976.

Solomon, H. *Geometric Probability.* Philadelphia, PA: SIAM, 1978.

Stoyan, D.; Kendall, W. S.; and Mecke, J. *Stochastic Geometry and Its Applications, with a Foreword by D. G. Kendall.* New York: Wiley, 1987.

Weisstein, E. W. "Books about Geometric Probability." http://www.treasure-troves.com/books/GeometricProbability.html.

Geometric Problems of Antiquity

The Greek problems of antiquity were a set of geometric problems whose solution was sought using only COMPASS and STRAIGHTEDGE:

1. CIRCLE SQUARING.
2. CUBE DUPLICATION.
3. TRISECTION of an ANGLE.

Only in modern times, more than 2,000 years after they were formulated, were all three ancient problems proved insoluble using only COMPASS and STRAIGHTEDGE.

Another ancient geometric problem not proved impossible until 1997 is ALHAZEN'S BILLIARD PROBLEM. As Ogilvy (1990) points out, constructing the general REGULAR POLYHEDRON was really a "fourth" unsolved problem of antiquity.

See also ALHAZEN'S BILLIARD PROBLEM, CIRCLE SQUARING, COMPASS, CONSTRUCTIBLE NUMBER, CONSTRUCTIBLE POLYGON, CUBE DUPLICATION, GEOMETRIC CONSTRUCTION, REGULAR POLYHEDRON, STRAIGHTEDGE, TRISECTION

References

Conway, J. H. and Guy, R. K. "Three Greek Problems." In *The Book of Numbers.* New York: Springer-Verlag, pp. 190–91, 1996.

Courant, R. and Robbins, H. "The Unsolvability of the Three Greek Problems." §3.3 in *What is Mathematics?: An Elementary Approach to Ideas and Methods, 2nd ed.* Oxford, England: Oxford University Press, pp. 117–18 and 134–40, 1996.

Ogilvy, C. S. *Excursions in Geometry.* New York: Dover, pp. 135–38, 1990.

Pappas, T. "The Impossible Trio." *The Joy of Mathematics.* San Carlos, CA: Wide World Publ./Tetra, pp. 130–32, 1989.

Jones, A.; Morris, S.; and Pearson, K. *Abstract Algebra and Famous Impossibilities.* New York: Springer-Verlag, 1991.

Stoschek, E. "Modul 41 Literatur." http://marvin.sn.schule.de/~inftreff/modul41/lit41.htm.

Stoschek, E. "Modul 41. Three Geometric Problems of Antiquity: Their Approximate Solutions in Automata Representation--Integrated Control Processors for Nanotechnology." http://marvin.sn.schule.de/~inftreff/modul41/task41.htm.

Geometric Progression

GEOMETRIC SEQUENCE

Geometric Realization

If the ABSTRACT SIMPLICIAL COMPLEX S is isomorphic with the VERTEX SCHEME of the SIMPLICIAL COMPLEX K, then K is said to be a geometric realization of S, and is uniquely determined up to a linear isomorphism.

See also ABSTRACT SIMPLICIAL COMPLEX, VERTEX SCHEME

References

Munkres, J. R. *Elements of Algebraic Topology.* Perseus Press, 1993.

Geometric Sequence

A geometric sequence is a SEQUENCE $\{a_k\}$, $k = 1, 2, ...,$ such that each term is given by a multiple r of the previous one. Another equivalent definition is that a sequence is geometric IFF it has a zero BIAS. If the multiplier is r, then the kth term is given by

$$a_k = ra_{k-1} = r^2 a_{k-2} = a_0 r^k.$$

Without loss of generality, take $a_0 = 1$, giving

$$a_k = r^k.$$

Geometric Series

A geometric series $\Sigma_k a_k$ is a series for which the ratio of each two consecutive terms a_{k+1}/a_k is a constant function of the summation index k. The more general case of the ratio a RATIONAL FUNCTION of the summation index k produces a series called a HYPERGEOMETRIC SERIES.

For the simplest case of the ratio $a_{k+1}/a_k = r$ equal to a constant r, the terms a_k are OF THE FORM $a_k = a_0 r^k$. Letting $a_0 = 1$, the GEOMETRIC SEQUENCE $\{a_k\}_{k=0}^n$ with constant $|r| < 1$ is given by

$$S_n = \sum_{k=0}^{n} a_k = \sum_{k=0}^{n} r^k \qquad (1)$$

is given by

$$S_n \equiv \sum_{k=0}^{n} r^k = 1 + r + r^2 + \ldots + r^n. \qquad (2)$$

Multiplying both sides by r gives

$$rS_n = r + r^2 + r^3 + \ldots + r^{n+1}, \qquad (3)$$

and subtracting (3) from (2) then gives

$$(1-r)S_n = (1 + r + r^2 + \ldots r^n)$$
$$- (r + r^2 + r^3 + \ldots + r^{n+1})$$
$$= 1 - r^{n+1}, \qquad (4)$$

so

$$S_n \equiv \sum_{k=0}^{n} r^k = \frac{1 - r^{n+1}}{1 - r}. \qquad (5)$$

For $-1 < r < 1$, the sum converges as $n \to \infty$, in which case

$$S \equiv S_\infty = \sum_{k=0}^{\infty} r^k = \frac{1}{1 - r} \qquad (6)$$

Similarly, if the sums are taken starting at $k = 1$ instead of $k = 0$,

$$\sum_{k=1}^{n} r^k = \frac{r(1 - r^n)}{1 - r} \qquad (7)$$

$$\sum_{k=1}^{\infty} r^k = \frac{r}{1 - r}, \qquad (8)$$

the latter of which is valid for $|r| < 1$.

See also ARITHMETIC SERIES, GABRIEL'S STAIRCASE, HARMONIC SERIES, HYPERGEOMETRIC SERIES, ST. IVES PROBLEM, WHEAT AND CHESSBOARD PROBLEM

References
Abramowitz, M. and Stegun, C. A. (Eds.). *Handbook of Mathematical Functions with Formulas, Graphs, and Mathematical Tables, 9th printing.* New York: Dover, p. 10, 1972.
Arfken, G. *Mathematical Methods for Physicists, 3rd ed.* Orlando, FL: Academic Press, pp. 278–79, 1985.
Beyer, W. H. *CRC Standard Mathematical Tables, 28th ed.* Boca Raton, FL: CRC Press, p. 8, 1987.
Courant, R. and Robbins, H. "The Geometric Progression." §1.2.3 in *What is Mathematics?: An Elementary Approach to Ideas and Methods, 2nd ed.* Oxford, England: Oxford University Press, pp. 13–4, 1996.
Pappas, T. "Perimeter, Area & the Infinite Series." *The Joy of Mathematics.* San Carlos, CA: Wide World Publ./Tetra, pp. 134–35, 1989.

Geometrization Conjecture
THURSTON'S GEOMETRIZATION CONJECTURE

Geometrography
A quantitative measure of the simplicity of a GEOMETRIC CONSTRUCTION which reduces geometric constructions to five steps. It was devised by È. Lemoine.

S_1 Place a STRAIGHTEDGE'S EDGE through a given POINT,
S_2 Draw a straight LINE,
C_1 Place a POINT of a COMPASS on a given POINT,
C_2 Place a POINT of a COMPASS on an indeterminate POINT on a LINE,
C_3 Draw a CIRCLE.

Geometrography seeks to reduce the number of operations (called the "SIMPLICITY") needed to effect a construction. If the number of the above operations are denoted m_1, m_2, n_1, n_2, and n_3, respectively, then the SIMPLICITY is $m_1 + m_2 + n_1 + n_2 + n_3$ and the symbol is $m_1 S_1 + m_2 S_2 + n_1 C_1 + n_2 C_2 + n_3 C_3$. It is apparently an unsolved problem to determine if a given GEOMETRIC CONSTRUCTION is of the smallest possible simplicity.

See also SIMPLICITY

References
De Temple, D. W. "Carlyle Circles and the Lemoine Simplicity of Polygonal Constructions." *Amer. Math. Monthly* **98**, 97–08, 1991.
Eves, H. *An Introduction to the History of Mathematics, 6th ed.* New York: Holt, Rinehart, and Winston, 1990.

Geometry
Geometry is the study of figures in a SPACE of a given number of dimensions and of a given type. The most common types of geometry are PLANE GEOMETRY (dealing with objects like the LINE, CIRCLE, TRIANGLE, and POLYGON), SOLID GEOMETRY (dealing with objects like the LINE, SPHERE, and POLYHEDRON), and SPHERICAL GEOMETRY (dealing with objects like the SPHERICAL TRIANGLE and SPHERICAL POLYGON). Geometry was part of the QUADRIVIUM taught in medieval universities.

Historically, the study of geometry proceeds from a small number of accepted truths (AXIOMS or POSTULATES), then builds up true statements using a systematic and rigorous step-by-step PROOF. However, there is much more to geometry than this relatively dry textbook approach, as evidenced by some of the beautiful and unexpected results of PROJECTIVE GEOMETRY (not to mention Schubert's powerful but questionable ENUMERATIVE GEOMETRY).

The late mathematician E. T. Bell has described geometry as follows (Coxeter and Greitzer 1967, p. 1): "With a literature much vaster than those of ALGEBRA and ARITHMETIC combined, and at least as extensive as that of ANALYSIS, geometry is a richer treasure house of more interesting and half-forgotten things, which a hurried generation has no leisure to enjoy, than any other division of mathematics." While the literature of ALGEBRA, ARITHMETIC, and ANALYSIS has grown extensively since Bell's day, the remainder of his commentary holds even more so today.

Formally, a geometry is defined as a complete locally homogeneous RIEMANNIAN METRIC. In $\mathbb{R}^2$, the possible geometries are Euclidean planar, hyperbolic planar, and elliptic planar. In $\mathbb{R}^3$, the possible geometries include Euclidean, hyperbolic, and elliptic, but also include five other types.

See also ABSOLUTE GEOMETRY, AFFINE GEOMETRY, CARTESIAN COORDINATES, COMBINATORIAL GEOMETRY, COMPUTATIONAL GEOMETRY, COORDINATE GEOMETRY, DIFFERENTIAL GEOMETRY, DISCRETE GEOMETRY, ENUMERATIVE GEOMETRY, FINSLER GEOMETRY, INVERSIVE GEOMETRY, KAWAGUCHI GEOMETRY, MINKOWSKI GEOMETRY, NIL GEOMETRY, NON-EUCLIDEAN GEOMETRY, ORDERED GEOMETRY, PLANE GEOMETRY, PROJECTIVE GEOMETRY, SOL GEOMETRY, SOLID GEOMETRY, SPHERICAL GEOMETRY, STOCHASTIC GEOMETRY, THURSTON'S GEOMETRIZATION CONJECTURE

References

Altshiller-Court, N. *College Geometry: A Second Course in Plane Geometry for Colleges and Normal Schools, 2nd ed., rev. enl.* New York: Barnes and Noble, 1952.

Bold, B. *Famous Problems of Geometry and How to Solve Them.* New York: Dover, 1964.

Brown, K. S. "Geometry." http://www.seanet.com/~ksbrown/igeometr.htm.

Cinderella, Inc. "Cinderella: The Interactive Geometry Software." http://www.cinderella.de/.

Coxeter, H. S. M. *Introduction to Geometry, 2nd ed.* New York: Wiley, 1969.

Coxeter, H. S. M. *The Beauty of Geometry: Twelve Essays.* New York: Dover, 1999.

Coxeter, H. S. M. and Greitzer, S. L. *Geometry Revisited.* Washington, DC: Math. Assoc. Amer., 1967.

Croft, H. T.; Falconer, K. J.; and Guy, R. K. *Unsolved Problems in Geometry.* New York: Springer-Verlag, 1994.

Davis, C.; Grünbaum, B.; and Scherk, F.A. *The Geometric Vein: The Coxeter Festschrift.* New York: Springer, 1981.

Eppstein, D. "Geometry Junkyard." http://www.ics.uci.edu/~eppstein/junkyard/.

Eppstein, D. "Many-Dimensional Geometry." http://www.ics.uci.edu/~eppstein/junkyard/highdim.html.

Eppstein, D. "Planar Geometry." http://www.ics.uci.edu/~eppstein/junkyard/2d.html.

Eppstein, D. "Three-Dimensional Geometry." http://www.ics.uci.edu/~eppstein/junkyard/3d.html.

Eves, H. W. *A Survey of Geometry, rev. ed.* Boston, MA: Allyn and Bacon, 1972.

Ghyka, M. C. *The Geometry of Art and Life, 2nd ed.* New York: Dover, 1977.

Hilbert, D. *The Foundations of Geometry, 2nd ed.* Chicago, IL: The Open Court Publishing Co., 1921.

Ivins, W. M. *Art and Geometry.* New York: Dover, 1964.

Johnson, R. A. *Modern Geometry: An Elementary Treatise on the Geometry of the Triangle and the Circle.* Boston, MA: Houghton Mifflin, 1929.

King, J. and Schattschneider, D. (Eds.). *Geometry Turned On: Dynamic Software in Learning, Teaching and Research.* Washington, DC: Math. Assoc. Amer., 1997.

Klee, V. "Some Unsolved Problems in Plane Geometry." *Math. Mag.* **52**, 131–45, 1979.

Klein, F. *Famous Problems of Elementary Geometry and Other Monographs.* New York: Dover, 1956.

Melzak, Z. A. *Invitation to Geometry.* New York: Wiley, 1983.

Meschkowski, H. *Unsolved and Unsolvable Problems in Geometry.* London: Oliver & Boyd, 1966.

Moise, E. E. *Elementary Geometry from an Advanced Standpoint, 3rd ed.* Reading, MA: Addison-Wesley, 1990.

Ogilvy, C. S. "Some Unsolved Problems of Modern Geometry." Ch. 11 in *Excursions in Geometry.* New York: Dover, pp. 143–53, 1990.

Playfair, J. *Elements of Geometry: Containing the First Six Books of Euclid, with a Supplement on the Circle and the Geometry of Solids to which are added Elements of Plane and Spherical Trigonometry.* New York: W. E. Dean.

Simon, M. *Über die Entwicklung der Elementargeometrie im XIX Jahrhundert.* Berlin, pp. 97–05, 1906.

Townsend, R. *Chapters on the Modern Geometry of the Point, Line, and Circle, 2 vols.* Dublin: Hodges, Smith and Co., 1863.

Uspenskii, V. A. *Some Applications of Mechanics to Mathematics.* New York: Blaisdell, 1961.

Weisstein, E. W. "Books about Geometry." http://www.treasure-troves.com/books/Geometry.html.

Woods, F. S. *Higher Geometry: An Introduction to Advanced Methods in Analytic Geometry.* New York: Dover, 1961.

Geometry of Position

PROJECTIVE GEOMETRY

Gergonne Line

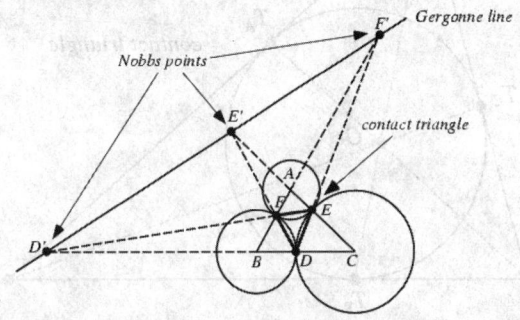

The perspective line for the CONTACT TRIANGLE ΔDEF and its TANGENTIAL TRIANGLE ΔABC. It is determined by the NOBBS POINTS D', E', and F'.

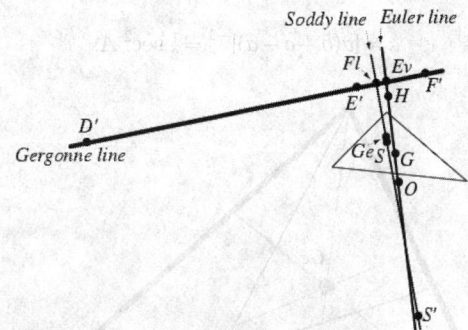

In addition to the NOBBS POINTS, the FLETCHER POINT and EVANS POINT also lie on the Gergonne line where it intersects the SODDY LINE and EULER LINE, respectively. The D and D' coordinates are given by

$$D = B + \frac{f}{e} C$$

$$D' = B - \frac{f}{e} C,$$

so $BDCD'$ form a HARMONIC RANGE. The equation of the Gergonne line is

$$\frac{\alpha}{d} + \frac{\beta}{e} + \frac{\gamma}{f} = 0.$$

See also CONTACT TRIANGLE, EULER LINE, EVANS POINT, FLETCHER POINT, NOBBS POINTS, SODDY LINE, TANGENTIAL TRIANGLE

References

Oldknow, A. "The Euler-Gergonne-Soddy Triangle of a Triangle." *Amer. Math. Monthly* **103**, 319–29, 1996.

Gergonne Point

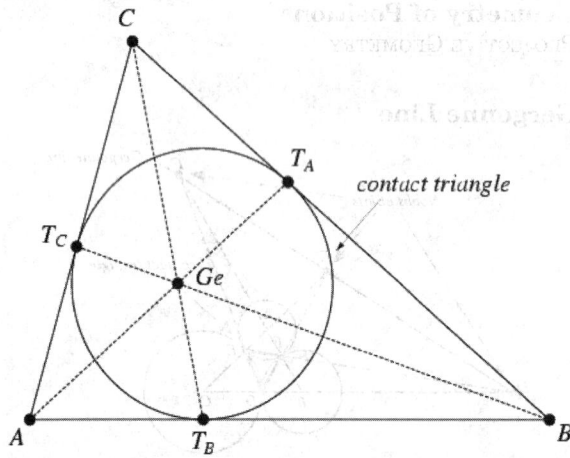

The common point Ge of the CONCURRENT lines from the CONTACT TRIANGLE TRIANGLE'S INCIRCLE to the opposite VERTICES. It has TRIANGLE CENTER FUNCTION

$$\alpha = [a(b+c-a)]^{-1} = \tfrac{1}{2} \sec^2 A.$$

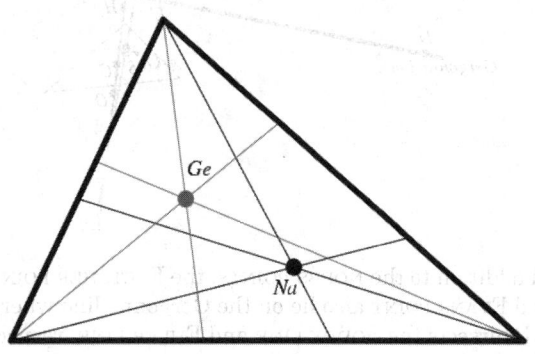

The Gergonne point Ge is the ISOTOMIC CONJUGATE POINT of the NAGEL POINT Na. The CONTACT TRIANGLE and TANGENTIAL TRIANGLE are perspective from the Gergonne point, and the Gergonne point of a triangle is the SYMMEDIAN POINT of its CONTACT TRIANGLE (Honsberger 1995).

See also ADAMS' CIRCLE, CONTACT TRIANGLE, GERGONNE LINE, NAGEL POINT

References

Altshiller-Court, N. *College Geometry: A Second Course in Plane Geometry for Colleges and Normal Schools*, 2nd ed. New York: Barnes and Noble, pp. 160–64, 1952.
Coxeter, H. S. M. and Greitzer, S. L. *Geometry Revisited.* New York: Random House, pp. 11–3, 1967.
Eves, H. W. *A Survey of Geometry*, rev. ed. Boston, MA: Allyn and Bacon, p. 83, 1972.
Gallatly, W. *The Modern Geometry of the Triangle*, 2nd ed. London: Hodgson, p. 22, 1913.
Honsberger, R. "The Gergonne Point." §7.4 (iv) in *Episodes in Nineteenth and Twentieth Century Euclidean Geometry*. Washington, DC: Math. Assoc. Amer., pp. 61–2, 1995.
Johnson, R. A. *Modern Geometry: An Elementary Treatise on the Geometry of the Triangle and the Circle.* Boston, MA: Houghton Mifflin, pp. 184 and 216, 1929.
Kimberling, C. "Gergonne Point." http://cedar.evansville.edu/~ck6/tcenters/class/gergonne.html.

Gergonne's Theorem

The internal (external) bisecting plane of a DIHEDRAL ANGLE of a TETRAHEDRON divides the opposite edge in the ratio of the areas of the adjacent faces.

References

Altshiller-Court, N. "Gergonne's Theorem." §235 in *Modern Pure Solid Geometry*. New York: Chelsea, p. 71, 1979.
Le Grand, Ferriot, Lambert, *et al.* "Questions Résolues: Démonstrations des deux théorèmes de géométrie énoncés à la page 196 de ce volume." *Ann. de math.* **3**, 317–23, 1812–813.

Germain Primes

SOPHIE GERMAIN PRIME

Gerono Lemniscate

EIGHT CURVE

Gergorin Circle Theorem

Gives a region in the COMPLEX PLANE containing all the EIGENVALUES of a COMPLEX SQUARE MATRIX. Define

$$R_i = \sum_{\substack{i=1 \\ j \neq i}}^{n} |a_{i,j}|, \qquad (1)$$

then each EIGENVALUE of the MATRIX of order n is in at least one of the disks

$$\{z : |z - a_{ii}| \leq R_i\}. \qquad (2)$$

G-Function

The theorem can be made stronger as follows. Let r be an INTEGER with $1 \le r \le n$, then each EIGENVALUE of is either in one of the disks Γ_1

$$\{z : |z - a_{jj}| \le S_j^{(r-1)}\}, \tag{3}$$

or in one of the regions

$$\left\{ z : \sum_{i=1}^{r} |z - a_{ii}| \le \sum_{i=1}^{r} R_i \right\}, \tag{4}$$

where $S_j^{(r-1)}$ is the sum of magnitudes of the $r-1$ largest off-diagonal elements in column j.

References

Brualdi, R. A. and Mellendorf, S. "Regions in the Complex Plane Containing the Eigenvalues of a Matrix." *Amer. Math. Monthly* **101**, 975–85, 1994.

Gradshteyn, I. S. and Ryzhik, I. M. *Tables of Integrals, Series, and Products, 6th ed.* San Diego, CA: Academic Press, pp. 1120–121, 2000.

Piziak, R. and Turner, D. "Exploring Gerschgorin Circles and Cassini Ovals." *Mathematica Educ.* **3**, 13–1, 1994.

Taussky-Todd, O. "A Recurring Theorem on Determinants." *Amer. Math. Monthly* **56**, 672–76, 1949.

G-Function

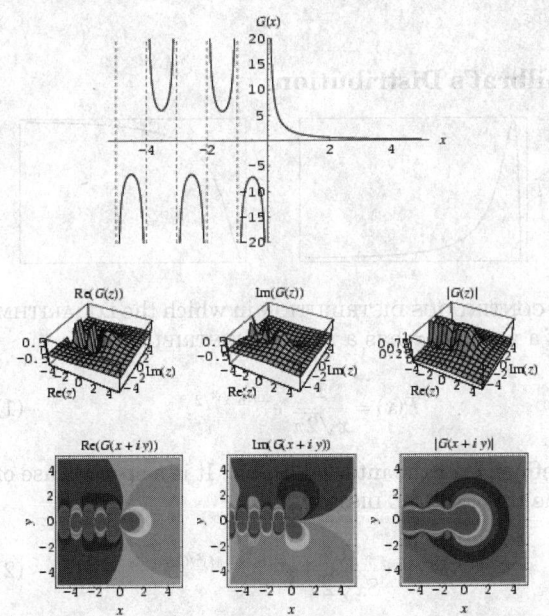

As defined by Erdélyi *et al.* (1981, p. 20), the G-function is given by

$$G(z) \equiv \psi_0(\tfrac{1}{2} + hz) - \psi_0(\tfrac{1}{2}z), \tag{1}$$

where $\psi_0(z)$ is the DIGAMMA FUNCTION. Integral representations are given by

$$G(z) = 2 \int_0^1 \frac{t^{z-1}}{1+t}\, dt \tag{2}$$

$$= 2 \int_0^\infty \frac{e^{-zt}}{1+e^{-t}}\, dt \tag{3}$$

for $\Re[z] > 0$. $G(z)$ is also given by the series

$$G(z) = 2 \sum_{n=0}^{\infty} \frac{(-1)^n}{z+n}, \tag{4}$$

and in terms of the HYPERGEOMETRIC FUNCTION by

$$G(z) = 2z^{-1}\,{}_2F_1(1, z;\ 1+z;\ -1). \tag{5}$$

It obeys the functional relations

$$G(1+z) = 2z^{-1} - G(z) \tag{6}$$

$$G(1-z) = 2\pi \csc(\pi z) - G(z) \tag{7}$$

$$G(mz) = \begin{cases} -\dfrac{2}{m} \displaystyle\sum_{r=0}^{m-1} (-1)^r \psi_0(z + \frac{r}{m}) & \text{for } m \text{ even} \\[2ex] \dfrac{1}{m} \displaystyle\sum_{r=0}^{m-1} (-1)^r G(z + \frac{r}{m}) & \text{for } m \text{ odd.} \end{cases} \tag{8}$$

See also BARNES' G-FUNCTION, DIGAMMA FUNCTION, MEIJER'S G-FUNCTION, RAMANUJAN g- AND G-FUNCTIONS

References

Erdélyi, A.; Magnus, W.; Oberhettinger, F.; and Tricomi, F. G. "The Function $G(z)$." §1.8 in *Higher Transcendental Functions, Vol. 1.* New York: Krieger, pp. 20 and 44–6, 1981.

Ghost

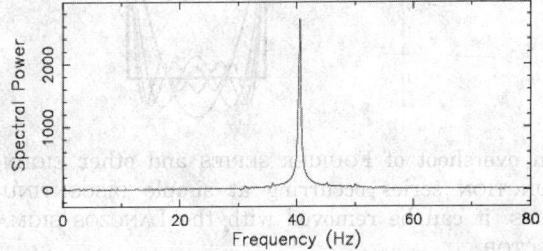

If the sampling of an interferogram is modulated at a definite frequency instead of being uniformly sampled, spurious spectral features called "ghosts" are produced (Brault 1985). Periodic ruling or sampling errors introduce a modulation superposed on top of the expected fringe pattern due to uniform stage translation. Because modulation is a multiplicative process, spurious features are generated in spectral space at the sum and difference of the true fringe and ghost fringe frequencies, thus throwing power out of its spectral band.

Ghosts are copies of the actual spectrum, but appear at reduced strength. The above shows the power spectrum for a pure sinusoidal signal sampled by translating a Fourier transform spectrometer mirror at constant speed. The small blips on either side of the main peaks are ghosts.

In order for a ghost to appear, the process producing it must exist for most of the interferogram. However, if the ruling errors are not truly sinusoidal but vary across the length of the screw, a longer travel path can reduce their effect.

See also JITTER

References

Brault, J. W. "Fourier Transform Spectroscopy." In *High Resolution in Astronomy: 15th Advanced Course of the Swiss Society of Astronomy and Astrophysics* (Ed. A. Benz, M. Huber, and M. Mayor). Geneva Observatory, Sauverny, Switzerland, 1985.

Gibbs Constant

WILBRAHAM-GIBBS CONSTANT

Gibbs Effect

GIBBS PHENOMENON

Gibbs Phenomenon

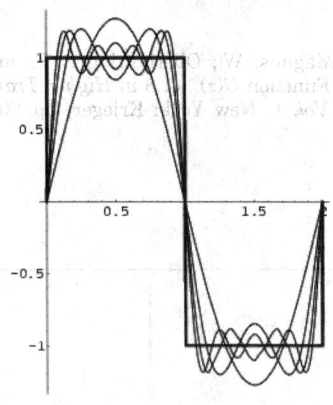

An overshoot of FOURIER SERIES and other EIGENFUNCTION series occurring at simple DISCONTINUITIES. it can be removed with the LANCZOS SIGMA FACTOR.

See also FOURIER SERIES

References

Arfken, G. "Gibbs Phenomenon." §14.5 in *Mathematical Methods for Physicists, 3rd ed.* Orlando, FL: Academic Press, pp. 783–87, 1985.
Foster, J. and Richards, F. B. "The Gibbs Phenomenon for Piecewise-Linear Approximation." *Amer. Math. Monthly* **98**, 47–9, 1991.
Gibbs, J. W. "Fourier Series." *Nature* **59**, 200 and 606, 1899.

Hewitt, E. and Hewitt, R. "The Gibbs-Wilbraham Phenomenon: An Episode in Fourier Analysis." *Arch. Hist. Exact Sci.* **21**, 129–60, 1980.
Jeffreys, H. and Jeffreys, B. S. "The Gibbs Phenomenon." §14.07 in *Methods of Mathematical Physics, 3rd ed.* Cambridge, England: Cambridge University Press, pp. 445–46, 1988.
Sansone, G. "Gibbs' Phenomenon." §2.10 in *Orthogonal Functions, rev. English ed.* New York: Dover, pp. 141–48, 1991.

Gift Wrap Theorem

No subspace of $\mathbb{R}^n$ can be homeomorphic to $\mathbb{S}^n$.

References

Dodson, C. T. J. and Parker, P. E. *A User's Guide to Algebraic Topology.* Dordrecht, Netherlands: Kluwer, p. 121, 1997.

Gigantic Prime

A PRIME with 10,000 or more decimal digits. As of Nov. 15, 1995, 127 were known.

See also TITANIC PRIME

References

Caldwell, C. "The Ten Largest Known Primes." http://www.utm.edu/research/primes/largest.html#largest.

Gilbrat's Distribution

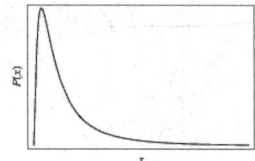
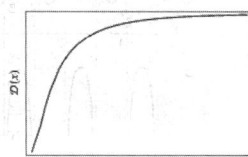

A CONTINUOUS DISTRIBUTION in which the LOGARITHM of a variable x has a NORMAL DISTRIBUTION,

$$P(x) = \frac{1}{x\sqrt{2\pi}}\, e^{-(\ln x)^2/2}, \tag{1}$$

defined over the interval $[0, \infty)$. It is a special case of the LOG NORMAL DISTRIBUTION

$$P(x) = \frac{1}{Sx\sqrt{2\pi}}\, e^{-(\ln x - M)^2/(2S^2)} \tag{2}$$

with $S = 1$ and $M = 0$, and so has distribution function

$$D(x) = \frac{1}{2}\left[1 + \operatorname{erf}\left(\frac{\ln x}{\sqrt{2}}\right)\right]. \tag{3}$$

The MEAN, VARIANCE, SKEWNESS, and KURTOSIS are

then given by

$$\mu = \sqrt{e} \qquad (4)$$

$$\sigma^2 = e(e-1) \qquad (5)$$

$$\gamma_1 = (e+2)\sqrt{e-1} \qquad (6)$$

$$\gamma_2 = e^4 + 2e^3 + 3e^2 - 3. \qquad (7)$$

See also LOG NORMAL DISTRIBUTION

Gilbreath's Conjecture

Let the DIFFERENCE of successive PRIMES be defined by $d_n \equiv p_{n+1} - p_n$, and d_n^k by

$$d_n^k \equiv \begin{cases} d_n & \text{for } k = 1 \\ |d_{n+1}^{k-1} - d_n^{k-1}| & \text{for } k > 1. \end{cases}$$

N. L. Gilbreath claimed that $d_1^k = 1$ for all k (Guy 1994). It has been verified for $k < 63{,}419$ and all PRIMES up to $\pi(10^{13})$, where $\pi(x)$ is the PRIME COUNTING FUNCTION.

See also PRIME DIFFERENCE FUNCTION

References

Gardner, M. "Patterns in Primes are a Clue to the Strong Law of Small Numbers." *Sci. Amer.* **243**, 18–8, Dec. 1980.
Guy, R. K. "Gilbreath's Conjecture." §A10 in *Unsolved Problems in Number Theory, 2nd ed.* New York: Springer-Verlag, pp. 25–6, 1994.
Kilgrove, R. B. and Ralston, K. E. "On a Conjecture Concerning the Primes." *Math. Tables Aids Comput.* **13**, 121–22, 1959.

Gill's Method

A formula for numerical solution of differential equations,

$$y_{n+1} = y_n + \tfrac{1}{6}[k_1 + (2-\sqrt{2})k_2 + (2+\sqrt{2})k_3 + k_4]$$
$$+ \mathcal{O}(h^5),$$

where

$$k_1 = hf(x_n, y_n)$$

$$k_2 = hf(x_n + \tfrac{1}{2}h, \ y_n + \tfrac{1}{2}k_1)$$

$$k_3 = hf[x_n + \tfrac{1}{2}h, \ y_n + \tfrac{1}{2}(-1+\sqrt{2})k_1 + (1 - \tfrac{1}{2}\sqrt{2})k_2]$$

$$k_4 = hf[x_n + h, \ y_n - \tfrac{1}{2}\sqrt{2}k_2 + (1 + \tfrac{1}{2}\sqrt{2})k_3].$$

See also ADAMS' METHOD, MILNE'S METHOD, PREDICTOR-CORRECTOR METHODS, RUNGE-KUTTA METHOD

References

Abramowitz, M. and Stegun, C. A. (Eds.). *Handbook of Mathematical Functions with Formulas, Graphs, and Mathematical Tables, 9th printing.* New York: Dover, p. 896, 1972.

Gingerbreadman Map

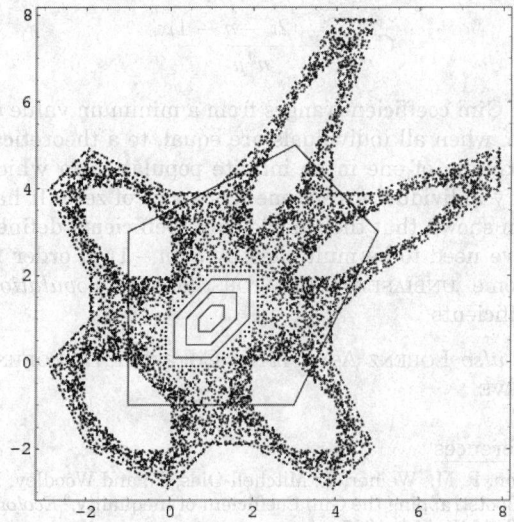

A 2-D piecewise linear MAP defined by

$$x_{n+1} = 1 - y_n + |x_n|$$

$$y_{n+1} = x_n.$$

The map is chaotic in the filled region above and stable in the six hexagonal regions. Each point in the interior hexagon defined by the vertices (0, 0), (1, 0), (2, 1), (2, 2), (1, 2), and (0, 1) has an orbit with period six (except the point (1, 1), which has period 1). Orbits in the other five hexagonal regions circulate from one to the other. There is a unique orbit of period five, with all others having period 30. The points having orbits of period five are (-1, 3), (-1, -1), (3, -1), (5, 3), and (3, 5), indicated in the above figure by the black line. However, there are infinitely many distinct periodic orbits which have an arbitrarily long period.

References

Devaney, R. L. "A Piecewise Linear Model for the Zones of Instability of an Area Preserving Map." *Physica D* **10**, 387–93, 1984.
Peitgen, H.-O. and Saupe, D. (Eds.). "A Chaotic Gingerbreadman." §3.2.3 in *The Science of Fractal Images.* New York: Springer-Verlag, pp. 149–50, 1988.

Gini Coefficient

This entry contributed by CHRISTIAN DAMGAARD

The Gini coefficient (or Gini ratio) G is a summary statistic of the LORENZ CURVE and a measure of inequality in a population. The Gini coefficient is most easily calculated from unordered size data as the "relative mean difference," i.e., the mean of the difference between every possible pair of individuals, divided by the mean size μ,

$$G = \frac{\sum_{i=1}^{n} \sum_{j=1}^{n} |x_i - x_j|}{2n^2 \mu}$$

Alternatively, if the data is ordered by increasing size of individuals, G is given by

$$G = \frac{\sum_{i=1}^{n} (2i - n - 1) x_i'}{n^2 \mu}.$$

The Gini coefficient ranges from a minimum value of zero, when all individuals are equal, to a theoretical maximum of one in an infinite population in which every individual except one has a size of zero. It has been shown that the sample Gini coefficients defined above need to be multiplied by $n/(n-1)$ in order to become UNBIASED ESTIMATORS for the *population* coefficients.

See also LORENZ ASYMMETRY COEFFICIENT, LORENZ CURVE

References

Dixon, P. M.; Weiner, J.; Mitchell-Olds, T.; and Woodley, R. "Bootstrapping the Gini Coefficient of Inequality." *Ecology* **68**, 1548–551, 1987.

Gini, C. "Variabilitá e mutabilita." 1912. Reprinted in *Memorie di metodologia statistica* (Ed. E. Pizetti and T. Salvemini.) Rome: Libreria Eredi Virgilio Veschi, 1955.

Glasser, G. J. "Variance Formulas for the Mean Difference and Coefficient of Concentration." *J. Amer. Stat. Assoc.* **57**, 648–54, 1962.

Sen, A. *On Economic Inequality.* Oxford, England: Clarendon Press, 1973.

Ginzburg-Landau Equation

The PARTIAL DIFFERENTIAL EQUATION

$$u_t = (1 + ia) u_{xx} + (1 + ic) u - (1 + id) |u|^2 u.$$

References

Katou, K. "Asymptotic Spatial Patterns on the Complex Time-Dependent Ginzburg-Landau Equation." *J. Phys. A: Math. Gen.* **19**, L1063-L1066, 1986.

Zwillinger, D. *Handbook of Differential Equations, 3rd ed.* Boston, MA: Academic Press, p. 133, 1997.

Girard's Spherical Excess Formula

Let a SPHERICAL TRIANGLE Δ have angles A, B, and C. Then the SPHERICAL EXCESS is given by

$$\Delta = A + B + C - \pi.$$

See also ANGULAR DEFECT, L'HUILIER'S THEOREM, SPHERICAL EXCESS, SPHERICAL TRIANGLE

References

Coxeter, H. S. M. *Introduction to Geometry, 2nd ed.* New York: Wiley, pp. 94–5, 1969.

Girard, A. *Invention nouvelle en algebra.* Amsterdam, Netherlands, 1629.

Zwillinger, D. (Ed.). *CRC Standard Mathematical Tables and Formulae.* Boca Raton, FL: CRC Press, p. 469, 1995.

Girko's Circular Law

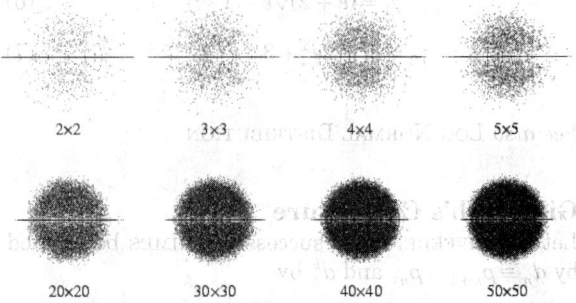

Let λ be (possibly complex) EIGENVALUES of a set of random $n \times n$ REAL MATRICES with entries independent and taken from a standard normal distribution. Then as $n \to \infty$, $\lambda / \sqrt{n}$ is uniformly distributed on the UNIT DISK in the COMPLEX PLANE. For small n, the distribution shows a concentration along the REAL LINE accompanied by a slight paucity above and below (with interesting embedded structure). However, as $n \to \infty$, the concentration about the line disappears and the distribution becomes truly uniform.

See also EIGENVALUE, MATRIX

References

Bai, Z. D. "Circular Law." *Ann. Prob.* **25**, 494–29, 1997.

Bai, Z. D. and Yin, Y. Q. "Limiting Behavior of the Norm Products of Random Matrices and Two Problems of Geman-Hwang." *Probab. Theory Related Fields* **73**, 555–69, 1986.

Edelman, A. and Kostlan, E. "How Many Zeros of a Random Polynomial are Real?" *Bull. Amer. Math. Soc.* **32**, 1–7, 1995.

Edelman, A. "The Probability that a Random Real Gaussian Matrix has k Real Eigenvalues, Related Distributions, and the Circular Law." *J. Multivariate Anal.* **60**, 203–32, 1997.

Geman, S. "The Spectral Radius of Large Random Matrices." *Ann. Probab.* **14**, 1318–328, 1986.

Girko, V. L. "Circular Law." *Theory Probab. Appl.* **29**, 694–06, 1984.

Girko, V. L. *Theory of Random Determinants.* Boston, MA: Kluwer, 1990.

Mehta, M. L. *Random Matrices, 2nd rev. enl. ed.* New York: Academic Press, 1991.

Girth

The length of the shortest GRAPH CYCLE (if any) in a GRAPH. Acyclic graphs are considered to have infinite girth (Skiena 1990, p. 191). The girth of a graph may be found using `Girth[g]` in the *Mathematica* add-on package `DiscreteMath`Combinatorica`` (which can be loaded with the command `<<DiscreteMath``). The following table gives examples of graphs with various girths.

girth	example
3	TETRAHEDRAL GRAPH, COMPLETE GRAPH K_n
4	CUBICAL GRAPH, UTILITY GRAPH
5	PETERSEN GRAPH
6	HEAWOOD GRAPH
7	McGEE GRAPH
8	LEVI GRAPH

See also CAGE GRAPH, GRAPH CIRCUMFERENCE, GRAPH CYCLE, MOORE GRAPH

References

Harary, F. *Graph Theory.* Reading, MA: Addison-Wesley, p. 13, 1994.
Skiena, S. "Girth." §5.3.2 in *Implementing Discrete Mathematics: Combinatorics and Graph Theory with Mathematica.* Reading, MA: Addison-Wesley, pp. 190-92, 1990.

Giuga Number

Any COMPOSITE NUMBER n with $p|(n/p-1)$ for all PRIME DIVISORS p of n. n is a Giuga number IFF

$$\sum_{k=1}^{n-1} k^{\phi(n)} \equiv -1 \pmod{n}$$

where ϕ is the TOTIENT FUNCTION and IFF

$$\sum_{p|n} \frac{1}{p} - \prod_{p|n} \frac{1}{p} \in \mathbb{N}.$$

n is a Giuga number IFF

$$nB_{\phi(n)} \equiv -1 \pmod{n},$$

where B_k is a BERNOULLI NUMBER and ϕ is the TOTIENT FUNCTION. Every counterexample to Giuga's conjecture is a contradiction to ARGOH'S CONJECTURE and vice versa. The smallest known Giuga numbers are 30 (3 factors), 858, 1722 (4 factors), 66198 (5 factors), 2214408306, 24423128562 (6 factors), 432749205173838, 14737133470010574, 5508433913 09130318 (7 factors),

244197000982499715087866346, 5540799146170708 01288578559178 (8 factors), ... (Sloane's A007850).

It is not known if there are an infinite number of Giuga numbers. All the above numbers have sum minus product equal to 1, and any Giuga number of higher order must have at least 59 factors. The smallest ODD Giuga number must have at least nine PRIME FACTORS.

See also ARGOH'S CONJECTURE, BERNOULLI NUMBER, PRIMARY PSEUDOPERFECT NUMBER, TOTIENT FUNCTION

References

Borwein, D.; Borwein, J. M.; Borwein, P. B.; and Girgensohn, R. "Giuga's Conjecture on Primality." *Amer. Math. Monthly* **103**, 40-0, 1996.
Butske, W.; Jaje, L. M.; and Mayernik, D. R. "The Equation $\Sigma_{p|N} 1/p + 1/N = 1$, Pseudoperfect Numbers, and Partially Weighted Graphs." *Math. Comput.* **69**, 407-20, 1999.
Sloane, N. J. A. Sequences A007850 in "An On-Line Version of the Encyclopedia of Integer Sequences." http://www.research.att.com/~njas/sequences/eisonline.html.

Giuga Sequence

A finite, increasing sequence of INTEGERS $\{n_1, \ldots, n_m\}$ such that

$$\sum_{i=1}^{m} \frac{1}{n_i} - \prod_{i=1}^{m} \frac{1}{n_i} \in \mathbb{N}.$$

A sequence is a Giuga sequence IFF it satisfies

$$n_i | (n_1 \cdots n_{i-1} \cdot n_{i+1} \cdot n_m - 1)$$

for $i = 1, \ldots, m$. There are no Giuga sequences of length 2, one of length 3 ($\{2, 3, 5\}$), two of length 4 ($\{2, 3, 7, 41\}$ and $\{2, 3, 11, 13\}$), 3 of length 5 ($\{2, 3, 7, 43, 1805\}$, $\{2, 3, 7, 83, 85\}$, and $\{2, 3, 11, 17, 59\}$), 17 of length 6, 27 of length 7, and hundreds of length 8. There are infinitely many Giuga sequences. It is possible to generate longer Giuga sequences from shorter ones satisfying certain properties.

See also CARMICHAEL SEQUENCE

References

Borwein, D.; Borwein, J. M.; Borwein, P. B.; and Girgensohn, R. "Giuga's Conjecture on Primality." *Amer. Math. Monthly* **103**, 40-0, 1996.

Giuga's Conjecture

If $n > 1$ and

$$n | 1^{n-1} + 2^{n-1} + \ldots + (n-1)^{n-1} + 1,$$

is n necessarily a PRIME? In other words, defining

$$s_n \equiv \sum_{k=1}^{n-1} k^{n-1},$$

does there exist a COMPOSITE n such that $s_n \equiv -1 \pmod{n}$? It is known that $s_n \equiv -1 \pmod{n}$ IFF for each prime divisor p of n, $(p-1)|(n/p-1)$ and $p|(n/p-1)$ (Giuga 1950, Borwein *et al.* 1996); therefore, any counterexample must be SQUAREFREE. A composite INTEGER n satisfies $s_n \equiv -1 \pmod{n}$ IFF it is both a CARMICHAEL NUMBER and a GIUGA NUMBER. Giuga showed that there are no exceptions to the conjecture up to 10^{1000}. This was later improved to 10^{1700} (Bedocchi 1985) and 10^{13800} (Borwein *et al.* 1996).

See also ARGOH'S CONJECTURE

References

Bedocchi, E. "The $\mathbb{Z}(\sqrt{14})$ Ring and the Euclidean Algorithm." *Manuscripta Math.* **53**, 199–16, 1985.

Borwein, D.; Borwein, J. M.; Borwein, P. B.; and Girgensohn, R. "Giuga's Conjecture on Primality." *Amer. Math. Monthly* **103**, 40–0, 1996.

Giuga, G. "Su una presumibile proprietà caratteristica dei numeri primi." *Ist. Lombardo Sci. Lett. Rend. A* **83**, 511–28, 1950.

Ribenboim, P. *The Book of Prime Number Records, 2nd ed.* New York: Springer-Verlag, pp. 20–1, 1989.

GL

GENERAL LINEAR GROUP

Glaisher

GLAISHER-KINKELIN CONSTANT

Glaisher Constant

GLAISHER-KINKELIN CONSTANT

Glaisher-Kinkelin Constant

N.B. A detailed online essay by S. Finch was the starting point for this entry.

Define

$$K(n) \equiv 0^0 1^1 2^2 3^3 \cdots (n-1)^{n-1} \qquad (1)$$

$$G(n) \equiv \frac{[\Gamma(n)]^n}{K(n)} = \begin{cases} 1 & \text{if } n = 0 \\ 0!1!2!\cdots(n-1)! & \text{if } n > 0. \end{cases} \qquad (2)$$

where $G(n)$ is BARNES' G-FUNCTION and $K(n)$ is the K-FUNCTION. Then

$$\lim_{n \to \infty} \frac{K(n+1)}{n^{n^2/2+n/2+1/12}e^{-n^2/4}} = A \qquad (3)$$

(Voros 1987) and

$$\lim_{n \to \infty} \frac{G(n)}{n^{n^2/2-1/12}(2\pi)^{n/2}e^{-3n^2/4}} = \frac{e^{1/12}}{A}, \qquad (4)$$

where

$$A = \exp[\tfrac{1}{12} - \zeta'(-1)] = 1.28242713\ldots \qquad (5)$$

is called the Glaisher-Kinkelin constant (Voros 1987) and $\zeta'(z)$ is the derivative of the RIEMANN ZETA FUNCTION (Kinkelin 1860, Glaisher 1877, 1878, 1893, 1894). The constant A is implemented in *Mathematica* 4.0 as `Glaisher`.

Glaisher (1877) also obtained

$$A = 2^{7/36}\pi^{-1/6}\exp\left\{\frac{1}{3} + \frac{2}{3}\int_0^{1/2}\ln[\Gamma(x+1)]\,dx\right\}. \qquad (6)$$

Glaisher (1894) showed that

$$1^{1/1}2^{1/2}3^{1/9}4^{1/16}5^{1/25}\cdots = \left(\frac{A^{12}}{2\pi e^\gamma}\right)^{\pi^2/6} \qquad (7)$$

$$1^{1/1}3^{1/9}5^{1/25}7^{1/49}9^{1/81}\cdots = \left(\frac{A^{12}}{2^{4/3}\pi e^\gamma}\right)^{\pi^2/8} \qquad (8)$$

$$\frac{1^{1/1}5^{1/125}9^{1/729}\cdots}{3^{1/27}7^{1/343}11^{1/1331}\cdots} = \left(\frac{A}{2^{5/32}\pi^{1/32}e^{3/32+\gamma/48+s/4}}\right)^{\pi^3}, \qquad (9)$$

where

$$s \equiv \frac{\zeta(3)}{3\cdot 4\cdot 5}\frac{1}{4^3} + \frac{\zeta(5)}{5\cdot 6\cdot 7}\frac{1}{4^5} + \frac{\zeta(7)}{7\cdot 8\cdot 9}\frac{1}{4^7}$$
$$+\cdots \qquad (10)$$

The constant appears in a number of sums and integrals, especially those involving GAMMA FUNCTIONS and ZETA FUNCTIONS (Wolfram 1999, p. 757).

See also BARNES' G-FUNCTION, HYPERFACTORIAL, K-FUNCTION

References

Finch, S. "Favorite Mathematical Constants." http://www.mathsoft.com/asolve/constant/glshkn/glshkn.html.

Glaisher, J. W. L. "On a Numerical Continued Product." *Messenger Math.* **6**, 71–6, 1877.

Glaisher, J. W. L. "On the Product $1^1 2^2 3^3 \cdots n^n$." *Messenger Math.* **7**, 43–7, 1878.

Glaisher, J. W. L. "On Certain Numerical Products." *Messenger Math.* **23**, 145–75, 1893.

Glaisher, J. W. L. "On the Constant which Occurs in the Formula for $1^1 2^2 3^3 \cdots n^n$." *Messenger Math.* **24**, 1–6, 1894.

Kinkelin. "Über eine mit der Gammafunktion verwandte Transcendente und deren Anwendung auf die Integralrechnung." *J. reine angew. Math.* **57**, 122–58, 1860.

Voros, A. "Spectral Functions, Special Functions and the Selberg Zeta Function." *Commun. Math. Phys.* **110**, 439–65, 1987.

Wolfram, S. *The Mathematica Book, 4th ed.* Cambridge, England: Cambridge University Press, pp. 756–57, 1999.

Glide

A product of a REFLECTION in a line and TRANSLATION along the same line.

See also REFLECTION, TRANSLATION

References

Addington, S. "The Four Types of Symmetry in the Plane." http://forum.swarthmore.edu/sum95/suzanne/symsusan.html.

Glide Reflection

GLIDE

Glissette

The LOCUS of a point P (or the envelope of a line) fixed in relation to a curve C which slides between fixed curves. For example, if C is a line segment and P a point on the line segment, then P describes an ELLIPSE when C slides so as to touch two ORTHOGONAL straight LINES. The glissette of the LINE SEGMENT C itself is, in this case, an ASTROID.

See also ROULETTE

References

Besant, W. H. *Notes on Roulettes and Glissettes, 2nd enl. ed.* Cambridge, England: Deighton, Bell & Co., 1890.

Lockwood, E. H. "Glissettes." Ch. 20 in *A Book of Curves.* Cambridge, England: Cambridge University Press, pp. 160–65, 1967.

Yates, R. C. "Glissettes." *A Handbook on Curves and Their Properties.* Ann Arbor, MI: J. W. Edwards, pp. 108–12, 1952.

Global

See also LOCAL

Global Analytic Continuation

Analytic continuation gives an equivalence relation between function elements, and the equivalence classes induced by this relation are called global analytic functions.

See also ANALYTIC CONTINUATION, DIRECT ANALYTIC CONTINUATION

References

Krantz, S. G. *The Elements of Advanced Mathematics.* Boca Raton, FL: CRC Press, 1995.

Krantz, S. G. "Global Analytic Continuation." §10.1.6 in *Handbook of Complex Analysis.* Boston, MA: Birkhäuser, pp. 129–30, 1999.

Global Extremum

A GLOBAL MINIMUM or GLOBAL MAXIMUM. It is impossible to construct an algorithm that will find a global extremum for an arbitrary function.

See also LOCAL EXTREMUM

Global Field

A global field is either a NUMBER FIELD, a FUNCTION FIELD on an ALGEBRAIC CURVE, or an extension of TRANSCENDENCE DEGREE one over a FINITE FIELD. From a modern point of view, a global field may refer to a FUNCTION FIELD on a complex ALGEBRAIC CURVE as well as one over a FINITE FIELD. A global field contains a canonical SUBRING, either the ALGEBRAIC INTEGERS or the POLYNOMIALS. By choosing a PRIME IDEAL in its SUBRING, a global field can be TOPOLOGICALLY COMPLETED to give a LOCAL FIELD. For example, the RATIONAL NUMBERS are a global field. By choosing a PRIME NUMBER p, the RATIONALS can be completed in the P-ADIC NORM to form the P-ADIC NUMBERS $\mathbb{Q}_p$.

A global field is called global because of the special case of a complex ALGEBRAIC CURVE, for which the field consists of global functions, (i.e., functions that are defined everywhere). These functions differ from functions defined near a point, whose completion is called a LOCAL FIELD. Under favorable conditions, the local information can be patched together to yield global information (e.g., the HASSE PRINCIPLE).

See also ALGEBRAIC CURVE, CLASS FIELD, FIELD, FUNCTION FIELD, HASSE PRINCIPLE, LOCAL FIELD, NUMBER FIELD, RIEMANN SURFACE

References

Cohn, H. *Advanced Number Theory.* New York: Dover, 1980.

Weil, A. Ch. 8 in *Basic Number Theory.* New York: Springer-Verlag, 1974.

Global Maximum

The largest overall value of a set, function, etc., over its entire range. It is impossible to construct an algorithm that will find a global maximum for an arbitrary function.

See also GLOBAL MINIMUM, LOCAL MAXIMUM, MAXIMUM

Global Minimum

The smallest overall value of a set, function, etc., over its entire range. It is impossible to construct an algorithm that will find a global minimum for an arbitrary function.

See also GLOBAL MAXIMUM, KUHN-TUCKER THEOREM, LOCAL MINIMUM, MINIMUM

Global Optimization

References

Floudas, C. A.; Pardalos, P. M.; Adjiman, C. S.; Esposito, W. R.; Gümüs, Z. H.; Harding, S. T.; Klepeis, J. L.; Meyer, C. A.; and Schweiger, C. A. *Handbook of Test Problems in Local and Global Optimization.* Dordrecht, Netherlands: Kluwer, 1999.

Törn, A. and Zilinskas, A. *Global Optimization.* New York: Springer-Verlag, 1989.

Globe

A SPHERE which acts as a model of a spherical (or ellipsoidal) celestial body, especially the Earth, and on which the outlines of continents, oceans, etc. are drawn.

See also LATITUDE, LONGITUDE, SPHERE

Glome

A 3-sphere

$$x^2 + y^2 + z^2 + w^2 = r^2$$

(as opposed to the usual 2-SPHERE). The term derives from the Latin 'glomus' meaning 'ball of string.'

See also HYPERSPHERE, SPHERE

Glove Problem

Let there be m doctors and $n \leq m$ patients, and let all mn possible combinations of examinations of patients by doctors take place. Then what is the minimum number of surgical gloves needed $G(m, n)$ so that no doctor must wear a glove contaminated by a patient and no patient is exposed to a glove worn by another doctor? In this problem, the gloves can be turned inside out and even placed on top of one another if necessary, but no "decontamination" of gloves is permitted. The optimal solution is

$$g(m,\ n) = \begin{cases} 2 & m = n = 2 \\ \frac{1}{2}(m+1) & n = 1,\ m = 2k+1 \\ \lceil \frac{1}{2}(m) + \frac{2}{3}\,n \rceil & \text{otherwise,} \end{cases}$$

where $\lceil x \rceil$ is the CEILING FUNCTION (Vardi 1991). The case $m = n = 2$ is straightforward since two gloves have a total of four surfaces, which is the number needed for $mn = 4$ examinations.

References

Gardner, M. *Aha! Insight.* New York: Scientific American, 1978.
Gardner, M. *Science Fiction Puzzle Tales.* New York: Crown, pp. 5, 67, and 104–50, 1981.
Hajnal, A. and Lovász, L. "An Algorithm to Prevent the Propagation of Certain Diseases at Minimum Cost." §10.1 in *Interfaces Between Computer Science and Operations Research* (Ed. J. K. Lenstra, A. H. G. Rinnooy Kan, and P. van Emde Boas). Amsterdam: Matematisch Centrum, 1978.
Orlitzky, A. and Shepp, L. "On Curbing Virus Propagation." Exercise 10.2 in Technical Memo. Bell Labs, 1989.
Vardi, I. "The Condom Problem." Ch. 10 in *Computational Recreations in Mathematica.* Redwood City, CA: Addison-Wesley, pp. 203–22, 1991.

Glue Vector

A VECTOR specifying how layers are stacked in a LAMINATED LATTICE.

Gnomon

A shape which, when added to a figure, yields another figure SIMILAR to the original.

References

Shanks, D. *Solved and Unsolved Problems in Number Theory,* 4th ed. New York: Chelsea, p. 123, 1993.

Gnomon Magic Square

A 3×3 array of numbers in which the elements in each 2×2 corner have the same sum.

See also MAGIC SQUARE

References

Stapleton, H. E. "The Gnomon as a Possible Link Between (a) One Type of Mesopotamian *Ziggurat* and (b) the Magic Square Numbers on which Jaribian Alchemy was Based." *Ambix: J. Soc. Study Alchemy and Early Chem.* **6**, 1–, 1957–958.

Gnomonic Number

A FIGURATE NUMBER OF THE FORM $g_n = 2n - 1$ which are the areas of square gnomons, obtained by removing a SQUARE of side $n - 1$ from a SQUARE of side n,

$$g_n = n^2 - (n-1)^2 = 2n - 1.$$

The gnomonic numbers are therefore equivalent to the ODD NUMBERS, and the first few are 1, 3, 5, 7, 9, 11, ... (Sloane's A005408). The GENERATING FUNCTION for the gnomonic numbers is

$$\frac{x(1+x)}{(x-1)^2} = x + 3x^2 + 5x^3 + 7x^4 + \dots.$$

See also FIGURATE NUMBER, ODD NUMBER

References

Sloane, N. J. A. Sequences A005408/M2400 in "An On-Line Version of the Encyclopedia of Integer Sequences." http://www.research.att.com/~njas/sequences/eisonline.html.

Gnomonic Projection

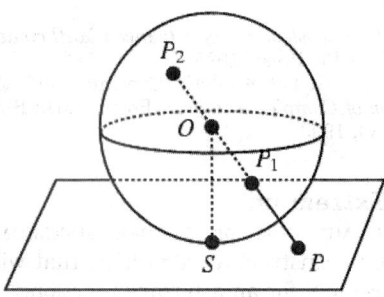

A nonconformal MAP PROJECTION obtained by projecting points P_1 (or P_2) on the surface of sphere from a sphere's center O to point P in a plane that is tangent to the south pole S (Coxeter 1969, p. 93). Since this projection obviously sends ANTIPODAL POINTS P_1 and P_2 to the same point P in the plane, it can only be used to project one HEMISPHERE as a time. In a gnomonic projection, ORTHODROMES are straight LINES.

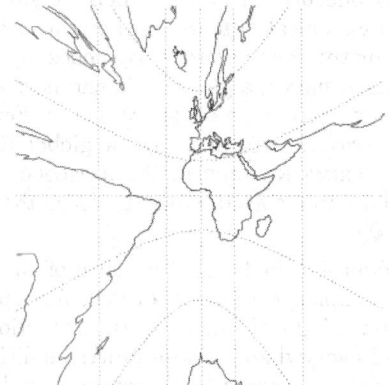

The transformation equations for a point at LATITUDE

ϕ and LONGITUDE λ are given by

$$x = \frac{\cos \phi \sin (\lambda - \lambda_0)}{\cos c} \qquad (1)$$

$$y = \frac{\cos \phi_1 \sin \phi - \sin \phi_1 \cos \phi \cos (\lambda - \lambda_0)}{\cos c}, \qquad (2)$$

where λ_0 is the central longitude, ϕ_1 is the central latitude, and c is the angular distance of the point (x, y) from the center of the projection, given by

$$\cos c = \sin \phi_1 \sin \phi + \cos \phi_1 \cos \phi \cos(\lambda - \lambda_0). \qquad (3)$$

The inverse FORMULAS are

$$\phi = \sin^{-1}\left(\cos \phi \sin \phi_1 + \frac{y \sin \theta \cos \theta \cos \phi_1}{\sqrt{x^2 + y^2}} \right), \qquad (4)$$

$$\lambda = \lambda_0$$
$$+ \tan^{-1}\left(\frac{x \sin \theta}{\sqrt{x^2 + y^2} \cos \phi_1 \cos \theta - y \sin \phi_1 \sin \theta} \right), \qquad (5)$$

where

$$\theta = \tan^{-1}(\sqrt{x^2 + y^2}). \qquad (6)$$

See also STEREOGRAPHIC PROJECTION

References

Coxeter, H. S. M. *Introduction to Geometry, 2nd ed.* New York: Wiley, pp. 93 and 289–90, 1969.
Coxeter, H. S. M. and Greitzer, S. L. *Geometry Revisited.* Washington, DC: Math. Assoc. Amer., pp. 150–53, 1967.
Snyder, J. P. *Map Projections--A Working Manual.* U. S. Geological Survey Professional Paper 1395. Washington, DC: U. S. Government Printing Office, pp. 164–68, 1987.

G-Number

EISENSTEIN INTEGER

Go

There are estimated to be about 4.63×10^{170} possible positions on a 19×19 board (Beeler *et al.*, Flammenkamp). The number of n-move Go games are 1, 362, 130683, 47046242, ... (Sloane's A007565).

References

Beeler, M. *et al.* Item 96 in Beeler, M.; Gosper, R. W.; and Schroeppel, R. *HAKMEM.* Cambridge, MA: MIT Artificial Intelligence Laboratory, Memo AIM-239, p. 35, Feb. 1972.
Bewersdorff, J. "Go und Mathematik." http://home.t-online.de/home/joerg.bewersdorff/go.htm.
Culin, S. "Pa-tok--Pebble Game." §75 in *Games of the Orient: Korea, China, Japan.* Rutland, VT: Charles E. Tuttle, pp. 91–01, 1965.
Kraitchik, M. "Go." §12.4 in *Mathematical Recreations.* New York: W. W. Norton, pp. 279–80, 1942.
Lasker, E. *Go and Go-Moku.* New York: Dover, 1960.

Sloane, N. J. A. Sequences A007565/M5447 in "An On-Line Version of the Encyclopedia of Integer Sequences." http://www.research.att.com/~njas/sequences/eisonline.html.
Warkentyne, K. "Ken's Go Page." http://nngs.cosmic.org/hmkw/.
Warkentyne, K. "The Web Go Page Index." http://nngs.cosmic.org/hmkw/golinks.html.

Goat Grazing Problem

GOAT PROBLEM

Goat Problem

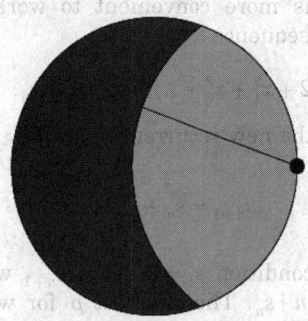

Let a circular field of unit radius be fenced in, and tie a goat to a point on the interior of the fence with a chain of length r. What length of chain must be used in order to allow the goat to graze exactly one half the area of the field?

The answer is obtained by using the equation for a CIRCLE-CIRCLE INTERSECTION

$$A = r^2 \cos^{-1}\left(\frac{d^2 + r^2 + R^2}{2dr} \right)$$
$$+ R^2 \cos^{-1}\left(\frac{d^2 + R^2 + -r^2}{2dR} \right)$$
$$- \frac{1}{2}\sqrt{(-d+r+R)(d+r-R)(d-r+R)(d+r+R)} \qquad (1)$$

with $R = d = 1$ and $A = \pi/2$ (i.e., half of πR^2). This leads to the equation

$$-\frac{1}{2}r\sqrt{4-r^2} + r^2 \cos^{-1}(\frac{1}{2}r) + \cos^{-1}(1 - \frac{1}{2}r^2) = \frac{1}{2}\pi, \qquad (2)$$

which cannot be solved exactly, but which has approximate solution

$$r \approx 1.15872847. \qquad (3)$$

See also CIRCLE-CIRCLE INTERSECTION, LENS

Göbel's Sequence

Consider the RECURRENCE RELATION

$$x_n = \frac{1 + x_0^2 + x_1^2 + \ldots + x_{n-1}^2}{n}, \qquad (1)$$

with $x_0 = 1$. The first few iterates of x_n are 1, 2, 3, 5,

10, 28, 154, ... (Sloane's A003504). The terms grow extremely rapidly, but are given by the asymptotic formula

$$x_n \approx (n^2 + 2n - 1 + 4n^{-1} - 21n^{-2} + 137n^{-3}$$
$$- \ldots)C^{2^n}, \qquad (2)$$

where

$$C = 1.0478314475764112295599090946274313755459\ldots \qquad (3)$$

(Zagier). It is more convenient to work with the transformed sequence

$$s_n = 2 + x_1^2 + x_2^2 + \ldots + x_{n-1}^2 = nx_n, \qquad (4)$$

which gives the new recurrence

$$s_{n+1} = s_n + \frac{s_n^2}{n^2} \qquad (5)$$

with initial condition $s_1 = 2$. Now, s_{n+1} will be nonintegral IFF $n \nmid s_n$. The smallest p for which $s_p \not\equiv 0$ (mod p) therefore gives the smallest nonintegral s_{p+1}. In addition, since $p \nmid s_p$, $x_p = s_p/p$ is also the smallest nonintegral x_p.

For example, we have the sequences $\{s_n \pmod{k}\}_{n=1}^k$:

$$2, \ 6 \equiv 2, \ \tfrac{5}{4} \equiv 0, \ 0, \ 0 \pmod 5 \qquad (6)$$

$$2, \ 6, \ 15 \equiv 1, \ \tfrac{5}{4} \equiv 0, \ 0, \ 0, \ 0 \pmod 7 \qquad (7)$$

$$2, \ 6, \ 15 \equiv 4, \ \tfrac{52}{9} \equiv 7, \ \tfrac{161}{16} \equiv 8, \ \tfrac{264}{5}$$
$$\equiv 0, \ 0, \ \ldots, \ 0 \pmod{11} \qquad (8)$$

Testing values of k shows that the first nonintegral x_n is x_{43}. Note that a direct verification of this fact is impossible since

$$x_{43} \approx 5.4093 \times 10^{178485291567} \qquad (9)$$

(calculated using the asymptotic formula) is much too large to be computed and stored explicitly.

A sequence even more striking for assuming integer values only for many terms is the 3-Göbel sequence

$$x_n = \frac{1 + x_0^3 + x_1^3 + \ldots + x_{n-1}^3}{n}. \qquad (10)$$

The first few terms of this sequence are 1, 2, 5, 45, 22815, ... (Sloane's A005166).

The Göbel sequences can be generalized to k powers by

$$x_n = \frac{1 + x_0^k + x_1^k + \ldots + x_{n-1}^k}{n}. \qquad (11)$$

See also SOMOS SEQUENCE

References

Guy, R. K. "The Strong Law of Small Numbers." *Amer. Math. Monthly* **95**, 697–12, 1988.

Guy, R. K. "A Recursion of Göbel." §E15 in *Unsolved Problems in Number Theory, 2nd ed.* New York: Springer-Verlag, pp. 214–15, 1994.

Sloane, N. J. A. Sequences A003504/M0728 and A005166/M1551 in "An On-Line Version of the Encyclopedia of Integer Sequences." http://www.research.att.com/~njas/sequences/eisonline.html.

Zaiger, D. "Solution: Day 5, Problem 3." http://www-groups.dcs.st-and.ac.uk/~john/Zagier/Solution5.3.html.

Goblet Illusion

An ILLUSION in which the eye alternately sees two black faces, or a white goblet.

References

Fineman, M. *The Nature of Visual Illusion.* New York: Dover, pp. 111 and 115, 1996.

Rubin, E. *Synoplevede Figurer.* Copenhagen, Denmark: Gyldendalske, 1915.

Gödel Number

A Gödel number is a unique number associated with a statement about arithmetic. It is formed as the PRODUCT of successive PRIMES raised to the POWER of the number corresponding to the individual symbols that comprise the sentence. For example, the statement $(\exists x)(x = sy)$ that reads "there EXISTS an x such that x is the immediate SUCCESSOR of y" is coded

$$(2^8)(3^4)(5^{13})(7^9)(11^8)(13^{13})(17^5)(19^7)(23^{16})(29^9),$$

where the numbers in the set (8, 4, 13, 9, 8, 13, 5, 7, 16, 9) correspond to the symbols that make up $(\exists x)(x = sy)$.

See also GÖDEL'S INCOMPLETENESS THEOREM

References

Hofstadter, D. R. *Gödel, Escher, Bach: An Eternal Golden Braid.* New York: Vintage Books, p. 18, 1989.

Gödel's Completeness Theorem

If T is a set of AXIOMS in a first-order language, and a statement p holds for any structure M satisfying T, then p can be formally deduced from T in some appropriately defined fashion.

See also GÖDEL'S INCOMPLETENESS THEOREM, LÖWENHEIM-SKOLEM THEOREM

References

Beth, E. W. *The Foundations of Mathematics.* Amsterdam, Netherlands: North-Holland, 1959.

Gödel's Incompleteness Theorem

Informally, Gödel's incompleteness theorem states that all CONSISTENT axiomatic formulations of NUMBER THEORY include undecidable propositions (Hofstadter 1989). This is sometimes called Gödel's first incompleteness theorem, and answers in the negative HILBERT'S PROBLEM asking whether mathematics is "complete" (in the sense that every statement in the language of NUMBER THEORY can be either proved or disproved). Formally, Gödel's theorem states, "To every ω-consistent recursive class κ of FORMULAS, there correspond recursive class-signs r such that neither (v Gen r) nor Neg(v Gen r) belongs to Flg(κ), where v is the FREE VARIABLE of r" (Gödel 1931).

A statement sometimes known as Gödel's second incompleteness theorem states that if NUMBER THEORY is consistent, then a proof of this fact does not exist using the methods of first-order PREDICATE CALCULUS. Stated more colloquially, any formal system that is interesting enough to formulate its own consistency can prove its own consistency IFF it is inconsistent.

Gerhard Gentzen showed that the consistency and completeness of arithmetic can be proved if "transfinite" induction is used. However, this approach does not allow proof of the consistency of all mathematics.

See also CONSISTENCY, GÖDEL'S COMPLETENESS THEOREM, HILBERT'S PROBLEMS, KREISEL CONJECTURE, NATURAL INDEPENDENCE PHENOMENON, NUMBER THEORY, RICHARDSON'S THEOREM, UNDECIDABLE

References

Barrow, J. D. *Pi in the Sky: Counting, Thinking, and Being.* Oxford, England: Clarendon Press, p. 121, 1993.
Erickson, G. W. and Fossa, J. A. *Dictionary of Paradox.* Lanham, MD: University Press of America, pp. 74–5, 1998.
Franzén, T. "Gödel on the Net." http://www.sm.luth.se/~torkel/eget/godel.html.
Gödel, K. "Uuml;ber Formal Unentscheidbare Sätze der *Principia Mathematica* und Verwandter Systeme, I." *Monatshefte für Math. u. Physik* **38**, 173–98, 1931.
Gödel, K. *On Formally Undecidable Propositions of Principia Mathematica and Related Systems.* New York: Dover, 1992.
Hofstadter, D. R. *Gödel, Escher, Bach: An Eternal Golden Braid.* New York: Vintage Books, p. 17, 1989.
Kolata, G. "Does Gödel's Theorem Matter to Mathematics?" *Science* **218**, 779–80, 1982.
Smullyan, R. M. *Gödel's Incompleteness Theorems.* New York: Oxford University Press, 1992.
Whitehead, A. N. and Russell, B. *Principia Mathematica.* New York: Cambridge University Press, 1927.

Gog Triangle

MONOTONE TRIANGLE

Golay-Rudin-Shapiro Sequence

RUDIN-SHAPIRO SEQUENCE

Goldbach Conjecture

Goldbach's original conjecture (sometimes called the "ternary" Goldbach conjecture), written in a June 7, 1742 letter to Euler, states that every INTEGER > 5 is the SUM of three PRIMES (Dickson 1957, p. 421). As re-expressed by Euler, an equivalent of this CONJECTURE (called the "strong" or "binary" Goldbach conjecture) asserts that all POSITIVE EVEN INTEGERS ≥ 4 can be expressed as the SUM of two PRIMES. According to Hardy (1999, p. 19), "It is comparatively easy to make clever guesses; indeed there are theorems, like 'Goldbach's Theorem', which have never been proved and which any fool could have guessed."

Schnirelman (1939) proved that every EVEN number can be written as the sum of not more than 300,000 PRIMES (Dunham 1990), which seems a rather far cry from a proof for *two* PRIMES! Pogorzelski (1977) claimed to have proven the Goldbach conjecture, but his proof is not generally accepted (Shanks 1993). The following table summarizes bounds n such that the strong Goldbach conjecture has been shown to be true for numbers $< n$.

bound	reference
1×10^4	Desboves 1885
1×10^5	Pipping 1938
1×10^8	Stein and Stein 1965ab
2×10^{10}	Granville *et al.* 1989
4×10^{11}	Sinisalo 1993
1×10^{14}	Deshouillers *et al.* 1998
4×10^{14}	Richstein 2000 (quoted in Peterson 2000)

The conjecture that all ODD numbers ≥ 9 are the SUM of three ODD PRIMES is called the "weak" Goldbach conjecture. Vinogradov proved that all ODD INTEGERS starting at some sufficiently large value are the SUM of three PRIMES (Guy 1994). The original "sufficiently large" $N \geq 3^{3^{15}} \approx e^{e^{16.573}} \approx 3.25 \times 10^{6,846,168}$ was subsequently reduced to $e^{e^{11.503}} \approx 3.33 \times 10^{43,000}$ by Chen and Wang (1989). Chen (1973, 1978) also showed that all sufficiently large EVEN NUMBERS are the sum of a PRIME and the PRODUCT of at most two PRIMES (Guy 1994, Courant and Robbins 1996).

It has been shown that if the weak Goldbach conjecture is false, then there are only a FINITE number of exceptions. A stronger version of the weak conjecture, namely that every odd number > 5 can be expressed as the sum of a prime plus twice a prime has been formulated by C. Eaton. This conjecture has been verified for $n \leq 10^9$ (Corbit).

Other variants of the Goldbach conjecture include the statements that every EVEN number ≥ 6 is the SUM of two ODD PRIMES, and every INTEGER > 17 the sum of exactly three distinct PRIMES. Let $R(n)$ be the number of representations of an EVEN INTEGER n as the sum of two PRIMES. Then the "extended" Goldbach conjecture states that

$$R(n) \sim 2 \prod_2 \prod_{\substack{k=2 \\ p_k \mid n}} \frac{p_k - 1}{p_k - 2} \int_2^x \frac{dx}{(\ln x)^2},$$

where $\prod_2$ is the TWIN PRIMES CONSTANT (Halberstam and Richert 1974).

If the Goldbach conjecture is true, then for every number m, there are PRIMES p and q such that

$$\phi(p) + \phi(q) = 2m,$$

where $\phi(x)$ is the TOTIENT FUNCTION (Guy 1994, p. 105).

Vinogradov (1937ab, 1954) proved that every sufficiently large ODD NUMBER is the sum of three PRIMES (Nagell 1951, p. 66), and Estermann (1938) proves that almost all EVEN NUMBERS are the sums of two PRIMES.

See also CHEN'S THEOREM, DE POLIGNAC'S CONJECTURE, GOLDBACH NUMBER, PRIME PARTITION, SCHNIRELMANN'S THEOREM, WARING'S PRIME NUMBER CONJECTURE

References

Ball, W. W. R. and Coxeter, H. S. M. *Mathematical Recreations and Essays, 13th ed.* New York: Dover, p. 64, 1987.

Caldwell, C. K. "Prime Links++: Resources in theory: conjectures: Goldbach." http://primes.utm.edu/links/theory/conjectures/Goldbach/.

Chen, J.-R. "On the Representation of a Large Even Number as the Sum of a Prime and the Product of at Most Two Primes.' *Sci. Sinica* **16**, 157–76, 1973.

Chen, J.-R. "On the Representation of a Large Even Number as the Sum of a Prime and the Product of at Most Two Primes, II." *Sci. Sinica* **21**, 421–30, 1978.

Chen, J.-R. and Wang, T.-Z. "On the Goldbach Problem." *Acta Math. Sinica* **32**, 702–18, 1989.

Corbit, D. sci.math posting. Nov 19, 1999.

Courant, R. and Robbins, H. *What is Mathematics?: An Elementary Approach to Ideas and Methods, 2nd ed.* Oxford, England: Oxford University Press, pp. 30–1, 1996.

Desboves, A. *Nouv. Ann. Math.* **14**, 293, 1855.

Deshouillers, J.-M.; te Riele, H. J. J.; and Saouter, Y. "New Experimental Results Concerning The Goldbach Conjecture." In *Algorithmic Number Theory: Proceedings of the 3rd International Symposium (ANTS-III) held at Reed College, Portland, OR, June 21–5, 1998* (Ed. J. P. Buhler). Berlin: Springer-Verlag, pp. 204–15, 1998.

Devlin, K. *Mathematics: The New Golden Age.* London: Penguin Books, 1988.

Dickson, L. E. "Goldbach's Empirical Theorem: Every Integer is a Sum of Two Primes." In *History of the Theory of*

Numbers, Vol. 1: Divisibility and Primality. New York: Chelsea, pp. 421–24, 1952.

Dunham, W. *Journey through Genius: The Great Theorems of Mathematics.* New York: Wiley, p. 83, 1990.

Estermann, T. "On Goldbach's Problem: Proof that Almost All Even Positive Integers are Sums of Two Primes." *Proc. London Math. Soc. Ser. 2* **44**, 307–14, 1938.

Granville, A.; van der Lune, J.; and te Riele, H. J. J. "Checking the Goldbach Conjecture on a Vector Computer." In *Number Theory and Applications: Proceedings of the NATO Advanced Study Institute held in Banff, Alberta, April 27-May 5, 1988* (Ed. R. A. Mollin). Dordrecht, Netherlands: Kluwer, pp. 423–33, 1989.

Guy, R. K. "Goldbach's Conjecture." §C1 in *Unsolved Problems in Number Theory, 2nd ed.* New York: Springer-Verlag, pp. 105–07, 1994.

Hardy, G. H. *Ramanujan: Twelve Lectures on Subjects Suggested by His Life and Work, 3rd ed.* New York: Chelsea, 1999.

Hardy, G. H. and Littlewood, J. E. "Some Problems of 'Partitio Numerorum.' III. On the Expression of a Number as a Sum of Primes." *Acta Math.* **44**, 1–0, 1922.

Hardy, G. H. and Littlewood, J. E. "Some Problems of Partitio Numerorum (V): A Further Contribution to the Study of Goldbach's Problem." *Proc. London Math. Soc. Ser. 2* **22**, 46–6, 1924.

Hardy, G. H. and Wright, E. M. *An Introduction to the Theory of Numbers, 5th ed.* Oxford, England: Clarendon Press, p. 19, 1979.

Halberstam, H. and Richert, H.-E. *Sieve Methods.* New York: Academic Press, 1974.

Nagell, T. *Introduction to Number Theory.* New York: Wiley, p. 66, 1951.

Peterson, I. "Prime Conjecture Verified to New Heights." *Sci. News* **158**, 103, Aug. 12, 2000.

Pipping, N. "Die Goldbachsche Vermutung und der Goldbach-Vinogradovsche Satz." *Acta. Acad. Aboensis, Math. Phys.* **11**, 4–5, 1938.

Pogorzelski, H. A. "Goldbach Conjecture." *J. reine angew. Math.* **292**, 1–2, 1977.

Richstein, J. To appear in *Math. Comput.*

Schnirelman, L. G. *Uspekhi Math. Nauk* **6**, 3–, 1939.

Shanks, D. *Solved and Unsolved Problems in Number Theory, 4th ed.* New York: Chelsea, pp. 30–1 and 222, 1985.

Sinisalo, M. K. "Checking the Goldbach Conjecture up to $4 \cdot 10^{11}$." *Math. Comput.* **61**, 931–34, 1993.

Stein, M. L. and Stein, P. R. "New Experimental Results on the Goldbach Conjecture." *Math. Mag.* **38**, 72–0, 1965a.

Stein, M. L. and Stein, P. R. "Experimental Results on Additive 2 Bases." *BIT* **38**, 427–34, 1965b.

Vinogradov, I. M. "Representation of an Odd Number as a Sum of Three Primes." *Comtes rendus (Doklady) de l'Académie des Sciences de l'U.R.S.S.* **15**, 169–72, 1937a.

Vinogradov, I. M. "Some Theorems Concerning the Theory of Primes." *Recueil Math.* **2**, 179–95, 1937b.

Vinogradov, I. M. *The Method of Trigonometrical Sums in the Theory of Numbers.* London: Interscience, p. 67, 1954.

Wang, Y. (Ed.). |it Goldbach Conjecture. Singapore: World Scientific, 1984.

Woon, M. S. C. On Partitions of Goldbach's Conjecture 4 Oct 2000. http://xxx.lanl.gov/abs/math.GM/0010027/.

Yuan, W. *Goldbach Conjecture.* Singapore: World Scientific, 1984.

Goldbach Number

A positive integer which is the sum of two ODD PRIMES is called a Goldbach number (Li 1999). Let $E(x)$ (the "exceptional set of Goldbach numbers") denote the number of even numbers not exceeding x which cannot be written as a sum of two primes. Then the GOLDBACH CONJECTURE is equivalent to proving that $E(x) = 2$ for every $x \geq 4$. Li (1999) proved that for sufficiently large x,

$$E(x) = \mathcal{O}(x^{0.921}).$$

See also GOLDBACH CONJECTURE

References

Chen, J. "The Exceptional Set of Goldbach Numbers (II)." *Sci. Sinica* **26**, 714–31, 1983.
Chen, J. and Liu, J. "The Exceptional Set of Goldbach Numbers (III)." *Chinese Quart. J. Math.* **4**, 1–5, 1989.
Chen, J. and Pan, C. "The Exceptional Set of Goldbach Numbers." *Sci. Sinica* **23**, 416–30, 1980.
Li, H. "The Exceptional Set of Goldbach Numbers." *Quart. J. Math. Oxford* **50**, 471–82, 1999.
Montgomery, H. L. and Vaughan, R. C. "The Exceptional Set of Goldbach's Problem." *Acta. Arith.* **27**, 353–70, 1975.

Goldbach's Theorem
GOLDBACH CONJECTURE

Golden Mean
GOLDEN RATIO

Golden Ratio

A number often encountered when taking the ratios of distances in simple geometric figures such as the PENTAGRAM, DECAGON and DODECAGON. It is denoted ϕ, or sometimes τ (which is an abbreviation of the Greek "tome," meaning "to cut"). ϕ is also known as the DIVINE PROPORTION, GOLDEN MEAN, and GOLDEN SECTION and is a PISOT-VIJAYARAGHAVAN CONSTANT. It has surprising connections with CONTINUED FRACTIONS and the EUCLIDEAN ALGORITHM for computing the GREATEST COMMON DIVISOR of two INTEGERS.

Given a RECTANGLE having sides in the ratio $1 : \phi$, ϕ is defined such that partitioning the original RECTANGLE into a SQUARE and new RECTANGLE results in a new RECTANGLE having sides with a ratio $1 : \phi$. Such a RECTANGLE is called a GOLDEN RECTANGLE, and successive points dividing a GOLDEN RECTANGLE into SQUARES lie on a LOGARITHMIC SPIRAL. This figure is known as a WHIRLING SQUARE.

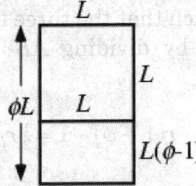

This means that

$$\frac{1}{\phi - 1} = \phi \qquad (1)$$

$$\phi^2 - \phi - 1 = 0. \qquad (2)$$

So, by the QUADRATIC EQUATION,

$$\phi = \tfrac{1}{2}(1 \pm \sqrt{1 + 4}) = \tfrac{1}{2}(1 + \sqrt{5}) \qquad (3)$$

$$= 1.6180339887498948482045868343656381177720\ldots \qquad (4)$$

(Sloane's A001622). The golden ratio is given by the INFINITE SERIES

$$\phi = \frac{13}{8} + \sum_{n=0}^{\infty} \frac{(-1)^{n+1}(2n + 1)!}{(n + 2)!\, n!\, 4^{2n+3}} \qquad (5)$$

(B. Roselle).

A geometric definition can be given in terms of the above figure. Let the ratio $x \equiv BC/AB$. The NUMERATOR and DENOMINATOR can then be taken as $\overline{AB} = a$ and $\overline{BC} = x$ without loss of generality. Now define the position of B by

$$\frac{AB}{BC} = \frac{BC}{AC}. \qquad (6)$$

Plugging in gives

$$\frac{1}{x} = \frac{x}{1 + x}, \qquad (7)$$

or

$$x^2 - x - 1 = 0, \qquad (8)$$

which can be solved using the QUADRATIC EQUATION to obtain

$$\phi \equiv x = \frac{1 + \sqrt{(-1)^2 - 4(1)(-1)}}{2} = \tfrac{1}{2}(1 + \sqrt{5}), \qquad (9)$$

where the plus sign has been taken to give the solution with $x > 1$.

ϕ is the "most" IRRATIONAL number because it has a CONTINUED FRACTION representation

$$\phi = [1, 1, 1, \ldots] \qquad (10)$$

(Sloane's A000012; Williams 1979, p. 52; Steinhaus 1983, p. 45). Another infinite representation in terms of a NESTED RADICAL is

$$\phi = \sqrt{1 + \sqrt{1 + \sqrt{1 + \sqrt{1 + \ldots}}}}. \qquad (11)$$

Ramanujan gave the curious CONTINUED FRACTION identities

$$\frac{1}{(\sqrt{\phi\sqrt{5}})e^{2\pi/5}} = 1 + \cfrac{e^{-2\pi}}{1 + \cfrac{e^{-4\pi}}{1 + \cfrac{e^{-6\pi}}{1 + \cfrac{e^{-8\pi}}{1 + \cfrac{e^{-10\pi}}{1 + \ldots}}}}} \qquad (12)$$

$$\cfrac{1}{\left\{\frac{\sqrt{5}}{1 + [5^{3/4}(\phi - 1)^{5/2} - 1] - \phi}\right\}e^{2\pi/\sqrt{5}}}$$
$$= 1 + \cfrac{e^{-2\pi\sqrt{5}}}{1 + \cfrac{e^{-4\pi\sqrt{5}}}{1 + \cfrac{e^{-6\pi\sqrt{5}}}{1 + \cfrac{e^{-8\pi\sqrt{5}}}{1 + \cfrac{e^{-10\pi\sqrt{5}}}{1 + \ldots}}}}} \qquad (13)$$

(Ramanathan 1984).

The SINE of certain complex numbers involving ϕ gives particularly simplex answers,

$$\sin(i \ln \phi) = \tfrac{1}{2} i \qquad (14)$$

$$\sin(\tfrac{1}{2}\pi - i \ln \phi) = \tfrac{1}{2}\sqrt{5} \qquad (15)$$

(Hoey). A curious approximation due to D. Barron is given by

$$\phi \approx \tfrac{1}{2} K^{\gamma - 19/7} \pi^{2/7 + \gamma}, \qquad (16)$$

where K is CATALAN'S CONSTANT and γ is the EULER-MASCHERONI CONSTANT, which is good to two digits.

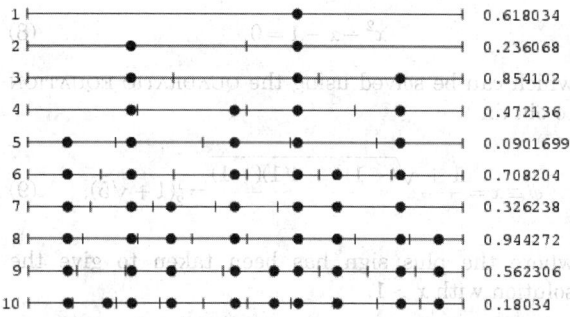

Steinhaus (1983, pp. 48–9) considers the distribution of the FRACTIONAL PARTS of $n\phi$ in the intervals bounded by 0, $1/n$, $2/n$, ..., $(n-1)/n$, 1, and notes that they are much more uniformly distributed than would be expected due to chance (i.e., frac($n\phi$) is close to an EQUIDISTRIBUTED SEQUENCE). In particular, the number of empty intervals for $n = 1, 2, ...$, are a mere 0, 0, 0, 0, 0, 0, 1, 0, 2, 0, 1, 1, 0, 2, 2, ... (Sloane's A036412). The values of n for which *no* bins are left blank are then given by 1, 2, 3, 4, 5, 6, 8, 10, 13, 16, 21, 34, 55, 89, 144, ... (Sloane's A036413). Steinhaus (1983) remarks that the highly uniform distribution has its roots in the CONTINUED FRACTION for ϕ.

The legs of a GOLDEN TRIANGLE are in a golden ratio to its base. In fact, this was the method used by Pythagoras to construct ϕ. Euclid used the following construction.

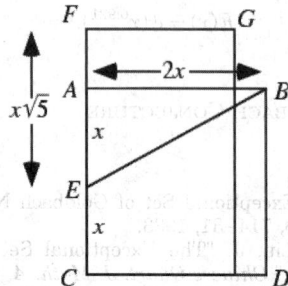

Draw the SQUARE $\square ABCD$, call E the MIDPOINT of AC, so that $AE = EC \equiv x$. Now draw the segment BE, which has length

$$x\sqrt{2^2 + 1^2} = x\sqrt{5}, \qquad (17)$$

and construct EF with this length. Now construct $FG = EF$, then

$$\phi = \frac{FC}{CD} = \frac{EF + CE}{CD} = \frac{x(\sqrt{5} + 1)}{2x} = \tfrac{1}{2}(\sqrt{5} + 1). \qquad (18)$$

The ratio of the CIRCUMRADIUS to the length of the side of a DECAGON is also ϕ,

$$\frac{R}{s} = \tfrac{1}{2}\csc\left(\frac{\pi}{10}\right) = \tfrac{1}{2}(1 + \sqrt{5}) = \phi. \qquad (19)$$

Similarly, the legs of a GOLDEN TRIANGLE (an ISOSCELES TRIANGLE with a VERTEX ANGLE of 36°) are in a golden ratio to the base. Bisecting a GAULLIST CROSS also gives a golden ratio (Gardner 1961, p. 102).

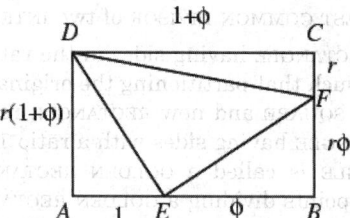

In the figure above, three TRIANGLES can be INSCRIBED in the RECTANGLE $\square ABCD$ of arbitrary aspect ratio $1 : r$ such that the three RIGHT TRIANGLES have equal areas by dividing AB and BC in the golden ratio. Then

$$K_{\triangle ADE} = \tfrac{1}{2} \cdot r(1 + \phi) \cdot 1 = \tfrac{1}{2} r\phi^2 \qquad (20)$$

$$K_{\triangle BEF} = \tfrac{1}{2} \cdot r\phi \cdot \phi = \tfrac{1}{2} r\phi^2 \qquad (21)$$

$$K_{\triangle CDF} = \tfrac{1}{2}(1 + \phi) \cdot r = \tfrac{1}{2} r\phi^2, \qquad (22)$$

which are all equal.

The golden ratio also satisfies the RECURRENCE RELATION

$$\phi^n = \phi^{n-1} + \phi^{n-2}, \tag{23}$$

so taking $n = 0$ gives

$$\phi = \phi^{-1} + 1. \tag{24}$$

The powers of the golden ratio also satisfy

$$\phi^n = F_n \phi + F_{n-1}, \tag{25}$$

where F_n is a FIBONACCI NUMBER (Wells 1986, p. 39). For the difference equations

$$\begin{cases} x_0 = 1 \\ x_n = 1 + \dfrac{1}{x_{n-1}} \quad \text{for } n = 1, 2, 3, \end{cases} \tag{26}$$

ϕ is also given by

$$\phi = \lim_{n \to \infty} x_n. \tag{27}$$

In addition,

$$\phi = \lim_{n \to \infty} \frac{F_n}{F_{n-1}}, \tag{28}$$

where F_n is the nth FIBONACCI NUMBER, as first proved by Scottish mathematician Robert Simson in 1753 (Wells 1986, p. 62).

The SUBSTITUTION MAP

$$0 \to 01 \tag{29}$$

$$1 \to 0 \tag{30}$$

gives

$$0 \to 01 \to 010 \to 01001 \to \dots, \tag{31}$$

giving rise to the sequence

$$01001010010010100101001001001 0 1 \dots \tag{32}$$

(Sloane's A003849). Here, the zeros occur at positions 1, 3, 4, 6, 8, 9, 11, 12, ... (Sloane's A000201), and the ones occur at positions 2, 5, 7, 10, 13, 15, 18, ... (Sloane's A001950). These are complementary BEATTY SEQUENCES generated by $\lfloor n\phi \rfloor$ and $\lfloor n\phi^2 \rfloor$. The sequence also has many connections with the FIBONACCI NUMBERS.

Salem showed that the set of PISOT-VIJAYARAGHAVAN CONSTANTS is closed, with ϕ the smallest accumulation point of the set (Le Lionnais 1983).

See also BERAHA CONSTANTS, DECAGON, FIVE DISKS PROBLEM, GOLDEN RATIO CONJUGATE, GOLDEN RECTANGLE, GOLDEN TRIANGLE, ICOSIDODECAHEDRON, NOBLE NUMBER, PENTAGON, PENTAGRAM, PHI NUMBER SYSTEM, PHYLLOTAXIS, PISOT-VIJAYARAGHAVAN CONSTANT, SECANT METHOD

References

Boyer, C. B. *History of Mathematics.* New York: Wiley, p. 56, 1968.

Coxeter, H. S. M. "The Golden Section, Phyllotaxis, and Wythoff's Game." *Scripta Mathematica* **19**, 135–43, 1953.

Dixon, R. *Mathographics.* New York: Dover, pp. 30–1 and 50, 1991.

Finch, S. "Favorite Mathematical Constants." http://www.mathsoft.com/asolve/constant/cntfrc/cntfrc.html.

Finch, S. "Favorite Mathematical Constants." http://www.mathsoft.com/asolve/constant/gold/gold.html.

Gardner, M. "Phi: The Golden Ratio." Ch. 8 in *The Second Scientific American Book of Mathematical Puzzles & Diversions, A New Selection.* New York: Simon and Schuster, pp. 89–03, 1961.

Gardner, M. "Notes on a Fringe-Watcher: The Cult of the Golden Ratio." *Skeptical Inquirer* **18**, 243–47, 1994.

Hambridge, J. *The Elements of Dynamic Stability.* New York: Dover, 1967.

Herz-Fischler, R. *A Mathematical History of the Golden Number.* New York: Dover, 1998.

Huntley, H. E. *The Divine Proportion.* New York: Dover, 1970.

Knott, R. "Fibonacci Numbers and the Golden Section." http://www.mcs.surrey.ac.uk/Personal/R.Knott/Fibonacci/fib.html.

Le Lionnais, F. *Les nombres remarquables.* Paris: Hermann, p. 40, 1983.

Markowsky, G. "Misconceptions About the Golden Ratio." *College Math. J.* **23**, 2–9, 1992.

Ogilvy, C. S. *Excursions in Geometry.* New York: Dover, pp. 122–34, 1990.

Olariu, A. Golden Section and the Art of Painting. 18 Aug 1999. http://xxx.lanl.gov/abs/physics/9908036/.

Pappas, T. "Anatomy & the Golden Section." *The Joy of Mathematics.* San Carlos, CA: Wide World Publ./Tetra, pp. 32–3, 1989.

Ramanathan, K. G. "On Ramanujan's Continued Fraction." *Acta. Arith.* **43**, 209–26, 1984.

Saaty, T. L. and Kainen, P. C. *The Four-Color Problem: Assaults and Conquest.* New York: Dover, p. 148, 1986.

Sloane, N. J. A. Sequences A000012/M0003, A000201/M2322, A001622/M4046, A001950/M1332, and A003849 in "An On-Line Version of the Encyclopedia of Integer Sequences." http://www.research.att.com/~njas/sequences/eisonline.html.

Steinhaus, H. *Mathematical Snapshots, 3rd ed.* New York: Dover, p. 45, 1999.

van Zanten, A. J. "The Golden Ratio in the Arts of Painting, Building, and Mathematics." *Nieuw Arch. Wisk.* **17**, 229–45, 1999.

Weisstein, E. W. "Books about Golden Ratio." http://www.treasure-troves.com/books/GoldenRatio.html.

Wells, D. *The Penguin Dictionary of Curious and Interesting Numbers.* Middlesex, England: Penguin Books, pp. 36–9, 1986.

Wells, D. *The Penguin Dictionary of Curious and Interesting Geometry.* London: Penguin, pp. 87–8, 1991.

Williams, R. "The Golden Proportion." §2– in *The Geometrical Foundation of Natural Structure: A Source Book of Design.* New York: Dover, pp. 52–3, 1979.

Zeising, A. *Neue Lehre von den Proportionen des menschlichen Körpers.*

Golden Ratio Conjugate

The quantity

$$\phi_C \equiv \frac{1}{\phi} = \phi - 1 = \frac{\sqrt{5} - 1}{2} \approx 0.6180339887, \tag{1}$$

where ϕ is the GOLDEN RATIO. The golden ratio conjugate is sometimes also called the SILVER RATIO. A quantity similar to the FEIGENBAUM CONSTANT can be found for the nth CONTINUED FRACTION representation

$$[a_0, a_1, a_2, \ldots]. \tag{2}$$

Taking the limit of

$$\delta_n \equiv \frac{\sigma_n - \sigma_{n-1}}{\sigma_n - \sigma_{n+1}} \tag{3}$$

gives

$$\delta \equiv \lim_{n \to \infty} = 1 + \phi = 2 + \phi_C. \tag{4}$$

See also GOLDEN RATIO, SILVER RATIO

Golden Rectangle

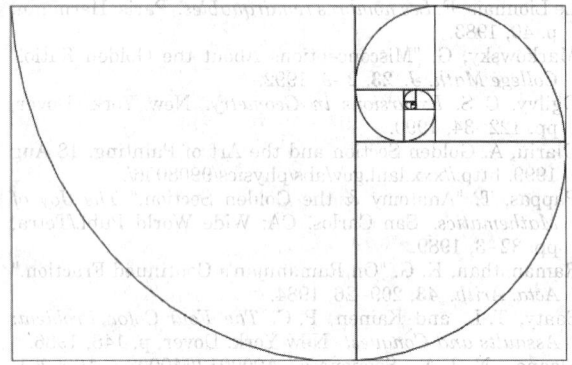

Given a RECTANGLE having sides in the ratio $1 : \phi$, the GOLDEN RATIO ϕ is defined such that partitioning the original RECTANGLE into a SQUARE and new RECTANGLE results in a new RECTANGLE having sides with a ratio $1 : \phi$. Such a RECTANGLE is called a golden rectangle, and successive points dividing a golden rectangle into SQUARES lie on a LOGARITHMIC SPIRAL (Wells 1986, p. 39). The spiral is not actually tangent at these points, however, but passes through them and intersects the adjacent side, as illustrated below.

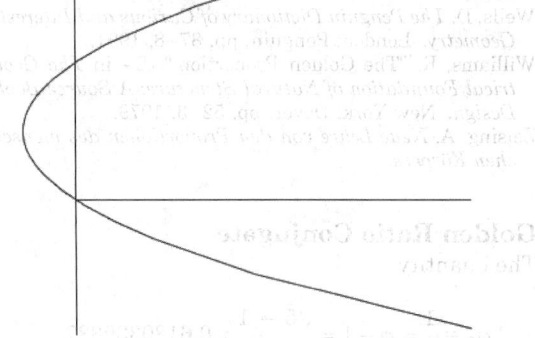

If the top left corner of the original square is

positioned at $(0, 0)$, the center of the spiral occurs at the position

$$x_0 = \sum_{n=0}^{\infty} \left(\frac{1}{\phi^{4n}} + \frac{1}{\phi^{4n+1}} - \frac{1}{\phi^{4n+2}} - \frac{1}{\phi^{4n+3}} \right)$$

$$= (1 + \phi^{-1} - \phi^{-2} - \phi^{-3}) \sum_{n=0}^{\infty} \frac{1}{\phi^{4n}} = \frac{2\phi + 1}{\phi + 2}$$

$$= \tfrac{1}{10}(5 + 3\sqrt{5}) \approx 1.17082 \tag{1}$$

$$y_0 = \sum_{n=0}^{\infty} \left(-\frac{1}{\phi^{4n}} + \frac{1}{\phi^{4n+1}} + \frac{1}{\phi^{4n+2}} - \frac{1}{\phi^{4n+3}} \right)$$

$$= (-1 + \phi^{-1} + \phi^{-2} - \phi^{-3}) \sum_{n=0}^{\infty} -\frac{1}{2 + \phi}$$

$$= \tfrac{1}{10}(\sqrt{5} - 5) \approx -0.276393, \tag{2}$$

and the parameters of the spiral $ae^{b\theta}$ are given by

$$a = (\tfrac{4}{5})^{1/4} \phi^{(\tan^{-1} 2)/\pi} \tag{3}$$

$$b = \frac{2 \ln \phi}{\pi} \approx 0.306349. \tag{4}$$

See also GOLDEN RATIO, GOLDEN TRIANGLE, LOGARITHMIC SPIRAL, RECTANGLE

References

Bicknell, M.; and Hoggatt, V. E. Jr. "Golden Triangles, Rectangles, and Cuboids." *Fib. Quart.* **7**, 73–1, 1969.

Cook, T. A. *The Curves of Life, Being an Account of Spiral Formations and Their Application to Growth in Nature, To Science and to Art.* New York: Dover, 1979.

Cundy, H. and Rollett, A. *Mathematical Models, 3rd ed.* Stradbroke, England: Tarquin Pub., p. 70, 1989.

Pappas, T. "The Golden Rectangle." *The Joy of Mathematics.* San Carlos, CA: Wide World Publ./Tetra, pp. 102–06, 1989.

Steinhaus, H. *Mathematical Snapshots, 3rd ed.* New York: Dover, pp. 45–7, 1999.

Wells, D. *The Penguin Dictionary of Curious and Interesting Geometry.* London: Penguin, p. 88, 1991.

Williams, R. *The Geometrical Foundation of Natural Structure: A Source Book of Design.* New York: Dover, p. 53, 1979.

Golden Root

GOLDEN RATIO

Golden Rule

The mathematical golden rule states that, for any FRACTION, both NUMERATOR and DENOMINATOR may be multiplied by the same number without changing the fraction's value.

See also DENOMINATOR, FRACTION, NUMERATOR

References

Conway, J. H. and Guy, R. K. *The Book of Numbers.* New York: Springer-Verlag, p. 151, 1996.

Golden Section

GOLDEN RATIO

Golden Theorem

QUADRATIC RECIPROCITY THEOREM

Golden Triangle

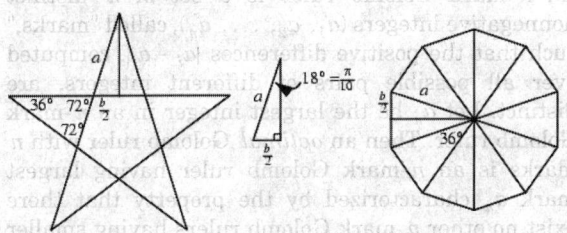

An ISOSCELES TRIANGLE with VERTEX angles 36°. Such TRIANGLES occur in the PENTAGRAM and DECAGON. The legs are in a GOLDEN RATIO to the base. For such a TRIANGLE,

$$\sin(18°) = \sin(\tfrac{1}{10}\pi) = \frac{\tfrac{1}{2}b}{l} \qquad (1)$$

$$b = 2a\sin(\tfrac{1}{10}\pi) = 2a\,\frac{\sqrt{5}-1}{4} = \tfrac{1}{2}a(\sqrt{5}-1) \qquad (2)$$

$$b + l = \tfrac{1}{2}a(\sqrt{5}+1) \qquad (3)$$

$$\frac{b+a}{a} = \frac{\sqrt{5}+1}{2} = \phi. \qquad (4)$$

Kimberling (1991) defines a second type of golden triangle in which the ratio of angles is $\phi : 1$, where ϕ is the GOLDEN RATIO.

See also DECAGON, GOLDEN RATIO, GOLDEN RECTANGLE, ISOSCELES TRIANGLE, PENTAGRAM

References

Bicknell, M.; and Hoggatt, V. E. Jr. "Golden Triangles, Rectangles, and Cuboids." *Fib. Quart.* **7**, 73–1, 1969.
Hoggatt, V. E. Jr. *The Fibonacci and Lucas Numbers.* Boston, MA: Houghton Mifflin, 1969.
Kimberling, C. "A New Kind of Golden Triangle." In *Applications of Fibonacci Numbers: Proceedings of the Fourth International Conference on Fibonacci Numbers and Their Applications,'* Wake Forest University (Ed. G. E. Bergum, A. N. Philippou, and A. F. Horadam). Dordrecht, Netherlands: Kluwer, pp. 171–76, 1991.
Pappas, T. "The Pentagon, the Pentagram & the Golden Triangle." *The Joy of Mathematics.* San Carlos, CA: Wide World Publ./Tetra, pp. 188–89, 1989.
Schoen, R. "The Fibonacci Sequence in Successive Partitions of a Golden Triangle." *Fib. Quart.* **20**, 159–63, 1982.

Goldschmidt Solution

The discontinuous solution of the SURFACE OF REVOLUTION AREA minimization problem for surfaces connecting two CIRCLES. When the CIRCLES are sufficiently far apart, the usual CATENOID is no longer stable and the surface will break and form two surfaces with the CIRCLES as boundaries.

See also CALCULUS OF VARIATIONS, SURFACE OF REVOLUTION

Göllnitz's Theorem

Let $A(n)$ denote the number of PARTITIONS of n into parts $\equiv 2,\ 5,\ 11 \pmod{12}$, let $B(n)$ denote the number of PARTITIONS of n into distinct parts $\equiv 2,\ 4,\ 5 \pmod 6$, and let $C(n)$ denote the number of PARTITIONS of n of the form

$$n = b_1 + b_2 + \ldots + b_t, \qquad (1)$$

where $b_i - b_{i+1} \geq 6$, with strict inequality if $b_i \equiv 0,\ 1$ or $3 \pmod 6$, and $b_t \neq 1,\ 3$. Then

$$A(n) = B(n) = C(n) \qquad (2)$$

(Andrews 1986, p. 101).

The values of $A(n) = B(n) = C(n)$ for $n = 1, 2, \ldots$ are 0, 1, 0, 1, 1, 1, 1, 1, 2, 2, 2, 2, 3, 3, 4, 4, 4, 5, 5, 6, 7, 7, 8, 9, ... (Sloane's A056970). For example, for $n = 24$, there are eight partitions satisfying these conditions, as summarized in the following table.

$A(24) = 8$	$B(24) = 8$	$C(24) = 8$
$17+5+2$	$22+2$	24
$14+5+5$	$20+4$	$22+2$
$14+2+2+2+2+2$	$17+5+2$	$20+4$
$11+11+2$	$16+8$	$19+5$
$11+5+2+2+2+2$	$14+10$	$18+6$
$5+5+5+5+2+2$	$14+8+2$	$17+$
$5+5+2+2+2+2+2$ $+2+2$	$11+8+5$	$16+8$
$2+2+2+2+2+2+2$ $+2+2+2+2+2$	$10+8+4+2$	$14+8+2$

The identity $A(n) = B(n)$ can be established using the identity

$$\sum_{n=0}^{\infty} B(n)q^n = \prod_{n=0}^{\infty} (1+q^{6n+2})(1+q^{6n+4})(1+q^{6n+5}) \qquad (3)$$

$$= \prod_{n=0}^{\infty} \frac{(1-q^{12n+4})(1-q^{12n+8})(1-q^{12n+10})}{(1-q^{6n+2})(1-q^{6n+4})(1-q^{6n+5})} \qquad (4)$$

$$= \prod_{n=0}^{\infty} \frac{1}{(1 - q^{12n+2})(1 - q^{12n+5})(1 - q^{12n+11})} \quad (5)$$

$$= \sum_{n=0}^{\infty} A(n)q^n \quad (6)$$

(Andrews 1986, p. 101). The assertion $B(n) = C(n)$ is significantly more difficult, and no simple proof is known. However, it can be established with the aid of computer algebra and the following refinement of the Göllnitz theorem.

Let $B(n, m)$ denote the number of partitions of n into m distinct parts $\equiv 2, 4, 5, 4, 5 \pmod 6$. Let $C(n, m)$ denote the number of partitions of n of the form

$$n = b_1 + b_2 + \ldots + b_n, \quad (7)$$

where $b_i - b_{i+1} \geq 6$, with strict inequality if $b_i = 0, 1, 3 \pmod 6$, where $b_s \neq 1, 3$, and m is the number of $b_i \equiv 2, 4, 5$ plus twice the number of $b_i \equiv 0, 1, 3$. Then $B(n, m) = C(n, m)$ for each n and m (Göllnitz 1967; Andrews 1986, p. 102).

See also SCHUR'S PARTITION THEOREM

References

Alladi, K. and Berkovich, A. A Double Bounded Key Identity for Göllnitz's (BIG) Partition Theorem. 1 Jul 2000. http://xxx.lanl.gov/abs/math.CO/0007001/.

Andrews, G. E. "Physics, Ramanujan, and Computer Algebra." In *Proc. Conf. Computer Algebra as a Tool for Researchers in Mathematics and Physics* (Ed. D. Chudnovsky and G. Chudnovsky). New York: Springer-Verlag.

Andrews, G. E. "Göllnitz's Theorem." §10.6 in *q-Series: Their Development and Application in Analysis, Number Theory, Combinatorics, Physics, and Computer Algebra.* Providence, RI: Amer. Math. Soc., pp. 101–04, 1986.

Göllnitz, H. "Partitionen mit Differenzenbedingungen." *J. reine angew. Math.* **225**, 154–90, 1967.

Sloane, N. J. A. Sequences A056970 in "An On-Line Version of the Encyclopedia of Integer Sequences." http://www.research.att.com/~njas/sequences/eisonline.html.

Göllnitz-Gordon Identities

$$\sum_{n=0}^{\infty} \frac{q^{n^2}(-q; q^2)_n}{(q^2; q^2)_n} = \frac{1}{(q; q^8)_\infty (q^4; q^8)_\infty (q^7; q^8)_\infty}$$

$$\sum_{n=0}^{\infty} \frac{q^{n(n+2)}(-q; q^2)_n}{(q^2; q^2)_n} = \frac{1}{(q^3; q^8)_\infty (q^4; q^8)_\infty (q^5; q^8)_\infty}.$$

References

Göllnitz, H. "Partitionen mit Differenzenbedingungen." *J. reine angew. Math.* **225**, 154–90, 1967.

Gordon, B. "Some Continued Fractions of the Rogers-Ramanujan Type." *Duke Math. J.* **32**, 741–48, 1965.

Gordon, B. and McIntosh, R. J. "Some Eighth Order Mock Theta Functions." To appear in *J. London Math. Soc.* 2000.

Selberg, A. "Über die Mock-Thetafunktionen siebenter Ordnung." *Arch. Math. og Naturvidenskab* **41**, 3–5, 1938.

Golomb Constant

GOLOMB-DICKMAN CONSTANT

Golomb Ruler

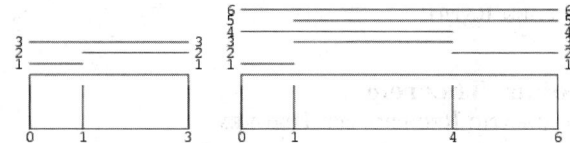

An n-mark Golomb ruler is a set of n distinct nonnegative integers $(a_1, a_2, \ldots, a_n)$, called "marks," such that the positive differences $|a_i - a_j|$, computed over all possible pairs of different integers, are distinct. Let a_n be the largest integer in an n-mark Golomb ruler. Then an *optimal* Golomb ruler with n marks is an n-mark Golomb ruler having largest mark a_n characterized by the property that there exist no other n-mark Golomb rulers having smaller a_n. In such a case, a_n is the called the "length" of the optimal n-mark ruler.

For example, the set $(0, 1, 3, 7)$ is 4-mark Golomb ruler since its differences are $(1 = 1-, 2 = 3-, 3 = 3-, 4 = 7-, 6 = 7-, 7 = 7-)$, all of which are distinct. However, the unique optimal Golomb 4-mark ruler is $(0, 1, 4, 6)$, which measures the distances $(1, 2, 3, 4, 5, 6)$ (and is therefore also a PERFECT RULER). As a further example, it turns out that the length of an optimal 6-mark Golomb ruler is 17. In fact, there are a total of four distinct 6-mark Golomb rulers, all of length 17, one of which is given by $(0, 1, 4, 10, 12, 17)$.

In general, the lengths of the optimal n-mark Golomb rulers for $n = 2, 3, 4, \ldots$ are 1, 3, 6, 11, 17, 25, 34, ... (Sloane's A003022, Vanderschel and Garry). Although the lengths of the optimal n-mark Golomb rulers are not known for $n \geq 23$, the known 21, 22, and 23-mark rulers were proved optimal by the Golomb ruler search project in 1998 and 1999. The number of inequivalent optimal n-mark Golomb rulers for $n = 2, 3, \ldots$ are 1, 1, 1, 2, 4, 5, 1, 1, 1, ... (Sloane's A036501), and the number of distances in an optimal n-mark ruler is given by the TRIANGULAR NUMBER $T_n = n(n-1)/2$, so for $n = 1, 2, \ldots$, the first few are 0, 1, 3, 6, 10, 15, ... (Sloane's A000217).

The following table gives the optimal Golomb rulers for small n. A more complete table is maintained by J. B. Shearer.

n	optimal rulers
2	(0, 1)
3	(0, 1, 3)
4	(0, 1, 4, 6)

5 (0, 1, 4, 9, 11), (0, 3, 4, 9, 11)

6 (0, 1, 4, 10, 12, 17), (0, 1, 4, 10, 15, 17), (0, 3, 5, 9, 16, 17),

   (0, 4, 6, 9, 16, 17)

7 (0, 1, 4, 10, 18, 23, 25), (0, 2, 3, 10, 16, 21, 25),

   (0, 2, 6, 9, 14, 24, 25), (0, 1, 7, 11, 20, 23, 25),

   (0, 3, 4, 12, 18, 23, 25)

8 (0, 1, 4, 9, 15, 22, 32, 34)

See also PERFECT DIFFERENCE SET, PERFECT RULER, RULER, TAYLOR'S CONDITION, WEIGHING

References

Atkinson, M. D.; Santoro, N.; and Urrutia, J. "Integer Sets with Distinct Sums and Differences and Carrier Frequency Assignments for Nonlinear Repeaters." *IEEE Trans. Comm.* **34**, 614–17, 1986.

Colbourn, C. J. and Dinitz, J. H. (Eds.). *CRC Handbook of Combinatorial Designs.* Boca Raton, FL: CRC Press, p. 315, 1996.

Dewdney, A. K. "Computer Recreations." *Sci. Amer.* **253**, 16, June 1985.

Dewdney, A. K. "Computer Recreations." *Sci. Amer.* **254**, 20, Mar. 1986.

distributed.net. "Project OGR." http://www.distributed.net/ogr/.

Golomb, S. W. "How to Number a Graph." In *Graph Theory and Computing* (Ed. R. C. Read). New York: Academic Press, pp. 23–7, 1972.

Guy, R. K. "Modular Difference Sets and Error Correcting Codes." §C10 in *Unsolved Problems in Number Theory, 2nd ed.* New York: Springer-Verlag, pp. 118–21, 1994.

Hewgill, G. "distributed.net OGR Project." http://www.hewgill.com/ogr/.

Kotzig, A. and Laufer, P. J. "Sum Triangles of Natural Numbers Having Minimum Top." *Ars. Combin.* **21**, 5–3, 1986.

Lam, A. W. and D. V. Sarwate, D. V. "On Optimum Time Hopping Patterns." *IEEE Trans. Comm.* **36**, 380–82, 1988.

Miller, L. "Golomb Rulers." http://www.cuug.ab.ca/~millerl/g3-records.html.

Robinson, J. P. and Bernstein, A. J. "A Class of Binary Recurrent Codes with Limited Error Propagation." *IEEE Trans. Inform. Th.* **13**, 106–13, 1967.

Shearer, J. B. "Golomb Rulers." http://www.research.ibm.com/people/s/shearer/grule.html.

Sloane, N. J. A. Sequences A000217/M2535, A003022/M2540, A036501, and A039953 in "An On-Line Version of the Encyclopedia of Integer Sequences." http://www.research.att.com/~njas/sequences/eisonline.html.

Sloane, N. J. A. and Plouffe, S. Figure M2540 in *The Encyclopedia of Integer Sequences.* San Diego, CA: Academic Press, 1995.

Vanderschel, D. and Garry, M. "In Search of the Optimal 20, 21, & 22 Mark Golomb Rulers." http://members.aol.com/golomb20/.

Golomb-Dickman Constant

N.B. A detailed online essay by S. Finch was the starting point for this entry.

Let Π be a PERMUTATION of n elements, and let α_i be the number of CYCLES of length i in this PERMUTATION. Picking Π at RANDOM gives

$$\left\langle \sum_{j=1}^{\infty} \alpha_j \right\rangle = \sum_{i=1}^{n} \frac{1}{i} = \ln n + \gamma + \mathcal{O}\left(\frac{1}{n}\right) \quad (1)$$

$$\mathrm{var}\left(\sum_{j=1}^{\infty} \alpha_j \right) = \sum_{i=1}^{n} \frac{i-1}{i^2} = \ln n + \gamma - \tfrac{1}{6}\pi^2 + \mathcal{O}\left(\frac{1}{n}\right) \quad (2)$$

$$\lim_{n \to \infty} P(\alpha_1 = 0) = \frac{1}{e} \quad (3)$$

(Shepp and Lloyd 1966, Wilf 1990). Goncharov (1942) showed that

$$\lim_{n \to \infty} P(\alpha_j = k) = \frac{1}{k!}\, e^{-1/j} j^{-k}, \quad (4)$$

which is a POISSON DISTRIBUTION, and

$$\lim_{n \to \infty} P\left[\left(\sum_{j=1}^{\infty} \alpha_j - \ln n \right)(\ln n)^{-1/2} \le x \right] = \Phi(x), \quad (5)$$

which is a NORMAL DISTRIBUTION, γ is the EULER-MASCHERONI CONSTANT, and $\Phi(x)$ is the NORMAL DISTRIBUTION FUNCTION.

Let

$$M(\alpha) \equiv \max_{\{}\ j : \alpha_j > 0\}, \quad (6)$$

i.e., the length of the longest cycle in Π. Then Golomb (1959) derived

$$\lambda \equiv \lim_{n \to \infty} \frac{\langle M(\alpha) \rangle}{n} = 0.6243299885\ldots, \quad (7)$$

which is known as the GOLOMB CONSTANT or Golomb-Dickman constant. Knuth (1981) asked for the constants b and c such that

$$\lim_{n \to \infty} n^b \left[\langle M(\alpha) \rangle - \lambda n - \tfrac{1}{2}\lambda \right] = c, \quad (8)$$

and Gourdon (1996) showed that

$$\langle M(\alpha) \rangle = \lambda\left(n + \tfrac{1}{2}\right) - \frac{e^{\gamma}}{24n} + \frac{\tfrac{1}{48}e^{\gamma} - \tfrac{1}{8}(-1)^n}{n^2}$$
$$+ \frac{\tfrac{17}{3840}e^{\gamma} + \tfrac{1}{8}(-1)^n + \tfrac{1}{6}j^{1+2n} + \tfrac{1}{6}j^{2+n}}{n^3}, \quad (9)$$

where

$$j \equiv e^{2\pi i/3}. \quad (10)$$

λ can be expressed in terms of the function $f(x)$ defined by $f(x) = 1$ for $1 \le x \le 2$ and

$$\frac{df}{dx} = -\frac{f(x-1)}{x-1} \quad (11)$$

for $x > 2$, by

$$\lambda = \int_1^\infty \frac{f(x)}{x^2}\, dx. \qquad (12)$$

Shepp and Lloyd (1966) derived

$$\lambda = \int_0^\infty \exp\left(-x - \int_x^\infty \frac{e^{-y}}{y}\, dy\right)$$

$$= \int_0^1 \exp\left(\int_0^x \frac{dy}{\ln y}\right) dx. \qquad (13)$$

Mitchell (1968) computed λ to 53 decimal places.

Surprisingly enough, there is a connection between λ and PRIME FACTORIZATION (Knuth and Pardo 1976, Knuth 1981, pp. 367–68, 395, and 611). Dickman (1930) investigated the probability $P(x, n)$ that the largest PRIME FACTOR p of a random INTEGER between 1 and n satisfies $p < n^x$ for $x \in (0, 1)$. He found that

$$F(x) \equiv \lim_{n \to \infty} P(x, n)$$

$$= \begin{cases} 1 & \text{if } x \geq 1 \\ \int_0^x F\left(\frac{t}{1-t}\right) \frac{dt}{t} & \text{if } 0 \leq x \leq 1. \end{cases} \qquad (14)$$

Dickman then found the average value of x such that $p = n^x$, obtaining

$$\mu \equiv \lim_{n \to \infty} \langle x \rangle = \lim_{n \to \infty} \left\langle \frac{\ln p}{\ln n} \right\rangle = \int_0^1 x\, \frac{dF}{dx}\, dx$$

$$= \int_0^1 F\left(\frac{1}{1-t}\right) dt = 0.62432999, \qquad (15)$$

which is λ.

References

Finch, S. "Favorite Mathematical Constants." http://www.mathsoft.com/asolve/constant/golomb/golomb.html.

Gourdon, X. 1996. http://www.mathsoft.com/asolve/constant/golomb/gourdon.html.

Knuth, D. E. *The Art of Computer Programming, Vol. 1: Fundamental Algorithms, 3rd ed.* Reading, MA: Addison-Wesley, 1997.

Knuth, D. E. *The Art of Computer Programming, Vol. 2: Seminumerical Algorithms, 3rd ed.* Reading, MA: Addison-Wesley, 1998.

Knuth, D. E. and Pardo, L. T. "Analysis of a Simple Factorization Algorithm." *Theor. Comput. Sci.* **3**, 321–48, 1976.

Mitchell, W. C. "An Evaluation of Golomb's Constant." *Math. Comput.* **22**, 411–15, 1968.

Purdom, P. W. and Williams, J. H. "Cycle Length in a Random Function." *Trans. Amer. Math. Soc.* **133**, 547–51, 1968.

Shepp, L. A. and Lloyd, S. P. "Ordered Cycle Lengths in Random Permutation." *Trans. Amer. Math. Soc.* **121**, 350–57, 1966.

Wilf, H. S. *Generatingfunctionology, 2nd ed.* New York: Academic Press, 1993.

Golygon

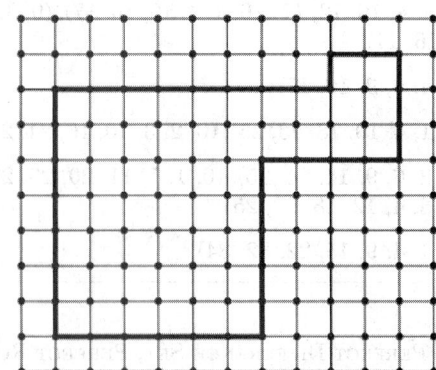

A PLANE path on a set of equally spaced LATTICE POINTS, starting at the ORIGIN, where the first step is one unit to the north or south, the second step is two units to the east or west, the third is three units to the north or south, etc., and continuing until the ORIGIN is again reached. No crossing or backtracking is allowed. The simplest golygon is $(0, 0)$, $(0, 1)$, $(2, 1)$, $(2, -2)$, $(-2, -2)$, $(-2, -7)$, $(-8, -7)$, $(-8, 0)$, $(0, 0)$. A golygon can be formed if there exists an EVEN INTEGER n such that

$$\pm 1 \pm 3 \pm \ldots \pm (n-1) = 0 \qquad (1)$$

$$\pm 2 \pm 4 \pm \ldots \pm n = 0 \qquad (2)$$

(Vardi 1991). Gardner proved that all golygons are OF THE FORM $n = 8k$. The number of golygons of length n (EVEN), with each initial direction counted separately, is the PRODUCT of the COEFFICIENT of $x^{n^2/8}$ in

$$(1+x)(1+x^3) \cdots (1+x^{n-1}), \qquad (3)$$

with the COEFFICIENT of $x^{n(n/2+1)/8}$ in

$$(1+x)(1+x^2) \cdots (1+x^{n/2}). \qquad (4)$$

The number of golygons $N(n)$ of length $8n$ for the first few n are 4, 112, 8432, 909288, ... (Sloane's A006718) and is asymptotic to

$$N(n) \sim \frac{3 \cdot 2^{8n-4}}{\pi n^2 (4n + 1)} \qquad (5)$$

(Sallows *et al.* 1991, Vardi 1991).

See also CANONICAL POLYGON, LATTICE PATH, LATTICE POLYGON

References

Dudeney, A. K. "An Odd Journey Along Even Roads Leads to Home in Golygon City." *Sci. Amer.* **263**, 118–21, July 1990.

Sallows, L. C. F. "New Pathways in Serial Isogons." *Math. Intell.* **14**, 55–7, 1992.

Sallows, L.; Gardner, M.; Guy, R. K.; and Knuth, D. "Serial Isogons of 90 Degrees." *Math Mag.* **64**, 315–24, 1991.

Sloane, N. J. A. Sequences A006718/M3707 in "An On-Line Version of the Encyclopedia of Integer Sequences." http://www.research.att.com/~njas/sequences/eisonline.html.

Smith, H. J. "Golygons." http://pweb.netcom.com/~hjsmith/
Golygons.html.
Vardi, I. "American Science." §5.3 in *Computational Recreations in Mathematica*. Redwood City, CA: Addison-Wesley, pp. 90–6, 1991.

Gomory's Theorem

Regardless of where one white and one black square are deleted from an ordinary 8×8 CHESSBOARD, the reduced board can always be covered exactly with 31 DOMINOES (of dimension 2×1).

See also CHESSBOARD

Gompertz Constant

$$G \equiv \int_0^\infty \frac{e^{-u}}{1+u} \, du = -e \, \text{ei}(-1)$$

$$= 0.596347362\ldots,$$

where $\text{ei}(x)$ is the EXPONENTIAL INTEGRAL. Stieltjes showed it has the CONTINUED FRACTION representation

$$G = \cfrac{1}{2-} \cfrac{1^2}{4-} \cfrac{2^2}{6-} \cfrac{3^2}{8-} \cdots$$

See also EXPONENTIAL INTEGRAL

References

Le Lionnais, F. *Les nombres remarquables*. Paris: Hermann, p. 29, 1983.

Gompertz Curve

The function defined by

$$y = ab^{q^x}.$$

It is used in actuarial science for specifying a simplified mortality law (Kenney and Keeping 1962, p. 241). Using $s(x)$ as the probability that a newborn will achieve age x, the Gompertz law is

$$s(x) = \exp[-m(c^x - 1)],$$

for $c > 1$, $x \geq 0$ (Gompertz 1832).

See also LAW OF GROWTH, LIFE EXPECTANCY, LOGISTIC GROWTH CURVE, MAKEHAM CURVE, POPULATION GROWTH

References

Bowers, N. L. Jr.; Gerber, H. U.; Hickman, J. C.; Jones, D. A.; and Nesbitt, C. J. *Actuarial Mathematics*. Itasca, IL: Society of Actuaries, p. 71, 1997.
Gompertz, B. "On the Nature of the Function Expressive of the Law of Human Mortality, and on a New Mode of Determining the Value of Life Contingencies." *Phil. Trans. Roy. Soc. London* **123**, 513–85, 1832.
Kenney, J. F. and Keeping, E. S. *Mathematics of Statistics, Pt. 1, 3rd ed.* Princeton, NJ: Van Nostrand, 1962.

Gon

GRADIAN

Gonal Number

POLYGONAL NUMBER

Good Binomial Coefficient

A BINOMIAL COEFFICIENT $\binom{N}{k}$ with $k \geq 2$ is called good if its LEAST PRIME FACTOR satisfies

$$\text{lpf}\binom{N}{k} > k$$

(Erdos *et al.* 1993). This is equivalent to the requirement that

$$\text{GCD}\left(\binom{N}{k}, k!\right) = 1.$$

The first few good binomial coefficients are therefore $\binom{3}{2}$, $\binom{5}{4}$, $\binom{6}{2}$, $\binom{7}{2}$, $\binom{7}{3}$, $\binom{7}{4}$, $\binom{7}{6}$, $\binom{10}{2}$, ... Good binomial coefficients are closely related to the ERDOS-SELFRIDGE FUNCTION $g(k)$, which gives the least integer $N > k + 1$ such that $\binom{N}{k}$ is good.

See also BINOMIAL COEFFICIENT, DEFICIENCY, ERDOS-SELFRIDGE FUNCTION, EXCEPTIONAL BINOMIAL COEFFICIENT

References

Erdos, P.; Lacampagne, C. B.; and Selfridge, J. L. "Estimates of the Least Prime Factor of a Binomial Coefficient." *Math. Comput.* **61**, 215–24, 1993.

Good Path

P-GOOD PATH

Good Prime

A PRIME p_n is called "good" if

$$p_n^2 > p_{n-i} p_{n+i}$$

for all $1 \leq i \leq n - 1$ (there is a typo in Guy 1994 in which the is are replaced by 1s). There are infinitely many good primes, and the first few are 5, 11, 17, 29, 37, 41, 53, ... (Sloane's A028388).

See also ANDRICA'S CONJECTURE, LANDAU'S PROBLEMS, PÓLYA CONJECTURE

References

Guy, R. K. "'Good' Primes and the Prime Number Graph." §A14 in *Unsolved Problems in Number Theory, 2nd ed.* New York: Springer-Verlag, pp. 32–3, 1994.
Sloane, N. J. A. Sequences A028388 in "An On-Line Version of the Encyclopedia of Integer Sequences." http://www.research.att.com/~njas/sequences/eisonline.html.

Goodman's Formula

A two-coloring of a COMPLETE GRAPH K_n of n nodes which contains exactly the number of MONOCHRO-

MATIC FORCED TRIANGLES and no more (i.e., a minimum of $R + B$ where R and B are the number of red and blue TRIANGLES) is called an EXTREMAL GRAPH. Goodman (1959) showed that for an extremal graph,

$$R + B = \begin{cases} \frac{1}{3} m(m-1)(m-2) & \text{for } n = 2m \\ \frac{2}{3} m(m-1)(4m+1) & \text{for } n = 4m+1 \\ \frac{2}{3} m(m+1)(4m-1) & \text{for } n = 4m+3. \end{cases}$$

Schwenk (1972) rewrote the equation in the form

$$R + B = \binom{n}{3} - \left\lfloor \frac{1}{2} n \left\lfloor \frac{1}{4}(n-1)^2 \right\rfloor \right\rfloor,$$

where $\binom{n}{k}$ is a BINOMIAL COEFFICIENT and $\lfloor x \rfloor$ is the FLOOR FUNCTION.

See also BLUE-EMPTY GRAPH, EXTREMAL GRAPH, MONOCHROMATIC FORCED TRIANGLE

References
Goodman, A. W. "On Sets of Acquaintances and Strangers at Any Party." *Amer. Math. Monthly* **66**, 778–83, 1959.
Schwenk, A. J. "Acquaintance Party Problem." *Amer. Math. Monthly* **79**, 1113–117, 1972.

Goodstein Sequence

Given a HEREDITARY REPRESENTATION of a number n in BASE b, let $B[b](n)$ be the NONNEGATIVE INTEGER which results if we syntactically replace each b by $b + 1$ (i.e., $B[b]$ is a base change operator that 'bumps the base' from b up to $b + 1$). The HEREDITARY REPRESENTATION of 266 in base 2 is

$$266 = 2^8 + 2^3 + 2$$

$$= 2^{2^2+1} + 2^{2+1} + 2,$$

so bumping the base from 2 to 3 yields

$$B[2](266) = 3^{3^3+1} + 3^{3+1} + 3.$$

Now repeatedly bump the base and subtract 1,

$$G_0(266) = 266 = 2^{2^2+1} + 2^{2+1} + 2$$

$$G_1(266) = B[2](266) - 1 = 3^{3^3+1} + 3^{3+1} + 2$$

$$G_2(266) = B[3](G_1) - 1 = 4^{4^4+1} + 4^{4+1} + 1$$

$$G_3(266) = B[4](G_2) - 1 = 5^{5^5+1} + 5^{5+1} + 1$$

$$G_4(266) = B[5](G_3) - 1 = 6^{6^6+1} + 6^{6+1} - 1$$

$$= 6^{6^6+1} + 5 \cdot 6^6 + 5 \cdot 6^5 + \ldots + 5 \cdot 6 + 5$$

$$G_5(266) = B[6](G_4) - 1$$

$$= 7^{7^7+1} + 5 \cdot 7^7 + 5 \cdot 7^5 + \ldots + 5 \cdot 7 + 4,$$

etc. Starting this procedure at an INTEGER n gives the Goodstein sequence $\{G_k(n)\}$. Amazingly, despite the apparent rapid increase in the terms of the sequence,

GOODSTEIN'S THEOREM states that $G_k(n)$ is 0 for any n and any sufficiently large k.

See also GOODSTEIN'S THEOREM, HEREDITARY REPRESENTATION

References
Goodstein, R. L. "On the Restricted Ordinal Theorem." *J. Symb. Logic* **9**, 33–1, 1944.
Henle, J. M. *An Outline of Set Theory.* New York: Springer-Verlag, 1986.

Goodstein's Theorem

For all n, there exists a k such that the kth term of the GOODSTEIN SEQUENCE $G_k(n) = 0$. In other words, every GOODSTEIN SEQUENCE converges to 0.

The secret underlying Goodstein's theorem is that the HEREDITARY REPRESENTATION of n in base b mimics an ordinal notation for ordinals less than some number. For such ordinals, the base bumping operation leaves the ordinal fixed whereas the subtraction of one decreases the ordinal. But these ordinals are well ordered, and this allows us to conclude that a Goodstein sequence eventually converges to zero.

Goodstein's theorem cannot be proved in PEANO ARITHMETIC (i.e., formal NUMBER THEORY).

See also NATURAL INDEPENDENCE PHENOMENON, PEANO ARITHMETIC

References
Goodstein, R. L. "On the Restricted Ordinal Theorem." *J. Symb. Logic* **9**, 33–1, 1944.
Henle, J. M. *An Outline of Set Theory.* New York: Springer-Verlag, 1986.

Googol

A LARGE NUMBER equal to 10^{100} (i.e., a 1 with 100 zeros following it). Written out explicitly,

10000000000000000000000000000000000000000000000000-
00000000000000000000000000000000000000000000000000-
0000000000000.

See also GOOGOLPLEX, LARGE NUMBER

References
Kasner, E. and Newman, J. R. *Mathematics and the Imagination.* Redmond, WA: Tempus Books, pp. 20–7, 1989.
Pappas, T. "Googol & Googolplex." *The Joy of Mathematics.* San Carlos, CA: Wide World Publ./Tetra, p. 76, 1989.

Googolplex

A LARGE NUMBER equal to $10^{10^{100}}$ (i.e., 1 with a GOOGOL number of 0s written after it).

See also GOOGOL, LARGE NUMBER

References
Kasner, E. and Newman, J. R. *Mathematics and the Imagination.* Redmond, WA: Tempus Books, pp. 23–7, 1989.

Pappas, T. "Googol & Googolplex." *The Joy of Mathematics.*
San Carlos, CA: Wide World Publ./Tetra, p. 76, 1989.

Gordian Distance

A metric characterizing the difference between two
knots K and K' in $\mathbb{S}^3$.

References

Murakami, H. "Some Metrics on Classical Knots." *Math.
Ann.* **270**, 35–5, 1985.

Gordon Function

Another name for the CONFLUENT HYPERGEOMETRIC
FUNCTION OF THE SECOND KIND, defined by

$$\left\{ \frac{\Gamma(1-c)}{\Gamma(1-a)} \left[e^{-\pi c} + \frac{\sin[\pi(a-c)]}{\sin(\pi a)} \right] {}_1F_1(a;\ c;\ z) \right.$$
$$\left. -2\ \frac{\Gamma(c-1)}{\Gamma(c-a)}\ z^{1-c} {}_1F_1(a-c+1;\ 2-c;\ z) \right\},$$

where $\Gamma(x)$ is the GAMMA FUNCTION and ${}_1F_1(a;\ b;\ z)$ is
the CONFLUENT HYPERGEOMETRIC FUNCTION OF THE
FIRST KIND.

See also CONFLUENT HYPERGEOMETRIC FUNCTION OF
THE SECOND KIND

References

Morse, P. M. and Feshbach, H. *Methods of Theoretical
Physics, Part I.* New York: McGraw-Hill, pp. 671–72,
1953.

Gordon Matrix

PRIME ARRAY

Gordon-Luecke Theorem

Two distinct knots cannot have the same exterior. Or,
equivalently, a knot is completely determined by its
KNOT EXTERIOR (Adams 1994, p. 261). The question
was first posed by Tietze in 1908, and finally proved
by Gordon and Luecke (1989).

See also KNOT EXTERIOR

References

Adams, C. C. "The Poincaré Conjecture, Dehn Surgery, and
the Gordon-Luecke Theorem." §9.3 in *The Knot Book: An
Elementary Introduction to the Mathematical Theory of
Knots.* New York: W. H. Freeman, pp. 257–63, 1994.
Gordon, C. and Luecke, J. "Knots Are Determined by Their
Complements." *J. Amer. Math. Soc.* **2**, 371–15, 1989.

Gorenstein Ring

An algebraic RING which appears in treatments of
duality in ALGEBRAIC GEOMETRY. Let A be a local
ARTINIAN RING with $m \subset A$ its maximal IDEAL. Then A
is a Gorenstein ring if the ANNIHILATOR of m has
DIMENSION 1 as a VECTOR SPACE over $K = A/m$.

See also CAYLEY-BACHARACH THEOREM

References

Eisenbud, D.; Green, M.; and Harris, J. "Cayley-Bacharach
Theorems and Conjectures." *Bull. Amer. Math. Soc.* **33**,
295–24, 1996.

Gosper Island

A modification of the KOCH SNOWFLAKE which has
FRACTAL DIMENSION

$$D = \frac{2 \ln 3}{\ln 7} = 1.12915\ldots.$$

The term "Gosper island" was used by Mandelbrot
(1977) because this curve bounds the space filled by
the PEANO-GOSPER CURVE; Gosper and Gardner use
the term FLOWSNAKE FRACTAL instead. Gosper islands
can TILE the PLANE.

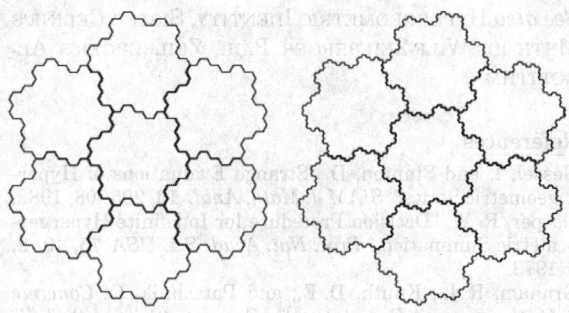

See also KOCH SNOWFLAKE, PEANO-GOSPER CURVE

References

Mandelbrot, B. B. *Fractals: Form, Chance, & Dimension.*
San Francisco, CA: W. H. Freeman, Plate 46, 1977.

Gosper's Algorithm

An ALGORITHM for finding closed form HYPERGEO-
METRIC IDENTITIES. The algorithm treats sums whose
successive terms have ratios which are RATIONAL
FUNCTIONS. Not only does it decide conclusively
whether there exists a hypergeometric sequence z_n
such that

$$t_n = z_{n+1} - z_n, \tag{1}$$

but actually produces z_n if it exists. If not, it produces
$\Sigma_{k=0}^{n-1} t_k$. An outline of the algorithm follows (Petkovsek 1996):

1. For the ratio $r(n) = t_{n+1}/t_n$ which is a RATIONAL FUNCTION of n.
2. Write

$$r(n) = \frac{a(n)}{b(n)} \frac{c(n+1)}{c(n)}, \qquad (2)$$

where $a(n)$, $b(n)$, and $c(n)$ are polynomials satisfying

$$\mathrm{GCD}(a(n), \, b(n+h)) = 1 \qquad (3)$$

for all nonnegative integers h.
3. Find a nonzero polynomial solution $x(n)$ of

$$a(n)x(n+1) - b(n-1)x(n) = c(n), \qquad (4)$$

if one exists.
4. Return $b(n-1)x(n)/c(n)t_n$ and stop.

Petkovsek *et al.* (1996) describe the algorithm as "one of the landmarks in the history of computerization of the problem of closed form summation." Gosper's algorithm is vital in the operation of ZEILBERGER'S ALGORITHM and the machinery of WILF-ZEILBERGER PAIRS.

See also HYPERGEOMETRIC IDENTITY, SISTER CELINE'S METHOD, WILF-ZEILBERGER PAIR, ZEILBERGER'S ALGORITHM

References

Gessel, I. and Stanton, D. "Strange Evaluations of Hypergeometric Series." *SIAM J. Math. Anal.* **13**, 295–08, 1982.
Gosper, R. W. "Decision Procedure for Indefinite Hypergeometric Summation." *Proc. Nat. Acad. Sci. USA* **75**, 40–2, 1978.
Graham, R. L.; Knuth, D. E.; and Patashnik, O. *Concrete Mathematics: A Foundation for Computer Science, 2nd ed.* Reading, MA: Addison-Wesley, 1994.
Koepf, W. "Algorithms for m-fold Hypergeometric Summation." *J. Symb. Comput.* **20**, 399–17, 1995.
Koepf, W. "Gosper's Algorithm." Ch. 5 in *Hypergeometric Summation: An Algorithmic Approach to Summation and Special Function Identities.* Braunschweig, Germany: Vieweg, pp. 61–9, 1998.
Lafron, J. C. "Summation in Finite Terms." In *Computer Algebra Symbolic and Algebraic Computation, 2nd ed.* (Ed. B. Buchberger, G. E. Collins, and R. Loos). New York: Springer-Verlag, 1983.
Paule, P. and Schorn, M. "A Mathematica Version of Zeilberger's Algorithm for Proving Binomial Coefficient Identities." *J. Symb. Comput.* **20**, 673–98, 1995.
Petkovsek, M.; Wilf, H. S.; and Zeilberger, D. "Gosper's Algorithm." Ch. 5 in $A = B$. Wellesley, MA: A. K. Peters, pp. 73–9, 1996.
Zeilberger, D. "The Method of Creative Telescoping." *J. Symb. Comput.* **11**, 195–04, 1991.

Gosper's Method
GOSPER'S ALGORITHM

Gossip Problem
GOSSIPING

Gossiping
This entry contributed by RONALD M. AARTS

Gossiping and broadcasting are two problems of information dissemination described for a group of individuals connected by a communication network. In gossiping, every person in the network knows a unique item of information and needs to communicate it to everyone else. In broadcasting, one individual has an item of information which needs to be communicated to everyone else (Hedetniemi *et al.* 1988).

A popular formulation assumes there are n people, each one of whom knows a scandal which is not known to any of the others. They communicate by telephone, and whenever two people place a call, they pass on to each other as many scandals as they know. How many calls are needed before everyone knows about all the scandals? Denoting the scandal-spreaders as A, B, C, and D, a solution for $n = 4$ is given by $\{A, B\}$, $\{C, D\}$, $\{A, C\}$, $\{B, D\}$. The $n = 4$ solution can then be generalized to $n > 4$ by adding the pair $\{A, X\}$ to the beginning and end of the previous solution, i.e., $\{A, E\}$, $\{A, B\}$, $\{C, D\}$, $\{A, C\}$, $\{B, D\}$, $\{A, E\}$.

Gossiping (which is also called total exchange or all-to-all communication) was originally introduced in discrete mathematics as a combinatorial problem in GRAPH THEORY, but it also has applications in communications and distributed memory multiprocessor systems (Bermond *et al.* 1998). Moreover, the gossip problem is implicit in a large class of parallel computation problems, such as linear system solving, the DISCRETE FOURIER TRANSFORM, and SORTING. Surveys are given in Hedetniemi *et al.* (1988) and Hromkovic *et al.* (1995).

Let $f(n)$ be the number of minimum calls necessary to complete gossiping among n people, where any pair of people may call each other. Then $f(1) = 0$, $f(2) = 1$, $f(3) = 3$, and

$$f(n) = 2n - 4$$

for $n \geq 4$. This result was proved by (Tijdeman 1971), as well as many others.

In the case of one-way communication ("polarized telephones"), e.g., where communication is done by letters or telegrams, the graph becomes a DIRECTED GRAPH and the minimum number of calls becomes

$$f(n) = 2n - 2$$

for $n \geq 4$ (Harary and Schwenk 1974).

References

Bermond, J.-C.; Gargano, L.; Rescigno, A. A.; and Vaccaro, U. "Fast Gossiping by Short Messages." *SIAM J. Comput.* **27**, 917–41, 1998.
Harary, F. and Schwenk, A. J. "The Communication Problem on Graphs and Digraphs." *J. Franklin Inst.* **297**, 491–95, 1974.

Hedetniemi, S. M.; Hedetniemi, S. T.; and Liestman, A. L. "A Survey of Gossiping and Broadcasting in Communication Networks." *Networks* **18**, 319–49, 1988.

Hromkovic, J.; Klasing, R.; Monien, B.; and Peine, R. "Dissemination of Information in Interconnection Networks (Broadcasting and Gossiping)." In *Combinatorial Network Theory* (Ed. F. Hsu and D.-A. Du). Norwell, MA: Kluwer, pp. 125–12, 1995.

Tijdeman, R. "On a Telephone Problem." *Nieuw Archief voor Wiskunde* **19**, 188–92, 1971.

Gould and Hsu Matrix Inversion Formula

Let (a_i) be a sequence of complex numbers and let the LOWER TRIANGULAR MATRICES $F = (F(n, k))$ and $G = (G(n, k))$ be defined as

$$F(n, k) = \frac{\prod_{j=k}^{n-1}(a_j + k)}{(n-k)!}$$

and

$$G(n, k) = (-1)^{n-k} \frac{a_k + k}{a_n + n} \frac{\prod_{j=k+1}^{n}(a_j + n)}{(n-k)!},$$

where the product over an EMPTY SET is 1. Then F and G are MATRIX INVERSES (Bhatnagar 1995, pp. 15–6 and 50–1). The KRATTENTHALER MATRIX INVERSION FORMULA is a generalization of this result.

See also KRATTENTHALER MATRIX INVERSION FORMULA

References

Bhatnagar, G. *Inverse Relations, Generalized Bibasic Series, and their U(n) Extensions.* Ph.D. thesis. Ohio State University, 1995.

Carlitz, L. "Some Inversion Relations." *Duke Math. J.* **40**, 803–01, 1972.

Chu, W. C. and Hsu, L. C. "Some New Applications of Gould-Hsu Inversions." *J. Combin. Inform. System Sci.* **14**, 1–, 1990.

Gessel, I. and Stanton, D. "Application of q-Lagrange Inversion to Basic Hypergeometric Series." *Trans. Amer. Math. Soc.* **277**, 173–01, 1983.

Gould, H. W. and Hsu, L. C. "Some New Inverse Series Relations." *Duke Math. J.* **40**, 885–91, 1973.

Riordan, J. *Combinatorial Identities.* New York: Wiley, 1979.

Gould Polynomial

The polynomials $G_n(x; a, b)$ given by the associated SHEFFER SEQUENCE with

$$f(t) = e^{at}(e^{bt} - 1),$$

where $b \neq 0$. The INVERSE FUNCTION (and therefore GENERATING FUNCTION) cannot be computed algebraically, but the GENERATING FUNCTION

$$\sum_{k=0}^{\infty} \frac{G_k(x; a, b)}{k!} t^k = e^{xf^{-1}(t)} \tag{1}$$

can be given in terms of the sum

$$f^{-1}(t) = \sum_{k=1}^{\infty} \frac{1}{b}\binom{-(b+ak)/b}{k-1} \frac{t^k}{k}. \tag{2}$$

This results in

$$G_n(x; a, b) = \frac{x}{x - an}\left(\frac{x - an}{b}\right)_n$$

where $(x)_n$ is a FALLING FACTORIAL. The first few are

$$G_0(x; a, b) = 1$$

$$G_1(x; a, b) = \frac{x}{b}$$

$$G_2(x; a, b) = -\frac{(2a + b - x)}{b^2}$$

$$G_3(x; a, b) = \frac{(3a + b - x)(3a + 2b - x)x}{b^3}$$

$$G_4(x; a, b)$$
$$= -\frac{(4a + b - x)(4a + 2b - x)(4a + 3b - x)x}{b^4}.$$

The binomial identity obtained from the SHEFFER SEQUENCE gives the generalized CHU-VANDERMONDE IDENTITY

$$\frac{x + y}{x + y - an}\binom{(x+y-an)/b}{n}$$
$$= \sum_{k=0}^{n} \frac{x}{x - ak} \frac{y}{y - a(n-k)} \binom{x - ak}{b}{k}\binom{y - a(n-k)}{a}{n-k} \tag{3}$$

(Roman 1984, p. 69).

In the special case $a = -b/2$, the function $f(t)$ simplifies to

$$f(t) = e^{bt/2} - e^{-bt/2} = 2\sinh(\tfrac{1}{2}bt), \tag{4}$$

which gives the GENERATING FUNCTION

$$\sum_{k=0}^{\infty} \frac{G_k(x; -\tfrac{1}{2}b, b)}{k!} t^k = \exp\left[\frac{2x\sinh^{-1}(\tfrac{1}{2}t)}{b}\right], \tag{5}$$

giving the polynomials

$$G_0(x; -b/2, b) = 1$$

$$G_1(x; -b/2, b) = \frac{x}{b}$$

$$G_2(x; -b/2, b) = \frac{x^2}{b^2}$$

$$G_3(x; -b/2, b) = -\frac{(b - 2x)x(b + 2x)}{4b^3}$$

$$G_4(x; -b/2, b) = -\frac{(b - x)x^2(b + x)}{b^4}.$$

See also CENTRAL FACTORIAL, FALLING FACTORIAL, SHEFFER SEQUENCE

References

Gould, H. W. "Note on a Paper of Sparre-Anderson." *Math. Scand.* **6**, 226–30, 1958.

Gould, H. W. "Stirling Number Representation Problems." *Proc. Amer. Math. Soc.* **11**, 447–51, 1960.

Gould, H. W. "A Series of Transformation for Finding Convolution Identities." *Duke Math. J.* **28**, 193–02, 1961.

Gould, H. W. "Note on a Paper of Klamkin Concerning Stirling Numbers." *Amer. Math. Monthly* **68**, 477–79, 1961.

Gould, H. W. "A New Convolution Formula and Some New Orthogonal Relations for the Inversion of Series." *Duke Math. J.* **29**, 393–04, 1962.

Gould, H. W. "Congruences Involving Sums of Binomial Coefficients and a Formula of Jensen." *Amer. Math. Monthly* **69**, 400–02, 1962.

Roman, S. "The Gould Polynomials and he Central Factorial Polynomials." §4.1.4 in *The Umbral Calculus.* New York: Academic Press, pp. 67–0, 1984.

Rota, G.-C.; Kahaner, D.; Odlyzko, A. "On the Foundations of Combinatorial Theory. VIII: Finite Operator Calculus." *J. Math. Anal. Appl.* **42**, 684–60, 1973.

Goursat Problem

For the HYPERBOLIC PARTIAL DIFFERENTIAL EQUATION

$$u_{xy} = F(x, y, u, p, q) \tag{1}$$

$$p = u_x \tag{2}$$

$$q = u_y \tag{3}$$

on a domain Ω, Goursat's problem asks to find a solution $u(x, y)$ of (3) from the BOUNDARY CONDITIONS

$$u(0, t) = \phi(t) \tag{4}$$

$$u(t, 1) = \psi(t) \tag{5}$$

$$\phi(1) = \phi(0) \tag{6}$$

for $0 \leq t \leq 1$ that is regular in Ω and continuous in the closure $\bar{\Omega}$, where ϕ and ψ are specified continuously differentiable functions.

The linear Goursat problem corresponds to the solution of the equation

$$\tilde{L}u = u_{xy} + au_x + bu_y + cu = f, \tag{7}$$

which can be effected using the so-called RIEMANN FUNCTION $R(x, y; \xi, \eta)$. The use of the RIEMANN FUNCTION to solve the linear Goursat problem is called the RIEMANN METHOD.

See also BOUNDARY VALUE PROBLEM, HYPERBOLIC PARTIAL DIFFERENTIAL EQUATION, FUNCTION, RIEMANN METHOD

References

Courant, R. and Hilbert, D. *Methods of Mathematical Physics, Vol. 2.* New York: Wiley, 1989.

Goursat, E. *Cours d'analyse mathématique, Vol. 3, Part 1.* Paris: Gauthier-Villars, 1923.

Hazewinkel, M. (Managing Ed.). *Encyclopaedia of Mathematics: An Updated and Annotated Translation of the Soviet "Mathematical Encyclopaedia."* Dordrecht, Netherlands: Reidel, p. 289, 1988.

Tricomi, F. G. *Integral Equations.* New York: Interscience, 1957.

Goursat's Surface

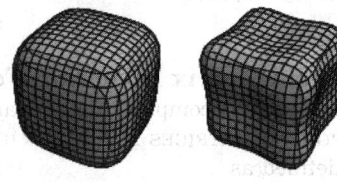

A general QUARTIC SURFACE defined by

$$x^4 + y^4 + z^4 + a(x^2 + y^2 + z^2)^2 + b(x^2 + y^2 + z^2) + c$$

(Gray 1997, p. 314). The above two images correspond to $a = b = 0$, $c = -1$, and $a = 0$, $b = -2$, $c = -1$, respectively.

The related surface

$$x^n + y^n + z^n = 1$$

for $n \geq 2$ an even integer is considered by Gray (1997, p. 292), and might appropriately be called a SUPERELLIPSOID.

See also CHMUTOV SURFACE, CUBE, SUPERELLIPSOID, TOOTH SURFACE

References

Banchoff, T. F. "Computer Graphics Tools for Rendering Algebraic Surfaces and for Geometry of Order." In *Geometric Analysis and Computer Graphics: Proceedings of a Workshop Held May 23–5, 1988* (Eds. P. Concus, R. Finn, D. A. Hoffman). New York: Springer-Verlag, pp. 31–7, 1991.

Goursat, E. "Eacute;tude des surfaces qui admettent tous les plans de symétrie d'un polyèdre régulier." *Ann. Sci. École Norm. Sup.* **4**, 159–000, 1897.

Gray, A. *Modern Differential Geometry of Curves and Surfaces with Mathematica, 2nd ed.* Boca Raton, FL: CRC Press, pp. 292 and 314, 1997.

Graceful Graph

A LABELED GRAPH which can be "gracefully numbered" is called a graceful graph. Label the nodes with distinct NONNEGATIVE INTEGERS. Then label the EDGES with the absolute differences between node values. If the EDGE numbers then run from 1 to e, the graph is gracefully numbered. In order for a graph to be graceful, it must be without loops or multiple EDGES.

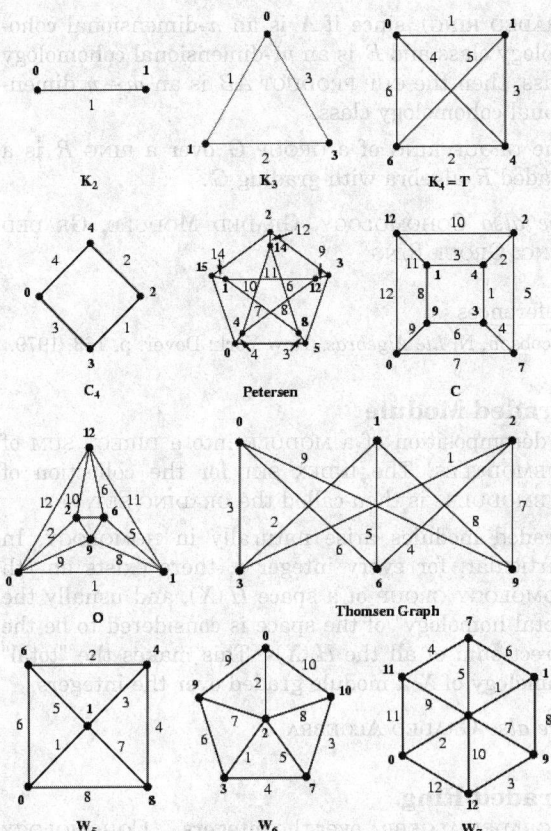

Golomb showed that the number of EDGES connecting the EVEN-numbered and ODD-numbered sets of nodes is $\lfloor (e+1/)2 \rfloor$, where e is the number of EDGES. In addition, if the nodes of a graph are all of EVEN ORDER, then the graph is graceful only if $\lfloor (e+1/)2 \rfloor$ is EVEN. The only ungraceful simple graphs with ≤ 5 nodes are shown below.

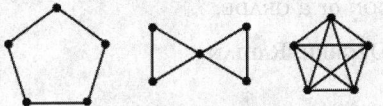

There are exactly $e!$ graceful graphs with e EDGES (Sheppard 1976), where $e!/2$ of these correspond to different labelings of the same graph. Golomb (1974) showed that all complete bipartite graphs are graceful. CATERPILLAR GRAPHS; COMPLETE GRAPHS K_2, K_3, $K_4 = W_4 = T$ (and only these; Golomb 1974); CYCLIC GRAPHS C_n when $n \equiv 0$ or $3 (\bmod 4)$, when the number of consecutive chords $k = 2$, 3, or $n - 3$ (Koh and Punnim 1982), or when they contain a P_k chord (Delorme *et al.* 1980, Koh and Yap 1985, Punnim and Pabhapote 1987); GEAR GRAPHS; PATH GRAPHS; the PETERSEN GRAPH; POLYHEDRAL GRAPHS $T = K_4 = W_4, C, O, D$, and I (Gardner 1983); STAR GRAPHS; the THOMSEN GRAPH (Gardner 1983); and WHEEL GRAPHS (Frucht 1988) are all graceful.

Some graceful graphs have only one numbering, but others have more than one. It is conjectured that all

trees are graceful (Bondy and Murty 1976), but this has only been proved for trees with ≤ 16 VERTICES. It has also been conjectured that all unicyclic graphs are graceful.

See also HARMONIOUS GRAPH, LABELED GRAPH

References

Abraham, J. and Kotzig, A. "All 2-Regular Graphs Consisting of 4-Cycles are Graceful." *Disc. Math.* **135**, 1–4, 1994.

Abraham, J. and Kotzig, A. "Extensions of Graceful Valuations of 2-Regular Graphs Consisting of 4-Gons." *Ars Combin.* **32**, 257–62, 1991.

Bloom, G. S. and Golomb, S. W. "Applications of Numbered Unidirected Graphs." *Proc. IEEE* **65**, 562–70, 1977.

Bolian, L. and Xiankun, Z. "On Harmonious Labellings of Graphs." *Ars Combin.* **36**, 315–26, 1993.

Bondy, J. A. and Murty, U. S. R. *Graph Theory with Applications.* New York: North Holland, p. 248, 1976.

Brualdi, R. A. and McDougal, K. F. "Semibandwidth of Bipartite Graphs and Matrices." *Ars Combin.* **30**, 275–87, 1990.

Cahit, I. "Are All Complete Binary Trees Graceful?" *Amer. Math. Monthly* **83**, 35–7, 1976.

Delorme, C.; Maheo, M.; Thuillier, H.; Koh, K. M.; and Teo, H. K. "Cycles with a Chord are Graceful." *J. Graph Theory* **4**, 409–15, 1980.

Frucht, R. W. and Gallian, J. A. "Labelling Prisms." *Ars Combin.* **26**, 69–2, 1988.

Gallian, J. A. "A Survey: Recent Results, Conjectures, and Open Problems in Labelling Graphs." *J. Graph Th.* **13**, 491–04, 1989.

Gallian, J. A. "Open Problems in Grid Labeling." *Amer. Math. Monthly* **97**, 133–35, 1990.

Gallian, J. A. "A Guide to the Graph Labelling Zoo." *Disc. Appl. Math.* **49**, 213–29, 1994.

Gallian, J. A.; Prout, J.; and Winters, S. "Graceful and Harmonious Labellings of Prism Related Graphs." *Ars Combin.* **34**, 213–22, 1992.

Gardner, M. "Golomb's Graceful Graphs." Ch. 15 in *Wheels, Life, and Other Mathematical Amusements.* New York: W. H. Freeman, pp. 152–65, 1983.

Golomb, S. W. "How to Number a Graph." In *Graph Theory and Computing* (Ed. R. C. Read). New York: Academic Press, pp. 23–7, 1972.

Golomb, S. W. "The Largest Graceful Subgraph of the Complete Graph." *Amer. Math. Monthly* **81**, 499–01, 1974.

Guy, R. "Monthly Research Problems, 1969–5." *Amer. Math. Monthly* **82**, 995–004, 1975.

Guy, R. "Monthly Research Problems, 1969–979." *Amer. Math. Monthly* **86**, 847–52, 1979.

Guy, R. K. "The Corresponding Modular Covering Problem. Harmonious Labelling of Graphs." §C13 in *Unsolved Problems in Number Theory, 2nd ed.* New York: Springer-Verlag, pp. 127–28, 1994.

Huang, J. H. and Skiena, S. "Gracefully Labelling Prisms." *Ars Combin.* **38**, 225–42, 1994.

Koh, K. M. and Punnim, N. "On Graceful Graphs: Cycles with 3-Consecutive Chords." *Bull. Malaysian Math. Soc.* **5**, 49–4, 1982.

Jungreis, D. S. and Reid, M. "Labelling Grids." *Ars Combin.* **34**, 167–82, 1992.

Koh, K. M. and Yap, K. Y. "Graceful Numberings of Cycles with a P_3-Chord." *Bull. Inst. Math. Acad. Sinica* **13**, 41–8, 1985.

Moulton, D. "Graceful Labellings of Triangular Snakes." *Ars Combin.* **28**, 3–3, 1989.

Punnim, N. and Pabhapote, N. "On Graceful Graphs: Cycles with a P_k-Chord, $k \geq 4$." *Ars Combin. A* **23**, 225–28, 1987.

Rosa, A. "On Certain Valuations of the Vertices of a Graph." In *Theory of Graphs, International Symposium, Rome, July 1966.* New York: Gordon and Breach, pp. 349–55, 1967.

Sheppard, D. A. "The Factorial Representation of Balanced Labelled Graphs." *Discr. Math.* **15**, 379–88, 1976.

Sierksma, G. and Hoogeveen, H. "Seven Criteria for Integer Sequences Being Graphic." *J. Graph Th.* **15**, 223–31, 1991.

Slater, P. J. "Note on k-Graceful, Locally Finite Graphs." *J. Combin. Th. Ser. B* **35**, 319–22, 1983.

Snevily, H. S. "New Families of Graphs That Have α-Labellings." Preprint.

Snevily, H. S. "Remarks on the Graceful Tree Conjecture." Preprint.

Xie, L. T. and Liu, G. Z. "A Survey of the Problem of Graceful Trees." *Qufu Shiyuan Xuebao* **1**, 8–5, 1984.

Graceful Permutation

A graceful permutation σ on n letters is a PERMUTATION such that

$$\{|\sigma(i) - \sigma(i+1)| : i = 1, 2, \ldots, n-1\}$$
$$= \{1, 2, \ldots, n-1\}.$$

For example, there are four graceful permutations on $\{1, 2, 3, 4\}$: $\{1, 4, 2, 3\}$, $\{2, 3, 1, 4\}$, $\{3, 2, 4, 1\}$, and $\{4, 1, 3, 2\}$. The number of graceful permutations on n letters for $n = 1, 2, \ldots$ are 1, 2, 4, 4, 8, 24, 32, 40, ... (Sloane's A006967).

References

Sloane, N. J. A. Sequences A006967/M3229 in "An On-Line Version of the Encyclopedia of Integer Sequences." http://www.research.att.com/~njas/sequences/eisonline.html.

Wilf, H. "On Crossing Numbers, and Some Unsolved Problems." In *Combinatorics, Geometry, and Probability: A Tribute to Paul Erdos. Papers from the Conference in Honor of Erdos' 80th Birthday Held at Trinity College, Cambridge, March 1993* (Ed. B. Bollobás and A. Thomason). Cambridge, England: Cambridge University Press, pp. 557–62, 1997.

Wilf, H. S. and Yoshimura, N. "Ranking Rooted Trees and a Graceful Application." In *Discrete Algorithms and Complexity (Proceedings of the Japan-US Joint Seminar June 4–, 1986, Kyoto, Japan)* (Ed. D. Johnson, T. Nishizeki, A. Nozaki and H. S. Wilf). Boston, MA: Academic Press, pp. 341–50, 1987.

Grade

GRADIAN

Graded Algebra

If A is a GRADED MODULE and there EXISTS a degree-preserving linear map $\phi : A \otimes A \to A$, then (A, ϕ) is called a graded algebra.

COHOMOLOGY is a graded algebra. In addition, the GRADING SET is MONOID having a compatibility relation such that if A is in the a grading of the algebra M, and B is in the b grading of the algebra M, then AB is in the ab grading of the algebra (where A and B are multiplied in M, and a and b are multiplied in the index monoid). For example, cohomology of a space is a graded algebra over the integers (i.e., a

GRADED RING), since if A is an n-dimensional cohomology class and B is an m-dimensional cohomology class, then the CUP PRODUCT AB is an $m + n$ dimensional cohomology class.

The GROUP RING of a GROUP G over a RING R is a graded R-algebra with grading G.

See also COHOMOLOGY, GRADED MODULE, GRADED RING, GROUP RING

References

Jacobson, N. *Lie Algebras.* New York: Dover, p. 163, 1979.

Graded Module

A decomposition of a MODULE into a DIRECT SUM of SUBMODULES. The INDEX SET for the collection of SUBMODULES is then called the GRADING SET.

Graded modules arise naturally in HOMOLOGY. In particular, for every integer i, there exists an ith HOMOLOGY GROUP of a space $H_i(X)$, and usually the "total homology" of the space is considered to be the direct sum of all the $H_i(X)$s. This makes the "total" homology of X a module graded over the integers.

See also GRADED ALGEBRA

Graded Ring

A GRADED ALGEBRA over the integers $\mathbb{Z}$. COHOMOLOGY of a space is a graded ring.

See also GRADED ALGEBRA

Gradian

A unit of angular measure in which the angle of an entire CIRCLE is 400 gradians. A RIGHT ANGLE is therefore 100 gradians. A gradian is sometimes also called a GON or a GRADE.

See also DEGREE, RADIAN

References

Harris, J. W. and Stocker, H. *Handbook of Mathematics and Computational Science.* New York: Springer-Verlag, p. 63, 1998.

Gradient

The gradient is a VECTOR operator denoted ∇ and sometimes also called DEL or NABLA. It is most often applied to a real function of three variables $f(u_1, u_2, u_3)$, and may be denoted

$$\nabla f \equiv \text{grad}(f). \tag{1}$$

For general CURVILINEAR COORDINATES, the gradient is given by

$$\nabla \phi = \frac{1}{h_1} \frac{\partial \phi}{\partial u_1} \hat{\mathbf{u}}_1 + \frac{1}{h_2} \frac{\partial \phi}{\partial u_2} \hat{\mathbf{u}}_2 + \frac{1}{h_3} \frac{\partial \phi}{\partial u_3} \hat{\mathbf{u}}_3, \tag{2}$$

which simplifies to

$$\nabla\phi(x,\ y,\ z)=\frac{\partial\phi}{\partial x}\ \hat{\mathbf{x}}+\frac{\partial\phi}{\partial y}\ \hat{\mathbf{y}}+\frac{\partial\phi}{\partial z}\ \hat{\mathbf{z}} \tag{3}$$

in CARTESIAN COORDINATES.

The direction of ∇f is the orientation in which the DIRECTIONAL DERIVATIVE has the largest value and $|\nabla f|$ is the value of that DIRECTIONAL DERIVATIVE. Furthermore, if $\nabla f\neq0$, then the gradient is PERPENDICULAR to the LEVEL CURVE through $(x_0,\ y_0)$ if $z=f(x,\ y)$ and PERPENDICULAR to the level surface through $(x_0,\ y_0,\ z_0)$ if $F(x,\ y,\ z)=0$.

In TENSOR notation, let

$$ds^2=g_\mu\,dx_\mu^2 \tag{4}$$

be the LINE ELEMENT in principal form. Then

$$\nabla_{\vec{e}_\alpha}\vec{e}_\beta=\nabla_\alpha\vec{e}_\beta=\frac{1}{\sqrt{g\alpha}}\frac{\partial}{\partial x_\alpha}\vec{e}_\beta. \tag{5}$$

For a MATRIX A,

$$\nabla|A\mathbf{x}|=\frac{(A\mathbf{x})^T A}{|A\mathbf{x}|}. \tag{6}$$

For expressions giving the gradient in particular coordinate systems, see CURVILINEAR COORDINATES.

See also CONVECTIVE DERIVATIVE, CURL, DIVERGENCE, LAPLACIAN, VECTOR DERIVATIVE

References
Arfken, G. "Gradient, ∇" and "Successive Applications of ∇." §1.6 and 1.9 in *Mathematical Methods for Physicists, 3rd ed.* Orlando, FL: Academic Press, pp. 33–7 and 47–1, 1985.

Gradient Descent Method
STEEPEST DESCENT METHOD

Gradient Four-Vector
The 4-dimensional version of the GRADIENT, encountered frequently in general relativity and special relativity, is

$$\nabla_\mu=\begin{bmatrix}\frac{1}{c}\frac{\partial}{\partial t}\\\frac{\partial}{\partial x}\\\frac{\partial}{\partial y}\\\frac{\partial}{\partial z}\end{bmatrix},$$

which can be written

$$(\nabla^\mu)^2\equiv\Box^2,$$

where $\Box^2$ is the D'ALEMBERTIAN.

See also D'ALEMBERTIAN, GRADIENT, TENSOR, VECTOR

References
Morse, P. M. and Feshbach, H. "The Differential Operator ∇." §1.4 in *Methods of Theoretical Physics, Part I.* New York: McGraw-Hill, pp. 31–4, 1953.

Gradient Theorem

$$\int_b^a(\nabla f)\cdot ds=f(b)-f(a),$$

where ∇ is the GRADIENT, and the integral is a LINE INTEGRAL. It is this relationship which makes the definition of a scalar potential function f so useful in gravitation and electromagnetism as a concise way to encode information about a VECTOR FIELD.

See also DIVERGENCE THEOREM, GREEN'S THEOREM, LINE INTEGRAL, POINCARÉ'S THEOREM

Grading Set
The INDEX SET for the collection of SUBMODULES in a GRADED MODULE.

See also GRADED MODULE

Graeco-Latin Square
EULER SQUARE

Graeco-Roman Square
EULER SQUARE

Graeffe Iteration
GRAEFFE'S METHOD

Graeffe's Method
A ROOT-finding method which was among the most popular methods for finding roots of UNIVARIATE POLYNOMIALS in the 19th and 20th centuries. It was invented independently by Graeffe, dandelin, and Lobachevsky (Householder 1959, Malajovich and Zubelli 1999). Graeffe's method has a number of drawbacks, among which are that its usual formulation leads to exponents exceeding the maximum allowed by floating-point arithmetic and also that it can map well-conditioned polynomials into ill-conditioned ones. However, these limitations are avoided in an efficient implementation by Malajovich and Zubelli (1999).

The method proceeds by multiplying a POLYNOMIAL $f(x)$ by $f(-x)$ and noting that

$$f(x)=(x-a_1)(x-a_2)\cdots(x-a_n) \tag{1}$$
$$f(-x)=(-1)^n(x+a_1)(x+a_2)\cdots(x+a_n) \tag{2}$$

so the result is

$$f(x)f(-x)=(-1)^n(x^2-a_1^2)(x^2-a_2^2)\cdots(x^2-a_n^2). \tag{3}$$

repeat v times, then write this in the form

$$y^n + b_1 y^{n-1} + \ldots + b_n = 0 \qquad (4)$$

where $y \equiv x^{2^\nu}$. Since the coefficients are given by NEWTON'S RELATIONS

$$b_1 = -(y_1 + y_2 + \ldots + y_n) \qquad (5)$$

$$b_2 = (y_1 y_2 + y_1 y_3 + \ldots + y_{n-1} y_n) \qquad (6)$$

$$b_n = (-1)^n y_1 y_2 \cdots y_n, \qquad (7)$$

and since the squaring procedure has separated the roots, the first term is larger than rest. Therefore,

$$b_1 \approx -y_1 \qquad (8)$$

$$b_2 \approx y_1 y_2 \qquad (9)$$

$$b_n \approx (-1)^n y_1 y_2 \cdots y_n, \qquad (10)$$

giving

$$y_1 \approx -b_1 \qquad (11)$$

$$y_2 \approx -\frac{b_2}{b_1} \qquad (12)$$

$$y_n \approx -\frac{b_n}{b_{n-1}}. \qquad (13)$$

Solving for the original roots gives

$$a_1 \approx \sqrt{-b_1} \qquad (14)$$

$$a_2 \approx \sqrt{-\frac{b_2}{b_1}} \qquad (15)$$

$$a_n \approx \sqrt{-\frac{b_n}{b_{n-1}}}. \qquad (16)$$

This method works especially well if all roots are real.

References

Bini, D. and Pan, V. Y. "Graeffe's, Chebyshev-Like, and Cardinal's Processes for Splitting a Polynomial Into Factors." *J. Complexity* **12**, 492–11, 1996.

Brodetsky, S. and Smeal, G. "on Graeffe's Method for Complex Roots of Algebraic Equations." *Proc. Cambridge Philos. Soc.* **22**, 83–7, 1924.

Dedieu, J.-P. "À Propos de la méthode de Dandelin-Graeffe." *C. R. Acad. Sci. Paris Sér. I Math* **309**, 1019–022, 1989.

Grau, A. A. "On the Reduction of Number Range in the Use of the Graeffe Process." *J. Assoc. Comput. Mach.* **10**, 538–44, 1963.

Householder, A. S. "dandelin, Lobacevskii, or Graeffe?" *Amer. Math. Monthly* **66**, 464–66, 1959.

Jana, P. and Sinha, B. "Fast Parallel Algorithms for Graeffe's Root Squaring." *Comput. Math. Appl.* **35**, 71–0, 1998.

Kármán, T. Von and Biot, M. a. "Squaring the Roots (Graeffe's Method)." §5.8.C in *Mathematical Methods in Engineering: an Introduction to the Mathematical Treatment of Engineering Problems.* New York: Mcgraw-Hill, pp. 194–96, 1940.

Malajovich, G. and Zubelli, J. P. "On the Geometry of Graeffe Iteration." *Informes de Matemática,* Série B-118, IMPA.

Malajovich, G. and Zubelli, J. P. Tangent Graeffe Iteration. 27 Aug 1999. http://xxx.lanl.gov/abs/math.AG/9908150/.

Ostrowski, A. "Recherches sur la méthode de Graeffe et les zéros des polynomes et des séries de Laurent." *Acta Math.* **72**, 99–55, 1940.

Ostrowski, A. "Recherches sur la méthode de Graeffe et les zéros des polynomes et des séries de Laurent. Chapitres III et IV." *Acta Math.* **72**, 157–57, 1940.

Pan, V. Y. "Solving a Polynomial Equation: Some History and Recent Progress." *SIAM Rev.* **39**, 187–20, 1997.

Whittaker, E. T. and Robinson, G. "The Root-Squaring Method of Dandelin, Lobachevsky, and Graeffe." §54 in *The Calculus of Observations: A Treatise on Numerical Mathematics, 4th ed.* New York: Dover, pp. 106–12, 1967.

Graham's Biggest Little Hexagon

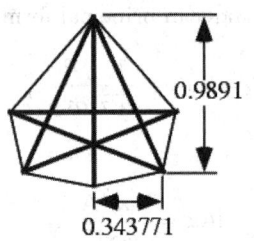

The largest possible (not necessarily regular) HEXAGON for which no two of the corners are more than unit distance apart. In the above figure, the heavy lines are all of unit length. The AREA of the hexagon is $A = 0.674981\ldots$, where A is the second-largest real ROOT of

$$4096A^{10} + 8192A^9 - 3008A^8 - 30{,}848A^7 + 21{,}056A^6$$
$$+ 146{,}496A^5 - 221{,}360A^4 + 1232A^3 + 144{,}464A^2$$
$$- 78{,}488A + 11{,}993$$
$$= 0.$$

Note that the sign of the A^9 is positive, not negative as erroneously given in Conway and Guy (1996).

See also CALABI'S TRIANGLE

References

Conway, J. H. and Guy, R. K. "Graham's Biggest Little Hexagon." In *The Book of Numbers.* New York: Springer-Verlag, pp. 206–07, 1996.

Graham, R. L. "The Largest Small Hexagon." *J. Combin. Th. Ser. A* **18**, 165–70, 1975.

Graham's Number

The smallest dimension n of a HYPERCUBE such that if the lines joining all pairs of corners are two-colored, a PLANAR COMPLETE GRAPH K_4 of one color will be forced. Stated colloquially, this is equivalent to considering every possible committee from some number of people n and enumerating every pair of committees. Now assign each pair of committees to one of two groups, and find the smallest n that will guarantee that there are four committees in which all pairs fall in the same group and all the people belong to an even number of committees (Hoffman 1998, p. 54).

An answer was proved to exist by R. L. Graham and B. L. Rothschild. However, although the actual answer is believed to be 6, the best bound proved is

$$64 \begin{cases} 3 \uparrow\uparrow\uparrow\uparrow\uparrow 3 \\ 3 \uparrow 3 \\ \vdots \\ 3 \uparrow 3 \end{cases}$$

where $\uparrow$ is stacked ARROW NOTATION. It is less than $3 \to 3 \to 3 \to 3$, where CHAINED ARROW NOTATION has been used.

See also ARROW NOTATION, CHAINED ARROW NOTATION, EXTREMAL GRAPH THEORY, RAMSEY THEORY, SKEWES NUMBER

References

Conway, J. H. and Guy, R. K. *The Book of Numbers.* New York: Springer-Verlag, pp. 61–2, 1996.
Gardner, M. "Mathematical Games." *Sci. Amer.* **237**, 18–8, Nov. 1977.
Hoffman, P. *The Man Who Loved Only Numbers: The Story of Paul Erdos and the Search for Mathematical Truth.* New York: Hyperion, pp. 18 and 54, 1998.

Gram Determinant

The DETERMINANT

$$G(f_1, f_2, \ldots, f_n)$$

$$= \begin{vmatrix} \int f_1^2 \, dt & \int f_1 f_2 \, dt & \cdots & \int f_1 f_n \, dt \\ \int f_2 f_1 \, dt & \int f_2^2 \, dt & \cdots & \int f_2 f_n \, dt \\ \vdots & \vdots & \ddots & \vdots \\ \int f_1 f_n \, dt & \int f_1 f_n \, dt & \cdots & \int f_n^2 \, dt \end{vmatrix}.$$

See also GRAM-SCHMIDT ORTHONORMALIZATION, WRONSKIAN

References

Andrews, G. E.; Askey, R.; and Roy, R. "Jacobi Polynomials and Gram Determinants." §6.3 in *Special Functions.* Cambridge, England: Cambridge University Press, pp. 293–97, 1999.
Sansone, G. *Orthogonal Functions, rev. English ed.* New York: Dover, p. 2, 1991.

Gram Matrix

Given m points with n-D vector coordinates $\mathbf{v}_i$, let M be the $n \times m$ matrix whose jth column consists of the coordinates of the vector $\mathbf{v}_j$, with $j = 1, \ldots, m$. Then define the $m \times m$ Gram matrix of dot products $a_{ij} = \mathbf{v}_i \cdot \mathbf{v}_j$ as

$$\mathsf{A} = \mathsf{M}^\mathsf{T} \mathsf{M},$$

where A^T denotes the TRANSPOSE. The Gram matrix determines the vectors $\mathbf{v}_i$ up to ISOMETRY.

Gram Series

$$G(x) = 1 + \sum_{k=1}^{\infty} \frac{(\ln x)^k}{kk!\zeta(k+1)},$$

where $\zeta(z)$ is the RIEMANN ZETA FUNCTION (Hardy 1999, p. 24). This approximation to the PRIME COUNTING FUNCTION is 10 times better than Li(x) for $x < 10^9$ but has been proven to be worse infinitely often by Littlewood (Ingham 1990). An equivalent formulation due to Ramanujan is

$$G(x) \equiv \frac{4}{\pi} \sum_{k=1}^{\infty} \frac{(-1)^{k-1} k}{B_{2k}(2k-1)} \left(\frac{\ln x}{2\pi} \right)^{2k-1} \sim \pi(x)$$

(Berndt 1994; Hardy 1999, p. 23), where B_{2k} is a BERNOULLI NUMBER. The integral analog, also found by Ramanujan, is

$$J(x) \equiv \int_0^{\infty} \frac{(\ln x)^t \, dt}{t\Gamma(t+1)\zeta(t+1)} \sim \pi(x)$$

(Berndt 1994; Hardy 1999, p. 23).
The Gram series is equivalent to the RIEMANN PRIME NUMBER FORMULA (Hardy 1999, pp. 24–5).

See also RIEMANN PRIME NUMBER FORMULA

References

Berndt, B. C. *Ramanujan's Notebooks, Part IV.* New York: Springer-Verlag, pp. 124–29, 1994.
Gram, J. P. "Undersøgelser angaaende Maengden af Primtal under en given Graeense." *K. Videnskab. Selsk. Skr.* **2**, 183–08, 1884.
Hardy, G. H. *Ramanujan: Twelve Lectures on Subjects Suggested by His Life and Work, 3rd ed.* New York: Chelsea, 1999.
Ingham, A. E. Ch. 5 in *The Distribution of Prime Numbers.* New York: Cambridge, 1990.
Ribenboim, P. *The New Book of Prime Number Records.* New York: Springer-Verlag, p. 225, 1996.
Vardi, I. *Computational Recreations in Mathematica.* Reading, MA: Addison-Wesley, p. 74, 1991.

Gram's Inequality

Let $f_1(x), \ldots, f_n(x)$ be REAL INTEGRABLE FUNCTIONS over the CLOSED INTERVAL $[a, b]$, then the DETERMINANT of their integrals satisfies

$$\begin{vmatrix} \int_a^b f_1^2(x)\,dx & \int_a^b f_1(x)f_2(x)\,dx & \cdots & \int_a^b f_1(x)f_n(x)\,dx \\ \int_a^b f_2(x)f_1(x)\,dx & \int_a^b f_2^2(x)\,dx & \cdots & \int_a^b f_2(x)f_n(x)\,dx \\ \vdots & \vdots & \ddots & \vdots \\ \int_a^b f_n(x)f_1(x)\,dx & \int_a^b f_n(x)f_2(x)\,dx & \cdots & \int_a^b f_n(x)f_n(x)\,dx \end{vmatrix}$$

$$\geq 0.$$

See also GRAM-SCHMIDT ORTHONORMALIZATION

References

Gradshteyn, I. S. and Ryzhik, I. M. *Tables of Integrals, Series, and Products, 6th ed.* San Diego, CA: Academic Press, p. 1100, 2000.

Gram-Charlier Series
EDGEWORTH SERIES

Gram-Schmidt Orthonormalization

A procedure which takes a nonorthogonal set of LINEARLY INDEPENDENT functions and constructs an ORTHOGONAL BASIS over an arbitrary interval with respect to an arbitrary WEIGHTING FUNCTION $w(x)$.

Given an original set of linearly independent functions $\{u_n\}_{n=0}^\infty$, let $\{\psi_n\}_{n=0}^\infty$ denote the orthogonalized (but not normalized) functions, $\{\phi_n\}_{n=0}^\infty$ denote the orthonormalized functions, and define

$$\psi_0(x) \equiv u_0(x) \tag{1}$$

$$\phi_0(x) \equiv \frac{\psi_0(x)}{\sqrt{\int \psi_0^2(x)w(x)\,dx}}. \tag{2}$$

Then take

$$\psi_1(x) = u_1(x) + a_{10}\phi_0(x), \tag{3}$$

where we require

$$\int \psi_1\phi_0 w\,dx = \int u_1\phi_0 w\,dx + a_{10}\int \phi_0^2 w\,dx = 0. \tag{4}$$

By definition,

$$\int \phi_0^2 w\,dx = 1, \tag{5}$$

so

$$a_{10} = -\int u_1\phi_0 w\,dx. \tag{6}$$

The first orthogonalized function is therefore

$$\psi_1 = u_1(x) - \left[\int u_1\phi_0 w\,dx\right]\phi_0, \tag{7}$$

and the corresponding normalized function is

$$\phi_1 = \frac{\psi_1(x)}{\sqrt{\int \psi_1^2 w\,dx}}. \tag{8}$$

By mathematical induction, it follows that

$$\phi_i(x) = \frac{\psi_i(x)}{\sqrt{\int \psi_i^2 w\,dx}}, \tag{9}$$

where

$$\psi_i(x) = u_i + a_{i0}\phi_0 + a_{i1}\phi_1 \ldots + a_{i,\,i-1}\phi_{i-1} \tag{10}$$

and

$$a_{ij} \equiv -\int u_i\phi_j w\,dx. \tag{11}$$

If the functions are normalized to N_j instead of 1, then

$$\int_a^b [\phi_j(x)]^2 w\,dx = N_j^2 \tag{12}$$

$$\phi_i(x) = N_i \frac{\psi_i(x)}{\sqrt{\int \psi_i^2 w\,dx}} \tag{13}$$

$$a_{ij} = -\frac{\int u_i\phi_j w\,dx}{N_j^2}. \tag{14}$$

ORTHOGONAL POLYNOMIALS are especially easy to generate using GRAM-SCHMIDT ORTHONORMALIZATION. Use the notation

$$\langle x_i | x_j \rangle \equiv \langle x_i | w | x_j \rangle \equiv \int_a^b x_i(x)x_j(x)w(x)\,dx, \tag{15}$$

where $w(x)$ is a WEIGHTING FUNCTION, and define the first few POLYNOMIALS,

$$p_0(x) \equiv 1 \tag{16}$$

$$p_1(x) = \left[x - \frac{\langle xp_0 | p_0 \rangle}{\langle p_0 | p_0 \rangle}\right]p_0. \tag{17}$$

As defined, p_0 and p_1 are ORTHOGONAL POLYNOMIALS, as can be seen from

$$\langle p_0 | p_1 \rangle = \left\langle \left[x - \frac{\langle xp_0 | p_0 \rangle}{\langle p_0 | p_0 \rangle}\right]p_0 \right\rangle = \langle xp_0 \rangle - \frac{\langle xp_0 | p_0 \rangle}{\langle p_0 | p_0 \rangle}\langle p_0 \rangle$$

$$= \langle xp_0 \rangle - \langle xp_0 \rangle = 0. \tag{18}$$

Now use the RECURRENCE RELATION

$$p_{i+1}(x) = \left[x - \frac{\langle xp_i | p_i \rangle}{\langle p_i | p_i \rangle}\right]p_i - \left[\frac{\langle p_i | p_i \rangle}{\langle p_{i-1} | p_{i-1} \rangle}\right]p_{i-1} \tag{19}$$

to construct all higher order POLYNOMIALS.

To verify that this procedure does indeed produce ORTHOGONAL POLYNOMIALS, examine

$$\langle p_{i+1}|p_i\rangle = \left\langle \left[x - \frac{\langle xp_i|p_i\rangle}{\langle p_i|p_i\rangle}\right]p_i\Big|p_i\right\rangle$$
$$- \left\langle \frac{\langle p_i|p_i\rangle}{\langle p_{i-1}|p_{i-1}\rangle}p_{i-1}\Big|p_i\right\rangle$$

$$= \langle xp_i|p_i\rangle - \frac{\langle xp_i|p_i\rangle}{\langle p_i|p_i\rangle}\langle p_i|p_i\rangle - \frac{\langle p_i|p_i\rangle}{\langle p_{i-1}|p_{i-1}\rangle}$$
$$\times \langle p_{i-1}|p_i\rangle$$

$$= -\frac{\langle p_i|p_i\rangle}{\langle p_{i-1}|p_{i-1}\rangle}\langle p_{i-1}|p_i\rangle$$

$$= -\frac{\langle p_i|p_i\rangle}{\langle p_{i-1}|p_{i-1}\rangle}\left[-\frac{\langle p_{i-1}|p_{j-1}\rangle}{\langle p_{j-2}|p_{j-2}\rangle}\langle p_{j-2}|p_{j-1}\rangle\right]$$

$$= \ldots = (-1)^j\frac{\langle p_j|p_j\rangle}{\langle p_0|p_0\rangle}\langle p_0|p_1\rangle = 0, \qquad (20)$$

since $\langle p_0|p_1\rangle = 0$. Therefore, all the POLYNOMIALS $p_i(x)$ are orthogonal.

Many common ORTHOGONAL POLYNOMIALS of mathematical physics can be generated in this manner. Unfortunately, the process turns out to be numerically unstable (Golub and van Loan 1989).

See also GRAM DETERMINANT, GRAM'S INEQUALITY, LATTICE REDUCTION, ORTHOGONAL POLYNOMIALS

References

Arfken, G. "Gram-Schmidt Orthogonalization." §9.3 in *Mathematical Methods for Physicists, 3rd ed.* Orlando, FL: Academic Press, pp. 516–20, 1985.

Cohen, H. *A Course in Computational Algebraic Number Theory.* New York: Springer-Verlag, 1993.

Golub, G. H. and van Loan, C. F. *Matrix Computations, 3rd ed.* Baltimore, MD: Johns Hopkins, 1989.

Pohst, M. and Zassenhaus, H. "Methods from the Geometry of Numbers." Ch. 3 in *Algorithmic Algebraic Number Theory.* Cambridge, England: Cambridge University Press, 1989.

Granny Knot

A COMPOSITE KNOT of seven crossings consisting of a KNOT SUM of TREFOILS. The granny knot has the same

ALEXANDER POLYNOMIAL $(x^2 - x + 1)^2$ as the SQUARE KNOT.

Graph

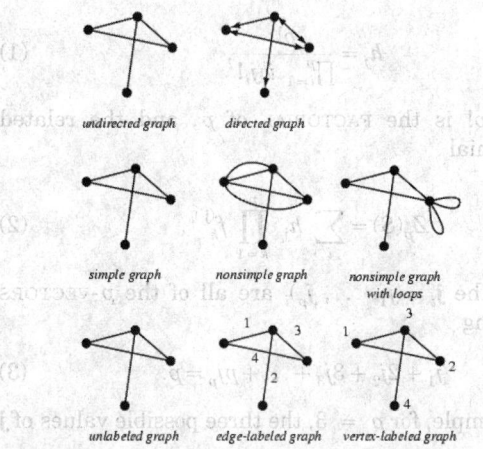

undirected graph   *directed graph*

simple graph   *nonsimple graph*   *nonsimple graph with loops*

unlabeled graph   *edge-labeled graph*   *vertex-labeled graph*

A mathematical object composed of points known as VERTICES or NODES and lines connecting some (possibly empty) SUBSET of them, known as EDGES. Formally, a graph is a binary relation on a set of vertices. If this relation is symmetric, the graph is said to be UNDIRECTED; otherwise, the graph is said to be DIRECTED. Graphs in which at most one edge connects any two nodes are said to be SIMPLE GRAPHS. Vertices are usually not allowed to be self-connected, but this restriction is sometimes relaxed to allow such "loops." The edges of a graph may be assigned specific values or labels, in which case the graph is called a LABELED GRAPH.

The study of graphs is known as GRAPH THEORY, and was first studied systematically by D. König in the 1930s (Gardner 1984, p. 91). As Gardner (1984, p. 91) notes, "The confusion of this term with the 'GRAPHS' of analytic geometry is regrettable, but the term has stuck."

Graphs are 1-D COMPLEXES, and there are always an EVEN NUMBER of ODD NODES in a graph. GRAPH SUMS, differences, powers, UNIONS, and PRODUCTS can be defined, as can GRAPH EIGENVALUES.

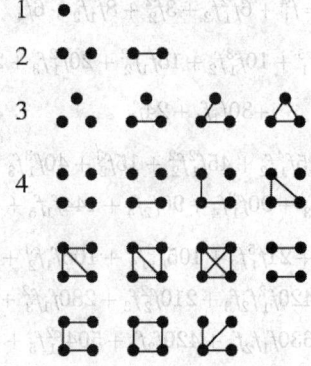

The number of nonisomorphic simple undirected

graphs with v NODES for $v = 1, 2, \ldots,$ are 1, 2, 4, 11, 34, 156, 1044, ... (Sloane's A000088; see above figure). The PÓLYA ENUMERATION THEOREM can be used to determine these numbers. In order to apply the PÓLYA ENUMERATION THEOREM, define the quantity

$$h_j = \frac{p!}{\prod_{i=1}^{p} i^j i^{j_i}!}, \tag{1}$$

where $p!$ is the FACTORIAL of p, and the related polynomial

$$Z_p(S) = \sum_i h_{\mathbf{j}_i} \prod_{k=1}^{p} f_k^{(\mathbf{j}_i)_k}, \tag{2}$$

where the $\mathbf{j}_i = (j_1, \ldots, j_p)_i$ are all of the p-VECTORS satisfying

$$j_1 + 2j_2 + 3j_3 + \ldots + pj_p = p. \tag{3}$$

For example, for $p = 3$, the three possible values of $\mathbf{j}$ are

$$\mathbf{j}_1 = (3, 0, 0), \text{ since } (1 \cdot 3) + (2 \cdot 0) + (3 \cdot 0) = 3,$$

$$\text{giving } h_{\mathbf{j}_1} = \frac{3!}{(1^3 3!)(2^0 0!)(3^0 0!)} = 1 \tag{4}$$

$$\mathbf{j}_2 = (1, 1, 0), \text{ since } (1 \cdot 1) + (2 \cdot 1) + (3 \cdot 0) = 3,$$

$$\text{giving } h_{\mathbf{j}_2} = \frac{3!}{(1^1 1!)(2^1 1!)(3^0 0!)} = 3, \tag{5}$$

$$\mathbf{j}_3 = (0, 0, 1), \text{ since } (1 \cdot 0) + (2 \cdot 0) + (3 \cdot 1) = 3$$

$$\text{giving } h_{\mathbf{j}_3} = \frac{3!}{(1^0 0!)(2^0 0!)(3^1 1!)} = 2. \tag{6}$$

Therefore,

$$Z_3(S) = f_1^3 + 3f_1 f_2 + 2f_3. \tag{7}$$

For small p, the first few values of $Z_p(S)$ are given by

$$Z_2(S) = f_1^2 + f_2 \tag{8}$$

$$Z_3(S) = f_1^3 + 3f_1 f_2 + 2f_3 \tag{9}$$

$$Z_4(S) = f_1^4 + 6f_1^2 f_2 + 3f_2^2 + 8f_1 f_3 + 6f_4 \tag{10}$$

$$Z_5(S) = f_1^5 + 10f_1^3 f_2 + 15f_1 f_2^2 + 20f_1^2 f_3 + 20f_2 f_3$$
$$+ 30f_1 f_4 + 24f_5 \tag{11}$$

$$Z_6(S) = f_1^6 + 15f_1^4 f_2 + 45f_1^2 f_2^2 + 15f_2^3 + 40f_1^3 f_3 + 120f_1 f_2 f_3$$
$$+ 40f_3^2 + 90f_1^2 f_4 + 90f_2 f_4 + 144f_1 f_5 + 120f_6 \tag{12}$$

$$Z_7(S) = f_1^7 + 21f_1^5 f_2 + 105f_1^3 f_2^2 + 105f_1 f_2^3 + 70f_1^4 f_3$$
$$+ 420f_1^2 f_2 f_3 + 210f_2^2 f_3 + 280f_1 f_3^2 + 210f_1^3 f_4$$
$$+ 630f_1 f_2 f_4 + 420f_3 f_4 + 504f_1^2 f_5 + 504f_2 f_5$$
$$+ 840f_1 f_6 + 720f_7. \tag{13}$$

Application of the PÓLYA ENUMERATION THEOREM then gives the formula

$$Z(R) = \frac{1}{p!} \sum_{(j)} h_j \prod_{n=0}^{\lfloor (p-1)/2 \rfloor} g_{2n+1}^{n j_{2n+1} + (2n+1)\binom{j_{2n+1}}{2}}$$

$$\times \prod_{n=1}^{\lfloor p/2 \rfloor} [(g_n g_{2n})^{n-1}]^{j_{2n}} g_{2n}^{2n\binom{j_{2n}}{2}}$$

$$\times \prod_{q=1}^{p} \prod_{r=q+1}^{p} g_{\mathrm{LCM}(q, r)}^{j_q j_r \mathrm{GCD}(q, r)}, \tag{14}$$

where $\lfloor x \rfloor$ is the FLOOR FUNCTION, $\binom{n}{m}$ is a BINOMIAL COEFFICIENT, LCM is the LEAST COMMON MULTIPLE, GCD is the GREATEST COMMON DIVISOR, and the SUM (j) is over all $\mathbf{j}_i$ satisfying the sum identity described above. The first few generating functions $Z_p(R)$ are

$$Z_2(R) = 2g_1 \tag{15}$$

$$Z_3(R) = g_1^3 + 3g_1 g_2 + 2g_3 \tag{16}$$

$$Z_4(R) = g_1^6 + 9g_1^2 g_2^2 + 8g_3^2 + 6g_2 g_4 \tag{17}$$

$$Z_5(R) = g_1^{10} + 10g_1^4 g_2^3 + 15g_1^2 g_2^4 + 20g_1 g_3^3 + 30g_2 g_4^2$$
$$+ 24g_5^2 + 20g_1 g_3 g_6 \tag{18}$$

$$Z_6(R) = g_1^{15} + 15g_1^7 g_2^4 + 60g_1^3 g_2^6 + 40g_1^3 g_3^4 + 40g_3^5$$
$$+ 180g_1 g_2 g_4^3 + 144g_5^3 + 120g_1 g_2 g_3^2 g_6$$
$$+ 120g_3 g_6^2 \tag{19}$$

$$Z_7(R) = g_1^{21} + 21g_1^{11} g_2^5 + 105g_1^5 g_2^8 + 105g_1^3 g_2^9 + 70g_1^6 g_3^5$$
$$+ 280g_3^7 + 210g_1^3 g_2 g_4^4 + 630g_1 g_2^2 g_4^4 + 504g_1 g_5^4$$
$$+ 420g_1^2 g_2^2 g_3^3 g_6 + 210\, g_1^2 g_2^2 g_3 g_6^2 + 840\, g_3^3 g_6^3$$
$$+ 720\, g_7^3 + 504\, g_1 g_5^2 g_{10} + 420\, g_2 g_3 g_4 g_{12}. \tag{20}$$

Letting $g_i = 1 + x^i$ then gives a POLYNOMIAL $S_i(x)$, which is a GENERATING FUNCTION for (i.e., the terms of x^i give) the number of graphs with i EDGES. The total number of graphs having i edges is $S_i(1)$. The first few $S_i(x)$ are

$$S_2 = 1 + x \tag{21}$$

$$S_3 = 1 + x + x^2 + x^3 \tag{22}$$

$$S_4 = 1 + x + 2x^2 + 3x^3 + 2x^4 + x^5 + x^6 \tag{23}$$

$$S_5 = 1 + x + 2x^2 + 4x^3 + 6x^4 + 6x^5 + 6x^6 + 4x^7 + 2x^8$$
$$+ x^9 + x^{10} \tag{24}$$

$$S_6 = 1 + x + 2x^2 + 5x^3 + 9x^4 + 15x^5 + 21x^6 + 24x^7$$
$$+ 24x^8 + 21x^9 + 15x^{10} + 9x^{11} + 5x^{12} + 2x^{13}$$
$$+ x^{14} + x^{15} \tag{25}$$

$$S_7 = 1 + x + 2x^2 + 5x^3 + 10x^4 + 21x^5 + 21x^6 + 24x^7$$
$$+ 41x^6 + 65x^7 + 97x^8 + 131x^9 + 148x^{10} + 148x^{11}$$
$$+ 131x^{12} + 97x^{13} + 65x^{14} + 41x^{15} + 21x^{16} + 10x^{17}$$
$$+ 5x^{18} + 2x^{19} + x^{20} + x^{21}, \qquad (26)$$

giving the number of graphs with n nodes as 1, 2, 4, 11, 34, 156, 1044, ... (Sloane's A000088). King and Palmer (cited in Read 1981) have calculated S_n up to $n = 24$, for which

$$S_{24} = 195, 704, 906, 302, 078, 447, 922, 174, 862, 416, \cdots$$
$$\cdots 726, 256, 004, 122, 075, 267, 063, 365, 754, 368. \quad (27)$$

See also BIPARTITE GRAPH, CATERPILLAR GRAPH, CAYLEY GRAPH, CIRCULANT GRAPH, COCKTAIL PARTY GRAPH, COMPARABILITY GRAPH, COMPLEMENT GRAPH, COMPLETE GRAPH, CONE GRAPH, CONNECTED GRAPH, COXETER GRAPH, CUBICAL GRAPH, DE BRUIJN GRAPH, DEGREE SEQUENCE, DIGRAPH, DIRECTED GRAPH, DODECAHEDRAL GRAPH, EULER GRAPH, EXTREMAL GRAPH, GEAR GRAPH, GRACEFUL GRAPH, GRAPH DIAMETER, GRAPH THEORY, HANOI GRAPH, HARARY GRAPH, HARMONIOUS GRAPH, HOFFMAN-SINGLETON GRAPH, ICOSAHEDRAL GRAPH, INTERVAL GRAPH, ISO-MORPHIC GRAPHS, LABELED GRAPH, LADDER GRAPH, LATTICE GRAPH, MATCHSTICK GRAPH, MINOR GRAPH, MOORE GRAPH, MULTIGRAPH, NULL GRAPH, OCTAHE-DRAL GRAPH, PATH GRAPH, PETERSEN GRAPH, PLANAR GRAPH, PSEUDOGRAPH, RANDOM GRAPH, REGULAR GRAPH, SEQUENTIAL GRAPH, SIMPLE GRAPH, STAR GRAPH, SUBGRAPH, SUPERGRAPH, SUPERREGULAR GRAPH, SYLVESTER GRAPH, TETRAHEDRAL GRAPH, THOMASSEN GRAPH, TOURNAMENT, TRIANGULAR GRAPH, TURAN GRAPH, TUTTE'S GRAPH, UNIVERSAL GRAPH, UTILITY GRAPH, WEB GRAPH, WHEEL GRAPH

References

Bogomolny, A. "Graph Puzzles." http://www.cut-the-knot.com/do_you_know/graphs2.html.
Fujii, J. N. *Puzzles and Graphs*. Washington, DC: National Council of Teachers, 1966.
Gardner, M. *The Sixth Book of Mathematical Games from Scientific American*. Chicago, IL: University of Chicago Press, p. 91, 1984.
Harary, F. "The Number of Linear, Directed, Rooted, and Connected Graphs." *Trans. Amer. Math. Soc.* **78**, 445–63, 1955.
Pappas, T. "Networks." *The Joy of Mathematics*. San Carlos, CA: Wide World Publ./Tetra, pp. 126–27, 1989.
Read, R. "The Graph Theorists Who Count--And What They Count." In *The Mathematical Gardner* (Ed. D. Klarner). Boston, MA: Prindle, Weber, and Schmidt, pp. 326–45, 1981.
Read, R. C. and Wilson, R. J. *Atlas of Graphs*. Oxford, England: Oxford University Press, 1998.
Sloane, N. J. A. Sequences A000088/M1253 in "An On-Line Version of the Encyclopedia of Integer Sequences." http://www.research.att.com/~njas/sequences/eisonline.html.
Sloane, N. J. A. and Plouffe, S. Figure M1253 in *The Encyclopedia of Integer Sequences*. San Diego: Academic Press, 1995.
Weisstein, E. W. "Graphs." MATHEMATICA NOTEBOOK GRAPHS.M.
Weisstein, E. W. "Books about Graph Theory." http://www.treasure-troves.com/books/GraphTheory.html.
Wilson, J. C. *On the Traversing of Geometrical Figures*. Oxford, England: Oxford University Press, 1905.

Graph (Function)

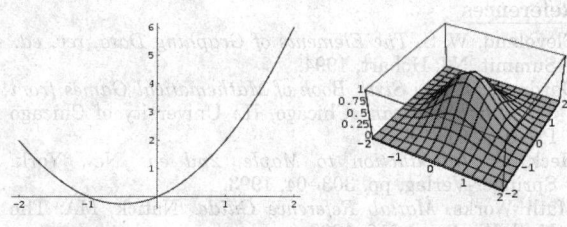

Given a FUNCTION $f(x_1, \ldots, x_n)$ defined on a DOMAIN U, the graph of f is defined as the set of points (which often form a CURVE or SURFACE) showing the values taken by f over U (or some portion of U). Technically, for real functions,

$$\text{graph } f(x) \equiv \{(x, f(x)) \in \mathbb{R}^2 : x \in U\}$$

$$\text{graph } f(x_1, \ldots, x_n) \equiv$$

$$\{(x_1, \ldots, x_n, f(x_1, \ldots, x_n)) \in \mathbb{R}^{n+1} : (x_1, \ldots, x_n) \in U\}.$$

A graph is sometimes also called a PLOT. Commenting on the unfortunate choice of the word "graph" in the completely different context of so-called GRAPH THE-ORY, Gardner (1984, p. 91) notes, "The confusion of this term with the 'graphs' of analytic geometry is regrettable, but the term has stuck."

2-D and 3-D graphs can be produced in *Mathematica* using the commands $\text{Plot}[f, \{x, xmin, xmin\}]$ and $\text{Plot3D}[f, \{x, xmin, xmin\}, \{y, ymin, ymax\}]$, respectively.

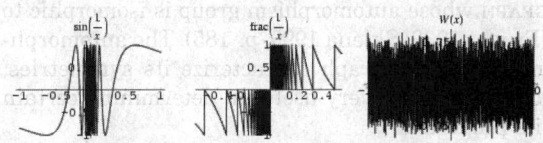

Several examples of continuous functions which are notoriously difficult to graph are shown above: $\sin(1/x)$, the FRACTIONAL PART $\text{frac}(1/x)$, and the WEIERSTRASS FUNCTION. Good routines for plotting graphs use adaptive algorithms which plot more points in regions where the function varies most rapidly (Wagon 1991, Math Works 1992, Heck 1993, Wickham-Jones 1994). Tupper (1996) has developed an algorithm that rigorously proves the pixels it generates are "on" if and only if there exists a mathematical point within the region of space represented by that pixel that is a solution to the relation being graphed. Although this method attempts to produce graphs that satisfy strict mathematical relationships, the problem of graphing is ultimately intractable, so no fixed algorithm can produce correct graphs for arbitrary relations.

See also CURVE, DATA CUBE, EXTREMUM, GRAPH, HISTOGRAM, MAXIMUM, MINIMUM

References

Cleveland, W. S. *The Elements of Graphing Data, rev. ed.* Summit, NJ: Hobart, 1994.

Gardner, M. *The Sixth Book of Mathematical Games from Scientific American.* Chicago, IL: University of Chicago Press, p. 91, 1984.

Heck, A. *Introduction to Maple, 2nd ed.* New York: Springer-Verlag, pp. 303–04, 1993.

Math Works. *Matlab Reference Guide.* Natick, MA: The Math Works, p. 216, 1992.

Tufte, E. R. *The Visual Display of Quantitative Information.* Cheshire, CN: Graphics Press, 1983.

Tufte, E. R. *Envisioning Information.* Cheshire, CN: Graphics Press, 1990.

Tupper, J. *Graphing Equations with Generalized Interval Arithmetic.* M.Sc. Thesis. Department of Computer Science. Toronto: University of Toronto, 1996. http://www.dgp.toronto.edu/~mooncake/msc.html.

Tupper, J. "GrafEq." http://www.peda.com/grafeq/.

Wagon, S. *Mathematica in Action.* New York: W. H. Freeman, pp. 24–5, 1991.

Weisstein, E. W. "Books about Graphing." http://www.treasure-troves.com/books/Graphing.html.

Wickham-Jones, T. *Computer Graphics with Mathematica.* Santa Clara, CA: TELOS, pp. 579–84, 1994.

Yates, R. C. "Sketching." *A Handbook on Curves and Their Properties.* Ann Arbor, MI: J. W. Edwards, pp. 188–05, 1952.

Graph Automorphism

An automorphism of a GRAPH is a GRAPH ISOMORPHISM with itself. The sets of automorphisms define a PERMUTATION GROUP. For every GROUP Γ, there exists a GRAPH whose automorphism group is isomorphic to Γ (Frucht 1939; Skiena 1990, p. 185). The automorphism groups of a graph characterize its symmetries, and are therefore very useful in determining certain of its properties.

The automorphism group of a GRAPH COMPLEMENT is the same as that for the original graph.

See also FRUCHT GRAPH, GRAPH ISOMORPHISM, ISOMORPHIC GRAPHS

References

Duijvestijn, A. J. W. "Algorithmic Calculation of the Order of the Automorphism Group of a Graph." Memorandum No. 221. Enschede, Netherlands: Twente Univ. Technology, 1978.

Frucht, R. "Herstellung von Graphen mit vorgegebener abstrakter Gruppe." *Compos. Math.* **6**, 239–50, 1939.

Lipton, R.; North, S.; and Sandberg, J. "A Method for Drawing Graphs." In *Proc. First ACM Symposium on Computation Geometry.* pp. 153–60, 1985.

Skiena, S. "Automorphism Groups." §5.2.2 in *Implementing Discrete Mathematics: Combinatorics and Graph Theory with Mathematica.* Reading, MA: Addison-Wesley, pp. 184–87, 1990.

Graph Cartesian Product

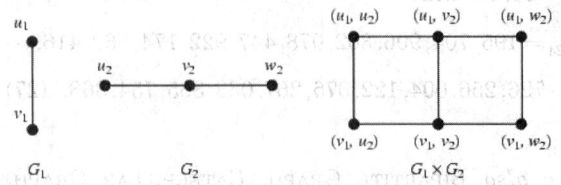

The Cartesian graph product $G = G_1 \square G_2$ of graphs G_1 and G_2 with disjoint point sets V_1 and V_2 and edge sets X_1 and X_2 is the graph with point set $V_1 \times V_2$ and $u = (u_1, u_2)$ adjacent with $v = (v_1, v_2)$ whenever [$u_1 = v_1$ and u_2 adj v_2] or [$u_2 = v_2$ and u_1 adj v_1] (Harary 1994, p. 22).

Graph Cartesian products can be computed using GraphProduct[*G1, G2*] in the *Mathematica* add-on package DiscreteMath`Combinatorica` (which can be loaded with the command < < DiscreteMath`).

See also GRAPH COMPOSITION, GRAPH PRODUCT, VIZING CONJECTURE

References

Clark, W. E. and Suen, S. "An Inequality Related to Vizing's Conjecture." *Electronic J. Combinatorics* **7**, No. 1, N4, 1–, 2000. http://www.combinatorics.org/Volume_7/v7i1toc.html#N4.

Harary, F. *Graph Theory.* Reading, MA: Addison-Wesley, 1994.

Hartnell, B. and Rall, D. "Domination in Cartesian Products: Vizing's Conjecture." In *Domination in Graphs--Advanced Topics* (Ed. T. W. Haynes, S. T. Hedetniemi, and P. J. Slater). New York: Dekker, pp. 163–89, 1998.

Sabidussi, G. "Graph Multiplication." *Math. Z.* **72**, 446–57, 1960.

Skiena, S. "Products of Graphs." §4.1.4 in *Implementing Discrete Mathematics: Combinatorics and Graph Theory with Mathematica.* Reading, MA: Addison-Wesley, pp. 133–35, 1990.

Vizing, V. G. "The Cartesian Product of Graphs." *Vycisl. Sistemy* **9**, 30–3, 1963.

Graph Categorical Product

This entry contributed by NICOLAS BRAY

The GRAPH PRODUCT denoted $G \times H$ and defined by the adjacency relations (g adj g' and h adj h').

See also GRAPH PRODUCT

Graph Center

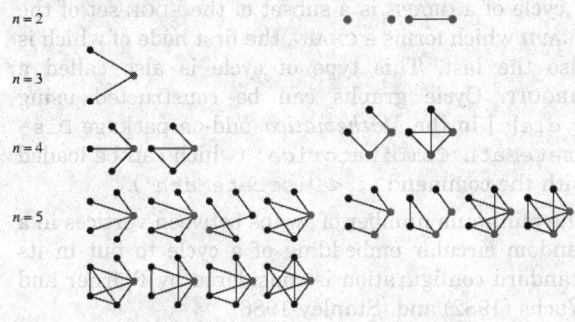

| | graphs with 1 center node | graphs with 2 center nodes |

The center of a GRAPH G is the set of vertices of GRAPH ECCENTRICITY equal to the GRAPH RADIUS (i.e., the set of CENTRAL POINTS). In the above illustration, center nodes are shown in red. The following table gives the number of n-node simple unlabeled graphs having k center nodes.

k	Sloane	$n = 1, 2, \ldots$
1	A052437	1, 0, 1, 2, 8, 29, 180, ...
2	A052438	0, 2, 0, 2, 4, 19, 84, ...
3	A052439	0, 0, 3, 0, 4, 18, 119, ...
4	A052340	0, 0, 0, 7, 0, 18, 118, ...
5	A052341	0, 0, 0, 0, 18, 0, 129, ...
6		0, 0, 0, 0, 0, 72, 0, ...
7		0, 0, 0, 0, 0, 0, 414, ...

See also BICENTERED TREE, CENTRAL POINT, CENTERED TREE, GRAPH ECCENTRICITY, GRAPH RADIUS

References

Harary, F. *Graph Theory.* Reading, MA: Addison-Wesley, p. 35, 1994.
Skiena, S. *Implementing Discrete Mathematics: Combinatorics and Graph Theory with Mathematica.* Reading, MA: Addison-Wesley, p. 107, 1990.
Sloane, N. J. A. Sequences A052437, A052438, A052439, A052340, and A052341 in "An On-Line Version of the Encyclopedia of Integer Sequences." http://www.research.att.com/~njas/sequences/eisonline.html.

Graph Circumference

The length of any longest cycle in a GRAPH.

See also GIRTH

References

Harary, F. *Graph Theory.* Reading, MA: Addison-Wesley, p. 13, 1994.

Skiena, S. *Implementing Discrete Mathematics: Combinatorics and Graph Theory with Mathematica.* Reading, MA: Addison-Wesley, p. 192, 1990.

Graph Coloring

The assignment of labels or colors to the edges or vertices of a graph. The most common types of graph colorings are EDGE COLORING and VERTEX COLORING.

See also EDGE COLORING, FOUR-COLOR THEOREM, K-COLORING, VERTEX COLORING

References

Jensen, T. R. and Toft, B. *Graph Coloring Problems.* New York: Wiley, 1994.
Morgenstern, C. and Shapiro, H. "Heuristics for Rapidly 4-Coloring Large Planar Graphs." *Algorithmica* **6**, 869–91, 1991.
Opsut, R. J. and Roberts, F. S. "On the Fleet Maintenance, Mobile Radio Frequency, Task Assignment, and Traffic Phasing Problems." In *The Theory and Applications of Graphs* (Ed. G. Chartrand, Y. Alavi, D. L. Goldsmith, L. Lesniak-Foster, and D. R. Lick). New York: Wiley, pp. 479–92, 1981.
Skiena, S. "Graph Coloring." §5.5 in *Implementing Discrete Mathematics: Combinatorics and Graph Theory with Mathematica.* Reading, MA: Addison-Wesley, pp. 210–16, 1990.
Wagon, S. "An April Fool's Hoax." *Mathematica in Educ. Res.* **7**, 46–2, 1998.
Wagon, S. "Coloring Planar Maps and Graphs." Ch. 24 in *Mathematica in Action, 2nd ed.* New York: Springer-Verlag, pp. 507–37, 1999.

Graph Complement

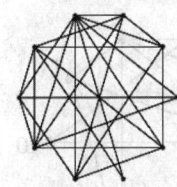 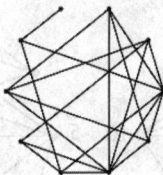 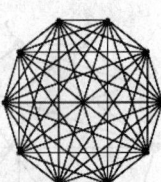

The complement of a graph G_n on n nodes is the graph G'_n (sometimes denoted $\bar{G}_n$) on the same nodes, but with the vertices in G_n omitted and the omitted vertices in G_n included. The GRAPH SUM $G_n + G'_n$ is therefore the COMPLETE GRAPH K_n. A graph complement can be given by the *Mathematica* command `GraphComplement[graph]` in the *Mathematica* add-on package `DiscreteMath`Combinatorica`` (which can be loaded with the command `<<DiscreteMath``).

See also COMPLETE GRAPH, GRAPH SUM, SELF-COMPLEMENTARY GRAPH

References

Skiena, S. "The Complement of a Graph." §3.2.3 in *Implementing Discrete Mathematics: Combinatorics and Graph Theory with Mathematica.* Reading, MA: Addison-Wesley, p. 93, 1990.

Graph Composition

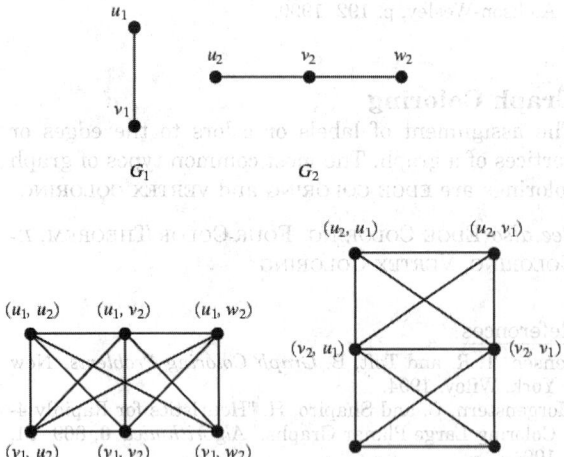

The composition $G = G_1[G_2]$ of graphs G_1 and G_2 with disjoint point sets V_1 and V_2 and edge sets X_1 and X_2 is the graph with point set $V_1 \times V_2$ and $u = (u_1, u_2)$ adjacent with $v = (v_1, v_2)$ whenever $[u_1 \text{ adj } v_1]$ or $[u_1 = v_1 \text{ and } u_2 \text{ adj } v_2]$ (Harary 1994, p. 22).

See also GRAPH PRODUCT

References

Harary, F. *Graph Theory*. Reading, MA: Addison-Wesley, p. 22, 1994.

Graph Contraction

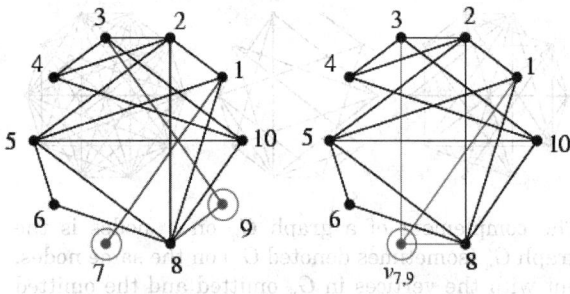

The contraction of an edge $\{v_i, v_j\}$ of a GRAPH is the graph obtained by replacing the two nodes v_1 and v_2 with a single node v such that v is adjacent to the union of the nodes to which v_1 and v_2 were originally adjacent. The figure above shows a random graph contracted on vertices v_7 and v_9. Graph contraction can be implemented using Contract[*g*, {*v1*, *v2*}] in the *Mathematica* add-on package DiscreteMath`-Combinatorica` (which can be loaded with the command < <DiscreteMath`).

References

Skiena, S. *Implementing Discrete Mathematics: Combinatorics and Graph Theory with Mathematica*. Reading, MA: Addison-Wesley, p. 91, 1990.

Graph Cycle

A cycle of a GRAPH is a subset of the EDGE-set of the GRAPH which forms a CHAIN, the first node of which is also the last. This type of cycle is also called a CIRCUIT. Cycle graphs can be constructed using Cycle[*n*] in the *Mathematica* add-on package DiscreteMath`Combinatorica` (which can be loaded with the command < <DiscreteMath`).

The minimum number of swaps between vertices in a random circular embedding of a cycle to put in its standard configuration is considered by Björner and Wachs (1982) and (Stanley 1986).

See also ACYCLIC DIGRAPH, CHAIN (GRAPH), CYCLE GRAPH, EULERIAN CIRCUIT, EULERIAN GRAPH, FOREST, HAMILTONIAN CIRCUIT, HAMILTONIAN GRAPH, WALK

References

Björner, A. and Wachs, M. "Bruhat Order of Coxeter Groups and Shellability." *Adv. Math.* **43**, 87–00, 1982.
Skiena, S. "Cycles in Graphs." §5.3 in *Implementing Discrete Mathematics: Combinatorics and Graph Theory with Mathematica*. Reading, MA: Addison-Wesley, pp. 188–02, 1990.
Stanley, R. P. *Enumerative Combinatorics, Vol. 1*. Cambridge, England: Cambridge University Press, 1999.

Graph Diameter

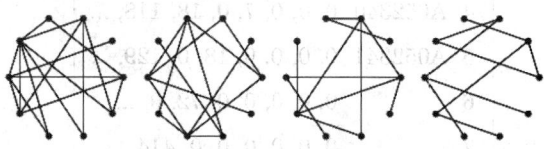

The length $\max_{u, v} d(u, v)$ of the "longest shortest path" (i.e., the longest GRAPH GEODESIC) between any two VERTICES (u, v) of a GRAPH. In other words, a graph's diameter is the largest number of vertices which must be traversed in order to travel from one vertex to another when paths which backtrack, detour, or loop are excluded from consideration. The above RANDOM GRAPHS on 10 vertices have diameters 3, 4, 5, and 7, respectively.

See also DIAMETER, GRAPH, GRAPH ECCENTRICITY, GRAPH GEODESIC, MOORE GRAPH, PERIPHERAL POINT

References

Harary, F. *Graph Theory*. Reading, MA: Addison-Wesley, p. 14, 1994.
Skiena, S. *Implementing Discrete Mathematics: Combinatorics and Graph Theory with Mathematica*. Reading, MA: Addison-Wesley, p. 107, 1990.

Graph Difference

The graph difference of graphs G and H is the graph with ADJACENCY MATRIX given by the difference of adjacency matrices of G and H. A graph difference is defined when the orders of G and H are the same, and can be computed using GraphDifference[*g*, *h*]

in the *Mathematica* add-on package `Discrete-Math`Combinatorica`` (which can be loaded with the command `< <DiscreteMath'`).

See also GRAPH SUM

References

Skiena, S. "Sum and Difference." §4.1.2 in *Implementing Discrete Mathematics: Combinatorics and Graph Theory with Mathematica.* Reading, MA: Addison-Wesley, p. 131, 1990.

Graph Eccentricity

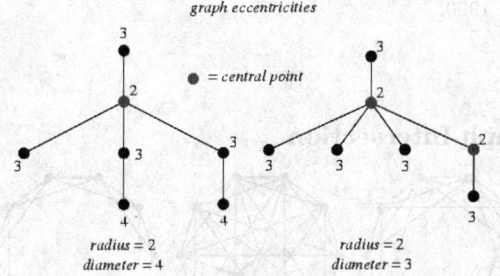

graph eccentricities

$\bullet$ = *central point*

radius = 2
diameter = 4

radius = 2
diameter = 3

The eccentricity of a node v in a CONNECTED GRAPH G is length $\max_u d(u, v)$ of the longest of all the shortest paths between v and every other point in G. The maximum eccentricity is the GRAPH DIAMETER. The minimum graph eccentricity is called the GRAPH RADIUS.

See also CENTRAL POINT, GRAPH CENTER, GRAPH DIAMETER, GRAPH RADIUS, PERIPHERAL POINT

References

Harary, F. *Graph Theory.* Reading, MA: Addison-Wesley, p. 35, 1994.
Skiena, S. *Implementing Discrete Mathematics: Combinatorics and Graph Theory with Mathematica.* Reading, MA: Addison-Wesley, p. 107, 1990.

Graph Eigenvalue

The eigenvalues of a GRAPH are defined as the EIGENVALUES of its ADJACENCY MATRIX. The set of eigenvalues of a GRAPH is called a GRAPH SPECTRUM.

See also GRAPH SPECTRUM

References

Biggs, N. L. *Algebraic Graph Theory, 2nd ed.* Cambridge, England: Cambridge University Press, 1993.
Cvetkovic, D.; Doob, M.; and Sachs, H. *Spectra of Graphs.* New York: Academic Press, 1980.
Skiena, S. *Implementing Discrete Mathematics: Combinatorics and Graph Theory with Mathematica.* Reading, MA: Addison-Wesley, p. 85, 1990.

Graph Embedding

A particular drawing of a GRAPH (with sometimes added constraint that the embedding be *planar*, i.e., has no crossing edges). The above figure shows the first several circular embeddings of the CUBICAL GRAPH.

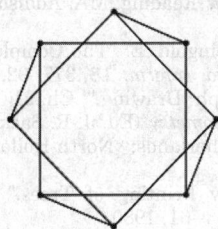

While the underlying object is independent of the embedding, a clever choice of embedding can lead to particularly illuminating diagrams. For example, the circular embedding of the CUBICAL GRAPH depicted above illustrates this graph's inherent symmetries.

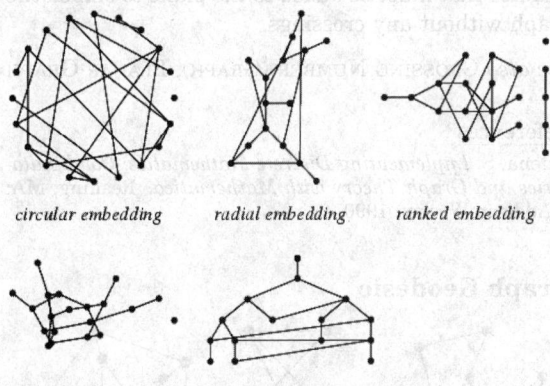

circular embedding   *radial embedding*   *ranked embedding*

spring embedding   *rooted embedding*

Skiena (1990) considers a number of different types of embeddings, including circular, ranked, radial, rooted, and spring.

See also EMBEDDING

References

Chung, F.; Leighton, T.; and Rosenberg, A. "Embeddings Graphs in Books: A Layout Problem with Applications to VLSI Design." *SIAM J. Algebraic Disc. Meth.* **8**, 33–8, 1987.

Di Battista, G.; Eades, P.; Tamassia, R.; and Tollis, I. G. *Graph Drawing: Algorithms for the Visualization of Graphs.* Englewood Cliffs, NJ: Prentice-Hall, 1998.

Eades, P. "A Heuristic for Graph Drawing." *Congr. Numer.* **42**, 149–60, 1984.

Eades, P.; Fogg, I.; and Kelly, D. *SPREMB: A System for Developing Graph Algorithms.* Technical Report. Department of Computer Science. St. Lucia, Queensland, Australia: University of Queensland, 1988.

Eades, P. and Tamassia, R. "Algorithms for Drawing Graphs: An Annotated Bibliography." Technical Report CS-89-9. Department of Computer Science. Providence, RI: Brown University, Feb. 1989.

Kamada, T. and Kawai, S. "An Algorithm for Drawing General Undirected Graphs." *Inform. Processing Lett.* **31**, 7–5, 1989.

Malitz, S. M. "Genus g Graphs Have Pagenumber $\mathcal{O}(\sqrt{g})$." In *Proc. 29th Sympos. Found. Computer Sci.* IEEE Press, pp. 458–68, 1988.

Reingold, E. and Tilford, J. "Tidier Drawings of Trees." *IEEE Trans. Software Engin.* **7**, 223–28, 1981.

Skiena, S. "Graph Embeddings." §3.3 in *Implementing Discrete Mathematics: Combinatorics and Graph Theory with Mathematica.* Reading, MA: Addison-Wesley, pp. 81 and 98–18, 1990.

Supowit, K. and Reingold, E. "The Complexity of Drawing Trees Nicely." *Acta. Inform.* **18**, 377–92, 1983.

Tamassia, R. "Graph Drawing." Ch. 21 in *Handbook of Computational Geometry* (Ed. J.-R. Sack and J. Urrutia). Amsterdam, Netherlands: North-Holland, pp. 937–71, 2000.

Vaucher, J. "Pretty Printing of Trees." *Software Pract. Experience* **10**, 553–61, 1980.

Wetherell, C. and Shannon, A. "Tidy Drawings of Trees." *IEEE Trans. Software Engin.* **5**, 514–20, 1979.

Graph Genus

The genus of a graph is the minimum number of handles that must be added to the plane to embed the graph without any crossings.

See also CROSSING NUMBER (GRAPH), PLANAR GRAPH

References

Skiena, S. *Implementing Discrete Mathematics: Combinatorics and Graph Theory with Mathematica.* Reading, MA: Addison-Wesley, 1990.

Graph Geodesic

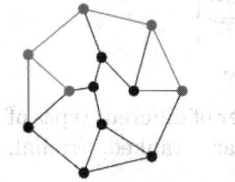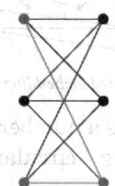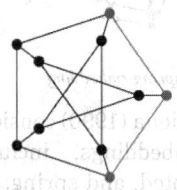

A shortest path between two VERTICES (u, v) of a GRAPH (Skiena 1990, p. 225). There may be more than one different shortest paths, all of the same length. Graph geodesics may be found using a BREADTH-FIRST TRAVERSAL (Moore 1959) or using DIJKSTRA'S ALGORITHM (Skiena 1990, p. 225). A graph geodesic can be found using ShortestPath[g, s, e] in the *Mathematica* add-on package DiscreteMath`Combinator-

ica` (which can be loaded with the command < <DiscreteMath`).

The length of the maximum graph geodesic in a given graph is called the GRAPH DIAMETER.

See also ALL-PAIRS SHORTEST PATH, GRAPH DIAMETER

References

Harary, F. *Graph Theory.* Reading, MA: Addison-Wesley, p. 14, 1994.

Moore, E. F. "The Shortest Path through a Maze." In *Proc. Internat. Symp. Switching Th., Part II.* Cambridge, MA: Harvard University Press, pp. 285–92, 1959.

Skiena, S. "Shortest Paths." §6.1 in *Implementing Discrete Mathematics: Combinatorics and Graph Theory with Mathematica.* Reading, MA: Addison-Wesley, pp. 225–53, 1990.

Graph Intersection

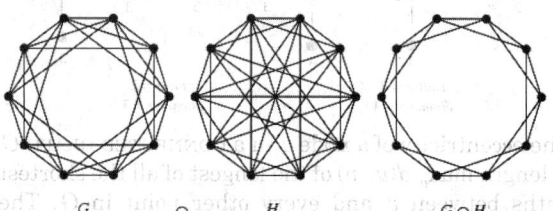

$$G \qquad H \qquad G \cap H$$

Let S be a set and $F = \{S_1, \ldots, S_p\}$ a nonempty family of distinct nonempty subsets of S whose union is $\cup_{i=1}^{p} S_i = S$. The intersection graph of F is denoted $\Omega(F)$ and defined by $V(\Omega(F)) = F$, with S_i and S_j adjacent whenever $i \neq j$ and $S_i \cap Sj \neq \varnothing$. Then a GRAPH G is an intersection graph on S if there exists a family F of subsets for which G and $\Omega(F)$ are ISOMORPHIC GRAPHS (Harary 1994, p. 19). Graph intersections can be computed using GraphIntersection[g, h] in the *Mathematica* add-on package DiscreteMath`Combinatorica` (which can be loaded with the command < <DiscreteMath`).

See also GRAPH UNION, INTERSECTION NUMBER

References

Harary, F. *Graph Theory.* Reading, MA: Addison-Wesley, 1994.

Skiena, S. "Unions and Intersections." §4.1.1 in *Implementing Discrete Mathematics: Combinatorics and Graph Theory with Mathematica.* Reading, MA: Addison-Wesley, pp. 129–31, 1990.

Graph Isomorphism

An isomorphism between two graphs is a one-to-one mapping between their two sets of vertices.

See also GRAPH AUTOMORPHISM, ISOMORPHIC GRAPHS

References

Du, D.-Z. and Ko, K.-I. *Theory of Computational Complexity.* New York; Wiley, p. 117, 2000.

Skiena, S. "Graph Isomorphism." §5.2 in *Implementing Discrete Mathematics: Combinatorics and Graph Theory with Mathematica.* Reading, MA: Addison-Wesley, pp. 181–87, 1990.

Graph Join

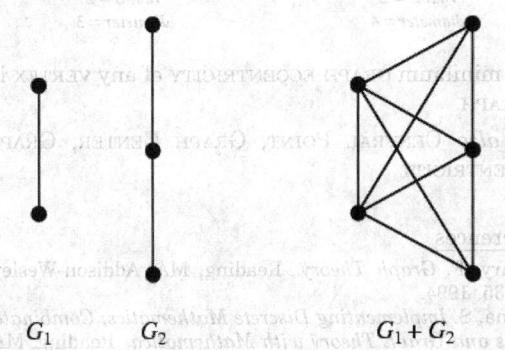

$$G_1 \qquad G_2 \qquad G_1 + G_2$$

The join $G = G_1 + G_2$ of graphs G_1 and G_2 with disjoint point sets V_1 and V_2 and edge sets X_1 and X_2 is the GRAPH UNION $G_1 \cup G_2$ together with all the edges joining V_1 and V_2 (Harary 1994, p. 21). Graph joins can be computed using GraphJoin[*G1, G2*] in the *Mathematica* add-on package DiscreteMath`-Combinatorica` (which can be loaded with the command < <DiscreteMath`).

A complete k-partite graph $k_{i, j, \ldots}$ is the graph join of empty graphs on $i, j, \ldots$ nodes. A WHEEL GRAPH is the join of a CYCLE GRAPH and the singleton graph. Finally, a STAR GRAPH is the join of an EMPTY GRAPH and the singleton graph (Skiena 1990, p. 132).

See also GRAPH SUM, GRAPH UNION

References

Harary, F. *Graph Theory.* Reading, MA: Addison-Wesley, 1994.

Skiena, S. "Joins of Graphs." §4.1.3 in *Implementing Discrete Mathematics: Combinatorics and Graph Theory with Mathematica.* Reading, MA: Addison-Wesley, pp. 131–32, 1990.

Graph Lexographic Product

This entry contributed by NICOLAS BRAY

The GRAPH PRODUCT denoted $G \cdot H$ and defined by the adjacency relations (g adj g') or ($g = g'$ and h adj h').

See also GRAPH PRODUCT

Graph Power

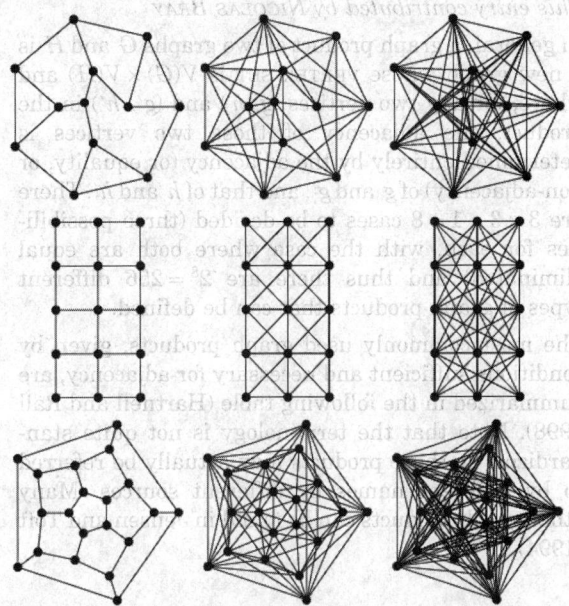

The kth power of a GRAPH G is a graph with the same set of vertices as G and an edge between two vertices IFF there is a path of length at most k between them (Skiena 1990, p. 229). Since a path of length two between vertices u and v exists for every vertex w such that $\{u, w\}$ and $\{w, v\}$ are edges in G, the square of the ADJACENCY MATRIX of G counts the number of such paths. Similarly, the (u, v)th element of the kth power of the ADJACENCY MATRIX of G gives the number of paths of length k between vertices u and v. The graph kth power is then defined as the graph whose adjacency matrix given by the sum of the first k powers of the ADJACENCY MATRIX,

$$\mathrm{adj}(G^k) = \sum_{i=1}^{k} [\mathrm{adj}(G)]^i,$$

which counts all paths of length up to k (Skiena 1990, p. 230).

Raising any graph to the power of its GRAPH DIAMETER gives a COMPLETE GRAPH. The square of any BICONNECTED GRAPH is HAMILTONIAN (Fleischner 1974, Skiena 1990, p. 231). Mukhopadhyay (1967) has considered "square root graphs," whose square gives a given graph G (Skiena 1990, p. 253).

See also ADJACENCY MATRIX, PÓSA'S THEOREM, SEYMOUR CONJECTURE

References

Fleischner, H. "The Square of Every Two-Connected Graph Is Hamiltonian." *J. Combin. Th. Ser. B* **16**, 29–4, 1974.

Mukhopadhyay, A. "The Square Root of a Graph." *J. Combin. Th.* **2**, 290–95, 1967.

Skiena, S. *Implementing Discrete Mathematics: Combinatorics and Graph Theory with Mathematica.* Reading, MA: Addison-Wesley, 1990.

Graph Product

This entry contributed by Nicolas Bray

In general, a graph product of two graphs G and H is a new graph whose VERTEX SET is $V(G) \times V(H)$ and where, for any two vertices (g, h) and (g', h') in the product, the adjacency of those two vertices is determined entirely by the adjacency (or equality, or non-adjacency) of g and g', and that of h and h'. There are $3 \times 3 - 1 = 8$ cases to be decided (three possibilities for each, with the case where both are equal eliminated) and thus there are $2^8 = 256$ different types of graph products that can be defined.

The most commonly used graph products, given by conditions sufficient and necessary for adjacency, are summarized in the following table (Hartnell and Rall 1998). Note that the terminology is not quite standardized, so these products may actually be referred to by different names by different sources. Many other graph products can be found in Jensen and Toft (1994).

graph product name	symbol	definition
GRAPH CARTESIAN PRODUCT	$G \square H$	$(g = g'$ and h adj $h')$ or $(g$ adj g' and $h = h')$
GRAPH CATEGORICAL PRODUCT	$G \times H$	$(g$ adj g' and h adj $h')$
GRAPH LEXOGRAPHIC PRODUCT	$G \cdot H$	$(g$ adj $g')$ or $(g = g'$ and h adj $h')$
GRAPH STRONG PRODUCT	$G \boxtimes H$	$(g = g'$ and h adj $h')$ or $(g$ adj g' and $h = h')$ or $(g$ adj g' and h adj $h')$

See also GRAPH CARTESIAN PRODUCT

References

Hartnell, B. and Rall, D. "Domination in Cartesian Products: Vizing's Conjecture." In *Domination in Graphs--Advanced Topics* (Ed. T. W. Haynes, S. T. Hedetniemi, and P. J. Slater). New York: Dekker, pp. 163–89, 1998.

Jensen, T. R. and Toft, B. *Graph Coloring Problems.* New York: Wiley, 1994.

Graph Radius

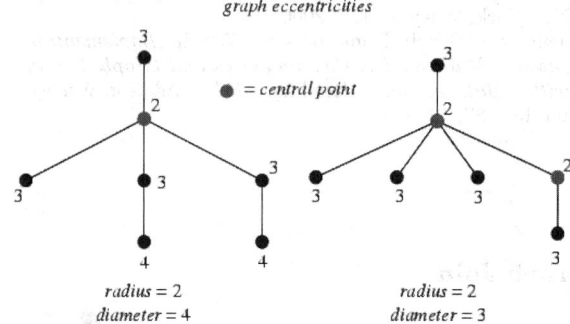

graph eccentricities

= *central point*

radius = 2
diameter = 4

radius = 2
diameter = 3

The minimum GRAPH ECCENTRICITY of any VERTEX in a GRAPH.

See also CENTRAL POINT, GRAPH CENTER, GRAPH ECCENTRICITY

References

Harary, F. *Graph Theory.* Reading, MA: Addison-Wesley, p. 35, 1994.

Skiena, S. *Implementing Discrete Mathematics: Combinatorics and Graph Theory with Mathematica.* Reading, MA: Addison-Wesley, p. 107, 1990.

Graph Section

A section of a GRAPH obtained by finding its intersection with a PLANE.

Graph Spectrum

The set of GRAPH EIGENVALUES is called the spectrum of the graph. The spectrum of a graph may be computed using `Spectrum[g]` in the *Mathematica* add-on package `DiscreteMath`Combinatorica` (which can be loaded with the command `<<DiscreteMath`).

Two nonisomorphic graphs can share the same spectrum, e.g., the GRAPH UNION $C_4 \cup K_1$ and STAR GRAPH S_5 (Skiena 1990, p. 85). The maximum degree of a CONNECTED GRAPH G is an eigenvalue of G IFF G is a REGULAR GRAPH.

See also GRAPH EIGENVALUE

References

Biggs, N. L. *Algebraic Graph Theory, 2nd ed.* Cambridge, England: Cambridge University Press, 1993.

Cvetkovic, D.; Doob, M.; and Sachs, H. *Spectra of Graphs.* New York: Academic Press, 1980.

Skiena, S. *Implementing Discrete Mathematics: Combinatorics and Graph Theory with Mathematica.* Reading, MA: Addison-Wesley, p. 85, 1990.

Wilf, H. "Graphs and Their Spectra: Old and New Results." *Congr. Numer.* **50**, 37–3, 1985.

Graph Strong Product

This entry contributed by NICOLAS BRAY

The GRAPH PRODUCT denoted $G \boxtimes H$ and defined by the adjacency relations ($g = g'$ and h adj h') or (g adj g' and $h = h'$) or (g adj g' and h adj h').

See also GRAPH PRODUCT

Graph Sum

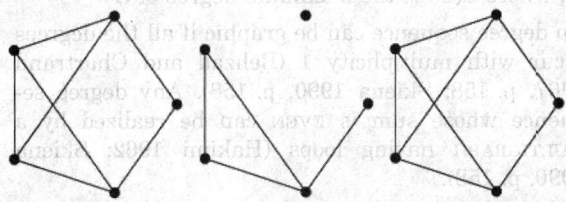

The graph sum of graphs G and H is the graph with ADJACENCY MATRIX given by the sum of adjacency matrices of G and H. A graph sum is defined when the orders of G and H are the same, and can be computed using `GraphSum[g, h]` in the *Mathematica* add-on package `DiscreteMath`Combinatorica` (which can be loaded with the command `<<DiscreteMath`).

See also GRAPH DIFFERENCE, GRAPH JOIN, GRAPH UNION

References

Skiena, S. "Sum and Difference." §4.1.2 in *Implementing Discrete Mathematics: Combinatorics and Graph Theory with Mathematica.* Reading, MA: Addison-Wesley, p. 131, 1990.

Graph Theory

The mathematical study of the properties of the formal mathematical structures called GRAPHS.

See also ADJACENCY MATRIX, ADJACENCY RELATION, ARTICULATION VERTEX, BLUE-EMPTY COLORING, BRIDGE, CHROMATIC NUMBER, CHROMATIC POLYNOMIAL, CIRCUIT RANK, CROSSING NUMBER (GRAPH), CYCLOMATIC NUMBER, DEGREE, DIJKSTRA'S ALGORITHM, ECCENTRICITY, EDGE COLORING, EDGE CONNECTIVITY, EULERIAN CIRCUIT, EULERIAN TRAIL, FACTOR (GRAPH), FLOYD'S ALGORITHM, GIRTH, GRAPH CYCLE, GRAPH DIAMETER, GRAPH RADIUS, GRAPH TWO-COLORING, GROUP THEORY, HAMILTONIAN CIRCUIT, HASSE DIAGRAM, HUB, INDEGREE, INTEGRAL DRAWING, ISTHMUS, JOIN (GRAPH), LOCAL DEGREE, MONOCHROMATIC FORCED TRIANGLE, OUTDEGREE, PARTY PROBLEM, PÓLYA ENUMERATION THEOREM, PÓLYA POLYNOMIAL, RAMSEY NUMBER, RE-ENTRANT CIRCUIT, SEPARATING EDGE, TAIT COLORING, TAIT CYCLE, TRAVELING SALESMAN PROBLEM, TREE, TUTTE'S THEOREM, UNICURSAL CIRCUIT, VERTEX COLORING, VERTEX DEGREE, WALK

References

Beinecke, L. W. and Wilson, R. J. (Eds.). *Graph Connections: Relationships Between Graph Theory and Other Areas of Mathematics.* Oxford, England: Oxford University Press, 1997.
Berge, C. *Graphs and Hypergraphs.* Amsterdam, Netherlands: North-Holland, 1976.
Berge, C. *The Theory of Graphs and Its Applications.* New York: Wiley, 1962.
Bogomolny, A. "Graphs." http://www.cut-the-knot.com/do_you_know/graphs.html.
Bollobás, B. *Graph Theory: An Introductory Course.* New York: Springer-Verlag, 1979.
Bollobás, B. *Modern Graph Theory.* New York: Springer-Verlag, 1998.
Caldwell, C. K. "Graph Theory Tutorials." http://www.utm.edu/departments/math/graph/.
Chartrand, G. *Introductory Graph Theory.* New York: Dover, 1985.
Emden-Weinert, T. "Graphs: Theory-Algorithms-Complexity." http://people.freenet.de/Emden-Weinert/graphs.html.
Foulds, L. R. *Graph Theory Applications.* New York: Springer-Verlag, 1992.
Chung, F. and Graham, R. *Erdos on Graphs: His Legacy of Unsolved Problems.* New York: A. K. Peters, 1998.
Gardner, M. "Graph Theory." Ch. 10 in *The Sixth Book of Mathematical Games from Scientific American.* Chicago, IL: University of Chicago Press, pp. 91–03, 1984.
Gould, R. (Ed.). *Graph Theory.* Menlo Park, CA: Benjamin-Cummings, 1988.
Grossman, I. and Magnus, W. *Groups and Their Graphs.* Washington, DC: Math. Assoc. Amer., 1965.
Harary, F. "Graphical Enumeration Problems." In *Graph Theory and Theoretical Physics* (Ed. F. Harary). London: Academic Press, pp. 1–1, 1967.
Harary, F. *Graph Theory.* Reading, MA: Addison-Wesley, 1994.
Hartsfield, N. and Ringel, G. *Pearls in Graph Theory: A Comprehensive Introduction, 2nd ed.* San Diego, CA: Academic Press, 1994.
Locke, S. C. "Graph Theory." http://www.math.fau.edu/locke/graphthe.htm.
Locke, S. C. "Graph Theory Books." http://www.math.fau.edu/locke/graphstx.htm.
Mehlhorn, K. and Näher, S. *LEDA: A Platform for Combinatorial and Geometric Computing.* Cambridge, England: Cambridge University Press, 1999.
Ore, Ø. *Graphs and Their Uses.* New York: Random House, 1963.
Read, R. C. and Wilson, R. J. *An Atlas of Graphs.* Oxford, England: Oxford University Press, 1998.
Ruskey, F. "Information on (Unlabelled) Graphs." http://www.theory.csc.uvic.ca/~cos/inf/grap/GraphInfo.html.
Saaty, T. L. and Kainen, P. C. *The Four-Color Problem: Assaults and Conquest.* New York: Dover, 1986.
Skiena, S. *Implementing Discrete Mathematics: Combinatorics and Graph Theory with Mathematica.* Redwood City, CA: Addison-Wesley, 1988.
Trudeau, R. J. *Introduction to Graph Theory.* New York: Dover, 1994.
Tutte, W. T. *Graph Theory as I Have Known It.* Oxford, England: Oxford University Press, 1998.
Weisstein, E. W. "Graphs." MATHEMATICA NOTEBOOK GRAPHS.M.
Weisstein, E. W. "Books about Graph Theory." http://www.treasure-troves.com/books/GraphTheory.html.
Woo, L. "Definitions of Graph Theory." http://www.simmons.edu/~woo/graphtheory/definition.html.

Graph Thickness

The thickness of a GRAPH G is the minimum number of PLANAR SUBGRAPHS of g whose GRAPH UNION is g (skiena 1990, p. 251).

References

Skiena, S. *Implementing Discrete Mathematics: Combinatorics and Graph Theory with Mathematica.* Reading, MA: Addison-Wesley, 1990.

Graph Two-Coloring

Assignment of each EDGE of a GRAPH to one of two color classes ("red" or "green").

See also BLUE-EMPTY GRAPH, MONOCHROMATIC FORCED TRIANGLE

Graph Union

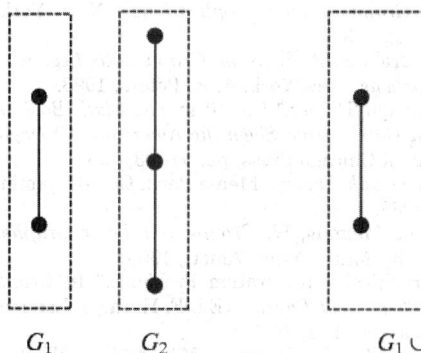

G_1 G_2 $G_1 \cup G_2$

The union $G = G_1 \cup G_2$ of graphs G_1 and G_2 with disjoint point sets V_1 and V_2 and edge sets X_1 and X_2 is the graph with $V = V_1 \cup V_2$ and $X = X_1 \cup X_2$ (Harary 1994, p. 21). Graph unions can be computed using `GraphUnion[g, h]` in the *Mathematica* add-on package `DiscreteMath`Combinatorica`` (which can be loaded with the command `<<DiscreteMath``).

See also GRAPH INTERSECTION, GRAPH JOIN

References

Harary, F. *Graph Theory.* Reading, MA: Addison-Wesley, 1994.
Skiena, S. "Unions and Intersections." §4.1.1 in *Implementing Discrete Mathematics: Combinatorics and Graph Theory with Mathematica.* Reading, MA: Addison-Wesley, pp. 129–31, 1990.

Graphic Sequence

A graphic sequence is a sequence of numbers which can be the DEGREE SEQUENCE of some GRAPH. A sequence can be checked to determine if it is graphic using `GraphicQ[g]` in the *Mathematica* add-on package `DiscreteMath`Combinatorica`` (which can be loaded with the command `<<DiscreteMath``).

Erdos and Gallai (1960) proved that a DEGREE SEQUENCE $\{d_1, \ldots, d_n\}$ is graphic IFF the sequence obeys the property

$$\sum_{i=1}^{r} d_i \leq r(r-1) + \sum_{i=r+1}^{n} \min(r, d_i)$$

for each integer $r < n$ (Skiena 1990, p. 157), and this condition also generalizes to DIRECTED GRAPHS. In addition, Hakimi (1962) and Havel (1955) showed that if a DEGREE SEQUENCE is graphic, then there exists a GRAPH G such that the node of highest degree is adjacent to the $\Delta(G)$ next highest degree vertices of G, where $\Delta(G)$ is the maximum degree of G.

No degree sequence can be graphic if all the degrees occur with multiplicity 1 (Behzad and Chartrand 1967, p. 158; Skiena 1990, p. 158). Any degree sequence whose sum is EVEN can be realized by a MULTIGRAPH having loops (Hakimi 1962; Skiena 1990, p. 158).

See also DEGREE SEQUENCE, GRAPHICAL PARTITION, VERTEX DEGREE

References

Behzad, M. and Chartrand, G. "No Graph is Perfect." *Amer. Math. Monthly* **74**, 962–63, 1967.
Eggleton, R. B. "Graphic Sequences and Graphic Polynomials." In *Infinite and Finite Sets* (Ed. A. Hajnal). Amsterdam, Netherlands: North-Holland, pp. 385–93, 1975.
Erdos, P. and Gallai, T. "Graphs with Prescribed Degrees of Vertices" [Hungarian]. *Mat. Lapok.* **11**, 264–74, 1960.
Fulkerson, D. R. "Upsets in Round Robin Tournaments." *Canad. J. Math.* **17**, 957–69, 1965.
Fulkerson, D. R.; Hoffman, A. J.; and McAndrew, M. H. "Some Properties of Graphs with Multiple Edges." *Canad. J. Math.* **17**, 166–77, 1965.
Hakimi, S. "On the Realizability of a Set of Integers as Degrees of the Vertices of a Graph." *SIAM J. Appl. Math.* **10**, 496–06, 1962.
Havel, V. "A Remark on the Existence of Finite Graphs" [Czech]. *Casopis Pest. Mat.* **80**, 477–80, 1955.
Ryser, H. J. "Combinatorial Properties of Matrices of Zeros and Ones." *Canad. J. Math.* **9**, 371–77, 1957.
Skiena, S. *Implementing Discrete Mathematics: Combinatorics and Graph Theory with Mathematica.* Reading, MA: Addison-Wesley, p. 157, 1990.

Graphical Partition

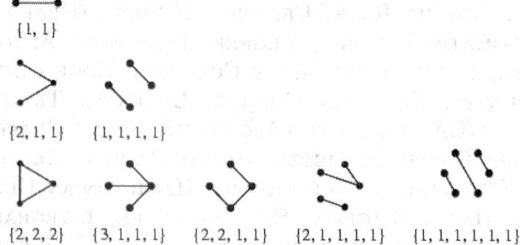

A partition $\{a_1, \ldots, a_n\}$ is called graphical if there exists a GRAPH G having DEGREE SEQUENCE $\{a_1, \ldots, a_n\}$. The number of graphical partitions on n-node graphs is therefore the same as the number of n-node graphs with no ISOLATED POINTS. A graphical partition of order p is one for which the sum of degrees is p. A p-graphical partition only exists for

EVEN p.

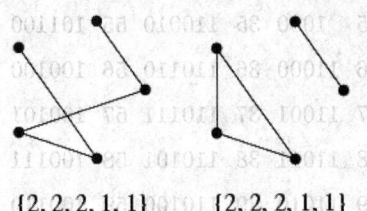

$\{2, 2, 2, 1, 1\}$ $\{2, 2, 2, 1, 1\}$

It is possible for two topologically distinct graphs to have the same DEGREE SEQUENCE.

For $n = 2, 4, 6, ...$, the numbers of graphical partitions $p_g(n)$ are 1, 2, 5, 9, 17, ... (Sloane's A000569).

Erdos and Richmond (1989) showed that

$$\liminf_{n \to \infty} \sqrt{2n} p_g(2n) \geq \frac{\pi}{\sqrt{6}}$$

and

$$\limsup_n p_g(2n) \leq 0.4258.$$

See also CUT, DEGREE SEQUENCE, SPECTRAL GRAPH PARTITIONING

References

Barnes, T. M. and Savage, C. D. "A Recurrence for Counting Graphical Partitions." *Electronic J. Combinatorics* **2**, R11 1–0, 1995. http://www.combinatorics.org/Volume_2/volume2.html#R11.

Barnes, T. M. and Savage, C. D. "Efficient Generation of Graphical Partitions." *Disc. Appl. Math.* **78**, 17–6, 1997.

Erdos, P. and Richmond, L. B. "On Graphical Partitions." Combinatorics and Optimization Research Report COPR 89–2. Waterloo, Ontario: University of Waterloo, pp. 1–3, 1989.

Harary, F. *Graph Theory.* Reading, MA: Addison-Wesley, p. 57, 1994.

Ruskey, F. "Information on Graphical Partitions." http://www.theory.csc.uvic.ca/~cos/inf/nump/GraphicalPartition.html.

Sloane, N. J. A. Sequences A000569 in "An On-Line Version of the Encyclopedia of Integer Sequences." http://www.research.att.com/~njas/sequences/eisonline.html.

Wilf, H. "On Crossing Numbers, and Some Unsolved Problems." In *Combinatorics, Geometry, and Probability: A Tribute to Paul Erdos. Papers from the Conference in Honor of Erdos' 80th Birthday Held at Trinity College, Cambridge, March 1993* (Ed. B. Bollobás and A. Thomason). Cambridge, England: Cambridge University Press, pp. 557–62, 1997.

Graphical Representation

FERRERS DIAGRAM

Graphoid

A graphoid consists of a set M of elements together with two collections $\mathfrak{C}$ and $\mathfrak{D}$ of nonempty subsets of M, called circuits and cocircuits respectively, such that

1. For any $C \in \mathfrak{C}$ and $D \in \mathfrak{D}$, $|C \cap D| \neq 1$,
2. No circuit properly contains another circuit and no cocircuit properly contains another cocircuit,
3. For any painting of M with colors exactly one element green and the rest either red or blue, there exists either (a) a circuit C containing the green element and no red elements, or (b) a cocircuit D containing the green element and no blue elements.

See also MATROID

References

Harary, F. *Graph Theory.* Reading, MA: Addison-Wesley, p. 41, 1994.

Grassmann Algebra

EXTERIOR ALGEBRA

Grassmann Coordinates

An $(m + 1)$-D SUBSPACE W of an $(n + 1)$-D VECTOR SPACE V can be specified by an $(m + 1) \times (n + 1)$ MATRIX whose rows are the coordinates of a BASIS of W. The set of all $\binom{n+1}{m+1}$ $(m + 1) \times (m + 1)$ MINORS of this MATRIX are then called the Grassmann (or sometimes Plücker; Stofli 1991) coordinates of w, where $\binom{a}{b}$ is a BINOMIAL COEFFICIENT. Hodge and Pedoe (1952) give a thorough treatment of Grassmann coordinates.

See also CHOW COORDINATES

References

Hodge, W. V. D. and Pedoe, D. *Methods of Algebraic Geometry.* Cambridge, England: Cambridge University Press, 1952.

Stofli, J. *Oriented Projective Geometry.* New York: Academic Press, 1991. Wilson, W. S.; Chern, S. S.; Abhyankar, S. S.; Lang, S.; and Igusa, J.-I. "Wei-Liang Chow." *Not. Amer. Math. Soc.* **43**, 1117–124, 1996.

Grassmann Manifold

A special case of a FLAG MANIFOLD. A Grassmann manifold is a certain collection of vector SUBSPACES of a VECTOR SPACE. In particular, $g_{n,k}$ is the Grassmann manifold of k-dimensional subspaces of the VECTOR SPACE $\mathbb{R}^n$. It has a natural MANIFOLD structure as an orbit-space of the STIEFEL MANIFOLD $v_{n,k}$ of orthonormal k-frames in $\mathbb{G}^n$. One of the main things about Grassmann manifolds is that they are classifying spaces for VECTOR BUNDLES.

Gray Code

An encoding of numbers so that adjacent numbers have a single DIGIT differing by 1. A BINARY Gray code with n DIGITS corresponds to a HAMILTONIAN PATH on an n-D HYPERCUBE (including direction reversals). The term Gray code is often used to refer to a "reflected" code, or more specifically still, the binary reflected Gray code.

To convert a BINARY number $d_1 d_2 \cdots d_{n-1} d_n$ to its corresponding binary reflected Gray code, start at the right with the digit d_n (the nth, or last, DIGIT). If the d_{n-1} is 1, replace d_n by $1 - d_n$; otherwise, leave it unchanged. Then proceed to d_{n-1}. Continue up to the first DIGIT d_1, which is kept the same since d_0 is assumed to be a 0. The resulting number $g_1 g_2 \cdots g_{n-1} g_n$ is the reflected binary Gray code.

To convert a binary reflected Gray code $g_1 g_2 \cdots g_{n-1} g_n$ to a BINARY number, start again with the nth digit, and compute

$$\sum_n \equiv \sum_{i=1}^{n-1} g_i \ (\mathrm{mod} \ 2).$$

If Σ_n is 1, replace g_n by $1 - g_n$; otherwise, leave it the unchanged. Next compute

$$\sum_{n-1} \equiv \sum_{i=1}^{n-2} g_i \ (\mathrm{mod} \ 2),$$

and so on. The resulting number $d_1 d_2 \cdots d_{n-1} d_n$ is the BINARY number corresponding to the initial binary reflected Gray code.

The code is called reflected because it can be generated in the following manner. Take the Gray code 0, 1. Write it forwards, then backwards: 0, 1, 1, 0. Then append 0s to the first half and 1s to the second half: 00, 01, 11, 10. Continuing, write 00, 01, 11, 10, 10, 11, 01, 00 to obtain: 000, 001, 011, 010, 110, 111, 101, 100, ... (Sloane's A014550). Each iteration therefore doubles the number of codes. The Gray codes corresponding to the first few nonnegative integers are given in the following table.

0	0	20	11110	40	111100
1	1	21	11111	41	111101
2	11	22	11101	42	111111
3	10	23	11100	43	111110
4	110	24	10100	44	111010
5	111	25	10101	45	111011
6	101	26	10111	46	111001
7	100	27	10110	47	111000
8	1100	28	10010	48	101000
9	1101	29	10011	49	101001
10	1111	30	10001	50	101011
11	1110	31	10000	51	101010
12	1010	32	110000	52	101110
13	1011	33	110001	53	101111
14	1001	34	110011	54	101101
15	1000	35	110010	55	101100
16	11000	36	110110	56	100100
17	11001	37	110111	57	100101
18	11011	38	110101	58	100111
19	11010	39	110100	59	100110

The binary reflected Gray code is closely related to the solutions of the TOWERS OF HANOI and BAGUENAUDIER, as well as to Hamiltonian circuits of hypercube graphs (Skiena 1990, p. 149).

See also BAGUENAUDIER, BINARY, HILBERT CURVE, RYSER FORMULA, THUE-MORSE SEQUENCE, TOWERS OF HANOI

References

Gardner, M. "The Binary Gray Code." Ch. 2 in *Knotted Doughnuts and Other Mathematical Entertainments.* New York: W. H. Freeman, 1986.

Gilbert, E. N. "Gray Codes and Paths on the n-Cube." *Bell System Tech. J.* **37**, 815–26, 1958.

Gray, F. "Pulse Code Communication." United States Patent Number 2,632,058. March 17, 1953.

Nijenhuis, A. and Wilf, H. *Combinatorial Algorithms for Computers and Calculators, 2nd ed.* New York: Academic Press, 1978.

Press, W. H.; Flannery, B. P.; Teukolsky, S. A.; and Vetterling, W. T. "Gray Codes." §20.2 in *Numerical Recipes in FORTRAN: The Art of Scientific Computing, 2nd ed.* Cambridge, England: Cambridge University Press, pp. 886–88, 1992.

Skiena, S. "Gray Code." §1.5.3 in *Implementing Discrete Mathematics: Combinatorics and Graph Theory with Mathematica.* Reading, MA: Addison-Wesley, pp. 42–3 and 149, 1990.

Sloane, N. J. A. Sequences A014550 in "An On-Line Version of the Encyclopedia of Integer Sequences." http://www.research.att.com/~njas/sequences/eisonline.html.

Vardi, I. *Computational Recreations in Mathematica.* Redwood City, CA: Addison-Wesley, pp. 111–12 and 246, 1991.

Wilf, H. S. *Combinatorial Algorithms: An Update.* Philadelphia, PA: SIAM, 1989.

Gray Graph

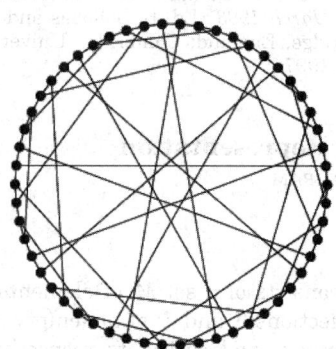

A CUBIC GRAPH on 54 vertices that is EDGE- but not

VERTEX-TRANSITIVE; the smallest known such example. It was discovered by Marion C. Gray in 1932, and was first published by Bouwer (1968). It has GIRTH 8, GRAPH DIAMETER 6, has |Aut *G*|=1296, and is the Levi graph of two dual, triangle-free, point-, line-, and flag-transitive, non-self-dual 27_3 configurations (Maruvic and Pisanski 2000). The symmetric embedding illustrated above is due to (Maruvic and Pisanski 2000). It can be constructed by taking three copies of the COMPLETE BIPARTITE GRAPH $K_{3,3}$ and, for a particular edge *e*, subdividing *e* in each of the three copies, joining the resulting three vertices to a new vertex, and repeating with each edge.

See also COMPLETE BIPARTITE GRAPH, CUBIC GRAPH, EDGE-TRANSITIVE GRAPH, VERTEX-TRANSITIVE GRAPH

References

Bondy, J. A. and Murty, U. S. R. *Graph Theory with Applications.* New York: North Holland, p. 235, 1976.
Bouwer, I. Z. "An Edge But Not Vertex Transitive Cubic Graph." *Bull. Canad. Math. Soc.* **11**, 533–35, 1968.
Bouwer, I. Z. "On Edge But Not Vertex Transitive Regular Graphs." *J. Combin. Th. B* **12**, 32–0, 1972.
Maruvic, D. and Pisanski, T. "The Gray Graph Revisited." *J. Graph Th.* **35**, 1–, 2000.
Pisanski, T. and Randic, M. "Bridged Between Geometry and Graph Theory." To appear.
Weisstein, E. W. "Graphs." MATHEMATICA NOTEBOOK GRAPHS.M.

Grazing Goat Problem

GOAT PROBLEM

Great Circle

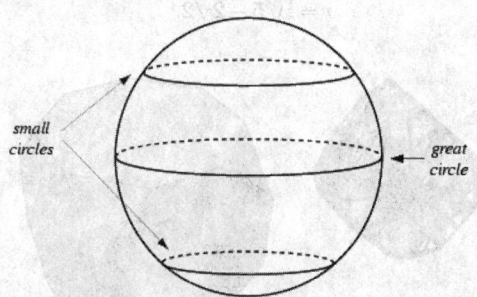

A great circle is a SECTION of a SPHERE which contains a DIAMETER of the SPHERE (Kern and Bland 1948, p. 87). Sections of the sphere that do not contain a diameter are called SMALL CIRCLES.

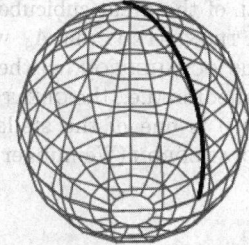

The shortest path between two points on a SPHERE,

also known as an ORTHODROME, is a segment of a great circle. To find the great circle (GEODESIC) distance between two points located at LATITUDE δ and LONGITUDE λ of (δ_1, λ_1) and (δ_2, λ_2) on a SPHERE of RADIUS *a*, convert SPHERICAL COORDINATES to CARTESIAN COORDINATES using

$$\mathbf{r}_i = a \begin{bmatrix} \cos \lambda_i \cos \delta_i \\ \sin \lambda_i \cos \delta_i \\ \sin \delta_i \end{bmatrix}. \tag{1}$$

(Note that the LATITUDE δ is related to the COLATITUDE ϕ of SPHERICAL COORDINATES by $\delta = 90° - \phi$, so the conversion to CARTESIAN COORDINATES replaces $\sin \phi$ and $\cos \phi$ by $\cos \delta$ and $\sin \delta$, respectively.) Now find the ANGLE α between $\mathbf{r}_1$ and $\mathbf{r}_2$ using the DOT PRODUCT,

$$\cos \alpha = \hat{\mathbf{r}}_1 \cdot \hat{\mathbf{r}}_2$$

$$= \cos \delta_1 \cos \delta_2 (\sin \lambda_1 \sin \lambda_2 + \cos \lambda_1 \cos \lambda_2)$$
$$+ \sin \delta_1 \sin \delta_2$$

$$= \cos \delta_1 \cos \delta_2 \cos(\lambda_1 - \lambda_2) + \sin \delta_1 \sin \delta_2. \tag{2}$$

The great circle distance is then

$$d = a \cos^{-1}[\cos \delta_1 \cos \delta_2 \cos(\lambda_1 - \lambda_2)$$
$$+ \sin \delta_1 \sin \delta_2]. \tag{3}$$

For the Earth, the *equatorial* RADIUS is $a \approx 6378$ km, or 3963 (statute) miles. Unfortunately, the FLATTENING of the Earth cannot be taken into account in this simple derivation, since the problem is considerably more complicated for a SPHEROID or ELLIPSOID (each of which has a RADIUS which is a function of LATITUDE). This leads to extremely complicated expressions for OBLATE SPHEROID GEODESICS and GEODESICS on other ELLIPSOIDS.

A great circle becomes a straight line in a GNOMONIC PROJECTION (Steinhaus 1983, pp. 220–21).

The equation of the great circle can be explicitly computed using the GEODESIC formalism. Writing

$$u = \lambda \tag{4}$$

$$v = \delta = \tfrac{1}{2} \pi - \phi \tag{5}$$

gives the *P*, *Q*, and *R* parameters of the GEODESIC (which are just combinations of the PARTIAL DERIVATIVES) as

$$P \equiv \left(\frac{\partial x}{\partial u}\right)^2 + \left(\frac{\partial y}{\partial u}\right)^2 + \left(\frac{\partial z}{\partial u}\right)^2 = a^2 \sin^2 v \tag{6}$$

$$Q \equiv \frac{\partial x}{\partial u} \frac{\partial x}{\partial v} + \frac{\partial y}{\partial u} \frac{\partial y}{\partial v} + \frac{\partial z}{\partial u} \frac{\partial z}{\partial v} = 0 \tag{7}$$

$$R \equiv \left(\frac{\partial x}{\partial v}\right)^2 + \left(\frac{\partial y}{\partial v}\right)^2 + \left(\frac{\partial z}{\partial v}\right)^2 = a^2. \tag{8}$$

The GEODESIC differential equation then becomes

$$\cos v \sin^4 v + 2 \cos v \sin^2 vv'^2 + \cos vv'^4 - \sin vv''$$
$$= 0. \tag{9}$$

However, because this is a special case of $Q = 0$ with P and R explicit functions of v only, the GEODESIC solution takes on the special form

$$v = c_1 \int \sqrt{\frac{R}{P^2 - c_1^2 P}} \, dv = c_1 \int \frac{dv}{a^2 \sin^4 v - c_1^2 \sin^2 v}$$

$$= \int \frac{dv}{\sin v \sqrt{\left(\dfrac{a}{c_1}\right)^2 \sin^2 v - 1}}$$

$$= -\tan^{-1}\left[\frac{\cos v}{\sqrt{\left(\dfrac{a}{c_1}\right)^2 - 1}}\right] + c_2 \tag{10}$$

(Gradshteyn and Ryzhik 2000, p. 174, eqn. 2.599.6), which can be rewritten as

$$v = -\sin^{-1}\left(\frac{\cot v}{\sqrt{\left(\dfrac{a}{c_1}\right)^2 - 1}}\right) + c_2. \tag{11}$$

It therefore follows that

$$(\sin c_2)a \sin v \cos u - (\cos c_2)a \sin v \sin u$$
$$-\frac{a \cos v}{\sqrt{\left(\dfrac{a}{c_1}\right)^2 - 1}}$$
$$= 0. \tag{12}$$

This equation can be written in terms of the CARTESIAN COORDINATES as

$$x \sin c_2 - y \cos c_2 - \frac{z}{\sqrt{\left(\dfrac{a}{c_1}\right)^2 - 1}} = 0, \tag{13}$$

which is simply a PLANE passing through the center of the SPHERE and the two points on the surface of the SPHERE.

See also GEODESIC, GREAT SPHERE, LOXODROME, MIKUSINSKI'S PROBLEM, OBLATE SPHEROID GEODESIC, ORTHODROME, POINT-POINT DISTANCE–2-D, PSEUDO-CIRCLE, SMALL CIRCLE, SPHERE

References

Gradshteyn, I. S. and Ryzhik, I. M. *Tables of Integrals, Series, and Products, 6th ed.* San Diego, CA: Academic Press, 2000.

Kern, W. F. and Bland, J. R. *Solid Mensuration with Proofs, 2nd ed.* New York: Wiley, 1948.

Steinhaus, H. *Mathematical Snapshots, 3rd ed.* New York: Dover, pp. 183 and 217, 1999.

Tietze, H. *Famous Problems of Mathematics: Solved and Unsolved Mathematics Problems from Antiquity to Modern Times.* New York: Graylock Press, pp. 24–5, 1965.

Weinstock, R. *Calculus of Variations, with Applications to Physics and Engineering.* New York: Dover, pp. 26–8 and 62–3, 1974.

Great Cubicuboctahedron

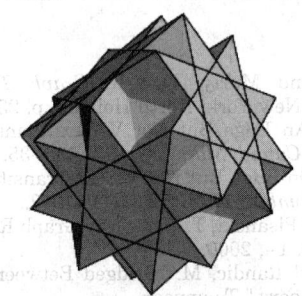

The UNIFORM POLYHEDRON U_{14} whose DUAL POLYHEDRON is the GREAT HEXACRONIC ICOSITETRAHEDRON. It has WYTHOFF SYMBOL $3\,4\,|\,\frac{4}{3}$ and is Wenninger model W_{77}. Its faces are $8\{3\} + 6\{4\} + 6\{\frac{8}{3}\}$. It is a FACETED version of the CUBE. The CIRCUMRADIUS of a great cubicuboctahedron with unit edge length is

$$r = \tfrac{1}{2}\sqrt{5 - 2\sqrt{2}}.$$

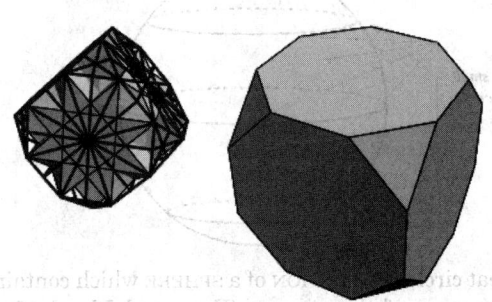

The CONVEX HULL of the great cubicuboctahedron is the Archimedean TRUNCATED CUBE A_9, whose dual is the SMALL TRIAKIS OCTAHEDRON, so the dual of the great cubicuboctahedron (i.e., the GREAT HEXACRONIC ICOSITETRAHEDRON) is one of the stellations of the SMALL TRIAKIS OCTAHEDRON (Wenninger 1983, p. 57).

References

Wenninger, M. J. *Dual Models.* Cambridge, England: Cambridge University Press, pp. 57–8, 1983.

Wenninger, M. J. *Polyhedron Models.* Cambridge, England: Cambridge University Press, pp. 118–19, 1989.

Great Deltoidal Hexecontahedron

The DUAL of the uniform GREAT RHOMBICOSIDODECA-HEDRON U_{67} and Wenninger dual W_{105}.

See also DUAL POLYHEDRON, GREAT RHOMBICOSIDO-DECAHEDRON (UNIFORM)

References
Wenninger, M. J. *Dual Models.* Cambridge, England: Cambridge University Press, p. 88, 1983.

Great Deltoidal Icositetrahedron

The DUAL of the uniform GREAT RHOMBICUBOCTAHE-DRON and Wenninger dual W_{85}.

References
Wenninger, M. J. *Dual Models.* Cambridge, England: Cambridge University Press, p. 59, 1983.

Great Dirhombicosidodecacron

The DUAL of the GREAT DIRHOMBICOSIDODECAHEDRON U_{75} and Wenninger dual W_{119}.

References
Wenninger, M. J. *Dual Models.* Cambridge, England: Cambridge University Press, p. 139, 1983.

Great Dirhombicosidodecahedron

The UNIFORM POLYHEDRON U_{75} whose DUAL is the GREAT DIRHOMBICOSIDODECACRON. This POLYHEDRON

is exceptional because it cannot be derived from SCHWARZ TRIANGLES and because it is the only UNI-FORM POLYHEDRON with more than six POLYGONS surrounding each VERTEX (four SQUARES alternating with two TRIANGLES and two PENTAGRAMS). This unique polyhedron has features in common with both snub forms and hemipolyhedra, and its octagrammic faces pass through the origin.

It has pseudo-WYTHOFF SYMBOL $\left|\frac{3}{2}\frac{5}{3}3\frac{5}{2}\right.$. Its faces are $40\{3\} + 60\{4\} + 24\{\frac{5}{2}\}$, and its CIRCUMRADIUS for unit edge length is

$$R = \tfrac{1}{2}\sqrt{2}.$$

See also UNIFORM POLYHEDRON

References
Wenninger, M. J. *Polyhedron Models.* Cambridge, England: Cambridge University Press, pp. 200–03, 1989.

Great Disdyakis Dodecahedron

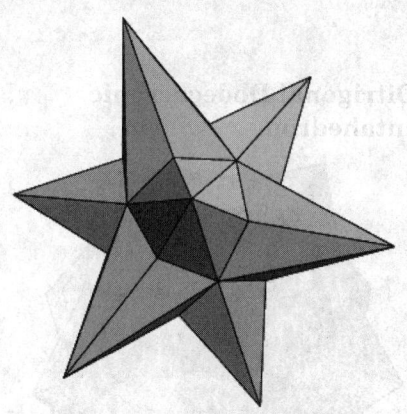

The DUAL of the GREAT TRUNCATED CUBOCTAHEDRON U_{20} and Wenninger dual W_{93}.

See also DUAL POLYHEDRON, GREAT TRUNCATED CUBOCTAHEDRON

References
Wenninger, M. J. *Dual Models.* Cambridge, England: Cambridge University Press, p. 92, 1983.

Great Disdyakis Triacontahedron

The DUAL of the GREAT TRUNCATED ICOSIDODECAHE-DRON U_{68} and Wenninger dual W_{108}.

See also DUAL POLYHEDRON, GREAT TRUNCATED ICOSIDODECAHEDRON

References

Wenninger, M. J. *Dual Models.* Cambridge, England: Cambridge University Press, p. 96, 1983.

Great Ditrigonal Dodecacronic Hexecontahedron

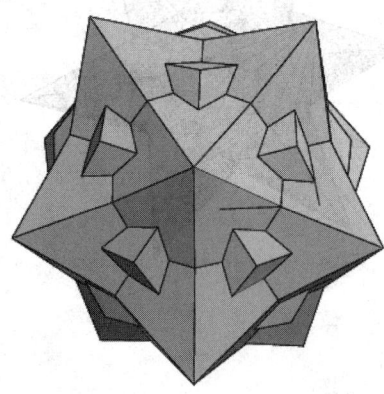

The DUAL of the GREAT DITRIGONAL DODECICOSIDODE-CAHEDRON U_{42} and Wenninger dual W_{81}.

See also DUAL POLYHEDRON, GREAT DITRIGONAL DODECICOSIDODECAHEDRON

References

Wenninger, M. J. *Dual Models.* Cambridge, England: Cambridge University Press, p. 62, 1983.

Great Ditrigonal Dodecicosidodecahedron

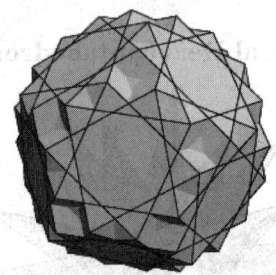

The UNIFORM POLYHEDRON U_{42} whose DUAL is the GREAT DITRIGONAL DODECACRONIC HEXECONTAHE-DRON. It has WYTHOFF SYMBOL $3\,5|\frac{5}{3}$. Its faces are $20\{3\} + 12\{5\} + 12\{\frac{10}{3}\}$, and its CIRCUMRADIUS for unit edge length is

$$R = \tfrac{1}{4}\sqrt{34 - 6\sqrt{5}}.$$

The CONVEX HULL of the great ditrigonal dodecicosi-dodecahedron is a regular DODECAHEDRON, whose dual is the ICOSAHEDRON, so the dual of the great ditrigonal dodecicosidodecahedron (the GREAT TRIAM-BIC ICOSAHEDRON) is one of the ICOSAHEDRON STELLA-TIONS (Wenninger 1983, p. 42).

References

Wenninger, M. J. *Dual Models.* Cambridge, England: Cambridge University Press, 1983.
Wenninger, M. J. *Polyhedron Models.* Cambridge, England: Cambridge University Press, p. 125, 1989.

Great Ditrigonal Icosidodecahedron

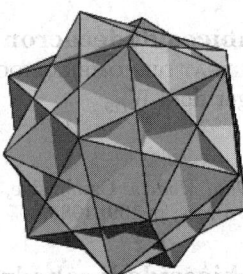

The UNIFORM POLYHEDRON U_{47} whose DUAL is the GREAT TRIAMBIC ICOSAHEDRON. It has WYTHOFF SYMBOL $\frac{3}{2}|3\,5$. Its faces are $20\{3\} + 12\{5\}$, and its CIRCUMRADIUS for unit edge length is

$$R = \tfrac{1}{2}\sqrt{3}.$$

The CONVEX HULL of the great triambic icosahedron is a regular DODECAHEDRON, whose dual is the ICOSA-HEDRON, so the dual of the great ditrigonal icosido-decahedron (the GREAT TRIAMBIC ICOSAHEDRON) is one of the ICOSAHEDRON STELLATIONS.

Great Dodecacronic Hexecontahedron

References

Wenninger, M. J. *Dual Models.* Cambridge, England: Cambridge University Press, p. 42, 1983.

Wenninger, M. J. *Polyhedron Models.* Cambridge, England: Cambridge University Press, pp. 135–36, 1989.

Great Dodecacronic Hexecontahedron

The DUAL of the GREAT DODECICOSIDODECAHEDRON U_{61} and Wenninger dual W_{99}.

See also DUAL POLYHEDRON, GREAT DODECICOSIDODECAHEDRON

References

Wenninger, M. J. *Dual Models.* Cambridge, England: Cambridge University Press, p. 88, 1983.

Great Dodecadodecahedron

DODECADODECAHEDRON

Great Dodecahedron

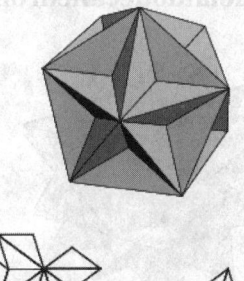

The KEPLER-POINSOT SOLID which is the DUAL of the SMALL STELLATED DODECAHEDRON. It is also UNIFORM POLYHEDRON U_{35} and Wenninger model W_{20}. Its SCHLÄFLI SYMBOL is $\{5, \frac{5}{2}\}$, and its WYTHOFF SYMBOL is $\frac{5}{2}|25$. Its faces are $12\{5\}$. Its CIRCUMRADIUS for unit edge length is

$$R = \tfrac{1}{2} 5^{1/4} \phi^{1/2} a = \tfrac{1}{4} 5^{1/4} \sqrt{2(1 + \sqrt{5})},$$

where ϕ is the GOLDEN RATIO. It can be constructed by CUMULATION of a unit edge-length ICOSAHEDRON by a pyramid with height $-\sqrt{\tfrac{1}{6}(7 - 3\sqrt{5})}$. This gives side of lengths

$$s_1 = \tfrac{1}{2}(\sqrt{5} - 1) = \phi - 1 \tag{1}$$

$$s_2 = 1 \tag{2}$$

The result solid has SURFACE AREA and VOLUME

$$S = 15\sqrt{5 - 2\sqrt{5}} \tag{3}$$

$$V = \tfrac{5}{4}(\sqrt{5} - 1). \tag{4}$$

Schläfli (1901, p. 134) did not recognize the great dodecahedron because it, like the SMALL STELLATED DODECAHEDRON, satisfies

$$N_0 - N_1 + N_2 = 12 - 30 + 12 = -6, \tag{5}$$

where N_0 is the number of vertices, N_1 the number of edges, and N_2 the number of faces (Coxeter 1973, p. 172), thus violating the POLYHEDRAL FORMULA.

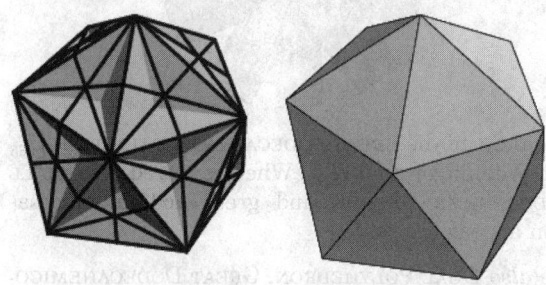

The CONVEX HULL of the great dodecahedron is a regular ICOSAHEDRON and the dual of the ICOSAHEDRON is the DODECAHEDRON, so the dual of the great dodecahedron (the SMALL STELLATED DODECAHEDRON) is one of the DODECAHEDRON STELLATIONS (Wenninger 1983, pp. 35 and 40)

See also DODECAHEDRON, GREAT ICOSAHEDRON, GREAT STELLATED DODECAHEDRON, KEPLER-POINSOT SOLID, SMALL STELLATED DODECAHEDRON, STELLATION

References

Coxeter, H. S. M. *Regular Polytopes, 3rd ed.* New York: Dover, 1973.

Cundy, H. and Rollett, A. "The Great Dodecahedron. $5^{5/2}$." §3.6.2 in *Mathematical Models, 3rd ed.* Stradbroke, England: Tarquin Pub., pp. 92–3, 1989.

Fischer, G. (Ed.). Plate 105 in *Mathematische Modelle/ Mathematical Models, Bildband/Photograph Volume.* Braunschweig, Germany: Vieweg, p. 104, 1986.

Schläfli, L. "Theorie der vielfachen Kontinuität." *Denkschriften der Schweizerischen naturforschenden Gessel.* **38**, 1–37, 1901.

Weisstein, E. W. "Polyhedra." Mathematica notebook POLYHEDRA.M.

Wenninger, M. J. *Dual Models.* Cambridge, England: Cambridge University Press, p. 39, 1983.

Wenninger, M. J. *Polyhedron Models.* Cambridge, England: Cambridge University Press, pp. 35 and 39, 1989.

Great Dodecahedron-Small Stellated Dodecahedron Compound

A polyhedron compound in which the great dodecahedron is interior to the small stellated dodecahedron.

See also Polyhedron Compound

Great Dodecahemicosacron

The dual of the great dodecahemicosahedron U_{65} and Wenninger dual W_{102}. When rendered, the small dodecahemicosacron and great dodecahemicosacron appear the same.

See also Dual Polyhedron, Great Dodecahemicosahedron, Uniform Polyhedron

References

Wenninger, M. J. *Dual Models.* Cambridge, England: Cambridge University Press, p. 107, 1983.

Great Dodecahemicosahedron

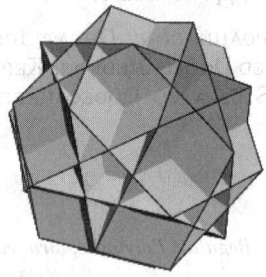

The uniform polyhedron U_{65} whose dual is the great dodecahemicosacron. It has Wythoff symbol $\frac{5}{4}$ 5|3. Its faces are $10\{6\} + 6\{5\} + 6\{\frac{5}{4}\}$. It is a faceted dodecadodecahedron. The circumradius for unit edge length is $R = 2$.

References

Wenninger, M. J. "Great Dodecahemicosahedron." Model 102 in *Polyhedron Models.* Cambridge, England: Cambridge University Press, p. 158, 1989.

Great Dodecahemidodecacron

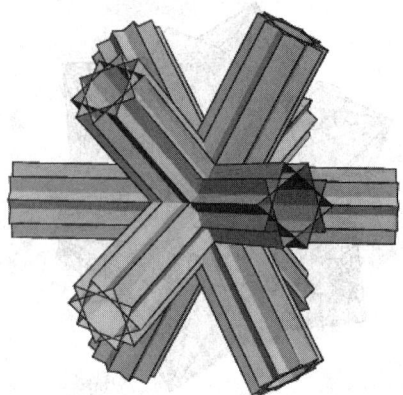

The dual of the great dodecahemidodecahedron U_{70} and Wenninger dual W_{107}. When rendered, the great dodecahemidodecacron and great icosihemidodecacron look the same, both consisting of a compound of six infinite $\{10/3\}$ prisms.

See also Dual Polyhedron, Great Dodecahemidodecahedron, Uniform Polyhedron

References

Wenninger, M. J. *Dual Models.* Cambridge, England: Cambridge University Press, p. 107, 1983.

Great Dodecahemidodecahedron

The uniform polyhedron U_{70} whose dual is the great dodecahemidodecacron. It has Wythoff symbol $\frac{5}{3}$ $\frac{5}{2}|\frac{5}{3}$. Its faces are $12\{\frac{5}{2}\} + 6\{\frac{10}{3}\}$. Its circumradius for unit edge length is

$$R = \phi^{-1},$$

where ϕ is the golden ratio.

References

Wenninger, M. J. *Polyhedron Models.* Cambridge, England: Cambridge University Press, p. 165, 1989.

Great Dodecicosacron

The DUAL of the GREAT DODECICOSAHEDRON and
Wenninger dual W_{101}.

References

Wenninger, M. J. *Dual Models*. Cambridge, England: Cambridge University Press, p. 67, 1983.

Great Dodecicosahedron

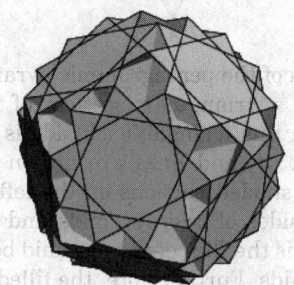

The UNIFORM POLYHEDRON U_{63} whose DUAL is the
GREAT DODECICOSACRON. It has WYTHOFF SYMBOL

$$3 \ \tfrac{5}{3} \Big| \begin{smallmatrix} \frac{3}{2} \\ \frac{5}{2} \end{smallmatrix}.$$

Its faces are $20\{6\} + 12\{\tfrac{10}{3}\}$. Its CIRCUMRADIUS for
unit edge length is

$$R = \tfrac{1}{4}\sqrt{34 - 6\sqrt{5}}.$$

References

Wenninger, M. J. *Polyhedron Models*. Cambridge, England: Cambridge University Press, pp. 156–57, 1989.

Great Dodecicosidodecahedron

The UNIFORM POLYHEDRON U_{61} whose DUAL is the
GREAT DODECACRONIC HEXECONTAHEDRON. Its WYTH-
OFF SYMBOL is $2 \ \tfrac{5}{2}|3$. Its faces are $20\{6\} + 12\{\tfrac{5}{2}\}$, and

its CIRCUMRADIUS for unit edge length is

$$R = \tfrac{1}{4}\sqrt{58 - 18\sqrt{5}}.$$

References

Wenninger, M. J. *Polyhedron Models*. Cambridge, England: Cambridge University Press, p. 148, 1989.

Great Hexacronic Icositetrahedron

The DUAL of the GREAT CUBICUBOCTAHEDRON and
Wenninger model W_{77}.

References

Wenninger, M. J. *Dual Models*. Cambridge, England: Cambridge University Press, p. 58, 1983.

Great Hexagonal Hexecontahedron

The DUAL of the GREAT SNUB DODECICOSIDODECAHE-
DRON and Wenninger dual W_{115}.

References

Wenninger, M. J. *Dual Models*. Cambridge, England: Cambridge University Press, p. 1356 1983.

Great Icosacronic Hexecontahedron

The DUAL of the GREAT ICOSICOSIDODECAHEDRON U_{48}
and Wenninger dual W_{88}.

See also DUAL POLYHEDRON, GREAT ICOSICOSIDODE-
CAHEDRON

References

Wenninger, M. J. *Dual Models*. Cambridge, England: Cambridge University Press, p. 65, 1983.

Great Icosahedron

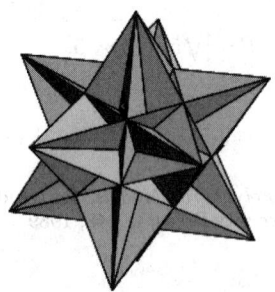

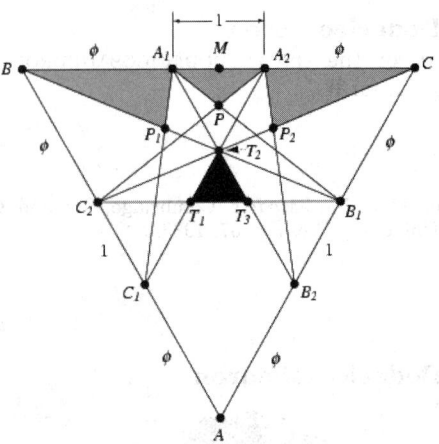

One of the KEPLER-POINSOT SOLIDS whose DUAL is the GREAT STELLATED DODECAHEDRON. It is also UNIFORM POLYHEDRON U_{53}, Wenninger model W_{22}, and has SCHLÄFLI SYMBOL $\{3, \frac{5}{2}\}$ and WYTHOFF SYMBOL $3\frac{5}{2}|\frac{5}{3}$. Its faces are $20\{3\} + 12\{\frac{5}{2}\} + 12\{\frac{10}{3}\}$.

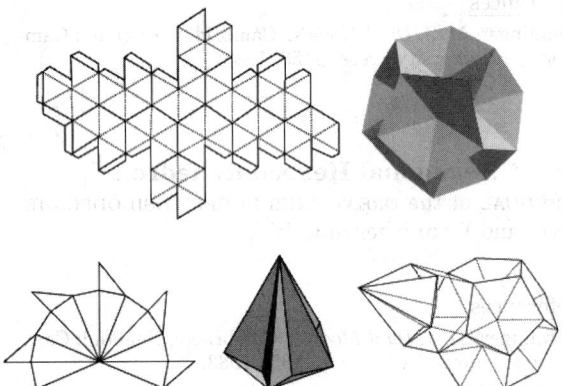

The great icosahedron can most easily be constructed by building a "squashed" dodecahedron (top right figure) from the corresponding net (top left figure). Then, using the net shown in the bottom left figure, build 12 PENTAGRAMMIC PYRAMIDS (bottom middle figure) and affix them into the dimples (bottom right). This method of construction is given in Cundy and Rollett (1989, pp. 98–9). If the edge lengths of the dodecahedron are unity, then the height of the pentagrammic pyramid (above the dodecahedron faces) is given by solving the equation for the SLANT HEIGHT of a PENTAGONAL PYRAMID

$$s = \sqrt{h^2 + \tfrac{1}{10}(5 + \sqrt{5})a^2} \qquad (1)$$

with $a = 1$, giving

$$h = \sqrt{\tfrac{1}{5}(5 + 2\sqrt{5})}. \qquad (2)$$

The distance from the center of the dodecahedron to the apex of a pyramid is then given by

$$H = h + r = \tfrac{1}{2}\sqrt{\tfrac{1}{2}(25 + 11\sqrt{5})}, \qquad (3)$$

where r is the INRADIUS of the DODECAHEDRON.

The dimensions of the pentagrammic pyramid can be by examining a triangular section of the great icosahedron. In this triangle, each side is divided in the ratios $\phi : 1 : \phi$, and lines are drawn as shown. Then the light shaded portions on the left and right correspond to sides of two pyramids and the center shaded portion is the "lip" of the pyramid between the first two pyramids. Furthermore, the filled portion of the diagram corresponds to one face of the ICOSAHEDRON inscribed in the great icosahedron. In the notation of the figure above,

$$|MP| = \tfrac{1}{10}\sqrt{15} \qquad (4)$$

$$|MT_2| = \tfrac{1}{2}\sqrt{3} \qquad (5)$$

$$|T_1 T_3| = \tfrac{1}{2}(\sqrt{5} - 1) = \phi - 1 \qquad (6)$$

$$|CP_2| = \sqrt{\tfrac{1}{5}(7 + 3\sqrt{5})} \qquad (7)$$

$$|PA_2| = \tfrac{1}{5}\sqrt{10}. \qquad (8)$$

The great icosahedron constructed from the DODECAHEDRON with unit edge lengths has edge lengths (where edges are interpreted to be broken where facial plane intersect) given by

$$s_1 = \tfrac{1}{5}\sqrt{10} \qquad (9)$$

$$s_2 = 1 \qquad (10)$$

$$s_3 = \tfrac{1}{2}(1 + \sqrt{5}) \qquad (11)$$

$$s_4 = \sqrt{\tfrac{1}{5}(7 + 3\sqrt{5})}. \qquad (12)$$

Its CIRCUMRADIUS is

$$R = \tfrac{1}{2}\sqrt{\tfrac{1}{2}(25 + 11\sqrt{5})}, \qquad (13)$$

and the SURFACE AREA and VOLUME are then

$$S = 3\sqrt{3}(5 + 4\sqrt{5}) \qquad (14)$$

$$V = \tfrac{1}{4}(25 + 9\sqrt{5}). \qquad (15)$$

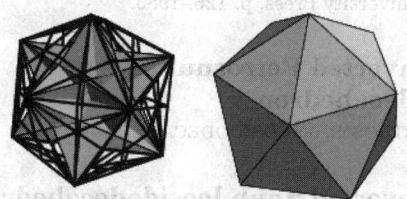

The CONVEX HULL of the great icosahedron is a regular ICOSAHEDRON and the dual of the ICOSAHEDRON is the DODECAHEDRON, so the dual of the great icosahedron is one of the DODECAHEDRON STELLATIONS (Wenninger 1983, p. 40)

See also GREAT DODECAHEDRON, GREAT STELLATED DODECAHEDRON, KEPLER-POINSOT SOLID, SMALL STELLATED DODECAHEDRON, TRUNCATED GREAT ICOSAHEDRON

References

Cundy, H. and Rollett, A. "The Great Icosahedron. $3^{5/2}$." §3.6.4 in *Mathematical Models, 3rd ed.* Stradbroke, England: Tarquin Pub., pp. 96–9, 1989.
Fischer, G. (Ed.). Plate 106 in *Mathematische Modelle/Mathematical Models, Bildband/Photograph Volume.* Braunschweig, Germany: Vieweg, p. 105, 1986.
Weisstein, E. W. "Polyhedra." MATHEMATICA NOTEBOOK POLYHEDRA.M.
Wenninger, M. J. *Dual Models.* Cambridge, England: Cambridge University Press, p. 40, 1983.
Wenninger, M. J. *Polyhedron Models.* Cambridge, England: Cambridge University Press, p. 154, 1989.

Great Icosahedron-Great Stellated Dodecahedron Compound

A POLYHEDRON COMPOUND of the GREAT ICOSAHEDRON and GREAT STELLATED DODECAHEDRON most easily constructed by adding the VERTICES OF THE FORMER to the latter.

See also GREAT ICOSAHEDRON, GREAT STELLATED DODECAHEDRON, POLYHEDRON COMPOUND

References

Cundy, H. and Rollett, A. "Great Icosahedron Plus Great Stellated Dodecahedron." §3.10.4 in *Mathematical Models, 3rd ed.* Stradbroke, England: Tarquin Pub., pp. 132–33, 1989.
Wenninger, M. J. *Dual Models.* Cambridge, England: Cambridge University Press, pp. 51–3 1983.

Great Icosicosidodecahedron

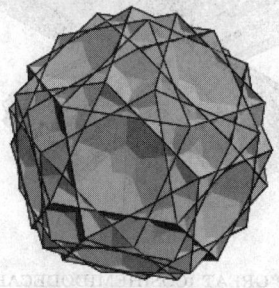

The UNIFORM POLYHEDRON U_{48} whose DUAL is the GREAT ICOSACRONIC HEXECONTAHEDRON. It has WYTHOFF SYMBOL $\frac{3}{2}$ 5|3. Its faces are $20\{3\} + 20\{6\} + 12\{5\}$. Its CIRCUMRADIUS for unit edge length is

$$R = \tfrac{1}{4}\sqrt{34 - 6\sqrt{5}}.$$

References

Wenninger, M. J. *Polyhedron Models.* Cambridge, England: Cambridge University Press, pp. 137–39, 1989.

Great Icosidodecahedron

A UNIFORM POLYHEDRON U_{54} whose DUAL is the GREAT RHOMBIC TRIACONTAHEDRON (also called the GREAT STELLATED TRIACONTAHEDRON). It is a STELLATED ARCHIMEDEAN SOLID. It has SCHLÄFLI SYMBOL $\left\{\frac{3}{\frac{5}{2}}\right\}$ and WYTHOFF SYMBOL $2|3\,\frac{5}{2}$. Its faces are $20\{3\} + 12\{\frac{5}{2}\}$. Its CIRCUMRADIUS for unit edge length is

$$R = \phi^{-1},$$

where ϕ is the GOLDEN RATIO.

References

Cundy, H. and Rollett, A. "Great Icosidodecahedron. $(3 \cdot \frac{5}{2})^{2}$" §3.9.2 in *Mathematical Models, 3rd ed.* Stradbroke, England: Tarquin Pub., p. 124, 1989.

Wenninger, M. J. *Polyhedron Models*. Cambridge, England: Cambridge University Press, p. 147, 1989.

Great Icosihemidodecacron

The DUAL of the GREAT ICOSIHEMIDODECAHEDRON U_{71} and Wenninger dual W_{106}. When rendered, the GREAT DODECAHEMIDODECACRON and great icosihemidodecacron look the same, both consisting of a compound of six infinite $\{10/3\}$ prisms.

See also DUAL POLYHEDRON, GREAT ICOSIHEMIDODECAHEDRON, UNIFORM POLYHEDRON

References

Wenninger, M. J. *Dual Models*. Cambridge, England: Cambridge University Press, p. 107, 1983.

Great Icosihemidodecahedron

The UNIFORM POLYHEDRON U_{71} whose DUAL is the GREAT ICOSIHEMIDODECACRON. It has WYTHOFF SYMBOL $\frac{3}{2}3|\frac{5}{3}$. Its faces are $20\{3\}+6\{\frac{10}{3}\}$. For unit edge length, its CIRCUMRADIUS is

$$R = \phi^{-1},$$

where ϕ is the GOLDEN RATIO.

References

Wenninger, M. J. *Polyhedron Models*. Cambridge, England: Cambridge University Press, p. 164, 1989.

Great Inverted Pentagonal Hexecontahedron

The DUAL of the GREAT INVERTED SNUB ICOSIDODECAHEDRON U_{69} and Wenninger dual W_{116}.

References

Wenninger, M. J. *Dual Models*. Cambridge, England: Cambridge University Press, p. 126, 1983.

Great Inverted Retrosnub Icosidodecahedron
GREAT RETROSNUB ICOSIDODECAHEDRON

Great Inverted Snub Icosidodecahedron

The UNIFORM POLYHEDRON U_{69} whose DUAL is the GREAT INVERTED PENTAGONAL HEXECONTAHEDRON. It has WYTHOFF SYMBOL $|2\,3\,\frac{5}{2}$. Its faces are $80\{3\}+12\{\frac{5}{2}\}$. For unit edge length, it has CIRCUMRADIUS

$$R = \frac{1}{2}\sqrt{\frac{8\cdot 2^{2/3}-16x+2^{1/3}x^2}{8\cdot 2^{2/3}-10x+2^{1/3}x^2}}$$

$$= 0.816080674799923,$$

where

$$x \equiv \left(49-27\sqrt{5}+3\sqrt{6}\sqrt{93-49\sqrt{5}}\right)^{1/3}.$$

References

Wenninger, M. J. *Polyhedron Models*. Cambridge, England: Cambridge University Press, p. 179, 1989.

Great Pentagonal Hexecontahedron

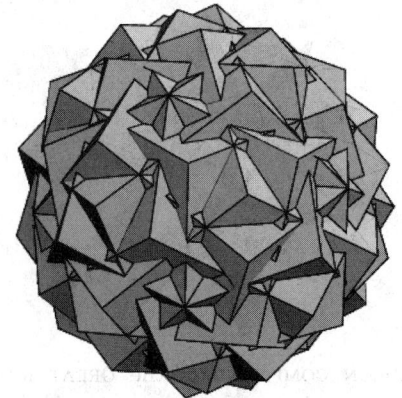

The DUAL of the GREAT SNUB ICOSIDODECAHEDRON U_{57} and Wenninger dual W_{113}.

See also DUAL POLYHEDRON, GREAT SNUB ICOSIDODECAHEDRON

References

Wenninger, M. J. *Dual Models*. Cambridge, England: Cambridge University Press, p. 123, 1983.

Great Pentagrammic Hexecontahedron

The DUAL of the GREAT RETROSNUB ICOSIDODECAHEDRON and Wenninger dual W_{117}.

References

Wenninger, M. J. *Dual Models*. Cambridge, England: Cambridge University Press, p. 128, 1983.

Great Pentakis Dodecahedron

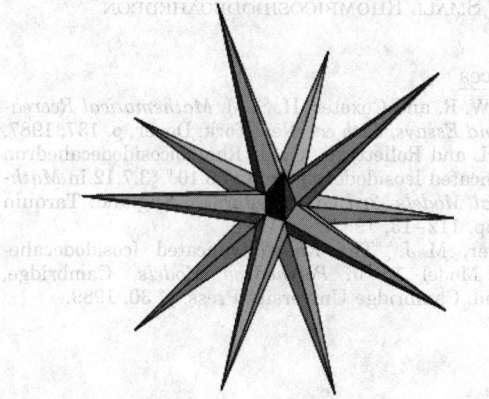

The DUAL of the SMALL STELLATED TRUNCATED DODECAHEDRON U_{58} and Wenninger dual W_{97}.

See also DUAL POLYHEDRON, SMALL STELLATED TRUNCATED DODECAHEDRON

References

Wenninger, M. J. *Dual Models*. Cambridge, England: Cambridge University Press, p. 70, 1983.

Great Quasitruncated Icosidodecahedron

GREAT TRUNCATED ICOSIDODECAHEDRON

Great Retrosnub Icosidodecahedron

The UNIFORM POLYHEDRON U_{74}, also called the GREAT INVERTED RETROSNUB ICOSIDODECAHEDRON, whose

DUAL is the GREAT PENTAGRAMMIC HEXECONTAHEDRON. It has WYTHOFF SYMBOL $|2\frac{3}{2}\frac{5}{3}$. Its faces are $80\{3\} + 12\{\frac{5}{2}\}$. For unit edge length, it has CIRCUMRADIUS

$$R = \tfrac{1}{2}\sqrt{\frac{2-x}{1-x}} \approx 0.5800015,$$

where x is the smaller NEGATIVE root of

$$x^3 + 2x^2 - \phi^{-2} = 0,$$

with ϕ the GOLDEN MEAN.

References

Wenninger, M. J. *Polyhedron Models*. Cambridge, England: Cambridge University Press, pp. 189–93, 1989.

Great Rhombic Triacontahedron

A ZONOHEDRON which is the DUAL of the GREAT ICOSIDODECAHEDRON and Wenninger model W_{94}. It is also called the GREAT STELLATED TRIACONTAHEDRON, and is one of the RHOMBIC DODECAHEDRON STELLATIONS.

See also DUAL POLYHEDRON, GREAT ICOSIDODECAHEDRON, RHOMBIC DODECAHEDRON STELLATIONS, ZONOHEDRON

References

Cundy, H. and Rollett, A. "Great Stellated Triacontahedron." V $(3.\frac{5}{2})^2$." §3.9.4 in *Mathematical Models, 3rd ed.* Stradbroke, England: Tarquin Pub., p. 126, 1989.
Wenninger, M. J. *Dual Models*. Cambridge, England: Cambridge University Press, pp. 54–5, 1983.

Great Rhombicosidodecahedron (Archimedean)

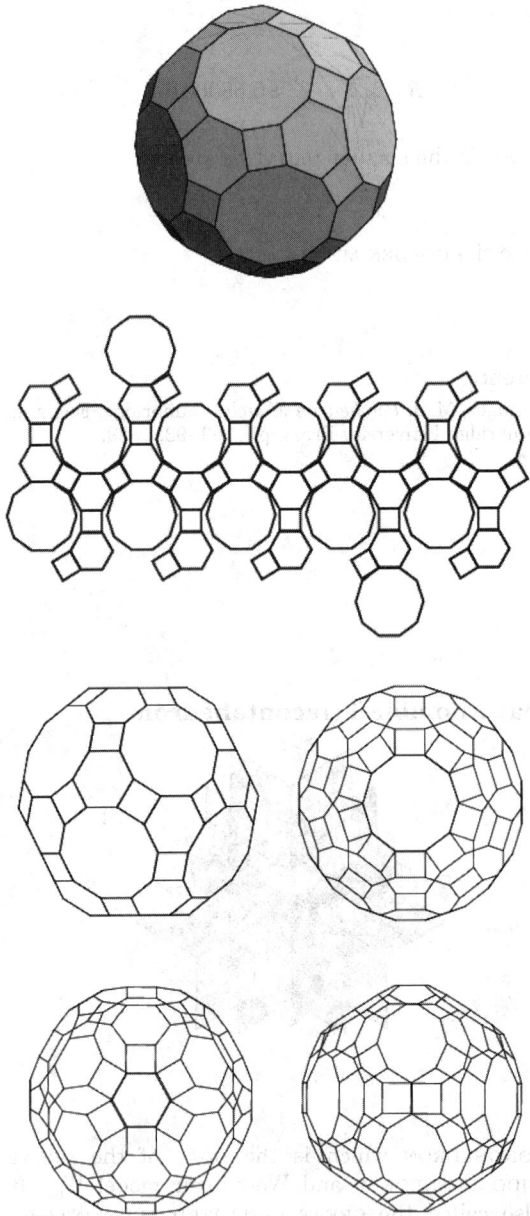

Its DUAL is the DISDYAKIS TRIACONTAHEDRON, also called the HEXAKIS ICOSAHEDRON. The INRADIUS of the dual, MIDRADIUS of the solid and dual, and CIRCUM-RADIUS of the solid for $a = 1$ are

$$r = \tfrac{1}{241}(105 + 6\sqrt{5})\sqrt{31 + 12\sqrt{5}} \approx 3.73665$$

$$\rho = \tfrac{1}{2}\sqrt{30 + 12\sqrt{5}} \approx 3.76938$$

$$R = \tfrac{1}{2}\sqrt{31 + 12\sqrt{5}} \approx 3.80239.$$

See also SMALL RHOMBICOSIDODECAHEDRON

References

Ball, W. W. R. and Coxeter, H. S. M. *Mathematical Recreations and Essays, 13th ed.* New York: Dover, p. 137, 1987.

Cundy, H. and Rollett, A. "Great Rhombicosidodecahedron or Truncated Icosidodecahedron. 4.6.10." §3.7.12 in *Mathematical Models, 3rd ed.* Stradbroke, England: Tarquin Pub., pp. 112–13, 1989.

Wenninger, M. J. "The Rhombitruncated Icosidodecahedron." Model 16 in *Polyhedron Models.* Cambridge, England: Cambridge University Press, p. 30, 1989.

Great Rhombicosidodecahedron (Uniform)

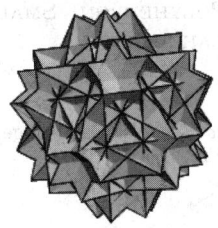

The UNIFORM POLYHEDRON U_{67}, also called the QUASIRHOMBICOSIDODECAHEDRON, whose DUAL is the GREAT DELTOIDAL HEXECONTAHEDRON. It has SCHLÄFLI SYMBOL r' $\left\{\begin{smallmatrix} 3 \\ 5 \\ 2 \end{smallmatrix}\right\}$. It has WYTHOFF SYMBOL $3 \frac{5}{2}|2$. Its faces are $20\{3\} + 30\{4\} + 12\{\frac{5}{2}\}$. For unit edge length, its CIRCUMRADIUS is

$$R = \tfrac{1}{2}\sqrt{11 - 4\sqrt{5}}.$$

References

Wenninger, M. J. *Polyhedron Models.* Cambridge, England: Cambridge University Press, pp. 162–63, 1989.

The 62-faced ARCHIMEDEAN SOLID A_2 with faces $30\{4\} + 20\{6\} + 12\{10\}$. It is also known as the rhombitruncated icosidodecahedron, and is sometimes improperly called the truncated icosidodecahedron, a name which is inappropriate since TRUNCATION would yield RECTANGULAR instead of SQUARE. The great rhombicosidodecahedron is also UNIFORM POLYHEDRON U_{28} and Wenninger model W_{16}. It has SCHLÄFLI SYMBOL t $\left\{\begin{smallmatrix} 3 \\ 5 \end{smallmatrix}\right\}$ and WYTHOFF SYMBOL 2 3 5|.

Great Rhombicuboctahedron (Archimedean)

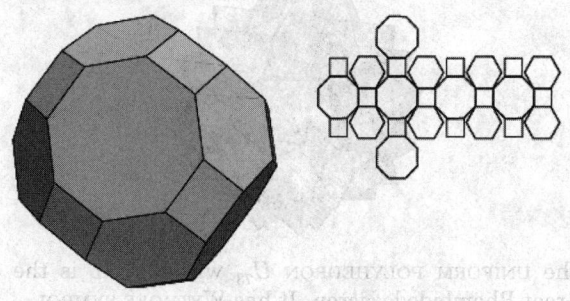

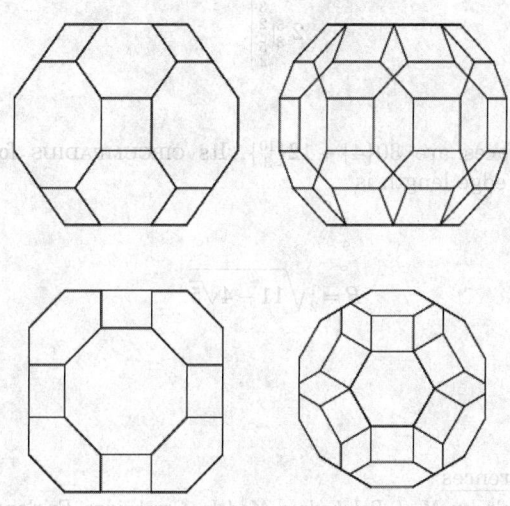

The 26-faced ARCHIMEDEAN SOLID A_3 consisting of faces $12\{4\} + 8\{6\} + 6\{8\}$. It is sometimes (improperly) called the truncated cuboctahedron, and is also called the rhombitruncated cuboctahedron. It is UNIFORM POLYHEDRON U_{11} and Wenninger model W_{15}. It has SCHLÄFLI SYMBOL $t \left\{{3 \atop 4}\right\}$ and WYTHOFF SYMBOL $2\ 3\ 4|$.

The SMALL CUBICUBOCTAHEDRON is a FACETED version of the great rhombicuboctahedron.

Its DUAL is the DISDYAKIS DODECAHEDRON, also called the HEXAKIS OCTAHEDRON. The INRADIUS r of the dual, MIDRADIUS ρ of the solid and dual, and CIRCUMRADIUS R of the solid for $a = 1$ are

$$r = \tfrac{3}{97}(14 + \sqrt{2})\sqrt{13 + 6\sqrt{2}} \approx 2.20974$$

$$\rho = \tfrac{1}{2}\sqrt{12 + 6\sqrt{2}} \approx 2.26303$$

$$R = \tfrac{1}{2}\sqrt{13 + 6\sqrt{2}} \approx 2.31761.$$

Additional quantities are

$$t = \tan(\tfrac{1}{8}\pi) = \sqrt{2} - 1$$

$$l = 2t = 2(\sqrt{2} - 1)$$

$$h = 1 + l\,\sin(\tfrac{1}{4}\pi) = 3 - \sqrt{2}.$$

The distances between the solid center and centroids of the square and octagonal faces are

$$r_4 = \tfrac{1}{2}(3 + \sqrt{2}) \tag{1}$$

$$r_8 = \tfrac{1}{2}(1 + 2\sqrt{2}). \tag{2}$$

The SURFACE AREA and VOLUME are

$$S = 12(2 + \sqrt{2} + \sqrt{3}) \tag{3}$$

$$V = 22 + 14\sqrt{2}. \tag{4}$$

See also ARCHIMEDEAN SOLID, GREAT RHOMBICUBOCTAHEDRON (UNIFORM), GREAT TRUNCATED CUBOCTAHEDRON, SMALL RHOMBICUBOCTAHEDRON, OCTATETRAHEDRON

References

Ball, W. W. R. and Coxeter, H. S. M. *Mathematical Recreations and Essays, 13th ed.* New York: Dover, p. 138, 1987.

Cundy, H. and Rollett, A. "Great Rhombicuboctahedron or Truncated Cuboctahedron. 4.6.8." §3.7.6 in *Mathematical Models, 3rd ed.* Stradbroke, England: Tarquin Pub., p. 106, 1989.

Wenninger, M. J. "The Rhombitruncated Cuboctahedron." Model 15 in *Polyhedron Models.* Cambridge, England: Cambridge University Press, p. 29, 1989.

Great Rhombicuboctahedron (Uniform)

The UNIFORM POLYHEDRON U_{17}, also known as the QUASIRHOMBICUBOCTAHEDRON, whose DUAL is the GREAT DELTOIDAL ICOSITETRAHEDRON. It has SCHLÄFLI SYMBOL $r' \left\{{3 \atop 4}\right\}$, WYTHOFF SYMBOL $\tfrac{3}{2}\ 4|2$, and is Wenninger model W_{85}. Its faces are $18\{4\} + 8\{3/2\}$. Its CIRCUMRADIUS for unit edge length is

$$R = \tfrac{1}{2}\sqrt{5 - 2\sqrt{2}}.$$

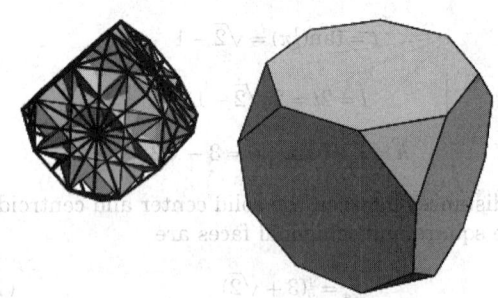

Great Rhombidodecahedron

The CONVEX HULL of the great cubicuboctahedron is the Archimedean TRUNCATED CUBE A_9, whose dual is the SMALL TRIAKIS OCTAHEDRON, so the dual of the great rhombicuboctahedron (i.e., the GREAT DELTOIDAL ICOSITETRAHEDRON) is one of the stellations of the SMALL TRIAKIS OCTAHEDRON (Wenninger 1983, p. 57).

The UNIFORM POLYHEDRON U_{73} whose DUAL is the Great Rhombidodecacron. It has WYTHOFF SYMBOL

$$2 \; \tfrac{5}{3} \; \tfrac{\tfrac{3}{2}}{\tfrac{5}{4}} \Big|$$

Its faces are $30\{4\} + 12\{\tfrac{10}{3}\}$. Its CIRCUMRADIUS for unit edge length is

$$R = \tfrac{1}{2}\sqrt{11 - 4\sqrt{5}}.$$

References

Wenninger, M. J. *Dual Models.* Cambridge, England: Cambridge University Press, pp. 57 and 59, 1983.
Wenninger, M. J. Model 85 in *Polyhedron Models.* Cambridge, England: Cambridge University Press, pp. 132–33, 1989.

References

Wenninger, M. J. *Polyhedron Models.* Cambridge, England: Cambridge University Press, pp. 168–70, 1989.

Great Rhombidodecacron

The DUAL of the GREAT RHOMBIDODECAHEDRON U_{73} and Wenninger dual W_{109}.

See also DUAL POLYHEDRON, GREAT RHOMBIDODECAHEDRON, UNIFORM POLYHEDRON

Great Rhombihexacron

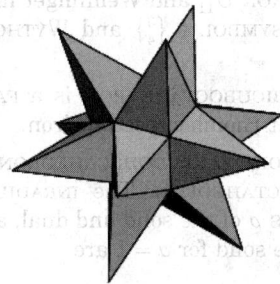

The DUAL of the GREAT RHOMBIHEXAHEDRON U_{21} and Wenninger dual W_{103}.

References

Wenninger, M. J. *Dual Models.* Cambridge, England: Cambridge University Press, p. 88, 1983.

References

Wenninger, M. J. *Dual Models.* Cambridge, England: Cambridge University Press, p. 60, 1983.

Great Rhombihexahedron

The UNIFORM POLYHEDRON U_{21} whose DUAL is the GREAT RHOMBIHEXACRON. It is Wenninger model W_{103}. Maeder gives its WYTHOFF SYMBOL as $\frac{4}{3}\frac{3}{2}2|$, and its faces as $6\{4\} + 3\{\frac{8}{3}\} + 3\{\frac{8}{5}\} + 6\{\frac{4}{3}\}$, while Wenninger (1989) gives the WYTHOFF SYMBOL as

$$2\ {\textstyle\frac{4}{3}}\ {\textstyle\begin{matrix}\frac{3}{2}\\[2pt]\frac{4}{2}\end{matrix}}\Big|$$

and its faces as $12\{4\} + 6\{\frac{8}{3}\}$. The CIRCUMRADIUS for a great rhombihexahedron of unit edge length is

$$R = {\textstyle\frac{1}{2}}\sqrt{5 - 2\sqrt{2}}.$$

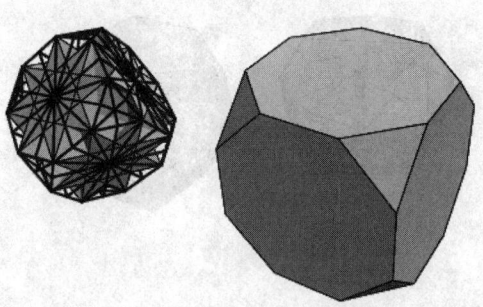

The CONVEX HULL of the great rhombihexahedron is the Archimedean TRUNCATED CUBE A_9, whose dual is the SMALL TRIAKIS OCTAHEDRON, so the dual of the great rhombihexahedron (i.e., the GREAT RHOMBIHEX-ACRON) is one of the stellations of the SMALL TRIAKIS OCTAHEDRON (Wenninger 1983, p. 57).

References

Maeder, R. E. Polyhedra.m and PolyhedraExamples *Mathematica* notebooks. http://www.inf.ethz.ch/department/TI/rm/programs.html.

Wenninger, M. J. *Dual Models*. Cambridge, England: Cambridge University Press, p. 57 and 160, 1983.

Wenninger, M. J. "Great Rhombihexahedron." Model 103 in *Polyhedron Models*. Cambridge, England: Cambridge University Press, pp. 159–60, 1989.

Great Snub Dodecicosidodecahedron

The UNIFORM POLYHEDRON U_{64} whose DUAL is the GREAT HEXAGONAL HEXECONTAHEDRON. It has WYTHOFF SYMBOL $|3\ \frac{5}{3}\frac{5}{2}$. Its faces are $80\{3\} + 24\{\frac{5}{2}\}$. Its CIRCUMRADIUS for unit edge length is

$$R = {\textstyle\frac{1}{2}}\sqrt{2}.$$

References

Wenninger, M. J. *Polyhedron Models*. Cambridge, England: Cambridge University Press, pp. 183–85, 1989.

Great Snub Icosidodecahedron

The UNIFORM POLYHEDRON U_{57} whose DUAL is the GREAT PENTAGONAL HEXECONTAHEDRON. It has WYTHOFF SYMBOL $|2\ 3\ \frac{5}{3}$. Its faces are $80\{3\} + 12\{\frac{5}{2}\}$. For unit edge length, it has CIRCUMRADIUS

$$R = {\textstyle\frac{1}{2}}\sqrt{\frac{2-x}{1-x}} \approx 0.6450202,$$

where x is the most NEGATIVE ROOT of

$$x^3 + 2x^2 - \phi^{-2} = 0,$$

with ϕ the GOLDEN RATIO.

References

Wenninger, M. J. *Polyhedron Models*. Cambridge, England: Cambridge University Press, pp. 186–88, 1989.

Great Sphere

The great sphere on the surface of a HYPERSPHERE is the 3-D analog of the GREAT CIRCLE on the surface of a SPHERE. Let $2h$ be the number of reflecting SPHERES, and let great spheres divide a HYPERSPHERE into g 4-D TETRAHEDRA. Then for the POLYTOPE with SCHLÄ-

FLI SYMBOL $\{p, q, r\}$,

$$\frac{64h}{g} = 12 - p - 2q - r + \frac{4}{p} + \frac{4}{r}.$$

See also GREAT CIRCLE

Great Stellapentakis Dodecahedron

The DUAL of the GREAT TRUNCATED ICOSAHEDRON U_{55} and Wenninger dual W_{95}.

See also DUAL POLYHEDRON, GREAT TRUNCATED ICOSAHEDRON

References
Wenninger, M. J. *Dual Models*. Cambridge, England: Cambridge University Press, p. 75, 1983.

Great Stellated Dodecahedron

One of the KEPLER-POINSOT SOLIDS. It is also UNIFORM POLYHEDRON U_{52}, Wenninger model W_{41}, and is the third DODECAHEDRON STELLATION (Wenninger 1989). Its DUAL is the GREAT ICOSAHEDRON. The great stellated dodecahedron has SCHLÄFLI SYMBOL $\{\frac{5}{2}, 3\}$ and WYTHOFF SYMBOL $3|2\,\frac{5}{2}$. Its faces are $12\{\frac{5}{2}\}$. Its CIRCUMRADIUS for unit edge length is

$$R = \tfrac{1}{2}\sqrt{3}\phi^{-1} = \tfrac{1}{4}\sqrt{3}(\sqrt{5}-1). \qquad (1)$$

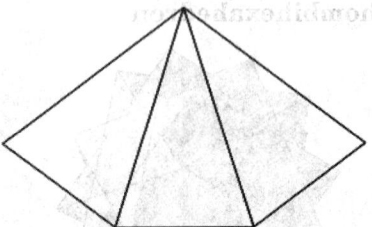

The easiest way to construct a great stellated dodecahedron is by CUMULATION, i.e., to making 20 TRIANGULAR PYRAMIDS with side length $\phi = (1+\sqrt{5})/2$ (the GOLDEN RATIO) times the base and attaching them to the sides of an ICOSAHEDRON. The height of these pyramids is then $\sqrt{\frac{1}{6}(7+3\sqrt{5})}$.

Cumulating a DODECAHEDRON to construct a great stellated dodecahedron produces a solid with edge lengths

$$s_1 = 1 \qquad (2)$$

$$s_2 = \phi = \tfrac{1}{2}(1+\sqrt{5}). \qquad (3)$$

The SURFACE AREA and VOLUME of such a great stellated dodecahedron are

$$S = 15\sqrt{5+2\sqrt{5}} \qquad (4)$$

$$V = \tfrac{5}{4}(3+\sqrt{5}). \qquad (5)$$

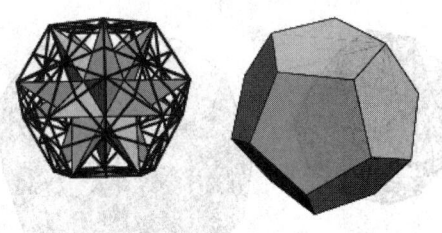

The CONVEX HULL of the great stellated dodecahedron is a regular DODECAHEDRON and the dual of the DODECAHEDRON is the ICOSAHEDRON, so the dual of the great stellated dodecahedron (i.e., the GREAT ICOSAHEDRON) is one of the ICOSAHEDRON STELLATIONS (Wenninger 1983, p. 40)

See also DODECAHEDRON, DODECAHEDRON STELLATIONS, GREAT DODECAHEDRON, GREAT ICOSAHEDRON, GREAT STELLATED TRUNCATED DODECAHEDRON, KEPLER-POINSOT SOLID, SMALL STELLATED DODECAHEDRON, STELLATION

References
Cundy, H. and Rollett, A. "Great Stellated Dodecahedron. $(\frac{5}{2})^3$." §3.6.3 in *Mathematical Models, 3rd ed.* Stradbroke, England: Tarquin Pub., pp. 94–5, 1989.
Fischer, G. (Ed.). Plate 104 in *Mathematische Modelle/ Mathematical Models, Bildband/Photograph Volume.* Braunschweig, Germany: Vieweg, p. 103, 1986.

Weisstein, E. W. "Polyhedra." MATHEMATICA NOTEBOOK POLYHEDRA.M.
Wenninger, M. J. *Dual Models.* Cambridge, England: Cambridge University Press, pp. 39–0, 1983.
Wenninger, M. J. *Polyhedron Models.* Cambridge, England: Cambridge University Press, pp. 35 and 40, 1989.

Great Stellated Triacontahedron

GREAT RHOMBIC TRIACONTAHEDRON

Great Stellated Truncated Dodecahedron

The UNIFORM POLYHEDRON U_{66}, also called the QUASI-TRUNCATED GREAT STELLATED DODECAHEDRON, whose DUAL is the GREAT TRIAKIS ICOSAHEDRON. It has SCHLÄFLI SYMBOL t' $\{\frac{5}{2}, 3\}$ and WYTHOFF SYMBOL $2\ 3|\frac{5}{3}$. Its faces are $20\{3\} + 12\{\frac{10}{3}\}$. Its CIRCUMRADIUS for unit edge length is

$$R = \tfrac{1}{4}\sqrt{74 - 30\sqrt{5}}.$$

References

Wenninger, M. J. *Polyhedron Models.* Cambridge, England: Cambridge University Press, p. 161, 1989.

Great Triakis Icosahedron

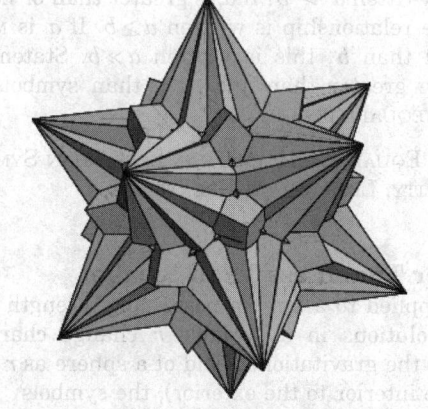

The DUAL of the GREAT STELLATED TRUNCATED DODE-CAHEDRON U_{66} and Wenninger dual W_{104}.

See also DUAL POLYHEDRON, GREAT STELLATED TRUNCATED DODECAHEDRON

References

Wenninger, M. J. *Dual Models.* Cambridge, England: Cambridge University Press, p. 77, 1983.

Great Triakis Octahedron

The DUAL of the STELLATED TRUNCATED HEXAHEDRON U_{19} and Wenninger dual W_{92}

See also DUAL POLYHEDRON, SMALL TRIAKIS OCTAHE-DRON, STELLATED TRUNCATED HEXAHEDRON

References

Wenninger, M. J. *Dual Models.* Cambridge, England: Cambridge University Press, p. 57, 1983.

Great Triambic Icosahedron

The DUAL of the GREAT DITRIGONAL ICOSIDODECAHE-DRON U_{47} and Wenninger model W_{87} whose appear-ance is the same as the MEDIAL TRIAMBIC ICOSAHEDRON (the dual of the DITRIGONAL DODECA-DODECAHEDRON), since internal vertices are hidden from view (Wenninger 1983, p. 42). The MEDIAL TRIAMBIC ICOSAHEDRON has hidden pentagrammic faces, while the great triambic icosahedron has hidden triangular faces (Wenninger 1983, pp. 45, 47, and 48–0).

The CONVEX HULL of the GREAT DITRIGONAL ICOSIDO-DECAHEDRON is a regular DODECAHEDRON, whose dual is the ICOSAHEDRON, so the dual of the GREAT DITRIGONAL ICOSIDODECAHEDRON (the great triambic icosahedron) is one of the ICOSAHEDRON STELLATIONS (Wenninger 1983, p. 42).

See also DUAL POLYHEDRON, GREAT DITRIGONAL ICOSIDODECAHEDRON, ICOSAHEDRON STELLATIONS, MEDIAL TRIAMBIC ICOSAHEDRON

References

Wenninger, M. J. *Dual Models.* Cambridge, England: Cambridge University Press, pp. 41 and 46, 1983.
Wenninger, M. J. "Ninth Stellation of the Icosahedron." §34 in *Polyhedron Models.* New York: Cambridge University Press, p. 55, 1989.

Great Truncated Cuboctahedron

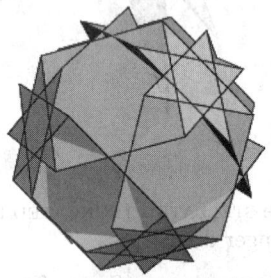

The UNIFORM POLYHEDRON U_{20}, also called the quasi-truncated cuboctahedron, whose DUAL is the GREAT DISDYAKIS DODECAHEDRON. Its faces consist of $8\{6\} + 12\{4\} + 6\{\frac{8}{3}\}$. It has SCHLÄFLI SYMBOL $t'\{\frac{3}{4}\}$ and WYTHOFF SYMBOL $\frac{4}{3}\,2\,3|$. Its CIRCUMRADIUS for unit edge length is

$$R = \tfrac{1}{2}\sqrt{13 - 6\sqrt{2}}.$$

References

Wenninger, M. J. *Polyhedron Models.* Cambridge, England: Cambridge University Press, pp. 145–46, 1989.

Great Truncated Icosahedron

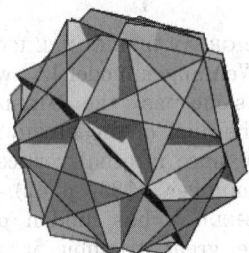

The UNIFORM POLYHEDRON U_{55}, also called the TRUNCATED GREAT ICOSAHEDRON, whose DUAL is the GREAT STELLAPENTAKIS DODECAHEDRON. It has SCHLÄFLI SYMBOL $t\{3, \frac{5}{2}\}$ and WYTHOFF SYMBOL $2\,\frac{5}{2}|3$. Its faces are $20\{6\} + 12\{\frac{5}{2}\}$. Its CIRCUMRADIUS for unit edge length is

$$R = \tfrac{1}{4}\sqrt{58 - 18\sqrt{5}}.$$

References

Wenninger, M. J. *Polyhedron Models.* Cambridge, England: Cambridge University Press, p. 148, 1989.

Great Truncated Icosidodecahedron

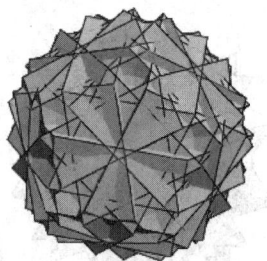

The UNIFORM POLYHEDRON U_{68}, also called the GREAT QUASITRUNCATED ICOSIDODECAHEDRON, whose DUAL is the GREAT DISDYAKIS TRIACONTAHEDRON. It has SCHLÄFLI SYMBOL $t'\{\frac{3}{\frac{5}{2}}\}$ and WYTHOFF SYMBOL $2\,3\,\frac{5}{3}|$. Its faces are $20\{6\} + 30\{4\} + 12\{\frac{10}{3}\}$. Its CIRCUMRADIUS for unit edge length is

$$R = \tfrac{1}{2}\sqrt{31 - 12\sqrt{5}}.$$

References

Wenninger, M. J. *Polyhedron Models.* Cambridge, England: Cambridge University Press, pp. 166–67, 1989.

Greater

A quantity a is said to be greater than b if a is larger than b, written $a > b$. If a is greater than or EQUAL to b, the relationship is written $a \geq b$. If a is MUCH GREATER than b, this is written $a \gg b$. Statements involving greater than and LESS than symbols are called INEQUALITIES.

See also EQUAL, GREATER THAN/LESS THAN SYMBOL, INEQUALITY, LESS, MUCH GREATER

Greater Than/Less Than Symbol

When applied to a system possessing a length R at which solutions in a variable r change character (such as the gravitational field of a sphere as r runs from the interior to the exterior), the symbols

$$r > \equiv \max(r, R)$$
$$r < \equiv \min(r, R)$$

are sometimes used.

See also EQUAL, GREATER, LESS

Greatest Common Denominator

GREATEST COMMON DIVISOR

Greatest Common Divisor

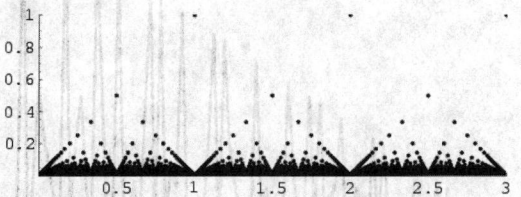

The greatest common divisor GCD(a, b) of two positive integers a and b, sometimes written (a, b), is the largest DIVISOR common to a and b. For example, GCD(3, 5) = 1, GCD(12, 60) = 12, and GCD(12, 90) = 6. The greatest common divisor GCD(a, b, c, ...) can also be defined for three or more positive integers as the largest divisor shared by all of them. The plot above shows GCD(1, b) with rational $b = m/n$.

The greatest common divisor of a and b is implemented in *Mathematica* as GCD[a, b, ...].

If d is the greatest common divisor of a and b, then d is the largest possible integer satisfying

$$a = dx \tag{1}$$

$$b = dy \tag{2}$$

with x and y positive integers. Therefore, there exists an INTEGER RELATION between a and b OF THE FORM

$$ay - bx = 0. \tag{3}$$

The EUCLIDEAN ALGORITHM can be used to find the greatest common divisor of two integers.

The notion can also be generalized to more general RINGS than simply the integers $\mathbb{Z}$. However, even for EUCLIDEAN RINGS, the notion of GCD of two elements of a ring is not the same as the GCD of two ideals of a ring. This is sometimes a source of confusion when studying rings other than $\mathbb{Z}$, such as polynomial rings in several variables.

To compute the GCD, write the PRIME FACTORIZATIONS of a and b,

$$a \equiv \prod_i p_i^{\alpha_i} \tag{4}$$

$$b \equiv \prod_i p_i^{\beta_i}, \tag{5}$$

where the p_is are all PRIME FACTORS of a and b, and if p_i does not occur in one factorization, then the corresponding exponent is taken as 0. Then the greatest common divisor GCD(a, b) is given by

$$\mathrm{GCD}(a, b) = \prod_i p_i^{\min(\alpha_i, \beta_i)}, \tag{6}$$

where min denotes the MINIMUM. For example, con-

sider GCD(12, 30).

$$12 = 2^2 \cdot 3^1 \cdot 5^0 \tag{7}$$

$$30 = 2^1 \cdot 3^1 \cdot 5^1, \tag{8}$$

so

$$\mathrm{GCD}(12, 30) = 2^1 \cdot 3^1 \cdot 5^0 = 6. \tag{9}$$

The GCD is DISTRIBUTIVE

$$\mathrm{GCD}(ma, mb) = m\,\mathrm{GCD}(a, b) \tag{10}$$

$$\mathrm{GCD}(ma, mb, mc) = m\,\mathrm{GCD}(a, b, c), \tag{11}$$

and ASSOCIATIVE

$$\mathrm{GCD}(a, b, c) = \mathrm{GCD}(\mathrm{GCD}(a, b), c)$$
$$= \mathrm{GCD}(a, \mathrm{GCD}(b, c)) \tag{12}$$

$$\mathrm{GCD}(ab, cd) = \mathrm{GCD}(a, c)\mathrm{GCD}(b, d)$$
$$\times \mathrm{GCD}\left(\frac{a}{\mathrm{GCD}(a, c)}, \frac{d}{\mathrm{GCD}(b, d)} \right)$$
$$\times \mathrm{GCD}\left(\frac{c}{\mathrm{GCD}(a, c)}, \frac{b}{\mathrm{GCD}(b, d)} \right). \tag{13}$$

If $a = a_1\mathrm{GCD}(a, b)$ and $b = b_1\mathrm{GCD}(a, b)$, then

$$\mathrm{GCD}(a, b) = \mathrm{GCD}(a_1\,\mathrm{GCD}(a, b), b_1\,\mathrm{GCD}(a, b))$$
$$= \mathrm{GCD}(a, b)\,\mathrm{GCD}(a_1, b_1), \tag{14}$$

so $\mathrm{GCD}(a_1, b_1) = 1$ and a_1 and b_1 are said to be RELATIVELY PRIME. The GCD is also IDEMPOTENT

$$\mathrm{GCD}(a, a) = a, \tag{15}$$

COMMUTATIVE

$$\mathrm{GCD}(a, b) = \mathrm{GCD}(b, a), \tag{16}$$

and satisfies the ABSORPTION LAW

$$\mathrm{LCM}(a, \mathrm{GCD}(a, b)) = a. \tag{17}$$

The probability that two INTEGERS picked at random are RELATIVELY PRIME is $[\zeta(2)]^{-1} = 6/\pi^2$, where $\zeta(z)$ is the RIEMANN ZETA FUNCTION. Polezzi (1997) observed that GCD(m, n) = k, where k is the number of LATTICE POINTS in the PLANE on the straight LINE connecting the VECTORS (0, 0) and (m, n) (excluding (m, n) itself). This observation is intimately connected with the probability of obtaining RELATIVELY PRIME integers, and also with the geometric interpretation of a REDUCED FRACTION y/x as a string through a LATTICE of points with ends at (1,0) and (x, y). The pegs it presses against (x_i, y_i) give alternate CONVERGENTS y_i/x_i of the CONTINUED FRACTION for y/x, while the other CONVERGENTS are obtained from the pegs it presses against with the initial end at (0, 1).

Knuth showed that

$$\gcd(2^p - 1, \, 2^q - 1) = 2^{\gcd(p, \, q)} - 1. \quad (18)$$

The extended greatest common divisor of two INTEGERS m and n can be defined as the greatest common divisor GCD(m, n) of m and n which also satisfies the constraint GCD(m, n) $= rm + sn$ for r and s given INTEGERS. It is used in solving LINEAR DIOPHANTINE EQUATIONS.

See also BÉZOUT NUMBERS, BÉZOUT'S THEOREM, DIRICHLET FUNCTION, EUCLID'S ORCHARD, EUCLIDEAN ALGORITHM, GAUSS'S LEMMA, LEAST COMMON MULTIPLE, LEAST PRIME FACTOR, ORCHARD-PLANTING PROBLEM, STAR OF DAVID THEOREM

References

Nagell, T. "Least Common Multiple and Greatest Common Divisor." §5 in *Introduction to Number Theory.* New York: Wiley, pp. 16–9, 1951.
Polezzi, M. "A Geometrical Method for Finding an Explicit Formula for the Greatest Common Divisor." *Amer. Math. Monthly* **104**, 445–46, 1997.
Séroul, R. "The Greatest Common Divisor." §2.4 in *Programming for Mathematicians.* Berlin: Springer-Verlag, pp. 9–1, 2000.

Greatest Common Divisor Theorem

Given m and n, it is possible to choose c and d such that $cm + dn$ is a common factor of m and n.

Greatest Common Factor

GREATEST COMMON DIVISOR

Greatest Dividing Exponent

The greatest dividing exponent gde(n, b) of a base b with respect to a number n is the largest integer value of k such that $b^k | n$, where $b^k \le n$. It is implemented as the *Mathematica* command IntegerExponent[n, b].

See also DIVIDE, EVEN PART, ODD PART

Greatest Integer Function

FLOOR FUNCTION

Greatest Lower Bound

INFIMUM

Greatest Prime Factor

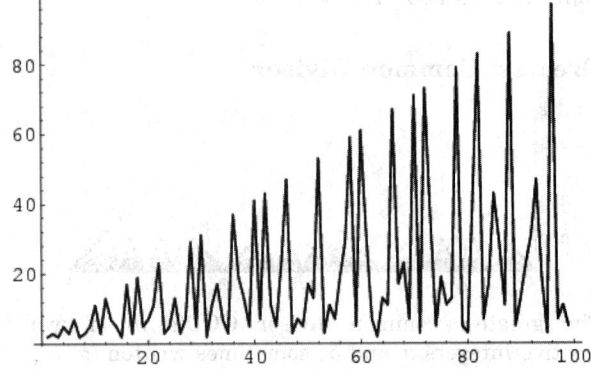

For an INTEGER $n \ge 2$, let gpf(x) denote the greatest prime factor of n, i.e., the number p_k in the factorization

$$n = p_1^{a_1} \ldots p_k^{a_k},$$

with $p_i < p_j$ for $i < j$. For $n = 2, 3, \ldots$, the first few are 2, 3, 2, 5, 3, 7, 2, 3, 5, 11, 3, 13, 7, 5, ... (Sloane's A006530). The greatest *multiple* prime factors for SQUAREFUL integers are 2, 2, 3, 2, 2, 3, 2, 2, 5, 3, 2, 2, 3, ... (Sloane's A046028).

The probability that the GREATEST PRIME FACTOR of a RANDOM integer n is greater than $\sqrt{n}$ is $\ln 2$ (Schroeppel 1972).

See also DICKMAN FUNCTION, DISTINCT PRIME FACTORS, FACTOR, LEAST COMMON MULTIPLE, LEAST PRIME FACTOR, MANGOLDT FUNCTION, PRIME FACTORS, TWIN PEAKS

References

Erdos, P. and Pomerance, C. "On the Largest Prime Factors of n and $n + 1$." *Aequationes Math.* **17**, 211–21, 1978.
Guy, R. K. "The Largest Prime Factor of n." §B46 in *Unsolved Problems in Number Theory, 2nd ed.* New York: Springer-Verlag, p. 101, 1994.
Heath-Brown, D. R. "The Largest Prime Factor of the Integers in an Interval." *Sci. China Ser. A* **39**, 449–76, 1996.
Mahler, K. "On the Greatest Prime Factor of $ax^m + by^n$." *Nieuw Arch. Wiskunde* **1**, 113–22, 1953.
Schroeppel, R. Item 29 in Beeler, M.; Gosper, R. W.; and Schroeppel, R. *HAKMEM.* Cambridge, MA: MIT Artificial Intelligence Laboratory, Memo AIM-239, p. 13, Feb. 1972.
Sloane, N. J. A. Sequences A006530/M0428 in "An On-Line Version of the Encyclopedia of Integer Sequences." http://www.research.att.com/~njas/sequences/eisonline.html.

Grebe Point

SYMMEDIAN POINT

Greedy Algorithm

An algorithm used to recursively construct a SET of objects from the smallest possible constituent parts.

Given a SET of k INTEGERS $(a_1, a_2, \ldots, a_k)$ with $a_1 < a_2 < \ldots < a_k$, a greedy algorithm can be used to find a

VECTOR of coefficients $(c_1, c_2, ..., c_k)$ such that

$$\sum_{i=1}^{k} c_i a_i = \mathbf{c} \cdot \mathbf{a} = n, \qquad (1)$$

where $\mathbf{c} \cdot \mathbf{a}$ is the DOT PRODUCT, for some given INTEGER n. This can be accomplished by letting $c_i = 0$ for $i = 1, ..., k-1$ and setting

$$c_k = \left\lfloor \frac{n}{a_k} \right\rfloor, \qquad (2)$$

where $\lfloor x \rfloor$ is the floor function. Now define the difference between the representation and n as

$$\Delta \equiv n - \mathbf{c} \cdot \mathbf{a}. \qquad (3)$$

If $\Delta = 0$ at any step, a representation has been found. Otherwise, decrement the NONZERO a_i term with least i, set all $a_j = 0$ for $j < i$, and build up the remaining terms from

$$c_j = \left\lceil \frac{\Delta_j}{a_k} \right\rceil \qquad (4)$$

for $j = i - 1, ..., 1$ until $\Delta = 0$ or all possibilities have been exhausted.

For example, MCNUGGET NUMBERS are numbers which are representable using only $(a_1, a_2, a_3) = (6, 9, 20)$. Taking $n = 62$ and applying the algorithm iteratively gives the sequence $(0, 0, 3)$, $(0, 2, 2)$, $(2, 1, 2)$, $(3, 0, 2)$, $(1, 4, 1)$, at which point $\Delta = 0$. 62 is therefore a MCNUGGET NUMBER with

$$62 = (1 \cdot 6) + (4 \cdot 9) + (1 \cdot 20). \qquad (5)$$

If *any* INTEGER n can be represented with $c_i = 0$ or 1 using a sequence $(a_1, a_2, ...)$, then this sequence is called a COMPLETE SEQUENCE.

A greedy algorithm can also be used to break down arbitrary fractions into UNIT FRACTIONS in a finite number of steps. For a FRACTION a/b, find the least INTEGER x_1 such that $1/x_1 \le a/b$, i.e.,

$$x_1 = \left\lceil \frac{b}{a} \right\rceil, \qquad (6)$$

where $\lceil x \rceil$ is the CEILING FUNCTION. Then find the least INTEGER x_2 such that $1/x_2 \le a/b - 1/x_1$. Iterate until there is no remainder. The ALGORITHM gives two or fewer terms for $1/n$ and $2/n$, three or fewer terms for $3/n$, and four or fewer for $4/n$.

See also COMPLETE SEQUENCE, INTEGER RELATION, LEVINE-O'SULLIVAN GREEDY ALGORITHM, MCNUGGET NUMBER, REVERSE GREEDY ALGORITHM, SQUARE NUMBER, SYLVESTER'S SEQUENCE, UNIT FRACTION

Greek Cross

An irregular DODECAHEDRON CROSS in the shape of a PLUS SIGN.

See also CROSS, DISSECTION, DODECAHEDRON, LATIN CROSS, PLUS SIGN, SAINT ANDREW'S CROSS

References
Wells, D. *The Penguin Dictionary of Curious and Interesting Geometry.* London: Penguin, p. 89, 1991.

Greek Problems
GEOMETRIC PROBLEMS OF ANTIQUITY

Green Space
A G-SPACE provides local notions of harmonic, hyperharmonic, and superharmonic functions. When there exists a nonconstant superharmonic function greater than 0, it is a called a Green space. Examples are $\mathbb{R}^n$ (for $n \ge 3$) and any bounded domain of $\mathbb{R}^n$.

See also G-SPACE

Green's Function
A Green's function is an integrating kernal which can be used to solve an inhomogeneous differential equation with boundary conditions. It serves roughly an analogous role in partial differential equations as does FOURIER ANALYSIS in the solution of ordinary differential equations.

As a special case, consider the 1-D DIFFERENTIAL OPERATOR

$$\tilde{L} = \tilde{D}^n + a_{n-1}(t)\tilde{D}^{n-1} + ... + a_1(t)\tilde{D} + a_0(t), \qquad (1)$$

with $a_i(t)$ CONTINUOUS for $i = 0, 1, ..., n-1$ on the interval I, and assume we wish to find the solution $y(t)$ to the equation

$$\tilde{L}y(t) = h(t), \qquad (2)$$

where $h(t)$ is a given CONTINUOUS FUNCTION on I. To solve equation (2), we look for a function $g : C^n(I) \mapsto C(I)$ such that $\tilde{L}(g(h)) = h$, where

$$y(t) = g(h(t)). \qquad (3)$$

This is a CONVOLUTION equation OF THE FORM

$$y = g \ast h, \qquad (4)$$

so the solution is

$$y(t) = \int_{t_0}^{t} g(t-x)h(x)\, dx, \qquad (5)$$

and the function $g(t)$ is called the Green's function for

$\tilde{L}$ on I. Now, note that if we take $h(t) = \delta(t)$, then

$$y(t) = \int_{t_0}^{t} g(t-x)\delta(x)\,dx = g(t), \qquad (6)$$

so the Green's function $g(t)$ can be defined by

$$\tilde{L}g(t) = \delta(t). \qquad (7)$$

However, the Green's function is determined uniquely only if some initial or boundary conditions are given.

For an arbitrary linear differential operator $\tilde{L}$ in 3-D, the Green's function $G(\mathbf{r}, \mathbf{r}')$ is defined by analogy with the 1-D case by

$$\tilde{L}G(\mathbf{r}, \mathbf{r}') = \delta(\mathbf{r} - \mathbf{r}'). \qquad (8)$$

The solution to $\tilde{L}\phi = f$ is then

$$\phi(\mathbf{r}) = \int G(\mathbf{r}, \mathbf{r}')f(\mathbf{r}')d^3\mathbf{r}'. \qquad (9)$$

Explicit expressions for $G(\mathbf{r}, \mathbf{r}')$ can often be found in terms of a basis of given eigenfunctions $\phi_n(\mathbf{r}_1)$ by expanding the Green's function

$$G(\mathbf{r}_1, \mathbf{r}_2) = \sum_{n=0}^{\infty} a_n(\mathbf{r}_2)\phi_n(\mathbf{r}_1) \qquad (10)$$

and DELTA FUNCTION,

$$\delta^3(\mathbf{r}_1 - \mathbf{r}_2) = \sum_{n=0}^{\infty} b_n\phi_n(\mathbf{r}_1). \qquad (11)$$

Multiplying both sides by $\phi_m(\mathbf{r}_2)$ and integrating over $\mathbf{r}_1$ space,

$$\int \phi_m(\mathbf{r}_2)\delta^3(\mathbf{r}_1 - \mathbf{r}_2)d^3\mathbf{r}_1$$

$$= \sum_{n=0}^{\infty} b_n \int \phi_m(\mathbf{r}_2)\phi_n(\mathbf{r}_1)d^3\mathbf{r}_1 \qquad (12)$$

$$\phi_m(\mathbf{r}_2) = \sum_{n=0}^{\infty} b_n\delta_{nm} = b_m, \qquad (13)$$

so

$$\delta^3(\mathbf{r}_1 - \mathbf{r}_2) = \sum_{n=0}^{\infty} \phi_n(\mathbf{r}_1)\phi_n(\mathbf{r}_2). \qquad (14)$$

By plugging in the differential operator, solving for the a_ns, and substituting into G, the original nonhomogeneous equation then can be solved.

The coefficient S of $\ln(1/r)$ in all normalized fundamental Green's function solutions

$$\phi(x, y; x_0, y_0)$$
$$= S(x, y; x_0, y_0) \ln(1/r) + T(x, y; x_0, y_0) \qquad (15)$$

with

$$r = \sqrt{(x-x_0)^2 + (y-y_0)^2} \qquad (16)$$

of the ELLIPTIC PARTIAL DIFFERENTIAL EQUATION

$$Ku = u_{xx} + v_{yy} + A(x, y)u_x + B(x, y)u_y + C(x, y)u$$
$$= 0 \qquad (17)$$

with analytic coefficients is an analytic function of four variables and is equal to the RIEMANN FUNCTION $S = R^*(\xi, \eta; \xi_0, \eta_0)$ of the conjugate equation

$$K^*v = v(\xi, \eta) - (av)(\xi) - (bv)(\eta) + cv = 0 \qquad (18)$$

which can be produced from $Ku = 0$ by the change of variables

$$\xi = x + iy \qquad (19)$$

$$\eta = x - iy \qquad (20)$$

$$\xi_0 = x_0 + iy_0 \qquad (21)$$

$$\eta_0 = x_0 - iy_0 \qquad (22)$$

$$4a(\xi, \eta) = A(x, y) + iB(x, y) \qquad (23)$$

$$4b(\xi, \eta) = A(x, y) - iB(x, y) \qquad (24)$$

$$4c(\xi, \eta) = C(x, y) \qquad (25)$$

(Garabedian 1964, Marichev 1990).

See also GREEN'S FUNCTION–HELMHOLTZ DIFFERENTIAL EQUATION, GREEN'S FUNCTION–POISSON'S EQUATION, RIEMANN METHOD

References

Arfken, G. "Nonhomogeneous Equation--Green's Function," "Green's Functions--One Dimension," and "Green's Functions--Two and Three Dimensions." §8.7 and §16.5–6.6 in *Mathematical Methods for Physicists, 3rd ed.* Orlando, FL: Academic Press, pp. 480–91 and 897–24, 1985.
Garabedian, P. R. *Partial Differential Equations.* New York: Wiley, 1964.
Marichev, O. I. "Funktionen vom hypergeometrischen Typ und einige Anwendungen auf Integral- under Differentialgleichungen." Ph.D. dissertation. Jena, Germany: Friedrich-Schiller-Universität, p. 266, 1990.

Green's Function—Helmholtz Differential Equation

The inhomogeneous HELMHOLTZ DIFFERENTIAL EQUATION is

$$\nabla^2\psi(\mathbf{r}) + k^2\psi(\mathbf{r}) = \rho(\mathbf{r}), \qquad (1)$$

where the Helmholtz operator is defined as $\tilde{L} \equiv \nabla^2 + k^2$. The Green's function is then defined by

$$(\nabla^2 + k^2)G(\mathbf{r}_1, \mathbf{r}_2) = \delta^3(\mathbf{r}_1 - \mathbf{r}_2). \qquad (2)$$

Define the basis functions ϕ_n as the solutions to the homogeneous HELMHOLTZ DIFFERENTIAL EQUATION

$$\nabla^2\phi_n(\mathbf{r}) + k_n^2\phi_n(\mathbf{r}) = 0. \qquad (3)$$

The Green's function can then be expanded in terms

of the ϕ_ns,

$$G(\mathbf{r}_1, \ \mathbf{r}_2) = \sum_{n=0}^{\infty} a_n(\mathbf{r}_2)\phi_n(\mathbf{r}_1), \qquad (4)$$

and the DELTA FUNCTION as

$$\delta^3(\mathbf{r}_1 - \mathbf{r}_2) = \sum_{n=0}^{\infty} \phi_n(\mathbf{r}_1)\phi_n(\mathbf{r}_2). \qquad (5)$$

Plugging (4) and (5) into (2) gives

$$\nabla^2 \left[\sum_{n=0}^{\infty} a_n(\mathbf{r}_2)\phi_n(\mathbf{r}_1) \right] + k^2 \sum_{n=0}^{\infty} a_n(\mathbf{r}_2)\phi_n(\mathbf{r}_1)$$

$$= \sum_{n=0}^{\infty} \phi_n(\mathbf{r}_1)\phi_n(\mathbf{r}_2). \qquad (6)$$

Using (3) gives

$$-\sum_{n=0}^{\infty} a_n(\mathbf{r}_2)k_n^2\phi_n(\mathbf{r}_1) + k^2 \sum_{n=0}^{\infty} a_n(\mathbf{r}_2)\phi_n(\mathbf{r}_1)$$

$$= \sum_{n=0}^{\infty} \phi_n(\mathbf{r}_1)\phi_n(\mathbf{r}_2) \qquad (7)$$

$$\sum_{n=0}^{\infty} a_n(\mathbf{r}_2)\phi_n(\mathbf{r}_1)(k^2 - k_n^2) = \sum_{n=0}^{\infty} \phi_n(\mathbf{r}_1)\phi_n(\mathbf{r}_2). \qquad (8)$$

This equation must hold true for each n, so

$$a_n(\mathbf{r}_2)\phi_n(\mathbf{r}_1)(k^2 - k_n^2) = \phi_n(\mathbf{r}_1)\phi_n(\mathbf{r}_2) \qquad (9)$$

$$a_n(\mathbf{r}_2) = \frac{\phi_n(\mathbf{r}_2)}{k^2 - k_n^2}, \qquad (10)$$

and (4) can be written

$$G(\mathbf{r}_1, \ \mathbf{r}_2) = \sum_{n=0}^{\infty} \frac{\phi_n(\mathbf{r}_1)\phi_n(\mathbf{r}_2)}{k^2 - k_n^2}. \qquad (11)$$

The general solution to (1) is therefore

$$\psi(\mathbf{r}_1) = \int G(\mathbf{r}_1, \ \mathbf{r}_2)\rho(\mathbf{r}_2)d^3\mathbf{r}_2$$

$$= \sum_{n=0}^{\infty} \int \frac{\phi_n(\mathbf{r}_1)\phi_n(\mathbf{r}_2)\rho(\mathbf{r}_2)}{k^2 - k_n^2} \, d^3\mathbf{r}_2. \qquad (12)$$

References

Arfken, G. *Mathematical Methods for Physicists, 3rd ed.* Orlando, FL: Academic Press, pp. 529–30, 1985.

Green's Function—Poisson's Equation

POISSON'S EQUATION equation is

$$\nabla^2 \phi = 4\pi\rho, \qquad (1)$$

where ϕ is often called a potential function and ρ a

density function, so the differential operator in this case is $\tilde{L} = \nabla^2$. As usual, we are looking for a Green's function $G(\mathbf{r}_1, \ \mathbf{r}_2)$ such that

$$\nabla^2 G(\mathbf{r}_1, \ \mathbf{r}_2) = \delta^3(\mathbf{r}_1, \ \mathbf{r}_2). \qquad (2)$$

But from LAPLACIAN,

$$\nabla^2 \left(\frac{1}{|\mathbf{r} - \mathbf{r}'|} \right) = -4\pi\delta^3(\mathbf{r} - \mathbf{r}'), \qquad (3)$$

so

$$G(\mathbf{r}, \ \mathbf{r}') = -\frac{1}{4\pi|\mathbf{r} - \mathbf{r}'|}, \qquad (4)$$

and the solution is

$$\phi(\mathbf{r}) = \int G(\mathbf{r}, \ \mathbf{r}')[4\pi\rho(\mathbf{r}')]d^3\mathbf{r}' = -\int \frac{\rho(\mathbf{r}')d^3\mathbf{r}'}{|\mathbf{r} - \mathbf{r}'|}. \qquad (5)$$

Expanding $G(\mathbf{r}_1, \ \mathbf{r}_2)$ in the SPHERICAL HARMONICS Y_l^m gives

$$G(\mathbf{r}_1, \ \mathbf{r}_2)$$

$$= \sum_{l=0}^{\infty} \sum_{m=-l}^{l} \frac{1}{2l+1} \frac{r_<^l}{r_>^{l+1}} Y_l^m(\theta_1, \ \phi_1)\bar{Y}_l^m(\theta_2, \ \phi_2), \qquad (6)$$

where $r_<$ and $r_>$ are GREATER THAN/LESS THAN SYMBOLS. this expression simplifies to

$$g(\mathbf{r}_1, \ \mathbf{r}_2) = \frac{1}{4\pi} \sum_{l=0}^{\infty} \frac{r_<^l}{r_>^{l+1}} p_l(\cos\gamma), \qquad (7)$$

where p_l are LEGENDRE POLYNOMIALS, and $\cos\gamma \equiv \mathbf{r}_1 \cdot \mathbf{r}_2$. Equations (6) and (7) give the addition theorem for LEGENDRE POLYNOMIALS.

In CYLINDRICAL COORDINATES, the Green's function is much more complicated,

$$G(\mathbf{r}_1, \ \mathbf{r}_2) = \frac{1}{2\pi^2} \sum_{m=-\infty}^{\infty} \int_0^{\infty} I_m(k\rho <)K_m$$

$$\times (k\rho >)e^{im(\phi_1 - \phi_2)}\cos[k(z_1 - z_2)] \, dk. \qquad (8)$$

where $I_m(x)$ and $K_m(x)$ are MODIFIED BESSEL FUNCTIONS OF THE FIRST and SECOND KINDS (Arfken 1985).

References

Arfken, G. *Mathematical Methods for Physicists, 3rd ed.* Orlando, FL: Academic Press, pp. 485–86, 905, and 912, 1985.

Green's Identities

Green's identities are a set of three vector derivative/integral identities which can be derived starting with the vector derivative identities

$$\nabla \cdot (\psi\nabla\phi) = \psi\nabla^2\phi + (\nabla\psi) \cdot (\nabla\phi) \qquad (1)$$

and

$$\nabla \cdot (\phi \nabla \psi) = \phi \nabla^2 \psi + (\nabla \phi)\cdot(\nabla \psi), \qquad (2)$$

where $\nabla\cdot$ is the DIVERGENCE, ∇ is the GRADIENT, ∇^2 is the LAPLACIAN, and $\mathbf{a} \cdot \mathbf{b}$ is the DOT PRODUCT. From the DIVERGENCE THEOREM,

$$\int_V (\nabla\cdot\mathbf{F})\, dV = \int_S \mathbf{F} \cdot d\mathbf{a}. \qquad (3)$$

Plugging (2) into (3),

$$\int_S \phi(\nabla\psi) \cdot d\mathbf{a} = \int_V [\phi\nabla^2\psi + (\nabla\phi)\cdot(\nabla\psi)]\, dV. \qquad (4)$$

This is Green's first identity.

Subtracting (2) from (1),

$$\nabla \cdot (\phi\nabla\psi - \psi\nabla\phi) = \phi\nabla^2\psi - \psi\nabla^2\phi. \qquad (5)$$

Therefore,

$$\int_V (\phi\nabla^2\psi - \psi\nabla^2\phi)\, dV = \int_S (\phi\nabla\psi - \psi\nabla\phi) \cdot d\mathbf{a}. \qquad (6)$$

This is Green's second identity.

Let u have continuous first PARTIAL DERIVATIVES and be HARMONIC inside the region of integration. Then Green's third identity is

$$u(x, y) = \frac{1}{2\pi} \oint_C \left[\ln\left(\frac{1}{r}\right)\frac{\partial u}{\partial n} - u\, \frac{\partial}{\partial n} \ln\left(\frac{1}{r}\right) \right] ds \qquad (7)$$

(Kaplan 1991, p. 361).

References

Kaplan, W. *Advanced Calculus, 4th ed.* Reading, MA: Addison-Wesley, 1991.

Green's Theorem

Green's theorem is a vector identity which is equivalent to the CURL THEOREM in the PLANE. Over a region D in the plane with boundary ∂D,

$$\int_{\partial D} f(x, y)\, dx + g(x, y)\, dy = \int\int_D \left(\frac{\partial g}{\partial x} - \frac{\partial f}{\partial y} \right) dx\, dy$$

$$\int_{\partial D} \mathbf{F} \cdot ds = \int\int_D (\nabla \times \mathbf{F}) \cdot \mathbf{k}\, dA.$$

If the region D is on the left when traveling around ∂D, then AREA of D can be computed using

$$A = \tfrac{1}{2} \int_{\partial D} x\, dy - y\, dx.$$

See also CURL THEOREM, DIVERGENCE THEOREM

References

Arfken, G. "Gauss's Theorem." §1.11 in *Mathematical Methods for Physicists, 3rd ed.* Orlando, FL: Academic Press, pp. 57–1, 1985.

Greene's Method

A method for predicting the onset of widespread CHAOS. It is based on the hypothesis that the dissolution of an invariant torus can be associated with the sudden change from stability to instability of nearly closed orbits (Tabor 1989, p. 163).

See also OVERLAPPING RESONANCE METHOD

References

Tabor, M. *Chaos and Integrability in Nonlinear Dynamics: An Introduction.* New York: Wiley, 1989.

Greenwood-Gleason Graph

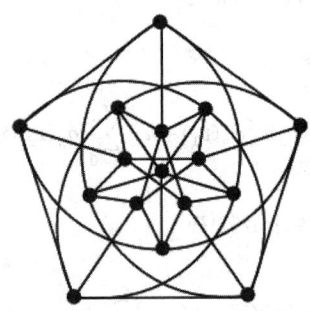

Kalbfleisch and Stanton (1968) showed that in a 3-edge coloring of the COMPLETE GRAPH K_{16} without monochromatic triangles, the subgraph induced by the edges of any one color is isomorphic to the graph illustrated above, known as the Greenwood-Gleason graph.

References

Bondy, J. A. and Murty, U. S. R. *Graph Theory with Applications.* New York: North Holland, p. 242, 1976.
Kalbfleisch, J. and Stanton, R. "On the Maximal Triangle-Free Edge-Chromatic Graph in Three Colors." *J. Combin. Th.* **5**, 9–0, 1968.

Gregory Number

A number

$$t_x = \tan^{-1}\left(\frac{1}{x}\right) = \cot^{-1} x,$$

where x is an INTEGER or RATIONAL NUMBER, $\tan^{-1} x$ is the INVERSE TANGENT, and $\cot^{-1} x$ is the INVERSE COTANGENT. Gregory numbers arise in the determination of MACHIN-LIKE FORMULAS. Every Gregory number t_x can be expressed uniquely as a sum of t_ns where the ns are STØRMER NUMBERS.

References

Conway, J. H. and Guy, R. K. "Gregory's Numbers" In *The Book of Numbers.* New York: Springer-Verlag, pp. 241–42, 1996.

Gregory's Formula

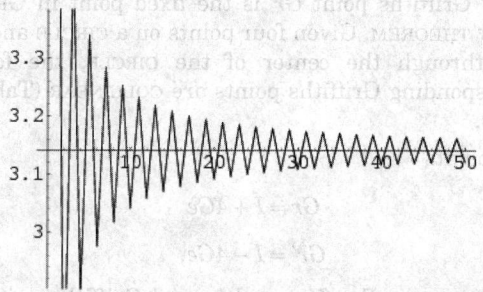

There are at least two formulas associated with Gregory. The first is a series PI FORMULA found by Gregory and Leibniz and obtained by plugging $x = 1$ into the LEIBNIZ SERIES,

$$\frac{\pi}{4} = 1 - \frac{1}{3} + \frac{1}{5} + \cdots$$

(Wells 1986, p. 50). The formula, also called the LEIBNIZ SERIES, converges very slowly, but its convergence can be accelerated using certain transformations, in particular

$$\pi = \sum_{k=1}^{\infty} \frac{3^k - 1}{4^k} \zeta(k+1),$$

where $\zeta(z)$ is the RIEMANN ZETA FUNCTION (Vardi 1991).

The second is the formula

$$\int_0^y p(u)\,du = \sum_{k \ge 0} \left\langle \frac{(e^{yt} - 1)^k |p(x)\rangle}{k!\,(e^t - 1)^k p(x),} \right.$$

discovered by Gregory in 1670 and reported to be the earliest formula in NUMERICAL INTEGRATION (Jordan 1950, Roman 1984).

See also LEIBNIZ SERIES, MACHIN'S FORMULA, MACHIN-LIKE FORMULAS, NUMERICAL INTEGRATION, PI FORMULAS

References

Jordan, C. *Calculus of Finite Differences, 3rd ed.* New York: Chelsea, p. 284, 1965.
Roman, S. *The Umbral Calculus.* New York: Academic Press, p. 59, 1984.
Vardi, I. *Computational Recreations in Mathematica.* Reading, MA: Addison-Wesley, pp. 157–58, 1991.
Wells, D. *The Penguin Dictionary of Curious and Interesting Numbers.* Middlesex, England: Penguin Books, p. 50, 1986.

Gregory-Newton Formula

NEWTON'S FORWARD DIFFERENCE FORMULA

Grelling's Paradox

A semantic PARADOX, also called the HETEROLOGICAL PARADOX, which arises by defining "heterological" to mean "a word which does not describe itself." The word "heterological" is therefore heterological IFF it is not.

See also RUSSELL'S PARADOX

References

Curry, H. B. *Foundations of Mathematical Logic.* New York: Dover, p. 6, 1977.
Erickson, G. W. and Fossa, J. A. *Dictionary of Paradox.* Lanham, MD: University Press of America, pp. 83–4, 1998.
Hofstadter, D. R. *Gödel, Escher, Bach: An Eternal Golden Braid.* New York: Vintage Books, pp. 20–1, 1989.

Grenz-Formel

An equation derived by Kronecker:

$$\sum_{x,\,y,\,z=-\infty}^{\infty}{}' \ (x^2 + y^2 + dz^2)^{-s}$$

$$= 4\zeta(s)\eta(s) + \frac{2\pi}{s-1}\frac{\zeta(2s-2)}{d^{s-1}} + \frac{2\pi^s}{\Gamma(s)}\,d^{(1-s)/2}$$

$$\times \sum_{n=1}^{\infty} n^{(s-1)/2} \sum_{u^2 | n} \frac{r\left(\dfrac{n}{u^2}\right)}{u^{2a-2}} \int_0^{\infty} e^{\pi\sqrt{nd}(y+y^{-1})} y^{s-2}\,dy,$$

where $r(n)$ is the SUM OF SQUARES FUNCTION, $\zeta(z)$ is the RIEMANN ZETA FUNCTION, $\eta(z)$ is the DIRICHLET ETA FUNCTION, $\Gamma(z)$ is the GAMMA FUNCTION, and the primed sum omits terms with zero DENOMINATOR (Selberg and Chowla 1967).

See also DIRICHLET ETA FUNCTION, EPSTEIN ZETA FUNCTION, SUM OF SQUARES FUNCTION

References

Borwein, J. M. and Borwein, P. B. *Pi & the AGM: A Study in Analytic Number Theory and Computational Complexity.* New York: Wiley, pp. 296–97, 1987.
Selberg, A. and Chowla, S. "On Epstein's Zeta-Function." *J. reine angew. Math.* **227**, 86–10, 1967.

Grid

This entry contributed by DANIEL SCOTT UZNANSKI

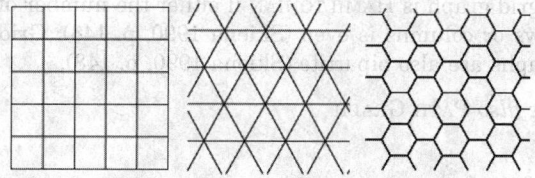

A grid usually refers to two or more infinite sets of evenly-spaced parallel lines at particular angles to each other in a plane, or the intersections of such lines. The two most common types of grid are orthogonal grids, with two sets of lines perpendicular to each other, and isometric grids, with three sets of lines at 60-degree angles to each other. It should be noted that in most grids with three or more sets of lines, every intersection includes one element of each set.

There are other types of planar grids, like hexagonal grids, which are formed by tessellating regular hexagons in the plane. These are often found in strategy and role-playing games because of the lack of single points of contact characteristic of isometric and orthogonal grids. The collection of cells created by a grid is often called a "BOARD" when these cells are used as resting places for pieces in a game.

Grids can be generalized into n-D space by using the centers of packed n-spheres or n-cubes as the points.

See also BOARD, FINITE ELEMENT METHOD, LATTICE POINT

References

Bern, M. W.; Flaherty, J. E.; and Luskin, M. (Eds.). *Grid Generation and Adaptive Algorithms.* New York: Springer-Verlag, 1999.
Liseikin, V. D. *Grid Generation Methods.* Berlin: Springer-Verlag, 1999.

Grid Graph

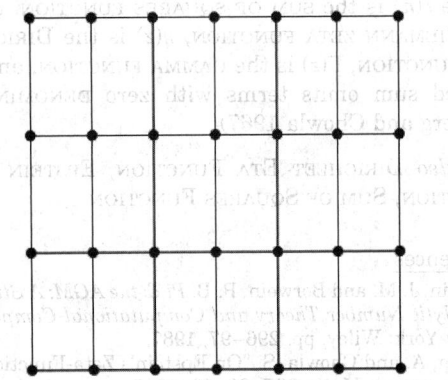

An $m \times n$ grid graph $G_{m,n}$ is the product of PATH GRAPHS on m and n vertices. A grid graph $G_{n,1}$ is called a PATH GRAPH. The grid graph $G_{2,2}$ is the CYCLE GRAPH C_4.

A grid graph is HAMILTONIAN if either the number of rows or columns is even (Skiena 1990, p. 148). Grid graphs are also bipartite (Skiena 1990, p. 148).

See also PATH GRAPH

References

Reddy, V. and Skiena, S. "Frequencies of Large Distances in Integer Lattices." Technical Report, Department of Computer Science. Stony Brook, NY: State University of New York, Stony Brook, 1989.
Skiena, S. "Grid Graphs." §4.2.4 in *Implementing Discrete Mathematics: Combinatorics and Graph Theory with Mathematica.* Reading, MA: Addison-Wesley, pp. 147–48, 1990.

Griffiths Points

"The" Griffiths point Gr is the fixed point in GRIFFITHS' THEOREM. Given four points on a CIRCLE and a line through the center of the CIRCLE, the four corresponding Griffiths points are COLLINEAR (Tabov 1995).

The points

$$Gr = I + 4Ge$$
$$Gr' = I - 4Ge,$$

are known as the first and second Griffiths points, where I is the INCENTER and Ge is the GERGONNE POINT (Oldknow 1996). The Griffiths points lie on the SODDY LINE.

See also GERGONNE POINT, GRIFFITHS' THEOREM, INCENTER, OLDKNOW POINTS, RIGBY POINTS, SODDY LINE

References

Oldknow, A. "The Euler-Gergonne-Soddy Triangle of a Triangle." *Amer. Math. Monthly* **103**, 319–29, 1996.
Tabov, J. "Four Collinear Griffiths Points." *Math. Mag.* **68**, 61–4, 1995.

Griffiths' Theorem

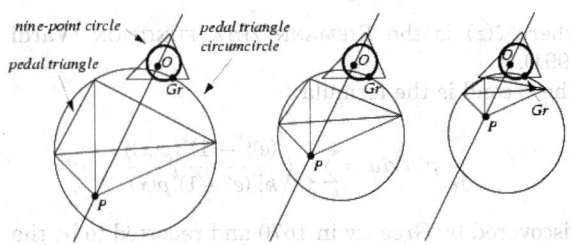

When a point P moves along a line through the CIRCUMCENTER of a given TRIANGLE Δ, the PEDAL CIRCLE of P with respect to Δ passes through a fixed point (the GRIFFITHS POINT) on the NINE-POINT CIRCLE of Δ.

See also CIRCUMCENTER, GRIFFITHS POINTS, NINE-POINT CIRCLE, PEDAL CIRCLE

Grimm's Conjecture

Grimm conjectured that if $n + 1, n + 2, ..., n + k$ are all COMPOSITE NUMBERS, then there are distinct PRIMES p_{i_j} such that $p_{i_j}|(n+j)$ for $1 \leq j \leq k$.

References

Guy, R. K. "Grimm's Conjecture." §B32 in *Unsolved Problems in Number Theory, 2nd ed.* New York: Springer-Verlag, p. 86, 1994.

Grinberg Formula

A formula satisfied by all HAMILTONIAN CIRCUITS with n nodes. Let f_j be the number of regions inside the circuit with j sides, and let g_j be the number of regions outside the circuit with j sides. If there are d interior diagonals, then there must be $d + 1$ regions

$$[\text{# regions in interior}] = d + 1 = f_2 + f_3 + \ldots + f_n. \quad (1)$$

Any region with j sides is bounded by j EDGES, so such regions contribute jf_j to the total. However, this counts each diagonal twice (and each EDGE only once). Therefore,

$$2f_2 + 3f_3 + \ldots nf_n = 2d + n. \quad (2)$$

Take (2) minus $2\times$(1),

$$f_3 + 2f_4 + 3f_5 + \ldots + (n-2)f_n = n - 2. \quad (3)$$

Similarly,

$$g_3 + 2g_4 + \ldots + (n-2)g_n = n - 2, \quad (4)$$

so

$$(f_3 - g_3) + 2(f_4 - g_4) + 3(f_5 - g_5) + \ldots + (n-2)(f_n - g_n)$$
$$= 0. \quad (5)$$

Gröbner Basis

A Gröbner basis for a system of POLYNOMIALS is an equivalence system that possesses useful properties, for example, that another polynomial f is a combination of those in the system IFF the remainder of f with respect to the system is 0. (Here, the division algorithm requires an ORDER of a certain type on the MONOMIALS.) Furthermore, the set of polynomials in a Gröbner basis have the same collection of roots as the original polynomials. For linear functions in any number of variables, a Gröbner basis is equivalent to GAUSSIAN ELIMINATION.

Gröbner bases are pervasive in the construction of symbolic algebra algorithms, and Gröbner bases with respect to LEXICOGRAPHIC ORDER are very useful for solving equations and for elimination of variables. The algorithm for computing Gröbner bases is known as BUCHBERGER'S ALGORITHM. The determination of a Gröbner basis is very roughly analogous to computing an ORTHONORMAL BASIS from a set of BASIS VECTORS and can be described roughly as a combination of GAUSSIAN ELIMINATION (for linear systems) and the EUCLIDEAN ALGORITHM (for UNIVARIATE POLYNOMIALS over a FIELD).

The time and memory required to calculate a Gröbner basis depend very much on the variable ordering, MONOMIAL ordering, and on which variables are regarded as constants. Gröbner bases are used implicitly in many routines in *Mathematica*, and can be called explicitly with the command GroebnerBasis[{*poly1, poly2, ...*}, {*x1, x2, ...*}].

See also BUCHBERGER'S ALGORITHM, COMMUTATIVE ALGEBRA, EUCLIDEAN ALGORITHM, GAUSSIAN ELIMINATION, MONOMIAL, ORTHONORMAL BASIS

References

Adams, W. W. and Loustaunau, P. *An Introduction to Gröbner Bases.* Providence, RI: Amer. Math. Soc., 1994.

Becker, T. and Weispfenning, V. *Gröbner Bases: A Computational Approach to Commutative Algebra.* New York: Springer-Verlag, 1993.

Boege, W.; Gebauer, R.; and Kredel, H. "Some Examples for Solving Systems of Algebraic Equations by Calculating Gröbner Bases." *J. Symb. Comput.* **1**, 83–8, 1986.

Buchberger, B. "Gröbner Bases: An Algorithmic Method in Polynomial Ideal Theory." Ch. 6 in *Multidimensional Systems Theory* (Ed. N. K. Bose). New York: van Nostrand Reinhold, 1982.

Cox, D.; Little, J.; and O'Shea, D. *Ideals, Varieties, and Algorithms: An Introduction to Algebraic Geometry and Commutative Algebra, 2nd ed.* New York: Springer-Verlag, 1996.

Eisenbud, D. *Commutative Algebra with a View toward Algebraic Geometry.* New York: Springer-Verlag, 1995.

Faugere, J. C.; Gianni, P.; Lazard, D.; and Mora, T. "Efficient Computation of Zero-Dimensional Groebner Bases by Change of Ordering." *J. Symb. Comput.* **16**, 329–44, 1993.

Harris, J. "Rearranging Expressions by Patterns." *Mathematica J.* **4**, 82–5, 1994.

Heck, A. "A Bird's-Eye View of Gröbner Bases." http://www.can.nl/CA_Library/Groebner/Tutorials/Heck/AI-HENP96.html.

Helzer, G. "Gröbner Bases." *Mathematica J.* **5**, 67–3, 1995.

Nakos, G. and Glinos, M. "Computing Gröbner Bases over the Integers." *Mathematica J.* **4**, 70–5, 1994.

Lichtblau, D. "Gröbner Bases in *Mathematica* 3.0." *Mathematica J.* **6**, 81–8, 1996.

Mishra, B. *Algorithmic Algebra.* New York: Springer-Verlag, 1993.

Robbiano, L. "Term Ordering on the Polynomial Ring." In *EUROCAL '85: European Conference on Computer Algebra, 1985 Linz, Austria, Vol. 2: Research Contributions* 0387159843 New York: Springer-Verlag, 1986.

Stoutemyer, D. "Which Polynomial Representation is Best? Surprises Abound!" In *Proceedings of the Third MAC-SYMA Users' Conference, Schenectady, NY.* pp. 221–43, 1984.

Trott, M. "Applying GroebnerBasis to Three Problems in Geometry." *Mathematica Educ. Res.* **6**, 15–8, 1997.

Wang, D. *Elimination Methods.* Berlin: Springer-Verlag, 1999.

Groemer Packing

A honeycomb-like packing that forms HEXAGONS.

See also GROEMER THEOREM

References

Stewart, I. "A Bundling Fool Beats the Wrap." *Sci. Amer.* **268**, 142–44, 1993.

Groemer Theorem

Given n CIRCLES and a PERIMETER p, the total AREA of the CONVEX HULL is

$$A_{\text{Convex Hull}} = 2\sqrt{3}(n-1) + p\left(1 - \tfrac{1}{2}\sqrt{3}\right) + \pi(\sqrt{3} - 1)$$

Furthermore, the actual AREA equals this value IFF the packing is a GROEMER PACKING. The theorem was proved in 1960 by Helmut Groemer.

See also CONVEX HULL

Gronwall's Theorem

Let $\sigma(n)$ be the DIVISOR FUNCTION. Then

$$\overline{\lim_{n \to \infty}} \frac{\sigma(n)}{n \ln \ln n} = e^{\gamma},$$

where γ is the EULER-MASCHERONI CONSTANT. Ramanujan independently discovered a less precise version of this theorem (Berndt 1994). Robin (1984) showed that the validity of the inequality

$$\sigma(n) < e^{\gamma} n \ln \ln n$$

for $n \geq 5041$ is equivalent to the RIEMANN HYPOTHESIS.

References

Berndt, B. C. *Ramanujan's Notebooks: Part I.* New York: Springer-Verlag, p. 94, 1985.
Gronwall, T. H. "Some Asymptotic Expressions in the Theory of Numbers." *Trans. Amer. Math. Soc.* **37**, 113–22, 1913.
Nicolas, J.-L. "On Highly Composite Numbers." In *Ramanujan Revisited: Proceedings of the Centenary Conference* (Ed. G. E. Andrews, B. C. Berndt, and R. A. Rankin). Boston, MA: Academic Press, pp. 215–44, 1988.
Robin, G. "Grandes Valeurs de la fonction somme des diviseurs et hypothèse de Riemann." *J. Math. Pures Appl.* **63**, 187–13, 1984.

Gross

A DOZEN DOZEN, or the SQUARE NUMBER 144.

See also 12, DOZEN, DUODECIMAL

Grössencharakter

In the original formulation, a quantity associated with ideal class groups. According to Chevalley's formulation, a Grössencharakter is a MULTIPLICATIVE CHARACTER of the group of ADÈLES that is trivial on the diagonally embedded $k^{\times}$, where k is a NUMBER FIELD.

See also ADÈLE, MULTIPLICATIVE CHARACTER

References

Hecke, E. *Math. Z.* **1**, 1918.
Hecke, E. *Math. Z.* **5**, 1920.
Iyanaga, S. and Kawada, Y. (Eds.). *Encyclopedic Dictionary of Mathematics.* Cambridge, MA: MIT Press, p. 24, 1980.
Knapp, A. W. "Group Representations and Harmonic Analysis, Part II." *Not. Amer. Math. Soc.* **43**, 537–49, 1996.
Tate, J. "Fourier Analysis in Number Fields and Hecke's Zeta Functions." Ch. 15 in *Algebraic Number Theory* (Ed. J. W. S. Cassels and A. Fröhlich). New York: Academic Press, 1950.

Grossman's Constant

Define the sequence $a_0 = 1$, $a_1 = x$, and

$$a_{n+2} = \frac{a_n}{1 + a_{n+1}}$$

for $n \geq 0$. Janssen and Tjaden (1987) showed that this sequence converges for exactly one value of x, $x = 0.73733830336929\ldots$, confirming Grossman's conjecture. However, no analytic form is known for this constant, either as the root of a function or as a combination of other constants.

See also FOIAS CONSTANT

References

Finch, S. "Favorite Mathematical Constants." http://www.mathsoft.com/asolve/constant/grssmn/grssmn.html.
Janssen, A. J. E. M. and Tjaden, D. L. A. Solution to Problem 86–. *Math. Intel.* **9**, 40–3, 1987.

Grothendieck's Constant

Let A be an $n \times n$ REAL SQUARE MATRIX and let x_i and y_j be real numbers with $|x_i|$, $|y_i| < 0$. Then Grothendieck showed that there exists a constant K independent of both A and n satisfying

$$\left| \sum_{1 \leq i, j \leq n} a_{ij} \langle x_i, y_j \rangle \right| \leq K \qquad (1)$$

in which the vectors x_i and y_j have a norm < 1 in any HILBERT SPACE. The Grothendieck constant is the smallest REAL NUMBER for which this inequality has been proven. Krivine (1977) showed that

$$1.676\ldots \leq K_G \leq 1.782\ldots, \qquad (2)$$

and has postulated that

$$K_G \equiv \frac{\pi}{2\ln(1 + \sqrt{2})} = 1.7822139\ldots, \qquad (3)$$

which is related to KHINTCHINE'S CONSTANT.

References

Krivine, J. L. "Sur la constante de Grothendieck." *C. R. A. S.* **284**, 8, 1977.
Le Lionnais, F. *Les nombres remarquables.* Paris: Hermann, p. 42, 1983.

Grothendieck's Theorem

Let E and F be paired spaces with S a family of absolutely convex bounded sets of F such that the sets of S generate F and, if $B_1, B_2 \in S$, then there exists a $B_3 \in S$ such that $B_3 \supset B_1$ and $B_3 \supset B_2$. Then E_S is complete IFF algebraic linear functional $f(y)$ of F that is weakly continuous on every $B \in S$ is expressed as $f(y) = \langle x, y \rangle$ for some $x \in E$. When E_S is not complete, the space of all linear functionals satisfying this condition gives the completion $\hat{E}_S$ of E_S.

See also MACKEY'S THEOREM

References

Iyanaga, S. and Kawada, Y. (Eds.). "Grothendieck's Theorem." §407L in *Encyclopedic Dictionary of Mathematics.* Cambridge, MA: MIT Press, p. 1274, 1980.

Ground Set

A PARTIALLY ORDERED SET is defined as an ordered pair $P = (X, \leq)$. Here, X is called the GROUND SET of P and $\leq$ is the PARTIAL ORDER of P.

See also PARTIAL ORDER, PARTIALLY ORDERED SET

Group

A group G is a finite or infinite set of elements together with a BINARY OPERATION which together satisfy the four fundamental properties of closure, associativity, the identity property, and the inverse property. The operation with respect to which a group is defined is often called the "group operation," and a set is said to be a group "under" this operation. Elements A, B, C, ... with binary operation between A and B denoted AB form a group if

1. Closure: If A and B are two elements in G, then the product AB is also in G.
2. Associativity: The defined multiplication is associative, i.e., for all A, B, $C \in G$, $(AB)C = A(BC)$.
3. Identity: There is an IDENTITY ELEMENT I (a.k.a. 1, E, or e) such that $IA = AI = A$ for every element $A \in G$.
4. Inverse: There must be an inverse or reciprocal of each element. Therefore, the set must contain an element $B = A^{-1}$ such that $AA^{-1} = A^{-1}A = I$ for each element of G.

A group is therefore a MONOID for which every element is invertible, and a group must contain at least one element.

The study of groups is known as GROUP THEORY. If there are a finite number of elements, the group is called a FINITE GROUP and the number of elements is called the ORDER of the group. A subset of a group that is CLOSED under the group operation and the inverse operation is called a SUBGROUP. SUBGROUPS are also groups, and many commonly encountered groups are in fact special subgroups of some more general larger group.

A basic example of a FINITE GROUP is the SYMMETRIC GROUP Σ_n, which is the group of PERMUTATIONS (or "under permutation") of n objects. The simplest infinite group is the set of INTEGERS under usual ADDITION. For continuous groups, one can consider the real numbers or the set of $n \times n$ invertible MATRICES. These last two are examples of LIE GROUPS.

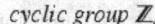

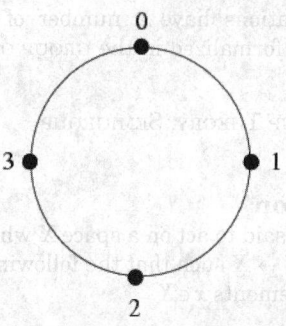

cyclic group $\mathbb{Z}_4$

One very common type of group is the CYCLIC GROUPS. This group is isomorphic to the group of integers (modulo n), is denoted Z_n, $\mathbb{Z}_n$, or $\mathbb{Z}/n\mathbb{Z}$, and is defined for every integer $n > 1$. It is CLOSED under addition, associative, and has unique inverses. The numbers from 0 to $n-1$ represent its elements, with the IDENTITY ELEMENT represented by 0, and the inverse of i is represented by $n-i$.

A map between two groups which preserves the identity and the group operation is called a HOMO-MORPHISM. If a homomorphism has an inverse which is also a homomorphism, then it is called an ISO-MORPHISM and the two groups are called isomorphic. Two groups which are isomorphic to each other are considered to be "the same" when viewed as abstract groups. For example, the group of rotations of a square, illustrated below, is the CYCLIC GROUP $\mathbb{Z}_4$.

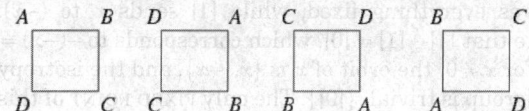

In general, a GROUP ACTION is when a group acts on a set, permuting its elements, so that the map from the group to the PERMUTATION GROUP of the set is a homomorphism. For example, the rotations of a square are a SUBGROUP of the PERMUTATIONS of its corners. One important GROUP ACTION for any group G is its action on itself by CONJUGATION. These are just some of the possible GROUP AUTOMORPHISMS. Another important kind of GROUP ACTION is a REPRE-SENTATION of a group, where the group acts on a VECTOR SPACE by INVERTIBLE LINEAR MAPS. When the FIELD of the VECTOR SPACE is the complex numbers, sometimes a representation is called a CG MODULE.

GROUP ACTIONS, and in particular representations, are very important in applications, not only to group theory, but also to physics and chemistry. Since a group can be thought of as an abstract mathematical object, the same group may arise in different contexts. It is therefore useful to think of a representation of the group as one particular incarnation of the group, which may also have other representations. An IRREDUCIBLE REPRESENTATION of a group is a representation for which there exists no UNITARY TRANSFORMATION which will transform the representation MATRIX into block diagonal form. The irreducible representations have a number of remarkable properties, as formalized in the GROUP ORTHOGONALITY THEOREM.

See also GROUP THEORY, SEMIGROUP

Group Action

A GROUP G is said to act on a space X when there is a map $\phi : G \times X \to X$ such that the following conditions hold for all elements $x \in X$.

1. $\phi(e, x) = x$ where e is the identity element of G.
2. $\phi(g, \phi(h, x)) = \phi(gh, x)$ for all $g, h \in G$.

In this case, G is called a TRANSFORMATION GROUP, X is a called a G-set, and ϕ is called the group action.

$$(5\ 7\ 9\ 3\ 4\ 6\ 8\ 2\ 0\ 1)$$

In a group action, a GROUP permutes the elements of X. The identity does nothing, while a composition of actions corresponds to the action of the composition. For example, as illustrated above, the SYMMETRIC GROUP S_{10} acts on the digits 0 to 9 by permutations.

For a given x, the set $\{gx\}$, where the group action moves x, is called the ORBIT of x. The SUBGROUP which fixes x is the ISOTROPY GROUP of x.

For example, the group $\mathbb{Z}_2 = \{[0], [1]\}$ acts on the real numbers by multiplication by $(-1)^n$. The identity leaves everything fixed, while $[1]$ sends x to $(-x)$. Note that $[1] \cdot [1] = [0]$, which corresponds to $-(-x) = x$. For $x \neq 0$, the orbit of x is $\{x, -x\}$, and the isotropy subgroup is trivial, $\{[0]\}$. The only FIXED POINT of this action is $x = 0$.

In a REPRESENTATION, a group acts by invertible LINEAR TRANSFORMATIONS of a VECTOR SPACE V. In fact, a representation is a GROUP HOMOMORPHISM from G to $GL(V)$, the GENERAL LINEAR GROUP of V. Some groups are described in a representation, such as the SPECIAL LINEAR GROUP, although they may have different representations.

Historically, the first group action studied was the action of the GALOIS GROUP on the roots of a POLYNOMIAL. However, there are numerous examples and applications of group actions in many branches of mathematics, including ALGEBRA, TOPOLOGY, GEOMETRY, NUMBER THEORY, and ANALYSIS, as well as the sciences, including chemistry and physics.

See also BLOCK (GROUP ACTION), EFFECTIVE ACTION, FREE ACTION, GALOIS GROUP, GROUP, ISOTROPY GROUP, MATRIX GROUP, ORBIT (GROUP), PRIMITIVE (GROUP ACTION), QUOTIENT SPACE (LIE GROUP), REPRESENTATION, TOPOLOGICAL GROUP, TRANSITIVE

References
Kawakubo, K. *The Theory of Transformation Groups.* Oxford, England: Oxford University Press, pp. 1–, 1987.

Group Convolution

The convolution of two COMPLEX-valued functions on a GROUP G is defined as

$$(a*b)(g) = \sum_{k \in G} a(k)b(k^{-1}g)$$

where the SUPPORT (set which is not zero) of each function is finite.

References
Weinstein, A. "Groupoids: Unifying Internal and External Symmetry." *Not. Amer. Math. Soc.* **43**, 744–52, 1996.

Group Direct Product

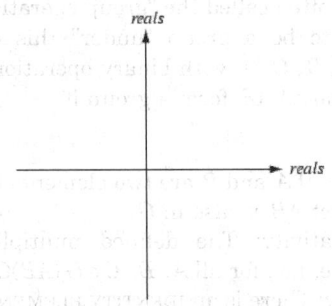

Given two GROUPS G and H, there are several ways to form a new group. The simplest is the direct product, denoted $G \times H$. As a set, the group direct product is the CARTESIAN PRODUCT of ordered pairs (g, h), and the group operation is componentwise, so

$$(g_1, h_1) \times (g_2, h_2) = (g_1 g_2, h_1 h_2).$$

For example, $\mathbb{R} \times \mathbb{R}$ is isomorphic to $\mathbb{R}^2$ under VECTOR ADDITION. In a similar fashion, one can take the direct product of any number of groups by taking the Cartesian product and operating componentwise.

Note that G is ISOMORPHIC to the SUBGROUP of elements g, e_H where e_H is the IDENTITY ELEMENT in H. Similarly, H can be realized as a SUBGROUP. The intersection of these two subgroups is the identity (e_G, e_H), and the two subgroups are NORMAL.

$$
\begin{array}{ccc}
X \to G & & X \to G \oplus H \to G \\
\downarrow & \Rightarrow & \downarrow \\
H & & H
\end{array}
$$

the universal property of a direct product;
X factors through $G \oplus H$.

Like the RING DIRECT PRODUCT, the group direct product has the UNIVERSAL PROPERTY that if any group X has a HOMOMORPHISM to G and a homomorphism to H, then these homomorphisms factor through $G \times H$ in a unique way.

If one has REPRESENTATIONS R_G of G and R_H of H, then there is a representation $R_G \otimes R_H$ sometimes called the EXTERNAL TENSOR PRODUCT, given by the TENSOR PRODUCT $\otimes$. In this case, the group CHARACTER satisfies

$$\chi(g \otimes h) = \chi R_G(g)\chi R_H(h).$$

See also CARTESIAN PRODUCT, EXTERNAL TENSOR PRODUCT, HOMOMORPHISM, REPRESENTATION, SUBGROUP, UNIVERSAL PROPERTY

References

Riesel, H. "The Direct Product of Two Given Groups." *Prime Numbers and Computer Methods for Factorization, 2nd ed.* Boston, MA: Birkhäuser, pp. 251–52, 1994.

Group Homomorphism

A group homomorphism is a map $f : G \to H$ between two groups such that

1. The group operation is preserved: $f(g_1 g_2) = f(g_1)f(g_2)$
2. The identity is mapped to the identity: $f(e_G) = e_H$,

where the product on the left-hand side is in G and on the right-hand side in H. Note that a homomorphism must preserve the inverse map because $f(g)f(g^{-1}) = f(gg^{-1}) = f(e_G) = e_H$, so $f(g)^{-1} = f(g^{-1})$.

In particular, the image of G is a SUBGROUP of H and the kernel, i.e., $f^{-1}(e_H)$ is a SUBGROUP of G. The kernel is actually a NORMAL SUBGROUP, as is the PREIMAGE of any NORMAL SUBGROUP of H. Hence, any homomorphism from a SIMPLE GROUP must be INJECTIVE.

See also HOMOMORPHISM, GROUP, NORMAL SUBGROUP, REPRESENTATION

Group Orthogonality Theorem

Let Γ be a representation for a GROUP of ORDER h, then

$$\sum_R \Gamma_i(R)_{mn}\Gamma_j(R)_{m'n'}{}^* = \frac{h}{\sqrt{l_i l_j}}\,\delta_{ij}\delta_{mm'}\delta_{nn'}.$$

The proof is nontrivial and may be found in Eyring *et al.* (1944).

See also CHARACTER (GROUP), GROUP, IRREDUCIBLE REPRESENTATION

References

Eyring, H.; Walker, J.; and Kimball, G. E. *Quantum Chemistry.* New York: Wiley, p. 371, 1944.

Group Representation

GROUP, IRREDUCIBLE REPRESENTATION, REPRESENTATION

Group Residue Theorem

If two groups are residual to a third, every group residual to one is residual to the other. The Gambier extension of this theorem states that if two groups are pseudoresidual to a third, then every group pseudoresidual to the first with an excess greater than or equal to the excess of the first minus the excess of the second is pseudoresidual to the second, with an excess ≥ 0.

References

Coolidge, J. L. *A Treatise on Algebraic Plane Curves.* New York: Dover, pp. 30–1, 1959.

Group Ring

The set of sums $\Sigma_x\, a_x x$ ranging over a multiplicative GROUP and a_i are elements of a FIELD with all but a finite number of $a_i = 0$. Group rings are GRADED ALGEBRAS.

See also GRADED ALGEBRA

Group Theory

The study of GROUPS. Gauss developed but did not publish parts of the mathematics of group theory, but Galois is generally considered to have been the first to develop the theory. Group theory is a powerful formal method for analyzing abstract and physical systems in which SYMMETRY is present and has surprising importance in physics, especially quantum mechanics.

See also FINITE GROUP, GROUP, HIGHER DIMENSIONAL GROUP THEORY, PLETHYSM, SYMMETRY

References

Alperin, J. L. and Bell, R. B. *Groups and Representations.* New York: Springer-Verlag, 1995.
Arfken, G. "Introduction to Group Theory." §4.8 in *Mathematical Methods for Physicists, 3rd ed.* Orlando, FL: Academic Press, pp. 237–76, 1985.
Burnside, W. *Theory of Groups of Finite Order, 2nd ed.* New York: Dover, 1955.
Burrow, M. *Representation Theory of Finite Groups.* New York: Dover, 1993.
Carmichael, R. D. *Introduction to the Theory of Groups of Finite Order.* New York: Dover, 1956.
Conway, J. H.; Curtis, R. T.; Norton, S. P.; Parker, R. A.; and Wilson, R. A. *Atlas of Finite Groups: Maximal Subgroups and Ordinary Characters for Simple Groups.* Oxford, England: Clarendon Press, 1985.
Cotton, F. A. *Chemical Applications of Group Theory, 3rd ed.* New York: Wiley, 1990.
Dixon, J. D. *Problems in Group Theory.* New York: Dover, 1973.
Farmer, D. *Groups and Symmetry.* Providence, RI: Amer. Math. Soc., 1995.
Grossman, I. and Magnus, W. *Groups and Their Graphs.* Washington, DC: Math. Assoc. Amer., 1965.
Hamermesh, M. *Group Theory and Its Application to Physical Problems.* New York: Dover, 1989.
Lomont, J. S. *Applications of Finite Groups.* New York: Dover, 1987.
Magnus, W.; Karrass, A.; and Solitar, D. *Combinatorial Group Theory: Presentations of Groups in Terms of Generators and Relations.* New York: Dover, 1976.
Mirman, R. *Group Theory: An Intuitive Approach.* River Edge, NJ: World Scientific, 1995.

Robinson, D. J. S. *A Course in the Theory of Groups, 2nd ed.* New York: Springer-Verlag, 1995.

Rose, J. S. *A Course on Group Theory.* New York: Dover, 1994.

Rotman, J. J. *An Introduction to the Theory of Groups, 4th ed.* New York: Springer-Verlag, 1995.

Scott, W. R. *Group Theory.* New York: Dover, 1987.

Weisstein, E. W. "Groups." MATHEMATICA NOTEBOOK GROUPS.M.

Weisstein, E. W. "Books about Group Theory." http://www.treasure-troves.com/books/GroupTheory.html.

Weyl, H. *The Classical Groups: Their Invariants and Representations.* Princeton, NJ: Princeton University Press, 1997.

Wybourne, B. G. *Classical Groups for Physicists.* New York: Wiley, 1974.

Groupoid

There are at least two definitions of "groupoid" currently in use.

The first type of groupoid is an algebraic structure on a SET with a BINARY OPERATOR. The only restriction on the operator is closure (i.e., applying the BINARY OPERATOR to two elements of a given set S returns a value which is itself a member of S). Associativity, commutativity, etc., are not required (Rosenfeld 1968, pp. 88–03). A groupoid can be empty. The numbers of nonisomorphic groupoids of this type having n elements are 1, 1, 10, 3330, 178981952, ... (Sloane's A001329), and the numbers of nonisomorphic and nonantiisomorphic groupoids are 1, 7, 1734, 89521056, ... (Sloane's A001424). An associative groupoid is called a SEMIGROUP.

The second type of groupoid is an algebraic structure first defined by Brandt (1926) and also known as a VIRTUAL GROUP. A groupoid with base B is a set G with mappings α and β from G onto B and a partially defined binary operation $(g, h) \mapsto gh$, satisfying the following four conditions:

1. gh is defined only when $\beta(g) = \alpha(h)$ for certain maps α and β from G onto $\mathbb{R}^2$ with $\alpha : (x, \gamma, y) \mapsto x$ and $\beta : (x, \gamma, y) \mapsto y$
2. ASSOCIATIVITY: If either $(gh)k$ or $g(hk)$ is defined, then so is the other and $(gh)k = g(hk)$.
3. For each g in G, there are left and right IDENTITY ELEMENTS λ_g and ρ_g such that $\lambda_g g = g = g\rho_g$.
4. Each g in G has an inverse g^{-1} for which $gg^{-1} = \lambda g$ and $g^{-1}g = \rho g$

(Weinstein 1996). A groupoid is a small CATEGORY with every morphism invertible.

See also BINARY OPERATOR, INVERSE SEMIGROUP, LIE ALGEBROID, LIE GROUPOID, MONOID, QUASIGROUP, SEMIGROUP, TOPOLOGICAL GROUPOID

References

Brandt, W. "Über eine Verallgemeinerung des Gruppengriffes." *Math. Ann.* **96**, 360–66, 1926.

Brown, R. "From Groups to Groupoids: A Brief Survey." *Bull. London Math. Soc.* **19**, 113–34, 1987.

Brown, R. *Topology: A Geometric Account of General Topology, Homotopy Types, and the Fundamental Groupoid.* New York: Halsted Press, 1988.

Higgins, P. J. *Notes on Categories and Groupoids.* London: Van Nostrand Reinhold, 1971.

Ramazan, B. "Groupoids Home Page." http://www.labomath.univ-orleans.fr/descriptions/ramazan/groupoides.html.

Rosenfeld, A. *An Introduction to Algebraic Structures.* New York: Holden-Day, 1968.

Sloane, N. J. A. Sequences A001329/M4760 and A001424 in "An On-Line Version of the Encyclopedia of Integer Sequences." http://www.research.att.com/~njas/sequences/eisonline.html.

Weinstein, A. "Groupoids: Unifying Internal and External Symmetry." *Not. Amer. Math. Soc.* **43**, 744–52, 1996.

Growth

A general term which refers to an increase (or decrease in the case of the oxymoron "negative growth") in a given quantity.

See also LAW OF GROWTH, LIFE EXPECTANCY, POPULATION GROWTH

Growth Function

BLOCK GROWTH

Growth Spiral

LOGARITHMIC SPIRAL

Grünbaum Graph

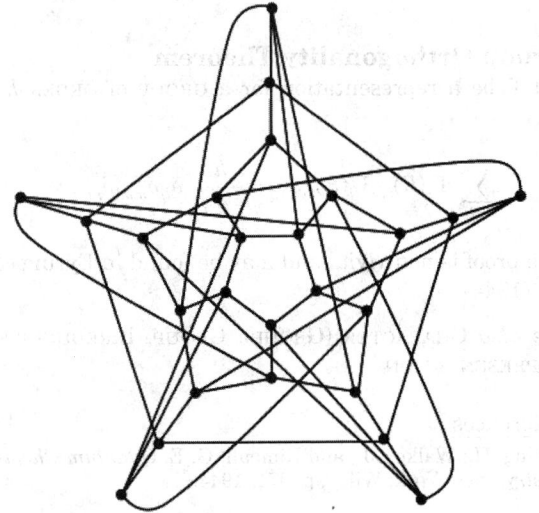

Grünbaum conjectured that for every $m > 1$, $n > 2$, there exists an m-regular, m-chromatic graph of GIRTH at least n. This result is trivial for $n = 2$ and $m = 2, 3$, but only two other such graphs are known: the Grünbaum graph illustrated above, and the CHVÁTAL GRAPH.

See also CHVÁTAL GRAPH

Grundy's Game

References

Bondy, J. A. and Murty, U. S. R. *Graph Theory with Applications.* New York: North Holland, pp. 241–42, 1976.

Grünbaum, B. "A Problem in Graph Coloring." *Amer. Math. Monthly* **77**, 1088–092, 1970.

Grundy's Game

A special case of NIM played by the following rules. Given a heap of size n, two players alternately select a heap and divide it into two unequal heaps. A player loses when he cannot make a legal move because all heaps have size 1 or 2. Flammenkamp gives a table of the extremal SPRAGUE-GRUNDY VALUES for this game. The first few values of Grundy's game are 0, 0, 0, 1, 0, 2, 1, 0, 2, ... (Sloane's A002188).

References

Sloane, N. J. A. Sequences A002188/M0044 in "An On-Line Version of the Encyclopedia of Integer Sequences." http://www.research.att.com/~njas/sequences/eisonline.html.

Grundy-Sprague Number

NIM-VALUE

G-Space

A G-space is a special type of HAUSDORFF SPACE. Consider a point x and a HOMEOMORPHISM of an open NEIGHBORHOOD V of x onto an OPEN SET of $\mathbb{R}^n$. Then a space is a G-space if, for any two such NEIGHBORHOODS v' and v'', the images of $v' \cup v''$ under the different HOMEOMORPHISMS are ISOMETRIC. If $n = 2$, the HOMEOMORPHISMS need only be conformal (but not necessarily orientation-preserving).

Hsiang (2000, p. 1) terms a space X with a topological (resp. differentiable, linear) transformation of a given GROUP G a topological (resp. differentiable, linear) G-space.

See also GREEN SPACE

References

Hsiang, W. Y. *Lectures on Lie Groups.* Singapore: World Scientific, p. 1, 2000.

G-Transform

The G-transform of a function $f(x)$ is defined by the integral

$$(Gf)(x) = \left(G_{pq}^{mn} \left| \begin{matrix} (a_p) \\ (b_q) \end{matrix} \right| f(t) \right)(x) \tag{1}$$

$$= \frac{1}{2\pi i} \int_\sigma \Gamma \left[\begin{matrix} (b_m) + s, & 1 - (a)_n - s \\ \left(a_p^{n+1}\right) + s, & 1 - \left(b_q^{m+1}\right) - s \end{matrix} \right] f^*(s) x^{-s} ds, \tag{2}$$

where G_{pq}^{mn} is MEIJER'S G-FUNCTION,

$$\Gamma \left[\begin{matrix} (b_m) + s, & 1 - (a_n) - s \\ \left(a_p^{n+1}\right) + s, & 1 - \left(b_q^{m+1}\right) - s \end{matrix} \right]$$

$$= \Gamma \left[\begin{matrix} b_1 + s, & \dots, & b_m + s, & 1 - a_1 - s, & \dots, & 1 - a_n - s \\ a_{n+1} + s, & \dots, & a_p + s, & 1 - b_{m+1} - s, & \dots, & 1 - b_q - s \end{matrix} \right] \tag{3}$$

$$= \frac{\prod_{j=1}^m \Gamma(b_j + s) \prod_{j=1}^n \Gamma(1 - a_j - s)}{\prod_{j=n+1}^p \Gamma(a_j + s) \prod_{j=m+1}^q \Gamma(1 - bj - s)}, \tag{4}$$

$f^*(s)$ is the MELLIN TRANSFORM of a function $f(x)$, σ is the CONTOUR $\sigma = \{1/2 - i\infty, \ 1/2 + i\infty\}$, $(a_n) = a_1, a_2, \dots, a_n$, $(a_p^{n+1}) = a_{n+1}, a_{n+2}, \dots, a_p$, $(b_m) = b_1, \dots b_m$, $(b_q^{m+1}) = b_{m+1}, \dots, b_q$, and the components of the vectors (a_p) and (b_q) are complex numbers satisfying the conditions $\Re[a_p] \neq 1/2, 3/2, 5/2, \dots$ and $\Re[b_q] \neq -1/2, -3/2, -5/2, \dots$.

See also MEIJER'S G-FUNCTION, W-TRANSFORM

References

Samko, S. G.; Kilbas, A. A.; and Marichev, O. I. "Definition of the G-Transform. The Spaces $\mathfrak{M}_{c,\gamma}^{-1}$ and $L_2^{(c,\gamma)}$ and Their Characterization." §36.1 in *Fractional Integrals and Derivatives.* Yverdon, Switzerland: Gordon and Breach, pp. 704–09, 1993.

Gudermannian Function

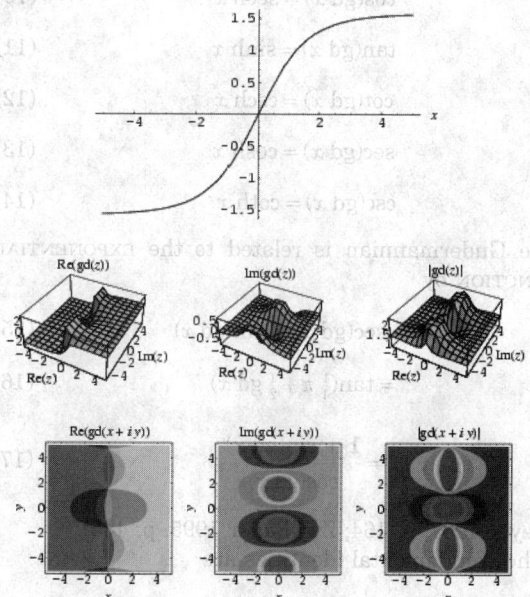

The ODD FUNCTION denoted either $\gamma(x)$ or gd(x) which arises in the inverse equations for the MERCATOR PROJECTION. $\phi(y) = $ gd(y) expresses the LATITUDE ϕ in terms of the vertical position y in this projection, so the Gudermannian function is defined by

$$\mathrm{gd}(x) \equiv \int_0^x \frac{dt}{\cosh t} \tag{1}$$

$$= \tan^{-1}(\sinh x) \tag{2}$$

$$2\tan^{-1}(e^x) - \tfrac{1}{2}\pi \tag{3}$$

The INVERSE FUNCTION of the Gudermannian function $y = \mathrm{gd}^{-1}\phi$ gives the vertical position y in the MERCATOR PROJECTION in terms of the LATITUDE ϕ, so

$$\mathrm{gd}^{-1}(x) \equiv \int_0^x \frac{dt}{\cos t} \tag{4}$$

$$= \ln[\tan(\tfrac{1}{4}\pi + \tfrac{1}{2}x)] \tag{5}$$

$$= \ln(\sec x + \tan x). \tag{6}$$

The derivatives of the function and its inverse are given by

$$\frac{d}{dx}\,\mathrm{gd}(x) = \mathrm{sech}\,x \tag{7}$$

$$\frac{d}{dx}\,\mathrm{gd}^{-1}(x) = \sec x. \tag{8}$$

The Gudermannian connects the TRIGONOMETRIC and HYPERBOLIC FUNCTIONS via

$$\sin(\mathrm{gd}\,x) = \tanh x \tag{9}$$

$$\cos(\mathrm{gd}\,x) = \mathrm{sech}\,x \tag{10}$$

$$\tan(\mathrm{gd}\,x) = \sinh x \tag{11}$$

$$\cot(\mathrm{gd}\,x) = \mathrm{csch}\,x \tag{12}$$

$$\sec(\mathrm{gd}\,x) = \cosh x \tag{13}$$

$$\csc(\mathrm{gd}\,x) = \coth x. \tag{14}$$

The Gudermannian is related to the EXPONENTIAL FUNCTION by

$$e^x = \sec(\mathrm{gd}\,x) = \tan(\mathrm{gd}\,x) \tag{15}$$

$$= \tan(\tfrac{1}{4}\pi + \tfrac{1}{2}\,\mathrm{gd}\,x) \tag{16}$$

$$= \frac{1 + \sin(\mathrm{gd}\,x)}{\cos(\mathrm{gd}\,x)} \tag{17}$$

(Beyer 1987, p. 164; Zwillinger 1995, p. 485). Other fundamental identities are

$$\tanh(\tfrac{1}{2}x) = \tan(\tfrac{1}{2}\,\mathrm{gd}\,x) \tag{18}$$

$$i\,\mathrm{gd}^{-1}\,x = \mathrm{gd}^{-1}(ix).$$

If $\mathrm{gd}(x + iy) = a + ib$, then

$$\tan a = \frac{\sinh x}{\cos y} \tag{19}$$

$$\tanh b = \frac{\sin y}{\cosh x} \tag{20}$$

$$\tanh x = \frac{\sin a}{\cosh b} \tag{21}$$

$$\tan y = \frac{\sin b}{\cosh a} \tag{22}$$

(Beyer 1987, p. 164; Zwillinger 1995, p. 485).

See also EXPONENTIAL FUNCTION, HYPERBOLIC FUNCTIONS, HYPERBOLIC SECANT, MERCATOR PROJECTION, SECANT, TRACTRIX, TRIGONOMETRIC FUNCTIONS

References

Beyer, W. H. "Gudermannian Function." *CRC Standard Mathematical Tables, 28th ed.* Boca Raton, FL: CRC Press, p. 164, 1987.
Zwillinger, D. (Ed.). "Gudermannian Function." §6.9 in *CRC Standard Mathematical Tables and Formulae.* Boca Raton, FL: CRC Press, pp. 484–86, 1995.

Guldinus Theorem
PAPPUS'S CENTROID THEOREM

Gumbel's Distribution
A special case of the FISHER-TIPPETT DISTRIBUTION with $a = 0$, $b = 1$. The MEAN, VARIANCE, SKEWNESS, and KURTOSIS are

$$\mu = \gamma$$

$$\sigma^2 = \tfrac{1}{6}\pi^2$$

$$\gamma_1 = \frac{12\sqrt{6}\zeta(3)}{\pi^3}$$

$$\gamma_2 = \tfrac{12}{5}.$$

where γ is the EULER-MASCHERONI CONSTANT, and $\zeta(3)$ is APÉRY'S CONSTANT.

See also FISHER-TIPPETT DISTRIBUTION

Guthrie's Problem
The problem of deciding if four colors are sufficient to color any map on a PLANE or SPHERE.

See also COLORING, FOUR-COLOR THEOREM

Gutschoven's Curve
KAPPA CURVE

Guy's Conjecture
Guy's conjecture, which has not yet been proven or disproven, states that the CROSSING NUMBER for a

COMPLETE GRAPH of order n is

$$\frac{1}{4}\left\lfloor\frac{n}{2}\right\rfloor\left\lfloor\frac{n-1}{2}\right\rfloor\left\lfloor\frac{n-2}{2}\right\rfloor\left\lfloor\frac{n-3}{2}\right\rfloor,$$

where $\lfloor x \rfloor$ is the FLOOR FUNCTION, which can be rewritten

$$\begin{cases} \frac{1}{64}n(n-2)^2(n-4) & \text{for } n \text{ even} \\ \frac{1}{64}(n-1)^2(n-3)^2 & \text{for } n \text{ odd.} \end{cases}$$

The first few values are 0, 0, 0, 0, 1, 3, 9, 18, 36, 60, ... (Sloane's A000241).

See also CROSSING NUMBER (GRAPH)

References
Sloane, N. J. A. Sequences A000241/M2772 in "An On-Line Version of the Encyclopedia of Integer Sequences." http://www.research.att.com/~njas/sequences/eisonline.html.

Gyrate Bidiminished Rhombicosidodecahedron

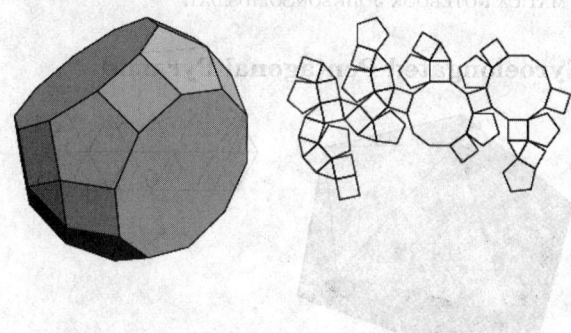

JOHNSON SOLID J_{82}.

References
Weisstein, E. W. "Johnson Solids." MATHEMATICA NOTEBOOK JOHNSONSOLIDS.M.
Weisstein, E. W. "Johnson Solid Netlib Database." MATHEMATICA NOTEBOOK JOHNSONSOLIDS.DAT.

Gyrate Rhombicosidodecahedron

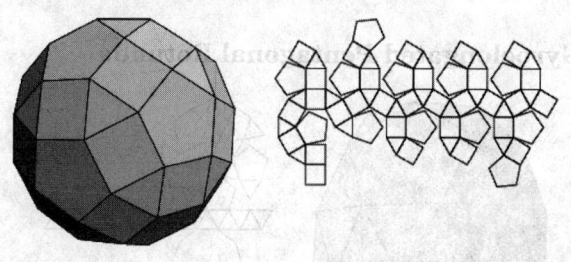

JOHNSON SOLID J_{72}.

References
Weisstein, E. W. "Johnson Solids." MATHEMATICA NOTEBOOK JOHNSONSOLIDS.M.
Weisstein, E. W. "Johnson Solid Netlib Database." MATHEMATICA NOTEBOOK JOHNSONSOLIDS.DAT.

Gyrobicupola

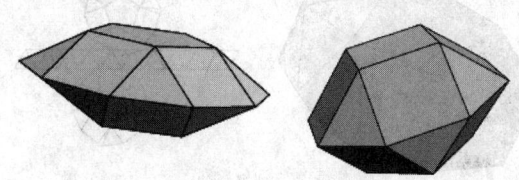

A BICUPOLA in which the bases are in opposite orientations.

See also BICUPOLA, PENTAGONAL GYROBICUPOLA, SQUARE GYROBICUPOLA

Gyrobifastigium

JOHNSON SOLID J_{26}, consisting of two joined triangular PRISMS.

Gyrobirotunda

A BIROTUNDA in which the bases are in opposite orientations.

Gyrocupolarotunda

A CUPOLAROTUNDA in which the bases are in opposite orientations.

See also ORTHOCUPOLAROTUNDA

Gyroelongated Cupola

A n-gonal CUPOLA adjoined to a $2n$-gonal ANTIPRISM.

See also GYROELONGATED PENTAGONAL CUPOLA, GYROELONGATED SQUARE CUPOLA, GYROELONGATED TRIANGULAR CUPOLA

Gyroelongated Dipyramid

GYROELONGATED PYRAMID, GYROELONGATED SQUARE DIPYRAMID

Gyroelongated Pentagonal Bicupola

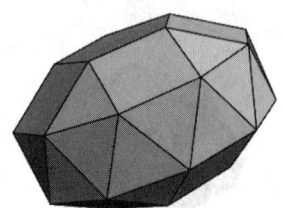
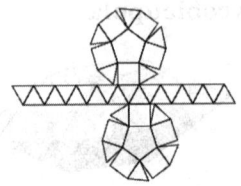

JOHNSON SOLID J_{46}, which consists of a PENTAGONAL ROTUNDA adjoined to a decagonal ANTIPRISM.

Gyroelongated Pentagonal Birotunda

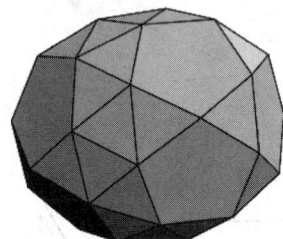
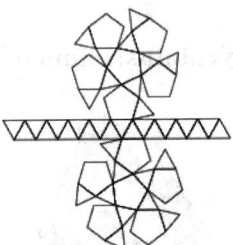

JOHNSON SOLID J_{48}.

References

Weisstein, E. W. "Johnson Solids." MATHEMATICA NOTEBOOK JOHNSONSOLIDS.M.

Weisstein, E. W. "Johnson Solid Netlib Database." MATHEMATICA NOTEBOOK JOHNSONSOLIDS.DAT.

Gyroelongated Pentagonal Cupola

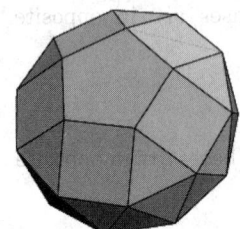
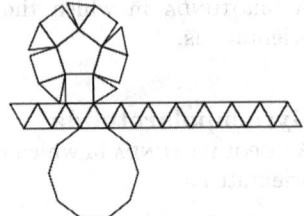

JOHNSON SOLID J_{24}.

References

Weisstein, E. W. "Johnson Solids." MATHEMATICA NOTEBOOK JOHNSONSOLIDS.M.

Weisstein, E. W. "Johnson Solid Netlib Database." MATHEMATICA NOTEBOOK JOHNSONSOLIDS.DAT.

Gyroelongated Pentagonal Cupolarotunda

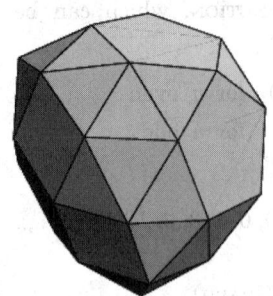
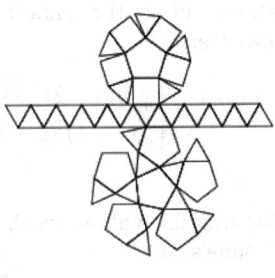

JOHNSON SOLID J_{47}.

References

Weisstein, E. W. "Johnson Solids." MATHEMATICA NOTEBOOK JOHNSONSOLIDS.M.

Weisstein, E. W. "Johnson Solid Netlib Database." MATHEMATICA NOTEBOOK JOHNSONSOLIDS.DAT.

Gyroelongated Pentagonal Pyramid

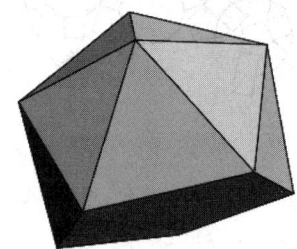
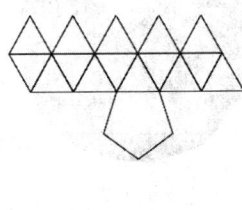

JOHNSON SOLID J_{11}.

References

Weisstein, E. W. "Johnson Solids." MATHEMATICA NOTEBOOK JOHNSONSOLIDS.M.

Weisstein, E. W. "Johnson Solid Netlib Database." MATHEMATICA NOTEBOOK JOHNSONSOLIDS.DAT.

Gyroelongated Pentagonal Rotunda

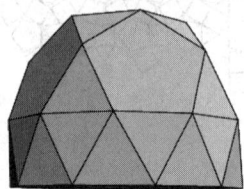
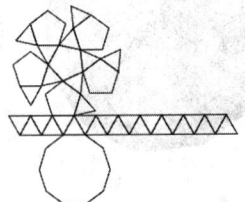

JOHNSON SOLID J_{25}.

References

Weisstein, E. W. "Johnson Solids." MATHEMATICA NOTEBOOK JOHNSONSOLIDS.M.

Weisstein, E. W. "Johnson Solid Netlib Database." MATHEMATICA NOTEBOOK JOHNSONSOLIDS.DAT.

Gyroelongated Pyramid

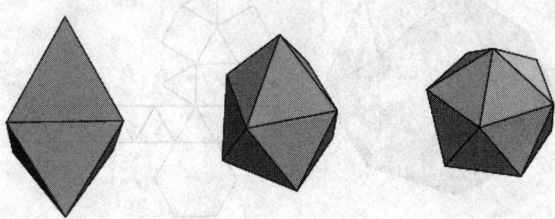

An n-gonal pyramid adjoined to the top of an n-gonal ANTIPRISM. In the 3-gonal gyroelongated pyramid, the pyramid and lateral antiprism are coplanar. However, the 4-gonal and 5-gonal gyroelongated pyramids correspond to JOHNSON SOLIDS J_{10} and J_{11}, respectively.

See also ANTIPRISM, ELONGATED PYRAMID, GYROELONGATED DIPYRAMID, GYROELONGATED PENTAGONAL PYRAMID, GYROELONGATED SQUARE DIPYRAMID, GYROELONGATED SQUARE PYRAMID

Gyroelongated Rotunda

GYROELONGATED PENTAGONAL ROTUNDA

Gyroelongated Square Bicupola

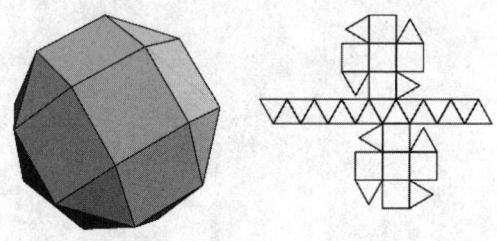

JOHNSON SOLID J_{45}.

References

Weisstein, E. W. "Johnson Solids." MATHEMATICA NOTEBOOK JOHNSONSOLIDS.M.

Weisstein, E. W. "Johnson Solid Netlib Database." MATHEMATICA NOTEBOOK JOHNSONSOLIDS.DAT.

Gyroelongated Square Cupola

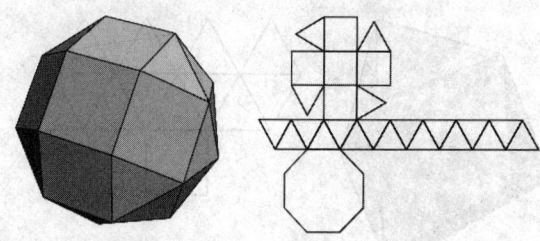

JOHNSON SOLID J_{23}.

References

Weisstein, E. W. "Johnson Solids." MATHEMATICA NOTEBOOK JOHNSONSOLIDS.M.

Weisstein, E. W. "Johnson Solid Netlib Database." MATHEMATICA NOTEBOOK JOHNSONSOLIDS.DAT.

Gyroelongated Square Dipyramid

One of the eight convex DELTAHEDRA built up from 16 equilateral triangles. It consists of two oppositely faced SQUARE PYRAMIDS rotated 45° to each other and separated by a 4-ANTIPRISM. It is JOHNSON SOLID J_{17}.

If the centroid is at the origin and the sides are of unit length, the equations of the 4-ANTIPRISM give height of the middle points as $\pm 2^{-5/4}$. Adding the height of the SQUARE PYRAMIDS gives apex heights of $\pm(2^{-5/4} + 2^{-1/2})$. The SURFACE AREA and VOLUME of the solid are

$$S = 4\sqrt{3}$$

$$V = \frac{2^{1/4}}{3}(1 + \sqrt{2} + 2^{1/4}).$$

See also ANTIPRISM, DELTAHEDRON, SNUB DISPHENOID, SQUARE PYRAMID

Gyroelongated Square Pyramid

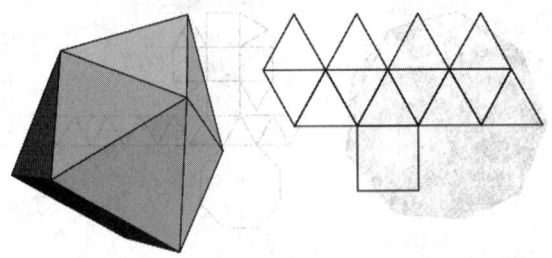

JOHNSON SOLID J_{10}.

References

Weisstein, E. W. "Johnson Solids." MATHEMATICA NOTEBOOK JOHNSONSOLIDS.M.
Weisstein, E. W. "Johnson Solid Netlib Database." MATHEMATICA NOTEBOOK JOHNSONSOLIDS.DAT.

Gyroelongated Triangular Bicupola

JOHNSON SOLID J_{44}.

One of the convex DELTAHEDRA, built up from 16 equilateral triangles. It consists of two opposing fused square pyramids, rotated 45° to each other and separated by a 4-ANTIPRISM. It is JOHNSON SOLID $J_{...}$

If the center is at the origin and the sides are of unit length, the equations of the 4 antiprism plus height of the middle points as $\pm 2^{-5/4}$. Adding the height of the square pyramids gives apex height of $(12^{1/4} + 2^{-5/4})$. The surface area and volume of the solid are

$$S = 4\sqrt{3}$$

$$V = \frac{2}{3}\left(1 + \sqrt{2} + 2\cdot4\right)$$

See also ANTIPRISM, DELTAHEDRON, SQUARE PYRAMID, SQUARE PYRAMID

References

Weisstein, E. W. "Johnson Solids." MATHEMATICA NOTEBOOK JOHNSONSOLIDS.M.
Weisstein, E. W. "Johnson Solid Netlib Database." MATHEMATICA NOTEBOOK JOHNSONSOLIDS.DAT.

Gyroelongated Triangular Cupola

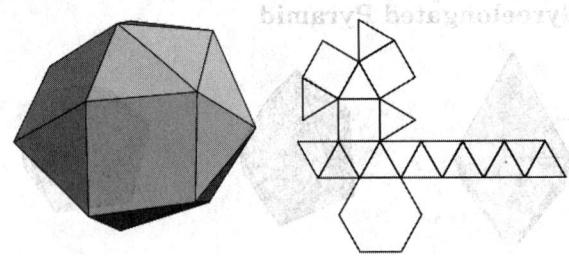

JOHNSON SOLID J_{22}.

References

Weisstein, E. W. "Johnson Solids." MATHEMATICA NOTEBOOK JOHNSONSOLIDS.M.
Weisstein, E. W. "Johnson Solid Netlib Database." MATHEMATICA NOTEBOOK JOHNSONSOLIDS.DAT.

Gyroid

An infinitely connected periodic MINIMAL SURFACE containing no straight lines.

See also MINIMAL SURFACE

References

Osserman, R. Frontispiece to *A Survey of Minimal Surfaces.* New York: Dover, 1986.

H

HA Measurement

INNER QUERMASS

Haar Condition

This entry contributed by RONALD M. AARTS

A set of VECTORS in n-space is said to satisfy the Haar condition if every set of n vectors is LINEARLY INDEPENDENT (Cheney 1999). Expressed otherwise, each selection of n vectors from such a set is a basis for n-space. A system of functions satisfying the Haar condition is sometimes termed a Tchebycheff system (Cheney 1999).

References

Cheney, E. W. *Introduction to Approximation Theory, 2nd ed.* Providence, RI: Amer. Math. Soc., 1999.

Haar Function

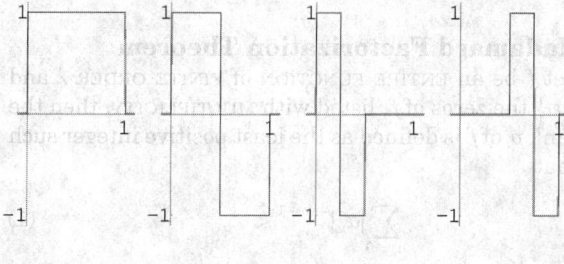

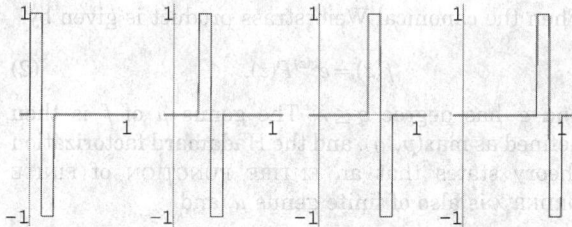

Define

$$\psi(x) \equiv \begin{cases} 1 & 0 \leq x \leq \frac{1}{2} \\ -1 & \frac{1}{2} \leq x \leq 1 \\ 0 & \text{otherwise} \end{cases} \quad (1)$$

and

$$\psi_{jk}(x) \equiv \psi(2^j x - k), \quad (2)$$

---

where the FUNCTIONS plotted above are

$$\psi_{00} = \psi(x)$$
$$\psi_{10} = \psi(2x)$$
$$\psi_{11} = \psi(2x - 1)$$
$$\psi_{20} = \psi(4x)$$
$$\psi_{21} = \psi(4x - 1)$$
$$\psi_{22} = \psi(4x - 2)$$
$$\psi_{23} = \psi(4x - 3).$$

Then a FUNCTION $f(x)$ can be written as a series expansion by

$$f(x) = c_0 + \sum_{j=0}^{\infty} \sum_{k=0}^{2^j - 1} c_{jk} \psi_{jk}(x). \quad (3)$$

The FUNCTIONS ψ_{jk} and ψ are all ORTHOGONAL in $[0, 1]$, with

$$\int_0^1 \phi(x)\phi_{jk}(x)\, dx = 0 \quad (4)$$

$$\int_0^1 \phi_{jk}(x)\phi_{lm}(x)\, dx = 0. \quad (5)$$

These functions can be used to define WAVELETS. Let a FUNCTION be defined on n intervals, with n a POWER of 2. Then an arbitrary function can be considered as an n-VECTOR $\mathbf{f}$, and the COEFFICIENTS in the expansion $\mathbf{b}$ can be determined by solving the MATRIX EQUATION

$$\mathbf{f} = \mathsf{W}_n \mathbf{b} \quad (6)$$

for $\mathbf{b}$, where W is the MATRIX of ψ basis functions. For example, the fourth-order Haar function WAVELET MATRIX is given by

$$W_4 = \begin{bmatrix} 1 & 1 & 1 & 0 \\ 1 & 1 & -1 & 0 \\ 1 & -1 & 0 & 1 \\ 1 & -1 & 0 & -1 \end{bmatrix}$$

$$= \begin{bmatrix} 1 & 1 & 0 & 0 \\ 1 & -1 & 0 & 0 \\ 0 & 0 & 1 & 1 \\ 0 & 0 & 1 & -1 \end{bmatrix} \begin{bmatrix} 1 & 0 & 0 & 0 \\ 0 & 0 & 1 & 0 \\ 0 & 1 & 0 & 0 \\ 0 & 0 & 0 & 1 \end{bmatrix} \begin{bmatrix} 1 & 1 & 0 & 0 \\ 1 & -1 & 0 & 0 \\ 0 & 0 & 1 & 0 \\ 0 & 0 & 0 & 1 \end{bmatrix}.$$

See also WAVELET, WAVELET MATRIX, WAVELET TRANSFORM

References

Haar, A. "Zur Theorie der orthogonalen Funktionensysteme." *Math. Ann.* **69**, 331–71, 1910.
Strang, G. "Wavelet Transforms Versus Fourier Transforms." *Bull. Amer. Math. Soc.* **28**, 288–05, 1993.

Haar Integral

The INTEGRAL associated with the HAAR MEASURE.

See also HAAR MEASURE

Haar Measure

Any locally compact Hausdorff topological group has a unique (up to scalars) NONZERO left invariant measure which is finite on compact sets. If the group is Abelian or compact, then this measure is also right invariant and is known as the Haar measure.

Haar Transform

A 1-D transform which makes use of the HAAR FUNCTIONS.

See also H-TRANSFORM, HAAR FUNCTION

References

Haar, A. "Zur Theorie der orthogonalen Funktionensysteme." *Math. Ann.* **69**, 331–71, 1910.

Haberdasher's Problem

With four cuts, DISSECT an EQUILATERAL TRIANGLE into a SQUARE. First proposed by Dudeney (1907) and discussed in Gardner (1961, p. 34), Stewart (1987, p. 169), and Wells (1991, pp. 61–2). The solution can be hinged so that the three pieces collapse into either the TRIANGLE or the SQUARE. Two of the hinges bisect sides of the triangle, while the third hinge and the corner of the large piece on the base cut the base in the approximate ratio $0.982 : 2 : 1.018$.

See also DISSECTION

References

Dudeney, H. E. *Amusements in Mathematics.* New York: Dover, p. 27, 1958.
Gardner, M. "Mathematical Games: About Henry Ernest Dudeney, A Brilliant Creator of Puzzles." *Sci. Amer.* **198**, 108–12, Jun. 1958.
Gardner, M. *The Second Scientific American Book of Mathematical Puzzles & Diversions: A New Selection.* New York: Simon and Schuster, 1961.
Stewart, I. *The Problems of Mathematics, 2nd ed.* Oxford, England: Oxford University Press, 1987.
Wells, D. *The Penguin Dictionary of Curious and Interesting Geometry.* London: Penguin, pp. 61–2, 1991.

Habiro Move

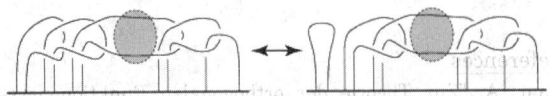

A KNOT MOVE illustrated above. Two knots cannot be distinguished using VASSILIEV INVARIANTS of order ≤

n IFF they are related by a sequence of such moves (Habiro 2000). There is a correspondence between the Habiro move and solution of the BAGUENAUDIER puzzle (Przytycki and Sikora 2000).

See also BAGUENAUDIER, KNOT MOVE

References

Habiro, K. "Claspers and Finite Type Invariants of Links." *Geom. Topol.* **4**, 1–3, 2000.
Przytycki, J. H. and Sikora, A. S. Topological Insights from the Chinese Rings. 21 Jul 2000. http://xxx.lanl.gov/abs/math.GT/0007134/.

Hadamard Design

A SYMMETRIC BLOCK DESIGN $(4n + 3, 2n + 1, n)$ which is equivalent to a HADAMARD MATRIX of order $4n + 4$. It is conjectured that Hadamard designs exist for all integers $n > 0$, but this has not yet been proven. This elusive proof (or disproof) remains one of the most important unsolved problems in COMBINATORICS.

See also HADAMARD MATRIX, SYMMETRIC BLOCK DESIGN

References

Dinitz, J. H. and Stinson, D. R. "A Brief Introduction to Design Theory." Ch. 1 in *Contemporary Design Theory: A Collection of Surveys* (Ed. J. H. Dinitz and D. R. Stinson). New York: Wiley, pp. 1–2, 1992.

Hadamard Factorization Theorem

Let f be an ENTIRE FUNCTION of FINITE ORDER λ and $\{a_j\}$ the zeros of f, listed with MULTIPLICITY, then the rank p of f is defined as the least positive integer such that

$$\sum_{a_n \neq 0} |a_n|^{-(p+1)} < \infty. \qquad (1)$$

Then the canonical Weierstrass product is given by

$$f(z) = e^{g(z)} P(z), \qquad (2)$$

and g has degree $q \leq \lambda$. The genus μ of f is then defined as $\max(p, q)$, and the Hadamard factorization theory states that an ENTIRE FUNCTION of FINITE ORDER λ is also of finite genus μ, and

$$\mu \leq \lambda. \qquad (3)$$

References

Krantz, S. G. "The Hadamard Factorization Theorem." §9.3.5 in *Handbook of Complex Analysis.* Boston, MA: Birkhäuser, pp. 121–22, 1999.

Hadamard Gap Theorem

OSTROWSKI-HADAMARD GAP THEOREM

Hadamard Matrix

A class of SQUARE MATRIX invented by Sylvester (1867) under the name of ANALLAGMATIC PAVEMENT. A Hadamard matrix is a SQUARE MATRIX containing only 1s and -1s such that when any two columns or rows are placed side by side, HALF the adjacent cells are the same SIGN and half the other (excepting from the count an L-shaped "half-frame" bordering the matrix on two sides which is composed entirely of 1s). When viewed as pavements, cells with 1s are colored black and those with -1s are colored white. Therefore, the $n \times n$ Hadamard matrix H_n must have $n(n-1)/2$ white squares (-1s) and $n(n+1)/2$ black squares (1s).

A Hadamard matrix of order n is a solution to HADAMARD'S MAXIMUM DETERMINANT PROBLEM, i.e., has the maximum possible DETERMINANT (in absolute value) of any $n \times n$ COMPLEX MATRIX with elements $|a_{ij}| \leq 1$ (Brenner 1972), namely $n^{n/2}$. An equivalent definition of the Hadamard matrices is given by

$$H_n H_n^T = n I_n, \tag{1}$$

where I_n is the $n \times n$ IDENTITY MATRIX. A Hadamard matrix of order $4n+4$ corresponds to a HADAMARD DESIGN $(4n+3, 2n+1, n)$.

Hadamard (1893) remarked that a NECESSARY condition for a Hadamard matrix to exist is that $n = 1, 2,$ or a positive multiple of 4 (Brenner 1972). PALEY'S THEOREM guarantees that there always exists a Hadamard matrix H_n when n is divisible by 4 and OF THE FORM $2^\epsilon(p^m + 1)$, where p is an ODD PRIME. In such cases, the MATRICES can be constructed using a PALEY CONSTRUCTION. The PALEY CLASS k is undefined for the following values of $m < 1000$: 92, 116, 156, 172, 184, 188, 232, 236, 260, 268, 292, 324, 356, 372, 376, 404, 412, 428, 436, 452, 472, 476, 508, 520, 532, 536, 584, 596, 604, 612, 652, 668, 712, 716, 732, 756, 764, 772, 808, 836, 852, 856, 872, 876, 892, 904, 932, 940, 944, 952, 956, 964, 980, 988, 996.

Sawade (1985) constructed H_{268}. It is conjectured (and verified up to $n < 428$) that H_n exists for all n DIVISIBLE by 4 (van Lint and Wilson 1993). However, the proof of this CONJECTURE remains an important problem in CODING THEORY. The number of Hadamard matrices of order 4_n are 1, 1, 1, 5, 3, 60, 487, ... (Sloane's A007299).

If H_n and H_m are known, then H_{nm} can be obtained by replacing all 1s in H_m by H_n and all -1s by $-H_n$. For $n \leq 100$, Hadamard matrices with $n = 12, 20, 28, 36, 44, 52, 60, 68, 76, 84, 92,$ and 100 cannot be built up from lower order Hadamard matrices.

$$H_2 = \begin{bmatrix} 1 & 1 \\ -1 & 1 \end{bmatrix} \tag{2}$$

$$H_4 = \begin{bmatrix} H_2 & H_2 \\ -H_2 & H_2 \end{bmatrix} = \begin{bmatrix} \begin{bmatrix} 1 & 1 \\ -1 & 1 \end{bmatrix} & \begin{bmatrix} 1 & 1 \\ -1 & 1 \end{bmatrix} \\ -\begin{bmatrix} 1 & 1 \\ -1 & 1 \end{bmatrix} & \begin{bmatrix} 1 & 1 \\ -1 & 1 \end{bmatrix} \end{bmatrix}$$

$$= \begin{bmatrix} 1 & 1 & 1 & 1 \\ -1 & 1 & -1 & 1 \\ -1 & -1 & 1 & 1 \\ 1 & -1 & -1 & 1 \end{bmatrix}. \tag{3}$$

H_8 can be similarly generated from H_4. Hadamard matrices can also be expressed in terms of the WALSH FUNCTIONS Cal and Sal

$$H_8 = \begin{bmatrix} \mathrm{Cal}(0, t) \\ \mathrm{Sal}(4, t) \\ \mathrm{Sal}(2, t) \\ \mathrm{Cal}(2, t) \\ \mathrm{Sal}(1, t) \\ \mathrm{Cal}(3, t) \\ \mathrm{Cal}(1, t) \\ \mathrm{Sal}(3, t) \end{bmatrix}. \tag{4}$$

Hadamard matrices can be used to make ERROR-CORRECTING CODES.

See also HADAMARD DESIGN, HADAMARD'S MAXIMUM DETERMINANT PROBLEM, INTEGER MATRIX, PALEY CONSTRUCTION, PALEY'S THEOREM, WALSH FUNCTION

References

Ball, W. W. R. and Coxeter, H. S. M. *Mathematical Recreations and Essays, 13th ed.* New York: Dover, pp. 107–09 and 274, 1987.

Beth, T.; Jungnickel, D.; and Lenz, H. *Design Theory.* New York: Cambridge University Press, 1986.

Brenner, J. and Cummings, L. "The Hadamard Maximum Determinant Problem." *Amer. Math. Monthly* **79**, 626–30, 1972.

Colbourn, C. J. and Dinitz, J. H. (Eds.). "Hadamard Matrices and Designs." Ch. 24 in *CRC Handbook of Combinatorial Designs.* Boca Raton, FL: CRC Press, pp. 370–77, 1996.

Gardner, M. "Mathematical Games: On the Remarkable Császár Polyhedron and Its Applications in Problem Solving." *Sci. Amer.* **232**, 102–07, May 1975.

Geramita, A. V. *Orthogonal Designs: Quadratic Forms and Hadamard Matrices.* New York: Dekker, 1979.

Golomb, S. W. and Baumert, L. D. "The Search for Hadamard Matrices." *Amer. Math. Monthly* **70**, 12–7, 1963.

Hadamard, J. "Résolution d'une question relative aux déterminants." *Bull. Sci. Math.* **17**, 30–1, 1893.

Hall, M. *Combinatorial Theory, 2nd ed.* New York: Wiley, 1998.

Hedayat, A. and Wallis, W. D. "Hadamard Matrices and Their Applications." *Ann. Stat.* **6**, 1184–238, 1978.

Kimura, H. "Classification of Hadamard Matrices of Order 28." *Disc. Math.* **133**, 171–80, 1994.

Kimura, H. "Classification of Hadamard Matrices of Order 28 with Hall Sets." *Disc. Math.* **128**, 257–69, 1994.

Kitis, L. "Paley's Construction of Hadamard Matrices." http://www.mathsource.com/cgi-bin/msitem?0205–60.

Ogilvie, G. A. "Solution to Problem 2511." *Math. Questions and Solutions* **10**, 74–6, 1868.

Paley, R. E. A. C. "On Orthogonal Matrices." *J. Math. Phys.* **12**, 311–20, 1933.

Ryser, H. J. *Combinatorial Mathematics.* Buffalo, NY: Math. Assoc. Amer., pp. 104–22, 1963.

Sawade, K. "A Hadamard Matrix of Order-268." *Graphs Combinatorics* **1**, 185–87, 1985.

Seberry, J. and Yamada, M. "Hadamard Matrices, Sequences, and Block Designs." Ch. 11 in *Contemporary Design Theory: A Collection of Surveys* (Ed. J. H. Dinitz and D. R. Stinson). New York: Wiley, pp. 431–60, 1992.

Sloane, N. J. A. Sequences A007299/M3736 in "An On-Line Version of the Encyclopedia of Integer Sequences." http://www.research.att.com/~njas/sequences/eisonline.html.

Spence, E. "Classification of Hadamard Matrices of Order 24 and 28." *Disc. Math* **140**, 185–43, 1995.

Sylvester, J. J. "Thoughts on Orthogonal Matrices, Simultaneous Sign-Successions, and Tessellated Pavements in Two or More Colours, with Applications to Newton's Rule, Ornamental Tile-Work, and the Theory of Numbers." *Phil. Mag.* **34**, 461–75, 1867.

Sylvester, J. J. "Problem 2511." *Math. Questions and Solutions* **10**, 74, 1868.

van Lint, J. H. and Wilson, R. M. *A Course in Combinatorics.* New York: Cambridge University Press, 1993.

Wallis, W. D.; Street, A. P.; and Wallis, J. S. *Combinatorics: Room Squares, Sum-free Sets, Hadamard Matrices.* New York: Springer-Verlag, 1972.

Williamson, J. "Hadamard's Determinant Theorem and the Sum of Four Squares." *Duke Math. J.* **11**, 65–1, 1944.

Williamson, J. "Note on Hadamard's Determinant Theorem." *Bull. Amer. Math. Soc.* **53**, 608–13, 1947.

Hadamard Transform

A FAST FOURIER TRANSFORM-like ALGORITHM which produces a hologram of an image.

Hadamard's Determinant Problem

HADAMARD'S MAXIMUM DETERMINANT PROBLEM

Hadamard's Inequality

Let $A = a_{ik}$ be an arbitrary $n \times n$ nonsingular MATRIX with REAL elements and DETERMINANT $|A|$, then

$$|A|^2 \leq \prod_{i=1}^{n} \left(\sum_{k=1}^{n} a_{ik}^2 \right).$$

See also HADAMARD'S THEOREM

References

Gradshteyn, I. S. and Ryzhik, I. M. *Tables of Integrals, Series, and Products, 6th ed.* San Diego, CA: Academic Press, p. 1110, 2000.

Hadamard's Maximum Determinant Problem

Find the largest possible DETERMINANT (in absolute value) for any $n \times n$ matrix whose elements are taken from some set. Hadamard (1893) proved that the DETERMINANT of any COMPLEX $n \times n$ matrix **A** with entries in the closed UNIT DISK $|a_{ij}| \leq 1$ satisfies

$$|\det A| \leq n^{n/2}, \tag{1}$$

with equality attained by the VANDERMONDE MATRIX of the n ROOTS OF UNITY (Faddeev and Sominskii 1965, p. 331; Brenner 1972). The first few values for $\max(\det A_n)$ for $n = 1, 2, \ldots$ are 1, 2, $3\sqrt{3}$, 16, $25\sqrt{5}$, 216, ..., and the squares of these are 1, 4, 27, 256, 3125, ... (Sloane's A000312). A matrix having such a maximal determinant is known as a HADAMARD MATRIX (Brenner 1972).

For real entries, Hadamard's bound can be improved for real matrices to

$$|\det A| \leq \frac{(n+1)^{(n+1)/2}}{2^n} \tag{2}$$

(Faddeev and Sominskii 1965, problem 523; Brenner 1972).

For an $n \times n$ BINARY MATRIX, i.e., a (0,1)-matrix, the largest possible determinants β_n for $n = 1, 2, \ldots$ are 1, 1, 2, 3, 5, 9, 32, 56, 144, 320, 1458, 3645, 9477, ... (Sloane's A003432). The numbers of distinct $n \times n$ binary matrices having the largest possible determinant are 1, 3, 3, 60, ... (Sloane's A051752).

n	matrices
1	$[1]$
2	$\begin{bmatrix} 1 & 0 \\ 0 & 1 \end{bmatrix} \begin{bmatrix} 1 & 0 \\ 1 & 1 \end{bmatrix} \begin{bmatrix} 1 & 1 \\ 0 & 1 \end{bmatrix}$
3	$\begin{bmatrix} 0 & 1 & 1 \\ 1 & 0 & 1 \\ 1 & 1 & 0 \end{bmatrix}, \begin{bmatrix} 1 & 0 & 1 \\ 1 & 1 & 0 \\ 0 & 1 & 1 \end{bmatrix}, \begin{bmatrix} 1 & 1 & 0 \\ 0 & 1 & 1 \\ 1 & 0 & 1 \end{bmatrix}$

For an $n \times n$ (−1, 1)-matrix, the largest possible determinants α_n for $n = 1, 2, \ldots$ are 1, 2, 4, 16, 48, 160, ... (Sloane's A003433; Ehrlich and Zeller 1962, Ehrlich 1964). The numbers of distinct $n \times n$ (−1, 1)-matrices having the largest possible determinant are 1, 4, 96, 384, ... α_n is related to the largest possible (0, 1)-matrix determinant β_{n-1} by

$$\alpha_n = 2^{n-1}\beta_{n-1} \tag{3}$$

(Williamson 1946, Brenner 1972).

n	matrices
1	$[1]$
2	$\begin{bmatrix} -1 & -1 \\ 1 & -1 \end{bmatrix}, \begin{bmatrix} -1 & 1 \\ -1 & -1 \end{bmatrix}, \begin{bmatrix} 1 & -1 \\ 1 & -1 \end{bmatrix}, \begin{bmatrix} 1 & 1 \\ -1 & 1 \end{bmatrix}$

For an $n \times n$ $(-1, 0, 1)$-matrix, the largest possible determinants γ_n are the same as α_n (Ehrlich 1964, Brenner 1972). The numbers of $n \times n$ $(-1, 0, 1)$-matrices having maximum determinants are 1, 4, 240, ... (Sloane's A051753).

See also DETERMINANT, HADAMARD MATRIX, INTEGER MATRIX

References

Brenner, J. and Cummings, L. "The Hadamard Maximum Determinant Problem." *Amer. Math. Monthly* **79**, 626–30, 1972.

Cohn, J. H. E. "Determinants with Elements ±1." *J. London Math. Soc.* **14**, 581–88, 1963.

Ehrlich, H. "Determinantenabschätzungen für binäre Matrizen." *Math. Z.* **83**, 123–32, 1964.

Ehrlich, H. and Zeller, K. "Binäre Matrizen." *Z. angew. Math. Mechanik* **42**, T20–1, 1962.

Faddeev, D. K. and Sominskii, I. S. *Problems in Higher Algebra.* San Francisco: W. H. Freeman, 1965.

Hadamard, J. "Résolution d'une question relative aux déterminants." *Bull. Sci. Math.* **17**, 30–1, 1893.

Hall, M. *Combinatorial Theory, 2nd ed.* New York: Wiley, 1998.

Kaplansky, I. "Never Too Late." *Amer. Math. Monthly* **102**, 259, 1995.

MacWilliams, F. J. and Sloane, N. J. A. *The Theory of Error-Correcting Codes.* Amsterdam, Netherlands: North-Holland, p. 54, 1978.

Sloane, N. J. A. Sequences A003432/M0720, A003433/M1291, A051752, and A051753 in "An On-Line Version of the Encyclopedia of Integer Sequences." http://www.research.att.com/~njas/sequences/eisonline.html.

Williamson, J. "Determinants Whose Elements are 0 and 1." *Amer. Math. Monthly* **53**, 427–34, 1946.

Yang, C. H. "Some Designs for Maximal $(+1, -1)$-Determinant of Order $n \equiv 2 \pmod 4$." *Math. Comput.* **20**, 147–48, 1966.

Yang, C. H. "A Construction for Maximal $(+1, -1)$-Matrix of Order 54." *Bull. Amer. Math. Soc.* **72**, 293, 1966.

Yang, C. H. "On Designs of Maximal $(+1, -1)$-Matrices of Order $n \equiv 2 \pmod 4$." *Math. Comput.* **22**, 174–80, 1968.

Yang, C. H. "On Designs of Maximal $(+1, -1)$-Matrices of Order $n \equiv 2 \pmod 4$ II." *Math. Comput.* **23**, 201–05, 1969.

Hadamard's Theorem

Let $|A|$ be an $n \times n$ DETERMINANT with COMPLEX (or REAL) elements a_{ij}, then $|A| \neq 0$ if

$$|a_{ii}| > \sum_{\substack{j=1 \\ j \neq i}}^{n} |a_{ij}|.$$

See also HADAMARD'S INEQUALITY

References

Gradshteyn, I. S. and Ryzhik, I. M. *Tables of Integrals, Series, and Products, 6th ed.* San Diego, CA: Academic Press, p. 1110, 2000.

Hadamard-Vallée Poussin Constants

N.B. A detailed online essay by S. Finch was the starting point for this entry.

The sum of RECIPROCALS of PRIMES diverges, but

$$\lim_{n \to \infty} \left[\sum_{k=1}^{\pi(n)} \frac{1}{p_k} - \ln(\ln n) \right] = \gamma + \sum_{k=1}^{\infty} \left[\ln\left(1 - \frac{1}{p_k}\right) + \frac{1}{p_k} \right]$$

$$\equiv C_1 = 0.2614972128..., \tag{1}$$

where $\pi(n)$ is the PRIME COUNTING FUNCTION and γ is the EULER-MASCHERONI CONSTANT (Le Lionnais 1983). Hardy and Wright (1985) show that, if $\omega(n)$ is the number of distinct PRIME FACTORS of n, then

$$\lim_{n \to \infty} \left[\frac{1}{n} \sum_{k=1}^{n} \omega(k) - \ln(\ln n) \right] = C_1. \tag{2}$$

Furthermore, if $\Omega(n)$ is the total number of PRIME FACTORS of n, then

$$\lim_{n \to \infty} \left[\frac{1}{n} \sum_{k=1}^{n} \Omega(k) - \ln(\ln n) \right] = C_1 + \sum_{k=1}^{\infty} \frac{1}{p_k(p_k - 1)}$$

$$= 1.0346538819... . \tag{3}$$

Similarly,

$$\lim_{n \to \infty} \left(\sum_{k=1}^{\pi(n)} \frac{\ln p_k}{p_k} - \ln n \right) = -\gamma - \sum_{j=2}^{\infty} \sum_{k=1}^{\infty} \frac{\ln p_k}{p_k^j} \equiv -C_2$$

$$= -1.3325822757... . \tag{4}$$

References

Finch, S. "Favorite Mathematical Constants." http://www.mathsoft.com/asolve/constant/hdmrd/hdmrd.html.

Hardy, G. H. and Wright, E. M. *An Introduction to the Theory of Numbers, 5th ed.* Oxford, England: Clarendon Press, 1985.

Le Lionnais, F. *Les nombres remarquables.* Paris: Hermann, p. 24, 1983.

Rosser, J. B. and Schoenfeld, L. "Approximate Formulas for Some Functions of Prime Numbers." *Ill. J. Math.* **6**, 64–4, 1962.

Hadwiger Number

References

Kostochka, A. V. "On Hadwiger Numbers of a Graph and Its Complement." In *Finite and Infinite Sets, Colloq. Math. Soc. János Bolyai, Vol. 37* (Ed. A. Hajnal, L. Lovász, and V. T. Sós). pp. 537–45, 1981.

Zelinka, B. "Hadwiger Number of Finite Graphs." *Math. Slov.* **26**, 23–0, 1976.

Hadwiger Problem

What is the largest number of subcubes (not necessarily different) into which a CUBE cannot be divided by plane cuts? The answer is 47.

See also CUBE DISSECTION, CUTTING

Hadwiger's Principal Theorem

The VECTORS $\pm\mathbf{a}_1, \ldots, \pm\mathbf{a}_n$ in a 3-space form a normalized EUTACTIC STAR IFF $T\mathbf{x} = \mathbf{x}$ for all $\mathbf{x}$ in the 3-space.

Hafner-Sarnak-McCurley Constant

N.B. A detailed online essay by S. Finch was the starting point for this entry.

Given two randomly chosen $n \times n$ INTEGER MATRICES, what is the probability $D(n)$ that the corresponding DETERMINANTS are RELATIVELY PRIME? Hafner *et al.* (1993) showed that

$$D(n) = \prod_{k=1}^{\infty}\left\{1 - \left[1 - \prod_{j=1}^{n}(1 - p_k^{-j})\right]^2\right\}, \qquad (1)$$

where p_n is the nth PRIME.

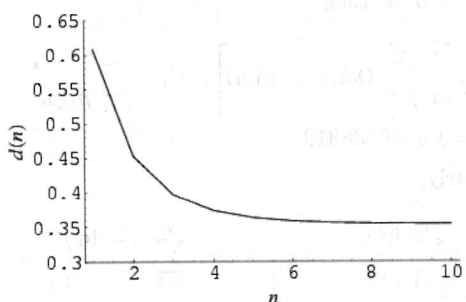

The case D_1 is just the probability that two random INTEGERS are RELATIVELY PRIME,

$$D(1) = \frac{6}{\pi^2} = 0.6079271019\ldots \qquad (2)$$

No analytic results are known for $n \geq 2$. Approximate values for the first few n are given by

$$D(2) \approx 0.453103 \qquad (3)$$

$$D(3) \approx 0.397276 \qquad (4)$$

$$D(4) \approx 0.373913 \qquad (5)$$

$$D(5) \approx 0.363321. \qquad (6)$$

Vardi (1991) computed the limit

$$\sigma \equiv \lim_{n \to \infty} D(n) = 0.3532363719\ldots. \qquad (7)$$

The speed of convergence is roughly $\sim 0.57^n$ (Flajolet and Vardi 1996).

See also INTEGER MATRIX, RELATIVELY PRIME

References

Finch, S. "Favorite Mathematical Constants." http://www.mathsoft.com/asolve/constant/hafner/hafner.html.

Flajolet, P. and Vardi, I. "Zeta Function Expansions of Classical Constants." Unpublished manuscript. 1996. http://pauillac.inria.fr/algo/flajolet/Publications/landau.ps.

Hafner, J. L.; Sarnak, P.; and McCurley, K. "Relatively Prime Values of Polynomials." In *Contemporary Mathematics* Vol. 143 (Ed. M. Knopp and M. Seingorn). Providence, RI: Amer. Math. Soc., 1993.

Vardi, I. *Computational Recreations in Mathematica.* Redwood City, CA: Addison-Wesley, 1991.

Hahn Polynomial

The orthogonal polynomials defined by

$$h_n^{(\alpha,\beta)}(x, N) = \frac{(-1)^n(N - x - n)_n(\beta + x + 1)_n}{n!}$$

$$\times {}_3F_2\left(\begin{matrix} -n, -x, \alpha + N - x \\ N - x - n, -\beta - x - n \end{matrix}; 1\right) \qquad (1)$$

$$= \frac{(-1)^n(N - n)_n(\beta + 1)_n}{n!}$$

$$\times {}_3F_2\left(\begin{matrix} -n, -x, \alpha + \beta + n + 1 \\ \beta + 1, \ 1 - N \end{matrix}; 1\right), \qquad (2)$$

where $(x)_n$ is the POCHHAMMER SYMBOL and ${}_3F_2(a, b, c; d, e; z)$ is a GENERALIZED HYPERGEOMETRIC FUNCTION (Koepf 1998). The first few are given by

$$h_0^{(\alpha,\beta)}(x, N) = 1$$

$$h_1^{(\alpha,\beta)}(x, N) = x(\alpha + \beta + 2) - (N - 1)(\beta + 1).$$

Koekoek and Swarttouw (1998) define another Hahn polynomial

$$Q_n(x; \alpha, \beta, N) = {}_3F_2\left(\begin{matrix} -n, n + \alpha + \beta + 1, -x \\ \alpha + 1, -N \end{matrix}; 1\right), \qquad (3)$$

the dual Hahn polynomial

$$R_n(\lambda(x); \gamma, \delta, N)$$

$$= {}_3F_2\left(\begin{matrix} -n, -x, x + \gamma + \delta + 1 \\ \gamma + 1, -N \end{matrix}; 1\right), \qquad (4)$$

the continuous Hahn polynomial

$$p_n(x; a, b, c, d) = i^n \frac{(a + c)_n(a + d)_n}{n!}$$

$$\times {}_3F_2\left(\begin{matrix} -n,\ n+a+b+c+d-1,\ a+ix \\ a+c,\ a+d \end{matrix};\ 1\right), \quad (5)$$

and the continuous dual Hahn polynomial

$$\frac{S_n(x^2;\ a,\ b,\ c)}{(a+b)_n(a+c)_n} = {}_3F_2\left(\begin{matrix} -n,\ a+ix,\ a-ix \\ a+b,\ a+c \end{matrix};\ 1\right), \quad (6)$$

for $n = 0, 1, ..., N$, and where

$$\lambda(x) = x(x + \gamma + \delta + 1). \quad (7)$$

References

Koekoek, R. and Swarttouw, R. F. "Continuous Dual Hahn," "Continuous Hahn," "Hahn," and "Dual Hahn." §1.3–.6 in *The Askey-Scheme of Hypergeometric Orthogonal Polynomials and its q*-Analogue. Delft, Netherlands: Technische Universiteit Delft, Faculty of Technical Mathematics and Informatics Report 98–7, pp. 29–6, 1998. ftp://www.twi.-tudelft.nl/publications/tech-reports/1998/DUT-TWI-98–7.ps.gz.

Koepf, W. *Hypergeometric Summation: An Algorithmic Approach to Summation and Special Function Identities.* Braunschweig, Germany: Vieweg, p. 115, 1998.

Hahn-Banach Theorem

A linear FUNCTIONAL defined on a SUBSPACE of a VECTOR SPACE V and which is dominated by a sublinear function defined on V has a linear extension which is also dominated by the sublinear function.

References

Casti, J. L. "The Hahn-Banach Theorem." Ch. 4 in *Five More Golden Rules: Knots, Codes, Chaos, and Other Great Theories of 20th-Century Mathematics.* New York: Wiley, pp. 155–05, 2000.

Zeidler, E. *Applied Functional Analysis: Applications to Mathematical Physics.* New York: Springer-Verlag, 1995.

Hailstone Number

Sequences of INTEGERS generated in the COLLATZ PROBLEM. For example, for a starting number of 7, the sequence is 7, 22, 11, 34, 17, 52, 26, 13, 40, 20, 10, 5, 16, 8, 4, 2, 1, 4, 2, 1, ... Such sequences are called hailstone sequences because the values typically rise and fall, somewhat analogously to a hailstone inside a cloud.

While a hailstone eventually becomes so heavy that it falls to ground, every starting INTEGER ever tested has produced a hailstone sequence that eventually drops down to the number 1 and then "bounces" into the small loop 4, 2, 1, ...

See also COLLATZ PROBLEM

References

Schwartzman, S. *The Words of Mathematics: An Etymological Dictionary of Mathematical Terms Used in English.* Washington, DC: Math. Assoc. Amer., 1994.

Hairy Ball Theorem

There does not exist an everywhere NONZERO tangent VECTOR FIELD on the 2-SPHERE $\mathbb{S}^2$. This implies that somewhere on the surface of the Earth, there is a point with zero horizontal wind velocity. The theorem can be generalized to the statement that the n-sphere $\mathbb{S}^n$ has a nonzero tangent vector field IFF n is ODD.

See also FIXED POINT THEOREM

References

Steinhaus, H. *Mathematical Snapshots, 3rd ed.* New York: Dover, pp. 279–81, 1999.

Wells, D. *The Penguin Dictionary of Curious and Interesting Geometry.* Middlesex, England: Penguin Books, p. 90, 1991.

Hajnal-Szemerédi Theorem

Every GRAPH with n vertices and maximum VERTEX DEGREE $\Delta(G) \le k$ is $(k+1)$-colorable with all color classes of size $\lfloor n/(k+1) \rfloor$ or $\lceil n/(k+1) \rceil$, where $\lfloor x \rfloor$ is the FLOOR FUNCTION and $\lceil x \rceil$ is the CEILING FUNCTION.

See also SEYMOUR CONJECTURE

References

Hajnal, A. and Szemerédi, E. "Proof of a Conjecture of Erdos." In *Combinatorial Theory and Its Applications, Vol. 2* (Ed. P. Erdos, A. Rényi, and V. T. Sós). Amsterdam, Netherlands: North-Holland, pp. 601–23, 1970.

Komlós, J.; Sárkozy, G. N.; and Szemerédi, E. "Proof of the Seymour Conjecture for Large Graphs." *Ann. Comb.* **2**, 43–0, 1998.

Hajós Number

The Hajós number $h(G)$ of a GRAPH G is the maximum k such that G contains a subdivision of the COMPLETE GRAPH K_k.

References

Erdos, P. and Fajtlowicz, S. "On the Conjecture of Hajós." *Combinatorica* **1**, 141–43, 1981.

Gutin, G.; Kostochka, A. V.; and Toft, B. "On the Hajós Number of Graphs." *Discr. Math.* **213**, 153–61, 2000.

Half

The UNIT FRACTION 1/2.

See also QUARTER, SQUARE ROOT, UNIT FRACTION

Half-Angle Formulas

Formulas expressing trigonometric functions of an angle $x/2$ in terms of functions of an angle x,

$$\sin\left(\tfrac{1}{2}x\right) = \pm\sqrt{\frac{1-\cos x}{2}} \quad (1)$$

$$\cos\left(\tfrac{1}{2}x\right) = \pm\sqrt{\frac{1 + \cos x}{2}} \qquad (2)$$

$$\tan\left(\tfrac{1}{2}x\right) = \frac{\sin x}{1 + \cos x} \qquad (3)$$

$$= \frac{1 - \cos x}{\sin x} \qquad (4)$$

$$= \frac{1 \pm \sqrt{1 + \tan^2 x}}{\tan x} \qquad (5)$$

$$= \frac{\tan x \sin x}{\tan x + \sin x}. \qquad (6)$$

The corresponding hyperbolic function double-angle formulas are

$$\sinh\left(\tfrac{1}{2}x\right) = \operatorname{sgn} x \sqrt{\frac{\cosh x - 1}{2}} \qquad (7)$$

$$\cosh\left(\tfrac{1}{2}x\right) = \sqrt{\frac{\cosh x + 1}{2}} \qquad (8)$$

$$\tanh\left(\tfrac{1}{2}x\right) = \frac{\sinh x}{\cosh x + 1} \qquad (9)$$

$$= \frac{\cosh x - 1}{\sinh x}. \qquad (10)$$

See also DOUBLE-ANGLE FORMULAS, HYPERBOLIC FUNCTIONS, MULTIPLE-ANGLE FORMULAS, PROSTHAPHAERESIS FORMULAS, TRIGONOMETRIC ADDITION FORMULAS, TRIGONOMETRIC FUNCTIONS, TRIGONOMETRY

Half-Closed Interval

half-closed interval [*a, b*)   *half-closed interval* (*a, b*]

An INTERVAL in which one endpoint is included but not the other. A half-closed interval is denoted [*a, b*) or (*a, b*] and is also called a HALF-OPEN INTERVAL. The non-standard notation [*a, b*[and]*a, b*] is sometimes also used.

See also CLOSED INTERVAL, INTERVAL, OPEN INTERVAL

Half-Normal Distribution

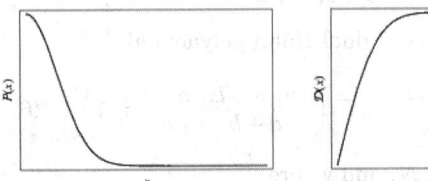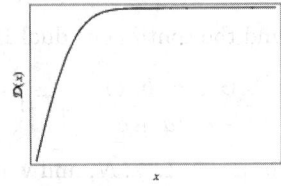

A NORMAL DISTRIBUTION with MEAN 0 and STANDARD DEVIATION $1/\theta$ limited to the domain $x \in [0, \infty)$.

$$P(x) = \frac{2\theta}{\pi} e^{-x^2\theta^2/\pi} \qquad (1)$$

$$D(x) = \operatorname{erf}\left(\frac{\theta x}{\sqrt{\pi}}\right). \qquad (2)$$

The MOMENTS are

$$\mu_1 = \frac{1}{\theta} \qquad (3)$$

$$\mu_2 = \frac{\pi}{2\theta^2} \qquad (4)$$

$$\mu_3 = \frac{\pi}{\theta^3} \qquad (5)$$

$$\mu_4 = \frac{3\pi^2}{4\theta^4}, \qquad (6)$$

so the MEAN, VARIANCE, SKEWNESS, and KURTOSIS are

$$\mu = \frac{1}{\theta} \qquad (7)$$

$$\sigma^2 = \frac{\pi - 2}{2\theta^2} \qquad (8)$$

$$\gamma_1 = 2\sqrt{\frac{2}{\pi}} \qquad (9)$$

$$\gamma_2 = 0. \qquad (10)$$

See also NORMAL DISTRIBUTION

Half-Open Interval

HALF-CLOSED INTERVAL

Half-Period Ratio

The ratio $\tau = \omega_1/\omega_2$ of the two half-periods ω_1 and ω_2 of an ELLIPTIC FUNCTION (Whittaker and Watson 1990, p. 475). The notation t is sometimes used instead of τ. The half-period ratio is most commonly encountered in the definition of the NOME q as

$$q(k) \equiv e^{\pi i \tau} = e^{-\pi K'(k)/K(k)} = e^{-\pi K\left(\sqrt{1-k^2}\right)/K(k)} \quad (1)$$

(Borwein and Borwein 1987, pp. 41, 109, and 114; Whittaker and Watson 1990, p. 463) where $K(k)$ is the complete ELLIPTIC INTEGRAL OF THE FIRST KIND, $m = k^2$ is the PARAMETER, k is the MODULUS, $K'(k) = K(k')$, and k' is the complementary MODULUS.

τ is defined such that the IMAGINARY PART $\Im[\tau] > 0$.

See also JACOBI THETA FUNCTIONS, MODULAR ANGLE, MODULUS (ELLIPTIC INTEGRAL), INVERSE NOME, NOME, PARAMETER

References

Borwein, J. M. and Borwein, P. B. *Pi & the AGM: A Study in Analytic Number Theory and Computational Complexity.* New York: Wiley, 1987.

Whittaker, E. T. and Watson, G. N. *A Course in Modern Analysis, 4th ed.* Cambridge, England: Cambridge University Press, 1990.

Half-Plane

This entry contributed by DANIEL SCOTT UZNANSKI

A half-plane is a planar region consisting of all points on one side of an infinite straight line, and no points on the other side.

See also HALF-SPACE, LOWER HALF-PLANE, PLANE, UPPER HALF-PLANE

Half-Space

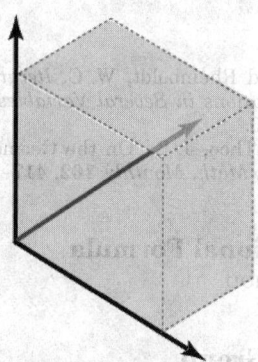

A half-space is that portion of an n-dimensional SPACE obtained by removing that part lying on one side of an $(n-1)$-dimensional hyperplane. For example, half a Euclidean space is given by the 3-dimensional region satisfying $x > 0$, $-\infty < y < \infty$, $-\infty < z < \infty$, while a HALF-PLANE is given by the 2-dimensional region satisfying $x > 0$, $-\infty < y < \infty$.

See also HALF-PLANE, SIEGEL'S UPPER HALF-SPACE

Half-Turn

A ROTATION through $180°$ (π radians).

See also ROTATION

References

Coxeter, H. S. M. and Greitzer, S. L. "Half-Turn." §4.3 in *Geometry Revisited.* Washington, DC: Math. Assoc. Amer., pp. 85–6, 1967.

Hall's Theorem

There exists a system of distinct representatives for a family of sets $S_1, S_2, ..., S_m$ IFF the union of any k of these sets contains at least k elements for all k from 1 to m (Harary 1994, p. 53).

References

Harary, F. *Graph Theory.* Reading, MA: Addison-Wesley, 1994.

Halley's Irrational Formula

A ROOT-finding ALGORITHM which makes use of a third-order TAYLOR SERIES

$$f(x) = f(x_n) + f'(x_n)(x - x_n) + \tfrac{1}{2} f''(x_n)(x - x_n)^2 + \dots \quad (1)$$

A ROOT of $f(x)$ satisfies $f(x) = 0$, so

$$0 \approx f(x_n) + f'(x_n)(x_{n+1} - x_n) + \tfrac{1}{2} f''(x_n)(x_{n+1} - x_n)^2. \quad (2)$$

Using the QUADRATIC EQUATION then gives

$$x_{n+1} = x_n + \frac{-f'(x_n) \pm \sqrt{[f'(x_n)]^2 - 2f(x_n)f''(x_n)}}{f''(x_n)}. \quad (3)$$

Picking the plus sign gives the iteration function

$$C_f(x) = x - \frac{1 - \sqrt{1 - \dfrac{2f(x)f''(x)}{[f'(x)]^2}}}{\dfrac{f''(x)}{f'(x)}}. \quad (4)$$

This equation can be used as a starting point for deriving HALLEY'S METHOD.

If the alternate form of the QUADRATIC EQUATION is used instead in solving (2), the iteration function becomes instead

$$C_f(x) = x - \frac{2f(x)}{f'(x) \pm \sqrt{[f'(x)]^2 - 2f(x)f''(x)}}. \quad (5)$$

This form can also be derived by setting $n = 2$ in LAGUERRE'S METHOD. Numerically, the SIGN in the DENOMINATOR is chosen to maximize its ABSOLUTE VALUE. Note that in the above equation, if $f''(x) = 0$, then NEWTON'S METHOD is recovered. This form of Halley's irrational formula has cubic convergence, and is usually found to be substantially more stable than NEWTON'S METHOD. However, it does run into difficulty when both $f(x)$ and $f'(x)$ or $f'(x)$ and $f''(x)$ are simultaneously near zero.

See also HALLEY'S METHOD, HOUSEHOLDER'S METHOD, LAGUERRE'S METHOD, NEWTON'S METHOD

References

Gourdon, X. and Sebah, P. "Newton's Iteration." http://xavier.gourdon.free.fr/Constants/Algorithms/newton.html.

Ortega, J. M. and Rheinboldt, W. C. *Iterative Solution of Nonlinear Equations in Several Variables.* Philadelphia, PA: SIAM, 2000.

Qiu, H. "A Robust Examination of the Newton-Raphson Method with Strong Global Convergence Properties." Master's Thesis. University of Central Florida, 1993.

Scavo, T. R. and Thoo, J. B. "On the Geometry of Halley's Method." *Amer. Math. Monthly* **102**, 417–26, 1995.

Halley's Method

Also known as the TANGENT HYPERBOLAS METHOD or HALLEY'S RATIONAL FORMULA. As in HALLEY'S IRRATIONAL FORMULA, take the second-order TAYLOR POLYNOMIAL

$$f(x) = f(x_n) + f'(x_n)(x - x_n) + \tfrac{1}{2}f''(x_n)(x - x_n)^2 + \dots. \quad (1)$$

A ROOT of $f(x)$ satisfies $f(x) = 0$, so

$$0 \approx f(x_n) + f'(x_n)(x_{n+1} - x_n) + \tfrac{1}{2}f''(x_n)(x_{n+1} - x_n)^2. \quad (2)$$

Now write

$$0 = f(x_n) + (x_{n+1} - x_n)$$
$$\times \left[f'(x_n) + \tfrac{1}{2}f''(x_n)(x_{n+1} - x_n) \right], \quad (3)$$

giving

$$x_{n+1} = x_n - \frac{f(x_n)}{f'(x_n) + \tfrac{1}{2}f''(x_n)(x_{n+1} - x_n)}. \quad (4)$$

Using the result from NEWTON'S METHOD,

$$x_{n+1} - x_n = -\frac{f(x_n)}{f'(x_n)}. \quad (5)$$

gives

$$x_{n+1} = x_n - \frac{2f(x_n)f'(x_n)}{2[f'(x_n)]^2 - f(x_n)f''(x_n)}, \quad (6)$$

so the iteration function is

$$H_f(x) = x - \frac{2f(x)f'(x)}{2[f'(x)]^2 - f(x)f''(x)}. \quad (7)$$

This satisfies $H_f(\alpha) = H_f''(\alpha) = 0$ where α is a ROOT, so it is third order for simple zeros. Curiously, the third derivative

$$H_f'''(\alpha) = -\left\{ \frac{f'''(\alpha)}{f'(\alpha)} - \frac{3}{2}\left[\frac{f''(\alpha)}{f'(\alpha)} \right]^2 \right\} \quad (8)$$

is the SCHWARZIAN DERIVATIVE. Halley's method may also be derived by applying NEWTON'S METHOD to $ff'^{-1/2}$. It may also be derived by using an OSCULATING CURVE OF THE FORM

$$y(x) = \frac{(x - x_n) + c}{a(x - x_n) + b}. \quad (9)$$

Taking derivatives,

$$f(x_n) = \frac{c}{b} \quad (10)$$

$$f'(x_n) = \frac{b - ac}{b^2} \quad (11)$$

$$f''(x_n) = \frac{2a(ac - b)}{b^3}, \quad (12)$$

which has solutions

$$a = -\frac{f''(x_n)}{2[f'(x_n)]^2 - f(x_n)f''(x_n)} \quad (13)$$

$$b = \frac{2f'(x_n)}{2[f'(x_n)]^2 - f(x_n)f''(x_n)} \quad (14)$$

$$c = \frac{2f(x_n)f'(x_n)}{2[f'(x_n)]^2 - f(x_n)f''(x_n)}, \quad (15)$$

so at a ROOT, $y(x_{n+1}) = 0$ and

$$x_{n+1} = x_n - c, \quad (16)$$

which is Halley's method.

See also HALLEY'S IRRATIONAL FORMULA, HOUSEHOLDER'S METHOD, LAGUERRE'S METHOD, NEWTON'S METHOD

References

Ortega, J. M. and Rheinboldt, W. C. *Iterative Solution of Nonlinear Equations in Several Variables.* Philadelphia, PA: SIAM, 2000.

Scavo, T. R. and Thoo, J. B. "On the Geometry of Halley's Method." *Amer. Math. Monthly* **102**, 417–26, 1995.

Halley's Rational Formula

HALLEY'S METHOD

Hall-Janko Group

The SPORADIC GROUP HJ, also denoted J_2.

See also JANKO GROUPS

Hall-Littlewood Polynomial

Let n be an integer such that $n \geq \lambda_1$, where $\lambda = (\lambda_1, \lambda_2, \dots)$ is a PARTITION of $n = |\lambda|$ if $\lambda_1 \geq \lambda_2 \geq \dots \geq 0$, where λ_i are a sequence of positive integers stabilizing 0 such that $\Sigma_i \lambda_i = n$. Also let $m_i(\lambda)$ be the number of parts of λ of size i. Then the PERMUTATION $w \in S_n$, where S_n is the symmetric group, acts on the variables $x_1, \dots, x_n$ by sending x_i to $x_{w(i)}$. Letting t be a COMPLEX NUMBER, the Hall-Littlewood polynomials are defined by

$$P_\lambda(x_1, \ldots, x_n; \ t)$$

$$= \frac{1}{\prod_{i \geq 0} \prod_{r=1}^{m_i(\lambda)} \frac{1-t^r}{1-t}} \sum_{w \in S_n} w \left(x_1^{\lambda_1} \cdots x_n^{\lambda_n} \prod_{i<j} \frac{x_i - tx_j}{x_i - x_j} \right).$$

These polynomials interpolate between the Schur functions (with $t = 0$) and the monomial symmetric functions (with $t = 1$; Fulman 1999).

References

Fulman, J. "The Rogers-Ramanujan Identities, the Finite General Linear Groups, and the Hall-Littlewood Polynomials." *Proc. Amer. Math. Soc.* **128**, 17–5, 1999.

Macdonald, I. G. *Symmetric Functions and Hall Polynomials, 2nd ed.* Oxford, England: Oxford University Press, p. 208, 1995.

Halm's Differential Equation

The second-order ORDINARY DIFFERENTIAL EQUATION

$$(1+x^2)^2 + y'' + \lambda y = 0$$

(Hille 1969, p. 357; Zwillinger 1997, p. 122).

References

Hille, E. *Lectures on Ordinary Differential Equations.* Reading, MA: Addison-Wesley, 1969.

Zwillinger, D. *Handbook of Differential Equations, 3rd ed.* Boston, MA: Academic Press, p. 122, 1997.

Halphen Constant

ONE-NINTH CONSTANT

Halphen's Transformation

A curve and its polar reciprocal with regard to the fixed CONIC have the same Halphen transformation.

References

Coolidge, J. L. *A Treatise on Algebraic Plane Curves.* New York: Dover, pp. 346–47, 1959.

Halting Problem

The determination of whether a TURING MACHINE will come to a halt given a particular input program. This problem is UNDECIDABLE, as first proved by Turing.

See also BUSY BEAVER, CHAITIN'S CONSTANT, TURING MACHINE, UNDECIDABLE

References

Chaitin, G. J. "Computing the Busy Beaver Function." §4.4 in *Open Problems in Communication and Computation* (Ed. T. M. Cover and B. Gopinath). New York: Springer-Verlag, pp. 108–12, 1987.

Davis, M. "What It a Computation." In *Mathematics Today: Twelve Informal Essays* (Ed. L. A. Steen). New York: Springer-Verlag, pp. 241–67, 1978.

Penrose, R. *The Emperor's New Mind: Concerning Computers, Minds, and the Laws of Physics.* Oxford, England: Oxford University Press, pp. 63–6, 1989.

Ham Sandwich Theorem

The volumes of any n n-D solids can always be simultaneously bisected by a $(n-1)$-D HYPERPLANE. Proving the theorem for $n = 2$ (where it is known as the PANCAKE THEOREM) is simple and can be found in Courant and Robbins (1978). The theorem was proved for $n > 3$ by Stone and Tukey (1942).

See also CUTTING, PANCAKE THEOREM

References

Chinn, W. G. and Steenrod, N. E. *First Concepts of Topology.* Washington, DC: Math. Assoc. Amer., 1966.

Courant, R. and Robbins, H. *What is Mathematics?: An Elementary Approach to Ideas and Methods.* Oxford, England: Oxford University Press, 1978.

Davis, P. J. and Hersh, R. *The Mathematical Experience.* Boston, MA: Houghton Mifflin, pp. 274–84, 1981.

Hunter, J. A. H. and Madachy, J. S. *Mathematical Diversions.* New York: Dover, pp. 67–9, 1975.

Steinhaus, H. "Sur la division des ensembles de l'espace par les plans et des ensembles plans par les cercles." *Fundamenta Math.* **33**, 245–63, 1945.

Steinhaus, H. *Mathematical Snapshots, 3rd ed.* New York: Dover, p. 145, 1999.

Stone, A. H. and Tukey, J. W. "Generalized 'Sandwich' Theorems." *Duke Math. J.* **9**, 356–59, 1942.

Hamburger Moment Problem

This entry contributed by RONALD M. AARTS

A NECESSARY and SUFFICIENT condition that there should exist at least one nondecreasing function $\alpha(t)$ such that

$$\mu_n = \int_{-\infty}^{\infty} t^n \, d\alpha(t)$$

for $n = 0, 1, 2, \ldots$, with all the integrals converging, is that sequence $\{\mu_n\}_0^\infty$ is positive (Widder 1941, p. 129).

References

Widder, D. V. *The Laplace Transform.* Princeton, NJ: Princeton University Press, 1941.

Hamel Basis

This entry contributed by KEVIN O'BRYANT

A basis for the real numbers $\mathbb{R}$, considered as a VECTOR SPACE over the rationals $\mathbb{Q}$, i.e., a set of real numbers $\{U_\alpha\}$ such that every real number β has a unique representation of the form

$$\beta = \sum_{i=1}^{n} r_i U_{\alpha i},$$

where r_i is rational and n depends on β.

The AXIOM OF CHOICE is equivalent to the statement: "Every VECTOR SPACE has a BASIS," and this is the only justification for the existence of a Hamel basis.

See also AXIOM OF CHOICE, BASIS, BASIS (VECTOR SPACE)

Hamilton's Equations

The equations defined by

$$\dot{q} = \frac{\partial H}{\partial p} \tag{1}$$

$$\dot{p} = -\frac{\partial H}{\partial q} \tag{2}$$

where $\dot{x} \equiv dx/dt$ and H is the so-called Hamiltonian, are called Hamilton's equations. These equations frequently arise in problems of celestial mechanics.

The vector form of these equations is

$$\dot{x}_i = H_{p_i}(t, \mathbf{x}, \mathbf{p}) \tag{3}$$

$$\dot{p}_i = H_{x_i}(t, \mathbf{x}, \mathbf{p}) \tag{4}$$

(Zwillinger 1997, p. 136; Iyanaga and Kawada 1980, p. 1005).

Another formulation related to Hamilton's equation is

$$p = \frac{\partial L}{\partial \dot{q}}, \tag{5}$$

where L is the so-called Lagrangian.

References

Iyanaga, S. and Kawada, Y. (Eds.). *Encyclopedic Dictionary of Mathematics.* Cambridge, MA: MIT Press, p. 1005, 1980.

Morse, P. M. and Feshbach, H. "Hamilton's Principle and Classical Dynamics." §3.2 in *Methods of Theoretical Physics, Part I.* New York: McGraw-Hill, pp. 280–01, 1953.

Zwillinger, D. *Handbook of Differential Equations, 3rd ed.* Boston, MA: Academic Press, 1997.

Hamilton's Rules

The rules for the MULTIPLICATION of QUATERNIONS.

See also QUATERNION

Hamilton-Connected Graph

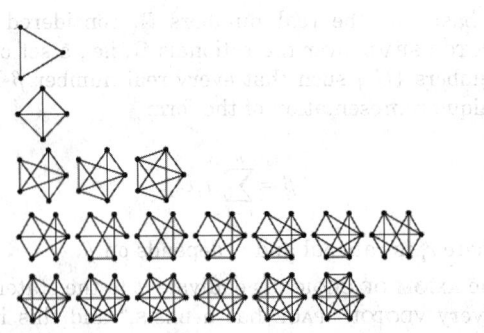

A graph G is Hamilton-connected if every two vertices of G are connected by a HAMILTONIAN PATH (Bondy and Murty 1976, p. 61). All COMPLETE GRAPHS are Hamilton-connected. The numbers of Hamilton-connected simple graphs on $n = 1, 2, \ldots$ nodes are 1, 1, 1, 1, 3, 13, 116, ... (Sloane's A057865).

See also HAMILTONIAN GRAPH, HAMILTONIAN PATH, HYPOTRACEABLE GRAPH, TRACEABLE GRAPH

References

Bondy, J. A. and Murty, U. S. R. *Graph Theory with Applications.* New York: North Holland, p. 61, 1976.

Sloane, N. J. A. Sequences A057865 in "An On-Line Version of the Encyclopedia of Integer Sequences." http://www.research.att.com/~njas/sequences/eisonline.html.

Hamiltonian Circuit

A GRAPH CYCLE (i.e., closed loop) through a GRAPH that visits each node exactly once (Skiena 1990, p. 196). A graph possessing a Hamiltonian circuit is said to be a HAMILTONIAN GRAPH. The Hamiltonian circuit is named after Sir William Rowan Hamilton, who devised a puzzle in which such a path along the EDGES of an ICOSAHEDRON was sought (the ICOSIAN GAME).

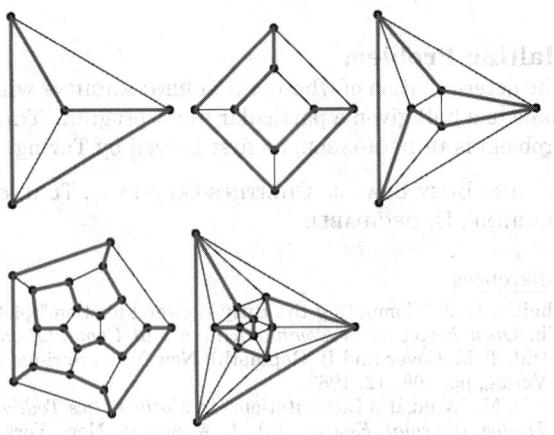

All PLATONIC SOLIDS have a Hamiltonian circuit, as illustrated above.

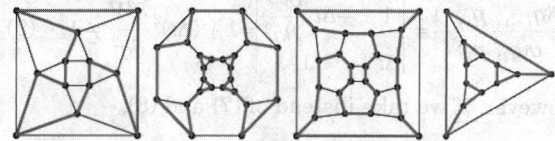

Although not explicitly stated by Gardner (1957), all ARCHIMEDEAN SOLIDS have Hamiltonian circuits as well, several of which are illustrated above. The Archimedean dual RHOMBIC DODECAHEDRON is Hamiltonian (Gardner 1984, p. 98). All PLANAR 4-connected graphs also have Hamiltonian circuits.

The number of Hamiltonian circuits on an n-HYPERCUBE is 2, 8, 96, 43008, ... (Sloane's A006069; Gardner 1986, pp. 23–4).

In general, the problem of finding a Hamiltonian circuit is NP-COMPLETE (Garey and Johnson 1983), so the only known way to determine whether a given general GRAPH has a Hamiltonian circuit is to undertake an exhaustive search.

See also CHVÁTAL'S THEOREM, DIRAC'S THEOREM, EULERIAN CIRCUIT, EULER GRAPH, GRINBERG FORMULA, HAMILTONIAN GRAPH, HAMILTONIAN PATH, ICOSIAN GAME, KOZYREV-GRINBERG THEORY, ORE'S THEOREM, PÓSA'S THEOREM, SMITH'S NETWORK THEOREM, TOUR, UNICURSAL CIRCUIT

References

Bollobás, B. *Graph Theory: An Introductory Course.* New York: Springer-Verlag, p. 12, 1979.

Chartrand, G. *Introductory Graph Theory.* New York: Dover, p. 68, 1985.

Gardner, M. "Mathematical Games: About the Remarkable Similarity between the Icosian Game and the Towers of Hanoi." *Sci. Amer.* **196**, 150–56, May 1957.

Gardner, M. *The Sixth Book of Mathematical Games from Scientific American.* Chicago, IL: University of Chicago Press, pp. 96–7, 1984.

Gardner, M. "The Binary Gray Code." In *Knotted Doughnuts and Other Mathematical Entertainments.* New York: W. H. Freeman, pp. 23–4, 1986.

Garey, M. R. and Johnson, D. S. *Computers and Intractability: A Guide to the Theory of NP-Completeness.* New York: W. H. Freeman, 1983.

Lederberg, J. "Hamilton Circuits of Convex Trivalent Polyhedra (up to 18 Vertices)." *Amer. Math. Monthly* **74**, 522–27, 1967.

Ore, O. "A Note on Hamiltonian Circuits." *Amer. Math. Monthly* **67**, 55, 1960.

Skiena, S. "Hamiltonian Cycles." §5.3.4 in *Implementing Discrete Mathematics: Combinatorics and Graph Theory with Mathematica.* Reading, MA: Addison-Wesley, pp. 196–98, 1990.

Sloane, N. J. A. Sequences A006069/M1903 in "An On-Line Version of the Encyclopedia of Integer Sequences." http://www.research.att.com/~njas/sequences/eisonline.html.

Hamiltonian Cycle

HAMILTONIAN CIRCUIT

Hamiltonian Graph

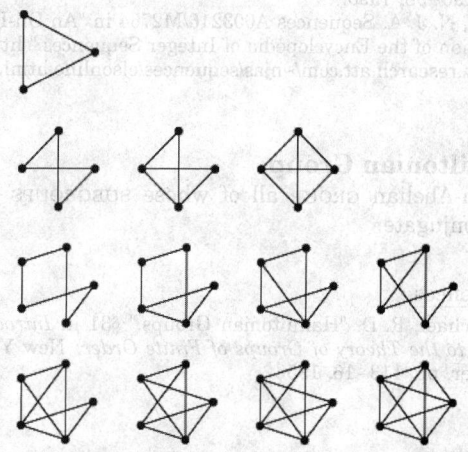

A GRAPH possessing a HAMILTONIAN CIRCUIT. By convention, the trivial graph on a single node is considered Hamiltonian, but the connected graph on two nodes is not. The numbers of simple Hamiltonian graphs on n nodes for $n = 1, 2, ...$ are then 1, 0, 1, 3, 8, 48, 383, ... (Sloane's A003216).

Testing whether a graph is Hamiltonian is an NP-COMPLETE PROBLEM (Skiena 1990, p. 196). An algorithm to test graphs is implemented as `HamiltonianQ[g]` in the *Mathematica* add-on package `DiscreteMath`Combinatorica`` (which can be loaded with the command `<<DiscreteMath`).

All Hamiltonian graphs are BICONNECTED, although the converse is not true (Skiena 1990, p. 197). If the sums of the degrees of nonadjacent vertices in a graph G is greater than the number of nodes n for all subsets of nonadjacent vertices, then G is Hamiltonian (Ore 1960; Skiena 1990, p. 197).

See also BARNETTE'S CONJECTURE, BICUBIC GRAPH, CHVÁTAL'S THEOREM, EULERIAN GRAPH, HAMILTONIAN CIRCUIT, HAMILTON-CONNECTED GRAPH, HAMILTONIAN PATH, HYPOHAMILTONIAN GRAPH, HYPOTRACEABLE GRAPH, ORE GRAPH, TAIT'S HAMILTONIAN GRAPH CONJECTURE, TUTTE CONJECTURE

References

Bollobás, B. *Graph Theory: An Introductory Course.* New York: Springer-Verlag, p. 12, 1979.

Chartrand, G. *Introductory Graph Theory.* New York: Dover, p. 68, 1985.

Chartrand, G.; Kapoor, S. F.; and Kronk, H. V. "The Many Facets of Hamiltonian Graphs." *Math. Student* **41**, 327–36, 1973.

Dolch, J. P. "Names of Hamiltonian Graphs." In *4th S-E Conf. Combin., Graph Theory, Computing. Congress. Numer.* **8**, 259–71, 1973.

Harary, F. and Palmer, E. M. *Graphical Enumeration.* New York: Academic Press, p. 219, 1973.

Ore, O. "A Note on Hamiltonian Circuits." *Amer. Math. Monthly* **67**, 55, 1960.

Skiena, S. "Hamiltonian Cycles." §5.3.4 in *Implementing Discrete Mathematics: Combinatorics and Graph Theory with Mathematica.*   Reading, MA: Addison-Wesley, pp. 196–98, 1990.

Sloane, N. J. A. Sequences A003216/M2764 in "An On-Line Version of the Encyclopedia of Integer Sequences." http://www.research.att.com/~njas/sequences/eisonline.html.

Hamiltonian Group

A non-Abelian GROUP all of whose SUBGROUPS are self-conjugate.

References

Carmichael, R. D. "Hamiltonian Groups." §31 in *Introduction to the Theory of Groups of Finite Order.* New York: Dover, pp. 113–16, 1956.

Hamiltonian Integer

A LINEAR COMBINATION of basis QUATERNIONS with integer coefficients.

See also QUATERNION

References

Ferguson, H. R. P.; Bailey, D. H.; and Arno, S. "Analysis of PSLQ, An Integer Relation Finding Algorithm." *Math. Comput.* **68**, 351–69, 1999.

Hamiltonian Map

Consider a 1-D Hamiltonian MAP OF THE FORM

$$H(p, q) = \tfrac{1}{2} p^2 + V(q), \qquad (1)$$

which satisfies HAMILTON'S EQUATIONS

$$\dot{q} = \frac{\partial H}{\partial p} \qquad (2)$$

$$\dot{p} = -\frac{\partial H}{\partial q}. \qquad (3)$$

Now, write

$$\dot{q}_i = \frac{(q_{i+1} - q_i)}{\Delta t}, \qquad (4)$$

where

$$q_i = q(t) \qquad (5)$$

$$q_{i+1} = q(t + \Delta t). \qquad (6)$$

Then the equations of motion become

$$q_{i+1} = q_i + p_i \Delta t \qquad (7)$$

$$p_{i+1} = p_i - \Delta t \left(\frac{\partial V}{\partial q_i} \right)_{q = qi} \qquad (8)$$

Note that equations (7) and (8) are not AREA-PRESERVING, since

$$\frac{\partial(q_{i+1}, p_{i+1})}{\partial(q_i, p_i)} = \begin{vmatrix} 1 & -\Delta t \dfrac{\partial^2 V}{\partial q_i^2} \\ \Delta t & 1 \end{vmatrix} = 1 + (\Delta t)^2 \frac{\partial^2 V}{\partial q_i^2} \neq 1. \quad (9)$$

However, if we take instead of (7) and (8),

$$q_{i+1} = q_i + p_i \Delta t \qquad (10)$$

$$p_{i+1} = p_i - \Delta t \left(\frac{\partial V}{\partial q_i} \right)_{q = q_{i+1}} \qquad (11)$$

$$\frac{\partial(q_{i+1}, p_{i+1})}{\partial(q_i, p_i)} = \begin{vmatrix} 1 & -\Delta t \dfrac{\partial}{\partial q_i} \left(\dfrac{\partial V}{\partial q} \right)_{q = q_{i+1}} \\ \Delta t & 1 \end{vmatrix}$$

$$= 1 + (\Delta t)^2 \frac{\partial^2 V}{\partial q_i^2} = 1, \qquad (12)$$

which is AREA-PRESERVING.

See also AREA-PRESERVING MAP

Hamiltonian Path

A path between two vertices of a GRAPH that visits each vertex exactly once. A Hamiltonian path that is also a GRAPH CYCLE is called a HAMILTONIAN CIRCUIT (or Hamiltonian cycle). Every TOURNAMENT has an ODD NUMBER of Hamiltonian paths (Rédei 1934; Szele 1943; Skiena 1990, p. 175).

The number of Hamiltonian paths on an n-HYPER-CUBE is 0, 0, 48, 48384, ... (Sloane's A006070; Gardner 1986, pp. 23–4).

See also HAMILTONIAN CIRCUIT, HAMILTONIAN GRAPH, TOURNAMENT

References

Gardner, M. "The Binary Gray Code." In *Knotted Doughnuts and Other Mathematical Entertainments.* New York: W. H. Freeman, pp. 23–4, 1986.
Rédei, L. "Ein Kombinatorischer Satz." *Acta Litt. Szeged.* **7**, 39–3, 1934.
Skiena, S. *Implementing Discrete Mathematics: Combinatorics and Graph Theory with Mathematica.* Reading, MA: Addison-Wesley, p. 175, 1990.
Sloane, N. J. A. Sequences A006070/M5295 in "An On-Line Version of the Encyclopedia of Integer Sequences." http://www.research.att.com/~njas/sequences/eisonline.html.
Szele, T. "Kombinatorische Untersuchungen über den gerichteten vollständigen Graphen." *Mat. Fiz. Lapok* **50**, 223–56, 1943.

Hamiltonian System

A system of variables which can be written in the form of HAMILTON'S EQUATIONS.

Hammer's X-Ray Problems

Let a homogeneous solid contain a convex hole K and take x-rays so that the "darkness" at each point on a photographic plate determines the length of the chord of K along the line of propagation of an x-ray. Then how many x-ray pictures must be taken to exactly reconstruct K if

1. The x-rays originate from a point source,
2. The x-rays originate from a source at infinity and so are parallel?

See also RADON TRANSFORM

References

Croft, H. T.; Falconer, K. J.; and Guy, R. K. "Hammer's X-Ray Problems." §A2 in *Unsolved Problems in Geometry.* New York: Springer-Verlag, pp. 11-4, 1991.

Hammer-Aitoff Equal-Area Projection

A MAP PROJECTION whose inverse is defined using the intermediate variable

$$z \equiv \sqrt{1 - \left(\tfrac{1}{4}x\right)^2 - \left(\tfrac{1}{2}y\right)^2}.$$

Then the longitude and latitude are given by

$$\lambda = 2 \tan^{-1}\left[\frac{zx}{2(2z^2 - 1)}\right]$$

$$\phi = \sin^{-1}(yz).$$

See also EQUAL-AREA PROJECTION

Hamming Code

A binary Hamming code H_r of length $n = 2^r - 1$ (with $r \geq 2$) is a linear code with parity-check matrix **H** whose columns consist of all nonzero binary vectors of length r, each used once. H_r is an ($n = 2^r - 1$, $k = 2^r - 1 - r$, $d = 3$) code. Hamming codes are PERFECT single ERROR-CORRECTING CODES.

See also ERROR-CORRECTING CODE, PERFECT CODE

References

MacWilliams, F. J. and Sloane, N. J. A. *The Theory of Error-Correcting Codes.* Amsterdam, Netherlands: North-Holland, 1977.

Hamming Function

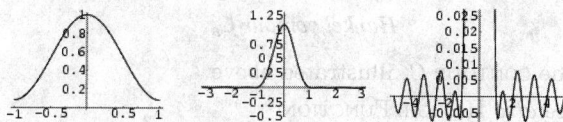

An APODIZATION FUNCTION chosen to minimize the height of the highest sidelobe (Hamming and Tukey, Blackman and Tukey 1959). The Hamming function is given by

$$A(x) = 0.54 + 0.46 \cos\left(\frac{\pi x}{a}\right), \qquad (1)$$

and its FULL WIDTH AT HALF MAXIMUM is $1.05543a$. The corresponding INSTRUMENT FUNCTION is

$$I(k) = \frac{a(1.08 - 0.64a^2k^2)\,\text{sinc}(2\pi ak)}{1 - 4a^2k^2}. \qquad (2)$$

This APODIZATION FUNCTION is close to the one produced by the requirement that the APPARATUS FUNCTION goes to 0 at $ka = 5/4$. From APODIZATION FUNCTION, a general symmetric apodization function $A(x)$ can be written as a FOURIER SERIES

$$A(x) = a_0 + 2 \sum_{n=1}^{\infty} a_n \cos\left(\frac{n\pi x}{b}\right), \qquad (3)$$

where the COEFFICIENTS satisfy

$$a_0 + 2 \sum_{n=1}^{\infty} a_n = 1. \qquad (4)$$

The corresponding apparatus function is

$$I(t) = 2b\{a_0\,\text{sinc}(2\pi kb)$$

$$+ \sum_{n=1}^{\infty} [\text{sinc}(2\pi kb + n\pi) + \text{sinc}(2\pi kb - n\pi)]\}. \qquad (5)$$

To obtain an APODIZATION FUNCTION with zero at $ka = 3/4$, use

$$a_0 + 2a_1 = 1, \qquad (6)$$

so

$$a_0\,\text{sinc}\left(\tfrac{5}{2}\pi\right) + a_1\left[\text{sinc}\left(\tfrac{7}{2}\pi\right) + \text{sinc}\left(\tfrac{3}{2}\pi\right)\right] = 0 \qquad (7)$$

$$(1 - 2a_1)\frac{2}{5\pi} - a_1\left(\frac{2}{7\pi} + \frac{2}{3\pi}\right) = (1 - 2a_1)\tfrac{1}{5} - a_1\left(\tfrac{1}{7} + \tfrac{1}{3}\right)$$

$$= 0 \qquad (8)$$

$$a_1\left(\tfrac{1}{7} + \tfrac{1}{3} + \tfrac{2}{5}\right) = \tfrac{1}{5} \qquad (9)$$

$$a_1 = \frac{\tfrac{1}{5}}{\tfrac{2}{5} + \tfrac{1}{7} + \tfrac{1}{3}} = \frac{7 \cdot 3}{2 \cdot 3 \cdot 7 + 3 \cdot 5 + 5 \cdot 7}$$

$$= \tfrac{21}{92} \approx 0.2283 \qquad (10)$$

$$a_0 = 1 - 2a_1 = \frac{92 - 2 \cdot 21}{92} = \frac{92 - 42}{92}$$

$$= \tfrac{50}{92} = \tfrac{25}{46} \approx 0.5435. \qquad (11)$$

The FWHM is 1.81522, the peak is 1.08, the peak NEGATIVE and POSITIVE sidelobes (in units of the peak) are -0.00689132 and 0.00734934, respectively.

See also APODIZATION FUNCTION, HANNING FUNCTION, INSTRUMENT FUNCTION

References

Blackman, R. B. and Tukey, J. W. "Particular Pairs of Windows." In *The Measurement of Power Spectra, From the Point of View of Communications Engineering.* New York: Dover, pp. 98–9, 1959.

Hamming, R. W. and Tukey, J. W. "Measuring Noise Color." Unpublished memorandum.

Handedness

Objects which are identical except for a mirror reflection are said to display handedness and to be CHIRAL.

See also AMPHICHIRAL, CHIRAL, ENANTIOMER, MIRROR IMAGE

Handkerchief Surface

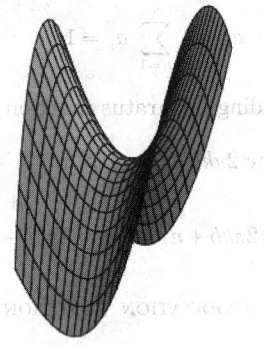

A surface given by the PARAMETRIC EQUATIONS

$$x(u, v) = u$$
$$y(u, v) = v$$
$$z(u, v) = \tfrac{1}{3}u^3 + uv^2 + 2(u^2 - v^2).$$

References

Gray, A. *Modern Differential Geometry of Curves and Surfaces with Mathematica, 2nd ed.* Boca Raton, FL: CRC Press, pp. 948–49, 1997.

Handle

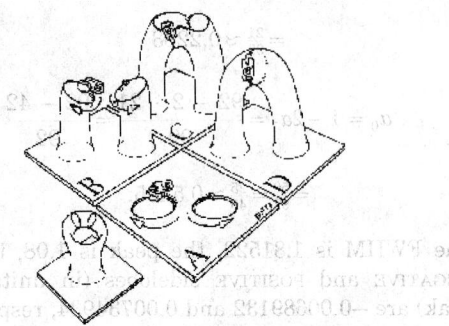

A handle is a topological structure which can be thought of as the object produced by puncturing a surface twice, attaching a ZIP around each puncture travelling in opposite directions, pulling the edges of the zips together, and then zipping up.

Handles are to MANIFOLDS as CELLS are to CW-COMPLEXES. If M is a MANIFOLD together with a $(k-1)$-SPHERE $\mathbb{S}^{k-1}$ embedded in its boundary with a trivial TUBULAR NEIGHBORHOOD, we attach a k-handle to M by gluing the tubular NEIGHBORHOOD of the $(k-1)$-SPHERE $\mathbb{S}^{k-1}$ to the TUBULAR NEIGHBORHOOD of the standard $(k-1)$-SPHERE $\mathbb{S}^{k-1}$ in the $\dim(M)$-dimensional DISK. In this way, attaching a k-handle is essentially just the process of attaching a fattened-up k-DISK to M along the $(k-1)$-SPHERE $\mathbb{S}^{k-1}$. The embedded DISK in this new MANIFOLD is called the k-handle in the UNION of M and the handle.

DYCK'S THEOREM states that HANDLES and cross-handles are equivalent in the presence of a CROSS-CAP.

See also CAP, CLASSIFICATION THEOREM OF SURFACES, CROSS-CAP, CROSS-HANDLE, HANDLEBODY, SURGERY, TUBULAR NEIGHBORHOOD

References

Francis, G. K. and Weeks, J. R. "Conway's ZIP Proof." *Amer. Math. Monthly* **106**, 393–99, 1999.

Handlebody

A handlebody of type (n, k) is an n-D MANIFOLD that is attained from the standard n-DISK by attaching only k-D HANDLES.

See also HANDLE, HEEGAARD SPLITTING, SURGERY

References

Rolfsen, D. *Knots and Links.* Wilmington, DE: Publish or Perish Press, p. 46, 1976.

Handsome Number

POWERFUL NUMBER

Hankel Contour

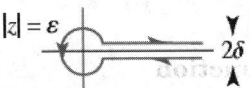

Hankel contour C_ϵ

The CONTOUR C_ϵ illustrated above.

See also HANKEL FUNCTION

References

Krantz, S. G. "The Hankel Contour and Hankel Functions." §13.2.4 in *Handbook of Complex Analysis.* Boston, MA: Birkhäuser, p. 159, 1999.

Hankel Function

There are two types of functions known as Hankel functions. The more common one is a COMPLEX

FUNCTION (also called a Bessel function of the third kind, or Weber Function) which is a LINEAR COMBINATION of BESSEL FUNCTIONS OF THE FIRST and SECOND KINDS. These are called the HANKEL FUNCTIONS OF THE FIRST and SECOND KINDS.

Another type of Hankel function is defined by the CONTOUR INTEGRAL

$$H_\epsilon(z) = \int_{C_\epsilon} \frac{(-w)^{z-1} e^{-w}}{1 - e^{-w}}\, dw$$

for $\Im[w] < 0$, $|\arg(-w)| < \pi$, $\epsilon \neq 2\pi k > 0$, where C_ϵ is a HANKEL CONTOUR. The RIEMANN ZETA FUNCTION can be expressed in terms of $H_\epsilon(z)$ as

$$\zeta(z) = -\frac{H_\epsilon(z)}{2i \sin(\pi z) \Gamma(z)}$$

for $0 < \epsilon < 2\pi$ and $\Re[z] > 1$, where $\Gamma(z)$ is the GAMMA FUNCTION (Krantz 1999, p. 160).

See also HANKEL CONTOUR, HANKEL FUNCTION OF THE FIRST KIND, HANKEL FUNCTION OF THE SECOND KIND, SPHERICAL HANKEL FUNCTION OF THE FIRST KIND, SPHERICAL HANKEL FUNCTION OF THE SECOND KIND, THIRD KIND

References

Arfken, G. "Hankel Functions." §11.4 in *Mathematical Methods for Physicists, 3rd ed.* Orlando, FL: Academic Press, pp. 604–10, 1985.
Hankel, H. "Die Cylinderfunctionen erster und zweiter Art." *Math. Ann.* **1**, 467–01, 1869.
Hankel, H. "Bestimmte Integrale mit Cylinderfunctionen." *Math. Ann.* **8**, 453–70, 1875.
Krantz, S. G. "The Hankel Contour and Hankel Functions." §13.2.4 in *Handbook of Complex Analysis.* Boston, MA: Birkhäuser, p. 159, 1999.
Morse, P. M. and Feshbach, H. *Methods of Theoretical Physics, Part I.* New York: McGraw-Hill, pp. 623–24, 1953.

Hankel Function of the First Kind

$$H_n^{(1)}(z) \equiv J_n(z) - iY_n(z),$$

where $J_n(z)$ is a BESSEL FUNCTION OF THE FIRST KIND and $Y_n(z)$ is a BESSEL FUNCTION OF THE SECOND KIND. Hankel functions of the first kind can be REPRESENTED AS A CONTOUR INTEGRAL over the UPPER HALF-PLANE using

$$H_n^{(1)}(z) = \frac{1}{i\pi} \int_{0[\text{upper half plane}]}^{\infty} \frac{e^{(z/2)(t-1/t)}}{t^{n+1}}\, dt.$$

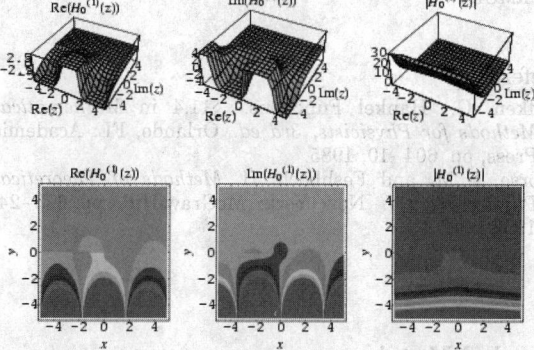

The plots above show the structure of $H_0^{(1)}(z)$ in the COMPLEX PLANE.

See also BESSEL FUNCTION OF THE FIRST KIND, BESSEL FUNCTION OF THE SECOND KIND, DEBYE'S ASYMPTOTIC REPRESENTATION, HANKEL FUNCTION OF THE SECOND KIND, MACDONALD FUNCTION, WATSON-NICHOLSON FORMULA, WEYRICH'S FORMULA

References

Arfken, G. "Hankel Functions." §11.4 in *Mathematical Methods for Physicists, 3rd ed.* Orlando, FL: Academic Press, pp. 604–10, 1985.
Morse, P. M. and Feshbach, H. *Methods of Theoretical Physics, Part I.* New York: McGraw-Hill, pp. 623–24, 1953.

Hankel Function of the Second Kind

$$H_n^{(2)}(z) \equiv J_n(z) - iY_n(z),$$

where $J_n(z)$ is a BESSEL FUNCTION OF THE FIRST KIND and $Y_n(z)$ is a BESSEL FUNCTION OF THE SECOND KIND. Hankel functions of the second kind can be REPRESENTED AS A CONTOUR INTEGRAL using

$$H_n^{(2)}(z) = \frac{1}{i\pi} \int_{-\infty[\text{lower half plane}]}^{0} \frac{e^{(z/2)(t-1/t)}}{t^{n+1}}\, dt.$$

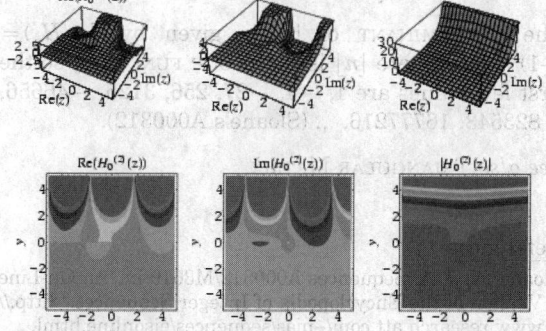

The plots above show the structure of $H_0^{(2)}(z)$ in the COMPLEX PLANE.

See also BESSEL FUNCTION OF THE FIRST KIND, BESSEL FUNCTION OF THE SECOND KIND, HANKEL FUNCTION OF THE FIRST KIND, WATSON-NICHOLSON FORMULA

References

Arfken, G. "Hankel Functions." §11.4 in *Mathematical Methods for Physicists, 3rd ed.* Orlando, FL: Academic Press, pp. 604–10, 1985.
Morse, P. M. and Feshbach, H. *Methods of Theoretical Physics, Part I.* New York: McGraw-Hill, pp. 623–24, 1953.

Hankel Matrix

A MATRIX H_n where the first row (and column) consists of the integers 1, 2, ..., n, the second row (and column) is given by 2, 3, ..., n, 0, and so on, with the nth row (and column) given by n,

$$\underbrace{0, \ldots, 0}_{n-1}.$$

A Hankel matrix can be given by `HankelMatrix[m, n]` in the *Mathematica* add-on package `LinearAlgebra'MatrixManipulation'` (which can be loaded with the command `<<LinearAlgebra'`). The first few such matrices are

$$H_2 = \begin{bmatrix} 1 & 2 \\ 2 & 0 \end{bmatrix}$$

$$H_3 = \begin{bmatrix} 1 & 2 & 3 \\ 2 & 3 & 0 \\ 3 & 0 & 0 \end{bmatrix}$$

$$H_4 = \begin{bmatrix} 1 & 2 & 3 & 4 \\ 2 & 3 & 4 & 0 \\ 3 & 4 & 0 & 0 \\ 4 & 0 & 0 & 0 \end{bmatrix}.$$

The elements of the Hankel matrix are given explicitly by

$$h_{ij} = \begin{cases} 0 & \text{if } i+j-1 > n \\ i+j-1 & \text{otherwise.} \end{cases}$$

The DETERMINANT of H_n is given by $\det(H_n) = (-1)^{\lfloor n/2 \rfloor} n^n$, where $\lfloor n \rfloor$ is the FLOOR FUNCTION, so the first few values are 1, -4, -27, 256, 3125, -46656, -823543, 16777216, ... (Sloane's A000312).

See also TRIANGULAR MATRIX

References

Sloane, N. J. A. Sequences A000312/M3619 in "An On-Line Version of the Encyclopedia of Integer Sequences." http://www.research.att.com/~njas/sequences/eisonline.html.

Hankel Transform

Equivalent to a 2-D FOURIER TRANSFORM with a radially symmetric KERNEL, and also called the FOURIER-BESSEL TRANSFORM.

$$g(u, v) = \mathscr{F}[f(r)] = \int_{-\infty}^{\infty} \int_{-\infty}^{\infty} f(r) e^{-2\pi i(ux+vy)} \, dx \, dy. \quad (1)$$

Let

$$x + iy = re^{i\theta} \quad (2)$$

$$u + iv = qe^{i\phi} \quad (3)$$

so that

$$x = r \cos \theta \quad (4)$$

$$y = r \sin \theta \quad (5)$$

$$r = \sqrt{x^2 + y^2} \quad (6)$$

$$u = q \cos \phi \quad (7)$$

$$v = q \sin \phi \quad (8)$$

$$q = \sqrt{u^2 + v^2}. \quad (9)$$

Then

$$g(q) = \int_0^{\infty} \int_0^{2\pi} f(r) e^{-2\pi i r q(\cos \phi \cos \theta + \sin \phi \sin \theta)} r \, dr \, d\theta$$

$$= \int_0^{\infty} \int_0^{2\pi} f(r) e^{-2\pi i r q \cos(\theta - \phi)} r \, dr \, d\theta$$

$$= \int_0^{\infty} \int_{-\phi}^{2\pi - \phi} f(r) e^{-2\pi i r q \cos \theta} r \, dr \, d\theta$$

$$= \int_0^{\infty} \int_0^{2\pi} f(r) e^{-2\pi i r q \cos \theta} r \, dr \, d\theta$$

$$= \int_0^{\infty} f(r) \left[\int_0^{2\pi} e^{-2\pi i r q \cos \theta} d\theta \right] r \, dr$$

$$= 2\pi \int_0^{\infty} f(r) J_0(2\pi q r) r \, dr, \quad (10)$$

where $J_0(z)$ is a zeroth order BESSEL FUNCTION OF THE FIRST KIND.

Therefore, the Hankel transform pairs are

$$g(q) = 2\pi \int_0^{\infty} f(x) J_0(2\pi q r) r \, dr \quad (11)$$

$$f(r) = 2\pi \int_0^{\infty} g(q) J_0(2\pi q r) q \, dq. \quad (12)$$

The following table gives Hankel transforms for a number of common functions (Bracewell 1999, p. 249). Here, $J_n(x)$ is a BESSEL FUNCTION and $\prod_a(r)$ is a RECTANGLE FUNCTION equal to 1 for $0 \le r \le a$ and 0 otherwise, and

$$M(x) = 2\pi \left[x^{-3} \int_0^x J_0(x)\, dx - x^{-2} J_0(x) \right] \qquad (13)$$

$$= \frac{\pi^2}{x^2} [J_1(x)\mathcal{H}_0(x) - J_0(x)\mathcal{H}_1(x)], \qquad (14)$$

where $J_n(x)$ is a BESSEL FUNCTION OF THE FIRST KIND, $\mathcal{H}_n(x)$ is a STRUVE FUNCTION and $\mathcal{L}_n(x)$ is a MODIFIED STRUVE FUNCTION.

$f(r)$	$g(q)$
$\prod_a(r)$	$\dfrac{aJ_1(2\pi aq)}{q}$
$\dfrac{\sin(2\pi ar)}{r}$	$\dfrac{\prod(q/(2a))}{\sqrt{a^2-q^2}}$
$\frac{1}{2}\delta(r-a)$	$\pi a J_0(2\pi aq)$
$M(ar)$	$a\Lambda\left(\dfrac{q}{2a}\right)$
$e^{-\pi r^2}$	$e^{-\pi q^2}$
$(a^2+r^2)^{-1/2}$	$\dfrac{e^{-2\pi aq}}{q}$
$(a^2+r^2)^{-1/3}$	$\dfrac{2\pi e^{-2\pi aq}}{a}$
$\dfrac{1}{a^2+r^2}$	$2\pi K_0(2\pi aq)$
$\dfrac{2a^2}{(a^2+r^2)^2}$	$4\pi^2 aq K_1(2\pi aq)$
$\dfrac{4a^4}{(a^2+r^2)^3}$	$4\pi^3 a^2 q^2 K_2(2\pi aq)$
$(a^2-r^2)\prod_a(r)$	$\dfrac{a^2}{\pi q^2}J_2(2\pi aq)$
$\dfrac{1}{r}$	$\dfrac{1}{q}$
e^{-ar}	$\dfrac{2\pi a}{(a^2+4\pi^2 q^2)^{3/2}}$
$\dfrac{e^{-ar}}{r}$	$\dfrac{2\pi}{\sqrt{a^2+4\pi^2 q^2}}$
$\dfrac{\delta(r)}{2\pi r}$	1
$r^2 e^{-\pi r^2}$	$\dfrac{e^{-\pi q^2}(1-\pi q^2)}{\pi}$
$-r^2 f(r)$	$\left(\dfrac{d^2 f}{dq^2}+\dfrac{1}{q}\dfrac{dF}{dq}\right)=\nabla^2 f$

See also BESSEL FUNCTION OF THE FIRST KIND, FOURIER TRANSFORM, LAPLACE TRANSFORM

References

Arfken, G. *Mathematical Methods for Physicists, 3rd ed.* Orlando, FL: Academic Press, p. 795, 1985.

Bracewell, R. "The Hankel Transform." *The Fourier Transform and Its Applications, 3rd ed.* New York: McGraw-Hill, pp. 244–50, 1999.
Oberhettinger, F. *Tables of Bessel Transforms.* New York: Springer-Verlag, 1972.
Samko, S. G.; Kilbas, A. A.; and Marichev, O. I. *Fractional Integrals and Derivatives.* Yverdon, Switzerland: Gordon and Breach, p. 23, 1993.

Hankel's Integral

$$J_m(x) = \frac{x^m}{2^{m-1}\sqrt{\pi}\,\Gamma\left(m+\frac{1}{2}\right)} \int_0^1 \cos(xt)$$
$$\times \left(1-t^2\right)^{m-1/2} dt,$$

where $J_m(x)$ is a BESSEL FUNCTION OF THE FIRST KIND and $\Gamma(z)$ is the GAMMA FUNCTION. Hankel's integral can be derived from SONINE'S INTEGRAL.

See also POISSON INTEGRAL, SONINE'S INTEGRAL

Hankel's Symbol
The symbol defined by

$$(v,\, n)$$

$$\equiv \frac{2^{-2n}\left\{(4v^2-1)(4v^2-3^2)\cdots\left[4v^2-(2n-1)^2\right]\right\}}{n!}$$
$$\qquad (1)$$

$$= \frac{(-1)^n \cos(\pi v)\Gamma\left(\frac{1}{2}+n-v\right)\Gamma\left(\frac{1}{2}+n+v\right)}{xn!}, \qquad (2)$$

where $\Gamma(z)$ is the GAMMA FUNCTION. If v is an integer, then this simplifies to

$$(v,\, n) = \frac{(-1)^{n+v}\Gamma\left(\frac{1}{2}+n-v\right)\Gamma\left(\frac{1}{2}+n+v\right)}{\pi n!}, \qquad (3)$$

given incorrectly by Erdélyi *et al.* (1981, p. 52).

See also KRAMP'S SYMBOL, POCHHAMMER SYMBOL

References

Erdélyi, A.; Magnus, W.; Oberhettinger, F.; and Tricomi, F. G. *Higher Transcendental Functions, Vol. 1.* New York: Krieger, p. 52, 1981.

Hann Function
HANNING FUNCTION

Hanning Function

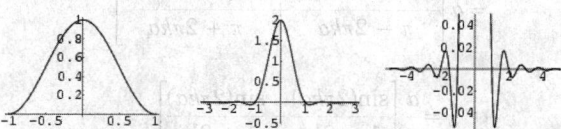

An APODIZATION FUNCTION, also called the HANN FUNCTION, frequently used to reduce ALIASING in

FOURIER TRANSFORMS. The illustrations above show the Hanning function, its INSTRUMENT FUNCTION, and a blowup of the INSTRUMENT FUNCTION sidelobes. It is named after the Austrian meteorologist Julius von Hann (Blackman and Tukey 1959, pp. 98–9). The Hanning function is given by

$$f(x) = \cos^2\left(\frac{\pi x}{2a}\right) = \frac{1}{2} - \frac{1}{2}\cos\left(\frac{\pi x}{a}\right). \qquad (1)$$

The INSTRUMENT FUNCTION for Hanning apodization can also be written

$$a\left[\operatorname{sinc}(2\pi ka) + \tfrac{1}{2}\operatorname{sinc}(2\pi ka - \pi) + \tfrac{1}{2}\operatorname{sinc}(2\pi ka + \pi)\right]. \qquad (2)$$

Its FULL WIDTH AT HALF MAXIMUM is a. It has APPARATUS FUNCTION

$$
\begin{aligned}
A(x) &= \int_{-a}^{a}\left[\frac{1}{2} - \frac{1}{2}\cos\left(\frac{\pi x}{a}\right)\right]e^{-2\pi ikx}\,dx \\
&= \frac{1}{2}\int_{-a}^{a}e^{-2\pi ikx}\,dx - \frac{1}{2}\int_{-a}^{a}e^{-2\pi ikx}\,dx \\
&\equiv \frac{1}{2}(A_1 + A_2).
\end{aligned} \qquad (3)
$$

The first integral is

$$I_1 = \int_{-a}^{a} e^{-2\pi ikx}\,dx = \frac{\sin(2\pi ka)}{\pi k} = 2a\,\operatorname{sinc}(2\pi ka). \qquad (4)$$

The second integral can be rewritten

$$
\begin{aligned}
I_2 &= \int_{-a}^{0}\cos\left(\frac{\pi x}{a}\right)e^{-2\pi ikx}\,dx + \int_{-a}^{0}\cos\left(\frac{\pi x}{a}\right)e^{-2\pi ikx}\,dx \\
&= \int_{0}^{a}\cos\left(\frac{\pi x}{a}\right)\left(e^{2\pi ikx} + e^{-2\pi ikx}\right)dx \\
&= 2\int_{0}^{a}\cos\left(\frac{\pi x}{a}\right)\cos(2\pi kx)\,dx \\
&= 2\left\{\frac{\sin\left(\frac{\pi}{a} - 2\pi k\right)x}{2\left(\frac{\pi}{a} - 2\pi k\right)} + \frac{\sin\left(\frac{\pi}{a} + 2\pi k\right)x}{2\left(\frac{\pi}{a} + 2\pi k\right)}\right\}_0^a \\
&= a\left[\frac{\sin(\pi - 2\pi ka)}{\pi - 2\pi ka} + \frac{\sin(\pi + 2\pi ka)}{\pi + 2\pi ka}\right] \\
&= \frac{a}{\pi}\left[\frac{\sin(2\pi ka)}{1 - 2ka} - \frac{\sin(2\pi ka)}{1 + 2ka}\right] \\
&= a[\operatorname{sinc}(\pi - 2\pi ka) + \operatorname{sinc}(\pi + 2\pi ka)]. \qquad (5)
\end{aligned}
$$

Combining (4) and (5) gives

$$A(x)$$
$$= a\left[\operatorname{sinc}(2\pi ka) + \tfrac{1}{2}\operatorname{sinc}(\pi - 2\pi ka) + \tfrac{1}{2}\operatorname{sinc}(\pi + 2\pi ka)\right]. \qquad (6)$$

To find the extrema, define $x \equiv 2\pi ka$ and rewrite (6) as

$$A(x) = a\left[\sin x + \tfrac{1}{2}\operatorname{sinc}(x - \pi) + \tfrac{1}{2}\operatorname{sinc}(x + \pi)\right]. \qquad (7)$$

Then solve

$$
\begin{aligned}
\frac{dA}{dx} &= \frac{\pi^2(-x^3\cos x + 3x^2\sin x + \pi^2 x\cos x - \pi^2\sin x)}{x^2(\pi^2 - x^2)^2} \\
&= 0 \qquad (8)
\end{aligned}
$$

to find the extrema. The roots are $x = 7.42023$ and 10.7061, giving a peak NEGATIVE sidelobe of -0.026708 and a peak POSITIVE sidelobe (in units of a) of 0.00843441. The peak in units of a is 1, and the full-width at half maximum is given by setting (7) equal to $1/2$ and solving for x, yielding

$$x_{1/2} = 2\pi k_{1/2}a = \pi. \qquad (9)$$

Therefore, with $L \equiv 2a$, the FULL WIDTH AT HALF MAXIMUM is

$$\text{FWHM} = 2k_{1/2} = \frac{1}{a} = \frac{2}{L}. \qquad (10)$$

See also APODIZATION FUNCTION, HAMMING FUNCTION

References

Blackman, R. B. and Tukey, J. W. "Particular Pairs of Windows." In *The Measurement of Power Spectra, From the Point of View of Communications Engineering*. New York: Dover, 1959.

Hanoi Graph

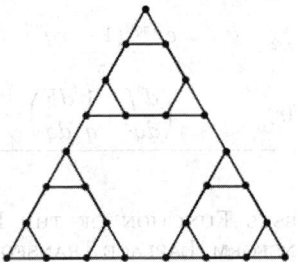

A GRAPH H_n arising in conjunction with the TOWERS OF HANOI problem. The above figure is the Hanoi graph H_3.

See also TOWERS OF HANOI

Hanoi Towers

Towers of Hanoi

Hansen Chain

An ADDITION CHAIN for which there is a SUBSET H of members such that each member of the chain uses the largest element of H which is less than the member.

See also ADDITION CHAIN, BRAUER CHAIN, HANSEN NUMBER

References

Guy, R. K. "Addition Chains. Brauer Chains. Hansen Chains." §C6 in *Unsolved Problems in Number Theory, 2nd ed.* New York: Springer-Verlag, pp. 111–13, 1994.

Hansen Number

A number n for which a shortest chain exists (which is also a HANSEN CHAIN) is called a Hansen number.

References

Guy, R. K. *Unsolved Problems in Number Theory, 2nd ed.* New York: Springer-Verlag, pp. 111–12, 1994.

Hansen's Problem

A SURVEYING PROBLEM: from the position of two known but inaccessible points A and B, determine the position of two unknown accessible points P and P' by bearings from A, B, P' to P and A, B, P to P'.

See also SURVEYING PROBLEMS

References

Dörrie, H. "Annex to a Survey." §40 in *100 Great Problems of Elementary Mathematics: Their History and Solutions.* New York: Dover, pp. 193–97, 1965.

Hansen-Bessel Formula

$$J_n(z) \frac{1}{2\pi} \int_{-\pi}^{\pi} e^{iz \cos t} e^{in(t-\pi/2)}\, dt$$

$$= \frac{i^{-n}}{\pi} \int_0^{\pi} e^{iz \cos t} \cos(nt)\, dt$$

$$= \frac{1}{\pi} \int_0^{\pi} \cos(z \sin t - nt)\, dt$$

for $n = 0, 1, 2, \ldots$, where $J_n(z)$ is a BESSEL FUNCTION OF THE FIRST KIND.

References

Iyanaga, S. and Kawada, Y. (Eds.). *Encyclopedic Dictionary of Mathematics.* Cambridge, MA: MIT Press, p. 1472, 1980.

Happy End Problem

The problem of determining the smallest number of points $g(n)$ in GENERAL POSITION in the plane (i.e., no three of which are COLLINEAR), which always determine a CONVEX POLYGON of n sides. The problem was so-named by Erdos when two investigators who first worked on the problem, E. Klein and G. Szekeres, became engaged and subsequently married (Hoffman 1998, p. 76).

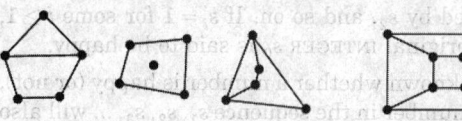

the three cases for 5 points　　　counterexample for 8 points

E. Klein proved that $g(4) = 5$ by showing that any arrangement of five points must fall into one of the three cases (left figure), and E. Makai proved $g(5) = 9$ after demonstrating that a counterexample could be found for eight points (right figure; Hoffman 1998, pp. 75–6). Erdos and Szekeres (1935) showed that $g(n)$ exists and derived the bound

$$2^{n-2} + 1 \leq g(n) \leq \binom{2n-4}{n-2} + 1, \qquad (1)$$

where $\binom{n}{k}$ is a BINOMIAL COEFFICIENT. For $n \geq 4$, this has since been reduced to

$$g(n) \leq \binom{2n-4}{n-2} \qquad (2)$$

by Chung and Graham (1998),

$$g(n) \leq \binom{2n-4}{n-2} + 7 - 2n \qquad (3)$$

by Kleitman and Pachter (1998), and

$$g(n) \leq \binom{2n-5}{n-2} + 2 \qquad (4)$$

by Tóth and Valtr (1998). For $g(6)$, these bounds give 71, 70, 65, and 37, respectively (Hoffman 1998, p. 78). The values of (4) for $n = 6, 7, \ldots$ are 37, 128, 464, 1718, ... (Sloane's A052473).

See also CONVEX HULL, CONVEX POLYGON

References

Chung, F. R. K. and Graham, R. L. "Forced Convex n-gons in the Plane." *Discr. Comput. Geom.* **19**, 367–71, 1998.

Erdos, P. and Szekeres, G. "A Combinatorial Problem in Geometry." *Compositio Math.* **2**, 463–70, 1935.

Hoffman, P. *The Man Who Loved Only Numbers: The Story of Paul Erdos and the Search for Mathematical Truth.* New York: Hyperion, pp. 75–8, 1998.

Kleitman, D. and Pachter, L. "Finding Convex Sets among Points in the Plane." *Discr. Comput. Geom.* **19**, 405–10, 1998.

Sloane, N. J. A. Sequences A052473 in "An On-Line Version of the Encyclopedia of Integer Sequences." http://www.research.att.com/~njas/sequences/eisonline.html.

Tóth, G. and Valtr, P. "Note on the Erdos-Szekeres Theorem." *Discr. Comput. Geom.* **19**, 457–59, 1998.

Happy Number

Let the sum of the SQUARES of the DIGITS of a POSITIVE INTEGER s_0 be represented by s_1. In a similar way, let the sum of the SQUARES of the DIGITS of s_1 be represented by s_2, and so on. If $s_i = 1$ for some $i \geq 1$, then the original INTEGER s_0 is said to be happy.

Once it is known whether a number is happy (or not), then any number in the sequence $s_1, s_2, s_3, \ldots$ will also be happy (or not). A number which is not happy is called UNHAPPY. Unhappy numbers have EVENTUALLY PERIODIC sequences of s_i which do not reach 1 (e.g., 4, 16, 37, 58, 89, 145, 42, 20, 4, ...).

Any PERMUTATION of the DIGITS of an UNHAPPY or happy number must also be unhappy or happy. This follows from the fact that ADDITION is COMMUTATIVE. The first few happy numbers are 1, 7, 10, 13, 19, 23, 28, 31, 32, 44, 49, 68, 70, 79, 82, 86, 91, 94, 97, 100, ... (Sloane's A007770). These are also the numbers whose 2-RECURRING DIGITAL INVARIANT sequences have period 1.

The first few happy primes are 7, 13, 19, 23, 31, 79, 97, 103, 109, 139, ... (Sloane's A035497).

See also KAPREKAR NUMBER, RECURRING DIGITAL INVARIANT , UNHAPPY NUMBER

References

Dudeney, H. E. Problem 143 in *536 Puzzles & Curious Problems.* New York: Scribner, pp. 43 and 258–59, 1967.
Guy, R. K. "Happy Numbers." §E34 in *Unsolved Problems in Number Theory, 2nd ed.* New York: Springer-Verlag, pp. 234–35, 1994.
Madachy, J. S. *Madachy's Mathematical Recreations.* New York: Dover, pp. 163–65, 1979.
Rivera, C. "Problems & Puzzles: Puzzle Happy Primes.-021." http://www.primepuzzles.net/puzzles/puzz_021.htm.
Schwartzman, S. *The Words of Mathematics: An Etymological Dictionary of Mathematical Terms Used in English.* Washington, DC: Math. Assoc. Amer., 1994.
Sloane, N. J. A. Sequences A007770 and A035497 in "An On-Line Version of the Encyclopedia of Integer Sequences." http://www.research.att.com/~njas/sequences/eisonline.html.
Weisstein, E. W. "Integer Sequences." MATHEMATICA NOTEBOOK INTEGERSEQUENCES.M.

Harada-Norton Group

The SPORADIC GROUP *HN*.

References

Wilson, R. A. "ATLAS of Finite Group Representation." http://for.mat.bham.ac.uk/atlas/html/HN.html.

Harary Graph

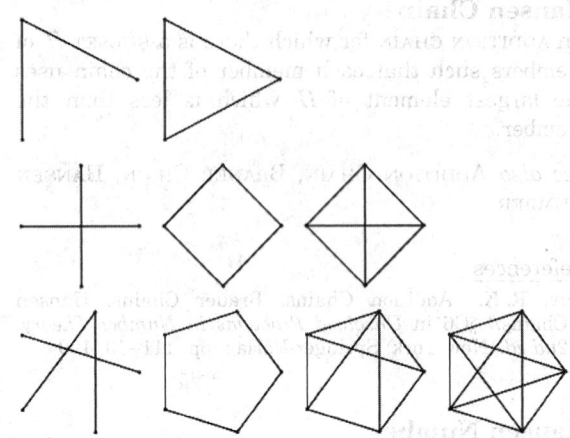

The smallest K-CONNECTED GRAPH $H_{k,n}$ with n VERTICES, having $\lceil kn/2 \rceil$ edges, where $\lceil x \rceil$ is the CEILING FUNCTION (Skiena 1990, p. 179). When n or k is even, $H_{k,n}$ is a CIRCULANT GRAPH. $H_{n-1,n}$ is the COMPLETE GRAPH K_n (Skiena 1990, p. 180).

See also K-CONNECTED GRAPH

References

Bondy, J. A. and Murty, U. S. R. *Graph Theory with Applications.* New York: North Holland, 1976.
Harary, F. "The Maximum Connectivity of a Graph." *Proc. Nat. Acad. Sci. USA* **48**, 1142–146, 1962.
Skiena, S. "Harary Graphs." §5.1.6 in *Implementing Discrete Mathematics: Combinatorics and Graph Theory with Mathematica.* Reading, MA: Addison-Wesley, pp. 179–80, 1990.

Harary-Read Number

POLYHEX

Harborth's Tiling

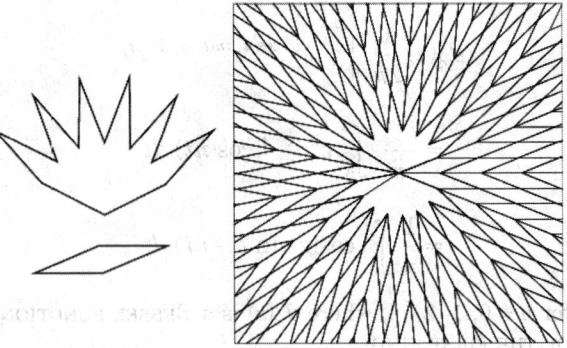

A TILING consisting of a RHOMBUS such that 17 rhombuses fit around a point and a second tile in the shape of six rhombuses stuck together. These two tiles can fill the plane in exactly four different ways.

Two tiles which tile the plane in n ways can be constructed using a rhombus of a shape such that $6n - 7$ pack around a point together with a complex piece made by sticking $2n - 2$ rhombuses together (Wells 1991).

References

Harborth, H. "Prescribed Numbers of Tiles and Tilings." *Math. Gaz.* **61**, 296–99, 1977.

Wells, D. *The Penguin Dictionary of Curious and Interesting Geometry.* Middlesex, England: Penguin Books, pp. 90–1, 1991.

Hard Hexagon Entropy Constant

N.B. A detailed online essay by S. Finch was the starting point for this entry.

A constant related to the HARD SQUARE ENTROPY CONSTANT. This constant is given by

$$\kappa_h \equiv \lim_{N \to \infty} [G(N)]^{1/N} = 1.395485972\ldots, \quad (1)$$

where $G(N)$ is the number of configurations of nonattacking KINGS on an $n \times n$ CHESSBOARD with regular hexagonal cells, where $N \equiv n^2$. Amazingly, κ_h is algebraic and given by

$$\kappa_h \equiv \kappa_1 \kappa_2 \kappa_3 \kappa_4, \quad (2)$$

where

$$\kappa_1 \equiv 4^{-1} 3^{5/4} 11^{-5/12} c^{-2} \quad (3)$$

$$\kappa_2 \equiv \left[1 - \sqrt{1-c} + \sqrt{2 + c + 2\sqrt{1 + c + c^2}} \right]^2 \quad (4)$$

$$\kappa_3 \equiv \left[-1 - \sqrt{1-c} + \sqrt{2 + c + 2\sqrt{1 + c + c^2}} \right]^2 \quad (5)$$

$$\kappa_4 \equiv \left[\sqrt{1-a} + \sqrt{2 + a + 2\sqrt{1 + a + a^2}} \right]^{-1/2} \quad (6)$$

$$a \equiv -\tfrac{124}{363} 11^{1/3} \quad (7)$$

$$b \equiv \tfrac{2501}{11979} 33^{1/2} \quad (8)$$

$$c \equiv \left\{ \tfrac{1}{4} + \tfrac{3}{8} a \left[(b+1)^{1/3} - (b-1)^{1/3} \right] \right\}^{1/3}. \quad (9)$$

(Baxter 1980, Joyce 1988).

References

Baxter, R. J. "Partition Function of the Eight-Vertex Lattice Model." *Ann. Phys.* **70**, 193–28, 1972.

Baxter, R. J. "Hard Hexagons: Exact Solution." *J. Physics A* **13**, 1023–030, 1980.

Baxter, R. J. *Exactly Solved Models in Statistical Mechanics.* New York: Academic Press, 1982.

Finch, S. "Favorite Mathematical Constants." http://www.mathsoft.com/asolve/constant/square/square.html.

Gaunt, D. S. "Hard-Sphere Lattice Gases. II. Plane-Triangular and Three-Dimensional Lattices." *J. Chem. Phys.* **46**, 3237–259, 1967.

Gaunt, D. S. and Fisher, M. E. "Hard-Sphere Lattice Gases. I. Plane-Square Lattice." *J. Chem. Phys.* **43**, 2840–863, 1965.

Joyce, G. S. "On the Hard Hexagon Model and the Theory of Modular Functions." *Phil. Trans. Royal Soc. London A* **325**, 643–02, 1988.

Joyce, G. S. "Exact Results for the Activity and Isothermal Compressibility of the Hard-Hexagon Model." *J. Phys. A: Math. Gen.* **21**, L983-L988, 1988.

Plouffe, S. "Hard Hexagons Constant." http://www.lacim.uqam.ca/piDATA/hardhex.html.

Hard Lefschetz Theorem

See also LEFSCHETZ THEOREMS

Hard Square Entropy Constant

N.B. A detailed online essay by S. Finch was the starting point for this entry.

Let $F(m, n)$ be the number of $m \times n$ BINARY MATRICES with no adjacent 1s (in either columns or rows). For $n = 1, 2, \ldots, F(n, n)$ is given by 2, 7, 63, 1234, ... (Sloane's A006506).

The hard square entropy constant is defined by

$$\kappa \equiv \lim_{n \to \infty} [F(n, n)]^{1/n^2} = 1.503048082\ldots.$$

The quantity $\ln \kappa$ arises in statistical physics (Baxter *et al.* 1980, Pearce and Seaton 1988), and is known as the entropy per site of hard squares. A related constant known as the HARD HEXAGON ENTROPY CONSTANT can also be defined.

See also BINARY MATRIX

References

Baxter, R. J.; Enting, I. G.; and Tsang, S. K. "Hard-Square Lattice Gas." *J. Statist. Phys.* **22**, 465–89, 1980.

Finch, S. "Favorite Mathematical Constants." http://www.mathsoft.com/asolve/constant/square/square.html.

Pearce, P. A. and Seaton, K. A. "A Classical Theory of Hard Squares." *J. Statist. Phys.* **53**, 1061–072, 1988.

Sloane, N. J. A. Sequences A006506/M1816 in "An On-Line Version of the Encyclopedia of Integer Sequences." http://www.research.att.com/~njas/sequences/eisonline.html.

Hardy Function

RIEMANN-SIEGEL FUNCTIONS

Hardy Space

If $0 < p < \infty$, then the Hardy space $H^p(D)$ is the class of functions holomorphic on the disk D and satisfying the growth condition

$$\|f\|_{H^p} \sup_{0 < r < 1} \left[\frac{1}{2\pi} \int_0^{2\pi} |f(re^{i\theta})|^p \, d\theta \right]^{1/p} < \infty,$$

where $\|f\|_{H^p}$ is the Hardy norm.

See also BERGMAN SPACE

References
Duren, P. L. *Theory of H^p Spaces.* New York: Academic Press, 1970.
Garnett, J. *Bounded Analytic Functions.* New York: Academic Press, 1981.
Koosis, P. *Introduction to H^p Spaces, 2nd ed.* Cambridge, England: Cambridge University Press, 1998.
Krantz, S. G. "Hardy Spaces." §12.3 in *Handbook of Complex Analysis.* Boston, MA: Birkhäuser, pp. 152–54, 1999.

Hardy Z-Function

RIEMANN-SIEGEL FUNCTIONS

Hardy's Inequality

Let $\{a_n\}$ be a NONNEGATIVE SEQUENCE and $f(x)$ a NONNEGATIVE integrable FUNCTION. Define

$$A_n = \sum_{k=1}^{n} a_k \qquad (1)$$

and

$$F(x) = \int_0^x f(t)\, dt \qquad (2)$$

and take $p > 1$. For sums,

$$\sum_{n=1}^{\infty} \left(\frac{A_n}{n}\right)^p < \left(\frac{p}{p-1}\right)^p \sum_{n=1}^{\infty} (a_n)^p \qquad (3)$$

(unless all $a_n = 0$), and for integrals,

$$\int_0^{\infty} \left[\frac{F(x)}{x}\right]^p dx < \left(\frac{p}{p-1}\right)^p \int_0^{\infty} [f(x)]^p\, dx \qquad (4)$$

(unless f is identically 0).

See also CARLEMAN'S INEQUALITY

References
Broadbent, T. A. A. "A Proof of Hardy's Convergence Theorem." *J. London Math. Soc.* **3**, 232–43, 1928.
Elliot, E. B. "A Simple Exposition of Some Recently Proved Facts as to Convergency." *J. London Math. Soc.* **1**, 93–6, 1926.
Grandjot, K. "On Some Identities Relating to Hardy's Convergence Theorem." *J. London Math. Soc.* **3**, 114–17, 1928.
Hardy, G. H. "Note on a Theorem of Hilbert." *Math. Z.* **6**, 314–17, 1920.
Hardy, G. H. "Notes on Some Points in the Integral Calculus. LX." *Messenger Math.* **54**, 150–56, 1925.
Hardy, G. H.; Littlewood, J. E.; and Pólya, G. "Hardy's Inequality." §9.8 in *Inequalities, 2nd ed.* Cambridge, England: Cambridge University Press, pp. 239–43, 1988.
Kaluza, T. and Szego, G. "Uuml;ber Reihen mit lauter positiven Gliedern." *J. London Math. Soc.* **2**, 266–72, 1927.
Knopp, K. "Über Reihen mit positiven Gliedern." *J. London Math. Soc.* **3**, 205–11, 1928.
Landau, E. "A Note on a Theorem Concerning Series of Positive Terms." *J. London Math. Soc.* **1**, 38–9, 1926.
Mitrinovic, D. S.; Pecaric, J. E.; and Fink, A. M. *Inequalities Involving Functions and Their Integrals and Derivatives.* New York: Kluwer, 1991.

Opic, B. and Kufner, A. *Hardy-Type Inequalities.* Essex, England: Longman, 1990.

Hardy's Rule

Let the values of a function $f(x)$ be tabulated at points x_i equally spaced by $h = x_{i+1} - x_i$, so $f_1 = f(x_1)$, $f_2 = f(x_2)$, ..., $f_7 = f(x_7)$. Then Hardy's rule approximating the integral of $f(x)$ is given by the NEWTON-COTES-like formula

$$\int_{x_1}^{x_7} f(x)\, dx = \frac{1}{100} h(28f_1 + 162f_2 + 220f_4 + 162f_6 + 28f_7).$$

See also BODE'S RULE, DURAND'S RULE, NEWTON-COTES FORMULAS, SHOVELTON'S RULE, SIMPSON'S 3/8 RULE, SIMPSON'S RULE, TRAPEZOIDAL RULE, WEDDLE'S RULE

References
King, A. E. "Approximate Integration. Note on Quadrature Formulae: Their Construction and Application to Actuarial Functions." *Trans. Faculty of Actuaries* **9**, 218–31, 1923.
Sheppard, W. F. "Some Quadrature-Formulæ." *Proc. London Math. Soc.* **32**, 258–77, 1900.
Whittaker, E. T. and Robinson, G. *The Calculus of Observations: A Treatise on Numerical Mathematics, 4th ed.* New York: Dover, p. 151, 1967.

Hardy-Littlewood Conjectures

The first Hardy-Littlewood conjecture is called the K-TUPLE CONJECTURE. It states that the asymptotic number of PRIME CONSTELLATIONS can be computed explicitly.

The second Hardy-Littlewood conjecture states that

$$\pi(x+y) - \pi(x) \le \pi(y)$$

for all x and y, where $\pi(x)$ is the PRIME COUNTING FUNCTION. Although it is not obvious, Richards (1974) proved that this conjecture is incompatible with the first Hardy-Littlewood conjecture.

See also PRIME CONSTELLATION, PRIME COUNTING FUNCTION

References
Richards, I. "On the Incompatibility of Two Conjectures Concerning Primes." *Bull. Amer. Math. Soc.* **80**, 419–38, 1974.
Riesel, H. *Prime Numbers and Computer Methods for Factorization, 2nd ed.* Boston, MA: Birkhäuser, pp. 61–2 and 68–9, 1994.

Hardy-Littlewood Constants

PRIME CONSTELLATION

Hardy-Littlewood k-Tuple Conjecture

PRIME PATTERNS CONJECTURE

Hardy-Littlewood Tauberian Theorem

Let $a_n \geq 0$ and suppose

$$\sum_{n=1}^{\infty} a_n e^{-an} \sim \frac{1}{a}$$

as $a \to 0^+$. Then

$$\sum_{n \leq x} a_n \sim x$$

as $x \to \infty$. This theorem is a step in the proof of the PRIME NUMBER THEOREM, but has subsequently been superseded by an approach due to Wiener (Hardy 1999, p. 34).

See also TAUBERIAN THEOREM

References

Berndt, B. C. *Ramanujan's Notebooks, Part IV.* New York: Springer-Verlag, pp. 118–19, 1994.

Hardy, G. H. *Ramanujan: Twelve Lectures on Subjects Suggested by His Life and Work,* 3rd ed. New York: Chelsea, pp. 34–5, 1999.

Hardy, G. H. and Littlewood, J. E. *Quart. J. Math.* **46**, 215–19, 1915.

Hardy, G. H. and Littlewood, J. E. *Acta Math.* **41**, 119–96, 1918.

Karamata. *Math. Z.* **32**, 319–20, 1930.

Hardy-Ramanujan Number

The smallest nontrivial TAXICAB NUMBER, i.e., the smallest number representable in two ways as a sum of two CUBES. It is given by

$$1729 = 1^3 + 12^3 = 9^3 + 10^3.$$

The number derives its name from the following story G. H. Hardy told about Ramanujan. "Once, in the taxi from London, Hardy noticed its number, 1729. He must have thought about it a little because he entered the room where Ramanujan lay in bed and, with scarcely a hello, blurted out his disappointment with it. It was, he declared, 'rather a dull number,' adding that he hoped that wasn't a bad omen. 'No, Hardy,' said Ramanujan, 'it is a very interesting number. It is the smallest number expressible as the sum of two [POSITIVE] cubes in two different ways'" (Hofstadter 1989, Kanigel 1991, Snow 1993; Hardy 1999, pp. 13 and 68).

See also DIOPHANTINE EQUATION–3RD POWERS, TAXICAB NUMBER

References

Guy, R. K. "Sums of Like Powers. Euler's Conjecture." §D1 in *Unsolved Problems in Number Theory,* 2nd ed. New York: Springer-Verlag, pp. 139–44, 1994.

Hardy, G. H. *Ramanujan: Twelve Lectures on Subjects Suggested by His Life and Work,* 3rd ed. New York: Chelsea, 1999.

Hofstadter, D. R. *Gödel, Escher, Bach: An Eternal Golden Braid.* New York: Vintage Books, p. 564, 1989.

Kanigel, R. *The Man Who Knew Infinity: A Life of the Genius Ramanujan.* New York: Washington Square Press, p. 312, 1991.

Snow, C. P. Foreword to Hardy, G. H. *A Mathematician's Apology, reprinted with a foreword by C. P. Snow.* New York: Cambridge University Press, p. 37, 1993.

Hardy-Ramanujan Theorem

Let $\omega(n)$ be the number of DISTINCT PRIME FACTORS of n. If $\Psi(x)$ tends steadily to infinity with x, then

$$\ln \ln x - \Psi(x)\sqrt{\ln \ln x} < \omega(n) < \ln \ln x + \Psi(x)\sqrt{\ln \ln x}$$

for ALMOST ALL numbers $n < x$. "ALMOST ALL" means here the frequency of those INTEGERS n in the interval $1 \leq n \leq x$ for which

$$|\omega(n) - \ln \ln x| > \Psi(x)\sqrt{\ln \ln x}$$

approaches 0 as $x \to \infty$.

See also DISTINCT PRIME FACTORS, ERDOS-KAC THEOREM

Harmonic

The word "harmonic" has several distinct meanings in mathematics, none of which is obviously related to the others. SIMPLE HARMONIC MOTION or "harmonic oscillation" refers to oscillations with a sinusoidal waveform. Such functions satisfy the differential equation

$$\frac{d^2x}{dt^2} + \omega^2 x = 0, \tag{1}$$

which has solution

$$x = A \cos(\omega t + \phi_1) + B \sin(\omega t + \phi_2). \tag{2}$$

The word HARMONIC ANALYSIS is therefore used to describe FOURIER ANALYSIS, which breaks an arbitrary function into a superposition of sinusoids.

In complex analysis, a HARMONIC FUNCTION refers to a real-valued function $f(x, y)$ which satisfies LAPLACE'S EQUATION

$$\nabla^2 f(x, y) = 0, \tag{3}$$

where ∇^2 is the LAPLACIAN. Although this definition is similar to that of harmonic oscillation, it omits the second term in the differential equation. The HELMHOLTZ DIFFERENTIAL EQUATION is obtained if it is added back in,

$$\nabla^2 f(x, y) + k^2 f(x, y) = 0. \tag{4}$$

For distances along a line segment, a HARMONIC RANGE is a set of four COLLINEAR points A, B, C, and D arranged such that

$$AB : BC = 2 : 1 \tag{5}$$

$$AD : DC = 6 : 3. \tag{6}$$

This use of the term probably arises from the use of "harmonics" to refer to ratios of notes in small integers producing an attractive sound, known in music theory as "harmony."

For a set of data points x_i, the HARMONIC MEAN is defined by

$$\frac{1}{H} \equiv \frac{1}{n} \sum_{i=1}^{n} \frac{1}{x_i}. \tag{7}$$

The connection of this use of "harmonic" with the preceding ones is not obvious.

See also HARMONIC FORM, HARMONIC FUNCTION, HARMONIC RANGE, SIMPLE HARMONIC MOTION

Harmonic Addition Theorem

To convert an equation OF THE FORM

$$f(\theta) = a \cos \theta + b \sin \theta \tag{1}$$

to the form

$$f(\theta) = c \cos(\theta + \delta), \tag{2}$$

expand (2) using the trigonometric addition formulas to obtain

$$f(\theta) = c \cos \theta \cos \delta - c \sin \theta \sin \delta. \tag{3}$$

Now equate the COEFFICIENTS of (1) and (3)

$$a = c \cos \delta \tag{4}$$

$$b = -c \sin \delta, \tag{5}$$

so

$$\tan \delta = -\frac{b}{a} \tag{6}$$

$$a^2 + b^2 = c^2, \tag{7}$$

and we have

$$\delta = \tan^{-1}\left(-\frac{b}{a}\right) \tag{8}$$

$$c = \sqrt{a^2 + b^2}. \tag{9}$$

Given two general sinusoidal functions with frequency ω:

$$\psi_1 = A_1 \sin(\omega t + \delta_1) \tag{10}$$

$$\psi_2 = A_2 \sin(\omega t + \delta_2), \tag{11}$$

their sum ψ can be expressed as a sinusoidal function with frequency ω

$$\begin{aligned}
\psi &\equiv \psi_1 + \psi_2 \\
&= A_1[\sin(\omega t) \cos \delta_1 + \sin \delta_1 \cos(\omega t)] \\
&\quad + A_2[\sin(\omega t) \cos \delta_2 + \sin \delta_2 \cos(\omega t)] \\
&= [A_1 \cos \delta_1 + A_2 \cos \delta_2] \sin(\omega t) \\
&\quad + [A_1 \sin \delta_1 + A_2 \sin \delta_2] \cos(\omega t). \tag{12}
\end{aligned}$$

Now, define

$$A \cos \delta \equiv A_1 \cos \delta_1 + A_2 \cos \delta_2 \tag{13}$$

$$A \sin \delta \equiv A_1 \sin \delta_1 + A_2 \sin \delta_2. \tag{14}$$

Then (12) becomes

$$A \cos \delta \sin(\omega t) + A \sin \delta \cos(\omega t) = A \sin(\omega t + \delta). \tag{15}$$

Square and add (13) and (14)

$$A^2 = A_1^2 + A_2^2 + 2A_1 A_2 \cos(\delta_2 - \delta_1). \tag{16}$$

Also, divide (14) by (13)

$$\tan \delta = \frac{A_1 \sin \delta_1 + A_2 \sin \delta_2}{A_1 \cos \delta_1 + A_2 \cos \delta_2}, \tag{17}$$

so

$$\psi = A \sin(\omega t + \delta), \tag{18}$$

where A and δ are defined by (16) and (17).

This procedure can be generalized to a sum of n harmonic waves, giving

$$\psi = \sum_{i=1}^{n} A_i \cos(\omega t + \delta_i) = A \cos(\omega t + \delta), \tag{19}$$

where

$$A^2 \equiv \sum_{i=1}^{n} \sum_{j=1}^{n} A_i A_j \cos(\delta_i - \delta_j) \tag{20}$$

$$= \sum_{i=1}^{n} A_i^2 + 2 \sum_{i=1}^{n} \sum_{j>1}^{n} A_i A_j \cos(\delta_i - \delta_j) \tag{21}$$

and

$$\tan \delta = \frac{\sum_{i=1}^{n} A_i \sin \delta_i}{\sum_{i=1}^{n} A_i \cos \delta_i} \tag{22}$$

Harmonic Analysis

FOURIER SERIES

Harmonic Brick

A right-angled PARALLELEPIPED with dimensions $a \times ab \times abc$, where a, b, and c are INTEGERS.

See also BRICK, DE BRUIJN'S THEOREM, EULER BRICK

Harmonic Conjugate Function

The harmonic conjugate to a given function $u(x, y)$ is a function $v(x, y)$ such that

$$f(x, y) = u(x, y) + iv(x, y)$$

is COMPLEX DIFFERENTIABLE (i.e., satisfies the CAUCHY-RIEMANN EQUATIONS). It is given by

$$v(z) = \int_{z_0}^{z} u_x \, dy - u_y \, dx + C,$$

where $u_x \equiv \partial u / \partial x$, $u_y \equiv \partial u / \partial y$, and C is a CONSTANT OF INTEGRATION.

Note that $u_x \, dy - u_y \, dx$ is a CLOSED FORM since u is HARMONIC, $u_{xx} + v_{yy} = 0$. The LINE INTEGRAL is WELL DEFINED on a SIMPLY CONNECTED domain because it is closed. However, on a domain which is not simply connected (such as the punctured disk), the harmonic conjugate may not exist.

See also CAUCHY-RIEMANN EQUATIONS, COMPLEX DIFFERENTIABLE, HARDY SPACE, HARMONIC FUNCTION, HILBERT TRANSFORM, SIMPLY CONNECTED

References
Rudin, W. *Real and Complex Analysis.* New York: McGraw-Hill, pp. 350–52, 1987.

Harmonic Conjugate Points

Given COLLINEAR points W, X, Y, and Z, Y and Z are harmonic conjugates with respect to W and X if

$$\frac{|WY|}{|YX|} = \frac{|WZ|}{|XZ|}.$$

The distances between such points are said to be in HARMONIC RATIO, and the LINE SEGMENT depicted above is called a HARMONIC SEGMENT. Harmonic points divide a LINE SEGMENT internally and externally in the same ratio. If $|WZ|=1$, then

$$|WY| = \frac{a(1 - a)}{1 + a}$$

$$|WX| = \frac{2a}{a + 1}.$$

Harmonic conjugate points are also defined for a TRIANGLE. If W and X have TRILINEAR COORDINATES $\alpha : \beta : \gamma$ and $\alpha' : \beta' : \gamma'$, then the TRILINEAR COORDINATES of the harmonic conjugates are

$$Y = \alpha + \alpha' : \beta + \beta' : \gamma + \gamma'$$

$$Z = \alpha - \alpha' : \beta - \beta' : \gamma - \gamma'$$

(Kimberling 1994).

See also HARMONIC RANGE, HARMONIC RATIO, POLAR, POLE (INVERSION)

References
Durell, C. V. *Modern Geometry: The Straight Line and Circle.* London: Macmillan, p. 65, 1928.
Kimberling, C. "Central Points and Central Lines in the Plane of a Triangle." *Math. Mag.* **67**, 163–87, 1994.
Lachlan, R. "Harmonic Ranges and Pencils." Ch. 4 in *An Elementary Treatise on Modern Pure Geometry.* London: Macmillian, pp. 24–6, 1893.
Ogilvy, C. S. *Excursions in Geometry.* New York: Dover, pp. 13–4, 1990.
Phillips, A. W. and Fisher, I. *Elements of Geometry.* New York: American Book Co., 1896.
Wells, D. *The Penguin Dictionary of Curious and Interesting Geometry.* New York: Viking Penguin, p. 92, 1992.

Harmonic Coordinates

Harmonic coordinates satisfy the condition

$$\Gamma^\lambda \equiv g^{\mu\nu} \Gamma_{\mu\nu}^\lambda = 0, \tag{1}$$

or equivalently,

$$\frac{\partial}{\partial x^\kappa} \left(\sqrt{g} g^{\lambda\kappa} \right) = 0. \tag{2}$$

It is always possible to choose such a system. Using the D'ALEMBERTIAN,

$$\square^2 \phi \equiv \left(g^{\lambda\kappa} \phi_{,\lambda} \right)_{;\kappa} = g^{\lambda\kappa} \frac{\partial^2 \phi}{\partial x^\lambda \partial x^\kappa} - \Gamma^\lambda \frac{\partial \phi}{\partial x^\lambda}. \tag{3}$$

But since $\Gamma^\lambda \equiv 0$ for harmonic coordinates, the result is a generalization of the harmonic equation

$$\nabla^2 \mathbf{x} = 0 \tag{4}$$

to

$$\square^2 x^\mu = 0. \tag{5}$$

See also D'ALEMBERTIAN

References
Weinberg, S. *Gravitation and Cosmology: Principles and Applications of the General Theory of Relativity.* New York: Wiley, 1972.

Harmonic Decomposition

A polynomial function of the elements of a VECTOR $\mathbf{x}$ can be uniquely decomposed into a sum of HARMONIC POLYNOMIALS times POWERS of $|\mathbf{x}|$.

See also HARMONIC FUNCTION

Harmonic Divisor Number

A number n for which the HARMONIC MEAN of the DIVISORS of n, i.e., $nd(n)/\sigma(n)$, is an INTEGER, where $d(n)$ is the number of POSITIVE integral DIVISORS of n and $\sigma(n)$ is the DIVISOR FUNCTION. For example, the divisors of $n = 140$ are 1, 2, 4, 5, 7, 10, 14, 20, 28, 35, 70, and 140, giving

$$d(140) = 12$$

$$\sigma(140) = 336$$

$$\frac{140 d(140)}{\sigma(140)} = \frac{140 \cdot 12}{336} = 5,$$

so 140 is a harmonic divisor number. Harmonic divisor numbers are also called ORE NUMBERS. Garcia (1954) gives the 45 harmonic divisor numbers less than 10^7. The first few are 1, 6, 140, 270, 672, 1638, ... (Sloane's A007340).

For distinct PRIMES p and q, harmonic divisor numbers are equivalent to EVEN PERFECT NUMBERS for numbers OF THE FORM $p^r q$. Mills (1972) proved that if there exists an ODD POSITIVE harmonic divisor number n, then n has a prime-POWER factor greater than 10^7.

Another type of number called "harmonic" is the HARMONIC NUMBER.

See also DIVISOR FUNCTION, HARMONIC NUMBER

References

Edgar, H. M. W. "Harmonic Numbers." *Amer. Math. Monthly* **99**, 783–89, 1992.

Garcia, M. "On Numbers with Integral Harmonic Mean." *Amer. Math. Monthly* **61**, 89–6, 1954.

Guy, R. K. "Almost Perfect, Quasi-Perfect, Pseudoperfect, Harmonic, Weird, Multiperfect and Hyperperfect Numbers." §B2 in *Unsolved Problems in Number Theory, 2nd ed.* New York: Springer-Verlag, pp. 45–3, 1994.

Mills, W. H. "On a Conjecture of Ore." *Proceedings of the 1972 Number Theory Conference.* University of Colorado, Boulder, pp. 142–46, 1972.

Ore, Ø. "On the Averages of the Divisors of a Number." *Amer. Math. Monthly* **55**, 615–19, 1948.

Pomerance, C. "On a Problem of Ore: Harmonic Numbers." Unpublished manuscript, 1973.

Sloane, N. J. A. Sequences A007340/M4299 in "An On-Line Version of the Encyclopedia of Integer Sequences." http://www.research.att.com/~njas/sequences/eisonline.html.

Sloane, N. J. A. and Plouffe, S. Figure M4299 in *The Encyclopedia of Integer Sequences.* San Diego: Academic Press, 1995.

Zachariou, A. and Zachariou, E. "Perfect, Semi-Perfect and Ore Numbers." *Bull. Soc. Math. Grèce (New Ser.)* **13**, 12–2, 1972.

Harmonic Equation

LAPLACE'S EQUATION

Harmonic Function

Any REAL FUNCTION $u(x, y)$ with continuous second PARTIAL DERIVATIVES which satisfies LAPLACE'S EQUATION,

$$\nabla^2 u(x, y) = 0, \qquad (1)$$

is called a harmonic function. Harmonic functions are called POTENTIAL FUNCTIONS in physics and engineer-

ing. Potential functions are extremely useful, for example, in electromagnetism, where they reduce the study of a 3-component VECTOR FIELD to a 1-component SCALAR FUNCTION. A scalar harmonic function is called a SCALAR POTENTIAL, and a vector harmonic function is called a VECTOR POTENTIAL.

To find a class of such functions in the PLANE, write the LAPLACE'S EQUATION in POLAR COORDINATES

$$u_{rr} + \frac{1}{r} u_r + \frac{1}{r^2} u_{\theta\theta} = 0, \qquad (2)$$

and consider only radial solutions

$$u_{rr} + \frac{1}{r} u_r = 0. \qquad (3)$$

This is integrable by quadrature, so define $v \equiv du/dr$,

$$\frac{dv}{dr} + \frac{1}{r} v = 0 \qquad (4)$$

$$\frac{dv}{v} = -\frac{dr}{r} \qquad (5)$$

$$\ln\left(\frac{v}{A}\right) = -\ln r \qquad (6)$$

$$\frac{v}{A} = \frac{1}{r} \qquad (7)$$

$$v = \frac{du}{dr} = \frac{A}{r} \qquad (8)$$

$$du = A \frac{dr}{r}, \qquad (9)$$

so the solution is

$$u = A \ln r. \qquad (10)$$

Ignoring the trivial additive and multiplicative constants, the general pure radial solution then becomes

$$u = \ln\left[(x-a)^2 + (y-b)^2\right]^{1/2}$$

$$= \tfrac{1}{2} \ln\left[(x-a)^2 + (y-b)^2\right]. \qquad (11)$$

Other solutions may be obtained by differentiation, such as

$$u = \frac{x-a}{(x-a)^2 + (y-b)^2} \qquad (12)$$

$$v = \frac{y-b}{(x-a)^2 + (y-b)^2}, \qquad (13)$$

$$u = e^x \sin y \qquad (14)$$

$$v = e^x \cos y, \qquad (15)$$

and

$$\tan^{-1}\left(\frac{y-b}{x-a}\right). \tag{16}$$

Harmonic functions containing azimuthal dependence include

$$u = r^n \cos(n\theta) \tag{17}$$

$$v = r^n \sin(n\theta). \tag{18}$$

The POISSON KERNEL

$$u(r,\, R,\, \theta,\, \phi) = \frac{R^2 - r^2}{R^2 - 2rR\,\cos(\theta - \phi) + r^2} \tag{19}$$

is another harmonic function.

See also CONFORMAL MAPPING, DIRICHLET PROBLEM, HARMONIC ANALYSIS, HARMONIC DECOMPOSITION, HARNACK'S INEQUALITY, HARNACK'S PRINCIPLE, KELVIN TRANSFORMATION, LAPLACE'S EQUATION, POISSON INTEGRAL, POISSON KERNEL, SCALAR POTENTIAL, SCHWARZ REFLECTION PRINCIPLE, SUBHARMONIC FUNCTION, VECTOR POTENTIAL

References

Ash, J. M. (Ed.). *Studies in Harmonic Analysis.* Washington, DC: Math. Assoc. Amer., 1976.
Axler, S.; Bourdon, P.; and Ramey, W. *Harmonic Function Theory.* Springer-Verlag, 1992.
Benedetto, J. J. *Harmonic Analysis and Applications.* Boca Raton, FL: CRC Press, 1996.
Cohn, H. *Conformal Mapping on Riemann Surfaces.* New York: Dover, 1980.
Krantz, S. G. "Harmonic Functions." §1.4.1 and Ch. 7 in *Handbook of Complex Analysis.* Boston, MA: Birkhäuser, pp. 16 and 89–01, 1999.
Weisstein, E. W. "Books about Potential Theory." http://www.treasure-troves.com/books/PotentialTheory.html.

Harmonic Homology

A PERSPECTIVE COLLINEATION with center O and axis o not incident is called a HOMOLOGY. A HOMOLOGY is said to be harmonic if the points A and A' on a line through O are harmonic conjugates with respect to O and $o \cdot a$. Every PERSPECTIVE COLLINEATION of period two is a harmonic homology.

See also HOMOLOGY (GEOMETRY), PERSPECTIVE COLLINEATION

References

Coxeter, H. S. M. *Introduction to Geometry, 2nd ed.* New York: Wiley, p. 248, 1969.

Harmonic Logarithm

For all INTEGERS n and NONNEGATIVE INTEGERS t, the harmonic logarithms $\lambda_n^{(t)}(x)$ of order t and degree n are defined as the unique functions satisfying

1. $\lambda_n^{(t)}(x) = (\ln x)^t$,
2. $\lambda_n^{(t)}(x)$ has no constant term except $\lambda_0^{(0)}(x) = 1$,
3. $\dfrac{d}{dx}\lambda_n^{(t)}(x) = \lfloor n \rceil \lambda_{n-1}^{(t)}(x)$,

where the "ROMAN SYMBOL" $\lfloor n \rceil$ is defined by

$$\lfloor n \rceil \equiv \begin{cases} n & \text{for } n \neq 0 \\ 1 & \text{for } n = 0 \end{cases} \tag{1}$$

(Roman 1992). This gives the special cases

$$\lambda_n^{(0)}(x) = \begin{cases} x^n & \text{for } n \geq 0 \\ 0 & \text{for } n < 0 \end{cases} \tag{2}$$

$$\lambda_n^{(1)}(x) = \begin{cases} x^n(\ln x - H_n) & \text{for } n \geq 0 \\ x^n & \text{for } n < 0, \end{cases} \tag{3}$$

where H_n is a HARMONIC NUMBER

$$H_n \equiv \sum_{k=1}^{n} \frac{1}{k}. \tag{4}$$

The harmonic logarithm has the INTEGRAL

$$\int \lambda_n^{(1)}(x)\, dx = \frac{1}{\lfloor n+1 \rceil}\lambda_{n+1}^{(1)}(x). \tag{5}$$

The harmonic logarithm can be written

$$\lambda_n^{(t)}(x) = \lfloor n \rceil!\, \tilde{D}^{-n}(\ln x)^t, \tag{6}$$

where $\tilde{D}$ is the DIFFERENTIAL OPERATOR, (so $\tilde{D}^{-n}$ is the nth INTEGRAL). Rearranging gives

$$\tilde{D}^k \lambda_n^{(t)}(x) = \left[\frac{\lfloor n \rceil!}{\lfloor n-k \rceil}\right]!\, \lambda_{n-k}^{(t)}(x). \tag{7}$$

This formulation gives an analog of the BINOMIAL THEOREM called the LOGARITHMIC BINOMIAL FORMULA. Another expression for the harmonic logarithm is

$$\lambda_n^{(t)}(x) = x^n \sum_{j=0}^{t} (-1)^j (t)_j c_n^{(j)} (\ln x)^{t-j}, \tag{8}$$

where $(t)_j = t(t-1)\cdots(t-j+1)$ is a POCHHAMMER SYMBOL and $c_n^{(j)}$ is a two-index HARMONIC NUMBER (Roman 1992).

See also LOGARITHM, ROMAN FACTORIAL

References

Loeb, D. and Rota, G.-C. "Formal Power Series of Logarithmic Type." *Advances Math.* **75**, 1–18, 1989.
Roman, S. "The Logarithmic Binomial Formula." *Amer. Math. Monthly* **99**, 641–48, 1992.

Harmonic Map

A map $u : M \to N$, between two COMPACT RIEMANNIAN MANIFOLDS, is a harmonic map if it is a critical point for the energy functional

$$\int_M |du|^2 \, d\mu_M.$$

The norm of the differential $|du|$ is given by the metric on M and N and $d\mu_M$ is the measure on M. Typically, the class of allowable maps lie in a fixed HOMOTOPY CLASS of maps.

The EULER-LAGRANGE DIFFERENTIAL EQUATION for the energy functional is a non-linear ELLIPTIC PARTIAL DIFFERENTIAL EQUATION. For example, when M is the circle, then the Euler-Lagrange equation is the same as the geodesic equation. Hence, u is a closed geodesic iff u is harmonic. The map from the circle to the equator of the standard 2-sphere is a harmonic map, and so are the maps that take the circle and map it around the equator n times, for any integer n. Note that these all lie in the same HOMOTOPY CLASS. A higher dimensional example is a MEROMORPHIC FUNCTION on a compact RIEMANN SURFACE, which is a harmonic map to the RIEMANN SPHERE.

A harmonic map may not always exist in a HOMOTOPY CLASS, and if it does it may not be unique. When N is negatively curved, a harmonic representative exists for each HOMOTOPY CLASS, and is also unique. For surfaces, the harmonic maps have been classified, and are precisely the holomorphic maps and the anti-holomorphic maps. Thus by HODGE'S THEOREM for surfaces, there are no non-trivial harmonic maps from the SPHERE to the TORUS.

A harmonic map between RIEMANNIAN MANIFOLDS can be viewed as a generalization of a GEODESIC when the domain DIMENSION is one, or of a HARMONIC FUNCTION when the range is a EUCLIDEAN SPACE.

See also BOCHNER IDENTITY, CALCULUS OF VARIATIONS, CURVATURE, EUCLIDEAN SPACE, EULER-LAGRANGE DIFFERENTIAL EQUATION, GEODESIC, HARMONIC FUNCTION, HODGE'S THEOREM, HOMOTOPY CLASS, RIEMANNIAN MANIFOLD, RIEMANN SURFACE

References

Burstal, F.; Lemaire, L.; and Rawnsley, J. "Harmonic Maps Bibliography."   http://www.bath.ac.uk/~masfeb/harmonic.html.
Eels, J. and Lemaire, L. "A Report on Harmonic Maps." *Bull. London Math. Soc.* **10**, 1–8, 1978.
Eels, J. and Lemaire, L. "Another Report on Harmonic Maps." *Bull. London Math. Soc.* **20**, 385–24, 1988.

Harmonic Mean

The harmonic mean $H(x_1, \ldots, x_n)$ of n points x_i (where $i = 1, \ldots, n$) is

$$\frac{1}{H} \equiv \frac{1}{n} \sum_{i=1}^n \frac{1}{x_i}. \tag{1}$$

The special cases of $n = 2$ and $n = 3$ are therefore given by

$$H(x_1, x_2) = \frac{2x_1 x_2}{x_1 + x_2} \tag{2}$$

$$H(x_1, x_2, x_3) = \frac{3x_1 x_2 x_3}{x_1 x_2 + x_1 x_3 + x_2 x_3}, \tag{3}$$

and so on.

The VOLUME-TO-SURFACE AREA ratio for a cylindrical container with height h and radius r and the MEAN CURVATURE of a general surface are related to the harmonic mean.

Hoehn and Niven (1985) show that

$$\begin{aligned} H(a_1 + c, \; a_2 + c, \; \ldots, \; a_n + c) \\ > c + H(a_1, \; a_2, \; \ldots, \; a_n) \end{aligned} \tag{4}$$

for any POSITIVE constant c.

See also ARITHMETIC MEAN, ARITHMETIC-GEOMETRIC MEAN, GEOMETRIC MEAN, HARMONIC-GEOMETRIC MEAN, HARMONIC RANGE, ROOT-MEAN-SQUARE

References

Abramowitz, M. and Stegun, C. A. (Eds.). *Handbook of Mathematical Functions with Formulas, Graphs, and Mathematical Tables, 9th printing.* New York: Dover, p. 10, 1972.
Hoehn, L. and Niven, I. "Averages on the Move." *Math. Mag.* **58**, 151–56, 1985.
Kenney, J. F. and Keeping, E. S. "Harmonic Mean." §4.13 in *Mathematics of Statistics, Pt. 1, 3rd ed.* Princeton, NJ: Van Nostrand, pp. 57–8, 1962.
Zwillinger, D. (Ed.). *CRC Standard Mathematical Tables and Formulae.* Boca Raton, FL: CRC Press, p. 602, 1995.

Harmonic Mean Index

The statistical INDEX

$$P_H \equiv \frac{\sum v_0}{\sum \dfrac{v_0 p_0}{p_n}} = \frac{\sum p_0 q_0}{\sum \dfrac{p_0^2 q_0}{p_n}},$$

where p_n is the price per unit in period n, q_n is the quantity produced in period n, and $v_n \equiv p_n q_n$ the value of the n units, and subscripts 0 indicate the reference year.

See also INDEX

References

Kenney, J. F. and Keeping, E. S. *Mathematics of Statistics, Pt. 1, 3rd ed.* Princeton, NJ: Van Nostrand, p. 69, 1962.

Harmonic Number

A number OF THE FORM

$$H_n = \sum_{k=1}^n \frac{1}{k}. \tag{1}$$

This can be expressed analytically as

$$H_n = \gamma + \psi_0(n+1), \tag{2}$$

where γ is the EULER-MASCHERONI CONSTANT and $\Psi(x) = \psi_0(x)$ is the DIGAMMA FUNCTION. The number formed by taking alternate signs in the sum also has an analytic solution

$$H'_n = \sum_{k=1}^{n} \frac{(-1)^{k+1}}{k} \qquad (3)$$

$$= \ln 2 + \tfrac{1}{2}(-1)^n \left[\psi_0\left(\tfrac{1}{2}n + \tfrac{1}{2}\right) - \psi_0\left(\tfrac{1}{2}n + 1\right) \right]. \qquad (4)$$

The first few harmonic numbers H_n are 1, 3/2, 11/6, 25/12, 137/60, ... (Sloane's A001008 and A002805). The harmonic numbers are implemented in *Mathematica* 4.0 as `HarmonicNumber[n]`.

The harmonic number H_n is never an INTEGER except for H_1, which can be proved by using the strong triangle inequality to show that the 2-ADIC VALUE of H_n is greater than 1 for $n > 1$. This result was proved in 1915 by Taeisinger, and the more general results that any number of consecutive terms not necessarily starting with 1 never sum to an integer was proved by Kürschák in 1918 (Hoffman 1998, p. 157).

The harmonic numbers have ODD NUMERATORS and EVEN DENOMINATORS. The nth harmonic number is given asymptotically by

$$H_n \sim \ln n + \gamma + \frac{1}{2n}, \qquad (5)$$

where γ is the EULER-MASCHERONI CONSTANT (Conway and Guy 1996). Gosper gave the interesting identity

$$\sum_{i=0}^{\infty} \frac{z^i H_i}{i!} = -e^z \sum_{k=1}^{\infty} \frac{(-z)^k}{kk!} = e^z [\ln z + \Gamma(0, z) + \gamma], \qquad (6)$$

where $\Gamma(0, z)$ is the incomplete GAMMA FUNCTION and γ is the EULER-MASCHERONI CONSTANT. Borwein and Borwein (1995) show that

$$\sum_{n=1}^{\infty} \frac{H_n^2}{(n+1)^2} = \tfrac{11}{4}\zeta(4) = \tfrac{11}{360}\pi^4 \qquad (7)$$

$$\sum_{n=1}^{\infty} \frac{H_n^2}{n^2} = \tfrac{17}{4}\zeta(4) = \tfrac{17}{360}\pi^4 \qquad (8)$$

$$\sum_{n=1}^{\infty} \frac{H_n}{n^3} = \tfrac{5}{4}\zeta(4) = \tfrac{1}{72}\pi^4, \qquad (9)$$

where $\zeta(z)$ is the RIEMANN ZETA FUNCTION. The first of these had been previously derived by de Doelder (1991), and the last by Euler (1775). These identities are corollaries of the identity

$$\frac{1}{\pi} \int_0^{\pi} x^2 \left\{ \ln\left[2\cos\left(\tfrac{1}{2}x\right)\right] \right\}^2 dx = \tfrac{11}{2}\zeta(4) = \tfrac{11}{180}\pi^4 \qquad (10)$$

(Borwein and Borwein 1995). Additional identities

due to Euler are

$$\sum_{n=1}^{\infty} \frac{H_n}{n^2} = 2\zeta(3) \qquad (11)$$

$$2\sum_{n=1}^{\infty} \frac{H_n}{n^m} = (m+2)\zeta(m+1)$$

$$-\sum_{n=1}^{m-2} \zeta(m-n)\zeta(n+1) \qquad (12)$$

for $m = 2, 3, ...$ (Borwein and Borwein 1995), where $\zeta(3)$ is APÉRY'S CONSTANT. These sums are related to so-called EULER SUMS.

There is an unexpected connection between the harmonic numbers and the RIEMANN HYPOTHESIS.

Harmonic numbers of order r can be defined by the relationship

$$H_n^{(r)} = \sum_{k=1}^{n} \frac{1}{k^r}. \qquad (13)$$

These number are built into *Mathematica* 4.0 as `HarmonicNumber[n, r]`. These numbers obey the unexpected identity

$$9H_8^{(n)} - 19H_9^{(n)} + 10H_{10}^{(n)} + \sum_{k=1}^{n-1} \left[H_8^{(n-k)} H_9^{(k)} - H_9^{(n-k)} H_9^{(k)} \right]$$

$$-H_8^{(n-k)} H_{10}^{(k)} + H_9^{(n-k)} H_{10}^{(k)} \right] = 0 \qquad (14)$$

(M. Trott).

Conway and Guy (1996) define the second harmonic number by

$$H_n^2 \equiv \sum_{i=1}^{n} H_i = (n+1)(H_{n+1} - 1)$$

$$= (n+1)(H_{n+1} - H_1), \qquad (15)$$

the third harmonic number by

$$H_n^3 \equiv \sum_{i=1}^{n} H_i^{(2)} = \binom{n+2}{2}(H_{n+2} - H_2), \qquad (16)$$

and the nth harmonic number by

$$H_n^k = \binom{n+k-1}{k-1}(H_{n+k-1} - H_{k-1}). \qquad (17)$$

A slightly different definition of a two-index harmonic number $c_n^{(j)}$ is given by Roman (1992) in connection with the HARMONIC LOGARITHM. Roman (1992) defines this by

$$c_n^{(0)} = \begin{cases} 1 & \text{for } n \geq 0 \\ 0 & \text{for } n < 0 \end{cases} \qquad (18)$$

$$c_0^{(j)} = \begin{cases} 1 & \text{for } j = 0 \\ 0 & \text{for } j \neq 0 \end{cases} \qquad (19)$$

plus the RECURRENCE RELATION

$$cn_n^{(j)} = c_n^{(j-1)} + nc_{n-1}^{(j)}. \tag{20}$$

For general $n > 0$ and $j > 0$, this is equivalent to

$$c_n^{(j)} = \sum_{i=1}^{n} \frac{1}{i} c_i^{(j-1)}, \tag{21}$$

and for $n > 0$, it simplifies to

$$c_n^{(j)} = \sum_{i=1}^{n} \binom{n}{i} (-1)^{i-1} i^{-j}. \tag{22}$$

For $n < 0$, the harmonic number can be written

$$c_n^{(j)} = (-1)^j \lfloor n \rceil! s(-n, j), \tag{23}$$

where $\lfloor n \rceil!$ is the ROMAN FACTORIAL and s is a STIRLING NUMBER OF THE FIRST KIND.

A separate type of number sometimes also called a "harmonic number" is a HARMONIC DIVISOR NUMBER (or ORE NUMBER).

See also APÉRY'S CONSTANT, EULER SUM, HARMONIC LOGARITHM, HARMONIC SERIES, ORE NUMBER, RAMANUJAN FUNCTION, UNIT FRACTION

References

Borwein, D. and Borwein, J. M. "On an Intriguing Integral and Some Series Related to $\zeta(4)$." *Proc. Amer. Math. Soc.* **123**, 1191–198, 1995.
Conway, J. H. and Guy, R. K. *The Book of Numbers.* New York: Springer-Verlag, pp. 143 and 258–59, 1996.
de Doelder, P. J. "On Some Series Containing $\Psi(x) - \Psi(y)$ and $(\Psi(x) - \Psi(y))^2$ for Certain Values of x and y." *J. Comp. Appl. Math.* **37**, 125–41, 1991.
Flajolet, P. and Salvy, B. "Euler Sums and Contour Integral Representation." *Experim. Math.* **7**, 15–5, 1998.
Graham, R. L.; Knuth, D. E.; and Patashnik, O. "Harmonic Numbers" and "Harmonic Summation." §6.3 and 6.4 in *Concrete Mathematics: A Foundation for Computer Science, 2nd ed.* Reading, MA: Addison-Wesley, pp. 272–82, 1994.
Hoffman, P. *The Man Who Loved Only Numbers: The Story of Paul Erdos and the Search for Mathematical Truth.* New York: Hyperion, 1998.
Roman, S. "The Logarithmic Binomial Formula." *Amer. Math. Monthly* **99**, 641–48, 1992.
Roman, S. *The Umbral Calculus.* New York: Academic Press, p. 99, 1984.
Sloane, N. J. A. Sequences A001008/M2885 and A002805/M1589 in "An On-Line Version of the Encyclopedia of Integer Sequences." http://www.research.att.com/~njas/sequences/eisonline.html.

Harmonic Progression
HARMONIC SERIES

Harmonic Range

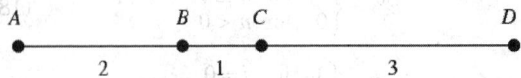

A set of four COLLINEAR points A, B, C, and D

arranged such that

$$AB : BC = 2 : 1$$

$$AD : DC = 6 : 3.$$

Hardy (1967) uses the term HARMONIC SYSTEM OF POINTS to refer to a harmonic range.

See also BIVALENT RANGE, EULER LINE, GERGONNE LINE, HARMONIC CONJUGATE POINTS, SODDY LINE

References

Casey, J. "Theory of Harmonic Section." §6.3 in *A Sequel to the First Six Books of the Elements of Euclid, Containing an Easy Introduction to Modern Geometry with Numerous Examples, 5th ed., rev. enl.* Dublin: Hodges, Figgis, & Co., pp. 87–4, 1888.
Durell, C. V. "Harmonic Ranges and Pencils." Ch. 6 in *Modern Geometry: The Straight Line and Circle.* London: Macmillan, pp. 65–7, 1928.
Graustein, W. C. "Harmonic Division." Ch. 4 in *Introduction to Higher Geometry.* New York: Macmillan, pp. 50–4, 1930.
Hardy, G. H. *A Course of Pure Mathematics, 10th ed.* Cambridge, England: Cambridge University Press, pp. 99 and 106, 1967.
Lachlan, R. "Harmonic Properties." §288–90 in *An Elementary Treatise on Modern Pure Geometry.* London: Macmillian, pp. 177 and 267–68, 1893.

Harmonic Ratio
HARMONIC RANGE

Harmonic Segment
HARMONIC CONJUGATE POINTS

Harmonic Series
The SUM

$$\sum_{k=1}^{\infty} \frac{1}{k} \tag{1}$$

is called the harmonic series. It can be shown to DIVERGE using the INTEGRAL TEST by comparison with the function $1/x$. The divergence, however, is very slow. The generalization of the harmonic series

$$\zeta(n) \equiv \sum_{k=1}^{\infty} \frac{1}{k^n} \tag{2}$$

is known as the RIEMANN ZETA FUNCTION. The sum

$$\sum_{k=1}^{\infty} \frac{1}{p_k} \tag{3}$$

taken over all PRIMES p_k also diverges (Wells 1986, p. 41) with asymptotic behavior

$$\sum_{k=1}^{x} \frac{1}{p_k} \sim \ln \ln x + \mathcal{O}(1) \qquad (4)$$

(Hardy 1999, p. 50).

Rather surprisingly, the ALTERNATING SERIES

$$\sum_{k=1}^{\infty} \frac{(-1)^{k-1}}{k} = \ln 2 \qquad (5)$$

converges to the natural logarithm of 2. An explicit formula for the partial sum of the alternating series is given by

$$\sum_{k=1}^{n} \frac{(-1)^{k-1}}{k}$$
$$= \ln 2 + \tfrac{1}{2}(-1)^n \left[\psi_0\left(\tfrac{1}{2} + \tfrac{1}{2}n\right) - \psi_0\left(1 + \tfrac{1}{2}n\right) \right]. \qquad (6)$$

Gardner (1984) notes that this series never reaches an integral sum.

The sum of the first few terms of the harmonic series is given analytically by the nth HARMONIC NUMBER

$$H_n = \sum_{j=1}^{n} \frac{1}{j} = \gamma + \psi_0(n+1), \qquad (7)$$

where γ is the EULER-MASCHERONI CONSTANT and $\Psi(x) = \psi_0(x)$ is the DIGAMMA FUNCTION. The number of terms needed to exceed 1, 2, 3, ... are 1, 4, 11, 31, 83, 227, 616, 1674, 4550, 12367, 33617, 91380, 248397, ... (Sloane's A004080). Using the analytic form shows that after 2.5×10^8 terms, the sum is still less than 20. Furthermore, to achieve a sum greater than 100, more than 1.509×10^{43} terms are needed! Written explicitly, the number of terms is 15,092,688,622,113,788,323,693,563,264,538,101,449, 859,497 (Gardner 1984, p. 167).

Progressions OF THE FORM

$$\frac{1}{a_1}, \frac{1}{a_1 + d}, \frac{1}{a_1 + 2d}, \dots \qquad (8)$$

are also sometimes called harmonic series (Beyer 1987).

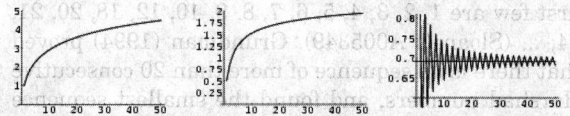

The partial sums of the harmonic series are plotted in the left figure above, together with two related series.

See also ARITHMETIC SERIES, BERNOULLI'S PARADOX, BOOK STACKING PROBLEM, EULER SUM, MERTENS CONSTANT, q-HARMONIC SERIES, ZIPF'S LAW

References

Arfken, G. *Mathematical Methods for Physicists, 3rd ed.* Orlando, FL: Academic Press, pp. 279–80, 1985.

Beyer, W. H. (Ed.). *CRC Standard Mathematical Tables, 28th ed.* Boca Raton, FL: CRC Press, p. 8, 1987.
Boas, R. P. and Wrench, J. W. "Partial Sums of the Harmonic Series." *Amer. Math. Monthly* **78**, 864–70, 1971.
Gardner, M. *The Sixth Book of Mathematical Games from Scientific American.* Chicago, IL: University of Chicago Press, pp. 165–72, 1984.
Hardy, G. H. *Ramanujan: Twelve Lectures on Subjects Suggested by His Life and Work, 3rd ed.* New York: Chelsea, 1999.
Hoffman, P. *The Man Who Loved Only Numbers: The Story of Paul Erdos and the Search for Mathematical Truth.* New York: Hyperion, p. 217, 1998.
Honsberger, R. "An Intriguing Series." Ch. 10 in *Mathematical Gems II.* Washington, DC: Math. Assoc. Amer., pp. 98–03, 1976.
Rosenbaum, B. "Solution to Problem E46." *Amer. Math. Monthly* **41**, 48, 1934.
Sloane, N. J. A. Sequences A004080 in "An On-Line Version of the Encyclopedia of Integer Sequences." http://www.research.att.com/~njas/sequences/eisonline.html.
Wells, D. *The Penguin Dictionary of Curious and Interesting Numbers.* Middlesex, England: Penguin Books, p. 41, 1986.

Harmonic System of Points

HARMONIC RANGE

Harmonic-Geometric Mean

Let

$$\alpha_{n+1} = \frac{2\alpha_n \beta_n}{\alpha_n + \beta_n}$$

$$\beta_{n+1} = \sqrt{\alpha_n \beta_n},$$

then

$$H(\alpha_0, \beta_0) \equiv \lim_{n \to \infty} a_n = \frac{1}{M(\alpha_0^{-1}, \beta_0^{-1})},$$

where M is the ARITHMETIC-GEOMETRIC MEAN.

See also ARITHMETIC MEAN, ARITHMETIC-GEOMETRIC MEAN, GEOMETRIC MEAN, HARMONIC MEAN

Harmonious Graph

A connected LABELED GRAPH with n EDGES in which all VERTICES can be labeled with distinct INTEGERS (mod n) so that the sums of the PAIRS of numbers at the ends of each EDGE are also distinct (mod n). The LADDER GRAPH, FAN, WHEEL GRAPH, PETERSEN GRAPH, TETRAHEDRAL GRAPH, DODECAHEDRAL GRAPH, and ICOSAHEDRAL GRAPH are all harmonious (Graham and Sloane 1980).

See also GRACEFUL GRAPH, LABELED GRAPH, POSTAGE STAMP PROBLEM, SEQUENTIAL GRAPH

References

Gallian, J. A. "Open Problems in Grid Labeling." *Amer. Math. Monthly* **97**, 133–35, 1990.
Gardner, M. *Wheels, Life, and other Mathematical Amusements.* New York: W. H. Freeman, p. 164, 1983.

Graham, R. L. and Sloane, N. "On Additive Bases and Harmonious Graphs." *SIAM J. Algebraic Discrete Math.* **1**, 382–04, 1980.

Guy, R. K. "The Corresponding Modular Covering Problem. Harmonious Labelling of Graphs." §C13 in *Unsolved Problems in Number Theory, 2nd ed.* New York: Springer-Verlag, pp. 127–28, 1994.

Harmonograph

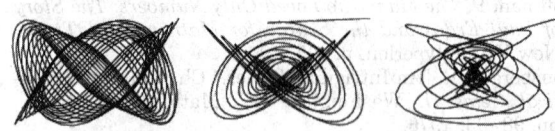

A device consisting of two coupled pendula, usually oscillating at right angles to each other, which are attached to a pen. The resulting damped SIMPLE HARMONIC MOTION can produce beautiful, complicated curves which eventually terminate in a point as the motion of the pendula is damped by friction. In the absence of friction, the figures produced by a harmonograph would be LISSAJOUS CURVES.

See also LISSAJOUS CURVE, SIMPLE HARMONIC MOTION, SPIROGRAPH

References

Cundy, H. and Rollett, A. "The Harmonograph." §5.5.4 in *Mathematical Models, 3rd ed.* Stradbroke, England: Tarquin Pub., pp. 244–48, 1989.

Wells, D. *The Penguin Dictionary of Curious and Interesting Geometry.* London: Penguin, pp. 92–3, 1991.

Harnack's Inequality

Let $D = D(z_0, R)$ be an OPEN DISK, and let u be a HARMONIC FUNCTION on D such that $u(z) \geq 0$ for all $z \in D$. Then for all $z \in D$, we have

$$0 \leq u(z) \leq \left(\frac{R}{R - |z - z_0|} \right)^2 u(z_0).$$

See also HARMONIC FUNCTION, HARNACK'S PRINCIPLE, LIOUVILLE'S CONFORMALITY THEOREM

References

Flanigan, F. J. "Harnack's Inequality." §2.5.1 in *Complex Variables: Harmonic and Analytic Functions.* New York: Dover, pp. 88–0, 1983.

Krantz, S. G. "The Harnack Inequality." §7.6.1 in *Handbook of Complex Analysis.* Boston, MA: Birkhäuser, p. 97, 1999.

Harnack's Principle

Let $u_1 \leq u_2 \leq \ldots$ be HARMONIC FUNCTIONS on a connected open set $U \subseteq \mathbb{C}$. Then either $u_j \to \infty$ uniformly on compact sets or there is a finite-values HARMONIC FUNCTION u on U such that $u_j \to u$ uniformly on compact sets.

See also HARMONIC FUNCTION, HARNACK'S INEQUALITY

References

Krantz, S. G. "Harnack's Principle." §7.6.2 in *Handbook of Complex Analysis.* Boston, MA: Birkhäuser, p. 97, 1999.

Harnack's Theorems

Let s_i be the orders of singular points on a curve (Coolidge 1959, p. 56). Harnack's first theorem states that a real irreducible curve of order n cannot have more than

$$\tfrac{1}{2}(n-1)(n-2) - \sum s_i(s_i - 1) + 1$$

circuits (Coolidge 1959, p. 57).

Harnack's second theorem states that there exists a curve of every order with the maximum number of circuits compatible with that order and with a certain number of double points, provided that number is not permissible for a curve of lower order (Coolidge 1959, p. 61).

References

Coolidge, J. L. *A Treatise on Algebraic Plane Curves.* New York: Dover, 1959.

Harry Dym Equation

The PARTIAL DIFFERENTIAL EQUATION

$$u_t = u_{xxx} u^3.$$

References

Calogero, F. and Degasperis, A. *Spectral Transform and Solitons: Tools to Solve and Investigate Nonlinear Evolution Equations.* New York: North-Holland, p. 53, 1982.

Zwillinger, D. *Handbook of Differential Equations, 3rd ed.* Boston, MA: Academic Press, p. 133, 1997.

Harshad Number

A POSITIVE INTEGER which is DIVISIBLE by the sum of its DIGITS, also called a Niven number (Kennedy *et al.* 1980) or a multidigital number (Kaprekar 1955). The first few are 1, 2, 3, 4, 5, 6, 7, 8, 9, 10, 12, 18, 20, 21, 24, ... (Sloane's A005349). Grundman (1994) proved that there is no sequence of more than 20 consecutive Harshad numbers, and found the smallest sequence of 20 consecutive Harshad numbers, each member of which has 44,363,342,786 digits.

Grundman (1994) defined an n-Harshad (or n-Niven) number to be a POSITIVE INTEGER which is DIVISIBLE by the sum of its digits in base $n \geq 2$. Cai (1996) showed that for $n = 2$ or 3, there exists an infinite family of sequences of consecutive n-Harshad numbers of length $2n$.

Define an all-Harshad (or all-Niven) number as a positive integer which is divisible by the sum of its digits in *all* bases $n \geq 2$. Then only 1, 2, 4, and 6 are all-Harshad numbers (A. Kertesz).

References

Cai, T. "On 2-Niven Numbers and 3-Niven Numbers." *Fib. Quart.* **34**, 118–20, 1996.

Cooper, C. N. and Kennedy, R. E. "Chebyshev's Inequality and Natural Density." *Amer. Math. Monthly* **96**, 118–24, 1989.

Cooper, C. N. and Kennedy, R. "On Consecutive Niven Numbers." *Fib. Quart.* **21**, 146–51, 1993.

Grundman, H. G. "Sequences of Consecutive *n*-Niven Numbers." *Fib. Quart.* **32**, 174–75, 1994.

Kaprekar, D. R. "Multidigital Numbers." *Scripta Math.* **21**, 27, 1955.

Kennedy, R. E. and Cooper, C. N. "On the Natural Density of the Niven Numbers." Abstract 816-1-19, *Abstracts Amer. Math. Soc.* **6**, 17, 1985.

Kennedy, R.; Goodman, R.; and Best, C. "Mathematical Discovery and Niven Numbers." *MATYC J.* **14**, 21–5, 1980.

Sloane, N. J. A. Sequences A005349/M0481 in "An On-Line Version of the Encyclopedia of Integer Sequences." http://www.research.att.com/~njas/sequences/eisonline.html.

Vardi, I. "Niven Numbers." §2.3 in *Computational Recreations in Mathematica.* Redwood City, CA: Addison-Wesley, pp. 19 and 28–1, 1991.

Wells, D. *The Penguin Dictionary of Curious and Interesting Numbers.* Middlesex, England: Penguin Books, p. 171, 1986.

Hart Circle

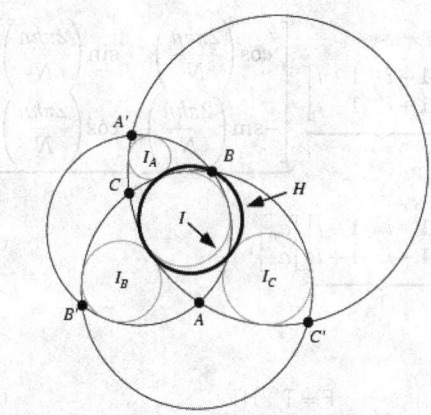

The CIRCLE H which touches the INCIRCLES I, I_A, I_B, and I_C of a CIRCULAR TRIANGLE ABC and its ASSOCIATED TRIANGLES. It is either externally tangent to I and internally tangent to incircles of the ASSOCIATED TRIANGLES I_A, I_B, and I_C (as in the above figure), or vice versa. The Hart circle has several properties which are analogous to the properties on the NINE-POINT CIRCLE of a linear triangle. There are eight Hart circles associated with a given CIRCULAR TRIANGLE.

The Hart circle of any CIRCULAR TRIANGLE and the Hart circles of the three ASSOCIATED TRIANGLES have a common tangent circle which touches the former in the opposite sense to that which it touches the latter (Lachlan 1893, p. 254). In addition, the CIRCUMCIRCLE of any CIRCULAR TRIANGLE is the Hart circle of the CIRCULAR TRIANGLE formed by the circumcircles of the inverse associated triangles (Lachlan 1893, p. 254).

See also ASSOCIATED TRIANGLES, CIRCLE, CIRCULAR TRIANGLE

References

Casey, J. "On the Equations and Properties--(1) of the System of Circles Touching Three Circles in a Plane; (2) of the System of Spheres Touching Four Spheres in Space; (3) of the System of Circles Touching Three Circles on a Sphere; (4) of the System of Conics Inscribed to a Conic, and Touching Three Inscribed Conics in a Plane." *Proc. Roy. Irish Acad.* **9**, 396–23, 1864–866.

Coolidge, J. L. *A Treatise on the Geometry of the Circle and Sphere.* New York: Chelsea, p. 43, 1971.

Johnson, R. A. *Modern Geometry: An Elementary Treatise on the Geometry of the Triangle and the Circle.* Boston, MA: Houghton Mifflin, pp. 127–28, 1929.

Lachlan, R. *An Elementary Treatise on Modern Pure Geometry.* London: Macmillian, pp. 254–57, 1893.

Larmor, A. "Contacts of Systems of Circles." *Proc. London Math. Soc.* **23**, 136–57, 1891.

Hart's Inversor

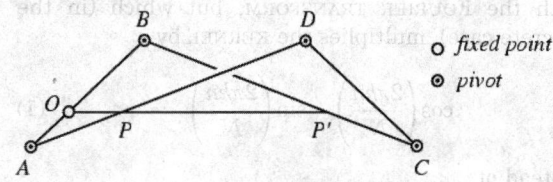

A LINKAGE which draws the inverse of a given curve. It can also convert circular to linear motion. The rods satisfy $AB = CD$ and $BC = DA$, and O, P, and P' remain COLLINEAR. Coxeter (1969, p. 428) shows that if $AO = \mu AB$, then

$$OP \times OP' = \mu(1 - \mu)(AD^2 - AB^2).$$

See also LINKAGE, PEAUCELLIER INVERSOR

References

Courant, R. and Robbins, H. *What is Mathematics?: An Elementary Approach to Ideas and Methods.* Oxford, England: Oxford University Press, p. 157, 1978.

Coxeter, H. S. M. *Introduction to Geometry, 2nd ed.* New York: Wiley, pp. 82–3, 1969.

Mannheim, A. "Sur l'inverseur de Hart." *Messenger Math.*, p. 151, Nov. 1896.

Rademacher, H. and Toeplitz, O. *The Enjoyment of Mathematics: Selections from Mathematics for the Amateur.* Princeton, NJ: Princeton University Press, pp. 124–29, 1957.

Hart's Theorem

Any one of the eight APOLLONIUS CIRCLES of three given CIRCLES is TANGENT to a CIRCLE H known as a

HART CIRCLE, as are the other three APOLLONIUS CIRCLES having (1) like contact with two of the given CIRCLES and (2) unlike contact with the third.

See also APOLLONIUS CIRCLES, HART CIRCLE

References

Casey, J. "On the Equations and Properties--(1) of the System of Circles Touching Three Circles in a Plane; (2) of the System of Spheres Touching Four Spheres in Space; (3) of the System of Circles Touching Three Circles on a Sphere; (4) of the System of Conics Inscribed to a Conic, and Touching Three Inscribed Conics in a Plane." *Proc. Roy. Irish Acad.* **9**, 396–23, 1864–866.

Casey, J. *A Sequel to the First Six Books of the Elements of Euclid, Containing an Easy Introduction to Modern Geometry with Numerous Examples,* 5th ed., rev. enl. Dublin: Hodges, Figgis, & Co., pp. 106–07, 1888.

Coolidge, J. L. *A Treatise on the Geometry of the Circle and Sphere.* New York: Chelsea, p. 43, 1971.

Johnson, R. A. *Modern Geometry: An Elementary Treatise on the Geometry of the Triangle and the Circle.* Boston, MA: Houghton Mifflin, pp. 127–28, 1929.

Lachlan, R. *An Elementary Treatise on Modern Pure Geometry.* London: Macmillian, pp. 254–57, 1893.

Larmor, A. "Contacts of Systems of Circles." *Proc. London Math. Soc.* **23**, 136–57, 1891.

Hartley Transform

An INTEGRAL TRANSFORM which shares some features with the FOURIER TRANSFORM, but which (in the discrete case), multiplies the KERNEL by

$$\cos\left(\frac{2\pi kn}{N}\right) - \sin\left(\frac{2\pi kn}{N}\right) \tag{1}$$

instead of

$$e - 2^{\pi i kn/N} = \cos\left(\frac{2\pi kn}{N}\right) - i\sin\left(\frac{2\pi kn}{N}\right). \tag{2}$$

The Hartley transform produces REAL output for a REAL input, and is its own inverse. It therefore can have computational advantages over the DISCRETE FOURIER TRANSFORM, although analytic expressions are usually more complicated for the Hartley transform.

The discrete version of the Hartley transform can be written explicitly as

$$\mathscr{H}[a] \equiv \frac{1}{\sqrt{N}}\sum_{n=0}^{N-1} a_n\left[\cos\left(\frac{2\pi kn}{N}\right) - \sin\left(\frac{2\pi kn}{N}\right)\right] \tag{3}$$

$$= \mathfrak{R}\mathscr{F}[a] - \mathfrak{I}\mathscr{F}[a], \tag{4}$$

where $\mathscr{F}$ denotes the FOURIER TRANSFORM. The Hartley transform obeys the CONVOLUTION property

$$\mathscr{H}[a*b]_k = \tfrac{1}{2}\left(A_k B_k - \bar{A}_k \bar{B}_k + A_k \bar{B}_k + \bar{A}_k B_k\right), \tag{5}$$

where

$$\bar{a}_0 \equiv a_0 \tag{6}$$

$$\bar{a}_{n/2} \equiv a_{n/2} \tag{7}$$

$$\bar{a}_k \equiv a_{n-k} \tag{8}$$

(Arndt). Like the FAST FOURIER TRANSFORM, there is a "fast" version of the Hartley transform. A decimation in time algorithm makes use of

$$\mathscr{H}_n^{\text{left}}[a] \equiv \mathscr{H}_{n/2}[a^{\text{even}}] + \mathscr{X}\mathscr{H}_{n/2}[a^{\text{odd}}] \tag{9}$$

$$\mathscr{H}_n^{\text{right}}[a] \equiv \mathscr{H}_{n/2}[a^{\text{even}}] - \mathscr{X}\mathscr{H}_{n/2}[a^{\text{odd}}], \tag{10}$$

where $\mathscr{X}$ denotes the sequence with elements

$$a_n \cos\left(\frac{\pi n}{N}\right) - \bar{a}_n \sin\left(\frac{\pi n}{N}\right). \tag{11}$$

A decimation in frequency algorithm makes use of

$$\mathscr{H}_n^{\text{even}}[a] = \mathscr{H}_{n/2}\left[a^{\text{left}} + a^{\text{right}}\right], \tag{12}$$

$$\mathscr{H}_n^{\text{odd}}[a] = \mathscr{H}_{n/2}\left[\mathscr{X}\left(a^{\text{left}} - a^{\text{right}}\right)\right]. \tag{13}$$

The DISCRETE FOURIER TRANSFORM

$$A_k \equiv \mathscr{F}[a] = \sum_{n=0}^{N-1} e^{-2\pi i kn/N} a_n \tag{14}$$

can be written

$$\begin{bmatrix} A_k \\ A_{-k} \end{bmatrix} = \sum_{n=0}^{N-1} \underbrace{\begin{bmatrix} e^{-2\pi i kn/N} & 0 \\ 0 & e^{-2\pi i kn/N} \end{bmatrix}}_{\mathsf{F}} \begin{bmatrix} a_n \\ a_n \end{bmatrix} \tag{15}$$

$$= \sum_{n=0}^{N-1} \underbrace{\frac{1}{2}\begin{bmatrix} 1-i & 1+i \\ 1+i & 1-i \end{bmatrix}}_{\mathsf{T}^{-1}} \underbrace{\begin{bmatrix} \cos\left(\frac{2\pi kn}{N}\right) & \sin\left(\frac{2\pi kn}{N}\right) \\ -\sin\left(\frac{2\pi kn}{N}\right) & \cos\left(\frac{2\pi kn}{N}\right) \end{bmatrix}}_{\mathsf{H}}$$

$$\times \underbrace{\frac{1}{2}\begin{bmatrix} 1+i & 1-i \\ 1-i & 1+i \end{bmatrix}}_{\mathsf{T}} \begin{bmatrix} a_n \\ a_n \end{bmatrix}, \tag{16}$$

so

$$\mathsf{F} = \mathsf{T}^{-1}\mathsf{H}\mathsf{T}. \tag{17}$$

See also DISCRETE FOURIER TRANSFORM, FAST FOURIER TRANSFORM, FOURIER TRANSFORM

References

Arndt, J. "The Hartley Transform (HT)." Ch. 2 in "Remarks on FFT Algorithms." http://www.jjj.de/fxt/.

Bracewell, R. N. *The Fourier Transform and Its Applications,* 3rd ed. New York: McGraw-Hill, 1999.

Bracewell, R. N. *The Hartley Transform.* New York: Oxford University Press, 1986.

Haruki's Theorem

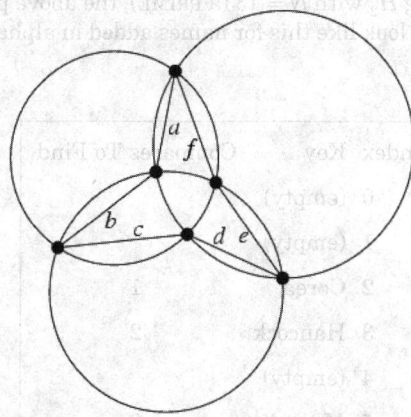

Given three circles, each intersecting the other two in two points, the line segments connecting their points of intersection satisfy

$$\frac{ace}{bdf} = 1$$

(Honsberger 1995).

See also CIRCULAR TRIANGLE, TRIQUETRA, VENN DIAGRAM

References

Honsberger, R. "Haruki's Cevian Theorem for Circles." §12.4 in *Episodes in Nineteenth and Twentieth Century Euclidean Geometry*. Washington, DC: Math. Assoc. Amer., pp. 144–46, 1995.

Hash Function

A hash function H projects a value from a set with many (or even an infinite number of) members to a value from a set with a fixed number of (fewer) members. Hash functions are not reversible. A hash function H might, for instance, be defined as $y = H(x) = \lfloor 10x \pmod 1 \rfloor$, where $x \in \mathbb{R}$, $y \in [0, 9]$, and $\lfloor x \rfloor$ is the FLOOR FUNCTION.

Hash functions can be used to determine if two objects are equal (possibly with a fixed average number of mistakes). Other common uses of hash functions are CHECKSUMS over a large amount of data (e.g., the CYCLIC REDUNDANCY CHECK [CRC]) and finding an entry in a database by a key value. The UNIX c-shell (csh) uses a hash table to store the location of executable programs. As a result adding new executables in a user's search path requires regeneration of the hash table using the rehash command before these programs can be executed without specifying the complete path.

To illustrate the use of hash functions in database lookups, consider a database consisting of an array containing an index n, a name, and a telephone number, with names listed in arbitrary order.

n	Name	Number
0	Parker	12345
1	(empty)	
2	Davis	43534
3	Harris	32452
4	Corea	46532
5	Hancock	96562
6	Brecker	37811
7	(empty)	
$N-1$	Marsalis	54323

To look up Hancock from this array, you would start at the beginning of the array, compare the names, then try the next until the names match. This very simple algorithm finds any entry in 1 to N steps, giving an average seek time of $N/2$. The seek time is therefore proportional to N. A much faster result can generally be achieved, if the database is sorted.

n	Name
0	Brecker
1	Corea
2	Davis
3	Hancock
4	Harris
5	Marsalis
6	Parker
7	(empty)
$N-1$	(empty)

An efficient algorithm on this sorted array first checks entry $N/2$, and then recursively uses bisection to check entries in intervals $[0, N/2 - 1]$ or $[N/2 + 1, N-1]$, depending wether the most recently looked-up name precedes or succeeds the name sought. The average seek time of this procedure this is proportional to $\ln N$.

The idea behind using a hash function here is that although the possible number of combinations of characters in a name is quite large, only a subset of them is usually found in practice (i.e., names such as "Kwqrst" are much less common than names like "Jones.") Therefore, when you insert an entry into the database at an index that can somehow be calculated using a key (which is also available at the time you search for it), you might be able to find it later at the first location you check.

Consider the following simple example in which the hash function H is simply the sum of ASCII codes of characters in a name (considered to be all in lower-case) computed mod $N = 13$.

Name	H
Brecker	6
Corea	2
Davis	2
Hancock	12
Harris	12
Marsalis	2
Parker	8

The above example illustrates that the hash function can give the same results for different keys. This difficulty is typically circumvented by introducing a second hash function H_2 whose results are designed to be completely different from that of H. For illustrative purposes, let H_2 be one plus the bitwise exclusive or of all codes in a name (again taken as all lower-case) mod $N - 1$. This gives the following table.

Name	H_2
Brecker	11
Corea	3
Davis	10
Hancock	4
Harris	8
Marsalis	3
Parker	8

A new index can then be calculated as the sum of the first index and H_2 (mod N) until an empty slot is found where new data can be stored. Note that when using H_2 as an offset to walk through the database, it is not, in general, guaranteed that any key will eventually reach any slot. However, for certain values of N, namely N a PRIME NUMBER, such behavior *is*

guaranteed, so N is always chosen to be PRIME. After computing H_2 with $N = 13$ (a PRIME), the above phone list would look like this for names added in alphabetic order.

Index	Key	Compares To Find
0	(empty)	
1	(empty)	
2	Corea	1
3	Hancock	2
4	(empty)	
5	Marsalis	2
6	Brecker	1
7	Harris	2
8	Parker	1
9	(empty)	
10	(empty)	
11	(empty)	
12	Davis	2

The average seek time for locating a name in this table depends on the kind of data, N, and the quality of the hash functions used. However, for reasonable choices of hash functions, it will be much smaller than $\ln N$.

See also COLLISION-FREE HASH FUNCTION, CRYPTOGRAPHIC HASH FUNCTION, CYCLIC REDUNDANCY CHECK, ONE-WAY HASH FUNCTION, HASH TABLE, UNIVERSAL HASH FUNCTION

Hash Table

A database accessed by one or more HASH FUNCTIONS.

See also HASH FUNCTION

HashLife

A LIFE ALGORITHM that achieves remarkable speed by storing subpatterns in a HASH FUNCTION table, and using them to skip forward, sometimes thousands of generations at a time. HashLife takes tremendous amounts of memory and can't show patterns at every step, but can quickly calculate the outcome of a pattern that takes millions of generations to complete.

See also HASH FUNCTION, LIFE

Hasse Diagram

A graphical rendering of a PARTIALLY ORDERED SET displayed via the COVER relation of the PARTIALLY ORDERED SET with an implied upward orientation. A point is drawn for each element of the POSET, and line segments are drawn between these points according to the following two rules:

1. If $x < y$ in the poset, then the point corresponding to x appears lower in the drawing than the point corresponding to y.
2. The line segment between the points corresponding to any two elements x and y of the poset is included in the drawing IFF x covers y or y covers x.

Hasse diagrams are also called UPWARD DRAWINGS.

A Hasse diagram of a GRAPH may be generated using `HasseDiagram[g]` in the *Mathematica* add-on package `DiscreteMath'Combinatorica'` (which can be loaded with the command `<<DiscreteMath'`).

References
Skiena, S. "Hasse Diagrams." §5.4.2 in *Implementing Discrete Mathematics: Combinatorics and Graph Theory with Mathematica*. Reading, MA: Addison-Wesley, p. 163 and 206–08, 1990.

Hasse Principle

A collection of equations satisfies the Hasse principle if, whenever one of the equations has solutions in $\mathbb{R}$ and all the $\mathbb{Q}_p$, then the equations have solutions in the RATIONALS $\mathbb{Q}$. Examples include the set of equations

$$ax^2 + bxy + cy^2 = 0$$

with a, b, and c INTEGERS, and the set of equations

$$x^2 + y^2 = a$$

for a rational. The trivial solution $x = y = 0$ is usually not taken into account when deciding if a collection of homogeneous equations satisfies the Hasse principle. The Hasse principle is sometimes called the local-global principle.

See also GLOBAL FIELD, LOCAL FIELD

Hasse's Algorithm
COLLATZ PROBLEM

Hasse's Conjecture

Define the ZETA FUNCTION of a VARIETY over a NUMBER FIELD by taking the product over all PRIME IDEALS of the ZETA FUNCTIONS of this VARIETY reduced modulo the PRIMES. Hasse conjectured that this product has a MEROMORPHIC continuation over the whole plane and a functional equation.

See also MEROMORPHIC FUNCTION, PRIME IDEAL

References
Lang, S. "Some History of the Shimura-Taniyama Conjecture." *Not. Amer. Math. Soc.* **42**, 1301–307, 1995.

Hasse's Resolution Modulus Theorem

The JACOBI SYMBOL $(a/y) = \chi(y)$ as a CHARACTER can be extended to the KRONECKER SYMBOL $(f(a)/y) = \chi^*(y)$ so that $\chi^*(y) = \chi(y)$ whenever $\chi(y) \neq 0$. When y is RELATIVELY PRIME to $f(a)$, then $\chi^*(y) \neq 0$, and for NONZERO values $\chi^*(y_1) = \chi^*(y_2)$ IFF $y_1 \equiv y_2 \bmod^+ f(a)$. In addition, $|f(a)|$ is the minimum value for which the latter congruence property holds in any extension symbol for $\chi(y)$.

See also CHARACTER (NUMBER THEORY), JACOBI SYMBOL, KRONECKER SYMBOL

References
Cohn, H. *Advanced Number Theory.* New York: Dover, pp. 35–6, 1980.

Hasse-Davenport Relation

Let F be a FINITE FIELD with q elements, and let F_s be a FIELD containing F such that $[F_s : F] = s$. Let χ be a nontrivial MULTIPLICATIVE CHARACTER of F and $\chi' = \chi \circ N_{F_s/F}$ a character of F_s. Then

$$(-g(\chi))^s = -g(\chi'),$$

where $g(x)$ is a GAUSSIAN SUM.

See also GAUSSIAN SUM, MULTIPLICATIVE CHARACTER

References
Ireland, K. and Rosen, M. "A Proof of the Hasse-Davenport Relation." §11.4 in *A Classical Introduction to Modern Number Theory, 2nd ed.* New York: Springer-Verlag, pp. 162–65, 1990.

Hasse-Minkowski Theorem

Two nonsingular forms are equivalent over the rationals IFF they have the same DETERMINANT and the same P-SIGNATURES for all p.

Hat

The hat is a CARET-shaped symbol most commonly used to denote a UNIT VECTOR (e.g., $\hat{\mathbf{v}}$) or an ESTIMATOR (e.g., $\hat{x}$). The symbol $\hat{x}$ is voiced "x-hat." The hat symbol is more commonly known as the circumflex (Bringhurst 1997, p. 274).

See also BAR, CARET, ESTIMATOR, MACRON, UNIT VECTOR

References
Bringhurst, R. *The Elements of Typographic Style, 2nd ed.* Point Roberts, WA: Hartley and Marks, 1997.

Hat-Box Theorem

ARCHIMEDES' HAT-BOX THEOREM

Haupt-Exponent

The smallest exponent e for which $b^e \equiv 1 \pmod 1$, where b and n are given numbers, is called the haupt-exponent (or sometimes "ORDER") of $b \pmod n$. The number of bases having a haupt-exponent e is $\phi(e)$, where $\phi(e)$ is the TOTIENT FUNCTION. Cunningham (1922) published the haupt-exponents for primes to 25409 and bases 2, 3, 5, 6, 7, 10, 11, and 12.

Haupt-exponents exists for n which are not factors of b. For example, the haupt-exponent of 10 (mod 7) is 6, since

$$10^6 \equiv 1 \pmod 7.$$

The haupt-exponent of 10 mod an integer n relatively prime to 10 gives the period of the DECIMAL EXPANSION of the reciprocal of n (Glaisher 1878, Lehmer 1941). For example, the haupt-exponent of 10 (mod 13) is 6, and

$$\tfrac{1}{13} = 0.\overline{0769230},$$

which has period 6. The haupt-exponent of 2 mod an integer n relatively prime to 2 gives the multiplicative order of 2 (mod $2n + 1$) (Golomb 1961).

The following table gives the first few haupt-exponents for bases $b \pmod p$ with $p = 1, 2, \ldots$.

b	Sloane	haupt-exponents
2	A002326	2, 4, 3, 6, 10, 12, 4, 8, 18, 6, 11, 20, 18, ...
3	A050975	1, 2, 4, 6, 2, 4, 5, 3, 6, 4, 16, 18, 4, 5, ...
4	A050976	1, 2, 3, 3, 5, 6, 2, 4, 9, 3, 11, 10, 9, 14, ...
5	A050977	1, 2, 1, 2, 6, 2, 6, 5, 2, 4, 6, 4, 16, 6, 9, ...
6	A050978	1, 2, 10, 12, 16, 9, 11, 5, 14, ...
7	A050979	1, 1, 2, 4, 1, 2, 3, 4, 10, 2, 12, 4, 2, 16, ...
8	A050980	2, 4, 1, 2, 10, 4, 4, 8, 6, 2, 11, 20, 6, 28, ...
9	A050981	1, 1, 2, 3, 1, 2, 5, 3, 3, 2, 8, 9, 2, 5, 11, ...
10	A002329	1, 6, 1, 2, 6, 16, 18, 6, 22, 3, 28, ...

See also COMPLETE RESIDUE SYSTEM, MULTIPLICATIVE ORDER, ORDER (MODULO), ORDER (POLYNOMIAL), PRIMITIVE ROOT

References

Cunningham, A. *Haupt-Exponents, Residue Indices, Primitive Roots.* London: F. Hodgson, 1922.

Glaisher, J. W. L. "Periods of Reciprocals of Integers Prime to 10." *Proc. Cambridge Philos. Soc.* **3**, 185–06, 1878.

Golomb, S. W. "Permutations by Cutting and Shuffling." *SIAM Rev.* **3**, 293–97, 1961.

Lehmer, D. H. "Guide to Tables in the Theory of Numbers." Bulletin No. 105. Washington, DC: National Research Council, pp. 7–2, 1941.

Nagell, T. "Exponent of an Integer Modulo n." §31 in *Introduction to Number Theory.* New York: Wiley, pp. 102–06, 1951.

Sloane, N. J. A. Sequences A0023260936, A0023294045, A050975, A050976, A050977, A050978, A050979, A050980, and A050981 in "An On-Line Version of the Encyclopedia of Integer Sequences." http://www.research.-att.com/~njas/sequences/eisonline.html.

Hausdorff

HAUSDORFF SPACE

Hausdorff Axioms

The axioms formulated by Hausdorff (1914) for his concept of a TOPOLOGICAL SPACE. These axioms describe the properties satisfied by subsets of elements x in a NEIGHBORHOOD SET E of x.

1. There corresponds to each point x at least one NEIGHBORHOOD $U(x)$, and each NEIGHBORHOOD $U(x)$ contains the point x.
2. If $U(x)$ and $V(x)$ are two NEIGHBORHOODS of the same point x, there must exist a NEIGHBORHOOD $W(x)$ that is a subset of both.
3. If the point y lies in $U(x)$, there must exist a NEIGHBORHOOD $U(y)$ that is a SUBSET of $U(x)$.
4. For two different points x and y, there are two corresponding NEIGHBORHOODS $U(x)$ and $U(y)$ with no points in common.

See also HAUSDORFF SPACE, TOPOLOGICAL SPACE

References

Hausdorff, F. *Grundzüge der Mengenlehre.* Leipzig, Germany: von Veit, 1914. Republished as *Set Theory, 2nd ed.* New York: Chelsea, 1962.

Hausdorff Dimension

Informally, SELF-SIMILAR objects with parameters N and s are described by a power law such as

$$N = s^d,$$

where

$$d = \frac{\ln N}{\ln s}$$

is the "DIMENSION" of the scaling law, known as the Hausdorff dimension.

Formally, let A be a SUBSET of a METRIC SPACE X. Then the Hausdorff dimension $D(A)$ of A is the

INFIMUM of $d \geq 0$ such that the d-dimensional HAUS-DORFF MEASURE of A is 0 (which need not be an INTEGER).

In many cases, the Hausdorff dimension correctly describes the correction term for a resonator with FRACTAL PERIMETER in Lorentz's conjecture. However, in general, the proper dimension to use turns out to be the MINKOWSKI-BOULIGAND DIMENSION (Schroeder 1991).

See also CAPACITY DIMENSION, FRACTAL, FRACTAL DIMENSION, MINKOWSKI-BOULIGAND DIMENSION, SELF-SIMILARITY

References

Duvall, P.; Keesling, J.; and Vince, A. "The Hausdorff Dimension of the Boundary of a Self-Similar Tile." *J. London Math. Soc.* **61**, 649–60, 2000.

Federer, H. *Geometric Measure Theory.* New York: Springer-Verlag, 1969.

Harris, J. W. and Stocker, H. "Hausdorff Dimension." §4.11.3 in *Handbook of Mathematics and Computational Science.* New York: Springer-Verlag, pp. 113–14, 1998.

Hausdorff, F. "Dimension und äußeres Maß." *Math. Ann.* **79**, 157–79, 1919.

Ott, E. "Appendix: Hausdorff Dimension." *Chaos in Dynamical Systems.* New York: Cambridge University Press, pp. 100–03, 1993.

Schroeder, M. *Fractals, Chaos, Power Laws: Minutes from an Infinite Paradise.* New York: W. H. Freeman, pp. 41–5, 1991.

Hausdorff Measure

Let X be a METRIC SPACE, A be a SUBSET of X, and d a number ≥ 0. The d-dimensional Hausdorff measure of A, $H^d(A)$, is the INFIMUM of POSITIVE numbers y such that for every $r > 0$, A can be covered by a countable family of closed sets, each of diameter less than r, such that the sum of the dth POWERS of their diameters is less than y. Note that $H^d(A)$ may be infinite, and d need not be an INTEGER.

References

Federer, H. *Geometric Measure Theory.* New York: Springer-Verlag, 1969.

Ott, E. *Chaos in Dynamical Systems.* Cambridge, England: Cambridge University Press, p. 103, 1993.

Rogers, C. A. *Hausdorff Measures, 2nd ed.* Cambridge, England: Cambridge University Press, 1999.

Hausdorff Moment Problem

MOMENT PROBLEM

Hausdorff Paradox

For $n \geq 3$, there exist no additive finite and invariant measures for the group of displacements in $\mathbb{R}^n$.

References

Le Lionnais, F. *Les nombres remarquables.* Paris: Hermann, p. 49, 1983.

Hausdorff Space

A TOPOLOGICAL SPACE in which any two points have disjoint NEIGHBORHOODS. A space that is Hausdorff is sometimes said to "have Hausdorff topology" or "be Hausdorff."

See also HAUSDORFF MEASURE, TOPOLOGICAL SPACE

References

Porter, J. R. *Extensions and Absolutes of Hausdorff Spaces.* New York: Springer-Verlag, 1987.

Hausdorff Topology

HAUSDORFF SPACE

Hausdorff-Besicovitch Dimension

CAPACITY DIMENSION

Hauy Construction

The construction of polyhedra using identical building blocks. The illustrations above show such constructions for the OCTAHEDRON and RHOMBIC DODECAHEDRON. In Book XIII of the *ELEMENTS*, Euclid used a Hauy construction to build the DODECAHEDRON (Wells 1991).

See also OCTAHEDRAL NUMBER, OCTAHEDRON, RHOMBIC DODECAHEDRAL NUMBER, RHOMBIC DODECAHEDRON

References

Hauy, R.-J. "Essai d'une théorie sur la structure des crystals appliquée à plusieurs genres de substances crystallisées." 1784.

Weisstein, E. W. "Häuy Construction." MATHEMATICA NOTEBOOK HAUY.M.

Wells, D. *The Penguin Dictionary of Curious and Interesting Geometry.* London: Penguin, p. 93, 1991.

Haversine

$$\text{hav}(x) \equiv \tfrac{1}{2} \text{vers}(x) = \tfrac{1}{2}(1 - \cos x),$$

where $\text{vers}(x)$ is the VERSINE and $\cos x$ is the COSINE. Using a trigonometric identity, the haversine is equal to

$$\mathrm{hav}(x) = \sin^2(\tfrac{1}{2}x).$$

See also COSINE, COVERSINE, EXSECANT, SPHERICAL TRIGONOMETRY, VERSINE

References

Abramowitz, M. and Stegun, C. A. (Eds.). *Handbook of Mathematical Functions with Formulas, Graphs, and Mathematical Tables, 9th printing.* New York: Dover, p. 78, 1972.
Smart, W. M. *Text-Book on Spherical Astronomy, 6th ed.* Cambridge, England: Cambridge University Press, p. 18, 1960.

h-Cobordism

An *h*-cobordism is a COBORDISM W between two MANIFOLDS M_1 and M_2 such that W is SIMPLY CONNECTED and the inclusion maps $M_1 \to W$ and $M_2 \to W$ are HOMOTOPY equivalences.

h-Cobordism Theorem

If W is a SIMPLY CONNECTED, COMPACT MANIFOLD with a boundary that has two components, M_1 and M_2, such that inclusion of each is a HOMOTOPY equivalence, then W is DIFFEOMORPHIC to the product $M_1 \times [0, 1]$ for $\dim(M_1) \geq 5$. In other words, if M and M' are two simply connected MANIFOLDS of DIMENSION ≥ 5 and there exists an *H*-COBORDISM W between them, then W is a product $M \times I$ and M is DIFFEOMORPHIC to M'.

The proof of the *h*-cobordism theorem can be accomplished using SURGERY. A particular case of the *h*-cobordism theorem is the POINCARÉ CONJECTURE in dimension $n \geq 5$. Smale proved this theorem in 1961.

See also DIFFEOMORPHISM, POINCARÉ CONJECTURE, SURGERY

References

Smale, S. "Generalized Poincaré's Conjecture in Dimensions Greater than Four." *Ann. Math.* **74**, 391–06, 1961.

Heads-Minus-Tails Distribution

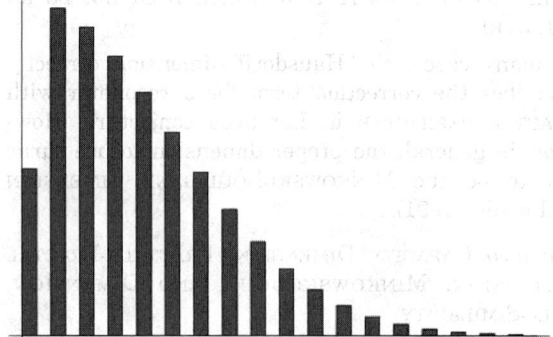

A fair COIN is tossed an even $2n$ number of times. Let $D \equiv |H - T|$ be the absolute difference in the number of heads and tails obtained. Then the probability distribution is given by

$$P(D = 2k) = \begin{cases} \left(\tfrac{1}{2}\right)^{2n} \dbinom{2n}{n} & k = 0 \\ 2\left(\tfrac{1}{2}\right)^{2n} \dbinom{2n}{n+k} & k = 1,\, 2,\, \ldots, \end{cases}$$

where $P(D = 2k - 1) = 0$. The most probable value of D is $D = 2$, and the expectation value is

$$\langle D_n \rangle = \frac{n \dbinom{2n}{n}}{2^{2n-1}}.$$

The generating function for $\langle D \rangle$ is given by

$$\sum \langle D_n \rangle x^{n-1} = (1 - x)^{-3/2} = 1 + \tfrac{3}{2}x + \tfrac{15}{8}x^2 + \tfrac{35}{16}x^3 + \ldots$$

(Sloane's A001803 and A046161; Abramowitz and Stegun 1972, Prévost 1933; Hughes 1995). These numbers also arise in 1-D RANDOM WALKS.

See also BERNOULLI DISTRIBUTION, COIN, COIN TOSSING, RANDOM WALK–1-D

References

Abramowitz, M. and Stegun, C. A. (Eds.). *Handbook of Mathematical Functions with Formulas, Graphs, and Mathematical Tables, 9th printing.* New York: Dover, p. 798, 1972.
Handelsman, M. B. Solution to Problem 436, "Distributing 'Heads' Minus 'Tails.'" *College Math. J.* **22**, 444–46, 1991.
Prévost, G. *Tables de Fonctions Sphériques.* Paris: Gauthier-Villars, pp. 156–57, 1933.
Hughes, B. D. Eq. (7.282) in *Random Walks and Random Environments, Vol. 1: Random Walks.* New York: Oxford University Press, p. 513, 1995.
Sloane, N. J. A. Sequences A001803/M2986 and A046161 in "An On-Line Version of the Encyclopedia of Integer Sequences." http://www.research.att.com/~njas/sequences/eisonline.html.

Heap

A SEQUENCE $\{a_n\}_{n=1}^N$ forms a (binary) heap if it satisfies $a_{\lfloor j/2 \rfloor} \leq a_j$ for $2 \leq j \leq N$, where $\lfloor x \rfloor$ is the FLOOR FUNCTION, which is equivalent to $a_i < a_{2i}$ and

$a_i < a_{2i+1}$ for $1 \leq i \leq (i-1)/2$. The first member must therefore be the smallest. A heap can be viewed as a labeled BINARY TREE in which the label of the ith node is smallest than the labels of any of its descendents (Skiena 1990, p. 35). Heaps support arbitrary insertion and seeking/deletion of the minimum value in $\mathcal{O}(\ln n)$ times per update (Skiena 1990, p. 38).

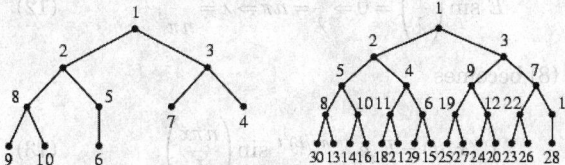

A list can be converted to a heap in $\mathcal{O}(n)$ times using an algorithm due to Floyd (1964). A binary heap can be generated from a PERMUTATION p using Heapify[p] in the *Mathematica* add-on package DiscreteMath`Combinatorica` (which can be loaded with the command <<DiscreteMath`). For example, given the RANDOM PERMUTATION $\{6, 2, 7, 9, 5, 3, 4, 8, 10, 1\}$, Floyd's algorithm gives the heap $\{1, 2, 3, 8, 5, 7, 4, 9, 10, 6\}$ (left figure). The right figure shows a heap containing 30 elements.

A PERMUTATION can be tested to see if it is a heap using the following *Mathematica* functions.

```
<<DiscreteMath`Combinatorica`;
  HeapQ[a_List?PermutationQ] := Module[{i, n
= Length[a]},
    And @@ Table[a[[Floor[i/2]]] < a[[i]], {i,
2, n}]
  ]
```

n	heaps
1	$\{1\}$
2	$\{1, 2\}$
3	$\{1, 2, 3\}$, $\{1, 3, 2\}$
4	$\{1, 2, 3, 4\}$, $\{1, 2, 4, 3\}$, $\{1, 3, 2, 4\}$

The numbers of heaps on $n = 1, 2, \ldots$ elements are 1, 1, 2, 3, 8, 20, 80, 896, 3360, ... (Sloane's A056971), the first few of which are summarized in the above table. The number of heaps of l levels (or equivalently, the number of heaps of $2^l - 1$ elements) is given by the RECURRENCE RELATION

$$S_l = \binom{2^l - 2}{2^{l-1} - 1} S_{l-1}^2$$

with $S_1 = 1$ (Skiena 1990, p. 36), the values of which for $l = 1, 2, \ldots$ are 1, 2, 80, 21964800, 74836825861835980800000, ... (Sloane's A056972).

See also BINARY TREE, COMPLETE BINARY TREE, HEAPSORT, PRIORITY QUEUE

References

Floyd, R. W. "Algorithm 245: Treesort 3." *Comm. ACM* **7**, 701, 1964.

Knuth, D. E. *The Art of Computer Programming, Vol. 3: Sorting and Searching, 2nd ed.* Reading, MA: Addison-Wesley, 1998.

Skiena, S. "Heaps." §1.4.4 in *Implementing Discrete Mathematics: Combinatorics and Graph Theory with Mathematica.* Reading, MA: Addison-Wesley, pp. 35–9, 1990.

Skiena, S. S. "Heaps." §1.4.4 in *The Algorithm Design Manual.* New York: Springer-Verlag, pp. 35–9, 1997.

Sloane, N. J. A. Sequences A056971 and A056972 in "An On-Line Version of the Encyclopedia of Integer Sequences." http://www.research.att.com/~njas/sequences/eisonline.html.

Heapsort

An $\mathcal{O}(n \lg n)$ SORTING ALGORITHM which is not quite as fast as QUICKSORT. It is a "sort-in-place" algorithm and requires no auxiliary storage, which makes it particularly concise and elegant to implement.

See also HEAP, QUICKSORT, SORTING

References

Knuth, D. E. *The Art of Computer Programming, Vol. 3: Sorting and Searching, 2nd ed.* Reading, MA: Addison-Wesley, pp. 144–48, 1998.

Press, W. H.; Flannery, B. P.; Teukolsky, S. A.; and Vetterling, W. T. "Heapsort." §8.3 in *Numerical Recipes in FORTRAN: The Art of Scientific Computing, 2nd ed.* Cambridge, England: Cambridge University Press, pp. 327–29, 1992.

Skiena, S. *Implementing Discrete Mathematics: Combinatorics and Graph Theory with Mathematica.* Reading, MA: Addison-Wesley, pp. 38–9, 1990.

Heart Surface

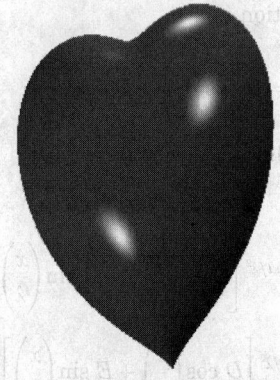

A heart-shaped surface given by the SEXTIC EQUATION

$$\left(2x^2 + 2y^2 + z^2 - 1\right)^3 - \tfrac{1}{10} x^2 z^3 - y^2 z^3 = 0.$$

See also ARCHIMEDEAN SPIRAL, BONNE PROJECTION, CARDIOID, PIRIFORM

References

Nordstrand, T. "Heart." http://www.uib.no/people/nfytn/hearttxt.htm.

Heat Conduction Equation

A PARTIAL DIFFERENTIAL diffusion equation OF THE FORM

$$\frac{\partial T}{\partial t} = \kappa \nabla^2 T. \tag{1}$$

Physically, the equation commonly arises in situations where κ is the thermal diffusivity and T the temperature.

The 1-D heat conduction equation is

$$\frac{\partial T}{\partial t} = \kappa \frac{\partial^2 T}{\partial x^2}. \tag{2}$$

This can be solved by SEPARATION OF VARIABLES using

$$T(x, t) = X(x)T(t). \tag{3}$$

Then

$$X \frac{dT}{dt} = \kappa T \frac{d^2 X}{dx^2}. \tag{4}$$

Dividing both sides by κXT gives

$$\frac{1}{\kappa T} \frac{dT}{dt} = \frac{1}{X} \frac{d^2 X}{dx^2} = -\frac{1}{\lambda^2}, \tag{5}$$

where each side must be equal to a constant. Anticipating the exponential solution in T, we have picked a negative separation constant so that the solution remains finite at all times and λ has units of length. The T solution is

$$T(t) = Ae^{-\kappa t/\lambda^2}, \tag{6}$$

and the X solution is

$$X(x) = C \cos\left(\frac{x}{\lambda}\right) + D \sin\left(\frac{x}{\lambda}\right). \tag{7}$$

The general solution is then

$$\begin{aligned}
T(x, t) &= T(t)X(x) \\
&= Ae^{-\kappa t/\lambda^2}\left[C \cos\left(\frac{x}{\lambda}\right) + D \sin\left(\frac{x}{\lambda}\right)\right] \\
&= e^{-\kappa t/\lambda^2}\left[D \cos\left(\frac{x}{\lambda}\right) + E \sin\left(\frac{x}{\lambda}\right)\right].
\end{aligned} \tag{8}$$

If we are given the boundary conditions

$$T(0, t) = 0 \tag{9}$$

and

$$T(L, t) = 0, \tag{10}$$

then applying (9) to (8) gives

$$D \cos\left(\frac{x}{\lambda}\right) = 0 \Rightarrow D = 0, \tag{11}$$

and applying (10) to (8) gives

$$E \sin\left(\frac{L}{\lambda}\right) = 0 \Rightarrow \frac{L}{\lambda} = n\pi \Rightarrow \lambda = \frac{L}{n\pi}, \tag{12}$$

so (8) becomes

$$T_n(x, t) = E_n e^{-\kappa(n\pi/L)^2 t} \sin\left(\frac{n\pi x}{L}\right). \tag{13}$$

Since the general solution can have any n,

$$T(x, t) = \sum_{n=1}^{\infty} c_n \sin\left(\frac{n\pi x}{L}\right) e^{-\kappa(n\pi/L)^2 t}. \tag{14}$$

Now, if we are given an initial condition $T(x, 0)$, we have

$$T(x, 0) = \sum_{n=1}^{\infty} c_n \sin\left(\frac{n\pi x}{L}\right). \tag{15}$$

Multiplying both sides by $\sin(m\pi x/L)$ and integrating from 0 to L gives

$$\begin{aligned}
\int_0^L & \sin\left(\frac{m\pi x}{L}\right) T(x, 0)\, dx \\
&= \int_0^L \sum_{n=1}^{\infty} c_n \sin\left(\frac{m\pi x}{L}\right) \sin\left(\frac{n\pi x}{L}\right) dx.
\end{aligned} \tag{16}$$

Using the ORTHOGONALITY of $\sin(nx)$ and $\sin(mx)$,

$$\begin{aligned}
\sum_{n=1}^{\infty} c_n \int_0^L & \sin\left(\frac{n\pi x}{L}\right) \sin\left(\frac{m\pi x}{L}\right) dx = \sum_{n=1}^{\infty} \frac{1}{2} \pi \delta_{mn} c_n \\
&= \frac{1}{2} \pi c_m = \int_0^L \sin\left(\frac{m\pi x}{L}\right) T(x, 0)\, dx,
\end{aligned} \tag{17}$$

so

$$c_n = \frac{2}{\pi} \int_0^L \sin\left(\frac{m\pi x}{L}\right) T(x, 0)\, dx. \tag{18}$$

If the boundary conditions are replaced by the requirement that the derivative of the temperature be zero at the edges, then (9) and (10) are replaced by

$$\left.\frac{\partial T}{\partial x}\right|_{(0,\, t)} = 0 \tag{19}$$

$$\left.\frac{\partial T}{\partial x}\right|_{(L,\, t)} = 0. \tag{20}$$

Following the same procedure as before, a similar

answer is found, but with sine replaced by cosine:

$$T(x,\ t) = \sum_{n=1}^{\infty} c_n \cos\left(\frac{n\pi x}{L}\right) e^{-\kappa(n\pi/L)^2 t}, \qquad (21)$$

where

$$c_n = \frac{2}{\pi} \int_0^L \cos\left(\frac{m\pi x}{L}\right) \frac{\partial T(x,\ 0)}{\partial x}\bigg|_{t=0}\ dx. \qquad (22)$$

Heat Conduction EquationDisk

To solve the HEAT CONDUCTION EQUATION on a 2-D disk of radius $R = 1$, try to separate the equation using

$$T(r,\ \theta,\ t) = R(r)\Theta(\theta)T(t). \qquad (1)$$

Writing the θ and r terms of the LAPLACIAN in SPHERICAL COORDINATES gives

$$\nabla^2 = \frac{d^2R}{dr^2} + \frac{2}{r}\frac{dR}{dr} + \frac{1}{r^2}\frac{d^2\Theta}{d\theta^2}, \qquad (2)$$

so the HEAT CONDUCTION EQUATION becomes

$$\frac{R\Theta}{\kappa}\frac{d^2T}{dt^2} = \frac{d^2R}{dr^2}\Theta T + \frac{2}{r}\frac{dR}{dr}\Theta T + \frac{1}{r^2}\frac{d^2\Theta}{d\theta^2}RT. \qquad (3)$$

Multiplying through by $r^2/R\Theta T$ gives

$$\frac{r^2}{\kappa T}\frac{d^2T}{dt^2} = \frac{r^2}{R}\frac{d^2R}{dr^2} + \frac{2r}{R}\frac{dR}{dr} + \frac{d^2\Theta}{d\theta^2}\frac{1}{\Theta}. \qquad (4)$$

The θ term can be separated.

$$\frac{d^2\Theta}{d\theta^2}\frac{1}{\Theta} = -n(n+1), \qquad (5)$$

which has a solution

$$\Theta(\theta) = A\cos\left[\sqrt{n(n+1)}\theta\right] + B\sin\left[\sqrt{n(n+1)}\theta\right]. \qquad (6)$$

The remaining portion becomes

$$\frac{r^2}{\kappa T}\frac{d^2T}{dt^2} = \frac{r^2}{R}\frac{d^2R}{dr^2} + \frac{2r}{R}\frac{dR}{dr} - n(n+1). \qquad (7)$$

Dividing by r^2 gives

$$\frac{1}{\kappa T}\frac{d^2T}{dt^2} = \frac{1}{R}\frac{d^2R}{dr^2} + \frac{2}{rR}\frac{dR}{dr} - \frac{n(n+1)}{r^2} = \frac{1}{\lambda^2}, \qquad (8)$$

where a NEGATIVE separation constant has been chosen so that the t portion remains finite

$$T(t) = Ce^{-\kappa t/\lambda^2}. \qquad (9)$$

The radial portion then becomes

$$\frac{1}{R}\frac{d^2R}{dr^2} + \frac{2}{rR}\frac{dR}{dr} - \frac{n(n+1)}{r^2} + \frac{1}{\lambda^2} = 0 \qquad (10)$$

$$r^2\frac{d^2R}{dr^2} + 2r\frac{dR}{dr} + \left[\frac{r^2}{\lambda^2} - n(n+1)\right]R = 0, \qquad (11)$$

which is the SPHERICAL BESSEL DIFFERENTIAL EQUATION. If the initial temperature is $T(r,\ 0) = 0$ and the boundary condition is $T(1,\ t) = 1$, the solution is

$$T(r,\ t) = 1 - 2\sum_{n=1}^{\infty}\frac{J_0(\alpha_n r)}{\alpha_n J_1(\alpha_n)}\ e^{\alpha_n^2 t}, \qquad (12)$$

where α_n is the nth POSITIVE zero of the BESSEL FUNCTION OF THE FIRST KIND $J_0(x)$.

Heaviside Calculus

The study, first developed by Boole, of SHIFT-INVARIANT OPERATORS which are polynomials in the DIFFERENTIAL OPERATOR $\check{D}$. Heaviside calculus can be used to solve any ORDINARY DIFFERENTIAL EQUATION OF THE FORM

$$p(\check{D})f(x) = g(x)$$

with $p(0) \neq 0$, and is frequently implemented using LAPLACE TRANSFORMS.

See also DIFFERENTIAL OPERATOR, LAPLACE TRANSFORM, SHIFT-INVARIANT OPERATOR

References

Rota, G.-C.; Kahaner, D.; Odlyzko, A. "On the Foundations of Combinatorial Theory. VIII: Finite Operator Calculus." *J. Math. Anal. Appl.* **42**, 684–60, 1973.

Heaviside Step Function

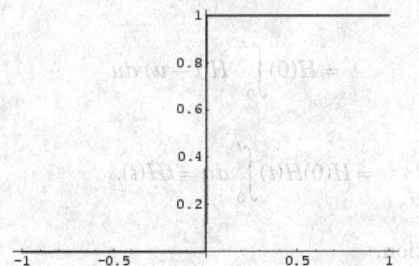

A discontinuous "step" function also called the unit step, and defined by

$$H(x) = \begin{cases} 0 & x < 0 \\ \frac{1}{2} & x = 0 \\ 1 & x > 0. \end{cases} \qquad (1)$$

It is related to the BOXCAR FUNCTION by

$$\prod(x) = H\left(x + \tfrac{1}{2}\right) - H\left(x - \tfrac{1}{2}\right) \qquad (2)$$

and can be defined in terms of the SGN function by

$$H(x) = \tfrac{1}{2}[1 + \text{sgn}(x)]. \tag{3}$$

The shorthand notation

$$H_c(x) \equiv H(x - c) \tag{4}$$

is sometimes also used. The Heaviside step function is given by the *Mathematica* command `UnitStep[x]`. The DERIVATIVE is given by

$$\frac{d}{dx} H(x) = \delta(x), \tag{5}$$

where $\delta(x)$ is the DELTA FUNCTION, and the step function is related to the RAMP FUNCTION $R(x)$ by

$$\frac{d}{dx} R(x) = -H(x) \tag{6}$$

$$R(x) = xH(x) \tag{7}$$

$$R(x) = H(x) * H(x), \tag{8}$$

where $*$ denotes CONVOLUTION.

Bracewell (1999) gives many identities, some of which include the following. Letting $*$ denote the CONVOLUTION,

$$H(x) * f(x) = \int_{-\infty}^{x} f(x')\, dx' \tag{9}$$

$$H(t) * H(t) = \int_{-\infty}^{\infty} H(u)H(t - u)\, du \tag{10}$$

$$= H(0) \int_{0}^{\infty} H(t - u)\, du$$

$$= H(0)H(t) \int_{0}^{t} du = tH(t). \tag{11}$$

In addition,

$$H(ax + b) = H\left(x + \frac{b}{a}\right)H(a) + H\left(-x - \frac{b}{a}\right)H(-a)$$

$$= \begin{cases} H\left(x + \dfrac{b}{a}\right) & a > 0 \\[2mm] H\left(-x - \dfrac{b}{a}\right) & a < 0. \end{cases} \tag{12}$$

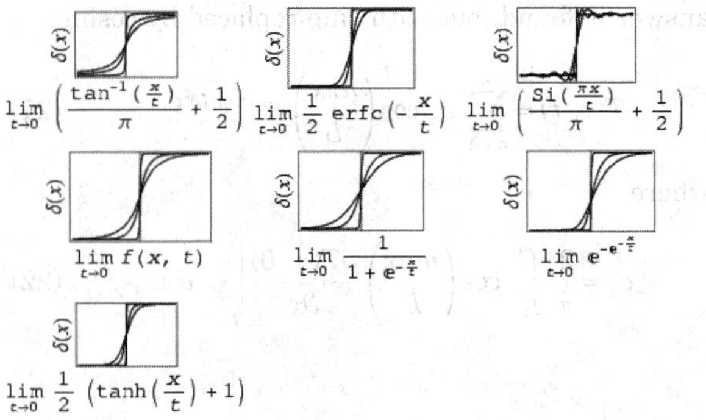

$$\lim_{t \to 0}\left(\frac{\tan^{-1}\left(\frac{x}{t}\right)}{\pi} + \frac{1}{2}\right) \quad \lim_{t \to 0}\frac{1}{2}\,\text{erfc}\left(-\frac{x}{t}\right) \quad \lim_{t \to 0}\left(\frac{\text{Si}\left(\frac{\pi x}{t}\right)}{\pi} + \frac{1}{2}\right)$$

$$\lim_{t \to 0} f(x, t) \quad \lim_{t \to 0}\frac{1}{1 + e^{-\frac{x}{t}}} \quad \lim_{t \to 0} e^{-e^{-\frac{x}{t}}}$$

$$\lim_{t \to 0}\frac{1}{2}\left(\tanh\left(\frac{x}{t}\right) + 1\right)$$

The Heaviside step function can be defined by the following limits,

$$H(x) = \lim_{t \to 0}\left[\frac{1}{2} + \frac{1}{\pi}\,\tan^{-1}\left(\frac{x}{t}\right)\right] \tag{13}$$

$$= \frac{1}{\sqrt{\pi}} \lim_{t \to 0} \int_{-x}^{\infty} t^{-1} e^{-u^2/t^2}\, du$$

$$= \frac{1}{2} \lim_{t \to 0} \text{erfc}\left(-\frac{x}{t}\right) \tag{14}$$

$$= \frac{1}{\pi} \lim_{t \to 0} \int_{-\infty}^{x} t^{-1} \text{sinc}\left(\frac{u}{t}\right) du$$

$$= \frac{1}{\pi} \lim_{t \to 0} \int_{-\infty}^{x} \frac{1}{u} \sin\left(\frac{u}{t}\right) du \tag{15}$$

$$= \frac{1}{2} + \frac{1}{\pi} \lim_{t \to 0} \text{si}\left(\frac{\pi x}{t}\right) \tag{16}$$

$$= \lim_{t \to 0}\begin{cases} \frac{1}{2} e^{x/t} & \text{for } x \leq 0 \\[2mm] 1 - \frac{1}{2} e^{-x/t} & \text{for } x \geq 0 \end{cases} \tag{17}$$

$$= \lim_{t \to 0} \frac{1}{1 + e^{-x/t}} \tag{18}$$

$$= \lim_{t \to 0} e^{e^{-x/t}} \tag{19}$$

$$= \frac{1}{2} \lim_{t \to 0}\left[1 + \tanh\left(\frac{x}{t}\right)\right] \tag{20}$$

$$= \lim_{t \to 0} \int_{-\infty}^{x} t^{-1}\Lambda\left(\frac{x - \frac{1}{2}t}{t}\right) dx, \tag{21}$$

where $\text{erfc}(x)$ is the ERFC function, $\text{si}(x)$ is the SINE INTEGRAL, $\text{sinc}\,x$ is the SINC FUNCTION, and $\Lambda(x)$ is the one-argument TRIANGLE FUNCTION. The first four of these are illustrated above for $t = 0.2$, 0.1, and 0.01.

Of course, any monotonic function with constant unequal horizontal asymptotes is a Heaviside step function under appropriate scaling and possible reflection. The FOURIER TRANSFORM of the Heaviside step function is given by

$$\mathscr{F}[H(x)] = \int_{-\infty}^{\infty} e^{-2\pi i k x} H(x)\, dx = \frac{1}{2}\left[\delta(k) - \frac{i}{\pi k}\right], \quad (22)$$

where $\delta(k)$ is the DELTA FUNCTION.

See also ABSOLUTE VALUE, BOXCAR FUNCTION, DELTA FUNCTION, FOURIER TRANSFORM–HEAVISIDE STEP FUNCTION, RAMP FUNCTION, RAMP FUNCTION, RECTANGLE FUNCTION, SGN, SQUARE WAVE, TRIANGLE FUNCTION

References

Bracewell, R. "Heaviside's Unit Step Function, $H(x)$." *The Fourier Transform and Its Applications, 3rd ed.* New York: McGraw-Hill, pp. 57–1, 1999.

Spanier, J. and Oldham, K. B. "The Unit-Step $u(x-a)$ and Related Functions." Ch. 8 in *An Atlas of Functions.* Washington, DC: Hemisphere, pp. 63–9, 1987.

Heawood Conjecture

The bound for the number of colors which are SUFFICIENT for MAP COLORING on a surface of GENUS g,

$$\gamma(g) = \left\lfloor \tfrac{1}{2}(7 + \sqrt{48g+1}) \right\rfloor$$

is the best possible, where $\lfloor x \rfloor$ is the FLOOR FUNCTION. $\gamma(g)$ is called the CHROMATIC NUMBER, and the first few values for $g = 0, 1, \ldots$ are 4, 7, 8, 9, 10, 11, 12, 12, 13, 13, 14, ... (Sloane's A000934).

The fact that $\gamma(g)$ is also NECESSARY was proved by Ringel and Youngs (1968) with two exceptions: the SPHERE (PLANE), and the KLEIN BOTTLE. When the FOUR-COLOR THEOREM was proved in 1976, the KLEIN BOTTLE was left as the only exception, in that the Heawood formula gives seven, but the correct bound is six (as demonstrated by the FRANKLIN GRAPH). The four most difficult cases to prove in the FOUR-COLOR THEOREM were $g = 59, 83, 158,$ and 257.

See also CHROMATIC NUMBER, FOUR-COLOR THEOREM, FRANKLIN GRAPH, MAP COLORING, SIX-COLOR THEOREM, TORUS COLORING

References

Bondy, J. A. and Murty, U. S. R. *Graph Theory with Applications.* New York: North Holland, p. 244, 1976.

Franklin, P. "A Six Color Problem." *J. Math. Phys.* **13**, 363–79, 1934.

Heawood, P. J. "Map Colour Theorem." *Quart. J. Math.* **24**, 332–38, 1890.

Ringel, G. *Map Color Theorem.* New York: Springer-Verlag, 1974.

Ringel, G. and Youngs, J. W. T. "Solution of the Heawood Map-Coloring Problem." *Proc. Nat. Acad. Sci. USA* **60**, 438–45, 1968.

Sloane, N. J. A. Sequences A000934/M3292 in "An On-Line Version of the Encyclopedia of Integer Sequences." http://www.research.att.com/~njas/sequences/eisonline.html.

Wagon, S. "Map Coloring on a Torus." §7.5 in *Mathematica in Action.* New York: W. H. Freeman, pp. 232–37, 1991.

Heawood Graph

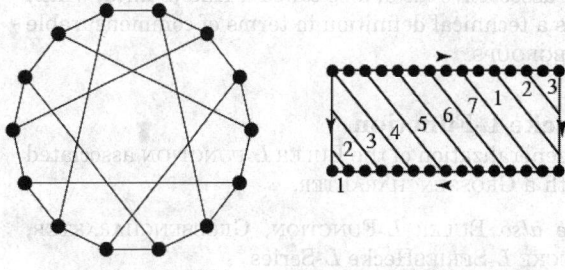

Heawood graph   *torus coloring*

The seven-color torus map on 14 nodes illustrated above. The Heawood graph is a CAGE GRAPH and is 4-transitive, but not 5-transitive (Harary 1994, p. 173). The Heawood graph is the point/line INCIDENCE GRAPH on the FANO PLANE (Royle).

See also CAGE GRAPH, FANO PLANE, SZILASSI POLYHEDRON, TORUS COLORING

References

Bondy, J. A. and Murty, U. S. R. *Graph Theory with Applications.* New York: North Holland, pp. 236 and 244, 1976.

Harary, F. *Graph Theory.* Reading, MA: Addison-Wesley, p. 173, 1994.

Royle, G. "Cubic Cages." http://www.cs.uwa.edu.au/~gordon/cages/.

Skiena, S. *Implementing Discrete Mathematics: Combinatorics and Graph Theory with Mathematica.* Reading, MA: Addison-Wesley, p. 192, 1990.

Weisstein, E. W. "Graphs." MATHEMATICA NOTEBOOK GRAPHS.M.

Wong, P. K. "Cages--A Survey." *J. Graph Th.* **6**, 1–2, 1982.

Hebesphenomegacorona

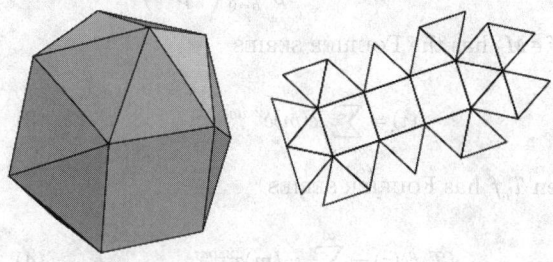

JOHNSON SOLID J_{89}.

References

Weisstein, E. W. "Johnson Solids." MATHEMATICA NOTEBOOK JOHNSONSOLIDS.M.

Weisstein, E. W. "Johnson Solid Netlib Database." MATHEMATICA NOTEBOOK JOHNSONSOLIDS.DAT.

Hecke Algebra

An associative RING, also called a HECKE RING, which has a technical definition in terms of commensurable SUBGROUPS.

Hecke L-Function

A generalization of the EULER L-FUNCTION associated with a GRÖSSENCHARAKTER.

See also EULER L-FUNCTION, GRÖSSENCHARAKTER, HECKE L-SERIESHecke L-Series

References

Knapp, A. W. "Group Representations and Harmonic Analysis, Part II." *Not. Amer. Math. Soc.* **43**, 537–49, 1996.

Hecke L-Series

See also HECKE L-FUNCTION

References

Koch, H. "Applications of Hecke L-Series." Ch. 8 in *Number Theory: Algebraic Numbers and Functions.* Providence, RI: Amer. Math. Soc., pp. 259–73, 2000.

Hecke Operator

A family of operators mapping each SPACE M_k of MODULAR FORMS onto itself. For a fixed integer k and any POSITIVE INTEGER n, the Hecke operator T_n is defined on the set M_k of entire modular forms of weight k by

$$(T_n f)(\tau) = n^{k-1} \sum_{d|n} d^{-k} \sum_{b=0}^{d-1} f\left(\frac{n\tau + bd}{d^2}\right). \tag{1}$$

For n a PRIME p, the operator collapses to

$$(T_p f)(\tau) = p^{k-1} f(p\tau) + \frac{1}{p} \sum_{b=0}^{p-1} \left(\frac{\tau + b}{p}\right). \tag{2}$$

If $f \in M_k$ has the FOURIER SERIES

$$f(\tau) = \sum_{m=0}^{\infty} c(m) e^{2\pi i m \tau}, \tag{3}$$

then $T_n f$ has FOURIER SERIES

$$(T_n f)(\tau) = \sum_{m=0}^{\infty} \gamma_n(m) e^{2\pi i m \tau}, \tag{4}$$

where

$$\gamma_n(m) = \sum_{d|(n,m)} d^{k-1} c\left(\frac{mn}{d^2}\right) \tag{5}$$

(Apostol 1997, p. 121).

If $(m, n) = 1$, the Hecke operators obey the composition property

$$T_m T_n = T_{mn}. \tag{6}$$

Any two Hecke operators $T(n)$ and $T(m)$ on M_k COMMUTE with each other, and moreover

$$T(m)T(n) = \sum_{d|(m, n)} d^{k-1} T\left(\frac{mn}{d^2}\right) \tag{7}$$

(Apostol 1997, pp. 126–27).

Each Hecke operator T_n has eigenforms when the dimension of M_k is 1, so for $k = 4$, 6, 8, 10, and 14, the eigenforms are the EISENSTEIN SERIES G_4, G_6, G_8, G_{10}, and G_{14}, respectively. Similarly, each T_n has eigenforms when the dimension of the set of CUSP FORMS $M_{k, 0}$ is 1, so for $k = 12$, 16, 18, 20, 22, and 26, the eigenforms are Δ, ΔG_4, ΔG_6, ΔG_8, ΔG_{10}, and ΔG_{14}, respectively, where Δ is the MODULAR DISCRIMINANT of the WEIERSTRASS ELLIPTIC FUNCTION (Apostol 1997, p. 130).

See also HECKE ALGEBRA, MODULAR FORM

References

Apostol, T. M. "The Hecke Operators." §6.7 in *Modular Functions and Dirichlet Series in Number Theory, 2nd ed.* New York: Springer-Verlag, pp. 120–22, 1997.

Hecke Ring

HECKE ALGEBRA

Hectogon

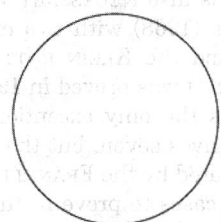

A 100-sided POLYGON, virtually indistinguishable in appearance from a CIRCLE except at very high magnification.

Hedgehog

An envelope parameterized by its GAUSS MAP. The PARAMETRIC EQUATIONS for a hedgehog are

$$x = p(\theta) \cos \theta + p'(\theta) \sin \theta$$

$$y = p(\theta) \sin \theta + p'(\theta) \cos \theta.$$

A plane convex hedgehog has at least four VERTICES where the CURVATURE has a stationary value. A plane convex hedgehog of constant width has at least six VERTICES (Martinez-Maure 1996).

References

Langevin, R.; Levitt, G.; and Rosenberg, H. "Hérissons et Multihérissons (Enveloppes paramétrées par leur applica-

tion de Gauss." Warsaw: Singularities, 245–53, 1985. Banach Center Pub. 20, PWN Warsaw, 1988.
Martinez-Maure, Y. "A Note on the Tennis Ball Theorem." *Amer. Math. Monthly* **103**, 338–40, 1996.

Heegaard Diagram

A diagram expressing how the gluing operation that connects the HANDLEBODIES involved in a HEEGAARD SPLITTING proceeds, usually by showing how the meridians of the HANDLEBODY are mapped.

See also HANDLEBODY, HEEGAARD SPLITTING

References

Rolfsen, D. *Knots and Links.* Wilmington, DE: Publish or Perish Press, p. 239, 1976.

Heegaard Splitting

A Heegaard splitting of a connected orientable 3-MANIFOLD M is any way of expressing M as the UNION of two (3,1)-HANDLEBODIES along their boundaries. The boundary of such a (3,1)-HANDLEBODY is an orientable SURFACE of some GENUS, which determines the number of HANDLES in the (3,1)-HANDLEBODIES. Therefore, the HANDLEBODIES involved in a Heegaard splitting are the same, but they may be glued together in a strange way along their boundary. A diagram showing how the gluing is done is known as a HEEGAARD DIAGRAM.

References

Adams, C. C. *The Knot Book: An Elementary Introduction to the Mathematical Theory of Knots.* New York: W. H. Freeman, p. 255, 1994.

Heegner Number

The values of $-d$ for which IMAGINARY QUADRATIC FIELDS $\mathbb{Q}(\sqrt{-d})$ are uniquely factorable into factors OF THE FORM $a + b\sqrt{-d}$). Here, a and b are half-integers, except for $d = 1$ and 2, in which case they are INTEGERS. The Heegner numbers therefore correspond to DISCRIMINANTS $-d$ which have CLASS NUMBER $h(-d)$ equal to 1, except for Heegner numbers -1 and -2, which correspond to $d = -4$ and -8, respectively.

The determination of these numbers is called GAUSS'S CLASS NUMBER PROBLEM, and it is now known that there are only nine Heegner numbers: -1, -2, -3, -7, -11, -19, -43, -67, and -163 (Sloane's A003173), corresponding to discriminants -4, -8, -3, -7, -11, -19, -43, -67, and -163, respectively.

Heilbronn and Linfoot (1934) showed that if a larger d existed, it must be 10^9. Heegner (1952) published a proof that only nine such numbers exist, but his proof was not accepted as complete at the time. Subsequent examination of Heegner's proof show it to be "essentially" correct (Conway and Guy 1996).

The Heegner numbers have a number of fascinating connections with amazing results in PRIME NUMBER theory. In particular, the J-FUNCTION provides stunning connections between e, π, and the ALGEBRAIC INTEGERS. They also explain why Euler's PRIME-GENERATING POLYNOMIAL $n^2 - n + 41$ is so surprisingly good at producing PRIMES.

See also CLASS NUMBER, DISCRIMINANT (BINARY QUADRATIC FORM), GAUSS'S CLASS NUMBER PROBLEM, J-FUNCTION, PRIME-GENERATING POLYNOMIAL, QUADRATIC FIELD, RAMANUJAN CONSTANT

References

Conway, J. H. and Guy, R. K. "The Nine Magic Discriminants." In *The Book of Numbers.* New York: Springer-Verlag, pp. 224–26, 1996.
Heegner, K. "Diophantische Analysis und Modulfunktionen." *Math. Z.* **56**, 227–53, 1952.
Heilbronn, H. A. and Linfoot, E. H. "On the Imaginary Quadratic Corpora of Class-Number One." *Quart. J. Math. (Oxford)* **5**, 293–01, 1934.
Sloane, N. J. A. Sequences A003173/M0827 in "An On-Line Version of the Encyclopedia of Integer Sequences." http://www.research.att.com/~njas/sequences/eisonline.html.

Heesch Number

The Heesch number of a closed plane figure is the maximum number of times that figure can be completely surrounded by copies of itself. The determination of the maximum possible (finite) Heesch number is known as HEESCH'S PROBLEM. The Heesch number of a TRIANGLE, QUADRILATERAL, regular HEXAGON, or any other shape that can TILE or TESSELLATE the plane, is infinity. Conversely, any shape with infinite Heesch number must tile the plane (Eppstein).

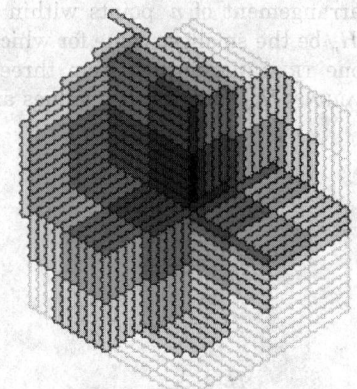

A tile invented by R. Ammann has Heesch number is three (Senechal 1995), and Mann has found an infinite family of tiles with Heesch number five (illustrated above), the largest (finite) number known.

See also HEESCH'S PROBLEM, TILING

References

Eppstein, D. "Heesch's Problem." http://www.ics.uci.edu/~eppstein/junkyard/heesch/.

Fontaine, A. "An Infinite Number of Plane Figures with Heesch Number Two." *J. Comb. Th. A* **57**, 151–56, 1991.

Friedman, E. "Heesch Tiles with Surround Numbers 3 and 4." http://www.stetson.edu/~efriedma/papers/heesch/heesch.html.

Grünbaum, B. and Sheppard, G. C. *Tilings and Patterns.* New York: W. H. Freeman, 1986.

Mann, C. "Heesch's Problem." http://www.math.unl.edu/~cmann/math/heesch/heesch.htm.

Raedschelders, P. "Heesch Tiles Based on Regular Polygons." *Combinatorics* **7**, 101–06, 1998.

Raedschelders, P. "Heesch-Tiles Based on *n*-gons." http://home.planetinternet.be/~praedsch/heersch.htm.

Senechal, M. *Quasicrystals and Geometry.* New York: Cambridge University Press, 1995.

Thompson, M. "Self-Surrounding Tiles." http://home.flash.net/~markthom/html/self-surrounding_tiles.html.

Heesch's Problem

How many times can a shape be completely surrounded by copies of itself without being able to TILE the entire plane, i.e., what is the maximum (finite) HEESCH NUMBER?

References

Eppstein, D. "Heesch's Problem." http://www.ics.uci.edu/~eppstein/junkyard/heesch/.

Height

The vertical length of an object from top to bottom.

See also LENGTH (SIZE), POLYNOMIAL HEIGHT, WIDTH (SIZE)

Heilbronn Triangle Problem

N.B. A detailed online essay by S. Finch was the starting point for this entry.

Given any arrangement of n points within a UNIT SQUARE, let H_n be the smallest value for which there is at least one TRIANGLE formed from three of the points with AREA $\leq H_n$. The first few values are

$$H_3 = \tfrac{1}{2}$$

$$H_4 = \tfrac{1}{2}$$

$$H_5 = \tfrac{1}{9}\sqrt{3}$$

$$H_6 = \tfrac{1}{8}$$

$$H_7 \geq \tfrac{1}{12}$$

$$H_8 \geq \tfrac{1}{4}(2-\sqrt{3})$$

$$H_9 \geq \tfrac{1}{21}$$

$$H_{10} \geq \tfrac{1}{32}(3\sqrt{17}-11)$$

$$H_{11} \geq \tfrac{1}{27}$$

$$H_{12} \geq \tfrac{1}{33}$$

$$H_{13} \geq 0.030$$

$$H_{14} \geq 0.022$$

$$H_{15} \geq 0.020$$

$$H_{16} \geq 0.0175.$$

Komlós *et al.* (1981, 1982) have shown that there are constants c such that

$$\frac{c\ln n}{n^2} \leq H_n \leq \frac{C}{n^{8/7}-\epsilon},$$

for any $\epsilon > 0$ and all sufficiently large n.

Using an EQUILATERAL TRIANGLE of unit AREA instead gives the constants

$$h_3 = 1$$

$$h_4 = \tfrac{1}{3}$$

$$h_5 = 3-2\sqrt{2}$$

$$h_6 = \tfrac{1}{8}.$$

References

Finch, S. "Favorite Mathematical Constants." http://www.mathsoft.com/asolve/constant/hlb/hlb.html.

Friedman, E. "The Heilbronn Problem." http://www.stetson.edu/~efriedma/heilbronn/.

Goldberg, M. "Maximizing the Smallest Triangle Made by N Points in a Square." *Math. Mag.* **45**, 135–44, 1972.

Guy, R. K. *Unsolved Problems in Number Theory, 2nd ed.* New York: Springer-Verlag, pp. 242–44, 1994.

Komlos, J.; Pintz, J.; and Szemerédi, E. "On Heilbronn's Triangle Problem." *J. London Math. Soc.* **24**, 385–96, 1981.

Komlos, J.; Pintz, J.; and Szemerédi, E. "A Lower Bound for Heilbronn's Triangle Problem." *J. London Math. Soc.* **25**, 13–4, 1982.

Roth, K. F. "Developments in Heilbronn's Triangle Problem." *Adv. Math.* **22**, 364–85, 1976.

Heine Differential Equation

The second-order ORDINARY DIFFERENTIAL EQUATION

$$y'' + \frac{1}{2}\left(\frac{1}{x-a_1}+\frac{2}{x-a_3}\right)y' + \frac{1}{4}$$
$$\times\left[\frac{A_0+A_1x+A_2x^2+A_3x^3}{(x-a_1)(x-a_2)^2(x-a_3)^2}\right]y$$
$$= 0$$

(Moon and Spencer 1961, p. 157; Zwillinger 1997, p. 123).

References

Moon, P. and Spencer, D. E. *Field Theory for Engineers.* New York: Van Nostrand, 1961.

Zwillinger, D. *Handbook of Differential Equations, 3rd ed.* Boston, MA: Academic Press, p. 123, 1997.

Heine Hypergeometric Series
Q-HYPERGEOMETRIC FUNCTION

Heine-Borel Theorem

If a CLOSED SET of points on a line can be covered by a set of intervals so that every point of the set is an interior point of at least one of the intervals, then there exist a finite number of intervals with the covering property.

The Heine-Borel theorem gives the BOLZANO-WEIER-STRASS THEOREM as a special case.

See also BOLZANO-WEIERSTRASS THEOREM

References

Baker, H. F. Cited in Lamb, H. *Proc. London Math. Soc.* **35**, 459–60, 1903.
Heine, E. "Die Elemente der Functionenlehre." *J. reine angew. Math.* **74**, 172–88, 1871.
Jeffreys, H. and Jeffreys, B. S. "The Heine-Borel Theorem" and "The Modified Heine-Borel Theorem." §1.0621–.0622 in *Methods of Mathematical Physics, 3rd ed.* Cambridge, England: Cambridge University Press, pp. 20–1, 1988.
Knopp, K. *Theory of Functions Parts I and II, Two Volumes Bound as One, Part I.* New York: Dover, p. 9, 1996.
Young, W. H. "Overlapping Intervals." *Proc. London Math. Soc.* **35**, 384–88, 1903.

Heisenberg Ferromagnet Equation

The system of PARTIAL DIFFERENTIAL EQUATIONS

$$\mathbf{S}_t = \mathbf{S} \times \mathbf{S}_{xx}.$$

References

Calogero, F. and Degasperis, A. *Spectral Transform and Solitons: Tools to Solve and Investigate Nonlinear Evolution Equations.* New York: North-Holland, p. 56, 1982.
Zwillinger, D. *Handbook of Differential Equations, 3rd ed.* Boston, MA: Academic Press, p. 138, 1997.

Heisenberg Group

The Heisenberg group H^n in n COMPLEX variables is the GROUP of all (z, t) with $z \in \mathbb{C}^n$ and $t \in \mathbb{R}$ having multiplication

$$(w, t)(z, t') = (w + z, t + t' + \Im[w^*z])$$

where w^* is the adjoint. The Heisenberg group is ISOMORPHIC to the group of MATRICES

$$\begin{bmatrix} 1 & z^T & \frac{1}{2}|z|^2 + it \\ 0 & 1 & z \\ 0 & 0 & 1 \end{bmatrix},$$

and satisfies

$$(z, t)^{-1} = (-z, -t).$$

Every finite-dimensional unitary representation is trivial on Z and therefore factors to a REPRESENTATION of the quotient $\mathbb{C}^n$.

See also NIL GEOMETRY

References

Knapp, A. W. "Group Representations and Harmonic Analysis, Part II." *Not. Amer. Math. Soc.* **43**, 537–49, 1996.

Heisenberg Space

The boundary of COMPLEX HYPERBOLIC 2-SPACE.

See also HYPERBOLIC SPACE

Held Group

The SPORADIC GROUP *He*.

References

Wilson, R. A. "ATLAS of Finite Group Representation." http://for.mat.bham.ac.uk/atlas/html/He.html.

Helen of Geometers

CYCLOID

Helicoid

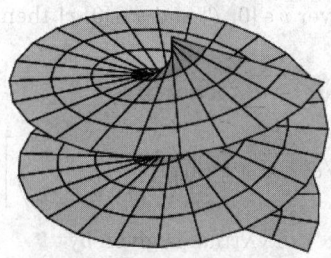

The MINIMAL SURFACE having a HELIX as its boundary. It is the only RULED MINIMAL SURFACE other than the PLANE (Catalan 1842, do Carmo 1986). For many years, the helicoid remained the only known example of a complete embedded MINIMAL SURFACE of finite topology with infinite CURVATURE. However, in 1992 a second example, known as HOFFMAN'S MINIMAL SURFACE and consisting of a helicoid with a HOLE, was discovered (*Sci. News* 1992). The helicoid is the only non-rotary surface which can glide along itself (Steinhaus 1983, p. 231).

The equation of a helicoid in CYLINDRICAL COORDINATES is

$$z = c\theta. \tag{1}$$

In CARTESIAN COORDINATES, it is

$$\frac{y}{x} = \tan\left(\frac{z}{c}\right). \tag{2}$$

It can be given in parametric form by

$$x = u \cos v \tag{3}$$

$$y = u \sin v \tag{4}$$

$$z = cv, \tag{5}$$

which has an obvious generalization to the ELLIPTIC HELICOID. Writing $z = -cu$ instead of $z = cv$ gives a CONE instead of a helicoid.

The FIRST FUNDAMENTAL FORM coefficients of the helicoid are given by

$$E = 1 \tag{6}$$

$$F = 0 \tag{7}$$

$$G^2 = c^2 + u^2, \tag{8}$$

and the SECOND FUNDAMENTAL FORM coefficients are

$$e = 0 \tag{9}$$

$$f = -\frac{c}{\sqrt{c^2 + u^2}} \tag{10}$$

$$g = 0, \tag{11}$$

giving AREA ELEMENT

$$dS = \sqrt{c^2 + u^2}\, du \wedge dv. \tag{12}$$

Integrating over $v \in [0, \theta]$ and $u \in [0, r]$ then gives

$$S = \int_0^\theta \int_0^r \sqrt{c^2 + u^2}\, du\, dv$$

$$= \tfrac{1}{2}\theta \left[r\sqrt{c^2 + r^2} + c^2 \ln\left(\frac{r + \sqrt{c^2 + r^2}}{c}\right)\right]. \tag{13}$$

The GAUSSIAN CURVATURE is given by

$$K = -\frac{c^2}{(c^2 + u^2)^2}, \tag{14}$$

and the MEAN CURVATURE is

$$H = 0 \tag{15}$$

making the helicoid a MINIMAL SURFACE.

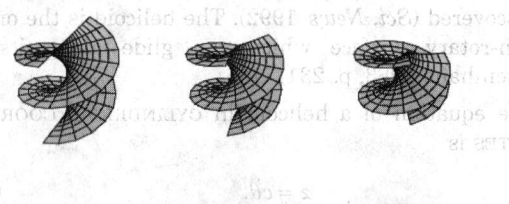

The helicoid can be continuously deformed into a CATENOID by the transformation

$$x(u, v) = \cos\alpha \sinh v \sin u + \sin\alpha \cosh v \cos u \tag{16}$$

$$y(u, v) = -\cos\alpha \sinh v \cos u + \sin\alpha \cosh v \sin u \tag{17}$$

$$z(u, v) = u\cos\alpha + v\sin\alpha, \tag{18}$$

where $\alpha = 0$ corresponds to a helicoid and $\alpha = \pi/2$ to a CATENOID.

If a twisted curve C (i.e., one with TORSION $\tau \neq 0$) rotates about a fixed axis A and, at the same time, is displaced parallel to A such that the speed of displacement is always proportional to the angular velocity of rotation, then C generates a GENERALIZED HELICOID.

See also CALCULUS OF VARIATIONS, CATENOID, CONE, ELLIPTIC HELICOID, GENERALIZED HELICOID, HELIX, HOFFMAN'S MINIMAL SURFACE, HYPERBOLIC HELICOID, MINIMAL SURFACE

References

Catalan E. "Sur les surfaces réglées dont l'aire est un minimum." *J. Math. Pure Appl.* **7**, 203–11, 1842.

do Carmo, M. P. "The Helicoid." §3.5B in *Mathematical Models from the Collections of Universities and Museums* (Ed. G. Fischer). Braunschweig, Germany: Vieweg, pp. 44–5, 1986.

Fischer, G. (Ed.). Plate 91 in *Mathematische Modelle / Mathematical Models, Bildband / Photograph Volume.* Braunschweig, Germany: Vieweg, p. 87, 1986.

Gray, A. *Modern Differential Geometry of Curves and Surfaces with Mathematica,* 2nd ed. Boca Raton, FL: CRC Press, pp. 449 and 644, 1997.

Kreyszig, E. *Differential Geometry.* New York: Dover, p. 88, 1991.

Meusnier, J. B. "Mémoire sur la courbure des surfaces." *Mém. des savans étrangers* **10** (lu 1776), 477–10, 1785.

Ogawa, A. "Helicatenoid." *Mathematica J.* **2**, 21, 1992.

Osserman, R. *A Survey of Minimal Surfaces.* New York: Dover, pp. 17–8, 1986.

Peterson, I. "Three Bites in a Doughnut." *Sci. News* **127**, 168, Mar. 16, 1985.

"Putting a Handle on a Minimal Helicoid." *Sci. News* **142**, 276, Oct. 24, 1992.

Steinhaus, H. *Mathematical Snapshots,* 3rd ed. New York: Dover, pp. 231–32, 1999.

Wells, D. *The Penguin Dictionary of Curious and Interesting Geometry.* London: Penguin, p. 94, 1991.

Wolfram, S. *The Mathematica Book,* 3rd ed. Champaign, IL: Wolfram Media, p. 164, 1996.

Helix

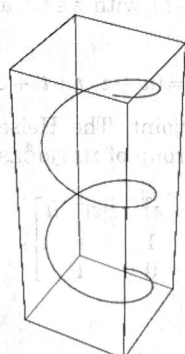

A helix is also called a CURVE OF CONSTANT SLOPE. It can be defined as a curve for which the TANGENT makes a constant ANGLE with a fixed line. The shortest path between two points on a cylinder (one not directly above the other) is a fractional turn of a helix, as can be seen by cutting the cylinder along one of its sides, flattening it out, and noting that a straight line connecting the points becomes helical upon re-wrapping (Steinhaus 1983, p. 229). It is for this reason that squirrels chasing one another up and around tree trunks follow helical paths.

Helices come in enantiomorphous left- (coils counterclockwise as it "goes away") and right-handed forms (coils clockwise). Standard screws, nuts, and bolts are all right-handed, as are both the helices in a double-stranded molecule of DNA (Gardner 1984, pp. 2–). Large helical structures in animals (such as horns) usually appear in both mirror-image forms, although the teeth of a male narwhal, usually only one which grows into a tusk, are *both* left-handed (Bonner 1951; Gardner 1984, p. 3; Thompson 1992). Gardner (1984) contains a fascinating discussion of helices in plants and animals, including an allusion to Shakespeare's *A Midsummer Night's Dream*.

The helix is a SPACE CURVE with PARAMETRIC EQUATIONS

$$x = r \cos t \tag{1}$$

$$y = r \sin t \tag{2}$$

$$z = ct, \tag{3}$$

where r is the radius of the helix and c is a constant giving the vertical separation of the helix's loops. The CURVATURE of the helix is given by

$$\kappa = \frac{r}{r^2 + c^2}, \tag{4}$$

and the LOCUS of the centers of CURVATURE of a helix is another helix. The ARC LENGTH is given by

$$s = \int \sqrt{x'^2 + y'^2 + z'^2}\, dt = \sqrt{r^2 + c^2}\, t. \tag{5}$$

The TORSION of a helix is given by

$$\tau = \frac{1}{r^2(r^2 + c^2)} \begin{vmatrix} -r \sin t & -r \cos t & r \sin t \\ r \cos t & -r \sin t & -r \cos t \\ c & 0 & 0 \end{vmatrix}$$

$$= \frac{c}{r^2 + c^2}, \tag{6}$$

so

$$\frac{\kappa}{\tau} = \frac{\dfrac{r}{r^2 + c^2}}{\dfrac{c}{r^2 + c^2}} = \frac{r}{c}, \tag{7}$$

which is a constant. In fact, LANCRET'S THEOREM states that a NECESSARY and SUFFICIENT condition for a curve to be a helix is that the ratio of CURVATURE

to TORSION be constant. The OSCULATING PLANE of the helix is given by

$$\begin{vmatrix} z_1 - r \cos t & z_2 - r \sin t & z_3 - ct \\ -r \sin t & r \cos t & c \\ -r \cos t & -r \sin t & 0 \end{vmatrix} = 0 \tag{8}$$

$$z_1 c \sin t - z_2 c \cos t + (z_3 - ct) r = 0. \tag{9}$$

The MINIMAL SURFACE of a helix is a HELICOID.

See also GENERALIZED HELIX, HELICOID, SPHERICAL HELIX, SPIRAL

References

Bonner, J. T. "The Horn of the Unicorn." *Sci. Amer.*, Mar. 1951.
Gardner, M. "The Helix." Ch. 1 in *The Sixth Book of Mathematical Games from Scientific American.* Chicago, IL: University of Chicago Press, pp. 1–, 1984.
Gray, A. "The Helix and Its Generalizations." §8.5 in *Modern Differential Geometry of Curves and Surfaces with Mathematica, 2nd ed.* Boca Raton, FL: CRC Press, pp. 198–00, 1997.
Isenberg, C. Plate 4.11 in *The Science of Soap Films and Soap Bubbles.* New York: Dover, 1992.
Pappas, T. "The Helix--Mathematics & Genetics." *The Joy of Mathematics.* San Carlos, CA: Wide World Publ./Tetra, pp. 166–68, 1989.
Steinhaus, H. *Mathematical Snapshots, 3rd ed.* New York: Dover, p. 229, 1999.
Thompson, D'A. W. *On Growth and Form, 2nd ed., compl. rev. ed.* New York: Cambridge University Press, 1992.
Wells, D. *The Penguin Dictionary of Curious and Interesting Geometry.* London: Penguin, p. 95, 1991.
Wolfram, S. *The Mathematica Book, 3rd ed.* Champaign, IL: Wolfram Media, p. 163, 1996.

Helly Number

Given a Euclidean n-space,

$$H^n \equiv n + 1.$$

See also EUCLIDEAN SPACE, HELLY'S THEOREM

Helly's Theorem

If F is a family of more than n bounded closed convex sets in Euclidean n-space $\mathbb{R}^n$, and if every H_n (where H_n is the HELLY NUMBER) members of F have at least one point in common, then all the members of F have at least one point in common.

See also CARATHÉODORY'S FUNDAMENTAL THEOREM, HELLY NUMBER

References

Eckhoff, J. "Helly, Radon, and Carathéodory Type Theorems." Ch. 2.1 in *Handbook of Convex Geometry* (Ed. P. M. Gruber and J. M. Wills). Amsterdam, Netherlands: North-Holland, pp. 389–48, 1993.

Helmholtz Differential Equation

An ELLIPTIC PARTIAL DIFFERENTIAL EQUATION given by

$$\nabla^2 \psi + k^2 \psi = 0, \tag{1}$$

where ψ is a SCALAR FUNCTION and ∇^2 is the scalar LAPLACIAN, or

$$\nabla^2 \mathbf{A} + k^2 \mathbf{A} = 0, \tag{2}$$

where $\mathbf{A}$ is a VECTOR FUNCTION and ∇^2 is the vector Laplacian (Moon and Spencer 1988, pp. 136–43).

When $k = 0$, the Helmholtz differential equation reduces to LAPLACE'S EQUATION. When $k^2 < 0$ (i.e., for imaginary k), the equation becomes the space part of the diffusion equation.

The Helmholtz differential equation can be solved by SEPARATION OF VARIABLES in only 11 coordinate systems, 10 of which (with the exception of CONFOCAL PARABOLOIDAL COORDINATES) are particular cases of the CONFOCAL ELLIPSOIDAL system: CARTESIAN, CONFOCAL ELLIPSOIDAL, CONFOCAL PARABOLOIDAL, CONICAL, CYLINDRICAL, ELLIPTIC CYLINDRICAL, OBLATE SPHEROIDAL, PARABOLOIDAL, PARABOLIC CYLINDRICAL, PROLATE SPHEROIDAL, and SPHERICAL COORDINATES (Eisenhart 1934). LAPLACE'S EQUATION (the Helmholtz differential equation with $k = 0$) is separable in the two additional BISPHERICAL COORDINATES and TOROIDAL COORDINATES.

If Helmholtz's equation is separable in a 3-D coordinate system, then Morse and Feshbach (1953, pp. 509–10) show that

$$\frac{h_1 h_2 h_3}{h_n^2} = f_n(u_n) g_n(u_i, u_j), \tag{3}$$

where $i \neq j \neq n$. The LAPLACIAN is therefore OF THE FORM

$$\nabla^2 = \frac{1}{h_1 h_2 h_3} \left\{ g_1(u_2, u_3) \frac{\partial}{\partial u_1} \left[f_1(u_1) \frac{\partial}{\partial u_1} \right] \right.$$

$$+ g_2(u_1, u_3) \frac{\partial}{\partial u_2} \left[f_2(u_2) \frac{\partial}{\partial u_2} \right]$$

$$\left. + g_3(u_1, u_2) \frac{\partial}{\partial u_3} \left[f_3(u_3) \frac{\partial}{\partial u_3} \right] \right\}, \tag{4}$$

which simplifies to

$$\nabla^2 = \frac{1}{h_1^2 f_1} \frac{\partial}{\partial u_1} \left[f_1(u_1) \frac{\partial}{\partial u_1} \right] + \frac{1}{h_2^2 f_2} \frac{\partial}{\partial u_2} \left[f_2(u_2) \frac{\partial}{\partial u_2} \right]$$

$$+ \frac{1}{h_3^2 f_3} \frac{\partial}{\partial u_3} \left[f_3(u_3) \frac{\partial}{\partial u_3} \right]. \tag{5}$$

Such a coordinate system obeys the ROBERTSON CONDITION, which means that the STÄCKEL DETERMINANT is of the form

$$S = \frac{h_1 h_2 h_3}{f_1(u_1) f_2(u_2) f_3(u_3)}. \tag{6}$$

See also LAPLACE'S EQUATION, POISSON'S EQUATION, SEPARATION OF VARIABLES, SPHERICAL BESSEL DIFFERENTIAL EQUATION, STÄCKEL DETERMINANT

References

Eisenhart, L. P. "Separable Systems in Euclidean 3-Space." *Physical Review* **45**, 427–28, 1934.

Eisenhart, L. P. "Separable Systems of Stäckel." *Ann. Math.* **35**, 284–05, 1934.

Eisenhart, L. P. "Potentials for Which Schroedinger Equations Are Separable." *Phys. Rev.* **74**, 87–9, 1948.

Moon, P. and Spencer, D. E. "Eleven Coordinate Systems" and "The Vector Helmholtz Equation." §1 and 5 in *Field Theory Handbook, Including Coordinate Systems, Differential Equations, and Their Solutions*, 2nd ed. New York: Springer-Verlag, pp. 1–8 and 136–43, 1988.

Morse, P. M. and Feshbach, H. *Methods of Theoretical Physics, Part I.* New York: McGraw-Hill, pp. 125–26, 271, and 509–10, 1953.

Zwillinger, D. (Ed.). *CRC Standard Mathematical Tables and Formulae.* Boca Raton, FL: CRC Press, p. 417, 1995.

Zwillinger, D. *Handbook of Differential Equations*, 3rd ed. Boston, MA: Academic Press, p. 129, 1997.

Helmholtz Differential Equation—Bipolar Coordinates

In BIPOLAR COORDINATES, the HELMHOLTZ DIFFERENTIAL EQUATION is not separable, but LAPLACE'S EQUATION is.

See also LAPLACE'S EQUATION–BIPOLAR COORDINATES

Helmholtz Differential Equation— Bispherical Coordinates

The HELMHOLTZ DIFFERENTIAL EQUATION is not separable in BISPHERICAL COORDINATES.

See also BISPHERICAL COORDINATES, HELMHOLTZ DIFFERENTIAL EQUATION, LAPLACE'S EQUATION–BISPHERICAL COORDINATES

Helmholtz Differential Equation— Cartesian Coordinates

In 2-D CARTESIAN COORDINATES, attempt SEPARATION OF VARIABLES by writing

$$F(x, y) = X(x) Y(y), \tag{1}$$

then the HELMHOLTZ DIFFERENTIAL EQUATION becomes

$$\frac{d^2 X}{dx^2} Y + \frac{d^2 Y}{dy^2} X + k^2 XY = 0. \tag{2}$$

Dividing both sides by XY gives

$$\frac{1}{X}\frac{d^2X}{dx^2} + \frac{1}{Y}\frac{d^2Y}{dy^2} + k^2 = 0. \tag{3}$$

This leads to the two coupled ordinary differential equations with a separation constant m^2,

$$\frac{1}{X}\frac{d^2X}{dx^2} = m^2 \tag{4}$$

$$\frac{1}{Y}\frac{d^2Y}{dy^2} = -(m^2 + k^2), \tag{5}$$

where X and Y could be interchanged depending on the boundary conditions. These have solutions

$$X = A_m e^{mx} + B_m e^{-mx} \tag{6}$$

$$Y = C_m e^{i\sqrt{m^2+k^2}\,y} + D_m e^{-i\sqrt{m^2+k^2}\,y}$$

$$= E_m \sin(\sqrt{m^2+k^2}\,y) + F_m \cos(\sqrt{m^2+k^2}\,y). \tag{7}$$

The general solution is then

$$F(x, y) = \sum_{m=1}^{\infty} (A_m e^{mx} + B_m e^{-mx})$$

$$\times [E_m \sin(\sqrt{m^2+k^2}\,y) + F_m \cos(\sqrt{m^2+k^2}\,y)]. \tag{8}$$

In 3-D CARTESIAN COORDINATES, attempt SEPARATION OF VARIABLES by writing

$$F(x, y, z) = X(x)Y(y)Z(z), \tag{9}$$

then the HELMHOLTZ DIFFERENTIAL EQUATION becomes

$$\frac{d^2X}{dx^2}\,YZ + \frac{d^2Y}{dy^2}\,XZ + \frac{d^2Z}{dz^2}\,XY + k^2XY = 0. \tag{10}$$

Dividing both sides by XYZ gives

$$\frac{1}{X}\frac{d^2X}{dx^2} + \frac{1}{Y}\frac{d^2Y}{dy^2} + \frac{1}{Z}\frac{d^2Z}{dz^2} + k^2 = 0. \tag{11}$$

This leads to the three coupled differential equations

$$\frac{1}{X}\frac{d^2X}{dx^2} = t^2 \tag{12}$$

$$\frac{1}{Y}\frac{d^2Y}{dy^2} = m^2 \tag{13}$$

$$\frac{1}{Z}\frac{d^2Z}{dz^2} = (k^2 + l^2 + m^2), \tag{14}$$

where X, Y, and Z could be permuted depending on boundary conditions. The general solution is therefore

$$F(x, y, z) = \sum_{l=1}^{\infty} \sum_{m=1}^{\infty} (A_l e^{lx} + B_l e^{-lx})(C_m e^{my} + D_m e^{-my})$$

$$\times (E_{lm} e^{-i\sqrt{k^2+l^2+m^2}\,z} + F_{lm} e^{i\sqrt{k^2+l^2+m^2}\,z}). \tag{15}$$

See also CARTESIAN COORDINATES, HELMHOLTZ DIFFERENTIAL EQUATION

References
Morse, P. M. and Feshbach, H. *Methods of Theoretical Physics, Part I.* New York: McGraw-Hill, pp. 501–02, 513–14 and 656, 1953.

Helmholtz Differential Equation— Circular Cylindrical Coordinates

In CYLINDRICAL COORDINATES, the SCALE FACTORS are $h_r = 1$, $h_\theta = r$, $h_z = 1$, so the LAPLACIAN is given by

$$\nabla^2 F = \frac{1}{r}\frac{\partial}{\partial r}\left(r\frac{\partial F}{\partial r}\right) + \frac{1}{r^2}\frac{\partial^2 F}{\partial \theta^2} + \frac{\partial^2 F}{\partial z^2}. \tag{1}$$

Attempt SEPARATION OF VARIABLES in the HELMHOLTZ DIFFERENTIAL EQUATION

$$\nabla^2 F + k^2 F = 0 \tag{2}$$

by writing

$$F(r, \theta, z) = R(r)\Theta(\theta)Z(z), \tag{3}$$

then combining (1) and (2) gives

$$\frac{d^2R}{dr^2}\,\Theta Z + \frac{1}{r}\frac{dR}{dr}\,\Theta Z + \frac{1}{r^2}\frac{d^2\Theta}{d\theta^2}\,RZ + \frac{d^2Z}{dz^2}\,R\Theta + k^2 R\Theta Z$$

$$= 0. \tag{4}$$

Now multiply by $r^2/(R\Theta Z)$,

$$\left(\frac{r^2}{R}\frac{d^2R}{dr^2} + \frac{r}{R}\frac{dR}{dr}\right) + \frac{1}{\Theta}\frac{d^2\Theta}{d\theta^2} + \frac{r^2}{Z}\frac{d^2Z}{dz^2} + k^2 r^2 = 0, \tag{5}$$

so the equation has been separated. Since the solution must be periodic in θ from the definition of the circular cylindrical coordinate system, the solution to the second part of (5) must have a NEGATIVE separation constant

$$\frac{1}{\Theta}\frac{d^2\theta}{d\theta^2} = -m^2, \tag{6}$$

which has a solution

$$\Theta(\theta) = C_m \cos(m\theta) + D_m \sin(m\theta). \tag{7}$$

Plugging (7) back into (5) gives

$$\frac{r^2}{R}\frac{d^2R}{dr^2} + \frac{r}{R}\frac{dR}{dr} - m^2 + \frac{r^2}{Z}\frac{d^2Z}{dz^2} + k^2 r^2 = 0, \tag{8}$$

and dividing through by r^2 results in

$$\frac{1}{R}\frac{d^2R}{dr^2} + \frac{1}{rR}\frac{dR}{dr} - \frac{m^2}{r^2} + \frac{1}{Z}\frac{d^2Z}{dz^2} + k^2 = 0. \qquad (9)$$

The solution to the second part of (9) must not be sinusoidal at $\pm\infty$ for a physical solution, so the differential equation has a POSITIVE separation constant

$$\frac{1}{Z}\frac{d^2Z}{dz^2} = n^2, \qquad (10)$$

and the solution is

$$Z(z) = E_n e^{-nx} + F_n e^{nx}. \qquad (11)$$

Plugging (11) back into (9) and multiplying through by R yields

$$\frac{d^2R}{dr^2} + \frac{1}{r}\frac{dR}{dr} + \left(n^2 + k^2 - \frac{m^2}{r^2}\right)R = 0 \qquad (12)$$

But this is just a modified form of the BESSEL DIFFERENTIAL EQUATION, which has a solution

$$R(r) = A_{mn}J_m(\sqrt{n^2 + k^2}\, r) + B_{mn}Y_m(\sqrt{n^2 + k^2}\, r), \quad (13)$$

where $J_n(x)$ and $Y_n(x)$ are BESSEL FUNCTIONS OF THE FIRST and SECOND KINDS, respectively. The general solution is therefore

$$F(r,\ \theta,\ z) = \sum_{m=o}^{\infty}\sum_{n=0}^{\infty}[A_{mn}J_m(\sqrt{k^2 + n^2}\, r)$$
$$+ B_{mn}Y_m(\sqrt{k^2 + n^2}\, r)]$$
$$\times[C_m\cos(m\theta) + D_m\sin(m\theta)](E_n e^{-nz} + F_n e^{nz}). \quad (14)$$

In the notation of Morse and Feshbach (1953), the separation functions are $f_1(r) = r, f_2(\theta) = 1, f_3(z) = 1$, so the STÄCKEL DETERMINANT is 1.

The HELMHOLTZ DIFFERENTIAL EQUATION is also separable in the more general case of k^2 OF THE FORM

$$k^2(r,\ \theta,\ z) = f(r) + \frac{g(\theta)}{r^2} + h(z) + k'^2. \qquad (15)$$

See also CYLINDRICAL COORDINATES, HELMHOLTZ DIFFERENTIAL EQUATION

References

Moon, P. and Spencer, D. E. *Field Theory Handbook, Including Coordinate Systems, Differential Equations, and Their Solutions, 2nd ed.* New York: Springer-Verlag, pp. 15–7, 1988.
Morse, P. M. and Feshbach, H. *Methods of Theoretical Physics, Part I.* New York: McGraw-Hill, pp. 514 and 656–57, 1953.

Helmholtz Differential Equation— Confocal Ellipsoidal Coordinates

Using the NOTATION of Byerly (1959, pp. 252–53), LAPLACE'S EQUATION can be reduced to

$$\nabla^2 F = (\mu^2 - v^2)\frac{\partial^2 F}{\partial\alpha^2} + (\lambda^2 - v^2)\frac{\partial^2 F}{\partial\beta^2} + (\lambda^2 - \mu^2)\frac{\partial^2 F}{\partial\gamma^2}$$
$$= 0, \qquad (1)$$

where

$$\alpha = c\int_c^{\lambda}\frac{d\lambda}{\sqrt{(\lambda^2 - b^2)(\lambda^2 - c^2)}}$$
$$= F\left(\frac{b}{c},\ \frac{\pi}{2}\right) - F\left(\frac{b}{c},\ \sin^{-1}\left(\frac{c}{\lambda}\right)\right) \qquad (2)$$

$$\beta = c\int_b^{\mu}\frac{d\mu}{\sqrt{(c^2 - \mu^2)(\mu^2 - b^2)}}$$
$$= F\left(\sqrt{1 - b^2 - c^2},\ \sin^{-1}\left(\sqrt{\frac{1 - \dfrac{b^2}{\mu^2}}{1 - \dfrac{b^2}{c^2}}}\right)\right) \qquad (3)$$

$$\gamma = c\int_0^{v}\frac{dv}{\sqrt{(b^2 - v^2)(c^2 - v^2)}}$$
$$= F\left(\frac{b}{c},\ \sin^{-1}\left(\frac{v}{b}\right)\right). \qquad (4)$$

In terms of α, β, and γ,

$$\lambda = c\ \mathrm{dc}\left(\alpha,\ \frac{b}{c}\right) \qquad (5)$$

$$\mu = b\ \mathrm{nd}\left(\beta,\ \sqrt{1 - \frac{b^2}{c^2}}\right) \qquad (6)$$

$$v = b\ \mathrm{sn}\left(\gamma,\ \frac{b}{c}\right). \qquad (7)$$

Equation (1) is not separable using a function OF THE FORM

$$F = L(\alpha)M(\beta)N(\gamma), \qquad (8)$$

but it is if we let

$$\frac{1}{L}\frac{d^2L}{d\alpha^2} = \sum a_k\lambda^k \qquad (9)$$

$$\frac{1}{M}\frac{d^2M}{d\beta^2} = \sum b_k\mu^k \qquad (10)$$

$$\frac{1}{N}\frac{d^2N}{d\gamma^2} = \sum c_k v^k. \qquad (11)$$

These give

$$a_0 = -b_0 = c_0 \tag{12}$$

$$a_2 = -b_2 = c_2, \tag{13}$$

and all others terms vanish. Therefore (1) can be broken up into the equations

$$\frac{d^2L}{d\alpha^2} = (a_0 + a_2\lambda^2)L \tag{14}$$

$$\frac{d^2M}{d\beta^2} = -(a_0 + a_2\mu^2)M \tag{15}$$

$$\frac{d^2N}{d\gamma^2} = (a_0 + a_2\nu^2)N. \tag{16}$$

For future convenience, now write

$$a_0 = -(b^2 + c^2)p \tag{17}$$

$$a_2 = m(m+1), \tag{18}$$

then

$$\frac{d^2L}{d\alpha^2} - [m(m+1)\lambda^2 - (b^2+c^2)p]L = 0 \tag{19}$$

$$\frac{d^2M}{d\beta^2} + [m(m+1)\mu^2 - (b^2+c^2)p]M = 0 \tag{20}$$

$$\frac{d^2N}{d\gamma^2} - [m(m+1)\nu^2 - (b^2+c^2)p]N = 0. \tag{21}$$

Now replace α, β, and γ to obtain

$$(\lambda^2 - b^2)(\lambda^2 - c^2)\frac{d^2L}{d\lambda^2} + \lambda(\lambda^2 - b^2 + \lambda^2 - c^2)\frac{dL}{d\lambda}$$

$$-[m(m+1)\lambda^2 - (b^2+c^2)p]L = 0 \tag{22}$$

$$(\mu^2 - b^2)(\mu^2 - c^2)\frac{d^2M}{d\mu^2} + \mu(\mu^2 - b^2 + \mu^2 - c^2)\frac{dM}{d\mu}$$

$$-[m(m+1)\mu^2 - (b^2+c^2)p]M = 0 \tag{23}$$

$$(\nu^2 - b^2)(\nu^2 - c^2)\frac{d^2N}{d\nu^2} + \nu(\nu^2 - b^2 + \nu^2 - c^2)\frac{dN}{d\nu}$$

$$-[m(m+1)\nu^2 - (b^2+c^2)p]N = 0. \tag{24}$$

Each of these is a LAMÉ'S DIFFERENTIAL EQUATION, whose solution is called an ELLIPSOIDAL HARMONIC. Writing

$$L(\lambda) = E_m^p(\lambda) \tag{25}$$

$$M(\lambda) = E_m^p(\mu) \tag{26}$$

$$N(\lambda) = E_m^p(\nu) \tag{27}$$

gives the solution to (1) as a product of ELLIPSOIDAL HARMONICS $E_m^p(x)$.

$$F = E_m^p(\lambda)E_m^p(\mu)E_m^p(\nu). \tag{28}$$

See also CONFOCAL ELLIPSOIDAL COORDINATES, HELMHOLTZ DIFFERENTIAL EQUATION

References

Arfken, G. "Confocal Ellipsoidal Coordinates (ξ_1, ξ_2, ξ_3)." §2.15 in *Mathematical Methods for Physicists, 2nd ed.* Orlando, FL: Academic Press, pp. 117–18, 1970.
Byerly, W. E. *An Elementary Treatise on Fourier's Series, and Spherical, Cylindrical, and Ellipsoidal Harmonics, with Applications to Problems in Mathematical Physics.* New York: Dover, pp. 251–58, 1959.
Moon, P. and Spencer, D. E. *Field Theory Handbook, Including Coordinate Systems, Differential Equations, and Their Solutions, 2nd ed.* New York: Springer-Verlag, pp. 43–4, 1988.
Morse, P. M. and Feshbach, H. *Methods of Theoretical Physics, Part I.* New York: McGraw-Hill, p. 663, 1953.

Helmholtz Differential Equation— Confocal Paraboloidal Coordinates

As shown by Morse and Feshbach (1953), the HELMHOLTZ DIFFERENTIAL EQUATION is separable in CONFOCAL PARABOLOIDAL COORDINATES.

See also CONFOCAL PARABOLOIDAL COORDINATES, HELMHOLTZ DIFFERENTIAL EQUATION

References

Moon, P. and Spencer, D. E. *Field Theory Handbook, Including Coordinate Systems, Differential Equations, and Their Solutions, 2nd ed.* New York: Springer-Verlag, pp. 47–8, 1988.
Morse, P. M. and Feshbach, H. *Methods of Theoretical Physics, Part I.* New York: McGraw-Hill, p. 664, 1953.

Helmholtz Differential Equation—Conical Coordinates

In CONICAL COORDINATES, LAPLACE'S EQUATION can be written

$$\frac{\partial^2 V}{\partial \alpha^2} + \frac{\partial^2 V}{\partial \beta^2} + (\mu^2 - \nu^2)\frac{\partial}{\partial \lambda}\left(\lambda^2 \frac{\partial V}{\partial \lambda}\right) = 0, \tag{1}$$

where

$$\alpha = \int_a^\mu \frac{d\mu}{\sqrt{(\mu^2 - a^2)(b^2 - \mu^2)}} \tag{2}$$

$$\beta = \int_0^\nu \frac{d\nu}{\sqrt{(a^2 - \nu^2)(b^2 - \nu^2)}} \tag{3}$$

(Byerly 1959). Letting

$$V = U(u)R(r) \tag{4}$$

breaks (1) into the two equations,

$$\frac{d}{dr}\left(r^2 \frac{dR}{dr}\right) = m(m+1)R \tag{5}$$

$$\frac{\partial^2 U}{\partial \alpha^2} + \frac{\partial^2 U}{\partial \beta^2} + m(m+1)(\mu^2 - \nu^2)U = 0 \qquad (6)$$

Solving these gives

$$R(r) = Ar^m + Br^{-m-1} \qquad (7)$$

$$U(u) = E_m^p(\mu)E_m^p(\nu), \qquad (8)$$

where E_m^p are ELLIPSOIDAL HARMONICS. The regular solution is therefore

$$V = Ar^m E_m^p(\mu)E_m^p(\nu), \qquad (9)$$

However, because of the cylindrical symmetry, the solution $E_m^p(\mu)E_m^p(\nu)$ is an mth degree SPHERICAL HARMONIC.

See also CONICAL COORDINATES, HELMHOLTZ DIFFERENTIAL EQUATION

References

Arfken, G. "Conical Coordinates (ξ_1, ξ_2, ξ_3)." §2.16 in *Mathematical Methods for Physicists, 2nd ed.* Orlando, FL: Academic Press, pp. 118–19, 1970.

Byerly, W. E. *An Elementary Treatise on Fourier's Series, and Spherical, Cylindrical, and Ellipsoidal Harmonics, with Applications to Problems in Mathematical Physics.* New York: Dover, p. 263, 1959.

Moon, P. and Spencer, D. E. *Field Theory Handbook, Including Coordinate Systems, Differential Equations, and Their Solutions, 2nd ed.* New York: Springer-Verlag, pp. 39–0, 1988.

Morse, P. M. and Feshbach, H. *Methods of Theoretical Physics, Part I.* New York: McGraw-Hill, pp. 514 and 659, 1953.

Helmholtz Differential Equation—Elliptic Cylindrical Coordinates

In ELLIPTIC CYLINDRICAL COORDINATES, the SCALE FACTORS are $h_u = h_v = \sqrt{\sinh^2 u + \sin^2 v}$, $h_z = 1$, and the separation functions are $f_1(u) = f_2(v) = f_3(z) = 1$, giving a STÄCKEL DETERMINANT of $S = (\sin^2 v + \sinh^2 u)$. The Helmholtz differential equation is

$$\frac{1}{\sinh^2 u + \sin^2 v}\left(\frac{\partial^2 F}{\partial u^2} + \frac{\partial^2 F}{\partial v^2}\right) + \frac{\partial^2 F}{\partial z^2} + k^2 F = 0. \quad (1)$$

Attempt SEPARATION OF VARIABLES by writing

$$F(u, v, z) = U(u)V(v)Z(z), \qquad (2)$$

then the HELMHOLTZ DIFFERENTIAL EQUATION becomes

$$\frac{Z}{\sinh^2 u + \sin^2 v}\left(V\frac{d^2 U}{du^2} + U\frac{d^2 V}{dv^2}\right) + UV\frac{d^2 Z}{dz^2}$$
$$+ k^2 UVZ$$
$$= 0. \qquad (3)$$

Now divide by UVZ to give

$$\frac{1}{\sinh^2 u + \sin^2 v}\left(\frac{1}{U}\frac{\partial^2 U}{\partial u^2} + \frac{1}{V}\frac{\partial^2 V}{\partial v^2}\right) + \frac{1}{Z}\frac{\partial^2 Z}{\partial z^2} + k^2$$
$$= 0. \qquad (4)$$

Separating the Z part,

$$\frac{1}{Z}\frac{d^2 Z}{dz^2} = -(k^2 + m^2) \qquad (5)$$

$$\frac{1}{\sinh^2 u + \sin^2 v}\left(\frac{1}{U}\frac{\partial^2 U}{\partial u^2} + \frac{1}{V}\frac{\partial^2 V}{\partial v^2}\right) = m^2 \qquad (6)$$

so

$$\frac{\partial^2 Z}{\partial z^2} = -(k^2 + m^2)Z, \qquad (7)$$

which has the solution

$$Z(z) = A_{km}\cos(\sqrt{k^2 + m^2}z) + B_{km}\sin(\sqrt{k^2 + m^2}z). \quad (8)$$

Rewriting (6) gives

$$\left(\frac{1}{U}\frac{d^2 U}{du^2} - m^2\sinh^2 u\right) + \left(\frac{1}{V}\frac{d^2 V}{dv^2} - m^2\sin^2 v\right)$$
$$= 0, \qquad (9)$$

which can be separated into

$$\frac{1}{U}\frac{d^2 U}{du^2} - m^2\sinh^2 u = c \qquad (10)$$

$$c + \frac{1}{V}\frac{d^2 V}{dv^2} - m^2\sin^2 v = 0, \qquad (11)$$

so

$$\frac{d^2 U}{du^2} - (c + m^2\sinh^2 u)U = 0 \qquad (12)$$

$$\frac{d^2 V}{dv^2} + (c - m^2\sin^2 v)V = 0. \qquad (13)$$

Now use

$$\sinh^2 u = \tfrac{1}{2}[\cosh(2u) - 1] \qquad (14)$$

$$\sin^2 v = \tfrac{1}{2}[1 - \cos(2v)] \qquad (15)$$

to obtain

$$\frac{d^2 U}{du^2} - \{c + \tfrac{1}{2}m^2[\cosh(2u) - 1]\}U = 0 \qquad (16)$$

$$\frac{d^2 V}{dv^2} + \{c - \tfrac{1}{2}m^2[1 - \cos(2v)]\}V = 0. \qquad (17)$$

Regrouping gives

$$\frac{d^2U}{du^2} - [(c - \tfrac{1}{2}m^2) + \tfrac{1}{2}m^2\cosh(2u)]U = 0 \qquad (18)$$

$$\frac{d^2V}{dv^2} + [(c - \tfrac{1}{2}m^2) + \tfrac{1}{2}m^2\cos(2v)]V = 0. \qquad (19)$$

Let $a \equiv c - m^2/2$ and $q \equiv -m^2/4$, then these become

$$\frac{d^2V}{dv^2} + [a - 2q\cos(2v)]V = 0 \qquad (20)$$

$$\frac{d^2U}{du^2} - [a - 2q\cosh(2u)]U = 0. \qquad (21)$$

Here, (20) is the MATHIEU DIFFERENTIAL EQUATION and (21) is the modified MATHIEU DIFFERENTIAL EQUATION. These solutions are known as MATHIEU FUNCTIONS.

See also ELLIPTIC CYLINDRICAL COORDINATES, HELMHOLTZ DIFFERENTIAL EQUATION, MATHIEU DIFFERENTIAL EQUATION, MATHIEU FUNCTION

References
Abramowitz, M. and Stegun, C. A. (Eds.). "Mathieu Functions." Ch. 20 in *Handbook of Mathematical Functions with Formulas, Graphs, and Mathematical Tables, 9th printing.* New York: Dover, pp. 721–46, 1972.

Moon, P. and Spencer, D. E. *Field Theory Handbook, Including Coordinate Systems, Differential Equations, and Their Solutions, 2nd ed.* New York: Springer-Verlag, pp. 17–9, 1988.

Morse, P. M. and Feshbach, H. *Methods of Theoretical Physics, Part I.* New York: McGraw-Hill, pp. 514 and 657, 1953.

Helmholtz Differential Equation—Oblate Spheroidal Coordinates

As shown by Morse and Feshbach (1953) and Arfken (1970), the HELMHOLTZ DIFFERENTIAL EQUATION is separable in OBLATE SPHEROIDAL COORDINATES.

See also HELMHOLTZ DIFFERENTIAL EQUATION, OBLATE SPHEROIDAL COORDINATES

References
Arfken, G. "Oblate Spheroidal Coordinates (u, v, φ)." §2.11 in *Mathematical Methods for Physicists, 2nd ed.* Orlando, FL: Academic Press, pp. 107–09, 1970.

Byerly, W. E. *An Elementary Treatise on Fourier's Series, and Spherical, Cylindrical, and Ellipsoidal Harmonics, with Applications to Problems in Mathematical Physics.* New York: Dover, pp. 242 and 245–47, 1959.

Moon, P. and Spencer, D. E. *Field Theory Handbook, Including Coordinate Systems, Differential Equations, and Their Solutions, 2nd ed.* New York: Springer-Verlag, pp. 33–4, 1988.

Morse, P. M. and Feshbach, H. *Methods of Theoretical Physics, Part I.* New York: McGraw-Hill, p. 662, 1953.

Helmholtz Differential Equation— Parabolic Coordinates

The SCALE FACTORS are $h_u = h_v = \sqrt{u^2 + v^2}$, $h_\theta = uv$ and the separation functions are $f_\theta(u) = u$, $f_2(v) = v$, $f_3(\theta) = 1$, given a STÄCKEL DETERMINANT of $S = u^2 + v^2$. The LAPLACIAN is

$$\frac{1}{u^2 + v^2}\left(\frac{1}{u}\frac{\partial F}{\partial u} + \frac{\partial^2 F}{\partial u^2} + \frac{1}{v}\frac{\partial F}{\partial v} + \frac{\partial^2 F}{\partial v^2}\right) + \frac{1}{u^2 v^2}\frac{\partial^2 F}{\partial \theta^2} + k^2 F = 0. \qquad (1)$$

Attempt SEPARATION OF VARIABLES by writing

$$F(u, v, \theta) \equiv U(u)V(v)\Theta(\theta), \qquad (2)$$

then the HELMHOLTZ DIFFERENTIAL EQUATION becomes

$$\frac{1}{u^2 + v^2}\left[V\Theta\left(\frac{1}{u}\frac{dU}{du} + \frac{d^2U}{du^2}\right) + U\Theta\left(\frac{1}{v}\frac{dV}{dv} + \frac{d^2V}{dv^2}\right)\right]$$
$$+ \frac{UV}{u^2 v^2}\frac{d^2\Theta}{d\theta^2} + k^2 UV\Theta = 0. \qquad (3)$$

Now multiply through by $u^2 v^2/(UV\Theta)$,

$$\frac{u^2 v^2}{u^2 + v^2}\left[\frac{1}{U}\left(\frac{1}{u}\frac{dU}{du} + \frac{d^2U}{du^2}\right) + \frac{1}{V}\left(\frac{1}{v}\frac{dV}{dv} + \frac{d^2V}{dv^2}\right)\right]$$
$$+ \frac{1}{\Theta}\frac{d^2\Theta}{d\theta^2} + k^2 u^2 v^2 = 0. \qquad (4)$$

Separating the Θ part gives

$$\frac{1}{\Theta}\frac{d^2\theta}{d\theta^2} = -m^2, \qquad (5)$$

which has solution

$$\Theta(\theta) = A_m\cos(m\theta) + B_m\sin(m\theta). \qquad (6)$$

Plugging (5) back into (4) and multiplying by $(u^2 + v^2)/(u^2 v^2)$ gives

$$\left[\frac{1}{U}\left(\frac{1}{u}\frac{dU}{du} + \frac{d^2U}{du^2}\right) + \frac{1}{V}\left(\frac{1}{v}\frac{dV}{dv} + \frac{d^2V}{dv^2}\right)\right]$$
$$- m^2\frac{u^2 + v^2}{u^2 v^2} + k^2(u^2 + v^2) \qquad (7)$$

Rewriting,

$$\left[\frac{1}{U}\left(\frac{1}{u}\frac{dU}{du} + \frac{d^2U}{du^2}\right) + \frac{1}{V}\left(\frac{1}{v}\frac{dV}{dv} + \frac{d^2V}{dv^2}\right)\right]$$
$$- m^2\left(\frac{1}{v^2} + \frac{1}{u^2}\right) + k^2(u^2 + v^2). \qquad (8)$$

This can be rearranged into two terms, each containing only u or v,

$$\left[\frac{1}{U}\left(\frac{1}{u}\frac{dU}{du}+\frac{d^2U}{du^2}\right)+k^2u^2-\frac{m^2}{u^2}\right]$$

$$+\left[\frac{1}{V}\left(\frac{1}{v}\frac{dV}{dv}+\frac{d^2V}{dv^2}\right)+k^2v^2-\frac{m^2}{v^2}\right] \quad (9)$$

and so can be separated by letting the first part equal c and the second equal $-c$, giving

$$\frac{d^2U}{du^2}+\frac{1}{u}\frac{dU}{du}+\left(k^2u^2-\frac{m^2}{u^2}-c\right)U=0 \quad (10)$$

$$\frac{d^2V}{dv^2}+\frac{1}{v}\frac{dV}{dv}+\left(k^2v^2-\frac{m^2}{v^2}+c\right)V=0. \quad (11)$$

See also HELMHOLTZ DIFFERENTIAL EQUATION, PARABOLIC COORDINATES

References

Arfken, G. "Parabolic Coordinates $(\xi,\ \eta,\ \phi)$." §2.12 in *Mathematical Methods for Physicists, 2nd ed.* Orlando, FL: Academic Press, pp. 109–11, 1970.

Moon, P. and Spencer, D. E. *Field Theory Handbook, Including Coordinate Systems, Differential Equations, and Their Solutions, 2nd ed.* New York: Springer-Verlag, p. 36, 1988.

Morse, P. M. and Feshbach, H. *Methods of Theoretical Physics, Part I.* New York McGraw-Hill, pp. 514–15 and 660, 1953.

Helmholtz Differential Equation—Parabolic Cylindrical Coordinates

In PARABOLIC CYLINDRICAL COORDINATES, the SCALE FACTORS are $h_u=h_v=\sqrt{u^2+v^2}$, $h_z=1$ and the separation functions are $f_1(u)=f_2(v)=f_3(z)=1$, giving STÄCKEL DETERMINANT of $s=u^2+v^2$. the HELMHOLTZ DIFFERENTIAL EQUATION is

$$\frac{1}{u^2+v^2}\left(\frac{\partial^2 f}{\partial u^2}+\frac{\partial^2 f}{\partial v^2}\right)+\frac{\partial^2 f}{\partial z^2}+k^2 f=0. \quad (1)$$

attempt SEPARATION OF VARIABLES by writing

$$f(u,\ v,\ z)\equiv u(u)v(v)z(z), \quad (2)$$

then the HELMHOLTZ DIFFERENTIAL EQUATION becomes

$$\frac{1}{u^2+v^2}\left(VZ\frac{d^2U}{du^2}+UZ\frac{d^2V}{dv^2}\right)+UV\frac{d^2Z}{dz^2}+k^2UVZ$$
$$=0. \quad (3)$$

Divide by UVZ,

$$\frac{1}{u^2+v^2}\left(\frac{1}{U}\frac{d^2U}{du^2}+\frac{1}{V}\frac{d^2V}{dv^2}\right)+\frac{1}{Z}\frac{d^2Z}{dz^2}+k^2=0. \quad (4)$$

Separating the Z part,

$$\frac{1}{Z}\frac{d^2Z}{dz^2}=-(k^2+m^2) \quad (5)$$

$$\frac{1}{u^2+v^2}\left(\frac{1}{U}\frac{d^2U}{du^2}+\frac{1}{V}\frac{d^2V}{dv^2}\right)-k^2=0. \quad (6)$$

$$\frac{1}{U}\frac{d^2U}{du^2}+\frac{1}{V}\frac{d^2V}{dv^2}-k^2(u^2+v^2)=0, \quad (7)$$

so

$$\frac{\partial^2 Z}{dz^2}=-(k^2+m^2)Z, \quad (8)$$

which has solution

$$Z(z)=A\cos(\sqrt{k^2+m^2}z)+B\sin(\sqrt{k^2+m^2}z), \quad (9)$$

and

$$\left(\frac{1}{U}\frac{d^2U}{du^2}-k^2u^2\right)+\left(\frac{1}{V}\frac{d^2V}{dv^2}-k^2v^2\right)=0. \quad (10)$$

This can be separated

$$\frac{1}{U}\frac{d^2U}{du^2}-k^2u^2=c \quad (11)$$

$$\frac{1}{V}\frac{d^2V}{dv^2}-k^2v^2=-c, \quad (12)$$

so

$$\frac{d^2U}{du^2}-(c+k^2u^2)U=0 \quad (13)$$

$$\frac{d^2V}{dv^2}-(c+k^2v^2)V=0. \quad (14)$$

These are the WEBER DIFFERENTIAL EQUATIONS, and the solutions are known as PARABOLIC CYLINDER FUNCTIONS.

See also HELMHOLTZ DIFFERENTIAL EQUATION, PARABOLIC CYLINDER FUNCTION, PARABOLIC CYLINDRICAL COORDINATES, WEBER DIFFERENTIAL EQUATIONS

References

Moon, P. and Spencer, D. E. *Field Theory Handbook, Including Coordinate Systems, Differential Equations, and Their Solutions, 2nd ed.* New York: Springer-Verlag, p. 36, 1988.

Morse, P. M. and Feshbach, H. *Methods of Theoretical Physics, Part I.* New York: McGraw-Hill, pp. 515 and 658, 1953.

Helmholtz Differential Equation—Polar Coordinates

In 2-D POLAR COORDINATES, attempt SEPARATION OF VARIABLES by writing

$$F(r,\ \theta)=R(r)\Theta(\theta), \quad (1)$$

then the HELMHOLTZ DIFFERENTIAL EQUATION becomes

$$\frac{d^2R}{dr^2}\Theta + \frac{1}{r}\frac{dR}{dr}\Theta + \frac{1}{r^2}\frac{d^2\Theta}{d\theta^2}R + k^2R\Theta = 0. \tag{2}$$

Divide both sides by $R\Theta$

$$\left(\frac{r^2}{R}\frac{d^2R}{dr^2} + \frac{r}{R}\frac{dR}{dr}\right) + \left(\frac{1}{\Theta}\frac{d^2\Theta}{d\theta^2} + k^2\right) = 0. \tag{3}$$

The solution to the second part of (3) must be periodic, so the differential equation is

$$\frac{d^2\Theta}{d\theta^2}\frac{1}{\Theta} = -(k^2 + m^2), \tag{4}$$

which has solutions

$$\Theta(\theta) = c_1e^{i\sqrt{k^2+m^2}\,\theta} + c_2e^{-i\sqrt{k^2+m^2}\,\theta}$$

$$= c_3\sin(\sqrt{k^2+m^2}\theta) + c_4\cos(\sqrt{k^2+m^2}\theta). \tag{5}$$

Plug (4) back into (3)

$$r^2R'' + rR' - m^2R = 0. \tag{6}$$

This is an EULER DIFFERENTIAL EQUATION with $\alpha \equiv 1$ and $\beta \equiv -m^2$. The roots are $r = \pm m$. So for $m = 0$, $r = 0$ and the solution is

$$R(r) = c_1 + c_2\ln r. \tag{7}$$

But since $\ln r$ blows up at $r = 0$, the only possible physical solution is $R(r) = c_1$. When $m > 0$, $r = \pm m$, so

$$R(r) = c_1r^m + c_2r^{-m}. \tag{8}$$

But since r^{-m} blows up at $r = 0$, the only possible physical solution is $R_m(r) = c_1r^m$. The solution for R is then

$$R_m(r) = c_mr^m \tag{9}$$

for $m = 0, 1, \ldots$ and the general solution is

$$F(r, \theta) = \sum_{m=0}^{\infty}[a_mr^m\sin(\sqrt{k^2+m^2}\theta)$$

$$+ b_mr^m\cos(\sqrt{k^2+m^2}\theta)]. \tag{10}$$

References

Morse, P. M. and Feshbach, H. *Methods of Theoretical Physics, Part I.* New York McGraw-Hill, pp. 502–04, 1953.

Helmholtz Differential Equation—Prolate Spheroidal Coordinates

As shown by Morse and Feshbach (1953) and Arfken (1970), the HELMHOLTZ DIFFERENTIAL EQUATION is separable in PROLATE SPHEROIDAL COORDINATES.

See also HELMHOLTZ DIFFERENTIAL EQUATION, PROLATE SPHEROIDAL COORDINATES

References

Arfken, G. "Prolate Spheroidal Coordinates (u, v, φ)." §2.10 in *Mathematical Methods for Physicists, 2nd ed.* Orlando, FL: Academic Press, pp. 103–07, 1970.

Byerly, W. E. *An Elementary Treatise on Fourier's Series, and Spherical, Cylindrical, and Ellipsoidal Harmonics, with Applications to Problems in Mathematical Physics.* New York: Dover, pp. 243–44, 1959.

Moon, P. and Spencer, D. E. *Field Theory Handbook, Including Coordinate Systems, Differential Equations, and Their Solutions, 2nd ed.* New York: Springer-Verlag, p. 30, 1988.

Morse, P. M. and Feshbach, H. *Methods of Theoretical Physics, Part I.* New York: McGraw-Hill, p. 661, 1953.

Helmholtz Differential Equation— Spherical Coordinates

In SPHERICAL COORDINATES, the SCALE FACTORS are $h_r = 1$, $h_\theta = r\sin\phi$, $h_\phi = r$, and the separation functions are $f_1(r) = r^2$, $f_2(\theta) = 1$, $f_3(\phi) = \sin\phi$, giving a STÄCKEL DETERMINANT of $S = 1$. The LAPLACIAN is

$$\nabla^2 \equiv \frac{1}{r^2}\frac{\partial}{\partial r}\left(r^2\frac{\partial}{\partial r}\right) + \frac{1}{r^2\sin^2\phi}\frac{\partial^2}{\partial\theta^2} + \frac{1}{r^2\sin\phi}\frac{\partial}{\partial\phi}$$

$$\times\left(\sin\phi\frac{\partial}{\partial\phi}\right). \tag{1}$$

To solve the HELMHOLTZ DIFFERENTIAL EQUATION in SPHERICAL COORDINATES, attempt SEPARATION OF VARIABLES by writing

$$F(r, \theta, \phi) = R(r)\Theta(\theta)\Phi(\phi). \tag{2}$$

Then the HELMHOLTZ DIFFERENTIAL EQUATION becomes

$$\frac{d^2R}{dr^2}\Phi\Theta + \frac{2}{r}\frac{dR}{dr}\Phi\Theta + \frac{1}{r^2\sin^2\phi}\frac{d^2\Theta}{d\theta^2}\Phi R$$

$$+ \frac{\cos\phi}{r^2\sin\phi}\frac{d\Phi}{d\phi}\Theta R + \frac{1}{r^2}\frac{d^2\Phi}{d\phi^2}\Theta R$$

$$= 0. \tag{3}$$

Now divide by $R\Theta\Phi$,

$$\frac{r^2\sin^2\phi}{\Phi R\Theta}\Phi\Theta\frac{d^2R}{dr^2} + \frac{2}{r}\frac{r^2\sin^2\phi}{\Phi R\Theta}\Phi\Theta\frac{dR}{dr}$$

$$+ \frac{1}{r^2\sin^2\phi}\frac{r^2\sin^2\phi}{\Phi R\Theta}\Phi R\frac{d^2\Theta}{d\theta^2}$$

$$+ \frac{\cos\phi}{r^2\sin\phi}\frac{r^2\sin^2\phi}{\Phi\Theta R}\frac{d\Phi}{d\phi}\Theta R$$

$$+ \frac{1}{r^2}\frac{r^2\sin^2\phi}{\Phi R\Theta}\frac{d^2\Phi}{d\phi^2}\Theta R = 0 \tag{4}$$

$$\left(\frac{r^2 \sin^2 \phi}{R} \frac{d^2R}{dr^2} + \frac{2r \sin^2 \phi}{R} \frac{dR}{dr}\right) + \left(\frac{1}{\Theta} \frac{d^2\Theta}{d\theta^2}\right)$$

$$+ \left(\frac{\cos\phi \sin\phi}{\Phi} \frac{d\Phi}{d\phi} + \frac{\sin^2\phi}{\Phi} \frac{d^2\Phi}{d\phi^2}\right) = 0. \quad (5)$$

The solution to the second part of (5) must be sinusoidal, so the differential equation is

$$\frac{d^2\Theta}{d\theta^2} \frac{1}{\Theta} = -m^2, \quad (6)$$

which has solutions which may be defined either as a COMPLEX function with $m = -\infty, ..., \infty$

$$\Theta(\theta) = A_m e^{im\theta}, \quad (7)$$

or as a sum of REAL sine and cosine functions with $m = -\infty, ..., \infty$

$$\Theta(\theta) = S_m \sin(m\theta) + C_m \cos(m\theta). \quad (8)$$

Plugging (6) back into (7),

$$\frac{r^2}{R} \frac{d^2R}{dr^2} + \frac{2r}{R} \frac{dR}{dr} - \frac{1}{\sin^2\phi}\left(m^2 + \frac{\cos\phi \sin\phi}{\Phi} \frac{d\Phi}{d\phi}\right)$$

$$+ \frac{\sin^2\phi}{\Phi} \frac{d^2\Phi}{d\phi^2}$$

$$= 0. \quad (9)$$

The radial part must be equal to a constant

$$\frac{r^2}{R} \frac{d^2R}{dr^2} + \frac{2r}{R} \frac{dR}{dr} = l(l+1) \quad (10)$$

$$r^2 \frac{d^2R}{dr^2} + 2r \frac{dR}{dr} = l(l+1)R. \quad (11)$$

But this is the EULER DIFFERENTIAL EQUATION, so we try a series solution OF THE FORM

$$R = \sum_{n=0}^{\infty} a_n r^{n+c} \quad (12)$$

Then

$$r^2 \sum_{n=0}^{\infty} (n+c)(n+c-1)a_n r^{n+c-2}$$

$$+ 2r \sum_{n=0}^{\infty} (n+c)a_n r^{n+c-1}$$

$$- l(l+1) \sum_{n=0}^{\infty} a_n r^{n+c} = 0 \quad (13)$$

$$\sum_{n=0}^{\infty} (n+c)(n+c-1)a_n r^{n+c} + 2 \sum_{n=0}^{\infty} (n+c)a_n r^{n+c}$$

$$- l(l+1) \sum_{n=0}^{\infty} a_n r^{n+c} = 0 \quad (14)$$

$$\sum_{n=0}^{\infty} [(n+c)(n+c-1) - l(l+1)]a_n r^{n+c} = 0. \quad (15)$$

This must hold true for all POWERS of r. For the r^c term (with $n = 0$),

$$c(c+1) = l(l+1), \quad (16)$$

which is true only if $c = l, -l-1$ and all other terms vanish. So $a_n = 0$ for $n \neq l, -l-1$. Therefore, the solution of the R component is given by

$$R_l(r) = A_l r^l + B_l r^{-l-1}. \quad (17)$$

Plugging (17) back into (9),

$$l(l+1) - \frac{m^2}{\sin^2\phi} + \frac{\cos\phi}{\sin\phi} \frac{1}{\Phi} \frac{d\Phi}{d\phi} + \frac{1}{\Phi} \frac{d^2\Phi}{d\phi^2} = 0 \quad (18)$$

$$\Phi'' \frac{\cos\phi}{\sin\phi} \Phi' + \left[l(l+1) - \frac{m^2}{\sin^2\phi}\right]\Phi = 0, \quad (19)$$

which is the associated LEGENDRE DIFFERENTIAL EQUATION for $x = \cos\phi$ and $m = 0, ..., l$. The general COMPLEX solution is therefore

$$\sum_{t=0}^{\infty} \sum_{m=-l}^{l} (A_l r^l + B_l r^{-l-1})P_l^m(\cos\phi)e^{-im\theta}$$

$$\equiv \sum_{t=0}^{\infty} \sum_{m=-1}^{l} (A_l r^l + B_l r^{-l-1})Y_l^m(\theta, \phi) \quad (20)$$

where

$$Y_l^m(\theta, \phi) \equiv P_l^m(\cos\phi)e^{-im\theta} \quad (21)$$

are the (COMPLEX) SPHERICAL HARMONICS. The general REAL solution is

$$\sum_{t=0}^{\infty} \sum_{m=0}^{l} (A_l r^l + B_l r^{-l-1})P_l^m(\cos\phi)$$

$$\times [S_m \sin(m\theta) + C_m \cos(m\theta)]. \quad (22)$$

Some of the normalization constants of P_l^m can be absorbed by S_m and C_m, so this equation may appear in the form

$$\sum_{t=0}^{\infty} \sum_{m=0}^{l} (A_l r^l + B_l r^{-l-1})P_l^m(\cos\phi)$$

$$\times [S_l^m \sin(m\theta) + C_l^m \cos(m\theta)]$$

$$\equiv \sum_{l=0}^{\infty} \sum_{m=0}^{l} (A_l r^l + B_l r^{-l-1})$$

$$\times [S_l^m Y_l^{m(o)}(\theta, \phi) + C_l^m Y_l^{m(e)}(\theta, \phi)], \quad (23)$$

where

$$Y_l^{m(0)}(\theta, \phi) \equiv P_l^m(\cos\theta)\sin(m\theta) \quad (24)$$

$$Y_l^{m(e)}(\theta, \phi) \equiv P_l^m(\cos\theta)\cos(m\theta) \quad (25)$$

are the EVEN and ODD (real) SPHERICAL HARMONICS. If azimuthal symmetry is present, then $\Theta(\theta)$ is constant and the solution of the Φ component is a LEGENDRE POLYNOMIAL $P_l(\cos \phi)$. The general solution is then

$$F(r, \phi) = \sum_{l=0}^{\infty} (A_l r^l B_l r^{-l-1}) P_l(\cos \phi). \tag{26}$$

Actually, the equation is separable under the more general condition that k^2 is OF THE FORM

$$k^2(r, \theta, \phi) = f(r) + \frac{g(\theta)}{r^2} + \frac{h(\phi)}{r^2 \sin \theta} + k'^2. \tag{27}$$

See also HELMHOLTZ DIFFERENTIAL EQUATION, SPHERICAL COORDINATES, SPHERICAL HARMONIC

References

Byerly, W. E. *An Elementary Treatise on Fourier's Series, and Spherical, Cylindrical, and Ellipsoidal Harmonics, with Applications to Problems in Mathematical Physics.* New York: Dover, p. 244, 1959.

Moon, P. and Spencer, D. E. *Field Theory Handbook, Including Coordinate Systems, Differential Equations, and Their Solutions, 2nd ed.* New York: Springer-Verlag, p. 27, 1988.

Morse, P. M. and Feshbach, H. *Methods of Theoretical Physics, Part I.* New York: McGraw-Hill, p. 514 and 658, 1953.

Helmholtz Differential Equation— Spherical Surface

On the surface of a SPHERE, attempt SEPARATION OF VARIABLES in SPHERICAL COORDINATES by writing

$$F(\theta, \phi) = \Theta(\theta)\Phi(\phi), \tag{1}$$

then the HELMHOLTZ DIFFERENTIAL EQUATION becomes

$$\frac{1}{\sin^2 \phi} \frac{d^2\Theta}{d\theta^2} \Phi + \frac{\cos \phi}{\sin \phi} \frac{d\Phi}{d\phi} \Theta + \frac{d^2\Phi}{d\phi^2} \Theta + k^2 \Theta \Phi = 0. \tag{2}$$

Dividing both sides by $\Phi\Theta$,

$$\left(\frac{\cos \phi \sin \phi}{\Phi} \frac{d\Phi}{d\phi} + \frac{\sin^2 \phi}{\Phi} \frac{d^2\Phi}{d\phi^2} \right) + \left(\frac{1}{\Theta} \frac{d^2\Theta}{d\theta^2} + k^2 \right) = 0, \tag{3}$$

which can now be separated by writing

$$\frac{d^2\Theta}{d\theta^2} \frac{1}{\Theta} = -(k^2 + m^2). \tag{4}$$

The solution to this equation must be periodic, so m must be an INTEGER. The solution may then be defined either as a COMPLEX function

$$\Theta(\theta) = A_m e^{i\sqrt{k^2+m^2}\,\theta} + B_m e^{-i\sqrt{k^2+m^2}\,\theta} \tag{5}$$

for $m = -\infty, ..., \infty$, or as a sum of REAL sine and cosine

functions

$$\Theta(\theta) = S_m \sin\left(\sqrt{k^2+m^2}\,\theta\right) + C_m \cos\left(\sqrt{k^2+m^2}\,\theta\right) \tag{6}$$

for $m = 0, ..., \infty$. Plugging (4) into (3) gives

$$\frac{\cos \phi \sin \phi}{\Phi} \frac{d\Phi}{d\phi} + \frac{\sin^2 \phi}{\Phi} \frac{d^2\Phi}{d\phi^2} + m^2 = 0 \tag{7}$$

$$\Phi'' + \frac{\cos \phi}{\sin \phi} \Phi' + \frac{m^2}{\sin^2 \phi} \Phi = 0, \tag{8}$$

which is the LEGENDRE DIFFERENTIAL EQUATION for $x = \cos \phi$ with

$$m^2 \equiv l(l+1), \tag{9}$$

giving

$$l^2 + l - m^2 = 0 \tag{10}$$

$$l = \tfrac{1}{2}(-1 \pm \sqrt{1+4m^2}). \tag{11}$$

Solutions are therefore LEGENDRE POLYNOMIALS with a COMPLEX index. The general COMPLEX solution is then

$$F(\theta, \phi) = \sum_{m=-\infty}^{\infty} P_l(\cos \phi)(A_m e^{im\theta} + B_m e^{-im\theta}), \tag{12}$$

and the general REAL solution is

$$F(\theta, \phi) = \sum_{m=0}^{\infty} P_l(\cos \phi) \times [S_m \sin(m\theta) + C_m \cos(m\theta)]. \tag{13}$$

Note that these solutions depend on only a single variable m. However, on the surface of a sphere, it is usual to express solutions in terms of the SPHERICAL HARMONICS derived for the 3-D spherical case, which depend on the two variables l and m.

Helmholtz Differential Equation— Toroidal Coordinates

The HELMHOLTZ DIFFERENTIAL EQUATION is not separable in TOROIDAL COORDINATES

See also HELMHOLTZ DIFFERENTIAL EQUATION, LAPLACE'S EQUATION–TOROIDAL COORDINATES, TOROIDAL COORDINATES

Helmholtz's Theorem

Any VECTOR FIELD $\mathbf{v}$ satisfying

$$[\nabla \cdot \mathbf{v}]_\infty = 0 \tag{1}$$

$$[\nabla \times \mathbf{v}]_\infty = 0 \tag{2}$$

may be written as the sum of an IRROTATIONAL part

and a SOLENOIDAL part,

$$\mathbf{v} = -\nabla\phi + \nabla \times \mathbf{A}, \qquad (3)$$

where for a VECTOR FIELD F,

$$\phi = -\int_V \frac{\nabla \cdot \mathbf{F}}{4\pi|\mathbf{r}' - \mathbf{r}|} d^3\mathbf{r}' \qquad (4)$$

$$\mathbf{A} = \int_V \frac{\nabla \times \mathbf{F}}{4\pi|\mathbf{r}' - \mathbf{r}|} d^3\mathbf{r}'. \qquad (5)$$

See also IRROTATIONAL FIELD, SOLENOIDAL FIELD, VECTOR FIELD

References

Arfken, G. "Helmholtz's Theorem." §1.15 in *Mathematical Methods for Physicists, 3rd ed.* Orlando, FL: Academic Press, pp. 78–4, 1985.
Gradshteyn, I. S. and Ryzhik, I. M. *Tables of Integrals, Series, and Products, 6th ed.* San Diego, CA: Academic Press, p. 1084, 2000.

Helson-Szego Measure

An absolutely continuous measure on ∂D whose density has the form $\exp(x + \bar{y})$, where x and y are real-valued functions in L^∞, $\|y\|_\infty < \pi/2$, exp is the EXPONENTIAL FUNCTION, and $\|y\|$ is the NORM.

Hemicylindrical Function

A function $S_n(z)$ which satisfies the RECURRENCE RELATION

$$S_{n-1}(z) - S_{n+1}(z) = 2S_n'(z)$$

together with

$$S_1(z) = -S_0'(z)$$

is called a hemicylindrical function.

References

Sonine, N. "Recherches sur les fonctions cylindriques et le développement des fonctions continues en séries." *Math. Ann.* **16**, 1– and 71–0, 1880.
Watson, G. N. "Hemi-Cylindrical Functions." §10.8 in *A Treatise on the Theory of Bessel Functions, 2nd ed.* Cambridge, England: Cambridge University Press, p. 353, 1966.

Hemisphere

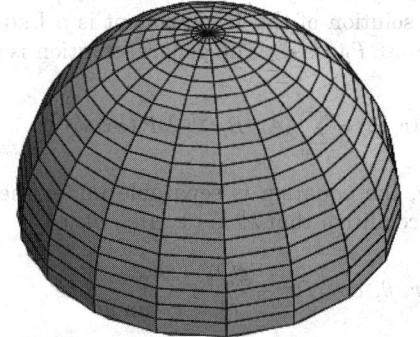

Half of a SPHERE cut by a PLANE passing through its CENTER. A hemisphere of RADIUS r can be given by the usual SPHERICAL COORDINATES

$$x = r\cos\theta\sin\phi \qquad (1)$$

$$y = r\sin\theta\sin\phi \qquad (2)$$

$$z = r\cos\phi, \qquad (3)$$

where $\theta \in [0, 2\pi)$ and $\phi \in [0, \pi/2]$. All CROSS SECTIONS passing through the z-AXIS are SEMICIRCLES.

The VOLUME of the hemisphere is

$$V = \pi\int_0^r (r^2 - z^2)\, dz = \tfrac{2}{3}\pi r^3. \qquad (4)$$

The weighted mean of z over the hemisphere is

$$\langle z \rangle = \pi\int_0^r z(r^2 - z^2)\, dz = \tfrac{1}{4}\pi r^2. \qquad (5)$$

The CENTROID is then given by

$$\bar{z} = \frac{\langle z \rangle}{V} = \tfrac{3}{8}r \qquad (6)$$

(Beyer 1987).

See also SEMICIRCLE, SPHERE

References

Beyer, W. H. (Ed.). *CRC Standard Mathematical Tables, 28th ed.* Boca Raton, FL: CRC Press, p. 133, 1987.

Hemispherical Function

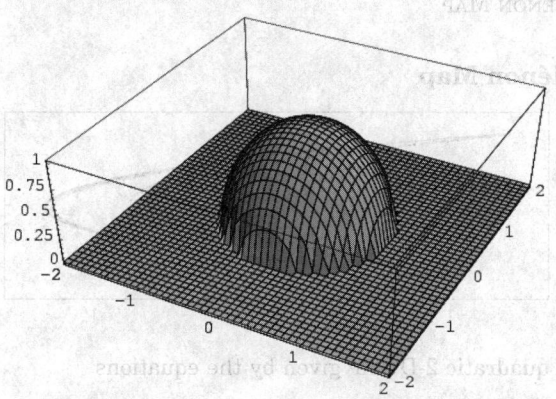

The hemisphere function is defined as

$$H(x,\ y) = \begin{cases} \sqrt{a - x^2 - y^2} & \text{for} \quad \sqrt{x^2 + y^2} \le a \\ 0 & \text{for} \quad \sqrt{x^2 + y^2} > a. \end{cases}$$

Watson (1966) defines a hemispherical function as a function S which satisfies the RECURRENCE RELATIONS

$$S_{n-1}(z) - S_{n+1}(z) = 2S_n'(z)$$

with

$$S_1(z) = -S_0'(z)$$

See also CYLINDER FUNCTION, CYLINDRICAL FUNCTION

References

Watson, G. N. *A Treatise on the Theory of Bessel Functions*, 2nd ed. Cambridge, England: Cambridge University Press, p. 353, 1966.

Hempel's Paradox

A purple cow is a confirming instance of the hypothesis that all crows are black.

References

Carnap, R. *Logical Foundations of Probability*. Chicago, IL: University of Chicago Press, pp. 224 and 469, 1950.
Erickson, G. W. and Fossa, J. A. *Dictionary of Paradox*. Lanham, MD: University Press of America, pp. 79–1, 1998.
Gardner, M. *The Scientific American Book of Mathematical Puzzles & Diversions*. New York: Simon and Schuster, pp. 52–4, 1959.
Goodman, N. Ch. 3 in *Fact, Fiction, and Forecast*. Cambridge, MA: Harvard University Press, 1955.
Hempel, C. G. "A Purely Syntactical Definition of Confirmation." *J. Symb. Logic* **8**, 122–43, 1943.
Hempel, C. G. "Studies in Logic and Confirmation." *Mind* **54**, 1–6, 1945.
Hempel, C. G. "Studies in Logic and Confirmation. II." *Mind* **54**, 97–21, 1945.
Hempel, C. G. "A Note on the Paradoxes of Confirmation." *Mind* **55**, 1946.
Hosiasson-Lindenbaum, J. "On Confirmation." *J. Symb. Logic* **5**, 133–48, 1940.
Whiteley, C. H. "Hempel's Paradoxes of Confirmation." *Mind* **55**, 156–58, 1945.

Hendecagon

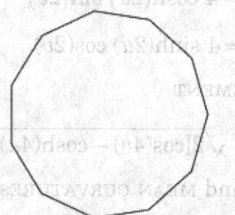

An 11-sided polygon, also variously known as the undecagon or unidecagon. The term "hendecagon" is preferable to the other two since it uses the Greek prefix and suffix instead of mixing a Roman prefix and Greek suffix. The regular 11-sided POLYGON has SCHLÄFLI SYMBOL $\{11\}$.

The hendecagon cannot be constructed using the classical Greek rules of GEOMETRIC CONSTRUCTION, but Conway and Guy (1996) give a NEUSIS CONSTRUCTION based on TRISECTION.

See also DECAGON, DODECAGON, TRIGONOMETRY VALUES PI/11

References

Conway, J. H. and Guy, R. K. *The Book of Numbers*. New York: Springer-Verlag, pp. 194–00, 1996.

Henneberg's Minimal Surface

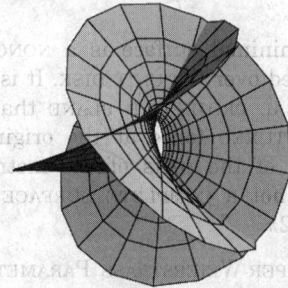

A MINIMAL SURFACE and double ALGEBRAIC SURFACE of 15th order and fifth class which can be given by PARAMETRIC EQUATIONS

$$x(u,\ v) = 2 \sinh u \cos v - \tfrac{2}{3} \sinh(3u) \cos(3v) \qquad (1)$$

$$y(u,\ v) = 2 \sinh u \sin v + \tfrac{2}{3} \sinh(3u) \sin(3v) \qquad (2)$$

$$z(u,\ v) = 2 \cosh(2u) \cos(2v). \qquad (3)$$

The coefficients of the FIRST FUNDAMENTAL FORM of this parameterization are given by

$$E = 8 \cosh^2 u [\cosh(4u) - \cos(4v)] \qquad (4)$$

$$F = 0 \qquad (5)$$

$$G = 8 \cosh^2 u [\cosh(4u) - \cos(4v)], \tag{6}$$

and the coefficients of the SECOND FUNDAMENTAL FORM are

$$e = -4 \cos(2v) \sinh(2u) \tag{7}$$

$$f = 4 \cosh(2u) \sin(2v) \tag{8}$$

$$g = 4 \sinh(2u) \cos(2v), \tag{9}$$

giving AREA ELEMENT

$$dS = 2 \sqrt{2[\cos(4v) - \cosh(4u)]} \tag{10}$$

and GAUSSIAN and MEAN CURVATURES are

$$K = \frac{\operatorname{sech}^4 u}{8[\cos(4v) - \cosh(4u)]} \tag{11}$$

$$H = 0. \tag{12}$$

The surface can also be obtained from the ENNEPER-WEIERSTRASS PARAMETERIZATION with

$$f = 2 - 2z^{-4} \tag{13}$$

$$g = z, \tag{14}$$

which gives a parameterization OF THE FORM

$$x = \frac{2(r^2 - 1)\cos\phi}{r} - \frac{2(r^6 - 1)\cos(3\phi)}{3r^3} \tag{15}$$

$$y = -\frac{6r^2(r^2 - 1)\sin\phi + 2(r^6 - 1)\sin(3\phi)}{3r^3} \tag{16}$$

$$z = \frac{2(r^4 + 1)\cos(2\phi)}{r^2} \tag{17}$$

Henneberg's minimal surface is a NONORIENTABLE SURFACE defined over the UNIT DISK. It is an immersion of the REAL PROJECTIVE PLANE that has been multiply PUNCTURED (once at the origin and four times at each of the roots of the metric). Consequently, it is not a COMPLETE SURFACE. The total curvature is -2π.

See also ENNEPER-WEIERSTRASS PARAMETERIZATION, MINIMAL SURFACE

References

Darboux, G. §226 in *Lecons sur la théorie générale des surfaces*. Paris: Gauthier-Villars, 1941.
Eisenhart, L. P. *A Treatise on the Differential Geometry of Curves and Surfaces*. New York: Dover, p. 267, 1960.
Gray, A. "Henneberg's Minimal Surface." *Modern Differential Geometry of Curves and Surfaces with Mathematica,* 2nd ed. Boca Raton, FL: CRC Press, pp. 691–92, 1997.
JavaView. "Classic Surfaces from Differential Geometry: Henneberg." http://www-sfb288.math.tu-berlin.de/vgp/javaview/demo/surface/common/PaSurface_Henneberg.html.
Nitsche, J. C. C. *Introduction to Minimal Surfaces.* Cambridge, England: Cambridge University Press, p. 144, 1989.

Hénon Attractor

HÉNON MAP

Hénon Map

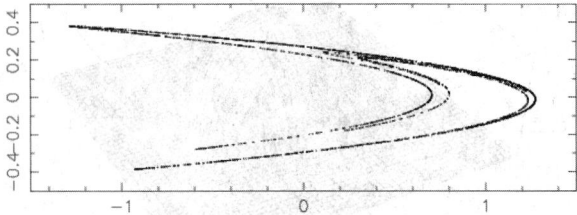

A quadratic 2-D MAP given by the equations

$$x_{n+1} = 1 - \alpha x_n^2 + y_n \tag{1}$$

$$y_{n+1} = \beta x_n \tag{2}$$

or

$$x_{n+1} = x_n \cos\alpha - (y_n - x_n^2)\sin\alpha \tag{3}$$

$$y_{n+1} = x_n \sin\alpha + (y_n - x_n^2)\cos\alpha. \tag{4}$$

The above map is for $\alpha = 1.4$ and $\beta = 0.3$. The Hénon map has CORRELATION EXPONENT 1.25 ± 0.02 (Grassberger and Procaccia 1983) and CAPACITY DIMENSION 1.261 ± 0.003 (Russell *et al.* 1980). Hitzl and Zele (1985) give conditions for the existence of periods 1 to 6.

See also BOGDANOV MAP, LOZI MAP, QUADRATIC MAP

References

Dickau, R. M. "The Hénon Attractor." http://forum.swarthmore.edu/advanced/robertd/henon.html.
Gleick, J. *Chaos: Making a New Science.* New York: Penguin Books, pp. 144–53, 1988.
Grassberger, P. and Procaccia, I. "Measuring the Strangeness of Strange Attractors." *Physica D* **9**, 189–08, 1983.
Hitzl, D. H. and Zele, F. "An Exploration of the Hénon Quadratic Map." *Physica D* **14**, 305–26, 1985.
Lauwerier, H. *Fractals: Endlessly Repeated Geometric Figures.* Princeton, NJ: Princeton University Press, pp. 128–33, 1991.
Morosawa, S.; Nishimura, Y.; Taniguchi, M.; and Ueda, T. "Dynamics of Generalized Hénon Maps." Ch. 7 in *Holomorphic Dynamics.* Cambridge, England: Cambridge University Press, pp. 225–62, 2000.
Peitgen, H.-O. and Saupe, D. (Eds.). "A Chaotic Set in the Plane." §3.2.2 in *The Science of Fractal Images.* New York: Springer-Verlag, pp. 146–48, 1988.
Russell, D. A.; Hanson, J. D.; and Ott, E. "Dimension of Strange Attractors." *Phys. Rev. Let.* **45**, 1175–178, 1980.
Wells, D. *The Penguin Dictionary of Curious and Interesting Geometry.* London: Penguin, pp. 95–7, 1991.

Hénon-Heiles Equation

A nonlinear *nonintegrable* HAMILTONIAN SYSTEM with

$$\ddot{x} = -\frac{\partial V}{\partial x} \tag{1}$$

$$\ddot{y} = -\frac{\partial V}{\partial y}, \qquad (2)$$

where the potential energy function is defined by the polar equation

$$V(r,\ \theta) = \tfrac{1}{2}r^2 + \tfrac{1}{3}r^3\sin(3\theta), \qquad (3)$$

giving Cartesian potential

$$V(x,\ y) = \tfrac{1}{2}\left(x^2 + y^2 + 2x^2y - \tfrac{2}{3}y^3\right). \qquad (4)$$

The total energy of the system is then given by

$$E = V(x,\ y) + \tfrac{1}{2}(\dot{x}^2 + \dot{y}^2), \qquad (5)$$

which is conserved during motion.

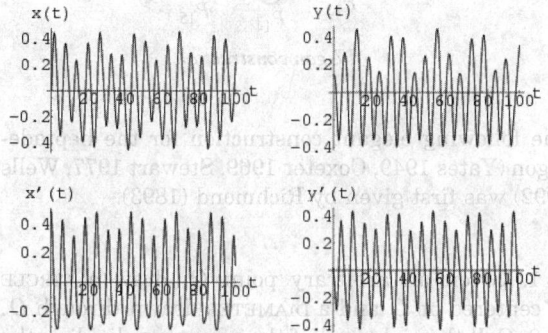

Integrating the above coupled ordinary differential equations from an arbitrary starting point with $x(t = 0) = 0$ and $E = 1/8$ gives the motion illustrated above. Computing the values of t at which $x = 0$ and plotting $y(t)$ vs. $\dot{y}(t)$ at these values gives a so-called SURFACE OF SECTION. The surfaces of section shown below correspond to $E = 1/12$ and $E = 1/8$.

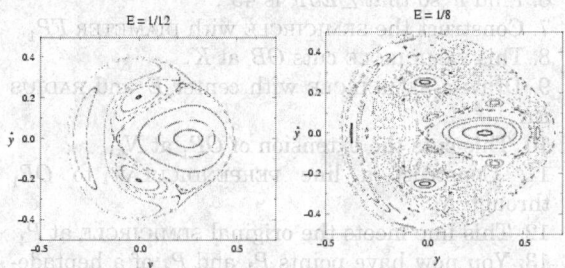

The Hamiltonian for a generalized Hénon-Heiles potential is

$$H = \tfrac{1}{2}(p_x^2 + p_y^2 + Ax^2 + By^2) + Dx^2y - \tfrac{1}{3}Cy^3. \qquad (6)$$

The equations of motion are integrable only for

1. $D/C = 0$,
2. $D/C = -1$, $A/B = 1$,
3. $D/C = -1/6$, and
4. $D/C = -1/16$, $A/B = 1/6$.

See also STANDARD MAP, SURFACE OF SECTION

References
Gleick, J. *Chaos: Making a New Science.* New York: Penguin Books, pp. 144–53, 1988.
Hénon, M. and Heiles, C. "The Applicability of the Third Integral of Motion: Some Numerical Experiments." *Astron. J.* **69**, 73–9, 1964.
Rasband, S. N. *Chaotic Dynamics of Nonlinear Systems.* New York: Wiley, pp. 171–72, 1990.
Tabor, M. "The Hénon-Heiles Hamiltonian." §4.1.b in *Chaos and Integrability in Nonlinear Dynamics: An Introduction.* New York: Wiley, pp. 121–22, 1989.

Henry VIII Prime

TRUNCATABLE PRIME

Hensel's Lemma

An important result in VALUATION THEORY which gives information on finding roots of POLYNOMIALS. Hensel's lemma is formally stated as follow. Let $(K,\ |\cdot|)$ be a complete NON-ARCHIMEDEAN FIELD, and let R be the corresponding VALUATION RING. Let $f(x)$ be a POLYNOMIAL whose COEFFICIENTS are in R and suppose a_0 satisfies

$$|f(a_0)| < |f'(a_0)|^2, \qquad (1)$$

where f' is the (formal) DERIVATIVE of f. Then there exists a unique element $a \in R$ such that $f(a) = 0$ and

$$|a - a_0| \le \left|\frac{f(a_0)}{f'(a_0)}\right|. \qquad (2)$$

Less formally, if $f(x)$ is a POLYNOMIAL with "INTEGER" COEFFICIENTS and $f(a_0)$ is "small" compared to $f'(a_0)$, then the equation $f(x) = 0$ has a solution "near" a_0. In addition, there are no other solutions near a_0, although there may be other solutions. The proof of the LEMMA is based around the Newton-Raphson method and relies on the non-Archimedean nature of the valuation.

Consider the following example in which Hensel's lemma is used to determine that the equation $x^2 = -1$ is solvable in the 5-adic numbers $\mathbb{Q}_5$ (and so we can embed the GAUSSIAN INTEGERS inside $\mathbb{Q}_5$ in a nice way). Let K be the 5-adic numbers $\mathbb{Q}_5$, let $f(x) = x^2 + 1$, and let $a_0 = 2$. Then we have $f(2) = 5$ and $f'(2) = 4$, so

$$|f(2)|_5 = \tfrac{1}{5} < |f'(2)|_5^2 = 1, \qquad (3)$$

and the condition is satisfied. Hensel's lemma then tells us that there is a 5-adic number a such that $a^2 + 1 = 0$ and

$$|a - 2|_5 \le \left|\tfrac{5}{4}\right|_5 = \tfrac{1}{5}. \qquad (4)$$

Similarly, there is a 5-adic number b such that $b^2 + 1 = 0$ and

$$|b - 3|_5 \le \left|\tfrac{10}{7}\right|_5 = \tfrac{1}{5}. \qquad (5)$$

Therefore, we have found both the square roots of -1 in $\mathbb{Q}_5$. It is possible to find the roots of any POLYNOMIAL using this technique.

See also P-ADIC NUMBER, VALUATION THEORY

References

Chevalley, C. C. "Hensel's Lemma." §3.2 in *Introduction to the Theory of Algebraic Functions of One Variable.* Providence, RI: Amer. Math. Soc., pp. 43–4, 1951.

Getz, J. "On Congruence Properties of the Partition Function." *Internat. J. Math. Math. Sci.* **23**, 493–96, 2000.

Koch, H. *Number Theory: Algebraic Numbers and Functions.* Providence, RI: Amer. Math. Soc., pp. 115–17, 2000.

Niven, I. M.; Zuckerman, H. S.; and Montgomery, H. L. *An Introduction to the Theory of Numbers, 5th ed.* New York: Wiley, 1991.

Henstock-Kurzweil Integral

HK INTEGRAL

Heptacontagon

A 70-sided POLYGON.

Heptadecagon

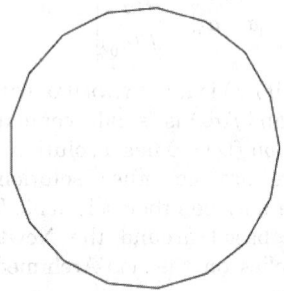

The REGULAR POLYGON of 17 sides is called the HEPTADECAGON, or sometimes the HEPTAKAIDECAGON. Gauss proved in 1796 (when he was 19 years old) that the heptadecagon is CONSTRUCTIBLE with a COMPASS and STRAIGHTEDGE. Gauss's proof appears in his monumental work *Disquisitiones Arithmeticae.* The proof relies on the property of irreducible POLYNOMIAL equations that ROOTS composed of a finite number of SQUARE ROOT extractions only exist when the order of the equation is a product OF THE FORM $2^a 3^b F_c \cdot F_d \cdots F_e$, where the F_n are distinct PRIMES OF THE FORM

$$F_n = 2^{2n} + 1,$$

known as FERMAT PRIMES. Constructions for the regular TRIANGLE (3^1), SQUARE (2^2), PENTAGON ($2^{2^1} + 1$), HEXAGON ($2^1 3^1$), etc., had been given by Euclid, but constructions based on the FERMAT PRIMES ≥ 17 were unknown to the ancients. The first explicit construction of a heptadecagon was given by Erchinger in

about 1800.

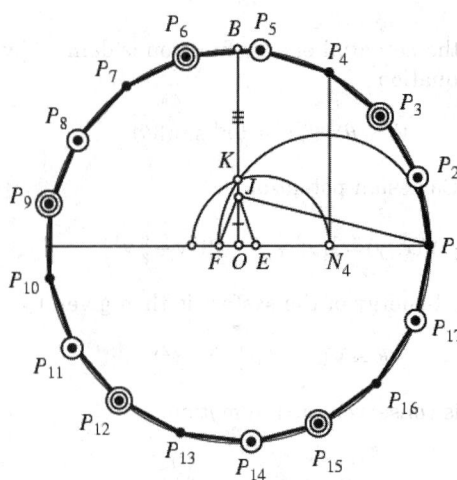

17-gon construction

The following elegant construction for the heptadecagon (Yates 1949, Coxeter 1969, Stewart 1977, Wells 1992) was first given by Richmond (1893).

1. Given an arbitrary point O, draw a CIRCLE centered on O and a DIAMETER drawn through O.
2. Call the right end of the DIAMETER dividing the CIRCLE into a SEMICIRCLE P_1.
3. Construct the DIAMETER PERPENDICULAR to the original DIAMETER by finding the PERPENDICULAR BISECTOR OB.
4. Construct J a QUARTER the way up OB.
5. Join JP_1 and find E so that $\angle OJE$ is a QUARTER of $\angle OJP_1$.
6. Find F so that $\angle EJF$ is $45°$.
7. Construct the SEMICIRCLE with DIAMETER FP_1.
8. This SEMICIRCLE cuts OB at K.
9. Draw a SEMICIRCLE with center E and RADIUS EK.
10. This cuts the extension of OP_1 at N_4.
11. Construct a line PERPENDICULAR to OP_1 through N_4.
12. This line meets the original SEMICIRCLE at P_4.
13. You now have points P_1 and P_4 of a heptadecagon.
14. Use P_1 and P_4 to get the remaining 15 points of the heptadecagon around the original CIRCLE by constructing P_1, P_4, P_7, P_{10}, P_{13}, P_{16} [filled circles], P_2, P_5, P_8, P_{11}, P_{14}, P_{17} [single-ringed filled circles], P_3, P_6, P_9, P_{12}, and P_{15} [double-ringed filled circles].
15. Connect the adjacent points P_i for $i = 1$ to 17, forming the heptadecagon.

This construction, when suitably streamlined, has SIMPLICITY 53. The construction of Smith (1920) has a greater SIMPLICITY of 58. Another construction due to

Tietze (1965) and reproduced in Hall (1970) has a SIMPLICITY of 50. However, neither Tietze (1965) nor Hall (1970) provides a proof that this construction is correct. Both Richmond's and Tietze's constructions require extensive calculations to prove their validity. De Temple (1991) gives an elegant construction involving the CARLYLE CIRCLES which has GEOMETRO-GRAPHY symbol $8S_1 + 4S_2 + 22C_1 + 11C_3$ and SIMPLICITY 45. The construction problem has now been automated to some extent (Bishop 1978).

See also 257-GON, 65537-GON, COMPASS, CONSTRUCTIBLE POLYGON, FERMAT NUMBER, FERMAT PRIME, REGULAR POLYGON, STRAIGHTEDGE, TRIGONOMETRY VALUES PI/17

References

Archibald, R. C. "The History of the Construction of the Regular Polygon of Seventeen Sides." *Bull. Amer. Math. Soc.* **22**, 239–46, 1916.

Archibald, R. C. "Gauss and the Regular Polygon of Seventeen Sides." *Amer. Math. Monthly* **27**, 323–26, 1920.

Ball, W. W. R. and Coxeter, H. S. M. *Mathematical Recreations and Essays, 13th ed.* New York: Dover, pp. 95–6, 1987.

Bishop, W. "How to Construct a Regular Polygon." *Amer. Math. Monthly* **85**, 186–88, 1978.

Bold, B. *Famous Problems of Geometry and How to Solve Them.* New York: Dover, pp. 63–9, 1982.

Conway, J. H. and Guy, R. K. *The Book of Numbers.* New York: Springer-Verlag, pp. 201 and 229–30, 1996.

Coxeter, H. S. M. *Introduction to Geometry, 2nd ed.* New York: Wiley, pp. 26–8, 1969.

De Temple, D. W. "Carlyle Circles and the Lemoine Simplicity of Polygonal Constructions." *Amer. Math. Monthly* **98**, 97–08, 1991.

Dickson, L. E. "Construction of the Regular Polygon of 17 Sides." §8.20 in *Monographs on Topics of Modern Mathematics Relevant to the Elementary Field* (Ed. J. W. A. Young). New York: Dover, pp. 372–73, 1955.

Dixon, R. "Gauss Extends Euclid." §1.4 in *Mathographics.* New York: Dover, pp. 52–4, 1991.

Dummit, D. S. and Foote, R. M. *Abstract Algebra, 2nd ed.* Englewood Cliffs, NJ: Prentice-Hall, 1998.

Gauss, C. F. §365 and 366 in *Disquisitiones Arithmeticae.* Leipzig, Germany, 1801. New Haven, CT: Yale University Press, 1965.

Hall, T. *Carl Friedrich Gauss: A Biography.* Cambridge, MA: MIT Press, 1970.

Hardy, G. H. and Wright, E. M. "Construction of the Regular Polygon of 17 Sides." §5.8 in *An Introduction to the Theory of Numbers, 5th ed.* Oxford, England: Clarendon Press, pp. 57–2, 1979.

Klein, F. *Famous Problems of Elementary Geometry and Other Monographs.* New York: Chelsea, 1956.

Ore, Ø. *Number Theory and Its History.* New York: Dover, 1988.

Rademacher, H. *Lectures on Elementary Number Theory.* New York: Blaisdell, 1964.

Richmond, H. W. "A Construction for a Regular Polygon of Seventeen Sides." *Quart. J. Pure Appl. Math.* **26**, 206–07, 1893.

Smith, L. L. "A Construction of the Regular Polygon of Seventeen Sides." *Amer. Math. Monthly* **27**, 322–23, 1920.

Stewart, I. "Gauss." *Sci. Amer.* **237**, 122–31, 1977.

Tietze, H. *Famous Problems of Mathematics.* New York: Graylock Press, 1965.

Wells, D. *The Penguin Dictionary of Curious and Interesting Geometry.* New York: Penguin, pp. 212–13, 1991.

Yates, R. C. *Geometrical Tools.* St. Louis, MO: Educational Publishers, 1949.

Heptagon

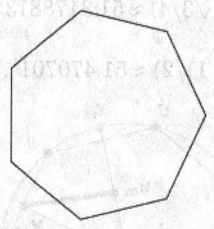

The regular seven-sided POLYGON, illustrated above, which has SCHLÄFLI SYMBOL {7}. According to Bankoff and Garfunkel (1973), "since the earliest days of recorded mathematics, the regular heptagon has been virtually relegated to limbo." Nevertheless, Thébault (1913) discovered many beautiful properties of the heptagon, some of which are discussed by Bankoff and Garfunkel (1973).

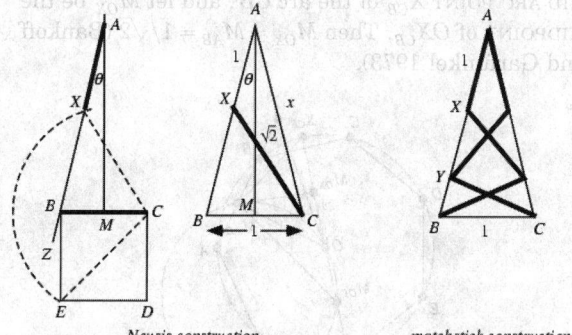

Neusis construction   *matchstick construction*

Although the regular heptagon is not a CONSTRUCTIBLE POLYGON using the classical rules of Greek GEOMETRIC CONSTRUCTION, it *is* constructible using a NEUSIS CONSTRUCTION (Johnson 1975; left figure above). To implement the construction, place a mark X on a ruler AZ, and then build a SQUARE of side length AX. Then construct the perpendicular bisector at M to BC, and draw an arc centered at C of radius CE. Now place the marked ruler so that it passes through B, X lies on the arc, and A falls on the perpendicular bisector. Then $2\theta = \angle BAC = \pi/7$, and two such triangles give the vertex angle $2\pi/7$ of a regular heptagon. Conway and Guy (1996) give a NEUSIS CONSTRUCTION for the heptagon. In addition, the regular heptagon can be constructed using seven identical toothpicks to form 1:3:3 triangles (Finlay 1959, Johnson 1975, Wells 1991; right figure above). Bankoff and Garfunkel (1973) discuss the heptagon, including a purported discovery of the NEUSIS CONSTRUCTION by Archimedes (Heath 1931). Madachy (1979) illustrates how to construct a heptagon by folding and knotting a strip of paper, and the regular heptagon can also be constructed using a CONCHOID OF NICOMEDES.

Although the regular heptagon not constructible using classical techniques, Dixon (1991) gives constructions for several angles very close to $360°/7$. While the ANGLE subtended by a side is $360°/7 \approx 51.428571°$, Dixon gives constructions containing angles of $2\sin^{-1}(\sqrt{3}/4) \approx 51.3178813°$, $\tan^{-1}(5/4) \approx 51.340192°$, and $30° + \sin^{-1}(\sqrt{3}-1)/2) \approx 51.470701°$.

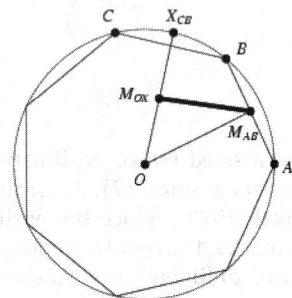

In the regular heptagon with unit CIRCUMRADIUS and center O, construct the MIDPOINT M_{AB} of AB and the MID-ARC POINT X_{CB} of the arc CB, and let M_{OX} be the MIDPOINT of OX_{CB}. Then $M_{OX} = M_{AB} = 1/\sqrt{2}$ (Bankoff and Garfunkel 1973).

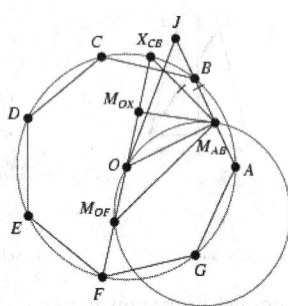

In the regular heptagon, construct the points X_{CB}, M_{AB}, and M_{OX} as above. Also construct the midpoint M_{OX} and construct J along the extension of $M_{AB}B$ such that $M_{AB}J = M_{AB}X_{CB}$. Note that the APOTHEM OM_{AB} of the heptagon has length $r = \cos(\pi/7)$. Then

1. The length $x = M_{AB}M_{OF}$ is equal to $\sqrt{2}r = \sqrt{2}\cos(\pi/7)$, and also to the largest root of

$$8x^6 - 20x^4 + 12x^2 - 1 = 0,$$

2. $M_{OJ} = \sqrt{6}/2$, and
3. $M_{AB}M_{OX}$ is tangent to the CIRCUMCIRCLE of $\Delta M_{OF}OM_{AB}$.

(Bankoff and Garfunkel 1973).

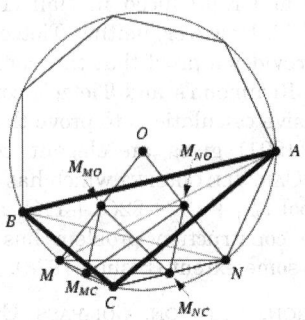

Construct a HEPTAGONAL TRIANGLE ΔABC in a regular heptagon with center O, and let BN and AM bisect $\angle ABC$ and $\angle BAC$, respectively, with M and N both lying on the circumcircle. Also define the midpoints M_{MO}, M_{NO}, M_{MC}, and M_{NC}. Then

$$MN = \tfrac{1}{2}M_{MO}M_{NO} = \tfrac{1}{2}M_{MC}M_{NC} \tag{1}$$

$$= \sqrt{2}M_{NO}M_{MC} \tag{2}$$

$$M_{MO}M_{MC} = M_{NO}M_{NC} = \tfrac{1}{2} \tag{3}$$

$$M_{MO}M_{NC} = \tfrac{1}{2}\sqrt{2} \tag{4}$$

(Bankoff and Garfunkel 1973).

See also CONCHOID OF NICOMEDES, EDMONDS' MAP, HEPTAGON THEOREM, HEPTAGONAL TRIANGLE, NEUSIS CONSTRUCTION, TRIGONOMETRY VALUES PI/7

References

Aaboe, A. *Episodes from the Early History of Mathematics.* Washington, DC: Math. Assoc. Amer., 1964.

Bankoff, L. and Garfunkel, J. "The Heptagonal Triangle." *Math. Mag.* **46**, 7–9, 1973.

Bold, B. *Famous Problems of Geometry and How to Solve Them.* New York: Dover, pp. 59–0, 1982.

Conway, J. H. and Guy, R. K. *The Book of Numbers.* New York: Springer-Verlag, pp. 194–00, 1996.

Courant, R. and Robbins, H. "The Regular Heptagon." §3.3.4 in *What is Mathematics?: An Elementary Approach to Ideas and Methods, 2nd ed.* Oxford, England: Oxford University Press, pp. 138–39, 1996.

Dixon, R. *Mathographics.* New York: Dover, pp. 35–0, 1991.

Finlay, A. H. "Zig-Zag Paths." *Math. Gaz.* **43**, 199, 1959.

Heath, T. L. *A Manual of Greek Mathematics.* Oxford, England: Clarendon Press, pp. 340–42, 1931.

Johnson, C. "A Construction for a Regular Heptagon." *Math. Gaz.* **59**, 17–1, 1975.

Madachy, J. S. *Madachy's Mathematical Recreations.* New York: Dover, pp. 59–1, 1979.

Bankoff, L. and Demir, H. "Solution to Problem E 1154." *Amer. Math. Monthly* **62**, 584–85, 1955. Wells, D. *The Penguin Dictionary of Curious and Interesting Geometry.* London: Penguin, p. 210, 1991.

Heptagon Theorem

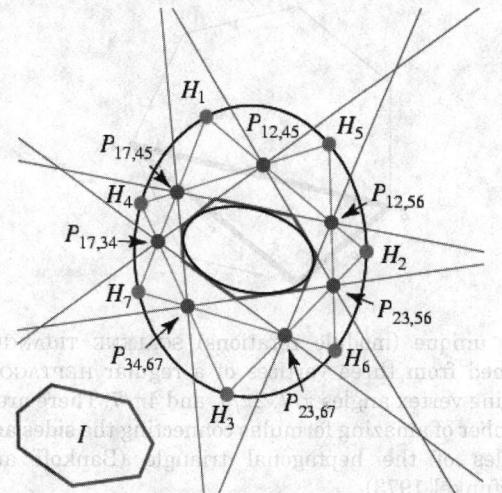

Let H be a heptagon with seven vertices given in cyclic order inscribed in a CONIC. Then the PASCAL LINES of the seven HEXAGONS obtained by omitting each vertex of H in turn and keeping the remaining vertices in the same cyclic order are the sides of a HEPTAGON I which circumscribes a CONIC. Moreover, the BRIANCHON POINTS of the seven HEXAGONS obtained by omitting the sides of I one at a time and keeping the remaining sides in the natural cyclic order are the vertices of the original HEPTAGON.

See also BRIANCHON POINT, CONIC SECTION, HEPTAGON, HEXAGON, PASCAL LINES

References

Evelyn, C. J. A.; Money-Coutts, G. B.; and Tyrrell, J. A. "The Heptagon Theorem." §2.1 in *The Seven Circles Theorem and Other New Theorems.* London: Stacey International, pp. 8–1, 1974.

Heptagonal Hexagonal Number

A number which is simultaneously a HEPTAGONAL NUMBER Hep_n and HEXAGONAL NUMBER Hex_m. Such numbers exist when

$$\tfrac{1}{2} n(5n-3) = m(2m-1). \qquad (1)$$

COMPLETING THE SQUARE and rearranging gives

$$(10n-3)^2 - 5(4m-1)^2 = 4. \qquad (2)$$

Substituting $x = 10n-3$ and $y = 4m-1$ gives the Pell-like quadratic Diophantine equation

$$x^2 - 5y^2 = 4, \qquad (3)$$

which has solutions $(x, y) = (3, 1), (7, 3), (18, 8), (47, 21), (123, 55), \dots$. The integer solutions in m and n are then given by $(n, m) = (1, 1), (221, 247), (71065, 79453), (22882613, 25583539), \dots$ (Sloane's A048902 and A048901), corresponding to the heptagonal hex-

agonal numbers 1, 121771, 12625478965, 1309034909945503, ... (Sloane's A048903).

See also HEPTAGONAL NUMBER, HEXAGONAL NUMBER

References

Sloane, N. J. A. Sequences A048901, A048902, and A048903 in "An On-Line Version of the Encyclopedia of Integer Sequences." http://www.research.att.com/~njas/sequences/eisonline.html.

Heptagonal Number

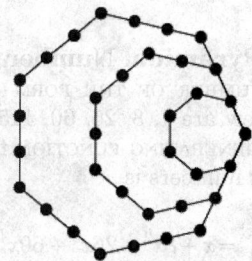

A FIGURATE NUMBER OF THE FORM $n(5n-3)/2$. The first few are 1, 7, 18, 34, 55, 81, 112, ... (Sloane's A000566). The GENERATING FUNCTION for the heptagonal numbers is

$$\frac{x(4x+1)}{(1-x)^3} = x + 7x^2 + 18x^3 + 34x^4 + \dots.$$

See also HEPTAGONAL HEXAGONAL NUMBER, HEPTAGONAL PENTAGONAL NUMBER, HEPTAGONAL SQUARE NUMBER, HEPTAGONAL TRIANGULAR NUMBER, OCTAGONAL HEPTAGONAL NUMBER

References

Sloane, N. J. A. Sequences A000566/M4358 in "An On-Line Version of the Encyclopedia of Integer Sequences." http://www.research.att.com/~njas/sequences/eisonline.html.

Heptagonal Pentagonal Number

A number which is simultaneously a HEPTAGONAL NUMBER H_n and PENTAGONAL NUMBER P_m. Such numbers exist when

$$\tfrac{1}{2} n(5n-3) = \tfrac{1}{2} m(3m-1). \qquad (1)$$

COMPLETING THE SQUARE and rearranging gives

$$3(10n-3)^2 - 5(6m-1)^2 = 22. \qquad (2)$$

Substituting $x = 10n-3$ and $y = 6m-1$ gives the Pell-like quadratic Diophantine equation

$$3x^2 - 5y^2 = 22, \qquad (3)$$

which has solutions $(x, y) = (3, 1), (7, 5), (17, 13), (53, 41), (133, 103), \dots$. The integer solutions in m and n are then given by $(n, m) = (1, 1), (42, 54), (2585, 3337), (160210, 206830), (9930417, 12820113) \dots$

(Sloane's A046198 and A046199), corresponding to the heptagonal pentagonal numbers 1, 4347, 16701685, 64167869935, 246532939589097, ... (Sloane's A048900).

See also HEPTAGONAL NUMBER, PENTAGONAL NUMBER

References

Sloane, N. J. A. Sequences A046198, A046199, and A048900 in "An On-Line Version of the Encyclopedia of Integer Sequences." http://www.research.att.com/~njas/sequences/eisonline.html.

Heptagonal Pyramidal Number

A PYRAMIDAL NUMBER OF THE FORM $n(n+1)(5n-2)/6$. The first few are 1, 8, 26, 60, 115, ... (Sloane's A002413). The GENERATING FUNCTION for the heptagonal pyramidal numbers is

$$\frac{x(4x+1)}{(x-1)^4} = x + 8x^2 + 26x^3 + 60x^4 + \ldots$$

References

Sloane, N. J. A. Sequences A002413/M4498 in "An On-Line Version of the Encyclopedia of Integer Sequences." http://www.research.att.com/~njas/sequences/eisonline.html.

Heptagonal Square Number

A number which is simultaneously a HEPTAGONAL NUMBER H_n and SQUARE NUMBER S_m. Such numbers exist when

$$\frac{1}{2} n(5n-3) = m^2. \tag{1}$$

COMPLETING THE SQUARE and rearranging gives

$$(10n-3)^2 - 40m^2 = 9. \tag{2}$$

Substituting $x = 10n-3$ and $y = 2m$ gives the Pell-like quadratic Diophantine equation

$$x^2 - 10y^2 = 9, \tag{3}$$

which has basic solutions $(x, y) = (7, 2)$, $(13, 4)$, and $(57, 18)$. Additional solutions can be obtained from the unit PELL EQUATION, and correspond to integer solutions when $(n, m) = (1, 1)$, $(6, 9)$, $(49, 77)$, $(961, 1519)$, ... (Sloane's A046195 and A046196), corresponding to the heptagonal square numbers 1, 81, 5929, 2307361, 168662169, 12328771225, ... (Sloane's A036354).

See also HEPTAGONAL NUMBER, SQUARE NUMBER

References

Sloane, N. J. A. Sequences A036354, A046195, and A046196 in "An On-Line Version of the Encyclopedia of Integer Sequences." http://www.research.att.com/~njas/sequences/eisonline.html.

Heptagonal Triangle

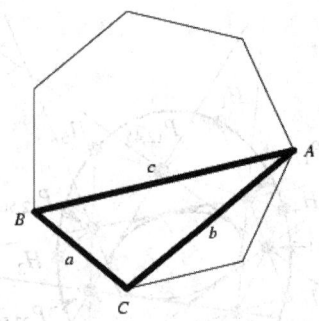

The unique (modulo rotations) SCALENE TRIANGLE formed from three vertices of a regular HEPTAGON, having vertex angles $\pi/7$, $2\pi/7$, and $4\pi/7$. There are a number of amazing formulas connecting the sides and angles of the heptagonal triangle (Bankoff and Garfunkel 1973).

The AREA of the TRIANGLE is

$$A = \tfrac{1}{4}\sqrt{7}R^2, \tag{1}$$

where R is the triangle's CIRCUMRADIUS. The sum of squares of sides of the heptagonal triangle is equal to $7R^2$ (Bankoff and Garfunkel 1973). The ratio $x = r/R$ of INRADIUS r to CIRCUMRADIUS R is given by the positive root of

$$8x^3 + 28x^2 + 14x - 7 = 0. \tag{2}$$

Also,

$$\frac{1}{a^2} + \frac{1}{b^2} + \frac{1}{c^2} = \frac{2}{R^2}. \tag{3}$$

The BROCARD ANGLE Ω satisfies

$$\cot \Omega = \sqrt{7}, \tag{4}$$

and the EXRADIUS r_a is equal to the radius of the NINE-POINT CIRCLE of $\triangle ABC$.

a is half the HARMONIC MEAN of the other two sides,

$$a = \frac{bc}{b+c} \tag{5}$$

$$b^2 - a^2 = ac, \tag{6}$$

and so on for all permutations of variables (Bankoff and Garfunkel 1973). Also,

$$\frac{b^2}{a^2} + \frac{c^2}{b^2} + \frac{a^2}{c^2} = 5. \tag{7}$$

If h_a, h_b, and h_c are the altitudes, then

$$h_a = h_b + h_c \tag{8}$$

$$h_a^2 + h_b^2 + h_c^2 = \tfrac{1}{2}(a^2 + b^2 + c^2). \tag{9}$$

If A', B', and C' are the feet of the altitudes, then

$$BA' \cdot A'C = \tfrac{1}{4} ac \qquad (10)$$

and so on (Bankoff and Garfunkel 1973). The internal angle bisectors of C and B are equal to the difference of the adjacent sides and the external angle bisector of A is equal to the sum of adjacent sides.

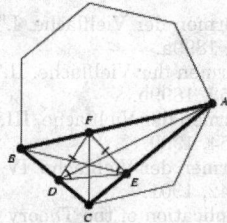

The triangle ΔDEF joining the feet of the angle bisectors of the heptagonal triangle is an ISOSCELES TRIANGLE with $DF = EF$.

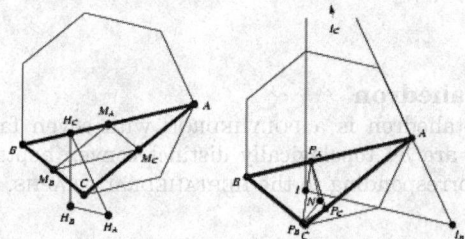

The ORTHIC TRIANGLE $\Delta H_A H_B H_C$ and MEDIAN TRIANGLE $M_A M_B M_C$ are congruent and perspective. In addition both are similar to ΔABC, to the PEDAL TRIANGLE $\Delta P_A P_B P_C$ of ΔABC with respect to the NINE-POINT CENTER N, and to the triangle $\Delta II_B I_C$ formed by the INCENTER I and the exterior angle bisectors I_B and I_C (Bankoff and Garfunkel 1973).

There are also a slew of curious trigonometric identities involving the angles of the heptagonal triangle:

$$\sin A \sin B \sin C = \tfrac{1}{8} \sqrt{7} \qquad (11)$$

$$\sin^2 A + \sin^2 B + \sin^2 C = \tfrac{7}{4} \qquad (12)$$

$$\sin(2A) + \sin(2B) + \sin(2C) = \tfrac{1}{2} \sqrt{7} \qquad (13)$$

$$\sin^2 A \sin^2 B \sin^2 C = \tfrac{7}{64} \qquad (14)$$

$$\sin^2 A \sin^2 B + \sin^2 A \sin^2 C + \sin^2 B \sin^2 C = \tfrac{7}{8} \qquad (15)$$

$$\cos A \cos B \cos C = -\tfrac{1}{8} \qquad (16)$$

$$\cos^2 A + \cos^2 B + \cos^2 C = \tfrac{5}{4} \qquad (17)$$

$$\cos^2 A \cos^2 B + \cos^2 A \cos^2 C + \cos^2 B \cos^2 C = \tfrac{3}{8} \qquad (18)$$

$$\cos(2A) + \cos(2B) + \cos(2C) = -\tfrac{1}{2} \qquad (19)$$

$$\sin A + \sin B + \sin C = \tfrac{1}{2}\sqrt{14} \qquad (20)$$

$$\tan A \tan B \tan C = -\sqrt{7} \qquad (21)$$

$$\cot A + \cot B + \cot C = \sqrt{7} \qquad (22)$$

$$\csc^2 A + \csc^2 B + \csc^2 C = 8 \qquad (23)$$

$$\sec^2 A + \sec^2 B + \sec^2 C = 24 \qquad (24)$$

$$\cot^2 A + \cot^2 B + \cot^2 C = 5 \qquad (25)$$

$$\tan^2 A + \tan^2 B + \tan^2 C = 21 \qquad (26)$$

$$\sec^4 A + \sec^4 B + \sec^4 C = 416 \qquad (27)$$

$$\cos^4 A + \cos^4 B + \cos^4 C = \tfrac{13}{16} \qquad (28)$$

$$\sin^4 A + \sin^4 B + \sin^4 C = \tfrac{21}{16} \qquad (29)$$

$$\csc^4 A + \csc^4 B + \csc^4 C = 32 \qquad (30)$$

$$\sec(2A) + \sec(2B) + \sec(2C) = -4 \qquad (31)$$

(Bankoff and Garfunkel 1973).

Finally, the heptagonal triangle satisfies the miscellaneous properties:

1. The first BROCARD POINT corresponds to the NINE-POINT CENTER and the second BROCARD POINT lies on the NINE-POINT CIRCLE.
2. $OH = R\sqrt{2}$, where O is the CIRCUMCENTER, H is the ORTHOCENTER, and R is the CIRCUMRADIUS.
3. $IH = (R^2 + 4r^2)/2$, where I is the INCENTER and r is the INRADIUS.
4. The two tangents from the ORTHOCENTER H to the CIRCUMCIRCLE of the heptagonal triangle are mutually perpendicular.
5. The center of the CIRCUMCIRCLE of the TANGENTIAL TRIANGLE corresponds with the symmetric point of O with respect to H.
6. The ALTITUDE from B is half the length of the internal bisector of the angle A.

See also HEPTAGON

References

Bankoff, L. and Garfunkel, J. "The Heptagonal Triangle." *Math. Mag.* **46**, 7–9, 1973.

Heptagonal Triangular Number

A number which is simultaneously a HEPTAGONAL NUMBER H_n and TRIANGULAR NUMBER T_m. Such numbers exist when

$$\tfrac{1}{2} n(5n - 3) = \tfrac{1}{2} m(m + 1). \qquad (1)$$

COMPLETING THE SQUARE and rearranging gives

$$(10n - 3)^2 - 5(2m + 1)^2 = 4. \qquad (2)$$

Substituting $x = 10n - 3$ and $y = 2m + 1$ gives the Pell-like quadratic Diophantine equation

$$x^2 - 5y^2 = 4, \qquad (3)$$

which has basic solutions $(x, y) = (3, 1)$, $(7, 3)$, and $(18, 8)$. Additional solutions can be obtained from the unit PELL EQUATION, and correspond to integer solutions when $(n, m) = (1, 1)$, $(5, 10)$, $(221, 493)$, $(1513, 3382)$, ... (Sloane's A046193 and A039835), corresponding to the heptagonal triangular numbers 1, 55, 121771, 5720653, 12625478965, ... (Sloane's A046194).

See also HEPTAGONAL NUMBER, TRIANGULAR NUMBER

References

Sloane, N. J. A. Sequences A039835, A046193, and A046194 in "An On-Line Version of the Encyclopedia of Integer Sequences." http://www.research.att.com/~njas/sequences/eisonline.html.

Heptagram

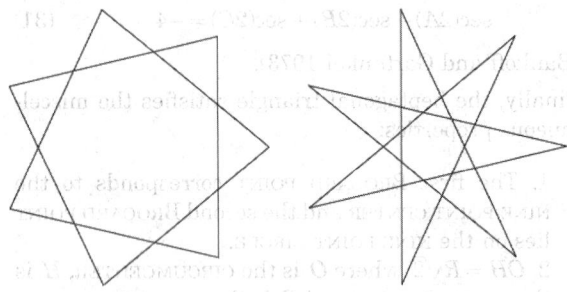

One of the two 7-sided STAR POLYGONS {7/2} and {7/3}, illustrated above.

See also HEPTAGON, STAR POLYGON

References

Steinhaus, H. *Mathematical Snapshots, 3rd ed.* New York: Dover, p. 211, 1999.

Heptahedral Graph

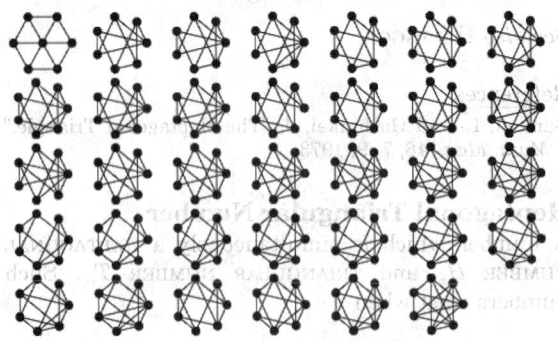

A POLYHEDRAL GRAPH on seven nodes. There are 34 nonisomorphic heptahedral graphs, as first enumerated by Kirkman (1862) and Hermes (1899ab, 1900, 1901; Federico 1969; Duijvestijn and Federico 1981).

See also HEPTAHEDRON, POLYHEDRAL GRAPH

References

Duijvestijn, A. J. W. and Federico, P. J. "The Number of Polyhedral (3-Connected Planar) Graphs." *Math. Comput.* **37**, 523–32, 1981.
Federico, P. J. "Enumeration of Polyhedra: The Number of 9-Hedra." *J. Combin. Th.* **7**, 155–61, 1969.
Grünbaum, B. *Convex Polytopes.* New York: Wiley, pp. 288 and 424, 1967.
Hermes, O. "Die Formen der Vielflache. I." *J. reine angew. Math.* **120**, 27–9, 1899a.
Hermes, O. "Die Formen der Vielflache. II." *J. reine angew. Math.* **120**, 305–53, 1899b.
Hermes, O. "Die Formen der Vielflache. III." *J. reine angew. Math.* **122**, 124–54, 1900.
Hermes, O. "Die Formen der Vielflache. IV." *J. reine angew. Math.* **123**, 312–42, 1901.
Kirkman, T. P. "Application of the Theory of the Polyhedra to the Enumeration and Registration of Results." *Proc. Roy. Soc. London* **12**, 341–80, 1862–863.
Pegg, E. Jr. "The 34 Convex Heptahedra and Their Characteristic Polynomials." http://www.mathpuzzle.com/charpoly.htm.

Heptahedron

A heptahedron is a POLYHEDRON with seven faces. There are 34 topologically distinct convex heptahedra, corresponding to the HEPTAHEDRAL GRAPHS.

The "regular" heptahedron is a one-sided surface made from four TRIANGLES and three QUADRILATERALS. It is topologically equivalent to the ROMAN SURFACE (Wells 1991). While all of the faces are regular and vertices equivalent, the heptahedron is self-intersecting and is therefore not considered an ARCHIMEDEAN SOLID.

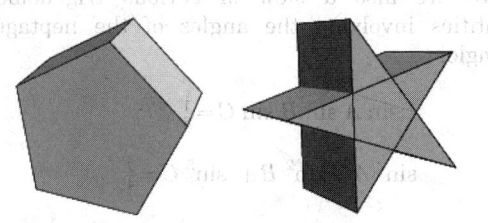

There are three semiregular heptahedra: the PENTAGONAL PRISM and PENTAGRAMMIC PRISM (illustrated above), and a FACETED version of the OCTAHEDRON (Holden 1991).

See also ARCHIMEDEAN SOLID, HEPTAHEDRAL GRAPH, OCTAHEDRON, POLYHEDRON, QUADRILATERAL, ROMAN SURFACE, SZILASSI POLYHEDRON

References

Holden, A. *Shapes, Space, and Symmetry.* New York: Dover, p. 95, 1991.
Wells, D. *The Penguin Dictionary of Curious and Interesting Geometry.* New York: Viking Penguin, p. 98, 1992.

Heptakaidecagon

HEPTADECAGON

Heptaparallelohedron

CUBOCTAHEDRON

Heptiamond

One of the 24 7-polyiamonds.

See also HEPTIAMOND TILING, POLYIAMOND

References

Gardner, M. *The Sixth Book of Mathematical Games from Scientific American.* Chicago, IL: University of Chicago Press, pp. 246, 248, and 250–51, 1984.

Heptiamond Tiling

See also HEPTIAMOND, HEXIAMOND TILING, OCTIAMOND TILING, PENTIAMOND TILING

References

Vichera, M. "Polyiamonds." http://alpha.ujep.cz/~vicher/puzzle/polyform/iamond/iamonds.htm.

Heptic Surface

An ALGEBRAIC SURFACE of degree 7.

See also ALGEBRAIC SURFACE

Heptomino

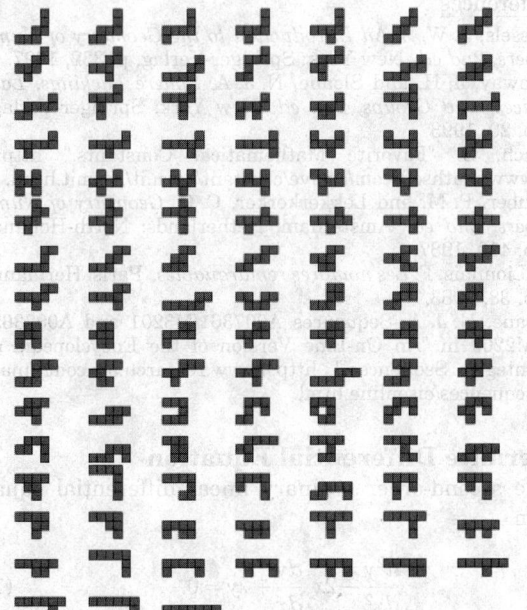

The heptominoes are the 7-POLYOMINOES. There are

108 FREE, 760 FIXED, and 196 one-sided heptominoes.

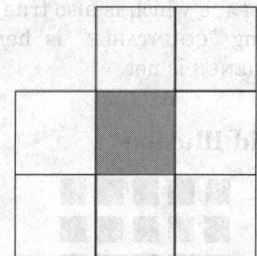

There is a single heptomino containing a hole (illustrated above), making heptominoes the smallest polyominoes for which the existence of a hole is possible.

See also DOMINO, HERSCHEL, HEXOMINO, OCTOMINO, PENTOMINO, PI HEPTOMINO, POLYOMINO, TETROMINO, TRIOMINO

Herbrand Function

References

Koch, H. *Number Theory: Algebraic Numbers and Functions.* Providence, RI: Amer. Math. Soc., p. 190, 2000.

Herbrand's Theorem

Let an ideal class be in $\mathscr{A}$ if it contains an IDEAL whose lth power is PRINCIPAL. Let i be an ODD INTEGER $1 \leq i \leq l$ and define j by $i + j = 1$. Then $\mathscr{A}_1 = \langle e \rangle$. If $i \geq 3$ and $l \nmid B_j$, then $\mathscr{A}_i = \langle e \rangle$.

See also IDEAL

References

Ireland, K. and Rosen, M. "Herbrand's Theorem." §15.3 in *A Classical Introduction to Modern Number Theory, 2nd ed.* New York: Springer-Verlag, pp. 241–48, 1990.

Hereditary Representation

The representation of a number as a sum of powers of a BASE b, followed by expression of each of the exponents as a sum of powers of b, etc., until the process stops. For example, the hereditary representation of 266 in base 2 is

$$266 = 2^8 + 2^3 + 2$$

$$= 2^{2^{2+1}} + 2^{2+1} + 2.$$

See also GOODSTEIN SEQUENCE, GOODSTEIN'S THEOREM

References

Henle, J. M. *An Outline of Set Theory.* New York: Springer-Verlag, 1986.

Heredity

A property of a SPACE which is also true of each of its SUBSPACES. Being "COUNTABLE" is hereditary, but having a given GENUS is not.

Hermann Grid Illusion

A regular 2-D arrangement of squares separated by vertical and horizontal "canals." Looking at the grid produces the illusion of gray spots in the white AREA between square VERTICES. The illusion was noted by Hermann (1870) while reading a book on sound by J. Tyndall.

References

Fineman, M. *The Nature of Visual Illusion*. New York: Dover, pp. 139–40, 1996.

Hermann's Formula

The MACHIN-LIKE FORMULA

$$\tfrac{1}{4}\pi = 2\tan^{-1}(\tfrac{1}{2}) - \tan^{-1}(\tfrac{1}{7}).$$

The other 2-term MACHIN-LIKE FORMULAS are EULER'S MACHIN-LIKE FORMULA, HUTTON'S FORMULA, and MACHIN'S FORMULA.

Hermann-Hering Illusion

The illusion in view by staring at the small black dot for a half minute or so, then switching to the white dot. The black squares appear stationary when staring at the white dot, but a fainter grid of moving squares also appears to be present.

Hermann-Mauguin Symbol

A symbol used to represent the POINT and SPACE GROUPS (e.g., $2/m\bar{3}$). Some symbols have abbreviated form. The equivalence between Hermann-Mauguin symbols (a.k.a. "crystallographic symbols"rpar; and SCHÖNFLIES SYMBOLS for the POINT GROUPS is given by Cotton (1990).

See also POINT GROUPS, SCHÖNFLIES SYMBOLS, SPACE GROUPS

References

Cotton, F. A. *Chemical Applications of Group Theory, 3rd ed.* New York: Wiley, p. 379, 1990.

Hermit Point

ISOLATED POINT

Hermite Constants

N.B. A detailed online essay by S. Finch was the starting point for this entry.

The Hermite constant is defined for DIMENSION n as the value

$$\gamma_n = \frac{\sup_f \min_{x_i} f(x_1, x_2, \ldots, x_n)}{[\text{discriminant}(f)]^{1/n}}$$

(Le Lionnais 1983). In other words, they are given by

$$\gamma_n = 4\left(\frac{\delta_n}{V_n}\right)^{2/n},$$

where δ_n is the maximum *lattice* PACKING DENSITY for HYPERSPHERE PACKING and V_n is the CONTENT of the n-HYPERSPHERE. The first few values of $(\gamma_n)^n$ are 1, 4/3, 2, 4, 8, 64/3, 64, 256, ... (Sloane's A007361 and A007362). Values for larger n are not known.

For sufficiently large n,

$$\frac{1}{2\pi e} \le \frac{\gamma_n}{n} \le \frac{1.744\ldots}{2\pi e}.$$

See also DISCRIMINANT, HYPERSPHERE PACKING, KISSING NUMBER, SPHERE PACKING

References

Cassels, J. W. S. *An Introduction to the Geometry of Numbers, 2nd ed.* New York: Springer-Verlag, p. 332, 1997.
Conway, J. H. and Sloane, N. J. A. *Sphere Packings, Lattices, and Groups, 2nd ed.* New York: Springer-Verlag, p. 20, 1993.
Finch, S. "Favorite Mathematical Constants." http://www.mathsoft.com/asolve/constant/hermit/hermit.html.
Gruber, P. M. and Lekkerkerker, C. G. *Geometry of Numbers, 2nd ed.* Amsterdam, Netherlands: North-Holland, p. 410, 1987.
Le Lionnais, F. *Les nombres remarquables.* Paris: Hermann, p. 38, 1983.
Sloane, N. J. A. Sequences A007361/M3201 and A007362/M2209 in "An On-Line Version of the Encyclopedia of Integer Sequences." http://www.research.att.com/~njas/sequences/eisonline.html.

Hermite Differential Equation

The second-order ordinary linear differential equation

$$\frac{d^2y}{dx^2} - 2x\,\frac{dy}{dx} + \lambda y = 0. \tag{1}$$

This differential equation has an irregular singularity at ∞. It can be solved using the series method

$$\sum_{n=0}^{\infty}(n+2)(n+1)a_{n+2}x^n - \sum_{n=1}^{\infty} 2na_nx^n + \sum_{n=0}^{\infty} \lambda a_nx^n$$

$$= 0 \qquad (2)$$

$$(2a_2 + \lambda a_0) + \sum_{n=1}^{\infty}[(n+2)(n+1)]a_{n+2} - 2na_n + \lambda a_n]x^n$$

$$= 0. \qquad (3)$$

Therefore,

$$a_2 = -\frac{\lambda a_0}{2} \qquad (4)$$

and

$$a_{n+2} = \frac{2n-\lambda}{(n+2)(n+1)} a_n \qquad (5)$$

for $n = 1, 2, \ldots$. Since (4) is just a special case of (5),

$$a_{n+2} = \frac{2n-\lambda}{(n+2)(n+1)} a_n \qquad (6)$$

for $n = 0, 1, \ldots$.
The linearly independent solutions are then

$$y_1$$

$$= a_0\left[1 - \frac{\lambda}{2!} x^2 - \frac{(4-\lambda)\lambda}{4!} x^4 - \frac{(8-\lambda)(4-\lambda)\lambda}{6!} x^6 - \cdots\right]$$

$$\qquad (7)$$

$$y_2 = a_1\left[x + \frac{(2-\lambda)}{3!} x^3 + \frac{(6-\lambda)(2-\lambda)}{5!} x^5 + \cdots\right]. \qquad (8)$$

These can be done in closed form as

$$y = a_0 \, {}_1F_1(-\tfrac{1}{4}\lambda; \tfrac{1}{2}; x^2) + a_1 x \, {}_1F_1(-\tfrac{1}{4}(\lambda-2); \tfrac{3}{2}; x^2) \qquad (9)$$

$$= a_0 \, {}_1F_1(-\tfrac{1}{4}\lambda; \tfrac{1}{2}; x^2) + a_2 H_{\lambda/2}(x), \qquad (10)$$

where ${}_1F_1(a; b; x)$ is a CONFLUENT HYPERGEOMETRIC FUNCTION OF THE FIRST KIND and $H_n(x)$ is a HERMITE POLYNOMIAL. In particular, for $\lambda = 0, 2, 4, \ldots$, the solutions can be written

$$y_{\lambda=0} = a_0 + \tfrac{1}{2}\sqrt{\pi} a_1 \, \mathrm{erfi}(x) \qquad (11)$$

$$y_{\lambda=2} = a_0\left[e^{x^2} - \sqrt{\pi} x \, \mathrm{erfi}(x)\right] + xa_1 \qquad (12)$$

$$y_{\lambda=4} = \tfrac{1}{4}\{2e^{x^2}xa_1 - (2x^2-1)[4a_0 + \sqrt{\pi} a_1 \, \mathrm{erfi}(x)]\}, \qquad (13)$$

where $\mathrm{erfi}(x)$ is the ERFI function.
If $\lambda = 0$, then Hermite's differential equation becomes

$$y'' - 2xy' = 0, \qquad (14)$$

which is OF THE FORM $P_2(x)y'' + P_1(x)y' = 0$ and so has solution

$$y = c_1 \int \frac{dx}{\exp\left(\int \frac{P_1}{P_2} dx\right)} + c_2$$

$$= c_1 \int \frac{dx}{\exp \int (-2x) \, dx} + c_2$$

$$= c_1 \int \frac{dx}{e^{-x^2}} + c_2 = c_1 \, \mathrm{erfi}(x) + c_2. \qquad (15)$$

Hermite Interpolation
HERMITE'S INTERPOLATING POLYNOMIAL

Hermite Polynomial

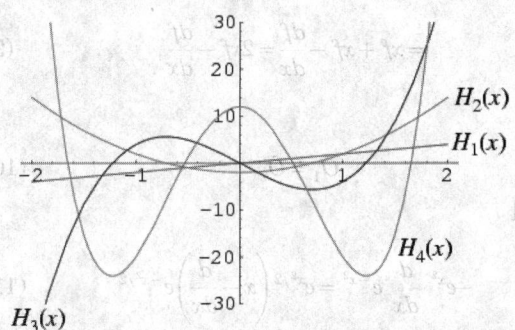

$H_2(x)$
$H_1(x)$
$H_4(x)$
$H_3(x)$

A set of ORTHOGONAL POLYNOMIALS $H_n(x)$, illustrated above for $x \in [0, 1]$ and $n = 1, 2, \ldots, 5$. Roman (1984, pp. 87–3) defines a generalized Hermite polynomial $H_n^{(v)}(x)$ of variance v.
The Hermite polynomials are a SHEFFER SEQUENCE with

$$g(t) = e^{t^2/4} \qquad (1)$$

$$f(t) = \tfrac{1}{2} t \qquad (2)$$

(Roman 1984, p. 30), giving the GENERATING FUNCTION

$$\exp(2xt - t^2) \equiv \sum_{n=0}^{\infty} \frac{H_n(x)t^n}{n!}. \qquad (3)$$

Using a TAYLOR SERIES shows that

$$H_n(x) = \left[\left(\frac{\partial}{\partial t}\right)^n \exp(2xt - t^2)\right]_{t=0}$$

$$= \left[e^{x^2}\left(\frac{\partial}{\partial t}\right)^n e^{-(x-t)^2}\right]_{t=0} \qquad (4)$$

Since $\partial f(x-t)/\partial t = -\partial f(x-t)/\partial x$,

$$H_n(x) = (-1)^n e^{x^2}\left[\left(\frac{\partial}{\partial x}\right)^n e^{-(x-t)^2}\right]_{t=0}$$

$$= (-1)^n e^{x^2} \frac{d^n}{dx^n} e^{-x^2}. \qquad (5)$$

Now define operators

$$\tilde{O}_1 \equiv -e^{x^2} \frac{d}{dx} e^{-x^2} \qquad (6)$$

$$\tilde{O}_2 \equiv e^{x^2/2} \left(x - \frac{d}{dx} \right) e^{-x^2/2}. \qquad (7)$$

It follows that

$$\tilde{O}_1 f = -e^{x^2} \frac{d}{dx} [fe^{-x^2}] = 2xf - \frac{df}{dx} \qquad (8)$$

$$\tilde{O}_2 f = e^{x^2/2} \left(x - \frac{d}{dx} \right) [fe^{-x^2/2}]$$

$$= xf + xf - \frac{df}{dx} = 2xf - \frac{df}{dx}, \qquad (9)$$

so

$$\tilde{O}_1 = \tilde{O}_2, \qquad (10)$$

and

$$-e^{x^2} \frac{d}{dx} e^{-x^2} = e^{x^2/2} \left(x - \frac{d}{dx} \right) e^{-x^2/2} \qquad (11)$$

(Arfken 1985, p. 720), which means the following definitions are equivalent:

$$\exp(2xt - t^2) \equiv \sum_{n=0}^{\infty} \frac{H_n(x) t^n}{n!} \qquad (12)$$

$$H_n(x) \equiv (-1)^n e^{x^2} \frac{d^n}{dx^n} e^{-x^2} \qquad (13)$$

$$H_n(x) \equiv e^{x^2/2} \left(x - \frac{d}{dx} \right)^n e^{-x^2/2} \qquad (14)$$

(Arfken 1985, pp. 712–13 and 720).
The Hermite polynomials may be written as

$$H_n(x) = (2x)^n \, {}_2F_0(-n/2, \, -(n-1)/2;; \, -1/x^2) \qquad (15)$$

(Koekoek and Swarttouw 1998), or

$$H_n(x) = 2^n U(-\tfrac{1}{2}n, \, \tfrac{1}{2}, \, x^2), \qquad (16)$$

where $U(a, b, x)$ is a CONFLUENT HYPERGEOMETRIC FUNCTION OF THE SECOND KIND. The Hermite polynomials are related to the derivative of the ERROR FUNCTION by

$$H_n(z) = (-1)^2 \frac{\sqrt{\pi}}{2} e^{z^2} \frac{d^{n+1}}{dz^{n+1}} \operatorname{erf}(z). \qquad (17)$$

They have a CONTOUR INTEGRAL representation

$$H_n(x) = \frac{n!}{2\pi i} \int e^{-t^2 + 2tx} t^{-n-1} \, dt. \qquad (18)$$

They are orthogonal in the range $(-\infty, \, \infty)$ with respect to the WEIGHTING FUNCTION e^{-x^2}

$$\int_{-\infty}^{\infty} H_m(x) H_n(x) e^{-x^2} \, dx = \delta_{mn} 2^n n! \, \sqrt{\pi}. \qquad (19)$$

The first few POLYNOMIALS are

$H_0(x) = 1$

$H_1(x) = 2x$

$H_2(x) = 4x^2 - 2$

$H_3(x) = 8x^3 - 12x$

$H_4(x) = 16x^4 - 48x^2 + 12$

$H_5(x) = 32x^5 - 160x^3 + 120x$

$H_6(x) = 64x^6 - 480x^4 + 720x^2 - 120$

$H_7(x) = 128x^7 - 1344x^5 + 3360x^3 - 1680x$

$H_8(x) = 256x^8 - 3584x^6 + 13440x^4 - 13440x^2 + 1680$

$H_9(x) = 512x^9 - 9216x^7 + 48348x^5 - 80640x^3 + 30240x$

$H_{10}(x) = 1024x^{10} - 23040x^8 + 161280x^6 - 403200x^4$
$\qquad\qquad + 302400x^2 - 30240.$

The Hermite polynomials obey the orthogonality conditions

$$\int_{-\infty}^{\infty} u_n(x) \frac{du_m}{dx} \, dx = \begin{cases} a \, \sqrt{\frac{n+1}{2}} & m = n+1 \\ -a \, \sqrt{\frac{n}{2}} & m = n-1 \\ 0 & \text{otherwise} \end{cases} \qquad (20)$$

$$\int_{-\infty}^{\infty} u_m(x) u_n(x) \, dx = \delta_{mn} \qquad (21)$$

$$\int_{-\infty}^{\infty} u_m(x) x u_n(x) \, dx = \begin{cases} \frac{1}{a} \sqrt{\frac{n+1}{2}} & m = n+1 \\ \frac{1}{a} \sqrt{\frac{n}{2}} & m = n-1 \\ 0 & \text{otherwise} \end{cases} \qquad (22)$$

$$\int_{-\infty}^{\infty} u_m(x) x^2 u_n(x) \, dx$$

$$= \begin{cases} \frac{2n+1}{2a^2} & m = n \\ \frac{\sqrt{(n+1)(n+2)}}{2a^2} & m = n+2 \\ 0 & m \neq n \neq n \pm 2 \end{cases} \qquad (23)$$

$$\int_{-\infty}^{\infty} e^{-x^2} H_\alpha H_\beta H_\gamma \, dx$$

$$= \sqrt{\pi} \, \frac{2^s \alpha! \beta! \gamma!}{(s-\alpha)!(s-\beta)!(s-\gamma)!}, \tag{24}$$

if $\alpha + \beta + \gamma = 2s$ is EVEN and $s \geq \alpha$, $s \geq \beta$, and $s \geq \gamma$. Otherwise, the last integral is 0 (Szego 1975, p. 390).

They also satisfy the RECURRENCE RELATIONS

$$H_{n+1}(x) = 2x H_n(x) - 2n H_{n-1}(x) \tag{25}$$

$$H'_n(x) = 2n H_{n-1}(x). \tag{26}$$

By solving the HERMITE DIFFERENTIAL EQUATION, the series

$$H_{2k}(x) = (-1)^k 2^k (2k-1)!!$$

$$\times \left[1 + \sum_{j=1}^{k} \frac{(-4k)(-4k+4)\cdots(-4k+4j-4)}{(2j)!} x^{2j} \right] \tag{27}$$

$$H_{2k+1}(x) = (-1)^k 2^{k+1} (2k+1)!!$$

$$\times \left[x + \sum_{j=1}^{k} \frac{(-4k)(-4k+4)\cdots(-4k+4j-4)}{(2j+1)!} x^{2j+1} \right] \tag{28}$$

are obtained, where the products in the numerators are equal to

$$(-4k)(-4k+4)\cdots(-4k+4j-4) = 4^j(-k)_j, \tag{29}$$

with $(x)_n$ the POCHHAMMER SYMBOL.
The DISCRIMINANT is

$$D_n = 2^{3n(n-1)/2} \prod_{k=1}^{n} k^k \tag{30}$$

(Szego 1975, p. 143), a normalized form of the HYPERFACTORIAL, the first few values of which are 1, 32, 55296, 7247757312, 92771293593600000, ... (Sloane's A054374). The table of RESULTANTS is given by $\{0\}$, $\{-8, 0\}$, $\{0, -2048, 0\}$, $\{192, 16384, 28311552, 0\}$, ... (Sloane's A054373).

Two interesting identities involving $H_n(x+y)$ are given by

$$\sum_{k=0}^{n} \binom{n}{k} H_k(x) H_{n-k}(y) = 2^{n/2} H_n(2^{-1/2}(x+y)) \tag{31}$$

and

$$\sum_{k=0}^{n} \binom{n}{k} H_k(x)(2y)^{n-k} = H_n(x+y) \tag{32}$$

(G. Colomer).

A set of associated functions is defined by

$$u_n(x) \equiv \sqrt{\frac{a}{\pi^{1/2} n! 2^n}} \, H_n(ax) e^{-a^2 x^2/2}. \tag{33}$$

A class of generalized Hermite POLYNOMIALS $\gamma_n^m(x)$ satisfying

$$e^{mxt - t^m} = \sum_{n=0}^{\infty} \gamma_n^m(x) t^n \tag{34}$$

was studied by Subramanyan (1990). A class of related POLYNOMIALS defined by

$$h_{n,m} = \gamma_n^m \left(\frac{2x}{m} \right) \tag{35}$$

and with GENERATING FUNCTION

$$e^{2xt - t^m} = \sum_{n=0}^{\infty} h_{n,m}(x) t^n \tag{36}$$

was studied by Djordjevic (1996). They satisfy

$$H_n(x) = n! h_{n,2}(x). \tag{37}$$

A modified version of the HERMITE POLYNOMIAL is sometimes defined by

$$He_n(x) \equiv H_n \left(\frac{x}{\sqrt{2}} \right). \tag{38}$$

See also MEHLER'S HERMITE POLYNOMIAL FORMULA, WEBER FUNCTIONS

References

Abramowitz, M. and Stegun, C. A. (Eds.). "Orthogonal Polynomials." Ch. 22 in *Handbook of Mathematical Functions with Formulas, Graphs, and Mathematical Tables, 9th printing.* New York: Dover, pp. 771–02, 1972.

Andrews, G. E.; Askey, R.; and Roy, R. "Hermite Polynomials." §6.1 in *Special Functions.* Cambridge, England: Cambridge University Press, pp. 278–82, 1999.

Arfken, G. "Hermite Functions." §13.1 in *Mathematical Methods for Physicists, 3rd ed.* Orlando, FL: Academic Press, pp. 712–21, 1985.

Chebyshev, P. L. "Sur un développement des fonctions à une seule variable." *Bull. ph.-math., Acad. Imp. Sc. St. Pétersbourg* **1**, 193–00, 1859.

Chebyshev, P. L. *Oeuvres, Vol. 1.* New York: Chelsea, pp. 49–08, 1987.

Djordjevic, G. "On Some Properties of Generalized Hermite Polynomials." *Fib. Quart.* **34**, 2–, 1996.

Hermite, C. "Sur un nouveau développement en série de fonctions." *Compt. Rend. Acad. Sci. Paris* **58**, 93–00 and 266–73, 1864. Reprinted in Hermite, C. *Oeuvres complètes, Vol. 2.* Paris, pp. 293–08, 1908.

Hermite, C. *Oeuvres complètes, Tome III.* Paris: Hermann, p. 432, 1912.

Iyanaga, S. and Kawada, Y. (Eds.). "Hermite Polynomials." Appendix A, Table 20.IV in *Encyclopedic Dictionary of Mathematics.* Cambridge, MA: MIT Press, pp. 1479–480, 1980.

Jeffreys, H. and Jeffreys, B. S. "The Parabolic Cylinder, Hermite, and Hh Functions" §23.08 in *Methods of Mathematical Physics, 3rd ed.* Cambridge, England: Cambridge University Press, pp. 620–22, 1988.

Koekoek, R. and Swarttouw, R. F. "Hermite." §1.13 in *The Askey-Scheme of Hypergeometric Orthogonal Polynomials and its q-Analogue.* Delft, Netherlands: Technische Universiteit Delft, Faculty of Technical Mathematics and Informatics Report 98–7, pp. 50–1, 1998. ftp://www.twi.tudelft.nl/publications/tech-reports/1998/DUT-TWI-98-7.ps.gz.

Roman, S. "The Hermite Polynomials." §4.2.1 in *The Umbral Calculus.* New York: Academic Press, pp. 87–3, 1984.

Rota, G.-C.; Kahaner, D.; Odlyzko, A. "Hermite Polynomials." §10 in "On the Foundations of Combinatorial Theory. VIII: Finite Operator Calculus." *J. Math. Anal. Appl.* **42**, 684–60, 1973.

Sansone, G. "Expansions in Laguerre and Hermite Series." Ch. 4 in *Orthogonal Functions, rev. English ed.* New York: Dover, pp. 295–85, 1991.

Sloane, N. J. A. Sequences A054373 and A054374 in "An On-Line Version of the Encyclopedia of Integer Sequences." http://www.research.att.com/~njas/sequences/eisonline.html.

Spanier, J. and Oldham, K. B. "The Hermite Polynomials $H_n(x)$." Ch. 24 in *An Atlas of Functions.* Washington, DC: Hemisphere, pp. 217–23, 1987.

Subramanyan, P. R. "Springs of the Hermite Polynomials." *Fib. Quart.* **28**, 156–61, 1990.

Szego, G. *Orthogonal Polynomials, 4th ed.* Providence, RI: Amer. Math. Soc., 1975.

Hermite Quadrature

HERMITE-GAUSS QUADRATURE

Hermite's Interpolating Polynomial

Let $l(x)$ be an nth degree POLYNOMIAL with zeros at $x_1, ..., x_n$. Then the fundamental Hermite interpolating polynomials of the first and second kinds are defined by

$$h_v^{(1)}(x) = \left[1 - \frac{l''(x_v)}{l'(x_v)}\right][l_v(x)]^2 \qquad (1)$$

and

$$h_v^{(2)}(x) = (x - x_v)[l_v(x)]^2 \qquad (2)$$

for $v = 1, 2, .., .n$. These polynomials have the properties

$$h_v^{(1)}(x_\mu) = \delta_{v\mu} \qquad (3)$$

$$h_v^{(1)'}(x_\mu) = 0 \qquad (4)$$

$$h_v^{(2)}(x_\mu) = 0 \qquad (5)$$

$$h_v^{(2)'}(x_\mu) = \delta_{v\mu}. \qquad (6)$$

for $\mu, v = 1, 2, ..., n$. Now let $f_1, ..., f_n$ and $f_1', ..., f_n'$ be values. Then the expansion

$$W_n(x) = \sum_{v=1}^n f_v h_v^{(1)}(x) + \sum_{v=1}^n f_v' h_v^{(2)}(x) \qquad (7)$$

gives the unique Hermite interpolating fundamental

polynomial for which

$$W_n(x_v) = f_v \qquad (8)$$

$$W_n'(x_v) = f_v'. \qquad (9)$$

If $f_v' = 0$, these are called STEP POLYNOMIALS. The fundamental polynomials satisfy

$$h_1(x) + ... + h_n(x) = 1 \qquad (10)$$

and

$$\sum_{v=1}^n x_v h_v^{(1)}(x) + \sum_{v=1}^n h_v^{(2)}(x) = x. \qquad (11)$$

Also, if $d\alpha(x)$ is an arbitrary distribution on the interval $[a, b]$, then

$$\int_a^b h_v^{(1)}(x)\, d\alpha(x) = \lambda_v \qquad (12)$$

$$\int_a^b h_v^{(1)'}(x)\, d\alpha(x) = 0 \qquad (13)$$

$$\int_a^b x h_v^{(1)}(x)\, d\alpha(x) = 0 \qquad (14)$$

$$\int_a^b h_v^{(2)}(x)\, d\alpha(x) = 0 \qquad (15)$$

$$\int_a^b h_v^{(2)'}(x)\, d\alpha(x) = \lambda_v \qquad (16)$$

$$\int_a^b x h_v^{(2)'}(x)\, d\alpha(x) = \lambda_v x_v, \qquad (17)$$

where λ_v are CHRISTOFFEL NUMBERS.

See also CHRISTOFFEL NUMBER, LAGRANGE INTERPOLATING POLYNOMIAL

References

Hildebrand, F. B. *Introduction to Numerical Analysis.* New York: McGraw-Hill, pp. 314–19, 1956.

Szego, G. *Orthogonal Polynomials, 4th ed.* Providence, RI: Amer. Math. Soc., pp. 330–32, 1975.

Hermite's Theorem

E IS TRANSCENDENTAL.

See also E, TRANSCENDENTAL NUMBER.

Hermite-Gauss Quadrature

Also called HERMITE QUADRATURE. A GAUSSIAN QUADRATURE over the interval $(-\infty, \infty)$ with WEIGHTING FUNCTION $W(x) = e^{-x^2}$ (Abramowitz and Stegun 1972, p. 890). The ABSCISSAS for quadrature order n are given by the roots of the HERMITE POLYNOMIALS $H_n(x)$, which occur symmetrically about 0. The WEIGHTS are

$$w_i = -\frac{A_{n+1}\gamma_n}{A_n H_n'(x_i)H_{n+1}(x_i)} = \frac{A_n}{A_{n-1}}\frac{\gamma_{n-1}}{H_{n-1}(x_1)H_n'(x_i)}, \quad (1)$$

where A_n is the COEFFICIENT of x^n in $H_n(x)$. For HERMITE POLYNOMIALS,

$$A_n = 2^n, \quad (2)$$

so

$$\left|\frac{A_{n+1}}{A_n}\right| = 2. \quad (3)$$

Additionally,

$$\gamma_n = \sqrt{\pi}\, 2^n n!, \quad (4)$$

so

$$w_i = -\frac{2^{n+1}n!\,\sqrt{\pi}}{H_{n+1}(x_i)H_n'(x_i)}$$

$$= \frac{2^n(n-1)!\,\sqrt{\pi}}{H_{n-1}(x_i)H_n'(x_i)}. \quad (5)$$

Using the RECURRENCE RELATION

$$H_n'(x) = 2nH_{n-1}(x) = 2xH_n(x) - H_{n+1}(x) \quad (6)$$

yields

$$H_n'(x_i) = 2nH_{n-1}(x_i) = -H_{n+1}(x_i) \quad (7)$$

and gives

$$w_i = \frac{2^{n+1}n!\,\sqrt{\pi}}{[H_n'(x_i)]^2} = \frac{2^{n+1}n!\,\sqrt{\pi}}{[H_{n+1}(x_i)]^2}. \quad (8)$$

The error term is

$$E = \frac{n!\,\sqrt{\pi}}{2^n(2n)!}\, f^{(2n)}(\xi). \quad (9)$$

Beyer (1987) gives a table of ABSCISSAS and weights up to $n = 12$.

n	x_i	w_i
2	$\pm$ 0.707107	0.886227
3	0	1.18164
	$\pm$ 1.22474	0.295409
4	$\pm$ 0.524648	0.804914
	$\pm$ 1.65068	0.0813128
5	0	0.945309
	$\pm$ 0.958572	0.393619
	$\pm$ 2.02018	0.0199532

The ABSCISSAS and weights can be computed analytically for small n.

n	x_i	w_i
2	$\pm\frac{1}{2}\sqrt{2}$	$\frac{1}{2}\sqrt{\pi}$
3	0	$\frac{2}{3}\sqrt{\pi}$
	$\pm\frac{1}{2}\sqrt{6}$	$\frac{1}{6}\sqrt{\pi}$
4	$\pm\sqrt{\frac{3-\sqrt{6}}{2}}$	$\frac{\sqrt{\pi}}{4(3-\sqrt{6})}$
	$\pm\sqrt{\frac{3+\sqrt{6}}{2}}$	$\frac{\sqrt{\pi}}{4(3+\sqrt{6})}$

References

Abramowitz, M. and Stegun, C. A. (Eds.). *Handbook of Mathematical Functions with Formulas, Graphs, and Mathematical Tables, 9th printing.* New York: Dover, p. 890, 1972.

Beyer, W. H. *CRC Standard Mathematical Tables, 28th ed.* Boca Raton, FL: CRC Press, p. 464, 1987.

Hildebrand, F. B. *Introduction to Numerical Analysis.* New York: McGraw-Hill, pp. 327–30, 1956.

HermiteH

HERMITE POLYNOMIAL

Hermite-Lindemann Theorem

Let α_i and A_1 be ALGEBRAIC NUMBERS such that the A_is differ from zero and the α_is differ from each other. Then the expression

$$A_1 e^{\alpha_1} + A_2 e^{\alpha_2} + A_3 e^{\alpha_3} + \ldots$$

cannot equal zero. The theorem was proved by Hermite (1873) in the special case of the A_is and α_is RATIONAL INTEGERS, and subsequently proved for algebraic numbers by Lindemann (1882). The proof was subsequently simplified by Weierstrass (1885) and Gordan (1893).

See also ALGEBRAIC NUMBER, CONSTANT PROBLEM, FOUR EXPONENTIALS CONJECTURE, INTEGER RELATION, LINDEMANN-WEIERSTRASS THEOREM, SIX EXPONENTIALS THEOREM

References

Dörrie, H. "The Hermite-Lindemann Transcendence Theorem." §26 in *100 Great Problems of Elementary Mathematics: Their History and Solutions.* New York: Dover, pp. 128–37, 1965.

Hermite, C. "Sur la fonction exponentielle." *Comptes rendus* **77**, 18–4, 1873.

Gordan, P. "Transcendenz von e und π." *Math. Ann.* **43**, 222–24, 1893.

Lindemann, F. "Über die Ludolph'sche Zahl." *Sitzungber. Königl. Preuss. Akad. Wissensch. zu Berlin* No. 2, pp. 679–82, 1888.

Weber, H. *Lehrbuch der Algebra, Vols. I-II.* New York: Chelsea, 1902.

Weierstrass, K. "Zu Hrn. Lindemann's Abhandlung: 'Über die Ludolph'sche Zahl'." *Sitzungber. Königl. Preuss. Akad. Wissensch. zu Berlin* No. 2, pp. 1067–086, 1885.

Hermitian Conjugate

Adjoint

Hermitian Form

A combination of variables x and y given by

$$a x\bar{x} + b x\bar{y} + \bar{b}\bar{x}y + c y\bar{y},$$

where $\bar{b}$, $\bar{x}$ and $\bar{y}$ are COMPLEX CONJUGATES.

Hermitian Inner Product

A Hermitian inner product on a COMPLEX VECTOR SPACE V is a complex-valued BILINEAR FORM on V which is ANTILINEAR in the second slot, and is positive definite. That is, it satisfies the following properties, where $\bar{z}$ denotes the COMPLEX CONJUGATE of z.

1. $\langle u+v,\ w\rangle = \langle u,\ w\rangle + \langle v,\ w\rangle$
2. $\langle u,\ v+w\rangle = \langle u,\ v\rangle + \langle u,\ w\rangle$
3. $\langle \alpha u,\ v\rangle = \alpha\langle u,\ v\rangle$
4. $\langle u,\ \alpha v\rangle = \bar{\alpha}\langle u,\ v\rangle$
5. $\langle u,\ v\rangle = \overline{\langle v,\ u\rangle}$
6. $\langle u,\ u\rangle \geq 0$, with equality only if $u=0$

The basic example is the form

$$h(z,\ w) = \sum z_i\bar{w}_i \qquad (1)$$

on $\mathbb{C}^n$, where $z = (z_1,\ \ldots,\ z_n)$ and $w = (w_1,\ \ldots,\ w_n)$. Note that by writing $z_k = x_k + iy_k$, it is possible to consider $\mathbb{C}^n \sim \mathbb{R}^{2n}$, in which case $\Re[h]$ is the Euclidean INNER PRODUCT and $\Im[h]$ is a nondegenerate alternating BILINEAR FORM, i.e., a SYMPLECTIC FORM. Explicitly, in $\mathbb{C}^2$, the standard Hermitian form is expressed below.

$$h((z_{11},\ z_{12}),\ (z_{21},\ z_{22})) = x_{11}, x_{21} + x_{12}x_{22} + y_{11}y_{21}$$

$$+ y_{12}y_{22} + i(x_{21}y_{11} - x_{11}y_{21} + x_{22}y_{12} - x_{12}y_{22}). \qquad (2)$$

A generic Hermitian inner product has its REAL PART symmetric positive definite, and its IMAGINARY PART symplectic by properties 5 and 6. A matrix $H = (h_{ij})$ defines an antilinear form, satisfying 1–, by $\langle e_i,\ e_j\rangle = h_{ij}$ IFF H is a HERMITIAN MATRIX. It is positive definite (satisfying 6) when $\Re[H]$ is a POSITIVE DEFINITE MATRIX. In matrix form,

$$\langle v,\ w\rangle = v^{\mathrm{T}}H\bar{w} \qquad (3)$$

and the canonical Hermitian inner product is when H is the IDENTITY MATRIX.

See also COMPLEX NUMBER, HERMITIAN METRIC, INNER PRODUCT, POSITIVE DEFINITE QUADRATIC FORM, SYMPLECTIC FORM, UNITARY BASIS, UNITARY GROUP, UNITARY MATRIX, VECTOR SPACE

Hermitian Matrix

A SQUARE MATRIX is called Hermitian if it is SELF-ADJOINT. Therefore, a Hermitian matrix is defined as one for which

$$A = A^* \qquad (1)$$

where A^* denotes the ADJOINT MATRIX. For example,

$$A = \begin{bmatrix} 1 & 1+i & 2i \\ 1-i & 5 & -3 \\ -2i & -3 & 0 \end{bmatrix} \qquad (2)$$

is a Hermitian matrix.

An INTEGER or REAL MATRIX is Hermitian iff it is SYMMETRIC. A matrix m can be tested to see if it is Hermitian using the *Mathematica* function

```
HermitianQ[m_List?MatrixQ]  :=  (m  = = =
Conjugate@Transpose@m)
```

Hermitian matrices have REAL EIGENVALUES whose EIGENVECTORS form a UNITARY BASIS. For REAL MATRICES, Hermitian is the same as SYMMETRIC.

Any MATRIX C which is not Hermitian can be expressed as the sum a Hermitian matrix and a SKEW HERMITIAN MATRIX using

$$C = \tfrac{1}{2}(C + C^*) + \tfrac{1}{2}(C - C^*). \qquad (3)$$

Let U be a UNITARY MATRIX and A be a Hermitian matrix. Then the ADJOINT MATRIX of a SIMILARITY TRANSFORMATION is

$$(UAU^{-1}) = [(UA)(U^{-1})]^* = (U^{-1})^*(UA)^*$$

$$= (U^*)^*(A^*U^*) = UAU^* = UAU^{-1}. \qquad (4)$$

The specific matrix

$$H(x,\ y,\ z) = \begin{bmatrix} z & x+iy \\ x-iy & -z \end{bmatrix} = xP_1 + yP_2 + zP_3, \qquad (5)$$

where P_i are PAULI SPIN MATRICES, is sometimes called "the" Hermitian matrix.

See also ADJOINT MATRIX, HERMITIAN OPERATOR, NORMAL MATRIX, PAULI SPIN MATRICES, SKEW HERMITIAN MATRIX, SYMMETRIC MATRIX

References

Arfken, G. "Hermitian Matrices, Unitary Matrices." §4.5 in *Mathematical Methods for Physicists, 3rd ed.* Orlando, FL: Academic Press, pp. 209–17, 1985.
Ayres, F. Jr. *Theory and Problems of Matrices.* New York: Schaum, pp. 13 and 117–18, 1962.

Hermitian Metric

A Hermitian metric on a COMPLEX VECTOR BUNDLE assigns a HERMITIAN INNER PRODUCT to every FIBER. The basic example is the TRIVIAL BUNDLE $\pi: U \times \mathbb{C}^k \to U$, where U is an OPEN SET in $\mathbb{R}^n$. Then a positive definite HERMITIAN MATRIX H defines a

Hermitian metric by

$$\langle v,\ w\rangle = v^{\mathrm{T}}\mathrm{H}\bar{w},$$

where $\bar{w}$ is the COMPLEX CONJUGATE of w. By a PARTITION OF UNITY, any COMPLEX VECTOR BUNDLE has a Hermitian metric.

In the special case of a COMPLEX MANIFOLD, the complexified TANGENT BUNDLE $TM \otimes \mathbb{C}$ may have a Hermitian metric, in which case its REAL PART is a RIEMANNIAN METRIC and its IMAGINARY PART is a nondegenerate ALTERNATING MULTILINEAR FORM ω. When ω is CLOSED, i.e., in this case a SYMPLECTIC FORM, then ω is a KÄHLER FORM.

On a HOLOMORPHIC VECTOR BUNDLE with a Hermitian metric h, there is a unique connection compatible with h and the complex structure. Namely, it must be $\nabla = \partial + \bar{\partial}$, where $\partial s = h^{-1}\partial hs$ in a TRIVIALIZATION.

See also COMPLEX GEOMETRY, COMPLEX MANIFOLD, COMPLEX VECTOR BUNDLE, HOLOMORPHIC VECTOR BUNDLE, KÄHLER FORM, KÄHLER MANIFOLD, RIEMANNIAN METRIC, SYMPLECTIC FORM, UNITARY GROUP

Hermitian Operator

A Hermitian OPERATOR $\bar{L}$ is one which satisfies

$$\int_a^b \bar{v}\bar{L}u\,dx = \int_a^b u\bar{L}\bar{v}\,dx. \tag{1}$$

where $\bar{z}$ denotes a COMPLEX CONJUGATE. As shown in STURM-LIOUVILLE THEORY, if $\bar{L}$ is SELF-ADJOINT and satisfies the boundary conditions

$$\bar{v}pu'|_{x=a} = \bar{v}pu'|_{x=b}, \tag{2}$$

then it is automatically Hermitian. Hermitian operators have REAL EIGENVALUES, ORTHOGONAL EIGENFUNCTIONS, and the corresponding EIGENFUNCTIONS form a COMPLETE set when $\bar{L}$ is second-order and linear.

In order to prove that EIGENVALUES must be REAL and EIGENFUNCTIONS ORTHOGONAL, consider

$$\bar{L}u_i + \lambda_i w u_i = 0. \tag{3}$$

Assume there is a second EIGENVALUE λ_j such that

$$\bar{L}u_j + \lambda_j w u_j = 0 \tag{4}$$

$$\bar{L}\bar{u}_j + \bar{\lambda}_j w \bar{u}_j = 0. \tag{5}$$

Now multiply (3) by $\bar{u}_j$ and (5) by u_i

$$\bar{u}_j \bar{L}u_i + \bar{u}_j \lambda w u_i = 0 \tag{6}$$

$$u_i \bar{L}\bar{u}_j + u_i \bar{\lambda}_j w \bar{u}_j = 0 \tag{7}$$

$$\bar{u}_j \bar{L}u_i - u_i \bar{L}\bar{u}_j = (\bar{\lambda}_j - \lambda_i) w u_i \bar{u}_j. \tag{8}$$

Now integrate

$$\int_a^b \bar{u}_j \bar{L}u_i - \int_a^b u_i \bar{L}\bar{u}_j = (\bar{\lambda}_j - \lambda_i)\int_a^b w u_i \bar{u}_j. \tag{9}$$

But because $\bar{L}$ is Hermitian, the left side vanishes.

$$(\bar{\lambda}_j - \lambda_i)\int_a^b w u_i \bar{u}_j = 0. \tag{10}$$

If EIGENVALUES λ_i and λ_j are not degenerate, then $\int_a^b w u_i \bar{u}_j = 0$, so the EIGENFUNCTIONS are ORTHOGONAL. If the EIGENVALUES are degenerate, the EIGENFUNCTIONS are not necessarily orthogonal. Now take $i = j$.

$$(\bar{\lambda}_i - \lambda_i)\int_a^b w u_i \bar{u}_i = 0. \tag{11}$$

The integral cannot vanish unless $u_i = 0$, so we have $\bar{\lambda}_i = \lambda_i$ and the EIGENVALUES are real.

For a Hermitian operator $\tilde{O}$,

$$\langle \phi|\tilde{O}\psi\rangle = \overline{\langle \phi|\tilde{O}\psi\rangle} = \langle \tilde{O}\phi|\psi\rangle. \tag{12}$$

In integral notation,

$$\int \overline{\tilde{A}\phi}\psi\,dx = \int \bar{\phi}\tilde{A}\psi\,dx. \tag{13}$$

Given Hermitian operators $\tilde{A}$ and $\tilde{B}$,

$$\langle \phi|\tilde{A}\tilde{B}\psi\rangle = \langle \tilde{A}\phi|\tilde{B}\psi\rangle = \langle \tilde{B}\tilde{A}\phi|\psi\rangle = \overline{\langle \phi|\tilde{B}\tilde{A}\psi\rangle}. \tag{14}$$

Because, for a Hermitian operator $\tilde{A}$ with EIGENVALUE a,

$$\langle \psi|\tilde{A}\psi\rangle = \langle \tilde{A}\psi|\psi\rangle \tag{15}$$

$$a\langle \psi|\psi\rangle = \bar{a}\langle \psi|\psi\rangle. \tag{16}$$

Therefore, either $\langle \psi|\psi\rangle = 0$ or $a = \bar{a}$. But $\langle \psi|\psi\rangle = 0$ IFF $\psi = 0$, so

$$\langle \psi|\psi\rangle \neq 0, \tag{17}$$

for a nontrivial EIGENFUNCTION. This means that $a = a^*$, namely that Hermitian operators produce REAL expectation values. Every observable must therefore have a corresponding Hermitian operator. Furthermore,

$$\langle \psi_n|\tilde{A}\psi_m\rangle = \langle \tilde{A}\psi_n|\psi_m\rangle \tag{18}$$

$$a_m\langle \psi_n|\psi_m\rangle = \bar{a}_n\langle \psi_n|\psi_m\rangle = a_n\langle \psi_n|\psi_m\rangle, \tag{19}$$

since $a_n = \bar{a}_n$. Then

$$(a_m - a_n)\langle \psi_n|\psi_m\rangle = 0 \tag{20}$$

For $a_m \neq a_n$ (i.e., $\psi_n \neq \psi_m$),

$$\langle \psi_n|\psi_m\rangle = 0. \tag{21}$$

For $a_m = a_n$ (i.e., $\psi_n = \psi_m$),

$$\langle \psi_n|\psi_m\rangle = \langle \psi_n|\psi_n\rangle \equiv 1. \tag{22}$$

Therefore,

$$\langle\psi_n|\psi_m\rangle = \delta_{nm}, \qquad (23)$$

so the basis of EIGENFUNCTIONS corresponding to a Hermitian operator are ORTHONORMAL.

Define the Hermitian conjugate operator $\tilde{A}^*$ by

$$\langle\tilde{A}\psi|\psi\rangle \equiv \langle\psi|\tilde{A}^*\psi\rangle. \qquad (24)$$

For a Hermitian operator, $\tilde{A} = \tilde{A}^*$. Furthermore, given two Hermitian operators $\tilde{A}$ and $\tilde{B}$,

$$\langle\psi_2|(\tilde{A}\tilde{B})^*\psi_1\rangle = \langle(\tilde{A}\tilde{B})\psi_2|\psi_1\rangle = \langle\tilde{B}\psi_2|\tilde{A}^*\psi_1\rangle$$
$$= \langle\psi_2|\tilde{B}^*\tilde{A}^*\psi_1\rangle, \qquad (25)$$

so

$$(\tilde{A}\tilde{B})^* = \tilde{B}^*\tilde{A}^*. \qquad (26)$$

By further iterations, this can be generalized to

$$(\tilde{A}\tilde{B}\cdots\tilde{Z})^* = \tilde{Z}^*\cdots\tilde{B}^*\tilde{A}^*. \qquad (27)$$

Given two Hermitian operators $\tilde{A}$ and $\tilde{B}$,

$$(\tilde{A}\tilde{B})^* = \tilde{B}^*\tilde{A}^* = \tilde{B}\tilde{A} = \tilde{A}\tilde{B} + [\tilde{B}, \tilde{A}], \qquad (28)$$

the operator $\tilde{A}\tilde{B}$ equals $(\tilde{A}\tilde{B})^*$, and is therefore Hermitian, only if

$$[\tilde{B}, \tilde{A}] = 0. \qquad (29)$$

Given an arbitrary operator $\tilde{A}$,

$$\langle\psi_1|(\tilde{A}+\tilde{A}^*)\psi_2\rangle = \langle(\tilde{A}^*+\tilde{A})\psi_1|\psi_2\rangle$$
$$= \langle(\tilde{A}+\tilde{A}^*)\psi_1|\psi_2\rangle, \qquad (30)$$

so $\tilde{A}+\tilde{A}^*$ is Hermitian.

$$\langle\psi_1|i(\tilde{A}-\tilde{A}^*)\psi_2\rangle = \langle-i(\tilde{A}^*-\tilde{A})\psi_1|\psi_2\rangle$$
$$= \langle i(\tilde{A}-\tilde{A}^*)\psi_1|\psi_2\rangle, \qquad (31)$$

so $i(\tilde{A}-\tilde{A}^*)$ is Hermitian. Similarly,

$$\langle\psi_1|(\tilde{A}\tilde{A}^*)\psi_2\rangle = \langle\tilde{A}^*\psi_1|\tilde{A}^*\psi_2\rangle = \langle(\tilde{A}\tilde{A}^*)\psi_1|\psi_2\rangle, \qquad (32)$$

so $\tilde{A}\tilde{A}^*$ is Hermitian.

See also ADJOINT, HERMITIAN MATRIX, SELF-ADJOINT, STURM-LIOUVILLE THEORY

References

Arfken, G. "Hermitian (Self-Adjoint) Operators." §9.2 in *Mathematical Methods for Physicists, 3rd ed.* Orlando, FL: Academic Press, pp. 504–06 and 510–16, 1985.

Heron Triangle

HERONIAN TRIANGLE

Heron's Formula

Gives the AREA of a TRIANGLE in terms of the lengths of the sides a, b, and c and the SEMIPERIMETER

$$s = \tfrac{1}{2}(a+b+c). \qquad (1)$$

Heron's formula then states

$$\Delta = \sqrt{s(s-a)(s-b)(s-c)}. \qquad (2)$$

Heron's formula may be stated beautifully using a CAYLEY-MENGER DETERMINANT as

$$-16\Delta^2 = \begin{vmatrix} 0 & a & b & c \\ a & 0 & c & b \\ b & c & 0 & a \\ c & b & a & 0 \end{vmatrix} \begin{vmatrix} 0 & 1 & 1 & 1 \\ 1 & 0 & c^2 & b^2 \\ 1 & c^2 & 0 & a^2 \\ 1 & b^2 & a^2 & 0 \end{vmatrix}. \qquad (3)$$

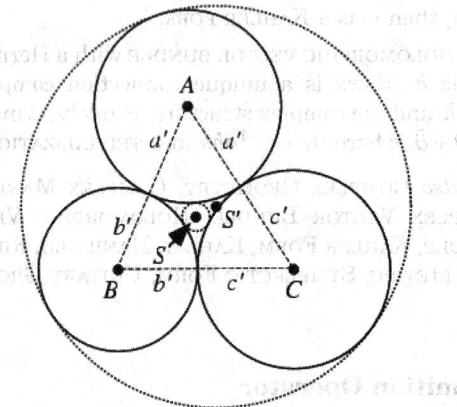

Expressing the side lengths a, b, and c in terms of the radii a', b', and c' of the mutually tangent circles centered on the TRIANGLE vertices (which define the SODDY CIRCLES),

$$a = b' + c' \qquad (4)$$
$$b = a' + c' \qquad (5)$$
$$c = a' + b', \qquad (6)$$

gives the particularly pretty form

$$\Delta = \sqrt{a'b'c'(a'+b'+c')}. \qquad (7)$$

Heron's proof (Dunham 1990) is ingenious but extremely convoluted, bringing together a sequence of apparently unrelated geometric identities and relying on the properties of CYCLIC QUADRILATERALS and RIGHT TRIANGLES. Heron's proof can be found in Proposition 1.8 of his work *Metrica* (ca. 100 BC–100 AD). This manuscript had been lost for centuries until a fragment was discovered in 1894 and a complete copy in 1896 (Dunham 1990, p. 118). More recently, writings of the Arab scholar Abu'l Raihan Muhammed al-Biruni have credited the formula to Heron's predecessor Archimedes prior to 212 BC (van der Waerden 1961, pp. 228 and 277; Coxeter and Greitzer 1967, p. 59; Kline 1972; Bell 1986, p. 58; Dunham 1990, p. 127).

A much more accessible algebraic proof proceeds from the LAW OF COSINES,

$$\cos A = \frac{b^2 + c^2 - a^2}{2bc}. \qquad (8)$$

Then

$$\sin A = \frac{\sqrt{-a^4 - b^4 - c^4 + 2b^2c^2 + 2c^2a^2 + 2a^2b^2}}{2bc}, \qquad (9)$$

giving

$$\Delta = \sqrt{s(s-a)(s-b)(s-c)} \qquad (10)$$

$$= \tfrac{1}{4}\sqrt{(2ab)^2 - (a^2 + b^2 - c^2)^2} \qquad (11)$$

$$= \tfrac{1}{2} bc \sin A \qquad (12)$$

$$= \tfrac{1}{4}\sqrt{(a+b+c)(-a+b+c)(a-b+c)(a+b-c)}$$

$$= \tfrac{1}{4}\sqrt{2(b^2c^2 + c^2a^2 + a^2b^2) - (a^4 + b^4 + c^4)} \qquad (13)$$

(Coxeter 1969). Heron's formula contains the PYTHA-GOREAN THEOREM as a degenerate case.

See also BRAHMAGUPTA'S FORMULA, BRETSCHNEIDER'S FORMULA, CAYLEY-MENGER DETERMINANT, HERO-NIAN TETRAHEDRON, HERONIAN TRIANGLE, SODDY CIRCLES, SSS THEOREM, TRIANGLE

References
Bell, E. T. *Men of Mathematics.* New York: Simon and Schuster, p. 58, 1986.
Brown, K. S. "Heron's FOrmula and Brahmagupta's Generalization." http://www.seanet.com/~ksbrown/kmath196.htm.
Coxeter, H. S. M. *Introduction to Geometry, 2nd ed.* New York: Wiley, p. 12, 1969.
Coxeter, H. S. M. and Greitzer, S. L. *Geometry Revisited.* Washington, DC: Math. Assoc. Amer., p. 59, 1967.
Dunham, W. "Heron's Formula for Triangular Area." Ch. 5 in *Journey through Genius: The Great Theorems of Mathematics.* New York: Wiley, pp. 113–32, 1990.
Kline, M. *Mathematical Thought from Ancient to Modern Times.* New York: Oxford University Press, 1972.
Pappas, T. "Heron's Theorem." *The Joy of Mathematics.* San Carlos, CA: Wide World Publ./Tetra, p. 62, 1989.
van der Waerden, B. L. *Science Awakening.* Oxford, England: Oxford University Press, pp. 228 and 277, 1961.

Heronian Mean

The Heronian mean of two numbers m and n is defined as

$$HM(a, b) = \tfrac{1}{3}(a + \sqrt{ab} + b),$$

which arises in the determination of the volume of a PYRAMIDAL FRUSTUM.

See also PYRAMIDAL FRUSTUM

References
Eves, H. *A Survey of Geometry, rev. ed.* Boston, MA: Allyn & Bacon, p. 7, 1965.

Heronian Tetrahedron

A TETRAHEDRON with RATIONAL sides, FACE AREAS, and VOLUME. The smallest examples have pairs of opposite sides (148, 195, 203), (533, 875, 888), (1183, 1479, 1804), (2175, 2296, 2431), (1825, 2748, 2873), (2180, 2639, 3111), (1887, 5215, 5512), (6409, 6625, 8484), and (8619, 10136, 11275).

See also HERON'S FORMULA, HERONIAN TRIANGLE

References
Guy, R. K. "Simplexes with Rational Contents." §D22 in *Unsolved Problems in Number Theory, 2nd ed.* New York: Springer-Verlag, pp. 190–92, 1994.

Heronian Triangle

A TRIANGLE with RATIONAL side lengths and RATIONAL AREA. Brahmagupta gave a parametric solution for *integer* Heronian triangles (the three side lengths and area can be multiplied by their LEAST COMMON MULTIPLE to make them all INTEGERS): side lengths $c(a^2 + b^2)$, $b(a^2 + c^2)$, and $(b+c)(a^2 - bc)$, giving SEMIPERIMETER

$$s = a^2(b + c) \qquad (1)$$

and AREA

$$\Delta = abc(a + b)(a^2 - bc). \qquad (2)$$

The first few integer Heronian triangles sorted by increasing maximal side lengths, are ((3, 4, 5), (5, 5, 6), (5, 5, 8), (6, 8, 10), (10, 10, 12), (5, 12, 13), (10, 13, 13), (9, 12, 15), (4, 13, 15), (13, 14, 15), (10, 10, 16), ... (Sloane's A055594, A055593, and A055592), having areas 6, 12, 12, 24, 48, 30, 60, 54, ... (Sloane's A055595). The first few integer Heronian SCALENE TRIANGLES, sorted by increasing maximal side lengths, are (3, 4, 5), (6, 8, 10), (5, 12, 13), (9, 12, 15), (4, 13, 15), (13, 14, 15), (9, 10, 17), ... (Sloane's A046128, A046129, and A046130), having areas 6, 24, 30, 54, 24, 84, 36, ... (Sloane's A046131).

Schubert (1905) claimed that Heronian triangles with two rational MEDIANS do not exist (Dickson 1952). This was shown to be incorrect by Buchholz and Rathbun (1997), who discovered the triangles given in the following table, where m_i are MEDIAN lengths and A is the area.

a	b	c	m_1	m_2	A
73	51	26	$\frac{35}{2}$	$\frac{97}{2}$	420
626	875	291	572	$\frac{433}{2}$	55440

4368	1241	3673	1657	$\frac{7975}{2}$	2042040
14791	14384	11257	$\frac{21177}{2}$	11001	75698280
28779	13816	15155	$\frac{3589}{2}$	21937	23931600
1823675	185629	1930456	$\frac{2048523}{2}$	$\frac{3751059}{2}$	142334216640

See also HERON'S FORMULA, MEDIAN (TRIANGLE), PYTHAGOREAN TRIPLE, TRIANGLE

References

Buchholz, R. H. *On Triangles with Rational Altitudes, Angle Bisectors or Medians.* Doctoral Dissertation. Newcastle, England: Newcastle University, 1989.

Buchholz, R. H. and Rathbun, R. L. "An Infinite Set of Heron Triangles with Two Rational Medians." *Amer. Math. Monthly* **104**, 107–15, 1997.

Dickson, L. E. *History of the Theory of Numbers, Vol. 2: Diophantine Analysis.* New York: Chelsea, pp. 199 and 208, 1952.

Fleenor, C. R. "Heronian Triangles with Consecutive Integer Sides." *J. Recr. Math.* **28**, 113–15, 1996–6.

Guy, R. K. "Simplexes with Rational Contents." §D22 in *Unsolved Problems in Number Theory, 2nd ed.* New York: Springer-Verlag, pp. 190–92, 1994.

Kraitchik, M. "Heronian Triangles." §4.13 in *Mathematical Recreations.* New York: W. W. Norton, pp. 104–08, 1942.

Rabinowitz, S. "Problem 2006: Heronian Properties." *J. Recr. Math.* **24**, 309, 1992.

Schubert, H. "Die Ganzzahligkeit in der algebraischen Geometrie." In *Festgabe 48 Versammlung d. Philologen und Schulmänner zu Hamburg.* Leipzig, Germany, pp. 1–6, 1905.

Sloane, N. J. A. Sequences A046128, A046129, A046130, A046131, A055592, A055593, A055594, and A055595 in "An On-Line Version of the Encyclopedia of Integer Sequences." http://www.research.att.com/~njas/sequences/eisonline.html.

Wells, D. G. *The Penguin Dictionary of Curious and Interesting Puzzles.* London: Penguin Books, p. 34, 1992.

Yiu, P. "Construction of Indecomposable Heronian Triangles." *Rocky Mountain J. Math.* **28**, 1189–202, 1998.

Herschel

A HEPTOMINO shaped like the astronomical symbol for Uranus (which was discovered by William Herschel).
See also HEPTOMINO

Herschfeld's Convergence Theorem

For real, NONNEGATIVE terms x_n and REAL p with $0 < p < 1$, the expression

$$\lim_{k \to \infty} x_0 + (x_1 + (x_2 + (\ldots + (x_k)^p)^p)^p)^p$$

converges IFF $(x_n)^{p^n}$ is bounded.

See also NESTED RADICAL

References

Herschfeld, A. "On Infinite Radicals." *Amer. Math. Monthly* **42**, 419–29, 1935.

Jones, D. J. "Continued Powers and a Sufficient Condition for Their Convergence." *Math. Mag.* **68**, 387–92, 1995.

Hesse's Theorem

If two pairs of opposite VERTICES of a COMPLETE QUADRILATERAL are pairs of CONJUGATE POINTS, then the third pair of opposite VERTICES is likewise a pair of CONJUGATE POINTS.

See also COMPLETE QUADRILATERAL

Hessenberg Matrix

A matrix OF THE FORM

$$\begin{bmatrix}
a_{11} & a_{12} & a_{13} & \cdots & a_{1(n-1)} & a_{1n} \\
a_{21} & a_{22} & a_{23} & \cdots & a_{2(n-1)} & a_{2n} \\
0 & a_{32} & a_{33} & \cdots & a_{3(n-1)} & a_{3n} \\
0 & 0 & a_{43} & \cdots & a_{4(n-1)} & a_{4n} \\
0 & 0 & 0 & \cdots & a_{5(n-1)} & a_{5n} \\
\vdots & \vdots & \vdots & \ddots & \vdots & \vdots \\
0 & 0 & 0 & 0 & a_{(n-1)(n-1)} & a_{(n-1)n} \\
0 & 0 & 0 & 0 & a_{n(n-1)} & a_{nn}
\end{bmatrix}.$$

See also TOEPLITZ MATRIX, TRIANGULAR MATRIX

References

Press, W. H.; Flannery, B. P.; Teukolsky, S. A.; and Vetterling, W. T. "Reduction of a General Matrix to Hessenberg Form." §11.5 in *Numerical Recipes in FORTRAN: The Art of Scientific Computing, 2nd ed.* Cambridge, England: Cambridge University Press, pp. 476–80, 1992.

Hessian Covariant

$$H \equiv |aa'a''|a_{x^{n-2}}a'_{x^{n-2}}a''_{x^{n-2}} = 0.$$

The nonsingular inflections of a curve are its nonsingular intersections with the Hessian.

References

Coolidge, J. L. *A Treatise on Algebraic Plane Curves.* New York: Dover, pp. 79, 95–8, and 151–61, 1959.

Hessian Determinant

The DETERMINANT

$$Hf(x, y) = \begin{vmatrix} \dfrac{\partial^2 f}{\partial x^2} & \dfrac{\partial^2 f}{\partial x \partial y} \\[2mm] \dfrac{\partial^2 f}{\partial y \partial x} & \dfrac{\partial^2 f}{\partial y^2} \end{vmatrix}$$

appearing in the SECOND DERIVATIVE TEST as $D \equiv Hf(x, y)$.

See also SECOND DERIVATIVE TEST

References

Gradshteyn, I. S. and Ryzhik, I. M. *Tables of Integrals, Series, and Products, 6th ed.* San Diego, CA: Academic Press, pp. 1112–113, 2000.

Heteroclinic Point

If intersecting stable and unstable MANIFOLDS (SEPARATRICES) emanate from FIXED POINTS of different families, they are called heteroclinic points.

See also HOMOCLINIC POINT, MANIFOLD, SEPARATRIX

Heterogeneous Numbers

Two numbers are heterogeneous if their PRIME FACTORS are distinct. For example, $6 = 2 \cdot 3$ and $24 = 2^3 \cdot 3$ are *not* heterogeneous since their factors are each $(2, 3)$.

See also DISTINCT PRIME FACTORS, HOMOGENEOUS NUMBERS

References

Le Lionnais, F. *Les nombres remarquables.* Paris: Hermann, p. 146, 1983.

Heterological Paradox

GRELLING'S PARADOX

Heteromecic Number

PRONIC NUMBER

Heteroscedastic

A set of STATISTICAL DISTRIBUTIONS having different VARIANCES.

See also HOMOSCEDASTIC, VARIANCE

Heterosquare

9	8	7
2	1	6
3	4	5

1	2	3	4
5	6	7	8
9	10	11	12
13	14	16	15

A heterosquare is an $n \times n$ ARRAY of the integers from 1 to n^2 such that the rows, columns, and diagonals have different sums. (By contrast, in a MAGIC SQUARE, they have the *same* sum.) There are no heterosquares of order two, but heterosquares of every ODD order exist. They can be constructed by placing consecutive INTEGERS in a SPIRAL pattern (Fults 1974, Madachy 1979).

An ANTIMAGIC SQUARE is a special case of a heterosquare for which the sums of rows, columns, and main diagonals form a SEQUENCE of consecutive integers.

See also ANTIMAGIC SQUARE, MAGIC SQUARE, TALISMAN SQUARE

References

Duncan, D. "Problem 86." *Math. Mag.* **24**, 166, 1951.
Fults, J. L. *Magic Squares.* Chicago, IL: Open Court, 1974.
Madachy, J. S. *Madachy's Mathematical Recreations.* New York: Dover, pp. 101–03, 1979.
Rivera, C. "Problems & Puzzles: Puzzle Primeful Heterosquares.-069." http://www.primepuzzles.net/puzzles/puzz_069.htm.
Weisstein, E. W. "Magic Squares." MATHEMATICA NOTEBOOK MAGICSQUARES.M.

Heuman Lambda Function

$$\Lambda_0(\phi|m) \equiv \frac{F(\phi|1-m)}{K(1-m)} + \frac{2}{\pi} K(m) Z(\phi|1-m),$$

where ϕ is the AMPLITUDE, m is the PARAMETER, Z is the JACOBI ZETA FUNCTION, and $F(\phi|m')$ and $K(m)$ are incomplete and complete ELLIPTIC INTEGRALS OF THE FIRST KIND.

See also ELLIPTIC INTEGRAL OF THE FIRST KIND, JACOBI ZETA FUNCTION

References

Abramowitz, M. and Stegun, C. A. (Eds.). *Handbook of Mathematical Functions with Formulas, Graphs, and Mathematical Tables, 9th printing.* New York: Dover, p. 595, 1972.
Tölke, F. "Jacobische Zeta- und Heumansche Lambda-Funktionen." §132 in *Praktische Funktionenlehre, dritter Band: Jacobische elliptische Funktionen, Legendresche elliptische Normalintegrale und spezielle Weierstraßsche Zeta- und Sigma Funktionen.* Berlin: Springer-Verlag, pp. 94–9, 1967.

Heun's Differential Equation

A natural extension of the RIEMANN P-DIFFERENTIAL EQUATION given by

$$\frac{d^2w}{dx^2} + \left(\frac{\gamma}{x} + \frac{\delta}{x-1} + \frac{\varepsilon}{x-a}\right) \frac{dw}{dx} + \frac{\alpha\beta x - q}{x(x-1)(x-a)} w$$
$$= 0$$

where

$$\alpha + \beta - \gamma - \delta - \varepsilon + 1 = 0.$$

See also RIEMANN P-DIFFERENTIAL EQUATION

References

Decarreau, A.; Dumont-Lepage, M.-C.; Maroni, P.; Robert, A.; and Ronveaux, A. "Formes canoniques des équations confluentes de l'équation de Heun." *Ann. Soc. Sci. de Bruxelles* **92**, 53–8, 1978.

Erdélyi, A.; Magnus, W.; Oberhettinger, F.; and Tricomi, F. G. *Higher Transcendental Functions, Vol. 3.* New York: Krieger, pp. 57-2, 1981.

Heun, K. "Zur Theorie der Riemann'schen Functionen Zweiter Ordnung mit Verzweigungspunkten." *Math. Ann.* **33**, 161-79.

Ronveaux, A. (Ed.). *Heun's Differential Equations.* Oxford, England: Oxford University Press, 1995.

Valent, G. "An Integral Transform Involving Heun Functions and a Related Eigenvalue Problem." *SIAM J. Math. Anal.* **17**, 688-03, 1986.

Whittaker, E. T. and Watson, G. N. *A Course in Modern Analysis, 4th ed.* Cambridge, England: Cambridge University Press, p. 576, 1990.

Zwillinger, D. *Handbook of Differential Equations, 3rd ed.* Boston, MA: Academic Press, p. 123, 1997.

Heuristic

(1) Based on or involving trial and error. (2) Convincing without being rigorous.

See also PARADOX, PROOF

Hex (Polyhex)

POLYHEX

Hex Game

A two-player GAME. There is a winning strategy for the first player if there is an even number of cells on each side; otherwise, there is a winning strategy for the second player.

References

Gardner, M. "The Game of Hex." Ch. 8 in *The Scientific American Book of Mathematical Puzzles & Diversions.* New York: Simon and Schuster, pp. 73-3, 1959.

Hex Number

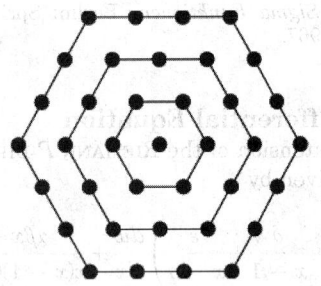

The CENTERED HEXAGONAL NUMBER given by

$$H_n = 1 + 6T_n = 2H_{n-1} - H_{n-2} + 6 = 3n^2 - 3n + 1,$$

where T_n is the nth TRIANGULAR NUMBER. The first few hex numbers are 1, 7, 19, 37, 61, 91, 127, 169, ... (Sloane's A003215). The GENERATING FUNCTION of the hex numbers is

$$\frac{x(x^2 + 4x + 1)}{(1-x)^3} = x + 7x^2 + 19x^3 + 37x^4 + \dots.$$

The first TRIANGULAR hex numbers are 1 and 91, and the first few SQUARE ones are 1, 169, 32761, 6355441,

... (Sloane's A006051). SQUARE hex numbers are obtained by solving the DIOPHANTINE EQUATION

$$3x^2 + 1 = y^2.$$

The only hex number which is SQUARE and TRIANGULAR is 1. There are no CUBIC hex numbers.

See also MAGIC HEXAGON, CENTERED PENTAGONAL NUMBER, CENTERED SQUARE NUMBER, STAR NUMBER, TALISMAN HEXAGON

References

Conway, J. H. and Guy, R. K. *The Book of Numbers.* New York: Springer-Verlag, p. 41, 1996.

Gardner, M. "Hexes and Stars." Ch. 2 in *Time Travel and Other Mathematical Bewilderments.* New York: W. H. Freeman, pp. 15-5, 1988.

Hindin, H. "Stars, Hexes, Triangular Numbers, and Pythagorean Triples." *J. Recr. Math.* **16**, 191-93, 1983-984.

Sloane, N. J. A. Sequences A003215/M4362 and A006051/M5409 in "An On-Line Version of the Encyclopedia of Integer Sequences." http://www.research.att.com/~njas/sequences/eisonline.html.

Hex Pyramidal Number

A FIGURATE NUMBER which is equal to the CUBIC NUMBER n^3. The first few are 1, 8, 27, 64, ... (Sloane's A000578).

References

Conway, J. H. and Guy, R. K. *The Book of Numbers.* New York: Springer-Verlag, pp. 42-4, 1996.

Sloane, N. J. A. Sequences A000578/M4499 in "An On-Line Version of the Encyclopedia of Integer Sequences." http://www.research.att.com/~njas/sequences/eisonline.html.

Hexa

POLYHEX

Hexabolo

A 6-POLYABOLO.

Hexacontagon

A 60-sided POLYGON.

Hexacronic Icositetrahedron

GREAT HEXACRONIC ICOSITETRAHEDRON, SMALL HEXACRONIC ICOSITETRAHEDRON

Hexad

A SET of six.

See also MONAD, QUARTET, QUINTET, TETRAD, TRIAD

Hexadecagon

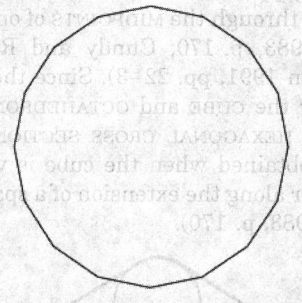

A 16-sided POLYGON, sometimes also called a HEXAKAIDECAGON. The regular hexadecagon is a CONSTRUCTIBLE POLYGON, and the INRADIUS r, CIRCUMRADIUS R, and area A of the regular hexadecagon of side length 1 are

$$r = \tfrac{1}{2}(1 + \sqrt{2} + \sqrt{2(2 + \sqrt{2})})$$

$$R = \sqrt{\tfrac{1}{2}(4 + 2\sqrt{2} + \sqrt{20 + 14\sqrt{2}})}$$

$$A = 4(1 + \sqrt{2} + \sqrt{2(2 + \sqrt{2})}).$$

See also POLYGON, REGULAR POLYGON, TRIGONOMETRY VALUES PI/16

Hexadecimal

The base 16 notational system for representing REAL NUMBERS. The digits used to represent numbers using hexadecimal NOTATION are 0, 1, 2, 3, 4, 5, 6, 7, 8, 9, A, B, C, D, E, and F. The following table gives the hexadecimal equivalents of the first few decimal numbers.

1	1	11	B	21	15
2	2	12	C	22	16
3	3	13	D	23	17
4	4	14	E	24	18
5	5	15	F	25	19
6	6	16	10	26	1A
7	7	17	11	27	1B
8	8	18	12	28	1C
9	9	19	13	29	1D
10	A	20	14	30	1E

The hexadecimal system is particularly important in computer programming, since four bits (each consisting of a one or zero) can be succinctly expressed using a single hexadecimal digit. Two hexadecimal digits represent numbers from 0 to 255, a common range used, for example, to specify colors. Thus, in the HTML language of the web, colors are specified using three pairs of hexadecimal digits RRGGBB, where *RR* is the amount of red, *GG* the amount of green, and *BB* the amount of blue.

In HEXADECIMAL, numbers with increasing digits are called METADROMES, those with nondecreasing digits are called PLAINDRONES, those with nonincreasing digits are called NIALPDROMES, and those with decreasing digits are called KATADROMES.

See also BASE (NUMBER), BINARY, DECIMAL, DIGIT, KATADROME, METADROME, NIALPDROME, OCTAL, PLAINDROME, QUATERNARY, TERNARY, VIGESIMAL

References

Gardner, M. *The Sixth Book of Mathematical Games from Scientific American.* Chicago, IL: University of Chicago Press, p. 105, 1984.
Weisstein, E. W. "Bases." MATHEMATICA NOTEBOOK BASES.M.

Hexaflexagon

A FLEXAGON made by folding a strip into adjacent EQUILATERAL TRIANGLES. The number of states possible in a hexaflexagon is the CATALAN NUMBER $C_4 = 42$.

See also FLEXAGON, FLEXATUBE, TETRAFLEXAGON

References

Cundy, H. and Rollett, A. *Mathematical Models, 3rd ed.* Stradbroke, England: Tarquin Pub., pp. 205–07, 1989.
Gardner, M. "Hexaflexagons." Ch. 1 in *The Scientific American Book of Mathematical Puzzles & Diversions.* New York: Simon and Schuster, pp. 1–4, 1959.
Gardner, M. "Tetraflexagons." Ch. 2 in *The Second Scientific American Book of Mathematical Puzzles & Diversions: A New Selection.* New York: Simon and Schuster, pp. 24–1, 1961.
Maunsell, F. G. "The Flexagon and the Hexaflexagon." *Math. Gazette* **38**, 213–14, 1954.
Wheeler, R. F. "The Flexagon Family." *Math. Gaz.* **42**, 1–, 1958.

Hexafrob

POLYHEX

Hexagon

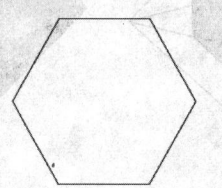

A six-sided POLYGON. In proposition IV.15, Euclid showed how to inscribe a regular hexagon in a CIRCLE. The INRADIUS r, CIRCUMRADIUS R, and AREA

A can be computed directly from the formulas for a general REGULAR POLYGON with side length s and $n = 6$ sides,

$$r = \tfrac{1}{2} s \cot\left(\frac{\pi}{6}\right) = \tfrac{1}{2}\sqrt{3}\, s \qquad (1)$$

$$R = \tfrac{1}{2} s \csc\left(\frac{\pi}{6}\right) = s \qquad (2)$$

$$A = \tfrac{1}{4} n s^2 \cot\left(\frac{\pi}{6}\right) = \tfrac{3}{2}\sqrt{3}\, s^2. \qquad (3)$$

Therefore, for a regular hexagon,

$$\frac{R}{r} = \sec\left(\frac{\pi}{6}\right) = \frac{2}{\sqrt{3}}, \qquad (4)$$

so

$$\frac{A_R}{A_r} = \left(\frac{R}{r}\right)^2 = \frac{4}{3}. \qquad (5)$$

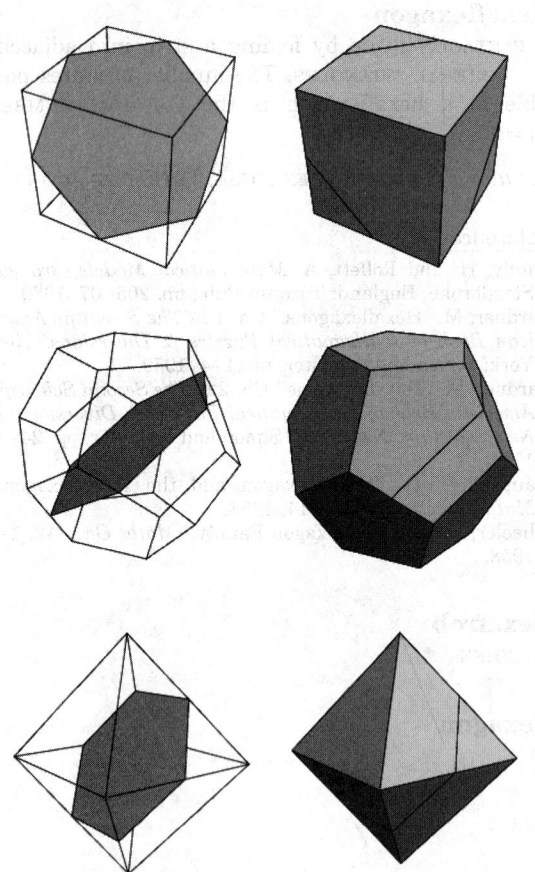

A PLANE PERPENDICULAR to a C_3 axis of a CUBE (Gardner 1960), DODECAHEDRON, or ICOSAHEDRON

cuts the solid in a regular HEXAGONAL CROSS SECTION (Holden 1991, pp. 22–3 and 27). For the CUBE, the PLANE passes through the MIDPOINTS of opposite sides (Steinhaus 1983, p. 170; Cundy and Rollett 1989, p. 157; Holden 1991, pp. 22–3). Since there are four such axes for the CUBE and OCTAHEDRON, there are four possible HEXAGONAL CROSS SECTIONS. A HEXAGON is also obtained when the cube is viewed from above a corner along the extension of a space diagonal (Steinhaus 1983, p. 170).

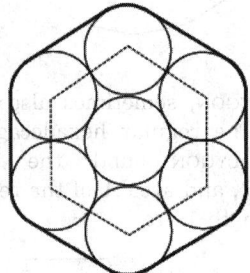

Take seven CIRCLES and close-pack them together in a hexagonal arrangement. The PERIMETER obtained by wrapping a band around the CIRCLE then consists of six straight segments of length d (where d is the DIAMETER) and 6 arcs with total length 1/6 of a CIRCLE. The PERIMETER is therefore

$$p = (12 + 2\pi)r = 2(6 + \pi)r. \qquad (6)$$

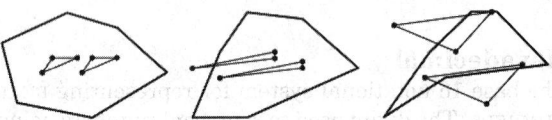

Given an arbitrary hexagon, take each three consecutive vertices, and mark the fourth point of the PARALLELOGRAM sharing these three vertices. Taking alternate points then gives two congruent triangles, as illustrated above (Wells 1991).

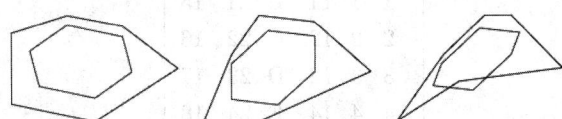

Given an arbitrary hexagon, connecting the centroids of each consecutive three sides gives a hexagon with equal and parallel sides known as the CENTROID HEXAGON (Wells 1991).

See also CENTROID HEXAGON, COSINE HEXAGON, CUBE, CYCLIC HEXAGON, DISSECTION, DODECAHEDRON, GRAHAM'S BIGGEST LITTLE HEXAGON, HEPTAGON THEOREM, HEXAGON POLYIAMOND, HEXAGRAM, LEMOINE HEXAGON, MAGIC HEXAGON, OCTAHEDRON, PAPPUS'S HEXAGON THEOREM, PASCAL'S THEOREM, TALISMAN HEXAGON, TUCKER HEXAGON

Hexagon Polyiamond

References

Cadwell, J. H. *Topics in Recreational Mathematics.* Cambridge, England: Cambridge University Press, 1966.

Coxeter, H. S. M. and Greitzer, S. L. "Hexagons." §3.7 in *Geometry Revisited.* Washington, DC: Math. Assoc. Amer., pp. 73–4, 1967.

Cundy, H. and Rollett, A. "Hexagonal Section of a Cube." §3.15.1 in *Mathematical Models, 3rd ed.* Stradbroke, England: Tarquin Pub., p. 157, 1989.

Dixon, R. *Mathographics.* New York: Dover, p. 16, 1991.

Gardner, M. "Mathematical Games: More About the Shapes that Can Be Made with Complex Dominoes." *Sci. Amer.* **203**, 186–98, Nov. 1960.

Holden, A. *Shapes, Space, and Symmetry.* New York: Dover, 1991.

Pappas, T. "Hexagons in Nature." *The Joy of Mathematics.* San Carlos, CA: Wide World Publ./Tetra, pp. 74–5, 1989.

Steinhaus, H. *Mathematical Snapshots, 3rd ed.* New York: Dover, 1999.

Wells, D. *The Penguin Dictionary of Curious and Interesting Geometry.* London: Penguin, pp. 53–4, 1991.

Hexagon Polyiamond

A 6-POLYIAMOND.

See also HEXAGON

References

Golomb, S. W. *Polyominoes: Puzzles, Patterns, Problems, and Packings, 2nd ed.* Princeton, NJ: Princeton University Press, p. 92, 1994.

Hexagon Tiling

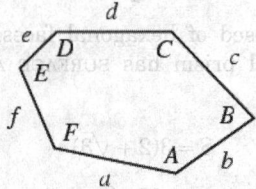

There are at least three aperiodic tilings of HEXAGONS, given by the following types:

$$
\begin{aligned}
A + B + C &= 360° & a &= d \\
A + B + D &= 360° & a &= d,\ c = e \\
A &= C = E & a &= b,\ c = d,\ e = f
\end{aligned}
\tag{1}
$$

(Gardner 1988). Note that the periodic hexagonal TESSELLATION is a degenerate case of all three tilings with

$$
A = B = C = D = E = F \quad a = b = c = d = e = f
\tag{2}
$$

Amazingly, the number of PLANE PARTITIONS $PL(a, b, c)$ contained in an $a \times b \times c$ box also gives the number of hexagon tilings by RHOMBI for a hexagon of side lengths a, b, c, a, b, c (David and Tomei 1989, Fulmek and Krattenthaler 2000). The asymptotic distribution of rhombi in a random hexa-

gon tiling by rhombi was given by Cohn *et al.* (1998). A variety of enumerations for various explicit positions of rhombi are given by Fulmek and Krattenthaler (1998, 2000).

See also PLANE PARTITION, TILING

References

Cohn, H.; Larsen, M.; and Propp, J. "The Shape of a Typical Boxed Plane Partition." *New York J. Math.* **4**, 137–66, 1998.

David, G. and Tomei, C. "The Problem of the Calissons." *Amer. Math. Monthly* **96**, 429–31, 1989.

Gardner, M. "Tilings with Convex Polygons." Ch. 13 in *Time Travel and Other Mathematical Bewilderments.* New York: W. H. Freeman, pp. 162–76, 1988.

Fulmek, M. and Krattenthaler, C. "The Number of Rhombus Tilings of a Symmetric Hexagon which Contains a Fixed Rhombus on the Symmetry Axis, I." *Ann. Combin.* **2**, 19–0, 1998.

Fulmek, M. and Krattenthaler, C. "The Number of Rhombus Tilings of a Symmetric Hexagon which Contains a Fixed Rhombus on the Symmetry Axes, II." *Europ. J. Combin.* **21**, 601–40, 2000.

Hexagon Triangle Picking

The mean area of a TRIANGLE picked inside a regular HEXAGON with unit area is $\bar{A} = 289/3888$ (Woolhouse 1867, Pfiefer 1989). This is a special case of a general POLYGON TRIANGLE PICKING result due to Alikoski (1939).

See also DISK TRIANGLE PICKING, POLYGON TRIANGLE PICKING, SQUARE TRIANGLE PICKING, SYLVESTER'S FOUR-POINT PROBLEM, TRIANGLE TRIANGLE PICKING

References

Alikoski, H. A. "Über das Sylversersche Vierpunktproblem." *Ann. Acad. Sci. Fenn.* **51**, No. 7, 1–0, 1939.

Pfiefer, R. E. "The Historical Development of J. J. Sylvester's Four Point Problem." *Math. Mag.* **62**, 309–17, 1989.

Solomon, H. *Geometric Probability.* Philadelphia, PA: SIAM, p. 114, 1978.

Woolhouse, W. S. B. "Question 2471" *Mathematical Questions, with Their Solutions, from the Educational Times, Vol. 8.* London: F. Hodgson and Son, pp. 100–05, 1867.

Hexagonal Close Packing

SPHERE PACKING

Hexagonal Number

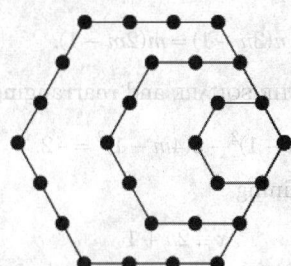

A FIGURATE NUMBER and 6-POLYGONAL NUMBER OF THE FORM $n(2n - 1)$. The first few are 1, 6, 15, 28, 45,

... (Sloane's A000384). The GENERATING FUNCTION of the hexagonal numbers

$$\frac{x(3x+1)}{(1-x)^3} = x + 6x^2 + 15x^3 + 28x^4 + \ldots.$$

Every hexagonal number is a TRIANGULAR NUMBER since

$$r(2r-1) = \tfrac{1}{2}(2r-1)[(2r-1)+1].$$

In 1830, Legendre (1979) proved that every number larger than 1791 is a sum of four hexagonal numbers, and Duke and Schulze-Pillot (1990) improved this to three hexagonal numbers for every sufficiently large integer. The numbers 11 and 26 can only be REPRESENTED AS a sum using the maximum possible of six hexagonal numbers:

$$11 = 1 + 1 + 1 + 1 + 1 + 6$$

$$26 = 1 + 1 + 6 + 6 + 6 + 6.$$

See also FIGURATE NUMBER, HEX NUMBER, HEPTAGONAL HEXAGONAL NUMBER, HEXAGONAL PENTAGONAL NUMBER, OCTAGONAL HEXAGONAL NUMBER, TRIANGULAR NUMBER

References

Duke, W. and Schulze-Pillot, R. "Representations of Integers by Positive Ternary Quadratic Forms and Equidistribution of Lattice Points on Ellipsoids." *Invent. Math.* **99**, 49–7, 1990.
Guy, R. K. "Sums of Squares." §C20 in *Unsolved Problems in Number Theory, 2nd ed.* New York: Springer-Verlag, pp. 136–38, 1994.
Legendre, A.-M. *Théorie des nombres, 4th ed., 2 vols.* Paris: A. Blanchard, 1979.
Sloane, N. J. A. Sequences A000384/M4108 in "An On-Line Version of the Encyclopedia of Integer Sequences." http://www.research.att.com/~njas/sequences/eisonline.html.

Hexagonal Pentagonal Number

A number which is simultaneously PENTAGONAL and HEXAGONAL. Let P_n denote the nth PENTAGONAL NUMBER and H_m the mth SQUARE NUMBER, then a number which is both pentagonal and hexagonal satisfies the equation $P_n = H_m$, or

$$\tfrac{1}{2}n(3n-1) = m(2m-1). \tag{1}$$

COMPLETING THE SQUARE and rearranging gives

$$(6n-1)^2 - 3(4m-1)^2 = -2. \tag{2}$$

Therefore, defining

$$x \equiv 2n+1 \tag{3}$$

$$y \equiv 2m \tag{4}$$

gives the Pell-like equation

$$x^2 - 3y^2 = -2 \tag{5}$$

The first few solutions are $(x, y) = (1, 1)$, $(5, 3)$, $(19, 11)$, $(71, 74)$, $(265, 153)$, $(989, 571)$, ... These give the solutions (n, m), $(1, 1)$, $(10/3, 3)$, $(12, 21/2)$, $(133/3, 77/2)$, $(165, 143)$, ..., of which the integer solutions are $(1, 1)$, $(165, 143)$, $(31977, 27693)$, $(6203341, 5372251)$, ... (Sloane's A046178 and A046179), corresponding to the pentagonal hexagonal numbers 1, 40755, 1533776805, 57722156241751, ... (Sloane's A046180).

See also HEXAGONAL NUMBER, PENTAGONAL NUMBER

References

Sloane, N. J. A. Sequences A046178, A046179, and A046180 in "An On-Line Version of the Encyclopedia of Integer Sequences." http://www.research.att.com/~njas/sequences/eisonline.html.

Hexagonal Prism

A PRISM composed of hexagonal faces. The regular right hexagonal prism has SURFACE AREA and VOLUME

$$S = 3(2 + \sqrt{3})$$

$$V = \tfrac{3}{2}\sqrt{3}.$$

See also HEXAGON, PRISM

Hexagonal Pyramid

A PYRAMID with a hexagonal base. The SLANT HEIGHT of a hexagonal pyramid is a special case of the formula for a regular n-gonal PYRAMID with $n = 6$, given by

$$s = \sqrt{h^2 + a^2}, \tag{1}$$

where h is the height and a is the length of a side of the base.

See also HEXAGON, PYRAMID

Hexagonal Pyramidal Number

A PYRAMIDAL NUMBER OF THE FORM $n(n+1)(4n-1)/6$, The first few are 1, 7, 22, 50, 95, ... (Sloane's A002412). The GENERATING FUNCTION of the hexagonal pyramidal numbers is

$$\frac{x(3x+1)}{(x-1)^4} = x + 7x^2 + 22x^3 + 50x^4 + \ldots.$$

References

Sloane, N. J. A. Sequences A002412/M4374 in "An On-Line Version of the Encyclopedia of Integer Sequences." http://www.research.att.com/~njas/sequences/eisonline.html.

Hexagonal Scalenohedron

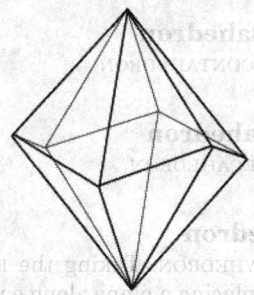

An irregular DODECAHEDRON which is also a TRAPEZOHEDRON.

See also DODECAHEDRON, TRAPEZOHEDRON

References

Cotton, F. A. *Chemical Applications of Group Theory, 3rd ed.* New York: Wiley, p. 63, 1990.

Hexagonal Square Number

Let H_n denote the nth HEXAGONAL NUMBER and S_m the mth SQUARE NUMBER, then a number which is both hexagonal and square satisfies the equation $H_n = S_m$, or

$$n(2n-1) = m^2. \tag{1}$$

COMPLETING THE SQUARE and rearranging gives

$$(4n-1)^2 - 8m^2 = 1. \tag{2}$$

Therefore, defining

$$x \equiv 4n - 1 \tag{3}$$

$$y \equiv 2m \tag{4}$$

gives the PELL EQUATION

$$x^2 - 2y^2 = 1. \tag{5}$$

The first few solutions are $(x, y) = (3, 2)$, $(17, 12)$, $(99, 70)$, $(577, 408)$, ... These give the solutions $(n, m) = (1, 1)$, $(9/2, 6)$, $(25, 35)$, $(289/2, 204)$, ..., giving the integer solutions (1, 1), (25, 35), (841, 1189), (28561,

40391), ... (Sloane's A008844 and A046176). The corresponding hexagonal square numbers are 1, 1225, 1413721, 1631432881, 1882672131025, ... (Sloane's A046177).

See also HEXAGONAL NUMBER, SQUARE NUMBER

References

Sloane, N. J. A. Sequences A008844, A046176, and A046177 in "An On-Line Version of the Encyclopedia of Integer Sequences." http://www.research.att.com/~njas/sequences/eisonline.html.

Hexagram

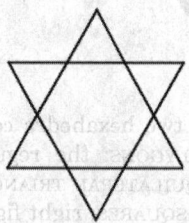

The STAR POLYGON $\{6/2\}$, also known as the STAR OF DAVID.

See also DISSECTION, PENTAGRAM, SOLOMON'S SEAL KNOT, STAR FIGURE, STAR OF LAKSHMI

Hexagrammum Mysticum Theorem

PASCAL'S THEOREM

Hexahedral Graph

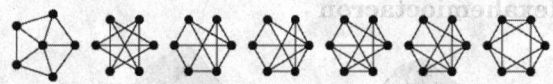

A POLYHEDRAL GRAPH on six vertices. There are seven topologically distinct hexahedral graphs (Gardner 1966, p. 233), of which three are the PENTAGONAL PYRAMID (first figure), TRIANGULAR PRISM (second figure), and OCTAHEDRON/square dipyramid/TRIANGULAR ANTIPRISM (last figure). The hexahedral graphs were first enumerated by Steiner (1828; Duijvestijn and Federico 1981).

See also HEXAHEDRON, POLYHEDRAL GRAPH

References

Duijvestijn, A. J. W. and Federico, P. J. "The Number of Polyhedral (3-Connected Planar) Graphs." *Math. Comput.* **37**, 523–32, 1981.

Gardner, M. *Martin Gardner's New Mathematical Diversions from Scientific American.* New York: Simon and Schuster, 1966.

Steiner, J. "Problème de situation." *Ann. de Math* **19**, 36, 1828. Reprinted in *Jacob Steiner's gesammelte Werke, Band I.* Bronx, NY: Chelsea, p. 227, 1971.

Hexahedron

A hexahedron is a POLYHEDRON with six faces. The regular hexahedron is the CUBE, although there are seven topologically different CONVEX hexahedra (Guy 1994, p. 189). Steiner (1828) was the first to enumerate the hexahedra (Duijvestijn and Federico 1981).

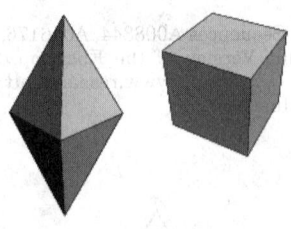

There are exactly two hexahedra composed of identical REGULAR POLYGONS: the regular TRIANGULAR DIPYRAMID (six EQUILATERAL TRIANGLES; left figure) and the CUBE (six SQUARES; right figure).

See also CUBE, HEXAHEDRAL GRAPH, POLYHEDRON, TRIANGULAR DIPYRAMID

References

Duijvestijn, A. J. W. and Federico, P. J. "The Number of Polyhedral (3-Connected Planar) Graphs." *Math. Comput.* **37**, 523–32, 1981.
Guy, R. K. *Unsolved Problems in Number Theory, 2nd ed.* New York: Springer-Verlag, 1994.
Steiner, J. "Problème de situation." *Ann. de Math.* **19**, 36, 1828. Reprinted in *Jacob Steiner's gesammelte Werke, Band I.* Bronx, NY: Chelsea, p. 227, 1971.

Hexahemioctacron

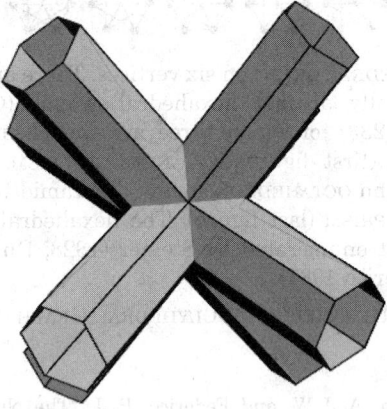

The DUAL POLYHEDRON of the CUBOHEMIOCTAHEDRON U_{15} and Wenninger dual W_{78}. When rendered, the OCTAHEMIOCTACRON and hexahemioctacron appear the same.

See also DUAL POLYHEDRON, CUBOHEMIOCTAHEDRON, OCTAHEMIOCTACRON, UNIFORM POLYHEDRON

References

Wenninger, M. J. *Dual Models.* Cambridge, England: Cambridge University Press, p. 104, 1983.

Hexahemioctahedron

$$(6n - 1)^2 - 3(4m - 1)^2 = -2.$$

The DUAL POLYHEDRON of the CUBOHEMIOCTAHEDRON $2n + 1$. When rendered, the OCTAHEMIOCTACRON and hexahemioctahedron appear the same.

See also DUAL POLYHEDRON, CUBOHEMIOCTAHEDRON, OCTAHEMIOCTACRON, UNIFORM POLYHEDRON

Hexakaidecagon

HEXADECAGON

Hexakis Icosahedron

DISDYAKIS TRIACONTAHEDRON

Hexakis Octahedron

DISDYAKIS DODECAHEDRON

Hexecontahedron

A 60-faced POLYHEDRON. Taking the RHOMBIC TRIACONTAHEDRON, placing a plane along each edge which is perpendicular to the plane of symmetry in which the edge lies, and taking the solid bounded by these planes gives a hexecontahedron (Steinhaus 1999).

See also DELTOIDAL HEXECONTAHEDRON, PENTAGONAL HEXECONTAHEDRON, PENTAKIS DODECAHEDRON, SMALL RHOMBICOSIDODECAHEDRON, SNUB DODECAHEDRON, TRIAKIS ICOSAHEDRON, TRUNCATED DODECAHEDRON, TRUNCATED ICOSAHEDRON

References

Steinhaus, H. *Mathematical Snapshots, 3rd ed.* New York: Dover, p. 210, 1999.

Hexiamond

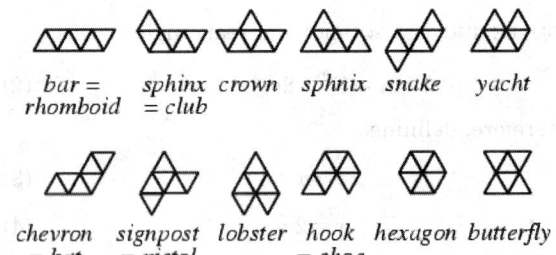

bar =   *sphinx*   *crown*   *sphnix*   *snake*   *yacht*
rhomboid   *= club*

chevron   *signpost*   *lobster*   *hook*   *hexagon*   *butterfly*
= but   *= pistol*   *= shoe*

hexiamonds

A POLYIAMOND composed of six equilateral triangles. The 12 hexiamonds are illustrated above. They are given the names BAR, CROOK, CROWN, SPHINX, SNAKE, YACHT, CHEVRON, SIGNPOST, LOBSTER, HOOK, HEXAGON, and BUTTERFLY.

See also POLYIAMOND, HEXAMOND TILING

References

Gardner, M. *The Sixth Book of Mathematical Games from Scientific American.* Chicago, IL: University of Chicago Press, pp. 174–75, 1984.

O'Beirne, T. H. "Pentominoes and Hexiamonds." *New Scientist* **12**, 379–80, 1961.

O'Beirne, T. H. "Some Hexiamond Solutions and an Introduction to a Set of 25 Remarkable Points." *New Scientists* **12**, 379–80, 1961.

O'Beirne, T. H. "Thirty-Six Triangles Make Six Hexiamonds Make One Triangle." *New Scientist* **12**, 706–07, 1961.

Zimpfer, H. *Die 12 Verhext.* Baden, Germany: privately printed, 1967.

Hexiamond Tiling

There are a number of tilings of various shapes by *all* the 12 order $n = 6$ polyiamonds, summarized in the following table. Several of these (starred in the table below) are also illustrated above (Beeler 1972). Beeler's numbers for the side 6 parallelogram of base 6 and side 4 trapezoid (156 and 76, respectively), differ from those quoted in Gardner (1984, p. 182) of 155 and 74, respectively.

Size	Solutions
side 9 Δ with inverted side 3 Δ hole	0
side 6 trapezoid with bases 3 and 9	0
two side 6 triangles	0
3 × 12 rhomboid	0
4 × 9 rhomboid*	37
side 4 trapezoid with bases 7 and 11*	76
side 6 parallelogram of base 6*	156
triangle of side 9 with 1, 2, 2 corners removed*	5885
trefoil*	several

The following table gives the number of solutions to various hexiamond tilings using fewer than 12 pieces. Those indicated with asterisks (*) have a solution illustrated above.

Size	Pieces	Solutions
2-hexagon		≥ 1
3-hexagon*	9	≥ 15
equilateral Δ		0
hexagonal ring		0
6-point star*	8	1
triangular ring		0
2 × 3 rhomboid		0
2 × 6 rhomboid*	4	1
3 × 3 rhomboid	3	0
3 × 4 rhomboid	4	many
3 × 5 rhomboid	5	many
3 × 6 rhomboid	6	many
3 × 7 rhomboid	7	many
3 × 8 rhomboid	8	many
3 × 9 rhomboid	9	many
3 × 10 rhomboid	10	many
3 × 11 rhomboid*	11	24
4 × 6 rhomboid	8	≥ 1
5 × 6 rhomboid	10	many

See also HEPTIAMOND TILING, HEXIAMOND, OCTIAMOND TILING, PENTIAMOND TILING, POLYHEX TILING, POLYIAMOND, POLYOMINO TILING

References

Beeler, M. Item 112 in Beeler, M.; Gosper, R. W.; and Schroeppel, R. *HAKMEM.* Cambridge, MA: MIT Artificial Intelligence Laboratory, Memo AIM-239, pp. 48–0, Feb. 1972.

Gardner, M. *The Sixth Book of Mathematical Games from Scientific American.* Chicago, IL: University of Chicago Press, pp. 176–81, 1984.

Vichera, M. "Polyiamonds." http://alpha.ujep.cz/~vicher/puzzle/polyform/iamond/iamonds.htm.

Hexlet

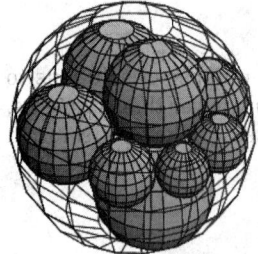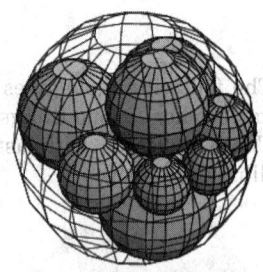

Consider two mutually tangent (externally) SPHERES A and B together with a larger sphere C inside which A and B are internally tangent. Then construct a chain of spheres each tangent externally to A, B and internally to C (so that C encloses the chain as well as the two original spheres). Surprisingly, every such chain closes into a "necklace" after six SPHERES, regardless of where the first SPHERE is placed.

This beautiful and amazing result due to Soddy (1937) is a special case of KOLLROS' THEOREM. It can be demonstrated using INVERSION of six identical spheres around an equal center sphere, all of which are sandwiched between two planes (Wells 1991, pp. 120 and 232). This result was given in a SANGAKU PROBLEM from Kanagawa Prefecture in 1822, more than a century before it was published by Soddy (Rothman 1998).

Moreover, the centers of the six spheres in the necklace and their six points of contact all lie in a plane. Furthermore, there are two planes which touch each of the six spheres, one on either side of the necklace. Finally, the radii r_i of the spheres are related by

$$\frac{1}{r_1} + \frac{1}{r_4} = \frac{1}{r_2} + \frac{1}{r_3} = \frac{1}{r_3} + \frac{1}{r_6}$$

(Rothman 1998).

Soddy's BOWL OF INTEGERS contains an infinite number of nested hexlets. The centers of a Soddy hexlet always lie on an ELLIPSE (Ogilvy 1990, p. 63).

See also BOWL OF INTEGERS, COXETER'S LOXODROMIC SEQUENCE OF TANGENT CIRCLES, DAISY, KOLLROS' THEOREM, SEVEN CIRCLES THEOREM, STEINER CHAIN, TANGENT SPHERES

References

Coxeter, H. S. M. "Interlocking Rings of Spheres." *Scripta Math.* **18**, 113–21, 1952.
Gosset, T. "The Hexlet." *Nature* **139**, 251–52, 1937.
Honsberger, R. *Mathematical Gems II.* Washington, DC: Math. Assoc. Amer., pp. 49–0, 1976.
Morley, F. "The Hexlet." *Nature* **139**, 72–3, 1937.
Ogilvy, C. S. *Excursions in Geometry.* New York: Dover, pp. 60–2, 1990.

Rothman, T. "Japanese Temple Geometry." *Sci. Amer.* **278**, 85–1, May 1998.
Soddy, F. "The Bowl of Integers and the Hexlet." *Nature* **139**, 77–9, 1937.
Soddy, F. "The Hexlet." *Nature* **139**, 154 and 252, 1937.
Wells, D. *The Penguin Dictionary of Curious and Interesting Geometry.* London: Penguin, pp. 120 and 231–32, 1991.

HexLife

An alternative LIFE game similar to Conway's, which is played on a hexagonal grid. No set of rules has yet emerged as uniquely interesting.

See also HIGHLIFE

Hexomino

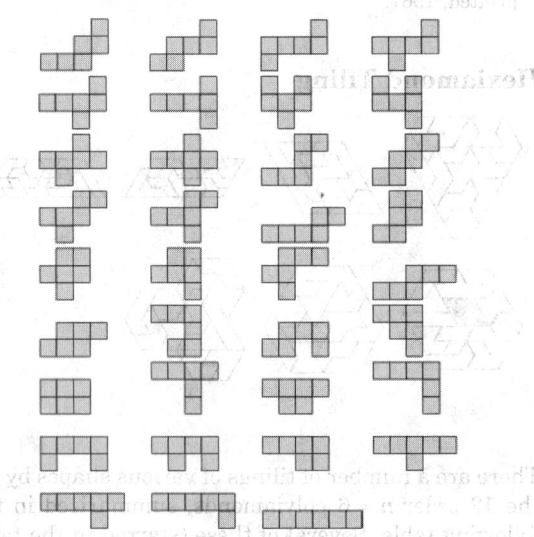

One of the 35 6-POLYOMINOES.

See also DOMINO, HEPTOMINO, OCTOMINO, PENTOMINO, POLYOMINO, TETROMINO, TRIOMINO

References

Pappas, T. "Triangular, Square & Pentagonal Numbers." *The Joy of Mathematics.* San Carlos, CA: Wide World Publ./Tetra, p. 214, 1989.

Heyting Algebra

An ALGEBRA which is a special case of a LOGOS.

See also LOGOS, TOPOS

H-Fractal

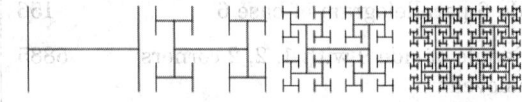

The FRACTAL illustrated above.

H-Function

References

Lauwerier, H. *Fractals: Endlessly Repeated Geometric Figures.* Princeton, NJ: Princeton University Press, pp. 1–, 1991.

Weisstein, E. W. "Fractals." MATHEMATICA NOTEBOOK FRACTAL.M.

H-Function

FOX'S *H*-FUNCTION

Hh Function

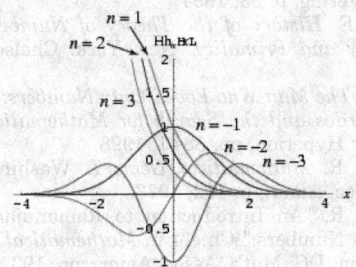

Let

$$Z(x) \equiv \frac{1}{\sqrt{2\pi}} e^{-x^2/2} \tag{1}$$

$$Q(x) \equiv \frac{1}{\sqrt{2\pi}} \int_x^\infty e^{-t^2/2} \, dt \tag{2}$$

$$= \frac{1}{2}\left[1 - \mathrm{erf}\left(\frac{x}{\sqrt{2}}\right)\right], \tag{3}$$

where $Z(x)$ and $Q(x)$ are closely related to the NORMAL DISTRIBUTION FUNCTION, then

$$\mathrm{Hh}_{-n}(x) = (-1)^{n-1} \sqrt{2\pi} \, Z^{(n-1)}(x) \tag{4}$$

$$\mathrm{Hh}_n(x) = \frac{(-1)^n}{n!} \mathrm{Hh}_{-1}(x) \frac{d^n}{dx^n}\left[\frac{Q(x)}{Z(x)}\right]. \tag{5}$$

The first few values are

$$\mathrm{Hh}_{-3}(x) = e^{-x^2/2}(x^2 - 1) \tag{6}$$

$$\mathrm{Hh}_{-2}(x) = e^{-x^2/2} x \tag{7}$$

$$\mathrm{Hh}_{-1}(x) = e^{-x^2/2} \tag{8}$$

$$\mathrm{Hh}_0(x) = 0 \tag{9}$$

$$\mathrm{Hh}_1(x) = e^{-x^2/2} - \sqrt{\frac{\pi}{2}} \, x \, \mathrm{erfc}\left(\frac{x}{\sqrt{2}}\right) \tag{10}$$

$$\mathrm{Hh}_2(x) = \frac{1}{4}\left[-2xe^{-x^2/2} + \sqrt{2\pi}(x^2 + 1)\mathrm{erfc}\left(\frac{x}{\sqrt{2}}\right)\right] \tag{11}$$

$$\mathrm{Hh}_3(x)$$

$$= \frac{1}{12}\left[2e^{-x^2/2}(x^2 + 2) - \sqrt{2\pi}\, x(x^2 + 3)\mathrm{erfc}\left(\frac{x}{\sqrt{2}}\right)\right]. \tag{12}$$

See also NORMAL DISTRIBUTION FUNCTION, TETRACHORIC FUNCTION

References

Jeffreys, H. and Jeffreys, B. S. "The Parabolic Cylinder, Hermite, and Hh Functions" et seq. §23.08–3.081 in *Methods of Mathematical Physics, 3rd ed.* Cambridge, England: Cambridge University Press, pp. 620–27, 1988.

Higher Arithmetic

An archaic term for NUMBER THEORY.

Higher Dimensional Group Theory

The term "higher dimensional group theory" was introduced by Brown (1982), and refers to a method for obtaining new homotopical information by generalizing to higher dimensions the fundamental group of a space with a base point.

See also GROUP THEORY, LOW-DIMENSIONAL TOPOLOGY

References

Brown, R. "Higher Dimensional Group Theory." In *Low-Dimensional Topology: Proceedings of a Conference on Topology in Low Dimension, Bangor, 1979* (Ed. R. Brown and T. L. Thickstun). Cambridge, England: Cambridge University Press, pp. 215–38, 1982.

Brown, R. "Higher Dimensional Group Theory." http://www.bangor.ac.uk/~mas010/hdaweb2.htm.

Higher Geometry

PROJECTIVE GEOMETRY

Highest Common Divisor

GREATEST COMMON DIVISOR

Highest Weight Theorem

A theorem proved by É. Cartan in 1913 which classifies the irreducible representations of COMPLEX semisimple LIE ALGEBRAS.

References

Knapp, A. W. "Group Representations and Harmonic Analysis, Part II." *Not. Amer. Math. Soc.* **43**, 537–49, 1996.

HighLife

An alternate set of LIFE rules similar to Conway's, but with the additional rule that six neighbors generate a birth. Most of the interest in this variant is due to the presence of a so-called replicator.

See also HEXLIFE, LIFE

Highly Abundant Number

HIGHLY COMPOSITE NUMBER

Highly Composite Number

A COMPOSITE NUMBER (also called a SUPERABUNDANT NUMBER) is a number n which has more FACTORS than any other number less than n. In other words, $\sigma(n)/n$ exceeds $\sigma(k)/k$ for all $k < n$, where $\sigma(n)$ is the DIVISOR FUNCTION. They were called highly composite numbers by Ramanujan, who found the first 100 or so, and superabundant numbers by Alaoglu and Erdos (1944).

There are an infinite number of highly composite numbers, and the first few are 2, 4, 6, 12, 24, 36, 48, 60, 120, 180, 240, 360, 720, 840, 1260, 1680, 2520, 5040, ... (Sloane's A002182). Ramanujan (1915) listed 102 up to 6746328388800 (but omitted 293, 318, 625, 600, and 29331862500). Robin (1983) gives the first 5000 highly composite numbers, and a comprehensive survey is given by Nicholas (1988).

If

$$N = 2^{a_2} 3^{a_3} \cdots p^{a_p} \tag{1}$$

is the PRIME FACTORIZATION of a highly composite number, then

1. The PRIMES 2, 3, ..., p form a string of consecutive PRIMES,

2. The exponents are nonincreasing, so $a_2 \geq a_3 \geq \ldots \geq a_p$, and

3. The final exponent a_p is always 1, except for the two cases $N = 4 = 2^2$ and $N = 36 = 2^2 \cdot 3^2$, where it is 2.

Let $Q(x)$ be the number of highly composite numbers $\leq x$. Ramanujan (1915) showed that

$$\lim_{x \to \infty} \frac{Q(x)}{\ln x} = \infty. \tag{2}$$

Erdos (1944) showed that there exists a constant

$c_1 > 0$ such that

$$Q(x) \geq (\ln x)^{1 + c_1} \tag{3}$$

Nicholas proved that there exists a constant $c_2 > 0$ such that

$$Q(x) \ll (\ln x)^{c_2}. \tag{4}$$

See also ABUNDANT NUMBER, ROUND NUMBER, ROUNDNESS, SMOOTH NUMBER

References

Alaoglu, L. and Erdos, P. "On Highly Composite and Similar Numbers." *Trans. Amer. Math. Soc.* **56**, 448–69, 1944.
Andree, R. V. "Ramanujan's Highly Composite Numbers." *Abacus* **3**, 61–2, 1986.
Berndt, B. C. *Ramanujan's Notebooks, Part IV.* New York: Springer-Verlag, p. 53, 1994.
Dickson, L. E. *History of the Theory of Numbers, Vol. 1: Divisibility and Primality.* New York: Chelsea, p. 323, 1952.
Hoffman, P. *The Man Who Loved Only Numbers: The Story of Paul Erdos and the Search for Mathematical Truth.* New York: Hyperion, pp. 88–1, 1998.
Honsberger, R. *Mathematical Gems I.* Washington, DC: Math. Assoc. Amer., p. 112, 1973.
Honsberger, R. "An Introduction to Ramanujan's Highly Composite Numbers." Ch. 14 in *Mathematical Gems III.* Washington, DC: Math. Assoc. Amer., pp. 193–07, 1985.
Kanigel, R. *The Man Who Knew Infinity: A Life of the Genius Ramanujan.* New York: Washington Square Press, p. 232, 1991.
Nicholas, J.-L. "On Highly Composite Numbers." In *Ramanujan Revisited: Proceedings of the Centenary Conference* (Ed. G. E. Andrews, B. C. Berndt, and R. A. Rankin). Boston, MA: Academic Press, pp. 215–44, 1988.
Ramanujan, S. "Highly Composite Numbers." *Proc. London Math. Soc.* **14**, 347–09, 1915.
Ramanujan, S. *Collected Papers.* New York: Chelsea, 1962.
Robin, G. "Méthodes d'optimalisation pour un problème de théories des nombres." *RAIRO Inform. Théor.* **17**, 239–47, 1983.
Séroul, R. "Highly Composite Numbers." §8.14 in *Programming for Mathematicians.* Berlin: Springer-Verlag, pp. 208–13, 2000.
Sloane, N. J. A. Sequences A002182/M1025 in "An On-Line Version of the Encyclopedia of Integer Sequences." http://www.research.att.com/~njas/sequences/eisonline.html.
Wells, D. *The Penguin Dictionary of Curious and Interesting Numbers.* New York: Penguin Books, p. 128, 1986.

Higman-Sims Group

The SPORADIC GROUP *HS*.

References

Wilson, R. A. "ATLAS of Finite Group Representation." http://for.mat.bham.ac.uk/atlas/html/HS.html.

Hilbert Basis

A Hilbert basis for the VECTOR SPACE of square summable sequences $(a_n) = a_1, a_2, \ldots$ is given by the standard basis e_i, where $e_i = \delta_{in}$, with δ_{in} the KRONECKER DELTA. Then

$$(a_n) = \sum a_i e_i,$$

with $\sum |a_i|^2 < \infty$. Although strictly speaking, the e_i are not a BASIS because there exist elements which are not a finite LINEAR COMBINATION, they are given the special term "Hilbert basis."

In general, a HILBERT SPACE V has a Hilbert basis e_i if the e_i are an ORTHONORMAL BASIS and every element $v \in V$ can be written

$$v = \sum_{i=1}^{\infty} a_i e_i$$

for some a_i with $\sum |a_i|^2 < \infty$.

See also BASIS (VECTOR SPACE), FOURIER SERIES, HILBERT SPACE, L2-SPACE, ORTHONORMAL SET

Hilbert Basis Theorem

If R is a NOETHERIAN RING, then $S = R[X]$ is also a NOETHERIAN RING.

See also ALGEBRAIC VARIETY, FUNDAMENTAL SYSTEM, NOETHERIAN RING, SYZYGY

References
Hilbert, D. "Über die Theorie der algebraischen Formen." *Math. Ann.* **36**, 473–34, 1890.

Hilbert Curve

A LINDENMAYER SYSTEM invented by Hilbert (1891) whose limit is a PLANE-FILLING CURVE which fills a square. Traversing the VERTICES of an n-D HYPERCUBE in GRAY CODE order produces a generator for the n-D Hilbert curve (Goetz). The Hilbert curve can be simply encoded with initial string "L", STRING REWRITING rules "L" -> "+RF-LFL-FR+RFR+FL-", and angle 90° (Peitgen and Saupe 1988, p. 278).

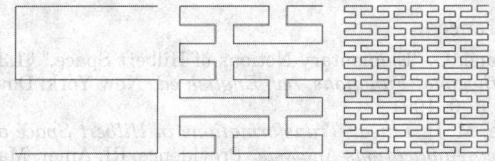

A related curve is the Hilbert II curve, shown above (Peitgen and Saupe 1988, p. 284). It is also a LINDENMAYER SYSTEM and the curve can be encoded with initial string "X", STRING REWRITING rules "X" -> "XFYFX+F+YFXFY-F-XFYFX", "Y" -> "YFX-FY-F-XFYFX+F+YFXFY", and angle 90°.

See also LINDENMAYER SYSTEM, PEANO CURVE, PLANE-FILLING CURVE, SIERPINSKI CURVE, SPACE-FILLING CURVE

References
Bogomolny, A. "Plane Filling Curves." http://www.cut-the-knot.com/do_you_know/hilbert.html.
Dickau, R. M. "Two-Dimensional L-Systems." http://forum.swarthmore.edu/advanced/robertd/lsys2d.html.
Dickau, R. M. "Three-Dimensional L-Systems." http://forum.swarthmore.edu/advanced/robertd/lsys3d.html.
Goetz, P. "Phil's Good Enough Complexity Dictionary." http://www.cs.buffalo.edu/~goetz/dict.html.
Hilbert, D. "Uuml;ber die stetige Abbildung einer Linie auf ein Flachenstück." *Math. Ann.* **38**, 459–60, 1891.
Peitgen, H.-O. and Saupe, D. (Eds.). *The Science of Fractal Images.* New York: Springer-Verlag, pp. 278 and 284, 1988.
Wagon, S. *Mathematica in Action.* New York: W. H. Freeman, pp. 198–06, 1991.
Weisstein, E. W. "Fractals." MATHEMATICA NOTEBOOK FRACTAL.M.
Wells, D. *The Penguin Dictionary of Curious and Interesting Geometry.* London: Penguin, pp. 100–01, 1991.

Hilbert Function

Let $\Gamma = \{p_1, \ldots, p_m\} \subset \mathbb{P}^2$ be a collection of m distinct points. Then the number of conditions imposed by Γ on forms of degree d is called the Hilbert function h_Γ of Γ. If curves X_1 and X_2 of degrees d and e meet in a collection Γ of $d \cdot e$ points, then for any k, the number $h_\Gamma(k)$ of conditions imposed by Γ on forms of degree k is independent of X_1 and X_2 and is given by

$$h_\Gamma(k) = \binom{k+2}{2} - \binom{k-d+2}{2} - \binom{k-e+2}{2} + \binom{k-d-e+2}{2},$$

where the BINOMIAL COEFFICIENT $\binom{a}{2}$ is taken as 0 if $a < 2$ (Cayley 1843).

References
Eisenbud, D.; Green, M.; and Harris, J. "Cayley-Bacharach Theorems and Conjectures." *Bull. Amer. Math. Soc.* **33**, 295–24, 1996.

Hilbert Hotel

Let a hotel have a DENUMERABLE set of rooms numbered 1, 2, 3, ... Then any finite number n of guests can be accommodated without evicting the current guests by moving the current guests from room i to room $i + n$. Furthermore, a DENUMERABLE number of guests can be similarly accommodated by moving the existing guests from i to $2i$, freeing up a DENUMERABLE number of rooms $2i - 1$.

See also CARDINAL NUMBER, DENUMERABLE SET

References
Erickson, G. W. and Fossa, J. A. *Dictionary of Paradox.* Lanham, MD: University Press of America, pp. 84–5, 1998.
Fadiman, C. *Fantasia Mathematica, Being a Set of Stories, Together with a Group of Oddments and Diversions, All Drawn from the Universe of Mathematics.* New York: Simon and Schuster, p. 286, 1958.

Gamow, G. *One, Two, Three, ... Infinity*. New York: Dover, 1988.

Hoffman, P. *The Man Who Loved Only Numbers: The Story of Paul Erdos and the Search for Mathematical Truth*. New York: Hyperion, p. 222, 1998.

Lauwerier, H. "Hilbert Hotel." In *Fractals: Endlessly Repeated Geometric Figures*. Princeton, NJ: Princeton University Press, p. 22, 1991.

Pappas, T. "Hotel Infinity." *The Joy of Mathematics*. San Carlos, CA: Wide World Publ./Tetra, p. 37, 1989.

Hilbert Matrix

A MATRIX H with elements

$$H_{ij} \equiv (i+j-1)^{-1}$$

for $i, j = 1, 2, ..., n$. Hilbert matrices are given by HilbertMatrix[m, n] in the *Mathematica* add-on package LinearAlgebra`MatrixManipulation` (which can be loaded with the command <<LinearAlgebra`). Although the MATRIX INVERSE is given analytically by

$$(H^{-1})_{ij} = \frac{(-1)^{i+j}}{i+j-1} \frac{(n+i-1)!(n+j-1)!}{[(i-1)!(j-1)!]^2(n-i)!(n-j)!},$$

Hilbert matrices are difficult to invert numerically. The DETERMINANTS for the first few values of H_n are given in the following table, and the numerical values for $n = 1, 2, ...$ are given by one divided by 1, 12, 2160, 6048000, 266716800000, ... (Sloane's A005249).

n	$\det(H)$
1	1
2	8.33333×10^{-2}
3	4.62963×10^{-4}
4	1.65344×10^{-7}
5	3.74930×10^{-12}
6	5.36730×10^{-18}

References

Choi, M.-D. "Tricks or Treats with the Hilbert Matrix." *Amer. Math. Monthly* **90**, 301–12, 1983.

Richardson, T. M. 1999. http://xxx.lanl.gov/abs/math.LA/9905079/.

Sloane, N. J. A. Sequences A005249/M4882 in "An On-Line Version of the Encyclopedia of Integer Sequences." http://www.research.att.com/~njas/sequences/eisonline.html.

Hilbert Number

GELFOND-SCHNEIDER CONSTANT

Hilbert Polynomial

Let Γ be an ALGEBRAIC CURVE in a projective space of DIMENSION n, and let p be the PRIME IDEAL defining Γ, and let $\chi(p, m)$ be the number of linearly independent forms of degree m modulo p. For large m, $\chi(p, m)$ is a POLYNOMIAL known as the Hilbert polynomial.

References

Iyanaga, S. and Kawada, Y. (Eds.). *Encyclopedic Dictionary of Mathematics*. Cambridge, MA: MIT Press, p. 36, 1980.

Hilbert Space

A Hilbert space is a VECTOR SPACE H with an INNER PRODUCT $\langle f, g \rangle$ such that the NORM defined by

$$|f| = \sqrt{\langle f, f \rangle}$$

turns H into a COMPLETE METRIC SPACE. If the INNER PRODUCT does not so define a NORM, it is instead known as an INNER PRODUCT SPACE.

Examples of FINITE-dimensional Hilbert spaces include

1. The REAL NUMBERS $\mathbb{R}^n$ with $\langle v, u \rangle$ the vector DOT PRODUCT of v and u.
2. The COMPLEX NUMBERS $\mathbb{C}^n$ with $\langle v, u \rangle$ the vector DOT PRODUCT of v and the COMPLEX CONJUGATE of u.

An example of an INFINITE-dimensional Hilbert space is L^2, the SET of all FUNCTIONS $f : \mathbb{R} \to \mathbb{R}$ such that the INTEGRAL of f^2 over the whole REAL LINE is FINITE. In this case, the INNER PRODUCT is

$$\langle f, g \rangle = \int_{-\infty}^{\infty} f(x)g(x)\,dx.$$

A Hilbert space is always a BANACH SPACE, but the converse need not hold.

See also BANACH SPACE, COMPLETE SET OF FUNCTIONS, HILBERT BASIS, $L2$-NORM, $L2$-SPACE, LIOUVILLE SPACE, PARALLELOGRAM LAW, VECTOR SPACE

References

Sansone, G. "Elementary Notions of Hilbert Space." §1.3 in *Orthogonal Functions, rev. English ed.* New York: Dover, pp. 5–0, 1991.

Stone, M. H. *Linear Transformations in Hilbert Space and Their Applications Analysis*. Providence, RI: Amer. Math. Soc., 1932.

Hilbert Symbol

For any two nonzero P-ADIC NUMBERS a and b, the Hilbert symbol is defined as

$$(a, b) = \begin{cases} 1 & \text{if } z^2 = ax^2 + by^2 \text{ has a nonzero solution} \\ -1 & \text{otherwise.} \end{cases}$$

If the p-adic field is not clear, it is said to be the

Hilbert symbol of a and b relative to k. The field can also be the reals ($p = \infty$). The Hilbert symbol satisfies the following formulas:

1. $(a, b) = (b, a)$.
2. $(a, c^2) = 1$ for any c.
3. $(a, -a) = 1$.
4. $(a, 1-a) = 1$.
5. $(a, b) = 1 \Rightarrow (aa', b) = (a', b)$.
6. $(a, b) = (a, -ab) = (a, (1-a)b)$.

The Hilbert symbol depends only the values of a and b modulo squares. So the symbol is a map $k^*/k^{*2} \times k^*/k^{*2} \to \{1, -1\}$.

Hilbert showed that for any two nonzero rational numbers a and b,

1. $(a, b)_v = 1$ for almost every prime v.
2. $\prod (a, b)_v = 1$ where v ranges over every prime, including $v = \infty$ corresponding to the reals.

See also Diophantine Equation–2nd Powers, Field, *P*-adic Number, Symmetric Bilinear Form (General Fields), Vector Space

References

Serre, J. P. *A Course in Arithmetic.* New York: Springer-Verlag, pp. 27–5, 1973.

Hilbert Transform

The Integral Transform

$$g(y) = \mathscr{H}[f(x)] = \frac{1}{\pi} \int_{-\infty}^{\infty} \frac{f(x)\, dx}{x - y}$$

$$f(x) = \mathscr{H}^{-1}[g(y)] = \frac{1}{\pi} \int_{-\infty}^{\infty} \frac{g(y)\, dy}{y - x},$$

where the Cauchy principal value is taken in each of the integrals.

In the following table, $\Pi(x)$ is the Rectangle Function, sinc x is the Sinc Function, $\delta(x)$ is the Delta Function, $\Pi(x)$ and $I_I(x)$ are Impulse Symbols, and $_1F_1(a; b; x)$ is a Confluent Hypergeometric Function of the First Kind.

$f(x)$	$g(y)$
$\sin x$	$\cos y$
$\cos x$	$-\sin y$
$\sin x$	$\cos y - 1$
x	y
$\Pi(x)$	$\dfrac{1}{\pi} \ln \left| \dfrac{y - \frac{1}{2}}{y + \frac{1}{2}} \right|$

$\dfrac{1}{1 + x^2}$	$-\dfrac{y}{1 + y^2}$
sinc$'$ x	$-\pi \,\text{sinc}\, y - \frac{1}{2}\pi\, \text{sinc}^2(\frac{1}{2}\pi y)$
$\delta(x)$	$-\dfrac{1}{\pi y}$
$\Pi(x)$	$\dfrac{y}{\pi(\frac{1}{4} - y^2)}$
$I_I(x)$	$-\dfrac{1}{2\pi(\frac{1}{4} - y^2)}$
e^{-x^2}	$-\dfrac{2y}{\sqrt{\pi}} \,_1F_1(a; b; x)$

See also Abel Transform, Fourier Transform, Integral Transform, Titchmarsh Theorem, Wiener-Lee Transform

References

Bracewell, R. "The Hilbert Transform." *The Fourier Transform and Its Applications, 3rd ed.* New York: McGraw-Hill, pp. 267–72, 1999.
Papoulis, A. "Hilbert Transforms." *The Fourier Integral and Its Applications.* New York: McGraw-Hill, pp. 198–01, 1962.

Hilbert's Axioms

The 21 assumptions which underlie the Geometry published in Hilbert's classic text *Grundlagen der Geometrie.* The eight Incidence Axioms concern collinearity and intersection and include the first of Euclid's Postulates. The four Ordering Axioms concern the arrangement of points, the five Congruence Axioms concern geometric equivalence, and the three Continuity Axioms concern continuity. There is also a single parallel axiom equivalent to Euclid's Parallel Postulate.

See also Congruence Axioms, Continuity Axioms, Incidence Axioms, Ordering Axioms, Parallel Postulate

References

Hilbert, D. *The Foundations of Geometry, 2nd ed.* Chicago, IL: Open Court, 1980.
Iyanaga, S. and Kawada, Y. (Eds.). "Hilbert's System of Axioms." §163B in *Encyclopedic Dictionary of Mathematics.* Cambridge, MA: MIT Press, pp. 544–45, 1980.

Hilbert's Constants

N.B. A detailed online essay by S. Finch was the starting point for this entry.

Extend HILBERT'S INEQUALITY by letting p, $q > 1$ and

$$\frac{1}{p} + \frac{1}{q} \geq 1, \tag{1}$$

so that

$$0 < \lambda = 2 - \frac{1}{p} - \frac{1}{q} \leq 1. \tag{2}$$

Levin (1937) and Steckin (1949) showed that

$$\sum_{m=1}^{\infty} \sum_{n=1}^{\infty} \frac{a_m b_n}{(m+n)^\lambda}$$
$$\leq \left\{ \pi \csc\left[\frac{\pi(q-1)}{\lambda q}\right] \right\}^\lambda \left[\sum_{m=1}^{\infty} (a_m)^p\right]^{1/p} \left[\sum_{n=1}^{\infty} (a_n)^q\right]^{1/q} \tag{3}$$

and

$$\int_0^\infty \int_0^\infty \frac{f(x)g(y)}{(x+y)^\lambda} \, dx \, dy < \pi \csc\left[\frac{\pi(q-1)}{p}\right]^\lambda$$
$$\times \left(\int_0^\infty [f(x)]^p \, dx\right)^{1/p} \left(\int_0^\infty [g(x)]^q \, dx\right)^{1/q}. \tag{4}$$

Mitrinovic *et al.* (1991) indicate that this constant is the best possible.

See also HILBERT'S INEQUALITY

References

Finch, S. "Favorite Mathematical Constants." http://www.mathsoft.com/asolve/constant/hilbert/hilbert.html.
Mitrinovic, D. S.; Pecaric, J. E.; and Fink, A. M. *Inequalities Involving Functions and Their Integrals and Derivatives.* Dordrecht, Netherlands: Kluwer, 1991.
Steckin, S. B. "On the Degree of Best Approximation to Continuous Functions." *Dokl. Akad. Nauk SSSR* **65**, 135–37, 1949.

Hilbert's Inequality

Given a POSITIVE SEQUENCE $\{a_n\}$,

$$\sqrt{\sum_{j=-\infty}^{\infty} \left|\sum_{\substack{n=-\infty \\ n \neq j}}^{\infty} \frac{a_n}{j-n}\right|^2} \leq \pi \sqrt{\sum_{n=-\infty}^{\infty} |a_n|^2},$$

where the a_ns are REAL and "square summable."

Another INEQUALITY known as Hilbert's applies to NONNEGATIVE sequences $\{a_n\}$ and $\{b_n\}$,

$$\sum_{m=1}^{\infty} \sum_{n=1}^{\infty} \frac{a_m b_n}{m+n} < \pi \csc\left(\frac{\pi}{p}\right) \left(\sum_{m=1}^{\infty} a_m^p\right)^{1/p} \left(\sum_{n=1}^{\infty} b_n^q\right)^{1/q}$$

unless all a_n or all b_n are 0. If $f(x)$ and $g(x)$ are NONNEGATIVE integrable functions, then the integral

form is

$$\int_0^\infty \int_0^\infty \frac{f(x)g(y)}{x+y} \, dx \, dy < \pi \csc\left(\frac{\pi}{p}\right)$$
$$\times \left(\int_0^\infty [f(x)]^p \, dx\right)^{1/p} \left(\int_0^\infty [g(x)]^q \, dx\right)^{1/q}.$$

The constant $\pi \csc(\pi/P)$ is the best possible, in the sense that counterexamples can be constructed for any smaller value.

References

Hardy, G. H.; Littlewood, J. E.; and Pólya, G. "Hilbert's Double Series Theorem" and "On Hilbert's Inequality." §9.1 and Appendix III in *Inequalities, 2nd ed.* Cambridge, England: Cambridge University Press, pp. 226–27 and 308–09, 1988.

Hilbert's Nullstellensatz

Let K be an algebraically closed field and let $\mathfrak{I}$ be an IDEAL in $K(x)$, where $x(x_1, x_2, \ldots, x_n$ is a finite set of indeterminates. Let $p \in K(x)$ be such that for any $(c_1, \ldots, c_n$ in K^n, if every element of $\mathfrak{I}$ vanishes when evaluated if we set each $(x_i = c_i)$, then p also vanishes. Then p^i lies in $\mathfrak{I}$ for some j. Colloquially, the theory of algebraically closed fields is a complete model.

See also ALGEBRAIC SET, IDEAL

References

Becker, T. and Weispfenning, V. "The Hilbert Nullstellensatz." §7.4 in *Gröbner Bases: A Computational Approach to Commutative Algebra.* New York: Springer-Verlag, pp. 312–23, 1993.
Hartshorne, R. *Algebraic Geometry.* New York: Springer-Verlag, 1977.

Hilbert's Problems

A set of (originally) unsolved problems in mathematics proposed by Hilbert. Of the 23 total, ten were presented at the Second International Congress in Paris in 1900. These problems were designed to serve as examples for the kinds of problems whose solutions would lead to the furthering of disciplines in mathematics.

1a. Is there a transfinite number between that of a DENUMERABLE SET and the numbers of the CONTINUUM? This question was answered by Gödel and Cohen to the effect that the answer depends on the particular version of SET THEORY assumed. 1b. Can the CONTINUUM of numbers be considered a WELL ORDERED SET? This question is related to Zermelo's AXIOM OF CHOICE. In 1963, the AXIOM OF CHOICE was demonstrated to be independent of all other AXIOMS in SET THEORY, so there appears to be no universally valid solution to this question either.

2. Can it be proven that the AXIOMS of logic are consistent? GÖDEL'S INCOMPLETENESS THEOREM indicated that the answer is "no," in the sense that any formal system interesting enough to formulate its own consistency can prove its own consistency IFF it is inconsistent.

3. Give two TETRAHEDRA which cannot be decomposed into congruent TETRAHEDRA directly or by adjoining congruent TETRAHEDRA. Max Dehn showed this could not be done in 1902 by inventing the theory of DEHN INVARIANTS, and W. F. Kagon obtained the same result independently in 1903.

4. Find GEOMETRIES whose AXIOMS are closest to those of EUCLIDEAN GEOMETRY if the ORDERING and INCIDENCE AXIOMS are retained, the CONGRUENCE AXIOMS weakened, and the equivalent of the PARALLEL POSTULATE omitted. This problem was solved by G. Hamel.

5. Can the assumption of differentiability for functions defining a continuous transformation GROUP be avoided? (This is a generalization of the CAUCHY FUNCTIONAL EQUATION.) Solved by John von Neumann in 1930 for bicompact groups. Also solved for the ABELIAN case, and for the solvable case in 1952 with complementary results by Montgomery and Zipin (subsequently combined by Yamabe in 1953). Andrew Glean showed in 1952 that the answer is also "yes" for all locally bicompact groups.

6. Can physics be axiomized?

7. Let $a \neq 1 \neq 0$ be ALGEBRAIC and β IRRATIONAL. Is α^β then TRANSCENDENTAL (Wells 1986, p. 45)? α^β is known to be transcendental for the special case of β an ALGEBRAIC NUMBER, as proved in 1934 by Aleksander Gelfond in a result now known as GELFOND'S THEOREM (Courant and Robins 1996). However, the case of general irrational β has not been resolved.

8. Prove the RIEMANN HYPOTHESIS. The CONJECTURE has still been neither proved nor disproved.

9. Construct generalizations of the RECIPROCITY THEOREM of NUMBER THEORY.

10. Does there exist a universal algorithm for solving DIOPHANTINE EQUATIONS? The impossibility of obtaining a general solution was proven by Julia Robinson and Martin Davis in 1970, following proof of the result that the relation $n = F_{2m}$ (where F_{2m} is a FIBONACCI NUMBER) is Diophantine by Yuri Matijasevich (Matiyasevich 1970; Davis 1973; Davis and Hersh 1973; Davis 1982; Matiyasevich 1993; Reid 1997, p. 107). More specifically, Matiyasevich showed that there is a polynomial P in n, m, and a number of other variables $x, y, z, \ldots$

having the property that $n = F_{2m}$ IFF there exist integers $x, y, z, \ldots$ such that $P(n, m, x, y, z, \ldots) = 0$.

11. Extend the results obtained for quadratic fields to arbitrary INTEGER algebraic fields.

12. Extend a theorem of Kronecker to arbitrary algebraic fields by explicitly constructing Hilbert class fields using special values. This calls for the construction of HOLOMORPHIC FUNCTIONS in several variables which have properties analogous to the exponential function and elliptic modular functions (Holzapfel 1995).

13. Show the impossibility of solving the general seventh degree equation by functions of two variables.

14. Show the finiteness of systems of relatively integral functions.

15. Justify Schubert's ENUMERATIVE GEOMETRY (Bell 1945).

16. Develop a topology of real algebraic curves and surfaces. The TANIYAMA-SHIMURA CONJECTURE postulates just this connection. See Gudkov and Utkin (1978), Ilyashenko and Yakovenko (1995), and Smale (2000).

17. Find a representation of definite form by SQUARES.

18. Build spaces with congruent POLYHEDRA.

19. Analyze the analytic character of solutions to variational problems.

20. Solve general BOUNDARY VALUE PROBLEMS.

21. Solve differential equations given a MONODROMY GROUP. More technically, prove that there always exists a FUCHSIAN SYSTEM with given singularities and a given MONODROMY GROUP. Several special cases had been solved, but a NEGATIVE solution was found in 1989 by B. Bolibruch (Anasov and Bolibruch 1994).

22. Uniformization.

23. Extend the methods of CALCULUS OF VARIATIONS.

See also GELFOND'S THEOREM, RIEMANN HYPOTHESIS, TANIYAMA-SHIMURA CONJECTURE, UNSOLVED PROBLEMS

References

Anasov, D. V. and Bolibruch, A. A. *The Riemann-Hilbert Problem.* Braunschweig, Germany: Vieweg, 1994.

Bell, E. T. *The Development of Mathematics, 2nd ed.* New York: McGraw-Hill, p. 340, 1945.

Borowski, E. J. and Borwein, J. M. (Eds.). "Hilbert Problems." Appendix 3 in *The Harper Collins Dictionary of Mathematics.* New York: Harper-Collins, p. 659, 1991.

Boyer, C. and Merzbach, U. "The Hilbert Problems." *History of Mathematics, 2nd ed.* New York: Wiley, pp. 610–14, 1991.

Browder, Felix E. (Ed.). *Mathematical Developments Arising from Hilbert Problems.* Providence, RI: Amer. Math. Soc., 1976.

Courant, R. and Robbins, H. *What is Mathematics?: An Elementary Approach to Ideas and Methods, 2nd ed.* Oxford, England: Oxford University Press, p. 107, 1996.

Davis, M. "Hilbert's Tenth Problem is Unsolvable." *Amer. Math. Monthly* **80**, 233–69, 1973.

Davis, M. and Hersh, R. "Hilbert's 10th Problem." *Sci. Amer.* **229**, 84–1, Nov. 1973.

Davis, M. "Hilbert's Tenth Problem is Unsolvable." Appendix 2 in *Computability and Unsolvability.* New York: Dover, 1999–35, 1982.

Gudkov, D. and Utkin, G. A. *Nine Papers on Hilbert's 16th Problem.* Providence, RI: Amer. Math. Soc., 1978.

Hilbert, D. "Mathematical Problems." *Bull. Amer. Math. Soc.* **8**, 437–79, 1901–902.

Holzapfel, R.-P. *The Ball and Some Hilbert Problems.* Boston, MA: Birkhäuser, 1995.

Ilyashenko, Yu. and Yakovenko, S. (Eds.). *Concerning the Hilbert 16th Problem.* Providence, RI: Amer. Math. Soc., 1995.

Itô, K. (Ed.). "Hilbert, David." §196 in *Encyclopedic Dictionary of Mathematics, 2nd ed., Vol. 2.* Cambridge, MA: MIT Press, pp. 736–37, 1987.

Joyce, D. E. "The Mathematical Problems of David Hilbert." http://aleph0.clarku.edu/~djoyce/hilbert/.

Matiyasevich, Yu. V. "Solution to of the Tenth Problem of Hilbert." *Mat. Lapok* **21**, 83–7, 1970.

Matijasevich, Yu. V. *Hilbert's Tenth Problem.* Cambridge, MA: MIT Press, 1993. http://www.informatik.uni-stuttgart.de/ifi/ti/personen/Matiyasevich/H10Pbook/.

Reid, C. *Julia: A Life in Mathematics.* Washington, DC: Math. Assoc. Amer., 1997.

Schroeppel, R. C. Transcription of Hilbert's Problems Lecture. http://www.cs.arizona.edu/~rcs/hilbert-speech.

Smale, S. "Mathematical Problems for the Next Century." In *Mathematics: Frontiers and Perspectives 2000* (Ed. V. Arnold, M. Atiyah, P. Lax, and B. Mazur). Providence, RI: Amer. Math. Soc., 2000.

Vsemirnov, M. "Welcome to Hilbert's Tenth Problem Page!" http://logic.pdmi.ras.ru/Hilbert10/.

Waldschmidt, M. "Schneider's Solution of Hilbert's Seventh Problem." §3.1 in *Transcendence Methods.* Queen's Papers in Pure and Applied Mathematics, No. 52. Kingston, Ontario, Canada: Queen's University, pp. 3.1–.4, 1979.

Weisstein, E. W. "Books about Hilbert's Problems." http://www.treasure-troves.com/books/HilbertsProblems.html.

Wells, D. *The Penguin Dictionary of Curious and Interesting Numbers.* Middlesex, England: Penguin Books, p. 45, 1986.

Hilbert's Theorem

Every MODULAR SYSTEM has a MODULAR SYSTEM BASIS consisting of a finite number of POLYNOMIALS. Stated another way, for every order n there exists a nonsingular curve with the maximum number of circuits and the maximum number for any one nest.

References
Coolidge, J. L. *A Treatise on Algebraic Plane Curves.* New York: Dover, p. 61, 1959.

Hilbert-Schmidt Norm

The Hilbert-Schmidt norm of a MATRIX A is defined as

$$|\mathsf{A}|_2 \equiv \sqrt{\sum_{ij} a_{ij}^2}.$$

Hilbert-Schmidt Theory

The study of linear integral equations of the Fredholm type with symmetric kernels

$$K(x,\, t) = K(t,\, x).$$

References
Arfken, G. "Hilbert-Schmidt Theory." §16.4 in *Mathematical Methods for Physicists, 3rd ed.* Orlando, FL: Academic Press, pp. 890–97, 1985.

Hill Determinant

A DETERMINANT which arises in the solution of the second-order ORDINARY DIFFERENTIAL EQUATION

$$x^2 \frac{d^2\psi}{dx^2} + x\, \frac{d\psi}{dx} + \left(\tfrac{1}{4} h^2 x^2 + \tfrac{1}{2} h^2 - b + \frac{h^2}{4x^2}\right)\psi = 0. \quad (1)$$

Writing the solution as a POWER SERIES

$$\psi = \sum_{n=-\infty}^{\infty} a_n x^{s+2n} \quad (2)$$

gives a RECURRENCE RELATION

$$h^2 a_{n+1} + [2h^2 - 4b + 16(n + \tfrac{1}{2} s)^2]a_n + h^2 a_{n-1} = 0. \quad (3)$$

The value of s can be computed using the Hill determinant

$$\Delta(s) = \begin{vmatrix} \ddots & \vdots & \vdots & \vdots & \vdots & \iddots \\ \cdots & \frac{(\sigma+2)-\alpha^2}{4-\alpha^2} & \frac{\beta^2}{4-\alpha^2} & 0 & 0 & \cdots \\ \cdots & 0 & -\frac{\beta^2}{\alpha^2} & -\frac{\sigma^2-\alpha^2}{\alpha^2} & \frac{\beta^2}{\alpha^2} & \cdots \\ \cdots & 0 & 0 & -\frac{\beta^2}{1-\alpha^2} & \frac{(\sigma-1)^2-a^2}{1-\alpha^2} & \cdots \\ \iddots & \vdots & \vdots & \vdots & \vdots & \ddots \end{vmatrix} \quad (4)$$

where

$$\sigma = \tfrac{1}{2} s \quad (5)$$

$$\alpha^2 = \tfrac{1}{4} b - \tfrac{1}{8} h^2 \quad (6)$$

$$\beta = \tfrac{1}{4} h, \quad (7)$$

and σ is the variable to solve for. The determinant can be given explicitly by the amazing formula

$$\Delta(s) = \Delta(0) - \frac{\sin^2(\pi s/2)}{\sin^2\left(\tfrac{1}{2}\pi\sqrt{b - \tfrac{1}{2} h^2}\right)}, \quad (8)$$

where

$$\Delta(0) = \begin{vmatrix} \ddots & \vdots & \vdots & \vdots & \vdots & \\ \cdots & 1 & \frac{h^2}{144+2h^2-4b} & 0 & 0 & \cdots \\ \cdots & \frac{h^2}{64+2h^2-4b} & 1 & \frac{h^2}{64+2h^2-4b} & 0 & \cdots \\ \cdots & 0 & \frac{h^2}{16+2h^2-4b} & 1 & \frac{h^2}{16+2h^2-4b} & \cdots \\ \cdots & 0 & 0 & \frac{h^2}{2h^2-4b} & 1 & \cdots \\ \cdots & 0 & 0 & 0 & \frac{h^2}{16+2h^2-4b} & \cdots \\ & \vdots & \vdots & \vdots & \vdots & \ddots \end{vmatrix}$$

(9)

leading to the implicit equation for s,

$$\sin^2\left(\tfrac{1}{2}\pi s\right) = \Delta(0)\sin^2\left(\tfrac{1}{2}\pi\sqrt{b-\tfrac{1}{2}h^2}\right).$$ (10)

See also HILL'S DIFFERENTIAL EQUATION

References

Hill, G. W. "On the Part of the Motion of Lunar Perigee Which is a Function of the Mean Motions of the Sum and Moon." *Acta Math.* **8**, 1–6, 1886.

Magnus, W. and Winkler, S. *Hill's Equation.* New York: Dover, 1979.

Morse, P. M. and Feshbach, H. *Methods of Theoretical Physics, Part I.* New York: McGraw-Hill, pp. 555–62, 1953.

Hill's Differential Equation

The second-order ORDINARY DIFFERENTIAL EQUATION

$$\frac{d^2y}{dx^2} + \left[\theta_0 + 2\sum_{n=1}^{\infty}\theta_n\cos(2nx)\right]y = 0,$$ (1)

where θ_n are fixed constants. A general solution can be given by taking the "DETERMINANT" of an infinite MATRIX.

If only the $n=0$ term is present, the equation have solution

$$y = C_1\sin(x\sqrt{\theta_0}) + C_2\cos(x\sqrt{\theta_0}).$$ (2)

If terms $n \le 1$ are included, the equation becomes the MATHIEU DIFFERENTIAL EQUATION, which has solution

$$y = C_1 C(a, -\tfrac{1}{2}b, x) + C_2 S\left(a, -\tfrac{1}{2}b, x\right).$$ (3)

If terms $n \le 2$ are included, it becomes the WHITTAKER-HILL DIFFERENTIAL EQUATION.

See also HILL DETERMINANT, WHITTAKER-HILL DIFFERENTIAL EQUATION

References

Hill, G. W. "On the Part of the Motion of Lunar Perigee Which is a Function of the Mean Motions of the Sun and Moon." *Acta Math.* **8**, 1–6, 1886.

Ince, E. L. *Ordinary Differential Equations.* New York: Dover, p. 384, 1956.

Magnus, W. and Winkler, S. *Hill's Equation.* New York: Dover, 1979.

Zwillinger, D. *Handbook of Differential Equations, 3rd ed.* Boston, MA: Academic Press, p. 123, 1997.

Hillam's Theorem

If $f : [a,\ b] \to [a,\ b]$ (where $[a,\ b]$ denotes the CLOSED INTERVAL from a to b on the REAL LINE) satisfies a LIPSCHITZ CONDITION with constant K, i.e., if

$$|f(x) - f(y)| \le K|x - y|$$

for all $x,\ y \in [a,\ b]$, then the iteration scheme

$$x_{n+1} = (1 - \lambda)x_n + \lambda f(x_n),$$

where $\lambda = 1/(K+1)$, converges to a FIXED POINT of f.

References

Falkowski, B.-J. "On the Convergence of Hillam's Iteration Scheme." *Math. Mag.* **69**, 299–03, 1996.

Geist, R.; Reynolds, R.; and Suggs, D. "A Markovian Framework for Digital Halftoning." *ACM Trans. Graphics* **12**, 136–59, 1993.

Hillam, B. P. "A Generalization of Krasnoselski's Theorem on the Real Line." *Math. Mag.* **48**, 167–68, 1975.

Krasnoselski, M. A. "Two Remarks on the Method of Successive Approximations." *Uspehi Math. Nauk (N. S.)* **10**, 123–27, 1955.

Hindu Check

CASTING OUT NINES

Hinge

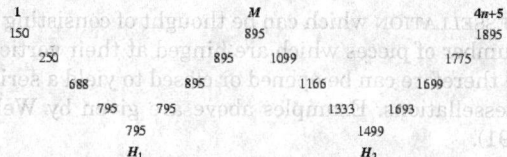

The upper and lower hinges are descriptive statistics of a set of N data values, where N is OF THE FORM $N = 4n + 5$ with $n = 0,\ 1,\ 2,\ \dots$. The hinges are obtained by ordering the data in increasing order $a_1, \dots, a_N$, and writing them out in the shape of a "w" as illustrated above. The values at the bottom legs are called the hinges H_1 and H_2 (and the central peak is the MEDIAN). In this ordering,

$$H_1 = a_{n+2} = a_{(N+3)/4}$$

$$M = a_{2n+3} = a_{(N+1)/2}$$

$$H_2 = a_{3n+4} = a_{(3N+1)/4}.$$

For N OF THE FORM $4n + 5$, the hinges are identical to the QUARTILES. The difference $H_2 - H_1$ is called the H-SPREAD.

See also H-SPREAD, HABERDASHER'S PROBLEM, MEDIAN (STATISTICS), ORDER STATISTIC, QUARTILE, TRIMEAN

References

Tukey, J. W. *Explanatory Data Analysis.* Reading, MA: Addison-Wesley, pp. 32–4, 1977.

Hinged Tessellation

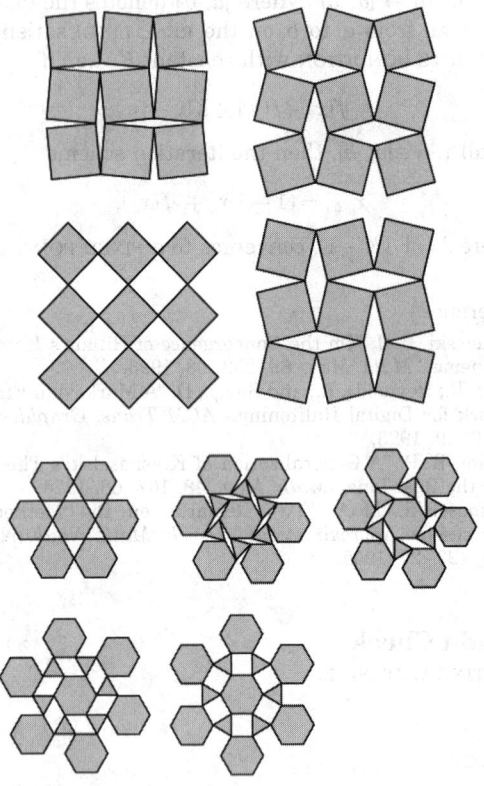

A TESSELLATION which can be thought of consisting of a number of pieces which are hinged at their vertices and therefore can be opened or closed to yield a series of tessellations. Examples above are given by Wells (1991).

See also BRACED SQUARE, TESSELLATION

References

Wells, D. *Hidden Connections, Double Meanings.* Cambridge, England: Cambridge University Press, 1988.
Wells, D. *The Penguin Dictionary of Curious and Interesting Geometry.* London: Penguin, pp. 101–03, 1991.

Hippias' Quadratrix

QUADRATRIX OF HIPPIAS

Hippopede

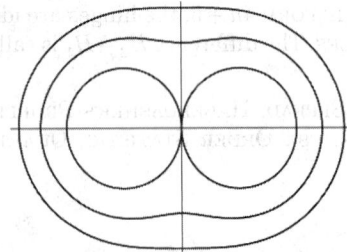

A curve also known as a HORSE FETTER and given by

the polar equation

$$r^2 = 4b(a - b \sin^2 \theta).$$

References

Lawrence, J. D. *A Catalog of Special Plane Curves.* New York: Dover, pp. 144–46, 1972.

Hi-Q

A triangular version of PEG SOLITAIRE with 15 holes and 14 pegs. Numbering hole 1 at the apex of the triangle and thereafter from left to right on the next lower row, etc., the following table gives possible ending holes for a single peg removed (Beeler 1972). Because of symmetry, only the first five pegs need be considered. Also because of symmetry, removing peg 2 is equivalent to removing peg 3 and flipping the board horizontally.

remove	possible ending pegs
1	1, 7 = 10, 13
2	2, 6, 11, 14
4	3 = 12, 4, 9, 15
5	13

References

Beeler, M. Item 76 in Beeler, M.; Gosper, R. W.; and Schroeppel, R. *HAKMEM.* Cambridge, MA: MIT Artificial Intelligence Laboratory, Memo AIM-239, p. 29, Feb. 1972.

Hirota Equation

The PARTIAL DIFFERENTIAL EQUATION

$$u_t + iau + ib(u_{xx} - 2\eta|u^2|u) + cu_x + d(u_{xxx} - 6\eta|u|^2) = 0.$$

References

Calogero, F. and Degasperis, A. *Spectral Transform and Solitons: Tools to Solve and Investigate Nonlinear Evolution Equations.* New York: North-Holland, p. 56, 1982.
Zwillinger, D. *Handbook of Differential Equations, 3rd ed.* Boston, MA: Academic Press, p. 133, 1997.

Hirota-Satsuma Equation

The system of PARTIAL DIFFERENTIAL EQUATIONS

$$u_t = \tfrac{1}{2} u_{xxx} + 3uu_x - 6ww_x$$

$$w_t = -w_{xxx} - 3uw_x.$$

Histogram

References

Weiss, J. "Periodic Fixed Points of Bäcklund Transformation and the Korteweg-de Vries Equation." *J. Math. Phys.* **27**, 2647–656, 1986.
Zwillinger, D. *Handbook of Differential Equations, 3rd ed.* Boston, MA: Academic Press, p. 138, 1997.

Histogram

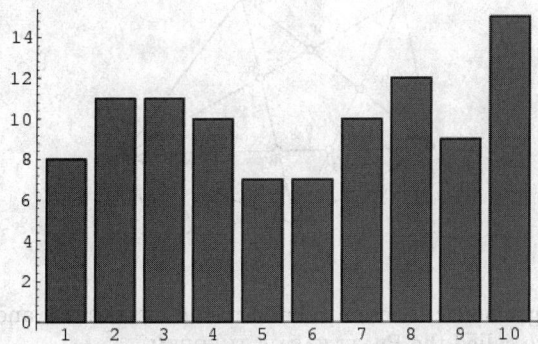

The grouping of data into BINS (spaced apart by the so-called CLASS INTERVAL) plotting the number of members in each bin versus the bin number. The above histogram shows the number of variates in bins with CLASS INTERVAL 1 for a sample of 100 real variates with a UNIFORM DISTRIBUTION from 0 and 10. Therefore, bin 1 gives the number of variates in the range 0–, bin 2 gives the number of variates in the range 1–, etc.

See also BAR CHART, BIN, CLASS INTERVAL, FREQUENCY DISTRIBUTION, FREQUENCY POLYGON, OGIVE, PIE CHART, SHEPPARD'S CORRECTION

Kenney, J. F. and Keeping, E. S. "Histograms." §2.4 in *Mathematics of Statistics, Pt. 1, 3rd ed.* Princeton, NJ: Van Nostrand, pp. 25–6, 1962.

Hitch

A KNOT that secures a rope to a post, ring, another rope, etc., but does not keep its shape by itself.

See also CLOVE HITCH, KNOT, LINK, LOOP (KNOT)

References

Owen, P. *Knots.* Philadelphia, PA: Courage, p. 17, 1993.

Hitting Set

VERTEX COVER

Hjelmslev's Theorem

When all the points P on one line are related by an ISOMETRY to all points P' on another, the MIDPOINTS of the segments PP' are either distinct and COLLINEAR or COINCIDENT.

HJLS Algorithm

An algorithm for finding INTEGER RELATIONS whose running time is bounded by a polynomial in the number of real variables (Ferguson and Bailey 1992). Unfortunately, it is numerically unstable and therefore requires extremely high numeric precision. The cause of this instability is not known, but is believed to derive from its reliance on GRAM-SCHMIDT ORTHONORMALIZATION (Ferguson and Bailey 1992), which is known to be numerically unstable (Golub and van Loan 1989).

Rössner, C. and Schnorr (1994) have developed a stable variation of HJLS (Ferguson *et al.* 1999).

See also FERGUSON-FORCADE ALGORITHM, INTEGER RELATION, LLL ALGORITHM, PSLQ ALGORITHM, PSOS ALGORITHM

References

Ferguson, H. R. P. and Bailey, D. H. "A Polynomial Time, Numerically Stable Integer Relation Algorithm." RNR Techn. Rept. RNR-91–32, Jul. 14, 1992.
Ferguson, H. R. P.; Bailey, D. H.; and Arno, S. "Analysis of PSLQ, An Integer Relation Finding Algorithm." *Math. Comput.* **68**, 351–69, 1999.
Golub, G. H. and van Loan, C. F. *Matrix Computations, 3rd ed.* Baltimore, MD: Johns Hopkins, 1996.
Hastad, J.; Just, B.; Lagarias, J. C.; and Schnorr, C. P. "Polynomial Time Algorithms for Finding Integer Relations Among Real Numbers." *SIAM J. Comput.* **18**, 859–81, 1988.
Rössner, C. and Schnorr, C. P. "A Stable Integer Relation Algorithm." Tech. Rep. TR-94–16. FB Mathematik/Informatik, Universität Frankfurt, 1–1, 1994.

HK Integral

A type of integral named after Henstock and Kurzweil. Every LEBESGUE INTEGRABLE function is HK integrable with the same value.

References

Shenitzer, A. and Steprans, J. "The Evolution of Integration." *Amer. Math. Monthly* **101**, 66–2, 1994.

H-Matrix

HADAMARD MATRIX

Hoax Number

A COMPOSITE NUMBER defined analogously to a SMITH NUMBER except that the SUM of the number's DIGITS equals the sum of the DIGITS of its *distinct* PRIME FACTORS (excluding 1). The first few hoax numbers are 22, 58, 84, 85, 94, 136, 160, 166, 202, 234, ... (Sloane's A019506), and the corresponding sums of digits are 4, 13, 12, 13, 13, 10, 7, 13, 4, 9, 7, ... (Sloane's A050223).

See also SMITH NUMBER

References

Sloane, N. J. A. Sequences A019506 and A050223 in "An On-Line Version of the Encyclopedia of Integer Se-

quences." http://www.research.att.com/~njas/sequences/
eisonline.html.

Hodge Conjecture

The Hodge conjecture asserts that, for particularly
nice types of spaces called PROJECTIVE ALGEBRAIC
VARIETIES, the pieces called HODGE CYCLES are
actually rational linear combinations of geometric
pieces called algebraic cycles.

See also HODGE CYCLE, PROJECTIVE ALGEBRAIC
VARIETY

References

Clay Mathematics Institute. "The Hodge Conjecture." http://
www.claymath.org/prize_problems/hodge.htm.
Deligne, P. "The Hodge Conjecture." http://www.clay-
math.org/prize_problems/hodge.pdf.
Grothendieck, A. "Hodge's General Conjecture Is False for
Trivial Reasons." *Topology* **8**, 299–03, 1969.
Hodge, W. V. D. "The Topological Invariants of Algebraic
Varieties." *Proc. Internat. Congress Math., Cambridge,
Mass., 1950, Vol. 1.* Providence, RI: Amer. Math. Soc.,
pp. 182–92, 1952.

Hodge Cycle

See also HODGE CONJECTURE

Hodge Diamond

See also HODGE STAR

Hodge Identities
KÄHLER IDENTITIES

Hodge Star

On an oriented n-D RIEMANNIAN MANIFOLD, the
Hodge star is a linear FUNCTION which converts
alternating DIFFERENTIAL K-FORMS to alternating
$(n-k)$-forms. If w is an alternating K-FORM, its
Hodge star is given by

$$w(v_1, \ldots, v_k) = (*w)(v_{k+1}, \ldots, v_n)$$

when $v_1, \ldots, v_n$ is an oriented orthonormal basis.

See also HODGE DIAMOND, STOKES' THEOREM

Hodge's Theorem

On a COMPACT oriented FINSLER MANIFOLD without
boundary, every COHOMOLOGY class has a UNIQUE
harmonic representation. The DIMENSION of the
SPACE of all harmonic forms of degree p is the pth
BETTI NUMBER of the MANIFOLD.

See also BETTI NUMBER, COHOMOLOGY, DIMENSION,
FINSLER MANIFOLD

References

Chern, S.-S. "Finsler Geometry is Just Riemannian Geometry without the Quadratic Restriction." *Not. Amer. Math.
Soc.* **43**, 959–63, 1996.

Hoehn's Theorem

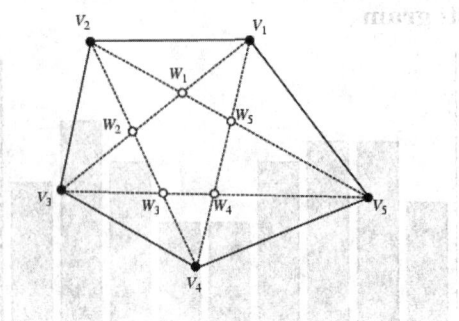

A geometric theorem related to the PENTAGRAM and
also called the PRATT-KASAPI THEOREM.

$$\frac{|V_1W_1|}{|W_2V_3|}\frac{|V_2W_2|}{|W_3V_4|}\frac{|V_3W_3|}{|W_4V_5|}\frac{|V_4W_4|}{|W_5V_1|}\frac{|V_5W_5|}{|W_1V_2|} = 1$$

$$\frac{|V_1W_2|}{|W_1V_3|}\frac{|V_2W_3|}{|W_2V_4|}\frac{|V_3W_4|}{|W_3V_5|}\frac{|V_4W_5|}{|W_4V_1|}\frac{|V_5W_1|}{|W_5V_2|} = 1.$$

In general, it is also true that

$$\frac{|V_iW_i|}{|W_{i+1}V_{i+2}|} = \frac{|V_iV_{i+1}V_{i+4}|}{|V_iV_{i+1}V_{i+2}V_{i+4}|}\frac{|V_iV_{i+1}V_{i+2}V_{i+3}|}{|V_{i+2}V_{i+3}V_{i+1}|}.$$

This type of identity was generalized to other figures
in the plane and their duals by Pinkernell (1996).

See also CEVA'S THEOREM, MENELAUS' THEOREM

References

Chou, S. C. *Mechanical Geometry Theorem Proving.* Dor-
drecht, Netherlands: Reidel, 1987.
Grünbaum, B. and Shepard, G. C. "Ceva, Menelaus, and the
Area Principle." *Math. Mag.* **68**, 254–68, 1995.
Hoehn, L. "A Menelaus-Type Theorem for the Pentagram."
Math. Mag. **68**, 254–68, 1995.
Pinkernell, G. M. "Identities on Point-Line Figures in the
Euclidean Plane." *Math. Mag.* **69**, 377–83, 1996.

Hoffman's Minimal Surface

A MINIMAL EMBEDDED SURFACE discovered in 1992
consisting of a HELICOID with a HOLE and HANDLE
(*Science News* 1992). It has the same topology as a
PUNCTURED sphere with a handle, and is only the
second complete embedded minimal surface of finite
topology and infinite total curvature discovered (the
HELICOID being the first).

A three-ended MINIMAL SURFACE of GENUS 1 is sometimes also called Hoffman's minimal surface (Peterson 1988).

See also HELICOID, MINIMAL SURFACE

Hoffman-Singleton Graph

References

Karcher, H.; Wei, F. S.; and Hoffman, D. "The Genus One Helicoid and the Minimal Surfaces that Led to Its Discovery." In *Global Analysis in Modern Mathematics. Proceedings of the Symposium in Honor of Richard Palais' Sixtieth Birthday held at the University of Maine, Orono, Maine, August 8–0, 1991, and at Brandeis University, Waltham, Massachusetts, August 12, 1992* (Ed. K. Uhlenbeck). Houston, TX: Publish or Perish Press, pp. 119–70, 1993.
Peterson, I. *Mathematical Tourist: Snapshots of Modern Mathematics.* New York: W. H. Freeman, pp. 57–9, 1988.
"Putting a Handle on a Minimal Helicoid." *Sci. News* **142**, 276, Oct. 24, 1992.

Hoffman-Singleton Graph

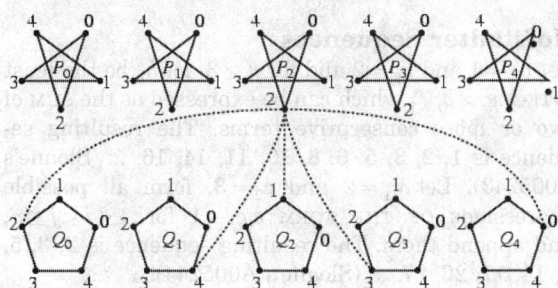

The only REGULAR GRAPH of VERTEX DEGREE 7, DIAMETER 2, and GIRTH 5. It is the unique (7, 5)-MOORE GRAPH (and is therefore also a (7,5)-CAGE GRAPH), and contains many copies of the PETERSEN GRAPH. It can be constructed from the 10 5-cycles illustrated above, with vertex i of P_j joined to vertex $i + jk$ (mod 5) of Q_k (Robertson 1969; Bondy and Murty 1976, p. 239; Wong 1982). (Note the correction of Wong's $j + jk$ to $i + jk$.)

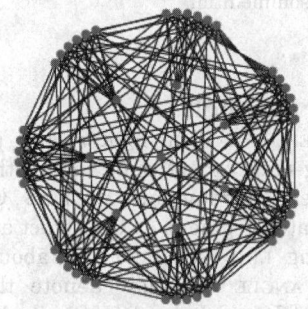

Other constructions are given by (Benson and Losey 1971; Biggs 1993, p. 163), and a RADIAL EMBEDDING is illustrated above.

See also CAGE GRAPH, HOFFMAN-SINGLETON THEOREM, MOORE GRAPH, PETERSEN GRAPH

References

Benson, C. T.; and Losey, N. E. "On a Graph of Hoffman and Singleton." *J. Combin. Th. Ser. B* **11**, 67–9, 1971.
Biggs, N. L. *Algebraic Graph Theory, 2nd ed.* Cambridge, England: Cambridge University Press, 1993.
Bondy, J. A. and Murty, U. S. R. *Graph Theory with Applications.* New York: North Holland, p. 235, 1976.
Hoffman, A. J. and Singleton, R. R. "On Moore Graphs of Diameter Two and Three." *IBM J. Res. Develop.* **4**, 497–04, 1960.
Robertson, N. *Graphs Minimal Under Girth, Valency, and Connectivity Constraints.* Dissertation. Waterloo, Ontario: University of Waterloo, 1969.
Weisstein, E. W. "Graphs." MATHEMATICA NOTEBOOK GRAPHS.M.
Wong, P. K. "Cages--A Survey." *J. Graph Th.* **6**, 1–2, 1982.

Hoffman-Singleton Theorem

Let G be a k-regular graph with GIRTH 5 and GRAPH DIAMETER 2. (Such a graph is a MOORE GRAPH). Then, $k = 2, 3, 7$, or 57. A proof of this theorem is difficult (Hoffman and Singleton 1960, Feit and Higman 1964, Damerell 1973, Bannai and Ito 1973), but can be found in Biggs (1993).

See also HOFFMAN-SINGLETON GRAPH, MOORE GRAPH

References

Bannai, E. and Ito, T. "On Moore Graphs." *J. Fac. Sci. Univ. Tokyo Ser. A* **20**, 191–08, 1973.
Biggs, N. L. Ch. 23 in *Algebraic Graph Theory, 2nd ed.* Cambridge, England: Cambridge University Press, 1993.
Damerell, R. M. "On Moore Graphs." *Proc. Cambridge Philos. Soc.* **74**, 227–36, 1973.
Feit, W. and Higman, G. "The Non-Existence of Certain Generalized Polygons." *J. Algebra* **1**, 114–31, 1964.
Hoffman, A. J. and Singleton, R. R. "On Moore Graphs of Diameter Two and Three." *IBM J. Res. Develop.* **4**, 497–04, 1960.

Hofstadter Figure-Figure Sequence

Define $F(1) = 1$ and $S(1) = 2$ and write

$$F(n) = F(n-1) + S(n-1),$$

where the sequence $\{S(n)\}$ consists of those integers not already contained in $\{F(n)\}$. For example, $F(2) = F(1) + S(1) = 3$, so the next term of $S(n)$ is $S(2) = 4$, giving $F(3) = F(2) + S(2) = 7$. The next integer is 5, so $S(3) = 5$ and $F(4) = F(3) + S(3) = 12$. Continuing in this manner gives the "figure" sequence $F(n)$ as 1, 3, 7, 12, 18, 26, 35, 45, 56, ... (Sloane's A005228) and the "space" sequence as 2, 4, 5, 6, 8, 9, 10, 11, 13, 14, ... (Sloane's A030124).

References

Hofstadter, D. R. *Gödel, Escher, Bach: An Eternal Golden Braid.* New York: Vintage Books, p. 73, 1989.
Sloane, N. J. A. Sequences A005228/M2629 and A030124 in "An On-Line Version of the Encyclopedia of Integer Sequences." http://www.research.att.com/~njas/sequences/eisonline.html.

Hofstadter G-Sequence

The sequence defined by $G(0) = 0$ and

$$G(n) = n - G(G(n-1)).$$

The first few terms are 1, 1, 2, 3, 3, 4, 4, 5, 6, 6, 7, 8, 8, 9, 9, ... (Sloane's A005206).

References

Hofstadter, D. R. *Gödel, Escher, Bach: An Eternal Golden Braid.* New York: Vintage Books, p. 137, 1989.
Sloane, N. J. A. Sequences A005206/M0436 in "An On-Line Version of the Encyclopedia of Integer Sequences." http://www.research.att.com/~njas/sequences/eisonline.html.

Hofstadter H-Sequence

The sequence defined by $H(0) = 0$ and

$$H(n) = n - H(H(H(n-1))).$$

The first few terms are 1, 1, 2, 3, 4, 4, 5, 5, 6, 7, 7, 8, 9, 10, 10, 11, 12, 13, 13, 14, ... (Sloane's A005374).

References

Hofstadter, D. R. *Gödel, Escher, Bach: An Eternal Golden Braid.* New York: Vintage Books, p. 137, 1989.
Sloane, N. J. A. Sequences A005374/M0449 in "An On-Line Version of the Encyclopedia of Integer Sequences." http://www.research.att.com/~njas/sequences/eisonline.html.

Hofstadter Male-Female Sequences

The pair of sequences defined by $F(0) = 1$, $M(0) = 0$, and

$$F(n) = n - M(F(n-1))$$

$$M(n) = n - F(M(n-1)).$$

The first few terms of the "male" sequence $M(n)$ are 0, 1, 2, 2, 3, 4, 4, 5, 6, 6, 7, 7, 8, 9, 9, ... (Sloane's A005379), and the first few terms of the "female" sequence $F(n)$ are 1, 2, 2, 3, 3, 4, 5, 5, 6, 6, 7, 8, 8, 9, 9, ... (Sloane's A005378).

References

Hofstadter, D. R. *Gödel, Escher, Bach: An Eternal Golden Braid.* New York: Vintage Books, p. 137, 1989.
Sloane, N. J. A. Sequences A005378/M0263 and A005379/M0278 in "An On-Line Version of the Encyclopedia of Integer Sequences." http://www.research.att.com/~njas/sequences/eisonline.html.

Hofstadter Point

The r-HOFSTADTER TRIANGLE of a given TRIANGLE ΔABC is perspective to ΔABC, and the PERSPECTIVE CENTER is called the Hofstadter point. The TRIANGLE CENTER FUNCTION is

$$\alpha = \frac{\sin(rA)}{\sin(r - rA)}.$$

As $r \to 0$, the TRIANGLE CENTER FUNCTION approaches

$$\alpha = \frac{A}{a},$$

and as $r \to 1$, the TRIANGLE CENTER FUNCTION approaches

$$\alpha = \frac{a}{A}.$$

See also HOFSTADTER TRIANGLE

References

Kimberling, C. "Hofstadter Points." *Nieuw Arch. Wiskunder* **12**, 109–14, 1994.
Kimberling, C. "Major Centers of Triangles." *Amer. Math. Monthly* **104**, 431–38, 1997.
Kimberling, C. "Hofstadter Points." http://cedar.evansville.edu/~ck6/tcenters/recent/hofstad.html.

Hofstadter Sequences

Let $b_1 = 1$ and $b_2 = 2$ and for $n \geq 3$, let b_n be the least INTEGER $> b_{n-1}$ which can be expressed as the SUM of two or more consecutive terms. The resulting sequence is 1, 2, 3, 5, 6, 8, 10, 11, 14, 16, ... (Sloane's A005243). Let $c_1 = 2$ and $c_2 = 3$, form all possible expressions OF THE FORM $c_i c_j - 1$ for $1 \leq i < j \leq n$, and append them. The resulting sequence is 2, 3, 5, 9, 14, 17, 26, 27, ... (Sloane's A005244).

See also HOFSTADTER-CONWAY \$10,000 SEQUENCE, HOFSTADTER'S Q-SEQUENCE, SUM-FREE SET

References

Guy, R. K. "Three Sequences of Hofstadter." §E31 in *Unsolved Problems in Number Theory, 2nd ed.* New York: Springer-Verlag, pp. 231–32, 1994.
Sloane, N. J. A. Sequences A005243/M0623 and A005244/M0705 in "An On-Line Version of the Encyclopedia of Integer Sequences." http://www.research.att.com/~njas/sequences/eisonline.html.

Hofstadter Triangle

For a NONZERO REAL NUMBER r and a TRIANGLE ΔABC, swing LINE SEGMENT BC about the vertex B towards vertex A through an ANGLE rB. Call the line along the rotated segment L. Construct a second line L' by rotating LINE SEGMENT BC about vertex C through an ANGLE rC. Now denote the point of intersection of L and L' by $A(r)$. Similarly, construct $B(r)$ and $C(r)$. The TRIANGLE having these points as vertices is called the Hofstadter r-triangle. Kimberling (1994) showed that the Hofstadter triangle is perspective to ΔABC, and calls PERSPECTIVE CENTER the HOFSTADTER POINT.

See also HOFSTADTER POINT

References

Kimberling, C. "Hofstadter Points." *Nieuw Arch. Wiskunde* **12**, 109–14, 1994.
Kimberling, C. "Hofstadter Points." http://cedar.evansville.edu/~ck6/tcenters/recent/hofstad.html.

Hofstadter's Q-Sequence

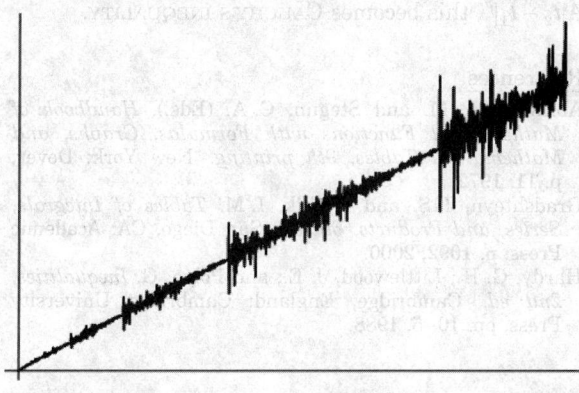

The INTEGER SEQUENCE given by

$$Q(n) = Q(n - Q(n-1)) + Q(n - Q(n-2)),$$

with $Q(1) = Q(2) = 1$. The first few values are 1, 1, 2, 3, 3, 4, 5, 5, 6, 6, ... (Sloane's A005185; illustrated above). These numbers are sometimes called Q-NUMBER.

There are currently no rigorous analyses or detailed predictions of the rather erratic behavior of $Q(n)$ (Guy 1994). It has, however, been demonstrated that the chaotic behavior of the Q-numbers shows some signs of order, namely that they exhibit approximate PERIOD DOUBLING, SELF-SIMILARITY and SCALING (Pinn 1998). These properties are shared with the related sequence

$$D(n) = D(D(n-1)) + D(n-1-D(n-2))$$

with $D(1) = D(2) = 1$, which exhibits exact PERIOD DOUBLING (Pinn 1998). The chaotic regions of $D(n)$ are separated by predictable smooth behavior.

See also HOFSTADTER-CONWAY $10,000 SEQUENCE, MALLOWS' SEQUENCE, PERIOD DOUBLING

References

Conolly, B. W. "Fibonacci and Meta-Fibonacci Sequences." In *Fibonacci and Lucas Numbers, and the Golden Section* (Ed. S. Vajda). New York: Halstead Press, pp. 127–38, 1989.

Dawson, R.; Gabor, G.; Nowakowski, R.; and Weins, D. "Random Fibonacci-Type Sequences." *Fib. Quart.* **23**, 169–76, 1985.

Guy, R. "Some Suspiciously Simple Sequences." *Amer. Math. Monthly* **93**, 186–91, 1986.

Guy, R. K. "Three Sequences of Hofstadter." §E31 in *Unsolved Problems in Number Theory, 2nd ed.* New York: Springer-Verlag, pp. 231–32, 1994.

Hofstadter, D. R. *Gödel, Escher Bach: An Eternal Golden Braid.* New York: Vintage Books, pp. 137–38, 1980.

Kubo, T. and Vakil, R. "On Conway's Recursive Sequence." *Disc. Math.* **152**, 225–52, 1996.

Mallows, C. L. "Conway's Challenge Sequence." *Amer. Math. Monthly* **98**, 5–0, 1991.

Pickover, C. A. "The Crying of Fractal Batrachion 1,489." Ch. 25 in *Keys to Infinity.* New York: W. H. Freeman, pp. 183–91, 1995.

Pinn, K. Order and Chaos is Hofstadter's $Q(n)$ Sequence. 1 Jul 1998. http://xxx.lanl.gov/abs/chao-dyn/9803012/. To appear in *Complexity*.

Pinn, K. A Chaotic Cousin of Conway's Recursive Sequence. 4 Aug 1998. http://xxx.lanl.gov/abs/cond-mat/9808031/.. Submitted to *J. Exper. Math.*

Sloane, N. J. A. Sequences A005185/M0438 in "An On-Line Version of the Encyclopedia of Integer Sequences." http://www.research.att.com/~njas/sequences/eisonline.html.

Tanny, S. M. "A Well-Behaved Cousin of the Hofstadter Sequence." *Disc. Math.* **105**, 227–39, 1992.

Hofstadter-Conway $10,000 Sequence

The INTEGER SEQUENCE defined by the RECURRENCE RELATION

$$a(n) = a(a(n-1)) + a(n - a(n-1))$$

with $a(1) = a(2) = 1$. The first few values are 1, 1, 2, 2, 3, 4, 4, 4, 5, 6, ... (Sloane's A004001). Plotting $a(n)/n$ against n gives the BATRACHION plotted below. Conway (1988) showed that $\lim_{n \to \infty} a(n)/n = 1/2$ and offered a prize of $10,000 to the discoverer of a value of n for which $|a(i)/i - 1/2| < 1/20$ for $i > n$. The prize was subsequently claimed by Mallows, after adjustment to Conway's "intended" prize of $1,000 (Schroeder 1991), who found $n = 1489$.

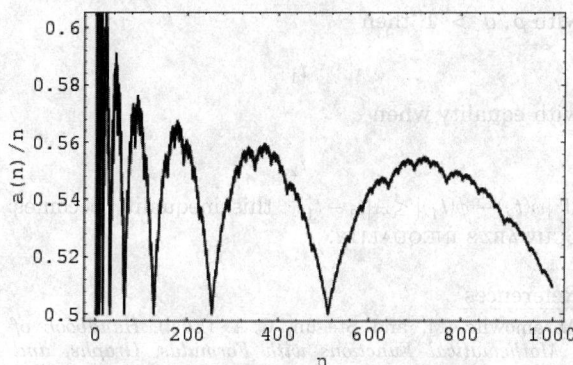

$a(n)/n$ takes a value of 1/2 for n OF THE FORM 2^k with $k = 1, 2, \ldots$. Pickover (1996) gives a table of analogous values of n corresponding to different values of $|a(n)/n - 1/2| < e$.

See also BLANCMANGE FUNCTION, HOFSTADTER'S Q-SEQUENCE, MALLOWS' SEQUENCE

References

Conolly, B. W. "Meta-Fibonacci Sequences." In *Fibonacci and Lucas Numbers, and the Golden Section* (Ed. S. Vajda). New York: Halstead Press, pp. 127–38, 1989.

Conway, J. "Some Crazy Sequences." Lecture at AT&T Bell Labs, July 15, 1988.

Guy, R. K. "Three Sequences of Hofstadter." §E31 in *Unsolved Problems in Number Theory, 2nd ed.* New York: Springer-Verlag, pp. 231–32, 1994.

Kubo, T. and Vakil, R. "On Conway's Recursive Sequence." *Disc. Math.* **152**, 225–52, 1996.

Mallows, C. L. "Conway's Challenge Sequence." *Amer. Math. Monthly* **98**, 5–0, 1991.

Pickover, C. A. "The Drums of Ulupu." In *Mazes for the Mind: Computers and the Unexpected.* New York: St. Martin's Press, 1993.

Pickover, C. A. "The Crying of Fractal Batrachion 1,489." Ch. 25 in *Keys to Infinity.* New York: W. H. Freeman, pp. 183–91, 1995.

Pinn, K. "A Chaotic Cousin of Conway's Recursive Sequence." *Exp. Math.* **9**, 55–6, 2000.

Schroeder, M. "John Horton Conway's 'Death Bet.'" *Fractals, Chaos, Power Laws.* New York: W. H. Freeman, pp. 57–9, 1991.

Sloane, N. J. A. Sequences A004001/M0276 in "An On-Line Version of the Encyclopedia of Integer Sequences." http://www.research.att.com/~njas/sequences/eisonline.html.

Hölder Condition

A function $\phi(t)$ satisfies the Hölder condition on two points t_1 and t_2 on an arc L when

$$|\phi(t_2) - \phi(t_1)| \le A|t_2 - t_1|^\mu,$$

with A and μ POSITIVE REAL constants.

See also LIPSCHITZ CONDITION

Hölder Integral Inequality

If

$$C(r)$$

with $p, q > 1$, then

$$t_1$$

with equality when

$$t_2$$

If $|\phi(t_2) - \phi(t_1)| \le A|t_2 - t_1|^\mu$, this inequality becomes SCHWARZ'S INEQUALITY.

References

Abramowitz, M. and Stegun, C. A. (Eds.). *Handbook of Mathematical Functions with Formulas, Graphs, and Mathematical Tables, 9th printing.* New York: Dover, p. 11, 1972.

Gradshteyn, I. S. and Ryzhik, I. M. *Tables of Integrals, Series, and Products, 6th ed.* San Diego, CA: Academic Press, p. 1099, 2000.

Hölder, O. "Über einen Mittelwertsatz." *Göttingen Nachr.*, 44, 1889.

Riesz, F. "Untersuchungen über Systeme integrierbarer Funktionen." *Math. Ann.* **69**, 456, 1910.

Riesz, F. "Su alcune disuguaglianze." *Boll. Un. Mat. It.* **7**, 77–9, 1928.

Sansone, G. *Orthogonal Functions, rev. English ed.* New York: Dover, pp. 32–3, 1991.

Hölder Sum Inequality

If

$$C(r)$$

with $p, q > 1$, then

$$\frac{1}{p} + \frac{1}{q} = 1$$

with equality when $q > 1$. If $|\phi(t_2) - \phi(t_1)| \le A|t_2 - t_1|^\mu$, this becomes CAUCHY'S INEQUALITY.

References

Abramowitz, M. and Stegun, C. A. (Eds.). *Handbook of Mathematical Functions with Formulas, Graphs, and Mathematical Tables, 9th printing.* New York: Dover, p. 11, 1972.

Gradshteyn, I. S. and Ryzhik, I. M. *Tables of Integrals, Series, and Products, 6th ed.* San Diego, CA: Academic Press, p. 1092, 2000.

Hardy, G. H.; Littlewood, J. E.; and Pólya, G. *Inequalities, 2nd ed.* Cambridge, England: Cambridge University Press, pp. 10–5, 1988.

Hölder's Inequalities

Let

$$\frac{1}{p} + \frac{1}{q} = 1 \tag{1}$$

with $p, q > 1$. Then Hölder's inequality for integrals states that

$$\int_a^b |f(x)g(x)| \, dx$$
$$\le \left[\int_a^b |f(x)|^p \, dx\right]^{1/p} \left[\int_a^b |g(x)|^q \, dx\right]^{1/q}, \tag{2}$$

with equality when

$$|g(x)| = c|f(x)|^{p-1}.$$

If $p = q = 2$, this inequality becomes SCHWARZ'S INEQUALITY.

Similarly, Hölder's inequality for sums states that

$$\sum_{k=1}^n |a_k b_k| \le \left(\sum_{k=1}^n |a_k|^p\right)^{1/p} \left(\sum_{k=1}^n |b_k|^q\right)^{1/q}, \tag{3}$$

with equality when $|b_k| = c|a_k|^{p-1}$. If $p = q = 2$, this becomes CAUCHY'S INEQUALITY.

See also CAUCHY'S INEQUALITY, SCHWARZ'S INEQUALITY

References

Abramowitz, M. and Stegun, C. A. (Eds.). *Handbook of Mathematical Functions with Formulas, Graphs, and Mathematical Tables, 9th printing.* New York: Dover, p. 11, 1972.

Gradshteyn, I. S. and Ryzhik, I. M. *Tables of Integrals, Series, and Products, 6th ed.* San Diego, CA: Academic Press, pp. 1092 and 1099, 2000.

Hardy, G. H.; Littlewood, J. E.; and Pólya, G. "Hölder's Inequality and Its Extensions." §2.7 and 2.8 in *Inequalities, 2nd ed.* Cambridge, England: Cambridge University Press, pp. 21–6, 1988.

Hölder, O. "Über einen Mittelwertsatz." *Göttingen Nachr.*, 38–7, 1889.

Riesz, F. "Untersuchungen über Systeme integrierbarer Funktionen." *Math. Ann.* **69**, 456, 1910.

Riesz, F. "Su alcune disuguaglianze." *Boll. Un. Mat. It.* **7**, 77–9, 1928.
Rogers, L. J. "An Extension of a Certain Theorem in Inequalities." *Messenger Math.* **17**, 145–50, 1888.
Sansone, G. *Orthogonal Functions, rev. English ed.* New York: Dover, pp. 32–3, 1991.

Holditch's Theorem

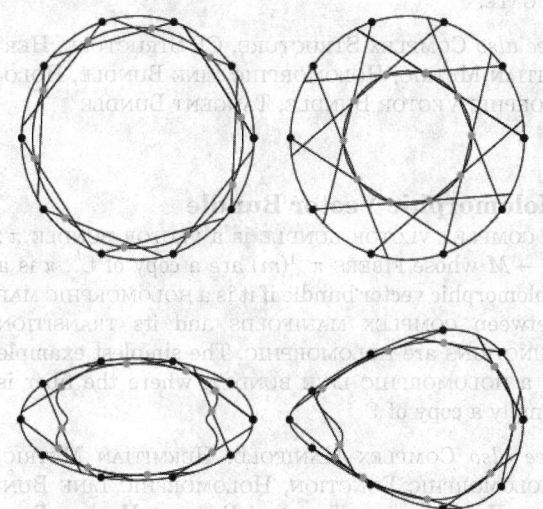

Let a CHORD of constant length be slid around a smooth, closed, convex curve C, and choose a point on the CHORD which divides it into segments of lengths p and q. This point will trace out a new closed curve C', as illustrated above. Provided certain conditions are met, the area between C and C' is given by πpq, as first shown by Holditch in 1858.

The Holditch curve for a CIRCLE of RADIUS R is another CIRCLE which, from the theorem, has RADIUS

$$r = \sqrt{R^2 - pq}.$$

References

Bender, W. "The Holditch Curve Tracer." *Math. Mag.* **54**, 128–29, 1981.
Broman, A. "Holditch's Theorem." *Math. Mag.* **54**, 99–08, 1981.
Kiliç, E. and Keles, S. "On Holditch's Theorem and Polar Inertia Momentum." *Comm. Fac. Sci. Univ. Ankara Ser. A_1 Math. Statist.* **43**, 41–7, 1996.
Weisstein, E. W. "Holditch's Theorem." MATHEMATICA NOTE-BOOK HOLDITCH.M.
Wells, D. *The Penguin Dictionary of Curious and Interesting Geometry.* London: Penguin, p. 103, 1991.

Hole

A hole in a mathematical object is a TOPOLOGICAL structure which prevents the object from being continuously shrunk to a point. When dealing with TOPOLOGICAL SPACES, a DISCONNECTIVITY is interpreted as a hole in the space. Examples of holes are things like the "donut hole" in the center of the TORUS, a domain removed from a plane, and the portion missing from EUCLIDEAN SPACE after cutting a KNOT out from it.

Singular HOMOLOGY GROUPS form a MEASURE of the hole structure of a SPACE, but they are one particular measure and they don't always detect all holes. HOMOTOPY GROUPS of a SPACE are another measure of holes in a SPACE, as well as BORDISM GROUPS, K-THEORY, COHOMOTOPY GROUPS, and so on.

There are many ways to measure holes in a space. Some holes are picked up by HOMOTOPY GROUPS that are not detected by HOMOLOGY GROUPS, and some holes are detected by HOMOLOGY GROUPS that are not picked up by HOMOTOPY GROUPS. (For example, in the TORUS, HOMOTOPY GROUPS "miss" the two-dimensional hole that is given by the TORUS itself, but the second HOMOLOGY GROUP picks that hole up.) In addition, HOMOLOGY GROUPS don't detect the varying hole structures of the complement of KNOTS in 3-space, but the first HOMOTOPY GROUP (the fundamental group) does.

See also BRANCH CUT, BRANCH POINT, CORK PLUG, CROSS-CAP, GENUS (SURFACE), PEG, PRINCE RUPERT'S CUBE, SINGULAR POINT (FUNCTION), SPHERICAL RING, TORUS

Holographic Projection

EQUAL-AREA PROJECTION

Holography

The mathematical study of a nonlinear equation $f(\varphi) = y$, where f maps from a HILBERT SPACE X to a HILBERT SPACE Y and $y \in Y$ which abstracts the construction of optical holograms.

References

Lannes, A. "Abstract Holography." *J. Math. Anal. Appl.* **74**, 530–59, 1980.

Holomorphic Function

A synonym for ANALYTIC FUNCTION, regular function, differentiable function, complex differentiable function, and holomorphic map (Krantz 1999, p. 16). The word derives from the Greek $o\lambda o\varsigma$ (*holos*), meaning "whole," and $\mu o\rho\varphi\eta$ (*morphe*), meaning "form" or "appearance."

Many mathematicians prefer the term "holomorphic function" (or "holomorphic map") to "analytic function" (Krantz 1999, p. 16), while "analytic" appears to be in widespread use among physicists, engineers, and in some older texts (Morse and Feshbach 1953, pp. 356–74; Knopp 1996, pp. 83–11; Whittaker and Watson 1990, p. 83).

See also ANALYTIC FUNCTION, COMPLEX DIFFERENTIABLE, HOLONOMIC FUNCTION, HOMEOMORPHIC, MEROMORPHIC FUNCTION

References

Knopp, K. "Analytic Continuation and Complete Definition of Analytic Functions." Ch. 8 in *Theory of Functions Parts I and II, Two Volumes Bound as One, Part I*. New York: Dover, pp. 83–11, 1996.

Krantz, S. G. "Holomorphic Functions." §1.3 in *Handbook of Complex Analysis*. Boston, MA: Birkhäuser, pp. 12–6, 1999.

Morse, P. M. and Feshbach, H. "Analytic Functions." §4.2 in *Methods of Theoretical Physics, Part I*. New York: McGraw-Hill, pp. 356–74, 1953.

Whittaker, E. T. and Watson, G. N. *A Course in Modern Analysis, 4th ed.* Cambridge, England: Cambridge University Press, 1990.

Holomorphic Line Bundle

A COMPLEX LINE BUNDLE is a VECTOR BUNDLE $\pi : E \to M$ whose FIBERS $\pi^{-1}(m)$ are a copy of $\mathbb{C}$. π is a holomorphic line bundle if it is a HOLOMORPHIC MAP between COMPLEX MANIFOLDS and its TRANSITION FUNCTIONS are HOLOMORPHIC.

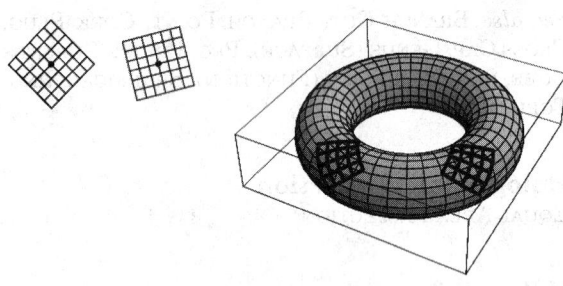

On a compact RIEMANN SURFACE, a DIVISOR $\Sigma n_i p_i$ determines a LINE BUNDLE. For example, consider $2p - q$ on X. Around p there is a COORDINATE CHART U given by the HOLOMORPHIC FUNCTION z_p with $z_p(p) = 0$. Similarly, z_q is a HOLOMORPHIC FUNCTION defining a disjoint chart V around q with $z_q(q) = 0$. Then letting $W = X - \{p, q\}$, the RIEMANN SURFACE is covered by $X = U \cup V \cup W$. The LINE BUNDLE corresponding to $2p - q$ is then defined by the following TRANSITION FUNCTIONS,

$$g_{UW}(x) = z_p(x)^2 \text{ defined for } x \in U \cap W$$

$$g_{VW}(x) = z_q(x)^{-1} \text{ defined for } x \in V \cap W.$$

See also CHERN CLASS, HERMITIAN METRIC, HOLOMORPHIC FUNCTION, HOLOMORPHIC TANGENT BUNDLE, HOLOMORPHIC VECTOR BUNDLE, LINE BUNDLE, RIEMANN-ROCH THEOREM, RIEMANN SURFACE, VECTOR BUNDLE

Holomorphic Map

HOLOMORPHIC FUNCTION

Holomorphic Tangent Bundle

The holomorphic tangent bundle to a COMPLEX MANIFOLD is given by its complexified tangent vectors which are of type $(1, 0)$. In a CHART $z = (z_1, \ldots, z_n)$, the bundle is spanned by the local SECTIONS $\partial / \partial z_k$. The antiholomorphic sections are spanned by $\partial / \partial \bar{z}_k$, of type $(0, 1)$, where $\bar{z}$ denotes the COMPLEX CONJUGATE.

See also COMPLEX STRUCTURE, CR-STRUCTURE, HERMITIAN METRIC, HOLOMORPHIC LINE BUNDLE, HOLOMORPHIC VECTOR BUNDLE, TANGENT BUNDLE

Holomorphic Vector Bundle

A COMPLEX VECTOR BUNDLE is a VECTOR BUNDLE $\pi : E \to M$ whose FIBERS $\pi^{-1}(m)$ are a copy of $\mathbb{C}^k$. π is a holomorphic vector bundle if it is a HOLOMORPHIC MAP between COMPLEX MANIFOLDS and its TRANSITION FUNCTIONS are HOLOMORPHIC. The simplest example is a HOLOMORPHIC LINE BUNDLE, where the fiber is simply a copy of $\mathbb{C}$.

See also COMPLEX MANIFOLD, HERMITIAN METRIC, HOLOMORPHIC FUNCTION, HOLOMORPHIC LINE BUNDLE, HOLOMORPHIC TANGENT BUNDLE, VECTOR BUNDLE

Holonomic Constant

A limiting value of a HOLONOMIC FUNCTION near a SINGULAR POINT. Holonomic constants include APÉRY'S CONSTANT, CATALAN'S CONSTANT, PÓLYA'S RANDOM WALK CONSTANTS for $d > 2$, and PI.

Holonomic Function

A solution of a linear homogeneous ORDINARY DIFFERENTIAL EQUATION with POLYNOMIAL COEFFICIENTS.

See also HOLOMORPHIC FUNCTION, HOLONOMIC CONSTANT

References

Koepf, W. *Hypergeometric Summation: An Algorithmic Approach to Summation and Special Function Identities.* Braunschweig, Germany: Vieweg, p. 2, 1998.

Zeilberger, D. "A Holonomic Systems Approach to Special Function Identities." *J. Comput. Appl. Math.* **32**, 321–48, 1990.

Holonomy

A general concept in CATEGORY THEORY involving the globalization of topological or differential structures. The term derives from the Greek $o\lambda o\varsigma$ (*holos*) "whole" and $\nu o\mu o\varsigma$ (*nomos*) "law, rule."

See also HOLONOMY GROUP, MONODROMY

Holonomy Group

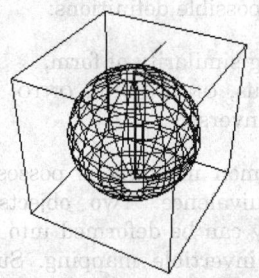

On a RIEMANNIAN MANIFOLD M, tangent vectors can be moved along a path by PARALLEL TRANSPORT, which preserves VECTOR ADDITION and SCALAR MULTIPLICATION. So a closed loop at a base point p, gives rise to an INVERTIBLE LINEAR MAP of TM_p, the tangent vectors at p. It is possible to compose closed loops by following one after the other, and to invert them by going backwards. Hence, the set of linear transformations arising from PARALLEL TRANSPORT along closed loops is a GROUP, called the holonomy group.

Since PARALLEL TRANSPORT preserves the RIEMANNIAN METRIC, the holonomy group is contained in the ORTHOGONAL GROUP $O(n)$. Moreover, if the manifold is ORIENTABLE, then it is contained in the SPECIAL ORTHOGONAL GROUP. A generic RIEMANNIAN METRIC on an ORIENTABLE MANIFOLD has holonomy group $SO(n)$, but for some special metrics it can be a subgroup, in which case the manifold is said to have special holonomy.

A KÄHLER MANIFOLD is a $2n$-dimensional MANIFOLD whose holonomy lies in the UNITARY GROUP $U(n) \subset O(2n)$. A CALABI-YAU MANIFOLD is a SIMPLY CONNECTED $2n$-dimensional manifold with holonomy in the SPECIAL UNITARY GROUP. A 4_n-dimensional manifold with holonomy group $Sp(n)$, the QUATERNIONIC UNITARY GROUP, is called a HYPER-KÄHLER MANIFOLD, and one with holonomy $Sp(n)Sp(1)$ is called a QUATERNION KÄHLER MANIFOLD. The possible groups that can arise as a holonomy group of the metric compatible LEVI-CIVITA CONNECTION were classified by Berger. The other possibilities for a nonproduct, nonsymmetric MANIFOLD are the LIE GROUPS G_2, Spin(7), and Spin(9).

On a FLAT MANIFOLD, two homotopic loops give the same linear transformation. Consequently, the holonomy group is a REPRESENTATION of the FUNDAMENTAL GROUP of M. In general though, the CURVATURE of M changes the PARALLEL TRANSPORT between homotopic loops. In fact, there is a formula for the difference as an integral of the curvature.

See also CALABI-YAU MANIFOLD, CONNECTION (PRINCIPAL BUNDLE), CONNECTION (VECTOR BUNDLE), CURVATURE FORM, HOMOGENEOUS SPACE, KÄHLER MANIFOLD, PARALLEL TRANSPORT, QUATERNION KÄHLER MANIFOLD, REPRESENTATION, TANGENT BUNDLE

References
Salamon, S. *Riemannian Geometry and Holonomy Groups.* Essex, England: Longman Group, 1989.

Holor

Moon, P. and Spencer, D. E. *Theory of Holors: A Generalization of Tensors.* Cambridge, England: Cambridge University Press, 1986.

Holyhedron

A polyhedron whose faces and holes are all finite-sided polygons and which contains at least one hole whose boundary shares no point with a face boundary. D. Wilson coined the term in 1997, although no actual holyhedron was known until 1999, when a holyhedron of GENUS approximately 54,000,000 was (apparently) constructed (Vinson 2000). J. H. Conway believes the construction to be correct, although he believes that the minimal GENUS should be closer to 100.

See also POLYHEDRON

References
Vinson, J. "On Holyhedra." *Disc. Comput. Geom.* **24**, 85–04, 2000.

Homalographic Projection

EQUAL-AREA PROJECTION

Home Plate

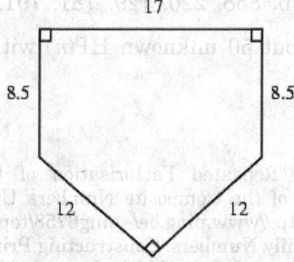

Home plate in the game of BASEBALL is an irregular PENTAGON. However, the Little League rulebook's specification of the shape of home plate (Kreutzer and Kerley 1990), illustrated above, is not physically realizable, since it requires the existence of a (12, 12, 17) RIGHT TRIANGLE, whereas

$$12^2 + 12^2 = 288 \neq 289 = 17^2$$

(Bradley 1996).

See also BASEBALL, BASEBALL COVER

References
Bradley, M. J. "Building Home Plate: Field of Dreams or Reality?" *Math. Mag.* **69**, 44–5, 1996.
Kreutzer, P. and Kerley, T. *Little League's Official How-to-Play Baseball Book.* New York: Doubleday, 1990.

Home Prime

The prime HP(n) reached starting from a number n, concatenating its prime factors, and repeating until a prime is reached. For example, for $n = 9$,

$$9 = 3 \cdot 3 \rightarrow 33 = 3 \cdot 11 \rightarrow 311,$$

so 311 is the home prime of 9. For $n = 2, 3, ...$, the first few are 2, 3, 211, 5, 23, 7, 3331113965338635107, 311, 773, ... (Sloane's A037274). Probabilistic arguments give exactly zero for the chance that the sequence of integers starting at a given number n contains no prime (J. H. Conway, Sloane), so a home prime should exist for every positive integer.

Since prime numbers have trivial home primes (themselves), we can restrict attention to composite numbers. The number of steps to arrive at a home prime for composite numbers 4, 6, 8, 9, ... are 1, 13, 2, 4, 1, 5, 4, 4, 1, 15, 1, ... (Sloane's A037271), and the primes they reach are 211, 23, 3331113965338635107, 311, 773, 223, ... (Sloane's A037272). The largest home prime for $n < 100$ is HP(49) = HP(77), although its value is not known. After 55 steps, the sequence reaches $3 \cdot 73 \cdot C105$, where $C105$ is the 105-digit composite number. This number was factored by P. Leyland in November 1999, and subsequently reached a number $C137$ in December 1999. In June 2000, Leyland factored this number as well, and proceeded a few steps to obtain a $C131$, which has not yet been factored. The next largest HP(n) for $n < 100$ is

HP(80) = 313, 169, 138, 727, 147, 145, 210, 044,

974, 146, 858, 220, 729, 781, 791, 489.

There are about 50 unknown HP(n) with $100 < n < 1000$ (Hoey).

References

De Geest, P. "Repeated Factorisation of Concatenated Primefactors of the Composite Numbers Up to 100 and Beyond..." http://www.ping.be/~ping6758/topic1.htm.
Heleen, J. "Family Numbers: Constructing Primes by Prime Factor Splitting." *J. Recr. Math.* **28**, 116–19, 1996–7.
Sloane, N. J. A. Sequences A037271, A037272, A037273 and A037274 in "An On-Line Version of the Encyclopedia of Integer Sequences." http://www.research.att.com/~njas/sequences/eisonline.html.

Homeoid

A shell bounded by two similar ELLIPSOIDS having a constant ratio of axes. Given a CHORD passing through a homeoid, the distance between inner and outer intersections is equal on both sides. Since a spherical shell is a symmetric case of a homeoid, this theorem is also true for spherical shells (CONCENTRIC CIRCLES in the PLANE), for which it is easily proved by symmetry arguments.

See also CHORD, ELLIPSOID

Homeomorphic

There are two possible definitions:

1. Possessing similarity of form,
2. Continuous, ONE-TO-ONE, ONTO, and having a continuous inverse.

The most common meaning is possessing intrinsic topological equivalence. Two objects are homeomorphic if they can be deformed into each other by a continuous, invertible mapping. Such a HOMEOMORPHISM ignores the space in which surfaces are embedded, so the deformation can be completed in a higher dimensional space than the surface was originally embedded. MIRROR IMAGES are homeomorphic, as are MÖBIUS STRIP with an EVEN number of half-twists, and MÖBIUS STRIP with an ODD number of half-twists.

In CATEGORY THEORY terms, homeomorphisms are ISOMORPHISMS in the CATEGORY of TOPOLOGICAL SPACES and CONTINUOUS MAPS.

See also HOMEOMORPHIC, HOMOMORPHIC, ISOGENY, POLISH SPACE

References

Krantz, S. G. "The Concept of Homeomorphism." §6.4.1 in *Handbook of Complex Analysis.* Boston, MA: Birkhäuser, p. 86, 1999.

Homeomorphic Type

The following three pieces of information completely determine the homeomorphic type of a surface (Massey 1967):

1. Orientability,
2. Number of boundary components,
3. EULER CHARACTERISTIC.

See also ALGEBRAIC TOPOLOGY, EULER CHARACTERISTIC

References

Massey, W. S. *Algebraic Topology: An Introduction.* New York: Springer-Verlag, 1996.

Homeomorphically Irreducible Tree

SERIES-REDUCED TREE

Homeomorphism

An EQUIVALENCE RELATION and one-to-one correspondence between points in two geometric figures or topological spaces which is continuous in both directions, also called a continuous transformation. A homeomorphism which also preserves distances is called an ISOMETRY. AFFINE TRANSFORMATIONS are another type of common geometric homeomorphism.

The similarity in meaning and form of the words "HOMOMORPHISM" and "homeomorphism" is unfortunate and a common source of confusion.

See also AFFINE TRANSFORMATION, HOMEOMORPHIC, HOMEOMORPHIC TYPE, HOMOMORPHISM, ISOMETRY, TOPOLOGICALLY CONJUGATE

References

Coxeter, H. S. M. and Greitzer, S. L. *Geometry Revisited.* Washington, DC: Math. Assoc. Amer., p. 101, 1967.

Krantz, S. G, "The Concept of Homeomorphism." §6.4.1 in *Handbook of Complex Analysis.* Boston, MA: Birkhäuser, p. 86, 1999.

Ore, Ø. *Graphs and Their Uses.* New York: Random House, 1963.

Homeomorphism Group

The homeomorphism group of a TOPOLOGICAL SPACE X is the set of all HOMEOMORPHISMS $f : X \to X$, which forms a GROUP by composition.

See also GROUP, INFINITE GROUP, TOPOLOGICAL SPACE

HOMFLY Polynomial

A 2-variable oriented KNOT POLYNOMIAL $P_L(a, z)$ motivated by the JONES POLYNOMIAL (Freyd *et al.* 1985). Its name is an acronym for the last names of its co-discoverers: Hoste, Ocneanu, Millett, Freyd, Lickorish, and Yetter (Freyd *et al.* 1985). Independent work related to the HOMFLY polynomial was also carried out by Prztycki and Traczyk (1987). HOMFLY polynomial is defined by the SKEIN RELATIONSHIP

$$a^{-1} P_{L+}(a, z) - a P_{L-}(a, z) = z P_{L_0}(a, z) \qquad (1)$$

(Doll and Hoste 1991), where v is sometimes written instead of a (Kanenobu and Sumi 1993) or, with a slightly different relationship, as

$$\alpha P_{L+}(\alpha, z) - \alpha^{-1} P_{L-}(\alpha, z) = z P_{L_0}(\alpha, z) \qquad (2)$$

(Kauffman 1991). It is also defined as $P_L(\ell, m)$ in terms of SKEIN RELATIONSHIP

$$\ell P_{L+} + \ell^{-1} P_{L-} + m P_{L_0} = 0 \qquad (3)$$

(Lickorish and Millett 1988). It can be regarded as a nonhomogeneous POLYNOMIAL in two variables or a homogeneous POLYNOMIAL in three variables. In three variables the SKEIN RELATIONSHIP is written

$$x P_{L+}(x, y, z) + y P_{L-}(x, y, z) + z P_{L_0}(x, y, z) = 0. \qquad (4)$$

It is normalized so that $P_{\text{unknot}} = 1$. Also, for n unlinked unknotted components,

$$P_L(x, y, z) = \left(-\frac{x + y}{z} \right)^{n-1}. \qquad (5)$$

This POLYNOMIAL usually detects CHIRALITY but does not detect the distinct ENANTIOMERS of the KNOTS 09–42, 10–48, 10–71, 10–91, 10–04, and 10–25 (Jones 1987). The HOMFLY polynomial of an oriented KNOT is the same if the orientation is reversed. It is a generalization of the JONES POLYNOMIAL $V(t)$, satisfying

$$V(t) = P(a = t, \ z = t^{1/2} - t^{-1/2}) \qquad (6)$$

$$V(t) = P(\ell = it^{-1}, \ m = i(t^{-1/2} - t^{1/2})). \qquad (7)$$

It is also a generalization of the ALEXANDER POLYNOMIAL $\nabla(z)$, satisfying

$$\nabla(z) = P(a = 1, \ z = t^{1/2} - t^{-1/2}). \qquad (8)$$

The HOMFLY POLYNOMIAL of the MIRROR IMAGE K^* of a KNOT K is given by

$$P_{K^*}(\ell, \ m) = P_K(\ell^{-1}, \ m), \qquad (9)$$

so P usually but not always detects CHIRALITY.

A split union of two links (i.e., bringing two links together without intertwining them) has HOMFLY polynomial

$$P(L_1 \cup L_2) = -(\ell + \ell^{-1}) m^{-1} P(L_1) P(L_2). \qquad (10)$$

Also, the composition of two links

$$P(L_1 \# L_2) = P(L_1) P(L_2), \qquad (11)$$

so the POLYNOMIAL of a COMPOSITE KNOT factors into POLYNOMIALS of its constituent knots (Adams 1994).

MUTANTS have the same HOMFLY polynomials. In fact, there are infinitely many distinct KNOTS with the same HOMFLY POLYNOMIAL (Kanenobu 1986). Examples include (05–01, 10–32), (08–08, 10–29) (08–16, 10–56), and (10–25, 10–56) (Jones 1987). Incidentally, these also have the same JONES POLYNOMIAL.

M. B. Thistlethwaite has tabulated the HOMFLY polynomial for KNOTS up to 13 crossings.

See also ALEXANDER POLYNOMIAL, JONES POLYNOMIAL, KNOT POLYNOMIAL

References

Adams, C. C. *The Knot Book: An Elementary Introduction to the Mathematical Theory of Knots.* New York: W. H. Freeman, pp. 171–72, 1994.

Doll, H. and Hoste, J. "A Tabulation of Oriented Links." *Math. Comput.* **57**, 747–61, 1991.

Freyd, P.; Yetter, D.; Hoste, J.; Lickorish, W. B. R.; Millett, K.; and Oceanu, A. "A New Polynomial Invariant of Knots and Links." *Bull. Amer. Math. Soc.* **12**, 239–46, 1985.

Jones, V. "Hecke Algebra Representations of Braid Groups and Link Polynomials." *Ann. Math.* **126**, 335–88, 1987.

Kanenobu, T. "Infinitely Many Knots with the Same Polynomial." *Proc. Amer. Math. Soc.* **97**, 158–61, 1986.

Kanenobu, T. and Sumi, T. "Polynomial Invariants of 2-Bridge Knots through 22 Crossings." *Math. Comput.* **60**, 771–78 and S17–S28, 1993.

Kauffman, L. H. *Knots and Physics.* Singapore: World Scientific, p. 52, 1991.

Lickorish, W. B. R. and Millett, B. R. "The New Polynomial Invariants of Knots and Links." *Math. Mag.* **61**, 1–3, 1988.

Morton, H. R. and Short, H. B. "Calculating the 2-Variable Polynomial for Knots Presented as Closed Braids." *J. Algorithms* **11**, 117–31, 1990.

Przytycki, J. and Traczyk, P. "Conway Algebras and Skein Equivalence of Links." *Proc. Amer. Math. Soc.* **100**, 744–48, 1987.

Stoimenow, A. "Jones Polynomials." http://guests.mpim-bonn.mpg.de/alex/ptab/j10.html.

Weisstein, E. W. "Knots and Links." MATHEMATICA NOTEBOOK KNOTS.M.

Homoclinic Point

A point where a stable and an unstable SEPARATRIX (invariant MANIFOLD) from the same fixed point or same family intersect. Therefore, the limits

$$\lim_{k \to \infty} f^k(X)$$

and

$$\lim_{k \to -\infty} f^k(X)$$

exist and are equal.

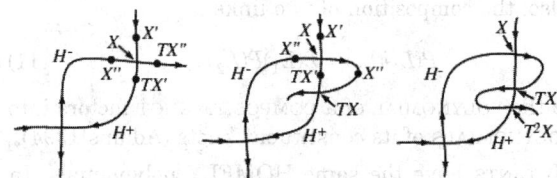

Refer to the above figure. Let X be the point of intersection, with X' ahead of X on one MANIFOLD and X'' ahead of X of the other. The mapping of each of these points TX' and TX'' must be ahead of the mapping of X, TX. The only way this can happen is if the MANIFOLD loops back and crosses itself at a new homoclinic point. Another loop must be formed, with T^2X another homoclinic point. Since T^2X is closer to the hyperbolic point than TX, the distance between T^2X and TX is less than that between X and TX. Area preservation requires the AREA to remain the same, so each new curve (which is closer than the previous one) must extend further. In effect, the loops become longer and thinner. The network of curves leading to a dense AREA of homoclinic points is known as a homoclinic tangle or tendril. Homoclinic points appear where CHAOTIC regions touch in a hyperbolic FIXED POINT.

A small DISK centered near a homoclinic point includes infinitely many periodic points of different periods. Poincaré showed that if there is a single homoclinic point, there are an infinite number. More specifically, there are infinitely many homoclinic points in each small disk (Nusse and Yorke 1996).

See also HETEROCLINIC POINT, MANIFOLD, SEPARATRIX

References
Nusse, H. E. and Yorke, J. A. "Basins of Attraction." *Science* **271**, 1376–380, 1996.

Tabor, M. *Chaos and Integrability in Nonlinear Dynamics: An Introduction.* New York: Wiley, p. 145, 1989.

Homogeneous Barycentric Coordinates

AREAL COORDINATES

Homogeneous Cartesian Coordinates

HOMOGENEOUS COORDINATES

Homogeneous Coordinates

Homogeneous coordinates (x_1, x_2, x_3) of a finite point (x, y) in the plane are any three numbers for which

$$\frac{x_1}{x_3} = x \tag{1}$$

$$\frac{x_2}{x_3} = y. \tag{2}$$

Coordinates $(x_1, x_2, 0)$ for which

$$\frac{x_2}{x_3} = \lambda \tag{3}$$

describe the POINT AT INFINITY in the direction of slope λ.

In homogeneous coordinates, the equation of a LINE

$$a_1 x + a_2 y + a_3 = 0 \tag{4}$$

is given by

$$a_1 x_1 + a_2 x_2 + a_3 x_3 = 0. \tag{5}$$

Two points expressed using homogeneous coordinates (a_1, a_2, a_3) and (b_1, b_2, b_3) are identical IFF

$$\begin{vmatrix} a_2 & a_3 \\ b_2 & b_3 \end{vmatrix} = \begin{vmatrix} a_3 & a_1 \\ b_3 & b_1 \end{vmatrix} = \begin{vmatrix} a_1 & a_2 \\ b_1 & b_2 \end{vmatrix} = 0. \tag{6}$$

Two lines expressed using homogeneous coordinates

$$a_1 x_1 + a_2 x_2 + a_3 x_3 = 0 \tag{7}$$

$$b_1 x_1 + b_2 x_2 + b_3 x_3 = 0 \tag{8}$$

are identical IFF

$$\begin{vmatrix} a_2 & a_3 \\ b_2 & b_3 \end{vmatrix} = \begin{vmatrix} a_3 & a_1 \\ b_3 & b_1 \end{vmatrix} = \begin{vmatrix} a_1 & a_2 \\ b_1 & b_2 \end{vmatrix} = 0. \tag{9}$$

The intersection of the two lines above is given by

$$x_1 = \begin{vmatrix} a_2 & a_3 \\ b_2 & b_3 \end{vmatrix} \tag{10}$$

$$x_2 = \begin{vmatrix} a_3 & a_1 \\ b_3 & b_1 \end{vmatrix} \tag{11}$$

$$x_3 = \begin{vmatrix} a_1 & a_2 \\ b_1 & b_2 \end{vmatrix}. \qquad (12)$$

See also TRILINEAR COORDINATES

References

Graustein, W. C. "Homogeneous Cartesian Coordinates. Linear Dependence of Points and Lines." Ch. 3 in *Introduction to Higher Geometry.* New York: Macmillan, pp. 29–9, 1930.

Homogeneous Function

A function which satisfies

$$f(tx, ty) = t^n f(x, y)$$

for a fixed n. MEANS, the WEIERSTRASS ELLIPTIC FUNCTION, and TRIANGLE CENTER FUNCTIONS are homogeneous functions. A transformation of the variables of a TENSOR changes the TENSOR into another whose components are linear homogeneous functions of the components of the original TENSOR.

See also EULER'S HOMOGENEOUS FUNCTION THEOREM

Homogeneous Ideal

A homogeneous ideal I in a GRADED RING $R = \oplus A_i$ is an IDEAL generated by a set of homogeneous elements, i.e., each one is contained in only one of the A_i. For example, the POLYNOMIAL RING $\mathbb{C}[x] = \oplus A_i$ is a GRADED RING, where $A_i = \{ax^i\}$. The IDEAL $I = \langle x^2 \rangle$, i.e., all polynomials with no constant or linear terms, is a homogeneous ideal in $\mathbb{C}[x]$. Another homogeneous ideal is $I = \langle x^2 + y^2 + z^2, \, xy + yz + zx, \, z^5 \rangle$ in $\mathbb{C}[x, y, z]$.

Given any finite set of polynomials in n variables, the process of homogenization converts them to homogeneous polynomials in $n+1$ variables. If $f = f(x_1, \ldots, x_n)$ is a polynomial of degree d then

$$f^h(x_0, x_1, \ldots, x_n) = x_0^d f(x_1/x_0, \ldots, x_n/x_0)$$

is the homogenization of f. Similarly, if I is an IDEAL in $\mathbb{C}[x_1, \ldots, x_n]$, then $I^h = \{f^h | f \in I\}$ is its homogenization and is a homogeneous ideal. For example, if $f = x_1^3 + 2x_1 x_2 - 3$ then $f^h = x_1^3 + 2x_0 x_1 x_2 - 3x_0^3$. Note that in general, if $I = \langle f_1, \ldots, f_k \rangle$ then I^h may have more elements than $\langle f_1^h, \ldots, f_k^h \rangle$. However, if $f_1, ..., f_k$ form a GRÖBNER BASIS using a graded monomial order, then $I^h = \langle f_1^h, \ldots, f_k^h \rangle$. A polynomial is easily dehomogenized by setting the extra variable $x_0 = 1$.

Here is a *Mathematica* function which takes a polynomial, in variables *vars*, and homogenizes it with the variable *x0*.

```
(*dg finds the degree of the polynomial f*)
    dg[f_?PolynomialQ,     {vars_?AtomQ}]    :=
Exponent[f, vars];
  dg[f_?PolynomialQ, vars_?ListQ] :=
```

```
Max[MapIndexed[(dg[#1, Rest[vars]] + #2 - 1
&), CoefficientList[f, First[vars]]]];  (*uses
dg = degree of polynomial above*)
    Homogenize[f_?PolynomialQ,    vars_?ListQ,
x0_?AtomQ] :=
  Expand[x0 ^ dg[f, vars] f /. Map[(#1 -> #1/
x0 &), vars]]
```

Here is a *Mathematica* function which dehomogenizes a polynomial in the variable *x0*.

```
Dehomogenize[f_?PolynomialQ, x0_?AtomQ] :=  f
/. x0 -> 1
```

The AFFINE VARIETY V corresponding to a homogeneous ideal has the property that $x \in V$ IFF $cx \in V$ for all COMPLEX c. Therefore, a homogeneous ideal defines an ALGEBRAIC VARIETY in COMPLEX PROJECTIVE SPACE.

See also ALGEBRAIC VARIETY, CATEGORY THEORY, COMMUTATIVE ALGEBRA, CONIC SECTION, IDEAL, PRIME IDEAL, PROJECTIVE VARIETY, SCHEME, ZARISKI TOPOLOGY

References

Hartshorne, R. *Algebraic Geometry.* New York: Springer-Verlag, 1977.

Homogeneous Numbers

Two numbers are homogeneous if they have identical PRIME FACTORS. An example of a homogeneous pair is (6, 72), both of which share PRIME FACTORS 2 and 3:

$$6 = 2 \cdot 3$$
$$72 = 2^3 \cdot 3^2.$$

See also HETEROGENEOUS NUMBERS, PRIME FACTORS, PRIME NUMBER

References

Le Lionnais, F. *Les nombres remarquables.* Paris: Hermann, p. 146, 1983.

Homogeneous Polynomial

A multivariate polynomial (i.e., a POLYNOMIAL in more than one variable) with all terms having the same degree. For example, $x^3 + xyz + y^2 z + z^3$ is a homogeneous polynomial of degree three. SYMMETRIC POLYNOMIALS are always homogeneous.

See also FORM (POLYNOMIAL), POLYNOMIAL, SYMMETRIC POLYNOMIAL

Homogeneous Space

A homogeneous space M is a SPACE with a TRANSITIVE GROUP ACTION by a LIE GROUP. Because a TRANSITIVE GROUP ACTION implies that there is only one ORBIT, M is ISOMORPHIC to the QUOTIENT SPACE G/H where H

is the ISOTROPY GROUP G_x. The choice of $x \in M$ does not affect the isomorphism type of G/G_x because all of the ISOTROPY GROUPS are CONJUGATE.

Many common spaces are homogeneous spaces, such as the HYPERSPHERE,

$$S^n \sim O(n+1)/O(n), \tag{1}$$

and the COMPLEX PROJECTIVE SPACE

$$\mathbb{C}\P^n \sim U(n+1)/U(n) \times U(1). \tag{2}$$

The real GRASSMANNIAN of k-dimensional SUBSPACES in $\mathbb{R}^{n+k}$ is

$$O(n+k)/O(n) \times O(k). \tag{3}$$

The projection $\pi : G \to G/H$ makes G a PRINCIPAL BUNDLE on G/H with FIBER H. For example, $\pi : SO(3) \to SO(3)/SO(2) \sim \mathbb{S}^2$ is a $SO(2)$ BUNDLE, i.e., a CIRCLE BUNDLE, on the sphere. The SUBGROUP

$$SO(2) = \begin{bmatrix} 1 & 0 & 0 \\ 0 & \cos t & -\sin t \\ 0 & \sin t & \cos t \end{bmatrix} \tag{4}$$

acts on the right, and does not affect the first column so $\pi(v_1 v_2 v_3) = v_1 \in \mathbb{S}^2$ is WELL DEFINED.

See also EFFECTIVE ACTION, FREE ACTION, GROUP, ISOTROPY GROUP, MATRIX GROUP, ORBIT (GROUP), QUOTIENT SPACE (LIE GROUP), REPRESENTATION, TOPOLOGICAL GROUP, TRANSITIVE

References
Kawakubo, K. *The Theory of Transformation Groups.* Oxford, England: Oxford University Press, pp. 41–9 and 89–4, 1987.

Homographic

Any two ranges $\{ABC \ldots\}$ and $\{A'B'C' \ldots\}$ which are situated on the same or different lines are said to be homographic when the CROSS-RATIO of any four points on one range is equal to the CROSS-RATIO of the corresponding points of the other range.

See also CROSS-RATIO, MÖBIUS TRANSFORMATION

References
Lachlan, R. "Homographic Ranges and Pencils." §433–39 in *An Elementary Treatise on Modern Pure Geometry.* London: Macmillian, pp. 279–82, 1893.

Homography

A CIRCLE-preserving transformation composed of an EVEN number of inversions.

See also ANTIHOMOGRAPHY

Homological Algebra

An abstract ALGEBRA concerned with results valid for many different kinds of SPACES. MODULES are the basic tools used in homological algebra.

See also MODULE

References
Enochs, E. E. and Jenda, O. M. G. *Relative Homological Algebra.* Berlin: de Gruyter, 2000.
Hilton, P. and Stammbach, U. *A Course in Homological Algebra, 2nd ed.* New York: Springer-Verlag, 1997.
Weibel, C. A. *An Introduction to Homological Algebra.* New York: Cambridge University Press, 1994.

Homological Projection

EQUAL-AREA PROJECTION

Homologous Points

The extremities of PARALLEL RADII of two CIRCLES are called homologous with respect to the SIMILITUDE CENTER collinear with them.

See also ANTIHOMOLOGOUS POINTS, INVARIABLE POINT, SIMILITUDE CENTER

References
Johnson, R. A. *Modern Geometry: An Elementary Treatise on the Geometry of the Triangle and the Circle.* Boston, MA: Houghton Mifflin, p. 19, 1929.

Homologous Triangles

PERSPECTIVE TRIANGLES

Homolographic Equal-Area Projection

MOLLWEIDE PROJECTION

Homology

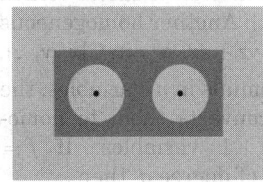 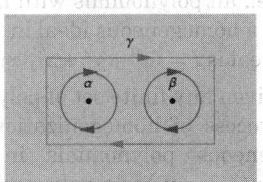

Homology is a concept which is used in many branches of algebra and topology. The basic example is degree one integral homology for a domain in $\mathbb{R}^2$. In this case, a HOMOLOGY CLASS is represented by a finite sum or difference of closed loops. For example, consider the loops in the twice PUNCTURED plane $\mathbb{R}^2 - \{(0, 0), (1, 0)\}$, illustrated above.

The equality $\alpha + \beta = \gamma$ holds in homology because the difference is the BOUNDARY of a COMPACTLY SUPPORTED region. The homology of a space is an algebraic object which reflects the topology. The algebraic tools used are called HOMOLOGICAL ALGEBRA, and in that language, the homology is a DERIVED FUNCTOR, the homology of a LONG EXACT SEQUENCE.

See also BOUNDARY (HOMOLOGY), COHOMOLOGY, DERIVED FUNCTOR, HOMOLOGY CLASS, HOMOLOGY (GEOMETRY), HOMOLOGY GROUP, INTERSECTION (HOMOLOGY), POINCARE DUALITY

Homology (Chain)

For every p, the kernel of $\partial_P : C_P \to C_{P-1}$ is called the group of cycles,

$$Z_P = \{c \in C_P : \partial(c) = 0\}. \tag{1}$$

The letter Z is short for the German word for cycle, "Zyklus." The image $\partial(C_{P+1})$ is contained in the group of cycles because $\partial \circ \partial = 0$, and is called the group of boundaries,

$$B_P = \{c \in C_P : \text{there exists } b \in C_{P+1} \text{ such that } \partial(b) \\ = c\}. \tag{2}$$

The quotients $H_P = Z_P/B_P$ are the HOMOLOGY GROUPS of the chain.

Given a SHORT EXACT SEQUENCE of CHAIN COMPLEXES

$$0 \to A_* \to B_* \to C_* \to 0, \tag{3}$$

there is a LONG EXACT SEQUENCE in homology.

$$\ldots \to H_P(A) \to H_P(B) \to H_P(C) \xrightarrow{\delta} H_{P-1}(A) \to \ldots . \tag{4}$$

In particular, a cycle a in A_P with $\partial a = 0$, is mapped to a cycle b in B_P. Similarly, a boundary $\partial a'$ in A_P gets mapped to a boundary $\partial b'$ in B_P. Consequently, the map between homologies $H_P(A) \to H_P(B)$ is well-defined. The only map which is not that obvious is δ, called the CONNECTING HOMOMORPHISM, which is well-defined by the SNAKE LEMMA.

Proofs of this nature are (with a modicum of humor) referred to as DIAGRAM CHASING.

See also CHAIN COMPLEX, CHAIN EQUIVALENCE, CHAIN HOMOMORPHISM, CHAIN HOMOTOPY, COCHAIN COMPLEX, HOMOLOGY, SNAKE LEMMA

References

Hilton, P. and Stammbach, U. *A Course in Homological Algebra.* New York: Springer-Verlag, pp. 117–18, 1997.
Munkres, J. *Elements of Algebraic Topology.* Reading, MA: Addison-Wesley, pp. 58 and 71–6, 1984.

Homology (Geometry)

A PERSPECTIVE COLLINEATION in which the center and axis are not incident. The term was first used by Poncelet (Cremona 1960, p. ix).

See also ELATION, HARMONIC HOMOLOGY, PERSPECTIVE COLLINEATION, PERSPECTIVE TRIANGLES

References

Cremona, L. *Elements of Projective Geometry, 3rd ed.* New York: Dover, 1960.
Desargues, G. (E*uvres de Desargues, réunies et analysées par M. Pudra, tome 1.* Paris, pp. 413–16, 1864.
Lambert, J. H. *Freie Perspective, 2nd ed.* Zürich, 1774.

Homology (Topology)

Historically, the term "homology" was first used in a topological sense by Poincaré. To him, it meant pretty much what is now called a COBORDISM, meaning that a homology was thought of as a relation between MANIFOLDS mapped into a MANIFOLD. Such MANIFOLDS form a homology when they form the boundary of a higher-dimensional MANIFOLD inside the MANIFOLD in question.

To simplify the definition of homology, Poincaré simplified the spaces he dealt with. He assumed that all the spaces he dealt with had a triangulation (i.e., they were "SIMPLICIAL COMPLEXES"). Then instead of talking about general "objects" in these spaces, he restricted himself to subcomplexes, i.e., objects in the space made up only on the simplices in the TRIANGULATION of the space. Eventually, Poincaré's version of homology was dispensed with and replaced by the more general SINGULAR HOMOLOGY. SINGULAR HOMOLOGY is the concept mathematicians mean when they say "homology."

In modern usage, however, the word homology is used to mean HOMOLOGY GROUP. For example, if someone says "X did Y by computing the homology of Z," they mean "X did Y by computing the HOMOLOGY GROUPS of Z." But sometimes homology is used more loosely in the context of a "homology in a SPACE," which corresponds to *singular* homology groups.

Singular homology groups of a SPACE measure the extent to which there are finite (compact) boundaryless GADGETS in that SPACE, such that these GADGETS are not the boundary of other finite (compact) GADGETS in that SPACE.

A generalized homology or cohomology theory must satisfy all of the EILENBERG-STEENROD AXIOMS with the exception of the DIMENSION AXIOM.

See also COHOMOLOGY, DIMENSION AXIOM, EILENBERG-STEENROD AXIOMS, GADGET, GRADED MODULE, HOMOLOGICAL ALGEBRA, HOMOLOGY GROUP, SIMPLICIAL COMPLEX, SIMPLICIAL HOMOLOGY, SINGULAR HOMOLOGY

References

Goldberg, S. I. *Curvature and Homology, enl. ed.* New York: Dover, 1998.

Homology Axis

PERSPECTIVE AXIS

Homology Center

PERSPECTIVE CENTER

Homology Class

A homology class in a singular homology theory is represented by a finite LINEAR COMBINATION of geometric subobjects with zero boundary. Such a linear combination is considered to be HOMOLOGOUS to zero if it is the boundary of something having dimension one greater. For instance, two points that can be connected by a path comprise the boundary for that path, so any two points in a component are homologous and represent the same homology class.

See also COHOMOLOGY, COHOMOLOGY CLASS, HOMOLOGY, HOMOLOGY GROUP, INTERSECTION (HOMOLOGY)

Homology Group

The term "homology group" usually means a singular homology group, which is an ABELIAN GROUP which partially counts the number of HOLES in a TOPOLOGICAL SPACE. In particular, singular homology groups form a MEASURE of the HOLE structure of a SPACE, but they are one particular measure and they don't always pick up everything.

In addition, there are "generalized homology groups" which are *not* singular homology groups.

See also HOMOLOGY (TOPOLOGY)

References

Munkres, J. R. *Elements of Algebraic Topology.* Perseus Press, 1993.

Homomorphic

Related to one another by a HOMOMORPHISM.

Homomorphism

A term used in CATEGORY THEORY to mean a general MORPHISM. The term derives from the Greek ομο (*omo*) "alike" and μορφωσις (*morphosis*), "to form" or "to shape." The similarity in meaning and form of the words "homomorphism" and "HOMEOMORPHISM" is unfortunate and a common source of confusion.

If G and H are GROUPS, then a group homomorphism of G into H is a function $\phi : G \to H$ which preserves the group operation, i.e., for all $g_1, g_2 \in G$,

$$(g_1 g_2)\phi = (g_1)\phi(g_2)\phi$$

(Yale 1988, p. 18).

See also GROUP HOMOMORPHISM, HOMEOMORPHISM, MORPHISM, RING HOMOMORPHISM

References

Yale, P. B. *Geometry and Symmetry.* New York: Dover, 1988.

Homomorphism (Ring)

See also RING

Homoscedastic

A set of STATISTICAL DISTRIBUTIONS having the same VARIANCE.

See also HETEROSCEDASTIC

Homothecy

A SIMILARITY TRANSFORMATION which preserves orientation, also called a homothety.

See also HOMOTHETIC, SIMILARITY TRANSFORMATION

References

Coxeter, H. S. M. *Introduction to Geometry, 2nd ed.* New York: Wiley, p. 68, 1969.

Croft, H. T.; Falconer, K. J.; and Guy, R. K. *Unsolved Problems in Geometry.* New York: Springer-Verlag, p. 3, 1991.

Homothetic

Two figures are homothetic if they are related by an EXPANSION or CONTRACTION. This means that they lie in the same plane and corresponding sides are PARALLEL; such figures have connectors of corresponding points which are CONCURRENT at a point known as the HOMOTHETIC CENTER. The HOMOTHETIC CENTER divides each connector in the same ratio k, known as the SIMILITUDE RATIO. For figures which are similar but do not have PARALLEL sides, a SIMILITUDE CENTER exists.

See also CONTRACTION (GEOMETRY), DIRECTLY SIMILAR, EXPANSION, HOMOTHECY, HOMOTHETIC CENTER, INVERSELY SIMILAR, PANTOGRAPH, PERSPECTIVE, SIMILAR, SIMILITUDE RATIO

References

Casey, J. *A Sequel to the First Six Books of the Elements of Euclid, Containing an Easy Introduction to Modern Geometry with Numerous Examples, 5th ed., rev. enl.* Dublin: Hodges, Figgis, & Co., p. 173, 1888.

Durell, C. V. *Modern Geometry: The Straight Line and Circle.* London: Macmillan, pp. 1–, 1928.

Johnson, R. A. *Modern Geometry: An Elementary Treatise on the Geometry of the Triangle and the Circle.* Boston, MA: Houghton Mifflin, 1929.

Lachlan, R. *An Elementary Treatise on Modern Pure Geometry.* London: Macmillian, p. 129, 1893.

Homothetic Center

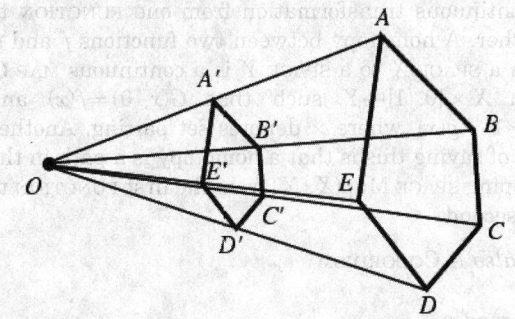

The meeting point of lines that connect corresponding points from HOMOTHETIC figures. In the above figure, O is the homothetic center of the HOMOTHETIC figures $ABCDE$ and $A'B'C'D'E'$. For figures which are similar but do not have PARALLEL sides, a SIMILITUDE CENTER exists (Johnson 1929, pp. 16–0).

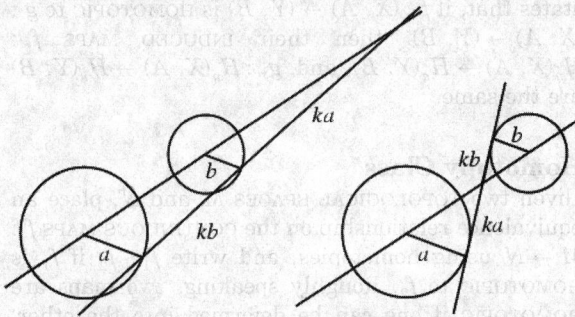

Given two nonconcentric CIRCLES, draw RADII PARALLEL and in the same direction. Then the line joining the extremities of the RADII passes through a fixed point on the line of centers which divides that line externally in the ratio of RADII. This point is called the external homothetic center, or external center of similitude (Johnson 1929, pp. 19–0 and 41).

If RADII are drawn PARALLEL but instead in opposite directions, the extremities of the RADII pass through a fixed point on the line of centers which divides that line internally in the ratio of RADII (Johnson 1929, pp. 19–0 and 41). This point is called the internal homothetic center, or internal center of similitude (Johnson 1929, pp. 19–0 and 41).

The position of the homothetic centers for two circles of radii r_i, centers (x_i, y_i), and segment angle θ are given by solving the simultaneous equations

$$y - y_2 = \frac{y_2 - y_1}{x_2 - x_1}(x - x_2)$$

$$y - y_2^{\pm} = \frac{y_2^{\pm} - y_1^{\pm}}{x_2^{\pm} - x_1^{\pm}}(x - x_2^{\pm})$$

for (x, y), where

$$x_i^{\pm} \equiv x_i + (-1)^i r_i \cos\theta$$

$$y_i^{\pm} \equiv y_i + (-1)^i r_i \sin\theta,$$

and the plus signs give the external homothetic center, while the minus signs give the internal homothetic center.

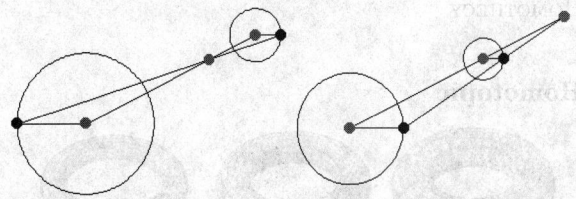

As the above diagrams show, as the angles of the parallel segments are varied, the positions of the homothetic centers remain the same. This fact provides a (slotted) LINKAGE for converting circular motion with one radius to circular motion with another.

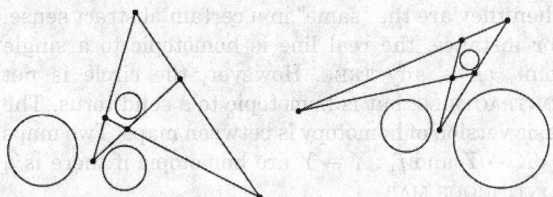

The six homothetic centers of three circles lie three by three on four lines (Johnson 1929, p. 120), which "enclose" the smallest circle.

The homothetic center of triangles is the PERSPECTIVE CENTER of HOMOTHETIC TRIANGLES. It is also called the SIMILITUDE CENTER (Johnson 1929, pp. 16–7).

See also APOLLONIUS' PROBLEM, HOMOTHETIC, PERSPECTIVE, SIMILITUDE CENTER

References

Johnson, R. A. *Modern Geometry: An Elementary Treatise on the Geometry of the Triangle and the Circle.* Boston, MA: Houghton Mifflin, 1929.

Lachlan, R. *An Elementary Treatise on Modern Pure Geometry.* London: Macmillian, p. 129, 1893.

Weisstein, E. W. "Plane Geometry." MATHEMATICA NOTEBOOK PLANEGEOMETRY.M.

Homothetic Position

Two similar figures with PARALLEL homologous LINES and connectors of HOMOLOGOUS POINTS CONCURRENT at the HOMOTHETIC CENTER are said to be in homothetic position. If two SIMILAR figures are in the same plane but the corresponding sides are not PARALLEL, there exists a self-HOMOLOGOUS POINT which occupies the same homologous position with respect to the two figures.

Homothetic Triangles

Nonconcurrent TRIANGLES with PARALLEL sides are always HOMOTHETIC. Homothetic triangles are always PERSPECTIVE TRIANGLES. Their PERSPECTIVE CENTER is called their HOMOTHETIC CENTER.

Homothety

HOMOTHECY

Homotopic

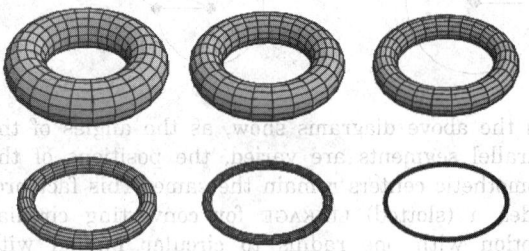

Two mathematical objects are said to be homotopic when they are the "same" in a certain abstract sense. For instance, the real line is homotopic to a single point, as is any TREE. However, the circle is not CONTRACTIBLE, but is homotopic to a solid torus. The basic version of homotopy is between maps. Two maps $f_0 : X \to Y$ and $f_1 : X \to Y$ are homotopic if there is a CONTINUOUS MAP

$$F : X \times [0, 1] \to Y$$

such that $F(x, 0) = f_0(x)$ and $F(x, 1) = f_1(x)$.

Whether or not two subsets are homotopic depends on the ambient space. For example, in the plane, the unit circle is homotopic to a point, but not in the PUNCTURED plane $\mathbb{R}^2 - 0$. The puncture can be thought of as an obstacle.

However, there is a way to compare two spaces via homotopy without ambient spaces. Two spaces X and Y are homotopy equivalent if there are maps $f : X \to Y$ and $g : X \to Y$ such that the composition $f \circ g$ is homotopic to the IDENTITY MAP of Y and $g \circ f$ is homotopic to the IDENTITY MAP of X. For example, the circle is not homotopic to a point, for then the constant map would be homotopic to the identity map of a circle, which is impossible because they have different DEGREES.

See also HOMEOMORPHISM, HOMOTOPY, HOMOTOPY CLASS, HOMOTOPY GROUP, HOMOTOPY TYPE, TOPOLOGICAL SPACE

Homotopy

A continuous transformation from one FUNCTION to another. A homotopy between two functions f and g from a SPACE X to a SPACE Y is a continuous MAP G from $X \times [0, 1] \mapsto Y$ such that $G(x, 0) = f(x)$ and $G(x, 1) = g(x)$, where $\times$ denotes set pairing. Another way of saying this is that a homotopy is a path in the mapping SPACE $\mathrm{Map}(X, Y)$ from the first FUNCTION to the second.

See also H-COBORDISM

References

Krantz, S. G. "The Concept of Homotopy" §10.3.2 in *Handbook of Complex Analysis*. Boston, MA: Birkhäuser, pp. 132–33, 1999.

Homotopy Axiom

One of the EILENBERG-STEENROD AXIOMS which states that, if $f : (X, A) \to (Y, B)$ is HOMOTOPIC to $g : (X, A) \to (Y, B)$, then their INDUCED MAPS $f_* : H_n(X, A) \to H_n(Y, B)$ and $g_* : H_n(X, A) \to H_n(Y, B)$ are the same.

Homotopy Class

Given two TOPOLOGICAL SPACES M and N, place an equivalence relationship on the CONTINUOUS MAPS $f : M \to N$ using homotopies, and write $f_1 \sim f_2$ if f_1 is HOMOTOPIC to f_2. Roughly speaking, two maps are HOMOTOPIC if one can be deformed into the other. This equivalence relation is transitive because these homotopy deformations can be composed (i.e., one can follow the other).

A simple example is the case of CONTINUOUS MAPS from one CIRCLE to another circle. Consider the number of ways an infinitely stretchable string can be tied around a tree trunk. The string forms the first circle, and the tree trunk's surface forms the second circle. For any integer n, the string can be wrapped around the tree n times, for positive n clockwise, and negative n counterclockwise. Each integer n corresponds to a homotopy class of maps from $\mathbb{S}^1$ to $\mathbb{S}^1$.

After the string is wrapped around the tree n times, it could be deformed a little bit to get another CONTINUOUS MAP, but it would still be in the same homotopy class, since it is HOMOTOPIC to the original map. Conversely, any map wrapped around n times can be deformed to any other.

See also HOMOTOPY, HOMOTOPY GROUP, TOPOLOGICAL SPACE

Homotopy Group

The homotopy groups generalize the FUNDAMENTAL GROUP to maps from higher dimensional spheres, instead of from the circle. The nth homotopy group of a TOPOLOGICAL SPACE X is the set of HOMOTOPY CLASSES of maps from the HYPERSPHERE to X, with a GROUP structure, and is denoted $\pi_n(X)$. The FUNDAMENTAL GROUP is $\pi_1(X)$, and, as in the case of π_1, the maps $\mathbb{S}^n \to X$ must pass through a BASEPOINT $p \in X$. For $n > 1$, the homotopy group $\pi_n(X)$ is an ABELIAN GROUP.

The group operations are not as simple as those for the FUNDAMENTAL GROUP. Consider two maps $a : \mathbb{S}^n \to X$ and $b : \mathbb{S}^n \to X$, which pass through $p \in X$. The product $a * b : \mathbb{S}^n \to X$ is given by mapping the equator to the BASEPOINT p. Then the northern hemisphere is mapped to the sphere by collapsing the equator to a point, and then it is mapped to X by a. The southern hemisphere is similarly mapped to X by b. The diagram above shows the product of two spheres.

The identity element is represented by the constant map $e(x) = p$. The choice of direction of a loop in the fundamental group corresponds to a ORIENTATION of $\mathbb{S}^n$ in a homotopy group. Hence the inverse of a map a is given by switching orientation for the sphere. By describing the sphere in $n + 1$ coordinates, switching the first and second coordinate changes the orientation of the sphere. Or as a HYPERSURFACE, $\mathbb{S}^n \subset \mathbb{R}^{n+1}$, switching orientation reverses the roles of inside and outside. The above diagram shows that $a * -a$ is homotopic to the constant map, i.e., the identity. It begins by expanding the equator in $a * -a$, and then the resulting map is contracted to the BASEPOINT.

As with the FUNDAMENTAL GROUP, the homotopy groups do not depend on the choice of BASEPOINT. But the higher homotopy groups are always ABELIAN. The above diagram shows an example of $a * b = b * a$. The BASEPOINT is fixed, and because $n > 1$ the map can be rotated. When $n = 1$, i.e., the FUNDAMENTAL

GROUP, it is impossible to rotate the map while keeping the BASEPOINT fixed.

A space with $\pi_i = 0$ for all $i \leq n$ is called n-connected. If X is $n-1$-connected, $n > 1$, then the HUREWICZ HOMOMORPHISM $\pi_n(X) \to H_n(X)$ from the nth-homotopy group to the nth-homology group is an ISOMORPHISM.

When $f : X \to Y$ is a CONTINUOUS MAP, then $f_* : \pi_n(X) \to \pi_n(Y)$ is defined by taking the images under f of the spheres in X. The pushforward is natural, i.e., $(f \circ g)_* = f_* \circ g_*$ whenever the composition of two maps is defined. In fact, given a FIBRATION,

$$F \to E \to B$$

where B is PATH-CONNECTED, there is a LONG EXACT SEQUENCE of homotopy groups

$$\ldots \to \pi_n(F) \to \pi_n(E) \to \pi_n(B) \to \pi_{n-1}(F) \to \ldots \to \pi_0(B)$$
$$= 0.$$

See also ABELIAN GROUP, COHOMOTOPY GROUP, FREUDENTHAL SUSPENSION THEOREM, FUNDAMENTAL GROUP, HOMOTOPY EXCISION, HUREWICZ HOMOMORPHISM, HYPERSPHERE, GROUP, RELATIVE HOMOTOPY GROUP, WEAK EQUIVALENCE

References

Dodson, C. T. J. and Parker, P. E. "Homotopy Groups" and "Tables of Homotopy Groups." §2.4 and Appendix D in *A User's Guide to Algebraic Topology.* Dordrecht, Netherlands: Kluwer, pp. 44–5 and 365–80, 1997.
Fulton, W. *Algebraic Topology: A First Course.* New York: Springer-Verlag, pp. 324–25, 1995.

Homotopy Theory

The branch of ALGEBRAIC TOPOLOGY which deals with HOMOTOPY GROUPS. Homotopy methods can be used to solve systems of polynomials by embedding the polynomials in a family of systems that define the deformation of the original problem into a simpler one whose solutions are known.

See also ALGEBRAIC TOPOLOGY, HOMOTOPY GROUP

References

Aubry, M. *Homotopy Theory and Models.* Boston, MA: Birkhäuser, 1995.

Honaker's Constant
PALINDROMIC PRIME

Honeycomb

A TESSELLATION in n-D, for $n \geq 3$. The only regular honeycomb in 3-D is {4, 3, 4}, which consists of eight cubes meeting at each VERTEX. The only quasiregular honeycomb (with regular cells and semiregular VERTEX FIGURES) has each VERTEX surrounded by eight

TETRAHEDRA and six OCTAHEDRA and is denoted $\left\{ {3 \atop 3,4} \right\}$.

Ball and Coxeter (1987) use the term "sponge" for a solid which can be parameterized by INTEGERS p, q, and n which satisfy the equation

$$2 \sin \left(\frac{\pi}{p} \right) \sin \left(\frac{\pi}{q} \right) = \cos \left(\frac{\pi}{n} \right).$$

The possible sponges are $\{p, q | n\} = \{6, 6|3\}$, $\{6, 4|4\}$, $\{4, 6|4\}$, $\{3, 6|6\}$, and $\{4, 4|\infty\}$.

There are many semiregular honeycombs, such as $\left\{ {3,3 \atop 4} \right\}$, in which each VERTEX consists of two OCTAHEDRA $\{3, 4\}$ and four CUBOCTAHEDRA $\left\{ {3 \atop 4} \right\}$.

See also HONEYCOMB CONJECTURE, MENGER SPONGE, SIERPINSKI SPONGE, TESSELLATION, TETRIX, TILING

References

Ball, W. W. R. and Coxeter, H. S. M. "Regular Sponges." In *Mathematical Recreations and Essays, 13th ed.* New York: Dover, pp. 152–53, 1987.
Bulatov, V. "Infinite Regular Polyhedra." http://www.physics.orst.edu/~bulatov/polyhedra/infinite/.
Coxeter, H. S. M. "Regular Honeycombs in Hyperbolic Space." *Proc. International Congress of Math., Vol. 3.* Amsterdam, Netherlands: pp. 155–69, 1954.
Coxeter, H. S. M. "Space Filled with Cubes," "Other Honeycombs," and "Polytopes and Honeycombs." §4.6, 4.7, and 7.4 in *Regular Polytopes, 3rd ed.* New York: Dover, pp. 68–2 and 126–28, 1973.
Cromwell, P. R. *Polyhedra.* New York: Cambridge University Press, p. 79, 1997.
Gott, J. R. III "Pseudopolyhedrons." *Amer. Math. Monthly* **73**, 497–04, 1967.
Wells, D. *The Penguin Dictionary of Curious and Interesting Geometry.* London: Penguin, pp. 104–06, 1991.
Williams, R. *The Geometrical Foundation of Natural Structure: A Source Book of Design.* New York: Dover, 1979.

Honeycomb Conjecture

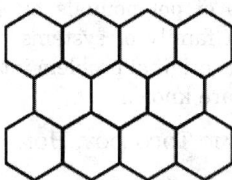

Any partition of the plane into regions of equal area has PERIMETER as least that of the regular hexagonal honeycomb TILING. Pappus refers to the problem in his fifth book, but the conjecture was finally proven by Hales (1999).

See also PERIMETER, TESSELLATION, TILING

References

Hales, T. C. The Honeycomb Conjecture. 8 Jun 1999. http://xxx.lanl.gov/abs/math.MG/9906042/.
Kepler, J. "L'étrenne ou la neige sexangulaire." C.N.R.S., 1975.
Mackenzie, D. "Proving the Perfection of the Honeycomb." *Science* **285**, 1338–339, 1999.

Thompson, D'A. W. *On Growth and Form, 2nd ed., compl. rev. ed.* New York: Cambridge University Press, 1992.
Weyl, H. *Symmetry.* Princeton, NJ: Princeton University Press, 1952.

Hoof

CYLINDRICAL WEDGE

Hook

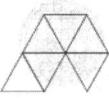

One of the 12 6-POLYIAMONDS.

See also POLYIAMOND

References

Golomb, S. W. *Polyominoes: Puzzles, Patterns, Problems, and Packings, 2nd ed.* Princeton, NJ: Princeton University Press, p. 92, 1994.

Hook Length Formula

A FORMULA for the number of YOUNG TABLEAUX associated with a given YOUNG DIAGRAM. In each box, write the sum of one plus the number of boxes horizontally to the right and vertically below the box (the "hook length"). The number of tableaux is then $n!$ divided by the product of all "hook lengths". The `NumberOfTableaux` in the *Mathematica* add-on package `DiscreteMath`Combinatorica`` (which can be loaded with the command `<<DiscreteMath``) function in *Mathematica* implements the hook length formula.

See also YOUNG DIAGRAM, YOUNG TABLEAU

References

Jones, V. "Hecke Algebra Representations of Braid Groups and Link Polynomials." *Ann. Math.* **126**, 335–88, 1987.
Skiena, S. *Implementing Discrete Mathematics: Combinatorics and Graph Theory with Mathematica.* Reading, MA: Addison-Wesley, 1990.

Hopf Algebra

Let a graded module A have a multiplication ϕ and a co-multiplication ψ. Then if ϕ and ψ have the unity of k as unity and $\psi : (A, \phi) \to (A, \phi) \otimes (A, \phi)$ is an algebra homomorphism, then (A, ϕ, ψ) is called a Hopf algebra.

Hopf Bifurcation

The BIFURCATION of a FIXED POINT to a LIMIT CYCLE (Tabor 1989).

References

Casti, J. L. "The Hopf Bifurcation Theorem." Ch. 2 in *Five More Golden Rules: Knots, Codes, Chaos, and Other Great Theories of 20th-Century Mathematics.* New York: Wiley, pp. 35–9, 2000.

Guckenheimer, J. and Holmes, P. *Nonlinear Oscillations, Dynamical Systems, and Bifurcations of Vector Fields, 3rd ed.* New York: Springer-Verlag, pp. 150–54, 1997.

Marsden, J. and McCracken, M. *Hopf Bifurcation and Its Applications.* New York: Springer-Verlag, 1976.

Tabor, M. *Chaos and Integrability in Nonlinear Dynamics: An Introduction.* New York: Wiley, p. 197, 1989.

Hopf Circle

Hopf Map

Hopf Fibration

Hopf Map

Hopf Link

The LINK 02-2-1 which has JONES POLYNOMIAL

$$V(t) = -t - t^{-1}$$

and HOMFLY POLYNOMIAL

$$P(z, \alpha) = z^{-1}(\alpha^{-1} - \alpha^{-3}) + z\alpha^{-1}.$$

It has BRAID WORD σ_1^2.

Hopf Map

The first example discovered of a MAP from a higher-dimensional SPHERE to a lower-dimensional SPHERE which is not null-HOMOTOPIC. Its discovery was a shock to the mathematical community, since it was believed at the time that all such maps were null-HOMOTOPIC, by analogy with HOMOLOGY GROUPS.

The Hopf map $f : \mathbb{S}^3 \to \mathbb{S}^2$ arises in many contexts, and can be generalized to a map $\mathbb{S}^7 \to \mathbb{S}^4$. For any point p in the sphere, its PREIMAGE $f^{-1}(p)$ is a circle $\mathbb{S}^1$ in $\mathbb{S}^3$. There are several descriptions of the Hopf map, also called the Hopf fibration.

As a SUBMANIFOLD of $\mathbb{R}^4$, the 3-SPHERE is

$$\mathbb{S}^3 = \{(X_1, X_2, X_3, X_4) : X_1^2 + X_2^2 + X_3^2 + X_4^2 = 1\} \quad (1)$$

and the 2-SPHERE is a SUBMANIFOLD of $\mathbb{R}^3$,

$$\mathbb{S}^2 = \{(x_1, x_2, x_3) : x_1^2 + x_2^2 + x_3^2 = 1\}. \quad (2)$$

The Hopf map takes points (X_1, X_2, X_3, X_4) on a 3-sphere to points on a 2-sphere (x_1, x_2, x_3)

$$x_1 = 2(X_1 X_2 + X_3 X_4) \quad (3)$$

$$x_2 = 2(X_1 X_4 - X_2 X_3) \quad (4)$$

$$x_3 = (X_1^2 + X_3^2) - (X_2^2 + X_4^2). \quad (5)$$

Every point on the 2-SPHERE corresponds to a CIRCLE called the HOPF CIRCLE on the 3-SPHERE.

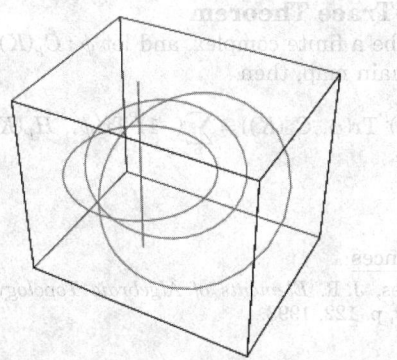

By STEREOGRAPHIC PROJECTION, the 3-sphere can be mapped to $\mathbb{R}^3$, where the point at infinity corresponds to the north pole. As a map, from $\mathbb{R}^3$, the Hopf map can be pretty complicated. The diagram above shows some of the preimages $f^{-1}(p)$, called HOPF CIRCLES. The straight red line is the circle through infinity.

By associating $\mathbb{R}^4$ with $\mathbb{C}^2$, the map is given by $f(z, w) = z/w$, which gives the map to the RIEMANN SPHERE.

The Hopf fibration is a FIBRATION

$$\mathbb{S}^1 \to \mathbb{S}^3 \to \mathbb{S}^2, \quad (6)$$

and is in fact a PRINCIPAL BUNDLE. The ASSOCIATED VECTOR BUNDLE

$$L = \mathbb{S}^3 \times \mathbb{C}/U(1), \quad (7)$$

where

$$((z, w), v) \sim ((e^{it}z, e^{it}w), e^{it}v) \quad (8)$$

is a complex LINE BUNDLE on $\mathbb{S}^2$. In fact, the set of line bundles on the sphere forms a group under TENSOR PRODUCT, and the bundle L generates all of them. That is, every line bundle on the sphere is $L^{\otimes k}$ for some k.

The sphere $\mathbb{S}^3$ is the LIE GROUP of unit QUATERNIONS, and can be identified with the SPECIAL UNITARY GROUP $SU(2)$, which is the SIMPLY CONNECTED double cover of $SO(3)$. The Hopf bundle is the quotient map $\mathbb{S}^2 \cong SU(2)/U(1)$.

See also FIBRATION, FIBER BUNDLE, HOMOGENEOUS SPACE, PRINCIPAL BUNDLE, STEREOGRAPHIC PROJECTION, VECTOR BUNDLE

References

Berger, M. Chs. 4 and 18 in *Geometry I.* New York: Springer-Verlag, 1987.

Kreminski, R. "Visualizing the Hopf Fibration." *Mathematica Educ. Res.* **6**, 9–4, 1997.

Penrose, R. and Rindler, W. *Spinors and Space-Time, Vol. 1: Two-Spinor Calculus and Relativistic Fields.* Cambridge, England: Cambridge University Press, 1987.

Ryder, L. H. *Quantum Field Theory, 2nd ed.* Cambridge, England: Cambridge University Press, 1996.

Whitehead, G. W. *Elements of Homotopy Theory.* New York: Springer Verlag, 1979.

Hopf Trace Theorem

Let K be a finite complex, and let $\phi : C_P(K) \to C_P(K)$ be a chain map, then

$$\sum_P (-1)^P \mathrm{Tr}(\phi, \, C_P(K)) = \sum_P (-1)^P \mathrm{Tr}(\phi_*, \, H_P(K)/T_P(K)).$$

References

Munkres, J. R. *Elements of Algebraic Topology.* Perseus Press, p. 122, 1993.

Hopf's Theorem

A NECESSARY and SUFFICIENT condition for a MEASURE which is quasi-invariant under a transformation to be equivalent to an invariant PROBABILITY MEASURE is that the transformation cannot (in a measure theoretic sense) compress the SPACE.

Horizontal

Oriented in position PERPENDICULAR to up-down, and therefore PARALLEL to a flat surface.

See also VERTICAL

Horizontal Cusp

SPINODE

Horizontal Cylinder

CYLINDRICAL SEGMENT

Horizontal Tank

CYLINDRICAL SEGMENT

Horizontally Convex Polyomino

ROW-CONVEX POLYOMINO

Horizontal-Vertical Illusion

VERTICAL-HORIZONTAL ILLUSION

Horn Angle

The configuration formed by two curves starting at a point, called the vertex V, in a common direction. Horn angles are concrete illustrations of NON-ARCHIMEDEAN GEOMETRIES.

See also NON-ARCHIMEDEAN GEOMETRY

References

Kasner, E. "The Recent Theory of the Horn Angle." *Scripta Math* **11**, 263–67, 1945.

Horn Cyclide

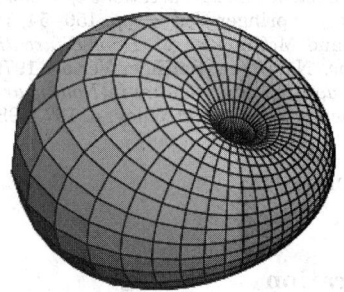

The INVERSION of a HORN TORUS. If the INVERSION CENTER lies on the TORUS, then the horn cyclide degenerates to a PARABOLIC HORN CYCLIDE.

See also CYCLIDE, HORN TORUS, INVERSION, PARABOLIC CYCLIDE, RING CYCLIDE, SPINDLE CYCLIDE, TORUS

Horn Function

The 34 distinct convergent hypergeometric series of order two enumerated by Horn (1931) and corrected by Borngässer (1933). There are 14 complete series for which $p = p' = q = q' = 2$,

$$F_1(\alpha, \, \beta, \, \beta', \, \gamma, \, x, \, y) = \sum_{m, \, n} \frac{(\alpha)_{m+n}(\beta)_m(\beta')_n}{(\gamma)_{m+n} m! n!} x^m y^n \quad (1)$$

$$F_2(\alpha, \, \beta, \, \beta', \, \gamma, \, \gamma', \, x, \, y) = \sum_{m, \, n} \frac{(\alpha)_{m+n}(\beta)_m(\beta')_n}{(\gamma)_m(\gamma')_n m! n!} x^m y^n \quad (2)$$

$$\begin{aligned} F_3(\alpha, \, \alpha', \, \beta, \, \beta', \, \gamma, \, x, \, y) \\ = \sum_{m, \, n} \frac{(\alpha)_m(\alpha')_n(\beta)_m(\beta')_n}{(\gamma)_{m+n} m! n!} x^m y^n \end{aligned} \quad (3)$$

$$F_4(\alpha, \, \beta, \, \gamma, \, \gamma', \, x, \, y) = \sum_{m, \, n} \frac{(\alpha)_{m+n}(\beta)_{m+n}}{(\gamma)_m(\gamma')_n m! n!} x^m y^n \quad (4)$$

$$G_1(\alpha, \, \beta, \, \beta', \, x, \, y) = \sum_{m, \, n} \frac{(\alpha)_{m+n}(\beta)_{n-m}(\beta')_{m-n}}{m! n!} x^m y^n \quad (5)$$

$$\begin{aligned} G_2(\alpha, \, \alpha', \, \beta, \, \beta', \, x, \, y) \\ = \sum_{m, \, n} \frac{(\alpha)_m(\alpha')_n(\beta)_{n-m}(\beta')_{m-n}}{m! n!} x^m y^n \end{aligned} \quad (6)$$

$$G_3(\alpha, \, \alpha', \, x, \, y) = \sum_{m, \, n} \frac{(\alpha)_{2n-m}(\alpha')_{2m-n}}{m! n!} x^m y^n \quad (7)$$

$$H_1(\alpha, \, \beta, \, \gamma, \, \delta, \, x, \, y) = \sum_{m, \, n} \frac{(\alpha)_{m-n}(\beta)_{m+n}(\gamma)_n}{(\delta)_m m! n!} x^m y^n \quad (8)$$

$$\begin{aligned} H_2(\alpha, \, \beta, \, \gamma, \, \delta, \, \epsilon, \, x, \, y) \\ = \sum_{m, \, n} \frac{(\alpha)_{m-n}(\beta)_m(\gamma)_n(\delta)_n}{(\epsilon)_m m! n!} x^m y^n \end{aligned} \quad (9)$$

$$H_3(\alpha, \beta, \gamma, x, y) = \sum_{m, n} \frac{(\alpha)_{2m+n}(\beta)_n}{(\gamma)_{m+n}m!n!} x^m y^n \qquad (10)$$

$$H_4(\alpha, \beta, \gamma, \delta, x, y) = \sum_{m, n} \frac{(\alpha)_{2m+n}(\beta)_n}{(\gamma)_m(\delta)_n m!n!} x^m y^n \qquad (11)$$

$$H_5(\alpha, \beta, \gamma, x, y) = \sum_{m, n} \frac{(\alpha)_{2m+n}(\beta)_{n-m}}{(\gamma)_n m!n!} x^m y^n \qquad (12)$$

$$H_6(\alpha, \beta, \gamma, x, y) = \sum_{m, n} \frac{(\alpha)_{2m-n}(\beta)_{n-m}(\gamma)_n}{m!n!} x^m y^n \qquad (13)$$

$$H_7(\alpha, \beta, \gamma, \delta, x, y) = \sum_{m, n} \frac{(\alpha)_{2m-n}(\beta)_n(\gamma)_n}{(\gamma)_m m!n!} x^m y^n \qquad (14)$$

(of which F_1, F_2, F_3, and F_4 are precisely APPELL HYPERGEOMETRIC FUNCTIONS), and 20 confluent series with $p \le p' = 2$, $q \le q' = 2$, and p, q not both 2,

$$\Phi_1(\alpha, \beta, \gamma, x, y) = \sum_{m, n} \frac{(\alpha)_{m+n}(\beta)_n}{(\gamma)_{m+n}m!n!} x^m y^n \qquad (15)$$

$$\Phi_2(\beta, \beta', \gamma, x, y) = \sum_{m, n} \frac{(\beta)_m(\beta')_m}{(\gamma)_{m+n}m!n!} x^m y^n \qquad (16)$$

$$\Phi_3(\beta, \gamma, x, y) = \sum_{m, n} \frac{(\beta)_m}{(\gamma)_{m+n}m!n!} x^m y^n \qquad (17)$$

$$\Psi_1(\alpha, \beta, \gamma, \gamma', x, y) = \sum_{m, n} \frac{(\alpha)_{m+n}(\beta)_m}{(\gamma)_m(\gamma')_n m!n!} x^m y^n \qquad (18)$$

$$\Psi_2(\alpha, \gamma, \gamma', x, y) = \sum_{m, n} \frac{(\alpha)_{m+n}}{(\gamma)_m(\gamma')_n m!n!} x^m y^n \qquad (19)$$

$$\Xi_1(\alpha, \alpha', \beta, \gamma, x, y) = \sum_{m, n} \frac{(\alpha)_m(\alpha')_n(\beta)_m}{(\gamma)_{m+n}m!n!} x^m y^n \qquad (20)$$

$$\Xi_2(\alpha, \beta, \gamma, x, y) = \sum_{m, n} \frac{(\alpha)_m(\beta)_n}{(\gamma)_{m+n}m!n!} x^m y^n \qquad (21)$$

$$\Gamma_1(\alpha, \beta, \beta', x, y) = \sum_{m, n} \frac{(\alpha)_m(\beta)_{n-m}(\beta')_{m-n}}{m!n!} x^m y^n \qquad (22)$$

$$\Gamma_2(\beta, \beta', x, y) = \sum_{m, n} \frac{(\beta)_{n-m}(\beta')_{m-n}}{m!n!} x^m y^n \qquad (23)$$

$$H_1(\alpha, \beta, \delta, x, y) = \sum_{m, n} \frac{(\alpha)_{m-n}(\beta)_{m+n}}{(\delta)_m m!n!} x^m y^n \qquad (24)$$

$$H_2(\alpha, \beta, \gamma, \delta, x, y) = \sum_{m, n} \frac{(\alpha)_{m-n}(\beta)_m(\gamma)_n}{(\delta)_m m!n!} x^m y^n \qquad (25)$$

$$H_3(\alpha, \beta, \delta, x, y) = \sum_{m, n} \frac{(\alpha)_{m-n}(\beta)_n}{(\delta)_m m!n!} x^m y^n \qquad (26)$$

$$H_4(\alpha, \gamma, \delta, x, y) = \sum_{m, n} \frac{(\alpha)_{m-n}(\gamma)_n}{(\delta)_m m!n!} x^m y^n \qquad (27)$$

$$H_5(\alpha, \delta, x, y) = \sum_{m, n} \frac{(\alpha)_{m-n}}{(\delta)_m m!n!} x^m y^n \qquad (28)$$

$$H_6(\alpha, \gamma, x, y) = \sum_{m, n} \frac{(\alpha)_{2m+n}}{(\gamma)_{m+n}m!n!} x^m y^n \qquad (29)$$

$$H_7(\alpha, \gamma, \delta, x, y) = \sum_{m, n} \frac{(\alpha)_{2m+n}}{(\gamma)_m(\delta)_n m!n!} x^m y^n \qquad (30)$$

$$H_8(\alpha, \beta, x, y) = \sum_{m, n} \frac{(\alpha)_{2m-n}(\beta)_{n-m}}{m!n!} x^m y^n \qquad (31)$$

$$H_9(\alpha, \beta, \delta, x, y) = \sum_{m, n} \frac{(\alpha)_{2m-n}(\beta)_n}{(\delta)_m m!n!} x^m y^n \qquad (32)$$

$$H_{10}(\alpha, \delta, x, y) = \sum_{m, n} \frac{(\alpha)_{2m-n}}{(\delta)_m m!n!} x^m y^n \qquad (33)$$

$$H_{11}(\alpha, \beta, \gamma, \delta, x, y) = \sum_{m, n} \frac{(\alpha)_{m-n}(\beta)_n(\gamma)_n}{(\delta)_m m!n!} x^m y^n \qquad (34)$$

(Erdélyi *et al.* 1981, pp. 224–26).

See also APPELL HYPERGEOMETRIC FUNCTION, KAMPÉ DE FÉRIET FUNCTION, LAURICELLA FUNCTIONS

References

Borngässer, L. *Über hypergeometrische Funktionen zweier Veränderlichen.* Dissertation. Darmstadt, Germany: University of Darmstadt, 1933.

Erdélyi, A.; Magnus, W.; Oberhettinger, F.; and Tricomi, F. G. "Horn's List" and "Convergence of the Series." §5.7.1 and 5.7.2 in *Higher Transcendental Functions, Vol. 1.* New York: Krieger, pp. 224–29, 1981.

Horn, J. "Hypergeometrische Funktionen zweier Veränderlichen." *Math. Ann.* **105**, 381–07, 1931.

Horn Torus

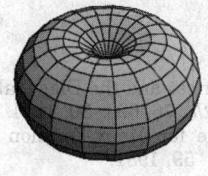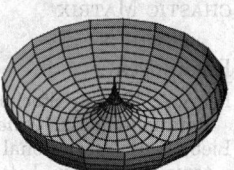

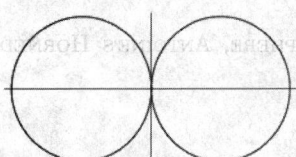

One of the three STANDARD TORI given by the PARAMETRIC EQUATIONS

$$x = (c + a \cos v)\cos u \qquad (1)$$

$$y = (c + a \cos v)\sin u \qquad (2)$$

$$z = a \sin v \qquad (3)$$

with $a = c$. The INVERSION of a horn torus is a HORN CYCLIDE (or PARABOLIC HORN CYCLIDE). The above figures show a horn torus (left), a cutaway (middle), and a CROSS SECTION of the horn torus through the xz-plane (right).

See also CYCLIDE, HORN CYCLIDE, RING TORUS, SPINDLE TORUS, STANDARD TORI, TORUS

References

Gray, A. "Tori." §13.4 in *Modern Differential Geometry of Curves and Surfaces with Mathematica, 2nd ed.* Boca Raton, FL: CRC Press, pp. 304–06, 1997.
Pinkall, U. "Cyclides of Dupin." §3.3 in *Mathematical Models from the Collections of Universities and Museums* (Ed. G. Fischer). Braunschweig, Germany: Vieweg, pp. 28–0, 1986.

Horn's Theorem

This entry contributed by FRED MANBY

Let

$$X = \{x_1 \geq x_2 \geq \cdots \geq x_n | x_i \in \mathbb{R}\} \qquad (1)$$

and

$$Y = \{y_1 \geq y_2 \geq \cdots \geq y_n | y_i \in \mathbb{R}\}. \qquad (2)$$

Then there exists an $n \times n$ HERMITIAN MATRIX with eigenvalues X and diagonal elements Y IFF

$$\sum_{i=1}^{t}(x_i - y_i) \geq 0 \quad \forall 1 \leq t \leq n \qquad (3)$$

and with equality for $t = n$. The theorem is sometimes also known as Schur's theorem.

See also HERMITIAN MATRIX, MAJORIZATION, STOCHASTIC MATRIX

References

Horn, A. "Doubly Stochastic Matrices and the Diagonal of a Rotation Matrix." *Amer. J. Math.* **76**, 620–30, 1954.
Lieb, E. H "Variational Principle for Many-Fermion Systems." *Phys. Rev. Lett.* **46**, 457–59, 1981.

Horned Sphere

ALEXANDER'S HORNED SPHERE, ANTOINE'S HORNED SPHERE

Horner's Method

A method for finding roots of a polynomial equation $f(x) = 0$. Now find an equation whose roots are the roots of this equation diminished by r, so

$$0 = f(x + r)$$
$$= f(r) + xf'(r) + \tfrac{1}{2}x^2 f''(r) + \tfrac{1}{3}x^3 f'''(r) + \ldots. \qquad (1)$$

The expressions for $f(r), f'(r), \ldots$ are then found as in the following example, where

$$f(x) \equiv Ax^5 + Bx^4 + Cx^3 + Dx^2 + Ex + F. \qquad (2)$$

Write the coefficients $A, B, \ldots, F$ in a horizontal row, and let a new letter shown as a denominator stand for the sum immediately above it so, in the following example, $P = Ar + B$. The result is the following table.

A	B	C	D	E	F
	$\dfrac{Ar}{P}$	$\dfrac{Pr}{Q}$	$\dfrac{Qr}{R}$	$\dfrac{Rr}{S}$	$\dfrac{Sr}{\omega}$
	$\dfrac{Ar}{T}$	$\dfrac{Tr}{U}$	$\dfrac{Ur}{R}$	$\dfrac{Vr}{\chi}$	
	$\dfrac{Ar}{W}$	$\dfrac{Wr}{X}$	$\dfrac{Xr}{\psi}$		
	$\dfrac{Ar}{Y}$	$\dfrac{Yr}{\phi}$			
	$\dfrac{Ar}{\theta}$				

Solving for the quantities θ, ϕ, ψ, χ, and ω gives

$$\theta = 5Ar + B = \frac{1}{4!}f^{(iv)}(r) \qquad (3)$$

$$\phi = 10Ar^2 + 4Br + C = \frac{1}{3!}f'''(r) \qquad (4)$$

$$\psi = 10Ar^3 + 6Br^2 + 3Cr + D = \frac{1}{2!}f''(r) \qquad (5)$$

$$\chi = 5Ar^4 + 4Br^3 + 3Cr^2 + 2Dr + E = f'(r) \qquad (6)$$

$$\omega = Ar^5 + Br^4 + Cr^3 + Dr^2 + Er + F = f(r), \qquad (7)$$

so the equation whose roots are the roots of $f(x) = 0$, each diminished by r, is

$$0 = Ax^5 + \theta x^4 + \phi x^3 + \psi x^2 + \chi x + \omega \qquad (8)$$

(Whittaker and Robinson 1967).

To apply the procedure, first determine the integer part of the root through whatever means are needed, then reduce the equation by this amount. This gives the second digit, by which the equation is once again reduced (after suitable multiplication by 10) to find the third digit, and so on.

1	−4	0	5(1	−10	−500	2000(3
	1	−3	−3	3	−21	−1563
	−3	−3	2	−7	−521	437
	1	−2		3	−12	
	−2	−5		−4	−523	
	1			3		
	−1			−1		

To see the method applied, consider the problem of finding the smallest positive root of

$$x^3 - 4x^2 + 5 = 0. \tag{9}$$

This root lies between 1 and 2, so diminish the equation by 1, resulting in the left table shown above. The resulting diminished equation is

$$x^3 - x^2 - 5x + 2 = 0, \tag{10}$$

and roots which are ten times the roots of this equation satisfy the equation

$$x^3 - 10x^2 - 500x + 2000 = 0. \tag{11}$$

The root of this equation between 1 and 10 lies between 3 and 4, so reducing the equation by 3 produces the right table shown above, giving the transformed equation

$$x^3 - x^2 - 533 + 437 = 0. \tag{12}$$

This procedure can be continued to yield the root as approximately 1.3819659.

Horner's process really boils down to the construction of a DIVIDED DIFFERENCE table (Whittaker and Robinson 1967).

See also DIVIDED DIFFERENCE, NEWTON'S METHOD

References
Boyer, C. B. and Merzbacher, U. C. *A History of Mathematics, 2nd ed.* New York: Wiley, pp. 202–04, 256, and 307, 1991.
Horner, W. G. *Philos. Trans.* **1**, 308, 1819.
Pena, J. M. and Sauer, T. *SIAM J. Numer. Anal.* **37**, 1186, 2000.
Ruffini, P. *Sopra la determinazione della radici.* Modena, Italy, 1804.
Ruffini, P. *Memorie di Mat. e di Fis. della Soc. Italiana delle Scienze.* Verona, Italy, 1813.
Séroul, R. "Evaluation of Polynomials: Horner's Method." §10.6 in *Programming for Mathematicians.* Berlin: Springer-Verlag, pp. 216–62, 2000.
Whittaker, E. T. and Robinson, G. "The Ruffini-Horner Method." §53 in *The Calculus of Observations: A Treatise on Numerical Mathematics, 4th ed.* New York: Dover, pp. 100–06, 1967.

Horner's Rule

A rule for POLYNOMIAL computation which both reduces the number of necessary multiplications and results in less numerical instability due to potential subtraction of one large number from another. The rule simply factors out POWERS of x,

giving

$$a_n x^n + a_{n-1} x^{n-1} + \ldots + a_0 = ((a_n x + a_{n-1})x + \ldots)x + a_0.$$

Horner's rule can be implemented to form a POLYNOMIAL from a list of coefficients in *Mathematica* as follows.

```
PolynomialFromCoefs[l_List, x_] := Fold[x#1 +
#2 &, 0, l]
```

See also POLYNOMIAL

References
Borwein, P. and Erdélyi, T. "Horner's Rule." §1.1.E.5 in *Polynomials and Polynomial Inequalities.* New York: Springer-Verlag, p. 8, 1995.
Knuth, D. E. *The Art of Computer Programming, Vol. 2: Seminumerical Algorithms, 3rd ed.* Reading, MA: Addison-Wesley, pp. 467–69, 1998.
Vardi, I. *Computational Recreations in Mathematica.* Reading, MA: Addison-Wesley, p. 9, 1991.

Horocycle

The LOCUS of a point which is derived from a fixed point Q by continuous parallel displacement.

References
Coxeter, H. S. M. *Introduction to Geometry, 2nd ed.* New York: Wiley, p. 300, 1969.

Horse Fetter

HIPPOPEDE

Horseshoe Map

SMALE HORSESHOE MAP

Horton Graph

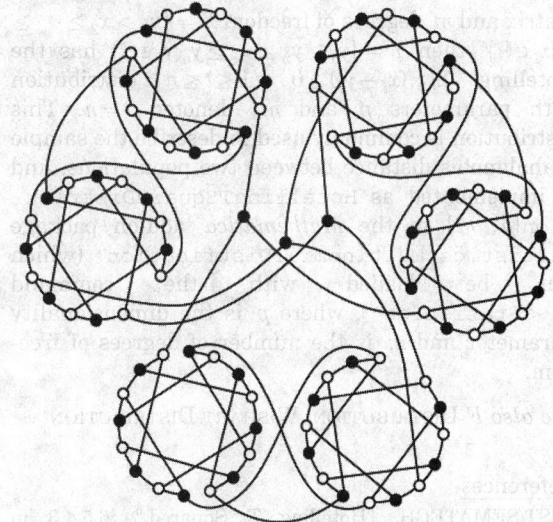

A graph on 93 nodes providing a counterexample to Tutte's conjecture that every 3-regular 3-connected bipartite graph is HAMILTONIAN. Two smaller counterexamples, each on 78 nodes, are now known (Ellingham 1981, 1982; Ellingham and Horton 1983; Owens 1983).

See also HAMILTONIAN GRAPH

References
Bondy, J. A. and Murty, U. S. R. *Graph Theory with Applications.* New York: North Holland, pp. 61 and 242, 1976.
Ellingham, M. N. "Non-Hamiltonian 3-Connected Cubic Partite Graphs." Research Report No. 28, Dept. of Math., Univ. Melbourne, Melbourne, 1981.
Ellingham, M. N. "Constructing Certain Cubic Graphs." In *Combinatorial Mathematics, IX: Proceedings of the Ninth Australian Conference held at the University of Queensland, Brisbane, August 24–8, 1981)* (Ed. E. J. Billington, S. Oates-Williams, and A. P. Street). Berlin: Springer-Verlag, pp. 252–74, 1982.
Ellingham, M. N. and Horton, J. D. "Non-Hamiltonian 3-Connected Cubic Bipartite Graphs." *J. Combin. Th. Ser. B* **34**, 350–53, 1983.
Owens, P. J. "Bipartite Cubic Graphs and a Shortness Exponent." *Disc. Math.* **44**, 327–30, 1983.

Hotelling T2 Distribution

A univariate distribution proportional to the *F*-DISTRIBUTION. If the vector **d** is Gaussian multivariate-distributed with zero mean and unit covariance matrix

$$\sum_{m,n} \frac{(\alpha)_{2m-n}(\beta)_n}{(\delta)_m m! n!} x^m y^n$$

and $H_{11}(\alpha, \beta, \gamma, \delta, x, y)$ is an

$$\sum_{m,n} \frac{(\alpha)_{m-n}(\beta)_n(\gamma)_n}{(\delta)_m m! n!} x^m y^n$$

matrix with a WISHART DISTRIBUTION with unit scale matrix and m degrees of freedom $X = \{x_1 > x_2 \geq \cdots \geq x_n | x_i \in \mathbb{R}\}$, then $Y = \{y_1 \geq y_2 \geq \cdots \geq y_n | y_i \in \mathbb{R}\}$ has the Hotelling $\sum_{i-1}^{t}(x_i - y_i) \geq 0 \quad \forall 1 \leq t \leq n$ distribution with parameters p and m, denoted $t = n$. This distribution is commonly used to describe the sample Mahalanobis distance between two populations, and is implemented as `HotellingTSquareDistribution[p, m]` in the *Mathematica* add-on package `Statistics'MultinomialDistribution'` (which can be loaded with the command `< <Statistics'`), where p is the dimensionality parameter and m is the number of degrees of freedom.

See also *F*-DISTRIBUTION, WISHART DISTRIBUTION

References
NIST/SEMATECH. "Hotelling *T* Squared." §6.5.4.3 in *NIST/Sematech Engineering Statistics Internet Handbook.* http://www.itl.nist.gov/div898/handbook/pmc/section5/pmc543.htm.

Hotelling T-Squared Distribution

A univariate distribution proportional to the *F*-DISTRIBUTION. If the vector **d** is Gaussian multivariate-distributed with zero mean and unit covariance matrix $N_p(0, I)$ and M is an $m \times p$ matrix with a WISHART DISTRIBUTION with unit scale matrix and m degrees of freedom $W_p(I, m)$, then $m\mathbf{d}^T M^{-1}\mathbf{d}$ has the Hotelling T^2 distribution with parameters p and m, denoted $T^2(p, m)$. This distribution is commonly used to describe the sample Mahalanobis distance between two populations, and is implemented as `HotellingTSquareDistribution[p, m]` in the *Mathematica* add-on package `Statistics'MultinomialDistribution'` (which can be loaded with the command `< <Statistics'`), where p is the dimensionality parameter and m is the number of degrees of freedom.

See also *F*-DISTRIBUTION, HOTELLING'S *T*-SQUARED TEST, WISHART DISTRIBUTION

References
NIST/SEMATECH. "Hotelling *T* Squared." §6.5.4.3 in *NIST/Sematech Engineering Statistics Internet Handbook.* http://www.itl.nist.gov/div898/handbook/pmc/section5/pmc543.htm.

Hotelling's T-Squared Test

See also HOTELLING *T*-SQUARED DISTRIBUTION

References
Winer, B. J. *Statistical Principles in Experimental Design.* New York: McGraw-Hill, 1962.

Hough Transform

A technique used to detect boundaries in digital images.

Householder's Method

A ROOT-finding algorithm based on the iteration formula

$$x_{n+1} = x_n - \frac{f(x_n)}{f'(x_n)}\left\{1 + \frac{f(x_n)f''(x_n)}{2[f'(x_n)]^2}\right\}.$$

This method, like NEWTON'S METHOD, has poor convergence properties near any point where the DERIVATIVE $f'(x) = 0$.

See also HALLEY'S IRRATIONAL FORMULA, HALLEY'S METHOD, NEWTON'S METHOD

References
Gourdon, X. and Sebah, P. "Newton's Iteration." http://xavier.gourdon.free.fr/Constants/Algorithms/newton.html.

Householder, A. S. *The Numerical Treatment of a Single Nonlinear Equation.* New York: McGraw-Hill, 1970.

Ortega, J. M. and Rheinboldt, W. C. *Iterative Solution of Nonlinear Equations in Several Variables.* Philadelphia, PA: SIAM, 2000.

Howe's Theorem

Let P be a PRIMITIVE POLYTOPE with eight vertices. Then there is a unimodular map that maps P to the polyhedron whose vertices are $(0, 0, 0)$, $(1, 0, 0)$, $(0, 1, 0)$, $(0, 0, 1)$, $(0, 1, 1)$, $(1, a, b)$, $(1, c, d)$, and $(1, a+c, b+d)$ with a, b, c, $d \in \mathbb{Z}$, a, b, c, $d \geq 0$, and $ad - bc = 1$. Furthermore, any primitive polyhedron with fewer than eight vertices can be embedded in one with eight vertices.

See also PRIMITIVE POLYTOPE

References

Khan, M. R. "A Counting Formula for Primitive Tetrahedra in Z^3." *Amer. Math. Monthly* **106**, 525–33, 1999.

Scarf, H. E. "Integral Polyhedra in Three Space." *Math. Oper. Res.* **10**, 403–38, 1985.

Howell Design

Let S be a set of $n + 1$ symbols, then a Howell design $H(s, 2n)$ on symbol set S is an $s \times s$ array H such that

1. Every cell of H is either empty or contains an unordered pair of symbols from S,
2. Every symbol of S occurs once in each row and column of H, and
3. Every unordered pair of symbols occurs in at most one cell of H.

References

Colbourn, C. J. and Dinitz, J. H. (Eds.). "Howell Designs." Ch. 26 in *CRC Handbook of Combinatorial Designs.* Boca Raton, FL: CRC Press, pp. 381–85, 1996.

H-Spread

The difference $H_2 - H_1$, where H_1 and H_2 are HINGES. It is the same as the INTERQUARTILE RANGE for $N = 5, 9, 13, \ldots$ points.

See also HINGE, INTERQUARTILE RANGE, STEP

References

Tukey, J. W. *Explanatory Data Analysis.* Reading, MA: Addison-Wesley, p. 44, 1977.

h-Statistic

An unbiased estimator for a MOMENT of a distribution.

See also κ-STATISTIC

H-Transform

A 2-D generalization of the HAAR TRANSFORM which is used for the compression of astronomical images. The algorithm consists of dividing the $2^N \times 2^N$ image into blocks of 2×2 pixels, calling the pixels in the block $a_{00}, a_{10}, a_{01},$ and a_{11}. For each block, compute the four coefficients

$$h_0 \equiv \tfrac{1}{2}(a_{11} + a_{10} + a_{01} + a_{00})$$

$$h_x \equiv \tfrac{1}{2}(a_{11} + a_{10} - a_{01} - a_{00})$$

$$h_y \equiv \tfrac{1}{2}(a_{11} - a_{10} + a_{01} - a_{00})$$

$$h_c \equiv \tfrac{1}{2}(a_{11} - a_{10} - a_{01} + a_{00}).$$

Construct a $2^{N-1} \times 2^{N-1}$ image from the h_0 values, and repeat until only one h_0 value remains. The H-transform can be performed in place and requires about $16N^2/3$ additions for an $N \times N$ image.

See also HAAR TRANSFORM

References

Capaccioli, M.; Held, E. V.; Lorenz, H.; Richter, G. M.; and Ziener, R. "Application of an Adaptive Filtering Technique to Surface Photometry of Galaxies. I. The Method Tested on NGC 3379." *Astron. Nachr.* **309**, 69–0, 1988.

Fritze, K.; Lange, M.; Möstle, G.; Oleak, H.; and Richter, G. M. "A Scanning Microphotometer with an On-Line Data Reduction for Large Field Schmidt Plates." *Astron. Nachr.* **298**, 189–96, 1977.

Richter, G. M. "The Evaluation of Astronomical Photographs with the Automatic Area Photometer." *Astron. Nachr.* **299**, 283–03, 1978.

White, R. L.; Postman, M.; and Lattanzi, M. G. "Compression of the Guide Star Digitised Schmidt Plates." In *Digitised Optical Sky Surveys: Proceedings of the Conference on "Digitised Optical Sky Surveys" held in Edinburgh, Scotland, 18–1 June 1991* (Ed. H. T. MacGillivray and E. B. Thompson). Dordrecht, Netherlands: Kluwer, pp. 167–75, 1992.

Hub

The central point in a WHEEL GRAPH W_n. The hub has DEGREE $n - 1$.

See also WHEEL GRAPH

References

Saaty, T. L. and Kainen, P. C. *The Four-Color Problem: Assaults and Conquest.* New York: Dover, p. 148, 1986.

Huffman Coding

A lossless data compression algorithm which uses a small number of bits to encode common characters. Huffman coding approximates the probability for each character as a POWER of 1/2 to avoid complications associated with using a nonintegral number of bits to encode characters using their actual probabilities.

Huffman coding works on a list of weights $\{w_i\}$ by building an EXTENDED BINARY TREE with minimum weighted PATH LENGTH and proceeds by finding the two smallest ws, w_1 and w_2, viewed as external nodes, and replacing them with an internal node of weight $w_1 + w_2$. The procedure is them repeated stepwise until the root node is reached. An individual external node can then encoded by a binary string of 0s (for left branches) and 1s (for right branches).

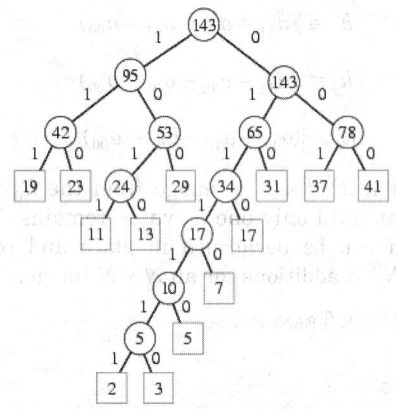

The procedure is summarized below for the weights 2, 3, 5, 7, 11, 13, 17, 19, 23, 29, 31, 37, and 41 given by the first 13 primes, and the resulting tree is shown above (Knuth 1997, pp. 402–03). As is clear from the diagram, the paths to the larger weights are shorter than those to the smaller weights. In this example, the number 13 would be encoded as 1010.

2	3	5	7	11	13	17	19	23	29	31	37	41
5	5	5	7	11	13	17	19	23	29	31	37	41
		10	7	11	13	17	19	23	29	31	37	41
			17	11	13	17	19	23	29	31	37	41
			17		24	17	19	23	29	31	37	41
					24	34	19	23	29	31	37	41
					24	34		42	29	31	37	41
						34		42	53	31	37	41
								42	53	65	37	41
								42	53	65		78
									95	65		78
									95			143
												238

The following *Mathematica* code can be used to construct the list of internal nodes and table of iterations.

```
HuffmanStep[10_List] := Module[
  {
```

```
    l = 10,
    s2 = Take[Select[Sort[10], Positive], 2]
  },
  l[[Take[Flatten[Position[l, #] & /@ s2], 2]]]
= 0;
  l[[Last[Position[l, 0]]]]] = Plus @@ s2;
  {l, s2}
] HuffmanList[l_List] := Module[{},
       Plus    @@@           Last  /@
NestWhileList[HuffmanStep[First[#]]       &,
HuffmanStep[l],      Length[Union[First[#]]]
> 2 &]
       ] HuffmanTable[l_List]  :=
NestWhileList[First[HuffmanStep[#]]   &,   l,
Length[Union[#]] > 2 &]
```

References

Huffman, D. A. "A Method for the Construction of Minimum-Redundancy Codes." *Proc. Inst. Radio Eng.* **40**, 1098–101, 1952.

Knuth, D. E. *The Art of Computer Programming, Vol. 1: Fundamental Algorithms, 3rd ed.* Reading, MA: Addison-Wesley, pp. 402–06, 1997.

Press, W. H.; Flannery, B. P.; Teukolsky, S. A.; and Vetterling, W. T. "Huffman Coding and Compression of Data." Ch. 20.4 in *Numerical Recipes in FORTRAN: The Art of Scientific Computing, 2nd ed.* Cambridge, England: Cambridge University Press, pp. 896–01, 1992.

Schwarz, E. S. "An Optimum Encoding with Minimum Longest Code and Total Number of Digits." *Information and Control* **7**, 37–4, 1964.

Hull

AFFINE HULL, CONVEX HULL

Hull Number

Let a set of vertices A in a CONNECTED GRAPH G be called convex if for every two vertices x, $y \in A$, the vertex set of every (x, y) GRAPH GEODESIC lies completely in A. Also define the convex hull $A \subseteq V(G)$ of a GRAPH G with vertex set $V(G)$ as the smallest CONVEX SET in G containing A. Then the smallest cardinality of a set A whose convex hull is $V(G)$ is called the hull number of G, denoted $h(G)$.

See also GEODETIC NUMBER

References

Chartrand, G. and Zhang, P. "On the Hull Number of a Graph." To appear in *Ars. Combin.*

Chartrand, G. and Zhang, P. "The Forcing Hull Number of a Graph." To appear in *J. Combin. Math. Comb. Comput.*

Chartrand, G. and Zhang, P. "The Geodetic Number of an Oriented Graph." *Europ. J. Combin.* **21**, 181–89, 2000.

Everett, M. G. and Seidman, S. B. "The Hull Number of a Graph." *Discr. Math.* **57**, 217–23, 1985.

Mulder, H. M. "The Expansion Procedure for Graphs." In *Contemporary Methods in Graph Theory* (Ed. R. Bodendiek). Mannheim, Germany: Wissenschaftsverlag, pp. 459–77, 1990.

Humbert's Theorem

The NECESSARY and SUFFICIENT condition that an ALGEBRAIC CURVE has an algebraic INVOLUTE is that the ARC LENGTH is a two-valued algebraic function of the coordinates of the extremities. Furthermore, this function is a ROOT of a QUADRATIC EQUATION whose COEFFICIENTS are rational functions of x and y.

See also ALGEBRAIC CURVE, INVOLUTE

References

Coolidge, J. L. *A Treatise on Algebraic Plane Curves.* New York: Dover, p. 195, 1959.

Hundkurve

TRACTRIX

Hundred

$100 = 10^2$. Madachy (1979) gives a number of algebraic equations using the digits 1 to 9 which evaluate to 100, such as

$$(7-5)^2 + 96 + 8 - 4 - 3 - 1 = 100$$
$$3^2 + 91 + 7 + 8 - 6 - 5 - 4 = 100$$
$$\sqrt{9} - 6 + 72 - (1)(3!) - 8 + 45 = 100$$
$$123 - 45 - 67 + 89 = 100,$$

and so on.

See also 10, BILLION, HUNDRED, LARGE NUMBER, MILLION, THOUSAND

References

Madachy, J. S. *Madachy's Mathematical Recreations.* New York: Dover, pp. 156–59, 1979.

Hunt's Surface

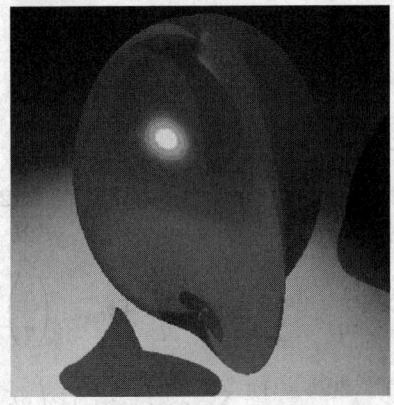

A SEXTIC SURFACE given by the implicit equation

$$4(x^2 + y^2 + z^2 - 13)^3 + 27(3x^2 + y^2 - 4z^2 - 12)^2 = 0.$$

References

Hunt, B. "Algebraic Surfaces." http://www.mathematik.uni-kl.de/~wwwagag/E/Galerie.html.

Nordstrand, T. "Hunt's Surface." http://www.uib.no/people/nfytn/hunttxt.htm.

Huntington Axiom

An axiom proposed by Huntington (1933) as part of his definition of a BOOLEAN ALGEBRA,

$$H(x, y) \equiv !(!x \vee y) \vee !(!x \vee !y) = x, \tag{1}$$

where $!x$ denotes NOT and $x \vee y$ denotes OR. Taken together, the three axioms consisting of (1), commutativity

$$x \vee y = y \vee x \tag{2}$$

and associativity

$$(x \vee y) \vee z = x \vee (y \vee z), \tag{3}$$

are equivalent to the axioms of BOOLEAN ALGEBRA. The Huntington operator can be defined in *Mathematica* by

```
Huntington := Function[{x, y}, ! (! x \[Or] y)
\[Or] ! (! x \[Or] ! y)]
```

That the Huntington axiom is a true statement in BOOLEAN ALGEBRA can be verified by examining its TRUTH TABLE.

x	y	$H(x, y)$
T	T	T
T	F	T
F	T	F
F	F	F

See also BOOLEAN ALGEBRA, ROBBINS ALGEBRA, ROBBINS AXIOM, WINKLER CONDITIONS, WOLFRAM AXIOM

References

Huntington, E. V. "New Sets of Independent Postulates for the Algebra of Logic, with Special Reference to Whitehead and Russell's *Principia Mathematica.*" *Trans. Amer. Math. Soc.* **35**, 274–04, 1933.
Huntington, E. V. "Boolean Algebra. A Correction." *Trans. Amer. Math. Soc.* **35**, 557–58, 1933.

Huntington Equation

An equation proposed by Huntington (1933) as part of his definition of a BOOLEAN ALGEBRA,

$$f : (X, A) \to (Y, B)$$

See also ROBBINS ALGEBRA, ROBBINS EQUATION

References

Huntington, E. V. "New Sets of Independent Postulates for the Algebra of Logic, with Special Reference to Whitehead and Russell's *Principia Mathematica.*" *Trans. Amer. Math. Soc.* **35**, 274–04, 1933.

Huntington, E. V. "Boolean Algebra. A Correction." *Trans. Amer. Math. Soc.* **35**, 557–58, 1933.

Hurwitz Equation

The DIOPHANTINE EQUATION

$$x_1^2 + x_2^2 + \ldots + x_n^2 = ax_1x_2\cdots x_n$$

which has no INTEGER solutions for $a > n$.

See also LAGRANGE NUMBER (DIOPHANTINE EQUATION)

References

Guy, R. K. "Markoff Numbers." §D12 in *Unsolved Problems in Number Theory, 2nd ed.* New York: Springer-Verlag, pp. 166–68, 1994.

Hurwitz Number

A number with a CONTINUED FRACTION whose terms are the values of one or more POLYNOMIALS evaluated on consecutive INTEGERS and then interleaved. This property is preserved by MÖBIUS TRANSFORMATIONS (Gosper 1972, p. 44).

References

Gosper, R. W. Item 101b in Beeler, M.; Gosper, R. W.; and Schroeppel, R. *HAKMEM.* Cambridge, MA: MIT Artificial Intelligence Laboratory, Memo AIM-239, pp. 39–4, Feb. 1972.

Hurwitz Polynomial

A POLYNOMIAL with REAL POSITIVE COEFFICIENTS and ROOTS which are either NEGATIVE or pairwise conjugate with NEGATIVE REAL PARTS.

Hurwitz Zeta Function

A generalization of the RIEMANN ZETA FUNCTION with a FORMULA

$$\zeta(s, a) \equiv \sum_{k=0}^{\infty} \frac{1}{(k+a)^s}, \tag{1}$$

where any term with $k + a = 0$ is excluded. The Hurwitz zeta function can also be given by the functional equation

$$\zeta\left(s, \frac{p}{q}\right) = 2\Gamma(1-s)$$

$$\times (2\pi q)^{s-1} \sum_{n=1}^{q} \sin\left(\frac{\pi s}{2} + \frac{2\pi np}{q}\right) \zeta\left(1-s, \frac{n}{q}\right) \tag{2}$$

(Apostol 1976, Miller and Adamchik), or the integral

$$\zeta(s, a) = \tfrac{1}{2} a^{-s} + \frac{a^{1-s}}{s-1}$$

$$+ 2 \int_0^{\infty} (a^2 + y^2)^{-s/2} \left\{ \sin\left[s \tan^{-1}\left(\frac{y}{a}\right)\right] \right\} \frac{dy}{e^{2\pi y} - 1}. \tag{3}$$

If $\Re|z| < 0$ and $0 < a \le 1$, then

$$\zeta(z, a) = \frac{2\Gamma(1-z)}{(2\pi)^{1-z}}$$

$$\times \left[\sin\left(\frac{\pi z}{2}\right) \sum_{n=1}^{\infty} \frac{\cos(2\pi an)}{n^{1-z}} + \cos\left(\frac{\pi z}{2}\right) \sum_{n=1}^{\infty} \frac{\sin(2\pi an)}{n^{1-z}} \right] \tag{4}$$

(Hurwitz 1882; Whittaker and Watson 1990, pp. 268–69). The Hurwitz zeta function satisfies

$$\zeta(0, a) = \tfrac{1}{2} - a \tag{5}$$

$$\frac{d}{ds} \zeta(0, a) = \ln[\Gamma(a)] - \tfrac{1}{2} \ln(2\pi) \tag{6}$$

$$\frac{d}{ds} \zeta(0, 0) = \tfrac{1}{2} \ln(2\pi), \tag{7}$$

where $\Gamma(z)$ is the GAMMA FUNCTION. In the limit,

$$\lim_{s \to 1} \zeta(s, a) - \frac{1}{s-1} = \frac{\Gamma'(a)}{\Gamma(a)} \tag{8}$$

(Whittaker and Watson 1990, p. 271; Allouche 1992). The POLYGAMMA FUNCTION $\psi_m(z)$ can be expressed in terms of the Hurwitz zeta function by

$$\psi_m(z) = (-1)^{m+1} m! \zeta(1+m, z). \tag{9}$$

For POSITIVE INTEGERS k, p, and $q > p$,

$$\zeta'\left(-2k+1, \frac{p}{q}\right)$$

$$= \frac{[\psi(2k) - \ln(2\pi q)]B_{2k}(p/q)}{2k} - \frac{[\psi(2k) - \ln(2\pi)]B_{2k}}{q^{2k} 2k}$$

$$+ \frac{(-1)^{k+1}\pi}{(2\pi q)^{2k}} \sum_{n=1}^{q-1} \sin\left(\frac{2\pi pn}{q}\right) \psi_{(2k-1)}\left(\frac{n}{q}\right)$$

$$+ \frac{(-1)^{k+1} 2(2k-1)!}{(2\pi q)^{2k}} \sum_{n=1}^{q-1} \cos\left(\frac{2\pi pn}{q}\right) \zeta'\left(2k, \frac{n}{q}\right)$$

$$+ \frac{\zeta'(-2k+1)}{q^2 k}, \tag{10}$$

where B_n is a BERNOULLI NUMBER, $B_n(x)$ a BERNOULLI POLYNOMIAL, $\psi_n(z)$ is a POLYGAMMA FUNCTION, and $\zeta(z)$ is a RIEMANN ZETA FUNCTION (Miller and Adamchik). Miller and Adamchik also give the closed-form expressions

$$\zeta'(-2k+1, \tfrac{1}{2}) = -\frac{B_{2k}\ln 2}{4^k k} - \frac{(2^{2k-1})\zeta'(-2k+1)}{2^{2k-1}} \quad (11)$$

$$\zeta'\left(-2k+1, \begin{array}{c}1/3\\2/3\end{array}\right)$$

$$= \mp \frac{(9^k-1)B_{2k}\pi}{\sqrt{3}(3^{2k-1}-1)8k} - \frac{B_{2k}\ln 3}{(3^{2k-1})4k} \quad (12)$$

$$\zeta'\left(-2k+1, \begin{array}{c}1/4\\3/4\end{array}\right)$$

$$= \mp\frac{(4^k+1)B_{2k}\pi}{4^{k+1}k} + \frac{(4^{k-1}-1)B_{2k}\ln 2}{2^{3k-1}k} \quad (13)$$

$$\zeta'\left(-2k+1, \begin{array}{c}1/6\\5/6\end{array}\right) = \mp\frac{(9^k-1)(2^{2k-1}+1)B_{2k}\pi}{\sqrt{3}(6^{2k-1})8k}$$

$$+\frac{B_{2k}(3^{2k-1}-1)\ln 2}{(6^{2k-1})4k} + \frac{B_{2k}(2^{2k-1}-1)\ln 3}{(6^{2k-1})4k}$$

$$\mp\frac{(-1)^k(2^{2k-1}+1)\psi_{2k-1}(\tfrac{1}{3})}{2\sqrt{3}(12\pi)^{2k-1}} \quad (14)$$

In these equations, $\zeta'(z_0, a)$ means $d\zeta(z, a)/dz|_{z=z_0}$, $\zeta'(z_0)$ means $d\zeta(z)/dz|_{z=z_0}$, and the upper and lower fractions on the left side of the equations correspond to the plus and minus signs, respectively, on the right side.

Gauss gave

$$\frac{\Gamma'(p/q)}{\Gamma(p/q)} = -\gamma - \ln(2q) - \tfrac{1}{2}\pi\cot\left(\frac{\pi p}{q}\right)$$

$$+2\sum_{0<n<q/2}\cos\left(\frac{2\pi pn}{q}\right)\ln\left[\sin\left(\frac{\pi n}{q}\right)\right] \quad (15)$$

(Allouche 1992, Knuth 1997, p. 94).

See also HURWITZ'S FORMULA, KHINTCHINE'S CONSTANT, POLYGAMMA FUNCTION, PSI FUNCTION, RIEMANN ZETA FUNCTION, ZETA FUNCTION

References

Adamchik, V. "A Class of Logarithmic Integrals." In *Proc. ISSAC'97, Maui, Hawaii* (Ed. W. W. Kuechlin). New York: ACM, 1997.

Adamchik, V. S. and Srivastava, H. M. "Some Series of the Zeta and Related Functions." *Analysis* 18, 131–44, 1998.

Allouche, J.-P. "Series and Infinite Products related to Binary Expansions of Integers." 1992. http://algo.inria.fr/seminars/sem92–3/allouche.ps.

Apostol, T. M. *Introduction to Analytic Number Theory.* New York: Springer-Verlag, 1995.

Berndt, B. C. "On the Hurwitz Zeta-Function." *Rocky Mountain J. Math.* 2, 151–57, 1972.

Cvijovic, D. and Klinowski, J. "Values of the Legendre Chi and Hurwitz Zeta Functions at Rational Arguments." *Math. Comput.* 68, 1623–630, 1999.

Elizalde, E.; Odintsov, A. D.; and Romeo, A. *Zeta Regularization Techniques with Applications.* River Edge, NJ: World Scientific, 1994.

Erdélyi, A.; Magnus, W.; Oberhettinger, F.; and Tricomi, F. G. "The Generalized Zeta Function." §1.10 in *Higher Transcendental Functions, Vol. 1.* New York: Krieger, pp. 24–7, 1981.

Hauss, M. *Verallgemeinerte Stirling, Bernoulli und Euler Zahlen, deren Anwendungen und schnell konvergente Reihen für Zeta Funktionen.* Aachen, Germany: Verlag Shaker, 1995.

Hurwitz. *Z. Math. Phys.* 27, 95, 1882.

Knopfmacher, J. "Generalised Euler Constants." *Proc. Edinburgh Math. Soc.* 21, 25–2, 1978.

Knuth, D. E. *The Art of Computer Programming, Vol. 1: Fundamental Algorithms, 3rd ed.* Reading, MA: Addison-Wesley, 1997.

Magnus, W. and Oberhettinger, F. *Formulas and Theorems for the Special Functions of Mathematical Physics, 3rd ed.* New York: Springer-Verlag, 1966.

Miller, J. and Adamchik, V. "Derivatives of the Hurwitz Zeta Function for Rational Arguments." *J. Comput. Appl. Math.* 100, 201–06, 1999. http://members.wri.com/victor/articles/hurwitz.html.

Prudnikov, A. P.; Marichev, O. I.; and Brychkov, Yu. A. "The Generalized Zeta Function $\zeta(s, x)$, Bernoulli Polynomials $B_n(x)$, Euler Polynomials $E_n(x)$, and Polylogarithms $\mathrm{Li}_\nu(x)$." §1.2 in *Integrals and Series, Vol. 3: More Special Functions.* Newark, NJ: Gordon and Breach, pp. 23–4, 1990.

Spanier, J. and Oldham, K. B. "The Hurwitz Function $\zeta(\nu; u)$." Ch. 62 in *An Atlas of Functions.* Washington, DC; Hemisphere, pp. 653–64, 1987.

Whittaker, E. T. and Watson, G. N. *A Course in Modern Analysis, 4th ed.* Cambridge, England: Cambridge University Press, pp. 268–69, 1950.

Hurwitz's Formula

$$\zeta(1-s, a) = \frac{\Gamma(s)}{(2\pi)^s}[e^{-\pi i s/2}F(a, s) + e^{\pi i s/2}F(-a, s)],$$

where $\zeta(z, a)$ is a HURWITZ ZETA FUNCTION, $\Gamma(z)$ is the GAMMA FUNCTION, and $F(a, s)$ is the PERIODIC ZETA FUNCTION.

See also GAMMA FUNCTION, HURWITZ ZETA FUNCTION, PERIODIC ZETA FUNCTION

References

Apostol, T. M. Theorem 12.6 in *Introduction to Analytic Number Theory.* New York: Springer-Verlag, 1995.

Apostol, T. M. *Modular Functions and Dirichlet Series in Number Theory, 2nd ed.* New York: Springer-Verlag, p. 71, 1997.

Hurwitz's Irrational Number Theorem

As Lagrange showed, any IRRATIONAL NUMBER α has an infinity of rational approximations p/q which satisfy

$$\left|\alpha - \frac{p}{q}\right| < \frac{1}{\sqrt{5}q^2}. \quad (1)$$

Furthermore, if there are no integers a, b, c, d with $|ad-bc|=1$ and $\alpha = \frac{a\alpha+b}{d\alpha+c}$ (corresponding to values of α

associated with the GOLDEN RATIO ϕ through their CONTINUED FRACTIONS), then

$$\left| \alpha - \frac{p}{q} \right| < \frac{1}{\sqrt{8}q^2}, \qquad (2)$$

and if values of α associated with the SILVER RATIO $1 + \sqrt{2}$ are also excluded, then

$$\left| \alpha - \frac{p}{q} \right| < \frac{5}{\sqrt{221}} \frac{1}{q^2}. \qquad (3)$$

In general, even tighter bounds OF THE FORM

$$\left| \alpha - \frac{p}{q} \right| < \frac{1}{L_n q^2} \qquad (4)$$

can be obtained for the best rational approximation possible for an arbitrary irrational number α, where the L_n are called LAGRANGE NUMBERS and get steadily larger for each "bad" set of irrational numbers which is excluded.

See also CONTINUED FRACTION, IRRATIONALITY MEASURE,

Hurwitz's Root Theorem

Let $\{f(x)\}$ be a SEQUENCE of ANALYTIC FUNCTIONS REGULAR in a region G, and let this sequence be UNIFORMLY CONVERGENT in every CLOSED SUBSET of G. If the ANALYTIC FUNCTION

$$\lim_{n \to \infty} f_n(x) = f(x)$$

does not vanish identically, then if $x = a$ is a zero of $f(x)$ of order k, a NEIGHBORHOOD $|x - a| < \delta$ of $x = a$ and a number N exist such that if $n > N$, $f_n(x)$ has exactly k zeros in $|x - a| < \delta$.

See also ARGUMENT PRINCIPLE, ROOT

References
Krantz, S. G. "Hurwitz's Theorem." §5.3.4 in *Handbook of Complex Analysis.* Boston, MA: Birkhäuser, p. 76, 1999.
Szego, G. *Orthogonal Polynomials, 4th ed.* Providence, RI: Amer. Math. Soc., p. 22, 1975.

Hurwitz-Radon Theorem

Determined the possible values of r and n for which there is an IDENTITY OF THE FORM

$$(x_1^2 + \ldots + x_r^2)(y_1^2 + \ldots + y_r^2) = z_1^2 + \ldots + z_n^2.$$

Hutton's Formula

The MACHIN-LIKE FORMULA

$$\tfrac{1}{4}\pi = 2\tan^{-1}\left(\tfrac{1}{3}\right) + \tan^{-1}\left(\tfrac{1}{7}\right).$$

The other two-term MACHIN-LIKE FORMULAS are

EULER'S MACHIN-LIKE FORMULA, HERMANN'S FORMULA, and MACHIN'S FORMULA.

Hutton's Method

LAMBERT'S METHOD

Hyperasymptotic Series

See also ASYMPTOTIC SERIES, SUPERASYMPTOTIC SERIES

References
Boyd, J. P. "The Devil's Invention: Asymptotic, Superasymptotic and Hyperasymptotic Series." *Acta Appl. Math.* **56**, 1–8, 1999.

Hyperbola

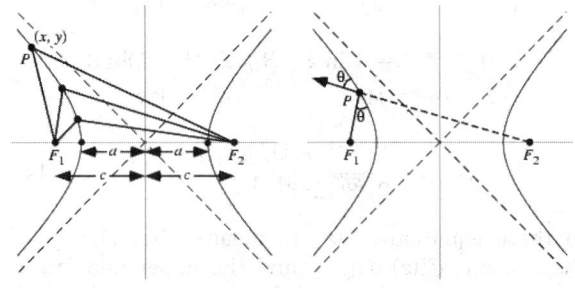

A hyperbola is a CONIC SECTION defined as the LOCUS of all points P in the PLANE the difference of whose distances $r_1 = F_1 P$ and $r_2 = F_2 P$ from two fixed points (the FOCI F_1 and F_2) separated by a distance $2c$ is a given POSITIVE constant k,

$$r_2 - r_1 = k \qquad (1)$$

(Hilbert and Cohn-Vossen 1999, p. 3). Letting P fall on the left x-intercept requires that

$$k = (c + a) - (c - a) = 2a, \qquad (2)$$

so the constant is given by $k = 2a$, i.e., twice the distance between the x-intercepts (left figure above). The hyperbola has the important property that a ray originating at a FOCUS F_1 reflects in such a way that the outgoing path lies along the line from the other FOCUS through the point of intersection (right figure above).

The special case of the RECTANGULAR HYPERBOLA, corresponding to a hyperbola with eccentricity $e = \sqrt{2}$, was first studied by Menaechmus. Euclid and Aristaeus wrote about the general hyperbola, but only studied one branch of it. The hyperbola was given its present name by Apollonius, who was the first to study both branches. The FOCUS and DIRECTRIX were considered by Pappus (MacTutor Archive). The hyperbola is the shape of an orbit of a body on an escape trajectory (i.e., a body with positive energy), such as some comets, about a fixed mass, such as the sun.

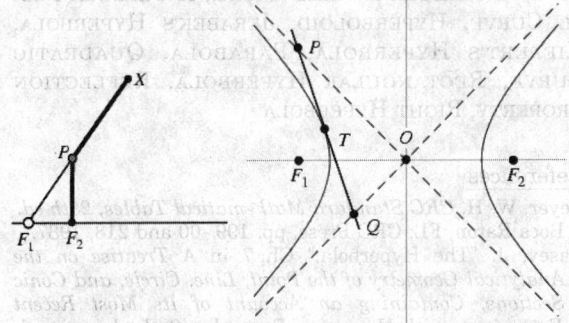

The hyperbola can be constructed by connecting the free end X of a rigid bar F_1X, where F_1 is a FOCUS, and the other FOCUS F_2 with a string F_2PX. As the bar AX is rotated about F_1 and P is kept taut against the bar (i.e., lies on the bar), the LOCUS of P is one branch of a hyperbola (left figure above; Wells 1991). A theorem of Apollonius states that for a line segment tangent to the hyperbola at a point T and intersecting the asymptotes at points P and Q, then $\overline{OP} \times \overline{OQ}$ is constant, and $PT = QT$ (right figure above; Wells 1991).

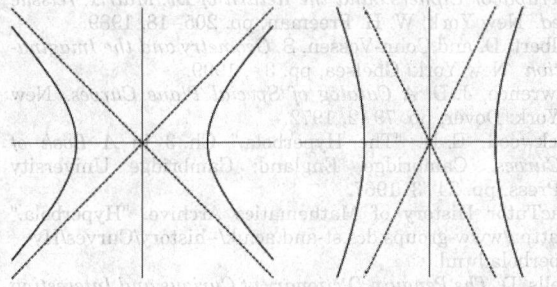

Let the point P on the hyperbola have Cartesian coordinates (x, y), then the definition of the hyperbola $r_2 - r_1 = 2a$ gives

$$\sqrt{(x-c)^2 + y^2} - \sqrt{(x+c)^2 + y^2} = 2a. \quad (3)$$

Rearranging and completing the square gives

$$x^2(c^2 - a^2) - a^2y^2 = a^2(c^2 - a^2), \quad (4)$$

and dividing both sides by $a^2(c^2 - a^2)$ results in

$$\frac{x^2}{a^2} - \frac{y^2}{c^2 - a^2} = 1. \quad (5)$$

By analogy with the definition of the ELLIPSE, define

$$b^2 \equiv c^2 - a^2, \quad (6)$$

so the equation for a hyperbola with SEMIMAJOR AXIS a parallel to the X-AXIS and SEMIMINOR AXIS b parallel to the Y-AXIS is given by

$$\frac{x^2}{a^2} - \frac{y^2}{b^2} = 1. \quad (7)$$

or, for a center at the point (x_0, y_0) instead of $(0, 0)$,

$$\frac{(x - x_0)^2}{a^2} - \frac{(y - y_0)^2}{b^2} = 1. \quad (8)$$

Unlike the ELLIPSE, no points of the hyperbola actually lie on the SEMIMINOR AXIS, but rather the ratio b/a determines the vertical scaling of the hyperbola. The ECCENTRICITY e of the hyperbola (which always satisfies $e > 1$) is then defined as

$$e \equiv \frac{c}{a} = \sqrt{1 + \frac{b^2}{a^2}}. \quad (9)$$

In the standard equation of the hyperbola, the center is located at (x_0, y_0), the FOCI are at $(x_0 \pm c, y_0)$, and the vertices are at $(x_0 \pm a, y_0)$. The so-called ASYMPTOTES (shown as the dashed lines in the above figures) can be found by substituting 0 for the 1 on the right side of the general equation (8),

$$y = \pm \frac{b}{a}(x - x_0) + y_0, \quad (10)$$

and therefore have SLOPES $\pm b/a$.

The special case $a = b$ (the left diagram above) is known as a RIGHT HYPERBOLA because the ASYMPTOTES are PERPENDICULAR.

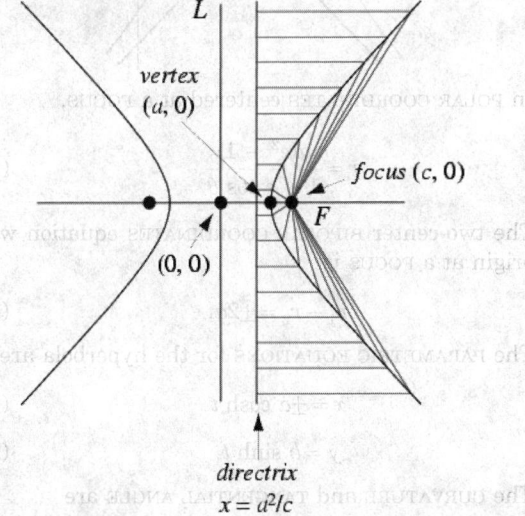

The hyperbola can also be defined as the LOCUS of points whose distance from the FOCUS F is proportional to the horizontal distance from a vertical line L known as the DIRECTRIX, where the ratio is > 1. Letting r be the ratio and d the distance from the center at which the directrix lies, then

$$d = \frac{a^2}{c}. \quad (11)$$

$$r = \frac{a}{c}. \tag{12}$$

Like noncircular ELLIPSES, hyperbolas have *two* distinct FOCI and two associated DIRECTRICES, each DIRECTRIX being PERPENDICULAR to the line joining the two foci (Eves 1965, p. 275).

The FOCAL PARAMETER of the hyperbola is

$$p = \frac{b^2}{\sqrt{a^2 + b^2}} \tag{13}$$

$$= \frac{c^2 - a^2}{c} \tag{14}$$

$$= \frac{a(e^2 - 1)}{e}. \tag{15}$$

In POLAR COORDINATES, the equation of a hyperbola centered at the ORIGIN (i.e., with $x_0 = y_0 = 0$) is

$$r^2 = \frac{a^2 b^2}{b^2 \cos^2 \theta - a^2 \sin^2 \theta}. \tag{16}$$

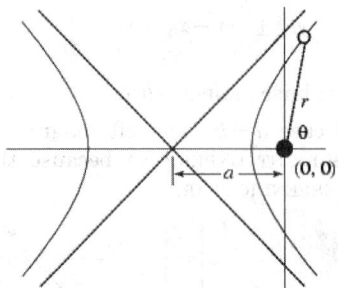

In POLAR COORDINATES centered at a FOCUS,

$$r = \frac{a(e^2 - 1)}{1 - e \cos \theta}. \tag{17}$$

The two-center BIPOLAR COORDINATES equation with origin at a FOCUS is

$$r_1 - r_2 = \pm 2a. \tag{18}$$

The PARAMETRIC EQUATIONS for the hyperbola are

$$x = \pm a \cosh t \tag{19}$$

$$y = b \sinh t. \tag{20}$$

The CURVATURE and TANGENTIAL ANGLE are

$$\kappa(t) = -[\cosh(2t)]^{-3/2} \tag{21}$$

$$\phi(t) = -\tan^{-1}(\tanh t). \tag{22}$$

The LOCUS of the apex of a variable CONE containing an ELLIPSE fixed in 3-space is a hyperbola through the FOCI of the ELLIPSE. In addition, the LOCUS of the apex of a CONE containing that hyperbola is the original ELLIPSE. Furthermore, the ECCENTRICITIES of the ELLIPSE and hyperbola are reciprocals.

See also CONIC SECTION, ELLIPSE, HYPERBOLA EVOLUTE, HYPERBOLA INVERSE CURVE, HYPERBOLA PEDAL CURVE, HYPERBOLOID, JERABEK'S HYPERBOLA, KIEPERT'S HYPERBOLA, PARABOLA, QUADRATIC CURVE, RECTANGULAR HYPERBOLA, REFLECTION PROPERTY, RIGHT HYPERBOLA

References

Beyer, W. H. *CRC Standard Mathematical Tables, 28th ed.* Boca Raton, FL: CRC Press, pp. 199–00 and 218, 1987.

Casey, J. "The Hyperbola." Ch. 7 in *A Treatise on the Analytical Geometry of the Point, Line, Circle, and Conic Sections, Containing an Account of Its Most Recent Extensions, with Numerous Examples, 2nd ed., rev. enl.* Dublin: Hodges, Figgis, & Co., pp. 250–84, 1893.

Courant, R. and Robbins, H. *What is Mathematics?: An Elementary Approach to Ideas and Methods, 2nd ed.* Oxford, England: Oxford University Press, pp. 75–6, 1996.

Coxeter, H. S. M. "Conics" §8.4 in *Introduction to Geometry, 2nd ed.* New York: Wiley, pp. 115–19, 1969.

Eves, H. *A Survey of Geometry, rev. ed.* Boston, MA: Allyn & Bacon, 1965.

Fukagawa, H. and Pedoe, D. "The One Hyperbola." §5.2 in *Japanese Temple Geometry Problems.* Winnipeg, Manitoba, Canada: Charles Babbage Research Foundation, pp. 51 and 136–38, 1989.

Gardner, M. "Hyperbolas." Ch. 15 in *Penrose Tiles and Trapdoor Ciphers...and the Return of Dr. Matrix, reissue ed.* New York: W. H. Freeman, pp. 205–18, 1989.

Hilbert, D. and Cohn-Vossen, S. *Geometry and the Imagination.* New York: Chelsea, pp. 3–, 1999.

Lawrence, J. D. *A Catalog of Special Plane Curves.* New York: Dover, pp. 79–2, 1972.

Lockwood, E. H. "The Hyperbola." Ch. 3 in *A Book of Curves.* Cambridge, England: Cambridge University Press, pp. 24–3, 1967.

MacTutor History of Mathematics Archive. "Hyperbola." http://www-groups.dcs.st-and.ac.uk/~history/Curves/Hyperbola.html.

Wells, D. *The Penguin Dictionary of Curious and Interesting Geometry.* London: Penguin, pp. 106–09, 1991.

Yates, R. C. "Conics." *A Handbook on Curves and Their Properties.* Ann Arbor, MI: J. W. Edwards, pp. 36–6, 1952.

Hyperbola Evolute

The EVOLUTE of a RECTANGULAR HYPERBOLA is the LAMÉ CURVE

$$(ax)^{2/3} - (by)^{2/3} = (a + b)^{2/3}.$$

From a point between the two branches of the EVOLUTE, two NORMALS can be drawn to the HYPERBOLA. However, from a point beyond the EVOLUTE, four NORMALS can be drawn.

Hyperbola Inverse Curve

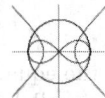

For a HYPERBOLA with $a = b$ with INVERSION CENTER

at the center, the INVERSE CURVE

$$x = \frac{2k \cos t}{a[3 - \cos(2t)]} \tag{1}$$

$$y = \frac{k \sin(2t)}{a[3 - \cos(2t)]} \tag{2}$$

is a LEMNISCATE.

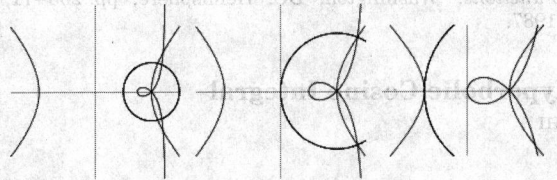

For an INVERSION CENTER at the VERTEX, the INVERSE CURVE

$$x = a + \frac{4k \cos t \sin^2\left(\frac{1}{2}t\right)}{a[5 - 4\cos t + \cos(2t) - 2\sin(2t)]} \tag{3}$$

$$y = a + \frac{k(\tan t - 1)}{a[(\sec t - 1)^2 + (\tan t - 1)^2]} \tag{4}$$

is a RIGHT STROPHOID.

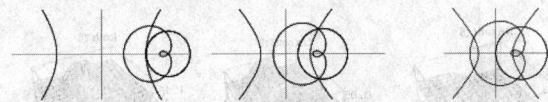

For an INVERSION CENTER at the FOCUS, the INVERSE CURVE

$$x = ae = \frac{k \cos t(1 - e \cos t)}{a(\cos t - e)^2} \tag{5}$$

$$y = \frac{\sqrt{e^2 - 1}\, k \sin(2t)}{2a(\cos t - e)^2} \tag{6}$$

is a LIMAÇON, where e is the ECCENTRICITY.

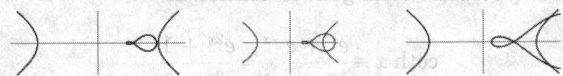

For a HYPERBOLA with $a = \sqrt{3}b$ and INVERSION CENTER at the VERTEX, the INVERSE CURVE

$$x = b + \frac{2k \cos t(\sqrt{3} - \cos t)}{b[9 - 4\sqrt{3}\cos t + \cos(2t) - 2\sin(2t)]} \tag{7}$$

$$y = b + \frac{k(\tan t - 1)}{b\left[(\sqrt{3}\sec t - 1)^2 + (\tan t - 1)^2\right]} \tag{8}$$

is a MACLAURIN TRISECTRIX.

References

Lawrence, J. D. *A Catalog of Special Plane Curves.* New York: Dover, p. 203, 1972.

Hyperbola Pedal Curve

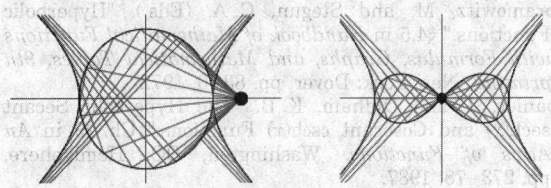

The PEDAL CURVE of a HYPERBOLA with the PEDAL POINT at the FOCUS is a CIRCLE (left figure; Hilbert and Cohn-Vossen 1999, p. 26). The PEDAL CURVE of a RECTANGULAR HYPERBOLA with PEDAL POINT at the center is a LEMNISCATE (right figure).

See also HYPERBOLA, PEDAL CURVE

References

Hilbert, D. and Cohn-Vossen, S. *Geometry and the Imagination.* New York: Chelsea, 1999.

Hyperbolic Automorphism

ANOSOV AUTOMORPHISM

Hyperbolic Cosecant

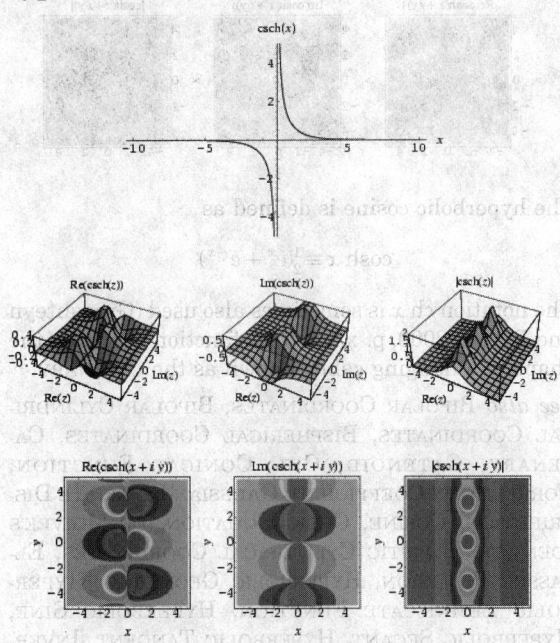

The hyperbolic cosecant is defined as

$$\operatorname{csch} x \equiv \frac{1}{\sinh x} = \frac{2}{e^2 - e^{-x}}.$$

See also BERNOULLI NUMBER, BIPOLAR COORDINATES, BIPOLAR CYLINDRICAL COORDINATES, COSECANT, HELMHOLTZ DIFFERENTIAL EQUATION–TOROIDAL COORDINATES, HYPERBOLIC SINE, POINSOT'S SPIRALS, SURFACE OF REVOLUTION, TOROIDAL FUNCTION

References

Abramowitz, M. and Stegun, C. A. (Eds.). "Hyperbolic Functions." §4.5 in *Handbook of Mathematical Functions with Formulas, Graphs, and Mathematical Tables, 9th printing.* New York: Dover, pp. 83–6, 1972.

Spanier, J. and Oldham, K. B. "The Hyperbolic Secant sech(x) and Cosecant csch(x) Functions." Ch. 29 in *An Atlas of Functions.* Washington, DC: Hemisphere, pp. 273–78, 1987.

Hyperbolic Cosine

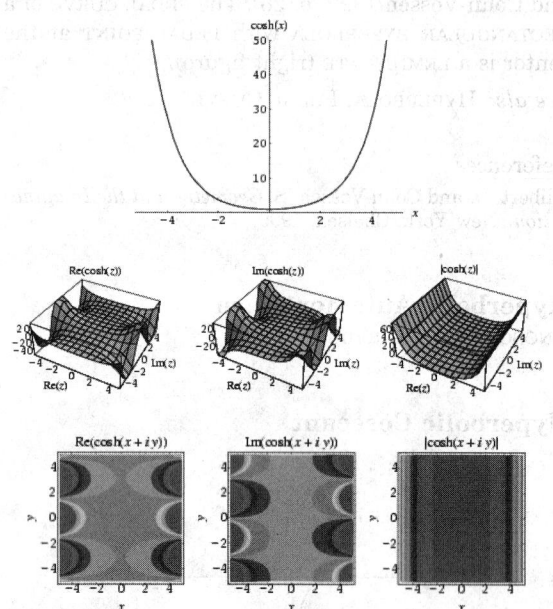

The hyperbolic cosine is defined as

$$\cosh x \equiv \tfrac{1}{2}(e^x + e^{-x}).$$

The notation ch x is sometimes also used (Gradshteyn and Ryzhik 2000, p. xxix). This function describes the shape of a hanging cable, known as the CATENARY.

See also BIPOLAR COORDINATES, BIPOLAR CYLINDRICAL COORDINATES, BISPHERICAL COORDINATES, CATENARY, CATENOID, CHI, CONICAL FUNCTION, CORRELATION COEFFICIENT–GAUSSIAN BIVARIATE DISTRIBUTION, COSINE, CUBIC EQUATION, DE MOIVRE'S IDENTITY, ELLIPTIC CYLINDRICAL COORDINATES, ELSASSER FUNCTION, HYPERBOLIC GEOMETRY, HYPERBOLIC LEMNISCATE FUNCTION, HYPERBOLIC SINE, HYPERBOLIC SECANT, HYPERBOLIC TANGENT, INVERSIVE DISTANCE, LAPLACE'S EQUATION–BIPOLAR COORDINATES, LAPLACE'S EQUATION–BISPHERICAL COORDINATES, LAPLACE'S EQUATION–TOROIDAL COORDINATES, LEMNISCATE FUNCTION, LORENTZ GROUP, MATHIEU DIFFERENTIAL EQUATION, MEHLER'S BESSEL FUNCTION FORMULA, MERCATOR PROJECTION, MODIFIED BESSEL FUNCTION OF THE FIRST KIND, OBLATE SPHEROIDAL COORDINATES, PROLATE SPHEROIDAL COORDINATES, PSEUDOSPHERE, RAMANUJAN COS/COSH IDENTITY, SINE-GORDON EQUATION, SURFACE OF REVOLUTION, TOROIDAL COORDINATES

References

Abramowitz, M. and Stegun, C. A. (Eds.). "Hyperbolic Functions." §4.5 in *Handbook of Mathematical Functions with Formulas, Graphs, and Mathematical Tables, 9th printing.* New York: Dover, pp. 83–6, 1972.

Gradshteyn, I. S. and Ryzhik, I. M. *Tables of Integrals, Series, and Products, 6th ed.* San Diego, CA: Academic Press, 2000.

Spanier, J. and Oldham, K. B. "The Hyperbolic Sine sinh(x) and Cosine cosh(x) Functions." Ch. 28 in *An Atlas of Functions.* Washington, DC: Hemisphere, pp. 263–71, 1987.

Hyperbolic Cosine Integral

CHI

Hyperbolic Cotangent

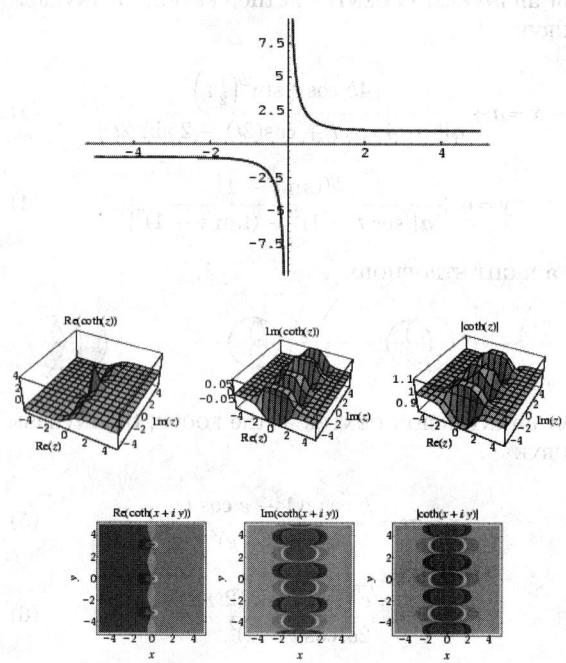

The hyperbolic cotangent is defined as

$$\coth x \equiv \frac{e^x + e^{-x}}{e^x - e^{-x}} = \frac{e^{2x} + 1}{e^{2x} - 1}.$$

The notation cth x is sometimes also used (Gradshteyn and Ryzhik 2000, p. xxix). The LAURENT SERIES of $\coth x$ is given by

$$\coth x = \frac{1}{x} + \tfrac{1}{3} x - \tfrac{1}{45} x^3 + \dots.$$

See also BERNOULLI NUMBER, BIPOLAR COORDINATES, BIPOLAR CYLINDRICAL COORDINATES, COTANGENT, HYPERBOLIC TANGENT, LAPLACE'S EQUATION–TOROIDAL COORDINATES, LEBESGUE CONSTANTS (FOURIER SERIES), PROLATE SPHEROIDAL COORDINATES, SURFACE OF REVOLUTION, TOROIDAL COORDINATES, TOROIDAL FUNCTION

References

Abramowitz, M. and Stegun, C. A. (Eds.). "Hyperbolic Functions." §4.5 in *Handbook of Mathematical Functions with Formulas, Graphs, and Mathematical Tables, 9th printing.* New York: Dover, pp. 83–6, 1972.

Gradshteyn, I. S. and Ryzhik, I. M. *Tables of Integrals, Series, and Products, 6th ed.* San Diego, CA: Academic Press, 2000.

Spanier, J. and Oldham, K. B. "The Hyperbolic Tangent tanh(*x*) and Cotangent coth(*x*) Functions." Ch. 30 in *An Atlas of Functions.* Washington, DC: Hemisphere, pp. 279–84, 1987.

Hyperbolic Cube

A hyperbolic version of the Euclidean CUBE.

See also HYPERBOLIC DODECAHEDRON, HYPERBOLIC ICOSAHEDRON, HYPERBOLIC OCTAHEDRON, HYPERBOLIC TETRAHEDRON

References

Rivin, I. "Hyperbolic Polyhedron Graphics." http://www.mathsource.com/cgi-bin/msitem22?0201–88.

Trott, M. "The Cover Image: Hyperbolic Platonic Bodies." §8.3.10 in *The Mathematica Guidebook, Vol. 2: Graphics in Mathematica.* New York: Springer-Verlag, 2000.

Hyperbolic Cylinder

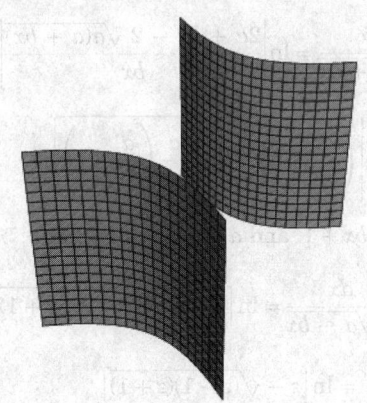

A QUADRATIC SURFACE given by the equation

$$\frac{x^2}{a^2} - \frac{y^2}{b^2} = -1.$$

See also ELLIPTIC PARABOLOID, PARABOLOID

References

Beyer, W. H. *CRC Standard Mathematical Tables, 28th ed.* Boca Raton, FL: CRC Press, pp. 210–11, 1987.

Hilbert, D. and Cohn-Vossen, S. *Geometry and the Imagination.* New York: Chelsea, p. 12, 1999.

Hyperbolic Disk

POINCARÉ HYPERBOLIC DISK

Hyperbolic Dodecahedron

A hyperbolic version of the Euclidean DODECAHEDRON.

See also HYPERBOLIC CUBE, HYPERBOLIC ICOSAHEDRON, HYPERBOLIC OCTAHEDRON, HYPERBOLIC TETRAHEDRON

References

Rivin, I. "Hyperbolic Polyhedron Graphics." http://www.mathsource.com/cgi-bin/msitem22?0201–88.

Trott, M. "The Cover Image: Hyperbolic Platonic Bodies." §8.3.10 in *The Mathematica Guidebook, Vol. 2: Graphics in Mathematica.* New York: Springer-Verlag, 2000.

Hyperbolic Fixed Point (Differential Equations)

A FIXED POINT for which the STABILITY MATRIX has EIGENVALUES $\lambda_1 < 0 < \lambda_2$, also called a SADDLE POINT.

See also ELLIPTIC FIXED POINT (DIFFERENTIAL EQUATIONS), FIXED POINT, STABLE IMPROPER NODE, STABLE SPIRAL POINT, STABLE STAR, UNSTABLE IMPROPER NODE, UNSTABLE NODE, UNSTABLE SPIRAL POINT, UNSTABLE STAR

References

Tabor, M. "Classification of Fixed Points." §1.4.b in *Chaos and Integrability in Nonlinear Dynamics: An Introduction.* New York: Wiley, pp. 22–5, 1989.

Hyperbolic Fixed Point (Map)

A FIXED POINT of a LINEAR TRANSFORMATION (MAP) for which the rescaled variables satisfy

$$(\delta - \alpha)^2 + 4\beta\gamma > 0.$$

See also ELLIPTIC FIXED POINT (MAP), LINEAR TRANSFORMATION, PARABOLIC FIXED POINT

Hyperbolic Functions

The hyperbolic functions sinh, cosh, tanh, csch, sech, coth (HYPERBOLIC SINE, HYPERBOLIC COSINE, etc.) share many properties with the corresponding CIRCULAR FUNCTIONS. The hyperbolic functions arise in many problems of mathematics and mathematical physics in which integrals involving $\sqrt{1+x^2}$ arise (whereas the CIRCULAR FUNCTIONS involve $\sqrt{1-x^2}$).

For instance, the HYPERBOLIC SINE arises in the gravitational potential of a cylinder and the calculation of the Roche limit. The HYPERBOLIC COSINE function is the shape of a hanging cable (the so-called CATENARY). The HYPERBOLIC TANGENT arises in the calculation of magnetic moment and rapidity of special relativity. All three appear in the Schwarzschild metric using external isotropic Kruskal coordinates in general relativity. The HYPERBOLIC SECANT arises in the profile of a laminar jet. The HYPERBOLIC COTANGENT arises in the Langevin function for magnetic polarization.

The hyperbolic functions are defined by

$$\sinh z \equiv \frac{e^z - e^{-z}}{2} = -\sinh(-z) \tag{1}$$

$$\cosh z \equiv \frac{e^z + e^{-z}}{2} = \cosh(-z) \tag{2}$$

$$\tanh z \equiv \frac{e^z - e^{-z}}{e^z + e^{-z}} = \frac{e^{2z} - 1}{e^{2z} + 1} \tag{3}$$

$$\operatorname{csch} z \equiv \frac{2}{e^z - e^{-z}} \tag{4}$$

$$\operatorname{sech} z \equiv \frac{2}{e^z + e^{-z}} \tag{5}$$

$$\coth z \equiv \frac{e^z + e^{-z}}{e^z - e^{-z}} = \frac{e^{2z} + 1}{e^{2z} - 1}. \tag{6}$$

For purely IMAGINARY arguments,

$$\sinh(iz) = i \sin z \tag{7}$$

$$\cosh(iz) = \cos z. \tag{8}$$

The hyperbolic functions satisfy many identities analogous to the trigonometric identities (which can be inferred using OSBORNE'S RULE) such as

$$\cosh^2 x - \sinh^2 x = 1 \tag{9}$$

$$\cosh x + \sinh x = e^x \tag{10}$$

$$\cosh x - \sinh x = e^{-x}. \tag{11}$$

See also Beyer (1987, p. 168). Some HALF-ANGLE FORMULAS are

$$\tanh\left(\frac{z}{2}\right) = \frac{\sinh x + i \sin y}{\cosh x + \cos y} \tag{12}$$

$$\coth\left(\frac{z}{2}\right) = \frac{\sinh x - i \sin y}{\cosh x - \cos y}. \tag{13}$$

Some DOUBLE-ANGLE FORMULAS are

$$\sinh(2x) = 2 \sinh x \cosh x \tag{14}$$

$$\cosh(2x) = 2\cosh^2 x - 1 = 1 + 2\sinh^2 x \tag{15}$$

Identities for COMPLEX arguments include

$$\sinh(x + iy) = \sinh x \cosh y + i \cosh x \sin y \tag{16}$$

$$\cosh(x + iy) = \cosh x \cos y + i \sinh x \sin y. \tag{17}$$

The ABSOLUTE SQUARES for COMPLEX arguments are

$$|\sinh(z)|^2 = \sinh^2 x + \sin^2 y \tag{18}$$

$$|\cosh(z)|^2 = \sinh^2 x + \cos^2 y. \tag{19}$$

Integrals involving hyperbolic functions include

$$\int \frac{dx}{x\sqrt{a + bx}} = \ln\left|\frac{\sqrt{a + bx} - \sqrt{a}}{\sqrt{a + bx} + \sqrt{a}}\right| \tag{20}$$

$$= \ln\left|\frac{(\sqrt{a + bx} - \sqrt{a})^2}{(a + bx) - a}\right|$$

$$= \ln\left|\frac{(a + bx) - 2\sqrt{a(a + bx)} + a}{bx}\right|. \tag{21}$$

If $b > 0$, then

$$\int \frac{dx}{x\sqrt{a + bx}} = \ln\left|\frac{2a + bx - 2\sqrt{a(a + bx)}}{bx}\right| \tag{22}$$

$$= \ln\left|\left(\frac{2a}{bx} + 1\right) - 2\sqrt{\frac{a}{bx}\left(\frac{a}{bx} + 1\right)}\right|. \tag{23}$$

Let $z \equiv 2a/bx + 1$, and $a/bx = (z-1)/2$ and

$$\int \frac{dx}{x\sqrt{a + bx}} = \ln\left[z - 2\sqrt{\tfrac{1}{2}(z-1)\tfrac{1}{2}(z+1)}\right]$$

$$= \ln\left[z - \sqrt{(z-1)(z+1)}\right] \tag{24}$$

$$= \ln\left(z - \sqrt{z^2 - 1}\right) = \cosh^{-1}(z) \tag{25}$$

$$= \cosh^{-1}\left(1 + \frac{2a}{bx}\right) \tag{26}$$

$$= 2\tanh\left(-\sqrt{\frac{a}{a+bx}}\right). \qquad (27)$$

See also DOUBLE-ANGLE FORMULAS, FIBONACCI HYPERBOLIC FUNCTIONS, HALF-ANGLE FORMULAS, HYPERBOLIC COSECANT, HYPERBOLIC COSINE, HYPERBOLIC COTANGENT, GENERALIZED HYPERBOLIC FUNCTIONS, HYPERBOLIC SECANT, HYPERBOLIC SINE, HYPERBOLIC TANGENT, INVERSE HYPERBOLIC FUNCTIONS, OSBORNE'S RULE

References

Abramowitz, M. and Stegun, C. A. (Eds.). "Hyperbolic Functions." §4.5 in *Handbook of Mathematical Functions with Formulas, Graphs, and Mathematical Tables, 9th printing.* New York: Dover, pp. 83–6, 1972.
Anderson, J. W. "Trigonometry in the Hyperbolic Plane." §5.7 in *Hyperbolic Geometry.* New York: Springer-Verlag, pp. 146–51, 1999.
Beyer, W. H. "Hyperbolic Function." *CRC Standard Mathematical Tables, 28th ed.* Boca Raton, FL: CRC Press, pp. 168–86 and 219, 1987.
Coxeter, H. S. M. and Greitzer, S. L. *Geometry Revisited.* Washington, DC: Math. Assoc. Amer., pp. 126–31, 1967.
Yates, R. C. "Hyperbolic Functions." *A Handbook on Curves and Their Properties.* Ann Arbor, MI: J. W. Edwards, pp. 113–18, 1952.

Hyperbolic Geometry

A NON-EUCLIDEAN GEOMETRY, also called LOBACHEVSKY-BOLYAI-GAUSS GEOMETRY, having constant SECTIONAL CURVATURE -1. This GEOMETRY satisfies all of EUCLID'S POSTULATES *except* the PARALLEL POSTULATE, which is modified to read: For any infinite straight LINE L and any POINT P not on it, there are *many other* infinitely extending straight LINES that pass through P and which do not intersect L.

In hyperbolic geometry, the sum of ANGLES of a TRIANGLE is less than 180°, and TRIANGLES with the same angles have the same areas. Furthermore, not all TRIANGLES have the same ANGLE sum (cf. the AAA THEOREM for TRIANGLES in Euclidean 2-space). There are no similar triangles in hyperbolic geometry. The best-known example of a hyperbolic space are SPHERES in Lorentzian 4-space. The POINCARÉ HYPERBOLIC DISK is a hyperbolic 2-space. Hyperbolic geometry is well understood in 2-D, but not in 3-D.

Geometric models of hyperbolic geometry include the KLEIN-BELTRAMI MODEL, which consists of an OPEN DISK in the Euclidean plane whose open chords correspond to hyperbolic lines. A 2-D model is the POINCARÉ HYPERBOLIC DISK. Felix Klein constructed an analytic hyperbolic geometry in 1870 in which a POINT is represented by a pair of REAL NUMBERS (x_1, x_2) with

$$x_1^2 + x_2^2 < 1$$

(i.e., points of an OPEN DISK in the COMPLEX PLANE)

and the distance between two points is given by

$$d(x, X) = a\cosh^{-1}\left[\frac{1 - x_1 X_1 - x_2 X_2}{\sqrt{1 - x_1^2 - x_2^2}\sqrt{1 - X_1^2 - X_2^2}}\right].$$

The geometry generated by this formula satisfies all of EUCLID'S POSTULATES except the fifth. The METRIC of this geometry is given by the CAYLEY-KLEIN-HILBERT METRIC,

$$g_{11} = \frac{a^2(1 - x_2^2)}{(1 - x_1^2 - x_2^2)^2}$$

$$g_{12} = \frac{a^2 x_1 x_2}{(1 - x_1^2 - x_2^2)^2}$$

$$g_{22} = \frac{a^2(1 - x_1^2)}{(1 - x_1^2 - x_2^2)^2}.$$

Hilbert extended the definition to general bounded sets in a EUCLIDEAN SPACE.

See also ELLIPTIC GEOMETRY, EUCLIDEAN GEOMETRY, HYPERBOLIC METRIC, KLEIN-BELTRAMI MODEL, NON-EUCLIDEAN GEOMETRY, PSEUDOSPHERE, SCHWARZ-PICK LEMMA

References

Anderson, J. W. *Hyperbolic Geometry.* New York: Springer-Verlag, 1999.
Dunham, W. *Journey through Genius: The Great Theorems of Mathematics.* New York: Wiley, pp. 57–0, 1990.
Eppstein, D. "Hyperbolic Geometry." http://www.ics.uci.edu/~eppstein/junkyard/hyper.html.
Stillwell, J. *Sources of Hyperbolic Geometry.* Providence, RI: Amer. Math. Soc., 1996.
Wells, D. *The Penguin Dictionary of Curious and Interesting Geometry.* London: Penguin, pp. 109–10, 1991.

Hyperbolic Helicoid

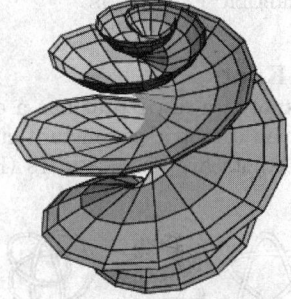

The surface with parametric equations

$$x = \frac{\sinh v \cos(\tau u)}{1 + \cosh u \cosh v} \qquad (1)$$

$$y = \frac{\sinh v \sin(\tau u)}{1 + \cosh u \cosh v} \qquad (1)$$

$$z = \frac{\cosh v \, \sinh(u)}{1 + \cosh u \cosh v}. \qquad (3)$$

where τ is a constant (the torsion).

See also HELICOID

References

JavaView. "Classic Surfaces from Differential Geometry: Hyperbolic Helicoid." http://www-sfb288.math.tu-berlin.de/vgp/javaview/demo/surface/common/PaSurface_HyperbolicHelicoid.html.

Hyperbolic Icosahedron

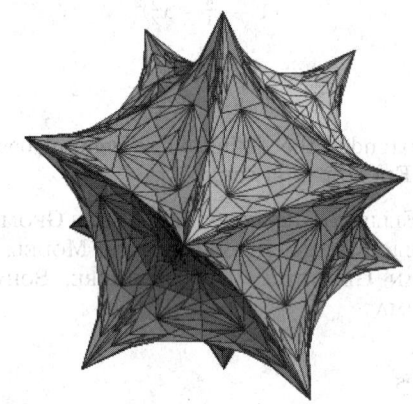

A hyperbolic version of the Euclidean ICOSAHEDRON.

See also HYPERBOLIC CUBE, HYPERBOLIC DODECAHEDRON, HYPERBOLIC OCTAHEDRON, HYPERBOLIC POLYHEDRON, HYPERBOLIC TETRAHEDRON

References

Trott, M. "The Cover Image: Hyperbolic Platonic Bodies." §8.3.10 in *The Mathematica Guidebook, Vol. 2: Graphics in Mathematica.* New York: Springer-Verlag, 2000.

Hyperbolic Inverse Functions

INVERSE HYPERBOLIC FUNCTIONS

Hyperbolic Knot

A hyperbolic knot is a KNOT that has a complement that can be given a metric of constant curvature -1. All hyperbolic knots are PRIME KNOTS (Hoste *et al.* 1998).

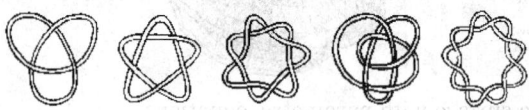

KNOTS which are not hyperbolic are either TORUS KNOTS or SATELLITE KNOTS, as proved by Thurston in 1978. Of the prime knots with 16 or fewer crossings, all but 32 are hyperbolic. Of these 32, 12 are torus knots and the remaining 20 are satellites of the TREFOIL KNOT (Hoste *et al.* 1998). The nonhyperbolic knots with nine or fewer crossings are all torus knots,

including 03–01 (the (3, 2)-TORUS KNOT), 05–01, 07–01, 08–19 (the (4, 3)-TORUS KNOT), and 09–01.

The following table gives the number of nonhyperbolic and hyperbolic knots of n crossing starting with $n = 3$.

type	Sloane	counts
torus	A051764	1, 0, 1, 0, 1, 1, 1, 1, 1, 0, 1, 1, 2, 1
satellite	A051765	0, 0, 0, 0, 0, 0, 0, 0, 0, 0, 2, 2, 6, 10
nonhyperbolic	A052407	1, 0, 1, 0, 1, 1, 1, 1, 1, 0, 3, 3, 8, 11
hyperbolic	A052408	0, 1, 1, 3, 6, 20, 48, 164, 551, 2176, 9985, 46969, 253285, 1388694

Almost all hyperbolic knots can be distinguished by their hyperbolic volumes (exceptions being 05–02 and a certain 12-crossing knot; see Adams 1994, p. 124). It has been conjectured that the smallest hyperbolic volume is 2.0298..., that of the FIGURE-OF-EIGHT KNOT. MUTANT KNOTS have the same hyperbolic knot volume.

The KNOT SYMMETRY group of a hyperbolic knot must be either a finite CYCLIC GROUP or a finite DIHEDRAL GROUP (Riley 1979, Kodama and Sakuma 1992, Hoste *et al.* 1998).

See also MUTANT KNOT, SATELLITE KNOT, TORUS KNOT

References

Adams, C. C. *The Knot Book: An Elementary Introduction to the Mathematical Theory of Knots.* New York: W. H. Freeman, pp. 119–27, 1994.

Adams, C.; Hildebrand, M.; and Weeks, J. "Hyperbolic Invariants of Knots and Links." *Trans. Amer. Math. Soc.* **326**, 1–6, 1991.

Hoste, J.; Thistlethwaite, M.; and Weeks, J. "The First 1,701,936 Knots." *Math. Intell.* **20**, 33–8, Fall 1998.

Kodama K. and Sakuma, M. "Symmetry Groups of Prime Knots Up to 10 Crossings." In *Knot 90, Proceedings of the International Conference on Knot Theory and Related Topics, Osaka, Japan, 1990* (Ed. A. Kawauchi.) Berlin: de Gruyter, pp. 323–40, 1992.

Riley, R. "An Elliptic Path from Parabolic Representations to Hyperbolic Structures." In *Topology of Low-Dimensional Manifolds, Proceedings, Sussex 1977* (Ed. R. Fenn). New York: Springer-Verlag, pp. 99–33, 1979.

Sloane, N. J. A. Sequences A051764, A051765, A052407, A052408 in "An On-Line Version of the Encyclopedia of Integer Sequences." http://www.research.att.com/~njas/sequences/eisonline.html.

Weisstein, E. W. "Knots and Links." MATHEMATICA NOTEBOOK KNOTS.M.

Hyperbolic Lemniscate Function

By analogy with the LEMNISCATE FUNCTIONS, hyperbolic lemniscate functions can also be defined

$$\operatorname{arcsinhlemn} x \equiv \int_0^x (1+t^4)^{1/2}\, dt \tag{1}$$

$$\operatorname{arccoshlemn} x \equiv \int_0^1 (1+t^4)^{1/2}\, dt. \tag{2}$$

Let $0 \le \theta \le \pi/2$ and $0 \le v \le 1$, and write

$$\frac{\theta\mu}{2} = \int_0^v \frac{dt}{\sqrt{1+t^2}}, \tag{3}$$

where μ is the constant obtained by setting $\theta = \pi/2$ and $v = 1$. Then

$$\mu = \frac{2}{\pi} K\left(\frac{1}{\sqrt{2}}\right), \tag{4}$$

where $K(k)$ is a complete ELLIPTIC INTEGRAL OF THE FIRST KIND, and Ramanujan showed

$$2\tan^{-1} v = \theta + \sum_{n=1}^{\infty} \frac{\sin(2n\theta)}{n\cosh(n\pi)}, \tag{5}$$

$$\tfrac{1}{8}\pi - \tfrac{1}{2}\tan^{-1}(v^2) = \sum_{n=0}^{\infty} \frac{(-1)^n \cos[(2n+1)\theta]}{(2n+1)\cosh\left[\tfrac{1}{2}(2n+1)\pi\right]} \tag{6}$$

and

$$\ln\left(\frac{1+v}{1-v}\right) = \ln\left[\tan\left(\tfrac{1}{4}\pi + \tfrac{1}{2}\theta\right)\right]$$
$$+ 4\sum_{n=0}^{\infty} \frac{(-1)^n \sin[(2n+1)\theta]}{(2n+1)[e^{(2n+1)\pi}-1]} \tag{7}$$

(Berndt 1994).

See also LEMNISCATE FUNCTION

References

Berndt, B. C. *Ramanujan's Notebooks, Part IV.* New York: Springer-Verlag, pp. 255–58, 1994.

Hyperbolic Map

A linear MAP $\mathbb{R}^n$ is hyperbolic if none of its EIGENVALUES has modulus 1. This means that $\mathbb{R}^n$ can be written as a DIRECT SUM of two A-invariant SUBSPACES E^s and E^u (where s stands for stable and u for unstable). This means that there exist constants $C > 0$ and $0 < \lambda < 1$ such that

$$\|A^n v\| \le C\lambda^n \|v\| \quad \text{if } v \in E^s$$

$$\|A^{-n} v\| \le C\lambda^n \|v\| \quad \text{if } v \in E^u$$

for $n = 0, 1, \ldots$.

See also PESIN THEORY

Hyperbolic Metric

The METRIC for the POINCARÉ HYPERBOLIC DISK, a model for HYPERBOLIC GEOMETRY. The hyperbolic metric is invariant under conformal maps of the disk onto itself.

See also HYPERBOLIC GEOMETRY, POINCARÉ HYPERBOLIC DISK

References

Bear, H. S. "Part Metric and Hyperbolic Metric." *Amer. Math. Monthly* **98**, 109–23, 1991.

Hyperbolic Octahedron

A hyperbolic version of the Euclidean OCTAHEDRON, which is a special case of the ASTROIDAL ELLIPSOID with $a = b = c = 1$. It is given by the PARAMETRIC EQUATIONS

$$x = (\cos u \cos v)^3$$
$$y = (\sin u \cos v)^3$$
$$z = \sin^3 v$$

for $u \in [-\pi/2,\ \pi/2]$ and $v \in [-\pi,\ \pi]$.
The FIRST FUNDAMENTAL FORM coefficients are

$$E = 9a^6 \cos^2 u \sin^2 u \cos^6 v \tag{1}$$

$$F = \tfrac{9}{4} a^6 \cos^5 v \sin v \sin(4u) \tag{2}$$

$$G = 9a^6 \cos^2 v \sin^2 v[\cos^2 v(\cos^6 u + \sin^6 u) + \sin^2 v], \tag{3}$$

the SECOND FUNDAMENTAL FORM coefficients are

$$e = \frac{24a^3 \cos^2 u \sin^2 u \csc(2u)\cos^3 v \sin v}{\sqrt{9 - \cos(4u) - [7 + \cos(4u)]\cos(2v)}} \tag{4}$$

$$f = 0 \tag{5}$$

$$g = \frac{24a^3 \cos^2 u \sin^2 u \csc(2u)\cos^3 v \sin v}{\sqrt{9 - \cos(4u) - [7 + \cos(4u)]\cos(2v)}}, \tag{6}$$

the AREA ELEMENT is

$$dA = \tfrac{9}{8} a^6 \cos^4 v \sin v \sin(2u)$$

$$\times \sqrt{9 - \cos(4u) - [7 + \cos(4u)\cos(2v)]}, \quad (7)$$

and the GAUSSIAN CURVATURE is

$$K = \frac{256 \sec^4 v}{9a^6 \{[7 + \cos(4u)]\cos(2v) + \cos(4u) - 9\}^2}. \quad (8)$$

The MEAN CURVATURE is given by a complicated expression.

See also ASTROIDAL ELLIPSOID, HYPERBOLIC CUBE, HYPERBOLIC DODECAHEDRON, HYPERBOLIC ICOSAHEDRON, HYPERBOLIC TETRAHEDRON

References

Gray, A. *Modern Differential Geometry of Curves and Surfaces with Mathematica, 2nd ed.* Boca Raton, FL: CRC Press, pp. 396–98, 1997.
Nordstrand, T. "Astroidal Ellipsoid." http://www.uib.no/people/nfytn/asttxt.htm.
Rivin, I. "Hyperbolic Polyhedron Graphics." http://www.mathsource.com/cgi-bin/msitem22?0201–88.
Trott, M. "The Cover Image: Hyperbolic Platonic Bodies." §8.3.10 in *The Mathematica Guidebook, Vol. 2: Graphics in Mathematica.* New York: Springer-Verlag, 2000.

Hyperbolic Paraboloid

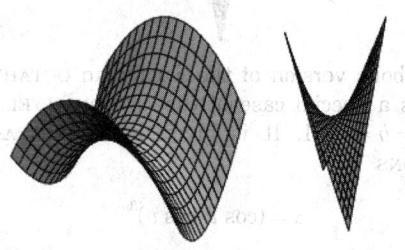

The QUADRATIC and DOUBLY RULED SURFACE given by the Cartesian equation

$$z = \frac{y^2}{b^2} - \frac{x^2}{a^2} \quad (1)$$

(left figure). An alternative form is

$$z = xy \quad (2)$$

(right figure; Fischer 1986), which has PARAMETRIC EQUATIONS

$$x(u, v) = u \quad (3)$$

$$y(u, v) = v \quad (4)$$

$$z(u, v) = uv \quad (5)$$

(Gray 1997, pp. 297–98).
The coefficients of the FIRST FUNDAMENTAL FORM are

$$E = 1 + v^2 \quad (6)$$

$$F = uv \quad (7)$$

$$G = 1 + u^2, \quad (8)$$

and the SECOND FUNDAMENTAL FORM coefficients are

$$e = 0 \quad (9)$$

$$f = (1 + u^2 + v^2)^{-1/2} \quad (10)$$

$$g = 0, \quad (11)$$

giving SURFACE AREA element

$$dS = \sqrt{1 + u^2 + v^2}. \quad (12)$$

The GAUSSIAN CURVATURE is

$$K = -(1 + u^2 + v^2)^{-2} \quad (13)$$

and the MEAN CURVATURE is

$$H = \frac{uv}{(1 + u^2 + v^2)^{3/2}}. \quad (14)$$

Three skew lines always define a one-sheeted HYPERBOLOID, except in the case where they are all parallel to a single PLANE but not to each other. In this case, they determine a hyperbolic paraboloid (Hilbert and Cohn-Vossen 1999, p. 15).

See also DOUBLY RULED SURFACE, ELLIPTIC PARABOLOID, PARABOLOID, RULED SURFACE, SADDLE, SKEW QUADRILATERAL

References

Beyer, W. H. *CRC Standard Mathematical Tables, 28th ed.* Boca Raton, FL: CRC Press, p. 227, 1987.
Fischer, G. (Ed.). *Mathematical Models from the Collections of Universities and Museums.* Braunschweig, Germany: Vieweg, pp. 3–, 1986.
Fischer, G. (Ed.). Plates 7– in *Mathematische Modelle/ Mathematical Models, Bildband/Photograph Volume.* Braunschweig, Germany: Vieweg, pp. 8–0, 1986.
Gray, A. "The Hyperbolic Paraboloid." *Modern Differential Geometry of Curves and Surfaces with Mathematica, 2nd ed.* Boca Raton, FL: CRC Press, pp. 297–98 and 449, 1997.
Hilbert, D. and Cohn-Vossen, S. *Geometry and the Imagination.* New York: Chelsea, 1999.
JavaView. "Classic Surfaces from Differential Geometry: Hyperbolic Paraboloid." http://www-sfb288.math.tu-berlin.de/vgp/javaview/demo/surface/common/PaSurface_HyperbolicParaboloid.html.
McCrea, W. H. *Analytical Geometry of Three Dimensions.* Edinburgh: Oliver and Boyd, 1947.
Meyer, W. "Spezielle algebraische Flächen." *Encylopädie der Math. Wiss. III,* **22B,** 1439–779.
Salmon, G. *Analytic Geometry of Three Dimensions.* New York: Chelsea, 1979.
Steinhaus, H. *Mathematical Snapshots, 3rd ed.* New York: Dover, p. 245, 1999.
Wells, D. *The Penguin Dictionary of Curious and Interesting Geometry.* London: Penguin, pp. 110–12, 1991.

Hyperbolic Partial Differential Equation

A PARTIAL DIFFERENTIAL EQUATION of second-order, i.e., one OF THE FORM

$$Au_{xx} + 2Bu_{xy} + Cu_{yy} + Du_x + Eu_y + F = 0, \quad (1)$$

is called hyperbolic if the MATRIX

$$Z \equiv \begin{bmatrix} A & B \\ B & C \end{bmatrix} \quad (2)$$

satisfies $\det(Z) < 0$. The WAVE EQUATION is an example of a hyperbolic partial differential equation. Initial-boundary conditions are used to give

$$u(x, y, t) = g(x, y, t) \quad \text{for } x \in \partial\Omega, \; t > 0 \quad (3)$$

$$u(x, y, 0) = v_0(x, y) \quad \text{in } \Omega \quad (4)$$

$$u_t(x, y, 0) = v_1(x, y) \quad \text{in } \Omega, \quad (5)$$

where

$$u_{xy} = f(u_x, u_t, x, y) \quad (6)$$

holds in Ω.

See also ELLIPTIC PARTIAL DIFFERENTIAL EQUATION, PARABOLIC PARTIAL DIFFERENTIAL EQUATION, PARTIAL DIFFERENTIAL EQUATION

Hyperbolic Plane

In the hyperbolic plane $\mathbb{H}^2$, a pair of LINES can be PARALLEL (diverging from one another in one direction and intersecting at an IDEAL POINT at infinity in the other), can intersect, or can be HYPERPARALLEL (diverge from each other in both directions).

See also EUCLIDEAN PLANE, RIEMANN SPHERE, RIGID MOTION

References

Anderson, J. W. "A Model for the Hyperbolic Plane." §1.1 in *Hyperbolic Geometry.* New York: Springer-Verlag, pp. 1–, 1999.

Hyperbolic Point

A point **p** on a REGULAR SURFACE $M \in \mathbb{R}^3$ is said to be hyperbolic if the GAUSSIAN CURVATURE $K(\mathbf{p}) < 0$ or equivalently, the PRINCIPAL CURVATURES κ_1 and κ_2, have opposite signs.

See also ANTICLASTIC, ELLIPTIC POINT, GAUSSIAN CURVATURE, HYPERBOLIC FIXED POINT (DIFFERENTIAL EQUATIONS), HYPERBOLIC FIXED POINT (MAP), PARABOLIC POINT, PLANAR POINT, SYNCLASTIC

References

Gray, A. *Modern Differential Geometry of Curves and Surfaces with Mathematica, 2nd ed.* Boca Raton, FL: CRC Press, p. 375, 1997.

Hyperbolic Polyhedron

A POLYHEDRON in a HYPERBOLIC GEOMETRY.

See also HYPERBOLIC CUBE, HYPERBOLIC DODECAHEDRON, HYPERBOLIC ICOSAHEDRON, HYPERBOLIC OCTAHEDRON, HYPERBOLIC TETRAHEDRON

References

Hodgson, C. D. and Riven, I. "A Characterization of Compact Convex Polyhedra in Hyperbolic 3-Space." *Invent. Math.* **111**, 77–11, 1993.
Kellerhals, R. "Shape and Size Through Hyperbolic Eyes." *Math. Intell.* **17**, 21–0, 1995.
Kellerhals, R. "Nichteuklidische Geometrie und Volumina hyperbolischer Polyeder." *Math. Semesterber.* **43**, 155–68, 1996.
Ratcliffe, J. G. *Foundations of Hyperbolic Manifolds.* New York: Springer-Verlag, 1994.
Rivin, I. "A Characterization of Ideal Polyhedra in Hyperbolic 3-Space." *Ann. Math.* **143**, 51–0, 1996.
Thurston, W. P. and Levy, S. (Eds.). *Three-Dimensional Geometry and Topology, Vol. 1.* Princeton, NJ: Princeton University Press, 1997.
Trott, M. "The Cover Image: Hyperbolic Platonic Bodies." §8.3.10 in *The Mathematica Guidebook, Vol. 2: Graphics in Mathematica.* New York: Springer-Verlag, 2000.

Hyperbolic Rotation

Also known as the a Lorentz transformation or Procrustian stretch, a hyperbolic transformation leaves each branch of the HYPERBOLA $x'y' = xy$ invariant and transforms CIRCLES into ELLIPSES with the same AREA.

$$x' = \mu^{-1}x$$

$$y' = \mu y.$$

See also CROSSED HYPERBOLIC ROTATION

References

Coxeter, H. S. M. and Greitzer, S. L. *Geometry Revisited.* Washington, DC: Math. Assoc. Amer., p. 101, 1967.

Hyperbolic Secant

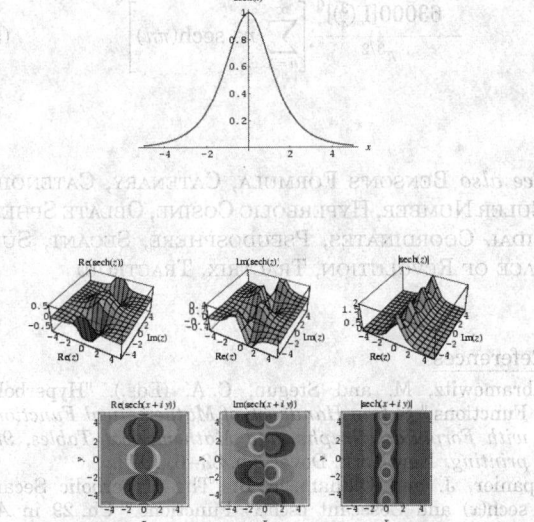

The hyperbolic secant is defined as

$$\text{sech } x \equiv \frac{1}{\cosh x} = \frac{2}{e^x + e^{-x}}, \tag{1}$$

where $\cosh x$ is the HYPERBOLIC COSINE. It has a MAXIMUM at $x = 0$ and inflection points at $x = \pm \text{sech}^{-1}(1\sqrt{2}) \approx 0.881374$.

Equating coefficients of θ^0, θ^4, and θ^8 in the RAMANUJAN COS/COSH IDENTITY

$$\left[1 + 2\sum_{n=1}^{\infty} \frac{\cos(n\theta)}{\cosh(n\pi)}\right]^{-2} + \left[1 + 2\sum_{n=1}^{\infty} \frac{\cosh(n\theta)}{\cosh(n\pi)}\right]^{-2}$$
$$= \frac{2\Gamma^4\left(\frac{3}{4}\right)}{\pi} \tag{2}$$

gives the amazing identities

$$\sum_{n=1}^{\infty} \text{sech}(\pi n) = \frac{1}{2}\left\{\frac{\sqrt{\pi}}{\left[\Gamma\left(\frac{3}{4}\right)\right]^2} - 1\right\} \tag{3}$$

$$\sum_{n=1}^{\infty} n^4 \, \text{sech}(\pi n) = \frac{18\left[\Gamma\left(\frac{3}{4}\right)\right]^2}{\sqrt{\pi}}\left[\sum_{n=1}^{\infty} n^2 \, \text{sech}(\pi n)\right]^2 \tag{4}$$

$$\sum_{n=1}^{\infty} n^8 \, \text{sech}(\pi n)$$
$$= \frac{168[\Gamma(\frac{3}{4})]^2}{\sqrt{\pi}}\left[\sum_{n=1}^{\infty} n^2 \, \text{sech}(\pi n)\right]\sum_{n=1}^{\infty} n^6 \, \text{sech}(\pi n)$$
$$- \frac{63000[\Gamma(\frac{3}{4})]^6}{\pi^{3/2}}\left[\sum_{n=1}^{\infty} n^2 \, \text{sech}(\pi n)\right]^4. \tag{5}$$

See also BENSON'S FORMULA, CATENARY, CATENOID, EULER NUMBER, HYPERBOLIC COSINE, OBLATE SPHEROIDAL COORDINATES, PSEUDOSPHERE, SECANT, SURFACE OF REVOLUTION, TRACTRIX, TRACTROID

References

Abramowitz, M. and Stegun, C. A. (Eds.). "Hyperbolic Functions." §4.5 in *Handbook of Mathematical Functions with Formulas, Graphs, and Mathematical Tables, 9th printing.* New York: Dover, pp. 83–6, 1972.
Spanier, J. and Oldham, K. B. "The Hyperbolic Secant sech(x) and Cosecant csch(x) Functions." Ch. 29 in *An Atlas of Functions.* Washington, DC: Hemisphere, pp. 273–78, 1987.

Hyperbolic Sine

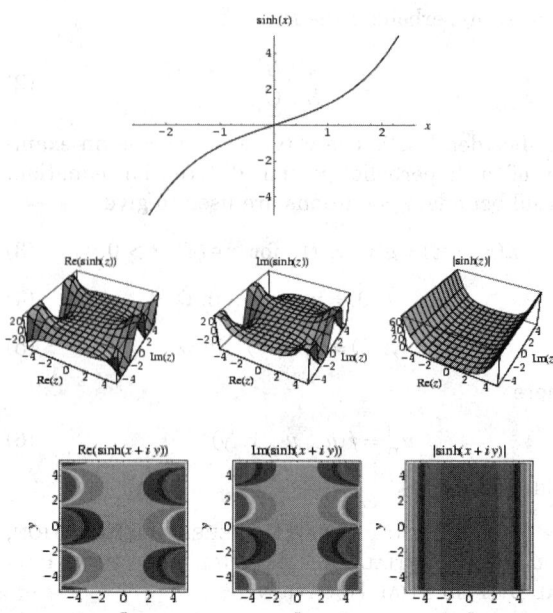

The hyperbolic sine is defined as

$$\sinh x \equiv \tfrac{1}{2}(e^x - e^{-x}).$$

The notation sh x is sometimes also used (Gradshteyn and Ryzhik 2000, p. xxix).

See also BETA EXPONENTIAL FUNCTION, BIPOLAR COORDINATES, BIPOLAR CYLINDRICAL COORDINATES, BISPHERICAL COORDINATES, CATENARY, CATENOID, CONICAL FUNCTION, CUBIC EQUATION, DE MOIVRE'S IDENTITY, DIXON-FERRAR FORMULA, ELLIPTIC CYLINDRICAL COORDINATES, ELSASSER FUNCTION, GUDERMANNIAN FUNCTION, HELICOID, HELMHOLTZ DIFFERENTIAL EQUATION–ELLIPTIC CYLINDRICAL COORDINATES, HYPERBOLIC COSECANT, LAPLACE'S EQUATION–BISPHERICAL COORDINATES, LAPLACE'S EQUATION–TOROIDAL COORDINATES, LEBESGUE CONSTANTS (FOURIER SERIES), LORENTZ GROUP, MERCATOR PROJECTION, MILLER CYLINDRICAL PROJECTION, MODIFIED BESSEL FUNCTION OF THE SECOND KIND, MODIFIED SPHERICAL BESSEL FUNCTION, MODIFIED STRUVE FUNCTION, NICHOLSON'S FORMULA, OBLATE SPHEROIDAL COORDINATES, PARABOLA INVOLUTE, PARTITION FUNCTION P, POINSOT'S SPIRALS, PROLATE SPHEROIDAL COORDINATES, RAMANUJAN'S TAU FUNCTION, SCHLÄFLI'S FORMULA, SHI, SINE, SINE-GORDON EQUATION, SURFACE OF REVOLUTION, TOROIDAL COORDINATES, TOROIDAL FUNCTION, TRACTRIX, WATSON'S FORMULA

References

Abramowitz, M. and Stegun, C. A. (Eds.). "Hyperbolic Functions." §4.5 in *Handbook of Mathematical Functions with Formulas, Graphs, and Mathematical Tables, 9th printing.* New York: Dover, pp. 83–6, 1972.

Gradshteyn, I. S. and Ryzhik, I. M. *Tables of Integrals, Series, and Products, 6th ed.* San Diego, CA: Academic Press, 2000.

Spanier, J. and Oldham, K. B. "The Hyperbolic Sine sinh(*x*) and Cosine cosh(*x*) Functions." Ch. 28 in *An Atlas of Functions.* Washington, DC: Hemisphere, pp. 263–71, 1987.

Hyperbolic Sine Integral

Shi

Hyperbolic Space

Hyperbolic Geometry

Hyperbolic Spiral

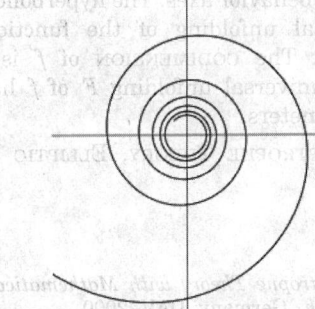

An Archimedean spiral with polar equation

$$r = \frac{a}{\theta}.$$

The hyperbolic spiral originated with Pierre Varignon in 1704 and was studied by Johann Bernoulli between 1710 and 1713, as well as by Cotes in 1722 (MacTutor Archive).

See also Archimedean Spiral, Spiral

References

Beyer, W. H. *CRC Standard Mathematical Tables, 28th ed.* Boca Raton, FL: CRC Press, p. 225, 1987.

Gray, A. *Modern Differential Geometry of Curves and Surfaces with Mathematica, 2nd ed.* Boca Raton, FL: CRC Press, p. 91, 1997.

Lawrence, J. D. *A Catalog of Special Plane Curves.* New York: Dover, pp. 186 and 188, 1972.

Lockwood, E. H. *A Book of Curves.* Cambridge, England: Cambridge University Press, p. 175, 1967.

MacTutor History of Mathematics Archive. "Hyperbolic Spiral." http://www-groups.dcs.st-and.ac.uk/~history/Curves/Hyperbolic.html.

Hyperbolic Spiral Inverse Curve

Taking the pole as the INVERSION CENTER, the HYPERBOLIC SPIRAL inverts to ARCHIMEDES' SPIRAL

$$r = a\theta.$$

Hyperbolic Spiral Roulette

The ROULETTE of the pole of a HYPERBOLIC SPIRAL rolling on a straight line is a TRACTRIX.

Hyperbolic Substitution

A substitution which can be used to transform integrals involving square roots into a more tractable form.

Form	Substitution
$\sqrt{x^2 + a^2}$	$x = a \sinh u$
$\sqrt{x^2 - a^2}$	$x = a \cosh u$

See also Integral, Trigonometric Substitution

Hyperbolic Tangent

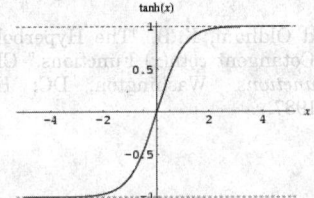

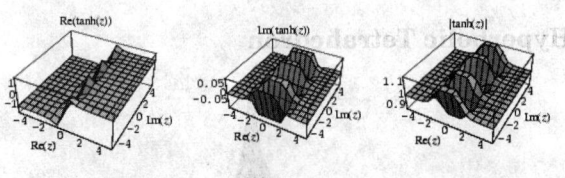

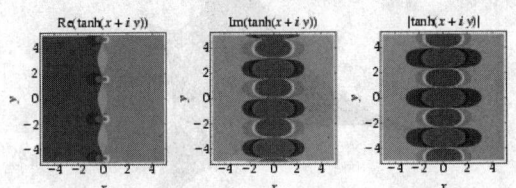

By way of analogy with the usual TANGENT

$$\tan x \equiv \frac{\sin x}{\cos x},$$

the hyperbolic tangent is defined as

$$\tanh x \equiv \frac{\sinh x}{\cosh x} = \frac{e^x - e^{-x}}{e^x + e^{-x}} = \frac{e^{2x} - 1}{e^{2x} + 1},$$

where sinh *x* is the HYPERBOLIC SINE and cosh *x* is the HYPERBOLIC COSINE. The notation th *x* is sometimes also used (Gradshteyn and Ryzhik 2000, p. xxix). The hyperbolic tangent can be written using a CONTINUED FRACTION as

$$\tanh x = \cfrac{x}{1 + \cfrac{x^2}{3 + \cfrac{x^3}{5 + \cdots}}}.$$

See also BERNOULLI NUMBER, CATENARY, CORRELATION COEFFICIENT–GAUSSIAN BIVARIATE DISTRIBUTION, FISHER'S z'-TRANSFORMATION, HYPERBOLIC COTANGENT, LORENTZ GROUP, MERCATOR PROJECTION, OBLATE SPHEROIDAL COORDINATES, PSEUDOSPHERE, SURFACE OF REVOLUTION, TANGENT, TRACTRIX, TRACTROID

References

Abramowitz, M. and Stegun, C. A. (Eds.). "Hyperbolic Functions." §4.5 in *Handbook of Mathematical Functions with Formulas, Graphs, and Mathematical Tables, 9th printing.* New York: Dover, pp. 83–6, 1972.
Gradshteyn, I. S. and Ryzhik, I. M. *Tables of Integrals, Series, and Products, 6th ed.* San Diego, CA: Academic Press, 2000.
Spanier, J. and Oldham, K. B. "The Hyperbolic Tangent $\tanh(x)$ and Cotangent $\coth(x)$ Functions." Ch. 30 in *An Atlas of Functions.* Washington, DC: Hemisphere, pp. 279–84, 1987.

Hyperbolic Tetrahedron

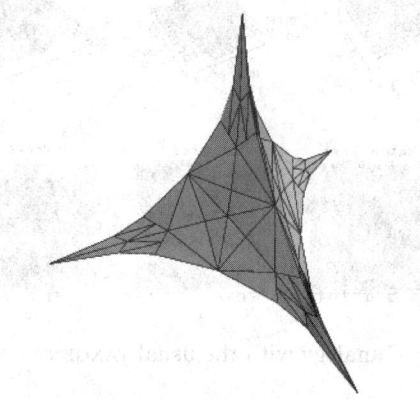

A hyperbolic version of the Euclidean TETRAHEDRON.

See also HYPERBOLIC CUBE, HYPERBOLIC DODECAHEDRON, HYPERBOLIC ICOSAHEDRON, HYPERBOLIC OCTAHEDRON, REULEAUX TETRAHEDRON

References

Rivin, I. "Hyperbolic Polyhedron Graphics." http://www.mathsource.com/cgi-bin/msitem22?0201–88.
Trott, M. "The Cover Image: Hyperbolic Platonic Bodies." §8.3.10 in *The Mathematica Guidebook, Vol. 2: Graphics in Mathematica.* New York: Springer-Verlag, 2000.

Hyperbolic Umbilic Catastrophe

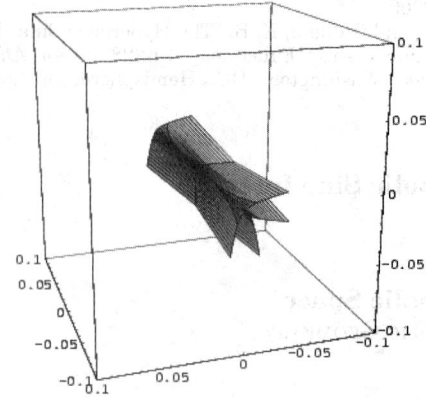

A CATASTROPHE which can occur for three control factors and two behavior axes. The hyperbolic umbilic is the universal unfolding of the function germ $f(x, y) = x^3 + y^3$. The CODIMENSION of f is 3, and therefore the universal unfolding F of f has three unfolding parameters.

See also CATASTROPHE THEORY, ELLIPTIC UMBILIC CATASTROPHE

References

Sanns, W. *Catastrophe Theory with Mathematica: A Geometric Approach.* Germany: DAV, 2000.

Hyperboloid

A QUADRATIC SURFACE which may be one- or two-sheeted. The one-sheeted hyperboloid is a SURFACE OF REVOLUTION obtained by rotating a HYPERBOLA about the perpendicular bisector to the line between the FOCI, while the two-sheeted hyperboloid is a SURFACE OF REVOLUTION obtained by rotating a HYPERBOLA about the line joining the FOCI (Hilbert and Cohn-Vossen 1991, p. 11).

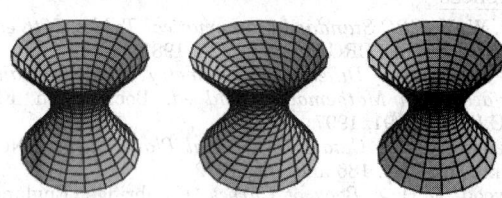

The one-sheeted circular hyperboloid is a DOUBLY RULED SURFACE. When oriented along the z-AXIS, the one-sheeted circular hyperboloid has CARTESIAN COORDINATES equation

$$\frac{x^2}{a^2} + \frac{y^2}{a^2} - \frac{z^2}{c^2} = 1, \tag{1}$$

and parametric equation

$$x = a\sqrt{1+u^2}\cos v \qquad (2)$$

$$y = a\sqrt{1+u^2}\sin v \qquad (3)$$

$$z = cu \qquad (4)$$

for $v \in [0, 2\pi)$ (left figure). Other parameterizations include

$$x(u, v) = a(\cos u \mp v \sin u) \qquad (5)$$

$$y(u, v) = a(\sin u \pm v \sin u) \qquad (6)$$

$$z(u, v) = \pm cv, \qquad (7)$$

(middle figure), or

$$x(u, v) = a \cosh v \cos u \qquad (8)$$

$$y(u, v) = a \cosh v \sin u \qquad (9)$$

$$z(u, v) = c \sinh v \qquad (10)$$

(right figure).

A hyperboloid of one sheet is also obtained as the envelope of a CUBE rotated about a space diagonal (Steinhaus 1983, pp. 171–72). Three skew lines always define a one-sheeted hyperboloid, except in the case where they are all parallel to a single PLANE but not to each other (Hilbert and Cohn-Vossen 1999, p. 15).

The VOLUME of a one-sheeted hyperboloid of height h, waist radius a, and top and bottom radii R is

$$V = \pi h a^2 \left(1 + \frac{h^2}{12 b^2}\right) \qquad (11)$$

$$= \tfrac{1}{3}\pi h(2a^2 + R^2), \qquad (12)$$

where

$$R^2 = a^2 \left(1 + \frac{h^2}{4b^2}\right) \qquad (13)$$

(Harris and Stocker 1998). An obvious generalization gives the one-sheeted ELLIPTIC HYPERBOLOID.

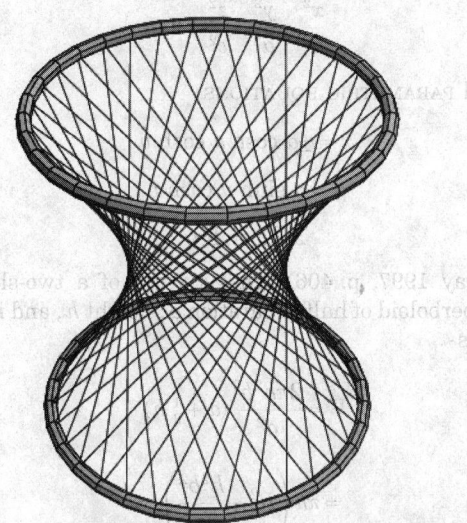

The hyperboloid of one sheet can be constructed by connecting two concentric vertically offset rings wire tilted wires, as illustrated above (Steinhaus 1983, pp. 242–43; Hilbert and Cohn-Vossen 1999, p. 11). Surprisingly, when the wires are fastened together so that rotation but not sliding is permitted, the framework can be expanded and collapsed as one ring is rotated relative to the other (Hilbert and Cohn-Vossen 1999, pp. 16–7 and 29–1).

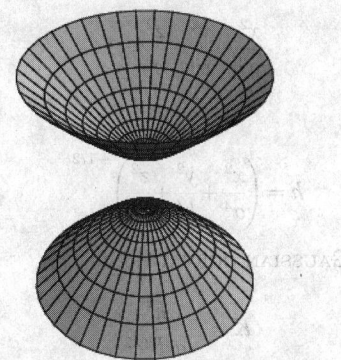

A two-sheeted circular hyperboloid oriented along the z-AXIS has CARTESIAN COORDINATES equation

$$\frac{x^2}{a^2} + \frac{y^2}{a^2} - \frac{z^2}{c^2} = -1. \qquad (14)$$

The PARAMETRIC EQUATIONS are

$$x = a \sinh u \cos v \qquad (15)$$

$$y = a \sinh u \sin v \qquad (16)$$

$$z = \pm c \cosh u \qquad (17)$$

for $v \in [0, 2\pi)$. Note that the plus and minus signs in z correspond to the upper and lower sheets. The two-sheeted circular hyperboloid oriented along the x-AXIS has Cartesian equation

$$\frac{x^2}{a^2} - \frac{y^2}{a^2} - \frac{z^2}{c^2} = 1 \qquad (18)$$

and PARAMETRIC EQUATIONS

$$x = \pm a \cosh u \cosh v \qquad (19)$$

$$y = a \sinh u \cosh v \qquad (20)$$

$$z = c \sinh v \qquad (21)$$

(Gray 1997, p. 406). The VOLUME of a two-sheeted hyperboloid of half-separation a, height h, and radius R is

$$V = \frac{2\pi h^2 b^2}{a^2}(a + \tfrac{1}{3}h) \qquad (22)$$

$$= \pi h \left(R^2 - \frac{h^2 b^2}{3a^2} \right), \qquad (23)$$

where

$$R^2 = \frac{h b^2}{a^2}(2a + h) \qquad (24)$$

(Harris and Stocket 1998). Again, an obvious generalization gives the two-sheeted ELLIPTIC HYPERBOLOID.

The SUPPORT FUNCTION of the hyperboloid of one sheet

$$\frac{x^2}{a^2} + \frac{y^2}{b^2} - \frac{z^2}{c^2} = 1 \qquad (25)$$

is

$$h = \left(\frac{x^2}{a^4} + \frac{y^2}{b^4} + \frac{z^2}{c^4} \right)^{-1/2}, \qquad (26)$$

and the GAUSSIAN CURVATURE is

$$K = -\frac{h^4}{a^2 b^2 c^2}. \qquad (27)$$

The SUPPORT FUNCTION of the hyperboloid of two sheets

$$\frac{x^2}{a^2} - \frac{y^2}{b^2} - \frac{z^2}{c^2} = 1 \qquad (28)$$

is

$$h = \left(\frac{x^2}{a^4} - \frac{y^2}{b^4} + \frac{z^2}{c^4} \right)^{-1/2}, \qquad (29)$$

and the GAUSSIAN CURVATURE is

$$K = \frac{h^4}{a^2 b^2 c^2} \qquad (30)$$

(Gray 1997, p. 414).

See also CATENOID, CONFOCAL QUADRICS, DOUBLY

RULED SURFACE, ELLIPSOID, ELLIPSOIDAL COORDINATES, ELLIPTIC HYPERBOLOID, HYPERBOLA, HYPERBOLOID EMBEDDING, PARABOLOID, RULED SURFACE

References

Beyer, W. H. *CRC Standard Mathematical Tables, 28th ed.* Boca Raton, FL: CRC Press, p. 227, 1987.
Fischer, G. (Ed.). Plates 67 and 69 in *Mathematische Modelle / Mathematical Models, Bildband / Photograph Volume.* Braunschweig, Germany: Vieweg, pp. 62 and 64, 1986.
Gray, A. "The Hyperboloid of Revolution." §20.5 in *Modern Differential Geometry of Curves and Surfaces with Mathematica, 2nd ed.* Boca Raton, FL: CRC Press, p. 470, 1997.
Harris, J. W. and Stocker, H. "Hyperboloid of Revolution." §4.10.3 in *Handbook of Mathematics and Computational Science.* New York: Springer-Verlag, p. 112, 1998.
Hilbert, D. and Cohn-Vossen, S. *Geometry and the Imagination.* New York: Chelsea, pp. 10–1, 1999.
JavaView. "Classic Surfaces from Differential Geometry: Hyperboloid." http://www-sfb288.math.tu-berlin.de/vgp/javaview/demo/surface/common/PaSurface_Hyperboloid.html.
Steinhaus, H. *Mathematical Snapshots, 3rd ed.* New York: Dover, 1999.
Wells, D. *The Penguin Dictionary of Curious and Interesting Geometry.* London: Penguin, pp. 112–13, 1991.

Hyperboloid Embedding

A 4-HYPERBOLOID has NEGATIVE CURVATURE, with

$$R^2 = x^2 + y^2 + z^2 - w^2 \qquad (1)$$

$$2x\,\frac{dx}{dw} + 2y\,\frac{dy}{dw} + 2z\,\frac{dz}{dw} - 2w = 0. \qquad (2)$$

Since

$$\mathbf{r} \equiv x\hat{\mathbf{x}} + y\hat{\mathbf{y}} + z\hat{\mathbf{z}} \qquad (3)$$

$$dw = \frac{x\,dx + y\,dy + z\,dz}{w} \equiv \frac{\mathbf{r} \cdot d\mathbf{r}}{\sqrt{r^2 - R^2}}. \qquad (4)$$

To stay on the surface of the HYPERBOLOID, the LINE ELEMENT is given by

$$ds^2 = dx^2 + dy^2 + dz^2 - dw^2$$

$$= dx^2 + dy^2 + dz^2 - \frac{r^2\,dr^2}{r^2 - R^2}$$

$$= dr^2 + r^2\,d\Omega^2 + \frac{dr^2}{1 - \frac{R^2}{r^2}}. \qquad (5)$$

Hypercomplex Number

There are at least two definitions of hypercomplex numbers. CLIFFORD ALGEBRAISTS call their higher dimensional numbers hypercomplex, even though they do not share all the properties of complex numbers and no classical function theory can be constructed over them.

According to van der Waerden (1985), a hypercomplex number is a number having properties departing from those of the REAL and COMPLEX NUMBERS. The most common examples are BIQUATERNIONS, EXTERIOR ALGEBRAS, GROUP algebras, MATRICES, OCTONIONS, and QUATERNIONS. One type of hypercomplex number due to Davenport (1996) and sometimes called "the" hypercomplex numbers are defined according to the multiplication table

$$ij = ji = k \tag{1}$$

$$jk = kj = -i \tag{2}$$

$$ki = ik = -j, \tag{3}$$

and therefore satisfy

$$i^2 = j^2 = -1 \tag{4}$$

$$k^2 = 1. \tag{5}$$

Unlike QUATERNIONS, multiplication of these hypercomplex numbers *is* commutative, and unlike real and complex numbers, not all nonzero hypercomplex numbers have a multiplicative inverse. An application of this sort of hypercomplex number can be found in the `julia_fractal` command in *POVRay*.

See also BIQUATERNION, CAYLEY NUMBER, CLIFFORD ALGEBRA, COMPLEX NUMBER, EXTERIOR ALGEBRA, GROUP, MATRIX, OCTONION, QUATERNION, REAL NUMBER, WEIERSTRASS'S THEOREM

References

Davenport, C. M. "A Commutative Hypercomplex Algebra with Associated Function Theory." In *Clifford Algebras with Numeric and Symbolic Computations* (Ed. R. Ab| amowicz, P. Lounesto, and J. M. Parra). Boston, MA: Birkhäuser, pp. 213–27, 1996.

Kantor, I. L. and Solodovnikov, A. S. *Hypercomplex Numbers : An Elementary Introduction to Algebras.* New York: Springer-Verlag, 1989.

van der Waerden, B. L. *A History of Algebra from al-Khwarizmi to Emmy Noether.* New York: Springer-Verlag, pp. 177–17, 1985.

Hypercube

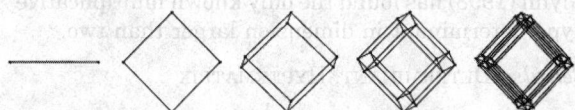

The generalization of a 3-CUBE to n-D, also called a MEASURE POLYTOPE. It is a regular POLYTOPE with mutually PERPENDICULAR sides, and is therefore an ORTHOTOPE. It is denoted γ_n and has SCHLÄFLI SYMBOL

$$\{4, \underbrace{3, 3}_{n-2}\}.$$

The number of k-cubes contained in an n-cube can be found from the COEFFICIENTS of $(2k + 1)^n$.

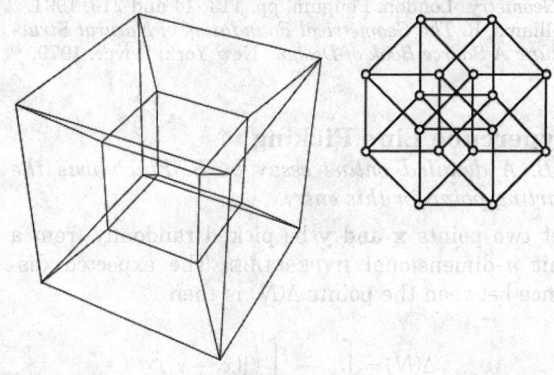

The 1-hypercube is a LINE SEGMENT, the 2-hypercube is the SQUARE, and the 3-hypercube is the CUBE. The hypercube in $\mathbb{R}^4$, called a TESSERACT, has the SCHLÄFLI SYMBOL $\{4, 3, 3\}$ and VERTICES $(\pm 1, \pm 1, \pm 1, \pm 1)$. The above figures show two visualizations of the TESSERACT. The figure on the left is a projection of the TESSERACT in 3-space (Gardner 1977; Williams 1979, p. 26), which also appears on the cover of Born (1926), and the figure on the right is the GRAPH of the TESSERACT symmetrically projected into the PLANE (Coxeter 1973). A TESSERACT has 16 VERTICES, 32 EDGES, 24 SQUARES, and eight CUBES. The dual of the 4-hypercube is the 16-CELL.

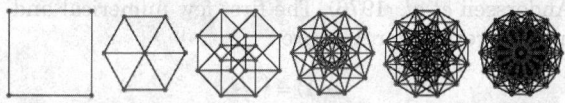

The above figures show the graphs for the n-hypercubes with $n = 2$ to 7. All hypercubes are HAMILTONIAN, and any HAMILTONIAN CIRCUIT of a labeled hypercube defines a GRAY CODE (Skiena 1990, p. 149).

See also CROSS POLYTOPE, CUBE, GLOME, HAMILTONIAN GRAPH, HYPERCUBE LINE PICKING, HYPERSPHERE, ORTHOTOPE, PARALLELEPIPED, POLYTOPE, SIMPLEX, TESSERACT

References

Born, M. *Problems of Atomic Dynamics.* Cambridge, MA: MIT Press, 1926.

Coxeter, H. S. M. *Regular Polytopes, 3rd ed.* New York: Dover, p. 123, 1973.

Dewdney, A. K. "Computer Recreations: A Program for Rotating Hypercubes Induces Four-Dimensional Dementia." *Sci. Amer.* **254**, 14–3, Mar. 1986.

Gardner, M. "Hypercubes." Ch. 4 in *Mathematical Carnival: A New Round-Up of Tantalizers and Puzzles from Scientific American.* New York: Vintage Books, pp. 41–4, 1977.

Pappas, T. "How Many Dimensions are There?" *The Joy of Mathematics.* San Carlos, CA: Wide World Publ./Tetra, pp. 204–05, 1989.

Skiena, S. "Hypercubes." §4.2.5 in *Implementing Discrete Mathematics: Combinatorics and Graph Theory with*

Mathematica. Reading, MA: Addison-Wesley, pp. 148–50, 1990.

Wells, D. *The Penguin Dictionary of Curious and Interesting Geometry.* London: Penguin, pp. 113–14 and 210, 1991.

Williams, R. *The Geometrical Foundation of Natural Structure: A Source Book of Design.* New York: Dover, 1979.

Hypercube Line Picking

N.B. A detailed online essay by S. Finch was the starting point for this entry.

Let two points **x** and **y** be picked randomly from a unit n-dimensional HYPERCUBE. The expected distance between the points $\Delta(N)$ is then

$$\Delta(N) = \underbrace{\int_0^1 \cdots \int_0^1}_{2n} [(x_1 - y_1)^2$$

$$+(x_2 - y_2)^2 + \ldots + (x_n + y_n)]^{1/2} \, dx_1 \cdots dx_n dy_1 \ldots dy_n.$$

(1)

This MULTIPLE INTEGRAL has been evaluated analytically only for small values of n. The case $\Delta(1)$ corresponds to the POINT-POINT DISTANCE between two random points in the interval [0, 1].

The function $\Delta(n)$ satisfies

$$\tfrac{1}{3} n^{1/2} \le \Delta(n) \le \left(\tfrac{1}{6} n\right)^{1/2} \sqrt{\frac{1}{3}\left[1 + 2\left(1 - \frac{3}{5n}\right)^{1/2}\right]}$$

(2)

(Anderssen *et al.* 1976). The first few numerical and analytic results for $\Delta(n)$ are

$$\Delta(1) = \tfrac{1}{3}$$

$$\Delta(2) = \tfrac{1}{15}[\sqrt{2} + 2 + 5\ln(1 + \sqrt{2})] = 0.521405433\ldots$$

$$\Delta(3) = \tfrac{1}{105}[4 + 17\sqrt{2} - 6\sqrt{3} + 21\ln(1 + \sqrt{2})$$

$$+ 42\ln(2 + \sqrt{3}) - 7\pi]$$

$$= 0.661707182\ldots$$

$$\Delta(4) = 0.77766\ldots$$

$$\Delta(5) = 0.87852\ldots$$

$$\Delta(6) = 0.96895\ldots$$

$$\Delta(7) = 1.05159\ldots$$

$$\Delta(8) = 1.12817\ldots$$

See also CUBE LINE PICKING, SQUARE TRIANGLE PICKING

References

Anderssen, R. S.; Brent, R. P.; Daley, D. J.; and Moran, A. P. "Concerning $\int_0^1 \cdots \int_0^1 \sqrt{x_1^2 + \ldots + x_k^2} \, dx_1 \cdots dx_k$ and

a Taylor Series Method." *SIAM J. Appl. Math.* **30**, 22–0, 1976.

Le Lionnais, F. *Les nombres remarquables.* Paris: Hermann, p. 30, 1983.

Robbins, D. "Average Distance between Two Points in a Box." *Amer. Math. Monthly* **85**, 278, 1978.

Trott, M. "The Area of a Random Triangle." *Mathematica J.* **7**, 189–98, 1998.

Hypercube Triangulation

References

Finch, S. "Unsolved Mathematics Problems: Triangulating an n-Dimensional Cube." http://www.mathsoft.com/asolve/simplex/simplex.html.

Hyperdeterminant

A technically defined extension of the ordinary DETERMINANT to "higher dimensional" HYPERMATRICES. Cayley (1845) originally coined the term, but subsequently used it to refer to an ALGEBRAIC INVARIANT of a multilinear form. The hyperdeterminant of the $2 \times 2 \times 2$ HYPERMATRIX $A = a_{ijk}$ (for $i, j, k = 0, 1$) is given by

$$\det(A) = (a_{000}^2 a_{111}^2 + a_{001}^2 a_{110}^2 + a_{010}^2 a_{101}^2 + a_{011}^2 a_{100}^2)$$

$$-2(a_{000}a_{001}a_{110}a_{111} + a_{000}a_{010}a_{101}a_{111} + a_{000}a_{011}a_{100}a_{111}$$

$$+ a_{001}a_{010}a_{101}a_{110} + a_{001}a_{011}a_{110}a_{100} + a_{010}a_{011}a_{101}a_{100})$$

$$+4(a_{000}a_{011}a_{101}a_{110} + a_{001}a_{010}a_{100}a_{111}).$$

The above hyperdeterminant vanishes IFF the following system of equations in six unknowns has a nontrivial solution,

$$a_{000}x_0 y_0 + a_{010}x_0 y_1 + a_{100}x_1 y_0 + a_{110}x_1 y_1 = 0$$

$$a_{001}x_0 y_0 + a_{011}x_0 y_1 + a_{101}x_1 y_0 + a_{111}x_1 y_1 = 0$$

$$a_{000}x_0 z_0 + a_{001}x_0 z_1 + a_{100}x_1 z_0 + a_{101}x_1 z_1 = 0$$

$$a_{010}x_0 z_0 + a_{011}x_0 z_1 + a_{110}x_1 z_0 + a_{111}x_1 z_1 = 0$$

$$a_{000}y_0 z_0 + a_{001}y_0 z_1 + a_{010}y_1 z_0 + a_{011}y_1 z_1 = 0$$

$$a_{100}y_0 z_0 + a_{101}y_0 z_1 + a_{110}y_1 z_0 + a_{111}y_1 z_1 = 0.$$

Glynn (1998) has found the only known multiplicative hyperdeterminant in dimension larger than two.

See also DETERMINANT, HYPERMATRIX

References

Cayley, A. "On the Theory of Linear Transformations." *Cambridge Math. J.* **4**, 193–09, 1845.

Gel'fand, I. M.; Kapranov, M. M.; and Zelevinsky, A. V. "Hyperdeterminants." *Adv. Math.* **96**, 226–63, 1992.

Glynn, D. G. "The Modular Counterparts of Cayley's Hyperdeterminant." *Bull. Austral. Math. Soc.* **57**, 479–97, 1998.

Schläfli, L. "Über die Resultante eine Systemes mehrerer algebraischer Gleichungen." *Denkschr. Kaiserl. Akad. Wiss., Math.-Naturwiss. Klasse* **4**, 1852.

Hyperedge

A connection between two or more vertices of a HYPERGRAPH. A hyperedge connecting just two vertices is simply a usual EDGE.

See also EDGE (GRAPH), HYPERGRAPH

Hyperellipse

$$y^{n/m} + c \left| \frac{x}{a} \right|^{n/m} - c = 0,$$

with $n/m > 2$. If $n/m < 2$, the curve is a HYPOELLIPSE.

See also ELLIPSE, HYPOELLIPSE, SUPERELLIPSE

References

von Seggern, D. *CRC Standard Curves and Surfaces.* Boca Raton, FL: CRC Press, p. 82, 1993.

Hyperelliptic Function

ABELIAN FUNCTION

Hyperelliptic Integral

ABELIAN INTEGRAL

Hyperfactorial

The function defined by

$$H(n) \equiv K(n+1) \equiv 1^1 2^2 3^3 \cdots n^n,$$

where $K(n)$ is the K-FUNCTION and the first few values for $n = 1, 2, \ldots$ are 1, 4, 108, 27648, 86400000, 4031078400000, 3319766398771200000, ... (Sloane's A002109), and these numbers are called hyperfactorials by Sloane and Plouffe (1995).

See also BARNES' G-FUNCTION, GLAISHER-KINKELIN CONSTANT, K-FUNCTION

References

Fletcher, A.; Miller, J. C. P.; Rosenhead, L.; and Comrie, L. J. *An Index of Mathematical Tables, Vol. 1, 2nd ed.* Reading, MA: Addison-Wesley, p. 50, 1962.
Graham, R. L.; Knuth, D. E.; and Patashnik, O. *Concrete Mathematics: A Foundation for Computer Science, 2nd ed.* Reading, MA: Addison-Wesley, p. 477, 1994.
Sloane, N. J. A. Sequences A002109/M3706 in "An On-Line Version of the Encyclopedia of Integer Sequences." http://www.research.att.com/~njas/sequences/eisonline.html.

Hyperfinite Set

One of the most useful tools in NONSTANDARD ANALYSIS is the concept of a hyperfinite set. To understand a hyperfinite set, begin with an arbitrary infinite set X whose members are not sets, and form the SUPERSTRUCTURE $S(X)$ over X. Assume that X includes the natural numbers as elements, let $\mathbb{N}$ denote the set of natural numbers as elements of X, and let $*S(X)$ be an ENLARGEMENT of $S(X)$. By the TRANSFER PRINCIPLE, the ordering $<$ on $\mathbb{N}$ extends to a strict linear ordering on $*\mathbb{N}$, which can be denoted with the symbol "$<$." Since $*S(X)$ is an enlargement of $S(X)$, it satisfies the CONCURRENCY PRINCIPLE, so that there is an element v of $*\mathbb{N}$ such that if $n \in \mathbb{N}$, then $n < v$. This follows because the relation $<$ is a CONCURRENT RELATION on the set of natural numbers.

Any member v of $*\mathbb{N}$ is called an infinite nonstandard natural number, and for any set $A \in *S(X)$, if A is in one-to-one correspondence with any element of $*\mathbb{N}$, then A is called a hyperfinite set in $*S(X)$. Because there are infinite nonstandard natural numbers in any enlargement $*S(X)$ of $S(X)$, there are hyperfinite sets that are not finite, in any such enlargement. Such hyperfinite sets can be used to study infinite structures satisfying various finiteness conditions.

References

Albeverio, S.; Fenstad, J.; Hoegh-Krohn, R.; and Lindstrøom, T. *Nonstandard Methods in Stochastic Analysis and Mathematical Physics.* New York: Academic Press, 1986.
Anderson, R. M. "Nonstandard Analysis with Applications to Economics." Ch. 39 in *Handbook of Mathematical Economics, Vol. 4* (Ed. W. Hildenbrand and H. Sonnenschein). New York: Elsevier, pp. 2145–208, 1991.
Dauben, J. W. *Abraham Robinson: The Creation of Nonstandard Analysis, A Personal and Mathematical Odyssey.* Princeton, NJ: Princeton University Press, 1998.
Davis, P. J. and Hersch, R. *The Mathematical Experience.* Boston, MA: Birkhäuser, 1981.
Insall, M. "Nonstandard Methods and Finiteness Conditions in Algebra" *Zeitschr. f. Math., Logik, und Grundlagen d. Math.* **37**, 525–32, 1991.
Keisler, H. J. *Elementary Calculus: An Infinitesimal Approach.* Boston, MA: PWS, 1986.
Lindstrøm, T. "An Invitation to Nonstandard Analysis." In *Nonstandard Analysis and Its Applications* (Ed. N. Cutland). New York: Cambridge University Press, 1988.
Robinson, A. *Non-Standard Analysis.* Princeton, NJ: Princeton University Press, 1996.
Stewart, I. "Non-Standard Analysis." In *From Here to Infinity: A Guide to Today's Mathematics.* Oxford, England: Oxford University Press, pp. 80–1, 1996.

Hypergame

A two-player game in which player 1 chooses any FINITE GAME and player 2 moves first. A PSEUDOPARADOX then arises as to whether the hypergame is itself a FINITE GAME.

See also FINITE GAME, GAME

Hypergeometric Differential Equation

$$x(x-1)\frac{d^2 y}{dx^2} + [(1+\alpha+\beta)x - \gamma]\frac{dy}{dx} + \alpha\beta y = 0.$$

It has REGULAR SINGULAR POINTS at 0, 1, and ∞. Every ORDINARY DIFFERENTIAL EQUATION of second-order with at most three REGULAR SINGULAR POINTS can be transformed into the hypergeometric differential equation.

See also CONFLUENT HYPERGEOMETRIC DIFFERENTIAL

EQUATION, CONFLUENT HYPERGEOMETRIC FUNCTION, GENERALIZED HYPERGEOMETRIC FUNCTION, HYPERGEOMETRIC FUNCTION

References

Bailey, W. N. *Generalised Hypergeometric Series.* Cambridge, England: University Press, pp. 1–, 1935.

Morse, P. M. and Feshbach, H. *Methods of Theoretical Physics, Part I.* New York: McGraw-Hill, pp. 542–43, 1953.

Zwillinger, D. *Handbook of Differential Equations, 3rd ed.* Boston, MA: Academic Press, p. 123, 1997.

Hypergeometric Distribution

Let there be n ways for a successful and m ways for an unsuccessful trial out of a total of $n+m$ possibilities. Take N samples and let x_i equal 1 if selection i is successful and 0 if it is not. Let x be the total number of successful selections,

$$x \equiv \sum_{i=1}^{N} x_i. \tag{1}$$

The probability of i successful selections is then

$$P(x=i) =$$

$$\frac{[\text{\# ways for } i \text{ successes}][\text{\# ways for } N-i \text{ unsuccesses}]}{[\text{total number of ways to select}]}$$

$$= \frac{\binom{n}{i}\binom{m}{N-i}}{\binom{n+m}{N}} = \frac{\frac{n!}{i!(n-i)!}\frac{m!}{(m+i-N)!(N-i)!}}{\frac{(n+m)!}{N!(N-n-m)!}}$$

$$= \frac{n!\,m!\,N!(N-m-n)!}{i!(n-i)!(m+i-N)!(N-i)!(n+m)!}. \tag{2}$$

The ith selection has an equal likelihood of being in any trial, so the fraction of acceptable selections p is

$$p \equiv \frac{n}{n+m} \tag{3}$$

$$P(x_i=1) = \frac{n}{n+m} \equiv p. \tag{4}$$

The expectation value of x is

$$\mu \equiv \langle x \rangle = \left\langle \sum_{i=1}^{N} x_i \right\rangle = \sum_{i=1}^{N} \langle x_i \rangle$$

$$= \sum_{i=1}^{N} \frac{n}{n+m} = \frac{nN}{n+m} = Np. \tag{5}$$

The VARIANCE is

$$\mathrm{var}(x) \equiv \sum_{i=1}^{N} \mathrm{var}(x_i) + \sum_{i=1}^{N}\sum_{\substack{j=1 \\ j \neq 1}}^{N} \mathrm{cov}(x_i, x_j). \tag{6}$$

Since x_i is a BERNOULLI variable,

$$\mathrm{var}(x_i) = p(1-p) = \frac{n}{n+m}\left(1 - \frac{n}{n+m}\right)$$

$$= \frac{n}{n+m}\left(1 - \frac{n}{n+m}\right)$$

$$= \frac{n}{n+m}\left(\frac{n+m-n}{n+m}\right) = \frac{nm}{(n+m)^2}, \tag{7}$$

so

$$\sum_{i=1}^{N} \mathrm{var}(x_i) = \frac{Nnm}{(n+m)^2}. \tag{8}$$

For $i<j$, the COVARIANCE is

$$\mathrm{cov}(x_i, x_j) = \langle x_i x_j \rangle - \langle x_i \rangle \langle x_j \rangle. \tag{9}$$

The probability that both i and j are successful for $i \neq j$ is

$$P(x_i=1,\, x_j=1) = P(x_i=1)P(x_j=1|x_i=1)$$

$$= \frac{n}{n+m}\frac{n-1}{n+m-1}$$

$$= \frac{n(n-1)}{(n+m)(n+m-1)}. \tag{10}$$

But since x_i and x_j are random BERNOULLI variables (each 0 or 1), their product is also a BERNOULLI variable. In order for $x_i x_j$ to be 1, both x_i and x_j must be 1,

$$\langle x_i x_j \rangle = P(x_i x_j = 1) = P(x_i=1,\, x_j=1)$$

$$= \frac{n}{n+m}\frac{n-1}{n+m-1}$$

$$= \frac{n(n-1)}{(n+m)(n+m-1)}. \tag{11}$$

Combining (11) with

$$\langle x_i \rangle \langle x_j \rangle = \frac{n}{n+m}\frac{n}{n+m} = \frac{n^2}{(n+m)^2}, \tag{12}$$

gives

$$\mathrm{cov}(x_i, x_j) = \frac{(n+m)(n^2-n) - n^2(n+m-1)}{(n+m)^2(n+m-1)}$$

$$= \frac{n^3 + mn^2 - n^2 - mn - n^3 - n^2m + n^2}{(n+m)^2(n+m-1)}$$

$$= -\frac{mn}{(n+m)^2(n+m-1)}. \tag{13}$$

There are a total of N^2 terms in a double summation

over N. However, $i = j$ for N of these, so there are a total of $N^2 - N = N(N - 1)$ terms in the COVARIANCE summation

$$\sum_{i=1}^{N} \sum_{\substack{j=1 \\ j \neq i}}^{n} \text{cov}(x_i, x_j) = -\frac{N(N - 1)mn}{(n + m)^2(n + m - 1)}. \quad (14)$$

Combining equations (6), (8), (11), and (14) gives the VARIANCE

$$\text{var}(x) = \frac{Nmn}{(n + m)^2} - \frac{N(N - 1)mn}{(n + m)^2(n + m - 1)}$$

$$= \frac{Nmn}{(n + m)^2}\left(1 - \frac{N - 1}{n + m - 1}\right)$$

$$= \frac{Nmn}{(n + m)^2}\left(\frac{N + m - 1 - N + 1}{n + m - 1}\right)$$

$$= \frac{Nmn(n + m - N)}{(n + m)^2(n + m - 1)}, \quad (15)$$

so the final result is

$$\langle x \rangle = Np \quad (16)$$

and, since

$$1 - p = \frac{m}{n + m} \quad (17)$$

and

$$np(1 - p) = \frac{mn}{(n + m)^2}, \quad (18)$$

we have

$$\sigma^2 = \text{var}(x) = Np(1 - p)\left(1 - \frac{N - 1}{n + m - 1}\right)$$

$$= \frac{mnN(m + n - N)}{(m + n)^2(m + n - 1)}. \quad (19)$$

The SKEWNESS is

$$\gamma_1 = \frac{q - p}{\sqrt{npq}}\sqrt{\frac{N - 1}{N - m}}\left(\frac{N - 2n}{N - 2}\right)$$

$$= \frac{(m - n)(m + n - 2N)}{m + n - 2}\sqrt{\frac{m + n - 1}{mnN(m + n - N)}}, \quad (20)$$

and the KURTOSIS is given by the complicated expression

$$\gamma_2 = \frac{F(m, n, N)}{mnN(-3 + m + n)(-2 + m + n)(-m - n + N)}, \quad (21)$$

where

$$F(m, n, N) = m^3 - m^5 + 3m^2n - 6m^3n + m^4n + 3mn^2$$

$$-12m^2n^2 + 8m^3n^2 + n^3 - 6mn^3 + 8m^2n^3$$

$$+mn^4 - n^5 - 6m^3N + 6m^4N + 18m^2nN$$

$$-6m^3nN + 18mn^2N - 24m^2n^2N - 6n^3N$$

$$-6mn^3N + 6n^4N + 6m^2N^2 - 6m^3N^2$$

$$-24mnN^2 + 12m^2nN^2 + 6n^2N^2$$

$$+12mn^2N^2 - 6n^3N^2. \quad (22)$$

The GENERATING FUNCTION is

$$\phi(t) = \frac{\dbinom{m}{N}}{\dbinom{n + m}{N}} \, {}_2F_1(-N, -n; m - N + 1; e^{it}), \quad (23)$$

where ${}_2F_1(a, b; c; z)$ is the HYPERGEOMETRIC FUNCTION.

If the hypergeometric distribution is written

$$h_n(x, s) = \frac{\dbinom{np}{x}\dbinom{nq}{s - x}}{\dbinom{n}{s}}, \quad (24)$$

then

$$\sum_{x=0}^{s} h_n(x, s)u^x = A \, {}_2F_1(-s, -np; nq - s + 1; u). \quad (25)$$

References

Beyer, W. H. *CRC Standard Mathematical Tables, 28th ed.* Boca Raton, FL: CRC Press, pp. 532–33, 1987.

Feller, W. "The Hypergeometric Series." §2.6 in *An Introduction to Probability Theory and Its Applications, Vol. 1, 3rd ed.* New York: Wiley, pp. 41–5, 1968.

Spiegel, M. R. *Theory and Problems of Probability and Statistics.* New York: McGraw-Hill, pp. 113–14, 1992.

Hypergeometric Function

A GENERALIZED HYPERGEOMETRIC FUNCTION ${}_pF_q(a_1, \ldots, a_p; b_1, \ldots, b_q; x)$ is a function which can be defined in the form of a HYPERGEOMETRIC SERIES, i.e., a series for which the ratio of successive terms can be written

$$\frac{c_{k+1}}{c_k} = \frac{P(k)}{Q(k)}$$

$$= \frac{(k + a_1)(k + a_2)\cdots(k + a_p)}{(k + b_1)(k + b_2)\cdots(k + b_q)(k + 1)}x. \quad (1)$$

(The factor of $k + 1$ in the DENOMINATOR is present for historical reasons of notation.)

The function $_2F_1(a, b; c; x)$ corresponding to $p = 2$, $q = 1$ is the first hypergeometric function to be studied (and, in general, arises the most frequently in physical problems), and so is frequently known as "the" hypergeometric equation or, more explicitly, Gauss's hypergeometric function (Gauss 1812; Barnes 1908). To confuse matters even more, the term "hypergeometric function" is less commonly used to mean CLOSED FORM, and "hypergeometric series" is sometimes used to mean hypergeometric function.

The hypergeometric functions are solutions to the HYPERGEOMETRIC DIFFERENTIAL EQUATION, which has a REGULAR SINGULAR POINT at the ORIGIN. To derive the hypergeometric function based on the HYPERGEOMETRIC DIFFERENTIAL EQUATION, plug

$$y = \sum_{n=0}^{\infty} A_n z^n \tag{2}$$

$$y' = \sum_{n=0}^{\infty} n A_n z^{n-1} \tag{3}$$

$$y'' = \sum_{n=0}^{\infty} n(n-1) A_n z^{n-2} \tag{4}$$

into

$$z(1-z)y'' + [c - (a+b+1)z]y' - aby = 0 \tag{5}$$

to obtain

$$\sum_{n=0}^{\infty} n(n-1)A_n z^{n-1} - \sum_{n=0}^{\infty} n(n-1)A_n z^n$$
$$+ c\sum_{n=0}^{\infty} nA_n z^{n-1} + (a+b+1)\sum_{n=0}^{\infty} nA_n z^n$$
$$- ab\sum_{n=0}^{\infty} A_n z^n = 0 \tag{6}$$

$$\sum_{n=2}^{\infty} n(n-1)A_n z^{n-1} - \sum_{n=0}^{\infty} n(n-1)A_n z^n$$
$$+ c\sum_{n=1}^{\infty} nA_n z^{n-1} - (a+b+1)\sum_{n=1}^{\infty} nA_n z^n$$
$$- ab\sum_{n=0}^{\infty} A_n z^n = 0 \tag{7}$$

$$\sum_{n=0}^{\infty} (n+1)nA_{n+1}z^n - \sum_{n=0}^{\infty} n(n-1)A_n z^n$$
$$+ c\sum_{n=0}^{\infty} (n+1)A_{n+1}z^n - (a+b+1)\sum_{n=0}^{\infty} nA_n z^n$$

$$- ab\sum_{n=0}^{\infty} A_n z^n = 0 \tag{8}$$

$$\sum_{n=0}^{\infty} [n(n+1)A_{n+1} - n(n-1)A_n + c(n+1)A_{n-1}$$
$$\times \sum_{n=0}^{\infty} \{(n+1)(n+c)A_{n+1}$$
$$- [n(n-1+a+b+1)+ab]A_n \}z^n = 0 \tag{9}$$

$$\sum_{n=0}^{\infty} \{(n+1)(n+c)A_{n+1}$$
$$- [n^2 + (a+b)n + ab]A_n \}z^n = 0, \tag{10}$$

so

$$A_{n+1} = \frac{(n+a)(n+b)}{(n+1)(n+c)} A_n \tag{11}$$

and

$$y = A_0 \left[1 + \frac{ab}{1!c} z + \frac{a(a+1)b(b+1)}{2!c(c+1)} z^2 + \ldots \right]. \tag{12}$$

This is the regular solution and is denoted

$$_2F_1(a, b; c; z) = 1 + \frac{ab}{1!c} z + \frac{a(a+1)b(b+1)}{2!c(c+1)} z^2 + \ldots$$
$$= \sum_{n=0}^{\infty} \frac{(a)_n (b)_n}{(c)_n} \frac{z^n}{n!}, \tag{13}$$

where $(a)_n$ are POCHHAMMER SYMBOLS. The hypergeometric series is convergent for REAL $-1 < z < 1$, and for $z = \pm 1$ if $c > a + b$. The complete solution to the HYPERGEOMETRIC DIFFERENTIAL EQUATION is

$$y = A \,_2F_1(a, b; c; z)$$
$$+ Bz^{1-c} \,_2F_1(a+1-c, b+1-c; 2-c; z). \tag{14}$$

Derivatives are given by

$$\frac{d \,_2F_1(a, b; c; z)}{dz} = \frac{ab}{c} \,_2F_1(a+1, b+1; c+1; z) \tag{15}$$

$$\frac{d^2 \,_2F_1(a, b; c; z)}{dz^2}$$
$$= \frac{a(a+1)b(b+1)}{c(c+1)} \,_2F_1(a+2, b+2; c+2; z) \tag{16}$$

(Magnus and Oberhettinger 1949, p. 8).

An integral giving the hypergeometric function is

$$_2F_1(a, b; c; z)$$
$$= \frac{\Gamma(c)}{\Gamma(b)\Gamma(c-b)} \int_0^1 \frac{t^{b-1}(1-t)^{c-b-1}}{(1-tz)^a} dt \tag{17}$$

as shown by Euler in 1748 (Bailey 1935, pp. 4–).

Barnes (1908) gave the CONTOUR INTEGRAL

$$_2F_1(a,\ b;\ c;\ z)$$

$$= \frac{1}{2\pi i}\int_{-i\infty}^{i\infty}\frac{\Gamma(a+s)\Gamma(b+s)\Gamma(-s)}{\Gamma(c-s)}(-z)^s\,ds,$$

$$(18)$$

where $|\arg(-z)| < \pi$ and the path is curved (if necessary) to separate the poles $s = -a-n$, $s = -b-n$, ... ($n = 0$, 1, ...) from the poles $s = 0$, 1 ... (Bailey 1935, pp. 4–; Whittaker and Watson 1990).

A hypergeometric function can be written using EULER'S HYPERGEOMETRIC TRANSFORMATIONS

$$t \to t \tag{19}$$

$$t \to 1-t \tag{20}$$

$$t \to (1-z-tz)^{-1} \tag{21}$$

$$t \to \frac{1-t}{1-tz} \tag{22}$$

in any one of four equivalent forms

$$_2F_1(a,\ b;\ c;\ z) = (1-z)^{-a}\,_2F_1(a,\ c-b;\ c;\ z/(z-1))$$

$$\text{nbsp;(23r}_p\text{ar}$$

$$= (1-z)^{-b}\,_2F_1(c-a,b;\ c;\ z/(z-1))$$

$$\text{nbsp;(24r}_p\text{ar}$$

$$= (1-z)^{c-a-b}\,_2F_1(c-a,\ c-b;\ c;\ z)$$

$$\text{nbsp;(25r}_p\text{ar}$$

It can also be written as a linear combination

$$_2F_1(a,\ b;\ c;\ z)$$

$$= \frac{\Gamma(c)\Gamma(c-a-b)}{\Gamma(c-a)\Gamma(c-b)}\,_2F_1(a,\ b;\ a+b+1-c;\ 1-z)$$

$$+\frac{\Gamma(c)\Gamma(a+b-c)}{\Gamma(a)\Gamma(b)}$$

$$\times (1-z)^{c-a-b}\,_2F_1(c-a,\ c-b;\ 1+c-a-b;\ 1-z) \tag{26}$$

(Barnes 1908; Bailey 1935, pp. 3–; Whittaker and Watson 1990, p. 291).

Kummer found all six solutions (not necessarily regular at the origin) to the HYPERGEOMETRIC DIFFERENTIAL EQUATION,

$$u_1(x) = {}_2F_1(a,\ b;\ c;\ z) \tag{27}$$

$$u_2(x) = {}_2F_1(a,\ b;\ a+b+1-c;\ 1-z) \tag{28}$$

$$u_3(x) = z^{-a}\,_2F_1(a,\ a+1-c;\ a+1-b;\ z^{-1}) \tag{29}$$

$$u_4(x) = z^{-b}\,_2F_1(b+1-c,\ b;\ b+1-a;\ z^{-1}) \tag{30}$$

$$u_5(x) = z^{1-c}\,_2F_1(b+1-c,\ a+1-c;\ 2-c;\ z) \tag{31}$$

$$u_6(x) = (1-z)^{c-a-b}\,_2F_1(c-a,\ c-b;\ c+1-a-b;\ 1-z) \tag{32}$$

(Abramowitz and Stegun 1972, p. 563).

Applying EULER'S HYPERGEOMETRIC TRANSFORMATIONS to the Kummer solutions then gives all 24 possible forms which are solutions to the HYPERGEOMETRIC DIFFERENTIAL EQUATION

$$u_1^{(1)}(x) = {}_2F_1(a,\ b;\ c;\ z) \tag{33}$$

$$u_1^{(2)}(x) = (1-z)^{c-a-b}\,_2F_1(c-a,\ c-b;\ c;\ z) \tag{34}$$

$$u_1^{(3)}(x) = (1-z)^{-a}\,_2F_1(a,\ c-b;\ c;\ z/(z-1)) \tag{35}$$

$$u_1^{(4)}(x) = (1-z)^{-b}\,_2F_1(c-a,\ b;\ c;\ z/(z-1)) \tag{36}$$

$$u_2^{(1)}(x) = {}_2F_1(a,\ b;\ a+b+1-c;\ 1-z) \tag{37}$$

$$u_2^{(2)}(x) = z^{1-c}\,_2F_1(a+1-c,\ b+1-c;\ a+b+1-c;\ 1-z) \tag{38}$$

$$u_2^{(3)}(x) = z^{-a}\,_2F_1(a,\ a+1-c;\ a+b+1-c;\ 1-z^{-1}) \tag{39}$$

$$u_2^{(4)}(x) = z^{-b}\,_2F_1(b+1-c,\ b;\ a+b+1-c;\ 1-z^{-1}) \tag{40}$$

$$u_3^{(1)}(x) = (-z)^{-a}\,_2F_1(a,\ a+1-c;\ a+1-b;\ z^{-1}) \tag{41}$$

$$u_3^{(2)}(x) = (-z)^{b-c}$$
$$\times (1-z)^{c-a-b}\,_2F_1(1-b,\ c-b;\ a+1-b;\ z^{-1}) \tag{42}$$

$$u_3^{(3)}(x) = (1-z)^{-a}\,_2F_1(a,\ c-b;\ a+1-b;(1-z)^{-1}) \tag{43}$$

$$u_3^{(4)}(x) = (-z)^{1-c}$$
$$\times (1-z)^{c-a-1}\,_2F_1(a+1-c,\ 1-b;\ a+1-b;(1-z)^{-1}) \tag{44}$$

$$u_4^{(1)}(x) = (-z)^{-b}\,_2F_1(b+1-c,\ b;\ b+1-a;\ z^{-1}) \tag{45}$$

$$u_4^{(2)} = (-z)^{a-c}$$
$$\times (1-z)^{c-a-b}\,_2F_1(1-a,\ c-a;\ b+1-a;\ z^{-1}) \tag{46}$$

$$u_4^{(3)}(x) = (1-z)^{-b}\,_2F_1(b,\ c-a;\ b+1-a;(1-z)^{-1}) \tag{47}$$

$$u_4^{(4)}(x) = (-z)^{1-c}$$
$$\times (1-z)^{c-b-1}\,_2F_1(b+1-c,\ 1-a;\ b+1-a;(1-z)^{-1}) \tag{48}$$

$$u_5^{(1)}(x) = z^{1-c}\,_2F_1(a+1-c,\ b+1-c;\ 2-c;\ z) \quad (49)$$

$$u_5^{(2)} = z^{1-c}(1-z)^{c-a-b}\,_2F_1(1-a,\ 1-b;\ 2-c;\ z) \quad (50)$$

$$u_5^{(3)}(x) = z^{1-c}(1-z)^{c-a-1}\,_2F_1(a+1-c,\ 1-b;\ 2-c;\ z/(z-1)) \quad (51)$$

$$u_5^{(4)}(x) = z^{1-c}(1-z)^{c-b-1}\,_2F_1(b+1-c,\ 1-a;\ 2-c;\ z/(z-1)) \quad (52)$$

$$u_6^{(1)}(x) = (1-z)^{c-a-b}\,_2F_1(c-a,\ c-b;\ c+1-a-b;\ 1-z) \quad (53)$$

$$u_6^{(2)}(x) = z^{1-c}(1-z)^{c-a-b}\,_2F_1(1-a,\ 1-b;\ c+1-a-b;\ 1-z) \quad (54)$$

$$u_6^{(3)}(x) = z^{a-c}(1-z)^{c-a-b}\,_2F_1(c-a,\ 1-a;\ c+1-a-b;\ 1-z^{-1}) \quad (55)$$

$$u_6^{(4)}(x) = z^{b-c}(1-z)^{c-a-b}\,_2F_1(c-b,\ 1-b;\ c+1-a-b;\ 1-z^{-1}) \quad (56)$$

(Kummer 1836; Erdélyi *et al.* 1981, pp. 105–06).

Goursat (1881) and Erdélyi *et al.* (1981) give many hypergeometric transformation formulas, including several cubic transformations.

Many functions of mathematical physics can be expressed as special cases of the hypergeometric functions. For example,

$$_2F_1(-l,\ l+1,\ 1;\ (1-z)/2) = P_l(z), \quad (57)$$

where $P_l(z)$ is a LEGENDRE POLYNOMIAL.

$$(1+z)^n = \,_2F_1(-n,\ b;\ b;\ -z) \quad (58)$$

$$\ln(1+z) = z\,_2F_1(1,\ 1;\ 2;\ -z) \quad (59)$$

Complete ELLIPTIC INTEGRALS and the RIEMANN *P*-SERIES can also be expressed in terms of $_2F_1(a,\ b;\ c;\ z)$. Special values include

$$_2F_1(a,\ b;\ a-b+1;\ -1)$$
$$= 2^{-a}\sqrt{\pi}\,\frac{\Gamma(1+a+b)}{\Gamma\left(1+\frac{1}{2}a-b\right)\Gamma\left(\frac{1}{2}+\frac{1}{2}a\right)} \quad (60)$$

$$_2F_1(1,\ -a;\ a;\ -1) = \frac{\sqrt{\pi}}{2}\,\frac{\Gamma(a)}{\Gamma\left(a+\frac{1}{2}\right)} + 1 \quad (61)$$

$$_2F_1\left(a,\ b;\ c;\ \tfrac{1}{2}\right) = 2\,_2^aF_1(a,\ c-b;\ c;\ -1) \quad (62)$$

$$_2F_1\left(a,\ b;\ \tfrac{1}{2}(a+b+1);\ \tfrac{1}{2}\right)$$
$$= \frac{\Gamma\left(\frac{1}{2}\right)\Gamma\left[\frac{1}{2}(1+a+b)\right]}{\Gamma\left[\frac{1}{2}(1+a)\right]\Gamma\left[\frac{1}{2}(1+b)\right]} \quad (63)$$

$$_2F_1\left(a,\ 1-a;\ c;\ \tfrac{1}{2}\right) = \frac{\Gamma\left(\frac{1}{2}c\right)\Gamma\left[\frac{1}{2}(c+1)\right]}{\Gamma\left[\frac{1}{2}(a+c)\right]\Gamma\left[\frac{1}{2}(1+c-a)\right]} \quad (64)$$

$$_2F_1(a,\ b;\ c;\ 1) = \frac{\Gamma(c)\Gamma(c-a-b)}{\Gamma(c-a)\Gamma(c-b)}. \quad (65)$$

KUMMER'S FIRST FORMULA gives

$$_2F_1\left(\tfrac{1}{2}+m-k,\ -n;\ 2m+1;\ 1\right)$$
$$= \frac{\Gamma(2m+1)\Gamma\left(m+\frac{1}{2}+k+n\right)}{\Gamma\left(m+\frac{1}{2}+k\right)\Gamma(2m+1+n)}, \quad (66)$$

where $m \neq -1/2,\ -1,\ -3/2,\ \dots$. Many additional identities are given by Abramowitz and Stegun (1972, p. 557).

Hypergeometric functions can be generalized to GENERALIZED HYPERGEOMETRIC FUNCTIONS

$$_nF_m(a_1,\ \dots,\ a_n;\ b_1,\ \dots,\ b_m;\ z). \quad (67)$$

A function OF THE FORM $_1F_1(a;\ b;\ z)$ is called a CONFLUENT HYPERGEOMETRIC FUNCTION OF THE FIRST KIND, and a function OF THE FORM $_0F_1(a;\ b;\ z)$ is called a CONFLUENT HYPERGEOMETRIC LIMIT FUNCTION.

See also APPELL HYPERGEOMETRIC FUNCTION, BARNES' LEMMA, BRADLEY'S THEOREM, CAYLEY'S HYPERGEOMETRIC FUNCTION THEOREM, CLAUSEN FORMULA, CLOSED FORM, CONFLUENT HYPERGEOMETRIC FUNCTION OF THE FIRST KIND, CONFLUENT HYPERGEOMETRIC FUNCTION OF THE SECOND KIND, CONFLUENT HYPERGEOMETRIC LIMIT FUNCTION, CONTIGUOUS FUNCTION, DARLING'S PRODUCTS, GENERALIZED HYPERGEOMETRIC FUNCTION, GOSPER'S ALGORITHM, HYPERGEOMETRIC IDENTITY, HYPERGEOMETRIC SERIES, JACOBI POLYNOMIAL, KUMMER'S FORMULAS, KUMMER'S QUADRATIC TRANSFORMATION, KUMMER'S RELATION, ORR'S THEOREM, PFAFF TRANSFORMATION, *Q*-HYPERGEOMETRIC FUNCTION, RAMANUJAN'S HYPERGEOMETRIC IDENTITY, SAALSCHÜTZIAN, SISTER CELINE'S METHOD, ZEILBERGER'S ALGORITHM

References

Abramowitz, M. and Stegun, C. A. (Eds.). "Hypergeometric Functions." Ch. 15 in *Handbook of Mathematical Functions with Formulas, Graphs, and Mathematical Tables, 9th printing.* New York: Dover, pp. 555–66, 1972.

Appell, P. and Kampé de Fériet, J. *Fonctions hypergéométriques et hypersphériques: polynomes d'Hermite.* Paris: Gauthier-Villars, 1926.

Arfken, G. "Hypergeometric Functions." §13.5 in *Mathematical Methods for Physicists, 3rd ed.* Orlando, FL: Academic Press, pp. 748–52, 1985.

Bailey, W. N. *Generalised Hypergeometric Series.* Cambridge, England: University Press, 1935.

Barnes, E. W. "A New Development in the Theory of the Hypergeometric Functions." *Proc. London Math. Soc.* **6**, 141–77, 1908.

Emmanuel, J. "Eacute;valuation rapide de fonctions hyper-géométriques." Report RT-0242. INRIA, Jul 2000. http://www.inria.fr.RRRT/RT-0242.html.

Erdélyi, A.; Magnus, W.; Oberhettinger, F.; and Tricomi, F. G. *Higher Transcendental Functions, Vol. 1.* New York: Krieger, 1981.

Exton, H. *Handbook of Hypergeometric Integrals: Theory, Applications, Tables, Computer Programs.* Chichester, England: Ellis Horwood, 1978.

Fine, N. J. *Basic Hypergeometric Series and Applications.* Providence, RI: Amer. Math. Soc., 1988.

Gasper, G. and Rahman, M. *Basic Hypergeometric Series.* Cambridge, England: Cambridge University Press, 1990.

Gauss, C. F. "Disquisitiones Generales Circa Seriem Infinitam $\left[\frac{\alpha\beta}{1\cdot\gamma}\right]x+\left[\frac{\alpha(\alpha+1)\beta(\beta+1)}{1\cdot 2\cdot\gamma(\gamma+1)}\right]x^2+\left[\frac{\alpha(\alpha+1)(\alpha+2)\beta(\beta+1)(\beta+2)}{1\cdot 2\cdot 3\cdot\gamma(\gamma+1)(\gamma+2)}\right]x^3+$ etc. Pars Prior." *Commentationes Societiones Regiae Scientiarum Gottingensis Recentiores, Vol. II.* 1812. Reprinted in *Gesammelte Werke, Bd. 3,* pp. 123–63 and 207–29, 1866.

Gessel, I. and Stanton, D. "Strange Evaluations of Hypergeometric Series." *SIAM J. Math. Anal.* **13,** 295–08, 1982.

Gosper, R. W. "Decision Procedures for Indefinite Hypergeometric Summation." *Proc. Nat. Acad. Sci. USA* **75,** 40–2, 1978.

Goursat, M. E. "Sur l'équation différentielle linéaire qui admet pour intégrale la série hypergéométrique." *Ann. Sci. École Norm. Super. Sup.* **10,** S3-S142, 1881.

Graham, R. L.; Knuth, D. E.; and Patashnik, O. *Concrete Mathematics: A Foundation for Computer Science, 2nd ed.* Reading, MA: Addison-Wesley, 1994.

Hardy, G. H. "A Chapter from Ramanujan's Note-Book." *Proc. Cambridge Philos. Soc.* **21,** 492–03, 1923.

Hardy, G. H. "Hypergeometric Series." Ch. 7 in *Ramanujan: Twelve Lectures on Subjects Suggested by His Life and Work, 3rd ed.* New York: Chelsea, pp. 101–12, 1999.

Iyanaga, S. and Kawada, Y. (Eds.). "Hypergeometric Functions and Spherical Functions." Appendix A, Table 18 in *Encyclopedic Dictionary of Mathematics.* Cambridge, MA: MIT Press, pp. 1460–468, 1980.

Kampé de Fériet, J. *La fonction hypergéométrique.* Paris: Gauthier-Villars, 1937.

Kohno, M. *Global Analysis in Linear Differential Equations.* Dordrecht, Netherlands: Kluwer, 1999.

Krattenthaler, C. "HYP and HYPQ." *J. Symb. Comput.* **20,** 737–44, 1995.

Kummer, E. E. "Über die Hypergeometrische Reihe." *J. reine angew. Math.* **15,** 39–3 and 127–72, 1836.

Magnus, W. and Oberhettinger, F. *Formulas and Theorems for the Special Functions of Mathematical Physics.* New York: Chelsea, 1949.

Morse, P. M. and Feshbach, H. *Methods of Theoretical Physics, Part I.* New York: McGraw-Hill, pp. 541–47, 1953.

Petkovsek, M.; Wilf, H. S.; and Zeilberger, D. *A = B.* Wellesley, MA: A. K. Peters, 1996.

Press, W. H.; Flannery, B. P.; Teukolsky, S. A.; and Vetterling, W. T. "Hypergeometric Functions." §6.12 in *Numerical Recipes in FORTRAN: The Art of Scientific Computing, 2nd ed.* Cambridge, England: Cambridge University Press, pp. 263–65, 1992.

Seaborn, J. B. *Hypergeometric Functions and Their Applications.* New York: Springer-Verlag, 1991.

Snow, C. *Hypergeometric and Legendre Functions with Applications to Integral Equations of Potential Theory.* Washington, DC: U. S. Government Printing Office, 1952.

Spanier, J. and Oldham, K. B. "The Gauss Function $F(a, b; c; x)$." Ch. 60 in *An Atlas of Functions.* Washington, DC: Hemisphere, pp. 599–07, 1987.

Thomae. *J. reine angew. Math.* **87,** 222–49, 1879.

Watson, G. N. "Ramanujan's Note Books." *J. London Math. Soc.* **6,** 137–53, 1931.

Weisstein, E. W. "Books about Hypergeometric Functions." http://www.treasure-troves.com/books/Hypergeometric-Functions.html.

Whittaker, E. T. and Watson, G. N. *A Course in Modern Analysis, 4th ed.* Cambridge, England: Cambridge University Press, 1990.

Hypergeometric Identity

A relation expressing a sum potentially involving BINOMIAL COEFFICIENTS, FACTORIALS, RATIONAL FUNCTIONS, and power functions in terms of a simple result. Thanks to results by Fasenmyer, Gosper, Zeilberger, Wilf, and Petkovsek, the problem of determining whether a given hypergeometric sum is expressible in simple closed form and, if so, finding the form, is now (subject to a mild restriction) completely solved. The algorithm which does so has been implemented in several computer algebra packages and is called ZEILBERGER'S ALGORITHM.

See also BINOMIAL SUMS, GENERALIZED HYPERGEOMETRIC FUNCTION, GOSPER'S ALGORITHM, HYPERGEOMETRIC SERIES, SISTER CELINE'S METHOD, WILFZEILBERGER PAIR, ZEILBERGER'S ALGORITHM

References

Koepf, W. "Hypergeometric Identities." Ch. 2 in *Hypergeometric Summation: An Algorithmic Approach to Summation and Special Function Identities.* Braunschweig, Germany: Vieweg, pp. 11–0, 1998.

Petkovsek, M.; Wilf, H. S.; and Zeilberger, D. *A = B.* Wellesley, MA: A. K. Peters, p. 18, 1996.

Hypergeometric Polynomial

JACOBI POLYNOMIAL

Hypergeometric Series

A hypergeometric series $\Sigma_k c_k$ is a series for which $c_0 = 1$ and the ratio of consecutive terms is a RATIONAL FUNCTION of the summation index k, i.e., one for which

$$\frac{c_{k+1}}{c_k} = \frac{P(k)}{Q(k)}, \qquad (1)$$

with $P(k)$ and $Q(k)$ POLYNOMIALS. In this case, c_k is called a HYPERGEOMETRIC TERM (Koepf 1998, p. 12). The functions generated by hypergeometric series are called HYPERGEOMETRIC FUNCTIONS or, more generally, GENERALIZED HYPERGEOMETRIC FUNCTIONS. If the polynomials are completely factored, the ratio of successive terms can be written

$$\frac{c_{k+1}}{c_k} = \frac{P(k)}{Q(k)}$$
$$= \frac{(k + a_1)(k + a_2)\cdots(k + a_p)}{(k + b_1)(k + b_2)\cdots(k + b_q)(k + 1)}x, \qquad (2)$$

where the factor of $k + 1$ in the DENOMINATOR is present for historical reasons of notation, and the resulting GENERALIZED HYPERGEOMETRIC FUNCTION is written

$$_pF_q\begin{bmatrix} a_1 & a_2 & \cdots & a_p \\ b_1 & b_2 & \cdots & b_q \end{bmatrix}; x = \sum_{k=0} c_k x^k. \qquad (3)$$

If $p = 2$ and $q = 1$, the function becomes a traditional HYPERGEOMETRIC FUNCTION $_2F_1(a, b; c; x)$.

Many sums can be written as GENERALIZED HYPERGEOMETRIC FUNCTIONS by inspections of the ratios of consecutive terms in the generating hypergeometric series.

See also BINOMIAL SUMS, GENERALIZED HYPERGEOMETRIC FUNCTION, GEOMETRIC SERIES, HYPERGEOMETRIC FUNCTION, HYPERGEOMETRIC IDENTITY, HYPERGEOMETRIC TERM

References

Koepf, W. *Hypergeometric Summation: An Algorithmic Approach to Summation and Special Function Identities.* Braunschweig, Germany: Vieweg, 1998.

Petkovsek, M.; Wilf, H. S.; and Zeilberger, D. "Hypergeometric Series," "How to Identify a Series as Hypergeometric," and "Software That Identifies Hypergeometric Series." §3.2–.4 in *A = B.* Wellesley, MA: A. K. Peters, pp. 34–2, 1996.

Hypergeometric Summation

The analytic summation of a HYPERGEOMETRIC SERIES. Powerful general techniques of hypergeometric summation include GOSPER'S ALGORITHM, SISTER CELINE'S METHOD, WILF-ZEILBERGER PAIRS, and ZEILBERGER'S ALGORITHM.

See also BINOMIAL SUMS, GOSPER'S ALGORITHM, SISTER CELINE'S METHOD, WILF-ZEILBERGER PAIR, ZEILBERGER'S ALGORITHM

References

Koepf, W. "Algorithms for m-fold Hypergeometric Summation." *J. Symb. Comput.* **20**, 399–17, 1995.

Hypergeometric Term

Given a HYPERGEOMETRIC SERIES $\Sigma_k c_k$, c_k is called a hypergeometric term (Koepf 1998, p. 12).

See also HYPERGEOMETRIC SERIES

References

Koepf, W. *Hypergeometric Summation: An Algorithmic Approach to Summation and Special Function Identities.* Braunschweig, Germany: Vieweg, 1998.

Hypergeometric0F1

CONFLUENT HYPERGEOMETRIC LIMIT FUNCTION

Hypergeometric0F1Regularized

CONFLUENT HYPERGEOMETRIC LIMIT FUNCTION

Hypergeometric1F1

CONFLUENT HYPERGEOMETRIC FUNCTION OF THE FIRST KIND

Hypergeometric2F1

HYPERGEOMETRIC FUNCTION

HypergeometricU

CONFLUENT HYPERGEOMETRIC FUNCTION OF THE SECOND KIND

Hypergraph

A hypergraph is a GRAPH in which generalized edges (called HYPEREDGES) may connect more than two nodes.

See also GRAPH, HYPEREDGE, MULTIGRAPH, PSEUDOGRAPH

References

Berge, C. *Graphs and Hypergraphs.* New York: Elsevier, 1973.

Berge, C. *Hypergraphs: The Theory of Finite Sets.* Amsterdam, Netherlands: North-Holland, 1989.

Hypergroup

A MEASURE ALGEBRA which has many properties associated with the convolution MEASURE ALGEBRA of a GROUP, but no algebraic structure is assumed for the underlying SPACE.

References

Bloom, W. R.; and Heyer, H. *The Harmonic Analysis of Probability Measures on Hypergroups.* Berlin: de Gruyter, 1995.

Jewett, R. I. "Spaces with an Abstract Convolution of Measures." *Adv. Math.* **18**, 1–01, 1975.

Hyper-Kähler Manifold

See also KÄHLER MANIFOLD

Hypermatrix

A generalization of the MATRIX to an $n_1 \times n_2 \times \cdots$ array of numbers.

See also HYPERDETERMINANT

References

Gel'fand, I. M.; Kapranov, M. M.; and Zelevinsky, A. V. "Hyperdeterminants." *Adv. Math.* **96**, 226–63, 1992.

Hyperparallel

Two lines in HYPERBOLIC GEOMETRY which diverge from each other in both directions.

See also ANTIPARALLEL, IDEAL POINT, PARALLEL

Hyperperfect Number

A number n is called k-hyperperfect if

$$n = 1 + k \sum_i d_i = 1 + k[\sigma(n) - n - 1],$$

where $\sigma(n)$ is the DIVISOR FUNCTION and the summation is over the PROPER DIVISORS with $1 < d_i < n$. Rearranging gives

$$k\sigma(n) = (k+1)n + k - 1.$$

Taking $k = 1$ gives the usual PERFECT NUMBERS.

If $k > 1$ is an odd integer, and $p = (3k + 1)/2$ and $q = 3k + 4 = 2p + 3$ are prime, then $p^2 q$ is k-hyperperfect. McCranie (2000) conjectures that all k-hyperperfect numbers for odd $k > 1$ are in fact of this form. Similarly, if p and q are distinct odd primes such that $k(p + q) = pq - 1$ for some integer k, then $n = pq$ is k-hyperperfect. Finally, if $k > 0$ and $p = k + 1$ is prime, then if $q = p^i - p + 1$ is prime for some $i > 1 <$ then $n = p^{i-1}q$ is k-hyperperfect (McCranie 2000).

The first few hyperperfect numbers (excluding PERFECT NUMBERS) are 21, 301, 325, 697, 1333, ... (Sloane's A007592). If PERFECT NUMBERS are included, the first few are 6, 21, 28, 301, 325, 496, ... (Sloane's A034897), whose corresponding values of k are 1, 2, 1, 6, 3, 1, 12, ... (Sloane's A034898). The following table gives the first few k-hyperperfect numbers for small values of k. McCranie (2000) has tabulated all hyperperfect numbers less than 10^{11}.

k	Sloane	k-hyperperfect number
1	A000396	6 ,28, 496, 8128, ...
2	A007593	21, 2133, 19521, 176661, ...
3		325, ...
4		1950625, 1220640625, ...
6	A028499	301, 16513, 60110701, ...
10		159841, ...
11		10693, ...
12	A028500	697, 2041, 1570153, 62722153, ...

See also PERFECT NUMBER

References

Guy, R. K. "Almost Perfect, Quasi-Perfect, Pseudoperfect, Harmonic, Weird, Multiperfect and Hyperperfect Numbers." §B2 in *Unsolved Problems in Number Theory, 2nd ed.* New York: Springer-Verlag, pp. 45-3, 1994.
McCranie, J. S.. "A Study of Hyperperfect Numbers." *J. Integer Sequences* **3**, No. 00.1.3, 2000. http://www.re-search.att.com/~njas/sequences/JIS/VOL3/mccranie.html.
Minoli, D. "Issues in Nonlinear Hyperperfect Numbers." *Math. Comput.* **34**, 639-45, 1980.
Roberts, J. *The Lure of the Integers.* Washington, DC: Math. Assoc. Amer., p. 177, 1992.
Sloane, N. J. A. Sequences A000396/M4186, A007592/M5113, A007593/M5121, A028499, A028500, A034897, and A034898 in "An On-Line Version of the Encyclopedia of Integer Sequences." http://www.research.att.com/~njas/sequences/eisonline.html.
te Riele, H. J. J. "Hyperperfect Numbers with Three Different Prime Factors." *Math. Comput.* **36**, 297-98, 1981.

Hyperplane

Let $a_1, a_2, ..., a_n$ be SCALARS not all equal to 0. Then the SET S consisting of all VECTORS

$$\mathbf{X} = \begin{bmatrix} x_1 \\ x_2 \\ \vdots \\ x_n \end{bmatrix}$$

in $\mathbb{R}^n$ such that

$$a_1 x_1 + a_2 x_2 + \ldots + a_n x_n = 0$$

is a SUBSPACE of $\mathbb{R}^n$ called a hyperplane.

More generally, a hyperplane is any CODIMENSION-1 vector SUBSPACE of a VECTOR SPACE. Equivalently, a hyperplane V in a VECTOR SPACE W is any SUBSPACE such that W/V is 1-dimensional. Equivalently, a hyperplane is the KERNEL of any NONZERO linear MAP from the VECTOR SPACE to the underlying FIELD.

Hyperreal Number

Hyperreal numbers are an extension of the REAL NUMBERS to include certain classes of infinite and infinitesimal numbers. A hyperreal number x is said to be finite IFF $|x| < n$ for some INTEGER n. x is said to be infinitesimal IFF $|x| < 1/n$ for all INTEGERS n.

See also AX-KOCHEN ISOMORPHISM THEOREM, NON-STANDARD ANALYSIS

References

Keisler, H. J. "The Hyperreal Line." In *Real Numbers, Generalizations of the Reals, and Theories of Continua* (Ed. P. Ehrlich). Norwell, MA: Kluwer, 1994.

Hyperspace

A SPACE having DIMENSION $n > 3$.

Hypersphere

The n-hypersphere (often simply called the n-sphere) is a generalization of the CIRCLE ($n = 2$) and SPHERE ($n = 3$) to dimensions $n \geq 4$. It is therefore defined as the set of n-tuples of points $(x_1, x_2, ..., x_n)$ such that

$$x_1^2 + x_2^2 + \ldots + x_n^2 = R^2, \tag{1}$$

where R is the RADIUS of the hypersphere. The CONTENT V_n (i.e., n-D VOLUME) of an n-hypersphere of RADIUS R is given by

$$V_n = \int_0^R S_n r^{n-1} \, dr = \frac{S_n R^n}{n}, \qquad (2)$$

where S_n is the hyper-SURFACE AREA of an n-sphere of unit radius. But, for a unit hypersphere, it must be true that

$$S_n \int_0^\infty e^{-r^2} r^{n-1} \, dr$$

$$= \underbrace{\int_{-\infty}^\infty \cdots \int_{-\infty}^\infty}_{n} e^{-(x_1^2 + \cdots + x_n^2)} \, dx_1 \cdots dx_m$$

$$= \left(\int_{-\infty}^\infty e^{-x^2} \, dx \right)^n. \qquad (3)$$

But the GAMMA FUNCTION can be defined by

$$\Gamma(m) = 2 \int_0^\infty e^{-r^2} r^{2m-1} \, dr, \qquad (4)$$

so

$$\tfrac{1}{2} S_n \Gamma\left(\tfrac{1}{2} n\right) = \left[\Gamma\left(\tfrac{1}{2}\right)\right]^n = (\pi^{1/2})^n \qquad (5)$$

$$S_n = \frac{2\pi^{n/2}}{\Gamma\left(\tfrac{1}{2} n\right)}. \qquad (6)$$

Special forms of $\Gamma\left(\tfrac{1}{2} n\right)$ for n an integer allow the above expression to be written as

$$S_n = \begin{cases} \dfrac{2^{(n+1)/2} \pi^{(n-1)/2}}{(n-2)!!} & \text{for } n \text{ odd} \\[2ex] \dfrac{2\pi^{n/2}}{\left(\tfrac{1}{2} n - 1\right)!} & \text{for } n \text{ even,} \end{cases} \qquad (7)$$

where $n!$ is a FACTORIAL and $n!!$ is a DOUBLE FACTORIAL.

Equation (6) gives the RECURRENCE RELATION

$$S_{n+2} = \frac{2\pi S_n}{n}. \qquad (8)$$

Using $\Gamma(n+1) = n\Gamma(n)$ then gives

$$V_n = \frac{S_n R^n}{n} = \frac{\pi^{n/2} R^n}{\left(\tfrac{1}{2} n\right)\Gamma\left(\tfrac{1}{2} n\right)} = \frac{\pi^{n/2} R^n}{\Gamma\left(1 + \tfrac{1}{2} n\right)} \qquad (9)$$

(Sommerville 1958, p. 136; Conway and Sloane 1993).

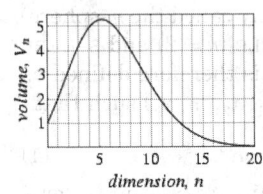
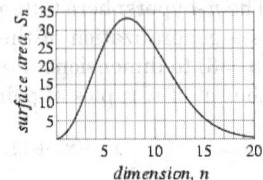

Strangely enough, the hyper-SURFACE AREA and CONTENT reach MAXIMA and then decrease towards

0 as n increases. The point of MAXIMAL hyper-SURFACE AREA satisfies

$$\frac{dS_n}{dn} = \frac{\pi^{n/2}\left[\ln \pi - \psi_0\left(\tfrac{1}{2} n\right)\right]}{\Gamma\left(\tfrac{1}{2} n\right)} = 0, \qquad (10)$$

where $\psi_0(x) \equiv \Psi(x)$ is the DIGAMMA FUNCTION. The point of MAXIMAL CONTENT satisfies

$$\frac{dV_n}{dn} = \frac{\pi^{n/2}\left[\ln \pi - \psi_0\left(1 + \tfrac{1}{2} n\right)\right]}{2\Gamma\left(1 + \tfrac{1}{2} n\right)} = 0. \qquad (11)$$

Neither can be solved analytically for n, but the numerical solutions are $n = 7.25695\ldots$ for hyper-SURFACE AREA and $n = 5.25695\ldots$ for CONTENT (Wells 1986, p. 67). As a result, the 7-D and 5-D hyperspheres have MAXIMAL hyper-SURFACE AREA and CONTENT, respectively (Le Lionnais 1983; Wells 1986, p. 60).

n	V_n	$V_{\text{sphere}}/V_{\text{cube}}$	S_n
0	1	1	0
1	2	1	2
2	π	$\tfrac{1}{4}\pi$	2π
3	$\tfrac{4}{3}\pi$	$\tfrac{1}{6}\pi$	4π
4	$\tfrac{1}{2}\pi^2$	$\tfrac{1}{32}\pi^2$	$2\pi^2$
5	$\tfrac{8}{15}\pi^2$	$\tfrac{1}{60}\pi^2$	$\tfrac{8}{3}\pi^2$
6	$\tfrac{1}{6}\pi^3$	$\tfrac{1}{384}\pi^3$	π^3
7	$\tfrac{16}{105}\pi^3$	$\tfrac{1}{840}\pi^3$	$\tfrac{16}{15}\pi^3$
8	$\tfrac{1}{24}\pi^4$	$\tfrac{1}{6144}\pi^4$	$\tfrac{1}{3}\pi^4$
9	$\tfrac{32}{945}\pi^4$	$\tfrac{1}{15120}\pi^4$	$\tfrac{32}{105}\pi^4$
10	$\tfrac{1}{120}\pi^5$	$\tfrac{1}{122880}\pi^5$	$\tfrac{1}{12}\pi^5$

In 4-D, the generalization of SPHERICAL COORDINATES is defined by

$$x_1 = R \sin\psi \sin\phi \cos\theta \qquad (12)$$

$$x_2 = R \sin\psi \sin\phi \sin\theta \qquad (13)$$

$$x_3 = R \sin\psi \cos\phi \qquad (14)$$

$$x_4 = R \cos\psi. \qquad (15)$$

The equation for a 4-sphere is

$$x_1^2 + x_2^2 + x_3^2 + x_4^2 = R^2, \qquad (16)$$

and the LINE ELEMENT is

$$ds^2 = R^2[d\psi^2 + \sin^2\psi(d\phi^2 + \sin^2\phi \, d\theta^2)]. \qquad (17)$$

By defining $r \equiv R \sin\psi$, the LINE ELEMENT can be rewritten

$$ds^2 = \frac{dr^2}{\left(1 - \frac{r^2}{R^2}\right)} + r^2(d\phi^2 + \sin^2 \phi \, d\theta^2). \qquad (18)$$

The hyper-SURFACE AREA is therefore given by

$$S_4 = \int_0^\pi R \, d\psi \int_0^\pi R \sin \psi \, d\phi \int_0^{2\pi} R \sin \psi \sin \phi \, d\phi$$

$$= 2\pi^2 R^3. \qquad (19)$$

See also CIRCLE, GLOME, HYPERCUBE, HYPERSPHERE PACKING, HYPERSPHERE POINT PICKING, MAZUR'S THEOREM, PEG, SPHERE, TESSERACT

References

Sommerville, D. M. Y. *An Introduction to the Geometry of n Dimensions.* New York: Dover, p. 136, 1958.
Conway, J. H. and Sloane, N. J. A. *Sphere Packings, Lattices, and Groups, 2nd ed.* New York: Springer-Verlag, p. 9, 1993.
Le Lionnais, F. *Les nombres remarquables.* Paris: Hermann, p. 58, 1983.
Peterson, I. *The Mathematical Tourist: Snapshots of Modern Mathematics.* New York: W. H. Freeman, pp. 96–01, 1988.

Hypersphere Packing

The analog of face-centered cubic packing is the densest lattice packing in 4- and 5-D. In 8-D, the densest lattice packing is made up of two copies of face-centered cubic. In 6- and 7-D, the densest lattice packings are CROSS SECTIONS of the 8-D case. In 24-D, the densest packing appears to be the LEECH LATTICE. For high dimensions (~ 1000-D), the densest known packings are nonlattice. The densest lattice packings in n-D have been rigorously proved to have PACKING DENSITY 1, $\pi/(2\sqrt{3})$, $\pi/(3\sqrt{2})$, $\pi^2/16$, $\pi^2/(15\sqrt{2})$, $\pi^3/(48\sqrt{3})$, $\pi^3/105$, and $\pi^4/384$ (Hilbert and Cohn-Vossen 1999, p. 47; Finch).

The densest known *non-lattice* packings of hyperspheres in dimensions up to 10 are given by Conway and Sloane (1995). However, there are no proofs that any packing in dimensions greater than 3 is optimal (Sloane 1998).

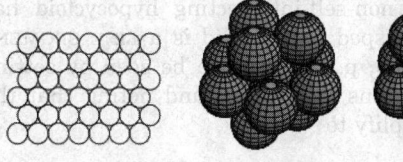

hexagonal close packing of circles   *hexagonal close packing of spheres*   *cubic close packing of spheres*

The largest number of UNIT CIRCLES which can touch a given UNIT CIRCLE is six. For SPHERES, the maximum number is 12. Newton considered this question long before a proof was published in 1874. The maximum number of hyperspheres that can touch another in n-D is the so-called KISSING NUMBER.

The following example illustrates the sometimes counterintuitive properties of hypersphere packings. Draw unit n-spheres in an n-D space centered at all ± 1 coordinates. Now place an additional HYPERSPHERE at the origin tangent to the other HYPERSPHERES. For values of n between 2 and 8, the central HYPERSPHERE is contained inside the HYPERCUBE with VERTICES at the centers of the other spheres. However, for $n = 9$, the central HYPERSPHERE just touches the HYPERCUBE of centers, and for $n > 9$, the central HYPERSPHERE is partially *outside* the HYPERCUBE.

This fact can be demonstrated by finding the distance from the origin to the center of one of the n HYPERSPHERES, which is given by

$$\underbrace{\sqrt{(\pm 1)^2 + \ldots + (\pm 1)^2}}_{n} = \sqrt{n}.$$

The radius of the central sphere is therefore $\sqrt{n} - 1$. Now, the distance from the origin to the center of a FACET bounding the HYPERCUBE is always 2 (two hypersphere radii), so the center HYPERSPHERE is tangent to the hypercube when $\sqrt{n} - 1 = 2$, or $n = 9$, and partially outside it for $n > 9$.

See also CIRCLE PACKING, ELLIPSOID PACKING, KEPLER CONJECTURE, KISSING NUMBER, LEECH LATTICE, PEG, SPHERE PACKING

References

Finch, S. "Favorite Mathematical Constants." http://www.mathsoft.com/asolve/constant/hermit/hermit.html.
Conway, J. H. and Sloane, N. J. A. *Disc. Comput. Geom.* **13**, 383–03, 1995.
Gardner, M. *Martin Gardner's New Mathematical Diversions from Scientific American.* New York: Simon and Schuster, pp. 89–0, 1966.
Hilbert, D. and Cohn-Vossen, S. *Geometry and the Imagination.* New York: Chelsea, p. 47, 1999.
Schnell, U. and Wills, J. M. "Densest Packings of More than Three *d*-Spheres are Nonplanar." *Disc. Comput. Geom.* **24**, 539–49, 2000.
Sloane, N. J. A. "Kepler's Conjecture Confirmed." *Nature* **395**, 435–36, 1998.

Hypersphere Point Picking

Marsaglia (1972) has given a simple method for selecting points with a uniform distribution on the surface of a 4-sphere. This is accomplished by picking two pairs of points (x_1, x_2) and (x_3, x_4), rejecting any points for which $x_1^2 + x_2^2 \geq 1$ and $x_3^2 + x_4^2 \geq 1$. Then the points

$$x = x_1 \qquad (1)$$

$$y = x_2 \qquad (2)$$

$$z = x_3 \sqrt{\frac{1 - x_1^2 - x_2^2}{x_2^3 + x_4^2}} \qquad (3)$$

$$w = x_4 \sqrt{\frac{1 - x_1^2 - x_2^2}{x_2^3 + x_4^2}} \qquad (4)$$

have a uniform distribution on the surface of the hypersphere. This extends the method of Marsaglia (1972) for SPHERE POINT PICKING.

See also SPHERE POINT PICKING

References

Hicks, J. S. ad Wheeling, R. F. "An Efficient Method for Generating Uniformly Distributed Points on the Surface of an *n*-Dimensional Sphere." *Comm. Assoc. Comput. Mach.* **2**, 13–5, 1959.
Marsaglia, G. "Choosing a Point from the Surface of a Sphere." *Ann. Math. Stat.* **43**, 645–46, 1972.

Hyperspherical Differential Equation

ULTRASPHERICAL DIFFERENTIAL EQUATION

Hypersurface

A generalization of an ordinary two-dimensional surface embedded in three-dimensional space to an $(n-1)$-dimensional surface embedded in n-dimensional space. A hypersurface is therefore the set of solutions to a single equation

$$f(x_1, \ldots, x_n) = 0$$

and so it has CODIMENSION one. For instance, the n-dimension HYPERSPHERE corresponds to the equation $x_1^2 + \ldots + x_n^2 = 1$.

See also HYPERSPHERE, SURFACE

Hypervolume

CONTENT

Hypocycloid

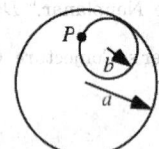

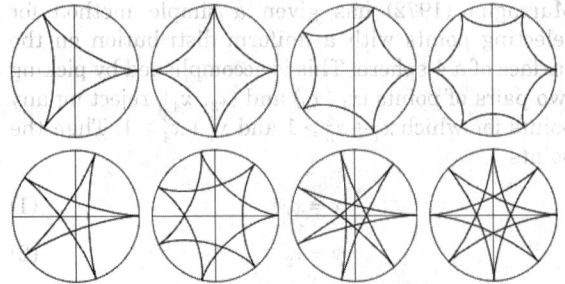

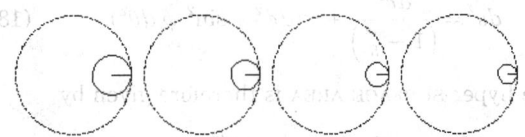

The curve produced by fixed point P on the CIRCUMFERENCE of a small CIRCLE of RADIUS b rolling around the inside of a large CIRCLE of RADIUS $a > b$. A hypocycloid is a HYPOTROCHOID with $h = b$. To derive the equations of the hypocycloid, call the ANGLE by which a point on the small CIRCLE rotates about its center ϑ, and the ANGLE from the center of the large CIRCLE to that of the small CIRCLE ϕ. Then

$$(a - b)\phi = b\vartheta, \qquad (1)$$

so

$$\vartheta = \frac{a - b}{b} \phi. \qquad (2)$$

Call $\rho \equiv a - 2b$. If $x(0) = \rho$, then the first point is at minimum radius, and the Cartesian parametric equations of the hypocycloid are

$$x = (a - b)\cos\phi - b\cos\vartheta$$
$$= (a - b)\cos\phi - b\cos\left(\frac{a - b}{b}\phi\right) \qquad (3)$$
$$y = (a - b)\sin\phi - b\sin\vartheta$$
$$= (a - b)\sin\phi + b\sin\left(\frac{a - b}{b}\phi\right). \qquad (4)$$

If $x(0) = a$ instead so the first point is at maximum radius (on the CIRCLE), then the equations of the hypocycloid are

$$x = (a - b)\cos\phi + b\cos\left(\frac{a - b}{b}\phi\right) \qquad (5)$$
$$y = (a - b)\sin\phi - b\sin\left(\frac{a - b}{b}\phi\right). \qquad (6)$$

An n-cusped non-self-intersecting hypocycloid has $a/b = n$. A 2-cusped hypocycloid is a LINE SEGMENT (Steinhaus 1983, p. 145), as can be seen by setting $a = b$ in equations (3) and (4) and noting that the equations simplify to

$$x = a\sin\phi \qquad (7)$$
$$y = 0. \qquad (8)$$

A 3-cusped hypocycloid is called a DELTOID or TRICUSPOID, and a 4-cusped hypocycloid is called an ASTROID. If a/b is rational, the curve closes on itself and has b cusps. If a/b is IRRATIONAL, the curve never closes and fills the entire interior of the CIRCLE.

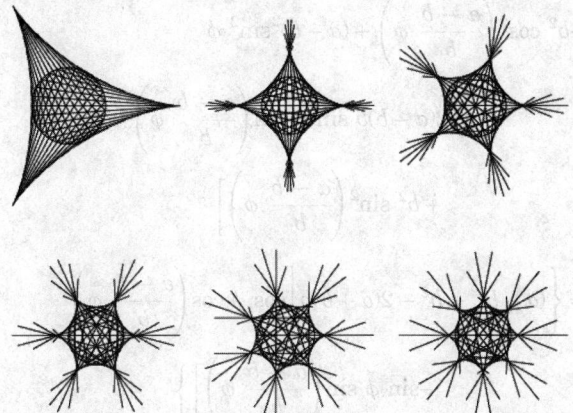

n-hypocycloids can also be constructed by beginning with the DIAMETER of a CIRCLE, offsetting one end by a series of steps while at the same time offsetting the other end by steps n times as large in the opposite direction and extending beyond the edge of the CIRCLE. After traveling around the CIRCLE once, an n-cusped hypocycloid is produced, as illustrated above (Madachy 1979).

Let r be the radial distance from a fixed point. For RADIUS OF TORSION ρ and ARC LENGTH s, a hypocycloid can given by the equation

$$s^2 + \rho^2 = 16r^2 \tag{9}$$

(Kreyszig 1991, pp. 63–4). A hypocycloid also satisfies

$$\sin^2 \psi = \frac{\rho^2}{a^2 - \rho^2} \frac{a^2 - r^2}{r^2}, \tag{10}$$

where

$$r \frac{dr}{d\theta} = \tan \psi \tag{11}$$

and ψ is the ANGLE between the RADIUS VECTOR and the TANGENT to the curve.

The ARC LENGTH of the hypocycloid can be computed as follows

$$x' = -(a-b)\sin \phi - (a-b)\sin\left(\frac{a-b}{b} \phi\right)$$

$$= (a-b)\left[\sin \phi + \sin\left(\frac{a-b}{b} \phi\right)\right] \tag{12}$$

$$y' = (a-b)\cos \phi - (a-b)\cos\left(\frac{a-b}{a} \phi\right)$$

$$= (a-b)\left[\cos \phi - \cos\left(\frac{a-b}{b} \phi\right)\right] \tag{13}$$

$$x'^2 + y'^2 = (a-b)^2\left[\sin^2 \phi + 2 \sin \phi \sin\left(\frac{a-b}{b} \phi\right)\right.$$

$$+\sin^2\left(\frac{a-b}{b} \phi\right) + \cos^2 \phi - 2 \cos \phi \cos\left(\frac{a-b}{b} \phi\right)$$

$$\left.+\cos^2\left(\frac{a-b}{b} \phi\right)\right]$$

$$= (a-b)^2\left\{2 + 2\left[\sin \phi \sin\left(\frac{a-b}{a} \phi\right)\right.\right.$$

$$\left.\left.-\cos \phi \cos\left(\frac{a-b}{b} \phi\right)\right]\right\}$$

$$= 2(a-b)^2\left[1 - \cos\left(\phi + \frac{a-b}{b} \phi\right)\right]$$

$$= 4(a-b)^2 \tfrac{1}{2}\left[1 - \cos\left(\frac{a}{b} \phi\right)\right]$$

$$= 4(a-b)^2\sin^2\left(\frac{a\phi}{2b}\right), \tag{14}$$

so

$$ds = \sqrt{x'^2 + y'^2} \, d\phi = 2(a-b)\sin\left(\frac{a\phi}{2b}\right) d\phi \tag{15}$$

for $\phi \le (b/2a)\pi$. Integrating,

$$s(\phi) = \int_0^\phi ds = 2(a-b)\left[-\frac{2b}{a} \cos\left(\frac{a\phi}{2b}\right)\right]_0^\phi$$

$$= \frac{4b(a-b)}{a}\left[-\cos\left(\frac{a}{2b} \phi\right) + 1\right]$$

$$= \frac{8b(a-b)}{a} \sin^2\left(\frac{a}{4b} \phi\right). \tag{16}$$

The length of a single cusp is then

$$s\left(2\pi \frac{b}{a}\right) = \frac{8b(a-b)}{a} \sin^2\left(\frac{\pi}{2}\right) = \frac{8b(a-b)}{a}. \tag{17}$$

If $n \equiv a/b$ is rational, then the curve closes on itself without intersecting after n cusps. For $n \equiv a/b$ and with $x(0) = a$, the equations of the hypocycloid become

$$x = \frac{1}{n}[(n-1)\cos \phi - \cos[(n-1)\phi] \, a, \tag{18}$$

$$y = \frac{1}{n}[(n-1)\sin \phi + \sin[(n-1)\phi] \, a, \tag{19}$$

and

$$s_n = n\,\frac{8b(bn - b)}{nb} = 8b(n-1) = \frac{8a(n-1)}{n}. \quad (20)$$

Compute

$$
\begin{aligned}
xy' - yx' &= \left[(a-b)\cos\phi + b\cos\!\left(\frac{a-b}{a}\,\phi\right)\right](b-a)\\
&\quad \times \left[\sin\phi + \sin\!\left(\frac{a-b}{b}\,\phi\right)\right]\\
&\quad - \left[(a-b)\sin\phi - b\sin\!\left(\frac{a-b}{b}\,\phi\right)\right](a-b)\\
&\quad \times \left[\cos\phi - \cos\!\left(\frac{a-b}{b}\,\phi\right)\right]\\
&= 2(a^2 - 3ab + 2b^2)\sin^2\!\left(\frac{a\phi}{2b}\right). \quad (21)
\end{aligned}
$$

The AREA of one cusp is then

$$
\begin{aligned}
A &= \frac{1}{2}\int_0^{2\pi b/a}(xy' - yx')\,d\phi\\
&= (a^2 - 3ab + 2b^2)\left[\frac{at - b\sin\!\left(\frac{at}{b}\right)}{2a}\right]_a^{2\pi b/a}\\
&= (a^2 - 3ab + 2b^2)\left[\frac{a\left(2\pi\frac{b}{a}\right)}{2a}\right]\\
&= \frac{b(a^2 - 3ab + 2b^2)}{a}\,\pi. \quad (22)
\end{aligned}
$$

If $n = a/b$ is rational, then after n cusps,

$$
\begin{aligned}
A_n &= n\pi\,\frac{b(a^2 - 3ab + 2b^2)}{a}\\
&= n\pi\,\frac{\frac{a}{n}\left(a^2 - 3a\,\frac{a}{n} + 2\,\frac{a^2}{n^2}\right)}{a}\\
&= \frac{n^2 - 3n + 2}{n^2}\,\pi a^2 = \frac{(n-1)(n-2)}{n^2}\,\pi a^2. \quad (23)
\end{aligned}
$$

The equation of the hypocycloid can be put in a form which is useful in the solution of CALCULUS OF VARIATIONS problems with radial symmetry. Consider the case $x(0) = \rho$, then

$$
r^2 = x^2 + y^2
$$
$$
= \left[(a-b)^2\cos^2\phi - 2(a-b)b\cos\phi\cos\!\left(\frac{a-b}{b}\,\phi\right)\right.
$$

$$
\begin{aligned}
&\left.+\,b^2\cos^2\!\left(\frac{a-b}{b}\,\phi\right) + (a-b)^2\sin^2\phi\right.\\
&\left.+\,2(a-b)b\sin\phi\sin\!\left(\frac{a-b}{b}\,\phi\right)\right.\\
&\left.+\,b^2\sin^2\!\left(\frac{a-b}{b}\,\phi\right)\right]\\
&= \left\{(a-b)^2 + b^2 - 2(a-b)b\left[\cos\phi\cos\!\left(\frac{a-b}{b}\,\phi\right)\right.\right.\\
&\left.\left.\quad -\sin\phi\sin\!\left(\frac{a-b}{b}\,\phi\right)\right]\right\}\\
&= (a-b)^2 + b^2 - 2(a-b)b\cos\!\left(\frac{a}{b}\,\phi\right). \quad (24)
\end{aligned}
$$

But $\rho = a - 2b$, so $b = (a-\rho)/2$, which gives

$$
\begin{aligned}
(a-b)^2 + b^2 &= \left[a - \tfrac{1}{2}(a-\rho)\right]^2 + \left[\tfrac{1}{2}(a-\rho)\right]^2\\
&= \left[\tfrac{1}{2}(a+\rho)\right]^2 + \left[\tfrac{1}{2}(a-\rho)\right]^2\\
&= \tfrac{1}{4}(a^2 + 2a\rho + \rho^2 + a^2 - 2a\rho + \rho^2)\\
&= \tfrac{1}{2}(a^2 + \rho^2) \quad (25)
\end{aligned}
$$

$$
\begin{aligned}
2(a-b)b &= 2\left[a - \tfrac{1}{2}(a-\rho)\right]\tfrac{1}{2}(a-\rho)\\
&= \tfrac{1}{2}(a+\rho)(a-\rho) = \tfrac{1}{2}(a^2 - \rho^2). \quad (26)
\end{aligned}
$$

Now let

$$
2\Omega t \equiv \frac{a}{b}\,\phi, \quad (27)
$$

so

$$
\phi = \frac{a-\rho}{a}\,\Omega t \quad (28)
$$

$$
\frac{\phi}{a-\rho} = \frac{\Omega t}{a}, \quad (29)
$$

then

$$
\begin{aligned}
r^2 &= \tfrac{1}{2}(a^2 + \rho^2) - \tfrac{1}{2}(a^2 - \rho^2)\cos\!\left(\frac{a}{b}\,\phi\right)\\
&= \tfrac{1}{2}(a^2 + \rho^2) - \tfrac{1}{2}(a^2 - \rho^2)\cos(2\Omega t). \quad (30)
\end{aligned}
$$

The POLAR ANGLE is

$$
\tan\theta \equiv \frac{y}{x} = \frac{(a-b)\sin\phi + b\sin\!\left(\frac{a-b}{a}\,\phi\right)}{(a-b)\cos\phi + b\cos\!\left(\frac{a-b}{a}\,\phi\right)}. \quad (31)
$$

But

$$b = \tfrac{1}{2}(a - \rho) \tag{32}$$

$$a - b = \tfrac{1}{2}(a + \rho) \tag{33}$$

$$\frac{a - b}{b} = \frac{a + \rho}{a - \rho}, \tag{34}$$

so

$$\tan\theta = \frac{\tfrac{1}{2}(a + \rho)\sin\phi + \tfrac{1}{2}(a - \rho)\sin\left(\frac{a + \rho}{z - \rho}\,\phi\right)}{\tfrac{1}{2}(a + \rho)\cos\phi - \tfrac{1}{2}(a - \rho)\cos\left(\frac{a + \rho}{a - \rho}\,\phi\right)}$$

$$= \frac{(a + \rho)\sin\left(\frac{a - \rho}{a}\,\Omega t\right) + (a - \rho)\sin\left(\frac{a + \rho}{a}\,\Omega t\right)}{(a + \rho)\cos\left(\frac{a - \rho}{a}\,\Omega t\right) - (a - \rho)\cos\left(\frac{a + \rho}{a}\,\Omega t\right)}$$

$$= \frac{a\left[\sin\left(\frac{a - \rho}{a}\,\Omega t\right) + \sin\left(\frac{a + \rho}{a}\,\Omega t\right)\right] + \rho\left[\sin\left(\frac{a - \rho}{a}\,\Omega t\right) - \sin\left(\frac{a + \rho}{a}\,\Omega t\right)\right]}{a\left[\cos\left(\frac{a - \rho}{a}\,\Omega t\right) - \cos\left(\frac{a + \rho}{a}\,\Omega t\right)\right] + \rho\left[\cos\left(\frac{a - \rho}{a}\,\Omega t\right) + \cos\left(\frac{a + \rho}{a}\,\Omega t\right)\right]}$$

$$= \frac{2a\,\sin(\Omega t)\cos\left(\frac{\rho}{q}\,\Omega t\right) - 2\rho\,\cos(\Omega t)\sin\left(\frac{\rho}{a}\,\Omega t\right)}{2a\,\sin(\Omega t)\sin\left(\frac{\rho}{q}\,\Omega t\right) + 2\rho\,\cos(\Omega t)\sin\left(\frac{\rho}{a}\,\Omega t\right)}$$

$$= \frac{a\,\tan(\Omega t) - \rho\,\tan\left(\frac{\rho}{a}\,\Omega t\right)}{a\,\tan(\Omega t)\tan\left(\frac{\rho}{a}\,\Omega t\right) + \rho}. \tag{35}$$

Computing

$$\tan\left(\theta + \frac{\rho}{a}\,\Omega t\right)$$

$$= \frac{\left[a\,\tan(\Omega t) - \rho\,\tan\left(\frac{\rho}{a}\,\Omega t\right) + \tan\left(\frac{\rho}{a}\,\Omega t\right)\right]\left[a\,\tan(\Omega t)\tan\left(\frac{\rho}{a}\,\Omega t\right) + \rho\right]}{\left[a\,\tan(\Omega t)\tan\left(\frac{\rho}{a}\,\Omega t\right) + \rho\right] - \left[a\,\tan(\Omega t) - \rho\,\tan\left(\frac{\rho}{a}\,\Omega t\right)\right]\tan\left(\frac{\rho}{a}\,\Omega t\right)}$$

$$= \frac{a\,\tan(\Omega t)\left[1 + \tan^2\left(\frac{\rho}{a}\,\Omega t\right)\right]}{\rho\left[1 + \tan^2\left(\frac{\rho}{a}\,\Omega t\right)\right]}$$

$$= \frac{a}{\rho}\,\tan(\Omega t), \tag{36}$$

then gives

$$\theta = \tan^{-1}\left[\frac{a}{\rho}\,\tan(\Omega t)\right] - \frac{\rho}{a}\,\Omega t. \tag{37}$$

Finally, plugging back in gives

$$\theta = \tan^{-1}\left[\frac{a}{\rho}\,\tan\left(\frac{a}{a - \rho}\,\phi\right)\right] - \frac{\rho}{a}\frac{a}{a - \rho}\,\phi$$

$$= \tan^{-1}\left[\frac{a}{\rho}\,\tan\left(\frac{a}{a - \rho}\,\phi\right)\right] - \frac{\rho}{a - \rho}\,\phi \tag{38}$$

This form is useful in the solution of the SPHERE WITH TUNNEL problem, which is the generalization of the BRACHISTOCHRONE PROBLEM, to find the shape of a tunnel drilled through a SPHERE (with gravity varying according to Gauss's law in a gravitational field such that the travel time between two points on the surface of the SPHERE under the force of gravity is minimized.

See also ASTROID, CYCLOID, DELTOID, EPICYCLOID

References
Bogomolny, A. "Cycloids." http://www.cut-the-knot.com/pythagoras/cycloids.html.
Kreyszig, E. *Differential Geometry.* New York: Dover, 1991.
Lawrence, J. D. *A Catalog of Special Plane Curves.* New York: Dover, pp. 171–73, 1972.
Lemaire, J. *Hypocycloïdes et epicycloïdes.* Paris: Albert Blanchard, 1967.
MacTutor History of Mathematics Archive. "Hypocycloid." http://www-groups.dcs.st-and.ac.uk/~history/Curves/Hypocycloid.html.
Madachy, J. S. *Madachy's Mathematical Recreations.* New York: Dover, pp. 225–31, 1979.
Steinhaus, H. *Mathematical Snapshots, 3rd ed.* New York: Dover, 1999.
Wagon, S. *Mathematica in Action.* New York: W. H. Freeman, pp. 50–2, 1991.
Yates, R. C. "Epi- and Hypo-Cycloids." *A Handbook on Curves and Their Properties.* Ann Arbor, MI: J. W. Edwards, pp. 81–5, 1952.

Hypocycloid Evolute

For $x(0) = a$,

$$x = \frac{a}{a - 2b}\left[(a - b)\cos\phi - b\,\cos\left(\frac{a - b}{b}\,\phi\right)\right]$$

$$y = \frac{a}{a - 2b}\left[(a - b)\sin\phi + b\,\sin\left(\frac{a - b}{b}\,\phi\right)\right].$$

If $a/b = n$, then

$$x = \frac{1}{n - 2}[(n - 1)\cos\phi - \cos[(n - 1)\phi]a$$

$$y = \frac{1}{n - 2}[(n - 1)\sin\phi - \sin[(n - 1)\phi]a.$$

This is just the original HYPOCYCLOID scaled by the factor $(n-2)/n$ and rotated by $1/(2n)$ of a turn.

Hypocycloid Involute

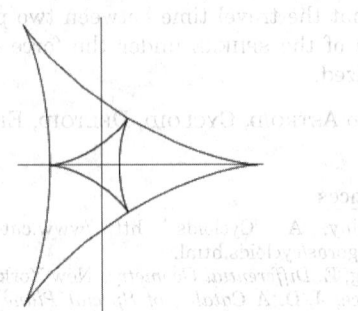

The HYPOCYCLOID

$$x = \frac{a}{a-2b}\left[(a-b)\cos\phi - b\cos\left(\frac{a-b}{b}\,\phi\right)\right]$$

$$y = \frac{a}{a-2b}\left[(a-b)\sin\phi + b\sin\left(\frac{a-b}{b}\,\phi\right)\right]$$

has INVOLUTE

$$x = \frac{a-2b}{a}\left[(a-b)\cos\phi + b\cos\left(\frac{a-b}{b}\,\phi\right)\right]$$

$$y = \frac{a-2b}{a}\left[(a-b)\sin\phi - b\sin\left(\frac{a-b}{b}\,\phi\right)\right],$$

which is another HYPOCYCLOID.

Hypocycloid Pedal Curve

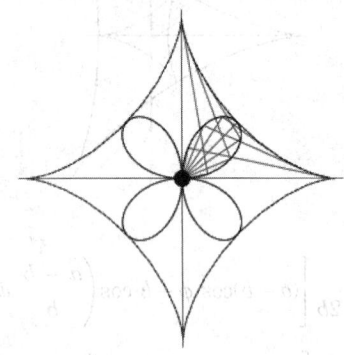

The PEDAL CURVE for a PEDAL POINT at the center is a ROSE.

Hypocycloid–3-Cusped
DELTOID

Hypocycloid–4-Cusped
ASTROID

Hypoellipse

$$y^{n/m} + c\left|\frac{x}{a}\right|^{n/m} - c = 0,$$

with $n/m < 2$. If $n/m > 2$, the curve is a HYPEREL-LIPSE.

See also ELLIPSE, HYPERELLIPSE, SUPERELLIPSE

References
von Seggern, D. *CRC Standard Curves and Surfaces.* Boca Raton, FL: CRC Press, p. 82, 1993.

Hypohamiltonian Graph

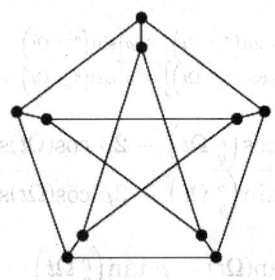

A graph G is hypohamiltonian if G is not HAMILTO-NIAN, but $G - v$ is HAMILTONIAN for every $v \in V$ (Bondy and Murty 1976, p. 61). The PETERSEN GRAPH, which has ten nodes and is illustrated above, is the smallest hypohamiltonian graph (Herz *et al.* 1967; Bondy and Murty 1976, p. 61). There are no hypohamiltonian graphs with 11 or 12 vertices. However, there exists a hypohamiltonian graph on p vertices for every $p \geq 13$ with the possible exceptions of $p = 14$, 17, 19. Thomassen (1973) found hypohamiltonian graphs on $p = 20$ and 25 vertices, which had previously been open.

A graph can be tested to see if it is hypohamiltonian using the following *Mathematica* function.

```
< <DiscreteMath`Combinatorica`;
HypohamiltonianQ[g_Graph]        :=     !
HamiltonianQ[g] &&     HamiltonianQ /@ And @@
(DeleteVertex[g, #] & /@ Range[V[g]])
```

See also HAMILTONIAN GRAPH, HYPOTRACEABLE GRAPH, TRACEABLE GRAPH

References
Bondy, J. A. and Murty, U. S. R. *Graph Theory with Applications.* New York: North Holland, p. 61, 1976.

Chvátal, V. "Flip-Flops in Hypohamiltonian Graphs." *Canad. Math. Bull.* **16**, 33–1, 1973.

Gaudin, T.; Herz, J.-C.; and Rossi, P. "Solution de problème no. 29." *Française Informat. Recherche Opérationnelle* **8**, 214–18, 1964.

Herz, J. C.; Duby, J. J.; and Vigué, F. "Recherche systématique des graphes hypohamiltoniens." In *Theory of Graphs: Internat. Sympos., Rome 1966* (Ed. P. Rosenstiehl). Paris: Gordon and Breach, pp. 153–59, 1967.

Lindgren, W. F. "An Infinite Class of Hypohamiltonian Graphs." *Amer. Math. Monthly* **74**, 1087–089, 1967.
Thomassen, C. "Hypohamiltonian and Hypotraceable Graphs." *Disc. Math.* **9**, 91–6, 1974.

Hypotenuse

The longest LEG of a RIGHT TRIANGLE (which is the side opposite the RIGHT ANGLE). The word derives from the Greek *hypo-* ("under") and *teinein* ("to stretch").

Hypothesis

A proposition that is consistent with known data, but has been neither verified nor shown to be false. It is synonymous with CONJECTURE.

See also BOURGET'S HYPOTHESIS, CHINESE HYPOTHESIS, CONTINUUM HYPOTHESIS, HYPOTHESIS TESTING, NESTED HYPOTHESIS, NULL HYPOTHESIS, POSTULATE, RAMANUJAN'S HYPOTHESIS, RIEMANN HYPOTHESIS, SCHINZEL'S HYPOTHESIS, SOUSLIN'S HYPOTHESIS

Hypothesis Testing

The use of statistics to determine the probability that a given hypothesis is true.

See also BONFERRONI CORRECTION, ESTIMATE, FISHER SIGN TEST, PAIRED *T*-TEST, PERMUTATION TESTS, STATISTICAL TEST, TYPE I ERROR, TYPE II ERROR, WILCOXON SIGNED RANK TEST

References

Good, P. *Permutation Tests: A Practical Guide to Resampling Methods for Testing Hypotheses, 2nd ed.* New York: Springer-Verlag, 2000.
Hoel, P. G.; Port, S. C.; and Stone, C. J. "Testing Hypotheses." Ch. 3 in *Introduction to Statistical Theory.* New York: Houghton Mifflin, pp. 52–10, 1971.
Iyanaga, S. and Kawada, Y. (Eds.). "Statistical Estimation and Statistical Hypothesis Testing." Appendix A, Table 23 in *Encyclopedic Dictionary of Mathematics.* Cambridge, MA: MIT Press, pp. 1486–489, 1980.
Shaffer, J. P. "Multiple Hypothesis Testing." *Ann. Rev. Psych.* **46**, 561–84, 1995.

Hypotraceable Graph

G is a hypotraceable graph if G has no HAMILTONIAN PATH (i.e., it is not a TRACEABLE GRAPH), but $G - v$ has a HAMILTONIAN PATH (i.e., is a TRACEABLE GRAPH) for every $v \in V$ (Bondy and Murty 1976, p. 61).

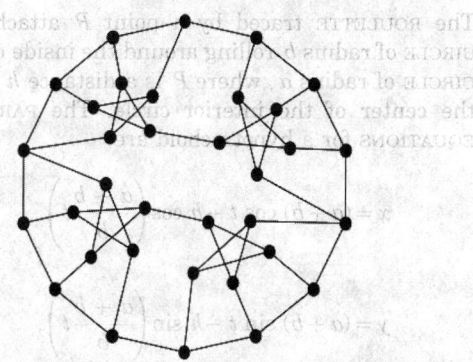

T. Gallai conjectured that there exist no hypotraceable graphs (there are none on seven or fewer nodes), but the THOMASSEN GRAPH, illustrated above, provides a counterexample (Bondy and Murty 1973, pp. 239–40). However, a hypotraceable graph with 40 vertices was found by Horton (Grünbaum 1973, Thomassen 1974). Thomassen (1974) showed that for $p = 34$, 37, 39, 40, and all $p \geq 42$, there exists a hypotraceable graph with p vertices. The smallest of these, the so-called THOMASSEN GRAPH, is illustrated above.

Walter (1969) gave an example of a connected graph in which the longest paths do not have a vertex in common, a property shared by hypotraceable graphs.

See also HAMILTON-CONNECTED GRAPH, THOMASSEN GRAPH, TRACEABLE GRAPH

References

Bondy, J. A. and Murty, U. S. R. *Graph Theory with Applications.* New York: North Holland, pp. 61 and 239–40, 1976.
Grünbaum, B. "Vertices Missed by Longest Paths or Circuits." Preprint, University of Washington, Seattle, May 1973.
Kapoor, S. F.; Kronk, H. V.; and Lick, D. R. "On Detours in Graphs." *Canad. Math. Bull.* **11**, 195–01, 1968.
Thomassen, C. "Hypohamiltonian and Hypotraceable Graphs." *Disc. Math.* **9**, 91–6, 1974.
Walter, H. "Über die Nichtexistenz eines Knotenpunktes, durch den alle längsten Wege eines Graphen gehen." *J. Combin. Th.* **6**, 1–, 1969.

Hypotrochoid

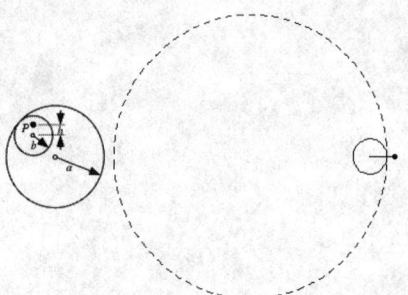

The ROULETTE traced by a point P attached to a CIRCLE of radius b rolling around the inside of a fixed CIRCLE of radius a, where P is a distance $h \leq b$ from the center of the interior circle. The PARAMETRIC EQUATIONS for a hypotrochoid are

$$x = (a+b)\cos t - h\cos\left(\frac{a+b}{b}t\right), \qquad (1)$$

$$y = (a+b)\sin t - h\sin\left(\frac{a+b}{b}t\right), \qquad (2)$$

Special cases include the HYPOCYCLOID with $h = b$, the ELLIPSE with $a = 2b$, and the ROSE with

$$a = \frac{2nh}{n+1} \qquad (3)$$

$$b = \frac{(n-1)h}{n+1}. \qquad (4)$$

See also EPITROCHOID, HYPOCYCLOID, SPIROGRAPH, TROCHOID

References

Lawrence, J. D. *A Catalog of Special Plane Curves.* New York: Dover, pp. 165–68, 1972.

MacTutor History of Mathematics Archive. "Hypotrochoid." http://www-groups.dcs.st-and.ac.uk/~history/Curves/Hypotrochoid.html.

Hypotrochoid Evolute

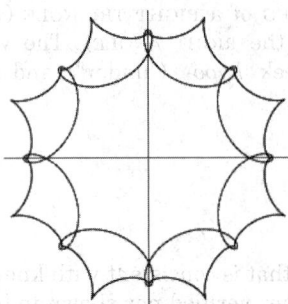

The EVOLUTE of the HYPOTROCHOID is illustrated above.

Hyzer's Illusion

FREEMISH CRATE

I

i

"The" IMAGINARY NUMBER i (also called the IMAGINARY UNIT) is defined as the SQUARE ROOT of -1, i.e., $i \equiv \sqrt{-1}$. Although there are two possible square roots of any number, the square roots of a negative number cannot be distinguished until one of the two is defined as the imaginary unit, at which point $+i$ and $-i$ can then be distinguished. Since either choice is possible, there is no ambiguity in defining i as "the" square root of -1.

In *Mathematica*, the imaginary number is implemented as I. For some reason engineers and physicists prefer the symbol J to i, probably because the symbol i (or I) is commonly used to denote current.

Numbers OF THE FORM iy, where y is a REAL NUMBER, are called IMAGINARY NUMBERS. Numbers OF THE FORM $z = x + iy$ where x and y are REAL NUMBERS are called COMPLEX NUMBERS, and when z is used to denote a COMPLEX NUMBER, it is sometimes (in older texts) called an "AFFIX."

The SQUARE ROOT of i is

$$\sqrt{i} = \pm \frac{i+1}{\sqrt{2}}, \tag{1}$$

since

$$\left[\frac{1}{\sqrt{2}}(i+1) \right]^2 = \tfrac{1}{2}(i^2 + 2i + 1) = i. \tag{2}$$

This can be immediately derived from the EULER FORMULA with $x = \pi/2$,

$$i = e^{i\pi/2} \tag{3}$$

$$\sqrt{i} = \sqrt{e^{i\pi/2}} = e^{i\pi/4} = \cos\left(\tfrac{1}{4}\pi\right) + i\sin\left(\tfrac{1}{4}\pi\right) = \frac{1+i}{\sqrt{2}}. \tag{4}$$

The PRINCIPAL VALUE of i^i is

$$i^i = \left(e^{i\pi/2}\right)^i = e^{i^2\pi/2} = e^{-\pi/2} = 0.207879\ldots \tag{5}$$

(Wells 1986, p. 26).

See also COMPLEX NUMBER, I, IMAGINARY IDENTITY, IMAGINARY NUMBER, REAL NUMBER, SURREAL NUMBER

References

Courant, R. and Robbins, H. *What is Mathematics?: An Elementary Approach to Ideas and Methods, 2nd ed.* Oxford, England: Oxford University Press, p. 89, 1996.
Nahin, P. J. *An Imaginary Tale: The Story of $\sqrt{-1}$.* Princeton, NJ: Princeton University Press, 1998.
Wells, D. *The Penguin Dictionary of Curious and Interesting Numbers.* Middlesex, England: Penguin Books, p. 26, 1986.

I

The double-struck capital letter I, $\mathbb{I}$, is a symbol sometimes used instead of $\mathbb{Z}$ for the RING of INTEGERS.

See also i, Z

Iamond

POLYIAMOND

Ice Fractal

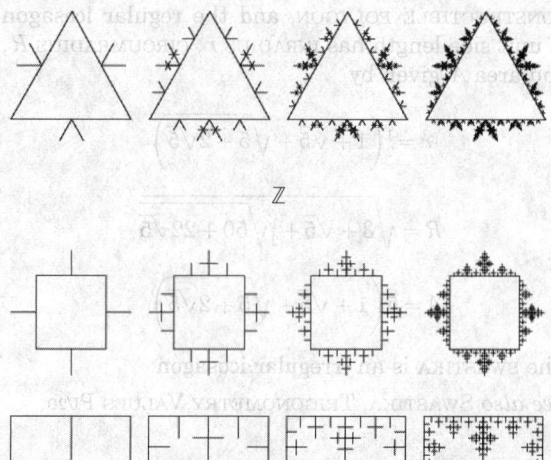

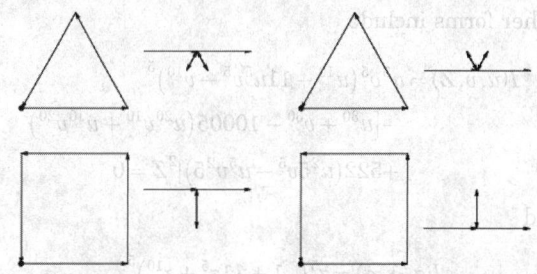

A FRACTAL (square, triangle, etc.) based on a simple generating motif. The above plots show the ice triangle, antitriangle, square, and antisquare. The base curves and motifs for the fractals illustrated above are shown below.

See also FRACTAL

References

Birch, M. W. "The Cross-Stitch Curve." *Eureka* **21**, 12–3, 1958.
Lauwerier, H. *Fractals: Endlessly Repeated Geometric Figures.* Princeton, NJ: Princeton University Press, p. 44, 1991.
Weisstein, E. W. "Fractals." MATHEMATICA NOTEBOOK FRACTAL.M.

Icosagon

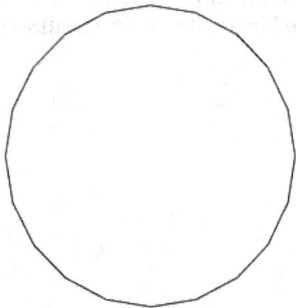

A 20-sided POLYGON. The regular icosagon is a CONSTRUCTIBLE POLYGON, and the regular icosagon of unit side length has INRADIUS r, CIRCUMRADIUS R, and area A given by

$$r = \tfrac{1}{2}\left(1 + \sqrt{5} + \sqrt{5 + 2\sqrt{5}}\right)$$

$$R = \sqrt{3 + \sqrt{5} + \tfrac{1}{2}\sqrt{50 + 22\sqrt{5}}}$$

$$A = 5\left(1 + \sqrt{5} + \sqrt{5 + 2\sqrt{5}}\right)_{.s}$$

The SWASTIKA is an irregular icosagon.

See also SWASTIKA, TRIGONOMETRY VALUES PI/20

Icosahedral Equation

Hunt (1996) gives the "dehomogenized" icosahedral equation as

$$\left[(z^{20} + 1) - 228(z^{15} - z^5) + 494z^{10}\right]^3$$
$$+ 1728 u z^5 (z^{10} + 11z^5 - 1)^5 = 0.$$

Other forms include

$$I(u, v, Z) = u^5 v^5 (u^{10} + 11 u^5 v^5 - v^{10})^5$$
$$- [u^{30} + v^{30} - 10005(u^{20}v^{10} + u^{10}v^{20})$$
$$+ 522(u^{25}v^5 - u^5v^25)]^2 Z = 0$$

and

$$I(z, 1, z) = z^5 (-1 + 11z^5 + z^{10})^5$$

$$- [1 + z^{30} - 10005(z^{10} + z^{20}) + 522(-z^5 + z^{25})]^2 z = 0.$$

References

Hunt, B. *The Geometry of Some Special Arithmetic Quotients.* New York: Springer-Verlag, p. 146, 1996.

Klein, F. "Sull' equazione dell' Icosaedro nella risoluzione delle equazioni del quinto grado [per funzioni ellittiche]." *Reale Istituto Lombardo, Rendiconto, Ser. 2* **10**, 1877.

Icosahedral Graph

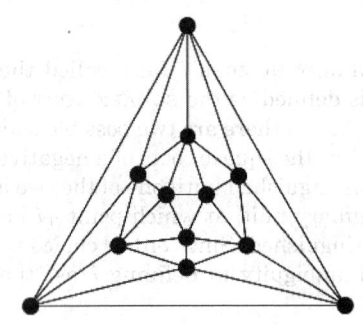

The PLATONIC GRAPH whose nodes have the connectivity of the ICOSAHEDRON. The icosahedral graph has 12 vertices, 30 edges, vertex connectivity 5, edge connectivity 5, GRAPH DIAMETER 3, GRAPH RADIUS 3, and GIRTH 3.

See also CUBICAL GRAPH, DODECAHEDRAL GRAPH, OCTAHEDRAL GRAPH, PLATONIC GRAPH, TETRAHEDRAL GRAPH

References

Bondy, J. A. and Murty, U. S. R. *Graph Theory with Applications.* New York: North Holland, p. 234, 1976.

Icosahedral Group

The POINT GROUP I_h of symmetries of the ICOSAHEDRON and DODECAHEDRON having order 60. The icosahedral group consists of the symmetry operations E, $12C_5$, $12C_5^2$, $20C_3$, $15C_2$, i, $12S_{10}$, $12S_{10}^3$, $20S_6$, and 15σ (Cotton 1990). The icosahedron group is a SUBGROUP of the SPECIAL ORTHOGONAL GROUP $SO(3)$.

See also BIPOLYHEDRAL GROUP, DODECAHEDRON, ICOSAHEDRON, OCTAHEDRAL GROUP, POINT GROUPS, POLYHEDRAL GROUP, SPECIAL ORTHOGONAL GROUP, TETRAHEDRAL GROUP

References

Cotton, F. A. *Chemical Applications of Group Theory, 3rd ed.* New York: Wiley, pp. 48–0, 1990.

Coxeter, H. S. M. "The Polyhedral Groups." §3.5 in *Regular Polytopes, 3rd ed.* New York: Dover, pp. 46–7, 1973.

Lomont, J. S. "Icosahedral Group." §3.10.E in *Applications of Finite Groups.* New York: Dover, p. 82, 1987.

Icosahedron

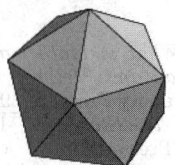 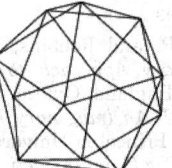

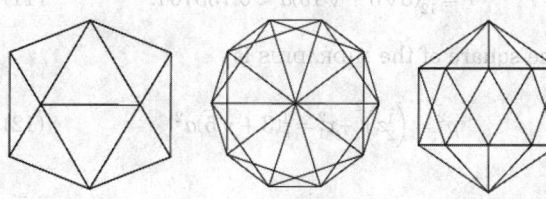

A PLATONIC SOLID P_5 having 12 VERTICES, 30 EDGES, and 20 equivalent EQUILATERAL TRIANGLE faces, $20\{3\}$. It is also UNIFORM POLYHEDRON U_{22} and Wenninger model W_4. It is described by the SCHLÄFLI SYMBOL $\{3,5\}$ and WYTHOFF SYMBOL $5|23$.

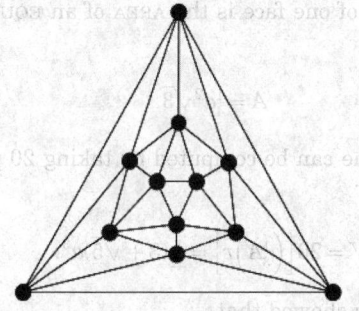

The icosahedron has the ICOSAHEDRAL GROUP I_h of symmetries. The connectivity of the vertices is given by the ICOSAHEDRAL GRAPH.

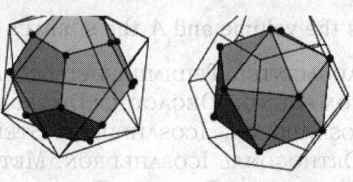

The DUAL POLYHEDRON of the icosahedron is the DODECAHEDRON, so the centers of the faces of an icosahedron form a DODECAHEDRON, and vice versa (Steinhaus 1983, pp. 199–01). There are 59 distinct icosahedra when each TRIANGLE is colored differently (Coxeter 1969).

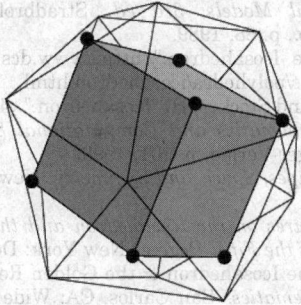

Taken eight at a time, the centers of the faces of an icosahedron comprise the vertices of a CUBE. This

leads to the beautiful CUBE 5-COMPOUND and is the basis for JESSEN'S ORTHOGONAL ICOSAHEDRON.

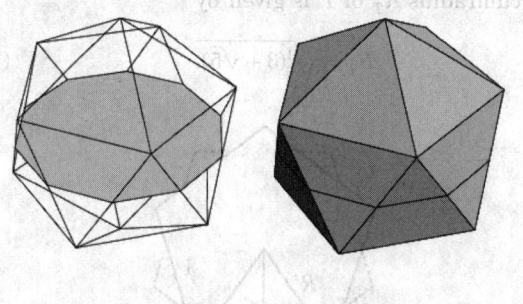

A plane PERPENDICULAR to a C_5 axis of an icosahedron cuts the solid in a regular DECAGONAL CROSS SECTION (Holden 1991, pp. 24–5).

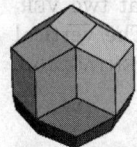  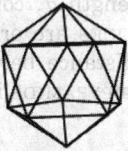

The long diagonals of the faces of the RHOMBIC TRIACONTAHEDRON give the edges of an icosahedron (Steinhaus 1983, pp. 209–10).

The following table gives polyhedra which can be constructed by CUMULATION of an icosahedron by pyramids of given heights h.

h	$(r+h)/h$	Result
$\frac{1}{6}\sqrt{3}(\sqrt{5}-3)$	$3(\sqrt{5}-2)$	GREAT DODECAHEDRON
$\frac{1}{15}\sqrt{15}$	$\frac{1}{5}(10-3\sqrt{5})$	SMALL TRIAMBIC ICOSAHEDRON
$\frac{1}{3}\sqrt{6}$	$1-3\sqrt{2}+\sqrt{10}$	60-faced star DELTAHEDRON
$\frac{1}{6}\sqrt{3}(3+\sqrt{5})$	3	GREAT STELLATED DODECAHEDRON

A construction for an icosahedron with side length $a = \sqrt{50-10\sqrt{5}}/5$ places the end vertices at $(0,0,\pm1)$ and the central vertices around two staggered CIRCLES of RADII $\frac{2}{5}\sqrt{5}$ and heights $\pm\frac{1}{5}\sqrt{5}$. By a suitable rotation, the VERTICES of an icosahedron of side length 2 can also be placed at $(0,\pm\phi,\pm1)$, $(\pm1,0,\pm\phi)$, and $(\pm\phi,\pm1,0)$, where ϕ is the GOLDEN RATIO. These points divide the EDGES of an OCTAHEDRON into segments with lengths in the ratio $\phi:1$. Another orientation of the icosahedron places two opposite triangular faces in an orientation parallel to the xy-plane. In this orientation, the distance h' from the top plane to the triangle T of vertices below it is $h' = \sqrt{3}/3$, equal to the circumradius of a face. The circumradius R_T of T is given by

$$R_T = \sqrt{\tfrac{1}{6}(3 + \sqrt{5})}. \tag{1}$$

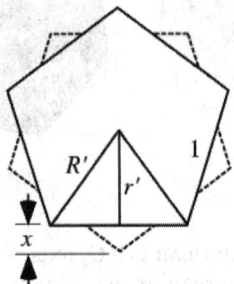

To derive the VOLUME of an icosahedron having edge length a, consider the orientation so that two VERTICES are oriented on top and bottom. The vertical distance between the top and bottom PENTAGONAL DIPYRAMIDS is then given by

$$z = \sqrt{\ell^2 - x^2}, \tag{2}$$

where

$$\ell = \tfrac{1}{2}\sqrt{3}a \tag{3}$$

is the height of an ISOSCELES TRIANGLE, and the SAGITTA $x = R' - r'$ of the pentagon is

$$x = \tfrac{1}{2}a\tfrac{1}{10}\sqrt{25 - 10\sqrt{5}}a, \tag{4}$$

giving

$$x^2 = \tfrac{1}{20}\sqrt{5 - 2\sqrt{5}}a^2. \tag{5}$$

Plugging (3) and (5) into (2) gives

$$z = \tfrac{1}{10}\sqrt{50 + 10\sqrt{5}}a, \tag{6}$$

which is identical to the radius of a PENTAGON of side a. The CIRCUMRADIUS is then

$$R = h + \tfrac{1}{2}z, \tag{7}$$

where

$$h = \tfrac{1}{10}\sqrt{50 - 10\sqrt{5}}a \tag{8}$$

is the height of a PENTAGONAL DIPYRAMID. Therefore,

$$R^2 = (h + \tfrac{1}{2}z)^2 = \tfrac{1}{8}(5 + \sqrt{5})a^2. \tag{9}$$

Taking the square root gives the CIRCUMRADIUS

$$R = \sqrt{\tfrac{1}{8}(5 + \sqrt{5})}a = \tfrac{1}{4}\sqrt{10 + 2\sqrt{5}}a \approx 0.95105a. \tag{10}$$

The INRADIUS is

$$r = \tfrac{1}{12}(3\sqrt{3} + \sqrt{15})a \approx 0.75576a. \tag{11}$$

The square of the MIDRADIUS is

$$\rho^2 = \left(\tfrac{1}{2}z\right)^2 + x_1^2 = \tfrac{1}{8}(3 + \sqrt{5})a^2, \tag{12}$$

so

$$\rho = \sqrt{\tfrac{1}{8}(3 + \sqrt{5})}a = \tfrac{1}{4}(1 + \sqrt{5})a \approx 0.80901a. \tag{13}$$

The DIHEDRAL ANGLE is

$$a = \cos^{-1}(-\tfrac{1}{3}\sqrt{5}) \approx 138.19°. \tag{14}$$

The AREA of one face is the AREA of an EQUILATERAL TRIANGLE

$$A = \tfrac{1}{4}a^2\sqrt{3}. \tag{15}$$

The volume can be computed by taking 20 pyramids of height r

$$V = 20\left[\left(\tfrac{1}{3}A\right)r\right] = \tfrac{5}{12}(3 + \sqrt{5})a^3. \tag{16}$$

Apollonius showed that

$$\frac{V_{\text{icosahedron}}}{V_{\text{dodecahedron}}} = \frac{A_{\text{icosahedron}}}{A_{\text{dodecahedron}}}, \tag{17}$$

where V is the volume and A the SURFACE AREA.

See also AUGMENTED TRIDIMINISHED ICOSAHEDRON, CUBE 5-COMPOUND, DECAGON, DODECAHEDRON, GREAT ICOSAHEDRON, ICOSAHEDRON STELLATIONS, JESSEN'S ORTHOGONAL ICOSAHEDRON, METABIDIMINISHED ICOSAHEDRON, RHOMBIC TRIACONTAHEDRON, TRIDIMINISHED ICOSAHEDRON, TRIGONOMETRY VALUES PI/5

References

Beyer, W. H. *CRC Standard Mathematical Tables, 28th ed.* Boca Raton, FL: CRC Press, p. 228, 1987.
Coxeter, H. S. M. *Introduction to Geometry, 2nd ed.* New York: Wiley, 1969.
Cundy, H. and Rollett, A. "Icosahedron 3^5." §3.5.5 in *Mathematical Models, 3rd ed.* Stradbroke, England: Tarquin Pub., p. 88, 1989.
Davie, T. "The Icosahedron." http://www.dcs.st-and.ac.uk/~ad/mathrecs/polyhedra/icosahedron.html.
Harris, J. W. and Stocker, H. "Icosahedron." §4.4.6 in *Handbook of Mathematics and Computational Science.* New York: Springer-Verlag, p. 101, 1998.
Holden, A. *Shapes, Space, and Symmetry.* New York: Dover, 1991.
Klein, F. *Lectures on the Icosahedron and the Solution of Equations of the Fifth Degree.* New York: Dover, 1956.
Pappas, T. "The Icosahedron & the Golden Rectangle." *The Joy of Mathematics.* San Carlos, CA: Wide World Publ./Tetra, p. 115, 1989.
Steinhaus, H. *Mathematical Snapshots, 3rd ed.* New York: Dover, pp. 199–01, 1999.

Wells, D. *The Penguin Dictionary of Curious and Interesting Geometry.* London: Penguin, p. 163, 1991.

Wenninger, M. J. "The Icosahedron." Model 4 in *Polyhedron Models.* Cambridge, England: Cambridge University Press, pp. 17–8, 1989.

Icosahedron Stellations

Applying the STELLATION process to the ICOSAHEDRON gives

$$20 + 30 + 60 + 20 + 60 + 120 + 12 + 30 + 60 + 60$$

cells of ten different shapes and sizes in addition to the ICOSAHEDRON itself. After application of five restrictions due to J. C. P. Miller to define which forms should be considered distinct, 59 stellations are found to be possible. Miller's restrictions are

1. The faces must lie in the twenty bounding planes of the icosahedron.
2. The parts of the faces in the twenty planes must be congruent, but those parts lying in one place may be disconnected.
3. The parts lying in one plane must have threefold rotational symmetry with or without reflections.
4. All parts must be accessible, i.e., lie on the outside of the solid.
5. Compounds are excluded that can be divided into two sets, each of which has the full symmetry of the whole.

Of these, 32 have full icosahedral symmetry and 27 are ENANTIOMERIC forms. Four are POLYHEDRON COMPOUNDS, one is a KEPLER-POINSOT SOLID, and one is the DUAL POLYHEDRON of an ARCHIMEDEAN SOLID.

n	name
1	ICOSAHEDRON
2	SMALL TRIAMBIC ICOSAHEDRON
3	OCTAHEDRON 5-COMPOUND
4	ECHIDNAHEDRON
11	GREAT ICOSAHEDRON
13	MEDIAL TRIAMBIC ICOSAHEDRON
13	GREAT TRIAMBIC ICOSAHEDRON
18	TETRAHEDRON 10-COMPOUND
36	TETRAHEDRON 5-COMPOUND

01 02 03

04 05 06

07 08 09

10 11 12

13 14 15

16 17 18

19   20   21

22   23   24

25   26   27

28   29   30

31   32   33

34   35   36

37   38   39

40   41   42

43   44   45

46   47   48

49   50   51

52   53   54

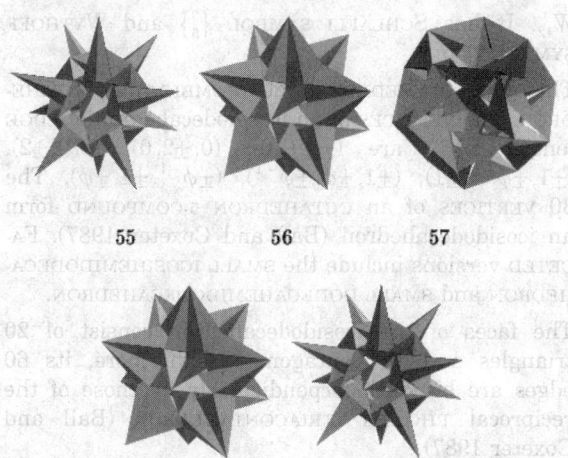

55   56   57

See also ARCHIMEDEAN SOLID STELLATION, DODECA-HEDRON STELLATIONS, STELLATION

References

Ball, W. W. R. and Coxeter, H. S. M. *Mathematical Recreations and Essays, 13th ed.* New York: Dover, pp. 146–47, 1987.

Bulatov, V. "Stellations of Icosahedron." http://www.physics.orst.edu/~bulatov/polyhedra/icosahedron/.

Coxeter, H. S. M.; Du Val, P.; Flather, H. T.; and Petrie, J. F. *The Fifty-Nine Icosahedra.* Stradbroke, England: Tarquin Publications, 1999.

Hart, G. "59 Stellations of the Icosahedron." http://www.georgehart.com/virtual-polyhedra/stellations-icosahedron-index.html.

Maeder, R. E. "Icosahedra." http://www.mathsource.com/cgi-bin/msitem?0206–42.

http://www.inf.ethz.ch/department/TI/rm/programs.html.

Maeder, R. E. "The Stellated Icosahedra." *Mathematica in Education* **3**, 1994. ftp://ftp.inf.ethz.ch/doc/papers/ti/scs/icosahedra94.ps.gz.

Maeder, R. E. "Stellated Icosahedra." http://www.mathconsult.ch/showroom/icosahedra/.

Weisstein, E. W. "Corrected version of Maeder's Icosahedra package." MATHEMATICA NOTEBOOK ICOSAHEDRA.M.

Wells, D. *The Penguin Dictionary of Curious and Interesting Geometry.* Middlesex, England: Penguin Books, pp. 77–8, 1991.

Wenninger, M. J. *Polyhedron Models.* New York: Cambridge University Press, pp. 41–5, 1989.

Wheeler, A. H. "Certain Forms of the Icosahedron and a Method for Deriving and Designating Higher Polyhedra." *Proc. Internat. Math. Congress* **1**, 701–08, 1924.

Icosian Game

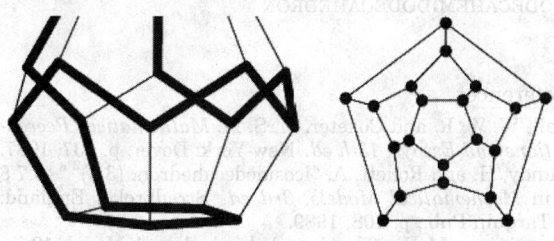

The problem of finding a HAMILTONIAN CIRCUIT along the edges of an DODECAHEDRON, i.e., a path such that every vertex is visited a single time, no edge is visited twice, and the ending point is the same as the starting point (left figure). The puzzle was distributed commercially as a pegboard with holes at the nodes of the DODECAHEDRAL GRAPH, illustrated above (right figure). The Icosian Game was invented in 1857 by William Rowan Hamilton. Hamilton sold it to a London game dealer in 1859 for 25 pounds, and the game was subsequently marketed in Europe in a number of forms (Gardner 1957).

See also HAMILTONIAN CIRCUIT, DODECAHEDRAL GRAPH, DODECAHEDRON

References

Ball, W. W. R. and Coxeter, H. S. M. *Mathematical Recreations and Essays, 13th ed.* New York: Dover, 1987.

Gardner, M. "Mathematical Games: About the Remarkable Similarity between the Icosian Game and the Towers of Hanoi." *Sci. Amer.* **196**, 150–56, May 1957.

Harary, F. *Graph Theory.* Reading, MA: Addison-Wesley, p. 4, 1994.

Herschel, A. S. "Sir Wm. Hamilton's Icosian Game." *Quart. J. Pure Applied Math.* **5**, 305, 1862.

MacTutor Archive. "Mathematical Games and Recreations." http://www-groups.dcs.st-and.ac.uk/~history/HistTo-Mathematical_games.html#49.

Skiena, S. *Implementing Discrete Mathematics: Combinatorics and Graph Theory with Mathematica.* Reading, MA: Addison-Wesley, p. 198, 1990.

Icosidodecadodecahedron

The UNIFORM POLYHEDRON U_{44} whose DUAL POLYHEDRON is the MEDIAL ICOSACRONIC HEXECONTAHEDRON. It has WYTHOFF SYMBOL $\frac{5}{3}5|3$. Its faces are $20\{6\} + 12\{\frac{5}{2}\} + 12\{5\}$. Its CIRCUMRADIUS for unit edge length is

$$R = \tfrac{1}{2}\sqrt{7}.$$

References

Wenninger, M. J. *Polyhedron Models.* Cambridge, England: Cambridge University Press, pp. 128–29, 1989.

Icosidodecagon

A 32-sided polygon. The regular icosidodecagon is a CONSTRUCTIBLE POLYGON, and the regular icosidodecahedron of side length 1 has INRADIUS r, CIRCUMRADIUS R, and AREA A

$$r = \frac{1}{2}\left[1 + \sqrt{2} + \sqrt{2(2+\sqrt{2})} + \sqrt{2(2+\sqrt{2})\left(2+\sqrt{2+\sqrt{2}}\right)}\right]$$

$$R = \sqrt{\frac{1}{2}(2+\sqrt{2})\left(2+\sqrt{2+\sqrt{2}}\right)\left(2+\sqrt{2+\sqrt{2+\sqrt{2}}}\right)}$$

$$A = 8\left[1 + \sqrt{2} + \sqrt{2(2+\sqrt{2})} + \sqrt{2(2+\sqrt{2})\left(2+\sqrt{2+\sqrt{2}}\right)}\right].$$

See also TRIGONOMETRY VALUES PI/32

Icosidodecahedron

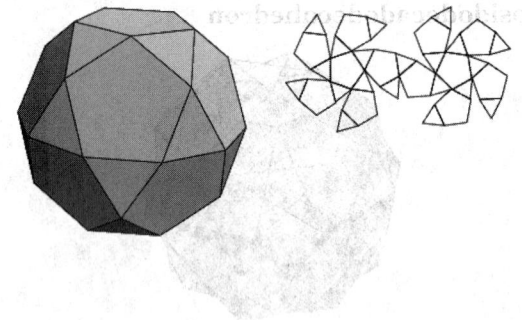

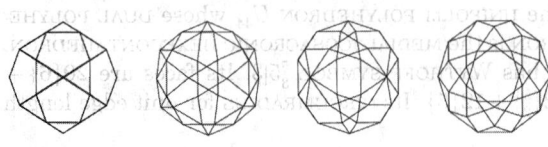

An icosidodecahedron is a 32-faced POLYHEDRON.

"The" icosidodecahedron is the 32-faced ARCHIMEDEAN SOLID A_4 with faces $20\{3\} + 12\{5\}$. It is one of the two convex QUASIREGULAR POLYHEDRA. It also UNIFORM POLYHEDRON U_{24} and Wenninger model

W_{12}. It has SCHLÄFLI SYMBOL $\left\{\begin{smallmatrix}3\\5\end{smallmatrix}\right\}$ and WYTHOFF SYMBOL $2|3\,5$.

The DUAL POLYHEDRON is the RHOMBIC TRIACONTAHEDRON. The VERTICES of an icosidodecahedron of EDGE length $2\phi^{-1}$ are $(\pm 2, 0, 0)$, $(0, \pm 2, 0)$, $(0, 0, \pm 2)$, $(\pm 1, \pm \phi^{-1}, \pm 1)$, $(\pm 1, \pm \phi, \pm \phi^{-1})$, $(\pm \phi^{-1}, \pm 1, \pm \phi)$. The 30 VERTICES of an OCTAHEDRON 5-COMPOUND form an icosidodecahedron (Ball and Coxeter 1987). FACETED versions include the SMALL ICOSIHEMIDODECAHEDRON and SMALL DODECAHEMIDODECAHEDRON.

The faces of the icosidodecahedron consist of 20 triangles and 12 pentagons. Furthermore, its 60 edges are bisected perpendicularly by those of the reciprocal RHOMBIC TRIACONTAHEDRON (Ball and Coxeter 1987).

The INRADIUS r of the dual, MIDRADIUS ρ of the solid and dual, and CIRCUMRADIUS R of the solid for $a = 1$ are

$$r = \frac{1}{8}(5 + 3\sqrt{5}) \approx 1.46353$$

$$\rho = \frac{1}{2}\sqrt{5 + 2\sqrt{5}} \approx 1.53884$$

$$R = \frac{1}{2}(1 + \sqrt{5}) = \phi \approx 1.61803.$$

The SURFACE AREA and VOLUME for an icosidodecahedron are given by

$$S = 5\sqrt{3} + 3\sqrt{5}\sqrt{5 + 2\sqrt{5}} \tag{1}$$

$$V = \frac{1}{6}45 + 17\sqrt{5} \tag{2}$$

The distance to the centers of the triangular and pentagonal faces are

$$r_3 = \sqrt{\frac{1}{6}\left(7 + 3\sqrt{5}\right)} \tag{3}$$

$$r_5 = \sqrt{\frac{1}{5}\left(5 + 2\sqrt{5}\right)}. \tag{4}$$

See also ARCHIMEDEAN SOLID, GREAT ICOSIDODECAHEDRON, ICOSIDODECAHEDRON, QUASIREGULAR POLYHEDRON, SMALL ICOSIHEMIDODECAHEDRON, SMALL DODECAHEMIDODECAHEDRON

References

Ball, W. W. R. and Coxeter, H. S. M. *Mathematical Recreations and Essays, 13th ed.* New York: Dover, p. 137, 1987.
Cundy, H. and Rollett, A. "Icosidodecahedron. $(3.5)^2$." §3.7.8 in *Mathematical Models, 3rd ed.* Stradbroke, England: Tarquin Pub., p. 108, 1989.
Wenninger, M. J. "The Icosidodecahedron." Model 12 in *Polyhedron Models.* Cambridge, England: Cambridge University Press, pp. 26 and 73, 1989.

Icosidodecahedron Stellation

The first stellation is a DODECAHEDRON-ICOSAHEDRON COMPOUND.

References

Wenninger, M. J. *Polyhedron Models.* Cambridge, England: Cambridge University Press, pp. 73–6, 1989.

Icosidodecahedron-Rhombic Triacontahedron Compound

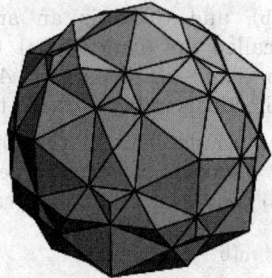

The POLYHEDRON COMPOUND of the ICOSIDODECAHE-DRON and its dual, the RHOMBIC TRIACONTAHEDRON. The compound can be constructed from an ICOSIDO-DECAHEDRON of unit edge length by midpoint CUMU-LATION with heights

$$h_3 = \frac{1}{4}\sqrt{\frac{1}{5}\left(\sqrt{7-3\sqrt{5}}\right)} \tag{1}$$

$$h_5 = \frac{1}{4}\sqrt{\left(\frac{1}{5}(5+2\sqrt{5})\right)}. \tag{2}$$

The resulting solid has edge lengths

$$s_1 = \frac{1}{4}\sqrt{\frac{1}{2}\left(5-\sqrt{5}\right)} \tag{3}$$

$$s_2 = \frac{1}{2} \tag{4}$$

$$s_3 = \frac{1}{4}\sqrt{5+2\sqrt{5}} \tag{5}$$

$$s_4 = \frac{1}{4}\left(1+\sqrt{5}\right), \tag{6}$$

CIRCUMRADIUS

$$R = \frac{1}{4}\sqrt{5\left(5+2\sqrt{5}\right)}, \tag{7}$$

SURFACE AREA S given by the largest positive root of

$$-612530859375 + 147622500000x + 36267750000x^2$$
$$-8164800000x^3 - 450360000x^4 + 82944000x^5$$
$$-230400x^6 - 184320x^7 + 4096x^8 = 0 \tag{8}$$

and VOLUME

$$V = \frac{5}{16}(27 + 10\sqrt{5}). \tag{9}$$

See also CUMULATION, ICOSIDODECAHEDRON, POLY-HEDRON COMPOUND, RHOMBIC TRIACONTAHEDRON

Icosidodecatruncated Icosidodecahedron

ICOSITRUNCATED DODECADODECAHEDRON

Icositetragon

A 24-sided POLYGON. The regular icositetragon is constructible. For side length 1, the INRADIUS r, CIRCUMRADIUS R, and AREA A are given by

$$r = \frac{1}{2}(2+\sqrt{2}+\sqrt{3}+\sqrt{6})$$

$$R = \frac{1}{2}\sqrt{16+10\sqrt{2}+8\sqrt{3}+6\sqrt{6}}$$

$$A = 6(2+\sqrt{2}+\sqrt{3}+\sqrt{6}).$$

See also TRIGONOMETRY VALUES PI/24

Icositetrahedron

A 24-faced POLYHEDRON.

See also DELTOIDAL ICOSITETRAHEDRON, PENTAGO-NAL ICOSITETRAHEDRON, SMALL RHOMBICUBOCTAHE-DRON, SMALL TRIAKIS OCTAHEDRON, SNUB CUBE, TETRAKIS HEXAHEDRON, TRUNCATED OCTAHEDRON

Icositruncated Dodecadodecahedron

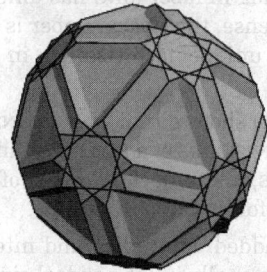

The UNIFORM POLYHEDRON U_{45} also called the ICOSI-

DODECATRUNCATED ICOSIDODECAHEDRON whose DUAL POLYHEDRON is the TRIDYAKIS ICOSAHEDRON. It has WYTHOFF SYMBOL $3\frac{5}{3}5|$. Its faces are $20\{6\} + 12\{10\} + 12\{\frac{10}{3}\}$. Its CIRCUMRADIUS for unit edge length is

$$R = 2.$$

References

Wenninger, M. J. *Polyhedron Models.* Cambridge, England: Cambridge University Press, pp. 130–31, 1989.

Ida Surface

A 3-D shadow of a 4-D KLEIN BOTTLE.

See also KLEIN BOTTLE

References

Peterson, I. *Islands of Truth: A Mathematical Mystery Cruise.* New York: W. H. Freeman, pp. 44–5, 1990.

Ideal

A subset $\Im$ of elements in a RING R which forms an additive GROUP and has the property that, whenever x belongs to R and y belongs to $\Im$, then xy and yx belong to $\Im$. For example, the set of EVEN INTEGERS is an ideal in the RING of INTEGERS $\mathbb{Z}$. Given an ideal $\Im$, it is possible to define a FACTOR RING R/I. Ideals are commonly denoted using a Gothic typeface.

An ideal may be viewed as a lattice and specified as the finite list of algebraic integers that form a basis for the lattice. Any two bases for the same lattice are equivalent. Ideals have multiplication, and this is basically the KRONECKER PRODUCT of the two bases. From the perspective of ALGEBRAIC GEOMETRY, ideals correspond to VARIETIES.

For any ideal $\Im$, there is an ideal $\Im_i$ such that

$$\Im\Im_i = \mathfrak{z},\tag{1}$$

where $\mathfrak{z}$ is a PRINCIPAL IDEAL, (i.e., an ideal of rank 1). Moreover there is a finite list of ideals $\Im_i$ such that this equation may be satisfied for every $\Im$. The size of this list is known as the CLASS NUMBER. In effect, the above relation imposes an EQUIVALENCE RELATION on ideals, and the number of ideals modulo this relation is the CLASS NUMBER. When the CLASS NUMBER is 1, the corresponding number RING has unique factorization and, in a sense, the class number is a measure of the failure of unique factorization in the original number ring.

Dedekind (1871) showed that every NONZERO ideal in the domain of INTEGERS of a FIELD is a unique product of PRIME IDEALS, and in fact all ideals of $\mathbb{Z}$ are of this form and therefore PRINCIPAL IDEALS.

Ideals can be added, multiplied and intersected. The union of ideals usually is not an ideal since it may not be closed under addition. From the perspective of ALGEBRAIC GEOMETRY, the addition of ideals corresponds to the intersection of VARIETIES and the intersection of ideals corresponds to the union of varieties. Also, the multiplication of ideals corresponds to the union of varieties.

Intersection and multiplication are different, for instance consider the ideal $\mathfrak{a} = (x)$ in $\mathbb{Z}[x,y]$. Then

$$\mathfrak{a}^2 = \mathfrak{a} \cdot \mathfrak{a} = \langle x^2 \rangle.\tag{2}$$

Sometimes they are the same. If $\mathfrak{b} = \langle y \rangle$, then

$$\mathfrak{a}\mathfrak{b} = \mathfrak{a} \cap \mathfrak{b} = \langle xy \rangle.\tag{3}$$

There is also an analog of division, the IDEAL QUOTIENT $(\mathfrak{a} : \mathfrak{b})$, and there is an analog of the RADICAL, also called the RADICAL $r(\mathfrak{a})$. Given a ring homomorphism $f : A \to B$, ideals in A EXTEND to ideals in B, while ideals in B CONTRACT to ideals in A.

The following formulas summarize operations on ideals, where $\mathfrak{r}^c$ denotes CONTRACT, $\mathfrak{r}^e$ denotes EXTENSION, and $(\mathfrak{a} : \mathfrak{b})$ denotes an IDEAL QUOTIENT.

$$\mathfrak{a}(\mathfrak{b} + \mathfrak{c}) = \mathfrak{a}\mathfrak{b} + \mathfrak{a}\mathfrak{c}\tag{4}$$

$$(\mathfrak{a} : \mathfrak{b})\mathfrak{b} \subset \mathfrak{a}\tag{5}$$

$$(\cap \mathfrak{a}_i : \mathfrak{b}) = \cap (\mathfrak{a}_i : \mathfrak{b})\tag{6}$$

$$(\mathfrak{a} : \sum \mathfrak{b}_i) = \cap (\mathfrak{a} : \mathfrak{b}_i)\tag{7}$$

$$\mathfrak{a} \subset r(\mathfrak{a})\tag{8}$$

$$r(r(\mathfrak{a})) = r(\mathfrak{a})\tag{9}$$

$$r(\mathfrak{a}\mathfrak{b}) = r(\mathfrak{a} \cap \mathfrak{b}) = r(\mathfrak{a}) \cap r(\mathfrak{b})\tag{10}$$

$$r(\mathfrak{a} + \mathfrak{b}) = r(r(\mathfrak{a}) + r(\mathfrak{b}))\tag{11}$$

$$\mathfrak{a} \subset \mathfrak{a}^{ec}\tag{12}$$

$$\mathfrak{b}^{ce} \subset \mathfrak{b}\tag{13}$$

$$\mathfrak{b}^c = \mathfrak{b}^{cec}\tag{14}$$

$$\mathfrak{a}^e = \mathfrak{a}^{ece}\tag{15}$$

$$(\mathfrak{a}_1 + \mathfrak{a}_2)^e = \mathfrak{a}_1^e + \mathfrak{a}_2^e\tag{16}$$

$$\mathfrak{b}_1^c + \mathfrak{b}_2^c \subset (\mathfrak{b}_1 + \mathfrak{b}_2)^c\tag{17}$$

$$(\mathfrak{a}_1 \cap \mathfrak{a}_2)^e \subset \mathfrak{a}_1^e \cap \mathfrak{a}_2^e\tag{18}$$

$$\mathfrak{b}_1^c \cap \mathfrak{b}_2^c = (\mathfrak{b}_1 \cap \mathfrak{b}_2)^c\tag{19}$$

$$\mathfrak{a}_1^e \mathfrak{a}_2^e = (\mathfrak{a}_1 \mathfrak{a}_2)^e\tag{20}$$

$$\mathfrak{b}_1^c \mathfrak{b}_2^c \subset (\mathfrak{b}_1 \mathfrak{b}_2)^c\tag{21}$$

$$(\mathfrak{a}_1 : \mathfrak{a}_2)^e \subset (\mathfrak{a}_1^e : \mathfrak{a}_2^e)\tag{22}$$

$$(\mathfrak{b}_1 : \mathfrak{b}_2)^c \subset (\mathfrak{b}_1^c : \mathfrak{b}_2^c)\tag{23}$$

$$r(\mathfrak{a})^e \subset r(\mathfrak{a}^e)\tag{24}$$

$$r(\mathfrak{b})^c = r(\mathfrak{b}^c) \qquad (25)$$

See also ALGEBRAIC GEOMETRY, CLASS NUMBER, CONTRACTION (IDEAL), DIVISOR THEORY, EXTENSION (IDEAL), HERBRAND'S THEOREM, HILBERT'S NULLSTELLENSATZ, HOMOGENEOUS IDEAL, IDEAL NUMBER, INTEGRAL DOMAIN, IDEAL QUOTIENT, JOSEPH IDEAL, MAXIMAL IDEAL, PRIME IDEAL, PRINCIPAL IDEAL, RADICAL, VARIETY

References

Atiyah, M. F. and MacDonald, I. G. *Introduction to Commutative Algebra.* Reading, MA: Addison-Wesley, pp. 6–0, 1969.

Dedekind, R. "Über die Theorie der ganzen algebraischen Zahlen." X. Supplement to *Vorlesungen über Zahlentheorie, 2nd ed.* Braunschweig, Germany: Vieweg, 1871.

Ferreirós, J. "Ideal Factors." §3.3.1 in *Labyrinth of Thought: A History of Set Theory and Its Role in Modern Mathematics.* Basel, Switzerland: Birkhäuser, pp. 95–7, 1999.

Halter-Koch, F. *Ideal Systems: An Introduction to Multiplicative Ideal Theory.* New York: Dekker, 1998.

Koch, H. "Dedekind's Theory of Ideals." Ch. 3 in *Number Theory: Algebraic Numbers and Functions.* Providence, RI: Amer. Math. Soc., pp. 65–02, 2000.

Malgrange, B. *Ideals of Differentiable Functions.* London: Oxford University Press, 1966.

Ideal (Partial Order)

An ideal I of a PARTIAL ORDER P is a subset of the elements of P which satisfy the property that if $y \in 1$ and $x < y$, then $x \in I$. For k disjoint chains in which the ith chain contains n_i elements, there are $(1 + n_1)(1 + n_2) \cdots (1 + n_k)$ ideals. The number of ideals of a n-element FENCE POSET is the FIBONACCI NUMBER F_n.

References

Ruskey, F. "Information on Ideals of Partially Ordered Sets." http://www.theory.csc.uvic.ca/~cos/inf/pose/Ideals.html.

Steiner, G. "An Algorithm to Generate the Ideals of a Partial Order." *Operat. Res. Let.* **5**, 317–20, 1986.

Ideal Function

DISTRIBUTION (GENERALIZED FUNCTION)

Ideal Number

A type of number involving the ROOTS OF UNITY which was developed by Kummer while trying to solve FERMAT'S LAST THEOREM. Although factorization over the INTEGERS is unique (the FUNDAMENTAL THEOREM OF ALGEBRA), factorization is *not* unique over the COMPLEX NUMBERS. Over the ideal numbers, however, factorization in terms of the COMPLEX NUMBERS becomes unique. Ideal numbers were so powerful that they were generalized by Dedekind into the more abstract IDEALS in general RINGS which are a key part of modern abstract ALGEBRA.

See also DIVISOR THEORY, FERMAT'S LAST THEOREM, IDEAL

References

Ferreirós, J. "Ideal Factors." §3.3.1 in *Labyrinth of Thought: A History of Set Theory and Its Role in Modern Mathematics.* Basel, Switzerland: Birkhäuser, pp. 95–7, 1999.

Ideal Point

A type of POINT AT INFINITY in which parallel lines in the HYPERBOLIC PLANE intersect at infinity in one direction, while diverging from one another in the other.

See also HYPERPARALLEL

Ideal Quotient

The ideal quotient $(\mathfrak{a} : \mathfrak{b})$ is an analog of division for IDEALS in a COMMUTATIVE RING R,

$$(\mathfrak{a} : \mathfrak{b}) = \{x \in R : x\mathfrak{b} \subset \mathfrak{a}\}.$$

The ideal quotient is always another ideal.

However, this operation is not exactly like division. For example, when R is the ring of integers, then $(\langle 12 \rangle : \langle 2 \rangle) = \langle 6 \rangle$, which is nice, while $(\langle 12 \rangle : \langle 5 \rangle) = \langle 12 \rangle$, which is not as nice.

See also ALGEBRAIC GEOMETRY, ALGEBRAIC NUMBER THEORY, IDEAL

Idele

The multiplicative subgroup of all elements in the product of the multiplicative groups $k_v^\times$ whose absolute value is 1 at all but finitely many v, where k is a number FIELD and v a PLACE.

See also ADÉLE

References

Knapp, A. W. "Group Representations and Harmonic Analysis, Part II." *Not. Amer. Math. Soc.* **43**, 537–49, 1996.

Idemfactor

DYADIC

Idempotent

An OPERATOR $\bar{A}$ such that $\bar{A}^2 = \bar{A}$ or an element of an ALGEBRA x such that $x^2 = x$.

See also AUTOMORPHIC NUMBER, BOOLEAN ALGEBRA, GROUP, IDEMPOTENT MATRIX, SEMIGROUP

Idempotent Matrix

A PERIODIC MATRIX with period 1, so that $A^2 = A$.

See also IDEMPOTENT, NILPOTENT MATRIX, PERIODIC MATRIX

Idempotent Number

The idempotent numbers are given by

$$B_{n,k}(1,2,3,\ldots) = \binom{n}{k} k^{n-k},$$

where $B_{n,k}$ is a BELL POLYNOMIAL and $\binom{n}{k}$ is a BINOMIAL COEFFICIENT. A table of the first few is given below.

k	$n=1$ A000027	$n=2$ A001788	$n=3$ A036216	$n=4$ A040075	$n=5$ A050982	$n=6$ A050988	$n=7$ A050989
1	1						
2	2	1					
3	3	6	1				
4	4	24	12	1			
5	5	80	90	20	1		
6	6	240	540	240	30	1	
7	7	672	2835	2240	525	42	1
8	8	1792	13608	17920	7000	1008	56
9	9	4608	61236	129024	78750	18144	1764
10	10	11520	262440	860160	787500	272160	41160

See also BELL POLYNOMIAL, LAH NUMBER

References

Comtet, L. *Advanced Combinatorics: The Art of Finite and Infinite Expansions, rev. enl. ed.* Dordrecht, Netherlands: Reidel, p. 91, 1974.

Roman, S. *The Umbral Calculus.* New York: Academic Press, p. 85, 1984.

Sloane, N. J. A. Sequences A000027/M0472, A001788/ M4161, A036216, A040075, A050982, A050988, and A050989 in "An On-Line Version of the Encyclopedia of Integer Sequences." http://www.research.att.com/~njas/sequences/eisonline.html.

Identical Congruence

FUNCTIONAL CONGRUENCE

Identity

An identity is a mathematical relationship equating one quantity to another (which may initially appear to be different).

See also ABEL'S DIFFERENTIAL EQUATION IDENTITY, ANDREWS-SCHUR IDENTITY, BAC-CAB IDENTITY, BEAUZAMY AND DÉGOT'S IDENTITY, BELTRAMI IDENTITY, BIANCHI IDENTITIES, BOCHNER IDENTITY, BRAHMAGUPTA IDENTITY, CASSINI'S IDENTITY, CAUCHY-LAGRANGE IDENTITY, CHRISTOFFEL-DARBOUX IDENTITY, CHU-VANDERMONDE IDENTITY, DE MOIVRE'S IDENTITY, DOUGALL-RAMANUJAN IDENTITY, EULER FOUR-SQUARE IDENTITY, EULER IDENTITY, EULER POLYNOMIAL IDENTITY, FERRARI'S IDENTITY, FIBONACCI IDENTITY, FROBENIUS TRIANGLE IDENTITIES, GREEN'S IDENTITIES, HYPERGEOMETRIC IDENTITY, IMAGINARY IDENTITY, JACKSON'S IDENTITY, JACOBI IDENTITIES, JACOBI'S DETERMINANT IDENTITY, JORDAN IDENTITY, LAGRANGE'S IDENTITY, LE CAM'S IDENTITY, LEIBNIZ IDENTITY, LIOUVILLE POLYNOMIAL IDENTITY, MATRIX POLYNOMIAL IDENTITY, MORGADO IDENTITY, NEWTON'S IDENTITIES, QUINTUPLE PRODUCT IDENTITY, RAMANUJAN 6-0- IDENTITY, RAMANUJAN COS/COSH IDENTITY, RAMANUJAN'S IDENTITY, RAMANUJAN'S SUM IDENTITY, REZNIK'S IDENTITY, ROGERS-RAMANUJAN IDENTITIES, SCHAAR'S IDENTITY, STREHL IDENTITIES, SYLVESTER'S DETERMINANT IDENTITY, TRINOMIAL IDENTITY, VISIBLE POINT VECTOR IDENTITY, WATSON QUINTUPLE PRODUCT IDENTITY, WORPITZKY'S IDENTITY

References

Petkovsek, M.; Wilf, H. S.; and Zeilberger, D. "Identities." §2.2 in *A = B.* Wellesley, MA: A. K. Peters, pp. 21–2, 1996.

Identity Element

The identity element I (also denoted E, e, or I) of a GROUP or related mathematical structure S is the unique element such that $IA = AI = A$ for every element $A \in S$. The symbol "E" derives from the German word for unity, "Einheit." An identity element is also called a unit element.

See also BINARY OPERATOR, GROUP, INVOLUTION (GROUP), MONOID

Identity Function

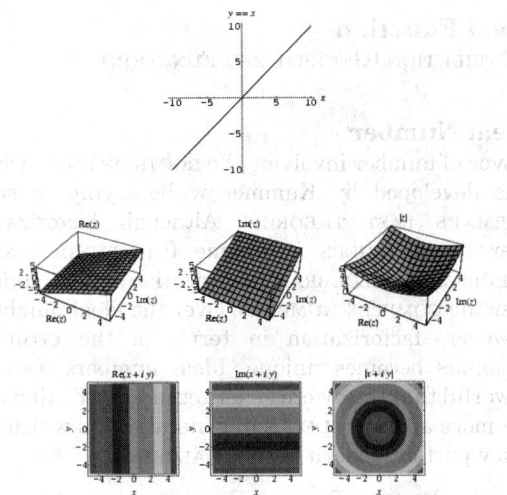

The function $f(x) = x$ which assigns every REAL

NUMBER x to the same REAL NUMBER x. It is identical to the IDENTITY MAP.

Identity Map

The MAP which assigns every member of a set A to the same element id_A. It is identical to the IDENTITY FUNCTION.

See also DONKIN'S THEOREM, IDENTITY FUNCTION, ZERO MAP

Identity Matrix

The identity matrix is a very special BINARY MATRIX denoted I (or I) and defined such that

$$I(\mathbf{X}) \equiv \mathbf{X} \tag{1}$$

for all VECTORS $\mathbf{X}$. The identity matrix is

$$I_{ij} = \delta_{ij} \tag{2}$$

for $i, j = 1, 2, ..., n$, where δ_{ij} is the KRONECKER DELTA. Written explicitly,

$$I = \begin{bmatrix} 1 & 0 & \cdots & 0 \\ 0 & 1 & \cdots & 0 \\ \vdots & \vdots & \ddots & \vdots \\ 0 & 0 & \cdots & 1 \end{bmatrix} \tag{3}$$

The notation E (an abbreviation for the German term, "Einheitsmatrix") is sometimes also used (Courant and Hilbert 1989, p. 7).

"Square root of identity" matrices can be defined for I_n by solving

$$\begin{bmatrix} a_{11} & a_{12} & \cdots & a_{1n} \\ a_{21} & a_{22} & \cdots & a_{2n} \\ \vdots & \vdots & \ddots & \vdots \\ a_{n1} & a_{n2} & \cdots & a_{nn} \end{bmatrix} \begin{bmatrix} a_{11} & a_{12} & \cdots & a_{1n} \\ a_{21} & a_{22} & \cdots & a_{2n} \\ \vdots & \vdots & \ddots & \vdots \\ a_{n1} & a_{n2} & \cdots & a_{nn} \end{bmatrix}$$

$$= \begin{bmatrix} 1 & 0 & \cdots & 0 \\ 0 & 1 & \cdots & 0 \\ \vdots & \vdots & \ddots & 0 \\ 0 & 0 & \cdots & 1 \end{bmatrix}. \tag{4}$$

For $n = 2$, the resulting matrices are

$$I_2^{1/2} = \begin{bmatrix} \pm 1 & 0 \\ 0 & \pm 1 \end{bmatrix}, \begin{bmatrix} \pm 1 & 0 \\ c & \mp 1 \end{bmatrix},$$

$$\begin{bmatrix} \pm 1 & b \\ 0 & \mp 1 \end{bmatrix}, \begin{bmatrix} -d & \dfrac{1 - d^2}{c^2} \\ c & d \end{bmatrix}. \tag{5}$$

"Cube root of identity" matrices can take on even more complicated forms. However, one simple class of such matrices is called K-MATRICES.

See also BINARY MATRIX, IDENTITY MATRIX, K-MATRIX, ZERO MATRIX

References

Ayres, F. Jr. *Theory and Problems of Matrices.* New York: Schaum, p. 10, 1962.
Courant, R. and Hilbert, D. *Methods of Mathematical Physics, Vol. 1.* New York: Wiley, 1989.

Identity Operator

The OPERATOR $\bar{I}$ which takes a REAL NUMBER to the same REAL NUMBER $\bar{I}r = r$.

See also IDENTITY FUNCTION, IDENTITY MAP

Identity Transformation

IDENTITY MAP

Identric Mean

This entry contributed by RONALD M. AARTS

The identric mean is defined by

$$I(a, b) = \frac{1}{e} \left(\frac{b^b}{a^a} \right)^{1/(b-a)}$$

for $a > 0$, $b > 0$, and $a \neq b$. The identric mean has been investigated intensively and many remarkable inequalities for $I(a, b)$ have been published (Bullen *et al.* 1988, Alzer 1993).

References

Alzer, H. "Some Gamma Function Inequalities." *Math. Comput.* **60**, 337–46, 1993.
Bullen, P. S.; Mitrinovic, D. S.; and Vasic, P. M. *Means and Their Inequalities.* Dordrecht, Netherlands: Reidel, 1988.

Idoneal Number

A POSITIVE value of D for which the fact that a number is a MONOMORPH (i.e., the number is expressible in only one way as $x^2 + Dy^2$ or $x^2 - Dy^2$ where x^2 is RELATIVELY PRIME to Dy^2) guarantees it to be a PRIME, POWER of a PRIME, or twice one of these. The numbers are also called EULER'S IDONEAL NUMBERS, or SUITABLE NUMBERS.

The 65 idoneal numbers found by Gauss and Euler and conjectured to be the *only* such numbers (Shanks 1969) are 1, 2, 3, 4, 5, 6, 7, 8, 9, 10, 12, 13, 15, 16, 18, 21, 22, 24, 25, 28, 30, 33, 37, 40, 42, 45, 48, 57, 58, 60, 70, 72, 78, 85, 88, 93, 102, 105, 112, 120, 130, 133, 165, 168, 177, 190, 210, 232, 240, 253, 273, 280, 312, 330, 345, 357, 385, 408, 462, 520, 760, 840, 1320, 1365, and 1848 (Sloane's A000926).

See also MONOMORPH

References

Shanks, D. "On Gauss's Class Number Problems." *Math. Comput.* **23**, 151–63, 1969.
Sloane, N. J. A. Sequences A000926/M0476 in "An On-Line Version of the Encyclopedia of Integer Sequences." http://www.research.att.com/~njas/sequences/eisonline.html.

Iff

If and only if (i.e., NECESSARY and SUFFICIENT). The terms "JUST IF" or "EXACTLY WHEN" are sometimes used instead. A iff B is written symbolically as $A \leftrightarrow B$. A iff B is also equivalent to $A \Rightarrow B$, together with $B \Rightarrow A$, where the symbol $\Rightarrow$ denotes "IMPLIES."

J. H. Conway believes that the word originated with P. Halmos and was transmitted through Kelley (1975). Halmos has stated, "To the best of my knowledge, I *did* invent the silly thing, but I wouldn't swear to it in a court of law. So there–give me credit for it anyway" (D. Asimov 1997).

See also EQUIVALENT, EXACTLY ONE, IMPLIES, NECESSARY, SUFFICIENT

References

Asimov, D. "Iff." math-fun@cs.arizona.edu posting, Sept. 19, 1997.
Kelley, J. L. *General Topology.* New York: Springer-Verlag, 1975.

Ill-Conditioned Matrix

A MATRIX is ill-conditioned if the CONDITION NUMBER is too large (and SINGULAR if it is INFINITE).

See also CONDITION NUMBER, SINGULAR MATRIX, SINGULAR VALUE DECOMPOSITION

References

Arfken, G. "Ill-Conditioned Systems." *Mathematical Methods for Physicists, 3rd ed.* Orlando, FL: Academic Press, pp. 233–34, 1985.

Ill Defined

A solution to a PARTIAL DIFFERENTIAL EQUATION that is *not* a continuous function of its values on the boundary is said to be ill defined. Otherwise, a solution is called WELL DEFINED.

The term "ill defined" is also used informally to mean AMBIGUOUS.

See also AMBIGUOUS, WELL DEFINED

Illumination Problem

In the early 1950s, Ernst Straus asked

1. Is every POLYGONAL region illuminable from every point in the region?
2. Is every POLYGONAL region illuminable from at least one point in the region?

Here, illuminable means that there is a path from every point to every other by repeated reflections. Tokarsky (1995) showed that unilluminable rooms exist in the plane and 3-D, but question (2) remains open. The smallest known counterexample to (1) in the PLANE has 26 sides.

See also ART GALLERY THEOREM

References

Croft, H. T.; Falconer, K. J.; and Guy, R. K. "Illumination Problems." §A5 in *Unsolved Problems in Geometry.* New York: Springer-Verlag, pp. 18–9, 1991.
Klee, V. "Is Every Polygonal Region Illuminable from Some Point?" *Math. Mag.* **52**, 180, 1969.
Tokarsky, G. W. "Polygonal Rooms Not Illuminable from Every Point." *Amer. Math. Monthly* **102**, 867–79, 1995.

Illusion

An object or drawing which appears to have properties which are physically impossible, deceptive, or counterintuitive.

See also BENHAM'S WHEEL, BLACK DOT ILLUSION, BULLSEYE ILLUSION, FREEMISH CRATE, GOBLET ILLUSION, HERMANN GRID ILLUSION, HERMANN-HERING ILLUSION, HYZER'S ILLUSION, IMPOSSIBLE FIGURE, IRRADIATION ILLUSION, KANIZSA TRIANGLE, MÜLLER-LYER ILLUSION, NECKER CUBE, ORBISON'S ILLUSION, PARALLELOGRAM ILLUSION, PENROSE STAIRWAY, POGGENDORFF ILLUSION, PONZO'S ILLUSION, RABBIT-DUCK ILLUSION, TRIBAR, TRIBOX, VERTICAL-HORIZONTAL ILLUSION, YOUNG GIRL-OLD WOMAN ILLUSION, ZÖLLNER'S ILLUSION

References

Ausbourne, B. "A Sensory Adventure." http://www.lainet.com/illusions/.
Ausbourne, B. "Optical Illusions: A Collection." http://www.lainet.com/~ausbourn/.
Ernst, B. *Optical Illusions.* New York: Taschen, 1996.
Fineman, M. *The Nature of Visual Illusion.* New York: Dover, 1996.
Gardner, M. "Optical Illusions." Ch. 1 in *Mathematical Circus: More Puzzles, Games, Paradoxes and Other Mathematical Entertainments from Scientific American.* New York: Knopf, pp. 3–5, 1979.
Gregory, R. L. *Eye and Brain, 5th ed.* Princeton, NJ: Princeton University Press, 1997.
Illusion Works. "Interactive Optical Illusions." http://www.illusionworks.com/.
Jablan, S. "Modularity in Art." http://www.mi.sanu.ac.yu/~jablans/d3.htm.
Landrigad, D. "Gallery of Illusions." http://dragon.uml.edu/psych/illusion/.html.
Luckiesh, M. *Visual Illusions: Their Causes, Characteristics, and Applications.* New York: Dover, 1965.
Pappas, T. "History of Optical Illusions." *The Joy of Mathematics.* San Carlos, CA: Wide World Publ./Tetra, pp. 172–73, 1989.
Robinson, J. O. *The Psychology of Visual Illusion.* New York: Dover, 1998.
Tolansky, S. *Optical Illusions.* New York: Pergamon Press, 1964.

Im

IMAGINARY PART

Image

RANGE (IMAGE)

Imaginary Axis

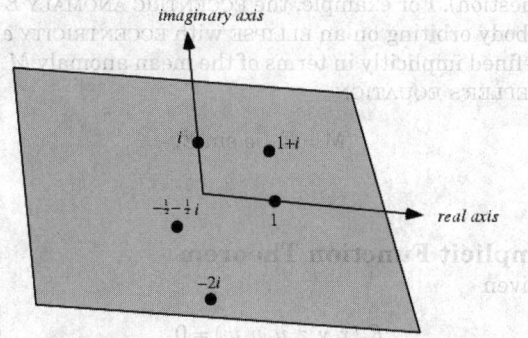

The axis in the COMPLEX PLANE corresponding to zero REAL PART, $\Re[z] = 0$.

See also COMPLEX PLANE, IMAGINARY LINE, REAL AXIS

Imaginary Identity

I

Imaginary Line

A "line" having imaginary coefficients in its equations which can arise in algebraic geometry.

See also IMAGINARY AXIS, LINE, REAL LINE

Imaginary Number

A COMPLEX NUMBER which has zero REAL PART, so that it can be written as a REAL NUMBER multiplied by the "IMAGINARY UNIT" I (equal to the SQUARE ROOT $\sqrt{-1}$).

See also COMPLEX NUMBER, GALOIS IMAGINARY, GAUSSIAN INTEGER, I, IMAGINARY PART, IMAGINARY UNIT, REAL NUMBER

References

Conway, J. H. and Guy, R. K. *The Book of Numbers.* New York: Springer-Verlag, pp. 211–16, 1996.

Imaginary Part

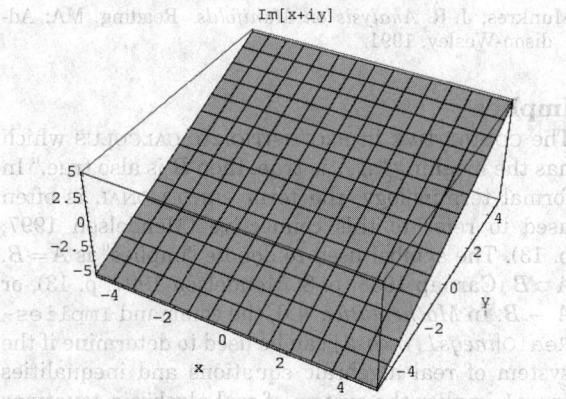

The imaginary part $\Im[z]$ of a COMPLEX NUMBER $z =$

$x + iy$ is the REAL NUMBER multiplying I, so $\Im[x + iy] = y$. In terms of z itself,

$$\Im[z] = \frac{z - \bar{z}}{2i},$$

where $\bar{z}$ is the COMPLEX CONJUGATE of z. The imaginary part is implemented in *Mathematica* as Im[z].

See also ABSOLUTE SQUARE, ARGUMENT (COMPLEX NUMBER), COMPLEX CONJUGATE, COMPLEX PLANE, MODULUS (COMPLEX NUMBER), REAL PART

References

Abramowitz, M. and Stegun, C. A. (Eds.). *Handbook of Mathematical Functions with Formulas, Graphs, and Mathematical Tables, 9th printing.* New York: Dover, p. 16, 1972.
Krantz, S. G. *Handbook of Complex Analysis.* Boston, MA: Birkhäuser, p. 2, 1999.

Imaginary Point

A pair of values x and y one or both of which is COMPLEX.

References

Woods, F. S. *Higher Geometry: An Introduction to Advanced Methods in Analytic Geometry.* New York: Dover, p. 2, 1961.

Imaginary Quadratic Field

A QUADRATIC FIELD $\mathbb{Q}(\sqrt{D})$ with $D < 0$.

See also JUGENDTRAUM, QUADRATIC FIELD

Imaginary Unit

The IMAGINARY NUMBER $i = \sqrt{-1}$, i.e., the SQUARE ROOT of -1. The imaginary unit is denoted and commonly referred to as "I." Although there are two possible square roots of any number, the square roots of a negative number cannot be distinguished until one of the two is defined as the imaginary unit, at which point $+i$ and $-i$ can then be distinguished. Since either choice is possible, there is no ambiguity in defining i as "the" square root of -1. In *Mathematica*, the imaginary unit is implemented as I.

See also COMPLEX NUMBER, I, IMAGINARY NUMBER, UNIT

Immanant

For an $n \times n$ matrix, let S denote any permutation e_1, e_2, ..., e_n of the set of numbers 1, 2, ..., n, and let $\chi^{(\lambda)}(S)$ be the character of the symmetric group corresponding to the partition (λ). Then the immanant $|a_{mn}|^{(\lambda)}$ is defined as

$$|a_{mn}|^{(\lambda)} = \sum \chi^{(\lambda)}(S) P_S$$

where the summation is over the $n!$ permutations of

the SYMMETRIC GROUP and

$$P_S = a_{1e_1} a_{2e_2} \cdots a_{ne_n}.$$

See also DETERMINANT, PERMANENT

References

Littlewood, D. E. and Richardson, A. R. "Group Characters and Algebra." *Philos. Trans. Roy. Soc. London A* **233**, 99–41, 1934.
Littlewood, D. E. and Richardson, A. R. "Immanants of Some Special Matrices." *Quart. J. Math. (Oxford)* **5**, 269–82, 1934.
Wybourne, B. G. "Immanants of Matrices." §2.19 in *Symmetry Principles and Atomic Spectroscopy.* New York: Wiley, pp. 12–3, 1970.

Immersed Minimal Surface
ENNEPER'S MINIMAL SURFACE

Immersion

A special nonsingular MAP from one MANIFOLD to another such that at every point in the domain of the map, the DERIVATIVE is an injective linear map. This is equivalent to saying that every point in the DOMAIN has a NEIGHBORHOOD such that, up to DIFFEOMORPHISMS of the TANGENT SPACE, the map looks like the inclusion map from a lower-dimensional EUCLIDEAN SPACE to a higher-dimensional EUCLIDEAN SPACE.

See also BOY SURFACE, EVERSION, SMALE-HIRSCH THEOREM, SUBMERSION

References

Boy, W. "Über die Curvatura integra und die Topologie geschlossener Flächen." *Math. Ann* **57**, 151–84, 1903.
Pinkall, U. "Models of the Real Projective Plane." Ch. 6 in *Mathematical Models from the Collections of Universities and Museums* (Ed. G. Fischer). Braunschweig, Germany: Vieweg, pp. 63–7, 1986.

Immersion Theorem
SMALE-HIRSCH THEOREM

Impartial Game

A GAME in which the possible moves are the same for each player in any position. All positions in all impartial GAMES form an additive ABELIAN GROUP. For impartial games in which the last player wins (normal form games), the nim-value of the sum of two GAMES is the nim-sum of their nim-values. If the last player loses, the GAME is said to be in misère form and the analysis is much more difficult.

See also FAIR GAME, GAME, PARTISAN GAME

Implicit Function

A function which is not defined explicitly, but rather is defined in terms of an algebraic relationship (which can not, in general, be "solved" for the function in question). For example, the ECCENTRIC ANOMALY E of a body orbiting on an ELLIPSE with ECCENTRICITY e is defined implicitly in terms of the mean anomaly M by KEPLER'S EQUATION

$$M = E - e \sin E.$$

Implicit Function Theorem
Given

$$F_1(x, y, z, u, v, w) = 0 \qquad (1)$$

$$F_2(x, y, z, u, v, w) = 0 \qquad (2)$$

$$F_3(x, y, z, u, v, w) = 0 \qquad (3)$$

if the JACOBIAN

$$JF(u, v, w) = \frac{\partial(F_1, F_2, F_3)}{\partial(u, v, w)} \neq 0, \qquad (4)$$

then u, v, and w can be solved for in terms of x, y, and z and PARTIAL DERIVATIVES of u, v, w with respect to x, y, and z can be found by differentiating implicitly.

More generally, let A be an OPEN SET in $\mathbb{R}^{n+k}$ and let $f : A \to \mathbb{R}^n$ be a C^τ FUNCTION. Write f in the form $f(x, y)$, where x and y are elements of $\mathbb{R}^k$ and $\mathbb{R}^n$. Suppose that (a, b) is a point in A such that $f(a, b) = 0$ and the DETERMINANT of the $n \times n$ MATRIX whose elements are the DERIVATIVES of the n component FUNCTIONS of f with respect to the n variables, written as y, evaluated at (a, b), is not equal to zero. The latter may be rewritten as

$$\text{rank}(Df(a, b)) = n. \qquad (5)$$

Then there exists a NEIGHBORHOOD B of a in $\mathbb{R}^k$ and a unique C^τ FUNCTION $g : B \to \mathbb{R}^n$ such that $g(a) = b$ and $f(x, g(x)) = 0$ for all $x \in B$.

See also CHANGE OF VARIABLES THEOREM, JACOBIAN

References

Munkres, J. R. *Analysis on Manifolds.* Reading, MA: Addison-Wesley, 1991.

Implies

The CONNECTIVE in PROPOSITIONAL CALCULUS which has the meaning "if A is true, then B is also true." In formal terminology, the term CONDITIONAL is often used to refer to this connective (Mendelson 1997, p. 13). The symbol used to denote "implies" is $A \Rightarrow B$, $A \supset B$ (Carnap 1958, p. 8; Mendelson 1997, p. 13), or $A \to B$. In *Mathematica* 4.0, the command Implies-RealQ[*ineqs1*, *ineqs2*] can be used to determine if the system of real algebraic equations and inequalities *ineqs1* implies the system of real algebraic equations and inequalities *ineqs2*.

$A \Rightarrow B$ is an abbreviation for $!A \vee B$, where $!A$ denotes NOT and $\vee$ denoted OR. $\Rightarrow$ is a binary operator that is implement in *Mathematica* as `Implies[A, B]`, and can *not* be extended to more than two arguments.

$A \Rightarrow B$ has the following TRUTH TABLE (Carnap 1958, p. 10; Mendelson 1997, p. 13).

A	B	$A \Rightarrow B$
T	T	T
T	F	F
F	T	T
F	F	T

If $A \Rightarrow B$ and $B \Rightarrow A$ (i.e, $A \Rightarrow B \wedge B \Rightarrow A$), then A and B are said to be EQUIVALENT, a relationship which is written symbolically as $A \Leftrightarrow B$, $A \rightleftharpoons B$, or $A \equiv B$ (Carnap 1958, p. 8).

See also CONNECTIVE, EQUIVALENT, EXISTS, FOR ALL, QUANTIFIER

References

Carnap, R. *Introduction to Symbolic Logic and Its Applications.* New York: Dover, p. 8, 1958.

Impossible Figure

A class of ILLUSION in which an object which is physically unrealizable is apparently depicted.

See also FREEMISH CRATE, HOME PLATE, ILLUSION, NECKER CUBE, PENROSE STAIRWAY, TRIBAR

References

Cowan, T. M. "The Theory of Braids and the Analysis of Impossible Figures." *J. Math. Psych.* **11**, 190–12, 1974.
Cowan, T. M. "Supplementary Report: Braids, Side Segments, and Impossible Figures." *J. Math. Psych.* **16**, 254–60, 1977.
Cowan, T. M. "Organizing the Properties of Impossible Figures." *Perception* **6**, 41–6, 1977.
Cowan, T. M. and Pringle, R. "An Investigation of the Cues Responsible for Figure Impossibility." *J. Exper. Psych. Human Perception Performance* **4**, 112–20, 1978.
Ernst, B. *Adventures with Impossible Figures.* Stradbroke, England: Tarquin, 1987.
Harris, W. F. "Perceptual Singularities in Impossible Pictures Represent Screw Dislocations." *South African J. Sci.* **69**, 10–3, 1973.
Fineman, M. *The Nature of Visual Illusion.* New York: Dover, pp. 119–22, 1996.
Jablan, S. "Impossible Figures." http://members.tripod.com/~modularity/impos.htm and "Are Impossible Figures Possible?" http://members.tripod.com/~modularity/kulpa.htm.
Kulpa, Z. "Are Impossible Figures Possible?" *Signal Processing* **5**, 201–20, 1983.
Kulpa, Z. "Putting Order in the Impossible." *Perception* **16**, 201–14, 1987.
Sugihara, K. "Classification of Impossible Objects." *Perception* **11**, 65–4, 1982.

Terouanne, E. "Impossible Figures and Interpretations of Polyhedral Figures." *J. Math. Psych.* **27**, 370–05, 1983.
Terouanne, E. "On a Class of 'Impossible' Figures: A New Language for a New Analysis." *J. Math. Psych.* **22**, 24–7, 1983.
Thro, E. B. "Distinguishing Two Classes of Impossible Objects." *Perception* **12**, 733–51, 1983.
Wilson, R. "Stamp Corner: Impossible Figures." *Math. Intell.* **13**, 80, 1991.

Impredicative

Definitions about a SET which depend on the entire SET.

Improper Divisor

A DIVISOR which is not a PROPER DIVISOR.

See also DIVISOR, PROPER DIVISOR

Improper Fraction

A FRACTION $p/q > 1$. A FRACTION with $p/q < 1$ is called a PROPER FRACTION. Therefore, the special cases 1/1, 2/2, 3/3, etc. are generally considered to be improper.

See also FRACTION, MIXED FRACTION, PROPER FRACTION

Improper Integral

An INTEGRAL which has either or both limits INFINITE or which has an INTEGRAND which approaches INFINITY at one or more points in the range of integration.

See also DEFINITE INTEGRAL, INDEFINITE INTEGRAL, INTEGRAL, PROPER INTEGRAL

References

Jeffreys, H. and Jeffreys, B. S. "Infinite and Improper Integrals." §1.104 in *Methods of Mathematical Physics, 3rd ed.* Cambridge, England: Cambridge University Press, pp. 33–4, 1988.
Press, W. H.; Flannery, B. P.; Teukolsky, S. A.; and Vetterling, W. T. "Improper Integrals." §4.4 in *Numerical Recipes in FORTRAN: The Art of Scientific Computing, 2nd ed.* Cambridge, England: Cambridge University Press, pp. 135–40, 1992.

Improper Node

A FIXED POINT for which the STABILITY MATRIX has equal nonzero EIGENVECTORS.

See also STABLE IMPROPER NODE, UNSTABLE IMPROPER NODE

Improper Rotation

The SYMMETRY OPERATION corresponding to a ROTATION followed by an INVERSION OPERATION, also called a ROTOINVERSION. This operation is denoted $\bar{n}$ for an improper rotation by $360°/n$ so the CRYSTALLOGRAPHY RESTRICTION gives only $\bar{1}$, $\bar{2}$, $\bar{3}$, $\bar{4}$, $\bar{6}$ for crystals. The

MIRROR PLANE symmetry operation is $(x,y,z) \rightarrow (x,y,-z)$, etc., which is equivalent to $\bar{2}$.

See also INVERSION OPERATION, ROTATION, SYMMETRY OPERATION

Impulse Pair

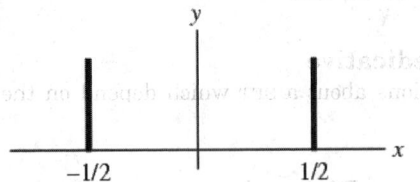

The even impulse pair is the FOURIER TRANSFORM of $\cos(\pi k)$,

$$\Pi(x) \equiv \tfrac{1}{2}\delta\left(x+\tfrac{1}{2}\right) + \tfrac{1}{2}\delta\left(x-\tfrac{1}{2}\right). \quad (1)$$

It satisfies

$$\Pi(x) * f(x) = \tfrac{1}{2}f\left(x+\tfrac{1}{2}\right) + \tfrac{1}{2}f\left(x-\tfrac{1}{2}\right), \quad (2)$$

where $*$ denotes CONVOLUTION, and

$$\int_{-\infty}^{\infty} \Pi(x)\,dx = 1. \quad (3)$$

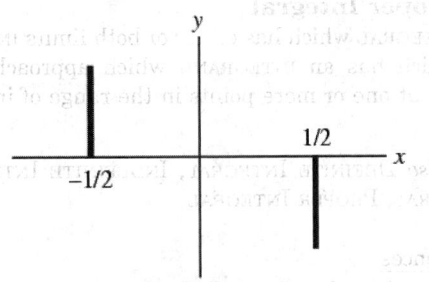

The odd impulse pair is the FOURIER TRANSFORM of $i\sin(\pi s)$,

$$\mathrm{I}_{\mathrm{I}}(x) \equiv \tfrac{1}{2}\delta\left(x+\tfrac{1}{2}\right) - \tfrac{1}{2}\delta\left(x-\tfrac{1}{2}\right). \quad (4)$$

Impulse Symbol

Bracewell's term for the DELTA FUNCTION.

See also DELTA FUNCTION, IMPULSE PAIR

References
Bracewell, R. *The Fourier Transform and Its Applications*, 3rd ed. New York: McGraw-Hill, 1999.

Inaccessible Cardinal

An inaccessible cardinal is a CARDINAL NUMBER which cannot be expressed in terms of a smaller number of smaller cardinals.

See also CARDINAL NUMBER

Inaccessible Cardinals Axiom

INACCESSIBLE CARDINAL, LEBESGUE MEASURABILITY PROBLEM

Inadmissible

A word or string which is not ADMISSIBLE.

In-and-Out Curve

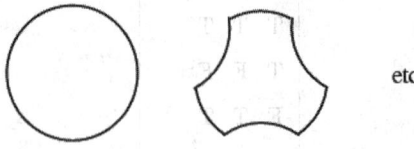

A curve created by starting with a circle, dividing it into six arcs, and flipping three alternating arcs. The process is then repeated an infinite number of times.

Incenter

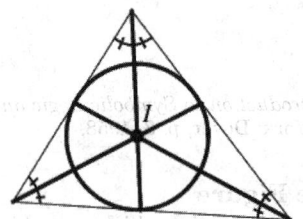

The center I of a TRIANGLE'S INCIRCLE. It can be found as the intersection of ANGLE BISECTORS, and it is the interior point for which distances to the sides of the triangle are equal. It has TRILINEAR COORDINATES 1:1:1 and homogeneous BARYCENTRIC COORDINATES (a,b,c). The distance between the incenter and CIRCUMCENTER is $\sqrt{R(R-2r)}$.

The incenter lies on the NAGEL LINE and SODDY LINE. The incenter lies on the EULER LINE only for an ISOSCELES TRIANGLE. For an EQUILATERAL TRIANGLE, the CIRCUMCENTER O, CENTROID G, NINE-POINT CENTER F, ORTHOCENTER H, and DE LONGCHAMPS POINT Z all coincide with I.

The incenter and EXCENTERS of a TRIANGLE are an ORTHOCENTRIC SYSTEM. The POWER of the incenter with respect to the CIRCUMCIRCLE is

$$p = \frac{a_1 a_2 a_3}{a_1 + a_2 + a_3}$$

(johnson 1929, p. 190). if the incenters of the TRIANGLES $\Delta A_1 H_2 H_3$, $\Delta A_2 H_3 A_1$, and $\Delta A_3 H_1 H_2$ are X_1, X_2, and X_3, then $X_2 X_3$ is equal and parallel to $I_2 I_3$, where H_i are the FEET of the ALTITUDES and I_i are the incenters of the TRIANGLES. Furthermore, X_1, X_2, X_3, are the reflections of I with respect to the sides of the TRIANGLE $\Delta I_1 I_2 I_3$ (Johnson 1929, p. 193).

See also CENTROID (ORTHOCENTRIC SYSTEM), CIRCUM-CENTER, CYCLIC QUADRILATERAL, EXCENTER, GERGONNE POINT, INCIRCLE, INRADIUS, ORTHOCENTER, NAGEL LINE

References

Carr, G. S. *Formulas and Theorems in Pure Mathematics*, 2nd ed. New York: Chelsea, p. 622, 1970.
Coxeter, H. S. M. and Greitzer, S. L. *Geometry Revisited.* Washington, DC: Math. Assoc. Amer., p. 10, 1967.
Dixon, R. *Mathographics.* New York: Dover, p. 58, 1991.
Johnson, R. A. *Modern Geometry: An Elementary Treatise on the Geometry of the Triangle and the Circle.* Boston, MA: Houghton Mifflin, pp. 182–94, 1929.
Kimberling, C. "Central Points and Central Lines in the Plane of a Triangle." *Math. Mag.* **67**, 163–87, 1994.
Kimberling, C. "Incenter." http://cedar.evansville.edu/~ck6/tcenters/class/incenter.html.
Wells, D. *The Penguin Dictionary of Curious and Interesting Geometry.* London: Penguin, pp. 115–16, 1991.

Incenter-Excenter Circle

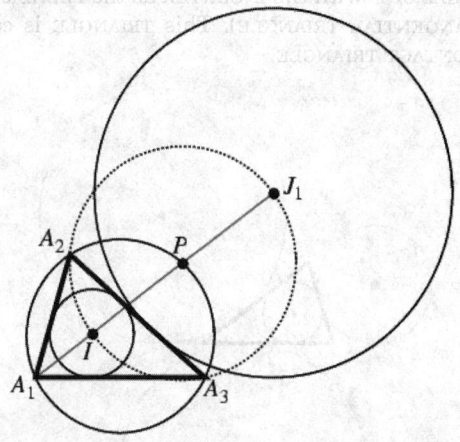

Given a triangle $\Delta A_1 A_2 A_3$, the points A_1, I, and J_1 lie on a line, where I is the INCENTER and J_1 is the EXCENTER corresponding to A_1. Furthermore, the CIRCLE with IJ_1 as the DIAMETER has P as its center, where P is the intersection of $A_1 J_1$ with the CIRCUMCIRCLE of $\Delta A_1 A_2 A_3$, and passes through A_2 and A_3. This CIRCLE has RADIUS

$$r = \tfrac{1}{2} a_1 \sec\left(\tfrac{1}{2}\alpha_1\right) = 2R \sin\left(\tfrac{1}{2}\alpha_1\right).$$

It arises because $IJ_1 J_2 J_3$ forms an ORTHOCENTRIC SYSTEM.

See also CIRCUMCIRCLE, EXCENTER, EXCENTER-EXCENTER CIRCLE, INCENTER, ORTHOCENTRIC SYSTEM

References

Johnson, R. A. *Modern Geometry: An Elementary Treatise on the Geometry of the Triangle and the Circle.* Boston, MA: Houghton Mifflin, p. 185, 1929.

Incidence Axioms

The eight of HILBERT'S AXIOMS which concern collinearity and intersection; they include the first four of EUCLID'S POSTULATES.

See also ABSOLUTE GEOMETRY, CONGRUENCE AXIOMS, CONTINUITY AXIOMS, EUCLID'S POSTULATES, HILBERT'S AXIOMS, ORDERING AXIOMS, PARALLEL POSTULATE

References

Hilbert, D. *The Foundations of Geometry*, 2nd ed. Chicago, IL: Open Court, 1980.
Iyanaga, S. and Kawada, Y. (Eds.). "Hilbert's System of Axioms." §163B in *Encyclopedic Dictionary of Mathematics.* Cambridge, MA: MIT Press, pp. 544–45, 1980.

Incidence Matrix

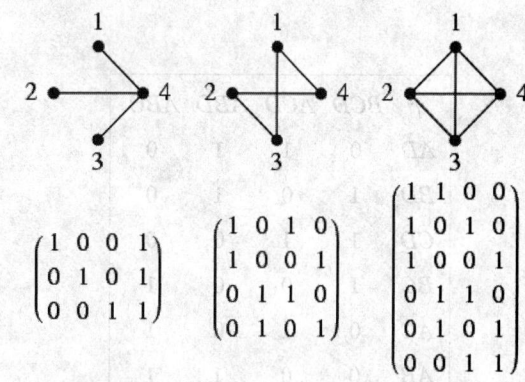

$$\begin{pmatrix} 1 & 0 & 0 & 1 \\ 0 & 1 & 0 & 1 \\ 0 & 0 & 1 & 1 \end{pmatrix} \quad \begin{pmatrix} 1 & 0 & 1 & 0 \\ 1 & 0 & 0 & 1 \\ 0 & 1 & 1 & 0 \\ 0 & 1 & 0 & 1 \end{pmatrix} \quad \begin{pmatrix} 1 & 1 & 0 & 0 \\ 1 & 0 & 1 & 0 \\ 1 & 0 & 0 & 1 \\ 0 & 1 & 1 & 0 \\ 0 & 1 & 0 & 1 \\ 0 & 0 & 1 & 1 \end{pmatrix}$$

The incidence matrix of a GRAPH gives the $(0,1)$-MATRIX which has a row for each vertex and column for each edge, and $(v,e) = 1$ IFF vertex v is incident upon edge e (Skiena 1990, p. 135). The physicist Kirchhoff (1847) was the first to define the incidence matrix. The incidence matrix of a graph can be computed using `IncidenceMatrix[g]` in the *Mathematica* add-on package `DiscreteMath`Combinatorica`` (which can be loaded with the command `<<DiscreteMath``).

The incidence matrix C of a graph and ADJACENCY MATRIX L of its LINE GRAPH are related by

$$\mathsf{L} = \mathsf{C}^{\mathsf{T}}\mathsf{C} - 2\mathsf{I},$$

where I is the IDENTITY MATRIX (Skiena 1990, p. 136). For a k-D POLYTOPE Π_k, the incidence matrix is defined by

$$\eta_{ij}^k = \begin{cases} 1 & \text{if } \Pi_{k-1}^i \text{ belongs to } \Pi_k^i \\ 0 & \text{if } \Pi_{k-1}^i \text{ does not belong } \Pi_k^i \end{cases}$$

The ith row shows which Π_ks surround Π_{k-1}^i, and the jth column shows which Π_{k-1}s bound Π_k^j. Incidence matrices are also used to specify PROJECTIVE PLANES. The incidence matrices for a TETRAHEDRON $ABCD$ are

η^0	1	A	B	C
	1	1	1	1

η^1	AD	BD	CD	BC	AC	AB
A	1	0	0	0	1	1
B	0	1	0	1	0	1
C	0	0	1	1	1	0
D	1	1	1	0	0	0

η^2	BCD	ACD	ABD	ABC
AD	0	1	1	0
BD	1	0	1	0
CD	1	1	0	0
BC	1	0	0	1
AC	0	1	0	1
AB	0	0	1	1

η^3	ABCD
BCD	1
ACD	1
ABD	1
ABC	1

See also ADJACENCY MATRIX, κ-CHAIN, κ-CIRCUIT, INTEGER MATRIX

References

Bruck, R. H. and Ryser, H. J. "The Nonexistence of Certain Finite Projective Planes." *Canad. J. Math.* **1**, 88–3, 1949.

Kirchhoff, G. "Über die Auflösung der Gleichungen, auf welche man bei der untersuchung der linearen verteilung galvanischer Ströme geführt wird." *Ann. Phys. Chem.* **72**, 497–08, 1847.

Skiena, S. *Implementing Discrete Mathematics: Combinatorics and Graph Theory with Mathematica.* Reading, MA: Addison-Wesley, pp. 135–36, 1990.

Incident

Two objects which touch each other are said to be incident.

See also CONCUR, TANGENT CURVES

Incircle

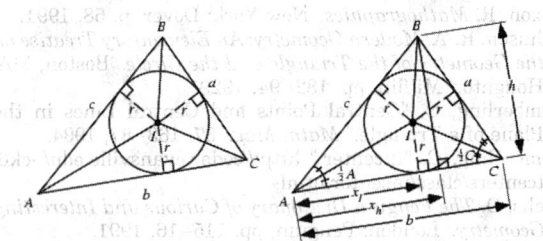

The INSCRIBED CIRCLE of a TRIANGLE ΔABC. The center I of the incircle is called the INCENTER and the RADIUS r the INRADIUS. The points of intersection of the incircle with T are the VERTICES of the PEDAL TRIANGLE of T with the INCENTER as the PEDAL POINT (cf. TANGENTIAL TRIANGLE). This TRIANGLE is called the CONTACT TRIANGLE.

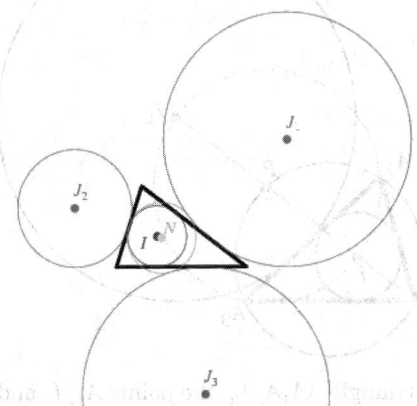

There are four CIRCLES that are tangent all three sides (or their extensions) of a given TRIANGLE: the incircle I and three EXCIRCLES J_1, J_2, and J_3. These four circles are, in turn, all touched by the NINE-POINT CIRCLE N.

The TRILINEAR COORDINATES of the INCENTER are $1 : 1 : 1$. The INRADIUS r and horizontal position of the INCENTER x_I for a given triangle with two angles A and C and adjacent side of length b is given by simultaneously solving the equations

$$\tan\left(\tfrac{1}{2}A\right) = \frac{r}{x_I} \tag{1}$$

$$\tan\left(\tfrac{1}{2}C\right) = \frac{r}{b - x_I}, \tag{2}$$

giving

$$r = \frac{\tan\left(\frac{1}{2}A\right)\tan\left(\frac{1}{2}C\right)}{\tan\left(\frac{1}{2}A\right) + \tan\left(\frac{1}{2}C\right)} b \qquad (3)$$

$$x_I = \frac{\tan\left(\frac{1}{2}C\right)}{\tan\left(\frac{1}{2}A\right) + \tan\left(\frac{1}{2}C\right)} b, \qquad (4)$$

whereas the ALTITUDE height h and horizontal position x_h of the ALTITUDE, are given by

$$h = \frac{\tan C}{\tan A + \tan C} b \qquad (5)$$

$$x_h = \frac{\tan A \tan C}{\tan A + \tan C} b. \qquad (6)$$

The AREA Δ of the TRIANGLE ΔABC is given by

$$\Delta = \Delta BIC + \Delta AIC + \Delta AIB$$

$$= \tfrac{1}{2}ar + \tfrac{1}{2}br + \tfrac{1}{2}cr = \tfrac{1}{2}(a+b+c)r = sr, \qquad (7)$$

where s is the SEMIPERIMETER, so the INRADIUS is

$$r = \frac{\Delta}{s} = \sqrt{\frac{(s-a)(s-b)(s-c)}{s}} \qquad (8)$$

Using the incircle of a TRIANGLE as the INVERSION CENTER, the sides of the TRIANGLE and its CIRCUMCIRCLE are carried into four equal CIRCLES (Honsberger 1976, p. 21). Pedoe (1995, p. xiv) gives a GEOMETRIC CONSTRUCTION for the incircle.

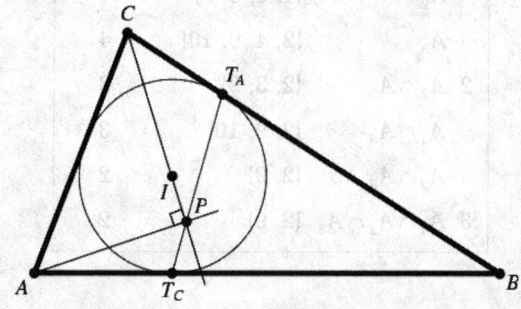

Let a triangle ΔABC have INCIRCLE with INCENTER I and let the incircle be tangent to ΔABC at T_A, T_C, (and T_B; not shown). Then the lines CI, $T_A T_C$, and the perpendicular to CI through A CONCUR in a point P (Honsberger 1995).

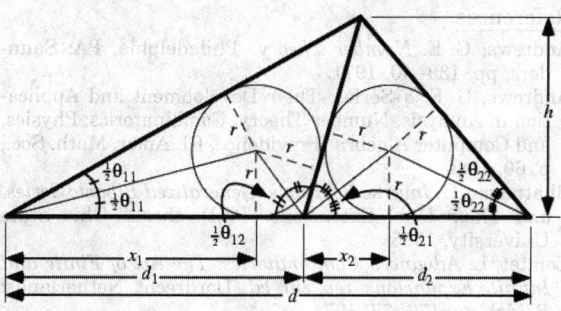

Given a triangle, draw a CEVIAN to one of the bases which divides it into two triangles having congruent incircles. The positions and sizes of these two circumcircles can then be determined by simultaneously solving the eight equations

$$x_1 = \frac{\tan\left(\frac{1}{2}\theta_{12}\right)}{\tan\left(\frac{1}{2}\theta_{11} + \tan\left(\frac{1}{2}\theta_{12}\right)\right)} d_1 \qquad (9)$$

$$x_2 = \frac{\tan\left(\frac{1}{2}\theta_{22}\right)}{\tan\left(\frac{1}{2}\theta_{21}\right) + \tan\left(\frac{1}{2}\theta_{22}\right)} d_2 \qquad (10)$$

$$a = \frac{\tan\left(\frac{1}{2}\theta_{11}\right)\tan\left(\frac{1}{2}\theta_{12}\right)}{\tan\left(\frac{1}{2}\theta_{11}\right) + \tan\left(\frac{1}{2}\theta_{12}\right)} d_1 \qquad (11)$$

$$a = \frac{\tan\left(\frac{1}{2}\theta_{21}\right)\tan\left(\frac{1}{2}\theta_{22}\right)}{\tan\left(\frac{1}{2}\theta_{21}\right) + \tan\left(\frac{1}{2}\theta_{22}\right)} d_2 \qquad (12)$$

$$h = \frac{\tan\theta_{11}\tan\theta_{12}}{\tan\theta_{11} + \tan\theta_{12}} d_1 \qquad (13)$$

$$h = \frac{\tan\theta_{21}\tan\theta_{22}}{\tan\theta_{21} + \tan\theta_{22}} d_2 \qquad (14)$$

$$d = d_1 + d_2 \qquad (15)$$

$$\pi = \theta_{12} + \theta_{21} \qquad (16)$$

for the eight variables d_1, d_2, θ_{12}, θ_{21}, a, x_1, x_2, and h, with θ_{11}, θ_{22}, and d given. Generalizing to n congruent circles gives the $4n$ equations

$$x_i = \frac{\tan\left(\frac{1}{2}\theta_{i2}\right)}{\tan\left(\frac{1}{2}\theta_{i1}\right) + \tan\left(\frac{1}{2}\theta_{i2}\right)} d_i \qquad (17)$$

$$a = \frac{\tan\left(\frac{1}{2}\theta_{i1}\right)\tan\left(\frac{1}{2}\theta_{i2}\right)}{\tan\left(\frac{1}{2}\theta_{i1}\right) + \tan\left(\frac{1}{2}\theta_{i2}\right)} d_i \qquad (18)$$

$$h = \frac{\tan\theta_{i1}\tan\theta_{i2}}{\tan\theta_{i1} + \tan\theta_{i2}} d_i \qquad (19)$$

for $i = 1, \ldots, n$,

$$\theta_{i2} + \theta_{i+1,1} = \pi \qquad (20)$$

for $i = 1, \ldots, n-1$, and

$$d = \sum_{i=1}^{n} d_i \qquad (21)$$

to be solved for the unknowns d_i and x_i (n of them), θ_{i1} and θ_{i2} ($n-2$ of each for $i = 2, \ldots, n-1$), and θ_{12}, θ_{n1}, a, and h, a total of $n + n + 2(n-2) + 4 = 4n$ unknowns.

Given an arbitrary TRIANGLE, let $n-1$ Cevians be drawn from one of its vertices so all of the n triangles so determined have equal incircles. Then the incircles determined by spanning 2, 3, ..., $n-1$ adjacent triangles are also equal (Wells 1991, p. 67).

See also Circumcircle, Congruent Incircles Point, Contact Triangle, Equal Incircles Theorem, Excircle, Incenter, Inradius, Japanese Theorem, Seven Circles Theorem, Tangent Circles, Tangential Triangle, Triangle Transformation Principle

References

Casey, J. *A Sequel to the First Six Books of the Elements of Euclid, Containing an Easy Introduction to Modern Geometry with Numerous Examples,* 5th ed., rev. enl. Dublin: Hodges, Figgis, & Co., pp. 53–5, 1888.

Coxeter, H. S. M. and Greitzer, S. L. "The Incircle and Excircles." §1.4 in *Geometry Revisited.* Washington, DC: Math. Assoc. Amer., pp. 10–3, 1967.

Honsberger, R. *Mathematical Gems II.* Washington, DC: Math. Assoc. Amer., 1976.

Honsberger, R. "An Unlikely Concurrence." §3.4 in *Episodes in Nineteenth and Twentieth Century Euclidean Geometry.* Washington, DC: Math. Assoc. Amer., pp. 31–2, 1995.

Johnson, R. A. *Modern Geometry: An Elementary Treatise on the Geometry of the Triangle and the Circle.* Boston, MA: Houghton Mifflin, pp. 182–94, 1929.

Lachlan, R. "The Inscribed and the Escribed Circles." §126–28 in *An Elementary Treatise on Modern Pure Geometry.* London: Macmillian, pp. 72–4, 1893.

Pedoe, D. *Circles: A Mathematical View,* rev. ed. Washington, DC: Math. Assoc. Amer., 1995.

Wells, D. *The Penguin Dictionary of Curious and Interesting Geometry.* London: Penguin, 1991.

Inclusion Map

Given a subset B of a set A, the injection $f : B \to A$ defined by $f(b) = b$ for all $b \in B$ is called the inclusion map.

See also Long Exact Sequence of a Pair Axiom

Inclusion-Exclusion Principle

Let $|A|$ denote the cardinality of set A, then it follows immediately that

$$|A \cup B| = |A| + |B| - |A \cap B|,$$

where $\cup$ denotes union, and $\cap$ denotes intersection.

This formula can be generalized in the following beautiful manner. Let $\mathcal{A} = \{A_i\}_{i=1}^p$ be a P-system of S consisting of sets $A_1, ..., A_p$, then

$$|A_1 \cup A_2 \cup ... \cup A_p| = \sum_{1 \le i \le p} |A_i| - \sum_{1 \le i_1 < i_2 \le p} |A_{i1} \cap A_{i2}|$$

$$+ \sum_{1 \le i_1 < i_2 < i_3 \le p} |A_{i1} \cap A_{i2} \cap A_{i3}| - ...$$

$$+ (-1)^{p-1} |A_{i1} \cap A_{i2} \cap ... \cap A_p|,$$

where the sums are taken over K-subsets of $\mathcal{A}$. This formula holds for infinite sets S as well as finite sets (Comtet 1972, p. 177).

The principle of inclusion-exclusion was used by Nicholas Bernoulli to solve the recontres problem of finding the number of derangements (Bhatnagar 1995, p. 8).

The following *Mathematica* programs give a list of the subsets appearing under each sum and the contribution each sum makes to the total.

```
<<                    DiscreteMath`Combinatorica`;
InclusionExclusionSubets[a_List]          :=
Module[{n, p = Length[a]},
    Table[Intersection @@ a[[#]] & /@
      KSubsets[Range[p], n],
    {n, p}]
    ]        InclusionExclusionTerms[a_List]  :=
Module[{n, p = Length[a]},
    Table[(-1)^(n - 1)Plus @@         Length /@
(Intersection @@ a[[#]] & /@
    KSubsets[Range[p], n]),
    {n, p}]
    ]
```

For example, for the three subsets $A_1 = \{2, 3, 7, 9, 10\}$, $A_2 = \{1, 2, 3, 9\}$, and $A_3 = \{2, 4, 9, 10\}$ of $S = \{1, 2, ..., 10\}$, the following table summarizes the terms appearing the sum.

#	term	set	length
1	A_1	$\{2, 3, 7, 9, 10\}$	5
	A_2	$\{1, 2, 3, 9\}$	4
	A_3	$\{2, 4, 9, 10\}$	4
2	$A_1 \cap A_2$	$\{2, 3, 9\}$	3
	$A_1 \cap A_3$	$\{2, 9, 10\}$	3
	$A_2 \cap A_3$	$\{2, 9\}$	2
3	$A_1 \cap A_2 \cap A_3$	$\{2, 9\}$	2

$|A_1 \cup A_2 \cup A_3|$ is therefore equal to $(5+4+4) - (3+3+2) + 2 = 7$, corresponding to the seven elements $A_1 \cup A_2 \cup A_3 = \{1, 2, 3, 4, 7, 9, 10\}$.

See also Bayes' Theorem

References

Andrews, G. E. *Number Theory.* Philadelphia, PA: Saunders, pp. 139–40, 1971.

Andrews, G. E. *q-Series: Their Development and Application in Analysis, Number Theory, Combinatorics, Physics, and Computer Algebra.* Providence, RI: Amer. Math. Soc., p. 60, 1986.

Bhatnagar, G. *Inverse Relations, Generalized Bibasic Series, and Their U(n) Extensions.* Ph.D. thesis. Ohio State University, 1995.

Comtet, L. *Advanced Combinatorics: The Art of Finite and Infinite Expansions,* rev. enl. ed. Dordrecht, Netherlands: Reidel, pp. 176–77, 1974.

da Silva. "Propriedades geraes." *J. de l'Ecole Polytechnique*, cah. 30.

de Quesada, C. A. "Daniel Augusto da Silva e la theoria delle congruenze binomie." *Ann. Sci. Acad. Polytech. Porto, Co 1 mbra* **4**, 166–92, 1909.

Knuth, D. E. *The Art of Computer Programming, Vol. 1: Fundamental Algorithms, 3rd ed.* Reading, MA: Addison-Wesley, pp. 178–79, 1997.

Sylvester, J. "Note sur la théorème de Legendre." *C. R. Acad. Sci. Paris* **96**, 463–65, 1883.

Inclusive Disjunction

A DISJUNCTION that remains true if either *or both* of its arguments are true. This is equivalent to the OR CONNECTIVE.

By contrast, the EXCLUSIVE DISJUNCTION is true if only one, but not both, of its arguments are true, and is false if neither or both are true, which is equivalent to the XOR connective.

See also DISJUNCTION, EXCLUSIVE DISJUNCTION, OR, XOR

Incommensurate

Two lengths are called incommensurate or incommensurable if their ratio cannot be expressed as a ratio of whole numbers. IRRATIONAL NUMBERS and TRANSCENDENTAL NUMBERS are incommensurate with the integers.

See also FRACTION, IRRATIONAL NUMBER, PYTHAGORAS'S CONSTANT, TRANSCENDENTAL NUMBER

Incomparable Rectangles

Two RECTANGLES, neither of which will fit inside the other, are said to be incomparable. This is equivalent to one rectangle being both longer and narrower. At least seven and at most eight mutually incomparable rectangles are needed to tile a given rectangle (Wells 1991).

See also RECTANGLE

References

Wells, D. *The Penguin Dictionary of Curious and Interesting Geometry.* London: Penguin, pp. 116–17, 1991.

Incomplete Beta Function

A generalization of the complete BETA FUNCTION defined by

$$B(z;a,b) \equiv \int_0^z u^{a-1}(1-u)^{b-1}du$$

$$= z^a \left[\frac{1}{a} + \frac{1-b}{a+1}z + \ldots + \frac{(1-b)\cdots(n-b)}{n!(a+n)}z^n + \ldots \right].$$

The symbol $B_z(a,b)$ is sometimes also used. The incomplete beta function $B(z;a,b)$ reduces to the use BETA FUNCTION $B(a,b)$ when $z = 1$,

$$B(1;a,b) = B(a,b)$$

The incomplete beta function is implemented in *Mathematica* as Beta[z, a, b].

See also BETA FUNCTION, REGULARIZED BETA FUNCTION

Incomplete Gamma Function

The "complete" GAMMA FUNCTION $\Gamma(x)$ can be generalized to the incomplete gamma function $\Gamma(a,x)$ such that $\Gamma(a) = \Gamma(a,0)$. This "upper" incomplete gamma function is given by

$$\Gamma(a,x) \equiv \int_x^\infty t^{a-1}e^{-t}dt. \tag{1}$$

For a an INTEGER n

$$\Gamma(n,x) = (n-1)!e^{-x} \sum_{s=0}^{n-1} \frac{x^s}{s!} = (n-1)!e^{-x}e_{n-1}(x), \tag{2}$$

where es is the EXPONENTIAL SUM FUNCTION. The lower incomplete gamma function is given by

$$\gamma(a,x) \equiv \int_0^x t^{a-1}e^{-t}dt$$

$$a^{-1}x^a e^{-x} {}_1F_1(1;1+a;x)$$

$$a^{-1}x^a {}_1F_1(a;1+a;-x), \tag{3}$$

where ${}_1F_1(a;b;x)$ is the CONFLUENT HYPERGEOMETRIC FUNCTION OF THE FIRST KIND. For a an INTEGER n,

$$\gamma(n,x) = (n-1)! \left(1 - e^{-x} \sum_{k=0}^{n-1} \frac{x^k}{k!} \right)$$

$$= (n-1)![1 - e^{-x}e_{n-1}(x)]. \tag{4}$$

The function $\Gamma(a,z)$ is denoted Gamma[a, z] and the function $\gamma(a,z)$ is denoted Gamma[a, 0, z] in *Mathematica*. By definition, the two incomplete functions satisfy

$$\Gamma(a,x) + \gamma(a,x) = \Gamma(a). \tag{5}$$

See also GAMMA FUNCTION, REGULARIZED GAMMA FUNCTION

Incompleteness

A formal theory is said to be incomplete if it contains fewer theorems than would be possible while still retaining CONSISTENCY.

See also CONSISTENCY, GÖDEL'S INCOMPLETENESS THEOREM

References

Chaitin, G. J. "G. J. Chaitin's Home Page." http://www.cs.auckland.ac.nz/CDMTCS/chaitin/.

Increasing Function

A function $f(x)$ increases on an INTERVAL I if $f(b) > f(a)$ for all $b > a$, where $a, b \in I$. Conversely, a function $f(x)$ decreases on an INTERVAL I if $f(b) < f(a)$ for all $b > a$ with $a, b \in I$.

If the DERIVATIVE $f'(x)$ of a CONTINUOUS FUNCTION $f(x)$ satisfies $f'(x) > 0$ on an OPEN INTERVAL (a, b), then $f(x)$ is increasing on (a, b). However, a function may increase on an interval without having a derivative defined at all points. For example, the function $x^{1/3}$ is increasing everywhere, including the origin $x = 0$, despite the fact that the DERIVATIVE is not defined at that point.

See also DECREASING FUNCTION, DERIVATIVE, NON-DECREASING FUNCTION, NONINCREASING FUNCTION

References

Jeffreys, H. and Jeffreys, B. S. "Increasing and Decreasing Functions." §1.065 in *Methods of Mathematical Physics, 3rd ed.* Cambridge, England: Cambridge University Press, p. 22, 1988.

Increasing Sequence

For a SEQUENCE $\{a_n\}$, if $a_{n+1} - a_n > 0$ for $n \geq x$, then a_n is increasing for $n \geq x$. Conversely, if $a_{n+1} - a_n < 0$ for $n \geq x$, then a_n is DECREASING for $n \geq x$.

If $a_n > 0$ and $a_{n+1}/a_n > 1$ for all $n \geq x$, then a_n is increasing for $n \geq x$. Conversely, if $a_n > 0$ and $a_{n+1}/a_n < 1$ for all $n \geq x$, then a_n is decreasing for $n \geq x$.

See also DECREASING SEQUENCE, SEQUENCE

Indecomposable

A P-FORM α is indecomposable if it cannot be written as the WEDGE PRODUCT of ONE-FORMS

$$\alpha = \beta_1 \wedge \ldots \wedge \beta_p.$$

A p-form that can be written as such a product is called DECOMPOSABLE.

See also DECOMPOSABLE, DIFFERENTIAL K-FORM

Indefinite Integral

An INTEGRAL

$$\int f(x)dx$$

without upper and lower limits, also called an ANTI-DERIVATIVE. The first FUNDAMENTAL THEOREM OF CALCULUS allows DEFINITE INTEGRALS to be computed in terms of indefinite integrals. If F is the indefinite integral for $f(x)$, then

$$\int_a^b f(x)dx = F(b) - F(a).$$

The question of which definite integrals can be expressed in terms of elementary function is not susceptible to any established theory. In fact, the problem belongs to transcendence theory, which appears to be "infinitely hard." For example, there are definite integrals that are equal to the EULER-MASCHERONI CONSTANT γ. However, the problem of deciding whether γ can be expressed in terms of the values at rational values of elementary functions involves the decision as to whether γ is rational or algebraic, which is not known.

See also ANTIDERIVATIVE, CALCULUS, DEFINITE INTEGRAL, FUNDAMENTAL THEOREMS OF CALCULUS, INTEGRAL

Indefinite Quadratic Form

A QUADRATIC FORM $Q(\mathbf{x})$ is indefinite if it is less than 0 for some values and greater than 0 for others. The QUADRATIC FORM, written in the form $(\mathbf{x}, \mathbf{Ax})$, is indefinite if EIGENVALUES of the MATRIX $\mathbf{A}$ are of both signs.

See also POSITIVE DEFINITE QUADRATIC FORM, POSITIVE SEMIDEFINITE QUADRATIC FORM

References

Gradshteyn, I. S. and Ryzhik, I. M. *Tables of Integrals, Series, and Products, 6th ed.* San Diego, CA: Academic Press, p. 1106, 2000.

Indefinite Summation Operator

The indefinite summation operator Δ^{-1} for discrete variables, is the equivalent of integration for continuous variables. If $\Delta Y(x) = y(x)$ then $\Delta^{-1}y(x) = Y(x)$.

Indegree

The number of inward directed EDGES from a given VERTEX in a DIRECTED GRAPH.

See also LOCAL DEGREE, OUTDEGREE

Independence Axiom

A rational choice between two alternatives should depend only on how they differ.

Independence Complement Theorem

If sets E and F are INDEPENDENT, then so are E and F', where F' is the complement of F (i.e., the set of all possible outcomes not contained in F). Let $\cup$ denote "or" and $\cap$ denote "and." Then

$$P(E) = P(EF \cup EF') \qquad (1)$$

$$= P(EF) + P(EF') - P(EF \cap EF'), \qquad (2)$$

where AB is an abbreviation for $A \cap B$. But E and F are independent, so

$$P(EF) = P(E)P(F). \qquad (3)$$

Also, since F and F' are complements, they contain no

common elements, which means that

$$P(EF \cap EF') = 0 \qquad (4)$$

for any E. Plugging (4) and (3) into (2) then gives

$$P(E) = P(E)P(F) + P(EF'). \qquad (5)$$

Rearranging,

$$P(EF') = P(E)[1 - P(F)] = P(E)P(F'), \qquad (6)$$

Q.E.D.

See also INDEPENDENT SET

Independence Number

The independence number $\alpha(G)$ of a graph is the cardinality of the largest INDEPENDENT SET. Formally,

$$\alpha(G) = \max(|U| : U \subset V \text{ independent})$$

for a GRAPH G, where $|U|$ denotes the CARDINALITY of the set U. The independence number of the DE BRUIJN GRAPH of order n is given by 1, 2, 3, 7, 13, 28, ... (Sloane's A006946).

By definition, the independence number of a graph G plus the number of elements in a minimal VERTEX COVER of G equals the number of vertices in the graph.

See also INDEPENDENT SET, VERTEX COVER

References

Skiena, S. "Maximum Independent Set" §5.6.3 in *Implementing Discrete Mathematics: Combinatorics and Graph Theory with Mathematica*. Reading, MA: Addison-Wesley, pp. 218–19, 1990.
Sloane, N. J. A. Sequences A006946/M0834 in "An On-Line Version of the Encyclopedia of Integer Sequences." http://www.research.att.com/~njas/sequences/eisonline.html.

Independent Equations

LINEARLY INDEPENDENT

Independent Events

Two events A and B are called independent if their probabilities satisfy $P(AB) = P(A)P(B)$ (Papoulis 1984, p. 40).

See also EVENT, INDEPENDENT STATISTICS

References

Papoulis, A. *Probability, Random Variables, and Stochastic Processes, 2nd ed.* New York: McGraw-Hill, 1984.

Independent Sequence

STRONGLY INDEPENDENT, WEAKLY INDEPENDENT

Independent Set

Two sets A and B are said to be independent if their INTERSECTION $A \cap B = \varnothing$, where $\varnothing$ is the EMPTY SET.

For example, $\{A, B, C\}$ and $\{D, E\}$ are independent, but $\{A, B, C\}$ and $\{C, D, E\}$ are not. Independent sets are also called DISJOINT or mutually exclusive.

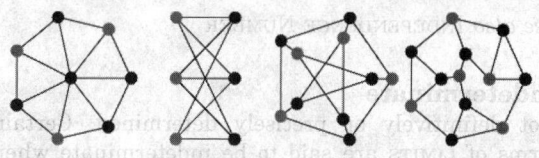

An independent set of a GRAPH G is a subset of the vertices such that no two vertices in the subset represent an edge of G. Given a VERTEX COVER of a GRAPH, all vertices not in the cover define an independent set (Skiena 1990, p. 218). The INDEPENDENCE NUMBER of a graph is the cardinality of the largest independent set. A maximum independent set of a graph can be computed using `MaximumIndependentSet[g]` in the *Mathematica* add-on package `DiscreteMath`Combinatorica`` (which can be loaded with the command $<<$ `DiscreteMath`). An independent set of edges can be defined similarly (Skiena 1990, p. 219). Gallai (1959) showed that the size of the minimum EDGE COVER plus the side of the maximum number of independent edges equals the number of vertices of a graph.

See also CLIQUE, DISJOINT SETS, EDGE COVER, EMPTY SET, INDEPENDENCE NUMBER, INTERSECTION, VENN DIAGRAM, VERTEX COVER

References

Gallai, T. "Über extreme Punkt- und Kantenmengen." *Ann. Univ. Sci. Budapest, Eotvos Sect. Math.* **2**, 133–38, 1959.
Skiena, S. "Maximum Independent Set" §5.6.3 in *Implementing Discrete Mathematics: Combinatorics and Graph Theory with Mathematica*. Reading, MA: Addison-Wesley, pp. 218–19, 1990.

Independent Statistics

Two variates A and B are statistically independent IFF the CONDITIONAL PROBABILITY $P(A|B)$ of A given B satisfies

$$P(A|B) = P(A), \qquad (1)$$

in which case the probability of A and B is just

$$P(A, B) = P(A \cap B) = P(A)P(B). \qquad (2)$$

Similarly, n events $A_1, A_2, ..., A_n$ are independent IFF

$$p\left(\bigcap_{i=1}^{n} A_i\right) = \prod_{i=1}^{n} P(A_i). \qquad (3)$$

Statistically independent variables are always UNCORRELATED, but the converse is not necessarily true.

See also BAYES' FORMULA, CONDITIONAL PROBABILITY, INDEPENDENT EVENTS, INDEPENDENCE COMPLEMENT THEOREM, UNCORRELATED

Independent Vertices

A set of VERTICES A of a GRAPH with EDGES V is independent if it contains no EDGES.

See also INDEPENDENCE NUMBER

Indeterminate

Not definitively or precisely determined. Certain forms of LIMITS are said to be indeterminate when merely knowing the limiting behavior of individual parts of the expression is not sufficient to actually determine the overall limit. For example, a LIMIT OF THE FORM $0/0$, i.e., $\lim_{x\to0} f(x)/g(x)$ where $\lim_{x\to0} f(x) = \lim_{x\to0} g(x) = 0$, is indeterminate since the value of the overall limit actually depends on the limiting behavior of the combination of the two functions (e.g. $\lim_{x\to0} x/x = 1$, while $\lim_{x\to0} x^2/x = 0$).

See also AMBIGUOUS, LIMIT, TRIVIAL, UNDEFINED, WELL DEFINED

Indeterminate Problems

DIOPHANTINE EQUATION

Index

The word "index" has a very large number of completely different meanings in mathematics. Most commonly, it is used in the context of an INDEX SET, where it means a quantity which can take on a set of values and is used to designate one out of a number of possible values associated with this value. For example, the subscript i in the symbol a_i could be called the index of a.

In a RADICAL $\sqrt{x}$, the quantity n is called the index.

The word index has a special meaning in economics, where it refers to a single quantity used to quantify the "average" value of a possibly complicated set of quantities. In this context, it is sometimes called an INDEX NUMBER.

In TOPOLOGY, INDEX THEORY refers to the study of topological invariants of MANIFOLDS.

See also INDEX LOWERING, INDEX RAISING, INDEX SET, MANIFOLD, MULTIPLICATIVE ORDER, STATISTICAL INDEX

Index (Extension Field)

DEGREE (EXTENSION FIELD)

Index (Modulo)

MULTIPLICATIVE ORDER

Index (Residue)

MULTIPLICATIVE ORDER

Index (Subgroup)

This entry contributed by NICOLAS BRAY

For a SUBGROUP H of a GROUP G, the index of H, denoted $(G:H)$, is the CARDINALITY of the set of LEFT COSETS of H in G (which is equal to the CARDINALITY of the set of RIGHT COSETS of H in G).

See also COSET, LAGRANGE'S GROUP THEOREM, LEFT COSET, RIGHT COSET

Index (Tensor)

See also INDEX LOWERING, INDEX RAISING

Index Law

EXPONENT LAWS

Index Lowering

The indices of a CONTRAVARIANT TENSOR A^j can be lowered, turning it into a COVARIANT TENSOR A_i, by multiplication by a so-called METRIC TENSOR, e.g.,

$$g_{ij}A^j = A_i.$$

See also CONTRAVARIANT TENSOR, COVARIANT TENSOR, INDEX RAISING, INDEX (TENSOR), TENSOR

Index Number

A STATISTIC which assigns a single number to several individual statistics in order to quantify trends. The best-known index in the United States is the consumer price index, which gives a sort of "average" value for inflation based on price changes for a group of selected products. The Dow Jones and NASDAQ indexes for the New York and American Stock Exchanges, respectively, are also index numbers.

Let p_n be the price per unit in period n, q_n be the quantity produced in period n, and $v_n \equiv p_n q_n$ be the value of the n units. Let q_a be the estimated relative importance of a product. There are several types of indices defined, among them those listed in the following table.

Index	Abbr.	Formula
BOWLEY INDEX	P_B	$\frac{1}{2}(P_L + P_P)$
FISHER INDEX	P_F	$\sqrt{P_L P_P}$
GEOMETRIC MEAN INDEX	P_G	$\left[\prod\left(\frac{p_n}{p_0}\right)^{v0}\right]^{1/\Sigma v_0}$

HARMONIC MEAN INDEX	P_H	$\dfrac{\Sigma p_0 q_0}{\Sigma \dfrac{p_0^2 q_0}{nm}}$
LASPEYRES' INDEX	P_L	$\dfrac{\Sigma p_n q_0}{\Sigma p_0 q_0}$
MARSHALL-EDGEWORTH INDEX	P_{ME}	$\dfrac{\Sigma p_n(q_0 + q_n)}{\Sigma(v_0 + v_n)}$
MITCHELL INDEX	P_M	$\dfrac{\Sigma p_n q_n}{\Sigma p_0 q_n}$
PAASCHE'S INDEX	P_P	$\dfrac{\Sigma p_n q_n}{\Sigma p_0 q_n}$
WALSH INDEX	P_W	$\dfrac{\Sigma \sqrt{q_0 q_n} p_n}{\Sigma \sqrt{q_0 q_a} p_n}$

See also BOWLEY INDEX, FISHER INDEX, GEOMETRIC MEAN INDEX, HARMONIC MEAN INDEX, LASPEYRES' INDEX, MARSHALL-EDGEWORTH INDEX, MITCHELL INDEX, PAASCHE'S INDEX, WALSH INDEX

References

Fisher, I. *The Making of Index Numbers: A Study of Their Varieties, Tests and Reliability, 3rd ed.* New York: Augustus M. Kelly, 1967.

Kenney, J. F. and Keeping, E. S. "Index Numbers." Ch. 5 in *Mathematics of Statistics, Pt. 1, 3rd ed.* Princeton, NJ: Van Nostrand, pp. 64–4, 1962.

Mudgett, B. D. *Index Numbers.* New York: Wiley, 1951.

Index Raising

The indices of a COVARIANT TENSOR A_j can be raised, forming a CONTRAVARIANT TENSOR A^i, by multiplication by a so-called METRIC TENSOR, e.g.,

$$g^{ij} A_j = A^i \tag{1}$$

See also CONTRAVARIANT TENSOR, COVARIANT TENSOR, INDEX LOWERING, INDEX (TENSOR), TENSOR

Index Set

A SET whose members index (label) members of another set. For example, in the set $A = \cup_{k \in K} A_k$, the set K is an index set of the set A.

See also SET

Index Theory

A branch of TOPOLOGY dealing with topological invariants of MANIFOLDS.

References

Roe, J. *Index Theory, Coarse Geometry, and Topology of Manifolds.* Providence, RI: Amer. Math. Soc., 1996.

Upmeier, H. *Toeplitz Operators and Index Theory in Several Complex Variables.* Boston, MA: Birkhäuser, 1996.

Indicator

References

Feller, W. *An Introduction to Probability Theory and Its Applications, Vol. 2, 3rd ed.* New York: Wiley, p. 104, 1971.

Indicatrix

A spherical image of a curve. The most common indicatrix is DUPIN'S INDICATRIX.

See also DUPIN'S INDICATRIX

Indicial Equation

The RECURRENCE RELATION obtained during application of the FROBENIUS METHOD of solving a second-order ordinary differential equation. The indicial equation (also called the CHARACTERISTIC EQUATION) is obtained by noting that, by definition, the lowest order term x^k (that corresponding to $n = 0$) must have a COEFFICIENT of zero. For an example of the construction of an indicial equation, see BESSEL DIFFERENTIAL EQUATION.

1. If the two ROOTS are equal, only one solution can be obtained.
2. If the two ROOTS differ by a noninteger, two solutions can be obtained.
3. If the two ROOTS differ by an INTEGER, the larger will yield a solution. The smaller may or may not.

References

Morse, P. M. and Feshbach, H. *Methods of Theoretical Physics, Part I.* New York: McGraw-Hill, pp. 532–34, 1953.

Indifference Principle

INSUFFICIENT REASON PRINCIPLE

Individual

One of the basic objects treated in a given formal language system. The term is sometimes also used as a synonym for URELEMENT.

See also URELEMENT

References

Carnap, R. *Introduction to Symbolic Logic and Its Applications.* New York: Dover, p. 4, 1958.

Induced Map

If $f : (X, A) \to (Y, B)$ is homotopic to $g : (X, A) \to (Y, B)$, then $f_* : H_n(X, A) \to H_n(Y, B)$ and $g_* : H_n(X, A) \to H_n(Y, B)$ are said to be the induced maps.

See also EILENBERG-STEENROD AXIOMS

Induced Norm

NATURAL NORM

Induced Representation

If a SUBGROUP H of G has a REPRESENTATION ϕ : $H \times W \to W$, then there is a unique induced representation of G on a VECTOR SPACE V. The original space W is contained in V, and in fact,

$$V = \oplus_{\sigma \in G/H} \sigma W,$$

where σW is a copy of W. The induced representation on V is denoted Ind_H^G.

Alternatively, the induced representation is the $\mathbb{C}G$-MODULE

$$\mathrm{Ind}_H^G \simeq \mathbb{C}G \otimes_{\mathbb{C}H} W. \qquad (1)$$

Also, it can be viewed as W-valued functions on G which commute with the H action.

$$\mathrm{Ind}_H^G \simeq \{f : G \to W : hf(g) = f(hg)\}. \qquad (2)$$

The induced representation is also determined by its UNIVERSAL PROPERTY:

$$\mathrm{Hom}_H(W, \mathrm{Res}\, U) = \mathrm{Hom}_G(\mathrm{Ind}\, W, U), \qquad (3)$$

where U is any representation of G. Also, the induced representation satisfies the following formulas.

1. $\mathrm{Ind} \oplus W_i = \otimes \mathrm{Ind}\, W_i$.
2. $U \otimes \mathrm{Ind}\, W = \mathrm{Ind}(\mathrm{Res}(U) \otimes W)$ for any REPRESENTATION U.
3. $\mathrm{Ind}_H^G(W) = \mathrm{Ind}_K^G(\mathrm{Ind}_H^K W)$ when $H \leq K \leq G$.

Some of the CHARACTERS of G can be calculated from the CHARACTERS of H, as induced representations, using FROBENIUS RECIPROCITY. ARTIN'S RECIPROCITY THEOREM says that the induced representations of CYCLIC SUBGROUPS of a FINITE GROUP G generates a LATTICE of finite index in the lattice of VIRTUAL CHARACTERS. BRAUER'S THEOREM says that the virtual characters are generated by the induced representations from P-ELEMENTARY SUBGROUPS.

See also ARTIN'S RECIPROCITY THEOREM, FROBENIUS RECIPROCITY, GROUP, IRREDUCIBLE REPRESENTATION, REPRESENTATION, RESTRICTION (REPRESENTATION), TENSOR PRODUCT (VECTOR SPACE), VECTOR SPACE

References
Fulton, W. and Harris, J. *Representation Theory.* New York: Springer-Verlag, 1991.

Induced Subgraph

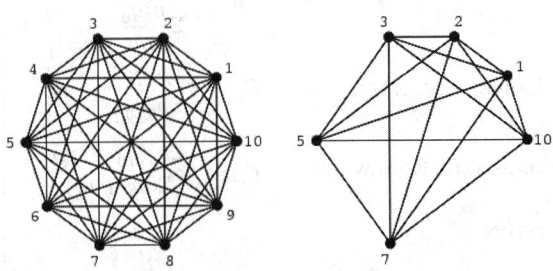

An induced subgraph is a subset of the edges of a GRAPH G together with any edges whose endpoints are both in this subset. The figure above illustrates the subgraph induced on the COMPLETE GRAPH K_5 by the vertex subset $\{1, 2, 3, 5, 7, 10\}$. An induced subgraph that is a COMPLETE GRAPH is called a CLIQUE. Any induced subgraph of a COMPLETE GRAPH forms a CLIQUE. An induced subgraph can be computed using `InduceSubgraph[g]` in the *Mathematica* add-on package `DiscreteMath`Combinatorica`` (which can be loaded with the command `<<DiscreteMath`$).

See also CLIQUE, SUBGRAPH

References
Skiena, S. "Induced Subgraphs." §3.2.2 in *Implementing Discrete Mathematics: Combinatorics and Graph Theory with Mathematica.* Reading, MA: Addison-Wesley, pp. 90–2, 1990.

Induction

The use of the INDUCTION PRINCIPLE in a PROOF. Induction used in mathematics is often called MATHEMATICAL INDUCTION.

See also PRINCIPLE OF STRONG INDUCTION, PRINCIPLE OF TRANSFINITE INDUCTION, PRINCIPLE OF WEAK INDUCTION

References
Buck, R. C. "Mathematical Induction and Recursive Definitions." *Amer. Math. Monthly* **70**, 128–35, 1963.
Séroul, R. "Reasoning by Induction." §2.14 in *Programming for Mathematicians.* Berlin: Springer-Verlag, pp. 22–5, 2000.

Induction Axiom

The fifth of PEANO'S AXIOMS, which states: If a SET S of numbers contains zero and also the successor of every number in S, then every number is in S.

See also PEANO'S AXIOMS

Induction Principle

The truth of an INFINITE sequence of propositions P_i for $i = 1, \ldots, \infty$ is established if (1) P_1 is true, and (2) P_k IMPLIES P_{k+1} for all k.

References

Courant, R. and Robbins, H. "The Principle of Mathematical Induction" and "Further Remarks on Mathematical Induction." §1.2.1 and 1.7 in *What is Mathematics?: An Elementary Approach to Ideas and Methods, 2nd ed.* Oxford, England: Oxford University Press, pp. 9–1 and 18–0, 1996.

Apostol, T. M. "The Principle of Mathematical Induction." §I 4.2 in *Calculus, 2nd ed., Vol. 1: One-Variable Calculus, with an Introduction to Linear Algebra.* Waltham, MA: Blaisdell, p. 34, 1967.

Inequality

A mathematical statement that one quantity is greater than or less than another. "a is less than b" is denoted $a < b$, and "a is greater than b" is denoted $a > b$. "a is less than or equal to b" is denoted $a \leq b$, and "a is greater than or equal to b" is denoted $a \geq b$. The symbols $a \ll b$ and $a \gg b$ are used to denote "a is much less than b" and "a is much greater than b," respectively.

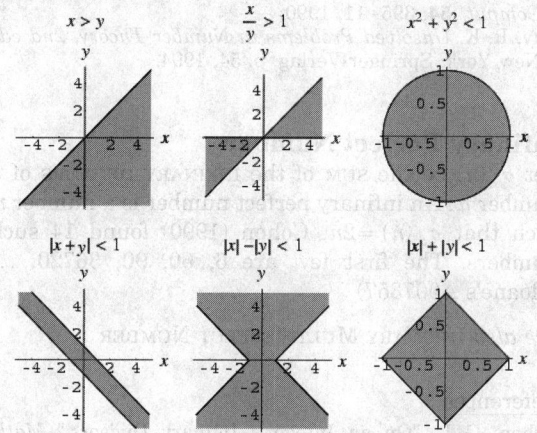

Solutions to the inequality $|x - a| < b$ consist of the set $\{x : a - b < x - a + b\}$, or equivalently $\{x : a - b < x < a + b\}$. Solutions to the inequality $|x - a| > b$ consist of the set $\{x : x - a > b\} \cup \{x : x - a < -b\}$. If a and b are both POSITIVE or both NEGATIVE and $a < b$, then $1/a > 1/b$. The portions of the xy-plane satisfying a number of specific inequalities are illustrated above.

In *Mathematica* 4.0, the command `InequalityInstance[ineqs, vars]` can be used to find a real solution of the system of real equations and inequalities *ineqs* in the variables *vars* or return the EMPTY SET if no such solution exists. Solution of inequalities can be performed using `[ineqs, vars]`, in the *Mathematica* add-on package `Algebra`Inequality-Solve`` (which can be loaded with the command `<<Algebra`) or directly using `CylindricalAlgebraicDecomposition[ineqs, vars]`.

See also Cylindrical Algebraic Decomposition, Equality, Exists, For All, Inequation, Quantifier, Strict Inequality

References

Abramowitz, M. and Stegun, C. A. (Eds.). *Handbook of Mathematical Functions with Formulas, Graphs, and Mathematical Tables, 9th printing.* New York: Dover, p. 16, 1972.

Beckenbach, E. F. and Bellman, Richard E. *An Introduction to Inequalities.* New York: Random House, 1961.

Beckenbach, E. F. and Bellman, Richard E. *Inequalities, 2nd rev. print.* Berlin: Springer-Verlag, 1965.

Hardy, G. H.; Littlewood, J. E.; and Pólya, G. *Inequalities, 2nd ed.* Cambridge, England: Cambridge University Press, 1952.

Kazarinoff, N. D. *Geometric Inequalities.* New York: Random House, 1961.

Mitrinovic, D. S. *Analytic Inequalities.* New York: Springer-Verlag, 1970.

Mitrinovic, D. S.; Pecaric, J. E.; and Fink, A. M. *Classical & New Inequalities in Analysis.* Dordrecht, Netherlands: Kluwer, 1993.

Mitrinovic, D. S.; Pecaric, J. E.; Fink, A. M. *Inequalities Involving Functions & Their Integrals & Derivatives.* Dordrecht, Netherlands: Kluwer, 1991.

Mitrinovic, D. S.; Pecaric, J. E.; and Volenec, V. *Recent Advances in Geometric Inequalities.* Dordrecht, Netherlands: Kluwer, 1989.

Weisstein, E. W. "Books about Inequalities." http://www.treasure-troves.com/books/Inequalities.html.

Inequation

While an equality

$$A = B$$

states that two mathematical expressions are equal, an inequation

$$A \neq B$$

states that two expressions are not equal.

See also Equation, Inequality, Strict Inequality

Inexact Differential

An infinitesimal which is not the differential of an actual function and which cannot be expressed as

$$dz = \left(\frac{\partial z}{\partial x}\right)_y dx + \left(\frac{\partial z}{\partial y}\right)_z dy,$$

the way an EXACT DIFFERENTIAL can. Inexact differentials are denoted with a bar through the d. The most common example of an inexact differential is the change in heat dQ encountered in thermodynamics.

See also Exact Differential, Pfaffian Form

References

Bringhurst, R. *The Elements of Typographic Style, 2nd ed.* Point Roberts, WA: Hartley and Marks, p. 277, 1997.

Zemansky, M. W. *Heat and Thermodynamics, 5th ed.* New York: McGraw-Hill, p. 38, 1968.

Inf

Infimum, Infimum Limit

Infimum

Portions of this entry contributed by JEROME R. BREITENBACH

The infimum is the greatest lower bound of a SET S, defined as a quantity m such that no member of the SET is less than m, but if ϵ is any POSITIVE quantity, however small, there is always one member that is less than $m + \epsilon$ (Jeffreys and Jeffreys 1988). When it exists (which is not required by this definition, e.g., $\mathbb{R}$ does not exist), the infimum is denoted $\inf S$ or $\inf_{x \in S} x$. The infimum can be computed using the *Mathematica* 4.0 command Infimum[f, $constr$, $vars$].

More formally, the infimum $\inf S$ for S a (nonempty) SUBSET of the extended reals $\overline{\mathbb{R}} = \mathbb{R} \cup \{\pm\infty\}$ is the largest value $y \in \overline{\mathbb{R}}$ such that for all $x \in S$ we have $x \geq y$. Using this definition, $\inf S$ *always* exists and, in particular, $\mathbb{R} = -\infty$.

Whenever an infimum exists, its value is unique.

See also INFIMUM LIMIT, LOWER BOUND, SUPREMUM

References

Croft, H. T.; Falconer, K. J.; and Guy, R. K. *Unsolved Problems in Geometry.* New York: Springer-Verlag, p. 2, 1991.
Jeffreys, H. and Jeffreys, B. S. "Upper and Lower Bounds." §1.044 in *Methods of Mathematical Physics, 3rd ed.* Cambridge, England: Cambridge University Press, p. 13, 1988.
Knopp, K. *Theory of Functions Parts I and II, Two Volumes Bound as One, Part I.* New York: Dover, p. 6, 1996.
Royden, H. L. *Real Analysis, 3rd ed.* New York: Macmillan, p. 31, 1988.
Rudin, W. *Real and Complex Analysis, 3rd ed.* New York: McGraw-Hill, p. 7, 1987.

Infimum Limit

Given a sequence of real numbers a_n, the infimum limit, also called the lower limit but more often simply pronounced 'lim-inf' and written liminf is the limit of

$$A_n = \inf_{k > n} a_k$$

as $n \to \infty$. Note that by definition, A_n is nondecreasing, and so either has a limit or tends to ∞. For example, suppose $a_n = (-1)^n / n$, then for n odd, $A_n = -1/n$, and for n even, $A_n = -1/(n+1)$. Another example is $a_n = \sin n$, in which case A_n is a constant sequence $A_n = -1$.
When $\limsup a_n = \liminf a_n$, the sequence converges to the real number

$$\lim a_n = \limsup a_n = \liminf a_n.$$

Otherwise, the sequence does not converge.

See also INFIMUM, LIMIT, LOWER LIMIT, SUPREMUM

Infinary Divisor

p^x is an infinary divisor of p^y (with $y > 0$) if $p^x |_{y-1} p^y$. This generalizes the concept of the K-ARY DIVISOR.

See also INFINARY PERFECT NUMBER, K-ARY DIVISOR

References

Cohen, G. L. "On an Integer's Infinary Divisors." *Math. Comput.* **54**, 395–11, 1990.
Cohen, G. and Hagis, P. "Arithmetic Functions Associated with the Infinary Divisors of an Integer." *Internat. J. Math. Math. Sci.* **16**, 373–83, 1993.
Guy, R. K. *Unsolved Problems in Number Theory, 2nd ed.* New York: Springer-Verlag, p. 54, 1994.

Infinary Multiperfect Number

Let $\sigma_\infty(n)$ be the SUM of the INFINARY DIVISORS of a number n. An infinary k-multiperfect number is a number n such that $\sigma_\infty(n) = kn$. Cohen (1990) found 13 infinary 3-multiperfects, seven 4-multiperfects, and two 5-multiperfects.

See also INFINARY PERFECT NUMBER

References

Cohen, G. L. "On an Integer's Infinary Divisors." *Math. Comput.* **54**, 395–11, 1990.
Guy, R. K. *Unsolved Problems in Number Theory, 2nd ed.* New York: Springer-Verlag, p. 54, 1994.

Infinary Perfect Number

Let $\sigma_\infty(n)$ be the SUM of the INFINARY DIVISORS of a number n. An infinary perfect number is a number n such that $\sigma_\infty(n) = 2n$. Cohen (1990) found 14 such numbers. The first few are 6, 60, 90, 36720, ... (Sloane's A007357).

See also INFINARY MULTIPERFECT NUMBER

References

Cohen, G. L. "On an Integer's Infinary Divisors." *Math. Comput.* **54**, 395–11, 1990.
Guy, R. K. *Unsolved Problems in Number Theory, 2nd ed.* New York: Springer-Verlag, p. 54, 1994.
Sloane, N. J. A. Sequences A007357/M4267 in "An On-Line Version of the Encyclopedia of Integer Sequences." http://www.research.att.com/~njas/sequences/eisonline.html.

Infinite

Greater than any assignable quantity of the sort in question. In mathematics, the concept of the infinite is made more precise through the notion of an INFINITE SET.

See also COUNTABLE SET, COUNTABLY INFINITE, FINITE, INFINITE SET, INFINITESIMAL, INFINITY

Infinite Group

A group having an infinite number of elements. Some infinite groups, such as the integers or rationals, are not CONTINUOUS GROUPS.

See also CONTINUOUS GROUP, FINITE GROUP

Infinite Product

N.B. A detailed online essay by S. Finch was the starting point for this entry.

A PRODUCT involving an INFINITE number of terms. Such products can converge. In fact, for POSITIVE a_n, the PRODUCT $\prod_{n=1}^{\infty} a_n$ converges to a NONZERO number IFF $\sum_{n=1}^{\infty} \ln a_n$ converges.

Infinite products can be used to define the COSINE

$$\cos x = \prod_{n=1}^{\infty} \left[1 - \frac{4x^2}{\pi^2 (2n-1)^2} \right], \tag{1}$$

GAMMA FUNCTION

$$\Gamma(z) = \left[z e^{\gamma z} \prod_{r=1}^{\infty} \left(1 + \frac{z}{r} \right) e^{-z/r} \right]^{-1}, \tag{2}$$

SINE, and SINC FUNCTION. They also appear in the POLYGON CIRCUMSCRIBING CONSTANT

$$k = \prod_{n=3}^{\infty} \frac{1}{\cos\left(\dfrac{\pi}{n}\right)}. \tag{3}$$

An interesting infinite product formula due to Euler which relates π and the nth PRIME p_n is

$$\pi = \frac{2}{\prod_{i=n}^{\infty} \left[1 + \dfrac{\sin\left(\frac{1}{2}\pi p_n\right)}{p_n} \right]} \tag{4}$$

$$= \frac{2}{\prod_{i=n}^{\infty} \left[1 + \dfrac{(-1)(p_n - 1)/2}{p_n} \right]} \tag{5}$$

(Blatner 1997). KNAR'S FORMULA gives a functional equation for the GAMMA FUNCTION $\Gamma(x)$ in terms of the infinite product

$$\Gamma(1+v) = 2^{2v} \prod_{m=1}^{\infty} \left[\pi^{-1/2} \Gamma\left(\tfrac{1}{2} + 2^{-m} v \right) \right]. \tag{6}$$

The class of products

$$\prod_{n=2}^{\infty} \frac{n^2 - 1}{n^2 + 1} = \pi \operatorname{csch} \pi \tag{7}$$

$$\prod_{n=2}^{\infty} \frac{n^3 - 1}{n^3 + 1} = \tfrac{2}{3} \tag{8}$$

$$\prod_{n=2}^{\infty} \frac{n^4 - 1}{n^4 + 1}$$

$$= -\tfrac{1}{2} \pi \sinh \pi \csc\left[(-1)^{1/4} \pi \right] \csc\left[(-1)^{3/4} \pi \right], \tag{9}$$

the first of which is given in Borwein and Corless (1999), can be done analytically.

The first few products

$$\prod_{k=1}^{\infty} \frac{(1 + k^{-1})^2}{1 + 2k^{-1}} = 2 \tag{10}$$

$$\prod_{k=1}^{\infty} \frac{(1 + k^{-1} + k^{-2})^2}{1 + 2k^{-1} + 3k^{-2}}$$

$$= \frac{3\sqrt{2} \cosh^2\left(\frac{1}{2} \pi \sqrt{3} \right) \operatorname{csch}\left(\pi \sqrt{2} \right)}{\pi} \tag{11}$$

$$\prod_{k=1}^{\infty} \frac{(1 + k^{-1} + k^{-2} + k^{-3})2}{1 + 2k^{-1} + 3k^{-2} + 4k^{-3}}$$

$$= \frac{\sinh^2 \pi \prod_{i=1}^{3} \Gamma(x_i)}{\pi^2}, \tag{12}$$

$$\prod_{k=1}^{\infty} \frac{(1 + k^{-1} + k^{-2} + k^{-3} + k^{-4})^2}{1 + 2k^{-1} + 3k^{-2} + 4k^{-3} + 5k^{-4}} = \prod_{i=1}^{4} \frac{\Gamma(y_i)}{\Gamma(z_i)} \tag{13}$$

where x_i, y_i, and z_i are the roots of

$$x^3 - 5x^2 + 10x - 10 = 0 \tag{14}$$

$$y^4 - 6y^3 + 15y^2 - 20y + 15 = 0, \tag{15}$$

and

$$z^4 - 5z^3 + 10z^2 - 10z + 5 = 0, \tag{16}$$

respectively, can also be done analytically. Note that (15) and (16) were unknown to Borwein and Corless (1999).

The product

$$\prod_{n=1}^{\infty} \left(1 + \frac{1}{n^p} \right) \tag{17}$$

has closed form expressions for small POSITIVE integral $p \geq 2$,

$$\prod_{n=1}^{\infty} \left(1 + \frac{1}{n^2} \right) = \frac{\sinh \pi}{\pi} \tag{18}$$

$$\prod_{n=1}^{\infty} \left(1 + \frac{1}{n^3} \right) = \frac{1}{\pi} \cosh\left(\tfrac{1}{2} \pi \sqrt{3} \right) \tag{19}$$

$$\prod_{n=1}^{\infty} \left(1 + \frac{1}{n^4} \right) = \frac{\cosh\left(\pi \sqrt{2} \right) - \cos\left(\pi \sqrt{2} \right)}{2\pi^2} \tag{20}$$

$$\prod_{n=1}^{\infty} \left(1 + \frac{1}{n^5} \right) = \left| \Gamma\left[\exp\left(\tfrac{2}{5} \pi i \right) \right] \Gamma\left[\exp\left(\tfrac{6}{5} \pi i \right) \right] \right|^{-2} \tag{21}$$

The D-ANALOG expression

$$[\infty!]_d = \prod_{n=3}^{\infty} \left(1 - \frac{2^d}{n^d} \right) \tag{22}$$

also has closed form expressions,

$$\prod_{n=3}^{\infty}\left(1-\frac{4}{n^2}\right)=\frac{1}{6} \tag{23}$$

$$\prod_{n=3}^{\infty}\left(1-\frac{8}{n^3}\right)=\frac{\sinh(\pi\sqrt{3})}{42\pi\sqrt{3}} \tag{24}$$

$$\prod_{n=3}^{\infty}\left(1-\frac{16}{n^4}\right)=\frac{\sinh(2\pi)}{120\pi} \tag{25}$$

$$\prod_{n=3}^{\infty}\left(1-\frac{32}{n^5}\right)=\left|\Gamma\left[\exp\left(\tfrac{1}{5}\pi i\right)\right]\Gamma\left[2\exp\left(\tfrac{7}{5}\pi i\right)\right]\right|^{-2} \tag{26}$$

General expressions for infinite products of this type include

$$\prod_{n=1}^{\infty}\left[1-\left(\frac{z}{n}\right)^{2N}\right]=\frac{\sin(\pi z)}{\pi z^{2N-1}}\prod_{k=1}^{N-1}\left|\Gamma\left(ze^{2\pi i(k-N)/(2N)}\right)\right|^{-2} \tag{27}$$

$$\prod_{n=1}^{\infty}\left[1+\left(\frac{z}{n}\right)^{2N}\right]=\frac{1}{z^{2N}}\prod_{k=1}^{N}\left|\Gamma\left(ze^{\pi i[2(k-N)-1]/(2N)}\right)\right|^{-2} \tag{28}$$

$$\prod_{n=1}^{\infty}\left[1+\left(\frac{z}{n}\right)^{2N+1}\right]=\frac{1}{\Gamma(1-z)z^{2N}}\prod_{k=1}^{N}\left|\Gamma\left(ze^{\pi i(2(k-N)-1/(2N+1))}\right)\right|^{-2} \tag{29}$$

$$\prod_{n=1}^{\infty}\left[1+\left(\frac{z}{n}\right)^{2N+1}\right]=\frac{1}{\Gamma(1+z)z^{2N}}\prod_{k=1}^{N}\left|\Gamma\left(ze^{2\pi i(k-N-1)/2N+1)}\right)\right|^{-2} \tag{30}$$

where $\Gamma(z)$ is the GAMMA FUNCTION and $|z|$ denotes the MODULUS (Kahovec). (27) and (28) can also be rewritten as

$$\prod_{n=1}^{\infty}\left[1-\left(\frac{z}{n}\right)^{2N}\right]=\frac{\sin(\pi z)}{\pi^3 z^2}\left(\frac{\sinh(\pi z)}{\pi z}\right)^{\mathrm{mod}(N+1,2)}$$

$$\times\prod_{k=1}^{\lceil N/2\rceil-1}\cosh^2\left[\pi z\sin\left(\frac{k\pi}{N}\right)\right]$$

$$-\cos^2\left[\pi z\cos\left(\frac{k\pi}{N}\right)\right] \tag{31}$$

$$\prod_{n=1}^{\infty}\left[1+\left(\frac{z}{n}\right)^{2N}\right]=\frac{1}{\pi^2 z^2}\left(\frac{\sinh(\pi z)}{\pi z}\right)^{\mathrm{mod}(N,2)}$$

$$\times\prod_{k=1}^{\lfloor N/2\rfloor}\cosh^2\left[\pi z\sin\left(\frac{(2k-1)\pi}{2N}\right)\right]$$

$$-\cos^2\left[\pi z\cos\left(\frac{(2k-1)\pi}{2N}\right)\right], \tag{32}$$

where $\lfloor x\rfloor$ is the FLOOR FUNCTION, $\lceil x\rceil$ is the CEILING FUNCTION, and $\mathrm{mod}(a,m)$ is the modulus of $a\ (\mathrm{mod}\ m)$ (Kahovec).

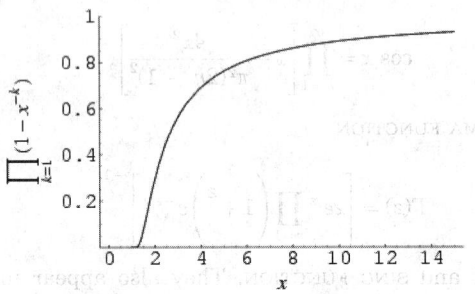

Infinite products OF THE FORM

$$\prod_{k=1}^{\infty}\left(1-\frac{1}{n^k}\right) \tag{33}$$

converge for $n\geq 2$. I am not aware of any analytic expressions, but the first few such products are numerically given by

$$\prod_{k=1}^{\infty}\left(1-\frac{1}{2^k}\right)\approx 0.28878809508660242128 \tag{34}$$

$$\prod_{k=1}^{\infty}\left(1-\frac{1}{3^k}\right)\approx 0.56012607792794894497 \tag{35}$$

$$\prod_{k=1}^{\infty}\left(1-\frac{1}{4^k}\right)\approx 0.68853753712033971546 \tag{36}$$

$$\prod_{k=1}^{\infty}\left(1-\frac{1}{5^k}\right)\approx 0.76033279587123242010. \tag{37}$$

A class of infinite products derived from the BARNES' G-FUNCTION is given by

$$\prod_{n=1}^{\infty}\left(1+\frac{z}{n}\right)^n e^{-z+z^2/(2n)}=\frac{G(z)}{(2\pi)^{z/2}}e^{[z(z+1)+\gamma z^2]/2}, \tag{38}$$

where γ is the EULER-MASCHERONI CONSTANT. The first few cases are

$$\prod_{n=1}^{\infty}\left(1+\frac{1}{n}\right)^n e^{1/(2n)-1}=\frac{e^{1+\gamma/2}}{\sqrt{2\pi}} \tag{39}$$

$$\prod_{n=1}^{\infty}\left(1+\frac{2}{n}\right)^n e^{4/(2n)-2}=\frac{e^{3+2\gamma}}{2\pi} \tag{40}$$

$$\prod_{n=1}^{\infty}\left(1+\frac{3}{n}\right)^n e^{9/(2n)-3}=\frac{e^{6+9\gamma/2}}{(2\pi)^{3/2}} \tag{41}$$

$$\prod_{n=1}^{\infty}\left(1+\frac{4}{n}\right)^n e^{16/(2n)-3} = \frac{e^{10+8\gamma}}{2\pi^2}. \tag{42}$$

The interesting identities

$$x\prod_{n=1}^{\infty}\frac{(1-x^{2n})^8}{(1-x^{2n-1})^8} = \sum_{n=1}^{\infty} 2^{3b(n)}\sigma_3(\mathrm{Od}(n))x^n \tag{43}$$

(Ewell 1995, 1999), where $b(n)$ is the exponent of the exact power of 2 dividing n, $\mathrm{Od}(n)$ is the ODD PART of n, $\sigma_k(n)$ is the DIVISOR FUNCTION of n, and $r_k(n)$ is the SUM OF SQUARES FUNCTION, and

$$\prod_{n=1}^{\infty}(1+x^{2n-1})^8 = \prod_{n=1}^{\infty}(1-x^{2n-1})^8 + 16x\prod_{n=1}^{\infty}(1+x^{2n})^8 \tag{44}$$

(Ewell 1998, 1999) arise is connection with the TAU FUNCTION.

See also ARTIN'S CONSTANT, BARNES' G-FUNCTION, COSINE, D-ANALOG, DEDEKIND ETA FUNCTION, DIRICHLET ETA FUNCTION, EULER IDENTITY, EULER-MASCHERONI CONSTANT, EULER'S PENTAGONAL NUMBER THEOREM, EULER PRODUCT, GAMMA FUNCTION, INFINITE SERIES, JACOBI TRIPLE PRODUCT, KNAR'S FORMULA, POLYGON CIRCUMSCRIBING CONSTANT, POLYGON INSCRIBING CONSTANT, POWER TOWER, Q-FUNCTION, Q-SERIES, RIEMANN ZETA FUNCTION, SINE, STEPHENS' CONSTANT

References

Abramowitz, M. and Stegun, C. A. (Eds.). *Handbook of Mathematical Functions with Formulas, Graphs, and Mathematical Tables, 9th printing.* New York: Dover, p. 75, 1972.

Arfken, G. "Infinite Products." §5.11 in *Mathematical Methods for Physicists, 3rd ed.* Orlando, FL: Academic Press, pp. 346-51, 1985.

Blatner, D. *The Joy of Pi.* New York: Walker, p. 119, 1997.

Borwein, J. M. and Corless, R. M. "Emerging Tools for Experimental Mathematics." *Amer. Math. Monthly* **106**, 899-09, 1999.

Ewell, J. A. "Arithmetical Consequences of a Sextuple Product Identity." *Rocky Mtn. J. Math.* **25**, 1287-293, 1995.

Ewell, J. A. "A Note on a Jacobian Identity." *Proc. Amer. Math. Soc.* **126**, 421-23, 1998.

Ewell, J. A. "New Representations of Ramanujan's Tau Function." *Proc. Amer. Math. Soc.* **128**, 723-26, 1999.

Finch, S. "Favorite Mathematical Constants." http://www.mathsoft.com/asolve/constant/infprd/infprd.html.

Hansen, E. R. *A Table of Series and Products.* Englewood Cliffs, NJ: Prentice-Hall, 1975.

Jeffreys, H. and Jeffreys, B. S. "Infinite Products." §1.14 in *Methods of Mathematical Physics, 3rd ed.* Cambridge, England: Cambridge University Press, pp. 52-3, 1988.

Kahovec, H. "Basic Infinite Products." http://www.mathsoft.com/asolve/constant/infprd/kahovec/ip.html.

Kahovec, H. "Proof of the Infinite Product Formulas." http://www.mathsoft.com/asolve/constant/infprd/kahovec/proof01.html.

Krantz, S. G. "The Concept of an Infinite Product." §8.1.6 in *Handbook of Complex Analysis.* Boston, MA: Birkhäuser, pp. 104-05, 1999.

Ritt, J. F. "Representation of Analytic Functions as Infinite Products." *Math. Z.* **32**, 1-, 1930.

Whittaker, E. T. and Watson, G. N. §7.5-.6 in *A Course in Modern Analysis, 4th ed.* Cambridge, England: Cambridge University Press, 1990.

Infinite Series

A SERIES with an INFINITE number of terms is called an infinite series. A (possibly infinite) series for which the ratio of each two consecutive terms a_{k+1}/a_k is a constant function of the summation index k. The more general case of the ratio a RATIONAL FUNCTION of the summation index k produces a series called a HYPERGEOMETRIC SERIES.

A particular infinite series identity is given by

$$\sum_{k=1,3,5,\ldots}^{\infty}\frac{e^{-kx}\sin(ky)}{k} = \frac{1}{2}\tan^{-1}\left(\frac{\sin y}{\sinh x}\right) \tag{1}$$

for $x > 0$. Apostol (1997, p. 25) gives the analytic sum

$$\sum_{n=1,3,5,\ldots}^{\infty}\frac{n^{4k+1}}{1+e^{n\pi}} = \frac{2^{4k+1}-1}{8k+4}B_{4k+2,} \tag{2}$$

where B_k is a BERNOULLI NUMBER.

Infinite series of the following type can also be computed analytically,

$$\left(\sum_{k=0}^{\infty}x^k\right)^p = (1-x)^{-p} \tag{3}$$

$$= \frac{1}{(p-1)!}\sum_{n=0}^{\infty}\frac{(n+p-1)!}{n!}x^n. \tag{4}$$

$$= \frac{1}{(p-1)!}\sum_{n=0}^{\infty}(n+1)_{p-1}x^n, \tag{5}$$

where $(n)_p$ is a POCHHAMMER SYMBOL.

An infinite series of the following form can be done in closed form.

$$\sum_{k=1}^{\infty}\frac{1}{[1+k^2\pi^2]^n} = \frac{p_n(e)}{2^{n+1}n!(e^2-1)^n}, \tag{6}$$

where $P_n(e^2)$ is an nth order polynomial in e^2. The first few polynomials are

$$P_1 = 1$$

$$P_2 = -e^4 + 8e^2 - 3$$

$$P_3 = -5e^6 + 41e^4 - 31e^2 + 11$$

$$P_4 = -33e^8 + 286e^6 - 344e^4 + 250e^2 - 63.$$

The related infinite series can also be done in closed form.

$$\sum_{k=1}^{\infty} \frac{1}{\left[1 + \left(k + \frac{1}{2} \right)^2 \pi^2 \right]^n}$$

$$= \frac{Q_n(e)}{2^{n+1} n! (e^2 + 1)^n} - \frac{4^n}{(4 + \pi^2)^n}, \qquad (7)$$

where $Q_n(e^2)$ is an nth order polynomial in e^2. The first few polynomials are

$$Q_1 = e^2 - 1$$

$$Q_2 = e^4 - 4e^2 - 1$$

$$Q_3 = 3e^6 - 17e^4 - 7e^2 - 3$$

$$Q_4 = 15e^8 - 94e^6 - 56e^4 - 58e^2 + 15$$

$$Q_5 = 105e^{10} - 657e^8 - 578e^6 - 982e^4 - 503 - 105.$$

See also ABSOLUTE CONVERGENCE, CONDITIONAL CONVERGENCE, CONVERGENT SERIES, DIVERGENT SERIES, GEOMETRIC SERIES, HYPERGEOMETRIC SERIES, INFINITE PRODUCT, SERIES

References
Apostol, T. M. *Modular Functions and Dirichlet Series in Number Theory, 2nd ed.* New York: Springer-Verlag, p. 25, 1997.
Bromwich, T. J. I'a. and MacRobert, T. M. "Alternating Series." §19 in *An Introduction to the Theory of Infinite Series, 3rd ed.* New York: Chelsea, pp. 55–7, 1991.
Gardner, M. "Limits of Infinite Series." Ch. 17 in *The Sixth Book of Mathematical Games from Scientific American.* Chicago, IL: University of Chicago Press, pp. 163–72, 1984.
Natanson, I. P. *Summation of Infinitely Small Quantities.* Boston, MA: Heath, 1963.
Rainville, E. D. *Infinite Series.* New York: Macmillan, 1967.

Infinite Set

A SET of S elements is said to be infinite if the elements of a PROPER SUBSET S' can be put into ONE-TO-ONE correspondence with the elements of S. An infinite set whose elements can be put into a ONE-TO-ONE correspondence with the set of INTEGERS is said to be COUNTABLY INFINITE; otherwise, it is called UNCOUNTABLY INFINITE.

See also ALEPH-0, ALEPH-1, CARDINAL NUMBER, COUNTABLY INFINITE, CONTINUUM, FINITE, INFINITE, INFINITY, ORDINAL NUMBER, TRANSFINITE NUMBER, UNCOUNTABLY INFINITE

References
Courant, R. and Robbins, H. *What is Mathematics?: An Elementary Approach to Ideas and Methods, 2nd ed.* Oxford, England: Oxford University Press, p. 77, 1996.

Infinite Sum

An infinite sum identity is given by

$$z^4 - 5z^3 + 10z^2 - 10z + 5 = 0,$$

for $\prod_{n=1}^{\infty} \left(1 + \frac{1}{n^p} \right).$

See also INFINITE PRODUCT

Infinitesimal

A quantity which yields 0 after the application of some LIMITING process. The understanding of infinitesimals was a major roadblock to the acceptance of CALCULUS and its placement on a firm mathematical foundation.

See also INFINITE, INFINITY, NONSTANDARD ANALYSIS

References
Bell, J. L. *A Primer of Infinitesimal Analysis.* Cambridge, England: Cambridge University Press, 1998.

Infinitesimal Analysis

An archaic term for CALCULUS.

Infinitesimal Matrix Change

Let B, A, and e be square matrices with e small, and define

$$\mathsf{B} \equiv \mathsf{A}(\mathsf{I} + \mathsf{e}), \qquad (1)$$

where I is the IDENTITY MATRIX. Then the inverse of B is approximately

$$\mathsf{B}\mathsf{B}^{-1} = (\mathsf{I} - \mathsf{e})\mathsf{A}^{-1}. \qquad (2)$$

This can be seen by multiplying

$$\mathsf{B}\mathsf{B}^{-1} = (\mathsf{A} + \mathsf{A}\mathsf{e})(\mathsf{A}^{-1} - \mathsf{e}\mathsf{A}^{-1})$$

$$= \mathsf{A}\mathsf{A}^{-1} - \mathsf{A}\mathsf{e}\mathsf{A}^{-1} + \mathsf{A}\mathsf{e}\mathsf{A}^{-1} - \mathsf{A}\mathsf{e}^2\mathsf{A}^{-1}$$

$$= \mathsf{I} - \mathsf{A}\mathsf{e}^2\mathsf{A}^{-1} \approx 1. \qquad (3)$$

Note that if we instead let $\mathsf{B}' \equiv \mathsf{A} + \mathsf{e}$, and look for an inverse OF THE FORM $\mathsf{B}'^{-1} = \mathsf{A}^{-1} + \mathsf{C}$, we obtain

$$\mathsf{B}\mathsf{B}'^{-1} = (\mathsf{A} + \mathsf{e})(\mathsf{A}^{-1} + \mathsf{C}) = \mathsf{A}\mathsf{A}^{-1} + \mathsf{A}\mathsf{C} + \mathsf{e}\mathsf{A}^{-1} + \mathsf{e}\mathsf{C}$$

$$= \mathsf{I} + \mathsf{A}\mathsf{C} + \mathsf{e}(\mathsf{C} + \mathsf{A}^{-1}) \equiv \mathsf{I}. \qquad (4)$$

In order to eliminate the e term, we require $\mathsf{C} = -\mathsf{A}^{-1}$. However, then $\mathsf{A}\mathsf{C} = -\mathsf{I}$, so $\mathsf{B}\mathsf{B}^{-1} = 0$ so there can be no inverse of this form.

The exact inverse of B' can be found as follows.

$$\mathsf{B}' = \mathsf{A}(\mathsf{I} + \mathsf{e}) = \mathsf{A}(\mathsf{I} + \mathsf{A}^{-1}\mathsf{e}), \qquad (5)$$

so

$$B'^{-1} = [A(I + A^{-1}e)]^{-1}. \qquad (6)$$

Using a general MATRIX INVERSE identity then gives

$$B'^{-1} = (I + A^{-1}e)^{-1} A^{-1}. \qquad (7)$$

Infinitesimal Rotation

An infinitesimal transformation of a VECTOR $\mathbf{r}$ is given by

$$\mathbf{r}' = (I + e)\mathbf{r}, \qquad (1)$$

where the MATRIX e is infinitesimal and I is the IDENTITY MATRIX. (Note that the infinitesimal transformation may not correspond to an inversion, since inversion is a discontinuous process.) The COMMUTATIVITY of infinitesimal transformations e_1 and e_2 is established by the equivalence of

$$(I + e_1)(I + e_2) = I^2 + e_1 I + I e_2 + e_1 e_2 \approx I + e_1 + e_2 \qquad (2)$$

$$(I + e_2)(I + e_1) = I^2 + e_2 I + I e_1 + e_2 e_1 \approx I + e_2 + e_1. \qquad (3)$$

Now let

$$A \equiv I + e, \qquad (4)$$

The inverse A^{-1} is then $I - e$, since

$$A A^{-1} = (I + e)(I - e) = I^2 - e^2 \approx I. \qquad (5)$$

Since we are defining our infinitesimal transformation to be a rotation, ORTHOGONALITY of ROTATION MATRICES requires that

$$A^T = A^{-1}, \qquad (6)$$

but

$$A^{-1} = I - e \qquad (7)$$

$$(I + e)^T = I^T + e^T = I + e^T, \qquad (8)$$

so $e = -e^T$ and the infinitesimal rotation is ANTISYMMETRIC. It must therefore have a MATRIX OF THE FORM

$$e = \begin{bmatrix} 0 & d\Omega_3 & -d\Omega_2 \\ -d\Omega_3 & 0 & d\Omega_1 \\ d\Omega_2 & -d\Omega_1 & 0 \end{bmatrix}. \qquad (9)$$

The differential change in a vector $\mathbf{r}$ upon application of the ROTATION MATRIX is then

$$d\mathbf{r} \equiv \mathbf{r}' - \mathbf{r} = (I + e)\mathbf{r} - \mathbf{r} = e\mathbf{r}. \qquad (10)$$

Writing in MATRIX form,

$$d\mathbf{r} = \begin{bmatrix} x \\ y \\ z \end{bmatrix} \begin{bmatrix} 0 & d\Omega_3 & -d\Omega_2 \\ -d\Omega_3 & 0 & d\Omega_1 \\ d\Omega_2 & -d\Omega_1 & 0 \end{bmatrix}$$

$$= \begin{bmatrix} y \ d\Omega_3 - z \ d\Omega_2 \\ z \ d\Omega_1 - x \ d\Omega_3 \\ x \ d\Omega_2 - y \ d\Omega_1 \end{bmatrix} \qquad (11)$$

$$= (y \ d\Omega_3 - z \ d\Omega_2)\hat{\mathbf{x}} + (z \ d\Omega_1 - x \ d\Omega_3)\hat{\mathbf{y}}$$

$$+ (x \ d\Omega_2 - y d\Omega_1)\hat{\mathbf{z}} = \mathbf{r} \times d\Omega. \qquad (12)$$

Therefore,

$$\left(\frac{d\mathbf{r}}{dt}\right)_{\text{rotation, body}} = \mathbf{r} \times \frac{d\Omega}{dt} = \mathbf{r} \times \omega, \qquad (13)$$

where

$$\omega \equiv \frac{d\Omega}{dt} = \hat{\mathbf{n}} \frac{d\phi}{dt}. \qquad (14)$$

The total rotation observed in the stationary frame will be a sum of the rotational velocity and the velocity in the rotating frame. However, note that an observer in the stationary frame will see a velocity opposite in direction to that of the observer in the frame of the rotating body, so

$$\left(\frac{d\mathbf{r}}{dt}\right)_{\text{space}} = \left(\frac{d\mathbf{r}}{dt}\right)_{\text{body}} + \omega \times \mathbf{r}. \qquad (15)$$

This can be written as an operator equation, known as the ROTATION OPERATOR, defined as

$$\left(\frac{d}{dt}\right)_{\text{space}} = \left(\frac{d}{dt}\right)_{\text{body}} + \omega \times. \qquad (16)$$

See also ACCELERATION, EULER ANGLES, ROTATION, ROTATION MATRIX, ROTATION OPERATOR

Infinitive Sequence

A sequence $\{x_n\}$ is called an infinitive sequence if, for every i, $x_n = i$ for infinitely many n. Write $a(i,j)$ for the jth index n for which $x_n = i$. Then as i and j range through N, the array $A = a(i,j)$, called the associative array of x, ranges through all of N.

See also FRACTAL SEQUENCE

References
Kimberling, C. "Fractal Sequences and Interspersions." *Ars Combin.* **45**, 157–68, 1997.

Infinitude of Primes

EUCLID'S THEOREMS

Infinity

An unbounded number greater than every REAL NUMBER, most often denoted as ∞. The symbol ∞ had been used as an alternative to M (1,000) in ROMAN NUMERALS until 1655, when John Wallis suggested it be used instead for infinity.

Infinity is a very tricky concept to work with, as evidenced by some of the counterintuitive results which follow from Georg Cantor's treatment of

INFINITE SETS. Informally, $1/\infty = 0$, a statement which can be made rigorous using the LIMIT concept,

$$\lim_{x \to \infty} \frac{1}{x} = 0.$$

Similarly,

$$\lim_{x \to 0^+} \frac{1}{x} = \infty,$$

where the notation 0^+ indicates that the LIMIT is taken from the POSITIVE side of the REAL LINE.

See also ALEPH, ALEPH-0, ALEPH-1, CARDINAL NUMBER, COMPLEX INFINITY, CONTINUUM, CONTINUUM HYPOTHESIS, HILBERT HOTEL, INFINITE, INFINITE SET, INFINITESIMAL, LINE AT INFINITY, L'HOSPITAL'S RULE, POINT AT INFINITY, TRANSFINITE NUMBER, UNCOUNTABLY INFINITE, ZERO

References
Conway, J. H. and Guy, R. K. *The Book of Numbers.* New York: Springer-Verlag, p. 19, 1996.
Courant, R. and Robbins, H. "The Mathematical Analysis of Infinity." §2.4 in *What is Mathematics?: An Elementary Approach to Ideas and Methods, 2nd ed.* Oxford, England: Oxford University Press, pp. 77–8, 1996.
Hardy, G. H. *Orders of Infinity, the 'infinitarcalcul' of Paul Du Bois-Reymond, 2nd ed.* Cambridge, England: Cambridge University Press, 1924.
Lavine, S. *Understanding the Infinite.* Cambridge, MA: Harvard University Press, 1994.
Maor, E. *To Infinity and Beyond: A Cultural History of the Infinite.* Boston, MA: Birkhäuser, 1987.
Moore, A. W. *The Infinite.* New York: Routledge, 1991.
Morris, R. *Achilles in the Quantum Universe: The Definitive History of Infinity.* New York: Henry Holt, 1997.
Owen, H. P. "Infinity in Theology and Metaphysics." In *The Encyclopedia of Philosophy, Vol. 4.* New York: Crowell Collier, pp. 190–93, 1967.
Péter, R. *Playing with Infinity.* New York: Dover, 1976.
Rucker, R. *Infinity and the Mind: The Science and Philosophy of the Infinite.* Princeton, NJ: Princeton University Press, 1995.
Smail, L. L. *Elements of the Theory of Infinite Processes.* New York: McGraw-Hill, 1923.
Thomson, J. "Infinity in Mathematics and Logic." In *The Encyclopedia of Philosophy, Vol. 4.* New York: Crowell Collier, pp. 183–90, 1967.
Vilenskin, N. Ya. *In Search of Infinity.* Boston, MA: Birkhäuser, 1995.
Weisstein, E. W. "Books about Infinity." http://www.trea-sure-troves.com/books/Infinity.html.
Wilson, A. M. *The Infinite in the Finite.* New York: Oxford University Press, 1996.
Zippin, L. *Uses of Infinity.* New York: Random House, 1962.

Inflection Point

A point on a curve at which the SIGN of the CURVATURE (i.e., the concavity) changes. The FIRST DERIVATIVE TEST can sometimes distinguish inflection points from EXTREMA for DIFFERENTIABLE functions $f(x)$.

See also CURVATURE, DIFFERENTIABLE, EXTREMUM, FIRST DERIVATIVE TEST, STATIONARY POINT

Information Dimension

Define the "information function" to be

$$I = -\sum_{i=1}^{N} P_i(\epsilon) \ln[P_i(\epsilon)], \tag{1}$$

where $P_i(\epsilon)$ is the NATURAL MEASURE, or probability that element i is populated, normalized such that

$$\sum_{i=1}^{N} P_i(\epsilon) = 1. \tag{2}$$

The information dimension is then defined by

$$d_{\text{inf}} \equiv -\lim_{\epsilon \to 0^+} \frac{I}{\ln(\epsilon)}$$

$$= \lim_{\epsilon \to 0^+} \sum_{i=1}^{N} \frac{P_i(\epsilon) \ln[P_i(\epsilon)]}{\ln(\epsilon)}. \tag{3}$$

If every element is equally likely to be visited, then $P_i(\epsilon)$ is independent of i, and

$$\sum_{i=1}^{N} P_i(\epsilon) = NP_i(\epsilon) = 1, \tag{4}$$

so

$$P_i(\epsilon) = \frac{1}{N}, \tag{5}$$

and

$$d_{\text{inf}} = \lim_{\epsilon \to 0^+} \frac{\sum_{i=1}^{N} \frac{1}{N} \ln\left(\frac{1}{N}\right)}{\ln \epsilon}$$

$$= \lim_{\epsilon \to 0^+} \frac{\ln(N^{-1})}{\ln \epsilon} = -\lim_{\epsilon \to 0^+} \frac{\ln N}{\ln \epsilon} = d_{\text{cap}}, \tag{6}$$

where d_{cap} is the CAPACITY DIMENSION.

See also CORRELATION EXPONENT

References
Balatoni, J. and Renyi, A. *Pub. Math. Inst. Hungarian Acad. Sci.* **1**, 9, 1956.
Farmer, J. D. "Chaotic Attractors of an Infinite-dimensional Dynamical System." *Physica D* **4**, 366–93, 1982.
Ott, E. *Chaos in Dynamical Systems.* New York: Cambridge University Press, p. 79, 1993.
Nayfeh, A. H. and Balachandran, B. *Applied Nonlinear Dynamics: Analytical, Computational, and Experimental Methods.* New York: Wiley, pp. 545–47, 1995.

Information Entropy

ENTROPY

Information Theory

The branch of mathematics dealing with the efficient and accurate storage, transmission, and representation of information.

See also CODING THEORY, COMPRESSION, ENTROPY

References

Goldman, S. *Information Theory.* New York: Dover, 1953.
Hankerson, D.; Harris, G. A.; and Johnson, P. D. Jr. *Introduction to Information Theory and Data Compression.* Boca Raton, FL: CRC Press, 1998.
Lee, Y. W. *Statistical Theory of Communication.* New York: Wiley, 1960.
Pierce, J. R. *An Introduction to Information Theory.* New York: Dover, 1980.
Reza, F. M. *An Introduction to Information Theory.* New York: Dover, 1994.
Singh, J. *Great Ideas in Information Theory, Language and Cybernetics.* New York: Dover, 1966.
Weisstein, E. W. "Books about Information Theory." http://www.treasure-troves.com/books/InformationTheory.html.
Zayed, A. I. *Advances in Shannon's Sampling Theory.* Boca Raton, FL: CRC Press, 1993.

Initial Ordinal

An ORDINAL NUMBER is called an initial ordinal if every smaller ordinal has a smaller CARDINALITY (Moore 1982, p. 248; Rubin 1967, p. 271). The ω_αs ordinal numbers are just the transfinite initial ordinals (Rubin 1967, p. 272).

This PROPER CLASS can be well ordered and put into one-to-one correspondence with the ORDINAL NUMBERS. For any two WELL ORDERED SETS that are ORDER ISOMORPHIC, there is only one order isomorphism between them. Let f be that isomorphism from the ordinals to the transfinite initial ordinals, then

$$\omega_\alpha = f(\alpha),$$

where $\omega_0 = \omega$.

See also ORDINAL NUMBER

References

Moore, G. H. *Zermelo's Axiom of Choice: Its Origin, Development, and Influence.* New York: Springer-Verlag, 1982.
Rubin, J. E. *Set Theory for the Mathematician.* New York: Holden-Day, 1967.

Initial Segment

Let $(A, \leq)$ be a WELL ORDERED SET. Then the set $\{a \in A : a < k\}$ for some $k \in A$ is called an initial segment of A (Rubin 1967, p. 161; Dauben 1990, pp. 196–97; Moore 1982, pp. 90–1). This term was first used by Cantor, who also proved that if $(A, \leq)$ and $(B, \leq)$ are WELL ORDERED SETS that are not ORDER ISOMORPHIC, then exactly one of the following statements is true:

1. A is ORDER ISOMORPHIC to an initial segment of B, or
2. B is ORDER ISOMORPHIC to an initial segment of A

(Dauben 1990, p. 198).

See also WELL ORDERED SET

References

Dauben, J. W. *Georg Cantor: His Mathematics and Philosophy of the Infinite.* Princeton, NJ: Princeton University Press, 1990.
Moore, G. H. *Zermelo's Axiom of Choice: Its Origin, Development, and Influence.* New York: Springer-Verlag, 1982.
Rubin, J. E. *Set Theory for the Mathematician.* New York: Holden-Day, 1967.

Initial Value Problem

An initial value problem is a problem that has its conditions specified at some time $t = t_0$. Usually, the problem is an ORDINARY DIFFERENTIAL EQUATION or a PARTIAL DIFFERENTIAL EQUATION. For example,

$$\begin{cases} \dfrac{\partial^2 u}{\partial t^2} - \nabla^2 u = f & \text{in } \Omega \\ u = u_0 & t = t_0 \\ u = u_1 & \text{on } \partial\Omega, \end{cases}$$

where $\partial\Omega$ denotes the boundary of Ω, is an initial value problem.

See also BOUNDARY CONDITIONS, BOUNDARY VALUE PROBLEM, PARTIAL DIFFERENTIAL EQUATION

References

Eriksson, K.; Estep, D.; Hansbo, P.; and Johnson, C. *Computational Differential Equations.* Lund, Sweden: Studentlitteratur, 1996.

Injection

ONE-TO-ONE

Injective

A MAP is injective when it is ONE-TO-ONE, i.e., f is injective when $x \neq y$ IMPLIES $f(x) \neq f(y)$.

See also ONE-TO-ONE, SURJECTIVE

Injective Patch

An injective patch is a PATCH such that $\mathbf{x}(u_1, v_1) = \mathbf{x}(u_2, v_2)$ implies that $u_1 = u_2$ and $v_1 = v_2$. An example of a PATCH which is injective but not REGULAR is the function defined by (u^3, v^3, uv) for $u, v \in (-1, 1)$. However, if $\mathbf{x} : U \to \mathbb{R}^n$ is an injective regular patch, then $\mathbf{x}$ maps U diffeomorphically onto $\mathbf{x}(U)$.

See also PATCH, REGULAR PATCH

References

Gray, A. *Modern Differential Geometry of Curves and Surfaces with Mathematica, 2nd ed.* Boca Raton, FL: CRC Press, p. 273, 1997.

Inner Automorphism Group

A particular type of AUTOMORPHISM GROUP which exists only for GROUPS. For a GROUP G, the inner automorphism group is defined by

$$\text{Inn}(G) = \{\sigma_a : a \in G\} \subset \text{Aut}(G)$$

where σ_a is an AUTOMORPHISM of G defined by

$$\sigma_a(x) = axa^{-1}.$$

See also AUTOMORPHISM, AUTOMORPHISM GROUP

Inner Product

DOT PRODUCT, HERMITIAN INNER PRODUCT, INTERIOR PRODUCT, L2-INNER PRODUCT

Inner Product Space

An inner product space is a VECTOR SPACE which has an INNER PRODUCT. If the INNER PRODUCT defines a NORM, then the inner product space is called a HILBERT SPACE.

See also HILBERT SPACE, INNER PRODUCT, NORM

Inner Quermass

The largest area of intersection of a solid body by a plane parallel to a given plane, also called the "HA measurement."

See also BRIGHTNESS, CROSS SECTION, SHADOW, STEREOLOGY

References

Bonnesen, T. "Om Minkowski's uligheder fur konvexer legemer." *Mat. Tidsskr.* **B**, 80, 1926.
Bonnesen, R. and Fenchel, W. *Theorie der Konvexer Körper.* New York: Chelsea, p. 140, 1971.
Croft, H. T.; Falconer, K. J.; and Guy, R. K. "What Can You Tell About a Convex Body from its Section." §A11 in *Unsolved Problems in Geometry.* New York: Springer-Verlag, pp. 24–5, 1991.
Klee, V. "Is a Body Spherical if All its HA Measurements are Constant?" *Amer. Math. Monthly* **76**, 539–42, 1969.
Zaks, J. "Nonspherical Bodies with Constant HA Measurements Exist." *Amer. Math. Monthly* **78**, 513–16, 1971.

Inradius

The radius of a TRIANGLE'S INCIRCLE or of a POLYHEDRON'S INSPHERE, denoted r (or sometimes ρ). For a TRIANGLE,

$$r = \frac{1}{2}\sqrt{\frac{(b+c-a)(c+a-b)(a+b-c)}{a+b+c}} \qquad (1)$$

$$= \frac{\Delta}{s} \qquad (2)$$

$$4R \sin\left(\tfrac{1}{2}A\right)\sin\left(\tfrac{1}{2}B\right)\sin\left(\tfrac{1}{2}C\right), \qquad (3)$$

where Δ is the AREA of the TRIANGLE, a, b, and c are the side lengths, s is the SEMIPERIMETER, R is the CIRCUMRADIUS, and A, B, and C are the angles opposite sides a, b, and c (Johnson 1929, p. 189). If two triangle side lengths a and b are known, together with the inradius r, then the length of the third side c can be found by solving (1) for c, resulting in a CUBIC EQUATION.

Equation (2) can be derived easily using TRILINEAR COORDINATES. Since the INCENTER is equally spaced from all three sides, its trilinear coordinates are 1:1:1, and its exact trilinear coordinates are $r:r:r$. The ratio k of the exact trilinears to the homogeneous coordinates is given by

$$k = \frac{2\Delta}{a+b+c} = \frac{\Delta}{s}. \qquad (4)$$

But since $k = r$ in this case,

$$r = k = \frac{\Delta}{s}, \qquad (5)$$

Q.E.D.

Other equations involving the inradius include

$$Rr = \frac{abc}{4s} \qquad (6)$$

$$\Delta^2 = rr_1r_2r_3 \qquad (7)$$

$$\cos A + \cos B + \cos C = 1 + \frac{r}{R} \qquad (8)$$

$$a^2 + b^2 + c^2 = 4rR + 8R^2, \qquad (9)$$

where r_i are the EXRADII (Johnson 1929, pp. 189–91). As shown in RIGHT TRIANGLE, the inradius of a RIGHT TRIANGLE side lengths a, b, and c is given by

$$r = \frac{ab}{a+b+c} \qquad (10)$$

$$= \sqrt{\tfrac{1}{2}(c-a)(c-b)} \qquad (11)$$

$$= \tfrac{1}{2}(a+b-c), \qquad (12)$$

where c is the HYPOTENUSE.

Let d be the distance between inradius r and CIRCUMRADIUS R, $d = \overline{rR}$. Then

$$R^2 - d^2 = 2Rr \qquad (13)$$

$$\frac{1}{R-d} + \frac{1}{R+d} = \frac{1}{r} \tag{14}$$

(Mackay 1886–7; Casey 1888, pp. 74–5). These and many other identities are given in Johnson (1929, pp. 186–90).

For a PLATONIC SOLID or ARCHIMEDEAN SOLID, the inradius of the solid is also the inradius of the DUAL POLYHEDRON. Expressing the MIDRADIUS ρ and CIRCUMRADIUS R in terms of the midradius gives

$$r = \frac{2}{\sqrt{\rho^2 + \frac{1}{4}a^2}} \tag{15}$$

$$r = \frac{R^2 - \frac{1}{4}a^2}{R} \tag{16}$$

for an ARCHIMEDEAN SOLID.

See also CARNOT'S THEOREM, CIRCUMRADIUS, JAPANESE THEOREM, MIDRADIUS

References

Casey, J. *A Sequel to the First Six Books of the Elements of Euclid, Containing an Easy Introduction to Modern Geometry with Numerous Examples, 5th ed., rev. enl.* Dublin: Hodges, Figgis, & Co., 1888.

Coxeter, H. S. M. and Greitzer, S. L. *Geometry Revisited.* Washington, DC: Math. Assoc. Amer., p. 10, 1967.

Johnson, R. A. *Modern Geometry: An Elementary Treatise on the Geometry of the Triangle and the Circle.* Boston, MA: Houghton Mifflin, 1929.

Mackay, J. S. "Historical Notes on a Geometrical Theorem and its Developments [18th Century]." *Proc. Edinburgh Math. Soc.* **5**, 62–8, 1886–887.

Mackay, J. S. "Formulas Connected with the Radii of the Incircle and Excircles of a Triangle." *Proc. Edinburgh Math. Soc.* **12**, 86–05.

Mackay, J. S. "Formulas Connected with the Radii of the Incircle and Excircles of a Triangle." *Proc. Edinburgh Math. Soc.* **13**, 103–04.

Inscribed

A geometric figure which touches only the sides (or interior) of another figure.

See also CIRCUMSCRIBED, INCENTER, INCIRCLE, INRADIUS

Inscribed Angle

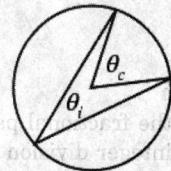

The ANGLE with VERTEX on a CIRCLE'S CIRCUMFERENCE formed by two points on a CIRCLE'S CIRCUMFER-ENCE. For ANGLES with the same endpoints,

$$\theta_c = 2\theta_i,$$

where θ_c is the CENTRAL ANGLE.

See also CENTRAL ANGLE

References

Pedoe, D. *Circles: A Mathematical View, rev. ed.* Washington, DC: Math. Assoc. Amer., pp. xxi-xxii, 1995.

Inside-Outside Theorem

Let $P(z)$ and $Q(z)$ be UNIVARIATE POLYNOMIALS in a complex variable z, and let the DEGREES of P and Q satisfy $\deg(Q) \geq \deg(P+2)$. Then

$$\int_\gamma \frac{P(z)}{Q(z)} dz = 2\pi i \sum_{a_i \in A} \operatorname*{Res}_{z=a_i} \frac{P(z)}{Q(z)} \tag{1}$$

$$= -2\pi i \sum_{b_i \in B} \operatorname*{Res}_{z=b_i} \frac{P(z)}{Q(z)}, \tag{2}$$

where γ is a simple closed clockwise-oriented CONTOUR, A is the set of ROOTS of Q inside of γ, and B is the set of ROOTS of Q outside of γ.

The first equality is an instance of the RESIDUE THEOREM. On the RIEMANN SPHERE, the simple closed CONTOUR γ splits the sphere into two regions. After the change of variables $w = 1/z$, the point zero is mapped to infinity and vice versa. What was the "inside" of γ becomes the outside of γ in the new coordinate. The second equality is the RESIDUE THEOREM applied to the MEROMORPHIC ONE-FORM $a = P/Q \, dz$ in the coordinate w, with a minus sign because γ travels clockwise after the coordinate change. The hypothesis on the degrees of P and Q ensure that α does not have a POLE at $z = \infty$.

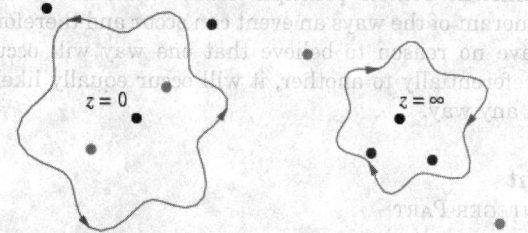

The above diagram shows two different points of view of the contour γ and the poles of the MEROMORPHIC ONE-FORM $P/Q \, dz$ on the RIEMANN SPHERE. The usual point of view is centered at $z = 0$, but the role of inside and outside is switched from the point of view of $z = \infty$. The poles inside are labeled blue and outside are green.

The theorem also follows from taking the CONTOUR INTEGRAL at infinity, i.e., a circle of large radius R. The hypothesis on the degree says that this integral tends to zero. Hence it must actually be zero, because at some point the circle contains all of the poles of

P/Q. This is a special case of the fact that on a COMPACT RIEMANN SURFACE, in this case the RIEMANN SPHERE, the sum of the RESIDUES of a MEROMORPHIC ONE-FORM is zero.

See also CONTOUR, CONTOUR INTEGRAL, JACOBIAN, RESIDUE (COMPLEX ANALYSIS), RESIDUE THEOREM, RIEMANN SPHERE, ROOT

Insphere

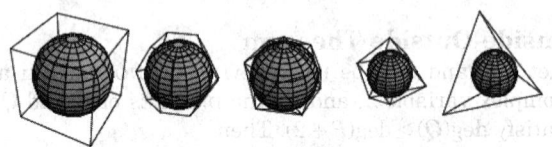

A SPHERE INSCRIBED in a given solid. The figures above depict the inspheres of the Platonic solids.

See also CIRCUMSPHERE, MIDSPHERE

Instrument Function

The finite FOURIER COSINE TRANSFORM of an APODIZATION FUNCTION, also known as an APPARATUS FUNCTION. The instrument function $I(x)$ corresponding to a given APODIZATION FUNCTION $A(x)$ is then given by

$$I(k) = \int_{-a}^{a} \cos(2\pi k x) A(x) dx.$$

See also APODIZATION FUNCTION, FOURIER COSINE TRANSFORM

Insufficient Reason Principle

A principle, also called the indifference principle, that was first enunciated by Johann Bernoulli. The insufficient reason principle states that, if we are ignorant of the ways an event can occur and therefore have no reason to believe that one way will occur preferentially to another, it will occur equally likely in any way.

Int

INTEGER PART

Integer

One of the numbers ..., -2, -1, 0, 1, 2, ... The SET of INTEGERS forms a RING which is denoted $\mathbb{Z}$. A given INTEGER n may be NEGATIVE ($a \equiv \mathbb{Z}^-$), NONNEGATIVE ($n \in \mathbb{Z}^*$), ZERO ($n = 0$), or POSITIVE ($n \in \mathbb{Z}^+ = \mathbb{N}$). The set of integers is denoted `Integers` in *Mathematica*, and a number x can be tested to see if it is an integer using the command `Element[x, Integers]`. Numbers that are integers are sometimes described as "integral" (instead of integer-valued), but this practice may lead to unnecessary confusions with the INTEGRALS of INTEGRAL CALCULUS.

The RING $\mathbb{Z}$ of integers has CARDINALITY of ALEPH-0. The GENERATING FUNCTION for the NONNEGATIVE INTEGERS is

$$f(x) = \frac{x}{(1-x)^2} = x + 2x^2 + 3x^3 + 4x^4 + \cdots.$$

There are several symbols used to perform operations having to do with conversion between REAL NUMBERS and integers. The symbol $\lfloor x \rfloor$ ("FLOOR x") means "the largest integer not greater than x," i.e., `int(x)` in computer parlance. The symbol $[x]$ means "the nearest integer to x" (NINT), i.e., `nint(x)` in computer parlance. The symbol $\lceil x \rceil$ ("CEILING x") means the smallest integer not smaller x," or `-int(-x)`, where `int(x)` is the INTEGER PART of x.

The German mathematician and logician Kronecker vociferously opposed the work of Georg Cantor on infinite sets and summarized his view that ARITHMETIC and ANALYSIS should be based on whole numbers only by saying, "God made the natural numbers; all else is the work of man" (Bell 1986, p. 477).

See also ALGEBRAIC INTEGER, ALMOST INTEGER, COMPLEX NUMBER, COUNTING NUMBER, CYCLOTOMIC INTEGER, EISENSTEIN INTEGER, FRACTIONAL PART, GAUSSIAN INTEGER, INTEGER PART, N, NATURAL NUMBER, NEGATIVE, POSITIVE, RADICAL INTEGER, REAL NUMBER, WHOLE NUMBER, Z, Z-, Z+, Z*, ZERO

References
Bell, E. T. *Men of Mathematics*. New York: Simon and Schuster, 1986.

Integer Array

See also INTEGER SEQUENCE

References
Kimberling, C. "Integer Sequences and Arrays." http:// cedar.evansville.edu/~ck6/integer/.

Integer Bowl

BOWL OF INTEGERS

Integer Cuboid

EULER BRICK

Integer Division

DIVISION in which the fractional part (remainder) is discarded is called integer division and is sometimes denoted \. Integer division can be defined as $a \backslash b \equiv \lfloor a/b \rfloor$ where "/" denotes normal division and $\lfloor x \rfloor$ is the FLOOR FUNCTION. For example,

$$10/3 = 3 + 1/3$$
$$10\backslash 3 = 3.$$

Integer Exponent
GREATEST DIVIDING EXPONENT

Integer Factorization
PRIME FACTORIZATION

Integer Function
A FUNCTION defined for all positive integers, sometimes also called an "arithmetical function" (Nagell 1951, p. 26).

See also COMPLEX MATRIX, REAL MATRIX

References
Nagell, T. "Arithmetical Functions." §9 in *Introduction to Number Theory*. New York: Wiley, pp. 26–9, 1951.

Integer Matrix
A MATRIX whose entries are all integers. Special cases which arise frequently are those having only $(1, -1)$ as entries (e.g., HADAMARD MATRIX), BINARY MATRICES having only $(0, 1)$ as entries (e.g., ADJACENCY MATRIX, FROBENIUS-KÖNIG THEOREM, GALE-RYSER THEOREM, HADAMARD'S MAXIMUM DETERMINANT PROBLEM, HARD SQUARE ENTROPY CONSTANT, IDENTITY MATRIX, INCIDENCE MATRIX, LAM'S PROBLEM), and those having $(-1, 0, 1)$ as entries (e.g., ALTERNATING SIGN MATRIX, C-MATRIX).

The ZERO MATRIX could be considered a degenerate case of an integer matrix.

See also ALTERNATING SIGN MATRIX, (-1,0,1)-MATRIX, (-1,1)-MATRIX, (0,1)-MATRIX, COMPLEX MATRIX, FROBENIUS-KÖNIG THEOREM, GALE-RYSER THEOREM, C-MATRIX, FIFTEEN THEOREM, GALE-RYSER THEOREM, HADAMARD'S MAXIMUM DETERMINANT PROBLEM, HADAMARD MATRIX, HAFNER-SARNAK-MCCURLEY CONSTANT, HARD SQUARE ENTROPY CONSTANT, IDENTITY MATRIX, INCIDENCE MATRIX, INTEGER-MATRIX FORM, INTERSPERSION, LAM'S PROBLEM, MORTAL, MORTALITY PROBLEM, REAL MATRIX, SMITH NORMAL FORM, SPECIAL MATRIX, UNIT MATRIX, ZERO MATRIX

Integer-Matrix Form
Let $Q(x) \equiv Q(x) = Q(x_1, x_2, \ldots, x_n)$ be an integer-valued n-ary QUADRATIC FORM, i.e., a POLYNOMIAL with integer COEFFICIENTS which satisfies $Q(x) > 0$ for REAL $x \neq 0$. Then $Q(x)$ can be represented by

$$Q(x) = \mathbf{x}^T A \mathbf{x},$$

where

$$A = \frac{1}{2} \frac{\partial^2 Q(x)}{\partial x_i \partial x_j}$$

is a POSITIVE SYMMETRIC MATRIX (Duke 1997). If A has POSITIVE entries, then $Q(x)$ is called an integer-matrix form. Conway *et al.* (1997) have proven that, if a POSITIVE integer-matrix quadratic form represents each of 1, 2, 3, 5, 6, 7, 10, 14, and 15, then it represents all POSITIVE INTEGERS.

See also FIFTEEN THEOREM

References
Conway, J. H.; Guy, R. K.; Schneeberger, W. A.; and Sloane, N. J. A. "The Primary Pretenders." *Acta Arith.* **78**, 307–13, 1997.
Duke, W. "Some Old Problems and New Results about Quadratic Forms." *Not. Amer. Math. Soc.* **44**, 190–96, 1997.

Integer Module
ABELIAN GROUP

Integer Part

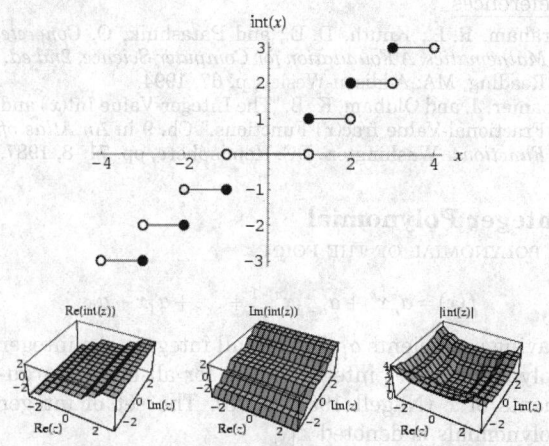

The function int x gives the integer part of x. In many computer languages, the function is denoted int(x). It is related to the FLOOR and CEILING FUNCTIONS $\lfloor x \rfloor$ and $\lceil x \rceil$ by

$$\text{int } x = \begin{cases} \lfloor x \rfloor & \text{for } x \geq \infty \\ \lceil x \rceil & \text{for } x < 0 \end{cases}$$

The integer part function satisfies

$$\text{int}(-x) = -\text{int}(x)$$

and is implemented in *Mathematica* as Integer-Part[x]. This definition is chosen so that int x + frac $x = x$, where frac x is the FRACTIONAL PART. Although Spanier and Oldham (1987) use the same definition as *Mathematica*, they mention the formula only very briefly and then say it will not be used further. Graham *et al.* (1994), and perhaps most other mathematicians, use the term "integer" part interchangeably with the FLOOR FUNCTION $\lfloor x \rfloor$.

Since usage concerning fractional part/value and integer part/value can be confusing, the following table gives a summary of names and notations used (D. W. Cantrell). Here, S&O indicates Spanier and Oldham (1987).

notation	name	S&O	Graham et al.	*Mathematica*
$\lfloor x \rfloor$	integer-value	Int(x)	floor or integer part	Floor[x]
$\text{sgn}(x)\lfloor \lvert x \rvert \rfloor$	integer-part	Ip(x)	no name	Integer-Part[x]
$x - \lfloor x \rfloor$	fractional-value	frac(x)	fractional part or $\{x\}$	no name
$\text{sgn}(x)(\lvert x \rvert - \lfloor \lvert x \rvert \rfloor)$	fractional-part	Fp(x)	no name	Fractional-Part[x]

See also CEILING FUNCTION, FLOOR FUNCTION, FRACTIONAL PART, INTEGER, NEAREST INTEGER FUNCTION

References

Graham, R. L.; Knuth, D. E.; and Patashnik, O. *Concrete Mathematics: A Foundation for Computer Science, 2nd ed.* Reading, MA: Addison-Wesley, p. 67, 1994.

Spanier, J. and Oldham, K. B. "The Integer-Value Int(x) and Fractional-Value frac(x) Functions." Ch. 9 in *An Atlas of Functions.* Washington, DC: Hemisphere, pp. 71-8, 1987.

Integer Polynomial

A POLYNOMIAL OF THE FORM

$$f(x) = a_n x^n + a_{n-1} x^{n-1} + \ldots + a_1 x + a_0$$

having coefficients a_i that are all integers. An integer polynomial gives integer values for all integer arguments of x (Nagell 1951, p. 73). The set of integer polynomials is denoted $\mathbb{Z}[x]$.

An integer polynomial is called primitive if the GREATEST COMMON DIVISOR $(a_0 + a_1, \ldots, a_n = 1.)$. Integer polynomials are sometimes called "integral polynomials," which is an unfortunately confusing choice of nomenclature.

See also INTEGER-REPRESENTING POLYNOMIAL, POLYNOMIAL, PRIME DIVISOR

References

Nagell, T. "Prime Divisors of Integral Polynomials" and "Divisibility of Integral Polynomials with Regard to a Prime Modulus." §25 and 29 in *Introduction to Number Theory.* New York: Wiley, pp. 73, 81-3, and 93-8, 1951.

Integer Relation

A set of REAL NUMBERS $x_1, \ldots, x_n$ is said to possess an integer relation if there exist integers a_i such that

$$a_1 x_1 + a_2 x_2 + \cdots + a_n x_n = 0,$$

with not all $a_i = 0$. For historical reasons, integer relation algorithms are sometimes called generalized Euclidean algorithms or multidimensional continued fraction algorithms.

An interesting example of such a relation is the 17-VECTOR $(1, x, x^2, \ldots, x^{16})$ with $x = 3^{1/4} - 2^{2/4}$, which has an integer relation $(1, 0, 0, 0, -3860, 0, 0, 0, -666, 0, 0, 0, -20, 0, 0, 0, 1)$, i.e.,

$$1 - 3860x^4 - 666x^8 - 20x^{12} + x^{16} = 0.$$

This is a special case of finding the polynomial of degree $n = rs$ satisfied by $x = 3^{1/r} - 2^{1/s}$.

Integer relation algorithms can be used to solve SUBSET SUM PROBLEMS, as well as to determine if a given numerical constant is equal to a root of a univariate polynomial of degree n or less (Bailey and Ferguson 1989, Ferguson and Bailey 1992).

One of the simplest cases of an integer relation between two numbers is the one inherent in the definition of the GREATEST COMMON DIVISOR. The well-known EUCLIDEAN ALGORITHM solves this problem, as well as the more general problem of an integer relation between two real numbers, yielding either an exact relation or an infinite sequence of approximate relations (Ferguson *et al.* 1999). Although attempts were made to generalize the algorithm to $n \geq 3$ by Hermite (1850), Jacobi (1868), Poincaré (1884), Perron (1907), Brun (1919, 1920, 1957), and Szekeres (1970), all such routines were known to fail in certain cases (Ferguson and Forcade 1979, Forcade 1981, Hastad *et al.* 1989). The first successful integer relation algorithm was developed by Ferguson and Forcade (1979) (Ferguson and Bailey 1992, Ferguson *et al.* 1999).

Algorithms for finding integer relations include the FERGUSON-FORCADE ALGORITHM, HJLS ALGORITHM, LLL ALGORITHM, PSLQ ALGORITHM, PSOS ALGORITHM, and the algorithm of Lagarias and Odlyzko (1985). Perhaps the simplest (and unfortunately most inefficient) such algorithm is the GREEDY ALGORITHM.

Plouffe's "Inverse Symbolic Calculator" site includes a huge database of 54 million REAL NUMBERS which are algebraically related to fundamental mathematical constants. The FERGUSON-FORCADE ALGORITHM has shown that there are no algebraic equations of degree ≤ 8 with integer coefficients having Euclidean norms below certain bounds for e/π, $e + \pi$, $\ln \pi$, γ, e^γ, γ/e, γ/π, and $\ln \gamma$, where E is the base for the NATURAL LOGARITHM, π is PI, and γ is the EULER-MASCHERONI CONSTANT (Bailey 1988).

Constant	Bound
e/π,	6.1030×10^{14}

$e + \pi$,	2.2753×10^{14}
$\ln \pi$,	8.7697×10^{9}
γ	3.5739×10^{9}
e^{γ},	1.6176×10^{17}
γ/e,	1.8440×10^{11}
γ/π	6.5403×10^{9}
$\ln \gamma$	2.6881×10^{10}

See also CONSTANT PROBLEM, FERGUSON-FORCADE ALGORITHM, GREEDY ALGORITHM, HERMITE-LINDE-MANN THEOREM, HJLS ALGORITHM, KNAPSACK PROBLEM, LATTICE REDUCTION, LINDEMANN-WEIERSTRASS THEOREM, LLL ALGORITHM, PSLQ ALGORITHM, PSOS ALGORITHM, RICHARDSON'S THEOREM, REAL NUMBER, SUBSET SUM PROBLEM

References

Bailey, D. H. and Ferguson, H. R. P. "Numerical Results on Relations Between Numerical Constants Using a New Algorithm." *Math. Comput.* **53**, 649–56, 1989.

Bailey, D. and Plouffe, S. "Recognizing Numerical Constants." http://www.cecm.sfu.ca/organics/papers/bailey/.

Bernstein, L. *The Jacobi-Perron Algorithm: Its Theory and Applications.* Berlin: Springer-Verlag, 1971.

Borwein, J. M. and Corless, R. M. "Emerging Tools for Experimental Mathematics." *Amer. Math. Monthly* **106**, 899–09, 1999.

Borwein, J. M. and Lisonek, P. "Applications of Integer Relation Algorithms." To appear in *Disc. Math.* http://www.cecm.sfu.ca/preprints/1997pp.html.

Brentjes, A. J. "Multi-Dimensional Continued Fraction Algorithms." Mathemat. Centre Tracts, No. 145. Amsterdam, Netherlands: Mathemat. Centrum, 1981.

Brun, V. "En generalisatiken av kjedebøøken, I." *Norske Vidensk. Skrifter I. Matemat. Naturvid. Klasse* **6**, 1–9, 1919.

Brun, V. "En generalisatiken av kjedebøøken, II." *Norske Vidensk. Skrifter I. Matemat. Naturvid. Klasse* **7**, 1–4, 1920.

Brun, V. "Algorithmes euclidiens pour trois et quatre nombres." In *Treizième Congrès des mathématiciens Scandinaves, tenu a Helsinki 18–3 août 1957.* Helsinki: Mercators Trycheri, pp. 46–4, 1958.

Centre for Experimental & Constructive Mathematics. "Integer Relations." http://www.cecm.sfu/projects/IntegerRelations/.

Ferguson, H. R. P. and Bailey, D. H. "A Polynomial Time, Numerically Stable Integer Relation Algorithm." RNR Techn. Rept. RNR-91–32, Jul. 14, 1992.

Ferguson, H. R. P.; Bailey, D. H.; and Arno, S. "Analysis of PSLQ, An Integer Relation Finding Algorithm." *Math. Comput.* **68**, 351–69, 1999.

Ferguson, H. R. P. and Forcade, R. W. "Generalization of the Euclidean Algorithm for Real Numbers to All Dimensions Higher than Two." *Bull. Amer. Math. Soc.* **1**, 912–14, 1979.

Forcade, R. W. "Brun's Algorithm." Unpublished manuscript, 1–7, Nov. 1981.

Hastad, J.; Just, B.; Lagarias, J. C.; and Schnorr, C. P. "Polynomial Time Algorithms for Finding Integer Relations Among Real Numbers." *SIAM J. Comput.* **18**, 859–81, 1988.

Hermite, C. "Extraits de lettres de M. Ch. Hermite à M. Jacobi sur différénts objets de la théorie de nombres." *J. reine angew. Math.* **3/4**, 261–15, 1850.

Jacobi, C. G. "Allgemeine Theorie der Kettenbruchahnlichen Algorithmen, in welche jede Zahl aus Drei vorhergehenden gebildet wird (Aus den hinterlassenen Papieren von C. G. Jacobi mitgetheilt durch Herrn E. Heine." *J. reine angew. Math.* **69**, 29–4, 1868.

Lagarias, J. C. and Odlyzko, A. M. "Solving Low-Density Subset Sum Problems." *J. ACM* **32**, 229–46, 1985.

Lenstra A. K.; Lenstra, H. W. Jr.; and Lovász, L. "Factoring Polynomials with Rational Coefficients." *Math. Ann.* **261**, 515–34, 1982.

Perron, O. "Grundlagen für eine Theorie des Jacobischen Kettenbruchalgorithmus." *Math. Ann.* **64**, 1–6, 1907.

Plouffe, S. "Inverse Symbolic Calculator." http://www.cecm.sfu.ca/projects/ISC/.

Poincaré, H. "Sur une généralisation des fractions continues." *Comptes Rendus Acad. Sci. Paris* **99**, 1014–016, 1884.

Szekeres, G. "Multidimensional Continued Fractions." *Ann. Univ. Sci. Budapest Eotvos Sect. Math.* **13**, 113–40, 1970.

Integer-Representing Polynomial

A polynomial that represents integers for all integer values of the variables. An INTEGER POLYNOMIAL is a special case of such a polynomial. In general, every integer representing polynomial $f(x)$ of degree n in the variable x can be written in the form

$$f(x) = A_0 + A_1 \binom{x}{1} + A_2 \binom{x}{2} + \ldots + A_n \binom{x}{n},$$

where $\binom{n}{k}$ is a BINOMIAL COEFFICIENT and $A_0, A_1, \ldots, A_n$ are integers (Nagell 1951, p. 121).

See also INTEGER POLYNOMIAL

References

Nagell, T. "Polynomials Representing Integers." §35 in *Introduction to Number Theory.* New York: Wiley, pp. 115–20 and 121, 1951.

Integer Sequence

A SEQUENCE whose terms are INTEGERS. The most complete printed references for such sequences are Sloane (1973) and its update, Sloane and Plouffe (1995). Sloane also maintains the sequences from both works together with many additional sequences in an on-line listing. In this listing, sequences are identified by a unique 6-DIGIT A-number. Sequences appearing in Sloane and Plouffe (1995) are ordered lexicographically and identified with a 4-DIGIT M-number, and those appearing in Sloane (1973) are identified with a 4-DIGIT N-number.

Sloane's huge (and enjoyable) database is accessible by either e-mail or web browser. To look up sequences by e-mail, send a message to either mailto:sequences@research.att.com or mailto:superseeker@research.att.com containing lines OF THE FORM `lookup 5 14 42 132 ...` (note that spaces *must* be used instead of commas). To use the browser version, point to http://www.research.att.com/~njas/sequences/eisonline.html.

Integer sequences can be analyzed by a variety techniques (Sloane and Plouffe 1995, p. 26), including the application a data compression algorithm (Bell *et al.* 1990) and computation of the DISCRETE FOURIER TRANSFORM (Loxton 1989). There are also a large number of transformations which relate integer sequences to one another, including the EULER TRANSFORM, EXPONENTIAL TRANSFORM, MÖBIUS TRANSFORM, and others (Bower, Sloane).

See also ARONSON'S SEQUENCE, COMBINATORICS, CONSECUTIVE NUMBER SEQUENCES, CONWAY SEQUENCE, EBAN NUMBER, EULER TRANSFORM, HOFSTADTER-CONWAY \$10,000 SEQUENCE, HOFSTADTER'S *Q*-SEQUENCE, INTEGER ARRAY, LEVINE-O'SULLIVAN SEQUENCE, LOOK AND SAY SEQUENCE, MALLOW'S SEQUENCE, MIAN-CHOWLA SEQUENCE, MÖBIUS TRANSFORMATION, MORSE-THUE SEQUENCE, NEWMAN-CONWAY SEQUENCE, NUMBER, PADOVAN SEQUENCE, PERRIN SEQUENCE, RATS SEQUENCE, SEQUENCE, SMARANDACHE SEQUENCES

References

Aho, A. V. and Sloane, N. J. A. "Some Doubly Exponential Sequences." *Fib. Quart.* **11**, 429–37, 1973.

Bell, T. C.; Cleary, J. G.; and Witten, I. H. *Text Compression.* Englewood Cliffs, NJ: 1990.

Bernstein, M. and Sloane, N. J. A. "Some Canonical Sequences of Integers." *Linear Algebra Appl.* **226//228**, 57–2, 1995.

Bower, C. G. "Further Transformations of Integer Sequences." http://www.research.att.com/~njas/sequences/transforms2.html.

Cameron, P. J. "Some Sequences of Integers." *Disc. Math.* **75**, 89–02, 1989.

Ding, C.; Helleseth, T.; and Niederreiter, H. (Eds.). *Sequences and Their Applications: Proceedings of SETA' 98.* New York: Springer-Verlag, 1999.

Erdos, P.; Sárközy, E.; and Szemerédi, E. "On Divisibility Properties of Sequences of Integers." In *Number Theory, Colloq. Math. Soc. János Bolyai, Vol. 2.* Amsterdam, Netherlands: North-Holland, pp. 35–9, 1970.

Guy, R. K. "Sequences of Integers." Ch. E in *Unsolved Problems in Number Theory, 2nd ed.* New York: Springer-Verlag, pp. 199–39, 1994.

Kimberling, C. "Integer Sequences and Arrays." http://cedar.evansville.edu/~ck6/integer/.

Krattenthaler, C. "RATE: A Mathematica Guessing Machine." http://radon.mat.univie.ac.at/People/kratt/rate/rate.html.

Loxton, J. H. "Spectral Studies of Automata." In *Irregularities of Partitions* (Ed. G. Halász and V. T. Sós). New York: Springer-Verlag, pp. 115–28, 1989.

Ostman, H. *Additive Zahlentheorie I, II.* Heidelberg, Germany: Springer-Verlag, 1956.

Petit, S. "Encyclopedia of Combinatorial Structures." http://algo.inria.fr/encyclopedia/.

Pomerance, C. and Sárközy, A. "Combinatorial Number Theory." In *Handbook of Combinatorics* (Ed. R. Graham, M. Grötschel, and L. Lovász). Amsterdam, Netherlands: North-Holland, 1994.

Ruskey, F. "The (Combinatorial) Object Server." http://www.theory.csc.uvic.ca/~cos/.

Sloane, N. J. A. *A Handbook of Integer Sequences.* Boston, MA: Academic Press, 1973.

Sloane, N. J. A. "Find the Next Term." *J. Recr. Math.* **7**, 146, 1974.

Sloane, N. J. A. "An On-Line Version of the Encyclopedia of Integer Sequences." *Elec. J. Combin.* **1**, F1 1–, 1994. http://www.combinatorics.org/Volume_1/volume1.html#F1.

Sloane, N. J. A. "An On-Line Version of the Encyclopedia of Integer Sequences." http://www.research.att.com/~njas/sequences/eisonline.html.

Sloane, N. J. A. "Some Important Integer Sequences." In *CRC Standard Mathematical Tables and Formulae.* (Ed. D. Zwillinger). Boca Raton, FL: CRC Press, 1995.

Sloane, N. J. A. "Transformation of Integer Sequences." http://www.research.att.com/~njas/sequences/transforms.html.

Sloane, N. J. A. and Plouffe, S. *The Encyclopedia of Integer Sequences.* San Diego, CA: Academic Press, 1995.

Stöhr, A. "Gelöste und ungelöste Fragen über Basen der natürlichen Zahlenreihe I, II." *J. reine angew. Math.* **194**, 40–5 and 111–40, 1955.

Turán, P. (Ed.). *Number Theory and Analysis: A Collection of Papers in Honor of Edmund Landau (1877–938).* New York: Plenum Press, 1969.

Weisstein, E. W. "Integer Sequences." MATHEMATICA NOTEBOOK INTEGERSEQUENCES.M.

Integers

INTEGER

Integrable

A function for which the INTEGRAL can be computed is said to be integrable.

See also DIFFERENTIABLE, INTEGRABLE (DIFFERENTIAL IDEAL), INTEGRAL, INTEGRATION, LOCALLY INTEGRABLE

Integrable (Differential Ideal)

A DIFFERENTIAL IDEAL is an IDEAL I in the RING of smooth FORMS on a MANIFOLD M. That is, it is closed under addition, scalar multiplication, and WEDGE PRODUCT with an arbitrary form. The IDEAL I is called integrable if, whenever $\alpha \in I$, then also $d\alpha \in I$, where d is the EXTERIOR DERIVATIVE.

For example, in $\mathbb{R}^3$, the IDEAL

$$I = \{a_1 y dx + a_2 dx \wedge dy + a_3 y dx \wedge dz + a_4 dx \wedge dy \wedge dz\},$$

$$(1)$$

where the a_i are arbitrary smooth functions, is an integrable differential ideal. However, if the second term were of the form $a_2 y dx \wedge dy$, then the ideal would not be integrable because it would not contain $d(y dx) = -dx \wedge dy$.

Given an integral differential ideal I on M, a SMOOTH MAP $f : X \to M$ is called integral if the PULLBACK of every form α vanishes on X, i.e., $f^* \alpha = 0$. In coordinates, an integral manifold solves a system of PARTIAL DIFFERENTIAL EQUATIONS. For example, using I above, a map $f = (f_1, f_2, f_3)$ from an OPEN SET in $\mathbb{R}^2$ is integral if

$$f_2 \frac{\partial f_1}{\partial x} = 0 \tag{2}$$

$$f_2 \frac{\partial f_1}{\partial y} = 0 \tag{3}$$

$$\frac{\partial f_1}{\partial x}\frac{\partial f_2}{\partial y} - \frac{\partial f_1}{\partial y}\frac{\partial f_2}{\partial x} = 0 \tag{4}$$

$$f_2 \left(\frac{\partial f_1}{\partial x}\frac{\partial f_3}{\partial y} - \frac{\partial f_1}{\partial y}\frac{\partial f_3}{\partial x} \right) = 0 \tag{5}$$

Conversely, any system of PARTIAL DIFFERENTIAL EQUATIONS can be expressed as an integrable differential ideal on a JET BUNDLE. For instance, $\partial f / \partial x = g$ on $\mathbb{R}$ corresponds to $I = \langle df - g dx \rangle$ on $\mathbb{R}^2 = \{(x, f)\}$.

See also DIFFERENTIAL κ-FORM, INTEGRABLE, JET BUNDLE, PARTIAL DIFFERENTIAL EQUATION, WEDGE PRODUCT

Integral

An integral is a mathematical object which can be interpreted as an AREA or a generalization of AREA. Integrals, together with DERIVATIVES, are the fundamental objects of CALCULUS. Other words for integral include ANTIDERIVATIVE and PRIMITIVE. The RIEMANN INTEGRAL is the simplest integral definition and the only one usually encountered in physics and elementary CALCULUS. In fact, according to Jeffreys and Jeffreys (1988, p. 29), "it appears that cases where these methods [i.e., generalizations of the Riemann integral] are applicable and Riemann's [definition of the integral] is not are too rare in physics to repay the extra difficulty." The RIEMANN INTEGRAL of the function $f(x)$ over x from a to b is written

$$\int_a^b f(x) dx. \tag{1}$$

Every definition of an integral is based on a particular MEASURE. For instance, the RIEMANN INTEGRAL is based on JORDAN MEASURE, and the LEBESGUE INTEGRAL is based on LEBESGUE MEASURE. The process of computing an integral is called INTEGRATION (a more archaic term for INTEGRATION is QUADRATURE), and the approximate computation of an integral is termed NUMERICAL INTEGRATION.

There are two classes of (Riemann) integrals: DEFINITE INTEGRALS such as (1), which have upper and lower limits, and INDEFINITE INTEGRALS, such as

$$\int f(x) dx \tag{2}$$

which are written without limits. The first FUNDAMENTAL THEOREM OF CALCULUS allows DEFINITE INTEGRALS to be computed in terms of INDEFINITE INTEGRALS, since if F is the INDEFINITE INTEGRAL for

$f(x)$, then

$$\int_a^b f(x) dx = F(b) - F(a). \tag{3}$$

WOLFRAM RESEARCH maintains a web site which will integrate many common (and not so common) functions. However, *Mathematica* 4.0 cannot solve some simple indefinite integrals such as

$$\int \left[\frac{d}{dx}\left(x\sqrt{\sin x}\right)\right] dx = \int \left(\frac{x \cos x}{2\sqrt{\sin x}} + \sqrt{\sin x} \right) dx \tag{4}$$

$$\int \left[\frac{d}{dx} \mathrm{Li}_2(x \ln x)\right] dx$$

$$= -\int \left[\frac{(\ln x + 1)\ln(1 - x \ln x)}{x \ln x}\right] dx, \tag{5}$$

where $\mathrm{Li}_2(x)$ is the DILOGARITHM. Consider integrals of this form

$$I(a) = \int_0^{\pi/2} \frac{dx}{1 + (\tan x)^a}, \tag{6}$$

can be done trivially by taking advantage of the trigonometric identity

$$\tan\left(\tfrac{1}{2}\pi - x\right) = \cot x \tag{7}$$

Letting $z \equiv (\tan x)^a$,

$$I(a) = \int_0^{\pi/4} \frac{dx}{1 + z} + \int_{\pi/4}^{\pi/2} \frac{dx}{1 + z}$$

$$= \int_0^{\pi/4} \frac{dx}{1 + z} + \int_0^{\pi/4} \frac{dx}{1 + \dfrac{1}{z}}$$

$$= \int_0^{\pi/4} \left(\frac{1}{1 + z} + \frac{1}{1 + \dfrac{1}{z}} \right) dx = \int_0^{\pi/4} dx$$

$$= \tfrac{1}{4}\pi \tag{8}$$

However, *Mathematica* 3.0 gives an incorrect answer of $\pi^{1 - 2\sqrt{3}} / \left(\sqrt{3} \cdot 4^{\sqrt{3}} \right)$ to

$$I(\sqrt{3}) = \int_0^{\pi/2} \frac{dx}{1 + (\tan x)^{\sqrt{3}}} = \tfrac{1}{4}\pi, \tag{9}$$

although integrals of this type remain unevaluated in *Mathematica* 4.0. Some care is therefore needed in the use of symbolic computer algebra packages for integration. This caveat is further illustrated by the example of the integral

$$\phi(\alpha) = \int_0^\pi \ln(1 - 2\alpha \cos x + \alpha^2)dx = 2\pi \ln|\alpha| \quad (10)$$

that has a simple analytic from for $|\alpha| > 1$ (Woods 1926) using the LEIBNIZ INTEGRAL RULE. However, *Mathematica* 4.0 gives a very complicated solution because it does not recognize the simple form above.

There are a wide range of methods available for NUMERICAL INTEGRATION. Good sources for such techniques include Press *et al.* (1992) and Hildebrand (1956). The most straightforward numerical integration technique uses the NEWTON-COTES FORMULAS (also called QUADRATURE FORMULAS), which approximate a function tabulated at a sequence of regularly spaced INTERVALS by various degree POLYNOMIALS. If the endpoints are tabulated, then the 2- and 3-point formulas are called the TRAPEZOIDAL RULE and SIMPSON'S RULE, respectively. The 5-point formula is called BODE'S RULE. A generalization of the TRAPEZOIDAL RULE is ROMBERG INTEGRATION, which can yield accurate results for many fewer function evaluations.

If the analytic form of a function is known (instead of its values merely being tabulated at a fixed number of points), the best numerical method of integration is called GAUSSIAN QUADRATURE. By picking the optimal ABSCISSAS at which to compute the function, Gaussian quadrature produces the most accurate approximations possible. However, given the speed of modern computers, the additional complication of the GAUSSIAN QUADRATURE formalism often makes it less desirable than the brute-force method of simply repeatedly calculating twice as many points on a regular grid until convergence is obtained. An excellent reference for GAUSSIAN QUADRATURE is Hildebrand (1956).

Here is a list of common INDEFINITE INTEGRALS:

$$\int x^r dx = \frac{x^{r+1}}{r+1} + C \quad (11)$$

$$\int \frac{dx}{x} = \ln|x| + C \quad (12)$$

$$\int a^x dx = \frac{a^x}{\ln a} + C \quad (13)$$

$$\int \sin x dx = -\cos x + C \quad (14)$$

$$\int \cos x dx = \sin x + C \quad (15)$$

$$\int \tan x dx = \ln|\sec x| + C \quad (16)$$

$$\int \csc x dx = \ln|\csc x - \cot x| + C \quad (17)$$

$$= \ln\left[\tan\left(\tfrac{1}{2}x\right)\right] + C \quad (18)$$

$$\tfrac{1}{2}\ln\left(\frac{1 - \cos x}{1 + \cos x}\right) + C \quad (19)$$

$$\int \sec x dx = \ln|\sec x + \tan x| + C \quad (20)$$

$$= \mathrm{gd}^{-1}(x) + C \quad (21)$$

$$\int \cot x dx = \ln|\sin x| + C \quad (22)$$

$$\int \sec^2 x dx = \tan x + C \quad (23)$$

$$\int \csc^2 x dx = -\cot x + C \quad (24)$$

$$\int \sec x \tan x dx = \sec x + C \quad (25)$$

$$\int \cos^{-1} x dx = x \cos^{-1} x - \sqrt{1 - x^2} + C \quad (26)$$

$$\int \sin^{-1} x dx = x \sin^{-1} x + \sqrt{1 - x^2} + C \quad (27)$$

$$\int \tan^{-1} x dx = x \tan^{-1} x - \tfrac{1}{2}\ln(1 + x^2) + C \quad (28)$$

$$\int \frac{dx}{\sqrt{a^2 - x^2}} = \sin^{-1}\left(\frac{x}{a}\right) + C \quad (29)$$

$$\int \frac{dx}{\sqrt{a^2 - x^2}} = -\cos^{-1}\left(\frac{x}{a}\right) + C \quad (30)$$

$$\int \frac{dx}{a^2 - x^2} = \left(\frac{1}{a}\right)\tan^{-1}\left(\frac{x}{a}\right) + C \quad (31)$$

$$\int \frac{dx}{a^2 + x^2} = -\frac{1}{a}\cot^{-1}\left(\frac{x}{a}\right) + C \quad (32)$$

$$\int \frac{dx}{x\sqrt{x^2 - a^2}} = -\frac{1}{a}\sec^{-1}\left(\frac{x}{a}\right) + C \quad (33)$$

$$\int \frac{dx}{x\sqrt{x^2 - a^2}} = -\frac{1}{a}\csc^{-1}\left(\frac{x}{a}\right) + C \quad (34)$$

$$\int \sin^2(ax)dx = \frac{x}{2} - \frac{1}{4a}\sin(2ax) + C \quad (35)$$

$$\int \mathrm{sn}\, u\, du = k^{-1}\ln(\mathrm{dn}\, u - k\, \mathrm{cn}\, u) + C \quad (36)$$

$$\int \mathrm{sn}^2 u\, du = \frac{u - E(u)}{k^2} + C \quad (37)$$

$$\int \mathrm{cn}\, u \, du = k^{-1} \sin^{-1}(k\, \mathrm{sn}\, u) + C \qquad (38)$$

$$\int \mathrm{dn}\, u \, du = \sin^{-1}(\mathrm{sn}\, u) + C, \qquad (39)$$

where $\sin x$ is the SINE; $\cos x$ is the COSINE; $\tan x$ is the TANGENT; $\csc x$ is the COSECANT; $\sec x$ is the SECANT; $\cot x$ is the COTANGENT; $\cos^{-1} x$ is the INVERSE COSINE; $\sin^{-1} x$ is the INVERSE SINE; $\tan^{-1} x$ is the INVERSE TANGENT; $\mathrm{sn}\, u$, $\mathrm{cn}\, u$, and $\mathrm{dn}\, u$ are JACOBI ELLIPTIC FUNCTIONS; $E(u)$ is a complete ELLIPTIC INTEGRAL OF THE SECOND KIND; and $\mathrm{gd}(x)$ is the GUDERMANNIAN FUNCTION.

To derive (16), let $u \equiv \cos x$, so $du = -\sin x\, dx$ and

$$\int \tan x = \int \frac{\sin u}{\cos x} dx = -\int \frac{du}{u}$$

$$= -\ln|u| + C = -\ln|\cos x| + C$$

$$= \ln|\cos x|^{-1} + C = \ln|\sec x| + C. \qquad (40)$$

To derive (17), let $u \equiv \csc x - \cot x$, so $du = (-\csc x \cot x + \csc^2 x) dx$ and

$$\int \csc x\, dx = \int \csc x \, \frac{\csc x - \cot x}{\csc x - \cot x} dx$$

$$= \int \frac{\csc^2 x + \cot x \csc x}{\csc x + \cot x} dx$$

$$= \int \frac{du}{u} = \ln|u| + C$$

$$= \ln|\csc x - \cot x| + C. \qquad (41)$$

To derive (20), let

$$u \equiv \sec x + \tan x, \qquad (42)$$

so

$$du = (\sec x \tan x + \sec^2 x) dx \qquad (43)$$

and

$$\int \sec x\, dx = \int \sec x \, \frac{\sec x + \tan x}{\sec x + \tan x} dx$$

$$= \int \frac{\sec^2 x + \sec x \tan x}{\sec x + \tan x} dx$$

$$= \int \frac{du}{u} = \ln|u| + C$$

$$= \ln|\sec x + \tan x| + C. \qquad (44)$$

To derive (22), let $u \equiv \sin x$, so $du = \cos x\, dx$ and

$$\int \cot x\, dx = \int \frac{\cos x}{\sin x} dx = \int \frac{du}{u}$$

$$= \ln|u| + C = \ln|\sin x| + C. \qquad (45)$$

Integral identities include

$$\frac{dx}{dy} = \frac{1}{\dfrac{dy}{dx}} \qquad (46)$$

$$\frac{d^2 x}{dy^2} = -\frac{d^2 y}{dx^2} \left(\frac{dy}{dx} \right)^{-3} \qquad (47)$$

$$\frac{d^3 x}{dy^3} = \left[3 \left(\frac{d^2 y}{dx^2} \right)^2 - \frac{d^3 y}{dx^3} \frac{dy}{dx} \right] \left(\frac{dy}{dx} \right)^{-5} \qquad (48)$$

Differentiating integrals leads to some useful and powerful identities, for instance

$$\frac{d}{dx} \int_a^x f(x')\, dx' = f(x), \qquad (49)$$

which is the first FUNDAMENTAL THEOREM OF CALCULUS. Other derivative-integral identities include

$$\frac{d}{dx} \int_x^b f(x')\, dx' = -f(x), \qquad (50)$$

the LEIBNIZ INTEGRAL RULE

$$\frac{d}{dx} \int_a^b f(x,t)\, dt = \int_a^b \frac{\partial}{\partial x} f(x,t)\, dt \qquad (51)$$

(Kaplan 1992, p. 275), and its generalization

$$\frac{d}{dx} \int_{u(x)}^{v(x)} f(x,t)\, dt$$

$$= v'(x) f(x, v(x)) - u'(x) f(x, u(x)) + \int_{u(x)}^{v(x)} \frac{\partial}{\partial x} f(x,t)\, dt \qquad (52)$$

(Leibniz 1992, p. 258). If $f(x,t)$ is singular or INFINITE, then

$$\frac{d}{dx} \int_a^x f(x,t)\, dx$$

$$= \frac{1}{x-a} \int_a^x \left[(x-a) \frac{\partial f}{\partial x} + (t-a) \frac{\partial f}{\partial x} + f \right] dt \qquad (53)$$

Other integral identities include

$$\int_0^x dt_n \int_0^{t_n} dt_{n-1} \cdots \int_0^{t_3} dt_2 \int_0^{t_2} f(t_1)\, dt_1$$

$$= \frac{1}{(n-1)!} \int_0^x (x-t)^{n-1} f(t)\, dt \qquad (54)$$

$$\frac{\partial}{\partial x_k} (x_j J_k) = \delta_{jk} J_k + x_j \frac{\partial}{\partial x_k} J_k = \mathbf{J} + \mathbf{r} \nabla \cdot \mathbf{J} \qquad (55)$$

$$\int_V \mathbf{J} d^3 \mathbf{r} = \int_V \frac{\partial}{\partial x_k} (x_i J_k) - \int_V \mathbf{r} \nabla \cdot \mathbf{J} d^3 \mathbf{r}$$

$$-\int_V \mathbf{r}\nabla \cdot \mathbf{J}d^3\mathbf{r} \tag{56}$$

and the amusing integral identity

$$\int_{-\infty}^{\infty} F(f(x))dx = \int_{-\infty}^{\infty} F(x)dx, \tag{57}$$

where F is any function and

$$f(x) = x - \sum_{n=0}^{\infty} \frac{a_n}{x + b_n} \tag{58}$$

as long as $a_n \geq 0$ and b_n is real (Glasser 1983).
Integrals OF THE FORM

$$\int_a^b f(x)dx \tag{59}$$

with one INFINITE LIMIT and the other NONZERO may be expressed as finite integrals over transformed functions. If $f(x)$ decreases at least as fast as $1/x^2$, then let

$$t \equiv \frac{1}{x} \tag{60}$$

$$dt = -\frac{dx}{x^2} \tag{61}$$

$$dx = -x^2 dt = -\frac{dt}{t^2}, \tag{62}$$

and

$$\int_a^b f(x)dx = -\int_{1/a}^{1/b} \frac{1}{t^2} f\left(\frac{1}{t}\right)dt = \int_{1/b}^{1/a} \frac{1}{t^2} f\left(\frac{1}{t}\right)dt. \tag{63}$$

If $f(x)$ diverges as $(x-a)^\gamma$ for $\gamma \in [0,1]$, let

$$x \equiv t^{1/(1-\gamma)} + a \tag{64}$$

$$dx = \frac{1}{1-\gamma} t^{1/(1-\gamma)-1}dt = \frac{1}{1-\gamma} t^{[1-(1-\gamma)]/(1-\gamma)}dt$$

$$= \frac{1}{\gamma-1} t^{\gamma/(1-\gamma)}dt \tag{65}$$

$$t = (x-a)^{1-\gamma}, \tag{66}$$

and

$$\int_a^b f(x)dx = \frac{1}{1-\gamma} = \int_0^{(b-a)^{1-\gamma}} t^{\gamma/(1-\gamma)} f\left(t^{1/(1-\gamma)} + a\right)dt. \tag{67}$$

If $f(x)$ diverges as $(x+b)^\gamma$ for $\gamma \in [0,1]$, let

$$x \equiv b - t^{1/(1-\gamma)} \tag{68}$$

$$dx = -\frac{1}{\gamma-1} t^{\gamma/(1-\gamma)}dt \tag{69}$$

$$t = (b-x)^{1-\gamma}, \tag{70}$$

and

$$\int_a^b f(x)dx = \frac{1}{1-\gamma} = \int_0^{(b-a)^{1-\gamma}} t^{\gamma/(1-\gamma)} f(b - t^{1/(1-\gamma)})dt. \tag{71}$$

If the integral diverges exponentially, then let

$$t \equiv e^{-x} \tag{72}$$

$$dt = -e^{-x}dx \tag{73}$$

$$x = -\ln t, \tag{74}$$

and

$$\int_a^\infty f(x)dx = \int_0^{e^{-a}} f(-\ln t)\frac{dt}{t}. \tag{75}$$

Integrals with rational exponents can often be solved by making the substitution $u = x^{1/n}$, where n is the LEAST COMMON MULTIPLE of the DENOMINATOR of the exponents.

Integration rules include

$$\int_a^a f(x)dx = 0 \tag{76}$$

$$\int_a^b f(x)dx = -\int_b^a f(x)dx. \tag{77}$$

For $c \in (a,b)$,

$$\int_a^b f(x)dx = \int_a^c f(x)dx + \int_c^b f(x)dx. \tag{78}$$

If g' is continuous on $[a, b]$ and f is continuous and has an antiderivative on an INTERVAL containing the values of $g(x)$ for $a \leq x \leq b$, then

$$\int_a^b f(g(x))g'(x)dx = \int_{g(a)}^{g(b)} f(u)du. \tag{79}$$

Liouville showed that the integrals

$$\int e^{-x^2}dx \quad \int \frac{e^x}{x}dx \quad \int \frac{\sin x}{x}dx \quad \int \frac{dx}{\ln x} \tag{80}$$

cannot be expressed as terms of a finite number of elementary functions. Other irreducibles include

$$\int x^x dx \quad \int x^{-x}dx \quad \int \sqrt{\sin x}\, dx. \tag{81}$$

Chebyshev proved that if U, V, and W are RATIONAL NUMBERS, then

$$\int x^U \left(A + Bx^V\right)^W dx \tag{82}$$

is integrable in terms of elementary functions IFF $(U+1)/V$, W, or $W + (U+1)/V$ is an INTEGER (Ritt 1948, Shanks 1993).

See also A-INTEGRABLE, ABELIAN INTEGRAL, CALCULUS, CHEBYSHEV-GAUSS QUADRATURE, CHEBYSHEV QUADRATURE, DARBOUX INTEGRAL, DEFINITE INTEGRAL, DENJOY INTEGRAL, DERIVATIVE, DOUBLE EXPONENTIAL INTEGRATION, EULER INTEGRAL, FUNDAMENTAL THEOREM OF GAUSSIAN QUADRATURE, GAUSS-JACOBI MECHANICAL QUADRATURE, GAUSSIAN QUADRATURE, HAAR INTEGRAL, HERMITE-GAUSS QUADRATURE, HERMITE QUADRATURE, HK INTEGRAL, INDEFINITE INTEGRAL, INTEGRATION, JACOBI-GAUSS QUADRATURE, JACOBI QUADRATURE, LAGUERRE-GAUSS QUADRATURE, LAGUERRE QUADRATURE, LEBESGUE INTEGRAL, LEBESGUE-STIELTJES INTEGRAL, LEGENDRE-GAUSS QUADRATURE, LEGENDRE QUADRATURE, LEIBNIZ INTEGRAL RULE, LOBATTO QUADRATURE, MECHANICAL QUADRATURE, MEHLER QUADRATURE, NEWTON-COTES FORMULAS, NUMERICAL INTEGRATION, PERRON INTEGRAL, QUADRATURE, RADAU QUADRATURE, RECURSIVE MONOTONE STABLE QUADRATURE, RIEMANN-STIELTJES INTEGRAL, ROMBERG INTEGRATION, RIEMANN INTEGRAL, STIELTJES INTEGRAL

References

Beyer, W. H. "Integrals." *CRC Standard Mathematical Tables, 28th ed.* Boca Raton, FL: CRC Press, pp. 233–96, 1987.

Bronstein, M. *Symbolic Integration I: Transcendental Functions.* New York: Springer-Verlag, 1996.

Glasser, M. L. "A Remarkable Property of Definite Integrals." *Math. Comput.* **40**, 561–63, 1983.

Gordon, R. A. *The Integrals of Lebesgue, Denjoy, Perron, and Henstock.* Providence, RI: Amer. Math. Soc., 1994.

Gradshteyn, I. S. and Ryzhik, I. M. *Tables of Integrals, Series, and Products, 6th ed.* San Diego, CA: Academic Press, 2000.

Hildebrand, F. B. *Introduction to Numerical Analysis.* New York: McGraw-Hill, pp. 319–23, 1956.

Jeffreys, H. and Jeffreys, B. S. *Methods of Mathematical Physics, 3rd ed.* Cambridge, England: Cambridge University Press, p. 29, 1988.

Kaplan, W. *Advanced Calculus, 4th ed.* Reading, MA: Addison-Wesley, 1992.

Piessens, R.; de Doncker, E.; Uberhuber, C. W.; and Kahaner, D. K. *QUADPACK: A Subroutine Package for Automatic Integration.* New York: Springer-Verlag, 1983.

Press, W. H.; Flannery, B. P.; Teukolsky, S. A.; and Vetterling, W. T. "Integration of Functions." Ch. 4 in *Numerical Recipes in FORTRAN: The Art of Scientific Computing, 2nd ed.* Cambridge, England: Cambridge University Press, pp. 123–58, 1992.

Ritt, J. F. *Integration in Finite Terms.* New York: Columbia University Press, p. 37, 1948.

Shanks, D. *Solved and Unsolved Problems in Number Theory, 4th ed.* New York: Chelsea, p. 145, 1993.

Woods, F. S. *Advanced Calculus: A Course Arranged with Special Reference to the Needs of Students of Applied Mathematics.* Boston, MA: Ginn, pp. 143–44, 1926.

Wolfram Research. "The Integrator." http://integrals.wolfram.com/.

Integral Brick

EULER BRICK

Integral Calculus

That portion of "the" CALCULUS dealing with INTEGRALS.

See also CALCULUS, DIFFERENTIAL CALCULUS, INTEGRAL

Integral Cohomology Class

See also COHOMOLOGY CLASS

Integral Cuboid

EULER BRICK

Integral Current

A RECTIFIABLE CURRENT whose boundary is also a RECTIFIABLE CURRENT.

Integral Curvature

Given a GEODESIC TRIANGLE (a triangle formed by the arcs of three GEODESICS on a smooth surface),

$$\int_{ABC} K\,da = A + B + C - \pi.$$

Given the EULER CHARACTERISTIC χ,

$$\iint K\,da = 2\pi\chi$$

so the integral curvature of a closed surface is not altered by a topological transformation.

See also GAUSS-BONNET FORMULA, GEODESIC TRIANGLE

Integral Domain

A RING that is COMMUTATIVE under multiplication, has an IDENTITY ELEMENT, and has no divisors of 0.

The INTEGERS form an integral domain.

See also FIELD, IDEAL, RING

References

Anderson, D. D. (Ed.). *Factorization in Integral Domains.* New York: Dekker, 1997.

Integral Drawing

A GRAPH drawn such that the EDGES have only INTEGER lengths. It is conjectured that every PLANAR GRAPH has an integral drawing.

References

Harborth, H. and Möller, M. "Minimum Integral Drawings of the Platonic Graphs." *Math. Mag.* **67**, 355–58, 1994.

Integral Equation

If the limits are fixed, an integral equation is called a Fredholm integral equation. If one limit is variable, it is called a Volterra integral equation. If the unknown function is only under the integral sign, the equation is said to be of the "first kind." If the function is both inside and outside, the equation is called of the "second kind." A Fredholm equation of the first kind is OF THE FORM

$$f(x) = \int_a^b K(x,t)\phi(t)dt. \tag{1}$$

A Fredholm equation of the second kind is OF THE FORM

$$\phi(x) = f(x) + \int_a^b K(x,t)\phi(t)dt. \tag{2}$$

A Volterra equation of the first kind is OF THE FORM

$$f(x) = \int_a^x K(x,t)\phi(t)dt. \tag{3}$$

A Volterra equation of the second kind is OF THE FORM

$$\phi(x) = f(x) + \int_a^x K(x,t)\phi(t)dt, \tag{4}$$

where the functions $K(x,t)$ are known as KERNELS. Integral equations may be solved directly if they are SEPARABLE. Otherwise, a NEUMANN SERIES must be used.

A KERNEL is separable if

$$K(x,t) = \lambda \sum_{j=1}^{n} M_j(x)N_j(t). \tag{5}$$

This condition is satisfied by all POLYNOMIALS and many TRANSCENDENTAL FUNCTIONS. a FREDHOLM INTEGRAL EQUATION OF THE SECOND KIND with separable KERNEL may be solved as follows:

$$\phi(x) = f(x) + \int_a^b K(x,t)\phi(t)dt$$

$$= f(x) + \lambda \sum_{j=1}^{n} M_j(x) \int_a^b N_j(t)\phi(t)dt$$

$$= f(x) + \lambda \sum_{j=1}^{n} c_j M_j(x), \tag{6}$$

where

$$c_j \equiv \int_a^b N_j(t)\phi(t)dt. \tag{7}$$

Now multiply both sides of (7) by $N_i(x)$ and integrate over dx.

$$\int_a^b \phi(x)N_i(x)dx$$

$$= \int_a^b f(x)N_i(x)dx + \lambda \sum_{j=1}^{n} c_j \int_a^b M_j(x)N_i(x)dx. \tag{8}$$

By (7), the first term is just c_i. Now define

$$b_i \equiv \int_a^b N_i(x)f(x)dx \tag{9}$$

$$a_{ij} = \int_a^b N_i(x)M_j(x)dx, \tag{10}$$

so (8) becomes

$$c_i = b_i + \lambda \sum_{j=1}^{n} a_{ij}c_j \tag{11}$$

Writing this in matrix form,

$$\mathbf{C} = \mathbf{B} + \lambda\mathbf{AC}, \tag{12}$$

so

$$(\mathsf{I} - \lambda\mathsf{A})\mathbf{C} = \mathbf{B} \tag{13}$$

$$\mathbf{C} = (\mathsf{I} - \lambda\mathsf{A})^{-1}\mathbf{B} \tag{14}$$

See also FREDHOLM INTEGRAL EQUATION OF THE FIRST KIND, FREDHOLM INTEGRAL EQUATION OF THE SECOND KIND, VOLTERRA INTEGRAL EQUATION OF THE FIRST KIND, VOLTERRA INTEGRAL EQUATION OF THE SECOND KIND

References

Corduneanu, C. *Integral Equations and Applications.* Cambridge, England: Cambridge University Press, 1991.

Davis, H. T. *Introduction to Nonlinear Differential and Integral Equations.* New York: Dover, 1962.

Kondo, J. *Integral Equations.* Oxford, England: Clarendon Press, 1992.

Lovitt, W. V. *Linear Integral Equations.* New York: Dover, 1950.

Mikhlin, S. G. *Integral Equations and Their Applications to Certain Problems in Mechanics, Mathematical Physics and Technology, 2nd rev. ed.* New York: Macmillan, 1964.

Mikhlin, S. G. *Linear Integral Equations.* New York: Gordon & Breach, 1961.

Pipkin, A. C. *A Course on Integral Equations.* New York: Springer-Verlag, 1991.

Porter, D. and Stirling, D. S. G. *Integral Equations: A Practical Treatment, from Spectral Theory to Applications.* Cambridge, England: Cambridge University Press, 1990.

Press, W. H.; Flannery, B. P.; Teukolsky, S. A.; and Vetterling, W. T. "Integral Equations and Inverse Theory." Ch. 18 in *Numerical Recipes in FORTRAN: The Art of Scientific Computing, 2nd ed.* Cambridge, England: Cambridge University Press, pp. 779–17, 1992.

Tricomi, F. G. *Integral Equations.* New York: Dover, 1957.

Weisstein, E. W. "Books about Integral Equations." http://www.treasure-troves.com/books/IntegralEquations.html.

Whittaker, E. T. and Robinson, G. "The Numerical Solution of Integral Equations." §183 in *The Calculus of Observa-*

tions: A Treatise on Numerical Mathematics, 4th ed. New York: Dover, pp. 376–81, 1967.

Integral Function

ENTIRE FUNCTION

Integral Geometry

See also GEOMETRIC PROBABILITY, STOCHASTIC GEOMETRY

Integral of Motion

A function of the coordinates which is constant along a trajectory in PHASE SPACE. The number of DEGREES OF FREEDOM of a DYNAMICAL SYSTEM such as the DUFFING differential equation can be decreased by one if an integral of motion can be found. In general, it is very difficult to discover integrals of motion.

Integral Polyhedron

PRIMITIVE POLYTOPE

Integral Polynomial

INTEGER POLYNOMIAL

Integral Sign

The symbol $\int$ used to denote an INTEGRAL $\int f(x)dx$. The symbol was invented by Leibniz and chosen to be a stylized script "S" to stand for "summation."

See also INTEGRAL, INTEGRATION UNDER THE INTEGRAL SIGN

Integral Test

Let Σu_k be a series with POSITIVE terms and let $f(x)$ be the function that results when k is replaced by x in the FORMULA for u_k. If f is decreasing and continuous for $x \geq 1$ and

$$\lim_{x \to \infty} f(x) = 0,$$

then

$$\sum_{k=1}^{\infty} u_k$$

and

$$\int_t^{\infty} f(x)dx$$

both converge or diverge, where $1 \leq t \leq \infty$. The test is also called the CAUCHY INTEGRAL TEST or MACLAURIN INTEGRAL TEST.

See also CONVERGENCE TESTS

References

Arfken, G. *Mathematical Methods for Physicists, 3rd ed.* Orlando, FL: Academic Press, pp. 283–284, 1985.

Integral Transform

A general integral transform is defined by

$$g(\alpha) = \int_a^b f(t)K(\alpha,t)dt,$$

where $K(\alpha, t)$ is called the KERNEL of the transform.

See also BUSCHMAN TRANSFORM, FOURIER TRANSFORM, FOURIER-STIELTJES TRANSFORM, G-TRANSFORM, H-TRANSFORM, HADAMARD TRANSFORM, HANKEL TRANSFORM, HARTLEY TRANSFORM, HOUGH TRANSFORM, KONTOROVICH-LEBEDEV TRANSFORM, MEHLER-FOCK TRANSFORM, MEIJER TRANSFORM, NARAIN G-TRANSFORM, OPERATIONAL MATHEMATICS, RADON TRANSFORM, STIELTJES TRANSFORM, W-TRANSFORM, WAVELET TRANSFORM, Z-TRANSFORM

References

Arfken, G. "Integral Transforms." Ch. 16 in *Mathematical Methods for Physicists, 3rd ed.* Orlando, FL: Academic Press, pp. 794–864, 1985.
Brychkov, Yu. A. and Prudnikov, A. P. *Integral Transforms of Generalized Functions.* New York: Gordon and Breach, 1989.
Carslaw, H. S. and Jaeger, J. C. *Operational Methods in Applied Mathematics.* New York: Dover, 1963.
Davies, B. *Integral Transforms and Their Applications, 2nd ed.* New York: Springer-Verlag, 1985.
Erdélyi, A.; Oberhettinger, M. F.; and Tricomi, F. G. *Tables of Integral Transforms. Based, in Part, on Notes Left by Harry Bateman and Compiled by the Staff of the Bateman Manuscript Project, 2 vols.* McGraw-Hill, 1954.
Krantz, S. G. "Transform Theory." Ch. 15 in *Handbook of Complex Analysis.* Boston, MA: Birkhäuser, pp. 195–217, 1999.
Marichev, O. I. *Handbook of Integral Transforms of Higher Transcendental Functions: Theory and Algorithmic Tables.* Chichester, England: Ellis Horwood, 1982.
Poularikas, A. D. (Ed.). *The Transforms and Applications Handbook.* Boca Raton, FL: CRC Press, 1995.
Weisstein, E. W. "Books about Integral Transforms." http://www.treasure-troves.com/books/IntegralTransforms.html.
Zayed, A. I. *Handbook of Function and Generalized Function Transformations.* Boca Raton, FL: CRC Press, 1996.

Integrand

The quantity being INTEGRATED, also called the KERNEL. For example, in $\int f(x)dx$, $f(x)$ is the integrand.

See also INTEGRAL, INTEGRATION

Integrating Factor

A FUNCTION by which an ORDINARY DIFFERENTIAL EQUATION is multiplied in order to make it integrable.

See also ORDINARY DIFFERENTIAL EQUATION

References

Morse, P. M. and Feshbach, H. *Methods of Theoretical Physics, Part I.* New York: McGraw-Hill, pp. 526–529, 1953.

Integration

The process of computing or obtaining an INTEGRAL. A more archaic term for integration is QUADRATURE.

See also CONTOUR INTEGRATION, INTEGRAL, INTEGRATION BY PARTS, MEASURE THEORY, NUMERICAL INTEGRATION

References

Shenitzer, A. and Steprans, S. J. "The Evolution of Integration." *Amer. Math. Monthly* **101**, 66–72, 1994.

Integration (Form)

A DIFFERENTIAL K-FORM can be integrated on an n-dimensional MANIFOLD. The basic example is an n-form α in the open unit ball in $\mathbb{R}^n$. Since α is a TOP-DIMENSIONAL FORM, it can be written $\alpha = f dx_1 \wedge \ldots \wedge dx_n$ and so

$$\int_B \alpha = \int_B f \, d\mu, \tag{1}$$

where the integral is the LEBESGUE INTEGRAL.

On a MANIFOLD M covered by COORDINATE CHARTS U_i, there is a PARTITION OF UNITY ρ_i such that

1. ρ_i is SUPPORTED in U_i and
2. $\Sigma \rho_i = 1$.

Then

$$\int_M \alpha = \sum \int_{U_i} \rho_i \alpha, \tag{2}$$

where the right-hand side is WELL DEFINED because each integration takes place in a COORDINATE CHART. The integral of the n-form α is WELL DEFINED because, under a change of coordinates $g : X \to Y$, the integral transforms according to the determinant of the JACOBIAN, while an n-form pulls back by the determinant of the JACOBIAN. Hence,

$$\int_X g^*(\alpha) = \int_X ||J|| f(g(x)) = \int_Y f(y) \tag{3}$$

is the same integral in either COORDINATE CHART.

For example, it is possible to integrate the 2-form

$$\alpha = z \, dx \wedge dy - y \, dx \wedge dz + x \, dy \wedge dz \tag{4}$$

on the SPHERE $\mathbb{S}^2$. Because a point has MEASURE ZERO, it is enough to integrate α on $\mathbb{S}^2 - (0, 0, 1)$, which can be covered by STEREOGRAPHIC PROJECTION $\phi : \mathbb{R}^2 \to \mathbb{S}^2 - (0, 0, 1)$. Since

$$\phi(x, y) = \left(\frac{2x}{1 + r^2}, \frac{2y}{1 + r^2}, \frac{1 - r^2}{1 + r^2} \right) \tag{5}$$

the PULLBACK MAP of α is

$$\phi^*(\alpha) = \frac{4}{\left(1 + r^2\right)^2} dx \wedge dy, \tag{6}$$

the integral of α on $\mathbb{S}^2$ is

$$\int \int \frac{4}{\left(1 + r^2\right)^2} 2\pi r \, d\theta = 4\pi. \tag{7}$$

Note that this computation is done more easily by STOKES' THEOREM, because $d\alpha = 3 dx \wedge dy \wedge dz$.

See also DE RHAM COHOMOLOGY, STOKES' THEOREM, SUBMANIFOLD, TOP-DIMENSIONAL FORM, VOLUME FORM

Integration by Parts

Integration by parts is a technique for performing definite integration $\int u \, dv$ by expanding the differential of a product of functions $d(uv)$ and expressing the original integral in terms of a known integral $\int v \, du$. A single integration by parts starts with

$$d(uv) = u \, dv + v \, du, \tag{1}$$

and integrates both sides,

$$\int d(uv) = uv = \int u \, dv + \int v \, du. \tag{2}$$

Rearranging gives

$$\int u \, dv = uv - \int v \, du, \tag{3}$$

so

$$\int_a^b u \, dv = [uv]_a^b - \int_{f(a)}^{f(b)} v \, du, \tag{4}$$

where $[f]_a^b = f(b) - f(a)$.

This procedure can also be applied n times to $\int f^{(n)}(x) g(x) dx$.

$$u = g(x) \quad dv = f^{(n)}(x) dx \tag{5}$$

$$du = g'(x) dx \quad v = f^{(n-1)}(x). \tag{6}$$

Therefore,

$$\int f^{(n)} g(x) dx = g(x) f^{(n-1)}(x) - \int f^{(n-1)}(x) g'(x) dx. \tag{7}$$

But

$$\int f^{(n-1)}(x) g'(x) dx = g'(x) f^{(n-2)}(x) - \int f^{(n-2)}(x) g''(x) dx \tag{8}$$

$$\int f^{(n-2)}(x)g''(x)dx$$

$$= g''(x)f^{(n-3)}(x) - \int f^{(n-3)}(x)g^{(3)}(x)dx, \qquad (9)$$

so

$$\int f^{(n)}(x)g(x)dx = g(x)f^{(n-1)}(x) - g'(x)f^{(n-2)}(x)$$

$$+ g(x)f^{(n-3)}(x) - \ldots + (-1)^n \int f(x)g^{(n)}(x)dx. \qquad (10)$$

Now consider this in the slightly different form $\int f(x)g(x)dx$. Integrate by parts a first time

$$u = f(x) \quad dv = g(x)dx \qquad (11)$$

$$du = f'(x)dx \quad v = \int g(x)dx, \qquad (12)$$

so

$$\int f(x)g(x)dx = f(x)\int g(x)dx - \int \left[\int g(x)dx\right]f'(x)dx. \qquad (13)$$

Now integrate by parts a second time,

$$u = f'(x) \quad dv = \int g(x)dx \qquad (14)$$

$$du = f''(x)dx \quad v = \int\int g(x)(dx)^2, \qquad (15)$$

so

$$\int f(x)g(x)dx = f(x)\int g(x)dx - f'(x)\int\int g(x)(dx)^2$$

$$+ \int \left[\int\int g(x)(dx)^2\right]f''(x)dx. \qquad (16)$$

Repeating a third time,

$$\int f(x)g(x)dx = f(x)\int g(x)dx - f'(x)\int\int g(x)(dx)^2$$

$$+ f''(x)\int\int\int g(x)(dx)^3 - \int \left[\int\int\int g(x)(dx)^3\right]f'''(x)dx. \qquad (17)$$

Therefore, after n applications,

$$\int f(x)g(x)dx = f(x)\int g(x)dx - f'(x)\int\int g(x)(dx)^2$$

$$+ f''(x)\int\int\int g(x)(dx)^3 - \ldots$$

$$+ (-1)^{n+1}f^{(n)}(x)\underbrace{\int \cdots \int}_{n+1} g(x)(dx)^{n+1}$$

$$+ (-1)^n \int \left[\underbrace{\int \cdots \int}_{n+1} g(x)(dx)^{n+1}\right]f^{(n+1)}(x)dx. \qquad (18)$$

If $f^{n+1}(x) = 0$ (e.g., for an nth degree POLYNOMIAL), the last term is 0, so the sum terminates after n terms and

$$\int f(x)g(x)dx = f(x)\int g(x)dx$$

$$- f'(x)\int\int g(x)(dx)^2 + f''(x)\int\int\int g(x)(dx)^3 - \ldots$$

$$+ (-1)^{n+1}f^{(n)}(x)\underbrace{\int \cdots \int}_{n+1} g(x)(dx)^{n+1}. \qquad (19)$$

See also INTEGRAL, INTEGRATION, SUMMATION BY PARTS

References

Abramowitz, M. and Stegun, C. A. (Eds.). *Handbook of Mathematical Functions with Formulas, Graphs, and Mathematical Tables, 9th printing.* New York: Dover, p. 12, 1972.

Integration Constant
CONSTANT OF INTEGRATION

Integration Lattice
A discrete subset of $\mathbb{R}^s$ which is CLOSED under addition and subtraction and which contains $\mathbb{Z}^s$ as a SUBSET.

See also LATTICE, POINT LATTICE

References

Sloan, I. H. and Joe, S. *Lattice Methods for Multiple Integration.* New York: Oxford University Press, 1994.

Integration Theory
MEASURE THEORY

Integration Under the Integral Sign
The use of the identity

$$\int_a^b dx \int_{\alpha_0}^\alpha f(x, \alpha)d\alpha = \int_{\alpha_0}^\alpha d\alpha \int_a^b f(x, \alpha)dx \qquad (1)$$

to compute an INTEGRAL. For example, consider

$$\int_0^1 x^\alpha dx = \frac{1}{\alpha + 1} \qquad (2)$$

for $\alpha > -1$. Multiplying by $d\alpha$ and integrating between a and b gives

$$\int_c^b d\alpha \int_0^1 x^\alpha dx = \int_a^b \frac{d\alpha}{\alpha + 1} = \ln \left(\frac{b + 1}{a + 1} \right). \qquad (3)$$

But the left-hand side is equal to

$$\int_0^1 d\alpha \int_a^b x^\alpha dx = \int_0^1 \frac{x^b - x^a}{\ln x} dx, \qquad (4)$$

so it follows that

$$\int_0^1 \frac{x^b - x^a}{\ln x} dx = \ln \left(\frac{b + 1}{a + 1} \right) \qquad (5)$$

(Woods 1926, pp. 145–146).

See also INTEGRAL, INTEGRAL SIGN, INTEGRATION, LEIBNIZ INTEGRAL RULE

References

Woods, F. S. "Integration Under the Integral Sign." §61 in *Advanced Calculus: A Course Arranged with Special Reference to the Needs of Students of Applied Mathematics.* Boston, MA: Ginn, pp. 145–146, 1926.

Intension

A definition of a SET by mentioning a defining property.

See also EXTENSION (SET)

References

Russell, B. "Definition of Number." *Introduction to Mathematical Philosophy.* New York: Simon and Schuster, 1971.

Interchange Graph

LINE GRAPH

Interest

Interest is a fee (or payment) made for the borrowing (or lending) of money. The two most common types of interest are SIMPLE INTEREST, for which interest is paid only on the initial PRINCIPAL, and COMPOUND INTEREST, for which interest earned can be re-invested to generate further interest.

See also COMPOUND INTEREST, CONVERSION PERIOD, PRESENT VALUE, RULE OF 72, SIMPLE INTEREST

References

Kellison, S. G. *Theory of Interest, 2nd ed.* Burr Ridge, IL: Richard D. Irwin, 1991.

Interior

That portion of a region lying "inside" a specified boundary. For example, the interior of the SPHERE is a BALL.

See also EXTERIOR

Interior Angle Bisector

ANGLE BISECTOR

Interior Product

The interior product is a dual notion of the EXTERIOR PRODUCT in an EXTERIOR ALGEBRA ΛV, where V is a VECTOR SPACE. Given an ORTHONORMAL BASIS $\{e_i\}$ of V, the forms

$$\{e_{i_1} \wedge \ldots \wedge e_{i_p}\} i_1 < \ldots < i_p \qquad (1)$$

are an ORTHONORMAL BASIS for $\Lambda^p V$. They define a metric on the EXTERIOR ALGEBRA, $\langle \alpha, \beta \rangle$. The interior product with a form γ is the ADJOINT of the EXTERIOR PRODUCT with γ. That is,

$$\langle \alpha \lrcorner \gamma, \beta \rangle = \langle \alpha, \beta \wedge \gamma \rangle \qquad (2)$$

for all β. For example,

$$e_1 \wedge e_2 \lrcorner e_3 = 0 \qquad (3)$$

and

$$e_1 \wedge e_2 \wedge e_3 \wedge e_4 \lrcorner e_1 \wedge e_4 = e_2 \wedge e_3, \qquad (4)$$

where the e_i are ORTHONORMAL, are two interior products.

An inner product on V gives an isomorphism $e : V \simeq V^*$ with the DUAL SPACE V^*. The interior product is the composition of this isomorphism with CONTRACTION.

See also CONTRACTION (TENSOR), EXTERIOR ALGEBRA, EXTERIOR PRODUCT, INNER PRODUCT, WEDGE PRODUCT

Intermediate Value Theorem

If f is continuous on a CLOSED INTERVAL $[a, b]$, and c is any number between $f(a)$ and $f(b)$ inclusive, then there is at least one number x in the CLOSED INTERVAL such that $f(x) = c$.

See also WEIERSTRASS INTERMEDIATE VALUE THEOREM

Internal Bisectors Problem

STEINER-LEHMUS THEOREM

Internal Contact

TANGENT INTERNALLY

Internal Knot

One of the "knots" $t_{p+1}, \ldots, t_{m-p-1}$ of a B-SPLINE with control points $\mathbf{P}_0, \ldots, \mathbf{P}_n$ and KNOT VECTOR

$$\mathbf{T} = \{t_0, t_1, \ldots, t_m\},$$

where

$$p \equiv m - n - 1.$$

See also B-SPLINE, KNOT VECTOR

Internal Path Length

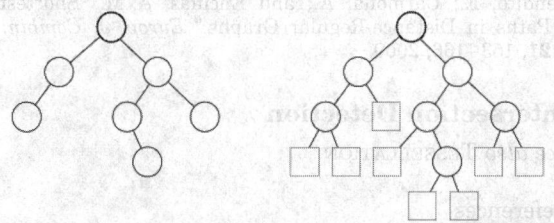

The sum I over all internal (circular) nodes of the paths from the root of an EXTENDED BINARY TREE to each node. For example, in the tree above, the external path length is 11 (Knuth 1997, p. 399–400). The internal and EXTERNAL PATH LENGTHS are related by

$$E = I + 2n,$$

where n is the number of internal nodes.

See also EXTENDED BINARY TREE, EXTERNAL PATH LENGTH

References

Knuth, D. E. *The Art of Computer Programming, Vol. 1: Fundamental Algorithms, 3rd ed.* Reading, MA: Addison-Wesley, 1997.

Internally Tangent

TANGENT INTERNALLY

Interpolation

The computation of points or values between ones that are known or tabulated using the surrounding points or values.

See also AITKEN INTERPOLATION, BESSEL'S INTERPOLATION FORMULA, EVERETT INTERPOLATION, EXTRAPOLATION, FINITE DIFFERENCE, GAUSS'S INTERPOLATION FORMULA, HERMITE INTERPOLATION, LAGRANGE INTERPOLATING POLYNOMIAL, NEWTON-COTES FORMULAS, NEWTON'S DIVIDED DIFFERENCE INTERPOLATION FORMULA, OSCULATING INTERPOLATION, THIELE'S INTERPOLATION FORMULA

References

Abramowitz, M. and Stegun, C. A. (Eds.). "Interpolation." §25.2 in *Handbook of Mathematical Functions with Formulas, Graphs, and Mathematical Tables, 9th printing.* New York: Dover, pp. 878–882, 1972.
Iyanaga, S. and Kawada, Y. (Eds.). "Interpolation." Appendix A, Table 21 in *Encyclopedic Dictionary of Mathematics.* Cambridge, MA: MIT Press, pp. 1482–1483, 1980.
Press, W. H.; Flannery, B. P.; Teukolsky, S. A.; and Vetterling, W. T. "Interpolation and Extrapolation." Ch. 3 in *Numerical Recipes in FORTRAN: The Art of Scientific Computing, 2nd ed.* Cambridge, England: Cambridge University Press, pp. 99–122, 1992.
Whittaker, E. T. and Robinson, G. "Interpolation with Equal Intervals of the Argument." Ch. 1 in *The Calculus of Observations: A Treatise on Numerical Mathematics, 4th ed.* New York: Dover, pp. 1–34, 1967.

Interquartile Range

Divide a set of data into two groups (high and low) of equal size at the MEDIAN if there is an EVEN number of data points, or two groups consisting of points on either side of the MEDIAN itself plus the MEDIAN if there is an ODD number of data points. Find the MEDIANS of the low and high groups, denoting these first and third quartiles by Q_1 and Q_3. The interquartile range is then defined by

$$\text{IQR} \equiv Q_3 - Q_1.$$

See also H-SPREAD, HINGE, MEDIAN (STATISTICS), QUARTILE

Interradius

MIDRADIUS

Intersecting Circles

CIRCLE-CIRCLE INTERSECTION

Intersecting Cylinders

STEINMETZ SOLID

Intersecting Lines

LINE-LINE INTERSECTION

Intersecting Spheres

SPHERE-SPHERE INTERSECTION

Intersection

The intersection of two SETS A and B is the SET of elements common to A and B. This is written $A \cap B$, and is pronounced "A intersection B" or "A cap B." The intersection of sets A_1 through A_n is written $\cap_{i=1}^{n} A_i$.

The intersection of two LINES AB and CD is written $AB \cap CD$. The intersection of two or more geometric objects is the point (points, lines, etc.) at which they CONCUR.

See also AND, CIRCLE-CIRCLE INTERSECTION, CIRCLE-LINE INTERSECTION, CONCUR, CONCURRENT, CONE-SPHERE INTERSECTION, CONIC SECTION, CYLINDRICAL SECTION, LINE-LINE INTERSECTION, SPHERE-SPHERE INTERSECTION, SPIRIC SECTION, STEINMETZ SOLID, TORIC SECTION, TOTAL INTERSECTION THEOREM, UNION, VENN DIAGRAM, VIVIANI'S CURVE

Intersection (Homology)

When two cycles intersect TRANSVERSALLY $X_1 \cap X_2 = Y$ on a SMOOTH MANIFOLD M, then Y is a cycle. Moreover, the homology class that Y represents depends only on the HOMOLOGY CLASS of X_1 and X_2. The sign of Y is determined by the orientations on M, X_1, and X_2.

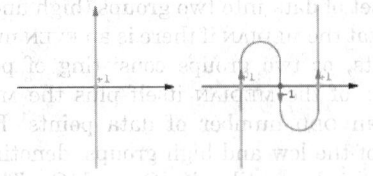

For example, two curves can intersect in one point on a surface transversally, since

$$\dim X_1 + \dim X_2 = 1 + 1 = 2 = \dim M - 0.$$

The curves can be deformed so that they intersect three times, but two of those intersections sum to zero since two intersect positively and one intersects negatively, i.e., with the ORIENTATION of the curves being the reverse orientation of the ambient space.

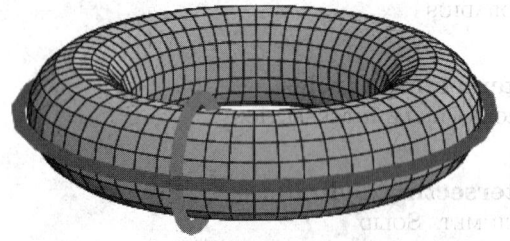

On the torus illustrated above, the cycles intersect in one point.

The binary operation of intersection makes homology on a MANIFOLD into a RING. That is, it plays the role of multiplication, which respects the grading. When $\alpha \in H_{n-p}$ and $\alpha \in H_{n-q}$, then $\alpha \cap \beta \in H_{n-(p+q)}$. In fact, intersection is the dual to the CUP PRODUCT in POINCARÉ DUALITY. That is, if $\alpha \in H^p$ is the POINCARÉ DUAL to $A \in H_{n-p}$ and $\beta \in H^q$ is the dual to $B \in H_{n-q}$ then $\alpha \wedge \beta \in H^{p+q}$ is the dual to $A \cap B \in H_{n-(p+q)}$.

Without the notion of TRANSVERSALITY, intersections are not well-defined in HOMOLOGY. On a more general space, even a manifold with singularities, the homology does not have a natural ring structure.

See also CODIMENSION, CUP PRODUCT, HOMOLOGY, MANIFOLD, ORIENTATION (MANIFOLD), ORIENTATION (VECTOR SPACE), POINCARE DUALITY, TRANSVERSAL INTERSECTION

Intersection Array

Given a DISTANCE-REGULAR GRAPH G with integers $b_i, c_i, i = 0, \ldots, d$ such that for any two vertices $x, y \in G$ at distance $i = d(x,y)$, there are exactly c_i neighbors of $y \in G_{i-1}(x)$ and b_i neighbors of $y \in G_{i+1}(x)$, the se-

quence

$$i(\gamma) = \{b_0, b_1, \ldots, b_{d-1}; c_1, \ldots, c_d\}$$

is called the intersection array of G.

References

Bendito, E.; Carmona, A.; and Encinas, A. M. "Shortest Paths in Distance-Regular Graphs." *Europ. J. Combin.* **21**, 153–166, 2000.

Intersection Detection

See also TESSELLATION

References

Skiena, S. S. "Intersection Detection" §8.6.8 in *The Algorithm Design Manual*. New York: Springer-Verlag, pp. 370–373, 1997.

Intersection Graph

GRAPH INTERSECTION

Intersection Number

The intersection number $\omega(G)$ of a given GRAPH G is the minimum number of elements in a set S such that G is an intersection graph on S.

See also GRAPH INTERSECTION

References

Harary, F. *Graph Theory*. Reading, MA: Addison-Wesley, 1994.

Interspersion

An ARRAY $A = a_{ij}$, $i, j \geq 1$ of POSITIVE INTEGERS is called an interspersion if

1. The rows of A comprise a PARTITION of the POSITIVE INTEGERS,
2. Every row of A is an INCREASING SEQUENCE,
3. Every column of A is a (possibly FINITE) INCREASING SEQUENCE,
4. If (u_j) and (v_j) are distinct rows of A and if p and q are any indices for which $u_p < v_q < u_{p+1}$, then $u_{p+1} < v_{q+1} < u_{p+2}$.

If an array $A = a_{ij}$ is an interspersion, then it is a DISPERSION. If an array $A = a(i,j)$ is an interspersion, then the sequence $\{x_n\}$ given by $\{x_n = i : n = (i,j)\}$ for some j is a FRACTAL SEQUENCE. Examples of interspersion are the STOLARSKY ARRAY and WYTHOFF ARRAY.

See also DISPERSION (SEQUENCE), FRACTAL SEQUENCE, STOLARSKY ARRAY

References

Kimberling, C. "Interspersions and Dispersions." *Proc. Amer. Math. Soc.* **117**, 313–321, 1993.
Kimberling, C. "Fractal Sequences and Interspersions." *Ars Combin.* **45**, 157–168, 1997.

Intersphere

MIDSPHERE

Interval

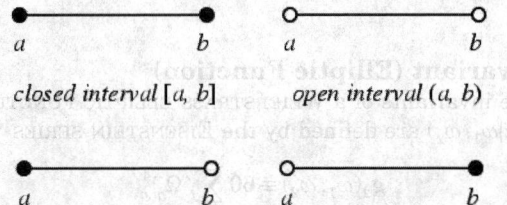

closed interval $[a, b]$ open interval (a, b)

half-closed interval $[a, b)$ half-closed interval $(a, b]$

A collection of points on a LINE SEGMENT. If the endpoints a and b are FINITE and are included, the interval is called CLOSED and is denoted $[a, b]$. If one of the endpoints is $\pm\infty$, then the interval still contains all of its LIMIT POINTS, so $[a, \infty)$ and $(-\infty, b]$ are also closed intervals. If the endpoints are not included, the interval is called OPEN and denoted (a, b). If one endpoint is included but not the other, the interval is denoted $[a, b)$ or $(a, b]$ and is called a HALF-CLOSED (or HALF-OPEN) interval.

The non-standard notation $]a, b[$ for an OPEN INTERVAL and $[a, b[$ or $]a, b]$ for a HALF-CLOSED INTERVAL is sometimes also used.

See also CLOSED INTERVAL, HALF-CLOSED INTERVAL, LIMIT POINT, OPEN INTERVAL, PENCIL

Interval Graph

A GRAPH $G = (V, E)$ is an interval graph if it captures the INTERSECTION RELATION for some set of INTERVALS on the REAL LINE. Formally, P is an interval graph provided that one can assign to each $v \in V$ an interval I_v such that $I_u \cap I_v$ is nonempty precisely when $uv \in E$. An interval graph on a list l can be generated using `IntervalGraph[l]` in the *Mathematica* add-on package `DiscreteMath`Combinatorica`` (which can be loaded with the command `<<DiscreteMath`$`).

STAR GRAPHS are interval graphs, but CYCLE GRAPHS are not (Skiena 1990, p. 164). Determining if a graph is an interval graph and realizing it can be done in $\mathcal{O}(n)$ time (Booth and Lueker 1976; Skiena 1990, p. 164).

See also COMPARABILITY GRAPH

References
Booth, K. S. and Lueker, G. S. "Testing for the Consecutive Ones Property, Interval Graphs, and Graph Planarity using PQ-Tree Algorithms." *J. Comput. System Sci.* **13**, 335–379, 1976.

Fishburn, P. C. *Interval Orders and Interval Graphs: A Study of Partially Ordered Sets.* New York: Wiley, 1985.

Gilmore, P. C. and Hoffman, A. J. "A Characterization of Comparability Graphs and of Interval Graphs." *Canad. J. Math.* **16**, 539–548, 1964.

Lekkerkerker, C. G. and Boland, J. C. "Representation of a Finite Graph by a Set of Intervals on the Real Line." *Fund. Math.* **51**, 45–64, 1962.

Skiena, S. *Implementing Discrete Mathematics: Combinatorics and Graph Theory with Mathematica.* Reading, MA: Addison-Wesley, pp. 163–164, 1990.

Interval Order

A POSET $P = (X, \leq)$ is an interval order if it is ISOMORPHIC to some set of INTERVALS on the REAL LINE ordered by left-to-right precedence. Formally, P is an interval order provided that one can assign to each $x \in X$ an INTERVAL $[x_L, x_R]$ such that $x_R < y_L$ in the REAL NUMBERS IFF $x < y$ in P.

See also PARTIALLY ORDERED SET

References
Fishburn, P. C. *Interval Orders and Interval Graphs: A Study of Partially Ordered Sets.* New York: Wiley, 1985.

Wiener, N. "A Contribution to the Theory of Relative Position." *Proc. Cambridge Philos. Soc.* **17**, 441–449, 1914.

Intrinsic Curvature

A CURVATURE such as GAUSSIAN CURVATURE which is detectable to the "inhabitants" of a surface and not just outside observers. An EXTRINSIC CURVATURE, on the other hand, is not detectable to someone who can't study the 3-dimensional space surrounding the surface on which he resides.

See also CURVATURE, EXTRINSIC CURVATURE, GAUSSIAN CURVATURE

Intrinsic Equation

An equation which specifies a CURVE in terms of intrinsic properties such as ARC LENGTH, RADIUS OF CURVATURE, and TANGENTIAL ANGLE instead of with reference to artificial coordinate axes. Intrinsic equations are also called NATURAL EQUATIONS.

See also CESÀRO EQUATION, NATURAL EQUATION, WHEWELL EQUATION

References
Yates, R. C. "Intrinsic Equations." *A Handbook on Curves and Their Properties.* Ann Arbor, MI: J. W. Edwards, pp. 123–126, 1952.

Intrinsic Variety

See also VARIETY

Intrinsically Linked

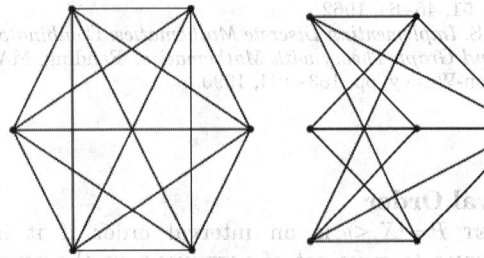

A GRAPH is intrinsically linked if any embedding of it in 3-D contains a nontrivial link. A GRAPH is intrinsically linked IFF it contains one of the seven PETERSEN GRAPHS (Robertson *et al.* 1993).

The COMPLETE GRAPH K_6 (left) is intrinsically linked because it contains at least two linked TRIANGLES. The COMPLETE K-PARTITE GRAPH $K_{3,3,1}$ (right) is also intrinsically linked.

See also COMPLETE GRAPH, COMPLETE K-PARTITE GRAPH, PETERSEN GRAPH

References

Adams, C. C. *The Knot Book: An Elementary Introduction to the Mathematical Theory of Knots.* New York: W. H. Freeman, pp. 217–221, 1994.
Robertson, N.; Seymour, P. D.; and Thomas, R. "Linkless Embeddings of Graphs in 3-Space." *Bull. Amer. Math. Soc.* **28**, 84–89, 1993.

Invaginatum

A negative-height (inward-pointing) PYRAMID used in CUMULATION. The term was introduced by B. Grünbaum.

See also CUMULATION, ELEVATUM

Invariable Point

Three concurrent homologous lines pass respectively through three fixed points on the SIMILITUDE CIRCLE which are known as the invariable points.

See also HOMOLOGOUS POINTS, SIMILITUDE CIRCLE

References

Johnson, R. A. *Modern Geometry: An Elementary Treatise on the Geometry of the Triangle and the Circle.* Boston, MA: Houghton Mifflin, 1929.

Invariant

A quantity which remains unchanged under certain classes of transformations. Invariants are extremely useful for classifying mathematical objects because they usually reflect intrinsic properties of the object of study.

See also ADIABATIC INVARIANT, ALEXANDER INVARIANT, ALGEBRAIC INVARIANT, ARF INVARIANT, GEOMETRIC INVARIANT THEORY, INTEGRAL OF MOTION, INVARIANT (ELLIPTIC FUNCTION), KNOT POLYNOMIAL

References

Hunt, B. "Invariants." Appendix B.1 in *The Geometry of Some Special Arithmetic Quotients.* New York: Springer-Verlag, pp. 282–290, 1996.
Olver, P. J. *Classical Invariant Theory.* Cambridge, England: Cambridge University Press, 1999.

Invariant (Elliptic Function)

The invariants of a WEIERSTRASS ELLIPTIC FUNCTION $\wp(z|\omega_1, \omega_2)$ are defined by the EISENSTEIN SERIES

$$g_2(\omega_1, \omega_2) \equiv 60 \sum_{m,n}{}' \Omega_{m,n}^{-4}$$

$$g_3(\omega_1, \omega_2) \equiv 140 \sum_{m,n}{}' \Omega_{m,n}^{-5}.$$

Here,

$$\Omega_{mn}(\omega_1, \omega_2) \equiv 2m\omega_1 - 2n\omega_2,$$

where ω_1 and ω_2 are the periods of the ELLIPTIC FUNCTION.
Writing $g_i(\tau) \equiv g_i(1, \tau)$,

$$g_2(\tau) \equiv g_2(1, \tau) = \omega_1^4(\omega_1, \omega_2) \tag{1}$$

$$g_3(\tau) \equiv g_3(1, \tau) = \omega_1^6(\omega_1, \omega_2), \tag{2}$$

and the invariants have the FOURIER SERIES

$$g_2(\tau) = \frac{4\pi^4}{4}\left[1 + 240 \sum_{k=1}^{\infty} \sigma_3(k)e^{2\pi i k\tau}\right] \tag{3}$$

$$g_3(\tau) = \frac{8\pi^6}{27}\left[1 - 504 \sum_{k=1}^{\infty} \sigma_5(k)e^{2\pi i k\tau}\right] \tag{4}$$

where $\tau \equiv \omega_2/\omega_2$ and $\sigma_k(n)$ is the DIVISOR FUNCTION (Apostol 1997).

See also DEDEKIND ETA FUNCTION, EISENSTEIN SERIES, MODULAR DISCRIMINANT, TAU FUNCTION, WEIERSTRASS ELLIPTIC FUNCTION

References

Apostol, T. M. "The Fourier Expansions of $g_2(\tau)$ and $g_3(\tau)$." §1.9 in *Modular Functions and Dirichlet Series in Number Theory,* 2nd ed. New York: Springer-Verlag, pp. 12–13, 1997.

Invariant Density

NATURAL INVARIANT

Invariant Factor

The polynomials in the DIAGONAL of the SMITH NORMAL FORM or RATIONAL CANONICAL FORM of a MATRIX are called its invariant factors.

See also RATIONAL CANONICAL FORM, SMITH NORMAL FORM

Invariant Factors

References

Ayres, F. Jr. "Smith Normal Form." Ch. 24 in *Theory and Problems of Matrices.* New York: Schaum, pp. 188–195, 1962.

Dummit, D. S. and Foote, R. M. *Abstract Algebra, 2nd ed.* Englewood Cliffs, NJ: Prentice-Hall, 1998.

Invariant Factors

The polynomials in the DIAGONAL of the SMITH NORMAL FORM of a MATRIX.

References

Ayres, F. Jr. "Smith Normal Form." Ch. 24 in *Theory and Problems of Matrices.* New York: Schaum, pp. 188–195, 1962.

Dummit, D. S. and Foote, R. M. *Abstract Algebra, 2nd ed.* Englewood Cliffs, NJ: Prentice-Hall, 1998.

Invariant Manifold

When stable and unstable invariant MANIFOLDS intersect, they do so in a HYPERBOLIC FIXED POINT (SADDLE POINT). The invariant MANIFOLDS are then called SEPARATRICES. A HYPERBOLIC FIXED POINT is characterized by two ingoing stable MANIFOLDS and two outgoing unstable MANIFOLDS. In integrable systems, incoming W^s and outgoing W^u MANIFOLDS all join up smoothly.

A stable invariant MANIFOLD W^s of a FIXED POINT Y^* is the set of all points Y_0 such that the trajectory passing through Y_0 tends to Y^* as $j \to \infty$.

An unstable invariant MANIFOLD W^u of a FIXED POINT Y^* is the set of all points Y_0 such that the trajectory passing through Y_0 tends to Y^* as $j \to -\infty$.

See also HOMOCLINIC POINT

Invariant Point

FIXED POINT (TRANSFORMATION)

Invariant Series

An invariant series of a GROUP G is a NORMAL SERIES

$$I = A_0 \lhd A_1 \lhd \ldots \lhd A_r = G$$

such that each $A_i \lhd G$, where $H \lhd G$ means that H is a NORMAL SUBGROUP of G.

See also COMPOSITION SERIES, NORMAL SERIES

References

Scott, W. R. *Group Theory.* New York: Dover, p. 36, 1987.

Invariant Subgroup

NORMAL SUBGROUP

Inverse Cosecant

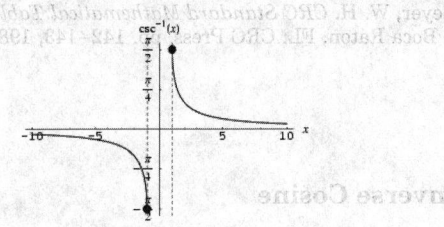

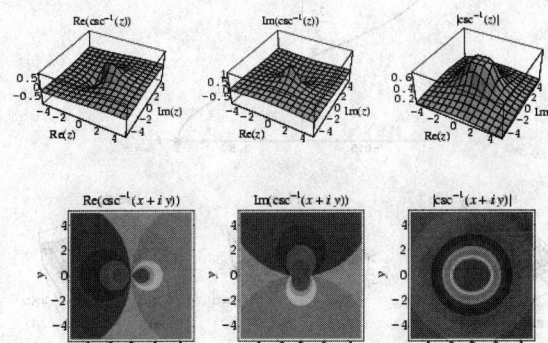

The function $\csc^{-1} x$, also denoted $\mathrm{arccsc}(x)$, where $\csc x$ is the COSECANT and the SUPERSCRIPT -1 denotes an INVERSE FUNCTION, *not* the multiplicative inverse. The inverse cosecant is implemented as `ArcCsc[x]` in *Mathematica*. The inverse cosecant satisfies

$$\csc^{-1} x = \sec^{-1}\left(\frac{x}{\sqrt{x^2-1}}\right) \qquad (1)$$

for POSITIVE or NEGATIVE x, and

$$\csc^{-1} x = \pi + \csc^{-1}(-x) \qquad (2)$$

for $x \geq 0$. The inverse cosecant has TAYLOR SERIES about infinity of

$$\csc^{-1} x = x^{-1} + \tfrac{1}{6}x^{-3} + \tfrac{3}{40}x^{-5} + \tfrac{5}{112}x^{-7} + \ldots \qquad (3)$$

The inverse cosecant is given in terms of other inverse trigonometric functions by

$$\csc^{-1} x = \cos^{-1}\left(\frac{\sqrt{x^2-1}}{x}\right) \qquad (4)$$

$$= \cot^{-1}\left(\sqrt{x^2-1}\right) \qquad (5)$$

$$= \tfrac{1}{2}\pi - \sec^{-1} x = -\tfrac{1}{2}\pi - \sec^{-1}(-x) \qquad (6)$$

$$= \sin^{-1}\left(\frac{1}{x}\right) \qquad (7)$$

for $x \geq 0$.

See also COSECANT, INVERSE SINE, SINE

References

Beyer, W. H. *CRC Standard Mathematical Tables, 28th ed.* Boca Raton, FL: CRC Press, pp. 142–143, 1987.

Inverse Cosine

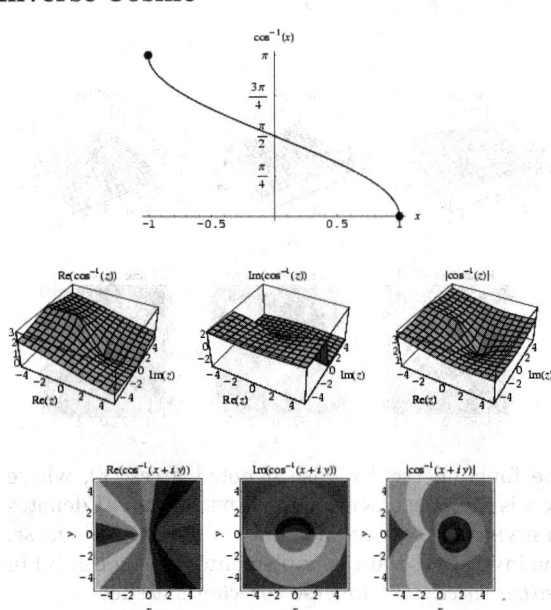

The function $\cos^{-1} x$, where $\cos x$ is the COSINE and the superscript -1 denotes the INVERSE FUNCTION, *not* the multiplicative inverse. The notation arccos x or Arccos$x$ is sometimes also used. The inverse cosine is implemented as ArcCos[x] in *Mathematica*. The inverse cosine satisfies

$$\cos^{-1} x = \pi - \cos^{-1}(-x) \tag{1}$$

for POSITIVE and NEGATIVE x, and

$$\cos^{-1} x = \begin{cases} \frac{1}{2}\pi - \cos^{-1}\left(\sqrt{1-x^2}\right) & \text{for } 0 \le x \le 1 \\ \frac{1}{2}\pi + \cos^{-1}\left(\sqrt{1-x^2}\right) & \text{for } -1 \le x \le 0. \end{cases} \tag{2}$$

The MACLAURIN SERIES for the inverse cosine with $-1 \le x \le 1$ is

$$\cos^{-1} x = \frac{1}{2}\pi - x - \frac{1}{6}x^3 - \frac{3}{40}x^5 - \frac{5}{112}x^7 - \frac{35}{1152}x^9 - \dots. \tag{3}$$

The inverse cosine is given in terms of other inverse trigonometric functions by

$$\cos^{-1} x = \cot^{-1}\left(\frac{x}{\sqrt{1-x^2}}\right) \tag{4}$$

$$= \frac{1}{2}\pi + \sin^{-1}(-x) = \frac{1}{2}\pi - \sin^{-1} x \tag{5}$$

$$= \frac{1}{2}\pi - \tan^{-1}\left(\frac{x}{\sqrt{1-x^2}}\right) \tag{6}$$

for POSITIVE or NEGATIVE x, and

$$\cos^{-1} x = \csc^{-1}\left(\frac{1}{\sqrt{1-x^2}}\right) \tag{7}$$

$$= \sec^{-1}\left(\frac{1}{x}\right) \tag{8}$$

$$= \sin^{-1}\left(\sqrt{1-x^2}\right) \tag{9}$$

$$= \tan^{-1}\left(\frac{\sqrt{1-x^2}}{x}\right) \tag{10}$$

for $x \ge 0$.

See also COSINE, INVERSE SECANT

References

Abramowitz, M. and Stegun, C. A. (Eds.). "Inverse Circular Functions." §4.4 in *Handbook of Mathematical Functions with Formulas, Graphs, and Mathematical Tables, 9th printing.* New York: Dover, pp. 79–83, 1972.

Beyer, W. H. *CRC Standard Mathematical Tables, 28th ed.* Boca Raton, FL: CRC Press, pp. 142–143 and 219, 1987.

Inverse Cotangent

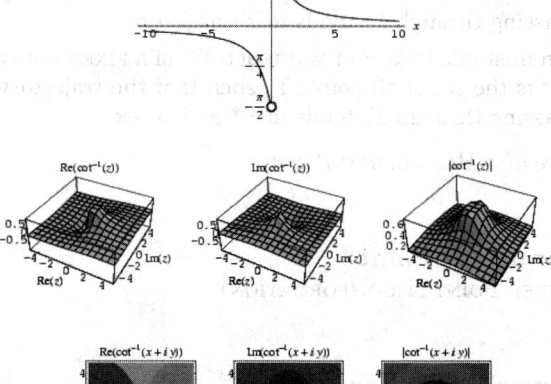

The function $\cot^{-1} x$, also denoted arccot(x), where $\cot x$ is the COTANGENT and the superscript -1 denotes an INVERSE FUNCTION and *not* the multiplicative inverse. The inverse cotangent is implemented as ArcCot[x] in *Mathematica*.

The MACLAURIN SERIES of the inverse cotangent is given by

$$\cot^{-1} x = \frac{1}{2}\pi - x + \frac{1}{3}x^3 - \frac{1}{5}x^5 + \frac{1}{7}x^7 - \frac{1}{9}x^9 + \dots, \tag{1}$$

and LAURENT SERIES by

$$\cot^{-1} x = x^{-1} - \tfrac{1}{3}x^{-3} + \tfrac{1}{5}x^{-5} - \tfrac{1}{7}x^{-7} + \tfrac{1}{9}x^{-9} + \dots. \quad (2)$$

Euler derived the INFINITE series

$$\cot^{-1} x = x\left[\frac{1}{x^2+1} + \frac{2}{3(x^2+1)^2} + \frac{2\cdot 4}{3\cdot 5(x^2+1)^3} + \dots\right] \quad (3)$$

(Wetherfield 1996).

The inverse cotangent satisfies

$$\cot^{-1} x = \tan^{-1}\left(\frac{1}{x}\right) \quad (4)$$

$$= -\cot^{-1}(-x) \quad (5)$$

for POSITIVE and NEGATIVE x, and

$$\cot^{-1} x = \cos^{-1}\left(\frac{x}{\sqrt{x^2+1}}\right) \quad (6)$$

$$= \tfrac{1}{2}\pi - \cot^{-1}\left(\frac{1}{x}\right) \quad (7)$$

$$= \csc^{-1}(\sqrt{x^2+1}) \quad (8)$$

$$= \sec^{-1}\left(\frac{\sqrt{x^2+1}}{x}\right) \quad (9)$$

$$= \sin^{-1}\left(\frac{1}{\sqrt{x^2+1}}\right) \quad (10)$$

$$= \tfrac{1}{2}\pi - \sin^{-1}\left(\frac{x}{\sqrt{x^2+1}}\right) \quad (11)$$

$$= \tfrac{1}{2}\pi + \tan^{-1}(-x) \quad (12)$$

$$= \tfrac{1}{2}\pi - \tan^{-1} x \quad (13)$$

for $x \geq 0$.

A number

$$t_x = \cot^{-1} x, \quad (14)$$

where x is an INTEGER or RATIONAL NUMBER, is sometimes called a GREGORY NUMBER. Lehmer (1938a) showed that $\cot^{-1}(a/b)$ can be expressed as a finite sum of inverse cotangents of INTEGER arguments

$$\cot^{-1}\left(\frac{a}{b}\right) = \sum_{i=1}^{k} (-1)^{i-1} \cot^{-1} n_i, \quad (15)$$

where

$$n_i = \left\lfloor \frac{a_i}{b_i} \right\rfloor, \quad (16)$$

with $\lfloor x \rfloor$ the FLOOR FUNCTION, and

$$a_{i+1} = a_i n + i + b_i \quad (17)$$

$$b_{i+1} = a_i - n_i b_i, \quad (18)$$

with $a_0 = a$ and $b_0 = b$, and where the recurrence is continued until $b_{k+1} = 0$. If an INVERSE TANGENT sum is written as

$$\tan^{-1} n = \sum_{k=1}^{} f_k \tan^{-1} n_k + f \tan^{-1} 1, \quad (19)$$

then equation (15) becomes

$$\cot^{-1} n = \sum_{k=1}^{} f_k \cot^{-1} n_k + c \cot^{-1} 1, \quad (20)$$

where

$$c = 2 - f - 2\sum_{k=1}^{} f_k. \quad (21)$$

Inverse cotangent sums can be used to generate MACHIN-LIKE FORMULAS.

An interesting inverse cotangent identity attributed to Charles Dodgson (Lewis Carroll) by Lehmer (1938b; Bromwich 1965, Castellanos 1988ab) is

$$\cot^{-1}(p+r) + \tan^{-1}(p+q) = \tan^{-1} p, \quad (22)$$

where

$$1 + p^2 = qr. \quad (23)$$

Other inverse cotangent identities include

$$2\cot^{-1}(2x) - \cot^{-1} x = \cot^{-1}(4x^3 + 3x) \quad (24)$$

$$3\cot^{-1}(3x) - \cot^{-1} x = \cot^{-1}\left(\frac{27x^4 + 18x^2 - 1}{8x}\right), \quad (25)$$

as well as many others (Bennett 1926, Lehmer 1938b).

See also COTANGENT, INVERSE TANGENT, MACHIN'S FORMULA, MACHIN-LIKE FORMULAS, TANGENT

References

Abramowitz, M. and Stegun, C. A. (Eds.). "Inverse Circular Functions." §4.4 in *Handbook of Mathematical Functions with Formulas, Graphs, and Mathematical Tables, 9th printing.* New York: Dover, pp. 79–83, 1972.

Bennett, A. A. "The Four Term Diophantine Arccotangent Relation." *Ann. Math.* **27**, 21–24, 1926.

Beyer, W. H. *CRC Standard Mathematical Tables, 28th ed.* Boca Raton, FL: CRC Press, pp. 142–143, 1987.

Bromwich, T. J. I. and MacRobert, T. M. *An Introduction to the Theory of Infinite Series, 3rd ed.* New York: Chelsea, 1991.

Castellanos, D. "The Ubiquitous Pi. Part I." *Math. Mag.* **61**, 67–98, 1988a.

Castellanos, D. "The Ubiquitous Pi. Part II." *Math. Mag.* **61**, 148–163, 1988b.

Lehmer, D. H. "A Cotangent Analogue of Continued Fractions." *Duke Math. J.* **4**, 323–340, 1938a.

Lehmer, D. H. "On Arccotangent Relations for π." *Amer. Math. Monthly* **45**, 657–664, 1938b.

Weisstein, E. W. "Arccotangent Series." MATHEMATICA NOTEBOOK COTSERIES.M.

Wetherfield, M. "The Enhancement of Machin's Formula by Todd's Process." *Math. Gaz.*, 333–344, July 1996.

Inverse Curve

Given a CIRCLE C with CENTER O and RADIUS k, then two points P and Q are inverse with respect to C if $OP \cdot OQ = k^2$. If P describes a curve C_1, then Q describes a curve C_2 called the inverse of C_1 with respect to the circle C (with INVERSION CENTER O). The PEAUCELLIER INVERSOR can be used to construct an inverse curve from a given curve.

If the POLAR equation of C is $r(\theta)$, then the inverse curve has polar equation

$$r = \frac{k^2}{r(\theta)}.$$

If $O = (x_0, y_0)$ and $P = (f(t), g(t))$, then the inverse has equations

$$x = x_0 + \frac{k^2(f - x_0)}{(f - x_0)^2 + (g - y_0)^2}$$

$$y = y_0 + \frac{k^2(g - y_0)}{(f - x_0)^2 + (g - y_0)^2}.$$

Curve	INVERSION CENTER	Inverse Curve
ARCHIMEDEAN SPIRAL	ORIGIN	ARCHIMEDEAN SPIRAL
CARDIOID	CUSP	PARABOLA
CIRCLE	any point	another CIRCLE
CISSOID OF DIOCLES	CUSP	PARABOLA
COCHLEOID	ORIGIN	QUADRATRIX OF HIPPIAS
EPISPIRAL	ORIGIN	ROSE
FERMAT'S SPIRAL	ORIGIN	LITUUS
HYPERBOLA	center	LEMNISCATE
HYPERBOLA	VERTEX	RIGHT STROPHOID
HYPERBOLA with $a = \sqrt{3}$	VERTEX	MACLAURIN TRISECTRIX
LEMNISCATE	center	HYPERBOLA
LITUUS	ORIGIN	FERMAT'S SPIRAL
LOGARITHMIC SPIRAL	ORIGIN	LOGARITHMIC SPIRAL
MACLAURIN TRISECTRIX	FOCUS	TSCHIRNHAUSEN'S CUBIC
PARABOLA	FOCUS	CARDIOID
PARABOLA	VERTEX	CISSOID OF DIOCLES
QUADRATRIX OF HIPPIAS		COCHLEOID
RIGHT STROPHOID	ORIGIN	the same RIGHT STROPHOID INVERSE CURVE
SINUSOIDAL SPIRAL	ORIGIN	SINUSOIDAL SPIRAL
TSCHIRNHAUSEN CUBIC		SINUSOIDAL SPIRAL

See also INVERSION, INVERSION CENTER, INVERSION CIRCLE, PEAUCELLIER INVERSOR, RECIPROCAL, RECIPROCATION

References

Welke, S. "Inversion of Elementary Algebraic Curves with Respect to a Circle." *Mathematica Educ. Res.* **4**, 16–22, 1995.
Wells, D. *The Penguin Dictionary of Curious and Interesting Geometry.* London: Penguin, p. 120, 1991.
Yates, R. C. "Inversion." *A Handbook on Curves and Their Properties.* Ann Arbor, MI: J. W. Edwards, pp. 127–134, 1952.

Inverse Elliptic Nome

INVERSE NOME

Inverse Filter

A linear DECONVOLUTION ALGORITHM.

Inverse Fourier Transform

FOURIER TRANSFORM

Inverse Function

Given a FUNCTION $f(x)$, its inverse $f^{-1}(x)$ is defined by

$$f(f^{-1}(x)) = f^{-1}(f(x)) \equiv x.$$

Therefore, $f(x)$ and $f^{-1}(x)$ are reflections about the line $y = x$.

See also COMPOSITION, INVERSE FUNCTION THEOREM, SERIES REVERSION

References

Jeffreys, H. and Jeffreys, B. S. "Inverse Functions." §1.066 in *Methods of Mathematical Physics, 3rd ed.* Cambridge, England: Cambridge University Press, pp. 22–23, 1988.

Inverse Function Theorem

Given a SMOOTH FUNCTION $f : \mathbb{R}^n \to \mathbb{R}^n$, if the JACOBIAN is invertible at 0, then there is a NEIGHBORHOOD U containing 0 such that $f : U \to f(U)$ is a DIFFEO-

MORPHISM. That is, there is a smooth inverse $f^{-1}: f(U) \to U$.

See also DIFFEOMORPHISM, IMPLICIT FUNCTION THEOREM, JACOBIAN

References

Rudin, W. *Principles of Mathematical Analysis, 3rd ed.* New York: McGraw-Hill, 1976.

Inverse Hyperbolic Cosecant

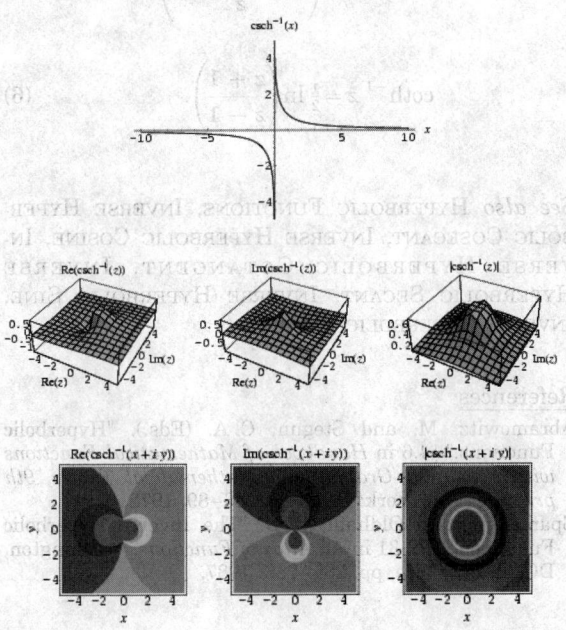

The INVERSE FUNCTION of the HYPERBOLIC COSECANT, denoted $\operatorname{csch}^{-1} z$. It can be defined for complex z by

$$\operatorname{csch}^{-1} z = \ln\left(\sqrt{1 + \frac{1}{z^2}} + \frac{1}{z}\right), \qquad (1)$$

or for real x by

$$\operatorname{csch}^{-1} x = \ln\left(\frac{1 \pm \sqrt{1 + x^2}}{x}\right). \qquad (2)$$

The inverse hyperbolic cosecant is implemented as ArcCsch[x] in *Mathematica*.

The inverse hyperbolic cosecant has TAYLOR SERIES

$$\operatorname{csch}^{-1} x = (\ln 2 - \ln x) + \tfrac{1}{4}x^2 - \tfrac{3}{32}x^4 + \tfrac{5}{96}x^6 + \dots \qquad (3)$$

$$\operatorname{csch}^{-1}\left(\frac{1}{x}\right) = x - \tfrac{1}{6}x^3 + \tfrac{3}{40}x^5 - \tfrac{5}{112}x^7 + \dots. \qquad (4)$$

See also HYPERBOLIC COSECANT, INVERSE HYPERBOLIC FUNCTIONS

Inverse Hyperbolic Cosine

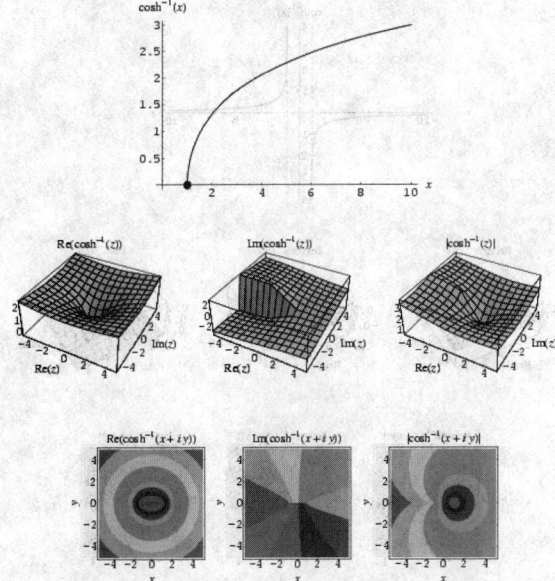

The INVERSE FUNCTION of the HYPERBOLIC COSINE, denoted $\cosh^{-1} z$. It can be defined for complex z by

$$\cosh^{-1} z = \ln\left(z + \sqrt{z+1}\sqrt{z-1}\right), \qquad (1)$$

and for real x by

$$\cosh^{-1} x = \ln\left(x \pm \sqrt{x^2 - 1}\right). \qquad (2)$$

The inverse cosine is implemented as ArcCosh[x] in *Mathematica*.

The inverse hyperbolic cosine has the TAYLOR SERIES

$$\cosh^{-1} x = \sqrt{2(x-1)}$$
$$\times \left[1 - \tfrac{1}{12}(x-1) + \tfrac{3}{160}(x-1)^2 - \tfrac{5}{896}(x-1)^3 + \dots\right] \qquad (3)$$

$$\cosh^{-1}\left(\frac{1}{x}\right) = (\ln 2 - \ln x) - \tfrac{1}{4}x^2 - \tfrac{3}{32}x^4 - \tfrac{5}{96}x^6 + \dots. \qquad (4)$$

See also HYPERBOLIC COSINE, INVERSE HYPERBOLIC FUNCTIONS

Inverse Hyperbolic Cotangent

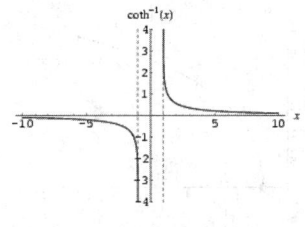

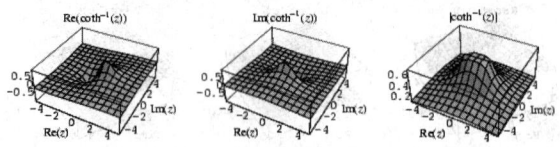

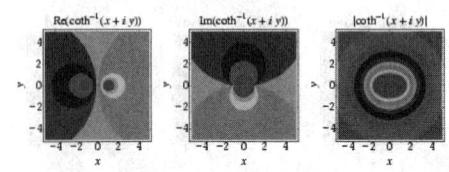

The INVERSE FUNCTION of the HYPERBOLIC COTAN-GENT, denoted $\coth^{-1} x$. It can be defined for complex z as

$$\coth^{-1} z = \frac{1}{2}\left[\ln\left(1 + \frac{1}{z}\right) - \ln\left(1 - \frac{1}{z}\right)\right], \quad (1)$$

and for real x as

$$\coth^{-1} x = \frac{1}{2}\ln\left(\frac{x+1}{x-1}\right). \quad (2)$$

The inverse hyperbolic cotangent is implemented as `ArcCoth[x]` in *Mathematica*.
It has the special values

$$\coth^{-1} 0 = -\tfrac{1}{2}i\pi \quad (3)$$

$$\coth^{-1} 1 = \infty \quad (4)$$

$$\coth^{-1} \infty = 0. \quad (5)$$

$$\coth^{-1} i = -\tfrac{1}{4}\pi i \quad (6)$$

and the MACLAURIN SERIES

$$\coth^{-1}\left(\frac{1}{x}\right) = x + \tfrac{1}{3}x^3 + \tfrac{1}{5}x^5 + \tfrac{1}{7}x^7 + \dots. \quad (7)$$

See also HYPERBOLIC COTANGENT, INVERSE HYPER-BOLIC FUNCTIONS, INVERSE HYPERBOLIC TANGENT

Inverse Hyperbolic Functions

The INVERSE of the HYPERBOLIC FUNCTIONS, denoted $\cosh^{-1} x$, $\coth^{-1} x$, $\operatorname{csch}^{-1} x$, $\operatorname{sech}^{-1} x$, $\sinh^{-1} x$, and $\tanh^{-1} x$. They are defined by

$$\sinh^{-1} z = \ln\left(z + \sqrt{z^2 + 1}\right) \quad (1)$$

$$\cosh^{-1} z = \ln\left(z \pm \sqrt{z^2 - 1}\right) \quad (2)$$

$$\tanh^{-1} z = \frac{1}{2}\ln\left(\frac{1+z}{1-z}\right) \quad (3)$$

$$\operatorname{csch}^{-1} z = \ln\left(\frac{1 \pm \sqrt{1+z^2}}{z}\right) \quad (4)$$

$$\operatorname{sech}^{-1} z = \ln\left(\frac{1 \pm \sqrt{1+z^2}}{z}\right) \quad (5)$$

$$\coth^{-1} z = \frac{1}{2}\ln\left(\frac{z+1}{z-1}\right). \quad (6)$$

See also HYPERBOLIC FUNCTIONS, INVERSE HYPER-BOLIC COSECANT, INVERSE HYPERBOLIC COSINE, IN-VERSE HYPERBOLIC COTANGENT, INVERSE HYPERBOLIC SECANT, INVERSE HYPERBOLIC SINE, INVERSE HYPERBOLIC TANGENT

References

Abramowitz, M. and Stegun, C. A. (Eds.). "Hyperbolic Functions." §4.6 in *Handbook of Mathematical Functions with Formulas, Graphs, and Mathematical Tables, 9th printing.* New York: Dover, pp. 86–89, 1972.
Spanier, J. and Oldham, K. B. "The Inverse Hyperbolic Functions." Ch. 31 in *An Atlas of Functions.* Washington, DC: Hemisphere, pp. 285–293, 1987.

Inverse Hyperbolic Secant

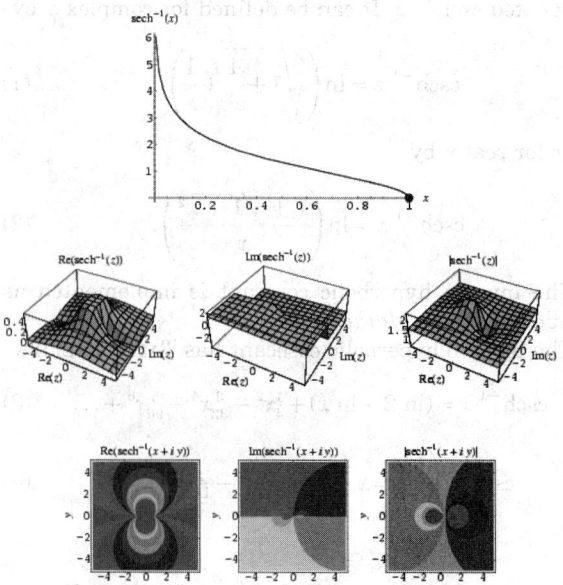

The INVERSE FUNCTION of the HYPERBOLIC SECANT,

denoted $\mathrm{sech}^{-1} x$. It can be defined for complex z as

$$\sec^{-1} z = \ln\left(\sqrt{\frac{1}{z}-1}\,\sqrt{\frac{1}{z}+1}+\frac{1}{z}\right), \qquad (1)$$

and for real x as

$$\mathrm{sech}^{-1} x = \ln\left(\frac{1\pm\sqrt{1-x^2}}{x}\right). \qquad (2)$$

The inverse hyperbolic secant is implemented as `ArcSech[x]` in *Mathematica*.
It has MACLAURIN SERIES

$$\mathrm{sech}^{-1} x = (\ln 2 - \ln x) - \tfrac{1}{4}x^2 - \tfrac{3}{32}x^4 - \tfrac{5}{96}x^6 - \tfrac{35}{1024}x^8 + \cdots$$

(Sloane's A052468 and A052469) and

$$\mathrm{sech}^{-1}\left(\frac{1}{x}\right) = i\left(\tfrac{1}{2}\pi - x - \tfrac{1}{6}x^3 - \tfrac{3}{40}x^5 + \cdots\right). \qquad (3)$$

See also HYPERBOLIC SECANT, INVERSE HYPERBOLIC FUNCTIONS

References
Sloane, N. J. A. Sequences A052468 and A052469 in "An On-Line Version of the Encyclopedia of Integer Sequences." http://www.research.att.com/~njas/sequences/eisonline.html.

Inverse Hyperbolic Sine

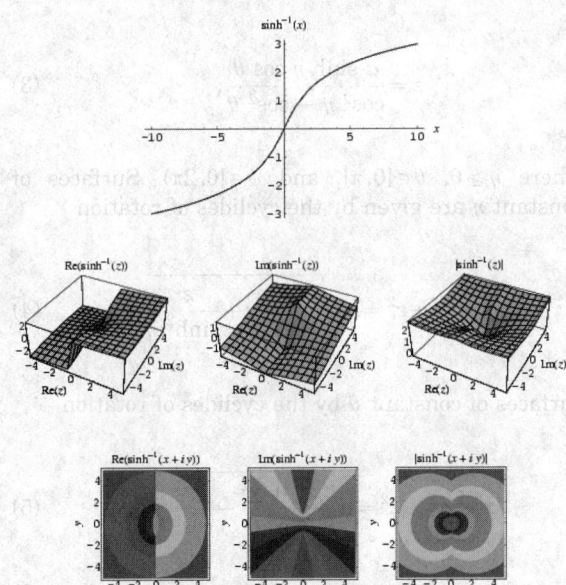

The INVERSE FUNCTION of the HYPERBOLIC SINE, denoted $\sinh^{-1} x$. It can be defined for complex z as

$$\sinh^{-1} z = \ln(z + \sqrt{1+z^2}).$$

The inverse hyperbolic sine is implemented as `Arc-Sinh[x]` in *Mathematica*.
It has a MACLAURIN SERIES

$$\sinh^{-1} x = x - \tfrac{1}{6}x^3 + \tfrac{3}{40}x^5 - \tfrac{5}{112}x^7 + \tfrac{35}{1152}x^9 + \cdots \qquad (1)$$

$$\sinh^{-1}\left(\frac{1}{x}\right) = (\ln 2 - \ln x) + \tfrac{1}{4}x^2 - \tfrac{3}{32}x^4 + \tfrac{5}{96}x^6 - \cdots. \qquad (2)$$

See also HYPERBOLIC SINE, INVERSE HYPERBOLIC FUNCTIONS

Inverse Hyperbolic Tangent

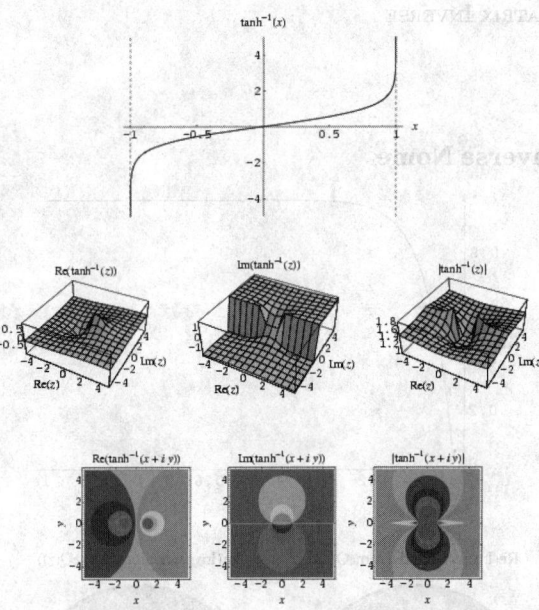

The INVERSE FUNCTION of the HYPERBOLIC TANGENT, denoted $\tanh^{-1} x$. It can be defined for complex z as

$$\tanh^{-1} z = \tfrac{1}{2}[\ln(1+z) - \ln(1-z)], \qquad (1)$$

and for real x as

$$\tanh^{-1} x = \frac{1}{2}\ln\left(\frac{1+x}{1-x}\right). \qquad (2)$$

The inverse hyperbolic tangent is implemented as `ArcTanh[x]` in *Mathematica*.
It has special values

$$\tanh^{-1} 0 = 0 \qquad (3)$$

$$\tanh^{-1} 1 = \infty \qquad (4)$$

$$\tanh^{-1} \infty = -\tfrac{1}{2}\pi i \qquad (5)$$

$$\tanh^{-1} i = \tfrac{1}{4}\pi i \qquad (6)$$

and MACLAURIN SERIES

$$\tanh^{-1} x = x + \tfrac{1}{3}x^3 + \tfrac{1}{5}x^5 + \tfrac{1}{7}x^7 + \tfrac{1}{9}x^9 + \ldots . \quad (7)$$

See also HYPERBOLIC TANGENT, INVERSE HYPERBOLIC COTANGENT, INVERSE HYPERBOLIC FUNCTIONS

Inverse Laplace Transform

BROMWICH INTEGRAL, LAPLACE TRANSFORM

Inverse Matrix

MATRIX INVERSE

Inverse Nome

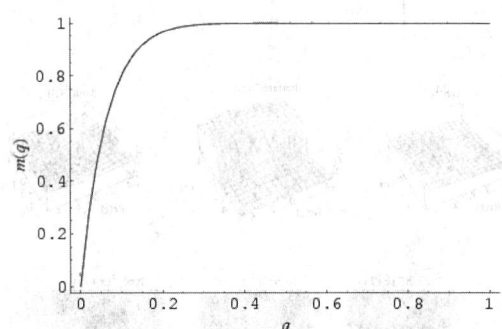

Re(InverseEllipticNomeQ(z))

Im(InverseEllipticNomeQ(z))

Solving the NOME q for the PARAMETER m gives

$$m(q) = \frac{\vartheta_2^4(0,q)}{\vartheta_3^4(0,q)},$$

where $\vartheta_i(z,q)$ is a JACOBI THETA FUNCTION. The inverse nome is implemented as InverseElliptic-NomeQ[q] in *Mathematica*. It satisfies

$$\lim_{q \to 0+} \frac{dm}{dq} = 16.$$

See also JACOBI THETA FUNCTIONS, NOME

Inverse Oblate Spheroidal Coordinates

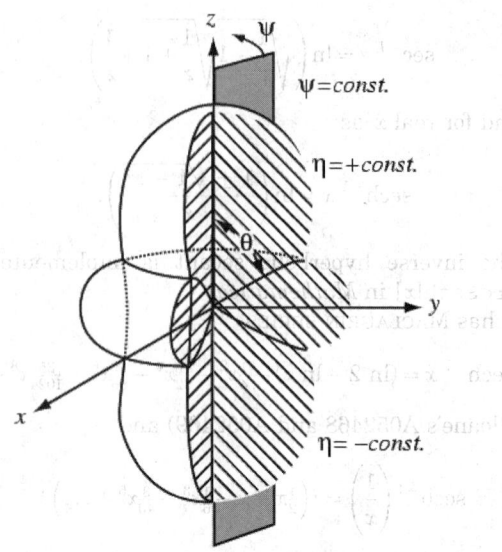

A system of coordinates obtained by INVERSION of the oblate spheroids and one-sheeted hyperboloids in OBLATE SPHEROIDAL COORDINATES. The inverse oblate spheroidal coordinates (η, θ, ψ) are given by the transformation equations

$$x = \frac{a \cosh \eta \sin \theta \cos \psi}{\cosh^2 \eta - \cos^2 \theta} \quad (1)$$

$$y = \frac{a \cosh \eta \sin \theta \sin \psi}{\cosh^2 \eta - \cos^2 \theta} \quad (2)$$

$$z = \frac{a \sinh \eta \cos \theta}{\cos^2 \eta - \cos^2 \theta}, \quad (3)$$

where $\eta \geq 0$, $\theta \in [0, \pi]$, and $\psi \in [0, 2\pi]$. Surfaces of constant η are given by the cyclides of rotation

$$x^2 + y^2 + z^2 = a \sqrt{\frac{x^2 + y^2}{\cosh^2 \eta} + \frac{z^2}{\sinh^2 \eta}}, \quad (4)$$

surfaces of constant θ by the cyclides of rotation

$$x^2 + y^2 + z^2 = a \sqrt{\frac{x^2 + y^2}{\sin^2 \theta} - \frac{z^2}{\cos^2 \theta}}, \quad (5)$$

and surfaces of constant ψ by the half-planes

$$\tan \psi = \frac{y}{x}. \quad (6)$$

The metric coefficients are given by

$$g_{\eta\eta} = \frac{a^2 (\cosh^2 \eta - \sin^2 \theta)}{(\cosh^2 \eta - \cos^2 \theta)} \qquad (7)$$

$$g_{\theta\theta} = \frac{a^2 (\cosh^2 \eta - \sin^2 \theta)}{(\cosh^2 \eta - \cos^2 \theta)} \qquad (8)$$

$$g_{\psi\psi} = \frac{a^2 \cosh^2 \eta \sin^2 \theta}{(\cosh^2 \eta - \cos^2 \theta)^2}. \qquad (9)$$

See also INVERSE PROLATE SPHEROIDAL COORDINATES, PROLATE SPHEROIDAL COORDINATES

References

Moon, P. and Spencer, D. E. "Inverse Oblate Spheroidal Coordinate (η, θ, ψ)." Fig. 4.06 in *Field Theory Handbook, Including Coordinate Systems, Differential Equations, and Their Solutions, 2nd ed.* New York: Springer-Verlag, pp. 119–121, 1988.

Inverse Permutation

An inverse permutation is a permutation in which each number and the number of the place which it occupies are exchanged. For example,

$$p_1 = \{3, 8, 5, 10, 9, 4, 6, 1, 7, 2\}$$

$$p_2 = \{8, 10, 1, 6, 3, 7, 9, 2, 5, 4\}$$

are inverse permutations, since the positions of 1, 2, 3, 4, 5, 6, 7, 8, 9, and 10 in p_1 are p_2, and the positions of 1, 2, 3, 4, 5, 6, 7, 8, 9, and 10 in p_2 are likewise p_1 (Muir 1960, p. 5). The inverse permutation of a given PERMUTATION can be computed using `InversePermutation[p]` in the *Mathematica* add-on package `DiscreteMath`Combinatorica`` (which can be loaded with the command `<<DiscreteMath``). Inverse permutations are sometimes also called conjugate or reciprocal permutations (Muir 1960, p. 4).

See also PERMUTATION, PERMUTATION INVERSION, SELF-CONJUGATE PARTITION

References

Muir, T. *A Treatise on the Theory of Determinants.* New York: Dover, 1960.

Inverse Points

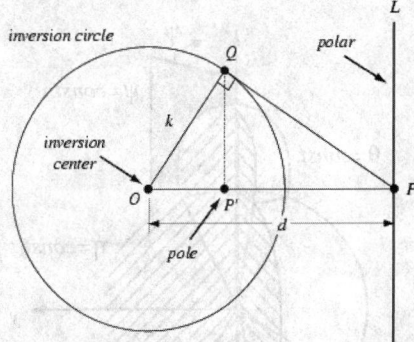

Points, also called polar reciprocals, which are transformed into each other through INVERSION about a given INVERSION CIRCLE C (or INVERSION SPHERE). The points P and P' are inverse points with respect to the INVERSION CIRCLE if

$$OP \cdot OP' = OQ^2 = k^2$$

(Wenninger 1983, p. 2). In this case, P' is called the POLE and the line L through P and perpendicular to OP is called the POLAR. In the above figure, the quantity k^2 is called the POWER of the point P relative to the circle C.

The point P' which is the inverse point of a given point P with respect to an INVERSION CIRCLE C may be constructed geometrically using a COMPASS only (Coxeter 1969, p. 78; Courant and Robbins 1996, pp. 144–145).

Inverse points can also be taken with respect to an INVERSION SPHERE, which is a natural extension of geometric INVERSION from the plane to 3-dimensional space.

See also GEOMETRIC CONSTRUCTION, INVERSION, INVERSION CIRCLE, INVERSION SPHERE, LIMITING POINT, POLAR, POLE (INVERSION), POWER (CIRCLE)

References

Courant, R. and Robbins, H. "Geometrical Construction of Inverse Points." §3.4.3 in *What is Mathematics?: An Elementary Approach to Ideas and Methods, 2nd ed.* Oxford, England: Oxford University Press, pp. 144–145, 1996.
Coxeter, H. S. M. *Introduction to Geometry, 2nd ed.* New York: Wiley, 1969.
Wenninger, M. J. *Dual Models.* Cambridge, England: Cambridge University Press, 1983.

Inverse Problem

References

Kozhanov, A. I. *Composite Type Equations and Inverse Problems.* Utrecht, Netherlands: VSP, 1999.
Prilepko, A. I.; Orlovsky, D. G.; and Vasin, I. A. *Methods for Solving Inverse Problems in Mathematical Physics.* New York: Dekker, 1999.

Inverse Prolate Spheroidal Coordinates

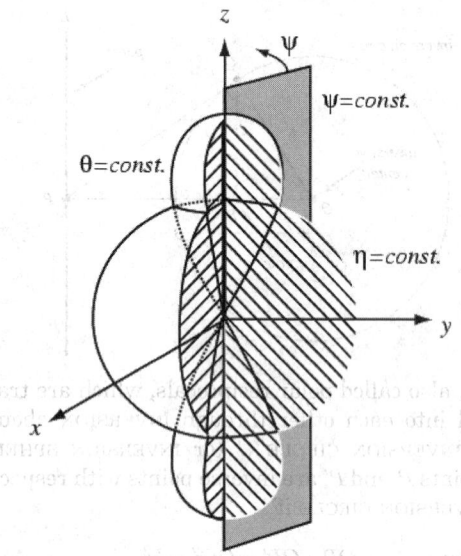

A system of coordinates obtained by INVERSION of the prolate spheroids and two-sheeted hyperboloids in PROLATE SPHEROIDAL COORDINATES. The inverse prolate spheroidal coordinates (η, θ, ψ) are given by the transformation equations

$$x = \frac{a \sinh \eta \sin \theta \cos \psi}{\cosh^2 \eta - \sin^2 \theta} \tag{1}$$

$$y = \frac{a \sinh \eta \sin \theta \sin \psi}{\cosh^2 \eta - \sin^2 \theta} \tag{2}$$

$$z = \frac{a \cosh \eta \cosh \theta}{\cosh^2 \eta - \sin^2 \theta}, \tag{3}$$

with $\eta \geq 0$, $\theta \in [0, \pi]$, and $\psi \in [0, 2\pi)$. Surfaces of constant η are given by the cyclides of rotation

$$x^2 + y^2 + z^2 = a \sqrt{\frac{x^2 + y^2}{\sinh^2 \eta} + \frac{z^2}{\cosh^2 \eta}}, \tag{4}$$

surfaces of constant θ by the cyclides of rotation

$$x^2 + y^2 + z^2 = a \sqrt{-\frac{x^2 + y^2}{\sin^2 \theta} + \frac{z^2}{\cosh^2 \theta}}, \tag{5}$$

and surfaces of constant ψ by the half-planes

$$\tan \psi = \frac{y}{x}. \tag{6}$$

The metric coefficients are given by

$$g_{\eta\eta} = \frac{a^2(\sinh^2 \eta + \sin^2 \theta)}{(\cosh^2 \eta - \sin^2 \theta)^2} \tag{7}$$

$$g_{\theta\theta} = \frac{a^2(\sinh^2 \eta + \sin^2 \theta)}{(\cosh^2 \eta - \sin^2 \theta)^2} \tag{8}$$

$$g_{\psi\psi} = \frac{a^2 \sinh^2 \eta \sinh^2 \theta}{(\cosh^2 \eta - \sin^2 \theta)^2}. \tag{9}$$

See also INVERSE OBLATE SPHEROIDAL COORDINATES, OBLATE SPHEROIDAL COORDINATES

References
Moon, P. and Spencer, D. E. "Inverse Prolate Spheroidal Coordinate (η, θ, ψ)." Fig. 4.05 in *Field Theory Handbook, Including Coordinate Systems, Differential Equations, and Their Solutions, 2nd ed.* New York: Springer-Verlag, pp. 115–118, 1988.

Inverse Proportion
INVERSELY PROPORTIONAL

Inverse Quadratic Interpolation
The use of three prior points in a ROOT-finding ALGORITHM to estimate the zero crossing.

Inverse Scattering Method
A method which can be used to solve the initial value problem for certain classes of nonlinear PARTIAL DIFFERENTIAL EQUATIONS. The method reduces the initial value problem to a linear INTEGRAL EQUATION in which time appears only implicitly. However, the solutions $u(x, t)$ and various of their derivatives must approach zero as $x \to \pm\infty$ (Infeld and Rowlands 2000).

See also ABLOWITZ-RAMANI-SEGUR CONJECTURE, BÄCKLUND TRANSFORMATION

References
Infeld, E. and Rowlands, G. "Inverse Scattering Method." §7.4 in *Nonlinear Waves, Solitons, and Chaos, 2nd ed.* Cambridge, England: Cambridge University Press, pp. 173–175, 2000.
Miura, R. M. (Ed.). *Bäcklund Transformations, the Inverse Scattering Method, Solitons, and Their Applications.* New York: Springer-Verlag, 1974.

Inverse Secant

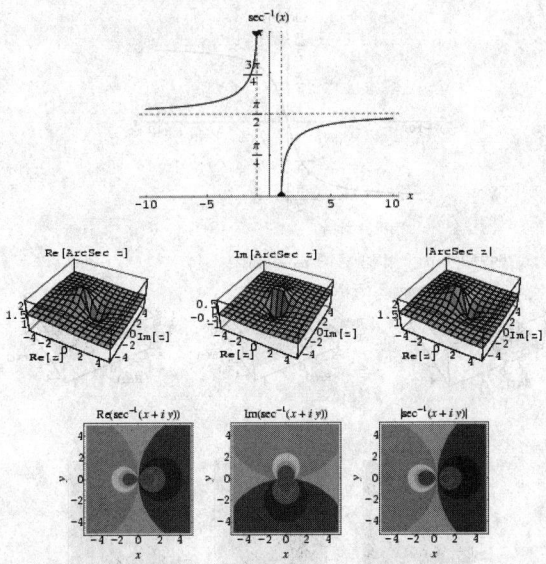

The function $\sec^{-1} x$, where $\sec x$ is the SECANT and the superscript -1 denotes the INVERSE FUNCTION, *not* the multiplicative inverse. The inverse secant is implemented as `ArcSec[x]` in *Mathematica*. The inverse secant satisfies

$$\sec^{-1} x = \csc^{-1}\left(\frac{x}{\sqrt{x^2 - 1}}\right) \qquad (1)$$

for POSITIVE or NEGATIVE x, and

$$\sec^{-1} x = \pi - \sec^{-1}(-x) \qquad (2)$$

for $x \geq 0$. The inverse secant has a TAYLOR SERIES about infinity of

$$\sec^{-1} x = \tfrac{1}{2}\pi - x^{-1} - \tfrac{1}{6}x^{-3} - \tfrac{3}{40}x^{-5} - \tfrac{5}{112}x^{-7} - \ldots. \qquad (3)$$

The inverse secant is given in terms of other inverse trigonometric functions by

$$\sec^{-1} x = \cos^{-1}\left(\frac{1}{x}\right) \qquad (4)$$

$$= \cot^{-1}\left(\frac{1}{\sqrt{x^2 - 1}}\right) \qquad (5)$$

$$= \tfrac{1}{2}\pi - \csc^{-1} x = -\tfrac{1}{2}\pi = \csc^{-1}(-x) \qquad (6)$$

$$= \sin^{-1}\left(\frac{\sqrt{x^2 - 1}}{x}\right) \qquad (7)$$

$$= \tan^{-1}(\sqrt{x^2 - 1}) \qquad (8)$$

for $x \geq 0$.

See also INVERSE COSECANT, SECANT

References

Beyer, W. H. *CRC Standard Mathematical Tables, 28th ed.* Boca Raton, FL: CRC Press, pp. 141–143, 1987.

Inverse Semigroup

This entry contributed by NICOLAS BRAY

A SEMIGROUP S is said to be an inverse semigroup if, for every a in S, there is a unique b (called the inverse of a) such that $a = aba$ and $b = bab$. This is equivalent to the condition that every element has at least one inverse and that the IDEMPOTENTS of S COMMUTE (Lawson 1999). Note that if b is an inverse of a, then ba is an IDEMPOTENT.

See also SEMIGROUP

References

Clifford, A. H. and Preston, G. B. *The Algebraic Theory of Semigroups, Vol. 1.* Providence, RI: Amer. Math. Soc., 1961.
Clifford, A. H. and Preston, G. B. *The Algebraic Theory of Semigroups, Vol. 2.* Providence, RI: Amer. Math. Soc., 1967.
Lawson, M. V. *Inverse Semigroups: The Theory of Partial Symmetries.* Singapore: World Scientific, 1999.
Lyapin, E. S. *Semigroups.* Providence, RI: Amer. Math. Soc., 1974.
Shevrin, L. N. "Inversion Semi-Group." In *Encyclopaedia of Mathematics: An Updated and Annotated Translation of the Soviet "Mathematical Encyclopaedia," Vol. 5* (Managing Ed. M. Hazewinkel). Dordrecht, Netherlands: Reidel, pp. 184–185, 1988.
Weinstein, A. "Groupoids: Unifying Internal and External Symmetry." *Not. Amer. Math. Soc.* **43**, 744–752, 1996.

Inverse Sine

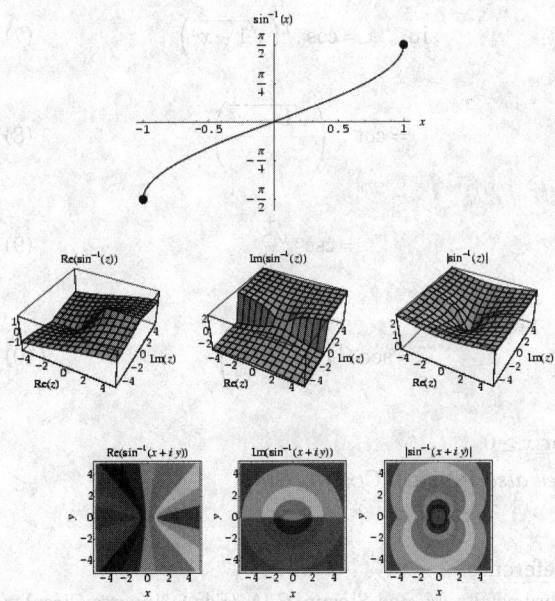

The function $\sin^{-1} x$, where $\sin x$ is the SINE and the superscript -1 denotes the INVERSE FUNCTION, *not* the multiplicative inverse. The notation $\arcsin x$ or $\text{Arcsin } x$ is sometimes also used. The inverse sine is implemented as `ArcSin[x]` in *Mathematica*. The inverse sine satisfies

$$\sin^{-1} x = -\sin^{-1}(-x) \tag{1}$$

for POSITIVE and NEGATIVE x, and

$$\sin^{-1} x = \begin{cases} \frac{1}{2}\pi - \sin^{-1}\left(\sqrt{1-x^2}\right) & \text{for } 0 \leq x \leq 1 \\ -\frac{1}{2}\pi + \sin^{-1}\left(\sqrt{1-x^2}\right) & \text{for } -1 \leq x \leq 0. \end{cases} \tag{2}$$

The MACLAURIN SERIES for the inverse sine with $-1 \leq x \leq 1$ is given by

$$\sin^{-1} x = x + \frac{1}{6}x^3 + \frac{3}{40}x^5 + \frac{5}{112}x^7 + \frac{35}{1152}x^9 + \cdots. \tag{3}$$

The inverse sine is given in terms of other inverse trigonometric functions by

$$\sin^{-1} x = \cos^{-1}(-x) - \frac{1}{2}\pi = \frac{1}{2}\pi - \cos^{-1} x \tag{4}$$

$$= \frac{1}{2}\pi - \cot^{-1}\left(\frac{x}{\sqrt{1-x^2}}\right) \tag{5}$$

$$= \tan^{-1}\left(\frac{x}{\sqrt{1-x^2}}\right) \tag{6}$$

for POSITIVE or NEGATIVE x, and

$$\sin^{-1} x = \cos^{-1}\left(\sqrt{1-x^2}\right) \tag{7}$$

$$= \cot^{-1}\left(\frac{\sqrt{1-x^2}}{x}\right) \tag{8}$$

$$= \csc^{-1}\frac{1}{x} \tag{9}$$

$$= \sec^{-1}\left(\frac{1}{\sqrt{1-x^2}}\right) \tag{10}$$

for $x \geq 0$.

See also INVERSE COSINE, SINE

References

Abramowitz, M. and Stegun, C. A. (Eds.). "Inverse Circular Functions." §4.4 in *Handbook of Mathematical Functions with Formulas, Graphs, and Mathematical Tables, 9th printing.* New York: Dover, pp. 79–83, 1972.

Beyer, W. H. *CRC Standard Mathematical Tables, 28th ed.* Boca Raton, FL: CRC Press, pp. 142–143 and 220, 1987.

Inverse Tangent

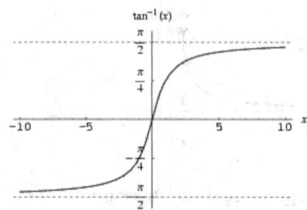

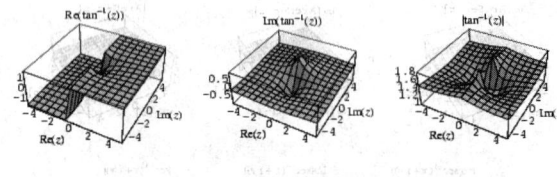

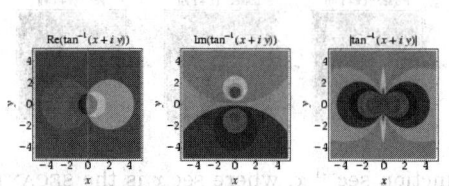

The inverse tangent is also called the arctangent and is denoted either $\tan^{-1} x$ or $\arctan x$, and is the INVERSE FUNCTION of the TANGENT $\tan x$. The inverse tangent is implemented as `ArcTan[x]` in *Mathematica*.

The ARGUMENT of a COMPLEX NUMBER $z = x + iy$ is often written as

$$\theta = \tan^{-1}\left(\frac{y}{x}\right), \tag{1}$$

where θ, sometimes also denoted ϕ, corresponds to the counterclockwise ANGLE from the POSITIVE REAL AXIS, i.e., the value of θ such that $x = \cos\theta$ and $y = \sin\theta$. This special kind of INVERSE TANGENT takes into account the quadrant in which z lies and is returned by the FORTRAN command `ATAN2(X,Y)` and the *Mathematica* command `ArcTan[x, y]`, and is often restricted to the range $-\pi < \theta \leq \pi$. In the degenerate case when $x = 0$,

$$\phi = \begin{cases} -\frac{1}{2}\pi & \text{if } y < 0 \\ \text{undefined} & \text{if } y = 0 \\ \frac{1}{2}\pi & \text{if } y > 0. \end{cases} \tag{2}$$

$\tan^{-1} x$ has the MACLAURIN SERIES for $-1 \leq x \leq 1$ of

$$\tan^{-1} x = \sum_{n=0}^{\infty} \frac{(-1)^n x^{2n+1}}{2n+1}$$

$$= x - \frac{1}{3}x^3 + \frac{1}{5}x^5 - \frac{1}{7}x^7 + \cdots. \tag{3}$$

A more rapidly converging form due to Euler is given by

$$\tan^{-1} x = \sum_{n=0}^{\infty} \frac{2^{2n}(n!)^2}{(2n+1)!} \frac{x^{2n+1}}{(1+x^2)^{n+1}} \quad (4)$$

(Castellanos 1988).

The inverse tangent satisfies

$$\tan^{-1} x = -\tan^{-1}(-x) \quad (5)$$

for POSITIVE and NEGATIVE x, and

$$\tan^{-1} x = \tfrac{1}{2}\pi - \tan^{-1}\left(\frac{1}{x}\right) \quad (6)$$

for $x \geq 0$. The inverse tangent is given in terms of other inverse trigonometric functions by

$$\tan^{-1} x = \tfrac{1}{2}\pi - \cos^{-1}\left(\frac{x}{\sqrt{x^2+1}}\right) \quad (7)$$

$$= \cot^{-1}(-x) - \tfrac{1}{2}\pi = \tfrac{1}{2}\pi - \cot^{-1} x \quad (8)$$

$$= \sin^{-1}\left(\frac{x}{\sqrt{x^2+1}}\right) \quad (9)$$

for POSITIVE or NEGATIVE x, and

$$\tan^{-1} x = \cos^{-1}\left(\frac{1}{\sqrt{x^2+1}}\right) \quad (10)$$

$$= \cot^{-1}\left(\frac{1}{x}\right) \quad (11)$$

$$= \csc^{-1}\left(\frac{\sqrt{x^2+1}}{x}\right) \quad (12)$$

$$= \sec^{-1}\left(\sqrt{x^2+1}\right) \quad (13)$$

for $x \geq 0$.

In terms of the HYPERGEOMETRIC FUNCTION,

$$\tan^{-1} x = x \,_2F_1\left(1, \tfrac{1}{2}; \tfrac{3}{2}; -x^2\right) \quad (14)$$

$$= \frac{x}{1+x^2} \,_2F_1\left(1, 1; \tfrac{3}{2}; \frac{x^2}{1+x^2}\right) \quad (15)$$

(Castellanos 1988). Castellanos (1986, 1988) also gives some curious formulas in terms of the FIBO-NACCI NUMBERS,

$$\tan^{-1} x = \sum_{n=0}^{\infty} \frac{(-1^n)f_{2n+1}t^{2n+1}}{5^n(2n+1)} \quad (16)$$

$$= 5 \sum_{n=0}^{\infty} \frac{(-1)^n f_{2n+1}^2}{(2n+1)\left(u+\sqrt{u^2+1}\right)^{2n+1}} \quad (17)$$

$$= \sum_{n=0}^{\infty} \frac{(-1)^n 5^{n+2} F_{2n+1}^3}{(2n+1)\left(v+\sqrt{v^2+5}\right)^{2n+1}}, \quad (18)$$

where

$$t \equiv \frac{2x}{1+\sqrt{\dfrac{4x^2}{5}}} \quad (19)$$

$$u \equiv \frac{5}{4x}\left(1+\sqrt{1+\frac{24}{25}x^2}\right), \quad (20)$$

and v is the largest POSITIVE ROOT of

$$8xv^4 - 100v^3 - 450xv^2 + 875v + 625x = 0. \quad (21)$$

The inverse tangent satisfies the addition FORMULA

$$\tan^{-1} x + \tan^{-1} y = \tan^{-1}\left(\frac{x+y}{1-xy}\right) \quad (22)$$

as well as the more complicated FORMULAS

$$\tan^{-1}\left(\frac{1}{a-b}\right)$$
$$= \tan^{-1}\left(\frac{1}{a}\right) + \tan^{-1}\left(\frac{b}{a^2-ab+1}\right) \quad (23)$$

$$\tan^{-1}\left(\frac{1}{a}\right) = 2\tan^{-1}\left(\frac{1}{2a}\right) - \tan^{-1}\left(\frac{1}{4a^3+3a}\right) \quad (24)$$

$$\tan^{-1}\left(\frac{1}{p}\right) = \tan^{-1}\frac{1}{p+q} + \tan^{-1}\left(\frac{q}{p^2+pq+1}\right), \quad (25)$$

the latter of which was known to Euler. The inverse tangent FORMULAS are connected with many interesting approximations to PI

$$\tan^{-1}(1+x)$$
$$= \tfrac{1}{4}\pi + \tfrac{1}{2}x - \tfrac{1}{4}x^2 + \tfrac{1}{12}x^3 + \tfrac{1}{40}x^5 + \tfrac{1}{48}x^6 + \tfrac{1}{112}x^7 + \dots \quad (26)$$

Euler gave

$$\tan^{-1} x = \frac{y}{x}\left(\frac{2}{3}y + \frac{2 \cdot 4}{3 \cdot 5}y^2 + \frac{2 \cdot 4 \cdot 6}{3 \cdot 5 \cdot 7}y^3 + \dots\right), \quad (27)$$

where

$$y \equiv \frac{x^2}{1+x^2} \quad (28)$$

The inverse tangent has CONTINUED FRACTION representations

$$\tan^{-1} x = \cfrac{x}{1 + \cfrac{x^2}{3 + \cfrac{4x^2}{5 + \cfrac{9x^2}{7 + \cfrac{16x^2}{9 + \dots}}}}} \quad (29)$$

$$= \cfrac{x}{1 + \cfrac{x^2}{3 - x^2 + \cfrac{9x^2}{5 - 3x^2 + \cfrac{25x^2}{7 - 5x^2 + \dots}}}} \qquad (30)$$

To find $\tan^{-1} x$ numerically, the following ARITH-METIC-GEOMETRIC MEAN-like ALGORITHM can be used. Let

$$a_0 = (1 + x^2)^{-1/2} \qquad (31)$$

$$b_0 = 1. \qquad (32)$$

Then compute

$$a_{i+1} = \tfrac{1}{2}(a_i + b_i) \qquad (33)$$

$$b_{i+1} = \sqrt{a_{i+1} b_i}, \qquad (34)$$

and the inverse tangent is given by

$$\tan^{-1} x = \lim_{n \to \infty} \frac{x}{\sqrt{1 + x^2} a_n} \qquad (35)$$

(Acton 1990).

An inverse tangent $\tan^{-1} n$ with integral n is called reducible if it is expressible as a finite sum OF THE FORM

$$\tan^{-1} n = \sum_k f_k \tan^{-1} n_k, \qquad (36)$$

where f_k are POSITIVE or NEGATIVE INTEGERS and n_i are INTEGERS $< n$. $\tan^{-1} m$ is reducible IFF all the PRIME FACTORS of $1 + m^2$ occur among the PRIME FACTORS of $1 + n^2$ for $n = 1, \dots, m - 1$. A second NECESSARY and SUFFICIENT condition is that the largest PRIME factor of $1 + m^2$ is less than $2m$. Equivalent to the second condition is the statement that every GREGORY NUMBER $t_x = \cot^{-1} x$ can be uniquely expressed as a sum in terms of t_ms for which m is a STØRMER NUMBER (Conway and Guy 1996). To find this decomposition, write

$$\arg(1 + in) = \arg \prod_{k=1} (1 + n_k i)^{f_k}, \qquad (37)$$

so the ratio

$$r = \frac{\prod_{k=1} (1 + n_k i)^{f_k}}{1 + in} \qquad (38)$$

is a RATIONAL NUMBER. Equation (38) can also be written

$$r^2 (1 + n^2) = \prod_{k=1} (1 + n_k^2)^{f_k}. \qquad (39)$$

Writing (36) in the form

$$\tan^{-1} n = \sum_k f_k \tan^{-1} n_k + f \tan^{-1} 1 \qquad (40)$$

allows a direct conversion to a corresponding INVERSE COTANGENT FORMULA

$$\cot^{-1} n = \sum_{k=1} f_k \cot^{-1} n_k + c \cot^{-1} 1, \qquad (41)$$

where

$$c = 2 - f - 2 \sum_{k=1} f_r. \qquad (42)$$

Todd (1949) gives a table of decompositions of $\tan^{-1} n$ for $n \leq 342$. Conway and Guy (1996) give a similar table in terms of STØRMER NUMBERS.

Arndt and Gosper give the remarkable inverse tangent identity

$$\sin \left(\sum_{k=1}^{2n+1} \tan^{-1} a_k \right)$$

$$= \frac{(-1)^n}{2n+1} \frac{\sum_{k=1}^{2n+1} \prod_{j=1}^{2n+1} \left[a_j - \tan \left(\frac{\pi(j-k)}{2n+1} \right) \right]}{\sqrt{\prod_{j=1}^{2n+1} \left(a_j^2 + 1 \right)}}. \qquad (43)$$

See also INVERSE COTANGENT, TANGENT

References

Abramowitz, M. and Stegun, C. A. (Eds.). "Inverse Circular Functions." §4.4 in *Handbook of Mathematical Functions with Formulas, Graphs, and Mathematical Tables, 9th printing.* New York: Dover, pp. 79–83, 1972.

Acton, F. S. "The Arctangent." In *Numerical Methods that Work, upd. and rev.* Washington, DC: Math. Assoc. Amer., pp. 6–10, 1990.

Arndt, J. "Completely Useless Formulas." http://www.jjj.de/hfloat/hfloatpage.html#formulas.

Beyer, W. H. *CRC Standard Mathematical Tables, 28th ed.* Boca Raton, FL: CRC Press, pp. 142–143 and 220, 1987.

Castellanos, D. "Rapidly Converging Expansions with Fibonacci Coefficients." *Fib. Quart.* **24**, 70–82, 1986.

Castellanos, D. "The Ubiquitous Pi. Part I." *Math. Mag.* **61**, 67–98, 1988.

Conway, J. H. and Guy, R. K. "Størmer's Numbers." *The Book of Numbers.* New York: Springer-Verlag, pp. 245–248, 1996.

Hildebrand, J. D. "Arctan() Appreciation Home Page!" http://www.undergrad.math.uwaterloo.ca/~jdhildeb/arctan.html.

Salamin, G. Item 137 in Beeler, M.; Gosper, R. W.; and Schroeppel, R. *HAKMEM.* Cambridge, MA: MIT Artificial Intelligence Laboratory, Memo AIM-239, pp. 67–68, Feb. 1972.

Todd, J. "A Problem on Arc Tangent Relations." *Amer. Math. Monthly* **56**, 517–528, 1949.

Inverse Tangent Integral

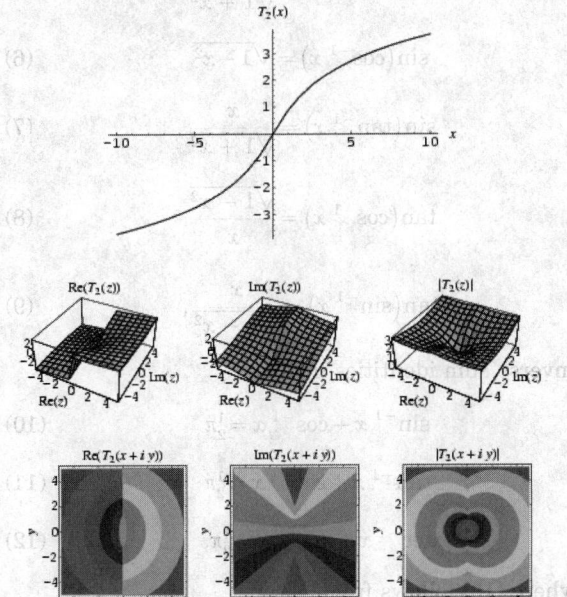

The inverse tangent integral $\mathrm{Ti}_2(x)$ is defined in terms of the DILOGARITHM $\mathrm{Li}_2(x)$ by

$$\mathrm{Li}_2(ix) = \tfrac{1}{4}\mathrm{Li}_2\left(-x^2\right) + i\,\mathrm{Ti}_2(x) \tag{1}$$

(Lewin 1958, p. 33). It has the series

$$\mathrm{Ti}_2(x) = \sum_{k=1}^{\infty}(-1)^{k-1}\frac{x^{2k-1}}{(2k-1)^2} \tag{2}$$

and gives in closed form the sum

$$\sum_{n=1}^{\infty}\frac{\sin[(4n-2)x]}{(2n-1)^2} = \mathrm{Ti}_2(\tan x) - x\ln(\tan x) \tag{3}$$

that was considered by Ramanujan (Lewin 1958, p. 39). The inverse tangent integral can be expressed in terms of the DILOGARITHM as

$$\mathrm{Ti}_2(x) = \frac{1}{2i}\left[\mathrm{Li}_2(ix) - \mathrm{Li}_2(-ix)\right], \tag{4}$$

in terms of LEGENDRE'S CHI-FUNCTION as

$$\mathrm{Ti}_2(x) = -i\chi_2(ix), \tag{5}$$

in terms of the LERCH TRANSCENDENT by

$$\mathrm{Ti}_2(x) = \tfrac{1}{4}x\Phi\left(-x^2, 2, \tfrac{1}{2}\right) \tag{6}$$

and as the integral

$$\mathrm{Ti}_2(x) = \int_0^x \frac{\tan^{-1}(x')}{x'}\,dx'. \tag{7}$$

$\mathrm{Ti}_2(x)$ has derivative

$$\frac{d\,\mathrm{Ti}_2(x)}{dx} = \frac{\tan^{-1}x}{x}. \tag{8}$$

It satisfies the identities

$$\mathrm{Ti}_2(x) - \mathrm{Ti}_2\left(\frac{1}{x}\right) = \tfrac{1}{2}\pi\,\mathrm{sgn}(x)\ln|x| \tag{9}$$

$$\tfrac{1}{2}\mathrm{Ti}_2\left(\frac{2x}{1-x^2}\right) = \mathrm{Ti}_2(x) + \mathrm{Ti}_2(-x,1) - \mathrm{Ti}_2(x,1), \tag{10}$$

where

$$\mathrm{Ti}_2(x,a) \equiv \int_0^x \frac{\tan^{-1}x'}{a+x'}\,dx' \tag{11}$$

is the generalized inverse tangent function. $\mathrm{Ti}_2(x)$ has the special value

$$\mathrm{Ti}_2(1) = K, \tag{12}$$

where K is CATALAN'S CONSTANT, and the functional relationships

$$3\,\mathrm{Ti}_2(1) - 2\,\mathrm{Ti}_2\left(\tfrac{1}{2}\right) - \mathrm{Ti}_2\left(\tfrac{1}{3}\right) - \tfrac{1}{2}\mathrm{Ti}_2\left(\tfrac{3}{4}\right) = \tfrac{1}{2}\pi\ln 2, \tag{13}$$

the two equivalent identities

$$3\,\mathrm{Ti}\left(2-\sqrt{3}\right) = 2\,\mathrm{Ti}_2(1) - \tfrac{1}{4}\pi\ln\left(2-\sqrt{3}\right) \tag{14}$$

$$\mathrm{Ti}_2\left(\tan\left(\tfrac{1}{12}\pi\right)\right)$$
$$= \tfrac{2}{3}\mathrm{Ti}_2\left(\tan\left(\tfrac{1}{4}\pi\right)\right) + \tfrac{1}{12}\pi\,\ln\left(\tan\left(\tfrac{1}{12}\pi\right)\right), \tag{15}$$

and

$$3\,\mathrm{Ti}\left(2+\sqrt{3}\right) = 2\,\mathrm{Ti}_2(1) + \tfrac{5}{4}\pi\,\ln\left(2+\sqrt{3}\right) \tag{16}$$

(Lewin 1958, p. 39). The triplication formula is given by

$$\tfrac{1}{3}\mathrm{Ti}_2\left(\frac{3x-x^3}{1-3x^2}\right) = \mathrm{Ti}_2(x) + \mathrm{Ti}_2\left(\frac{1-x\sqrt{3}}{\sqrt{3}+x}\right)$$
$$-\mathrm{Ti}_2\left(\frac{1+x\sqrt{3}}{\sqrt{3}-x}\right) + \tfrac{1}{6}\pi\,\ln\left(\frac{\left(\sqrt{3}+x\right)\left(1+x\sqrt{3}\right)}{\left(1-x\sqrt{3}\right)\left(\sqrt{3}-x\right)}\right), \tag{17}$$

which leads to

$$\mathrm{Ti}_2\left(\tan\left(\tfrac{1}{24}\pi\right)\right) - \mathrm{Ti}_2\left(\tan\left(\tfrac{5}{24}\pi\right)\right) + \tfrac{2}{3}\mathrm{Ti}_2\left(\tan\left(\tfrac{1}{8}\pi\right)\right)$$
$$+\tfrac{1}{6}\pi\,\ln\left(\frac{\tan\left(\tfrac{5}{24}\pi\right)}{\tan\left(\tfrac{1}{8}\pi\right)}\right) = 0 \tag{18}$$

and the algebraic form

$$\mathrm{Ti}_2\left(\frac{\sqrt{3}-\sqrt{2}}{\sqrt{2}+1}\right) - \mathrm{Ti}_2\left(\frac{\sqrt{3}-\sqrt{2}}{\sqrt{2}-1}\right) + \tfrac{2}{3}\mathrm{Ti}_2\left(\sqrt{2}-1\right)$$

$$= \tfrac{1}{6}\pi \, \ln\left(\frac{\sqrt{2}-1}{(\sqrt{3}-\sqrt{2})(\sqrt{2}+1)}\right) \tag{19}$$

(Lewin 1958, p. 41).

See also DILOGARITHM, LEGENDRE'S CHI-FUNCTION, LERCH TRANSCENDENT

References

Lewin, L. "The Inverse Tangent Integral" and "The Generalized Inverse Tangent Integral." Chs. 2–3 in *Dilogarithms and Associated Functions.* London: Macdonald, pp. 33–90, 1958.

Lewin, L. *Polylogarithms and Associated Functions.* Amsterdam, Netherlands: North-Holland, p. 45, 1981.

Nielsen, N. "Der Eulersche Dilogarithmus und seine Verallgemeinerungen." *Nova Acta (Leopold)* **90**, 121–212, 1909.

Inverse Trigonometric Functions

INVERSE FUNCTIONS of the TRIGONOMETRIC FUNCTIONS written $\cos^{-1} x$, $\cot^{-1} x$, $\csc^{-1} x$, $\sec^{-1} x$, $\sin^{-1} x$, and $\tan^{-1} x$. As noted by Feynman (1997), the notation $f^{-1}x$ is unfortunate because it conflicts with the common interpretation of a superscripted quantity as indicating a power, i.e., $f^{-1}x = (1/f)x = x/f$.

The inverse trigonometric functions are generally defined on the following domains.

Function	Domain
$\sin^{-1} x$	$-\tfrac{1}{2}\pi \leq y \leq \tfrac{1}{2}\pi$
$\cos^{-1} x$	$0 \leq y \leq \pi$
$\tan^{-1} x$	$-\tfrac{1}{2}\pi < y < \tfrac{1}{2}\pi$
$\csc^{-1} x$	$0 \leq y \leq \tfrac{1}{2}\pi$ or $\pi \leq y \leq \dfrac{3\pi}{2}$
$\sec^{-1} x$	$0 \leq y \leq \pi$
$\cot^{-1} x$	$0 \leq y \leq \tfrac{1}{2}\pi$ or $-\pi \leq y \leq -\tfrac{1}{2}\pi$

Inverse-forward identities are

$$\tan^{-1}(\cot x) = \tfrac{1}{2}\pi - x \tag{1}$$

$$\sin^{-1}(\cos x) = \tfrac{1}{2}\pi - x \tag{2}$$

$$\sec^{-1}(\csc x) = \tfrac{1}{2}\pi - x, \tag{3}$$

and forward-inverse identities are

$$\cos(\sin^{-1} x) = \sqrt{1-x^2} \tag{4}$$

$$\cos(\tan^{-1} x) = \frac{1}{\sqrt{1+x^2}} \tag{5}$$

$$\sin(\cos^{-1} x) = \sqrt{1-x^2} \tag{6}$$

$$\sin(\tan^{-1} x) = \frac{x}{\sqrt{1+x^2}} \tag{7}$$

$$\tan(\cos^{-1} x) = \frac{\sqrt{1-x^2}}{x} \tag{8}$$

$$\tan(\sin^{-1} x) = \frac{x}{\sqrt{1-x^2}}. \tag{9}$$

Inverse sum identities include

$$\sin^{-1} x + \cos^{-1} x = \tfrac{1}{2}\pi \tag{10}$$

$$\tan^{-1} x + \cot^{-1} x = \tfrac{1}{2}\pi \tag{11}$$

$$\sec^{-1} x + \csc^{-1} x = \tfrac{1}{2}\pi, \tag{12}$$

where (10) follows from

$$x = \sin(\sin^{-1} x) = \cos\left(\tfrac{1}{2}\pi - \sin^{-1} x\right). \tag{13}$$

Complex inverse identities in terms of LOGARITHMS include

$$\sin^{-1}(z) = -i \, \ln\left(iz \pm \sqrt{1-z^2}\right) \tag{14}$$

$$\cos^{-1}(z) = -i \, \ln\left(z \pm i\sqrt{1-z^2}\right) \tag{15}$$

$$\tan^{-1}(z) = -i \, \ln\left(\frac{1+iz}{\sqrt{1+z^2}}\right) \tag{16}$$

$$= \tfrac{1}{2}i \, \ln\left(\frac{1-iz}{1+iz}\right). \tag{17}$$

See also INVERSE COSECANT, INVERSE COSINE, INVERSE COTANGENT, INVERSE SECANT, INVERSE SINE, INVERSE TANGENT

References

Abramowitz, M. and Stegun, C. A. (Eds.). "Inverse Circular Functions." §4.4 in *Handbook of Mathematical Functions with Formulas, Graphs, and Mathematical Tables, 9th printing.* New York: Dover, pp. 79–83, 1972.

Feynman, R. P. and Leighton, R. "He Fixes Radios by Thinking!" In *'Surely You're Joking, Mr. Feynman!': Adventures of a Curious Character.* New York: W. W. Norton, p. 12, 1997.

Spanier, J. and Oldham, K. B. "Inverse Trigonometric Functions." Ch. 35 in *An Atlas of Functions.* Washington, DC: Hemisphere, pp. 331–341, 1987.

InverseEllipticNomeQ

INVERSE NOME

InverseJacobiCD
JACOBI ELLIPTIC FUNCTIONS

InverseJacobiCN
JACOBI ELLIPTIC FUNCTIONS

InverseJacobiCS
JACOBI ELLIPTIC FUNCTIONS

InverseJacobiDC
JACOBI ELLIPTIC FUNCTIONS

InverseJacobiDN
JACOBI ELLIPTIC FUNCTIONS

InverseJacobiDS
JACOBI ELLIPTIC FUNCTIONS

InverseJacobiNC
JACOBI ELLIPTIC FUNCTIONS

InverseJacobiND
JACOBI ELLIPTIC FUNCTIONS

InverseJacobiNS
JACOBI ELLIPTIC FUNCTIONS

InverseJacobiSC
JACOBI ELLIPTIC FUNCTIONS

InverseJacobiSD
JACOBI ELLIPTIC FUNCTIONS

InverseJacobiSN
JACOBI ELLIPTIC FUNCTIONS

Inversely Proportional
Two quantities y and x are said to be inversely proportional (or "in inverse proportion") if y is given by a constant multiple of $1/x$, i.e., $y = c/x$ for c a constant. This relationship is commonly written $y \propto x^{-1}$.

See also DIRECTLY PROPORTIONAL, PROPORTIONAL

Inversely Similar

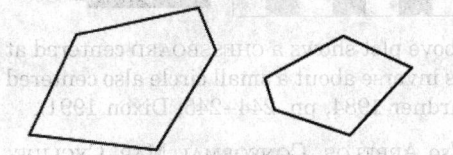

inversely similar

Two figures are said to be SIMILAR when all corre-
sponding ANGLES are equal, and are inversely similar when all corresponding ANGLES are equal and described in the opposite rotational sense.

See also DIRECTLY SIMILAR, HOMOTHETIC, SIMILAR

References
Lachlan, R. "Properties of Two Figures Inversely Similar." §220–222 in *An Elementary Treatise on Modern Pure Geometry*. London: Macmillian, pp. 138–139, 1893.

InverseWeierstrassP
WEIERSTRASS ELLIPTIC FUNCTION

Inversion

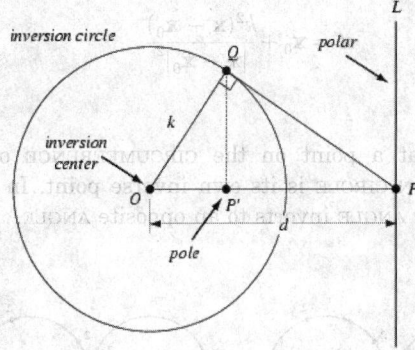

Inversion is the process of transforming points P to a corresponding set of points P' known as their INVERSE POINTS. Two points P and P' are said to be inverses with respect to an INVERSION CIRCLE having INVERSION CENTER $O = (x_0, y_0)$ and INVERSION RADIUS k if P' is the foot of the altitude of ΔOQP, where Q is a point on the circle such that $OQ \perp PQ$. The analogous notation of inversion can be carried to in 3-dimensional space with respect to an INVERSION SPHERE.

If P and P' are inverse points, then the line L through P and perpendicular to OP is sometimes called a "POLAR" with respect to point P, known as the "POLE". In addition, the curve to which a given curve is transformed under inversion is called its INVERSE CURVE (or more simply, its "inverse"). This sort of inversion was first systematically investigated by Jakob Steiner.

From similar triangles, it immediately follows that the inverse points P and P' obey

$$\frac{OP}{k} = \frac{k}{OP'}, \tag{1}$$

or

$$k^2 = OP \times OP' \tag{2}$$

(Coxeter 1969, p. 78), where the quantity k^2 is known as the POWER (Coxeter 1969, p. 81).

The general equation for the inverse of the point (x, y) relative to the INVERSION CIRCLE with INVERSION

CENTER (x_0, y_0) and INVERSION RADIUS k is given by

$$x' = x_0 + \frac{k^2(x - x_0)}{(x - x_0)^2 + (y - y_0)^2} \qquad (3)$$

$$y' = y_0 + \frac{k^2(y - y_0)}{(x - x_0)^2 + (y - y_0)^2}. \qquad (4)$$

In vector form,

$$\mathbf{x}' = \mathbf{x}_0 + \frac{k^2(\mathbf{x} - \mathbf{x}_0)}{|\mathbf{x} - \mathbf{x}_0|^2}. \qquad (5)$$

Note that a point on the CIRCUMFERENCE of the INVERSION CIRCLE is its own inverse point. In addition, any ANGLE inverts to an opposite ANGLE.

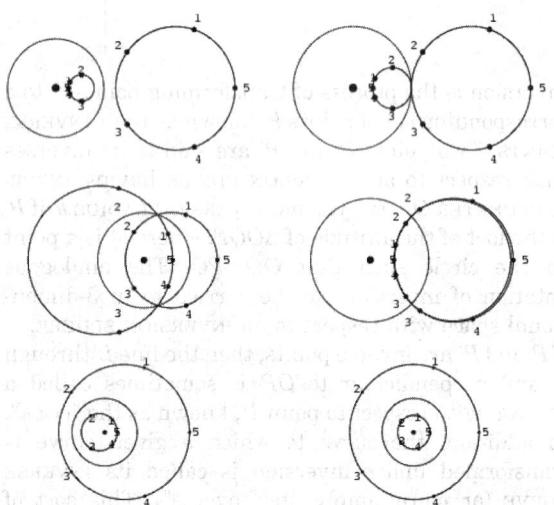

Treating LINES as CIRCLES of INFINITE RADIUS, all CIRCLES invert to CIRCLES (Lachlan 1893, p. 221). Furthermore, any two nonintersecting circles can be inverted into concentric circles by taking the INVERSION CENTER at one of the two so-called LIMITING POINTS of the two circles (Coxeter 1969), and any two circles can be inverted into themselves or into two equal circles (Casey 1888, pp. 97–98). ORTHOGONAL CIRCLES invert to ORTHOGONAL CIRCLES (Coxeter 1969). The INVERSION CIRCLE itself, circles orthogonal to it, and lines through the INVERSION CENTER are invariant under inversion. Furthermore, inversion is a CONFORMAL MAP, so angles are preserved.

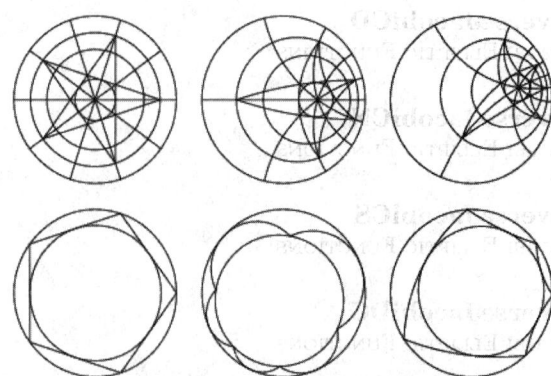

The property that inversion transforms circles and lines to circles or lines (and that inversion is conformal) makes it an extremely important tool of plane analytic geometry. By picking a suitable inversion circle, it is often possible to transform one geometric configuration into another simpler one in which a proof is more easily effected. The illustration above shows examples of the results of geometric inversion.

The inverse of a CIRCLE of RADIUS a with CENTER (x, y) with respect to an inversion circle with INVERSION CENTER (x_0, y_0) and INVERSION RADIUS k is another CIRCLE with CENTER

$$x' = x_0 + s(x - x_0) \qquad (6)$$

$$y' = y_0 + s(y - y_0) \qquad (7)$$

and RADIUS

$$r' = |s|a, \qquad (8)$$

where

$$s \equiv \frac{k^2}{(x - x_0)^2 + (y - y_0)^2 - a^2}. \qquad (9)$$

These equations can also be naturally extended to inversion with respect to a sphere in 3-dimensional space.

The above plot shows a CHESSBOARD centered at $(0, 0)$ and its inverse about a small circle also centered at $(0, 0)$ (Gardner 1984, pp. 244–245; Dixon 1991).

See also ARBELOS, CONFORMAL MAP, CYCLIDE, HEXLET, INVERSE CURVE, INVERSE POINTS, INVERSION CIRCLE, INVERSION OPERATION, INVERSION RADIUS, INVERSION SPHERE, INVERSIVE DISTANCE, INVERSIVE

GEOMETRY, LIMITING POINT, MIDCIRCLE, PAPPUS CHAIN, PEAUCELLIER INVERSOR, PERMUTATION INVERSION, POLAR, POLE (INVERSION), POWER (CIRCLE), RADICAL LINE, STEINER CHAIN, STEINER'S PORISM

References

Casey, J. "Theory of Inversion." §6.4 in *A Sequel to the First Six Books of the Elements of Euclid, Containing an Easy Introduction to Modern Geometry with Numerous Examples, 5th ed., rev. enl.* Dublin: Hodges, Figgis, & Co., pp. 95–112, 1888.

Coolidge, J. L. "Inversion." §1.2 in *A Treatise on the Geometry of the Circle and Sphere.* New York: Chelsea, pp. 21–30, 1971.

Courant, R. and Robbins, H. "Geometrical Transformations. Inversion." §3.4 in *What is Mathematics?: An Elementary Approach to Ideas and Methods, 2nd ed.* Oxford, England: Oxford University Press, pp. 140–146, 1996.

Coxeter, H. S. M. "Inversion in a Circle" and "Inversion of Lines and Circles." §6.1 and 6.3 in *Introduction to Geometry, 2nd ed.* New York: Wiley, pp. 77–83, 1969.

Coxeter, H. S. M. and Greitzer, S. L. "An Introduction to Inversive Geometry." Ch. 5 in *Geometry Revisited.* Washington, DC: Math. Assoc. Amer., pp. 103–131, 1967.

Darboux, G. *Leçons sur les systemes orthogonaux et les coordonnées curvilignes.* Paris: Gauthier-Villars, 1910.

Dixon, R. "Inverse Points and Mid-Circles." §1.6 in *Mathographics.* New York: Dover, pp. 62–73, 1991.

Durell, C. V. "Inversion." Ch. 10 in *Modern Geometry: The Straight Line and Circle.* London: Macmillan, pp. 105–120, 1928.

Fukagawa, H. and Pedoe, D. "Problems Soluble by Inversion." §1.8 in *Japanese Temple Geometry Problems.* Winnipeg, Manitoba, Canada: Charles Babbage Research Foundation, pp. 17–22 and 93–99, 1989.

Gardner, M. *The Sixth Book of Mathematical Games from Scientific American.* Chicago, IL: University of Chicago Press, 1984.

Jeans, J. H. *The Mathematical Theory of Electricity and Magnetism, 5th ed.* Cambridge, England: The University Press, 1925.

Johnson, R. A. *Modern Geometry: An Elementary Treatise on the Geometry of the Triangle and the Circle.* Boston, MA: Houghton Mifflin, pp. 43–57, 1929.

Kelvin, W. T. and Tait, P. G. *Principles of Mechanics and Dynamics, Vol. 2.* New York: Dover, p. 62, 1962.

Lachlan, R. "The Theory of Inversion." Ch. 14 in *An Elementary Treatise on Modern Pure Geometry.* London: Macmillian, pp. 218–236, 1893.

Liouville, J. "Note au sujet de l'article précédent." *J. math. pures appl.* **12**, 265–290, 1847.

Lockwood, E. H. "Inversion." Ch. 23 in *A Book of Curves.* Cambridge, England: Cambridge University Press, pp. 176–181, 1967.

Maxwell, J. C. *A Treatise on Electricity and Magnetism, Vol. 1, unabridged 3rd ed.* New York: Dover, 1954.

Maxwell, J. C. *A Treatise on Electricity and Magnetism, Vol. 2, unabridged 3rd ed.* New York: Dover, 1954.

Morley, F. and Morley, F. V. *Inversive Geometry.* Boston, MA: Ginn, 1933.

Ogilvy, C. S. *Excursions in Geometry.* New York: Dover, pp. 25–31, 1990.

Schmidt, H. *Die Inversion und ihre Anwendung.* Munich, Germany: Oldenbourg, 1950.

Thomson, W. "Extrait d'un lettre de M. William Thomson a M. Liouville." *J. math. pures appl.* **10**, 364–367, 1845.

Thomson, W. "Extrait de deux lettres adressées à M. Liouville." *J. math. pures appl.* **12**, 256, 1847.

Wangerin, A. S. 147 in *Theorie des Potentials und der Kugelfunktionen, Bd. II.* Berlin: de Gruyter, 1921.

Weber, E. *Electromagnetic Fields.* New York: Wiley, p. 244, 1950.

Weisstein, E. W. "Plane Geometry." MATHEMATICA NOTEBOOK PLANEGEOMETRY.M.

Wells, D. *The Penguin Dictionary of Curious and Interesting Geometry.* London: Penguin, pp. 119–121, 1991.

Inversion Center

The point that INVERSION OF A CURVE is performed with respect to.

See also INVERSE POINTS, INVERSION CIRCLE, INVERSION RADIUS, INVERSIVE DISTANCE, LIMITING POINT, POLAR, POLE (INVERSION), POWER (CIRCLE)

Inversion Circle

The CIRCLE with respect to which an INVERSE CURVE is computed or relative to which INVERSE POINTS are computed. In 3-D, INVERSE POINTS can be computed relative to an INVERSION SPHERE.

See also INVERSE POINTS, INVERSION CENTER, INVERSION RADIUS, INVERSION SPHERE, INVERSIVE DISTANCE, MIDCIRCLE, POLAR, POLE (INVERSION), POWER (CIRCLE)

Inversion Number

In DETERMINANT EXPANSION BY MINORS, the minimal number of TRANSPOSITIONS of adjacent columns in a SQUARE MATRIX needed to turn the matrix representing a permutation of $\{1, 2, \ldots, n\}$ into the IDENTITY MATRIX.

See also DETERMINANT EXPANSION BY MINORS, TRANSPOSITION

References

Bressoud, D. and Propp, J. "How the Alternating Sign Matrix Conjecture was Solved." *Not. Amer. Math. Soc.* **46**, 637–646.

Inversion Operation

The SYMMETRY OPERATION $(x, y, z) \to (-x, -y, -z)$. When used in conjunction with a ROTATION, it becomes an IMPROPER ROTATION.

Inversion Poset

A relation between permutations p and q that exists if there is a sequence of TRANSPOSITIONS such that each transposition increases the number of inversions (Stanton and White 1986; Skiena 1990, p. 162).

See also PERMUTATION

References

Skiena, S. *Implementing Discrete Mathematics: Combinatorics and Graph Theory with Mathematica.* Reading, MA: Addison-Wesley, 1990.

Stanton, D. W. and White, D. E. *Constructive Combinatorics.* New York: Springer-Verlag, 1986.

Inversion Radius

The RADIUS used in performing an INVERSION with respect to an INVERSION CIRCLE.

See also INVERSE POINTS, INVERSION CENTER, INVERSION CIRCLE, INVERSIVE DISTANCE, POLAR, POLE (INVERSION), POWER (CIRCLE)

Inversion Semigroup

INVERSE SEMIGROUP

Inversion Sphere

The SPHERE with respect to which INVERSE POINTS are computed (i.e., with respect to which geometrical INVERSION is performed). For example, the CYCLIDES are inversions in a sphere of TORI. The center of the inversion sphere is called the INVERSION CENTER, and its radius is called the INVERSION RADIUS. When DUAL POLYHEDRA are being considered, the inversion sphere is commonly called the MIDSPHERE (or intersphere, or reciprocating sphere).

In 2-D, the inversion sphere collapses to an INVERSION CIRCLE.

See also CYCLIDE, INVERSE POINTS, INVERSION, INVERSION CENTER, INVERSION CIRCLE, INVERSION RADIUS, INVERSIVE DISTANCE, MIDCIRCLE, MIDSPHERE, POLAR, POLE (INVERSION), POWER (CIRCLE)

Inversion Statistic

See also WEIGHTED INVERSION STATISTIC

References

Milne, S. and Degenhardt, S. "Weighted Inversion Statistics and Their Symmetry Group." To appear in *J. Combin. Th. Ser. A.* http://www.math.ohio-state.edu/~milne/preprints.html.

Inversion Vector

The number of elements greater than i to the left of i in a PERMUTATION gives the ith element of the inversion vector (Skiena 1990, p. 27). A PERMUTATION p can be converted to an inversion vector using `ToInversionVector[p]` in the *Mathematica* add-on package `DiscreteMath`Combinatorica`` (which can be loaded with the command `<<DiscreteMath``), and an inversion vector v can be converted to a PERMUTATION using `ToInversionVector[v]`.

See also PERMUTATION INVERSION

References

Skiena, S. "Inversion Vectors." §1.3.1 in *Implementing Discrete Mathematics: Combinatorics and Graph Theory with Mathematica.* Reading, MA: Addison-Wesley, pp. 27–28, 1990.
Thompkins, C. B. *Machine Attacks on Problems Whose Variables are Permutations.* Providence, RI: Amer. Math. Soc., p. 203, 1956.

Inversive Distance

The inversive distance is the NATURAL LOGARITHM of the ratio of two concentric circles into which the given circles can be inverted. Let c be the distance between the centers of two nonintersecting CIRCLES of RADII a and $b < a$. Then the inversive distance is

$$\delta = \cosh^{-1}\left|\frac{a^2 + b^2 - c^2}{2ab}\right|$$

(Coxeter and Greitzer 1967).

The inversive distance between the SODDY CIRCLES is given by

$$\delta = 2\cosh^{-1} 2,$$

and the CIRCUMCIRCLE and INCIRCLE of a TRIANGLE with CIRCUMRADIUS R and INRADIUS r are at inversive distance

$$\delta = 2\sinh^{-1}\left(\frac{1}{2}\sqrt{\frac{r}{R}}\right)$$

(Coxeter and Greitzer 1967, pp. 130–131).

References

Coxeter, H. S. M. and Greitzer, S. L. *Geometry Revisited.* Washington, DC: Math. Assoc. Amer., pp. 123–124 and 127–131, 1967.

Inversive Geometry

The GEOMETRY resulting from the application of the INVERSION operation. It can be especially powerful for solving apparently difficult problems such as STEINER'S PORISM and APOLLONIUS' PROBLEM.

See also HEXLET, INVERSE CURVE, INVERSION, PEAUCELLIER INVERSOR, POLAR, POLE (INVERSION), POWER (CIRCLE), RADICAL LINE

References

Coxeter, H. S. M. and Greitzer, S. L. "An Introduction to Inversive Geometry." Ch. 5 in *Geometry Revisited.* Washington, DC: Math. Assoc. Amer., pp. 103–131, 1967.
Ogilvy, C. S. "Inversive Geometry" and "Applications of Inversive Geometry." Chs. 3–4 in *Excursions in Geometry.* New York: Dover, pp. 24–55, 1990.
Morley, F. and Morley, F. V. *Inversive Geometry.* Boston, MA: Ginn, 1933.

Inverted Funnel

FUNNEL, SINCLAIR'S SOAP FILM PROBLEM

Inverted Snub Dodecadodecahedron

The UNIFORM POLYHEDRON U_{60} whose DUAL POLYHEDRON is the MEDIAL INVERTED PENTAGONAL HEXECONTAHEDRON. It has WYTHOFF SYMBOL $|2\frac{5}{3}5$. Its faces are $12\{\frac{5}{3}\} + 60\{3\} + 12\{5\}$. It has CIRCUMRADIUS for unit edge length of

$$R \approx 0.8516302.$$

References

Wenninger, M. J. *Polyhedron Models*. Cambridge, England: Cambridge University Press, pp. 180–182, 1989.

Invertible Knot

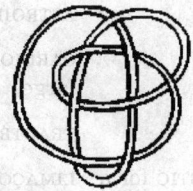

A knot which can be deformed via an AMBIENT ISOTOPY into itself but with the orientation reversed. No noninvertible knots were known until Trotter (1964) discovered an infinite family, the smallest of which had nine crossings. The simplest noninvertible knot is 08-017, illustrated above. The following table gives the numbers of noninvertible and invertible knots of n crossings.

type	Sloane	counts
noninvertible	A052403	0, 0, 0, 0, 0, 0, 0, 1, 2, 33, 187, 1144, 6919, 38118, 226581, 1309875, ...
invertible	A052402	0, 0, 1, 1, 2, 3, 7, 20, 47, 132, 365, 1032, 3069, 8854, 26712, 78830, ...

No general technique is known for determining if a KNOT is invertible. Burde and Zieschang (1985) give a tabulation from which it is possible to extract the noninvertible knots up to 10 crossings.

See also AMPHICHIRAL KNOT

References

Burde, G. and Zieschang, H. *Knots*. Berlin: de Gruyter, 1985.
Hoste, J.; Thistlethwaite, M.; and Weeks, J. "The First 1,701,936 Knots." *Math. Intell.* **20**, 33–48, Fall 1998.
Sloane, N. J. A. Sequences A052402 and A052403 in "An On-Line Version of the Encyclopedia of Integer Sequences." http://www.research.att.com/~njas/sequences/eisonline.html.
Trotter, H. F. "Noninvertible Knots Exist." *Topology* **2**, 275–280, 1964.

Invertible Linear Map

An invertible linear transformation $T : V \to W$ is a map between VECTOR SPACES V and W with an inverse map which is also a LINEAR TRANSFORMATION. When T is given by MATRIX MULTIPLICATION, i.e., $T(v) = \mathsf{A}v$, then T is invertible IFF A is a INVERTIBLE MATRIX. Note that the dimensions of V and W must be the same.

See also INVERTIBLE MATRIX, LINEAR TRANSFORMATION, MATRIX, VECTOR SPACE

Invertible Linear Transformation

INVERTIBLE LINEAR MAP

Invertible Matrix

NONSINGULAR MATRIX

Invertible Polynomial Map

A POLYNOMIAL MAP $\phi_{\mathbf{f}}$, with $\mathbf{f} = (f_1, \ldots, f_n) \in (K[X_1, \ldots, X_n])^m$ in a FIELD K is called invertible if there exist $g_1, \ldots, g_m \in K[X_1, \ldots, x_n]$ such that $g_i(f_1, \ldots, f_n) = X_i$ for $1 \leq n \leq n$ so that $\phi_{\mathbf{g}} \circ \phi_{\mathbf{f}} = \mathrm{id}_{k^n}$ (Becker and Weispfenning 1993, p. 330). GRÖBNER BASES provide a means to decide for given $\mathbf{f}$ whether or not $\phi_{\mathbf{f}}$ is invertible.

See also JACOBIAN CONJECTURE, POLYNOMIAL MAP

References

Becker, T. and Weispfenning, V. *Gröbner Bases: A Computational Approach to Commutative Algebra*. New York: Springer-Verlag, p. 330, 1993.

Involuntary

A LINEAR TRANSFORMATION of period two. Since a LINEAR TRANSFORMATION has the form,

$$\lambda' = \frac{\alpha\lambda + \beta}{\gamma\lambda + \delta}, \tag{1}$$

applying the transformation a second time gives

$$\lambda'' + \frac{\alpha\lambda' + \beta}{\gamma\lambda' + \delta} = \frac{(\alpha^2 + \beta\gamma)\lambda + \beta(\alpha + \delta)}{(\alpha + \delta)\gamma\lambda + \beta\gamma + \delta^2}, \tag{2}$$

For an involuntary, $\lambda'' = \lambda$, so

$$\gamma(\alpha + \delta)\lambda^2 + (\delta^2 - \alpha^2)\lambda - (\alpha + \delta)\beta = 0. \tag{3}$$

Since each COEFFICIENT must vanish separately,

$$\alpha\gamma + \gamma\delta = 0 \tag{4}$$

$$\delta^2 - \alpha^2 = 0 \tag{5}$$

$$\alpha\beta + \beta\delta = 0. \tag{6}$$

The first equation gives $\delta = \pm\alpha$. Taking $\delta = \alpha$ would require $\gamma = \beta = 0$, giving $\lambda = \lambda'$, the identity transformation. Taking $\delta = -\alpha$ gives $\delta = -\alpha$, so

$$\lambda' = \frac{\alpha\lambda + \beta}{\gamma\lambda - \alpha} \tag{7}$$

the general form of an INVOLUTION.

See also CROSS-RATIO, INVOLUTION (LINE)

References

Woods, F. S. *Higher Geometry: An Introduction to Advanced Methods in Analytic Geometry.* New York: Dover, pp. 14–15, 1961.

Involute

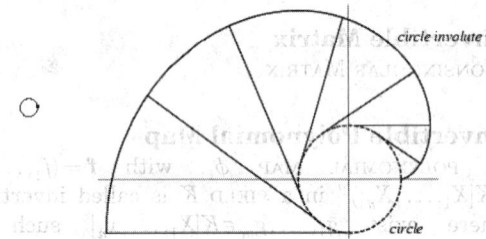

Attach a string to a point on a curve. Extend the string so that it is tangent to the curve at the point of attachment. Then wind the string up, keeping it always taut. The LOCUS of points traced out by the end of the string is the involute of the original curve, and the original curve is called the EVOLUTE of its involute. Although a curve has a unique EVOLUTE, it has infinitely many involutes corresponding to different choices of initial point. An involute can also be thought of as any curve ORTHOGONAL to all the TANGENTS to a given curve.

The equation of the involute is

$$\mathbf{r}_i = \mathbf{r} - s\hat{\mathbf{T}}, \tag{1}$$

where $\hat{\mathbf{T}}$ is the TANGENT VECTOR

$$\hat{\mathbf{T}} = \frac{\dfrac{d\mathbf{r}}{dt}}{\left|\dfrac{d\mathbf{r}}{dt}\right|} \tag{2}$$

and s is the ARC LENGTH

$$s = \int ds = \int \frac{ds}{dt} dt = \int \frac{\sqrt{ds^2}}{dt} dt = \int \sqrt{f'^2 + g'^2}\, dt. \tag{3}$$

This can be written for a parametrically represented function $(f(t), g(t))$ as

$$x(t) = f - \frac{sf'}{\sqrt{f'^2 + g'^2}} \tag{4}$$

$$y(t) = g - \frac{sg'}{\sqrt{f'^2 + g'^2}}. \tag{5}$$

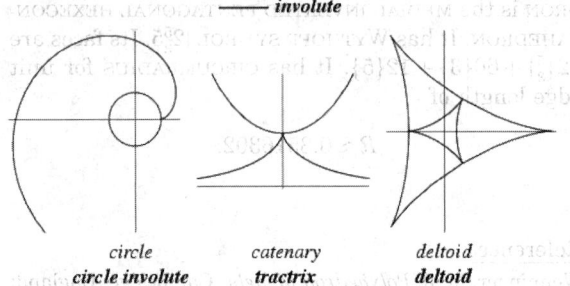

circle		catenary		deltoid	
circle involute		**tractrix**		**deltoid**	

The following table lists the involutes of some common curves, some of which are illustrated above.

Curve	Involute
ASTROID	ASTROID 1/2 as large
CARDIOID	CARDIOID 3 times as large
CATENARY	TRACTRIX
CIRCLE CATACAUSTIC for a point source	LIMAÇON
CIRCLE	CIRCLE INVOLUTE (a SPIRAL)
CYCLOID	equal CYCLOID
DELTOID	DELTOID 1/3 as large
ELLIPSE	ELLIPSE INVOLUTE
EPICYCLOID	reduced EPICYCLOID
HYPOCYCLOID	similar HYPOCYCLOID
LOGARITHMIC SPIRAL	equal LOGARITHMIC SPIRAL
NEILE'S PARABOLA	PARABOLA
NEPHROID	CAYLEY'S SEXTIC
NEPHROID	NEPHROID 2 times as large

See also ENVELOPE, EVOLUTE, HUMBERT'S THEOREM, ROULETTE

Involution

References

Cundy, H. and Rollett, A. "Roulettes and Involutes." §2.6 in *Mathematical Models, 3rd ed.* Stradbroke, England: Tarquin Pub., pp. 46–55, 1989.

Dixon, R. "String Drawings." Ch. 2 in *Mathographics.* New York: Dover, pp. 75–78, 1991.

Gray, A. "Involutes." §5.4 in *Modern Differential Geometry of Curves and Surfaces with Mathematica, 2nd ed.* Boca Raton, FL: CRC Press, pp. 103–107, 1997.

Lawrence, J. D. *A Catalog of Special Plane Curves.* New York: Dover, pp. 40–42 and 202, 1972.

Lockwood, E. H. "Evolutes and Involutes." Ch. 21 in *A Book of Curves.* Cambridge, England: Cambridge University Press, pp. 166–171, 1967.

Pappas, T. "The Involute." *The Joy of Mathematics.* San Carlos, CA: Wide World Publ./Tetra, p. 187, 1989.

Yates, R. C. "Involutes." *A Handbook on Curves and Their Properties.* Ann Arbor, MI: J. W. Edwards, pp. 135–137, 1952.

Involution

An OPERATOR of period 2, i.e., an OPERATOR * which satisfies $((a)^*)^* = a$.

Involution (Group)

An element of order 2 in a GROUP (i.e., an element A of a GROUP such that $A^2 = I$, where I is the IDENTITY ELEMENT).

See also GROUP, IDENTITY ELEMENT

Involution (Line)

Pairs of points of a line, the product of whose distances from a FIXED POINT is a given constant. This is more concisely defined as a PROJECTIVITY of period two.

If $\{AA', BB', CC'\}$ is a range in involution, then the ranges $\{AA', BC\}$ and $\{A'A, B'C'\}$ are EQUICROSS, and conversely.

See also EQUICROSS, INVOLUTORY

References

Casey, J. *A Sequel to the First Six Books of the Elements of Euclid, Containing an Easy Introduction to Modern Geometry with Numerous Examples, 5th ed., rev. enl.* Dublin: Hodges, Figgis, & Co., p. 133, 1888.

Lachlan, R. "Theory of Involutions" and "Involution." Ch. 5 and §426–427 in *An Elementary Treatise on Modern Pure Geometry.* London: Macmillian, pp. 272–274, 1893.

Involution (Operator)

An OPERATOR of period 2, i.e., an OPERATOR $\bar{3}$ which satisfies

$$\sin^{-1} x = x + \tfrac{1}{6}x^3 + \tfrac{3}{40}x^5 + \tfrac{5}{112}x^7 + \tfrac{35}{1152}x^9 + \dots.$$

Involution (Permutation)

An involution of a SET S is a PERMUTATION of S which does not contain any CYCLES of length > 2 (i.e., it consists exclusively of fixed points and TRANSPOSITIONS). Involutions are in one-to-one correspondence with self-conjugate permutations (i.e, permutations that are their own INVERSE PERMUTATION). For example, the unique permutation involution on 1 element is $\{1\}$, the two involution permutations on 2 elements are $\{1, 2\}$ and $\{2, 1\}$, and the four involution permutations on 3 elements are $\{1, 2, 3\}$, $\{1, 3, 2\}$, $\{2, 1, 3\}$, and $\{3, 2, 1\}$. A PERMUTATION p can be tested to determine if it is a permutation using `InvolutionQ[p]` in the *Mathematica* add-on package `DiscreteMath`Combinatorica`` (which can be loaded with the command `< <DiscreteMath``).

The PERMUTATION MATRICES of an involution are SYMMETRIC. The number of involutions on n elements is the same as the number of distinct YOUNG TABLEAUX on n elements (Skiena 1990, p. 32).

In general, the number of involution permutations on n letters is given by the formula

$$I(n) = 1 + \sum_{k=0}^{\lfloor (n-2)/2 \rfloor} \frac{1}{(k+1)!} \binom{n-2i}{2}, \tag{1}$$

where $\binom{n}{k}$ is a BINOMIAL COEFFICIENT (Muir 1960, p. 5), or alternatively by

$$I(n) = n! \sum_{k=0}^{\lfloor n \rfloor} \frac{1}{2^k k! (n-2k)!} \tag{2}$$

(Skiena 1990, p. 32). Although the number of involutions on n symbols cannot be expressed as a fixed number of hypergeometric terms (Petkovsek *et al.* 1996, p. 160), it can be written in terms of the CONFLUENT HYPERGEOMETRIC FUNCTION OF THE SECOND KIND $U(a, b, z)$ as

$$I(n) = (-i)^n 2^{n/2} U\left(-\tfrac{1}{2}n, \tfrac{1}{2}, -\tfrac{1}{2}\right). \tag{3}$$

Breaking this up into n even and odd gives

$$I(n) = \begin{cases} (-2)^k U\left(-k, \tfrac{1}{2}, -\tfrac{1}{2}\right) & \text{for } n = 2k \\ (-2)^k U\left(-k, \tfrac{3}{2}, -\tfrac{1}{2}\right) & \text{for } n = 2k+1 \end{cases} \tag{4}$$

The number of involutions $I(n)$ of a SET containing the first n integers is given by the RECURRENCE RELATION

$$I(n) = I(n-1) + (n-1)I(n-2) \tag{5}$$

(Muir 1960, pp. 3–7; Skiena 1990, p. 32). For $n = 1, 2, \dots$, the first few values of $I(n)$ are 1, 2, 4, 10, 26, 76, ... (Sloane's A000085).

See also CYCLE (PERMUTATION), INVERSE PERMUTATION, PERMUTATION, PERMUTATION MATRIX, YOUNG TABLEAU

References

Knuth, D. E. *The Art of Computer Programming, Vol. 3: Sorting and Searching, 2nd ed.* Reading, MA: Addison-Wesley, 1998.

Muir, T. "On Self-Conjugate Permutations." *Proc. Royal Soc. Edinburgh* **17**, 7–22, 1889.

Muir, T. *A Treatise on the Theory of Determinants.* New York: Dover, 1960.

Petkovsek, M.; Wilf, H. S.; and Zeilberger, D. *A = B.* Wellesley, MA: A. K. Peters, 1996.

Ruskey, F. "Information on Involutions." http://www.theory.csc.uvic.ca/~cos/inf/perm/Involutions.html.

Skiena, S. "Involutions." §1.4.1 in *Implementing Discrete Mathematics: Combinatorics and Graph Theory with Mathematica.* Reading, MA: Addison-Wesley, pp. 32–33, 1990.

Sloane, N. J. A. Sequences A000085/M1221 in "An On-Line Version of the Encyclopedia of Integer Sequences." http://www.research.att.com/~njas/sequences/eisonline.html.

Involution (Transformation)

A TRANSFORMATION of period 2.

Involution Principle

GARSIA-MILNE INVOLUTION PRINCIPLE

Involutory

A LINEAR TRANSFORMATION of period two. Since a LINEAR TRANSFORMATION has the form,

$$\lambda' = \frac{\alpha\lambda + \beta}{\gamma\lambda + \delta}, \tag{1}$$

applying the transformation a second time gives

$$\lambda'' = \frac{\alpha\lambda' + \beta}{\gamma\lambda' + \delta} = \frac{(\alpha^2 + \beta\gamma)\lambda + (\beta\alpha + \delta)}{(\alpha + \delta)\gamma\lambda + \beta\gamma + \delta^2}. \tag{2}$$

For an involutory, $\lambda'' = \lambda$, so

$$\gamma(\alpha + \delta)\lambda^2 + (\delta^2 - \alpha^2)\lambda - (\alpha + \delta)\beta = 0. \tag{3}$$

Since each COEFFICIENT must vanish separately,

$$\gamma(\alpha + \delta) = 0 \tag{4}$$

$$\delta^2 - \alpha^2 = 0 \tag{5}$$

$$\beta(\alpha + \delta) = 0. \tag{6}$$

Equation (5) requires $\delta = \pm\alpha$. Taking $\delta = \alpha$ in turn requires that $\gamma = \beta = 0$, giving $\lambda = \lambda'$, i.e., the IDENTITY MAP, while taking $\delta = -\alpha$ gives $\delta = -\alpha$, so

$$\lambda' = \frac{\alpha\lambda + \beta}{\gamma\lambda - \alpha}, \tag{7}$$

which is the general form of an INVOLUTION.

See also CROSS-RATIO, INVOLUTION (LINE)

References

Woods, F. S. *Higher Geometry: An Introduction to Advanced Methods in Analytic Geometry.* New York: Dover, pp. 14–15, 1961.

Involutory Matrix

A SQUARE MATRIX A such that $A^2 = I$, where I is the IDENTITY MATRIX. An involutory matrix is its own MATRIX INVERSE.

References

Ayres, F. Jr. *Theory and Problems of Matrices.* New York: Schaum, p. 11, 1962.

Irradiation Illusion

The ILLUSION shown above which was discovered by Helmholtz in the 19th century. Despite the fact that the two above figures are identical in size, the white hole looks bigger than the black one in this ILLUSION.

See also ILLUSION

References

Pappas, T. "Irradiation Optical Illusion." *The Joy of Mathematics.* San Carlos, CA: Wide World Publ./Tetra, p. 199, 1989.

Irrational Number

A number which cannot be expressed as a FRACTION p/q for any INTEGERS p and q. The most famous irrational number is $\sqrt{2}$, sometimes called PYTHAGORAS'S CONSTANT. Legend has it that the Pythagorean philosopher Hippasus used geometric methods to demonstrate the irrationality of $\sqrt{2}$ while at sea and, upon notifying his comrades of his great discovery, was immediately thrown overboard by the fanatic Pythagoreans. Other examples include $\sqrt{3}$, e, π, etc.

Every TRANSCENDENTAL NUMBER is irrational. Numbers OF THE FORM $n^{1/m}$ are irrational unless n is the mth POWER of an INTEGER. Numbers OF THE FORM $\log_n m$, where log is the LOGARITHM, are irrational if m and n are INTEGERS, one of which has a PRIME factor which the other lacks. e^r is irrational for rational $r \neq 0$. $\cos r$ is irrational for every nonnegative rational number r (Niven 1956, Stevens 1999), and $\cos(\theta)$ (for θ measured in degrees) is irrational for every rational $0° < \theta < 90°$ with the exception of $\theta = 60°$ (Niven 1956). $\tan r$ is irrational for every rational $r \neq 0$ (Stevens 1999).

The irrationality of e was proven by Lambert in 1761; for the general case, see Hardy and Wright (1979,

p. 46). π^n is irrational for POSITIVE integral n. The irrationality of PI itself was proven by Lambert in 1760; for the general case, see Hardy and Wright (1979, p. 47). APÉRY'S CONSTANT $\zeta(3)$ (where $\zeta(z)$ is the RIEMANN ZETA FUNCTION) was proved irrational by Apéry (Apéry 1979, van der Poorten 1979). In addition, T. Rivoal (2000) recently proved that there are infinitely many integers n such that $\zeta(2n+1)$ is irrational.

From GELFOND'S THEOREM, a number OF THE FORM a^b is TRANSCENDENTAL (and therefore irrational) if a is ALGEBRAIC $\neq 0$, 1 and b is irrational and ALGEBRAIC. This establishes the irrationality of e^π (since $(-1)^{-i} = (e^{i\pi})^{-i} = e^\pi)$), $2^{\sqrt{2}}$, and $e\pi$. Nesterenko (1996) proved that $\pi + e^\pi$ is irrational. In fact, he proved that π, e^π and $\Gamma(1/4)$ are ALGEBRAICALLY INDEPENDENT, but it was not previously known that $\pi + e^\pi$ was irrational.

Given a POLYNOMIAL equation

$$x^m + c_{m-1}x^{m-1} + \ldots + c_0, \qquad (1)$$

where c_i are INTEGERS, the roots x_i are either integral or irrational. If $\cos(2\theta)$ is irrational, then so are $\cos\theta$, $\sin\theta$, and $\tan\theta$.

Irrationality has not yet been established for 2^e, π^e, $\pi^{\sqrt{2}}$, or γ (where γ is the EULER-MASCHERONI CONSTANT).

QUADRATIC SURDS are irrational numbers which have periodic CONTINUED FRACTIONS.

HURWITZ'S IRRATIONAL NUMBER THEOREM gives bounds OF THE FORM

$$\left| \alpha - \frac{p}{q} < \frac{1}{l_n q^2} \right| \qquad (2)$$

for the best rational approximation possible for an arbitrary irrational number α, where the l_n are called LAGRANGE NUMBERS and get steadily larger for each "bad" set of irrational numbers which is excluded.

The SERIES

$$\sum_{n=1}^{\infty} \frac{\sigma_k(n)}{n!}, \qquad (3)$$

where $\sigma_k(n)$ is the DIVISOR FUNCTION, is irrational for $k = 1$ and 2, and the series

$$\sum_{n=1}^{\infty} \frac{1}{2^n - 1} = \sum_{n=1}^{\infty} \frac{d(n)}{2^n}, \qquad (4)$$

where $d(n)$ is the number of divisors of n, is also irrational (Guy 1994).

See also ALGEBRAIC INTEGER, ALGEBRAIC NUMBER, ALMOST INTEGER, DIRICHLET FUNCTION, E, FERGUSON-FORCADE ALGORITHM, GELFOND'S THEOREM, HURWITZ'S IRRATIONAL NUMBER THEOREM, NEAR NOBLE NUMBER, NOBLE NUMBER, PI, PYTHAGORAS'S CONSTANT, PYTHAGORAS'S THEOREM, Q-HARMONIC SERIES, QUADRATIC IRRATIONAL NUMBER, RATIONAL NUMBER, SEGRE'S THEOREM, TRANSCENDENTAL NUMBER

References

Apéry, R. "Irrationalité de $\zeta(2)$ et $\zeta(3)$." *Astérisque* **61**, 11–13, 1979.

Courant, R. and Robbins, H. "Incommensurable Segments, Irrational Numbers, and the Concept of Limit." §2.2 in *What is Mathematics?: An Elementary Approach to Ideas and Methods, 2nd ed.* Oxford, England: Oxford University Press, pp. 58–61, 1996.

Guy, R. K. "Some Irrational Series." §B14 in *Unsolved Problems in Number Theory, 2nd ed.* New York: Springer-Verlag, p. 69, 1994.

Hardy, G. H. and Wright, E. M. *An Introduction to the Theory of Numbers, 5th ed.* Oxford, England: Clarendon Press, 1979.

Manning, H. P. *Irrational Numbers and Their Representation by Sequences and Series.* New York: Wiley, 1906.

Nagell, T. "Irrational Numbers" and "Irrationality of the numbers e and π." §12–13 in *Introduction to Number Theory.* New York: Wiley, pp. 38–40, 1951.

Nesterenko, Yu. "Modular Functions and Transcendence Problems." *C. R. Acad. Sci. Paris Sér. I Math.* **322**, 909–914, 1996.

Nesterenko, Yu. V. "Modular Functions and Transcendence Questions." *Mat. Sb.* **187**, 65–96, 1996.

Niven, I. M. *Irrational Numbers.* New York: Wiley, 1956.

Niven, I. M. *Numbers: Rational and Irrational.* New York: Random House, 1961.

Pappas, T. "Irrational Numbers & the Pythagoras Theorem." *The Joy of Mathematics.* San Carlos, CA: Wide World Publ./Tetra, pp. 98–99, 1989.

Rivoal, T. "Irrationalité d'une infinité de valeurs de la fonction Zeta aux entiers impairs." Preprint 2000–9. http://www.math.unicaen.fr/~leclerc/publi_labo/2000/index2000.html.

Stevens, J. "Zur Irrationalität von π." *Mitt. Math. Ges. Hamburg* **18**, 151–158, 1999.

van der Poorten, A. "A Proof that Euler Missed... Apéry's Proof of the Irrationality of $\zeta(3)$." *Math. Intel.* **1**, 196–203, 1979.

Weisstein, E. W. "Books about Irrational Numbers." http://www.treasure-troves.com/books/IrrationalNumbers.html.

Irrationality Measure

N.B. A detailed online essay by S. Finch was the starting point for this entry.

Let x be a REAL NUMBER, and let R be the SET of POSITIVE REAL NUMBERS for which

$$\left| x - \frac{p}{q} \right| < q^{-r} \qquad (1)$$

has (at most) finitely many solutions p/q for p and q INTEGERS. Then the irrationality measure, sometimes called the Liouville-Roth constant, is defined as the threshold at which LIOUVILLE'S APPROXIMATION THEOREM kicks in and x is no longer approximable by RATIONAL NUMBERS,

$$r(x) \equiv \inf_{r \in R} r. \qquad (2)$$

There are three regimes:

$$\begin{cases} r(x) = 1 & x \text{ is rational} \\ r(x) = 2 & x \text{ is algebraic} \\ r(x) \geq 3 & x \text{ is transcendental} \end{cases} \qquad (3)$$

Exact values include

$$r(L) = \infty$$

$$r(e) = 2,$$

where L is LIOUVILLE'S CONSTANT. The best known upper bounds for other common constants are summarized in the following table, where $\zeta(3)$ is APÉRY'S CONSTANT, $Ln_q(2)$ and $h_q(1)$ are Q-HARMONIC SERIES, and the lower bounds are 2.

constant x	upper bound	reference
π	8.0161	Hata (1992)
π^2	6.3489	Hata (1992)
$\ln 2$	4.13	
$\zeta(3)$	7.377956	Hata (2000)
$Ln_q(2)$	4.80	Amdeberhan and Zeilberger (1998)
$h_q(1)$	4.80	Amdeberhan and Zeilberger (1998)

See also LIOUVILLE'S APPROXIMATION THEOREM, ROTH'S THEOREM, THUE-SIEGEL-ROTH THEOREM

References

Amdeberhan, T. and Zeilberger, D. "q-Apéry Irrationality Proofs by q-WZ Pairs." *Adv. Appl. Math.* **20**, 275–283, 1998.

Borwein, J. M. and Borwein, P. B. *Pi & the AGM: A Study in Analytic Number Theory and Computational Complexity.* New York: Wiley, 1987.

Finch, S. "Favorite Mathematical Constants." http://www.mathsoft.com/asolve/constant/lvlrth/lvlrth.html.

Hardy, G. H. and Wright, E. M. *An Introduction to the Theory of Numbers, 5th ed.* Oxford: Clarendon Press, 1979.

Hata, M. "Legendre Type Polynomials and Irrationality Measures." *J. reine angew. Math.* **407**, 99–125, 1990.

Hata, M. "Improvement in the Irrationality Measures of π and π^2." *Proc. Japan. Acad. Ser. A Math. Sci.* **68**, 283–286, 1992.

Hata, M. "Rational Approximations to π and Some Other Numbers." *Acta Arith.* **63** 335–349, 1993.

Hata, M. "A Note on Beuker's Integral." *J. Austral. Math. Soc.* **58**, 143–153, 1995.

Hata, M. "A New Irrationality Measure for $\zeta(3)$." *Acta Arith.* **92**, 47–57, 2000.

Stark, H. M. *An Introduction to Number Theory.* Cambridge, MA: MIT Press, 1978.

Irrationality Sequence

A sequence of POSITIVE INTEGERS $\{a_n\}$ such that $\Sigma \, 1/(a_n b_n)$ is IRRATIONAL for all integer sequences $\{b_n\}$. Erdos showed that $\{2^{2^n}\} = \{1, 2, 4, 16, 256, \dots, \}$ (Sloane's A001146) is an irrationality sequence.

References

Guy, R. K. "Irrationality Sequence." §E24 in *Unsolved Problems in Number Theory, 2nd ed.* New York: Springer-Verlag, p. 225, 1994.

Sloane, N. J. A. Sequences A001146/M1297 in "An On-Line Version of the Encyclopedia of Integer Sequences." http://www.research.att.com/~njas/sequences/eisonline.html.

Irreducible Matrix

A SQUARE MATRIX which is not REDUCIBLE is said to be irreducible.

See also REDUCIBLE MATRIX

Irreducible Polynomial

A POLYNOMIAL is said to be irreducible if it cannot be factored into nontrivial polynomials over the same FIELD.

For example, in the FIELD of rational polynomials $\mathbb{Q}[x]$ (i.e., polynomials $f(x)$ with rational coefficients), a $f(x)$ is said to be irreducibility if there do not exist two nonconstant polynomials $g(x)$ and $h(x)$ in x with rational coefficients such that

$$f(x) = g(x)h(x)$$

(Nagell 1951, p. 160). Similarly, in the FINITE FIELD GF(2), $x^2 + x + 1$ is irreducible, but $x^2 + 1$ is not, since $(x + 1)(x + 1) = x^2 + 2x + 1 \equiv x^2 + 1 \pmod 2$. A polynomial can be tested to see if it is primitive using the *Mathematica* function

```
IrreducibleQ[p_,n_] := SameQ[Factor[p, Modulus->n], p]
```

In general, the number of irreducible polynomials of degree n over the FINITE FIELD GF(q) is given by

$$L_q(n) = \frac{1}{n} \sum_{d|n} \mu\left(\frac{n}{d}\right) q^d,$$

where $\mu(n)$ is the MÖBIUS FUNCTION.

The number of irreducible polynomials of degree n over GF(2) is equal to the number of n-bead fixed aperiodic NECKLACES of two colors and the number of binary LYNDON WORDS of length n. The first few values for $n = 1, 2, \dots$ are 2, 1, 2, 3, 6, 9, 18, $\dots$ (Sloane's A001037). The following table lists the irreducible polynomials (mod 2) of degrees 1 through 5.

n	irreducible polynomials
1	$1, x$
2	$1 + x + x^2$
3	$1 + x + x^3, 1 + x^2 + x^3$
4	$1 + x + x^4, 1 + x + x^2 + x^3 + x^4, 1 + x^3 + x^4$
5	$1 + x^2 + x^5, 1 + x + x^2 + x^3 + x^5, 1 + x^3 + x^5,$
	$1 + x + x^3 + x^4 + x^5, 1 + x^2 + x^3 + x^4 + x^5,$
	$1 + x + x^2 + x^4 + x^5$

See also FIELD, FINITE FIELD, LYNDON WORD, NECKLACE, POLYNOMIAL, PRIMITIVE POLYNOMIAL

References

Marsh, R. *Tables of Irreducible Polynomials of GF(2) through Degree 19.* Washington, DC: U. S. Dept. Commerce., 1957.

Nagell, T. "Irreducibility of the Cyclotomic Polynomial." §47 in *Introduction to Number Theory.* New York: Wiley, pp. 160–164, 1951.

Ruskey, F. "Information on Primitive and Irreducible Polynomials." http://www.theory.csc.uvic.ca/~cos/inf/neck/PolyInfo.html.

Sloane, N. J. A. Sequences A001037/M0116 in "An On-Line Version of the Encyclopedia of Integer Sequences." http://www.research.att.com/~njas/sequences/eisonline.html.

Sloane, N. J. A. and Plouffe, S. Figure M0564 in *The Encyclopedia of Integer Sequences.* San Diego: Academic Press, 1995.

Irreducible Representation

An irreducible representation of a GROUP is a REPRESENTATION that has no nontrivial invariant subspaces. For example, the ORTHOGONAL GROUP $O(n)$ has an irreducible representation on $\mathbb{R}^n$.

Any representation of a finite or SEMISIMPLE LIE GROUP breaks up into a DIRECT SUM of irreducible representations. But in general, this is not the case, e.g., $(\mathbb{R}, +)$ has a representation on $\mathbb{R}^2$ by

$$\phi(a) = \begin{bmatrix} 1 & a \\ 0 & 1 \end{bmatrix},$$

i.e., $\phi(a)(x, y) = (x + ay, y)$. But the subspace $y = 0$ is fixed, hence ϕ is not irreducible, but there is no complementary invariant subspace.

The irreducible representation has a number of remarkable properties, as formalized in the GROUP ORTHOGONALITY THEOREM. Let the ORDER of a GROUP be h, and the dimension of the ith representation (the order of each constituent matrix) be l_i (a POSITIVE INTEGER). Let any operation be denoted R, and let the mth row and nth column of the matrix corresponding to a matrix R in the ith IRREDUCIBLE REPRESENTATION be $\Gamma_i(R)_{mn}$. The following properties can be derived from the GROUP ORTHOGONALITY THEOREM,

$$\sum_R \Gamma_i(R)_{mn} \Gamma_j(R)^*_{m'n'} = \frac{h}{\sqrt{l_i l_j}} \delta_{ij} \delta_{mm'} \delta_{nn'}. \quad (1)$$

1. The DIMENSIONALITY THEOREM:

$$h = \sum_i l_i^2 = l_1^2 + l_2^2 + l_3^2 + \ldots = \sum_i \chi_i^2(I), \quad (2)$$

where each l_i must be a POSITIVE INTEGER and χ is the CHARACTER (trace) of the representation.

2. The sum of the squares of the CHARACTERS in any IRREDUCIBLE REPRESENTATION i equals h,

$$h = \sum_R \chi_i^2(R). \quad (3)$$

3. ORTHOGONALITY of different representations

$$\sum_R \chi_i(R) \chi_i(R) = 0 \quad \text{for } i \neq j. \quad (4)$$

4. In a given representation, reducible or irreducible, the CHARACTERS of all MATRICES belonging to operations in the same class are identical (but differ from those in other representations).

5. The number of IRREDUCIBLE REPRESENTATIONS of a GROUP is equal to the number of CONJUGACY CLASSES in the GROUP. This number is the dimension of the Γ MATRIX (although some may have zero elements).

6. A one-dimensional representation with all 1s (totally symmetric) will always exist for any GROUP.

7. A 1-D representation for a GROUP with elements expressed as MATRICES can be found by taking the CHARACTERS of the MATRICES.

8. The number a_i of IRREDUCIBLE REPRESENTATIONS χ_i present in a reducible representation c is given by

$$a_i = \frac{1}{h} \sum_R \chi(R) \chi_i(R), \quad (5)$$

where h is the ORDER of the GROUP and the sum must be taken over all elements in each class. Written explicitly,

$$a_i = \frac{1}{h} \sum_R \chi(R) \chi_i'(R) n_R, \quad (6)$$

where χ_i' is the CHARACTER of a single entry in the CHARACTER TABLE and n_R is the number of elements in the corresponding CONJUGACY CLASS.

Irreducible representations can be indicated using MULLIKEN SYMBOLS.

See also CHARACTER (GROUP), CHARACTER TABLE, FINITE GROUP, GROUP, GROUP ORTHOGONALITY THEOREM, ITÔ'S THEOREM, MULLIKEN SYMBOLS, REPRESENTATION, REPRESENTATION (LIE ALGEBRA),

SEMISIMPLE LIE GROUP UNITARY TRANSFORMATION, VECTOR SPACE, WEDDERBURN'S THEOREM

References

Fulton, W. and Harris, J. *Representation Theory.* New York:Springer-Verlag, 1991.

Jacobson, N. *Lie Algebras.* New York: Dover, 1979.

Huang, J.-S. "Irreducible Representations." §2.3 in *Lectures on Representation Theory.* Singapore: World Scientific, pp. 11–14, 1999.

Knapp, A. *Lie Groups Beyond an Introduction.* Boston, MA: Birkhäuser, 1996.

Irreducible Semiperfect Number

PRIMITIVE PSEUDOPERFECT NUMBER

Irreducible Tensor

Given a general second RANK TENSOR A_{ij} and a METRIC g_{ij}, define

$$\theta \equiv A_{ij}g^{ij} = A_i^i \tag{1}$$

$$\omega^i \equiv \epsilon^{ijk}A_{jk} \tag{2}$$

$$\sigma_{ij} \equiv \tfrac{1}{2}(A_{ij} + A_{ji}) - \tfrac{1}{3}g_{ij}A_k^k, \tag{3}$$

where δ_{ij} is the KRONECKER DELTA and ϵ^{ijk} is the LEVI-CIVITA symbol. Then

$$\sigma_{ij} + \tfrac{1}{3}\theta g_{ij} + \tfrac{1}{2}\epsilon_{ijk}\omega^k$$

$$= \left[\tfrac{1}{2}(A_{ij} + A_{ji}) - \tfrac{1}{3}g_{ij}A_k^k\right] + \tfrac{1}{3}A_k^k g_{ij} + \tfrac{1}{2}\epsilon_{ijk}\left[\epsilon^{\lambda\mu k}A_{\lambda\mu}\right]$$

$$= \tfrac{1}{2}(A_{ij} + A_{ji}) + \tfrac{1}{2}\left(\delta_i^\lambda \delta_j^\mu - \delta_i^\mu \delta_j^\lambda\right)A_{\lambda\mu}$$

$$= \tfrac{1}{2}(A_{ij} + A_{ji}) + \tfrac{1}{2}(A_{ij} - A_{ji}) = A_{ij}, \tag{4}$$

where θ, ω^i, and σ_{ij} are TENSORS of RANK 0, 1, and 2.

See also TENSOR

References

Varshalovich, D. A.; Moskalev, A. N.; and Khersonskii, V. K. "Irreducible Tensors." Ch. 3 in *Quantum Theory of Angular Momentum.* Singapore: World Scientific, pp. 61–71, 1988.

Irreducible Variety

An ALGEBRAIC VARIETY is called irreducible if it cannot be written as the union of nonempty algebraic varieties. For example, the set of solutions to $xy = 0$ is reducible because it is the union of the solutions to $x = 0$ and the solutions to $y = 0$.

See also ALGEBRAIC SET, ALGEBRAIC VARIETY, PROJECTIVE VARIETY

Irredundant Ramsey Number

Let G_1, G_2, ..., G_t be a t-EDGE coloring of the COMPLETE GRAPH K_n, where for each $i = 1, 2, ..., t$, G_i is the spanning SUBGRAPH of K_n consisting of all EDGES colored with the ith color. The irredundant Ramsey number $s(q_1, ..., q_t)$ is the smallest INTEGER n such that for any t-EDGE coloring of K_n, the COMPLEMENT GRAPH $\overline{G_i}$ has an irredundant set of size q_i for at least one $i = 1, ..., t$. Irredundant Ramsey numbers were introduced by Brewster *et al.* (1989) and satisfy

$$s(q_1, \cdots q_t) \leq R(q_1, \ldots q_t).$$

For a summary, see Mynhardt (1992).

s	Bounds	Reference
$s(3,3)$	6	Brewster *et al.* 1989
$s(3,4)$	8	Brewster *et al.* 1989
$s(3,5)$	12	Brewster *et al.* 1989
$s(3,6)$	15	Brewster *et al.* 1990
$s(3,7)$	18	Chen and Rousseau 1995, Cockayne *et al.* 1991
$s(4,4)$	13	Cockayne *et al.* 1992
$s(3,3,3)$	13	Cockayne and Mynhardt 1994

References

Brewster, R. C.; Cockayne, E. J.; and Mynhardt, C. M. "Irredundant Ramsey Numbers for Graphs." *J. Graph Theory* **13**, 283–290, 1989.

Brewster, R. C.; Cockayne, E. J.; and Mynhardt, C. M. "The Irredundant Ramsey Number $s(3,6)$." *Quaest. Math.* **13**, 141–157, 1990.

Chen, G. and Rousseau, C. C. "The Irredundant Ramsey Number $s(3,7)$." *J. Graph. Th.* **19**, 263–270, 1995.

Cockayne, E. J.; Exoo, G.; Hattingh, J. H.; and Mynhardt, C. M. "The Irredundant Ramsey Number $s(4,4)$." *Util. Math.* **41**, 119–128, 1992.

Cockayne, E. J.; Hattingh, J. H.; and Mynhardt, C. M. "The Irredundant Ramsey Number $s(3,7)$." *Util. Math.* **39**, 145–160, 1991.

Cockayne, E. J. and Mynhardt, C. M. "The Irredundant Ramsey Number $s(3,3,3) = 13$." *J. Graph. Th.* **18**, 595–604, 1994.

Hattingh, J. H. "On Irredundant Ramsey Numbers for Graphs." *J. Graph Th.* **14**, 437–441, 1990.

Mynhardt, C. M. "Irredundant Ramsey Numbers for Graphs: A Survey." *Congres. Numer.* **86**, 65–79, 1992.

Irreflexive

A RELATION R on a SET S is irreflexive provided that no element is related to itself; in other words, xRx for no x in S.

See also RELATION

Irregular Pair

If p divides the NUMERATOR of the BERNOULLI NUMBER B_{2k} for $0 < 2k < p - 1$, then $(p, 2k)$ is called an irregular pair. For $p < 30000$, the irregular pairs of various forms are $p = 16843$ for $(p, p - 3)$, $p = 37$ for $(p, p - 5)$, none for $(p, p - 7)$, and $p = 67,877$ for $(p, p - 9)$.

See also BERNOULLI NUMBER, IRREGULAR PRIME

References

Johnson, W. "Irregular Primes and Cyclotomic Invariants." *Math. Comput.* **29**, 113–120, 1975.

Irregular Prime

PRIMES for which Kummer's theorem on the unsolvability of FERMAT'S LAST THEOREM does not apply. An irregular prime p divides the NUMERATOR of one of the BERNOULLI NUMBERS B_0, B_2, ..., B_{p-3}, as shown by Kummer in 1850. The FERMAT EQUATION has no solutions for REGULAR PRIMES.

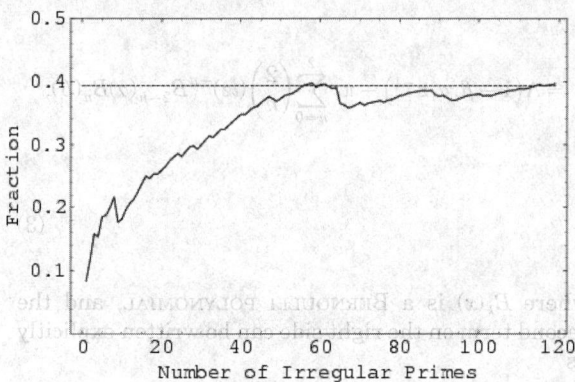

An INFINITE number of irregular primes exist, as proven in 1915 by Jensen. The first few irregular primes are 37, 59, 67, 101, 103, 131, 149, 157, ... (Sloane's A000928). Of the 283,145 PRIMES less than 4×10^6, 111,597 (or 39.41%) are irregular. The conjectured FRACTION is $1 - e^{-1/2} \approx 39.35\%$ (Ribenboim 1996, p. 415).

See also BERNOULLI NUMBER, FERMAT'S LAST THEOREM, IRREGULAR PAIR, REGULAR PRIME

References

Buhler, J.; Crandall, R.; Ernvall, R.; and Metsänkylä, T. "Irregular Primes and Cyclotomic Invariants to Four Million." *Math. Comput.* **60**, 151–153, 1993.

Hardy, G. H. and Wright, E. M. *An Introduction to the Theory of Numbers, 5th ed.* Oxford, England: Clarendon Press, p. 202, 1979.

Hoffman, P. *The Man Who Loved Only Numbers: The Story of Paul Erdos and the Search for Mathematical Truth.* New York: Hyperion, p. 192, 1998.

Johnson, W. "Irregular Primes and Cyclotomic Invariants." *Math. Comput.* **29**, 113–120, 1975.

Ribenboim, P. *The New Book of Prime Number Records.* New York: Springer-Verlag, pp. 325–329 and 414–425, 1996.

Sloane, N. J. A. Sequences A000928/M5260 in "An On-Line Version of the Encyclopedia of Integer Sequences." http://www.research.att.com/~njas/sequences/eisonline.html.

Stewart, C. L. "A Note on the Fermat Equation." *Mathematika* **24**, 130–132, 1977.

Irregular Singularity

Consider a second-order ORDINARY DIFFERENTIAL EQUATION

$$y'' + P(x)y' + Q(x)y = 0.$$

If $P(x)$ and $Q(x)$ remain FINITE at $x = x_0$, then x_0 is called an ORDINARY POINT. If either $P(x)$ or $Q(x)$ diverges as $x \to x_0$, then x_0 is called a singular point. If $P(x)$ diverges more quickly than $1/(x - x_0)$, so $(x - x_0)P(x)$ approaches INFINITY as $x \to x_0$, or $Q(x)$ diverges more quickly than $1/(x - x_0)^2 Q$ so that $(x - x_0)^2 Q(x)$ goes to INFINITY as $x \to x_0$, then x_0 is called an IRREGULAR SINGULARITY (or ESSENTIAL SINGULARITY).

See also ORDINARY POINT, REGULAR SINGULAR POINT, SINGULAR POINT (DIFFERENTIAL EQUATION)

References

Arfken, G. "Singular Points." §8.4 in *Mathematical Methods for Physicists, 3rd ed.* Orlando, FL: Academic Press, pp. 451–453 and 461–463, 1985.

Irrotational Field

A VECTOR FIELD **v** for which the CURL vanishes,

$$\nabla \times \mathbf{v} = \mathbf{0}.$$

See also BELTRAMI FIELD, CONSERVATIVE FIELD, POINCARÉ'S THEOREM, SOLENOIDAL FIELD, VECTOR FIELD

Isarithm

EQUIPOTENTIAL CURVE

ISBN

Publisher	Digits
Addison-Wesley	0–201
Amer. Math. Soc.	0–821
Birkhäuser Basel	3–7643
Birkhäuser Boston	0–8176
Cambridge University Press	0–521
CRC Press	0–8493
Dover	0–486
McGraw-Hill	0–070

Oxford University Press	0–198
Springer-Verlag Berlin	3–540
Springer-Verlag New York	0–387
Tarquin Publications	0–906212
Wiley	0–471

The International Standard Book Number (ISBN) is a 10-digit CODE which is used to uniquely identify a book. The digits d_i are arranged in four groups, which are sometimes (but not always) separated by hyphens. The first group is a single digit which codes country or language in which a publisher is incorporated: 0 for English, 2 for French, 3 for German, 4 for Japanese, 8 for Indian publishers, etc. The next group of digits specifies the publisher, and may range in length from two to seven digits, with fewer digits used for larger publishers. Some publishers with offices in more than one country (at least when different languages are spoken in those countries) have multiple publisher codes and initial digits.

The third group of digits specifies an individual book, and may be from one to six digits in length. The actual number is eight minus the number of digits in the publisher group, so that small publishers may have only 10 books, while large ones can have up to a millions books. The last digit d_{10} is a check digit which may be in the range 0–9 or X (where X is the ROMAN NUMERAL for 10). The check digit is computed from the equation

$$10d_1 + 9d_2 + 8d_3 + \ldots + 2d_9 + d_{10} \equiv 0 \pmod{11}.$$

For example, the number for this book is 0–8493–9640–9, and

$$10 \cdot 0 + 9 \cdot 8 + 8 \cdot 4 + 7 \cdot 9 + 6 \cdot 3 + 5 \cdot 9$$

$$+4 \cdot 6 + 3 \cdot 4 + 2 \cdot 0 + 1 \cdot 9 = 275 = 25 \cdot 11 \equiv 0 \pmod{11}.$$

as required.

The ISBN is error-detecting, but not error-correcting (unless it is known that only a single digit is erroneous). The ISBN detects any single-digit error, as well as any two-digit error resulting from transposing two digits.

See also CODE, CODING THEORY, UPC

References

Hill, R. *First Course in Coding Theory.* Oxford, England: Oxford University Press, 1986.
Press, W. H.; Flannery, B. P.; Teukolsky, S. A.; and Vetterling, W. T. *Numerical Recipes in FORTRAN: The Art of Scientific Computing, 2nd ed.* Cambridge, England: Cambridge University Press, p. 894, 1992.

Iseki's Formula

Let $\Re[z] > 0$, $0 \le \alpha, \beta \le 1$, and

$$\Lambda(\alpha, \beta, z) \equiv \sum_{r=0}^{\infty} [\lambda((r+\alpha)z - i\beta) + \lambda((r+1-\alpha)z + i\beta)], \tag{1}$$

where

$$\lambda(x) \equiv -\ln(1 - e^{-2\pi x}) = \sum_{m=1}^{\infty} \frac{e^{-2\pi m x}}{m}. \tag{2}$$

Then if either $0 \le \alpha \le 1$ and $0 < \beta < 1$, or $0 < \alpha < 1$ and $0 \le \beta \le 1$,

$$\Lambda(\alpha, \beta, z)$$

$$= \Lambda(1 - \beta, \alpha, z^{-1}) - \pi z \sum_{n=0}^{2} \binom{2}{n} (iz)^{-n} B_{2-n}(\alpha) B_n(\beta), \tag{3}$$

where $B_k(x)$ is a BERNOULLI POLYNOMIAL, and the second term on the right side can be written explicitly as

$$-\pi z \left(\alpha^2 \alpha + \tfrac{1}{6}\right) + \frac{\pi}{z}\left(\beta^2 - \beta + \tfrac{1}{6}\right) + 2\pi i \left(\alpha - \tfrac{1}{2}\right)(\beta - h). \tag{4}$$

See also DEDEKIND ETA FUNCTION

References

Apostol, T. M. "Iseki's Transformation Formula" and "Deduction of Dedekind's Functional Equation from Iseki's Formula." §3.5–3.6 in *Modular Functions and Dirichlet Series in Number Theory, 2nd ed.* New York: Springer-Verlag, pp. 53–61, 1997.
Iseki, S. "The Transformation Formula for the Dedekind Modular Function and Related Functional Equations." *Duke Math. J.* **24**, 653–662, 1957.

I-Signature

SIGNATURE (RECURRENCE RELATION)

Island

islands

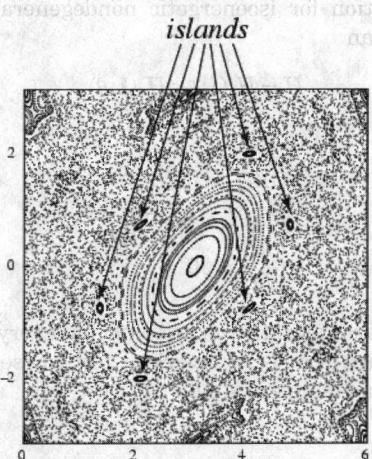

If an integrable QUASIPERIODIC system is slightly perturbed so that it becomes nonintegrable, only a finite number of n-CYCLES remain as a result of MODE LOCKING. One will be elliptical and one will be hyperbolic.

Surrounding the ELLIPTIC FIXED POINT is a region of stable ORBITS which circle it, as illustrated above in the STANDARD MAP with $K = 1.5$. As the map is iteratively applied, the island is mapped to a similar structure surrounding the next point of the elliptic cycle. The map thus has a chain of islands, with the FIXED POINT alternating between ELLIPTIC (at the center of the islands) and HYPERBOLIC (between islands). Because the unperturbed system goes through an INFINITY of rational values, the perturbed system must have an INFINITE number of island chains.

See also MODE LOCKING, ORBIT (MAP), QUASIPERIODIC FUNCTION

Isobaric Polynomial

A POLYNOMIAL in which the sum of SUBSCRIPTS is the same in each term.

See also HOMOGENEOUS POLYNOMIAL

Isochronous Curve

SEMICUBICAL PARABOLA, TAUTOCHRONE PROBLEM

Isoclinal

ISOCLINAL LINE, ISOCLINAL PLANE, ISOCLINE

Isoclinal Line

A line making equal angles with the edges of a TRIHEDRON is called an isoclinal line of the TRIHEDRON.

See also ISOCLINAL PLANE, TRIHEDRON

References

Altshiller-Court, N. "Isoclinal Lines and Planes." §2.3 in *Modern Pure Solid Geometry*. New York: Chelsea, pp. 32–37, 1979.

Isoclinal Plane

A PLANE making equal angles with the three edges of a TRIHEDRON.

See also ISOCLINAL LINE, TETRAHEDRON

References

Altshiller-Court, N. "Isoclinal Lines and Planes." §2.3 in *Modern Pure Solid Geometry*. New York: Chelsea, pp. 32–37, 1979.

Isocline

A graphical method of solving an ORDINARY DIFFERENTIAL EQUATION OF THE FORM

$$\frac{dy}{dx} = f(x, y)$$

by plotting a series of curves $f(x, y) = [\text{const}]$, then drawing a curve PERPENDICULAR to each curve such that it satisfies the initial condition. This curve is the solution to the ORDINARY DIFFERENTIAL EQUATION.

See also ISOCLINAL LINE, ISOCLINAL PLANE

References

Kármán, T. von and Biot, M. A. *Mathematical Methods in Engineering: An Introduction to the Mathematical Treatment of Engineering Problems*. New York: McGraw-Hill, pp. 3 and 7, 1940.

Isoclinic Groups

Two GROUPS G and H are said to be isoclinic if there are isomorphisms $G/Z(G) \to H/Z(H)$ and $G' \to H'$, where $Z(G)$ is the CENTER of the group, which identify the two commutator maps.

References

Conway, J. H.; Curtis, R. T.; Norton, S. P.; Parker, R. A.; and Wilson, R. A. "Isoclinism." §6.7 in *Atlas of Finite Groups: Maximal Subgroups and Ordinary Characters for Simple Groups*. Oxford, England: Clarendon Press, pp. xxiii-xxiv, 1985.

Isodynamic Points

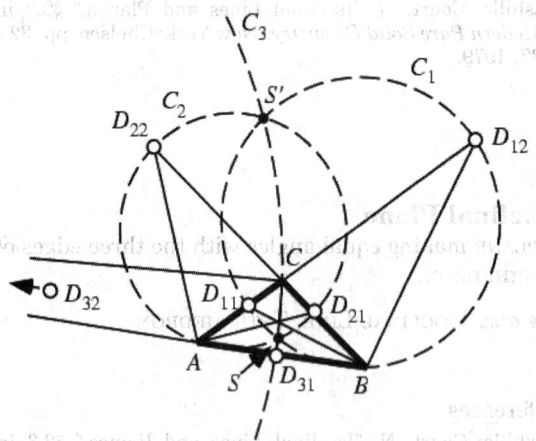

The first and second isodynamic points of a TRIANGLE ΔABC can be constructed by drawing the triangle's ANGLE BISECTORS and EXTERIOR ANGLE BISECTORS. Each pair of bisectors intersects a side of the triangle (or its extension) in two points D_{i1} and D_{i2}, for $i = 1, 2, 3$. The three CIRCLES having $D_{11}D_{12}$, $D_{21}D_{22}$, and $D_{31}D_{32}$ as DIAMETERS are the APOLLONIUS CIRCLES C_1, C_2, and C_3. The points S and S' in which the three APOLLONIUS CIRCLES intersect are the first and second isodynamic points, respectively.

S and S' have TRIANGLE CENTER FUNCTIONS

$$\alpha = \sin\left(A \pm \tfrac{1}{3}\pi\right),$$

respectively. The ANTIPEDAL TRIANGLES of both points are EQUILATERAL and have AREAS

$$\Delta' = 2\Delta\left[\cot\omega\cot\left(\tfrac{1}{3}\pi\right)\right],$$

where ω is the BROCARD ANGLE.

The isodynamic points are ISOGONAL CONJUGATES of the FERMAT POINTS. They lie on the BROCARD AXIS. The distances from either isodynamic point to the VERTICES are inversely proportional to the sides. The PEDAL TRIANGLE of either isodynamic point is an EQUILATERAL TRIANGLE. An INVERSION with either isodynamic point as the INVERSION CENTER transforms the triangle into an EQUILATERAL TRIANGLE.

The CIRCLE which passes through both the isodynamic points and the CENTROID of a TRIANGLE is known as the PARRY CIRCLE.

See also APOLLONIUS CIRCLES, BROCARD AXIS, CENTROID (TRIANGLE), FERMAT POINTS, PARRY CIRCLE

References
Gallatly, W. *The Modern Geometry of the Triangle, 2nd ed.* London: Hodgson, p. 106, 1913.
Johnson, R. A. *Modern Geometry: An Elementary Treatise on the Geometry of the Triangle and the Circle.* Boston, MA: Houghton Mifflin, pp. 295–297, 1929.
Kimberling, C. "Central Points and Central Lines in the Plane of a Triangle." *Math. Mag.* **67**, 163–187, 1994.

Isoenergetic Nondegeneracy
The condition for isoenergetic nondegeneracy for a Hamiltonian

$$H = H_0(\mathbf{I}) + \epsilon H_1(\mathbf{I}, \theta)$$

is

$$\left| \begin{matrix} \dfrac{\partial^2 H_0}{\partial I_i \partial I_j} & \dfrac{\partial H_0}{\partial I_i} \\ \dfrac{\partial H_0}{\partial I_j} & 0 \end{matrix} \right| \neq 0,$$

which guarantees the EXISTENCE on every energy level surface of a set of invariant tori whose complement has a small MEASURE.

References
Tabor, M. *Chaos and Integrability in Nonlinear Dynamics: An Introduction.* New York: Wiley, pp. 113–114, 1989.

Isogeny
A rational homomorphism $\varphi G \to G'$ defined over a FIELD is called an isogeny when $\dim G = \dim G'$. Two GROUPS G and G' are then called isogenous if there exists a third group G'' and isogenies $G'' \to G$ and $G'' \to G'$.

See also HOMEOMORPHIC

References
Iyanaga, S. and Kawada, Y. (Eds.). *Encyclopedic Dictionary of Mathematics.* Cambridge, MA: MIT Press, p. 47, 1980.

Isogonal Conjugate

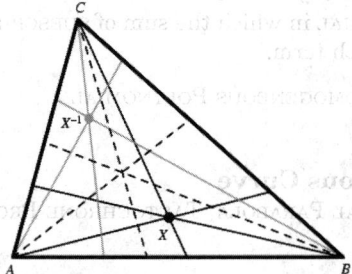

The isogonal conjugate X^{-1} of a point X in the plane of the TRIANGLE ΔABC is constructed by reflecting the lines AX, BX, and CX about the ANGLE BISECTORS at A, B, and C. The three reflected lines then CONCUR at the isogonal conjugate (Honsberger 1995, pp. 55–56). The TRILINEAR COORDINATES of the isogonal conjugate of the point with coordinates

$$\alpha : \beta : \gamma$$

are

$$\alpha^{-1} : \beta^{-1} : \gamma^{-1}.$$

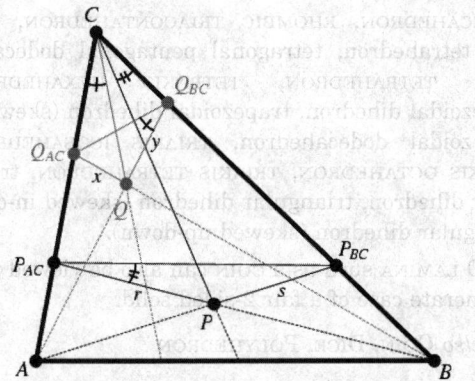

In the above figure with P and Q isogonal conjugates,

$$\frac{x}{y} = \frac{s}{r} \qquad (1)$$

(Honsberger 1995, pp. 54–55).

Isogonal conjugation maps the interior of a TRIANGLE onto itself. This mapping transforms lines onto CONIC SECTIONS that CIRCUMSCRIBE the TRIANGLE. The type of CONIC SECTION is determined by whether the line d meets the CIRCUMCIRCLE C'.

1. If d does not intersect C', the isogonal transform is an ELLIPSE;
2. If d is tangent to C', the transform is a PARABOLA;
3. If d cuts C', the transform is a HYPERBOLA, which is a RECTANGULAR HYPERBOLA if the line passes through the CIRCUMCENTER

(Casey 1893, Vandeghen 1965).

The isogonal conjugate of a point on the CIRCUMCIRCLE is a POINT AT INFINITY (and conversely). The sides of the PEDAL TRIANGLE of a point are PERPENDICULAR to the connectors of the corresponding VERTICES with the isogonal conjugate. The isogonal conjugate of a set of points is the LOCUS of their isogonal conjugate points.

The product of ISOTOMIC and isogonal conjugation is a COLLINEATION which transforms the sides of a TRIANGLE to themselves (Vandeghen 1965).

See also ANTIPEDAL TRIANGLE, COLLINEATION, ISOGONAL LINE, ISOTOMIC CONJUGATE POINT, LINE AT INFINITY, SYMMEDIAN

References
Barrow, D. F. "A Theorem about Isogonal Conjugates." *Amer. Math. Monthly* **20**, 251–253, 1913.

Casey, J. "Theory of Isogonal and Isotomic Points, and of Antiparallel and Symmedian Lines." Supp. Ch. §1 in *A Sequel to the First Six Books of the Elements of Euclid, Containing an Easy Introduction to Modern Geometry with Numerous Examples, 5th ed., rev. enl.* Dublin: Hodges, Figgis, & Co., pp. 165–173, 1888.

Casey, J. *A Treatise on the Analytical Geometry of the Point, Line, Circle, and Conic Sections, Containing an Account of Its Most Recent Extensions with Numerous Examples, 2nd rev. enl. ed.* Dublin: Hodges, Figgis, & Co., 1893.

Coolidge, J. L. *A Treatise on the Geometry of the Circle and Sphere.* New York: Chelsea, p. 49, 1971.

Coxeter, H. S. M. and Greitzer, S. L. *Geometry Revisited.* Washington, DC: Math. Assoc. Amer., p. 93, 1967.

Honsberger, R. *Episodes in Nineteenth and Twentieth Century Euclidean Geometry.* Washington, DC: Math. Assoc. Amer., pp. 53–57, 1995.

Johnson, R. A. *Modern Geometry: An Elementary Treatise on the Geometry of the Triangle and the Circle.* Boston, MA: Houghton Mifflin, pp. 153–158, 1929.

Lachlan, R. §10 in *An Elementary Treatise on Modern Pure Geometry.* London: Macmillian, pp. 55–57, 1893.

Vandeghen, A. "Some Remarks on the Isogonal and Cevian Transforms. Alignments of Remarkable Points of a Triangle." *Amer. Math. Monthly* **72**, 1091–1094, 1965.

Isogonal Line

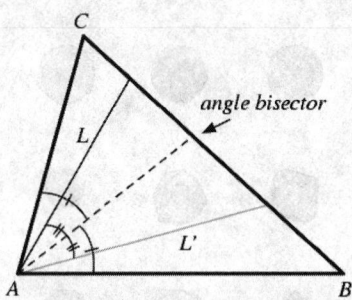

The line L' through a TRIANGLE VERTEX obtained by reflecting an initial line L (also through a VERTEX) about the ANGLE BISECTOR. If three lines from the VERTICES of a TRIANGLE $\triangle ABC$ are CONCURRENT at $X = L_1 L_2 L_3$, then their isogonal lines are also CONCURRENT, and the point of concurrence $X' = L_1' L_2' L_3'$ is called the ISOGONAL CONJUGATE point.

See also ISOGONAL CONJUGATE

References
Johnson, R. A. *Modern Geometry: An Elementary Treatise on the Geometry of the Triangle and the Circle.* Boston, MA: Houghton Mifflin, pp. 153–157, 1929.

Isogonic Centers
FERMAT POINTS

Isograph
The substitution of $re^{i\theta}$ for z in a POLYNOMIAL $p(z)$. $p(z)$ is then plotted as a function of θ for a given r in the COMPLEX PLANE. By varying r so that the curve passes through the ORIGIN, it is possible to determine a value for one ROOT of the POLYNOMIAL.

Isohedral Tiling
Let $S(T)$ be the group of symmetries which map a MONOHEDRAL TILING T onto itself. The TRANSITIVITY

CLASS of a given tile T is then the collection of all tiles to which T can be mapped by one of the symmetries of $S(T)$. If T has k TRANSITIVITY CLASSES, then T is said to be k-isohedral. Berglund (1993) gives examples of k-isohedral tilings for $k = 1$, 2, and 4.

See also ANISOHEDRAL TILING

References

Berglund, J. "Is There a k-Anisohedral Tile for $k \geq 5$?" *Amer. Math. Monthly* **100**, 585–588, 1993.
Grünbaum, B. and Shephard, G. C. "The 81 Types of Isohedral Tilings of the Plane." *Math. Proc. Cambridge Philos. Soc.* **82**, 177–196, 1977.

Isohedron

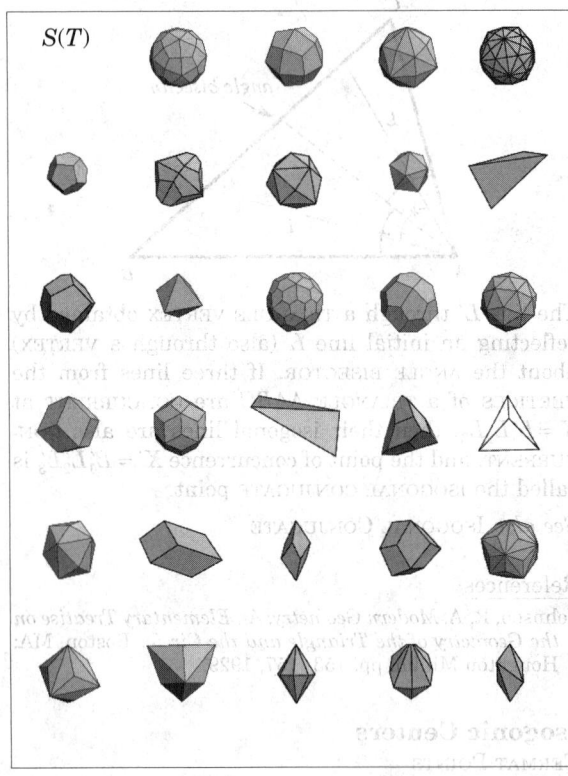

A convex POLYHEDRON with symmetries acting transitively on its faces. Every isohedron has an EVEN number of faces (Grünbaum 1960). The isohedra make fair DICE, and there are 30 of them, many of which are PLATONIC SOLIDS, ARCHIMEDEAN SOLIDS, or duals of ARCHIMEDEAN SOLIDS.

The 30 isohedra are the CUBE, DISDYAKIS DODECAHEDRON, DELTOIDAL HEXECONTAHEDRON, DELTOIDAL ICOSITETRAHEDRON, DISDYAKIS TRIACONTAHEDRON, DODECAHEDRON, dyakis dodecahedron, hexakis tetra-

hedron, ICOSAHEDRON, isosceles tetrahedron, octahedral pentagonal dodecahedron, OCTAHEDRON, PENTAGONAL HEXECONTAHEDRON, PENTAGONAL ICOSITETRAHEDRON, PENTAKIS DODECAHEDRON, RHOMBIC DODECAHEDRON, RHOMBIC TRIACONTAHEDRON, scalene tetrahedron, tetragonal pentagonal dodecahedron, TETRAHEDRON, TETRAKIS HEXAHEDRON, trapezoidal dihedron, trapezoidal dihedron (skewed), trapezoidal dodecahedron, TRIAKIS ICOSAHEDRON, TRIAKIS OCTAHEDRON, TRIAKIS TETRAHEDRON, triangular dihedron, triangular dihedron (skewed in-out), triangular dihedron (skewed up-down).

A 2-D LAMINA such as a COIN can also be viewed as a degenerate case of a fair 2-sided solid.

See also COIN, DICE, POLYHEDRON

References

Grünbaum, B. "On Polyhedra in E^3 Having All Faces Congruent." *Bull. Research Council Israel* **8F**, 215–218, 1960.
Grünbaum, B. and Shepard, G. C. "Spherical Tilings with Transitivity Properties." In *The Geometric Vein: The Coxeter Festschrift* (Ed. C. Davis, B. Grünbaum, and F. Shenk). New York: Springer-Verlag, 1982.
Pegg, E. Jr. "Fair Dice." http://www.mathpuzzle.com/Fairdice.htm.
Weisstein, E. W. "Fair Dice." MATHEMATICA NOTEBOOK FAIRDICE.M.

Isolated Point

An isolated point on a curve, also known as an ACNODE or HERMIT POINT, is a point which has no other points in its NEIGHBORHOOD.

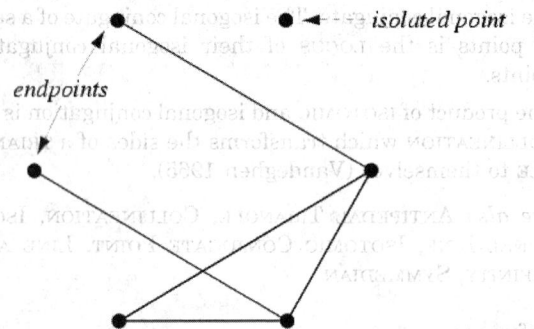

An isolated point of a GRAPH is a node of degree 1 (Harary 1994, p. 15). The number of n-node graphs with no isolated points are 0, 1, 2, 7, 23, 122, 888, ... (Sloane's A002494), the first few of which are illustrated below.

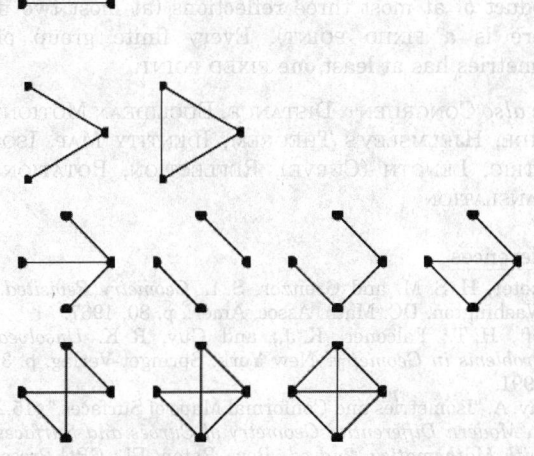

An isolated point of a DISCRETE SET S is a member of S (Krantz 1999, p. 63).

See also ENDPOINT, NEIGHBORHOOD

References

Harary, F. *Graph Theory*. Reading, MA: Addison-Wesley, 1994.
Krantz, S. G. "Discrete Sets and Isolated Points." §4.6.2 in *Handbook of Complex Analysis*. Boston, MA: Birkhäuser, pp. 63–64, 1999.
Sloane, N. J. A. Sequences A002494/M1762 in "An On-Line Version of the Encyclopedia of Integer Sequences." http://www.research.att.com/~njas/sequences/eisonline.html.

Isolated Singular Point

ISOLATED SINGULARITY

Isolated Singularity

An isolated singularity is a SINGULARITY for which there exists a (small) REAL NUMBER ϵ such that there are no other SINGULARITIES within a NEIGHBORHOOD of radius ϵ centered about the SINGULARITY. Isolated singularities are also known as conic double points.

The types of isolated singularities possible for CUBIC SURFACES have been classified (Schläfli 1864, Cayley 1869, Bruce and Wall 1979) and are summarized in the following table from Fischer (1986).

Name	Symbol	Normal Form	COXETER DIAGRAM
conic double point	C_2	$x^2+y^2+z^2$	A_1
biplanar double point	B_3	$x^2+y^2+z^3$	A_2
biplanar double point	B_4	$x^2+y^2+z^4$	A_3
biplanar double point	B_5	$x^2+y^2+z^5$	A_4
biplanar double point	B_6	$x^2+y^2+z^6$	A_5
uniplanar double point	U_6	$x^2+z(y^2+z^2)$	D_4
uniplanar double point	U_7	$x^2+z(y^2+z^3)$	D_5
uniplanar double point	U_8	$x^2+y^3+z^4$	E_6
elliptic cone point	–	$xy^2-4z^3-g_2x^2y+g_3x^3$	$\tilde{E}_6$

See also CUBIC SURFACE, RATIONAL DOUBLE POINT, SINGULAR POINT (FUNCTION)

References

Bruce, J. and Wall, C. T. C. "On the Classification of Cubic Surfaces." *J. London Math. Soc.* **19**, 245–256, 1979.
Cayley, A. "A Memoir on Cubic Surfaces." *Phil. Trans. Roy. Soc.* **159**, 231–326, 1869.
Fischer, G. (Ed.). *Mathematical Models from the Collections of Universities and Museums*. Braunschweig, Germany: Vieweg, pp. 12–13, 1986.
Krantz, S. G. *Handbook of Complex Analysis*. Boston, MA: Birkhäuser, p. 41, 1999.
Morse, P. M. and Feshbach, H. *Methods of Theoretical Physics, Part I*. New York: McGraw-Hill, pp. 380–381, 1953.
Schläfli, L. "On the Distribution of Surfaces of Third Order into Species." *Phil. Trans. Roy. Soc.* **153**, 193–247, 1864.

Isolating Integral

An integral of motion which restricts the PHASE SPACE available to a DYNAMICAL SYSTEM.

Isometric

A METRIC SPACE X is isometric to a METRIC SPACE Y if there is a BIJECTION f between X and Y that preserves distances. That is, $d(a,b)=d(f(a),f(b))$. In the context of RIEMANNIAN GEOMETRY, two manifolds M and N are isometric if there is a DIFFEOMORPHISM such that the RIEMANNIAN METRIC from one pulls back to the metric on the other. Since the GEODESICS define a distance, a RIEMANNIAN METRIC makes the MANIFOLD M a METRIC SPACE. An isometry between Riemannian manifolds is also an isometry between the two manifolds, considered as metric spaces.

Isometric spaces are considered isomorphic. For instance, the circle of radius one around the origin is isometric to the circle of radius one around $(0,3)$.

See also ISOMETRIC LATITUDE, ISOMETRY, METRIC SPACE, RIEMANNIAN METRIC, TOPOLOGICAL SPACE

Isometric Latitude

An AUXILIARY LATITUDE which is directly proportional to the spacing of parallels of LATITUDE from the equator on an ellipsoidal MERCATOR PROJECTION. It is defined by

$$\psi = \ln \left| \tan \left(\tfrac{1}{4}\pi + \tfrac{1}{2}\phi \right) \left(\frac{1 - e\sin\phi}{1 + e\sin\phi} \right)^{e/2} \right|, \quad (1)$$

where the symbol τ is sometimes used instead of ψ. The isometric latitude is related to the CONFORMAL LATITUDE by

$$\psi = \ln \tan \left(\tfrac{1}{4}\pi + \tfrac{1}{2}\chi \right). \quad (2)$$

The inverse is found by iterating

$$\phi = 2\tan^{-1} \left[\exp(\psi) \left(\frac{1 + e\sin\phi}{1 - e\sin\phi} \right)^{e/2} \right] - \tfrac{1}{2}\pi, \quad (3)$$

with the first trial as

$$\phi_0 = 2\tan^{-1}\left(e^{\psi}\right) - \tfrac{1}{2}\pi. \quad (4)$$

See also LATITUDE

References
Adams, O. S. "Latitude Developments Connected with Geodesy and Cartography with Tables, Including a Table for Lambert Equal-Area Meridional Projections." Spec. Pub. No. 67. U. S. Coast and Geodetic Survey, 1921.
Snyder, J. P. *Map Projections--A Working Manual.* U. S. Geological Survey Professional Paper 1395. Washington, DC: U. S. Government Printing Office, p. 15, 1987.

Isometry

A BIJECTIVE MAP between two METRIC SPACES that preserves distances, i.e.,

$$d(f(x), f(y)) = d(x, y),$$

where f is the MAP and $d(a, b)$ is the DISTANCE function. Isometries are sometimes also called congruence transformations. Two figures that can be transformed into each other by an isometry are said to be CONGRUENT (Coxeter and Greitzer 1967, p. 80).

An isometry of the PLANE is a linear transformation which preserves length. Isometries include ROTATION, TRANSLATION, REFLECTION, GLIDES, and the IDENTITY MAP. If an isometry has more than one FIXED POINT, it must be either the identity transformation or a reflection. Every isometry of period two (two applications of the transformation preserving lengths in the original configuration) is either a reflection or a half-turn rotation. Every isometry in the plane is the product of at most three reflections (at most two if there is a FIXED POINT). Every finite group of isometries has at least one FIXED POINT.

See also CONGRUENT, DISTANCE, EUCLIDEAN MOTION, GLIDE, HJELMSLEV'S THEOREM, IDENTITY MAP, ISOMETRIC, LENGTH (CURVE), REFLECTION, ROTATION, TRANSLATION

References
Coxeter, H. S. M. and Greitzer, S. L. *Geometry Revisited.* Washington, DC: Math. Assoc. Amer., p. 80, 1967.
Croft, H. T.; Falconer, K. J.; and Guy, R. K. *Unsolved Problems in Geometry.* New York: Springer-Verlag, p. 3, 1991.
Gray, A. "Isometries and Conformal Maps of Surfaces." §15.2 in *Modern Differential Geometry of Curves and Surfaces with Mathematica, 2nd ed.* Boca Raton, FL: CRC Press, pp. 346–351, 1997.

Isomorphic

The term "isomorphic" means "having the same form," and is used in many branches of mathematics to identify mathematical objects which have the same structural properties. Objects which may be represented (or "embedded") differently but which have the same essential structure are often said to be "identical up to an isomorphism." The statement "A is isomorphic to B" is denoted $A \cong B$ (Harary 1994, p. 161).

See also ISOMORPHIC GRAPHS, ISOMORPHIC GROUPS, ORDER ISOMORPHIC, ISOMORPHIC POSETS, ISOMORPHISM

References
Harary, F. *Graph Theory.* Reading, MA: Addison-Wesley, 1994.

Isomorphic Graphs

Two GRAPHS which contain the same number of VERTICES connected in the same way are said to be isomorphic. Formally, two graphs G and H with VERTICES $V_n = \{1, 2, \ldots, n\}$ are said to be isomorphic if there is a PERMUTATION p of V_n such that $\{u, v\}$ is in the set of EDGES $E(G)$ IFF $\{p(u), p(v)\}$ is in the set of EDGES $E(H)$.

Determining if two GRAPHS are isomorphic is thought to be an NP-HARD PROBLEM (Skiena 1990, p. 181), although this has not been proved. However, a polynomial-time algorithm is known when the maximum VERTEX DEGREE is bounded by a constant (Luks 1980; Skiena 1990, p. 181). The equivalence or nonequivalence of two graphs can be ascertained using `IsomorphicQ[g1, g2]` in the *Mathematica* add-on package `DiscreteMath`Combinatorica`` (which can be loaded with the command `<<DiscreteMath``).

See also GRAPH, GRAPH AUTOMORPHISM, GRAPH ISOMORPHISM, GRAPH THEORY, ULAM'S CONJECTURE

References

Chartrand, G. "Isomorphic Graphs." §2.2 in *Introductory Graph Theory*. New York: Dover, pp. 32–40, 1985.

Corneil, D. G. and Gottlieb, C. C. "An Efficient Algorithm for Graph Isomorphism." *J. ACM* **17**, 51–64, 1970.

Cvetkovic, D. M.; Doob, M.; and Sachs, H. *Spectra of Graphs: Theory and Applications, 3rd rev. enl. ed.* New York: Wiley, 1998.

Harary, F. *Graph Theory*. Reading, MA: Addison-Wesley, pp. 10–11, 1994.

Luks, E. M. "Isomorphism of Bounded Valence can be Tested in Polynomial Time." In *Proc. 21st Annual Symposium on Foundations of Computing*. IEEE Press, pp. 42–49, 1980.

Schmidt, D. C. and Druffel, L. E. "A Fast Backtracking Algorithm to Test Directed Graphs for Isomorphism Using Distance Matrices." *J. ACM* **23**, 433–445, 1976.

Skiena, S. "Graph Isomorphism." §5.2 in *Implementing Discrete Mathematics: Combinatorics and Graph Theory with Mathematica*. Reading, MA: Addison-Wesley, pp. 181–187, 1990.

Isomorphic Groups

Two GROUPS are isomorphic if the correspondence between them is ONE-TO-ONE and the "multiplication" table is preserved. For example, the POINT GROUPS C_2 and D_1 are isomorphic GROUPS, written $C_2 \cong D_1$ or $C_2 \rightleftharpoons D_1$ (Shanks 1993).

See also JORDAN-HÖLDER THEOREM

References

Shanks, D. *Solved and Unsolved Problems in Number Theory, 4th ed.* New York: Chelsea, 1993.

Isomorphic Posets

Two POSETS are said to be isomorphic if their "structures" are entirely analogous. Formally, POSETS $P = (X, \leq)$ and $Q = (X, \leq')$ are isomorphic if there is a BIJECTION f from X to X' such that $x \leq x'$ precisely when $f(x) \leq' f(x')$.

Isomorphism

Isomorphism is a very general concept which appears in several areas of mathematics. The word derives from the Greek ισο (*iso*), meaning "equal," and μορφωσις (*morphosis*), meaning "to form" or "to shape." Formally, an isomorphism is BIJECTIVE MORPHISM. Informally, an isomorphism is a map which preserves sets and relations among elements. "A is isomorphic to B" is written $A \cong B$. Unfortunately, this symbol is also used to denote geometric CONGRUENCE.

A space isomorphism is a VECTOR SPACE in which addition and scalar multiplication are preserved. An isomorphism of a TOPOLOGICAL SPACE is called a HOMEOMORPHISM.

Two groups G_1 and G_2 with binary operators $+$ and $\times$ are isomorphic if there exists a map $f : G_1 \rightarrow G_2$ which satisfies

$$f(x + y) = f(x) \times f(y).$$

An isomorphism preserves the identities and inverses of a GROUP. An isomorphism of a GROUP onto itself is called an AUTOMORPHISM.

See also AUTOMORPHISM, AX-KOCHEN ISOMORPHISM THEOREM, HOMEOMORPHISM, ISOMORPHIC GRAPHS, ISOMORPHIC GROUPS, MORPHISM

Isoperimetric Inequality

Let a PLANE figure have AREA A and PERIMETER p. Then

$$Q \equiv \frac{4\pi A}{p^2} \leq 1,$$

where Q is known as the ISOPERIMETRIC QUOTIENT. The equation becomes an EQUALITY only for a CIRCLE.

See also ISOPERIMETRIC QUOTIENT

References

Osserman, R. "Isoperimetric Inequalities." Appendix 3, §3 in *A Survey of Minimal Surfaces*. New York: Dover, pp. 147–148, 1986.

Solomon, H. *Geometric Probability*. Philadelphia, PA: SIAM, p. 35, 1978.

Isoperimetric Point

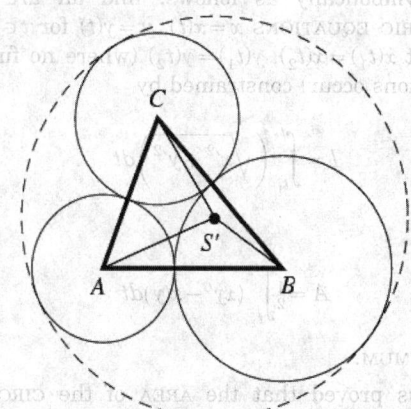

The point S' which makes the PERIMETERS of the TRIANGLES $\Delta BS'C$, $\Delta CS'A$, and $\Delta AS'B$ equal. The isoperimetric point exists IFF the largest ANGLE of the triangle satisfies

$$\max(A, B, C) < 2 \sin^{-1}\left(\tfrac{4}{5}\right) \approx 1.85459 \text{ rad} \approx 106.26°,$$

or equivalently

$$a + b + c > 4R + r,$$

where a, b, and c are the side lengths of ΔABC, r is the INRADIUS, and R is the CIRCUMRADIUS. The isoperimetric point is also the center of the outer SODDY CIRCLE of ΔABC and has TRIANGLE CENTER

FUNCTION

$$\alpha = 1 - \frac{2\Delta}{a(b+c-a)} = \sec\left(\tfrac{1}{2}A\right)\cos\left(\tfrac{1}{2}B\right)\cos\left(\tfrac{1}{2}C\right) - 1.$$

See also EQUAL DETOUR POINT, PERIMETER, SODDY CIRCLES

References

Kimberling, C. "Central Points and Central Lines in the Plane of a Triangle." *Math. Mag.* **67**, 163–187, 1994.

Kimberling, C. "Isoperimetric Point and Equal Detour Point." http://cedar.evansville.edu/~ck6/tcenters/recent/isoper.html.

Kimberling, C. and Wagner, R. W. "Problem E 3020 and Solution." *Amer. Math. Monthly* **93**, 650–652, 1986.

Veldkamp, G. R. "The Isoperimetric Point and the Point(s) of Equal Detour." *Amer. Math. Monthly* **92**, 546–558, 1985.

Isoperimetric Problem

Find a closed plane curve of a given PERIMETER which encloses the greatest AREA. The solution is a CIRCLE. If the class of curves to be considered is limited to smooth curves, the isoperimetric problem can be stated symbolically as follows: find an arc with PARAMETRIC EQUATIONS $x = x(t)$, $y = y(t)$ for $t \in |t_1, t_2|$ such that $x(t_1) = x(t_2)$, $y(t_1) = y(t_2)$ (where no further intersections occur) constrained by

$$l = \int_{t_1}^{t_2}\left(\sqrt{x'^2 + y'^2}\right)dt$$

such that

$$A = \tfrac{1}{2}\int_{t_1}^{t_2}(xy' - x'y)dt$$

is a MAXIMUM.

Zenodorus proved that the AREA of the CIRCLE is larger than that of any POLYGON having the same PERIMETER, but the problem was not rigorously solved until Steiner published several proofs in 1841 (Wells 1991).

See also CIRCLE, DIDO'S PROBLEM, DOUBLE BUBBLE, ISOPERIMETRIC QUOTIENT, ISOPERIMETRIC THEOREM, ISOVOLUME PROBLEM, PERIMETER

References

Bogomolny, A. "Isoperimetric Theorem and Inequality." http://www.cut-the-knot.com/do_you_know/isoperimetric.html.

Isenberg, C. "The Maximum Area Contained by a Given Circumference." Appendix V in *The Science of Soap Films and Soap Bubbles.* New York: Dover, pp. 171–173, 1992.

Steinhaus, H. *Mathematical Snapshots, 3rd ed.* New York: Dover, pp. 149–150, 1999.

Wells, D. *The Penguin Dictionary of Curious and Interesting Geometry.* London: Penguin, pp. 122–124, 1991.

Isoperimetric Quotient

Portions of this entry contributed by HERMANN KREMER

The isoperimetric quotient of a closed curve is defined as the ratio of the curve area to the area of a circle with same perimeter as the curve,

$$Q \equiv \frac{4\pi A}{p^2}, \tag{1}$$

where A is the area of the plane figure and p is its PERIMETER. The ISOPERIMETRIC INEQUALITY gives $Q \leq 1$, with equality only in the case of the CIRCLE.

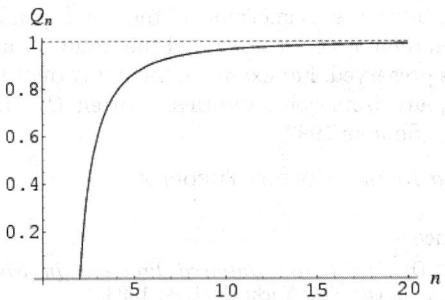

For a regular n-gon with INRADIUS r, the area is given by

$$A = nr^2 \tan\left(\frac{\pi}{n}\right), \tag{2}$$

edge length by

$$a = 2r \tan\left(\frac{\pi}{n}\right), \tag{3}$$

and the perimeter is given by

$$p = na = 2nr \tan\left(\frac{\pi}{n}\right). \tag{4}$$

Thus,

$$Q_n = \frac{\pi}{n \tan\left(\dfrac{\pi}{n}\right)}, \tag{5}$$

which converges to 1 for $n \to \infty$.

See also ISOPERIMETRIC INEQUALITY

References

Croft, H. T.; Falconer, K. J.; and Guy, R. K. *Unsolved Problems in Geometry.* New York: Springer-Verlag, p. 23, 1991.

Isoperimetric Theorem

Of all convex n-gons of a given PERIMETER, the one which maximizes AREA is the regular n-gon.

See also ISOPERIMETRIC INEQUALITY, ISOPERIMETRIC PROBLEM

Isopleth

EQUIPOTENTIAL CURVE

Isoptic Curve

For a given curve C, consider the locus of the point P from where the TANGENTS from P to C meet at a fixed given ANGLE. This is called an isoptic curve of the given curve.

Curve	Isoptic
CYCLOID	curtate or prolate CYCLOID
EPICYCLOID	EPITROCHOID
HYPOCYCLOID	HYPOTROCHOID
PARABOLA	HYPERBOLA
SINUSOIDAL SPIRAL	SINUSOIDAL SPIRAL

See also ORTHOPTIC CURVE

References

Lawrence, J. D. *A Catalog of Special Plane Curves.* New York: Dover, pp. 58–59 and 206, 1972.
Yates, R. C. "Isoptic Curves." *A Handbook on Curves and Their Properties.* Ann Arbor, MI: J. W. Edwards, pp. 138–140, 1952.

Isosceles Tetrahedron

A nonregular TETRAHEDRON in which each pair of opposite EDGES are equal such that all triangular faces are congruent. A TETRAHEDRON is isosceles IFF the sum of the face angles at each VERTEX is 180°, and IFF its INSPHERE and CIRCUMSPHERE are concentric.

The only way for all the faces of a general TETRAHEDRON to have the same PERIMETER or to have the same AREA is for them to be fully congruent, in which case the tetrahedron is isosceles. If the CIRCUMCENTER and the INCENTER of a general TETRAHEDRON coincide, then the TETRAHEDRON is isosceles (Altshiller-Court 1930, p. 97).

See also CIRCUMSPHERE, INSPHERE, ISOSCELES TRIANGLE, TETRAHEDRON

References

Altshiller-Court, N. "The Isosceles Tetrahedron." §4.6b in *Modern Pure Solid Geometry.* New York: Chelsea, pp. 94–101 and 300, 1979.

Biddle, D. Problem 14684. *Math. Questions and Solutions from the Educational Times* **75**, 133–136, 1901.
Biddle, D. *Mathesis*, p. 91, 1931.
Brown, B. H. "Theorem of Bang. Isosceles Tetrahedra." *Amer. Math. Monthly* **33**, 224–226, 1926.
Gentry, E. "Exercices sur le tétraèdre." *Nouvelles ann. de math.* **37**, 223–225, 1878.
Honsberger, R. "A Theorem of Bang and the Isosceles Tetrahedron." Ch. 9 in *Mathematical Gems II.* Washington, DC: Math. Assoc. Amer., pp. 90–97, 1976.
Jacobi, C. F. A. In Swinden, J. H. *Elemente.* p. 457, 1834.
Lemoine, E. "Quelques théorèmes sur les tétraèdres dont les arêtes opposées sont égales deux a deux, et solution de la question 1272." *Nouvelle ann. de math.* **39**, 133–138, 1880.
Lemoine, E. *Z. Math. u. Physik* **29**, 321, 1884.
Monge, G. *Corresp. sur l'École Polytech.*, pp. 1–6, 1809.
Monge, G. Arts. 7 and 8. *Ann. de math.* **1**, 355, 1810–1811.
Morley, F. "Problem 12032." *Math. Questions and Solutions from the Educational Times* **61**, 26–27, 1894.

Isosceles Trapezoid

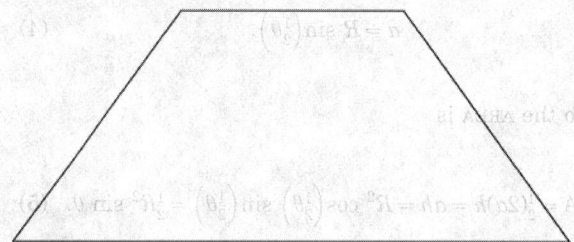

A TRAPEZOID in which the base angles are equal.

See also TRAPEZOID

References

Harris, J. W. and Stocker, H. *Handbook of Mathematics and Computational Science.* New York: Springer-Verlag, p. 83, 1998.

Isosceles Triangle

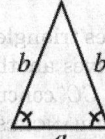

A TRIANGLE with two equal sides (and two equal ANGLES). The name derives from the Greek *iso* (same) and *skelos* (LEG). The height of the above isosceles triangle can be found from the PYTHAGOREAN THEOREM as

$$h = \sqrt{b^2 - \tfrac{1}{4}a^2}. \tag{1}$$

The AREA is therefore given by

$$A = \tfrac{1}{2}ah = \tfrac{1}{2}a\sqrt{b^2 - \tfrac{1}{4}a^2}. \qquad (2)$$

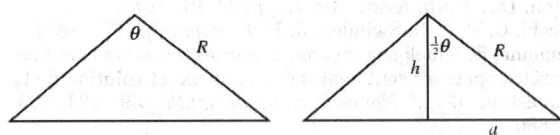

There is a surprisingly simple relationship between the AREA and VERTEX ANGLE θ. As shown in the above diagram, simple TRIGONOMETRY gives

$$h = R\cos\left(\tfrac{1}{2}\theta\right) \qquad (3)$$

$$a = R\sin\left(\tfrac{1}{2}\theta\right), \qquad (4)$$

so the AREA is

$$A = \tfrac{1}{2}(2a)h = ah = R^2\cos\left(\tfrac{1}{2}\theta\right)\sin\left(\tfrac{1}{2}\theta\right) = \tfrac{1}{2}R^2\sin\theta. \quad (5)$$

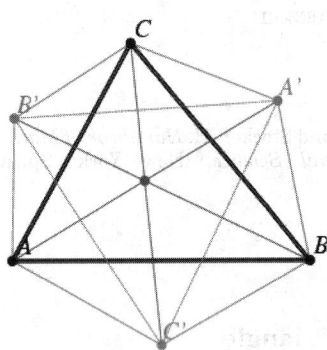

Erecting similar isosceles triangles on the edges of an initial triangle $\triangle ABC$ gives another triangle $\triangle A'B'C'$ such that AA', BB', and CC' concur. The triangles are therefore PERSPECTIVE TRIANGLES.

No set of $n > 6$ points in the PLANE can determine only ISOSCELES TRIANGLES.

See also ACUTE TRIANGLE, EQUILATERAL TRIANGLE, INTERNAL BISECTORS PROBLEM, ISOSCELES TETRAHEDRON, ISOSCELIZER, KIEPERT'S PARABOLA, OBTUSE TRIANGLE, POINT PICKING, PONS ASINORUM, RIGHT TRIANGLE, SCALENE TRIANGLE, STEINER-LEHMUS THEOREM

Isoscelizer

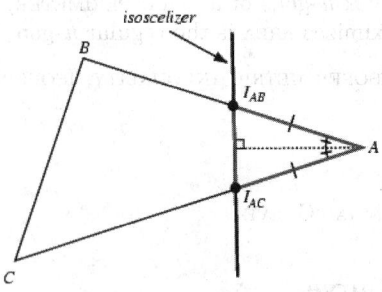

An isoscelizer of an (interior) ANGLE A in a TRIANGLE $\triangle ABC$ is a LINE through points $I_{AB}I_{AC}$ where I_{AB} lies on AB and I_{AC} on AC such that $\triangle AI_{AB}I_{AC}$ is an ISOSCELES TRIANGLE. An isoscelizer is therefore a line perpendicular to an ANGLE BISECTOR, and if the angle is A, the line is known as an A-isoscelizer. There are obviously an infinite number of isoscelizers for any given angle. Isoscelizers were invented by P. Yff in 1963.

Through any point P draw the line parallel to BC as well as the corresponding ANTIPARALLEL. Then the A-isoscelizer through P bisects the angle formed by the parallel and the antiparallel. Another way of saying this is that an isoscelizer is a line which is both parallel and antiparallel to itself (P. Yff).

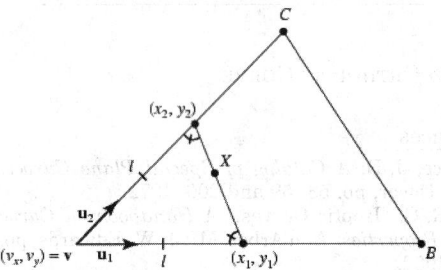

Let $\mathbf{u}_1 = (u_{1x}, u_{1y})$ and $\mathbf{u}_2 = (u_{2x}, u_{2y})$ be the unit vectors from a given vertex $\mathbf{v} = (v_x, v_y)$, let $X = (x, y)$ be a point in the interior of a triangle through which an isoscelizer passes, and the side lengths of the isosceles triangle be l. Then setting the POINT-LINE DISTANCE from the vector (u_1, u_2) to the point $\mathbf{x}$ equal to 0 gives

$$(y_2 - y_1)(x_0 - x_1) - (x_2 - x_1)(y_0 - y_1) = 0 \qquad (1)$$

$$l(u_{2y} - u_{1y})[(x - v_x) - lu_{1x}]$$

$$-l(u_{2x} - u_{1x})[(y - v_y) - lu_{1y}] = 0 \qquad (2)$$

$$l = \frac{(x - v_x)(u_{2y} - u_{1y}) - (y - v_y)(u_{2x} - u_{1x})}{u_{1x}u_{2y} - u_{2x}u_{1y}}. \qquad (3)$$

See also ANGLE BISECTOR, ANTIPARALLEL, CONGRUENT ISOSCELIZERS POINT, ISOSCELES TRIANGLE, YFF CENTER OF CONGRUENCE, YFF CENTRAL TRIANGLE

Isospectral Manifolds

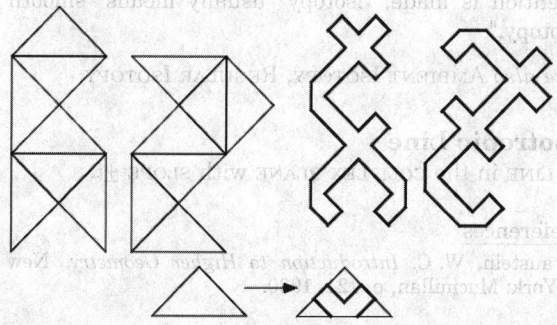

DRUMS that sound the same, i.e., have the same eigenfrequency spectrum. Two drums with differing AREA, PERIMETER, or GENUS can always be distinguished. However, Kac (1966) asked if it was possible to construct differently shaped drums which have the same eigenfrequency spectrum. This question was answered in the affirmative by Gordon *et al.* (1992). Two such isospectral manifolds are shown in the right figure above (Cipra 1992).

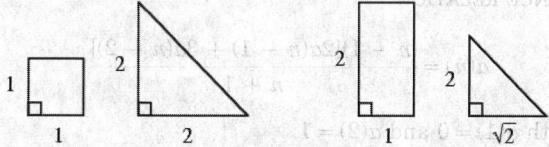

Furthermore, pairs of separate drums (having the same total area) can be constructed which have the same eigenfrequency spectrum when played together (illustrated above). Therefore, you cannot hear the shape of a two-piece band (Zwillinger 1995, p. 426).

References

Chapman, S. J. "Drums That Sound the Same." *Amer. Math. Monthly* **102**, 124–138, 1995.

Cipra, B. "You Can't Hear the Shape of a Drum." *Science* **255**, 1642–1643, 1992.

Gordon, C.; Webb, D.; and Wolpert, S. "Isospectral Plane Domains and Surfaces via Riemannian Orbifolds." *Invent. Math.* **110**, 1–22, 1992.

Gordon, C.; Webb, D.; and Wolpert, S. "You Cannot Hear the Shape of a Drum." *Bull. Amer. Math. Soc.* **27**, 134–138, 1992.

Kac, M. "Can One Hear the Shape of a Drum?" *Amer. Math. Monthly* **73**, 1–23, 1966.

Zwillinger, D. (Ed.). "Eigenvalues." §5.8 in *CRC Standard Mathematical Tables and Formulae.* Boca Raton, FL: CRC Press, pp. 425–426, 1995.

Isothermal Parameterization

A parameterization is isothermal if, for $\zeta \equiv u + iv$ and

$$\phi_k(\zeta) = \frac{\partial x_k}{\partial u} - i\frac{\partial x_k}{\partial v},$$

the identity

$$\phi_1^2(\zeta) + \phi_2^2(\zeta) + \phi_3^2(\zeta) = 0$$

holds.

See also MINIMAL SURFACE, TEMPERATURE

References

Osserman, R. "Isothermal Parameters." §4 in *A Survey of Minimal Surfaces.* New York: Dover, pp. 27–33, 1986.

Isotomic Conjugate Point

The point of concurrence Q of the ISOTOMIC LINES relative to a point P. The isotomic conjugate $\alpha' : \beta' : \gamma'$ of a point with TRILINEAR COORDINATES $\alpha : \beta : \gamma$ is

$$(a^2\alpha)^{-1} : (b^2\beta)^{-1} : (c^2\gamma)^{-1}. \tag{1}$$

The isotomic conjugate of a LINE d having trilinear equation

$$l\alpha + m\beta + n\gamma \tag{2}$$

is a CONIC SECTION circumscribed on the TRIANGLE $\triangle ABC$ (Casey 1893, Vandeghen 1965). The isotomic conjugate of the LINE AT INFINITY having trilinear equation

$$a\alpha + b\beta + c\gamma = 0 \tag{3}$$

is STEINER'S ELLIPSE

$$\frac{\beta'\gamma'}{a} + \frac{\gamma'\alpha'}{b} + \frac{\alpha'\beta'}{c} = 0 \tag{4}$$

(Vandeghen 1965). The type of CONIC SECTION to which d is transformed is determined by whether the line d meets STEINER'S ELLIPSE E.

1. If d does not intersect E, the isotomic transform is an ELLIPSE.
2. If d is tangent to E, the transform is a PARABOLA.
3. If d cuts E, the transform is a HYPERBOLA, which is a RECTANGULAR HYPERBOLA if the line passes through the isotomic conjugate of the ORTHOCENTER

(Casey 1893, Vandeghen 1965).

There are four points which are isotomically self-conjugate: the CENTROID M and each of the points of intersection of lines through the VERTICES PARALLEL to the opposite sides. The isotomic conjugate of the EULER LINE is called JERABEK'S HYPERBOLA (Casey 1893, Vandeghen 1965).

Vandeghen (1965) calls the transformation taking points to their isotomic conjugate points the CEVIAN TRANSFORM. The product of isotomic and ISOGONAL is a COLLINEATION which transforms the sides of a TRIANGLE to themselves (Vandeghen 1965).

See also CEVIAN TRANSFORM, GERGONNE POINT, ISOGONAL CONJUGATE, JERABEK'S HYPERBOLA, NAGEL POINT, STEINER'S ELLIPSE

References

Casey, J. "Theory of Isogonal and Isotomic Points, and of Antiparallel and Symmedian Lines." Supp. Ch. §1 in *A Sequel to the First Six Books of the Elements of Euclid, Containing an Easy Introduction to Modern Geometry with Numerous Examples, 5th ed., rev. enl.* Dublin: Hodges, Figgis, & Co., pp. 165–173, 1888.

Casey, J. *A Treatise on the Analytical Geometry of the Point, Line, Circle, and Conic Sections, Containing an Account of Its Most Recent Extensions with Numerous Examples, 2nd rev. enl. ed.* Dublin: Hodges, Figgis, & Co., 1893.

Eddy, R. H. and Fritsch, R. "The Conics of Ludwig Kiepert: A Comprehensive Lesson in the Geometry of the Triangle." *Math. Mag.* **67**, 188–205, 1994.

Johnson, R. A. *Modern Geometry: An Elementary Treatise on the Geometry of the Triangle and the Circle.* Boston, MA: Houghton Mifflin, pp. 157–159, 1929.

Vandeghen, A. "Some Remarks on the Isogonal and Cevian Transforms. Alignments of Remarkable Points of a Triangle." *Amer. Math. Monthly* **72**, 1091–1094, 1965.

Isotomic Lines

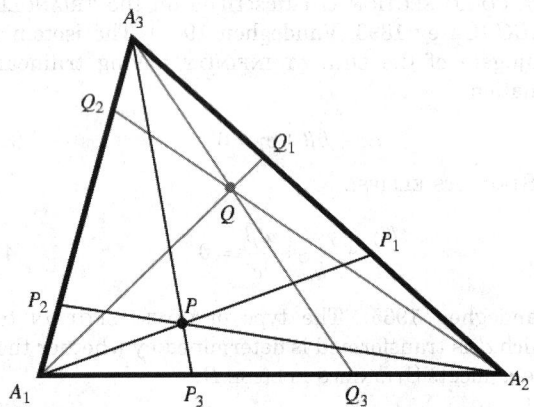

Given a point P in the interior of a TRIANGLE $\Delta A_1 A_2 A_3$, draw the CEVIANS through P from each VERTEX which meet the opposite sides at P_1, P_2, and P_3. Now, mark off point Q_1 along side $A_2 A_3$ such that $A_3 P_1 = A_2 Q_1$, etc., i.e., so that Q_i and P_i are equidistance from the MIDPOINT of $A_j A_k$. The lines $A_1 Q_1$, $A_2 Q_2$, and $A_3 Q_3$ then coincide in a point Q known as the ISOTOMIC CONJUGATE POINT.

See also CEVIAN, ISOTOMIC CONJUGATE POINT, MIDPOINT

Isotone Map

A MAP which is monotone increasing and therefore order-preserving.

Isotope

To rearrange without cutting or pasting.

Isotopy

A HOMOTOPY from one embedding of a MANIFOLD M in N to another such that at every time, it is an embedding. The notion of isotopy is category independent, so notions of topological, piecewise-linear, smooth, isotopy (and so on) exist. When no explicit mention is made, "isotopy" usually means "smooth isotopy."

See also AMBIENT ISOTOPY, REGULAR ISOTOPY

Isotropic Line

A LINE in the COMPLEX PLANE with SLOPE $\pm i$.

References

Graustein, W. C. *Introduction to Higher Geometry.* New York: Macmillan, p. 121, 1930.

Isotropic Tensor

A TENSOR which has the same components in all rotated coordinate systems. All rank-0 TENSORS (SCALARS) are isotropic, but no rank-1 TENSORS (VECTORS) are. The unique rank-2 isotropic tensor is the KRONECKER DELTA. The number of isotropic tensors of rank 0, 1, 2, ... are 1, 0, 1, 1, 3, 6, 15, 36, 91, 232, ... (Sloane's A005043). These numbers are called the Motzkin sum numbers and are given by the RECURRENCE RELATION

$$a(n) = \frac{(n-1)[2a(n-1) + 3a(n-2)]}{n+1}$$

with $a(1) = 0$ and $a(2) = 1$.

Starting at rank 5, SYZYGIES play a role in restricting the number of isotropic tensors. In particular, SYZYGIES occur at rank 5, 7, 8, and all higher ranks.

See also KRONECKER DELTA, SCALAR, SYZYGY, TENSOR, VECTOR

References

Jeffreys, H. and Jeffreys, B. S. "Isotropic Tensors." §3.03 in *Methods of Mathematical Physics, 3rd ed.* Cambridge, England: Cambridge University Press, pp. 87–89, 1988.

Kearsley, E. A. and Fong, J. T. ""Linearly Independent Sets of Isotropic Cartesian Tensors of Ranks up to Eight." *J. Res. Nat. Bureau Standards* **79B**, 49–58, 1975.

Sloane, N. J. A. Sequences A005043/M2587 in "An On-Line Version of the Encyclopedia of Integer Sequences." http://www.research.att.com/~njas/sequences/eisonline.html.

Smith G. F. "On Isotropic Tensors and Rotation Tensors of Dimension m and Order n." *Tensor, N. S.* **19**, 79–88, 1968.

Isotropy Group

Some elements of a GROUP G ACTING on a space X may fix a point x. These group elements form a SUBGROUP called the isotropy group, defined by

$$G_x = \{g \in G \text{ such that } gx = x\}.$$

For example, consider the group $SO(3)$ of all rotations of a sphere $\mathbb{S}^2$. Let x be the north pole $(0,0,1)$. Then a rotation which does not change x must turn about the usual axis, leaving the north pole and the south pole fixed. These rotations correspond to the action of the circle group $\mathbb{S}^1$ on the equator.

When two points x and y are on the same ORBIT, say $y = gx$, then the isotropy groups are CONJUGATE SUBGROUPS. More precisely, $G_y = gG_xg^{-1}$. In fact, any subgroup conjugate to G_x occurs as an isotropy group G_y to some point y on the same orbit as x.

See also EFFECTIVE ACTION, FREE ACTION, GROUP ACTION, MATRIX GROUP, ORBIT (GROUP), QUOTIENT SPACE (LIE GROUP), REPRESENTATION, TOPOLOGICAL GROUP, TRANSITIVE

References

Kawakubo, K. *The Theory of Transformation Groups.* Oxford, England: Oxford University Press, pp. 4 and 49–52, 1987.

Isovolume Problem

Find the surface enclosing the maximum VOLUME per unit SURFACE AREA, $I \equiv V/S$. The solution is a SPHERE, which has

$$I_{\text{sphere}} = \frac{\frac{4}{3}\pi r^3}{4\pi r^2} = \frac{1}{3}r.$$

The fact that a sphere solves the isovolume problem was only proved as recently as 1882 by Schwarz (Haas 2000).

See also DIDO'S PROBLEM, DOUBLE BUBBLE, ISOPERIMETRIC PROBLEM, SPHERE, SURFACE AREA, VOLUME

References

Bogomolny, A. "Isoperimetric Theorem and Inequality." http://www.cut-the-knot.com/do_you_know/isoperimetric.html.

Haas, J. "General Double Bubble Conjecture in $\mathbb{R}^3$ Solved." *Focus: The Newsletter of the Math. Assoc. Amer.*, No. 5, pp. 4–5, May/June 2000.

Isenberg, C. "The Maximum Volume Contained by a Closed Surface of Fixed Area." Appendix VI in *The Science of Soap Films and Soap Bubbles.* New York: Dover, pp. 174–177, 1992.

Steinhaus, H. *Mathematical Snapshots, 3rd ed.* New York: Dover, p. 214, 1999.

Isthmus

BRIDGE

Iterated Exponential

POWER TOWER

Iterated Function System

A finite set of contraction maps ω_i for $i = 1, 2, \ldots, N$, each with a contractivity factor $s < 1$, which map a compact METRIC SPACE onto itself. It is the basis for FRACTAL image compression techniques.

See also BARNSLEY'S FERN, SELF-SIMILARITY

References

Barnsley, M. F. "Fractal Image Compression." *Not. Amer. Math. Soc.* **43**, 657–662, 1996.

Barnsley, M. *Fractals Everywhere, 2nd ed.* Boston, MA: Academic Press, 1993.

Barnsley, M. F. and Demko, S. G. "Iterated Function Systems and the Global Construction of Fractals." *Proc. Roy. Soc. London, Ser. A* **399**, 243–275, 1985.

Barnsley, M. F. and Hurd, L. P. *Fractal Image Compression.* Wellesley, MA: A. K. Peters, 1993.

Diaconis, P. M. and Shashahani, M. "Products of Random Matrices and Computer Image Generation." *Contemp. Math.* **50**, 173–182, 1986.

Fisher, Y. *Fractal Image Compression.* New York: Springer-Verlag, 1995.

Hutchinson, J. "Fractals and Self-Similarity." *Indiana Univ. J. Math.* **30**, 713–747, 1981.

Wagon, S. "Iterated Function Systems." §5.2 in *Mathematica in Action.* New York: W. H. Freeman, pp. 149–156, 1991.

Iterated Radical

NESTED RADICAL

Iteration

The repeated application of a transformation.

See also ITERATED FUNCTION SYSTEM, ITERATION SEQUENCE, POWER TOWER

References

Chang, G. and Sederberg, T. W. *Over and Over Again.* Washington, DC: Math. Assoc. Amer., 1997.

Iteration Sequence

A SEQUENCE $\{a_j\}$ of POSITIVE INTEGERS is called an iteration sequence if there EXISTS a strictly INCREASING SEQUENCE $\{s_k\}$ of POSITIVE INTEGERS such that $a_1 = s_1 \geq 2$ and $a_j = s_{a_{j-1}}$ for $j = 2, 3, \ldots$. A NECESSARY and SUFFICIENT condition for $\{a_j\}$ to be an iteration sequence is

$$a_j \geq 2a_{j-1} - a_{j-2}$$

for all $j \geq 3$.

References

Kimberling, C. "Interspersions and Dispersions." *Proc. Amer. Math. Soc.* **117**, 313–321, 1993.

Itô's Lemma

Let $W(u)$ be a WIENER PROCESS. Then

$$V_t - V_0 = \int_0^t f_x(W(u), u)\,dW(u) - \int_0^t f_\tau(W(u), u)\,du$$

$$+ \tfrac{1}{2}\int_0^t f_{xx}(W(u), u)\,du,$$

where $V_t = f(W(t), \tau)$ for $0 \leq \tau \equiv T - t \leq T$, and $f \in C^{2,1}((0, \infty) \times [0, T])$.

See also WIENER PROCESS

References

Karatsas, I. and Shreve, S. *Brownian Motion and Stochastic Calculus, 2nd ed.* New York: Springer-Verlag, 1997.

Price, J. F. "Optional Mathematics is Not Optional." *Not. Amer. Math. Soc.* **43**, 964–971, 1996.

Itô's Theorem

The dimension d of any IRREDUCIBLE REPRESENTATION of a GROUP G must be a DIVISOR of the index of each maximal normal ABELIAN SUBGROUP of G.

See also ABELIAN GROUP, IRREDUCIBLE REPRESENTATION, SUBGROUP

References

Lomont, J. S. *Applications of Finite Groups.* New York: Dover, p. 55, 1993.

Iverson Bracket

Let S be a mathematical statement, then the Iverson bracket is defined by

$$[S] \equiv \begin{cases} 0 & \text{if } S \text{ is false} \\ 1 & \text{if } S \text{ is true.} \end{cases}$$

This notation conflicts with the brackets sometimes used to denote the FLOOR FUNCTION. (However, because of the elegant symmetry of the FLOOR FUNCTION and CEILING FUNCTION symbols $\lfloor x \rfloor$ and $\lceil x \rceil$, the use of $[x]$ to denote the FLOOR FUNCTION should be deprecated.) The Iverson bracket is implemented in *Mathematica* 4.1 as `Boole[S]`.

See also CEILING FUNCTION, FLOOR FUNCTION

References

Graham, R. L.; Knuth, D. E.; and Patashnik, O. *Concrete Mathematics: A Foundation for Computer Science.* Reading, MA: Addison-Wesley, p. 24, 1990.

Iverson, K. E. *A Programming Language.* New York: Wiley, p. 11, 1962.

Iwasawa's Theorem

Every finite-dimensional LIE ALGEBRA of characteristic $p \neq 0$ has a FAITHFUL finite-dimensional representation.

See also ADO'S THEOREM, LIE ALGEBRA

References

Jacobson, N. *Lie Algebras.* New York: Dover, pp. 204–205, 1979.

J

j

The symbol used by engineers and some physicists to denote $\imath$, the IMAGINARY NUMBER $\sqrt{-1}$. j is probably preferred over i because the symbol i (or I) is commonly used to denote current.

Jack Polynomial

References

Lasalle, M. "Some Combinatorial Conjectures for Jack Polynomials." *Ann. Combin.* **2**, 61–3, 1998.

Jackknife

See also BOOTSTRAP METHODS, PERMUTATION TESTS, RESAMPLING STATISTICS

Jackson's Difference Fan

If, after constructing a DIFFERENCE TABLE, no clear pattern emerges, turn the paper through an ANGLE of $60°$ and compute a new table. If necessary, repeat the process. Each ROTATION reduces POWERS by 1, so the sequence $\{k^n\}$ multiplied by any POLYNOMIAL in n is reduced to 0s by a k-fold difference fan.

References

Conway, J. H. and Guy, R. K. "Jackson's Difference Fans." In *The Book of Numbers*. New York: Springer-Verlag, pp. 84–5, 1996.

Jackson's Identity

The Q-HYPERGEOMETRIC FUNCTION identity

$$_r\phi_s'\left[\begin{matrix} a, q\sqrt{a}, -q\sqrt{a}, 1/b, 1/c, 1/d, 1/e, 1/f \\ \sqrt{a}, -\sqrt{a}, abq, acq, adq, aeq, afq \end{matrix}\right]$$
$$= \frac{(aq)_q^m (aqde)_q^m (adec)_q^m (aqcd)_q^m}{(aqc)_q^m (aqd)_q^m (aqe)_q^m (aqcde)_q^m},$$

where

$$a^2 bcdefq = 1,$$

$_r\phi_s'$ is a Q-HYPERGEOMETRIC FUNCTION, and one of b, c, d, e, or f is equal to q^m (Hardy 1999, pp. 108–09). This identity includes the DOUGALL-RAMANUJAN IDENTITY as a special case.

See also DOUGALL-RAMANUJAN IDENTITY, Q-HYPERGEOMETRIC FUNCTION

References

Bailey, W. N. *Generalised Hypergeometric Series.* Cambridge, England: Cambridge University Press, pp. 66–2, 1935.

Hardy, G. H. *Ramanujan: Twelve Lectures on Subjects Suggested by His Life and Work*, 3rd ed. New York: Chelsea, pp. 109–10, 1959.
Jackson, F. H. "Summation of q-Hypergeometric Series." *Messenger Math.* **50**, 101–12, 1921.

Jackson's Theorem

Jackson's theorem is a statement about the error $E_n(f)$ of the best uniform approximation to a REAL FUNCTION $f(x)$ on $[-1, 1]$ by REAL POLYNOMIALS of degree at most n. Let $f(x)$ be of bounded variation in $[-1, 1]$ and let M' and V' denote the least upper bound of $|f(x)|$ and the total variation of $f(x)$ in $[-1, 1]$, respectively. Given the function

$$F(x) = F(-1) + \int_{-1}^{x} f(x)dx, \tag{1}$$

then the coefficients

$$a_n = \tfrac{1}{2}(2n+1) \int_{-1}^{1} F(x)P_n(x)dx \tag{2}$$

of its LEGENDRE SERIES, where $P_n(x)$ is a LEGENDRE POLYNOMIAL, satisfy the inequalities

$$|a_n| < \begin{cases} \dfrac{6}{\sqrt{\pi}}(M'+V')n^{-3/2} & \text{for } n \geq 1 \\[2mm] \dfrac{4}{\sqrt{\pi}}(M'+V')n^{-3/2} & \text{for } n \geq 1 \end{cases} \tag{3}$$

Moreover, the LEGENDRE SERIES of $F(x)$ converges uniformly and absolutely to $F(x)$ in $[-1, 1]$.

Bernstein strengthened Jackson's theorem to

$$2nE_{2n}(\alpha) \leq \frac{4n}{\pi(2n+1)} < \frac{2}{\pi} = 0.6366. \tag{4}$$

A specific application of Jackson's theorem shows that if

$$\alpha(x) = |x|, \tag{5}$$

then

$$E_n(\alpha) \leq \frac{6}{n}. \tag{6}$$

See also LEGENDRE SERIES, PICONE'S THEOREM

References

Cheney, E. W. *Introduction to Approximation Theory*, 2nd ed. Providence, RI: Amer. Math. Soc., 1999.
Jackson, D. *The Theory of Approximation.* New York: Amer. Math. Soc., p. 76, 1930.
Rivlin, T. J. *An Introduction to the Approximation of Functions.* New York: Dover, 1981.
Sansone, G. *Orthogonal Functions*, rev. English ed. New York: Dover, pp. 205–08, 1991.

Jacobi Algorithm

A method which can be used to solve a TRIDIAGONAL MATRIX equation with largest absolute values in each row and column dominated by the diagonal element. Each diagonal element is solved for, and an approximate value plugged in. The process is then iterated until it converges. This algorithm is a stripped-down version of the JACOBI METHOD of matrix diagonalization.

See also JACOBI METHOD, TRIDIAGONAL MATRIX

References

Acton, F. S. *Numerical Methods That Work, 2nd printing.* Washington, DC: Math. Assoc. Amer., pp. 161–63, 1990.

Jacobi Differential Equation

$$\left(1-x^2\right)y'' + \left[\beta - \alpha - (\alpha + \beta + 2)x\right]y'$$
$$+ n(n + \alpha + \beta + 1)y = 0 \tag{1}$$

or

$$\frac{d}{dx}\left[(1-x)^{\alpha+1}(1+x)^{\beta+1}y'\right] + n(n + \alpha + \beta + 1)$$
$$\times (1-x)^{\alpha}(1+x)^{\beta}y = 0. \tag{2}$$

The solutions are JACOBI POLYNOMIALS $P_n^{(\alpha,\beta)}(x)$ or, in terms of hypergeometric functions, as

$$y(x) = C_1\,{}_2F_1\!\left(-n, n + 1 + \alpha + \beta, 1 + \alpha, \tfrac{1}{2}(x-1)\right)$$
$$+ 2^{\alpha}(x-1)^{-\alpha}C_2\,{}_2F_1\!\left(-n - \alpha, n + 1 + \beta, 1 - \alpha, \tfrac{1}{2}(1-x)\right). \tag{3}$$

The equation (2) can be transformed to

$$\frac{d^2y}{dx^2} + \left[\frac{1}{4}\frac{1-\alpha^2}{(1-x)^2} + \frac{1}{4}\frac{1-\beta^2}{(1+x)^2}\right.$$
$$\left. + \frac{n(n + \alpha + \beta + 1) + \frac{1}{2}(\alpha + 1)(\beta + 1)}{1 - x^2}\right]u = 0, \tag{4}$$

where

$$u(x) = (1-x)^{(\alpha+1)/2}(1+x)^{(\beta+1)/2}P_n^{(\alpha,\beta)}(x), \tag{5}$$

and

$$\frac{d^2u}{d\theta^2} + \left[\frac{\frac{1}{4} - \alpha^2}{4\sin^2\left(\frac{1}{2}\theta\right)} + \frac{\frac{1}{4} - \beta^2}{4\cos^2\left(\frac{1}{2}\theta\right)} + \left(n + \frac{\alpha + \beta + 1}{2}\right)^2\right]u$$
$$= 0, \tag{6}$$

where

$$u(\theta) = \sin^{\alpha+1/2}\left(\tfrac{1}{2}\theta\right)\cos^{\beta+1/2}\left(\tfrac{1}{2}\theta\right)P_n^{(\alpha,\beta)}(\cos\theta). \tag{7}$$

Zwillinger (1997, p. 123) gives a related differential

equation he terms Jacobi's equation

$$x(1-x)y'' + [\gamma - (\alpha + 1)x]y' + n(\alpha + n)y = 0 \tag{8}$$

(Iyanaga and Kawada 1980, p. 1480), which has solution

$$y = C_1\,{}_2F_1(-n, n + \alpha, \gamma, x)$$
$$-(-1)^{-\gamma}x^{1-\gamma}C_2\,{}_2F_1(1 - n - \gamma, 1 + n + \alpha - \gamma, 2 - \gamma, x). \tag{9}$$

Zwillinger (1997, p. 120; duplicated twice) also gives another types of ordinary differential equation called a Jacobi equation,

$$(a_1 + b_1x + c_1y)(xy' - y) - (a_2 + b_2x + c_2y)y'$$
$$+ (a_3 + b_3x + c_3y) = 0 \tag{10}$$

(Ince 1956, p. 22).

In the CALCULUS OF VARIATIONS, the PARTIAL DIFFERENTIAL EQUATION

$$\frac{d}{dx}\Omega_{\eta'} - \Omega_{\eta} = \frac{d}{dx}\left(f_{y'y}\eta + f_{y'y'}\eta'\right) - \left(f_{yy}\eta + f_{yy'}\eta'\right) = 0, \tag{11}$$

where

$$\Omega(x, \eta, \eta') \equiv \tfrac{1}{2}(f_{yy}\eta^2 + 2f_{yy'}\eta\eta' + f_{y'y'}\eta'^2) \tag{12}$$

is called the Jacobi differential equation.

References

Bliss, G. A. *Calculus of Variations.* Chicago, IL: Open Court, pp. 162–63, 1925.
Ince, E. L. *Ordinary Differential Equations.* New York: Dover, p. 22, 1956.
Iyanaga, S. and Kawada, Y. (Eds.). *Encyclopedic Dictionary of Mathematics.* Cambridge, MA: MIT Press, p. 1480, 1980.
Zwillinger, D. *Handbook of Differential Equations, 3rd ed.* Boston, MA: Academic Press, p. 120, 1997.

Jacobi Differential Equation (Calculus of Variations)

$$u(x) = (1-x)^{(\alpha+1)/2}(1+x)^{(\beta+1)/2}P_n^{(\alpha,\beta)}(x),$$

where

$$\frac{d^2u}{d\theta^2} + \left[\frac{\frac{1}{4} - \alpha^2}{4\sin^2\left(\frac{1}{2}\theta\right)} + \frac{\frac{1}{4} - \beta^2}{4\cos^2\left(\frac{1}{2}\theta\right)} + \left(n + \frac{\alpha + \beta + 1}{2}\right)^2\right]u$$
$$= 0,$$

This equations arises in the CALCULUS OF VARIATIONS.

References

Bliss, G. A. *Calculus of Variations.* Chicago, IL: Open Court, pp. 162–63, 1925.

Jacobi Elliptic Functions

The Jacobi elliptic functions are standard forms of ELLIPTIC FUNCTIONS. The three basic functions are denoted $\mathrm{cn}(u,k)$, $\mathrm{dn}(u,k)$, and $\mathrm{sn}(u,k)$, where k is known as the MODULUS. The arise from the inversion of the ELLIPTIC INTEGRAL OF THE FIRST KIND,

$$u = F(\phi, k) = \int_o^\phi \frac{dt}{\sqrt{1 - k^2 \sin^2 t}}, \tag{1}$$

where $0 < k^2 < 1$, $k = \mathrm{mod}\ u$ is the MODULUS, and $\phi = \mathrm{am}(u,k) = \mathrm{am}(u)$ is the AMPLITUDE, giving

$$\phi = F^{-1}(u,k) = \mathrm{am}(u,k) = \mathrm{am}(u). \tag{2}$$

From this, it follows that

$$\sin\phi = \sin(\mathrm{am}(u,k)) = \sin(\mathrm{am}\ u) = \mathrm{sn}(u,k) = \mathrm{sn}(u) \tag{3}$$

$$\cos\phi = \cos(\mathrm{am}(u,k)) = \cos(\mathrm{am}\ u) = \mathrm{cn}(u,k) = \mathrm{cn}(u) \tag{4}$$

$$\sqrt{1 - k^2 \sin^2\phi} = \sqrt{1 - k^2 \sin^2(\mathrm{am}(u,k))}$$
$$= \sqrt{1 - k^2\ \mathrm{sn}^2\ u} = \mathrm{dn}(u,k) = \mathrm{dn}(u). \tag{5}$$

These functions are doubly periodic generalizations of the trigonometric functions satisfying

$$\mathrm{sn}(u,0) = \sin u \tag{6}$$

$$\mathrm{cn}(u,0) = \cos u \tag{7}$$

$$\mathrm{dn}(u,0) = 1. \tag{8}$$

In terms of JACOBI THETA FUNCTIONS,

$$\mathrm{sn}(u,k) = \frac{\vartheta_3}{\vartheta_4} \frac{\vartheta_1(u\vartheta_3^{-2})}{\vartheta_4(u\vartheta_3^{-2})} \tag{9}$$

$$\mathrm{cn}(u,k) = \frac{\vartheta_4}{\vartheta_2} \frac{\vartheta_2(u\vartheta_3^{-2})}{\vartheta_4(u\vartheta_3^{-2})} \tag{10}$$

$$\mathrm{dn}(u,k) = \frac{\vartheta_4}{\vartheta_3} \frac{\vartheta_3(u\vartheta_3^{-2})}{\vartheta_4(u\vartheta_3^{-2})} \tag{11}$$

(Whittaker and Watson 1990, p. 492), where $\vartheta_i \equiv \vartheta_i(0)$ (Whittaker and Watson 1990, p. 464). Ratios of Jacobi elliptic functions are denoted by combining the first letter of the NUMERATOR elliptic function with the first of the DENOMINATOR elliptic function. The multiplicative inverses of the elliptic functions are denoted by reversing the order of the two letters. These combinations give a total of 12 functions: cd, cn, cs, dc, dn, ds, nc, nd, ns, sc, sd, and sn. The AMPLITUDE ϕ is defined in terms of sn u by

$$y = \sin\phi = \mathrm{sn}(u,k). \tag{12}$$

The k argument is often suppressed for brevity so, for example, $\mathrm{sn}(u,k)$ can be written as sn u.

The Jacobi elliptic functions are periodic in $K(k)$ and $K'(k)$ as

$$\mathrm{sn}(u + 2mK + 2niK', k) = (-1)^m\ \mathrm{sn}(u,k) \tag{13}$$

$$\mathrm{cn}(u + 2mK + 2niK', k) = (-1)^{m+n}\ \mathrm{cn}(u,k) \tag{14}$$

$$\mathrm{dn}(u + 2mK + 2niK', k) = (-1)^n\ \mathrm{dn}(u,k), \tag{15}$$

where $K(k)$ is the complete ELLIPTIC INTEGRAL OF THE FIRST KIND, $K'(k) \equiv K(k')$, and $k' \equiv \sqrt{1 - k^2}$ (Whittaker and Watson 1990, p. 503).

The cn x, dn x, and sn x functions may also be defined as solutions to the differential equations

$$\frac{d^2y}{dx^2} = -(1 + k^2)y + 2k^2 y^3 \tag{16}$$

$$\frac{d^2y}{dx^2} = -(1 - 2k^2)y - 2k^2 y^3 \tag{17}$$

$$\frac{d^2y}{dx^2} = (2 - k^2)y - 2y^3. \tag{18}$$

The standard Jacobi elliptic functions satisfy the identities

$$\mathrm{sn}^2\ u + \mathrm{cn}^2\ u = 1 \tag{19}$$

$$k^2\ \mathrm{sn}^2\ u + \mathrm{dn}^2\ u = 1 \tag{20}$$

$$k^2\ \mathrm{cn}^2\ u + k'^2 = \mathrm{dn}^2\ u \tag{21}$$

$$\mathrm{cn}^2\ u + k'^2\ \mathrm{sn}^2\ u = \mathrm{dn}^2\ u. \tag{22}$$

Special values include

$$\mathrm{cn}(0,k) = \mathrm{cn}(0) = 1 \tag{23}$$

$$\mathrm{cn}(K(k),k) = \mathrm{cn}(K(k)) = 0 \tag{24}$$

$$\mathrm{dn}(0,k) = \mathrm{dn}(0) = 1 \tag{25}$$

$$\mathrm{dn}(K(k),k) = \mathrm{dn}(K(k)) = k' \equiv \sqrt{1 - k^2}, \tag{26}$$

$$\mathrm{sn}(0,k) = \mathrm{sn}(0) = 0 \tag{27}$$

$$\mathrm{sn}(K(k),k) = \mathrm{sn}(K(k)) = 1, \tag{28}$$

where $K = K(k)$ is a complete ELLIPTIC INTEGRAL OF THE FIRST KIND and $k' = \sqrt{1 - k^2}$ is the complementary MODULUS (Whittaker and Watson 1990, pp. 498–99), and

$$\mathrm{cn}(u,1) = \mathrm{sech}\ u \tag{29}$$

$$\mathrm{dn}(u,1) = \mathrm{sech}\ u \tag{30}$$

$$\mathrm{sn}(u,1) = \tanh u. \tag{31}$$

In terms of integrals,

$$u = \int_0^{\mathrm{sn}\ u} (1 - t^2)^{-1/2}(1 - k^2 t^2)^{-1/2} dt \tag{32}$$

$$= \int_{\mathrm{ns}\ u}^\infty (t^2 - 1)^{-1/2}(t^2 - l^2)^{-1/2} dt \tag{33}$$

$$= \int_{\mathrm{cn}\ u}^1 (1 - t^2)^{-1/2}(k'^2 + k^2 t^2)^{-1/2} dt \tag{34}$$

$$= \int_1^{\text{nc } u} \left(t^2 - 1\right)^{-1/2} \left(k'^2 t^2 + k^2\right)^{-1/2} dt \tag{35}$$

$$= \int_{\text{dn } u}^1 \left(1 - t^2\right)^{-1/2} \left(t^2 - k'^2\right)^{-1/2} dt \tag{36}$$

$$= \int_1^{\text{nd } u} \left(t^2 - 1\right)^{-1/2} \left(1 - k'^2 t^2\right)^{-1/2} dt \tag{37}$$

$$= \int_0^{\text{sc } u} \left(1 + t^2\right)^{-1/2} \left(1 + k'^2 t^2\right)^{-1/2} dt \tag{38}$$

$$= \int_{\text{cs } u}^\infty \left(t^2 + 1\right)^{-1/2} \left(t^2 + k'^2\right)^{-1/2} dt \tag{39}$$

$$= \int_0^{\text{sd } u} \left(1 - k'^2 t^2\right)^{-1/2} \left(1 + k^2 t^2\right)^{-1/2} dt \tag{40}$$

$$= \int_{\text{ds } u}^\infty \left(t^2 - k'^2\right)^{-1/2} \left(t^2 + k^2\right)^{-1/2} dt \tag{41}$$

$$= \int_1^{\text{cd } u} \left(1 - t^2\right)^{-1/2} \left(1 - k^2 t^2\right)^{-1/2} dt \tag{42}$$

$$= \int_{\text{dc } u}^1 \left(t^2 - 1\right)^{-1/2} \left(t^2 - k^2\right)^{-1/2} dt \tag{43}$$

(Whittaker and Watson 1990, p. 494).

Jacobi elliptic functions addition formulas include

$$\text{sn}(u+v) = \frac{\text{sn } u \text{ cn } v \text{ dn } v + \text{sn } v \text{ cn } u \text{ dn } u}{1 - k^2 \text{ sn}^2 u \text{ sn}^2 v} \tag{44}$$

$$\text{cn}(u+v) = \frac{\text{cn } u \text{ cn } v - \text{sn } u \text{ sn } v \text{ dn } u \text{ dn } v}{1 - k^2 \text{ sn}^2 u \text{ sn}^2 v} \tag{45}$$

$$\text{dn}(u+v) = \frac{\text{dn } u \text{ dn } v - k^2 \text{sn } u \text{ sn } v \text{ cn } u \text{ cn } v}{1 - k^2 \text{ sn}^2 u \text{ sn}^2 v}. \tag{46}$$

Extended to integral periods,

$$\text{sn}(u+K) = \frac{\text{cn } u}{\text{dn } u} \tag{47}$$

$$\text{cn}(u+K) = \frac{k' \text{ sn } u}{\text{dn } u} \tag{48}$$

$$\text{dn}(u+K) = \frac{k'}{\text{dn } u} \tag{49}$$

$$\text{sn}(u+2K) = -\text{sn } u \tag{50}$$

$$\text{cn}(u+2K) = -\text{cn } u \tag{51}$$

$$\text{dn}(u+2K) = \text{dn } u \tag{52}$$

For COMPLEX arguments,

$$\text{sn}(u+iv) = \frac{\text{sn}(u,k) \text{ dn}(v,k')}{1 - \text{dn}^2(u,k) \text{ sn}^2(v,k')}$$
$$+ \frac{i \text{ cn}(u,k) \text{ dn}(u,k) \text{ sn}(v,k') \text{ cn}(v,k')}{1 - \text{dn}^2(u,k) \text{ sn}^2(v,k')} \tag{53}$$

$$\text{cn}(u+iv) = \frac{\text{cn}(u,k) \text{ cn}(v,k')}{1 - \text{dn}^2(u,k) \text{ sn}^2(v,k')}$$
$$+ \frac{i \text{ sn}(u,k) \text{ dn}(u,k) \text{ sn}(v,k') \text{ dn}(v,k')}{1 - \text{dn}^2(u,k) \text{ sn}^2(v,k')} \tag{54}$$

$$\text{dn}(u+iv) = \frac{\text{dn}(u,k) \text{ cn}(v,k') \text{ dn}(v,k')}{1 - \text{dn}^2(u,k) \text{ sn}^2(v,k')}$$
$$+ \frac{ik^2 \text{ sn}(u,k) \text{ cn}(u,k) \text{ sn}(v,k')}{1 - \text{dn}^2(u,k) \text{ sn}^2(v,k')} \tag{55}$$

DERIVATIVES of the Jacobi elliptic functions include

$$\frac{d \text{ sn } u}{du} = \text{cn } u \text{ dn } u \tag{56}$$

$$\frac{d \text{ cn } u}{du} = -\text{sn } u \text{ dn } u \tag{57}$$

$$\frac{d \text{ dn } u}{du} = -k^2 \text{ sn } u \text{ cn } u \tag{58}$$

(Hille 1969, p. 66; Zwillinger 1997, p. 136).

Double-period formulas involving the Jacobi elliptic functions include

$$\text{sn}(2u) = \frac{2 \text{ sn } u \text{ cn } u \text{ dn } u}{1 - k^2 \text{ sn}^4 u} \tag{59}$$

$$\text{cn}(2u) = \frac{1 - 2 \text{ sn}^2 u + k^2 \text{ sn}^4 u}{1 - k^2 \text{ sn}^4 u} \tag{60}$$

$$\text{dn}(2u) = \frac{1 - 2k^2 \text{ sn}^2 u + k^2 \text{ sn}^4 u}{1 - k^2 \text{ sn}^4 u}. \tag{61}$$

Half-period formulas involving the Jacobi elliptic functions include

$$\text{sn}\left(\tfrac{1}{2}K\right) = \frac{1}{\sqrt{1+k'}} \tag{62}$$

$$\text{cn}\left(\tfrac{1}{2}K\right) = \sqrt{\frac{k'}{1+k'}} \tag{63}$$

$$\text{dn}\left(\tfrac{1}{2}K\right) = \sqrt{k'}. \tag{64}$$

Squared formulas include

$$\text{sn}^2 u = \frac{1 - \text{cn}(2u)}{1 + \text{dn}(2u)} \tag{65}$$

$$\text{cn}^2 u = \frac{\text{dn}(2u) + \text{cn}(2u)}{1 + \text{dn}(2u)} \tag{66}$$

$$\mathrm{dn}^2 u = \frac{\mathrm{dn}(2u) + \mathrm{cn}(2u)}{1 + \mathrm{cn}(2u)}. \qquad (67)$$

See also AMPLITUDE, ELLIPTIC FUNCTION, JACOBI DIFFERENTIAL EQUATION, JACOBI'S IMAGINARY TRANSFORMATION, JACOBI FUNCTION OF THE SECOND KIND, JACOBI THETA FUNCTIONS, WEIERSTRASS ELLIPTIC FUNCTION

References

Abramowitz, M. and Stegun, C. A. (Eds.). "Jacobian Elliptic Functions and Theta Functions." Ch. 16 in *Handbook of Mathematical Functions with Formulas, Graphs, and Mathematical Tables, 9th printing.* New York: Dover, pp. 567–81, 1972.

Bellman, R. E. *A Brief Introduction to Theta Functions.* New York: Holt, Rinehart and Winston, 1961.

Hille, E. *Lectures on Ordinary Differential Equations.* Reading, MA: Addison-Wesley, 1969.

Morse, P. M. and Feshbach, H. *Methods of Theoretical Physics, Part I.* New York: McGraw-Hill, p. 433, 1953.

Press, W. H.; Flannery, B. P.; Teukolsky, S. A.; and Vetterling, W. T. "Elliptic Integrals and Jacobi Elliptic Functions." §6.11 in *Numerical Recipes in FORTRAN: The Art of Scientific Computing, 2nd ed.* Cambridge, England: Cambridge University Press, pp. 254–63, 1992.

Spanier, J. and Oldham, K. B. "The Jacobian Elliptic Functions." Ch. 63 in *An Atlas of Functions.* Washington, DC: Hemisphere, pp. 635–52, 1987.

Tölke, F. "Jacobische elliptische Funktionen und zugehörige logarithmische Ableitungen," "Umkehrfunktionen der Jacobischen elliptischen Funktionen und elliptische Normalintegrale erster Gattung. Elliptische Amplitudenfunktionen sowie Legendresche *F*- und *E*-Funktion. Elliptische Normalintegrale zweiter Gattung. Jacobische Zeta- und Heumansche Lambda-Funktionen," and "Normalintegrale dritter Gattung. Legendresche Π-Funktion. Zurückführung des allgemeinen elliptischen Integrals auf Normalintegrale erster, zweiter, und dritter Gattung." Chs. 5– in *Praktische Funktionenlehre, dritter Band: Jacobische elliptische Funktionen, Legendresche elliptische Normalintegrale und spezielle Weierstraßsche Zeta- und Sigma Funktionen.* Berlin: Springer-Verlag, pp. 1–44, 1967.

Tölke, F. *Praktische Funktionenlehre, vierter Band: Elliptische Integralgruppen und Jacobische elliptische Funktionen im Komplexen.* Berlin: Springer-Verlag, 1967.

Whittaker, E. T. and Watson, G. N. *A Course in Modern Analysis, 4th ed.* Cambridge, England: Cambridge University Press, 1990.

Jacobi Function of the First Kind

JACOBI POLYNOMIAL

Jacobi Function of the Second Kind

$$Q_n^{(\alpha,\beta)}(x) = 2^{-n-1}(x-1)^{-\alpha}(x+1)^{-\beta}$$
$$\times \int_{-1}^{1} (1-t)^{n+\alpha}(1+t)^{n+\beta}(x-t)^{-n-1}\,dt.$$

In the exceptional case $n = 0$, $\alpha + \beta + 1 = 0$, a non-constant solution is given by

$$Q^{(\alpha)}(x) = \ln(x+1) + \pi^{-1}\sin(\pi\alpha)(x-1)^{-\alpha}(x+1)^{-\beta}$$

$$\times \int_{-1}^{1} \frac{(1-t)^{\alpha}(1+t)^{\beta}}{x-t}\ln(1+t)\,dt.$$

See also JACOBI DIFFERENTIAL EQUATION, JACOBI POLYNOMIAL

References

Szego, G. "Jacobi Polynomials." Ch. 4 in *Orthogonal Polynomials, 4th ed.* Providence, RI: Amer. Math. Soc., pp. 73–9, 1975.

Jacobi Identities

"The" Jacobi identity is a relationship

$$[A, [B, C]] + [B, [C, A]] + [C, [A, B]] = 0, \qquad (1)$$

between three elements A, B, and C, where $[A, B]$ is the COMMUTATOR. The elements of a LIE ALGEBRA satisfy this identity.

Relationships between the Q-FUNCTIONS Q_i are also known as Jacobi identities:

$$Q_1 Q_2 Q_3 = 1, \qquad (2)$$

equivalent to the JACOBI TRIPLE PRODUCT (Borwein and Borwein 1987, p. 65) and

$$Q_2^8 = 16q Q_1^8 + Q_3^8, \qquad (3)$$

where

$$q \equiv e^{-\pi K'(k)/K(k)} \qquad (4)$$

$K = K(k)$ is the complete ELLIPTIC INTEGRAL OF THE FIRST KIND, and $K'(k) = K(k') = K\left(\sqrt{1-k^2}\right)$. Using WEBER FUNCTIONS

$$f_1 = q^{-1/24} Q_3 \qquad (5)$$

$$f_2 = 2^{1/2} q^{1/12} Q_1 \qquad (6)$$

$$f = q^{-1/24} Q_2, \qquad (7)$$

(5) and (6) become

$$f_1 f_2 f = \sqrt{2} \qquad (8)$$

$$f^8 = f_1^8 + f_2^8 \qquad (9)$$

(Borwein and Borwein 1987, p. 69).

See also COMMUTATOR, JACOBI TRIPLE PRODUCT, PARTITION FUNCTION Q, Q-FUNCTION, WEBER FUNCTIONS

References

Borwein, J. M. and Borwein, P. B. *Pi & the AGM: A Study in Analytic Number Theory and Computational Complexity.* New York: Wiley, 1987.

Hardy, G. H. and Wright, E. M. *An Introduction to the Theory of Numbers, 5th ed.* Oxford, England: Clarendon Press, 1979.

Schafer, R. D. *An Introduction to Nonassociative Algebras.* New York: Dover, p. 3, 1996.

Jacobi Matrix

JACOBI ROTATION MATRIX, JACOBIAN

Jacobi Method

A method of diagonalizing a MATRIX A using JACOBI ROTATION MATRICES P_{pq}. It consists of a sequence of ORTHOGONAL SIMILARITY TRANSFORMATIONS OF THE FORM

$$\mathsf{A}' = \mathsf{P}_{pq}^T \mathsf{A} \mathsf{P}_{pq},$$

each of which eliminates one off-diagonal element. Each application of P_{pq} affects only rows and columns of A, and the sequence of such matrices is chosen so as to eliminate the off-diagonal elements.

See also JACOBI ALGORITHM, JACOBI ROTATION MATRIX

References

Gentle, J. E. "Givens Transformations (Rotations)." §3.2.5 in *Numerical Linear Algebra for Applications in Statistics.* Berlin: Springer-Verlag, pp. 99–02, 1998.
Press, W. H.; Flannery, B. P.; Teukolsky, S. A.; and Vetterling, W. T. "Jacobi Transformation of a Symmetric Matrix." §11.1 in *Numerical Recipes in FORTRAN: The Art of Scientific Computing, 2nd ed.* Cambridge, England: Cambridge University Press, pp. 456–62, 1992.

Jacobi Polynomial

Also known as the HYPERGEOMETRIC POLYNOMIALS, they occur in the study of ROTATION GROUPS and in the solution to the equations of motion of the symmetric top. They are solutions to the JACOBI DIFFERENTIAL EQUATION. Plugging

$$y = \sum_{v=0}^{\infty} a_v (x-1)^v \qquad (1)$$

into the differential equation gives the RECURRENCE RELATION

$$[\gamma - v(v+\alpha+\beta+1)]a_v - 2(v+1)(v+\alpha+1)a_{v+1} = 0 \quad (2)$$

for $v = 0, 1, \ldots,$ where

$$\gamma \equiv n(n+\alpha+\beta+1). \qquad (3)$$

Solving the RECURRENCE RELATION gives

$$P_n^{(\alpha+\beta)}(x) = \frac{(-1)^n}{2^n n!} (1-x)^{-\alpha} (1+x)^{-\beta} \frac{d^n}{dx^n}$$

$$\times \left[(1-x)^{\alpha+n} (1+x)^{\beta+n} \right] \qquad (4)$$

for $\alpha, \beta > -1$. They form a complete orthogonal system in the interval $[-1, 1]$ with respect to the weighting function

$$w_n(x) = (1-x)^\alpha (1+x)^\beta, \qquad (5)$$

and are normalized according to

$$P_n^{(\alpha,\beta)}(1) = \binom{n+\alpha}{n}, \qquad (6)$$

where $\binom{n}{k}$ is a BINOMIAL COEFFICIENT. Jacobi polynomials can also be written

$$P_n^{\alpha,\beta} = \frac{\Gamma(2n+\alpha+\beta+1)}{n!\Gamma(n+\alpha+\beta+1)} G_n\left(\alpha+\beta+1, \beta+1, \tfrac{1}{2}(x+1)\right), \qquad (7)$$

where $\Gamma(z)$ is the GAMMA FUNCTION and

$$G_n(p,q,x) \equiv \frac{n!\Gamma(n+p)}{\Gamma(2n+p)} P_n^{(p-q,q-1)}(2x-1). \qquad (8)$$

Jacobi polynomials are ORTHOGONAL satisfying

$$\int_{-1}^1 P_m^{(\alpha,\beta)} P_n^{(\alpha,\beta)} (1-x)^\alpha (1+x)^\beta dx$$

$$= \frac{2^{\alpha+\beta+1}}{2n+\alpha+\beta+1} \frac{\Gamma(n+\alpha+1)\Gamma(n+\beta+1)}{n!\Gamma(n+\alpha+\beta+1)} \delta_{mn}. \quad (9)$$

The COEFFICIENT of the term x^n in $P_n^{(\alpha,\beta)}(x)$ is given by

$$A_n = \frac{\Gamma(2n+\alpha+\beta+1)}{2^n n!\Gamma(n+\alpha+\beta+1)}. \qquad (10)$$

They satisfy the RECURRENCE RELATION

$$2(n+1)(n+\alpha+\beta+1)(2n+\alpha+\beta)P_{n+1}^{(\alpha,\beta)}(x)$$

$$= \left[(2n+\alpha+\beta+1)(\alpha^2-\beta^2) + (2n+\alpha+\beta)_3 x \right] P_n^{(\alpha,\beta)}(x)$$

$$- 2(n+\alpha)(n+\beta)(2n+\alpha+\beta+2)P_{n-1}^{(\alpha,\beta)}(x), \qquad (11)$$

where $(m)_n$ is the RISING FACTORIAL

$$(m)_n \equiv m(m+1)\cdots(m+n-1) = \frac{(m+n-1)!}{(m-1)!}. \qquad (12)$$

The DERIVATIVE is given by

$$\frac{d}{dx}\left[P_n^{(\alpha,\beta)}(x) \right] = \tfrac{1}{2}(n+\alpha+\beta+1)P_{n-1}^{(\alpha+1,\beta+1)}(x). \qquad (13)$$

The ORTHOGONAL POLYNOMIALS with WEIGHTING FUNCTION $(b-x)^\alpha (x-a)^\beta$ on the CLOSED INTERVAL $[a, b]$ can be expressed in the form

$$[\text{const.}] P_n^{(\alpha,\beta)}\left(2\frac{x-a}{b-a} - 1 \right) \qquad (14)$$

(Szego 1975, p. 58).

Special cases with $\alpha = \beta$ are

$$P_{2v}^{(\alpha,\alpha)}(x) = \frac{\Gamma(2v+\alpha+1)\Gamma(v+1)}{\Gamma(v+\alpha+1)\Gamma(2v+1)} P_v^{(\alpha,-1/2)}(2x^2-1) \quad (15)$$

$$= (-1)^v \frac{\Gamma(2v+\alpha+1)\Gamma(v+1)}{\Gamma(v+\alpha+1)\Gamma(2v+1)} P_v^{(-1/2,\alpha)}(1-2x^2) \quad (16)$$

$$P_{2v+1}^{(\alpha,\alpha)}(x) = \frac{\Gamma(2v+\alpha+2)\Gamma(v+1)}{\Gamma(v+\alpha+1)\Gamma(2v+2)} x P_v^{(\alpha,1/2)}(2x^2-1)$$

(17)

$$= (-1)^v \frac{\Gamma(2v+\alpha+2)\Gamma(v+1)}{\Gamma(v+\alpha+1)\Gamma(2v+2)} x P_v^{(1/2,\alpha)}(1-2x^2). \quad (18)$$

Further identities are

$$P_n^{(\alpha+1,\beta)}(x) = \frac{2}{2n+\alpha+\beta+2}$$

$$\times \frac{(n+\alpha+1)P_n^{(\alpha,\beta)} - (n+1)P_{n+1}^{(\alpha,\beta)}(x)}{1-x}$$

(19)

$$P_n^{(\alpha+\beta+1)}(x) = \frac{2}{2n+\alpha+\beta+2}$$

$$\times \frac{(n+\beta+1)P_n^{(\alpha,\beta)}(x) + (n+1)P_{n+1}^{(\alpha,\beta)}(x)}{1+x}$$

(20)

$$\sum_{v=0}^{n} \frac{2v+\alpha+\beta+1}{2^{\alpha+\beta+1}}$$

$$\times \frac{\Gamma(v+1)\Gamma(v+\alpha+\beta+1)}{\Gamma(v+\alpha+1)\Gamma(v+\beta+1)} P_v^{(\alpha,\beta)}(x)Q_v^{(\alpha,\beta)}(y)$$

$$= \frac{1}{2} \frac{(y-1)^{\sim\alpha}(y+1)^{\sim\beta}}{y-x} + \frac{2^{\sim\alpha\sim\beta}}{2n+\alpha+\beta+2}$$

$$\times \frac{\Gamma(n+2)\Gamma(n+\alpha+\beta+2)}{\Gamma(n+\alpha+1)\Gamma(n+\beta+1)}$$

$$\times \frac{P_{n+1}^{(\alpha,\beta)}(x)Q_n^{(\alpha,\beta)}(y) - P_n^{(\alpha,\beta)}(x)Q_{n+1}^{\alpha,\beta}(y)}{x-y}$$

(21)

(Szego 1975, p. 79).

The KERNEL POLYNOMIAL is

$$K_n^{(\alpha,\beta)}(x,y) = \frac{2^{\sim\alpha\sim\beta}}{2n+\alpha+\beta+2}$$

$$\times \frac{\Gamma(n+2)\Gamma(n+\alpha+\beta+2)}{\Gamma(n+\alpha+1)\Gamma(n+\beta+1)}$$

$$\times \frac{P_{n+1}^{(\alpha,\beta)}(x)P_n^{(\alpha,\beta)}(y) - P_n^{(\alpha,\beta)}(x)P_{n+1}^{(\alpha,\beta)}(y)}{x-y} \quad (22)$$

(Szego 1975, p. 71).

The DISCRIMINANT is

$$D_n^{(\alpha,\beta)} = 2^{-n(n-1)} \prod_{v=1}^{n} v^{v-2n+2}(v+\alpha)^{v-1}(v+\beta)^{v-1}$$

$$\times (n+v+\alpha+\beta)^{n-v} \quad (23)$$

(Szego 1975, p. 143).

For $\alpha = \beta = 0$, $P_n^{(0,0)}(x)$ reduces to a LEGENDRE POLY-NOMIAL. The GEGENBAUER POLYNOMIAL

$$G_n(p,q,x) = \frac{n!\Gamma(n+p)}{\Gamma(2n+p)} P_n^{(p-q,q-1)}(2x-1) \quad (24)$$

and CHEBYSHEV POLYNOMIAL OF THE FIRST KIND can also be viewed as special cases of the Jacobi polynomials. In terms of the HYPERGEOMETRIC FUNCTION,

$$P_n^{(\alpha,\beta)}(x) = \binom{n+\alpha}{n} {}_2F_1(-n,n+\alpha+\beta;\ \alpha+1;\ \tfrac{1}{2}(1-x))$$

(25)

$$= \frac{(\alpha+1)_n}{n!} {}_2F_1\left(-n,n+\alpha+\beta;\alpha+1;\tfrac{1}{2}(1-x)\right) \quad (26)$$

$$= \binom{n+\alpha}{n}\left(\frac{x+1}{2}\right)^2$$

$$\times {}_2F_1\left(-n,-n-\beta;\alpha+1;\frac{x-1}{x+1}\right), \quad (27)$$

where $(\alpha)_n$ is the POCHHAMMER SYMBOL (Koekoek 1998).

Let N_1 be the number of zeros in $x \in (-1,1)$, N_2 the number of zeros in $x \in (-\infty,-1)$, and N_3 the number of zeros in $x \in (1,\infty)$. Define Klein's symbol

$$E(u) = \begin{cases} 0 & \text{if } u \le 0 \\ \lfloor u \rfloor & \text{if } u \text{ positive and nonintegral} \quad (28) \\ u-1 & \text{if } u = 1,2\ldots, \end{cases}$$

where $\lfloor x \rfloor$ is the FLOOR FUNCTION, and

$$X(\alpha,\beta) = E\left[\tfrac{1}{2}(|2n+\alpha+\beta+1|-|\alpha|-|\beta|+1)\right] \quad (29)$$

$$Y(\alpha,\beta) = E\left[\tfrac{1}{2}(-|2n+\alpha+\beta+1|+|\alpha|-|\beta|+1)\right] \quad (30)$$

$$Z(\alpha,\beta) = E\left[\tfrac{1}{2}(-|2n+\alpha+\beta+1|-|\alpha|+|\beta|+1)\right]. \quad (31)$$

If the cases $\alpha = -1, -2, \ldots, -n$, $\beta = -1, -2, \ldots, -n$, and $n+\alpha+\beta = -1, -2, \ldots, -n$ are excluded, then the number of zeros of $P_n^{(\alpha,\beta)}$ in the respective intervals are

$$N_1(\alpha,\beta)$$

$$= \begin{cases} 2\left\lfloor \tfrac{1}{2}(X+1) \right\rfloor & \text{for } (-1)^n \binom{n+\alpha}{n}\binom{n+\beta}{n} > 0 \\ 2\left\lfloor \tfrac{1}{2}X \right\rfloor + 1 & \text{for } (-1)^n \binom{n+\alpha}{n}\binom{n+\beta}{n} < 0 \end{cases} \quad (32)$$

$$N_2(\alpha,\beta)$$

$$= \begin{cases} 2\left\lfloor \tfrac{1}{2}(Y+1) \right\rfloor & \text{for } \binom{2n+\alpha+\beta}{n}\binom{n+\beta}{n} > 0 \\ 2\left\lfloor \tfrac{1}{2}Y \right\rfloor + 1 & \text{for } \binom{2n+\alpha+\beta}{n}\binom{n+\beta}{n} < 0 \end{cases} \quad (33)$$

$$N_3(\alpha,\beta)$$

$$= \begin{cases} 2\left\lfloor \frac{1}{2}(Z+1) \right\rfloor & \text{for } \binom{2n+\alpha+\beta}{n}\binom{n+\alpha}{n} > 0 \\ 2\left\lfloor \frac{1}{2}Z \right\rfloor + 1 & \text{for } \binom{2n+\alpha+\beta}{n}\binom{n+\alpha}{n} < 0 \end{cases} \quad (34)$$

(Szegö 1975, pp. 144–46).

The first few POLYNOMIALS are

$$P_0^{(\alpha,\beta)}(x) = 1 \quad (35)$$

$$P_1^{(\alpha,\beta)}(x) = \tfrac{1}{2}[2(\alpha+1) + (\alpha+\beta+2)(x-1)] \quad (36)$$

$$P_2^{(\alpha,\beta)}(x) = \tfrac{1}{8}[4(\alpha+1)^{(2)} + 4(\alpha+\beta+3)(\alpha+2)(x-1)$$
$$+ (\alpha+\beta+3)(2)(x-1)^2], \quad (37)$$

where $(m)_n$ is a RISING FACTORIAL (Abramowitz and Stegun 1972, p. 793).

See Abramowitz and Stegun (1972, pp. 782–93) and Szegö (1975, Ch. 4) for additional identities.

See also CHEBYSHEV POLYNOMIAL OF THE FIRST KIND, GEGENBAUER POLYNOMIAL, JACOBI FUNCTION OF THE SECOND KIND, RISING FACTORIAL, ZERNIKE POLYNOMIAL

References

Abramowitz, M. and Stegun, C. A. (Eds.). "Orthogonal Polynomials." Ch. 22 in *Handbook of Mathematical Functions with Formulas, Graphs, and Mathematical Tables, 9th printing.* New York: Dover, pp. 771–02, 1972.

Andrews, G. E.; Askey, R.; and Roy, R. "Jacobi Polynomials and Gram Determinants" and "Generating Functions for Jacobi Polynomials." §6.3 and 6.4 in *Special Functions.* Cambridge, England: Cambridge University Press, pp. 293–06, 1999.

Iyanaga, S. and Kawada, Y. (Eds.). "Jacobi Polynomials." Appendix A, Table 20.V in *Encyclopedic Dictionary of Mathematics.* Cambridge, MA: MIT Press, p. 1480, 1980.

Koekoek, R. and Swarttouw, R. F. "Jacobi." §1.8 in *The Askey-Scheme of Hypergeometric Orthogonal Polynomials and its q-Analogue.* Delft, Netherlands: Technische Universiteit Delft, Faculty of Technical Mathematics and Informatics Report 98–7, pp. 38–4, 1998. ftp://www.twi.-tudelft.nl/publications/tech-reports/1998/DUT-TWI-98–7.ps.gz.

Roman, S. "The Theory of the Umbral Calculus I." *J. Math. Anal. Appl.* **87**, 58–15, 1982.

Szegö, G. "Jacobi Polynomials." Ch. 4 in *Orthogonal Polynomials, 4th ed.* Providence, RI: Amer. Math. Soc., 1975.

Jacobi Quadrature

JACOBI-GAUSS QUADRATURE

Jacobi Rotation Matrix

A MATRIX used in the JACOBI TRANSFORMATION method of diagonalizing MATRICES. The Jacobi rotation matrix P_{pq} contains 1s along the DIAGONAL, except for the two elements $\cos\phi$ in rows and columns p and q. In addition, all off-diagonal elements are zero except the elements $\sin\phi$ and $-\sin\phi$. The rotation angle ϕ for an initial matrix A is chosen such that

$$\cot(2\phi) = \frac{a_{qq} - a_{pp}}{2a_{pq}}.$$

Then the corresponding Jacobi rotation matrix which annihilates the off-diagonal element a_{pq} is

$$\mathsf{P}_{pq} \equiv \begin{bmatrix} 1 & & & & & & & 0 \\ & \ddots & & \vdots & & \vdots & & \iddots \\ & & \cos\phi & \cdots & 0 & \cdots & \sin\phi & \\ \cdots & & 0 & \cdots & 1 & \cdots & 0 & \cdots \\ & & -\sin\phi & \cdots & 0 & \cdots & \cos\phi & \\ & \iddots & & \vdots & & \vdots & & \ddots \\ 0 & & & & & & & 1 \end{bmatrix}$$

See also JACOBI TRANSFORMATION

References

Gentle, J. E. "Givens Transformations (Rotations)." §3.2.5 in *Numerical Linear Algebra for Applications in Statistics.* Berlin: Springer-Verlag, pp. 99–02, 1998.

Press, W. H.; Flannery, B. P.; Teukolsky, S. A.; and Vetterling, W. T. "Jacobi Transformation of a Symmetric Matrix." §11.1 in *Numerical Recipes in FORTRAN: The Art of Scientific Computing, 2nd ed.* Cambridge, England: Cambridge University Press, pp. 456–62, 1992.

Jacobi Symbol

The product of LEGENDRE SYMBOLS (n/p_i) for each of the PRIME FACTORS p_i such that $m = \prod_i p_i$, denoted (n/m) or $\left(\frac{n}{m}\right)$. When m is a PRIME, the Jacobi symbol reduces to the LEGENDRE SYMBOL. (The Legendre symbol is equal to ± 1 depending on whether m is a QUADRATIC RESIDUE modulo m.) Analogously to the Legendre symbol, the Jacobi symbol is commonly generalized to have value

$$\left(\frac{n}{m}\right) = 0 \text{ if } m|n, \quad (1)$$

giving

$$\left(\frac{n}{n}\right) = 0 \quad (2)$$

as a special case. Note that the Jacobi symbol is *not defined* for $m \le 0$ or m EVEN. The Jacobi symbol is implemented in *Mathematica* as `JacobiSymbol[n, m]`.

Use of the Jacobi symbol provides the generalization of the QUADRATIC RECIPROCITY THEOREM

$$\left(\frac{m}{n}\right)\left(\frac{n}{m}\right) = (-1)^{(m-1)(n-1)/4} \quad (3)$$

for m and n RELATIVELY PRIME ODD INTEGERS with $n \ge 3$ (Nagell 1951, pp. 147–48). Written another way,

$$\left(\frac{m}{n}\right) = (-1)^{(m-1)(n-1)/4}\left(\frac{n}{m}\right) \tag{4}$$

or

$$\left(\frac{n}{m}\right) = \begin{cases} \left(\dfrac{m}{n}\right) & \text{for } m \text{ or } n \equiv 1 \ (\text{mod } 4) \\[2mm] -\left(\dfrac{m}{n}\right) & \text{for } m, n \equiv 3 \ (\text{mod } 4) \end{cases} . \tag{5}$$

The Jacobi symbol satisfies the same rules as the Legendre symbol

$$\left(\frac{n}{m}\right)\left(\frac{n}{m'}\right) = \left(\frac{n}{(mm')}\right) \tag{6}$$

$$\left(\frac{n}{m}\right)\left(\frac{n'}{m}\right) = \left(\frac{(nn')}{m}\right) \tag{7}$$

$$\left(\frac{n^2}{m}\right) = \left(\frac{n}{m^2}\right) = 1 \quad \text{if } (m, n) = 1 \tag{8}$$

$$\left(\frac{n}{m}\right) = \left(\frac{n'}{m}\right) \quad \text{if } n \equiv n' \ (\text{mod } m) \tag{9}$$

$$\left(\frac{-1}{m}\right) = (-1)^{(m-1)/2} = \begin{cases} 1 & \text{for } m \equiv 1 \ (\text{mod } 4) \\ -1 & \text{for } m \equiv -1 \ (\text{mod } 4) \end{cases} \tag{10}$$

$$\left(\frac{2}{m}\right) = (-1)^{(m^2-1)/8} = \begin{cases} 1 & \text{for } m \equiv \pm 1 \ (\text{mod } 8) \\ -1 & \text{for } m \equiv \pm 3 \ (\text{mod } 8) \end{cases} \tag{11}$$

Bach and Shallit (1996) show how to compute the Jacobi symbol in terms of the SIMPLE CONTINUED FRACTION of a RATIONAL NUMBER n/m.

See also KRONECKER SYMBOL, LEGENDRE SYMBOL, QUADRATIC RESIDUE

References

Bach, E. and Shallit, J. *Algorithmic Number Theory, Vol. 1: Efficient Algorithms.* Cambridge, MA: MIT Press, pp. 343–44, 1996.

Guy, R. K. "Quadratic Residues. Schur's Conjecture." §F5 in *Unsolved Problems in Number Theory, 2nd ed.* New York: Springer-Verlag, pp. 244–45, 1994.

Nagell, T. "Jacobi's Symbol and the Generalization of the Reciprocity Law." §42 in *Introduction to Number Theory.* New York: Wiley, pp. 145–49, 1951.

Riesel, H. "Jacobi's Symbol." *Prime Numbers and Computer Methods for Factorization, 2nd ed.* Boston, MA: Birkhäuser, pp. 281–84, 1994.

Jacobi Tensor

$$J^{\mu}_{\nu\alpha\beta} = J^{\mu}_{\nu\beta\alpha} \equiv \frac{1}{2}\left(R^{\mu}_{\alpha\nu\beta} + R^{\mu}_{\beta\nu\alpha}\right),$$

where R is the RIEMANN TENSOR.

See also RIEMANN TENSOR

Jacobi Theta Function

THETA FUNCTIONS

Jacobi Theta Functions

The Jacobi theta functions are the elliptic analogs of the EXPONENTIAL FUNCTION, and may be used to express the JACOBI ELLIPTIC FUNCTIONS. The theta functions are quasi-doubly periodic, and are most commonly denoted $\vartheta_n(z, q)$ in modern texts, although the notations $\Theta_n(z, q)$ and $\theta_n(z, q)$ (Borwein and Borwein 1987) are sometimes also used. Whittaker and Watson (1990, p. 487) gives a table summarizing notations used by various earlier writers. The theta functions are given in *Mathematica* by EllipticTheta[n, z, q].

The theta functions may be expressed in terms of the NOME q, denoted $\vartheta_n(z, q)$, or the HALF-PERIOD RATIO τ, denoted $\vartheta_n(z|\tau)$, where $|q| < 1$ and q and τ are related by

$$q \equiv e^{i\pi\tau}. \tag{1}$$

Let the many-valued function q^λ be interpreted to stand for $e^{\lambda\pi i\tau}$. Then for a complex number z, the Jacobi theta functions are defined as

$$\vartheta_1(z, q) \equiv \sum_{n=-\infty}^{\infty} (-1)^{n-1/2} q^{(n+1/2)^2} e^{(2n+1)iz} \tag{2}$$

$$\vartheta_2(z, q) \equiv \sum_{n=-\infty}^{\infty} q^{(n+1/2)^2} e^{(2n+1)iz} \tag{3}$$

$$\vartheta_3(z, q) \equiv \sum_{n=-\infty}^{\infty} q^{n^2} e^{2niz} \tag{4}$$

$$\vartheta_4(z, q) \equiv \sum_{n=-\infty}^{\infty} (-1)^n q^{n^2} e^{2niz}. \tag{5}$$

Writing the doubly infinite sums as singly infinite sums gives the slightly less symmetrical forms

$$\vartheta_1(z, q) = 2\sum_{n=0}^{\infty} (-1)^n q^{(n+1/2)^2} \sin[(2n+1)z] \tag{6}$$

$$= 2q^{1/4}\sum_{n=0}^{\infty} (-1)^n q^{n(n+1)} \sin[(2n+1)z] \tag{7}$$

$$\vartheta_2(z, q) = 2\sum_{n=0}^{\infty} q^{(n+1/2)^2} \cos[(2n+1)z] \tag{8}$$

$$= 2q^{1/4}\sum_{n=0}^{\infty} q^{n(n+1)} \cos[(2n+1)z] \tag{9}$$

$$\vartheta_3(z, q) = 1 + 2\sum_{n=0}^{\infty} q^{n^2} \cos(2nz) \tag{10}$$

$$\vartheta_4(z,q) = 1 + 2\sum_{n=0}^{\infty} (-1)^n q^{n^2} \cos(2nz) \qquad (11)$$

(Whittaker and Watson 1990, p. 463–64). Explicitly writing out the series gives

$$\vartheta_1(z,q) = 2q^{1/4}\sin z - 2q^{9/4}\sin(3z) + 2q^{25/4}\sin(5z) \\ + \ldots \qquad (12)$$

$$\vartheta_2(z,q) = 2q^{1/4}\cos z + 2q^{9/4}\cos(3z) + 2q^{25/4}\cos(5z) \\ + \ldots \qquad (13)$$

$$\vartheta_3(z,q) = 1 + 2q\cos(2z) + 2q^4\cos(4z) + 2q^9\cos(6z) \\ + \ldots \qquad (14)$$

$$\vartheta_4(z,q) = 1 - 2q\cos(2z) + 2q^4\cos(4z) - 2q^9\cos(6z) \\ + \ldots \qquad (15)$$

(Borwein and Borwein 1987, p. 52; Whittaker and Watson 1990, p. 464). $\vartheta_1(z,q)$ is an ODD FUNCTION of z, while the other three are even functions of z.

The following table illustrates the quasi-double periodicity of the Jacobi theta functions.

ϑ_i	$\vartheta_i(z+\pi)/\vartheta_i(z)$	$\vartheta_i(z+\tau\pi)/\vartheta_i(z)$
ϑ_1	-1	$-N$
ϑ_2	-1	N
ϑ_3	1	N
ϑ_4	1	$-N$

Here,

$$N \equiv q^{-1}e^{-2iz}. \qquad (16)$$

The quasi-periodicity can be established as follows for the specific case of ϑ_4,

$$\vartheta_4(z+\pi,q) = \sum_{n=-\infty}^{\infty} (-1)^n q^{n^2} e^{2niz} e^{2ni\pi}$$

$$= \sum_{n=-\infty}^{\infty} (-1)^n q^{n^2} e^{2niz} = \vartheta_4(z,q) \qquad (17)$$

$$\vartheta_4(z+\pi\tau,q) = \sum_{n=-\infty}^{\infty} (-1)^n q^{n^2} e^{2ni\pi t} e^{2niz}$$

$$= \sum_{n=-\infty}^{\infty} (-1)^n q^{n^2} q^{2n} e^{2niz}$$

$$= -q^{-1}e^{-2iz} \sum_{n=-\infty}^{\infty} (-1)^{n+1} q^{(n+1)^2} q^{2(n+1)iz}$$

$$= -q^{-1}e^{-2iz} \sum_{n=-\infty}^{\infty} (-1)^n q^{n^2} q^{2niz}$$

$$= -q^{-1}e^{-2iz} \vartheta_4(z,q). \qquad (18)$$

The Jacobi theta functions can be written in terms of each other:

$$\vartheta_1(z,q) = -ie^{iz+\pi i\tau/4}\vartheta_4\left(z+\tfrac{1}{4}\pi\tau,q\right) \qquad (19)$$

$$\vartheta_2(z,q) = \vartheta_1\left(z+\tfrac{1}{2}\pi,q\right) \qquad (20)$$

$$\vartheta_3(z,q) = \vartheta_4\left(z+\tfrac{1}{2}\pi,q\right) \qquad (21)$$

Any Jacobi theta function of given arguments can be expressed in terms of any other two Jacobi theta functions with the same arguments.

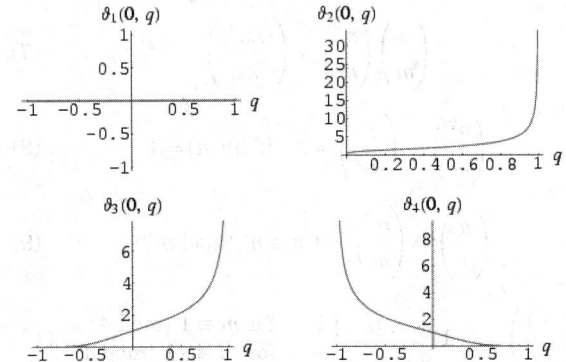

Define

$$\vartheta_i(q) \equiv \vartheta_i(z=0,q) \qquad (22)$$

to be the Jacobi theta functions with argument $z=0$, plotted above. Then the doubly infinite sums (2) to (5) take on the particularly simple forms

$$\vartheta_1(q) = 0 \qquad (23)$$

$$\vartheta_2(q) = \sum_{n=-\infty}^{\infty} q^{(n+1/2)^2} \qquad (24)$$

$$\vartheta_3(q) = \sum_{n=-\infty}^{\infty} q^{n^2} \qquad (25)$$

$$\vartheta_4(q) = \sum_{n=-\infty}^{\infty} (-1)^n q^{n^2} \qquad (26)$$

(Borwein and Borwein 1987, p. 33).

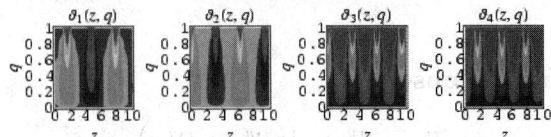

The plots above show the Jacobi theta functions plotted as a function of argument z and NOME q restricted to real values.

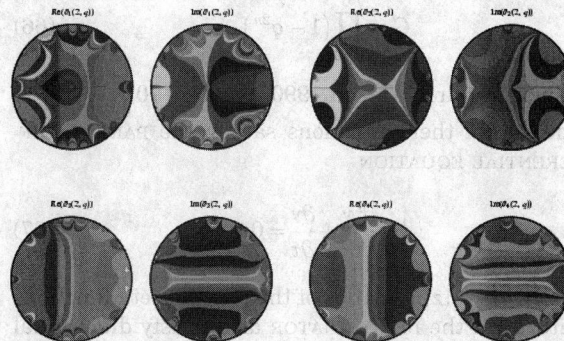

Particularly beautiful plots are obtained by examining the REAL and IMAGINARY PARTS of $\vartheta_i(z,q)$ for fixed z in the complex plane for $|q| < 1$, illustrated above.

The Jacobi theta functions satisfy an almost bewilderingly large number of identities involving the four functions, their derivatives, multiples of their arguments, and sums of their arguments. Among the unusual identities given by Whittaker and Watson (1990) are

$$\vartheta_3(z,q) = \vartheta_3(2z,q^4) + \vartheta_2(2z,q^4) \tag{27}$$

$$\vartheta_3(z,q) = \vartheta_3(2z,q^4) - \vartheta_2(2z,q^4) \tag{28}$$

(Whittaker and Watson 1990, p. 464) and

$$\frac{\vartheta_k'(z+\pi)}{\vartheta_k(z+\pi)} = \frac{\vartheta_k'(z)}{\vartheta_k(z)} \tag{29}$$

$$\frac{\vartheta_k'(z+\pi\gamma)}{\vartheta_k(z+\pi\gamma)} = -2i + \frac{\vartheta_k'(z)}{\vartheta_k(z)} \tag{30}$$

(Whittaker and Watson 1990, p. 465), for $k = 1, ..., 4$, where $\vartheta_k(z) \equiv \vartheta_k(z,q)$ and $\vartheta_i \equiv \vartheta_i(0,q)$. A class of identities involving the squares of Jacobi theta functions are

$$\vartheta_1^2(z)\vartheta_4^2 = \vartheta_3^2(z)\vartheta_2^2 - \vartheta_2^2(z)\vartheta_3^2 \tag{31}$$

$$\vartheta_2^2(z)\vartheta_4^2 = \vartheta_4^2(z)\vartheta_2^2 - \vartheta_1^2(z)\vartheta_3^2 \tag{32}$$

$$\vartheta_3^2(z)\vartheta_4^2 = \vartheta_4^2(z)\vartheta_3^2 - \vartheta_1^2(z)\vartheta_2^2 \tag{33}$$

$$\vartheta_4^2(z)\vartheta_4^2 = \vartheta_3^2(z)\vartheta_3^2 - \vartheta_2^2(z)\vartheta_2^2 \tag{34}$$

(Whittaker and Watson 1990, p. 466). Taking $z = 0$ in (34) gives the special case

$$\vartheta_4^4 = \vartheta_3^4 - \vartheta_2^4, \tag{35}$$

which is the only identity of this type.
In addition,

$$\vartheta_3(x) = \sum_{n=-\infty}^{\infty} x^{n^2} = 1 + 2x + 2x^4 + 2x^9 + \ldots \tag{36}$$

$$\vartheta_3^2(x) = 1$$
$$+ 4\left(\frac{x}{1-x} - \frac{x^3}{1-x^3} + \frac{x^5}{1-x^5} - \frac{x^7}{1-x^7} + \ldots\right) \tag{37}$$

$$\vartheta_3^4(x) = 1$$
$$+ 8\left(\frac{x}{1-x} + \frac{2x^2}{1-x^2} + \frac{3x^3}{1-x^3} + \frac{4x^4}{1-x^4} + \ldots\right) \tag{38}$$

The Jacobi theta functions obey addition rules such as

$$\vartheta_1(y+z)\vartheta_1(y-z)\vartheta_4^2 = \vartheta_3^2(y)\vartheta_2^2(z) - \vartheta_2^2(y)\vartheta_3^2(z)$$
$$= \vartheta_1^2(y)\vartheta_4^2(z) - \vartheta_4^2(y)\vartheta_1^2(z) \tag{39}$$

$$\vartheta_2(y+z)\vartheta_2(y-z)\vartheta_4^2 = \vartheta_4^2(y)\vartheta_2^2(z) - \vartheta_1^2(y)\vartheta_3^2(z)$$
$$= \vartheta_2^2(y)\vartheta_4^2(y) - \vartheta_3^2(y)\vartheta_1^2(z) \tag{40}$$

$$\vartheta_3(y+z)\vartheta_3(y-z)\vartheta_4^2 = \vartheta_4^2(y)\vartheta_3^2(z) - \vartheta_1^2(y)\vartheta_2^2(z)$$
$$= \vartheta_3^2(y)\vartheta_4^2(z) - \vartheta_2^2(y)\vartheta_1^2(z) \tag{41}$$

$$\vartheta_4(y+z)\vartheta_4(y-z)\vartheta_4^2 = \vartheta_3^2(y)\vartheta_3^2(z) - \vartheta_2^2(y)\vartheta_2^2(z)$$
$$= \vartheta_4^2(y)\vartheta_4^2(z) - \vartheta_1^2(y)\vartheta_1^2(z) \tag{42}$$

(Whittaker and Watson 1990, p. 487), and

$$\vartheta_3(y+z)\vartheta_3(y-z)\vartheta_2^2 = \vartheta_3^2(y)\vartheta_2^2(z) + \vartheta_4^2(y)\vartheta_1^2(z)$$
$$= \vartheta_2^2(y)\vartheta_3^2(z) + \vartheta_1^2(y)\vartheta_4^2(z)$$

$$\vartheta_3(y+z)\vartheta_3(y-z)\vartheta_3^2 = \vartheta_1^2(y)\vartheta_1^2(z) + \vartheta_3^2(y)\vartheta_3^2(z)$$
$$= \vartheta_2^2(y)\vartheta_2^2(z) - \vartheta_4(y)\vartheta_4^2(z)$$

$$\vartheta_4(y+z)\vartheta_4(y-z)\vartheta_2^2 = \vartheta_4^2(y)\vartheta_2^2(z) + \vartheta_3^2(y)\vartheta_1^2(z)$$
$$= \vartheta_2^2(y)\vartheta_4^2(z) + \vartheta_1^2(y)\vartheta_3^2(z) \tag{43}$$

$$\vartheta_4(y+z)\vartheta_4(y-z)\vartheta_3^2 = \vartheta_4^2(y)\vartheta_3^2(z) + \vartheta_2^2(y)\vartheta_1^2(z)$$
$$= \vartheta_3^2(y)\vartheta_4^2(z) + \vartheta_1^2(y)\vartheta_2^2(z) \tag{44}$$

(Whittaker and Watson 1990, p. 488).

$$\vartheta_1(y\pm z)\vartheta_2(y\mp z)\vartheta_3\vartheta_4$$
$$= \vartheta_1(y)\vartheta_2(y)\vartheta_3(z)\vartheta_4(z) \pm \vartheta_3(y)\vartheta_4(y)\vartheta_1(z)\vartheta_2(z) \tag{45}$$

$$\vartheta_1(y\pm z)\vartheta_3(y\mp z)\vartheta_2\vartheta_4$$
$$= \vartheta_1(y)\vartheta_3(y)\vartheta_2(z)\vartheta_4(z) \pm \vartheta_2(y)\vartheta_4(y)\vartheta_1(z)\vartheta_3(z) \tag{46}$$

$$\vartheta_1(y\pm z)\vartheta_4(y\mp z)\vartheta_2\vartheta_3$$
$$= \vartheta_1(y)\vartheta_4(y)\vartheta_2(z)\vartheta_3(z) \pm \vartheta_2(y)\vartheta_3(y)\vartheta_1(z)\vartheta_4(z) \tag{47}$$

$$\vartheta_2(y\pm z)\vartheta_3(y\mp z)\vartheta_2\vartheta_3$$
$$= \vartheta_2(y)\vartheta_3(y)\vartheta_2(z)\vartheta_3(z) \mp \vartheta_1(y)\vartheta_4(y)\vartheta_1(z)\vartheta_4(z) \tag{48}$$

$$\vartheta_2(y \pm z)\vartheta_4(y \mp z)\vartheta_2\vartheta_4$$
$$= \vartheta_2(y)\vartheta_4(y)\vartheta_2(z)\vartheta_4(z) \mp \vartheta_1(y)\vartheta_3(y)\vartheta_1(z)\vartheta_3(z) \quad (49)$$

$$\vartheta_3(y \pm z)\vartheta_4(y \pm z)\vartheta_3\vartheta_4$$
$$= \vartheta_3(y)\vartheta_4(y)\vartheta_3(z)\vartheta_4(z) \mp \vartheta_1(y)\vartheta_2(y)\vartheta_1(z)\vartheta_2(z) \quad (50)$$

(Whittaker and Watson 1990, p. 488).

There are also a series of DUPLICATION FORMULAS

$$\vartheta_3(2z)\vartheta_3^3 = \vartheta_3^4(z) + \vartheta_1^4(z) \quad (51)$$

$$\vartheta_2(2z)\vartheta_2\vartheta_4^2 = \vartheta_2^2(z)\vartheta_4^2(z) - \vartheta_1^2(z)\vartheta_3^2(z) \quad (52)$$

$$\vartheta_3(2z)\vartheta_3\vartheta_4^2 = \vartheta_3^2(z)\vartheta_4^2(z) - \vartheta_1^2(z)\vartheta_2^2(z) \quad (53)$$

$$\vartheta_4(2z)\vartheta_4^3 = \vartheta_3^4(z) - \vartheta_2^4(z) \quad (54)$$

$$= \vartheta_4^4(z) - \vartheta_1^4(z) \quad (55)$$

$$\vartheta_1(2z)\vartheta_2\vartheta_3\vartheta_4 = 2\vartheta_1(z)\vartheta_2(z)\vartheta_3(z)\vartheta_4(z) \quad (56)$$

(Whittaker and Watson 1990, p. 488).

Ratios of Jacobi theta function derivatives to the functions themselves have the simple forms

$$\frac{\vartheta_1'(z)}{\vartheta_1(z)} = \cot z + 4\sum_{n=1}^{\infty} \frac{q^{2n}}{1-q^{2n}}\sin(2nz) \quad (57)$$

$$\frac{\vartheta_2'(z)}{\vartheta_2(z)} = -\tan z + 4\sum_{n=1}^{\infty}(-1)^n \frac{q^{2n}}{1-q^{2n}}\sin(2nz) \quad (58)$$

$$\frac{\vartheta_3'(z)}{\vartheta_3(z)} = 4\sum_{n=1}^{\infty}(-1)^n \frac{q^n}{1-q^{2n}}\sin(2nz) \quad (59)$$

$$\frac{\vartheta_4'(z)}{\vartheta_4(z)} = \sum_{n=1}^{\infty} \frac{q^{2n-1}\sin(2z)}{1-2q^{2n-1}\cos(2z)+q^{4n-2}} \quad (60)$$

$$= \sum_{n=1}^{\infty} \frac{4q^n \sin(2nz)}{1-q^{2n}} \quad (61)$$

(Whittaker and Watson 1990, p. 489).

The Jacobi theta functions can be expressed as products instead of sums by

$$\vartheta_1(z) = 2Gq^{1/4}\sin z \prod_{n=1}^{\infty}\left[1-2q^{2n}\cos(2z)+q^{4n}\right] \quad (62)$$

$$\vartheta_2(z) = 2Gq^{1/4}\cos z \prod_{n=1}^{\infty}\left[1+2q^{2n}\cos(2z)+q^{4n}\right] \quad (63)$$

$$\vartheta_3(z) = G \prod_{n=1}^{\infty}\left[1+2q^{2n-1}\cos(2z)+q^{4n-2}\right] \quad (64)$$

$$\vartheta_4(z) = G \prod_{n=1}^{\infty}\left[1-2q^{2n-1}\cos(2z)+q^{4n-2}\right], \quad (65)$$

where

$$G \equiv \prod_{n=1}^{\infty}(1-q^{2n}) \quad (66)$$

(Whittaker and Watson 1990, pp. 469–70).

The Jacobi theta functions satisfy the PARTIAL DIF-FERENTIAL EQUATION

$$\tfrac{1}{4}\pi i\frac{\partial^2 y}{\partial z^2} + \frac{\partial y}{\partial \tau} = 0, \quad (67)$$

where $y \equiv \vartheta_i(z|\tau)$. Ratios of the Jacobi theta functions with ϑ_4 in the DENOMINATOR also satisfy differential equations

$$\frac{d}{dz}\left[\frac{\vartheta_1(z)}{\vartheta_4(z)}\right] = \vartheta_4^2 \frac{\vartheta_2(z)\vartheta_3(z)}{\vartheta_4^2(z)} \quad (68)$$

$$\frac{d}{dz}\left[\frac{\vartheta_2(z)}{\vartheta_4(z)}\right] = -\vartheta_3^2 \frac{\vartheta_1(z)\vartheta_3(z)}{\vartheta_4^2(z)} \quad (69)$$

$$\frac{d}{dz}\left[\frac{\vartheta_3(z)}{\vartheta_4(z)}\right] = \vartheta_2^2 \frac{\vartheta_1(z)\vartheta_2(z)}{\vartheta_4^2(z)} \quad (70)$$

JACOBI'S IMAGINARY TRANSFORMATION expresses $\vartheta_i(z/\tau|-1/\tau)$ in terms of $\vartheta_i(z|\tau)$. There are a large number of beautiful identities involving Jacobi theta functions of arguments w, x, y, and z and w', x', y', and z', related by

$$2w' = -w+x+y+z \quad (71)$$

$$2x' = w-x-y+z \quad (72)$$

$$2y' = w+x-y+z \quad (73)$$

$$2z' = w+x+y-z \quad (74)$$

(Whittaker and Watson 1990, pp. 467–69, 488, and 490). Using the notation

$$\vartheta_i(w+\pi/2,q)\vartheta_j(x+\pi/2,q)\vartheta_k(y,q)\vartheta_l(z,q) \equiv [ijkl] \quad (75)$$

$$\vartheta_i(w',q)\vartheta_j(x',q)\vartheta_k(y'+\pi/2,q)\vartheta_l(z'+\pi/2,q) \equiv ijkl, \quad (76)$$

gives a whopping 288 identities of the form

$$\pm[a_1 a_2 a_3 a_4] \pm [b_1 b_2 b_3 b_4] = \pm a_1' a_2' a_3' a_4' \pm b_1' b_2' b_3' b_4'. \quad (77)$$

The complete ELLIPTIC INTEGRALS OF THE FIRST and SECOND KINDS can be expressed using Jacobi theta functions. Let

$$\xi \equiv \frac{\vartheta_1(z)}{\vartheta_4(z)}, \quad (78)$$

and plug into (68)

$$\left(\frac{d\xi}{dz}\right)^2 = \left(\vartheta_2^2 - \xi^2\vartheta_3^2\right)\left(\vartheta_3^2 - \xi^2\vartheta_2^2\right). \quad (79)$$

Now write

$$\xi \frac{\vartheta_3}{\vartheta_2} \equiv y \tag{80}$$

and

$$z\vartheta_3^2 \equiv u. \tag{81}$$

Then

$$\left(\frac{dy}{du}\right)^2 = (1-y^2)(1-k^2y^2), \tag{82}$$

where the MODULUS is defined by

$$k = k(q) = \frac{\vartheta_2^2(q)}{\vartheta_3^2(q)}. \tag{83}$$

Define also the complementary MODULUS

$$k' = k'(q) = \frac{\vartheta_4^2(-q)}{\vartheta_3^2(q)}. \tag{84}$$

Now, since

$$\vartheta_2^4 + \vartheta_4^4 = \vartheta_3^4, \tag{85}$$

we have shown

$$k^2 + k'^2 = 1. \tag{86}$$

The solution to the equation is

$$y = \frac{\vartheta_3}{\vartheta_2} \frac{\vartheta_1(u\vartheta_3^{-2}|r)}{\vartheta_4(u\vartheta_3^{-2}|r)} \equiv \mathrm{sn}(u, k), \tag{87}$$

which is a JACOBI ELLIPTIC FUNCTION with periods

$$4K(k) = 2\pi\vartheta_3^2(q) \tag{88}$$

and

$$2iK'(k) = \pi r\vartheta_3^2(q). \tag{89}$$

Here, K is the complete ELLIPTIC INTEGRAL OF THE FIRST KIND,

$$K(k) = \tfrac{1}{2}\pi\vartheta_3^2(q). \tag{90}$$

The Jacobi theta functions provide analytic solutions to many tricky problems in mathematics and mathematical physics. For example, the Jacobi theta functions are related to the SUM OF SQUARES FUNCTION $r_2(n)$ giving the number of representations of n by two squares via

$$\vartheta_3^2(q) = \sum_{n=0}^{\infty} r_2(n)q^n \tag{91}$$

$$\vartheta_4^2(q) = \sum_{n=0}^{\infty} (-1)^n r_2(n)q^n \tag{92}$$

(Borwein and Borwein 1987, p. 34). The general QUINTIC EQUATION is solvable in terms of Jacobi theta functions, and these functions also provide a uni-

formly convergent form of the GREEN'S FUNCTION for a rectangular region (Oberhettinger and Magnus 1949). Finally, Jacobi theta functions can be used to uniformize all elliptic and hyperelliptic curves, the classical example being

$$y^2 - x(x^4 - 1) = 0, \tag{93}$$

with

$$x = -\frac{\vartheta_3(0|\tfrac{1}{2}\tau)}{\vartheta_4(0|\tfrac{1}{2}\tau)} \tag{94}$$

$$y = \frac{i\vartheta_3^{1/2}(0|\tfrac{1}{2}\tau)\vartheta_2^2(0|\tfrac{1}{2}\tau)}{\vartheta_4^{5/2}(0|\tfrac{1}{2}\tau)}. \tag{95}$$

See also BLECKSMITH-BRILLHART-GERST THEOREM, ELLIPTIC FUNCTION, ETA FUNCTION, EULER'S PENTAGONAL NUMBER THEOREM, HALF-PERIOD RATIO, JACOBI ELLIPTIC FUNCTIONS, JACOBI TRIPLE PRODUCT, LANDEN'S FORMULA, MOCK THETA FUNCTION, MODULAR EQUATION, MODULAR TRANSFORMATION, MORDELL INTEGRAL, NEVILLE THETA FUNCTIONS, NOME, POINCARÉ-FUCHS-KLEIN AUTOMORPHIC FUNCTION, QUINTUPLE PRODUCT IDENTITY, RAMANUJAN THETA FUNCTIONS, SCHRÖTER'S FORMULA, SUM OF SQUARES FUNCTION, THETA FUNCTIONS, WEBER FUNCTIONS

References

Abramowitz, M. and Stegun, C. A. (Eds.). *Handbook of Mathematical Functions with Formulas, Graphs, and Mathematical Tables, 9th printing.* New York: Dover, pp. 576–79, 1972.

Bellman, R. E. *A Brief Introduction to Theta Functions.* New York: Holt, Rinehart and Winston, 1961.

Berndt, B. C. "Theta-Functions and Modular Equations." Ch. 25 in *Ramanujan's Notebooks, Part IV.* New York: Springer-Verlag, pp. 138–44, 1994.

Borwein, J. M. and Borwein, P. B. "Theta Functions and the Arithmetic-Geometric Mean Iteration." Ch. 2 in *Pi & the AGM: A Study in Analytic Number Theory and Computational Complexity.* New York: Wiley, pp. 33–1, 1987.

Euler, L. *Opera Omnia, Vol. 20.* Leipzig, Germany, 1912.

Hermite, C. *Oeuvres Mathématiques.* Paris, 1905–917.

Jacobi, C. G. J. *Fundamentia Nova Theoriae Functionum Ellipticarum.* Königsberg, Germany: Regiomonti, Sumtibus fratrum Borntraeger, 1829. Reprinted in *Gesammelte Mathematische Werke, Vol. 1*, pp. 497–38.

Klein, F. *Vorlesungen über die Theorie der elliptischen Modulfunctionen, 2 vols.* Leipzig, Germany: Teubner, 1890–2.

Kronecker, L. *J. reine angew. Math.* **102**, 260–72, 1887.

Morse, P. M. and Feshbach, H. *Methods of Theoretical Physics, Part I.* New York: McGraw-Hill, pp. 430–32, 1953.

Oberhettinger, F. and Magnus, W. *Anwendung der Elliptischen Funktionen in Physik und Technik.* Berlin: Springer-Verlag, 1949.

Tannery, J. and Molk, J. *Elements de la Theorie des Fonctions Elliptiques, 4 vols.* Paris: Gauthier-Villars, 1893–902.

Tölke, F. "Theta-Funktionen" and "Logarithmen der Theta-Funktionen." Chs. 1– in *Praktische Funktionenlehre, zweiter Band: Theta-Funktionen und spezielle Weier-straßsche Funktionen.* Berlin: Springer-Verlag, pp. 1–3, 1966.

Tölke, F. *Praktische Funktionenlehre, fünfter Band: Allge-meine Weierstraßsche Funktionen und Ableitungen nach dem Parameter. Integrale der Theta-Funktionen und Bi-linear-Entwicklungen.* Berlin: Springer-Verlag, 1968.

Weber, H. *Elliptische Funktionen und algebraische Zahlen.* Brunswick, Germany, 1891.

Whittaker, E. T. and Watson, G. N. *A Course in Modern Analysis, 4th ed.* Cambridge, England: Cambridge University Press, 1990.

Jacobi Transformation

Jacobi Method

Jacobi Triple Product

The Jacobi triple product is the beautiful identity

$$\prod_{n=1}^{\infty}\left(1-x^{2n}\right)\left(1+x^{2n-1}z^2\right)\left(1+\frac{x^{2n-1}}{z^2}\right)$$

$$=\sum_{m=-\infty}^{\infty}x^{m^2}z^{2m}. \qquad (1)$$

In terms of the Q-FUNCTION, (1) is written

$$Q_1 Q_2 Q_3 = 1, \qquad (2)$$

which is one of the two JACOBI IDENTITIES. In Q-SERIES notation, the Jacobi triple product identity is written

$$(q, -xq, -1/x; q)_{\infty} = \sum_{k=-\infty}^{\infty} x^k q^{(k^2+k)/2} \qquad (3)$$

for $0 < |q| < 1$ and $x \neq 0$ (Gasper and Rahman 1990, p. 12; Leininger and Milne 1997). Another form of the identity is

$$\sum_{n=-\infty}^{\infty}(-1)^n a^n q^{(n^2-n)/2}$$

$$= \prod_{n=1}^{\infty}\left(1 - aq^{n-1}\right)\left(1 - a^{-1}q^n\right)\left(1 - q^n\right) \qquad (4)$$

(Hirschhorn 1999).

Dividing (4) by $1-a$ and letting $a \to 1$ gives the limiting case

$$(q, q)_{\infty}^3 = \sum_{n=0}^{\infty}(-1)^n(2n+1)q^{n(n+1)/2} \qquad (5)$$

$$= \frac{1}{2}\sum_{n=-\infty}^{\infty}(-1)^n(2n+1)q^{n(n+1)/2} \qquad (6)$$

(Jacobi 1829; Hardy and Wright 1979; Leininger and Milne 1997; Hardy 1999, p. 87; Hirschhorn 1999).

For the special case of $z = 1$, (1) becomes

$$\varphi(x) \equiv G(1) = \prod_{n=1}^{\infty}\left(1 + x^{2n-1}\right)^2\left(1 - x^{2n}\right)$$

$$= \sum_{m=-\infty}^{\infty}x^{m^2} = 1 + 2\sum_{m=1}^{\infty}x^{m^2}, \qquad (7)$$

where $\varphi(x)$ is the one-variable RAMANUJAN THETA FUNCTION. In terms of the two-variable RAMANUJAN THETA FUNCTION $f(a,b)$, the Jacobi triple product is equivalent to

$$f(a,b) = (-a; ab)_{\infty}(-b; ab)_{\infty}(ab; ab)_{\infty} \qquad (8)$$

(Berndt *et al.*).

One method of proof for the Jacobi identity proceeds by defining the function

$$F(z) \equiv \prod_{n=1}^{\infty}\left(1 + x^{2n-1}z^2\right)\left(1 + \frac{x^{2n-1}}{z^2}\right)$$

$$= \left(1 + xz^2\right)\left(1 + \frac{x}{z^2}\right)\left(1 + x^3z^2\right)\left(1 + \frac{x^3}{z^2}\right)\left(1 + x^5z^2\right)$$

$$\times \left(1 + \frac{x^5}{z^2}\right)\cdots, \qquad (9)$$

Then

$$F(xz) = \left(1 + x^3z^2\right)\left(1 + \frac{1}{xz^2}\right)\left(1 + x^5z^2\right)\left(1 + \frac{x}{z^2}\right)$$

$$\times \left(1 + x^7z^2\right)\left(1 + \frac{x^3}{z^2}\right)\cdots. \qquad (10)$$

Taking (10) ÷ (9),

$$\frac{F(xz)}{F(z)} = \left(1 + \frac{1}{xz^2}\right)\left(\frac{1}{1 + xz^2}\right)$$

$$= \frac{xz^2 + 1}{xz^2}\frac{1}{1 + xz^2} = \frac{1}{xz^2}, \qquad (11)$$

which yields the fundamental relation

$$xz^2 F(xz) = F(z). \qquad (12)$$

Now define

$$G(z) \equiv F(z)\prod_{n=1}^{\infty}\left(1 - x^{2n}\right) \qquad (13)$$

$$G(xz) = F(xz)\prod_{n=1}^{\infty}\left(1 - x^{2n}\right). \qquad (14)$$

Using (12), (14) becomes

$$G(xz) = \frac{F(z)}{xz^2}\prod_{n=1}^{\infty}\left(1 - x^{2n}\right) = \frac{G(z)}{xz^2}, \qquad (15)$$

so

$$G(z) = xz^2 G(xz). \qquad (16)$$

Expand G in a LAURENT SERIES. Since G is an EVEN FUNCTION, the LAURENT SERIES contains only even terms.

$$G(z) = \sum_{m=-\infty}^{\infty} a_m z^{2m}. \qquad (17)$$

Equation (16) then requires that

$$\sum_{m=-\infty}^{\infty} a_m z^{2m} = xz^2 \sum_{m=-\infty}^{\infty} a_m (xz)^{2m}$$

$$= \sum_{m=-\infty}^{\infty} a_m x^{2m+1} z^{2m+2}. \qquad (18)$$

This can be re-indexed with $m' \equiv m - 1$ on the left side of (18)

$$\sum_{m=-\infty}^{\infty} a_m z^{2m} = \sum_{m=-\infty}^{\infty} a_m x^{2m-1} z^{2m}, \qquad (19)$$

which provides a RECURRENCE RELATION

$$a_m = a_{m-1} x^{2m-1}, \qquad (20)$$

so

$$a_1 = a_0 x \qquad (21)$$

$$a_2 = a_1 x^3 = a_0 x^{3+1} = a_0 x^4 = a_0 x^{2^2} \qquad (22)$$

$$a_3 = a_2 x^5 = a_0 x^{5+4} = a_0 x^9 = a_0 x^{3^2}. \qquad (23)$$

The exponent grows greater by $(2m-1)$ for each increase in m of 1. It is given by

$$\sum_{n=1}^{m} (2m-1) = 2 \frac{m(m+1)}{2} - m = m^2. \qquad (24)$$

Therefore,

$$a_m = a_0 x^{m^2}. \qquad (25)$$

This means that

$$G(z) = a_0 \sum_{m=-\infty}^{\infty} x^{m^2} z^{2m}. \qquad (26)$$

The COEFFICIENT a_0 must be determined by going back to (9) and (13) and letting $z = 1$. Then

$$F(1) = \prod_{n=1}^{\infty} \left(1 + x^{2n-1}\right)\left(1 + x^{2n-1}\right)$$

$$= \prod_{n=1}^{\infty} \left(1 + x^{2n-1}\right)^2 \qquad (27)$$

$$G(1) = F(1) \prod_{n=1}^{\infty} \left(1 - x^{2n}\right)$$

$$= \prod_{n=1}^{\infty} \left(1 + x^{2n-1}\right)^2 \prod_{n=1}^{\infty} \left(1 - x^{2n}\right)$$

$$= \prod_{n=1}^{\infty} \left(1 + x^{2n-1}\right)^2 \left(1 - x^{2n}\right), \qquad (28)$$

since multiplication is ASSOCIATIVE. It is clear from this expression that the a_0 term must be 1, because all other terms will contain higher POWERS of x. Therefore,

$$a_0 = 1, \qquad (29)$$

so we have the Jacobi triple product,

$$G(z) = \prod_{n=1}^{\infty} \left(1 - x^{2n}\right)\left(1 + x^{2n-1} z^2\right)\left(1 + \frac{x^{2n-1}}{z^2}\right)$$

$$= \sum_{m=-\infty}^{\infty} x^{m^2} z^{2m}. \qquad (30)$$

See also EULER IDENTITY, JACOBI IDENTITIES, PARTITION FUNCTION Q, Q-FUNCTION, QUINTUPLE PRODUCT IDENTITY, RAMANUJAN PSI SUM, RAMANUJAN THETA FUNCTIONS, SCHRÖTER'S FORMULA, THETA FUNCTIONS

References

Andrews, G. E. *q*-Series: Their Development and Application in Analysis, Number Theory, Combinatorics, Physics, and Computer Algebra. Providence, RI: Amer. Math. Soc., pp. 63–4, 1986.

Berndt, B. C.; Huang, S.-S.; Sohn, J.; and Son, S. H. "Some Theorems on the Rogers-Ramanujan Continued Fraction in Ramanujan's Lost Notebook." To appears in *Trans. Amer. Math. Soc.*

Borwein, J. M. and Borwein, P. B. "Jacobi's Triple Product and Some Number Theoretic Applications." Ch. 3 in *Pi & the AGM: A Study in Analytic Number Theory and Computational Complexity.* New York: Wiley, pp. 62–01, 1987.

Gasper, G. and Rahman, M. *Basic Hypergeometric Series.* Cambridge, England: Cambridge University Press, 1990.

Hardy, G. H. and Wright, E. M. *An Introduction to the Theory of Numbers, 5th ed.* Oxford, England: Clarendon Press, 1979.

Hirschhorn, M. D. "Another Short Proof of Ramanujan's Mod 5 Partition Congruences, and More." *Amer. Math. Monthly* **106**, 580–83, 1999.

Jacobi, C. G. J. *Fundamentia Nova Theoriae Functionum Ellipticarum.* Regiomonti, Sumtibus fratrum Borntraeger, p. 90, 1829.

Leininger, V. E. and Milne, S. C. "Expansions for $(q)_\infty^{n^2+n}$ and Basic Hypergeometric Series in $U(n)$." Preprint. http://www.math.ohio-state.edu/~milne/preprints.html.

Whittaker, E. T. and Watson, G. N. *A Course in Modern Analysis, 4th ed.* Cambridge, England: Cambridge University Press, p. 470, 1990.

Jacobi Zeta Function
Denoted zn(u, k) or $Z(u)$.

$$Z(\phi|m) \equiv E(\phi|m) - \frac{E(m)F(\phi|m)}{K(m)},$$

where ϕ is the AMPLITUDE, m is the PARAMETER, and $F(\phi|m)$ and $K(m)$ are ELLIPTIC INTEGRALS OF THE FIRST KIND, and $e(m)$ is an ELLIPTIC INTEGRAL OF THE SECOND KIND. See Gradshteyn and Ryzhik (2000, p. xxxi) for expressions in terms of THETA FUNCTIONS. The Jacobi zeta functions is implemented in *Mathematica* as JacobiZeta[*phi*, *m*].

See also ELLIPTIC INTEGRAL OF THE FIRST KIND, ELLIPTIC INTEGRAL OF THE SECOND KIND, HEUMAN LAMBDA FUNCTION, ZETA FUNCTION

References
Abramowitz, M. and Stegun, C. A. (Eds.). *Handbook of Mathematical Functions with Formulas, Graphs, and Mathematical Tables, 9th printing.* New York: Dover, p. 595, 1972.

Gradshteyn, I. S. and Ryzhik, I. M. *Tables of Integrals, Series, and Products, 6th ed.* San Diego, CA: Academic Press, 2000.

Tölke, F. "Jacobische Zeta- und Heumansche Lambda-Funktionen." §132 in *Praktische Funktionenlehre, dritter Band: Jacobische elliptische Funktionen, Legendresche elliptische Normalintegrale und spezielle Weierstraßsche Zeta- und Sigma Funktionen.* Berlin: Springer-Verlag, pp. 94–9, 1967.

Jacobi's Curvature Theorem
The principal normal indicatrix of a closed SPACE CURVE with nonvanishing curvature bisects the AREA of the unit sphere if it is embedded.

Jacobi's Determinant Identity
Let

$$A = \begin{bmatrix} B & D \\ E & C \end{bmatrix} \tag{1}$$

$$A^{-1} = \begin{bmatrix} W & X \\ Y & Z \end{bmatrix}, \tag{2}$$

where B and W are $k \times k$ MATRICES. Then

$$(\det Z)(\det A) = \det B. \tag{3}$$

The proof follows from equating determinants on the two sides of the block matrices

$$\begin{bmatrix} B & D \\ E & C \end{bmatrix} \begin{bmatrix} I & X \\ O & Z \end{bmatrix} = \begin{bmatrix} B & O \\ E & I \end{bmatrix}, \tag{4}$$

where I is the IDENTITY MATRIX and O is the ZERO MATRIX.

References
Gantmacher, F. R. *The Theory of Matrices, Vol. 1.* New York: Chelsea, p. 21, 1960.

Horn, R. A. and Johnson, C. R. *Matrix Analysis.* Cambridge, England: Cambridge University Press, p. 21, 1985.

Jacobi's Imaginary Transformation
Transformations which relate elliptic functions to other elliptic functions of the same type but having different arguments. In the case of the JACOBI ELLIPTIC FUNCTIONS sn u, cn u, and dn u, the transformations are

$$\text{sn}(iu, k) = i\frac{\text{sn}(u, k')}{\text{cn}(u, k')} \tag{1}$$

$$\text{cn}(iu, k) = \frac{1}{\text{cn}(u, k')} \tag{2}$$

$$\text{dn}(iu, k) = \frac{\text{dn}(u, k')}{\text{cn}(u, k')}, \tag{3}$$

where k is the MODULUS, and $k' = \sqrt{1-k^2}$ is the COMPLEMENTARY MODULUS (Abramowitz and Stegun 1972; Whittaker and Watson 1990, p. 505).

In the case of the JACOBI THETA FUNCTIONS, Jacobi's imaginary transformation gives

$$\vartheta_1(z|\tau) = -i(-i\tau)^{-1/2} e^{i\tau z^2/\pi} \vartheta_1(z\tau'|\tau') \tag{4}$$

$$\vartheta_2(z|\tau) = (-i\tau)^{-1/2} e^{i\tau z^2/\pi} \vartheta_4(z\tau'|\tau') \tag{5}$$

$$\vartheta_3(z|\tau) = (-i\tau)^{-1/2} e^{i\tau z^2/\pi} \vartheta_3(z\tau'|\tau') \tag{6}$$

$$\vartheta_4(z|\tau) = (-i\tau)^{-1/2} e^{i\tau z^2/\pi} \vartheta_2(z\tau'|\tau), \tag{7}$$

where

$$\tau' \equiv -\frac{1}{\tau'} \tag{8}$$

and $(-i\tau)^{-1/2}$ is interpreted as satisfying $|\arg(-i\tau)| < \pi/2$ (Whittaker and Watson 1990, p. 475). These transformations were first obtained by Jacobi (1828), but Poisson (1827) had previously obtained a formula equivalent to one of the four, and from which the other three follow from elementary algebra (Whittaker and Watson 1990, p. 475).

See also JACOBI ELLIPTIC FUNCTIONS, JACOBI THETA FUNCTIONS

References
Abramowitz, M. and Stegun, C. A. (Eds.). *Handbook of Mathematical Functions with Formulas, Graphs, and Mathematical Tables, 9th printing.* New York: Dover, pp. 592 and 595, 1972.

Borwein, J. M. and Borwein, P. B. *Pi & the AGM: A Study in Analytic Number Theory and Computational Complexity.* New York: Wiley, p. 73, 1987.

Jacobi, C. G. J. "Suite des notices sur les fonctions elliptiques." *J. reine angew. Math.* **3**, 403–04, 1828. Reprinted in *Gesammelte Werke, Vol. 1.* Providence, RI: Amer. Math. Soc., pp. 264–65, 1969.

Landsberg, G. "Zur Theorie der Gaussschen Summen und der linearen Transformation der Thetafunctionen." *J. reine angew. Math.* **111**, 234–53, 1893.

Poisson, S. *Mém. de l'Acad. des Sci.* **6**, 592, 1827.

Whittaker, E. T. and Watson, G. N. "Jacobi's Imaginary Transformation." §21.51 in *A Course in Modern Analysis, 4th ed.* Cambridge, England: Cambridge University Press, pp. 474–76 and 505, 1990.

Jacobi's Theorem

Let M_r be an r-rowed MINOR of the nth order DETERMINANT $|\mathsf{A}|$ associated with an $n \times n$ MATRIX $\mathsf{A} = a_{ij}$ in which the rows $i_1, i_2, ..., i_r$ are represented with columns $k_1, k_2, ..., k_r$. Define the complementary minor to M_r as the $(n-k)$-rowed MINOR obtained from $|\mathsf{A}|$ by deleting all the rows and columns associated with M_r and the signed complementary minor $M^{(r)}$ to M_r to be

$$M^{(r)} = (-1)^{i_1 + i_2 + ... + i_r + k_1 + k_2 + ... + k_r}$$
$$\times [\text{complementary minor to } M_r].$$

Let the MATRIX of cofactors be given by

$$\Delta = \begin{vmatrix} A_{11} & A_{12} & \cdots & A_{1n} \\ A_{21} & A_{22} & \cdots & A_{2n} \\ \vdots & \vdots & \ddots & \vdots \\ A_{n1} & A_{n2} & \cdots & A_{nn} \end{vmatrix},$$

with M_r and M'_r the corresponding r-rowed minors of $|\mathsf{A}|$ and Δ, then it is true that

$$M'_r = |\mathsf{A}|^{r-1} M^{(r)}.$$

References

Gradshteyn, I. S. and Ryzhik, I. M. *Tables of Integrals, Series, and Products, 6th ed.* San Diego, CA: Academic Press, pp. 1109–100, 2000.

JacobiAmplitude

AMPLITUDE

Jacobian

Given a set $\mathbf{y} = \mathbf{f}(\mathbf{x})$ of n equations in n variables $x_1, ..., x_n$, written explicitly as

$$\mathbf{y} \equiv \begin{bmatrix} f_1(\mathbf{x}) \\ f_2(\mathbf{x}) \\ \vdots \\ f_n(\mathbf{x}) \end{bmatrix}, \tag{1}$$

or more explicitly as

$$\begin{cases} y_1 = f_1(x_1, ..., x_n) \\ \vdots \\ y_n = f_n(x_1, ..., x_n), \end{cases} \tag{2}$$

the Jacobian matrix, sometimes simply called "the Jacobian" (Simon and Blume 1994) is defined by

$$\mathsf{J}(x_1, ..., x_n) = \begin{bmatrix} \dfrac{\partial y_1}{\partial x_1} & \cdots & \dfrac{\partial y_1}{\partial x_n} \\ \vdots & \ddots & \vdots \\ \dfrac{\partial y_n}{\partial x_1} & \cdots & \dfrac{\partial y_n}{\partial x_n} \end{bmatrix}. \tag{3}$$

The Jacobian matrix can be computed using the *Mathematica* command

```
JacobianMatrix[fns_List, vars_List] :=
  Outer[D, fns, vars]
```

The DETERMINANT of J is the Jacobian determinant (confusingly, often called "the Jacobian" as well) and is denoted

$$J = \left| \frac{\partial(y_1, ..., y_n)}{\partial(x_1, ..., x_n)} \right|. \tag{4}$$

It can be computed using the *Mathematica* command

```
JacobianDeterminant[fns_List, vars_List] :=
Module[
    {
      nf = Length[fns],
      nv = Length[vars],
      j = JacobianMatrix[fns, vars]
    },
    Which[
      nf > nv, Sqrt[Det[Transpose[j].j]],
      nf == nv, Det[j],
      nf < nv, Sqrt[Det[j.Transpose[j]]]
    ]
]
```

Taking the differential

$$d\mathbf{y} = \mathbf{y_x} d\mathbf{x} \tag{5}$$

shows that J is the DETERMINANT of the MATRIX $\mathbf{y_x}$, and therefore gives the ratios of n-D volumes (CONTENTS) in y and x,

$$dy_1 \cdots dy_n = \left| \frac{\partial(y_1, ..., y_n)}{\partial(x_1, ..., x_n)} \right| dx_1 \cdots dx_n. \tag{6}$$

The concept of the Jacobian can also be applied to n functions in more than n variables. For example, considering $f(u, v, w)$ and $g(u, v, w)$, the Jacobians

$$\frac{\partial(f, g)}{\partial(u, v)} = \begin{vmatrix} f_u & f_v \\ g_u & g_v \end{vmatrix} \tag{7}$$

$$\frac{\partial(f, g)}{\partial(u, w)} = \begin{vmatrix} f_u & f_w \\ g_u & g_w \end{vmatrix} \tag{8}$$

can be defined (Kaplan 1984, p. 99).

For the case of $n = 3$ variables, the Jacobian takes the special form

$$Jf(x_1, x_2, x_3) \equiv \left| \frac{\partial \mathbf{y}}{\partial x_1} \cdot \frac{\partial \mathbf{y}}{\partial x_2} \times \frac{\partial \mathbf{y}}{\partial x_3} \right|, \tag{9}$$

where $\mathbf{a} \cdot \mathbf{b}$ is the DOT PRODUCT and $\mathbf{b} \times \mathbf{c}$ is the CROSS PRODUCT, which can be expanded to give

$$\left| \frac{\partial(y_1, y_2, y_3)}{\partial(x_1, x_2, x_3)} \right| = \begin{vmatrix} \dfrac{\partial y_1}{\partial x_1} & \dfrac{\partial y_1}{\partial x_2} & \dfrac{\partial y_1}{\partial x_3} \\ \dfrac{\partial y_2}{\partial x_1} & \dfrac{\partial y_2}{\partial x_2} & \dfrac{\partial y_2}{\partial x_3} \\ \dfrac{\partial y_3}{\partial x_1} & \dfrac{\partial y_3}{\partial x_2} & \dfrac{\partial y_3}{\partial x_3} \end{vmatrix}. \quad (10)$$

See also CHANGE OF VARIABLES THEOREM, CURVILINEAR COORDINATES, IMPLICIT FUNCTION THEOREM

References

Kaplan, W. *Advanced Calculus, 3rd ed.* Reading, MA: Addison-Wesley, pp. 98–9, 123, and 238–45, 1984.
Simon, C. P. and Blume, L. E. *Mathematics for Economists.* New York: W. W. Norton, 1994.

Jacobian Conjecture

If $\det[F'(x)] = 1$ for a POLYNOMIAL MAP F (where det is the DETERMINANT), then F is BIJECTIVE with polynomial inverse (i.e., F is an INVERTIBLE POLYNOMIAL MAP).

See also INVERTIBLE POLYNOMIAL MAP, POLYNOMIAL MAP

References

Becker, T. and Weispfenning, V. *Gröbner Bases: A Computational Approach to Commutative Algebra.* New York: Springer-Verlag, p. 330, 1993.
Smale, S. "Mathematical Problems for the Next Century." In *Mathematics: Frontiers and Perspectives 2000* 0821820702 (Ed. V. Arnold, M. Atiyah, P. Lax, and B. Mazur). Providence, RI: Amer. Math. Soc., 2000.

Jacobian Curve

The Jacobian of a linear net of curves of order n is a curve of order $3(n-1)$. It passes through all points common to all curves of the net. It is the LOCUS of points where the curves of the net touch one another and of singular points of the curve.

See also CAYLEYIAN CURVE, HESSIAN COVARIANT, STEINERIAN CURVE

References

Coolidge, J. L. *A Treatise on Algebraic Plane Curves.* New York: Dover, p. 149, 1959.

Jacobian Determinant

JACOBIAN

Jacobian Group

The Jacobian group of a 1-D linear series is given by intersections of the base curve with the JACOBIAN CURVE of itself and two curves cutting the series.

References

Coolidge, J. L. *A Treatise on Algebraic Plane Curves.* New York: Dover, p. 283, 1959.

Jacobian Matrix

JACOBIAN

Jacobi-Anger Expansion

$$e^{iz \cos \theta} = \sum_{n=-\infty}^{\infty} i^n J_n(z) e^{in\theta},$$

where $J_n(z)$ is a BESSEL FUNCTION OF THE FIRST KIND. The identity can also be written

$$e^{iz \cos \theta} = J_0(z) + 2 \sum_{n=1}^{\infty} i^n J_n(z) \cos(n\theta).$$

This expansion represents an expansion of plane waves into a series of cylindrical waves.

See also BESSEL FUNCTION OF THE FIRST KIND

JacobiCD

JACOBI ELLIPTIC FUNCTIONS

JacobiCN

JACOBI ELLIPTIC FUNCTIONS

JacobiCS

JACOBI ELLIPTIC FUNCTIONS

JacobiDC

JACOBI ELLIPTIC FUNCTIONS

JacobiDN

JACOBI ELLIPTIC FUNCTIONS

JacobiDS

JACOBI ELLIPTIC FUNCTIONS

Jacobi-Gauss Quadrature

Also called JACOBI QUADRATURE or MEHLER QUADRATURE. A GAUSSIAN QUADRATURE over the interval $[-1, 1]$ with WEIGHTING FUNCTION

$$W(x) = (1-x)^\alpha (1+x)^\beta. \quad (1)$$

The ABSCISSAS for quadrature order n are given by the roots of the JACOBI POLYNOMIALS $P_n^{(\alpha, \beta)}(x)$. The weights are

$$w_i = -\frac{A_{n+1} \gamma_n}{A_n P_n^{(\alpha, \beta)'}(x_i) P_{n+1}^{(\alpha, \beta)}(x_i)}$$

$$= \frac{A_n}{A_{n-1}} \frac{\gamma_{n-1}}{P_{n-1}^{(\alpha, \beta)}(x_i) P_n^{(\alpha, \beta)'}(x_i)}, \quad (2)$$

where A_n is the COEFFICIENT of x^n in $P_n^{(\alpha,\beta)}(x)$. For JACOBI POLYNOMIALS,

$$A_n \frac{\Gamma(2n+\alpha+\beta+1)}{2^n n! \Gamma(n+\alpha+\beta+1)}, \tag{3}$$

where $\Gamma(z)$ is a GAMMA FUNCTION. Additionally,

$$\gamma_n = \frac{1}{2^{2n}(n!)^2} \frac{2^{2n+\alpha+\beta+1} n!}{2n+\alpha+\beta+1}$$
$$\times \frac{\Gamma(n+\alpha+1)\Gamma(n+\beta+1)}{\Gamma(n+\alpha+\beta+1)}, \tag{4}$$

so

$$w_i = \frac{2n+\alpha+\beta+2}{n+\alpha+\beta+1} \frac{\Gamma(n+\alpha+1)\Gamma(n+\beta+1)}{\Gamma(n+\alpha+\beta+1)}$$
$$\times \frac{2^{2n+\alpha+\beta+1} n!}{V_n'(x_i) V_{n+1}(x_i)} \tag{5}$$

$$= \frac{\Gamma(n+\alpha+1)\Gamma(n+\beta+1)}{\Gamma(n+\alpha+\beta+1)} \frac{2^{2n+\alpha+\beta+1} n!}{\left(1-x_i^2\right)\left[V_n'(x_i)\right]^2}, \tag{6}$$

where

$$V_m \equiv P_n^{(\alpha,\beta)}(x) \frac{2^n n!}{(-1)^n}. \tag{7}$$

The error term is

$$E_n = \frac{\Gamma(n+\alpha+1)\Gamma(n+\beta+1)\Gamma(n+\alpha+\beta+1)}{(2n+\alpha+\beta+1)\left[\Gamma(2n+\alpha+\beta+1)\right]^2}$$
$$\times \frac{2^{2n+\alpha+\beta+1} n!}{(2n)!} f^{(2n)}(\xi) \tag{8}$$

(Hildebrand 1959).

References

Hildebrand, F. B. *Introduction to Numerical Analysis.* New York: McGraw-Hill, pp. 331–34, 1956.

JacobiNC
JACOBI ELLIPTIC FUNCTIONS

JacobiND
JACOBI ELLIPTIC FUNCTIONS

JacobiNS
JACOBI ELLIPTIC FUNCTIONS

JacobiP
JACOBI POLYNOMIAL

JacobiSC
JACOBI ELLIPTIC FUNCTIONS

JacobiSD
JACOBI ELLIPTIC FUNCTIONS

JacobiSN
JACOBI ELLIPTIC FUNCTIONS

JacobiZeta
JACOBI ZETA FUNCTION

Jacobson Canonical Form

Let A be a matrix with the elementary divisors of its characteristic matrix expressed as powers of its irreducible polynomials in the field $F[\lambda]$, and consider an elementary divisor $[p(\lambda)]^q$. If $q > 1$, then

$$C_q(p) = \begin{bmatrix} C(p) & M & 0 & \cdots & 0 & 0 \\ 0 & C(p) & M & \cdots & 0 & 0 \\ \vdots & \ddots & \ddots & \ddots & \vdots & \vdots \\ 0 & 0 & 0 & \cdots & C(p) & M \\ 0 & 0 & 0 & \cdots & 0 & C(p) \end{bmatrix},$$

where M is a matrix of the same order as $C(p)$ having the element 1 in the lower left-hand corner and zeros everywhere else.

Ayres, F. Jr. *Theory and Problems of Matrices.* New York: Schaum, pp. 205–06, 1962.

Jacobson Radical

A special ideal in a COMMUTATIVE RING R. The Jacobson radical is the intersection of the maximal ideals in R. It could be the zero ideal, as in the case of the integers.

See also ALGEBRAIC GEOMETRY, ALGEBRAIC NUMBER THEORY, IDEAL, NILRADICAL, RADICAL (IDEAL)

Jacobsthal Number

The Jacobsthal numbers are the numbers obtained by the U_ns in the LUCAS SEQUENCE with $P = 1$ and $Q = -2$, corresponding to $a = 2$ and $b = -1$. They and the Jacobsthal-Lucas numbers (the V_ns) satisfy the RECURRENCE RELATION

$$J_n = J_{n-1} + 2J_{n-2}. \tag{1}$$

The Jacobsthal numbers satisfy $J_0 = 0$ and $J_1 = 1$ and are 0, 1, 1, 3, 5, 11, 21, 43, 85, 171, 341, ... (Sloane's A001045). The Jacobsthal-Lucas numbers satisfy $j_0 = 2$ and $j_1 = 1$ and are 2, 1, 5, 7, 17, 31, 65, 127, 257, 511, 1025, ... (Sloane's A014551). The properties of these numbers are summarized in Horadam (1996). They are given by the closed form expressions

$$J_n = \sum_{r=0}^{[(n-1)/2]} \binom{n-1-r}{r} 2^r \tag{2}$$

$$j_n = \sum_{r=0}^{\lfloor n/2 \rfloor} \frac{n}{n-r} \binom{n-r}{r} 2^r, \tag{3}$$

where $\lfloor x \rfloor$ is the FLOOR FUNCTION and $\binom{n}{k}$ is a BINOMIAL COEFFICIENT. The Binet forms are

$$J_n = \tfrac{1}{3}(a^n - b^n) = \tfrac{1}{3}[2^n - (-1)^n] \tag{4}$$

$$j_n = a^n + b^n = 2^n + (-1)^n. \tag{5}$$

The GENERATING FUNCTIONS are

$$\sum_{i=1}^{\infty} J_i x^{i-1} = \left(1 - x - 2x^2\right)^{-1} \tag{6}$$

$$\sum_{i=1}^{\infty} j_i x^{i-1} = \left(1 + 4x\right)\left(1 - x - 2x^2\right)^{-1}. \tag{7}$$

The Simson FORMULAS are

$$J_{n+1}J_{n-1} - J_n^2 = (-1)^n 2^{n-1} \tag{8}$$

$$j_{n+1}j_{n-1} - j_n^2 = 9(-1)^{n-1}2^{n-1} = -9\left(J_{n+1}J_{n-1} - J_n^2\right). \tag{9}$$

Summation FORMULAS include

$$\sum_{i=2}^{n} J_i = \tfrac{1}{2}\left(J_{n+2} - 3\right). \tag{10}$$

$$\sum_{i=1}^{n} j_i = \tfrac{1}{2}\left(j_{n+2} - 5\right). \tag{11}$$

Interrelationships are

$$j_n J_n = J_{2n} \tag{12}$$

$$j_n = J_{n+1} + 2J_{n-1} \tag{13}$$

$$9J_n = j_{n+1} + 2j_{n-1} \tag{14}$$

$$j_{n+1} + j_n = 3\left(J_{n+1} + J_n\right) = 3 \cdot 2^n \tag{15}$$

$$j_{n+1} - j_n = 3\left(J_{n+1} - J_n\right) + 4(-1)^{n+1}$$
$$= 2^n + 2(-1)^{n+1} \tag{16}$$

$$j_{n+1} - 2j_n = 3\left(2J_n - J_{n+1}\right) = 3(-1)^{n+1} \tag{17}$$

$$2j_{n+1} + j_{n-1} = 3\left(2J_{n+1} + J_{n-1}\right) + 6(-1)^{n+1} \tag{18}$$

$$j_{n+r} + j_{n-r} = 3\left(J_{n+r} + J_{n-r}\right) + 4(-1)^{n-r} \tag{19}$$

$$= 2^{n-r}\left(2^{2r} + 1\right) + 2(-1)^{n-r} \tag{20}$$

$$j_{n+r} - j_{n-r} = 3\left(J_{n+r} - J_{n-r}\right) = 2^{n-r}\left(2^{2r} - 1\right) \tag{21}$$

$$j_n = 3J_n + 2(-1)^n \tag{22}$$

$$3J_n + j_n = 2^{n+1} \tag{23}$$

$$J_n + j_n = 2J_{n+1} \tag{24}$$

$$j_{n+2}j_{n-2} - j_n^2 = -9\left(J_{n+2}J_{n-2} - J_n\right)^2 = 9(-1)^n 2^{n-2} \tag{25}$$

$$J_m j_n + J_n j_m = 2J_{m+n} \tag{26}$$

$$j_m j_n + 9J_m J_n = 2j_{m+n} \tag{27}$$

$$j_n^2 + 9J_n^2 = 2j_{2n} \tag{28}$$

$$J_m j_n - J_n j_m = (-1)^n 2^{n+1} J_{m-n} \tag{29}$$

$$j_m j_n - 9J_m J_n = (-1)^n 2^{n+1} j_{m-n} \tag{30}$$

$$j_n^2 - 9J_n^2 = (-1)^n 2^{n+2} \tag{31}$$

(Horadam 1996).

References

Horadam, A. F. "Jacobsthal and Pell Curves." *Fib. Quart.* **26**, 79–3, 1988.

Horadam, A. F. "Jacobsthal Representation Numbers." *Fib. Quart.* **34**, 40–4, 1996.

Sloane, N. J. A. Sequences A001045/M2482 and A014551 in "An On-Line Version of the Encyclopedia of Integer Sequences." http://www.research.att.com/~njas/sequences/eisonline.html.

Jacobsthal Polynomial

The Jacobsthal polynomials are the POLYNOMIALS obtained by setting $p(x) = 1$ and $q(x) = 2x$ in the LUCAS POLYNOMIAL SEQUENCE. The first few Jacobsthal polynomials are

$$J_1(x) = 1$$

$$J_2(x) = 1$$

$$J_3(x) = 1 + 2x$$

$$J_4(x) = 1 + 4x$$

$$J_5(x) = 4x^2 + 6x + 1,$$

and the first few Jacobsthal-Lucas polynomials are

$$j_1(x) = 1$$

$$j_2(x) = 4x + 1$$

$$j_3(x) = 6x + 1$$

$$j_4(x) = 8x^2 + 8x + 1$$

$$j_5(x) = 20x^2 + 10x + 1.$$

Jacobsthal and Jacobsthal-Lucas polynomials satisfy

$$J_n(1) = J_n$$

$$j_n(1) = j_n$$

where J_n is a JACOBSTHAL NUMBER and j_n is a JACOBSTHAL-LUCAS NUMBER.

Jacobsthal-Lucas Number
JACOBSTHAL NUMBER

Jacobsthal-Lucas Polynomial
JACOBSTHAL POLYNOMIAL

Jaco-Shalen-Johannson Torus Decomposition

Irreducible orientable COMPACT 3-MANIFOLDS have a canonical (up to ISOTOPY) minimal collection of disjointly EMBEDDED incompressible TORI such that each component of the 3-MANIFOLD removed by the TORI is either "atoroidal" or "Seifert-fibered."

Janko Groups

The SPORADIC GROUPS J_1, J_2, J_3 and J_4. The Janko group J_2 is also known as the HALL-JANKO GROUP.

See also SPORADIC GROUP

References

Ivanov, A. A. and Meierfrankenfeld, U. "A Computer-Free Construction of J_4." *J. Algebra* **219**, 113–72, 1999.
Wilson, R. A. "ATLAS of Finite Group Representation." http://for.mat.bham.ac.uk/atlas/html/contents.html#spo.

Japanese Temple Problem

SANGAKU PROBLEM

Japanese Theorem

Let a convex CYCLIC POLYGON be TRIANGULATED in any manner, and draw the INCIRCLE to each TRIANGLE so constructed. Then the sum of the INRADII is a constant independent of the TRIANGULATION chosen. This theorem can be proved using CARNOT'S THEOREM. In the above figures, for example, the INRADII of the left triangulation are 0.142479, 0.156972, 0.232307, 0.498525, and the INRADII of the right triangulation are 0.157243, 0.206644, 0.312037, 0.354359, giving a sum of 1.03028 in each case.

According to an ancient custom of Japanese mathematicians, this theorem was a SANGAKU PROBLEM inscribed on tablets hung in a Japanese temple to honor the gods and the author in 1800 (Johnson 1929).

The converse is also true: if the sum of INRADII does not depend on the TRIANGULATION of a POLYGON, then the POLYGON is CYCLIC.

See also CARNOT'S THEOREM, CYCLIC POLYGON, INCIRCLE, INRADIUS, SANGAKU PROBLEM, TRIANGULATION

References

Hayashi, T. "Sur un soi-disant théorème chinois." *Mathesis* **6**, 257–60, 1906.
Honsberger, R. *Mathematical Gems III.* Washington, DC: Math. Assoc. Amer., pp. 24–6, 1985.
Johnson, R. A. *Modern Geometry: An Elementary Treatise on the Geometry of the Triangle and the Circle.* Boston, MA: Houghton Mifflin, p. 193, 1929.
Lambert, T. "The Delaunay Triangulation Maximizes the Mean Inradius." *Proc. Sixth Canadian Conf. Comput. Geometry.* Saskatoon, Saskatchewan, Canada, pp. 201–06, Aug. 1994.
Weisstein, E. W. "Plane Geometry." MATHEMATICA NOTEBOOK PlaneGeometry.M.
Wells, D. *The Penguin Dictionary of Curious and Interesting Geometry.* London: Penguin, p. 125, 1991.

Japanese Triangulation Theorem

JAPANESE THEOREM

Jarnick's Inequality

Given a CONVEX plane region with AREA A and PERIMETER p, then

$$|N - A| < p,$$

where N is the number of enclosed LATTICE POINTS.

See also LATTICE POINT, NOSARZEWSKA'S INEQUALITY

j-Conductor

FREY CURVE

Jeep Problem

Maximize the distance a jeep can penetrate into the desert using a given quantity of fuel. The jeep is allowed to go forward, unload some fuel, and then return to its base using the fuel remaining in its tank. At its base, it may refuel and set out again. When it reaches fuel it has previously stored, it may then use it to partially fill its tank. This problem is also called the EXPLORATION PROBLEM (Ball and Coxeter 1987).

Given $n + f$ (with $0 \le f < 1$) drums of fuel at the edge of the desert and a jeep capable of holding one drum (and storing fuel in containers along the way), the maximum one-way distance which can be traveled (assuming the jeep travels one unit of distance per drum of fuel expended) is

$$d = \frac{f}{2n+1} + \sum_{i=1}^{n} \frac{1}{2i-1}$$

$$= \frac{f}{2n+1} + \tfrac{1}{2}\left[\gamma + 2\ln 2 + \psi_0\left(\tfrac{1}{2}+n\right)\right],$$

where γ is the EULER-MASCHERONI CONSTANT and $\psi_n(z)$ the POLYGAMMA FUNCTION.

For example, the farthest a jeep with $n = 1$ drum can travel is obviously 1 unit. However, with $n = 2$ drums of gas, the maximum distance is achieved by filling up the jeep's tank with the first drum, traveling 1/3 of a unit, storing 1/3 of a drum of fuel there, and then returning to base with the remaining 1/3 of a tank. At the base, the tank is filled with the second drum. The jeep then travels 1/3 of a unit (expending 1/3 of a drum of fuel), refills the tank using the 1/3 of a drum of fuel stored there, and continues an additional 1 unit of distance on a full tank, giving a total distance of 4/3. The solutions for $n = 1, 2, \ldots$ drums are 1, 4/3, 23/15, 176/105, 563/315, ..., which can also be written as $a(n)/b(n)$, where

$$a(n) = \left(\frac{1}{1} + \frac{1}{3} + \ldots + \frac{1}{2n-1} \right) \mathrm{LCM}(1,3,5,\ldots,2n-1)$$

$$b(n) = \mathrm{LCM}(1,3,5,\ldots,2n-1)$$

(Sloane's A025550 and A025547).

See also HARMONIC NUMBER

References

Alway, G. C. "Crossing the Desert." *Math. Gaz.* **41**, 209, 1957.
Ball, W. W. R. and Coxeter, H. S. M. *Mathematical Recreations and Essays, 13th ed.* New York: Dover, p. 32, 1987.
Bellman, R. Exercises 54–5 *Dynamic Programming.* Princeton, NJ: Princeton University Press, p. 103, 1955.
Fine, N. J. "The Jeep Problem." *Amer. Math. Monthly* **54**, 24–1, 1947.
Gale, D. "The Jeep Once More or Jeeper by the Dozen." *Amer. Math. Monthly* **77**, 493–01, 1970.
Gardner, M. *The Second Scientific American Book of Mathematical Puzzles & Diversions: A New Selection.* New York: Simon and Schuster, pp. 152 and 157–59, 1961.
Haurath, A.; Jackson, B.; Mitchem, J.; and Schmeichel, E. "Gale's Round-Trip Jeep Problem." *Amer. Math. Monthly* **102**, 299–09, 1995.
Helmer, O. "A Problem in Logistics: The Jeep Problem." Project Rand Report No. Ra 15015, Dec. 1947.
Phipps, C. G. "The Jeep Problem, A More General Solution." *Amer. Math. Monthly* **54**, 458–62, 1947.
Sloane, N. J. A. Sequences A025550 and A025547 in "An On-Line Version of the Encyclopedia of Integer Sequences." http://www.research.att.com/~njas/sequences/eisonline.html.

Jenkins' Theorem

This entry contributed by RONALD M. AARTS

A theorem in the theory of univalent CONFORMAL MAPPINGS of families of domains on a RIEMANN SURFACE, containing an inequality for the coefficients of the mapping functions, as well as conditions to be satisfied by the function so that the inequality becomes an equality. Jenkins' theorem is an exact expression and generalization of TEICHMÜLLER'S PRINCIPLE (Jenkins 1958, Jenkins 1964).

See also CONFORMAL MAPPING, TEICHMÜLLER'S PRINCIPLE

References

Jenkins, J. A. *Univalent Functions and Conformal Mapping.* New York: Springer-Verlag, 1958.
Jenkins, J. A. "Some Area Theorems and a Special Coefficient Theorem." *Illinois J. Math.* **8**, 80–9, 1964.

Jenkins-Traub Method

A complicated POLYNOMIAL ROOT-finding algorithm which is used in the *IMSL* ® (IMSL, Houston, TX) library and which Press *et al.* (1992) describe as "practically a standard in black-box POLYNOMIAL ROOT-finders."

References

IMSL, Inc. *IMSL Math/Library User's Manual.* Houston, TX: IMSL, Inc.
Press, W. H.; Flannery, B. P.; Teukolsky, S. A.; and Vetterling, W. T. *Numerical Recipes in FORTRAN: The Art of Scientific Computing, 2nd ed.* Cambridge, England: Cambridge University Press, p. 369, 1992.
Ralston, A. and Rabinowitz, P. §8.9–.13 in *A First Course in Numerical Analysis, 2nd ed.* New York: McGraw-Hill, 1978.

Jensen Polynomial

Let $f(x)$ be a real ENTIRE FUNCTION OF THE FORM

$$f(x) = \sum_{k=0}^{\infty} \gamma_k \frac{x^k}{k!},$$

where the γ_ks are POSITIVE and satisfy TURÁN'S INEQUALITIES

$$\gamma_k^2 - \gamma_{k-1}\gamma_{k+1} \geq 0$$

for $k = 1, 2, \ldots$. The Jensen polynomial $g(t)$ associated with $f(x)$ is then given by

$$g_n(t) = \sum_{k=0}^{n} \binom{n}{k} \gamma_k t^k,$$

where $\binom{a}{b}$ is a BINOMIAL COEFFICIENT.

References

Csordas, G.; Varga, R. S.; and Vincze, I. "Jensen Polynomials with Applications to the Riemann ζ-Function." *J. Math. Anal. Appl.* **153**, 112–35, 1990.

Jensen's Formula

Portions of this entry contributed by RONALD M. AARTS

A relation connecting the values of a MEROMORPHIC FUNCTION inside a disk with its boundary values on the circumference and with its zeros and poles (Jensen 1899, Levin 1980). Let f be holomorphic on a NEIGHBORHOOD of the CLOSED DISK $\bar{D}(0,r)$ and $f(0) \neq 0$, $a_1, \ldots, a_k$ be the zeros of f in the OPEN DISK $D(0,r)$ counted according to their multiplicities, and assume that $f \neq 0$ on $\partial D(0,r)$. Then

$$\ln|f(0)| + \sum_{j=1}^{k} \ln\left|\frac{r}{a_j}\right| = \frac{1}{2\pi} \int_0^{2\pi} \ln|f(re^{i\theta})| \, d\theta$$

(Krantz 1999, p. 118).

See also CONTOUR INTEGRAL, JENSEN'S INEQUALITY, MAHLER MEASURE

References

Borwein, P. and Erdélyi, T. "Jensen's Formula." §4.2.E.10c in *Polynomials and Polynomial Inequalities.* New York: Springer-Verlag, p. 187, 1995.

Jensen, J. L. "Sur un nouvel et important théorème de la théorie des fonctions." *Acta Math.* **22**, 359–64, 1899.

Krantz, S. G. "Jensen's Formula." §9.1.2 in *Handbook of Complex Analysis.* Boston, MA: Birkhäuser, pp. 117–18, 1999.

Levin, B. Ya. *Distribution of Zeros of Entire Functions.* Providence, RI: Amer. Math. Soc., 1980.

Jensen's Inequality

For a REAL CONTINUOUS CONCAVE FUNCTION

$$\frac{\sum f(x_i)}{n} \le f\left(\frac{\sum x_i}{n}\right) \qquad (1)$$

if f is concave down,

$$\frac{\sum f(x_i)}{n} \ge f\left(\frac{\sum x_i}{n}\right) \qquad (2)$$

if f is concave up, and

$$\frac{\sum f(x_i)}{n} = f\left(\frac{\sum x_i}{n}\right) \qquad (3)$$

IFF $x_1 = x_2 = \ldots = x_n$. A special case is

$$\sqrt{x_1 x_2 \cdots x_n} \le \frac{x_1 + x_2 + \ldots + x_n}{n}, \qquad (4)$$

with equality IFF $x_1 = x_2 = \ldots = x_n$.

See also CONCAVE FUNCTION, JENSEN'S FORMULA

References

Gradshteyn, I. S. and Ryzhik, I. M. *Tables of Integrals, Series, and Products, 6th ed.* San Diego, CA: Academic Press, p. 1101, 2000.

Hardy, G. H.; Littlewood, J. E.; and Pólya, G. "Some Theorems Concerning Monotonic Functions." §3.14 in *Inequalities, 2nd ed.* Cambridge, England: Cambridge University Press, pp. 83–4, 1988.

Jensen, J. L. W. V. "Sur les fonctions convexes et les inégalités entre les valeurs moyennes." *Acta Math.* **30**, 175–93, 1906.

Krantz, S. G. "Jensen's Inequality." §9.1.3 in *Handbook of Complex Analysis.* Boston, MA: Birkhäuser, p. 118, 1999.

Jensen's Theorem

This entry contributed by RONALD M. AARTS

For fixed $v = (v_1, \ldots, v_m)$, the function

$$\|v\|_p = \left[\sum_{i=1}^{m} |v_i|^p\right]^{1/p}$$

is a DECREASING FUNCTION of p (Cheney 1999).

References

Cheney, E. W. *Introduction to Approximation Theory, 2nd ed.* Providence, RI: Amer. Math. Soc., 1999.

Jerabek's Hyperbola

The ISOGONAL CONJUGATE of the EULER LINE. It passes through the vertices of a TRIANGLE, the ORTHOCENTER, CIRCUMCENTER, the SYMMEDIAN POINT, and the ISOGONAL CONJUGATE points of the NINE-POINT CENTER and DE LONGCHAMPS POINT.

See also CIRCUMCENTER, DE LONGCHAMPS POINT, EULER LINE, ISOGONAL CONJUGATE, SYMMEDIAN POINT, NINE-POINT CENTER, ORTHOCENTER

References

Casey, J. *A Treatise on the Analytical Geometry of the Point, Line, Circle, and Conic Sections, Containing an Account of Its Most Recent Extensions with Numerous Examples, 2nd rev. enl. ed.* Dublin: Hodges, Figgis, & Co., 1893.

Pinkernell, G. M. "Cubic Curves in the Triangle Plane." *J. Geom.* **55**, 141–61, 1996.

Vandeghen, A. "Some Remarks on the Isogonal and Cevian Transforms. Alignments of Remarkable Points of a Triangle." *Amer. Math. Monthly* **72**, 1091–094, 1965.

Jerk

The jerk **j** is defined as the time DERIVATIVE of the VECTOR ACCELERATION **a**,

$$\mathbf{j} \equiv \frac{d\mathbf{a}}{dt}$$

See also ACCELERATION, VELOCITY

Jessen's Orthogonal Icosahedron

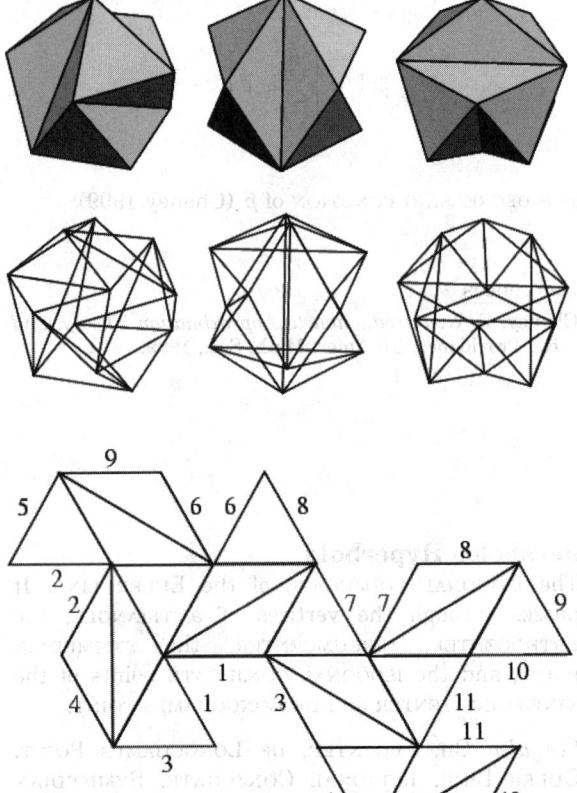

A SHAKY POLYHEDRON constructed by replacing six pairs of adjacent triangles in an ICOSAHEDRON (whose edges form a SKEW QUADRILATERAL) with pairs of ISOSCELES TRIANGLES sharing a common base. The polyhedron can be constructed by dividing the sides of the ICOSAHEDRON in the GOLDEN RATIO (as used in the construction of the ICOSAHEDRON along the edges of the OCTAHEDRON), but reversing the long and short segments.

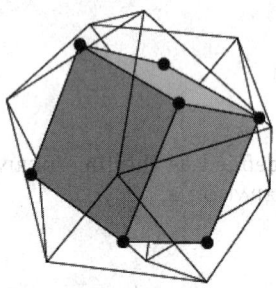

The centers of the eight EQUILATERAL TRIANGLES which remain are then the vertices of a CUBE. The polyhedron can be deformed infinitesimally by pinching the angles between the isosceles triangles whose bases act as hinges. If the polyhedron is constructed using paper and tape instead of entirely rigid faces, it

is possible to collapse the isosceles triangles onto one another, resulting in an OCTAHEDRON.

See also FLEXIBLE POLYHEDRON, RIGID POLYHEDRON, RIGIDITY THEOREM, SHAKY POLYHEDRON

References

Goldberg, M. "Unstable Polyhedral Structures." *Math. Mag.* **51**, 165–70, 1978.
Jessen, B. "Orthogonal Icosahedron." *Nordisk Mat. Tidskr.* **15**, 90–6, 1967.
Weisstein, E. W. "Polyhedra." MATHEMATICA NOTEBOOK POLYHEDRA.M.
Wells, D. *The Penguin Dictionary of Curious and Interesting Geometry.* London: Penguin, p. 161, 1991.

j-Function

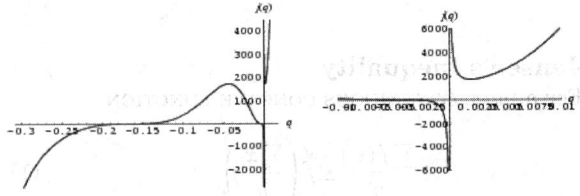

The j-function is defined as

$$j(q) \equiv 1728 J(\sqrt{q}), \qquad (1)$$

where

$$J(q) \equiv \frac{4}{27} \frac{\left[1 - \lambda(q) + \lambda^2(q)\right]^3}{\lambda^2(q)[1 - \lambda(q)]^2} \qquad (2)$$

is KLEIN'S ABSOLUTE INVARIANT, $\lambda(q)$ the ELLIPTIC LAMBDA FUNCTION

$$\lambda(q) \equiv k^2(q) = \left[\frac{\vartheta_2(q)}{\vartheta_3(q)}\right]^4, \qquad (3)$$

ϑ_i a JACOBI THETA FUNCTION, and $1728 = 12^3$. This function can also be specified in terms of the WEBER FUNCTIONS f, f_1, f_2, γ_2, and γ_3 as

$$j(z) = \frac{\left[f^{24}(z) - 16\right]^3}{f^{24}(z)} \qquad (4)$$

$$= \frac{\left[f_1^{24}(z) + 16\right]^3}{f_1^{24}(z)} \qquad (5)$$

$$= \frac{\left[f_2^{24}(z) + 16\right]^3}{f_2^{24}(z)} \qquad (6)$$

$$= \gamma_2^3(z) \qquad (7)$$

$$= \gamma_3^2(z) + 1728 \qquad (8)$$

(Weber 1902, p. 179; Atkin and Morain 1993).
The j-function is a MEROMORPHIC FUNCTION on the UPPER HALF-PLANE which is invariant with respect to

the SPECIAL LINEAR GROUP $SL(2, Z)$. It has a FOURIER SERIES

$$j(q) = \sum_{n=-\infty}^{\infty} c_n q^n, \qquad (9)$$

for the NOME

$$q \equiv e^{2\pi i \tau} \qquad (10)$$

with $\Im[\tau] > 0$. The coefficients in the expansion of the j-function satisfy:

1. $c_n = 0$ for $n < -1$ and $c_{-1} = 1$,
2. all c_ns are INTEGERS with fairly limited growth with respect to n, and
3. $j(q)$ is an ALGEBRAIC NUMBER, sometimes a RATIONAL NUMBER, and sometimes even an INTEGER at certain very special values of q (or τ).

The latter result is the end result of the massive and beautiful theory of COMPLEX multiplication and the first step of Kronecker's so-called "JUGENDTRAUM."

Then all of the COEFFICIENTS in the LAURENT SERIES

$$j(q) = \frac{1}{q} + 744 + 196884q + 21493760q^2 + 864299970q^3$$

$$+ 20245856256q^4 + 333202640600q^5 + \ldots \quad (11)$$

(Sloane's A000521) are POSITIVE INTEGERS (Rankin 1977, Apostol 1997). Berwick calculated the first seven $c(n)$ in 1916, Zuckerman found the first 24 in 1939, and van Wijngaarden gave the first 100 in 1963.

Some remarkable sum formulas involving $j(q)$ for $\tau \in H$, where H is the UPPER HALF-PLANE, and $c(n)$ include

$$\left[504 \sum_{n=0}^{\infty} \sigma_5(n) q^n \right]^2 = \left[j(q) - 12^3 \right] \sum_{n=1}^{\infty} \tau(n) x^n, \quad (12)$$

where $\sigma_k(n)$ is the DIVISOR FUNCTION and $\sigma_5(0) = -1/504$. In addition,

$$(504)^2 \sum_{k=0}^{n} \sigma_5(k) \sigma_5(n-k)$$

$$= \tau(n+1) - 984\tau(n) + \sum_{k=1}^{n-1} c(k)\tau(n-k) \quad (13)$$

$$\frac{65520}{691} [\sigma_{11}(n) - \tau(n)]$$

$$= \tau(n+1) + 24\tau(n) + \sum_{k=1}^{n-1} c(k)\tau(n-k), \quad (14)$$

where $\tau(n)$ is the TAU FUNCTION (Lehmer 1942; Apostol 1997, p. 92). The latter leads immediately to the remarkable congruence

$$\tau(n) \equiv \sigma_{11}(n) \pmod{691}. \quad (15)$$

Lehmer (1942) showed that

$$(n+1)c(n) \equiv 0 \pmod{24} \quad (16)$$

for all $n \geq 1$, and Lehner (1949) and Apostol (1997, pp. 22, 74, and 90–1) demonstrated that

$$c(2n) \equiv 0 \pmod{2^{11}} \quad (17)$$

$$c(3n) \equiv 0 \pmod{3^5} \quad (18)$$

$$c(5n) \equiv 0 \pmod{5^2} \quad (19)$$

$$c(7n) \equiv 0 \pmod{7} \quad (20)$$

$$c(11n) \equiv 0 \pmod{11}. \quad (21)$$

More generally,

$$c(2^\alpha n) \equiv 0 \pmod{2^{3\alpha+8}} \quad (22)$$

$$c(3^\alpha n) \equiv 0 \pmod{3^{2\alpha+3}} \quad (23)$$

$$c(5^\alpha n) \equiv 0 \pmod{5^{\alpha+1}} \quad (24)$$

$$c(7^\alpha n) \equiv 0 \pmod{7^\alpha} \quad (25)$$

(Lehner 1949; Apostol 1997, p. 91). Congruences of this type cannot exist for 13, but Newman (1958) showed

$$c(13np) + c(13n)c(13p) + p^{-1}c\left(\frac{13n}{p}\right) \equiv 0 \pmod{13},$$

$$(26)$$

where $p^{-1}p \equiv 1 \pmod{13}$ and $c(x) = 0$ if x is not an integer (Apostol 1997, p. 91). Congruences for $c(kn)$ have been generalized by Atkin and O'Brien (1967).

An asymptotic formula for $c(n)$ was discovered by Petersson (1932), and subsequently independently rediscovered by Rademacher (1938):

$$c(n) \sim \frac{e^{4\pi\sqrt{n}}}{\sqrt{2} n^{3/4}}. \quad (27)$$

Let d be a POSITIVE SQUAREFREE INTEGER, and define

$$t \equiv \begin{cases} i\sqrt{d} & \text{for } d \equiv 1 \text{ or } 2 \pmod 4 \\ \frac{1}{2}\left(1 + i\sqrt{d}\right) & \text{for } d \equiv 3 \pmod 4. \end{cases} \quad (28)$$

Then the NOME is

$$q \equiv e^{i\pi\tau} = \begin{cases} e^{2\pi i (i\sqrt{d})} & \text{for } d \equiv 1 \text{ or } 2 \pmod 4 \\ e^{2\pi i (1+i\sqrt{d})/2} & \text{for } d \equiv 3 \pmod 4 \end{cases}$$

$$= \begin{cases} e^{-2\pi\sqrt{d}} & \text{for } d \equiv 1 \text{ or } 2 \pmod 4 \\ -e^{-\pi\sqrt{d}} & \text{for } d \equiv 3 \pmod 4. \end{cases} \quad (29)$$

It then turns out that $j(q)$ is an ALGEBRAIC INTEGER of degree $h(-d)$, where $h(-d)$ is the CLASS NUMBER of the DISCRIMINANT $-d$ of the QUADRATIC FIELD $\mathbb{Q}(\sqrt{n})$ (Silverman 1986). The first term in the LAURENT

SERIES is then $q^{-1} = e^{-2\pi\sqrt{n}}$ or $-e^{-\pi\sqrt{n}}$, and all the later terms are POWERS of q^{-1}, which are small numbers. The larger n, the faster the series converges. If $h(-d) = 1$, then $j(q)$ is a ALGEBRAIC INTEGER of degree 1, i.e., just a plain INTEGER. Furthermore, the INTEGER is a perfect CUBE.

The numbers whose LAURENT SERIES give INTEGERS are those with CLASS NUMBER 1. But these are precisely the HEEGNER NUMBERS -1, -2, -3, -7, -11, -19, -43, -67, -163. The greater (in ABSOLUTE VALUE) the HEEGNER NUMBER d, the closer to an INTEGER is the expression $e^{\pi\sqrt{-n}}$, since the initial term in $j(q)$ is the largest and subsequent terms are the smallest. The best approximations with $h(-d) = 1$ are therefore

$$e^{\pi\sqrt{43}} \approx 960^3 + 744 - 2.2 \times 10^{-4} \tag{30}$$

$$e^{\pi\sqrt{67}} \approx 5280^3 + 744 - 1.3 \times 10^{-6} \tag{31}$$

$$e^{\pi\sqrt{163}} \approx 640320^3 + 744 - 7.5 \times 10^{-13}. \tag{32}$$

The exact values of $j(q)$ corresponding to the HEEGNER NUMBERS are

$$j(-e^{-\pi}) = 12^3 \tag{33}$$

$$j\left(e^{-2\pi\sqrt{2}}\right) = 20^3 \tag{34}$$

$$j\left(-e^{-\pi\sqrt{3}}\right) = 0^3 \tag{35}$$

$$j\left(-e^{-\pi\sqrt{7}}\right) = -15^3 \tag{36}$$

$$j\left(-e^{-\pi\sqrt{11}}\right) = -32^3 \tag{37}$$

$$j\left(-e^{-\pi\sqrt{19}}\right) = -96^3 \tag{38}$$

$$j\left(-e^{-\pi\sqrt{43}}\right) = -960^3 \tag{39}$$

$$j\left(-e^{-\pi\sqrt{67}}\right) = -5280^3 \tag{40}$$

$$j\left(-e^{-\pi\sqrt{163}}\right) = -640320^3. \tag{41}$$

(The number 5280 is particularly interesting since it is also the number of feet in a mile.) The ALMOST INTEGER generated by the last of these, $e^{\pi\sqrt{163}}$ (corresponding to the field $\mathbb{Q}(\sqrt{-163})$ and the IMAGINARY QUADRATIC FIELD of maximal discriminant), is sometimes known as the RAMANUJAN CONSTANT. However, this attribution is historically fallacious since this amazing property of $e^{\pi\sqrt{163}}$ was first noted by Hermite (1859) and does not seem to appear in any of the works of Ramanujan.

$e^{\pi\sqrt{22}}$, $e^{\pi\sqrt{37}}$, and $e^{\pi\sqrt{58}}$ are also ALMOST INTEGERS. These correspond to binary quadratic forms with discriminants -88, -148, and -232, all of which have CLASS NUMBER two and were noted by Ramanujan (Berndt 1994).

It turns out that the j-function also is important in the CLASSIFICATION THEOREM for finite simple groups, and that the factors of the orders of the SPORADIC GROUPS, including the celebrated MONSTER GROUP, are also related.

See also ALMOST INTEGER, HEEGNER NUMBER, IMAGINARY QUADRATIC FIELD, KLEIN'S ABSOLUTE INVARIANT, RAMANUJAN CONSTANT, WEBER FUNCTIONS

References

Apostol, T. M. "The Fourier Expansions of $\Delta(\tau)$ and $J(\tau)$" and "Congruences for the Coefficients of the Modular Function j." §1.15 and Ch. 4 in *Modular Functions and Dirichlet Series in Number Theory, 2nd ed.* New York: Springer-Verlag, pp. 20–2 and 74–3, 1997.

Atkin, A. O. L. and Morain, F. "Elliptic Curves and Primality Proving." *Math. Comput.* **61**, 29–8, 1993.

Atkin, A. O. L. and O'Brien, J. N. "Some Properties of $p(n)$ and $c(n)$ Modulo Powers of 13." *Trans. Amer. Math. Soc.* **126**, 442–59, 1967.

Berndt, B. C. *Ramanujan's Notebooks, Part IV.* New York: Springer-Verlag, pp. 90–1, 1994.

Borwein, J. M. and Borwein, P. B. *Pi & the AGM: A Study in Analytic Number Theory and Computational Complexity.* New York: Wiley, pp. 117–18, 1987.

Cohn, H. *Introduction to the Construction of Class Fields.* New York: Dover, p. 73, 1994.

Conway, J. H. and Guy, R. K. "The Nine Magic Discriminants." In *The Book of Numbers.* New York: Springer-Verlag, pp. 224–26, 1996.

Hermite, C. "Sur la théorie des équations modulaires." *C. R. Acad. Sci. (Paris)* **49**, 16–4, 110–18, and 141–44, 1859 *Oeuvres complètes, Tome II.* Paris: Hermann, p. 61, 1912.

Lehmer, D. H. "Properties of the Coefficients of the Modular Invariant $J(\tau)$." *Amer. J. Math.* **64**, 488–02, 1942.

Lehner, J. "Divisibility Properties of the Fourier Coefficients of the Modular Invariant $j(\tau)$." *Amer. J. Math.* **71**, 136–48, 1949.

Lehner, J. "Further Congruence Properties of the Fourier Coefficients of the Modular Invariant $j(\tau)$." *Amer. J. Math.* **71**, 373–86, 1949.

Morain, F. "Implementation of the Atkin-Goldwasser-Kilian Primality Testing Algorithm." Rapport de Recherche 911, INRIA, Oct. 1988.

Newman, M. "Congruences for the Coefficients of Modular Forms and for the Coefficients of $j(\tau)$." *Proc. Amer. Math. Soc.* **9**, 609–12, 1958.

Petersson, H. "Über die Entwicklungskoeffizienten der automorphen formen." *Acta Math.* **58**, 169–15, 1932.

Rademacher, H. "The Fourier Coefficients of the Modular Invariant $j(\tau)$." *Amer. J. Math.* **60**, 501–12, 1938.

Rankin, R. A. *Modular Forms.* New York: Wiley, 1985.

Rankin, R. A. *Modular Forms and Functions.* Cambridge, England: Cambridge University Press, p. 199, 1977.

Serre, J. P. *Cours d'arithmétique.* Paris: Presses Universitaires de France, 1970.

Silverman, J. H. *The Arithmetic of Elliptic Curves.* New York: Springer-Verlag, p. 339, 1986.

Sloane, N. J. A. Sequences A000521/M5477 in "An On-Line Version of the Encyclopedia of Integer Sequences." http://www.research.att.com/~njas/sequences/eisonline.html.

Weber, H. *Lehrbuch der Algebra, Vols. I-II.* New York: Chelsea, 1979.

Weisstein, E. W. "j-Function." MATHEMATICA NOTEBOOK JFUNCTION.M.

Jinc Function

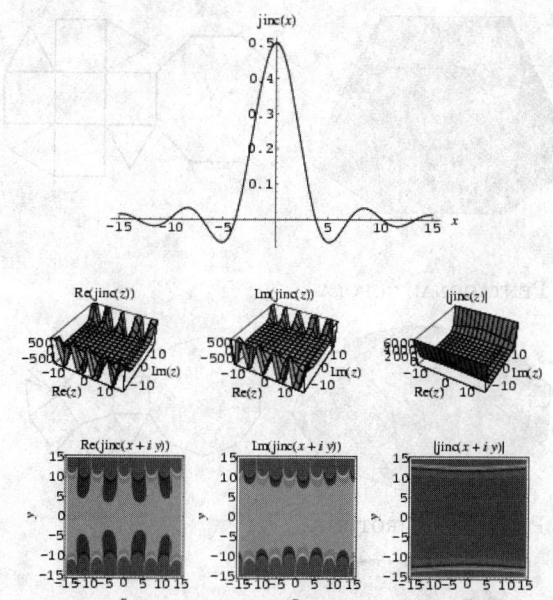

The jinc function is defined as

$$\mathrm{jinc}(x) \equiv \frac{J_1(x)}{x},$$

where $J_1(x)$ is a BESSEL FUNCTION OF THE FIRST KIND, and satisfies $\lim_{x\to 0} \mathrm{jinc}(x) = 1/2$. The DERIVATIVE of the jinc function is given by

$$\mathrm{jinc}'(x) = -\frac{J_2(x)}{x}.$$

The function is sometimes normalized by multiplying by a factor of 2 so that $\mathrm{jinc}(0) = 1$ (Siegman 1986, p. 729).

See also BESSEL FUNCTION OF THE FIRST KIND, SINC FUNCTION

References

Bracewell, R. *The Fourier Transform and Its Applications, 3rd ed.* New York: McGraw-Hill, p. 64, 1999.
Siegman, A. E. *Lasers.* Sausalito, CA: University Science Books, 1986.

j-Invariant

An invariant of an ELLIPTIC CURVE given in the form

$$y^2 = x^3 + ax + b$$

which is closely related to the DISCRIMINANT and defined by

$$j(E) \equiv \frac{2^8 3^3 a^3}{4a^3 + 27b^2}.$$

The determination of j as an ALGEBRAIC INTEGER in the QUADRATIC FIELD $\mathbb{Q}(j)$ is discussed by Greenhill (1891), Weber (1902), Berwick (1928), Watson (1938),

Gross and Zaiger (1985), and Dorman (1988). The norm of j in $\mathbb{Q}(j)$ is the CUBE of an INTEGER in $\mathbb{Z}$.

See also DISCRIMINANT (ELLIPTIC CURVE), ELLIPTIC CURVE, FREY CURVE

References

Berwick, W. E. H. "Modular Invariants Expressible in Terms of Quadratic and Cubic Irrationalities." *Proc. London Math. Soc.* **28**, 53–9, 1928.
Dorman, D. R. "Special Values of the Elliptic Modular Function and Factorization Formulae." *J. reine angew. Math.* **383**, 207–20, 1988.
Greenhill, A. G. "Table of Complex Multiplication Moduli." *Proc. London Math. Soc.* **21**, 403–22, 1891.
Gross, B. H. and Zaiger, D. B. "On Singular Moduli." *J. reine angew. Math.* **355**, 191–20, 1985.
Stepanov, S. A. "The j-Invariant." §7.2 in *Codes on Algebraic Curves.* New York: Kluwer, pp. 178–80, 1999.
Watson, G. N. "Ramanujans Vermutung über Zerfällungsanzahlen." *J. reine angew. Math.* **179**, 97–28, 1938.
Weber, H. *Lehrbuch der Algebra, Vols. I-II.* New York: Chelsea, 1979.

Jitter

A SAMPLING phenomenon produced when a waveform is not sampled uniformly at an interval t each time, but rather at a series of slightly shifted intervals $t + \Delta t_i$ such that the average $\langle \Delta t_i \rangle = 0$.

See also GHOST, SAMPLING

Joachimsthal's Equation

Using CLEBSCH-ARONHOLD NOTATION, an algebraic curve satisfies

$$\xi_1^n a_y^n + \xi_1^{n-1}\xi_2 a_y^{n-1}a_x + \tfrac{1}{2}n(n-1)\xi_1^{n-2}\xi_2^2 a_y^{n-2}a_x^2 + \ldots$$

$$+ n\xi_1\xi_2^{n-1}a_y a_x^{n-1} + \xi_2^n a_x^n = 0.$$

References

Coolidge, J. L. *A Treatise on Algebraic Plane Curves.* New York: Dover, p. 89, 1959.

Johnson Bound

A bound on error-correcting codes.

Johnson Circle

The CIRCUMCIRCLE in JOHNSON'S THEOREM.

See also JOHNSON'S THEOREM

Johnson Solid

The Johnson solids are the CONVEX POLYHEDRA having regular faces and equal edge lengths (with the exception of the completely regular PLATONIC SOLIDS, the "SEMIREGULAR" ARCHIMEDEAN SOLIDS, and the two infinite families of PRISMS and ANTIPRISMS). There are 28 simple (i.e., cannot be dissected into two other regular-faced polyhedra by a plane) regular-faced polyhedra in addition to the PRISMS and ANTIPRISMS (Zalgaller 1969), and Johnson (1966) proposed and Zalgaller (1969) proved that there exist exactly 92 Johnson solids in all.

There is a near-Johnson solid which can be constructed by inscribing regular nonagons inside the eight triangular faces of a regular octahedron, then joining the free edges to the 24 triangles and finally the remaining edges of the triangles to six squares, with one square for each octahedral vertex. It turns out that the triangles are not quite equilateral, making the edges that bound the squares a slightly different length from that of the enneagonal edge. However, because the differences in edge lengths are so small, the flexing of an average model allows the solid to be constructed with all edges equal (Olshevsky).

A database of solids and VERTEX NETS of these solids is maintained on the Bell Laboratories Netlib server, but a few errors exist in several entries. A concatenated and corrected version of the files is given by Weisstein, together with *Mathematica* code to display the solids and nets. The following table summarizes the names of the Johnson solids and gives their images and nets.

1. SQUARE PYRAMID

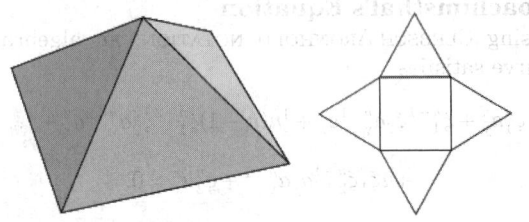

2. PENTAGONAL PYRAMID

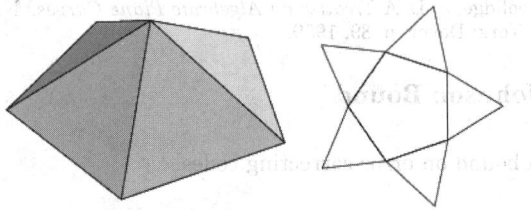

3. TRIANGULAR CUPOLA

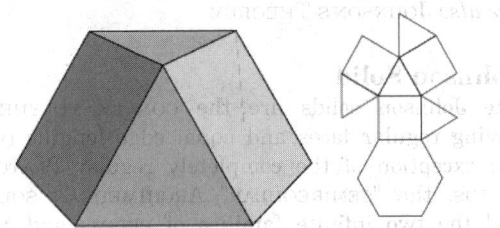

4. SQUARE CUPOLA

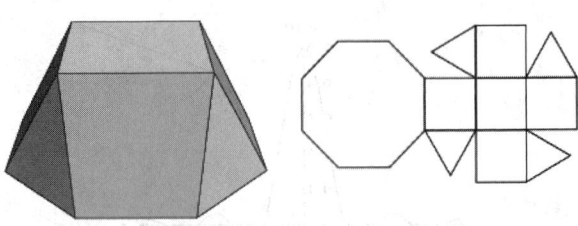

5. PENTAGONAL CUPOLA

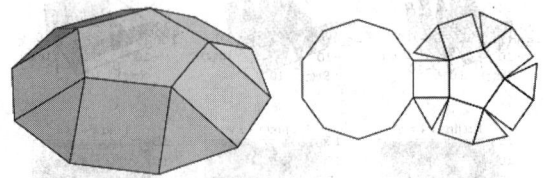

6. PENTAGONAL ROTUNDA

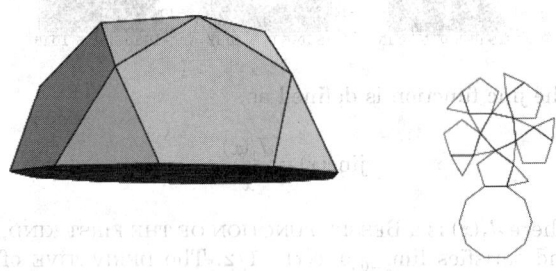

7. ELONGATED TRIANGULAR PYRAMID

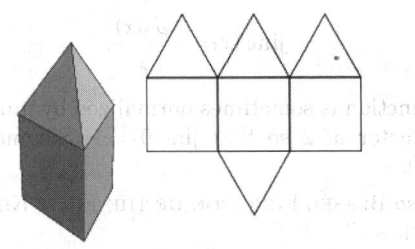

8. ELONGATED SQUARE PYRAMID

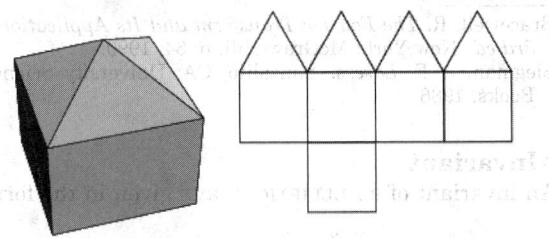

9. ELONGATED PENTAGONAL PYRAMID

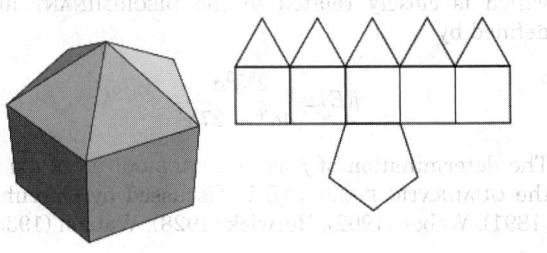

10. GYROELONGATED SQUARE PYRAMID

11. GYROELONGATED PENTAGONAL PYRAMID

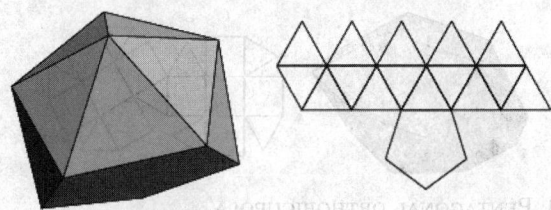

12. TRIANGULAR DIPYRAMID

13. PENTAGONAL DIPYRAMID

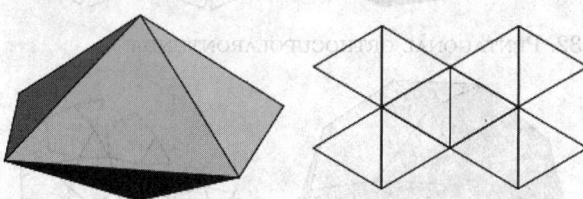

14. ELONGATED TRIANGULAR DIPYRAMID

15. ELONGATED SQUARE DIPYRAMID

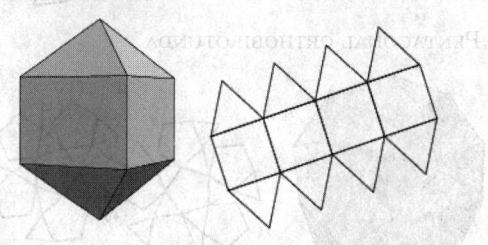

16. ELONGATED PENTAGONAL DIPYRAMID

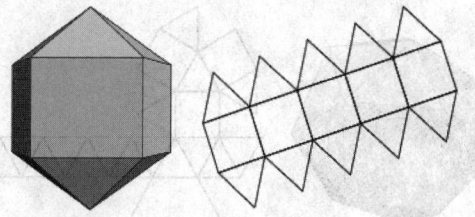

17. GYROELONGATED SQUARE DIPYRAMID

18. ELONGATED TRIANGULAR CUPOLA

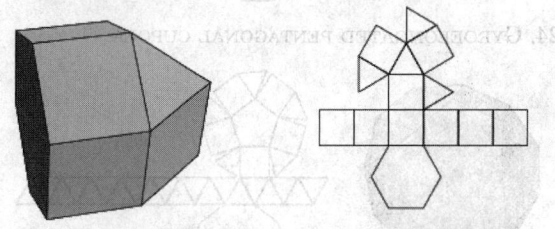

19. ELONGATED SQUARE CUPOLA

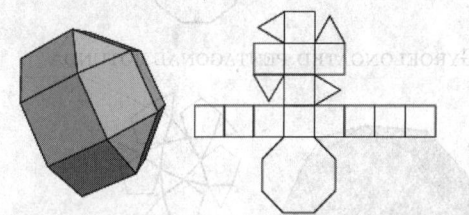

20. ELONGATED PENTAGONAL CUPOLA

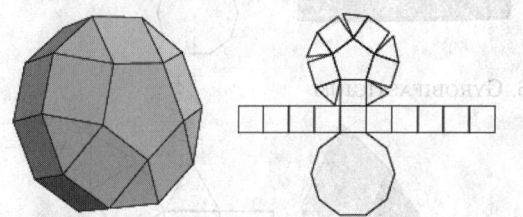

21. ELONGATED PENTAGONAL ROTUNDA

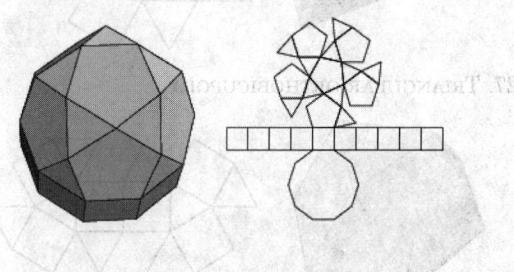

22. GYROELONGATED TRIANGULAR CUPOLA

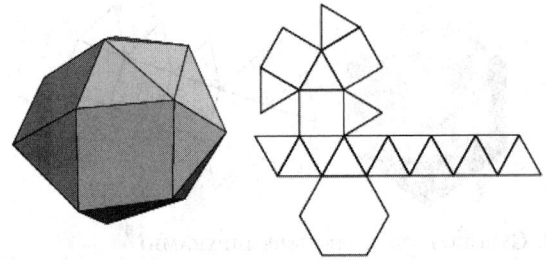

23. GYROELONGATED SQUARE CUPOLA

24. GYROELONGATED PENTAGONAL CUPOLA

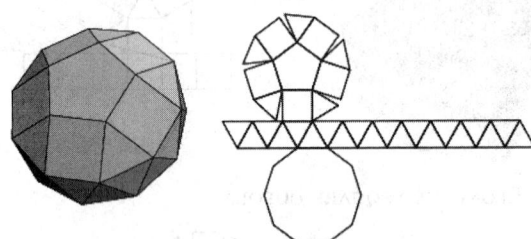

25. GYROELONGATED PENTAGONAL ROTUNDA

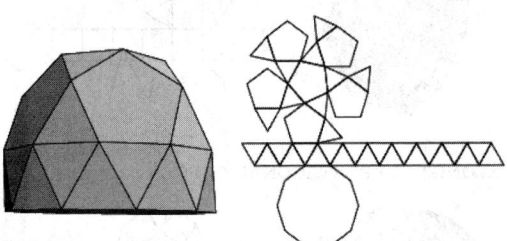

26. GYROBIFASTIGIUM

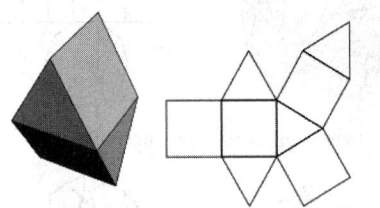

27. TRIANGULAR ORTHOBICUPOLA

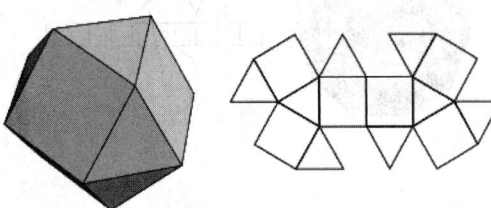

28. SQUARE ORTHOBICUPOLA

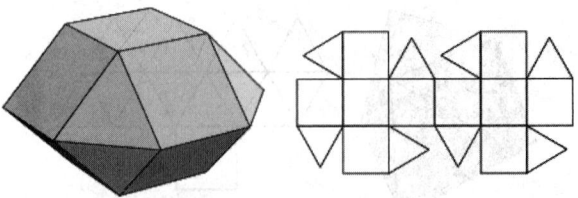

29. SQUARE GYROBICUPOLA

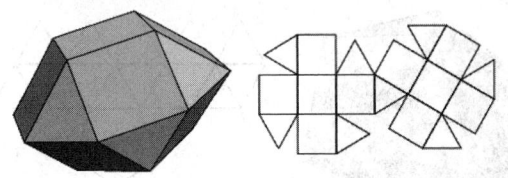

30. PENTAGONAL ORTHOBICUPOLA

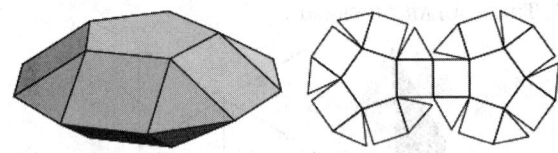

31. PENTAGONAL GYROBICUPOLA

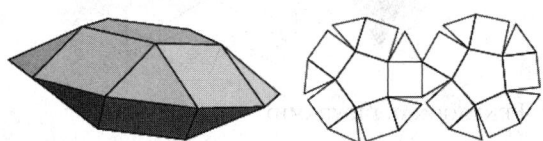

32. PENTAGONAL ORTHOCUPOLARONTUNDA

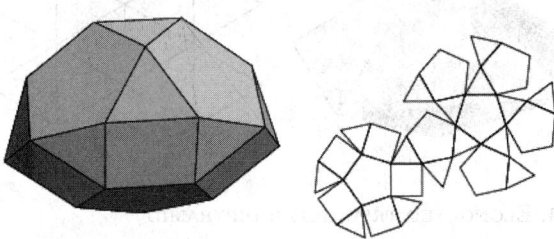

33. PENTAGONAL GYROCUPOLAROTUNDA

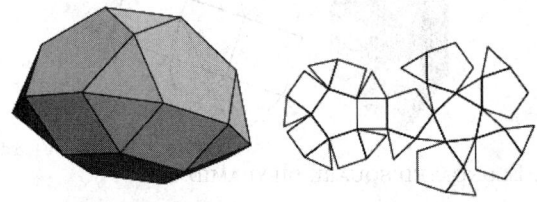

34. PENTAGONAL ORTHOBIROTUNDA

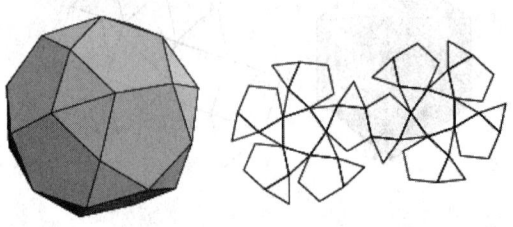

35. ELONGATED TRIANGULAR ORTHOBICUPOLA

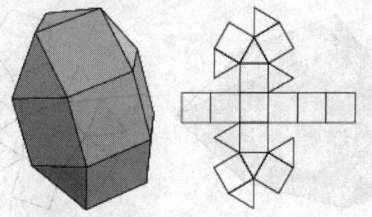

36. ELONGATED TRIANGULAR GYROBICUPOLA

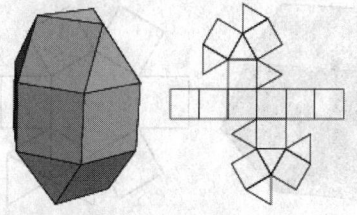

37. ELONGATED SQUARE GYROBICUPOLA

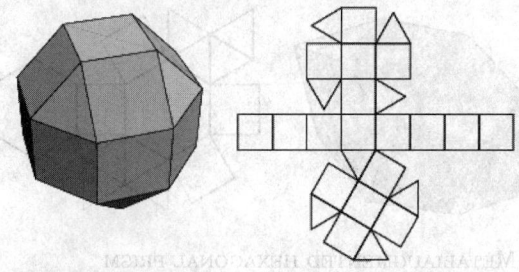

38. ELONGATED PENTAGONAL ORTHOBICUPOLA

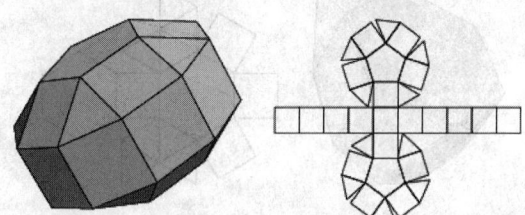

39. ELONGATED PENTAGONAL GYROBICUPOLA

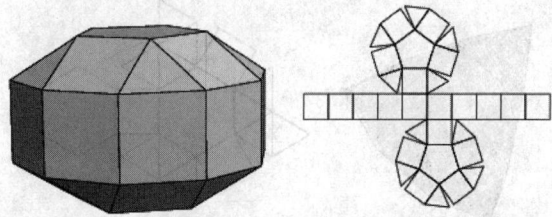

40. ELONGATED PENTAGONAL ORTHOCUPOLAROTUNDA

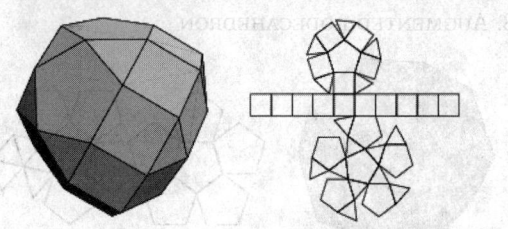

41. ELONGATED PENTAGONAL GYROCUPOLAROTUNDA

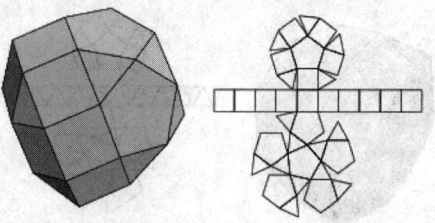

42. ELONGATED PENTAGONAL ORTHOBIROTUNDA

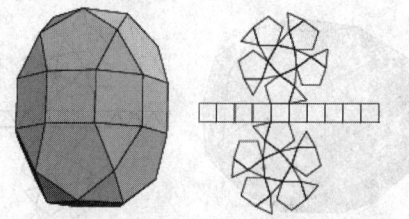

43. ELONGATED PENTAGONAL GYROBIROTUNDA

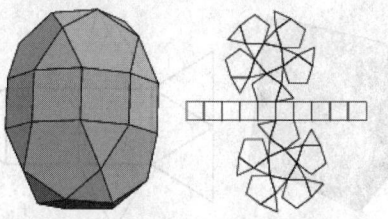

44. GYROELONGATED TRIANGULAR BICUPOLA

45. GYROELONGATED SQUARE BICUPOLA

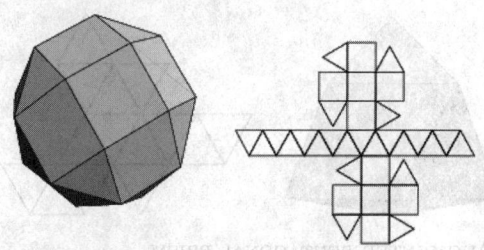

46. GYROELONGATED PENTAGONAL BICUPOLA

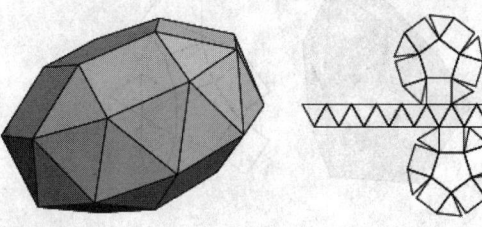

47. GYROELONGATED PENTAGONAL CUPOLAROTUNDA

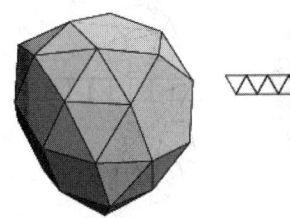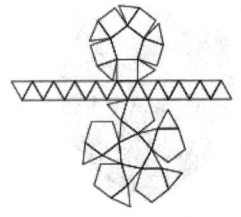

48. GYROELONGATED PENTAGONAL BIROTUNDA

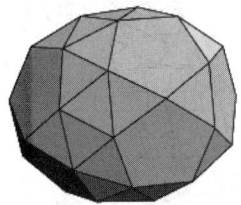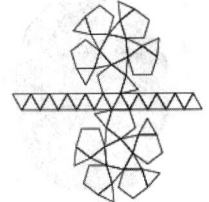

49. AUGMENTED TRIANGULAR PRISM

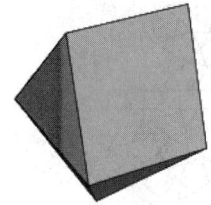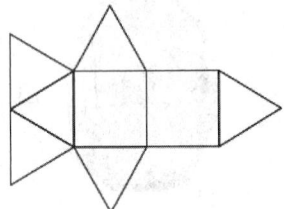

50. BIAUGMENTED TRIANGULAR PRISM

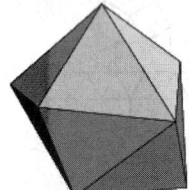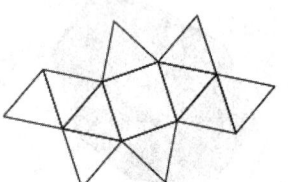

51. TRIAUGMENTED TRIANGULAR PRISM

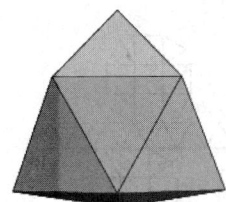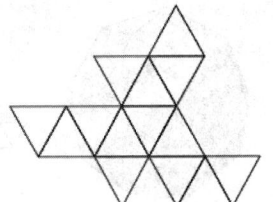

52. AUGMENTED PENTAGONAL PRISM

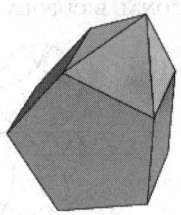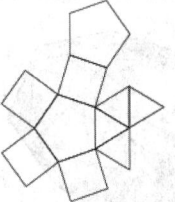

53. BIAUGMENTED PENTAGONAL PRISM

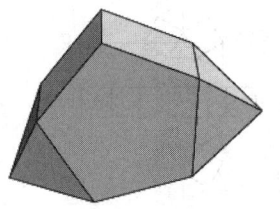

54. AUGMENTED HEXAGONAL PRISM

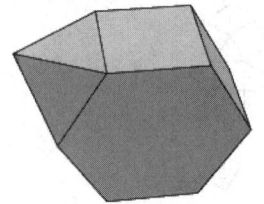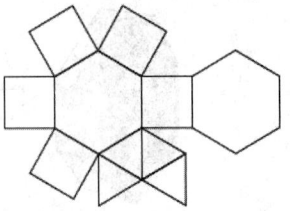

55. PARABIAUGMENTED HEXAGONAL PRISM

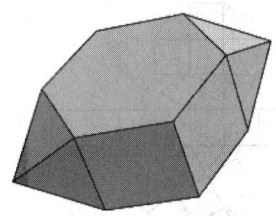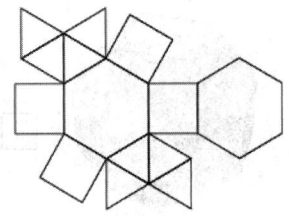

56. METABIAUGMENTED HEXAGONAL PRISM

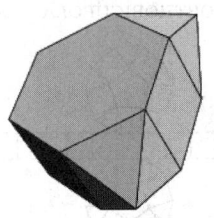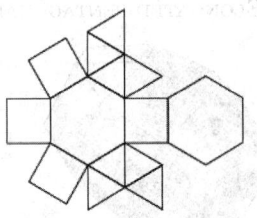

57. TRIAUGMENTED HEXAGONAL PRISM

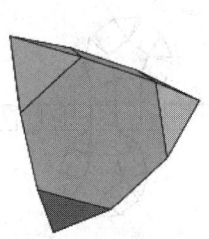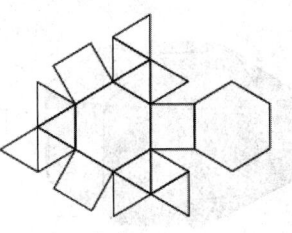

58. AUGMENTED DODECAHEDRON

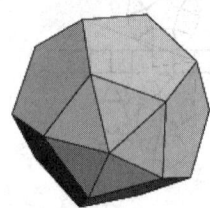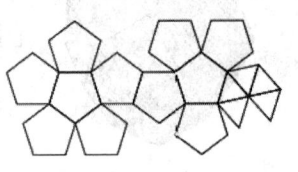

59. PARABIAUGMENTED DODECAHEDRON

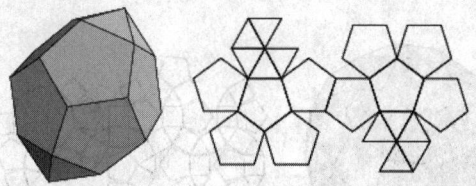

60. METABIAUGMENTED DODECAHEDRON

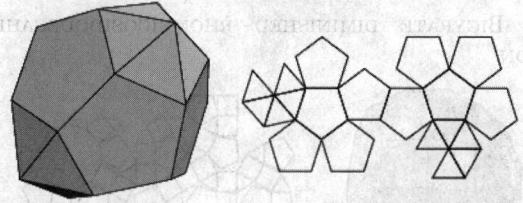

61. TRIAUGMENTED DODECAHEDRON

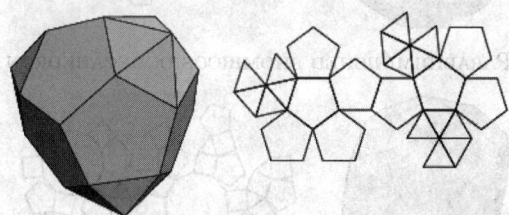

62. METABIDIMINISHED ICOSAHEDRON

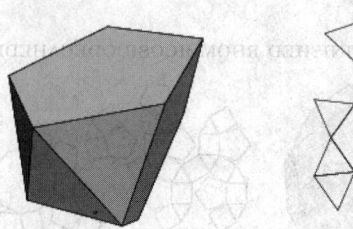

63. TRIDIMINISHED ICOSAHEDRON

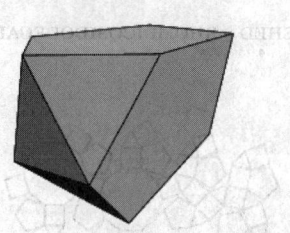

64. AUGMENTED TRIDIMINISHED ICOSAHEDRON

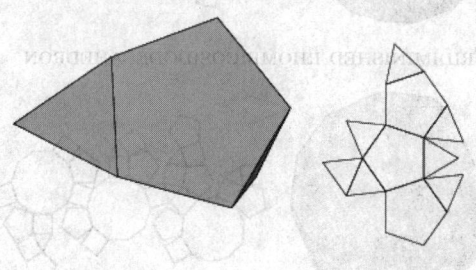

65. AUGMENTED TRUNCATED TETRAHEDRON

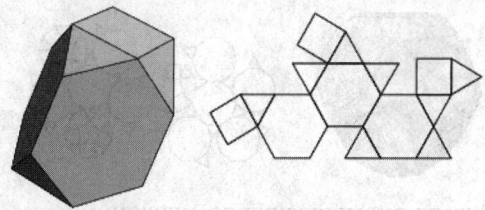

66. AUGMENTED TRUNCATED CUBE

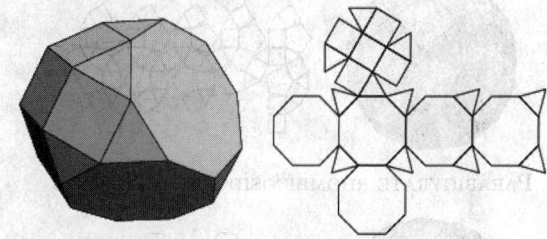

67. BIAUGMENTED TRUNCATED CUBE

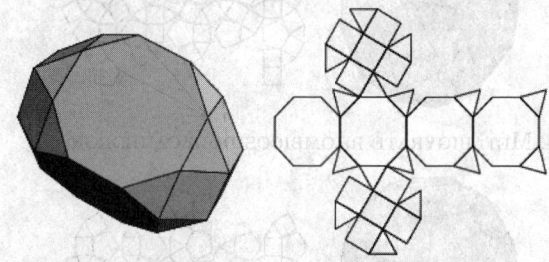

68. AUGMENTED TRUNCATED DODECAHEDRON

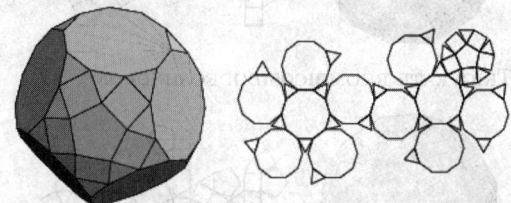

69. PARABIAUGMENTED TRUNCATED DODECAHEDRON

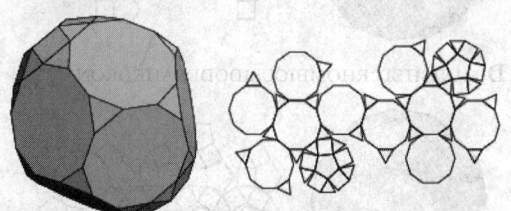

70. METABIAUGMENTED TRUNCATED DODECAHEDRON

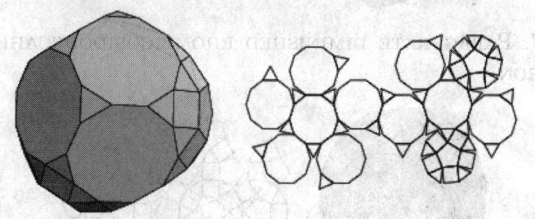

71. TRIAUGMENTED TRUNCATED DODECAHEDRON

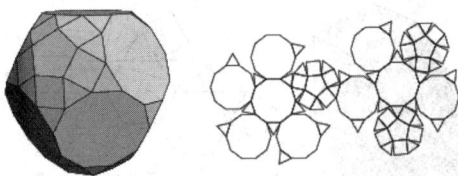

72. GYRATE RHOMBICOSIDODECAHEDRON

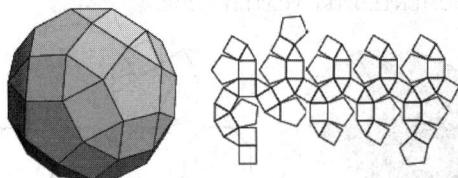

73. PARABIGYRATE RHOMBICOSIDODECAHEDRON

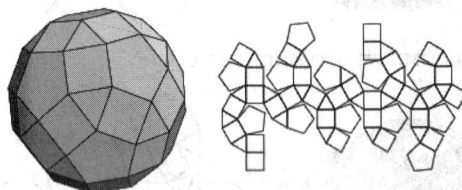

74. METABIGYRATE RHOMBICOSIDODECAHEDRON

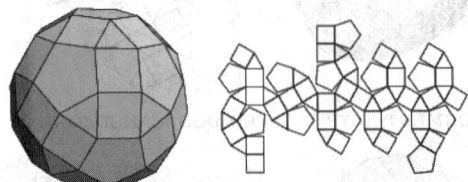

75. TRIGYRATE RHOMBICOSIDODECAHEDRON

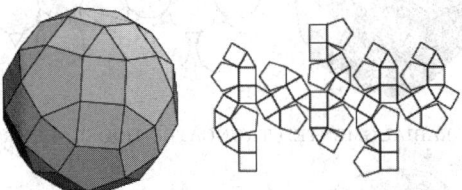

76. DIMINISHED RHOMBICOSIDODECAHEDRON

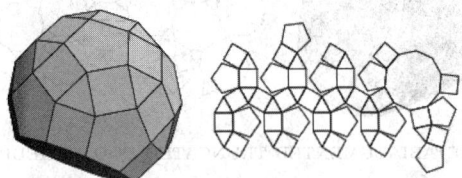

77. PARAGYRATE DIMINISHED RHOMBICOSIDODECAHE-
DRON

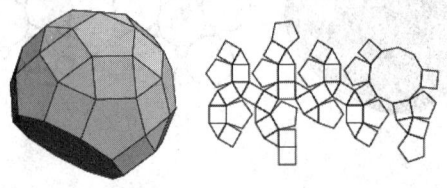

78. METAGYRATE DIMINISHED RHOMBICOSIDODECAHE-
DRON

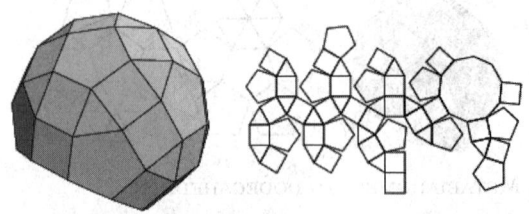

79. BIGYRATE DIMINISHED RHOMBICOSIDODECAHE-
DRON

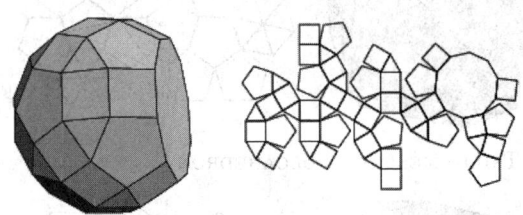

80. PARABIDIMINISHED RHOMBICOSIDODECAHEDRON

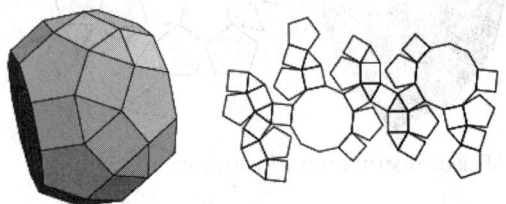

81. METABIDIMINISHED RHOMBICOSIDODECAHEDRON

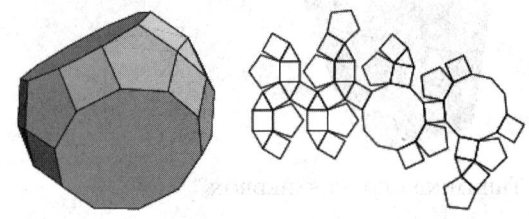

82. GYRATE BIDIMINISHED RHOMBICOSIDODECAHE-
DRON

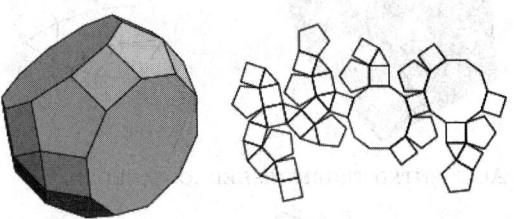

83. TRIDIMINISHED RHOMBICOSIDODECAHEDRON

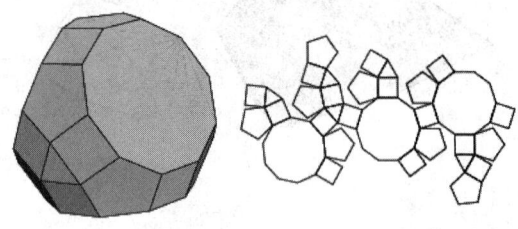

84. SNUB DISPHENOID

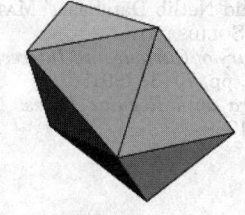 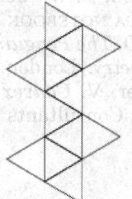

85. SNUB SQUARE ANTIPRISM

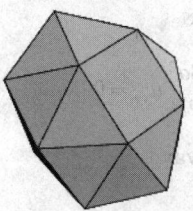 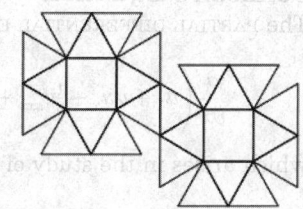

86. SPHENOCORONA

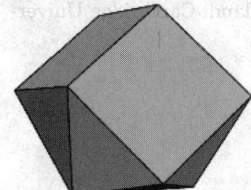 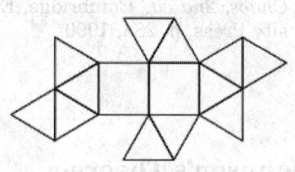

87. AUGMENTED SPHENOCORONA

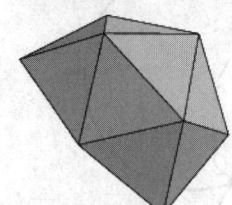 

88. SPHENOMEGACORONA

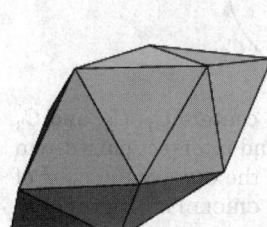 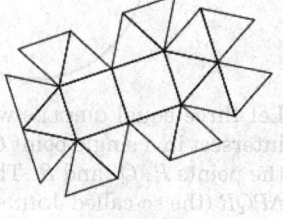

89. HEBESPHENOMEGACORONA

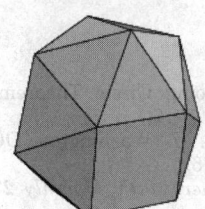 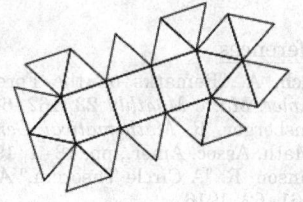

90. DISPHENOCINGULUM

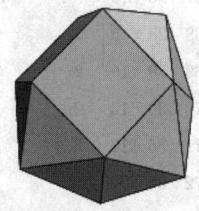 

91. BILUNABIROTUNDA

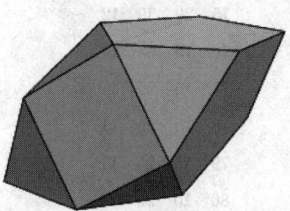 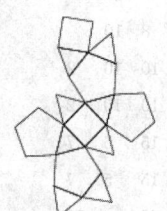

92. TRIANGULAR HEBESPHENOROTUNDA

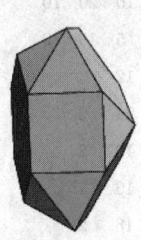 

The number of constituent n-gons ($\{n\}$) for each Johnson solid are given in the following table.

J_n	$\{3\}$	$\{4\}$	$\{5\}$	$\{6\}$	$\{8\}$	$\{10\}$	J_n	$\{3\}$	$\{4\}$	$\{5\}$	$\{6\}$	$\{8\}$	$\{10\}$
1	4		1				47	35		5	7		
2	5		1				48	40		12			
3	4	3		1			49	6	2				
4	4	5			1		50	10	1				
5	5	5	1			1	51	14					
6	10		6			1	52	4	4		2		
7	4	3					53	8	3		2		
8	4	5					54	4	5			2	
9	5	5	1				55	4	5			2	
10	12	1					56	8	4			2	
11	15		1				57	12	3			2	
12	6						58	5		11			
13	10						59	10		10			
14	6	3					60	10		10			
15	8	4					61	15		9			
16	10	5					62	10		2			
17	16						63	5		3	3		
18	4	9		1			64	4		7	3		
19	4	13			1		65	8	3		3		

20	5	15	1		1
21	10	10	6		1
22	16	3		1	
23	20	5			1
24	25	5	1		1
25	30		6		1
26	4	4			
27	8	6			
28	8	10			
29	8	10			
30	10	10	2		
31	10	10	2		
32	15	5	7		
33	15	5	7		
34	20		12		
35	8	12			
36	8	12			
37	8	18			
38	10	20	2		
39	10	20	2		
40	15	15	7		
41	15	15	7		
42	20	10	12		
43	20	10	12		
44	20	6			
45	24	10			
46	30	10	2		

66	12	5		5
67	16	10		4
68	25	5	1	11
69	30	10	2	10
70	30	10	2	10
71	35	15	3	9
72	20	30	12	
73	20	30	12	
74	20	30	12	
75	20	30	12	
76	15	25	11	1
77	15	25	11	1
78	15	25	11	1
79	15	25	11	1
80	10	20	10	2
81	10	20	10	2
82	10	20	10	2
83	5	15	9	3
84	12			
85	24	2		
86	12	2		
87	16	1		
88	16	2		
89	18	3		
90	20	4		
91	8	2	4	
92	13	3	3	1

See also ANTIPRISM, ARCHIMEDEAN SOLID, CONVEX POLYHEDRON, KEPLER-POINSOT SOLID, POLYHEDRON, PLATONIC SOLID, PRISM, UNIFORM POLYHEDRON

References

Bell Laboratories. http://netlib.bell-labs.com/netlib/polyhedra/.
Bulatov, V. "Johnson Solids." http://www.physics.orst.edu/~bulatov/polyhedra/johnson/.
Cromwell, P. R. *Polyhedra.* New York: Cambridge University Press, pp. 86-2, 1997.
Hart, G. "NetLib Polyhedra DataBase." http://www.george-hart.com/virtual-polyhedra/netlib-info.html.
Holden, A. *Shapes, Space, and Symmetry.* New York: Dover, 1991.
Hume, A. *Exact Descriptions of Regular and Semi-Regular Polyhedra and Their Duals.* Computer Science Technical Report #130. Murray Hill, NJ: AT&T Bell Laboratories, 1986.
Johnson, N. W. "Convex Polyhedra with Regular Faces." *Canad. J. Math.* **18**, 169-00, 1966.
Pedagoguery Software. `Poly.` http://www.peda.com/poly/.
Pugh, A. "Further Convex Polyhedra with Regular Faces." Ch. 3 in *Polyhedra: A Visual Approach.* Berkeley, CA: University of California Press, pp. 28-5, 1976.
Weisstein, E. W. "Johnson Solids." MATHEMATICA NOTEBOOK JOHNSONSOLIDS.M.
Weisstein, E. W. "Johnson Solid Netlib Database." MATHEMATICA NOTEBOOK JOHNSONSOLIDS.DAT.
Wells, D. *The Penguin Dictionary of Curious and Interesting Geometry.* London: Penguin, pp. 70-1, 1991.
Zalgaller, V. *Convex Polyhedra with Regular Faces.* New York: Consultants Bureau, 1969.

Johnson's Equation

The PARTIAL DIFFERENTIAL EQUATION

$$\frac{\partial}{\partial x}\left(u_1 + uu_x + \tfrac{1}{2}u_{xxx} + \frac{u}{2t}\right) + \frac{3a^2}{2t^2}u_{yy} = 0$$

which arises in the study of water waves.

References

Infeld, E. and Rowlands, G. *Nonlinear Waves, Solitons, and Chaos, 2nd ed.* Cambridge, England: Cambridge University Press, p. 284, 1990.

Johnson's Theorem

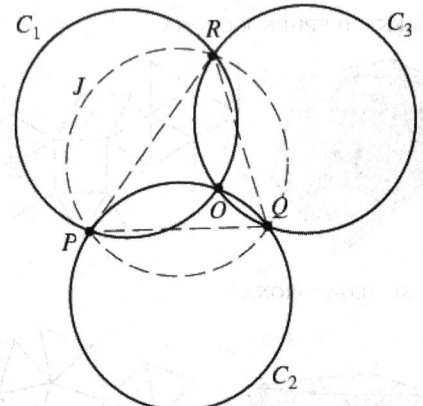

Let three equal CIRCLES with centers C_1, C_2, and C_3 intersect in a single point O and intersect pairwise in the points P, Q, and R. Then the CIRCUMCIRCLE J of ΔPQR (the so-called JOHNSON CIRCLE) is congruent to the original three.

See also CIRCUMCIRCLE, JOHNSON CIRCLE

References

Emch, A. "Remarks on the Foregoing Circle Theorem." *Amer. Math. Monthly* **23**, 162-64, 1916.
Honsberger, R. *Mathematical Gems II.* Washington, DC: Math. Assoc. Amer., pp. 18-1, 1976.
Johnson, R. "A Circle Theorem." *Amer. Math. Monthly* **23**, 161-62, 1916.
Wells, D. *The Penguin Dictionary of Curious and Interesting Geometry.* London: Penguin, pp. 125-26, 1991.

Join (Graph)

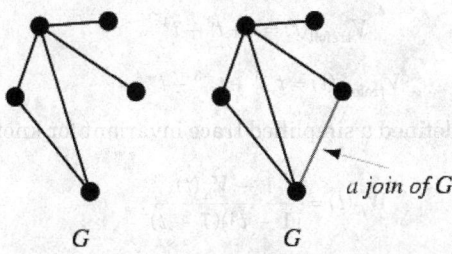

G G

Let x and y be distinct nodes of G which are not joined by an EDGE. Then the graph G/xy which is formed by adding the EDGE (x, y) to G is called a join of G.

Join (Spaces)

Let X and Y be TOPOLOGICAL SPACES. Then their join is the factor space

$$X * Y = (X \times Y \times I)/\sim,$$

where $\sim$ is the EQUIVALENCE RELATION

$$(x, y, t) \sim (x', y', t') \Leftrightarrow \begin{cases} t = t' = 0 \text{ and } x = x' \\ \qquad\qquad\text{or} \\ t = t' = 1 \text{ and } y = y. \end{cases}$$

See also CONE (SPACE), SUSPENSION

References

Rolfsen, D. *Knots and Links*. Wilmington, DE: Publish or Perish Press, p. 6, 1976.

Joint Denial

The term used in PROPOSITIONAL CALCULUS for the NOR CONNECTIVE. The notation $A \downarrow B$ is used for this connective.

See also ALTERNATIVE DENIAL, NAND

References

Mendelson, E. *Introduction to Mathematical Logic, 4th ed.* London: Chapman & Hall, p. 26, 1997.

Joint Distribution Function

A joint distribution function is a DISTRIBUTION FUNCTION $D(x, y)$ in two variables defined by

$$D(x, y) \equiv P(X \leq x, Y \leq y) \tag{1}$$

$$D_x(x) \equiv \lim_{y \to \infty} D(x, y) \tag{2}$$

$$D_y(y) \equiv \lim_{x \to \infty} D(x, y) \tag{3}$$

so that the joint probability function satisfies

$$D[(x, y) \in C] = \iint_{(X,Y) \in C} P(X, Y) dX dY \tag{4}$$

$$D(x \in A, y \in B) = \int_{Y \in B} \int_{X \in A} P(X, Y) dX dY \tag{5}$$

$$D(x, y) = P\{X \in (-\infty, x], Y \in (-\infty, y]\}$$

$$= \int_{-\infty}^{x} \int_{-\infty}^{y} P(X, Y) dX dY \tag{6}$$

$$D(a \leq x \leq a + da, b \leq y \leq b + db)$$

$$= \int_{b}^{b+db} \int_{a}^{a+da} P(X, Y) dX dY \approx P(a, b) da\, db. \tag{7}$$

Two random variables X and Y are independent IFF

$$D(x, y) = D_x(x) D_y(y) \tag{8}$$

for all x and y and

$$P(x, y) = \frac{\partial^2 D(x, y)}{\partial x \partial y}. \tag{9}$$

A multiple distribution function is OF THE FORM

$$D(x_1, \ldots, x_n) \equiv P(X_1 \leq x_1, \ldots, X_n \leq x_n). \tag{10}$$

See also DISTRIBUTION FUNCTION

References

Grimmett, G. and Stirzaker, D. *Probability and Random Processes, 2nd ed.* New York: Oxford University Press, 1992.

Joint Probability Density Function

JOINT DISTRIBUTION FUNCTION

Joint Theorem

GAUSSIAN JOINT VARIABLE THEOREM

Joke Number

HOAX NUMBER, SMITH NUMBER

Jonah Formula

A formula for the generalized CATALAN NUMBER $_p d_{qi}$. The general formula is

$$\binom{n - q}{k - 1} = \sum_{i=1}^{k} {}_p d_{qi} \binom{n - pi}{k - i},$$

where $\binom{n}{k}$ is a BINOMIAL COEFFICIENT, although Jonah's original formula corresponded to $p = 2$, $q = 0$ (Hilton and Pederson 1991).

See also BINOMIAL COEFFICIENT, CATALAN NUMBER

References

Hilton, P. and Pederson, J. "Catalan Numbers, Their Generalization, and Their Uses." *Math. Intel.* **13**, 64–5, 1991.

Jones Polynomial

The second KNOT POLYNOMIAL discovered. Unlike the first-discovered ALEXANDER POLYNOMIAL, the Jones polynomial can sometimes distinguish handedness (as can its more powerful generalization, the HOMFLY POLYNOMIAL). Jones polynomials are LAURENT POLYNOMIALS in t assigned to an $\mathbb{R}^3$ KNOT. The Jones polynomials are denoted $V_L(t)$ for LINKS, $V_K(t)$ for KNOTS, and normalized so that

$$V_{\text{unknot}}(t) = 1. \tag{1}$$

For example, the Jones polynomial of the TREFOIL KNOT is given by

$$V_{\text{trefoil}}(t) = t + t^3 - t^4. \tag{2}$$

If a LINK has an ODD number of components, then V_L is a LAURENT POLYNOMIAL over the INTEGERS; if the number of components is EVEN, $V_L(t)$ is $t^{1/2}$ times a LAURENT POLYNOMIAL. The Jones polynomial of a KNOT SUM $L_1 \# L_2$ satisfies

$$V_{L_1 \# L_2} = \left(V_{L_1}\right)\left(V_{L_2}\right). \tag{3}$$

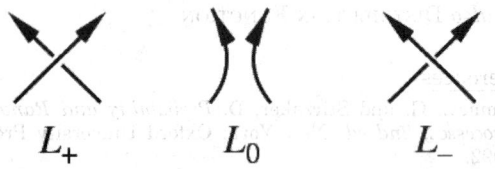

$$L_+ \qquad L_0 \qquad L_-$$

The SKEIN RELATIONSHIP for under- and overcrossings is

$$t^{-1}V_{L_+} - tV_{L_-} = \left(t^{1/2} - t^{-1/2}\right)V_{L_0}. \tag{4}$$

Combined with the link sum relationship, this allows Jones polynomials to be built up from simple knots and links to more complicated ones.

Some interesting identities from Jones (1985) follow. For any LINK L,

$$V_L(-1) = \Delta_L(-1), \tag{5}$$

where Δ_L is the ALEXANDER POLYNOMIAL, and

$$V_L(1) = (-2)^{p-1}, \tag{6}$$

where p is the number of components of L. For any KNOT K,

$$V_K\left(e^{2\pi i/3}\right) = 1 \tag{7}$$

and

$$\frac{d}{dt}V_K(1) = 0 \tag{8}$$

Let K^* denote the MIRROR IMAGE of a KNOT K. Then

$$V_{K^*}(t) = V_K\left(t^{-1}\right). \tag{9}$$

For example, the right-hand and left-hand TREFOIL

KNOTS have polynomials

$$V_{\text{trefoil}}(t) = t + t^3 - t^4 \tag{10}$$

$$V_{\text{trefoil}^*}(t) = t^{-1} + t^{-3} - t^{-4}. \tag{11}$$

Jones defined a simplified trace invariant for knots by

$$W_K(t) = \frac{1 - V_K(t)}{(1 - t^3)(1 - t)}. \tag{12}$$

The ARF INVARIANT of W_K is given by

$$\text{Arf}(K) = W_K(i) \tag{13}$$

(Jones 1985), where i is $\sqrt{-1}$. A table of the W polynomials is given by Jones (1985) for knots of up to eight crossings, and by Jones (1987) for knots of up to 10 crossings. (Note that in these papers, an additional polynomial which Jones calls V is also tabulated, but it is not the conventionally defined Jones polynomial.)

Jones polynomials were subsequently generalized to the two-variable HOMFLY POLYNOMIALS, the relationship being

$$V(t) = P\left(a = t, x = t^{1/2} - t^{-1/2}\right) \tag{14}$$

$$V(t) = P\left(\ell = it, m = i\left(t^{-1/2} - t^{1/2}\right)\right). \tag{15}$$

They are related to the KAUFFMAN POLYNOMIAL F by

$$V(t) = F\left(-t^{-3/4}, t^{-1/4} + t^{1/4}\right). \tag{16}$$

Jones (1987) gives a table of BRAID WORDS and W polynomials for knots up to 10 crossings. Jones polynomials for KNOTS up to nine crossings are given in Adams (1994) and for oriented links up to nine crossings by Doll and Hoste (1991). All PRIME KNOTS with 10 or fewer crossings have distinct Jones polynomials. It is not known if there is a nontrivial knot with Jones polynomial 1. The Jones polynomial of an (m, n)-TORUS KNOT is

$$\frac{t(m-1)(n-1)/2(1 - t^{m+1} - t^{n+1} + t^{m+n})}{1 - t^2} \tag{17}$$

Let k be one component of an oriented LINK L. Now form a new oriented LINK L^* by reversing the orientation of k. Then

$$V_{L^*} = t^{-3\lambda}V(L), \tag{18}$$

where V is the Jones polynomial and λ is the LINKING NUMBER of k and $L - k$. No such result is known for HOMFLY POLYNOMIALS (Lickorish and Millett 1988).

Birman and Lin (1993) showed that substituting the POWER SERIES for e^x as the variable in the Jones polynomial yields a POWER SERIES whose COEFFICIENTS are VASSILIEV INVARIANTS.

Let L be an oriented connected LINK projection of n crossings, then

$$n \geq \text{span } V(L), \qquad (19)$$

with equality if L is ALTERNATING and has no REMOVABLE CROSSING (Lickorish and Millett 1988).

There exist distinct KNOTS with the same Jones polynomial. Examples include (05–01, 10–32), (08–08, 10–29), (08–16, 10–56), (10–25, 10–56), (10–22, 10–35), (10–41, 10–94), (10–43, 10–91), (10–59, 10–06), (10–60, 10–83), (10–71, 10–04), (10–73, 10–86), (10–81, 10–09), and (10–37, 10–55) (Jones 1987). Incidentally, the first four of these also have the same HOMFLY POLYNOMIAL.

Witten (1989) gave a heuristic definition in terms of a topological quantum field theory, and Sawin (1996) showed that the "quantum group" $U_q(sl_2)$ gives rise to the Jones polynomial.

See also ALEXANDER POLYNOMIAL, HOMFLY POLYNOMIAL, KAUFFMAN POLYNOMIAL F, KNOT, LINK, VASSILIEV INVARIANT

References

Adams, C. C. *The Knot Book: An Elementary Introduction to the Mathematical Theory of Knots.* New York: W. H. Freeman, 1994.

Birman, J. S. and Lin, X.-S. "Knot Polynomials and Vassiliev's Invariants." *Invent. Math.* **111**, 225–70, 1993.

Doll, H. and Hoste, J. "A Tabulation of Oriented Links." *Math. Comput.* **57**, 747–61, 1991.

El-Misiery, A. "An Algorithm for Calculating Jones Polynomials." *Appl. Math. Comput.* **74**, 249–59, 1996.

Jones, V. "A Polynomial Invariant for Knots via von Neumann Algebras." *Bull. Am. Math. Soc.* **12**, 103–11, 1985.

Jones, V. "Hecke Algebra Representations of Braid Groups and Link Polynomials." *Ann. Math.* **126**, 335–88, 1987.

Khovanov, M. A Categorification of the Jones Polynomial. 30 Aug 1999. http://xxx.lanl.gov/abs/math.QA/9908171/.

Khovanov, M. "A Categorification of the Jones Polynomial." *Duke Math. J.* **101**, 359–26, 2000.

Lickorish, W. B. R. and Millett, B. R. "The New Polynomial Invariants of Knots and Links." *Math. Mag.* **61**, 1–3, 1988.

Murasugi, K. "Jones Polynomials and Classical Conjectures in Knot Theory." *Topology* **26**, 297–07, 1987.

Murasugi, K. and Kurpita, B. I. *A Study of Braids.* Dordrecht, Netherlands: Kluwer, 1999.

Praslov, V. V. and Sossinsky, A. B. *Knots, Links, Braids and 3-Manifolds: An Introduction to the New Invariants in Low-Dimensional Topology.* Providence, RI: Amer. Math. Soc., 1996.

Sawin, S. "Links, Quantum Groups, and TQFTS." *Bull. Amer. Math. Soc.* **33**, 413–45, 1996.

Stoimenow, A. "Jones Polynomials." http://guests.mpim-bonn.mpg.de/alex/ptab/j10.html.

Thistlethwaite, M. "A Spanning Tree Expansion for the Jones Polynomial." *Topology* **26**, 297–09, 1987.

Weisstein, E. W. "Knots and Links." MATHEMATICA NOTEBOOK KNOTS.M.

Witten, E. "Quantum Field Theory and the Jones Polynomial." *Comm. Math. Phys.* **121**, 351–99, 1989.

Jonquière's Function

POLYGAMMA FUNCTION

Jordan Algebra

A NONASSOCIATIVE ALGEBRA named after physicist Pascual Jordan which satisfies

$$xy = yx \qquad (1)$$

and

$$(xx)(xy) = x((xx)y)). \qquad (2)$$

The latter is equivalent to the so-called JORDAN IDENTITY

$$(xy)x^2 = x(yx^2) \qquad (3)$$

(Schafer 1996, p. 4). An ASSOCIATIVE ALGEBRA A with associative product xy can be made into a Jordan algebra A^+ by the JORDAN PRODUCT

$$x \cdot y = \tfrac{1}{2}(xy + yx). \qquad (4)$$

Division by 2 gives the nice identity $x \cdot x = xx$, but it must be omitted in characteristic $p = 2$.

Unlike the case of a LIE ALGEBRA, not every Jordan algebra is isomorphic to a SUBALGEBRA of some A^+. Jordan algebras which are isomorphic to a subalgebra are called SPECIAL JORDAN ALGEBRAS, while those that are not are called EXCEPTIONAL JORDAN ALGEBRAS.

See also ANTICOMMUTATOR, NONASSOCIATIVE ALGEBRA

References

Jacobson, N. *Structure and Representations of Jordan Algebras.* Providence, RI: Amer. Math. Soc., 1968.

Jordan, P. "Über eine Klasse nichtassoziativer hyperkomplexer Algebren." *Nachr. Ges. Wiss. Göttingen*, 569–75, 1932.

Schafer, R. D. *An Introduction to Nonassociative Algebras.* New York: Dover, pp. 4–, 1996.

Jordan Basis

Given a matrix A, a Jordan basis satisfies

$$Ab_{i,1} = \lambda_i b_{i,1}$$

and

$$Ab_{i,j} = \lambda_i b_{i,j} + b_{i,j-1},$$

and provides the means by which any COMPLEX MATRIX A can be written in JORDAN CANONICAL FORM.

See also JORDAN BLOCK, JORDAN CANONICAL FORM

Jordan Block

A matrix, also called a canonical box matrix, having zeros everywhere except along the DIAGONAL and SUPERDIAGONAL, with each element of the DIAGONAL consisting of a single number λ, and each element of the SUPERDIAGONAL consisting of a 1. For example,

$$\begin{bmatrix} \lambda & 1 & 0 & \cdots & 0 & 0 \\ 0 & \lambda & 1 & \ddots & 0 & 0 \\ 0 & 0 & \lambda & \ddots & 0 & 0 \\ 0 & 0 & 0 & \ddots & 0 & 0 \\ \vdots & \ddots & \ddots & \ddots & \ddots & 1 \\ 0 & 0 & 0 & \cdots & 0 & \lambda \end{bmatrix}$$

(Ayres 1962, p. 206). A JORDAN CANONICAL FORM consists of one or more Jordan blocks.

The convention that 1s be along the SUBDIAGONAL instead of the SUPERDIAGONAL is sometimes adopted instead (Faddeeva 1958, p. 50).

See also DIAGONAL MATRIX, JORDAN CANONICAL FORM, SUBDIAGONAL

References

Ayres, F. Jr. *Theory and Problems of Matrices.* New York: Schaum, p. 206, 1962.
Faddeeva, V. N. *Computational Methods of Linear Algebra.* New York: Dover, p. 50, 1958.
Golub, G. H. and van Loan, C. F. *Matrix Computations, 3rd ed.* Baltimore, MD: Johns Hopkins University Press, p. 317, 1996.

Jordan Canonical Form

A BLOCK MATRIX in which the blocks consist of CANONICAL BOX MATRICES with possibly differing constants λ_i, also called classical canonical form. For example,

$$\begin{bmatrix} \lambda_1 & 1 & 0 & \cdots & 0 \\ 0 & \lambda_1 & 1 & \ddots & 0 \\ 0 & 0 & \lambda_1 & \ddots & 0 \\ \vdots & & \ddots & \ddots & 1 \\ 0 & 0 & 0 & \cdots & \lambda_1 \\ & & & & & \ddots \\ & & & & & & \lambda_k & 1 & 0 & \cdots & 0 \\ & & & & & & 0 & \lambda_k & 1 & \ddots & 0 \\ & & & & & & 0 & 0 & \lambda_k & \ddots & 0 \\ & & & & & & \vdots & \ddots & \ddots & \ddots & 1 \\ & & & & & & 0 & 0 & 0 & \cdots & \lambda_k \end{bmatrix}$$

(Ayres 1962, p. 206). A specific example is given by

$$\begin{bmatrix} 5 & 1 & 0 & 0 & 0 & 0 \\ 0 & 5 & 0 & 0 & 0 & 0 \\ 0 & 0 & 5 & 0 & 0 & 0 \\ 0 & 0 & 0 & 1-2i & 1 & 0 \\ 0 & 0 & 0 & 0 & 1-2i & 1 \\ 0 & 0 & 0 & 0 & 0 & 1-2i \end{bmatrix},$$

which has three JORDAN BLOCKS.

Any COMPLEX MATRIX A can be written in Jordan canonical form by finding a JORDAN BASIS b_{ij} for each JORDAN BLOCK. In fact, any matrix with coefficients in an algebraically closed FIELD can be put into Jordan canonical form. The dimensions of the blocks corresponding to the EIGENVALUE λ can be recovered by the sequence

$$a_i = \dim \mathrm{Null}(\mathsf{A} - \lambda)^i.$$

The convention that the submatrices have 1s on the SUBDIAGONAL instead of the SUPERDIAGONAL is also used sometimes (Faddeeva 1958, p. 50).

See also JORDAN BASIS, JORDAN BLOCK, JORDAN MATRIX DECOMPOSITION

References

Ayres, F. Jr. *Theory and Problems of Matrices.* New York: Schaum, p. 206, 1962.
Faddeeva, V. N. *Computational Methods of Linear Algebra.* New York: Dover, p. 50, 1958.

Jordan Curve

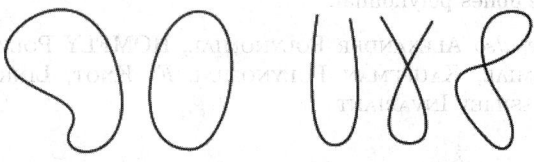

Jordan curves   *non-Jordan curves*

A Jordan curve is a plane curve which is topologically equivalent to (a HOMEOMORPHIC image of) the UNIT CIRCLE, i.e., it is SIMPLE and CLOSED.

It is not known if every Jordan curve contains all four VERTICES of some SQUARE, but it has been proven true for "sufficiently smooth" curves and closed convex curves (Schnirelman 1944; Steinhaus 1990, p. 104). For every TRIANGLE T and Jordan curve J, J has an INSCRIBED TRIANGLE similar to T.

See also CARATHÉODORY'S THEOREM, CLOSED CURVE, JORDAN CURVE THEOREM, SQUARE INSCRIBING, SIMPLE CURVE, UNIT CIRCLE

References

Krantz, S. G. "Closed Curves." §2.1.2 in *Handbook of Complex Analysis.* Boston, MA: Birkhäuser, pp. 19-0, 1999.
Schnirelman, L. G. "On Certain Geometrical Properties of Closed Curves." *Uspehi Matem. Nauk* **10**, 34-4, 1944.
Steinhaus, H. *Mathematical Snapshots, 3rd ed.* New York: Dover, 1999.

Jordan Curve Theorem

If J is a simple closed curve in $\mathbb{R}^2$, then $\mathbb{R}^2 - J$ has two components (an "inside" and "outside"), with J the BOUNDARY of each.

See also JORDAN CURVE, SCHÖNFLIES THEOREM

References

Knopp, K. *Theory of Functions Parts I and II, Two Volumes Bound as One, Part I.* New York: Dover, p. 14, 1996.
Rolfsen, D. *Knots and Links.* Wilmington, DE: Publish or Perish Press, p. 9, 1976.

Jordan Decomposition Theorem

Let $V \neq (0)$ be a finite dimensional VECTOR SPACE over the COMPLEX NUMBERS, and let A be a linear operator on V. Then V can be expressed as a DIRECT SUM of cyclic subspaces.

References

Gohberg, I. and Goldberg, S. "A Simple Proof of the Jordan Decomposition Theorem for Matrices." *Amer. Math. Monthly* **103**, 157–59, 1996.

Jordan Identity

The identity

$$(xy)x^2 = x(yx^2)$$

satisfied by elements x and y in a JORDAN ALGEBRA.

See also JORDAN ALGEBRA

References

Schafer, R. D. *An Introduction to Nonassociative Algebras.* New York: Dover, p. 4, 1996.

Jordan Matrix Decomposition

The Jordan matrix decomposition is the decomposition of a square matrix M into the form

$$\mathsf{M} = \mathsf{S}\mathsf{J}\mathsf{S}^{-1}, \tag{1}$$

where M and J are SIMILAR MATRICES, J is a matrix of JORDAN CANONICAL FORM, and S^{-1} is the MATRIX INVERSE of S. In other words, M is a SIMILARITY TRANSFORMATION of a matrix J in JORDAN CANONICAL FORM. The proof that any square matrix can be brought into JORDAN CANONICAL FORM is rather complicated (Turnbull and Aitken 1932; Faddeeva 1958, p. 49; Halmos 1958, p. 112).

Jordan decomposition is also associated with the MATRIX EQUATION $\mathsf{AX} = \mathsf{XB}$ and the special case $\mathsf{A} = \mathsf{B}$.

The Jordan matrix decomposition is implemented in *Mathematica* as `JordanDecomposition[m]`, and returns a list $\{s, j\}$. Note that *Mathematica* takes the CANONICAL BOX MATRICES in the JORDAN CANONICAL FORM to have 1s along the SUPERDIAGONAL instead of the SUBDIAGONAL. For example, a Jordan decomposition of

$$\mathsf{M} = \begin{bmatrix} 2 & 4 & -6 & 0 \\ 4 & 6 & -3 & -4 \\ 0 & 0 & 4 & 0 \\ 0 & 4 & -6 & 2 \end{bmatrix} \tag{2}$$

is given by

$$\mathsf{S} = \begin{bmatrix} 1 & -\frac{1}{4} & 0 & 1 \\ 0 & \frac{1}{4} & 3 & 1 \\ 0 & 0 & 2 & 0 \\ 1 & 0 & 0 & 1 \end{bmatrix} \tag{3}$$

$$\mathsf{J} = \begin{bmatrix} 2 & 1 & 0 & 0 \\ 0 & 2 & 0 & 0 \\ 0 & 0 & 4 & 0 \\ 0 & 0 & 0 & 6 \end{bmatrix}, \tag{4}$$

See also JORDAN CANONICAL FORM, MATRIX DECOMPOSITION, SIMILAR MATRICES

References

Faddeeva, V. N. "The Jordan Canonical Form." §4 in *Computational Methods of Linear Algebra.* New York: Dover, pp. 49–4 and 235, 1958.
Frazer, R. A.; Duncan, W. J.; and Collar, A. R. "Collinearity Transformation of a Numerical Matrix to a Canonical Form." §3.16 in *Elementary Matrices and Some Applications to Dynamics and Differential Equations.* Cambridge, England: Cambridge University Press, pp. 93–5, 1955.
Golub, G. H. and van Loan, C. F. *Matrix Computations, 3rd ed.* Baltimore, MD: Johns Hopkins University Press, p. 317, 1996.
Halmos, P. R. *Finite-Dimensional Vector Spaces, 2nd ed.* Princeton, NJ: Van Nostrand, p. 112, 1958.
Turnbull, H. W. and Aitken, A. C. Chs. 5– in *An Introduction to the Theory of Canonical Matrices.* London: Blackie and Sons, 1932.

Jordan Measure

Let the set M correspond to a bounded, NONNEGATIVE function f on an interval $0 \le f(x) \le c$ for $x \in [a, b]$. The Jordan measure, when it exists, is the common value of the outer and inner Jordan measures of M.

The outer Jordan measure is the greatest lower bound of the areas of the covering of M, consisting of finite unions of RECTANGLES. The inner Jordan measure of M is the difference between the AREA $c(a - b)$ of the RECTANGLE S with base $[a, b]$ and height c, and the outer measure of the complement of M in S.

References

Shenitzer, A. and Steprans, J. "The Evolution of Integration." *Amer. Math. Monthly* **101**, 66–2, 1994.

Jordan Measure Decomposition

If μ is a REAL MEASURE (i.e., a MEASURE that takes on real values), then one can decompose it according to where it is positive and negative. The positive variation is defined by

$$\mu^+ = \tfrac{1}{2}(|\mu| + \mu), \tag{1}$$

where $|\mu|$ is the TOTAL VARIATION MEASURE. Similarly, the negative variation is

$$\mu^- = \tfrac{1}{2}(|\mu| - \mu). \tag{2}$$

Then the Jordan decomposition of μ is defined as

$$\mu = \mu^+ - \mu^-. \tag{3}$$

When μ already is a positive measure then $\mu = \mu^+$.

More generally, if μ is ABSOLUTELY CONTINUOUS, i.e.,

$$\mu(E) = \int_E f \, dx, \tag{4}$$

then so are μ^+ and μ^-. The positive and negative variations can also be written as

$$\mu^+(E) = \int_E f^+ \, dx \tag{5}$$

and

$$\mu^-(E) = \int_E f^- \, dx, \tag{6}$$

where $f = f^+ - f^-$ is the decomposition of f into its positive and negative parts.

The Jordan decomposition has a so-called minimum property. In particular, given any positive measure λ, the measure μ has another decomposition

$$\mu = (\mu^+ + \lambda) - (\mu^- + \lambda). \tag{7}$$

The Jordan decomposition is minimal with respect to these changes. One way to say this is that any decomposition $\mu = \lambda_1 - \lambda_2$ must have $\lambda_1 \geq \mu^+$ and $\lambda_2 \geq \mu^-$.

See also MEASURE, POLAR REPRESENTATION (MEASURE), TOTAL VARIATION MEASURE

References
Rudin, W. *Real and Complex Analysis.* New York: McGraw-Hill, p. 119, 1987.

Jordan Polygon
SIMPLE POLYGON

Jordan Product
The Jordan product of quantities x and y is defined by

$$x \cdot y = \tfrac{1}{2}(xy + yx).$$

See also ANTICOMMUTATOR, JORDAN ALGEBRA

Jordan's Inequality

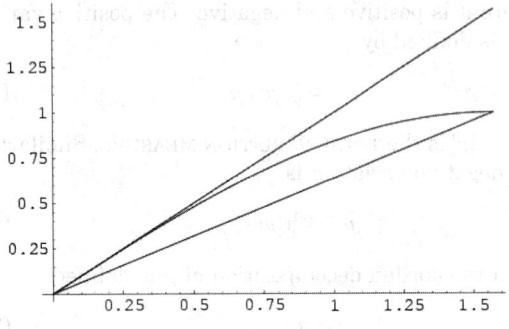

For $0 \leq x \leq \pi/2$

$$\frac{2}{\pi}x \leq \sin x \leq x.$$

References
Yuefeng, F. "Jordan's Inequality." *Math. Mag.* **69**, 126, 1996.

Jordan's Lemma
Jordan's lemma shows the value of the INTEGRAL

$$I \equiv \int_{-\infty}^{\infty} f(x)e^{iax} \, dx \tag{1}$$

along the REAL AXIS is 0 for "nice" functions which satisfy $\lim_{R \to \infty} |f(Re^{i\theta})| = 0$. This is established using a CONTOUR INTEGRAL I_R which satisfies

$$\lim_{R \to \infty} |I_R| \leq \frac{\pi}{a} \lim_{R \to \infty} \epsilon = 0. \tag{2}$$

To derive the lemma, write

$$x \equiv Re^{i\theta} = R(\cos \theta + i \sin \theta) \tag{3}$$

$$dx = iRe^{i\theta} d\theta \tag{4}$$

and define the CONTOUR INTEGRAL

$$I_R = \int_0^{\pi} f(Re^{i\theta})e^{iaR \cos \theta - aR \sin \theta} iRe^{i\theta} d\theta \tag{5}$$

Then

$$|I_R| = R \int_0^{\pi} |f(Re^{i\theta})| |e^{iaR \cos \theta}| |e^{-aR \sin \theta}| |i| |e^{i\theta}| d\theta$$

$$= R \int_0^{\pi} |f(Re^{i\theta})| e^{-aR \sin \theta} d\theta.$$

$$= 2R \int_0^{\pi/2} |f(Re^{i\theta})| e^{-aR \sin \theta} d\theta. \tag{6}$$

Now, if $\lim_{R \to \infty} |f(Re^{i\theta})| = 0$, choose an ϵ such that $|f(Re^{i\theta})| \leq \epsilon$, so

$$|I_R| \leq 2R\epsilon \int_0^{\pi/2} e^{-aR \sin \theta} d\theta. \tag{7}$$

But, for $\theta \in [0, \pi/2]$,

$$\frac{2}{\pi}\theta \leq \sin \theta, \tag{8}$$

so

$$|I_R| \leq 2R\epsilon \int_0^{\pi/2} e^{-2aR\theta/\pi} d\theta$$

$$=2\epsilon R\,\frac{1-e^{-aR}}{\dfrac{2aR}{\pi}}=\frac{\pi\epsilon}{a}\left(1-e^{-aR}\right). \tag{9}$$

As long as $\lim_{R\to\infty}|f(z)|=0$, Jordan's lemma

$$\lim_{R\to\infty}|I_R|\le\frac{\pi}{a}\lim_{R\to\infty}\epsilon=0 \tag{10}$$

then follows.

See also CONTOUR INTEGRATION

References

Arfken, G. *Mathematical Methods for Physicists, 3rd ed.* Orlando, FL: Academic Press, pp. 406–08, 1985.

Jordan's Symmetric Group Theorem

A primitive subgroup of the SYMMETRIC GROUP S_n is equal to either the ALTERNATING GROUP A_n or S_n whenever it contains at least one PERMUTATION which is a q-cycle for some prime $q\le n-3$.

References

Dixon, J. D. "The Probability of Generating the Symmetric Group." *Math. Z.* **110**, 199–05, 1969.
Wielandt, H. *Finite Permutation Groups.* New York: Academic Press, 1964.

Jordan-Hölder Theorem

The composition QUOTIENT GROUPS belonging to two COMPOSITION SERIES of a FINITE GROUP G are, apart from their sequence, ISOMORPHIC in pairs. In other words, if

$$I\subset H_s\subset\ldots\subset H_2\subset H_1\subset G$$

is one COMPOSITION SERIES and

$$I\subset K_t\subset\ldots\subset K_2\subset K_1\subset G$$

is another, then $t=s$, and corresponding to any composition quotient group K_j/K_{j+1}, there is a composition QUOTIENT GROUP H_i/H_{i+1} such that

$$\frac{K_j}{K_{j+1}}\cong\frac{H_i}{H_{i+1}}.$$

This theorem was proven in 1869–889.

See also COMPOSITION SERIES, FINITE GROUP, ISOMORPHIC GROUPS

References

Lomont, J. S. *Applications of Finite Groups.* New York: Dover, p. 26, 1993.
Scott, W. R. §2.5.8 in *Group Theory.* New York: Dover, p. 37, 1987.

Joseph Ideal

See also IDEAL

References

Huang, J.-S. "Joseph Ideals and Minimal Representations." §12.3 in *Lectures on Representation Theory.* Singapore: World Scientific, pp. 169–71, 1999.

Josephus Problem

Given a group of n men arranged in a CIRCLE under the edict that every mth man will be executed going around the CIRCLE until only one remains, find the position $L(n,m)$ in which you should stand in order to be the last survivor (Ball and Coxeter 1987). The list giving the place in the execution sequence of the first, second, etc. man can be given by Josephus[m, n] in the *Mathematica* add-on package DiscreteMath`Combinatorica` (which can be loaded with the command < <DiscreteMath`). To obtain the ordered list of men who are consecutively slaughtered, InversePermutation in the *Mathematica* add-on package DiscreteMath`Combinatorica` (which can be loaded with the command < <DiscreteMath`) can be applied to the output of Josephus.

The following array gives the original position of the last survivor out of a group of $n=1,2,\ldots$, if every mth man is killed:

1									
2	1								
3	3	2							
4	1	1	2						
5	3	4	1	2					
6	5	1	5	1	4				
7	7	4	2	6	3	5			
8	1	7	6	3	1	4	4		
9	3	1	8	7	2	3	8		
10	5	4	5	3	3	9	1	7	8

(Sloane's A032434). The survivor for $m=2$ can be given analytically by

$$L(n,2)=1+2n-2^{1+\lfloor\lg n\rfloor},$$

where $\lfloor n\rfloor$ is the FLOOR FUNCTION and LG is the LOGARITHM to base 2. The first few solutions are therefore 1, 1, 3, 1, 3, 5, 7, 1, 3, 5, 7, 9, 11, 13, 15, 1, ... (Sloane's A006257).

The original position of the second-to-last survivor is given in the following table for $n=2,3,\ldots$:>

1	1							
2	1	1						
3	1	1	2					
4	3	2	1	2				
5	1	1	5	1	4			
6	3	1	2	1	3	4		
7	1	4	6	3	1	3	4	
8	3	1	1	2	7	1	3	7
9	5	4	5	3	3	8	1	6	4

(Sloane's A032435).

The original position of the second-to-last survivor is given in the following table for $n = 2, 3, ...:>$

1	1	1					
2	1	1	1				
3	1	2	1	2			
4	1	1	3	1	2		
5	3	1	2	1	1	2	
6	1	4	3	3	1	1	2
7	3	1	1	2	4	1	1	2
8	1	4	1	3	3	5	1	1	4

(Sloane's A032436).

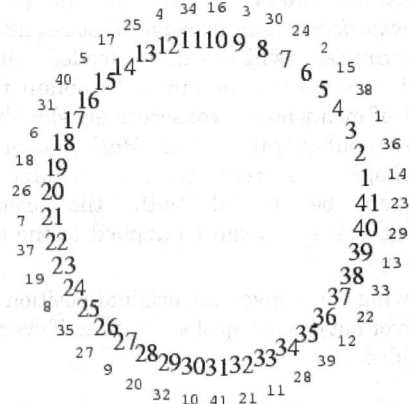

The original Josephus problem consisted of a CIRCLE of 41 men with every third man killed ($n = 41, m = 3$). In order for the lives of the last two men to be spared, they must be placed at positions 31 (last) and 16 (second-to-last). The complete list in order of execution is 3, 6, 9, 12, 15, 18, 21, 24, 27, 30, 33, 36, 39, 1, 5, 10, 14, 19, 23, 28, 32, 37, 41, 7, 13, 20, 26, 34, 40, 8, 17, 29, 38, 11, 25, 2, 22, 4, 35, 16, 31.

Another version of the problem considers a CIRCLE of two groups (say, "A" and "B") of 15 men each (giving a total of 30 men), with every ninth man cast overboard. To save all the members of the "A" group, the men must be placed at positions 1, 2, 3, 4, 10, 11, 13, 14, 15, 17, 20, 21, 25, 28, 29. Written out explicitly, the order is

AAAABBBBBBAABAAABABBAABBBABBBAAB.

This sequence of letters can be remembered with the aid of the MNEMONIC "From numbers' aid and art, never will fame depart." Consider the vowels only, assign $a = 1, e = 2, i = 3, o = 4, u = 5$, and alternately add a number of letters corresponding to a vowel value, so 4A (o), 5B (u), 2A (e), etc. (Mott-Smith 1954, §149, pp. 94 and 209–10; Ball and Coxeter 1987).

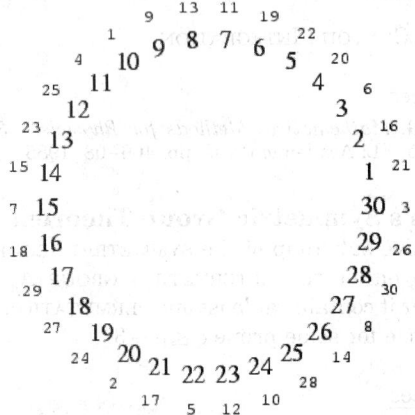

If instead every *tenth* man is thrown overboard, the men from the "A" group must be placed in positions 1, 2, 4, 5, 6, 12, 13, 16, 17, 18, 19, 21, 25, 28, 29. Written out explicitly,

AABAAABBBBBAABBAAAABABBBBABBBAAB

which can be constructed using the Latin MNEMONIC "Rex paphi cum gente bona dat signa serena" (Ball and Coxeter 1987).

Mott-Smith (1954, §153, pp. 96 and 212) discusses a card game called "Out and Under" in which cards at the top of a deck are alternately discarded and placed at the bottom. This is a Josephus problem with parameter $m = 2$, and Mott-Smith hints at the above closed-form solution.

See also KIRKMAN'S SCHOOLGIRL PROBLEM, NECKLACE

References

Bachet, C. G. Problem 23 in *Problèmes plaisans et délectables, 2nd ed.* p. 174, 1624.

Ball, W. W. R. and Coxeter, H. S. M. *Mathematical Recreations and Essays, 13th ed.* New York: Dover, pp. 32–6, 1987.

Graham, R. L.; Knuth, D. E.; and Patashnik, O. *Concrete Mathematics: A Foundation for Computer Science, 2nd ed.* Reading, MA: Addison-Wesley, 1994.

Knuth, D. E. *The Art of Computer Programming, Vol. 1: Fundamental Algorithms, 3rd ed.* Reading, MA: Addison-Wesley, 1997.

Knuth, D. E. *The Art of Computer Programming, Vol. 3: Sorting and Searching, 2nd ed.* Reading, MA: Addison-Wesley, 1998.

Kraitchik, M. "Josephus' Problem." §3.13 in *Mathematical Recreations.* New York: W. W. Norton, pp. 93–4, 1942.

Mott-Smith, G. "Decimation Puzzles." Ch. 9, §149–54 in *Mathematical Puzzles for Beginners and Enthusiasts,*

2nd rev. ed. New York: Dover, pp. 94–7 and 209–14, 1954.

Odlyzko, A. M. and Wilf, H. S. "Functional Iteration and the Josephus Problem." *Glasgow Math. J.* **33**, 235–40, 1991.

Skiena, S. "Josephus' Problem." §1.4.3 in *Implementing Discrete Mathematics: Combinatorics and Graph Theory with Mathematica.* Reading, MA: Addison-Wesley, pp. 34–5, 1990.

Sloane, N. J. A. Sequences A0062572216, A032434, A032435, and A032436 in "An On-Line Version of the Encyclopedia of Integer Sequences." http://www.research.-att.com/~njas/sequences/eisonline.html.

Smith, H. J. "Josephus Permutation Problems." http://pweb.netcom.com/~hjsmith/Josephus.html.

Joyce Sequence

The sequence of numbers giving the number of digits in n^{n^n}. The sequence n^{n^n} for $n = 1, 2, \ldots$ is 1, 16, 7625597484987, ... (Sloane's A002488; Rossier 1948), so the Joyce sequence is 1, 2, 13, 155, 2185, 36306, ... (Sloane's A054382). Laisant (1906) found the term $j(9)$, and Uhler (1947) published the logarithm of this number to 250 decimal places (Wells 1986, p. 208).

The sequence is named in honor of the following excerpt from the "Ithaca" chapter of James Joyce's *Ulysses*: "Because some years previously in 1886 when occupied with the problem of the quadrature of the circle he had learned of the existence of a number computed to a relative degree of accuracy to be of such magnitude and of so many places, e.g., the 9th power of the 9th power of 9, that, the result having been obtained, 33 closely printed volumes of 1000 pages each of innumerable quires and reams of India paper would have to be requisitioned in order to contain the complete tale of its printed integers of units, tens, hundreds, thousands, tens of thousands, hundreds of thousands, millions, tens of millions, hundreds of millions, billions, the nucleus of the nebula of every digit of every series containing succinctly the potentiality of being raised to the utmost kinetic elaboration of any power of any of its powers."

References

Joyce, J. "Ithaca" Chapter in *Ulysses*. New York: Random House, 1986.

Rossier, P. "Grands nombres." *Elemente der Math.* **3**, 20, 1948.

Sloane, N. J. A. Sequences A002488/M5031 and A054382 in "An On-Line Version of the Encyclopedia of Integer Sequences." http://www.research.att.com/~njas/sequences/eisonline.html.

Wells, D. *The Penguin Dictionary of Curious and Interesting Numbers.* Middlesex, England: Penguin Books, p. 208, 1986.

Jug

THREE JUG PROBLEM

Jugendtraum

The German mathematician Kronecker proved that all the Galois extensions of the RATIONALS Q with ABELIAN Galois groups are SUBFIELDS of cyclotomic fields $Q(\mu_n)$, where μ_n is the group of nth ROOTS OF UNITY. He then sought to find a similar function whose division values would generate the Abelian extensions of an arbitrary NUMBER FIELD. He discovered that the J-FUNCTION works for IMAGINARY QUADRATIC FIELDS K, but the completion of this problem, known as Kronecker's Jugendtraum ("dream of youth"), for more general FIELDS remains one of the great unsolved problems in NUMBER THEORY.

See also IMAGINARY QUADRATIC FIELD, J-FUNCTION

References

Shimura, G. *Introduction to the Arithmetic Theory of Automorphic Functions.* Princeton, NJ: Princeton University Press, 1981.

Juggling

The throwing and catching of multiple objects such that at least one is always in the air. Some aspects of juggling turn out to be quite mathematical. The best examples are the two-handed asynchronous juggling sequences known as "SITESWAPS."

See also SITESWAP

References

Buhler, J.; Eisenbud, D.; Graham, R.; and Wright, C. "Juggling Drops and Descents." *Amer. Math. Monthly* **101**, 507–19, 1994.

Donahue, B. "Jugglers Now Juggle Numbers to Compute New Tricks for Ancient Art." *New York Times,* pp. B5 and B10, Apr. 16, 1996.

Juggling Information Service. "Siteswaps." http://www.juggling.org/help/siteswap/.

Julia Fractal

JULIA SET

Julia Set

Let $R(z)$ be a RATIONAL FUNCTION

$$R(z) \equiv \frac{P(z)}{Q(z)}, \tag{1}$$

where $z \in \mathbb{C}^*$, $z \in \mathbb{C}^*$ is the RIEMANN SPHERE $\mathbb{C} \cup \{\infty\}$, and P and Q are POLYNOMIALS without common divisors. The "filled-in" Julia set J_R is the set of points z which do not approach infinity after $R(z)$ is repeatedly applied (corresponding to a STRANGE ATTRACTOR). The true Julia set J is the boundary of the filled-in set (the set of "exceptional points"). There are two types of Julia sets: connected sets (FATOU SET) and Cantor sets (FATOU DUST).

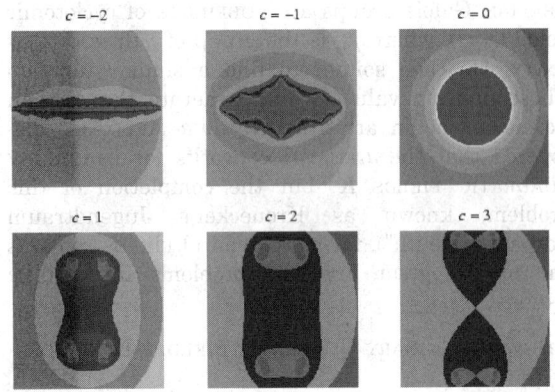

Quadratic Julia sets are generated by the quadratic mapping

$$z_{n+1} = z_n^2 + c \tag{2}$$

for fixed c. For almost every c, this transformation generates a FRACTAL. Examples are shown above for various values of c. The resulting object is *not* a fractal for $c = -2$ (Dufner *et al.* 1998, pp. 224–26) and $c = 0$ (Dufner *et al.* 1998, pp. 125–26), although it does not seem to be known if these two are the *only* such exceptional values.

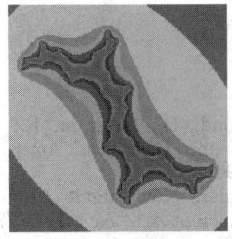

The special case of c on the boundary of the MANDELBROT SET is called a DENDRITE FRACTAL (top left figure, computed using $c = i$), $c = -0.123 + 0.745i$ is called DOUADY'S RABBIT FRACTAL (left figure), $c = -0.75$ is called the SAN MARCO FRACTAL (middle figure), and $c = -0.391 - 0.587i$ is the SIEGEL DISK FRACTAL (right figure). Julia sets can be rendered in *Mathematica* using the following code.

```
JuliaSet[n_:50,c_,rmax_:3.,{{x1_,x2_},{y1_,-
y2_}},opts___]:=
   DensityPlot[-Length[
     FixedPointList[#^2+c&,x+I y,n,SameTest-
>(Abs[#2]>rmax&)]],
```

```
         {x,x1,x2},{y,y1,y2},opts,PlotPoints-
>200,Mesh->False,
      Frame->False,AspectRatio->Automatic
   ]
```

The equation for the quadratic Julia set is a CONFORMAL MAPPING, so angles are preserved. Let J be the JULIA SET, then $x' \mapsto x$ leaves J invariant. If a point P is on J, then all its iterations are on J. The transformation has a two-valued inverse. If $b = 0$ and y is started at 0, then the map is equivalent to the LOGISTIC MAP. The set of all points for which J is connected is known as the MANDELBROT SET.

For a Julia set J_c with $c \ll 1$, the CAPACITY DIMENSION is

$$d_{\text{capacity}} = 1 + \frac{|c|^2}{4 \ln 2} + \mathcal{O}(|c|^3). \tag{3}$$

For small c, J_c is also a JORDAN CURVE, although its points are not COMPUTABLE.

See also DENDRITE FRACTAL, DOUADY'S RABBIT FRACTAL, FATOU DUST, FATOU SET, MANDELBROT SET, NEWTON'S METHOD, SAN MARCO FRACTAL, SIEGEL DISK FRACTAL, STRANGE ATTRACTOR

References

Dickau, R. M. "Julia Sets." http://forum.swarthmore.edu/advanced/robertd/julias.html.
Dickau, R. M. "Another Method for Calculating Julia Sets." http://forum.swarthmore.edu/advanced/robertd/inversejulia.html.
Douady, A. "Julia Sets and the Mandelbrot Set." In *The Beauty of Fractals: Images of Complex Dynamical Systems* (Ed. H.-O. Peitgen and D. H. Richter). Berlin: Springer-Verlag, p. 161, 1986.
Dufner, J.; Roser, A.; and Unseld, F. *Fraktale und Julia-Mengen.* Harri Deutsch, 1998.
Lauwerier, H. *Fractals: Endlessly Repeated Geometric Figures.* Princeton, NJ: Princeton University Press, pp. 124–26, 138–48, and 177–79, 1991.
Mendes-France, M. "Nevertheless." *Math. Intell.* **10**, 35, 1988.
Peitgen, H.-O. and Saupe, D. (Eds.). "The Julia Set," "Julia Sets as Basin Boundaries," "Other Julia Sets," and "Exploring Julia Sets." §3.3.2 to 3.3.5 in *The Science of Fractal Images.* New York: Springer-Verlag, pp. 152–63, 1988.
Schroeder, M. *Fractals, Chaos, Power Laws.* New York: W. H. Freeman, p. 39, 1991.
Wagon, S. "Julia Sets." §5.4 in *Mathematica in Action.* New York: W. H. Freeman, pp. 163–78, 1991.
Wells, D. *The Penguin Dictionary of Curious and Interesting Geometry.* London: Penguin, pp. 126–27, 1991.

Jump

A point of DISCONTINUITY, also called a LEAP.

See also DISCONTINUITY, JUMP ANGLE, JUMPING CHAMPION

References

Jeffreys, H. and Jeffreys, B. S. "Leap at a Discontinuity." §1.094 in *Methods of Mathematical Physics, 3rd ed.*

Cambridge, England: Cambridge University Press, p. 26, 1988.

Jump Angle

A GEODESIC TRIANGLE with oriented boundary yields a curve which is piecewise DIFFERENTIABLE. Furthermore, the TANGENT VECTOR varies continuously at all but the three corner points, where it changes suddenly. The angular difference of the tangent vectors at these corner points are called the jump angles.

See also ANGULAR DEFECT, GAUSS-BONNET FORMULA

Jumping Champion

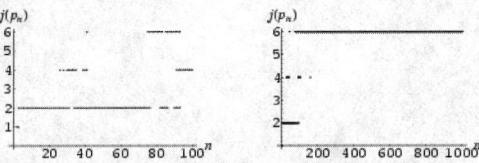

An integer $j(n)$ is called a JUMPING CHAMPION if $j(n)$ is the most frequently occurring difference between consecutive PRIMES $\leq n$ (Odlyzko *et al.*). This term was coined by J. H. Conway in 1993. There are occasionally several jumping champions in a range. The scatter plots above show the jumping champions for small n, and the ranges of number having given jumping champion sets are summarized in the following table.

$j(n)$		n
1		3
1, 2		5
2	7–00, 103–06, 109–12, ...	
2, 4	101–02, 107–08, 113–30, ...	
4		131–38, ...
2, 4, 6	179–80, 467–90, ...	
2, 6	379–88, 421–32, ...	
6		389–20, ...

Odlyzko *et al.* give a table of jumping champions for $n \leq 1000$, consisting mainly of 2, 4, and 6. 6 is the jumping champion up to about $n \approx 1.74 \times 10^{35}$, at which point 30 dominates. At $n \approx 10^{425}$, 210 becomes champion, and subsequent PRIMORIALS are conjectured to take over at larger and larger n. Erdos and Straus (1980) proved that the jumping champions tend to infinity under the assumption of a quantitative form of the k-tuples conjecture.

Wolf gives a table of approximate values $\tilde{n}$ at which the PRIMORIAL (p_n) will become a champion. An estimate for $\tilde{n}$ is given by

$$\tilde{n} = n^{n + o(n)}.$$

See also PRIME DIFFERENCE FUNCTION, PRIME GAPS, PRIME NUMBER, PRIMORIAL

References

Erdos, P.; and Straus, E. G. "Remarks on the Differences Between Consecutive Primes." *Elem. Math.* **35**, 115–18, 1980.
Guy, R. K. *Unsolved Problems in Number Theory, 2nd ed.* New York: Springer-Verlag, 1994.
Nelson, H. "Problem 654." *J. Recr. Math.* **11**, 231, 1978–979.
Odlyzko, A.; Rubinstein, M.; and Wolf, M. "Jumping Champions." http://www.research.att.com/~amo/doc/recent.html.

Jumping Octahedron

A bistable eight-sided polyhedron discovered by Wunderlich and Schwabe (1986).

See also FLEXIBLE POLYHEDRON, MULTISTABLE, RIGID POLYHEDRON

References

Cromwell, P. R. *Polyhedra.* New York: Cambridge University Press, pp. 222–23, 1997.
Wunderlich, W. and Schwabe, C. "Eine Familie von geschlossen gleichflachigen Polyedern, die fast beweglich sind." *Elem. Math.* **41**, 88–8, 1986.

Jung's Theorem

Every finite set of points with SPAN d has an enclosing CIRCLE with RADIUS no greater than $\sqrt{3}d/3$.

In 3-D, a generalization of the theorem states that every set of points with SPAN d has an enclosing SPHERE with RADIUS no greater than $\sqrt{6}d/4$ (Smarandache 1992, 1996).

See also SPAN (GEOMETRY)

References

Le Lionnais, F. *Les nombres remarquables.* Paris: Hermann, p. 28, 1983.
Rademacher, H. and Toeplitz, O. *The Enjoyment of Mathematics: Selections from Mathematics for the Amateur.* Princeton, NJ: Princeton University Press, pp. 103–10, 1957.
Smarandache, F. "A Generalization in Space of Jung's Theorem." *Gazeta Matematica (Bucharest)*, No. 9--12, 352, 1992.
Smarandache, F. "A Generalization in Space of Jung's Theorem." In *Collected Papers, Vol. 1.* Bucharest, Romania: Tempus, pp. 223–24, 1996.
Wells, D. *The Penguin Dictionary of Curious and Interesting Geometry.* London: Penguin, p. 128, 1991.

Just If

IFF

Just One

EXACTLY ONE

Just Rigid

A FRAMEWORK is called "just rigid" if it is RIGID, but ceases to be so when any single bar is removed. Lamb (1928, pp. 93–4) proved that a NECESSARY (but not SUFFICIENT) condition that a graph be just rigid is that

$$E = 2V - 3,$$

where E is the number of edges (bars) and V is the node of vertices (i.e., pivots; Coxeter and Greitzer 1967, p. 56).

See also RIGID GRAPH

References

Coxeter, H. S. M. and Greitzer, S. L. *Geometry Revisited.* Washington, DC: Math. Assoc. Amer., p. 56, 1967.
Lamb, H. *Statics, Including Hydrostatics and the Elements of the Theory of Elasticity,* 3rd ed. London: Cambridge University Press, 1928.

K

Kabon Triangles

The largest number $N(n)$ of nonoverlapping TRIANGLES which can be produced by n straight LINE SEGMENTS. The first few terms are 1, 2, 5, 7, 11, 15, 21, ... (Sloane's A006066).

References

Sloane, N. J. A. Sequences A006066/M1334 in "An On-Line Version of the Encyclopedia of Integer Sequences." http://www.research.att.com/~njas/sequences/eisonline.html.

Kac Formula

The expected number of REAL zeros E_n of a RANDOM POLYNOMIAL of degree n if the coefficients are independent and distributed normally is given by

$$E_n = \frac{1}{\pi} \int_{-\infty}^{\infty} \sqrt{\frac{1}{(t^2-1)^2} - \frac{(n+1)^2 t^{2n}}{(t^{2n+2}-1)^2}}\, dt \qquad (1)$$

$$= \frac{4}{\pi} \int_0^1 \sqrt{\frac{1}{(1-t^2)^2} - \frac{(n+1)^2 t^{2n}}{(1-t^{2n+2})^2}}\, dt. \qquad (2)$$

(Kac 1943, Edelman and Kostlan 1995). Another form of the equation is given by

$$E_n = \frac{1}{\pi} \int_{-\infty}^{\infty} \sqrt{\left[\frac{\partial^2}{\partial x\, \partial y} \ln \frac{1-(xy)^{n+1}}{1-xy}\right]_{x=y=t}}\, dt \qquad (3)$$

(Kostlan 1993, Edelman and Kostlan 1995). As $n \to \infty$,

$$E_n = \frac{2}{\pi} \ln n + C_1 + \frac{2}{\pi n} + \mathcal{O}(n^{-2}), \qquad (4)$$

where

$$C_1 = \frac{2}{\pi}\left\{\ln 2 + \int_0^\infty \left[\sqrt{\frac{1}{x^2} - \frac{4e^{-2x}}{(1-e^{-2x})^2}} - \frac{1}{x+1}\right] dx\right\}$$

$$= 0.6257358072\ldots. \qquad (5)$$

The initial term was derived by Kac (1943).

See also RANDOM POLYNOMIAL

References

Edelman, A. and Kostlan, E. "How Many Zeros of a Random Polynomial are Real?" *Bull. Amer. Math. Soc.* **32**, 1–37, 1995.
Kac, M. "On the Average Number of Real Roots of a Random Algebraic Equation." *Bull. Amer. Math. Soc.* **49**, 314–320, 1943.
Kac, M. "A Correction to 'On the Average Number of Real Roots of a Random Algebraic Equation'." *Bull. Amer. Math. Soc.* **49**, 938, 1943.
Kostan, E. "On the Distribution of Roots in a Random Polynomial." Ch. 38 in *From Topology to Computation: Proceedings of the Smalefest* (Ed. M. W. Hirsch,

J. E. Marsden, and M. Shub). New York: Springer-Verlag, pp. 419–431, 1993.

Kac Matrix

The $(n+1) \times (n+1)$ TRIDIAGONAL MATRIX (also called the CLEMENT MATRIX) defined by

$$S_n = \begin{bmatrix} 0 & n & 0 & 0 & \cdots & 0 \\ 1 & 0 & n-1 & 0 & \cdots & 0 \\ 0 & 2 & 0 & n-2 & \cdots & 0 \\ \vdots & \vdots & & \ddots & \ddots & \vdots \\ 0 & 0 & 0 & n-1 & 0 & 1 \\ 0 & 0 & 0 & 0 & n & 0 \end{bmatrix}$$

The EIGENVALUES are $2k - n$ for $k = 0, 1, \ldots, n$.

Kadomtsev-Petviashvili Equation

The PARTIAL DIFFERENTIAL EQUATION

$$\tfrac{3}{4}\mathcal{U}_y + \mathcal{W}_x = 0, \qquad (1)$$

where

$$\mathcal{W}_y + \mathcal{U}_t - \tfrac{1}{4}\mathcal{U}_{xxx} + \tfrac{3}{2}\mathcal{U}\mathcal{U}_x = 0 \qquad (2)$$

(Krichever and Novikov 1980; Novikov 1999). Zwillinger (1997, p. 131) and Calogero and Degasperis (1982, p. 54) give the equation as

$$\frac{\partial}{\partial x}(u_t + u_{xxx} - 6uu_x) \pm u_{yy} = 0. \qquad (3)$$

The modified Kadomtsev-Petviashvili equation is given by

$$u_{xt} = u_{xxx} + 3u_{yy} - 6u_x^2 u_{xx} - 6u_y u_{xx} \qquad (4)$$

(Clarkson 1986; Zwillinger 1997, p. 133).

See also KADOMTSEV-PETVIASHVILI-BURGERS EQUATION, KORTEWEG-DE VRIES EQUATION, KRICHEVER-NOVIKOV EQUATION

References

Baker, H. F. *Abelian Functions: Abel's Theorem and the Allied Theory, Including the Theory of the Theta Functions.* New York: Cambridge University Press, p. xix, 1995.
Calogero, F. and Degasperis, A. *Spectral Transform and Solitons: Tools to Solve and Investigate Nonlinear Evolution Equations.* New York: North-Holland, 1982.
Clarkson, P. A. "The Painlevé Property, a Modified Boussinesq Equation and a Modified Kadomtsev-Petviashvili Equation." *Physica D* **19**, 447–450, 1986.
Krichever, I. M. and Novikov, S. P. "Holomorphic Bundles over Algebraic Curves, and Nonlinear Equations." *Russ. Math. Surv.* **35**, 53–80, 1980. English translation of *Uspekhi Mat. Nauk* **35**, 47–68, 1980.
Novikov, D. P. "Algebraic-Geometric Solutions of the Krichever-Novikov Equation." *Theoret. Math. Phys.* **121**, 1567–15773, 1999.
Zwillinger, D. *Handbook of Differential Equations, 3rd ed.* Boston, MA: Academic Press, p. 131, 1997.

Kadomtsev-Petviashvili-Burgers Equation

The so-called generalized Kadomtsev-Petviashvili-Burgers equation is the PARTIAL DIFFERENTIAL EQUATION

$$\frac{\partial}{\partial x}\left(u_t + \frac{Ju}{2t} + J_1 u u_x + J_2 u_{xx} + J_3 u_{xxx}\right) + J_4(t)u_{yy} = 0$$

(Brugarino 1986; Zwillinger 1997, p. 131).

See also KADOMTSEV-PETVIASHVILI EQUATION

References

Brugarino, T. "Similarity Solutions of the Generalized Kadomtsev-Petviashvili-Burgers Equation." *Nuovo Cimento B* **92**, 142–156, 1986.
Infeld, E. and Rowlands, G. "An Example: The Kadomtsev-Petviashvili Equation." §7.10.4 in *Nonlinear Waves, Solitons, and Chaos, 2nd ed.* Cambridge, England: Cambridge University Press, pp. 196–199, 2000.
Zwillinger, D. *Handbook of Differential Equations, 3rd ed.* Boston, MA: Academic Press, p. 131, 1997.

Kähler Form

A CLOSED TWO-FORM ω on a COMPLEX MANIFOLD M which is also the negative IMAGINARY PART of a HERMITIAN METRIC $h = g - i\omega$ is called a Kähler form. In this case, M is called a KÄHLER MANIFOLD and g, the REAL PART of the HERMITIAN METRIC, is called a KÄHLER METRIC. The Kähler form combines the metric and the COMPLEX STRUCTURE, indeed

$$g(X, Y) = \omega(X, JY), \tag{1}$$

where J is the ALMOST COMPLEX STRUCTURE induced by multiplication by i. Since the Kähler form comes from a HERMITIAN METRIC, it is preserved by J, i.e., since $h(X, Y) = h(JX, JY)$. The equation $d\omega = 0$ implies that the metric and the complex structure are related. It gives M a KÄHLER STRUCTURE, and has many implications.

On $\mathbb{C}^2$, the Kähler form can be written as

$$\omega = -\tfrac{1}{2}i(dz_1 \wedge \overline{dz_1} + dz_2 \wedge \overline{dz_2}) = dx_1 \wedge dy_1$$
$$+ dx_2 \wedge dy_2, \tag{2}$$

where $z_n = x_n + iy_n$. In general, the Kähler form can be written in coordinates

$$\omega = \sum g_{i\bar{k}}\, dz_i \wedge d\bar{z}_k, \tag{3}$$

where $g_{i\bar{k}}$ is a HERMITIAN METRIC, the REAL PART of which is the KÄHLER METRIC. Locally, a Kähler form can be written as $\partial\bar{\partial}f$, where f is a function called a KÄHLER POTENTIAL. The Kähler form is a real (1, 1)-COMPLEX FORM.

Since the Kähler form ω is closed, it represents a COHOMOLOGY CLASS in DE RHAM COHOMOLOGY. On a COMPACT MANIFOLD, it cannot be EXACT because $\omega^n/n! \neq 0$ is the volume form determined by the metric. In the special case of a PROJECTIVE VARIETY, the Kähler form represents an INTEGRAL COHOMOLOGY CLASS. That is, it integrates to an integer on any one-dimensional submanifold, i.e., an ALGEBRAIC CURVE. The KODAIRA EMBEDDING THEOREM says that if the Kähler form represents an INTEGRAL COHOMOLOGY CLASS on a compact manifold, then it must be a PROJECTIVE VARIETY. There exist Kähler forms which are not projective algebraic, but it is an open question whether or not any KÄHLER MANIFOLD can be deformed to a PROJECTIVE VARIETY (in the compact case).

A Kähler form satisfies WIRTINGER'S INEQUALITY,

$$|\omega(X, Y)| \leq |X \wedge Y|, \tag{4}$$

where the right-hand side is the volume of the parallelogram formed by the tangent vectors X and Y. Corresponding inequalities hold for the EXTERIOR POWERS of ω. Equality holds IFF X and Y form a complex subspace. Therefore, ω is a CALIBRATION FORM, and the complex submanifolds of a Kähler manifold are CALIBRATED SUBMANIFOLDS. In particular, the complex submanifolds are locally volume minimizing in a Kähler manifold. For example, the graph of a holomorphic function is a locally area-minimizing surface in $\mathbb{C}^2 \simeq \mathbb{R}^4$.

See also CALABI-YAU SPACE, CALIBRATION FORM, COMPLEX MANIFOLD, COMPLEX PROJECTIVE SPACE, DOLBEAULT COHOMOLOGY, KÄHLER IDENTITIES, KÄHLER MANIFOLD, KÄHLER METRIC, KÄHLER POTENTIAL, KÄHLER STRUCTURE, KODAIRA EMBEDDING THEOREM, PROJECTIVE VARIETY, SYMPLECTIC FORM, WIRTINGER'S INEQUALITY

References

Griffiths, P. and Harris, J. *Principles of Algebraic Geometry.* New York: Wiley, pp. 106–126, 1994.
Weil, A. *Introduction à l'étude des variétès Kähleriennes.* Publications de l'Institut de Mathématiques de l'Université de Nancago, VI, Actualites Scientifiques et Industrielles, no. 1267. Paris: Hermann, 1958.
Wells, R. O. *Differential Analysis on Complex Manifolds.* New York: Springer-Verlag, 1980.

Kähler Identities

A collection of identities which hold on a KÄHLER MANIFOLD, also called the Hodge identities. Let ω be a KÄHLER FORM, $d = \partial + \bar{\partial}$ be the EXTERIOR DERIVATIVE, where $\bar{\partial}$ is the DEL BAR OPERATOR, $[A, B] = AB - BA$ be the COMMUTATOR of two differential operators, and A^* denote the FORMAL ADJOINT of A. The following operators also act on DIFFERENTIAL FORMS on a KÄHLER MANIFOLD:

$$L(\alpha) = \alpha \wedge \omega \tag{1}$$

$$\Lambda(\alpha) = L^*(\alpha) = \alpha \lrcorner \omega \tag{2}$$

$$d_c = -JdJ, \tag{3}$$

where J is the ALMOST COMPLEX STRUCTURE, $J^2 = -I$,

and $\lrcorner$ denotes the INTERIOR PRODUCT. Then

$$[L, \bar{\partial}] = [L, \partial] = 0 \qquad (4)$$

$$[\Lambda, \bar{\partial}^*] = [\Lambda, \partial^*] = 0 \qquad (5)$$

$$[L, \bar{\partial}^*] = -i\partial \qquad (6)$$

$$[L, \partial^*] = i\bar{\partial} \qquad (7)$$

$$[\Lambda, \bar{\partial}] = -i\partial^* \qquad (8)$$

$$[\Lambda, \partial] = i\bar{\partial}. \qquad (9)$$

In addition,

$$d^*d_c = -d_c d^* = d^*Ld^* = -d_c \Lambda d_c \qquad (10)$$

$$dd_c{}^* = -d_c{}^*d = d_c{}^*Ld_c{}^* = -d\Lambda d \qquad (11)$$

$$\partial\bar{\partial}^* = -\bar{\partial}^*\partial = -i\bar{\partial}^*L\bar{\partial}^* = -i\partial\Lambda\partial \qquad (12)$$

$$\bar{\partial}\partial^* = -\partial^*\bar{\partial} = i\partial^*L\partial^* = i\bar{\partial}\Lambda\bar{\partial}. \qquad (13)$$

These identities have many implications. For instance, the two operators

$$\Delta_d = dd^* + d^*d \qquad (14)$$

and

$$\Delta_{\bar{\partial}} = \bar{\partial}\bar{\partial}^* + \bar{\partial}^*\bar{\partial} \qquad (15)$$

(called Laplacians because they are elliptic operators) satisfy $\Delta_d = 2\Delta_{\bar{\partial}}$. At this point, assume that M is also a COMPACT MANIFOLD. Along with HODGE'S THEOREM, this equality of Laplacians proves the HODGE DECOMPOSITION. The operators L and Λ commute with these Laplacians. By HODGE'S THEOREM, they act on cohomology, which is represented by HARMONIC FORMS. Moreover, defining

$$H = [L, \Lambda] = \sum(p + q - n)\Pi^{p, q}, \qquad (16)$$

where $\Pi^{p, q}$ is projection onto the (p, q)-DOLBEAULT COHOMOLOGY, they satisfy

$$[L, \Lambda] = H \qquad (17)$$

$$[H, L] = -2L \qquad (18)$$

$$[H, \Lambda] = 2\Lambda. \qquad (19)$$

In other words, these operators provide a REPRESENTATION of the SPECIAL LINEAR LIE ALGEBRA $\mathfrak{sl}_2(\mathbb{C})$ on the complex cohomology of a compact Kähler manifold. In effect, this is the content of the HARD LEFSCHETZ THEOREM.

See also CALIBRATED MANIFOLD, COMPLEX MANIFOLD, COMPLEX PROJECTIVE SPACE, HARD LEFSCHETZ THEOREM, HODGE'S THEOREM, KÄHLER FORM, KÄHLER MANIFOLD, KÄHLER POTENTIAL, KÄHLER STRUCTURE, PROJECTIVE VARIETY, RIEMANNIAN METRIC, SYMPLECTIC MANIFOLD

References

Griffiths, P. and Harris, J. *Principles of Algebraic Geometry.* New York: Wiley, p. 111–122, 1994.

Weil, A. *Introduction à l'étude des variétès Kähleriennes.* Publications de l'Institut de Mathématiques de l'Université de Nancago, VI, Actualites Scientifiques et Industrielles, no. 1267. Paris: Hermann, p. 44, 1958.

Wells, R. O. *Differential Analysis on Complex Manifolds.* New York: Springer-Verlag, pp. 191–195, 1980.

Kähler Manifold

A COMPLEX MANIFOLD for which the EXTERIOR DERIVATIVE of the fundamental form Ω associated with the given HERMITIAN METRIC vanishes, so $d\Omega = 0$. In other words, it is a complex manifold with a KÄHLER STRUCTURE. It has a KÄHLER FORM, so it is also a SYMPLECTIC MANIFOLD. It has a KÄHLER METRIC, so it is also a RIEMANNIAN MANIFOLD.

The simplest example of a Kähler manifold is a RIEMANN SURFACE, which is a COMPLEX MANIFOLD of dimension 1. In this case, the IMAGINARY PART of any HERMITIAN METRIC must be a CLOSED FORM since all 2-forms are CLOSED on a two real dimensional MANIFOLD.

See also CALIBRATED MANIFOLD, COMPLEX MANIFOLD, COMPLEX PROJECTIVE SPACE, HYPER-KÄHLER MANIFOLD, KÄHLER FORM, KÄHLER IDENTITIES, KÄHLER METRIC, KÄHLER POTENTIAL, KÄHLER STRUCTURE, PROJECTIVE VARIETY, QUATERNION KÄHLER MANIFOLD RIEMANNIAN METRIC, SYMPLECTIC MANIFOLD

References

Amorós, J. *Fundamental Groups of Compact Kähler Manifolds.* Providence, RI: Amer. Math. Soc., 1996.

Goldberg, S. I. *Curvature and Homology, enl. ed.* New York: Dover, 1998.

Griffiths, P. and Harris, J. *Principles of Algebraic Geometry.* New York: Wiley pp. 106–126, 1994.

Iyanaga, S. and Kawada, Y. (Eds.). "Kähler Manifolds." §232 in *Encyclopedic Dictionary of Mathematics.* Cambridge, MA: MIT Press, pp. 732–734, 1980.

Weil, A. *Introduction à l'étude des variétès Kähleriennes.* Publications de l'Institut de Mathématiques de l'Université de Nancago, VI, Actualites Scientifiques et Industrielles, no. 1267. Paris: Hermann, 1958.

Wells, R. O. *Differential Analysis on Complex Manifolds.* New York: Springer-Verlag, 1980.

Kähler Metric

A Kähler metric is a RIEMANNIAN METRIC g on a COMPLEX MANIFOLD which gives M a KÄHLER STRUCTURE, i.e., it is a KÄHLER MANIFOLD with a KÄHLER FORM. However, the term "Kähler metric" can also refer to the corresponding HERMITIAN METRIC $h = g - i\omega$, where ω is the KÄHLER FORM, defined by $\omega(X, Y) = g(JX, Y)$. Here, the operator J is the ALMOST COMPLEX STRUCTURE, a linear map on tangent vectors satisfying $J^2 = -I$, induced by multiplication by i. In coordinates $z_k = x_k + iy_k$, the operator J satisfies $J(\partial/\partial x_k) = \partial/\partial y_k$ and $J(\partial/\partial y_k) = -\partial/\partial x_k$.

The operator J depends on the COMPLEX STRUCTURE, and on a KÄHLER MANIFOLD, it must preserve the Kähler metric. For a metric to be Kähler, one additional condition must also be satisfied, namely that it can be expressed in terms of the metric and the complex structure. Near any point p, there exists holomorphic coordinates $z_k = x_k + iy_k$ such that the metric has the form

$$g = \sum dx_k \otimes dx_k + dy_k \otimes dy_k + \mathcal{O}(|z|^2),$$

where $\otimes$ denotes the TENSOR PRODUCT; that is, it vanishes up to order two at p. Hence, any geometric equation in $\mathbb{C}^n$ involving only the first derivatives can be defined on a Kähler manifold. Note that a generic metric can be written to vanish up to order two, but not necessarily in holomorphic coordinates, using a GAUSSIAN COORDINATE SYSTEM.

See also CALIBRATED MANIFOLD, COMPLEX MANIFOLD, COMPLEX PROJECTIVE SPACE, KÄHLER FORM, KÄHLER IDENTITIES, KÄHLER MANIFOLD, KÄHLER POTENTIAL, KÄHLER STRUCTURE, PROJECTIVE VARIETY, RIEMANNIAN METRIC, SYMPLECTIC MANIFOLD

References

Griffiths, P. and Harris, J. *Principles of Algebraic Geometry.* New York: Wiley, pp. 106–126, 1994.

Weil, A. *Introduction à l'étude des variétès Kähleriennes.* Publications de l'Institut de Mathématiques de l'Université de Nancago, VI, Actualites Scientifiques et Industrielles, no. 1267. Paris: Hermann, 1958.

Wells, R. O. *Differential Analysis on Complex Manifolds.* New York: Springer-Verlag, 1980.

Kähler Potential

The Kähler potential is a real-valued function f on a KÄHLER MANIFOLD for which the KÄHLER FORM ω can be written as $\omega = i\partial\bar{\partial}f$. Here, the operators

$$\partial = \sum \frac{\partial}{\partial z_k} dz_k \qquad (1)$$

and

$$\bar{\partial} = \sum \frac{\partial}{\partial \bar{z}_k} d\bar{z}_k \qquad (2)$$

are called the del and DEL BAR OPERATOR, respectively.

For example, in $\mathbb{C}^n$, the function $f = |z|^2/2$ is a Kähler potential for the standard Kähler form, because

$$i\partial\bar{\partial}(\tfrac{1}{2}|z|^2) = \tfrac{1}{2}i\partial\bar{\partial}\sum z_k\bar{z}_k$$

$$= \tfrac{1}{2}i\partial\sum z_k\,d\bar{z}_k$$

$$= \tfrac{1}{2}i\sum dz_k \wedge d\bar{z}_k = \omega.$$

See also CALIBRATED MANIFOLD, COMPLEX MANIFOLD, COMPLEX PROJECTIVE SPACE, KÄHLER FORM, KÄHLER IDENTITIES, KÄHLER MANIFOLD, KÄHLER METRIC, KÄHLER STRUCTURE, PROJECTIVE VARIETY, RIEMANNIAN METRIC, SYMPLECTIC MANIFOLD

References

Griffiths, P. and Harris, J. *Principles of Algebraic Geometry.* New York: Wiley, pp. 106–126, 1994.

Weil, A. *Introduction à l'étude des variétès Kähleriennes.* Publications de l'Institut de Mathématiques de l'Université de Nancago, VI, Actualites Scientifiques et Industrielles, no. 1267. Paris: Hermann, 1958.

Wells, R. O. *Differential Analysis on Complex Manifolds.* New York: Springer-Verlag, 1980.

Kähler Structure

A Kähler structure on a COMPLEX MANIFOLD M combines a RIEMANNIAN METRIC on the underlying REAL MANIFOLD with the COMPLEX STRUCTURE. Such a structure brings together geometry and complex analysis, and the main examples come from ALGEBRAIC GEOMETRY. When M has n complex dimensions, then it has $2n$ real dimensions. A Kähler structure is related to the UNITARY GROUP $U(n)$, which embeds in $SO(2n)$ as the orthogonal matrices that preserve the ALMOST COMPLEX STRUCTURE (multiplication by 'i'). In a COORDINATE CHART, the COMPLEX STRUCTURE of M defines a multiplication by i and the metric defines orthogonality for tangent vectors. On a Kähler manifold, these two notions (and their derivatives) are related.

The following are elements of a Kähler structure, with each condition SUFFICIENT for a Kähler structure to exist.

1. A KÄHLER METRIC. Near any point p, there exists holomorphic coordinates $z_k = x_k + iy_k$ such that the metric has the form

$$g = \sum dx_k \otimes dx_k + dy_k \otimes dy_k + \mathcal{O}(|z|^2), \qquad (1)$$

where $\otimes$ denotes the TENSOR PRODUCT; that is, it vanishes up to order two at p. Hence any geometric equation in $\mathbb{C}^n$ involving only the first derivatives can be defined on a KÄHLER MANIFOLD. Note that a generic metric can be written to vanish up to order two, but not necessarily in holomorphic coordinates, using a GAUSSIAN COORDINATE SYSTEM.

2. A KÄHLER FORM ω is a real CLOSED nondegenerate TWO-FORM, i.e., a SYMPLECTIC FORM, for which $\omega(X, JX) > 0$ for nonzero tangent vectors X. Moreover, it must also satisfy $\omega(JX, JY) = \omega(X, Y)$, where J is the ALMOST COMPLEX STRUCTURE

induced by multiplication by i. That is,

$$J\left(\frac{\partial}{\partial x_k}\right) = \frac{\partial}{\partial y_k} \qquad (2)$$

and

$$J\left(\frac{\partial}{\partial y_k}\right) = -\frac{\partial}{\partial x_k}. \qquad (3)$$

Locally, a Kähler form can be written as $\partial \bar{\partial} f$, where f is a function called a KÄHLER POTENTIAL. The Kähler form is a real (1, 1)-FORM.

3. A HERMITIAN METRIC $h = g - i\omega$ where the REAL PART is a KÄHLER METRIC, as in item (1) above, and where the IMAGINARY PART is a KÄHLER FORM, as in item (2).

4. A metric for which the ALMOST COMPLEX STRUCTURE J is PARALLEL. Since PARALLEL TRANSPORT is always an isometry, a HERMITIAN METRIC is well-defined by PARALLEL TRANSPORT along paths from a base point. The HOLONOMY GROUP is contained in the UNITARY GROUP.

It is easy to see that a complex SUBMANIFOLD of a KÄHLER MANIFOLD inherits its Kähler structure, and so must also be Kähler. The main source of examples are PROJECTIVE VARIETIES, complex submanifolds of COMPLEX PROJECTIVE SPACE which are solutions to algebraic equations.

There are several deep consequences of the Kähler condition. For example, the KÄHLER IDENTITIES, the HODGE DECOMPOSITION of COHOMOLOGY, and the LEFSCHETZ THEOREMS depend on the Kähler condition for compact manifolds.

See also CALIBRATED MANIFOLD, COMPLEX MANIFOLD, COMPLEX PROJECTIVE SPACE, COMPLEX STRUCTURE, KÄHLER FORM, KÄHLER IDENTITIES, KÄHLER MANIFOLD, KÄHLER METRIC, KÄHLER POTENTIAL, PROJECTIVE VARIETY, RIEMANN SURFACE, SYMPLECTIC MANIFOLD

References

Griffiths, P. and Harris, J. *Principles of Algebraic Geometry.* New York: Wiley, pp. 106–126, 1994.
Weil, A. *Introduction à l'étude des variétès Kähleriennes.* Publications de l'Institut de Mathématiques de l'Université de Nancago, VI, Actualites Scientifiques et Industrielles, no. 1267. Paris: Hermann, 1958.
Wells, R. O. *Differential Analysis on Complex Manifolds.* New York: Springer-Verlag, 1980.

Kakeya Needle Problem

What is the plane figure of least AREA in which a line segment of width 1 can be freely rotated (where translation of the segment is also allowed)? When the figure is restricted to be convex, Cunningham and Schoenberg (1965) found there is still *no* minimum AREA, although Wells (1991) states that Kakeya

discovered that the smallest convex region is an EQUILATERAL TRIANGLE of unit height. The smallest *simple convex* domain in which one can put a segment of length 1 which will coincide with itself when rotated by 180° is

$$\frac{1}{24}(5 - 2\sqrt{2})\pi = 0.284258\ldots$$

(Le Lionnais 1983).

For a general convex shape, Besicovitch (1928) proved that there is *no* MINIMUM AREA. This can be seen by rotating a line segment inside a DELTOID, star-shaped 5-oid, star-shaped 7-oid, etc. Another iterative construction which tends to as small an area as desired is called a PERRON TREE (Falconer 1990, Wells 1991).

See also CURVE OF CONSTANT WIDTH, LEBESGUE MINIMAL PROBLEM, PERRON TREE, REULEAUX POLYGON, REULEAUX TRIANGLE

References

Ball, W. W. R. and Coxeter, H. S. M. *Mathematical Recreations and Essays, 13th ed.* New York: Dover, pp. 99–101, 1987.
Besicovitch, A. S. "On Kakeya's Problem and a Similar One." *Math. Z.* **27**, 312–320, 1928.
Besicovitch, A. S. "The Kakeya Problem." *Amer. Math. Monthly* **70**, 697–706, 1963.
Cunningham, F. Jr. and Schoenberg, I. J. "On the Kakeya Constant." *Canad. J. Math.* **17**, 946–956, 1965.
Falconer, K. J. *The Geometry of Fractal Sets, 1st pbk. ed., with corrections.* Cambridge, England: Cambridge University Press, 1990.
Le Lionnais, F. *Les nombres remarquables.* Paris: Hermann, p. 24, 1983.
Ogilvy, C. S. *A Calculus Notebook.* Boston, MA: Prindle, Weber, & Schmidt, 1968.
Ogilvy, C. S. *Excursions in Geometry.* New York: Dover, pp. 147–153, 1990.
Pál, J. "Ein Minimumproblem für Ovale." *Math. Ann.* **88**, 311–319, 1921.
Plouffe, S. "Kakeya Constant." http://www.lacim.uqam.ca/piDATA/kakeya.txt.
Steinhaus, H. *Mathematical Snapshots, 3rd ed.* New York: Dover, pp. 151–152, 1999.
Wagon, S. *Mathematica in Action.* New York: W. H. Freeman, pp. 50–52, 1991.
Wells, D. *The Penguin Dictionary of Curious and Interesting Geometry.* London: Penguin, pp. 128–129, 1991.

Kakeya Set

KAKEYA NEEDLE PROBLEM

Kakutani's Fixed Point Theorem

Every correspondence that maps a compact convex subset of a locally convex space into itself with a closed graph and convex nonempty images has a fixed point.

See also FIXED POINT THEOREM

Kakutani's Problem

COLLATZ PROBLEM

Kalman Filter

An ALGORITHM in CONTROL THEORY introduced by R. Kalman in 1960 and refined by Kalman and R. Bucy. It is an ALGORITHM which makes optimal use of imprecise data on a linear (or nearly linear) system with Gaussian errors to continuously update the best estimate of the system's current state.

See also WIENER FILTER

References

Casti, J. L. "The Kalman Filter." Ch. 1 in *Five More Golden Rules: Knots, Codes, Chaos, and Other Great Theories of 20th-Century Mathematics.* New York: Wiley, pp. 101–154, 2000.
Chui, C. K. and Chen, G. *Kalman Filtering: With Real-Time Applications, 2nd ed.* Berlin: Springer-Verlag, 1991.
Grewal, M. S. *Kalman Filtering: Theory & Practice.* Englewood Cliffs, NJ: Prentice-Hall, 1993.
Kalman, H. E. "Transversal Filters." *Proc. I.R.E.* **28**, 302–310, 1940.

KAM Theorem

KOLMOGOROV-ARNOLD-MOSER THEOREM

Kampé de Fériet Function

A SPECIAL FUNCTION generalizes the GENERALIZED HYPERGEOMETRIC FUNCTION to two variables and includes the APPELL HYPERGEOMETRIC FUNCTION $F_1(\alpha;\ \beta,\ \beta';\ \gamma;\ x,\ y)$ as a special case. The Kampe de Feriet function can represent derivatives of GENERALIZED HYPERGEOMETRIC FUNCTIONS with respect to their parameters, as well as indefinite integrals of two and three MEIJER'S G-FUNCTIONS. Exton and Krupnikov (1998) have derived a large collection of formulas involving this function.

Kampé de Fériet functions are written in the notation

$$F_{q,\ s,\ u}^{p,\ r,\ t}\left(\begin{array}{c|c|c}c_p & a_r & \alpha_t \\ d_q & b_s & \beta_u\end{array}\ x,\ y\right). \tag{1}$$

Special cases include

$$F_{1,\ 0,\ 0}^{1,\ 1,\ 1}\left(\begin{array}{c|c|c}1/2 & 1/2 & -1/2 \\ 3/2 & - & -\end{array}\ x,\ y\right)$$
$$=\frac{1}{\sqrt{x}}\ E\left(\sin^{-1}(\sqrt{x}),\ \sqrt{y/x}\right) \tag{2}$$

$$F_{1,\ 0,\ 0}^{1,\ 1,\ 1}\left(\begin{array}{c|c|c}1/2 & 1/2 & -1/2 \\ 3/2 & - & -\end{array}\ x,\ y\right)$$
$$=\frac{1}{\sqrt{x}}\ F\left(\sin^{-1}(\sqrt{x}),\ \sqrt{y/x}\right) \tag{3}$$

for $x \neq 0$ and $|x|,\ |y| \leq 1$, where $E(x,\ k)$ is the incomplete ELLIPTIC INTEGRAL OF THE SECOND KIND and $F(x,\ k)$ is the incomplete ELLIPTIC INTEGRAL OF THE FIRST KIND, as well as

$$F_{1,\ 0,\ 0}^{1,\ 1,\ 1}\left(\begin{array}{c|c|c}1/2 & 1 & 1/2 \\ 1 & - & -\end{array}\ x,\ y\right)=\frac{2}{\pi}\ \Pi(1;\ x,\ \sqrt{y}) \tag{4}$$

for $|x|,\ |y| < 1$, where $\Pi(n;\ x,\ k)$ is the incomplete ELLIPTIC INTEGRAL OF THE THIRD KIND (Exton and Krupnikov 1998, p. 1). Additional identities are given by

$$F_{q,\ s,\ u}^{1+p,\ r,\ t}\left(\begin{array}{c|c|c}0, c_p & a_r & \alpha_t \\ d_q & b_s & \beta_u\end{array}\ x,\ y\right)=1 \tag{5}$$

$$F_{q,\ s,\ u}^{p,\ r,\ t}\left(\begin{array}{c|c|c}c_p & a_r & \alpha_t \\ d_q & b_s & \beta_u\end{array}\ x,\ 0\right)=F_{q+s}^{p+r}\left(\begin{array}{c|c}c_p,\ a_r \\ d_q,\ d_s\end{array}\ x\right) \tag{6}$$

$$F_{q,\ s,\ u}^{p,\ r,1+\ t}\left(\begin{array}{c|c|c}c_p & a_r & 0,\ \alpha_t \\ d_q & b_s & \beta_u\end{array}\ x,\ y\right)=F_{q+s}^{p+r}\left(\begin{array}{c|c}c_p,\ a_r \\ d_q,\ d_s\end{array}\ x\right) \tag{7}$$

(Exton and Krupnikov 1998, p. 3).

See also APPELL HYPERGEOMETRIC FUNCTION, FOX'S H-FUNCTION, GENERALIZED HYPERGEOMETRIC FUNCTION, HORN FUNCTION, LAURICELLA FUNCTIONS, MACROBERT'S E-FUNCTION, MEIJER'S G-FUNCTION

References

Appell, P. *Sur le fonctions hypergéométriques de plusieurs variables.* Paris: Gauthier-Villars, 1925.
Appell, P. and Kampé de Fériet, J. *Fonctions hypergéométriques et hypersphériques: polynomes d'Hermite.* Paris: Gauthier-Villars, 1926.
Exton, H. "The Kampé de Fériet Function." §1.3.2 in *Handbook of Hypergeometric Integrals: Theory, Applications, Tables, Computer Programs.* Chichester, England: Ellis Horwood, pp. 24–25, 1978.
Exton, H. *Multiple Hypergeometric Functions and Applications.* Chichester, England: Ellis Horwood, 1976.
Exton, H. and Krupnikov, E. D. *A Register of Computer-Oriented Reduction Identities for the Kampé de Fériet Function.* Draft manuscript. Novosibirsk, 1998.
Kampé de Fériet, J. *La fonction hypergéométrique.* Paris: Gauthier-Villars, 1937.
Srivastava, H. M., Karlsson, P. W. *Multiple Gaussian Hypergeometric Series.* Chichester, England: Ellis Horwood, 1985.

Kampyle of Eudoxus

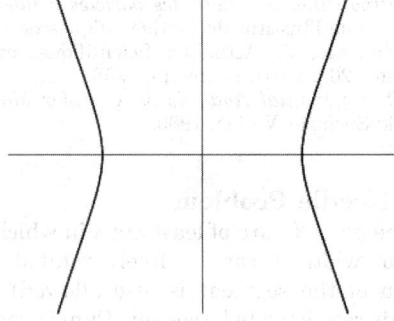

A curve studied by Eudoxus in relation to the classical problem of CUBE DUPLICATION. It is given

by the polar equation

$$r \cos^2 \theta = a,$$

and the PARAMETRIC EQUATIONS

$$x = a \sec t$$

$$y = a \tan t \sec t$$

with $t \in [-\pi/2, \ \pi/2]$.

References

Lawrence, J. D. *A Catalog of Special Plane Curves.* New York: Dover, pp. 141–143, 1972.

MacTutor History of Mathematics Archive. "Kampyle of Eudoxus." http://www-groups.dcs.st-and.ac.uk/~history/Curves/Kampyle.html.

Kanizsa Triangle

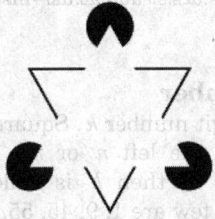

An optical ILLUSION, illustrated above, in which the eye perceives a white upright EQUILATERAL TRIANGLE where none is actually drawn.

See also ILLUSION

References

Bradley, D. R. and Petry, H. M. "Organizational Determinants of Subjective Contour." *Amer. J. Psychology* **90**, 253–262, 1977.

Fineman, M. *The Nature of Visual Illusion.* New York: Dover, pp. 26, 137, and 156, 1996.

Kantorovich Inequality

Suppose $x_1 < x_2 < \ldots < x_n$ are given POSITIVE numbers. Let $\lambda_1, \ldots, \lambda_n \geq 0$ and $\Sigma_{j=1}^n \lambda_j = 1$. Then

$$\left(\sum_{j=1}^n \lambda_j x_j \right) \left(\sum_{j=1}^n \lambda_j x_j^{-1} \right) \leq A^2 G^{-2}, \tag{1}$$

where

$$A = \tfrac{1}{2}(x_1 + x_n) \tag{2}$$

$$G = \sqrt{x_1 x_n} \tag{3}$$

are the ARITHMETIC and GEOMETRIC MEAN, respectively, of the first and last numbers. The Kantorovich inequality is central to the study of convergence properties of descent methods in optimization (Luenberger 1984).

See also ARITHMETIC MEAN, GEOMETRIC MEAN

References

Bauer, F. L. "A Further Generalization of the Kantorovich Inequality." *Numer. Math.* **3**, 117–119, 1961.

Greub, W. and Rheinboldt, W. "On a Generalization of an Inequality of L. V. Kantorovich." *Proc. Amer. Math. Soc.* **10**, 407–413, 1959.

Henrici, P. "Two Remarks of the Kantorovich Inequality." *Amer. Math. Monthly* **68**, 904–906, 1961.

Kantorovic, L. V. "Functional Analysis and Applied Mathematics" [Russian]. *Uspekhi Mat. Nauk* **3**, 89–185, 1948.

Luenberger, D. G. *Linear and Nonlinear Programming, 2nd ed.* Reading, MA: Addison-Wesley, pp. 217–219, 1984.

Newman, M. "Kantorovich's Inequality." *J. Res. National Bur. Standards* **64B**, 33–34, 1960.

Pólya, G. and Szego, G. *Aufgaben und Lehrsätze der Analysis.* Berlin, 1925.

Pták, V. "The Kantorovich Inequality." *Amer. Math. Monthly* **102**, 820–821, 1995.

Schopf, A. H. "On the Kantorovich Inequality." *Numer. Math.* **2**, 344–346, 1960.

Strang, W. G. "On the Kantorovich Inequality." *Proc. Amer. Math. Soc.* **11**, 468, 1960.

Kaplan-Yorke Conjecture

There are several versions of the Kaplan-Yorke conjecture, with many of the higher dimensional ones remaining unsettled. The original Kaplan-Yorke conjecture (Kaplan and Yorke 1979) proposed that, for a two-dimensional mapping, the CAPACITY DIMENSION D equals the KAPLAN-YORKE DIMENSION D_{KY},

$$D = D_{KY} = d_{\text{Lya}} = 1 + \frac{\sigma_1}{\sigma_2},$$

where σ_1 and σ_2 are the LYAPUNOV CHARACTERISTIC EXPONENTS. This was subsequently proven to be true in 1982. A later conjecture held that the KAPLAN-YORKE DIMENSION is generically equal to a probabilistic dimension which appears to be identical to the INFORMATION DIMENSION (Frederickson *et al.* 1983). This conjecture is partially verified by Ledrappier (1981). For invertible 2-D maps, $\nu = \sigma = D$, where ν is the CORRELATION EXPONENT, σ is the INFORMATION DIMENSION, and D is the CAPACITY DIMENSION (Young 1984).

See also CAPACITY DIMENSION, KAPLAN-YORKE DIMENSION, LYAPUNOV CHARACTERISTIC EXPONENT, LYAPUNOV DIMENSION

References

Chen, Z. M. "A Note on Kaplan-Yorke-Type Estimates on the Fractal Dimension of Chaotic Attractors." *Chaos, Solitons, and Fractals* **3**, 575–582, 1994.

Frederickson, P.; Kaplan, J. L.; Yorke, E. D.; and Yorke, J. A. "The Liapunov Dimension of Strange Attractors." *J. Diff. Eq.* **49**, 185–207, 1983.

Kaplan, J. L. and Yorke, J. A. In *Functional Differential Equations and Approximations of Fixed Points* (Ed. H.-O. Peitgen and H.-O. Walther). Berlin: Springer-Verlag, p. 204, 1979.

Ledrappier, F. "Some Relations Between Dimension and Lyapunov Exponents." *Commun. Math. Phys.* **81**, 229–238, 1981.

Worzbusekros, A. "Remark on a Conjecture of Kaplan and Yorke." *Proc. Amer. Math. Soc.* **85**, 381–382, 1982.
Young, L. S. "Dimension, Entropy, and Lyapunov Exponents in Differentiable Dynamical Systems." *Phys. A* **124**, 639–645, 1984

Kaplan-Yorke Dimension

$$D_{KY} \equiv j + \frac{\sigma_1 + \ldots + \sigma_j}{|\sigma_{j+1}|},$$

where $\sigma_1 \leq \sigma_n$ are LYAPUNOV CHARACTERISTIC EXPONENTS and j is the largest INTEGER for which

$$\lambda_1 + \ldots + \lambda_j \geq 0.$$

If $\nu = \sigma = D$, where ν is the CORRELATION EXPONENT, σ the INFORMATION DIMENSION, and D the HAUSDORFF DIMENSION, then

$$D \leq D_{KY}$$

(Grassberger and Procaccia 1983).

References

Grassberger, P. and Procaccia, I. "Measuring the Strangeness of Strange Attractors." *Physica D* **9**, 189–208, 1983.

Kaplan-Yorke Map

$$x_{n+1} = 2x_n$$

$$y_{n+1} = ay_n + \cos(4\pi x_n),$$

where x_n, y_n are computed mod 1. (Kaplan and Yorke 1979). The Kaplan-Yorke map with $\alpha = 0.2$ has CORRELATION EXPONENT 1.42 ± 0.02 (Grassberger Procaccia 1983) and CAPACITY DIMENSION 1.43 (Russell *et al.* 1980).

References

Grassberger, P. and Procaccia, I. "Measuring the Strangeness of Strange Attractors." *Physica D* **9**, 189–208, 1983.
Kaplan, J. L. and Yorke, J. A. In *Functional Differential Equations and Approximations of Fixed Points* (Ed. H.-O. Peitgen and H.-O. Walther). Berlin: Springer-Verlag, p. 204, 1979.
Russell, D. A.; Hanson, J. D.; and Ott, E. "Dimension of Strange Attractors." *Phys. Rev. Let.* **45**, 1175–1178, 1980.

Kappa Curve

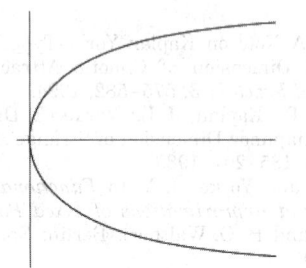

A curve also known as GUTSCHOVEN'S CURVE which was first studied by G. van Gutschoven around 1662

(MacTutor Archive). It was also studied by Newton and, some years later, by Johann Bernoulli. It is given by the Cartesian equation

$$(x^2 + y^2)y^2 = a^2 x^2, \tag{1}$$

by the polar equation

$$r = a \cot \theta, \tag{2}$$

and the PARAMETRIC EQUATIONS

$$x = a \cos t \cot t \tag{3}$$

$$y = a \cos t. \tag{4}$$

References

Lawrence, J. D. *A Catalog of Special Plane Curves.* New York: Dover, pp. 136 and 139–141, 1972.
MacTutor History of Mathematics Archive. "Kappa Curve." http://www-groups.dcs.st-and.ac.uk/~history/Curves/Kappa.html.

Kaprekar Number

Consider an n-digit number k. Square it and add the right n digits to the left n or $n-1$ digits. If the resultant sum is k, then k is called a Kaprekar number. The first few are 1, 9, 45, 55, 99, 297, 703, ... (Sloane's A006886).

$$9^2 = 81 \quad 8 + 1 = 9$$

$$297^2 = 88,209 \quad 88 + 209 = 297.$$

See also DIGITAL ROOT, DIGITADDITION, HAPPY NUMBER, KAPREKAR ROUTINE, NARCISSISTIC NUMBER, RECURRING DIGITAL INVARIANT

References

Iannucci, D. E.. "The Kaprekar Numbers." *J. Integer Sequences* **3**, No. 00.1.2, 2000. http://www.research.att.com/~njas/sequences/JIS/VOL3/iann2a.html.
Sloane, N. J. A. Sequences A006886/M4625 in "An On-Line Version of the Encyclopedia of Integer Sequences." http://www.research.att.com/~njas/sequences/eisonline.html.
Wells, D. *The Penguin Dictionary of Curious and Interesting Numbers.* Middlesex, England: Penguin Books, p. 73, 1986.

Kaprekar Routine

A routine discovered in 1949 by D. R. Kaprekar for 4-digit numbers, but which can be generalized to k-digit numbers. To apply the Kaprekar routine to a number n, arrange the digits in descending (n') and ascending (n'') order. Now compute $K(n) \equiv n' - n''$ and iterate. The algorithm reaches 0 (a degenerate case), a constant, or a cycle, depending on the number of digits in k and the value of n.

For a 3-digit number n in base 10, the Kaprekar routine reaches the number 495 in at most six iterations. In base r, there is a unique number $((r-2)/2, r-1, r/2)_r$ to which n converges in at most $(r+2)/2$ iterations IFF r is EVEN. For any 4-digit number n in base-10, the routine terminates on the number 6174 after seven or fewer steps (where it enters the 1-cycle $K(6174) = 6174$).

2. 0, 0, 9, 21, {(45), (49)}, ...,

3. 0, 0, (32, 52), 184, (320, 580, 484), ...,

4. 0, 30, {201, (126, 138)}, (570, 765), {(2550), (3369), (3873)}, ...,

5. 8, (48, 72), 392, (1992, 2616, 2856, 2232), (7488, 10712, 9992, 13736, 11432), ...,

6. 0, 105, (430, 890, 920, 675, 860, 705), {5600, (4305, 5180)}, {(27195), (33860), (42925), (16840, 42745, 35510)}, ...,

7. 0, (144, 192), (1068, 1752, 1836), (9936, 15072, 13680, 13008, 10608), (55500, 89112, 91800, 72012, 91212, 77388), ...,

8. 21, 252, {(1589, 3178, 2723), (1022, 3122, 3290, 2044, 2212)}, {(17892, 20475), (21483, 25578, 26586, 21987)}...,

9. (16, 48), (320, 400), {(2256, 5312, 3856),(3712, 5168, 5456)}, {41520,(34960, 40080, 55360, 49520, 42240)}, ...,

10. 0, 495, 6174, {(53955, 59994), (61974, 82962, 75933, 63954), (62964, 71973, 83952, 74943)}, ...,

See also 196-ALGORITHM, KAPREKAR NUMBER, RATS SEQUENCE

References

Eldridge, K. E. and Sagong, S. "The Determination of Kaprekar Convergence and Loop Convergence of All 3-Digit Numbers." *Amer. Math. Monthly* **95**, 105–112, 1988.

Kaprekar, D. R. "An Interesting Property of the Number 6174." *Scripta Math.* **15**, 244–245, 1955.

Trigg, C. W. "All Three-Digit Integers Lead to..." *The Math. Teacher*, **67**, 41–45, 1974.

Young, A. L. "A Variation on the 2-digit Kaprekar Routine." *Fibonacci Quart.* **31**, 138–145, 1993.

Kaps-Rentrop Methods

A generalization of the RUNGE-KUTTA METHOD for solution of ORDINARY DIFFERENTIAL EQUATIONS, also called ROSENBROCK METHODS.

See also RUNGE-KUTTA METHOD

References

Press, W. H.; Flannery, B. P.; Teukolsky, S. A.; and Vetterling, W. T. *Numerical Recipes in FORTRAN: The Art of Scientific Computing, 2nd ed.* Cambridge, England: Cambridge University Press, pp. 730–735, 1992.

Kapteyn Series

A series OF THE FORM

$$\sum_{n=0}^{\infty} \alpha_n J_{v+n}[(v+n)z],$$

where $J_n(z)$ is a BESSEL FUNCTION OF THE FIRST KIND. Examples include Kapteyn's original series

$$\frac{1}{1-z} = 1 + 2\sum_{n=0}^{\infty} J_n(nz)$$

and

$$\frac{z^2}{2(1-z^2)} = \sum_{n=0}^{\infty} J_{2n}(2nz).$$

See also BESSEL FUNCTION OF THE FIRST KIND, LEMON, NEUMANN SERIES (BESSEL FUNCTION)

References

Iyanaga, S. and Kawada, Y. (Eds.). *Encyclopedic Dictionary of Mathematics.* Cambridge, MA: MIT Press, p. 1473, 1980.

Karamata's Tauberian Theorem

References

Widder, D. V. Ch. 5 in *The Laplace Transform.* Princeton, NJ: Princeton University Press, 1941.

Karatsuba Multiplication

It is possible to perform MULTIPLICATION of LARGE NUMBERS in (many) fewer operations than the usual brute-force technique of "long multiplication." As discovered by Karatsuba and Ofman (1962), MULTIPLICATION of two n-DIGIT numbers can be done with a BIT COMPLEXITY of less than n^2 using identities OF THE FORM

$$(a + b \cdot 10^n)(c + d \cdot 10^n)$$
$$= ac + [(a+b)(c+d) - ac - bd]10^n + bd \cdot 10^{2n}. \quad (1)$$

Proceeding recursively then gives BIT COMPLEXITY $\mathcal{O}(n^{\lg 3})$, where $\lg 3 = 1.58\ldots < 2$ (Borwein *et al.* 1989). The best known bound is $\mathcal{O}(n \lg n \lg n)$ steps for $n \gg 1$ (Schönhage and Strassen 1971, Knuth 1981). However, this ALGORITHM is difficult to implement, but a procedure based on the FAST FOURIER TRANSFORM is straightforward to implement and gives BIT COMPLEXITY $\mathcal{O}((\lg n)^{2+\epsilon}n)$ (Brigham 1974, Borodin and Munro 1975, Knuth 1981, Borwein *et al.* 1989).

As a concrete example, consider MULTIPLICATION of two numbers each just two "digits" long in base w,

$$N_1 = a_0 + a_1 w \quad (2)$$

$$N_2 = b_0 + b_1 w, \tag{3}$$

then their PRODUCT is

$$P \equiv N_1 N_2 = a_0 b_0 + (a_0 b_1 + a_1 b_0)w + a_1 b_1 w^2$$
$$= p_0 + p_1 w + p_2 w^2. \tag{4}$$

Instead of evaluating products of individual digits, now write

$$q_0 = a_0 b_0 \tag{5}$$

$$q_1 = (a_0 + a_1)(b_0 + b_1) \tag{6}$$

$$q_2 = a_1 b_1. \tag{7}$$

The key term is q_1, which can be expanded, regrouped, and written in terms of the p_j as

$$q_1 = p_1 + p_0 + p_2. \tag{8}$$

However, since $p_0 = q_0$, and $p_2 = q_2$, it immediately follows that

$$p_0 = q_0 \tag{9}$$

$$p_1 = q_1 - q_0 - q_2 \tag{10}$$

$$p_2 = q_2, \tag{11}$$

so the three "digits" of p have been evaluated using three multiplications rather than four. The technique can be generalized to multidigit numbers, with the trade-off being that more additions and subtractions are required.

Now consider four-"digit" numbers

$$N_1 = a_0 + a_1 w + a_2 w^2 + a_3 w^3, \tag{12}$$

which can be written as a two-"digit" number represented in the base w^2,

$$N_1 = (a_0 + a_1 w) + (a_2 + a_3 w) * w^2. \tag{13}$$

The "digits" in the new base are now

$$a_0' = a_0 + a_1 w \tag{14}$$

$$a_1' = a_2 + a_3 w, \tag{15}$$

and the Karatsuba algorithm can be applied to N_1 and N_2 in this form. Therefore, the Karatsuba algorithm is not restricted to multiplying two-digit numbers, but more generally expresses the multiplication of two numbers in terms of multiplications of numbers of half the size. The asymptotic speed the algorithm obtains by recursive application to the smaller required subproducts is $\mathcal{O}(n^{\lg 3})$ (Knuth 1981).

When this technique is recursively applied to multidigit numbers, a point is reached in the recursion when the overhead of additions and subtractions makes it more efficient to use the usual $\mathcal{O}(n^2)$ MULTIPLICATION algorithm to evaluate the partial products. The most efficient overall method therefore relies on a combination of Karatsuba and conventional multiplication.

See also COMPLEX MULTIPLICATION, MULTIPLICATION, STRASSEN FORMULAS

References

Borodin, A. and Munro, I. *The Computational Complexity of Algebraic and Numeric Problems.* New York: American Elsevier, 1975.

Borwein, J. M.; Borwein, P. B.; and Bailey, D. H. "Ramanujan, Modular Equations, and Approximations to Pi, or How to Compute One Billion Digits of Pi." *Amer. Math. Monthly* **96**, 201–219, 1989.

Brigham, E. O. *The Fast Fourier Transform.* Englewood Cliffs, NJ: Prentice-Hall, 1974.

Brigham, E. O. *Fast Fourier Transform and Applications.* Englewood Cliffs, NJ: Prentice-Hall, 1988.

Cook, S. A. *On the Minimum Computation Time of Functions.* Ph.D. Thesis. Cambridge, MA: Harvard University, pp. 51–77, 1966.

Hollerbach, U. "Fast Multiplication & Division of Very Large Numbers." `sci.math.research` posting, Jan. 23, 1996.

Karatsuba, A. and Ofman, Yu. "Multiplication of Many-Digital Numbers by Automatic Computers." *Doklady Akad. Nauk SSSR* **145**, 293–294, 1962. Translation in *Physics-Doklady* **7**, 595–596, 1963.

Knuth, D. E. *The Art of Computing, Vol. 2: Seminumerical Algorithms, 3rd ed.* Reading, MA: Addison-Wesley, pp. 278–286, 1998.

Schönhage, A. and Strassen, V. "Schnelle Multiplikation Grosser Zahlen." *Computing* **7**, 281–292, 1971.

Toom, A. L. "The Complexity of a Scheme of Functional Elements Simulating the Multiplication of Integers." *Dokl. Akad. Nauk SSSR* **150**, 496–498, 1963. English translation in *Soviet Mathematics* **3**, 714–716, 1963.

Zuras, D. "More on Squaring and Multiplying Large Integers." *IEEE Trans. Comput.* **43**, 899–908, 1994.

Karnaugh Map

In combinatorial logic minimization, a device known as a Karnaugh map is frequently used. It is similar to a TRUTH TABLE, but the various variables are represented along two axes, and are arranged in such a way that only one input bit changes in going from one square to an adjacent square.

See also TRUTH TABLE

k-ary Divisor

Let a DIVISOR d of n be called a 1-ary divisor if $d \perp nd$ (i.e., d is RELATIVELY PRIME to n/d). Then d is called a k-ary divisor of n, written $d|_k n$, if the GREATEST COMMON $(k-1)$-ary divisor of d and (n/d) is 1.

In this notation, $d\|n$ is written $d|_0 n$, and $d\|n$ is written $d|_1 n$. p^x is an INFINARY DIVISOR of p^y (with $y > 0$) if $p^x|_{y-1} p^y$.

See also BIUNITARY DIVISOR, DIVISOR, GREATEST COMMON DIVISOR, INFINARY DIVISOR, UNITARY DIVISOR

References

Cohen, G. L. "On an Integer's Infinary Divisors." *Math. Comput.* **54**, 395–411, 1990.

Guy, R. K. *Unsolved Problems in Number Theory, 2nd ed.*
New York: Springer-Verlag, p. 54, 1994.
Suryanarayana, D. "The Number of k-ary Divisors of an
Integer." *Monatschr. Math.* **72**, 445–450, 1968.

Katadrome

A katadrome is a number whose HEXADECIMAL digits
are in strict descending order. The first few are 1, 2, 3,
4, 5, 6, 7, 8, 9, 10, 11, 12, 13, 14, 15, 16, 32, 33, 48, 49,
... (Sloane's A023797), corresponding to 1, 2, 3, 4, 5, 6,
7, 8, 9, A, B, C, D, E, F, 10, 20, 21, 30, 31, ...

See also DIGIT, HEXADECIMAL, METADROME, NIALP-
DROME, PLAINDROME

References

Sloane, N. J. A. Sequences A023797 in "An On-Line Version
of the Encyclopedia of Integer Sequences." http://www.re-
search.att.com/~njas/sequences/eisonline.html.
Weisstein, E. W. "Integer Sequences." MATHEMATICA NOTE-
BOOK INTEGERSEQUENCES.M.

Katona's Problem

Find the minimum number $f(n)$ of SUBSETS in a
SEPARATING FAMILY for a SET of n elements, where a
SEPARATING FAMILY is a SET of SUBSETS in which each
pair of adjacent elements is found separated, each in
one of two DISJOINT SUBSETS. For example, the 26
letters of the alphabet can be separated by a family of
nine:

$$(abcdefghi) \quad (jklmnopqr) \quad (stuvwxyz)$$
$$(abcjklstu) \quad (defmnovwx) \quad (ghipqryz).$$
$$(adgjmpsvy) \quad (behknqtwz) \quad (cfilorux)$$

The problem was posed by Katona (1973) and solved
by C. Mao-Cheng in 1982,

$$f(n) = \min\left\{2p + 3\left\lceil \log_3\left(\frac{n}{2^p}\right)\right\rceil : p = 0, 1, 2\right\},$$

where $\lceil x \rceil$ is the CEILING FUNCTION. $f(n)$ is nondecreasing, and the values for $n = 1, 2, ...$ are 0, 2, 3, 4,
5, 5, 6, 6, 6, 7, ... (Sloane's A007600). The values at
which $f(n)$ increases are 1, 2, 3, 4, 5, 7, 10, 13, 19, 28,
37, ... (Sloane's A007601), so $f(26) = 9$, as illustrated
in the preceding example.

See also SEPARATING FAMILY

References

Honsberger, R. "Cai Mao-Cheng's Solution to Katona's
Problem on Families of Separating Subsets." Ch. 18 in
Mathematical Gems III. Washington, DC: Math. Assoc.
Amer., pp. 224–239, 1985.
Katona, G. O. H. "Combinatorial Search Problem." In *A
Survey of Combinatorial Theory* (Ed. J. N. Srivasta, F.
Harary, C. R. Rao, G.-C. Rota, and S. S. Shrikhande).
Amsterdam, Netherlands: North-Holland, pp. 285–308,
1973.
Sloane, N. J. A. Sequences A007600/M0456 and A007601/
M0525 in "An On-Line Version of the Encyclopedia of

Integer Sequences." http://www.research.att.com/~njas/
sequences/eisonline.html.

Kauffman Polynomial F

A semi-oriented 2-variable KNOT POLYNOMIAL defined
by

$$F_L(a, z) = a^{-w(L)}\langle|L|\rangle, \tag{1}$$

where L is an oriented LINK DIAGRAM, $w(L)$ is the
WRITHE of L, $|L|$ is the unoriented diagram corre-
sponding to L, and $\langle L \rangle$ is the BRACKET POLYNOMIAL.
It was developed by Kauffman by extending the BLM/
HO POLYNOMIAL Q to two variables, and satisfies

$$F(1, x) = Q(x). \tag{2}$$

The Kauffman POLYNOMIAL is a generalization of the
JONES POLYNOMIAL $V(t)$ since it satisfies

$$V(t) = F(-t^{-3/4}, t^{-1/4} + t^{1/4}), \tag{3}$$

but its relationship to the HOMFLY POLYNOMIAL is
not well understood. In general, it has more terms
than the HOMFLY POLYNOMIAL, and is therefore
more powerful for discriminating KNOTS. It is a semi-
oriented POLYNOMIAL because changing the orientation only changes F by a POWER of a. In particular,
suppose L^* is obtained from L by reversing the
orientation of component k, then

$$F_{L^*} = a^{4\lambda}F_L, \tag{4}$$

where λ is the LINKING NUMBER of k with $L - k$
(Lickorish and Millett 1988). F is unchanged by
MUTATION.

$$F_{L_1 + L_2} = F(L_1)F(L_2) \tag{5}$$

$$F_{L_1 \cup L_2} = [(a^{-1} + a)x^{-1} - 1]F_{L_1}F_{L_2}. \tag{6}$$

M. B. Thistlethwaite has tabulated the Kauffman 2-
variable POLYNOMIAL for KNOTS up to 13 crossings.

See also KAUFFMAN POLYNOMIAL X

References

Lickorish, W. B. R. and Millett, B. R. "The New Polynomial
Invariants of Knots and Links." *Math. Mag.* **61**, 1–23,
1988.
Stoimenow, A. "Kauffman Polynomials." http://guests.mpim-
bonn.mpg.de/alex/ptab/k10.html.
Weisstein, E. W. "Knots and Links." MATHEMATICA NOTE-
BOOK KNOTS.M.

Kauffman Polynomial X

A 1-variable KNOT POLYNOMIAL denoted X or $\mathscr{L}$.

$$\mathscr{L}_L(A) \equiv (-A^3)^{-w(L)}\langle L \rangle, \tag{1}$$

where $\langle L \rangle$ is the BRACKET POLYNOMIAL and $w(L)$ is
the WRITHE of L. This POLYNOMIAL is invariant under
AMBIENT ISOTOPY, and relates MIRROR IMAGES by

$$\mathcal{L}_{L^*} = \mathcal{L}_L(A^{-1}). \qquad (2)$$

It is identical to the JONES POLYNOMIAL with the change of variable

$$\mathcal{L}(t^{-1/4}) = V(t). \qquad (3)$$

The X POLYNOMIAL of the MIRROR IMAGE K^* is the same as for K but with A replaced by A^{-1}.

See also KAUFFMAN POLYNOMIAL F

References

Kauffman, L. H. *Knots and Physics.* Singapore: World Scientific, p. 33, 1991.

Kaup's Equation

The system of PARTIAL DIFFERENTIAL EQUATIONS

$$f_x = 2fgc(x - t)$$
$$g_t = 2fgc(x - t).$$

References

Dodd, R. and Fordy, A. "The Prolongation Structures of Quasi-Polynomial Flows." *Proc. Roy. Soc. A* **385**, 389–429, 1983.
Zwillinger, D. *Handbook of Differential Equations, 3rd ed.* Boston, MA: Academic Press, p. 138, 1997.

k-Automatic Set

AUTOMATIC SET

k-Balanced

A GENERALIZED HYPERGEOMETRIC FUNCTION

$${}_pF_q\begin{bmatrix} \alpha_1, & \alpha_2, & \ldots, & \alpha_p \\ \beta_1, & \beta_2, & \ldots, & \beta_q \end{bmatrix}; z ,$$

is said to be k-balanced if

$$\sum_{i=1}^{q} \beta_i = k + \sum_{i=1}^{p} \alpha_i.$$

See also GENERALIZED HYPERGEOMETRIC FUNCTION, NEARLY-POISED, SAALSCHÜTZIAN, WELL-POISED

References

Koepf, W. *Hypergeometric Summation: An Algorithmic Approach to Summation and Special Function Identities.* Braunschweig, Germany: Vieweg, p. 43, 1998.

k-Chain

Any sum of a selection of Π_ks, where Π_k denotes a k-D POLYTOPE.

See also K-CIRCUIT, POLYTOPE

k-Circuit

A K-CHAIN whose bounding $(K\text{-}1)$-CHAIN vanishes.

See also K-CHAIN

k-Coloring

A k-coloring of a GRAPH G is an assignment of one of k possible colors to each vertex of G (i.e, a VERTEX COLORING) such that no two adjacent vertices receive the same color.

See also CHROMATIC NUMBER, CHROMATIC POLYNOMIAL, COLORING, EDGE COLORING, VERTEX COLORING

References

Saaty, T. L. and Kainen, P. C. *The Four-Color Problem: Assaults and Conquest.* New York: Dover, p. 13, 1986.

k-Connected Graph

A graph G is said to be k-connected if there does not exist a set of $k - 1$ vertices whose removal disconnects the graph, i.e., the VERTEX CONNECTIVITY of G is $\geq k$ (Skiena 1990, p. 177). Therefore, a CONNECTED GRAPH is 1-connected, and a BICONNECTED GRAPH is 2-connected (Skiena 1990, p. 177).

The following table gives the numbers of k-connected graphs for n-node graphs. Note that there is a unique n-connected n-node graph, namely, the COMPLETE GRAPH K_n. The WHEEL GRAPH is the basic 3-connected graph (Tutte 1961; Skiena 1990, p. 179).

k	k-connected graphs on 1, 2, ... nodes
1	1, 1, 2, 6, 21, 112, 853, ...
2	0, 1, 1, 3, 10, 56, 468, ...
3	0, 0, 1, 1, 3, 17, 136, ...
4	0, 0, 0, 1, 1, 4, 25, ...
5	0, 0, 0, 0, 1, 1, 4, ...
6	0, 0, 0, 0, 0, 1, 1, ...
7	0, 0, 0, 0, 0, 0, 1, ...
8	0, 0, 0, 0, 0, 0, 0, ...

See also BARNETTE'S CONJECTURE, BICONNECTED GRAPH, CONNECTED GRAPH, DISCONNECTED GRAPH, HARARY GRAPH, K-EDGE-CONNECTED GRAPH, MENGER'S N-ARC THEOREM, POLYHEDRAL GRAPH

References

Harary, F. *Graph Theory.* Reading, MA: Addison-Wesley, p. 45, 1994.
Skiena, S. *Implementing Discrete Mathematics: Combinatorics and Graph Theory with Mathematica.* Reading, MA: Addison-Wesley, 1990.
Sloane, N. J. A. Sequences A000719/M1452, A052442, A052443, A052444, and A052445 in "An On-Line Version

of the Encyclopedia of Integer Sequences." http://www.re-search.att.com/~njas/sequences/eisonline.html.

Tutte, W. T. "A Theory of 3-Connected Graphs." *Indag. Math.* **23**, 441–455, 1961.

k-Edge-Connected Graph

A graph is k-edge-connected if there does not exist a set of k edges whose removal disconnects the graph (Skiena 1990, p. 177). The maximum edge connectivity of a given graph is the smallest degree of any node, since deleting these edges disconnects the graph. Complete bipartite graphs have maximum edge connectivity. The following table gives the numbers of k-edge-connected graphs for n-node graphs.

k	Sloane	$n = 1, 2, \ldots$
0	A000719	0, 1, 2, 5, 13, 44, 191, ...
1	A052446	0, 1, 1, 3, 10, 52, 351, ...
2	A052447	0, 0, 1, 2, 8, 41, 352, ...
3	A052448	0, 0, 0, 1, 2, 15, 121, ...
4		0, 0, 0, 0, 1, 3, 25, ...
5		0, 0, 0, 0, 0, 1, 3, ...
6		0, 0, 0, 0, 0, 0, 1, ...

See also κ-CONNECTED GRAPH

References

Harary, F. *Graph Theory.* Reading, MA: Addison-Wesley, p. 45, 1994.

Skiena, S. *Implementing Discrete Mathematics: Combinatorics and Graph Theory with Mathematica.* Reading, MA: Addison-Wesley, 1990.

Sloane, N. J. A. Sequences A000719/M1452, A052446, A052447, and A052448 in "An On-Line Version of the Encyclopedia of Integer Sequences." http://www.research.att.com/~njas/sequences/eisonline.html.

Kei

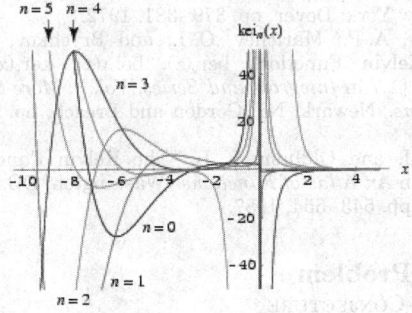

The IMAGINARY PART of

$$e^{-v\pi i/2}K_v(xe^{\pi i/4}) = \ker_v(x) + i\,\mathrm{kei}_v(x),$$

where $K_v(z)$ is a MODIFIED BESSEL FUNCTION OF THE SECOND KIND.

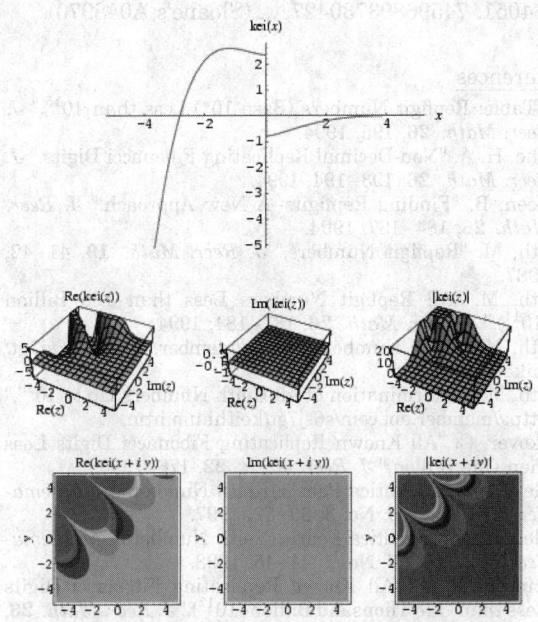

The special case $v = 0$ gives the plots shown above.

See also BEI, BER, KER, KELVIN FUNCTIONS

References

Abramowitz, M. and Stegun, C. A. (Eds.). "Kelvin Functions." §9.9 in *Handbook of Mathematical Functions with Formulas, Graphs, and Mathematical Tables, 9th printing.* New York: Dover, pp. 379–381, 1972.

Prudnikov, A. P.; Marichev, O. I.; and Brychkov, Yu. A. "The Kelvin Functions $\mathrm{ber}_v(x)$, bei $v(x)$, $\ker_v(x)$ and $\mathrm{kei}_v(x)$." §1.7 in *Integrals and Series, Vol. 3: More Special Functions.* Newark, NJ: Gordon and Breach, pp. 29–30, 1990.

Keith Number

A Keith number is an n-digit INTEGER N such that if a Fibonacci-like sequence (in which each term in the sequence is the sum of the n previous terms) is formed with the first n terms taken as the decimal digits of the number N, then N itself occurs as a term in the sequence. For example, 197 is a Keith number since it generates the sequence 1, 9, 7, 17, 33, 57, 107, 197, ... (Keith). Keith numbers are also called REPFIGIT NUMBERS.

There is no known general technique for finding Keith numbers except by exhaustive search. Keith numbers are much rarer than the PRIMES, with only 52 Keith numbers with < 15 digits: 14, 19, 28, 47, 61, 75, 197, 742, 1104, 1537, 2208, 2580, 3684, 4788, 7385, 7647, 7909, ... (Sloane's A007629). The number of Keith numbers having $n = 1, 2, \ldots$ digits are 0, 6, 2, 9, 7, 10, 2, 3, 2, 0, 2, 4, 2, 3, 3, 3, 5, 3, 5, ... (Sloane's A050235; Keith), so there are only 71 less than 10^{19}. It is not known if there are an INFINITE number of Keith numbers.

The known prime Keith numbers are 19, 47, 61, 197, 1084051, 74596893730427, ... (Sloane's A048970).

References

--. "Table: Repfigit Numbers (Base 10*) Less than 10^{15}." *J. Recr. Math.* **26**, 195, 1994.
Esche, H. A. "Non-Decimal Replicating Fibonacci Digits." *J. Recr. Math.* **26**, 193–194, 1994.
Heleen, B. "Finding Repfigits--A New Approach." *J. Recr. Math.* **26**, 184–187, 1994.
Keith, M. "Repfigit Numbers." *J. Recr. Math.* **19**, 41–42, 1987.
Keith, M. "All Repfigit Numbers Less than 100 Billion (10^{11})." *J. Recr. Math.* **26**, 181–184, 1994.
Keith, M. "Keith Numbers." http://member.aol.com/s6sj7gt/mikekeit.htm.
Keith, M. "Determination of All Keith Numbers Up to 10^{19}." http://member.aol.com/s6sj7gt/keithnum.htm.
Pickover, C. "All Known Replicating Fibonacci Digits Less then One Billion." *J. Recr. Math.* **22**, 176, 1990.
Piele, D. "Mathematica Pearls: Keith Numbers." *Mathematica Res. Educ.* **6**, No. 3, 50–52, 1997.
Piele, D. "Mathematica Pearls: Keith Numbers." *Mathematica Res. Educ.* **7**, No. 1, 44–45, 1998.
Robinson, N. M. "All Known Replicating Fibonacci Digits Less than One Thousand Billion (10^{12})." *J. Recr. Math.* **26**, 188–191, 1994.
Sherriff, K. "Computing Replicating Fibonacci Digits." *J. Recr. Math.* **26**, 191–193, 1994.
Sloane, N. J. A. Sequences A007629, A048970, and A050235 in "An On-Line Version of the Encyclopedia of Integer Sequences." http://www.research.att.com/~njas/sequences/eisonline.html.
Weisstein, E. W. "Integer Sequences." MATHEMATICA NOTEBOOK INTEGERSEQUENCES.M.

Keller's Conjecture

Keller conjectured that tiling an n-D space with n-D HYPERCUBES of equal size yields an arrangement in which at least two hypercubes have an entire $(n-1)$-D "side" in common. The CONJECTURE has been proven true for $n = 1$ to 6, but disproven for $n \geq 10$.

References

Cipra, B. "If You Can't See It, Don't Believe It." *Science* **259**, 26–27, 1993.
Cipra, B. *What's Happening in the Mathematical Sciences, Vol. 1.* Providence, RI: Amer. Math. Soc., p. 24, 1993.

Kelvin Differential Equation

The second-order complex ORDINARY DIFFERENTIAL EQUATION

$$x^2 y'' + xy' - (ix^2 + v^2)y = 0 \tag{1}$$

(Abramowitz and Stegun 1972, p. 379; Zwillinger 1997, p. 123), whose solutions can be given in terms of the KELVIN functions

$$y = \mathrm{ber}_v\, x + i\, \mathrm{bei}_v \tag{2}$$

$$= \mathrm{ber}_{-v}\, x + i\, \mathrm{bei}_{-v} \tag{3}$$

$$= \mathrm{ker}_v\, x + i\, \mathrm{kei}_v \tag{4}$$

$$= \mathrm{ker}_{-v}\, x + i\, \mathrm{kei}_{-v} \tag{5}$$

(Abramowitz and Stegun 1972, p. 379).

See also KELVIN FUNCTIONS

References

Abramowitz, M. and Stegun, C. A. (Eds.). "Kelvin Functions." §9.9 in *Handbook of Mathematical Functions with Formulas, Graphs, and Mathematical Tables, 9th printing.* New York: Dover, pp. 379–381, 1972.
Zwillinger, D. *Handbook of Differential Equations, 3rd ed.* Boston, MA: Academic Press, p. 123, 1997.

Kelvin Functions

Kelvin defined the Kelvin functions BEI and BER according to

$$\mathrm{ber}_v(x) + i\, \mathrm{bei}_v(x) = J_v(xe^{3\pi i/4}) \tag{1}$$

$$= e^{v\pi i} J_v(xe^{-\pi i/4}), \tag{2}$$

$$= e^{v\pi i/2} I_v(xe^{\pi i/4}) \tag{3}$$

$$= e^{3v\pi i/2} I_v(xe^{-3\pi i/4}), \tag{4}$$

where $J_v(x)$ is a BESSEL FUNCTION OF THE FIRST KIND and $I_v(x)$ is a MODIFIED BESSEL FUNCTION OF THE FIRST KIND. These functions satisfy the KELVIN DIFFERENTIAL EQUATION.

Similarly, the functions KEI and KER by

$$\mathrm{ker}_v(x) + i\, \mathrm{kei}_v(x) = e^{-v\pi i/2} K_v(xe^{\pi i/4}), \tag{5}$$

where $K_v(x)$ is a MODIFIED BESSEL FUNCTION OF THE SECOND KIND. For the special case $v = 0$,

$$J_0\left(i\sqrt{i}x\right) = J_0\left(\tfrac{1}{2}\sqrt{2}(i-1)x\right) \equiv \mathrm{ber}(x) + i\, \mathrm{bei}(x). \tag{6}$$

See also BEI, BER, KEI, KELVIN DIFFERENTIAL EQUATION, KER

References

Abramowitz, M. and Stegun, C. A. (Eds.). "Kelvin Functions." §9.9 in *Handbook of Mathematical Functions with Formulas, Graphs, and Mathematical Tables, 9th printing.* New York: Dover, pp. 379–381, 1972.
Prudnikov, A. P.; Marichev, O. I.; and Brychkov, Yu. A. "The Kelvin Functions $\mathrm{ber}_v(x)$, $\mathrm{bei}\,v(x)$, $\mathrm{ker}_v(x)$ and $\mathrm{kei}_v(x)$." §1.7 in *Integrals and Series, Vol. 3: More Special Functions.* Newark, NJ: Gordon and Breach, pp. 29–30, 1990.
Spanier, J. and Oldham, K. B. "The Kelvin Functions." Ch. 55 in *An Atlas of Functions.* Washington, DC: Hemisphere, pp. 543–554, 1987.

Kelvin Problem

KELVIN'S CONJECTURE

Kelvin Transformation

Let D be a DOMAIN in $\mathbb{R}^n$ for $n \geq 3$. Then the transformation

$$v(x_1', \ldots, x_n') = \left(\frac{a}{r'}\right)^{n-2} u\left(\frac{a^2 x_1'}{r'^2}, \ldots, \frac{a^2 x_n'}{r'^2}\right)$$

onto a domain D', where

$$r'^2 = x_1'^2 + \ldots + x_n'^2$$

is called a Kelvin transformation. If $u(x_1, \ldots, x_n)$ is a HARMONIC FUNCTION on D, then $v(x_1', \ldots, x_n')$ is also HARMONIC on D'.

See also HARMONIC FUNCTION

References

Itô, K. (Ed.). "Harmonic Functions and Subharmonic Functions: Invariance of Harmonicity." §193B in *Encyclopedic Dictionary of Mathematics, 2nd ed.* Cambridge, MA: MIT Press, p. 725, 1980.

Kelvin's Conjecture

What space-filling arrangement of similar polyhedral cells of equal volume has minimal SURFACE AREA? Kelvin (Thomson 1887) proposed that the solution was the 14-sided TRUNCATED OCTAHEDRON. The isoperimetric quotient for the TRUNCATED OCTAHEDRON is given by

$$Q = \frac{36\pi V^3}{S^2} = \frac{36\pi(8\sqrt{2})^2}{(6 + 12\sqrt{3})^3}$$

$$= \frac{64\pi}{3(1 + 2\sqrt{3})^3} \approx 0.753367.$$

Despite one hundred years of failed attempts and Weyl's (1952) opinion that the TRUNCATED OCTAHEDRON could not be improved upon, Weaire and Phelan (1994) discovered a space-filling unit cell consisting of six 14-sided polyhedra and two 12-sided polyhedra that has 0.3% less SURFACE AREA.

See also SPACE-FILLING POLYHEDRON, TRUNCATED OCTAHEDRON

References

Gray, J. "Parsimonious Polyhedra." *Nature* **367**, 598–599, 1994.
Matzke, E. *Amer. J. Botany* **32**, 130, 1946.
Princen, H. M. and Levinson, P. *J. Colloid Interface Sci.* **120**, 172, 1987.
Ross, S. *Amer. J. Phys.* **46**, 513, 1978.
Thomson, W. *Philos. Mag.* **25**, 503, 1887.
Weaire, D. *Philos. Mag. Let.* **69**, 99, 1994.
Weaire, D. and Phelan, R. "A Counter-Example to Kelvin's Conjecture on Minimal Surfaces." *Philos. Mag. Let.* **69**, 107–110, 1994.
Weaire, D. *The Kelvin Problem: Foam Structures of Minimal Surface Area.* London: Taylor and Francis, 1996.
Weyl, H. *Symmetry.* Princeton, NJ: Princeton University Press, 1952.
Williams, R. *Science* **161**, 276, 1968.

Kempe Linkage

A double rhomboid LINKAGE which gives rectilinear motion from circular without an inversion.

See also PEAUCELLIER INVERSOR

References

Rademacher, H. and Toeplitz, O. *The Enjoyment of Mathematics: Selections from Mathematics for the Amateur.* Princeton, NJ: Princeton University Press, pp. 126–127, 1957.

Kepler Conjecture

In 1611, Kepler proposed that close packing (cubic or hexagonal) is the densest possible SPHERE PACKING (has the greatest η), and this assertion is known as the Kepler conjecture. Finding the densest (not necessarily periodic) packing of spheres is known as the KEPLER PROBLEM.

Buckminster Fuller (1975) claimed to have a proof, but it was really a description of face-centered cubic packing, not a proof of its optimality (Sloane 1998). A second putative proof of the Kepler conjecture was put forward by W.-Y. Hsiang (Cipra 1991, Hsiang 1992, Hsiang 1993, Cipra 1993), but was subsequently determined to be flawed (Conway *et al.* 1994, Hales 1994, Sloane 1998). According to J. H. Conway, nobody who has read Hsiang's proof has any doubts about its validity: it is nonsense.

Soon thereafter, Hales (1997a) published a detailed plan describing how the Kepler conjecture might be proved using a significantly different approach from earlier attempts and making extensive use of computer calculations. Hales subsequently completed a full proof, which appears in a series of papers totaling more than 250 pages (Cipra 1998) The proof relies extensively on methods from the theory of global optimization, linear programming, and interval arithmetic. The computer files containing the computer code and data files for combinatorics, interval arithmetic, and linear programs require over 3 gigabytes of space for storage.

See also DODECAHEDRAL CONJECTURE, KEPLER PROBLEM, KISSING NUMBER, SPHERE PACKING

References

Buckminster Fuller, R. *Synergetics.* London: Macmillan, 1975.
Cipra, B. "Gaps in a Sphere Packing Proof?" *Science* **259**, 895, 1993.
Cipra, B. "Packing Challenge Mastered at Last." *Science* **281**, 1267, 1998.
Cipra, B. "Music of the Spheres." *Science* **251**, 1028, 1991.
Conway, J. H.; Hales, T. C.; Muder, D. J.; and Sloane, N. J. A. "On the Kepler Conjecture." *Math. Intel.* **16**, 5, Spring 1994.
Eppstein, D. "Sphere Packing and Kissing Numbers." http://www.ics.uci.edu/~eppstein/junkyard/spherepack.html.
Ferguson, S. P. "Sphere Packings. V." http://www.math.lsa.umich.edu/~samf/MyStuff/Research/draft.ps.gz.

Ferguson, S. P. and Hales, T. C. "A Formulation of the Kepler Conjecture." http://www.math.lsa.umich.edu/~hales/countdown/form.ps.

Hales, T. C. "The Kepler Conjecture." http://www.math.lsa.umich.edu/~hales/countdown/.

Hales, T. C. "An Overview of the Kepler Conjecture." http://www.math.lsa.umich.edu/~hales/countdown/sphere0.ps.

Hales, T. C. "Recent Progress on the Kepler Conjecture." http://www.math.lsa.umich.edu/~hales/countdown/recent.ps.

Hales, T. C. "The Sphere Packing Problem." *J. Comput. Appl. Math.* **44**, 41–76, 1992.

Hales, T. C. "Remarks on the Density of Sphere Packings in 3 Dimensions." *Combinatori* **13**, 181–197, 1993.

Hales, T. C. "The Status of the Kepler Conjecture." *Math. Intel.* **16**, 47–58, Summer 1994.

Hales, T. C. "Sphere Packings. I." *Disc. Comput. Geom.* **17**, 1–51, 1997a. http://www.math.lsa.umich.edu/~hales/countdown/sphere1.ps.

Hales, T. C. "Sphere Packings. II." *Disc. Comput. Geom.* **18**, 135–149, 1997b. http://www.math.lsa.umich.edu/~hales/countdown/sphere2.ps.

Hales, T. C. "Sphere Packings. III." http://www.math.lsa.umich.edu/~hales/countdown/sphere3.ps.

Hales, T. C. "Sphere Packings. IV." http://www.math.lsa.umich.edu/~hales/countdown/sphere4.ps.

Hales, T. C. "Sphere Packings. VI." http://www.math.lsa.umich.edu/~hales/countdown/sphere6.ps.

Hsiang, W.-Y. "On Soap Bubbles and Isoperimetric Regions in Noncompact Symmetrical Spaces. 1." *Tôhoku Math. J.* **44**, 151–175, 1992.

Hsiang, W.-Y. "On the Sphere Packing Problem and the Proof of Kepler's Conjecture." *Int. J. Math.* **4**, 739–831, 1993.

Hsiang, W.-Y. "A Rejoinder to Hales's Article." *Math. Intel.* **17**, 35–42, Winter 1995.

Sloane, N. J. A. "Kepler's Conjecture Confirmed." *Nature* **395**, 435–436, 1998.

Zong, C. and Talbot, J. *Sphere Packings.* New York: Springer-Verlag, 1999.

Kepler Problem

Finding the densest not necessarily periodic SPHERE PACKING.

See also KEPLER CONJECTURE, SPHERE PACKING

Kepler Solid

KEPLER-POINSOT SOLID

Kepler's Equation

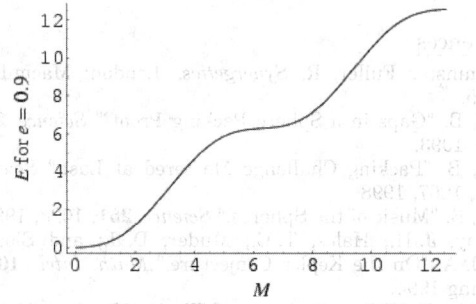

Kepler's equation gives the relation between the polar coordinates of a celestial body (like a planet) and the time elapsed from a given initial point. Kepler's equation is of fundamental importance in celestial mechanics, but cannot be directly inverted in terms of simple functions in order to determine where the planet will be at a given time.

Let M be the mean anomaly (a parameterization of time) and E the ECCENTRIC ANOMALY (a parameterization of polar angle) of a body orbiting on an ELLIPSE with ECCENTRICITY e, then

$$M = E - e \sin E. \qquad (1)$$

For M not a multiple of π, Kepler's equation has a unique solution, but is a TRANSCENDENTAL EQUATION and so cannot be inverted and solved directly for E given an arbitrary M. However, many algorithms have been derived for solving the equation as a result of its importance in celestial mechanics.

Writing a E as a POWER SERIES in e gives

$$E = M + \sum_{n=1}^{\infty} a_n e^n, \qquad (2)$$

where the coefficients are given by the LAGRANGE INVERSION THEOREM as

$$a_n = \frac{1}{2^{n-1} n!} \sum_{k=0}^{\lfloor n/2 \rfloor} (-1)^k \binom{n}{k} \\ \times (n - 2k)^{n-1} \sin[(n - 2k)M] \qquad (3)$$

(Wintner 1941, Moulton 1970, Henrici 1974, Finch). Surprisingly, this series diverges for

$$e > 0.6627434193\ldots, \qquad (4)$$

a value known as the LAPLACE LIMIT. In fact, E converges as a GEOMETRIC SERIES with ratio

$$r = \frac{e}{1 + \sqrt{1 + e^2}} \exp\left(\sqrt{1 + e^2}\right) \qquad (5)$$

(Finch).

There is also a series solution in BESSEL FUNCTIONS OF THE FIRST KIND,

$$E = M + \sum_{n=1}^{\infty} \frac{2}{n} J_n(ne) \sin(nM). \qquad (6)$$

This series converges for all $e < 1$ like a GEOMETRIC SERIES with ratio

$$r = \frac{e}{1 + \sqrt{1 - e^2}} \exp\left(\sqrt{1 - e^2}\right). \qquad (7)$$

The equation can also be solved by letting ψ be the ANGLE between the planet's motion and the direction PERPENDICULAR to the RADIUS VECTOR. Then

$$\tan \psi = \frac{e \sin E}{\sqrt{1 - e^2}}. \qquad (8)$$

Alternatively, we can define e in terms of an intermediate variable ϕ

$$e \equiv \sin \phi, \qquad (9)$$

then

$$\sin\left[\tfrac{1}{2}(v - E)\right] = \sqrt{\frac{r}{p}} \sin\left(\tfrac{1}{2}\phi\right) \sin v \qquad (10)$$

$$\sin\left[\tfrac{1}{2}(v + E)\right] = \sqrt{\frac{r}{p}} \cos\left(\tfrac{1}{2}\phi\right) \sin v. \qquad (11)$$

Iterative methods such as the simple

$$E_{i+1} = M + e \sin E_i \qquad (12)$$

with $E_0 = 0$ work well, as does NEWTON'S METHOD,

$$E_{i+1} = E_i + \frac{M + e \sin E_i - E_i}{1 - e \cos E_i}. \qquad (13)$$

In solving Kepler's equation, Stieltjes required the solution to

$$e^x(x - 1) = e^{-x}(x + 1), \qquad (14)$$

which is 1.1996678640257734... (Goursat 1959, Le Lionnais 1983).

See also ECCENTRIC ANOMALY

References

Danby, J. M. *Fundamentals of Celestial Mechanics, 2nd ed., rev. ed.* Richmond, VA: Willmann-Bell, 1988.

Dörrie, H. "The Kepler Equation." §81 in *100 Great Problems of Elementary Mathematics: Their History and Solutions.* New York: Dover, pp. 330–334, 1965.

Finch, S. "Favorite Mathematical Constants." http://www.mathsoft.com/asolve/constant/lpc/lpc.html.

Goldstein, H. *Classical Mechanics, 2nd ed.* Reading, MA: Addison-Wesley, pp. 101–102 and 123–124, 1980.

Goursat, E. *A Course in Mathematical Analysis, Vol. 2.* New York: Dover, p. 120, 1959.

Henrici, P. *Applied and Computational Complex Analysis, Vol. 1: Power Series-Integration-Conformal Mapping-Location of Zeros.* New York: Wiley, 1974.

Ioakimidis, N. I. and Papadakis, K. E. "A New Simple Method for the Analytical Solution of Kepler's Equation." *Celest. Mech.* **35**, 305–316, 1985.

Ioakimidis, N. I. and Papadakis, K. E. "A New Class of Quite Elementary Closed-Form Integrals Formulae for Roots of Nonlinear Systems." *Appl. Math. Comput.* **29**, 185–196, 1989.

Le Lionnais, F. *Les nombres remarquables.* Paris: Hermann, p. 36, 1983.

Marion, J. B. and Thornton, S. T. "Kepler's Equations." §7.8 in *Classical Dynamics of Particles & Systems, 3rd ed.* San Diego, CA: Harcourt Brace Jovanovich, pp. 261–266, 1988.

Moulton, F. R. *An Introduction to Celestial Mechanics, 2nd rev. ed.* New York: Dover, pp. 159–169, 1970.

Montenbruck, O. and Pfleger, T. "Mathematical Treatment of Kepler's Equation." §4.3 in *Astronomy on the Personal Computer, 4th ed.* Berlin: Springer-Verlag, pp. 62–63 and 65–68, 2000.

Siewert, C. E. and Burniston, E. E. "An Exact Analytical Solution of Kepler's Equation." *Celest. Mech.* **6**, 294–304, 1972.

Wintner, A. *The Analytic Foundations of Celestial Mechanics.* Princeton, NJ: Princeton University Press, 1941.

Kepler's Folium

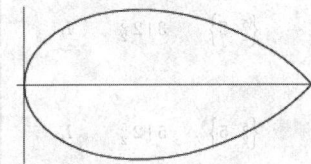 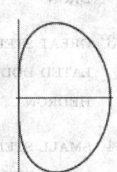

The plane curve with implicit equation

$$[(x - b)^2 + y^2][x(x - b) + y^2] = 4a(x - b)y^2.$$

References

Gray, A. *Modern Differential Geometry of Curves and Surfaces with Mathematica, 2nd ed.* Boca Raton, FL: CRC Press, p. 93, 1997.

Kepler-Poinsot Solid

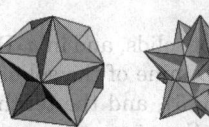   

The Kepler-Poinsot solids are the four regular CONCAVE POLYHEDRA with intersecting facial planes. They are composed of regular CONCAVE POLYGONS and were unknown to the ancients. Kepler discovered two and described them in his work *Harmonice Mundi* in 1619. These two were subsequently rediscovered by Poinsot, who also discovered the other two, in 1809. As shown by Cauchy, they are stellated forms of the DODECAHEDRON and ICOSAHEDRON.

The Kepler-Poinsot solids, illustrated above, are known as the GREAT DODECAHEDRON, GREAT ICOSAHEDRON, GREAT STELLATED DODECAHEDRON, and SMALL STELLATED DODECAHEDRON. These names probably originated with Arthur Cayley, who first used them in 1859. Cauchy (1813) proved that these four exhaust all possibilities for regular star polyhedra (Ball and Coxeter 1987).

A table listing these solids, their DUALS, and COMPOUNDS is given below. Like the five Platonic solids, duals of the Kepler-Poinsot solids are themselves Kepler-Poinsot solids (Wenninger 1983, pp. 39 and 43–45).

n	solid	UNIFORM POLYHEDRON	SCHLÄFLI SYMBOL	WYTHOFF SYMBOL	POINT GROUP
1	GREAT DODECA-HEDRON	U_{35}	$\left\{5, \frac{5}{2}\right\}$	$\frac{5}{2} \mid 25$	I_h
2	GREAT ICOSAHE-DRON	U_{53}	$\left\{3, \frac{5}{2}\right\}$	$3\,\frac{5}{2} \mid \frac{5}{3}$	I_h
3	GREAT STEL-LATED DODECA-HEDRON	U_{52}	$\left\{\frac{5}{2}, 3\right\}$	$3 \mid 2\,\frac{5}{2}$	I_h
4	SMALL STEL-LATED DODECA-HEDRON	U_{34}	$\left\{\frac{5}{2}, 5\right\}$	$5 \mid 2\,\frac{5}{2}$	I_h

The polyhedra $\left\{\frac{5}{2}, 5\right\}$ and $\left\{5, \frac{5}{2}\right\}$ fail to satisfy the POLYHEDRAL FORMULA

$$V - E + F = 2,$$

where V is the number of vertices, E the number of edges, and F the number of faces, despite the fact that the formula holds for all ordinary polyhedra (Ball and Coxeter 1987). This unexpected result led none less than Schläfli (1860) to erroneously conclude that they could not exist.

In 4-D, there are 10 Kepler-Poinsot solids, and in n-D with $n \geq 5$, there are none. In 4-D, nine of the solids have the same VERTICES as $\{3, 3, 5\}$, and the tenth has the same as $\{5, 3, 3\}$. Their SCHLÄFLI SYMBOLS are $\left\{\frac{5}{2}, 5, 3\right\}$, $\left\{3, 5, \frac{5}{2}\right\}$, $\left\{5, \frac{5}{2}, 5\right\}$, $\left\{\frac{5}{2}, 3, 5\right\}$, $\left\{5, 3, \frac{5}{2}\right\}$, $\left\{\frac{5}{2}, 5, \frac{5}{2}\right\}$, $\left\{5, \frac{5}{2}, 3\right\}$, $\left\{3, \frac{5}{2}, 5\right\}$, $\left\{\frac{5}{2}, 3, 3\right\}$, and $\left\{3, 3, \frac{5}{2}\right\}$. Coxeter *et al.* (1954) have investigated star "Archimedean" polyhedra.

See also ARCHIMEDEAN SOLID, DELTAHEDRON, JOHNSON SOLID, PLATONIC SOLID, POLYHEDRON COMPOUND, UNIFORM POLYHEDRON

References

Ball, W. W. R. and Coxeter, H. S. M. *Mathematical Recreations and Essays, 13th ed.* New York: Dover, pp. 144–146, 1987.

Cauchy, A. L. "Recherches sur les polyèdres." *J. de l'École Polytechnique* **9**, 68–86, 1813.

Cayley, A. "On Poinsot's Four New Regular Solids." *Philos. Mag.* **17**, 123–127 and 209, 1859.

Coxeter, H. S. M.; Longuet-Higgins, M. S.; and Miller, J. C. P. "Uniform Polyhedra." *Phil. Trans. Roy. Soc. London Ser. A* **246**, 401–450, 1954.

Pappas, T. "The Kepler-Poinsot Solids." *The Joy of Mathematics.* San Carlos, CA: Wide World Publ./Tetra, p. 113, 1989.

Quaisser, E. "Regular Star-Polyhedra." Ch. 5 in *Mathematical Models from the Collections of Universities and Museums* (Ed. G. Fischer). Braunschweig, Germany: Vieweg, pp. 56–62, 1986.

Schläfli. *Quart. J. Math.* **3**, 66–67, 1860.

Wells, D. *The Penguin Dictionary of Curious and Interesting Geometry.* London: Penguin, pp. 130–131, 1991.

Wenninger, M. J. *Dual Models.* Cambridge, England: Cambridge University Press, pp. 39–41, 1983.

Ker

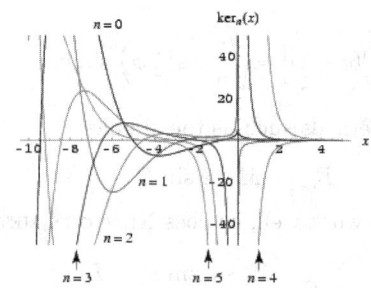

The REAL PART of

$$e^{-\nu\pi i/2} K_\nu(x e^{\pi i/4}) = \mathrm{ker}_\nu(x) + i\,\mathrm{kei}_\nu(x),$$

where $K_\nu(x)$ is a MODIFIED BESSEL FUNCTION OF THE SECOND KIND.

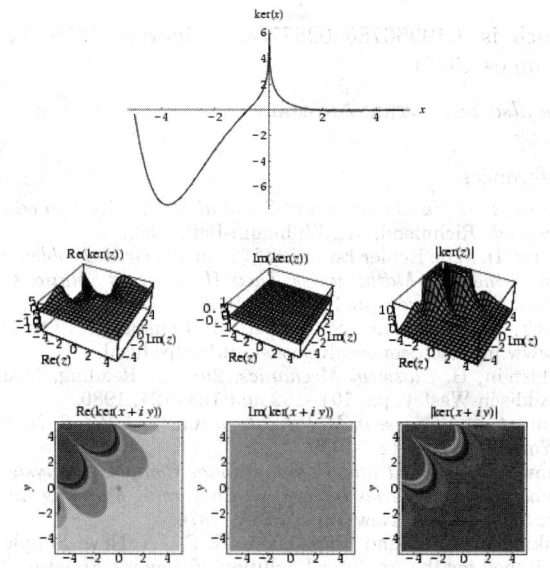

The special case $\nu = 0$ gives the plots shown above.

See also BEI, BER, KEI, KELVIN FUNCTIONS

References

Abramowitz, M. and Stegun, C. A. (Eds.). "Kelvin Functions." §9.9 in *Handbook of Mathematical Functions with Formulas, Graphs, and Mathematical Tables, 9th printing.* New York: Dover, pp. 379–381, 1972.

Prudnikov, A. P.; Marichev, O. I.; and Brychkov, Yu. A. "The Kelvin Functions $\mathrm{ber}_\nu(x)$, $\mathrm{bei}\,\nu(x)$, $\mathrm{ker}_\nu(x)$ and $\mathrm{kei}_\nu(x)$." §1.7 in *Integrals and Series, Vol. 3: More Special Functions.* Newark, NJ: Gordon and Breach, pp. 29–30, 1990.

Keratoid Cusp

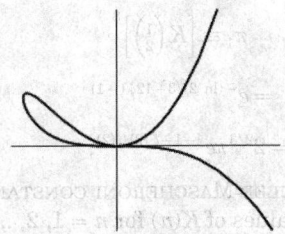

The PLANE CURVE given by the Cartesian equation

$$y^2 = x^2 y + x^5.$$

References

Cundy, H. and Rollett, A. *Mathematical Models, 3rd ed.* Stradbroke, England: Tarquin Pub., p. 72, 1989.

Kernel (Integral)

The function $K(\alpha, t)$ in an INTEGRAL or INTEGRAL TRANSFORM

$$g(\alpha) = \int_a^b f(t) K(\alpha,\, t)\, dt.$$

Whittaker and Robinson (1967, p. 376) use the term nucleus for kernel.

See also BERGMAN KERNEL, INTEGRAL, POISSON KERNEL

References

Whittaker, E. T. and Robinson, G. *The Calculus of Observations: A Treatise on Numerical Mathematics, 4th ed.* New York: Dover, p. 376, 1967.

Kernel (Linear Algebra)

NULLSPACE

Kernel Polynomial

The function

$$K_n(x_0,\, x) = \overline{K_n(x,\, x_0)} = K_n(\bar{x},\, \bar{x}_0)$$

which is useful in the study of many POLYNOMIALS.

References

Szego, G. *Orthogonal Polynomials, 4th ed.* Providence, RI: Amer. Math. Soc., 1975.

Kervaire's Characterization Theorem

Let G be a GROUP, then there exists a piecewise linear KNOT K^{n-2} in $\mathbb{S}^n$ for $n \geq 5$ with $G = \pi_1(\mathbb{S}^n - K)$ IFF G satisfies

1. G is finitely presentable,
2. The Abelianization of G is infinite cyclic,
3. The normal closure of some single element is all of G,
4. $H_2(G) = 0$; the second homology of the group is trivial.

References

Rolfsen, D. *Knots and Links.* Wilmington, DE: Publish or Perish Press, pp. 350–351, 1976.

Ket

A CONTRAVARIANT VECTOR, denoted $|\psi\rangle$. The ket is DUAL to the COVARIANT BRA one-forms $\langle\psi|$. Taken together, the BRA and ket form an ANGLE BRACKET (bra + ket = bracket) $\langle\psi|\psi\rangle$. The ket is commonly encountered in quantum mechanics.

See also ANGLE BRACKET, BRA, BRACKET PRODUCT, CONTRAVARIANT VECTOR, COVARIANT VECTOR, DIFFERENTIAL K-FORM, ONE-FORM

References

Dirac, P. A. M. "Bra and Ket Vectors." §6 in *Principles of Quantum Mechanics, 4th ed.* Oxford, England: Oxford University Press, pp. 16 and 18–22, 1982.

k-Factor

A k-factor of a GRAPH is a k-regular SUBGRAPH of order n. k-factors are a generalization of complete matchings. A PERFECT MATCHING is a 1-factor (Skiena 1990, p. 244).

See also MATCHING

References

Skiena, S. *Implementing Discrete Mathematics: Combinatorics and Graph Theory with Mathematica.* Reading, MA: Addison-Wesley, 1990.

k-Factorable Graph

A GRAPH G is k-factorable if it is the union of disjoint K-FACTORS (Skiena 1990, p. 244).

See also K-FACTOR

References

Skiena, S. *Implementing Discrete Mathematics: Combinatorics and Graph Theory with Mathematica.* Reading, MA: Addison-Wesley, 1990.

k-Form

DIFFERENTIAL K-FORM

K-Function

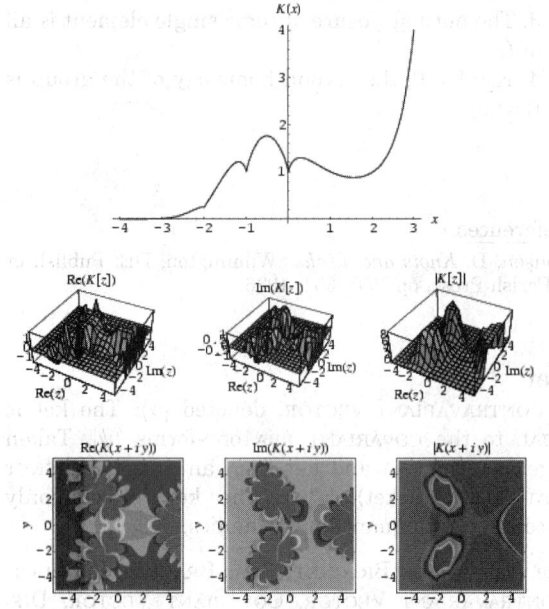

For positive integer n, the K-function is defined by

$$K(n) \equiv 0^0 1^1 2^2 3^3 \cdots (n-1)^{n-1} \tag{1}$$

and is related to the BARNES' G-FUNCTION by

$$K(n) = \frac{[\Gamma(n)]^{n-1}}{G(n)}, \tag{2}$$

where $G(n)$ is defined by

$$G(n) = \begin{cases} 1 & \text{if } n = 0 \\ 0! 1! 2! \cdots (n-2)! & \text{if } n > 0. \end{cases} \tag{3}$$

The K-function is given by the integral

$$K(z) = (2\pi)^{-(z-1)/2} \exp\left[\binom{z}{2} + \int_0^{z-1} \ln(t!) \, dt\right] \tag{4}$$

and the closed-form expression

$$K(z) = \exp[\zeta'(-1, z) - \zeta'(-1)], \tag{5}$$

where $\zeta(z)$ is the RIEMANN ZETA FUNCTION, $\zeta'(z)$ its DERIVATIVE, $\zeta(a, z)$ is the HURWITZ ZETA FUNCTION, and

$$\zeta'(a, z) \equiv \left[\frac{d\zeta(s, z)}{ds}\right]_{s=a}. \tag{6}$$

$K(z)$ also has a STIRLING-like series

$$K(z+1) = (2^{1/3} \pi_1 z)^{1/12} z^{\binom{z+1}{2}}$$

$$\times \exp\left(\tfrac{1}{4} z^2 + \tfrac{1}{12} - \frac{B_4}{2 \cdot 3 \cdot 4 z^2} - \frac{B_6}{4 \cdot 5 \cdot 6 z^4} - \cdots\right), \tag{7}$$

where

$$\pi_1 \equiv \left[K\left(\tfrac{1}{2}\right)\right]^8 \tag{8}$$

$$= e^{-(\ln 2)/3 - 12\zeta'(-1)} \tag{9}$$

$$= 2^{2/3} \pi e^{\gamma - 1 - \zeta'(2)/\zeta(2)}, \tag{10}$$

and γ is the EULER-MASCHERONI CONSTANT (Gosper). The first few values of $K(n)$ for $n = 1, 2, \ldots$ are 1, 1, 1, 4, 108, 27648, 86400000, 4031078400000, ... (Sloane's A002109). These numbers are called HYPERFACTOR-IALS by Sloane and Plouffe (1995).

See also BARNES' G-FUNCTION, GLAISHER-KINKELIN CONSTANT, HYPERFACTORIAL, STIRLING'S SERIES

References

Sloane, N. J. A. Sequences A002109/M3706 in "An On-Line Version of the Encyclopedia of Integer Sequences." http://www.research.att.com/~njas/sequences/eisonline.html.

Whittaker, E. T. and Watson, G. N. *A Course in Modern Analysis,* 4th ed. Cambridge, England: Cambridge University Press, p. 264, 1990.

K-Graph

The GRAPH obtained by dividing a set of VERTICES $\{1, \ldots, n\}$ into $k-1$ pairwise disjoint subsets with VERTICES of degree $n_1, \ldots, n_{k-1}$, satisfying

$$n = n_1 + \ldots + n_{k-1},$$

and with two VERTICES joined IFF they lie in distinct VERTEX sets. Such GRAPHS are denoted $K_{n_1, \ldots, n_k}$.

See also BIPARTITE GRAPH, COMPLETE GRAPH, COMPLETE K-PARTITE GRAPH, K-PARTITE GRAPH

Khinchin

KHINTCHINE'S CONSTANT

Khinchin Constant

KHINTCHINE'S CONSTANT

Khintchine's Constant

N.B. A detailed online essay by S. Finch was the starting point for this entry.

Let

$$x = [a_0, a_1, \ldots] = a_0 + \cfrac{1}{a_1 + \cfrac{1}{a_2 + \cfrac{1}{a_3 + \cdots}}} \tag{1}$$

be the SIMPLE CONTINUED FRACTION of a REAL NUMBER x, where the numbers a_i are the PARTIAL QUOTIENTS. Khintchine (1934) considered the limit of the GEOMETRIC MEAN

$$G_n(x) = (a_1 a_2 \cdots a_n)^{1/n} \tag{2}$$

as $n \to \infty$. Amazingly enough, this limit is a constant *independent* of x—except if x belongs to a set of MEASURE 0-given by

$$K = 2.685452001\ldots \tag{3}$$

(Sloane's A002210), as proved in Kac (1959). The constant is built into *Mathematica* 4.0 as `Khinchin`.

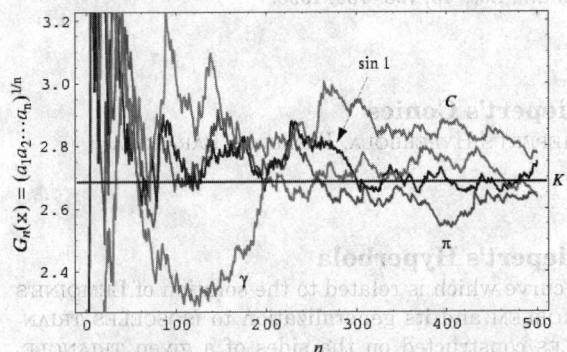

The values $G_n(x)$ are plotted above for $n = 1$ to 500 and $x = \pi$, $1/\pi$, $\sin 1$, the EULER-MASCHERONI CONSTANT γ, and the COPELAND-ERDOS CONSTANT. REAL NUMBERS x for which $\lim_{n\to\infty} G_n(x) \neq K$ include $x = e$, $\sqrt{2}$, $\sqrt{3}$, and the GOLDEN RATIO ϕ, plotted below.

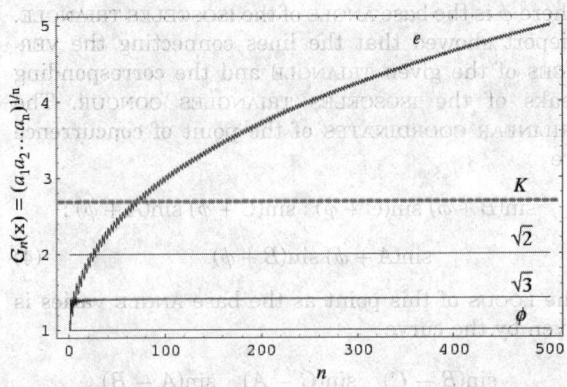

The CONTINUED FRACTION for K is $[2, 1, 2, 5, 1, 1, 2, 1, 1, \ldots]$ (Sloane's A002211; Havermann). It is not known if K is IRRATIONAL, let alone TRANSCENDENTAL. Bailey *et al.* (1995) have computed K to 7350 DIGITS.

Explicit expressions for K include

$$K = \prod_{n=1}^{\infty} \left[1 + \frac{1}{n(n+2)} \right]^{\ln n / \ln 2} \tag{4}$$

$$\ln 2 \ln K = \tfrac{1}{12} \pi^2 + \tfrac{1}{2}(\ln 2)^2 + \int_0^\pi \frac{\ln(\theta|\cot \theta|)\, d\theta}{\theta} \tag{5}$$

$$\ln K = \frac{1}{\ln 2} \sum_{m=1}^{\infty} \frac{h_{m-1}}{m} [\zeta(2m) - 1], \tag{6}$$

where $\zeta(z)$ is the RIEMANN ZETA FUNCTION and

$$h_m = \sum_{j=1}^{m} \frac{(-1)^{j-1}}{j} \tag{7}$$

(Shanks and Wrench 1959). Gosper gave

$$\ln K = \frac{1}{\ln 2} \sum_{j=2}^{\infty} \frac{(-1)^j (2 - 2^j)\zeta'(j)}{j}, \tag{8}$$

where $\zeta'(z)$ is the DERIVATIVE of the RIEMANN ZETA FUNCTION. An extremely rapidly converging sum also due to Gosper is

$$\ln K = \frac{1}{\ln 2} \sum_{k=0}^{\infty} \left\{ -\ln(k+1)[\ln(k+3) \right.$$
$$- 2\ln(k+2) + \ln(k+1)]$$
$$- \frac{(-1)^k (2 - 2^{k+2})}{k+2}$$
$$\times \left[\frac{\ln(k+1)}{(k+1)^{k+2}} - \zeta'(k+2, k+2) \right]$$
$$\left. + \ln(k+1) \left[\sum_{s=1}^{k+2} \frac{(-1)^s (2 - 2^s)}{(k+1)^s s} \right] \right\}, \tag{9}$$

where $\zeta(s, a)$ is the HURWITZ ZETA FUNCTION.

Khintchine's constant is also given by the integral

$$\ln 2 \ln\left(\tfrac{1}{2}K\right) = \int_0^1 \frac{1}{x(1+x)} \ln\left[\frac{\pi x(1 - x^2)}{\sin(\pi x)} \right] dx. \tag{10}$$

If P_n/Q_n is the nth CONVERGENT of the CONTINUED FRACTION of x, then

$$\lim_{n\to\infty}(Q_n)^{1/n} = \lim_{n\to\infty} \left(\frac{P_n}{x} \right)^{1/n} = e^{\pi^2/(12 \ln 2)} \approx 3.27582 \tag{11}$$

for almost all REAL x (Lévy 1936, Finch). This number is sometimes called the LÉVY CONSTANT, and the argument of the exponential is sometimes called the KHINTCHINE-LÉVY CONSTANT.

Define the following quantity in terms of the kth partial quotient q_k,

$$M(s, n, x) = \left(\frac{1}{n} \sum_{k=1}^{n} q_k^s \right)^{1/s}. \tag{12}$$

Then

$$\lim_{n\to\infty} M(1, n, x) = \infty \tag{13}$$

for almost all real x (Khintchine, Knuth 1981, Finch), and

$$M(1, n, x) \sim \mathcal{O}(\ln n). \tag{14}$$

Furthermore, for $s < 1$, the limiting value

$$\lim_{n \to \infty} M(s, n, x) = K(s) \qquad (15)$$

exists and is a constant $K(s)$ with probability 1 (Rockett and Szüsz 1992, Khintchine 1997).

See also CONTINUED FRACTION, CONVERGENT, KHINTCHINE-LÉVY CONSTANT, LÉVY CONSTANT, PARTIAL QUOTIENT, SIMPLE CONTINUED FRACTION

References

Bailey, D. H.; Borwein, J. M.; and Crandall, R. E. "On the Khintchine Constant." *Math. Comput.* **66**, 417–431, 1997.

Finch, S. "Favorite Mathematical Constants." http://www.mathsoft.com/asolve/constant/khntchn/khntchn.html.

Havermann, H. "Simple Continued Fraction Expansion of Khinchin's Constant." http://members.home.net/hahaj/cfk.html.

Kac, M. *Statistical Independence and Probability, Analysts and Number Theory.* Providence, RI: Math. Assoc. Amer., 1959.

Khinchin, A. Ya. *Continued Fractions.* New York: Dover, 1997.

Knuth, D. E. Exercise 24 in *The Art of Computer Programming, Vol. 2: Seminumerical Algorithms, 3rd ed.* Reading, MA: Addison-Wesley, p. 604, 1998.

Le Lionnais, F. *Les nombres remarquables.* Paris: Hermann, p. 46, 1983.

Lehmer, D. H. "Note on an Absolute Constant of Khintchine." *Amer. Math. Monthly* **46**, 148–152, 1939.

Phillipp, W. "Some Metrical Theorems in Number Theory." *Pacific J. Math.* **20**, 109–127, 1967.

Plouffe, S. "Plouffe's Inverter: Table of Current Records for the Computation of Constants." http://www.lacim.u-qam.ca/pi/records.html.

Rockett, A. M. and Szüsz, P. *Continued Fractions.* Singapore: World Scientific, 1992.

Shanks, D. and Wrench, J. W. "Khintchine's Constant." *Amer. Math. Monthly* **66**, 148–152, 1959.

Sloane, N. J. A. Sequences A002210/M1564 and A002211/M0118 in "An On-Line Version of the Encyclopedia of Integer Sequences." http://www.research.att.com/~njas/sequences/eisonline.html.

Vardi, I. "Khinchin's Constant." §8.4 in *Computational Recreations in Mathematica.* Reading, MA: Addison-Wesley, pp. 163–171, 1991.

Wolfram, S. *The Mathematica Book, 4th ed.* Cambridge, England: Cambridge University Press, pp. 756–757, 1999.

Wrench, J. W. "Further Evaluation of Khintchine's Constant." *Math. Comput.* **14**, 370–371, 1960.

Khintchine-Lévy Constant

A constant related to KHINTCHINE'S CONSTANT and defined by

$$KL \equiv \frac{\pi^2}{12 \ln 2} = 1.1865691104\ldots.$$

See also KHINTCHINE'S CONSTANT, LÉVY CONSTANT

References

Plouffe, S. "Khintchine-Levy Constant." http://www.lacim.u-qam.ca/piDATA/klevy.txt.

Khovanski's Theorem

If $f_1, \ldots, f_m : \mathbb{R}^n \to \mathbb{R}$ are exponential polynomials, then $\{x \in \mathbb{R}^n : f_1(x) = \cdots f_n(x) = 0\}$ has finitely many connected components.

References

Marker, D. "Model Theory and Exponentiation." *Not. Amer. Math. Soc.* **43**, 753–759, 1996.

Kiepert's Conics

KIEPERT'S HYPERBOLA, KIEPERT'S PARABOLA

Kiepert's Hyperbola

A curve which is related to the solution of LEMOINE'S PROBLEM and its generalization to ISOSCELES TRIANGLES constructed on the sides of a given TRIANGLE. The VERTICES of the constructed TRIANGLES are

$$A' = -\sin \phi : \sin(C + \phi) : \sin(B + \phi) \qquad (1)$$

$$B' = \sin(C + \phi) : -\sin \phi : \sin(A + \phi) \qquad (2)$$

$$C' = \sin(B + \phi) : \sin(A + \phi) : -\sin \phi, \qquad (3)$$

where ϕ is the base ANGLE of the ISOSCELES TRIANGLE. Kiepert showed that the lines connecting the VERTICES of the given TRIANGLE and the corresponding peaks of the ISOSCELES TRIANGLES CONCUR. The TRILINEAR COORDINATES of the point of concurrence are

$$\sin(B + \phi) \sin(C + \phi) : \sin(C + \phi) \sin(A + \phi) :$$

$$\sin(A + \phi) \sin(B + \phi). \qquad (4)$$

The LOCUS of this point as the base ANGLE varies is given by the curve

$$\frac{\sin(B - C)}{\alpha} + \frac{\sin(C - A)}{\beta} + \frac{\sin(A - B)}{\gamma}$$

$$= \frac{bc(b^2 - c^2)}{\alpha} + \frac{ca(c^2 - a^2)}{\beta} + \frac{ab(a^2 - b^2)}{\gamma} = 0. \qquad (5)$$

Writing the TRILINEAR COORDINATES as

$$\alpha_i = d_i s_i, \qquad (6)$$

where d_i is the distance to the side opposite α_i of length s_i and using the POINT-LINE DISTANCE FORMULA with (x_0, y_0) written as (x, y),

$$d_i = \frac{|(y_{i+2} - y_{i+1})(x - x_{i+1})}{s_i}$$

$$- \frac{(x_{i+2} - x_{i+1})(y - y_{i+1})|}{s_i}, \qquad (7)$$

where $y_4 \equiv y_1$ and $y_5 \equiv y_2$ gives the FORMULA

$$\sum_{i=1}^{3} s_{i+1}s_{i+2}(s_{i+1}^2 - s_{i+2}^2)$$

$$\times \frac{s_i}{(y_{i+2} - y_{i+1})(x - x_{i+1}) - (x_{i+2} - x_{i+1})(y - y_{i+1})} = 0 \quad (8)$$

$$\sum_{i=1}^{3} \frac{(s_{i+1}^2 - s_{i+2}^2)}{(y_{i+2} - y_{i+1})(x - x_{i+1}) - (x_{i+2} - x_{i+1})(y - y_{i+1})}$$
$$= 0. \quad (9)$$

Bringing this equation over a common DENOMINATOR then gives a quadratic in x and y, which is a CONIC SECTION (in fact, a HYPERBOLA). The curve can also be written as $\csc(A + t) : \csc(B + t) : \csc(C + t)$, as t varies over $[-\pi/4, \pi/4]$.

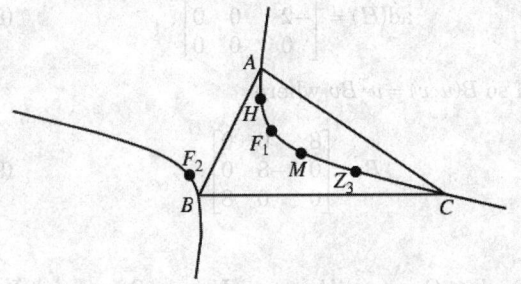

Kiepert's hyperbola passes through the triangle's CENTROID M ($\phi = 0$), ORTHOCENTER H ($\phi = \pi/2$), VERTICES A ($\phi = -\alpha$ if $\alpha \le \pi/2$ and $\phi = \pi - \alpha$ if $\alpha > \pi/2$), B ($\phi = -\beta$), C ($\phi = -\gamma$), FERMAT POINTS F_1 ($\phi = \pi/3$) and F_2 ($\phi = -\pi/3$), ISOGONAL CONJUGATE of the BROCARD MIDPOINT ($\phi = \omega$), and BROCARD'S THIRD POINT Z_3 ($\phi = \omega$), where ω is the BROCARD ANGLE (Eddy and Fritsch 1994, p. 193).

The ASYMPTOTES of Kiepert's hyperbola are the SIMSON LINES of the intersections of the BROCARD AXIS with the CIRCUMCIRCLE. Kiepert's hyperbola is a RECTANGULAR HYPERBOLA. In fact, all nondegenerate conics through the VERTICES and ORTHOCENTER of a TRIANGLE are RECTANGULAR HYPERBOLAS the centers of which lie halfway between the FERMAT POINTS and on the NINE-POINT CIRCLE. The LOCUS of centers of these HYPERBOLAS is the NINE-POINT CIRCLE.

The ISOGONAL CONJUGATE curve of Kiepert's hyperbola is the BROCARD AXIS. The center of the INCIRCLE of the TRIANGLE constructed from the MIDPOINTS of the sides of a given TRIANGLE lies on Kiepert's hyperbola of the original TRIANGLE.

See also BROCARD ANGLE, BROCARD AXIS, BROCARD POINTS, CENTROID (TRIANGLE), CIRCUMCIRCLE, FERMAT POINTS, ISOGONAL CONJUGATE, ISOSCELES TRIANGLE, KIEPERT'S PARABOLA, LEMOINE'S PROBLEM, NINE-POINT CIRCLE, ORTHOCENTER, SIMSON LINE

References

Casey, J. *A Treatise on the Analytical Geometry of the Point, Line, Circle, and Conic Sections, Containing an Account of Its Most Recent Extensions with Numerous Examples,* 2nd rev. enl. ed. Dublin: Hodges, Figgis, & Co., 1893.
Eddy, R. H. and Fritsch, R. "The Conics of Ludwig Kiepert: A Comprehensive Lesson in the Geometry of the Triangle." *Math. Mag.* **67**, 188–205, 1994.
Kelly, P. J. and Merriell, D. "Concentric Polygons." *Amer. Math. Monthly* **71**, 37–41, 1964.
Mineuer, A. "Sur les asymptotes de l'hyperbole de Kiepert." *Mathesis* **49**, 30–33, 1935.
Rigby, J. F. "A Concentrated Dose of Old-Fashioned Geometry." *Math. Gaz.* **57**, 296–298, 1953.
Vandeghen, A. "Some Remarks on the Isogonal and Cevian Transforms. Alignments of Remarkable Points of a Triangle." *Amer. Math. Monthly* **72**, 1091–1094, 1965.

Kiepert's Parabola

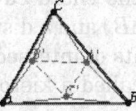

Let three similar ISOSCELES TRIANGLES $\triangle A'BC$, $\triangle AB'C$, and $\triangle ABC'$ be constructed on the sides of a TRIANGLE $\triangle ABC$. Then $\triangle ABC$ and $\triangle A'B'C''$ are PERSPECTIVE TRIANGLES, and the ENVELOPE of their PERSPECTIVE AXIS as the vertex angle of the erected triangles is varied is a PARABOLA known as Kiepert's parabola. It has equation

$$\frac{\sin A(\sin^2 B - \sin^2 C)}{u} + \frac{\sin B(\sin^2 C - \sin^2 A)}{v}$$
$$+ \frac{\sin C(\sin^2 A - \sin^2 B)}{w} = 0 \quad (1)$$

$$\frac{a(b^2 - c^2)}{u} + \frac{b(c^2 - a^2)}{v} + \frac{c(a^2 - b^2)}{w} = 0, \quad (2)$$

where $[u, v, w]$ are the TRILINEAR COORDINATES for a line tangent to the parabola.

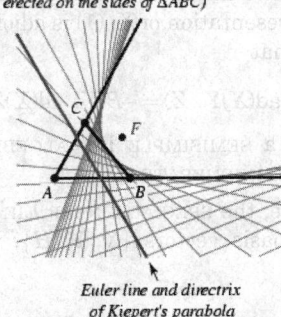

perspective axes of $\triangle ABC$ and $\triangle A'B'C'$ (constructed from isosceles triangles erected on the sides of $\triangle ABC$)

Euler line and directrix of Kiepert's parabola

Kiepert's parabola is tangent to the sides of the TRIANGLE (or their extensions), the line at infinity, and the LEMOINE LINE. The FOCUS has TRIANGLE CENTER FUNCTION

$$\alpha = \csc(B - C). \qquad (3)$$

The EULER LINE of a triangle is the DIRECTRIX of Kiepert's parabola. In fact, the DIRECTRICES of all parabolas inscribed in a TRIANGLE pass through the ORTHOCENTER. The BRIANCHON POINT for Kiepert's parabola is the STEINER POINT of $\triangle ABC$.

See also BRIANCHON POINT, ENVELOPE, EULER LINE, ISOSCELES TRIANGLE, LEMOINE LINE, PARABOLA, STEINER POINTS

Kieroid

Let the center B of a CIRCLE of RADIUS a move along a line BA. Let O be a fixed point located a distance c away from AB. Draw a SECANT LINE through O and D, the MIDPOINT of the chord cut from the line DE (which is parallel to AB) and a distance b away. Then the LOCUS of the points of intersection of OD and the CIRCLE P_1 and P_2 is called a kieroid.

Special Case	Curve
$b = 0$	CONCHOID OF NICOMEDES
$b = a$	CISSOID plus asymptote
$b = a = -c$	STROPHOID plus ASYMPTOTE

References
Yates, R. C. "Kieroid." *A Handbook on Curves and Their Properties.* Ann Arbor, MI: J. W. Edwards, pp. 141–142, 1952.

Killing Form

The Killing form is an INNER PRODUCT on a finite dimensional LIE ALGEBRA $\mathfrak{g}$ defined by

$$B(X, Y) = \mathrm{Tr}(\mathrm{ad}(X)\,\mathrm{ad}(Y)) \qquad (1)$$

in the ADJOINT REPRESENTATION, where $\mathrm{ad}(X)$ is the adjoint representation of X. (1) is adjoint-invariant in the sense that

$$B(\mathrm{ad}(X)Y, Z) = -B(Y, \mathrm{ad}(X)Z). \qquad (2)$$

When $\mathfrak{g}$ is a SEMISIMPLE LIE ALGEBRA, the Killing form is NONDEGENERATE.

For example, the SPECIAL LINEAR LIE ALGEBRA $\mathfrak{sl}_2(\mathbb{C})$ has three basis vectors $\{X, Y, H\}$, where $[X, Y] = 2H$:

$$X = \begin{bmatrix} 0 & 1 \\ 1 & 0 \end{bmatrix} \qquad (3)$$

$$Y = \begin{bmatrix} 0 & -1 \\ 1 & 0 \end{bmatrix} \qquad (4)$$

$$H = \begin{bmatrix} 1 & 0 \\ 0 & -1 \end{bmatrix}. \qquad (5)$$

The other brackets are given by $[X, H] = 2Y$ and $[Y, H] = 2X$. In the adjoint representation, with the ordered basis $\{X, Y, H\}$, these elements are represented by

$$\mathrm{ad}(X) = \begin{bmatrix} 0 & 0 & 0 \\ 0 & 0 & 2 \\ 0 & 2 & 0 \end{bmatrix} \qquad (6)$$

$$\mathrm{ad}(Y) = \begin{bmatrix} 0 & 0 & 2 \\ 0 & 0 & 0 \\ -2 & 0 & 0 \end{bmatrix} \qquad (7)$$

$$\mathrm{ad}(H) = \begin{bmatrix} 0 & -2 & 0 \\ -2 & 0 & 0 \\ 0 & 0 & 0 \end{bmatrix}, \qquad (8)$$

and so $B(u, v) = u^{\mathrm{T}} B v$ where

$$B = \begin{bmatrix} 8 & 0 & 0 \\ 0 & -8 & 0 \\ 0 & 0 & 8 \end{bmatrix}. \qquad (9)$$

See also CARTAN MATRIX, INNER PRODUCT, LIE ALGEBRA, SEMISIMPLE LIE ALGEBRA, SIGNATURE (MATRIX), SPECIAL LINEAR LIE ALGEBRA, WEYL GROUP

References
Fulton, W. and Harris, J. *Representation Theory.* New York: Springer-Verlag, 1991.

Huang, J.-S. "The Killing Form." §4.4 in *Lectures on Representation Theory.* Singapore: World Scientific, pp. 33–36, 1999.

Jacobson, N. *Lie Algebras.* New York: Dover, 1979.

Knapp, A. *Lie Groups Beyond an Introduction.* Boston, MA: Birkhäuser, 1996.

Killing Vectors

If any set of points is displaced by $X^i\,dx_i$ where all distance relationships are unchanged (i.e., there is an ISOMETRY), then the VECTOR FIELD is called a Killing vector.

$$g_{ab} = \frac{\partial x'^c}{\partial x^a} \frac{\partial x'^d}{\partial x^b} g_{cd}(x'), \qquad (1)$$

so let

$$x'^a = x^a + \epsilon x^a \qquad (2)$$

$$\frac{\partial x'^a}{\partial x^b} = \delta^a_b + \epsilon x^a{}_{,b} \qquad (3)$$

$$g_{ab}(x) = (\delta^c_a + \epsilon x^c{}_{,a})(\delta^d_b + \epsilon x^d{}_{,b})g_{cd}(x^e + \epsilon X^e)$$

$$= (\delta^c_a + \epsilon x^c{}_{,a})(\delta^d_b + \epsilon x^d{}_{,b})[g_{cd}(x) + \epsilon X^e g_{cd}(x)_{,e} + \dots]$$

$$= g_{ab}(x) + \epsilon[g_{ad}X^d_{,b} + g_{bd}X^d_{,a} + X^e g_{ab,e}] + \mathcal{O}(\epsilon^2)$$

$$= g_{ab} + \mathcal{L}_X g_{ab}$$

$$= g'_{ab}, \tag{4}$$

where $\mathcal{L}$ is the LIE DERIVATIVE.

An ordinary derivative can be replaced with a COVARIANT DERIVATIVE in a LIE DERIVATIVE, so we can take as the definition

$$g_{ab; c} = 0 \tag{5}$$

$$g_{ab}g^{bc} = \delta^c_a, \tag{6}$$

which gives KILLING'S EQUATION

$$\mathcal{L}_X g_{ab} = X_{a; b} + X_{b; a} = 2X_{(a; b)} = 0, \tag{7}$$

where $X_{(a; b)}$ denotes the SYMMETRIC TENSOR part and $X_{a; b}$ is a COVARIANT DERIVATIVE.

A Killing vector X^b satisfies

$$g^{bc}X_{c; ab} - R_{ab}X^b = 0 \tag{8}$$

$$X_{a; bc} = R_{abcd}X^d \tag{9}$$

$$X^{a; b}{}_{;b} + R^a_c X^c = 0, \tag{10}$$

where R_{ab} is the RICCI TENSOR and R_{abcd} is the RIEMANN TENSOR.

A 2-sphere with METRIC

$$ds^2 = d\theta^2 + \sin^2\theta \, d\phi^2 \tag{11}$$

has three Killing vectors, given by the angular momentum operators

$$\tilde{L}_x = -\cos\phi \frac{\partial}{\partial\theta} + \cot\theta \sin\phi \frac{\partial}{\partial\phi} \tag{12}$$

$$\tilde{L}_y = \sin\phi \frac{\partial}{\partial\theta} + \cot\theta \cos\phi \frac{\partial}{\partial\phi} \tag{13}$$

$$\tilde{L}_z = \frac{\partial}{\partial\phi}. \tag{14}$$

The Killing vectors in Euclidean 3-space are

$$x^1 = \frac{\partial}{\partial x} \tag{15}$$

$$x^2 = \frac{\partial}{\partial y} \tag{16}$$

$$x^3 = \frac{\partial}{\partial z} \tag{17}$$

$$x^4 = y\frac{\partial}{\partial z} - z\frac{\partial}{\partial y} \tag{18}$$

$$x^5 = z\frac{\partial}{\partial x} - x\frac{\partial}{\partial z} \tag{19}$$

$$x^6 = x\frac{\partial}{\partial y} - y\frac{\partial}{\partial x}. \tag{20}$$

In MINKOWSKI SPACE, there are 10 Killing vectors

$$X^\mu_i = a^\mu_i \quad \text{for } i = 1, 2, 3, 4 \tag{21}$$

$$X^0_k = 0 \tag{22}$$

$$X^l_k = \epsilon^{lkm}x_m \quad \text{for } k = 1, 2, 3 \tag{23}$$

$$X^k_\mu = \delta^{[0, k]}_\mu \quad \text{for } k = 1, 2, 3. \tag{24}$$

The first group is TRANSLATION, the second ROTATION, and the final corresponds to a "boost."

See also KILLING'S EQUATION, LIE DERIVATIVE

Killing's Equation

The equation defining KILLING VECTORS.

$$\mathcal{L}_X g_{ab} = X_{a; b} + X_{b; a} = 2X_{(a; b)} = 0,$$

where $\mathcal{L}$ is the LIE DERIVATIVE and $X_{b; a}$ is a COVARIANT DERIVATIVE.

See also KILLING VECTORS, LIE DERIVATIVE

References
Schafer, R. D. *An Introduction to Nonassociative Algebras.* New York: Dover, pp. 23–26, 1996.

Kilroy Curve

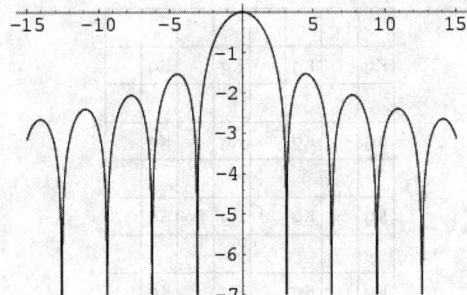

The curve defined by the Cartesian equation

$$f(x) = \ln\left|\frac{\sin x}{x}\right| = \ln|\text{sinc } x|.$$

The Kilroy curve arises in the study of spread spectra plotted on a logarithmic (decibel) scale, and is so named because it resembles Kilroy looking over a wall.

See also SINC FUNCTION

Kimberling Sequence

A sequence generated by beginning with the POSITIVE INTEGERS, then iteratively applying the following algorithm:

1. In iteration i, discard the ith element,
2. Alternately write the $i+k$ and $i-k$th elements until $k=i$,
3. Write the remaining elements in order.

The first few iterations are therefore

$$\begin{array}{ccccccccccc}
\boxed{1} & 2 & 3 & 4 & 5 & 6 & 7 & 8 & 9 & 10 & 11 \\
2 & \boxed{3} & 4 & 5 & 6 & 7 & 8 & 9 & 10 & 11 & 12 \\
4 & 2 & \boxed{5} & 6 & 7 & 8 & 9 & 10 & 11 & 12 & 13 \\
6 & 2 & 7 & \boxed{4} & 8 & 9 & 10 & 11 & 12 & 13 & 14 \\
8 & 7 & 9 & 2 & \boxed{10} & 6 & 11 & 12 & 13 & 14 & 15
\end{array}$$

The diagonal elements form the sequence 1, 3, 5, 4, 10, 7, 15, ... (Sloane's A007063).

See also PERFECT SHUFFLE, SHUFFLE

References
Guy, R. K. "The Kimberling Shuffle." §E35 in *Unsolved Problems in Number Theory, 2nd ed.* New York: Springer-Verlag, pp. 235–236, 1994.
Kimberling, C. "Problem 1615." *Crux Math.* **17**, 44, 1991.
Sloane, N. J. A. Sequences A007063/M2387 in "An On-Line Version of the Encyclopedia of Integer Sequences." http://www.research.att.com/~njas/sequences/eisonline.html.

Kimberling Shuffle
KIMBERLING SEQUENCE

King Walk
DELANNOY NUMBER

Kings Problem

The problem of determining how many nonattacking kings can be placed on an $n \times n$ CHESSBOARD. For $n = 8$, the solution is 16, as illustrated above (Madachy 1979). In general, the solutions are

$$K(n) = \begin{cases} \frac{1}{4}n^2 & n \text{ even} \\ \frac{1}{4}(n+1)^2 & n \text{ odd} \end{cases} \quad (1)$$

(Madachy 1979), giving the sequence of doubled squares 1, 1, 4, 4, 9, 9, 16, 16, ... (Sloane's A008794).

This sequence has GENERATING FUNCTION

$$\frac{1+x^2}{(1-x^2)^2(1-x)}$$
$$= 1 + x + 4x^2 + 4x^3 + 9x^4 + 9x^5 + \dots \quad (2)$$

The minimum number of kings needed to attack or occupy all squares on an 8×8 CHESSBOARD is nine, illustrated above (Madachy 1979).

See also BISHOPS PROBLEM, CHESS, HARD HEXAGON ENTROPY CONSTANT, KNIGHTS PROBLEM, QUEENS PROBLEM, ROOKS PROBLEM

References
Madachy, J. S. *Madachy's Mathematical Recreations.* New York: Dover, p. 39, 1979.
Sloane, N. J. A. Sequences A008794 in "An On-Line Version of the Encyclopedia of Integer Sequences." http://www.research.att.com/~njas/sequences/eisonline.html.

Kinney's Set

A set of plane MEASURE 0 that contains a CIRCLE of every RADIUS.

References
Falconer, K. J. *The Geometry of Fractal Sets.* New York: Cambridge University Press, 1985.
Fejzic, H. "On Thin Sets of Circles." *Amer. Math. Monthly* **103**, 582–585, 1996.
Kinney, J. R. "A Thin Set of Circles." *Amer. Math. Monthly* **75**, 1077–1081, 1968.

Kinoshita-Terasaka Knot

The KNOT with BRAID WORD

$$\sigma_1^3 \sigma_3^2 \sigma_2 \sigma_3^{-1} \sigma_1^{-2} \sigma_2 \sigma_1^{-1} \sigma_3^{-1} \sigma_2^{-1}.$$

Its JONES POLYNOMIAL is

$$t^{-4}(-1 + 2t - 2t^2 + 2t^3 + t^6 - 2t^7 + 2t^8 - 2t^9 + t^{10}),$$

the *same* as for CONWAY'S KNOT. It has the same ALEXANDER POLYNOMIAL as the UNKNOT.

See also CONWAY'S KNOT, KNOT, UNKNOT

References
Kinoshita, S. and Terasaka, H. "On Unions of Knots." *Osaka Math. J.* **9**, 131–153, 1959.

Kinoshita-Terasaka Mutants

 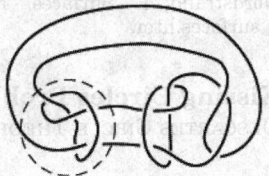

References

Adams, C. C. *The Knot Book: An Elementary Introduction to the Mathematical Theory of Knots.* New York: W. H. Freeman, pp. 49–50, 1994.

Kirby Calculus

The manipulation of DEHN SURGERY descriptions by a certain set of operations.

See also DEHN SURGERY

References

Adams, C. C. *The Knot Book: An Elementary Introduction to the Mathematical Theory of Knots.* New York: W. H. Freeman, p. 263, 1994.

Kirby's List

A list of problems in low-dimensional TOPOLOGY maintained by R. C. Kirby. The list currently runs about 380 pages.

References

Kirby, R. "Problems in Low-Dimensional Topology." http://www.math.berkeley.edu/~kirby/.

Kirkman Points

The 60 PASCAL LINES of a HEXAGON inscribed in a conic intersect three at a time through 20 STEINER POINTS, and also three at a time in 60 points known as Kirkman points. Each STEINER POINT lines together with three Kirkman points on a total of 20 lines known as CAYLEY LINES. There is a dual relationship between the 60 Kirkman points and the 60 PASCAL LINES.

See also CAYLEY LINES, PASCAL LINES, PASCAL'S THEOREM, PLÜCKER LINES, SALMON POINTS, STEINER POINTS

References

Johnson, R. A. *Modern Geometry: An Elementary Treatise on the Geometry of the Triangle and the Circle.* Boston, MA: Houghton Mifflin, pp. 236–237, 1929.
Kirkman, T. P. *Cambridge Dublin Math. J.* **5**, 185.
Lachlan, R. *An Elementary Treatise on Modern Pure Geometry.* London: Macmillian, p. 116, 1893.
Salmon, G. "Notes: Pascal's Theorem, Art. 267" in *A Treatise on Conic Sections, 6th ed.* New York: Chelsea, pp. 379–382, 1960.
Wells, D. *The Penguin Dictionary of Curious and Interesting Geometry.* London: Penguin, p. 172, 1991.

Kirkman Triple System

A Kirkman triple system of order $v = 6n + 3$ is a STEINER TRIPLE SYSTEM with parallelism (Ball and Coxeter 1987), i.e., one with the following additional stipulation: the set of $b = (2n + 1)(3n + 1)$ triples is partitioned into $(3n + 1)$ components such that each component is a $(2n + 1)$-subset of triples and each of the v elements appears exactly once in each component. The STEINER TRIPLE SYSTEMS of order 3 and 9 are Kirkman triple systems with $n = 0$ and 1. Solution to KIRKMAN'S SCHOOLGIRL PROBLEM requires construction of a Kirkman triple system of order $n = 2$.

Ray-Chaudhuri and Wilson (1971) showed that there exists at least one Kirkman triple system for every NONNEGATIVE order n. Earlier editions of Ball and Coxeter (1987) gave constructions of Kirkman triple systems with $9 \le v < 99$. For $n = 1$, there is a single unique (up to an isomorphism) solution, while there are 7 different systems for $n = 2$ (Mulder 1917, Cole 1922, Ball and Coxeter 1987).

See also STEINER TRIPLE SYSTEM

References

Abel, R. J. R. and Furino, S. C. "Kirkman Triple Systems." §I.6.3 in *The CRC Handbook of Combinatorial Designs* (Ed. C. J. Colbourn and J. H. Dinitz). Boca Raton, FL: CRC Press, pp. 88–89, 1996.
Ball, W. W. R. and Coxeter, H. S. M. *Mathematical Recreations and Essays, 13th ed.* New York: Dover, pp. 287–289, 1987.
Kirkman, T. P. "On a Problem in Combinations." *Cambridge and Dublin Math. J.* **2**, 191–204, 1847.
Lindner, C. C. and Rodger, C. A. *Design Theory.* Boca Raton, FL: CRC Press, 1997.
Mulder, P. *Kirkman-Systemen.* Groningen Dissertation. Leiden, Netherlands, 1917.
Ray-Chaudhuri, D. K. and Wilson, R. M. "Solution of Kirkman's Schoolgirl Problem." *Combinatorics, Proc. Sympos. Pure Math., Univ. California, Los Angeles, Calif., 1968* **19**, 187–203, 1971.
Ryser, H. J. *Combinatorial Mathematics.* Buffalo, NY: Math. Assoc. Amer., pp. 101–102, 1963.

Kirkman's Schoolgirl Problem

In a boarding school there are fifteen schoolgirls who always take their daily walks in rows of threes. How can it be arranged so that each schoolgirl walks in the same row with every other schoolgirl exactly once a week? Solution of this problem is equivalent to constructing a KIRKMAN TRIPLE SYSTEM of order $n = 2$. The following table gives one of the 7 distinct (up to permutations of letters) solutions to the problem.

Sun	Mon	Tue	Wed	Thu	Fri	Sat
ABC	ADE	AFG	AHI	AJK	ALM	ANO
DHL	BIK	BHJ	BEG	CDF	BEF	BDG

EJN	CMO	CLN	CMN	BLO	CIJ	CHK
FIO	FHN	DIM	DJO	EHM	DKN	EIL
GKM	GJL	EKO	FKL	GIN	GHO	FJM

(The table of Dörrie 1965 contains four omissions in which the $a_1 = B$ and $a_2 = C$ entries for Wednesday and Thursday are written simply as a.)

See also JOSEPHUS PROBLEM, KIRKMAN TRIPLE SYSTEM, STEINER TRIPLE SYSTEM

References

Abel, R. J. R. and Furino, S. C. "Kirkman Triple Systems." §I.6.3 in *The CRC Handbook of Combinatorial Designs* (Ed. C. J. Colbourn and J. H. Dinitz). Boca Raton, FL: CRC Press, pp. 88–89, 1996.

Ball, W. W. R. and Coxeter, H. S. M. *Mathematical Recreations and Essays, 13th ed.* New York: Dover, pp. 287–289, 1987.

Carpmael. *Proc. London Math. Soc.* **12**, 148–156, 1881.

Cole, F. N. "Kirkman Parades." *Bull. Amer. Math. Soc.* **28**, 435–437, 1922.

Dörrie, H. §5 in *100 Great Problems of Elementary Mathematics: Their History and Solutions.* New York: Dover, pp. 14–18, 1965.

Frost, A. "General Solution and Extension of the Problem of the 15 School Girls." *Quart. J. Pure Appl. Math.* **11**, 26–37, 1871.

Kirkman, T. P. "On a Problem in Combinatorics." *Cambridge and Dublin Math. J.* **2**, 191–204, 1847.

Kirkman, T. P. *Lady's and Gentleman's Diary*. 1850.

Kraitchik, M. §9.3.1 in *Mathematical Recreations.* New York: W. W. Norton, pp. 226–227, 1942.

Peirce, B. "Cyclic Solutions of the School-Girl Puzzle." *Astron. J.* **6**, 169–174, 1859–1861.

Ryser, H. J. *Combinatorial Mathematics.* Buffalo, NY: Math. Assoc. Amer., pp. 101–102, 1963.

Woolhouse. *Lady's and Gentleman's Diary*. 1862–1863.

Kiss Surface

The QUINTIC SURFACE given by the equation

$$\tfrac{1}{2}x^5 + \tfrac{1}{2}x^4 - (y^2 + z^2) = 0.$$

See also QUINTIC SURFACE

References

Nordstrand, T. "Surfaces." http://www.uib.no/people/nfytn/surfaces.htm.

Kissing Circles Problem

DESCARTES CIRCLE THEOREM, SODDY CIRCLES

Kissing Number

The number of equivalent HYPERSPHERES in n-D which can touch an equivalent HYPERSPHERE without any intersections, also sometimes called the NEWTON NUMBER, CONTACT NUMBER, COORDINATION NUMBER, or LIGANCY. Newton correctly believed that the kissing number in 3-D was 12, but the first proofs were not produced until the 19th century (Conway and Sloane 1993, p. 21) by Bender (1874), Hoppe (1874), and Günther (1875). More concise proofs were published by Schütte and van der Waerden (1953) and Leech (1956). After packing 12 spheres around the central one (which can be done, for example, by arranging the spheres so that their points of tangency with the central sphere correspond to the vertices of an ICOSAHEDRON), there is a significant amount of free space left (above figure), although not enough to fit a 13th sphere.

Exact values for *lattice packings* are known for $n = 1$ to 9 and $n = 24$ (Conway and Sloane 1992, Sloane and Nebe). Odlyzko and Sloane (1979) found the exact value for 24-D.

The arrangement of n points on the surface of a sphere, corresponding to the placement of n identical spheres around a central sphere (not necessarily of the same radius) is called a SPHERICAL PACKING.

The following table gives the largest known kissing numbers in DIMENSION D for lattice (L) and non-lattice (NL) packings (if a nonlattice packing with higher number exists). In nonlattice packings, the kissing number may vary from sphere to sphere, so the largest value is given below (Conway and Sloane 1993, p. 15). A more extensive and up-to-date tabulation is maintained by Sloane and Nebe.

D	L	NL	D	L	NL
1	2		13	≥ 918	$\geq 1{,}130$

2	6	14	$\geq 1,422$	$\geq 1,582$
3	12	15	$\geq 2,340$	
4	24	16	$\geq 4,320$	
5	40	17	$\geq 5,346$	
6	72	18	$\geq 7,398$	
7	126	19	$\geq 10,668$	
8	240	20	$\geq 17,400$	
9	272	≥ 306	21	$\geq 27,720$
10	≥ 336	≥ 500	22	$\geq 49,896$
11	≥ 438	≥ 582	23	$\geq 93,150$
12	≥ 756	≥ 840	24	$196,560$

The lattices having maximal packing numbers in 12- and 24-D have special names: the COXETER-TODD LATTICE and LEECH LATTICE, respectively. The general form of the lower bound of n-D lattice densities given by

$$\eta \geq \frac{\zeta(n)}{2^{n-1}},$$

where $\zeta(n)$ is the RIEMANN ZETA FUNCTION, is known as the MINKOWSKI-HLAWKA THEOREM.

See also COXETER-TODD LATTICE, HERMITE CONSTANTS, HYPERSPHERE PACKING, KEPLER CONJECTURE, LEECH LATTICE, MINKOWSKI-HLAWKA THEOREM, SPHERE PACKING

References
Bender, C. "Bestimmung der grössten Anzahl gleich Kugeln, welche sich auf eine Kugel von demselben Radius, wie die übrigen, auflegen lassen." *Archiv Math. Physik (Grunert)* **56**, 302–306, 1874.
Conway, J. H. and Sloane, N. J. A. "The Kissing Number Problem" and "Bounds on Kissing Numbers." §1.2 and Ch. 13 in *Sphere Packings, Lattices, and Groups, 2nd ed.* New York: Springer-Verlag, pp. 21–24 and 337–339, 1993.
Edel, Y.; Rains, E. M.; Sloane, N. J. A. "On Kissing Numbers in Dimensions 32 to 128." *Electronic J. Combinatorics* **5**, No. 1, R22, 1–5, 1998. http://www.combinatorics.org/Volume_5/v5i1toc.html.
Günther, S. "Ein stereometrisches Problem." *Archiv Math. Physik* **57**, 209–215, 1875.
Hoppe, R. "Bemerkung der Redaction." *Archiv Math. Physik. (Grunert)* **56**, 307–312, 1874.
Kuperberg, G. "Average Kissing Numbers for Sphere Packings." Preprint.
Kuperberg, G. and Schramm, O. "Average Kissing Numbers for Non-Congruent Sphere Packings." *Math. Res. Let.* **1**, 339–344, 1994.
Leech, J. "The Problem of Thirteen Spheres." *Math. Gaz.* **40**, 22–23, 1956.
Odlyzko, A. M. and Sloane, N. J. A. "New Bounds on the Number of Unit Spheres that Can Touch a Unit Sphere in n Dimensions." *J. Combin. Th. A* **26**, 210–214, 1979.
Schütte, K. and van der Waerden, B. L. "Das Problem der dreizehn Kugeln." *Math. Ann.* **125**, 325–334, 1953.
Sloane, N. J. A. Sequences A001116/M1585 in "An On-Line Version of the Encyclopedia of Integer Sequences." http://www.research.att.com/~njas/sequences/eisonline.html.
Sloane, N. J. A. and Nebe, G. "Table of Highest Kissing Numbers Presently Known." http://www.research.att.com/~njas/lattices/kiss.html.
Stewart, I. *The Problems of Mathematics, 2nd ed.* Oxford, England: Oxford University Press, pp. 82–84, 1987.
Zong, C. and Talbot, J. *Sphere Packings.* New York: Springer-Verlag, 1999.

Kite

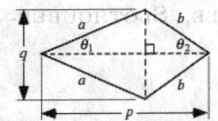  

A planar convex QUADRILATERAL consisting of two adjacent sides of length a and the other two sides of length b. The RHOMBUS is a special case of the kite, and the LOZENGE is a special case of the RHOMBUS. The AREA of a kite is given by

$$A = \tfrac{1}{2} pq,$$

where p and q are the lengths of the DIAGONALS, which are PERPENDICULAR.

See also LOZENGE, PARALLELOGRAM, PENROSE TILES, QUADRILATERAL, RHOMBUS

References
Harris, J. W. and Stocker, H. "Kite." §3.6.9 in *Handbook of Mathematics and Computational Science.* New York: Springer-Verlag, p. 86, 1998.

Kittell Graph

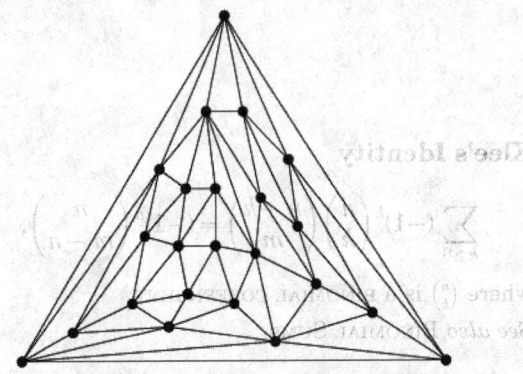

A planar 23-node graph which tangles the Kempe chains in Kempe's algorithm and thus provides an example of how Kempe's supposed proof of the FOUR-COLOR THEOREM fails.

See also ERRERA GRAPH, FOUR-COLOR THEOREM

References

Kittell, I. "A Group of Operations on a Partially Colored Map." *Bull. Amer. Math. Soc.* **41**, 407–413, 1935.

Wagon, S. *Mathematica in Action, 2nd ed.* New York: Springer-Verlag, pp. 533–534, 1999.

Klarner's Theorem

An $a \times b$ RECTANGLE can be packed with $1 \times n$ strips IFF $n|a$ or $n|b$.

See also BOX-PACKING THEOREM, CONWAY PUZZLE, DE BRUIJN'S THEOREM, RECTANGLE, SLOTHOUBER-GRAATSMA PUZZLE

References

Honsberger, R. *Mathematical Gems II.* Washington, DC: Math. Assoc. Amer., p. 88, 1976.

Klarner-Rado Sequence

The thinnest sequence which contains 1, and whenever it contains x, also contains $2x$, $3x+2$, and $6x+3$: 1, 2, 4, 5, 8, 9, 10, 14, 15, 16, 17, ... (Sloane's A005658).

See also DOUBLE-FREE SET

References

Guy, R. K. "Klarner-Rado Sequences." §E36 in *Unsolved Problems in Number Theory, 2nd ed.* New York: Springer-Verlag, p. 237, 1994.

Klarner, D. A. and Rado, R. "LINEAR COMBINATIONS of Sets of Consecutive Integers." *Amer. Math. Monthly* **80**, 985–989, 1973.

Sloane, N. J. A. Sequences A005658/M0969 in "An On-Line Version of the Encyclopedia of Integer Sequences." http://www.research.att.com/~njas/sequences/eisonline.html.

Klee's Identity

$$\sum_{k \geq 0} (-1)^k \binom{n}{k} \binom{n+k}{m} = (-1)^n \binom{n}{m-n},$$

where $\binom{n}{k}$ is a BINOMIAL COEFFICIENT.

See also BINOMIAL SUMS

References

Riordan, J. *Combinatorial Identities.* New York: Wiley, p. 13, 1979.

Rota, G.-C.; Kahaner, D.; Odlyzko, A. "On the Foundations of Combinatorial Theory. VIII: Finite Operator Calculus." *J. Math. Anal. Appl.* **42**, 684–760, 1973.

Klein Bottle

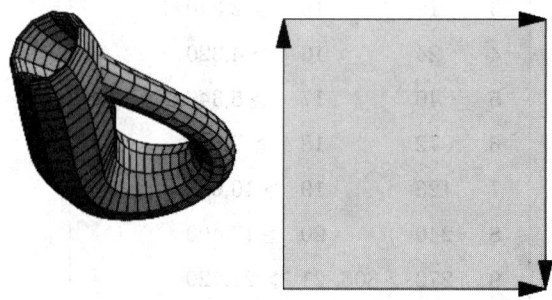

Klein bottle

A closed NONORIENTABLE SURFACE of EULER CHARACTERISTIC 0 (Dodson and Parker 1997, p. 125) that has no inside or outside. It can be constructed by gluing both pairs of opposite edges of a RECTANGLE together giving one pair a half-twist, but can be physically realized only in 4-D, since it must pass through itself without the presence of a HOLE. Its TOPOLOGY is equivalent to a pair of CROSS-CAPS with coinciding boundaries (Francis and Weeks 1999). It can be cut in half along its length to make two MÖBIUS STRIPS (Dodson and Parker 1997, p. 88), but can also be cut into a *single* MÖBIUS STRIP (Gardner 1984, pp. 14 and 17).

The above picture is an IMMERSION of the Klein bottle in $\mathbb{R}^3$ (3-space). There is also another possible IMMERSION called the "figure-8" IMMERSION (Geometry Center).

The equation for the usual IMMERSION is given by the implicit equation

$$(x^2+y^2+z^2+2y-1)[(x^2+y^2+z^2+2y-1)^2-8z^2]$$

$$+16xz(x^2+y^2+z^2-2y-1)=0 \qquad (1)$$

(Stewart 1991). Nordstrand gives the parametric form

$$x = \cos u \left[\cos\left(\tfrac{1}{2}u\right)\left(\sqrt{2}+\cos v\right) + \sin\left(\tfrac{1}{2}u\right)\sin v \cos v \right] \tag{2}$$

$$y = \sin u \left[\cos\left(\tfrac{1}{2}u\right)\left(\sqrt{2}+\cos v\right) + \sin\left(\tfrac{1}{2}u\right)\sin v \cos v \right] \tag{3}$$

$$z = -\sin\left(\tfrac{1}{2}u\right)\left(\sqrt{2}+\cos v\right) + \cos\left(\tfrac{1}{2}u\right)\sin v \cos v. \tag{4}$$

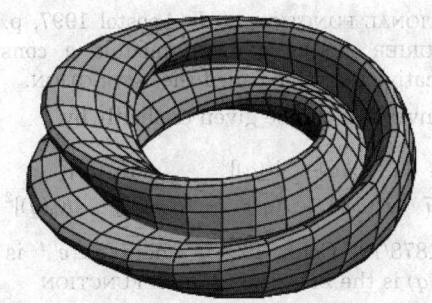

The "figure-8" form of the Klein bottle is obtained by rotating a figure eight about an axis while placing a twist in it, and is given by PARAMETRIC EQUATIONS

$$x(u, v) = \left[a + \cos\left(\tfrac{1}{2}u\right) \sin(v) - \sin\left(\tfrac{1}{2}u\right) \sin(2v)\right]\cos(u) \quad (5)$$

$$y(u, v) = \left[a + \cos\left(\tfrac{1}{2}u\right) \sin(v) - \sin\left(\tfrac{1}{2}u\right) \sin(2v)\right]\sin(u) \quad (6)$$

$$z(u, v) = \sin\left(\tfrac{1}{2}u\right) \sin(v) + \cos\left(\tfrac{1}{2}u\right) \sin(2v) \quad (7)$$

for $u \in [0, 2\pi)$, $v \in [0, 2\pi)$, and $a > 2$ (Gray 1997).

The image of the CROSS-CAP map of a TORUS centered at the ORIGIN is a Klein bottle (Gray 1997, p. 339). The MÖBIUS SHORTS are topologically equivalent to a Klein bottle with a hole (Gramain 1984, Stewart 2000).

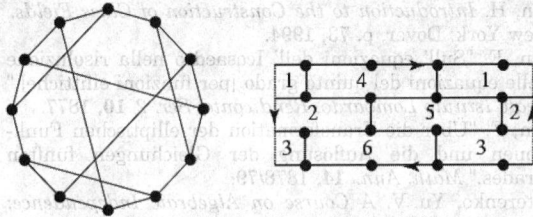

Franklin graph　　　　　coloring of the Klein bottle

Any set of regions on the Klein bottle can be colored using six colors only (Franklin 1934, Saaty and Kainen 1986), providing the sole exception to the HEAWOOD CONJECTURE (Bondy and Murty 1976, p. 244).

See also CROSS-CAP, ETRUSCAN VENUS SURFACE, FRANKLIN GRAPH, HEAWOOD CONJECTURE, IDA SURFACE, MAP COLORING, MÖBIUS SHORTS, MÖBIUS STRIP

References

Bondy, J. A. and Murty, U. S. R. *Graph Theory with Applications.* New York: North Holland, p. 244, 1976.
Dickson, S. "Klein Bottle Graphic." http://www.mathsource.com/cgi-bin/msitem22?0201-801.
Dodson, C. T. J. and Parker, P. E. *A User's Guide to Algebraic Topology.* Dordrecht, Netherlands: Kluwer, 1997.
Francis, G. K. and Weeks, J. R. "Conway's ZIP Proof." *Amer. Math. Monthly* **106**, 393-399, 1999.
Franklin, P. "A Six Colour Problem." *J. Math. Phys.* **13**, 363-369, 1934.
Gardner, M. "Klein Bottles and Other Surfaces." Ch. 2 in *The Sixth Book of Mathematical Games from Scientific American.* Chicago, IL: University of Chicago Press, pp. 9-18, 1984.
Gramain, A. *Topology of Surfaces.* Moscow, ID: BCS Associates, 1984.
Gray, A. "The Klein Bottle" and "A Different Klein Bottle." §14.4 and 14.5 in *Modern Differential Geometry of Curves and Surfaces with Mathematica, 2nd ed.* Boca Raton, FL: CRC Press, pp. 327-330, 1997.
Hilbert, D. and Cohn-Vossen, S. *Geometry and the Imagination.* New York: Chelsea, pp. 308-311, 1999.
JavaView. "Classic Surfaces from Differential Geometry: Klein Bottle." http://www-sfb288.math.tu-berlin.de/vgp/javaview/demo/surface/common/PaSurface_KleinBottle.html.
Nordstrand, T. "The Famed Klein Bottle." http://www.uib.no/people/nfytn/kleintxt.htm.
Pappas, T. "The Moebius Strip & the Klein Bottle." *The Joy of Mathematics.* San Carlos, CA: Wide World Publ./Tetra, pp. 44-46, 1989.
Saaty, T. L. and Kainen, P. C. *The Four-Color Problem: Assaults and Conquest.* New York: Dover, p. 45, 1986.
Stewart, I. *Game, Set and Math.* New York: Viking Penguin, 1991.
Stewart, I. "Mathematical Recreations: Reader Feedback." *Sci. Amer.* **283**, 101, Sep. 2000.
Wells, D. *The Penguin Dictionary of Curious and Interesting Geometry.* London: Penguin, pp. 131-132, 1991.
Wolfram Research, Inc. "Algebraic Construction of a Klein Bottle." http://library.wolfram.com/demos/v4/KleinBottleFormula.nb.

Klein Bottle Dissection

Every MÖBIUS STRIP DISSECTION of unequal squares can be glued along its edge to produce a dissection of the Klein bottle. There are no other ways to tile a Klein bottle with six or fewer squares, the situation is unknown for seven or eight squares, but it is known that other types of dissections do exists for nine squares (Stewart 1997).

See also CYLINDER DISSECTION, MÖBIUS STRIP DISSECTION, PERFECT SQUARE DISSECTION, TORUS DISSECTION

References

Stewart, I. "Squaring the Square." *Sci. Amer.* **277**, 94-96, July 1997.

Klein Four-Group

VIERGRUPPE

Klein Quartic

A 3-holed TORUS. In 1879, Felix Klein discovered that the surface has a 366-fold symmetry, the maximum possible for a surface of its type.

See also QUARTIC SURFACE

References

Levy, S. (Ed.). *The Eightfold Way: The Beauty of the Klein Quartic.* New York: Cambridge University Press, 1999.

Klein's Absolute Invariant

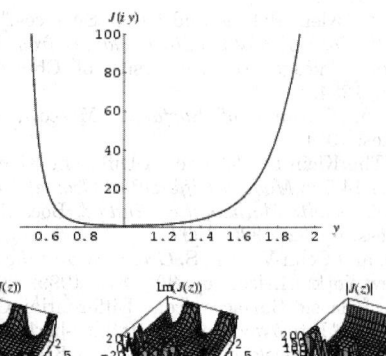

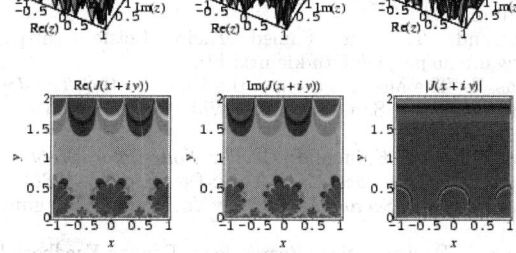

Let ω_1 and ω_2 be periods of a DOUBLY PERIODIC FUNCTION, with $\tau = \omega_2 / \omega_1$ the HALF-PERIOD RATIO a number with $\Im[\tau] \neq 0$. Then Klein's absolute invariant (also called Klein's modular function) is defined as

$$J(\omega_1, \ \omega_2) \equiv \frac{g_2^3(\omega_1, \ \omega_2)}{\Delta(\omega_1, \ \omega_2)}, \tag{1}$$

where g_2 and g_3 are the invariants of the WEIERSTRASS ELLIPTIC FUNCTION with MODULAR DISCRIMINANT

$$\Delta \equiv g_2^3 - 27 g_3^2 \tag{2}$$

(Klein 1877). If $\tau \in H$, where H is the UPPER HALF-PLANE, then

$$J(\tau) \equiv J(1, \ \tau) = J(\omega_1, \ \omega_2) \tag{3}$$

is a function of the ratio τ only, as are g_2, g_3, and Δ. Furthermore, $g_2(\tau), g_3(\tau), \Delta(\tau)$, and $J(\tau)$ are analytic in H (Apostol 1997, p. 15).

$J(\tau)$ is invariant under a UNIMODULAR TRANSFORMATION, so

$$J\!\left(\frac{a\tau + b}{c\tau + d}\right) = J(\tau), \tag{4}$$

and $J(\tau)$ is a MODULAR FUNCTION. $J(\tau)$ takes on the special values

$$J(\rho = e^{2\pi i/3}) = 0 \tag{5}$$

$$J(i) = 1 \tag{6}$$

$$J(i\infty) = \infty. \tag{7}$$

Every RATIONAL FUNCTION of J is a MODULAR FUNCTION, and every MODULAR FUNCTION can be expressed

as a RATIONAL FUNCTION of J (Apostol 1997, p. 40). The FOURIER SERIES of $J(\tau)$, modulo a constant multiplicative factor, is called the J-FUNCTION.

Klein's invariant can be given explicitly by

$$J(q) \equiv \frac{4}{27} \frac{[1 - \lambda(q) + \lambda^2(q)]^3}{\lambda^2(q)[1 - \lambda(q)]^2} = \frac{[E_4(q)]^3}{[E_4(q)]^3 - [E_6(q)]^2} \tag{8}$$

(Klein 1878/79, Cohn 1994), where $q \equiv e^{i\pi t}$ is the NOME, $\lambda(q)$ is the ELLIPTIC LAMBDA FUNCTION

$$\lambda(q) \equiv k^2(q) = \left[\frac{\vartheta_2(q)}{\vartheta_3(q)}\right]^4, \tag{9}$$

$\vartheta_i(q)$ is a JACOBI THETA FUNCTION, and the $E_i(q)$ are RAMANUJAN-EISENSTEIN SERIES.

See also ELLIPTIC LAMBDA FUNCTION, J-FUNCTION, JACOBI THETA FUNCTIONS, LAMBDA ELLIPTIC FUNCTION, PI, RAMANUJAN-EISENSTEIN SERIES

References

Apostol, T. M. "Klein's Modular Function $J(\tau)$," "Invariance of J Under Unimodular Transformation," "The Fourier Expansions of $\Delta(\tau)$ and $J(\tau)$," "Special Values of J," and "Modular Functions as Rational Functions of J." §1.12–1.13, 1.15, and 2.5–2.6 in *Modular Functions and Dirichlet Series in Number Theory, 2nd ed.* New York: Springer-Verlag, pp. 15–18, 20–22, and 39–40, 1997.
Borwein, J. M. and Borwein, P. B. *Pi & the AGM: A Study in Analytic Number Theory and Computational Complexity.* New York: Wiley, pp. 115 and 179, 1987.
Cohn, H. *Introduction to the Construction of Class Fields.* New York: Dover, p. 73, 1994.
Klein, F. "Sull' equazioni dell' Icosaedro nella risoluzione delle equazioni del quinto grado [per funzioni ellittiche]." *Reale Istituto Lombardo, Rendiconto, Ser. 2* **10**, 1877.
Klein, F. "Über die Transformation der elliptischen Funktionen und die Auflösung der Gleichungen fünften Grades." *Math. Ann.* **14**, 1878/79.
Nesterenko, Yu. V. *A Course on Algebraic Independence: Lectures at IHP 1999.* http://www.math.jussieu.fr/~nesteren/.
Weisstein, E. W. "j-Function." MATHEMATICA NOTEBOOK JFUNCTION.M.

Klein's Equation

If a real ALGEBRAIC CURVE has no singularities except nodes and CUSPS, BITANGENTS, and INFLECTION POINTS, then

$$n + 2\tau_2' + \iota' = m + 2\delta_2' + \kappa',$$

where n is the order, τ' is the number of conjugate tangents, ι' is the number of REAL inflections, m is the class, δ' is the number of REAL conjugate points, and κ' is the number of REAL CUSPS. This is also called KLEIN'S THEOREM.

See also PLÜCKER'S EQUATION

References

Coolidge, J. L. *A Treatise on Algebraic Plane Curves.* New York: Dover, p. 114, 1959.

Klein's Modular Function
KLEIN'S ABSOLUTE INVARIANT

Klein's Theorem
KLEIN'S EQUATION

Klein-Beltrami Model
The Klein-Beltrami model of HYPERBOLIC GEOMETRY consists of an OPEN DISK in the Euclidean plane whose open chords correspond to hyperbolic lines. Two lines l and m are then considered parallel if their chords fail to intersect and are PERPENDICULAR under the following conditions,

1. If at least one of l and m is a diameter of the DISK, they are hyperbolically perpendicular IFF they are perpendicular in the Euclidean sense.
2. If neither is a diameter, l is perpendicular to m IFF the Euclidean line extending l passes through the pole of m (defined as the point of intersection of the tangents to the disk at the "endpoints" of m).

There is an isomorphism between the POINCARÉ HYPERBOLIC DISK model and the Klein-Beltrami model. Consider a Klein disk in Euclidean 3-space with a SPHERE of the same radius seated atop it, tangent at the ORIGIN. If we now project chords on the disk orthogonally upward onto the SPHERE's lower HEMISPHERE, they become arcs of CIRCLES orthogonal to the equator. If we then stereographically project the SPHERE's lower HEMISPHERE back onto the plane of the Klein disk from the north pole, the equator will map onto a disk somewhat larger than the Klein disk, and the chords of the original Klein disk will now be arcs of CIRCLES orthogonal to this larger disk. That is, they will be Poincaré lines. Now we can say that two Klein lines or angles are congruent IFF their corresponding Poincaré lines and angles under this isomorphism are congruent in the sense of the Poincaré model.

See also HYPERBOLIC GEOMETRY, POINCARÉ HYPERBOLIC DISK

Klein-Erdos-Szekeres Problem
HAPPY END PROBLEM

Klein-Gordon Equation
The PARTIAL DIFFERENTIAL EQUATION

$$\frac{1}{c^2}\frac{\partial^2 \psi}{\partial t^2} = \frac{\partial^2 \psi}{\partial x^2} - \mu^2 \psi \tag{1}$$

that arises in mathematical physics.

The quasilinear Klein-Gordon equation is given by

$$u_{tt} - \alpha^2 u_{xx} + \gamma^2 u = \beta u^3 \tag{2}$$

(Nayfeh 1972, p. 76; Zwillinger 1997, p. 133), and the

nonlinear Klein-Gordon equation by

$$\sum_{i=1}^{n} u_{x_i x_i} + \lambda u^p = 0 \tag{3}$$

(Matsumo 1987; Zwillinger 1997, p. 133).

See also LIOUVILLE'S EQUATION, SINE-GORDON EQUATION, WAVE EQUATION

References
Matsumo, Y. "Exact Solution for the Nonlinear Klein-Gordon and Liouville Equations in Four-Dimensional Euclidean Space." *J. Math. Phys.* **28**, 2317–2322, 1987.
Morse, P. M. and Feshbach, H. *Methods of Theoretical Physics, Part I.* New York: McGraw-Hill, p. 272, 1953.
Nayfeh, A. H. *Perturbation Methods.* New York: Wiley, 1973.
Zwillinger, D. *Handbook of Differential Equations, 3rd ed.* Boston, MA: Academic Press, pp. 129 and 133, 1997.

Klein-Gordon-Maxwell Equation
The system of PARTIAL DIFFERENTIAL EQUATIONS

$$\nabla^2 s - (|\mathbf{a}|^2 + 1)s = 0$$

$$\nabla^2 \mathbf{a} - \nabla(\nabla \cdot \mathbf{a}) - s^2 \mathbf{a} = \mathbf{a}.$$

References
Deumens, E. "The Klein-Gordon-Maxwell Nonlinear System of Equations." *Physica D* **18**, 371–373, 1986.
Zwillinger, D. *Handbook of Differential Equations, 3rd ed.* Boston, MA: Academic Press, p. 138, 1997.

Kleinian Group
A finitely generated discontinuous group of linear fractional transformation acting on a domain in the COMPLEX PLANE.

References
Iyanaga, S. and Kawada, Y. (Eds.). *Encyclopedic Dictionary of Mathematics.* Cambridge, MA: MIT Press, p. 425, 1980.
Kra, I. *Automorphic Forms and Kleinian Groups.* Reading, MA: W. A. Benjamin, 1972.

KleinInvariantJ
KLEIN'S ABSOLUTE INVARIANT

Kloosterman's Sum

$$S(u, v, n) \equiv \sum_{n} \exp\left[\frac{2\pi i(uh + v\bar{h})}{n}\right], \tag{1}$$

where h runs through a complete set of residues RELATIVELY PRIME to n, and $\bar{h}$ is defined by

$$h\bar{h} \equiv 1 \pmod{n}. \tag{2}$$

If $(n, n) = 1$ (if n and (n') are RELATIVELY PRIME), then

$$S(u, v, n)S(u, v', n') = S(u, vn'^2 + v'n^2, nn'). \quad (3)$$

Kloosterman's sum essentially solves the problem introduced by Ramanujan of representing sufficiently large numbers by QUADRATIC FORMS $ax_1^2 + bx_2^2 + cx_3^2 + dx_4^2$. Weil improved on Kloosterman's estimate for Ramanujan's problem with the best possible estimate

$$|S(u, v, n)| \leq 2\sqrt{n} \quad (4)$$

(Duke 1997).

See also GAUSSIAN SUM

References

Duke, W. "Some Old Problems and New Results about Quadratic Forms." *Not. Amer. Math. Soc.* **44**, 190–196, 1997.

Hardy, G. H. and Wright, E. M. *An Introduction to the Theory of Numbers, 5th ed.* Oxford, England: Clarendon Press, p. 56, 1979.

Katz, N. M. *Gauss Sums, Kloosterman Sums, and Monodromy Groups.* Princeton, NJ: Princeton University Press, 1987.

Kloosterman, H. D. "On the Representation of Numbers in the Form $ax^2 + by^2 + cz^2 + dt^2$." *Acta Math.* **49**, 407–464, 1926.

Ramanujan, S. "On the Expression of a Number in the Form $ax^2 + by^2 + cz^2 + du^2$." *Collected Papers.* New York: Chelsea, 1962.

k-Matrix

A k-matrix is a kind of CUBE ROOT of the IDENTITY MATRIX (distinct from the IDENTITY MATRIX) which is defined by the COMPLEX MATRIX

$$k = \begin{bmatrix} 0 & 0 & -i \\ i & 0 & 0 \\ 0 & 1 & 0 \end{bmatrix}.$$

It satisfies

$$k^3 = I$$

where I is the IDENTITY MATRIX.

See also COMPLEX MATRIX, CUBE ROOT, IDENTITY MATRIX, QUATERNION

K-Means Clustering Algorithm

An algorithm for partitioning (or clustering) N data points into K disjoint subsets $\mathscr{S}_j$ containing N_j data points so as to minimize the sum-of-squares criterion

$$J = \sum_{j=1}^{K} \sum_{n \in \mathscr{S}_j} \|x_n - \mu_j\|^2,$$

where x_n is a vector representing the nth data point and μ_j is the CENTROID of the data points in $\mathscr{S}_j$. In general, the algorithm does not achieve a GLOBAL MINIMUM of J over the assignments. In fact, since the algorithm uses discrete assignment rather than a set of continuous parameters, the "minimum" it reaches cannot even be properly called a LOCAL MINIMUM. Despite these limitations, the algorithm is used fairly frequently as a result of its ease of implementation.

The algorithm consists of a simple re-estimation procedure as follows. First, the data points are assigned at random to the K sets. Then the centroid is computed for each set. These two steps are alternated until a stopping criterion is met, i.e., when there is no further change in the assignment of the data points.

See also GLOBAL MINIMUM, LOCAL MINIMUM, MINIMUM

References

Bishop, C. M. *Neural Networks for Pattern Recognition.* Oxford, England: Oxford University Press, 1995.

Knapsack Problem

Given a SUM and a set of WEIGHTS, find the WEIGHTS which were used to generate the SUM. The values of the weights are then encrypted in the sum. This system relies on the existence of a class of knapsack problems which can be solved trivially (those in which the weights are separated such that they can be "peeled off" one at a time using a GREEDY-like algorithm), and transformations which convert the trivial problem to a difficult one and vice versa. Modular multiplication is used as the TRAPDOOR ONE-WAY FUNCTION. The simple knapsack system was broken by Shamir in 1982, the Graham-Shamir system by Adleman, and the iterated knapsack by Ernie Brickell in 1984.

See also SUBSET SUM PROBLEM , TRAPDOOR ONE-WAY FUNCTION

References

Coppersmith, D. "Knapsack Used in Factoring." §4.6 in *Open Problems in Communication and Computation* (Ed. T. M. Cover and B. Gopinath). New York: Springer-Verlag, pp. 117–119, 1987.

Honsberger, R. *Mathematical Gems III.* Washington, DC: Math. Assoc. Amer., pp. 163–166, 1985.

Knar's Formula

The INFINITE PRODUCT identity

$$\Gamma(1+v) = 2^{2v} \prod_{m=1}^{\infty} \left[\pi^{-1/2} \Gamma\left(\tfrac{1}{2} + 2^{-m}v\right) \right],$$

where $\Gamma(x)$ is the GAMMA FUNCTION.

See also GAMMA FUNCTION, INFINITE PRODUCT

References

Erdélyi, A.; Magnus, W.; Oberhettinger, F.; and Tricomi, F. G. *Higher Transcendental Functions, Vol. 1.* New York: Krieger, p. 6, 1981.

Kneser-Sommerfeld Formula

Let $J_\nu(z)$ be a BESSEL FUNCTION OF THE FIRST KIND, $N_\nu(z)$ a NEUMANN FUNCTION, and $j_{\nu,n}(z)$ the zeros of $z^{-\nu}J_\nu(z)$ in order of ascending REAL PART. Then for $0 < x < X < 1$ and $\Re[z] > 0$,

$$\frac{\pi J_\nu(xz)}{4J_\nu(z)}[J_\nu(z)N_\nu(Xz) - N_\nu(z)J_\nu(Xz)]$$

$$= \sum_{n=1}^{\infty} \frac{J_\nu(j_{\nu,n}x)J_\nu(j_{\nu,n}X)}{(z^2 - j_{\nu,n}^2)J_{\nu,n}^{\prime 2}(j_{\nu,n})}.$$

References

Iyanaga, S. and Kawada, Y. (Eds.). *Encyclopedic Dictionary of Mathematics.* Cambridge, MA: MIT Press, p. 1474, 1980.

Knight's Tour

A knight's tour of a CHESSBOARD (or any other grid) is a sequence of moves by a knight CHESS piece (which may only make moves which simultaneously shift one square along one axis and two along the other) such that each square of the board is visited exactly once (i.e., a HAMILTONIAN CIRCUIT). If the final position is a knight's move away from the first position, the tour is called re-entrant. The above figures shows six knight's tours on an 8×8 CHESSBOARD, all but the first of which are re-entrant. The final tour has the additional property that it is a SEMIMAGIC SQUARE with row and column sums of 260 and main diagonal sums of 348 and 168 (Steinhaus 1983, p. 30).

BACKTRACKING algorithms (in which the knight is allowed to move as far as possible until it comes to a blind alley, at which point it backs up some number of steps and then tries a different path) can be used to find knight's tours, but such methods can be very slow. Warnsdorff (1823) proposed an algorithm that finds a path without any backtracking by computing ratings for "successor" steps at each position. Here, successors of a position are those squares that have not yet been visited and can be reached by a single move from the given position. The rating is highest for the successor whose number of successors is least. In this way, squares tending to be isolated are visited first and therefore prevented from being isolated (Roth). The time needed for this algorithm grows roughly linearly with the number of squares of the chessboard, but unfortunately computer implementation show that this algorithm runs into blind alleys for chessboards bigger than 76×76, despite the fact that it works well on smaller boards (Roth).

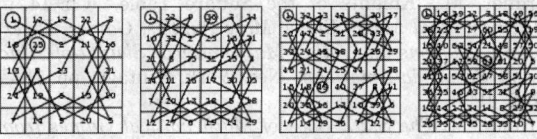

Recently, Conrad *et al.* (1994) discovered another linear time algorithm and proved that it solves the problem for all $n \geq 5$. The Conrad *et al.* algorithm works by decomposition of the chessboard into smaller chessboards (not necessarily square) for which explicit solutions are known. This algorithm is rather complicated because it has to deal with many special cases, but has been implemented in *Mathematica* by A. Roth. Example tours are illustrated above for $n \times n$ boards with $n = 5$ to 8.

Löbbing and Wegener (1996) computed the number of cycles covering the directed knight's graph for an 8×8 CHESSBOARD. They obtained α^2, where $\alpha = 2,849,759,680$, i.e., $8,121,130,233,753,702,400$. They also computed the number of undirected tours, obtaining an incorrect answer $33,439,123,484,294$ (which is not divisible by 4 as it must be), and so are currently redoing the calculation.

The following results are given by Kraitchik (1942). The number of possible tours on a $4k \times 4k$ board for $k = 3, 4, \ldots$ are 8, 0, 82, 744, 6378, 31088, 189688, 1213112, ... (Kraitchik 1942, p. 263). There are 14 tours on the 3×7 rectangle, two of which are symmetrical. There are 376 tours on the 3×8 rectangle, none of which is closed. There are 16 symmetric tours on the 3×9 rectangle and 8 closed tours on the 3×10 rectangle. There are 58 symmetric tours on the 3×11 rectangle and 28 closed tours on the 3×12 rectangle. There are five doubly symmetric tours on the 6×6 square. There are 1728 tours on the 5×5 square, 8 of which are symmetric. The longest "uncrossed" knight's tours on an $n \times n$ board for $n = 3, 4, \ldots$ are 2, 5, 10, 17, 24, 35, ... (Sloane's A003192).

See also CHESS, HAMILTONIAN CIRCUIT, KINGS PROBLEM, KNIGHTS PROBLEM, MAGIC TOUR, QUEENS PROBLEM, TOUR

References

Ahrens, W. *Mathematische Unterhaltungen und Spiele.* Leipzig, Germany: Teubner, p. 381, 1910.
Ball, W. W. R. and Coxeter, H. S. M. *Mathematical Recreations and Essays, 13th ed.* New York: Dover, pp. 175–186, 1987.

Chartrand, G. "The Knight's Tour." §6.2 in *Introductory Graph Theory.* New York: Dover, pp. 133–135, 1985.

Conrad, A.; Hindrichs, T.; Morsy, H.; and Wegener, I. "Solution of the Knight's Hamiltonian Path Problem on Chessboards." *Discr. Appl. Math.* **50**, 125–134, 1994.

Dudeney, H. E. *Amusements in Mathematics.* New York: Dover, pp. 102–103, 1970.

Euler, L. "Solution d'une question curieuse qui ne paroit soumise a aucune analyse." *Mémoires de l'Académie Royale des Sciences et Belles Lettres de Berlin, Année 1759* **15**, 310–337, 1766.

Gardner, M. "Knights of the Square Table." Ch. 14 in *Mathematical Magic Show: More Puzzles, Games, Diversions, Illusions and Other Mathematical Sleight-of-Mind from Scientific American.* New York: Vintage, pp. 188–202, 1978.

Gardner, M. *The Sixth Book of Mathematical Games from Scientific American.* Chicago, IL: University of Chicago Press, pp. 98–100, 1984.

Guy, R. K. "The *n* Queens Problem." §C18 in *Unsolved Problems in Number Theory, 2nd ed.* New York: Springer-Verlag, pp. 133–135, 1994.

Jelliss, G. "Knight's Tour Notes." http://homepages.stayfree.co.uk/gpj/ktn.htm.

Jelliss, G. "Magic Knight's Tours." http://homepages.stayfree.co.uk/gpj/mkt.htm.

Kraitchik, M. "The Problem of the Knights." Ch. 11 in *Mathematical Recreations.* New York: W. W. Norton, pp. 257–266, 1942.

Madachy, J. S. *Madachy's Mathematical Recreations.* New York: Dover, pp. 87–89, 1979.

Roget, P. M. *Philos. Mag.* **16**, 305–309, 1840.

Roth, A. "The Problem of the Knight: A Fast and Simple Algorithm." http://www.mathsource.com/cgi-bin/msitem?0202-127.

Ruskey, F. "Information on the *n* Knight's Tour Problem." http://www.theory.csc.uvic.ca/~cos/inf/misc/Knight.html.

Skiena, S. *Implementing Discrete Mathematics: Combinatorics and Graph Theory with Mathematica.* Reading, MA: Addison-Wesley, p. 166, 1990.

Sloane, N. J. A. Sequences A003192/M1369 and A006075/M3224 in "An On-Line Version of the Encyclopedia of Integer Sequences." http://www.research.att.com/~njas/sequences/eisonline.html.

Steinhaus, H. *Mathematical Snapshots, 3rd ed.* New York: Dover, p. 30, 1999.

van der Linde, A. *Geschichte und Literatur des Schachspiels, Vol. 2.* Berlin: Springer-Verlag, pp. 101–111, 1874.

Vandermonde, A.-T. "Remarques sur les Problèmes de Situation." *L'Histoire de l'Académie des Sciences avec les Mémoires, Année 1771.* Paris: Mémoirs, pp. 566–574 and Plate I, 1774.

Volpicelli, P. "Soluzione completa e generale, mediante la geometria di situazione, del problema relativo alle corse del cavallo sopra qualunque scacchiere." *Atti della Reale Accad. dei Lincei* **25**, 87–162, 1872.

Warnsdorff, H. C. von *Des Rösselsprungs einfachste und allgemeinste Lösung.* Schmalkalden, 1823.

Wegener, I. and Löbbing, M. "The Number of Knight's Tours Equals 33,439,123,484,294--Counting with Binary Decision Diagrams." *Electronic J. Combinatorics* **3**, R5 1–4, 1996. http://www.combinatorics.org/Volume_3/volume3.html#R5.

Knights of the Round Table

NECKLACE

Knights Problem

The problem of determining how many nonattacking knights $K(n)$ can be placed on an $n \times n$ CHESSBOARD. For $n = 8$, the solution is 32 (illustrated above). In general, the solutions are

$$K(n) = \begin{cases} \frac{1}{2} n^2 & n > 2 \text{ even} \\ \frac{1}{2}(n^2 + 1) & n > 1 \text{ odd,} \end{cases}$$

giving the sequence 1, 4, 5, 8, 13, 18, 25, ... (Sloane's A030978, Dudeney 1970, p. 96; Madachy 1979).

The minimal number of knights needed to occupy or attack every square on an $n \times n$ CHESSBOARD is given by 1, 4, 4, 4, 5, 8, 10, ... (Sloane's A006075). The number of such solutions are given by 1, 1, 2, 3, 8, 22, 3, ... (Sloane's A006076).

See also BISHOPS PROBLEM, CHESS, KINGS PROBLEM, KNIGHT'S TOUR, QUEENS PROBLEM, ROOKS PROBLEM

References

Dudeney, H. E. "The Knight-Guards." §319 in *Amusements in Mathematics.* New York: Dover, p. 95, 1970.

Madachy, J. S. *Madachy's Mathematical Recreations.* New York: Dover, pp. 38–39, 1979.

Moser, L. "King Paths on a Chessboard." *Math. Gaz.* **39**, 54, 1955.

Sloane, N. J. A. Sequences A006075/M3224, A006076/M0884, and A030978 in "An On-Line Version of the Encyclopedia of Integer Sequences." http://www.research.att.com/~njas/sequences/eisonline.html.

Sloane, N. J. A. and Plouffe, S. Figure M3224 in *The Encyclopedia of Integer Sequences.* San Diego: Academic Press, 1995.

Vardi, I. *Computational Recreations in Mathematica.* Redwood City, CA: Addison-Wesley, pp. 196–197, 1991.

Wilf, H. S. "The Problem of Kings." *Electronic J. Combinatorics* **2**, 3 1–7, 1995. http://www.combinatorics.org/Volume_2/volume2.html#3.

Knödel Numbers

For every $k \geq 1$, let C_k be the set of COMPOSITE NUMBERS $n > k$ such that if $1 < a < n$, $GCD(a, n) = 1$ (where GCD is the GREATEST COMMON DIVISOR), then $a^{n-k} \equiv 1 \pmod{n}$. C_1 is the set of CARMICHAEL NUMBERS. Makowski (1962/1963) proved that there are infinitely many members of C_k for $k \geq 2$.

k	Sloane	C_k
1	A002997	561, 1105, 1729, 2465, 2821, 6601, 8911, ...
2	A050990	4, 6, 8, 10, 12, 14, 22, 24, 26, 30, ...
3	A050991	9, 15, 21, 33, 39, 51, 57, 63, 69, 87, ...
4	A050992	6, 8, 12, 16, 20, 24, 28, 40, 44, 48, ...
5	A050993	25, 65, 85, 145, 165, 185, 205, ...

See also CARMICHAEL NUMBER, D-NUMBER, GREATEST COMMON DIVISOR

References

Makowski, A. "Generalization of Morrow's D-Numbers." *Simon Stevin* **36**, 71, 1962/1963.
Ribenboim, P. *The Book of Prime Number Records, 2nd ed.* New York: Springer-Verlag, p. 101, 1989.
Sloane, N. J. A. Sequences A002997/M5462, A050990, A050991, A050992, and A050993 in "An On-Line Version of the Encyclopedia of Integer Sequences." http://www.research.att.com/~njas/sequences/eisonline.html.

Knot

A knot is defined as a closed, non-self-intersecting curve embedded in 3-D. A knot is a single component LINK. Knot theory was given its first impetus when Lord Kelvin proposed a theory that atoms were vortex loops, with different chemical elements consisting of different knotted configurations (Thompson 1867). P. G. Tait then cataloged possible knots by trial and error. Much progress has been made in the intervening years.

Klein proved that knots cannot exist in an EVEN-numbered dimensional space ≥ 4. It has since been shown that a knot cannot exist in *any* dimension ≥ 4. Two distinct knots cannot have the same KNOT COMPLEMENT (Gordon and Luecke 1989), but two LINKS can! (Adams 1994, p. 261). Schubert (1949) showed that every knot can be uniquely decomposed (up to the order in which the decomposition is performed) as a KNOT SUM of a class of knots known as PRIME KNOTS, which cannot themselves be further decomposed. Combining PRIME KNOTS gives no new knot types for knots of three to five crossing, but one additional COMPOSITE KNOT each for knots of six and seven crossings.

Knots are most commonly cataloged based on the minimum number of crossings present (the so-called CROSSING NUMBER. Thistlethwaite has used DOWKER NOTATION to enumerate the number of PRIME KNOTS of up to 13 crossings, and ALTERNATING KNOTS up to 14 crossings. In this compilation, MIRROR IMAGES are counted as a single knot type. Hoste *et al.* (1998) subsequently tabulated all prime knots up to 16 crossings. Hoste and Weeks are currently begun compiling a list of 17-crossing knots (Hoste *et al.* 1998).

The following table gives the number of distinct PRIME, ALTERNATING, NONALTERNATING, TORUS, and SATELLITE KNOTS, in addition to the number of chiral noninvertible c, + amphichiral noninvertible, − amphichiral noninvertible, chiral invertible i, and fully amphichiral and invertible knots a for $n = 3$ to 16 (Hoste *et al.* 1998).

n	prime	alt.	nonalt.	torus	sat.
Sloane	A002863	A002864	A051763	A051764	A051765
3	1	1	0	1	0
4	1	1	0	0	0
5	2	2	0	1	0
6	3	3	0	0	0
7	7	7	0	1	0
8	21	18	3	1	0
9	49	41	8	1	0
10	165	123	42	1	0
11	552	367	185	1	0
12	2176	1288	888	0	0
13	9988	4878	5110	1	2
14	46972	19536	27436	1	2
15	253293	85263	168030	2	6
16	1388705	379799	1008906	1	10

n	c	+	−	i	a
Sloane	A051766	A051767	A051768	A051769	A052400
3	0	0	0	1	0
4	0	0	0	0	1
5	0	0	0	2	0
6	0	0	0	2	1
7	0	0	0	7	0
8	0	0	1	16	4
9	2	0	0	47	0
10	27	0	6	125	7
11	187	0	0	365	0

12	1103	1	40	1015	17
13	6919	0	0	3069	0
14	37885	6	227	8813	41
15	226580	0	1	26712	0
16	1308449	65	1361	78717	113

A pictorial enumeration of PRIME KNOTS of up to 10 crossings appears in Rolfsen (1976, Appendix C). Note, however, that in this table, the PERKO PAIR 10-161 and 10-162 are actually identical, and the uppermost crossing in 10-144 should be changed (Jones 1987). The kth knot having n crossings in this (arbitrary) ordering of knots is given the symbol n_k. Another possible representation for knots uses the BRAID GROUP. A knot with $n + 1$ crossings is a member of the BRAID GROUP n.

There is no general ALGORITHM to determine if a tangled curve is a knot or if two given knots are interlocked. Haken (1961) and Hemion (1979) have given ALGORITHMS for rigorously determining if two knots are equivalent, but they are too complex to apply even in simple cases (Hoste *et al.* 1998).

If a knot is AMPHICHIRAL, the "amphichirality" is $A = 1$, otherwise $A = 0$ (Jones 1987). ARF INVARIANTS are designated a. BRAID WORDS are denoted b (Jones 1987). CONWAY'S KNOT NOTATION C for knots up to 10 crossings is given by Rolfsen (1976). Hyperbolic volumes are given (Adams, Hildebrand, and Weeks 1991; Adams 1994). The BRAID INDEX i is given by Jones (1987). ALEXANDER POLYNOMIALS Δ are given in Rolfsen (1976), but with the POLYNOMIALS for 10-083 and 10-086 reversed (Jones 1987). The ALEXANDER POLYNOMIALS are normalized according to Conway, and given in abbreviated form $[a_1, a_2, \ldots$ for $a_1 + a_2(x^{-1} + x) + \ldots$.

The JONES POLYNOMIALS W for knots of up to 10 crossings are given by Jones (1987), and the JONES POLYNOMIALS V can be either computed from these, or taken from Adams (1994) for knots of up to 9 crossings (although most POLYNOMIALS are associated with the wrong knot in the first printing). The JONES POLYNOMIALS are listed in the abbreviated form $\{n\}a_0a_1\ldots$ for $t^{-n}(a_0 + a_1t + \ldots)$, and correspond either to the knot depicted by Rolfsen or its MIRROR IMAGE, whichever has the lower POWER of t^{-1}. The HOMFLY POLYNOMIAL $P(\ell, m)$ and KAUFFMAN POLYNOMIAL $F(A, x)$ are given in Lickorish and Millett (1988) for knots of up to 7 crossings.

M. B. Thistlethwaite has tabulated the HOMFLY POLYNOMIAL and KAUFFMAN POLYNOMIAL F for KNOTS of up to 13 crossings.

03-001 04-001 05-001 05-002 06-001 06-002 06-003 07-001
07-002 07-003 07-004 07-005 07-006 07-007 08-001 08-002
08-003 08-004 08-005 08-006 08-007 08-008 08-009 08-010
08-011 08-012 08-013 08-014 08-015 08-016 08-017 08-018
08-019 08-020 08-021 09-001 09-002 09-003 09-004 09-005
09-006 09-007 09-008 09-009 09-010 09-011 09-012 09-013
09-014 09-015 09-016 09-017 09-018 09-019 09-020 09-021
09-022 09-023 09-024 09-025 09-026 09-027 09-028 09-029
09-030 09-031 09-032 09-033 09-034 09-035 09-036 09-037
09-038 09-039 09-040 09-041 09-042 09-043 09-044 09-045
09-046 09-047 09-048 09-049 10-001 10-002 10-003 10-004
10-005 10-006 10-007 10-008 10-009 10-010 10-011 10-012
10-013 10-014 10-015 10-016 10-017 10-018 10-019 10-020
10-021 10-022 10-023 10-024 10-025 10-026 10-027 10-028
10-029 10-030 10-031 10-032 10-033 10-034 10-035 10-036
10-037 10-038 10-039 10-040 10-041 10-042 10-043 10-044
10-045 10-046 10-047 10-048 10-049 10-050 10-051 10-052
10-053 10-054 10-055 10-056 10-057 10-058 10-059 10-060
10-061 10-062 10-063 10-064 10-065 10-066 10-067 10-068
10-069 10-070 10-071 10-072 10-073 10-074 10-075 10-076
10-077 10-078 10-079 10-080 10-081 10-082 10-083 10-084
10-085 10-086 10-087 10-088 10-089 10-090 10-091 10-092
10-093 10-094 10-095 10-096 10-097 10-098 10-099 10-100
10-101 10-102 10-103 10-104 10-105 10-106 10-107 10-108
10-109 10-110 10-111 10-112 10-113 10-114 10-115 10-116
10-117 10-118 10-119 10-120 10-121 10-122 10-123 10-124
10-125 10-126 10-127 10-128 10-129 10-130 10-131 10-132
10-133 10-134 10-135 10-136 10-137 10-138 10-139 10-140
10-141 10-142 10-143 10-144 10-145 10-146 10-147 10-148
10-149 10-150 10-151 10-152 10-153 10-154 10-155 10-156
10-157 10-158 10-159 10-160 10-161 10-162 10-163 10-164
10-165 10-166

See also ALEXANDER POLYNOMIAL, ALEXANDER'S HORNED SPHERE, AMBIENT ISOTOPY, AMPHICHIRAL KNOT, ANTOINE'S NECKLACE, BEND (KNOT), BENNEQUIN'S CONJECTURE, BORROMEAN RINGS, BRAID GROUP, BRUNNIAN LINK, BURAU REPRESENTATION, CHEFALO KNOT, CLOVE HITCH, COLORABLE, CONWAY'S KNOT, CROOKEDNESS, DEHN'S LEMMA, DOWKER NOTATION, FIGURE-OF-EIGHT KNOT, GRANNY KNOT, HITCH, INVERTIBLE KNOT, JONES POLYNOMIAL, KINOSHITA-TERASAKA KNOT, KNOT POLYNOMIAL, KNOT SUM, LINKING NUMBER, LOOP (KNOT), MARKOV'S THEOREM, MENASCO'S THEOREM, MILNOR'S CONJECTURE, NASTY KNOT, ORIENTED KNOT, PRETZEL KNOT, PRIME KNOT, REIDEMEISTER MOVES, RIBBON KNOT, RUNNING KNOT, SATELLITE KNOT, SCHÖNFLIES THEOREM, SHORTENING, SIGNATURE (KNOT), SKEIN RELATIONSHIP, SLICE KNOT, SLIP KNOT, SMITH CONJECTURE, SOLOMON'S SEAL KNOT, SPAN (LINK), SPLITTING, SQUARE KNOT, STEVEDORE'S KNOT, STICK NUMBER, STOPPER KNOT, TAIT'S KNOT CONJECTURES, TAME KNOT, TANGLE, TORSION NUMBER, TORUS KNOT, TREFOIL KNOT, UNKNOT, UNKNOTTING NUMBER, VASSILIEV INVARIANT, WHITEHEAD LINK

References

Adams, C. C. *The Knot Book: An Elementary Introduction to the Mathematical Theory of Knots.* New York: W. H. Freeman, pp. 280–286, 1994.

Adams, C.; Hildebrand, M.; and Weeks, J. "Hyperbolic Invariants of Knots and Links." *Trans. Amer. Math. Soc.* **1**, 1–56, 1991.

Alexander, J. W. and Briggs, G. B. "On Types of Knotted Curves." *Ann. Math.* **28**, 562–586, 1927.

Aneziris, C. N. *The Mystery of Knots: Computer Programming for Knot Tabulation.* Singapore: World Scientific, 1999.

Ashley, C. W. *The Ashley Book of Knots.* New York: McGraw-Hill, 1996.

Bogomolny, A. "Knots...." http://www.cut-the-knot.com/do_you_know/knots.html.

Bruzelius, L. "Knots and Splices." http://pc-78-120.udac.se:8001/WWW/Nautica/Bibliography/Knots&Splices.html.

Caudron, A. "Classification des noeuds et des enlacements." Prepublication Math. d'Orsay. Orsay, France: Université Paris-Sud, 1980.

Cerf, C. "Atlas of Oriented Knots and Links." *Topology Atlas Invited Contributions* **3**, No. 2, 1–32, 1998. http://at.yorku.ca/t/a/i/c/31.htm.

Conway, J. H. "An Enumeration of Knots and Links." In *Computational Problems in Abstract Algebra* (Ed. J. Leech). Oxford, England: Pergamon Press, pp. 329–358, 1970.

Eppstein, D. "Knot Theory." http://www.ics.uci.edu/~eppstein/junkyard/knot.html.

Eppstein, D. "Knot Theory." http://www.ics.uci.edu/~eppstein/junkyard/knot/.

Erdener, K.; Candy, C.; and Wu, D. "Verification and Extension of Topological Knot Tables." ftp://chs.cusd.claremont.edu/pub/knot/FinalReport.sit.hqx.

Gordon, C. and Luecke, J. "Knots are Determined by their Complements." *J. Amer. Math. Soc.* **2**, 371–415, 1989.

Haken, W. "Theorie der Normalflachen." *Acta Math.* **105**, 245–375, 1961.

Hemion, G. "On the Classification of Homeomorphisms of 2-Manifolds and the Classification of 3-Manifolds." *Acta Math.* **142**, 123–155, 1979.

Hoste, J.; Thistlethwaite, M.; and Weeks, J. "The First 1,701,936 Knots." *Math. Intell.* **20**, 33–48, Fall 1998.

Kauffman, L. *Knots and Applications.* River Edge, NJ: World Scientific, 1995.

Kauffman, L. *Knots and Physics.* Teaneck, NJ: World Scientific, 1991.

Kirkman, T. P. "The Enumeration, Description, and Construction of Knots Fewer than Ten Crossings." *Trans. Roy. Soc. Edinburgh* **32**, 1885, 281–309.

Kirkman, T. P. "The 634 Unifilar Knots of Ten Crossings Enumerated and Defined." *Trans. Roy. Soc. Edinburgh* **32**, 483–506, 1885.

Korpegård, J. "The Knotting Dictionary of Kännet." http://www.korpegard.nu/jan/knots.html.

Lickorish, W. B. R. and Millett, B. R. "The New Polynomial Invariants of Knots and Links." *Math. Mag.* **61**, 1–23, 1988.

Listing, J. B. "Vorstudien zur Topologie." Göttingen Studien, University of Göttingen, Germany, 1848.

Little, C. N. "On Knots, with a Census of Order Ten." *Trans. Connecticut Acad. Sci.* **18**, 374–378, 1885.

Livingston, C. *Knot Theory.* Washington, DC: Math. Assoc. Amer., 1993.

Murasugi, K. and Kurpita, B. I. *A Study of Braids.* Dordrecht, Netherlands: Kluwer, 1999.

Neuwirth, L. "The Theory of Knots." *Sci. Amer.* **140**, 84–96, Jun. 1979.

Perko, K. "Invariants of 11-Crossing Knots." Prepublications Math. d'Orsay. Orsay, France: Université Paris-Sub, 1980.

Perko, K. "Primality of Certain Knots." *Topology Proc.* **7**, 109–118, 1982.

Praslov, V. V. and Sossinsky, A. B. *Knots, Links, Braids and 3-Manifolds: An Introduction to the New Invariants in Low-Dimensional Topology.* Providence, RI: Amer. Math. Soc., 1996.

Przytycki, J. "A History of Knot Theory from Vandermonde to Jones." *Proc. Mexican Nat. Congress Math.*, Nov. 1991.

Reidemeister, K. *Knotentheorie.* Berlin: Springer-Verlag, 1932.

Rolfsen, D. "Table of Knots and Links." Appendix C in *Knots and Links.* Wilmington, DE: Publish or Perish Press, pp. 280–287, 1976.

Schubert, H. *Sitzungsber. Heidelberger Akad. Wiss., Math.-Naturwiss. Klasse, 3rd Abhandlung.* 1949.

Sloane, N. J. A. Sequences A002863/M0851 in "An On-Line Version of the Encyclopedia of Integer Sequences." http://www.research.att.com/~njas/sequences/eisonline.html.

Sloane, N. J. A. and Plouffe, S. Figure M0851 in *The Encyclopedia of Integer Sequences.* San Diego: Academic Press, 1995.

Suber, O. "Knots on the Web." http://www.earlham.edu/~peters/knotlink.htm.

Tait, P. G. "On Knots I, II, and III." *Scientific Papers, Vol. 1.* Cambridge, England: University Press, pp. 273–347, 1898.

Thistlethwaite, M. B. "Knot Tabulations and Related Topics." In *Aspects of Topology in Memory of Hugh Dowker 1912–1982* (Ed. I. M. James and E. H. Kronheimer). Cambridge, England: Cambridge University Press, pp. 2–76, 1985.

Thistlethwaite, M. B. ftp://chs.cusd.claremont.edu/pub/knot/Thistlethwaite_Tables/.

Thistlethwaite, M. B. "Morwen's Home Page." http://www.math.utk.edu/~morwen/.

Thompson, W. T. "On Vortex Atoms." *Philos. Mag.* **34**, 15–24, 1867.

Weisstein, E. W. "Knots." MATHEMATICA NOTEBOOK KNOTS.M.

Wells, D. *The Penguin Dictionary of Curious and Interesting Geometry.* London: Penguin, pp. 132–135, 1991.

Weisstein, E. W. "Books about Knot Theory." http://www.treasure-troves.com/books/KnotTheory.html.

Knot Complement

Let $\mathbb{R}^3$ be the space in which a KNOT K sits. Then the space "around" the knot, i.e., everything but the knot itself, is denoted $\mathbb{R}^3 - K$ and is called the knot complement of K (Adams 1994, p. 84).

If a knot complement is hyperbolic (in the sense that it admits a complete Riemannian metric of constant GAUSSIAN CURVATURE -1), then this metric is unique (Prasad 1973, Hoste *et al.* 1998).

See also COMPLEMENT, COMPRESSIBLE SURFACE, KNOT, KNOT EXTERIOR

References

Adams, C. C. "Knot Complements and Three-Manifolds." §9.1 in *The Knot Book: An Elementary Introduction to the Mathematical Theory of Knots.* New York: W. H. Freeman, pp. 243–246, 1994.

Cipra, B. "To Have and Have Knot: When are Two Knots Alike?" *Science* **241**, 1291–1292, 1988.

Gordon, C. and Luecke, J. "Knots are Determined by their Complements." *J. Amer. Math. Soc.* **2**, 371–415, 1989.

Hoste, J.; Thistlethwaite, M.; and Weeks, J. "The First 1,701,936 Knots." *Math. Intell.* **20**, 33–48, Fall 1998.

Prasad, G. "Stong Rigidity of Q-Rank 1 Lattices." *Invent. Math.* **21**, 255–286, 1973.

Knot Curve

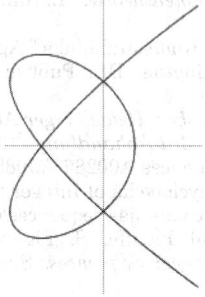

$$(x^2 - 1)^2 = y^2(3 + 2y).$$

References

Cundy, H. and Rollett, A. *Mathematical Models, 3rd ed.* Stradbroke, England: Tarquin Pub., p. 72, 1989.

Knot Determinant

The determinant of a knot is $|\Delta(-1)|$, where $\Delta(z)$ is the ALEXANDER POLYNOMIAL.

Knot Diagram

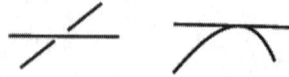

A picture of a projection of a KNOT onto a PLANE. Usually, only double points are allowed (no more than two points are allowed to be superposed), and the double or crossing points must be "genuine crossings" which are transverse in the plane. This means that double points must look like the above left diagram, and not the above right one. Also, it is usually demanded that a knot diagram contain the information if the crossings are overcrossings or undercrossings so that the original knot can be reconstructed. The knot diagram of the TREFOIL KNOT is illustrated below.

KNOT POLYNOMIALS can be computed from knot diagrams. Such POLYNOMIALS often (but not always) allow the knots corresponding to given diagrams to be uniquely identified.

See also NUGATORY CROSSING, REDUCED KNOT DIAGRAM, REIDEMEISTER MOVES

References

Hoste, J.; Thistlethwaite, M.; and Weeks, J. "The First 1,701,936 Knots." *Math. Intell.* **20**, 33–48, Fall 1998.

Knot Exterior

The exterior of a knot K is the complement of an open solid TORUS knotted like K. The removed open solid TORUS is called a TUBULAR NEIGHBORHOOD (Adams 1994, p. 258).

See also KNOT COMPLEMENT, GORDON-LUECKE THEOREM, TUBULAR NEIGHBORHOOD

References

Adams, C. C. *The Knot Book: An Elementary Introduction to the Mathematical Theory of Knots.* New York: W. H. Freeman, 1994.

Knot Invariant

A knot invariant is a function from the set of all KNOTS to any other set such that the function does not change as the knot is changed (up to isotopy). In other words, a knot invariant always assigns the same value to equivalent knots (although different knots may have the same knot invariant). Standard knot invariants include the FUNDAMENTAL GROUP of the KNOT COMPLEMENT, numerical knot invariants (such as VASSILIEV INVARIANTS), polynomial invariants (KNOT POLYNOMIALS such as the ALEXANDER POLYNOMIAL, JONES POLYNOMIAL, KAUFFMAN POLYNOMIAL F, and KAUFFMAN POLYNOMIAL X), and torsion invariants (such as the TORSION NUMBER).

See also ARF INVARIANT, KNOT, KNOT POLYNOMIAL, LINK INVARIANT, TORSION NUMBER, VASSILIEV INVARIANT

References

Aneziris, C. N. "The Knot INvariants." Ch. 5 in *The Mystery of Knots: Computer Programming for Knot Tabulation.* Singapore: World Scientific, pp. 35–42, 1999.

Knot Linking

In general, it is possible to link two n-D HYPERSPHERES in $(n + 2)$-D space in an infinite number of inequivalent ways. In dimensions greater than $n + 2$ in the piecewise linear category, it is true that these spheres are themselves unknotted. However, they may still form nontrivial links. In this way, they are something like higher dimensional analogs of two 1-spheres in 3-D. The following table gives the number of nontrivial ways that two n-D HYPERSPHERES can be linked in k-D.

D of spheres	D of space	Distinct Linkings
23	40	239
31	48	959
102	181	3

| 102 | 182 | 10438319 |
| 102 | 183 | 3 |

Two 10-D HYPERSPHERES link up in 12, 13, 14, 15, and 16-D, then unlink in 17-D, link up again in 18, 19, 20, and 21-D. The proof of these results consists of an "easy part" (Zeeman 1962) and a "hard part" (Ravenel 1986). The hard part is related to the calculation of the (stable and unstable) HOMOTOPY GROUPS of SPHERES.

References
Bing, R. H. *The Geometric Topology of 3-Manifolds*. Providence, RI: Amer. Math. Soc., 1983.
Ravenel, D. *Complex Cobordism and Stable Homotopy Groups of Spheres*. New York: Academic Press, 1986.
Rolfsen, D. *Knots and Links*. Wilmington, DE: Publish or Perish Press, p. 7, 1976.
Zeeman. "Isotopies and Knots in Manifolds." In *Topology of 3-Manifolds and Related Topics* (Ed. M. K. Fort). Englewood Cliffs, NJ: Prentice-Hall, 1962.

Knot Move
An operation on a knot or link diagram which preserves its crossing number. Thistlethwaite used 13 different moves in generating a list of 16-crossing alternating knots (Hoste *et al.* 1998). While these moves eliminate all duplicate knots up to 13 crossings with only a single exception, there are 9,868 duplicates in his list of 1,018,774 16-crossing knots (Hoste *et al.* 1998).

See also FLYPE, HABIRO MOVE, MARKOV MOVES, PASS MOVE, PERKO MOVE, POKE MOVE, REIDEMEISTER MOVES, SLIDE MOVE, TWIST MOVE

References
Hoste, J.; Thistlethwaite, M.; and Weeks, J. "The First 1,701,936 Knots." *Math. Intell.* **20**, 33–48, Fall 1998.

Knot Polynomial
A knot invariant in the form of a POLYNOMIAL such as the ALEXANDER POLYNOMIAL, BLM/HO POLYNOMIAL, BRACKET POLYNOMIAL, CONWAY POLYNOMIAL, HOMFLY POLYNOMIAL, JONES POLYNOMIAL, KAUFFMAN POLYNOMIAL F, KAUFFMAN POLYNOMIAL X, and VASSILIEV INVARIANT.

See also KNOT, LINK

References
Lickorish, W. B. R. and Millett, K. C. "The New Polynomial Invariants of Knots and Links." *Math. Mag.* **61**, 3–23, 1988.

Knot Problem
The problem of deciding if two KNOTS in 3-space are equivalent such that one can be continuously deformed into another.

Knot Shadow
A KNOT DIAGRAM which does not specify whether crossings are under- or overcrossings.

Knot Sum
Two oriented knots (or links) can be summed by placing them side by side and joining them by straight bars so that orientation is preserved in the sum. This operation is denoted #, so the knot sum of knots K_1 and K_2 is written

$$K_1 \# K_2 = K_2 \# K_1.$$

The KNOT SUM of any number of knots cannot be the UNKNOT unless each knot in the sum is the UNKNOT (Schubert 1949; Steinhaus 1983, p. 265).

See also CONNECTED SUM

References
Schubert, H. *Sitzungsber. Heidelberger Akad. Wiss., Math.-Naturwiss. Klasse, 3rd Abhandlung.* 1949.
Steinhaus, H. *Mathematical Snapshots, 3rd ed.* New York: Dover, 1999.

Knot Symmetry
A symmetry of a knot K is a homeomorphism of $\mathbb{R}^3$ which maps K onto itself. More succinctly, a knot symmetry is a homeomorphism of the pair of spaces $(\mathbb{R}^3, K)$. Hoste *et al.* (1998) consider four types of symmetry based on whether the symmetry preserves or reverses orienting of $\mathbb{R}^3$ and K,

1. preserves $\mathbb{R}^3$, preserves K (identity operation),
2. preserves $\mathbb{R}^3$, reverses K,
3. reverses $\mathbb{R}^3$, preserves K,
4. reverses $\mathbb{R}^3$, reverses K.

This then gives the five possible classes of symmetry summarized in the table below.

class	symmetries	knot symmetries
c	1	chiral, noninvertible
$+$	1, 3	+ amphichiral, noninvertible
$-$	1, 4	− amphichiral, noninvertible
i	1, 2	chiral, invertible
a	1, 2, 3, 4	+ and − amphichiral, invertible

In the case of HYPERBOLIC KNOTS, the symmetry group must be finite and either CYCLIC or DIHEDRAL (Riley 1979, Kodama and Sakuma 1992, Hoste *et al.* 1998). The classification is slightly more complicated for nonhyperbolic knots. Furthermore, all knots with ≤ 8 crossings are either amphichiral or invertible

(Hoste *et al.* 1998). Any symmetry of a prime alternating link must be visible up to flypes in any alternating diagram of the link (Bonahon and Siebermann, Menasco and Thistlethwaite 1993, Hoste *et al.* 1998).

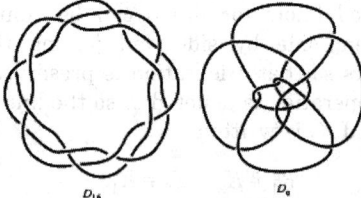

The following tables (Hoste *et al.* 1998) give the numbers of n-crossing knots belonging to cyclic symmetry groups $\mathbb{Z}_k$ (Sloane's A052411 for $\mathbb{Z}_1$ and A052412 for $\mathbb{Z}_2$) and dihedral symmetry groups D_k (Sloane's A052415 through A052422). Of knots with 16 or fewer crossings, there are only one each having symmetry groups $\mathbb{Z}_3$, D_{14}, and D_{16} (above left). There are only two knots with symmetry group D_9, one hyperbolic (above right), and one a satellite knot. In addition, there are 2, 4, and 10 satellite knots having 14-, 15-, and 16-crossings, respectively, which belong to the dihedral group D_∞.

n	D_1	D_2	D_3	D_4	D_5	D_6	D_7	D_8	D_9	D_{10}	D_{14}	D_{16}
1	0	0	0	0	0	0	0	0	0	0	0	0
2	0	0	0	0	0	0	0	0	0	0	0	0
3	0	0	0	0	0	0	0	0	0	0	0	0
4	0	0	1	0	0	0	0	0	0	0	0	0
5	0	1	0	0	0	0	0	0	0	0	0	0
6	0	2	0	1	0	0	0	0	0	0	0	0
7	0	4	0	2	0	0	0	0	0	0	0	0
8	4	12	0	3	0	0	0	1	0	0	0	0
9	13	23	3	4	0	3	0	0	0	0	0	0
10	66	62	1	5	0	1	0	0	0	1	0	0
11	217	134	2	11	0	0	0	0	0	0	0	0
12	728	309	6	18	0	8	1	2	0	0	0	0
13	2391	647	1	21	2	3	1	2	0	0	0	0
14	7575	1463	4	31	2	2	0	0	0	0	1	0
15	23517	3065	50	53	3	12	0	2	1	4	0	0
16	73263	6791	15	89	0	10	1	8	1	1	0	1

n	$\mathbb{Z}_1$	$\mathbb{Z}_2$	$\mathbb{Z}_3$	$\mathbb{Z}_4$
1	0	0	0	0
2	0	0	0	0
3	0	0	0	0
4	0	0	0	0
5	0	0	0	0
6	0	0	0	0
7	0	0	0	0
8	0	0	0	0
9	2	0	0	0
10	24	3	0	0
11	173	14	0	0
12	1047	57	0	0
13	6709	210	0	0
14	37177	712	0	2
15	224311	2268	1	0
16	1301492	7011	0	11

See also AMPHICHIRAL KNOT, CHIRAL KNOT, KNOT

References

Bonahon, F. and Siebermann, L. "The Classification of Algebraic Links." Unpublished manuscript.

Hoste, J.; Thistlethwaite, M.; and Weeks, J. "The First 1,701,936 Knots." *Math. Intell.* **20**, 33–48, Fall 1998.

Kodama K. and Sakuma, M. "Symmetry Groups of Prime Knots Up to 10 Crossings." In *Knot 90, Proceedings of the International Conference on Knot Theory and Related Topics, Osaka, Japan, 1990* (Ed. A. Kawauchi.) Berlin: de Gruyter, pp. 323–340, 1992.

Menasco, W. and Thistlethwaite, M. "The Classification of Alternating Links." *Ann. Math.* **138**, 113–171, 1993.

Riley, R. "An Elliptic Path from Parabolic Representations to Hyperbolic Structures." In *Topology of Low-Dimensional Manifolds, Proceedings, Sussex 1977* (Ed. R. Fenn). New York: Springer-Verlag, pp. 99–133, 1979.

Sloane, N. J. A. Sequences A052411, A052412, A052415, A052416, A052417, A052418, A052420, and A052422 in "An On-Line Version of the Encyclopedia of Integer Sequences." http://www.research.att.com/~njas/sequences/eisonline.html.

Knot Theory

The mathematical study of KNOTS. Knot theory considers questions such as the following:

1. Given a tangled loop of string, is it really knotted or can it, with enough ingenuity and/or luck, be untangled without having to cut it?
2. More generally, given two tangled loops of string, when are they deformable into each other?
3. Is there an effective algorithm (or any algorithm to speak of) to make these determinations?

Although there has been almost explosive growth in the number of important results proved since the discovery of the JONES POLYNOMIAL, there are still many "knotty" problems and conjectures whose answers remain unknown.

See also KNOT, LINK

Knot Vector

B-SPLINE

Knuth Number

The numbers defined by the RECURRENCE RELATION

$$K_{n+1} = 1 + \min(2K_{\lfloor n/2 \rfloor}, 3K_{\lfloor n/3 \rfloor}),$$

with $K_0 = 1$. The first few values for $n = 0, 1, 2, \dots$ are 1, 3, 3, 4, 7, 7, 7, 9, 9, 10, 13, ... (Sloane's A007448).

References

Graham, R. L.; Knuth, D. E.; and Patashnik, O. *Concrete Mathematics: A Foundation for Computer Science, 2nd ed.* Reading, MA: Addison-Wesley, 1994.

Sloane, N. J. A. Sequences A0074482276 in "An On-Line Version of the Encyclopedia of Integer Sequences." http://www.research.att.com/~njas/sequences/eisonline.html.

Köbe Function

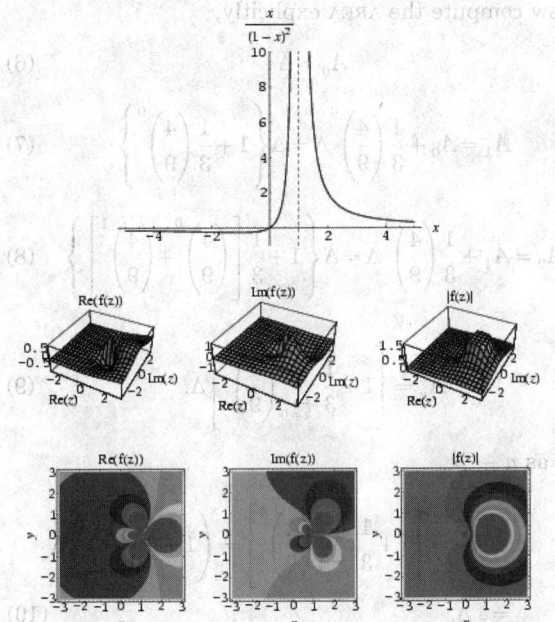

The function

$$f_\theta(z) \equiv \frac{z}{(1 + e^{i\theta}z)^2} \tag{1}$$

defined on the UNIT DISK $|z| < 1$. For $\theta \in [0, 2\pi)$, the Köbe function is a SCHLICHT FUNCTION

$$f(z) = z + \sum_{j=2}^{\infty} a_j z^j \tag{2}$$

with $|a_j| = j$ for all j (Krantz 1999, p. 149). For $\theta = 0$,

$$f_0(z) = \frac{z}{(z-1)^2}, \tag{3}$$

illustrated above.

See also KÖBE'S ONE-FOURTH THEOREM, SCHLICHT FUNCTION

References

Bombieri, E. "On the Local Maximum of the Koebe Function." *Invent. Math.* **4**, 26–67, 1967.

Krantz, S. G. *Handbook of Complex Analysis.* Boston, MA: Birkhäuser, p. 149, 1999.

Pederson, R. and Schiffer, M. "A Proof of the Bieberbach Conjecture for the Fifth Coefficient." *Arch. Rational Mech. Anal.* **45**, 161–193, 1972.

Stewart, I. *From Here to Infinity: A Guide to Today's Mathematics.* Oxford, England: Oxford University Press, pp. 164–165, 1996.

Köbe's One-Fourth Theorem

If f is a SCHLICHT FUNCTION and $D(z_0, r)$ is the OPEN DISK of radius r centered at z_0, then

$$f(D(0, 1)) \supseteq D(0, 1/4),$$

where $\supseteq$ denotes a (not necessarily proper) SUPERSET (Krantz 1999, p. 150).

See also KÖBE FUNCTION, SCHLICHT FUNCTION

References

Krantz, S. G. "The Köbe 1/4 Theorem." §12.1.5 in *Handbook of Complex Analysis.* Boston, MA: Birkhäuser, pp. 150–151, 1999.

Le Lionnais, F. *Les nombres remarquables.* Paris: Hermann, p. 24, 1983.

Koch Antisnowflake

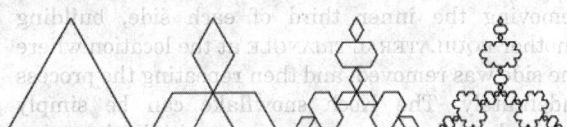

A FRACTAL derived from the KOCH SNOWFLAKE. The base curve and motif for the fractal are illustrated below.

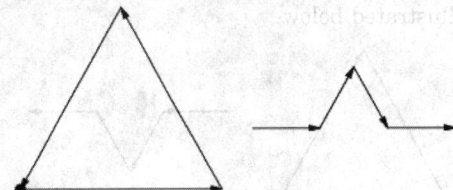

The AREA after the nth iteration is

$$A_n = A_{n-1} - \frac{1}{3} \frac{\ell_{n-1}}{a} \frac{\Delta}{3^n},$$

where Δ is the area of the original EQUILATERAL

TRIANGLE, so from the derivation for the KOCH SNOWFLAKE,

$$A \equiv \lim_{n \to \infty} A_n = (1 - \tfrac{3}{5})\Delta = \tfrac{2}{5}\Delta.$$

See also EXTERIOR SNOWFLAKE, FLOWSNAKE FRACTAL, KOCH SNOWFLAKE, PENTAFLAKE, SIERPINSKI CURVE

References

Cundy, H. and Rollett, A. *Mathematical Models, 3rd ed.* Stradbroke, England: Tarquin Pub., pp. 66–67, 1989.
Lauwerier, H. *Fractals: Endlessly Repeated Geometric Figures.* Princeton, NJ: Princeton University Press, pp. 36–37, 1991.
Weisstein, E. W. "Fractals." MATHEMATICA NOTEBOOK FRACTAL.M.
Wells, D. *The Penguin Dictionary of Curious and Interesting Geometry.* London: Penguin, p. 136, 1991.

Koch Island

KOCH SNOWFLAKE

Koch Snowflake

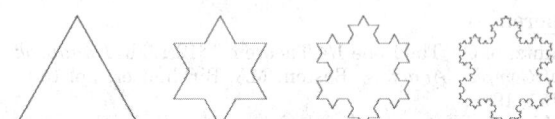

A FRACTAL, also known as the KOCH ISLAND, which was first described by Helge von Koch in 1904. It is built by starting with an EQUILATERAL TRIANGLE, removing the inner third of each side, building another EQUILATERAL TRIANGLE at the location where the side was removed, and then repeating the process indefinitely. The Koch snowflake can be simply encoded as a LINDENMAYER SYSTEM with initial string "F-F-F", STRING REWRITING rule "F" -> "F+F-F+F", and angle 60°. The zeroth through third iterations of the construction are shown above. The fractal can also be constructed using a base curve and motif, illustrated below.

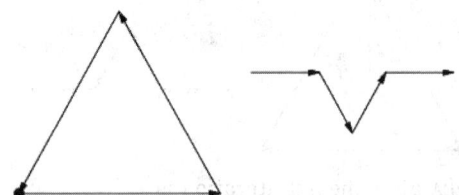

Let N_n be the number of sides, L_n be the length of a single side, ℓ_n be the length of the PERIMETER, and A_n the snowflake's AREA after the nth iteration. Further, denote the AREA of the initial $n = 0$ TRIANGLE Δ, and the length of an initial $n = 0$ side 1. Then

$$N_n = 3 \cdot 4^n \tag{1}$$

$$L_n = \left(\tfrac{1}{3}\right)^n = 3^{-n} \tag{2}$$

$$\ell_n \equiv N_n L_n = 3\left(\tfrac{4}{3}\right)^n \tag{3}$$

$$A_n = A_{n-1} + \tfrac{1}{4} N_n L_n^2 \Delta = A_{n-1} + \frac{3 \cdot 4^n}{4}\left(\frac{1}{3}\right)^{2n}\Delta$$

$$= A_{n-1} + \frac{3 \cdot 4^{n-1}}{9^n}\Delta = A_{n-1} + \frac{3 \cdot 4^{n-1}}{9 \cdot 9^{n-1}}\Delta$$

$$= A_{n-1} + \tfrac{1}{3}\left(\tfrac{4}{9}\right)^{n-1}\Delta. \tag{4}$$

The CAPACITY DIMENSION is then

$$d_{\text{cap}} = -\lim_{n \to \infty} \frac{\ln N_n}{\ln L_n} = -\lim_{n \to \infty} \frac{\ln(3 \cdot 4)^n}{\ln(3^{-n})}$$

$$= \lim_{n \to \infty} \frac{\ln 3 + n \ln 4}{n \ln 3} = \frac{\ln 4}{\ln 3} = \frac{2 \ln 2}{\ln 3}$$

$$= 1.261859507\ldots . \tag{5}$$

Now compute the AREA explicitly,

$$A_0 = \Delta \tag{6}$$

$$A_1 = A_0 + \frac{1}{3}\left(\frac{4}{9}\right)^0 \Delta = \Delta\left\{1 + \frac{1}{3}\left(\frac{4}{9}\right)^0\right\} \tag{7}$$

$$A_2 = A_1 + \frac{1}{3}\left(\frac{4}{9}\right)^1 \Delta = \Delta\left\{1 + \frac{1}{3}\left[\left(\frac{4}{9}\right)^0 + \left(\frac{4}{9}\right)^1\right]\right\} \tag{8}$$

$$A_n = \left[1 + \frac{1}{3}\sum_{k=0}^{n}\left(\frac{4}{9}\right)^k\right]\Delta, \tag{9}$$

so as $n \to \infty$,

$$A \equiv A_\infty = \left[1 + \frac{1}{3}\sum_{k=1}^{\infty}\left(\frac{4}{9}\right)^k\right] = \left(1 + \frac{1}{3}\frac{1}{1 - \frac{4}{9}}\right)\Delta$$

$$= \tfrac{8}{5}\Delta. \tag{10}$$

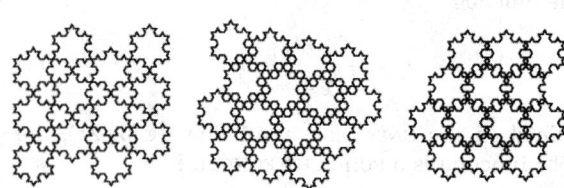

Some beautiful TILINGS, a few examples of which are illustrated above, can be made with iterations toward

Koch snowflakes.

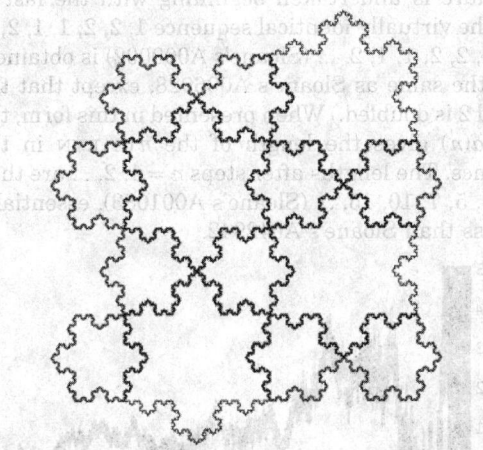

In addition, two sizes of Koch snowflakes in AREA ratio 1:3 TILE the PLANE, as shown above (Mandelbrot).

Another beautiful modification of the Koch snowflake involves inscribing the constituent triangles with filled-in triangles, possibly rotated at some angle. Some sample results are illustrated above for 3 and 4 iterations.

See also CESÀRO FRACTAL, EXTERIOR SNOWFLAKE, GOSPER ISLAND, KOCH ANTISNOWFLAKE, PEANO-GOSPER CURVE, PENTAFLAKE, SIERPINSKI SIEVE

References

Bulaevsky, J. "The Koch Curve Fractal." http://www.best.com/~ejad/java/fractals/koch.shtml.

Cesàro, E. "Remarques sur la courbe de von Koch." *Atti della R. Accad. della Scienze fisiche e matem. Napoli* **12**, No. 15, 1905. Reprinted as §228 in *Opere scelte, a cura dell'Unione matematica italiana e col contributo del Consiglio nazionale delle ricerche, Vol. 2: Geometria, analisi, fisica matematica.* Rome: Edizioni Cremonese, pp. 464–479, 1964.

Cundy, H. and Rollett, A. *Mathematical Models, 3rd ed.* Stradbroke, England: Tarquin Pub., pp. 65–66, 1989.

Dickau, R. M. "Two-Dimensional L-Systems." http://forum.swarthmore.edu/advanced/robertd/lsys2d.html.

Dixon, R. *Mathographics.* New York: Dover, pp. 175–177 and 179, 1991.

Gardner, M. *The Sixth Book of Mathematical Games from Scientific American.* Chicago, IL: University of Chicago Press, p. 227, 1984.

Harris, J. W. and Stocker, H. "Koch's Curve" and "Koch's Snowflake." §4.11.5–4.11.6 in *Handbook of Mathematics and Computational Science.* New York: Springer-Verlag, pp. 114–115, 1998.

King, B. W. "Snowflake Curves." *Math. Teacher* **57**, 219–222, 1964.

Koch, von. *Acta Math.* **30**, 145, 1906.

Koch, von. *Archiv för Matemat., Astron. och Fysik.*, pp. 681–702, 1914.

Lauwerier, H. *Fractals: Endlessly Repeated Geometric Figures.* Princeton, NJ: Princeton University Press, pp. 28–29 and 32–36, 1991.

Pappas, T. "The Snowflake Curve." *The Joy of Mathematics.* San Carlos, CA: Wide World Publ./Tetra, pp. 78 and 160–161, 1989.

Peitgen, H.-O.; Jürgens, H.; and Saupe, D. *Chaos and Fractals: New Frontiers of Science.* New York: Springer-Verlag, 1992.

Peitgen, H.-O. and Saupe, D. (Eds.). "The von Koch Snowflake Curve Revisited." §C.2 in *The Science of Fractal Images.* New York: Springer-Verlag, pp. 275–279, 1988.

Schneider, J. E. "A Generalization of the Von Koch Curves." *Math. Mag.* **38**, 144–147, 1965.

Wagon, S. *Mathematica in Action.* New York: W. H. Freeman, pp. 185–195, 1991.

Weisstein, E. W. "Fractals." MATHEMATICA NOTEBOOK FRACTAL.M.

Wells, D. *The Penguin Dictionary of Curious and Interesting Geometry.* London: Penguin, pp. 135–136, 1991.

Kochansky's Approximation

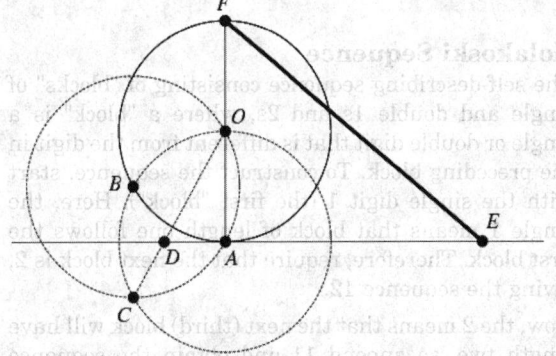

The approximation for PI given by

$$\pi \approx \sqrt{\frac{40}{3} - 2\sqrt{3}} = \tfrac{1}{3}\sqrt{120 - 18\sqrt{3}} = 3.141533\ldots.$$

In the above figure, let $OA = AF = 1$, and construct the circle centered at $A = (0, 0)$ of radius 1. This intersects O at point $B = (-\sqrt{3}/2, 1/2)$. Now construct the circle about B with radius 1. The circles A and B intersect in $C = (-\sqrt{3}/2, -1/2)$, and the line

CO intersects the perpendicular to OA through A in the point $D = (-\sqrt{3}/3,\ 0)$. Now construct the point $E = (3 - \sqrt{3}/3,\ 0)$ to be a distance 3 along DA. The line segment EF is then of length

$$\sqrt{2^2 + \left(3 - \tfrac{1}{2}\sqrt{3}\right)^2} = \sqrt{\frac{40}{3} - 2\sqrt{3}}.$$

This construction was given by the Polish Jesuit priest Kochansky (Steinhaus 1983).

See also GEOMETRIC CONSTRUCTION, PI

References

Bold, B. *Famous Problems of Geometry and How to Solve Them.* New York: Dover, p. 44, 1982.
Kochansky. *Acta Eruditorum.* 1685.
Steinhaus, H. *Mathematical Snapshots, 3rd ed.* New York: Dover, p. 143, 1999.

Kodaira Embedding Theorem

A theorem which states that if a KÄHLER FORM represents an INTEGRAL COHOMOLOGY CLASS on a COMPACT MANIFOLD, then it must be a PROJECTIVE VARIETY.

See also KÄHLER FORM

Koenigs-Poincaré Theorem

Let G denote the group of GERMS of holomorphic diffeomorphisms of $(\mathbb{C},\ 0)$. Then if $|\lambda| \neq 1$, then G_λ is a conjugacy class, i.e., all $f \in G_\lambda$ are linearizable.

References

Marmi, S. An Introduction to Small Divisors Problems 27 Sep 2000. http://xxx.lanl.gov/abs/math.DS/0009232/.

Kolakoski Sequence

The self-describing sequence consisting of "blocks" of single and double 1s and 2s, where a "block" is a single or double digit that is different from the digit in the preceding block. To construct the sequence, start with the single digit 1 (the first "block"). Here, the single 1 means that block of length one follows the first block. Therefore, require that the next block is 2, giving the sequence 12.

Now, the 2 means that the next (third) block will have length two, so append 11 and obtain the sequence 1211. We have added two 1s, so the fourth and fifth blocks have length one each, giving 12112 and then 121121. As a result of adding 21, we obtain 121121221. As a result of adding 221, we obtain 12112122122112, and so on, giving the sequence 1, 2, 1, 1, 2, 1, 2, 2, 1, 2, 2, 1, 1, 2, ... (Sloane's A006928). The sequence after successive iterations is given by 1, 12, 1211, 121121, 121121221, ..., and the lengths of this sequence after steps $n = 1, 2, ...$ are given by 1, 2, 4, 6, 9, 14, 22, ... (Sloane's A042942).

If the sequence is started with 1, 2, 2 and the above procedure is undertaken beginning with the last 2, then the virtually identical sequence 1, 2, 2, 1, 1, 2, 1, 2, 2, 1, 2, 2, 1, 1, 2, ... (Sloane's A000002) is obtained. (It is the same as Sloane's A006928, except that the second 2 is doubled.) When presented in this form, the term $a(n)$ gives the length of the nth RUN in the sequence. The lengths after steps $n = 1, 2, ...$ are then 1, 2, 3, 5, 7, 10, 15, ... (Sloane's A001083), essentially one less than Sloane's A042942.

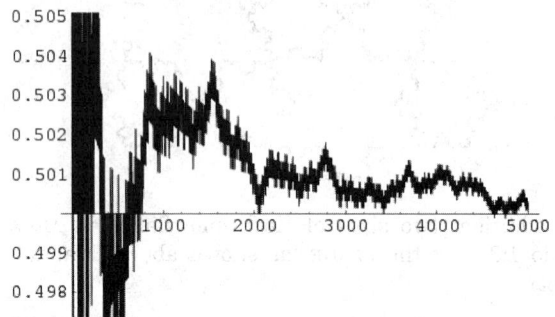

The question of whether the number of 1s is "asymptotically" equal to the number of 2s is unsettled, although the above plot (which shows the fraction of 1s as a function of number of digits) is certainly consistent with 1 and 2 being equidistributed.

See also RUN

References

Dekking, F. M. "What Is the Long Range Order in the Kolakoski Sequence?" *Reports of the Faculty of Technical Mathematics and Informatics,* No. 95–100. Delft, Netherlands: Delft University of Technology, 1995.
Kimberling, C. "Integer Sequences and Arrays." http://cedar.evansville.edu/~ck6/integer/.
Kimberling, C. "Unsolved Problems and Rewards." http://cedar.evansville.edu/~ck6/integer/unsolved.html.
Kolakoski, W. "Problem 5304: Self Generating Runs." *Amer. Math. Monthly* **72**, 674, 1965.
Kolakoski, W. "Problem 5304." *Amer. Math. Monthly* **73**, 681–682, 1966.
Lagarias, J. C. "Number Theory and Dynamical Systems." In *The Unreasonable Effectiveness of Number Theory* (Ed. S. A. Burr). Providence, RI: Amer. Math. Soc., pp. 35–72, 1992.
Paun, G. and Salomaa, A. "Self-Reading Sequences." *Amer. Math. Monthly* **103**, 166–168, 1996.
Sellke. Problem 324 in *Statistica Neerlandica* **50**, 222–223, 1996.
Sloane, N. J. A. Sequences A000002/M0190, A001083, and A006298/M0070, A042942 in "An On-Line Version of the Encyclopedia of Integer Sequences." http://www.research.att.com/~njas/sequences/eisonline.html.
Vardi, I. *Computational Recreations in Mathematica.* Redwood City, CA: Addison-Wesley, p. 233, 1991.

Kollros' Theorem

For every ring containing p SPHERES, there exists a ring of q SPHERES, each touching each of the p SPHERES, where

$$\frac{1}{p} + \frac{1}{q} = \frac{1}{3}.$$

The HEXLET is a special case with $p = 3$.

See also HEXLET, SPHERE

References

Honsberger, R. *Mathematical Gems II.* Washington, DC: Math. Assoc. Amer., p. 50, 1976.

Kolmogorov Complexity

The complexity of a pattern parameterized as the shortest ALGORITHM required to reproduce it. Also known as ALGORITHMIC COMPLEXITY.

References

Goetz, P. "Phil's Good Enough Complexity Dictionary." http://www.cs.buffalo.edu/~goetz/dict.html.

Kolmogorov Constant

The exponent 5/3 in the spectrum of homogeneous turbulence, $k^{-5/3}$.

References

Le Lionnais, F. *Les nombres remarquables.* Paris: Hermann, p. 41, 1983.

Kolmogorov Criterion

STRONG LAW OF LARGE NUMBERS

Kolmogorov Entropy

Also known as METRIC ENTROPY. Divide PHASE SPACE into D-dimensional HYPERCUBES of CONTENT ϵ^D. Let $P_{i_0, \ldots, i_n}$ be the probability that a trajectory is in HYPERCUBE i_0 at $t = 0$, i_1 at $t = T$, i_2 at $t = 2T$, etc. Then define

$$K_n = h_K = -\sum_{i_0, \ldots, i_n} P_{i_0, \ldots, i_n} \ln P_{i_0, \ldots, i_n}, \tag{1}$$

where $K_{N+1} - K_N$ is the information needed to predict which HYPERCUBE the trajectory will be in at $(n+1)T$ given trajectories up to nT. The Kolmogorov entropy is then defined by

$$K \equiv \lim_{T \to 0} \lim_{\epsilon \to 0^+} \lim_{N \to \infty} \frac{1}{NT} \sum_{n=0}^{N-1} (K_{n+1} - K_n). \tag{2}$$

The Kolmogorov entropy is related to LYAPUNOV CHARACTERISTIC EXPONENTS by

$$h_K = \int_P \sum_{\sigma_i > 0} \sigma_i \, d\mu. \tag{3}$$

See also HYPERCUBE, LYAPUNOV CHARACTERISTIC EXPONENT

References

Ott, E. *Chaos in Dynamical Systems.* New York: Cambridge University Press, p. 138, 1993.
Schuster, H. G. *Deterministic Chaos: An Introduction, 3rd ed.* New York: Wiley, p. 112, 1995.

Kolmogorov-Arnold-Moser Theorem

A theorem outlined in 1954 by Kolmogorov which was subsequently proved in the 1960s by Arnold and Moser (Tabor 1989, p. 105). It gives conditions under which CHAOS is restricted in extent. Moser's 1962 proof was valid for TWIST MAPS

$$\theta' = \theta + 2\pi f(I) + g(\theta, I) \tag{1}$$

$$I' = I + f(\theta, I). \tag{2}$$

In 1963, Arnold produced a proof for Hamiltonian systems

$$H = H_0(\mathbf{I}) + \epsilon H_1(\mathbf{I}). \tag{3}$$

The original theorem required perturbations $\epsilon \sim 10^{-48}$, although this has since been significantly increased. Arnold's proof required C^∞, and Moser's original proof required C^{333}. Subsequently, Moser's version has been reduced to C^6, then $C^{2+\epsilon}$, although counterexamples are known for C^2. Conditions for applicability of the KAM theorem are:

1. small perturbations,
2. smooth perturbations, and
3. sufficiently irrational WINDING NUMBER.

Moser considered an integrable Hamiltonian function H_0 with a TORUS T_0 and set of frequencies ω having an incommensurate frequency vector ω^* (i.e., $\omega \cdot \mathbf{k} \neq 0$ for all INTEGERS k_i). Let H_0 be perturbed by some periodic function H_1. The KAM theorem states that, if H_1 is small enough, then for almost every ω^* there exists an invariant TORUS $T(\omega^*)$ of the perturbed system such that $T(\omega^*)$ is "close to" $T_0(\omega^*)$. Moreover, the TORI $T(\omega^*)$ form a set of POSITIVE measures whose complement has a measure which tends to zero as $|H_1| \to 0$. A useful paraphrase of the KAM theorem is, "For sufficiently small perturbation, almost all TORI (excluding those with rational frequency vectors) are preserved." The theorem thus explicitly excludes TORI with rationally related frequencies, that is, $n-1$ conditions of the form

$$\omega \cdot \mathbf{k} = 0. \tag{4}$$

These TORI are destroyed by the perturbation. For a system with two DEGREES OF FREEDOM, the condition of closed orbits is

$$\sigma = \frac{\omega_1}{\omega_2} = \frac{r}{s}. \tag{5}$$

For a QUASIPERIODIC ORBIT, σ is IRRATIONAL. KAM shows that the preserved TORI satisfy the irrationality condition

$$\left| \frac{\omega_1}{\omega_2} - \frac{r}{s} \right| > \frac{K(\epsilon)}{s^{2.5}} \qquad (6)$$

for all r and s, although not much is known about $K(\epsilon)$.

The KAM theorem broke the deadlock of the small divisor problem in classical perturbation theory, and provides the starting point for an understanding of the appearance of CHAOS. For a HAMILTONIAN SYSTEM, the ISOENERGETIC NONDEGENERACY condition

$$\left| \frac{\partial^2 H_0}{\partial I_j \, \partial I_j} \right| \neq 0 \qquad (7)$$

guarantees preservation of most invariant TORI under small perturbations $\epsilon \ll 1$. The Arnold version states that

$$\left| \sum_{k=1}^{n} m_k \omega_k \right| > K(\epsilon) \left(\sum_{k=1}^{n} |m_k| \right)^{-n-1} \qquad (8)$$

for all $m_k \in \mathbb{Z}$. This condition is less restrictive than Moser's, so fewer points are excluded.

See also CHAOS, HAMILTONIAN SYSTEM, QUASIPERIODIC FUNCTION, TORUS

References

Tabor, M. *Chaos and Integrability in Nonlinear Dynamics: An Introduction.* New York: Wiley, 1989.

Kolmogorov-Sinai Entropy
KOLMOGOROV ENTROPY, METRIC ENTROPY

Kolmogorov-Smirnov Test
A goodness-of-fit test for any STATISTICAL DISTRIBUTION. The test relies on the fact that the value of the sample cumulative density function is asymptotically normally distributed.

To apply the Kolmogorov-Smirnov test, calculate the cumulative frequency (normalized by the sample size) of the observations as a function of class. Then calculate the cumulative frequency for a true distribution (most commonly, the NORMAL DISTRIBUTION). Find the greatest discrepancy between the observed and expected cumulative frequencies, which is called the "D-STATISTIC." Compare this against the critical D-STATISTIC for that sample size. If the calculated D-STATISTIC is greater than the critical one, then reject the NULL HYPOTHESIS that the distribution is of the expected form. The test is an R-ESTIMATE.

See also ANDERSON-DARLING STATISTIC, D-STATISTIC, KUIPER STATISTIC, NORMAL DISTRIBUTION, R-ESTIMATE

References

Boes, D. C.; Graybill, F. A.; and Mood, A. M. *Introduction to the Theory of Statistics, 3rd ed.* New York: McGraw-Hill, 1974.
DeGroot, M. H. Ch. 9 in *Probability and Statistics, 3rd ed.* Reading, MA: Addison-Wesley, 1991.
Knuth, D. E. §3.3.1B in *The Art of Computer Programming, Vol. 2: Seminumerical Algorithms, 3rd ed.* Reading, MA: Addison-Wesley, pp. 45–52, 1998.
Press, W. H.; Flannery, B. P.; Teukolsky, S. A.; and Vetterling, W. T. "Kolmogorov-Smirnov Test." In *Numerical Recipes in FORTRAN: The Art of Scientific Computing, 2nd ed.* Cambridge, England: Cambridge University Press, pp. 617–620, 1992.

König's Theorem
If an ANALYTIC FUNCTION has a single simple POLE at the RADIUS OF CONVERGENCE of its POWER SERIES, then the ratio of the coefficients of its POWER SERIES converges to that POLE.

See also POLE

References

König, J. "Über eine Eigenschaft der Potenzreihen." *Math. Ann.* **23**, 447–449, 1884.

König-Egeváry Theorem
A theorem on BIPARTITE GRAPHS.

See also BIPARTITE GRAPH, FROBENIUS-KÖNIG THEOREM

Königsberg Bridge Problem

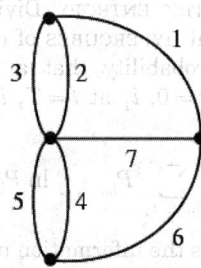

The Königsberg bridges cannot all be traversed in a single trip without doubling back. This problem was solved by Euler (1736), and represented the beginning of GRAPH THEORY.

See also CIRCUIT, EULERIAN CIRCUIT, GRAPH THEORY, UNICURSAL CIRCUIT

References

Biggs, N. L.; Lloyd, E. K.; and Wilson, R. J. *Graph Theory 1736–1936.* Oxford, England: Oxford University Press, 1976.
Bogomolny, A. "Graphs." http://www.cut-the-knot.com/do_you_know/graphs.html.
Chartrand, G. "The Königsberg Bridge Problem: An Introduction to Eulerian Graphs." §3.1 in *Introductory Graph Theory.* New York: Dover, pp. 51–66, 1985.

Euler, L. "Solutio problematis ad geometriam situs pertinentis." *Comment. Acad. Sci. U. Petrop.* **8**, 128–140, 1736. Reprinted in *Opera Omnia Ser. I-7*, pp. 1–10, 1766.

Harary, F. *Graph Theory.* Reading, MA: Addison-Wesley, pp. 1–2, 1994.

Kraitchik, M. §8.4.1 in *Mathematical Recreations.* New York: W. W. Norton, pp. 209–211, 1942.

Newman, J. "Leonhard Euler and the Königsberg Bridges." *Sci. Amer.* **189**, 66–70, 1953.

Pappas, T. "Königsberg Bridge Problem & Topology." *The Joy of Mathematics.* San Carlos, CA: Wide World Publ./Tetra, pp. 124–125, 1989.

Skiena, S. *Implementing Discrete Mathematics: Combinatorics and Graph Theory with Mathematica.* Reading, MA: Addison-Wesley, p. 192, 1990.

Steinhaus, H. *Mathematical Snapshots, 3rd ed.* New York: Dover, pp. 256–259, 1999.

Wilson, R. J. "An Eulerian Trail through Königsberg." *J. Graph Th.* **10**, 265–275, 1986.

Kontorovich-Lebedev Transform

The forward and inverse Kontorovich-Lebedev transforms are defined by

$$K_{ix}[f(t)] = \int_0^\infty K_{ix}(t)f(t)\, dt$$

$$K_{ix}^{-1}[g(t)] = \frac{2}{\pi^2 x} \int_0^\infty t\, \sinh(\pi t) K_{it}(x)g(t)\, dt,$$

respectively, where $K_\nu(z)$ is a MODIFIED BESSEL FUNCTION OF THE SECOND KIND with imaginary index $\nu = ix$.

References

Samko, S. G.; Kilbas, A. A.; and Marichev, O. I. *Fractional Integrals and Derivatives.* Yverdon, Switzerland: Gordon and Breach, p. 753, 1993.

Kontsevich Integral

This entry contributed by SERGEI DUZHIN AND S. CHMUTOV

Kontsevich's integral is a far-reaching generalization of the GAUSS INTEGRAL for the LINKING NUMBER, and provides a tool to construct the UNIVERSAL VASSILIEV INVARIANT of a knot. In fact, any VASSILIEV KNOT INVARIANT can be derived from it.

To construct the Kontsevich integral, represent the three-dimensional space $\mathbb{R}^3$ as a DIRECT PRODUCT of a complex line $\mathbb{C}$ with coordinate z and a real line $\mathbb{R}$ with coordinate t. The integral is defined for MORSE KNOTS, i.e., knots K embedded in $\mathbb{R}^3 = \mathbb{C}_z \times \mathbb{R}_t$ in such a way that the coordinate t is a MORSE FUNCTION on K, and its values belong to the GRADED COMPLETION $\bar{\mathcal{A}}$ of the ALGEBRA OF CHORD DIAGRAMS $\mathcal{A}$.

The Kontsevich integral $Z(K)$ of the knot K is defined as

$$Z(K) = \sum_{m=0}^\infty \frac{1}{(2\pi i)^m} \int\limits_{\substack{t_{\min} < t_1 < \ldots < t_m < t_{\max} \\ t_j \text{ are noncritical}}} \sum_{P=\{(z_j, z_j')\}} (-1)^{\downarrow} Dp$$

$$\times \bigwedge_{j=1}^m \frac{dz_j - dz_j'}{z_j - z_j'}, \tag{1}$$

where the ingredients of this formula have the following meanings. The real numbers $t_{\min}$ and $t_{\max}$ are the minimum and the maximum of the function t on K.

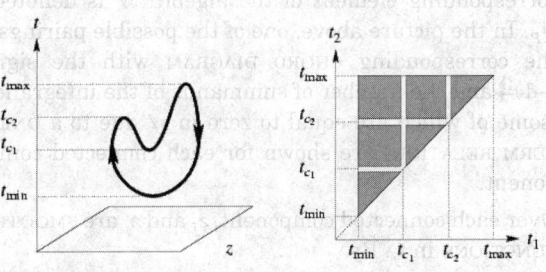

The integration domain is the m-dimensional simplex $t_{\min} < t_1 < \ldots < t_m < t_{\max}$ divided by the critical values into a certain number of connected components. For example, for the embedding of the unknot and $m = 2$ (left figure), the corresponding integration domain has six connected components, illustrated in the right figure above.

The number of summands in the integrand is constant in each connected component of the integration domain, but can be different for different components. In each plane $\{t = t_j\} \subset \mathbb{R}^3$, choose an unordered pair of distinct points (z_j, t_j) and (z_j', t_j) on K so that $z_j(t_j)$ and $z_j'(t_j)$ are continuous functions. Denote by $P = \{(z_j, z_j')\}$ the set of such pairs for $j = 1, \ldots, m$, then the integrand is the sum over all choices of P. In the example above, for the component $\{t_{\min} < t_1 < t_{c_1}, t_{c_2} < t_2 < t_{\max}\}$, we have only one possible pair of points on the levels $\{t = t_1\}$ and $\{t = t_2\}$. Therefore, the sum over P for this component consists of only one summand. In contrast, in the component $\{t_{\min} < t_1 < t_{c_1}, t_{c_1} < t_2 < t_{c_2}\}$, we still have only one possibility for the level $\{t = t_1\}$, but the plane $\{t = t_2\}$ intersects our knot K in four points. So we have $\binom{4}{2} = 6$ possible pairs (z_2, z_2'), and the total number of summands is six (see the picture below).

For a pairing P the symbol "$\downarrow$" denotes the number of points (z_j, t_j) or (z_j', t_j) in P where the coordinate t decreases along the ORIENTATION of K.

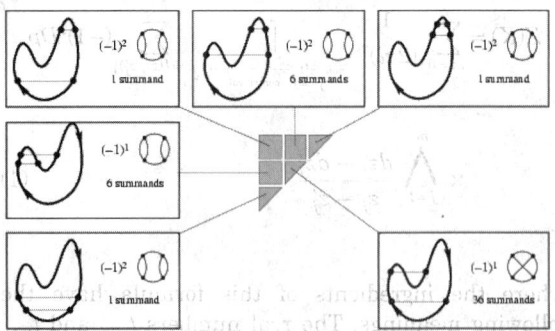

Fix a pairing P, consider the knot K as an oriented circle, and connect the points (z_j, t_j) and (z'_j, t_j) by a chord to obtain a chord diagram with m chords. The corresponding element of the algebra $\mathscr{A}$ is denoted D_P. In the picture above, one of the possible pairings, the corresponding CHORD DIAGRAM with the sign $(-1)^{\downarrow}$, and the number of summands of the integrand (some of which are equal to zero in $\mathscr{A}$ due to a ONE-TERM RELATION) are shown for each connected component.

Over each connected component, z_j and z'_j are SMOOTH FUNCTIONS in t_j. By

$$\bigwedge_{j=1}^{m} \frac{dz_j - dz'_j}{z_j - z'_j}$$

we mean the PULLBACK of this form to the integration domain of variables $t_1, ..., t_m$. The integration domain is considered with the ORIENTATION of the space $\mathbb{R}^m$ defined by the natural order of the coordinates $t_1, ..., t_m$.

By convention, the term in the Kontsevich integral corresponding to $m = 0$ is the (only) CHORD DIAGRAM of order 0 with coefficient one. It represents the unit of the algebra $\mathscr{A}$.

The Kontsevich integral is convergent thanks to ONE-TERM RELATIONS. It is invariant under DEFORMATIONS of the knot in the class of MORSE KNOTS. Unfortunately, the Kontsevich integral is not invariant under deformations that change the number of critical points of the function t. However, the formula shows how the integral changes under such deformations:

$$Z\left(\;\raisebox{-1ex}{\includegraphics{}}\;\right) = Z(H) \cdot Z\left(\;\raisebox{-1ex}{\includegraphics{}}\;\right).$$

In the above equation, the graphical arguments of Z represent two embeddings of an arbitrary knot, differing only in the illustrated fragment,

H is the *hump* (i.e, the UNKNOT embedded in $\mathbb{R}^3$ in the specified way; illustrated above), and the product is the product in the completed algebra $\bar{\mathscr{A}}$ of CHORD DIAGRAMS. The last equality allows the definition of the UNIVERSAL VASSILIEV INVARIANT by the formula

$$I(K) = \frac{Z(K)}{Z(H)^{c/2}}, \tag{2}$$

where c denotes the number of critical points of K and quotient means division in the algebra $\bar{\mathscr{A}}$ according to the rule $(1 + a)^{-1} = 1 - a + a^2 - a^3 + \dots$. The UNIVERSAL VASSILIEV INVARIANT $I(K)$ is invariant under an arbitrary DEFORMATION of K.

Consider a function w on the set of CHORD DIAGRAMS with m chords satisfying ONE- AND FOUR-TERM RELATIONS (a WEIGHT SYSTEM). Applying this function to the UNIVERSAL VASSILIEV INVARIANT $w(I(K))$, we get a numerical knot invariant. This invariant will be a VASSILIEV INVARIANT of order m, and any VASSILIEV INVARIANT can be obtained in this way.

The Kontsevich integral behaves in a nice way with respect to the natural operations on knots, such as mirror reflection, changing the orientation of the knot, and mutation of knots. In a proper normalization it is multiplicative under the CONNECTED SUM of knots:

$$I'(K_1 \# K_2) = I'(K_1)I'(K_2), \tag{3}$$

where $I'(K) = Z(H)I(K)$. For any knot K the coefficients in the expansion of $Z(K)$ over an arbitrary basis consisting of CHORD DIAGRAMS are rational (Kontsevich 1993, Le and Murakami 1996).

The task of computing the Kontsevich integral is very difficult. The explicit expression of the universal Vassiliev invariant $I(K)$ is currently known only for the UNKNOT,

$$I(O) = \exp\left(\sum_{n=0}^{\infty} b_{2n} w_{2n}\right) \tag{4}$$

$$= 1 + \left(\sum_{n=0}^{\infty} b_{2n} w_{2n}\right) + \frac{1}{2}\left(\sum_{n=0}^{\infty} b_{2n} w_{2n}\right)^2 + \dots. \tag{5}$$

(Bar-Natan *et al.* 1997). Here, b_{2n} are MODIFIED BERNOULLI NUMBERS, i.e., the coefficients of the TAYLOR SERIES

$$\sum_{n=0}^{\infty} b_{2n} x^{2n} = \frac{1}{2} \ln\left(\frac{e^{x/2} - e^{-x/2}}{\frac{1}{2}x}\right) \tag{6}$$

$(b_2 = 1/48,\ b_4 = -1/5760,\ \ldots;$ Sloane's A057868), and w_{2n} are the *wheels*, i.e., diagrams of the form

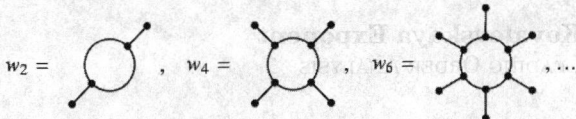

$$w_2 = \bigcirc \quad,\quad w_4 = \bigcirc \!\!\!\!\!\! \quad,\quad w_6 = \bigcirc \!\!\!\!\!\! \quad,\ \ldots$$

The linear combination is understood as an element of the ALGEBRA OF CHINESE CHARACTERS $\mathscr{B}$, which is isomorphic to the ALGEBRA OF CHORD DIAGRAMS $\mathscr{A}$. Expressed through CHORD DIAGRAMS, the beginning of this series looks as follows:

$$1 - \frac{1}{24}\bigotimes - \frac{1}{5760}\oplus + \frac{1}{1152}\bigotimes + \frac{1}{2880}\bigotimes + \ldots$$

The Kontsevich integral was invented by Kontsevich (1993), and detailed expositions can be found in Arnol'd (1994), Bar-Natan (1995), and Chmutov and Duzhin (2000).

See also CHORD DIAGRAM, GAUSS INTEGRAL, MORSE KNOT, VASSILIEV INVARIANT

References

Arnol'd, V. I. "Vassiliev's Theory of Discriminants and Knots." In *First European Congress of Mathematics, Vol. 1 (Paris, 1992)* 3764327987 (Ed. A. Joseph, F. Mignot, F. Murat, B. Prum, and R. Rentschler). Basel, Switzerland: Birkhäuser, pp. 3–29, 1994.

Bar-Natan, D.; Garoufalidis, S.; Rozansky, L.; and Thurston, D. "Wheels, Wheeling, and the Kontsevich Integral of the Unknot." Preprint, 1997.

Bar-Natan, D. "On the Vassiliev Knot Invariants." *Topology* **34** 423–472, 1995.

Chmutov, S. V. and Duzhin, S. V. "The Kontsevich Integral." To appear in *Acta Appl. Math.*, 2000. ftp://ftp.botik.ru/pub/local/zmr/ki.ps.gz.

Kontsevich, M. "Vassiliev's Knot Invariants." *Adv. Soviet Math.* **16**, Part 2, 137–150, 1993.

Le, T. Q. T. and Murakami, J. "The Universal Vassiliev-Kontsevich Invariant for Framed Oriented Links." *Compos. Math.* **102**, 42–64, 1996.

Sloane, N. J. A. Sequences A057868 in "An On-Line Version of the Encyclopedia of Integer Sequences." http://www.research.att.com/~njas/sequences/eisonline.html.

Vassiliev, V. A. "Cohomology of Knot Spaces." In *Theory of Singularities and Its Applications* (Ed. V. I. Arnold). *Adv. Soviet Math.* **1**, 23–69, 1990.

Kontsevich's Integral

See also VASSILIEV INVARIANT

Korselt's Criterion

n DIVIDES $a^n - a$ for all INTEGERS a IFF n is SQUAREFREE and $(p-1)|n/p-1$ for all PRIME DIVISORS p of n. CARMICHAEL NUMBERS satisfy this CRITERION.

See also CARMICHAEL NUMBER

References

Borwein, D.; Borwein, J. M.; Borwein, P. B.; and Girgensohn, R. "Giuga's Conjecture on Primality." *Amer. Math. Monthly* **103**, 40–50, 1996.

Korteweg de Vries Equation

The PARTIAL DIFFERENTIAL EQUATION

$$K_0 = 1$$

See also KADOMTSEV-PETVIASHVILI EQUATION, KRICHEVER-NOVIKOV EQUATION

References

Baker, H. F. *Abelian Functions: Abel's Theorem and the Allied Theory, Including the Theory of the Theta Functions.* New York: Cambridge University Press, p. xix, 1995.

Segal, G. "The Geometry of the KdV Equation." *Int. J. Math. Phys. A* **6**, 2859–2869, 1991.

Zwillinger, D. (Ed.). *CRC Standard Mathematical Tables and Formulae.* Boca Raton, FL: CRC Press, p. 417, 1995.

Korteweg-de Vries Equation

The PARTIAL DIFFERENTIAL EQUATION

$$u_t + u_{xxx} - 6uu_x = 0 \tag{1}$$

(Lamb 1980; Zwillinger 1997, p. 131), often abbreviated "KdV."

The so-called generalized KdV equation is given by

$$u_t + uu_x - u_{xxxxx} = 0 \tag{2}$$

(Boyd 1986; Zwillinger 1997, p. 131). The so-called deformed KdV equation is given by

$$u_t + \frac{\partial}{\partial x}\left(u_{xx} - 2\eta u^3 - \frac{3}{2}\frac{uu_x^2}{\eta + u^2}\right) = 0 \tag{3}$$

(Dodd and Fordy 1983; Zwillinger 1997, p. 133), and the modified KdV equation is given by

$$u_t + u_{xxx} \pm 6u^2 u_x = 0 \tag{4}$$

(Calogero and Degasperis 1982, p. 51; Tabor 1990, p. 304; Zwillinger 1997, p. 133), or

$$u_t + u_{xxx} - \frac{1}{8}u_x^3 + u_x(Ae^u + B + Ce^{-u}) = 0 \tag{5}$$

(Dodd and Fordy 1983; Zwillinger 1997, p. 133).

The cylindrical KdV equation is given by

$$u_t + u_{xxx} - 6uu_x + \frac{u}{2t} = 0 \tag{6}$$

(Calogero and Degasperis 1982, p. 50; Zwillinger 1997, p. 131), and the spherical KdV by

$$u_t + u_{xxx} - 6uu_x + \frac{u}{t} = 0 \tag{7}$$

(Calogero and Degasperis 1982, p. 51; Zwillinger 1997, p. 132).

See also Kadomtsev-Petviashvili Equation, Korteweg-de Vries-Burger Equation, Krichever-Novikov Equation, Regularized Long-Wave Equation, Soliton

References

Baker, H. F. *Abelian Functions: Abel's Theorem and the Allied Theory, Including the Theory of the Theta Functions.* New York: Cambridge University Press, p. xix, 1995.

Boyd, J. P. "Solitons from Sine Waves: Analytical and Numerical Methods of Non-Integrable Solitary and Cnoidal Waves." *Physica D* **21**, 227–246, 1986.

Calogero, F. and Degasperis, A. *Spectral Transform and Solitons: Tools to Solve and Investigate Nonlinear Evolution Equations.* New York: North-Holland, 1982.

Dodd, R. and Fordy, A. "The Prolongation Structures of Quasi-Polynomial Flows." *Proc. Roy. Soc. A* **385**, 389–429, 1983.

Gardner, C. S. "The Korteweg-de Vries Equation and Generalizations, IV. The Korteweg-de Vries Equation as a Hamiltonian System." *J. Math. Phys.* **12**, 1548–1551, 1971.

Gardner, C. S.; Greene, C. S.; Kruskal, M. D.; and Miura, R. M. "Method for Solving the Korteweg-de Vries Equation." *Phys. Rev. Lett.* **19**, 1095–1097, 1967.

Infeld, E. and Rowlands, G. *Nonlinear Waves, Solitons, and Chaos, 2nd ed.* Cambridge, England: Cambridge University Press, 2000.

Korteweg, D. J. and de Vries, F. "On the Change of Form of Long Waves Advancing in a Rectangular Canal, and on a New Type of Long Stationary Waves." *Philos. Mag.* **39**, 422–443, 1895.

Lamb, G. L. Jr. Ch. 4 in *Elements of Soliton Theory.* New York: Wiley, 1980.

Miles, J. W. "The Korteweg-de Vries Equation, A Historical Essay." *J. Fluid Mech.* **106**, 131–147, 1981.

Russell, J. S. "Report on Waves." *Report of the 14th Meeting of the British Association for the Advancement of Science.* London: Jon Murray, pp. 311–390, 1844.

Segal, G. "The Geometry of the KdV Equation." *Int. J. Math. Phys. A* **6**, 2859–2869, 1991.

Tabor, M. "Nonlinear Evolution Equations and Solitons." Ch. 7 in *Chaos and Integrability in Nonlinear Dynamics: An Introduction.* New York: Wiley, pp. 278–321, 1989.

Zakharov, V. E. and Faddeev, L. D. "Korteweg-de Vries Equation, A Completely Integrable System." *Funct. Anal. Appl.* **5**, 280–287, 1971.

Zwillinger, D. (Ed.). *CRC Standard Mathematical Tables and Formulae.* Boca Raton, FL: CRC Press, p. 417, 1995.

Zwillinger, D. *Handbook of Differential Equations, 3rd ed.* Boston, MA: Academic Press, p. 131, 1997.

Korteweg-de Vries-Burger Equation

The Partial Differential Equation

$$u_t + 2uu_x - vu_{xx} + \mu u_{xxx} = 0.$$

See also Korteweg-de Vries Equation

References

Canosa, J. and Gazdag, J. "The Korteweg-de Vries-Burgers Equation." *J. Comput. Phys.* **23**, 393–403, 1977.

Zwillinger, D. *Handbook of Differential Equations, 3rd ed.* Boston, MA: Academic Press, p. 131, 1997.

Kovalevskaya Exponent

Leading Order Analysis

Kovalevskaya Top Equations

The system of Ordinary Differential Equations

$$\frac{d\mathbf{m}}{dt} = \lambda\mathbf{m} \times \mathbf{m} + \gamma \times \mathbf{1}$$

$$\frac{d\gamma}{dt} = \lambda\gamma \times \mathbf{m}.$$

References

Haine, L. and Horozov, E. "A Lax Pair for Kowalevski's Top." *Physica D* **29**, 173–180, 1987.

Zwillinger, D. *Handbook of Differential Equations, 3rd ed.* Boston, MA: Academic Press, p. 136, 1997.

Kozyrev-Grinberg Theory

A theory of Hamiltonian circuits.

See also Grinberg Formula, Hamiltonian Circuit

k-Partite Graph

A k-partite graph is a Graph whose Vertices can be partitioned into k Disjoint Sets so that no two vertices within the same set are adjacent.

See also Complete K-Partite Graph, K-Graph

References

Saaty, T. L. and Kainen, P. C. *The Four-Color Problem: Assaults and Conquest.* New York: Dover, p. 12, 1986.

Kramers Equation

The Partial Differential Equation

$$P_t = P_{xx} - uP_x + \frac{\partial}{\partial x}\{[u - F(x)]P\}.$$

References

Duck, P. W.; Marshall, T. W.; and Watson, E. J. "First-Passage Times for the Uhlenbeck-Ornstein Process." *J. Phys. A: Math. Gen.* **19**, 3545–3558, 1986.

Zwillinger, D. *Handbook of Differential Equations, 3rd ed.* Boston, MA: Academic Press, p. 130, 1997.

Kramers Rate

The characteristic escape rate from a stable state of a potential in the absence of signal.

See also Stochastic Resonance

Kramp's Symbol

References

Bulsara, A. R. and Gammaitoni, L. "Tuning in to Noise." *Phys. Today* **49**, 39–45, March 1996.

Kramp's Symbol

The symbol defined by

$$c^{a/b} \equiv c(c+b)(c+2b) \cdots [c+(a-1)b] \tag{1}$$

$$= b^a \left(\frac{c}{b}\right)_a \tag{2}$$

$$= \frac{b^a \Gamma\left(a + \dfrac{c}{b}\right)}{\Gamma\left(\dfrac{c}{b}\right)}, \tag{3}$$

where $(a)_n$ is the POCHHAMMER SYMBOL and $\Gamma(z)$ is the GAMMA FUNCTION. Note that the definition by Erdélyi *et al.* (1981, p. 52) incorrectly gives the PREFACTOR of (3) as b^{a-1}.

See also HANKEL'S SYMBOL, POCHHAMMER SYMBOL

References

Erdélyi, A.; Magnus, W.; Oberhettinger, F.; and Tricomi, F. G. *Higher Transcendental Functions, Vol. 1.* New York: Krieger, p. 52, 1981.

Krattenthaler Matrix Inversion Formula

Let (a_i) and (b_i) be sequences of complex numbers such that $b_j \neq b_k$ for $j \neq k$, and let the LOWER TRIANGULAR MATRICES $F = (F(n, k))$ and $G = (G(n, k))$ be defined as

$$F(n, k) = \frac{\prod_{j=k}^{n-1}(a_j + k)}{\prod_{j=k+1}^{n}(b_j - b_k)}$$

and

$$G(n, k) = \frac{a_k + b_k}{a_n + b_n} \frac{\prod_{j=k+1}^{n}(a_j + b_n)}{\prod_{j=k}^{n-1}(b_j - b_n)},$$

where the product over an EMPTY SET is 1. Then F and G are MATRIX INVERSES (Bhatnagar 1995, pp. 16–17). This result simplifies to the GOULD AND HSU MATRIX INVERSION FORMULA when $b_k = k$, to Carlitz's q-analog for $b_k = q^k$ (Carlitz 1972), and to Bressoud's matrix theorem for $b_k = q^{-k} + aq^k$ and $a_k = -(aq^{-j}/b) - bq^j$ (Bressoud 1983).

The formula can be extended to a summation theorem which generalizes Gosper's bibasic sum (Gasper and Rahman 1990, p. 240; Bhatnagar 1995, p. 19).

See also GOULD AND HSU MATRIX INVERSION FORMULA

References

Bhatnagar, G. *Inverse Relations, Generalized Bibasic Series, and their U(n) Extensions.* Ph.D. thesis. Ohio State University, 1995.
Bressoud, D. M. "A Matrix Inverse." *Proc. Amer. Math. Soc.* **88**, 446–448, 1983.
Carlitz, L. "Some Inversion Relations." *Duke Math. J.* **40**, 803–901, 1972.
Gasper, G. and Rahman, M. *Basic Hypergeometric Series.* Cambridge, England: Cambridge University Press, 1990.
Krattenthaler, C. "Operator Methods and Lagrange Inversions: A Unified Approach to Lagrange Formulas." *Trans. Amer. Math. Soc.* **305**, 431–465, 1988.
Riordan, J. *Combinatorial Identities.* New York: Wiley, 1979.

Krawtchouk Polynomial

Let $\alpha(x)$ be a STEP FUNCTION with the JUMP

$$j(x) = \binom{N}{x} p^x q^{N-x} \tag{1}$$

at $x = 0, 1, ..., N$, where $p > 0$, $q > 0$, and $p + q = 1$. Then the Krawtchouk polynomial is defined by

$$k_n^{(p)}(x, N) = \sum_{v=0}^{n} (-1)^{n-v} \binom{N-x}{n-v} \binom{x}{v} p^{n-v} q^v, \tag{2}$$

$$= (-1)^n \binom{N}{n} p^n {}_2F_1(-n, -x; -N; 1/p) \tag{3}$$

$$= \frac{(-1)^n p^n}{n!} \frac{\Gamma(N-x+1)}{\Gamma(N-x-n+1)}$$

$$\times {}_2F_1(-n, -x; N-x-n+1; (p-1)/p). \tag{4}$$

for $n = 0, 1, ..., N$. The first few Krawtchouk polynomials are

$$k_0^{(p)}(x, N) = 1$$

$$k_1^{(p)}(x, N) = -Np + x$$

$$k_2^{(p)}(x, N) = \tfrac{1}{2}[N^2 p^2 + x(2p + x - 1) - Np(p + 2x)].$$

Koekoek and Swarttouw (1998) define the Krawtchouk polynomial without the leading coefficient as

$$K_n(x; p, N) = {}_2F_1(-n, -x; -N; 1/p). \tag{5}$$

The Krawtchouk polynomials have WEIGHT FUNCTION

$$w = \frac{N! p^x q^{N-x}}{\Gamma(1+x)\Gamma(N+1-x)}, \tag{6}$$

where $\Gamma(x)$ is the GAMMA FUNCTION, RECURRENCE RELATION

$$(n+1)k_{n+1}^{(p)}(x, N) + pq(N-n+1)k_{n-1}^{(p)}(x, N)$$

$$= [x - n - (N-2)]k_n^{(p)}(x, N), \tag{7}$$

and squared norm

$$\frac{N!}{n!(N-n)!}(pq)^n. \tag{8}$$

It has the limit

$$\lim_{n\to\infty}\left(\frac{2}{Npq}\right)^{n/2} n!k_n^{(p)}(Np+\sqrt{2Npq}\,s,\,N)=H_n(s), \tag{9}$$

where $H_n(x)$ is a HERMITE POLYNOMIAL.

The Krawtchouk polynomials are a special case of the MEIXNER POLYNOMIALS OF THE FIRST KIND.

See also MEIXNER POLYNOMIAL OF THE FIRST KIND, ORTHOGONAL POLYNOMIALS

References

Koekoek, R. and Swarttouw, R. F. "Krawtchouk." §1.10 in *The Askey-Scheme of Hypergeometric Orthogonal Polynomials and its q-Analogue.* Delft, Netherlands: Technische Universiteit Delft, Faculty of Technical Mathematics and Informatics Report 98–17, pp. 46–47, 1998. ftp://www.twi.tudelft.nl/publications/tech-reports/1998/DUT-TWI-98-17.ps.gz.

Koepf, W. *Hypergeometric Summation: An Algorithmic Approach to Summation and Special Function Identities.* Braunschweig, Germany: Vieweg, p. 115, 1998.

Nikiforov, A. F.; Uvarov, V. B.; and Suslov, S. S. *Classical Orthogonal Polynomials of a Discrete Variable.* New York: Springer-Verlag, 1992.

Szego, G. *Orthogonal Polynomials, 4th ed.* Providence, RI: Amer. Math. Soc., pp. 35–37, 1975.

Zelenkov, V. "Krawtchouk Polynomial Home Page." http://www.isir.minsk.by/~zelenkov/physmath/kr_polyn/.

Kreisel Conjecture

A CONJECTURE in DECIDABILITY theory which postulates that, if there is a uniform bound to the lengths of shortest proofs of instances of $S(n)$, then the universal generalization is necessarily provable in PEANO ARITHMETIC. The CONJECTURE was proven true by M. Baaz in 1988 (Baaz and Pudlák 1993).

See also DECIDABLE

References

Baaz, M. and Pudlák P. "Kreisel's Conjecture for $L\exists_1$." In *Arithmetic, Proof Theory, and Computational Complexity, Papers from the Conference Held in Prague, July 2–5, 1991* (Ed. P. Clote and J. Krajicek). New York: Oxford University Press, pp. 30–60, 1993.

Dawson, J. "The Gödel Incompleteness Theorem from a Length of Proof Perspective." *Amer. Math. Monthly* **86**, 740–747, 1979.

Kreisel, G. "On the Interpretation of Nonfinitistic Proofs, II." *J. Symbolic Logic* **17**, 43–58, 1952.

Krichever-Novikov Equation

The PARTIAL DIFFERENTIAL EQUATION

$$\frac{u_t}{u_x}=\frac{1}{4}\frac{u_{xxx}}{u_x}-\frac{3}{8}\frac{u_{xx}^2}{u_x^2}+\frac{3}{2}\frac{p(u)}{u_x^2},$$

where

$$p(u)=\tfrac{1}{4}(4u^3-g_2u-g_3).$$

The special cases $p(u)=(u-e_1)^2(u-e_2)$ and $p(u)=u^3$ can be reduced to the KORTEWEG-DE VRIES EQUATION by a change of variables.

See also KADOMTSEV-PETVIASHVILI EQUATION, KORTEWEG-DE VRIES EQUATION

References

Krichever, I. M. and Novikov, S. P. "Holomorphic Bundles over Algebraic Curves, and Nonlinear Equations." *Russ. Math. Surv.* **35**, 53–80, 1980. English translation of *Uspekhi Mat. Nauk* **35**, 47–68, 1980.

Mokhov, O. I. "Canonical Hamiltonian Representation of the Krichever-Novikov Equation." *Math. Notes* **50**, 939–945, 1991. English translation of *Mat. Zametki* **50**, 87–96, 1991.

Novikov, D. P. "Algebraic-Geometric Solutions of the Krichever-Novikov Equation." *Theoret. Math. Phys.* **121**, 1567–15773, 1999.

Sokolov, V. V. "Hamiltonian Property of the Krichever-Novikov Equation." *Dokl. Akad. Nauk SSSR* **277**, 48–50, 1984.

Svinolupov, S. I.; Sokolov, V. V.; and Yamilov, R. I. "Bäcklund Transformations for Integrable Evolution Equations." *Dokl. Akad. Nauk SSSR* **271**, 802–805, 1983. English translation of *Sov. Math. Dokl.* **28**, 165–168, 1983.

Kronecker Decomposition Theorem

Every FINITE ABELIAN GROUP can be written as a GROUP DIRECT PRODUCT of CYCLIC GROUPS of PRIME POWER ORDERS. In fact, the number of nonisomorphic ABELIAN FINITE GROUPS $a(n)$ of any given ORDER n is given by writing n as

$$n=\prod_i p_i^{\alpha_i},$$

where the p_i are distinct PRIME FACTORS, then

$$a(n)=\prod_i P(\alpha_i),$$

where $P(n)$ is the PARTITION FUNCTION. This gives 1, 1, 1, 2, 1, 1, 1, 3, 2, ... (Sloane's A000688).

See also ABELIAN GROUP, FINITE GROUP, ORDER (GROUP), PARTITION FUNCTION P

References

Sloane, N. J. A. Sequences A000688/M0064 in "An On-Line Version of the Encyclopedia of Integer Sequences." http://www.research.att.com/~njas/sequences/eisonline.html.

Kronecker Delta

The simplest interpretation of the Kronecker delta is as the discrete version of the DELTA FUNCTION defined by

$$\delta_{ij}\equiv\begin{cases}0 & \text{for } i\neq j\\ 1 & \text{for } i=j.\end{cases} \tag{1}$$

It has the COMPLEX GENERATING FUNCTION

$$\delta_{mn} = \frac{1}{2\pi i} \int z^{m-n-1} \, dz, \qquad (2)$$

where m and n are INTEGERS. In 3-space, the Kronecker delta satisfies the identities

$$\delta_{ii} = 3 \qquad (3)$$

$$\delta_{ij}\epsilon_{ijk} = 0 \qquad (4)$$

$$\epsilon_{ipq}\epsilon_{jpg} = 2\delta_{ij} \qquad (5)$$

$$\epsilon_{ijk}\epsilon_{pqk} = \delta_{ip}\delta_{jq} - \delta_{iq}\delta_{jp}, \qquad (6)$$

where EINSTEIN SUMMATION is implicitly assumed, $i, j = 1, 2, 3$, and ϵ_{ijk} is the PERMUTATION SYMBOL. Technically, the Kronecker delta is a TENSOR defined by the relationship

$$\delta_l^k \frac{\partial x_i'}{\partial x_k} \frac{\partial x_l}{\partial x_j'} = \frac{\partial x_i'}{\partial x_k} \frac{\partial x_k}{\partial x_j'} = \frac{\partial x_i'}{\partial x_j'} \qquad (7)$$

Since, by definition, the coordinates x_i and x_j are independent for $i \neq j$,

$$\frac{\partial x_i'}{\partial x_j'} = \delta_j^i, \qquad (8)$$

so

$$\delta_j'^i = \frac{\partial x_i'}{\partial x_k} \frac{\partial x_l}{\partial x_j'} \, \partial_l^k, \qquad (9)$$

and δ_j^i is really a mixed second-RANK TENSOR. It satisfies

$$\delta_{ab}^{jk} = \epsilon_{abi}\epsilon^{jki} = \delta_a^j \delta_b^k - \delta_a^k \delta_b^j \qquad (10)$$

$$\delta_{abjk} = g_{aj}g_{bk} - g_{ak}g_{bj} \qquad (11)$$

$$\epsilon_{aij}\epsilon^{bij} = \delta_{ai}^{bi} = 2\delta_a^b. \qquad (12)$$

The generalization of the Kronecker delta viewed as a tensor is called the PERMUTATION TENSOR.

See also DELTA FUNCTION, PERMUTATION SYMBOL, PERMUTATION TENSOR

Kronecker Product

MATRIX DIRECT PRODUCT

Kronecker Symbol

An extension of the JACOBI SYMBOL (n/m) to all INTEGERS. It is variously written as (n/m) or $\left(\frac{n}{m}\right)$ (Cohn 1980) or $(n|m)$ (Dickson 1957). The Kronecker symbol can be computed using the normal rules for the JACOBI SYMBOL

$$\left(\frac{ab}{cd}\right) = \left(\frac{a}{cd}\right)\left(\frac{b}{cd}\right) = \left(\frac{ab}{c}\right)\left(\frac{ab}{d}\right)$$

$$= \left(\frac{a}{c}\right)\left(\frac{b}{c}\right)\left(\frac{a}{d}\right)\left(\frac{b}{d}\right) \qquad (1)$$

plus additional rules for $m = -1$,

$$(n/-1) = \begin{cases} -1 & \text{for } n < 0 \\ 1 & \text{for } n > 0, \end{cases} \qquad (2)$$

and $m = 2$. The definition for $(n/2)$ is variously written as

$$(n/2) = \begin{cases} 0 & \text{for } n \text{ even} \\ 1 & \text{for } n \text{ odd}, \ n \equiv \pm 1 \pmod 8 \\ -1 & \text{for } n \text{ odd}, \ n \equiv \pm 3 \pmod 8 \end{cases} \qquad (3)$$

or

$$(n/2) \equiv \begin{cases} 0 & \text{for } 4|n \\ 1 & \text{for } n \equiv 1 \pmod 8 \\ -1 & \text{for } n \equiv 5 \pmod 8 \\ \text{undefined} & \text{otherwise} \end{cases} \qquad (4)$$

(Cohn 1980). Cohn's form "undefines" $(n/2)$ for SINGLY EVEN NUMBERS $n \equiv 2 \pmod 4$ and $n \equiv -1, 3 \pmod 8$, probably because no other values are needed in applications of the symbol involving the DISCRIMINANTS d of QUADRATIC FIELDS, where $m > 0$ and d always satisfies $d \equiv 0, 1 \pmod 4$.

The KRONECKER SYMBOL is a REAL CHARACTER modulo n, and is, in fact, essentially the only type of REAL PRIMITIVE CHARACTER (Ayoub 1963).

See also CHARACTER (NUMBER THEORY), CLASS NUMBER, DIRICHLET L-SERIES, JACOBI SYMBOL, LEGENDRE SYMBOL, PRIMITIVE CHARACTER, QUADRATIC RESIDUE

References

Ayoub, R. G. *An Introduction to the Analytic Theory of Numbers.* Providence, RI: Amer. Math. Soc., 1963.
Cohn, H. *Advanced Number Theory.* New York: Dover, p. 35, 1980.
Dickson, L. E. "Kronecker's Symbol." §48 in *Introduction to the Theory of Numbers.* New York: Dover, p. 77, 1957.

Kronecker's Algorithm

A POLYNOMIAL FACTORIZATION algorithm that proceeds by considering the vector of coefficients of a polynomial P, calculating $b_i = P(i)/a_i$, constructing the LAGRANGE INTERPOLATING POLYNOMIALS from the conditions $A(i) = a_i$ and $B(i) = b_i$, and checking to see which are factorizations.

See also POLYNOMIAL FACTORIZATION

References

Hausmann, B. A. "A New Simplification of Kronecker's Method of Factorization of Polynomials." *Amer. Math. Monthly* **47**, 574–576, 1937.

Séroul, R. "Kronecker's Factorization Algorithm." §10.14.2 in *Programming for Mathematicians*. Berlin: Springer-Verlag, pp. 288–289, 2000.

Kronecker's Approximation Theorem

If θ is a given IRRATIONAL NUMBER, then the sequence of numbers $\{n\theta\}$, where $\{x\} \equiv x - \lfloor x \rfloor$, is DENSE in the unit interval. Explicitly, given any α, $0 \le \alpha \le 1$, and given any $\epsilon > 0$, there exists a POSITIVE INTEGER k such that

$$|\{k\theta\} - \alpha| < \epsilon.$$

Therefore, if $h = \lfloor k\theta \rfloor$, it follows that $|k\theta - h - \alpha| < \epsilon$. The restriction on α can be removed as follows. Given *any* real α, any irrational θ, and any $\epsilon > 0$, there exist integers h and k with $k > 0$ such that

$$|k\theta - h - \alpha| < \epsilon.$$

See also RATIONAL APPROXIMATION

References
Apostol, T. M. "Kronecker's Approximation Theorem: The One-Dimensional Case" and "Extension of Kronecker's Theorem to Simultaneous Approximation." §7.4 and 7.5 in *Modular Functions and Dirichlet Series in Number Theory, 2nd ed.* New York: Springer-Verlag, pp. 148–155, 1997.

Kronecker's Constant

MERTENS CONSTANT

Kronecker's Polynomial Theorem

An algebraically soluble equation of ODD PRIME degree which is irreducible in the natural FIELD possesses either

1. Only a single REAL ROOT, or
2. All REAL ROOTS.

See also ABEL'S IRREDUCIBILITY THEOREM, ABEL'S LEMMA, SCHÖNEMANN'S THEOREM

References
Dörrie, H. *100 Great Problems of Elementary Mathematics: Their History and Solutions.* New York: Dover p. 127, 1965.

Krull Dimension

If R is a RING (commutative with 1), the height of a PRIME IDEAL p is defined as the SUPREMUM of all n so that there is a chain $p_0 \subset \cdots p_{n-1} \subset p_n = p$ where all p_i are distinct PRIME IDEALS. Then, the Krull dimension of R is defined as the SUPREMUM of all the heights of all its PRIME IDEALS.

See also PRIME IDEAL

References
Eisenbud, D. *Commutative Algebra with a View Toward Algebraic Geometry.* New York: Springer-Verlag, 1995.
Macdonald, I. G. and Atiyah, M. F. *Introduction to Commutative Algebra.* Reading, MA: Addison-Wesley, 1969.

Kruskal's Algorithm

An ALGORITHM for finding a GRAPH's spanning TREE of minimum length.

See also KRUSKAL'S TREE THEOREM

References
Gardner, M. *Mathematical Magic Show: More Puzzles, Games, Diversions, Illusions and Other Mathematical Sleight-of-Mind from Scientific American.* New York: Vintage, pp. 248–249, 1978.

Kruskal's Tree Theorem

A theorem which plays a fundamental role in computer science because it is one of the main tools for showing that certain orderings on TREES are well-founded. These orderings play a crucial role in proving the termination of rewriting rules and the correctness of the Knuth-Bendix equational completion procedures.

See also KRUSKAL'S ALGORITHM, NATURAL INDEPENDENCE PHENOMENON, TREE

References
Gallier, J. "What's so Special about Kruskal's Theorem and the Ordinal Gamma[0]? A Survey of Some Results in Proof Theory." *Ann. Pure and Appl. Logic* **53**, 199–260, 1991.

KS Entropy

METRIC ENTROPY

k-Statistic

The ith k-statistic k_i is an UNBIASED ESTIMATOR of the CUMULANT κ_i of a given DISTRIBUTION, i.e., k_i is defined so that $\langle k_i \rangle = \kappa_i$, where $\langle x \rangle$ denotes the EXPECTATION VALUE of x (Kenney and Keeping 1951, p. 189). For a SAMPLE SIZE n, the first few k-statistics are given by

$$k_1 = \mu \tag{1}$$

$$k_2 = \frac{n}{n-1} m_2 \tag{2}$$

$$k_3 = \frac{n^2}{(n-1)(n-2)} m_3 \tag{3}$$

$$k_4 = \frac{n^2[(n+1)m_4 - 3(n-1)m_2^2]}{(n-1)(n-2)(n-3)}, \tag{4}$$

where μ is the sample MEAN, m_2 is the SAMPLE VARIANCE, and m_i is the sample ith CENTRAL MOMENT

(Kenney and Keeping 1951, pp. 109–110, 163–165, and 189; Kenney and Keeping 1962).

The k-statistics can be obtained by defining the sums of the rth powers of the data points as

$$s_r \equiv \sum_{i=1}^{n} X_i^r, \qquad (5)$$

then the CENTRAL MOMENTS m_i are given in terms of the s_r by

$$m_2 = -\frac{s_1^2}{n^2} + \frac{s_2}{n} \qquad (6)$$

$$m_3 = \frac{2s_1^3}{n^3} - \frac{3s_1 s_2}{n^2} + \frac{s_3}{n} \qquad (7)$$

$$m_4 = -\frac{3s_1^4}{n^4} + \frac{6s_1^2 s_2}{n^3} - \frac{4s_1 s_3}{n^2} + \frac{s_4}{n}. \qquad (8)$$

Taking the raw expectations of these equations and expressing the answers in terms of moments m_i using

$$m_i = \frac{s_i}{n} \qquad (9)$$

then gives the expectation values of the *observed* central moments m_i in terms of the population central moments as

$$\langle m_2 \rangle = \frac{n-1}{n} \mu_2 \qquad (10)$$

$$\langle m_3 \rangle = \frac{(n-1)(n-2)}{n^2} \mu_3 \qquad (11)$$

$$\langle m_4 \rangle = \frac{(n-1)[(n^2 - 3n + 3)\mu_4 + 3(2n - 3)\mu_2^2]}{n^3}, \qquad (12)$$

together with

$$\langle m_2^2 \rangle = \frac{(n-1)[(n-1)\mu_4 + (n^2 - 2n + 3)\mu_2^2]}{n^3} \qquad (13)$$

(Kenney and Keeping 1951, p. 189). Solving for the population central moments μ_i in terms of the expectation values of the observed central moments then gives the formulas for the k-statistics, e.g., (10) becomes

$$\mu_2 = \frac{n}{n-1} \langle m_2 \rangle, \qquad (14)$$

so

$$k_2 = \frac{n}{n-1} m_2 \qquad (15)$$

is an UNBIASED ESTIMATOR for $\kappa_2 = \mu_2$.

In terms of the power sums, the k-statistics can then be written as

$$k_2 = \frac{ns_2 - s_1^2}{n(n-1)} \qquad (16)$$

$$k_3 = \frac{2s_1^3 - 3ns_1 s_2 + n^2 s_3}{n(n-1)(n-2)} \qquad (17)$$

$$k_4 = \frac{-6s_1^4 + 12ns_1^2 s_2 - 3n(n-1)s_2^2 - 4n(n+1)s_1 s_3 + n^2(n+1)s_4}{n(n-1)(n-2)(n-3)}. \qquad (18)$$

The VARIANCE $\mathrm{var}(k_2)$ of k_2 is given by the second central expectation of k_2 which, when expressed in terms of CUMULANTS, becomes

$$\mathrm{var}(k_2) = \frac{\kappa_4}{n} + \frac{2\kappa_2^2}{n-1}. \qquad (19)$$

The UNBIASED ESTIMATOR of $\mathrm{var}(k_2)$ is

$$\widehat{\mathrm{var}(k_2)} = \frac{2k_2^2 n + (n-1)k_4}{n(n+1)} \qquad (20)$$

(Kenney and Keeping 1951, p. 189).

The VARIANCE of k_3 can be expressed in terms of CUMULANTS by

$$\mathrm{var}(k_3) = \frac{\kappa_6}{n} + \frac{9\kappa_2 \kappa_4}{n-1} + \frac{9\kappa_3^2}{n-1} + \frac{6n\kappa_2^3}{(n-1)(n-2)}, \qquad (21)$$

and the UNBIASED ESTIMATOR for $\mathrm{var}(k_3)$ is

$$\widehat{\mathrm{var}(k_3)} = \frac{6k_2^3 n(n-1)}{(n-2)(n+1)(n+3)} \qquad (22)$$

(Kenney and Keeping 1951, p. 190).

For a *finite* population, let a SAMPLE SIZE n be taken from a population size N. Then UNBIASED ESTIMATORS M_1 for the population MEAN μ, M_2 for the population VARIANCE μ_2, G_1 for the population SKEWNESS γ_1, and G_2 for the population KURTOSIS γ_2 are

$$M_1 = \mu \qquad (23)$$

$$M_2 = \frac{N-n}{n(N-1)} \mu_2 \qquad (24)$$

$$G_1 = \frac{N-2n}{N-2} \sqrt{\frac{N-1}{n(N-n)}} \gamma_1 \qquad (25)$$

$$G_2 = \frac{(N-1)(N^2 - 6Nn + N + 6n^2)\gamma_2}{n(N-2)(N-3)(N-n)} - \frac{6N(Nn + N - n^2 - 1)}{n(N-2)(N-3)(N-n)} \qquad (26)$$

(Church 1926, p. 357; Carver 1930; Irwin and Kendall 1944; Kenney and Keeping 1951, p. 143), where γ_1 is the sample SKEWNESS and γ_2 is the sample KURTOSIS.

See also CUMULANT, GAUSSIAN DISTRIBUTION, *H*-STATISTIC, KURTOSIS, MEAN, MOMENT, SKEWNESS, STATISTIC, UNBIASED ESTIMATOR, VARIANCE

References

Carver, H. C. (Ed.). "Fundamentals of the Theory of Sampling." *Ann. Math. Stat.* **1**, 101–121, 1930.
Church, A. E. R. "On the Means and Squared Standard-Deviations of Small Samples from Any Population." *Biometrika* **18**, 321–394, 1926.
Irwin, J. O. and Kendall, M. G. "Sampling Moments of Moments for a Finite Population." *Ann. Eugenics* **12**, 138–142, 1944.
Kenney, J. F. and Keeping, E. S. *Mathematics of Statistics, Pt. 2, 2nd ed.* Princeton, NJ: Van Nostrand, 1951.
Kenney, J. F. and Keeping, E. S. "The *k*-Statistics." §7.9 in *Mathematics of Statistics, Pt. 1, 3rd ed.* Princeton, NJ: Van Nostrand, pp. 99–100, 1962.

k-Subset

A *k*-subset is a SUBSET of a set on *n* elements containing exactly *k* elements. The number of *k*-subsets on *n* elements is therefore given by the BINOMIAL COEFFICIENT $\binom{n}{k}$. For example, there are $\binom{3}{2} = 3$ 2-subsets of $\{1, 2, 3\}$, namely $\{1, 2\}$, $\{1, 3\}$, and $\{2, 3\}$. The *k*-subsets on a list can be enumerated using KSubsets[*list*, *k*] in the *Mathematica* add-on package DiscreteMath`Combinatorica` (which can be loaded with the command << DiscreteMath`).

The total number of distinct *k*-subsets on a set of *n* elements (i.e., the number of SUBSETS) is given by

$$\sum_{k=0}^{n} \binom{n}{k} = 2^n.$$

See also BINOMIAL COEFFICIENT, COMBINATION, *P*-SYSTEM, PERMUTATION, SUBSET

References

Nijenhuis, A. and Wilf, H. *Combinatorial Algorithms for Computers and Calculators, 2nd ed.* New York: Academic Press, 1978.
Skiena, S. "Generating *k*-Subsets." §1.5.5 in *Implementing Discrete Mathematics: Combinatorics and Graph Theory with Mathematica.* Reading, MA: Addison-Wesley, pp. 44–46, 1990.

K-Theory

A branch of mathematics which brings together ideas from ALGEBRAIC GEOMETRY, LINEAR ALGEBRA, and NUMBER THEORY. In general, there are two main types of *K*-theory: topological and algebraic.

Topological *K*-theory is the "true" *K*-theory in the sense that it came first. Topological *K*-theory has to do with VECTOR BUNDLES over TOPOLOGICAL SPACES. Elements of a *K*-theory are STABLE EQUIVALENCE classes of VECTOR BUNDLES over a TOPOLOGICAL SPACE. You can put a RING structure on the collection of STABLY EQUIVALENT bundles by defining ADDITION through the WHITNEY SUM, and MULTIPLICATION through the TENSOR PRODUCT of VECTOR BUNDLES. This defines "the reduced real topological *K*-theory of a space."

"The reduced *K*-theory of a space" refers to the same construction, but instead of REAL VECTOR BUNDLES, COMPLEX VECTOR BUNDLES are used. Topological *K*-theory is significant because it forms a generalized COHOMOLOGY theory, and it leads to a solution to the vector fields on spheres problem, as well as to an understanding of the *J*-homeomorphism of HOMOTOPY THEORY.

Algebraic *K*-theory is somewhat more involved. Swan (1962) noticed that there is a correspondence between the CATEGORY of suitably nice TOPOLOGICAL SPACES (something like regular HAUSDORFF SPACES) and C*-ALGEBRAS. The idea is to associate to every SPACE the C*-ALGEBRA of CONTINUOUS MAPS from that SPACE to the REALS.

A VECTOR BUNDLE over a SPACE has sections, and these sections can be multiplied by CONTINUOUS FUNCTIONS to the REALS. Under Swan's correspondence, VECTOR BUNDLES correspond to modules over the C*-ALGEBRA of CONTINUOUS FUNCTIONS, the MODULES being the modules of sections of the VECTOR BUNDLE. This study of MODULES over C*-ALGEBRA is the starting point of algebraic *K*-theory.

The QUILLEN-LICHTENBAUM CONJECTURE connects algebraic *K*-theory to Étale cohomology.

See also C*-ALGEBRA

References

Atiyah, M. F. *K-Theory.* New York: Benjamin, 1967.
Bass, H.; Kuku, A. O.; and Pedrini, C. *Proceedings of the Workshop and Symposium: Algebraic K-Theory and Its Applications, ICTP, Trieste, Italy, 1–19 Sept. 1997.* Singapore: World Scientific, 1999.
Raskind, W. and Weibel, C. (Eds.). *Algebraic K-Theory: AMS-IMS-SIAM Joint Summer Research Conference on Algebraic K-Theory, July 13–24, 1997, University of Washington, Seattle.* Providence, RI: Amer. Math. Soc., 1997.
Srinivas, V. *Algebraic K-Theory, 2nd ed.* Boston, MA: Birkhäuser, 1995.
Swan, R. G. "Vector Bundles and Projective Modules." *Trans. Amer. Math. Soc.* **105**, 264–277, 1962.

k-Tuple Conjecture

The first of the HARDY-LITTLEWOOD CONJECTURES. The *k*-tuple conjecture states that the asymptotic number of PRIME CONSTELLATIONS can be computed explicitly. In particular, unless there is a trivial divisibility condition that stops p, $p + a_1$, ..., $p + a_k$ from consisting of PRIMES infinitely often, then such PRIME CONSTELLATIONS will occur with an asymptotic density which is computable in terms of a_1, ..., a_k. Let $0 < m_1 < m_2 < \ldots < m_k$, then the *k*-tuple conjecture predicts that the number of PRIMES $p \leq x$ such that

$p + 2m_1, p + 2m_2, ..., p + 2m_k$ are all PRIME is

$$P(x; m_1, m_2, ..., m_k)$$

$$\sim C(m_1, m_2, ..., m_k) \int_2^x \frac{dt}{\ln^{k+1} t}, \tag{1}$$

where

$$C(m_1, m_2, ..., m_k)$$

$$= 2k \prod_q \frac{1 - \dfrac{w(q; m_1, m_2, ..., m_k)}{q}}{\left(1 - \dfrac{1}{q}\right)^{k+1}}, \tag{2}$$

the product is over ODD PRIMES q, and

$$w(q; m_1, m_2, ..., m_k) \tag{3}$$

denotes the number of distinct residues of $0, m_1, ..., m_k \pmod q$ (Halberstam and Richert 1974, Odlyzko). If $k = 1$, then this becomes

$$C(m) = 2 \prod_q \frac{q(q-2)}{(q-1)^2} \prod_{q|m} \frac{q-1}{q-2}. \tag{4}$$

This conjecture is generally believed to be true, but has not been proven (Odlyzko *et al.*). The following special case of the conjecture is sometimes known as the PRIME PATTERNS CONJECTURE. Let S be a FINITE set of INTEGERS. Then it is conjectured that there exist infinitely many k for which $\{k + s : s \in S\}$ are all PRIME IFF S does not include all the RESIDUES of any PRIME. The TWIN PRIME CONJECTURE is a special case of the prime patterns conjecture with $S = \{0, 2\}$. This conjecture also implies that there are arbitrarily long ARITHMETIC PROGRESSIONS of PRIMES.

See also ARITHMETIC PROGRESSION, DIRICHLET'S THEOREM, HARDY-LITTLEWOOD CONJECTURES, K-TUPLE CONJECTURE, PRIME ARITHMETIC PROGRESSION, PRIME CONSTELLATION, PRIME QUADRUPLET, PRIME PATTERNS CONJECTURE, TWIN PRIME CONJECTURE, TWIN PRIMES

References

Brent, R. P. "The Distribution of Small Gaps Between Successive Primes." *Math. Comput.* **28**, 315–324, 1974.

Brent, R. P. "Irregularities in the Distribution of Primes and Twin Primes." *Math. Comput.* **29**, 43–56, 1975.

Halberstam, E. and Richert, H.-E. *Sieve Methods.* New York: Academic Press, 1974.

Hardy, G. H. and Littlewood, J. E. "Some Problems of 'Partitio Numerorum.' III. On the Expression of a Number as a Sum of Primes." *Acta Math.* **44**, 1–70, 1922.

Odlyzko, A.; Rubinstein, M.; and Wolf, M. "Jumping Champions."

Riesel, H. *Prime Numbers and Computer Methods for Factorization, 2nd ed.* Boston, MA: Birkhäuser, pp. 66–68, 1994.

Kuen Surface

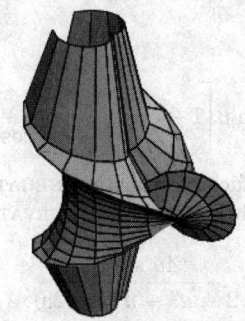

A special case of ENNEPER'S NEGATIVE CURVATURE SURFACES which can be given parametrically by

$$x = \frac{2(\cos u + u \sin u) \sin v}{1 + u^2 \sin^2 v} \tag{1}$$

$$= \frac{2\sqrt{1 + u^2} \cos(u - \tan^{-1} u) \sin v}{1 + u^2 \sin^2 v} \tag{2}$$

$$y = \frac{2(\sin u + u \cos u) \sin v}{1 + u^2 \sin^2 v} \tag{3}$$

$$= \frac{2\sqrt{1 + u^2} \sin(u - \tan^{-1} u) \sin v}{1 + u^2 \sin^2 v} \tag{4}$$

$$z = \ln\left[\tan\left(\tfrac{1}{2}v\right)\right] + \frac{2 \cos v}{1 + u^2 \sin^2 v} \tag{5}$$

for $v \in [0, \pi)$, $u \in [0, 2\pi)$ (Reckziegel *et al.* 1986; Gray 1997, p. 496).

The coefficients of the FIRST FUNDAMENTAL FORM are

$$E = \frac{16u^2 \sin^2 v}{[2 + u^2 - u^2 \cos^2(2v)]^2} \tag{6}$$

$$F = 0 \tag{7}$$

$$G = \csc^2 v - \frac{16u^2 \sin^2 v}{[2 + u^2 - u^2 \cos^2(2v)]^2}, \tag{8}$$

the SECOND FUNDAMENTAL FORM coefficients are

$$e = \frac{4u[2 - u^2 + u^2 \cos^2(2v)] \sin v}{[2 + u^2 - u^2 \cos^2(2v)]^2}, \tag{9}$$

$$f = 0 \tag{10}$$

$$g = \frac{4u[2 - u^2 + u^2 \cos^2(2v)] \csc v}{[2 + u^2 - u^2 \cos^2(2v)]^2}, \tag{11}$$

and the surface area element is

$$dS = \frac{4u[2 - u^2 + u^2 \cos^2(2v)]}{[2 + u^2 - u^2 \cos^2(2v)]^2}. \tag{12}$$

The GAUSSIAN and MEAN CURVATURES are

$$K = -1 \tag{13}$$

$$H = -\frac{\csc v}{4u}$$

$$+\frac{1}{4} u \sin v \left[1 + \frac{8}{2 - u^2 + u^2 \cos(2v)} \right], \tag{14}$$

so the Kuen surface has constant NEGATIVE GAUSSIAN CURVATURE, and the PRINCIPAL CURVATURES are

$$\kappa_1 = \frac{4u \sin v}{2 - u^2 + u^2 \cos(2v)} \tag{15}$$

$$\kappa_2 = \frac{[2 - u^2 + u^2 \cos(2v)] \csc v}{4u} \tag{16}$$

(Gray 1997, p. 496).

See also ENNEPER'S NEGATIVE CURVATURE SURFACES

References

--. Cover of *La Gaceta de la Real Sociedad Matemática Española* **2**, 1999.

Fischer, G. (Ed.). Plate 86 in *Mathematische Modelle/ Mathematical Models, Bildband/Photograph Volume.* Braunschweig, Germany: Vieweg, p. 82, 1986.

Gray, A. "Kuen's Surface." §21.6 in *Modern Differential Geometry of Curves and Surfaces with Mathematica, 2nd ed.* Boca Raton, FL: CRC Press, pp. 496–497, 1997.

JavaView. "Classic Surfaces from Differential Geometry: Kuen." http://www-sfb288.math.tu-berlin.de/vgp/java-view/demo/surface/common/PaSurface_Kuen.html.

Kuen, T. "Ueber Flächen von constantem Krümmungs-maass." *Sitzungsber. d. königl. Bayer. Akad. Wiss. Math.-phys. Classe,* Heft II, 193–206, 1884.

Nordstrand, T. "Kuen's Surface." http://www.uib.no/people/nfytn/kuentxt.htm.

Reckziegel, H. "Kuen's Surface." §3.4.4.2 in *Mathematical Models from the Collections of Universities and Museums* (Ed. G. Fischer). Braunschweig, Germany: Vieweg, p. 38, 1986.

Kuhn-Tucker Theorem

A theorem in nonlinear programming which states that if a regularity condition holds and f and the functions h_j are convex, then a solution x^0 which satisfies the conditions h_j for a VECTOR of multipliers λ is a GLOBAL MINIMUM. The Kuhn-Tucker theorem is a generalization of LAGRANGE MULTIPLIERS. FARKAS'S LEMMA is key in proving this theorem.

See also FARKAS'S LEMMA, LAGRANGE MULTIPLIER

Kuiper Statistic

A statistic defined to improve the KOLMOGOROV-SMIRNOV TEST in the TAILS.

See also ANDERSON-DARLING STATISTIC

References

Press, W. H.; Flannery, B. P.; Teukolsky, S. A.; and Vetter-ling, W. T. *Numerical Recipes in FORTRAN: The Art of*

Scientific Computing, 2nd ed. Cambridge, England: Cambridge University Press, p. 621, 1992.

Kulikowski's Theorem

For every POSITIVE INTEGER n, there exists a SPHERE which has exactly n LATTICE POINTS on its surface. The SPHERE is given by the equation

$$(x - a)^2 + (y - b)^2 + (z - \sqrt{2})^2 = c^2 + 2,$$

where a and b are the coordinates of the center of the so-called SCHINZEL CIRCLE

$$\begin{cases} \left(x - \frac{1}{2}\right)^2 + y^2 = \frac{1}{4} 5^{k-1} & \text{for } n = 2k \text{ even} \\ \left(x - \frac{1}{3}\right)^2 + y^2 = \frac{1}{9} 5^{2k} & \text{for } n = 2k + 1 \text{ odd} \end{cases}$$

and c is its RADIUS.

See also CIRCLE LATTICE POINTS, LATTICE POINT, SCHINZEL'S THEOREM

References

Honsberger, R. "Circles, Squares, and Lattice Points." Ch. 11 in *Mathematical Gems I.* Washington, DC: Math. Assoc. Amer., pp. 117–127, 1973.

Kulikowski, T. "Sur l'existence d'une sphère passant par un nombre donné aux coordonnées entières." *L'Enseignement Math. Ser. 2* **5**, 89–90, 1959.

Schinzel, A. "Sur l'existence d'un cercle passant par un nombre donné de points aux coordonnées entières." *L'Enseignement Math. Ser. 2* **4**, 71–72, 1958.

Sierpinski, W. "Sur quelques problèmes concernant les points aux coordonnées entières." *L'Enseignement Math. Ser. 2* **4**, 25–31, 1958.

Sierpinski, W. "Sur un problème de H. Steinhaus concernant les ensembles de points sur le plan." *Fund. Math.* **46**, 191–194, 1959.

Sierpinski, W. *A Selection of Problems in the Theory of Numbers.* New York: Pergamon Press, 1964.

Kullback-Leibler Distance

RELATIVE ENTROPY

Kummer Extension

References

Koch, H. "Kummer Extensions." §6.8 in *Number Theory: Algebraic Numbers and Functions.* Providence, RI: Amer. Math. Soc., pp. 195–199, 2000.

Kummer Group

A GROUP of LINEAR FRACTIONAL TRANSFORMATIONS which transform the arguments of Kummer solutions to the HYPERGEOMETRIC DIFFERENTIAL EQUATION into each other. Define

$$A(z) = 1 - z$$

$$B(z) = 1/z,$$

then the elements of the group are $\{I, A, B, AB, BA, ABA = BAB\}$.

Kummer Surface

The Kummer surfaces are a family of QUARTIC SURFACES given by the algebraic equation

$$(x^2 + y^2 + z^2 - \mu^2 w^2)^2 - \lambda pqrs = 0, \tag{1}$$

where

$$\lambda \equiv \frac{3\mu^2 - 1}{3 - \mu^2}, \tag{2}$$

p, q, r, and s are the TETRAHEDRAL COORDINATES

$$p = w - z - \sqrt{2}x \tag{3}$$

$$q = w - z + \sqrt{2}x \tag{4}$$

$$r = w + z + \sqrt{2}y \tag{5}$$

$$s = w + z - \sqrt{2}y, \tag{6}$$

and w is a parameter which, in the above plots, is set to $w = 1$. The above plots correspond to $\mu^2 = 1/3$

$$(3x^2 + 3y^2 + 3z^2 + 1)^2 = 0, \tag{7}$$

(double sphere), 2/3, 1

$$x^4 - 2x^2y^2 + y^4 + 4x^2z + 4y^2z + 4x^2z^2 + 4y^2z^2 = 0 \tag{8}$$

(ROMAN SURFACE), $\sqrt{2}$, $\sqrt{3}$

$$[(z-1)^2 - 2x^2][y^2 - (z+1)^2] = 0 \tag{9}$$

(four planes), 2, and 5. The case $0 \le \mu^2 \le 1/3$ corresponds to four real points.

The following table gives the number of ORDINARY DOUBLE POINTS for various ranges of μ^2, corresponding to the preceding illustrations.

$0 \le \mu^2 \le \frac{1}{3}$	4	12
$\mu^2 = \frac{1}{3}$		
$\frac{1}{3} \le \mu^2 < 1$	4	12
$\mu^2 = 1$		
$1 < \mu^2 < 3$	16	0
$\mu^2 = 3$		
$\mu^2 > 3$	16	0

The Kummer surfaces can be represented parametrically by hyperelliptic THETA FUNCTIONS. Most of the Kummer surfaces admit 16 ORDINARY DOUBLE POINTS, the maximum possible for a QUARTIC SURFACE. A special case of a Kummer surface is the TETRAHEDROID.

Nordstrand gives the implicit equations as

$$x^4 + y^4 + z^4 - x^2 - y^2 - z^2 - x^2y^2 - x^2z^2 - y^2z^2 + 1 = 0 \tag{10}$$

or

$$x^4 + y^4 + z^4 + a(x^2 + y^2 + z^2) + b(x^2y^2 + x^2z^2 + y^2z^2)$$
$$+ cxyz - 1 = 0. \tag{11}$$

See also QUARTIC SURFACE, ROMAN SURFACE, TETRAHEDROID

References

Endraß, S. "Flächen mit vielen Doppelpunkten." *DMV-Mitteilungen* **4**, 17–20, Apr. 1995.

Endraß, S. "Kummer Surfaces." http://enriques.mathematik.uni-mainz.de/kon/docs/Ekummer.shtml.

Fischer, G. (Ed.). *Mathematical Models from the Collections of Universities and Museums.* Braunschweig, Germany: Vieweg, pp. 14–19, 1986.

Fischer, G. (Ed.). Plates 34–37 in *Mathematische Modelle/Mathematical Models, Bildband/Photograph Volume.* Braunschweig, Germany: Vieweg, pp. 33–37, 1986.

Gray, A. *Modern Differential Geometry of Curves and Surfaces with Mathematica, 2nd ed.* Boca Raton, FL: CRC Press, p. 313, 1997.

Guy, R. K. *Unsolved Problems in Number Theory, 2nd ed.* New York: Springer-Verlag, p. 183, 1994.

Hudson, R. *Kummer's Quartic Surface.* Cambridge, England: Cambridge University Press, 1990.

Kummer, E. "Über die Flächen vierten Grades mit sechszehn singulären Punkten." *Ges. Werke* **2**, 418–432.

Kummer, E. "Über Strahlensysteme, deren Brennflächen Flächen vierten Grades mit sechszehn singulären Punkten sind." *Ges. Werke* **2**, 418–432.

Nordstrand, T. "Kummer's Surface." http://www.uib.no/people/nfytn/kummtxt.htm.

Kummer's Conjecture

A conjecture concerning PRIMES.

Kummer's Differential Equation

CONFLUENT HYPERGEOMETRIC DIFFERENTIAL EQUATION

Kummer's Formulas

Kummer's first formula is

$${}_2F_1\left(\tfrac{1}{2} + m - k, -n; 2m + 1; 1\right)$$
$$= \frac{\Gamma(2m+1)\Gamma\left(m + \tfrac{1}{2} + k + n\right)}{\Gamma(m + \tfrac{1}{2} + k)\Gamma(2m + 1 + n)}, \tag{1}$$

where $_2F_1(a, b; c; z)$ is the HYPERGEOMETRIC FUNCTION with $m \neq -1/2, -1, -3/2, \ldots$, and $\Gamma(z)$ is the GAMMA FUNCTION. The identity can be written in the more symmetrical form as

$$_2F_1(a, b; c; -1) = \frac{\Gamma\left(\frac{1}{2}b + 1\right)\Gamma(b - a + 1)}{\Gamma(b + 1)\Gamma\left(\frac{1}{2}b - a + 1\right)}, \quad (2)$$

where $a - b + c = 1$ and b is a positive integer (Bailey 1935, p. 35; Petkovsek *et al.* 1996; Koepf 1998, p. 32; Hardy 1999, p. 106). If b is a negative integer, the identity takes the form

$$_2F_1(a, b; c; -1) = 2\cos\left(\frac{1}{2}\pi b\right)\frac{\Gamma(|b|)\Gamma(b - a + 1)}{\Gamma\left(\frac{1}{2}b - a + 1\right)} \quad (3)$$

(Petkovsek *et al.* 1996).

Kummer's second formula is

$$_1F_1\left(\frac{1}{2} + m; 2m + 1; z\right) = M_{0,m}(z)$$

$$= z^{m+1/2}\left[1 + \sum_{p=1}^{\infty}\frac{z^{2p}}{2^{4p}p!(m + 1)(m + 2)\cdots(m + p)}\right], \quad (4)$$

where $_1F_1(a; b; z)$ is the CONFLUENT HYPERGEOMETRIC FUNCTION and $m \neq -1/2, -1, -3/2, \ldots$.

See also CONFLUENT HYPERGEOMETRIC FUNCTION, HYPERGEOMETRIC FUNCTION

References

Bailey, W. N. *Generalised Hypergeometric Series.* Cambridge, England: Cambridge University Press, 1935.
Hardy, G. H. *Ramanujan: Twelve Lectures on Subjects Suggested by His Life and Work, 3rd ed.* New York: Chelsea, 1999.
Koepf, W. *Hypergeometric Summation: An Algorithmic Approach to Summation and Special Function Identities.* Braunschweig, Germany: Vieweg, 1998.
Petkovsek, M.; Wilf, H. S.; and Zeilberger, D. *A = B.* Wellesley, MA: A. K. Peters, pp. 42–43 and 126, 1996.

Kummer's Function

CONFLUENT HYPERGEOMETRIC FUNCTION

Kummer's Quadratic Transformation

A transformation of a HYPERGEOMETRIC FUNCTION,

$$_2F_1\left(\alpha, \beta; 2\beta; \frac{4z}{(1 + z)^2}\right)$$

$$= (1 + z)^{2\alpha}{}_2F_1\left(\alpha, \alpha + \frac{1}{2} - \beta; \beta + \frac{1}{2}; z^2\right).$$

Kummer's Relation

An identity which relates HYPERGEOMETRIC FUNCTIONS,

$$_2F_1\left(2a, 2b; a + b + \frac{1}{2}; x\right)$$

$$= {}_2F_1(a, b; a + b + \frac{1}{2}, 4x(1 - x)).$$

Kummer's Series

HYPERGEOMETRIC FUNCTION

Kummer's Series Transformation

Let $\Sigma_{k=0}^{\infty} a_k = a$ and $\Sigma_{k=0}^{\infty} c_k = c$ be convergent series such that

$$\lim_{k \to \infty}\frac{a_k}{c_k} = \lambda \neq 0.$$

Then

$$a = \lambda c + \sum_{k=0}^{\infty}\left(1 - \lambda\,\frac{c_k}{a_k}\right)a_k.$$

References

Abramowitz, M. and Stegun, C. A. (Eds.). *Handbook of Mathematical Functions with Formulas, Graphs, and Mathematical Tables, 9th printing.* New York: Dover, p. 16, 1972.

Kummer's Test

Given a SERIES of POSITIVE terms u_i and a sequence of finite POSITIVE constants a_i, let

$$\rho \equiv \lim_{n \to \infty}\left(a_n\,\frac{u_n}{u_{n+1}} - a_{n+1}\right).$$

1. If $\rho > 0$, the series converges.
2. If $\rho < 0$, the series diverges.
3. If $\rho = 0$, the series may converge or diverge.

The test is a general case of BERTRAND'S TEST, the ROOT TEST, GAUSS'S TEST, and RAABE'S TEST. With $a_n = n$ and $a_{n+1} = n + 1$, the test becomes RAABE'S TEST.

See also CONVERGENCE TESTS, RAABE'S TEST

References

Arfken, G. *Mathematical Methods for Physicists, 3rd ed.* Orlando, FL: Academic Press, pp. 285–286, 1985.
Jingcheng, T. "Kummer's Test Gives Characterizations for Convergence or Divergence of All Series." *Amer. Math. Monthly* **101**, 450–452, 1994.
Samelson, H. "More on Kummer's Test." *Amer. Math. Monthly* **102**, 817–818, 1995.

Kummer's Theorem

The identity

$$_2F_1(x, -x; x+n+1; -1) = \frac{\Gamma(x+n+1)\Gamma\left(\frac{1}{2}n+1\right)}{\Gamma\left(x+\frac{1}{2}n+1\right)\Gamma(n+1)},$$

or equivalently

$$_2F_1(\alpha, \beta; 1+\alpha-\beta; -1) = \frac{\Gamma(1+\alpha-\beta)\Gamma\left(1+\frac{1}{2}\alpha\right)}{\Gamma(1+\alpha)\Gamma\left(1+\frac{1}{2}\alpha-\beta\right)},$$

where $_2F_1(a, b; c; z)$ is a HYPERGEOMETRIC FUNCTION and $\Gamma(z)$ is the GAMMA FUNCTION. This formula was first stated by Kummer (1836, p. 53).

See also SAALSCHÜTZ'S THEOREM

References

Bailey, W. N. "Kummer's Theorem." §2.3 in *Generalised Hypergeometric Series.* Cambridge, England: Cambridge University Press, pp. 9–10, 1935.
Kummer, E. E. "Ueber die hypergeometrische Reihe." *J. für Math.* **15**, 39–83, 1836.

Kupershmidt Equation

The PARTIAL DIFFERENTIAL EQUATION

$$u_t = u_{xxxxx} + \frac{5}{2}u_{xxx}u + \frac{25}{4}u_{xx}u_x + \frac{5}{4}u^2u_x.$$

References

Fuchssteiner, B.; Oevel, W.; and Wiwianka, W. "Computer-Algebra Methods for Investigation of Hereditary Operators of High Order Soliton Equations." *Comput. Phys. Commun.* **44**, 47–55, 1987.
Zwillinger, D. *Handbook of Differential Equations, 3rd ed.* Boston, MA: Academic Press, p. 133, 1997.

Kuramoto-Sivashinsky Equation

The PARTIAL DIFFERENTIAL EQUATION

$$u_1 + \nabla^4 u + \nabla^2 u + \frac{1}{2}\left|\nabla^2 u\right|^2 = 0,$$

where ∇^2 is the LAPLACIAN and ∇^4 is the BIHARMONIC OPERATOR.

References

Michelson, D. "Steady Solutions of the Kuramoto-Sivashinsky Equation." *Physica D* **19**, 89–111, 1986.
Zwillinger, D. *Handbook of Differential Equations, 3rd ed.* Boston, MA: Academic Press, p. 131, 1997.

Kuratowski Reduction Theorem

Every nonplanar graph is a SUPERGRAPH of an expansion of the UTILITY GRAPH $UG = K_{3,3}$ (i.e., the COMPLETE BIPARTITE GRAPH on two sets of three vertices) or the COMPLETE GRAPH K_5. This theorem was also proven earlier by Pontryagin (1927–1928), and later by Frink and Smith (1930). Kennedy *et al.*

(1985) give a detailed history of the theorem, and there exists a generalization known as the ROBERTSON-SEYMOUR THEOREM.

See also COMPLETE BIPARTITE GRAPH, COMPLETE GRAPH, PLANAR GRAPH, ROBERTSON-SEYMOUR THEOREM, UTILITY GRAPH

References

Harary, F. "Kuratowski's Theorem." In *Graph Theory.* Reading, MA: Addison-Wesley, pp. 108–113, 1994.
Kennedy, J. W.; Quintas, L. V.; and Syslo, M. M. "The Theorem on Planar Graphs." *Historia Math.* **12**, 356–368, 1985.
Kuratowski, C. "Sur l'operation A de l'analysis situs." *Fund. Math.* **3**, 182–199, 1922.
Kuratowski, C. "Sur le problème des courbes gauches en topologie." *Fund. Math.* **15**, 217–283, 1930.
Skiena, S. *Implementing Discrete Mathematics: Combinatorics and Graph Theory with Mathematica.* Reading, MA: Addison-Wesley, p. 247, 1990.
Thomassen, C. "Kuratowski's Theorem." *J. Graph Th.* **5**, 225–241, 1981.
Thomassen, C. "A Link Between the Jordan Curve Theorem and the Kuratowski Planarity Criterion." *Amer. Math. Monthly* **97**, 216–218, 1990.

Kuratowski's Closure-Component Problem

Let X be an arbitrary TOPOLOGICAL SPACE. Denote the CLOSURE of a SUBSET A of X by A^- and the COMPLEMENT of A by A'. Then at most 14 different SETS can be derived from A by repeated application of closure and complementation (Berman and Jordan 1975, Fife 1991). The problem was first proved by Kuratowski (1922) and popularized by Kelley (1955).

See also KURATOWSKI REDUCTION THEOREM

References

Anusiak, J. and Shum, K. P. "Remarks on Finite Topological Spaces." *Colloq. Math.* **23**, 217–223, 1971.
Aull, C. E. "Classification of Topological Spaces." *Bull. de l'Acad. Pol. Sci. Math. Astron. Phys.* **15**, 773–778, 1967.
Baron, S. Advanced Problem 5569. *Amer. Math. Monthly* **75**, 199, 1968.
Beeler *et al.* Item 105 in Beeler, M.; Gosper, R. W.; and Schroeppel, R. *HAKMEM.* Cambridge, MA: MIT Artificial Intelligence Laboratory, Memo AIM-239, p. 45, Feb. 1972.
Berman, J. and Jordan, S. L. "The Kuratowski Closure-Complement Problem." *Amer. Math. Monthly* **82**, 841–842, 1975.
Buchman, E. "Problem E 3144." *Amer. Math. Monthly* **93**, 299, 1986.
Chagrov, A. V. "Kuratowski Numbers, Application of Functional Analysis in Approximation Theory." Kalinin: Kalinin Gos. Univ., pp. 186–190, 1982.
Chapman, T. A. "A Further Note on Closure and Interior Operators." *Amer. Math. Monthly* **69**, 524–529, 1962.
Fife, J. H. "The Kuratowski Closure-Complement Problem." *Math. Mag.* **64**, 180–182, 1991.
Fishburn, P. C. "Operations on Binary Relations." *Discrete Math.* **21**, 7–22, 1978.
Graham, R. L.; Knuth, D. E.; and Motzkin, T. S. "Complements and Transitive Closures." *Discrete Math.* **2**, 17–29, 1972.
Hammer, P. C. "Kuratowski's Closure Theorem." *Nieuw Arch. Wisk.* **8**, 74–80, 1960.

Herda, H. H. and Metzler, R. C. "Closure and Interior in Finite Topological Spaces." *Colloq. Math.* **15**, 211–216, 1966.

Kelley, J. L. *General Topology.* Princeton: Van Nostrand, p. 57, 1955.

Koenen, W. "The Kuratowski Closure Problem in the Topology of Convexity." *Amer. Math. Monthly* **73**, 704–708, 1966.

Kuratowski, C. "Sur l'operation A de l'analysis situs." *Fund. Math.* **3**, 182–199, 1922.

Langford, E. "Characterization of Kuratowski 14-Sets." *Amer. Math. Monthly* **78**, 362–367, 1971.

Levine, N. "On the Commutativity of the Closure and Interior Operators in Topological Spaces." *Amer. Math. Monthly* **68**, 474–477, 1961.

Moser, L. E. "Closure, Interior, and Union in Finite Topological Spaces." *Colloq. Math.* **38**, 41–51, 1977.

Munkres, J. R. *Topology: A First Course.* Englewood Cliffs, NJ: Prentice-Hall, 1975.

Peleg, D. "A Generalized Closure and Complement Phenomenon." *Discrete Math.* **50**, 285–293, 1984.

Shum, K. P. "On the Boundary of Kuratowski 14-Sets in Connected Spaces." *Glas. Mat. Ser. III* **19**, 293–296, 1984.

Shum, K. P. "The Amalgamation of Closure and Boundary Functions on Semigroups and Partially Ordered Sets." In *Proceedings of the Conference on Ordered Structures and Algebra of Computer Languages.* Singapore: World Scientific, pp. 232–243, 1993.

Smith, A. Advanced Problem 5996. *Amer. Math. Monthly* **81**, 1034, 1974.

Soltan, V. P. "On Kuratowski's Problem." *Bull. Acad. Polon. Sci. Ser. Sci. Math.* **28**, 369–375, 1981.

Soltan, V. P. "Problems of Kuratowski Type." *Mat. Issled.* **65**, 121–131 and 155, 1982.

Steen, L. A. and Seebach, J. A. Jr. *Counterexamples in Topology.* New York: Dover, 1996.

Kuratowski's Theorem

KURATOWSKI REDUCTION THEOREM

Kurschák's Theorem

The AREA of the DODECAGON ($n = 12$) inscribed in a UNIT CIRCLE with $R = 1$ is

$$A = \tfrac{1}{2} n R^2 \sin\left(\frac{2\pi}{n}\right) = 3. \tag{1}$$

See also DODECAHEDRON

References

Wells, D. *The Penguin Dictionary of Curious and Interesting Geometry.* London: Penguin, p. 137, 1991.

Kurschák's Tile

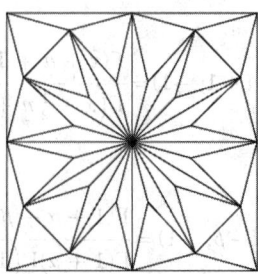

An attractive tiling of the SQUARE composed of two types of triangular tiles. It consists of 16 EQUILATERAL TRIANGLES and 32 15°-15°-150° ISOSCELES TRIANGLES arranged in the shape of a DODECAGON.

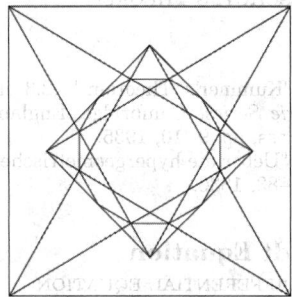

The composition of Kürschák's tile is motivated by drawing inward-pointing EQUILATERAL TRIANGLES on each side of a UNIT SQUARE and then connecting adjacent vertices to form a smaller SQUARE rotated 45° with respect to the original SQUARE. Joining the midpoints of the square together with the intersections of the EQUILATERAL TRIANGLES then gives a DODECAGON (Wells 1991) with CIRCUMRADIUS

$$R = \sin\left(\frac{\pi}{12}\right) = \tfrac{1}{4}(\sqrt{6} - \sqrt{2}).$$

See also DODECAGON, EQUILATERAL TRIANGLE, ISOSCELES TRIANGLE

References

Alexanderson, G. L. and Seydel, K. "Kürschák's Tile." *Math. Gaz.* **62**, 192–196, 1978.

Honsberger, R. *Mathematical Gems III.* Washington, DC: Math. Assoc. Amer., pp. 30–32, 1985.

Schoenberg, I. *Mathematical Time Exposures.* Washington, DC: Math. Assoc. Amer., p. 7, 1982.

Weisstein, E. W. "Kürschák's Tile." MATHEMATICA NOTEBOOK KURSCHAKSTILE.M.

Wells, D. *The Penguin Dictionary of Curious and Interesting Geometry.* London: Penguin, pp. 136–137, 1991.

Kurtosis

The degree of peakedness of a distribution, also called the "excess" or "excess coefficient." Kurtosis is a normalized form of the fourth CENTRAL MOMENT of a distribution. There are several flavors of kurtosis commonly encountered, including FISHER KURTOSIS (denoted γ_2 or b_2) and PEARSON KURTOSIS (denoted β_2 or α_4). If not specifically qualified, then term "kurtosis" is generally taken to refer to FISHER KURTOSIS. A distribution with a high peak ($\gamma_2 > 0$) is called LEPTOKURTIC, a flat-topped curve ($\gamma_2 < 0$) is called PLATYKURTIC, and the normal distribution ($\gamma_2 = 0$) is called MESOKURTIC.

Let μ_i denote the ith CENTRAL MOMENT. Then the FISHER KURTOSIS is defined by

$$\gamma_2 \equiv \frac{\mu_4}{\mu_2^2} - 3 = \frac{\mu_4}{\sigma^4} - 3, \tag{1}$$

where σ^2 is the VARIANCE. Similarly, the PEARSON KURTOSIS is defined by

$$\beta_2 \equiv \frac{\mu_4}{\mu_2^2} = \frac{\mu_4}{\sigma^4}. \tag{2}$$

An ESTIMATOR for the FISHER KURTOSIS γ_2 is given by

$$\hat{g}_2 = \frac{k_4}{k_2^2}, \tag{3}$$

where the ks are K-STATISTIC. For a normal distribution, the variance of this estimator is

$$\mathrm{var}(g_2) \approx \frac{24}{N}. \tag{4}$$

The following table lists the FISHER KURTOSIS for a number of common distributions.

distribution	FISHER KURTOSIS
BERNOULLI DISTRIBUTION	$\dfrac{1}{1-p} + \dfrac{1}{p} - 6$
BETA DISTRIBUTION	$\dfrac{6[a^3 + a^2(1 - 2b) + b^2(1 + b) - 2ab(2 + b)]}{ab(2 + a + b)(3 + a + b)}$
BINOMIAL DISTRIBUTION	$\dfrac{6p^2 - 6p + 1}{np(1 - p)}$
CHI-SQUARED DISTRIBUTION	$\dfrac{12}{r}$
EXPONENTIAL DISTRIBUTION	6
FISHER-TIPPETT DISTRIBUTION	$\dfrac{12}{5}$
GAMMA DISTRIBUTION	$\dfrac{6}{a}$
GEOMETRIC DISTRIBUTION	$5 - p + \dfrac{1}{1 - p}$

HALF-NORMAL DISTRIBUTION	$\dfrac{8(\pi - 3)}{(\pi - 2)^2}$
LAPLACE DISTRIBUTION	3
LOG NORMAL DISTRIBUTION	$e^{4S^2} + 2e^{3S^2} + 3e^{2S^2} - 6$
MAXWELL DISTRIBUTION	$-\dfrac{4}{3}$
NEGATIVE BINOMIAL DISTRIBUTION	$\dfrac{6 - p(6 - p)}{r(1 - p)}$
NORMAL DISTRIBUTION	0
POISSON DISTRIBUTION	$\dfrac{1}{v}$
RAYLEIGH DISTRIBUTION	$\dfrac{6\pi(4 - \pi) - 16}{(\pi - 4)^2}$
STUDENT'S T-DISTRIBUTION	$\dfrac{6}{n - 4}$
continuous UNIFORM DISTRIBUTION	$-\dfrac{6}{5}$
discrete UNIFORM DISTRIBUTION	$\dfrac{6(n^2 + 1)}{5(n^2 - 1)}$

See also FISHER KURTOSIS, MEAN, PEARSON KURTOSIS, SKEWNESS, STANDARD DEVIATION

References

Abramowitz, M. and Stegun, C. A. (Eds.). *Handbook of Mathematical Functions with Formulas, Graphs, and Mathematical Tables, 9th printing.* New York: Dover, p. 928, 1972.

Darlington, R. B. "Is Kurtosis Really Peakedness?" *Amer. Statist.* **24**, 19–22, 1970.

Dodge, Y. and Rousson, V. "The Complications of the Fourth Central Moment." *Amer. Statist.* **53**, 267–269, 1999.

Kenney, J. F. and Keeping, E. S. "Kurtosis." §7.12 in *Mathematics of Statistics, Pt. 1, 3rd ed.* Princeton, NJ: Van Nostrand, pp. 102–103, 1962.

Moors, J. J. A. "The Meaning of Kurtosis: Darlington Reexamined." *Amer. Statist.* **40**, 283–284, 1986.

Press, W. H.; Flannery, B. P.; Teukolsky, S. A.; and Vetterling, W. T. "Moments of a Distribution: Mean, Variance, Skewness, and So Forth." §14.1 in *Numerical Recipes in FORTRAN: The Art of Scientific Computing, 2nd ed.* Cambridge, England: Cambridge University Press, pp. 604–609, 1992.

Rupert, D. "What is Kurtosis? An Influence Function Approach." *Amer. Statist.* **41**, 1–5, 1987.

L

L1-Norm

A VECTOR NORM defined for a VECTOR

$$\mathbf{x} = \begin{bmatrix} x_1 \\ x_2 \\ \vdots \\ x_n \end{bmatrix},$$

with COMPLEX entries by

$$\|\mathbf{x}\|_1 = \sum_{r=1}^{n} |x_r|.$$

The vector norm $\|\mathbf{x}\|_1$ is implemented as Vector-Norm[m, 1] in the *Mathematica* add-on package LinearAlgebra`MatrixMultiplication` (which can be loaded with the command << LinearAlgebra`).

See also $L1$-SPACE, $L2$-NORM, L-INFINITY-NORM, VEC-TOR NORM

References

Gradshteyn, I. S. and Ryzhik, I. M. *Tables of Integrals, Series, and Products, 6th ed.* San Diego, CA: Academic Press, pp. 1114–125, 2000.

L1-Space

See also $L1$-NORM

L2-Function

Informally, an L^2-function is a function $f : X \to \mathbb{R}$ that is SQUARE INTEGRABLE, i.e.,

$$\|f\|^2 = \int_X |f|^2 \, d\mu$$

with respect to the MEASURE μ, exists (and is finite), in which case $\|f\|$ is its $L2$-NORM. Here X is a MEASURE SPACE and the integral is the LEBESGUE INTEGRAL. The collection of L^2 functions on X is called $L^2(X)$ (ell-two) of $L2$-SPACE, which is a HILBERT SPACE.

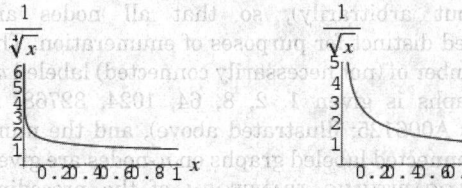

On the unit interval $(0, 1)$, the functions $f(x) = 1/x^p$ are in L^2 for $p < 1/2$. However, the function $f(x) = x^{-1/2}$ is not in L^2 since

$$\int_0^1 (x^{-1/2})^2 \, dx = \int_0^1 \frac{dx}{x}$$

does not exist.

More generally, there are L^2-COMPLEX FUNCTIONS obtained by replacing the ABSOLUTE VALUE of a REAL NUMBER in the definition with the NORM of the COMPLEX NUMBER. In fact, this generalizes to functions from a MEASURE SPACE X to any NORMED SPACE.

L^2-functions play an important role in many areas of ANALYSIS. They also arise in physics, and especially quantum mechanics, where probabilities are given as the integral of the absolute square of a wavefunction ψ. In this and in the context of energy density, L^2-functions arise due to the requirement that these quantities remain finite.

See also HILBERT SPACE, LEBESGUE INTEGRAL, LP-SPACE, $L2$-SPACE, MEASURE, MEASURE SPACE, SQUARE INTEGRABLE

L2-Inner Product

The L^2-inner product of two REAL FUNCTIONS f and g on a MEASURE SPACE X with respect to the MEASURE μ is given by

$$\langle f, g \rangle_{L^2} = \int_X fg \, d\mu,$$

sometimes also called the bracket product, where the symbol $\langle f, g \rangle$ are called ANGLE BRACKETS. If the functions are COMPLEX, the generalization of the HERMITIAN INNER PRODUCT

$$\int_X f\bar{g} \, d\mu$$

is used.

See also ANGLE BRACKET, BRA, HILBERT SPACE, KET, LEBESGUE INTEGRAL, $L2$-FUNCTION, $L2$-SPACE

L2-Norm

A VECTOR NORM defined for a VECTOR

$$\mathbf{x} = \begin{bmatrix} x_1 \\ x_2 \\ \vdots \\ x_n \end{bmatrix}, \tag{1}$$

with COMPLEX entries by

$$\|\mathbf{x}\|_2 = \sqrt{\sum_{r=1}^{n} |x_r|^2}. \tag{2}$$

This discrete norm for a vector is sometimes called the l_2-norm, while the L_2-norm (denoted with an upper-case L) is reserved for application with a function $\phi(x)$, where it is defined by

$$\|\phi\|^2 \equiv \phi \cdot \phi \equiv \langle \phi | \phi \rangle \equiv \int |\phi(x)|^2 \, dx, \tag{3}$$

with $\langle f | g \rangle$ denoting an ANGLE BRACKET.

The L_2-norm $\|\mathbf{x}\|_2$ is also called the Euclidean norm, and is implemented as VectorNorm[m, 2] in the *Mathematica* add-on package LinearAlgebra`MatrixMultiplication` (which can be loaded with the command < < LinearAlgebra`).

See also ANGLE BRACKET, COMPLETE SET OF FUNCTIONS, L1-NORM, L2-SPACE, L-INFINITY-NORM, PARALLELOGRAM LAW, VECTOR NORM

References

Gradshteyn, I. S. and Ryzhik, I. M. *Tables of Integrals, Series, and Products, 6th ed.* San Diego, CA: Academic Press, pp. 1114–125, 2000.

L2-Space

On a MEASURE SPACE X, the set of SQUARE INTEGRABLE L2-FUNCTIONS is an L^2-space. Taken together with the L2-INNER PRODUCT (a.k.a. BRACKET PRODUCT) with respect to a MEASURE μ,

$$\langle f, g \rangle \equiv \int_X fg \, d\mu \tag{1}$$

the L^2-space forms a HILBERT SPACE. The functions in an L^2-space satisfy

$$\langle \phi | \psi \rangle \equiv \int \bar{\psi} \phi \, dx \tag{2}$$

and

$$\overline{\langle \phi | \psi \rangle} = \langle \psi | \phi \rangle \tag{3}$$

$$\langle \phi | \lambda_1 \psi_1 + \lambda_2 \psi_2 \rangle = \lambda_1 \langle \phi | \psi_1 \rangle + \lambda_2 \langle \phi | \psi_2 \rangle \tag{4}$$

$$\langle \lambda_1 \phi_1 + \lambda_2 \phi_2 | \psi \rangle = \bar{\lambda}_1 \langle \phi_1 | \psi \rangle + \bar{\lambda}_2 \langle \phi_2 | \psi \rangle \tag{5}$$

$$\langle \psi | \psi \rangle \in \mathbb{R} \geq 0 \tag{6}$$

$$|\langle \psi_1 | \psi_2 \rangle|^2 \leq \langle \psi_1 | \psi_1 \rangle \langle \psi_2 | \psi_2 \rangle. \tag{7}$$

The inequality (7) is called SCHWARZ'S INEQUALITY.

The basic example is when $X = \mathbb{R}$ with LEBESGUE MEASURE. Another important example is when X is the positive integers, in which case it is denoted as l^2, or "little ell-two." These are the square summable SERIES.

Strictly speaking, L^2-space really consists of EQUIVALENCE CLASSES of functions. Two functions represent the same L^2-function if the set where they differ has measure zero. It is not hard to see that this makes $\langle f, g \rangle$ an inner product, because $\langle f, f \rangle = 0$ if and only if $f = 0$ ALMOST EVERYWHERE. A good way to think of an L^2-function is as a density function, so only its integral on sets with positive measure matter.

In practice, this does not cause much trouble, except that some care has to be taken with boundary conditions in DIFFERENTIAL EQUATIONS. The problem is that for any particular point p, the value $f(p)$ isn't WELL DEFINED for an L^2-function f.

If an L^2-function in EUCLIDEAN SPACE can be represented by a continuous function f, then f is the only continuous representative. In such a case, it is not harmful to consider the L^2-function as the continuous function f. Also, it is often convenient to think of $L^2(\mathbb{R}^n)$ as the COMPLETION of the CONTINUOUS functions with respect to the L2-NORM.

See also BRACKET PRODUCT, COMPLETION, HILBERT SPACE, L2-NORM, $L$$P$-SPACE, L-FUNCTION, LEBESGUE INTEGRAL, LEBESGUE MEASURE, MEASURE, MEASURE SPACE, RIESZ-FISCHER THEOREM, SCHWARZ'S INEQUALITY

Labeled Graph

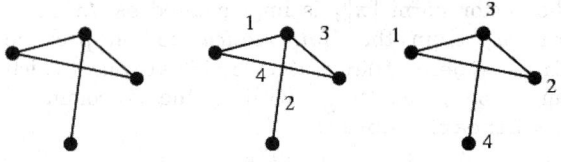

unlabeled graph   *edge-labeled graph*   *vertex-labeled graph*

A labeled graph $G = (V, E)$ is a finite series of VERTICES V with a set of EDGES E of 2-SUBSETS of V. Given a VERTEX set $V_n = \{1, 2, \ldots, n\}$, the number of vertex-labeled graphs is given by $2^{n(n-1)/2}$. Two graphs G and H with VERTICES $V_n = \{1, 2, \ldots, n\}$ are said to be ISOMORPHIC if there is a PERMUTATION p of V_n such that $\{u, v\}$ is in the set of EDGES $E(G)$ IFF $\{p(u), p(v)\}$ is in the set of EDGES $E(H)$.

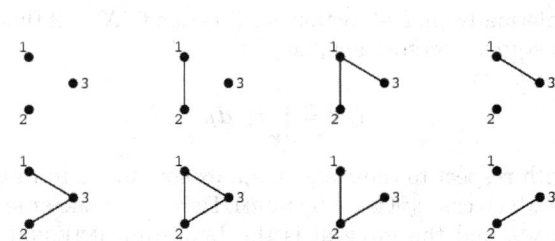

The term "labeled graph" when used without qualification means a graph with each node labeled differently (but arbitrarily), so that all nodes are considered distinct for purposes of enumeration. The *total* number of (not necessarily connected) labeled n-node graphs is given 1, 2, 8, 64, 1024, 32768, ... (Sloane's A006125; illustrated above), and the numbers of connected labeled graphs on n-nodes are given by the LOGARITHMIC TRANSFORM of the preceding sequence, 1, 1, 4, 38, 728, 26704, ... (Sloane's A001187; Sloane and Plouffe 1995, p. 19).

See also 15 PUZZLE, A-CORDIAL GRAPH, CONNECTED GRAPH, CORDIAL GRAPH, EDGE-GRACEFUL GRAPH, ELEGANT GRAPH, EQUITABLE GRAPH, GRACEFUL GRAPH, GRAPH, H-CORDIAL GRAPH, HARMONIOUS GRAPH, LABELED TREE, MAGIC GRAPH, ORIENTED

GRAPH, SUPER-EDGE-GRACEFUL GRAPH, TAYLOR'S
CONDITION, UNLABELED GRAPH, WEIGHTED TREE

References

Cahit, I. "Homepage for the Graph Labelling Problems and
　New Results." http://www.emu.edu.tr/~cahit/COR-
　DIAL.htm.
Gallian, J. A. "Graph Labeling." *Elec. J. Combin.* DS6, 1–2,
　Apr. 15, 1999. http://www.combinatorics.org/Surveys/.
Gilbert, E. N. "Enumeration of Labeled Graphs." *Canad. J.
　Math.* **8**, 405–11, 1956.
Harary, F. "Labeled Graphs." *Graph Theory.* Reading, MA:
　Addison-Wesley, pp. 10 and 178–80, 1994.
Sloane, N. J. A. Sequences A001187/M3671 and A006125/
　M1897 in "An On-Line Version of the Encyclopedia of
　Integer Sequences." http://www.research.att.com/~njas/
　sequences/eisonline.html.
Sloane, N. J. A. and Plouffe, S. *The Encyclopedia of Integer
　Sequences.* San Diego, CA: Academic Press, 1995.

Labeled Tree

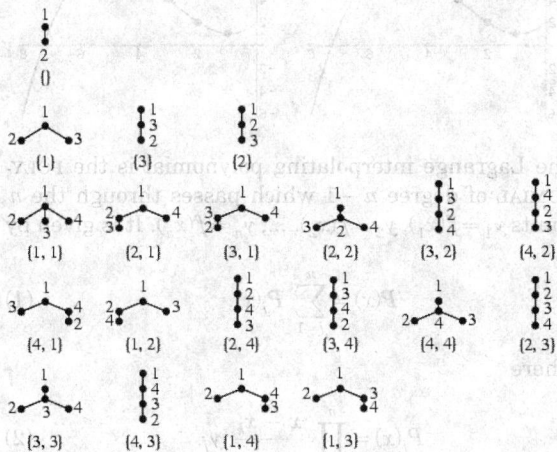

{1}　{3}　{2}

{1, 1}　{2, 1}　{3, 1}　{2, 2}　{3, 2}　{4, 2}

{4, 1}　{1, 2}　{2, 4}　{3, 4}　{4, 4}　{2, 3}

{3, 3}　{4, 3}　{1, 4}　{1, 3}

A TREE with its nodes labeled. The number of labeled
trees on n nodes is n^{n-2}, the first few values of which
are 1, 1, 3, 16, 125, 1296, ... (Sloane's A000272).
Cayley (1889) provided the first proof of the number
of labeled trees (Skiena 1990, p. 151), and a constructive proof was subsequently provided by Prüfer
(1918). Prüfer's result gives an encoding for labeled
trees known as PRÜFER CODE (indicated underneath
the trees above, where the trees are depicted using an
embedding with root at the node labeled 1).
The probability that a random labeled tree is CEN-
TERED is asymptotically equal to 1/2 (Szekeres 1983;
Skiena 1990, p. 167).

See also LABELED GRAPH, PRÜFER CODE, TREE

References

Biggs, N. L.; Lloyd, E. K.; and Wilson, R. J. *Graph Theory
　1736–936.* Oxford, England: Oxford University Press,
　p. 51, 1976.
Cayley, A. "A Theorem on Trees." *Quart. J. Math.* **23**, 376–
　78, 1889.
Prüfer, H. "Neuer Beweis eines Satzes über Permutationen."
　Arch. Math. Phys. **27**, 742–44, 1918.

Riordan, J. *An Introduction to Combinatorial Analysis.* New
　York: Wiley, p. 128, 1980.
Skiena, S. *Implementing Discrete Mathematics: Combinato-
　rics and Graph Theory with Mathematica.* Reading, MA:
　Addison-Wesley, 1990.
Sloane, N. J. A. Sequences A000272/M3027 in "An On-Line
　Version of the Encyclopedia of Integer Sequences." http://
　www.research.att.com/~njas/sequences/eisonline.html.
Szekeres, G. *Distribution of Labeled Trees by Diameter.* New
　York: Springer-Verlag, pp. 392–97, 1983.
van Lint, J. H. and Wilson, R. M. *A Course in Combinato-
　rics.* New York: Cambridge University Press, 1992.

Lacunarity

Quantifies deviation from translational invariance by
describing the distribution of gaps within a set at
multiple scales. The more lacunar a set, the more
heterogeneous the spatial arrangement of gaps.

Lacunary Function

This entry contributed by JONATHAN DEANE

A function that has a NATURAL BOUNDARY.

See also NATURAL BOUNDARY

References

Ash, R. B. Ch. 3 in *Complex Variables.* New York: Academic
　Press, 1971.

Ladder

ASTROID, CROSSED LADDERS PROBLEM, CROSSED LAD-
DERS THEOREM, LADDER GRAPH

Ladder Graph

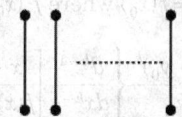

A GRAPH consisting of two rows of paired nodes each
connected by an EDGE. Its complement is the COCK-
TAIL PARTY GRAPH.

See also COCKTAIL PARTY GRAPH

Lagerstrom Differential Equation

The second-order ORDINARY DIFFERENTIAL EQUATION

$$y'' + \frac{k}{x}\,y' + \epsilon y' y = 0.$$

References

Rosenblat, S. and Shepherd, J. "On the Asymptotic Solution
　of the Lagerstrom Model Equation." *SIAM J. Appl. Math.*
　29, 110–20, 1975.
Zwillinger, D. *Handbook of Differential Equations, 3rd ed.*
　Boston, MA: Academic Press, p. 124, 1997.

Lagrange Bracket

Let F and G be infinitely differentiable functions of x, u, and p. Then the Lagrange bracket is defined by

$$[F, G] = \sum_{\nu=1}^{n} \left[\frac{\partial F}{\partial p_\nu} \left(\frac{\partial G}{\partial x_p} + p_\nu \frac{\partial G}{\partial u} \right) - \frac{\partial G}{\partial p_\nu} \left(\frac{\partial F}{\partial x_\nu} + p_\nu \frac{\partial F}{\partial u} \right) \right]. \tag{1}$$

The Lagrange bracket satisfies

$$[F, G] = -[G, F] \tag{2}$$

$$[[F, G], H] + [[G, H], F] + [[H, F], G]$$
$$= \frac{\partial F}{\partial u} [G, H] + \frac{\partial G}{\partial u} [H, F] + \frac{\partial H}{\partial u} [F, G]. \tag{3}$$

If F and G are functions of x and p only, then the Lagrange bracket $[F, G]$ collapses the Poisson bracket (F, G).

See also Lie Bracket, Poisson Bracket

References
Iyanaga, S. and Kawada, Y. (Eds.). *Encyclopedic Dictionary of Mathematics.* Cambridge, MA: MIT Press, p. 1004, 1980.

Lagrange-Bürmann Expansion
Lagrange Inversion Theorem

Lagrange-Bürmann Theorem
Lagrange Inversion Theorem

Lagrange Expansion
Let $y = f(x)$ and $y_0 = f(x_0)$ where $f'(x_0) \neq 0$, then

$$x = x_0 + \sum_{k=1}^{\infty} \frac{(y - y_0)^k}{k!} \left\{ \frac{d^{k-1}}{dx^{k-1}} \left[\frac{x - x_0}{f(x) - y_0} \right]^k \right\}_{x=x_0}$$

$$g(x) = g(x_0) + \sum_{k=1}^{\infty} \frac{(y - y_0)^k}{k!}$$
$$\times \left\{ \frac{d^{k-1}}{dx^{k-1}} \left[g'(x) \left(\frac{x - x_0}{f(x) - y_0} \right)^k \right] \right\}_{x=x_0}.$$

Expansions of this form were first considered by Lagrange (1770; Lagrange 1868, pp. 680–93).

See also Bürmann's Theorem, Maclaurin Series, Taylor Series, Teixeira's Theorem

References
Abramowitz, M. and Stegun, C. A. (Eds.). *Handbook of Mathematical Functions with Formulas, Graphs, and Mathematical Tables, 9th printing.* New York: Dover, p. 14, 1972.
Goursat, E. *A Course in Mathematical Analysis, Vol. 2, Pt. 1.* New York: Dover, p. 106, 1959.

Lagrange, J. L. "Nouvelle méthode pour résoudre les problèmes indéterminés en nombres entiers." *Mém. de l'Acad. Roy. des Sci. et Belles-Lettres de Berlin* **24**, 1770. Reprinted in *Oeuvres de Lagrange, tome 2, section deuxième: Mémoires extraits des recueils de l'Academie royale des sciences et Belles-Lettres de Berlin.* Paris: Gauthier-Villars, pp. 655–26, 1868.
Whittaker, E. T. and Watson, G. N. "Lagrange's Theorem." §7.32 in *A Course in Modern Analysis, 4th ed.* Cambridge, England: Cambridge University Press, p. 132, 1990.

Lagrange Interpolating Polynomial

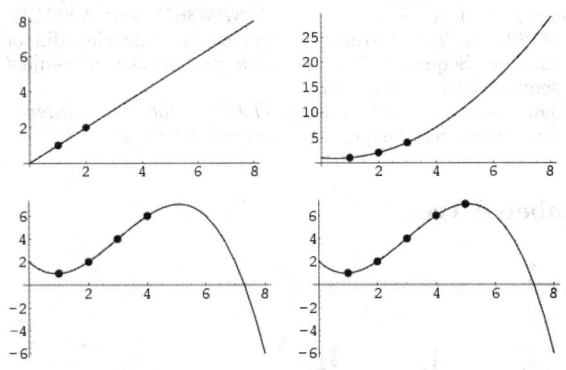

The Lagrange interpolating polynomial is the Polynomial of degree $n-1$ which passes through the n points $y_1 = f(x_1)$, $y_2 = f(x_2)$, ..., $y_n = f(x_n)$. It is given by

$$P(x) = \sum_{j=1}^{n} P_j(x), \tag{1}$$

where

$$P_j(x) = \prod_{\substack{k=1 \\ k \neq j}}^{n} \frac{x - x_k}{x_j - x_k} \, y_j. \tag{2}$$

Written explicitly,

$$P(x) = \frac{(x - x_2)(x - x_3) \cdots (x - x_n)}{(x_1 - x_2)(x_1 - x_3) \cdots (x_1 - x_n)} y_1$$
$$+ \frac{(x - x_1)(x - x_3) \cdots (x - x_n)}{(x_2 - x_1)(x_2 - x_3) \cdots (x_2 - x_n)} y_2 + \cdots$$
$$+ \frac{(x - x_1)(x - x_2) \cdots (x - x_{n-1})}{(x_n - x_1)(x_n - x_2) \cdots (x_n - x_{n-1})} y_n. \tag{3}$$

The formula was first published by Waring (1779), rediscovered by Euler in 1783, and published by Lagrange in 1795 (Jeffreys and Jeffreys 1988). For $n = 3$ points,

$$P(x) = \frac{(x - x_2)(x - x_3)}{(x_1 - x_2)(x_1 - x_3)} y_1 + \frac{(x - x_1)(x - x_3)}{(x_2 - x_1)(x_2 - x_3)} y_2$$
$$+ \frac{(x - x_1)(x - x_2)}{(x_3 - x_1)(x_3 - x_2)} y_3 \tag{4}$$

$$P'(x) = \frac{2x - x_2 - x_3}{(x_1 - x_2)(x_1 - x_3)} y_1 + \frac{2x - x_1 - x_3}{(x_2 - x_1)(x_2 - x_3)} y_2$$

$$+ \frac{2x - x_1 - x_2}{(x_3 - x_1)(x_3 - x_2)} y_3 \qquad (5)$$

Note that the function $P(x)$ passes through the points (x_i, y_i), as can be seen for the case $n = 3$,

$$P(x_1) = \frac{(x_1 - x_2)(x_1 - x_3)}{(x_1 - x_2)(x_1 - x_3)} y_1 + \frac{(x_1 - x_1)(x_1 - x_3)}{(x_2 - x_1)(x_2 - x_3)} y_2$$

$$+ \frac{(x_1 - x_1)(x_1 - x_2)}{(x_3 - x_1)(x_3 - x_2)} y_3 = y_1 \qquad (6)$$

$$P(x_2) = \frac{(x_2 - x_2)(x_2 - x_3)}{(x_1 - x_2)(x_1 - x_3)} y_1 + \frac{(x_2 - x_1)(x_2 - x_3)}{(x_2 - x_1)(x_2 - x_3)} y_2$$

$$+ \frac{(x_2 - x_1)(x_2 - x_2)}{(x_3 - x_1)(x_3 - x_2)} y_3 = y_2 \qquad (7)$$

$$P(x_3) = \frac{(x_3 - x_2)(x_3 - x_3)}{(x_1 - x_2)(x_1 - x_3)} y_1 + \frac{(x_3 - x_1)(x_3 - x_3)}{(x_2 - x_1)(x_2 - x_3)} y_2$$

$$+ \frac{(x_3 - x_1)(x_3 - x_2)}{(x_3 - x_1)(x_3 - x_2)} y_3 = y_3 . \qquad (8)$$

Generalizing to arbitrary n,

$$P(x_j) = \sum_{k=1}^{n} P_k(x_j) = \sum_{k=1}^{n} \delta_{jk} y_k = y_j . \qquad (9)$$

The Lagrange interpolating polynomials can also be written using what Szego (1975) called Lagrange's fundamental interpolating polynomials. Let

$$\pi(x) \equiv \prod_{k=1}^{n} (x - x_k), \qquad (10)$$

$$\pi(x_j) \equiv \prod_{k=1}^{n} (x_j - x_k), \qquad (11)$$

$$\pi'(x_j) = \left[\frac{d\pi}{dx} \right]_{x=x_j} = \prod_{\substack{k=1 \\ k \neq j}}^{n} (x_j - x_k) \qquad (12)$$

so that $\pi(x)$ is an nth degree POLYNOMIAL with zeros at $x_1, ..., x_n$. Then define the fundamental polynomials by

$$\pi_\nu(x) = \frac{\pi(x)}{\pi'(x_\nu)(x - x_\nu)}, \qquad (13)$$

which satisfy

$$\pi_\nu(x_\mu) = \delta_{\nu\mu}, \qquad (14)$$

where $\delta_{\nu\mu}$ is the KRONECKER DELTA. Now let $y_1 = P(x_1), ..., y_n = P(x_n)$, then the expansion

$$P(x) = \sum_{k=1}^{n} \pi_k(x) y_k = \sum_{k=1}^{n} \frac{\pi(x)}{(x - x_k)\pi'(x_k)} y_k \qquad (15)$$

gives the unique Lagrange interpolating polynomial assuming the values y_k at x_k. More generally, let $d\alpha(x)$ be an arbitrary distribution on the interval $[a, b]$, $\{p_n(x)\}$ the associated ORTHOGONAL POLYNOMIALS, and $l_1(x), ..., l_n(x)$ the fundamental POLYNOMIALS corresponding to the set of zeros of a polynomial $P_n(x)$. Then

$$\int_a^b l_\nu(x) l_\mu(x) \, d\alpha(x) = \lambda_\mu \delta_{\nu\mu} \qquad (16)$$

for $\nu, \mu = 1, 2, ..., n$, where λ_ν are CHRISTOFFEL NUMBERS.

Lagrange interpolating polynomials give no error estimate. A more conceptually straightforward method for calculating them is NEVILLE'S ALGORITHM.

See also AITKEN INTERPOLATION, HERMITE'S INTERPOLATING POLYNOMIAL, LEBESGUE CONSTANTS (LAGRANGE INTERPOLATION), NEVILLE'S ALGORITHM, NEWTON'S DIVIDED DIFFERENCE INTERPOLATION FORMULA

References

Abramowitz, M. and Stegun, C. A. (Eds.). *Handbook of Mathematical Functions with Formulas, Graphs, and Mathematical Tables, 9th printing.* New York: Dover, pp. 878–79 and 883, 1972.

Beyer, W. H. (Ed.). *CRC Standard Mathematical Tables, 28th ed.* Boca Raton, FL: CRC Press, p. 439, 1987.

Jeffreys, H. and Jeffreys, B. S. "Lagrange's Interpolation Formula." §9.011 in *Methods of Mathematical Physics, 3rd ed.* Cambridge, England: Cambridge University Press, p. 260, 1988.

Pearson, K. *Tracts for Computers* **2**, 1920.

Press, W. H.; Flannery, B. P.; Teukolsky, S. A.; and Vetterling, W. T. "Polynomial Interpolation and Extrapolation" and "Coefficients of the Interpolating Polynomial." §3.1 and 3.5 in *Numerical Recipes in FORTRAN: The Art of Scientific Computing, 2nd ed.* Cambridge, England: Cambridge University Press, pp. 102–04 and 113–16, 1992.

Séroul, R. "Lagrange Interpolation." §10.9 in *Programming for Mathematicians.* Berlin: Springer-Verlag, pp. 269–73, 2000.

Szego, G. *Orthogonal Polynomials, 4th ed.* Providence, RI: Amer. Math. Soc., pp. 329 and 332, 1975.

Waring, E. *Philos. Trans.* **69**, 59–7, 1779.

Whittaker, E. T. and Robinson, G. "Lagrange's Formula of Interpolation." §17 in *The Calculus of Observations: A Treatise on Numerical Mathematics, 4th ed.* New York: Dover, pp. 28–0, 1967.

Lagrange Interpolation

LAGRANGE INTERPOLATING POLYNOMIAL

Lagrange Inversion Theorem

Let z be defined as a function of w in terms of a parameter α by

$$z = w + \alpha \phi(z).$$

Then any function of z can be expressed as a POWER SERIES in α which converges for sufficiently small α and has the form

$$F(z) = F(w) + \frac{\alpha}{1} \phi(w)F'(w) + \frac{\alpha^2}{1 \cdot 2} \frac{\partial}{\partial w}\{[\phi(w)]^2 F'(w)\}$$

$$+ \ldots + \frac{\alpha^{n+1}}{(n+1)!} \frac{\partial^n}{\partial w^n}\{[\phi(w)]^{n+1} F'(w)\} + \ldots$$

See also BÜRMANN'S THEOREM, SCHUR-JABOTINSKY THEOREM

References

Goursat, E. *Functions of a Complex Variable, Vol. 2, Pt. 1.* New York: Dover, 1959.

Henrici, P. "An Algebraic Proof of the Lagrange-Burmann Formula." *J. Math. Anal. Appl.* **8**, 218–24, 1964.

Henrici, P. "The Lagrange-Bürmann Theorem." §1.9 in *Applied and Computational Complex Analysis, Vol. 1: Power Series-Integration-Conformal Mapping-Location of Zeros.* New York: Wiley, pp. 55–5, 1988.

Joni, S. A. "Lagrange Inversion in Higher Dimensions and Umbral Operators." *J. Linear Multi-Linear Algebra* **6**, 111–21, 1978.

Moulton, F. R. *An Introduction to Celestial Mechanics, 2nd rev. ed.* New York: Dover, p. 161, 1970.

Popoff, M. "Sur le reste de la série de Lagrange." *Comptes Rendus Herbdom. Séances de l'Acad. Sci.* **53**, 795–98, 1861.

Roman, S. "The Lagrange Inversion Formula." §5.2. in *The Umbral Calculus.* New York: Academic Press, pp. 138–40, 1984.

Whittaker, E. T. and Watson, G. N. "Lagrange's Theorem." §7.32 in *A Course in Modern Analysis, 4th ed.* Cambridge, England: Cambridge University Press, pp. 132–33, 1990.

Williamson, B. "Remainder in Lagrange's Series." §119 in *An Elementary Treatise on the Differential Calculus, 9th ed.* London: Longmans, pp. 158–59, 1895.

Lagrange Multiplier

Used to find the EXTREMUM of $f(x_1, x_2, \ldots, x_n)$ subject to the constraint $g(x_1, x_2, \ldots, x_n) = C$, where f and g are functions with continuous first PARTIAL DERIVATIVES on the OPEN SET containing the curve $g(x_1, x_2, \ldots, x_n) = 0$, and $\nabla g \neq 0$ at any point on the curve (where ∇ is the GRADIENT). For an EXTREMUM to exist,

$$df = \frac{\partial f}{\partial x_1} dx_1 + \frac{\partial f}{\partial x_2} dx_2 + \ldots + \frac{\partial f}{\partial x_n} dx_n = 0. \quad (1)$$

But we also have

$$dg = \frac{\partial g}{\partial x_1} dx_1 + \frac{\partial g}{\partial x_2} dx_2 + \ldots + \frac{\partial g}{\partial x_n} dx_n = 0. \quad (2)$$

Now multiply (2) by the as yet undetermined parameter λ and add to (1),

$$\left(\frac{\partial f}{\partial x_1} + \lambda \frac{\partial q}{\partial x_1}\right) dx_1 + \left(\frac{\partial f}{\partial x_2} + \lambda \frac{\partial q}{\partial x_2}\right) dx_2$$

$$+ \ldots + \left(\frac{\partial f}{\partial x_n} + \lambda \frac{\partial q}{\partial x_n}\right) dx_n = 0. \quad (3)$$

Note that the differentials are all independent, so we can set any combination equal to 0, and the remainder must still give zero. This requires that

$$\frac{\partial f}{\partial x_k} + \lambda \frac{\partial g}{\partial x_k} = 0 \quad (4)$$

for all $k = 1, \ldots, n$. The constant λ is called the Lagrange multiplier. For multiple constraints, $g_1 = 0, g_2 = 0, \ldots,$

$$\nabla f = \lambda_1 \nabla g_1 + \lambda_2 \nabla g_2 + \ldots. \quad (5)$$

See also KUHN-TUCKER THEOREM

References

Arfken, G. "Lagrange Multipliers." §17.6 in *Mathematical Methods for Physicists, 3rd ed.* Orlando, FL: Academic Press, pp. 945–50, 1985.

Lagrange Number (Diophantine Equation)

Given a FERMAT DIFFERENCE EQUATION (a quadratic DIOPHANTINE EQUATION)

$$x^2 - r^2 y^2 = 4$$

with r a QUADRATIC SURD, assign to each solution $x|y$ the Lagrange number

$$z \equiv \tfrac{1}{2}(x + yr).$$

The product and quotient of two Lagrange numbers are also Lagrange numbers. Furthermore, every Lagrange number is a POWER of the smallest Lagrange number with an integral exponent.

See also PELL EQUATION

References

Dörrie, H. *100 Great Problems of Elementary Mathematics: Their History and Solutions.* New York: Dover, pp. 94–5, 1965.

Lagrange Number (Rational Approximation)

HURWITZ'S IRRATIONAL NUMBER THEOREM gives the best rational approximation possible for an arbitrary irrational number $-\beta$ as

$$\phi$$

The $\sqrt{8}$ are called Lagrange numbers and get steadily larger for each "bad" set of irrational numbers which is excluded.

n	Exclude	$\sqrt{8}$

1	none	$\sqrt{2}$
2	$\dfrac{\sqrt{221}}{5}$	$\sqrt{9 - \dfrac{4}{3}}$,
3	m	$f(x) = f(x_0) + (x - x_0)f'(x_0)$ $+ \dfrac{(x - x_0)^2}{2!} f''(x_0) + \dots$

Lagrange numbers are OF THE FORM

$$+ \frac{(x - x_0)^n}{n!} f^{(n)}(x_0) + R_n,$$

where m is a MARKOV NUMBER. The Lagrange numbers form a SPECTRUM called the LAGRANGE SPECTRUM.

See also HURWITZ'S IRRATIONAL NUMBER THEOREM, IRRATIONALITY MEASURE, LIOUVILLE'S APPROXIMATION THEOREM, MARKOV NUMBER, ROTH'S THEOREM, SPECTRUM SEQUENCE, THUE-SIEGEL-ROTH THEOREM.

References
Conway, J. H. and Guy, R. K. *The Book of Numbers*. New York: Springer-Verlag, pp. 187–89, 1996.

Lagrange Polynomial
LAGRANGE INTERPOLATING POLYNOMIAL

Lagrange Remainder
Given a TAYLOR SERIES

$$f(x) = f(x_0) + (x - x_0)f'(x_0) + \frac{(x - x_0)^2}{2!} f''(x_0) + \dots$$

$$+ \frac{(x - x_0)^n}{n!} f^{(n)}(x_0) + R_n, \tag{1}$$

the error R_n after n terms is given by

$$R_n = \int_{x_0}^{x} f^{(n+1)}(t) \frac{(x - t)^n}{n!} \, dt. \tag{2}$$

Using the MEAN-VALUE THEOREM, this can be bounded by

$$R_n = \frac{f^{(n+1)}(x^*)}{(n + 1)!} (x - x_0)^{n+1} \tag{3}$$

for some $x^* \in (x_0, x)$ (Abramowitz and Stegun 1972, p. 880).

Note that the Lagrange remainder R_n is also sometimes taken to refer to the remainder when terms up to the $(n - 1)$st power are taken in the TAYLOR SERIES,

and that a notation in which $h \to x - x_0$, $x^* \to a + \theta h$, and $x - x^* \to 1 - \theta$ is sometimes used (Blumenthal 1926; Whittaker and Watson 1990, pp. 95–6).

See also CAUCHY REMAINDER, SCHLÖMILCH REMAINDER, TAYLOR SERIES

References
Abramowitz, M. and Stegun, C. A. (Eds.). *Handbook of Mathematical Functions with Formulas, Graphs, and Mathematical Tables, 9th printing.* New York: Dover, 1972.
Beesack, P. R. "A General Form of the Remainder in Taylor's Theorem." *Amer. Math. Monthly* **73**, 64–7, 1966.
Blumenthal, L. M. "Concerning the Remainder Term in Taylor's Formula." *Amer. Math. Monthly* **33**, 424–26, 1926.
Firey, W. J. "Remainder Formulae in Taylor's Theorem." *Amer. Math. Monthly* **67**, 903–05, 1960.
Fulks, W. *Advanced Calculus.* New York: Wiley, p. 137, 1961.
Nicholas, C. P. "Taylor's Theorem in a First Course." *Amer. Math. Monthly* **58**, 559–62, 1951.
Poffald, E. I. "The Remainder in Taylor's Formula." *Amer. Math. Monthly* **97**, 205–13, 1990.
Whittaker, E. T. and Watson, G. N. "Forms of the Remainder in Taylor's Series." §5.41 in *A Course in Modern Analysis, 4th ed.* Cambridge, England: Cambridge University Press, pp. 95–6, 1990.

Lagrange Resolvent
A quantity involving primitive cube ROOTS OF UNITY which can be used to solve the CUBIC EQUATION.

References
Faucette, W. M. "A Geometric Interpretation of the Solution of the General Quartic Polynomial." *Amer. Math. Monthly* **103**, 51–7, 1996.

Lagrange's Continued Fraction Theorem
The REAL ROOTS of quadratic expressions with integral COEFFICIENTS have periodic CONTINUED FRACTIONS, as first proved by Lagrange.

See also CONTINUED FRACTION

Lagrange's Equation
The PARTIAL DIFFERENTIAL EQUATION

$$(1 + f_y^2)f_{xx} + 2f_x f_y f_{xy} + (1 + f_x^2)f_{yy} = 0,$$

whose solutions are called MINIMAL SURFACES. This corresponds to the MEAN CURVATURE H equalling 0 over the surface.

D'ALEMBERT'S EQUATION

$$y = xf(y') + g(y')$$

is sometimes also known as Lagrange's equation (Zwillinger 1997, pp. 120 and 265–68).

See also D'ALEMBERT'S EQUATION, MEAN CURVATURE, MINIMAL SURFACE

References

do Carmo, M. P. "Minimal Surfaces." §3.5 in *Mathematical Models from the Collections of Universities and Museums* (Ed. G. Fischer). Braunschweig, Germany: Vieweg, pp. 41–3, 1986.

Zwillinger, D. "Lagrange's Equation." §II.A.69 in *Handbook of Differential Equations, 3rd ed.* Boston, MA: Academic Press, pp. 120 and 265–68, 1997.

Lagrange's Four-Square Theorem

A theorem also known as BACHET'S CONJECTURE which was stated but not proven by Diophantus. It states that every POSITIVE INTEGER can be written as the SUM of at most four SQUARES. Although the theorem was proved by Fermat using infinite descent, the proof was suppressed. Euler was unable to prove the theorem. The first published proof was given by Lagrange in 1770 and made use of the EULER FOUR-SQUARE IDENTITY.

Lagrange proved that $g(2) = 4$, where 4 may be reduced to 3 except for numbers OF THE FORM $4^n(8k + 7)$, as proved by Legendre in 1798 (Nagell 1951, p. 194; Wells 1986, pp. 48 and 56; Hardy 1999, p. 12; Savin 2000).

See also DIOPHANTINE EQUATION–2ND POWERS, EULER FOUR-SQUARE IDENTITY, FERMAT'S POLYGONAL NUMBER THEOREM, FIFTEEN THEOREM, LEBESGUE IDENTITY, SUM OF SQUARES FUNCTION, VINOGRADOV'S THEOREM, WARING'S PROBLEM

References

Hardy, G. H. *Ramanujan: Twelve Lectures on Subjects Suggested by His Life and Work, 3rd ed.* New York: Chelsea, 1999.

Hardy, G. H. and Wright, E. M. "The Four-Square Theorem." §20.5 in *An Introduction to the Theory of Numbers, 5th ed.* Oxford, England: Clarendon Press, pp. 302–03, 1979.

Landau, E. *Vorlesungen über Zahlentheorie, Vol. 1.* New York: Chelsea, pp. 114–22, 1970.

Nagell, T. "Bachet's Theorem." §55 in *Introduction to Number Theory.* New York: Wiley, pp. 191–95, 1951.

Niven, I. M.; Zuckerman, H. S.; and Montgomery, H. L. *An Introduction to the Theory of Numbers, 5th ed.* New York: Wiley, 1991.

Savin, A. "Shape Numbers." *Quantum* **11**, 14–8, 2000.

Séroul, R. "Sums of Four Squares." §8.13 in *Programming for Mathematicians.* Berlin: Springer-Verlag, pp. 207–08, 2000.

Wells, D. *The Penguin Dictionary of Curious and Interesting Numbers.* Middlesex, England: Penguin Books, p. 48, 1986.

Lagrange's Group Theorem

This entry contributed by NICOLAS BRAY

Also known as Lagrange's lemma. The most general form of Lagrange's theorem states that for a GROUP G, a SUBGROUP H of G, and a subgroup K of H, $(G : K) = (G : H)(H : K)$, where the products are taken as cardinalities (thus the theorem holds even for INFINITE GROUPS) and (G_H) denotes the INDEX. A fre-

quently stated corollary (which follows from taking $K = \{e\}$, where e is the IDENTITY ELEMENT) is that the order of G is equal to the product of the order of H and the INDEX of H.

The corollary is easily proven in the case of G being a FINITE GROUP, as the LEFT COSETS of H form a partition of G, and so the number of blocks in the partition (which is $(G : H)$) multiplied by the number of elements in each partition (which is just the order of H).

For a FINITE GROUP G, this corollary gives that the order of H must divide the order of G. Then, because the order of an element x of G is the order of the cyclic subgroup generated by x, we must have that the order of any element of G divides the order of G.

The converse of Lagrange's theorem is not, in general, true (Gallian 1993, 1994).

References

Birkhoff, G. and Mac Lane, S. *A Survey of Modern Algebra, 5th ed.* New York: Macmillan, p. 111, 1996.

Gallian, J. A. "On the Converse of Lagrange's Theorem." *Math. Mag.* **63**, 23, 1993.

Gallian, J. A. *Contemporary Abstract Algebra, 3rd ed.* Lexington, MA: D. C. Heath, 1994.

Herstein, I. N. *Abstract Algebra, 3rd ed.* New York: Macmillan, p. 66, 1996.

Hogan, G. T. "More on the Converse of Lagrange's Theorem." *Math. Mag.* **69**, 375–76, 1996.

Shanks, D. *Solved and Unsolved Problems in Number Theory, 4th ed.* New York: Chelsea, p. 86, 1993.

Lagrange's Identity

The algebraic identity

$$\left(\sum_{k=1}^{n} a_k b_k \right)^2$$

$$= \left(\sum_{k=1}^{n} a_k^2 \right) \left(\sum_{k=1}^{n} b_k^2 \right) - \sum_{1 \le k < j \le n} (a_k b_j - a_j b_k)^2 \quad (1)$$

(Mitrinovic 1970, p. 41). In determinant form,

$$(\mathbf{a}_1 \times \cdots \times \mathbf{a}_{n-1}) \cdot (\mathbf{b}_1 \times \cdots \times \mathbf{b}_{n-1})$$

$$= \begin{vmatrix} \mathbf{a}_1 \cdot \mathbf{b}_1 & \cdots & \mathbf{a}_1 \cdot \mathbf{b}_{n-1} \\ \vdots & \ddots & \vdots \\ \mathbf{a}_{n-1} \cdot \mathbf{b}_1 & \cdots & \mathbf{a}_{n-1} \cdot \mathbf{b}_{n-1} \end{vmatrix}, \quad (2)$$

where $|A|$ is the DETERMINANT of A. Lagrange's identity is a special case of the BINET-CAUCHY IDENTITY, and CAUCHY'S INEQUALITY in n-D follows from it. It can be coded in *Mathematica* as follow.

```
<<DiscreteMath`Combinatorica`;
CauchyLagrangeId[n_] := Module[
  {aa = Array[a, n], bb = Array[b, n]},
  Plus @@ (aa^2)Plus @@ (bb^2) ==
  Plus @@ ((a[#1]b[#2] - a[#2]b[#1])^2 & @@@
    KSubsets[Range[n], 2]) +
```

```
(aa.bb)^2
]
```

Plugging in gives the $n = 2$ and $n = 3$ identities

$$(a_1^2 + a_2^2)(b_1^2 + b_2^2) = (a_1 b_1 + a_2 b_2)^2 + (a_1 b_2 - a_2 b_1)^2 \quad (3)$$

$$(a_1^2 + a_2^2 + a_3^2)(b_1^2 + b_2^2 + b_3^2) = (a_1 b_1 + a_2 b_2 + a_3 b_3)^2$$

$$+ [(a_1 b_2 - a_2 b_1)^2 + (a_1 b_3 - a_3 b_1)^2 + (a_2 b_3 - a_3 b_2)^2]. \quad (4)$$

See also BINET-CAUCHY IDENTITY, CAUCHY'S INEQUALITY, VECTOR TRIPLE PRODUCT, VECTOR QUADRUPLE PRODUCT

References

Gradshteyn, I. S. and Ryzhik, I. M. *Tables of Integrals, Series, and Products, 6th ed.* San Diego, CA: Academic Press, p. 1093, 2000.
Mitrinovic, D. S. *Analytic Inequalities.* New York: Springer-Verlag, 1970.

Lagrange's Inequality
CAUCHY'S INEQUALITY

Lagrange's Lemma
LAGRANGE'S FOUR-SQUARE THEOREM

Lagrange Spectrum

A SPECTRUM formed by the LAGRANGE NUMBERS. The only ones less than three are the LAGRANGE NUMBERS, but the last gaps end at FREIMAN'S CONSTANT. REAL NUMBERS larger than FREIMAN'S CONSTANT are in the MARKOV SPECTRUM.

See also FREIMAN'S CONSTANT, LAGRANGE NUMBER (RATIONAL APPROXIMATION), MARKOV SPECTRUM, SPECTRUM SEQUENCE

References

Conway, J. H. and Guy, R. K. *The Book of Numbers.* New York: Springer-Verlag, pp. 187–89, 1996.

Lagrangian Coefficient

COEFFICIENTS which appear in LAGRANGE INTERPOLATING POLYNOMIALS where the points are equally spaced along the ABSCISSA.

Lagrangian Derivative
CONVECTIVE DERIVATIVE

Laguerre Differential Equation

$$xy'' + (1 - x)y' + \lambda y = 0. \quad (1)$$

The Laguerre differential equation is a special case of the more general "associated Laguerre differential equation"

$$xy'' + (v + 1 - x)y' + \lambda y = 0 \quad (2)$$

(Iyanaga and Kawada 1980, p. 1481; Zwillinger 1997, p. 124) with $v = 0$. The general solution is

$$t = C_1 U(-\lambda, \ 1 + v, \ x) + C_2 L_\lambda^v(x), \quad (3)$$

where $U(a, \ b, \ x)$ is a CONFLUENT HYPERGEOMETRIC FUNCTION OF THE FIRST KIND and $L_\lambda^v(x)$ is an associated LAGUERRE POLYNOMIAL.

Note that in the special case $\lambda = 0$, the associated Laguerre differential equation is OF THE FORM

$$y''(x) + P(x)y'(x) = 0, \quad (4)$$

so the solution can be found using an INTEGRATING FACTOR

$$\mu = \exp\left(\int P(x)\, dx\right) = \exp\left(\int \frac{v + 1 - x}{x}\, dx\right)$$

$$= \exp[(v + 1) \ln x - x] = x^{v+1} e^{-x}, \quad (5)$$

as

$$y = C_1 \int \frac{dx}{\mu} + C_2 = C_1 \int \frac{e^x}{x^{v+1}}\, dx + C_2 \quad (6)$$

$$= C_2 - C_1 x^{-v} E_{1+v}(-x), \quad (7)$$

where $E_n(x)$ is the E_N-FUNCTION.

The associated Laguerre differential equation has a REGULAR SINGULAR POINT at 0 and an IRREGULAR SINGULARITY at ∞. It can be solved using a series expansion,

$$x \sum_{n=2}^{\infty} n(n-1)a_n x^{n-2} + (v+1) \sum_{n=1}^{\infty} n a_n x^{n-1}$$

$$-x \sum_{n=1}^{\infty} n a_n x^{n-1} + \lambda \sum_{n=0}^{\infty} a_n x^n = 0 \quad (8)$$

$$\sum_{n=2}^{\infty} n(n-1)a_n x^{n-1} + (v+1) \sum_{n=1}^{\infty} n a_n x^{n-1}$$

$$-\sum_{n=1}^{\infty} n a_n x^n + \lambda \sum_{n=0}^{\infty} a_n x^n = 0 \quad (9)$$

$$\sum_{n=1}^{\infty} (n+1)n a_{n+1} x^n + (v+1) \sum_{n=0}^{\infty} (n+1)a_{n+1} x^n$$

$$-\sum_{n=1}^{\infty} n a_n x^n + \lambda \sum_{n=0}^{\infty} a_n x^n = 0 \quad (10)$$

$$[(n+1)a_1 + \lambda a_0]$$

$$+ \sum_{n=1}^{\infty} \{[(n+1)n + (\nu+1)(n+1)]a_{n+1} - na_n + \lambda a_n\}x^n$$

$$= 0 \tag{11}$$

$$[(n+1)a_1 + \lambda a_0]$$

$$+ \sum_{n=1}^{\infty} [(n+1)(n+\nu+1)a_{n+1} + (\lambda-n)a_n]x^n = 0. \tag{12}$$

This requires

$$a_1 = -\frac{\lambda}{\nu+1}a_0 \tag{13}$$

$$a_{n+1} = \frac{n-\lambda}{(n+1)(n+\nu+1)}a_n \tag{14}$$

for $n > 1$. Therefore,

$$a_{n+1} = \frac{n-\lambda}{(n+1)(n+\nu+1)}a_n \tag{15}$$

for $n = 1, 2, ...,$ so

$$y = a_0 \left[1 - \frac{\lambda}{\nu+1}x - \frac{\lambda(1-\lambda)}{2(\nu+1)(\nu+2)}x^2 \right.$$

$$\left. - \frac{\lambda(1-\lambda)(2-\lambda)}{2 \cdot 3(\nu+1)(\nu+2)(\nu+3)} + \cdots \right]. \tag{16}$$

If λ is a POSITIVE INTEGER, then the series terminates and the solution is a POLYNOMIAL, known as an associated LAGUERRE POLYNOMIAL (or, if $\nu = 0$, simply a LAGUERRE POLYNOMIAL).

See also LAGUERRE POLYNOMIAL

References

Iyanaga, S. and Kawada, Y. (Eds.). *Encyclopedic Dictionary of Mathematics.* Cambridge, MA: MIT Press, p. 1481, 1980.

Zwillinger, D. *Handbook of Differential Equations*, 3rd ed. Boston, MA: Academic Press, p. 120, 1997.

Laguerre-Gauss Quadrature

Also called GAUSS-LAGUERRE QUADRATURE or LAGUERRE QUADRATURE. A GAUSSIAN QUADRATURE over the interval $[0, \infty)$ with WEIGHTING FUNCTION $W(x) = e^{-x}$ (Abramowitz and Stegun 1972, p. 890). The ABSCISSAS for quadrature order n are given by the ROOTS of the LAGUERRE POLYNOMIALS $L_n(x)$. The weights are

$$w_i = -\frac{A_{n+1}\gamma_n}{A_n L_n'(x_i)L_{n+1}(x_i)} = \frac{A_n}{A_{n-1}} \frac{\gamma_{n-1}}{L_{n-1}(x_i)L_n'(x_i)}, \tag{1}$$

where A_n is the COEFFICIENT of x^n in $L_n(x)$. For LAGUERRE POLYNOMIALS,

$$A_n = \frac{(-1)^n}{n!}, \tag{2}$$

where $n!$ is a FACTORIAL, so

$$\frac{A_{n+1}}{A_n} = -\frac{1}{n+1} \tag{3}$$

$$\frac{A_n}{A_{n-1}} = -\frac{1}{n}. \tag{4}$$

Additionally,

$$\gamma_n = \int_0^{\infty} W(x)[L_n(x)]^2 \, dx = 1, \tag{5}$$

so

$$w_i = \frac{1}{(n+1)L_n'(x_i)L_{n+1}(x_i)} = -\frac{1}{nL_{n-1}(x_i)L_n'(x_i)}. \tag{6}$$

Using the RECURRENCE RELATION

$$xL_n'(x) = nL_n(x) - nL_{n-1}(x)$$

$$= (x-n-1)L_n(x) + (n+1)L_{n+1}(x) \tag{7}$$

which, since x_i is a root of $L_n(x)$, gives

$$nL_n(x) = (x-n-1)L_n(x) = 0, \tag{8}$$

so (7) becomes

$$x_i L_n'(x_i) = -nL_{n-1}(x_i) = (n+1)L_{n+1}(x_i) \tag{9}$$

gives

$$w_i = \frac{1}{x_i[L_n'(x_i)]^2} = \frac{x_i}{(n+1)^2[L_{n+1}(x_i)]^2}. \tag{10}$$

The error term is

$$E = \frac{(n!)^2}{(2n)!}f^{(2n)}(\xi) \tag{11}$$

(Abramowitz and Stegun 1972, p. 890).

Beyer (1987) gives a table of ABSCISSAS and weights up to $n = 6$.

n	x_i	w_i
2	0.585786	0.853553
	3.41421	0.146447
3	0.415775	0.711093
	2.29428	0.278518
	6.28995	0.0103893
4	0.322548	0.603154
	1.74576	0.357419
	4.53662	0.0388879

	9.39507	0.000539295
5	0.26356	0.521756
	1.4134	0.398667
	3.59643	0.0759424
	7.08581	0.00361176
	12.6408	0.00002337

The ABSCISSAS and weights can be computed analytically for small n.

n	x_i	w_i
2	$2 - \sqrt{2}$	$\frac{1}{4}(2 + \sqrt{2})$
	$2 + \sqrt{2}$	$\frac{1}{4}(2 - \sqrt{2})$

For the associated Laguerre polynomial $L_n^\beta(x)$ with WEIGHTING FUNCTION $w(x) = x^\beta e^{-x}$,

$$A_n = \frac{(-1)^n}{n!} \tag{12}$$

is the coefficient of x^n in $L_n^\beta(x)$ and

$$\gamma_n = \int_0^\infty x^\beta e^{-x} [L_n^\beta(x)]^2 \, dx = \frac{\Gamma(n + \beta + 1)}{n!}, \tag{13}$$

where $\Gamma(z)$ is the GAMMA FUNCTION. The weights are then

$$w_i = \frac{\Gamma(n + \beta) x_i}{n!(n + \beta)[L_{n-1}^\beta(x_i)]^2} = \frac{\Gamma(n + \beta + 1) x_i}{n!(n + 1)^2 [L_{n+1}^\beta(x_i)]^2}, \tag{14}$$

and the error term is

$$E_n = \frac{n! \Gamma(n + \beta + 1)}{(2n)!} f^{(2n)}(\xi). \tag{15}$$

See also GAUSSIAN QUADRATURE

References
Abramowitz, M. and Stegun, C. A. (Eds.). *Handbook of Mathematical Functions with Formulas, Graphs, and Mathematical Tables, 9th printing.* New York: Dover, pp. 890 and 923, 1972.

Beyer, W. H. *CRC Standard Mathematical Tables, 28th ed.* Boca Raton, FL: CRC Press, p. 463, 1987.

Chandrasekhar, S. *Radiative Transfer.* New York: Dover, pp. 64–5, 1960.

Hildebrand, F. B. *Introduction to Numerical Analysis.* New York: McGraw-Hill, pp. 325–27, 1956.

LaguerreL
LAGUERRE POLYNOMIAL

Laguerre Polynomial

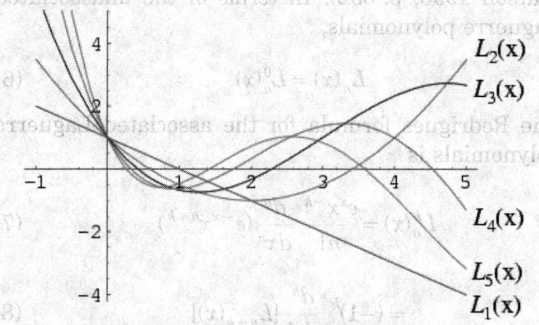

Solutions $L_n(x)$ to the LAGUERRE DIFFERENTIAL EQUATION with $\nu = 0$ are called Laguerre polynomials, illustrated above for $x \in [0, 1]$ and $n = 1, 2, ..., 5$. The Rodrigues formula for the Laguerre polynomials is

$$L_n(x) = \frac{e^x}{n!} \frac{d^n}{dx^n}(x^n e^{-x}) \tag{1}$$

and the GENERATING FUNCTION for Laguerre polynomials is

$$g(x, z) = \frac{\exp\left(-\dfrac{zz}{1 - z}\right)}{1 - z} = 1 + (-x + 1)z$$

$$+ \left(\frac{1}{2}x^2 - 2x + 1\right)z^2 + \left(-\frac{1}{6}x^3 + \frac{3}{2}x^2 - 3x + 1\right)z^3 + \dots. \tag{2}$$

A CONTOUR INTEGRAL is given by

$$L_n(x) = \frac{1}{2\pi i} \int \frac{e^{-xz/(1-z)}}{(1 - z)z^{n+1}} \, dz. \tag{3}$$

The Laguerre polynomials satisfy the RECURRENCE RELATIONS

$$(n + 1)L_{n+1}(x) = (2n + 1 - x)L_n(x) - nL_{n-1}(x) \tag{4}$$

(Petkovsek *et al.* 1996) and

$$xL_n'(x) = nL_n(x) - nL_{n-1}(x). \tag{5}$$

The first few Laguerre polynomials are

$$L_0(x) = 1$$

$$L_1(x) = -x + 1$$

$$L_2(x) = \frac{1}{2}(x^2 - 4x + 2)$$

$$L_3(x) = \frac{1}{6}(-x^3 + 9x^2 - 18x + 6).$$

Solutions to the associated LAGUERRE DIFFERENTIAL EQUATION with $\nu \neq 0$ are called associated Laguerre polynomials $L_n^k(x)$ or, in older literature, Sonine polynomials (Sonine 1880, p. 41; Whittaker and Watson 1990, p. 352). In terms of the unassociated Laguerre polynomials,

$$L_n(x) = L_n^0(x). \tag{6}$$

The Rodrigues formula for the associated Laguerre polynomials is

$$L_n^k(x) = \frac{e^x x^{-k}}{n!} \frac{d^n}{dx^n}(e^{-x} x^{n+k}) \tag{7}$$

$$= (-1)^k \frac{d^k}{dx^k}[L_{n+k}(x)] \tag{8}$$

$$= \frac{(-1)^n x^{-(k+1)/2}}{n!} e^{x/2} W_{k/2+n+1/2,\, k/2}(x) \tag{9}$$

$$= \sum_{m=0}^{n} (-1)^m \frac{(n+k)!}{(n-m)!(k+m)!m!} x^m, \tag{10}$$

where $W_{k,\,m}(x)$ is a WHITTAKER FUNCTION. The associated Laguerre polynomials are a SHEFFER SEQUENCE with

$$g(t) = (1-t)^{-k-1} \tag{11}$$

$$f(t) = \frac{t}{t-1}, \tag{12}$$

giving the GENERATING FUNCTION

$$g(x, z) = \frac{\exp\left(-\dfrac{zz}{1-z}\right)}{(1-z)^{k+1}}$$

$$= 1 + (k+1-x)z + \tfrac{1}{2}[x^2 - 2(k+2)x + (k+1)(k+2)]z^2$$

$$+ \dots. \tag{13}$$

where the usual factor of $n!$ in the denominator has been suppressed (Roman 1984, p. 31). Many interesting properties of the associated Laguerre polynomials follow from the fact that $f^{-1}(t) = f(t)$ (Roman 1984, p. 31).

The associated Laguerre polynomials are given explicitly by the formula

$$L_n^{(k)}(x) = \frac{1}{n!} \sum_{i=0}^{n} \frac{n!}{i!} \binom{k+n}{n-i,} (-x)^i, \tag{14}$$

where $\binom{n}{k}$ is a BINOMIAL COEFFICIENT, and have Sheffer identity

$$\frac{1}{n!} L_n^{(k)}(x+y) = \sum_{i=0}^{n} \binom{n}{i} \frac{1}{i!} L_i^{(k)}(x) \frac{1}{(n-i)!} L_{n-i}^{(-1)}(y) \tag{15}$$

(Roman 1984, p. 31). The associated Laguerre polynomial can also be written as

$$L_n^{(k)}(x) = \frac{(k+1)_n}{n!}\, {}_1F_1(-n;\ k+1;\ x), \tag{16}$$

where $(a)_n$ is the POCHHAMMER SYMBOL and ${}_1F_1(a;\ b;\ x)$ is a CONFLUENT HYPERGEOMETRIC FUNCTION (Koekoek and Swarttouw 1998).

The associated Laguerre polynomials are orthogonal over $[0,\ \infty)$ with respect to the WEIGHTING FUNCTION $x^k e^{-x}$.

$$\int_0^\infty e^{-x} x^k L_n^k(x) L_m^k(x)\, dx = \frac{(n+k)!}{n!}\, \delta_{mn}, \tag{17}$$

where δ_{mn} is the KRONECKER DELTA. They also satisfy

$$\int_0^\infty e^{-x} x^{k+1} [L_n^k(x)]^2\, dx = \frac{(n+k)!}{n!}(2n+k+1). \tag{18}$$

RECURRENCE RELATIONS include

$$\sum_{v=0}^{n} L_v^{(k)}(x) = L_n^{(k+1)}(x) \tag{19}$$

and

$$L_n^{(k)}(x) = L_n^{(k+1)}(x) - L_{n-1}^{(k+1)}(x). \tag{20}$$

The DERIVATIVE is given by

$$\frac{d}{dx} L_n^{(k)}(x) = -L_{n-1}^{(k+1)}(x)$$

$$= x^{-1}[nL_n^{(k)}(x) - (n+k)L_{n-1}^{(k)}(x). \tag{21}$$

An interesting identity is

$$\sum_{n=0}^{\infty} \frac{L_n^{(k)}(x)}{\Gamma(n+k+1)}\, w^n = e^w (xw)^{-k/2} J_k(2\sqrt{xw}), \tag{22}$$

where $\Gamma(z)$ is the GAMMA FUNCTION and $J_k(z)$ is the BESSEL FUNCTION OF THE FIRST KIND (Szego 1975, p. 102). An integral representation is

$$e^{-x} x^{k/2} L_n^{(k)}(x) = \frac{1}{n!} \int_0^\infty e^{-t} t^{n+k/2} J_k(2\sqrt{tx})\, dt \tag{23}$$

for $n = 0, 1, \dots$ and $k > -1$. The DISCRIMINANT is

$$D_n^{(k)} = \prod_{v=1}^{n} v^{v-2n+2}(v+k)^{v-1} \tag{24}$$

(Szego 1975, p. 143). The KERNEL POLYNOMIAL is

$$K_n^{(k)}(x, y) = \frac{n+1}{\Gamma(k+1)}$$

$$\times \binom{n+k}{n}^{-1}$$

$$\times \frac{L_n^{(k)}(x)L_{n+1}^{(k)}(y) - L_{n+1}^{(k)}(x)L_n(k)(y)}{x-y}, \quad (25)$$

where $\binom{n}{k}$ is a BINOMIAL COEFFICIENT (Szego 1975, p. 101).

The first few associated Laguerre polynomials are

$$L_0^k(x) = 1$$
$$L_1^k(x) = -x + k + 1$$
$$L_2^k(x) = \tfrac{1}{2}[x^2 - 2(k+2)x + (k+1)(k+2)]$$
$$L_3^k(x) = \tfrac{1}{6}[-x^3 + 3(k+3)x^2 - 3(k+2)(k+3)x$$
$$+ (k+1)(k+2)(k+3)].$$

See also LAGUERRE DIFFERENTIAL EQUATION, SONINE POLYNOMIAL

References

Abramowitz, M. and Stegun, C. A. (Eds.). "Orthogonal Polynomials." Ch. 22 in *Handbook of Mathematical Functions with Formulas, Graphs, and Mathematical Tables, 9th printing.* New York: Dover, pp. 771–02, 1972.
Andrews, G. E.; Askey, R.; and Roy, R. "Laguerre Polynomials." §6.2 in *Special Functions.* Cambridge, England: Cambridge University Press, pp. 282–93, 1999.
Arfken, G. "Laguerre Functions." §13.2 in *Mathematical Methods for Physicists, 3rd ed.* Orlando, FL: Academic Press, pp. 721–31, 1985.
Chebyshev, P. L. "Sur le développement des fonctions à une seule variable." *Bull. Ph.-Math., Acad. Imp. Sc. St. Pétersbourg* **1**, 193–00, 1859.
Chebyshev, P. L. *Oeuvres, Vol. 1.* New York: Chelsea, pp. 499–08, 1987.
Iyanaga, S. and Kawada, Y. (Eds.). "Laguerre Functions." Appendix A, Table 20.VI in *Encyclopedic Dictionary of Mathematics.* Cambridge, MA: MIT Press, p. 1481, 1980.
Koekoek, R. and Swarttouw, R. F. "Laguerre." §1.11 in *The Askey-Scheme of Hypergeometric Orthogonal Polynomials and its q-Analogue.* Delft, Netherlands: Technische Universiteit Delft, Faculty of Technical Mathematics and Informatics Report 98–7, pp. 47–9, 1998. ftp://www.twi.tudelft.nl/publications/tech-reports/1998/DUT-TWI-98-7.ps.gz.
Laguerre, E. de. "Sur l'intégrale $\int_x^{+\infty} x^{-1}e^{-x}\,dx$." *Bull. Soc. math. France* **7**, 72–1, 1879. Reprinted in *Oeuvres, Vol. 1.* New York: Chelsea, pp. 428–37, 1971.
Petkovsek, M.; Wilf, H. S.; and Zeilberger, D. *A = B.* Wellesley, MA: A. K. Peters, pp. 61–2, 1996.
Roman, S. "The Laguerre Polynomials." §3.1 i *The Umbral Calculus.* New York: Academic Press, pp. 108–13, 1984.
Rota, G.-C.; Kahaner, D.; Odlyzko, A. "Laguerre Polynomials." §11 in "On the Foundations of Combinatorial Theory. VIII: Finite Operator Calculus." *J. Math. Anal. Appl.* **42**, 684–60, 1973.
Sansone, G. "Expansions in Laguerre and Hermite Series." Ch. 4 in *Orthogonal Functions, rev. English ed.* New York: Dover, pp. 295–85, 1991.
Sonine, N. J. "Sur les fonctions cylindriques et le développement des fonctions continues en séries." *Math. Ann.* **16**, 1–0, 1880.

Spanier, J. and Oldham, K. B. "The Laguerre Polynomials $L_n(x)$." Ch. 23 in *An Atlas of Functions.* Washington, DC: Hemisphere, pp. 209–16, 1987.
Szego, G. *Orthogonal Polynomials, 4th ed.* Providence, RI: Amer. Math. Soc., 1975.
Whittaker, E. T. and Watson, G. N. Ch. 16, Ex. 8 in *A Course in Modern Analysis, 4th ed.* Cambridge, England: Cambridge University Press, p. 352, 1990.

Laguerre Quadrature

A GAUSSIAN QUADRATURE-like FORMULA for numerical estimation of integrals. It fits exactly all POLYNOMIALS of degree $2m - 1$.

References

Chandrasekhar, S. *Radiative Transfer.* New York: Dover, p. 61, 1960.

Laguerre's Method

A ROOT-finding algorithm which converges to a COMPLEX ROOT from any starting position.

$$P_n(x) = (x - x_1)(x - x_2)\cdots(x - x_n) \quad (1)$$

$$\ln|P_n(x)| = \ln|x - x_1| + \ln|x - x_2| + \ldots + \ln|x - x_n| \quad (2)$$

$$P_n'(x) = (x - x_2)\cdots(x - x_n) + (x - x_1)\cdots(x - x_n) + \ldots$$

$$= P_n(x)\left(\frac{1}{x - x_1} + \ldots + \frac{1}{x - x_n}\right) \quad (3)$$

$$\frac{d\ln|P_n(x)|}{dx} = \frac{1}{x - x_1} + \frac{1}{x - x_2} + \ldots + \frac{1}{x - x_n}$$

$$= \frac{P_n'(x)}{P_n(x)} \equiv G(x) \quad (4)$$

$$-\frac{d^2\ln|P_n(x)|}{dx^2} = \frac{1}{(x - x_1)^2} + \frac{1}{(x - x_2)^2} + \ldots + \frac{1}{(x - x_n)^2}$$

$$= \left[\frac{P_n'(x)}{P_n(x)}\right]^2 - \frac{P_n''(x)}{P_n(x)} \equiv H(x). \quad (5)$$

Now let $a \equiv x - x_1$ and $b \equiv x - x_1$. Then

$$G \equiv \frac{1}{a} + \frac{n-1}{b} \quad (6)$$

$$H \equiv \frac{1}{a^2} + \frac{n-1}{b^2}, \quad (7)$$

so

$$a = \frac{n}{\max[G \pm \sqrt{(n-1)(nH - G^2)}]}. \quad (8)$$

Setting $n = 2$ gives HALLEY'S IRRATIONAL FORMULA.

See also HALLEY'S IRRATIONAL FORMULA, HALLEY'S METHOD, NEWTON'S METHOD, ROOT

References

Press, W. H.; Flannery, B. P.; Teukolsky, S. A.; and Vetterling, W. T. *Numerical Recipes in FORTRAN: The Art of Scientific Computing, 2nd ed.* Cambridge, England: Cambridge University Press, pp. 365–66, 1992.

Ralston, A. and Rabinowitz, P. §8.9–.13 in *A First Course in Numerical Analysis, 2nd ed.* New York: McGraw-Hill, 1978.

Laguerre's Repeated Fraction

The CONTINUED FRACTION

$$\frac{(x+1)^n - (x-1)^n}{(x+1)^n + (x-1)^n} = \frac{n}{x+} \frac{n^2-1}{3x+} \frac{n^2-2^2}{5x+\ldots}.$$

References

Hardy, G. H. *Ramanujan: Twelve Lectures on Subjects Suggested by His Life and Work, 3rd ed.* New York: Chelsea, pp. 13 and 21, 1959.

Watson, G. N. "Ramanujan's Note Books." *J. London Math. Soc.* **6**, 137–53, 1931.

Watson, G. N. "The Mock Theta Functions (II)." *Proc. London Math. Soc.* **42**, 274–04, 1937.

Lah Number

The numbers

$$B_{n,k}(1!,\ 2!,\ 3!,\ \ldots) = \binom{n-1}{k-1}\frac{n!}{k!},$$

where $B_{n,k}$ is a BELL POLYNOMIAL.

See also BELL POLYNOMIAL, IDEMPOTENT NUMBER

References

Comtet, L. *Advanced Combinatorics: The Art of Finite and Infinite Expansions, rev. enl. ed.* Dordrecht, Netherlands: Reidel, p. 156, 1974.

Roman, S. *The Umbral Calculus.* New York: Academic Press, p. 86, 1984.

Rota, G.-C.; Kahaner, D.; Odlyzko, A. "On the Foundations of Combinatorial Theory. VIII: Finite Operator Calculus." *J. Math. Anal. Appl.* **42**, 684–60, 1973.

Laisant's Recurrence Formula

The RECURRENCE RELATION

$$(n-1)A_{n+1} = (n^2-1)A_n + (n+1)A_{n-1} + 4(-1)^n$$

with $A(1) = A(2) = 1$ which solves the MARRIED COUPLES PROBLEM.

See also MARRIED COUPLES PROBLEM

Lakshmi Star

STAR OF LAKSHMI

L-Algebraic Number

An L-algebraic number is a number $\theta \in (0,\ 1)$ which satisfies

$$\sum_{k=0}^{n} c_k L(\theta^k) = 0, \tag{1}$$

where $L(x)$ is the ROGERS L-FUNCTION and c_k are integers not all equal to 0 (Gordon and Mcintosh 1997). Loxton (1991, p. 289) gives a slew of similar identities having rational coefficients

$$\sum_{k=0}^{n} \frac{e_k}{k} L(\theta^k) = 0 \tag{2}$$

instead of integers.

The only known L-algebraic numbers of order 1 are

$$L(0) = 0 \tag{3}$$

$$L(1-\rho) = \tfrac{2}{5} \tag{4}$$

$$L\left(\tfrac{1}{2}\right) = \tfrac{1}{2} \tag{5}$$

$$L(\rho) = \tfrac{3}{5} \tag{6}$$

$$L(1) = 1 \tag{7}$$

(Loxton 1991, pp. 287 and 289; Bytsko 1999), where $\rho = (\sqrt{5}-1)/2$.

The only known rational L-algebraic numbers are $1/2$ and $1/3$:

$$L\left(\tfrac{1}{64}\right) - 2L\left(\tfrac{1}{8}\right) - 6L\left(\tfrac{1}{4}\right) + 2L(1) = 0 \tag{8}$$

$$L\left(\tfrac{1}{9}\right) - 6L\left(\tfrac{1}{3}\right) + 2L(1) = 0 \tag{9}$$

(Lewin 1982, pp. 317–18; Gordon and McIntosh 1997).

There are a number of known quadratic L-algebraic numbers. Watson (1937) found

$$L(\alpha) - L(\alpha^2) = \tfrac{1}{42}\pi^2 \tag{10}$$

$$2L(\beta) + L(\beta^2) = \tfrac{5}{21}\pi^2 \tag{11}$$

$$2L(\gamma) + L(\gamma^2) = \tfrac{4}{21}\pi^2, \tag{12}$$

where α, $-\beta$, and $-1/\gamma$ are the roots of

$$x^3 + 2x^2 - 1 = 0, \tag{13}$$

so that

$$\alpha = \tfrac{1}{2}\sec\left(\tfrac{2}{7}\pi\right) \tag{14}$$

$$\beta = \tfrac{1}{2}\sec\left(\tfrac{1}{7}\pi\right) \tag{15}$$

$$\gamma = 2\cos\left(\tfrac{3}{7}\pi\right) \tag{16}$$

(Loxton 1991, pp. 287–88).

Higher order algebraic identities include

$$5L(\delta^3) - 5L(\delta) + L(1) = 0, \tag{17}$$

$$L(\delta^{12}) - 2L(\delta^6) - 6L(\delta^4) + 4L(\delta^3) + 3L(\delta^2) + 4L(\delta)$$
$$-4L(1) = 0 \tag{18}$$

$$3L(\kappa^3) - 9L(\kappa^2) - 9K(\kappa) + 7L(1) = 0 \tag{19}$$

$$3L(\lambda^6) - 6L(\lambda^3) - 27L(\lambda^2) + 18L(\lambda)2L(1) = 0 \tag{20}$$

$$3L(\mu^6) - 6L(\mu^3) - 27L(\mu^2) + 18L(\mu) - 2L(1) = 0 \tag{21}$$

$$2L(a^3) - 2L(a^2) - 11L(a) + 3L(1) = 0 \tag{22}$$

$$2L(b^6) - 4L(b^3) - 15L(b^2) + 22L(b) - 6L(1) = 0 \tag{23}$$

$$2L(c^6) - 4L(c^3) - 15L(c^2) + 22L(c) - 4L(1) = 0,$$

where

$$\delta = \tfrac{1}{2}\left(\sqrt{3 + 2\sqrt{5}} - 1\right) \tag{24}$$

$$\kappa = \tfrac{1}{2}\sec\left(\tfrac{1}{9}\pi\right) \tag{25}$$

$$\lambda = \tfrac{1}{2}\sec\left(\tfrac{2}{9}\pi\right) \tag{26}$$

$$\mu = 2\cos\left(\tfrac{4}{9}\pi\right) \tag{27}$$

$$a = 2\sqrt{3}\cos\left(\frac{5\pi}{18}\right) - 2 \tag{28}$$

$$b = 2\sqrt{3}\cos\left(\frac{11\pi}{18}\right) + 2 \tag{29}$$

$$c = 2\sqrt{3}\cos\left(\frac{7\pi}{18}\right) - 1 \tag{30}$$

(Gordon and McIntosh 1997).

See also DILOGARITHM, ROGERS *L*-FUNCTION

References

Bytsko, A. G. Two-Term Dilogarithm Identities Related to Conformal Field Theory. 9 Nov 1999. http://xxx.lanl.gov/abs/math-ph/9911012/.

Gordon, B. and McIntosh, R. J. "Algebraic Dilogarithm Identities." *Ramanujan J.* **1**, 431–48, 1997.

Lewin, L. "The Dilogarithm in Algebraic Fields." *J. Austral. Soc. Ser. A* **33**, 302–30, 1982.

Lewin, L. (Ed.). *Structural Properties of Polylogarithms.* Providence, RI: Amer. Math. Soc., 1991.

Loxton, J. H. "Special Values of the Dilogarithm Function." *Acta Arith.* **43**, 155–66, 1984.

Loxton, J. H. "Partition Identities and the Dilogarithm." Ch. 13 in *Structural Properties of Polylogarithms* (Ed. L. Lewin). Providence, RI: Amer. Math. Soc., pp. 287–99, 1991.

Watson, G. N. *Quart. J. Math. Oxford Ser.* **8**, 39, 1937.

Lal's Constant

Let $P(N)$ denote the number of PRIMES OF THE FORM $n^2 + 1$ for $1 \le n \le N$, then

$$P(N) \sim 0.68641\,\mathrm{li}(N), \tag{1}$$

where $\mathrm{li}(N)$ is the LOGARITHMIC INTEGRAL (Shanks 1960, pp. 321–32). Let $Q(N)$ denote the number of PRIMES OF THE FORM $n^4 + 1$ for $1 \le n \le N$, then

$$Q(N) \sim \tfrac{1}{4}s_1\,\mathrm{li}(N) = 0.66974\,\mathrm{li}(N) \tag{2}$$

(Shanks 1961, 1962). Let $R(N)$ denote the number of pairs of PRIMES $(n-1)^2 + 1$ and $(n+1)^2 + 1$ for $n \le N - 1$, then

$$R(N) \sim 0.487621\,\mathrm{li}_2(N), \tag{3}$$

where

$$\mathrm{li}_2(N) \equiv \int_2^N \frac{dn}{(\ln n)^2} \tag{4}$$

(Shanks 1960, pp. 201–03). Finally, let $S(N)$ denote the number of pairs of PRIMES $(n-1)^4 + 1$ and $(n+1)^4 + 1$ for $n \le N - 1$, then

$$S(N) \sim \lambda\,\mathrm{li}_2(N) \tag{5}$$

(Lal 1967), where λ is called Lal's constant. Shanks (1967) showed that $\lambda \approx 0.79220$.

References

Lal, M. "Primes of the Form $n^4 + 1$." *Math. Comput.* **21**, 245–47, 1967.

Shanks, D. "On the Conjecture of Hardy and Littlewood Concerning the Number of Primes of the Form $n^2 + a$." *Math. Comput.* **14**, 321–32, 1960.

Shanks, D. "On Numbers of the Form $n^4 + 1$." *Math. Comput.* **15**, 186–89, 1961.

Shanks, D. Corrigendum to "On the Conjecture of Hardy and Littlewood Concerning the Number of Primes of the Form $n^2 + a$." *Math. Comput.* **16**, 513, 1962.

Shanks, D. "Lal's Constant and Generalization." *Math. Comput.* **21**, 705–07, 1967.

Laman's Theorem

Let a GRAPH G have exactly $2n - 3$ EDGES, where n is the number of VERTICES in G. Then G is "generically" RIGID in $\mathbb{R}^2$ iff $e' \le 2n' - 3$ for every SUBGRAPH of G having n' VERTICES and e' EDGES.

See also RIGID GRAPH

References

Laman, G. "On Graphs and Rigidity of Plane Skeletal Structures." *J. Engineering Math.* **4**, 331–40, 1970.

Lambda Calculus

Developed by Alonzo Church and Stephen Kleene to address the COMPUTABLE NUMBER problem. In the lambda calculus, λ is defined as the ABSTRACTION OPERATOR. Three theorems of lambda calculus are λ-conversion, α-conversion, and η-conversion.

See also ABSTRACTION OPERATOR, COMPUTABLE NUMBER

References

Hankin, C. *Lambda Calculi: A Guide for Computer Scientists.* Oxford, England: Oxford University Press, 1995.

Penrose, R. *The Emperor's New Mind: Concerning Computers, Minds, and the Laws of Physics.* Oxford, England: Oxford University Press, pp. 66–0, 1989.

Lambda Elliptic Function

ELLIPTIC LAMBDA FUNCTION

Lambda Function

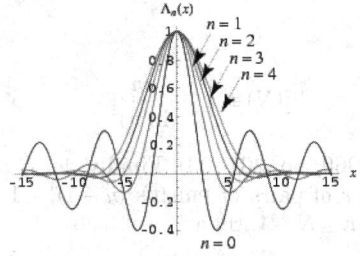

The lambda function defined by Jahnke and Emden (1945) is

$$\Lambda_\nu(z) \equiv \Gamma(\nu+1) \frac{J_\nu(z)}{\left(\frac{1}{2}z\right)^\nu} \qquad (1)$$

where $J_n(z)$ is a BESSEL FUNCTION OF THE FIRST KIND and $\Gamma(x)$ is the GAMMA FUNCTION. $\Lambda_0(z) = J_0(z)$, and taking $\nu = 1$ gives the special case

$$\Lambda_1(z) \equiv \frac{J_1(z)}{\frac{1}{2}z} = 2\,\mathrm{jinc}(z), \qquad (2)$$

where $\mathrm{jinc}(z)$ is the JINC FUNCTION.

A two-variable lambda function is defined as

$$\lambda(x,\, y) \equiv \int_0^y \frac{\Gamma(t+1)\,dt}{x^t}, \qquad (3)$$

where $\Gamma(z)$ is the GAMMA FUNCTION (McLachlan *et al.* 1950, p. 9; Prudnikov *et al.* 1990, p. 798; Gradshteyn and Ryzhik 2000, p. 1109).

The MANGOLDT FUNCTION is sometimes called the lambda function.

See also AIRY FUNCTIONS, DIRICHLET LAMBDA FUNCTION, ELLIPTIC LAMBDA FUNCTION, JINC FUNCTION, MANGOLDT FUNCTION, MU FUNCTION, NU FUNCTION

References

Gradshteyn, I. S. and Ryzhik, I. M. "The Functions $\nu(x)$, $\nu(x,\, a)$, $\mu(x,\, \beta)$, $\mu(x,\, \beta,\, \alpha)$, $\lambda(x,\, y)$." §9.64 in *Tables of Integrals, Series, and Products, 6th ed.* San Diego, CA: Academic Press, p. 1109, 2000.

Jahnke, E. and Emde, F. *Tables of Functions with Formulae and Curves, 4th ed.* New York: Dover, 1945.

McLachlan, N. W. *et al. Supplément au formulaire pour le calcul symbolique.* Paris: L'Acad. des Sciences de Paris, Fasc. 113, p. 9, 1950.

Prudnikov, A. P.; Marichev, O. I.; and Brychkov, Yu. A. *Integrals and Series, Vol. 3: More Special Functions.* Newark, NJ: Gordon and Breach, 1990.

Lambda Group

MODULAR GROUP LAMBDA

Lambda Modular Function

ELLIPTIC LAMBDA FUNCTION

Lambert Azimuthal Equal-Area Projection

A special case of a CYLINDRICAL EQUAL-AREA PROJECTION with standard parallel of $\phi_s = 0°$.

$$x = k' \cos\phi \sin(\lambda - \lambda_0) \qquad (1)$$

$$y = k'[\cos\phi_1 \sin\phi - \sin\phi_1 \cos\phi \cos(\lambda - \lambda_0)], \qquad (2)$$

where

$$k' = \sqrt{\frac{2}{1 + \sin\phi_1 \sin\phi + \cos\phi_1 \cos\phi \cos(\lambda - \lambda_0)}}. \qquad (3)$$

The inverse FORMULAS are

$$\phi = \sin^{-1}\left(\cos c \sin\phi_1 + \frac{y \sin c \cos\phi_1}{\rho}\right) \qquad (4)$$

$$\lambda = \lambda_0 + \tan^{-1}\left(\frac{x \sin c}{\rho \cos\phi_1 \cos c - y \sin\phi_1 \sin c}\right), \qquad (5)$$

where

$$\rho = \sqrt{x^2 + y^2} \qquad (6)$$

$$c = 2 \sin^{-1}\left(\frac{1}{2}\rho\right). \qquad (7)$$

See also AZIMUTHAL PROJECTION, BALTHASART PROJECTION, BEHRMANN CYLINDRICAL EQUAL-AREA PROJECTION, CYLINDRICAL EQUAL-AREA PROJECTION, EQUAL-AREA PROJECTION, GALL ORTHOGRAPHIC PROJECTION, PETERS PROJECTION, TRISTAN EDWARDS PROJECTION

Lambert Conformal Conic Projection

References

Snyder, J. P. *Map Projections--A Working Manual.* U. S. Geological Survey Professional Paper 1395. Washington, DC: U. S. Government Printing Office, pp. 182–90, 1987.

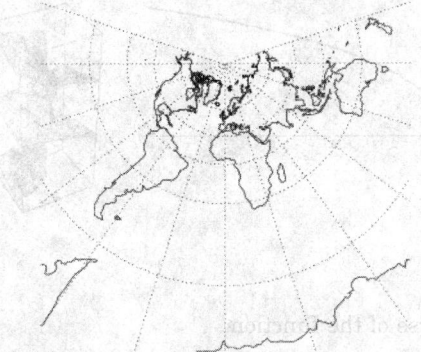

Let λ be the longitude, λ_0 the reference longitude, ϕ the latitude, ϕ_0 the reference latitude, and ϕ_1 and ϕ_2 the standard parallels. Then the transformation of SPHERICAL COORDINATES to the plane via the Lambert conformal conic projection is given by

$$x = \rho \sin[n(\lambda - \lambda_0)] \tag{1}$$

$$y = \rho_0 - \rho \cos[n(\lambda - \lambda_0)], \tag{2}$$

where

$$\rho = F \cot^n\left(\tfrac{1}{4}\pi + \tfrac{1}{2}\phi\right) \tag{3}$$

$$\rho_0 = F \cot^n\left(\tfrac{1}{4}\pi + \tfrac{1}{2}\phi_0\right) \tag{4}$$

$$F = \frac{\cos\phi_1 \tan^n\left(\tfrac{1}{4}\pi + \tfrac{1}{2}\phi_1\right)}{n} \tag{5}$$

$$n = \frac{\ln(\cos\phi_1 \sec\phi_2)}{\ln\left[\tan\left(\tfrac{1}{4}\pi + \tfrac{1}{2}\phi_2\right)\cot\left(\tfrac{1}{4}\pi + \tfrac{1}{2}\phi_1\right)\right]}. \tag{6}$$

The inverse formulas are

$$\phi = 2\tan^{-1}\left[\left(\frac{F}{\rho_0}\right)^{1/n}\right] - \tfrac{1}{2}\pi \tag{7}$$

$$\lambda = \lambda_0 + \frac{\theta}{n}, \tag{8}$$

where

$$\rho = \mathrm{sgn}(n)\sqrt{x^2 + (\rho_0 - y)^2} \tag{9}$$

$$\theta = \tan^{-1}\left(\frac{x}{\rho_0 - y}\right), \tag{10}$$

with F, ρ_0, and n as defined above.

See also CONFORMAL PROJECTION, CONIC PROJECTION

References

Snyder, J. P. *Map Projections--A Working Manual.* U. S. Geological Survey Professional Paper 1395. Washington, DC: U. S. Government Printing Office, pp. 104–10, 1987.

Lambert Cylindrical Equal-Area Projection

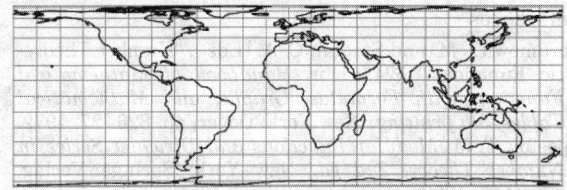

A CYLINDRICAL EQUAL-AREA PROJECTION with standard parallel $\phi_s = 0°$.

See also CYLINDRICAL EQUAL-AREA PROJECTION

Lambert Series

A series OF THE FORM

$$F(x) \equiv \sum_{n=1}^{\infty} a_n \frac{x_n}{1 - x_n} \tag{1}$$

for $|x| < 1$. Then

$$F(x) = \sum_{n=1}^{\infty} a_n \sum_{m=1}^{\infty} x^{mn} = \sum_{N=1}^{\infty} b_N x^N, \tag{2}$$

where

$$b_N \equiv \sum_{n|N} a_n. \tag{3}$$

Some beautiful series of this type include

$$\sum_{n=1}^{\infty} \frac{\mu(n)x^n}{1 - x^n} = x \tag{4}$$

$$\sum_{n=1}^{\infty} \frac{\phi(n)x^n}{1 - x^n} = \frac{x}{(1 - x)^2} \tag{5}$$

$$\sum_{n=1}^{\infty} \frac{x^n}{1 - x^n} = \sum_{n=1}^{\infty} d(n)x^n \tag{6}$$

$$\sum_{n=1}^{\infty} \frac{n^k x^n}{1 - x^n} = \sum_{n=1}^{\infty} \sigma_k(n)x^n \tag{7}$$

$$\sum_{n=1}^{\infty} \frac{4(-1)^{n+1} x^n}{1 - x^n} = \sum_{n=1}^{\infty} r(n)x^n \tag{8}$$

$$\sum_{n=1}^{\infty} \frac{\lambda(n)x^n}{1 - x^n} = \sum_{n=1}^{\infty} x^{n^2}, \tag{9}$$

where $\mu(n)$ is the MÖBIUS FUNCTION, $\phi(n)$ is the TOTIENT FUNCTION, $d(n) = \sigma_0(n)$ is the number of divisors of n, $\sigma_k(n)$ is the DIVISOR FUNCTION, $r(n)$ is the number of representations of n in the form $n = A^2 + B^2$ where A and B are rational integers (Hardy and Wright 1979), and $\lambda(n)$ is the LAMBDA FUNCTION.

See also DIVISOR FUNCTION, LAMBDA FUNCTION, MÖBIUS FUNCTION, MÖBIUS TRANSFORM, TOTIENT FUNCTION

References

Abramowitz, M. and Stegun, C. A. (Eds.). "Number Theoretic Functions." §24.3.1 in *Handbook of Mathematical Functions with Formulas, Graphs, and Mathematical Tables, 9th printing.* New York: Dover, pp. 826–27, 1972.

Apostol, T. M. *Modular Functions and Dirichlet Series in Number Theory, 2nd ed.* New York: Springer-Verlag, pp. 24–5, 1997.

Erdos, P. "On Arithmetical Properties of Lambert Series." *J. Indian Math. Soc.* **12**, 63–6, 1948.

Hardy, G. H. and Wright, E. M. *An Introduction to the Theory of Numbers, 5th ed.* Oxford, England: Clarendon Press, pp. 257–58, 1979.

Lambert's Method

A ROOT-finding method also called BAILEY'S METHOD and HUTTON'S METHOD If $g(x) = x^d - r$, then

$$H_g(x) = \frac{(d-1)x^d + (d+1)r}{(d+1)x^d + (d-1)r} \, x.$$

References

Scavo, T. R. and Thoo, J. B. "On the Geometry of Halley's Method." *Amer. Math. Monthly* **102**, 417–26, 1995.

Lambert's Transcendental Equation

An equation proposed by Lambert (1758) and studied by Euler in 1779 (Euler 1921).

$$x^\alpha - x^\beta = (\alpha - \beta)vx^{\alpha+\beta}.$$

When $\alpha \to \beta$, the equation becomes

$$\ln x = vx^\beta,$$

which has the solution

$$x = \exp\left[-\frac{W(-\beta v)}{\beta}\right],$$

where $W(x)$ is LAMBERT'S W-FUNCTION.

See also LAMBERT'S W-FUNCTION

References

Corless, R. M.; Gonnet, G. H.; Hare, D. E. G.; Jeffrey, D. J.; and Knuth, D. E. "On the Lambert W Function." *Adv. Comput. Math.* **5**, 329–59, 1996.

de Bruijn, N. G. *Asymptotic Methods in Analysis.* Amsterdam, Netherlands: North-Holland, pp. 27–8, 1961.

Euler, L. "De Serie Lambertina Plurismique Eius Insignibus Proprietatibus." *Leonhardi Euleri Opera Omnia, Ser. 1. Opera Mathematica,* Bd. 6, 1921.

Lambert, J. H. "Observations variae in Mathesin Puram." *Acta Helvitica, physico-mathematico-anatomico-botanico-medica* **3**, 128–68, 1758.

Lambert's W-Function

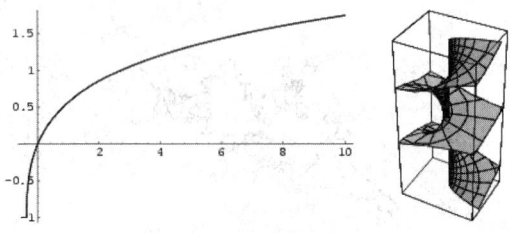

The inverse of the function

$$f(W) = We^W, \tag{1}$$

also called the omega function. The plots above show the function along the REAL AXIS (left figure) and its RIEMANN SURFACE (right figure). The principal value of the Lambert W-function is implemented in *Mathematica* as `ProductLog[z]`. Different branches of the function are available as `ProductLog[k, z]`, where k is any integer and $k = 0$ corresponds to the principal value.

Lambert's W-function can be used to analytically express the value of the POWER TOWER $h(x) = x \uparrow\uparrow \infty = x^{x^{\cdot^{\cdot}}}$, where x^{x^x} is an abbreviation for $x^{(x^x)}$, as

$$h(x) = -\frac{W(-\ln x)}{\ln x}. \tag{2}$$

$W(1)$ is called the OMEGA CONSTANT and can be considered a sort of "GOLDEN RATIO" of exponentials since

$$\exp[-W(1)] = W(1), \tag{3}$$

giving

$$\ln\left[\frac{1}{W(1)}\right] = W(1). \tag{4}$$

Lambert's W-Function has the series expansion

$$W(x) = \sum_{n=1}^{\infty} \frac{(-1)^{n-1} n^{n-2}}{(n-1)!} x^n = x - x^2 + \tfrac{3}{2} x^3 - \tfrac{8}{3} x^4$$
$$+ \tfrac{125}{24} x^5 - \tfrac{54}{5} x^6 + \tfrac{16807}{720} x^7 + \ldots \tag{5}$$

The LAGRANGE INVERSION THEOREM gives the equivalent series expansion

$$W_0(z) = \sum_{n=1}^{\infty} \frac{(-n)^{n-1}}{n!} z^n, \tag{6}$$

where $n!$ is a FACTORIAL. However, this series oscillates between ever larger POSITIVE and NEGATIVE values for REAL $z \gtrsim 0.4$, and so cannot be used for practical numerical computation. An asymptotic FORMULA which yields reasonably accurate results for $z \gtrsim 3$ is

$$W(z) = \operatorname{Ln} z - \ln \operatorname{Ln} z + \sum_{k=0}^{\infty} \sum_{m=0}^{\infty} c_{km} (\ln \operatorname{Ln} z)^{m+1}$$

$$\times (\operatorname{Ln} z)^{-k-m-1}$$

$$= L_1 - L_2 + \frac{L_2}{L_1} + \frac{L_2(-2+L_2)}{2L_1^2} + \frac{L_2(6 - 9L_2 + 2L_2^2)}{6L_1^3}$$

$$+ \frac{L_2(-12 + 36L_2 - 22L_2^2 + 3L_2^3)}{12L_1^4}$$

$$+ \frac{L_2(60 - 300L_2 + 350L_2^2 - 125L_2^3 + 12L_2^4)}{60L_1^5}$$

$$+ \mathcal{O}\left[\left(\frac{L_2}{L_1}\right)^6\right], \tag{7}$$

where

$$L_1 = \operatorname{Ln} z \tag{8}$$

$$L_2 = \ln \operatorname{Ln} z \tag{9}$$

(Corless *et al.* 1996), correcting a typographical error in de Bruijn (1961). Another expansion due to Gosper is the DOUBLE SUM

$$W(x) = a + \sum_{n=0}^{\infty} \left\{ \sum_{k=0}^{n} \frac{S_1(n, k)}{\left[\ln\left(\frac{x}{a}\right) - a\right]^{k-1} (n - k + 1)!} \right\}$$

$$\times \left[1 - \frac{\ln\left(\frac{x}{a}\right)}{a}\right]^n, \tag{10}$$

where S_1 is a *nonnegative* STIRLING NUMBER OF THE FIRST KIND and a is a first approximation which can be used to select between branches. Lambert's W-function is two-valued for $-1/e \leq x < 0$. For $W(x) \geq -1$, the function is denoted $W_0(x)$ or simply $W(x)$, and this is called the principal branch. For $W(x) \leq -1$, the function is denoted $W_{-1}(x)$. The DERIVATIVE of W is

$$W'(x) = \frac{1}{[1 + W(x)] \exp[W(x)]} = \frac{W(x)}{x[1 + W(x)]} \tag{11}$$

for $x \neq 0$. For the principal branch when $z > 0$,

$$\ln W(z) = \ln z - W(z) \tag{12}$$

See also ABEL POLYNOMIAL, DIGIT-SHIFTING CONSTANTS, LAMBERT'S TRANSCENDENTAL EQUATION, OMEGA CONSTANT, POWER TOWER

References

--. "Time for a New Elementary Function?" *FOCUS: Newsletter Math. Assoc. Amer.* **20**, 2, Feb. 2000.

Borwein, J. M. and Corless, R. M. "Emerging Tools for Experimental Mathematics." *Amer. Math. Monthly* **106**, 899–09, 1999.

Briggs, K. "W-ology, or, Some Exactly Solvable Growth Models." http://epidem13.plantsci.cam.ac.uk/~kbriggs/W-ology.html.

Corless, R. M.; Jeffrey, D. J.; and Knuth, D. E. "A Sequence of Series for the Lambert W Function." In *Proc. ISSAC '97, Maui, Hawaii* (Ed. W. W. Küchlin). New York: ACM, pp. 197–04, 1997.

Corless, R. M.; Gonnet, G. H.; Hare, D. E. G.; Jeffrey, D. J.; and Knuth, D. E. "On the Lambert W Function." *Adv. Comput. Math.* **5**, 329–59, 1996.

Corless, R. M.; Gonnet, G. H.; Hare, D. E. G.; and Jeffrey, D. J. "Lambert's W Function in Maple." *Maple Technical Newsletter* **9**, 12–2, Spring 1993.

de Bruijn, N. G. *Asymptotic Methods in Analysis.* Amsterdam, Netherlands: North-Holland, pp. 27–8, 1961.

Euler, L. "De serie Lambertina Plurimisque eius insignibus proprietatibus." *Acta Acad. Scient. Petropol.* **2**, 29–1, 1783. Reprinted in Euler, L. *Opera Omnia I6: Commentationes Algebraicae.* pp. 350–69.

Fritsch, F. N.; Shafer, R. E.; and Crowley, W. P. "Algorithm 443: Solution of the Transcendental Equation $we^w = x$." *Comm. ACM* **16**, 123–24, 1973.

Jeffrey, D. J.; Hare, D. E. G.; and Corless, R. M. "Unwinding the Branches of the Lambert W Function." *Math. Scientist* **21**, 1–, 1996.

Jeffrey, D. J.; Corless, R. M.; Hare, D. E. G.; and Knuth, D. E. "Sur l'inversion de yâ ey au moyen des nombres de Stirling associes. " *Comptes Rendus Acad. Sci. Paris* **320**, 1449–452, 1995.

Pólya, G. and Szego, G. *Problems and Theorems in Analysis I.* Berlin: Springer-Verlag, 1998.

Lamé Curve

There are two curves commonly known as the Lamé curve: the ELLIPSE EVOLUTE and the SUPERELLIPSE.

See also ELLIPSE EVOLUTE, SUPERELLIPSE

Lamé Function

ELLIPSOIDAL HARMONIC

Lamé's Differential Equation

The ORDINARY DIFFERENTIAL EQUATION

$$(x^2 - b^2)(x^2 - c^2) \frac{d^2 z}{dx^2} + x(x^2 - b^2 + x^2 - c^2) \frac{dz}{dx}$$

$$-[m(m+1)x^2 - (b^2 + c^2)p]z = 0. \tag{1}$$

(Byerly 1959, p. 255). The solution is denoted $E_m^p(x)$ and is known as a LAMÉ FUNCTION or an ELLIPSOIDAL HARMONIC. Whittaker and Watson (1990, pp. 554–55) give the alternative forms

$$4\Delta_\lambda \frac{d}{d\lambda}\left[\Delta\lambda \frac{d\Lambda}{d\lambda}\right] = [n(n+1)\lambda + C]\Lambda \tag{2}$$

$$\frac{d^2\Lambda}{d\lambda^2} + \left[\frac{\frac{1}{2}}{a^2+\lambda} + \frac{\frac{1}{2}}{b^2+\lambda} + \frac{\frac{1}{2}}{c^2} \right] \frac{d\Lambda}{d\lambda}$$

$$= \frac{[n(n+1)\lambda + C]\Lambda}{4\Delta_\lambda} \tag{3}$$

$$\frac{d^2\Lambda}{du^2} = \left[n(n+1)\wp(u) + C - \tfrac{1}{3}n(n+1)(a^2+b^2+c^2) \right]\Lambda \tag{4}$$

$$\frac{d^2\Lambda}{dz^2} = n(n+1)k^2\,\mathrm{sn}^2(z,k) + A\Lambda \tag{5}$$

(Whittaker and Watson 1990, pp. 554–55; Ward 1997; Zwillinger 1997, p. 124). Here, $\wp$ is a WEIER-STRASS ELLIPTIC FUNCTION, $\mathrm{sn}(z,k)$ is a JACOBI ELLIPTIC FUNCTION, and

$$\Lambda(\theta) \equiv \prod_{q=1}^{m}(\theta - \theta_q) \tag{6}$$

$$\Delta_\lambda \equiv \sqrt{(a^2+\lambda)(b^2+\lambda)(c^2+\lambda)} \tag{7}$$

$$A \equiv \frac{C - \tfrac{1}{3}n(n+1)(a^2+b^2+c^2) + e_3 n(n+1)}{e_1 - e_3}. \tag{8}$$

Two other equations named after Lamé are given by

$$y'' + \frac{1}{2}\left[\frac{1}{x-a_1} + \frac{1}{x-a_2} + \frac{1}{x-a_3} \right]y'$$

$$+ \frac{1}{4}\left[\frac{A_0 + A_1 x}{(x-a_1)(x-a_2)(x-a_3)} \right]y = 0 \tag{9}$$

and

$$y'' + \frac{1}{2}\left[\frac{1}{x} + \frac{1}{x-a_2} + \frac{1}{x-a_3} \right]y'$$

$$+ \frac{1}{4}\left[\frac{(a_2^2+a_3^2)q - p(p+1)x + \kappa x^2}{x(x-a_2)(x-a_3)} \right]y = 0 \tag{10}$$

(Moon and Spencer 1961, p. 157; Zwillinger 1997, p. 124).

See also ELLIPSOIDAL WAVE EQUATION, LAMÉ'S DIFFERENTIAL EQUATION TYPES, WANGERIN DIFFERENTIAL EQUATION

References

Byerly, W. E. *An Elementary Treatise on Fourier's Series, and Spherical, Cylindrical, and Ellipsoidal Harmonics, with Applications to Problems in Mathematical Physics.* New York: Dover, 1959.

Moon, P. and Spencer, D. E. *Field Theory for Engineers.* New York: Van Nostrand, 1961.

Ward, R. S. "The Nahn Equations, Finite-Gap Potentials and Lamé Functions." *J. Phys. A: Math. Gen.* **20**, 2679–683, 1987.

Whittaker, E. T. and Watson, G. N. *A Course in Modern Analysis, 4th ed.* Cambridge, England: Cambridge University Press, 1990.

Zwillinger, D. *Handbook of Differential Equations, 3rd ed.* Boston, MA: Academic Press, p. 124, 1997.

Lamé's Differential Equation Types

Whittaker and Watson (1990, pp. 539–40) write Lamé's differential equation for ELLIPSOIDAL HARMONICS of the four types as

$$4\delta(\theta)\frac{d}{d\theta}\left[f(\theta)\frac{d\lambda(\theta)}{d\theta} \right] = [2m(2m+1)\theta + c]\lambda(\theta) \tag{1}$$

$$4\delta(\theta)\frac{d}{d\theta}\left[f(\theta)\frac{d\lambda(\theta)}{d\theta} \right]$$

$$= [(2m+1)(2m+2)\theta + c]\lambda(\theta) \tag{2}$$

$$4\delta(\theta)\frac{d}{d\theta}\left[f(\theta)\frac{d\lambda(\theta)}{d\theta} \right]$$

$$= [(2m+2)(2m+3)\theta + c]\lambda(\theta) \tag{3}$$

$$4\delta(\theta)\frac{d}{d\theta}\left[f(\theta)\frac{d\lambda(\theta)}{d\theta} \right]$$

$$= [(2m+3)(2m+4)\theta + c]\lambda(\theta), \tag{4}$$

where

$$\delta(\theta) \equiv \sqrt{(a^2+\theta)(b^2+\theta)(c^2+\theta)} \tag{5}$$

$$\lambda(\theta) \equiv \prod_{q=1}^{m}(\theta - \theta_q). \tag{6}$$

See also LAMÉ'S DIFFERENTIAL EQUATION

References

Whittaker, E. T. and Watson, G. N. *A Course in Modern Analysis, 4th ed.* Cambridge, England: Cambridge University Press, 1990.

Lamé's Theorem

If a is the smallest INTEGER for which there is a smaller INTEGER b such that a and b generate a EUCLIDEAN ALGORITHM remainder sequence with n steps, then a is the FIBONACCI NUMBER F_{n+2}. Furthermore, the number of steps in the EUCLIDEAN ALGORITHM never exceeds 5 times the number of digits in the smaller number.

See also EUCLIDEAN ALGORITHM

References

Honsberger, R. "A Theorem of Gabriel Lamé." Ch. 7 in *Mathematical Gems II.* Washington, DC: Math. Assoc. Amer., pp. 54–7, 1976.

Lamina

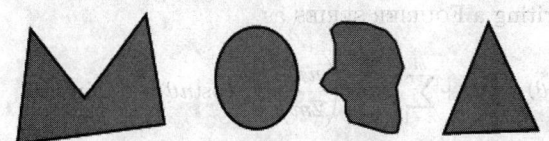

A 2-D planar closed surface L which has a mass M and a surface density $\sigma(x, y)$ (in units of mass per areas squared) such that

$$M = \int_L \sigma(x, y)\, dx\, dy.$$

The CENTER OF MASS of a lamina is called its CENTROID.

See also CENTROID (GEOMETRIC), CROSS SECTION, SOLID

Laminated Lattice

A LATTICE which is built up of layers of n-D lattices in $(n+1)$-D space. The VECTORS specifying how layers are stacked are called GLUE VECTORS.

See also GLUE VECTOR, LATTICE

References

Conway, J. H. and Sloane, N. J. A. "Laminated Lattices." Ch. 6 in *Sphere Packings, Lattices, and Groups, 2nd ed.* New York: Springer-Verlag, pp. 157–80, 1993.

Lamp Paradox

THOMPSON LAMP PARADOX

Lam's Problem

Given a 111×111 BINARY MATRIX, fill 11 spaces in each row in such a way that all columns also have 11 spaces filled. Furthermore, each pair of rows must have exactly one filled space in the same column. This problem is equivalent to finding a PROJECTIVE PLANE of order 10. Using a computer program, Lam *et al.* (1989) showed that no such arrangement exists.

Lam's problem is equivalent to finding nine orthogonal LATIN SQUARES of order 10.

See also BINARY MATRIX, LATIN SQUARE, PROJECTIVE PLANE

References

--. *Science.* 1507–508, Dec. 20, 1988.
Beezer, R. "Graeco-Latin Squares." http://buzzard.ups.edu/squares.html.
Browne, M. W. "Is a Math Proof a Proof If No One Can Check It?" *New York Times*, Sec. 3, p. 1, col. 1, Dec. 20, 1988.
Lam, C. W. H.; Thiel, L.; and Swiercz, S. "The Nonexistence of Finite Projective Planes of Order 10." *Canad. J. Math.* **41**, 1117–123, 1989.
Petersen, I. "Search Yields Math Proof No One Can Check." *Science News* **134**, 406, Dec. 24 & 31, 1988.

Lancret Equation

$$ds_N^2 = ds_T^2 + ds_B^2,$$

where N is the NORMAL VECTOR, T is the TANGENT, and B is the BINORMAL VECTOR.

Lancret's Theorem

A NECESSARY and SUFFICIENT condition for a curve to be a HELIX is that the ratio of CURVATURE to TORSION be constant.

Lanczos Algorithm

An algorithm for computing the eigenvalues and eigenvectors for large symmetric sparse matrices.

References

Chung, F. R. K. *Spectral Graph Theory.* Providence, RI: Amer. Math. Soc., 1997.
Demmel, J. "CS 267: Notes for Lecture 23, April 9, 1999. Graph Partitioning, Part 2." http://www.cs.berkeley.edu/~demmel/cs267/lecture20/lecture20.html.

Lanczos Approximation

An approximation for the GAMMA FUNCTION $\Gamma(z+1)$ with $z > 0$ is given by

$$\Gamma(z+1) = \sqrt{2\pi}$$

$$\times \left(z + \sigma + \tfrac{1}{2}\right)^{z+1/2} e^{-(z+\sigma+1/2)} \sum_{k=0}^{\infty} g_k H_k(z), \tag{1}$$

where σ is an arbitrary constant such that $\Re[z + \sigma + 1/2] > 0$,

$$g_k = \frac{e^{\sigma} \varepsilon_k (-1)^k}{\sqrt{2\pi}} \sum_{r=0}^{k} (-1)^r \binom{k}{r} (k)_r \left(\frac{e}{r + \sigma + \frac{1}{2}}\right)^{r+1/2} \tag{2}$$

where $(k)_r$ is a POCHHAMMER SYMBOL and

$$\varepsilon_k = \begin{cases} 1 & \text{for } k = 0 \\ 2 & \text{otherwise,} \end{cases} \tag{3}$$

and

$$H_k(z) = \frac{1}{(z+1)_k (z+1)_{-k}} \tag{4}$$

$$= \frac{(-1)^k (-z)^k}{(z+1)_k}, \tag{5}$$

with $H_0(z) = 1$ (Lanczos 1964; Luke 1969, p. 30). g_k satisfies

$$\sum_{k=0}^{\infty} g_k = 1, \tag{6}$$

and if z is a POSITIVE INTEGER, then g_k satisfies the identity

$$\sum_{k=0}^{n} \frac{(-1)^k (-n)_k}{(n+1)_k} g_k = \frac{e^{n+\sigma+1/2} n!}{\sqrt{2\pi}(n+\sigma+1/2)^{n+1/2}} \quad (7)$$

(Luke 1969, p. 30).

A similar result is given by

$$\ln[\Gamma(z)] = \left(z - \tfrac{1}{2}\right)\ln z - z + \tfrac{1}{2}\ln(2\pi)$$

$$+ \tfrac{1}{2}\left[\frac{c_1}{z+1} + \frac{c_2}{2(z+1)(z+2)} + \dots\right] \quad (8)$$

where

$$c_n = \int_0^1 (x)_n (2x-1)\, dx, \quad (9)$$

with $(x)_n$ a POCHHAMMER SYMBOL. The first few values of c_n are

$$c_1 = \tfrac{1}{6}$$

$$c_2 = \tfrac{1}{3}$$

$$c_3 = \tfrac{59}{60}$$

$$c_4 = \tfrac{58}{15}$$

$$c_5 = \tfrac{533}{28}$$

(Sloane's A054379 and A054380; Whittaker and Watson 1990, p. 253). Note that Whittaker and Watson incorrectly give c_4 as 227/60.

Yet another related result gives

$$\ln[\Gamma(z)] = \left(z - \tfrac{1}{2}\right)\ln z - z + \tfrac{1}{2}\ln(2\pi)$$

$$+ \tfrac{1}{2}\left[\frac{1}{2\cdot 3}\sum_{r=1}^{\infty}\frac{1}{(z+r)^2} + \frac{2}{3\cdot 4}\sum_{r=1}^{\infty}\frac{1}{(z+r)^3}\right.$$

$$\left. + \frac{3}{4\cdot 5}\sum_{r=1}^{\infty}\frac{1}{(z+r)^4} + \dots\right] \quad (10)$$

(Whittaker wand Watson 1990, p. 261).

See also GAMMA FUNCTION

References

Lanczos, C. *J. Soc. Indust. Appl. Math. Ser. B: Numer. Anal.* **1**, 86–6, 1964.

Luke, Y. L. "An Expansion for $\Gamma(z+1)$." §2.10.3 in *The Special Functions and their Approximations, Vol. 1.* New York: Academic Press, pp. 29–1, 1969.

Sloane, N. J. A. Sequences A054379 and A054379 in "An On-Line Version of the Encyclopedia of Integer Sequences." http://www.research.att.com/~njas/sequences/eisonline.html.

Whittaker, E. T. and Watson, G. N, *A Course in Modern Analysis, 4th ed.* Cambridge, England: Cambridge University Press, 1990.

Lanczos Sigma Factor

Writing a FOURIER SERIES as

$$f(\theta) = \tfrac{1}{2}a_0 + \sum_{n=1}^{m}\operatorname{sin} c\left(\frac{n\pi}{2m}\right)[a_n\cos(n\theta) + b_n\sin(n\theta)],$$

where m is the last term and the sinc x terms are the Lanczos σ factor, removes the GIBBS PHENOMENON (Acton 1990).

See also FOURIER SERIES, GIBBS PHENOMENON, SINC FUNCTION

References

Acton, F. S. *Numerical Methods That Work, 2nd printing.* Washington, DC: Math. Assoc. Amer., p. 228, 1990.

Landau Constant

N.B. A detailed online essay by S. Finch was the starting point for this entry.

Let F be the set of COMPLEX analytic functions f defined on an open region containing the closure of the unit disk $D = \{z : |z| < 1\}$ satisfying $f(0) = 0$ and $df/dz(0) = 1$. For each f in F, let (f) be the SUPREMUM of all numbers r such that $f(D)$ contains a disk of radius r. Then

$$L \equiv \inf\{l(f) : f \in F\}.$$

This constant is called the Landau constant, or the BLOCH-LANDAU CONSTANT. Robinson (1938, unpublished) and Rademacher (1943) derived the bounds

$$\tfrac{1}{2} < L \le \frac{\Gamma\left(\tfrac{1}{3}\right)\Gamma\left(\tfrac{5}{6}\right)}{\Gamma\left(\tfrac{1}{6}\right)} = 0.5432588\dots,$$

where $\Gamma(z)$ is the GAMMA FUNCTION, and conjectured that the second inequality is actually an equality,

$$L = \frac{\Gamma\left(\tfrac{1}{3}\right)\Gamma\left(\tfrac{5}{6}\right)}{\Gamma\left(\tfrac{1}{6}\right)} = 0.5432588\dots.$$

See also BLOCH CONSTANT

References

Finch, S. "Favorite Mathematical Constants." http://www.mathsoft.com/asolve/constant/bloch/bloch.html.

Rademacher, H. "On the Bloch-Landau Constant." *Amer. J. Math.* **65**, 387–90, 1943.

Landau-Kolmogorov Constants

N.B. A detailed online essay by S. Finch was the starting point for this entry.

Let $\|f\|$ be the SUPREMUM of $|f(x)|$, a real-valued function f defined on $(0, \infty)$. If f is twice differentiable and both f and f'' are bounded, Landau (1913) showed that

$$\|f'\| \le 2\|f\|^{1/2}\|f''\|^{1/2}, \qquad (1)$$

where the constant 2 is the best possible. Schoenberg (1973) extended the result to the nth derivative of f defined on $(0, \infty)$ if both f and $f^{(n)}$ are bounded,

$$\|f^{(k)}\| \le C(n, k)\|f\|^{1-k/n}\|f(n)\|^{k/n}. \qquad (2)$$

An explicit FORMULA for $C(n, k)$ is not known, but particular cases are

$$C(3, 1) = \left(\frac{243}{8}\right)^{1/3} \qquad (3)$$

$$C(3, 2) = 24^{1/3} \qquad (4)$$

$$C(4, 1) = 4.288\ldots \qquad (5)$$

$$C(4, 2) = 5.750\ldots \qquad (6)$$

$$C(4, 3) = 3.708\ldots. \qquad (7)$$

Let $\|f\|$ be the SUPREMUM of $|f(x)|$, a real-valued function f defined on $(-\infty, \infty)$. If f is twice differentiable and both f and f'' are bounded, Hadamard (1914) showed that

$$\|f'\| \le \sqrt{2}\|f\|^{1/2}\|f''\|^{1/2}, \qquad (8)$$

where the constant $\sqrt{2}$ is the best possible. Kolmogorov (1962) determined the best constants $C(n, k)$ for

$$\|f^{(k)}\| \le C(n, k)\|f\|^{1-k/n}\|f^{(n)}\|^{k/n} \qquad (9)$$

in terms of the FAVARD CONSTANTS

$$a_n = \frac{4}{\pi} \sum_{j=0}^{\infty} \left[\frac{(-1)^j}{2j+1}\right]^{n+1} \qquad (10)$$

by

$$C(n, k) = a_{n-k} a_n^{-1+k/n}. \qquad (11)$$

Special cases derived by Shilov (1937) are

$$C(3, 1) = \left(\frac{9}{8}\right)^{1/3} \qquad (12)$$

$$C(3, 2) = 3^{1/3} \qquad (13)$$

$$C(4, 1) = \left(\frac{512}{375}\right)^{1/4} \qquad (14)$$

$$C(4, 2) = \sqrt{\frac{6}{5}} \qquad (15)$$

$$C(4, 3) = \left(\frac{24}{5}\right)^{1/4} \qquad (16)$$

$$C(5, 1) = \left(\frac{1953125}{1572864}\right)^{1/5} \qquad (17)$$

$$C(5, 2) = \left(\frac{125}{72}\right)^{1/5}. \qquad (18)$$

For a real-valued function f defined on $(-\infty, \infty)$, define

$$\|f\|\sqrt{\int_{-\infty}^{\infty} [f(x)]^2 \, dx}. \qquad (19)$$

If f is n differentiable and both f and $f^{(n)}$ are bounded, Hardy *et al.* (1934) showed that

$$\|f^{(k)}\| \le \|f\|^{1-k/n}\|f^{(n)}\|^{k/n}, \qquad (20)$$

where the constant 1 is the best possible for all n and $0 < k < n$.

For a real-valued function f defined on $(0, \infty)$, define

$$\|f\| = \sqrt{\int_0^{\infty} [f(x)]^2 \, dx}. \qquad (21)$$

If f is twice differentiable and both f and f'' are bounded, Hardy *et al.* (1934) showed that

$$\|f'\| \le \sqrt{2}\|f\|^{1/2}\|f^{(n)}\|^{1/2}, \qquad (22)$$

where the constant $\sqrt{2}$ is the best possible. This inequality was extended by Ljubic (1964) and Kupcov (1975) to

$$\|f^{(k)}\| \le C(n, k)\|f\|^{1-k/n}\|f^{(n)}\|^{k/n} \qquad (23)$$

where $C(n, k)$ are given in terms of zeros of POLYNOMIALS. Special cases are

$$C(3, 1) = C(3, 2) = 3^{1/2}[2(2^{1/2} - 1)]^{-1/3}$$
$$= 1.84420\ldots \qquad (24)$$

$$C(4, 1) = C(4, 3) = \sqrt{\frac{3^{1/4} + 3^{-3/4}}{a}}$$
$$= 2.27432\ldots \qquad (25)$$

$$C(4, 2) = \sqrt{\frac{2}{b}} = 2.97963\ldots \qquad (26)$$

$$C(4, 3) = \left(\frac{24}{5}\right)^{1/4} \qquad (27)$$

$$C(5, 1) = C(5, 4) = 2.70247\ldots \qquad (28)$$

$$C(5, 2) = C(5, 3) = 4.37800\ldots, \qquad (29)$$

where a is the least POSITIVE ROOT of

$$x^8 - 6x^4 - 8x^2 + 1 = 0 \qquad (30)$$

and b is the least POSITIVE ROOT of

$$x^4 - 2x^2 - 4x + 1 = 0 \qquad (31)$$

(Franco *et al.* 1985, Neta 1980). The constants $C(n, 1)$ are given by

$$C(n, 1) = \sqrt{\frac{(n - 1)^{1/n} + (n + 1)^{-1+1/n}}{c}}, \qquad (32)$$

where c is the least POSITIVE ROOT of

$$\int_0^c \int_0^\infty \frac{dx \, dy}{(x^{2n} - yx^2 + 1)\sqrt{y}} = \frac{\pi^2}{2n}. \qquad (33)$$

An explicit FORMULA of this type is not known for $k > 1$.

The cases $p = 1, 2, \infty$ are the only ones for which the best constants have exact expressions (Kwong and Zettl 1992, Franco *et al.* 1983).

References

Finch, S. "Favorite Mathematical Constants." http://www.mathsoft.com/asolve/constant/lk/lk.html.

Franco, Z. M.; Kaper, H. G.; Kwong, M. N.; and Zettl, A. "Bounds for the Best Constants in Landau's Inequality on the Line." *Proc. Roy. Soc. Edinburgh* **95A**, 257–62, 1983.

Franco, Z. M.; Kaper, H. G.; Kwong, M. N.; and Zettl, A. "Best Constants in Norm Inequalities for Derivatives on a Half Line." *Proc. Roy. Soc. Edinburgh* **100A**, 67–4, 1985.

Hardy, G. H.; Littlewood, J. E.; and Pólya, G. *Inequalities.* Cambridge, England: Cambridge University Press, 1934.

Kolmogorov, A. "On Inequalities Between the Upper Bounds of the Successive Derivatives of an Arbitrary Function on an Infinite Integral." *Amer. Math. Soc. Translations, Ser. 1* **2**, 233–43, 1962.

Kupcov, N. P. "Kolmogorov Estimates for Derivatives in $L_2(0, \infty)$." *Proc. Steklov Inst. Math.* **138**, 101–25, 1975.

Kwong, M. K. and Zettl, A. *Norm Inequalities for Derivatives and Differences.* New York: Springer-Verlag, 1992.

Landau, E. "Einige Ungleichungen für zweimal different-zierbare Funktionen." *Proc. London Math. Soc. Ser. 2* **13**, 43–9, 1913.

Landau, E. "Die Ungleichungen für zweimal differentzier-bare Funktionen." *Danske Vid. Selsk. Math. Fys. Medd.* **6**, 1–9, 1925.

Ljubic, J. I. "On Inequalities Between the Powers of a Linear Operator." *Amer. Math. Soc. Trans. Ser. 2* **40**, 39–4, 1964.

Neta, B. "On Determinations of Best Possible Constants in Integral Inequalities Involving Derivatives." *Math. Comput.* **35**, 1191–193, 1980.

Schoenberg, I. J. "The Elementary Case of Landau's Problem of Inequalities Between Derivatives." *Amer. Math. Monthly* **80**, 121–58, 1973.

Landau-Lifshitz Equation

The system of PARTIAL DIFFERENTIAL EQUATIONS

$$U_t = U \cdot U_{xx} + U \cdot AU.$$

References

Fuchssteiner, B. "On the Hierarchy of the Landau-Lifshitz Equation." *Physica D* **13**, 387–94, 1984.

Zwillinger, D. *Handbook of Differential Equations*, 3rd ed. Boston, MA: Academic Press, p. 138, 1997.

Landau-Ramanujan Constant

N.B. A detailed online essay by S. Finch was the starting point for this entry.

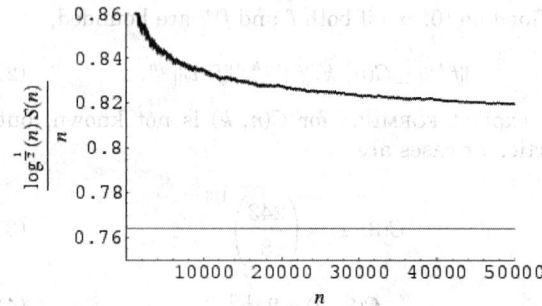

Let $S(x)$ denote the number of POSITIVE INTEGERS not exceeding x which can be expressed as a sum of two squares, then

$$\lim_{x \to \infty} \frac{\sqrt{\ln x}}{x} S(x) = K, \qquad (1)$$

as proved by Landau (1908). Ramanujan independently stated the theorem in the slightly different form that the number of numbers between A and x which are either squares of sums of two squares is

$$S(x) = K \int_A^x \frac{dt}{\sqrt{\ln t}} + \theta(x), \qquad (2)$$

where $K \approx 0.764$ and $\theta(x)$ is very small compared with the previous integral (Hardy 1999, p. 8; Moree and Cazaran 1999). However, the convergence to the constant K is very slow.

The exact value for

$$K = 0.764223653\ldots \qquad (3)$$

(sometimes denoted λ) is given by

$$K = \frac{1}{\sqrt{2}} \prod_{\substack{p \text{ prime} \\ \equiv 3(\text{mod } 4)}} \left(1 - \frac{1}{p^2}\right)^{-1/2} \qquad (4)$$

(Landau 1908; Le Lionnais 1983, p. 31; Berndt 1994; Hardy 1999; Moree and Cazaran 1999). An equivalent formula is given by

$$K = \frac{\pi}{4} \prod_{\substack{p \text{ prime} \\ \equiv 1(\text{mod } 4)}} \left(1 - \frac{1}{p^2}\right)^{-1/2}. \qquad (5)$$

Flajolet and Vardi (1996) give a beautiful FORMULA with fast convergence

$$K = \frac{1}{\sqrt{2}} \prod_{n=1}^\infty \left[\left(1 - \frac{1}{2^n}\right) \frac{\zeta(2^n)}{\beta(2^n)}\right]^{1/(2^n+1)}, \qquad (6)$$

where

$$\beta(s) \equiv \frac{1}{4^s}\left[\zeta\left(s,\tfrac{1}{4}\right) - \varsigma\left(s,\tfrac{3}{4}\right)\right] \tag{7}$$

is the DIRICHLET BETA FUNCTION, and $\zeta(z, a)$ is the HURWITZ ZETA FUNCTION. Landau proved the even stronger fact

$$\lim_{x\to\infty} \frac{(\ln x)^{3/2}}{Kx}\left[S(x)\frac{Kx}{\sqrt{\ln x}}\right] = C, \tag{8}$$

where

$$C \equiv \frac{1}{2}\left[1 - \ln\left(\frac{\pi e^\gamma}{L}\right)\right] - \frac{1}{4}\frac{d}{ds}\left[\ln\prod_{\substack{p \text{ prime} \\ p=4k+3}} \frac{1}{p^{-2s}}\right]_{s=1}$$

$$= 0.581948659\ldots. \tag{9}$$

Here,

$$L = 5.2441151086\ldots \tag{10}$$

is the ARC LENGTH of a LEMNISCATE with $a = 1$ (the LEMNISCATE CONSTANT to within a factor of 2 or 4), and γ is the EULER-MASCHERONI CONSTANT.

Landau's method of proof can be extended to show that

$$B(x) \sim K\frac{x}{\sqrt{\ln x}} \tag{11}$$

has an ASYMPTOTIC SERIES

$$B(x) = K\frac{x}{\sqrt{\ln x}}$$
$$\times\left[1 + \frac{C_1}{\ln x} + \frac{C_2}{(\ln x)^2} + \ldots + \frac{C_n}{(\ln x)^n} + \mathcal{O}\left(\frac{1}{(\ln x)^{n+1}}\right)\right], \tag{12}$$

where n can be arbitrarily large and the C_j are constants (Moree and Cazaran 1999).

See also SQUARE NUMBER

References

Berndt, B. C. *Ramanujan's Notebooks, Part IV.* New York: Springer-Verlag, pp. 60-6, 1994.

Berndt, B. C. and Rankin, R. A. Ch. 2 in *Ramanujan: Letters and Commentary.* Providence, RI: Amer. Math. Soc, 1995.

Finch, S. "Favorite Mathematical Constants." http://www.mathsoft.com/asolve/constant/lr/lr.html.

Flajolet, P. and Vardi, I. "Zeta Function Expansions of Classical Constants." Unpublished manuscript. 1996. http://pauillac.inria.fr/algo/flajolet/Publications/landau.ps.

Hardy, G. H. *Ramanujan: Twelve Lectures on Subjects Suggested by His Life and Work, 3rd ed.* New York: Chelsea, pp. 9-0, 55, and 60-4, 1999.

Landau, E. "Über die Einteilung der positiven ganzen Zahlen in vier Klassen nach der Mindeszahl der zu ihrer additiven Zusammensetzung erforderlichen Quadrate." *Arch. Math. Phys.* **13**, 305-12, 1908.

Landau, E. *Handbuch der Lehre von der Verteilung der Primzahlen, Bd. II, 2nd ed.* New York: Chelsea, pp. 641-69, 1953.

Le Lionnais, F. *Les nombres remarquables.* Paris: Hermann, 1983.

Moree, P. and Cazaran, J. "On a Claim of Ramanujan in His First Letter to Hardy." *Expos. Math.* **17**, 289-12, 1999.

Selberg, A. *Collected Papers, Vol. II.* Berlin: Springer-Verlag, pp. 183-85, 1991.

Shanks, D. "The Second-Order Term in the Asymptotic Expansion of $B(x)$." *Math. Comput.* **18**, 75-6, 1964.

Shanks, D. "Non-Hypotenuse Numbers." *Fibonacci Quart.* **13**, 319-21, 1975.

Shanks, D. and Schmid, L. P. "Variations on a Theorem of Landau. I." *Math. Comput.* **20**, 551-69, 1966.

Shiu, P. "Counting Sums of Two Squares: The Meissel-Lehmer Method." *Math. Comput.* **47**, 351-60, 1986.

Stanley, G. K. "Two Assertions Made by Ramanujan." *J. London Math. Soc.* **3**, 232-37, 1928.

Stanley, G. K. Corrigendum to "Two Assertions Made by Ramanujan." *J. London Math. Soc.* **4**, 32, 1929.

Wolfram Research, Inc. "Computing the Landau-Ramanujan Constant." http://library.wolfram.com/demos/v4/LandauRamanujan.nb.

Landau's Problems

The four "unattackable" problems mentioned by Landau in the 1912 Fifth Congress of Mathematicians in Cambridge. The four were

1. The GOLDBACH CONJECTURE,
2. TWIN PRIME CONJECTURE,
3. The conjecture that there exists a PRIME p such that $n^2 < p < (n+1)^2$ for every n (Hardy and Wright 1979, p. 415; Ribenboim 1996, pp. 397-98), and
4. The conjecture that there are infinitely many PRIMES p OF THE FORM $p = n^2 + 1$ (Hardy and Wright 1979, p. 19; Ribenboim 1996, pp. 206-08).

The first few PRIMES p which are OF THE FORM $p = n^2 + 1$ are given by 2, 5, 17, 37, 101, 197, 257, 401, ... (Sloane's A002496). These correspond to $n = 1, 2, 4, 6, 10, 14, 16, 20, ...$ (Sloane's A005574; Hardy and Wright 1979, p. 19).

Although it is not know if there always exists a PRIME p such that $n^2 < p < (n+1)^2$, Chen (1975) has shown that a number P which is either a PRIME or SEMIPRIME does always satisfy this inequality. Moreover, there is always a prime between $n - n^\theta$ and n where $\theta = 23/42$ (Iwaniec and Pintz 1984; Hardy and Wright 1979, p. 415). The smallest PRIMES between n^2 and $(n+1)^2$ for $n = 1, 2, ...,$ are 2, 5, 11, 17, 29, 37, 53, 67, 83, ... (Sloane's A007491).

See also GOLDBACH CONJECTURE, GOOD PRIME, PRIME NUMBER, TWIN PRIME CONJECTURE

References

Chen, J. R. "On the Distribution of Almost Primes in an Interval." *Sci. Sinica* **18**, 611-27, 1975.

Hardy, G. H. and Wright, W. M. "Unsolved Problems Concerning Primes." §2.8 and Appendix §3 in *An Introduction to the Theory of Numbers, 5th ed.* Oxford, England: Oxford University Press, pp. 19 and 415–16, 1979.

Iwaniec, H. and Pintz, J. "Primes in Short Intervals." *Monatsh. f. Math.* **98**, 115–43, 1984.

Ogilvy, C. S. *Tomorrow's Math: Unsolved Problems for the Amateur, 2nd ed.* Oxford, England: Oxford University Press, p. 116, 1972.

Ribenboim, P. *The New Book of Prime Number Records, 3rd ed.* New York: Springer-Verlag, pp. 132–34 and 206–08, 1996.

Sloane, N. J. A. Sequences A002496/M1506, A005574/M1010, and A007491/Min "An On-Line Version of the Encyclopedia of Integer Sequences." http://www.research.att.com/~njas/sequences/eisonline.html.

Landau Symbol

Let $f(z)$ be a function $\neq 0$ in an interval containing $z = 0$. Let $g(z)$ be another function also defined in this interval such that $g(z)/f(z) \to 0$ as $z \to 0$. Then $g(z)$ is said to be $o(f(z))$.

See also ASYMPTOTIC NOTATION

Landen's Formula

$$\frac{\vartheta_3(z, t)\vartheta_4(z, t)}{\vartheta_4(2z, 2t)} = \frac{\vartheta_3(0, t)\vartheta_4(0, t)}{\vartheta_4(0, 2t)} = \frac{\vartheta_2(z, t)\vartheta_4(z, t)}{\vartheta_1(2z, 2t)},$$

where ϑ_i are JACOBI THETA FUNCTIONS. This transformation was used by Gauss to show that ELLIPTIC INTEGRALS could be computed using the ARITHMETIC-GEOMETRIC MEAN.

See also JACOBI THETA FUNCTIONS

Landen's Identity

The DILOGARITHM identity

$$\text{Li}_2(-x) = -\text{Li}_2\left(\frac{x}{1+x}\right) - \tfrac{1}{2}[\ln(1+x)]^2.$$

See also DILOGARITHM

References

Gordon, B. and McIntosh, R. J. "Algebraic Dilogarithm Identities." *Ramanujan J.* **1**, 431–48, 1997.

Landen, J. *Mathematical Memoirs Respecting a Variety of Subjects, with an Appendix Containing Tables of Theorems, Vol. 1.* London: printed for the author, p. 112, 1780–789.

Landen's Transformation

If $x \sin \alpha = \sin(2\beta - \alpha)$, then

$$(1+x)\int_0^\alpha \frac{d\phi}{\sqrt{1 - x^2 \sin^2 \phi}}$$
$$= 2\int_0^\beta \frac{d\phi}{\sqrt{1 - \dfrac{4x}{(1+x)^2}\sin^2 \phi}}.$$

See also ELLIPTIC INTEGRAL OF THE FIRST KIND, GAUSS'S TRANSFORMATION

References

Abramowitz, M. and Stegun, C. A. (Eds.). "Ascending Landen Transformation" and "Landen's Transformation." §16.14 and 17.5 in *Handbook of Mathematical Functions with Formulas, Graphs, and Mathematical Tables, 9th printing.* New York: Dover, pp. 573–74 and 597–98, 1972.

Lane-Emden Differential Equation

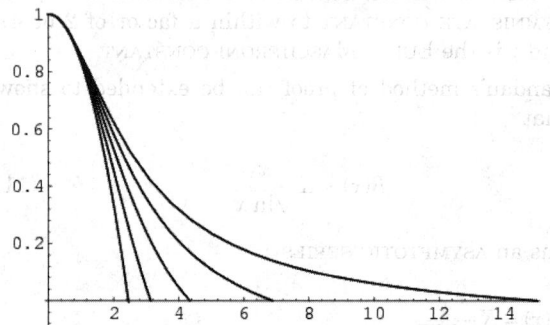

A second-order ORDINARY DIFFERENTIAL EQUATION arising in the study of stellar interiors, also called the polytropic differential equations. It is given by

$$\frac{1}{\xi^2}\frac{d}{d\xi}\left(\xi^2 \frac{d\theta}{d\xi}\right) + \theta^n = 0 \tag{1}$$

$$\frac{1}{\xi^2}\left(2\xi\frac{d\theta}{d\xi} + \xi^2 \frac{d^2\theta}{d\xi^2}\right) + \theta^n = \frac{d^2\theta}{d\xi^2} + \frac{2}{\xi}\frac{d\theta}{d\xi} + \theta^n = 0 \tag{2}$$

(Zwillinger 1997, pp. 124 and 126). It has the BOUNDARY CONDITIONS

$$\theta(0) = 1 \tag{3}$$

$$\left[\frac{d\theta}{d\xi}\right]_{\xi=0} = 0. \tag{4}$$

Solutions $\theta(\xi)$ for $n = 0$, 1, 2, 3, and 4 are shown above. The cases $n = 0$, 1, and 5 can be solved analytically (Chandrasekhar 1967, p. 91); the others must be obtained numerically.

For $n = 0$ ($(\gamma = \infty)$), the LANE-EMDEN DIFFERENTIAL EQUATION is

$$\frac{1}{\xi^2}\frac{d}{d\xi}\left(\xi^2\frac{d\theta}{d\xi}\right)+1=0 \tag{5}$$

(Chandrasekhar 1967, pp. 91–2). Directly solving gives

$$\frac{d}{d\xi}\left(\xi^2\frac{d\theta}{d\xi}\right)+1=-\xi^2 \tag{6}$$

$$\int d\left(\xi^2\frac{d\theta}{d\xi^2}\right)=-\int \xi^2\,d\xi \tag{7}$$

$$\xi^2\frac{d\theta}{d\xi}=c_1-\tfrac{1}{3}\xi^3 \tag{8}$$

$$\frac{d\theta}{d\xi}=\frac{c_1-\tfrac{1}{3}\xi^3}{\xi^2} \tag{9}$$

$$\theta(\xi)=\int d\theta=\int \frac{c_1-\tfrac{1}{3}\xi^3}{\xi^2}\,d\xi \tag{10}$$

$$\theta(\xi)=\theta_0-c_1\xi^{-1}-\tfrac{1}{6}\xi^2. \tag{11}$$

The BOUNDARY CONDITION $\theta(0)=1$ then gives $\theta_0=1$ and $c_1=0$, so

$$\theta_1(\xi)=1-\tfrac{1}{6}\xi^2, \tag{12}$$

and $\theta_1(\xi)$ is PARABOLIC.

For $n=1$ ($\gamma=2$), the differential equation becomes

$$\frac{1}{\xi^2}\frac{d}{d\xi}\left(\xi^2\frac{d\theta}{d\xi}\right)+\theta=0 \tag{13}$$

$$\frac{d}{d\xi}\left(\xi^2\frac{d\theta}{d\xi}\right)+\theta\xi^2=0, \tag{14}$$

which is the SPHERICAL BESSEL DIFFERENTIAL EQUATION

$$\frac{d}{dr}\left(r^2\frac{dR}{dr}\right)+[k^2r^2-n(n+1)]R=0 \tag{15}$$

with $k=1$ and $n=0$, so the solution is

$$\theta(\xi)=Aj_0(\xi)+Bn_0(\xi). \tag{16}$$

Applying the BOUNDARY CONDITION $\theta(0)=1$ gives

$$\theta_2(\xi)=j_0(\xi)=\frac{\sin\xi}{\xi}, \tag{17}$$

where $j_0(x)$ is a SPHERICAL BESSEL FUNCTION OF THE FIRST KIND (Chandrasekhar 1967, pp. 92).

For $n=5$, make Emden's transformation

$$\theta=Ax^\omega z \tag{18}$$

$$\omega=\frac{2}{n-1}, \tag{19}$$

which reduces the Lane-Emden equation to

$$\frac{d^2z}{dt^2}+(2\omega-1)\frac{dz}{dt}+\omega(\omega-1)z+A^{n-1}z^n=0 \tag{20}$$

(Chandrasekhar 1967, p. 90). After further manipulation (not reproduced here), the equation becomes

$$\frac{d^2z}{dt^2}=\tfrac{1}{4}z(1-z^4) \tag{21}$$

and then, finally,

$$\theta_5(\xi)=\left(1+\tfrac{1}{3}\xi^2\right)^{-1/2}. \tag{22}$$

References

Chandrasekhar, S. *An Introduction to the Study of Stellar Structure.* New York: Dover, pp. 84–82, 1967.

Iyanaga, S. and Kawada, Y. (Eds.). *Encyclopedic Dictionary of Mathematics.* Cambridge, MA: MIT Press, p. 908, 1980.

Seshadi, R. and Na, T. Y. *Group Invariance in Engineering Boundary Value Problems.* New York: Springer-Verlag, p. 193, 1985.

Zwillinger, D. *Handbook of Differential Equations, 3rd ed.* Boston, MA: Academic Press, pp. 124 and 126, 1997.

Langford's Problem

Arrange copies of the n digits $1, ..., n$ such that there is one digit between the 1s, two digits between the 2s, etc. For example, the unique (modulo reversal) $n=3$ solution is 231213, and the unique (again modulo reversal) $n=4$ solution is 23421314. Solutions to Langford's problem exist only if $n\equiv 0, 3\pmod 4$, so the next solutions occur for $n=7$. There are 26 of these, as exhibited by Lloyd (1971). In lexicographically smallest order (i.e., small digits come first), the first few Langford sequences are 231213, 23421314, 14156742352637, 14167345236275, 15146735423627, ... (Sloane's A050998).

The number of solutions for $n=3, 4, 5, ...$ (modulo reversal of the digits) are 1, 1, 0, 0, 26, 150, 0, 0, 17792, 108144, ... (Sloane's A014552). No formula is known for the number of solutions of a given order $n\not\equiv 0, 3\pmod 4$.

References

Davies, R. O. "On Langford's Problem. II." *Math. Gaz.* **43**, 253–55, 1959.

Gardner, M. *Mathematical Magic Show: More Puzzles, Games, Diversions, Illusions and Other Mathematical Sleight-of-Mind from Scientific American.* New York: Vintage, pp. 70 and 77–8, 1978.

Langford, C. D. "Problem." *Math. Gaz.* **42**, 228, 1958.

Lloyd, P. R. Correspondence to the Editor. *Math. Gaz.* **55**, 73, 1971.

Lorimer, P. "A Method of Constructing Skolem and Langford Sequences." *Southeast Asian Bull. Math.* **6**, 115–19, 1982.

Miller, J. "Langford's Problem." http://www.lclark.edu/~miller/langford.html.
Miller, J. "Langford's Problem Bibliography." http://www.lclark.edu/~miller/langford/langford-biblio.html.
Simpson, J. E. "Langford Sequences: Perfect and Hooked." *Disc > Math.* **44**, 97–04, 1983.
Priday, C. J. "On Langford's Problem. I." *Math. Gaz.* **43**, 250–53, 1959.
Sloane, N. J. A. Sequences A014552 and A050998 in "An On-Line Version of the Encyclopedia of Integer Sequences." http://www.research.att.com/~njas/sequences/eisonline.html.

Langlands Conjectures

LANGLANDS PROGRAM

Langlands Program

A grand unified theory of mathematics which includes the search for a generalization of ARTIN RECIPROCITY (known as LANGLANDS RECIPROCITY) to non-Abelian Galois extensions of NUMBER FIELDS. In a January 1967 letter to André Weil, Langlands proposed that the mathematics of algebra (Galois representations) and analysis (AUTOMORPHIC FORMS) are intimately related, and that congruences over FINITE FIELDS are related to infinite-dimensional representation theory. In particular, Langlands conjectured that the transformations behind general reciprocity laws could be represented by means of MATRICES (Mackenzie 2000).

In 1998, three mathematicians proved Langlands' conjectures for LOCAL FIELDS, and in a November 1999 lecture at the Institute for Advanced Study at Princeton University, L. Lafforgue presented a proof of the conjectures for FUNCTION FIELDS. This leaves only the case of NUMBER FIELDS as unresolved (Mackenzie 2000).

Langlands was a co-recipient of the 1996 Wolf Prize for the web of conjectures underlying this program.

See also ARTIN RECIPROCITY, AUTOMORPHIC FORM, ENDOSCOPY, LANGLANDS RECIPROCITY, RECIPROCITY THEOREM, TANIYAMA-SHIMURA CONJECTURE

References

American Mathematical Society. "Langlands and Wiles Share Wolf Prize." *Not. Amer. Math. Soc.* **43**, 221–22, 1996.
Knapp, A. W. "Group Representations and Harmonic Analysis from Euler to Langlands." *Not. Amer. Math. Soc.* **43**, 410–15, 1996.
Mackenzie, D. "Fermat's Last Theorem's Cousin." *Science* **287**, 792–93, 2000.

Langlands Reciprocity

The conjecture that the ARTIN L-FUNCTION of any n-D GALOIS GROUP representation is an L-FUNCTION obtained from the GENERAL LINEAR GROUP $GL_1(\mathbb{A})$.

See also ARTIN L-FUNCTION

References

Knapp, A. W. "Group Representations and Harmonic Analysis, Part II." *Not. Amer. Math. Soc.* **43**, 537–49, 1996.

Langton's Ant

A CELLULAR AUTOMATON for which the COHEN-KUNG THEOREM guarantees that the ant's trajectory is unbounded.

See also CELLULAR AUTOMATON, COHEN-KUNG THEOREM

References

Stewart, I. "The Ultimate in Anty-Particles." *Sci. Amer.* **271**, 104–07, 1994.

Laplace-Beltrami Operator

A self-adjoint elliptic differential operator defined somewhat technically as

$$\Delta = d\delta + \delta d,$$

where d is the EXTERIOR DERIVATIVE and d and δ are adjoint to each other with respect to the INNER PRODUCT.

References

Iyanaga, S. and Kawada, Y. (Eds.). *Encyclopedic Dictionary of Mathematics.* Cambridge, MA: MIT Press, p. 628, 1980.

Laplace Distribution

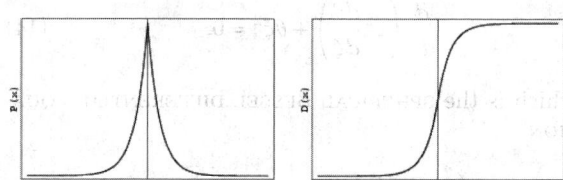

Also called the DOUBLE EXPONENTIAL DISTRIBUTION. It is the distribution of differences between two independent variates with identical EXPONENTIAL DISTRIBUTIONS (Abramowitz and Stegun 1972, p. 930).

$$P(x) = \frac{1}{2b} e^{-|x-\mu|/b} \tag{1}$$

$$D(x) = \tfrac{1}{2}[1 + \operatorname{sgn}(x-\mu)(1 - e^{-|x-\mu|/b})]. \tag{2}$$

The MOMENTS about the MEAN μ_n are related to the MOMENTS about 0 by

$$\mu_n = \sum_{j=0}^{n} \binom{n}{j}(-1)^{n-j}\mu_j'\mu^{n-j}, \tag{3}$$

where $\binom{n}{k}$ is a BINOMIAL COEFFICIENT, so

$$\mu_n = \sum_{j=0}^{n} \sum_{k=0}^{\lfloor j/2 \rfloor} (-1)^{n-j} \binom{n}{j} \binom{j}{2k} b^{2k} \mu^{n-2k} \Gamma(2k+1)$$

$$= \begin{cases} n!b^n & \text{for } n \text{ even} \\ 0 & \text{for } n \text{ odd,} \end{cases} \qquad (4)$$

where $\lfloor x \rfloor$ is the FLOOR FUNCTION and $\Gamma(2k+1)$ is the GAMMA FUNCTION. The MOMENTS can also be computed using the CHARACTERISTIC FUNCTION,

$$\phi(t) \equiv \int_{-\infty}^{\infty} e^{itx} P(x)\,dx = \frac{1}{2b} \int_{-\infty}^{\infty} e^{itx} e^{-|x-\mu|/b}\,dx. \qquad (5)$$

Using the FOURIER TRANSFORM OF THE EXPONENTIAL FUNCTION

$$\mathscr{F}[e^{-2\pi k_0 |x|}] = \frac{1}{\pi} \frac{k_0}{k^2 + k_0^2} \qquad (6)$$

gives

$$\phi(t) = \frac{e^{i\mu t}}{2b} \frac{\frac{2}{b}}{t^2 + \left(\frac{1}{b}\right)^2} = \frac{e^{i\mu t}}{1 + b^2 t^2} \qquad (7)$$

(Abramowitz and Stegun 1972, p. 930). The MOMENTS are therefore

$$\mu_n = (-i)^n \phi(0) = (-i)^n \left[\frac{d^n \phi}{dt^n}\right]_{t=0}. \qquad (8)$$

The MEAN, VARIANCE, SKEWNESS, and KURTOSIS are

$$\mu = \mu \qquad (9)$$
$$\sigma^2 = 2b^2 \qquad (10)$$
$$\gamma_1 = 0 \qquad (11)$$
$$\gamma_2 = 3. \qquad (12)$$

References

Abramowitz, M. and Stegun, C. A. (Eds.). *Handbook of Mathematical Functions with Formulas, Graphs, and Mathematical Tables, 9th printing.* New York: Dover, 1972.
Papoulis, A. *Probability, Random Variables, and Stochastic Processes, 2nd ed.* New York: McGraw-Hill, p. 104, 1984.

Laplace-Everett Formula
EVERETT'S FORMULA

Laplace Limit
The value $e = 0.6627434193\ldots$ (Sloane's A033259) for which Laplace's formula for solving KEPLER'S EQUATION begins diverging. The constant is defined as the value e at which the function

$$f(x) = \frac{x \exp(\sqrt{1+x^2})}{1 + \sqrt{1+x^2}}$$

equals $f(\lambda) = 1$. The CONTINUED FRACTION of e is given by [0, 1, 1, 1, 27, 1, 1, 1, 8, 2, 154, ...] (Sloane's A033260). The positions of the first occurrences of n in the CONTINUED FRACTION of e are 2, 10, 35, 13, 15, 32, 101, 9, ... (Sloane's A033261). The incrementally largest terms in the CONTINUED FRACTION are 1, 27, 154, 1601, 2135, ... (Sloane's A033262), which occur at positions 2, 5, 11, 19, 1801, ... (Sloane's A033263).

See also ECCENTRIC ANOMALY, KEPLER'S EQUATION

References

Finch, S. "Favorite Mathematical Constants." http://www.mathsoft.com/asolve/constant/lpc/lpc.html.
Plouffe, S. "Laplace Limit Constant." http://www.lacim.u-qam.ca/piDATA/laplace.txt.
Sloane, N. J. A. Sequences A033259, A033260, A033261, A033262, and A033263 in "An On-Line Version of the Encyclopedia of Integer Sequences." http://www.research.att.com/~njas/sequences/eisonline.html.

Laplace-Mehler Integral

$$p_n(\cos\theta) = \frac{1}{\pi} \int_0^{2\pi} (\cos\theta + i\sin\theta\cos\phi)^n \, d\phi$$

$$= \frac{\sqrt{2}}{\pi} \int_0^\theta \frac{\cos\left[\left(n + \frac{1}{2}\right)\phi\right]}{\sqrt{\cos\phi - \cos\theta}} \, d\phi$$

$$= \frac{\sqrt{2}}{\pi} \int_\theta^\pi \frac{\sin\left[\left(n + \frac{1}{2}\right)\phi\right]}{\sqrt{\cos\theta - \cos\phi}} \, d\phi.$$

References

Iyanaga, S. and Kawada, Y. (Eds.). *Encyclopedic Dictionary of Mathematics.* Cambridge, MA: MIT Press, p. 1463, 1980.

Laplace's Equation
The scalar form of Laplace's equation is the PARTIAL DIFFERENTIAL EQUATION

$$\nabla^2 \psi = 0. \qquad (1)$$

Note that the operator ∇^2 is commonly written as Δ by mathematicians (Krantz 1999, p. 16). Laplace's equation is a special case of the HELMHOLTZ DIFFERENTIAL EQUATION

$$\nabla^2 \psi + k^2 \psi = 0 \qquad (2)$$

with $k = 0$, or POISSON'S EQUATION

$$\nabla^2 \psi = -4\pi\rho \qquad (3)$$

with $\rho = 0$. The vector Laplace's equation is given by

$$\nabla^2 \mathbf{F} = 0. \qquad (4)$$

A FUNCTION ψ which satisfies Laplace's equation is said to be HARMONIC. A solution to Laplace's equation has the property that the average value over a spherical surface is equal to the value at the center of the SPHERE (GAUSS'S HARMONIC FUNCTION THEOREM). Solutions have no local maxima or minima. Because Laplace's equation is linear, the superposition of any two solutions is also a solution.

A solution to Laplace's equation is uniquely determined if (1) the value of the function is specified on all boundaries (DIRICHLET BOUNDARY CONDITIONS) or (2) the normal derivative of the function is specified on all boundaries (NEUMANN BOUNDARY CONDITIONS).

Coordinate System	Variables	Solution Functions
CARTESIAN	$X(x)Y(y)Z(z)$	EXPONENTIAL FUNCTIONS, CIRCULAR FUNCTIONS, HYPERBOLIC FUNCTIONS
CIRCULAR CYLINDRICAL	$R(r)\Theta(\theta)Z(z)$	BESSEL FUNCTIONS, EXPONENTIAL FUNCTIONS, CIRCULAR FUNCTIONS
CONICAL		ELLIPSOIDAL HARMONICS, POWER
ELLIPSOIDAL	$\Lambda(\lambda)M(\mu)N(\nu)$	ELLIPSOIDAL HARMONICS
ELLIPTIC CYLINDRICAL	$U(u)V(v)Z(z)$	MATHIEU FUNCTION, CIRCULAR FUNCTIONS
OBLATE SPHEROIDAL	$\Lambda(\lambda)M(\mu)N(\nu)$	LEGENDRE POLYNOMIAL, CIRCULAR FUNCTIONS
PARABOLIC		BESSEL FUNCTIONS, CIRCULAR FUNCTIONS
PARABOLIC CYLINDRICAL		PARABOLIC CYLINDER FUNCTIONS, BESSEL FUNCTIONS, CIRCULAR FUNCTIONS
PARABOLOIDAL	$U(u)V(v)\Theta(\theta)$	CIRCULAR FUNCTIONS
PROLATE SPHEROIDAL	$\Lambda(\lambda)M(\mu)N(\nu)$	LEGENDRE POLYNOMIAL, CIRCULAR FUNCTIONS
SPHERICAL	$R(r)\Theta(\theta)\Phi(\phi)$	LEGENDRE POLYNOMIAL, POWER, CIRCULAR FUNCTIONS

Laplace's equation can be solved by SEPARATION OF VARIABLES in all 11 coordinate systems that the HELMHOLTZ DIFFERENTIAL EQUATION can. The form these solutions take is summarized in the table above. In addition to these 11 coordinate systems, separation can be achieved in two additional coordinate systems by introducing a multiplicative factor. In these coordinate systems, the separated form is

$$\psi = \frac{X_1(u_1)X_2(u_2)X_3(u_3)}{R(u_1, u_2, u_3)}, \qquad (5)$$

and setting

$$\frac{h_1 h_2 h_3}{h_i^2} = g_i(u_{i+1}, u_{i+2}) f_i(u_i) R^2, \qquad (6)$$

where h_i are SCALE FACTORS, gives the Laplace's equation

$$\sum_{i=1}^{3} \frac{1}{h_i^2 X_i} \left[\frac{1}{f_i} \frac{d}{du_i} \left(f_i \frac{dX_i}{du_i} \right) \right]$$
$$= \sum_{i=1}^{3} \frac{1}{h_i^2 R} \left[\frac{1}{f_i} \frac{\partial}{\partial u_i} \left(f_i \frac{\partial R}{\partial u_i} \right) \right]. \qquad (7)$$

If the right side is equal to $-k_1^2/F(u_1, u_2, u_3)$, where k_1 is a constant and F is any function, and if

$$h_1 h_2 h_3 = S f_1 f_2 f_3 R^2 F, \qquad (8)$$

where S is the STÄCKEL DETERMINANT, then the equation can be solved using the methods of the HELMHOLTZ DIFFERENTIAL EQUATION. The two systems where this is the case are BISPHERICAL and TOROIDAL, bringing the total number of separable systems for Laplace's equation to 13 (Morse and Feshbach 1953, pp. 665–66).

In 2-D BIPOLAR COORDINATES, Laplace's equation is separable, although the HELMHOLTZ DIFFERENTIAL EQUATION is not.

Zwillinger (1997, p. 128) calls

$$(a_0 x + b_0)y^{(n)} + (a_1 x + b_1)y^{(n-1)} + \cdots + (a_n x + b_n)y$$
$$= 0 \qquad (9)$$

the Laplace equations.

See also BOUNDARY CONDITIONS, HARMONIC EQUATION, HARMONIC FUNCTION, HELMHOLTZ DIFFERENTIAL EQUATION, PARTIAL DIFFERENTIAL EQUATION, POISSON'S EQUATION, SEPARATION OF VARIABLES, STÄCKEL DETERMINANT

References

Abramowitz, M. and Stegun, C. A. (Eds.). *Handbook of Mathematical Functions with Formulas, Graphs, and Mathematical Tables, 9th printing.* New York: Dover, p. 17, 1972.

Byerly, W. E. *An Elementary Treatise on Fourier's Series, and Spherical, Cylindrical, and Ellipsoidal Harmonics, with Applications to Problems in Mathematical Physics.* New York: Dover, 1959.

Eisenhart, L. P. "Separable Systems in Euclidean 3-Space." *Physical Review* **45**, 427–28, 1934.

Eisenhart, L. P. "Separable Systems of Stäckel." *Ann. Math.* **35**, 284–05, 1934.

Eisenhart, L. P. "Potentials for Which Schroedinger Equations Are Separable." *Phys. Rev.* **74**, 87–9, 1948.

Krantz, S. G. "The Laplace Equation." §7.1.1 in *Handbook of Complex Analysis.* Boston, MA: Birkhäuser, pp. 16 and 89, 1999.

Moon, P. and Spencer, D. E. "Recent Investigations of the Separation of Laplace's Equation." *Proc. Amer. Math. Soc.* **4**, 302, 1953.

Moon, P. and Spencer, D. E. "Eleven Coordinate Systems." §1 in *Field Theory Handbook, Including Coordinate Systems, Differential Equations, and Their Solutions, 2nd ed.* New York: Springer-Verlag, pp. 1–8, 1988.

Morse, P. M. and Feshbach, H. *Methods of Theoretical Physics, Part I.* New York: McGraw-Hill, pp. 125–26 and 271, 1953.

Valiron, G. *The Geometric Theory of Ordinary Differential Equations and Algebraic Functions.* Brookline, MA: Math. Sci. Press, pp. 306–15, 1950.

Zwillinger, D. (Ed.). *CRC Standard Mathematical Tables and Formulae.* Boca Raton, FL: CRC Press, p. 417, 1995.

Zwillinger, D. *Handbook of Differential Equations, 3rd ed.* Boston, MA: Academic Press, p. 128, 1997.

Laplace's Equation—Bipolar Coordinates

In 2-D BIPOLAR COORDINATES, LAPLACE'S EQUATION is

$$\frac{(\cosh v - \cos u)^2}{a^2}\left(\frac{\partial F^2}{\partial u^2} + \frac{\partial F^2}{\partial v^2}\right) = 0, \qquad (1)$$

which simplifies to

$$\frac{\partial F^2}{\partial u^2} + \frac{\partial F^2}{\partial v^2} = 0, \qquad (2)$$

so LAPLACE'S EQUATION is separable, although the HELMHOLTZ DIFFERENTIAL EQUATION is not.

See also BIPOLAR COORDINATES, LAPLACE'S EQUATION

Laplace's Equation—Bispherical Coordinates

In BISPHERICAL COORDINATES, LAPLACE'S EQUATION becomes

$$\nabla^2 f = \frac{\sin u}{(\cosh v - \cos u)^3}\left[\frac{\partial}{\partial u}\left(\frac{\sin u}{\cosh v - \cos u}\frac{\partial f}{\partial u}\right)\right.$$

$$+\frac{\partial}{\partial v}\left(\frac{\sin u}{\cosh v - \cos u}\frac{\partial f}{\partial v}\right) + \frac{\partial}{\partial \phi}$$

$$\left.\times\left(\frac{\csc u}{\cosh v - \cos u}\frac{\partial f}{\partial \phi}\right)\right]. \qquad (1)$$

Attempt SEPARATION OF VARIABLES by plugging in the trial solution

$$f(u, v, \phi) = \sqrt{\cosh v - \cos u}\,U(u)V(v)\Psi(\psi), \qquad (2)$$

then divide the result by $\csc^2 u(\cosh v - \cos u)^{5/2} U(u)V(v)\Phi(\phi)$ to obtain

$$-\tfrac{1}{4}\sinh^2 u + \cos u \sin u\frac{U'(u)}{U(u)} + \sin^2 u\frac{U''(u)}{U(u)}$$

$$+\sin^2 u\frac{V''(v)}{V(v)} + \frac{\Phi''(\phi)}{\Phi(\phi)} = 0. \qquad (3)$$

The function $\Phi(\phi)$ then separates with

$$\frac{\Phi''(\phi)}{\Phi(\phi)} = -m^2, \qquad (4)$$

giving solution

$$\Psi(\psi) = \genfrac{}{}{0pt}{}{\sin}{\cos}(m\phi) = \sum_{k=1}^{\infty}[A_k\sin(m\psi) + B_k\cos(m\psi)]. \qquad (5)$$

Plugging $\Psi(\psi)$ back in and dividing by $\sin^2 u$ gives

$$\cot u\frac{U'(u)}{U(u)} + \frac{U''(u)}{U(u)} - \frac{m^2}{\sin^2 u} - \frac{1}{4} + \frac{V''(v)}{V(v)} = 0. \qquad (6)$$

The function $V(v)$ then separates with

$$\frac{V''(v)}{V(v)} = -n^2, \qquad (7)$$

giving solution

$$V(v) = \genfrac{}{}{0pt}{}{\sin}{\cos}(nv) = \sum_{k=1}^{\infty}[C_k\sin(nv) + D_k\cos(nv)]. \qquad (8)$$

Plugging $V(v)$ back in and multiplying by $V(v)$ gives

$$U''(u) + \cot u\,U'(u) - \left[\frac{m^2}{\sin^2 u} + \left(n^2 + \tfrac{1}{4}\right)\right]U(u) = 0, \qquad (9)$$

so LAPLACE'S EQUATION is partially separable in BISPHERICAL COORDINATES. However, the HELMHOLTZ DIFFERENTIAL EQUATION cannot be separated in this manner.

See also BISPHERICAL COORDINATES, LAPLACE'S EQUATION

References

Arfken, G. "Bispherical Coordinates (ξ, η, ϕ)." §2.14 in *Mathematical Methods for Physicists, 2nd ed.* Orlando, FL: Academic Press, pp. 115–17, 1970.

Morse, P. M. and Feshbach, H. *Methods of Theoretical Physics, Part I.* New York: McGraw-Hill, pp. 665–66, 1953.

Laplace's Equation—Spherical Coordinates

Laplace's Equation–Spherical

HELMHOLTZ DIFFERENTIAL EQUATION–SPHERICAL COORDINATES

Laplace's Equation—Toroidal Coordinates

In TOROIDAL COORDINATES, LAPLACE'S EQUATION becomes

$$\nabla^2 f = \frac{\sinh u}{(\cosh u - \cos v)^3} \left[\frac{\partial}{\partial u} \left(\frac{\sinh u}{\cosh u - \cos v} \frac{\partial f}{\partial u} \right) \right.$$

$$+ \frac{\partial}{\partial v} \left(\frac{\sinh u}{\cosh u - \cos v} \frac{\partial f}{\partial v} \right) + \frac{\partial}{\partial \phi}$$

$$\left. \times \left(\frac{\operatorname{csch} u}{\cosh u - \cos v} \frac{\partial f}{\partial \phi} \right) \right] \qquad (1)$$

Attempt SEPARATION OF VARIABLES by plugging in the trial solution

$$f(u, v, \phi) = \sqrt{\cosh u - \cos u}\, U(u) V(v) \Psi(\psi), \qquad (2)$$

then divide the result by $\operatorname{csch}^2 u (\cosh u - \cos v)^{5/2} U(u) V(v) \Phi(\phi)$ to obtain

$$\tfrac{1}{4} \sinh^2 u + \cosh u \sinh u \frac{U'(u)}{U(u)} + \sin^2 u \frac{U''(u)}{U(u)}$$

$$+ \sinh^2 u \frac{V''(v)}{V(v)} + \frac{\Phi''(\phi)}{\Phi(\phi)} = 0. \qquad (3)$$

The function $\Phi(\phi)$ then separates with

$$\frac{\Phi''(\phi)}{\Phi(\phi)} = -m^2, \qquad (4)$$

giving solution

$$\Psi(\psi) = \frac{\sin}{\cos}(m\phi) = \sum_{k=1}^{\infty} [A_k \sin(m\psi) + B_k \cos(m\psi)]. \qquad (5)$$

Plugging $\Psi(\psi)$ back in and dividing by $\sinh^2 u$ gives

$$\coth u \frac{U'(u)}{U(u)} + \frac{U''(u)}{U(u)} - \frac{m^2}{\sinh^2 u} + \frac{1}{4} + \frac{V''(v)}{V(v)} = 0. \qquad (6)$$

The function $V(v)$ then separates with

$$\frac{V''(v)}{V(v)} = -n^2, \qquad (7)$$

giving solution

$$V(v) = \frac{\sin}{\cos}(nv) = \sum_{k=1}^{\infty} [C_k \sin(nv) + D_k \cos(nv)]. \qquad (8)$$

Plugging $V(v)$ back in and multiplying by $V(v)$ gives

$$U''(u) + \coth u U'(u) - \left[\frac{m^2}{\sinh^2 u} + \left(n^2 - \tfrac{1}{4} \right) \right] U(u)$$

$$= 0, \qquad (9)$$

which can also be written

$$\frac{1}{\sinh u} \frac{d}{du} \left(\sinh u \frac{dU}{du} \right) - \left[\frac{m^2}{\sinh^2 u} + \left(n^2 - \tfrac{1}{4} \right) \right] U$$

$$= 0 \qquad (10)$$

(Arfken 1970, pp. 114–15). LAPLACE'S EQUATION is partially separable, although the HELMHOLTZ DIFFERENTIAL EQUATION is not.

See also LAPLACE'S EQUATION, LAPLACIAN, TOROIDAL COORDINATES

References

Arfken, G. "Toroidal Coordinates (ξ, η, ϕ)." §2.13 in *Mathematical Methods for Physicists, 2nd ed.* Orlando, FL: Academic Press, pp. 112–15, 1970.

Byerly, W. E. *An Elementary Treatise on Fourier's Series, and Spherical, Cylindrical, and Ellipsoidal Harmonics, with Applications to Problems in Mathematical Physics.* New York: Dover, pp. 264–66, 1959.

Morse, P. M. and Feshbach, H. *Methods of Theoretical Physics, Part I.* New York: McGraw-Hill, p. 666, 1953.

Laplace Series

A function $f(\theta, \phi)$ expressed as a double sum of SPHERICAL HARMONICS is called a Laplace series. Taking f as a COMPLEX FUNCTION,

$$f(\theta, \phi) = \sum_{l=0}^{\infty} \sum_{m=-l}^{l} a_{lm} Y_l^m(\theta, \phi). \qquad (1)$$

Now multiply both sides by $\bar{Y}_{l'}^{m'} \sin \theta$ and integrate over $d\theta$ and $d\phi$.

$$\int_0^{2\pi} \int_0^{\pi} f(\theta, \phi) \bar{Y}_{l'}^{m'} \sin \theta \, d\theta \, d\phi$$

$$= \sum_{l=0}^{\infty} \sum_{m=-l}^{l} a_{lm} \int_0^{2\pi} \int_0^{\pi} \bar{Y}_{l'}^{m'}(\theta, \phi) Y_l^m(\theta, \phi) \sin \theta \, d\theta \, d\phi.$$

$$(2)$$

Now use the ORTHOGONALITY of the SPHERICAL HARMONICS

$$\int_0^{2\pi} \int_0^{\pi} Y_l^m(\theta, \phi) \bar{Y}_{l'}^{m'} \sin \theta \, d\theta \, d\phi = \delta_{mm'} \delta_{ll'}, \qquad (3)$$

so (2) becomes

$$\int_0^{2\pi} \int_0^{\pi} f(\theta, \ \phi) \bar{Y}_{l'}^{m'} \sin \theta \ d\theta \ d\phi = \sum_{l=0}^{\infty} \sum_{m=-1}^{l} a_{lm} \delta_{mm'} \delta_{ll'}$$
$$= a_{lm}, \tag{4}$$

where δ_{mn} is the KRONECKER DELTA.

For a REAL series, consider

$$f(\theta, \ \phi) = \sum_{l=0}^{\infty} \sum_{m=-1}^{l} [C_l^m \cos(m\phi)$$
$$+ S_l^m \sin(m\phi)] P_l^m (\cos \theta). \tag{5}$$

Proceed as before, using the orthogonality relationships

$$\int_0^{2\pi} \int_0^{\pi} P_l^m (\cos \theta) \cos(m\phi) P_{l'}^{m'} (\cos \theta) \cos(m'\phi)$$
$$\times \sin(\theta) \ d\theta \ d\phi = - \frac{2\pi(l+m)!}{(2l+1)(l-m)!} \delta_{mm'} \delta_{ll'} \tag{6}$$

$$\int_0^{2\pi} \int_0^{\pi} P_l^m (\cos \theta) \sin(m\phi) P_{l'}^{m'} (\cos \theta) \sin(m'\phi)$$
$$\times \sin \theta \ d\theta \ d\phi = - \frac{2\pi(l+m)!}{(2l+1)(l-m)!} \delta_{mm'} \delta_{ll'}. \tag{7}$$

So C_l^m and S_l^m are given by

$$C_l^m = - \frac{(2l+1)(l-m)!}{2\pi(l+m)!} \int_0^{2\pi} \int_0^{\pi} f(\theta, \ \phi)$$
$$\times P_l^m \cos \theta \cos(m\phi) \sin \theta \ d\theta \ d\phi \tag{8}$$

$$S_l^m = - \frac{(2l+1)(l-m)!}{2\pi(l+m)!} \int_0^{2\pi} \int_0^{\pi} f(\theta, \ \phi)$$
$$\times P_l^m \cos \theta \sin(m\phi) \sin \theta \ d\theta \ d\phi. \tag{9}$$

Laplace's Integral

$$P_n(x) = \frac{1}{\pi} \int_0^{\pi} \frac{du}{\left(x + \sqrt{x^2 - 1} \cos u\right)^{n+1}} \ du$$
$$= \frac{1}{\pi} \int_0^{\pi} \left(x + \sqrt{x^2 - 1} \cos u\right)^n \ du.$$

It can be evaluated in terms of the HYPERGEOMETRIC FUNCTION.

Laplace's Problem

BUFFON-LAPLACE NEEDLE PROBLEM

Laplace-Stieltjes Transform

An integral transform which is often written as an ordinary LAPLACE TRANSFORM involving the DELTA FUNCTION. The LAPLACE TRANSFORM and DIRICHLET SERIES are special cases of the Laplace-Stieltjes transform (Apostol 1997, p. 162).

See also DIRICHLET SERIES, LAPLACE TRANSFORM

References

Abramowitz, M. and Stegun, C. A. (Eds.). *Handbook of Mathematical Functions with Formulas, Graphs, and Mathematical Tables, 9th printing.* New York: Dover, p. 1029, 1972.
Apostol, T. M. *Modular Functions and Dirichlet Series in Number Theory, 2nd ed.* New York: Springer-Verlag, p. 162, 1997.
Morse, P. M. and Feshbach, H. *Methods of Theoretical Physics, Part I.* New York: McGraw-Hill, 1953.
Widder, D. V. *The Laplace Transform.* Princeton, NJ: Princeton University Press, 1941.

Laplace Transform

The Laplace transform is an INTEGRAL TRANSFORM perhaps second only to the FOURIER TRANSFORM in its utility in solving physical problems. Due to its useful properties, the Laplace transform is particularly useful in solving linear ORDINARY DIFFERENTIAL EQUATIONS such as those arising in the analysis of electronic circuits.

The (one-sided) Laplace transform $\mathscr{L}$ (not to be confused with the LIE DERIVATIVE) is defined by

$$\mathscr{L}(s) = \mathscr{L}[f(t)] \equiv \int_0^{\infty} f(t)e^{-st} \ dt, \tag{1}$$

where $f(t)$ is defined for $t \geq 0$. The one-sided Laplace transform is implemented in *Mathematica* as LaplaceTransform[*expr*, *t*, *s*].

A two-sided Laplace transform is sometimes also defined by

$$\mathscr{L}(s) = \mathscr{L}|f(t)| = \int_{-\infty}^{\infty} f(t)e^{-st} \ dt. \tag{2}$$

The Laplace transform existence theorem states that, if $f(t)$ is PIECEWISE CONTINUOUS function on every finite interval in $[0, \ \infty)$ satisfying

$$|f(t)| \leq Me^{at} \tag{3}$$

for all $t \in [0, \ \infty)$, then $\mathscr{L}[f(t)]$ exists for all $s > a$. The Laplace transform is also UNIQUE, in the sense that, given two functions $F_1(t)$ and $F_2(t)$ with the same transform so that

$$\mathscr{L}[F_1(t)] = \mathscr{L}[F_2(t)] \equiv f(s), \tag{4}$$

then LERCH'S THEOREM guarantees that the integral

$$\int_0^a N(t) \ dt = 0 \tag{5}$$

vanishes for all $a > 0$ for a NULL FUNCTION defined by

$$N(t) \equiv F_1(t) - F_2(t). \tag{6}$$

The Laplace transform is LINEAR since

$$\mathscr{L}[af(t) + bg(t)] = \int_0^{\infty} [af(t) + bg(t)]e^{-st} \ dt$$

$$= a \int_0^\infty f(t)e^{-st}\, dt + b \int_0^\infty g(t)e^{-st}\, dt$$

$$= a\mathscr{L}[f(t)] + b\mathscr{L}[g(t)]. \tag{7}$$

The inverse Laplace transform is given by the BROMWICH INTEGRAL (see also DUHAMEL'S CONVOLUTION PRINCIPLE). A table of several important Laplace transforms follows.

$f(t)$	$\mathscr{L}[f(t)]$	Range
1	$\dfrac{1}{s}$	$s > 0$
t	$\dfrac{1}{s^2}$	$s > 0$
t^n	$\dfrac{n!}{s^{n+1}}$	$n \in \mathbb{Z} > 0$
t^a	$\dfrac{\Gamma(a+1)}{s^{a+1}}$	$a > 0$
e^{at}	$\dfrac{1}{s-a}$	$s > a$
$\cos(\omega t)$	$\dfrac{s}{s^2+\omega^2}$	$s > 0$
$\sin(\omega t)$	$\dfrac{\omega}{s^2+\omega^2}$	
$\cosh(\omega t)$	$\dfrac{s}{s^2-\omega^2}$	$s > |a|$
$\sinh(\omega t)$	$\dfrac{\omega}{s^2-\omega^2}$	$s > |a|$
$e^{at}\sin(bt)$	$\dfrac{b}{(s-a)^2+b^2}$	$s > a$
$e^{at}\cos(bt)$	$\dfrac{s-a}{(s-a)^2+b^2}$	$s > a$
$\delta(t-c)$	e^{-cs}	$c > 0$
$H_c(t)$	$\dfrac{e^{-cs}}{s}$	$s > 0$
$J_0(t)$	$\dfrac{1}{\sqrt{s^2+1}}$	
$J_n(t)$	$\dfrac{{}_2F_1\left(\frac{1}{2}(n+1), \frac{1}{2}(n+2); n+1; -s^{-2}\right)}{2^n s^{n+1}}$	

In the above table, $J_0(t)$ is the zeroth order BESSEL FUNCTION OF THE FIRST KIND, $\delta(t)$ is the DELTA FUNCTION, and $H_c(t)$ is the HEAVISIDE STEP FUNCTION. The Laplace transform has many important properties.

The Laplace transform of a CONVOLUTION is given by

$$\mathscr{L}[f(t) * g(t)] = \mathscr{L}(f(t))\mathscr{L}(g(t)) \tag{8}$$

$$\mathscr{L}^{-1}[F(s)G(s)] = \mathscr{L}^{-1}(F(s)) * \mathscr{L}^{-1}(G(s)). \tag{9}$$

Now consider DIFFERENTIATION. Let $f(t)$ be continuously differentiable $n-1$ times in $[0, \infty)$. If $|f(t)| \leq Me^{at}$, then

$$\mathscr{L}[f^{(n)}(t)] = s^n \mathscr{L}(f(t)) - s^{n-1}f(0) - s^{n-2}f'(0) - \dots \\ - f^{(n-1)}(0). \tag{10}$$

This can be proved by INTEGRATION BY PARTS,

$$\mathscr{L}[f'(t)] = \lim_{a \to \infty} \int_0^a e^{-st}f'(t)\, dt$$

$$= \lim_{a \to \infty} \left\{ [e^{-st}f(t)]_0^a + s\int_0^a e^{-st}f(t)\, dt \right\}$$

$$= \lim_{a \to \infty} \left[e^{-sa}f(a) - f(0) + s\int_0^a e^{-st}f(t)\, dt \right]$$

$$= s\mathscr{L}[f(t)] - f(0). \tag{11}$$

Continuing for higher order derivatives then gives

$$\mathscr{L}[f''(t)] = s^2 \mathscr{L}[f(t)] - sf(0) - f'(0). \tag{12}$$

This property can be used to transform differential equations into algebraic equations, a procedure known as the HEAVISIDE CALCULUS, which can then be inverse transformed to obtain the solution. For example, applying the Laplace transform to the equation

$$f''(t) + a_1 f'(t) + a_0 f(t) = 0 \tag{13}$$

gives

$$\{s^2 \mathscr{L}[f(t)] - sf(0) - f'(0)\} + a_1\{s\mathscr{L}[f(t)] - f(0)\} \\ + a_0 \mathscr{L}[f(t)] = 0 \tag{14}$$

$$\mathscr{L}[f(t)](s^2 + a_1 s + a_0) - sf(0) - f'(0) - a_1 f(0) = 0, \tag{15}$$

which can be rearranged to

$$\mathscr{L}[f(t)] = \frac{sf(0) + [f'(0) + a_1 f(0)]}{s^2 + a_1 s + a_0}. \tag{16}$$

If this equation can be inverse Laplace transformed, then the original differential equation is solved.

Consider EXPONENTIATION. If $\mathscr{L}[f(t)] = F(s)$ for $s > \alpha$, then $\mathscr{L}(e^{at}f(t)) = F(s-a)$ for $s > a + \alpha$.

$$F(s-a) = \int_0^\infty f(t)e^{-(s-a)t}\, dt = \int_0^\infty [f(t)e^{at}]e^{-st}\, dt$$

$$= \mathscr{L}[e^{at}f(t)]. \tag{17}$$

Consider INTEGRATION. If $f(t)$ is PIECEWISE CONTINUOUS and $|f(t)| \leq Me^{at}$, then

$$\mathscr{L}\left[\int_0^t f(t)\,dt\right] = \frac{1}{s}\,\mathscr{L}[f(t)]. \qquad (18)$$

The inverse transform is known as the Bromwich Integral, or sometimes the Fourier-Mellin Integral.

See also Bromwich Integral, Fourier-Mellin Integral, Fourier Transform, Integral Transform, Laplace-Stieltjes Transform, Operational Mathematics

References

Abramowitz, M. and Stegun, C. A. (Eds.). "Laplace Transforms." Ch. 29 in *Handbook of Mathematical Functions with Formulas, Graphs, and Mathematical Tables, 9th printing.* New York: Dover, pp. 1019–030, 1972.

Arfken, G. *Mathematical Methods for Physicists, 3rd ed.* Orlando, FL: Academic Press, pp. 824–63, 1985.

Churchill, R. V. *Operational Mathematics.* New York: McGraw-Hill, 1958.

Doetsch, G. *Introduction to the Theory and Application of the Laplace Transformation.* Berlin: Springer-Verlag, 1974.

Franklin, P. *An Introduction to Fourier Methods and the Laplace Transformation.* New York: Dover, 1958.

Jaeger, J. C. and Newstead, G. H. *An Introduction to the Laplace Transformation with Engineering Applications.* London: Methuen, 1949.

Henrici, P. *Applied and Computational Complex Analysis, Vol. 2: Special Functions, Integral Transforms, Asymptotics, Continued Fractions.* New York: Wiley, pp. 322–50, 1991.

Krantz, S. G. "The Laplace Transform." §15.3 in *Handbook of Complex Analysis.* Boston, MA: Birkhäuser, pp. 212–14, 1999.

Morse, P. M. and Feshbach, H. *Methods of Theoretical Physics, Part I.* New York: McGraw-Hill, pp. 467–69, 1953.

Oberhettinger, F. *Tables of Laplace Transforms.* New York: Springer-Verlag, 1973.

Prudnikov, A. P.; Brychkov, Yu. A.; and Marichev, O. I. *Integrals and Series, Vol. 4: Direct Laplace Transforms.* New York: Gordon and Breach, 1992.

Prudnikov, A. P.; Brychkov, Yu. A.; and Marichev, O. I. *Integrals and Series, Vol. 5: Inverse Laplace Transforms.* New York: Gordon and Breach, 1992.

Spiegel, M. R. *Theory and Problems of Laplace Transforms.* New York: McGraw-Hill, 1965.

Weisstein, E. W. "Books about Laplace Transforms." http://www.treasure-troves.com/books/LaplaceTransforms.html.

Widder, D. V. *The Laplace Transform.* Princeton, NJ: Princeton University Press, 1941.

Zwillinger, D. (Ed.). *CRC Standard Mathematical Tables and Formulae.* Boca Raton, FL: CRC Press, pp. 231 and 543, 1995.

Laplacian

The Laplacian operator for a scalar function ϕ is defined by

$$\nabla^2 \phi = \frac{1}{h_1\,h_2\,h_3}\left[\frac{\partial}{\partial u_1}\left(\frac{h_2 h_3}{h_1}\frac{\partial}{\partial u_1}\right)\right.$$

$$\left.+\frac{\partial}{\partial u_2}\left(\frac{h_1 h_3}{h_2}\frac{\partial}{\partial u_2}\right)+\frac{\partial}{\partial u_3}\left(\frac{h_1 h_2}{h_3}\frac{\partial}{\partial u_3}\right)\right]\phi \qquad (1)$$

in vector notation, where the h_i are the scale factors of the coordinate system. In tensor notation, the Laplacian is written

$$\nabla^2 \phi = (g^{\lambda\kappa}\phi_{;\lambda})_{;\kappa} = g^{\lambda\kappa}\frac{\partial^2 \phi}{\partial x^\lambda \partial x^\kappa} - \Gamma^\lambda\,\frac{\partial\phi}{\partial x^\lambda}$$

$$= \frac{1}{\sqrt{g}}\frac{\partial}{\partial x^j}\left(\sqrt{g}\,g^{ij}\,\frac{\partial\phi}{\partial x^i}\right), \qquad (2)$$

where $g_{;\kappa}$ is a covariant derivative and

$$\Gamma^\lambda \equiv \tfrac{1}{2}g^{\mu\nu}g^{\lambda\kappa}\left(\frac{\partial g_{\kappa\mu}}{\partial x^\nu}+\frac{\partial g_{\kappa\nu}}{\partial x^\mu}-\frac{\partial g_{\mu\nu}}{\partial x^\kappa}\right). \qquad (3)$$

Note that the operator ∇^2 is commonly written as Δ by mathematicians (Krantz 1999, p. 16).

The following table gives the form of the Laplacian in several common coordinate systems.

coordinate system	∇^2
Cartesian coordinates	$\dfrac{\partial^2}{\partial x^2}+\dfrac{\partial^2}{\partial y^2}+\dfrac{\partial^2}{\partial z^2}$
Cylindrical coordinates	$\dfrac{1}{r}\dfrac{\partial}{\partial r}\left(r\dfrac{\partial f}{\partial r}\right)+\dfrac{1}{r^2}\dfrac{\partial^2 f}{\partial \theta^2}+\dfrac{\partial^2 f}{\partial z^2}$
Parabolic coordinates	$\dfrac{1}{uv(u^2+v^2)}\left[\dfrac{\partial}{\partial u}\left(uv\dfrac{\partial f}{\partial u}\right)+\dfrac{\partial}{\partial v}\left(uv\dfrac{\partial f}{\partial v}\right)\right]$ $+\dfrac{1}{u^2 v^2}\dfrac{\partial^2 f}{\partial \theta^2}$
Parabolic cylindrical coordinates	$\dfrac{1}{u^2+v^2}\left(\dfrac{\partial^2 f}{\partial u^2}+\dfrac{\partial^2 f}{\partial v^2}\right)+\dfrac{\partial^2 f}{\partial z^2}$
Spherical coordinates	$\dfrac{1}{r^2}\dfrac{\partial}{\partial r}\left(r^2\dfrac{\partial}{\partial r}\right)+\dfrac{1}{r^2 \sin^2\phi}\dfrac{\partial^2}{\partial \theta^2}$ $+\dfrac{1}{r^2 \sin\phi}\dfrac{\partial}{\partial \phi}\left(\sin\phi\dfrac{\partial}{\partial \phi}\right)$

The finite difference form is

$$\nabla^2 \psi(x,\,y,\,z) = \frac{1}{h^2}[\psi(x+h,\,y,\,z)+\psi(x-h,\,y,\,z)$$

$$+\psi(x,\,y+h,\,z)+\psi(x,\,y-h,\,z)+\psi(x,\,y,\,z+h)$$

$$+\psi(x,\,y,\,z-h)-6\psi(x,\,y,\,z)]. \qquad (4)$$

For a pure radial function $g(r)$,

$$\nabla^2 g(r) \equiv \nabla\cdot[\nabla g(r)]$$

$$= \nabla \cdot \left[\frac{\partial g(r)}{\partial r} \, \hat{\mathbf{r}} + \frac{1}{r} \, \frac{\partial g(r)}{\partial \theta} \, \hat{\boldsymbol{\theta}} + \frac{1}{r \sin \theta} \, \frac{\partial g(r)}{\partial \phi} \, \hat{\boldsymbol{\phi}} \right]$$

$$= \nabla \cdot \left(\hat{\mathbf{r}} \, \frac{dg}{dr} \right). \tag{5}$$

Using the VECTOR DERIVATIVE identity

$$\nabla \cdot (f\mathbf{A}) = f(\nabla \cdot \mathbf{A}) + (\nabla f) \cdot (\mathbf{A}), \tag{6}$$

so

$$\nabla^2 g(r) \equiv \nabla \cdot [\nabla g(r)] = \frac{dg}{dr} \, \nabla \cdot \hat{\mathbf{r}} + \nabla \left(\frac{dg}{dr} \right) \cdot \hat{\mathbf{r}}$$

$$= \frac{2}{r} \frac{dg}{dr} + \frac{d^2 g}{dr^2}. \tag{7}$$

Therefore, for a radial power law,

$$\nabla^2 r^n = \frac{2}{r} \, nr^{n-1} + n(n-1)r^{n-2} = [2n + n(n-1)]r^{n-2}$$

$$= n(n+1)r^{n-2}. \tag{8}$$

A vector Laplacian can also be defined for a VECTOR **A** by

$$\nabla^2 \mathbf{A} = \nabla(\nabla \cdot \mathbf{A}) - \nabla \times (\nabla \times \mathbf{A}) \tag{9}$$

in vector notation. The notation ☆ is sometimes also used for a vector Laplacian (Moon and Spencer 1988, p. 3). In tensor notation, **A** is written A_μ, and the identity becomes

$$\nabla^2 A_\mu = A_{\mu;\lambda}^{;\lambda} = (g^{\lambda\kappa} A_{\mu;\lambda})_{;\kappa}$$

$$= g^{\lambda}{}_{\kappa;\kappa} A_{\mu;\lambda} + g^{\lambda\kappa} A_{\mu;\ \lambda\kappa}. \tag{10}$$

Similarly, a TENSOR Laplacian can be given by

$$\nabla^2 A_{\alpha\beta} = A_{\alpha\beta;\lambda}^{;\lambda} \tag{11}$$

An identity satisfied by the Laplacian is

$$\nabla^2 |\mathbf{x} \mathsf{A}| = \frac{|\mathsf{A}|_2^2 - |(\mathbf{x}\mathsf{A})\mathsf{A}^T|^2}{|\mathbf{x}\mathsf{A}|^3}, \tag{12}$$

where $|\mathsf{A}|_2$ is the HILBERT-SCHMIDT NORM, **x** is a row VECTOR, and A^T is the MATRIX TRANSPOSE of A.

To compute the LAPLACIAN of the inverse distance function $1/r$, where $r \equiv |\mathbf{r} - \mathbf{r}'|$, and integrate the LAPLACIAN over a volume,

$$\int_V \nabla^2 \left(\frac{1}{|\mathbf{r} - \mathbf{r}'|} \right) d^3 \mathbf{r}. \tag{13}$$

This is equal to

$$\int_V \nabla^2 \frac{1}{r} \, d^3 \mathbf{r} = \int_V \nabla \cdot \left(\nabla \frac{1}{r} \right) d^3 \mathbf{r} = \int_S \left(\nabla \frac{1}{r} \right) \cdot d\mathbf{a}$$

$$= \int_S \frac{\partial}{\partial r} \left(\frac{1}{r} \right) \hat{\mathbf{r}} \cdot d\mathbf{a} = \int_S -\frac{1}{r^2} \, \hat{\mathbf{r}} \cdot d\mathbf{a}$$

$$= -4\pi \frac{R^2}{r^2}, \tag{14}$$

where the integration is over a small SPHERE of RADIUS R. Now, for $r > 0$ and $R \to 0$, the integral becomes 0. Similarly, for $r = R$ and $R \to 0$, the integral becomes -4π. Therefore,

$$\nabla^2 \left(\frac{1}{|\mathbf{r} - \mathbf{r}'|} \right) = -4\pi \delta^3(\mathbf{r} - \mathbf{r}'), \tag{15}$$

where $\delta(\mathbf{x})$ is the DELTA FUNCTION.

The tensor Laplacian is given by

$$\nabla \cdot (\nabla \psi) = \frac{1}{g^{1/2}} (g^{1/2} g^{ik} \psi_{,k})_{,i}, \tag{16}$$

where g_{ij} is the METRIC TENSOR, $g = \det(g_{ij})$, and $A_{,k}$ is the COMMA DERIVATIVE (Arfken 1985, p. 185).

See also ANTILAPLACIAN, D'ALEMBERTIAN, HELMHOLTZ DIFFERENTIAL EQUATION, LAPLACE'S EQUATION, VECTOR LAPLACIAN

References

Arfken, G. *Mathematical Methods for Physicists, 3rd ed.* Orlando, FL: Academic Press, 1985.
Krantz, S. G. *Handbook of Complex Analysis.* Boston, MA: Birkhäuser, p. 16, 1999.
Moon, P. and Spencer, D. E. *Field Theory Handbook, Including Coordinate Systems, Differential Equations, and Their Solutions, 2nd ed.* New York: Springer-Verlag, 1988.

Laplacian Determinant Expansion by Minors

DETERMINANT EXPANSION BY MINORS

Laplacian Expansion

DETERMINANT EXPANSION BY MINORS

Laplacian Matrix

The Laplacian matrix $L(G)$ of a graph G, where $G = (N, E)$ is an undirected, unweighted graph without self edges (i, i) or multiple edges from one node to another, is an $|N| \times |N|$ SYMMETRIC MATRIX with one row and column for each node. It is defined as follows,

$$L_{ij}(G) = \begin{cases} \text{degree of node } i & \text{if } i = j \\ -1 & \text{if } i \neq j \text{ and } \exists \text{ edge}(i, j) \\ 0 & \text{otherwise.} \end{cases}$$

A normalized version of the Laplacian matrix, denoted $\mathscr{L}$, is similar defined by

$$\mathcal{L}_{ij}(G) = \begin{cases} 1 & \text{if } i=j \text{ and } d_j \neq 0 \\ -\dfrac{1}{\sqrt{d_i d_j}} & \text{if } i \text{ and } j \text{ are adjacent} \\ 0 & \text{otherwise.} \end{cases}$$

See also ALGEBRAIC CONNECTIVITY, FIEDLER VECTOR, SPECTRAL GRAPH PARTITIONING

References

Bendito, E.; Carmona, A.; and Encinas, A. M. "Shortest Paths in Distance-Regular Graphs." *Europ. J. Combin.* **21**, 153–66, 2000.

Chung, F. R. K. *Spectral Graph Theory.* Providence, RI: Amer. Math. Soc., 1997.

Demmel, J. "CS 267: Notes for Lecture 23, April 9, 1999. Graph Partitioning, Part 2." http://www.cs.berkeley.edu/~demmel/cs267/lecture20/lecture20.html.

Large Number

There are a wide variety of large numbers which crop up in mathematics. Some are contrived, but some actually arise in proofs. Often, it is possible to prove existence theorems by deriving some potentially huge upper limit which is frequently greatly reduced in subsequent versions (e.g., GRAHAM'S NUMBER, KOLMOGOROV-ARNOLD-MOSER THEOREM, MERTENS CONJECTURE, SKEWES NUMBER, WANG'S CONJECTURE).

Large decimal numbers beginning with 10^9 are named according to two mutually conflicting nomenclatures: the American system (in which the prefix stands for n in 10^{3+3n}) and the British system (in which the prefix stands for n in 10^{6n}). However, it should be noted that in more recent years, the "American" system is now widely used in England as well as in the United States. The following table gives the names assigned to various POWERS of 10 (Woolf 1982).

American	British	power of 10
MILLION	million	10^6
BILLION	milliard	10^9
TRILLION	billion	10^{12}
QUADRILLION		10^{15}
QUINTILLION	trillion	10^{18}
SEXTILLION		10^{21}
SEPTILLION	quadrillion	10^{24}
OCTILLION		10^{27}
NONILLION	quintillion	10^{30}
DECILLION		10^{33}
UNDECILLION	sexillion	10^{36}
DUODECILLION		10^{39}
TREDECILLION	septillion	10^{42}
QUATTUORDECILLION		10^{45}
QUINDECILLION	octillion	10^{48}
SEXDECILLION		10^{51}
SEPTENDECILLION	nonillion	10^{54}
OCTODECILLION		10^{57}
NOVEMDECILLION	decillion	10^{60}
VIGINTILLION		10^{63}
	undecillion	10^{66}
	duodecillion	10^{72}
	tredecillion	10^{78}
	quattuordecillion	10^{84}
	quindecillion	10^{90}
	sexdecillion	10^{96}
	septendecillion	10^{102}
	octodecillion	10^{108}
	novemdecillion	10^{114}
	vigintillion	10^{120}
centillion		10^{303}
	centillion	10^{600}

See also 10, ACKERMANN NUMBER, ARROW NOTATION, BARNES' G-FUNCTION, BILLION, CIRCLE NOTATION, EDDINGTON NUMBER, ERDOS-MOSER EQUATION, FRIVOLOUS THEOREM OF ARITHMETIC, GÖBEL'S SEQUENCE, GOOGOL, GOOGOLPLEX, GRAHAM'S NUMBER, HUNDRED, HYPERFACTORIAL, JUMPING CHAMPION, LAW OF TRULY LARGE NUMBERS, MEGA, MEGISTRON, MILLION, MONSTER GROUP, MOSER, N-PLEX, POWER TOWER, SKEWES NUMBER, SMALL NUMBER, STEINHAUS-MOSER NOTATION, STRONG LAW OF LARGE NUMBERS, SUPERFACTORIAL, THOUSAND, WEAK LAW OF LARGE NUMBERS, ZILLION

References

Conway, J. H. and Guy, R. K. *The Book of Numbers.* New York: Springer-Verlag, pp. 59–2, 1996.

Crandall, R. E. "The Challenge of Large Numbers." *Sci. Amer.* **276**, 74–9, Feb. 1997.

Davis, P. J. *The Lore of Large Numbers.* New York: Random House, 1961.

Knuth, D. E. "Mathematics and Computer Science: Coping with Finiteness. Advances in Our Ability to Compute Are Bringing Us Substantially Closer to Ultimate Limitations." *Science* **194**, 1235–242, 1976.

Munafo, R. "Large Numbers." http://www.mrob.com/largenum.html.

Spencer, J. "Large Numbers and Unprovable Theorems." *Amer. Math. Monthly* **90**, 669–75, 1983.
Woolf, H. B. (Ed. in Chief). *Webster's New Collegiate Dictionary.* Springfield, MA: Merriam, p. 782, 1980.

Large Prime

GIGANTIC PRIME, LARGE NUMBER, TITANIC PRIME

Largest Prime Factor

GREATEST PRIME FACTOR

Laspeyres' Index

The statistical INDEX

$$P_L \equiv \frac{\sum P_n q_0}{\sum p_0 q_0},$$

where p_n is the price per unit in period n and q_n is the quantity produced in period n.

See also INDEX

References

Kenney, J. F. and Keeping, E. S. *Mathematics of Statistics, Pt. 1, 3rd ed.* Princeton, NJ: Van Nostrand, pp. 65–7, 1962.

Latent Root

EIGENVALUE

Latent Vector

EIGENVECTOR

Latin Cross

An irregular DODECAHEDRON CROSS in the shape of a dagger †. The six faces of a CUBE can be cut along seven EDGES and unfolded into a Latin cross (i.e., the Latin cross is the NET of the CUBE). Similarly, eight hypersurfaces of a HYPERCUBE can be cut along 17 SQUARES and unfolded to form a 3-D Latin cross.

Another cross also called the Latin cross is illustrated above. It is a GREEK CROSS with flared ends, and is also known as the crux immissa or cross patée.

See also CROSS, DISSECTION, DODECAHEDRON, GREEK CROSS, MALTESE CROSS

Latin-Graeco Square

EULER SQUARE

Latin Rectangle

A $k \times n$ Latin rectangle is a $k \times n$ MATRIX with elements $a_{ij} \in \{1, 2, \ldots, n\}$ such that entries in each row and column are distinct. If $k = n$, the special case of a LATIN SQUARE results. A normalized Latin rectangle has first row $\{1, 2, \ldots, n\}$ and first column $\{1, 2, \ldots, k\}$. Let $L(k, n)$ be the number of normalized $k \times n$ Latin rectangles, then the total number of $k \times n$ Latin rectangles is

$$N(k, n) = \frac{n!(n - 1)!L(k, n)}{(n - k)!}$$

(McKay and Rogoyski 1995), where $n!$ is a FACTORIAL. Kerewala (1941) found a RECURRENCE RELATION for $L(3, n)$, and Athreya, Pranesachar, and Singhi (1980) found a summation FORMULA for $L(4, n)$.

The asymptotic value of $L(o(n^{6/7}), n)$ was found by Godsil and McKay (1990). The numbers of $k \times n$ Latin rectangles are given in the following table from McKay and Rogoyski (1995). The entries $L(1, n)$ and $L(n, n)$ are omitted, since

$$L(1, n) = 1$$

$$L(n, n) = L(n - 1, n),$$

but $L(1, 1)$ and $L(2, 1)$ are included for clarity. The values of $L(k, n)$ are given as a "wrap-around" series by Sloane's A001009.

n	k	$L(k, n)$
1	1	1
2	1	1
3	2	1
4	2	3
4	3	4
5	2	11
5	3	46
5	4	56
6	2	53
6	3	1064
6	4	6552
6	5	9408
7	2	309
7	3	35792
7	4	1293216

7	5	11270400
7	6	16942080
8	2	2119
8	3	1673792
8	4	420909504
8	5	27206658048
8	6	335390189568
8	7	535281401856
9	2	16687
9	3	103443808
9	4	207624560256
9	5	112681643083776
9	6	12952605404381184
9	7	224382967916691456
9	8	377597570964258816
10	2	148329
10	3	8154999232
10	4	147174521059584
10	5	746988383076286464
10	6	870735405591003709440
10	7	1771144296983054185922560
10	8	4292039421591854273003520
10	9	7580721483160132811489280

References

Athreya, K. B.; Pranesachar, C. R.; and Singhi, N. M. "On the Number of Latin Rectangles and Chromatic Polynomial of $L(K_{r,s})$." *Europ. J. Combin.* **1**, 9–7, 1980.

Colbourn, C. J. and Dinitz, J. H. (Eds.). *CRC Handbook of Combinatorial Designs.* Boca Raton, FL: CRC Press, 1996.

Godsil, C. D. and McKay, B. D. "Asymptotic Enumeration of Latin Rectangles." *J. Combin. Th. Ser. B* **48**, 19–4, 1990.

Kerawla, S. M. "The Enumeration of Latin Rectangle of Depth Three by Means of Difference Equation" [sic]. *Bull. Calcutta Math. Soc.* **33**, 119–27, 1941.

McKay, B. D. and Rogoyski, E. "Latin Squares of Order 10." *Electronic J. Combinatorics* **2**, N3 1–, 1995. http://www.combinatorics.org/Volume_2/volume2.html#N3.

Ryser, H. J. "Latin Rectangles." §3.3 in *Combinatorial Mathematics.* Buffalo, NY: Math. Assoc. of Amer., pp. 35–7, 1963.

Sloane, N. J. A. Sequences A001009 in "An On-Line Version of the Encyclopedia of Integer Sequences." http://www.research.att.com/~njas/sequences/eisonline.html.

Latin Square

An $n \times n$ Latin square is a LATIN RECTANGLE with $k = n$. Specifically, a Latin square consists of n sets of the numbers 1 to n arranged in such a way that no orthogonal (row or column) contains the same two numbers. The numbers of Latin squares of order $n = 1, 2, \ldots$ are 1, 2, 12, 576, 161280, ... (Sloane's A002860). For example, the two Latin squares of order two are given by

$$\begin{bmatrix} 1 & 2 \\ 2 & 1 \end{bmatrix}, \begin{bmatrix} 2 & 1 \\ 1 & 2 \end{bmatrix}, \tag{1}$$

the 12 Latin squares of order three are given by

$$\begin{bmatrix} 1 & 2 & 3 \\ 2 & 3 & 1 \\ 3 & 1 & 2 \end{bmatrix}, \begin{bmatrix} 1 & 2 & 3 \\ 3 & 1 & 2 \\ 2 & 3 & 1 \end{bmatrix}, \begin{bmatrix} 1 & 3 & 2 \\ 2 & 1 & 3 \\ 3 & 2 & 1 \end{bmatrix}, \begin{bmatrix} 1 & 3 & 2 \\ 3 & 2 & 1 \\ 2 & 1 & 3 \end{bmatrix},$$

$$\begin{bmatrix} 2 & 1 & 3 \\ 1 & 3 & 2 \\ 3 & 2 & 1 \end{bmatrix}, \begin{bmatrix} 2 & 1 & 3 \\ 3 & 2 & 1 \\ 1 & 3 & 2 \end{bmatrix}, \begin{bmatrix} 2 & 3 & 1 \\ 1 & 2 & 3 \\ 3 & 1 & 2 \end{bmatrix}, \begin{bmatrix} 2 & 3 & 1 \\ 3 & 1 & 2 \\ 1 & 2 & 3 \end{bmatrix},$$

$$\begin{bmatrix} 3 & 2 & 1 \\ 1 & 3 & 2 \\ 2 & 1 & 3 \end{bmatrix}, \begin{bmatrix} 3 & 2 & 1 \\ 2 & 1 & 3 \\ 1 & 3 & 2 \end{bmatrix}, \begin{bmatrix} 3 & 1 & 2 \\ 1 & 2 & 3 \\ 2 & 3 & 1 \end{bmatrix}, \begin{bmatrix} 3 & 1 & 2 \\ 2 & 3 & 1 \\ 1 & 2 & 3 \end{bmatrix}, \tag{2}$$

and two of the whopping 576 Latin squares of order 4 are given by

$$\begin{bmatrix} 1 & 2 & 3 & 4 \\ 2 & 1 & 4 & 3 \\ 3 & 4 & 1 & 2 \\ 4 & 3 & 2 & 1 \end{bmatrix} \text{ and } \begin{bmatrix} 1 & 2 & 3 & 4 \\ 3 & 4 & 1 & 2 \\ 4 & 3 & 2 & 1 \\ 2 & 1 & 4 & 3 \end{bmatrix}. \tag{3}$$

A pair of Latin squares is said to be orthogonal if the n^2 pairs formed by juxtaposing the two arrays are all distinct. For example, the two Latin squares

$$\begin{bmatrix} 3 & 2 & 1 \\ 2 & 1 & 3 \\ 1 & 3 & 2 \end{bmatrix} \quad \begin{bmatrix} 2 & 3 & 1 \\ 1 & 2 & 3 \\ 3 & 1 & 2 \end{bmatrix} \tag{4}$$

are orthogonal.

A normalized, or reduced, Latin square is a Latin square with the first row and column given by $\{1, 2, \ldots, n\}$. General FORMULAS for the number of *normalized* $n \times n$ Latin squares $L(n, n)$ are given by Nechvatal (1981), Gessel (1987), and Shao and Wei (1992). The *total* number of Latin squares $N(n, n)$ of order n can then be computed from

$$N(n, n) = n!(n - 1)!L(n, n). \tag{5}$$

The numbers of normalized Latin squares of order $n = 1, 2, \ldots,$ are 1, 1, 1, 4, 56, 9408, ... (Sloane's A000315). McKay and Rogoyski (1995) give the number of normalized LATIN RECTANGLES $L(k, n)$ for $n = 1, \ldots, 10$, as well as estimates for $L(n, n)$ with $n = 11, 12, \ldots, 15$.

n	$L(n, n)$
11	5.36×10^{33}
12	1.62×10^{44}
13	2.51×10^{56}
14	2.33×10^{70}
15	1.5×10^{86}

See also 36 OFFICER PROBLEM, EULER SQUARE, KIRKMAN TRIPLE SYSTEM, LAM'S PROBLEM, PARTIAL LATIN SQUARE, QUASIGROUP, SOMA

References

Colbourn, C. J. and Dinitz, J. H. *CRC Handbook of Combinatorial Designs.* Boca Raton, FL: CRC Press, 1996.
Gessel, I. "Counting Latin Rectangles." *Bull. Amer. Math. Soc.* **16**, 79–3, 1987.
Hunter, J. A. H. and Madachy, J. S. *Mathematical Diversions.* New York: Dover, pp. 33–4, 1975.
Kraitchik, M. "Latin Squares." §7.11 in *Mathematical Recreations.* New York: W. W. Norton, p. 178, 1942.
Lindner, C. C. and Rodger, C. A. *Design Theory.* Boca Raton, FL: CRC Press, 1997.
McKay, B. D. and Rogoyski, E. "Latin Squares of Order 10." *Electronic J. Combinatorics* **2**, N3 1–, 1995. http://www.combinatorics.org/Volume_2/volume2.html#N3.
Nechvatal, J. R. "Asymptotic Enumeration of Generalised Latin Rectangles." *Util. Math.* **20**, 273–92, 1981.
Rohl, J. S. *Recursion via Pascal.* Cambridge, England: Cambridge University Press, pp. 162–65, 1984.
Ryser, H. J. "Latin Rectangles." §3.3 in *Combinatorial Mathematics.* Buffalo, NY: Math. Assoc. Amer., pp. 35–7, 1963.
Shao, J.-Y. and Wei, W.-D. "A Formula for the Number of Latin Squares." *Disc. Math.* **110**, 293–96, 1992.
Sloane, N. J. A. Sequences A002860/M2051 and A000315/M3690 in "An On-Line Version of the Encyclopedia of Integer Sequences." http://www.research.att.com/~njas/sequences/eisonline.html.

Latitude

The latitude of a point on a SPHERE is the elevation of the point from the PLANE of the equator. The latitude δ is related to the COLATITUDE (the polar angle in SPHERICAL COORDINATES) by $\delta = \phi - 90°$. More generally, the latitude of a point on an ELLIPSOID is the ANGLE between a LINE PERPENDICULAR to the surface of the ELLIPSOID at the given point and the PLANE of the equator (Snyder 1987).

The equator therefore has latitude $0°$, and the north and south poles have latitude $\pm 90°$, respectively. Latitude is also called GEOGRAPHIC LATITUDE or GEODETIC LATITUDE in order to distinguish it from several subtly different varieties of AUXILIARY LATITUDES.

The shortest distance between any two points on a SPHERE is the so-called GREAT CIRCLE distance, which can be directly computed from the latitudes and LONGITUDES of the two points.

See also AUXILIARY LATITUDE, COLATITUDE, CONFORMAL LATITUDE, GREAT CIRCLE, ISOMETRIC LATITUDE, LATITUDE, LONGITUDE, SPHERICAL COORDINATES

References

Snyder, J. P. *Map Projections--A Working Manual.* U. S. Geological Survey Professional Paper 1395. Washington, DC: U. S. Government Printing Office, p. 13, 1987.

Lattice

A lattice is a system K such that $\forall A \in K, A \subset A$, and if $A \subset B$ and $B \subset A$, then $A = B$, where $\subset$ means "is included in." Lattices offer a natural way to formalize and study the ordering of objects using a general concept known as the POSET (partially ordered set). The study of lattices is called LATTICE THEORY. Note that this type of lattice is distinct from the regular array of points known as a POINT LATTICE (or informally as a mesh or grid).

The following inequalities hold for any lattice:

$$(x \wedge y) \vee (x \wedge z) \leq x \wedge (y \vee z)$$

$$x \vee (y \wedge z) \leq (x \vee y) \wedge (x \vee z)$$

$$(x \wedge y) \vee (y \wedge z) \vee (z \wedge x) \leq (x \vee y) \wedge (y \vee z) \wedge (z \vee x)$$

$$(x \wedge y) \vee (x \wedge z) \leq x \wedge (y \vee (x \wedge z))$$

(Grätzer 1971, p. 35). The first three are the distributive inequalities, and the last is the modular identity.

See also DISTRIBUTIVE LATTICE, INTEGRATION LATTICE, LATTICE THEORY, MODULAR LATTICE, POINT LATTICE, TORIC VARIETY

Lattice Algebraic System

A generalization of the concept of SET UNIONS and INTERSECTIONS.

Lattice Animal

A distinct (including reflections and rotations) arrangement of adjacent squares on a grid, also called a FIXED POLYOMINO.

See also ANIMAL, PERCOLATION THEORY, POLYOMINO

References

Delest, M.-P. and Viennot, G. "Algebraic Languages and Polyominoes [sic] Enumeration." *Theoret. Comput. Sci.* **34**, 169–06, 1984.
Read, R. C. "Contributions to the Cell Growth Problem." *Canad. J. Math.* **14**, 1–0, 1962.

Lattice Basis Reduction

LATTICE REDUCTION

Lattice Distribution

A DISCRETE DISTRIBUTION of a random variable such that every possible value can be represented in the form $a + bn$, where a, $b \neq 0$ and n is an INTEGER.

References

Abramowitz, M. and Stegun, C. A. (Eds.). *Handbook of Mathematical Functions with Formulas, Graphs, and Mathematical Tables, 9th printing.* New York: Dover, p. 927, 1972.

Lattice Graph

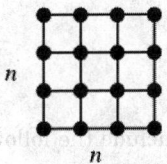

The lattice graph with n nodes on a side is denoted $L(n)$.

See also TRIANGULAR GRAPH

Lattice Groups

In the plane, there are 17 lattice groups, eight of which are pure translation. In $\mathbb{R}^3$, there are 32 POINT GROUPS and 230 SPACE GROUPS. In $\mathbb{R}^4$, there are 4783 space lattice groups.

See also POINT GROUPS, SPACE GROUPS, WALLPAPER GROUPS

Lattice Invariant

INVARIANT (ELLIPTIC FUNCTION)

Lattice Path

A path composed of connected horizontal and vertical line segments, each passing between adjacent LATTICE POINTS. A lattice path is therefore a SEQUENCE of points $P_0, P_1, ..., P_n$ with $n \geq 0$ such that each P_i is a LATTICE POINT and P_{i+1} is obtained by offsetting one unit east (or west) or one unit north (or south).

The number of paths of length $a + b$ from the ORIGIN $(0,0)$ to a point (a, b) which are restricted to east and north steps is given by the BINOMIAL COEFFICIENT $\binom{a+b}{a}$.

See also BALLOT PROBLEM, DYCK PATH, FABER POLYNOMIAL, GOLYGON, KINGS PROBLEM, LATTICE POINT, P-GOOD PATH, RANDOM WALK, STAIRCASE WALK

References

Dickau, R. M. "Shortest-Path Diagrams." http://forum.swarthmore.edu/advanced/robertd/manhattan.html.
Hilton, P. and Pederson, J. "Catalan Numbers, Their Generalization, and Their Uses." *Math. Intel.* **13**, 64–5, 1991.

Mohanty, S. G. *Lattice Path Counting and Applications.* New York: Academic Press, 1979.
Moser, L. and Zayachkowski, H. S. "Lattice Paths with Diagonal Steps." *Scripta Math.* **26**, 223–29, 1963.
Narayana, T. V. *Lattice Path Combinatorics with Statistical Applications.* Toronto, Ontario, Canada: University of Toronto Press, 1979.

Lattice Point

A POINT at the intersection of two or more grid lines in a POINT LATTICE.

See also POINT LATTICE

Lattice Polygon

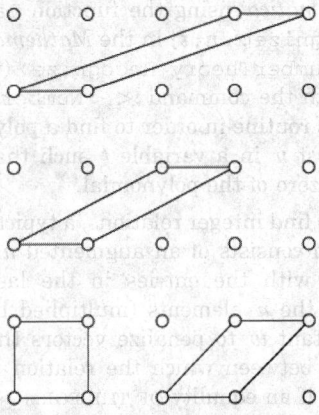

A POLYGON whose vertices are points of a POINT LATTICE. Regular lattice n-gons exists only for $n = 3$, 4, and 6 (Schoenberg 1937, Klamkin and Chrestenson 1963, Maehara 1993). A lattice n-gon in the plane can be equiangular to a regular polygon only for $n = 4$ and 8 (Scott 1987, Maehara 1993).

Maehara (1993) presented a NECESSARY and SUFFICIENT condition for a polygon to be angle-equivalent to a lattice polygon in $\mathbb{R}^n$. In addition, Maehara (1993) proved that $\cos^2(\Sigma_{\theta \in S} \theta)$ is a RATIONAL NUMBER for any collection S of interior angles of a lattice polygon.

See also BAR GRAPH POLYGON, CANONICAL POLYGON, CONVEX POLYGON, CONVEX POLYOMINO, FERRERS GRAPH POLYGON, GOLYGON, POINT LATTICE, POLYOMINO, SELF-AVOIDING POLYGON, STACK POLYGON, STAIRCASE POLYGON, THREE-CHOICE POLYGON

References

Beeson, M. J. "Triangles with Vertices on Lattice Points." *Amer. Math. Monthly* **99**, 243–52, 1992.
Jensen, I. Size and Area of Square Lattice Polygons. 28 Mar 2000. http://xxx.lanl.gov/abs/cond-mat/0003442/.
Klamkin, M. and Chrestenson, H. E. "Polygon Imbedded in a Lattice." *Amer. Math. Monthly* **70**, 51–1, 1963.
Maehara, H. "Angles in Lattice Polygons." *Ryukyu Math. J.* **6**, 9–9, 1993.
Schoenberg, I. J. "Regular Simplices and Quadratic Forms." *J. London Math. Soc.* **12**, 48–5, 1937.
Scott, P. R. "Equiangular Lattice Polygons and Semiregular Lattice Polyhedra." *College Math. J.* **18**, 300–06, 1987.

LatticeReduce

LLL ALGORITHM

Lattice Reduction

The process of finding a reduced set of basis vectors for a given LATTICE having certain special properties. Lattice reduction algorithms are used in a number of modern number theoretical applications, including in the discovery of a SPIGOT ALGORITHM for PI. Although determining the shortest basis is possibly an NP-COMPLETE PROBLEM, algorithms such as the LLL ALGORITHM can find a short basis in polynomial time with guaranteed worst-case performance.

The LLL ALGORITHM of lattice reduction is implemented in *Mathematica* using the function LatticeReduce. Recognize[x, n, t] in the *Mathematica* add-on package NumberTheory`Recognize` (which can be loaded with the command < <NumberTheory`) also calls this routine in order to find a polynomial of degree at most n in a variable t such that x is an approximate zero of the polynomial.

When used to find integer relations, a typical input to the algorithm consists of an augmented $n \times n$ IDENTITY MATRIX with the entries in the last column consisting of the n elements (multiplied by a large positive constant w to penalize vectors that do not sum to zero) between which the relation is sought. For example, if an equality OF THE FORM

$$a_1 x + a_2 y + a_3 z = 0$$

is known to exist, then doing a lattice reduction on the matrix

$$m = \begin{bmatrix} 1 & 0 & 0 & wx \\ 0 & 1 & 0 & wy \\ 0 & 0 & 1 & wz \end{bmatrix}$$

will produce a new matrix in which one or more entries in the last column being close to zero. This row then gives the coefficients $\{a_1, a_2, a_3, 0\}$ of the identity. An example lattice reduction calculation is illustrated in both Borwein and Corless (1999) and Borwein and Lisonek.

See also GRAM-SCHMIDT ORTHONORMALIZATION, INTEGER RELATION, LLL ALGORITHM, PSLQ ALGORITHM

References

Borwein, J. M. and Corless, R. M. "Emerging Tools for Experimental Mathematics." *Amer. Math. Monthly* **106**, 899–09, 1999.

Borwein, J. M. and Lisonek, P. "Applications of Integer Relation Algorithms." To appear in *Disc. Math.* http://www.cecm.sfu.ca/preprints/1997pp.html.

Cohen, H. *A Course in Computational Algebraic Number Theory.* New York: Springer-Verlag, 1993.

Coster, M. J.; Joux, A.; LaMacchia, B. A.; Odlyzko, A. M.; Schnorr, C. P.; and Stern, J. "Improved Low-Density Subset Sum Algorithms." *Comput. Complex.* **2**, 111–28, 1992.

Hastad, J.; Just, B.; Lagarias, J. C.; and Schnorr, C. P. "Polynomial Time Algorithms for Finding Integer Relations Among Real Numbers." *SIAM J. Comput.* **18**, 859–81, 1988.

Lagarias, J. C.; Lenstra, H. W. Jr.; and Schnorr, C. P. "Korkin-Zolotarev Bases and Successive Minima of a Lattice and Its Reciprocal Lattice." *Combinatorica* **10**, 333–48, 1990.

Schnorr, C. P. "A More Efficient Algorithm for Lattice Basis Reduction." *J. Algorithms* **9**, 47–2, 1988.

Schnorr, C. P. and Euchner, M. "Lattice Basis Reduction: Improved Practical Algorithms and Solving Subset Sum Problems." In *Fundamentals of Computation Theory (Gosen 1991).* Berlin: Springer-Verlag, pp. 68–5, 1991.

Lattice Sum

Cubic lattice sums include the following:

$$b_2(2s) \equiv \sum_{i,\,j=-\infty}^{\infty}{}' \frac{(-1)^{i+j}}{(i^2+j^2)^s} \tag{1}$$

$$b_3(2s) \equiv \sum_{i,\,j,\,k=-\infty}^{\infty}{}' \frac{(-1)^{i+j+k}}{(i^2+j^2+k^2)^s} \tag{2}$$

$$b_n(2s) \equiv \sum_{k_1,\,...,\,k_n=-\infty}^{\infty}{}' \frac{(-1)^{k_1+...+k_n}}{(k_1^2+...+k_n^2)^s} \tag{3}$$

where the prime indicates that summation over the original $(0,0)$, $(0,0,0)$, ... is excluded (Borwein and Borwein 1986, p. 288).

As shown in Borwein and Borwein (1987, pp. 288–01), these have closed forms for even n

$$b_2(2s) = -4\beta(s)\eta(s) \tag{4}$$

$$b_4(2s) = -8\eta(s)\eta(s-1) \tag{5}$$

$$b_8(2s) = -16\zeta(s)\eta(s-3), \quad \text{for } \Re[s] > 1 \tag{6}$$

where $\beta(z)$ is the DIRICHLET BETA FUNCTION, $\eta(z)$ is the DIRICHLET ETA FUNCTION, and $\zeta(z)$ is the RIEMANN ZETA FUNCTION. The lattice sums evaluated at $s = 1$ are called the MADELUNG CONSTANTS. An additional form for $b_2(2s)$ is given by

$$b_2(2s) = \sum_{n=1}^{\infty} \frac{(-1)^n r_2(n)}{n^s} \tag{7}$$

for $\Re[s] > 1/3$, where $r_2(n)$ is the SUM OF SQUARES FUNCTION, i.e., the number of representations of n by two squares (Borwein and Borwein 1986, p. 291). Borwein and Borwein (1986) prove that $b_8(2)$ converges (the closed form for $b_8(2s)$ above does not apply for $s = 1$), but its value has not been computed. A number of other related DOUBLE SERIES can be evaluated analytically.

For hexagonal sums, Borwein and Borwein (1987, p. 292) give

$$h_2(2s) \equiv \frac{4}{3} \sum_{m,\,n=-\infty}^{\infty}$$

$$\times \frac{\sin[(n+1)\theta]\sin[(m+1)\theta] - \sin(n\theta)\sin[(m-1)\theta]}{\left[\left(n+\frac{1}{2}m\right)^2 + 3\left(\frac{1}{2}m\right)^2\right]^s}, \tag{8}$$

where $\theta = 2\pi/3$. This MADELUNG CONSTANT is expressible in closed form for $s = 1$ as

$$h_2(2) = \pi \ln 3\sqrt{3}. \tag{9}$$

Other interesting analytic lattice sums are given by

$$\sum_{k,\,m,\,n=-\infty}^{\infty} \frac{(-1)^{k+m+n}}{\left[\left(k+\frac{1}{6}\right)^2\left(m+\frac{1}{6}\right)^2\left(n+\frac{1}{6}\right)^2\right]^s}$$

$$= 12^s \beta(2s-1), \tag{10}$$

giving the special case

$$\sum_{k,\,m,\,n=-\infty}^{\infty} \frac{(-1)^{k+m+n}}{\left[\left(k+\frac{1}{6}\right)^2\left(m+\frac{1}{6}\right)^2\left(n+\frac{1}{6}\right)^2\right]^{1/2}} = \sqrt{3} \tag{11}$$

(Borwein and Borwein 1986, p. 303), and

$$\sum_{k,\,m,\,n=-\infty}^{\infty} \frac{(-1)^{k+m+n+1}}{(|k|+|m|+|n|)^s} = 2\eta(s) + 4\eta(s-2) \tag{12}$$

(Borwein and Borwein 1986, p. 305).

See also BENSON'S FORMULA, DOUBLE SERIES, MADELUNG CONSTANTS

References

Borwein, D. and Borwein, J. M. "On Some Trigonometric and Exponential Lattice Sums." *J. Math. Anal.* **188**, 209–18, 1994.

Borwein, D.; Borwein, J. M.; and Shail, R. "Analysis of Certain Lattice Sums." *J. Math. Anal.* **143**, 126–37, 1989.

Borwein, D.; Borwein, J. M.; and Taylor, K. F. "Convergence of Lattice Sums and Madelung's Constant." *J. Math. Phys.* **26**, 2999–009, 1985.

Borwein, D. and Borwein, J. M. "A Note on Alternating Series in Several Dimensions." *Amer. Math. Monthly* **93**, 531–39, 1986.

Borwein, J. M. and Borwein, P. B. *Pi & the AGM: A Study in Analytic Number Theory and Computational Complexity.* New York: Wiley, 1987.

Finch, S. "Favorite Mathematical Constants." http://www.mathsoft.com/asolve/constant/mdlung/mdlung.html.

Glasser, M. L. and Zucker, I. J. "Lattice Sums." In *Perspectives in Theoretical Chemistry: Advances and Perspectives, Vol. 5* (Ed. H. Eyring).

Lattice Theory

Lattice theory is the study of sets of objects known as LATTICES. It is an outgrowth of the study of BOOLEAN ALGEBRAS, and provides a framework for unifying the study of classes or ordered sets in mathematics. The study of lattice theory was given a great boost by a series of papers and subsequent textbook written by Birkhoff (1967).

See also BOOLEAN ALGEBRA, LATTICE

References

Birkhoff, G. *Lattice Theory, 3rd ed.* Providence, RI: Amer. Math. Soc., 1967.

Grätzer, G. *Lattice Theory: First Concepts and Distributive Lattices.* San Francisco, CA: W. H. Freeman, 1971.

Grätzer, G. *General Lattice Theory, 2nd ed.* Boston, MA: Birkhäuser, 1998.

Priestly, H. A. and Davey, B. A. *Introduction to Lattices and Order.* Cambridge, England: Cambridge University Press, 1990.

Weisstein, E. W. "Books about Lattice Theory." http://www.treasure-troves.com/books/LatticeTheory.html.

Latus Rectum

Twice the SEMILATUS RECTUM of a CONIC SECTION.

See also PARABOLA, SEMILATUS RECTUM

References

Coxeter, H. S. M. *Introduction to Geometry, 2nd ed.* New York: Wiley, pp. 116–18, 1969.

Laurent Polynomial

A Laurent polynomial with COEFFICIENTS in the FIELD $\mathbb{F}$ is an algebraic object that is typically expressed in the form

$$\ldots + a_{-n}t^{-n} + a_{-(n-1)}t^{-(n-1)} + \ldots$$

$$+ a_{-1}t^{-1} + a_0 + a_1 t + \ldots + a_n t^n + \ldots,$$

where the a_i are elements of $\mathbb{F}$, and only finitely many of the a_i are NONZERO. A Laurent polynomial is an algebraic object in the sense that it is treated as a POLYNOMIAL except that the indeterminant "t" can also have NEGATIVE POWERS.

Expressed more precisely, the collection of Laurent polynomials with COEFFICIENTS in a FIELD $\mathbb{F}$ form a RING, denoted $\mathbb{F}[t, t^{-1}]$, with RING operations given by componentwise addition and multiplication according to the relation

$$at^n \cdot bt^m = abt^{n+m}$$

for all n and m in the INTEGERS. Formally, this is equivalent to saying that $\mathbb{F}[t, t^{-1}]$ is the GROUP RING of the INTEGERS and the FIELD $\mathbb{F}$. This corresponds to $\mathbb{F}[t]$ (the POLYNOMIAL ring in one variable for $\mathbb{F}$) being the GROUP RING or MONOID ring for the MONOID of natural numbers and the FIELD $\mathbb{F}$.

See also POLYNOMIAL, PRINCIPAL PART

References

Lang, S. *Undergraduate Algebra, 2nd ed.* New York: Springer-Verlag, 1990.

Laurent Series

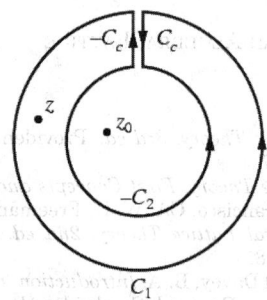

Let there be two circular contours C_2 and C_1, with the radius of C_1 larger than that of C_2. Let z_0 be interior to C_1 and C_2, and z be between C_1 and C_2. Now create a cut line C_c between C_1 and C_2, and integrate around the path $C \equiv C_1 + C_c - C_2 - C_c$, so that the plus and minus contributions of C_c cancel one another, as illustrated above. From the CAUCHY INTEGRAL FORMULA,

$$f(z) = \frac{1}{2\pi i} \int_C \frac{f(z')}{z' - z} \, dz'$$

$$= \frac{1}{2\pi i} \int_{C_1} \frac{f(z')}{z' - z} \, dz' + \frac{1}{2\pi i} \int_{C_c} \frac{f(z')}{z' - z} \, dz'$$

$$- \frac{1}{2\pi i} \int_{C_1} \frac{f(z')}{z' - z} - \frac{1}{2\pi i} \int_{C_c} \frac{f(z')}{z' - z} \, dz'$$

$$= \frac{1}{2\pi i} \int_{C_1} \frac{f(z')}{z' - z} \, dz' - \frac{1}{2\pi i} \int_{C_2} \frac{f(z')}{z' - z} \, dz'. \quad (1)$$

Now, since contributions from the cut line in opposite directions cancel out,

$$f(z) = \frac{1}{2\pi i} \int_{C_1} \frac{f(z')}{(z' - z_0) - (z - z_0)} \, dz'$$

$$- \frac{1}{2\pi i} \int_{C_2} \frac{f(z')}{(z' - z_0) - (z - z_0)} \, dz'$$

$$= \frac{1}{2\pi i} \int_{C_1} \frac{f(z')}{(z' - z_0)\left(1 - \frac{z - z_0}{z' - z_0}\right)} \, dz'$$

$$- \frac{1}{2\pi i} \int_{C_2} \frac{f(z')}{(z - z_0)\left(\frac{z' - z_0}{z - z_0} - 1\right)} \, dz'$$

$$= \frac{1}{2\pi i} \int_{C_1} \frac{f(z')}{(z' - z_0)\left(1 - \frac{z - z_0}{z' - z_0}\right)} \, dz'$$

$$- \frac{1}{2\pi i} \int_{C_2} \frac{f(z')}{(z - z_0)\left(1 - \frac{z' - z_0}{z - z_0}\right)} \, dz'. \quad (2)$$

For the first integral, $|z' - z_0| > |z - z_0|$. For the second, $|z' - z_0| < |z - z_0|$. Now use the TAYLOR EXPANSION

(valid for $|t| < 1$)

$$\frac{1}{1 - t} = \sum_{n=0}^{\infty} t^n \quad (3)$$

to obtain

$$f(z) = \frac{1}{2\pi i} \left[\int_{C_1} \frac{f(z')}{z' - z_0} \sum_{n=0}^{\infty} \left(\frac{z - z_0}{z - z_0}\right)^n dz' \right.$$

$$\left. + \int_{C_2} \frac{f(z')}{z - z_0} \sum_{n=0}^{\infty} \left(\frac{z' - z_0}{z - z_0}\right)^n dz' \right]$$

$$= \frac{1}{2\pi i} \sum_{n=0}^{\infty} (z - z_0)^n \int_{C_1} \frac{f(z')}{(z' - z_0)^{n+1}} \, dz'$$

$$+ \frac{1}{2\pi i} \sum_{n=0}^{\infty} (z - z_0)^{-n-1} \int_{C_2} (z' - z_0)^n f(z') \, dz'$$

$$= \frac{1}{2\pi i} \sum_{n=0}^{\infty} (z - z_0)^n \int_{C_1} \frac{f(z')}{(z' - z_0)^{n+1}} \, dz'$$

$$+ \frac{1}{2\pi i} \sum_{n=1}^{\infty} (z - z_0)^{-n} \int_{C_2} (z' - z_0)^{n+1} f(z') \, dz', \quad (4)$$

where the second term has been re-indexed. Re-indexing again,

$$f(z) = \frac{1}{2\pi i} \sum_{n=0}^{\infty} (z - z_0)^n \int_{C_1} \frac{f(z')}{(z' - z_0)^{n+1}} \, dz'$$

$$+ \frac{1}{2\pi i} \sum_{n=-\infty}^{-1} (z - z_0) \int_{C_2} \frac{f(z')}{(z' - z_0)^{n+1}} \, dz'. \quad (5)$$

Now, use the CAUCHY INTEGRAL THEOREM, which requires that any CONTOUR INTEGRAL of a function which encloses no POLES has value 0. But $1/(z' - z_0)^{n+1}$ is never singular inside C_2 for $n \geq 0$, and $1/(z' - z_0)^{n+1}$ is never singular inside C_1 for $n \leq -1$. Similarly, there are no POLES in the closed cut $C_c - C_c$. We can therefore replace C_1 and C_2 in the above integrals by C without altering their values, so

$$f(z) = \frac{1}{2\pi i} \sum_{n=0}^{\infty} (z - z_0)^n \int_C \frac{f(z')}{(z' - z_0)^{n+1}} \, dz'$$

$$+ \frac{1}{2\pi i} \sum_{n=-\infty}^{-1} (z - z_0)^n \int_C \frac{f(z')}{(z' - z_0)^{n+1}} \, dz'$$

$$= \frac{1}{2\pi i} \sum_{n=-\infty}^{\infty} (z - z_0)^n \int_C \frac{f(z')}{(z' - z_0)^{n+1}} \, dz'$$

$$\equiv \sum_{n=-\infty}^{\infty} a_n (z - z_0)^n. \quad (6)$$

The only requirement on C is that it encloses z, so we are free to choose any contour γ that does so. The

RESIDUES a_n are therefore defined by

$$a_n \equiv \frac{1}{2\pi i} \int_\gamma \frac{f(z')}{(z'-z_0)^{n+1}} \, dz'. \tag{7}$$

See also MACLAURIN SERIES, PRINCIPAL PART, RESI-DUE (COMPLEX ANALYSIS), TAYLOR SERIES

References

Arfken, G. "Laurent Expansion." §6.5 in *Mathematical Methods for Physicists, 3rd ed.* Orlando, FL: Academic Press, pp. 376–84, 1985.

Knopp, K. "The Laurent Expansion." Ch. 10 in *Theory of Functions Parts I and II, Two Volumes Bound as One, Part I.* New York: Dover, pp. 117–22, 1996.

Krantz, S. G. "Laurent Series." §4.2.1 in *Handbook of Complex Analysis.* Boston, MA: Birkhäuser, p. 43, 1999.

Morse, P. M. and Feshbach, H. "Derivatives of Analytic Functions, Taylor and Laurent Series." §4.3 in *Methods of Theoretical Physics, Part I.* New York: McGraw-Hill, pp. 374–98, 1953.

Lauricella Functions

This entry contributed by RONALD M. AARTS

Lauricella functions are generalizations of the Gauss hypergeometric functions to multiple variables. Four such generalizations were investigated by Lauricella (1893), and more fully by Appell and Kampé de Fériet (1926, p. 117). Let n be the number of variables, then the Lauricella functions are defined by

$$F_A^{(n)}(a, b_1, \ldots, b_n; c_1, \ldots, c_n; x_1, \ldots x_n)$$
$$= \sum \frac{(a, m_1 + \ldots + m_n)(b_1, m_1) \cdots (b_n, m_n) x_1^{m_1} \cdots x_n^{m_n}}{(c_1, m_1) \cdots (c_n, m_n) m_1! \cdots m_n!} \tag{1}$$

$$F_B^{(n)}(a_1, \ldots, a_n, b_1, \ldots, b_n; c; x_1, \ldots, x_n)$$
$$= \sum \frac{(a_1, m_1) \cdots (a_n, m_n)(b_1, m_1) \cdots (b_n, m_n) x_1^{m_1} \cdots x_n^{m_n}}{(c, m_1 + \ldots m_n) m_1! \cdots m_n!} \tag{2}$$

$$F_C^{(n)}(a, b; c_1, \ldots, c_n; x_1, \ldots, x_n)$$
$$= \sum \frac{(a_1, m_1 + \ldots m_n)(b, m_1 + \ldots m_n) x_1^{m_1} \cdots x_n^{m_n}}{(c_1, m_1) \cdots (c_n, m_n) m_1! \cdots m_n!} \tag{3}$$

$$F_D^{(n)}(a, b_1, \ldots, b_n; c; x_1, \ldots, x_n)$$
$$= \sum \frac{(a, m_1 + \ldots + m_n)(b_1, m_1) \cdots (b_n, m_n) x_1^{m_1} \cdots x_n^{m_n}}{(c, m_1 + \ldots m_n) m_1! \cdots m_n!}. \tag{4}$$

If $n = 2$, then these functions reduce to the APPELL HYPERGEOMETRIC FUNCTIONS F_2, F_3, F_4, and F_1, respectively. If $n = 1$, all four become the Gauss hypergeometric function $_2F_1$ (Exton 1978, p. 29).

See also APPELL HYPERGEOMETRIC FUNCTION, GEN-ERALIZED HYPERGEOMETRIC FUNCTION, HORN FUNC-TION, KAMPÉ DE FÉRIET FUNCTION

References

Appell, P. and Kampé de Fériet, J. *Fonctions hypergéo-métriques et hypersphériques: polynomes d'Hermite.* Paris: Gauthier-Villars, 1926.

Erdélyi, A. "Hypergeometric Functions of Two Variables." *Acta Math.* **83**, 131–64, 1950.

Exton, H. Ch. 5 in *Multiple Hypergeometric Functions and Applications.* New York: Wiley, 1976.

Exton, H. "The Lauricella Functions and Their Confluent Forms," "Convergence," and "Systems of Partial Differen-tial Equations." §1.4.1–.4.3 in *Handbook of Hypergeo-metric Integrals: Theory, Applications, Tables, Computer Programs.* Chichester, England: Ellis Horwood, pp. 29–1, 1978.

Lauricella, G. "Sulla funzioni ipergeometriche a più varia-bili." *Rend. Circ. Math. Palermo* **7**, 111–58, 1893.

Law

A law is a mathematical statement which always holds true. Whereas "laws" in physics are generally experimental observations backed up by theoretical underpinning, laws in mathematics are generally THEOREMS which can formally be proven true under the stated conditions. However, the term is also sometimes used in the sense of an empirical observa-tion, e.g., BENFORD'S LAW.

See also ABSORPTION LAW, BENFORD'S LAW, CONTRA-DICTION LAW, DE MORGAN'S DUALITY LAW, DE MOR-GAN'S LAWS, ELLIPTIC CURVE GROUP LAW, EXCLUDED MIDDLE LAW, EXPONENT LAWS, GIRKO'S CIRCULAR LAW, LAW OF COSINES, LAW OF SINES, LAW OF TANGENTS, LAW OF TRULY LARGE NUMBERS, MOR-RIE'S LAW, PARALLELOGRAM LAW, PLATEAU'S LAWS, QUADRATIC RECIPROCITY LAW, STRONG LAW OF LARGE NUMBERS, STRONG LAW OF SMALL NUMBERS, SYLVESTER'S INERTIA LAW, TRICHOTOMY LAW, VEC-TOR TRANSFORMATION LAW, WEAK LAW OF LARGE NUMBERS, ZIPF'S LAW

Law of Anomalous Numbers

BENFORD'S LAW

Law of Cancellation

CANCELLATION LAW

Law of Cosines

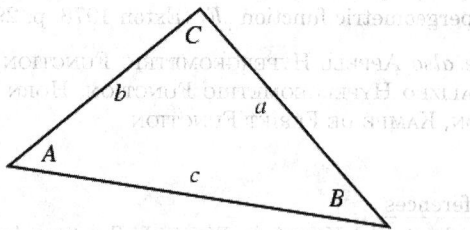

Let a, b, and c be the lengths of the legs of a TRIANGLE opposite ANGLES A, B, and C. Then the law of cosines states

$$c^2 = a^2 + b^2 - 2ab \cos C. \qquad (1)$$

This law can be derived in a number of ways. The definition of the DOT PRODUCT incorporates the law of cosines, so that the length of the VECTOR from $\mathbf{X}$ to $\mathbf{Y}$ is given by

$$|\mathbf{X} - \mathbf{Y}|^2 = (\mathbf{X} - \mathbf{Y}) \cdot (\mathbf{X} - \mathbf{Y}) \qquad (2)$$

$$= \mathbf{X} \cdot \mathbf{X} - 2\mathbf{X} \cdot \mathbf{Y} + \mathbf{Y} \cdot \mathbf{Y} \qquad (3)$$

$$= |\mathbf{X}|^2 + |\mathbf{Y}|^2 - 2|\mathbf{X}||\mathbf{Y}|\cos \theta, \qquad (4)$$

where θ is the ANGLE between $\mathbf{X}$ and $\mathbf{Y}$.

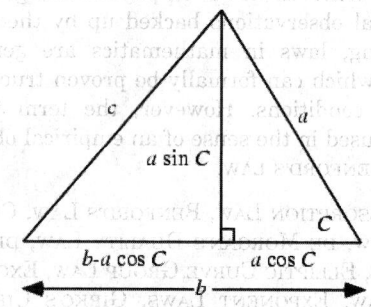

The formula can also be derived using a little geometry and simple algebra. From the above diagram,

$$c^2 = (a \sin C)^2 + (b - a \cos C)^2$$

$$= a^2 \sin^2 C + b^2 - 2ab \cos C + a^2 \cos^2 C$$

$$= a^2 + b^2 - 2ab \cos C. \qquad (5)$$

The law of cosines for the sides of a SPHERICAL TRIANGLE states that

$$\cos a = \cos b \cos c + \sin b \sin c \cos A \qquad (6)$$

$$\cos b = \cos c \cos a + \sin c \sin a \cos B \qquad (7)$$

$$\cos c = \cos a \cos b + \sin a \sin b \cos C \qquad (8)$$

(Beyer 1987). The law of cosines for the angles of a

SPHERICAL TRIANGLE states that

$$\cos A = -\cos B \cos C + \sin B \sin C \cos a \qquad (9)$$

$$\cos B = -\cos C \cos A + \sin C \sin A \cos b \qquad (10)$$

$$\cos C = -\cos A \cos B + \sin A \sin B \cos c \qquad (11)$$

(Beyer 1987).
For similar triangles, a generalized law of cosines is given by

$$aa' = bb' + cc' - (bc' + b'c)\cos A \qquad (12)$$

(Lee 1997). Furthermore, consider an arbitrary TETRAHEDRON $A_1 A_2 A_3 A_4$ with triangles $T_1 = \Delta A_2 A_3 A_4$, $T_2 = \Delta A_1 A_3 A_4$, $T_3 = \Delta A_1 A_2 A_4$, and $T_4 = A_1 A_2 A_3$. Let the areas of these triangles be s_1, s_2, s_3, and s_4, respectively, and denote the DIHEDRAL ANGLE with respect to T_i and T_j for $i \neq j = 1$, 2, 3, 4 by θ_{ij}. Then

$$s_k = \sum_{\substack{j \neq k \\ 1 \leq i \leq 4}} s_i \cos \theta_{ki}, \qquad (13)$$

which gives the law of cosines in a tetrahedron,

$$s_k^2 = \sum_{\substack{i \neq k \\ 1 \leq j \leq 4}} s_j^2 - 2 \sum_{\substack{i, j \neq k \\ 1 \leq i, j \leq 4}} s_i s_j \cos \theta_{ij} \qquad (14)$$

(Lee 1997). A corollary gives the nice identity

$$s_1 s_1' = s_2 s_2' + s_3 s_3' + s_4 s_4' - (s_2 s_3' + s_2' s_3)\cos \theta_{23}$$

$$- (s_3 s_4' + s_3' s_4)\cos \theta_{34} - (s_2 s_4' + s_2' s_4)\cos \theta_{24} \qquad (15)$$

See also LAW OF SINES, LAW OF TANGENTS

References

Abramowitz, M. and Stegun, C. A. (Eds.). *Handbook of Mathematical Functions with Formulas, Graphs, and Mathematical Tables, 9th printing.* New York: Dover, p. 79, 1972.
Beyer, W. H. *CRC Standard Mathematical Tables, 28th ed.* Boca Raton, FL: CRC Press, pp. 148–49, 1987.
Lee, J. R. "The Law of Cosines in a Tetrahedron." *J. Korea Soc. Math. Ed. Ser. B: Pure Appl. Math.* **4**, 1–, 1997.

Law of Exponents
EXPONENT LAWS

Law of Growth

An exponential growth law OF THE FORM

$$y = ar^x$$

characterizing a quantity which increases at a fixed rate proportionally to itself.

See also GROWTH, LOGISTIC GROWTH CURVE, POPULATION GROWTH

Law of Indices

References
Kenney, J. F. and Keeping, E. S. "The Law of Growth." §4.12 in *Mathematics of Statistics, Pt. 1, 3rd ed.* Princeton, NJ: Van Nostrand, pp. 56–7, 1962.

Law of Indices

EXPONENT LAWS

Law of Large Numbers

STRONG LAW OF LARGE NUMBERS, WEAK LAW OF LARGE NUMBERS

Law of Sines

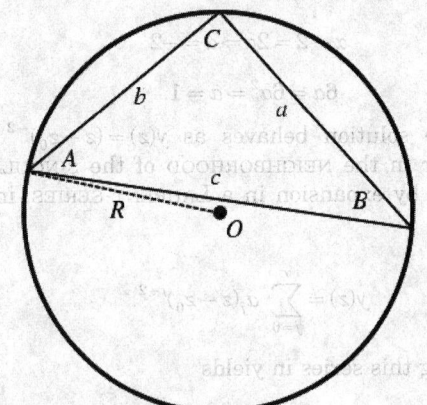

Let a, b, and c be the lengths of the LEGS of a TRIANGLE opposite ANGLES A, B, and C. Then the law of sines states that

$$\frac{a}{\sin A} = \frac{b}{\sin B} = \frac{c}{\sin C} = 2R, \tag{1}$$

where R is the radius of the CIRCUMCIRCLE. Other related results include the identities

$$a(\sin B - \sin C) + b(\sin C - \sin A) + c(\sin A - \sin B)$$
$$= 0 \tag{2}$$

$$a = b \cos C + c \cos B, \tag{3}$$

the LAW OF COSINES

$$\cos A = \frac{c^2 + b^2 - a^2}{2bc}, \tag{4}$$

and the LAW OF TANGENTS

$$\frac{a + b}{a - b} = \frac{\tan\left[\frac{1}{2}(A + B)\right]}{\tan\left[\frac{1}{2}(A - B)\right]}. \tag{5}$$

The law of sines for oblique SPHERICAL TRIANGLES

states that

$$\frac{\sin a}{\sin A} = \frac{\sin b}{\sin B} = \frac{\sin c}{\sin C}. \tag{6}$$

See also LAW OF COSINES, LAW OF TANGENTS

References
Abramowitz, M. and Stegun, C. A. (Eds.). *Handbook of Mathematical Functions with Formulas, Graphs, and Mathematical Tables, 9th printing.* New York: Dover, p. 79, 1972.
Beyer, W. H. *CRC Standard Mathematical Tables, 28th ed.* Boca Raton, FL: CRC Press, p. 148, 1987.
Coxeter, H. S. M. and Greitzer, S. L. "The Extended Law of Sines." §1.1 in *Geometry Revisited.* Washington, DC: Math. Assoc. Amer., pp. 1–, 1967.

Law of Small Numbers

STRONG LAW OF SMALL NUMBERS

Law of Tangents

Let a TRIANGLE have sides of lengths a, b, and c and let the ANGLES opposite these sides by A, B, and C. The law of tangents states

$$\frac{a - b}{a + b} = \frac{\tan\left[\frac{1}{2}(A - B)\right]}{\tan\left[\frac{1}{2}(A + B)\right]}.$$

An analogous result for oblique SPHERICAL TRIANGLES states that

$$\frac{\tan\left[\frac{1}{2}(a - b)\right]}{\tan\left[\frac{1}{2}(a + b)\right]} = \frac{\tan\left[\frac{1}{2}(A - B)\right]}{\tan\left[\frac{1}{2}(A + B)\right]}.$$

See also LAW OF COSINES, LAW OF SINES

References
Abramowitz, M. and Stegun, C. A. (Eds.). *Handbook of Mathematical Functions with Formulas, Graphs, and Mathematical Tables, 9th printing.* New York: Dover, p. 79, 1972.
Beyer, W. H. *CRC Standard Mathematical Tables, 28th ed.* Boca Raton, FL: CRC Press, pp. 145 and 149, 1987.

Law of Truly Large Numbers

With a large enough sample, any outrageous thing is likely to happen (Diaconis and Mosteller 1989). Littlewood (1953) considered an event which occurs one in a million times to be "surprising." Taking this definition, close to 100,000 surprising events are "expected" each year in the United States alone and, in the world at large, "we can be absolutely sure that we will see incredibly remarkable events" (Diaconis and Mosteller 1989).

See also COINCIDENCE, FRIVOLOUS THEOREM OF ARITHMETIC, STRONG LAW OF LARGE NUMBERS, STRONG LAW OF SMALL NUMBERS

References

Diaconis, P. and Mosteller, F. "Methods of Studying Coincidences." *J. Amer. Statist. Assoc.* **84**, 853–61, 1989.
Littlewood, J. E. *Littlewood's Miscellany.* Cambridge, England: Cambridge University Press, 1986.

Lax-Milgram Theorem

Let ϕ be a bounded COERCIVE bilinear FUNCTIONAL on a HILBERT SPACE H. Then for every bounded linear FUNCTIONAL f on H, there exists a unique $x_f \in H$ such that

$$f(x) = \phi(x, x_f)$$

for all $x \in H$.

References

Debnath, L. and Mikusinski, P. *Introduction to Hilbert Spaces with Applications.* San Diego, CA: Academic Press, 1990.
Zeidler, E. *Applied Functional Analysis: Applications to Mathematical Physics.* New York: Springer-Verlag, 1995.

Lax Pair

A pair of linear OPERATORS L and A associated with a given PARTIAL DIFFERENTIAL EQUATION which can be used to solve the equation. However, it turns out to be very difficult to find the L and A corresponding to a given equation, so it is actually simpler to postulate a given L and A and determine to which PARTIAL DIFFERENTIAL EQUATION they correspond (Infeld and Rowlands 2000).

See also PARTIAL DIFFERENTIAL EQUATION

References

Infeld, E. and Rowlands, G. "Integrable Equations in Two Space Dimensions as Treated by the Zakharov-Shabat Method." §7.10 in *Nonlinear Waves, Solitons, and Chaos, 2nd ed.* Cambridge, England: Cambridge University Press, pp. 192–99, 2000.

Layer

P-LAYER

LCM

LEAST COMMON MULTIPLE

Leading Digit Phenomenon

BENFORD'S LAW

Leading Order Analysis

A procedure for determining the behavior of an nth order ORDINARY DIFFERENTIAL EQUATION at a REMOVABLE SINGULARITY without actually solving the equation. Consider

$$\frac{d^n y}{dz^n} = F\left(\frac{d^{n-1}y}{dz^{n-1}}, \ldots, \frac{dy}{dx}, y, z\right), \tag{1}$$

where F is ANALYTIC in z and rational in its other arguments. Proceed by making the substitution

$$y(z) \equiv a(z - z_0)^\alpha \tag{2}$$

with $\alpha < 1$. For example, in the equation

$$\frac{d^2 y}{dz^2} = 6y^2 + Ay, \tag{3}$$

making the substitution gives

$$a\alpha(\alpha - 1)(z - z_0)^{\alpha - 2} = 6a^2(z - z_0)^{2\alpha} + Aa(az - z_0)^\alpha. \tag{4}$$

The most singular terms (those with the most NEGATIVE exponents) are called the "dominant balance terms," and must balance exponents and COEFFICIENTS at the SINGULARITY. Here, the first two terms are dominant, so

$$\alpha - 2 = 2\alpha \Rightarrow \alpha = -2 \tag{5}$$

$$6a = 6a^2 \Rightarrow a = 1, \tag{6}$$

and the solution behaves as $y(z) = (z - z_0)^{-2}$. The behavior in the NEIGHBORHOOD of the SINGULARITY is given by expansion in a LAURENT SERIES, in this case,

$$y(z) = \sum_{j=0}^{\infty} a_j (z - z_0)^{j-2}. \tag{7}$$

Plugging this series in yields

$$\sum_{j=0}^{\infty} a_j (j-2)(j-3)(z - z_0)^{j-4}$$

$$= 6 \sum_{j=0}^{\infty} \sum_{k=0}^{\infty} a_j a_k (z - z_0)^{j+k-4} + A \sum_{j=0}^{\infty} a_j (z - z_0)^{j-2}. \tag{8}$$

This gives RECURRENCE RELATIONS, in this case with a_6 arbitrary, so the $(z - z_0)^6$ term is called the resonance or KOVALEVSKAYA EXPONENT. At the resonances, the COEFFICIENT will always be arbitrary. If no resonance term is present, the POLE present is not ordinary, and the solution must be investigated using a PSI FUNCTION.

See also PSI FUNCTION

References

Tabor, M. *Chaos and Integrability in Nonlinear Dynamics: An Introduction.* New York: Wiley, p. 330, 1989.

Leaf (Foliation)

Let M^n be an n-MANIFOLD and let $\mathsf{F} = \{F_\alpha\}$ denote a PARTITION of M into DISJOINT path-connected SUB-

SETS. Then if F is a FOLIATION of M, each F_x is called a leaf and is not necessarily closed or compact.

See also FOLIATION

References

Rolfsen, D. *Knots and Links.* Wilmington, DE: Publish or Perish Press, p. 284, 1976.

Leaf (Tree)

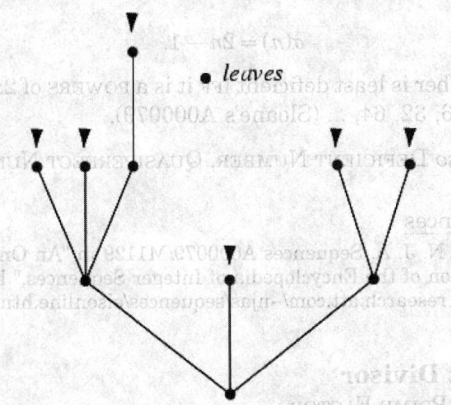

leaves

An unconnected end of a TREE (i.e., a node of VERTEX DEGREE 1). The following tables gives the total numbers of leaves for various classes of graphs on $n = 1, 2, \ldots$ nodes. For ROOTED TREES, the ROOT NODE is not counted as a leaf.

graph type	Sloane	leaf count for $n = 1, 2, \ldots$ nodes
GRAPH	A055540	0, 2, 4, 14, 38, 153, 766, ...
TREE	A003228	0, 2, 2, 5, 9, 21, 43, 101, ...
LABELED TREE	A055541	0, 2, 6, 36, 320, 3750, ...
ROOTED TREE	A003227	1, 1, 3, 8, 22, 58, 160, 434, 1204, ...

See also BRANCH, CHILD, FORK, ROOT NODE, TREE

References

Robinson, R. W. and Schwenk, A. J. "The Distribution of Degrees in a Large Random Tree." *Discr. Math.* **12**, 359–72, 1975.

Sloane, N. J. A. Sequences A003227/M2744, A003228/M0351, A055540, and A055541 in "An On-Line Version of the Encyclopedia of Integer Sequences." http://www.research.att.com/~njas/sequences/eisonline.html.

Leakage

ALIASING

Leap

JUMP

Least Bound

SUPREMUM

Least Common Multiple

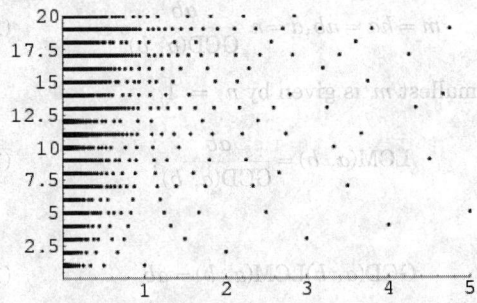

The least common multiple of two numbers a and b, denoted LCM(a, b) or $[a, b]$, is the smallest number m for which there exist positive integers n_a and n_b such that

$$n_a a = n_b b = m. \tag{1}$$

The least common multiple LCM$(a, b, c, \ldots)$ of more than two numbers is similarly defined. The plot above shows LCM$(1, r)$ for rational $r = m/n$, which is equivalent to the NUMERATOR of the reduced form of m/n.

The least common multiple of $a, b, c, \ldots$, is denoted LCM$[a, b, c, \ldots]$ in *Mathematica*.

The least common multiple of two numbers a and b can be obtained by finding the PRIME FACTORIZATION of each

$$a = p_1^{a_1} \cdots p_n^{a_n} \tag{2}$$

$$b = p_1^{b_1} \cdots p_n^{b_n}, \tag{3}$$

where the p s are all PRIME FACTORS of a and b, and if p_i does not occur in one factorization, then the corresponding exponent is taken as 0. The least common multiple is then given by

$$\text{LCM}(a, b) = \prod_{i=1}^{n} p_i^{\max(a_i, b_i)}. \tag{4}$$

For example, consider LCM$(12, 30)$.

$$12 = 2^2 \cdot 3^1 \cdot 5^0 \tag{5}$$

$$30 = 2^1 \cdot 3^1 \cdot 5^1, \tag{6}$$

so

$$LCM(12, 30) = 2^2 \cdot 3^1 \cdot 5^1 = 60. \tag{7}$$

Let m be a common multiple of a and b so that

$$m = ha = kb. \tag{8}$$

Write $a = a_1 \, \mathrm{GCD}(a, b)$ and $b = b_1 \, \mathrm{GCD}(a, b)$, where a_1 and b_1 are RELATIVELY PRIME by definition of the GREATEST COMMON DIVISOR $GCD(a_1, b_1) = 1$. Then $ha_1 = kb_1$, and from the DIVISION LEMMA (given that ha_1 is DIVISIBLE by b_1 and $\mathrm{GCD}(b_1, a_1) = 1$), we have h is DIVISIBLE by b_1, so

$$h = nb_1 \tag{9}$$

$$m = ha = nb_1 a = n\,\frac{ab}{\mathrm{GCD}(a, b)}. \tag{10}$$

The smallest m is given by $n = 1$,

$$\mathrm{LCM}(a, b) = \frac{ab}{\mathrm{GCD}(a, b)}, \tag{11}$$

so

$$\mathrm{GCD}(a, b)\mathrm{LCM}(a, b) = ab \tag{12}$$

The LCM is IDEMPOTENT

$$\mathrm{LCM}(a, a) = a \tag{13}$$

COMMUTATIVE

$$\mathrm{LCM}(a, b) = \mathrm{LCM}(b, a), \tag{14}$$

ASSOCIATIVE

$$\mathrm{LCM}(a, b, c) = \mathrm{LCM}(\mathrm{LCM}(a, b), c)$$
$$= \mathrm{LCM}(a, \mathrm{LCM}(b, c)), \tag{15}$$

DISTRIBUTIVE

$$\mathrm{LCM}(ma, mb, mc) = m\,\mathrm{LCM}(a, b, c), \tag{16}$$

and satisfies the ABSORPTION LAW

$$\mathrm{GCD}(a, \mathrm{LCM}(a, b)) = a. \tag{17}$$

It is also true that

$$\mathrm{LCM}(ma, mb) = \frac{\mathrm{GCD}(ma)\mathrm{GCD}(mb)}{\mathrm{GCD}(ma, mb)} = m\,\frac{ab}{\mathrm{GCD}(a, b)}$$
$$= m\,\mathrm{LCM}(a, b). \tag{18}$$

See also GREATEST COMMON DIVISOR, MANGOLDT FUNCTION, RELATIVELY PRIME

References

Guy, R. K. "Density of a Sequence with L.C.M. of Each Pair Less than x." §E2 in *Unsolved Problems in Number Theory*, 2nd ed. New York: Springer-Verlag, pp. 200–01, 1994.

Nagell, T. "Least Common Multiple and Greatest Common Divisor." §5 in *Introduction to Number Theory*. New York: Wiley, pp. 16–9, 1951.

Least Common Multiple Matrix

Let $S = \{x_1, \ldots, x_n\}$ be a set of n distinct POSITIVE INTEGERS. Then the matrix $[S]_n$ having the LEAST COMMON MULTIPLE $\mathrm{LCM}(x_i, x_j)$ of x_i and x_j as its i, jth

entry is called the least common multiple matrix on S.

See also BOURQUE-LIGH CONJECTURE

References

Hong, S. "On the Bourque-Ligh Conjecture of Least Common Multiple Matrices." *J. Algebra* **218**, 216–28, 1999.

Least Deficient Number

A number for which

$$\sigma(n) = 2n - 1.$$

A number is least deficient IFF it is a POWERS of 2: 1, 2, 4, 8, 16, 32, 64, ... (Sloane's A000079).

See also DEFICIENT NUMBER, QUASIPERFECT NUMBER

References

Sloane, N. J. A. Sequences A000079/M1129 in "An On-Line Version of the Encyclopedia of Integer Sequences." http://www.research.att.com/~njas/sequences/eisonline.html.

Least Divisor

LEAST PRIME FACTOR

Least Period

The smallest n for which a point x_0 is a PERIODIC POINT of a function f so that $f^n(x_0) = x_0$. For example, for the FUNCTION $f(x) = -x$, all points x have period 2 (including $x = 0$). However, $x = 0$ has a *least* period of 1. The analogous concept exists for a PERIODIC SEQUENCE, but not for a PERIODIC FUNCTION. The least period is also called the exact period.

Least Prime Factor

Let $n > 1$ be any integer and let $\mathrm{LD}(n)$ be the least integer greatest than 1 that divides n. Then $\mathrm{LD}(n)$ is a prime number, and if n is not prime, then $[\mathrm{LD}(n)]^2 \le n$ (Séroul 2000, p. 7).

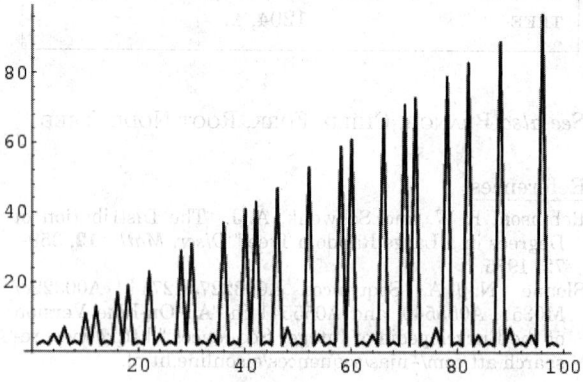

For an INTEGER $n \ge 2$, let $\mathrm{lpf}(x)$ denote the LEAST PRIME FACTOR of n, i.e., the number p_1 in the factorization

$$n = p_1^{a_1} \cdots p_k^{a_k},$$

with $p_i < p_j$ for $i < j$. For $n = 2, 3, \ldots$, the first few are 2, 3, 2, 5, 2, 7, 2, 3, 2, 11, 2, 13, 2, 3, ... (Sloane's A020639). The above plot of the least prime factor function can be seen to resemble a jagged terrain of mountains, which leads to the appellation of "TWIN PEAKS" to a PAIR of INTEGERS (x, y) such that

1. $x < y$,
2. $\mathrm{lpf}(x) = \mathrm{lpf}(y)$,
3. For all z, $x < z < y$ IMPLIES $\mathrm{lpf}(z) < \mathrm{lpf}(x)$.

The least *multiple* prime factors for SQUAREFUL integers are 2, 2, 3, 2, 2, 3, 2, 2, 5, 3, 2, 2, 2, ... (Sloane's A046027).

Erdos *et al.* (1993) consider the least prime factor of the BINOMIAL COEFFICIENTS, and define what they term GOOD BINOMIAL COEFFICIENTS and EXCEPTIONAL BINOMIAL COEFFICIENTS. They also conjecture that

$$\mathrm{lpf}\binom{N}{k} \leq \max(N/k, 29). \tag{1}$$

See also ALLADI-GRINSTEAD CONSTANT, DISTINCT PRIME FACTORS, ERDOS-SELFRIDGE FUNCTION, EUCLID-MULLIN SEQUENCE, EXCEPTIONAL BINOMIAL COEFFICIENT, FACTOR, GOOD BINOMIAL COEFFICIENT, GREATEST PRIME FACTOR, LEAST COMMON MULTIPLE, MANGOLDT FUNCTION, PRIME FACTORS, TWIN PEAKS

References

Erdos, P.; Lacampagne, C. B.; and Selfridge, J. L. "Estimates of the Least Prime Factor of a Binomial Coefficient." *Math. Comput.* **61**, 215–24, 1993.

Séroul, R. "The Lowest Divisor Function." §8.4 in *Programming for Mathematicians.* Berlin: Springer-Verlag, pp. 9–1 and 165–67, 2000.

Sloane, N. J. A. Sequences A020639 and A046027 in "An On-Line Version of the Encyclopedia of Integer Sequences." http://www.research.att.com/~njas/sequences/eisonline.html.

Least Squares Fitting

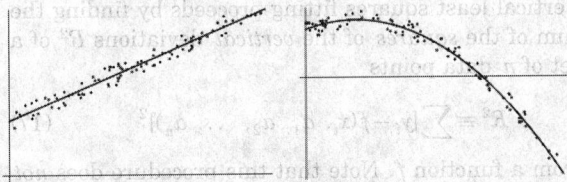

A mathematical procedure for finding the best fitting curve to a given set of points by minimizing the sum of the squares of the offsets ("the residuals"rpar; of the points from the curve. The sum of the *squares* of the offsets is used instead of the offset absolute values because this allows the residuals to be treated as a continuous differentiable quantity. However, because squares of the offsets are used, outlying points can have a disproportionate effect on the fit, a property which may or may not be desirable depending on the problem at hand.

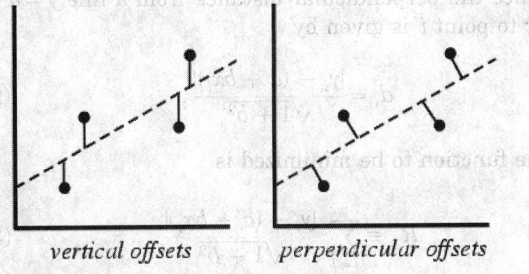

vertical offsets   *perpendicular offsets*

In practice, the *vertical* offsets from a line are almost always minimized instead of the *perpendicular* offsets. This allows uncertainties of the data points along the x- and y-axes to be incorporated simply, and also provides a much simpler analytic form for the fitting parameters than would be obtained using a fit based on perpendicular distances. In addition, the fitting technique can be easily generalized from a best-fit *line* to a best-fit *polynomial* when sums of vertical distances are used (which is not the case using perpendicular distances). For a reasonable number of noisy data points, the difference between vertical and perpendicular fits is quite small.

The *linear* least squares fitting technique is the simplest and most commonly applied form of LINEAR REGRESSION and provides a solution to the problem of finding the best fitting *straight* line through a set of points. In fact, if the functional relationship between the two quantities being graphed is known to within additive or multiplicative constants, it is common practice to transform the data in such a way that the resulting line *is* a straight line, say by plotting T vs. $\sqrt{\ell}$ instead of T vs. ℓ in the case of analyzing the period T of a pendulum as a function of its length l. For this reason, standard forms for EXPONENTIAL, LOGARITHMIC, and POWER laws are often explicitly computed. The formulas for linear least squares fitting were independently derived by Gauss and Legendre.

For NONLINEAR LEAST SQUARES FITTING to a number of unknown parameters, linear least squares fitting may be applied iteratively to a linearized form of the function until convergence is achieved. Depending on the type of fit and initial parameters chosen, the nonlinear fit may have good or poor convergence properties. If uncertainties (in the most general case, error ellipses) are given for the points, points can be weighted differently in order to give the high-quality points more weight.

The residuals of the best-fit line for a set of n points using *unsquared perpendicular* distances d_i of points (x_i, y_i) are given by

$$R_\perp \equiv \sum_{i=1}^{n} d_i. \tag{1}$$

Since the perpendicular distance from a line $y = a + bx$ to point i is given by

$$d_i = \frac{|y_i - (a + bx_i)|}{\sqrt{1 + b^2}}, \tag{2}$$

the function to be minimized is

$$R_\perp \equiv \sum_{i=1}^{n} \frac{|y_i - (a + bx_i)|}{\sqrt{1 + b^2}}. \tag{3}$$

Unfortunately, because the absolute value function does not have continuous derivatives, minimizing $R_\perp$ is not amenable to analytic solution. However, if the *square* of the perpendicular distances

$$R_\perp^2 \equiv \sum_{i=1}^{n} \frac{[y_i - (a + bx_i)]^2}{1 + b^2} \tag{4}$$

is minimized instead, the problem can be solved in closed form. $R_\perp^2$ is a minimum when (suppressing the indices)

$$\frac{\partial R_\perp^2}{\partial a} = \frac{2}{1 + b^2} \sum [y - (a + bx)](-1) = 0 \tag{5}$$

and

$$\frac{\partial R_\perp^2}{\partial b} = \frac{2}{1 + b^2} \sum [y - (a + bx)](-x)$$
$$+ \sum \frac{[y - (a + bx)]^2(-1)(2b)}{(1 + b^2)^2}$$
$$= 0. \tag{6}$$

The former gives

$$a = \frac{\sum y - b \sum x}{n} = \bar{y} - b\bar{x}, \tag{7}$$

and the latter

$$(1 + b^2) \sum [y - (a + bx)]x + b \sum [y - (a + bx)]^2 = 0. \tag{8}$$

But

$$[y - (a + bx)]^2 = y^2 - 2(a + bx)y + (a + bx)^2$$
$$= y^2 - 2ay - 2bxy + a^2 + 2abx + b^2x^2, \tag{9}$$

so (8) becomes

$$(1 + b^2)\left(\sum xy - a \sum x - b \sum x^2 \right)$$
$$+ b\left(\sum y^2 - 2a \sum y - 2b \right.$$
$$\times \sum xy + a^2 \sum 1 + 2ab \sum x + b^2 \sum x^2 \Big) = 0 \tag{10}$$

$$[(1 + b^2)(-b) + b(b^2)] \sum x^2 + [(1 + b^2) - 2b^2] \sum xy$$

$$+ b \sum y^2 + [-a(1 + b^2) + 2ab^2] \sum x - 2ab \sum y$$
$$+ ba^2 \sum 1 = 0 \tag{11}$$

$$-b \sum x^2 + (1 - b^2) \sum xy + b \sum y^2 + a(b^2 - 1) \sum x$$
$$-2ab \sum y + ba^2 n = 0. \tag{12}$$

Plugging (7) into (12) then gives

$$-b \sum x^2 + (1 - b^2) \sum xy + b \sum y^2 + \frac{1}{n}(b^2 - 1)$$
$$\times \left(\sum y - b \sum x \right) \sum x$$
$$-\frac{2}{n}\left(\sum y - b \sum x \right)b \sum y$$
$$+\frac{1}{n}b\left(\sum y - b \sum x \right)^2$$
$$= 0 \tag{13}$$

After a fair bit of algebra, the result is

$$b^2 + \frac{\sum y^2 - \sum x^2 + \frac{1}{n}\left[(\sum x)^2 - (\sum y)^2 \right]}{\frac{1}{n} \sum x \sum y - \sum xy} b - 1$$
$$= 0. \tag{14}$$

So define

$$B \equiv \frac{1}{2} \frac{\left[\sum y^2 - \frac{1}{n}(\sum y)^2 \right] - \left[\sum x^2 - \frac{1}{n}(\sum x)^2 \right]}{\frac{1}{n} \sum x \sum y - \sum xy}$$
$$= \frac{1}{2} \frac{(\sum y^2 - n\bar{y}^2) - (\sum x^2 - n\bar{x}^2)}{n\bar{x}\bar{y} - \sum xy}, \tag{15}$$

and the QUADRATIC FORMULA gives

$$b = -B \pm \sqrt{B^2 + 1}, \tag{16}$$

with a found using (7). Note the rather unwieldy form of the best-fit parameters in the formulation. In addition, minimizing $R_\perp^2$ for a second- or higher-order POLYNOMIAL leads to polynomial equations having *higher* order, so this formulation cannot be extended.

Vertical least squares fitting proceeds by finding the sum of the *squares* of the *vertical* deviations R^2 of a set of n data points

$$R^2 \equiv \sum [y_i - f(x_i, a_1, a_2, \ldots, a_n)]^2 \tag{17}$$

from a function f. Note that this procedure does *not* minimize the actual deviations from the line (which would be measured perpendicular to the given function). In addition, although the *unsquared* sum of distances might seem a more appropriate quantity to minimize, use of the absolute value results in discontinuous derivatives which cannot be treated analytically. The square deviations from each point are therefore summed, and the resulting residual is then minimized to find the best fit line. This procedure results in outlying points being given disproportionately large weighting.

The condition for R^2 to be a minimum is that

$$\frac{\partial(R^2)}{\partial a_i} = 0 \qquad (18)$$

for $i = 1, ..., n$. For a linear fit,

$$f(a, b) = a + bx, \qquad (19)$$

so

$$R^2(a, b) \equiv \sum_{i=1}^{n}[y_i - (a + bx_i)]^2 \qquad (20)$$

$$\frac{\partial(R^2)}{\partial a} = -2\sum_{i=1}^{n}[y_i - (a + bx_i)] = 0 \qquad (21)$$

$$\frac{\partial(R^2)}{\partial b} = -2\sum_{i=1}^{n}[y_i - (a + bx_i)]x_i = 0. \qquad (22)$$

These lead to the equations

$$na + b\sum x = \sum y \qquad (23)$$

$$a\sum x + b\sum x^2 = \sum xy, \qquad (24)$$

where the subscripts have been dropped for conciseness. In MATRIX form,

$$\begin{bmatrix} n & \sum x \\ \sum x & \sum x^2 \end{bmatrix}\begin{bmatrix} a \\ b \end{bmatrix} = \begin{bmatrix} \sum y \\ \sum xy \end{bmatrix}, \qquad (25)$$

so

$$\begin{bmatrix} a \\ b \end{bmatrix} = \begin{bmatrix} n & \sum x \\ \sum x & \sum x^2 \end{bmatrix}^{-1}\begin{bmatrix} \sum y \\ \sum xy \end{bmatrix}. \qquad (26)$$

The 2×2 MATRIX INVERSE is

$$\begin{bmatrix} a \\ b \end{bmatrix} = \frac{1}{n\sum x^2 - \left(\sum x\right)^2}$$

$$\times \begin{bmatrix} \sum y \sum x^2 - \sum x \sum xy \\ n\sum xy - \sum x \sum y \end{bmatrix}, \qquad (27)$$

so

$$a = \frac{\sum y \sum x^2 - \sum x \sum xy}{n\sum x^2 - \left(\sum x\right)^2} \qquad (28)$$

$$= \frac{\bar{y}\sum x^2 - \bar{x}\sum xy}{\sum x^2 - n\bar{x}^2} \qquad (29)$$

$$b = \frac{n\sum xy - \sum x \sum y}{n\sum x^2 - \left(\sum x\right)^2} \qquad (30)$$

$$= \frac{\sum xy - n\bar{x}\bar{y}}{\sum x^2 - n\bar{x}^2} \qquad (31)$$

(Kenney and Keeping 1962). These can be rewritten

in a simpler form by defining the sums of squares

$$ss_{xx} = \sum_{i=1}^{n}(x_i - \bar{x})^2 = \left(\sum x^2\right) - n\bar{x}^2 \qquad (32)$$

$$ss_{yy} = \sum_{i=1}^{n}(y_i - \bar{y})^2 = \left(\sum y^2\right) - n\bar{y}^2 \qquad (33)$$

$$ss_{xy} = \sum_{i=1}^{n}(x_i - \bar{x})(y_i - \bar{y}) = \left(\sum xy\right) - n\bar{x}\bar{y}, \qquad (34)$$

which are also written as

$$\sigma_x^2 = ss_{xx} \qquad (35)$$

$$\sigma_y^2 = ss_{yy} \qquad (36)$$

$$cov(x, y) = ss_{xy}. \qquad (37)$$

Here, $cov(x, y)$ is the COVARIANCE and σ_x^2 and σ_y^2 are variances. Note that the quantities $\sum xy$ and $\sum x^2$ can also be interpreted as the DOT PRODUCTS

$$\sum x^2 = \mathbf{x} \cdot \mathbf{x} \qquad (38)$$

$$\sum xy = \mathbf{x} \cdot \mathbf{y}. \qquad (39)$$

In terms of the sums of squares, the REGRESSION COEFFICIENT b is given by

$$b = \frac{cov(x, y)}{\sigma_x^2} = \frac{ss_{xy}}{ss_{xx}}, \qquad (40)$$

and a is given in terms of b using (24) as

$$a = \bar{y} - b\bar{x}. \qquad (41)$$

The overall quality of the fit is then parameterized in terms of a quantity known as the CORRELATION COEFFICIENT, defined by

$$r^2 = \frac{ss_{xy}^2}{ss_{xx}ss_{yy}}, \qquad (42)$$

which gives the proportion of ss_{yy} which is accounted for by the regression.

The STANDARD ERRORS for a and b are

$$SE(a) = s\sqrt{\frac{1}{n} + \frac{\bar{x}^2}{ss_{xx}}} \qquad (43)$$

$$SE(b) = \frac{s}{\sqrt{ss_{xx}}}. \qquad (44)$$

Let $\hat{y}_i$ be the vertical coordinate of the best-fit line with x-coordinate x_i, so

$$\hat{y}_i \equiv a + bx_i, \qquad (45)$$

then the error between the actual vertical point y_i and the fitted point is given by

$$e_i \equiv y_i - \hat{y}_i \qquad (46)$$

Now define s^2 as an estimator for the variance in e_i,

$$s^2 = \sum_{i=1}^{n} \frac{e_i^2}{n-2}. \qquad (47)$$

Then s can be given by

$$s = \sqrt{\frac{\text{ss}_{yy} - b\,\text{ss}_{xy}}{n-2}} = \sqrt{\frac{\text{ss}_{yy} - \frac{\text{ss}_{xy}^2}{\text{ss}_{xx}}}{n-2}} \qquad (48)$$

(Acton 1966, pp. 32–5; Gonick and Smith 1993, pp. 202–04).

Generalizing from a straight line (i.e., first degree polynomial) to a kth degree POLYNOMIAL

$$y = a_0 + a_1 x + \ldots + a_k x^k, \qquad (49)$$

the residual is given by

$$R^2 \equiv \sum_{i=1}^{n} [y_i - (a_0 + a_1 x_i + \ldots + a_k x_i^k)]^2. \qquad (50)$$

The PARTIAL DERIVATIVES (again dropping superscripts) are

$$\frac{\partial(R^2)}{\partial a_0} = -2 \sum [y - (a_0 + a_1 x + \ldots + a_k x^k)] = 0 \qquad (51)$$

$$\frac{\partial(R^2)}{\partial a_1} = -2 \sum [y - (a_0 + a_1 x + \ldots + a_k x^k)]x = 0 \qquad (52)$$

$$\frac{\partial(R^2)}{\partial a_k} = -2 \sum [y - (a_0 + a_1 x + \ldots + a_k x^k)]x^k = 0. \qquad (53)$$

These lead to the equations

$$a_0 n + a_1 \sum x + \ldots + a_k \sum x^k = \sum y \qquad (54)$$

$$a_0 \sum x + a_1 \sum x^2 + \ldots + a_k \sum x^{k+1} = \sum xy \qquad (55)$$

$$a_0 \sum x^k + a_1 \sum x^{k+1} + \ldots + a_k \sum x^{2k} = \sum x^k y \qquad (56)$$

or, in MATRIX form

$$\begin{bmatrix} n & \sum x & \cdots & \sum x^k \\ \sum x & \sum x^2 & \cdots & \sum x^{k+1} \\ \vdots & \vdots & \ddots & \vdots \\ \sum x^k & \sum x^{k+1} & \cdots & \sum x^{2k} \end{bmatrix} \begin{bmatrix} a_0 \\ a_1 \\ \vdots \\ a_k \end{bmatrix}$$
$$= \begin{bmatrix} \sum y \\ \sum xy \\ \vdots \\ \sum x^k y \end{bmatrix}. \qquad (57)$$

This is a VANDERMONDE MATRIX. We can also obtain the MATRIX for a least squares fit by writing

$$\begin{bmatrix} 1 & x_1 & \cdots & x_1^k \\ 1 & x_2 & \cdots & x_2^k \\ \vdots & \vdots & \ddots & \vdots \\ 1 & x_n & \cdots & x_n^k \end{bmatrix} \begin{bmatrix} a_0 \\ a_1 \\ \vdots \\ a_k \end{bmatrix} = \begin{bmatrix} y_1 \\ y_2 \\ \vdots \\ y_n \end{bmatrix}. \qquad (58)$$

Premultiplying both sides by the TRANSPOSE of the first MATRIX then gives

$$\begin{bmatrix} 1 & 1 & \cdots & 1 \\ x_1 & x_2 & \cdots & x_n \\ \vdots & \vdots & \ddots & \vdots \\ x_1^k & x_2^k & \cdots & x_n^k \end{bmatrix} \begin{bmatrix} 1 & x_1 & \cdots & x_1^k \\ 1 & x_2 & \cdots & x_2^k \\ \vdots & \vdots & \ddots & \vdots \\ 1 & x_n & \cdots & x_n^k \end{bmatrix} \begin{bmatrix} a_0 \\ a_1 \\ \vdots \\ a_k \end{bmatrix}$$
$$= \begin{bmatrix} 1 & 1 & \cdots & 1 \\ x_1 & x_2 & \cdots & x_n \\ \vdots & \vdots & \ddots & \vdots \\ x_1^k & x_2^k & \cdots & x_n^k \end{bmatrix} \begin{bmatrix} y_1 \\ y_2 \\ \vdots \\ y_n \end{bmatrix}, \qquad (59)$$

so

$$\begin{bmatrix} n & \sum x & \cdots & \sum x^n \\ \sum x & \sum x^2 & \cdots & \sum x^{n+1} \\ \vdots & \vdots & \ddots & \vdots \\ \sum x^n & \sum x^{n+1} & \cdots & \sum x^{2n} \end{bmatrix} \begin{bmatrix} a_0 \\ a_1 \\ \vdots \\ a_k \end{bmatrix}$$
$$= \begin{bmatrix} \sum y \\ \sum xy \\ \vdots \\ \sum x^k y \end{bmatrix}. \qquad (60)$$

As before, given m points (x_i, y_i) and fitting with POLYNOMIAL COEFFICIENTS a_0, ..., a_n gives

$$\begin{bmatrix} y_1 \\ y_2 \\ \vdots \\ y_m \end{bmatrix} = \begin{bmatrix} 1 & x_1 & x_1^2 & \cdots & x_1^n \\ 1 & x_2 & x_2^2 & \cdots & x_2^n \\ \vdots & \vdots & \vdots & \ddots & \vdots \\ 1 & x_m & x_m^2 & \cdots & x_m^n \end{bmatrix} \begin{bmatrix} a_0 \\ a_1 \\ \vdots \\ a_n \end{bmatrix}, \qquad (61)$$

In MATRIX notation, the equation for a polynomial fit is given by

$$\mathbf{y} = \mathsf{X}\mathbf{a}. \qquad (62)$$

This can be solved by premultiplying by the MATRIX TRANSPOSE X^T,

$$\mathsf{X}^T \mathbf{y} = \mathsf{X}^T \mathsf{X} \mathbf{a}. \qquad (63)$$

This MATRIX EQUATION can be solved numerically, or can be inverted directly if it is well formed, to yield the solution vector

$$\mathbf{a} = (\mathsf{X}^T \mathsf{X})^{-1} \mathsf{X}^T \mathbf{y}. \qquad (64)$$

Setting $m = 1$ in the above equations reproduces the linear solution.

See also CORRELATION COEFFICIENT, INTERPOLATION, LEAST SQUARES FITTING–EXPONENTIAL, LEAST SQUARES FITTING–LOGARITHMIC, LEAST SQUARES FITTING–POWER LAW, MOORE-PENROSE GENERALIZED MATRIX INVERSE, NONLINEAR LEAST SQUARES FITTING, REGRESSION COEFFICIENT, SPLINE

References

Acton, F. S. *Analysis of Straight-Line Data.* New York: Dover, 1966.

Bevington, P. R. *Data Reduction and Error Analysis for the Physical Sciences.* New York: McGraw-Hill, 1969.

Chatterjee, S.; Hadi, A.; and Price, B. "Simple Linear Regression." Ch. 2 in *Regression Analysis by Example, 3rd ed.* New York: Wiley, pp. 21–0, 2000.

Gauss, C. F. "Theoria combinationis obsevationum erroribus minimis obnoxiae." *Werke, Bd. 4*, p. 1.

Gonick, L. and Smith, W. *The Cartoon Guide to Statistics.* New York: Harper Perennial, 1993.

Kenney, J. F. and Keeping, E. S. "Linear Regression, Simple Correlation, and Contingency." Ch. 8 in *Mathematics of Statistics, Pt. 2, 2nd ed.* Princeton, NJ: Van Nostrand, pp. 199–37, 1951.

Kenney, J. F. and Keeping, E. S. "Linear Regression and Correlation." Ch. 15 in *Mathematics of Statistics, Pt. 1, 3rd ed.* Princeton, NJ: Van Nostrand, pp. 252–85, 1962.

Lancaster, P. and Salkauskas, K. *Curve and Surface Fitting: An Introduction.* London: Academic Press, 1986.

Laplace, P. S. Ch. 4 in *Théorie anal. des prob., Livre 2.* 1812.

Lawson, C. and Hanson, R. *Solving Least Squares Problems.* Englewood Cliffs, NJ: Prentice-Hall, 1974.

Nash, J. C. *Compact Numerical Methods for Computers: Linear Algebra and Function Minimisation, 2nd ed.* Bristol, England: Adam Hilger, pp. 21–4, 1990.

Press, W. H.; Flannery, B. P.; Teukolsky, S. A.; and Vetterling, W. T. "Fitting Data to a Straight Line" "Straight-Line Data with Errors in Both Coordinates," and "General Linear Least Squares." §15.2, 15.3, and 15.4 in *Numerical Recipes in FORTRAN: The Art of Scientific Computing, 2nd ed.* Cambridge, England: Cambridge University Press, pp. 655–75, 1992.

Whittaker, E. T. and Robinson, G. "The Method of Least Squares." Ch. 9 in *The Calculus of Observations: A Treatise on Numerical Mathematics, 4th ed.* New York: Dover, pp. 209–, 1967.

York, D. "Least-Square Fitting of a Straight Line." *Canad. J. Phys.* **44**, 1079–086, 1966.

Least Squares Fitting—Exponential

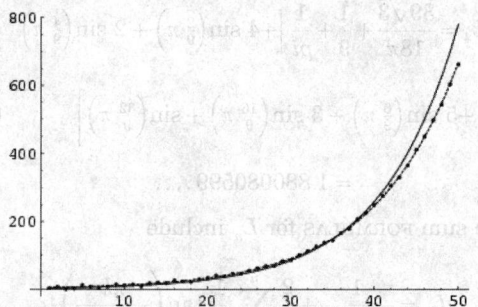

To fit a functional form

$$y = Ae^{Bx}, \tag{1}$$

take the LOGARITHM of both sides

$$\ln y = \ln A + Bx. \tag{2}$$

The best-fit values are then

$$a = \frac{\sum \ln y \sum x^2 - \sum x \sum x \ln y}{n \sum x^2 - \left(\sum x\right)^2} \tag{3}$$

$$b = \frac{n \sum x \ln y - \sum x \sum \ln y}{n \sum x^2 - \left(\sum x\right)^2}, \tag{4}$$

where $B \equiv b$ and $A \equiv \exp(a)$.

This fit gives greater weights to small y values so, in order to weight the points equally, it is often better to minimize the function

$$\sum y(\ln y - a - bx)^2. \tag{5}$$

Applying LEAST SQUARES FITTING gives

$$a \sum y + b \sum xy = \sum y \ln y \tag{6}$$

$$a \sum xy + b \sum x^2 y = \sum xy \ln y \tag{7}$$

$$\begin{bmatrix} \sum y & \sum xy \\ \sum xy & \sum x^2 y \end{bmatrix} \begin{bmatrix} a \\ b \end{bmatrix} = \begin{bmatrix} \sum y \ln y \\ \sum xy \ln y \end{bmatrix}. \tag{8}$$

Solving for a and b,

$$a = \frac{\sum(x^2 y) \sum(y \ln y) - \sum(xy) \sum(xy \ln y)}{\sum y \sum(x^2 y) - \left(\sum xy\right)^2} \tag{9}$$

$$b = \frac{\sum y \sum(xy \ln y) - \sum(xy) \sum(y \ln y)}{\sum y \sum(x^2 y) - \left(\sum xy\right)^2}. \tag{10}$$

In the plot above, the short-dashed curve is the fit computed from (3) and (4) and the long-dashed curve is the fit computed from (9) and (10).

See also LEAST SQUARES FITTING, LEAST SQUARES FITTING–LOGARITHMIC, LEAST SQUARES FITTING–POWER LAW

Least Squares Fitting—Logarithmic

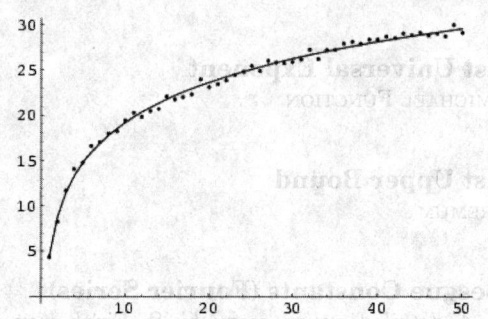

Given a function OF THE FORM

$$y = a + b \ln x, \tag{1}$$

the COEFFICIENTS can be found from LEAST SQUARES

FITTING as

$$b = \frac{n \sum (y \ln x) - \sum y \sum (\ln x)}{n \sum \left[(\ln x)^2 \right] - \left[\sum (\ln x) \right]^2} \qquad (2)$$

$$a = \frac{\sum y - b \sum (\ln x)}{n}. \qquad (3)$$

See also LEAST SQUARES FITTING, LEAST SQUARES FITTING–EXPONENTIAL, LEAST SQUARES FITTING–POWER LAW

Least Squares Fitting—Power Law

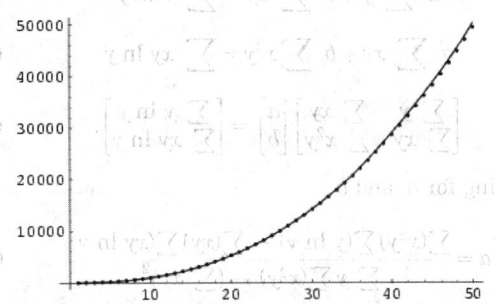

Given a function OF THE FORM

$$y = Ax^B, \qquad (1)$$

LEAST SQUARES FITTING gives the COEFFICIENTS as

$$b = \frac{n \sum (\ln x \ln y) - \sum (\ln x) \sum (\ln y)}{n \sum [(\ln x)^2] - (\sum \ln x)^2} \qquad (2)$$

$$a = \frac{\sum (\ln y) - b \sum (\ln x)}{n}, \qquad (3)$$

where $B \equiv b$ and $A \equiv \exp(a)$.

See also LEAST SQUARES FITTING, LEAST SQUARES FITTING–EXPONENTIAL, LEAST SQUARES FITTING–LOGARITHMIC

Least Universal Exponent
CARMICHAEL FUNCTION

Least Upper Bound
SUPREMUM

Lebesgue Constants (Fourier Series)
N.B. A detailed online essay by S. Finch was the starting point for this entry.

Assume a function f is integrable over the interval $[-\pi, \pi]$ and $S_n(f, x)$ is the nth partial sum of the FOURIER SERIES of f, so that

$$a_k = \frac{1}{\pi} \int_{-\pi}^{\pi} f(t) \cos(kt) \, dt \qquad (1)$$

$$b_k = \frac{1}{\pi} \int_{-\pi}^{\pi} f(t) \sin(kt) \, dt \qquad (2)$$

and

$$S_n(f, x) = \frac{1}{2} a_0 + \left\{ \sum_{k=1}^{n} [a_k \cos(kx) + b_k \sin(kx)] \right\}. \qquad (3)$$

If

$$|f(x)| \le 1 \qquad (4)$$

for all x, then

$$S_n(f, x) \le \frac{1}{\pi} \int_0^{\pi} \left| \frac{\sin \left[\frac{1}{2}(2n + 1)\theta \right]}{\sin \left(\frac{1}{2} \theta \right)} \right| d\theta = L_n, \qquad (5)$$

and L_n is the smallest possible constant for which this holds for all continuous f. The first few values of L_n are

$$L_0 = 1 \qquad (6)$$

$$L_1 = \frac{1}{3} + \frac{2\sqrt{3}}{\pi} = 1.435991124\ldots \qquad (7)$$

$$L_2 = \frac{1}{5} + \frac{\sqrt{25 - 2\sqrt{5}}}{\pi} = 1.642188435\ldots \qquad (8)$$

$$L_3 = \frac{1}{7} + \frac{1}{\pi} \left[4 \sin\left(\frac{2}{7}\pi\right) - 2 \sin\left(\frac{4}{7}\pi\right) + \frac{16}{3} \sin\left(\frac{6}{7}\pi\right) \right.$$

$$\left. -2 \sin\left(\frac{8}{7}\pi\right) + \frac{2}{3} \sin\left(\frac{12}{7}\pi\right) + \frac{4}{3} \sin\left(\frac{18}{7}\pi\right) \right]$$

$$= 1.778322861\ldots. \qquad (9)$$

$$L_4 = \frac{39\sqrt{3}}{18\pi} + \frac{1}{9} + \frac{1}{pi} \left[+4 \sin\left(\frac{2}{9}\pi\right) + 2 \sin\left(\frac{4}{9}\pi\right) \right.$$

$$\left. +5 \sin\left(\frac{8}{9}\pi\right) + 3 \sin\left(\frac{16}{9}\pi\right) + \sin\left(\frac{32}{9}\pi\right) \right] \qquad (10)$$

$$= 1.880080599\ldots. $$

Some sum FORMULAS for L_n include

$$L_n = \frac{1}{2n + 1} + \frac{2}{\pi} \sum_{k=1}^{n} \frac{1}{k} \tan\left(\frac{\pi k}{2n + 1} \right)$$

$$= \frac{16}{\pi^2} \sum_{k=1}^{\infty} \sum_{j=1}^{(2n+1)k} \frac{1}{4k^2 - 1} \frac{1}{2j - 1} \qquad (11)$$

(Zygmund 1959) and integral FORMULAS include

$$L_n = 4 \int_0^{\infty} \frac{\tanh[(2n + 1)x]}{\tanh x} \frac{dx}{\pi^2 + 4x^2}$$

$$= \frac{4}{\pi^2} \int_0^\infty \frac{\sinh[(2n+1)x]}{\sinh x} \ln \left\{ \coth \left[\tfrac{1}{2}(2n+1)x \right] \right\} dx \tag{12}$$

(Hardy 1942). For large n,

$$\frac{4}{\pi^2} \ln n < L_n < 3 + \frac{4}{\pi^2} \ln n. \tag{13}$$

This result can be generalized for an r-differentiable function satisfying

$$\left| \frac{d^r f}{dx^r} \right| \leq 1 \tag{14}$$

for all x. In this case,

$$|f(x) - S_n(f, x)| \leq L_{n,r} = \frac{4}{\pi^2} \frac{\ln n}{n^r} + \mathcal{O}\left(\frac{1}{n^r}\right), \tag{15}$$

where

$$L_{n,r} = \begin{cases} \dfrac{1}{\pi} \displaystyle\int_{-\pi}^{\pi} \left| \displaystyle\sum_{k=n+1}^{\infty} \frac{\sin(kx)}{k^r} \right| dx & \text{for } r \geq 1 \text{ odd} \\[4mm] \dfrac{1}{\pi} \displaystyle\int_{-\pi}^{\pi} \left| \displaystyle\sum_{k=n+1}^{\infty} \frac{\cos(kx)}{k^r} \right| dx & \text{for } r \geq 1 \text{ even} \end{cases} \tag{16}$$

(Kolmogorov 1935, Zygmund 1959).

Watson (1930) showed that

$$\lim_{n \to \infty} \left[L_n - \frac{4}{\pi^2} \ln(2n+1) \right] = c, \tag{17}$$

where

$$c = \frac{8}{\pi^2} \left(\sum_{k=1}^{\infty} \frac{\ln k}{4k^2 - 1} \right) - \frac{4}{\pi^2} \frac{\Gamma'\left(\frac{1}{2}\right)}{\Gamma\left(\frac{1}{2}\right)} \tag{18}$$

$$= \frac{8}{\pi^2} \left[\sum_{j=0}^{\infty} \frac{\lambda(2j+2) - 1}{2j+1} \right] + \frac{4}{\pi^2} (2\ln 2 + \gamma) \tag{19}$$

$$= 0.9894312738..., \tag{20}$$

where $\Gamma(z)$ is the GAMMA FUNCTION, $\lambda(z)$ is the DIRICHLET LAMBDA FUNCTION, and γ is the EULER-MASCHERONI CONSTANT.

References

Finch, S. "Favorite Mathematical Constants." http://www.mathsoft.com/asolve/constant/lbsg/lbsg.html.

Hardy, G. H. "Note on Lebesgue's Constants in the Theory of Fourier Series." *J. London Math. Soc.* **17**, 4–3, 1942.

Kolmogorov, A. N. "Zur Grössenordnung des Restgliedes Fourierscher reihen differenzierbarer Funktionen." *Ann. Math.* **36**, 521–26, 1935.

Watson, G. N. "The Constants of Landau and Lebesgue." *Quart. J. Math. Oxford* **1**, 310–18, 1930.

Zygmund, A. G. *Trigonometric Series, 2nd ed., Vols. 1–*. Cambridge, England: Cambridge University Press, 1959.

Lebesgue Constants (Lagrange Interpolation)

N.B. A detailed online essay by S. Finch was the starting point for this entry.

Define the nth Lebesgue constant for the LAGRANGE INTERPOLATING POLYNOMIAL by

$$\Lambda_n(X) \equiv \max_{-1 \leq x \leq 1} \sum_{k=1}^{n} \left| \prod_{j \neq k} \frac{x - x_j}{x_k - x_j} \right|. \tag{1}$$

It is true that

$$\Lambda_n > \frac{4}{\pi^2} \ln n - 1. \tag{2}$$

The efficiency of a Lagrange interpolation is related to the rate at which Λ_n increases. Erdos (1961) proved that there exists a POSITIVE constant such that

$$\Lambda_n > \frac{2}{\pi} \ln n - C \tag{3}$$

for all n. Erdos (1961) further showed that

$$\Lambda_n < \frac{2}{\pi} \ln n + 4, \tag{4}$$

so (3) cannot be improved upon.

References

Erdos, P. "Problems and Results on the Theory of Interpolation, II." *Acta Math. Acad. Sci. Hungary* **12**, 235–44, 1961.

Finch, S. "Favorite Mathematical Constants." http://www.mathsoft.com/asolve/constant/lbsg/lbsg.html.

Lebesgue Covering Dimension

An important DIMENSION and one of the first dimensions investigated. It is defined in terms of covering sets, and is therefore also called the COVERING DIMENSION. Another name for the Lebesgue covering dimension is the TOPOLOGICAL DIMENSION.

A SPACE has Lebesgue covering dimension m if for every open COVER of that space, there is an open COVER that refines it such that the refinement has order at most $m + 1$. Consider how many elements of the cover contain a given point in a base space. If this has a maximum over all the points in the base space, then this maximum is called the order of the cover. If a SPACE does not have Lebesgue covering dimension m for any m, it is said to be infinite dimensional.

Results of this definition are:

1. Two homeomorphic spaces have the same dimension,
2. $\mathbb{R}^n$ has dimension n,
3. A TOPOLOGICAL SPACE can be embedded as a closed subspace of a EUCLIDEAN SPACE IFF it is LOCALLY COMPACT, HAUSDORFF, SECOND COUNTA-

BLE, and is finite-dimensional (in the sense of the LEBESGUE DIMENSION), and

4. Every compact metrizable m-dimensional TOPOLOGICAL SPACE can be embedded in $\mathbb{R}^{2m+1}$.

See also LEBESGUE MINIMAL PROBLEM

References

Dieudonne, J. A. *A History of Algebraic and Differential Topology.* Boston, MA: Birkhäuser, 1994.

Iyanaga, S. and Kawada, Y. (Eds.). *Encyclopedic Dictionary of Mathematics.* Cambridge, MA: MIT Press, p. 414, 1980.

Munkres, J. R. *Topology: A First Course.* Englewood Cliffs, NJ: Prentice-Hall, 1975.

Lebesgue Decomposition (Measure)

Any COMPLEX MEASURE λ decomposes into an ABSOLUTELY CONTINUOUS measure λ_a and a SINGULAR MEASURE λ_c, with respect to some positive measure μ. This is the LEBESGUE DECOMPOSITION

$$\lambda = \lambda_a + \lambda_c.$$

See also ABSOLUTELY CONTINUOUS, COMPLEX MEASURE, FUNDAMENTAL THEOREMS OF CALCULUS, LEBESGUE MEASURE, POLAR REPRESENTATION (MEASURE), RADON-NIKODYM THEOREM, SINGULAR MEASURE

References

Rudin, W. *Real and Complex Analysis.* New York: McGraw-Hill, p. 121, 1987.

Lebesgue Dimension

LEBESGUE COVERING DIMENSION

Lebesgue Identity

$$(a^2 + b^2 + c^2 + d^2)^2$$

$$= (a^2 + b^2 - c^2 - d^2)^2 + (2ac + 2bd)^2 + (2ad - 2bc)^2$$

(Nagell 1951, pp. 194–95).

See also DIOPHANTINE EQUATION–2ND POWERS, EULER FOUR-SQUARE IDENTITY

References

Nagell, T. *Introduction to Number Theory.* New York: Wiley, 1951.

Lebesgue Integrable

A real-valued function f defined on the reals $\mathbb{R}$ is called Lebesgue integrable if there exists a SEQUENCE of STEP FUNCTIONS $\{f_n\}$ such that the following two conditions are satisfied:

1. $\sum_{n=1}^{\infty} \int |f_n| < \infty$,
2. $f(x) = \sum_{n=1}^{\infty} f_n(x)$ for every $x \in \mathbb{R}$ such that $\sum_{n=1}^{\infty} \int |f_n| < \infty$.

Here, the above integral denotes the ordinary RIEMANN INTEGRAL. Note that this definition avoids explicit use of the LEBESGUE MEASURE.

See also INTEGRAL, LEBESGUE INTEGRAL, RIEMANN INTEGRAL, STEP FUNCTION

Lebesgue Integral

The LEBESGUE INTEGRAL is defined in terms of upper and lower bounds using the LEBESGUE MEASURE of a SET. It uses a LEBESGUE SUM $S_n = \eta_i \mu(E_i)$ where η_i is the value of the function in subinterval i, and $\mu(E_i)$ is the LEBESGUE MEASURE of the SET E_i of points for which values are approximately η_i. This type of integral covers a wider class of functions than does the RIEMANN INTEGRAL.

The Lebesgue integral of a function f over a MEASURE SPACE X is written

$$\int_X f,$$

or sometimes

$$\int_X f \, d\mu$$

to emphasize that the integral is taken with respect to the MEASURE μ.

See also A-INTEGRABLE, COMPLETE FUNCTIONS, INTEGRAL, MEASURE, MEASURE SPACE

References

Kestelman, H. "Lebesgue Integral of a Non-Negative Function" and "Lebesgue Integrals of Functions Which Are Sometimes Negative." Chs. 5– in *Modern Theories of Integration, 2nd rev. ed.* New York: Dover, pp. 113–60, 1960.

Papoulis, A. *Probability, Random Variables, and Stochastic Processes, 2nd ed.* New York: McGraw-Hill, p. 141, 1984.

Lebesgue Measurability Problem

A problem related to the CONTINUUM HYPOTHESIS which was solved by Solovay (1970) using the INACCESSIBLE CARDINALS AXIOM. It has been proven by Shelah and Woodin (1990) that use of this AXIOM is essential to the proof.

See also CONTINUUM HYPOTHESIS, INACCESSIBLE CARDINALS AXIOM, LEBESGUE MEASURE

References

Shelah, S. and Woodin, H. "Large Cardinals Imply that Every Reasonable Definable Set of Reals is Lebesgue Measurable." *Israel J. Math.* **70**, 381–94, 1990.

Solovay, R. M. "A Model of Set-Theory in which Every Set of Reals is Lebesgue Measurable." *Ann. Math.* **92**, 1–6, 1970.

Lebesgue Measure

An extension of the classical notions of length and AREA to more complicated sets. Given an open set $S \equiv \Sigma_k(a_k, b_k)$ containing DISJOINT intervals,

$$\mu_L(S) \equiv \sum_k (b_k - a_k).$$

Given a CLOSED SET $S' \equiv [a, b] - \Sigma_k(a_k, b_k)$,

$$\mu_L(S') \equiv (b - a) - \sum_k (b_k - a_k).$$

A unit LINE SEGMENT has Lebesgue measure 1; the CANTOR SET has Lebesgue measure 0. The MINKOWSKI MEASURE of a bounded, CLOSED SET is the same as its Lebesgue measure (Ko 1995).

See also CANTOR SET, MEASURE, RIESZ-FISCHER THEOREM

References

Croft, H. T.; Falconer, K. J.; and Guy, R. K. *Unsolved Problems in Geometry.* New York: Springer-Verlag, p. 4, 1991.
Kestelman, H. "Lebesgue Measure." Ch. 3 in *Modern Theories of Integration, 2nd rev. ed.* New York: Dover, pp. 67–1, 1960.
Ko, K.-I. "A Polynomial-Time Computable Curve whose Interior has a Nonrecursive Measure." *Theoret. Comput. Sci.* **145**, 241–70, 1995.

Lebesgue Minimal Problem

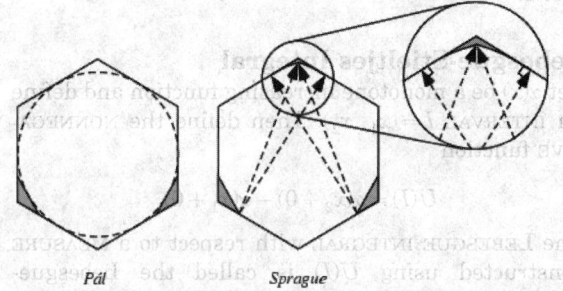

Pál   Sprague

Find the plane LAMINA of least AREA A which is capable of covering any plane figure of unit GENERALIZED DIAMETER. A UNIT CIRCLE is too small, but a HEXAGON circumscribed on the UNIT CIRCLE is larger than necessary. Pál (1920) showed that the hexagon can be reduced by cutting off two EQUILATERAL TRIANGLES on the corners of the hexagon which are tangent to the hexagon's INCIRCLE (Wells 1991; left figure above). Sprague subsequently demonstrated that an additional small curvilinear region could be removed (Wells 1991; right figure above). These

constructions give upper bounds.

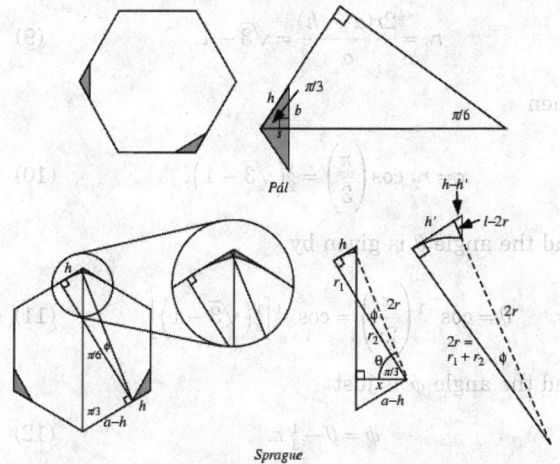

Pál

Sprague

The HEXAGON having INRADIUS $r = 1/2$ (giving a DIAMETER of 1) has side length

$$a = 2r \tan\left(\frac{\pi}{n}\right) = \tfrac{1}{3}\sqrt{3}, \tag{1}$$

and the area of this HEXAGON is

$$A_1 = nr^2 \tan\left(\frac{\pi}{n}\right) = \tfrac{1}{2}\sqrt{3} \approx 0.866025. \tag{2}$$

In the above figure, the SAGITTA is given by

$$s = r \tan\left(\frac{\pi}{n}\right)\tan\left(\frac{\pi}{2n}\right) = \tfrac{1}{6}\left(2\sqrt{3} - 3\right), \tag{3}$$

and the other distances by

$$b = s \tan\left(\frac{\pi}{3}\right) = \sqrt{3}s \tag{4}$$

$$h = \sqrt{s^2 + b^2} = 2s, \tag{5}$$

so the area of one of the equilateral triangles removed in Pál's reduction is

$$A_\Delta = bs = \sqrt{3}s^2 = \tfrac{1}{12}\left(7\sqrt{3} - 12\right) \approx 0.0773505, \tag{6}$$

so the area left after removing two of these triangles is

$$A_2 \equiv A_1 - 2A_\Delta = \tfrac{2}{3}\left(3 - \sqrt{3}\right) \approx 0.845299. \tag{7}$$

Computing the area of the region removed in Sprague's construction is more involved. First, use similar triangles

$$\frac{a - h}{h} = \frac{r_2}{r_1} \tag{8}$$

together with $r_1 + r_2 = r$ to obtain

$$r_2 = \frac{2r(a - h)}{a} = \sqrt{3} - 1. \tag{9}$$

Then

$$x = r_2 \cos\left(\frac{\pi}{3}\right) = \tfrac{1}{2}\left(\sqrt{3} - 1\right), \tag{10}$$

and the angle θ is given by

$$\theta = \cos^{-1}\left(\frac{x}{2r}\right) = \cos^{-1}\left[\tfrac{1}{2}\left(\sqrt{3} - 1\right)\right], \tag{11}$$

and the angle ϕ is just

$$\phi = \theta - \tfrac{1}{3}\pi. \tag{12}$$

The distance h' is

$$h' = 2r \tan \phi \tag{13}$$

$$l = 2r \sec \phi, \tag{14}$$

and the area between the triangle and sector is

$$dA_3^{(1)} = rh - \tfrac{1}{2}(2r)^2\phi = 2r^2(\tan \phi - \phi) = \tfrac{1}{2}(\tan \phi - \phi)$$

$$\approx 0.000554738. \tag{15}$$

The area of the small triangle is

$$dA_3^{(2)} = \tfrac{1}{2}(l - 2r)(h - h')$$

$$= \tfrac{1}{6}(\sec \phi - 1)(2\sqrt{3} - 3 - 3\tan \phi)$$

$$\approx 0.0000264307, \tag{16}$$

so the total area remaining is

$$A_3 = A_2 - 2(dA_3^{(1)} - dA_3^{(2)}) = 0.844137. \tag{17}$$

It is also known that a lower bound for the AREA is given by

$$A > \tfrac{1}{8}\pi + \tfrac{1}{4}\sqrt{3} \approx 0.825712 \tag{18}$$

(Ogilvy 1990).

See also AREA, BORSUK'S CONJECTURE, GENERALIZED DIAMETER, KAKEYA NEEDLE PROBLEM

References

Ball, W. W. R. and Coxeter, H. S. M. *Mathematical Recreations and Essays, 13th ed.* New York: Dover, p. 99, 1987.
Coxeter, H. S. M. "Lebesgue's Minimal Problem." *Eureka* **21**, 13, 1958.
Grünbaum, B. "Borsuk's Problem and Related Questions." *Proc. Sympos. Pure Math, Vol. 7.* Providence, RI: Amer. Math. Soc., pp. 271–84, 1963.
Kakeya, S. "Some Problems on Maxima and Minima Regarding Ovals." *Sci. Reports Tôhoku Imperial Univ., Ser. 1 (Math., Phys., Chem.)* **6**, 71–8, 1917.
Ogilvy, C. S. *Tomorrow's Math: Unsolved Problems for the Amateur, 2nd ed.* New York: Oxford University Press, 1972.
Ogilvy, C. S. *Excursions in Geometry.* New York: Dover, pp. 142–44, 1990.
Pál, J. "Ueber ein elementares Variationsproblem." *Det Kgl. Danske videnkabernes selskab, Math.-fys. meddelelser* **3**, Nr. 2, 1–5, 1920.
Wells, D. *The Penguin Dictionary of Curious and Interesting Geometry.* London: Penguin, p. 138, 1991.
Yaglom, I. M. and Boltyanskii, V. G. *Convex Figures.* New York: Holt, Rinehart, & Winston, pp. 18 and 100, 1961.

Lebesgue-Radon Integral

LEBESGUE-STIELTJES INTEGRAL

Lebesgue's Dominated Convergence Theorem

Suppose that $\{f_n\}$ is a sequence of MEASURABLE FUNCTIONS, that $f_n \to f$, as $n \to \infty$, and that $|f_n| \le g$ for all n, where g is integrable. Then f is integrable, and

$$\int f \, d\mu = \lim_{n \to \infty} \int f_n \, d\mu.$$

See also ALMOST EVERYWHERE CONVERGENCE, MEASURE THEORY, POINTWISE CONVERGENCE

References

Browder, A. *Mathematical Analysis: An Introduction.* New York: Springer-Verlag, 1996.

Lebesgue Singular Integrals

$$\mathscr{U}_n(f) = \int_a^b f(x)K_n(x) \, dx,$$

where $\{K_n(x)\}$ is a SEQUENCE of CONTINUOUS FUNCTIONS.

Lebesgue-Stieltjes Integral

Let $\alpha(x)$ be a monotone increasing function and define an INTERVAL $I = (x_1, x_2)$. Then define the NONNEGATIVE function

$$U(I) = \alpha(x_2 + 0) - \alpha(x_1 + 0).$$

The LEBESGUE INTEGRAL with respect to a MEASURE constructed using $U(I)$ is called the Lebesgue-Stieltjes integral, or sometimes the LEBESGUE-RADON INTEGRAL.

References

Iyanaga, S. and Kawada, Y. (Eds.). *Encyclopedic Dictionary of Mathematics.* Cambridge, MA: MIT Press, p. 326, 1980.

Lebesgue Sum

$$S_n = \sum_i \eta_i \mu(E_i),$$

where $\mu(E_i)$ is the MEASURE of the SET E_i of points on the x-AXIS for which $f(x) \approx \eta_i$.

Le Cam's Identity

Let S_n be the sum of n random variates X_i with a BERNOULLI DISTRIBUTION with $P(X_i = 1) = p_i$. Then

$$\sum_{k=0}^{\infty} \left| P(S_n = k) - \frac{e^{-\lambda} \lambda^k}{k!} \right| < 2 \sum_{i=1}^{n} p_i^2,$$

where

$$\lambda \equiv \sum_{i=1}^{n} p_i.$$

See also BERNOULLI DISTRIBUTION

References

Cox, D. A. "Introduction to Fermat's Last Theorem." *Amer. Math. Monthly* **101**, 3–4, 1994.

Leech Lattice

A 24-D Euclidean lattice. An AUTOMORPHISM of the Leech lattice modulo a center of two leads to the CONWAY GROUP Co_1. Stabilization of the 1- and 2-D sublattices leads to the CONWAY GROUPS Co_2 and Co_3, the HIGMAN-SIMS GROUP HS and the McLAUGHLIN GROUP McL.

The Leech lattice appears to be the densest HYPERSPHERE PACKING in 24-D, and results in each HYPERSPHERE touching 195,560 others. The number of vectors with norm n in the Leech lattice (i.e., its "theta series"rpar; is given by

$$\theta(n) = \tfrac{65520}{691}[\sigma_{11}(n) - \tau(n)], \tag{1}$$

where σ_{11} is the DIVISOR FUNCTION giving the sum of the 11th powers of the DIVISORS of n and $\tau(n)$ is the TAU FUNCTION (Conway and Sloane 1993, p. 135). The first few values for $n = 1, 2, \ldots$ are 0, 196560, 16773120, 398034000, ... (Sloane's A008408). This is an immediate consequence of the theta function for Leech's lattice being a weight 12 MODULAR FORM and having no vectors of norm two. $\theta(n)$ has the generating function

$$f(q) = [E_2(q)]^3 - 720q^2 \prod_{m=1}^{\infty} (1 - q^{2m})^{24} \tag{2}$$

$$= \left(1 + 240 \sum_{m=1}^{\infty} \sigma_3(m) q^{2m} \right)^3 - 720q^2 \prod_{m=1}^{\infty} (1 - q^{2m})^{24} \tag{3}$$

$$1 + 196560q^4 + 16773120q^6 + 3980034000q^8 + \cdots, \tag{4}$$

where $E_2(q)$ is the RAMANUJAN-EISENSTEIN SERIES

which is the theta series of the E_8 lattice (Sloane's A004009).

See also BARNES-WALL LATTICE, CONWAY GROUPS, COXETER-TODD LATTICE, EISENSTEIN SERIES, HIGMAN-SIMS GROUP, HYPERSPHERE, HYPERSPHERE PACKING, KISSING NUMBER, McLAUGHLIN GROUP, TAU FUNCTION

References

Conway, J. H. and Sloane, N. J. A. "The 24-Dimensional Leech Lattice Λ_{24}," "A Characterization of the Leech Lattice," "The Covering Radius of the Leech Lattice," "Twenty-Three Constructions for the Leech Lattice," "The Cellular of the Leech Lattice," "Lorentzian Forms for the Leech Lattice." §4.11, Ch. 12, and Chs. 23–6 in *Sphere Packings, Lattices, and Groups, 2nd ed.* New York: Springer-Verlag, pp. 131–35, 331–36, and 478–26, 1993.
Leech, J. "Notes on Sphere Packings." *Canad. J. Math.* **19**, 251–67, 1967.
Sloane, N. J. A. Sequences A004009/M5416 and A008408 in "An On-Line Version of the Encyclopedia of Integer Sequences." http://www.research.att.com/~njas/sequences/eisonline.html.
Wilson, R. A. "Vector Stabilizers and Subgroups of Leech Lattice Groups." *J. Algebra* **127**, 387–08, 1989.

Lefschetz Number

If K is a finite complex and $h : |K| \to |K|$ is a continuous map, then

$$\Lambda(h) = \sum (-1)^p \mathrm{Tr}(h_*, H_p(K)/T_p(K))$$

is the Lefschetz number of the map h.

See also EULER NUMBER (FINITE COMPLEX)

References

Munkres, J. R. *Elements of Algebraic Topology.* Perseus Press, p. 125, 1993.

Lefschetz Theorems

Each DOUBLE POINT assigned to an irreducible ALGEBRAIC CURVE whose GENUS is NONNEGATIVE imposes exactly one condition.

See also HARD LEFSCHETZ THEOREM

References

Coolidge, J. L. *A Treatise on Algebraic Plane Curves.* New York: Dover, p. 104, 1959.

Lefshetz Fixed Point Formula

Let K be a finite complex, let $h : |K| \to |K|$ be a continuous map. If $\Lambda(h) \neq 0$, then h has a fixed point.

See also LEFSHETZ TRACE FORMULA

References

Munkres, J. R. "Application: The Lefschetz Fixed-Point Theorem." §22 in *Elements of Algebraic Topology.* Perseus Press, pp. 121–28, 1993.

Lefshetz Trace Formula

A formula which counts the number of FIXED POINTS for a topological transformation.

Left Coset

Consider a countable SUBGROUP H with ELEMENTS h_i and an element x not in H, then xh_i for $i = 1, 2, \ldots$ are the left cosets of the SUBGROUP H with respect to x.

See also COSET, RIGHT COSET

Left Half-Plane

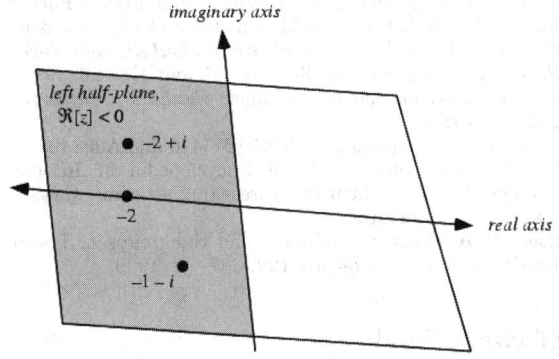

The portion of the COMPLEX PLANE $z = x + iy$ with REAL PART $\Re[z] < 0$.

See also COMPLEX PLANE, LOWER HALF-PLANE, RIGHT HALF-PLANE, UPPER HALF-PLANE

Left-Handed Coordinate System

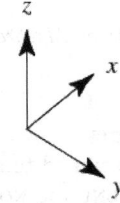

left-handed coordinate system

A three-dimensional COORDINATE SYSTEM in which the axes do not satisfy the RIGHT-HAND RULE.

See also CROSS PRODUCT, RIGHT-HAND RULE, RIGHT-HANDED COORDINATE SYSTEM

Leg

A leg of a TRIANGLE is one of its sides. For a RIGHT TRIANGLE, the term "leg" generally refers to a side other than the one opposite the RIGHT ANGLE, which is termed the HYPOTENUSE.

See also HYPOTENUSE, TRIANGLE

Legendre Addition Theorem

SPHERICAL HARMONIC ADDITION THEOREM

Legendre Differential Equation

The second-order ORDINARY DIFFERENTIAL EQUATION

$$(1 - x^2)\frac{d^2y}{dx^2} - 2x\frac{dy}{dx} + l(l+1)y = 0, \tag{1}$$

which can be rewritten

$$\frac{d}{dx}\left[(1 - x^2)\frac{dy}{dx}\right] + l(l+1)y = 0. \tag{2}$$

The above form is a special case of the associated Legendre differential equation with $m = 0$. The Legendre differential equation has REGULAR SINGULAR POINTS at -1, 1, and ∞.

If the variable x is replaced by $\cos\theta$, then the Legendre differential equation becomes

$$\frac{d^2y}{d\theta^2} + \frac{\cos\theta}{\sin\theta}\frac{dy}{d\theta} + l(l+1)y = 0, \tag{3}$$

as is derived below for the associated Legendre differential equation with $m = 0$.

Since the Legendre differential equation is a second-order ORDINARY DIFFERENTIAL EQUATION, it has two linearly independent solutions. A solution $P_l(x)$ which is regular at the origin is called a LEGENDRE FUNCTION OF THE FIRST KIND, while a solution $Q_l(x)$ which is singular at the origin is called a LEGENDRE FUNCTION OF THE SECOND KIND. If l is an integer, the function of the first kind reduces to a polynomial known as the LEGENDRE POLYNOMIAL.

The Legendre differential equation can be solved using the standard method of making a series expansion,

$$y = \sum_{n=0}^{\infty} a_n x^n \tag{4}$$

$$y' = \sum_{n=0}^{\infty} n a_n x^{n-1} \tag{5}$$

$$y'' = \sum_{n=0}^{\infty} n(n-1) a_n x^{n-2}. \tag{6}$$

Plugging in,

$$(1 - x^2)\sum_{n=0}^{\infty} n(n-1)a_n x^{n-2} - 2x\sum_{n=0}^{\infty} n a_n x^{n-1}$$

$$+ l(l+1)\sum_{n=0}^{\infty} a_n x^n = 0 \tag{7}$$

$$\sum_{n=0}^{\infty} n(n-1)a_n x^{n-2} - \sum_{n=0}^{\infty} n(n-1)a_n x^n$$

$$-2x \sum_{n=0}^{\infty} n a_n x^{n-1} + l(l+1) \sum_{n=0}^{\infty} a_n x^n = 0 \qquad (8)$$

$$\sum_{n=0}^{\infty} n(n-1) a_n x^{n-2} - \sum_{n=0}^{\infty} n(n-1) a_n x^n$$

$$-2 \sum_{n=0}^{\infty} n a_n x^n + l(l+1) \sum_{n=0}^{\infty} a_n x^n = 0 \qquad (9)$$

$$\sum_{n=0}^{\infty} (n+2)(n+1) a_{n+2} x^n - \sum_{n=0}^{\infty} n(n-1) a_n x^n$$

$$-2 \sum_{n=0}^{\infty} n a_n x^n + l(l+1) \sum_{n=0}^{\infty} a_n x^n = 0 \qquad (10)$$

$$\sum_{n=0}^{\infty} \{(n+1)(n+2) a_{n+2} + [-n(n-1)$$

$$-2n + l(l+1)] a_n \} = 0, \qquad (11)$$

so each term must vanish and

$$(n+1)(n+2) a_{n+2} + [-n(n+1) + l(l+1)] a_n = 0 \qquad (12)$$

$$a_{n+2} = \frac{n(n+1) - l(l+1)}{(n+1)(n+2)} a_n$$

$$= -\frac{[l+(n+1)](l-n)}{(n+1)(n+2)} a_n. \qquad (13)$$

Therefore,

$$a_2 = -\frac{l(l+1)}{1 \cdot 2} a_0 \qquad (14)$$

$$a_4 = -\frac{(l-2)(l+3)}{3 \cdot 4} a_2$$

$$= (-1)^2 \frac{[(l-2)l][(l+1)(l+3)]}{1 \cdot 2 \cdot 3 \cdot 4} a_0 \qquad (15)$$

$$a_6 = -\frac{(l-4)(l+5)}{5 \cdot 6} a_4$$

$$= (-1)^3 \frac{[(l-4)(l-2)l][(l+1)(l+3)(l+5)]}{1 \cdot 2 \cdot 3 \cdot 4 \cdot 5 \cdot 6} a_0; \qquad (16)$$

so the EVEN solution is

$$y_1(x) = 1 + \sum_{n=1}^{\infty} (-1)^n$$

$$\times \frac{[(l-2n+2)\ldots(l-2)l][(l+1)(l+3)\ldots(l+2n-1)]}{(2n)!} x^{2n}. \qquad (17)$$

Similarly, the ODD solution is

$$y_2(x) = x + \sum_{n=1}^{\infty} (-1)^n$$

$$\times \frac{[(l-2n+1)\cdots(l-3)(l-1)][(l+2)(l+4)\cdots(l+2n)]}{(2n+1)!} x^{2m+1}. \qquad (18)$$

If l is an EVEN INTEGER, the series $y_1(x)$ reduces to a POLYNOMIAL of degree l with only EVEN POWERS of x and the series $y_2(x)$ diverges. If l is an ODD INTEGER, the series $y_2(x)$ reduces to a POLYNOMIAL of degree l with only ODD POWERS of x and the series $y_1(x)$ diverges. The general solution for an INTEGER l is then given by the LEGENDRE POLYNOMIALS

$$P_n(x) = c_n \begin{cases} y_1(x) & \text{for } l \text{ even} \\ y_2(x) & \text{for } l \text{ odd}, \end{cases} \qquad (19)$$

where c_n is chosen so as to yield the normalization $P_n(1) = 1$.

The *associated* Legendre differential equation is

$$\frac{d}{dx}\left[(1-x^2)\frac{dy}{dx}\right] + \left[l(l+1) - \frac{m^2}{1-x^2}\right] y = 0, \qquad (20)$$

which can be written

$$(1-x^2)\frac{d^2y}{dx^2} - 2x \frac{dy}{dx} + \left[l(l+1) - \frac{m^2}{1-x^2}\right] y = 0 \qquad (21)$$

(Abramowitz and Stegun 1972; Zwillinger 1997, p. 124). The solutions $P_l^m(x)$ to this equation are called the associated Legendre polynomials (if l is an integer), or associated Legendre functions of the first kind (if l is not an integer). The complete solution is

$$y = C_1 P_l^m(x) + C_2 Q_l^m(x), \qquad (22)$$

where $Q_l^m(x)$ is a LEGENDRE FUNCTION OF THE SECOND KIND.

The associated Legendre differential equation is often written in a form obtained by setting $x \equiv \cos\theta$. Using the identities

$$\frac{dy}{dx} = \frac{dy}{d(\cos\theta)} = -\frac{1}{\sin\theta}\frac{dy}{d\theta} \qquad (23)$$

$$x\frac{dy}{dx} = -\frac{\cos\theta}{\sin\theta}\frac{dy}{d\theta}, \qquad (24)$$

$$\frac{d^2y}{dx^2} = \frac{1}{\sin\theta}\frac{d}{d\theta}\left(\frac{1}{\sin\theta}\frac{dy}{d\theta}\right)$$

$$= \frac{1}{\sin\theta}\left(\frac{-\cos\theta}{\sin^2\theta}\right)\frac{dy}{d\theta} + \frac{1}{\sin^2\theta}\frac{d^2y}{d\theta^2}, \qquad (25)$$

and

$$1 - x^2 = 1 - \cos^2\theta = \sin^2\theta, \qquad (26)$$

therefore gives

$$(1-x^2)\frac{d^2y}{dx^2} = \sin^2\theta \frac{1}{\sin\theta}\left(\frac{-\cos\theta}{\sin^2\theta}\right)\frac{dy}{d\theta} + \frac{1}{\sin^2\theta}\frac{d^2y}{d\theta^2}$$

$$= \frac{d^2y}{d\theta^2} - \frac{\cos\theta}{\sin\theta}\frac{dy}{d\theta}. \qquad (27)$$

Plugging (23) into (27) and the result back into (21) gives

$$\left(\frac{d^2y}{d\theta^2} - \frac{\cos\theta}{\sin\theta}\frac{dy}{d\theta}\right) + 2\frac{\cos\theta}{\sin\theta}\frac{dy}{d\theta} + \left[l(l+1) - \frac{m^2}{\sin^2\theta}\right]y$$
$$= 0 \qquad (28)$$

$$\frac{d^2y}{d\theta^2} + \frac{\cos\theta}{\sin\theta}\frac{dy}{d\theta} + \left[l(l+1) - \frac{m^2}{\sin^2\theta}\right]y = 0. \qquad (29)$$

Moon and Spencer (1961, p. 155) call

$$(1-x^2)y'' - 2xy' - \left[k^2a^2(x^2-1) - p(p+1) - \frac{q^2}{x^2-1}\right]y$$
$$= 0 \qquad (30)$$

The Legendre wave function (Zwillinger 1997, p.124).

See also Legendre Function of the First Kind, Legendre Function of the Second Kind, Legendre Polynomial

References

Abramowitz, M. and Stegun, C. A. (Eds.). *Handbook of Mathematical Functions with Formulas, Graphs, and Mathematical Tables, 9th printing.* New York: Dover, p. 332, 1972.
Moon, P. and Spencer, D. E. *Field Theory for Engineers.* New York: Van Nostrand, 1961.
Zwillinger, D. (Ed.). *CRC Standard Mathematical Tables and Formulae.* Boca Raton, FL: CRC Press, 1995.

Legendre Duplication Formula

Gamma functions of argument $2z$ can be expressed in terms of Gamma functions of smaller arguments. From the definition of the Beta function,

$$B(m, n) = \frac{\Gamma(m)\Gamma(n)}{\Gamma(m+n)} = \int_0^1 u^{m-1}(1-u)^{n-1}\, du. \qquad (1)$$

Now, let $m = n \equiv z$, then

$$\frac{\Gamma(z)\Gamma(z)}{\Gamma(2z)} = \int_0^1 u^{z-1}(1-u)^{z-1}\, du \qquad (2)$$

and $u \equiv (1+x)/2$, so $du = dx/2$ and

$$\frac{\Gamma(z)\Gamma(z)}{\Gamma(2z)} = \int_0^1 \left(\frac{1+x}{2}\right)^{z-1}\left(1 - \frac{1+x}{2}\right)^{z-1}(\tfrac{1}{2}\, dx)$$

$$= \frac{1}{2}\int_0^1 \left(\frac{1+x}{2}\right)^{z-1}\left(\frac{1+x}{2}\right)^{z-1}\, dx$$

$$= \frac{1}{2^{1+2(z-1)}}\int_0^1 (1-x^2)^{z-1}\, dx$$

$$= 2^{1-2x}\int_0^1 (1-x^2)^{z-1}\, dx. \qquad (3)$$

Now, use the Beta function identity

$$B(m, n) = 2\int_0^1 x^{2z-1}(1-x^2)^{z-1}\, dx \qquad (4)$$

to write the above as

$$\frac{\Gamma(z)\Gamma(z)}{\Gamma(2z)} = 2^{1-2z}B(\tfrac{1}{2}, z) = 2^{1-2z}\frac{\Gamma(\tfrac{1}{2})\Gamma(z)}{\Gamma(z+\tfrac{1}{2})}. \qquad (5)$$

Solving for $\Gamma(2x)$,

$$\Gamma(2z) = \frac{\Gamma(z)\Gamma(z+\tfrac{1}{2})2^{2z-1}}{\Gamma(\tfrac{1}{2})} = \frac{\Gamma(z)\Gamma(z+\tfrac{1}{2})2^{2z-1}}{\sqrt{\pi}}$$

$$= (2\pi)^{-1/2}2^{2z-1/2}\Gamma(z)\Gamma(z+\tfrac{1}{2}), \qquad (6)$$

since $\Gamma(\tfrac{1}{2}) = \sqrt{\pi}$.

See also Gamma Function, Gauss Multiplication Formula

References

Abramowitz, M. and Stegun, C. A. (Eds.). *Handbook of Mathematical Functions with Formulas, Graphs, and Mathematical Tables, 9th printing.* New York: Dover, p. 256, 1972.
Arfken, G. *Mathematical Methods for Physicists, 3rd ed.* Orlando, FL: Academic Press, pp. 561–62, 1985.
Erdélyi, A.; Magnus, W.; Oberhettinger, F.; and Tricomi, F. G. *Higher Transcendental Functions, Vol. 1.* New York: Krieger, p. 5, 1981.
Morse, P. M. and Feshbach, H. *Methods of Theoretical Physics, Part I.* New York: McGraw-Hill, pp. 424–25, 1953.

Legendre Function of the First Kind

The (associated) Legendre function of the first kind $P_n^m(z)$ is the solution to the Legendre differential equation which is regular at the origin. For m, n integers and z real, the Legendre function of the first kind simplifies to a polynomial, called the Legendre polynomial. The associated Legendre function of first kind is given by the *Mathematica* command LegendreP[n, m, z], and the unassociated function by LegendreP[n, z].

See also Legendre Differential Equation, Legendre Function of the Second Kind, Legendre Polynomial

Legendre Function of the Second Kind

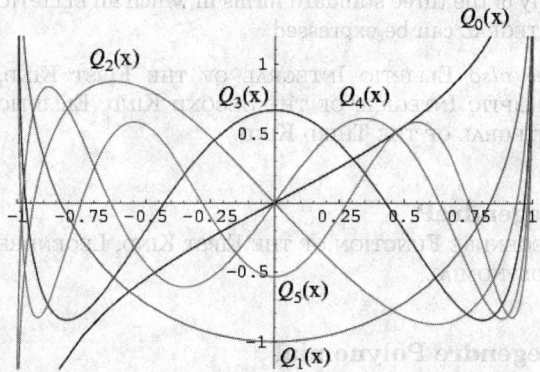

The second solution $Q_1(x)$ to the LEGENDRE DIFFER-ENTIAL EQUATION. The Legendre functions of the second kind satisfy the same RECURRENCE RELATION as the LEGENDRE POLYNOMIALS. The Legendre functions of the second kind are implemented in *Mathematica* as LegendreQ[l, x]. The first few are

$$Q_0(x) = \frac{1}{2} \ln\left(\frac{1+x}{1-x}\right)$$

$$Q_1(x) = \frac{x}{2} \ln\left(\frac{1+x}{1-x}\right) - 1$$

$$Q_2(x) = \frac{3x^2 - 1}{4} \ln\left(\frac{1+x}{1-x}\right) - \frac{3x}{2}$$

$$Q_3(x) = \frac{5x^3 - 3x}{4} \ln\left(\frac{1+x}{1-x}\right) - \frac{5x^2}{2} + \frac{2}{3}.$$

The associated Legendre functions of the second kind $Q_l^m(x)$ are the second solution to the associated Legendre differential equation, and are implemented in *Mathematica* as LegendreQ[l, m, x] $Q_v^\mu(x)$ has DERIVATIVE about 0 of

$$\left[\frac{dQ_v^\mu(x)}{dx}\right]_{x=0} = \frac{2^\mu \sqrt{\pi} \cos[\frac{1}{2}\pi(v+\mu)]\Gamma(\frac{1}{2}v + \frac{1}{2}\mu + 1)}{\Gamma(\frac{1}{2}v - \frac{1}{2}\mu + \frac{1}{2})}$$

(Abramowitz and Stegun 1972, p. 334). The LOGA-RITHMIC DERIVATIVE is

$$\left[\frac{d \ln Q_\lambda^\mu(z)}{dz}\right]_{z=0}$$

$$= 2\exp\{\tfrac{1}{2}\pi i \operatorname{sgn}(\Im[z])\} \frac{[\frac{1}{2}(\lambda + \mu)]![\frac{1}{2}(\lambda - \mu)]!}{[\frac{1}{2}(\lambda + \mu - 1)]![\frac{1}{2}(\lambda - \mu - 1)]!}$$

(Binney and Tremaine 1987, p. 654).

See also LEGENDRE DIFFERENTIAL EQUATION, LEGENDRE FUNCTION OF THE FIRST KIND, LEGENDRE POLYNOMIAL

References

Abramowitz, M. and Stegun, C. A. (Eds.). "Legendre Functions." Ch. 8 in *Handbook of Mathematical Functions with Formulas, Graphs, and Mathematical Tables, 9th printing.* New York: Dover, pp. 331–39, 1972.

Arfken, G. "Legendre Functions of the Second Kind, $Q_n(x)$." *Mathematical Methods for Physicists, 3rd ed.* Orlando, FL: Academic Press, pp. 701–07, 1985.

Binney, J. and Tremaine, S. "Associated Legendre Functions." Appendix 5 in *Galactic Dynamics.* Princeton, NJ: Princeton University Press, pp. 654–55, 1987.

Morse, P. M. and Feshbach, H. *Methods of Theoretical Physics, Part I.* New York: McGraw-Hill, pp. 597–00, 1953.

Snow, C. *Hypergeometric and Legendre Functions with Applications to Integral Equations of Potential Theory.* Washington, DC: U. S. Government Printing Office, 1952.

Spanier, J. and Oldham, K. B. "The Legendre Functions $P_v(x)$ and $Q_v(x)$." Ch. 59 in *An Atlas of Functions.* Washington, DC: Hemisphere, pp. 581–97, 1987.

Legendre-Gauss Quadrature

Also called "the" GAUSSIAN QUADRATURE or LEGENDRE QUADRATURE. A GAUSSIAN QUADRATURE over the interval $[-1, 1]$ with WEIGHTING FUNCTION $W(x) = 1$. The ABSCISSAS for quadrature order n are given by the roots of the LEGENDRE POLYNOMIALS $P_n(x)$, which occur symmetrically about 0. The weights are

$$w_i = -\frac{A_{n+1}\gamma_n}{A_n P_n'(x_i)P_{n+1}(x_i)} = \frac{A_n}{A_{n-1}} \frac{\gamma_{n-1}}{P_{n-1}(x_i)P_n'(x_i)}, \quad (1)$$

where A_n is the COEFFICIENT of x^n in $P_n(x)$. For LEGENDRE POLYNOMIALS,

$$A_n = \frac{(2n)!}{2^n(n!)^2}, \quad (2)$$

so

$$\frac{A_{n+1}}{A_n} = \frac{[2(n+1)]!}{2^{n+1}[(n+1)!]^2} \frac{2^n(n!)^2}{(2n)!}$$

$$= \frac{(2n+1)(2n+2)}{2(n+1)^2} = \frac{2n+1}{n+1}. \quad (3)$$

Additionally,

$$\gamma_n = \frac{2}{2n+1}, \quad (4)$$

so

$$w_i = -\frac{2}{(n+1)P_{n+1}(x_i)P_n'(x_i)} = \frac{2}{nP_{n-1}(x_i)P_n'(x_i)}. \quad (5)$$

Using the RECURRENCE RELATION

$$(1 - x^2)P_n'(x) = nxP_n(x) + nP_{n-1}(x)$$

$$= (n+1)xP_n(x) - (n+1)P_{n+1}(x) \quad (6)$$

gives

$$w_i = -\frac{2}{(1 - x^2)[P_n'(x_i)]^2} = \frac{2(1 - x_i^2)}{(n + 1)^2[P_{n+1}(x_i)]^2}. \quad (7)$$

The error term is

$$E = \frac{2^{2n+1}(n!)^4}{(2n + 1)[(2n)!]^3} f^{(2n)}(\xi). \quad (8)$$

Beyer (1987) gives a table of ABSCISSAS and weights up to $n = 16$, and Chandrasekhar (1960) up to $n = 8$ for n EVEN.

n	x_i	w_i
2	± 0.57735	1.000000
3	0	0.888889
	± 0.774597	0.555556
4	± 0.339981	0.652145
	± 0.861136	0.347855
5	0	0.568889
	± 0.538469	0.478629
	± 0.90618	0.236927

The ABSCISSAS and weights can be computed analytically for small n.

n	x_i	w_i
2	$\pm\frac{1}{3}\sqrt{3}$	1
3	0	$\frac{8}{9}$
	$\pm\frac{1}{5}\sqrt{15}$	$\frac{5}{9}$
4	$\pm\frac{1}{35}\sqrt{525 - 70\sqrt{30}}$	$\frac{1}{36}(18 + \sqrt{30})$
	$\pm\frac{1}{35}\sqrt{525 + 70\sqrt{30}}$	$\frac{1}{36}(18 - \sqrt{30})$
5	0	$\frac{128}{225}$
	$\pm\frac{1}{21}\sqrt{245 - 14\sqrt{70}}$	$\frac{1}{900}(322 + 13\sqrt{70})$
	$\pm\frac{1}{21}\sqrt{245 + 14\sqrt{70}}$	$\frac{1}{900}(322 - 13\sqrt{70})$

References
Beyer, W. H. *CRC Standard Mathematical Tables, 28th ed.* Boca Raton, FL: CRC Press, pp. 462–63, 1987.

Chandrasekhar, S. *Radiative Transfer.* New York: Dover, pp. 56–2, 1960.

Hildebrand, F. B. *Introduction to Numerical Analysis.* New York: McGraw-Hill, pp. 323–25, 1956.

Legendre-Jacobi Elliptic Integral

Any of the three standard forms in which an ELLIPTIC INTEGRAL can be expressed.

See also ELLIPTIC INTEGRAL OF THE FIRST KIND, ELLIPTIC INTEGRAL OF THE SECOND KIND, ELLIPTIC INTEGRAL OF THE THIRD KIND

LegendreP

LEGENDRE FUNCTION OF THE FIRST KIND, LEGENDRE POLYNOMIAL

Legendre Polynomial

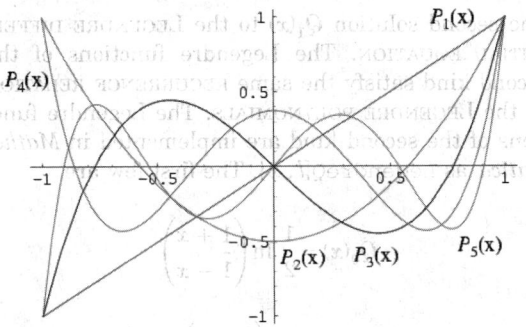

The Legendre polynomials, sometimes called Legendre functions of the first kind, Legendre coefficients, or ZONAL HARMONICS (Whittaker and Watson 1990, p. 302), are solutions to the LEGENDRE DIFFERENTIAL EQUATION. If l is an INTEGER, they are POLYNOMIALS. The Legendre polynomials $P_n(x)$ are illustrated above for $x \in [0, 1]$ and $n = 1, 2, ..., 5$.

The Legendre polynomials are a special case of the ULTRASPHERICAL FUNCTIONS with $\alpha = 1/2$, a special case of the JACOBI POLYNOMIALS $P_n^{(\alpha, \beta)}$ with $\alpha = \beta = 0$, and can be written as a HYPERGEOMETRIC FUNCTION using Murphy's formula

$$P_n(x) = P_n^{(0, 0)}(x) = {}_2F_1(-n, n + 1; 1; \tfrac{1}{2}(1 - x)) \quad (1)$$

(Bailey 1933; Bailey 1935, p. 101; Koekoek and Swarttouw 1998).

The Rodrigues formula provides the GENERATING FUNCTION

$$P_l(x) = \frac{l}{2^l l!} \frac{d^l}{dx^l}(x^2 - 1)^l, \quad (2)$$

which yields upon expansion

$$P_l(x) = \frac{1}{2^l} \sum_{k=0}^{\lfloor l/2 \rfloor} \frac{(-1)^k (2l - 2k)!}{k!(l - k)!(l - 2k)!} x^{l-2k} \quad (3)$$

$$= \frac{1}{2^l} \sum_{k=0}^{\lfloor l/2 \rfloor} (-1)^k \binom{l}{k} \binom{2l - 2k}{l} x^{l-2k} \quad (4)$$

where $\lfloor r \rfloor$ is the FLOOR FUNCTION. Additional sum formulas include

$$P_l(x) = \frac{1}{2^l} \sum_{k=0}^{l} \binom{l}{k}^2 (x-1)^{l-k}(x+1)^k \qquad (5)$$

$$= \sum_{k=0}^{l} \binom{l}{k}\binom{-l-1}{k}\left(\frac{1-x}{2}\right)^k \qquad (6)$$

(Koepf 1998, p. 1). In terms of HYPERGEOMETRIC FUNCTIONS, these can be written

$$P_n(x) = \left(\frac{x-1}{2}\right)^n {}_2F_1(-n, -n; 1; (x+1)/(x-1)) \qquad (7)$$

$$P_n(x) = \binom{2n}{n}\frac{x^n}{2^n} {}_2F_1(-n/2, (1-n)/2; 1/2-n; x^{-2}) \qquad (8)$$

$$P_n(x) = {}_2F_1(-n, n+1; 1; (1-x)/2) \qquad (9)$$

(Koepf 1998, p. 3).

A GENERATING FUNCTION for $P_n(x)$ is given by

$$g(t, x) = (1 - 2xt + t^2)^{-1/2} = \sum_{n=0}^{\infty} P_n(x)t^n. \qquad (10)$$

Take $\partial g/\partial t$,

$$-\frac{1}{2}(1 - 2xt + t^2)^{-3/2}(-2x + 2t) = \sum_{n=0}^{\infty} nP_n(x)t^{n-1}. \qquad (11)$$

Multiply (11) by $2t$,

$$-t(1 - 2xt + t^2)^{-3/2}(-2x + 2t) = \sum_{n=0}^{\infty} 2nP_n(x)t^n \qquad (12)$$

and add (10) and (12),

$$(1 - 2xt + t^2)^{-3/2}[(2xt - 2t^2) + (1 - 2xt + t^2)]$$
$$= \sum_{n=0}^{\infty}(2n+1)P_n(x)t^n \qquad (13)$$

This expansion is useful in some physical problems, including expanding the Heyney-Greenstein phase function and computing the charge distribution on a SPHERE. Another GENERATING FUNCTION is given by

$$\sum_{n=0}^{\infty} \frac{P_n(x)}{n!} z^n = e^{xz} J_0(z\sqrt{1-x^2}), \qquad (14)$$

where $J_0(x)$ is a zeroth order BESSEL FUNCTION OF THE FIRST KIND (Koepf 1998, p. 2).

The Legendre polynomials satisfy the RECURRENCE RELATION

$$(l+1)P_{l+1}(x) - (2l+1)xP_l(x) + lP_{l-1}(x) = 0 \qquad (15)$$

(Koepf 1998, p. 2).

The Legendre polynomials are orthogonal over $(-1, 1)$ with WEIGHTING FUNCTION 1 and satisfy

$$\int_{-1}^{1} P_n(x)P_m(x)\, dx = \frac{2}{2n+1}\delta_{mn}, \qquad (16)$$

where δ_{mn} is the KRONECKER DELTA.

A COMPLEX GENERATING FUNCTION is

$$P_l(x) = \frac{1}{2\pi i}\int (1 - 2zx + z^2)^{-1/2}z^{-l-1}\, dz, \qquad (17)$$

and the Schläfli integral is

$$P_l(x) = \frac{(-1)^l}{2^l}\frac{1}{2\pi i}\int \frac{(1-z^2)^l}{(z-x)^{l+1}}\, dz. \qquad (18)$$

Additional integrals (Byerly 1959, p. 172) include

$$\int_0^1 P_m(x)\, dx$$
$$= \begin{cases} 0 & m \text{ even} \neq 0 \\ (-1)^{(m-1)/2}\dfrac{m!!}{m(m+1)(m-1)!!} & m \text{ odd} \end{cases} \qquad (19)$$

$$\int_0^1 P_m(x)P_n(x)\, dx$$
$$= \begin{cases} 0 & m, n \text{ both even or odd } m \neq n \\ (-1)^{(m+n+1)/2} \\ \quad \times \dfrac{m!\,n!}{2^{m+n+1}(m-n)(m+n+1)(\frac{1}{2}m)!\{[\frac{1}{2}(n-1)]!\}^2} \\ \qquad m \text{ even, } n \text{ odd} \\ \dfrac{1}{2n+1} \\ \qquad m = n. \end{cases} \qquad (20)$$

Integrals with weighting functions x and x^2 are given by

$$\int_{-1}^{1} xP_L(x)P_N(x)\, dx = \begin{cases} \frac{2(L+1)}{(2L+1)(2L+3)} & N = L+1 \\ \frac{2L}{(2L-1)(2L+1)} & N = L-1 \end{cases} \qquad (21)$$

$$\int_{-1}^{1} x^2 P_L(x)P_N(x)\, dx$$
$$= \begin{cases} \frac{2(L+1)(L+2)}{(2L+1)(2L+3)(2L+5)} & N = L+2 \\ \frac{2(L^2+2L-1)}{(2L-1)(2L+1)(2L+3)} & N = L \\ \frac{2L(L-1)}{(2L-3)(2L-1)(2L+1)} & N = L-2 \end{cases} \qquad (22)$$

(Arfken 1985, p. 700). An additional identity is

$$1 - [P_n(x)]^2 = \sum_{v=1}^{n} \frac{1-x^2}{1-x_v^2}\left[\frac{P_n(x)}{P_n'(x_v)(x-x_v)}\right]^2, \qquad (23)$$

where x_v is the vth root of $P_n(x)$ (Szego 1975, p. 348).

The first few Legendre polynomials are

$$P_0(x) = 1$$
$$P_1(x) = x$$
$$P_2(x) = \tfrac{1}{2}(3x^2 - 1)$$
$$P_3(x) = \tfrac{1}{2}(5x^3 - 3x)$$
$$P_4(x) = \tfrac{1}{8}(35x^4 - 30x^2 + 3)$$
$$P_5(x) = \tfrac{1}{8}(63x^5 - 70x^3 + 15x)$$
$$P_6(x) = \tfrac{1}{16}(231x^6 - 315x^4 + 105x^2 - 5).$$

The first few POWERS in terms of Legendre polynomials are

$$x = P_1$$
$$x^2 = \tfrac{1}{3}[P_0(x) + 2P_2(x)]$$
$$x^3 = \tfrac{1}{5}[3P_1(x) + 2P_3(x)]$$
$$x^4 = \tfrac{1}{35}[7P_0(x) + 20P_2(x) + 8P_4(x)]$$
$$x^5 = \tfrac{1}{63}[27P_1(x) + 28P_3(x) + 8P_5(x)]$$
$$x^6 = \tfrac{1}{231}[33P_0(x) + 110P_2(x) + 72P_4(x) + 16P_6(x)].$$

For Legendre polynomials and POWERS up to exponent 12, see Abramowitz and Stegun (1972, p. 798).

The Legendre POLYNOMIALS can also be generated using GRAM-SCHMIDT ORTHONORMALIZATION in the OPEN INTERVAL $(-1, 1)$ with the WEIGHTING FUNCTION 1.

$$P_0(x) = 1 \qquad (24)$$

$$P_1(x) = \left[x - \frac{\int_{-1}^{1} x \, dx}{\int_{-1}^{1} dx} \right] \cdot 1$$

$$= x - \frac{\tfrac{1}{2}[x^2]_{-1}^{1}}{[x]_{-1}^{1}} = x - \frac{\tfrac{1}{2}(1-1)}{1-(-1)} = x \qquad (25)$$

$$P_2(x) = \left[x - \frac{\int_{-1}^{1} x^3 \, dx}{\int_{-1}^{1} x^2 \, dx} \right] - \left[\frac{\int_{-1}^{1} x^2 \, dx}{\int_{-1}^{1} dx} \right] \cdot 1$$

$$= \left[x - \frac{\tfrac{1}{4}[x^4]_{-1}^{1}}{\tfrac{1}{3}[x^3]_{-1}^{1}} \right] x - \frac{\tfrac{1}{3}[x^3]_{-1}^{1}}{[x]_{-1}^{1}} = x^2 - \tfrac{1}{3} \qquad (26)$$

$$P_3(x) = \left[x - \frac{\int_{-1}^{1} x(x^2 - \tfrac{1}{3})^2 \, dx}{\int_{-1}^{1} (x^2 - \tfrac{1}{3})^2 \, dx} \right] (x^2 - \tfrac{1}{3})$$

$$- \left[\frac{\int_{-1}^{1} (x^2 - \tfrac{1}{3})^2 \, dx}{\int_{-1}^{1} x^2 \, dx} \right] x$$

$$= x \left[x^2 - \tfrac{1}{3} - \frac{(\tfrac{1}{5} - \tfrac{2}{9} + \tfrac{1}{9})x}{\tfrac{1}{3}} \right]$$

$$= x^3 - \tfrac{1}{3}x - 3(\tfrac{1}{5} - \tfrac{1}{9})$$

$$= x^3 - x(\tfrac{1}{3} + \tfrac{3}{5} - \tfrac{1}{3}) = x^3 - \tfrac{3}{5}x. \qquad (27)$$

Normalizing so that $P_n(1) = 1$ gives the expected Legendre polynomials.

The "shifted" Legendre polynomials are a set of functions analogous to the Legendre polynomials, but defined on the interval $(0, 1)$. They obey the ORTHOGONALITY relationship

$$\int_0^1 \bar{P}_m(x) \bar{P}_n(x) \, dx = \frac{1}{2n+1} \, \delta_{mn}. \qquad (28)$$

The first few are

$$\bar{P}_0(x) = 1$$
$$\bar{P}_1(x) = 2x - 1$$
$$\bar{P}_2(x) = 6x^2 - 6x + 1$$
$$\bar{P}_3(x) = 20x^3 - 30x^2 + 12x - 1.$$

The associated Legendre polynomials $P_l^m(x)$ are solutions to the associated LEGENDRE DIFFERENTIAL EQUATION, where l is a POSITIVE INTEGER and $m = 0, \ldots, l$. They can be given in terms of the unassociated polynomials by

$$P_l^m(x) = (-1)^m (1 - x^2)^{m/2} \frac{d^m}{dx^m} P_l(x)$$

$$= \frac{(-1)^m}{2^l l!} (1 - x^2)^{m/2} \frac{d^{l+m}}{dx^{l+m}} (x^2 - 1)^l, \qquad (29)$$

where $P_l(x)$ are the unassociated LEGENDRE POLYNOMIALS. Note that some authors (e.g., Arfken 1985, p. 668) omit the CONDON-SHORTLEY PHASE $(-1)^m$, while others include it (e.g., Abramowitz and Stegun 1972, Press *et al.* 1992, and the LegendreP[l, m, z] command of *Mathematica*). Abramowitz and Stegun (1972, p. 332) use the notation

$$P_{lm}(X) \equiv (-1)^m P_m^l(x) \qquad (30)$$

to distinguish these two cases.

Associated polynomials are sometimes called FERRERS' FUNCTIONS (Sansone 1991, p. 246). If $m = 0$, they reduce to the unassociated POLYNOMIALS. The associated Legendre functions are part of the SPHERICAL HARMONICS, which are the solution of LAPLACE'S EQUATION in SPHERICAL COORDINATES. They are ORTHOGONAL over $[-1, 1]$ with the WEIGHTING FUNCTION 1

$$\int_{-1}^{1} P_l^m(x) P_{l'}^m(x) \, dx = \frac{2}{2l+1} \frac{(l+m)!}{(l-m)!} \delta_{ll'}, \qquad (31)$$

and ORTHOGONAL over $[-1, 1]$ with respect to m with the WEIGHTING FUNCTION $(1 - x^2)^{-2}$

$$\int_{-1}^{1} P_l^m(x) P_l^{m'}(x) \frac{dx}{1-x^2} = \frac{(l+m)!}{m(l-m)!} \delta_{mm'}. \qquad (32)$$

The associated Legendre polynomials also obey the following RECURRENCE RELATIONS

$$(l-m)P_l^m(x)$$
$$= x(2l-1)P_{l-1}^m(x) - (l+m-1)P_{l-2}^m(x). \qquad (33)$$

Letting $x \equiv \cos\theta$ (commonly denoted μ in this context),

$$\frac{dP_l^m(\mu)}{d\theta} = \frac{l\mu P_l^m(\mu) - (l+m)P_{l-1}^m(\mu)}{\sqrt{1-\mu^2}} \qquad (34)$$

$$(2l+1)\mu P_l^m(\mu)$$
$$= (l+m)P_{l-1}^m(\mu) + (l-m+1)P_{l+1}^m(\mu). \qquad (35)$$

An identity relating associated POLYNOMIALS with NEGATIVE m to the corresponding functions with POSITIVE m is

$$P_l^{-m}(x) = (-1)^m \frac{(l-m)!}{(l+m)!} P_l^m(x). \qquad (36)$$

Additional identities are

$$P_l^l(x) = (-1)^l (2l-1)!!(1-x^2)^{1/2} \qquad (37)$$

$$P_{l+1}^l(x) = x(2l+1)P_l^l(x). \qquad (38)$$

Written in terms of x and using the convention without a leading factor of $(-1)^m$ (Arfken 1985, p. 669), the first few associated Legendre polynomials are

$$P_0^0(x) = 1$$

$$P_1^0(x) = x$$

$$P_1^1(x) = -(1-x^2)^{1/2}$$

$$P_2^0(x) = \frac{1}{2}(3x^2 - 1)$$

$$P_2^1(x) = -3x(1-x^2)^{1/2}$$

$$P_2^2(x) = 3(1-x^2)$$

$$P_3^0(x) = \frac{1}{2}x(5x^2 - 3)$$

$$P_3^1(x) = \frac{3}{2}(1-5x^2)(1-x^2)^{1/2}$$

$$P_3^2(x) = 15x(1-x^2)$$

$$P_3^3(x) = -15(1-x^2)^{3/2}$$

$$P_4^0(x) = \frac{1}{8}(35x^4 - 30x^2 + 3)$$

$$P_4^1(x) = \frac{5}{2}x(3 - 7x^2)(1-x^2)^{1/2}$$

$$P_4^2(x) = \frac{15}{2}(7x^2 - 1)(1-x^2)$$

$$P_4^3(x) = -105x(1-x^2)^{3/2}$$

$$P_4^4(x) = 105(1-x^2)^2$$

$$P_5^0(x) = \frac{1}{8}x(63x^4 - 70x^2 + 15).$$

Written in terms $x = \cos\theta$ (commonly written $\mu = \cos\theta$), the first few become

$$P_0^0(\cos\theta) = 1$$

$$P_1^0(\cos\theta) = \cos\theta$$

$$P_1^1(\cos\theta) = -\sin\theta$$

$$P_2^0(\cos\theta) = \frac{1}{2}(3\cos^2\theta - 1)$$

$$P_2^1(\cos\theta) = -3\sin\theta\cos\theta$$

$$P_2^2(\cos\theta) = 3\sin^2\theta$$

$$P_3^0(\cos\theta) = \frac{1}{2}\cos\theta(5\cos^2\theta - 3)$$

$$P_3^1(\cos\theta) = -\frac{3}{2}(5\cos^2\theta - 1)\sin\theta$$

$$P_3^2(\cos\theta) = 15\cos\theta\sin^2\theta$$

$$P_3^3(\cos\theta) = -15\sin^3\theta.$$

The derivative about the origin is

$$\left[\frac{dP_\nu^\mu(x)}{dx}\right]_{x=0} = \frac{2^{\mu+1}\sin[\frac{1}{2}\pi(\nu+\mu)]\Gamma(\frac{1}{2}\nu + \frac{1}{2}\mu + 1)}{\pi^{-1/2}\Gamma(\frac{1}{2}\nu - \frac{1}{2}\mu + \frac{1}{2})} \qquad (39)$$

(Abramowitz and Stegun 1972, p. 334), and the logarithmic derivative is

$$\left[\frac{d\ln P_\lambda^\mu(z)}{dz}\right]_{z=0}$$
$$= 2\tan[\frac{1}{2}\pi(\lambda+\mu)]$$
$$\times \frac{[\frac{1}{2}(\lambda+\mu)]![\frac{1}{2}(\lambda-\mu)]!}{[\frac{1}{2}(\lambda+\mu-1)]![\frac{1}{2}(\lambda-\mu-1)]!} \qquad (40)$$

(Binney and Tremaine 1987, p. 654).

See also CONDON-SHORTLEY PHASE, CONICAL FUNCTION, KINGS PROBLEM, LAPLACE'S INTEGRAL, LAPLACE-MEHLER INTEGRAL, LEGENDRE FUNCTION OF THE FIRST KIND, LEGENDRE FUNCTION OF THE SECOND KIND, SUPER CATALAN NUMBER, TOROIDAL FUNCTION, TURÁN'S INEQUALITIES, ULTRASPHERICAL POLYNOMIAL, ZONAL HARMONIC

References

Abramowitz, M. and Stegun, C. A. (Eds.). "Legendre Functions" and "Orthogonal Polynomials." Ch. 22 in Chs. 8 and 22 in *Handbook of Mathematical Functions with Formulas, Graphs, and Mathematical Tables, 9th printing.* New York: Dover, pp. 331–39 and 771–02, 1972.

Arfken, G. "Legendre Functions." Ch. 12 in *Mathematical Methods for Physicists, 3rd ed.* Orlando, FL: Academic Press, pp. 637–11, 1985.

Bailey, W. N. "On the Product of Two Legendre Polynomials." *Proc. Cambridge Philos. Soc.* **29**, 173–77, 1933.

Bailey, W. N. *Generalised Hypergeometric Series.* Cambridge, England: Cambridge University Press, 1935.

Binney, J. and Tremaine, S. "Associated Legendre Functions." Appendix 5 in *Galactic Dynamics.* Princeton, NJ: Princeton University Press, pp. 654–55, 1987.

Byerly, W. E. "Zonal Harmonics." Ch. 5 in *An Elementary Treatise on Fourier's Series, and Spherical, Cylindrical, and Ellipsoidal Harmonics, with Applications to Problems in Mathematical Physics.* New York: Dover, pp. 144–94, 1959.

Iyanaga, S. and Kawada, Y. (Eds.). "Legendre Function" and "Associated Legendre Function." Appendix A, Tables 18.II and 18.III in *Encyclopedic Dictionary of Mathematics.* Cambridge, MA: MIT Press, pp. 1462–468, 1980.

Koekoek, R. and Swarttouw, R. F. "Legendre / Spherical." §1.8.3 in *The Askey-Scheme of Hypergeometric Orthogonal Polynomials and its q-Analogue.* Delft, Netherlands: Technische Universiteit Delft, Faculty of Technical Mathematics and Informatics Report 98–7, p. 44, 1998. ftp://www.twi.tudelft.nl/publications/tech-reports/1998/DUT-TWI-98-7.ps.gz.

Koepf, W. *Hypergeometric Summation: An Algorithmic Approach to Summation and Special Function Identities.* Braunschweig, Germany: Vieweg, 1998.

Lagrange, R. *Polynomes et fonctions de Legendre.* Paris: Gauthier-Villars, 1939.

Legendre, A. M. "Sur l'attraction des Sphéroides." *Mém. Math. et Phys. présentés à l'Ac. r. des. sc. par divers savants* **10**, 1785.

Morse, P. M. and Feshbach, H. *Methods of Theoretical Physics, Part I.* New York: McGraw-Hill, pp. 593–97, 1953.

Press, W. H.; Flannery, B. P.; Teukolsky, S. A.; and Vetterling, W. T. *Numerical Recipes in FORTRAN: The Art of Scientific Computing, 2nd ed.* Cambridge, England: Cambridge University Press, p. 252, 1992.

Sansone, G. "Expansions in Series of Legendre Polynomials and Spherical Harmonics." Ch. 3 in *Orthogonal Functions, rev. English ed.* New York: Dover, pp. 169–94, 1991.

Snow, C. *Hypergeometric and Legendre Functions with Applications to Integral Equations of Potential Theory.* Washington, DC: U. S. Government Printing Office, 1952.

Spanier, J. and Oldham, K. B. "The Legendre Polynomials $P_n(x)$" and "The Legendre Functions $P_\nu(x)$ and $Q_\nu(x)$." Chs. 21 and 59 in *An Atlas of Functions.* Washington, DC: Hemisphere, pp. 183–92 and 581–97, 1987.

Szego, G. *Orthogonal Polynomials, 4th ed.* Providence, RI: Amer. Math. Soc., 1975.

Legendre Polynomial of the Second Kind

LEGENDRE FUNCTION OF THE SECOND KIND

LegendreQ

LEGENDRE FUNCTION OF THE SECOND KIND

Legendre Quadrature

LEGENDRE-GAUSS QUADRATURE

Legendre Relation

Let $E(k)$ and $K(k)$ be complete ELLIPTIC INTEGRALS OF THE FIRST and SECOND KINDS, with $E'(k)$ and $K'(k)$ the complementary integrals. Then

$$E(k)K'(k) + E'(k)K(k) - K(k)K'(k) = \tfrac{1}{2}\pi.$$

References

Abramowitz, M. and Stegun, C. A. (Eds.). *Handbook of Mathematical Functions with Formulas, Graphs, and Mathematical Tables, 9th printing.* New York: Dover, p. 591, 1972.

Legendre's Chi-Function

Portions of this entry contributed by Joe Keane.

The function defined by

$$\chi_\nu(z) = \sum_{k=0}^{\infty} \frac{z^{2k+1}}{(2k+1)^\nu} \tag{1}$$

for integral $\nu = 2, 3, \ldots$. It is related to the POLYLOGARITHM by

$$\chi_\nu(z) = \tfrac{1}{2}[\mathrm{Li}_\nu(z) - \mathrm{Li}_\nu(-z)] \tag{2}$$

$$= \mathrm{Li}_\nu(z) - 2^{-\nu}\mathrm{Li}_\nu(z^2) \tag{3}$$

and to the LERCH TRANSCENDENT by

$$\chi_\nu(z) = 2^{-\nu}z\Phi(z^2, n, \tfrac{1}{2}). \tag{4}$$

It takes the special values

$$\chi_2(i) = iK \tag{5}$$

$$\chi_2(\sqrt{2}-1) = \tfrac{1}{16}\pi^2 - \tfrac{1}{4}[\ln(\sqrt{2}+1)]^2 \tag{6}$$

$$\chi_2(\tfrac{1}{2}(\sqrt{5}-1)) = \tfrac{1}{12}\pi^2 - \tfrac{3}{4}[\ln(\tfrac{1}{2}(\sqrt{5}+1))]^2 \tag{7}$$

$$\chi_2(\sqrt{5}-2) = \tfrac{1}{24}\pi^2 - \tfrac{3}{4}[\ln(\tfrac{1}{2}(\sqrt{5}+1))]^2 \tag{8}$$

$$\chi_2(-1) = -\tfrac{1}{8}\pi^2 \tag{9}$$

$$\chi_2(1) = \tfrac{1}{8}\pi^2, \tag{10}$$

where i is the imaginary unit and K is CATALAN'S CONSTANT (Lewin, p. 19). Other special values include

$$\chi_n(1) = \lambda(n) \tag{11}$$

$$\chi_n(1) = i\beta(n), \tag{12}$$

where $\lambda(n)$ is the DIRICHLET LAMBDA FUNCTION and $\beta(n)$ is the DIRICHLET BETA FUNCTION.

See also LERCH TRANSCENDENT, POLYLOGARITHM

References

Cvijovic, D. and Klinowski, J. "Closed-Form Summation of Some Trigonometric Series." *Math. Comput.* **64**, 205–10, 1995.

Edwards, J. *A Treatise on the Integral Calculus, Vol. 2.* New York: Chelsea, p. 290, 1955.

Legendre, A. M. *Exercices de calcul intégral, tome 1.* p. 247, 1811.

Lewin, L. "Legendre's Chi-Function." §1.8 in *Dilogarithms and Associated Functions.* London: Macdonald, pp. 17–9, 1958.

Lewin, L. *Polylogarithms and Associated Functions.* Amsterdam, Netherlands: North-Holland, pp. 282–83, 1981.

Nielsen, N. "Der Eulersche Dilogarithmus und seine Verallgemeinerungen." *Nova Acta (Leopold)* **90**, 121–12, 1909.

Legendre's Constant

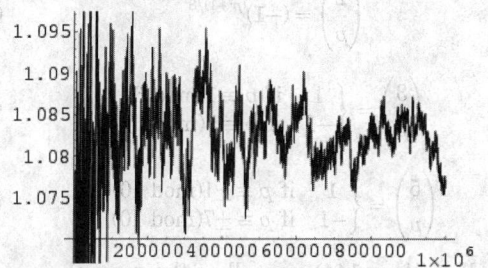

The number 1.08366 in Legendre's guess at the PRIME NUMBER THEOREM

$$\pi(n) = \frac{n}{\ln n - A(n)}$$

with $\lim_{n \to \infty} A(n) \approx 1.08366$. This expression is correct to leading term only, since it is actually true that this limit approaches 1 (Rosser and Schoenfeld 1962, Panaitopol 1999).

See also PRIME COUNTING FUNCTION

References

Le Lionnais, F. *Les nombres remarquables.* Paris: Hermann, p. 147, 1983.

Panaitopol, L. "Several Approximations of $\pi(x)$." *Math. Ineq. Appl.* **2**, 317–24, 1999.

Ribenboim, P. *The New Book of Prime Number Records.* New York: Springer-Verlag, 1996.

Rosser, J. B. and Schoenfeld, L. "Approximate Formulas for Some Functions of Prime Numbers." *Ill. J. Math.* **6**, 64–4, 1962.

Wagon, S. *Mathematica in Action.* New York: W. H. Freeman, pp. 28–9, 1991.

Legendre Series

Because the LEGENDRE FUNCTIONS OF THE FIRST KIND form a COMPLETE ORTHOGONAL BASIS, any FUNCTION may be expanded in terms of them

$$f(x) = \sum_{n=0}^{\infty} a_n P_n(x). \tag{1}$$

Now, multiply both sides by $P_m(x)$ and integrate

$$\int_{-1}^{1} P_m(x) f(x) \, dx = \sum_{n=0}^{\infty} a_n \int_{-1}^{1} P_n(x) P_m(x) \, dx. \tag{2}$$

But

$$\int_{-1}^{1} P_n(x) P_m(x) \, dx = \frac{2}{2m+1} \delta_{mn}, \tag{3}$$

where δ_{mn} is the KRONECKER DELTA, so

$$\int_{-1}^{1} P_m(x) f(x) \, dx = \sum_{n=0}^{\infty} a_n \frac{2}{2m+1} \delta_{mn}$$

$$= \frac{2}{2m+1} a_m \tag{4}$$

and

$$a_m = \frac{2m+1}{2} \int_{-1}^{1} P_m(x) f(x) \, dx. \tag{5}$$

See also FOURIER SERIES, JACKSON'S THEOREM, LEGENDRE POLYNOMIAL, MACLAURIN SERIES, PICONE'S THEOREM, TAYLOR SERIES

Legendre's Factorization Method

A PRIME FACTORIZATION ALGORITHM in which a sequence of TRIAL DIVISORS is chosen using a QUADRATIC SIEVE. By using QUADRATIC RESIDUES of N, the QUADRATIC RESIDUES of the factors can also be found.

See also PRIME FACTORIZATION ALGORITHMS, QUADRATIC RESIDUE, QUADRATIC SIEVE, TRIAL DIVISOR

Legendre's Formula

Counts the number of POSITIVE INTEGERS less than or equal to a number x which are not divisible by any of the first a PRIMES,

$$\phi(x, a) = \lfloor x \rfloor - \sum \left\lfloor \frac{x}{p_i} \right\rfloor + \sum \left\lfloor \frac{x}{p_i p_j} \right\rfloor - \sum \left\lfloor \frac{x}{p_i p_j p_k} \right\rfloor + \ldots, \tag{1}$$

where $\lfloor x \rfloor$ is the FLOOR FUNCTION. Taking $a = x$ gives

$$\phi(x, x) = \pi(x) - \pi(\sqrt{x}) + 1$$

$$= \lfloor x \rfloor - \sum_{p_i \leq \sqrt{x}} \left\lfloor \frac{x}{pi} \right\rfloor + \sum_{p_i < p_j \leq \sqrt{x}} \left\lfloor \frac{x}{p_i p_j} \right\rfloor$$

$$- \sum_{p_i < p_j < p_k \leq \sqrt{x}} \left\lfloor \frac{x}{p_i p_j p_k} \right\rfloor + \ldots, \tag{2}$$

where $\pi(n)$ is the PRIME COUNTING FUNCTION. Legendre's formula holds since one more than the number of PRIMES in a range equals the number of INTEGERS minus the number of composites in the interval.

Legendre's formula satisfies the RECURRENCE RELATION

$$\phi(x, a) = \phi(x, a-1) - \phi\left(\frac{x}{p_a}, a-1\right). \tag{3}$$

Let $m_k \equiv p_1 p_2 \cdots p_k$, then

$$\phi(m_k,\, k) = \lfloor m_k \rfloor - \sum \left\lfloor \frac{m_k}{p_i} \right\rfloor + \sum \left\lfloor \frac{m_k}{p_i p_j} \right\rfloor - \cdots$$

$$= m_k - \sum \frac{m_k}{p_i} + \sum \frac{m_k}{p_i p_j} - \cdots$$

$$= m_k \left(1 - \frac{1}{p-1}\right)\left(1 - \frac{1}{p^2}\right) \cdots \left(1 - \frac{1}{p_k}\right)$$

$$= \prod_{i=1}^{k}(p_i - 1) = \phi(m_k), \tag{4}$$

where $\phi(n)$ is the TOTIENT FUNCTION, and

$$\phi(sm_k + t,\, k) = s\phi(m_k) + \phi(t,\, k), \tag{5}$$

where $0 \leq t \leq m_k$. If $t > m_k/2$, then

$$\phi(t,\, k) = \phi(m_k) - \phi(m_k - t - 1,\, k). \tag{6}$$

Note that $\phi(n,\, n)$ is not practical for computing $\pi(n)$ for large arguments. A more efficient modification is MEISSEL'S FORMULA.

See also LEHMER'S FORMULA, MAPES' METHOD, MEISSEL'S FORMULA, PRIME COUNTING FUNCTION

References

Séroul, R. "Legendre's Formula" and "Implementation of Legendre's Formula." §8.7.1 and 8.7.2 in *Programming for Mathematicians*. Berlin: Springer-Verlag, pp. 175–79, 2000.

Legendre's Quadratic Reciprocity Law
QUADRATIC RECIPROCITY LAW

Legendre Sum
LEGENDRE'S FORMULA

Legendre Symbol
The Legendre symbol is a number theoretic function $\left(\frac{m}{n}\right)$ which is defined to be equal to ± 1 depending on whether m is a QUADRATIC RESIDUE modulo n. The definition is sometimes generalized to have value 0 if $m|n$,

$$\left(\frac{m}{n}\right) = (m|n)$$

$$\equiv \begin{cases} 0 & \text{if } m|n \\ 1 & \text{if } m \text{ is a quadratic residue modulo } n \\ -1 & \text{if } m \text{ is a quadratic nonresidue modulo } n. \end{cases}$$

$$\tag{1}$$

If n is an ODD PRIME, then the JACOBI SYMBOL reduces to the Legendre symbol. The Legendre symbol is implemented in *Mathematica* via the JACOBI SYMBOL, `JacobiSymbol[n, m]`.

The Legendre symbol obeys the identity

$$\left(\frac{ab}{p}\right) = \left(\frac{a}{p}\right)\left(\frac{b}{p}\right). \tag{2}$$

Particular identities include

$$\left(\frac{-1}{p}\right) = (-1)^{(p-1)/2} \tag{3}$$

$$\left(\frac{2}{p}\right) = (-1)^{(p^2-1)/8} \tag{4}$$

$$\left(\frac{3}{p}\right) = \begin{cases} 1 & \text{if } p \equiv 1 \pmod 6 \\ -1 & \text{if } p \equiv 5 \pmod 6 \end{cases} \tag{5}$$

$$\left(\frac{5}{p}\right) = \begin{cases} 1 & \text{if } p \equiv \pm 1 \pmod{10} \\ -1 & \text{if } p \equiv \pm 7 \pmod{10} \end{cases} \tag{6}$$

(Nagell 1951, p. 144), as well as the general

$$\left(\frac{q}{p}\right) = \left(\frac{p}{q}\right)(-1)^{[(p-1)/2][(q-1)/2]}. \tag{7}$$

See also JACOBI SYMBOL, KRONECKER SYMBOL, QUADRATIC RECIPROCITY THEOREM, QUADRATIC RESIDUE

References

Guy, R. K. "Quadratic Residues. Schur's Conjecture." §F5 in *Unsolved Problems in Number Theory, 2nd ed.* New York: Springer-Verlag, pp. 244–45, 1994.
Hardy, G. H. and Wright, E. M. "Quadratic Residues." §6.5 in *An Introduction to the Theory of Numbers, 5th ed.* Oxford, England: Clarendon Press, pp. 67–8, 1979.
Nagell, T. "Euler's Criterion and Legendre's Symbol." §38 in *Introduction to Number Theory.* New York: Wiley, pp. 133–36, 1951.
Shanks, D. *Solved and Unsolved Problems in Number Theory, 4th ed.* New York: Chelsea, pp. 33–4 and 40–2, 1993.

Legendre Transform
The Legendre transform of a sequence $\{c_k\}$ is the sequence $\{a_k\}$ with terms given by

$$a_n = \sum_{k=0}^{n} c_k \binom{n}{k}\binom{n+k}{k},$$

where $\binom{n}{k}$ is a BINOMIAL COEFFICIENT (Jin and Dickinson 2000). Strehl (1994) and Schmidt (1995) showed that

$$\sum_{k=0}^{n}\binom{n}{k}^2\binom{n+k}{k}^2 = \sum_{k=0}^{n}\binom{n}{k}\binom{n+k}{k}\sum_{j=0}^{k}\binom{k}{j}^3.$$

Legendre Transformation

References

Jin, Y. and Dickinson, H. "Apéry Sequences and Legendre Transforms." *J. Austral. Math. Soc. Ser. A* **68**, 349–56, 2000.
Schmidt, A. L. "Legendre Transforms and Apéry's Sequences." *J. Austral. Math. Soc. Ser. A* **58**, 358–75, 1995.
Strehl, V. "Binomial Identities--Combinatorial and Algorithmic Aspects. Trends in Discrete Mathematics." *Disc. Math.* **136**, 309–46, 1994.

Legendre Transformation

Given a function of two variables

$$df = \frac{\partial f}{\partial x}\, dx + \frac{\partial f}{\partial y}\, dy \equiv u\, dx + v\, dy, \tag{1}$$

change the differentials from dx and dy to du and dy with the transformation

$$g \equiv f - ux \tag{2}$$

$$dg = df - u\, dx - x\, du = u\, dx + v\, dy - u\, dx - x\, du$$

$$= v\, dy - x\, du. \tag{3}$$

Then

$$x \equiv -\frac{\partial g}{\partial u}. \tag{4}$$

$$v \equiv \frac{\partial g}{\partial y}. \tag{5}$$

Lehmer Continued Fraction

A CONTINUED FRACTION OF THE FORM

$$b_0 + \cfrac{e_1}{b_1 + \cfrac{e_2}{b_2 + \cfrac{e_3}{b_3 + \ldots}}}$$

where $(b_i, e_{i+1}) = (1, 1)$ or $(2, -1)$ for $x \in [1, 2)$ an IRRATIONAL NUMBER (Lehmer 1994, Dajani and Kraaikamp 1999).

See also CONTINUED FRACTION

References

Dajani, K. and Kraaikamp, C. "The Mother of All Continued Fractions." http://www.math.uu.nl/publications/preprints/1106.ps.gz.
Lehmer, J. "Semiregular Continued Fractions whose Partial Denominators are 1 or 2." In *The Mathematical Legacy of Wilhelm Magnus: Groups, Geometry, and Special Functions. Conference on the Legacy of Wilhelm Magnus May 1–, 1992 (Brooklyn, NY)* (Ed. W. Abikoff, J. S. Birman, and K. Kuiken). Providence, RI: Amer. Math. Soc., 1994.

Lehmer Method

LEHMER-SCHUR METHOD

Lehmer Number

A number generated by a generalization of a LUCAS SEQUENCE. Let α and β be COMPLEX NUMBERS with

$$\alpha + \beta = \sqrt{R} \tag{1}$$

$$\alpha\beta = Q, \tag{2}$$

where Q and R are RELATIVELY PRIME NONZERO INTEGERS and α/β is a ROOT OF UNITY. Then the Lehmer numbers are

$$U_n(\sqrt{R},\, Q) = \frac{\alpha^n - \beta^n}{\alpha - \beta}, \tag{3}$$

and the companion numbers

$$V_n\left(\sqrt{R},\, Q\right) = \begin{cases} \dfrac{\alpha^n + \beta^n}{\alpha + \beta} & \text{for } n \text{ odd} \\[2mm] \alpha^n + \beta^n & \text{for } n \text{ even} \end{cases} \tag{4}$$

References

Lehmer, D. H. "An Extended Theory of Lucas' Functions." *Ann. Math.* **31**, 419–48, 1930.
Ribenboim, P. *The Book of Prime Number Records, 2nd ed.* New York: Springer-Verlag, pp. 61 and 70, 1989.
Shorey, T. N. and Stewart, C. L. "On Divisors of Fermat, Fibonacci, Lucas and Lehmer Numbers, 2." *J. London Math. Soc.* **23**, 17–3, 1981.
Stewart, C. L. "On Divisors of Fermat, Fibonacci, Lucas and Lehmer Numbers." *Proc. London Math. Soc.* **35**, 425–47, 1977.
Williams, H. C. "The Primality of $N = 2A3^n - 1$." *Canad. Math. Bull.* **15**, 585–89, 1972.

Lehmer-Schur Method

An ALGORITHM which isolates ROOTS in the COMPLEX PLANE by generalizing 1-D bracketing.

References

Acton, F. S. *Numerical Methods That Work, 2nd printing.* Washington, DC: Math. Assoc. Amer., pp. 196–98, 1990.

Lehmer's Conjecture

LEHMER'S MAHLER MEASURE PROBLEM

Lehmer's Constant

N.B. A detailed online essay by S. Finch was the starting point for this entry.

Lehmer (1938) showed that every POSITIVE IRRATIONAL NUMBER x has a unique infinite continued cotangent representation OF THE FORM

$$x = \cot\left[\sum_{k=0}^{\infty} (-1)^k \cot^{-1} b_k\right],$$

where the b_ks are NONNEGATIVE and

$$b_k \geq (b_{k-1})^2 + b_{k-1} + 1.$$

The case for which the convergence is slowest occurs when the inequality is replaced by equality, giving $c_0 = 0$ and

$$c_k = (c_{k-1})^2 + c_{k-1} + 1$$

for $k \geq 1$. The first few values are c_k are 0, 1, 3, 13, 183, 33673, ... (Sloane's A024556), resulting in the constant

$$\xi = \cot(\cot^{-1} 0 - \cot^{-1} 1 + \cot^{-1} 3 - \cot^{-1} 13$$

$$+ \cot^{-1} 183 - \cot^{-1} 33673 + \cot^{-1} 1133904603$$

$$- \cot^{-1} 1285739649838492213 + \ldots + (-1)^k c_k \ldots)$$

$$= \cot\left(\tfrac{1}{4}\pi + \cot^{-1} 3 - \cot^{-1} 13\right.$$

$$+ \cot^{-1} 183 - \cot^{-1} 33673 + \cot^{-1} 1133904603$$

$$\left. - \cot^{-1} 1285739649838492213 + \ldots + (-1)^k c_k \ldots\right)$$

$$= 0.59263271\ldots$$

(Sloane's A030125). ξ is not an ALGEBRAIC NUMBER of degree less than 4, but Lehmer's approach cannot show whether or not ξ is TRANSCENDENTAL.

See also ALGEBRAIC NUMBER, TRANSCENDENTAL NUMBER

References
Finch, S. "Favorite Mathematical Constants." http://www.mathsoft.com/asolve/constant/lehmer/lehmer.html.
Le Lionnais, F. *Les nombres remarquables.* Paris: Hermann, p. 29, 1983.
Lehmer, D. H. "A Cotangent Analogue of Continued Fractions." *Duke Math. J.* **4**, 323–40, 1938.
Plouffe, S. "The Lehmer Constant." http://www.lacim.u-qam.ca/piDATA/lehmer.txt.
Sloane, N. J. A. Sequences A024556 and A030125 in "An On-Line Version of the Encyclopedia of Integer Sequences." http://www.research.att.com/~njas/sequences/eisonline.html.

Lehmer's Formula

A FORMULA related to MEISSEL'S FORMULA.

$$\pi(x) = \lfloor x \rfloor - \sum_{i=1}^{a} \left\lfloor \frac{x}{p_i} \right\rfloor + \sum_{1 \leq i \leq j \leq a} \left\lfloor \frac{x}{p_i p_j} \right\rfloor - \ldots$$

$$+ \tfrac{1}{2}(b + a - 2)(b - a + 1) - \sum_{a \leq i \leq b} \pi\left(\frac{x}{p_i}\right)$$

$$- \sum_{i=a+1}^{c} \sum_{j=i}^{b_i} \left[\pi\left(\frac{x}{p_i p_j}\right) - (j - 1) \right],$$

where $\lfloor x \rfloor$ is the FLOOR FUNCTION,

$$a \equiv \pi(x^{1/4})$$

$$b \equiv \pi(x^{1/2})$$

$$b_i \equiv \left(\pi \sqrt{x/p_i}\right)$$

$$c \equiv \pi(x^{1/3}),$$

and $\pi(n)$ is the PRIME COUNTING FUNCTION.

References
Riesel, H. "Lehmer's Formula." *Prime Numbers and Computer Methods for Factorization, 2nd ed.* Boston, MA: Birkhäuser, pp. 13–4, 1994.

Lehmer's Mahler Measure Problem
Portions of this entry contributed by KEVIN O'BRYANT

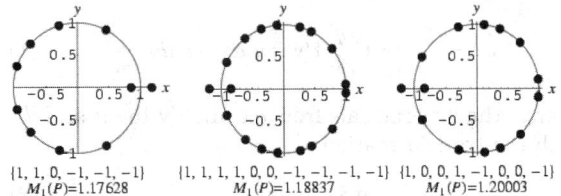

$\{1, 1, 0, -1, -1, -1\}$ $\{1, 1, 1, 1, 0, 0, -1, -1, -1, -1\}$ $\{1, 0, 0, 1, -1, 0, 0, -1\}$
$M_1(P) = 1.17628$ $M_1(P) = 1.18837$ $M_1(P) = 1.20003$

An UNSOLVED PROBLEM in mathematics attributed to Lehmer that concerns the minimum MAHLER MEASURE $M_1(P)$ for a UNIVARIATE POLYNOMIAL $P(x)$ that is not a product of CYCLOTOMIC POLYNOMIALS. Lehmer conjectured that if $P(x)$ is such a polynomial with integer coefficients, then

$$M_1(P) \geq M_1(1 - x + x^3 - x^4 + x^5 - x^6 + x^7 - x^9 + x^{10})$$

$$= m^*, \tag{1}$$

where $m^* \approx 1.1762$ is the largest positive root of this polynomial. The roots of this polynomial, plotted in the left figure above, are very special, since 8 of the 10 lie on the UNIT CIRCLE in the COMPLEX PLANE. The roots of the polynomials (represented by half their coefficients) giving the two next smallest known Mahler measures are also illustrated above (Mossinghoff, p. S11).

The best current bound is that of Smyth (1971), who showed that $M(F) > \theta_1$, where F is a nonzero nonreciprocal polynomial that is not a product of CYCLOTOMIC POLYNOMIALS (Everest 1999), and $\theta_1 \approx 1.324$ is the real root of $x^3 - x - 1 = 0$. Generalizations of Smyth's result have been constructed by Lloyd-Smith (1985) and Dubickas (1997).

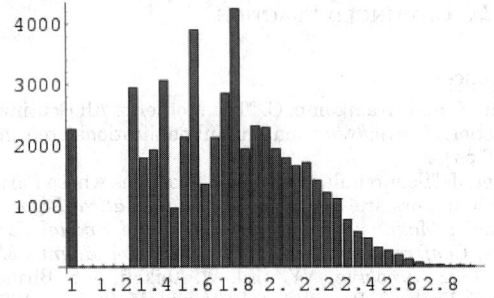

In general, the smallest MAHLER MEASURES occur for polynomials with integers coefficients that are small in absolute value. The histogram above shows the distribution of measures for random $(-1, 0, 1)$-polynomials of random orders 1 to 10. Mossinghoff (1998) gives a table of the smallest known Mahler measures for polynomial degrees up to $d = 24$.

See also MAHLER MEASURE

References

Boyd, D. W. "Reciprocal Polynomials Having Small Measure." *Math. Comput.* **35**, 1361–377, 1980.

Boyd, D. W. "Reciprocal Polynomials Having Small Measure. II." *Math. Comput.* **53**, 355–57 and S1-S5, 1989.

Dubickas, A. "Algebraic Conjugates Outside the Unit Circle." In *New Trends in Probability and Statistics, Vol. 4: Analytic and Probabilistic Methods in Number Theory. Proceedings of the 2nd International Conference held in Honor of J. Kubilius on His 75th Birthday in Palanga, September 23–7, 1996* (Ed. A. Laurincikas, E. Manstavicius, and V. Stakenas). Utrecht, Netherlands: VSP, pp. 11–1, 1997.

Everest, G. Ch. 1 in *Heights of Polynomials and Entropy in Algebraic Dynamics.* London: Springer-Verlag, 1999.

Lloyd-Smith, C. W. "Algebraic Numbers Near the Unit Circle." *Acta Arith.* **45**, 43–7, 1985.

Mossinghoff, M. J. "Polynomials with Small Mahler Measure." *Math. Comput.* **67**, 1697–705 and S11-S14, 1998.

Smyth, C. J. "On the Product of the Conjugates Outside the Unit Circle of an Algebraic Integer." *Bull. London Math. Soc.* **3**, 169–75, 1971.

Lehmer's Phenomenon

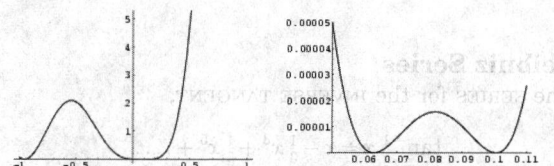

The appearance of nontrivial zeros (i.e., those along the CRITICAL STRIP with $\Re[z] = 1/2$) of the RIEMANN ZETA FUNCTION $\zeta(z)$ very close together. An example is the pair of zeros $\zeta\left(\frac{1}{2} + (7005 + t)i\right)$ given by $t_1 \approx 0.0606918$ and $t_2 \approx 0.100055$, illustrated above in the plot of $\left|\zeta(\frac{1}{2} + (7005 + t)i)\right|^2$.

See also CRITICAL STRIP, RIEMANN ZETA FUNCTION

References

Csordas, G.; Odlyzko, A. M.; Smith, W.; and Varga, R. S. "A New Lehmer Pair of Zeros and a New Lower Bound for the de Bruijn-Newman Constant." *Elec. Trans. Numer. Analysis* **1**, 104–11, 1993.

Csordas, G.; Smith, W.; and Varga, R. S. "Lehmer Pairs of Zeros, the de Bruijn-Newman Constant and the Riemann Hypothesis." *Constr. Approx.* **10**, 107–29, 1994.

Csordas, G.; Smith, W.; and Varga, R. S. "Lehmer Pairs of Zeros and the Riemann ζ-Function." In *Mathematics of Computation 1943–993: A Half-Century of Computational Mathematics* (Vancouver, BC, 1993). *Proc. Sympos. Appl. Math.* **48**, 553–56, 1994.

Wagon, S. *Mathematica in Action.* New York: W. H. Freeman, pp. 357–58, 1991.

Lehmer's Problem

LEHMER'S MAHLER MEASURE PROBLEM, LEHMER'S TOTIENT PROBLEM

Lehmer's Theorem

FERMAT'S LITTLE THEOREM CONVERSE

Lehmer's Totient Problem

Do there exist any COMPOSITE NUMBERS n such that $\phi(n)|(n-1)$, where $\phi(n)$ is the TOTIENT FUNCTION? No such numbers are known. In 1932, Lehmer showed that such an n must be ODD and SQUAREFREE, and that the number of distinct PRIME FACTORS $d(7) \geq 7$. This was subsequently extended to $d(n) \geq 11$. The best current results are $n > 10^{20}$ and $d(n) \geq 14$ (Cohen and Hagis 1980), if $30 \nmid n$, then $d(n) \geq 26$ (Wall 1980), and if $3|n$ then $d(n) \geq 213$ and $n \geq 5.5 \times 10^{570}$ (Lieuwens 1970).

See also LEHMER'S MAHLER MEASURE PROBLEM, TOTIENT FUNCTION

References

Cohen, G. L. and Hagis, P. Jr. "On the Number of Prime Factors of n is $\phi(n)|(n-1)$." *Nieuw Arch. Wisk.* **28**, 177–85, 1980.

Lieuwens, E. "Do There Exist Composite Numbers for Which $k\phi(M) = M - 1$ Holds?" *Nieuw. Arch. Wisk.* **18**, 165–69, 1970.

Ribenboim, P. *The Book of Prime Number Records, 2nd ed.* New York: Springer-Verlag, pp. 27–8, 1989.

Wall, D. W. "Conditions for $\phi(N)$ to Properly Divide $N - 1$." In *A Collection of Manuscripts Related to the Fibonacci Sequence* (Ed. V. E. Hoggatt and M. V. E. Bicknell-Johnson). San Jose, CA: Fibonacci Assoc., pp. 205–08, 1980.

Lehmus' Theorem

STEINER-LEHMUS THEOREM

Leibniz Criterion

Also known as the ALTERNATING SERIES TEST. Given a SERIES

$$\sum_{n=1}^{\infty} (-1)^{n+1} a_n$$

with $a_n > 0$, if a_n is monotonic decreasing as $n \to \infty$ and

$$\lim_{n \to \infty} a_n = 0$$

then the series CONVERGES.

Leibniz Harmonic Triangle

$$\frac{1}{1}$$

$$\frac{1}{2} \quad \frac{1}{2}$$

$$\frac{1}{3} \quad \frac{1}{6} \quad \frac{1}{3}$$

$$\frac{1}{4} \quad \frac{1}{12} \quad \frac{1}{12} \quad \frac{1}{4}$$

$$\frac{1}{5} \quad \frac{1}{20} \quad \frac{1}{30} \quad \frac{1}{20} \quad \frac{1}{5}$$

(Sloane's A003506). In the Leibniz harmonic triangle, each FRACTION is the sum of numbers below it, with the initial and final entry on each row one over the corresponding entry in PASCAL'S TRIANGLE. The DE-NOMINATORS in the second diagonals are 6, 12, 20, 30, 42, 56, ... (Sloane's A007622).

See also CATALAN'S TRIANGLE, CLARK'S TRIANGLE, EULER'S TRIANGLE, LOSSNITSCH'S TRIANGLE, NUMBER TRIANGLE, PASCAL'S TRIANGLE, SEIDEL-ENTRINGER-ARNOLD TRIANGLE

References

Sloane, N. J. A. Sequences A003506 and A007622/M4096 in "An On-Line Version of the Encyclopedia of Integer Sequences." http://www.research.att.com/~njas/sequences/eisonline.html.

Leibniz Identity

$$\frac{d^n}{dx^n}(uv) = \frac{d^n u}{dx^n} v + \binom{n}{1} \frac{d^{n-1}u}{dx^{n-1}} \frac{dv}{dx} + \ldots + \binom{n}{r}$$

$$\times \frac{d^{n-r}u}{dx^{n-r}} \frac{d^n v}{dx^r} + \ldots + u \frac{d^n v}{dx^n}.$$

where $\binom{n}{k}$ is a BINOMIAL COEFFICIENT. This can also be written explicitly as

$$D^n f(t)g(t) = \sum_{k=0}^{n} \binom{n}{k} D^k f(t) D^{n-k} g(t)$$

(Roman 1980).

See also FAÁ DI BRUNO'S FORMULA

References

Abramowitz, M. and Stegun, C. A. (Eds.). *Handbook of Mathematical Functions with Formulas, Graphs, and Mathematical Tables, 9th printing.* New York: Dover, p. 12, 1972.
Roman, S. "The Formula of Faa di Bruno." *Amer. Math. Monthly* **87**, 805–09, 1980.

Leibniz Integral Rule

$$\frac{\partial}{\partial z} \int_{a(z)}^{b(z)} f(x, z) \, dx$$

$$= \int_{a(z)}^{b(z)} \frac{\partial f}{\partial z} \, dx + f(b(z), z) \frac{\partial b}{\partial z} - f(a(z), z) \frac{\partial a}{\partial z}.$$

The differentiation of a definite integral whose limits are functions of the differential variable. The rule can be used to evaluate certain unusual definite integrals such as

$$\phi(\alpha) = \int_0^{\pi} \ln(1 - 2\alpha \cos x + \alpha^2) \, dx = 2\pi \ln|\alpha|$$

for $|\alpha| > 1$ (Woods 1926). Although the symbolic mathematics program *Mathematica* gives *an* analy-

tic solution to this integral, it gives the solution in a much more complicated form.

Feynman (1997) recalled seeing the method in Woods (1926) and remarked "So because I was self-taught using that book, I had peculiar methods for doing integrals," and "I used that one damn tool again and again."

See also DERIVATIVE, INTEGRAL, INTEGRATION UNDER THE INTEGRAL SIGN

References

Abramowitz, M. and Stegun, C. A. (Eds.). *Handbook of Mathematical Functions with Formulas, Graphs, and Mathematical Tables, 9th printing.* New York: Dover, p. 11, 1972.
Beyer, W. H. *CRC Standard Mathematical Tables, 28th ed.* Boca Raton, FL: CRC Press, p. 232, 1987.
Feynman, R. P. and Leighton, R. "A Different Set of Tools." In *'Surely You're Joking, Mr. Feynman!': Adventures of a Curious Character.* New York: W. W. Norton, pp. 69–2, 1997.
Kaplan, W. "Integrals Depending on a Parameter--Leibnitz's Rule.' §4.9 in *Advanced Calculus, 4th ed.* Reading, MA: Addison-Wesley, pp. 256–58, 1992.
Woods, F. S. "Differentiation of a Definite Integral." §60 in *Advanced Calculus: A Course Arranged with Special Reference to the Needs of Students of Applied Mathematics.* Boston, MA: Ginn, pp. 141–44, 1926.

Leibniz Series

The SERIES for the INVERSE TANGENT,

$$\tan^{-1} x = x - \frac{1}{3}x^3 + \frac{1}{5}x^5 + \ldots.$$

Plugging in $x = 1$ gives GREGORY'S FORMULA

$$\frac{1}{4}\pi = 1 - \frac{1}{3} + \frac{1}{5} - \frac{1}{7} + \frac{1}{9} - \ldots.$$

This series is intimately connected with the number of representations of n by k squares $r_k(n)$, and also with GAUSS'S CIRCLE PROBLEM (Hilbert and Cohn-Vossen 1999, pp. 27–9).

See also GAUSS'S CIRCLE PROBLEM, GREGORY'S FOR-MULA, SUM OF SQUARES FUNCTION

References

Hilbert, D. and Cohn-Vossen, S. *Geometry and the Imagination.* New York: Chelsea, p. 37, 1999.
Wells, D. *The Penguin Dictionary of Curious and Interesting Numbers.* Middlesex, England: Penguin Books, p. 50, 1986.

Lelong's Theorem

References

Morosawa, S.; Nishimura, Y.; Taniguchi, M.; and Ueda, T. "Lelong's Theorem." §8.2 in *Holomorphic Dynamics.* Cambridge, England: Cambridge University Press, pp. 270–76, 2000.

Lemarié's Wavelet

A wavelet used in multiresolution representation to analyze the information content of images. The WAVELET is defined by

$$H(\omega) = \left[2(1-u)^4 \frac{315 - 420u + 126u^2 - 4u^3}{315 - 420v + 126v^2 - 4v^3} \right]^{1/2},$$

where

$$u \equiv \sin^2\left(\tfrac{1}{2}\,\omega\right)$$

$$v \equiv \sin^2 \omega$$

(Mallat 1989).

See also WAVELET

References

Mallat, S. G. "A Theory for Multiresolution Signal Decomposition: The Wavelet Representation." *IEEE Trans. Pattern Analysis Machine Intel.* **11**, 674–93, 1989.

Mallat, S. G. "Multiresolution Approximation and Wavelet Orthonormal Bases of $L^2(\mathbb{R})$." *Trans. Amer. Math. Soc.* **315**, 69–7, 1989.

Lemma

A short THEOREM used in proving a larger THEOREM. Related concepts are the AXIOM, PORISM, POSTULATE, PRINCIPLE, and THEOREM.

See also ABEL'S LEMMA, ARCHIMEDES' LEMMA, BARNES' LEMMA, BLICHFELDT'S LEMMA, BOREL-CANTELLI LEMMA, BURNSIDE'S LEMMA, DANIELSON-LANCZOS LEMMA, DEHN'S LEMMA, DILWORTH'S LEMMA, DIRICHLET'S LEMMA, DIVISION LEMMA, FARKAS'S LEMMA, FATOU'S LEMMA, FUNDAMENTAL LEMMA OF CALCULUS OF VARIATIONS, GAUSS'S LEMMA, HENSEL'S LEMMA, ITÔ'S LEMMA, JORDAN'S LEMMA, LAGRANGE'S LEMMA, NEYMAN-PEARSON LEMMA, POINCARÉ'S HOLOMORPHIC LEMMA, POINCARÉ'S LEMMA, PÓLYA-BURNSIDE LEMMA, RIEMANN-LEBESGUE LEMMA, SCHUR'S LEMMA, SCHUR'S REPRESENTATION LEMMA, SCHWARZ-PICK LEMMA, SPIJKER'S LEMMA, ZORN'S LEMMA

Lemma That Is Not Burnside's

CAUCHY-FROBENIUS LEMMA, PÓLYA ENUMERATION THEOREM

Lemniscate

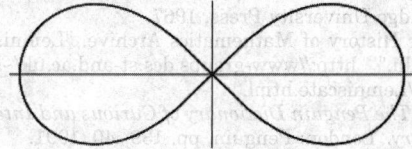

A polar curve also called LEMNISCATE OF BERNOULLI which is the LOCUS of points the product of whose

distances from two fixed points (called the FOCI) a distance $2a$ away is the constant a^2. Letting the FOCI be located at $(\pm a,\ 0)$, the Cartesian equation is

$$[(x-a)^2 + y^2][(x+a)^2 + y^2] = a^4, \qquad (1)$$

which can be rewritten

$$x^4 + y^4 + 2x^2 y^2 = 2a^2(x^2 - y^2). \qquad (2)$$

Letting $a' \equiv \sqrt{2}a$, the POLAR COORDINATES are given by

$$r^2 = a^2 \cos(2\theta). \qquad (3)$$

An alternate form is

$$r^2 = a^2 \sin(2\theta) \qquad (4)$$

The PARAMETRIC EQUATIONS for the lemniscate are

$$x = \frac{a \cos t}{1 + \sin^2 t}. \qquad (5)$$

$$y = \frac{a \sin t \cos t}{1 + \sin^2 t}. \qquad (6)$$

The bipolar equation of the lemniscate is

$$rr' = \tfrac{1}{2}\,a^2, \qquad (7)$$

and in PEDAL COORDINATES with the PEDAL POINT at the center, the equation is

$$pa^2 = r^3. \qquad (8)$$

The two-center BIPOLAR COORDINATES equation with origin at a FOCUS is

$$r_1 r_2 = c^2. \qquad (9)$$

The lemniscate can also be generated as the ENVELOPE of circles centered on a RECTANGULAR HYPERBOLA and passing through the center of the HYPERBOLA (Wells 1991).

Jakob Bernoulli published an article in *Acta Eruditorum* in 1694 in which he called this curve the *lemniscus* (Latin for "a pendant ribbon"). Jakob Bernoulli was not aware that the curve he was describing was a special case of CASSINI OVALS which had been described by Cassini in 1680. The general properties of the lemniscate were discovered by G. Fagnano in 1750 (MacTutor Archive). Gauss's and Euler's investigations of the ARC LENGTH of the curve led to later work on ELLIPTIC FUNCTIONS.

The lemniscate is the INVERSE CURVE of the HYPERBOLA with respect to its center.

The CURVATURE of the lemniscate is

$$\kappa = \frac{3\sqrt{2}\cos t}{\sqrt{3 - \cos(2t)}}. \tag{10}$$

The ARC LENGTH is more problematic. Using the polar form,

$$ds^2 = dr^2 + r^2 \, d\theta^2 \tag{11}$$

so

$$ds = \sqrt{1 + \left(r\frac{d\theta}{dr}\right)^2} \, dr. \tag{12}$$

But we have

$$2r \, dr = 2a^2 \sin(2\theta) \, d\theta \tag{13}$$

$$r\frac{dr}{d\theta} = \frac{r^2}{a^2 \sin(2\theta)} \tag{14}$$

$$\left(r\frac{d\theta}{dr}\right)^2 = \frac{r^4}{a^4 \sin^2(2\theta)} = \frac{r^4}{a^4[1 - \cos^2(2\theta)]}$$

$$= \frac{r^4}{a^4 - r^4}, \tag{15}$$

so

$$ds = \sqrt{1 + \frac{r^4}{a^4 - r^4}} \, dr = \sqrt{\frac{a^4}{a^4 - r^4}} \, dr = \frac{a^2}{\sqrt{a^4 - r^4}} \, dr$$

$$= \frac{dr}{\sqrt{1 - \left(\frac{r}{a}\right)^4}}, \tag{16}$$

and

$$L = \int_0^a ds = 2\int_0^a \frac{ds}{dr} \, dr = 2\int_0^a \frac{dr}{\sqrt{1 - \left(\frac{r}{a}\right)^4}}. \tag{17}$$

Let $t \equiv r/a$, so $dt = dr/a$, and

$$L = 2a\int_0^1 (1 - t^4)^{-1/2} \, dt \tag{18}$$

which, as shown in LEMNISCATE FUNCTION, is given analytically by

$$L = \sqrt{2}aK\left(\frac{1}{\sqrt{2}}\right) = \frac{\Gamma^2\left(\frac{1}{4}\right)}{2^{3/2}\sqrt{\pi}} \, a. \tag{19}$$

If $a = 1$, then

$$L = 5.2441151086\ldots \tag{20}$$

which is related to GAUSS'S CONSTANT M by

$$L = \frac{2\pi}{M}. \tag{21}$$

The quantity $L/2$ or $L/4$ is called the LEMNISCATE CONSTANT and plays a role for the lemniscate analogous to that of π for the CIRCLE.

The AREA of one loop of the lemniscate is

$$A = \frac{1}{2}\int r^2 \, d\theta = \frac{1}{2}a^2 \int_{-\pi/4}^{\pi/4} \cos(2\theta) \, d\theta$$

$$= \frac{1}{4}a^2[\sin(2\theta)]_{-\pi/4}^{+\pi/4}$$

$$= \frac{1}{2}a^2[\sin(2\theta)]_0^{\pi/4} = \frac{1}{2}a^2\left[\sin\left(\frac{\pi}{2}\right) - \sin 0\right] = \frac{1}{2}a^2. \tag{22}$$

See also LEMNISCATE FUNCTION, LICHTENFELS MINIMAL SURFACE

References

Ayoub, R. "The Lemniscate and Fagnano's Contributions to Elliptic Integrals." *Arch. Hist. Exact Sci.* **29**, 131–49, 1984.

Beyer, W. H. *CRC Standard Mathematical Tables, 28th ed.* Boca Raton, FL: CRC Press, p. 220, 1987.

Borwein, J. M. and Borwein, P. B. *Pi & the AGM: A Study in Analytic Number Theory and Computational Complexity.* New York: Wiley, 1987.

Gray, A. "Lemniscates of Bernoulli." §3.2 in *Modern Differential Geometry of Curves and Surfaces with Mathematica, 2nd ed.* Boca Raton, FL: CRC Press, pp. 52–3, 1997.

Lawrence, J. D. *A Catalog of Special Plane Curves.* New York: Dover, pp. 120–24, 1972.

Le Lionnais, F. *Les nombres remarquables.* Paris: Hermann, p. 37, 1983.

Lockwood, E. H. *A Book of Curves.* Cambridge, England: Cambridge University Press, 1967.

MacTutor History of Mathematics Archive. "Lemniscate of Bernoulli." http://www-groups.dcs.st-and.ac.uk/~history/Curves/Lemniscate.html.

Wells, D. *The Penguin Dictionary of Curious and Interesting Geometry.* London: Penguin, pp. 139–40, 1991.

Yates, R. C. "Lemniscate." *A Handbook on Curves and Their Properties.* Ann Arbor, MI: J. W. Edwards, pp. 143–47, 1952.

Lemniscate (Mandelbrot Set)

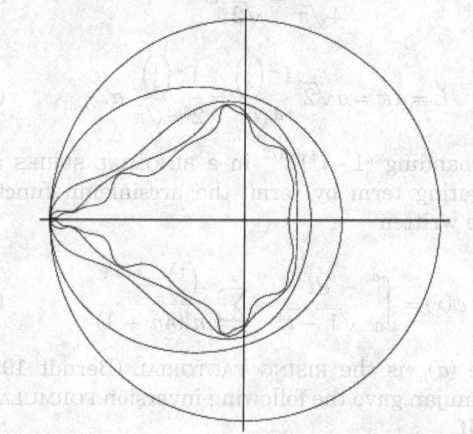

A curve on which points of a MAP z_n (such as the MANDELBROT SET) diverge to a given value r_{max} at the same rate. A common method of obtaining lemniscates is to define an INTEGER called the COUNT which is the largest n such that $|z_n| < r$ where r is usually taken as $r = 2$. Successive COUNTS then define a series of lemniscates, which are called EQUIPOTENTIAL CURVES by Peitgen and Saupe (1988).

See also COUNT, MANDELBROT SET

References

Peitgen, H.-O. and Saupe, D. (Eds.). *The Science of Fractal Images.* New York: Springer-Verlag, pp. 178–79, 1988.

Lemniscate Case

The case of the WEIERSTRASS ELLIPTIC FUNCTION with invariants $g_2 = 1$ and $g_3 = 0$.

See also EQUIANHARMONIC CASE, WEIERSTRASS ELLIPTIC FUNCTION, PSEUDOLEMNISCATE CASE

References

Abramowitz, M. and Stegun, C. A. (Eds.). "Lemniscate Case ($g_2 = 1$, $g_3 = 0$)." §18.14 in *Handbook of Mathematical Functions with Formulas, Graphs, and Mathematical Tables, 9th printing.* New York: Dover, pp. 658–62, 1972.

Lemniscate Constant

Let

$$L = \frac{1}{\sqrt{2\pi}} \left[\Gamma\left(\tfrac{1}{4}\right) \right]^2 = 5.2441151086\ldots$$

be the ARC LENGTH of a LEMNISCATE with $a = 1$. Then the lemniscate constant is the quantity $L/2$ (Abramowitz and Stegun 1972), or $L/4 = 1.311028777\ldots$ (Todd 1975, Le Lionnais 1983). Todd (1975) cites T. Schneider (1937) as proving L to be a TRANSCENDENTAL NUMBER.

See also LEMNISCATE

References

Abramowitz, M. and Stegun, C. A. (Eds.). *Handbook of Mathematical Functions with Formulas, Graphs, and Mathematical Tables, 9th printing.* New York: Dover, 1972.

Borwein, J. M. and Borwein, P. B. *Pi & the AGM: A Study in Analytic Number Theory and Computational Complexity.* New York: Wiley, 1987.

Finch, S. "Favorite Mathematical Constants." http://www.mathsoft.com/asolve/constant/gauss/gauss.html.

Le Lionnais, F. *Les nombres remarquables.* Paris: Hermann, p. 37, 1983.

Todd, J. "The Lemniscate Constant." *Comm. ACM* **18**, 14–9 and 462, 1975.

Lemniscate Function

The lemniscate functions arise in rectifying the ARC LENGTH of the LEMNISCATE. The lemniscate functions were first studied by Jakob Bernoulli and Giulio Fagnano. A historical account is given by Ayoub (1984), and an extensive discussion by Siegel (1969). The lemniscate functions were the first functions defined by inversion of an integral, which was first done by Gauss.

$$L = 2a \int_0^1 (1 - t^4)^{-1/2} \, dt. \tag{1}$$

Define the functions

$$\phi(x) \equiv \text{arcsinlemn } x \int_0^x (1 - t^4)^{-1/2} \, dt \tag{2}$$

$$\phi'(x) \equiv \text{arccoslemn } x = \int_x^1 (1 - t^4)^{-1/2} \, dt, \tag{3}$$

where

$$\varpi \equiv \frac{L}{a}, \tag{4}$$

and write

$$x = \text{sinlemn } \phi \tag{5}$$

$$x = \text{coslemn } \phi'. \tag{6}$$

There is an identity connecting ϕ and ϕ' since

$$\phi(x) + \phi'(x) = \frac{L}{2a} = \tfrac{1}{2} \varpi, \tag{7}$$

so

$$\text{sinlemn } \phi = \text{coslemn}\left(\tfrac{1}{2} \varpi - \phi\right). \tag{8}$$

These functions can be written in terms of JACOBI ELLIPTIC FUNCTIONS,

$$u = \int_0^{\text{sd}(u, \, k)} \left[(1 - k'^2 y^2)(1 + k^2 y^2)\right]^{-1/2} \, dy. \tag{9}$$

Now, if $k = k' = 1/\sqrt{2}$, then

$$u = \int_0^{\mathrm{sd}(u,\,1/\sqrt{2})} \left[\left(1 - \tfrac{1}{2}y^2\right)\left(1 + \tfrac{1}{2}y^2\right)\right]^{-1/2} dy$$

$$= \int_0^{\mathrm{sd}(u,\,1/\sqrt{2})} \left(1 - \tfrac{1}{4}y^4\right)^{-1/2} dy. \tag{10}$$

Let $t \equiv y/\sqrt{2}$ so $dy = \sqrt{2}\,dt$,

$$u = \sqrt{2} \int_0^{\mathrm{sd}(u,\,1/\sqrt{2})/\sqrt{2}} (1 - t^4)^{-1/2}\,dt \tag{11}$$

$$\frac{u}{\sqrt{2}} = \int_0^{\mathrm{sd}(u,\,1/\sqrt{2})/\sqrt{2}} (1 - t^4)^{-1/2}\,dt \tag{12}$$

$$u = \int_0^{\mathrm{sd}(u\sqrt{2},\,1/\sqrt{2})/\sqrt{2}} (1 - t^4)^{-1/2}\,dt \tag{13}$$

and

$$\mathrm{sinlemn}\,\phi = \frac{1}{\sqrt{2}}\,\mathrm{sd}\left(\phi\sqrt{2},\,\frac{1}{\sqrt{2}}\right). \tag{14}$$

Similarly,

$$u = \int_{\mathrm{cn}(u,\,k)}^1 (1 - t^2)^{-1/2}(k'^2 + k^2 t^2)^{-1/2}\,dt$$

$$= \int_{\mathrm{cn}(u,\,1/\sqrt{2})}^1 (1 - t^2)^{-1/2}\left(\tfrac{1}{2} + \tfrac{1}{2}t^2\right)^{-1/2}\,dt$$

$$= \sqrt{2} \int_{\mathrm{cn}(u,\,1/\sqrt{2})}^1 (1 - t^4)^{-1/2}\,dt \tag{15}$$

$$\frac{u}{\sqrt{2}} = \int_{\mathrm{cn}(u,\,1/\sqrt{2})}^1 (1 - t^4)^{-1/2}\,dt \tag{16}$$

$$u = \int_{\mathrm{cn}(u\sqrt{2},\,1/\sqrt{2})}^1 (1 - t^4)^{-1/2}\,dt, \tag{17}$$

and

$$\mathrm{coslemn}\,\phi = \mathrm{cn}\left(\phi\sqrt{2},\,\frac{1}{\sqrt{2}}\right). \tag{18}$$

We know

$$\mathrm{coslemn}\left(\tfrac{1}{2}\varpi\right) = \mathrm{cn}\left(\tfrac{1}{2}\varpi\sqrt{2},\,\frac{1}{\sqrt{2}}\right) = 0. \tag{19}$$

But it is true that

$$\mathrm{cn}(K,\,k) = 0, \tag{20}$$

so

$$K\left(\frac{1}{\sqrt{2}}\right) = \tfrac{1}{2}\sqrt{2}\varpi = \frac{1}{\sqrt{2}}\varpi \tag{21}$$

$$\frac{\Gamma^2\left(\tfrac{1}{4}\right)}{4\sqrt{\pi}} = \frac{1}{\sqrt{2}}\varpi \tag{22}$$

$$L = a\varpi = a\sqrt{2}\,\frac{\Gamma^2\left(\tfrac{1}{4}\right)}{4\sqrt{\pi}} = \frac{\Gamma^2\left(\tfrac{1}{4}\right)}{2^{3/2}\sqrt{\pi}}\,a. \tag{23}$$

By expanding $(1 - t^4)^{-1/2}$ in a BINOMIAL SERIES and integrating term by term, the arcsinlemn function can be written

$$\phi(x) = \int_0^v \frac{dt}{\sqrt{1 - t^4}} = \sum_{n=0}^{\infty} \frac{\left(\tfrac{1}{2}\right)_n}{n!(4n+1)} x^{4n+1}, \tag{24}$$

where $(a)_n$ is the RISING FACTORIAL (Berndt 1994). Ramanujan gave the following inversion FORMULA for $\phi(x)$. If

$$\frac{\theta\mu}{\sqrt{2}} = \sum_{n=0}^{\infty} \frac{\left(\tfrac{1}{2}\right)_n}{n!(4n+1)} x^{4n+1}, \tag{25}$$

where

$$\mu = \frac{\Gamma^2\left(\tfrac{1}{4}\right)}{2\pi^{3/2}} \tag{26}$$

is the constant obtained by letting $x = 1$ and $\theta = \pi/2$, and

$$v = 2^{-1/2}\mathrm{sd}(\mu\theta), \tag{27}$$

then

$$\frac{\mu^2}{2x^2} = \csc^2\theta - \frac{1}{\pi} - 8\sum_{n=1}^{\infty} \frac{n\cos(2n\theta)}{e^{2\pi n} - 1} \tag{28}$$

(Berndt 1994). Ramanujan also showed that if $0 < \theta < \pi/2$, then

$$-\frac{\mu}{\sqrt{2}} \sum_{n=0}^{\infty} \frac{\left(\tfrac{1}{2}\right)_n}{n!(4n-1)} v^{4n-1}$$

$$= \cot\theta + \frac{\theta}{\pi} + 4\sum_{n=1}^{\infty} \frac{\sin(2n\theta)}{2^{2\pi n} - 1}, \tag{29}$$

$$\ln v + \tfrac{1}{6}\pi - \tfrac{1}{2}\ln 2 + \sum_{n=0}^{\infty} \frac{\left(\tfrac{1}{4}\right)_n}{\left(\tfrac{3}{4}\right)_n} \frac{v^{4n}}{4n}$$

$$= \ln(\sin\theta) + \frac{\theta^2}{2\pi} - 2\sum_{n=1}^{\infty} \frac{\cos(2n\theta)}{n(e^{2\pi n} - 1)}, \tag{30}$$

$$\tfrac{1}{2}\tan^{-1} v = \sum_{n=0}^{\infty} \frac{\sin[(2n+1)\theta]}{(2n+1)\cosh\left[\tfrac{1}{2}(2n+1)\pi\right]}, \tag{31}$$

$$\tfrac{1}{4}\cos^{-1}(v^2) = \sum_{n=0}^{\infty} \frac{(-1)^n\cos[(2n+1)\theta]}{(2n+1)\cosh\left[\tfrac{1}{2}(2n+1)\pi\right]}, \tag{32}$$

and

$$\frac{\sqrt{2}}{4\mu} \sum_{n=0}^{\infty} \frac{2^{2n}(n!)^2}{(2n+1)!(4n+3)} v^{4n+3}$$

$$= \frac{\pi\theta}{8} - \sum_{n=0}^{\infty} \frac{(-1)^n \sin[(2n+1)\theta]}{(2n+1)^2 \cosh\left[\frac{1}{2}(2n+1)\pi\right]} \quad (33)$$

(Berndt 1994).

A generalized version of the lemniscate function can be defined by letting $0 \le \theta \le \pi/2$ and $0 \le v \le 1$. Write

$$\frac{2}{3}\theta\mu = \int_0^v \frac{dt}{\sqrt{1-t^6}}, \quad (34)$$

where μ is the constant obtained by setting $\theta = \pi/2$ and $v = 1$. Then

$$\mu = \frac{\sqrt{\pi}}{\Gamma\left(\frac{2}{3}\right)\Gamma\left(\frac{5}{6}\right)}, \quad (35)$$

and Ramanujan showed

$$\frac{4\mu^2}{9v^2} = \csc^2\theta - \frac{2}{\pi\sqrt{3}} + 8 \sum_{n=1}^{\infty} \frac{(-1)^{n-1}n\cos(2n\theta)}{e^{\pi n\sqrt{3}} - (-1)^n} \quad (36)$$

(Berndt 1994).

See also ELLIPTIC FUNCTION, ELLIPTIC INTEGRAL, HYPERBOLIC LEMNISCATE FUNCTION

References

Ayoub, R. "The Lemniscate and Fagnano's Contributions to Elliptic Integrals." *Arch. Hist. Exact Sci.* **29**, 131–49, 1984.

Berndt, B. C. *Ramanujan's Notebooks, Part IV.* New York: Springer-Verlag, pp. 245, and 247–55, 258–60, 1994.

Siegel, C. L. *Topics in Complex Function Theory, Vol. 1.* New York: Wiley, 1969.

Lemniscate Inverse Curve

The INVERSE CURVE of a LEMNISCATE in a CIRCLE centered at the origin and touching the LEMNISCATE where it crosses the x-AXIS produces a RECTANGULAR HYPERBOLA (Wells 1991).

See also RECTANGULAR HYPERBOLA

References

Wells, D. *The Penguin Dictionary of Curious and Interesting Geometry.* London: Penguin, p. 209, 1991.

Lemniscate of Bernoulli

LEMNISCATE

Lemniscate of Gerono

EIGHT CURVE

Lemoine Axis

LEMOINE LINE

Lemoine Circle

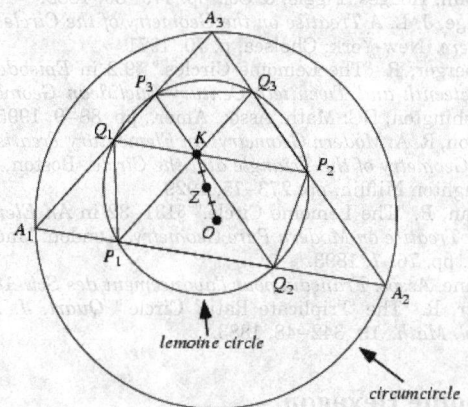

Draw lines P_1Q_1, P_2Q_2, and P_3Q_3 through the SYMMEDIAN POINT K and parallel to the sides of the triangle $\Delta A_1A_2A_3$. The points where the parallel lines intersect the sides of $\Delta A_1A_2A_3$ then lie on a CIRCLE known as the Lemoine circle, or sometimes the TRIPLICATE-RATIO CIRCLE (Tucker 1883). This circle has center at the MIDPOINT Z of OK, where O is the CIRCUMCENTER, and RADIUS

$$\frac{1}{2}\sqrt{R^2 + r_c^2} = \frac{1}{2}R \sec \omega,$$

where R is the CIRCUMRADIUS, r_c is RADIUS of the COSINE CIRCLE, and ω is the BROCARD ANGLE of the original triangle (Johnson 1929, p. 274). The Lemoine circle and BROCARD CIRCLE are concentric, and the triangles ΔQ_1P_3K, ΔKQ_3P_2, and ΔP_1KQ_2 are similar to $\Delta A_1A_3A_2$ (Tucker 1883).

The Lemoine circle divides any side into segments proportional to the squares of the sides

$$\overline{A_2P_2} : \overline{P_2Q_3} : \overline{Q_3A_3} = a_3^2 : a_1^2 : a_2^2$$

Furthermore, the chords cut from the sides by the Lemoine circle are proportional to the squares of the sides.

The COSINE CIRCLE is sometimes called the second Lemoine circle. The Lemoine circle is a special case of a TUCKER CIRCLE.

See also COSINE CIRCLE, LEMOINE HEXAGON, LEMOINE LINE, SYMMEDIAN POINT, TAYLOR CIRCLE, TUCKER CIRCLES

References

Casey, J. "On the Equations and Properties--(1) of the System of Circles Touching Three Circles in a Plane; (2) of the System of Spheres Touching Four Spheres in Space; (3) of the System of Circles Touching Three Circles on a Sphere; (4) of the System of Conics Inscribed to a Conic, and Touching Three Inscribed Conics in a Plane." *Proc. Roy. Irish Acad.* **9**, 396–23, 1864–866.

Casey, J. "Lemoine's, Tucker's, and Taylor's Circle." Supp. Ch. §3 in *A Sequel to the First Six Books of the Elements of Euclid, Containing an Easy Introduction to Modern*

Geometry with Numerous Examples, 5th ed., rev. enl. Dublin: Hodges, Figgis, & Co., pp. 179–89, 1888.

Coolidge, J. L. *A Treatise on the Geometry of the Circle and Sphere.* New York: Chelsea, p. 70, 1971.

Honsberger, R. "The Lemoine Circles." §9.2 in *Episodes in Nineteenth and Twentieth Century Euclidean Geometry.* Washington, DC: Math. Assoc. Amer., pp. 88–9, 1995.

Johnson, R. A. *Modern Geometry: An Elementary Treatise on the Geometry of the Triangle and the Circle.* Boston, MA: Houghton Mifflin, pp. 273–75, 1929.

Lachlan, R. "The Lemoine Circle." §131–32 in *An Elementary Treatise on Modern Pure Geometry.* London: Macmillian, pp. 76–7, 1893.

Lemoine. *Assoc. Français pour l'avancement des Sci.* 1873.

Tucker, R. "The 'Triplicate Ratio' Circle." *Quart. J. Pure Appl. Math.* **19**, 342–48, 1883.

Lemoine Hexagon

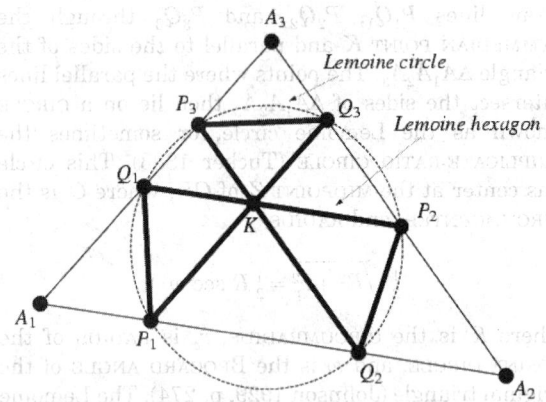

The closed self-intersecting cyclic hexagon formed by joining the adjacent PARALLELS in the construction of the LEMOINE CIRCLE. The sides of this hexagon have the property that, in addition to $Q_1P_2 \| A_1A_2$, $Q_2P_3 \| A_2A_3$, and $Q_3P_2 \| A_1A_3$, the remaining sides Q_1P_1, Q_2P_2, and Q_3P_3 are ANTIPARALLEL to A_2A_3, A_1A_3, and A_1A_2, respectively. The Lemoine hexagon is a special case of a TUCKER HEXAGON.

See also COSINE HEXAGON, LEMOINE CIRCLE, TUCKER HEXAGON

Lemoine Line

The Lemoine line, also called the LEMOINE AXIS, is the perspectivity axis of a TRIANGLE and its TANGENTIAL TRIANGLE, and also the TRILINEAR POLAR of the CENTROID of the triangle vertices. It is also the POLAR of K with regard to its CIRCUMCIRCLE, and is PERPENDICULAR to the BROCARD AXIS.

The centers of the APOLLONIUS CIRCLES L_1, L_2, and L_3 are COLLINEAR on the LEMOINE LINE. This line is PERPENDICULAR to the BROCARD AXIS OK and is the RADICAL AXIS of the CIRCUMCIRCLE and the BROCARD CIRCLE. It has equation

$$\frac{\alpha}{a} + \frac{\beta}{b} + \frac{\gamma}{c}$$

in terms of TRILINEAR COORDINATES (Oldknow 1996).

See also APOLLONIUS CIRCLES, BROCARD AXIS, CENTROID (TRIANGLE), CIRCUMCIRCLE, COLLINEAR, LEMOINE CIRCLE, SYMMEDIAN POINT, POLAR, RADICAL AXIS, SYMMEDIAN, TANGENTIAL TRIANGLE, TRILINEAR POLAR

References

Johnson, R. A. *Modern Geometry: An Elementary Treatise on the Geometry of the Triangle and the Circle.* Boston, MA: Houghton Mifflin, p. 295, 1929.

Oldknow, A. "The Euler-Gergonne-Soddy Triangle of a Triangle." *Amer. Math. Monthly* **103**, 319–29, 1996.

Lemoine Point
SYMMEDIAN POINT

Lemoine's Problem

Given the vertices of the three EQUILATERAL TRIANGLES placed on the sides of a TRIANGLE T, construct T. The solution can be given using KIEPERT'S HYPERBOLA.

See also KIEPERT'S HYPERBOLA

Lemon

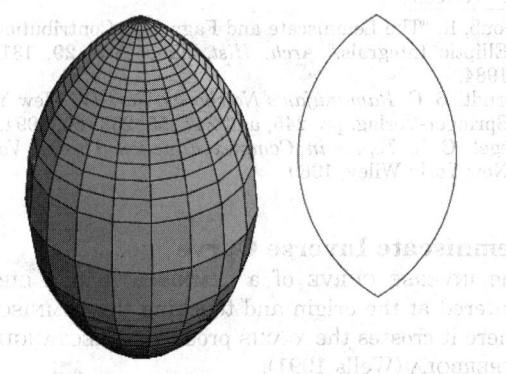

A SURFACE OF REVOLUTION defined by Kepler. It consists of less than half of a circular ARC rotated about an axis passing through the endpoints of the ARC. The equations of the upper and lower boundaries in the xz plane are

$$z_{\pm} = \pm\sqrt{R^2 - (x + r)^2}$$

for $R > r$ and $x \in [-(R-r), R-r]$. The CROSS SECTION of a lemon is a LENS. The lemon is the inside surface of a SPINDLE TORUS. The American football is shaped like a lemon.

See also APPLE, LENS, OVAL, PROLATE SPHEROID, SPINDLE TORUS

Length (Curve)

References
JavaView. "Classic Surfaces from Differential Geometry: Football/Barrel." http://www-sfb288.math.tu-berlin.de/vgp/javaview/demo/surface/common/PaSurface_Football-Barrel.html.

Length (Curve)

Let $\gamma(t)$ be a smooth curve in a MANIFOLD M from x to y with $\gamma(0) = x$ and $\gamma(1) = y$. Then $\gamma'(t) \in T_{\gamma(t)}$ where T_x is the TANGENT SPACE of M at x. The length of γ with respect to the Riemannian structure is given by

$$\int_0^1 \|\gamma'(t)\|_{\gamma(t)} \, dt.$$

See also ARC LENGTH, DISTANCE

Length (Number)

The length of a number n in base b is the number of DIGITS in the base-b numeral for n, given by the formula

$$L(n, b) = \lfloor \log_b(n) \rfloor + 1,$$

where $\lfloor x \rfloor$ is the FLOOR FUNCTION.

The MULTIPLICATIVE PERSISTENCE of an n-DIGIT is sometimes also called its length.

See also CONCATENATION, DIGIT, FIGURES, MULTIPLICATIVE PERSISTENCE

Length (Partial Order)

For a PARTIAL ORDER, the size of the longest CHAIN is called the length.

See also WIDTH (PARTIAL ORDER)

Length (Size)

The longest dimension of a 3-D object.

See also HEIGHT, WIDTH (SIZE)

Length Distribution Function

A function giving the distribution of the interpoint distances of a curve. It is defined by

$$p(r) = \frac{1}{N} \sum_{ij} \delta_{rij} = r.$$

See also RADIUS OF GYRATION

References
Pickover, C. A. *Keys to Infinity.* New York: Wiley, pp. 204–06, 1995.

Length-Preserving Transformation
ISOMETRY

Lengyel's Constant

N.B. A detailed online essay by S. Finch was the starting point for this entry.

Let L denote the partition lattice of the SET $\{1, 2, \ldots, n\}$. The MAXIMUM element of L is

$$M = \{\{1, 2, \ldots, n\}\} \tag{1}$$

and the MINIMUM element is

$$m = \{\{1\}, \{2\}, \ldots, \{n\}\}. \tag{2}$$

Let Z_n denote the number of chains of any length in L containing both M and m. Then Z_n satisfies the RECURRENCE RELATION

$$Z_n = \sum_{k=1}^{n-1} s(n, k) Z_k, \tag{3}$$

where $s(n, k)$ is a STIRLING NUMBER OF THE SECOND KIND. Lengyel (1984) proved that the QUOTIENT

$$r(n) = \frac{Z_n}{(n!)^2 (2 \ln 2)^{-n} n^{1 - (\ln 2)/3}} \tag{4}$$

is bounded between two constants as $n \to \infty$, and Flajolet and Salvy (1990) improved the result of Babai and Lengyel (1992) to show that

$$\Lambda \equiv \lim_{n \to \infty} r(n) = 1.0986858055\ldots. \tag{5}$$

References
Babai, L. and Lengyel, T. "A Convergence Criterion for Recurrent Sequences with Application to the Partition Lattice." *Analysis* **12**, 109–19, 1992.
Finch, S. "Favorite Mathematical Constants." http://www.mathsoft.com/asolve/constant/lngy/lngy.html.
Flajolet, P. and Salvy, B. "Hierarchal Set Partitions and Analytic Iterates of the Exponential Function." Unpublished manuscript, 1990.
Lengyel, T. "On a Recurrence Involving Stirling Numbers." *Europ. J. Comb.* **5**, 313–21, 1984.
Plouffe, S. "The Lengyel Constant." http://www.lacim.u-qam.ca/piDATA/lengyel.txt.

Lens

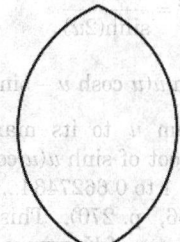

A figure composed of two equal and symmetrically placed circular ARCS. It is also known as the FISH BLADDER (Pedoe 1995, p. xii) or VESICA PISCIS. The latter term is often used for the particular lens formed by the intersection of two unit CIRCLES whose centers are offset by a unit distance (Rawles 1997). In this case, the height of the lens is given by letting $d = r = R = 1$ in the equation for a CIRCLE-CIRCLE INTERSECTION

$$a = \frac{1}{d} \sqrt{4d^2R^2 - (d^2 - r^2 + R^2)^2}, \qquad (1)$$

giving $a = \sqrt{3}$. The AREA of the VESICA PISCIS is given by plugging $d = R$ into the CIRCLE-CIRCLE INTERSECTION area equation with $r = R$,

$$A = 2R^2 \cos^{-1}\left(\frac{d}{2R}\right) - \tfrac{1}{2}d\sqrt{4R^2 - d^2}, \qquad (2)$$

giving

$$A = \tfrac{1}{6}\left(4\pi - 3\sqrt{3}\right) \approx 1.22837. \qquad (3)$$

Renaissance artists frequently surrounded images of Jesus with the vesica piscis (Rawles 1997). An asymmetrical lens is produced by a CIRCLE-CIRCLE INTERSECTION for unequal CIRCLES.

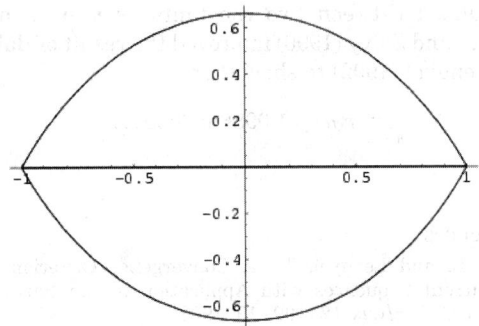

A lens-shaped region also arises in the study of BESSEL FUNCTIONS. Letting $z = e^{i\theta}$, the inequality

$$\left| \frac{z \exp(1 - z^2)}{1 + \sqrt{1 - z^2}} \right| \leq 1$$

holds in the region illustrated above. This region can be parameterized in terms of a variable u as

$$r^2 = \frac{2u}{\sinh(2u)} \qquad (4)$$

$$\sin^2 \theta = \sinh u(u \cosh u - \sinh u). \qquad (5)$$

As u increases from u to its maximum value of 1.19967874... (the root of $\sinh u(u \cosh u - \sinh u) = 0$), r decreases from 1 to 0.6627434... (Plummer 1960, p. 47; Watson 1966, p. 270). This curve is very important in the theory of KAPTEYN SERIES.

See also CIRCLE, CIRCLE-CIRCLE INTERSECTION, DOUBLE BUBBLE, FLOWER OF LIFE, GOAT PROBLEM, KAPTEYN SERIES, LEMON, LUNE, REULEAUX TRIANGLE, SECTOR, SEED OF LIFE, SEGMENT, VENN DIAGRAM

References
Pedoe, D. *Circles: A Mathematical View, rev. ed.* Washington, DC: Math. Assoc. Amer., 1995.
Plummer, H. *An Introductory Treatise of Dynamical Astronomy.* New York: Dover, 1960.
Rawles, B. *Sacred Geometry Design Sourcebook: Universal Dimensional Patterns.* Nevada City, CA: Elysian Pub., p. 11, 1997.
Watson, G. N. *A Treatise on the Theory of Bessel Functions, 2nd ed.* Cambridge, England: Cambridge University Press, 1966.

Lens Space

A lens space $L(p, q)$ is the 3-MANIFOLD obtained by gluing the boundaries of two solid TORI together such that the meridian of the first goes to a (p, q)-curve on the second, where a (p, q)-curve has p meridians and q longitudes.

References
Adams, C. C. "The Three-Sphere and Lens Spaces." §9.2 in *The Knot Book: An Elementary Introduction to the Mathematical Theory of Knots.* New York: W. H. Freeman, pp. 246–56, 1994.
Rolfsen, D. *Knots and Links.* Wilmington, DE: Publish or Perish Press, 1976.

Lenstra Elliptic Curve Method

A method of factoring INTEGERS using ELLIPTIC CURVES.

References
Montgomery, P. L. "Speeding up the Pollard and Elliptic Curve Methods of Factorization." *Math. Comput.* **48**, 243–64, 1987.

Léon Anne's Theorem

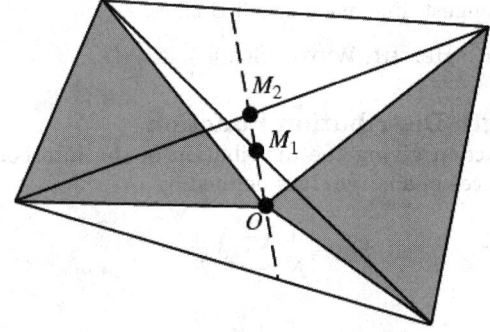

Pick a point O in the interior of a QUADRILATERAL which is not a PARALLELOGRAM. Join this point to each of the four VERTICES, then the LOCUS of points O for which the sum of opposite TRIANGLE areas is half the QUADRILATERAL AREA is the line joining the MIDPOINTS M_1 and M_2 of the DIAGONALS.

See also DIAGONAL (POLYGON), MIDPOINT, QUADRILATERAL

References

Honsberger, R. *More Mathematical Morsels.* Washington, DC: Math. Assoc. Amer., pp. 174–75, 1991.

Leonardo's Paradox

In the depiction of a row of identical columns parallel to the plane of a PERSPECTIVE drawing, the outer columns should appear wider even though they are farther away.

See also PERSPECTIVE, VANISHING POINT, ZEEMAN'S PARADOX

References

Dixon, R. *Mathographics.* New York: Dover, p. 82, 1991.

Leptokurtic

A distribution with a high peak so that the KURTOSIS satisfies $\gamma_2 > 0$.

See also KURTOSIS

LerchPhi

LERCH TRANSCENDENT

Lerch's Theorem

If there are two functions $F_1(t)$ and $F_2(t)$ with the same integral transform

$$\mathcal{T}[F_1(t)] = \mathcal{T}[F_2(t)] \equiv f(s), \qquad (1)$$

then a NULL FUNCTION can be defined by

$$\delta_0(t) \equiv F_1(t) - F_2(t) \qquad (2)$$

so that the integral

$$\int_0^a \delta_0(t)\, dt = 0 \qquad (3)$$

vanishes for all $a > 0$.

See also NULL FUNCTION

Lerch Transcendent

A generalization of the HURWITZ ZETA FUNCTION and POLYLOGARITHM function. Many sums of reciprocal POWERS can be expressed in terms of it. It is defined by

$$\Phi(z, s, a) \equiv \sum_{k=0}^{\infty} \frac{z^k}{(a+k)^s}, \qquad (1)$$

where any term with $a + k = 0$ is excluded. The Lerch transcendent is given by the *Mathematica* command `LerchPhi[z, s, a]`.

The Lerch transcendent can be used to express the DIRICHLET BETA FUNCTION

$$\beta(s) \equiv \sum_{k=0}^{\infty} (-1)^k (2k+1)^{-s} 2^{-s} \Phi\left(-1,\, s,\, \tfrac{1}{2}\right), \qquad (2)$$

the integral of the FERMI-DIRAC DISTRIBUTION

$$\int_0^{\infty} \frac{k^s}{e^{k-\mu} + 1}\, dk = e^{\mu} \Gamma(s+1) \Phi(-e^{\mu},\, s+1, 1), \qquad (3)$$

where $\Gamma(z)$ is the GAMMA FUNCTION, and to evaluate the DIRICHLET L-SERIES.

See also DIRICHLET BETA FUNCTION, DIRICHLET L-SERIES, FERMI-DIRAC DISTRIBUTION, HURWITZ ZETA FUNCTION, LEGENDRE'S CHI-FUNCTION, POLYLOGARITHM

References

Erdélyi, A.; Magnus, W.; Oberhettinger, F.; and Tricomi, F. G. "The Function $\Psi(z,\, s,\, v) = \sum_{n=0}^{\infty}(v+n)^{-s}z^n$." §1.11 in *Higher Transcendental Functions, Vol. 1.* New York: Krieger, pp. 27–1, 1981.

Less

A quantity a is said to be less than b if a is smaller than b, written $a < b$. If a is less than or EQUAL to b, the relationship is written $a \leq b$. If a is MUCH LESS than b, this is written $a \ll b$. Statements involving GREATER than and less than symbols are called INEQUALITIES.

See also EQUAL, GREATER, INEQUALITY, MUCH GREATER, MUCH LESS

Lester Circle

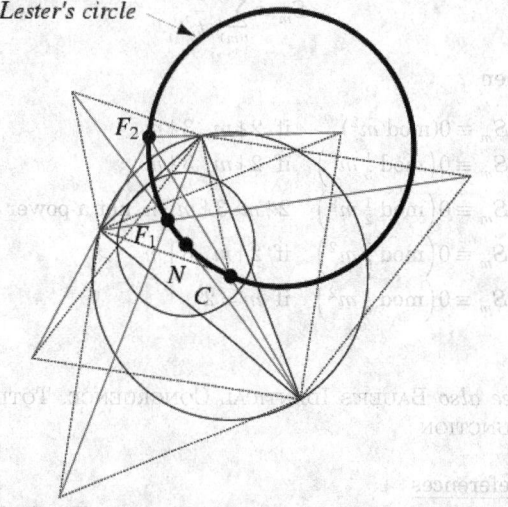

The CIRCUMCENTER C, NINE-POINT CENTER N, and the first and second FERMAT POINTS F_1 and F_2 of a triangle lie on a circle known as the Lester circle.

See also CIRCUMCENTER, FERMAT POINTS, NINE-POINT CENTER

References

Kimberling, C. "Lester Circle." *Math. Teacher* **89**, 26, 1996.
Lester, J. "Triangles III: Complex Triangle Functions." *Aequationes Math.* **53**, 4–5, 1997.
Trott, M. "Applying `GroebnerBasis` to Three Problems in Geometry." *Mathematica Educ. Res.* **6**, 15–8, 1997.
Trott, M. "A Proof of Lester's Circle Theorem." http://library.wolfram.com/demos/v3/GeometryProof.nb.

L-Estimate

A ROBUST ESTIMATION based on LINEAR COMBINATIONS of ORDER STATISTICS. Examples include the MEDIAN and TUKEY'S TRIMEAN.

See also M-ESTIMATE, R-ESTIMATE

References

Press, W. H.; Flannery, B. P.; Teukolsky, S. A.; and Vetterling, W. T. "Robust Estimation." §15.7 in *Numerical Recipes in FORTRAN: The Art of Scientific Computing, 2nd ed.* Cambridge, England: Cambridge University Press, pp. 694–00, 1992.

Letter-Value Display

A method of displaying simple statistical parameters including HINGES, MEDIAN, and upper and lower values.

References

Tukey, J. W. *Explanatory Data Analysis.* Reading, MA: Addison-Wesley, p. 33, 1977.

Leudesdorf Theorem

Let $t(m)$ denote the set of the $\phi(m)$ numbers less than and RELATIVELY PRIME to m, where $\phi(n)$ is the TOTIENT FUNCTION. Then if

$$S_m \equiv \sum_{t(m)} \frac{1}{t},$$

then

$$
\begin{cases}
S_m \equiv 0 (\mathrm{mod}\ m^2) & \text{if } 2 \nmid m,\ 3 \nmid m \\
S_m \equiv 0 \left(\mathrm{mod}\ \frac{1}{3} m^2\right) & \text{if } 2 \nmid m,\ 3 \nmid m \\
S_m \equiv 0 \left(\mathrm{mod}\ \frac{1}{2} m^2\right) & 2 \nmid m,\ 3 \nmid m,\ m \text{ not a power of 2} \\
S_m \equiv 0 \left(\mathrm{mod}\ \frac{1}{6} m^2\right) & \text{if } 2 \nmid m,\ 3 \nmid m \\
S_m \equiv 0 \left(\mathrm{mod}\ \frac{1}{4} m^2\right) & \text{if } m = 2^a.
\end{cases}
$$

See also BAUER'S IDENTICAL CONGRUENCE, TOTIENT FUNCTION

References

Hardy, G. H. and Wright, E. M. "A Theorem of Leudesdorf." §8.7 in *An Introduction to the Theory of Numbers, 5th ed.* Oxford, England: Clarendon Press, pp. 100–02, 1979.

Level Curve

A LEVEL SET in 2-D.

See also CONTOUR PLOT, EQUIPOTENTIAL CURVE, LEVEL SURFACE

Level Set

The level set of c is the SET of points

$$\{(x_1, \ldots, x_n) \in U : f(x_1, \ldots, x_n) = c\} \in \mathbb{R}^n,$$

and is in the DOMAIN of the function. If $n = 2$, the level set is a plane curve (a LEVEL CURVE). If $n = 3$, the level set is a surface (a level surface).

See also CONTOUR PLOT, EQUIPOTENTIAL CURVE, LEVEL CURVE, LEVEL SURFACE

References

Gray, A. "Level Surfaces in $\mathbb{R}^3$." §12.7 in *Modern Differential Geometry of Curves and Surfaces with Mathematica, 2nd ed.* Boca Raton, FL: CRC Press, pp. 291–93, 1997.

Level Surface

A LEVEL SET in 3-D.

Levenberg-Marquardt Method

Levenberg-Marquardt is a popular alternative to the Gauss-Newton method of finding the minimum of a function $F(x)$ that is a sum of squares of nonlinear functions,

$$F(x) = \frac{1}{2} \sum_{i=1}^{m} [f_i(x)]^2.$$

Let the JACOBIAN of $f_i(x)$ be denoted $J_i(x)$, then the Levenberg-Marquardt method searches in the direction given by the solution **p** to the equations

$$(J_k^{\mathrm{T}} J) + \lambda_k \mathsf{I}) p_k = -J_k^{\mathrm{T}} f_k,$$

where λ_k are nonnegative scalars and I is the IDENTITY MATRIX. The method has the nice property that, for some scalar Δ related to λ_k, the vector p_k is the solution of the constrained subproblem of minimizing $\|J_k p + f_k\|_2^2 / 2$ subject to $\|p\|_2 \le \Delta$ (Gill *et al.* 1981, p. 136).

The method is used by the *Mathematica* 4.0 command `FindMinimum[f, {x, x0}]` when given the `Method->LevenbergMarquardt` option.

See also MINIMUM, OPTIMIZATION

References

Gill, P. R.; Murray, W.; and Wright, M. H. "The Levenberg-Marquardt Method." §4.7.3 in *Practical Optimization.* London: Academic Press, pp. 136–37, 1981.
Levenberg, K. "A Method for the Solution of Certain Problems in Least Squares." *Quart. Appl. Math.* **2**, 164–68, 1944.
Marquardt, D. "An Algorithm for Least-Squares Estimation of Nonlinear Parameters." *SIAM J. Appl. Math.* **11**, 431–41, 1963.

Leviathan Number

The number $(10^{666})!$, where 666 is the BEAST NUMBER and $n!$ denotes a FACTORIAL. The number of trailing zeros in the Leviathan number is $25 \times 10^{664} - 143$ (Pickover 1995).

See also 666, APOCALYPSE NUMBER, APOCALYPTIC NUMBER, BEAST NUMBER

References

Pickover, C. A. *Keys to Infinity*. New York: Wiley, pp. 97–02, 1995.

Levi-Civita Connection

On a RIEMANNIAN MANIFOLD M, there is a canonical CONNECTION called the Levi-Civita connection (pronounced le-ve shi-vit-), sometimes also known as the Riemannian connection or COVARIANT DERIVATIVE. As a CONNECTION on the TANGENT BUNDLE, it provides a well-defined method for differentiating VECTOR FIELDS, forms, or any other kind of TENSOR. The theorem asserting the existence of the Levi-Civita connection, which is the unique TORSION-free CONNECTION ∇ on the TANGENT BUNDLE TM compatible with the metric, is called the FUNDAMENTAL THEOREM OF RIEMANNIAN GEOMETRY.

These properties can be described as follows. Let X, Y, and Z be any VECTOR FIELDS, and $\langle , \rangle$ denote the METRIC. Recall that vector fields act as DERIVATIONS on the ring of smooth functions by the DIRECTIONAL DERIVATIVE, and that this action extends to an action on vector fields. The notation $[X, Y]$ is the COMMUTATOR of vector fields, $XY - YX$. The Levi-Civita connection is torsion-free, meaning

$$\nabla_X \nabla_Y Z - \nabla_Y \nabla_X Z = \nabla_{[X, Y]} Z, \tag{1}$$

and is compatible with the metric

$$X(Y, Z) = \langle \nabla_X Y, Z \rangle + \langle Y, \nabla_X Z \rangle. \tag{2}$$

In coordinates, the Levi-Civita connection can be described using the CHRISTOFFEL SYMBOLS OF THE SECOND KIND $\Gamma^k_{i, j}$. In particular, if $e_i = \partial/\partial x_i$, then

$$\Gamma^k_{i, j} = \langle \nabla_{e_i} e_j, e_k \rangle, \tag{3}$$

or in other words,

$$\nabla_{e_i} e_j = \sum_k \Gamma^k_{i, j} e_k. \tag{4}$$

As a CONNECTION on the TANGENT BUNDLE TM, it induces a connection on the DUAL BUNDLE T^*M and on all their TENSOR PRODUCTS $TM^k \otimes TM^{*l}$. Also, given a SUBMANIFOLD N it restricts to TN to give the Levi-Civita connection from the restriction of the metric to N.

The Levi-Civita connection can be used to describe many intrinsic geometric objects. For instance, a path $c : \mathbb{R} \to M$ is a geodesic IFF $\nabla_{\dot{c}(t)} \dot{c}(t) = 0$ where $\dot{c}$ is the

path's TANGENT VECTOR. On a more general path c, the equation $\nabla_{\dot{c}(t)} v(t) = 0$ defines PARALLEL TRANSPORT for a VECTOR FIELD v along c. The SECOND FUNDAMENTAL FORM **II** of a submanifold N is given by $\pi_Q \circ \nabla_{TN}$ where TN is the TANGENT BUNDLE of N and π_Q is projection onto the NORMAL BUNDLE Q. The CURVATURE of M is given by $\nabla \circ \nabla$.

See also CHRISTOFFEL SYMBOL, CONNECTION, COVARIANT DERIVATIVE, CURVATURE, FUNDAMENTAL THEOREM OF RIEMANNIAN GEOMETRY, GEODESIC, PRINCIPAL BUNDLE, RIEMANNIAN MANIFOLD, RIEMANNIAN METRIC

References

Carmo, M. *Differential Geometry of Curves and Surfaces.* Englewood Cliffs, NJ: Prentice-Hall, pp. 441–42, 1976.
Gallot, S.; Hulin, D.; and Lafontaine, J. §II.B in *Riemannian Geometry.* New York: Springer-Verlag, 1980.
Lee, J. M. *Riemannian Manifolds: An Introduction to Curvature.* New York: Springer-Verlag, pp. 65–1, 1997.
Sternberg, S. *Differential Geometry.* New York: Chelsea, 1983.

Levi-Civita Density
PERMUTATION SYMBOL

Levi-Civita Symbol
PERMUTATION SYMBOL

Levi-Civita Tensor
PERMUTATION TENSOR

Levi Graph

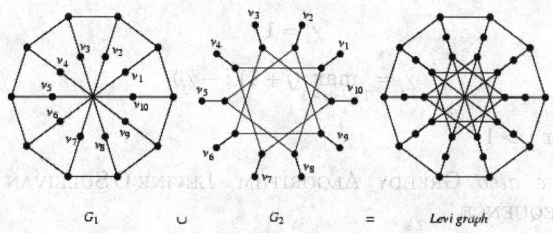

$$G_1 \quad \cup \quad G_2 \quad = \quad \text{Levi graph}$$

The unique 8-CAGE GRAPH (right figure) consisting of the union of the two leftmost subgraphs illustrated above. It has 45 nodes, 15 edges, and all nodes have degree 3. The Levi graph is a GENERALIZED POLYGON which is the point/line INCIDENCE GRAPH of the generalized quadrangle W_2. The graph is a 4-arc transitive cubic graph, was first discovered by Tutte (1947), and is also called the Tutte-Coxeter graph

(Bondy and Murty 1976, p. 237).

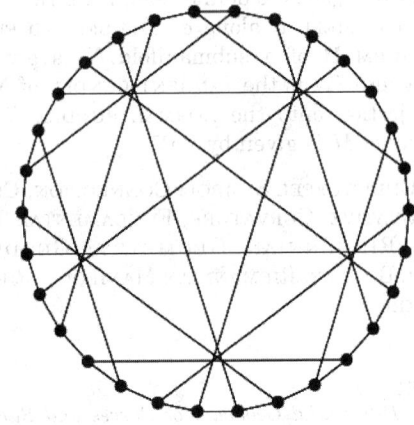

An alternative embedding is illustrated above.

See also CAGE GRAPH

References

Bondy, J. A. and Murty, U. S. R. *Graph Theory with Applications.* New York: North Holland, p. 276, 1976.
Coxeter, H. S. M. "The Chords of the Non-Ruled Quadratic in PG(3,3)." *Canad. J. Math.* **10**, 484–88, 1958.
Coxeter, H. S. M. "Twelve Points in PG(5,3) with 95040 Self-Transformations." *Proc. Roy. Soc. London Ser. A* **247**, 279–93, 1958.
Harary, F. *Graph Theory.* Reading, MA: Addison-Wesley, pp. 174–75, 1994.
Royle, G. "Cubic Cages." http://www.cs.uwa.edu.au/~gordon/cages/.
Tutte, W. T. "A Family of Cubical Graphs." *Proc. Cambridge Philos. Soc.*, 459–74, 1947.
Tutte, W. T. *Connectivity in Graphs.* Toronto, Ontario: University of Toronto Press, 1966.
Tutte, W. T. "The Chords of the Non-Ruled Quadratic in PG(3,3)." *Canad. J. Math.* **10**, 481–83, 1958.
Weisstein, E. W. "Graphs." MATHEMATICA NOTEBOOK GRAPHS.M.
Wong, P. K. "Cages--A Survey." *J. Graph Th.* **6**, 1–2, 1982.

Levine-O'Sullivan Greedy Algorithm

For a sequence $\{\chi_i\}$, the Levine-O'Sullivan greedy algorithm is given by

$$\chi_1 = 1$$

$$\chi_i = \max_{1 \le j \le i-1} (j+1)(i - \chi_j)$$

for $i > 1$.

See also GREEDY ALGORITHM, LEVINE-O'SULLIVAN SEQUENCE

References

Levine, E. and O'Sullivan, J. "An Upper Estimate for the Reciprocal Sum of a Sum-Free Sequence." *Acta Arith.* **34**, 9–4, 1977.

Levine-O'Sullivan Sequence

The sequence generated by the LEVINE-O'SULLIVAN GREEDY ALGORITHM: 1, 2, 4, 6, 9, 12, 15, 18, 21, 24, 28, 32, 36, 40, 45, 50, 55, 60, 65, ... (Sloane's A014011). The reciprocal sum of this sequence is conjectured to bound the reciprocal sum of all A-SEQUENCE.

References

Finch, S. "Favorite Mathematical Constants." http://www.mathsoft.com/asolve/constant/erdos/erdos.html.
Levine, E. and O'Sullivan, J. "An Upper Estimate for the Reciprocal Sum of a Sum-Free Sequence." *Acta Arith.* **34**, 9–4, 1977.
Sloane, N. J. A. Sequences A014011 in "An On-Line Version of the Encyclopedia of Integer Sequences." http://www.research.att.com/~njas/sequences/eisonline.html.

Lévy Constant

Let p_n/q_n be the nth CONVERGENT of a REAL NUMBER x. Then almost all REAL NUMBERS satisfy

$$L \equiv \lim_{n \to \infty} (q_n)^{1/n} = e^{\pi^2/(12 \ln 2)} = 3.27582291872\ldots$$

See also CONTINUED FRACTION, KHINTCHINE'S CONSTANT, KHINTCHINE-LÉVY CONSTANT

References

Le Lionnais, F. *Les nombres remarquables.* Paris: Hermann, p. 51, 1983.

Lévy Distribution

$$\mathscr{F}[P_N(k)] = \mathscr{F}[\exp(-N|k|^\beta)],$$

where $\mathscr{F}$ is the FOURIER TRANSFORM of the probability $P_N(k)$ for N-step addition of random variables. Lévy showed that $\beta \in (0, 2)$ for $P(x)$ to be NONNEGATIVE. The Lévy distribution has infinite variance and sometimes infinite mean. The case $\beta = 1$ gives a CAUCHY DISTRIBUTION, while $\beta = 2$ gives a GAUSSIAN DISTRIBUTION.

See also CAUCHY DISTRIBUTION, GAUSSIAN DISTRIBUTION, LÉVY FLIGHT

Lévy Dragon

LÉVY FRACTAL

Lévy Flight

RANDOM WALK trajectories which are composed of self-similar jumps. They are described by the LÉVY DISTRIBUTION.

See also LÉVY DISTRIBUTION

References

Shlesinger, M.; Zaslavsky, G. M.; and Frisch, U. (Eds.). *Lévy Flights and Related Topics in Physics.* New York: Springer-Verlag, 1995.

Lévy Fractal

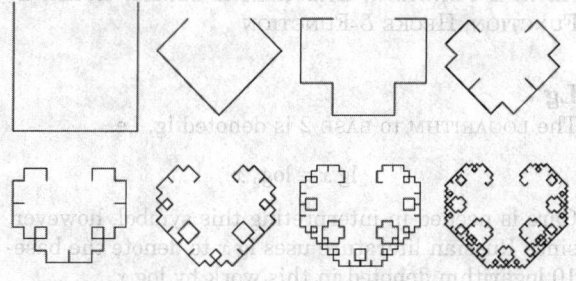

A FRACTAL curve, also called the C-CURVE (Gosper 1972). The base curve and motif are illustrated below.

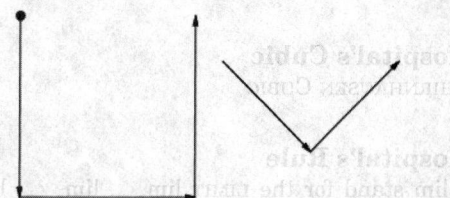

Duvall and Keesling (1999) proved that the HAUS-DORFF DIMENSION of the boundary of the Lévy fractal is rigorously greater than one, obtaining an estimate of 1.934007183.

See also LÉVY TAPESTRY

References

Dixon, R. *Mathographics.* New York: Dover, pp. 182–83, 1991.

Duvall, P. and Keesling, J. The Hausdorff Dimension of the Boundary of the Lévy Dragon. 22 Jul 1999. http://xxx.lanl.gov/abs/math.DS/9907145/.

Gosper, R. W. Item 135 in Beeler, M.; Gosper, R. W.; and Schroeppel, R. *HAKMEM.* Cambridge, MA: MIT Artificial Intelligence Laboratory, Memo AIM-239, pp. 65–6, Feb. 1972.

Lauwerier, H. *Fractals: Endlessly Repeated Geometric Figures.* Princeton, NJ: Princeton University Press, pp. 45–8, 1991.

Lévy, P. "Les courbes planes ou gauches et les surfaces composées de parties semblales au tout." *J. l'École Polytech.,* 227–47 and 249–91, 1938.

Lévy, P. "Plane or Space Curves and Surfaces Consisting of Parts Similar to the Whole." In *Classics on Fractals* (Ed. G. A. Edgar). Reading, MA: Addison-Wesley, pp. 181–39, 1993.

Weisstein, E. W. "Fractals." MATHEMATICA NOTEBOOK FRACTAL.M.

Lévy Function

BROWN FUNCTION

Lévy Process

References

Sato, K.-I. *Lévy Processes and Infinitely Divisible Distributions.* Cambridge, England: Cambridge University Press, 1999.

Lévy Tapestry

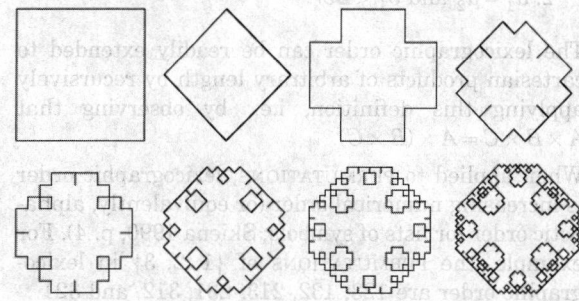

The FRACTAL curve illustrated above, with base curve and motif illustrated below.

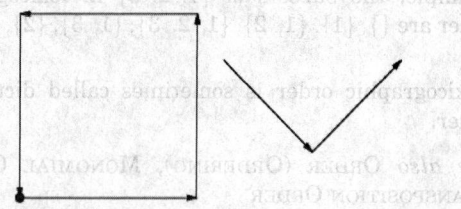

See also LÉVY FRACTAL

References

Lauwerier, H. *Fractals: Endlessly Repeated Geometric Figures.* Princeton, NJ: Princeton University Press, pp. 45–8, 1991.

Weisstein, E. W. "Fractals." MATHEMATICA NOTEBOOK FRACTAL.M.

Lewis Regulator

The ORDINARY DIFFERENTIAL EQUATION

$$y'' + (1 - |y|)y' + y = 0.$$

References

Hagerdorn, P. *Non-Linear Oscillations.* Oxford, England: Clarendon Press, p. 152, 1982.

Zwillinger, D. *Handbook of Differential Equations, 3rd ed.* Boston, MA: Academic Press, p. 124, 1997.

Lew k-Gram

Diagrams invented by Lewis Carroll which can be used to determine the number of minimal MINIMAL COVERS of n numbers with k members.

References

Macula, A. J. "Lewis Carroll and the Enumeration of Minimal Covers." *Math. Mag.* **68**, 269–74, 1995.

Lexicographic Order

An ordering for the Cartesian product $\times$ of any two sets A and B with order relations $<A$ and $<B$, respectively, such that if (a_1, b_1) and (a_2, b_2) both belong to $A \times B$, then $(a_1, b_1) < (a_2, b_2)$ IFF either

1. $a_1 < Aa_2$, or
2. $a_1 = a_2$ and $b_1 < Bb_2$.

The lexicographic order can be readily extended to cartesian products of arbitrary length by recursively applying this definition, i.e., by observing that $A \times B \times C = A \times (B \times C)$.

When applied to PERMUTATIONS, lexicographic order is increasing numerical order (or equivalently, alphabetic order for lists of symbols; Skiena 1990, p. 4). For example, the PERMUTATIONS of $\{1, 2, 3\}$ in lexicographic order are 123, 132, 213, 231, 312, and 321.

When applied to subsets, two subsets are ordered by their smallest elements (Skiena 1990, p. 44). For example, the subsets of $\{1, 2, 3\}$ in lexicographic order are $\{\}$, $\{1\}$, $\{1, 2\}$, $\{1, 2, 3\}$, $\{1, 3\}$, $\{2\}$, $\{2, 3\}$, $\{3\}$.

Lexicographic order is sometimes called dictionary order.

See also ORDER (ORDERING), MONOMIAL ORDER, TRANSPOSITION ORDER

References

Ruskey, F. "Information on Combinations of a Set." http://www.theory.csc.uvic.ca/~cos/inf/comb/CombinationsInfo.html.
Séroul, R. *Programming for Mathematicians*. Berlin: Springer-Verlag, p. 23, 2000.
Skiena, S. "Lexicographically Ordered Permutations" and "Lexicographically Ordered Subsets." §1.1.1 and 1.5.4 in *Implementing Discrete Mathematics: Combinatorics and Graph Theory with Mathematica*. Reading, MA: Addison-Wesley, pp. 3– and 43–4, 1990.

Lexis Ratio

$$L \equiv \frac{\sigma}{\sigma_B},$$

where σ is the VARIANCE in a set of s LEXIS TRIALS and σ_B is the VARIANCE assuming BERNOULLI TRIALS. If $L < 1$, the trials are said to be SUBNORMAL, and if $L > 1$, the trials are said to be SUPERNORMAL.

See also BERNOULLI TRIAL, LEXIS TRIALS, SUBNORMAL, SUPERNORMAL

Lexis Trials

n sets of s trials each, with the probability of success p constant in each set.

$$\text{var}\left(\frac{x}{n}\right) = spq + s(s-1)\sigma_p^2,$$

where σ_p^2 is the VARIANCE of p_i.

See also BERNOULLI TRIAL, LEXIS RATIO

L-Function

ARTIN L-FUNCTION, DIRICHLET L-SERIES, EULER L-FUNCTION, HECKE L-FUNCTION

Lg

The LOGARITHM to BASE 2 is denoted lg, i.e.,

$$\lg x \equiv \log_2 x.$$

Care is needed in interpreting this symbol, however, since Russian literature uses $\lg x$ to denote the base-10 logarithm denoted in this work by $\log x$.

See also BASE (LOGARITHM), E, LN, LOGARITHM, NAPIERIAN LOGARITHM, NATURAL LOGARITHM

L'Hospital's Cubic

TSCHIRNHAUSEN CUBIC

L'Hospital's Rule

Let lim stand for the LIMIT $\lim_{x \to c}$, $\lim_{x \to c^-}$, $\lim_{x \to c^+}$, $\lim_{x \to \infty}$, or $\lim_{x \to -\infty}$, and suppose that $\lim f(x)$ and $\lim g(x)$ are both ZERO or are both $\pm\infty$. If

$$\lim \frac{f'(x)}{g'(x)}$$

has a finite value or if the LIMIT is $\pm\infty$, then

$$\lim \frac{f(x)}{g(x)} = \lim \frac{f'(x)}{g'(x)}.$$

L'Hospital's rule occasionally fails to yield useful results, as in the case of the function $\lim_{u \to \infty} u(u^2 + 1)^{-1/2}$. Repeatedly applying the rule in this case gives expressions which oscillate and never converge,

$$\lim_{u \to \infty} \frac{u}{(u^2 + 1)^{1/2}} = \lim_{u \to \infty} \frac{1}{u(u^2 + 1)^{-1/2}}$$

$$= \lim_{u \to \infty} \frac{(u^2 + 1)^{1/2}}{u} = \lim_{u \to \infty} \frac{u(u^2 + 1)^{-1/2}}{1}$$

$$= \lim_{u \to \infty} \frac{u}{(u^2 + 1)^{1/2}}.$$

(The actual LIMIT is 1.)

References

Abramowitz, M. and Stegun, C. A. (Eds.). *Handbook of Mathematical Functions with Formulas, Graphs, and Mathematical Tables, 9th printing*. New York: Dover, p. 13, 1972.
L'Hospital, G. de *L'analyse des infiniment petits pour l'intelligence des lignes courbes*. 1696.

L'Huilier's Theorem

Let a SPHERICAL TRIANGLE have sides of length a, b, and c, and SEMIPERIMETER s. Then the SPHERICAL EXCESS E is given by

$$\tan\left(\tfrac{1}{4}E\right)$$

$$= \sqrt{\tan\left(\tfrac{1}{2}s\right)\tan\left[\tfrac{1}{2}(s-a)\right]\tan\left[\tfrac{1}{2}(s-b)\right]\tan\left[\tfrac{1}{2}(s-c)\right]}.$$

See also GIRARD'S SPHERICAL EXCESS FORMULA, SPHERICAL EXCESS, SPHERICAL TRIANGLE

References

Beyer, W. H. *CRC Standard Mathematical Tables, 28th ed.* Boca Raton, FL: CRC Press, p. 148, 1987.

Zwillinger, D. (Ed.). *CRC Standard Mathematical Tables and Formulae.* Boca Raton, FL: CRC Press, p. 469, 1995.

Liar's Paradox

The paradox of a man who states "I am lying." If he *is* lying, then he is telling the truth, and vice versa. Another version of this paradox is the EPIMENIDES PARADOX. Such paradoxes are often analyzed by creating so-called "metalanguages" to separate statements into different levels on which truth and falsity can be assessed independently. For example, Bertrand Russell noted that, "The man who says, 'I am telling a lie of order n' *is* telling a lie, but a lie of order $n + 1$" (Gardner 1984, p. 222).

See also EPIMENIDES PARADOX, EUBULIDES PARADOX

References

Beth, E. W. *The Foundations of Mathematics.* Amsterdam, Netherlands: North-Holland, p. 485, 1959.

Bochenski, I. M. §23 and 25 in *Formale Logik.* Munich, Germany, 1956.

Church, A. "Paradoxes, Logical." In *The Dictionary of Philosophy, rev. enl. ed.* (Ed. D. D. Runes). New York: Rowman and Littlefield, p. 224, 1984.

Curry, H. B. *Foundations of Mathematical Logic.* New York: Dover, pp. 5–, 1977.

Erickson, G. W. and Fossa, J. A. *Dictionary of Paradox.* Lanham, MD: University Press of America, pp. 108–11, 1998.

Fraenkel, A. A. and Bar-Hillel, Y. *Foundations of Set Theory.* Amsterdam, Netherlands, p. 11, 1958.

Gardner, M. *The Sixth Book of Mathematical Games from Scientific American.* Chicago, IL: University of Chicago Press, p. 222, 1984.

Kleene, S. C. *Introduction to Metamathematics.* Princeton, NJ: Van Nostrand, p. 39, 1964.

Prior, A. N. "Epimenides the Cretan." *J. Symb. Logic* **23**, 261–66, 1958.

Tarski, A. "The Semantic Conception of Truth and the Foundations of Semantics." *Philos. Phenomenol. Res.* **4**, 341–76, 1944.

Tarski, A. "Der Wahrheitsbegriff in den formalisierten Sprachen." *Studia Philos.* **1**, 261–05, 1936.

Weyl, H. *Philosophy of Mathematics and Natural Science.* Princeton, NJ, p. 228, 1949.

Lichnerowicz Conditions

Second and higher derivatives of the METRIC TENSOR g_{ab} need not be continuous across a surface of discontinuity, but g_{ab} and $g_{ab,\,c}$ must be continuous across it.

Lichnerowicz Formula

$$D^*D\psi = \nabla^*\nabla\psi + \tfrac{1}{4}R\psi - \tfrac{1}{2}F_L^+(\psi),$$

where D is the Dirac operator $D : \Gamma(W^+) \to \Gamma(W^-)$, ∇ is the COVARIANT DERIVATIVE on SPINORS, R is the CURVATURE SCALAR, and F_L^+ is the self-dual part of the curvature of L.

See also LICHNEROWICZ-WEITZENBOCK FORMULA

References

Donaldson, S. K. "The Seiberg-Witten Equations and 4-Manifold Topology." *Bull. Amer. Math. Soc.* **33**, 45–0, 1996.

Lichnerowicz-Weitzenbock Formula

$$D^*D\psi = \nabla^*\nabla\psi + \tfrac{1}{4}R\psi,$$

where D is the Dirac operator $D : \Gamma(S^+) \to \Gamma(S^-)$, ∇ is the COVARIANT DERIVATIVE on SPINORS, and R is the CURVATURE SCALAR.

See also LICHNEROWICZ FORMULA

References

Donaldson, S. K. "The Seiberg-Witten Equations and 4-Manifold Topology." *Bull. Amer. Math. Soc.* **33**, 45–0, 1996.

Lichtenfels Minimal Surface

A MINIMAL SURFACE that contains LEMNISCATES as geodesics which is given by the parametric equations

$$x = \Re\left[\sqrt{2}\cos\left(\tfrac{1}{2}\zeta\right)\sqrt{\cos\left(\tfrac{2}{3}\zeta\right)}\right] \tag{1}$$

$$y = \Re\left[-\sqrt{2}\sin\left(\tfrac{1}{3}\zeta\right)\sqrt{\cos\left(\tfrac{2}{3}\zeta\right)}\right] \tag{2}$$

$$z = \Re\left[\tfrac{1}{3}\sqrt{2}i\int_0^\zeta \frac{d\zeta}{\sqrt{\cos\left(\tfrac{2}{3}\zeta\right)}}\right] \tag{3}$$

$$= \Re\left[-i\sqrt{2}\,F\left(\sqrt{\tfrac{1}{3}}\zeta,\,2\right)\right], \tag{4}$$

where $F(x, x)$ is an incomplete ELLIPTIC INTEGRAL OF THE FIRST KIND and $\zeta = u + iv$ is a COMPLEX NUMBER. A given LEMNISCATE is the intersection of the surface with the xy-plane. The surface is periodic in the direction of the axis with period

$$\omega = 2 \int_0^1 \frac{dt}{\sqrt{1 - t^2}\sqrt{1 - \frac{1}{2}t^2}} = 2K\left(\frac{1}{2}\right), \quad (5)$$

where $K(x)$ is a complete ELLIPTIC INTEGRAL OF THE FIRST KIND.

See also LEMNISCATE, MINIMAL SURFACE

References

do Carmo, M. P. "Minimal Surfaces with a Lemniscate as a Geodesic." §3.5F in *Mathematical Models from the Collections of Universities and Museums* (Ed. G. Fischer). Braunschweig, Germany: Vieweg, p. 47, 1986.
Lichtenfels, O. von. "Notiz über eine transcendente Minimalfläche." *Sitzungsber. Kaiserl. Akad. Wiss. Wien* **94**, 41–4, 1889.

Lie Algebra

A NONASSOCIATIVE ALGEBRA obeyed by objects such as the LIE BRACKET and POISSON BRACKET. Elements f, g, and h of a Lie algebra satisfy

$$[f, f] = 0 \quad (1)$$

$$[f + g, h] = [f, h] + [g, h], \quad (2)$$

and

$$[f, [g, h]] + [g, [h, f]] + [h, [f, g]] = 0 \quad (3)$$

(the JACOBI IDENTITY). The relation $[f, f] = 0$ implies

$$[f, g] = -[g, f]. \quad (4)$$

For characteristic not equal to two, these two relations are equivalent.

The binary operation of a Lie algebra is the bracket

$$[fg, h] = f[g, h] + g[f, h]. \quad (5)$$

An ASSOCIATIVE ALGEBRA A with associative product xy can be made into a Lie algebra A^- by the Lie product

$$[x, y] = xy - yx. \quad (6)$$

Every Lie algebra L is isomorphic to a SUBALGEBRA of some A^- where the associative algebra A may be taken to be the linear operators over a VECTOR SPACE V (the POINCARÉ-BIRKHOFF-WITT THEOREM; Jacobson 1979, pp. 159–60). If L is finite dimensional, then V can be taken to be finite dimensional (ADO'S THEOREM for characteristic $p = 0$; IWASAWA'S THEOREM for characteristic $p \neq 0$).

The classification of finite dimensional simple Lie algebras over an algebraically closed field of characteristic 0 can be accomplished by (1) determining matrices called CARTAN MATRICES corresponding to indecomposable simple systems of roots and (2) determining the simple algebras associated with these matrices (Jacobson 1979, p. 128). This is one of the major results in Lie algebra theory, and is frequently accomplished with the aid of diagrams called DYNKIN DIAGRAMS.

See also ADO'S THEOREM, DERIVATION ALGEBRA, DYNKIN DIAGRAM, JACOBI IDENTITIES, LIE ALGEBROID, LIE BRACKET, IWASAWA'S THEOREM, POINCARÉ-BIRKHOFF-WITT THEOREM, POISSON BRACKET, REDUCED ROOT SYSTEM, ROOT SYSTEM, WEYL GROUP

References

Humphrey, J. E. *Introduction to Lie Algebras and Representation Theory.* New York: Springer-Verlag, 1972.
Jacobson, N. *Lie Algebras.* New York: Dover, 1979.
Schafer, R. D. *An Introduction to Nonassociative Algebras.* New York: Dover, p. 3, 1996.
Weisstein, E. W. "Books about Lie Algebra." http://www.treasure-troves.com/books/LieAlgebra.html.

Lie Algebroid

The infinitesimal algebraic object associated with a LIE GROUPOID. A Lie algebroid over a MANIFOLD B is a VECTOR BUNDLE A over B with a LIE ALGEBRA structure $[,]$ (LIE BRACKET) on its SPACE of smooth sections together with its ANCHOR ρ.

See also LIE ALGEBRA

References

Weinstein, A. "Groupoids: Unifying Internal and External Symmetry." *Not. Amer. Math. Soc.* **43**, 744–52, 1996.

Liebmann's Theorem

A SPHERE is rigid.

See also SPHERE

References

Gray, A. *Modern Differential Geometry of Curves and Surfaces with Mathematica, 2nd ed.* Boca Raton, FL: CRC Press, p. 483 and 653–54, 1997.
O'Neill, B. *Elementary Differential Geometry, 2nd ed.* New York: Academic Press, p. 262, 1997.

Lie Bracket

The commutation operation

$$[a, b] = ab - ba$$

corresponding to the LIE PRODUCT.

See also LAGRANGE BRACKET, POISSON BRACKET

Lie Commutator

LIE PRODUCT

Lie Derivative

The Lie derivative of TENSOR T_{ab} with respect to the VECTOR FIELD X is defined by

$$\mathscr{L}_X T_{ab} \equiv \lim_{\delta x \to 0} \frac{T'_{ab}(x') - T_{ab}(x)}{\delta x}. \qquad (1)$$

Explicitly, it is given by

$$\mathscr{L}_X T_{ab} = T_{ab} X^d_{,b} + T_{bd} X^d_{,a} + T_{ab,e} X^e, \qquad (2)$$

where $X_{,a}$ is a COMMA DERIVATIVE. The Lie derivative of a METRIC TENSOR g_{ab} with respect to the VECTOR FIELD X is given by

$$\mathscr{L}_X g_{ab} = X_{a;\,b} + X_{b;\,a} = 2X_{(a;\,b)}, \qquad (3)$$

where $X_{(a,\,b)}$ denotes the SYMMETRIC TENSOR part and $X_{a;\,b}$ is a COVARIANT DERIVATIVE.

See also COVARIANT DERIVATIVE, KILLING'S EQUATION, KILLING VECTORS, LIE DERIVATIVE (SPINOR)

Lie Derivative (Spinor)

The Lie derivative of a SPINOR ψ is defined by

$$\mathscr{L}_X \psi(x) = \lim_{t \to 0} \frac{\tilde{\psi}_t(x) - \psi(x)}{t},$$

where $\tilde{\psi}_t$ is the image of ψ by a one-parameter group of isometries with X its generator. For a VECTOR FIELD X^a and a COVARIANT DERIVATIVE ∇_a, the Lie derivative of ψ is given explicitly by

$$\mathscr{L}_X \psi = X^a \nabla_a \psi - \tfrac{1}{8}(\nabla_a X_b - \nabla_b X_a)\gamma^a \gamma^b \psi,$$

where γ^a and γ^b are DIRAC MATRICES (Choquet-Bruhat and DeWitt-Morette 2000).

See also COVARIANT DERIVATIVE, DIRAC MATRICES, LIE DERIVATIVE, SPINOR

References

Choquet-Bruhat, Y. and DeWitt-Morette, C. *Analysis, Manifolds and Physics, Part II: 92 Applications, rev. ed.* Amsterdam, Netherlands: North-Holland, 2000.

Lie Group

A Lie group is a DIFFERENTIABLE MANIFOLD obeying the group properties and that satisfies the additional condition that the group operations are continuous.

The simplest examples of Lie groups are one-dimensional. Under addition, the REAL LINE is a Lie group. After picking a specific point to be the IDENTITY ELEMENT, the CIRCLE is also a Lie group. Another point on the circle at angle θ from the identity then acts by rotating the circle by the angle θ. In general, a Lie group may have a more complicated group structure, such as the ORTHOGONAL GROUP $O(n)$ (i.e., the $n \times n$ orthogonal matrices), or the GENERAL LINEAR GROUP $GL(n)$ (i.e., the $n \times n$ invertible matrices). The LORENTZ GROUP is also a Lie group.

The TANGENT SPACE at the identity of a Lie group always has the structure of a LIE ALGEBRA, and this LIE ALGEBRA determines the local structure of the Lie

group via the EXPONENTIAL MAP. For example, the function e^{it} gives the EXPONENTIAL MAP from the circle's tangent space (i.e., the reals), to the circle, thought of as a the UNIT CIRCLE in $\mathbb{C}$. A more difficult example is the exponential map e^A from SKEW SYMMETRIC $n \times n$ matrices to the SPECIAL ORTHOGONAL GROUP $SO(n)$, the subset of $O(n)$ with determinant 1.

The topology of a Lie group is fairly restricted. For example, there always exists a nonvanishing VECTOR FIELD. This structure has allowed complete classification of the finite dimensional SEMISIMPLE LIE GROUPS and their representations.

See also COMPACT GROUP, CONTINUOUS GROUP, GROUP, DIFFERENTIABLE MANIFOLD, LIE ALGEBRA, LIE GROUPOID, LIE-TYPE GROUP, LORENTZ GROUP, NIL GEOMETRY, ORTHOGONAL GROUP, SEMISIMPLE LIE GROUP, SOL GEOMETRY, TANGENT SPACE, VECTOR FIELD

References

Arfken, G. "Infinite Groups, Lie Groups." *Mathematical Methods for Physicists, 3rd ed.* Orlando, FL: Academic Press, pp. 251–52, 1985.
Chevalley, C. *Theory of Lie Groups.* Princeton, NJ: Princeton University Press, 1946.
Hsiang, W. Y. *Lectures on Lie Groups.* Singapore: World Scientific, 2000.
Knapp, A. W. *Lie Groups Beyond an Introduction.* Boston, MA: Birkhäuser, 1996.
Lipkin, H. J. *Lie Groups for Pedestrians, 2nd ed.* Amsterdam, Netherlands: North-Holland, 1966.

Lie Groupoid

A GROUPOID G over B for which G and B are differentiable manifolds and α, β, and multiplication are differentiable maps. Furthermore, the derivatives of α and β are required to have maximal RANK everywhere. Here, α and β are maps from G onto $\mathbb{R}^2$ with $\alpha : (x,\,\gamma,\,y) \mapsto z$ and $\beta : (x,\,\gamma,\,y) \mapsto y$

See also LIE ALGEBROID, NILPOTENT LIE GROUP, SEMISIMPLE LIE GROUP, SOLVABLE LIE GROUP

References

Weinstein, A. "Groupoids: Unifying Internal and External Symmetry." *Not. Amer. Math. Soc.* **43**, 744–52, 1996.

Liénard's Differential Equation

The second-order ORDINARY DIFFERENTIAL EQUATION

$$y'' + f(x)y' + y = 0.$$

References

Villari, G. "Periodic Solutions of Liénard's Equation." *J. Math. Anal. Appl.* **86**, 379–86, 1982.
Zwillinger, D. *Handbook of Differential Equations, 3rd ed.* Boston, MA: Academic Press, p. 124, 1997.

Lie Product

The multiplication operation corresponding to the LIE BRACKET.

Lie-Type Group

A finite analog of LIE GROUPS. The Lie-type groups include the CHEVALLEY GROUPS $[PSL(n, q), PSU(n, q), PSp(2n, q), P\Omega^\epsilon(n, q)]$, TWISTED CHEVALLEY GROUPS, and the TITS GROUP.

See also CHEVALLEY GROUPS, FINITE GROUP, LIE GROUP, LINEAR GROUP, ORTHOGONAL GROUP, SIMPLE GROUP, SYMPLECTIC GROUP, TITS GROUP, TWISTED CHEVALLEY GROUPS, UNITARY GROUP

References

Wilson, R. A. "ATLAS of Finite Group Representation." http://for.mat.bham.ac.uk/atlas/html/contents.html#lie.

Life

The most well-known CELLULAR AUTOMATON, invented by John Conway and popularized in Martin Gardner's *Scientific American* column starting in October 1970. The game was originally played (i.e., successive generations were produced) by hand with counters, but implementation on a computer greatly increased the ease of exploring patterns.

The Life CELLULAR AUTOMATON is run by placing a number of filled cells on a 2-D grid. Each generation then switches cells on or off depending on the state of the cells that surround it. The rules are defined as follows. All eight of the cells surrounding the current one are checked to see if they are on or not. Any cells that are on are counted, and this count is then used to determine what will happen to the current cell.

1. Death: if the count is less than 2 or greater than 3, the current cell is switched off.
2. Survival: if (a) the count is exactly 2, or (b) the count is exactly 3 and the current cell is on, the current cell is left unchanged.
3. Birth: if the current cell is off and the count is exactly 3, the current cell is switched on.

Hensel gives a JAVA APPLET implementing the Game of Life on his web page. Weisstein gives an extensive alphabetical tabulation of life forms and terms.

A pattern which does not change from one generation to the next is known as a still life , and is said to have period 1. Conway originally believed that no pattern could produce an infinite number of cells, and offered a \$50 prize to anyone who could find a counterexample before the end of 1970 (Gardner 1983, p. 216). Many counterexamples were subsequently found, including guns and puffer trains.

A Life pattern which has no father pattern is known as a Garden of Eden (for obvious biblical reasons). The first such pattern was not found until 1971, and

at least 3 are now known. It is not, however, known if a pattern exists which has a father pattern , but no grandfather pattern (Gardner 1983, p. 249).

Rather surprisingly, Gosper and J. H. Conway independently showed that Life can be used to generate a UNIVERSAL TURING MACHINE (Berlekamp *et al.* 1982, Gardner 1983, pp. 250–53).

Similar CELLULAR AUTOMATON games with different rules are HEXLIFE and HIGHLIFE. HASHLIFE is a life ALGORITHM that achieves remarkable speed by storing subpatterns in a hash table, and using them to skip forward, sometimes thousands of generations at a time.

See also CELLULAR AUTOMATON, HASHLIFE, HEXLIFE, HIGHLIFE

References

Berlekamp, E. R.; Conway, J. H.; and Guy, R. K. "What Is Life." Ch. 25 in *Winning Ways for Your Mathematical Plays, Vol. 2: Games in Particular*. London: Academic Press, 1982.

Flammenkamp, A. "Game of Life." http://www.uni-bielefeld.de/~achim/gol.html.

"The Game of Life." *Math Horizons*. p. 9, Spring 1994.

Gardner, M. "The Game of Life, Parts I-III." Chs. 20–2 in *Wheels, Life, and other Mathematical Amusements*. New York: W. H. Freeman, 1983.

Hensel, A. "PC Life Distribution." http://www.mindspring.com/~alanh/lifep.zip.

Hensel, A. "Conway's Game of Life." Includes a Java applet for the Game of Life. http://www.mindspring.com/~alanh/life/.

Koenig, H. "Game of Life Information." http://www.halcyon.com/hkoenig/LifeInfo/LifeInfo.html.

Poundstone, W. *The Recursive Universe: Cosmic Complexity and the Limits of Scientific Knowledge*. New York: Morrow, 1985.

Resnick, M. and Silverman, B. "A Zoo of Life Forms." http://lcs.www.media.mit.edu/groups/el/projects/emergence/life-zoo.html.

Toffoli, T. and Margolus, N. *Cellular Automata Machines: A New Environment for Modeling*. Cambridge, MA: MIT Press, 1987.

Wainwright, R. T. "LifeLine." http://members.aol.com/life1ine/life/lifepage.htm.

Wainwright, R. T. *LifeLine: A Quarterly Newsletter for Enthusiasts of John Conway's Game of Life*. Nos. 1–1, 1971–973.

Weisstein, E. W. "Eric's Treasure Trove of Life." http://www.treasure-troves.com/life/.

Life Expectancy

An l_x table is a tabulation of numbers which is used to calculate life expectancies.

x	n_x	d_x	l_x	q_x	L_x	T_x	e_x
0	1000	200	1.00	0.20	0.90	2.70	2.70
1	800	100	0.80	0.12	0.75	1.80	2.25
2	700	200	0.70	0.29	0.60	1.05	1.50

3	500	300	0.50	0.60	0.35	0.45	0.90
4	200	200	0.20	1.00	0.10	0.10	0.50
5	0	0	0.00	–	0.00	0.00	–
Σ		1000	2.70				

x: Age category ($x = 0, 1, ..., k$). These values can be in any convenient units, but must be chosen so that no observed lifespan extends past category $k-1$.

n_x: Census size, defined as the number of individuals in the study population who survive to the beginning of age category x. Therefore, $n_0 = N$ (the total population size) and $n_k = 0$.

d_x: $n_x - n_{x+1}$; $\Sigma_{i=0}^k d_i = n_0$. Crude death rate, which measures the number of individuals who die within age category x.

l_x: $= n_x/n_0$. Survivorship, which measures the *proportion* of individuals who survive to the beginning of age category x.

q_x: $= dx/n_x$; $q_{k-1} = 1$. Proportional death rate, or "risk," which measures the proportion of individuals surviving to the beginning of age category x who die within that category.

L_x: $= (l_x + l_{x+1})/2$. Midpoint survivorship, which measures the proportion of individuals surviving to the *midpoint* of age category x. Note that the simple averaging formula must be replaced by a more complicated expression if survivorship is nonlinear within age categories. The sum $\Sigma_{i=0}^k L_x$ gives the total number of age categories lived by the entire study population.

T_x: $T_{x-1} - L_{x-1}$; $T_0 = \Sigma_{i=0}^k L_x$. Measures the total number of age categories left to be lived by all individuals who survive to the beginning of age category x.

e_x: $= T_x/l_x$; $e_{k-1} = 1/2$. Life expectancy, which is the mean number of age categories remaining until death for individuals surviving to the beginning of age category x.

For all x, $e_{x+1} + 1 > e_x$. This means that the total expected lifespan increases monotonically. For instance, in the table above, the one-year-olds have an average age at death of $2.25 + 1 = 3.25$, compared to 2.70 for newborns. In effect, the age of death of older individuals is a distribution conditioned on the fact that they have survived to their present age.

It is common to study survivorship as a semilog plot of l_x vs. x, known as a SURVIVORSHIP CURVE. A so-called $l_x m_x$ table can be used to calculate the mean generation time of a population. Two $l_x m_x$ tables are illustrated below.

Population 1

x	l_x	m_x	$l_x m_x$	$x l_x m_x$
0	1.00	0.00	0.00	0.00
1	0.70	0.50	0.35	0.35
2	0.50	1.50	0.75	1.50
3	0.20	0.00	0.00	0.00
4	0.00	0.00	0.00	0.00
			$R_0 = 1.10$	$\Sigma = 1.85$

$$T = \frac{\sum xl_x m_x}{\sum l_x m_x} = \frac{1.85}{1.10} = 1.68$$

$$r = \frac{\ln R_0}{T} = \frac{\ln 1.10}{1.68} = 0.057.$$

Population 2

x	l_x	m_x	$l_x m_x$	$x l_x m_x$
0	1.00	0.00	0.00	0.00
1	0.70	0.00	0.00	0.00
2	0.50	2.00	1.00	2.00
3	0.20	0.50	0.10	0.30
4	0.00	0.00	0.00	0.00
			$R_0 = 1.10$	$\Sigma = 2.30$

$$T = \frac{\sum xl_x m_x}{\sum l_x m_x} = \frac{2.30}{1.10} = 2.09$$

$$r = \frac{\ln R_0}{T} = \frac{\ln 1.10}{2.09} = 0.046.$$

x: Age category ($x = 0, 1, ..., k$). These values can be in any convenient units, but must be chosen so that no observed lifespan extends past category $k-1$ (as in an l_x table).

l_x: $= n_x/n_0$. Survivorship, which measures the *proportion* of individuals who survive to the beginning of age category x (as in an l_x table).

m_x: The average number of offspring produced by an individual in age category x *while in that age category*. $\Sigma_{i=0}^k m_x$ therefore represents the average lifetime number of offspring produced by an individual of maximum lifespan.

$l_x m_x$: The average number of offspring produced by an individual within age category x weighted by the probability of surviving to the beginning of that age category. $\Sigma_{i=0}^k l_x m_x$ therefore represents the average lifetime number of offspring produced by a member of the study population. It is called the net reproductive rate per generation and is often denoted R_0.

$xl_x m_x$: A column weighting the offspring counted in the previous column by their parents' age when they were born. Therefore, the ratio $T = \Sigma(xl_x m_x)/\Sigma(l_x m_x)$ is the mean generation time of the population.

The MALTHUSIAN PARAMETER r measures the reproductive rate per unit time and can be calculated as $r = (\ln R_0)/T$. For an exponentially increasing population, the population size $N(t)$ at time t is then given by

$$N(t) = N_0 e^{rt}.$$

In the above two tables, the populations have identical reproductive rates of $R_0 = 1.10$. However, the shift toward later reproduction in population 2 increases the generation time, thus slowing the rate of POPULATION GROWTH. Often, a slight delay of reproduction decreases POPULATION GROWTH more strongly than does even a fairly large reduction in reproductive rate.

See also GOMPERTZ CURVE, LOGISTIC GROWTH CURVE, MAKEHAM CURVE, MALTHUSIAN PARAMETER, POPULATION GROWTH, SURVIVORSHIP CURVE

References
Steinhaus, H. *Mathematical Snapshots, 3rd ed.* New York: Dover, pp. 294–95, 1999.

Lift
Given a MAP f from a SPACE X to a SPACE Y and another MAP g from a SPACE Z to a SPACE Y, a lift is a MAP h from X to Z such that $gh = f$. In other words, a lift of f is a MAP h such that the diagram (shown below) commutes.

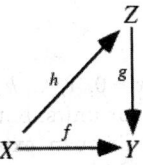

If f is the identity from Y to Y, a MANIFOLD, and if g is the BUNDLE PROJECTION from the TANGENT BUNDLE to Y, the lifts are precisely VECTOR FIELDS. If g is a bundle projection from any FIBER BUNDLE to Y, then lifts are precisely sections. If f is the identity from Y to Y, a MANIFOLD, and g a projection from the orientation double cover of Y, then lifts exist IFF Y is an orientable MANIFOLD.

If f is a MAP from a CIRCLE to Y, an n-MANIFOLD, and g the bundle projection from the FIBER BUNDLE of alternating K-FORMS on Y, then lifts always exist IFF Y is orientable. If f is a MAP from a region in the

COMPLEX PLANE to the COMPLEX PLANE (complex analytic), and if g is the exponential MAP, lifts of f are precisely LOGARITHMS of f.

See also LIFTING PROBLEM

Lifting Problem
Given a MAP f from a SPACE X to a SPACE Y and another MAP g from a SPACE Z to a SPACE Y, does there exist a MAP h from X to Z such that $gh = f$? If such a map h exists, then h is called a LIFT of f.

See also EXTENSION PROBLEM, LIFT

Ligancy
KISSING NUMBER

Likelihood
The hypothetical PROBABILITY that an event which has already occurred would yield a specific outcome. The concept differs from that of a probability in that a probability refers to the occurrence of future events, while a likelihood refers to past events with known outcomes.

See also LIKELIHOOD RATIO, MAXIMUM LIKELIHOOD, NEGATIVE LIKELIHOOD RATIO, PROBABILITY

Likelihood Ratio
A quantity used to test NESTED HYPOTHESES. Let H' be a NESTED HYPOTHESIS with n' DEGREES OF FREEDOM within H (which has n DEGREES OF FREEDOM), then calculate the MAXIMUM LIKELIHOOD of a given outcome, first given H', then given H. Then

$$LR = \frac{[\text{likelihood } H']}{[\text{likelyhood } H]}.$$

Comparison of this ratio to the critical value of the CHI-SQUARED DISTRIBUTION with $n - n'$ DEGREES OF FREEDOM then gives the SIGNIFICANCE of the increase in LIKELIHOOD.

The term likelihood ratio is also used (especially in medicine) to test nonnested complementary hypotheses as follows,

$$LR = \frac{[\text{true positive rate}]}{[\text{false positive rate}]} = \frac{[\text{sensitivity}]}{1 - [\text{specificity}]}.$$

See also NEGATIVE LIKELIHOOD RATIO, SENSITIVITY, SPECIFICITY

Limaçon of Pascal
LIMAÇON

Limaçon

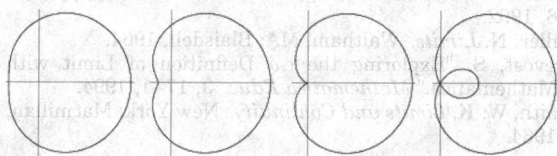

The limaçon is a polar curve OF THE FORM

$$r = b + a \cos \theta$$

also called the LIMAÇON OF PASCAL. It was first investigated by Dürer, who gave a method for drawing it in *Underweysung der Messung* (1525). It was rediscovered by Étienne Pascal, father of Blaise Pascal, and named by Gilles-Personne Roberval in 1650 (MacTutor Archive). The word "limaçon" comes from the Latin *limax*, meaning "snail."

If $b \geq 2a$, we have a convex limaçon. If $2a > b > a$, we have a dimpled limaçon. If $b = a$, the limaçon degenerates to a CARDIOID. If $b < a$, we have limaçon with an inner loop. If $b = a/2$, it is a TRISECTRIX (but *not* the MACLAURIN TRISECTRIX) with inner loop of AREA

$$A_{\text{inner loop}} = \frac{1}{4} a^2 \left(\pi - 3 \sqrt{\frac{3}{2}} \right),$$

and AREA between the loops of

$$A_{\text{between loops}} = \frac{1}{4} a^2 \left(\pi + 3 \sqrt{3} \right)$$

(MacTutor Archive).

The limaçon can be generated by specifying a fixed point P, then drawing a sequences of circles with centers on a given circle which all pass through P. The ENVELOPE of these curves is a limaçon. If the fixed point is on the CIRCUMFERENCE of the circle, then the ENVELOPE is a CARDIOID.

The limaçon is an ANALLAGMATIC CURVE, and is also the CATACAUSTIC of a CIRCLE when the RADIANT POINT is a finite (NONZERO) distance from the CIRCUMFERENCE, as shown by Thomas de St. Laurent in 1826 (MacTutor Archive). The limaçon is the CONCHOID of a CIRCLE with respect to a point on its CIRCUMFERENCE (Wells 1991).

See also CARDIOID

References

Beyer, W. H. *CRC Standard Mathematical Tables, 28th ed.* Boca Raton, FL: CRC Press, pp. 220–21, 1987.

Baudoin, P. *Les ovales de Descartes et le limaçon de Pascal.* Paris: Vuibert, 1938.
Lawrence, J. D. *A Catalog of Special Plane Curves.* New York: Dover, pp. 113–17, 1972.
Lockwood, E. H. "The Limaçon." Ch. 5 in *A Book of Curves.* Cambridge, England: Cambridge University Press, pp. 44–1, 1967.
MacTutor History of Mathematics Archive. "Limacon of Pascal." http://www-groups.dcs.st-and.ac.uk/~history/Curves/Limacon.html.
Steinhaus, H. *Mathematical Snapshots, 3rd ed.* New York: Dover, pp. 154–55, 1999.
Wells, D. *The Penguin Dictionary of Curious and Interesting Geometry.* London: Penguin, pp. 140–41, 1991.
Yates, R. C. "Limacon of Pascal." *A Handbook on Curves and Their Properties.* Ann Arbor, MI: J. W. Edwards, pp. 148–51, 1952.

Limaçon Evolute

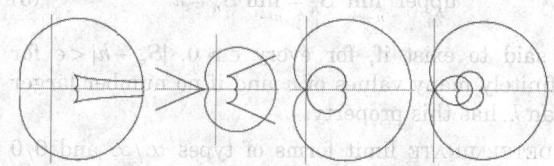

The CATACAUSTIC of a CIRCLE for a RADIANT POINT is the limaçon evolute. It has PARAMETRIC EQUATIONS

$$x = \frac{a[4a^2 + 4b^2 + 9ab \cos t - ab \cos(3t)]}{4(2a^2 + b^2 + 3ab \cos t)}$$

$$y = \frac{a^2 b \sin^3 t}{2a^2 + b^2 + 3ab \cos t}.$$

Limb

A limb of a TREE at a vertex v is the union of one or more BRANCHES at v in the tree. v is then called the base of the limb.

See also BRANCH, TREE

References

Lu, T. "The Enumeration of Trees with and without Given Limbs." *Disc. Math.* **154**, 153–65, 1996.
Schwenk, A. "Almost All Trees are Cospectral." In *New Directions in the Theory of Graphs* (Ed. F. Harary). New York: Academic Press, pp. 275–07, 1973.

Lim Inf

INFIMUM LIMIT

Limit

A function $f(z)$ is said to have a limit $\lim_{z \to a} f(z) = c$ if, for all $\epsilon > 0$, there exists a $\delta > 0$ such that $|f(z) - c| < \epsilon$ whenever $0 < |z - a| < \delta$. This form of definition is sometimes called an EPSILON-DELTA DEFINITION. Limits may be taken from below

$$\lim_{z \to a^-} = \lim_{x \uparrow a} \tag{1}$$

or from above

$$\lim_{z \to a^+} = \lim_{z \downarrow a}. \tag{2}$$

if the two are equal, then "the" limit is said to exist

$$\lim_{z \to a} = \lim_{z \to a^-} = \lim_{z \to a^+}. \tag{3}$$

A LOWER LIMIT h

$$\text{lower} \lim_{n \to \infty} S_n = \underline{\lim_{n \to \infty}} S_n = h \tag{4}$$

is said to exist if, for every $\epsilon > 0$, $|S_n - h| < \epsilon$ for infinitely many values of n and if no number less than h has this property.

An UPPER LIMIT k

$$\text{upper} \lim_{n \to \infty} S_n = \overline{\lim_{n \to \infty}} S_n = k \tag{5}$$

is said to exist if, for every $\epsilon > 0$, $|S_n - h| < \epsilon$ for infinitely many values of n and if no number larger than k has this property.

INDETERMINATE limit forms of types ∞/∞ and $0/0$ can often be computed with L'HOSPITAL'S RULE. Types $0 \cdot \infty$ can be converted to the form $0/0$ by writing

$$f(x)g(x) = \frac{f(x)}{1/g(x)}. \tag{6}$$

Types 0^0, ∞^0, and 1^∞ are treated by introducing a dependent variable

$$y = f(x)^{g(x)} \tag{7}$$

so that

$$\ln y = g(x)\ln[f(x)], \tag{8}$$

then calculating $\lim \ln y$. The original limit then equals $e^{\lim \ln y}$,

$$L = \lim f(x)^{g(x)} = e^{\lim \ln y} \tag{9}$$

The INDETERMINATE form $\infty - \infty$ is also frequently encountered.

See also CENTRAL LIMIT THEOREM, CONTINUOUS, DERIVATIVE, DISCONTINUITY, INDETERMINATE, INFIMUM LIMIT, L'HOSPITAL'S RULE, LIMIT COMPARISON TEST, LIMIT TEST, LOWER LIMIT, PINCHING THEOREM, SQUEEZING THEOREM, SUPREMUM LIMIT, UPPER LIMIT

References

Courant, R. and Robbins, H. "Limits. Infinite Geometrical Series." §2.2.3 in *What is Mathematics?: An Elementary Approach to Ideas and Methods, 2nd ed.* Oxford, England: Oxford University Press, pp. 63–6, 1996.

Gruntz, D. *On Computing Limits in a Symbolic Manipulation System.* Doctoral thesis. Zürich: Swiss Federal Institute of Technology, 1996.

Hight, D. W. *A Concept of Limits.* New York: Prentice-Hall, 1966.

Kaplan, W. "Limits and Continuity." §2.4 in *Advanced Calculus, 4th ed.* Reading, MA: Addison-Wesley, pp. 82–6, 1992.

Miller, N. *Limits.* Waltham, MA: Blaisdell, 1964.

Prevost, S. "Exploring the ϵ-δ Definition of Limit with Mathematica." *Mathematica Educ.* **3**, 17–1, 1994.

Smith, W. K. *Limits and Continuity.* New York: Macmillan, 1964.

Limit Comparison Test

Let Σa_k and Σb_k be two SERIES with POSITIVE terms and suppose

$$\lim_{k \to \infty} \frac{a_k}{b_k} = \rho.$$

If ρ is finite and $\rho > 0$, then the two SERIES both CONVERGE or DIVERGE.

See also CONVERGENCE TESTS, LIMIT, LIMIT TEST

Limit Cycle

An attracting set to which orbits or trajectories converge and upon which trajectories are periodic.

See also HOPF BIFURCATION

Limiting Point

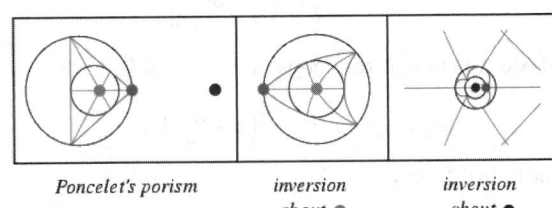

Poncelet's porism   *inversion about ●*   *inversion about ●*

A point about which INVERSION of two circles produced CONCENTRIC CIRCLES. Every pair of distinct circles has two limiting points.

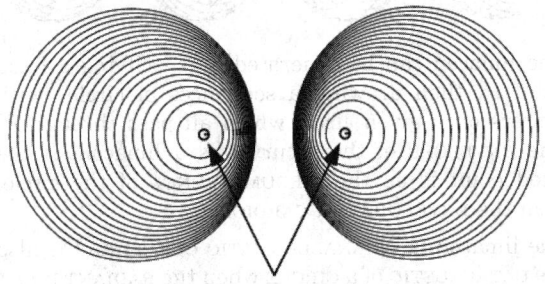

point circles (limit points)

The limiting points correspond to the POINT CIRCLES of a COAXAL SYSTEM, and the limiting points of a COAXAL SYSTEM are INVERSE POINTS with respect to any circle of the system.

To find the limiting point of two circles of radii r and R with centers separated by a distance d, set up a coordinate system centered on the circle of radius R and with the other circle centered at $(d, 0)$. Then the equation for the position of the center of the inverted circles with inversion center $(x_0, 0)$,

$$x' = x_0 + \frac{k^2(x - x_0)}{(x - x_0)^2 + (y - y_0)^2 - a^2}, \tag{1}$$

becomes

$$x_1' = x_0 + \frac{k^2(d - x_0)}{(d - x_0)^2 - r^2} \tag{2}$$

$$x_2' = x_0 + \frac{k^2(0 - x_0)}{(0 - x_0)^2 - R^2} \tag{3}$$

for the first and second circles, respectively. Setting $x_1' = x_2'$ gives

$$\frac{d - x_0}{(d - x_0)^2 - r^2} = \frac{-x_0}{x_0^2 - R^2}, \tag{4}$$

and solving using the quadratic equation gives the positions of the limiting points as

$$x' = \frac{d^2 - r^2 + R^2 \pm \sqrt{(d^2 - r^2 + R^2)^2 - 4d^2R^2}}{2d}. \tag{5}$$

See also COAXAL SYSTEM, CONCENTRIC CIRCLES, INVERSE POINTS, INVERSION CENTER, POINT CIRCLE

References

Casey, J. *A Sequel to the First Six Books of the Elements of Euclid, Containing an Easy Introduction to Modern Geometry with Numerous Examples, 5th ed., rev. enl.* Dublin: Hodges, Figgis, & Co., p. 43, 1888.
Durell, C. V. *Modern Geometry: The Straight Line and Circle.* London: Macmillan, pp. 123 and 130, 1928.

Limit Ordinal

An ORDINAL NUMBER $\alpha > 0$ is called a limit ordinal IFF it has no immediate PREDECESSOR, i.e., if there is no ORDINAL NUMBER β such that $\beta + 1 = \alpha$ (Ciesielski 1997, p. 46; Moore 1982, p. 60; Rubin 1967, p. 182; Suppes 1972, p. 196). The first limit ordinal is ω.

See also ORDINAL NUMBER, SUCCESSOR

References

Ciesielski, K. *Set Theory for the Working Mathematician.* Cambridge, England: Cambridge University Press, 1997.
Moore, G. H. *Zermelo's Axiom of Choice: Its Origin, Development, and Influence.* New York: Springer-Verlag, 1982.
Rubin, J. E. *Set Theory for the Mathematician.* New York: Holden-Day, 1967.
Suppes, P. *Axiomatic Set Theory.* New York: Dover, 1972.

Limit Point

A number x such that for all $\epsilon > 0$, there exists a member of the SET y different from x such that $|y - x| < \epsilon$. The topological definition of limit point P of A is that P is a point such that every OPEN SET around it intersects A.

See also ACCUMULATION POINT, CLOSED SET, OPEN SET

References

Jeffreys, H. and Jeffreys, B. S. *Methods of Mathematical Physics, 3rd ed.* Cambridge, England: Cambridge University Press, pp. 9–0, 1988.
Lauwerier, H. *Fractals: Endlessly Repeated Geometric Figures.* Princeton, NJ: Princeton University Press, pp. 25–6, 1991.

Limit Test

If $\lim a_n \neq 0$ or this LIMIT does not exist as n tends to infinity, then the INFINITE SERIES Σa_n does not CONVERGE. For example, $\sum_{n=1}^{\infty} (-1)^n$ does not converge by the limit test. The limit test is inconclusive when the limit is zero.

See also CONVERGENT SERIES, CONVERGENCE TESTS, LIMIT, LIMIT COMPARISON TEST, SEQUENCE, SERIES

Limit Theorem

CENTRAL LIMIT THEOREM, LEBESGUE'S DOMINATED CONVERGENCE THEOREM LINDEBERG-FELLER CENTRAL LIMIT THEOREM, MONOTONE CONVERGENCE THEOREM, POINTWISE CONVERGENCE

Lim Sup

SUPREMUM LIMIT

Lindeberg Condition

A SUFFICIENT condition on the LINDEBERG-FELLER CENTRAL LIMIT THEOREM. Given random variates X_1, $X_2, ...,$ let $\langle X_i \rangle = 0$, the VARIANCE σ_i^2 of X_i be finite, and VARIANCE of the distribution consisting of a sum of X_is

$$S_n \equiv X_1 + X_2 + ... + X_n \tag{1}$$

be

$$s_n^2 \equiv \sum_{i=1}^{n} \sigma_i^2. \tag{2}$$

In the terminology of Zabell (1995), let

$$\Lambda_n(\epsilon) \equiv \sum_{k=1}^{n} \left\langle \left(\frac{X_k}{s_n}\right)^2 : \frac{|X_k|}{s_n} \geq \epsilon \right\rangle, \tag{3}$$

where $\langle f : g \rangle$ denotes the EXPECTATION VALUE of f restricted to outcomes g, then the Lindeberg condition is

$$\lim_{n \to \infty} \Lambda_n(\epsilon) = 0 \tag{4}$$

for all $\epsilon > 0$ (Zabell 1995).

In the terminology of Feller (1971), the Lindeberg condition assumed that for each $t > 0$,

$$\frac{1}{s_n^2} \sum_{k=1}^{n} \int_{|y| \geq ts_n} y^2 F_k\{dy\} \to 0, \tag{5}$$

or equivalently

$$\frac{1}{s_n^2} \sum_{k=1}^{n} \int_{|y| < ts_n} y^2 F_k\{dy\} \to 1. \qquad (6)$$

Then the distribution

$$S_n^* = \frac{X_1 + \ldots + X_n}{s_n} \qquad (7)$$

tends to the NORMAL DISTRIBUTION with zero expectation and unit variance (Feller 1971, p. 256). The Lindeberg condition (5) guarantees that the individual variances σ_k^2 are small compared to their sum s_n^2 in the sense that for given $\epsilon > 0$ for for all SUFFICIENTLY LARGE n, $\sigma_k/s_n < \epsilon$ for $k = 1, \ldots, n$ (Feller 1971, p. 256).

See also CENTRAL LIMIT THEOREM, FELLER-LÉVY CONDITION

References

Feller, W. "Uuml;ber den zentralen Grenzwertsatz der Wahrscheinlichkeitsrechnung." *Math. Zeit.* **40**, 521–59, 1935.

Feller, W. "Über den zentralen Grenzwertsatz der Wahrscheinlichkeitsrechnung, II." *Math. Zeit.* **42**, 301–12, 1935.

Feller, W. *An Introduction to Probability Theory and Its Applications, Vol. 2, 3rd ed.* New York: Wiley, pp. 257–58, 1971.

Lindeberg, J. W. "Eine neue Herleitung des Exponentialgesetzes in der Wahrscheinlichkeitsrechnung." *Math. Zeit.* **15**, 211–35, 1922.

Trotter, H. F. "An Elementary Proof of the Central Limit Theorem." *Arch. Math.* **10**, 226–34, 1959.

Wallace, D. L. "Asymptotic Approximations to Distributions." *Ann. Math. Stat.* **29**, 635–54, 1958.

Zabell, S. L. "Alan Turing and the Central Limit Theorem." *Amer. Math. Monthly* **102**, 483–94, 1995.

Lindeberg-Feller Central Limit Theorem

If the random variates X_1, X_2, ... satisfy the LINDEBERG CONDITION, then for all $a < b$,

$$\lim_{n \to \infty} P\left(a < \frac{S_n}{s_n} < b\right) = \Phi(b) - \Phi(a),$$

where Φ is the NORMAL DISTRIBUTION FUNCTION.

See also BERRY-ESSÉEN THEOREM, CENTRAL LIMIT THEOREM, FELLER-LÉVY CONDITION, NORMAL DISTRIBUTION FUNCTION

References

Feller, W. "Über den zentralen Genzwertsatz der Wahrscheinlichkeitsrechnung." *Math. Z.* **40**, 521–59, 1935.

Feller, W. *An Introduction to Probability Theory and Its Applications, Vol. 1, 3rd ed.* New York: Wiley, p. 229, 1968.

Lindeberg, J. W. "Eine neue Herleitung des Exponentialgesetzes in der Wahrschienlichkeitsrechnung." *Math. Z.* **15**, 211–25, 1922.

Zabell, S. L. "Alan Turing and the Central Limit Theorem." *Amer. Math. Monthly* **102**, 483–94, 1995.

Lindelof's Theorem

The SURFACE OF REVOLUTION generated by the external CATENARY between a fixed point a and its conjugate on the ENVELOPE of the CATENARY through the fixed point is equal in AREA to the surface of revolution generated by its two Lindelof TANGENTS, which cross the axis of rotation at the point a and are calculable from the position of the points and CATENARY.

See also CATENARY, ENVELOPE, SURFACE OF REVOLUTION

Lindemann-Weierstrass Theorem

If $\alpha_1, \ldots, \alpha_n$ are linearly independent over $\mathbb{Q}$, then $e^{\alpha_1}, \ldots, e^{\alpha_n}$ are ALGEBRAICALLY INDEPENDENT over $\mathbb{Q}$. The Lindemann-Weierstrass theorem is implied by SCHANUEL'S CONJECTURE (Chow 1999).

See also ALGEBRAICALLY INDEPENDENT, HERMITE-LINDEMANN THEOREM, SCHANUEL'S CONJECTURE

References

Baker, A. Theorem 2.1 in *Transcendental Number Theory.* Cambridge, England: Cambridge University Press, 1990.

Chow, T. Y. "What is a Closed-Form Number?" *Amer. Math. Monthly* **106**, 440–48, 1999.

Lindenmayer System

A STRING REWRITING system which can be used to generate FRACTALS with DIMENSION between 1 and 2. The term L-system is often used as an abbreviation.

See also ARROWHEAD CURVE, DRAGON CURVE EXTERIOR SNOWFLAKE, FRACTAL, HILBERT CURVE, KOCH SNOWFLAKE, PEANO CURVE, PEANO-GOSPER CURVE, SIERPINSKI CURVE, STRING REWRITING

References

Bulaevsky, J. "*L*-System Based Fractals." http://www.best.com/~ejad/java/fractals/lsystems.shtml.

Bulaevsky, J. "A Process to Generate Fractals." http://www.best.com/~ejad/java/fractals/process.shtml.

Dickau, R. M. "Two-dimensional L-systems." http://forum.swarthmore.edu/advanced/robertd/lsys2d.html.

Prusinkiewicz, P. and Hanan, J. *Lindenmayer Systems, Fractal, and Plants.* New York: Springer-Verlag, 1989.

Prusinkiewicz, P. and Lindenmayer, A. *The Algorithmic Beauty of Plants.* New York: Springer-Verlag, 1990.

Stevens, R. T. *Fractal Programming in C.* New York: Holt, 1989.

Wagon, S. "Recursion via String Rewriting." §6.2 in *Mathematica in Action.* New York: W. H. Freeman, pp. 190–96, 1991.

Line

Euclid defined a line as a "breadthless length," and a straight line as a line which "lies evenly with the points on itself" (Kline 1956, Dunham 1990). Lines are intrinsically 1-dimensional objects, but may be embedded in higher dimensional SPACES. An infinite line passing through points A and B is denoted $\breve{A}B$. A

LINE SEGMENT terminating at these points is denoted $\overline{AB}$. A line is sometimes called a STRAIGHT LINE or, more archaically, a RIGHT LINE (Casey 1893), to emphasize that it has no curves anywhere along its length.

Harary (1994) called an edge of a graph a "line."

Consider first lines in a 2-D PLANE. The line with x-INTERCEPT a and y-INTERCEPT b is given by the *intercept form*

$$\frac{x}{a} + \frac{y}{b} = 1. \tag{1}$$

The line through (x_1, y_1) with SLOPE m is given by the *point-slope form*

$$y - y_1 = m(x - x_1). \tag{2}$$

The line with y-intercept b and slope m is given by the *slope-intercept form*

$$y = mx + b. \tag{3}$$

The line through (x_1, y_1) and (x_2, y_2) is given by the *two point form*

$$y - y_1 = \frac{y_2 - y_1}{x_2 - x_1}(x - x_1). \tag{4}$$

Other forms are

$$a(x - x_1) + b(y - y_1) = 0 \tag{5}$$

$$ax + by + c = 0 \tag{6}$$

$$\begin{vmatrix} x & y & 1 \\ x_1 & y_1 & 1 \\ x_2 & y_2 & 1 \end{vmatrix} = 0. \tag{7}$$

A line in 2-D can also be REPRESENTED AS a VECTOR. The VECTOR along the line

$$ax + by = 0 \tag{8}$$

is given by

$$t \begin{bmatrix} -b \\ a \end{bmatrix}, \tag{9}$$

where $t \in \mathbb{R}$. Similarly, VECTORS OF THE FORM

$$t \begin{bmatrix} a \\ b \end{bmatrix} \tag{10}$$

are PERPENDICULAR to the line. Three points lie on a line if

$$\begin{vmatrix} x_1 & y_1 & 1 \\ x_2 & y_2 & 1 \\ x_3 & y_3 & 1 \end{vmatrix} = 0. \tag{11}$$

The ANGLE between lines

$$A_1 x + B_1 y + C_1 = 0 \tag{12}$$

$$A_2 x + B_2 y + C_2 = 0 \tag{13}$$

is

$$\tan \theta = \frac{A_1 B_2 - A_2 B_1}{A_1 A_2 + B_1 B_2}. \tag{14}$$

The line joining points with TRILINEAR COORDINATES $\alpha_1 : \beta_1 : \gamma_1$ and $\alpha_2 : \beta_2 : \gamma_2$ is the set of point $\alpha : \beta : \gamma$ satisfying

$$\begin{vmatrix} \alpha & \beta & \gamma \\ \alpha_1 & \beta_1 & \gamma_1 \\ \alpha_2 & \beta_2 & \gamma_2 \end{vmatrix} = 0 \tag{15}$$

$$(\beta_1 \gamma_2 - \gamma_1 \beta_2)\alpha + (\gamma_1 \alpha_2 - \alpha_1 \gamma_2)\beta + (\alpha_1 \beta_2 - \beta_1 \alpha_2)\gamma = 0. \tag{16}$$

Three lines CONCUR if their TRILINEAR COORDINATES satisfy

$$l_1 \alpha + m_1 \beta + n_1 \gamma = 0 \tag{17}$$

$$l_2 \alpha + m_2 \beta + n_2 \gamma = 0 \tag{18}$$

$$l_3 \alpha + m_3 \beta + n_3 \gamma = 0, \tag{19}$$

in which case the point is

$$m_2 n_3 - n_2 m_3 : n_2 l_3 - l_2 n_3 : l_2 m_3 - m_2 l_3, \tag{20}$$

or if the COEFFICIENTS of the lines

$$A_1 x + B_1 y + C_1 = 0 \tag{21}$$

$$A_2 x + B_2 y + C_2 = 0 \tag{22}$$

$$A_3 x + B_3 y + C_3 = 0 \tag{23}$$

satisfy

$$\begin{vmatrix} A_1 & B_1 & C_1 \\ A_2 & B_2 & C_2 \\ A_3 & B_3 & C_3 \end{vmatrix} = 0. \tag{24}$$

Two lines CONCUR if their TRILINEAR COORDINATES satisfy

$$\begin{vmatrix} l_1 & m_1 & n_1 \\ l_2 & m_2 & n_2 \\ l_3 & m_3 & n_3 \end{vmatrix} = 0. \tag{25}$$

The line through P_1 is the direction (a_1, b_1, c_1) and the line through P_2 in direction (a_2, b_2, c_2) intersect IFF

$$\begin{vmatrix} x_2 - x_1 & y_2 - y_1 & z_2 - z_1 \\ a_1 & b_1 & c_1 \\ a_2 & b_2 & c_2 \end{vmatrix} = 0. \tag{26}$$

The line through a point $\alpha' : \beta' : \gamma'$ PARALLEL to

$$l\alpha + m\beta + n\gamma = 0 \tag{27}$$

is

$$\begin{vmatrix} \alpha & \beta & \gamma \\ \alpha' & \beta' & \gamma' \\ bn-cm & cl-an & am-bl \end{vmatrix} = 0. \qquad (28)$$

The lines

$$l\alpha + m\beta + n\gamma = 0 \qquad (29)$$

$$l'\alpha + m'\beta + n'\gamma = 0 \qquad (30)$$

are PARALLEL if

$$a(mn' - nm') + b(nl' - ln') + c(lm' - ml') = 0 \qquad (31)$$

for all (a, b, c), and PERPENDICULAR if

$$2abc(ll' + mm' + nn') - (mn' + m'm)\cos A$$

$$-(nl' + n'l)\cos B - (lm' + l'm)\cos C = 0 \qquad (32)$$

for all (a, b, c) (Sommerville 1924). The line through a point $\alpha' : \beta' : \gamma'$ PERPENDICULAR to (32) is given by

$$\begin{vmatrix} \alpha & \beta & \gamma \\ \alpha' & \beta' & \gamma' \\ l - m\cos C & m - n\cos A & n - l\cos B \\ -n\cos B & -l\cos C & -m\cos A \end{vmatrix} = 0. \qquad (33)$$

In 3-D SPACE, the line passing through the point (x_0, y_0, z_0) and PARALLEL to the NONZERO VECTOR

$$\mathbf{v} = \begin{bmatrix} a \\ b \\ c \end{bmatrix} \qquad (34)$$

has PARAMETRIC EQUATIONS

$$x = x_0 + at \qquad (35)$$

$$y = y_0 + bt \qquad (36)$$

$$z = z_0 + ct, \qquad (37)$$

written concisely as

$$\mathbf{x} = \mathbf{x}_0 + \mathbf{v}t. \qquad (38)$$

Similarly, the line in 3-D passing through (x_1, y_1) and (x_2, y_2) has parametric vector equation

$$\mathbf{x} = \mathbf{x}_1 + (\mathbf{x}_2 - \mathbf{x}_1)t, \qquad (39)$$

where this parametrization corresponds to $\mathbf{x}(t = 0) = \mathbf{x}_1$ and $\mathbf{x}(t = 1) = \mathbf{x}_2$.

See also Asymptote, Branch Line, Brocard Line, Cayley Lines, Collinear, Concur, Critical Line, Desargues' Theorem, Erdos-Anning Theorem, Euler Line, Flow Line, Gergonne Line, Imaginary Line, Isogonal Line, Isotropic Line, Lemoine Line, Line-Line Intersection, Line-Plane Intersection, Line Segment, Ordinary Line, Pascal Lines, Pedal Line, Pencil, Philo Line, Point, Point-Line Distance–2-D, Point-Line Distance–3-D, Plane, Plücker Lines, Polar Line, Power Line, Radical Line, Range (Line Segment), Ray, Real Line, Rhumb Line, Secant Line, Simson Line, Skew Lines, Soddy Line, Solomon's Seal Lines, Steiner Set, Steiner's Theorem, Sylvester's Line Problem, Symmedian, Tangent Line, Transversal Line, Trilinear Line, World Line

References

Casey, J. "The Right Line." Ch. 2 in *A Treatise on the Analytical Geometry of the Point, Line, Circle, and Conic Sections, Containing an Account of Its Most Recent Extensions, with Numerous Examples,* 2nd ed., rev. enl. Dublin: Hodges, Figgis, & Co., pp. 30–5, 1893.
Dunham, W. *Journey through Genius: The Great Theorems of Mathematics.* New York: Wiley, p. 32, 1990.
Harary, F. *Graph Theory.* Reading, MA: Addison-Wesley, 1994.
Kern, W. F. and Bland, J. R. "Lines and Planes in Space." §4 in *Solid Mensuration with Proofs,* 2nd ed. New York: Wiley, pp. 9–2, 1948.
Kline, M. "The Straight Line." *Sci. Amer.* **156**, 105–14, Mar. 1956.
MacTutor History of Mathematics Archive. "Straight Line." http://www-groups.dcs.st-and.ac.uk/~history/Curves/Straight.html.
Sommerville, D. M. Y. *Analytical Conics.* London: G. Bell, p. 186, 1924.
Spanier, J. and Oldham, K. B. "The Linear Function $bx + c$ and Its Reciprocal." Ch. 7 in *An Atlas of Functions.* Washington, DC: Hemisphere, pp. 53–2, 1987.

Linear Algebra

The study of linear sets of equations and their transformation properties. Linear algebra allows the analysis of ROTATIONS in space, LEAST SQUARES FITTING, solution of coupled differential equations, determination of a circle passing through three given points, as well as many other problems in mathematics, physics, and engineering.

The MATRIX and DETERMINANT are extremely useful tools of linear algebra. One central problem of linear algebra is the solution of the matrix equation

$$\mathsf{A}\mathbf{x} = \mathbf{b}$$

for $\mathbf{x}$. While this can, in theory, be solved using a MATRIX INVERSE

$$\mathbf{x} = \mathsf{A}^{-1}\mathbf{b},$$

other techniques such as GAUSSIAN ELIMINATION are numerically more robust.

See also Control Theory, Cramer's Rule, Determinant, Gaussian Elimination, Linear Transformation, Matrix, Vector

References

Axler, S. *Linear Algebra Done Right,* 2nd ed. New York: Springer-Verlag, 1997.
Ayres, F. Jr. *Theory and Problems of Matrices.* New York: Schaum, 1962.
Banchoff, T. and Wermer, J. *Linear Algebra Through Geometry,* 2nd ed. New York: Springer-Verlag, 1992.
Bellman, R. E. *Introduction to Matrix Analysis,* 2nd ed. New York: McGraw-Hill, 1970.

BLAS. "BLAS (Basic Linear Algebra Subprograms)." http://www.netlib.org/blas/.

Carlson, D.; Johnson, C. R.; Lay, D. C.; Porter, A. D.; Watkins, A. E.; and Watkins, W. (Eds.). *Resources for Teaching Linear Algebra.* Washington, DC: Math. Assoc. Amer., 1997.

Faddeeva, V. N. *Computational Methods of Linear Algebra.* New York: Dover, 1958.

Golub, G. and van Loan, C. *Matrix Computations, 3rd ed.* Baltimore, MD: Johns Hopkins University Press, 1996.

Halmos, P. R. *Linear Algebra Problem Book.* Providence, RI: Math. Assoc. Amer., 1995.

Lang, S. *Introduction to Linear Algebra, 2nd ed.* New York: Springer-Verlag, 1997.

LAPACK. "LAPACK--Linear Algebra PACKage." http://www.netlib.org/lapack/.

Lipschutz, S. *Schaum's Outline of Theory and Problems of Linear Algebra, 2nd ed.* New York: McGraw-Hill, 1991.

Lumsdaine, J. and Siek, J. "The Matrix Template Library: Generic Components for High Performance Scientific Computing." http://www.lsc.nd.edu/research/mtl/.

Marcus, M. and Minc, H. *Introduction to Linear Algebra.* New York: Dover, 1988.

Marcus, M. and Minc, H. *A Survey of Matrix Theory and Matrix Inequalities.* New York: Dover, 1992.

Marcus, M. *Matrices and Matlab: A Tutorial.* Englewood Cliffs, NJ: Prentice-Hall, 1993.

Mirsky, L. *An Introduction to Linear Algebra.* New York: Dover, 1990.

Muir, T. *A Treatise on the Theory of Determinants.* New York: Dover, 1960.

Nash, J. C. *Compact Numerical Methods for Computers: Linear Algebra and Function Minimisation, 2nd ed.* Bristol, England: Adam Hilger, 1990.

Petard, H. *Problems in Linear Algebra, preliminary ed.* New York: W.A. Benjamin, 1967.

Strang, G. *Linear Algebra and its Applications, 3rd ed.* Philadelphia, PA: Saunders, 1988.

Strang, G. *Introduction to Linear Algebra.* Wellesley, MA: Wellesley-Cambridge Press, 1993.

Strang, G. and Borre, K. *Linear Algebra, Geodesy, & GPS.* Wellesley, MA: Wellesley-Cambridge Press, 1997.

Weisstein, E. W. "Books about Linear Algebra." http://www.treasure-troves.com/books/LinearAlgebra.html.

Zhang, F. *Matrix Theory: Basic Results and Techniques.* New York: Springer-Verlag, 1999.

Linear Algebraic Group

A linear algebraic group is a GROUP which is also an AFFINE VARIETY. In particular, its elements satisfy polynomial equations. For example, $GL(n)$, the GENERAL LINEAR GROUP, is a linear algebraic group because an INVERTIBLE MATRIX is given by n^2 entries that satisfy the polynomial $\det a_n = 1$. The group operations are required to be given by REGULAR RATIONAL FUNCTIONS. The linear algebraic groups are similar to the LIE GROUPS, except that linear algebraic groups may be defined over any FIELD, including those of positive CHARACTERISTIC.

See also AFFINE VARIETY, ALGEBRAIC GROUP, FORMAL GROUP, GROUP, GROUP SCHEME, LIE ALGEBRA, LIE GROUP, VARIETY

Linear Approximation

A linear approximation to a function $f(x)$ at a point x_0 can be computed by taking the first term in the TAYLOR SERIES

$$f(x_0 + \Delta x) = f(x_0) + f'(x_0)\Delta x + \ldots .$$

See also MACLAURIN SERIES, TAYLOR SERIES

Linear Code

A linear code over a FINITE FIELD with q elements F_q is a linear SUBSPACE $C \subset F_q^n$. The vectors forming the SUBSPACE are called code words. When code words are chosen such that the distance between them is maximized, the code is called error-correcting since slightly garbled vectors can be recovered by choosing the nearest code word.

See also CODE, CODING THEORY, ERROR-CORRECTING CODE, GRAY CODE, HUFFMAN CODING, ISBN, UPC

Linear Combination

A sum of the elements from some set with constant coefficients placed in front of each. For example, a linear combination of the VECTORS $\mathbf{x}$, $\mathbf{y}$, and $\mathbf{z}$ is given by

$$a\mathbf{x} + b\mathbf{y} + c\mathbf{z},$$

where a, b, and c are constants.

See also BASIS, BASIS (VECTOR SPACE), SPAN (VECTOR SPACE)

Linear Congruence Equation

A linear congruence equation

$$ax \equiv b \pmod{m} \tag{1}$$

is solvable IFF the CONGRUENCE

$$b \equiv 0 \pmod{d} \tag{2}$$

is solvable, where $d \equiv \mathrm{GCD}(a, m)$ is the GREATEST COMMON DIVISOR. Let one solution to the original equation be $x_0 < m/d$. Then the solutions are $x = x_0$, $x_0 + m/d, x_0 + 2m/d, ..., x_0 + (d-1)m/d$. If $d = 1$, then there is only one solution $< m$. The solution of a linear congruence can be found in *Mathematica* using `Solve[ax == b && Modulus == m, x]`.

Solution to a linear congruence equation is equivalent to finding the value of a fractional CONGRUENCE, for which a greedy-type algorithm exists. In particular, (1) can be rewritten as

$$x \equiv \frac{b}{a} \pmod{m} \tag{3}$$

which can also be written

$$\frac{x}{b} \equiv \frac{1}{a} \pmod{m}. \tag{4}$$

In this form, the solution x can be found as `Mod[by,`

m] of the solution y returned by the *Mathematica* command $y = \texttt{PowerMod}[a, -1, m]$.

See also CHINESE REMAINDER THEOREM, CONGRUENCE, CONGRUENCE EQUATION, QUADRATIC CONGRUENCE EQUATION

References

Nagell, T. "Linear Congruences." §23 in *Introduction to Number Theory*. New York: Wiley, pp. 76–8, 1951.

Linear Congruence Method

A METHOD for generating RANDOM (PSEUDORANDOM) numbers using the linear RECURRENCE RELATION

$$X_{n+1} = aX_n + c \pmod{m},$$

where a and c must assume certain fixed values and X_0 is an initial number known as the SEED.

See also PSEUDORANDOM NUMBER, RANDOM NUMBER, SEED

References

Brunner, D. and Uhl, A. "Optimal Multipliers for Linear Congruential Pseudo Random Number Generators with Prime Moduli: Parallel Computation and Properties." *BIT. Numer. Math.* **39**, 193–09, 1999.
Pickover, C. A. "Computers, Randomness, Mind, and Infinity." Ch. 31 in *Keys to Infinity*. New York: W. H. Freeman, pp. 233–47, 1995.

Linear Diophantine Equation

DIOPHANTINE EQUATION

Linear Equation

An algebraic equation OF THE FORM

$$y = ax + b$$

involving only a constant and a first-order (linear) term.

See also LINE, POLYNOMIAL, QUADRATIC EQUATION

Linear Equation System

When solving a system of n linear equations with $k > n$ unknowns, use MATRIX operations to solve the system as far as possible. Then solve for the first $(k - n)$ components in terms of the last n components to find the solution space.

Linear Extension

A linear extension of a PARTIALLY ORDERED SET P is a PERMUTATION of the elements $p_1, p_2, \ldots$ of P such that $i < j$ IMPLIES $p_i < p_j$. For example, the linear extensions of the PARTIALLY ORDERED SET $((1, 2), (3, 4))$ are 1234, 1324, 1342, 3124, 3142, and 3412, all of which have 1 before 2 and 3 before 4.

References

Brightwell, G. and Winkler, P. "Counting Linear Extensions." *Order* **8**, 225–42, 1991.
Bubley, R. and Dyer, M. "Faster Random Generation of Linear Extensions." In *Proc. Ninth Annual ACM-SIAM Symposium on Discrete Algorithms, San Francisco, Calif.,* pp. 350–54, 1998.
Preusse, G. and Ruskey, F. "Generating Linear Extensions Fast." *SIAM J. Comput.* **23**, 373–86, 1994.
Ruskey, F. "Information on Linear Extension." http://www.theory.csc.uvic.ca/~cos/inf/pose/LinearExt.html.
Varol, Y. and Rotem, D. "An Algorithm to Generate All Topological Sorting Arrangements." *Comput. J.* **24**, 83–4, 1981.

Linear Fractional Transformation

A transformation OF THE FORM

$$w = f(z) = \frac{az + b}{cz + d}, \tag{1}$$

where $a, b, c, d \in \mathbb{C}$ and

$$ad - bc \neq 0, \tag{2}$$

is a CONFORMAL MAPPING called a linear fractional transformation. The transformation can be extended to the entire extended COMPLEX PLANE $\mathbb{C}^* = \mathbb{C} \cup \{\infty\}$ by defining

$$f\left(-\frac{d}{c}\right) = \infty \tag{3}$$

$$f(\infty) = \frac{a}{c} \tag{4}$$

(Apostol 1997, p. 26). The linear fractional transformation is linear in both w and z, and analytic everywhere except for a simple POLE at $z = -d/c$.

Every linear fractional transformation except $f(z) = z$ has one or two FIXED POINTS. The linear fractional transformation sends CIRCLES and lines to CIRCLES or lines. Linear fractional transformations preserve symmetry. The CROSS-RATIO is invariant under a linear fractional transformation. A linear fractional transformation is a composition of translations, rotations, magnifications, and inversions.

To determine a particular linear fractional transformation, specify the map of three points which preserve orientation. A particular linear fractional transformation is then uniquely determined. To determine a general linear fractional transformation, pick two symmetric points α and α_S. Define $\beta \equiv f(\alpha)$, restricting β as required. Compute β_S. $f(\alpha_S)$ then equals β_S since the linear fractional transformation preserves symmetry (the SYMMETRY PRINCIPLE). Plug in α and α_S into the general linear fractional transformation and set equal to β and β_S. Without loss of generality, let $c = 1$ and solve for a and b in terms of β. Plug back into the general expression to obtain a linear fractional transformation.

See also Cayley Transform, Möbius Transform, Modular Group Gamma, Schwarz's Lemma, Symmetry Principle, Unimodular Transformation

References

Anderson, J. W. "The Group of Möbius Transformations." §2.1 in *Hyperbolic Geometry*. New York: Springer-Verlag, pp. 19–5, 1999.

Apostol, T. M. "Möbius Transformations." Ch. 2.1 in *Modular Functions and Dirichlet Series in Number Theory*, 2nd ed. New York: Springer-Verlag, pp. 26–8, 1997.

Krantz, S. G. "Linear Fractional Transformations." §6.3 in *Handbook of Complex Analysis*. Boston, MA: Birkhäuser, pp. 81–6, 1999.

Mathews, J. "The Moebius Transformation." http://www.ecs.fullerton.edu/~mathews/fofz/mobius/.

Linear Function

A linear function is a function f which satisfies

$$f(x+y) = f(x) + f(y)$$

and

$$f(\alpha x) = \alpha f(x)$$

for all x and y in the DOMAIN, and all SCALARS α.

See also Bilinear Function, Function, Vector Space

Linear Functional

A linear functional on a REAL VECTOR SPACE V is a function $T : V \to \mathbb{R}$, which satisfies the following properties.

1. $T(v+w) = T(v) + T(w)$, and
2. $T(\alpha v) = \alpha T(v)$.

When V is a COMPLEX VECTOR SPACE, then T is a linear map into the COMPLEX NUMBERS.

DISTRIBUTIONS are a special case of linear functionals, and have a rich theory surrounding them.

See also Distribution (Generalized Function), Dual Space, Functional, Vector Space

Linear Group

See also General Linear Group, Lie-Type Group, Projective General Linear Group, Projective Special Linear Group, Special Linear Group

References

Hsiang, W. Y. "Linear Groups and Linear Representations." Lec. 1 in *Lectures on Lie Groups*. Singapore: World Scientific, pp. 1–9, 2000.

Wilson, R. A. "ATLAS of Finite Group Representation." http://for.mat.bham.ac.uk/atlas/html/contents.html#lin.

Linear Group Theorem

Any linear system of point-groups on a curve with only ordinary singularities may be cut by ADJOINT CURVES.

References

Coolidge, J. L. *A Treatise on Algebraic Plane Curves*. New York: Dover, pp. 122 and 251, 1959.

Linear Map

LINEAR TRANSFORMATION

Linear Operator

An operator $\tilde{L}$ is said to be linear if, for every pair of functions f and g and SCALAR t,

$$\tilde{L}(f+g) = \tilde{L}f + \tilde{L}g$$

and

$$\tilde{L}(tf) = t\tilde{L}f.$$

See also Linear Transformation, Operator

Linear Ordinary Differential Equation

Ordinary Differential Equation–First-Order, Ordinary Differential Equation–Second-Order

Linear Programming

The problem of maximizing a linear function over a convex polyhedron, also known as OPERATIONS RESEARCH, OPTIMIZATION THEORY, or CONVEX OPTIMIZATION THEORY. Linear programming is extensively used in economics and engineering. Examples from economics include Leontief's input-output model, the determination of shadow prices, etc., while an example of an engineering application would be maximizing profit in a factory that manufactures a number of different products from the same raw material using the same resources.

Linear programming can be solved using the SIMPLEX METHOD (Wood and Dantzig 1949, Dantzig 1949) which runs along EDGES of the visualization solid to find the best answer. In 1979, L. G. Khachian found a $\mathcal{O}(x^5)$ POLYNOMIAL-time ALGORITHM. A much more efficient POLYNOMIAL-time ALGORITHM was found by Karmarkar (1984). This method goes through the middle of the solid and then transforms and warps, and offers many advantages over the simplex method. Karmarkar's method is patented, so it has not received much detailed discussion.

See also Criss-Cross Method, Ellipsoidal Calculus, Kuhn-Tucker Theorem, Lagrange Multiplier, Optimization, Optimization Theory, Stochastic Optimization, Vertex Enumeration

References

Bellman, R. and Kalaba, R. *Dynamic Programming and Modern Control Theory.* New York: Academic Press, 1965.

Dantzig, G. B. "Programming of Interdependent Activities. II. Mathematical Model." *Econometrica* **17**, 200–11, 1949.

Dantzig, G. B. *Linear Programming and Extensions.* Princeton, NJ: Princeton University Press, 1963.

Karloff, H. *Linear Programming.* Boston, MA: Birkhäuser, 1991.

Karmarkar, N. "A New Polynomial-Time Algorithm for Linear Programming." *Combinatorica* **4**, 373–95, 1984.

Pappas, T. "Projective Geometry & Linear Programming." *The Joy of Mathematics.* San Carlos, CA: Wide World Publ./Tetra, pp. 216–17, 1989.

Press, W. H.; Flannery, B. P.; Teukolsky, S. A.; and Vetterling, W. T. "Linear Programming and the Simplex Method." §10.8 in *Numerical Recipes in FORTRAN: The Art of Scientific Computing, 2nd ed.* Cambridge, England: Cambridge University Press, pp. 423–36, 1992.

Sultan, A. *Linear Programming: An Introduction with Applications.* San Diego, CA: Academic Press, 1993.

Tokhomirov, V. M. "The Evolution of Methods of Convex Optimization." *Amer. Math. Monthly* **103**, 65–1, 1996.

Weisstein, E. W. "Books about Linear Programming." http://www.treasure-troves.com/books/LinearProgramming.html.

Wood, M. K. and Dantzig, G. B. "Programming of Interdependent Activities. I. General Discussion." *Econometrica* **17**, 193–99, 1949.

Yudin, D. B. and Nemirovsky, A. S. *Problem Complexity and Method Efficiency in Optimization.* New York: Wiley, 1983.

Linear Recurrence Sequence

RECURRENCE SEQUENCE

Linear Regression

The fitting of a straight LINE through a given set of points according to some specified goodness-of-fit criterion. The most common form of linear regression is LEAST SQUARES FITTING.

See also LEAST SQUARES FITTING, MULTIPLE REGRESSION, NONLINEAR LEAST SQUARES FITTING

References

Edwards, A. L. *An Introduction to Linear Regression and Correlation.* San Francisco, CA: W. H. Freeman, 1976.

Edwards, A. L. *Multiple Regression and the Analysis of Variance and Covariance.* San Francisco, CA: W. H. Freeman, 1979.

Linear Space

VECTOR SPACE

Linear Stability

Consider the general system of two first-order ORDINARY DIFFERENTIAL EQUATIONS

$$\dot{x} = f(x, y) \tag{1}$$
$$\dot{y} = g(x, y). \tag{2}$$

Let x_0 and y_0 denote FIXED POINTS with $\dot{x} = \dot{y} = 0$, so

$$f(x_0, y_0) = 0 \tag{3}$$
$$g(x_0, y_0) = 0. \tag{4}$$

Then expand about (x_0, y_0) so

$$\delta\dot{x} = f_x(x_0, y_0)\delta x + f_y(x_0, y_0)\delta y + f_{xy}(x_0, y_0)\delta x \delta y \\ + \cdots \tag{5}$$

$$\delta\dot{y} = g_x(x_0, y_0)\delta x + g_y(x_0, y_0)\delta y + g_{xy}(x_0, y_0)\delta x \delta y \\ + \cdots \tag{6}$$

To first-order, this gives

$$\frac{d}{dt}\begin{bmatrix} \delta x \\ \delta y \end{bmatrix} = \begin{bmatrix} f_x(x_0, y_0) & f_y(x_0, y_0) \\ g_x(x_0, y_0) & g_y(x_0, y_0) \end{bmatrix}\begin{bmatrix} \delta x \\ \delta y \end{bmatrix}, \tag{7}$$

where the 2×2 MATRIX is called the STABILITY MATRIX.

In general, given an n-D MAP $\mathbf{x}' = T(\mathbf{x})$, let $\mathbf{x}_0$ be a FIXED POINT, so that

$$T(\mathbf{x}_0) = \mathbf{x}_0. \tag{8}$$

Expand about the fixed point,

$$T(\mathbf{x}_0 + \delta\mathbf{x}) = T(\mathbf{x}_0) + \frac{\partial T}{\partial \mathbf{x}}\delta\mathbf{x} + \mathcal{O}(\delta\mathbf{x})^2$$

$$\equiv T(\mathbf{x}_0) + \delta T, \tag{9}$$

so

$$\delta T = \frac{\partial T}{\partial \mathbf{x}}\delta\mathbf{x} \equiv A\delta\mathbf{x}. \tag{10}$$

The map can be transformed into the principal axis frame by finding the EIGENVECTORS and EIGENVALUES of the MATRIX A

$$(A - \lambda I)\delta\mathbf{x} = 0, \tag{11}$$

so the DETERMINANT

$$|A - \lambda I| = 0. \tag{12}$$

The mapping is

$$\delta\mathbf{x}'_{\text{princ}} = \begin{bmatrix} \lambda_1 & \cdots & 0 \\ \vdots & \ddots & \vdots \\ 0 & \cdots & \lambda_n \end{bmatrix}. \tag{13}$$

When iterated a large number of times,

$$\delta T'_{\text{princ}} \to 0 \tag{14}$$

only if $|\Re(\lambda_i)| < 1$ for $i = 1, ..., n$ but $\to \infty$ if any $|\lambda_i| > 1$. Analysis of the EIGENVALUES (and EIGENVECTORS) of A therefore characterizes the type of FIXED POINT. The condition for stability is $|\Re(\lambda_i)| < 1$ for $i = 1, ..., n$.

See also FIXED POINT, LYAPUNOV FUNCTION, NONLINEAR STABILITY, STABILITY MATRIX

References

Tabor, M. "Linear Stability Analysis." §1.4 in *Chaos and Integrability in Nonlinear Dynamics: An Introduction.* New York: Wiley, pp. 20–1, 1989.

Linear Transformation

A linear transformation between two VECTOR SPACES V and W is a MAP $T : V \to W$ such that the following hold:

1. $T(\mathbf{v}_1 + \mathbf{v}_2) = T(\mathbf{v}_1) T(\mathbf{v}_2)$ for any VECTORS $\mathbf{v}_1$ and $\mathbf{v}_2$ in V, and
2. $T(\alpha \mathbf{v}) = \alpha T(\mathbf{v})$ for any SCALAR α.

A linear transformation may not be INJECTIVE or ONTO. When V and W have the same DIMENSION, it is possible for T to be invertible, meaning there exists a T^{-1} such that $TT^{-1} = I$. It is always the case that $T(0) = 0$. Also, a linear transformation always maps LINES to LINES (or to zero).

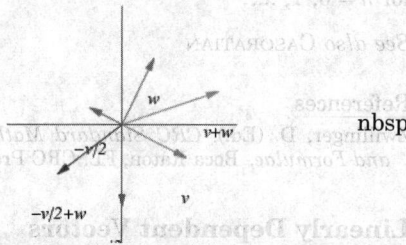

nbsp

The main example of a linear transformation is given by MATRIX MULTIPLICATION. Given an $n \times m$ MATRIX A, define $T(\mathbf{v}) = A\mathbf{v}$, where $\mathbf{v}$ is written as a COLUMN VECTOR (with m coordinates). For example, consider

$$A = \begin{bmatrix} 0 & 1 \\ -1 & 3 \\ 4 & 0 \end{bmatrix}, \tag{1}$$

then T is a linear transformation from $\mathbb{R}^2$ to $\mathbb{R}^3$, defined by,

$$T(x, y) = (y, -2x + 2y, x). \tag{2}$$

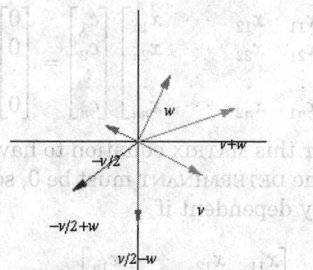

Another example is $T(x, y) = (1.4x - y, 0.8x)$. The homotopy from the identity transformation to T is illustrated above.

When V and W are FINITE dimensional, a general linear transformation can be written as a matrix multiplication only after specifying a BASIS for V and

W. When V and W have an INNER PRODUCT, and their BASES, $\{v_1, \cdots, v_m\}$ and $\{w, \cdots, w_n\}$, are ORTHONORMAL, it is easy to write the corresponding matrix $A = (a_{ij})$. In particular, $a_{ij} = \langle w_i, T(v_j) \rangle$. Note that when using the standard basis for $\mathbb{R}^n$ and $\mathbb{R}^m$, the jth column corresponds to the image of the jth standard basis vector.

When V and W are INFINITE dimensional, then it is possible for a linear transformation to not be CONTINUOUS. For example, let V be the space of polynomials in one variable, and T be the DERIVATIVE. Then $T(x^3) = nx^{n-1}$, which is not CONTINUOUS because $x^n/n \to 0$ while $T(x^n/n)$ does not converge.

Linear 2-D transformations have a simple classification. Consider the 2-D linear transformation

$$\rho x_1' = a_{11} x_1 + a_{12} x_2 \tag{3}$$
$$\rho x_2' = a_{21} x_1 + a_{22} x_2. \tag{4}$$

Now rescale by defining $\lambda \equiv x_1/x_2$ and $\lambda' \equiv x_1'/x_2'$. Then the above equations become

$$\lambda' = \frac{\alpha \lambda + \beta}{\gamma \lambda + \delta} \tag{5}$$

where $\alpha \delta - \beta \gamma \neq 0$ and α, β, γ and δ are defined in terms of the old constants. Solving for λ gives

$$\lambda = \frac{\delta \lambda' - \beta}{-\gamma \lambda' + \alpha}, \tag{6}$$

so the transformation is ONE-TO-ONE. To find the FIXED POINTS of the transformation, set $\lambda = \lambda'$ to obtain

$$\gamma \lambda^2 + (\delta - \alpha) \lambda - \beta = 0. \tag{7}$$

This gives two fixed points which may be distinct or coincident. The fixed points are classified as follows.

variables	type
$(\delta - \alpha)^2 + 4\beta\gamma > 0$	HYPERBOLIC FIXED POINT
$(\delta - \alpha)^2 + 4\beta\gamma < 0$	ELLIPTIC FIXED POINT
$(\delta - \alpha)^2 + 4\beta\gamma = 0$	PARABOLIC FIXED POINT

See also BASIS (VECTOR SPACE), ELLIPTIC FIXED POINT (MAP), GENERAL LINEAR GROUP, HYPERBOLIC FIXED POINT (MAP), INVERTIBLE LINEAR MAP, INVOLUTORY, LINEAR OPERATOR, MATRIX, MATRIX MULTIPLICATION, PARABOLIC FIXED POINT, VECTOR SPACE

References

Woods, F. S. *Higher Geometry: An Introduction to Advanced Methods in Analytic Geometry.* New York: Dover, pp. 13–5, 1961.

Linear Weighted Moment

L-MOMENT

Linearly Dependent Curves

Two curves ϕ and ψ satisfying

$$\phi + \psi = 0$$

are said to be linearly dependent. Similarly, n curves ϕ_i, $i = 1, ..., n$ are said to be linearly dependent if

$$\sum_{i=1}^{n} \phi_i = 0.$$

See also BERTINI'S THEOREM, STUDY'S THEOREM

References

Coolidge, J. L. *A Treatise on Algebraic Plane Curves.* New York: Dover, pp. 32–4, 1959.

Linearly Dependent Functions

The n functions $f_1(x)$, $f_2(x)$, ..., $f_n(x)$ are linearly dependent if, for some $c_1, c_2, ..., c_n \in \mathbb{R}$ not all zero,

$$c_i f_i(x) = 0 \tag{1}$$

(where EINSTEIN SUMMATION is used) for all x in some interval I. If the functions are not linearly dependent, they are said to be linearly independent. Now, if the functions $\in \mathbb{R}^{n-1}$, we can differentiate (1) up to $n-1$ times. Therefore, linear dependence also requires

$$c_i f_i' = 0 \tag{2}$$

$$c_i f_i'' = 0 \tag{3}$$

$$c_i f_i^{(n-1)} = 0, \tag{4}$$

where the sums are over $i = 1, ..., n$. These equations have a nontrivial solution IFF the DETERMINANT

$$\begin{vmatrix} f_1 & f_2 & \cdots & f_n \\ f_1' & f_2' & \cdots & f_2' \\ \vdots & \vdots & \ddots & \vdots \\ f_1^{(n-1)} & f_2^{(n-1)} & \cdots & f_n^{(n-1)} \end{vmatrix} = 0, \tag{5}$$

where the DETERMINANT is conventionally called the WRONSKIAN and is denoted $W(f_1, f_2, ..., f_n)$. If the WRONSKIAN $\neq 0$ for any value c in the interval I, then the only solution possible for (2) is $c_i = 0$ ($i = 1, ..., n$), and the functions are linearly independent. If, on the other hand, $W = 0$ for a range, the functions are linearly dependent in the range. This is equivalent to stating that if the vectors $\mathbf{V}[f_1(c)], ..., \mathbf{V}[f_n(c)]$ defined by

$$\mathbf{V}[f_i(x)] = \begin{bmatrix} f_i(x) \\ f_i'(x) \\ f_i''(x) \\ \vdots \\ f_i^{n-1}(x) \end{bmatrix} \tag{6}$$

are linearly independent for at least one $c \in I$, then the functions f_i are linearly independent in I.

References

Sansone, G. "Linearly Independent Functions." §1.2 in *Orthogonal Functions, rev. English ed.* New York: Dover, pp. 2–, 1991.

Linearly Dependent Sequences

Sequences $x_n^{(1)}$, $x_n^{(2)}$, ..., $x_n^{(k)}$ are linearly independent if constants $c_1, c_2, ..., c_k$ (not all zero) exist such that

$$\sum_{i=1}^{k} c_i x_n^{(i)} = 0$$

for $n = 0, 1, ...$

See also CASORATIAN

References

Zwillinger, D. (Ed.). *CRC Standard Mathematical Tables and Formulae.* Boca Raton, FL: CRC Press, p. 229, 1995.

Linearly Dependent Vectors

n VECTORS $\mathbf{X}_1, \mathbf{X}_2, ..., \mathbf{X}_n$ are linearly dependent IFF there exist SCALARS $c_1, c_2, ..., c_n$, not all zero, such that

$$c_i \mathbf{X}_i = 0, \tag{1}$$

where EINSTEIN SUMMATION is used and $i = 1, ..., n$. If no such SCALARS exist, then the vectors are said to be linearly independent. In order to satisfy the CRITERION for linear dependence,

$$c_1 \begin{bmatrix} x_{11} \\ x_{12} \\ \vdots \\ x_{n1} \end{bmatrix} + c_2 \begin{bmatrix} x_{12} \\ x_{22} \\ \vdots \\ x_{n2} \end{bmatrix} + \cdots + c_n \begin{bmatrix} x_{1n} \\ x_{2n} \\ \vdots \\ x_{nn} \end{bmatrix} = \begin{bmatrix} 0 \\ 0 \\ \vdots \\ 0 \end{bmatrix} \tag{2}$$

$$\begin{bmatrix} x_{11} & x_{12} & \cdots & x_{1n} \\ x_{21} & x_{22} & \cdots & x_{2n} \\ \vdots & \vdots & \ddots & \vdots \\ x_{n1} & x_{n2} & \cdots & x_{nn} \end{bmatrix} \begin{bmatrix} c_1 \\ c_2 \\ \vdots \\ c_n \end{bmatrix} = \begin{bmatrix} 0 \\ 0 \\ \vdots \\ 0 \end{bmatrix}. \tag{3}$$

In order for this MATRIX equation to have a nontrivial solution, the DETERMINANT must be 0, so the VECTORS are linearly dependent if

$$\begin{bmatrix} x_{11} & x_{12} & \cdots & x_{1n} \\ x_{21} & x_{22} & \cdots & x_{2n} \\ \vdots & \vdots & \ddots & \vdots \\ x_{n1} & x_{n2} & \cdots & x_{nn} \end{bmatrix} = 0, \tag{4}$$

and linearly independent otherwise.

Let $\mathbf{p}$ and $\mathbf{q}$ be n-D VECTORS. Then the following three conditions are equivalent (Gray 1997).

1. **p** and **q** are linearly dependent.

2. $\begin{vmatrix} \mathbf{p} \cdot \mathbf{p} & \mathbf{p} \cdot \mathbf{q} \\ \mathbf{q} \cdot \mathbf{p} & \mathbf{q} \cdot \mathbf{q} \end{vmatrix} = 0.$

3. The $2 \times n$ MATRIX $\begin{bmatrix} \mathbf{p} \\ \mathbf{q} \end{bmatrix}$ has rank less than two.

References

Gray, A. *Modern Differential Geometry of Curves and Surfaces with Mathematica, 2nd ed.* Boca Raton, FL: CRC Press, pp. 272–73, 1997.

Linearly Independent

Two or more functions, equations, or vectors $f_1, f_2, \ldots$, which are not linearly dependent, i.e., cannot be expressed in the form

$$a_1 f_1 + a_2 f_2 + \cdots + a_n f_n = 0$$

with $a_1, a_2, \ldots$ constants which are not all zero are said to be linearly independent.

See also LINEARLY DEPENDENT CURVES, LINEARLY DEPENDENT FUNCTIONS, LINEARLY DEPENDENT VECTORS, MAXIMALLY LINEARLY INDEPENDENT

Linearly Ordered Set

TOTAL ORDER

Line at Infinity

The straight line on which all POINTS AT INFINITY lie. The line at infinity is given in terms of TRILINEAR COORDINATES by

$$a\alpha + b\beta + c\gamma = 0,$$

which follows from the fact that a REAL TRIANGLE will have POSITIVE AREA, and therefore that

$$2\Delta = a\alpha + b\beta + c\gamma > 0.$$

Instead of the three reflected segments concurring for the ISOGONAL CONJUGATE of a point X on the CIRCUMCIRCLE of a TRIANGLE, they become parallel (and can be considered to meet at infinity). As X varies around the CIRCUMCIRCLE, X^{-1} varies through a line called the line at infinity. Every line is PERPENDICULAR to the line at infinity.

Poncelet was the first to systematically employ the line at infinity (Graustein 1930).

See also POINT AT INFINITY

References

Lachlan, R. §10 in *An Elementary Treatise on Modern Pure Geometry.* London: Macmillian, p. 6, 1893.
Graustein, W. C. *Introduction to Higher Geometry.* New York: Macmillan, p. 30, 1930.
Wells, D. *The Penguin Dictionary of Curious and Interesting Geometry.* London: Penguin, pp. 141–42, 1991.

Line Bisector

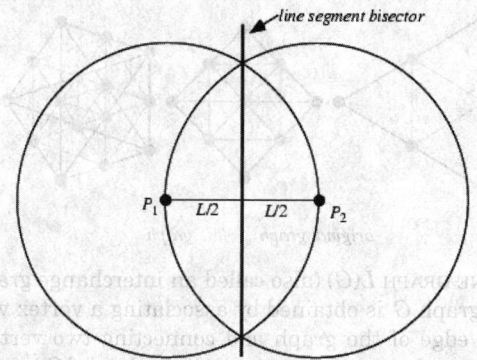

The line bisecting a given LINE SEGMENT $P_1 P_2$ can be constructed geometrically, as illustrated above.

References

Courant, R. and Robbins, H. "How to Bisect a Segment and Find the Center of a Circle with the Compass Alone." §3.4.4 in *What is Mathematics?: An Elementary Approach to Ideas and Methods, 2nd ed.* Oxford, England: Oxford University Press, pp. 145–46, 1996.
Dixon, R. *Mathographics.* New York: Dover, p. 22, 1991.

Line Bundle

A line bundle is a special case of a VECTOR BUNDLE in which the fiber is either $\mathbb{R}$, in the case of a real line bundle, or $\mathbb{C}$, in the case of a complex line bundle.

See also MANIFOLD, PRINCIPAL BUNDLE, TRIVIAL BUNDLE, VECTOR BUNDLE

Line-Circle Intersection

CIRCLE-LINE INTERSECTION

Line Connectivity

EDGE CONNECTIVITY

Line Element

Also known as the first FUNDAMENTAL FORM

$$ds^2 = g_{ab} dx^a dx^b.$$

In the principal axis frame for 3-D,

$$ds^2 = g_{aa}(dx^a)^2 + g_{bb}(dx^b)^2 + g_{cc}(dx^c)^2.$$

At ORDINARY POINTS on a surface, the line element is positive definite.

See also AREA ELEMENT, FUNDAMENTAL FORMS, VOLUME ELEMENT

Line Graph

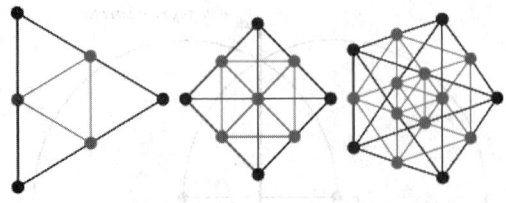

original graph   *line graph*

A LINE GRAPH $L(G)$ (also called an interchange graph) of a graph G is obtained by associating a vertex with each edge of the graph and connecting two vertices with an edge IFF the corresponding edges of G meet at one or both endpoints. In the three examples above, the original graphs are the COMPLETE GRAPHS K_3, K_4, and K_5.

The line graph of a GRAPH with n nodes, e edges, and vertex degrees d_i contains $n' = e$ nodes and

$$e' = \frac{1}{2} \sum_{i=1}^{n} d_i^2 - e$$

edges (Skiena 1990, p. 137). The INCIDENCE MATRIX C of a graph and ADJACENCY MATRIX L of its line graph are related by

$$\mathsf{L} = \mathsf{C}^T\mathsf{C} - 2\mathsf{I},$$

where I is the IDENTITY MATRIX (Skiena 1990, p. 136).

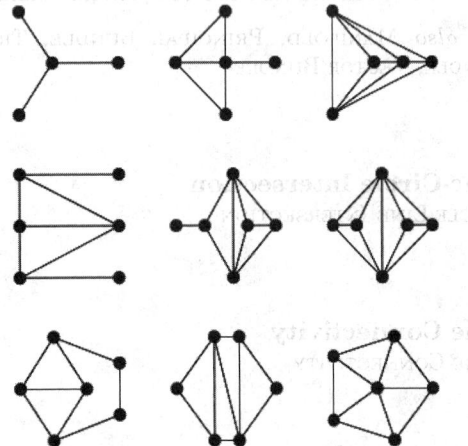

A graph is a line graph IFF if does not contain any of the above graphs as SUBGRAPHS (van Rooij and Wilf 1965; Beineke 1968; Skiena 1990, p. 138). Of the nine, one has four nodes (the STAR GRAPH $S_4 = K_{1,3}$), two have five nodes, and six have six nodes (including the WHEEL GRAPH W_6).

The only CONNECTED GRAPH that is isomorphic to its line graph is a CYCLE GRAPH C_n (Skiena 1990, p. 137). Whitney (1932) showed that, with the exception of K_3 and $K_{1,3}$, any two CONNECTED GRAPHS with isomorphic line graphs are isomorphic (Skiena 1990, p. 138).

The line graph of an EULERIAN GRAPH is both Eulerian and HAMILTONIAN (Skiena 1990, p. 138). More information about cycles of line graphs is given by Harary and Nash-Williams (1965) and Chartrand (1968).

See also TOTAL GRAPH

References

Beineke, L. W. "Derived Graphs and Digraphs." In *Beiträge zur Graphentheorie* (Ed. H. Sachs, H. Voss, and H. Walther). Leipzig, Germany: Teubner, pp. 17–3, 1968.
Chartrand, G. "On Hamiltonian Line Graphs." *Trans. Amer. Math. Soc.* **134**, 559–66, 1968.
Harary, F. *Graph Theory.* Reading, MA: Addison-Wesley, 1994.
Harary, F. and Nash-Williams, C. J. A. "On Eulerian and Hamiltonian Graphs and Line Graphs." *Canad. Math. Bull.* **8**, 701–09, 1965.
Saaty, T. L. and Kainen, P. C. "Line Graphs." §4– in *The Four-Color Problem: Assaults and Conquest.* New York: Dover, pp. 108–12, 1986.
Skiena, S. "Line Graph." §4.1.5 in *Implementing Discrete Mathematics: Combinatorics and Graph Theory with Mathematica.* Reading, MA: Addison-Wesley, pp. 128 and 135–39, 1990.
van Rooij, A. and Wilf, H. "The Interchange Graph of a Finite Graph." *Acta Math. Acad. Sci. Hungar.* **16**, 263–69, 1965.
Whitney, H. "Congruent Graphs and the Connectivity of Graphs." *Amer. J. Math.* **54**, 150–68, 1932.

Line Integral

The line integral of a VECTOR FIELD $\mathbf{F}(\mathbf{x})$ on a curve σ is defined by

$$\int_\sigma \mathbf{F} \cdot d\mathbf{s} = \int_a^b \mathbf{F}(\sigma(t)) \cdot \sigma'(t)\,dt, \quad (1)$$

where $\mathbf{a} \cdot \mathbf{b}$ denotes a DOT PRODUCT. In Cartesian coordinates, the line integral can be written

$$\int_\sigma \mathbf{F} \cdot d\mathbf{s} = \int_C F_1\,dx + F_2\,dy + F_3\,dz, \quad (2)$$

where

$$\mathbf{F} \equiv \begin{bmatrix} F_1(\mathbf{x}) \\ F_2(\mathbf{x}) \\ F_3(\mathbf{x}) \end{bmatrix} \quad (3)$$

For z COMPLEX and $\gamma : z = z(t)$ a path in the COMPLEX PLANE parameterized by $t \in [a, b]$,

$$\int_\gamma f\,dz = \int_a^b f(z(t))z'(t)\,dt. \quad (4)$$

POINCARÉ'S THEOREM states that if $\nabla \times \mathbf{F} = 0$ in a simply connected neighborhood $U(\mathbf{x})$ of a point $\mathbf{x}$, then in this neighborhood, $\mathbf{F}$ is the GRADIENT of a SCALAR FIELD $\phi(\mathbf{x})$,

$$\mathbf{F}(x) = -\nabla \phi(\mathbf{x}) \quad (5)$$

for $\mathbf{x} \in U(\mathbf{x})$, where ∇ is the gradient operator. Consequently, the GRADIENT THEOREM gives

$$\int_\sigma \mathbf{F} \cdot ds = \phi(\mathbf{x}_1) - \phi(\mathbf{x}_2) \qquad (6)$$

for any path σ located completely within $U(\mathbf{x})$, starting at $\mathbf{x}_1$ and ending at $\mathbf{x}_2$.

This means that if $\nabla \times \mathbf{F} = 0$ (i.e., $\mathbf{F}(\mathbf{x})$ is an IRROTATIONAL FIELD in some region), then the line integral is path-independent in this region. If desired, a Cartesian path can therefore be chosen between starting and ending point to give

$$\int_{(a,\,b,\,c)}^{(x,\,y,\,z)} F_1\,dx + F_2\,dy + F_3\,dz$$

$$= \int_{(a,\,b,\,c)}^{(x,\,b,\,c)} F_1\,dx + \int_{(x,\,b,\,c)}^{(x,\,y,\,c)} F_2\,dy + \int_{(x,\,y,\,c)}^{(x,\,y,\,z)} F_3\,dz. \qquad (7)$$

If $\nabla \cdot \mathbf{F} = 0$ (i.e., $\mathbf{F}(\mathbf{x})$ is a DIVERGENCELESS FIELD, a.k.a. SOLENOIDAL FIELD), then there exists a VECTOR FIELD $\mathbf{A}$ such that

$$\mathbf{F} = \nabla \times \mathbf{A}, \qquad (8)$$

where $\mathbf{A}$ is uniquely determined up to a gradient field (and which can be chosen so that $\nabla \cdot \mathbf{A} = 0$).

See also CONSERVATIVE FIELD, CONTOUR INTEGRAL, GRADIENT THEOREM, IRROTATIONAL FIELD, PATH INTEGRAL, POINCARÉ'S THEOREM

References

Krantz, S. G. "The Complex Line Integral." §2.1.6 in *Handbook of Complex Analysis.* Boston, MA: Birkhäuser, p. 22, 1999.

Line-Line Intersection

The INTERSECTION of two LINES L_1 and L_2 in 2-D with, L_1 containing the points (x_1, y_1) and (x_2, y_2), and L_2

containing the points (x_3, y_3) and (x_4, y_4), is given by

$$x = \frac{\begin{vmatrix} \begin{vmatrix} x_1 & y_1 \\ x_2 & y_2 \end{vmatrix} & \begin{vmatrix} x_1 & 1 \\ x_2 & 1 \end{vmatrix} \\ \begin{vmatrix} x_3 & y_3 \\ x_4 & y_4 \end{vmatrix} & \begin{vmatrix} x_3 & 1 \\ x_4 & 1 \end{vmatrix} \end{vmatrix}}{\begin{vmatrix} \begin{vmatrix} x_1 & 1 \\ x_2 & 1 \end{vmatrix} & \begin{vmatrix} y_1 & 1 \\ y_2 & 1 \end{vmatrix} \\ \begin{vmatrix} x_3 & 1 \\ x_4 & 1 \end{vmatrix} & \begin{vmatrix} y_3 & 1 \\ y_4 & 1 \end{vmatrix} \end{vmatrix}} = \frac{\begin{vmatrix} \begin{vmatrix} x_1 & y_1 \\ x_2 & y_2 \end{vmatrix} & x_1 - x_2 \\ \begin{vmatrix} x_3 & y_3 \\ x_4 & y_4 \end{vmatrix} & x_3 - x_4 \end{vmatrix}}{\begin{vmatrix} x_1 - x_2 & y_1 - y_2 \\ x_3 - x_4 & y_3 - y_4 \end{vmatrix}} \qquad (1)$$

$$y = \frac{\begin{vmatrix} \begin{vmatrix} x_1 & y_1 \\ x_2 & y_2 \end{vmatrix} & \begin{vmatrix} y_1 & 1 \\ y_2 & 1 \end{vmatrix} \\ \begin{vmatrix} x_3 & y_3 \\ x_4 & y_4 \end{vmatrix} & \begin{vmatrix} y_3 & 1 \\ y_4 & 1 \end{vmatrix} \end{vmatrix}}{\begin{vmatrix} \begin{vmatrix} x_1 & 1 \\ x_2 & 1 \end{vmatrix} & \begin{vmatrix} y_1 & 1 \\ y_2 & 1 \end{vmatrix} \\ \begin{vmatrix} x_3 & 1 \\ x_4 & 1 \end{vmatrix} & \begin{vmatrix} y_3 & 1 \\ y_4 & 1 \end{vmatrix} \end{vmatrix}} = \frac{\begin{vmatrix} \begin{vmatrix} x_1 & y_1 \\ x_2 & y_2 \end{vmatrix} & y_1 - y_2 \\ \begin{vmatrix} x_3 & y_3 \\ x_4 & y_4 \end{vmatrix} & y_3 - y_4 \end{vmatrix}}{\begin{vmatrix} x_1 - x_2 & y_1 - y_2 \\ x_3 - x_4 & y_3 - y_4 \end{vmatrix}}. \qquad (2)$$

In 3-D, let the two lines pass through points given by the vectors $(\mathbf{p}_1, \mathbf{q}_1)$ and $(\mathbf{p}_2, \mathbf{q}_2)$ and define

$$\mathbf{v}_1 = \frac{\mathbf{q}_1 - \mathbf{p}_1}{|\mathbf{q}_1 - \mathbf{p}_1|} \qquad (3)$$

$$\mathbf{v}_2 = \frac{\mathbf{q}_2 - \mathbf{p}_2}{|\mathbf{q}_2 - \mathbf{p}_2|} \qquad (4)$$

$$\mathbf{v}_{12} = \mathbf{v}_1 \times \mathbf{v}_2 \qquad (5)$$

$$s_1 = \det(\mathbf{p}_2 - \mathbf{p}_1 \quad \mathbf{v}_2 \quad \mathbf{v}_{12}) \qquad (6)$$

$$s_2 = \det(\mathbf{p}_2 - \mathbf{p}_1 \quad \mathbf{v}_1 \quad \mathbf{v}_{12}). \qquad (7)$$

Then the point of intersection $\mathbf{p}$ of the two lines is given by

$$\mathbf{p} = \tfrac{1}{2}(\mathbf{p}_1 + \mathbf{v}_1 s_1 + \mathbf{p}_2 + \mathbf{v}_2 s_2) \qquad (8)$$

(Glassner).

See also CONCUR, CONCURRENT, INTERSECTION, LINE, LINE-PLANE INTERSECTION

References

Glassner, A. S. (Ed.). *Graphics Gems.*

Line Line Picking

POINT-POINT DISTANCE–1-D

Line of Curvature

A curve on a surface whose tangents are always in the direction of PRINCIPAL CURVATURE. The equation of the lines of curvature can be written

$$\begin{vmatrix} g_{11} & g_{12} & g_{22} \\ b_{11} & b_{12} & b_{22} \\ du^2 & -du\,dv & dv^2 \end{vmatrix} = 0,$$

where g and b are the COEFFICIENTS of the first and second FUNDAMENTAL FORMS.

See also DUPIN'S THEOREM, FUNDAMENTAL FORMS, PRINCIPAL CURVATURES

Line-Plane Intersection

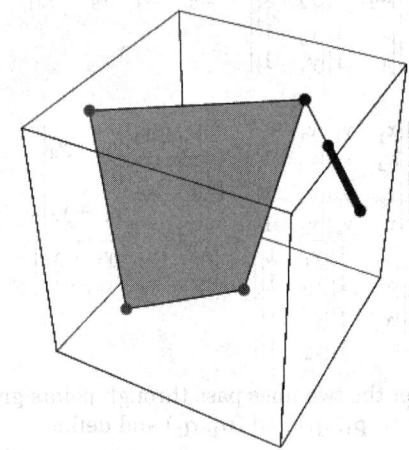

The PLANE determined by the points $\mathbf{x}_1$, $\mathbf{x}_2$, and $\mathbf{x}_3$ and the LINE passing through the points $\mathbf{x}_4$ and $\mathbf{x}_5$ intersect in a point which can be determined by solving the four simultaneous equations

$$\begin{vmatrix} x & y & z & 1 \\ x_1 & y_1 & z_1 & 1 \\ x_2 & y_2 & z_2 & 1 \\ x_3 & y_3 & z_3 & 1 \end{vmatrix} = 0 \tag{1}$$

$$x = x_4 + (x_4 - x_5)t \tag{2}$$

$$y = y_4 + (y_4 - y_5)t \tag{3}$$

$$z = z_4 + (z_4 - z_5)t \tag{4}$$

for x, y, z, and t, giving

$$t = \frac{\begin{vmatrix} 1 & 1 & 1 & 1 \\ x_1 & x_2 & x_3 & x_4 \\ y_1 & y_2 & y_3 & y_4 \\ z_1 & z_2 & z_3 & z_4 \end{vmatrix}}{\begin{vmatrix} 1 & 1 & 1 & 0 \\ x_1 & x_2 & x_3 & x_5 - x_4 \\ y_1 & y_2 & y_3 & y_5 - y_4 \\ z_1 & z_2 & z_3 & z_5 - z_4 \end{vmatrix}}. \tag{5}$$

This value can then be plugged back in to (2), (3), and (4) to give the point of intersection (x, y, z).

See also LINE, LINE-LINE INTERSECTION, PLANE

Line Segment

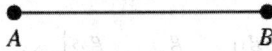

A closed interval corresponding to a FINITE portion of an infinite LINE. Line segments are generally labeled

with two letters corresponding to their endpoints, say A and B, and then written AB. The length of the line segment is indicated with an overbar, so the length of the line segment AB would be written $\overline{AB}$.

Curiously, the number of points in a line segment (ALEPH-1) is equal to that in an entire 1-D SPACE (a LINE), and also to the number of points in an n-D SPACE, as first recognized by Georg Cantor.

See also ALEPH-1, COLLINEAR, CONTINUUM, LINE, RANGE (LINE SEGMENT), RAY

References
Lachlan, R. *An Elementary Treatise on Modern Pure Geometry.* London: Macmillian, pp. 14–6, 1893.

Line Space
LIOUVILLE SPACE

L-Infinity-Norm
A VECTOR NORM defined for a VECTOR

$$\mathbf{x} = \begin{bmatrix} x_1 \\ x_2 \\ \vdots \\ x_n \end{bmatrix},$$

with COMPLEX entries by

$$\|\mathbf{x}\|_\infty = \max_i |x_i|.$$

The vector norm $|\mathbf{x}|_\infty$ is implemented as Vector-Norm[m, Infinity] in the *Mathematica* add-on package LinearAlgebra`MatrixMultiplication` (which can be loaded with the command <<LinearAlgebra`).

See also $L1$-NORM, $L2$-NORM, VECTOR NORM

References
Gradshteyn, I. S. and Ryzhik, I. M. *Tables of Integrals, Series, and Products, 6th ed.* San Diego, CA: Academic Press, pp. 1114–125, 2000.

L-Infinity-Space
The SPACE called L^∞ (ell-infinity) generalizes the L^p-SPACES to $p = \infty$. No integration is used to define them, and instead, the norm on L^∞ is given by the ESSENTIAL SUPREMUM.

More precisely,

$$\|f\|_\infty = \text{ess sup}|f|$$

is the norm which makes L^∞ a BANACH SPACE. It is the space of all essentially bounded functions. The space of bounded continuous functions is *not* DENSE in L^∞.

See also BANACH SPACE, COMPLETION, DENSE, ESSENTIAL SUPREMUM, L^p-SPACE, $L2$-SPACE, MEASURE, MEASURABLE FUNCTION, MEASURE SPACE

Link

Formally, a link is one or more disjointly embedded CIRCLES in 3-space. More informally, a link is an assembly of KNOTS with mutual entanglements. Kuperberg (1994) has shown that a nontrivial KNOT or link in $\mathbb{R}^3$ has four COLLINEAR points (Eppstein). Doll and Hoste (1991) list POLYNOMIALS for oriented links of nine or fewer crossings.

A listing of the first few simple links follows, arranged by CROSSING NUMBER. The numbers of nontrivial 2-component links of 0, 1, 2, ... crossings are 1, 0, 1, 0, 1, 1, 3, 8, 16, 61, ... (Sloane's A048952). The numbers of nontrivial 3-component links of 6, 7, ... crossings are 3, 1, 10, 21, ... (Sloane's A048953). The number of nontrivial 4-component links of 8, 9, ... crossings are 3, 1, ...

00–2–1 02–2–1 04–2–1 05–2–1 06–2–1 06–2–2 06–2–3 07–2–1 07–2–2 07–2–3 07–2–4 07–2–5 07–2–6 07–2–7 07–2–8 08–2–1 08–2–2 08–2–3 08–2–4 08–2–5 08–2–6 08–2–7 08–2–8 08–2–9 08–2–0 08–2–1 08–2–2 08–2–3 08–2–4 08–2–5 08–2–6 09–2–1 09–2–2 09–2–3 09–2–4 09–2–5 09–2–6 09–2–7 09–2–8 09–2–9 09–2–0 09–2–1 09–2–2 09–2–3 09–2–4 09–2–5 09–2–6 09–2–7 09–2–8 09–2–9 09–2–0 09–2–1 09–2–2 09–2–3 09–2–4 09–2–5 09–2–6 09–2–7 09–2–8 09–2–9 09–2–0 09–2–1 09–2–2 09–2–3 09–2–4 09–2–5 09–2–6 09–2–7 09–2–8 09–2–9 09–2–0 09–2–1 09–2–2 09–2–3 09–2–4 09–2–5 09–2–6 09–2–7 09–2–8 09–2–9 09–2–0 09–2–1 06–3–1 06–3–2 06–3–3 07–3–1 08–3–1 08–3–2 08–3–3 08–3–4 08–3–5 08–3–6 08–3–7 08–3–8 08–3–9 08–3–0 09–3–1 09–3–2 09–3–3 09–3–4 09–3–5 09–3–6 09–3–7 09–3–8 09–3–9 09–3–0 09–3–1 09–3–2 09–3–3 09–3–4 09–3–5 09–3–6 09–3–7 09–3–8 09–3–9 09–3–0 09–3–1 08–4–1 08–4–2 08–4–3 09–4–1

See also ANDREWS-CURTIS LINK, BORROMEAN RINGS, BRUNNIAN LINK, HOPF LINK, KNOT, ORIENTED LINK, WHITEHEAD LINK

References

Cerf, C. "Atlas of Oriented Knots and Links." *Topology Atlas Invited Contributions* **3**, No. 2, 1–2, 1998. http://at.yorku.ca/t/a/i/c/31.htm.
Doll, H. and Hoste, J. "A Tabulation of Oriented Links." *Math. Comput.* **57**, 747–61, 1991.
Eppstein, D. "Colinear Points on Knots." http://www.ics.uci.edu/~eppstein/junkyard/knot-colinear.html.
Kuperberg, G. "Quadrisecants of Knots and Links." *J. Knot Theory Ramifications* **3**, 41–0, 1994.
Rolfsen, D. *Knots and Links.* Wilmington, DE: Publish or Perish Press, 1976.
Sloane, N. J. A. Sequences A048952 and A048953 in "An On-Line Version of the Encyclopedia of Integer Sequences." http://www.research.att.com/~njas/sequences/eisonline.html.
Weisstein, E. W. "Knots." MATHEMATICA NOTEBOOK KNOTS.M.

Link (Simplicial Complex)

The set $\overline{\text{St}}\,v - \text{St}\,v$, where $\overline{\text{St}}\,v$ is a CLOSED STAR and St v is a STAR, is called the link of v in a SIMPLICIAL COMPLEX K and is denoted Lkv (Munkres 1993, p. 11).

See also CLOSED STAR, SIMPLICIAL COMPLEX, STAR

References

Munkres, J. R. *Elements of Algebraic Topology.* Perseus Press, 1993.

Link Complement

KNOT COMPLEMENT

Link Diagram

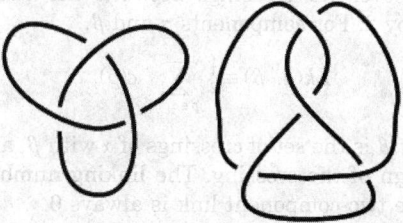

A planar diagram depicting a LINK (or KNOT) as a sequence of segments with gaps representing undercrossings and solid lines overcrossings. In such a diagram, only two segments should ever cross at a single point. Link diagrams for the TREFOIL KNOT and FIGURE-OF-EIGHT KNOT are illustrated above.

Link Invariant

A link invariant is a function from the set of all LINKS to any other set such that the function does not change as the link is changed (up to isotopy). In other words, a link invariant always assigns the same value to equivalent links (although different knots may have the same link invariant). When the link has a single component and therefore generates to a KNOT, the invariant is called a KNOT INVARIANT.

See also KNOT, KNOT INVARIANT, LINK

Linkage

Sylvester, Kempe and Cayley developed the geometry associated with the theory of linkages in the 1870s. Kempe proved that every finite segment of an algebraic curve can be generated by a linkage in the manner of WATT'S CURVE.

See also HART'S INVERSOR, KEMPE LINKAGE, PANTOGRAPH, PEAUCELLIER INVERSOR, SARRUS LINKAGE, WATT'S PARALLELOGRAM

References

Chuan, J. C. "Machine." http://www.math.ntnu.edu.tw/~jcchuan/demo/gear/machine.html.
Cundy, H. and Rollett, A. *Mathematical Models, 3rd ed.* Stradbroke, England: Tarquin Pub., 1989.
Kempe, A. B. *How to Draw a Straight Line: A Lecture on Linkages.* 1977.

King, H. C. Configuration Spaces of Linkages in $\mathbb{R}^n$ 23 Nov 1998. http://xxx.lanl.gov/abs/math.GT/9811138/.

King, H. C. Semiconfiguration Spaces of Planar Linkages. 20 Oct 1998. http://xxx.lanl.gov/abs/math.GT/9810130/.

McCarthy, J. M. "Geometric Design of Linkages." http://www.eng.uci.edu/~mccarthy/.

Rademacher, H. and Toeplitz, O. "Producing Rectilinear Motion by Means of Linkages." §18 in *The Enjoyment of Mathematics: Selections from Mathematics for the Amateur.* Princeton, NJ: Princeton University Press, pp. 119–29, 1957.

Linking Number

A LINK INVARIANT defined for a two-component oriented LINK as the sum of $+1$ crossings and -1 crossing over all crossings between the two links divided by 2. For components α and β,

$$Lk(\alpha, \beta) \equiv \frac{1}{2} \sum_{p \in \alpha \sqcap \beta} \epsilon(p),$$

where $\alpha \sqcap \beta$ is the set of crossings of α with β, and $\epsilon(p)$ is the sign of the crossing. The linking number of a splittable two-component link is always 0.

See also CALUGAREANU THEOREM, GAUSS INTEGRAL, JONES POLYNOMIAL, LINK, TWIST, WRITHE

References

Pohl, W. F. "The Self-Linking Number of a Closed Space Curve." *J. Math. Mech.* **17**, 975–85, 1968.

Rolfsen, D. *Knots and Links.* Wilmington, DE: Publish or Perish Press, pp. 132–33, 1976.

Links Curve

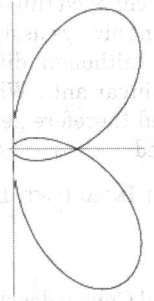

The curve given by the Cartesian equation

$$(x^2 + y^2 - 3x)^2 = 4x^2(2 - x).$$

The origin of the curve is a TACNODE.

References

Cundy, H. and Rollett, A. *Mathematical Models, 3rd ed.* Stradbroke, England: Tarquin Pub., p. 72, 1989.

Linnik's Constant

The constant L in LINNIK'S THEOREM. Heath-Brown (1992) has shown that $L \leq 5.5$, and Schinzel, Sierpinski, and Kanold (Ribenboim 1989) have conjectured that $L = 2$.

References

Finch, S. "Favorite Mathematical Constants." http://www.mathsoft.com/asolve/constant/linnik/linnik.html.

Guy, R. K. *Unsolved Problems in Number Theory, 2nd ed.* New York: Springer-Verlag, p. 13, 1994.

Heath-Brown, D. R. "Zero-Free Regions for Dirichlet L-Functions and the Least Prime in an Arithmetic Progression." *Proc. London Math. Soc.* **64**, 265–38, 1992.

Ribenboim, P. *The Book of Prime Number Records, 2nd ed.* New York: Springer-Verlag, 1989.

Linnik's Theorem

Let $p(d, a)$ be the smallest PRIME in the arithmetic progression $\{a + kd\}$ for k an INTEGER > 0. Let

$$p(d) \equiv \max p(d, a)$$

such that $1 \leq a < d$ and $(a, d) = 1$. Then there exists a $d_0 \geq 2$ and an $L > 1$ such that $p(d) < d^L$ for all $d > d_0$. L is known as LINNIK'S CONSTANT.

References

Linnik, U. V. "On the Least Prime in an Arithmetic Progression. I. The Basic Theorem." *Mat. Sbornik N. S.* **15 (57)**, 139–78, 1944.

Linnik, U. V. "On the Least Prime in an Arithmetic Progression. II. The Deuring-Heilbronn Phenomenon" *Mat. Sbornik N. S.* **15 (57)**, 347–68, 1944.

Lin's Method

An ALGORITHM for finding ROOTS for QUARTIC EQUATIONS with COMPLEX ROOTS.

References

Acton, F. S. *Numerical Methods That Work, 2nd printing.* Washington, DC: Math. Assoc. Amer., pp. 198–99, 1990.

Lin-Tsien Equation

The PARTIAL DIFFERENTIAL EQUATION

$$2u_{tx} + u_x u_{xx} - u_{yy} = 0.$$

References

Ames, W. F. and Nucci, W. N. "Analysis of Fluid Equations by Group Methods." *J. Eng. Mech.* **20**, 181–87, 1985.

Zwillinger, D. *Handbook of Differential Equations, 3rd ed.* Boston, MA: Academic Press, p. 131, 1997.

Linus Sequence

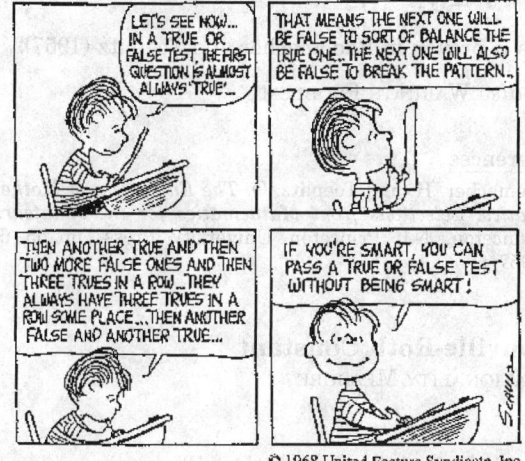

© 1968 United Feature Syndicate, Inc.

The sequence composed of 1s and 2s obtained by starting with the number 1, and picking subsequent elements to avoid repeating the longest possible substring. The first few terms are 1, 2, 1, 1, 2, 2, 1, 2, 1, 1, 2, 1, 2, 2, ... (Sloane's A006345). The SALLY SEQUENCE gives the length of the run that was avoided.

See also SALLY SEQUENCE

References

Sloane, N. J. A. Sequences A006345/M0126 in "An On-Line Version of the Encyclopedia of Integer Sequences." http://www.research.att.com/~njas/sequences/eisonline.html.

Sloane, N. J. A. and Plouffe, S. Figure M0126 in *The Encyclopedia of Integer Sequences.* San Diego: Academic Press, 1995.

Liouville Function

The function

$$\lambda(n) = (-1)^{r(n)}, \tag{1}$$

where $r(n)$ is the number of not necessarily distinct PRIME FACTORS of n, with $r(1) = 0$. The first few values of $\lambda(n)$ are 1, -1, -1, 1, -1, 1, -1, -1, 1, 1, -1, -1, ... The Liouville function is connected

with the RIEMANN ZETA FUNCTION by the equation

$$\frac{\zeta(2s)}{\zeta(s)} = \sum_{n=1}^{\infty} \frac{\lambda(n)}{n^s} \tag{2}$$

(Lehman 1960).

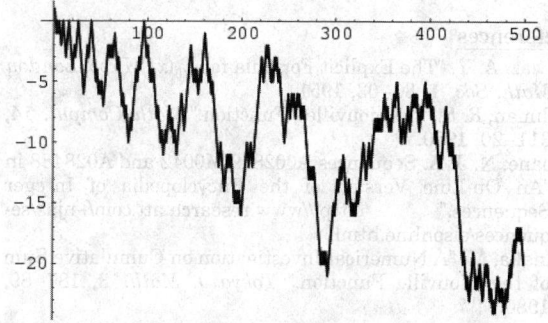

The CONJECTURE that the SUMMATORY FUNCTION

$$L(n) \equiv \sum_{k=1}^{n} \lambda(n) \tag{3}$$

satisfies $L(n) \leq 0$ for $n \geq 2$ is called the PÓLYA CONJECTURE and has been proved to be false. The first n for which $L(n) = 0$ are for $n = 2, 4, 6, 10, 16, 26, 40, 96, 586, 906150256, \ldots$ (Sloane's A028488), and $n = 906150257$ is, in fact, the first counterexample to the PÓLYA CONJECTURE (Tanaka 1980). However, it is unknown if $L(x)$ changes sign infinitely often (Tanaka 1980). The first few values of $L(n)$ are 1, 0, -1, 0, -1, 0, -1, -2, -1, 0, -1, -2, -3, -2, -1, 0, -1, -2, -3, -4, ... (Sloane's A002819). $L(n)$ also satisfies

$$\sum_{n=1}^{x} L\left(\frac{x}{n}\right) = \lfloor \sqrt{x} \rfloor, \tag{4}$$

where $\lfloor x \rfloor$ is the FLOOR FUNCTION (Lehman 1960). Lehman (1960) also gives the formulas

$$L(x) = \sum_{m=1}^{x/w} \mu(m)$$
$$\times \left\{ \left\lfloor \sqrt{\frac{x}{m}} \right\rfloor - \sum_{k=1}^{v-1} \lambda(k) \left(\left\lfloor \frac{x}{km} \right\rfloor - \left\lfloor \frac{x}{mv} \right\rfloor \right) \right\}$$
$$- \sum_{l=x/w-1}^{x/v} L\left(\frac{x}{l}\right) \sum_{\substack{m|l \\ m=1}}^{x/w} \mu(m) \tag{5}$$

and

$$L(x) = \sum_{k=1}^{g} M\left(\frac{x}{k^2}\right) + \sum_{l=1}^{x/g^2} \mu(l) \left\lfloor \sqrt{\frac{x}{l}} \right\rfloor - M\left(\frac{x}{g^2}\right)$$
$$\times \left\lfloor \sqrt{\frac{x}{g^2}} \right\rfloor, \tag{6}$$

where k, l, and m are variables ranging over the POSITIVE INTEGERS, $\mu(n)$ is the MÖBIUS FUNCTION, $M(x)$ is MERTENS FUNCTION, and v, w, and x are POSITIVE real numbers with $v < w < x$.

See also PÓLYA CONJECTURE, PRIME FACTORS, RIEMANN ZETA FUNCTION

References

Fawaz, A. Y. "The Explicit Formula for $L_0(x)$." *Proc. London Math. Soc.* **1**, 86–03, 1951.
Lehman, R. S. "On Liouville's Function." *Math. Comput.* **14**, 311–20, 1960.
Sloane, N. J. A. Sequences A002819/M0042 and A028488 in "An On-Line Version of the Encyclopedia of Integer Sequences." http://www.research.att.com/~njas/sequences/eisonline.html.
Tanaka, M. "A Numerical Investigation on Cumulative Sum of the Liouville Function." *Tokyo J. Math.* **3**, 187–89, 1980.

Liouville Measure

$$\prod_i dp_i\, dq_i,$$

where p_i and q_i are momenta and positions of particles.

See also LIOUVILLE'S PHASE SPACE THEOREM, PHASE SPACE

Liouville Number

A Liouville number is a TRANSCENDENTAL NUMBER which has very close RATIONAL NUMBER approximations. An IRRATIONAL NUMBER β is a Liouville number if, for any n, there exist an infinite number of pairs of INTEGERS p and q such that

$$0 < \left| \beta - \frac{p}{q} \right| < \frac{1}{q^n}.$$

LIOUVILLE'S CONSTANT is an example of a Liouville number. Mahler (1953) proved that π is not a Liouville number.

See also LIOUVILLE'S CONSTANT, LIOUVILLE'S APPROXIMATION THEOREM, ROTH'S THEOREM, TRANSCENDENTAL NUMBER

References

Apostol, T. M. *Modular Functions and Dirichlet Series in Number Theory, 2nd ed.* New York: Springer-Verlag, p. 147, 1997.
Mahler, K. "On the Approximation of π." *Nederl. Akad. Wetensch. Proc. Ser. A.* **56**/*Indagationes Math.* **15**, 30–2, 1953.

Liouville Polynomial Identity

$$6(x_1^2 + x_2^2 + x_3^2 + x_4^2) = (x_1 + x_2)^4 + (x_1 + x_3)^4 + (x_2 + x_3)^4$$

$$+ (x_1 + x_4)^4 + (x_2 + x_4)^4 + (x_3 + x_4)^4 + (x_1 - x_2)^4$$

$$+ (x_1 - x_3)^4 + (x_2 - x_3)^4 + (x_1 - x_4)^4 + (x_2 - x_4)^4$$

$$+ (x_3 - x_4)^4.$$

This is proven in Rademacher and Toeplitz (1957).

See also WARING'S PROBLEM

References

Rademacher, H. and Toeplitz, O. *The Enjoyment of Mathematics: Selections from Mathematics for the Amateur.* Princeton, NJ: Princeton University Press, pp. 55–6, 1957.

Liouville-Roth Constant

IRRATIONALITY MEASURE

Liouville's Approximation Theorem

For any ALGEBRAIC NUMBER x of degree $n \geq 2$, a RATIONAL approximation $x = p/q$ must satisfy

$$\left| x - \frac{p}{q} \right| > \frac{1}{q^{n+1}}$$

for sufficiently large q. Writing $r \equiv n + 1$ leads to the definition of the IRRATIONALITY MEASURE of a given number. Apostol (1997) states the theorem in the slightly modified form that for all integers p and q with $q > 0$, there exists a positive constant $C(x)$ depending only on x such that

$$\left| x - \frac{p}{q} \right| > \frac{C(x)}{q^n}.$$

See also DIRICHLET'S APPROXIMATION THEOREM, IRRATIONALITY MEASURE, LAGRANGE NUMBER (RATIONAL APPROXIMATION), LIOUVILLE'S CONSTANT, LIOUVILLE NUMBER, MARKOV NUMBER, ROTH'S THEOREM, THUE-SIEGEL-ROTH THEOREM

References

Apostol, T. M. "Liouville's Approximation Theorem." §7.3 in *Modular Functions and Dirichlet Series in Number Theory, 2nd ed.* New York: Springer-Verlag, pp. 146–48, 1997.
Courant, R. and Robbins, H. "Liouville's Theorem and the Construction of Transcendental Numbers." §2.6.2 in *What is Mathematics?: An Elementary Approach to Ideas and Methods, 2nd ed.* Oxford, England: Oxford University Press, pp. 104–07, 1996.

Liouville's Boundedness Theorem

A bounded ENTIRE FUNCTION in the COMPLEX PLANE $\mathbb{C}$ is constant. The FUNDAMENTAL THEOREM OF ALGEBRA follows as a simple corollary.

See also COMPLEX PLANE, ENTIRE FUNCTION, FUNDAMENTAL THEOREM OF ALGEBRA

References

Knopp, K. *Theory of Functions Parts I and II, Two Volumes Bound as One, Part II.* New York: Dover, p. 74, 1996.

Krantz, S. G. "Entire Functions and Liouville's Theorem." §3.1.3 in *Handbook of Complex Analysis.* Boston, MA: Birkhäuser, pp. 31–2, 1999.

Morse, P. M. and Feshbach, H. *Methods of Theoretical Physics, Part I.* New York: McGraw-Hill, pp. 381–82, 1953.

Liouville's Conformality Theorem

In SPACE, the only CONFORMAL MAPPINGS are inversions, SIMILARITY TRANSFORMATIONS, and CONGRUENCE TRANSFORMATIONS. Or, restated, every ANGLE-preserving transformation is a SPHERE-preserving transformation.

See also CONFORMAL MAP

Liouville's Conic Theorem

The lengths of the TANGENTS from a point P to a CONIC C are proportional to the CUBE ROOTS of the RADII OF CURVATURE of C at the corresponding points of contact.

See also CONIC SECTION

Liouville's Constant

$$L \equiv \sum_{n=1}^{\infty} 10^{-n!}$$

$$= 0.110001000000000000000001\ldots$$

(Sloane's A012245). Liouville's constant is a decimal fraction with a 1 in each decimal place corresponding to a FACTORIAL $n!$, and ZEROS everywhere else. Liouville (1844) constructed an infinite class of TRANSCENDENTAL NUMBERS using CONTINUED FRACTIONS, but the above number was the first decimal constant to be proven TRANSCENDENTAL (Liouville 1850). However, Cantor subsequently proved that "almost all" real numbers are in fact transcendental. Liouville's constant nearly satisfies

$$10x^6 - 75x^3 - 190x + 21 = 0,$$

but plugging $x = L$ into this equation gives $-0.0000000059\ldots$ instead of 0.

See also LIOUVILLE NUMBER

References

Apostol, T. M. *Modular Functions and Dirichlet Series in Number Theory, 2nd ed.* New York: Springer-Verlag, p. 147, 1997.

Conway, J. H. and Guy, R. K. "Liouville's Number." In *The Book of Numbers.* New York: Springer-Verlag, pp. 239–41, 1996.

Courant, R. and Robbins, H. "Liouville's Theorem and the Construction of Transcendental Numbers." §2.6.2 in *What is Mathematics?: An Elementary Approach to Ideas and*

Methods, 2nd ed. Oxford, England: Oxford University Press, pp. 104–07, 1996.

Liouville, J. "Sur des classes très étendues de quantités dont la valeur n'est ni algébrique, ni même reductible à des irrationelles algébriques." *C. R. Acad. Sci. Paris* **18**, 883–85 and 993–95, 1844.

Liouville, J. "Sur des classes très-étendues de quantités dont la valeur n'est ni algébrique, ni même réductible à des irrationelles algébriques." *J. Math. pures appl.* **15**, 133–42, 1850.

Sloane, N. J. A. Sequences A012245 in "An On-Line Version of the Encyclopedia of Integer Sequences." http://www.research.att.com/~njas/sequences/eisonline.html.

Wells, D. *The Penguin Dictionary of Curious and Interesting Numbers.* Middlesex, England: Penguin Books, p. 26, 1986.

Liouville's Elliptic Function Theorem

An ELLIPTIC FUNCTION with no POLES in a FUNDAMENTAL CELL is a constant.

See also ELLIPTIC FUNCTION, FUNDAMENTAL CELL, POLE

References

Whittaker, E. T. and Watson, G. N. *A Course in Modern Analysis, 4th ed.* Cambridge, England: Cambridge University Press, p. 431, 1990.

Liouville's Equation

The second-order ORDINARY DIFFERENTIAL EQUATION

$$y'' + g(y)y'^2 + f(x)y' = 0 \tag{1}$$

is called Liouville's equation (Goldstein and Braun 1973; Zwillinger 1997, p. 124), as are the PARTIAL DIFFERENTIAL EQUATIONS

$$\sum_{i=1}^{n} u_{x_i x_i} + e^{\lambda u} = 0 \tag{2}$$

(Matsumo 1987; Zwillinger 1997, p. 133) and

$$u_{xt} = e^{\eta u} \tag{3}$$

(Calogero and Degasperis 1982, p. 60; Zwillinger 1997, p. 133).

See also KLEIN-GORDON EQUATION

References

Calogero, F. and Degasperis, A. *Spectral Transform and Solitons: Tools to Solve and Investigate Nonlinear Evolution Equations.* New York: North-Holland, p. 60, 1982.

Goldstein, M. E. and Braun, W. H. *Advanced Methods for the Solution of Differential Equations.* NASA SP-316. Washington, DC: U.S. Government Printing Office, p. 98, 1973.

Matsumo, Y. "Exact Solution for the Nonlinear Klein-Gordon and Liouville Equations in Four-Dimensional Euclidean Space." *J. Math. Phys.* **28**, 2317–322, 1987.

Zwillinger, D. *Handbook of Differential Equations, 3rd ed.* Boston, MA: Academic Press, pp. 124 and 133, 1997.

Liouville Space

Also known as LINE SPACE or "extended" HILBERT SPACE, it is the SET DIRECT PRODUCT of two HILBERT SPACES.

See also HILBERT SPACE, SET DIRECT PRODUCT

Liouville's Phase Space Theorem

States that for a nondissipative HAMILTONIAN SYSTEM, phase space density (the AREA between phase space contours) is constant. This requires that, given a small time increment dt,

$$q_1 = q(t_0 + dt) = q_0 + \frac{\partial H(q_0, p_0, t)}{\partial p_0} dt + \mathcal{O}(dt^2) \quad (1)$$

$$p_1 \equiv p(t_0 + dt) = p_0 - \frac{\partial H(q_0, p_0, t)}{\partial q_0} dt + \mathcal{O}(dt^2), \quad (2)$$

the JACOBIAN be equal to one:

$$\frac{\partial(q_1, p_1)}{\partial(q_0, p_0)} = \begin{vmatrix} \dfrac{\partial q_1}{\partial q_0} & \dfrac{\partial p_1}{\partial q_0} \\ \dfrac{\partial q_1}{\partial p_0} & \dfrac{\partial p_1}{\partial p_0} \end{vmatrix}$$

$$= \begin{vmatrix} 1 + \dfrac{\partial^2 H}{\partial q_0 \partial p_0} dt & -\dfrac{\partial^2 H}{\partial q_0^2} dt \\ \dfrac{\partial^2 H}{\partial p_0^2} dt & 1 - \dfrac{\partial^2 H}{\partial q_0 \partial p_0} dt \end{vmatrix} + \mathcal{O}(dt^2)$$

$$= 1 + \mathcal{O}(dt^2). \quad (3)$$

Expressed in another form, the integral of the LIOUVILLE MEASURE,

$$\prod_{i=1}^{N} \int dp_i \, dq_i, \quad (4)$$

is a constant of motion. SYMPLECTIC MAPS of HAMILTONIAN SYSTEMS must therefore be AREA preserving (and have DETERMINANTS equal to 1).

See also LIOUVILLE MEASURE, PHASE SPACE

References

Chavel, I. *Riemannian Geometry: A Modern Introduction.* New York: Cambridge University Press, 1994.

Liouville's Principle

Let F be a differential field with constant field K. For $f \in F$, suppose that the equation $g' = f$ (i.e., $g = \int f$) has a solution $g \in G$, where G is an elementary extension of F having the same constant FIELD K. Then there exist $v_0, v_1, ..., v_m \in F$ and constants $c_1, ..., c_m \in K$ such that

$$f = v_0' + \sum_{i=1}^{m} c_i \frac{v_i'}{v_i},$$

In other words, such that

$$\int f = v_0 + \sum_{i=1}^{m} c_i \ln v_i.$$

See also ELEMENTARY FUNCTION

References

Geddes, K. O.; Czapor, S. R.; and Labahn, G. "Liouville's Principle." §12.4 in *Algorithms for Computer Algebra.* Amsterdam, Netherlands: Kluwer, pp. 523–29, 1992.

Liouville's Sphere-Preserving Theorem

LIOUVILLE'S CONFORMALITY THEOREM

Liouvillian Number

A member of the smallest algebraically closed SUBFIELD $\mathbb{L}$ of $\mathbb{C}$ which is CLOSED under the exponentiation and logarithm operations.

See also ELEMENTARY NUMBER

References

Chow, T. Y. "What is a Closed-Form Number." *Amer. Math. Monthly* **106**, 440–48, 1999.
Richardson, D. "The Elementary Constant Problem." In *Proc. Internat. Symp. on Symbolic and Algebraic Computation, Berkeley, July 27–9, 1992* (Ed. P. S. Wang). ACM Press, 1992.
Ritt, J. *Integration in Finite Terms: Liouville's Theory of Elementary Models.* New York: Columbia University Press, 1948.

Lipschitz Condition

A function $f(x)$ satisfies the Lipschitz condition of order α at $x = 0$ if

$$|f(h) - f(0)| \leq B|h|^\beta$$

for all $|h| < \epsilon$, where B and β are independent of h, $\beta > 0$, and α is an UPPER BOUND for all β for which a finite B exists.

See also HILLAM'S THEOREM, HÖLDER CONDITION, LIPSCHITZ FUNCTION

References

Jeffreys, H. and Jeffreys, B. S. "The Lipschitz Condition." §1.15 in *Methods of Mathematical Physics, 3rd ed.* Cambridge, England: Cambridge University Press, p. 53, 1988.

Lipschitz Function

A function f such that

$$|f(x) - f(y)| \leq C|x - y|$$

for all x and y, where C is a constant independent of x and y, is called a Lipschitz function. For example, any function with a bounded first derivative must be Lipschitz.

Lipschitz's Integral

See also LIPSCHITZ CONDITION

References

Morgan, F. "What Is a Surface?" *Amer. Math. Monthly* **103**, 369–76, 1996.

Lipschitz's Integral

$$\int_0^\infty e^{-ax} J_0(bx)\,dx = \frac{1}{\sqrt{a^2 + b^2}},$$

where $J_0(z)$ is the zeroth order BESSEL FUNCTION OF THE FIRST KIND.

References

Bowman, F. *Introduction to Bessel Functions.* New York: Dover, p. 58, 1958.

Lissajous Curve

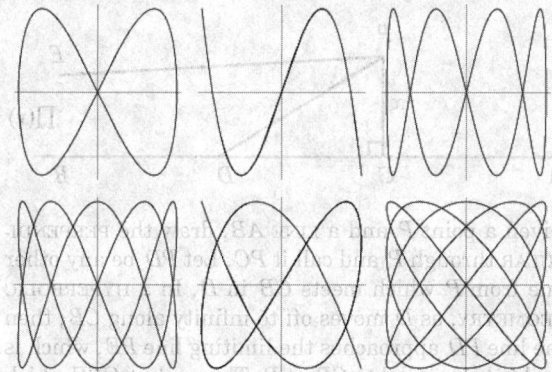

Lissajous curves are the family of curves described by the PARAMETRIC EQUATIONS

$$x(t) = A\cos(\omega_x t - \delta_x) \tag{1}$$

$$y(t) = B\cos(\omega_y t - \delta_y), \tag{2}$$

sometimes also written in the form

$$x(t) = a\sin(nt + c) \tag{3}$$

$$y(t) = b\sin t. \tag{4}$$

They are sometimes known as BOWDITCH CURVES after Nathaniel Bowditch, who studied them in 1815. They were studied in more detail (independently) by Jules-Antoine Lissajous in 1857 (MacTutor Archive). Lissajous curves have applications in physics, astronomy, and other sciences. The curves close IFF ω_x/ω_y is RATIONAL.

Lissajous curves are a special case of the HARMONOGRAPH with damping constants $\beta_1 = \beta_2 = 0$.

See also HARMONOGRAPH

References

Cundy, H. and Rollett, A. "Lissajous's Figures." §5.5.3 in *Mathematical Models, 3rd ed.* Stradbroke, England: Tarquin Pub., pp. 242–44, 1989.

Gray, A. *Modern Differential Geometry of Curves and Surfaces with Mathematica, 2nd ed.* Boca Raton, FL: CRC Press, pp. 70–1, 1997.
Lawrence, J. D. *A Catalog of Special Plane Curves.* New York: Dover, pp. 178–79 and 181–83, 1972.
MacTutor History of Mathematics Archive. "Lissajous Curves." http://www-groups.dcs.st-and.ac.uk/~history/Curves/Lissajous.html.
Wells, D. *The Penguin Dictionary of Curious and Interesting Geometry.* London: Penguin, p. 142, 1991.

Lissajous Figure

LISSAJOUS CURVE

List

An DATA STRUCTURE consisting of an ordered SET of elements, each of which may be a number, another list, etc. A list is usually denoted $(a_1, a_2, \ldots, a_n)$ or $\{a_1, a_2, \ldots, a_n\}$, and may also be interpreted as a VECTOR. Multiplicity matters in a list, so $(1, 1, 2)$ and $(1, 2)$ are not equivalent.

See also MULTISET, QUEUE, SET, STACK, STRING, VECTOR

Little Moment Problem

MOMENT PROBLEM

Lituus

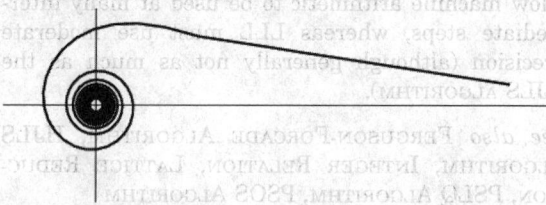

An ARCHIMEDEAN SPIRAL with $m = -2$, having polar equation

$$r^2\theta = a^2.$$

Lituus means a "crook," in the sense of a bishop's crosier. The lituus curve originated with Cotes in 1722. Maclaurin used the term lituus in his book *Harmonia Mensurarum* in 1722 (MacTutor Archive). The lituus is the locus of the point P moving such that the AREA of a circular SECTOR remains constant.

References

Beyer, W. H. *CRC Standard Mathematical Tables, 28th ed.* Boca Raton, FL: CRC Press, p. 221, 1987.
Gray, A. *Modern Differential Geometry of Curves and Surfaces with Mathematica, 2nd ed.* Boca Raton, FL: CRC Press, p. 91, 1997.
Lawrence, J. D. *A Catalog of Special Plane Curves.* New York: Dover, pp. 186 and 188, 1972.
Lockwood, E. H. *A Book of Curves.* Cambridge, England: Cambridge University Press, p. 175, 1967.
MacTutor History of Mathematics Archive. "Lituus." http://www-groups.dcs.st-and.ac.uk/~history/Curves/Lituus.html.

Lituus Inverse Curve

The INVERSE CURVE of the LITUUS is an ARCHIMEDEAN SPIRAL with $m = 2$, which is FERMAT'S SPIRAL.

See also ARCHIMEDEAN SPIRAL, FERMAT'S SPIRAL, LITUUS

LLL Algorithm

A LATTICE REDUCTION algorithm, named after discoverers Lenstra, Lenstra, and Lovasz (1982), that produces a lattice basis of "short" vectors. It was noticed by Lenstra *et al.* (1928) that the algorithm could be used to obtain factors of univariate polynomials, which amounts to the determination of INTEGER RELATIONS. However, this application of the algorithm, which later came to be one of its primary applications, was not stressed in the original paper.

The *Mathematica* command LatticeReduce[*matrix*] implements the LLL algorithm to perform LATTICE REDUCTION. *Mathematica*'s implementation requires the input to consist of rational numbers, so Rationalize may need to be called first.

More recently, other algorithms such as PSLQ, which can be significant faster than LLL, have been developed for finding INTEGER RELATIONS. PSLQ achieves its performance because of clever techniques that allow machine arithmetic to be used at many intermediate steps, whereas LLL must use moderate precision (although generally not as much as the HJLS ALGORITHM).

See also FERGUSON-FORCADE ALGORITHM, HJLS ALGORITHM, INTEGER RELATION, LATTICE REDUCTION, PSLQ ALGORITHM, PSOS ALGORITHM

References

Borwein, J. M. and Corless, R. M. "Emerging Tools for Experimental Mathematics." *Amer. Math. Monthly* **106**, 899–09, 1999.
Borwein, J. M. and Lisonek, P. "Applications of Integer Relation Algorithms." To appear in *Disc. Math.* http://www.cecm.sfu.ca/preprints/1997pp.html.
Cohen, H. *A Course in Computational Algebraic Number Theory.* New York: Springer-Verlag, 1993.
Lenstra, A. K.; Lenstra, H. W.; and Lovasz, L. "Factoring Polynomials with Rational Coefficients." *Math. Ann.* **261**, 515–34, 1982.
Matthews, K. "Keith Matthews' LLL Page." http://www.maths.uq.edu.au/~krm/lll.html.
Mignotte, M. *Mathematics for Computer Algebra.* New York: Springer-Verlag, 1991.

L-Moment

A type of statistic which can be useful for determining asymmetry and tailedness of a population.

See also MOMENT, ORDER STATISTIC

References

Hosking, J. R. M. "*L*-Moments: Analysis and Estimation of Distributions Using Linear Combinations of Order Statistics." *J. Roy. Stat. Soc. B* **52**, 105–24, 1990.

Ln

The LOGARITHM to BASE E, also called the NATURAL LOGARITHM, is denoted ln, i.e.,

$$\ln x \equiv \log_e x.$$

See also BASE (LOGARITHM), e, LG, LOGARITHM, NAPIERIAN LOGARITHM, NATURAL LOGARITHM

Lobachevsky-Bolyai-Gauss Geometry

HYPERBOLIC GEOMETRY

Lobachevsky's Formula

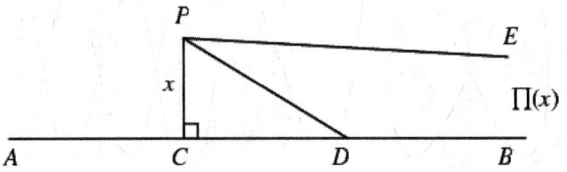

Given a point P and a LINE AB, draw the PERPENDICULAR through P and call it PC. Let PD be any other line from P which meets CB in D. In a HYPERBOLIC GEOMETRY, as D moves off to infinity along CB, then the line PD approaches the limiting line PE, which is said to be parallel to CB at P. The angle $\angle CPE$ which PE makes with PC is then called the ANGLE OF PARALLELISM for perpendicular distance x, and is given by

$$\prod(x) = 2\tan^{-1}(e^{-x}),$$

which is called Lobachevsky's formula.

See also ANGLE OF PARALLELISM, HYPERBOLIC GEOMETRY

References

Manning, H. P. *Introductory Non-Euclidean Geometry.* New York: Dover, p. 58, 1963.

Lobatto Quadrature

Also called RADAU QUADRATURE (Chandrasekhar 1960). A GAUSSIAN QUADRATURE with WEIGHTING FUNCTION $W(x) = 1$ in which the endpoints of the interval $[-1, 1]$ are included in a total of n ABSCISSAS, giving $r = n - 2$ free abscissas. ABSCISSAS are symmetrical about the origin, and the general FORMULA is

$$\int_{-1}^{1} f(x)\, dx = w_1 f(-1) + w_n f(1) + \sum_{i=2}^{n-1} w_i f(x_i). \qquad (1)$$

The free ABSCISSAS x_i for $i = 2, \ldots, n - 1$ are the roots of the POLYNOMIAL $P'_{n-1}(x)$, where $P(x)$ is a LEGENDRE POLYNOMIAL. The weights of the free abscissas are

$$w_i = -\frac{2n}{(1 - x_i^2)P''_{n-1}(x_i)P'_m(x_i)} \quad (2)$$

$$= \frac{2}{n(n-1)[P_{n-1}(x_i)]^2}, \quad (3)$$

and of the endpoints are

$$w_{1,n} = \frac{2}{n(n-1)}. \quad (4)$$

The error term is given by

$$E = -\frac{n(n-1)^3 2^{2n-1}[(n-2)!]^4}{(2n-1)[(2n-1)!]^3}f^{(2n-2)}(\xi), \quad (5)$$

for $\xi \in (-1, 1)$. Beyer (1987) gives a table of parameters up to $n = 11$ and Chandrasekhar (1960) up to $n = 9$ (although Chandrasekhar's $\mu_{3,4}$ for $m = 5$ is incorrect).

n	x_i	w_i
3	0	1.33333
	± 1	0.333333
4	± 0.447214	0.833333
	± 1	0.166667
5	0	0.711111
	± 0.654654	0.544444
	± 1	0.100000
6	± 0.285232	0.554858
	± 0.765055	0.378475
	± 1	0.0666667

The ABSCISSAS and weights can be computed analytically for small n.

n	x_i	w_i
3	0	$\frac{4}{3}$
	± 1	$\frac{1}{3}$
4	$\pm\frac{1}{5}\sqrt{5}$	$\frac{1}{6}$
	± 1	$\frac{5}{6}$
5	0	$\frac{32}{45}$
	$\pm\frac{1}{7}\sqrt{21}$	$\frac{49}{90}$
	± 1	$\frac{1}{10}$

See also CHEBYSHEV QUADRATURE, RADAU QUADRATURE

References

Abramowitz, M. and Stegun, C. A. (Eds.). *Handbook of Mathematical Functions with Formulas, Graphs, and Mathematical Tables, 9th printing.* New York: Dover, pp. 888–90, 1972.
Beyer, W. H. *CRC Standard Mathematical Tables, 28th ed.* Boca Raton, FL: CRC Press, p. 465, 1987.
Chandrasekhar, S. *Radiative Transfer.* New York: Dover, pp. 63–4, 1960.
Hildebrand, F. B. *Introduction to Numerical Analysis.* New York: McGraw-Hill, pp. 343–45, 1956.
Hunter, D. and Nikolov, G. "On the Error Term of Symmetric Gauss-Lobatto Quadrature Formulae for Analytic Functions." *Math. Comput.* **69**, 269–82, 2000.
Ueberhuber, C. W. *Numerical Computation 2: Methods, Software, and Analysis.* Berlin: Springer-Verlag, p. 105, 1997.

Lobster

One of the 12 6-POLYIAMONDS.

See also POLYIAMOND

References

Golomb, S. W. *Polyominoes: Puzzles, Patterns, Problems, and Packings, 2nd ed.* Princeton, NJ: Princeton University Press, p. 92, 1994.

Local

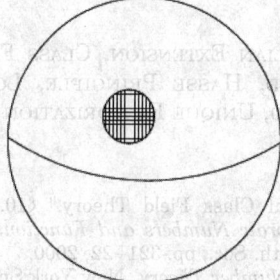

A mathematical property P holds locally if P is true near every point. In many different areas of mathematics, this notion is very useful. For instance, the sphere, and more generally a MANIFOLD, is locally Euclidean. For every point on the sphere, there is a NEIGHBORHOOD which is the same as a piece of EUCLIDEAN SPACE.

The description of local as "near every point" has a different interpretation in algebra. For instance, given a RING R and a PRIME IDEAL p, there is the LOCAL RING R_p, which often is simpler to study. It is possible to understand the original ring better by patching together the information from the local rings.

What ties all the notions of local together is the concept of a topology, a collection of open sets. For a SUBMANIFOLD of Euclidean space, or for the set of ideals of a ring, the topology is chosen as is appropriate.

A property P holds locally on a TOPOLOGICAL SPACE if every point has a NEIGHBORHOOD on which P holds. This concept is useful on any topological space.

See also GLOBAL, LOCAL FIELD, LOCAL RING, MANIFOLD, TOPOLOGICAL SPACE

Local Cell

The POLYHEDRON resulting from letting each SPHERE in a SPHERE PACKING expand uniformly until it touches its neighbors on flat faces.

See also LOCAL DENSITY, SPHERE PACKING

Local Class Field Theory

The study of NUMBER FIELDS by embedding them in a LOCAL FIELD is called local class field theory. Information about an equation in a LOCAL FIELD may give information about the equation in a GLOBAL FIELD, such as the rational numbers or a NUMBER FIELD (e.g., the HASSE PRINCIPLE).

Local class field theory is termed "local" because the local fields are LOCALIZED at a PRIME IDEAL in the RING of ALGEBRAIC INTEGERS. The methods of using CLASS FIELDS have developed over the years, from the LEGENDRE SYMBOL, to the CHARACTERS of ABELIAN EXTENSIONS of a number field, and is applied to LOCAL FIELDS.

See also ABELIAN EXTENSION, CLASS FIELD, FIELD, GLOBAL FIELD, HASSE PRINCIPLE, LOCAL FIELD, NUMBER FIELD, UNIQUE FACTORIZATION

References
Koch, H. "Local Class Field Theory." §10.3 in *Number Theory: Algebraic Numbers and Functions.* Providence, RI: Amer. Math. Soc., pp. 321–22, 2000.
Weil, A. *Basic Number Theory.* New York:Springer-Verlag, Chapter VII, 1974.

Local Degree

The degree of a VERTEX of a GRAPH is the number of EDGES which touch the VERTEX, also called the LOCAL DEGREE. The VERTEX degree of a point A in a GRAPH, denoted $\rho(A)$, satisfies

$$\sum_{i=1}^{n} \rho(A_i) = 2E,$$

where E is the total number of EDGES. Directed graphs have two types of degrees, known as the INDEGREE and OUTDEGREE.

See also INDEGREE, OUTDEGREE

Local Density

Let each SPHERE in a SPHERE PACKING expand uniformly until it touches its neighbors on flat faces. Call the resulting POLYHEDRON the LOCAL CELL. Then the local density is given by

$$\rho \equiv \frac{V_{\text{sphere}}}{V_{\text{local cell}}}.$$

When the LOCAL CELL is a regular DODECAHEDRON, then

$$\rho_{\text{dodecahedron}} = \frac{\pi\sqrt{5+\sqrt{5}}}{15\sqrt{10}(\sqrt{5}-2)} = 0.7547\ldots.$$

See also LOCAL CELL, LOCAL DENSITY CONJECTURE, SPHERE PACKING

Local Density Conjecture

The CONJECTURE that the maximum LOCAL DENSITY is given by $\rho_{\text{dodecahedron}}$.

See also DODECAHEDRAL CONJECTURE, LOCAL DENSITY

Local Extremum

A LOCAL MINIMUM or LOCAL MAXIMUM.

See also EXTREMUM, GLOBAL EXTREMUM

Local Field

A FIELD which is complete with respect to a discrete VALUATION is called a local field if its FIELD of RESIDUE CLASSES is FINITE. The HASSE PRINCIPLE is one of the chief applications of local field theory.

See also FUNCTION FIELD, HASSE PRINCIPLE, NUMBER FIELD, VALUATION

References
Iyanaga, S. and Kawada, Y. (Eds.). "Local Fields." §257 in *Encyclopedic Dictionary of Mathematics.* Cambridge, MA: MIT Press, pp. 811–15, 1980.

Local-Global Principle

HASSE PRINCIPLE

Local Group Theory

The study of a FINITE GROUP G using the LOCAL SUBGROUPS of G. Local group theory plays a critical role in the CLASSIFICATION THEOREM.

See also SYLOW THEOREMS

Local Maximum

The largest value of a set, function, etc., within some local neighborhood.

See also GLOBAL MAXIMUM, LOCAL MINIMUM, MAXIMUM, PEANO SURFACE

Local Minimum

The smallest value of a set, function, etc., within some local neighborhood.

See also GLOBAL MINIMUM, LOCAL MAXIMUM, MINIMUM

Local Ring

A NOETHERIAN RING R with a JACOBSON RADICAL which has only a single MAXIMAL IDEAL. One property of a local ring R is that the SUBSET $R - m$ is precisely the set of UNITS, where m is the MAXIMAL IDEAL. This follows because, in a ring, any nonunit belongs to at least one MAXIMAL IDEAL.

See also JACOBSON RADICAL, MAXIMAL IDEAL, NOETHERIAN RING, RESIDUE FIELD, UNIT (RING)

References

Iyanaga, S. and Kawada, Y. (Eds.). "Local Rings." §281D in *Encyclopedic Dictionary of Mathematics*. Cambridge, MA: MIT Press, pp. 890–91, 1980.

Local Subgroup

A normalizer of a nontrivial SYLOW P-SUBGROUP of a GROUP G.

See also LOCAL GROUP THEORY

Local Surface

PATCH

Locally Compact

A TOPOLOGICAL SPACE X is locally compact if every point has a NEIGHBORHOOD which is itself contained in a COMPACT SET. Many familiar topological spaces are locally compact, including the EUCLIDEAN SPACE. Of course, any COMPACT SET is locally compact. Some common spaces are not locally compact, such as infinite dimensional BANACH SPACES. For instance, the $L2$-SPACE of SQUARE INTEGRABLE functions is not locally compact.

See also COMPACT SET, LOCALLY COMPACT GROUP, NEIGHBORHOOD, TOPOLOGICAL SPACE

Locally Convex Space

LOCALLY PATHWISE-CONNECTED

Locally Finite Complex

A SIMPLICIAL COMPLEX K is said to be locally finite if each vertex of K belongs only to finitely many SIMPLICES of K.

References

Munkres, J. R. *Elements of Algebraic Topology*. Perseus Press, 1993.

Locally Finite Space

A locally finite SPACE is one for which every point of a given space has a NEIGHBORHOOD that meets only finitely many elements of the COVER.

Locally Integrable

A function is called locally integrable if, around every point in the domain, there is a NEIGHBORHOOD on which the function is INTEGRABLE. The space of locally integrable functions is denoted L^1_{loc}. Any integrable function is also locally integrable. One possibility for a nonintegrable function which is locally integrable is if it does not decay at infinity. For instance, $f(x) = 1$ is locally integrable on $\mathbb{R}$, as is any CONTINUOUS FUNCTION.

See also FRECHET SPACE, INTEGRABLE, LEBESGUE INTEGRABLE, $L1$-SPACE

Locally Pathwise-Connected

A SPACE X is locally pathwise-connected if for every NEIGHBORHOOD around every point in X, there is a smaller, PATHWISE-CONNECTED NEIGHBORHOOD.

See also ARCWISE-CONNECTED, PATHWISE-CONNECTED

Locally Pathwise-Connected Space

A SPACE X is locally pathwise-connected if for every NEIGHBORHOOD around every point in X, there is a smaller, PATHWISE-CONNECTED NEIGHBORHOOD.

Lochs' Theorem

For a real number $x \in (0, 1)$, let m be the number of terms in the CONVERGENT to a CONTINUED FRACTION that are required to represent n decimal places of x. Then for almost all x,

$$\lim_{n \to \infty} \frac{m}{n} = \frac{6 \ln 2 \ln 10}{\pi^2} = 0.97027014\ldots$$

(Lochs 1964). Therefore, the CONTINUED FRACTION is only slightly more efficient at representing real numbers than is the decimal expansion. The set of x for which this statement does not hold is of measure 0.

See also CONTINUED FRACTION

References

Kintchine, A. "Zur metrischen Kettenbruchtheorie." *Compos. Math.* **3**, 276–85, 1936.
Lévy, P. "Sur le developpement en fraction continue d'un nombre choisi au hasard." *Compos. Math.* **3**, 286–03, 1936.

Lochs, G. *Abh. Hamburg Univ. Math. Sem.* **27**, 142–44, 1964.

Perron, O. *Die Lehre von Kettenbrüchen, 3. verb. und erweiterte Aufl.* Stuttgart, Germany: Teubner, 1954–7.

Loculus of Archimedes

STOMACHION

Locus

The set of all points (usually forming a curve or surface) satisfying some condition. For example, the locus of points in the plane equidistant from a given point is a CIRCLE, and the set of points in 3-space equidistant from a given point is a SPHERE.

References

Casey, J. *A Sequel to the First Six Books of the Elements of Euclid, Containing an Easy Introduction to Modern Geometry with Numerous Examples, 5th ed., rev. enl.* Dublin: Hodges, Figgis, & Co., pp. 5–, 1888.

Log

COMMON LOGARITHM, LOGARITHM, NATURAL LOGARITHM

Log Likelihood Procedure

A method for testing NESTED HYPOTHESES. To apply the procedure, given a specific model, calculate the LIKELIHOOD of observing the actual data. Then compare this likelihood to a nested model (i.e., one in which fewer parameters are allowed to vary independently).

Log Normal Distribution

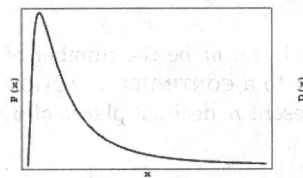 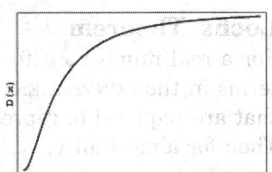

A CONTINUOUS DISTRIBUTION in which the LOGARITHM of a variable has a NORMAL DISTRIBUTION. It is a general case of GILBRAT'S DISTRIBUTION, to which the log normal distribution reduces with $S = 1$ and $M = 0$. The probability density and cumulative distribution functions for the log normal distribution are

$$P(x) = \frac{1}{Sx\sqrt{2\pi}} e^{-(\ln x - M)^2/(2S^2)} \quad (1)$$

$$D(x) = \frac{1}{2}\left[1 + \operatorname{erf}\left(\frac{\ln x - M}{S\sqrt{2}}\right)\right], \quad (2)$$

where $\operatorname{erf}(x)$ is the ERF function. This distribution is normalized, since letting $y \equiv \ln x$ gives $dy = dx/x$ and

$x = e^y$, so

$$\int_0^\infty P(x)\,dx = \frac{1}{S\sqrt{2\pi}} \int_{-\infty}^\infty e^{-(y-M)^2/2s^2}\,dy = 1. \quad (3)$$

The RAW MOMENTS are

$$\mu_1' = e^{M+S^2/2} \quad (4)$$

$$\mu_2' = e^{2(M+S)^2} \quad (5)$$

$$\mu_3' = e^{3M+9S^2/2} \quad (6)$$

$$\mu_4' = e^{4M+8S^2}, \quad (7)$$

and the CENTRAL MOMENTS are

$$\mu_2 = e^{2M+S^2}(e^{S^2} - 1) \quad (8)$$

$$\mu_3 = e^{3M+3S^2/2}(e^{S^2} - 1)^2(e^{S^2} + 2) \quad (9)$$

$$\mu_4 = e^{4M+2S^2}(e^{S^2} - 1)^2(e^{4S^2} + 2e^{3S^2} + 3e^{2S^2} - 3). \quad (10)$$

Therefore, the MEAN, VARIANCE, SKEWNESS, and KURTOSIS are given by

$$\mu = e^{M+S^2/2} \quad (11)$$

$$\sigma^2 = e^{S^2+2M}(e^{S^2} - 1) \quad (12)$$

$$\gamma_1 = \sqrt{e^{S^2} - 1}\,(2 + e^{S^2}) \quad (13)$$

$$\gamma_2 = e^{4S^2} + 2e^{3S^2} + 3e^{2S^2} - 6. \quad (14)$$

These can be found by direct integration

$$\mu = \frac{1}{S\sqrt{2\pi}} \int_0^\infty e^{-(\ln x - M)^2/(2S^2)}\,dx$$

$$= \frac{1}{S\sqrt{2\pi}} \int_{-\infty}^\infty e^{-(y-M)^2/2S^2}e^y\,dy$$

$$= e^{M+S^2/2}, \quad (15)$$

and similarly for σ^2.

Examples of variates which have approximately log normal distributions include the size of silver particles in a photographic emulsion, the survival time of bacteria in disinfectants, the weight and blood pressure of humans, and the number of words written in sentences by George Bernard Shaw.

See also GILBRAT'S DISTRIBUTION, WEIBULL DISTRIBUTION

References

Aitchison, J. and Brown, J. A. C. *The Lognormal Distribution, with Special Reference to Its Use in Economics.* New York: Cambridge University Press, 1957.

Balakrishnan, N. and Chen, W. W. S. *Handbook of Tables for Order Statistics from Lognormal Distributions with Applications.* Amsterdam, Netherlands: Kluwer, 1999.

Crow, E. L. and Shimizu, K. (Ed.). *Lognormal Distributions: Theory and Applications.* New York: Dekker, 1988.

Kenney, J. F. and Keeping, E. S. *Mathematics of Statistics, Pt. 2, 2nd ed.* Princeton, NJ: Van Nostrand, p. 123, 1951.

Logarithm

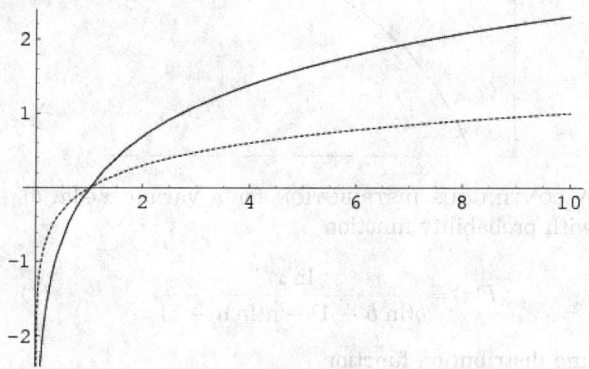

The logarithm $\log_b x$ for a BASE b and a number x is defined to be the INVERSE FUNCTION of taking x to the POWER b. Therefore, for any x and b,

$$x = b^{\log_b x}, \tag{1}$$

or equivalently,

$$x = \log_b(b^x). \tag{2}$$

Whereas power of trigonometric functions are denoted using notations like $\sin^k x$, $\ln^k x$ is less commonly used in favor of the notation $(\ln x)^k$.

For any BASE, the logarithm function has a SINGULARITY at $x = 0$. In the above plot, the solid curve is the logarithm to BASE e (the NATURAL LOGARITHM), and the dotted curve is the logarithm to BASE 10 (LOG).

Logarithms are used in many areas of science and engineering in which quantities vary over a large range. For example, the decibel scale for the loudness of sound, the Richter scale of earthquake magnitudes, and the astronomical scale of stellar brightnesses are all logarithmic scales.

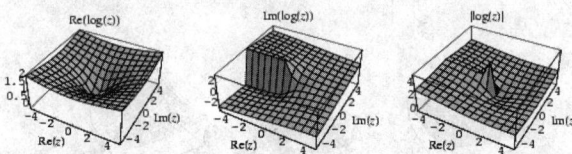

The logarithm can also be defined for COMPLEX arguments, as shown above. If the logarithm is taken as the forward function, the function taking the BASE to a given POWER is then called the ANTILOGARITHM.

For $x = \log N$, $\lfloor x \rfloor$ is called the CHARACTERISTIC and $x - \lfloor x \rfloor$ is called the MANTISSA. Division and multiplication identities follow from these

$$xy = b^{\log_b x} b^{\log_b y} = b^{\log_b x + \log_b y}, \tag{3}$$

from which it follows that

$$\log_b(xy) = \log_b x + \log_b y \tag{4}$$

$$\log_b\left(\frac{x}{y}\right) = \log_b x - \log_b y \tag{5}$$

$$\log_b x^n = n \log_b x. \tag{6}$$

There are a number of properties which can be used to change from one logarithm BASE to another

$$a = a^{\log_a b / \log_a b} = (a^{\log_a b})^{1/\log_a b} = b^{1/\log_a b} \tag{7}$$

$$\log_b a = \frac{1}{\log_a b} \tag{8}$$

$$\log_b x = \log_b\left(y^{\log_y x}\right) = \log_y x \, \log_b y \tag{9}$$

$$\log_b x = \frac{\log_n x}{\log_n b} \tag{10}$$

$$a^x = b^{x/\log_a b} = b^{x \log_b a}. \tag{11}$$

The logarithm BASE e is called the NATURAL LOGARITHM and is denoted $\ln x$ (LN). The logarithm BASE 10 is denoted $\log x$ (LOG), (although mathematics texts often use $\log x$ to mean $\ln x$). The logarithm BASE 2 is denoted $\lg x$ (LG).

An interesting property of logarithms follows from looking for a number y such that

$$\log_b(x + y) = -\log_b(x - y) \tag{12}$$

$$x + y = \frac{1}{x - y} \tag{13}$$

$$x^2 - y^2 = 1 \tag{14}$$

$$y = \sqrt{x^2 - 1}, \tag{15}$$

so

$$\log_b\left(x + \sqrt{x^2 - 1}\right) = -\log_b\left(x - \sqrt{x^2 - 1}\right). \tag{16}$$

Numbers OF THE FORM $\log_a b$ are IRRATIONAL if a and b are INTEGERS, one of which has a PRIME factor which the other lacks. A. Baker made a major step forward in TRANSCENDENTAL NUMBER theory by proving the transcendence of sums of numbers OF THE FORM $\alpha \ln \beta$ for α and β ALGEBRAIC NUMBERS.

See also ANTILOGARITHM, BASE (LOGARITHM), COLOGARITHM, E, EXPONENTIAL FUNCTION, HARMONIC LOGARITHM, LG, LN, LOG, LOGARITHMIC SERIES, LOGARITHMIC NUMBER, NAPIERIAN LOGARITHM, NATURAL LOGARITHM, POWER

References

Abramowitz, M. and Stegun, C. A. (Eds.). "Logarithmic Function." §4.1 in *Handbook of Mathematical Functions with Formulas, Graphs, and Mathematical Tables, 9th printing.* New York: Dover, pp. 67–9, 1972.

Beyer, W. H. *CRC Standard Mathematical Tables, 28th ed.* Boca Raton, FL: CRC Press, p. 221, 1987.

Conway, J. H. and Guy, R. K. "Logarithms." *The Book of Numbers.* New York: Springer-Verlag, pp. 248–52, 1996.

Beyer, W. H. "Logarithms." *CRC Standard Mathematical Tables, 28th ed.* Boca Raton, FL: CRC Press, pp. 159–60, 1987.

Pappas, T. "Earthquakes and Logarithms." *The Joy of Mathematics.* San Carlos, CA: Wide World Publ./Tetra, pp. 20–1, 1989.

Spanier, J. and Oldham, K. B. "The Logarithmic Function ln(x)." Ch. 25 in *An Atlas of Functions.* Washington, DC: Hemisphere, pp. 225–32, 1987.

Logarithmic Binomial Formula

Logarithmic Binomial Theorem

Logarithmic Binomial Theorem

For all integers n and $|x| < a$,

$$\lambda_n^{(t)}(x+a) = \sum_{k=0}^{\infty} \left\lfloor \begin{matrix} n \\ k \end{matrix} \right\rceil \lambda_{n-k}^t(a) x^k,$$

where $\lambda_n^{(t)}$ is the Harmonic Logarithm and $\left\lfloor \begin{matrix} n \\ k \end{matrix} \right\rceil$ is a Roman Coefficient. For $t=0$, the logarithmic binomial theorem reduces to the classical Binomial Theorem for Positive n, since $\lambda_1^{(0)}(a) - c^{n-k}$ for $n \geq k$, $\lambda_{n-k}^{(0)}(a) = 0$ for $n < k$, and $\left\lfloor \begin{matrix} n \\ k \end{matrix} \right\rceil = \binom{n}{k}$ when $n \geq k \geq 0$. Similarly, taking $t=1$ and $n<0$ gives the Negative Binomial Series. Roman (1992) gives expressions obtained for the case $t=1$ and $n \geq 0$ which are not obtainable from the Binomial Theorem.

See also Harmonic Logarithm, Roman Coefficient

References

Roman, S. "The Logarithmic Binomial Formula." *Amer. Math. Monthly* **99**, 641–48, 1992.

Logarithmic Derivative

The logarithmic derivative of a function f is defined as the Derivative of the Logarithm of a function. For example, the Digamma Function is defined as the logarithmic derivative of the Gamma Function,

$$\Psi(z) = \frac{d}{dz} \ln \Gamma(z).$$

See also Derivative, Digamma Function, Logarithm, Polygamma Function

References

Zwillinger, D. (Ed.). "Logarithmic Derivative." §6.11.8 in *CRC Standard Mathematical Tables and Formulae.* Boca Raton, FL: CRC Press, p. 496, 1995.

Logarithmic Distribution

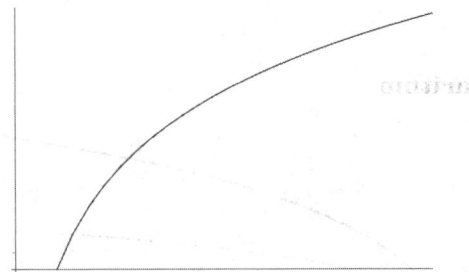

A Continuous Distribution for a variate $x \in [a, b]$ with probability function

$$P(x) = \frac{\ln x}{b(\ln b - 1) - a(\ln a - 1)} \tag{1}$$

and distribution function

$$D(x) = \frac{a(1 - \ln a) - x(1 - \ln x)}{a(1 - \ln a) - b(1 - \ln b)}. \tag{2}$$

The moments about zero are given by

$$\mu_n' = \frac{a^{n+1}[1 - (n+1)\ln a] - b^{n+1}[1 - (n+1)\ln b]}{(n+1)^2[a(1 - \ln a) - b(1 - \ln b)]}, \tag{3}$$

giving Mean

$$\mu = \frac{a^2(1 - 2\ln a) - b^2(1 - 2\ln b)}{4[a(1 - \ln a) - b(1 - \ln b)]}. \tag{4}$$

The Variance, Skewness, and Kurtosis are complicated expressions involving the μ_n'.

Logarithmic Integral

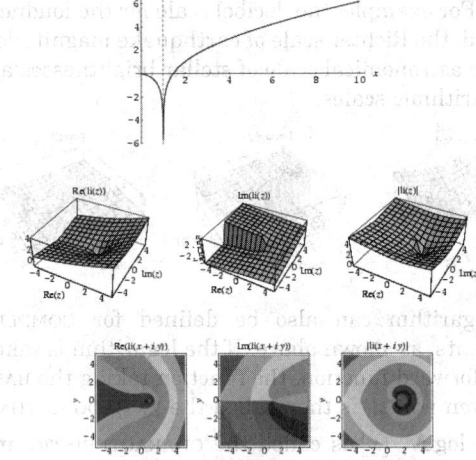

The logarithmic integral is defined by

$$\mathrm{li}(x) \equiv \int_0^x \frac{du}{\ln u}. \tag{1}$$

This function is implemented in *Mathematica* as `LogIntegral[x]`. The logarithmic integral obeys the identity

$$\mathrm{li}(x^m) = \gamma + \ln \ln x - \ln m + \sum_{n=1}^{\infty} \frac{(\ln x)^n}{n \cdot n! m^n} \qquad (2)$$

(Bromwich and MacRobert 1991, p. 334; Hardy 1999, p. 25).

The form of this function appearing in the PRIME NUMBER THEOREM is defined so that $\mathrm{Li}(2) = 0$:

$$\mathrm{Li}(x) \equiv \int_2^x \frac{du}{\ln u} \qquad (3)$$

$$= \mathrm{li}(x) - \mathrm{li}(2) \approx \mathrm{li}(x) - 1.04516 \qquad (4)$$

$$= \mathrm{ei}(\ln x), \qquad (5)$$

where $\mathrm{ei}(x)$ is the EXPONENTIAL INTEGRAL. (Note that the NOTATION $\mathrm{Li}_n(z)$ is also used for the POLYLOGARITHM.) Nielsen (1965, pp. 3 and 11) showed and Ramanujan independently discovered (Berndt 1994) that

$$\int_\mu^x \frac{dt}{\ln t} = \gamma + \ln \ln x + \sum_{k=1}^{\infty} \frac{(\ln x)^k}{k! k}, \qquad (6)$$

where γ is the EULER-MASCHERONI CONSTANT and μ is SOLDNER'S CONSTANT. Another FORMULA due to Ramanujan which converges more rapidly is

$$\int_\mu^x \frac{dt}{\ln t} = \gamma + \ln \ln x$$
$$+ \sqrt{x} \sum_{n=0}^{\infty} \frac{(-1)^{n-1}(\ln x)^n}{n! 2^{n-1}} \sum_{k=0}^{[(n-1)/2]} \frac{1}{2k+1} \qquad (7)$$

(Berndt 1994).

See also POLYLOGARITHM, PRIME CONSTELLATION, PRIME NUMBER THEOREM, SKEWES NUMBER

References

Berndt, B. C. *Ramanujan's Notebooks, Part IV.* New York: Springer-Verlag, pp. 126–31, 1994.
Bromwich, T. J. I'a and MacRobert, T. M. *An Introduction to the Theory of Infinite Series, 3rd ed.* New York: Chelsea, p. 334, 1991.
de Morgan, A. *The Differential and Integral Calculus, Containing Differentiation, Integration, Development, Series, Differential Equations, Differences, Summation, Equations of Differences, Calculus of Variations, Definite Integrals,--With Applications to Algebra, Plane Geometry, Solid Geometry, and Mechanics.* London: Robert Baldwin, p. 662, 1839.
Hardy, G. H. *Ramanujan: Twelve Lectures on Subjects Suggested by His Life and Work, 3rd ed.* New York: Chelsea, 1999.
Koosis, P. *The Logarithmic Integral I.* Cambridge, England: Cambridge University Press, 1998.
Nielsen, N. "Theorie des Integrallograrithmus und Verwandter Transzendenten." Part II in *Die Gammafunktion.* New York: Chelsea, 1965.

Vardi, I. *Computational Recreations in Mathematica.* Reading, MA: Addison-Wesley, p. 151, 1991.
Hardy, G. H. *Ramanujan: Twelve Lectures on Subjects Suggested by His Life and Work, 3rd ed.* New York: Chelsea, p. 45, 1999.
Le Lionnais, F. *Les nombres remarquables.* Paris: Hermann, p. 39, 1983.
Soldner. *Abhandlungen* **2**, 333, 1812.

Logarithmic Number

A COEFFICIENT of the MACLAURIN SERIES of

$$\frac{1}{\ln(1+x)} = \frac{1}{x} + \frac{1}{2} - \frac{1}{12}x + \frac{1}{24}x^2 - \frac{19}{720}x^3 + \frac{3}{160}x^4 + \cdots$$

(Sloane's A002206 and A002207), the multiplicative inverse of the MERCATOR SERIES function $\ln(1+x)$.

See also MERCATOR SERIES

References

Sloane, N. J. A. Sequences A002206/M5066 and A002207/M2017 in "An On-Line Version of the Encyclopedia of Integer Sequences." http://www.research.att.com/~njas/sequences/eisonline.html.

Logarithmic Series

$$\sum_{k=1}^{\infty} (-1)^k \ln k = \frac{1}{2} \ln\left(\frac{1}{2}\pi\right)$$

$$\sum_{k=1}^{\infty} \ln k = \frac{1}{2} \ln(2\pi).$$

See also LOGARITHM

References

Bromwich, T. J. I'a. and MacRobert, T. M. *An Introduction to the Theory of Infinite Series, 3rd ed.* New York: Chelsea, p. 351, 1991.
Hardy, G. H. *Ramanujan: Twelve Lectures on Subjects Suggested by His Life and Work, 3rd ed.* New York: Chelsea, p. 37, 1999.

Logarithmic Spiral

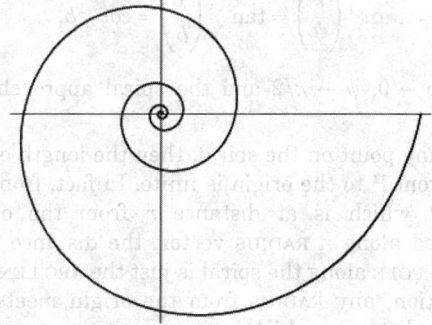

A curve whose equation in POLAR COORDINATES is

given by

$$r = ae^{b\theta}, \qquad (1)$$

where r is the distance from the ORIGIN, θ is the angle from the x-AXIS, and a and b are arbitrary constants. The logarithmic spiral is also known as the GROWTH SPIRAL, EQUIANGULAR SPIRAL, and SPIRA MIRABILIS. It can be expressed parametrically using

$$\cos\theta = \frac{1}{\sqrt{1 - \tan^2\theta}} = \frac{1}{\sqrt{1 + \frac{y^2}{x^2}}} = \frac{x}{\sqrt{x^2 + y^2}} = \frac{x}{r}, \qquad (2)$$

which gives

$$x = r\cos\theta = a\cos\theta e^{b\theta} \qquad (3)$$

$$y = x\tan\theta = r\sin\theta = a\sin\theta e^{b\theta}. \qquad (4)$$

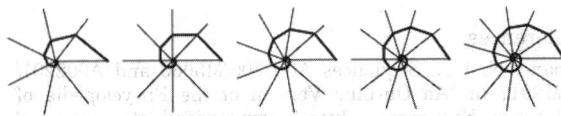

The logarithmic spiral can be constructed from equally spaced rays by starting at a point along one ray, and drawing the perpendicular to a neighboring ray. As the number of rays approached infinity, the sequence of segments approaches the smooth logarithmic spiral (Hilton *et al.* 1997, pp. 2–).

The logarithmic spiral was first studied by Descartes in 1638 and Jakob Bernoulli. Bernoulli was so fascinated by the spiral that he had one engraved on his tombstone (although the engraver did not draw it true to form) together with the words "eadem mutata resurgo" ("I shall arise the same though changed"rpar;. Torricelli worked on it independently and found the length of the curve (MacTutor Archive).

The rate of change of RADIUS is

$$\frac{dr}{d\theta} = abe^{b\theta} = br, \qquad (5)$$

and the ANGLE between the tangent and radial line at the point (r, θ) is

$$\psi = \tan^{-1}\left(\frac{r}{\frac{dr}{d\theta}}\right) = \tan^{-1}\left(\frac{1}{b}\right) = \cot^{-1}b. \qquad (6)$$

So, as $b \to 0$, $\psi \to \pi/2$ and the spiral approaches a CIRCLE.

If P is any point on the spiral, then the length of the spiral from P to the origin is finite. In fact, from the point P which is at distance r from the origin measured along a RADIUS vector, the distance from P to the POLE along the spiral is just the ARC LENGTH. In addition, any RADIUS from the origin meets the spiral at distances which are in GEOMETRIC PROGRESSION (MacTutor Archive).

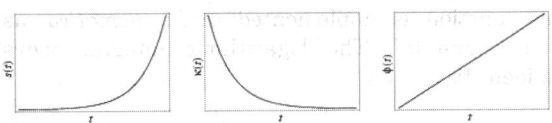

The ARC LENGTH, CURVATURE, and TANGENTIAL ANGLE of the logarithmic spiral are

$$s = \int ds = \int \sqrt{x'^2 + y'^2}\, dt = \frac{a\sqrt{1 + b^2}}{b} e^{b\theta}$$

$$= \frac{r\sqrt{1 + b^2}}{b} \qquad (7)$$

$$\kappa = \frac{x'y'' - y'x''}{(x'^2 + y'^2)^{3/2}} = \left(a\sqrt{1 + b^2}\, e^{b\theta}\right)^{-1} \qquad (8)$$

$$\phi = \int \kappa(s)\, ds = \theta. \qquad (9)$$

The CESÀRO EQUATION is

$$\kappa = \frac{1}{bs}. \qquad (10)$$

On the surface of a SPHERE, the analog is a LOXODROME. This SPIRAL is related to FIBONACCI NUMBERS and the GOLDEN RATIO.

See also GOLDEN RECTANGLE, LOGARITHMIC SPIRAL CAUSTIC CURVE, LOGARITHMIC SPIRAL EVOLUTE, LOGARITHMIC SPIRAL INVERSE CURVE, LOGARITHMIC SPIRAL PEDAL CURVE, LOGARITHMIC SPIRAL RADIAL CURVE, MICE PROBLEM, SPIRAL, WHIRL

References

Boyadzhiev, K. N. "Spirals and Conchospirals in the Flight of Insects." *Coll. Math. J.* **30**, 23–1, 1999.

Cook, T. A. *The Curves of Life, Being an Account of Spiral Formations and Their Application to Growth in Nature, To Science and to Art.* New York: Dover, 1979.

Gray, A. "Logarithmic Spirals." *Modern Differential Geometry of Curves and Surfaces with Mathematica, 2nd ed.* Boca Raton, FL: CRC Press, pp. 40–2, 1997.

Hilton, P.; Holton, D.; and Pedersen, J. *Mathematical Reflections in a Room with Many Mirrors.* New York: Springer-Verlag, 1997.

Lawrence, J. D. *A Catalog of Special Plane Curves.* New York: Dover, pp. 184–86, 1972.

Lockwood, E. H. "The Equiangular Spiral." Ch. 11 in *A Book of Curves.* Cambridge, England: Cambridge University Press, pp. 98–09, 1967.

MacTutor History of Mathematics Archive. "Equiangular Spiral." http://www-groups.dcs.st-and.ac.uk/~history/Curves/Equiangular.html.

Steinhaus, H. *Mathematical Snapshots, 3rd ed.* New York: Dover, pp. 132–36, 1999.

Thompson, D'Arcy W. *Science and the Classics.* Oxford, England: Oxford University Press, pp. 114–47, 1940.

Wells, D. *The Penguin Dictionary of Curious and Interesting Geometry.* London: Penguin, pp. 67–8, 1991.

Logarithmic Spiral Caustic Curve

The CAUSTIC of a LOGARITHMIC SPIRAL, where the pole is taken as the RADIANT POINT, is an equal LOGARITHMIC SPIRAL.

Logarithmic Spiral Evolute

In POLAR COORDINATES $r = r(\theta)$, the RADIUS OF CURVATURE is given by

$$R = \frac{(r^2 + r_\theta^2)^{3/2}}{r^2 + 2r^2r_\theta^2 - rr_{\theta\theta}}, \tag{1}$$

so plugging in the equation of the LOGARITHMIC SPIRAL and its derivatives

$$r = ae^{b\theta} \tag{2}$$

$$r_\theta = abe^{b\theta} \tag{3}$$

$$r_{\theta\theta} = ab^2 e^{b\theta} \tag{4}$$

gives

$$R = a\sqrt{1 + b^2}\, e^{b\theta}. \tag{5}$$

To find the VELOCITY VECTOR, compute

$$\begin{bmatrix} x \\ y \end{bmatrix} = \begin{bmatrix} ae^{b\theta} & \cos\theta \\ ae^{b\theta} & \sin\theta \end{bmatrix}$$

$$\begin{bmatrix} x' \\ y' \end{bmatrix} = \begin{bmatrix} abe^{b\theta} & \cos\theta - ae^{b\theta}\sin\theta \\ abe^{b\theta} & \sin\theta + ae^{b\theta}\cos\theta \end{bmatrix}$$

$$= ae^{b\theta} \begin{bmatrix} b\cos\theta - \sin\theta \\ b\sin\theta + \cos\theta \end{bmatrix}, \tag{6}$$

so

$$|\mathbf{r}'| = ae^{b\theta}\sqrt{(b\cos\theta - \sin\theta)^2 + (b\sin\theta + \cos\theta)^2}$$

$$= ae^{b\theta}\sqrt{1 + b^2}, \tag{7}$$

and the TANGENT VECTOR is given by

$$\hat{\mathbf{T}} = \frac{\mathbf{r}'}{|\mathbf{r}'|} = \frac{1}{ae^{b\theta}\sqrt{1 + b^2}} \begin{bmatrix} ae^{b\theta}\cos\theta \\ ae^{b\theta}\sin\theta \end{bmatrix}$$

$$= \frac{1}{\sqrt{1 + b^2}} \begin{bmatrix} \cos\theta \\ \sin\theta \end{bmatrix}. \tag{8}$$

The coordinates of the EVOLUTE are therefore

$$\xi = -abe^{b\theta}\sin\theta \tag{9}$$

$$\eta = -abe^{b\theta}\cos\theta. \tag{10}$$

Therefore, the EVOLUTE is another logarithmic spiral with $a' \equiv ab$, as first shown by Johann Bernoulli.

In some cases, the EVOLUTE is identical to the original, as can be demonstrated by making the substitution to the new variable

$$\theta \equiv \phi - \tfrac{1}{2}\pi \pm 2n\pi. \tag{11}$$

Then the above equations become

$$\xi = -abe^{b(\phi - \pi/2 \pm 2n\pi)}\sin(\phi - \pi/2 \pm 2n\pi)$$

$$= abe^{b\phi}e^{b(-\pi/2 \pm 2n\pi)}\cos\phi \tag{12}$$

$$\eta = abe^{b(\phi - \pi/2 \pm 2n\pi)}\cos(\phi - \pi/2 \pm 2n\pi)$$

$$= abe^{b\phi}e^{b(-\pi/2 \pm 2n\pi)}\sin\phi, \tag{13}$$

which are equivalent to the form of the original equation if

$$be^{b\left(-\frac{1}{2}\pi \pm 2n\pi\right)} = 1 \tag{14}$$

$$\ln b + b\left(-\tfrac{1}{2}\pi \pm 2n\pi\right) = 0 \tag{15}$$

$$\frac{\ln b}{b} = \tfrac{1}{2}\pi \mp 2n\pi = -\left(2n - \tfrac{1}{2}\right)\pi, \tag{16}$$

where only solutions with the minus sign in $\mp$ exist. Solving gives the values summarized in the following table.

n	b_n	$\psi = \cot^{-1}b_n$
1	0.2744106319...	74°39′18.53″
2	0.1642700512...	80°40′16.80″
3	0.1218322508...	83°03′13.53″
4	0.0984064967...	84°22′47.53″
5	0.0832810611...	85°14′21.60″
6	0.0725974881...	85°50′51.92″
7	0.0645958183...	86°18′14.64″
8	0.0583494073...	86°39′38.20″
9	0.0533203211...	86°56′52.30″
10	0.0491732529...	87°11′05.45″

References

Lauwerier, H. *Fractals: Endlessly Repeated Geometric Figures.* Princeton, NJ: Princeton University Press, pp. 60–4, 1991.

Logarithmic Spiral Inverse Curve

The INVERSE CURVE of the LOGARITHMIC SPIRAL

$$r = e^{a\theta}$$

with INVERSION CENTER at the origin and inversion radius k is the LOGARITHMIC SPIRAL

$$r = ke^{-a\theta}.$$

Logarithmic Spiral Pedal Curve

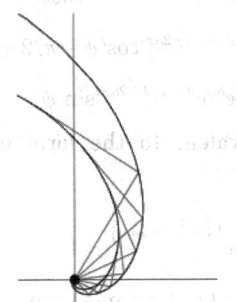

The PEDAL CURVE of a LOGARITHMIC SPIRAL with parametric equation

$$f = e^{at} \cos t \tag{1}$$
$$g = e^{at} \sin t \tag{2}$$

for a PEDAL POINT at the pole is an identical LOGARITHMIC SPIRAL

$$x = \frac{(a \sin t + \cos t)e^{at}}{1 + a^2} \tag{3}$$

$$y = \frac{(\sin t - a \cos t)e^{at}}{1 + a^2} \tag{4}$$

so

$$r = \sqrt{x^2 + y^2} = \frac{e^{at}}{\sqrt{1 + a^2}}. \tag{5}$$

Logarithmic Spiral Radial Curve

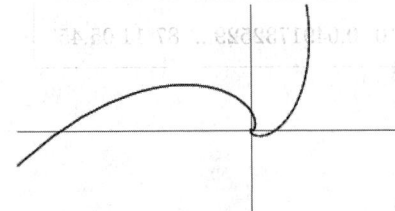

The RADIAL CURVE of the LOGARITHMIC SPIRAL is another LOGARITHMIC SPIRAL.

Logarithmic Transform

The inverse transform

$$\sum_{n=1}^{\infty} \frac{a_n x^n}{n!} = \ln\left(1 + \sum_{n=1}^{\infty} \frac{b_n x^n}{n!}\right)$$

of the EXPONENTIAL TRANSFORM

$$1 + \sum_{n=1}^{\infty} \frac{b_n x^n}{n!} = \exp\left(\sum_{n=1}^{\infty} \frac{a_n x^n}{n!}\right)$$

which relate sequences $a_1, a_2, \ldots$ and $b_1, b_2, \ldots$.

See also EXPONENTIAL TRANSFORM

References

Sloane, N. J. A. and Plouffe, S. *The Encyclopedia of Integer Sequences.* San Diego, CA: Academic Press, pp. 19–0, 1995.

Logarithmically Concave Function

A function $f(x)$ is logarithmically concave on the interval $[a, b]$ if $f > 0$ and $\ln f(x)$ is CONCAVE on $[a, b]$. The definition can also be extended to $\mathbb{R}^k \to (0, \infty)$ functions (Dharmadhikari and Joag-Dev 1988, p. 18).

See also CONCAVE FUNCTION, LOGARITHMICALLY CONVEX FUNCTION

References

Dharmadhikari, S. and Joag-Dev, K. *Unimodality, Convexity, and Applications.* Boston, MA: Academic Press, 1988.

Logarithmically Convex Function

A function $f(x)$ is logarithmically convex on the interval $[a, b]$ if $f > 0$ and $\ln f(x)$ is CONVEX on $[a, b]$. If $f(x)$ and $g(x)$ are logarithmically convex on the interval $[a, b]$, then the functions $f(x) + g(x)$ and $f(x)g(x)$ are also logarithmically convex on $[a, b]$. The definition can also be extended to $\mathbb{R}^k \to (0, \infty)$ functions (Dharmadhikari and Joag-Dev 1988, p. 18).

See also CONVEX FUNCTION, LOGARITHMICALLY CONCAVE FUNCTION

References

Dharmadhikari, S. and Joag-Dev, K. *Unimodality, Convexity, and Applications.* Boston, MA: Academic Press, 1988.
Gradshteyn, I. S. and Ryzhik, I. M. *Tables of Integrals, Series, and Products, 6th ed.* San Diego, CA: Academic Press, p. 1100, 2000.

Logconcave Function

LOGARITHMICALLY CONCAVE FUNCTION

Logconvex Function

LOGARITHMICALLY CONVEX FUNCTION

LogGamma

GAMMA FUNCTION

Logic

The formal mathematical study of the methods, structure, and validity of mathematical deduction and proof.

In Hilbert's day, formal logic sought to devise a complete, consistent formulation of mathematics such that propositions could be formally stated and proved using a small number of symbols with WELL DEFINED meanings. The difficulty of formal logic was demonstrated in the monumental *Principia Mathematica* (1925) of Whitehead and Russell's , in which hundred of pages of symbols were required before the statement $1 + 1 = 2$ could be deduced. In 1931, Gödel unexpectedly showed that Hilbert's goal to be impossible, and this proved only the first of a number of difficult and counterintuitive results which have since been demonstrated.

A very simple form of logic is the study of "TRUTH TABLES" and digital logic circuits in which one or more outputs depend on a combination of circuit elements (AND, OR, NAND, NOR, NOT, XOR, etc.; "gates") and the input values. In such a circuit, values at each point can take on values of only TRUE (1) or FALSE (0). DE MORGAN'S DUALITY LAW is a useful principle for the analysis and simplification of such circuits.

A generalization of this simple type of logic in which possible values are TRUE, FALSE, and "undecided" is called THREE-VALUED LOGIC. A further generalization called FUZZY LOGIC treats "truth" as a continuous quantity ranging from 0 to 1.

See also ABSORPTION LAW, ALETHIC, BOOLEAN ALGEBRA, BOOLEAN CONNECTIVE, BOUND, CALIBAN PUZZLE, CONTRADICTION LAW, DE MORGAN'S DUALITY LAW, DE MORGAN'S LAWS, DEDUCIBLE, EXCLUDED MIDDLE LAW, FREE, FUZZY LOGIC, GÖDEL'S INCOMPLETENESS THEOREM, KHOVANSKI'S THEOREM, LOGICAL PARADOX, LOGOS, LÖWENHEIM-SKOLEM THEOREM, METAMATHEMATICS, MODEL THEORY, QUANTIFIER, SENTENCE, TARSKI'S THEOREM, TAUTOLOGY, THREE-VALUED LOGIC, TOPOS, TRUTH TABLE, TURING MACHINE, UNIVERSAL TURING MACHINE, VENN DIAGRAM, WILKIE'S THEOREM

References

Adamowicz, Z. and Zbierski, P. *Logic of Mathematics: A Modern Course of Classical Logic.* New York: Wiley, 1997.

Bogomolny, A. "Falsity Implies Anything." http://www.cut-the-knot.com/do_you_know/falsity.html.

Carnap, R. *Introduction to Symbolic Logic and Its Applications.* New York: Dover, 1958.

Church, A. *Introduction to Mathematical Logic, Vol. 1.* Princeton, NJ: Princeton University Press, 1996.

Enderton, H. B. *A Mathematical Introduction to Logic.* New York: Academic Press, 1972.

Enderton, H. B. *Elements of Set Theory.* New York: Academic Press, 1977.

Heijenoort, J. van. *From Frege to Gödel: A Sourcebook in Mathematical Logic, 1879–931.* Cambridge, MA: Cambridge University Press, 1967.

Gödel, K. *On Formally Undecidable Propositions of Principia Mathematica and Related Systems.* New York: Dover, 1992.

Jeffrey, R. C. *Formal Logic: Its Scope and Limits.* New York: McGraw-Hill, 1967.

Kac, M. and Ulam, S. M. *Mathematics and Logic: Retrospect and Prospects.* New York: Dover, 1992.

Kleene, S. C. *Introduction to Metamathematics.* Princeton, NJ: Van Nostrand, 1971.

Smullyan, R. M. *First-Order Logic.* New York: Dover.

Weisstein, E. W. "Books about Logic." http://www.treasure-troves.com/books/Logic.html.

Whitehead, A. N. and Russell, B. *Principia Mathematica, 2nd ed.* Cambridge, England: Cambridge University Press, 1962.

Logical And

AND

Logical Connective

CONNECTIVE

Logical Not

NEGATION SIGN, NOT

Logical Or

OR

Logical Paradox

PARADOX

LogIntegral

Logarithmic Integral

Logistic Distribution

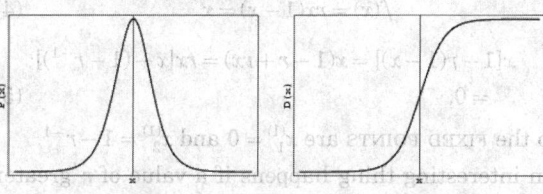

$$P(x) = \frac{e^{(x-m)/b}}{|b|[1 + e^{(x-m)/b}]^2} \qquad (1)$$

$$D(x) = \frac{1}{1 + e^{(m-x)/|b|}}, \qquad (2)$$

and the MEAN, VARIANCE, SKEWNESS, and KURTOSIS are

$$\mu = m \qquad (3)$$

$$\sigma^2 = \tfrac{1}{3}\pi^2\beta^2 \qquad (4)$$

$$\gamma_1 = 0 \qquad (5)$$

$$\gamma_2 = \tfrac{6}{5}. \qquad (6)$$

See also LOGISTIC EQUATION, LOGISTIC GROWTH CURVE

References
von Seggern, D. *CRC Standard Curves and Surfaces*. Boca Raton, FL: CRC Press, p. 250, 1993.

Logistic Equation

The logistic equation (sometimes called the VERHULST MODEL since it was first published in 1845 by the Belgian P.-F. Verhulst) is defined by

$$x_{n+1} = rx_n(1 - x_n), \tag{1}$$

where r (sometimes also denoted μ) is a POSITIVE constant (the "biotic potential"). Let an initial point x_0 lie in the interval $[0, 1]$. Now find appropriate conditions on r which keep points in the interval. The maximum value x_{n+1} can take is found from

$$\frac{dx_{n+1}}{dx_n} = r(1 - 2x_n) = 0, \tag{2}$$

so the largest value of x_{n+1} occurs for $x_n = 1/2$. Plugging this in, $\max(x_{n+1}) = r/4$. Therefore, to keep the MAP in the desired region, we must have $r \in (0, 4]$. The JACOBIAN is

$$J = \left| \frac{dx_{n+1}}{dx_n} \right| = |r(1 - 2x_n)|, \tag{3}$$

and the MAP is stable at a point x_0 if $J(x_0) < 1$.

Now find the FIXED POINTS of the MAP, which occur when $x_{n+1} = x_n$. For convenience, drop the n subscript on x_n

$$f(x) = rx(1 - x) = x \tag{4}$$

$$x[1 - r(1 - x)] = x(1 - r + rx) = rx[x - (1 - r^{-1})]$$
$$= 0, \tag{5}$$

so the FIXED POINTS are $x_1^{(1)} = 0$ and $x_2^{(1)} = 1 - r^{-1}$.

An interesting thing happens if a value of r greater than 3 is chosen. The map becomes unstable and we get a PITCHFORK BIFURCATION with two stable orbits of period two corresponding to the two stable FIXED POINTS of $f^2(x)$. The fixed points of order two must satisfy $x_{n+2} = x_n$, so

$$x_{n+2} = rx_{n+1}(1 - x_{n+1})$$
$$= r[rx_n(1 - x_n)][1 - rx_n(1 - x_n)]$$
$$= r^2 x_n(1 - x_n)(1 - rx_n + rx_n^2) = x_n. \tag{6}$$

For convenience, drop the n subscripts and rewrite

$$x\{r^2[1 - x(1 + r) + 2rx^2 - rx^3] - 1\} = 0 \tag{7}$$

$$x[-r^3 x^3 + 2r^3 x^2 - r^2(1 + r)x + (r^2 - 1)] = 0 \tag{8}$$

$$-r^3 x[x - (1 - r^{-1})][x^2 - (1 + r^{-1})x + r^{-1}(1 + r^{-1})]$$
$$= 0. \tag{9}$$

Notice that we have found the first-order FIXED POINTS as well, since two iterations of a first-order

FIXED POINT produce a trivial second-order FIXED POINT. The true 2-CYCLES are given by solutions to the quadratic part

$$x_{\pm}^{(2)} = \tfrac{1}{2}[(1 + r^{-1}) \pm \sqrt{(1 + r^{-1})^2 - 4r^{-1}(1 + r^{-1})}]$$

$$= \tfrac{1}{2}[(1 + r^{-1}) \pm \sqrt{1 + 2r^{-1} + r^{-2} - 4r^{-1} - 4r^{-2}}]$$

$$= \tfrac{1}{2}[(1 + r^{-1}) \pm \sqrt{1 - 2r^{-1} - 3r^{-2}}]$$

$$= \tfrac{1}{2}[(1 + r^{-1}) \pm r^{-1}\sqrt{(r - 3)(r + 1)}]. \tag{10}$$

These solutions are only REAL for $r \geq 3$, so this is where the 2-CYCLE begins. Note that the 2-cycle can also be found by computing the DISCRIMINANT of

$$\frac{f^2(x) - x}{f(x) - x} = r^2 x^2 - r(1 + r)x + (1 + r) = 0, \tag{11}$$

which is

$$\frac{(1 + r)(3 - r)}{r^2}. \tag{12}$$

When this equals 0, two roots coincide, so $r_2 = 3$ is the onset of period doubling.

Now look for the onset of the 3-CYCLE. To eliminate the 1-CYCLES, consider

$$\frac{f^3(x) - x}{f(x) - x} = 0. \tag{13}$$

This gives

$$1 + r + r^2 - (r^4 + 2r^3 + 2r^2 + r)x$$
$$+ (2r^5 + 3r^4 + 3r^3 + r^2)x^2$$
$$-(r^6 + 5r^5 + 3r^4 + r^3)x^3 + (3r^6 + 4r^5 + r^4)x^4$$
$$-(3r^6 - r^5)x^5 + r^6 x^6 = 0. \tag{14}$$

The ROOTS of this equation are all IMAGINARY for r less than some cutoff r_3, at which point two of them convert to REAL roots. The value of r_3 can be found by computing the DISCRIMINANT of (14),

$$D = \frac{(r^2 - 5r + 7)^2(r^2 - 2r - 7)^3(1 + r + r^2)^2}{r^{30}}. \tag{15}$$

When the DISCRIMINANT is zero, two roots coincide. This happens at $r_3 = 1 + 2\sqrt{2}$, so the 3-CYCLE starts at r_3.

To find the onset of the 4-CYCLE, eliminate the 2- and 1-CYCLES by considering

$$\frac{f^4(x) - x}{f^2(x) - x} = 0. \tag{16}$$

This gives

$$1 + r^2 + (-r^2 - r^3 - r^4 - r^5)x$$

$$+(2r^3 + r^4 + 4r^5 + r^6 + 2r^7)x^2$$

$$+(-r^3 - 5r^5 - 4r^6 - 5r^7 - 4r^8 - r^9)x^3$$

$$+(2r^5 + 6r^6 + 4r^7 + 14r^8 + 5r^9 + 3r^{10})x^4$$

$$+(-4r^6 - r^7 - 18r^8 - 12r^9 - 12r^{10} - 3r^{11})x^5$$

$$+(r^6 + 10r^8 + 17r^9 + 18r^{10} + 15r^{11} + r^{12})x^6$$

$$+(-2r^8 - 14r^9 - 12r^{10} - 30r^{11} - 6r^{12})x^7$$

$$+(6r^9 + 3r^{10} + 30r^{11} + 15r^{12})x^8$$

$$+(-r^9 - 15r^{11} - 20r^{12})x^9 + (3r^{11} + 15r^{12})x^{10} \quad (17)$$

The value of r_4 can be found by computing the DISCRIMINANT of (17),

$$D = \frac{(r^2 + 1)^3(r^2 - 4r + 5)^3}{r^{132}}$$

$$\times (r^6 - 6r^5 + 3r^4 + 28r^3 - 9r^2 - 54r - 135), \quad (18)$$

which has roots at $r_4 = 1 + \sqrt{6}$, as well as at the 2nd root of

$$r^6 - 6r^5 + 3r^4 + 28r^3 - 9r^2 - 54r - 135 = 0.$$

The 4-CYCLE therefore starts at $r_4 = 1 + \sqrt{6} = 3.449489\ldots$.

The onset of 5-cycles can be found analogously, and gives a messy 22nd-order polynomial in r whose real positive roots are 3.73817, 3.90557, and 3.99026.

In general, the set of $n + 1$ equations which can be solved to give the onset of an arbitrary n-cycle (Saha and Strogatz 1995) is

$$\begin{cases} x_2 = rx_1(1 - x_1) \\ x_3 = rx_2(1 - x_2) \\ \vdots \\ x_n = rx_{n-1}(1 - x_{n-1}) \\ x_1 = rx_n(1 - x_n) \\ r^n \prod_{k=1}^n (1 - 2x_k) = 1. \end{cases} \quad (19)$$

The first n of these give $f(x)$, $f^2(x)$, ..., $f^n(x)$, and the last uses the fact that the onset of period n occurs by a TANGENT BIFURCATION, so the nth DERIVATIVE is 1. For small n, these can be solved exactly, but the complexity rapidly increases with n.

For $n = 2$, the solutions (x_1, x_2, r) are given by $(0, 0, \pm 1)$ and $(2/3, 2/3, 3)$, so the first BIFURCATION occurs at $r_2 = 3$.

For $n = 3$,

$$\frac{d[f^3(x)]}{dx} = \frac{d[f^3(x)]}{d[f^2(x)]} \frac{d[f^2(x)]}{d[f(x)]} \frac{d[f(x)]}{dx}$$

$$= \frac{d[f(z)]}{dz} \frac{d[f(y)]}{dy} \frac{d[f(x)]}{dx}$$

$$= r^3(1 - 2z)(1 - 2y)(1 - 2x). \quad (20)$$

Solving the resulting CUBIC EQUATION using computer algebra gives

$$r = 1 + 2\sqrt{2} \quad (21)$$

and x_1, x_2, x_3 the 2nd, 4th, and 5th roots of the sextic

$$343x^6 - 980x^5 + 868x^4 - 134x^3 - 161x^2 + 70x - 7 = 0, \quad (22)$$

giving numerical roots

$$x_1 \approx 0.514355 \quad (23)$$

$$x_2 \approx 0.956318 \quad (24)$$

$$x_3 \approx 0.159929 \quad (25)$$

$$r \approx 3.828427. \quad (26)$$

Saha and Strogatz (1995) give a simplified algebraic treatment for the 3-cycle which involves solving

$$r^3(1 - 2\alpha + 4\beta - 8\gamma) = 1, \quad (27)$$

together with three other simultaneous equations, where

$$\alpha \equiv x_1 + x_2 + x_3 \quad (28)$$

$$\beta \equiv x_1 x_2 + x_1 x_3 + x_2 x_3 \quad (29)$$

$$\gamma \equiv x_1 x_2 x_3. \quad (30)$$

Further simplifications still are provided in Bechhoeffer (1996) and Gordon (1996), but neither of these techniques generalizes easily to higher CYCLES. Bechhoeffer (1996) expresses the three additional equations as

$$2\alpha = 3 + r^{-1} \quad (31)$$

$$4\beta = \frac{3}{2} + 5r^{-1} + \frac{3}{2}r^{-2} \quad (32)$$

$$8\gamma = -\frac{1}{2} + \frac{7}{2}r^{-1} + \frac{5}{2}r^{-2} + \frac{5}{2}r^{-3}, \quad (33)$$

giving

$$r^2 - 2r - 7 = 0. \quad (34)$$

This has the positive solution found previously, $r_3 = 1 + 2\sqrt{2}$.

Gordon (1996) derives not only the value for the onset of the 3-CYCLE, but also an upper bound for the r-values supporting stable period-3 orbits. This value is obtained by solving the CUBIC EQUATION

$$s^3 - 11s^2 + 37s - 108 = 0 \quad (35)$$

for s, then

$$r' = 1 + \sqrt{s} \quad (36)$$

$$= 1 + \sqrt{\frac{11}{3} + \left(\frac{1915}{54} + \frac{5}{2}\sqrt{201}\right)^{1/3} + \left(\frac{1915}{54} - \frac{5}{2}\sqrt{201}\right)^{1/3}}$$

$$= 3.841499007543\ldots \quad (37)$$

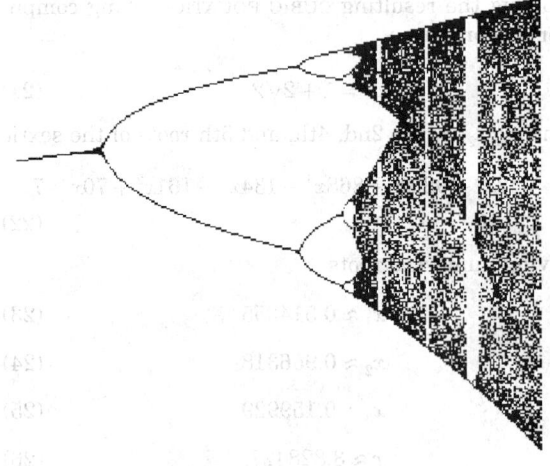

The illustration above shows the logistic map. A table of the CYCLE type and value of r_n at which the cycle 2^n appears is given below.

n	cycle (2^n)	r_n
1	2	3
2	4	3.449490
3	8	3.544090
4	16	3.564407
5	32	3.568750
6	64	3.56969
7	128	3.56989
8	256	3.569934
9	512	3.569943
10	1024	3.5699451
11	2048	3.569945557
∞	ACC. PT.	3.569945672

For additional values, see Rasband (1990, p. 23). Note that the table in Tabor (1989, p. 222) is incorrect, as is the $n = 2$ entry in Lauwerier (1991). The period doubling BIFURCATIONS come faster and faster (8, 16, 32, ...), then suddenly break off. Beyond a certain point known as the ACCUMULATION POINT, periodicity gives way to CHAOS, as illustrated below. In the middle of the complexity, a window suddenly appears with a regular period like 3 or 7 as a result of MODE LOCKING. The period-3 BIFURCATION occurs at $r = 1 + 2\sqrt{2} = 3.828427$, and PERIOD DOUBLINGS then begin again with CYCLES of 6, 12, ...and 7, 14, 28, ..., and then once again break off to CHAOS.

It is relatively easy to show that the logistic map is chaotic on an invariant Cantor set for $r > 2 + \sqrt{5} \approx 4.236$ (Devaney 1989, pp. 31–0; Gulik 1992, pp. 112–26; Holmgren 1996, pp. 69–5), but in fact, it is also chaotic for all $r > 4$ (Robinson 1995, pp. 33–7; Kraft 1999).

The logistic equation has CORRELATION EXPONENT 0.500 ± 0.005 (Grassberger and Procaccia 1983), CAPACITY DIMENSION 0.538 (Grassberger 1981), and INFORMATION DIMENSION 0.5170976 (Grassberger and Procaccia 1983).

See also BIFURCATION, FEIGENBAUM CONSTANT, LOGISTIC DISTRIBUTION, LOGISTIC EQUATION $R = 4$, LOGISTIC GROWTH CURVE, PERIOD THREE THEOREM, QUADRATIC MAP

References

Bechhoeffer, J. "The Birth of Period 3, Revisited." *Math. Mag.* **69**, 115–18, 1996.

Beck, C.; and Schlögl, F. *Thermodynamics of Chaotic Systems.* Cambridge, England: Cambridge University Press, 1993.

Bogomolny, A. "Chaos Creation (There is Order in Chaos)." http://www.cut-the-knot.com/blue/chaos.html.

Costa, U. M. S. and Lyra, M. L. *Phys. Rev. E* **56**, 245, 1997.

Devaney, R. *An Introduction to Chaotic Dynamical Systems,* 2nd ed. Redwood City, CA: Addison-Wesley, 1989.

Dickau, R. M. "Bifurcation Diagram." http://forum.swarthmore.edu/advanced/robertd/bifurcation.html.

Gleick, J. *Chaos: Making a New Science.* New York: Penguin Books, pp. 69–0, 1988.

Gordon, W. B. "Period Three Trajectories of the Logistic Map." *Math. Mag.* **69**, 118–20, 1996.

Grassberger, P. "On the Hausdorff Dimension of Fractal Attractors." *J. Stat. Phys.* **26**, 173–79, 1981.

Grassberger, P. and Procaccia, I. "Measuring the Strangeness of Strange Attractors." *Physica D* **9**, 189–08, 1983.

Gulick, D. *Encounters with Chaos.* New York: McGraw-Hill, 1992.

Holmgren, R. *A First Course in Discrete Dynamical Systems,* 2nd ed. New York: Springer-Verlag, 1996.

Kraft, R. L. "Chaos, Cantor Sets, and Hyperbolicity for the Logistic Maps." *Amer. Math. Monthly* **106**, 400–08, 1999.

Latora, V.; Rapisarda, A.; Tsallis, C.; and Baranger, M. The Rate of Entropy Increase at the Edge of Chaos. 1999. http://xxx.lanl.gov/abs/cond-mat/9907412/.

Lauwerier, H. *Fractals: Endlessly Repeated Geometrical Figures.* Princeton, NJ: Princeton University Press, pp. 119–22, 1991.

May, R. M. "Simple Mathematical Models with Very Complicated Dynamics." *Nature* **261**, 459–67, 1976.

Peitgen, H.-O.; Jürgens, H.; and Saupe, D. *Chaos and Fractals: New Frontiers of Science.* New York: Springer-Verlag, pp. 585–53, 1992.

Rasband, S. N. *Chaotic Dynamics of Nonlinear Systems.* New York: Wiley, p. 23, 1990.

Robinson, C. *Stability, Symbolic Dynamics, and Chaos.* Boca Raton, FL: CRC Press, 1995.

Russell, D. A.; Hanson, J. D.; and Ott, E. "Dimension of Strange Attractors." *Phys. Rev. Let.* **45**, 1175–178, 1980.

Saha, P. and Strogatz, S. H. "The Birth of Period Three." *Math. Mag.* **68**, 42–7, 1995.

Strogatz, S. H. *Nonlinear Dynamics and Chaos.* Reading, MA: Addison-Wesley, 1994.

Tabor, M. *Chaos and Integrability in Nonlinear Dynamics: An Introduction.* New York: Wiley, 1989.

Tsallis, C.; Plastino, A. R.; and Zheng, W.-M. *Chaos, Solitons & Fractals* **8**, 885, 1997.

Trott, M. "Numerical Computations." §1.2.1 in *The Mathematica Guidebook, Vol. 1: Programming in Mathematica.* New York: Springer-Verlag, 2000.

Wagon, S. "The Dynamics of the Quadratic Map." §4.4 in *Mathematica in Action.* New York: W. H. Freeman, pp. 117–40, 1991.

Logistic Equation r = 4

With $r = 4$, the LOGISTIC EQUATION becomes

$$x_{n+1} = 4x_n(1 - x_n), \tag{1}$$

which is equivalent to the TENT MAP with $\mu = 1$. Now let

$$x \equiv \sin^2(\tfrac{1}{2}\pi y) = \tfrac{1}{2}[1 - \cos(\pi y)] \tag{2}$$

$$\sqrt{x} = \sin\left(\tfrac{1}{2}\pi y\right) \tag{3}$$

$$y = \frac{2}{\pi} \sin^{-1}(\sqrt{x}), \tag{4}$$

so

$$\frac{dy}{dx} = \frac{2}{\pi} \frac{1}{\sqrt{1-x}} \tfrac{1}{2} x^{-1/2} = \frac{1}{\pi\sqrt{x(1-x)}}. \tag{5}$$

Manipulating (2) gives

$$\sin^2\left(\tfrac{1}{2}\pi y_{n+1}\right)$$
$$= 4\tfrac{1}{2}[1 - \cos(\pi y_n)]\left\{1 - \tfrac{1}{2}\left[1 - \tfrac{1}{2}(1 - \cos(\pi y_n))\right]\right\}$$
$$= 2[1 - \cos(\pi y = 1 - \cos^2(\pi y_n)\sin^2(\pi y_n), \tag{6}$$

so

$$\tfrac{1}{2}\pi y_{n+1} = \pm y_n + s\pi \tag{7}$$

$$y_{n+1} = \pm 2y_n + \tfrac{1}{2}s. \tag{8}$$

But $y \in [0, 1]$. Taking $y_n \in [0, 1/2]$, then $s = 0$ and

$$y_{n+1} = 2y_n. \tag{9}$$

For $y \in [1/2, 1]$, $s = 1$ and

$$y_{n+1} = 2 - 2y_n. \tag{10}$$

Combining gives

$$y_{n+1} = \begin{cases} 2y_n & \text{for } y_n \in \left[0, \tfrac{1}{2}\right] \\ 2 - 2y_n & \text{for } y_n \in \left[\tfrac{1}{2}, 1\right], \end{cases} \tag{11}$$

which can be written

$$y_{n+1} = 1 - 2\left|x_n - \tfrac{1}{2}\right|, \tag{12}$$

which is just the TENT MAP with $\mu = 1$, whose NATURAL INVARIANT in y is

$$\rho(y) = 1. \tag{13}$$

Transforming back to x therefore gives

$$\rho(x) = \left|\frac{dy}{dx}\right|\rho(y(x)) = \frac{2}{\pi} \frac{1}{\sqrt{1-x}} \tfrac{1}{2} x^{-1/2}$$

$$= \frac{1}{\pi\sqrt{x(1-x)}}. \tag{14}$$

This can also be derived from

$$\rho(x) = \lim_{N \to \infty} \frac{1}{N} \sum_{i=1}^{N} \delta(x_i - x) = \frac{1}{\pi\sqrt{x(1-x)}}, \tag{15}$$

where $\delta(x)$ is the DELTA FUNCTION.

See also LOGISTIC EQUATION, TENT MAP

References

Jaffe, S. "The Logistic Equation: Computable Chaos." http://www.mathsource.com/cgi-bin/msitem?0204–13.

Whittaker, J. V. "An Analytical Description of Some Simple Cases of Chaotic Behavior." *Amer. Math. Monthly* **98**, 489–04, 1991.

Logistic Growth Curve

The POPULATION GROWTH law which arises frequently in biology and is given by the differential equation

$$\frac{dN}{dt} = \frac{r(K - N)}{K}, \tag{1}$$

where r is the MALTHUSIAN PARAMETER and K is the so-called CARRYING CAPACITY (i.e., the maximum sustainable population). Rearranging and integrating both sides gives

$$\int_{N_0}^{N} \frac{dN}{K - N} = \frac{r}{K} \int_{0}^{t} dt \tag{2}$$

$$\ln\left(\frac{N_0 - K}{N - K}\right) = \frac{r}{K} t \tag{3}$$

$$N(t) = K + (N_0 - K)e^{-rt/K}. \tag{4}$$

The curve

$$y = \frac{a}{1 + bq^x} \tag{5}$$

is sometimes also known as the logical curve.

See also GOMPERTZ CURVE, LAW OF GROWTH, LIFE EXPECTANCY, LOGISTIC EQUATION, MAKEHAM CURVE, MALTHUSIAN PARAMETER, POPULATION GROWTH

References

Pearl, R. Ch. 18 in *The Biology of Population Growth.* New York: Knopf, 1978.

Logistic Map
LOGISTIC EQUATION

Logit Transformation

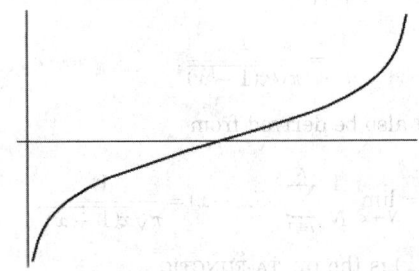

The function

$$z = f(x) = \ln\left(\frac{x}{1-x}\right).$$

This function has an inflection point at $x = 1/2$, where

$$f''(x) = \frac{2x-1}{x^2(x-1)^2} = 0.$$

Applying the logit transformation to values obtained by iterating the LOGISTIC EQUATION generates a sequence of RANDOM NUMBERS having distribution

$$P_z = \frac{1}{\pi(e^{x/2} + e^{-x/2})},$$

which is very close to a GAUSSIAN DISTRIBUTION.

References
Collins, J.; Mancilulli, M.; Hohlfeld, R.; Finch, D.; Sandri, G.; and Shtatland, E. "A Random Number Generator Based on the Logit Transform of the Logistic Variable." *Computers in Physics* **6**, 630–32, 1992.
Pickover, C. A. *Keys to Infinity.* New York: Wiley, pp. 244–45, 1995.

Logos
A generalization of a HEYTING ALGEBRA which replaces BOOLEAN ALGEBRA in "intuitionistic" LOGIC.

See also TOPOS

Log-Series Distribution
The terms in the series expansion of $\ln(1-\theta)$ about $\theta = 0$ are proportional to this distribution.

$$P(n) = -\frac{\theta^n}{n \ln(1-\theta)} \qquad (1)$$

$$D(n) \equiv \sum_{i=1}^{n} P(i) = \frac{\theta^{1+n}\Phi(\theta, 1, 1+n) + \ln(1-\theta)}{\ln(1-\theta)}, \qquad (2)$$

where Φ is the LERCH TRANSCENDENT. The MEAN, VARIANCE, SKEWNESS, and KURTOSIS

$$\mu = \frac{\theta}{(\theta-1)\ln(1-\theta)} \qquad (3)$$

$$\sigma^2 = -\frac{\theta[\theta + \ln(1-\theta)]}{(\theta-1)^2[\ln(1-\theta)]^2} \qquad (4)$$

$$\gamma_1 = \frac{2\theta^2 + 3\theta\ln(1-\theta) + (1+\theta)\ln^2(1-\theta)}{\ln(1-\theta)[\theta + \ln(1-\theta)]\sqrt{-\theta[\theta + \ln(1-\theta)]}}$$
$$\times \ln(1-\theta) \qquad (5)$$

$$\gamma_2 = \frac{6\theta^3 + 12\theta^2\ln(1-\theta) + \theta(7+4\theta)\ln^2(1-\theta)}{\theta[\theta + \ln(1-\theta)]^2}$$
$$+ \frac{(1+4\theta+\theta^2)\ln^3(1-\theta)}{\theta[\theta + \ln(1-\theta)]^2}. \qquad (6)$$

Log-Weibull Distribution
FISHER-TIPPETT DISTRIBUTION

Lommel Differential Equation
A generalization of the BESSEL DIFFERENTIAL EQUATION

$$z^2 \frac{d^2y}{dz^2} + z\frac{dy}{dz} - (z^2 + v^2)y = kz^{\mu+1}$$

(Watson 1966, p. 345; Zwillinger 1997, p. 125; Gradshteyn and Ryzhik 2000, p. 986). A further generalization gives

$$z^2 \frac{d^2y}{dz^2} + z\frac{dy}{dz} - (z^2 + v^2)y = \pm kz^{\mu+1}.$$

The solutions are LOMMEL FUNCTIONS.

See also LOMMEL FUNCTION

References
Gradshteyn, I. S. and Ryzhik, I. M. *Tables of Integrals, Series, and Products, 6th ed.* San Diego, CA: Academic Press, p. 986, 2000.
Watson, G. N. *A Treatise on the Theory of Bessel Functions, 2nd ed.* Cambridge, England: Cambridge University Press, 1966.
Zwillinger, D. *Handbook of Differential Equations, 3rd ed.* Boston, MA: Academic Press, p. 125, 1997.

Lommel Function
There are several functions called "Lommel functions." One type of Lommel function is the solution to the LOMMEL DIFFERENTIAL EQUATION with a PLUS SIGN, given by

$$y = ks_{\mu,v}(z), \qquad (1)$$

where

$$s_{\mu, \nu}^{(+)}(z) \equiv \frac{1}{2\pi}\left[Y_\nu(z) \int_0^z z^\mu J_\nu(z)\, dz - J_\nu(z) \int_0^z z^\mu Y_\nu(z)\, dz \right]. \tag{2}$$

Here, $J_\nu(z)$ and $Y_\nu(z)$ are BESSEL FUNCTIONS OF THE FIRST and SECOND KINDS (Watson 1966, p. 346). If a minus sign precedes k, then the solution is

$$s_{\mu, \nu}^{-} \equiv I_\nu(z) \int_z^{c_1} z^\mu K_\nu(z)\, dz - J_\nu(z) \int_{c_2}^z z^\mu I_\nu(z)\, dz, \tag{3}$$

where $K_\nu(z)$ and $I_\nu(z)$ are MODIFIED BESSEL FUNCTIONS OF THE FIRST and SECOND KINDS.

Lommel functions of two variables are related to the BESSEL FUNCTION OF THE FIRST KIND and arise in the theory of diffraction and, in particular, Mie scattering (Watson 1966, p. 537),

$$U_n(w, z) = \sum_{m=0}^\infty (-1)^m \left(\frac{w}{z}\right)^{n+2m} J_{n+2m}(z) \tag{4}$$

$$V_n(w, z) = \sum_{m=0}^\infty (-1)^m \left(\frac{w}{z}\right)^{-n-2m} J_{-n-2m}(z). \tag{5}$$

See also LOMMEL DIFFERENTIAL EQUATION, LOMMEL POLYNOMIAL

References

Chandrasekhar, S. *Radiative Transfer.* New York: Dover, p. 369, 1960.

Prudnikov, A. P.; Marichev, O. I.; and Brychkov, Yu. A. "The Lommel Functions $s_{\mu, \nu}(x)$ and $S_{\mu, \nu}(x)$." §1.5 in *Integrals and Series, Vol. 3: More Special Functions.* Newark, NJ: Gordon and Breach, pp. 28–9, 1990.

Watson, G. N. *A Treatise on the Theory of Bessel Functions, 2nd ed.* Cambridge, England: Cambridge University Press, 1966.

Lommel Polynomial

$$R_{m, \nu}(z) =$$

$$\frac{\Gamma(\nu + m)}{\Gamma(\nu)(z/2)^m} {}_2F_3(\tfrac{1}{2}(1-m), -\tfrac{1}{2}m;\ \nu, -m,\ 1 - \nu - m;\ z^2)$$

$$\times \frac{\pi z}{2\sin(\nu\pi)} [J_{\nu+m}(z)J_{-\nu+1}(z) + (-1)^m J_{-\nu-m}(z)J_{\nu-1}(z)],$$

where $\Gamma(z)$ is a GAMMA FUNCTION, $J_n(x)$ is a BESSEL FUNCTION OF THE FIRST KIND, and ${}_2F_3(a, b;\ c, d, e;\ z)$ is a GENERALIZED HYPERGEOMETRIC FUNCTION.

See also LOMMEL FUNCTION

References

Iyanaga, S. and Kawada, Y. (Eds.). *Encyclopedic Dictionary of Mathematics.* Cambridge, MA: MIT Press, p. 1477, 1980.

Lommel's Integrals

$$(\beta^2 - \alpha^2) \int x J_n(\alpha x) J_n(\beta x)\, dx$$

$$= x[\alpha J_n'(\alpha x)J_n(\beta x) - \beta J_n'(\beta x)J_n(\alpha x)]$$

$$\int x J_n^2(\alpha x)\, dx = \tfrac{1}{2} x^2 [J_n^2(\alpha x) + J_{n-1}(\alpha x)J_{n+1}(\alpha x)],$$

where $J_n(x)$ is a BESSEL FUNCTION OF THE FIRST KIND.

References

Bowman, F. *Introduction to Bessel Functions.* New York: Dover, p. 101, 1958.

Long Cross

DAGGER

Long Division

```
     7              72              726
17 )123456.     17 )123456.     17 )123456.
   -119            -119            -119
    44              44              44
                   -34             -34
                   105             105
                                  -102
                                    36
```

```
     7262.1              7262.11...
17 )123456.0        17 )123456.00
   -119                 -119
    44                   44
   -34                  -34
    105                  105
   -102                 -102
     36                   36
    -34                  -34
     20                   20
                         -17
                          30
```

Long division is an algorithm for dividing two numbers, obtaining the QUOTIENT one DIGIT at a time. The above example shows how the division of $123456/17$ is performed to obtain the result $7262.11...$.

See also DIVISION

References

Beck, G. "Long Multiplication and Division." MATHEMATICA NOTEBOOK LONGDIVISION.NB.

Longest Increasing Scattered Subsequence

The longest increasing scattered subsequence is the longest subsequence of increasing terms, where intervening nonincreasing terms may be dropped. Finding the largest scattered subsequence is a much harder problem. The longest increasing scattered subsequence of a PARTITION can be found using LongestIncreasingSubsequence[p] in the *Mathematica* add-on package DiscreteMath'Combinatorica` (which can be loaded with the command `< <DiscreteMath`). For example, the longest increasing scattered subsequence of the PERMUTATION {6, 3, 4, 8, 10, 5, 7, 1, 9, 2} is {3, 4, 5, 7, 9}, whereas the longest contiguous subsequence is {3, 4, 8, 10}.

Any sequence of $n^2 + 1$ distinct integers must contain either an increasing or decreasing scattered subsequence of length $n + 1$ (Erdos and Szekeres 1935; Skiena 1990, p. 75).

See also LONGEST INCREASING SUBSEQUENCE, PERMUTATION

References

Erdos, P. and Szekeres, G. "A Combinatorial Problem in Geometry." *Compos. Math.* **2**, 464–70, 1935.
Schensted, C. "Longest Increasing and Decreasing Subsequences." *Canad. J. Math.* **13**, 179–91, 1961.
Skiena, S. "Longest Increasing Subsequences." §2.3.6 in *Implementing Discrete Mathematics: Combinatorics and Graph Theory with Mathematica.* Reading, MA: Addison-Wesley, pp. 73–5, 1990.

Longest Increasing Subsequence

The longest increasing subsequence of a given sequence is the subsequence of increasing terms containing the largest number of elements. For example, the longest increasing subsequence of the PERMUTATION {6, 3, 4, 8, 10, 5, 7, 1, 9, 2} is {3, 4, 8, 10}.

See also LONGEST INCREASING SCATTERED SUBSEQUENCE

References

Skiena, S. "Longest Increasing Subsequences." §2.3.6 in *Implementing Discrete Mathematics: Combinatorics and Graph Theory with Mathematica.* Reading, MA: Addison-Wesley, pp. 73–5, 1990.

Long Exact Sequence

See also LONG EXACT SEQUENCE OF A PAIR AXIOM

Long Exact Sequence of a Pair Axiom

One of the EILENBERG-STEENROD AXIOMS. It states that, for every pair (X, A), there is a natural long exact sequence

$$\ldots \to H_n(A) \to H_n(X) \to H_n(X, A) \to H_{n-1}(A)$$
$$\to \ldots , \tag{1}$$

where the MAP $H_n(A) \to H_n(X)$ is induced by the INCLUSION MAP $A \to X$ and $H_n(X) \to H_n(X, A)$ is induced by the INCLUSION MAP $(X, \phi) \to (X, A)$. The MAP $H_n(X, A) \to H_{n-1}(A)$ is called the BOUNDARY MAP.

See also EILENBERG-STEENROD AXIOMS

Longimeter

A longimeter is a transparent sheet of plastic with a regular grid of lines inclined at an angle of 30° to the sides of the sheet. By counting the number of squares occupied by a linear feature on a map (such as a river) for six different rotations of the sheet, the length of the feature can be determined.

See also COASTLINE PARADOX

References

Steinhaus, H. *Mitteilungen der Sächsischen Akad.* **82**, 120–30, 1930.
Steinhaus, H. *Przeglad Geogr.* **21**, 1947.
Steinhaus, H. *Comptes Rendus Soc. des Sciences et des Lettres de Wroclaw, Sér. B,* 1949.
Steinhaus, H. *Mathematical Snapshots, 3rd ed.* New York: Dover, pp. 105–10, 1999.

Longitude

The azimuthal coordinate on the surface of a SPHERE (θ in SPHERICAL COORDINATES) or on a SPHEROID (in PROLATE or OBLATE SPHEROIDAL COORDINATES). Longitude is defined such that $0° = 360°$. Lines of constant longitude are generally called MERIDIANS. The other angular coordinate on the surface of a SPHERE is called the LATITUDE.

The shortest distance between any two points on a SPHERE is the so-called GREAT CIRCLE distance, which can be directly computed from the LATITUDE and longitudes of two points.

See also GREAT CIRCLE, LATITUDE, MERIDIAN, OBLATE SPHEROIDAL COORDINATES, PROLATE SPHEROIDAL COORDINATES

Longitudinal Data

Data resulting from the observation of a population on a number of variables over time. Whenever observations are made more than once, the data is considered to be longitudinal.

References

Bijleveld, C. C. J. H.; van der Kamp, L. J. T.; Mooijaart, A.; van der Kloot, W. A.; van der Leeden, R.; and van der Burg, E. *Longitudinal Data Analysis: Designs, Models and Methods.* London: Sage, 1998.

Long Prime

FULL REPTEND PRIME

Look and Say Sequence

The INTEGER SEQUENCE beginning with a single digit in which the next term is obtained by describing the previous term. Starting with 1, the sequence would be defined by "1, one 1, two 1s, one 2 one 1," etc., and the result is 1, 11, 21, 1211, 111221, 312211, 13112221, 1113213211, ... (Sloane's A005150).

Starting the sequence instead with the digit d for $2 \leq d \leq 9$ gives d, $1d$, $111d$, $311d$, $13211d$, $111312211d$, $31131122211d$, $1321132132211d$, ... The sequences for $d = 2$ and 3 are Sloane's A006751 and A006715. n

terms of the look and say sequence (given as lists of digits) starting with digit d can be implemented in *Mathematica* as follows.

```
RunLengthEncode[x_List]  :=  (Through[{First,
Length}[#]]        &)        /@        Split[x]
LookAndSay[n_Integer?Positive,    d_:1]    :=
NestList[Flatten[Reverse                   /@
RunLengthEncode[#]] &, {d}, n - 1]
```

The number of DIGITS in the nth term the sequence for $1 \le d \le 9$ is given by the sequence 1, 2, 2, 4, 6, 6, 8, 10, 14, 20, 26, 34, 46, 62, ... (Sloane's A005341), which is asymptotic to $C\lambda^n$, where C is a constant and

$$\lambda = 1.303577269034296\ldots$$

(Sloane's A014715) is CONWAY'S CONSTANT, given by the unique positive real root of the POLYNOMIAL

$$\begin{aligned}
0 = &\, x^{71} - x^{69} - 2x^{68} - x^{67} + 2x^{66} + 2x^{65} + x^{64} - x^{63} - x^{62} \\
&- x^{61} - x^{60} - x^{59} + 2x^{58} + 5x^{57} + 3x^{56} - 2x^{55} - 10x^{54} \\
&- 3x^{53} - 2x^{52} + 6x^{51} + 6x^{50} + x^{49} + 9x^{48} - 3x^{47} \\
&- 7x^{46} - 8x^{45} - 8x^{44} + 10x^{43} + 6x^{42} + 8x^{41} - 4x^{40} \\
&- 12x^{39} + 7x^{38} - 7x^{37} + 7x^{36} + x^{35} - 3x^{34} + 10x^{33} \\
&+ x^{32} - 6x^{31} - 2x^{30} - 10x^{29} - 3x^{28} + 2x^{27} + 9x^{26} \\
&- 3x^{25} + 14x^{24} - 8x^{23} - 7x^{21} + 9x^{20} - 3x^{19} - 4x^{18} \\
&- 10x^{17} - 7x^{16} + 12x^{15} + 7x^{14} + 2x^{13} - 12x^{12} - 4x^{11} \\
&- 2x^{10} - 5x^9 + x^7 - 7x^6 + 7x^5 - 4x^4 + 12x^3 - 6x^2 \\
&+ 3x - 6.
\end{aligned}$$

In fact, the constant is even more general than this, applying to *all* starting sequences (i.e., even those starting with arbitrary starting digits), with the exception of 22, a result which follows from the COSMOLOGICAL THEOREM. Conway discovered that strings sometimes factor as a concatenation of two strings whose descendants never interfere with one another. A string with no nontrivial splittings is called an "element," and other strings are called "compounds." Every string of 1s, 2s, and 3s eventually "decays" into a compound of 92 special elements, named after the chemical elements.

See also CONWAY'S CONSTANT, COSMOLOGICAL THEOREM, RUN-LENGTH ENCODING

References

Conway, J. H. "The Weird and Wonderful Chemistry of Audioactive Decay." *Eureka* **45**, 5–8, 1985.
Conway, J. H. "The Weird and Wonderful Chemistry of Audioactive Decay." §5.11 in *Open Problems in Communications and Computation.* (Ed. T. M. Cover and B. Gopinath). New York: Springer-Verlag, pp. 173–88, 1987.
Conway, J. H. and Guy, R. K. "The Look and Say Sequence." In *The Book of Numbers.* New York: Springer-Verlag, pp. 208–09, 1996.
Hilgemeier, M. "Die Gleichniszahlen-Reihe." *Bild der Wissensch.* **12**, 19, 1986.
Hilgemeier, M. "'One Metaphor Fits All': A Fractal Voyage with Conway's Audioactive Decay." Ch. 7 in Pickover,

C. A. (Ed.). *Fractal Horizons: The Future Use of Fractals.* New York: St. Martin's Press, 1996.
Sloane, N. J. A. Sequences A005150/M4780, A005341/M0321, A006715/M2965, and A006751/M2052 in "An On-Line Version of the Encyclopedia of Integer Sequences." http://www.research.att.com/~njas/sequences/eisonline.html.
Vardi, I. *Computational Recreations in Mathematica.* Reading, MA: Addison-Wesley, pp. 13–4, 1991.

Loop

A path whose initial and final points coincide in a fixed point p known as the BASEPOINT.

Loop (Algebra)

A QUASIGROUP with an IDENTITY ELEMENT e such that $xe = x$ and $ex = x$ for any x in the QUASIGROUP. All GROUPS are loops.

See also GROUP, QUASIGROUP

References

Albert, A. A. (Ed.). *Studies in Modern Algebra.* Washington, DC: Math. Assoc. Amer., 1963.

Loop (Graph)

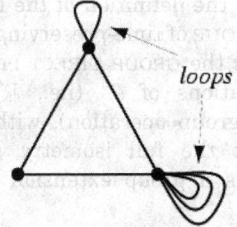

loops

A degenerate edge of a graph which joins a vertex to itself, also called a self-loop. A SIMPLE GRAPH cannot contain any loops, but a PSEUDOGRAPH can contain both multiple edges and loops.

See also PSEUDOGRAPH, SIMPLE GRAPH

References

Skiena, S. *Implementing Discrete Mathematics: Combinatorics and Graph Theory with Mathematica.* Reading, MA: Addison-Wesley, p. 82, 1990.

Loop (Knot)

A KNOT or HITCH which holds its form rigidly.

References

Owen, P. *Knots.* Philadelphia, PA: Courage, p. 35, 1993.

Loop Space

Let Y^X be the set of continuous mappings $f : X \to Y$. Then the TOPOLOGICAL SPACE for Y^X supplied with a compact-open topology is called a MAPPING SPACE, and if $Y = I$ is taken as the interval $(0, 1)$, then $Y^I = \Omega(Y)$ is called a loop space (or SPACE OF CLOSED PATHS).

See also MACHINE, MAPPING SPACE, MAY-THOMASON UNIQUENESS THEOREM

References

Brylinski, J.-L. *Loop Spaces, Characteristic Classes and Geometric Quantization.* Boston, MA: Birkhäuser, 1993.

Iyanaga, S. and Kawada, Y. (Eds.). *Encyclopedic Dictionary of Mathematics.* Cambridge, MA: MIT Press, p. 658, 1980.

Lopez Minimal Surface

See also MINIMAL SURFACE

Lorentz Group

The Lorentz group is the GROUP L of time-preserving linear ISOMETRIES of MINKOWSKI SPACE $\mathbb{R}^4$ with the pseudo-Riemannian metric

$$dr^2 = -dt^2 + dx^2 + dy^2 + dz^2.$$

It is also the GROUP of ISOMETRIES of 3-D HYPERBOLIC SPACE. It is time-preserving in the sense that the unit time VECTOR $(1, 0, 0, 0)$ is sent to another VECTOR (t, x, y, z) such that $t > 0$.

A consequence of the definition of the Lorentz group is that the full GROUP of time-preserving isometries of MINKOWSKI $\mathbb{R}^4$ is the GROUP DIRECT PRODUCT of the group of translations of $\mathbb{R}^4$ (i.e., $\mathbb{R}^4$ itself, with addition as the group operation), with the Lorentz group, and that the full isometry group of the MINKOWSKI $\mathbb{R}^4$ is a group extension of $\mathbb{Z}_2$ by the product $L \otimes \mathbb{R}^4$.

The Lorentz group is invariant under space rotations and LORENTZ TRANSFORMATIONS.

See also LORENTZ TENSOR, LORENTZ TRANSFORMATION

References

Arfken, G. "Homogeneous Lorentz Group." §4.13 in *Mathematical Methods for Physicists, 3rd ed.* Orlando, FL: Academic Press, pp. 271–75, 1985.

Lorentz Tensor

The TENSOR in the LORENTZ TRANSFORMATION given by

$$\mathsf{L} \equiv \begin{bmatrix} \gamma & -\gamma\beta & 0 & 0 \\ -\gamma\beta & \gamma & 0 & 0 \\ 0 & 0 & 1 & 0 \\ 0 & 0 & 0 & 1 \end{bmatrix}, \tag{1}$$

where beta and gamma are defined by

$$\beta \equiv \frac{v}{c} \tag{2}$$

$$\gamma \equiv \frac{1}{\sqrt{1 - \beta^2}}. \tag{3}$$

See also LORENTZ GROUP, LORENTZ TRANSFORMATION

Lorentz Transformation

A 4-D transformation satisfied by all FOUR-VECTORS a^v,

$$a'^\mu = \Lambda^\mu_v a^v. \tag{1}$$

In the theory of special relativity, the Lorentz transformation replaces the GALILEAN TRANSFORMATION as the valid transformation law between reference frames moving with respect to one another at constant VELOCITY. Let x^v be the POSITION FOUR-VECTOR with $x^0 = ct$, and let the relative motion be along the x^1 axis with VELOCITY v. Then (1) becomes

$$x'^\mu = \Lambda^\mu_v x^v, \tag{2}$$

where the LORENTZ TENSOR is given by

$$\mathsf{L} = \begin{bmatrix} \Lambda^0_0 & \Lambda^0_1 & \Lambda^0_2 & \Lambda^0_3 \\ \Lambda^1_0 & \Lambda^1_1 & \Lambda^1_2 & \Lambda^1_3 \\ \Lambda^2_0 & \Lambda^2_1 & \Lambda^2_2 & \Lambda^2_3 \\ \Lambda^3_0 & \Lambda^3_1 & \Lambda^3_2 & \Lambda^3_3 \end{bmatrix} \equiv \begin{bmatrix} \gamma & -\gamma\beta & 0 & 0 \\ -\gamma\beta & \gamma & 0 & 0 \\ 0 & 0 & 1 & 0 \\ 0 & 0 & 0 & 1 \end{bmatrix}. \tag{3}$$

Here,

$$\beta \equiv \frac{v}{c} \tag{4}$$

$$\gamma \equiv \frac{1}{\sqrt{1 - \beta^2}}. \tag{5}$$

Written explicitly, the transformation between x^v and $x^{v'}$ coordinate is

$$x^{0'} = \gamma(x^0 - \beta x^1) \tag{6}$$

$$x^{1'} = \gamma(x^1 - \beta x^0) \tag{7}$$

$$x^{2'} = x^2 \tag{8}$$

$$x^{3'} = x^3. \tag{9}$$

The DETERMINANT of the upper left 2×2 MATRIX in (3) is

$$D = (\gamma)^2 - (-\gamma\beta)^2 = \gamma^2(1 - \beta^2) = \frac{\gamma^2}{\gamma^2} = 1, \tag{10}$$

so

$$\mathsf{L}^{-1} = \begin{bmatrix} (\Lambda^{-1})^0_0 & (\Lambda^{-1})^0_1 & (\Lambda^{-1})^0_2 & (\Lambda^{-1})^0_3 \\ (\Lambda^{-1})^1_0 & (\Lambda^{-1})^1_1 & (\Lambda^{-1})^1_2 & (\Lambda^{-1})^1_3 \\ (\Lambda^{-1})^2_0 & (\Lambda^{-1})^2_1 & (\Lambda^{-1})^2_2 & (\Lambda^{-1})^2_3 \\ (\Lambda^{-1})^3_0 & (\Lambda^{-1})^3_1 & (\Lambda^{-1})^3_2 & (\Lambda^{-1})^3_3 \end{bmatrix}$$

$$\equiv \begin{bmatrix} \gamma & \gamma\beta & 0 & 0 \\ \gamma\beta & \gamma & 0 & 0 \\ 0 & 0 & 1 & 0 \\ 0 & 0 & 0 & 1 \end{bmatrix}. \qquad (11)$$

A Lorentz transformation along the x^1-axis can also be written

$$\begin{bmatrix} x^{0'} \\ x^{1'} \\ x^{2'} \\ x^{3'} \end{bmatrix} \begin{bmatrix} \cosh\theta & -\sinh\theta & 0 & 0 \\ -\sinh\theta & \cosh\theta & 0 & 0 \\ 0 & 0 & 1 & 0 \\ 0 & 0 & 0 & 1 \end{bmatrix} \begin{bmatrix} x^0 \\ x^1 \\ x^2 \\ x^3 \end{bmatrix}. \qquad (12)$$

where θ is called the rapidity,

$$x^0 \equiv ct, \qquad (13)$$

and

$$\tanh\theta \equiv \beta \equiv \frac{v}{c} \qquad (14)$$

$$\cosh\theta \equiv \gamma \equiv \frac{1}{\sqrt{1 - \beta^2}} \qquad (15)$$

$$\sinh\theta = \gamma\beta. \qquad (16)$$

See also HYPERBOLIC ROTATION, LORENTZ GROUP, LORENTZ TENSOR

References

Fraundorf, P. "Accel-1D: Frame-Dependent Relativity at UM-StL." http://www.umsl.edu/~fraundor/a1toc.html.
Griffiths, D. J. *Introduction to Electrodynamics.* Englewood Cliffs, NJ: Prentice-Hall, pp. 412–14, 1981.
Morse, P. M. and Feshbach, H. "The Lorentz Transformation, Four-Vectors, Spinors." §1.7 in *Methods of Theoretical Physics, Part I.* New York: McGraw-Hill, pp. 93–07, 1953.

Lorentzian Distribution

CAUCHY DISTRIBUTION

Lorentzian Function

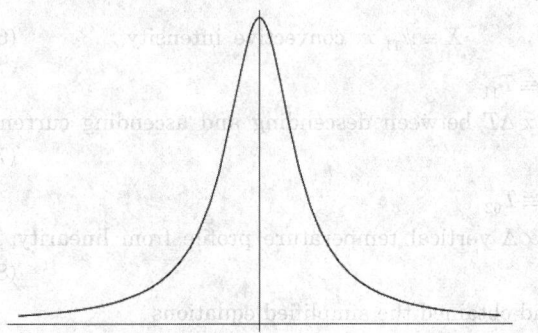

The Lorentzian function is the singly peaked function

given by

$$L(x) = \frac{1}{\pi} \frac{\frac{1}{2}\Gamma}{(x - x_0)^2 + \left(\frac{1}{2}\Gamma\right)^2}. \qquad (1)$$

It is normalized to that

$$\int_{-\infty}^{\infty} L(x) = 1. \qquad (2)$$

It has a maximum at $x = x_0$, where

$$L'(x) = -\frac{16(x - x_0)\Gamma}{\pi[4(x - x_0)^2 + \Gamma^2]} = 0. \qquad (3)$$

Its value at the maximum is

$$L(x_0) = \frac{2}{\pi\Gamma}. \qquad (4)$$

It is equal to half its maximum at

$$x = \left(x_0 \pm \tfrac{1}{2}\Gamma\right), \qquad (5)$$

and so has FULL WIDTH AT HALF MAXIMUM Γ. The function has inflection points at

$$L''(x) = 16\,\Gamma\,\frac{12(x - x_0)^2 - \Gamma^2}{\pi[4(x - x_0)^2 + \Gamma^2]} = 0, \qquad (6)$$

giving

$$x_1 = x_0 - \tfrac{1}{6}\sqrt{3}\,\Gamma, \qquad (7)$$

where

$$L(x_1) = \frac{3}{2\pi\Gamma}. \qquad (8)$$

The Lorentzian function gives the shape of certain types of spectral lines and is the distribution function in the CAUCHY DISTRIBUTION. The Lorentzian function has FOURIER TRANSFORM

$$\mathcal{F}\left[\frac{1}{\pi}\frac{\frac{1}{2}\Gamma}{(x - x_0)^2 + (\frac{1}{2}\Gamma)^2}\right] = e^{-2\pi i k x_0 - \Gamma\pi|k|}. \qquad (9)$$

See also CAUCHY DISTRIBUTION, DAMPED EXPONENTIAL COSINE INTEGRAL, FOURIER TRANSFORM–LORENTZIAN FUNCTION

Lorentzian Inner Product

The standard Lorentzian inner product on $\mathbb{R}^4$ is given by

$$-dx_0^2 + dx_1^2 + dx_2^2 + dx_3^2, \qquad (1)$$

i.e, for vectors **v** and **w**,

$$\langle \mathbf{v}, \mathbf{w} \rangle = -v_0 w_0 + v_1 w_1 + v_2 w_2 + v_3 w_3. \qquad (2)$$

The Lorentzian inner product is used in special relativity as a measurement, replacing distances, which is independent of reference frame. The variables x_1, x_2, and x_3 can be thought of as space variables, and the x_0 variable as the time variable. Sometimes, the time variable is labelled t instead of x_0 and when used in special relativity, $x_0 = ct$, where c is the speed of light. The formula (1) uses the convention that units are chosen so that the speed of light has the value $c = 1$ in order to simplify formulas.

For a vector $\mathbf{v}$, the sign of $\langle \mathbf{v}, \mathbf{v} \rangle$ determines the type of $\mathbf{v}$. If it is positive, then $\mathbf{v}$ is a space-like vector. If it is zero, then $\mathbf{v}$ is called a null vector, or light-like vector. If it is negative, then $\mathbf{v}$ is called a time-like vector. After a change of variables, it is possible to rewrite the Lorentzian inner product as above where t is in the direction of a given time-like vector v with $\langle \mathbf{v}, \mathbf{v} \rangle = -1$. Such a change of variables corresponds to a change in reference frame. Altogether, these form the LORENTZ GROUP, also called the ORTHOGONAL GROUP $O(3, 1)$.

See also ORTHOGONAL GROUP

Lorenz Asymmetry Coefficient

This entry contributed by CHRISTIAN DAMGAARD

The Lorenz asymmetry coefficient is a summary statistic of the Lorenz curve that measures the degree of asymmetry of a LORENZ CURVE. The Lorenz asymmetry coefficient is defined as

$$S \equiv F(\mu) + L(\mu), \tag{1}$$

where the functions F and L are defined as for the Lorenz curve. If $S > 1$, then the point where the LORENZ CURVE is parallel with the line of equality is above the axis of symmetry. Correspondingly, if $S < 1$, then the point where the LORENZ CURVE is parallel to the line of equality is below the axis of symmetry.

The sample statistic S can be calculated from ordered size data using the following equations

$$\delta = \frac{\mu - x'_m}{x'_{m+1} - x'_m} \tag{2}$$

$$F(\mu) = \frac{m + \delta}{n} \tag{3}$$

$$L(\mu) = \frac{L_m + \delta x'_{m+1}}{L_n}, \tag{4}$$

where m is the number of individuals with a size less than μ.

See also GINI COEFFICIENT, LORENZ CURVE

References
Damgaard, C. and Weiner, J. "Describing Inequality in Plant Size or Fecundity." *Ecology* **81**, 1139–142, 2000.

Lorenz Attractor

The Lorenz attractor is a STRANGE ATTRACTOR that arises in a simplified system of equations describing the 2-D flow of fluid of uniform depth H, with an imposed temperature difference ΔT, under gravity g, with buoyancy α, thermal diffusivity κ, and kinematic viscosity v. The full equations are

$$\frac{\partial}{\partial t}(\nabla^2 \phi) = \frac{\partial \psi}{\partial z}\frac{\partial}{\partial x}(\nabla^2 \psi) - \frac{\partial \psi}{\partial x}\frac{\partial}{\partial z}(\nabla^2 \psi) + v\nabla^2(\nabla^2 \psi)$$
$$+ g\alpha \frac{dT}{dx} \tag{1}$$

$$\frac{\partial T}{\partial t} = \frac{\partial T}{\partial z}\frac{\partial \psi}{\partial x} - \frac{\partial \theta}{\partial x}\frac{\partial \psi}{\partial z} + \kappa\nabla^2 T + \frac{\Delta T}{H}\frac{\partial \psi}{\partial x}. \tag{2}$$

Here, ψ is the "stream function," as usual defined such that

$$u = \frac{\partial \psi}{\partial x}, \quad v = \frac{\partial \psi}{\partial x}. \tag{3}$$

In the early 1960s, Lorenz accidentally discovered the chaotic behavior of this system when he found that, for a simplified system, periodic solutions OF THE FORM

$$\psi = \psi_0 \sin\left(\frac{\pi a x}{H}\right)\sin\left(\frac{\pi z}{H}\right) \tag{4}$$

$$\theta = \theta_0 \cos\left(\frac{\pi a x}{H}\right)\sin\left(\frac{\pi z}{H}\right) \tag{5}$$

grew for Rayleigh numbers larger than the critical value, $Ra > Ra_c$. Furthermore, vastly different results were obtained for very small changes in the initial values, representing one of the earliest discoveries of the so-called BUTTERFLY EFFECT.

Lorenz included the following terms in his system of equations,

$$X \equiv \psi_{11} \propto \text{ convective intensity} \tag{6}$$

$Y \equiv T_{11}$
$$\propto \Delta T \text{ between descending and ascending currents} \tag{7}$$

$Z \equiv T_{02}$
$$\propto \Delta \text{ vertical temperature profile from linearity,} \tag{8}$$

and obtained the simplified equations

$$\dot{X} = \sigma(Y - X) \tag{9}$$

$$\dot{Y} = -XZ + rX - Y \qquad (10)$$

$$\dot{Z} = XY - bZ, \qquad (11)$$

now known as the LORENZ EQUATIONS, where $\dot{X} = dX/dt$, $\dot{Y} = dY/dt$, $\dot{Z} = dZ/dt$, and

$$\sigma \equiv \frac{v}{\kappa} = \text{Prandtl number} \qquad (12)$$

$$r \equiv \frac{Ra}{Ra_c} = \text{normalized Rayleigh number} \qquad (13)$$

$$b \equiv \frac{4}{1 + a^2} = \text{geometric factor.} \qquad (14)$$

Lorenz took $b \equiv 8/3$ and $\sigma \equiv 10$.

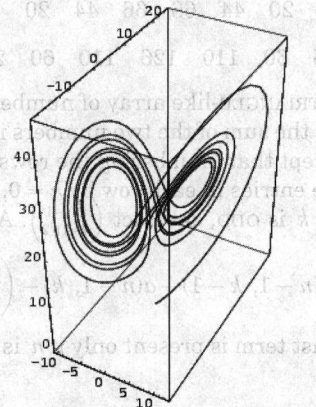

The CRITICAL POINTS at (0, 0, 0) correspond to no convection, and the CRITICAL POINTS at

$$\left(\sqrt{b(r-1)}, \ \sqrt{b(r-1)}, \ r-1 \right) \qquad (15)$$

and

$$\left(-\sqrt{b(r-1)}, -\sqrt{b(r-1)}, \ r-1 \right) \qquad (16)$$

correspond to steady convection. This pair is stable only if

$$r = \frac{\sigma(\sigma + b + 3)}{\sigma - b - 1}, \qquad (17)$$

which can hold only for POSITIVE r if $\sigma > b + 1$. The Lorenz attractor has a CORRELATION EXPONENT of 2.05 ± 0.01 and CAPACITY DIMENSION 2.06 ± 0.01 (Grassberger and Procaccia 1983). For more details, see Lichtenberg and Lieberman (1983, p. 65) and Tabor (1989, p. 204).

See also BUTTERFLY EFFECT, LORENZ EQUATIONS, RÖSSLER MODEL

References

Gleick, J. *Chaos: Making a New Science.* New York: Penguin Books, pp. 27–1, 1988.

Grassberger, P. and Procaccia, I. "Measuring the Strangeness of Strange Attractors." *Physica D* **9**, 189–08, 1983.
Lichtenberg, A. and Lieberman, M. *Regular and Stochastic Motion.* New York: Springer-Verlag, 1983.
Lorenz, E. N. "Deterministic Nonperiodic Flow." *J. Atmos. Sci.* **20**, 130–41, 1963.
Lorenz, E. N. "On the Prevalence of Aperiodicity in Simple Systems." In *Global Analysis: Proceedings of the Biennial Seminar of the Canadian Mathematical Congress Held at the University of Calgary, Alberta., June 12–7* (Ed. M. Grmela and J. E. Marsden). New York: Springer-Verlag, pp. 53–5, 1979.
Peitgen, H.-O.; Jürgens, H.; and Saupe, D. *Chaos and Fractals: New Frontiers of Science.* New York: Springer-Verlag, pp. 697–08, 1992.
Smale, S. "Mathematical Problems for the Next Century." In *Mathematics: Frontiers and Perspectives 2000* 0821820702 (Ed. V. Arnold, M. Atiyah, P. Lax, and B. Mazur). Providence, RI: Amer. Math. Soc., 2000.
Sparrow, C. *The Lorenz Equations: Bifurcations, Chaos, and Strange Attractors.* New York: Springer-Verlag, 1982.
Stewart, I. "The Lorenz Attractor Exists." *Nature* **406**, 948–49, 2000.
Tabor, M. *Chaos and Integrability in Nonlinear Dynamics: An Introduction.* New York: Wiley, 1989.
Viana, M. "What's New on Lorenz Strange Attractors." *Math. Intell.* **22**, 6–9.
Wells, D. *The Penguin Dictionary of Curious and Interesting Geometry.* London: Penguin, pp. 142–43, 1991.

Lorenz Curve

This entry contributed by CHRISTIAN DAMGAARD

The Lorenz curve is used in economics and ecology to describe inequality in wealth or size. The Lorenz curve is a function of the cumulative proportion of ordered individuals mapped onto the corresponding cumulative proportion of their size. Given a sample of n ordered individuals with x_i' the size of individual i and $x_1' < x_2' < \ldots < x_n'$, then the sample Lorenz curve is the polygon joining the points $(h/n, \ L_h/L_n)$, where $h = 0, 1, 2, \ldots n$, $L_0 = 0$, and $L_h = \Sigma_{i=1}^h x_i'$. Alternatively, the Lorenz curve can be expressed as

$$L(y) = \frac{\int_0^y x \, dF(x)}{\mu},$$

where $F(y)$ is the cumulative distribution function of ordered individuals and μ is the average size.

If all individuals are the same size, the Lorenz curve is a straight diagonal line, called the line of equality. If there is any inequality in size, then the Lorenz curve falls below the line of equality. The total amount of inequality can be summarized by the GINI COEFFICIENT (also called the Gini ratio), which is the ratio between the area enclosed by the line of equality and the Lorenz curve, and the total triangular area under the line of equality. The degree of asymmetry around the axis of symmetry is measured by the so-called LORENZ ASYMMETRY COEFFICIENT.

See also GINI COEFFICIENT, LORENZ ASYMMETRY COEFFICIENT

References

Dagum, C. "The Generation and Distribution of Income, the Lorenz Curve and the Gini Ratio." *Écon. Appl.* **33**, 327–67, 1980.

Kotz, S.; Johnson, N. L.; and Read, C. B. *Encyclopedia of Statistical Science.* New York: Wiley, 1983.

Lorenz, M. O. "Methods for Measuring the Concentration of Wealth." *Amer. Stat. Assoc.* **9**, 209–19, 1905.

Weiner, J. and Solbrig, O. T. "The Meaning and Measurement of Size Hierarchies in Plant Populations." *Oecologia* **61**, 334–36, 1984.

Lorenz Equations

The system of ordinary differential equations

$$\dot{X} = \sigma(Y - X) \tag{1}$$

$$\dot{Y} = rX - Y - XZ \tag{2}$$

$$\dot{Z} = XY - bZ, \tag{3}$$

See also LORENZ ATTRACTOR

References

Sparrow, C. *The Lorenz Equations: Bifurcations, Chaos, and Strange Attractors.* New York: Springer-Verlag, 1982.

Zwillinger, D. *Handbook of Differential Equations, 3rd ed.* Boston, MA: Academic Press, p. 137, 1997.

Lorenz System

LORENZ ATTRACTOR, LORENZ EQUATIONS

Lorraine Cross

GAULLIST CROSS

Lo Shu

8	1	6
3	5	7
4	9	2

The unique MAGIC SQUARE of order three. The Lo Shu is an ASSOCIATIVE MAGIC SQUARE, but not a PANMAGIC SQUARE.

See also ASSOCIATIVE MAGIC SQUARE, MAGIC SQUARE, PANMAGIC SQUARE

References

Gardner, M. *The Sixth Book of Mathematical Games from Scientific American.* Chicago, IL: University of Chicago Press, pp. 19 and 24, 1984.

Hunter, J. A. H. and Madachy, J. S. *Mathematical Diversions.* New York: Dover, pp. 23–4, 1975.

Kraitchik, M. *Mathematical Recreations.* New York: W. W. Norton, pp. 146–47, 1942.

Wells, D. *The Penguin Dictionary of Curious and Interesting Numbers.* Middlesex, England: Penguin Books, pp. 75–6, 1986.

Lossnitsch's Triangle

```
                    1
                 1     1
              1     1     1
           1     2     2     1
        1     2     4     2     1
     1     3     6     6     3     1
  1     3     9    10     9     3     1
1     4    12    19    19    12     4     1
1  4   16   28   38   28   16   4   1
1  5   20   44   66   66   44   20   5   1
1  5   25   60  110  126  110   60   25   5   1
```

A PASCAL'S TRIANGLE-like array of numbers for which each term is the sum of the two numbers immediately above it, except that, numbering the rows by $n = 0, 1, 2, \ldots$ and the entries in each row by $k = 0, 1, 2, \ldots$, if n is EVEN and k is ODD, subtract $\binom{n/2-1}{(k-1)/2}$. Analytically,

$$a(n, k) = a(n-1, k-1) + a(n-1, k) - \binom{n/2-1}{(k-1)/2},$$

where the last term is present only if n is EVEN and k is ODD.

References

Lossnitsch, S. M. "Die Isometrie-Arten ... Paraffin-Reihe." *Chem. Ber.* **30**, 1917–926, 1897.

Sloane, N. J. A. http://www.research.att.com/~njas/sequences/classic.html#LOSS.

Sloane, N. J. A. Sequences A034851 in "An On-Line Version of the Encyclopedia of Integer Sequences." http://www.research.att.com/~njas/sequences/eisonline.html.

Los' Theorem

Let I be a set, and let $\mathfrak{U}$ be an ULTRAFILTER on I, let ϕ be a formula of a given language L, and let $\{\mathbb{A}_i : i \in I\}$ be any collection of structures which is indexed by the set I. Denote by $[x]_{\mathfrak{U}}$ the EQUIVALENCE CLASS of x under $\mathfrak{U}$, for any element x of the product $\prod_{i \in I} \mathbb{A}_i$. Then the ULTRAPRODUCT $(\prod_{i \in I} \mathbb{A})/\mathfrak{U}$ satisfies ϕ via a valuation $s = [(x_i)_{i \in I}]_{\mathfrak{U}}$ in $(\prod_{i \in I} \mathbb{A})/\mathfrak{U}$ if and only if Tarski's recursive definition of SATISFACTION holds,

$$\left\{ i \in I : \mathbb{A}_i \vDash_{x_i} \phi \right\} \in \mathfrak{U}.$$

See also NONSTANDARD ANALYSIS, TRANSFER PRINCIPLE

Lost in a Forest Problem

References

Bell, J. L. and Slomson, A. B. *Models and Ultraproducts: An Introduction.* Amsterdam, Netherlands: North-Holland, 1971.

Hurd, A. E. and Loeb, P. A. *An Introduction to Nonstandard Real Analysis.* Orlando, FL: Academic Press, 1985.

Lost in a Forest Problem

The problem of finding the strategy to guarantee reaching the boundary of a given region ("forest") in the shortest distance (i.e., a strategy having the best worst-case performance). For example, one simple strategy would consist of walking in a straight line in a random direction until encountering a boundary. Although this straightforward approach is indeed the best for some simple geometries, other approaches (e.g., walking in a spiral, alternating left and right turns after traveling some fixed distance, etc.) might be optimal for forests with more complicated boundaries.

References

Bellman, R. "Minimization Problem." *Bull. Amer. Math. Soc.* **62**, 270, 1956.

Berzsenyi, G. "Lost in a Forest (A Problem Area Initiated by the Late Richard E. Bellman)." *Quantum*, p. 41, Nov./Dec. 1995.

Finch, S. "Unsolved Mathematics Problems: Lost in a Forest." http://www.mathsoft.com/asolve/forest/forest.html.

Lotka-Volterra Equations

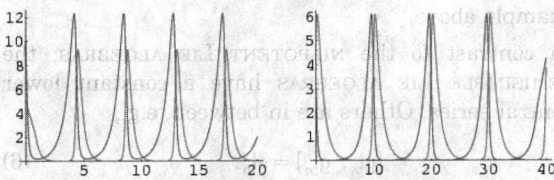

An ecological model which assumes that a population x increases at a rate $dx = Ax\, dt$, but is destroyed at a rate $dx = -Bxy\, dt$. Population y decreases at a rate $dy = -Cy\, dt$, but increases at $dy = Dxy\, dt$, giving the coupled differential equations

$$\frac{dx}{dt} = Ax - Bxy \tag{1}$$

$$\frac{dy}{dt} = -Cy + Dxy. \tag{2}$$

Critical points occur when $dx/dt = dy/dt = 0$, so

$$A - By = 0 \tag{3}$$

$$-C + Dx = 0. \tag{4}$$

The sole STATIONARY POINT is therefore located at $(x, y) = (C/D, A/B)$.

References

Boyce, W. E. and DiPrima, R. C. *Elementary Differential Equations and Boundary Value Problems, 5th ed.* New York: Wiley, p. 494, 1992.

Zwillinger, D. *Handbook of Differential Equations, 3rd ed.* Boston, MA: Academic Press, p. 135, 1997.

Lovász Number

Let $\vartheta(G)$ be the Lovász number of a GRAPH of G. Then

$$\omega(G) \le \vartheta(\bar{G}) \le \chi(G),$$

where $\omega(G)$ is the CLIQUE NUMBER and $\chi(G)$ is the minimum number of colors needed to color the VERTICES of G. This is the SANDWICH THEOREM.

See also CLIQUE NUMBER, COLORING, SANDWICH THEOREM

References

Knuth, D. E. "The Sandwich Theorem." *Electronic J. Combinatorics* **1**, A1 1–8, 1994. http://www.combinatorics.org/Volume_1/volume1.html#A1.

Love Transform

The INTEGRAL TRANSFORM

$$(Kf)(x) = \int_{-\infty}^{\infty} \frac{(x-t)_+^{c-1}}{\Gamma(c)} \, {}_2F_1\left(a, b; c; 1 - \frac{t}{x}\right) f(t)\, dt,$$

where $\Gamma(x)$ is the GAMMA FUNCTION, ${}_2F_1(a, b; c; z)$ is a HYPERGEOMETRIC FUNCTION, where y_+^z denotes the TRUNCATED POWER FUNCTION.

References

Samko, S. G.; Kilbas, A. A.; and Marichev, O. I. *Fractional Integrals and Derivatives.* Yverdon, Switzerland: Gordon and Breach, p. 23, 1993.

Low-Dimensional Topology

Low-dimensional topology usually deals with objects that are 2-, 3-, or 4-dimensional in nature. Properly speaking, low-dimensional topology should be part of DIFFERENTIAL TOPOLOGY, but the general machinery of ALGEBRAIC and DIFFERENTIAL TOPOLOGY gives only limited information. This fact is particularly noticeable in dimensions three and four, and so alternative specialized methods have evolved.

See also ALGEBRAIC TOPOLOGY, DIFFERENTIAL TOPOLOGY, HIGHER DIMENSIONAL GROUP THEORY, TOPOLOGY

References

Boroczky, K. Jr.; Neumann, W.; and Stipsicz, A. (Eds.). *Low Dimensional Topology.* Budapest, Hungary: János Bolyai Mathematical Society, 1999.

Brown, R. and Thickstun, T. L. (Eds.). *Low-Dimensional Topology: Proceedings of a Conference on Topology in Low*

Dimension, Bangor, 1979. Cambridge, England: Cambridge University Press, 1982.

Stillwell, J. *Classical Topology and Combinatorial Group Theory, 2nd ed.* New York: Springer-Verlag, 1993.

Löwenheim-Skolem Theorem

A fundamental result in MODEL THEORY which states that if a countable theory has a model, then it has a countable model. Furthermore, it has a model of every CARDINALITY greater than or equal to $\aleph_0$ (ALEPH-0). This theorem established the existence of "nonstandard" models of arithmetic.

See also ALEPH-0, CARDINALITY, GÖDEL'S COMPLETENESS THEOREM, MODEL THEORY

References

Berry, G. D. W. *Symposium on the Ontological Significance of the Löwenheim-Skolem Theorem, Academic Freedom, Logic, and Religion.* Philadelphia, PA: Amer. Philos. Soc., pp. 39–5, 1953.

Beth, E. W. "A Topological Proof of the Theorem of Löwenheim-Skolem-Gödel." *Nederl. Akad. Wetensch., Ser. A* **54**, 436–44, 1951.

Beth, E. W. "Some Consequences of the Theorem of Löwenheim-Skolem-Gödel-Malcev." *Nederl. Akad. Wetensch., Ser. A* **56**, 66–1, 1953.

Chang, C. C. and Keisler, H. J. *Model Theory, 3rd enl. ed.* New York: Elsevier, 1990.

Church, A. §45 and 49 in *Introduction to Mathematical Logic.* Princeton, NJ: Princeton University Press, 1996.

Curry, H. B. *Foundations of Mathematical Logic, 2nd rev. ed.* New York: Dover, pp. 6–, 95–6, and 121, 1977.

Fraenkel, A. A. and Bar-Hillel, Y. *Foundations of Set Theory.* Amsterdam, Netherlands, p. 105, 1958.

Myhill, J. *Symposium on the Ontological Significance of the Löwenheim-Skolem Theorem, Academic Freedom, Logic, and Religion.* Philadelphia, PA: Amer. Philos. Soc., pp. 57–0, 1953.

Quine, W. V. "Completeness of Quantification Theory: Löwenheim's Theorem." Appendix to *Methods of Logic, rev. ed.* New York: pp. 253–60, 1959.

Quine, W. V. "Interpretation of Sets of Conditions." *J. Symb. Logic* **19**, 97–02, 1954.

Rasiowa, H. and Sikorski, R. "A Proof of the Löwenheim-Skolem Theorem." *Fund. Math.* **38**, 230–32, 1952.

Skolem, T. "Sur la portée du théorème de Löwenheim-Skolem." *Les Entretiens de Zurich sur les fondements et la méthode des sciences mathématiques (December 6–, 1938),* pp. 25–2, 1941.

Vaught, R. L. "Applications of the Löwenheim-Skolem-Tarski Theorem to Problems of Completeness and Decidability." *Nederl. Akad. Wetensch., Ser. A* **57**, 467–72, 1954.

Lower Bound

A function f is said to have a lower bound c if $c \le f(x)$ for all x in its DOMAIN. The GREATEST LOWER BOUND is called the INFIMUM.

See also INEQUALITY, INFIMUM, SUPREMUM, UPPER BOUND

Lower Central Series (Lie Algebra)

The lower central series of a LIE ALGEBRA $\mathfrak{g}$ is the sequence of subalgebras recursively defined by

$$\mathfrak{g}_{k+1} = [\mathfrak{g}, \mathfrak{g}_k], \qquad (1)$$

with $\mathfrak{g}_0 = \mathfrak{g}$. The sequence of subspaces is always decreasing with respect to inclusion or dimension, and becomes stable when $\mathfrak{g}$ is finite dimensional. The notation $[\mathfrak{a}, \mathfrak{b}]$ means the linear span of elements of the form $[A, B]$, where $A \in \mathfrak{a}$ and $B \in \mathfrak{b}$.

When the lower central series ends in the zero subspace, the Lie algebra is called NILPOTENT. For example, consider the LIE ALGEBRA of strictly UPPER TRIANGULAR MATRICES, then

$$\mathfrak{g}_0 = \begin{bmatrix} 0 & a_{12} & a_{13} & a_{14} & a_{15} \\ 0 & 0 & a_{23} & a_{24} & a_{25} \\ 0 & 0 & 0 & a_{34} & a_{35} \\ 0 & 0 & 0 & 0 & a_{45} \\ 0 & 0 & 0 & 0 & 0 \end{bmatrix} \qquad (2)$$

$$\mathfrak{g}_1 = \begin{bmatrix} 0 & 0 & a_{13} & a_{14} & a_{15} \\ 0 & 0 & 0 & a_{24} & a_{25} \\ 0 & 0 & 0 & 0 & a_{35} \\ 0 & 0 & 0 & 0 & 0 \\ 0 & 0 & 0 & 0 & 0 \end{bmatrix} \qquad (3)$$

$$\mathfrak{g}_2 = \begin{bmatrix} 0 & 0 & 0 & a_{14} & a_{15} \\ 0 & 0 & 0 & 0 & a_{25} \\ 0 & 0 & 0 & 0 & 0 \\ 0 & 0 & 0 & 0 & 0 \\ 0 & 0 & 0 & 0 & 0 \end{bmatrix} \qquad (4)$$

$$\mathfrak{g}_3 = \begin{bmatrix} 0 & 0 & 0 & 0 & a_{15} \\ 0 & 0 & 0 & 0 & 0 \\ 0 & 0 & 0 & 0 & 0 \\ 0 & 0 & 0 & 0 & 0 \\ 0 & 0 & 0 & 0 & 0 \end{bmatrix}, \qquad (5)$$

and $\mathfrak{g}_4 = 0$. By definition, $\mathfrak{g}^k \subset \mathfrak{g}_k$, where $\mathfrak{g}^k$ is the term in the COMMUTATOR SERIES, as can be seen by the example above.

In contrast to the NILPOTENT LIE ALGEBRAS, the SEMISIMPLE LIE ALGEBRAS have a constant lower central series. Others are in between, e.g.,

$$[\mathfrak{gl}_n, \mathfrak{gl}_n] = \mathfrak{sl}_n, \qquad (6)$$

which is semisimple, because the TRACE satisfies

$$\mathrm{Tr}(AB) = \mathrm{Tr}(BA). \qquad (7)$$

Here, $\mathfrak{gl}_n$ is a general linear Lie algebra and $\mathfrak{sl}_n$ is the SPECIAL LINEAR LIE ALGEBRA.

Here are some *Mathematica* functions for determining the lower central series, when given a list of matrices which is a basis for $\mathfrak{g}$.

```
MatrixBasis[a_-
List]:=Partition[#1,Length[a[[1]]]]&/@
    LatticeReduce[Flatten/@a]
            LieCommutator[a_,b_]:=a.b-b.a
NextLCS[gold_List,{}]={};
            NextLCS[gold_List,g_List]:=
MatrixBasis[Flatten[Outer[LieCommutator,gold,-
```

```
g,1],1]]                    kthLCS[g_List,
k_Integer]:=Nest[NextLCS[g,#1]&,g,k]
```

For example,

```
gl5=Flatten[Table[ReplacePart[
                                    Ta-
ble[0,{i,5},{j,5}],1,{k,1}],{k,5},{1,5}],1];
   sl5=kthLCS[gl5,1]
```

See also COMMUTATOR SERIES (LIE ALGEBRA), LIE ALGEBRA, LIE GROUP, LOWER CENTRAL SERIES (GROUP), NILPOTENT LIE GROUP, REPRESENTATION (LIE ALGEBRA), REPRESENTATION (NILPOTENT LIE GROUP), UNIPOTENT

Lower Denjoy Sum
LOWER SUM

Lower Factorial
FALLING FACTORIAL

Lower Half-Disk

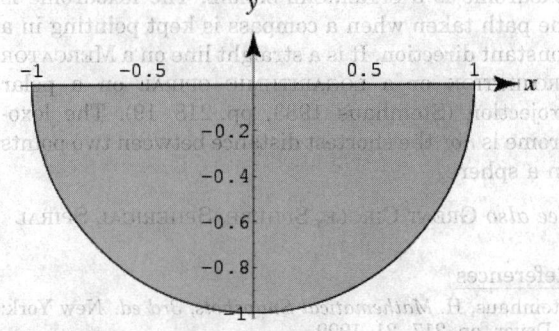

The unit lower half-disk is the portion of the COMPLEX PLANE satisfying $\{|z| \leq 1, \Im[z] < 0\}$.

See also DISK, REAL AXIS, SEMICIRCLE, UNIT DISK, LOWER HALF-PLANE, UPPER HALF-DISK

Lower Half-Plane

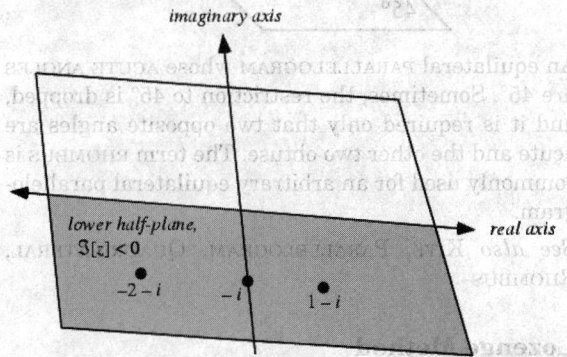

The portion of the COMPLEX PLANE $\{x+iy : x, y \in$

$(-\infty, \infty)\}$ satisfying $y = \Im[z] < 0$, i.e., $\{x+iy : x \in (-\infty, \infty), y \in (\infty, 0)\}$.

See also COMPLEX PLANE, HALF-PLANE, LEFT HALF-PLANE, LOWER HALF-DISK, RIGHT HALF-PLANE, UPPER HALF-PLANE

Lower Integral

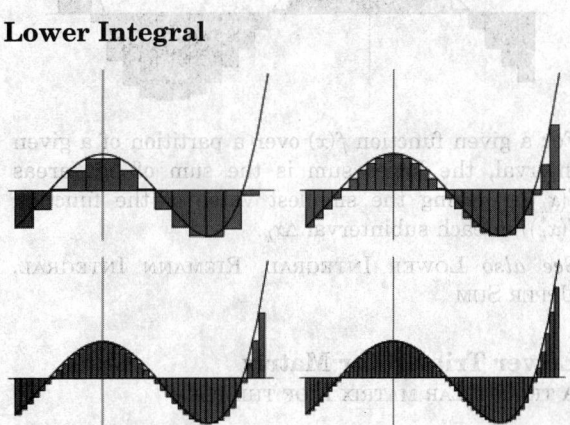

The limit of a LOWER SUM, when it exists, as the MESH SIZE approaches 0.

See also LOWER SUM, RIEMANN INTEGRAL, UPPER INTEGRAL

Lower Limit

Let the least term h of a SEQUENCE be a term which is smaller than all but a finite number of the terms which are equal to h. Then h is called the lower limit of the SEQUENCE.

A lower limit of a SERIES

$$\text{lower } \lim_{n \to \infty} S_n = \varliminf_{n \to \infty} S_n = h$$

is said to exist if, for every $\epsilon > 0$, $|S_n - h| < \epsilon$ for infinitely many values of n and if no number less than h has this property.

See also INFIMUM LIMIT, LIMIT, SUPREMUM LIMIT, UPPER LIMIT

References

Bromwich, T. J. I'a and MacRobert, T. M. "Upper and Lower Limits of a Sequence." §5.1 in *An Introduction to the Theory of Infinite Series, 3rd ed.* New York: Chelsea, p. 40 1991.

Lower Sum

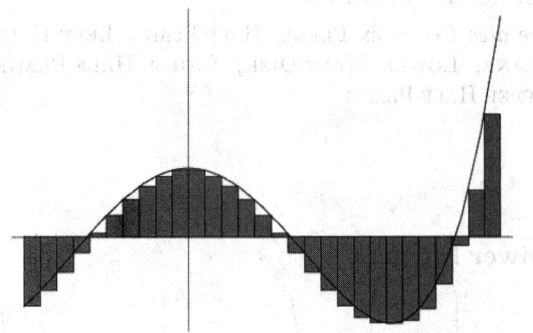

For a given function $f(x)$ over a partition of a given interval, the lower sum is the sum of box areas $f(x_k^*)\Delta x_k$ using the smallest value of the function $f(x_k^*)$) in each subinterval Δx_k.

See also LOWER INTEGRAL, RIEMANN INTEGRAL, UPPER SUM

Lower Triangular Matrix

A TRIANGULAR MATRIX **L** OF THE FORM

$$L_{ij} = \begin{cases} a_{ij} & \text{for } i \geq j \\ 0 & \text{for } i < j. \end{cases}$$

Written explicitly,

$$\mathbf{L} = \begin{bmatrix} a_{11} & 0 & \cdots & 0 \\ a_{21} & a_{22} & \cdots & 0 \\ \vdots & \vdots & \ddots & 0 \\ a_{n1} & a_{n2} & \cdots & a_{nn} \end{bmatrix}$$

A lower triangular matrix with elements `f[i,j]` below the diagonal can be formed using `LowerDiagonalMatrix[f, n]` in the *Mathematica* add-on package `LinearAlgebra`MatrixMultiplication` (which can be loaded with the command `<<LinearAlgebra`).

See also TRIANGULAR MATRIX, UPPER TRIANGULAR MATRIX

References
Ayres, F. Jr. *Theory and Problems of Matrices.* New York: Schaum, p. 10, 1962.

Lower-Trimmed Subsequence

The lower-trimmed subsequence of $x = \{x_n\}$ is the sequence $V(x)$ obtained by subtracting 1 from each x_n and then removing all 0s. If x is a FRACTAL SEQUENCE, then $V(x)$ is a FRACTAL SEQUENCE. If x is a SIGNATURE SEQUENCE, then $V(x) = x$.

See also SIGNATURE SEQUENCE, UPPER-TRIMMED SUBSEQUENCE

References
Kimberling, C. "Fractal Sequences and Interspersions." *Ars Combin.* **45**, 157–68, 1997.

Lowest Divisor Function

LEAST PRIME FACTOR

Lowest Terms Fraction

REDUCED FRACTION

Löwner's Differential Equation

The ORDINARY DIFFERENTIAL EQUATION

$$y' = -y \frac{1 + \kappa(x)y}{1 - \kappa(x)y}.$$

References
Iyanaga, S. and Kawada, Y. (Eds.). *Encyclopedic Dictionary of Mathematics.* Cambridge, MA: MIT Press, p. 1345, 1980.
Zwillinger, D. *Handbook of Differential Equations, 3rd ed.* Boston, MA: Academic Press, p. 120, 1997.

Loxodrome

A path, also known as a RHUMB LINE, which cuts a MERIDIAN on a given surface at any constant ANGLE but a RIGHT ANGLE. If the surface is a SPHERE, the loxodrome is a SPHERICAL SPIRAL. The loxodrome is the path taken when a compass is kept pointing in a constant direction. It is a straight line on a MERCATOR PROJECTION or a LOGARITHMIC SPIRAL on a polar projection (Steinhaus 1983, pp. 218–19). The loxodrome is *not* the shortest distance between two points on a sphere.

See also GREAT CIRCLE, SPHERE, SPHERICAL SPIRAL

References
Steinhaus, H. *Mathematical Snapshots, 3rd ed.* New York: Dover, pp. 217–21, 1999.

Lozenge

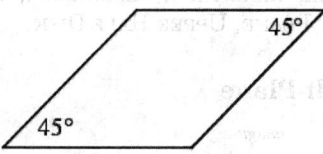

An equilateral PARALLELOGRAM whose ACUTE ANGLES are 45°. Sometimes, the restriction to 45° is dropped, and it is required only that two opposite angles are acute and the other two obtuse. The term RHOMBUS is commonly used for an arbitrary equilateral parallelogram.

See also KITE, PARALLELOGRAM, QUADRILATERAL, RHOMBUS

Lozenge Method

A method for constructing MAGIC SQUARES of ODD order.

See also MAGIC SQUARE

Lozi Map

A 2-D map similar to the HÉNON MAP which is given by the equations

$$x_{n+1} = 1 - \alpha|x_n| + y_n$$

$$y_{n+1} = \beta x_n.$$

See also HÉNON MAP

References
Dickau, R. M. "Lozi Attractor." http://forum.swarthmore.edu/advanced/robertd/lozi.html.
Peitgen, H.-O.; Jürgens, H.; and Saupe, D. §12.1 in *Chaos and Fractals: New Frontiers of Science.* New York: Springer-Verlag, p. 672, 1992.

Lp'-Balance Theorem

If every component L of $X/O_{p'}(X)$ satisfies the "Schreler property," then

$$L_{p'}(Y) \le L_{p'}(X)$$

for every p-local SUBGROUP Y of X, where $L_{p'}$ is the P-LAYER.

See also P-LAYER, SUBGROUP

L-Polyomino

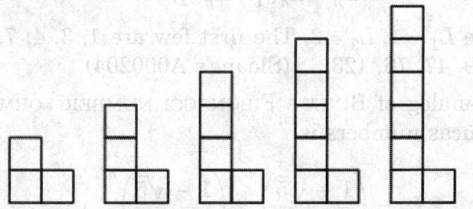

The order $n \ge 2$ L-polyomino consists of a vertical line of n SQUARES with a single additional SQUARE attached at the bottom.

See also L-POLYOMINO, SKEW POLYOMINO, SQUARE, SQUARE POLYOMINO, STRAIGHT POLYOMINO

Lp-Space

The set of L^p-functions generalizes $L2$-SPACE. Instead of SQUARE INTEGRABLE, the MEASURABLE FUNCTION f must be p-integrable for f to be in L^p.

On a MEASURE SPACE X, the L^p norm of a function f is

$$\|f\|_{Lp} = \left(\int_X |f|^p \right)^{1/p}.$$

The L^p-functions are the functions for which this integral converges. For $p \ne 2$, the space of L^p-functions is a BANACH SPACE which is not a HILBERT SPACE.

The L^p-space on $\mathbb{R}^n$, and in most other cases, is the COMPLETION of the continuous functions with COMPACT SUPPORT using the L^p norm. As in the case of an $L2$-SPACE, an L^p-function is really an equivalence class of functions which agree ALMOST EVERYWHERE. It is possible for a sequence of functions f_n to converge in L^p but not in $L^{p'}$ for some other p', e.g., $f_n = (1+x^2)^{-1/2-1/n}$ converges in $L^2(\mathbb{R})$ but not $L^1(\mathbb{R})$. However, if a sequence converges in L^p and in $L^{p'}$, then its limit must be the same in both spaces.

For $p > 1$, the DUAL SPACE to L^p is given by integrating against functions in L^q, where $1/p + 1/q = 1$. This makes sense because of HÖLDER'S INEQUALITY FOR INTEGRALS. In particular, the only L^p-space which is SELF-DUAL is L^2.

While the use of L^p functions is not as common as L^2, they are very important in ANALYSIS and PARTIAL DIFFERENTIAL EQUATIONS. For instance, some OPERATORS are only BOUNDED in L^p for some $p > 2$.

See also BANACH SPACE, COMPLETION, HILBERT SPACE, LEBESGUE INTEGRAL, L^p-SPACE, $L2$-SPACE, MEASURE, MEASURE SPACE

LQ Decomposition

The orthogonal decomposition of a matrix into lower trapezoidal matrices.

References
Ferguson, H. R. P.; Bailey, D. H.; and Arno, S. "Analysis of PSLQ, An Integer Relation Finding Algorithm." *Math. Comput.* **68**, 351–69, 1999.

L-Series

DIRICHLET L-SERIES, ROGERS L-FUNCTION

L-System

LINDENMAYER SYSTEM

Lubbock's Formula

$$f_0 + f_{1/m} + f_{2/m} + \ldots + f_r$$

$$m(f_0 + f_1 + \ldots + f_r) - \tfrac{1}{2}(m-2)(f_r + f_0)$$

$$-\frac{m^2-1}{12m}(\Delta f_{r-1} - \Delta f_0) - \frac{m^2-1}{24m}(\Delta^2 f_{r-2} - \Delta^2 f_0)$$

$$-\frac{(m^2-1)(19m^2-1)}{720m^3}(\Delta^3 f_{r-3} - \Delta^3 f_0)$$

$$-\frac{(m^2-1)(9m^2-1)}{480m^3}(\Delta^4 f_{r-4} - \Delta^4 f_0).$$

References

Lubbock, J. W. *Cambridge Philos. Trans.* **3**, 323, 1829.

Whittaker, E. T. and Robinson, G. "Lubbock's Formula of Summation." §74 in *The Calculus of Observations: A Treatise on Numerical Mathematics*, 4th ed. New York: Dover, pp. 149–50, 1967.

Lucas Correspondence

The correspondence which relates the HANOI GRAPH to the ISOMORPHIC GRAPH of the ODD BINOMIAL COEFFICIENTS in PASCAL'S TRIANGLE, where the adjacencies are determined by adjacency (either horizontal or diagonal) in PASCAL'S TRIANGLE. The proof of the correspondence is given by the LUCAS CORRESPONDENCE THEOREM.

See also BINOMIAL COEFFICIENT, HANOI GRAPH, PASCAL'S TRIANGLE

References

Poole, David G. "The Towers and Triangles of Professor Claus (or, Pascal Knows Hanoi)." *Math. Mag.* **67**, 323–44, 1994.

Lucas Correspondence Theorem

Let p be PRIME and

$$r = r_m p^m + \ldots + r_1 p + r_0 \quad (0 \le r_i < p) \quad (1)$$

$$k = k_m p^m + \ldots + k_1 p + k_0 \quad (0 \le k_i < p), \quad (2)$$

then

$$\binom{r}{k} = \prod_{i=0}^{m} \binom{r_i}{k_i} \pmod{p}. \quad (3)$$

This is proved in Fine (1947).

References

Fine, N. J. "Binomial Coefficients Modulo a Prime." *Amer. Math. Monthly* **54**, 589–92, 1947.

Lucas-Lehmer Residue

LUCAS-LEHMER TEST

Lucas-Lehmer Test

A MERSENNE NUMBER M_p is prime IFF M_p divides s_{p-2}, where $s_0 \equiv 4$ and

$$s_i \equiv s_{i-1}^2 - 2 \pmod{2^p - 1} \quad (1)$$

for $i \ge 1$. The first few terms of this series are 4, 14, 194, 37634, 1416317954, ... (Sloane's A003010). The remainder when s_{p-2} is divided by M_p is called the LUCAS-LEHMER RESIDUE for p. The LUCAS-LEHMER RESIDUE is 0 IFF M_p is PRIME. This test can also be extended to arbitrary INTEGERS.

A generalized version of the Lucas-Lehmer test lets

$$N + 1 = \prod_{j=1}^{n} q_j^{\beta_j}, \quad (2)$$

with q_j the distinct PRIME FACTORS, and β_j their respective POWERS. If there exists a LUCAS SEQUENCE U_v such that

$$\text{GCD}(U_{(N+1)/q_j}, N) = 1 \quad (3)$$

for $j = 1, \ldots, n$ and

$$U_{N+1} \equiv 0 \pmod{N}, \quad (4)$$

then N is a PRIME. The test is particularly simple for MERSENNE NUMBERS, yielding the conventional Lucas-Lehmer test.

See also LUCAS SEQUENCE, MERSENNE NUMBER, RABIN-MILLER STRONG PSEUDOPRIME TEST

References

Sloane, N. J. A. Sequences A003010/M3494 in "An On-Line Version of the Encyclopedia of Integer Sequences." http://www.research.att.com/~njas/sequences/eisonline.html.

Lucas' Married Couples Problem

MARRIED COUPLES PROBLEM

Lucas Number

The numbers produced by the V recurrence in the LUCAS SEQUENCE with $(P, Q) = (1, -1)$ are called Lucas numbers. They are the companions to the FIBONACCI NUMBERS F_n and satisfy the same recurrence

$$L_n = L_{n-1} + L_{n-2}, \quad (1)$$

where $L_1 = 1$, $L_2 = 3$. The first few are 1, 3, 4, 7, 11, 18, 29, 47, 76, 123, ... (Sloane's A000204).

The analog of BINET'S FIBONACCI NUMBER FORMULA for Lucas numbers is

$$L_n = \left(\frac{1 + \sqrt{5}}{2}\right)^n + \left(\frac{1 - \sqrt{5}}{2}\right)^n. \quad (2)$$

Another formula is

$$L_n = [\phi^n], \quad (3)$$

where ϕ is the GOLDEN RATIO and $[x]$ denotes the NINT function. Given L_n,

$$L_{n+1} = \left\lfloor \frac{L_n(1 + \sqrt{5}) + 1}{2} \right\rfloor, \quad (4)$$

where $\lfloor x \rfloor$ is the FLOOR FUNCTION,

$$L_n^2 - L_{n-1}L_{n+1} = 5(-1)^n, \quad (5)$$

and

$$\sum_{k=0}^{n} L_k^2 = L_n L_{n+1} - 2. \quad (6)$$

The Lucas numbers obey the negation formula

$$L_{-n} = (-1)^n L_n, \tag{7}$$

the addition formula

$$L_{m+n} = \tfrac{1}{2}(5 F_m F_n + L_m L_n), \tag{8}$$

where F_n is a FIBONACCI NUMBER, the subtraction formula

$$L_{m-n} = \tfrac{1}{2}(-1)(L_m L_n - 5 F_m F_n), \tag{9}$$

the fundamental identity

$$L_n^2 - 5 F_n^2 = 4(-1)^n, \tag{10}$$

conjugation relation

$$L_n = F_{n-1} + F_{n+1}, \tag{11}$$

successor relation

$$L_{n+1} = \tfrac{1}{2}(5 F_n + L_n), \tag{12}$$

double-angle formula

$$L_{2n} = \tfrac{1}{2}(5 F_n^2 + L_n^2), \tag{13}$$

multiple-angle recurrence

$$L_{kn} = L_k L_{k(n-1)} - (-1)^k L_{k(n-2)}, \tag{14}$$

multiple-angle formulas

$$L_{kn} = \frac{1}{2^{k-1}} \sum_{i=0}^{\lfloor k/2 \rfloor} \binom{k}{2i} 5^i F_n^{2i} L_n^{k-2i} \tag{15}$$

$$= \sum_{i=0}^{\lfloor k/2 \rfloor} \frac{k}{k-i} \binom{k-i}{i} (-1)^{i(n+1)} L_n^{k-2i} \tag{16}$$

$$= \begin{cases} \sum_{i=0}^{k/2} \frac{k}{k-i} \binom{k-i}{i} (-1)^{in} 5^{k/2-i} F_n^{k-2i} & \text{for } k \text{ even} \\ L_n \sum_{i=0}^{\lfloor k/2 \rfloor} \binom{k-1-i}{i} (-1)^{in} 5^{\lfloor k/2 \rfloor - i} F_n^{k-1-2i} & \text{for } k \text{ odd} \end{cases} \tag{17}$$

$$= \sum_{i=0}^{k} \binom{k}{i} L_i F_n^i F_{n-1}^{k-i}, \tag{18}$$

product expansions

$$F_m L_n = F_{m+n} + (-1)^n F_{m-n} \tag{19}$$

and

$$F_m F_n = \tfrac{1}{5}[L_{m+n} - (-1)^n L_{m-n}], \tag{20}$$

square expansion,

$$L_n^2 = L_{2n} - 2(-1)^n, \tag{21}$$

and power expansion

$$L_n^k = \frac{1}{2} \sum_{i=0}^{k} \binom{k}{i} (-1)^{in} L_{(k-2i)n}. \tag{22}$$

The Lucas numbers satisfy the power recurrence

$$\sum_{j=0}^{t+1} (-1)^{j(j+1)/2} \begin{bmatrix} t+1 \\ j \end{bmatrix}_F L_{n-j}^t = 0, \tag{23}$$

where $\begin{bmatrix} a \\ b \end{bmatrix}_F$ is a FIBONACCI COEFFICIENT, the reciprocal sum

$$\sum_{k=1}^{n} \frac{(-1)^k}{L_k L_{k+a}} = \frac{F_n}{F_a} \sum_{k=1}^{a} \frac{(-1)^k}{L_k L_{k+n}}, \tag{24}$$

the convolution

$$\sum_{k=0}^{n} L_k L_{n-k} = (n+2)L_n + F_n, \tag{25}$$

the partial fraction decomposition

$$-\frac{5}{L_{n+a} L_{n+b} L_{n+c}} = \frac{A}{L_{n+a}} + \frac{B}{L_{n+b}} + \frac{C}{L_{n+c}}, \tag{26}$$

where

$$A = \frac{(-1)^{n-a}}{F_{b-a} F_{c-a}} \tag{27}$$

$$B = \frac{(-1)^{n-b}}{F_{c-b} F_{a-b}} \tag{28}$$

$$C = \frac{(-1)^{n-c}}{F_{a-c} F_{b-c}}, \tag{29}$$

and the summation formula

$$\sum_{k=0}^{n} x^k L_{ak+b} = \frac{g(n+1) - g(0)}{1 - L_a x + (-1)^a x^2}, \tag{30}$$

where

$$g(n) = (-1)^a L_{a(n-1)+b} x^{n+1} - L_{an+b} x^n. \tag{31}$$

Let p be a PRIME > 3 and k be a POSITIVE INTEGER. Then L_{2p^k} ends in a 3 (Honsberger 1985, p. 113). Analogs of the Cesàro identities for FIBONACCI NUMBERS are

$$\sum_{k=0}^{n} \binom{n}{k} L_k = L_{2n} \tag{32}$$

$$\sum_{k=0}^{n} \binom{n}{k} 2^k L_k = L_{3n}, \tag{33}$$

where $\binom{n}{k}$ is a BINOMIAL COEFFICIENT.

$L_n | F_m$ (L_n DIVIDES F_m) IFF n DIVIDES into m an EVEN number of times. $L_n | L_m$ IFF n divides into m an ODD number of times. $2^n L_n$ always ends in 2 (Honsberger 1985, p. 137).

Defining

$$D_n \equiv \begin{vmatrix} 3 & i & 0 & 0 & \cdots & 0 & 0 \\ i & 1 & i & 0 & \cdots & 0 & 0 \\ 0 & i & 1 & i & \cdots & 0 & 0 \\ 0 & 0 & i & 1 & \cdots & 0 & 0 \\ \vdots & \vdots & \vdots & \vdots & \ddots & \vdots & \vdots \\ 0 & 0 & 0 & 0 & \cdots & 1 & i \\ 0 & 0 & 0 & 0 & 0 & i & 1 \end{vmatrix} = L_{n+1} \qquad (34)$$

gives

$$D_n = D_{n-1} + D_{n-2} \qquad (35)$$

(Honsberger 1985, pp. 113–14).

The number of ways of picking a set (including the EMPTY SET) from the numbers 1, 2, ..., n without picking two consecutive numbers (where 1 and n are now consecutive) is L_n (Honsberger 1985, p. 122).

The only SQUARE NUMBERS in the Lucas sequence are 1 and 4, as proved by John H. E. Cohn (Alfred 1964). The only TRIANGULAR Lucas numbers are 1, 3, and 5778 (Ming 1991). The only Lucas CUBIC NUMBER is 1. The first few Lucas PRIMES L_n occur for $n = 2, 4, 5, 7, 8, 11, 13, 16, 17, 19, 31, 37, 41, 47, 53, 61, 71, 79, 113, 313, 353, ...$ (Dubner and Keller 1999, Sloane's A001606).

See also FIBONACCI NUMBER

References

Alfred, Brother U. "On Square Lucas Numbers." *Fib. Quart.* **2**, 11–2, 1964.

Borwein, J. M. and Borwein, P. B. *Pi & the AGM: A Study in Analytic Number Theory and Computational Complexity.* New York: Wiley, pp. 94–01, 1987.

Brillhart, J.; Montgomery, P. L.; and Solverman, R. D. "Tables of Fibonacci and Lucas Factorizations." *Math. Comput.* **50**, 251–60 and S1–S15, 1988.

Brown, J. L. Jr. "Unique Representation of Integers as Sums of Distinct Lucas Numbers." *Fib. Quart.* **7**, 243–52, 1969.

Dubner, H. and Keller, W. "New Fibonacci and Lucas Primes." *Math. Comput.* **68**, 417–27 and S1–S12, 1999.

Guy, R. K. "Fibonacci Numbers of Various Shapes." §D26 in *Unsolved Problems in Number Theory, 2nd ed.* New York: Springer-Verlag, pp. 194–95, 1994.

Hilton, P.; Holton, D.; and Pedersen, J. "Fibonacci and Lucas Numbers." Ch. 3 in *Mathematical Reflections in a Room with Many Mirrors.* New York: Springer-Verlag, pp. 61–5, 1997.

Hilton, P. and Pedersen, J. "Fibonacci and Lucas Numbers in Teaching and Research." *J. Math. Informatique* **3**, 36–7, 1991–992.

Hoggatt, V. E. Jr. *The Fibonacci and Lucas Numbers.* Boston, MA: Houghton Mifflin, 1969.

Honsberger, R. "A Second Look at the Fibonacci and Lucas Numbers." Ch. 8 in *Mathematical Gems III.* Washington, DC: Math. Assoc. Amer., 1985.

Leyland, P. ftp://sable.ox.ac.uk/pub/math/factors/lucas.Z.

Ming, L. "On Triangular Lucas Numbers." *Applications of Fibonacci Numbers, Vol. 4* (Ed. G. E. Bergum, A. N. Philippou, and A. F. Horadam). Dordrecht, Netherlands: Kluwer, pp. 231–40, 1991.

Sloane, N. J. A. Sequences A000204/M2341 and A001606/M0961 in "An On-Line Version of the Encyclopedia of Integer Sequences." http://www.research.att.com/~njas/sequences/eisonline.html.

Lucas Polynomial

The w POLYNOMIALS obtained by setting $p(x) = x$ and $q(x) = 1$ in the LUCAS POLYNOMIAL SEQUENCE. The first few are

$$F_1(x) = x$$

$$F_2(x) = x^2 + 2$$

$$F_3(x) = 3x^3 + 3x$$

$$F_4(x) = x^4 + 4x^2 + 2$$

$$F_5(x) = x^5 + 5x^3 + 5x.$$

The corresponding W POLYNOMIALS are called FIBONACCI POLYNOMIALS. The Lucas polynomials satisfy

$$L_n(1) = L_n,$$

where the L_ns are LUCAS NUMBERS.

See also FIBONACCI POLYNOMIAL, LUCAS NUMBER, LUCAS POLYNOMIAL SEQUENCE

Lucas Polynomial Sequence

A pair of generalized POLYNOMIALS which generalize the LUCAS SEQUENCE to POLYNOMIALS is given by

$$W_n^k(x) = \frac{\Delta^k(x)[a^n(x) - (-1)^k b^n(x)]}{\Delta(x)} \qquad (1)$$

$$w_n^k(x) = \Delta^k(x)\left[a^n(x) + (-1)^k b^n(x)\right], \qquad (2)$$

where

$$a(x) + b(x) = p(x) \qquad (3)$$

$$a(x)b(x) = -q(x) \qquad (4)$$

$$a(x) - b(x) = \sqrt{p^2(x) + 4q(x)} \equiv \Delta(x) \qquad (5)$$

(Horadam 1996). Setting $n = 0$ gives

$$W_0^k(x) = \Delta^k(x)\frac{1 - (-1)^k}{\Delta(x)} \qquad (6)$$

$$w_0^k(x) = \Delta^k(x)[1 + (-1)^k], \qquad (7)$$

giving

$$W_0^0(x) = 0 \qquad (8)$$

$$w_0^0(x) = 2. \qquad (9)$$

The sequences most commonly considered have $k = 0$, giving

$$W_n(x) \equiv W_n^0(x) = \frac{a^n(x) - b^n(x)}{a(x) - b(x)} \qquad (10)$$

$$= \frac{\left[p(x) + \sqrt{p^2(x) + 4q(x)}\right]^n + \left[p(x) - \sqrt{p^2(x) + 4q(x)}\right]^n}{2^n \sqrt{p^2(x) + 4q^2(x)}} \tag{11}$$

$$w_n(x) \equiv w_n^0(x) = a^n(x) + b^n(x) \tag{12}$$

$$\frac{\left[p(x) + \sqrt{p^2(x) + 4q(x)}\right]^n + \left[p(x) + \sqrt{p^2(x) + 4q(x)}\right]^n}{2^n} \tag{13}$$

The w polynomials satisfy the RECURRENCE RELATION

$$w_n(x) = p(x)w_{n-1}(x) + q(x)w_{n-2}(x). \tag{14}$$

Special cases of the W and w polynomials are given in the following table.

$p(x)$	$q(x)$	Polynomial 1	Polynomial 2
x	1	FIBONACCI $F_n(x)$	LUCAS $L_n(x)$
$2x$	1	PELL $P_n(x)$	PELL-LUCAS $Q_n(x)$
1	$2x$	JACOBSTHAL $J_n(x)$	JACOBSTHAL $j_n(x)$
$3x$	-2	FERMAT $\mathscr{F}_n(x)$	FERMAT-LUCAS $f_n(x)$
$2x$	-1	CHEBYSHEV POLYNOMIAL OF THE SECOND KIND $U_{n-1}(x)$	CHEBYSHEV POLYNOMIAL OF THE FIRST KIND $2T_n(x)$

See also CHEBYSHEV POLYNOMIAL OF THE FIRST KIND, CHEBYSHEV POLYNOMIAL OF THE SECOND KIND, FERMAT POLYNOMIAL, FIBONACCI POLYNOMIAL, JACOBSTHAL POLYNOMIAL, LUCAS POLYNOMIAL, LUCAS SEQUENCE, PELL POLYNOMIAL

References

Horadam, A. F. "Extension of a Synthesis for a Class of Polynomial Sequences." *Fib. Quart.* **34**, 68–4, 1996.

Lucas Pseudoprime

When P and Q are INTEGERS such that $D = P^2 - 4Q \neq 0$, define the LUCAS SEQUENCE $\{U_k\}$ by

$$U_k = \frac{a^k - b^k}{a - b}$$

for $k \geq 0$, with a and b the two ROOTS of $x^2 - Px + Q = 0$. Then define a Lucas pseudoprime as an ODD COMPOSITE number n such that $n \nmid Q$, the JACOBI SYMBOL $(D/n) = -1$, and $n \mid U_{n+1}$.

There are no EVEN Lucas pseudoprimes (Bruckman 1994). The first few Lucas pseudoprimes are 705, 2465, 2737, 3745, ... (Sloane's A005845).

See also EXTRA STRONG LUCAS PSEUDOPRIME, LUCAS SEQUENCE, PSEUDOPRIME, STRONG LUCAS PSEUDOPRIME

References

Bruckman, P. S. "Lucas Pseudoprimes are Odd." *Fib. Quart.* **32**, 155–57, 1994.
Ribenboim, P. "Lucas Pseudoprimes (lpsp(P, Q))." §2.X.B in *The New Book of Prime Number Records, 3rd ed.* New York: Springer-Verlag, p. 129, 1996.
Sloane, N. J. A. Sequences A005845/M5469 in "An On-Line Version of the Encyclopedia of Integer Sequences." http://www.research.att.com/~njas/sequences/eisonline.html.

Lucas Sequence

Let P, Q be POSITIVE INTEGERS. The ROOTS of

$$x^2 - Px + Q = 0 \tag{1}$$

are

$$a \equiv \tfrac{1}{2}\left(P + \sqrt{D}\right) \tag{2}$$

$$b \equiv \tfrac{1}{2}\left(P - \sqrt{D}\right), \tag{3}$$

where

$$D \equiv P^2 - 4Q, \tag{4}$$

so

$$a + b = P \tag{5}$$

$$ab = \tfrac{1}{4}(P^2 - D) = Q \tag{6}$$

$$a - b = \sqrt{D}. \tag{7}$$

Then define

$$U_n(P, Q) \equiv \frac{a^n - b^n}{a - b} \tag{8}$$

$$V_n(P, Q) \equiv a^n + b^n. \tag{9}$$

The first few values are therefore

$$U_0(P, Q) = 0 \tag{10}$$

$$U_1(P, Q) = 1 \tag{11}$$

$$V_0(P, Q) = 2 \tag{12}$$

$$V_1(P, Q) = P. \tag{13}$$

The sequences

$$U(P, Q) = \{U_n(P, Q) : n \geq 1\} \tag{14}$$

$$V(P, Q) = \{V_n(P, Q) : n \geq 1\} \tag{15}$$

are called Lucas sequences, where the definition is usually extended to include

$$U_{-1} = \frac{a^{-1} - b^{-1}}{a - b} = \frac{-1}{ab} = -\frac{1}{Q}. \tag{16}$$

For $(P, Q) = (1, -1)$, the U_n are the FIBONACCI NUM-

BERS and V_n are the LUCAS NUMBERS. For $(P, Q) = (2, -1)$, the PELL NUMBERS and Pell-Lucas numbers are obtained. $(P, Q) = (1, -2)$ produces the JACOBSTHAL NUMBERS and Pell-Jacobsthal Numbers.

The Lucas sequences satisfy the general RECURRENCE RELATIONS

$$U_{m+n} = \frac{a^{m+n} - b^{m+n}}{a - b}$$

$$= \frac{(a^m - b^m)(a^n + b^n)}{a - b} - \frac{a^n b^n (a^{m-n} - b^{m-n})}{a - b}$$

$$= U_m V_n - a^n b^n U_{m-n} \tag{17}$$

$$V_{m+n} = a^{m+n} + b^{m+n}$$

$$= (a^m + b^m)(a^n + b^n) - a^n b^n (a^{m-n} + b^{m-n})$$

$$= V_m V_n - a^n b^n V_{m-n}. \tag{18}$$

Taking $n = 1$ then gives

$$U_m(P, Q) = PU_{m-1}(P, Q) - QU_{m-2}(P, Q) \tag{19}$$

$$V_m(P, Q) = PV_{m-1}(P, Q) - QV_{m-2}(P, Q). \tag{20}$$

Other identities include

$$U_{2n} = U_n V_n \tag{21}$$

$$U_{2n+1} = U_{n+1} V_n - Q^n \tag{22}$$

$$V_{2n} = V_n^2 - 2(ab)^n = V_n^2 - 2Q^n \tag{23}$$

$$V_{2n+1} = V_{n+1} V_n - PQ^n. \tag{24}$$

These formulas allow calculations for large n to be decomposed into a chain in which only four quantities must be kept track of at a time, and the number of steps needed is $\sim \lg n$. The chain is particularly simple if n has many 2s in its factorization.

The Us in a Lucas sequence satisfy the CONGRUENCE

$$U_{p^{n-1}[p-(D/p)]} \equiv 0 \pmod{p^n} \tag{25}$$

if

$$\mathrm{GCD}(2QcD, p) = 1, \tag{26}$$

where

$$P^2 - 4Q^2 = c^2 D. \tag{27}$$

This fact is used in the proof of the general LUCAS-LEHMER TEST.

See also FIBONACCI NUMBER, JACOBSTHAL NUMBER, LUCAS-LEHMER TEST, LUCAS NUMBER, LUCAS POLYNOMIAL SEQUENCE, PELL NUMBER, RECURRENCE SEQUENCE, SYLVESTER CYCLOTOMIC NUMBER

References

Dickson, L. E. "Recurring Series; Lucas' u_n, v_n." Ch. 17 in *History of the Theory of Numbers, Vol. 1: Divisibility and Primality.* New York: Chelsea, pp. 393–11, 1952.
Ribenboim, P. *The Little Book of Big Primes.* New York: Springer-Verlag, pp. 35–3, 1991.

Lucas's Theorem

Let $n \geq 3$ be a SQUAREFREE integer, and $\Phi_n(z)$ a CYCLOTOMIC POLYNOMIAL. Then

$$\Phi_n(z) = U_n^2(z) - (-1)^{(n-1)/2} nz V_n^2(z), \tag{1}$$

where $U_n(z)$ and $V_n(z)$ are INTEGER POLYNOMIALS of degree $\phi(n)/2$ and $\phi(n)/2 - 1$, respectively. This identity can be expressed as

$$\begin{cases} \Phi_n((-1)^{(n-1)/2} z) = C_n^2(z) - nz D_n^2(z) & \text{for } n \text{ odd} \\ \Phi_{n/2}(-z^2) = C_n^2(z) - nz D_n^2(z) & n = 4k + 2 \\ \Phi_1(-z^2) = C_2^2(z) - 2z D_2^2(z) & \text{for } n = 2, \end{cases} \tag{2}$$

with $C_n(z)$ and $D_n(z)$ SYMMETRIC POLYNOMIALS. The following table gives the first few $C_n(z)$ and $D_n(z)$s (Riesel 1994, pp. 443–56).

n	$C_n(z)$	$D_n(z)$
2	$z + 1$	1
3	$z + 1$	1
5	$z^2 + 3z + 1$	$z + 1$
6	$z^2 + 3z + 1$	$z + 1$
7	$z^3 + 3z^2 + 3z + 1$	$z^2 + z + 1$
10	$z^4 + 5z^3 + 7z^2 + 5z + 1$	$z^3 + 2z^2 + 2z + 1$

See also CYCLOTOMIC POLYNOMIAL, GAUSS'S CYCLOTOMIC FORMULA

References

Brent, R. P. "On Computing Factors of Cyclotomic Polynomials." *Math. Comput.* **61**, 131–49, 1993.
Kraitchik, M. *Recherches sue la théorie des nombres, tome I.* Paris: Gauthier-Villars, pp. 126–28, 1924.
Riesel, H. "Lucas's Formula for Cyclotomic Polynomials." In tables at end of *Prime Numbers and Computer Methods for Factorization, 2nd ed.* Boston, MA: Birkhäuser, pp. 443–56, 1994.

Lucky Number

Write out all the ODD numbers: 1, 3, 5, 7, 9, 11, 13, 15, 17, 19, ... The first ODD number > 1 is 3, so strike out every third number from the list: 1, 3, 7, 9, 13, 15, 19, ... The first ODD number greater than 3 in the list is 7, so strike out every seventh number: 1, 3, 7, 9, 13, 15, 21, 25, 31, ...

Numbers remaining after this procedure has been carried out completely are called lucky numbers. The first few are 1, 3, 7, 9, 13, 15, 21, 25, 31, 33, 37, ... (Sloane's A000959). Many asymptotic properties of the PRIME NUMBERS are shared by the lucky numbers. The asymptotic density is $1/\ln N$, just as the PRIME

NUMBER THEOREM, and the frequency of TWIN PRIMES and twin lucky numbers are similar. A version of the GOLDBACH CONJECTURE also seems to hold.

It therefore appears that the SIEVING process accounts for many properties of the PRIMES.

See also GOLDBACH CONJECTURE, LUCKY NUMBER OF EULER, PRIME NUMBER, PRIME NUMBER THEOREM, SIEVE

References

Gardner, M. "Mathematical Games: Tests Show whether a Large Number can be Divided by a Number from 2 to 12." *Sci. Amer.* **207**, 232, Sep. 1962.

Gardner, M. "Lucky Numbers and 2187." *Math. Intell.* **19**, 26, 1997.

Guy, R. K. "Lucky Numbers." §C3 in *Unsolved Problems in Number Theory, 2nd ed.* New York: Springer-Verlag, pp. 108–09, 1994.

Ogilvy, C. S. and Anderson, J. T. *Excursions in Number Theory.* New York: Dover, pp. 100–02, 1988.

Peterson, I. "MathTrek: Martin Gardner's Luck Number." http://www.sciencenews.org/sn_arc97/9_6_97/mathland.htm.

Sloane, N. J. A. Sequences A000959/M2616 in "An On-Line Version of the Encyclopedia of Integer Sequences." http://www.research.att.com/~njas/sequences/eisonline.html.

Ulam, S. M. *A Collection of Mathematical Problems.* New York: Interscience Publishers, p. 120, 1960.

Wells, D. G. *The Penguin Dictionary of Curious and Interesting Numbers.* London: Penguin, p. 32, 1986.

Lucky Number of Euler

A number p such that the PRIME-GENERATING POLYNOMIAL

$$n^2 - n + p$$

is PRIME for $n = 0, 1, \ldots, p-2$. Such numbers are related to the COMPLEX QUADRATIC FIELD in which the RING of INTEGERS is factorable. Specifically, the Lucky numbers of Euler (excluding the trivial case $p = 3$) are those numbers p such that the QUADRATIC FIELD $\mathbb{Q}(\sqrt{1-4p})$ has CLASS NUMBER 1 (Rabinowitz 1913, Le Lionnais 1983, Conway and Guy 1996).

As established by Stark (1967), there are only nine numbers $-d$ such that $h(-d) = 1$ (the HEEGNER NUMBERS -2, -3, -7, -11, -19, -43, -67, and -163), and of these, only 7, 11, 19, 43, 67, and 163 are of the required form. Therefore, the only Lucky numbers of Euler are 2, 3, 5, 11, 17, and 41 (Le Lionnais 1983, Sloane's A014556), and there does not exist a better PRIME-GENERATING POLYNOMIAL of Euler's form.

See also CLASS NUMBER, HEEGNER NUMBER, PRIME-GENERATING POLYNOMIAL

References

Conway, J. H. and Guy, R. K. "The Nine Magic Discriminants." In *The Book of Numbers.* New York: Springer-Verlag, pp. 224–26, 1996.

Le Lionnais, F. *Les nombres remarquables.* Paris: Hermann, pp. 88 and 144, 1983.

Rabinowitz, G. "Eindeutigkeit der Zerlegung in Primzahlfaktoren in quadratischen Zahlkörpern." *Proc. Fifth Internat. Congress Math. (Cambridge)* **1**, 418–21, 1913.

Sloane, N. J. A. Sequences A014556 in "An On-Line Version of the Encyclopedia of Integer Sequences." http://www.research.att.com/~njas/sequences/eisonline.html.

Stark, H. M. "A Complete Determination of the Complex Quadratic Fields of Class Number One." *Michigan Math. J.* **14**, 1–7, 1967.

LUCY

A nonlinear DECONVOLUTION technique used in deconvolving images from the Hubble Space Telescope before corrective optics were installed.

See also DECONVOLUTION, MAXIMUM ENTROPY METHOD

LU Decomposition

A procedure for decomposing an $N \times N$ matrix A into a product of a LOWER TRIANGULAR MATRIX L and an UPPER TRIANGULAR MATRIX U,

$$LU = A. \tag{1}$$

LU decomposition is implemented in *Mathematica* as `LUDecomposition[m]`.

Written explicitly for a 3×3 MATRIX, the decomposition is

$$\begin{bmatrix} l_{11} & 0 & 0 \\ l_{21} & l_{22} & 0 \\ l_{31} & l_{32} & l_{33} \end{bmatrix} \begin{bmatrix} u_{11} & u_{12} & u_{13} \\ 0 & u_{22} & u_{23} \\ 0 & 0 & u_{33} \end{bmatrix} = \begin{bmatrix} a_{11} & a_{12} & a_{13} \\ a_{21} & a_{22} & a_{23} \\ a_{31} & a_{32} & a_{33} \end{bmatrix} \tag{2}$$

$$\begin{bmatrix} l_{11}u_{11} & l_{11}u_{12} & l_{11}u_{13} \\ l_{21}u_{11} & l_{21}u_{12}+l_{22}u_{22} & l_{21}u_{13}+l_{22}u_{23} \\ l_{31}u_{11} & l_{31}u_{12}+l_{32}u_{22} & l_{31}u_{13}+l_{32}u_{23}+l_{33}u_{23} \end{bmatrix}$$

$$= \begin{bmatrix} a_{11} & a_{12} & a_{13} \\ a_{21} & a_{22} & a_{23} \\ a_{31} & a_{32} & a_{33} \end{bmatrix}. \tag{3}$$

This gives three types of equations

$$i < j \quad l_{i1}u_{1j} + l_{i2}u_{2j} + \ldots + l_{ii}u_{ij} = a_{ij} \tag{4}$$

$$i = j \quad l_{i1}u_{1j} + l_{i2}u_{2j} + \ldots + l_{ii}u_{jj} = a_{ij} \tag{5}$$

$$i > j \quad l_{i1}u_{1j} + l_{i2}u_{2j} + \ldots + l_{ij}u_{jj} = a_{ij}. \tag{6}$$

This gives N^2 equations for $N^2 + N$ unknowns (the decomposition is not unique), and can be solved using CROUT'S METHOD. To solve the MATRIX equation

$$A\mathbf{x} = (LU)\mathbf{x} = L(U\mathbf{x}) = \mathbf{b}, \tag{7}$$

first solve $L\mathbf{y} = \mathbf{b}$ for $\mathbf{y}$. This can be done by forward substitution

$$y_1 = \frac{b_1}{l_{11}} \tag{8}$$

$$y_i = \frac{1}{l_{ii}} \left(b_i - \sum_{j=1}^{i-1} l_{ij} y_j \right) \qquad (9)$$

for $i = 2, ..., N$. Then solve $\mathsf{U}\mathbf{x} = \mathbf{y}$ for $\mathbf{x}$. This can be done by back substitution

$$x_N = \frac{y_N}{u_{NN}} \qquad (10)$$

$$x_i = \frac{1}{u_{ii}} \left(y_i - \sum_{j=i+1}^{N} u_{ij} x_j \right) \qquad (11)$$

for $i = N-1, ..., 1$.

See also LOWER TRIANGULAR MATRIX, MATRIX DE-COMPOSITION, CHOLESKY DECOMPOSITION, QR DE-COMPOSITION, TRIANGULAR MATRIX, UPPER TRIANGULAR MATRIX

References

Press, W. H.; Flannery, B. P.; Teukolsky, S. A.; and Vetterling, W. T. "LU Decomposition and Its Applications." §2.3 in *Numerical Recipes in FORTRAN: The Art of Scientific Computing, 2nd ed.* Cambridge, England: Cambridge University Press, pp. 34–2, 1992.

Ludolph's Constant
PI

Ludwig's Inversion Formula

Expresses a function in terms of its RADON TRANS-FORM,

$$f(x, y) = \mathcal{R}^{-1}(\mathcal{R}f)(x, y)$$

$$= \frac{1}{\pi} \frac{1}{2\pi} \int_{-\infty}^{\infty} \frac{\frac{\partial}{\partial p}(\mathcal{R}f)(p, \alpha)}{x \cos \alpha + y \sin \alpha - p} \, dp \, d\alpha.$$

See also RADON TRANSFORM

Ludwig's Law
FIBONACCI NUMBER

Lukács Theorem

Let $\rho(x)$ be an mth degree POLYNOMIAL which is NONNEGATIVE in $[-1, 1]$. Then $\rho(x)$ can be represented in the form

$$\begin{cases} [A(x)]^2 + (1-x^2)[B(x)]^2 & \text{for } m \text{ even} \\ (1+x)[C(x)]^2 + (1-x)[D(x)]^2 & \text{for } m \text{ odd,} \end{cases}$$

where $A(x)$, $B(x)$, $C(x)$, and $D(x)$ are REAL POLYNOMIALS whose degrees do not exceed m.

References

Szego, G. *Orthogonal Polynomials, 4th ed.* Providence, RI: Amer. Math. Soc., p. 4, 1975.

Lune

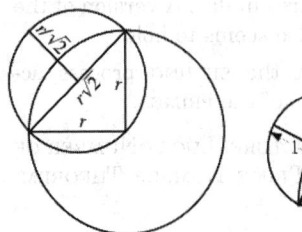 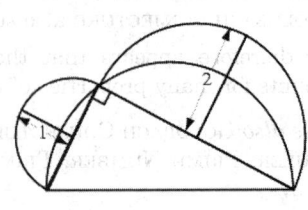

A figure bounded by two circular ARCS of unequal RADII. Hippocrates of Chios SQUARED the above left lune, as well as two others, in the fifth century BC. Two more SQUARABLE lunes were found by T. Clausen in the 19th century (Dunham 1990 attributes these discoveries to Euler in 1771). In the 20th century, N. G. Tschebatorew and A. W. Dorodnow proved that these are the only five squarable lunes (Shenitzer and Steprans 1994). The left lune above is squared as follows,

$$A_{\text{half small circle}} = \frac{1}{2} \pi \left(\frac{r}{\sqrt{2}} \right)^2 = \frac{1}{4} \pi r^2$$

$$A_{\text{lens}} = A_{\text{quarter big circle}} - A_{\text{triangle}}$$
$$= \frac{1}{4} \pi r^2 - \frac{1}{2} r^2$$

$$A_{\text{lune}} = A_{\text{half small circle}} - A_{\text{lens}} = \frac{1}{2} r^2$$
$$= A_{\text{triangle}},$$

so the lune and TRIANGLE have the same AREA. In the right figure, $A_1 + A_2 = A_\Delta$.

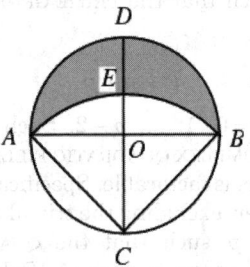

For the above lune,

$$A_{\text{lune}} = 2A_{\Delta OBC}.$$

See also ANNULUS, ARC, CIRCLE, SALINON, SPHERICAL LUNE

References

Dunham, W. "Hippocrates' Quadrature of the Lune." Ch. 1 in *Journey through Genius: The Great Theorems of Mathematics.* New York: Wiley, pp. 1–0, 1990.

Heath, T. L. *A History of Greek Mathematics, Vol. 1: From Thales to Euclid.* New York: Dover, p. 185, 1981.

Pappas, T. "Lunes." *The Joy of Mathematics.* San Carlos, CA: Wide World Publ./Tetra, pp. 72–3, 1989.

Shenitzer, A. and Steprans, J. "The Evolution of Integration." *Amer. Math. Monthly* **101**, 66–2, 1994.

Wells, D. *The Penguin Dictionary of Curious and Interesting Geometry.* London: Penguin, pp. 143–44, 1991.

Lunule

LUNE

Lüroth's Theorem

If x and y are nonconstant rational functions of a parameter, the curve so defined has GENUS 0. Furthermore, x and y may be expressed rationally in terms of a parameter which is rational in them.

References

Coolidge, J. L. *A Treatise on Algebraic Plane Curves.* New York: Dover, p. 246, 1959.

Lusin Area Integral

If $\Omega \subseteq \mathbb{C}$ is a DOMAIN and $\varphi : \Omega \to \mathbb{C}$ is a ONE-TO-ONE ANALYTIC FUNCTION, then $\varphi(\Omega)$ is a DOMAIN, and

$$\text{area}(\varphi(\Omega)) = \int_\Omega |\varphi'(z)|^2 \, dx \, dy$$

(Krantz 1999, p. 150).

See also AREA INTEGRAL

References

Krantz, S. G. "The Lusin Area Integral." §12.1.3 in *Handbook of Complex Analysis.* Boston, MA: Birkhäuser, p. 150, 1999.

Lusin's Theorem

Let $f(x)$ be a finite and MEASURABLE FUNCTION in $(-\infty, \infty)$, and let ϵ be freely chosen. Then there is a function $g(x)$ such that

1. $g(x)$ is continuous in $(-\infty, \infty)$,
2. The MEASURE of $\{x : f(x) \neq g(x)\}$ is $< \epsilon$,
3. $M(|g|; R_1) \leq M(|f|; R_1)$,

where $M(f; S)$ denotes the upper bound of the aggregate of the values of $f(P)$ as P runs through all values of S.

References

Kestelman, H. §4.4 in *Modern Theories of Integration, 2nd rev. ed.* New York: Dover, pp. 30 and 109–12, 1960.

Lusternik-Schnirelmann Theorem

LYUSTERNIK-SCHNIRELMANN THEOREM

LUX Method

A method for constructing MAGIC SQUARES of SINGLY EVEN order $n \geq 6$.

See also MAGIC SQUARE

Lyapunov Characteristic Exponent

The Lyapunov characteristic exponent [LCE] gives the rate of exponential divergence from perturbed initial conditions. To examine the behavior of an orbit

around a point $\mathbf{X}^*(t)$, perturb the system and write

$$\mathbf{X}(t) = \mathbf{X}^*(t) + U(t), \tag{1}$$

where $U(t)$ is the average deviation from the unperturbed trajectory at time t. In a CHAOTIC region, the LCE σ is independent of $\mathbf{X}^*(0)$. It is given by the OSEDELEC THEOREM, which states that

$$\sigma_i = \lim_{t \to \infty} \frac{1}{t} \ln |\mathbf{U}(t)|. \tag{2}$$

For an n-dimensional mapping, the Lyapunov characteristic exponents are given by

$$\sigma_i = \lim_{N \to \infty} \ln |\lambda_i(N)| \tag{3}$$

for $i = 1, ..., n$, where λ_i is the LYAPUNOV CHARACTERISTIC NUMBER.

One Lyapunov characteristic exponent is always 0, since there is never any divergence for a perturbed trajectory in the direction of the unperturbed trajectory. The larger the LCE, the greater the rate of exponential divergence and the wider the corresponding SEPARATRIX of the CHAOTIC region. For the STANDARD MAP, an analytic estimate of the width of the CHAOTIC zone by Chirikov (1979) finds

$$\delta I = Be^{-AK - 1/2}. \tag{4}$$

Since the Lyapunov characteristic exponent increases with increasing K, some relationship likely exists connecting the two. Let a trajectory (expressed as a MAP) have initial conditions (x_0, y_0) and a nearby trajectory have initial conditions $(x', y') = (x_0 + dx, y_0 + dy)$. The distance between trajectories at iteration k is then

$$dk = \|(x' - x_0, y' - y_0)\|, \tag{5}$$

and the mean exponential rate of divergence of the trajectories is defined by

$$\sigma_1 = \lim_{k \to \infty} \frac{1}{k} \ln \left(\frac{d_k}{d_0} \right). \tag{6}$$

For an n-dimensional phase space (MAP), there are n Lyapunov characteristic exponents $\sigma_1 \geq \sigma_2 \geq \ldots > \sigma_n$. However, because the largest exponent σ_1 will dominate, this limit is practically useful only for finding the largest exponent. Numerically, since d_k increases exponentially with k, after a few steps the perturbed trajectory is no longer nearby. It is therefore necessary to renormalize frequently every t steps. Defining

$$r_{kr} \equiv \frac{d_{kr}}{d_0}, \tag{7}$$

one can then compute

$$\sigma_1 = \lim_{k \to \infty} \frac{1}{nr} \sum_{k=1}^{n} \ln r_{kr}. \qquad (8)$$

Numerical computation of the second (smaller) Lyapunov exponent may be carried by considering the evolution of a 2-D surface. It will behave as

$$e^{(\sigma_1 + \sigma_2)t}, \qquad (9)$$

so σ_2 can be extracted if σ_1 is known. The process may be repeated to find smaller exponents.

For HAMILTONIAN SYSTEMS, the LCEs exist in additive inverse pairs, so if σ is an LCE, then so is $-\sigma$. One LCE is always 0. For a 1-D oscillator (with a 2-D phase space), the two LCEs therefore must be $\sigma_1 = \sigma_2 = 0$, so the motion is QUASIPERIODIC and cannot be CHAOTIC. For higher order HAMILTONIAN SYSTEMS, there are always at least two 0 LCEs, but other LCEs may enter in plus-and-minus pairs l and $-l$. If they, too, are both zero, the motion is integrable and not CHAOTIC. If they are NONZERO, the POSITIVE LCE l results in an exponential separation of trajectories, which corresponds to a CHAOTIC region. Notice that it is not possible to have all LCEs NEGATIVE, which explains why convergence of orbits is never observed in HAMILTONIAN SYSTEMS.

Now consider a dissipative system. For an arbitrary n-D phase space, there must always be one LCE equal to 0, since a perturbation along the path results in no divergence. The LCEs satisfy $\Sigma_i \, \sigma_i < 0$. Therefore, for a 2-D phase space of a dissipative system, $\sigma_1 = 0$, $\sigma_2 < 0$. For a 3-D phase space, there are three possibilities:

1. (Integrable): $\sigma_1 = 0$, $\sigma_2 = 0$, $\sigma_3 < 0$,
2. (Integrable): $\sigma_1 = 0$, σ_2, $\sigma_3 < 0$.
3. (CHAOTIC): $\sigma_1 = 0$, $\sigma_2 > 0$, $\sigma_3 < -\sigma_2 < 0$.

See also CHAOS, HAMILTONIAN SYSTEM, LYAPUNOV CHARACTERISTIC NUMBER, OSEDELEC THEOREM

References

Chirikov, B. V. "A Universal Instability of Many-Dimensional Oscillator Systems." *Phys. Rep.* **52**, 264–79, 1979.

Ramasubramanian, K. and Sriram, M. S. A Comparative Study of Computation of Lyapunov Spectra with Different Algorithms 1999. http://xxx.lanl.gov/abs/chao-dyn/9909029/.

Trott, M. "Numerical Computations." §1.2.1 in *The Mathematica Guidebook, Vol. 1: Programming in Mathematica.* New York: Springer-Verlag, 2000.

Lyapunov Characteristic Number

Given a LYAPUNOV CHARACTERISTIC EXPONENT σ_i, the corresponding Lyapunov characteristic number λ_i is defined as

$$\lambda_i \equiv e^{\sigma_i}. \qquad (1)$$

For an n-dimensional linear MAP,

$$\mathbf{X}_{n+1} = M\mathbf{X}_n. \qquad (2)$$

The Lyapunov characteristic numbers $\lambda_1, \ldots, \lambda_n$ are the EIGENVALUES of the MAP MATRIX. For an arbitrary MAP

$$x_{n+1} = f_1(x_n, y_n) \qquad (3)$$

$$y_{n+1} = f_2(x_n, y_n), \qquad (4)$$

the Lyapunov numbers are the EIGENVALUES of the limit

$$\lim_{n \to \infty} [J(x_n, y_n)J(x_{n-1}, y_{n-1}) \cdots J(x_1, y_1)]^{1/n}, \qquad (5)$$

where $J(x, y)$ is the JACOBIAN

$$J(x, y) \equiv \begin{vmatrix} \dfrac{\partial f_1(x, y)}{\partial x} & \dfrac{\partial f_1(x, y)}{\partial y} \\ \dfrac{\partial f_2(x, y)}{\partial x} & \dfrac{\partial f_2(x, y)}{\partial y} \end{vmatrix}. \qquad (6)$$

If λ_i for all i, the system is not CHAOTIC. If $\lambda \neq 0$ and the MAP is AREA-PRESERVING (HAMILTONIAN), the product of EIGENVALUES is 1.

See also ADIABATIC INVARIANT, CHAOS, LYAPUNOV CHARACTERISTIC EXPONENT

Lyapunov Condition

If the third MOMENT exists for a STATISTICAL DISTRIBUTION of x_i and the LEBESGUE INTEGRAL is given by

$$r_n^3 = \sum_{i=1}^{n} \int_{-\infty}^{\infty} |x|^3 \, dF_i(x),$$

then if

$$\lim_{n \to \infty} \frac{r_n}{s_n} = 0,$$

the CENTRAL LIMIT THEOREM holds.

See also CENTRAL LIMIT THEOREM

Lyapunov Dimension

For a 2-D MAP with $\sigma_2 > \sigma_1$,

$$d_{\mathrm{Lya}} = 1 - \frac{\sigma_1}{\sigma_2},$$

where σ_n are the LYAPUNOV CHARACTERISTIC EXPONENTS.

See also CAPACITY DIMENSION, KAPLAN-YORKE CONJECTURE

References

Frederickson, P.; Kaplan, J. L.; Yorke, E. D.; and Yorke, J. A. "The Liapunov Dimension of Strange Attractors." *J. Diff. Eq.* **49**, 185–07, 1983.

Nayfeh, A. H. and Balachandran, B. *Applied Nonlinear Dynamics: Analytical, Computational, and Experimental Methods.* New York: Wiley, p. 549, 1995.

Lyapunov Function

This entry contributed by MARTIN KELLER-RESSEL

A Lyapunov function is a SCALAR FUNCTION $V(y)$ defined on a region D that is continuous, positive definite (i.e., $V(0) = 0$, $V(y) > 0$ for all $y \neq 0$), and has continuous first-order PARTIAL DERIVATIVES at every point of D. The derivative of V with respect to the system $y' = f(y)$, written as $V^*(y)$ is defined as the DOT PRODUCT

$$V^*(y) = \nabla V(y) \cdot F(y).$$

The existence of a Lyapunov function for which $V^*(y) \leq 0$ on some region D containing the origin, guarantees the stability of the zero solution of $y' = f(y)$, while the existence of a Lyapunov function for which $V^*(y)$ is negative definite (i.e., $V^*(0) = 0$, $V^*(y) < 0$ for all $y \neq 0$) on some region D containing the origin guarantees the asymptotical stability of the zero solution of $y' = f(y)$

For example, given the system

$$y' = z$$
$$z' = -y - 2z$$

and the Lyapunov function $V(y, z) = (y^2 + z^2)/2$, we obtain

$$V^*(y, z) = yz + z(-y - 2z) = -2z^2,$$

which is nonnegative on every region containing the origin, and thus the zero solution is stable.

See also LINEAR STABILITY, NONLINEAR STABILITY

References

Boyce, W. E. and DiPrima, R. C. *Elementary Differential Equations and Boundary Value Problems, 5th ed.* New York: Wiley, pp. 502–12, 1992.
Brauer, F. and Nohel, J. A. *The Qualitative Theory of Ordinary Differential Equations: An Introduction.* New York: Dover, 1989.
Hahn, W. *Theory and Application of Liapunov's Direct Method.* Englewood Cliffs, NJ: Prentice-Hall, 1963.
Jordan, D. W. and Smith, P. *Nonlinear Ordinary Differential Equations.* Oxford, England: Clarendon Press, p. 283, 1977.
Kalman, R. E. and Bertram, J. E. "Control System Analysis and Design Via the 'Second Method' of Liapunov, I. Continuous-Time Systems." *J. Basic Energ. Trans. ASME* **82**, 371–93, 1960.
Oguztöreli, M. N.; Lakshmikantham, V.; and Leela, S. "An Algorithm for the Construction of Liapunov Functions." *Nonlinear Anal.* **5**, 1195–212, 1981.
Zwillinger, D. "Liapunov Functions." §120 in *Handbook of Differential Equations, 3rd ed.* Boston, MA: Academic Press, pp. 429–32, 1997.

Lyapunov's First Theorem

A NECESSARY and SUFFICIENT condition for all the EIGENVALUES of a REAL $n \times n$ matrix A to have NEGATIVE REAL PARTS is that the equation

$$A^T V + VA = -1$$

has as a solution where V is an $n \times n$ matrix and $(\mathbf{x}, V\mathbf{x})$ is a POSITIVE DEFINITE QUADRATIC FORM.

References

Gradshteyn, I. S. and Ryzhik, I. M. *Tables of Integrals, Series, and Products, 6th ed.* San Diego, CA: Academic Press, p. 1122, 2000.

Lyapunov's Second Theorem

If all the EIGENVALUES of a REAL MATRIX A have REAL PARTS, then to an arbitrary negative definite quadratic form $(\mathbf{x}, W\mathbf{x})$ with $\mathbf{x} = \mathbf{x}(t)$ there corresponds a positive definite quadratic form $(\mathbf{x}, V\mathbf{x})$ such that if one takes

$$\frac{dx}{dt} = A\mathbf{A}\mathbf{x},$$

then $(\mathbf{x}, V\mathbf{x})$ and $(\mathbf{x}, W\mathbf{x})$ satisfy

$$\frac{d}{dt}(\mathbf{x}, V\mathbf{x}) = (\mathbf{x}, W\mathbf{x}).$$

References

Gradshteyn, I. S. and Ryzhik, I. M. *Tables of Integrals, Series, and Products, 6th ed.* San Diego, CA: Academic Press, p. 1122, 2000.

Lyndon Word

A Lyndon word is an aperiodic notation for representing a NECKLACE.

See also DE BRUIJN SEQUENCE, IRREDUCIBLE POLYNOMIAL, NECKLACE

References

Ruskey, F. "Information on Necklaces, Lyndon Words, de Bruijn Sequences." http://www.theory.csc.uvic.ca/~cos/inf/neck/NecklaceInfo.html.
Sloane, N. J. A. Sequences A001037/M0116 in "An On-Line Version of the Encyclopedia of Integer Sequences." http://www.research.att.com/~njas/sequences/eisonline.html.

Lyons Group

The SPORADIC GROUP Ly.

See also SPORADIC GROUP

References

Wilson, R. A. "ATLAS of Finite Group Representation." http://for.mat.bham.ac.uk/atlas/html/Ly.html.

Lyusternik-Schnirelmann Theorem

If a sphere is covered by three closed sets, then one of them must contain a pair of ANTIPODAL POINTS.

References

Dodson, C. T. J. and Parker, P. E. *A User's Guide to Algebraic Topology*. Dordrecht, Netherlands: Kluwer, pp. 122 and 284, 1997.

M

MacDonald Function

A modified HANKEL FUNCTION.

Macdonald Polynomial

See also N! THEOREM

References

Haiman, M. "Macdonald Polynomials and Geometry." In *New Perspectives in Algebraic Combinatorics* (Ed. L. J. Billera, A. Björner, C. Greene, R. E. Simion, and R. P. Stanley). Cambridge, England: Cambridge University Press, pp. 207–54, 1999.

Macdonald, I. G. *Symmetric Functions and Hall Polynomials, 2nd ed.* Oxford, England: Oxford University Press, 1995.

Zabrocki, M. "Macdonald Polynomials." http://www.lacim.uqam.ca/~zabrocki/MPWP.html.

Macdonald's Constant-Term Conjecture

Macdonald's constant term conjectures are related to ROOT SYSTEMS of LIE ALGEBRAS (Macdonald 1982, Andrews 1986). They can be regarded as generalizations of DYSON'S CONJECTURE (Dyson 1962), its q-analog due to Andrews, and Mehta's conjecture (Mehta 1991). The simplest of these states that if R is a ROOT SYSTEM, then the constant term in $\Pi_{\alpha \in R}(1 - e^{\alpha})^{k}$, where k is a NONNEGATIVE INTEGER, is $\Pi_{i=1}^{l}\binom{kd_l}{k}$, where the d_l are fixed integer parameters of the ROOT SYSTEM R corresponding to the fundamental invariants of the WEYL GROUP W of R (Andrews 1986, p. 41).

Opdam (1989) proved the $q = 1$ case for all root systems. The general conjecture had remained "almost proved" for some time, since the infinite families were accomplished by Zeilberger-Bressoud (A_n), Kadell (B_n, D_n) Gustafson (BC_n, C_n), while the exceptional cases were done by Zeilberger and (independently) Habsieger (G_2), Zeilberger (G_2 dual), and Garvan and Gonnet (F_4 and F_4 dual), using Zeilberger's method. This left only the three root systems (E_6, E_7, E_8) which were infeasible to address using existing computers. In the meanwhile, however, Cherednik (1993) proved the constant term conjectures for all root systems using a methodology not dependent on classification.

A special case of the constant-term conjecture is given by the assertion that the constant term in

$$\prod_{1 < i \neq j \leq n}\left(1 - \frac{x_i}{x_j}\right)^{k} \tag{1}$$

is $(nk)!/(k!)^n$. Another special case asserts that the constant term in

$$\left[\prod_{i \leq \leq n}(x_i;\ q)_a(q/x_i;\ q)_a\right]$$
$$\times \prod_{1 \leq i \leq j \leq n}(x_i x_j;\ q)_b\left(\frac{q}{x_i x_j};\ q\right)_b\left(\frac{x_i}{x_j};\ q\right)_b\left(\frac{qx_j}{x_i};\ q\right)_b \tag{2}$$

is

$$\frac{(q;\ q)_{nb}}{[(q;\ q)_b]^n}\prod_{1 \leq j \leq n-1}\frac{(q;\ q)_{2a+2jb}(q;\ q)_{2jb}}{(q;\ q)_{a+(n+j-1)n}(q;\ q)_{a+jb}} \tag{3}$$

(Andrews 1986, p. 41).

See also DYSON'S CONJECTURE, ROOT SYSTEM, WEYL GROUP

References

Andrews, G. E. "The Macdonald Conjectures." §4.5 in *q-Series: Their Development and Application in Analysis, Number Theory, Combinatorics, Physics, and Computer Algebra.* Providence, RI: Amer. Math. Soc., pp. 40–2, 1986.

Cherednik, I. "The Macdonald Constant-Term Conjecture." *Duke Math. J.* **70**, 165–77, 1993 and *Internat. Math. Res. Not.*, No. 6, 165–77, 1993.

Dyson, F. "Statistical Theory of the Energy Levels of Complex Systems. I." *J. Math. Phys.* **3**, 140–56, 1962.

Macdonald, I. G. "Some Conjectures for Root Systems." *SIAM J. Math. Anal.* **13**, 988–007, 1982.

Mehta, M. L. *Random Matrices, 2nd ref. enl. ed.* New York: Academic Press, 1991.

Opdam, E M. "Some Applications of Hypergeometric Shift Operators." *Invent. Math.* **98**, 1–8, 1989.

Macdonald's Plane Partition Conjecture

Macdonald's plane partition conjecture proposes a formula for the number of CYCLICALLY SYMMETRIC PLANE PARTITIONS (CSPPs) of a given integer whose YOUNG DIAGRAMS fit inside an $n \times n \times n$ box. Macdonald gave a product representation for the power series whose coefficients q^n were the number of such partitions of n.

Let $D(\pi)$ be the set of all integer points (i, j, k) in the first OCTANT such that a PLANE PARTITION $\pi = (a_{ij})$ is defined and $1 \leq k \leq a_{ij}$. Then π is said to be cyclically symmetric if $D(\pi)$ is invariant under the mapping $(i, j, k) \rightarrow (j, k, i)$. Let $M(m, n)$ be the number of cyclically symmetric partitions of n such that none of i, j, a_{ij} exceed m. Let $\mathfrak{B}_m$ be the box containing all integer points (i, j, k) such that $1 \leq i, j, k \leq m$, then $M(m, n)$ is the number of cyclically symmetric plane partitions of n such that $D(\pi) \subseteq \mathfrak{B}_m$. Now, let $\mathfrak{C}_m$ be the set of all the orbits in $\mathfrak{B}_m$. Finally, for each point $p = (i, j, k)$ in $\mathfrak{B}_m$, let its height

$$ht(p) = i + j + k - 2 \tag{1}$$

and for each ξ in $\mathfrak{C}_m$, let $|\xi|$ be the number of points in ξ (either 1 or 3) and write

$$ht(\xi) = \sum_{p \in \xi} ht(p). \tag{2}$$

Then Macdonald conjectured that

$$\sum_{n \geq 0} M(m, n) q^n = \prod_{\xi \in \mathfrak{C}_m} \frac{1 - q^{|\xi| + ht(\xi)}}{1 - q^{ht(\xi)}} \tag{3}$$

$$= \prod_{i=1}^{m} \left[\frac{1 - q^{3i-1}}{1 - q^{3i-2}} \prod_{j=i}^{m} \frac{1 - q^{3(m+i+j-1)}}{1 - q^{3(2i+j-1)}} \right], \tag{4}$$

(Mills *et al.* 1982, Macdonald 1995), where the latter form is due to Andrews (1979).

Andrews (1979) proved the $q = 1$ case, giving the total number of CSPPs fitting inside an $n \times n \times n$ box. The general case was proved by Mills *et al.* (1982).

See also CYCLICALLY SYMMETRIC PLANE PARTITION, DYSON'S CONJECTURE, PLANE PARTITION, ROOT SYSTEM, ZEILBERGER-BRESSOUD THEOREM

References

Andrews, G. E. "Plane Partitions (III): The Weak Macdonald Conjecture." *Invent. Math.* **53**, 193–25, 1979.
Andrews, G. E. "Macdonald's Conjecture and Descending Plane Partitions." In *Combinatorics, Representation Theory and Statistical Methods in Groups* (Ed. T. V. Narayana, R. M. Mathsen, and J. G. Williams). New York: Dekker, pp. 91–06, 1980.
Bressoud, D. *Proofs and Confirmations: The Story of the Alternating Sign Matrix Conjecture.* Cambridge, England: Cambridge University Press, 1999.
Bressoud, D. and Propp, J. "How the Alternating Sign Matrix Conjecture was Solved." *Not. Amer. Math. Soc.* **46**, 637–46.
Macdonald, I. G. "Some conjectures for Root Systems." *SIAM J. Math. Anal.* **13**, 988–007, 1982.
Macdonald, I. G. *Symmetric Functions and Hall Polynomials, 2nd ed.* Oxford, England: Oxford University Press, 1995.
Mills, W. H.; Robbins, D. P.; and Rumsey, H. Jr. "Proof of the Macdonald Conjecture." *Invent. Math.* **66**, 73–7, 1982.
Morris, W. G. *Constant Term Identities for Finite and Affine Root Systems: Conjectures and Theorems.* Ph.D. thesis. Madison, WI: University of Wisconsin, 1982.

Machine

A method for producing infinite LOOP SPACES and spectra.

See also GADGET, LOOP SPACE, MAY-THOMASON UNIQUENESS THEOREM, TURING MACHINE

Machin-Like Formulas

Machin-like formulas have the form

$$m \cot^{-1} u + n \cot^{-1} v = \tfrac{1}{4} k\pi, \tag{1}$$

where u, v, and k are POSITIVE INTEGERS and m and n are NONNEGATIVE INTEGERS. Some such FORMULAS can be found by converting the INVERSE TANGENT decompositions for which $c_n \neq 0$ in the table of Todd (1949) to INVERSE COTANGENTS. However, this gives

only Machin-like formulas in which the smallest term is ± 1.

Machin-like formulas can be derived by writing

$$\cot^{-1} z = \frac{1}{2i} \ln \left(\frac{z + i}{z - i} \right) \tag{2}$$

and looking for a_k and u_k such that

$$\sum_k a_k \cot^{-1} u_k = \tfrac{1}{4} \pi, \tag{3}$$

so

$$\prod_k \left(\frac{u_k + i}{u_k - i} \right)^{a_k} = e^{2\pi i/4} = i. \tag{4}$$

Machin-like formulas exist IFF (4) has a solution in INTEGERS. This is equivalent to finding INTEGER values such that

$$(1 - i)^k (u + i)^m (v + i)^n \tag{5}$$

is REAL (Borwein and Borwein 1987, p. 345). An equivalent formulation is to find all integral solutions to one of

$$1 + x^2 = 2y^n \tag{6}$$

$$1 + x^2 = y^n \tag{7}$$

for $n = 3, 5, \ldots$.

There are only four such FORMULAS,

$$\tfrac{1}{4}\pi = 4 \tan^{-1} \left(\tfrac{1}{5} \right) - \tan^{-1} \left(\tfrac{1}{239} \right) \tag{8}$$

$$\tfrac{1}{4}\pi = \tan^{-1} \left(\tfrac{1}{2} \right) + \tan^{-1} \left(\tfrac{1}{3} \right) \tag{9}$$

$$\tfrac{1}{4}\pi = 2 \tan^{-1} \left(\tfrac{1}{2} \right) - \tan^{-1} \left(\tfrac{1}{7} \right) \tag{10}$$

$$\tfrac{1}{4}\pi = 2 \tan^{-1} \left(\tfrac{1}{3} \right) + \tan^{-1} \left(\tfrac{1}{7} \right), \tag{11}$$

known as MACHIN'S FORMULA, EULER'S MACHIN-LIKE FORMULA, HERMANN'S FORMULA, and HUTTON'S FORMULA. These follow from the identities

$$\left(\frac{5 + i}{5 - i} \right)^4 \left(\frac{239 + i}{239 - i} \right)^{-1} = i \tag{12}$$

$$\left(\frac{2 + i}{2 - i} \right) \left(\frac{3 + i}{3 - i} \right) = i \tag{13}$$

$$\left(\frac{2 + i}{2 - i} \right)^2 \left(\frac{7 + i}{7 - i} \right)^{-1} = i \tag{14}$$

$$\left(\frac{3 + i}{3 - i} \right)^2 \left(\frac{7 + i}{7 - i} \right) = i. \tag{15}$$

Machin-like formulas with two terms can also be

generated which do not have integral arc cotangent arguments such as Euler's

$$\tfrac{1}{4}\pi = 5\tan^{-1}\left(\tfrac{1}{7}\right) + 2\tan^{-1}\left(\tfrac{3}{79}\right) \tag{16}$$

(Wetherfield 1996), and which involve inverse SQUARE ROOTS, such as

$$\frac{\pi}{2} = 2\tan^{-1}\left(\frac{1}{\sqrt{2}}\right) + \tan^{-1}\left(\frac{1}{\sqrt{8}}\right). \tag{17}$$

Three-term Machin-like formulas include GAUSS'S MACHIN-LIKE FORMULA

$$\tfrac{1}{4}\pi = 12\cot^{-1} 18 + 8\cot^{-1} 57 - 5\cot^{-1} 239, \tag{18}$$

STRASSNITZKY'S FORMULA

$$\tfrac{1}{4}\pi = \cot^{-1} 2 + \cot^{-1} 5 + \cot^{-1} 8, \tag{19}$$

and the following,

$$\tfrac{1}{4}\pi = 6\cot^{-1} 8 + 2\cot^{-1} 57 + \cot^{-1} 239 \tag{20}$$

$$\tfrac{1}{4}\pi = 4\cot^{-1} 5 - 1\cot^{-1} 70 + \cot^{-1} 99 \tag{21}$$

$$\tfrac{1}{4}\pi = 1\cot^{-1} 2 + 1\cot^{-1} 5 + \cot^{-1} 8 \tag{22}$$

$$\tfrac{1}{4}\pi = 8\cot^{-1} 10 - 1\cot^{-1} 239 - 4\cot^{-1} 515 \tag{23}$$

$$\tfrac{1}{4}\pi = 5\cot^{-1} 7 + 4\cot^{-1} 53 + 2\cot^{-1} 4443. \tag{24}$$

The first is due to Størmer, the second due to Rutherford, and the third due to Dase.

Using trigonometric identities such as

$$\cot^{-1} x = 2\cot^{-1}(2x) - \cot^{-1}(4x^3 + 3x), \tag{25}$$

it is possible to generate an infinite sequence of Machin-like formulas. Systematic searches therefore most often concentrate on formulas with particularly "nice" properties (such as "efficiency").

The efficiency of a FORMULA is the time it takes to calculate π with the POWER SERIES for arctangent

$$\pi = a_1\cot(b_1) + a_2\cot(b_2) + \ldots, \tag{26}$$

and can be roughly characterized using Lehmer's "measure" formula

$$e \equiv \sum \frac{1}{\log_{10} b_i}. \tag{27}$$

The number of terms required to achieve a given precision is roughly proportional to e, so lower e-values correspond to better sums. The best currently known efficiency is 1.51244, which is achieved by the 6-term series

$$\tfrac{1}{4}\pi = 183\cot^{-1} 239 + 32\cot^{-1} 1023 - 68\cot^{-1} 5832$$
$$+ 12\cot^{-1} 110443 - 12\cot^{-1} 4841182$$
$$- 100\cot^{-1} 6826318 \tag{28}$$

discovered by C.-L. Hwang (1997). Hwang (1997) also discovered the remarkable identities

$$\tfrac{1}{4}\pi = P\cot^{-1} 2 - M\cot^{-1} 3 + L\cot^{-1} 5 + K\cot^{-1} 7$$
$$+ (N + K + L - 2M + 3P - 5)\cot^{-1} 8$$
$$+ (2N + M - P + 2 - L)\cot^{-1} 18$$
$$- (2P - 3 - M + L + K - N)\cot^{-1} 57 - N\cot^{-1} 239, \tag{29}$$

where $K, L, M, N,$ and P are POSITIVE INTEGERS, and

$$\tfrac{1}{4}\pi = (N + 2)\cot^{-1} 2 - N\cot^{-1} 3$$
$$- (N + 1)\cot^{-1} N. \tag{30}$$

The following table gives the number $N(n)$ of Machin-like formulas of n terms in the compilation by Wetherfield and Hwang. Except for previously known identities (which are included), the criteria for inclusion are the following:

1. first term < 8 digits: measure < 1.8.
2. first term = 8 digits: measure < 1.9.
3. first term = 9 digits: measure < 2.0.
4. first term = 10 digits: measure < 2.0.

n	$N(n)$	min e
1	1	0
2	4	1.85113
3	106	1.78661
4	39	1.58604
5	90	1.63485
6	120	1.51244
7	113	1.54408
8	18	1.65089
9	4	1.72801
10	78	1.63086
11	34	1.6305
12	188	1.67458
13	37	1.71934
14	5	1.75161
15	24	1.77957
16	51	1.81522

17	5	1.90938
18	570	1.87698
19	1	1.94899
20	11	1.95716
21	1	1.98938
Total	1500	1.51244

See also EULER'S MACHIN-LIKE FORMULA, GAUSS'S MACHIN-LIKE FORMULA, GREGORY NUMBER, HERMANN'S FORMULA, HUTTON'S FORMULA, INVERSE COTANGENT, MACHIN'S FORMULA, PI, STØRMER NUMBER, STRASSNITZKY'S FORMULA

References

Ball, W. W. R. and Coxeter, H. S. M. *Mathematical Recreations and Essays, 13th ed.* New York: Dover, pp. 347–59, 1987.
Berstel, J.; Pin, J.-E.; and Pocchiola, M. *Mathématiques et Informatique.* New York: McGraw-Hill, 1991.
Borwein, J. M. and Borwein, P. B. *Pi & the AGM: A Study in Analytic Number Theory and Computational Complexity.* New York: Wiley, 1987.
Castellanos, D. "The Ubiquitous Pi. Part I." *Math. Mag.* **61**, 67–8, 1988.
Conway, J. H. and Guy, R. K. *The Book of Numbers.* New York: Springer-Verlag, pp. 241–48, 1996.
Hwang, C.-L. "More Machin-Type Identities." *Math. Gaz.* **81**, 120–21, 1997.
Lehmer, D. H. "On Arccotangent Relations for π." *Amer. Math. Monthly* **45**, 657–64, 1938.
Lewin, L. *Polylogarithms and Associated Functions.* New York: North-Holland, 1981.
Lewin, L. *Structural Properties of Polylogarithms.* Providence, RI: Amer. Math. Soc., 1991.
Nielsen, N. *Der Euler'sche Dilogarithms.* Leipzig, Germany: Halle, 1909.
Séroul, R. "Machin Formulas." §9.3 in *Programming for Mathematicians.* Berlin: Springer-Verlag, pp. 240–52, 2000.
Størmer, C. "Sur l'Application de la Théorie des Nombres Entiers Complexes à la Solution en Nombres Rationnels x_1, x_2, ..., c_1, c_2, ..., k de l'Equation...." *Archiv for Mathematik og Naturvidenskab* **B 19**, 75–5, 1896.
Todd, J. "A Problem on Arc Tangent Relations." *Amer. Math. Monthly* **56**, 517–28, 1949.
Weisstein, E. W. "Machin-Like Formulas." MATHEMATICA NOTEBOOK MACHINFORMULAS.M.
Wetherfield, M. "The Enhancement of Machin's Formula by Todd's Process." *Math. Gaz.* **80**, 333–44, 1996.
Wetherfield, M. "Machin Revisited." *Math. Gaz.* **81** 121–23, 1997.

Machin's Formula

$$\frac{1}{4}\pi = 4\tan^{-1}\left(\frac{1}{5}\right) - \tan^{-1}\left(\frac{1}{239}\right).$$

There are a whole class of MACHIN-LIKE FORMULAS with various numbers of terms (although only four such formulas with only two terms). The properties of these formulas are intimately connected with COTANGENT identities.

See also 239, GREGORY NUMBER, MACHIN-LIKE FORMULAS, PI

Mackey's Theorem

Let E and F be paired spaces with S a family of absolutely convex bounded sets of F such that the sets of S generate F and, if B_1, $B_2 \in S$, there exists a $B_3 \in S$ such that $B_3 \supset B_1$ and $B_3 \supset B_2$. Then the dual space of E_S is equal to the union of the weak completions of λB, where $\lambda > 0$ and $B \in S$.

See also GROTHENDIECK'S THEOREM

References

Iyanaga, S. and Kawada, Y. (Eds.). "Mackey's Theorem." §407M in *Encyclopedic Dictionary of Mathematics.* Cambridge, MA: MIT Press, p. 1274, 1980.

Mac Lane's Theorem

A theorem which treats constructions of FIELDS of CHARACTERISTIC p.

See also CHARACTERISTIC (FIELD), FIELD

Maclaurin-Bézout Theorem

The Maclaurin-Bézout theorem says that two curves of degree n intersect in n^2 points, so two CUBICS intersect in nine points. This means that $n(n+3)/2$ points do not always uniquely determine a single curve of order n.

See also CRAMÉR-EULER PARADOX

Maclaurin-Cauchy Theorem

If $f(x)$ is positive and decreases to 0, then an EULER CONSTANT

$$\gamma_f = \lim_{n\to\infty}\left[\sum_{k=1}^{n} f(k) - \int_a^n f(x)\,dx\right]$$

can be defined. If $f(x) = 1/x$, then

$$\gamma = \lim_{n\to\infty}\left(\sum_{k=1}^{n}\frac{1}{k} - \int_1^n\frac{dx}{x}\right) = \lim_{n\to\infty}\left(\sum_{k=1}^{n}\frac{1}{k} - \ln n\right),$$

where γ is the EULER-MASCHERONI CONSTANT.

Maclaurin Integral Test

INTEGRAL TEST

Maclaurin Polynomial

MACLAURIN SERIES

Maclaurin Series

A series expansion of a function about 0,

$$f(x) = f(0) + f'(0)x + \frac{f''(0)}{2!} x^2 + \frac{f^{(3)}(0)}{3!} x^3 + \dots$$

$$+ \frac{f^{(n)}(0)}{n!} x^n + \dots, \tag{1}$$

named after the Scottish mathematician Maclaurin. Maclaurin series for common functions include

$$\frac{1}{1-x} = 1 + x + x^2 + x^3 + x^4 + x^5 + \dots$$

$$\text{for } -1 < x < 1 \tag{2}$$

$$cn(x, k) = 1 - \frac{1}{2}x^2 + \frac{1}{24}(1 + 4k^2)x^4 + \dots \tag{3}$$

$$\cos x = 1 - \frac{1}{2}x^2 + \frac{1}{24}x^4 - \frac{1}{720}x^6 - \dots$$

$$\text{for } -\infty < x < \infty \tag{4}$$

$$\cos^{-1} x = \frac{1}{2}\pi - x - \frac{1}{6}x^3 - \frac{3}{40}x^5 - \frac{5}{112}x^7 - \dots$$

$$\text{for } -1 < x < 1 \tag{5}$$

$$\cosh x = 1 + \frac{1}{2}x^2 + \frac{1}{24}x^4 + \frac{1}{720}x^6 + \frac{1}{40,320}x^8 + \dots \tag{6}$$

$$\cosh^{-1}(1+x) = \sqrt{2x}\left(1 - \frac{1}{2}x + \frac{3}{160}x^2 - \frac{5}{896}x^3 + \dots\right) \tag{7}$$

$$\cot x = x^{-1} - \frac{1}{3}x - \frac{1}{45}x^3 - \frac{2}{945}x^5 - \frac{1}{4725}x^7 - \dots \tag{8}$$

$$\cot^{-1} x = \frac{1}{2}\pi - x + \frac{1}{3}x^3 - \frac{1}{5}x^5 + \frac{1}{7}x^7 - \frac{1}{9}x^9 + \dots \tag{9}$$

$$\cot^{-1}\left(\frac{1}{x}\right) = x - \frac{1}{3}x^3 + \frac{1}{5}x^5 - \frac{1}{7}x^7 + \frac{1}{9}x^9 + \dots \tag{10}$$

$$\coth x = x^{-1} + \frac{1}{3}x - \frac{1}{45}x^4 + \frac{2}{945}x^5 - \frac{1}{4725}x^7 + \dots \tag{11}$$

$$\coth^{-1}(1+x) = \frac{1}{2}\ln 2 - \frac{1}{2}\ln x + \frac{1}{4}x - \frac{1}{16}x^2 + \dots \tag{12}$$

$$\csc x = x^{-1} + \frac{1}{6}x + \frac{7}{360}x^3 + \frac{31}{15120}x^5 + \dots \tag{13}$$

$$\operatorname{csch} x = x^{-1} - \frac{1}{6}x + \frac{7}{360}x^3 - \frac{31}{15120}x^5 + \dots \tag{14}$$

$$\operatorname{csch}^{-1} x = \ln 2 - \ln x + \frac{1}{4}x^2 - \frac{3}{32}x^4 + \frac{5}{96}x^6 - \dots \tag{15}$$

$$dn(x, k) = 1 - \frac{1}{2}k^2x^2 + \frac{1}{24}k^2(4 + k^2)x^4 + \dots \tag{16}$$

$$\operatorname{erf} x = \frac{1}{\sqrt{\pi}}\left(2x - \frac{2}{3}x^3 + \frac{1}{5}x^5 - \frac{1}{21}x^7 + \dots\right) \tag{17}$$

$$e^x = 1 + x + \frac{1}{2}x^2 + \frac{1}{6}x^3 + \frac{1}{24}x^4 + \dots$$

$$\text{for } -\infty < x < \infty \tag{18}$$

$${}_2F_1(\alpha, \beta, \gamma; x)$$

$$= 1 + \frac{\alpha\beta}{1\gamma} x + \frac{\alpha(\alpha+1)\beta(\beta+1)}{2\gamma(\gamma+1)} x^2 + \dots \tag{19}$$

$$\ln(1+x) = x - \frac{1}{2}x^2 + \frac{1}{3}x^3 - \frac{1}{4}x^4 + \dots$$

$$\text{for } -1 < x < 1 \tag{20}$$

$$\ln\left(\frac{1+x}{1-x}\right) = 2x + \frac{2}{3}x^3 + \frac{2}{5}x^5 + \frac{2}{7}x^7 + \dots$$

$$\text{for } -1 < x < 1 \tag{21}$$

$$\sec x = 1 + \frac{1}{2}x^2 + \frac{5}{24}x^4 + \frac{61}{720}x^6 + \frac{277}{8064}x^8 + \dots \tag{22}$$

$$\operatorname{sech} x = 1 - \frac{1}{2}x^2 + \frac{5}{24}x^4 - \frac{61}{720}x^6 + \frac{277}{8064}x^8 - \dots \tag{23}$$

$$\operatorname{sech}^{-1} x = \ln 2 - \ln x - \frac{1}{4}x^2 - \frac{3}{32}x^4 - \dots \tag{24}$$

$$\sin x = x - \frac{1}{6}x^3 + \frac{1}{120}x^5 - \frac{1}{5040}x^7 + \dots$$

$$\text{for } -\infty < x < \infty \tag{25}$$

$$\sin^{-1} x = x + \frac{1}{6}x^3 + \frac{3}{40}x^5 + \frac{5}{112}x^7 + \frac{35}{112}x^9 + \dots \tag{26}$$

$$\sinh x = x + \frac{1}{6}x^3 + \frac{1}{120}x^5 + \frac{1}{5040}x^7 + \frac{1}{362,880}x^9 + \dots \tag{27}$$

$$\sinh^{-1} x = x - \frac{1}{6}x^3 + \frac{3}{40}x^5 - \frac{5}{112}x^7 + \frac{35}{1152}x^9 - \dots \tag{28}$$

$$sn(x, k) = x - \frac{1}{6}(1+k^2)x^3 + \frac{1}{120}(1 + 14k^2 + k^4)x^5 + \dots \tag{29}$$

$$\tan x = x + \frac{1}{3}x^3 + \frac{2}{15}x^5 + \frac{17}{315}x^7 + \frac{62}{2835}x^9 + \dots \tag{30}$$

$$\tan^{-1} x = x - \frac{1}{3}x^3 + \frac{1}{5}x^5 - \frac{1}{7}x^7 + \dots$$

$$\text{for } -1 < x < 1 \tag{31}$$

$$\tan^{-1}(1+x) = \frac{1}{4}\pi + \frac{1}{2}x - \frac{1}{4}x^2 + \frac{1}{12}x^3 + \frac{1}{40}x^5 + \dots \tag{32}$$

$$\tanh x = x - \frac{1}{3}x^3 + \frac{2}{15}x^5 - \frac{17}{315}x^7 + \frac{62}{2835}x^9 - \dots \tag{33}$$

$$\tanh^{-1} x = x + \frac{1}{3}x^3 + \frac{1}{5}x^5 + \frac{1}{7}x^7 + \frac{1}{9}x^9 + \dots \tag{34}$$

The explicit forms for some of these are

$$\frac{1}{1-x} = \sum_{n=0}^{\infty} x^n \tag{35}$$

$$\cos x = \sum_{n=0}^{\infty} \frac{(-1)^n}{(2n)!} x^{2n} \tag{36}$$

$$\cosh x = \sum_{n=0}^{\infty} \frac{1}{(2n)!} x^{2n} \tag{37}$$

$$\csc x = \sum_{n=0}^{\infty} \frac{(-1)^{n+1} 2(2^{2n-1} - 1)B_{2n}}{(2n)!} x^{2n-1} \tag{38}$$

$$e^x = \sum_{n=0}^{\infty} \frac{1}{n!} x^n \qquad (39)$$

$$\ln(1+x) = \sum_{n=1}^{\infty} \frac{(-1)^{n+1}}{n} x^n \qquad (40)$$

$$\ln\left(\frac{1+x}{1-x}\right) = \sum_{n=1}^{\infty} \frac{2}{(2n-1)} x^{2n-1} \qquad (41)$$

$$\sec x = \sum_{n=0}^{\infty} \frac{(-1)^n E_{2n}}{(2n)!} x^{2n} \qquad (42)$$

$$\sin x = \sum_{n=0}^{\infty} \frac{(-1)^n}{(2n+1)!} x^{2n+1} \qquad (43)$$

$$\sinh x = \sum_{n=0}^{\infty} \frac{1}{(2n+1)!} x^{2n+1} \qquad (44)$$

$$\tan x = \sum_{n=0}^{\infty} \frac{(-1)^n 2^{2n+2}(2^{2n+2}-1)B_{2n+2}}{(2n+2)!} x^{2n+1} \qquad (45)$$

$$\tan^{-1} x = \sum_{n=1}^{\infty} \frac{(-1)^{n+1}}{(2n-1)} x^{2n-1} \qquad (46)$$

$$\tanh^{-1} x = \sum_{n=1}^{\infty} \frac{1}{2n-1} x^{2n-1}, \qquad (47)$$

where B_n are BERNOULLI NUMBERS and E_n are EULER NUMBERS.

See also ALCUIN'S SEQUENCE, LAGRANGE EXPANSION, LAGRANGE REMAINDER, LEGENDRE SERIES, TAYLOR SERIES

References

Beyer, W. H. (Ed.). *CRC Standard Mathematical Tables, 28th ed.* Boca Raton, FL: CRC Press, pp. 299–00, 1987.

Maclaurin Trisectrix

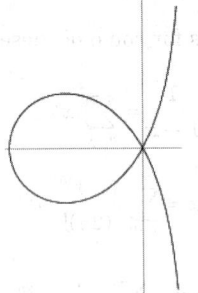

A curve first studied by Colin Maclaurin in 1742. It was studied to provide a solution to one of the GEOMETRIC PROBLEMS OF ANTIQUITY, in particular TRISECTION of an ANGLE, whence the name trisectrix.

The Maclaurin trisectrix is an ANALLAGMATIC CURVE, and the origin is a CRUNODE.
The Maclaurin trisectrix has CARTESIAN equation

$$y^2 = \frac{x^2(x+3a)}{a-x}, \qquad (1)$$

or the PARAMETRIC EQUATIONS

$$x = a \frac{t^2-3}{t^2+1} \qquad (2)$$

$$y = a \frac{t(t^2-3)}{t^2+1}. \qquad (3)$$

The ASYMPTOTE has equation $x = a$, and the center of the loop is at $(-2a, 0)$. If P is a point on the loop so that the line CP makes an ANGLE of 3α with the negative Y-AXIS, then the line OP will make an ANGLE of α with the negative Y-AXIS.

The Maclaurin trisectrix is sometimes defined instead as

$$x(x^2+y^2) = a(y^2 - 3x^2) \qquad (4)$$

$$y^2 = \frac{x^2(3a+x)}{a-x} \qquad (5)$$

$$r = \frac{2a \sin(3\theta)}{\sin(2\theta)}. \qquad (6)$$

Another form of the equation is the POLAR EQUATION

$$r = a \sec\left(\tfrac{1}{3}\theta\right), \qquad (7)$$

where the origin is inside the loop and the crossing point is on the NEGATIVE X-AXIS.

The tangents to the curve at the origin make angles of $\pm 60°$ with the X-AXIS. The AREA of the loop is

$$A_{\text{loop}} = 3\sqrt{3}a^2, \qquad (8)$$

and the NEGATIVE x-intercept is $(-3a, 0)$ (MacTutor Archive).

The Maclaurin trisectrix is the PEDAL CURVE of the PARABOLA where the PEDAL POINT is taken as the reflection of the FOCUS in the DIRECTRIX.

See also RIGHT STROPHOID, TSCHIRNHAUSEN CUBIC

References

Lawrence, J. D. *A Catalog of Special Plane Curves.* New York: Dover, pp. 103–06, 1972.
MacTutor History of Mathematics Archive. "Trisectrix of Maclaurin." http://www-groups.dcs.st-and.ac.uk/~history/Curves/Trisectrix.html.

Maclaurin Trisectrix Inverse Curve

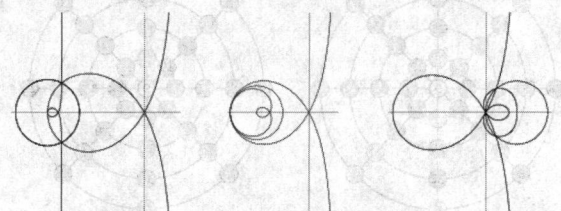

The INVERSE CURVE of the MACLAURIN TRISECTRIX with INVERSION CENTER at the NEGATIVE x-intercept is a TSCHIRNHAUSEN CUBIC.

MacMahon's Prime Number of Measurement

PRIME NUMBER OF MEASUREMENT

MacRobert's E-Function

$E(p;\alpha_r:\rho_s:x)$

$$\equiv \frac{\Gamma(\alpha_{q+1})}{\Gamma(\rho_1 - \alpha_1)\Gamma(\rho_2 - \alpha_2)\cdots\Gamma(\rho_q - \alpha_q)}$$

$$\times \prod_{\mu=1}^{q}\int_0^\infty \lambda_\mu^{\rho_\mu - \alpha_\mu - 1}(1 + \lambda_\mu)^{-\rho_\mu}\,d\lambda_\mu$$

$$\times \prod_{\nu=2}^{p-q-1}\int_0^\infty e^{-\lambda_{q+\nu}}\lambda_{q+\nu}^{\alpha_{q+\nu}-1}\,d\lambda_{q+\nu}$$

$$\times \int_0^\infty e^{-\lambda_p}\lambda_p^{\alpha_p - 1}\left[1 + \frac{\lambda_{q+2}\lambda_{q+3}\cdots\lambda_p}{(1+\lambda_1)\cdots(1+\lambda_q)x}\right]^{-\alpha_q + 1}d\lambda_p,$$

where $\Gamma(z)$ is the GAMMA FUNCTION and other details are discussed by Gradshteyn and Ryzhik (2000).

See also FOX'S H-FUNCTION, KAMPÉ DE FÉRIET FUNCTION, MEIJER'S G-FUNCTION

References

Erdélyi, A.; Magnus, W.; Oberhettinger, F.; and Tricomi, F. G. "Definition of the E-Function." §5.2 in *Higher Transcendental Functions, Vol. 1.* New York: Krieger, pp. 203–06, 1981.

Gradshteyn, I. S. and Ryzhik, I. M. *Tables of Integrals, Series, and Products, 6th ed.* San Diego, CA: Academic Press, pp. 896–03 and 1071–072, 2000.

MacRobert, T. M. "Induction Proofs of the Relations between Certain Asymptotic Expansions and Corresponding Generalised Hypergeometric Series." *Proc. Roy. Soc. Edinburgh* **58**, 1–3, 1937–8.

MacRobert, T. M. "Some Formulæ for the E-Function." *Philos. Mag.* **31**, 254–60, 1941.

Macron

A macron is a BAR placed over a single symbol or character, such as $\bar{x}$. The symbol $\bar{z}$ is sometimes used to denote the following operations.

1. The COMPLEX CONJUGATE.
2. NEGATION of a logical expression.
3. Infrequently, ADJOINT operator.

A bar placed over multiple symbols or characters is called a VINCULUM.

See also BAR, HAT, VINCULUM

References

Bringhurst, R. *The Elements of Typographic Style, 2nd ed.* Point Roberts, WA: Hartley and Marks, p. 281, 1997.

Madelung Constants

The quantities obtained from cubic, hexagonal, etc., LATTICE SUMS, evaluated at $s = 1$, are called Madelung constants. For cubic LATTICE SUMS, they are expressible in closed form for EVEN indices,

$$b_2(2) = -4\beta(1)\eta(1) = -4\frac{\pi}{4}\ln 2 = -\pi\ln 2 \tag{1}$$

$$b_4(2) = -8\eta(1)\eta(0) = -8\ln 2 \cdot \tfrac{1}{2} = -4\ln 2, \tag{2}$$

where $\beta(n)$ is the DIRICHLET BETA FUNCTION and $\eta(n)$ is the DIRICHLET ETA FUNCTION. $b_3(1)$ is given by BENSON'S FORMULA,

$$-b_3(1) = \sideset{}{'}\sum_{i,j,k=-\infty}^{\infty} \frac{(-1)^{i+j+k+1}}{\sqrt{i^2 + j^2 + k^2}}$$

$$= 12\pi \sum_{m,n=1,3,\ldots}^{\infty} \text{sech}^2\left(\tfrac{1}{2}\pi\sqrt{m^2 + n^2}\right), \tag{3}$$

where the prime indicates that summation over $(0, 0, 0)$ is excluded. $b_3(1)$ is sometimes called "the" Madelung constant, corresponds to the Madelung constant for a 3-D NaCl crystal, and is numerically equal to $-1.74756\ldots$.

For hexagonal LATTICE SUM, $h_2(2)$ is expressible in closed form as

$$h_2(2) = \pi\ln 3\sqrt{3}. \tag{4}$$

See also BENSON'S FORMULA, LATTICE SUM

References

Borwein, J. M. and Borwein, P. B. *Pi & the AGM: A Study in Analytic Number Theory and Computational Complexity.* New York: Wiley, 1987.

Buhler, J. and Wagon, S. "Secrets of the Madelung Constant." *Mathematica in Education and Research* **5**, 49–5, Spring 1996.

Crandall, R. E. and Buhler, J. P. "Elementary Function Expansions for Madelung Constants." *J. Phys. Ser. A: Math. and Gen.* **20**, 5497–510, 1987.

Finch, S. "Favorite Mathematical Constants." http://www.mathsoft.com/asolve/constant/mdlung/mdlung.html.

Maeder's Owl Minimal Surface
BOUR'S MINIMAL SURFACE

Maehly's Procedure
A method for finding ROOTS which defines

$$P_j(x) = \frac{P(x)}{(x - x_1) \cdots (x - x_j)}, \quad (1)$$

so the derivative is

$$P_j'(x) = \frac{P'(x)}{(x - x_1) \cdots (x - x_j)}$$
$$- \frac{P(x)}{(x - x_1) \cdots (x - x_j)} \sum_{i=1}^{j} (x - x_i)^{-1} \quad (2)$$

One step of NEWTON'S METHOD can then be written as

$$x_{k+1} = x_k - \frac{P(x_k)}{P'(x_k) - P(x_k) \sum_{i=1}^{j}(x_k - x_i)^{-1}}. \quad (3)$$

Magic Circles

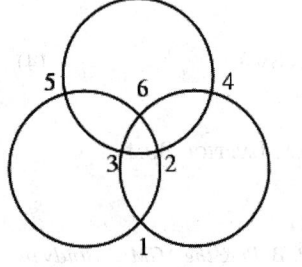 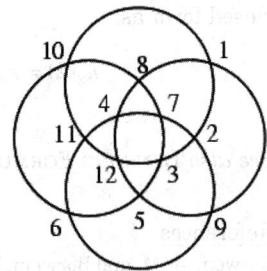

A set of n magic circles is a numbering of the intersections of the n CIRCLES such that the sum over all intersections is the same constant for all circles. The above sets of three and four magic circles have magic constants 14 and 39 (Madachy 1979).

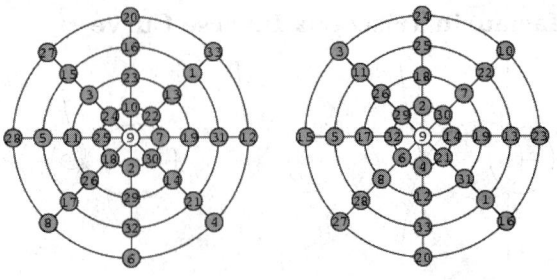

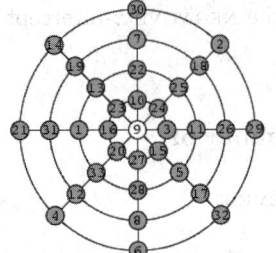

Another type of magic circle arranges the number 1, 2, ..., n in a number of rings, which each ring containing the same number of elements and corresponding elements being connected with radial lines. One of the numbers (which is subsequently ignored) is placed at the center. In a magic circle arrangement, the rings have equal sums and this sum is also equal to the sum of elements along each diameter (excluding the central number). Three magic circles using the numbers 1 to 33 are illustrated above. (Hung).

See also MAGIC GRAPH, MAGIC SQUARE

References
Madachy, J. S. *Madachy's Mathematical Recreations.* New York: Dover, p. 86, 1979.

Magic Constant
The number

$$M_2(n) = \frac{1}{n} \sum_{k=1}^{n^2} k = \tfrac{1}{2} n(n^2 + 1)$$

to which the n numbers in any horizontal, vertical, or *main* diagonal line must sum in a MAGIC SQUARE. The first few values are 1, 5, 15, 34, 65, 111, 175, 260, ... (Sloane's A006003). The magic constant for an nth order magic square starting with an INTEGER A and with entries in an increasing ARITHMETIC SERIES with difference D between terms is

$$M_2(n;\ A,\ D) = \tfrac{1}{2} n \left[2a + D(n^2 - 1)\right]$$

(Hunter and Madachy 1975, Madachy 1979). In a PANMAGIC SQUARE, in addition to the main diagonals, the broken diagonals also sum to $M_2(n)$.

For a MAGIC CUBE, MAGIC TESSERACT, etc., the magic d-D constant is

$$M_d(n) = \frac{1}{n^{d-1}} \sum_{k=1}^{n^d} k = \tfrac{1}{2} n \left(n^d + 1\right).$$

The first few magic constants are summarized in the following table.

n	$M_2(n)$	$M_3(n)$	$M_4(n)$
Sloane	A006003	A027441	A021003
1	1	1	1
2	5	9	17
3	15	42	123
4	34	130	514
5	65	315	1565

There is a corresponding multiplicative magic constant for MULTIPLICATION MAGIC SQUARES.

A similar magic constant $M_n^{(j)}$ of degree k is defined for MAGIC SERIES and MULTIMAGIC SERIES as $1/n$ times the sum of the first n^2 kth powers,

$$M_n^{(k)} = \frac{1}{n} \sum_{i=1}^{n^2} i^k = \frac{H_{n^2}^{(-p)}}{n},$$

where $H_n^{(k)}$ is a HARMONIC NUMBER of order k. The following table gives the first few values.

n	$k=1$	$k=2$	$k=3$	$k=4$
Sloane	A006003	A052459	A052460	A052461
1	1	1	1	1
2	5	15	50	177
3	15	95	675	5111
4	34	374	4624	60962
5	65	1105	21125	430729

See also MAGIC CUBE, MAGIC GEOMETRIC CONSTANTS, MAGIC HEXAGON, MAGIC SERIES, MAGIC SQUARE, MULTIMAGIC SERIES, MULTIPLICATION MAGIC SQUARE, PANMAGIC SQUARE

References

Hunter, J. A. H. and Madachy, J. S. "Mystic Arrays." Ch. 3 in *Mathematical Diversions.* New York: Dover, pp. 23–4, 1975.
Madachy, J. S. *Madachy's Mathematical Recreations.* New York: Dover, p. 86, 1979.
Sloane, N. J. A. Sequences A006003/M3849, A021003, A027441, A052459, A052460, and A052461 in "An On-Line Version of the Encyclopedia of Integer Sequences."

http://www.research.att.com/~njas/sequences/eisonline.html.

Magic Cube

An $n \times n \times n$ 3-D version of the MAGIC SQUARE in which the n^2 rows, n^2 columns, n^2 pillars (or "files"), and four space diagonals each sum to a single number $M_3(n)$ known as the MAGIC CONSTANT. If the CROSS SECTION diagonals also sum to $M_3(n)$, the magic cube is called a PERFECT MAGIC CUBE; if they do not, the cube is called a SEMIPERFECT MAGIC CUBE, or sometimes an ANDREWS CUBE (Gardner 1988). A pandiagonal cube is a perfect or SEMIPERFECT MAGIC CUBE which is magic not only along the main space diagonals, but also on the broken space diagonals.

A magic cube using the numbers 1, 2, ..., n^3, if it exists, has MAGIC CONSTANT

$$M_3(n) = \tfrac{1}{2} n \left(n^3 + 1\right).$$

For $n = 1, 2, ...,$ the magic constants are 1, 9, 42, 130, 315, 651, ... (Sloane's A027441).

4	12	26
11	25	6
27	5	10

20	7	15
9	14	19
13	21	8

18	23	1
22	3	17
2	16	24

60	37	12	21
13	20	61	36
56	41	8	25
1	32	49	48

7	26	55	42
50	47	2	31
11	22	59	38
62	35	14	19

57	40	9	24
16	17	64	33
53	44	5	28
4	29	52	45

6	27	54	43
51	46	3	30
10	23	58	39
63	34	15	18

The above SEMIPERFECT MAGIC CUBES of orders three (Hunter and Madachy 1975, p. 31; Ball and Coxeter 1987, p. 218) and four (Ball and Coxeter 1987, p. 220) have magic constants 42 and 130, respectively. There is a trivial SEMIPERFECT MAGIC CUBE of order one, but no semiperfect cubes of orders two or three exist. Semiperfect cubes of ODD order with $n \geq 5$ and DOUBLY EVEN order can be constructed by extending the methods used for MAGIC SQUARES.

Semiperfect pandiagonal cubes exist for all orders $8n$ and all ODD $n > 8$ (Ball and Coxeter 1987). A perfect pandiagonal magic cube has been constructed by Planck (1950), cited in Gardner (1988).

See also BIMAGIC CUBE, MAGIC CONSTANT, MAGIC GRAPH, MAGIC HEXAGON, MAGIC SQUARE, MAGIC TESSERACT, PERFECT MAGIC CUBE, SEMIPERFECT MAGIC CUBE

References

Adler, A. and Li, S.-Y. R. "Magic Cubes and Prouhet Sequences." *Amer. Math. Monthly* **84**, 618–27, 1977.
Andrews, W. S. *Magic Squares and Cubes, 2nd rev. ed.* New York: Dover, 1960.

Ball, W. W. R. and Coxeter, H. S. M. *Mathematical Recreations and Essays, 13th ed.* New York: Dover, pp. 216–24, 1987.

Barnard, F. A. P. "Theory of Magic Squares and Cubes." *Mem. Nat. Acad. Sci.* **4**, 209–70, 1888.

Benson, W. H. and Jacoby, O. *Magic Cubes: New Recreations.* New York: Dover, 1981.

Gardner, M. *Sci. Amer.*, Jan. 1976.

Gardner, M. "Magic Squares and Cubes." Ch. 17 in *Time Travel and Other Mathematical Bewilderments.* New York: W. H. Freeman, pp. 213–25, 1988.

Hirayama, A. and Abe, G. *Researches in Magic Squares.* Osaka, Japan: Osaka Kyoikutosho, 1983.

Hunter, J. A. H. and Madachy, J. S. "Mystic Arrays." Ch. 3 in *Mathematical Diversions.* New York: Dover, p. 31, 1975.

Lei, A. "Magic Cube and Hypercube." http://www.cs.ust.hk/~philipl/magic/mcube2.html.

Madachy, J. S. *Madachy's Mathematical Recreations.* New York: Dover, pp. 99–00, 1979.

Pappas, T. "A Magic Cube." *The Joy of Mathematics.* San Carlos, CA: Wide World Publ./Tetra, p. 77, 1989.

Planck, C. *Theory of Path Nasiks.* Rugby, England: Privately Published, 1905.

Rosser, J. B. and Walker, R. J. "The Algebraic Theory of Diabolical Squares." *Duke Math. J.* **5**, 705–28, 1939.

Sloane, N. J. A. Sequences A027441 in "An On-Line Version of the Encyclopedia of Integer Sequences." http://www.research.att.com/~njas/sequences/eisonline.html.

Trenkler, M. "A Construction of Magic Cubes." *Math. Gaz.* **84**, 36–1, 2000.

Wynne, B. E. "Perfect Magic Cubes of Order 7." *J. Recr. Math.* **8**, 285–93, 1975–976.

Magic Geometric Constants

N.B. A detailed online essay by S. Finch was the starting point for this entry.

Let E be a compact connected subset of d-dimensional EUCLIDEAN SPACE. Gross (1964) and Stadje (1981) proved that there is a unique REAL NUMBER $a(E)$ such that for all $x_1, x_2, ..., x_n \in E$, there exists $y \in E$ with

$$\frac{1}{n} \sum_{j=1}^{n} \sqrt{\sum_{k=1}^{d} \left(x_{j,k} - y_k\right)^2} = a(E). \tag{1}$$

The magic constant $m(E)$ of E is defined by

$$m(E) = \frac{a(E)}{\text{diam}(E)}, \tag{2}$$

where

$$\text{diam}(E) \equiv \max_{u, v \in E} \sqrt{\sum_{k=1}^{d} (u_k - v_k)^2}. \tag{3}$$

These numbers are also called DISPERSION NUMBERS and RENDEZVOUS VALUES. For any E, Gross (1964) and Stadje (1981) proved that

$$\tfrac{1}{2} \leq m(E) < 1. \tag{4}$$

If I is a subinterval of the LINE and D is a circular DISK in the PLANE, then

$$m(I) = m(D) = \tfrac{1}{2}. \tag{5}$$

If C is a CIRCLE, then

$$m(C) = \frac{2}{\pi} = 0.6366\ldots \tag{6}$$

An expression for the magic constant of an ELLIPSE in terms of its SEMIMAJOR and SEMIMINOR AXES lengths is not known. Nikolas and Yost (1988) showed that for a REULEAUX TRIANGLE T

$$0.6675276 \leq m(T) \leq 0.6675284. \tag{7}$$

Denote the MAXIMUM value of $m(E)$ in n-D space by $M(n)$. Then

$M(1)$	$\tfrac{1}{2}$
$M(2)$	$m(T) \leq M(2) \leq \dfrac{2 + \sqrt{3}}{3\sqrt{3}} < 0.7182336$
$M(d)$	$\dfrac{d}{d+1} \leq M(d) \leq \dfrac{[\Gamma(\tfrac{1}{2}d)]^2 2^{d-2}\sqrt{2d}}{\Gamma(d - \tfrac{1}{2})\sqrt{(d+1)\pi}} < \sqrt{\dfrac{d}{d+1}}$

where $\Gamma(z)$ is the GAMMA FUNCTION (Nikolas and Yost 1988).

An unrelated quantity characteristic of a given MAGIC SQUARE is also known as a MAGIC CONSTANT.

References

Finch, S. "Favorite Mathematical Constants." http://www.mathsoft.com/asolve/constant/magic/magic.html.

Cleary, J.; Morris, S. A.; and Yost, D. "Numerical Geometry--Numbers for Shapes." *Amer. Math. Monthly* **95**, 260–75, 1986.

Croft, H. T.; Falconer, K. J.; and Guy, R. K. *Unsolved Problems in Geometry.* New York: Springer-Verlag, 1994.

Gross, O. *The Rendezvous Value of Metric Space.* Princeton, NJ: Princeton University Press, pp. 49–3, 1964.

Nikolas, P. and Yost, D. "The Average Distance Property for Subsets of Euclidean Space." *Arch. Math. (Basel)* **50**, 380–84, 1988.

Stadje, W. "A Property of Compact Connected Spaces." *Arch. Math. (Basel)* **36**, 275–80, 1981.

Magic Graph

An edge-magic graph is a LABELED GRAPH with e EDGES labeled with distinct elements $\{1, 2, ..., e\}$ so that the sum of the EDGE labels at each VERTEX is the same.

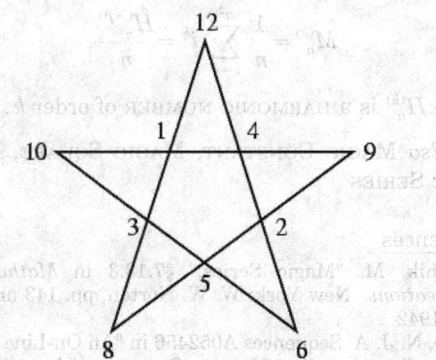

A vertex-magic graph labeled VERTICES which give the same sum along every straight line segment. No magic pentagrams can be formed with the number 1, 2, ..., 10 (Trigg 1960; Langman 1962, pp. 80–3; Dongre 1971; Richards 1975; Buckley and Rubin 1977–8; Trigg 1998), but 168 almost magic pentagrams (in which the sums are the same for four of the five lines) can. The figure above show a magic pentagram with sums 24 built using the labels 1, 2, 3, 4, 5, 6, 8, 9, 10, and 12 (Madachy 1979).

See also ANTIMAGIC GRAPH, LABELED GRAPH, MAGIC CIRCLES, MAGIC CONSTANT, MAGIC CUBE, MAGIC HEXAGON, MAGIC SQUARE

References

Buckley, M. R. W. and Rubin, F. Solution to Problem 385. "Do Pentacles Exists?" *J. Recr. Math.* **10**, 288–89, 1977–8.

Doob, M. "Characterization of Regular Magic Graphs." *J. Comb. Th. B* **25**, 94–04, 1978.

Dongre, N. M. "More About Magic Star Polygons." *Amer. Math. Monthly* **78**, 1025, 1971.

Gallian, J. A. "Graph Labeling." *Elec. J. Combin.* DS6, 1–2, Apr. 15, 1999. http://www.combinatorics.org/Surveys/.

Hartsfield, N. and Ringel, G. *Pearls in Graph Theory: A Comprehensive Introduction.* San Diego, CA: Academic Press, 1990.

Heinz, H. "Magic Stars." http://www.geocities.com/CapeCanaveral/Launchpad/4057/magicstar.htm.

Jezný, S. and Trenkler, M. "Characterization of Magic Graphs." *Czech. Math. J.* **33**, 435–38, 1983.

Jeurissen, R. H. "Magic Graphs, a Characterization." *Europ. J. Combin.* **9**, 363–68, 1988.

Langman, H. *Play Mathematics.* New York: Hafner, 1962.

Madachy, J. S. *Madachy's Mathematical Recreations.* New York: Dover, pp. 98–9, 1979.

Richards, I. "Impossibility." *Math. Mag.* **48**, 249–62, Nov. 1975.

Rivera, C. "Problems & Puzzles: Puzzle The Prime-Magical Pentagram.-013." http://www.primepuzzles.net/puzzles/puzz_013.htm.

Trigg, C. W. "Solution of Problem 113." *Pi Mu Epsilon J.* **3**, 119–20, Fall 1960.

Trigg, C. W. "Ten Elements on a Pentagram." *Eureka (Canada)* **3**, 5–, Jan. 1977.

Trigg, C. W. "Almost Magic Pentagrams." *J. Recr. Math.* **29**, 8–1, 1998.

Wynne, B. E. "Perfect Magic Icosapentacles." *J. Recr. Math.* **9**, 241–48, 1976–7.

Magic Hexagon

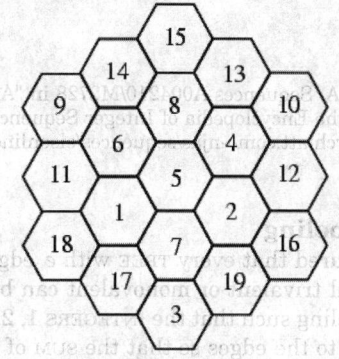

An arrangement of close-packed HEXAGONS containing the numbers 1, 2, ..., $H_n = 3n(n-1)+1$, where H_n is the nth HEX NUMBER, such that the numbers along each straight line add up to the same sum. In the above magic hexagon, each line (those of lengths 3, 4, and 5) adds up to 38. This is the only magic hexagon of the counting numbers for any size hexagon, as proved by Trigg (Gardner 1984, p. 24). It was discovered by C. W. Adams, who worked on the problem from 1910 to 1957.

Trigg showed that the magic constant for an order n hexagon would be

$$\frac{9(n^4 - 2n^3 + 2n^2 - n) + 2}{2(2n - 1)},$$

which requires $5/(2n-1)$ to be an integer for a solution to exist. But this is an integer for only $n = 1$ (the trivial case of a single hexagon) and Adam's $n = 3$ (Gardner 1984, p. 24).

See also HEX NUMBER, HEXAGON, MAGIC GRAPH, MAGIC SQUARE, TALISMAN HEXAGON

References

Abraham, K. *Philadelphia Evening Bulletin.* July 19, 1963, p. 18 and July 30, 1963.

Beeler, M. *et al.* Item 49 in Beeler, M.; Gosper, R. W.; and Schroeppel, R. *HAKMEM.* Cambridge, MA: MIT Artificial Intelligence Laboratory, Memo AIM-239, p. 18, Feb. 1972.

Gardner, M. "Permutations and Paradoxes in Combinatorial Mathematics." *Sci. Amer.* **209**, 112–19, Aug. 1963.

Gardner, M. *The Sixth Book of Mathematical Games from Scientific American.* Chicago, IL: University of Chicago Press, pp. 22–4, 1984.

Honsberger, R. *Mathematical Gems I.* Washington, DC: Math. Assoc. Amer., pp. 69–6, 1973.

Madachy, J. S. *Madachy's Mathematical Recreations.* New York: Dover, pp. 100–01, 1979.

Trigg, C. W. "A Unique Magic Hexagon." *Recr. Math. Mag.*, Jan. 1964.

Vickers, T. *Math. Gaz.*, p. 291, 1958.

Magic Integer

References

Sloane, N. J. A. Sequences A004210/M2728 in "An On-Line Version of the Encyclopedia of Integer Sequences." http://www.research.att.com/~njas/sequences/eisonline.html.

Magic Labeling

It is conjectured that every TREE with e edges whose nodes are all trivalent or monovalent can be given a "magic" labeling such that the INTEGERS 1, 2, ..., e can be assigned to the edges so that the SUM of the three meeting at a node is constant.

See also MAGIC CONSTANT, MAGIC CUBE, MAGIC GRAPH, MAGIC HEXAGON, MAGIC SQUARE

References

Guy, R. K. "Unsolved Problems Come of Age." *Amer. Math. Monthly* **96**, 903–09, 1989.

Magic Number

DIGITAL ROOT, MAGIC CONSTANT

Magic Pentagram

MAGIC GRAPH

Magic Series

A set n distinct numbers taken from the interval $[1, n^2]$ form a magic series if their sum is the nth MAGIC CONSTANT

$$M_n = \tfrac{1}{2}n(n^2+1)$$

(Kraitchik 1942, p. 143). The numbers of magic series of orders $n = 1, 2, ...$, are 1, 2, 8, 86, 1394, ... (Sloane's A052456). The following table gives the first few magic series of small order.

n	magic series
1	{1}
2	{1, 4}, {2, 3}
3	{1, 5, 9}, {1, 6, 8}, {2, 4, 9}, {2, 5, 8}, {2, 6, 7}, {3, 4, 8}, {3, 5, 7}, {4, 5, 6}

If the sum of the kth powers of these number is the MAGIC CONSTANT of degree k for all $k \in [1, p]$, then they are said to form a pth order MULTIMAGIC SERIES. Here, the magic constant $M_n^{(j)}$ of degree k is defined as $1/n$ times the sum of the first n^2 kth powers,

$$M_n^{(k)} = \frac{1}{n}\sum_{i=1}^{n^2} i^k = \frac{H_{n^2}^{(-p)}}{n},$$

where $H_n^{(k)}$ is a HARMONIC NUMBER of order k.

See also MAGIC CONSTANT, MAGIC SQUARE, MULTIMAGIC SERIES

References

Kraitchik, M. "Magic Series." §7.13.3 in *Mathematical Recreations.* New York: W. W. Norton, pp. 143 and 183–86, 1942.

Sloane, N. J. A. Sequences A052456 in "An On-Line Version of the Encyclopedia of Integer Sequences." http://www.research.att.com/~njas/sequences/eisonline.html.

Magic Square

8	1	6
3	5	7
4	9	2

16	2	3	13
5	11	10	8
9	7	6	12
4	14	15	1

17	24	1	8	15
23	5	7	14	16
4	6	13	20	22
10	12	19	21	3
11	18	25	2	9

32	29	4	1	24	21
30	31	2	3	22	23
12	9	17	20	28	25
10	11	18	19	26	27
13	16	36	33	5	8
14	15	34	35	6	7

30	39	48	1	10	19	28
38	47	7	9	18	27	29
46	6	8	17	26	35	37
5	14	16	25	34	36	45
13	15	24	33	42	44	4
21	23	32	41	43	3	12
22	31	40	49	2	11	20

64	2	3	61	60	6	7	57
9	55	54	12	13	51	50	16
17	47	46	20	21	43	42	24
40	26	27	37	36	30	31	33
32	34	35	29	28	38	39	25
41	23	22	44	45	19	18	48
49	15	14	52	53	11	10	56
8	58	59	5	4	62	63	1

A (normal) magic square consists of the distinct POSITIVE INTEGERS 1, 2, ..., n^2 such that the sum of the n numbers in any horizontal, vertical, or *main* diagonal line is always the same MAGIC CONSTANT

$$M_2(n) = \frac{1}{n}\sum_{k=1}^{n^2} k = \tfrac{1}{2}n(n^2+1).$$

The unique normal square of order three was known to the ancient Chinese, who called it the LO SHU. A version of the order 4 magic square with the numbers 15 and 14 in adjacent middle columns in the bottom row is called DÜRER'S MAGIC SQUARE. Magic squares of order 3 through 8 are shown above.

The MAGIC CONSTANT for an nth order magic square starting with an INTEGER A and with entries in an increasing ARITHMETIC SERIES with difference D between terms is

$$M_2(n;\ A,\ D) = \tfrac{1}{2}n[2a + D(n^2-1)]$$

(Hunter and Madachy 1975). If every number in a magic square is subtracted from n^2+1, another magic square is obtained called the complementary magic square. Squares which are magic under multiplication instead of addition can be constructed and are known as MULTIPLICATION MAGIC SQUARES. In addition, squares which are magic under both addition *and* multiplication can be constructed and are known as ADDITION-MULTIPLICATION MAGIC SQUARES (Hunter and Madachy 1975).

A square that fails to be magic only because one or both of the main diagonal sums do not equal the MAGIC CONSTANT is called a SEMIMAGIC SQUARE. If *all* diagonals (including those obtained by wrapping around) of a magic square sum to the MAGIC CONSTANT, the square is said to be a PANMAGIC SQUARE (also called a DIABOLIC SQUARE or PANDIAGONAL SQUARE). If replacing each number n_i by its square n_i^2 produces another magic square, the square is said to be a BIMAGIC SQUARE (or DOUBLY MAGIC SQUARE). If a square is magic for n_i, n_i^2, and n_i^3, it is called a TREBLY MAGIC SQUARE. If all pairs of numbers symmetrically opposite the center sum to $n^2 + 1$, the square is said to be an ASSOCIATIVE MAGIC SQUARE.

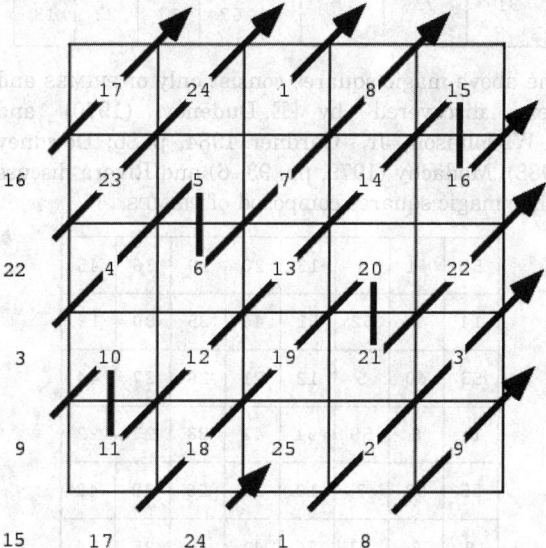

Kraitchik (1942) gives general techniques of constructing EVEN and ODD squares of order n. For n ODD, a very straightforward technique known as the Siamese method can be used, as illustrated above (Kraitchik 1942, pp. 148–49). It begins by placing a 1 in any location (in the center square of the top row in the above example), then incrementally placing subsequent numbers in the square one unit above and to the right. The counting is wrapped around, so that falling off the top returns on the bottom and falling off the right returns on the left. When a square is encountered which is already filled, the next number is instead placed *below* the previous one and the method continues as before. The method, also called de la Loubere's method, is purported to have been first reported in the West when de la Loubere returned to France after serving as ambassador to Siam.

A generalization of this method uses an "ordinary vector" (x, y) which gives the offset for each noncolliding move and a "break vector" (u, v) which gives the offset to introduce upon a collision. The standard

Siamese method therefore has ordinary vector $(1, -1)$ and break vector $(0, 1)$. In order for this to produce a magic square, each break move must end up on an unfilled cell. Special classes of magic squares can be constructed by considering the absolute sums $|u+v|$, $|(u-x)+(v-y)|$, $|u-v|$, and $|(u-x)-(v-y)| = |u+y-x-v|$. Call the set of these numbers the sumdiffs (sums and differences). If all sumdiffs are RELATIVELY PRIME to n and the square is a magic square, then the square is also a PANMAGIC SQUARE. This theory originated with de la Hire. The following table gives the sumdiffs for particular choices of ordinary and break vectors.

Ordinary Vector	Break Vector	Sumdiffs	Magic Squares	Panmagic Squares
$(1, -1)$	$(0, 1)$	$(1, 3)$	$2k+1$	none
$(1, -1)$	$(0, 2)$	$(0, 2)$	$6k \pm 1$	none
$(2, 1)$	$(1, -2)$	$(1, 2, 3, 4)$	$6k \pm 1$	none
$(2, 1)$	$(1, -1)$	$(0, 1, 2, 3)$	$6k \pm 1$	$6k \pm 1$
$(2, 1)$	$(1, 0)$	$(0, 1, 2)$	$2k+1$	none
$(2, 1)$	$(1, 2)$	$(0, 1, 2, 3)$	$6k \pm 1$	none

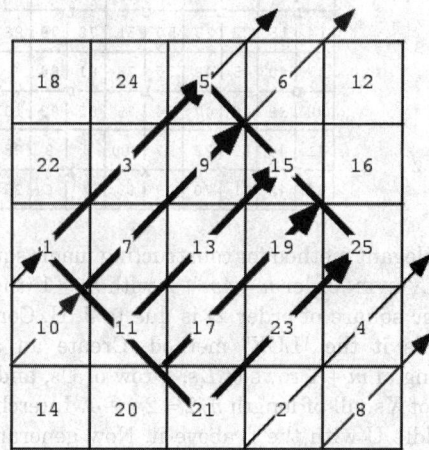

A second method for generating magic squares of ODD order has been discussed by J. H. Conway under the name of the "lozenge" method. As illustrated above, in this method, the ODD numbers are built up along diagonal lines in the shape of a DIAMOND in the central part of the square. The EVEN numbers which were missed are then added sequentially along the continuation of the diagonal obtained by wrapping around the square until the wrapped diagonal reaches its initial point. In the above square, the first diagonal therefore fills in 1, 3, 5, 2, 4, the second diagonal fills in 7, 9, 6, 8, 10, and so on.

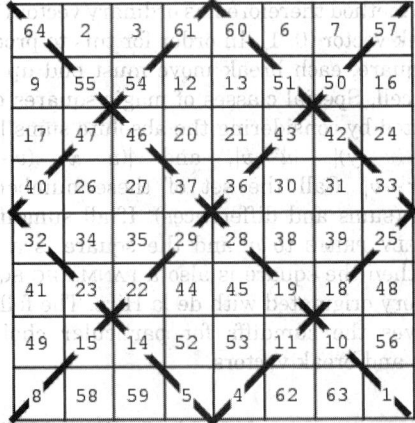

An elegant method for constructing magic squares of DOUBLY EVEN order $n = 4m$ is to draw xs through each 4×4 subsquare and fill all squares in sequence. Then replace each entry a_{ij} on a crossed-off diagonal by $(n^2 + 1) - a_{ij}$ or, equivalently, reverse the order of the crossed-out entries. Thus in the above example for $n = 8$, the crossed-out numbers are originally 1, 4, ..., 61, 64, so entry 1 is replaced with 64, 4 with 61, etc.

68	65	96	93	4	1	32	29	60	57
66	67	94	95	2	3	30	31	58	59
92	89	20	17	28	25	56	53	64	61
90	91	18	19	26	27	54	55	62	63
16	13	24	21	49	52	80	77	88	85
14	15	22	23	50	51	78	79	86	87
37	40	45	48	76	73	81	84	9	12
38	39	46	47	74	75	82	83	10	11
41	44	69	72	97	100	5	8	33	36
43	42	71	70	99	98	7	6	35	34

A very elegant method for constructing magic squares of SINGLY EVEN order $n = 4m + 2$ with $m \geq 1$ (there is no magic square of order 2) is due to J. H. Conway, who calls it the "LUX" method. Create an array consisting of $m + 1$ rows of Ls, 1 row of Us, and $m - 1$ rows of Xs, all of length $n/2 = 2m + 1$. Interchange the middle U with the L above it. Now generate the magic square of order $2m + 1$ using the Siamese method centered on the array of letters (starting in the center square of the top row), but fill each set of four squares surrounding a letter sequentially according to the order prescribed by the letter. That order is illustrated on the left side of the above figure, and the completed square is illustrated to the right. The "shapes" of the letters L, U, and X naturally suggest the filling order, hence the name of the algorithm.

It is an unsolved problem to determine the number of magic squares of an arbitrary order, but the number of distinct magic squares (excluding those obtained by rotation and reflection) of order $n = 1, 2, \ldots$ are 1, 0, 1, 880, 275305224, ... (Sloane's A006052; Madachy 1979, p. 87). The 880 squares of order four were enumerated by Frenicle de Bessy in the seventeenth century, and are illustrated in Berlekamp *et al.* (1982, pp. 778–83). The number of 6×6 squares is not known, but Pinn and Wieczerkowski (1998) estimated it to be $(1.7745 \pm 0.0016) \times 10^{19}$ using Monte Carlo simulation and methods from statistical mechanics.

67	1	43
13	37	61
31	73	7

3	61	19	37
43	31	5	41
7	11	73	29
67	17	23	13

The above magic squares consist only of PRIMES and were discovered by E. Dudeney (1970) and A. W. Johnson, Jr. (Gardner 1984, p. 86; Dewdney 1988). Madachy (1979, pp. 93–6) and Rivera discuss other magic squares composed of PRIMES.

52	61	4	13	20	29	36	45
14	3	62	51	46	35	30	19
53	60	5	12	21	28	37	44
11	6	59	54	43	38	27	22
55	58	7	10	23	26	39	42
9	8	57	56	41	40	25	24
50	63	2	15	18	31	34	47
16	1	64	49	48	33	32	17

Benjamin Franklin constructed the above 8×8 PAN-MAGIC SQUARE having MAGIC CONSTANT 260. Any half-row or half-column in this square totals 130, and the four corners plus the middle total 260. In addition, bent diagonals (such as 52–5–4–0–7–3–6) also total 260 (Madachy 1979, p. 87).

1480028159	1480028153	1480028201
1480028213	1480028171	1480028129
1480028141	1480028189	1480028183

In addition to other special types of magic squares, a 3×3 square whose entries are consecutive PRIMES, illustrated above, has been discovered by H. Nelson (Rivera).

According to a 1913 proof of J. N. Murray (cited in Gardner 1984, pp. 86–7), the smallest magic square composed of consecutive primes *starting with 3 and including the number 1* is of order 12.

Variations on magic squares can also be constructed using letters (either in defining the square or as entries in it), such as the ALPHAMAGIC SQUARE and TEMPLAR MAGIC SQUARE.

Various numerological properties have also been associated with magic squares. Pivari associates the squares illustrated above with Saturn, Jupiter, Mars, the Sun, Venus, Mercury, and the Moon, respectively. Attractive patterns are obtained by connecting consecutive numbers in each of the squares (with the exception of the Sun magic square).

See also ADDITION-MULTIPLICATION MAGIC SQUARE ALPHAMAGIC SQUARE, ANTIMAGIC SQUARE, ASSOCIATIVE MAGIC SQUARE, BIMAGIC SQUARE, BORDER SQUARE, DÜRER'S MAGIC SQUARE, EULER SQUARE, FRANKLIN MAGIC SQUARE, GNOMON MAGIC SQUARE, HETEROSQUARE, LATIN SQUARE, MAGIC CIRCLES, MAGIC CONSTANT, MAGIC CUBE, MAGIC HEXAGON, MAGIC LABELING, MAGIC SERIES, MAGIC TESSERACT, MAGIC TOUR, MULTIMAGIC SQUARE, MULTIPLICATION MAGIC SQUARE, PANMAGIC SQUARE, SEMIMAGIC SQUARE, TALISMAN SQUARE, TEMPLAR MAGIC SQUARE, TRIMAGIC SQUARE

References

Abe, G. "Unsolved Problems on Magic Squares." *Disc. Math.* **127**, 3–3, 1994.

Alejandre, S. "Suzanne Alejandre's Magic Squares." http://forum.swarthmore.edu/alejandre/magic.square.html.

Andrews, W. S. *Magic Squares and Cubes, 2nd rev. ed.* New York: Dover, 1960.

Andrews, W. S. and Sayles, H. A. "Magic Squares Made with Prime Numbers to have the Lowest Possible Summations." *Monist* **23**, 623–30, 1913.

Ball, W. W. R. and Coxeter, H. S. M. "Magic Squares." Ch. 7 in *Mathematical Recreations and Essays, 13th ed.* New York: Dover, 1987.

Barnard, F. A. P. "Theory of Magic Squares and Cubes." *Memoirs Natl. Acad. Sci.* **4**, 209–70, 1888.

Benson, W. H. and Jacoby, O. *New Recreations with Magic Squares.* New York: Dover, 1976.

Berlekamp, E. R.; Conway, J. H; and Guy, R. K. *Winning Ways for Your Mathematical Plays, Vol. 2: Games in Particular.* London: Academic Press, 1982.

Chabert, J.-L. (Ed.). "Magic Squares." Ch. 2 in *A History of Algorithms: From the Pebble to the Microchip.* New York: Springer-Verlag, pp. 49–1, 1999.

Danielsson, H. "Magic Squares." http://www.magic-squares.de/magic.html.

Dewdney, A. K. "Computer Recreations: How to Pan for Primes in Numerical Gravel." *Sci. Amer.* **259**, pp. 120–23, July 1988.

Dudeney, E. *Amusements in Mathematics.* New York: Dover, 1970.

Fults, J. L. *Magic Squares.* Chicago, IL: Open Court, 1974.

Gardner, M. "Magic Squares." Ch. 12 in *The Second Scientific American Book of Mathematical Puzzles & Diversions: A New Selection.* New York: Simon and Schuster, pp. 130–40, 1961.

Gardner, M. *The Sixth Book of Mathematical Games from Scientific American.* Chicago, IL: University of Chicago Press, 1984.

Gardner, M. "Magic Squares and Cubes." Ch. 17 in *Time Travel and Other Mathematical Bewilderments.* New York: W. H. Freeman, pp. 213–25, 1988.

Grogono, A. W. "Magic Squares by Grog." http://www.grogono.com/magic/.

Hawley, D. "Magic Squares." http://www.nrich.maths.org.uk/mathsf/journalf/aug98/art1/.

Heinz, H. "Magic Squares." http://www.geocities.com/CapeCanaveral/Launchpad/4057/magicsquare.htm.

Hirayama, A. and Abe, G. *Researches in Magic Squares.* Osaka, Japan: Osaka Kyoikutosho, 1983.

Horner, J. "On the Algebra of Magic Squares, I., II., and III." *Quart. J. Pure Appl. Math.* **11**, 57–5, 123–31, and 213–24, 1871.

Hunter, J. A. H. and Madachy, J. S. "Mystic Arrays." Ch. 3 in *Mathematical Diversions.* New York: Dover, pp. 23–4, 1975.

Kraitchik, M. "Magic Squares." Ch. 7 in *Mathematical Recreations.* New York: Norton, pp. 142–92, 1942.

Lei, A. "Magic Square, Cube, Hypercube." http://www.cs.ust.hk/~philipl/magic/.

Madachy, J. S. "Magic and Antimagic Squares." Ch. 4 in *Madachy's Mathematical Recreations.* New York: Dover, pp. 85–13, 1979.

Moran, J. *The Wonders of Magic Squares.* New York: Vintage, 1982.

Pappas, T. "Magic Squares," "The "Special" Magic Square," "The Pyramid Method for Making Magic Squares," "Ancient Tibetan Magic Square," "Magic "Line"," and "A Chinese Magic Square." *The Joy of Mathematics.* San Carlos, CA: Wide World Publ./Tetra, pp. 82–7, 112, 133, 169, and 179, 1989.

Peterson, I. "Ivar Peterson's MathLand: More than Magic Squares." http://www.maa.org/mathland/mathland_10_14.html.

Pinn, K. and Wieczerkowski, C. "Number of Magic Squares from Parallel Tempering Monte Carlo." *Int. J. Mod. Phys. C* **9**, 541–47, 1998. http://xxx.lanl.gov/abs/cond-mat/9804109/

Pivari, F. "Nice Examples." http://www.geocities.com/CapeCanaveral/Lab/3469/examples.html.

Pivari, F. "Simple Magic Square Checker and GIF Maker." http://www.geocities.com/CapeCanaveral/Lab/3469/squaremaker.html.

Rivera, C. "Problems & Puzzles: Puzzle Magic Squares with Consecutive Primes.-003." http://www.primepuzzles.net/puzzles/puzz_003.htm.

Rivera, C. "Problems & Puzzles: Puzzle Prime-Magical Squares.-004." http://www.primepuzzles.net/puzzles/puzz_004.htm.

Sloane, N. J. A. Sequences A006052/M5482 in "An On-Line Version of the Encyclopedia of Integer Sequences." http://www.research.att.com/~njas/sequences/eisonline.html.

Suzuki, M. "Magic Squares." http://www.pse.che.tohoku.ac.jp/~msuzuki/MagicSquare.html.

Weisstein, E. W. "Magic Squares." Mathematica notebook MagicSquares.M.

Weisstein, E. W. "Books about Magic Squares." http://www.treasure-troves.com/books/MagicSquares.html.

Wells, D. *The Penguin Dictionary of Curious and Interesting Numbers.* Middlesex, England: Penguin Books, p. 75, 1986.

Magic Star

Magic Graph

Magic Tesseract

A magic tesseract is a 4-D generalization of the 2-D MAGIC SQUARE and the 3-D MAGIC CUBE. A magic tesseract has MAGIC CONSTANT

$$M_4(n) = \tfrac{1}{2} n(n^4 + 1),$$

so for $n = 1, 2, \ldots$, the magic tesseract constants are 1, 17, 123, 514, 1565, 3891, ... (Sloane's A021003).

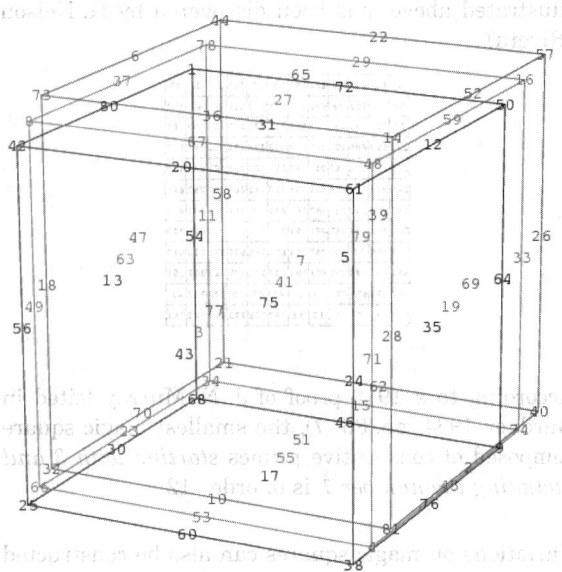

Berlekamp *et al.* (1982, p. 783) give a magic TESSERACT. J. Hendricks has constructed magic tesseracts of orders three, four, five (Hendricks 1999a, pp. 128–29), and six (Heinz). M. Houlton has used Hendricks' techniques to construct magic tesseracts of orders 5, 7, and 9.

There are 58 distinct magic tesseracts of order three, modulo rotations and reflections (Heinz, Hendricks 1999), one of which is illustrated above. Each of the 27 rows (e.g., 1–2–0), columns (e.g., 1–0–2), pillars (e.g., 1–4–8), and files (e.g., 1–8–4) sum to the magic constant 123.

Hendricks (1968) has constructed a pan-4-agonal magic tesseract of order 4. No pan-4-agonal magic tesseract of order five is known, and Andrews (1960) and Schroeppel (1972) state that no such tesseract can exist.

The smallest perfect magic tesseract is of order 16, having MAGIC CONSTANT 524,296, and has been constructed by Hendricks (Peterson 1999).

n-dimensional magic hypercubes of order 3 are known for $n = 5, 6, 7,$ and 8 (Hendricks). Hendricks has also constructed a perfect 16th order magic tesseract (where perfect means that all hyperplanes are perfect).

See also Magic Cube, Magic Square

References

Adler, A. "Magic N-Cubes Form a Free Monoid." *Electronic J. Combinatorics* **4**, No. 1, R15, 1–, 1997. http://www.combinatorics.org/Volume_4/v4i1toc.html#R15.

Andrews, W. S. *Magic Squares and Cubes, 2nd rev. ed.* New York: Dover, 1960.

Berlekamp, E. R.; Conway, J. H; and Guy, R. K. *Winning Ways for Your Mathematical Plays, Vol. 2: Games in Particular.* London: Academic Press, 1982.

Heinz, H. "John Hendricks: Inlaid Magic Tesseract." http://www.geocities.com/~harveyh/Hendricks.htm#Inlaid Magic Tesseract.

Hendricks, J. R. "The Five and Six Dimensional Magic Hypercubes of Order 3." *Canad. Math. Bull.* **5**, 171–89, 1952.

Hendricks, J. R. "A Pan-4-agonal Magic Tesseract." *Amer. Math. Monthly* **75**, 384, 1968.

Hendricks, J. R. "Magic Tesseracts and *N*-Dimensional Magic Hypercubes." *J. Recr. Math.* **6**, 193–01, 1973.

Hendricks, J. R. Erratum to 'Magic Tesseracts and *N*-Dimensional Magic Hypercubes." *J. Recr. Math.* **7**, 80, 1974.

Hendricks, J. R. "Ten Magic Tesseracts of Order Three." *J. Recr. Math.* **18**, 125–34, 1985–986.

Hendricks, J. R. *Magic Squares to Tesseracts by Computer.* Published by the author, 1999a.

Hendricks, J. R. *All Third Order Magic Tesseracts.* Published by the author, 1999b.

Hendricks, J. R. *Perfect n-Dimensional Hypercubes of Order 2^n.* Published by the author, 1999c.

Peterson, I. "Ivar Peterson's MathTrek: Magic Tesseracts." http://www.maa.org/mathland/mathtrek_10_18_99.html .

Schroeppel, R. Item 51 in Beeler, M.; Gosper, R. W.; and Schroeppel, R. *HAKMEM.* Cambridge, MA: MIT Artificial Intelligence Laboratory, Memo AIM-239, p. 18, Feb. 1972.

Sloane, N. J. A. Sequences A021003 in "An On-Line Version of the Encyclopedia of Integer Sequences." http://www.research.att.com/~njas/sequences/eisonline.html.

Trenkler, M. "Magic *p*-Dimensional Cubes of Order $n \not\equiv 2$ (mod 4)." *Acta Arith.* **92**, 189–04, 2000.

Trenkler, M. "A Construction of Magic Cubes." *Math. Gaz.* **84**, 36–1, 2000.

Trenkler, M. "Magic *p*-Dimensional Cubes." Submitted to *Acta Arith.*, 2000.

Magic Tour

Let a chess piece make a TOUR on an $n \times n$ CHESSBOARD whose squares are numbered from 1 to n^2 along the path of the chess piece. Then the TOUR is called a magic tour if the resulting arrangement of numbers is a MAGIC SQUARE. If the first and last squares traversed are connected by a move, the tour is said to be closed (or "re-entrant"); otherwise it is open. The MAGIC CONSTANT for the 8×8 CHESSBOARD is 260.

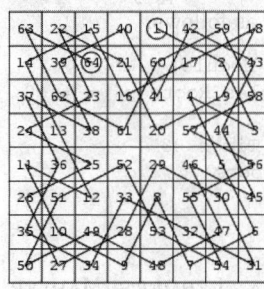 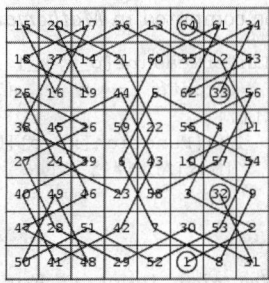

Magic KNIGHT'S TOURS are not possible on $n \times n$ boards for n ODD, and are believed to be impossible for $n = 8$. The "most magic" knight tour known on the 8×8 board is the SEMIMAGIC SQUARE illustrated in the above left figure (Ball and Coxeter 1987, p. 185) having main diagonal sums of 348 and 168. Combining two half-knights' tours one above the other as in the above right figure does, however, give a MAGIC

SQUARE (Ball and Coxeter 1987, p. 185).

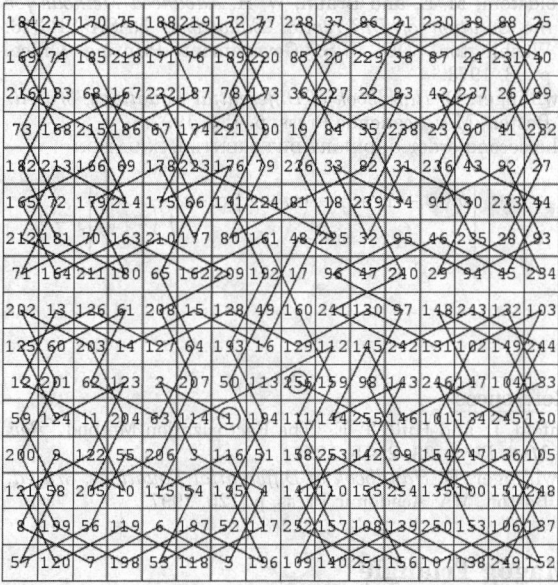

The above illustration shows a 16×16 closed magic KNIGHT'S TOUR (Madachy 1979).

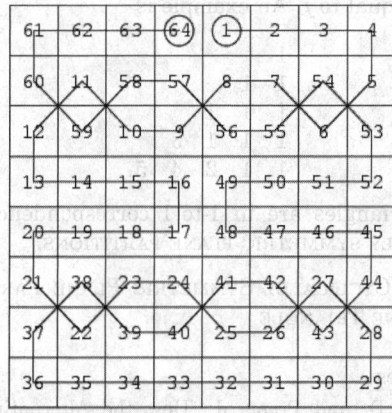

A magic tour for king moves is illustrated above (Coxeter 1987, p. 186).

See also CHESSBOARD, KNIGHT'S TOUR, MAGIC SQUARE, SEMIMAGIC SQUARE, TOUR

References

Ball, W. W. R. and Coxeter, H. S. M. *Mathematical Recreations and Essays, 13th ed.* New York: Dover, pp. 185–87, 1987.

Madachy, J. S. *Madachy's Mathematical Recreations.* New York: Dover, pp. 87–9, 1979.

Magnetic Pole Differential Equation

The second-order ORDINARY DIFFERENTIAL EQUATION

$$y'' + g(y)y'^2 + f(x)y' = 0.$$

References

Goldstein, M. E. and Braun, W. H. *Advanced Methods for the Solution of Differential Equations.* NASA SP-316. Washington, DC: U.S. Government Printing Office, p. 98, 1973.

Zwillinger, D. *Handbook of Differential Equations, 3rd ed.* Boston, MA: Academic Press, p. 124, 1997.

The second-order ORDINARY DIFFERENTIAL EQUATION

$$y'' - \left[\frac{m(m+1) + \frac{1}{4} - \left(m + \frac{1}{2}\right) \cos x}{\sin^2 x} + \left(\lambda + \frac{1}{2}\right) \right] y = 0.$$

References

Infeld, L. and Hull, T. E. "The Factorization Method." *Rev. Mod. Phys.* **23**, 21–8, 1951.

Zwillinger, D. *Handbook of Differential Equations, 3rd ed.* Boston, MA: Academic Press, p. 125, 1997.

Magog Triangle

A NUMBER TRIANGLE of order n with entries 1 to n such that entries are nondecreasing across rows and down columns and all entries in column j are less than or equal to j. An example is

$$\begin{array}{ccccc}
1 & & & & \\
1 & 1 & & & \\
1 & 1 & 1 & & \\
1 & 1 & 1 & 3 & \\
1 & 1 & 2 & 4 & 5.
\end{array}$$

Magog triangles are in 1-to-1 correspondence with CYCLICALLY SYMMETRIC PLANE PARTITIONS.

See also CYCLICALLY SYMMETRIC PLANE PARTITION, MONOTONE TRIANGLE

References

Bressoud, D. and Propp, J. "How the Alternating Sign Matrix Conjecture was Solved." *Not. Amer. Math. Soc.* **46**, 637–46.

Mahler-Lech Theorem

Let K be a FIELD of CHARACTERISTIC 0 (e.g., the rationals $\mathbb{Q}$) and let $\{u_n\}$ be a SEQUENCE of elements of K which satisfies a difference equation OF THE FORM

$$0 = c_0 u_n + c_1 u_{n+1} + \ldots + c_k u_{n+k},$$

where the COEFFICIENTS c_i are fixed elements of K. Then, for any $c \in K$, we have *either* $u_n = c$ for only finitely many values of n, or $u_n = c$ for the values of n in some ARITHMETIC PROGRESSION.

The proof involves embedding certain FIELDS inside the P-ADIC NUMBERS $\mathbb{Q}_p$ for some PRIME p, and using properties of zeros of POWER SERIES over $\mathbb{Q}_p$ (STRASSMAN'S THEOREM).

See also ARITHMETIC PROGRESSION, P-ADIC NUMBER, STRASSMAN'S THEOREM

Mahler Measure

This entry contributed by KEVIN O'BRYANT

For a polynomial $P(x_1, x_2, \ldots, x_k)$, the Mahler measure of P is defined by

$$M_k(P)$$

$$\equiv \exp\left[\int_0^1 \ldots \int_0^1 \ln\left|P\left(e^{2\pi i t_1}, \ldots, e^{2\pi i t_k}\right)\right| \, dt_1 \cdots dt_k \right]. \quad (1)$$

Using JENSEN'S FORMULA, it can be shown that for $P(x) = a \prod_{i=1}^n (x - \alpha_i)$,

$$M_1(P) = |a| \prod_{i=1}^n \max\{1, |\alpha_i|\} \quad (2)$$

(Borwein and Erdélyi 1995, p. 271).

Specific cases are given by

$$M_1(ax + b) = \max\{|a|, |b|\} \quad (3)$$

$$M_2(1 + x + y) = M_1(\max\{1, |1 + x|\}) \quad (4)$$

$$M_2(1 + x + y - xy) = M_1(\max\{|1 - x|, |1 + x|\}) \quad (5)$$

(Borwein and Erdélyi 1995, p. 272).

A product of CYCLOTOMIC POLYNOMIALS has Mahler measure 1. LEHMER'S MAHLER MEASURE PROBLEM conjectures that a particular univariate polynomial has the smallest possible Mahler measure other than 1.

The Mahler measure for a univariate polynomial can be computed in *Mathematica* as follows.

```
MahlerMeasure[p_, x_] := Module[
    {roots = x /. {ToRules[Roots[p == 0,
x]]}},
    Abs[Function[x, p][0]] Times @@
    (Max[Abs[#], 1] & /@ roots)
]
```

See also JENSEN'S FORMULA, LEHMER'S MAHLER MEASURE PROBLEM

References

Borwein, P. and Erdélyi, T. "Mahler's Measure." §5.3.E.4 in *Polynomials and Polynomial Inequalities.* New York: Springer-Verlag, pp. 271–72, 1995.

Graham, E. *Heights of Polynomials and Entropy in Algebraic Dynamics.* London: Springer-Verlag, 1999.

Mahler Polynomial

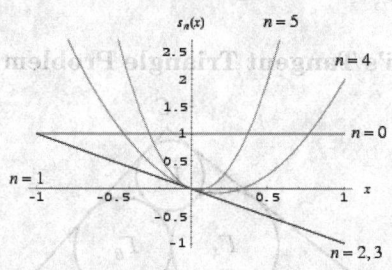

Polynomials $s_n(x)$ which form the SHEFFER SEQUENCE for

$$f^{-1}(t) = 1 + t - e^t,$$

where $f^{-1}(t)$ is the INVERSE FUNCTION of $f(t)$, and have GENERATING FUNCTION

$$\sum_{k=0}^{\infty} \frac{s_k(x)}{k!} t^k = e^{x(1+t-e^t)}.$$

The first few are

$$s_0(x) = 1$$
$$s_1(x) = 0$$
$$s_2(x) = -x$$
$$s_3(x) = -x$$
$$s_4(x) = 3x^2 - x$$
$$s_5(x) = 10x^2 - x.$$

References

Erdélyi, A.; Magnus, W.; Oberhettinger, F.; and Tricomi, F. G. *Higher Transcendental Functions, Vol. 3.* New York: Krieger, p. 254, 1981.

Roman, S. *The Umbral Calculus.* New York: Academic Press, 1984.

Mahler's Measure

For a POLYNOMIAL ,

$$c \in K$$

It is related to JENSEN'S INEQUALITY.

See also JENSEN'S INEQUALITY

Mainardi-Codazzi Equations

PETERSON-MAINARDI-CODAZZI EQUATIONS

Main Diagonal

DIAGONAL

Majorant

A function used to study ORDINARY DIFFERENTIAL EQUATIONS.

Major Axis

SEMIMAJOR AXIS

Majorization

This entry contributed by SERGE BELONGIE

Let $x = (x_1, x_2, \ldots, x_n)$ and $y = (y_1, y_2, \ldots, y_n)$ be nonincreasing sequences of real numbers. Then x majorizes y if, for each $k = 1, 2, \ldots, n$,

$$\sum_{i=1}^{k} x_i \geq \sum_{i=1}^{k} y_i,$$

with equality if $k = n$. Note that some caution is needed when consulting the literature, since the direction of the inequality is not consistent from reference to reference. An order-free characterization along the lines of HORN'S THEOREM is also readily available.

If P is a doubly stochastic matrix, then $y = Px$ iff y is majorized by x. Intuitively, if x majorizes y, then y is more "mixed" than x. HORN'S THEOREM relates the eigenvalues of a HERMITIAN MATRIX A to its diagonal entries using majorization. Given two vectors λ, $\mathbf{v} \in \mathbb{R}^n$, then λ majorizes $\mathbf{v}$ iff there exists a HERMITIAN MATRIX A with eigenvalues λ_i and diagonal entries v_i.

See also BIRKHOFF'S THEOREM, HORN'S THEOREM, SCHUR CONVEXITY

References

Bhatia, R. *Matrix Analysis.* New York: Springer-Verlag, 1997.

Horn, R. A. and Johnson, C. R. *Matrix Analysis, Repr. with Corrections.* Cambridge, England: Cambridge University Press, 1987.

Marshall, A. W. and Olkin, I. *Inequalities: The Theory of Majorizations and Its Applications.* New York: Academic Press, 1979.

Nielsen, M. A. "Conditions for a Class of Entanglement Transformations." *Phys. Rev. Lett.* **83**, 436–39, 1999.

Major Triangle Center

A TRIANGLE CENTER $\alpha : \beta : \gamma$ is called a major center if the TRIANGLE CENTER FUNCTION $\alpha = f(a, b, c, A, B, C)$ is a function of ANGLE A alone, and therefore β and γ of B and C alone, respectively.

See also REGULAR TRIANGLE CENTER, TRIANGLE CENTER

References

Kimberling, C. "Major Centers of Triangles." *Amer. Math. Monthly* **104**, 431–38, 1997.

Makeham Curve

The function defined by

$$y \equiv ks^x b^{q^x}$$

which is used in actuarial science for specifying a simplified mortality law (Kenney and Keeping 1962, pp. 241–42). Using $s(x)$ as the probability that a newborn will achieve age x, the Makeham law (1860) uses

$$s(x) = \exp(-Ax - B(c^x - 1))$$

for $B > 0$, $A \geq -B$, $c > 1$, $x \geq 0$.

See also GOMPERTZ CURVE, LAW OF GROWTH, LIFE EXPECTANCY, LOGISTIC GROWTH CURVE, POPULATION GROWTH

References

Bowers, N. L. Jr.; Gerber, H. U.; Hickman, J. C.; Jones, D. A.; and Nesbitt, C. J. *Actuarial Mathematics.* Itasca, IL: Society of Actuaries, p. 71, 1997.

Kenney, J. F. and Keeping, E. S. *Mathematics of Statistics, Pt. 1, 3rd ed.* Princeton, NJ: Van Nostrand, 1962.

Makeham, W. M. "On the Law of Mortality and the Construction of Annuity Tables." *J. Inst. Actuaries and Assur. Mag.* **8**, 301–10, 1860.

Makeham, W. M. "On an Application of the Theory of the Composition of Decremental Forces." *J. Inst. Actuaries and Assur. Mag.* **18**, 317–22, 1874.

Malfatti Circles

Three circles packed inside a RIGHT TRIANGLE which are each tangent to the other two and to two sides of the TRIANGLE. Although these circles were for many years thought to provide the solutions to MALFATTI'S RIGHT TRIANGLE PROBLEM, they were subsequently shown *never* to provide the solution.

See also APOLLONIAN GASKET, MALFATTI'S RIGHT TRIANGLE PROBLEM, SODDY CIRCLES

Malfatti Points

AJIMA-MALFATTI POINTS

Malfatti's Right Triangle Problem

In 1803, Malfatti asked for the three columns (of possibly different sizes) which, when carved out of a right triangular prism, would have the largest possible total CROSS SECTION. This is equivalent to finding the maximum total AREA of three CIRCLES which can be packed inside a RIGHT TRIANGLE of any shape without overlapping. Malfatti gave the solution as three CIRCLES (the MALFATTI CIRCLES) tangent to each other and to two sides of the TRIANGLE. In 1930, it was shown that the MALFATTI CIRCLES were not always the best solution. Then Goldberg (1967) showed that, even worse, they are *never* the best solution. Wells (1991) illustrates specific cases where alternative solutions are clearly optimal.

See also CIRCLE PACKING, MALFATTI'S TANGENT TRIANGLE PROBLEM

References

Eves, H. *A Survey of Geometry, rev. ed.* Boston, MA: Allyn & Bacon, p. 245, 1965.

Goldberg, M. "On the Original Malfatti Problem." *Math. Mag.* **40**, 241–47, 1967.

Ogilvy, C. S. *Excursions in Geometry.* New York: Dover, pp. 145–47, 1990.

Rothman, T. "Japanese Temple Geometry." *Sci. Amer.* **278**, 85–1, May 1998.

Wells, D. *The Penguin Dictionary of Curious and Interesting Geometry.* London: Penguin, 1991.

Malfatti's Tangent Triangle Problem

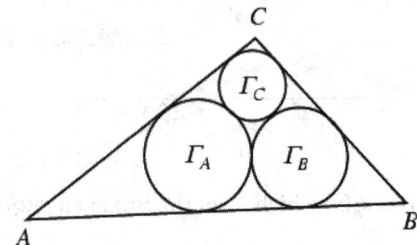

Draw within a given TRIANGLE three CIRCLES, each of which is TANGENT to the other two and to two sides of the TRIANGLE. Denote the three CIRCLES so constructed Γ_A, Γ_B, and Γ_C. Then Γ_A is tangent to AB and AC, Γ_B is tangent to BC and BA, and Γ_C is tangent to AC and BC.

See also AJIMA-MALFATTI POINTS, MALFATTI'S RIGHT TRIANGLE PROBLEM

References

Casey, J. *A Sequel to the First Six Books of the Elements of Euclid, Containing an Easy Introduction to Modern Geometry with Numerous Examples, 5th ed., rev. enl.* Dublin: Hodges, Figgis, & Co., pp. 154–55, 1888.

Dörrie, H. "Malfatti's Problem." §30 in *100 Great Problems of Elementary Mathematics: Their History and Solutions.* New York: Dover, pp. 147–51, 1965.

Forder, H. G. *Higher Course Geometry.* Cambridge, England: Cambridge University Press, pp. 244–45, 1931.

Fukagawa, H. and Pedoe, D. "The Malfatti Problem." *Japanese Temple Geometry Problems (San Gaku).* Winnipeg: The Charles Babbage Research Centre, pp. 106–20, 1989.

F. Gabriel-Marie. *Exercices de géométrie.* Tours, France: Maison Mame, pp. 710–12, 1912.

Gardner, M. *Fractal Music, Hypercards, and More Mathematical Recreations from Scientific American Magazine.* New York: W. H. Freeman, pp. 163–65, 1992.

Goldberg, M. "On the Original Malfatti Problem." *Math. Mag.* **40**, 241–47, 1967.

Hart. *Quart. J.* **1**, p. 219.

Lob, H. and Richmond, H. W. "On the Solution of Malfatti's Problem for a Triangle." *Proc. London Math. Soc.* **2**, 287–04, 1930.

Ogilvy, C. S. *Excursions in Geometry.* New York: Dover, pp. 145–47, 1990.

Rouché, E. and de Comberousse, C. *Traité de géométrie plane.* Paris: Gauthier-Villars, pp. 311–14, 1900.

Woods, F. S. *Higher Geometry.* New York: Dover, pp. 206–09, 1961.

Malliavin Calculus

An infinite-dimensional DIFFERENTIAL CALCULUS on the WIENER SPACE. Also called STOCHASTIC CALCULUS OF VARIATIONS.

Mallows' Sequence

An INTEGER SEQUENCE given by the RECURRENCE RELATION

$$a(n) = a(a(n-2)) + a(n - a(n-2))$$

with $a(1) = a(2) = 1$. The first few values are 1, 1, 2, 3, 3, 4, 5, 6, 6, 7, 7, 8, 9, 10, 10, 11, 12, 12, 13, 14, ... (Sloane's A005229).

See also HOFSTADTER-CONWAY $10,000 SEQUENCE, HOFSTADTER'S Q-SEQUENCE

References

Mallows, C. L. "Conway's Challenge Sequence." *Amer. Math. Monthly* **98**, 5–0, 1991.
Sloane, N. J. A. Sequences A005229/M0441 in "An On-Line Version of the Encyclopedia of Integer Sequences." http://www.research.att.com/~njas/sequences/eisonline.html.

Malmstén's Differential Equation

The ORDINARY DIFFERENTIAL EQUATION

$$y'' + \frac{r}{z} y' = \left(A z^m + \frac{s}{z^2} \right) y.$$

References

Watson, G. N. *A Treatise on the Theory of Bessel Functions*, 2nd ed. Cambridge, England: Cambridge University Press, pp. 99–00, 1966.

Malmstén's Formula

The integral representation of $\ln[\Gamma(z)]$ by

$$\ln[(z)] = \int_1^z \psi_0(z') \, dz'$$

$$= \int_0^\infty \left[(z-1) - \frac{1 - e^{-(z-1)t}}{1 - e^{-t}} \right] \frac{e^{-t}}{t} \, dt,$$

where $\Gamma(z)$ is the GAMMA FUNCTION and $\psi_0(z)$ is the DIGAMMA FUNCTION.

See also BINET'S LOG GAMMA FORMULAS, GAMMA FUNCTION

References

Erdélyi, A.; Magnus, W.; Oberhettinger, F.; and Tricomi, F. G. *Higher Transcendental Functions, Vol. 1.* New York: Krieger, pp. 20–1, 1981.

Maltese Cross

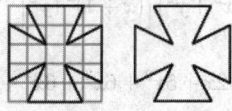

An irregular DODECAHEDRON CROSS shaped like a + sign but whose points flange out at the end: ✠. The conventional proportions as computed on a 5×5 grid as illustrated above.

See also CROSS, DISSECTION, DODECAHEDRON, MALTESE CROSS CURVE

References

Frederickson, G. "Maltese Crosses." Ch. 14 in *Dissections: Plane and Fancy.* New York: Cambridge University Press, pp. 157–62, 1997.

Maltese Cross Curve

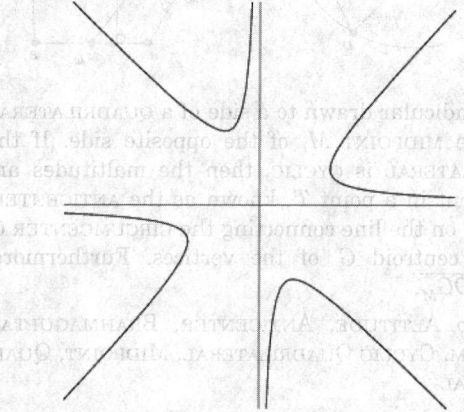

The plane curve with Cartesian equation

$$xy(x^2 - y^2) = x^2 + y^2$$

and polar equation

$$r^2 = \frac{1}{\cos\theta \sin\theta(\cos^2\theta - \sin^2\theta)}$$

(Cundy and Rollett 1989, p. 71), so named for its resemblance to the MALTESE CROSS.

See also MALTESE CROSS

References

Cundy, H. and Rollett, A. *Mathematical Models, 3rd ed.* Stradbroke, England: Tarquin Pub., p. 71, 1989.

Malthusian Parameter

The parameter α in the exponential POPULATION GROWTH equation

$$N_1(t) = N_0 e^{\alpha t}.$$

See also LIFE EXPECTANCY, POPULATION GROWTH

Maltitude

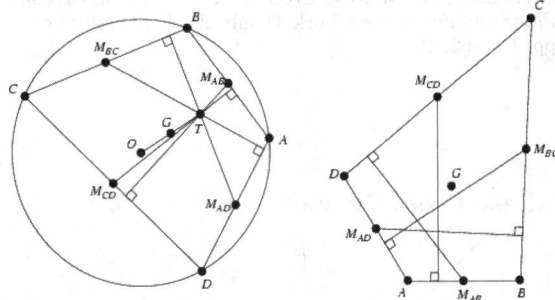

A perpendicular drawn to a side of a QUADRILATERAL from the MIDPOINT M_i of the opposite side. If the QUADRILATERAL is CYCLIC, then the maltitudes are concurrent in a point T, known as the ANTICENTER, which is on the line connecting the CIRCUMCENTER O an the centroid G of the vertices. Furthermore, $\overline{OM} = 2\overline{OG_M}$.

See also ALTITUDE, ANTICENTER, BRAHMAGUPTA'S THEOREM, CYCLIC QUADRILATERAL, MIDPOINT, QUADRILATERAL

References

Honsberger, R. *Episodes in Nineteenth and Twentieth Century Euclidean Geometry.* Washington, DC: Math. Assoc. Amer., pp. 36–7, 1995.
Wells, D. *The Penguin Dictionary of Curious and Interesting Geometry.* London: Penguin, p. 146, 1991.

Mandelbar Set

A FRACTAL set analogous to the MANDELBROT SET or its generalization to a higher power with the variable z replaced by its COMPLEX CONJUGATE $\bar{z}$.

See also MANDELBROT SET

Mandelbrot Set

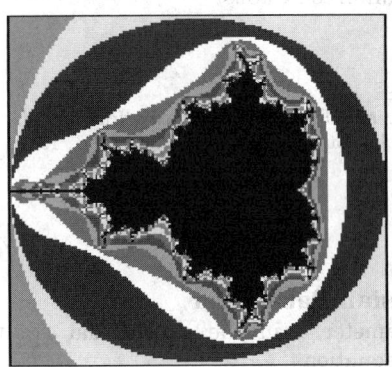

The set obtained by the QUADRATIC RECURRENCE

$$z_{n+1} = z_n^2 + C, \qquad (1)$$

where points C for which the orbit $z_0 = 0$ does not tend to infinity are in the SET. It marks the set of points in the COMPLEX PLANE such that the corre-

sponding JULIA SET is CONNECTED and not COMPUTABLE. The Mandelbrot set was originally called a MU MOLECULE by Mandelbrot.

J. Hubbard and A. Douady proved that the Mandelbrot set is CONNECTED. Shishikura (1994) proved that the boundary of the Mandelbrot set is a FRACTAL with HAUSDORFF DIMENSION 2. However, it is not yet known if the Mandelbrot set is pathwise-connected. If it is pathwise-connected, then Hubbard and Douady's proof implies that the Mandelbrot set is the image of a CIRCLE and can be constructed from a DISK by collapsing certain arcs in the interior (Douady 1986).

The AREA of the set is known to lie between 1.5031 and 1.5702; it is estimated as 1.50659....

Decomposing the COMPLEX coordinate $z = x + iy$ and $z_0 = a + ib$ gives

$$x' = x^2 - y^2 + a \qquad (2)$$

$$y' = 2xy + b. \qquad (3)$$

In practice, the limit is approximated by

$$\lim_{n \to \infty} |z_n| \approx \lim_{n \to n_{max}} |z_n| < r_{max}. \qquad (4)$$

Beautiful computer-generated plots can be created by coloring nonmember points depending on how quickly they diverge to r_{max}. A common choice is to define an INTEGER called the COUNT to be the largest n such that $|z_n| < r$, where r is usually taken as $r = 2$, and to color points of different COUNT different colors. The boundary between successive COUNTS defines a series of "LEMNISCATES," called EQUIPOTENTIAL CURVES by Peitgen and Saupe (1988), $|L_n(C)| = r$ which have distinctive shapes. The first few LEMNISCATES are

$$L_1(C) = C \qquad (5)$$

$$L_2(C) = C(C + 1) \qquad (6)$$

$$L_3(C) = C + \left(C + C^2\right)^2 \qquad (7)$$

$$L_4(C) = C + \left[\left(C + C^2\right)^2\right]^2. \qquad (8)$$

When written in CARTESIAN COORDINATES, the first three of these are

$$r^2 = x^2 + y^2 \qquad (9)$$

$$r^2 = \left(x^2 + y^2\right)\left[(x+1)^2 + y^2\right] \qquad (10)$$

$$\begin{aligned} r^2 = \left(x^2 + y^2\right)&\left(1 + 2x + 5x^2 + 6x^3 + 6x^4 + 4x^5 + x^6 \right. \\ &-3y^2 - 2xy^2 + 8x^2y^2 + 8x^3y^2 + 3x^4y^2 + 2y^4 + 4xy^4 \\ &\left. +3x^2y^4 + y^6\right) \end{aligned} \qquad (11)$$

which are a CIRCLE, an OVAL, and a PEAR CURVE. In fact, the second LEMNISCATE L_2 can be written in terms of a new coordinate system with $x' \equiv x - 1/2$ as

$$\left[\left(x'-\tfrac{1}{2}\right)^2+y^2\right]\left(\left(x'+\tfrac{1}{2}\right)^2+y^2\right)=r^2, \qquad (12)$$

which is just a CASSINI OVAL with $a = 1/2$ and $b^2 = r$. The LEMNISCATES grow increasingly convoluted with higher COUNT and approach the Mandelbrot set as the COUNT tends to infinity.

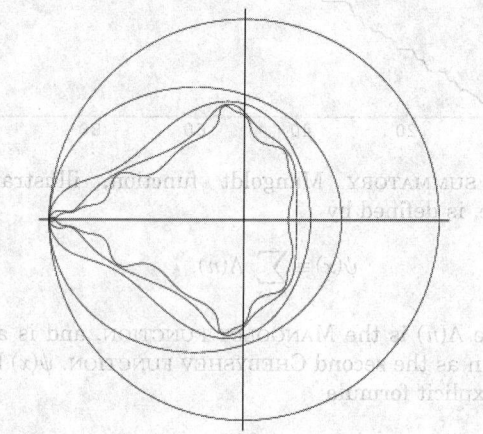

The kidney bean-shaped portion of the Mandelbrot set is bordered by a CARDIOID with equations

$$4x = 2\cos t - \cos(2t) \qquad (13)$$

$$4y = 2\sin t - \sin(2t). \qquad (14)$$

The adjoining portion is a CIRCLE with center at $(-1, 0)$ and RADIUS $1/4$. One region of the Mandelbrot set containing spiral shapes is known as SEA HORSE VALLEY because the shape resembles the tail of a sea horse.

Generalizations of the Mandelbrot set can be constructed by replacing z_n^2 with z_n^k or $(\bar{z}_n)^k$, where k is a POSITIVE INTEGER and $\bar{z}$ denotes the COMPLEX CONJUGATE of z. The following figures show the FRACTALS obtained for $k = 2$, 3, and 4 (Dickau). The plots on the right have z replaced with $\bar{z}$ and are sometimes called "MANDELBAR SETS."

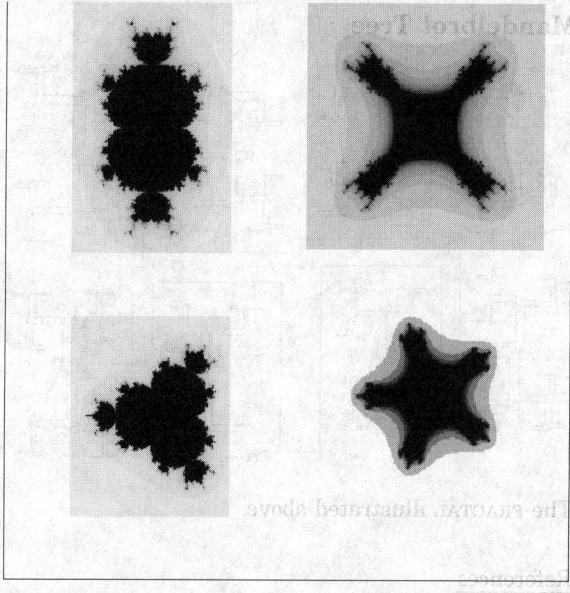

See also CACTUS FRACTAL, FRACTAL, JULIA SET, LEMNISCATE (MANDELBROT SET), MANDELBAR SET, QUADRATIC MAP, RANDELBROT SET, SEA HORSE VALLEY

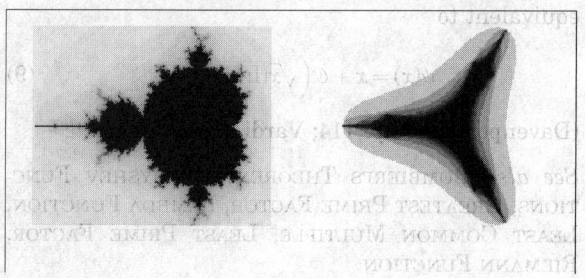

References

Alfeld, P. "The Mandelbrot Set." http://www.math.utah.edu/~alfeld/math/mandelbrot/mandelbrot.html.

Branner, B. "The Mandelbrot Set." In *Chaos and Fractals: The Mathematics Behind the Computer Graphics, Proc. Sympos. Appl. Math., Vol. 39* (Ed. R. L. Devaney and L. Keen). Providence, RI: Amer. Math. Soc., 75–05, 1989.

Devaney, R. "The Mandelbrot Set and the Farey Tree, and the Fibonacci Sequence." *Amer. Math. Monthly* **106**, 289–02, 1999.

Dickau, R. M. "Mandelbrot (and Similar) Sets." http://forum.swarthmore.edu/advanced/robertd/mandelbrot.html.

Douady, A. "Julia Sets and the Mandelbrot Set." In *The Beauty of Fractals: Images of Complex Dynamical Systems* (Ed. H.-O. Peitgen and D. H. Richter). Berlin: Springer-Verlag, p. 161, 1986.

Eppstein, D. "Area of the Mandelbrot Set." http://www.ics.uci.edu/~eppstein/junkyard/mand-area.html.

Fisher, Y. and Hill, J. "Bounding the Area of the Mandelbrot Set." Submitted.

Hill, J. R. "Fractals and the Grand Internet Parallel Processing Project." Ch. 15 in *Fractal Horizons: The Future Use of Fractals.* New York: St. Martin's Press, pp. 299–23, 1996.

Lauwerier, H. *Fractals: Endlessly Repeated Geometric Figures.* Princeton, NJ: Princeton University Press, pp. 148–51 and 179–80, 1991.

Lei, T. (Ed.) *The Mandelbrot Set, Theme and Variations.* Cambridge, England: Cambridge University Press, 2000.

Munafo, R. "Mu-Ency--The Encyclopedia of the Mandelbrot Set." http://www.mrob.com/muency.html.

Peitgen, H.-O. and Saupe, D. (Eds.). *The Science of Fractal Images.* New York: Springer-Verlag, pp. 178–79, 1988.

Shishikura, M. "The Boundary of the Mandelbrot Set has Hausdorff Dimension Two." *Astérisque*, No. 222, **7**, 389–05, 1994.

Wells, D. *The Penguin Dictionary of Curious and Interesting Geometry.* London: Penguin, pp. 146–48, 1991.

Mandelbrot Tree

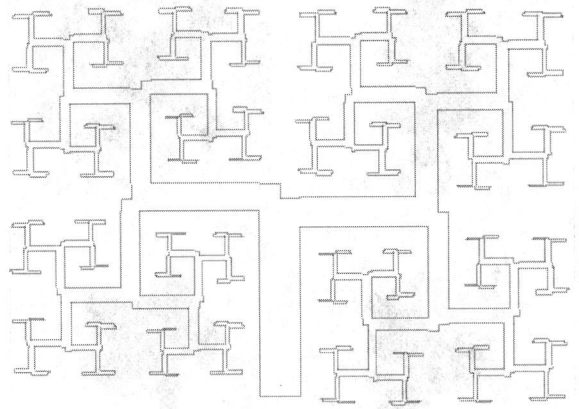

The FRACTAL illustrated above.

References

Lauwerier, H. *Fractals: Endlessly Repeated Geometric Figures.* Princeton, NJ: Princeton University Press, pp. 71–3, 1991.

Weisstein, E. W. "Fractals." MATHEMATICA NOTEBOOK FRACTAL.M.

Mangoldt Function

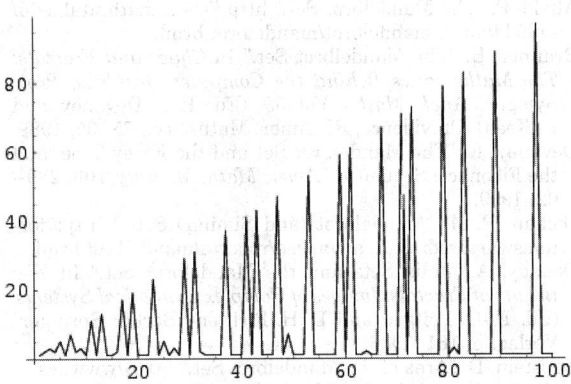

The function defined by

$$\Lambda(n) \equiv \begin{cases} \ln p & \text{if } n = p^k \text{ for } p \text{ a prime} \\ 0 & \text{otherwise,} \end{cases} \quad (1)$$

sometimes also called the lambda function. $\exp(\Lambda(n))$ is also given by $[1, 2, ..., n]/[1, 2, ..., n-1]$, where $[a, b, c, ...]$ denotes the LEAST COMMON MULTIPLE. The first few values of $\exp((n))$ for $n = 1, 2, ...$, plotted above, are 1, 2, 3, 2, 5, 1, 7, 2, ... (Sloane's A014963). The Mangoldt function is related to the RIEMANN ZETA FUNCTION $\zeta(z)$ by

$$-\frac{\zeta'(s)}{\zeta(s)} = \sum_{n=1}^{\infty} \frac{\Lambda(n)}{n^s}, \quad (2)$$

where $\Re[s] > 1$ (Hardy 1999, p. 28; Krantz 1999,

p. 161).

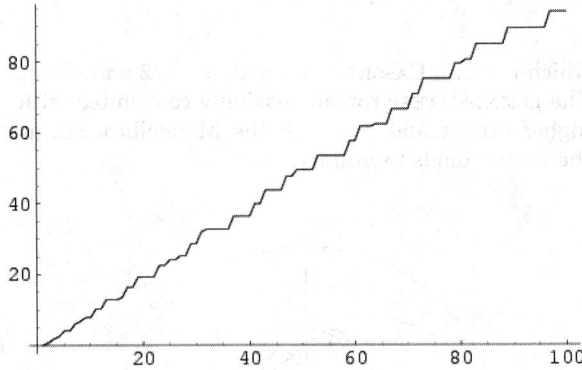

The SUMMATORY Mangoldt function, illustrated above, is defined by

$$\psi(x) \equiv \sum_{n \leq x} \Lambda(n), \quad (3)$$

where $\Lambda(n)$ is the MANGOLDT FUNCTION, and is also known as the second CHEBYSHEV FUNCTION. $\psi(x)$ has the explicit formula

$$\psi(x) = x - \sum_{\rho} \frac{x^{\rho}}{\rho} - \ln(2\pi) - \tfrac{1}{2}\ln(1-x^2), \quad (4)$$

where the second SUM is over all complex zeros ρ of the RIEMANN ZETA FUNCTION $\zeta(s)$, i.e., those in the CRITICAL STRIP so $0 < \Re[\rho] < 1$, and interpreted as

$$\lim_{t \to \infty} \sum_{|\Im(\rho)| < t} \frac{x^{\rho}}{\rho}. \quad (5)$$

Vardi (1991, p. 155) also gives the interesting formula

$$\ln([x]!) = \psi(x) + \psi\left(\tfrac{1}{2}x\right) + \psi\left(\tfrac{1}{3}x\right) + ..., \quad (6)$$

where $[x]$ is the NINT function and $n!$ is a FACTORIAL. Vallée Poussin's version of the PRIME NUMBER THEOREM states that

$$\psi(x) = x + \mathcal{O}\left(xe^{-a\sqrt{\ln x}}\right) \quad (7)$$

for some a (Davenport 1980, Vardi 1991). The PRIME NUMBER THEOREM is equivalent to the statement that

$$\psi(x) = x + o(x) \quad (8)$$

as $x \to \infty$ (Dusart 1999). The RIEMANN HYPOTHESIS is equivalent to

$$\psi(x) = x + \mathcal{O}\left(\sqrt{x}(\ln x)^2\right) \quad (9)$$

(Davenport 1980, p. 114; Vardi 1991).

See also BOMBIERI'S THEOREM, CHEBYSHEV FUNCTIONS, GREATEST PRIME FACTOR, LAMBDA FUNCTION, LEAST COMMON MULTIPLE, LEAST PRIME FACTOR, RIEMANN FUNCTION

References

Costa Pereira, N. "Estimates for the Chebyshev Function $\psi(x) - \theta(x)$." *Math. Comp.* **44**, 211–21, 1985.

Costa Pereira, N. "Corrigendum: Estimates for the Chebyshev Function $\psi(x) - \theta(x)$." *Math. Comp.* **48**, 447, 1987.

Costa Pereira, N. "Elementary Estimates for the Chebyshev Function $\psi(x)$ and for the Möbius Function $M(x)$." *Acta Arith.* **52**, 307–37, 1989.

Davenport, H. *Multiplicative Number Theory, 2nd ed.* New York: Springer-Verlag, p. 110, 1980.

Dusart, P. "Inégalités explicites pour $\psi(X)$, $\theta(X)$, $\pi(X)$ et les nombres premiers." *C. R. Math. Rep. Acad. Sci. Canad* **21**, 53–9, 1999.

Hardy, G. H. *Ramanujan: Twelve Lectures on Subjects Suggested by His Life and Work, 3rd ed.* New York: Chelsea, p. 28, 1999.

Krantz, S. G. "The Lambda Function" and "Relation of the Zeta Function to the Lambda Function." §13.2.10 and 13.2.11 in *Handbook of Complex Analysis.* Boston, MA: Birkhäuser, p. 161, 1999.

Rosser, J. B. and Schoenfeld, L. "Sharper Bounds for Chebyshev Functions $\theta(x)$ and $\psi(x)$." *Math. Comput.* **29**, 243–69, 1975.

Schoenfeld, L. "Sharper Bounds for Chebyshev Functions $\theta(x)$ and $\psi(x)$. II," *Math. Comput.* **30**, 337–60, 1976.

Sloane, N. J. A. Sequences A014963 in "An On-Line Version of the Encyclopedia of Integer Sequences." http://www.research.att.com/~njas/sequences/eisonline.html.

Vardi, I. *Computational Recreations in Mathematica.* Reading, MA: Addison-Wesley, pp. 146–47, 152–53, and 249, 1991.

Manhattan Distance

The distance between two points (x, y) and (u, v) given by the METRIC

$$d = |x - u| + |y - v|$$

(Skiena 1990, p. 227).

See also METRIC

References

Skiena, S. *Implementing Discrete Mathematics: Combinatorics and Graph Theory with Mathematica.* Reading, MA: Addison-Wesley, pp. 172 and 227, 1990.

Manifold

A manifold is a TOPOLOGICAL SPACE which is LOCALLY EUCLIDEAN (i.e., around every point, there is a NEIGHBORHOOD which is topologically the same as the OPEN UNIT BALL in $\mathbb{R}^n$). To illustrate this idea, consider the ancient belief that the Earth was flat as contrasted with the modern evidence that it is round. This discrepancy arises essentially from the fact that on the small scales that we see, the Earth does indeed look flat (although the Greeks did notice that the last part of a ship to disappear over the horizon was the mast). In general, any object which is nearly "flat" on small scales is a manifold, and so manifolds constitute a generalization of objects we could live on in which we would encounter the round/flat Earth problem, as first codified by Poincaré. More formally, any object that can be "charted" is a manifold.

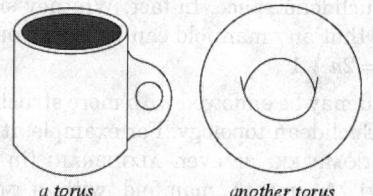

a torus   *another torus*

As a TOPOLOGICAL SPACE, a manifold can be COMPACT or not compact, and CONNECTED or disconnected. Typically, by "manifold," one means a manifold without boundary. However, an author will sometimes be more precise and use the term OPEN MANIFOLD (for a noncompact manifold without boundary) or CLOSED MANIFOLD (for a COMPACT MANIFOLD without boundary).

If a manifold contains its own boundary, it is called, not surprisingly, a "MANIFOLD WITH BOUNDARY." The closed unit ball in $\mathbb{R}^n$ is a manifold with boundary, and its boundary is the unit sphere. The concept can be generalized to manifolds with corners. By definition, every point on a manifold has a neighborhood together with a HOMEOMORPHISM of that neighborhood with an OPEN BALL in $\mathbb{R}^n$. In addition, a manifold must have a SECOND COUNTABLE TOPOLOGY. Unless otherwise indicated, a manifold is assumed to have finite DIMENSION n, for n a positive integer.

DIFFERENTIABLE MANIFOLDS are manifolds for which overlapping charts "relate smoothly" to each other, meaning that the inverse of one followed by the other is an infinitely differentiable map from EUCLIDEAN SPACE to itself. Manifolds arise naturally in a variety of mathematical and physical applications as "global objects." For example, in order to precisely describe all the configurations of a robot arm or all the possible positions and momenta of a rocket, an object is needed to store all of these parameters. The objects that crop up are manifolds. From the geometric perspective, manifolds represent the profound idea having to do with global versus local properties.

The basic example of a manifold is EUCLIDEAN SPACE, and many of its properties carry over to manifolds. In addition, any smooth boundary of a subset of Euclidean space, like the circle or the sphere, is a manifold. Manifolds are therefore of interest in the study of GEOMETRY, TOPOLOGY, and ANALYSIS.

One of the goals of topology is to find ways of distinguishing manifolds. For instance, a circle is topologically the same as any closed loop, no matter how different these two manifolds may appear. Similarly, the surface of a coffee mug with a handle is topologically the same as the surface of the donut, and this type of surface is called a (one-handled) TORUS.

A SUBMANIFOLD is a subset of a manifold which is itself a manifold, but has smaller dimension. For example, the equator of a sphere is a submanifold. Many common examples of manifolds are submanifolds of Euclidean space. In fact, Whitney showed in the 1930s that any manifold can be EMBEDDED in $\mathbb{R}^N$, where $N = 2n + 1$.

A manifold may be endowed with more structure than a locally Euclidean topology. For example, it could be SMOOTH, COMPLEX, or even ALGEBRAIC (in order of specificity). A smooth manifold with a METRIC is called a RIEMANNIAN MANIFOLD, and one with a SYMPLECTIC STRUCTURE is called a SYMPLECTIC MANIFOLD. Finally, a COMPLEX MANIFOLD with a KÄHLER STRUCTURE is called a KÄHLER MANIFOLD.

See also ALGEBRAIC MANIFOLD, COBORDANT MANIFOLD, COMPACT MANIFOLD, COMPLEX MANIFOLD, CONNECTED SUM DECOMPOSITION, COORDINATE CHART, DIFFERENTIABLE MANIFOLD, EUCLIDEAN SPACE, FLAG MANIFOLD, GRASSMANN MANIFOLD, HEEGAARD SPLITTING, ISOSPECTRAL MANIFOLDS, JACO-SHALEN-JOHANNSON TORUS DECOMPOSITION, KÄHLER MANIFOLD, LIE GROUP, MANIFOLD WITH BOUNDARY, POINCARÉ CONJECTURE, POISSON MANIFOLD, PRIME MANIFOLD, RIEMANNIAN MANIFOLD, SET, SMOOTH MANIFOLD, SPACE, STIEFEL MANIFOLD, STRATIFIED MANIFOLD, SUBMANIFOLD, SURGERY, SYMPLECTIC MANIFOLD, TANGENT BUNDLE, TANGENT VECTOR (MANIFOLD), THURSTON'S GEOMETRIZATION CONJECTURE, TOPOLOGICAL MANIFOLD, TOPOLOGICAL SPACE, TRANSITION FUNCTION, WHITEHEAD MANIFOLD, WIEDERSEHEN MANIFOLD

References

Conlon, L. *Differentiable Manifolds: A First Course.* Boston, MA: Birkhäuser, 1993.
Ferreirós, J. "A New Fundamental Notion: Riemann's Manifolds." Ch. 2 in *Labyrinth of Thought: A History of Set Theory and Its Role in Modern Mathematics.* Basel, Switzerland: Birkhäuser, pp. 39–0, 1999.

Mannheim's Theorem

The four planes determined by the four altitudes of a TETRAHEDRON and the orthocenters of the corresponding faces pass through the MONGE POINT of the TETRAHEDRON.

See also MONGE POINT, TETRAHEDRON

References

Altshiller-Court, N. "The Monge Point." §4.2c in *Modern Pure Solid Geometry.* New York: Chelsea, pp. 69–1, 1979.
Mannheim, A. *J. de math. élémentaires*, p. 225, 1895.
Thompson, H. F. "A Geometrical Proof of a Theorem Connected with the Tetrahedron." *Proc. Edinburgh Math. Soc.* **17**, 51–3, 1908–909.

Mann's Theorem

This entry contributed by KEVIN O'BRYANT

A theorem widely circulated as the "α-β conjecture" and proved by Mann (1942). It states that if A and B are sets of integers each containing 0, then

$$\sigma(A \oplus B) \geq \min\{1, \ \sigma(A) + \sigma(B)\}.$$

Here, $A \oplus B$ denotes the DIRECT SUM, i.e., $A \oplus B = \{a + b : a \in A, \ b \in B\}$, and σ is the SCHNIRELMANN DENSITY.

Mann's theorem is best possible in the sense that $A = B = \{0, 1, 11, 12, 13, \ldots\}$ satisfies $\sigma(A \oplus B) = \sigma(A) + \sigma(B)$.

Mann's theorem implies SCHNIRELMANN'S THEOREM as follows. Let $P = \{0, 1\} \cup \{p : p \text{ prime}\}$, then Mann's theorem proves that $\sigma(P + P + P + P) > 2\sigma(P + P)$, so as more and more copies of the primes are included, the SCHNIRELMANN density increases at least linearly, and so reaches 1 with at most $2 \cdot 1/(\sigma(P + P))$ copies of the primes. Since the only sets with SCHNIRELMANN density 1 are the sets containing all positive integers, SCHNIRELMANN'S THEOREM follows.

See also SCHNIRELMANN DENSITY, SCHNIRELMANN'S THEOREM

References

Garrison, B. K. "A Nontransformation Proof of Mann's Density Theorem." *J. reine angew. Math.* **245**, 41–6, 1970.
Khinchin, A. Y. "The Landau-Schnirelmann Hypothesis and Mann's Theorem." Ch. 2 in *Three Pearls of Number Theory.* New York: Dover, pp. 18–6, 1998.
Mann, H. B. "A Proof of the Fundamental Theorem on the Density of Sets of Positive Integers." *Ann. Math.* **43**, 523–27, 1942.

MANOVA

MANOVA ("multiple analysis of variance") is a procedure for testing the equality of mean vectors of more than two populations. The technique is analogous to ANOVA for univariate data, except that groups are compared on multiple response variables simultaneously. While F-tests can be used in the uniseriate case to assess the hypothesis under consideration, there is no single test statistic in the multivariate case that is optimal in all situations (Everitt and Wykes 1999, p. 125).

See also ANOVA

References

Bijleveld, C. C. J. H.; van der Kamp, L. J. T.; Mooijaart, A.; van der Kloot, W. A.; van der Leeden, R.; and van der Burg, E. *Longitudinal Data Analysis: Designs, Models and Methods.* London: Sage, 1998.
Everitt, B. S. and Wykes, T. *Dictionary of Statistics for Psychologists.* London: Arnold, p. 125, 1999.

Mantissa

For a REAL NUMBER x, the mantissa is defined as the POSITIVE FRACTIONAL PART $x - \lfloor x \rfloor = \operatorname{frac}(x)$, where $\lfloor x \rfloor$ denotes the FLOOR FUNCTION.

See also CHARACTERISTIC (REAL NUMBER), FLOOR FUNCTION, SCIENTIFIC NOTATION

Many-to-One

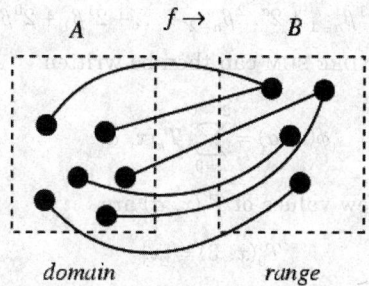

many-to-one

A FUNCTION f which may (but does not necessarily) associate a given member of the RANGE of f with more than one member of the DOMAIN of f. For example, TRIGONOMETRIC FUNCTIONS such as $\sin x$ are many-to-one since $\sin x = \sin(2\pi + x) = \sin(4\pi + x) = \cdots$.

See also DOMAIN, ONE-TO-ONE, RANGE (IMAGE)

Many Valued Logic

References

Rescher, N. *Many Valued Logic*. Ashgate, 1993.

Map

A way of associating unique objects to every point in a given SET. So a map from $A \mapsto B$ is an object f such that for every $A \in B$, there is a unique object $f(a) \in B$. The terms FUNCTION and MAPPING are synonymous with map.

the following table gives several common types of complex maps.

Mapping	FORMULA	Domain
Inversion	$f(z) = \dfrac{1}{z}$	
Magnification	$f(z) = az$	$a \in \mathbb{R} \neq 0$
Magnification + Rotation	$f(z) = az$	$a \in \mathbb{C} \neq 0$
MÖBIUS TRANSFORMATION	$f(z) = \dfrac{az + b}{cz + d}$	$a, b, c, d \in \mathbb{C}$
ROTATION	$f(z) = e^{i\theta}z$	$\theta \in \mathbb{R}$
TRANSLATION	$f(z) = z + a$	$a \in \mathbb{C}$

See also 2X MOD 1 MAP, ARNOLD'S CAT MAP, BAKER'S MAP, BOUNDARY MAP, CONFORMAL MAP, FUNCTION, GAUSS MAP, GINGERBREADMAN MAP, HARMONIC MAP, HÉNON MAP, IDENTITY MAP, INCLUSION MAP, KAPLAN-YORKE MAP, LOGISTIC MAP, MANDELBROT SET, MAP PROJECTION, PULLBACK MAP, QUADRATIC MAP,

SYMPLECTIC MAP, TANGENT MAP, TENT MAP, TRANSFORMATION, ZASLAVSKII MAP

References

Arfken, G. "Mapping." §6.6 in *Mathematical Methods for Physicists, 3rd ed.* Orlando, FL: Academic Press, pp. 384–92, 1985.

Map-Airy Distribution

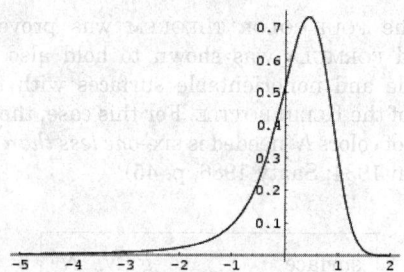

A probability distribution having density

$$P(x) = 2e^{-2x^3/3}\left[x\,\mathrm{Ai}(x^2) - \mathrm{Ai}'(x^2)\right],$$

where $\mathrm{Ai}(x)$ is the AIRY FUNCTION and $\mathrm{Ai}'(x) = d\mathrm{Ai}(x)/dx$. The corresponding distribution function is

$$D(x) = \frac{1}{3} - 2x^5\,\frac{{}_2F_2\left(\frac{7}{6}, \frac{5}{3}, \frac{7}{3}, \frac{8}{3}; -\frac{4}{3}x^3\right)}{15 \cdot 3^{2/3}\Gamma\left(\frac{5}{3}\right)}$$

$$-x^4\,\frac{{}_2F_2\left(\frac{5}{6}, \frac{4}{3}, \frac{5}{3}, \frac{7}{3}; -\frac{4}{3}x^3\right)}{6 \cdot 3^{1/3}\left(\frac{4}{3}\right)}$$

$$+x^2\,\frac{{}_2F_2\left(\frac{1}{6}, \frac{2}{3}, \frac{1}{3}, \frac{5}{3}; -\frac{4}{3}x^3\right)}{3^{2/3}\Gamma\left(\frac{2}{3}\right)}$$

$$+2x\,\frac{{}_2F_2\left(-\frac{1}{6}, \frac{1}{3}, -\frac{1}{3}, \frac{4}{3}; -\frac{4}{3}x^3\right)}{3^{1/3}\Gamma\left(\frac{1}{3}\right)}$$

(M. Trott). The density is normalized with

$$\int_{-\infty}^{\infty} A(x)\,dx = 1.$$

The MEAN is 0, but the second moment μ_2 is undefined.

See also AIRY FUNCTIONS

References

Banderier, C.; Flajolet, P.; Schaeffer, G.; and Soria, M. "Planar Maps and Airy Phenomena." Preprint.

Map Coloring

Given a map with GENUS $g > 0$, Heawood showed in 1890 that the maximum number N_u of colors necessary to color a map (the CHROMATIC NUMBER) on an unbounded surface is

$$N_u \equiv \left\lfloor \tfrac{1}{2}\left(7 + \sqrt{48g+1}\right)\right\rfloor = \left\lfloor \tfrac{1}{2}\left(7 + \sqrt{49-24\chi}\right)\right\rfloor,$$

where $\lfloor x \rfloor$ is the FLOOR FUNCTION, g is the GENUS, and χ is the EULER CHARACTERISTIC. This is the HEAWOOD CONJECTURE. In 1968, for any orientable surface other than the SPHERE (or equivalently, the PLANE) and any nonorientable surface other than the KLEIN BOTTLE, N_u was shown to be not merely a maximum, but the actual number needed (Ringel and Youngs 1968).

When the FOUR-COLOR THEOREM was proven, the Heawood FORMULA was shown to hold also for all orientable and nonorientable surfaces with the exception of the KLEIN BOTTLE. For this case, the actual number of colors N needed is six—*one less than $N_u = 7$* (Franklin 1934; Saaty 1986, p. 45).

surface	g	N_u	N
KLEIN BOTTLE	1	7	6
MÖBIUS STRIP	$\frac{1}{2}$	6	6
PLANE	0	4	4
PROJECTIVE PLANE	$\frac{1}{2}$	6	6
SPHERE	0	4	4
TORUS	1	7	7

See also CHROMATIC NUMBER, FOUR-COLOR THEOREM, HEAWOOD CONJECTURE, SIX-COLOR THEOREM, TORUS COLORING

References

Ball, W. W. R. and Coxeter, H. S. M. *Mathematical Recreations and Essays, 13th ed.* New York: Dover, pp. 237–38, 1987.
Barnette, D. *Map Coloring, Polyhedra, and the Four-Color Problem.* Washington, DC: Math. Assoc. Amer., 1983.
Franklin, P. "A Six Colour Problem." *J. Math. Phys.* **13**, 363–69, 1934.
Franklin, P. *The Four-Color Problem.* New York: Scripta Mathematica, Yeshiva College, 1941.
Ore, Ø. *The Four-Color Problem.* New York: Academic Press, 1967.
Ringel, G. and Youngs, J. W. T. "Solution of the Heawood Map-Coloring Problem." *Proc. Nat. Acad. Sci. USA* **60**, 438–45, 1968.
Saaty, T. L. and Kainen, P. C. *The Four-Color Problem: Assaults and Conquest.* New York: Dover, 1986.

Mapes' Method

A method for computing the PRIME COUNTING FUNCTION. Define the function

$$T_k(x, a) = (-1)^{\beta_0+\beta_1+\ldots+\beta_{a-1}} \left\lfloor \frac{x}{p_1^{\beta_0} p_2^{\beta_1} \cdots p_a^{\beta_{a-1}}} \right\rfloor, \quad (1)$$

where $\lfloor x \rfloor$ is the FLOOR FUNCTION and the β_i are the binary digits (0 or 1) in

$$k = 2^{a-1}\beta_{a-1} + 2^{a-2}\beta_{a-2} + \ldots + 2^1\beta_1 + 2^0\beta_0. \quad (2)$$

The LEGENDRE SUM can then be written

$$\phi(x, a) = \sum_{k=0}^{2^a-1} T_k(x, a). \quad (3)$$

The first few values of $T_k(x, a)$ are

$$T_0(x, 3) = \lfloor x \rfloor \quad (4)$$

$$T_1(x, 3) = -\left\lfloor \frac{x}{p_1} \right\rfloor \quad (5)$$

$$T_2(x, 3) = -\left\lfloor \frac{x}{p_2} \right\rfloor \quad (6)$$

$$T_3(x, 3) = \left\lfloor \frac{x}{p_1 p_2} \right\rfloor \quad (7)$$

$$T_4(x, 3) = \left\lfloor \frac{x}{p_3} \right\rfloor \quad (8)$$

$$T_5(x, 3) = \left\lfloor \frac{x}{p_1 p_3} \right\rfloor \quad (9)$$

$$T_6(x, 3) = \left\lfloor \frac{x}{p_2 p_3} \right\rfloor \quad (10)$$

$$T_7(x, 3) = -\left\lfloor \frac{x}{p_1 p_2 p_3} \right\rfloor. \quad (11)$$

Mapes' method takes time $\sim x^{0.7}$, which is slightly faster than the LEHMER-SCHUR METHOD.

See also LEHMER-SCHUR METHOD, PRIME COUNTING FUNCTION

References

Mapes, D. C. "Fast Method for Computing the Number of Primes Less than a Given Limit." *Math. Comput.* **17**, 179–85, 1963.
Riesel, H. "Mapes' Method." *Prime Numbers and Computer Methods for Factorization, 2nd ed.* Boston, MA: Birkhäuser, p. 23, 1994.

Map Folding

A general FORMULA giving the number of distinct ways of folding an $N = m \times n$ rectangular map is not known. A distinct folding is defined as a permutation of N numbered cells reading from the top down. Lunnon (1971) gives values up to $n = 28$.

n	$1 \times n$	$2 \times n$	$3 \times n$	$4 \times n$	$5 \times n$
1	1	1			

2	2	8	
3	6	60	1368
4	16	1980	300608
5	59	19512	18698669
6	144	15552	

The limiting ratio of the number of $1 \times (n+1)$ strips to the number of $1 \times n$ strips is given by

$$\lim_{n \to \infty} \frac{[1 \times (n+1)]}{[1 \times n]} \in [3.3868, 3.9821].$$

See also STAMP FOLDING

References

Gardner, M. "The Combinatorics of Paper Folding." Ch. 7 in *Wheels, Life, and Other Mathematical Amusements.* New York: W. H. Freeman, pp. 60–3, 1983.
Koehler, J. E. "Folding a Strip of Stamps." *J. Combin. Th.* **5**, 135–52, 1968.
Lunnon, W. F. "A Map-Folding Problem." *Math. Comput.* **22**, 193–99, 1968.
Lunnon, W. F. "Multi-Dimensional Strip Folding." *Computer J.* **14**, 75–9, 1971.

Mapping (Function)

MAP

Mapping Space

Let Y^X be the set of continuous mappings $f : X \to Y$. Then the TOPOLOGICAL SPACE for Y^X supplied with a compact-open topology is called a mapping space.

See also LOOP SPACE

References

Iyanaga, S. and Kawada, Y. (Eds.). "Mapping Spaces." §204B in *Encyclopedic Dictionary of Mathematics.* Cambridge, MA: MIT Press, p. 658, 1980.

Map Projection

A projection which maps a SPHERE (or SPHEROID) onto a PLANE. Map projections are generally classified into groups according to common properties (cylindrical vs. conical, conformal vs. area-preserving, etc.), although such schemes are generally not mutually exclusive. Early compilers of classification schemes include Tissot (1881), Close (1913), and Lee (1944). However, the categories given in Snyder (1987) remain the most commonly used today, and Lee's terms authalic and aphylactic are not commonly encountered.

No projection can be simultaneously CONFORMAL and AREA-PRESERVING.

See also AIRY PROJECTION, ALBERS EQUAL-AREA

CONIC PROJECTION, AXONOMETRY, AZIMUTHAL EQUIDISTANT PROJECTION, AZIMUTHAL PROJECTION, BALTHASART PROJECTION, BEHRMANN CYLINDRICAL EQUAL-AREA PROJECTION, BONNE PROJECTION, CASSINI PROJECTION, CHROMATIC NUMBER, CONIC EQUIDISTANT PROJECTION, CONIC PROJECTION, CYLINDRICAL EQUAL-AREA PROJECTION, CYLINDRICAL EQUIDISTANT PROJECTION, CYLINDRICAL PROJECTION, ECKERT IV PROJECTION, ECKERT VI PROJECTION, FOUR-COLOR THEOREM, GALL ISOGRAPHIC PROJECTION, GALL ORTHOGRAPHIC PROJECTION, GNOMONIC PROJECTION, GUTHRIE'S PROBLEM, HAMMER-AITOFF EQUAL-AREA PROJECTION, LAMBERT AZIMUTHAL EQUAL-AREA PROJECTION, LAMBERT CONFORMAL CONIC PROJECTION, MAP COLORING, MERCATOR PROJECTION, MILLER CYLINDRICAL PROJECTION, MOLLWEIDE PROJECTION, ORTHOGRAPHIC PROJECTION, PETERS PROJECTION, POLYCONIC PROJECTION, PSEUDOCYLINDRICAL PROJECTION, RECTANGULAR PROJECTION, SINUSOIDAL PROJECTION, SIX-COLOR THEOREM, STEREOGRAPHIC PROJECTION, TRISTAN EDWARDS PROJECTION, VAN DER GRINTEN PROJECTION, VERTICAL PERSPECTIVE PROJECTION

References

Anderson, P. B. "Reciprocal Links." http://www.series2000.com/users/pbander/.
Close, C. F. *Text-Book of Topographical and Geographical Surveying, 2nd ed.* London: H. M. Stationary Office, 1913.
Craig, T. *A Treatise on Projections.* Washington, DC: U.S. Government Printing Office, 1882.
Dana, P. H. "Map Projections." http://www.colorado.edu/geography/gcraft/notes/mapproj/mapproj_f.html.
Hinks, A. R. *Map Projections, 2nd rev. ed.* Cambridge, England: Cambridge University Press, 1921.
Lee, L. P. "The Nomenclature and Classification of Map Projections." *Empire Survey Review* **7**, 190–00, 1944.
Mulcahy, K. "The Map Projection Home Page." http://everest.hunter.cuny.edu/mp/.
Maling, D. H. *Coordinate Systems and Map Projections, 2nd ed, rev.* Woburn, MA: Butterworth-Heinemann, 1993.
Snyder, J. P. *Flattening the Earth: Two Thousand Years of Map Projections.* Chicago, IL: University of Chicago Press, 1993.
Snyder, J. P. *Map Projections--A Working Manual.* U. S. Geological Survey Professional Paper 1395. Washington, DC: U. S. Government Printing Office, 1987.
Tissot, A. *Mémoir sur la représentation des surfaces et les projections des cartes géographiques.* Paris: Gauthier-Villars, 1881.
Weisstein, E. W. "Books about Cartography." http://www.treasure-troves.com/books/Cartography.html.

Marcus's Theorem

A COMPACT MANIFOLD admits a LORENTZIAN STRUCTURE IFF its EULER CHARACTERISTIC vanishes. Therefore, every noncompact manifold admits a LORENTZIAN STRUCTURE.

See also EULER CHARACTERISTIC, LORENTZIAN STRUCTURE

References

Dodson, C. T. J. and Parker, P. E. "Marcus's Theorem." §9.5 in *A User's Guide to Algebraic Topology.* Dordrecht, Netherlands: Kluwer, pp. 289–91, 1997.

Marginal Analysis

Let $R(x)$ be the revenue for a production x, $C(x)$ the cost, and $P(x)$ the profit. Then

$$P(x) = R(x) - C(x),$$

and the marginal profit for the x_0th unit is defined by

$$P'(x_0) = R'(x_0) - C'(x_0),$$

where $P'(x)$, $R'(x)$, and $C'(x)$ are the DERIVATIVES of $P(x)$, $R(x)$, and $C(x)$, respectively.

See also DERIVATIVE

Marginal Probability

Let S be partitioned into $r \times s$ disjoint sets E_i and F_j where the general subset is denoted $E_i \cap F_j$. Then the marginal probability of E_i is

$$P(E_i) = \sum_{j=1}^{s} P(E_i \cap F_j).$$

See also CONDITIONAL PROBABILITY, DISTRIBUTION FUNCTION, JOINT DISTRIBUTION FUNCTION, PROBABILITY FUNCTION

Markoff Chain

MARKOV CHAIN

Markoff Number

MARKOV NUMBER

Markoff's Formulas

Formulas obtained from differentiating NEWTON'S FORWARD DIFFERENCE FORMULA,

$$f'(a_0 + ph) = \frac{1}{h} \left[\Delta_0 + \tfrac{1}{2}(2p-1)\Delta_0^2 \right.$$

$$\left. + \tfrac{1}{6}(3p^2 - 6p + 2)\Delta_0^3 + \ldots + \frac{d}{dp}\binom{p}{n}\Delta_0^n \right] + R_n',$$

where

$$R_n' = h^n f^{(n+1)}(\xi) \frac{d}{dp}\binom{p}{n+1} + h^{n+1}\binom{p}{n+1}$$

$$\times \frac{d}{dp} f^{(n+1)}(\xi), \tag{1}$$

$\binom{n}{k}$ is a BINOMIAL COEFFICIENT, and $a_0 < \xi < a_n$. Abramowitz and Stegun (1972) and Beyer (1987)

give derivatives $h^n f_0^{(n)}$ in terms of Δ^k and derivatives in terms of δ^k and ∇^k.

See also FINITE DIFFERENCE

References

Abramowitz, M. and Stegun, C. A. (Eds.). *Handbook of Mathematical Functions with Formulas, Graphs, and Mathematical Tables, 9th printing.* New York: Dover, p. 883, 1972.

Beyer, W. H. *CRC Standard Mathematical Tables, 28th ed.* Boca Raton, FL: CRC Press, pp. 449–50, 1987.

Markov Algorithm

An ALGORITHM which constructs allowed mathematical statements from simple ingredients.

Markov Chain

A collection of random variables $\{X_t\}$ (where the index t runs through 0, 1, ...) having the property that, given the present, the future is conditionally independent of the past. In other words,

$$P(X_t = j | X_0 = i_0, X_1 = i_1, \ldots X_{t-1} = i_{t-1})$$

$$= P(X_t = j | X_{t-1} = i_{t-1}).$$

If a MARKOV SEQUENCE of random variates x_n take the discrete values $a_1, ..., a_N$, then

$$P\left(x_n = a_{i_n} | x_{n-1} = a_{i_{n-1}}, \ldots, x_1 = a_1\right)$$

$$= P\left(x_n = a_{i_n} | x_{n-1} = a_{i_{n-1}}\right),$$

and the sequence x_n is called a Markov chain (Papoulis 1984, p. 532).

A SIMPLE RANDOM WALK is an example of a Markov chain.

See also MARKOV SEQUENCE, MONTE CARLO METHOD, RANDOM WALK

References

Gamerman, D. *Markov Chain Monte Carlo: Stochastic Simulation for Bayesian Inference.* Boca Raton, FL: CRC Press, 1997.

Gilks, W. R.; Richardson, S.; and Spiegelhalter, D. J. (Eds.). *Markov Chain Monte Carlo in Practice.* Boca Raton, FL: Chapman & Hall, 1996.

Grimmett, G. and Stirzaker, D. *Probability and Random Processes, 2nd ed.* Oxford, England: Oxford University Press, 1992.

Harary, F. *Graph Theory.* Reading, MA: Addison-Wesley, p. 6, 1994.

Kallenberg, O. *Foundations of Modern Probability.* New York: Springer-Verlag, 1997.

Kemeny, J. G. and Snell, J. L. *Finite Markov Chains.* New York: Springer-Verlag, 1976.

Papoulis, A. "Brownian Movement and Markoff Processes." Ch. 15 in *Probability, Random Variables, and Stochastic Processes, 2nd ed.* New York: McGraw-Hill, pp. 515–53, 1984.

Stewart, W. J. *Introduction to the Numerical Solution of Markov Chains.* Princeton, NJ: Princeton University Press, 1995.

Markov Matrix

STOCHASTIC MATRIX

Markov Moves

A type I move (CONJUGATION) takes $AB \to BA$ for A, $B \in B_n$ where B_n is a BRAID GROUP.

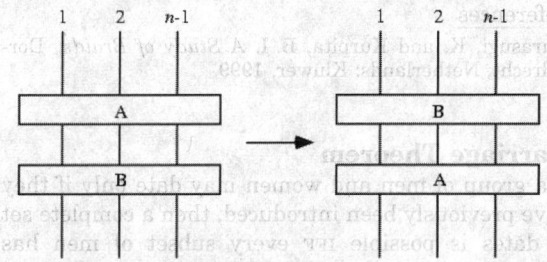

A type II move (STABILIZATION) takes $A \to Ab_n$ or $A \to Ab_n^{-1}$ for $A \in B_n$ and b_n, Ab_n, and $Ab_n^{-1} \in B_{n+1}$.

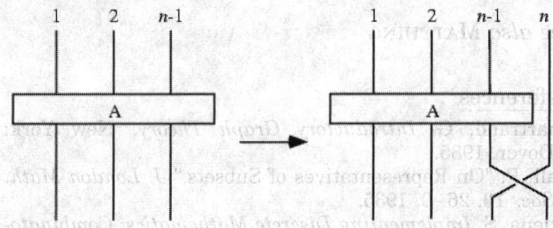

See also BRAID GROUP, CONJUGATION, KNOT MOVE, REIDEMEISTER MOVES, STABILIZATION

Markov Number

The Markov numbers m are the union of the solutions (x, y, z) to the DIOPHANTINE EQUATION

$$x^2 + y^2 + z^2 = 3xyz,$$

and are related to LAGRANGE NUMBERS L_n by

$$L_n = \sqrt{9 - \frac{4}{n^2}}.$$

The first few solutions are $(x, y, z) = (1, 1, 1)$, $(1, 1, 2)$, $(1, 2, 5)$, $(1, 5, 13)$, $(2, 5, 29)$, ... All solutions can be generated from the first two of these since the equation is a quadratic in each of the variables, so one integer solution leads to a second, and it turns out that all solutions (other than the first two singular ones) have distinct values of x, y, and z, and share two of their three values with three other solutions (Guy 1994, p. 166). The Markov numbers are then given by 1, 2, 5, 13, 29, 34, ... (Sloane's A002559).

The Markov numbers for triples (x, y, z) in which one term is 5 are 1, 2, 13, 29, 194, 433, ... (Sloane's A030452), whose terms are given by the RECURRENCE

$$a(n) = 15a(n-2) - a(n-4), \qquad (1)$$

with $a(0) = 1$, $a(1) = 2$, $a(2) = 13$, and $a(3) = 29$. The solutions can be arranged in an infinite tree with two smaller branches on each trunk. It is not known if two different regions can have the same label. Strangely, the regions adjacent to 1 have alternate FIBONACCI NUMBERS 1, 2, 5, 13, 34, ..., and the regions adjacent to 2 have alternate PELL NUMBERS 1, 5, 29, 169, 985, ...

Let $M(N)$ be the number of TRIPLES with $x \leq y \leq z \leq N$, then

$$M(n) = C(\ln N)^2 + \mathcal{O}((\ln N)^{1+\epsilon}),$$

where $C \approx 0.180717105$ (Guy 1994, p. 166).

See also HURWITZ EQUATION, HURWITZ'S IRRATIONAL NUMBER THEOREM, IRRATIONALITY MEASURE, LAGRANGE NUMBER (RATIONAL APPROXIMATION) LIOUVILLE'S APPROXIMATION THEOREM, ROTH'S THEOREM, SEGRE'S THEOREM, THUE-SIEGEL-ROTH THEOREM

References

Conway, J. H. and Guy, R. K. *The Book of Numbers.* New York: Springer-Verlag, pp. 187–89, 1996.
Descombes, R. "Problèmes d'approximation diophantienne." *Enseign. Math.* **6**, 18–6, 1960.
Guy, R. K. "Don't Try to Solve These Problems." *Amer. Math. Monthly* **90**, 35–1, 1983.
Guy, R. K. "Markoff Numbers." §D12 in *Unsolved Problems in Number Theory, 2nd ed.* New York: Springer-Verlag, pp. 166–68, 1994.
Sloane, N. J. A. Sequences A002559/M1432 and A030452 in "An On-Line Version of the Encyclopedia of Integer Sequences." http://www.research.att.com/~njas/sequences/eisonline.html.

Markov Process

A random process whose future probabilities are determined by its most recent values. A STOCHASTIC PROCESS $x(t)$ is called Markov if for every n and

$$t_1 < t_2 \ldots < t_n$$

we have

$$P(x(t_n) \leq x_n | x(t_{n-1}), \ldots, x(t_1))$$

$$= P(x(t_n) \leq x_n | x(t_{n-1})).$$

This is equivalent to

$$P(x(t_n) \leq x_n | x(t) \text{ for all } t \leq t_{n-1})$$

$$= P(x(t_n) \leq x_n | x(t_{n-1}))$$

(Papoulis 1984, p. 535).

See also DOOB'S THEOREM

References

Bharucha-Reid, A. T. *Elements of the Theory of Markov Processes and Their Applications.* New York: McGraw-Hill, 1960.
Papoulis, A. "Brownian Movement and Markoff Processes." Ch. 15 in *Probability, Random Variables, and Stochastic Processes, 2nd ed.* New York: McGraw-Hill, pp. 515–53, 1984.

Markov Sequence

A sequence X_1, X_2, ... of random variates is called Markov (or Markoff) if, for any n,

$$F(X_n | X_{n-1}, X_{n-2}, \ldots, X_1) = F(X_n | X_{n-1}),$$

i.e., if the conditional distribution F of X_n assuming $X_{n-1}, X_{n-2}, ..., X_1$ equals the conditional distribution F of X_n assuming only X_{n-1} (Papoulis 1984, pp. 528–29). The transitional densities of a Markov sequence satisfy the CHAPMAN-KOLMOGOROV EQUATION.

See also CHAPMAN-KOLMOGOROV EQUATION, MARKOV CHAIN

References

Papoulis, A. "Markoff Sequences." §15- in *Probability, Random Variables, and Stochastic Processes, 2nd ed.* New York: McGraw-Hill, pp. 528–35, 1984.

Markov's Inequality

If x takes only NONNEGATIVE values, then

$$P(x \geq a) \leq \frac{\langle x \rangle}{a}.$$

To prove the theorem, write

$$\langle x \rangle = \int_0^\infty xf(x)\,dx = \int_0^a xf(x)\,dx + \int_a^\infty xf(x)\,dx.$$

Since $P(x)$ is a probability density, it must be ≥ 0. We have stipulated that $x \geq 0$, so

$$\langle x \rangle = \int_0^a xf(x)\,dx + \int_a^\infty xf(x)\,dx$$

$$\geq \int_0^\infty xf(x)\,dx \geq \int_0^\infty af(x)\,dx$$

$$= a \int_0^\infty f(x)\,dx = aP(x \geq a),$$

Q.E.D.

Markov Spectrum

A SPECTRUM containing the REAL NUMBERS larger than FREIMAN'S CONSTANT.

See also FREIMAN'S CONSTANT, SPECTRUM SEQUENCE

References

Conway, J. H. and Guy, R. K. *The Book of Numbers.* New York: Springer-Verlag, pp. 188–89, 1996.

Markov's Theorem

Published by A. A. Markov in 1935, Markov's theorem states that equivalent BRAIDS expressing the same LINK are mutually related by successive applications of two types of MARKOV MOVES. Markov's theorem is difficult to apply in practice, so it is difficult to establish the equivalence or nonequivalence of LINKS having different BRAID representations.

See also BRAID, LINK, MARKOV MOVES

References

Murasugi, K. and Kurpita, B. I. *A Study of Braids.* Dordrecht, Netherlands: Kluwer, 1999.

Marriage Theorem

If a group of men and women may date only if they have previously been introduced, then a complete set of dates is possible IFF every subset of men has collectively been introduced to at least as many women, and vice versa (Hall 1935; Chartrand 1985, p. 121; Skiena 1990, p. 240).

See also MATCHING

References

Chartrand, G. *Introductory Graph Theory.* New York: Dover, 1985.
Hall, P. "On Representatives of Subsets." *J. London Math. Soc.* **10**, 26–0, 1935.
Skiena, S. *Implementing Discrete Mathematics: Combinatorics and Graph Theory with Mathematica.* Reading, MA: Addison-Wesley, 1990.

Married Couples Problem

Also called the MÉNAGE PROBLEM. In how many ways can n married couples be seated around a circular table in such a manner than there is always one man between two women and none of the men is next to his own wife? The solution (Ball and Coxeter 1987, p. 50) uses DISCORDANT PERMUTATIONS and can be given in terms of LAISANT'S RECURRENCE FORMULA

$$(n-1)A_{n+1} = (n^2-1)A_n + (n+1)A_{n-1} + 4(-1)^n, \quad (1)$$

with $A_1 = A_2 = 1$. A closed form expression due to Touchard (1934) is

$$A_n = \sum_{k=0}^n \frac{2n}{2n-k} \binom{2n-k}{k} (n-k)!(-1)^k, \quad (2)$$

where $\binom{n}{k}$ is a BINOMIAL COEFFICIENT (Vardi 1991). The sum can be evaluated explicitly as

$$A_n = \frac{n\pi I_{-n}(2)\csc(n\pi)}{e^2}$$

$$- \frac{4(-1)^n}{n^2-1}\,_2F_2(1, \tfrac{3}{2}; 2-n, 2-n; 2+n; -4), \quad (3)$$

where $_2F_2(a, b; c, d; x)$ is a GENERALIZED HYPERGEOMETRIC FUNCTION.

The first few values of A_n are -1, 1, 0, 2, 13, 80, 579, ... (Sloane's A000179), which are sometimes called MÉNAGE NUMBERS. The desired solution is then $2n!A_n$. The numbers A_n can be considered a special case of a restricted ROOKS PROBLEM.

See also DISCORDANT PERMUTATION, LAISANT'S RECURRENCE FORMULA, ROOKS PROBLEM

References

Ball, W. W. R. and Coxeter, H. S. M. *Mathematical Recreations and Essays, 13th ed.* New York: Dover, p. 50, 1987.

Comtet, L. "The 'Problème des Ménages'." §4.3 in *Advanced Combinatorics: The Art of Finite and Infinite Expansions, rev. enl. ed.* Dordrecht, Netherlands: Reidel, pp. 182–85, 1974.

Dörrie, H. §8 in *100 Great Problems of Elementary Mathematics: Their History and Solutions.* New York: Dover, pp. 27–3, 1965.

Halmos, P. R.; Vaughan, H. E. "The Marriage Problem." *Amer. J. Math.* **72**, 214–15, 1950.

Lucas, E. *Théorie des Nombres.* Paris: A. Blanchard, pp. 215 and 491–95, 1979.

MacMahon, P. A. *Combinatory Analysis, Vol. 1.* London: Cambridge University Press, pp. 253–56, 1915.

Newman, D. J. "A Problem in Graph Theory." *Amer. Math. Monthly* **65**, 611, 1958.

Sloane, N. J. A. Sequences A000179/M2062 in "An On-Line Version of the Encyclopedia of Integer Sequences." http://www.research.att.com/~njas/sequences/eisonline.html.

Touchard, J. "Sur un problème de permutations." *C. R. Acad. Sci. Paris* **198**, 631–33, 1934.

Vardi, I. *Computational Recreations in Mathematica.* Reading, MA: Addison-Wesley, p. 123, 1991.

Marshall-Edgeworth Index

The statistical INDEX

$$P_{\text{ME}} \equiv \frac{\sum p_n(q_0 + q_n)}{\sum (v_0 + v_n)},$$

where p_n is the price per unit in period n, q_n is the quantity produced in period n, and $v_n \equiv p_n q_n$ is the value of the n units.

See also INDEX

References

Kenney, J. F. and Keeping, E. S. *Mathematics of Statistics, Pt. 1, 3rd ed.* Princeton, NJ: Van Nostrand, pp. 66–7, 1962.

Martingale

A sequence of random variates X_0, X_1, ... with finite means such that the conditional expectation of X_{n+1} given X_0, X_1, X_2, ..., X_n is equal to X_n, i.e.,

$$\langle x_{n+1} | X_0, \ldots, X_n \rangle = X_n$$

(Feller 1971, p. 210). The term was first used to describe a type of wagering in which the bet is doubled or halved after a loss or win, respectively.

The concept of martingales is due to Lévy, and it was developed extensively by Doob.

A 1-D RANDOM WALK with steps equally likely in either direction ($p = q = 1/2$) is an example of a martingale.

See also ABSOLUTELY FAIR, GAMBLER'S RUIN, RANDOM WALK–1-D, SAINT PETERSBURG PARADOX

References

Doob, J. L. *Stochastic Processes.* New York: Wiley, 1953.

Feller, W. "Martingales." §6.12 in *An Introduction to Probability Theory and Its Applications, Vol. 2, 3rd ed.* New York: Wiley, pp. 210–15, 1971.

Lévy, P. *Calcul de probabilités.* Paris: Gauthier-Villars, 1925.

Lévy, P. *Théorie de l'addition des variables aléatoires.* Paris: Gauthier-Villars, 1954.

Lévy, P. *Processus stochastiques et mouvement Brownien, 2nd ed.* Paris: Gauthier-Villars, 1965.

Loève, M. *Probability Theory, 3rd ed.* Princeton, NJ: Van Nostrand, 1963.

Mascheroni Constant

EULER-MASCHERONI CONSTANT

Mascheroni Construction

A geometric construction done with a movable COMPASS alone. All constructions possible with a COMPASS and STRAIGHTEDGE are possible with a *movable* COMPASS alone, as was proved by Mascheroni (1797). Mascheroni's results are now known to have been anticipated largely by Mohr (1672).

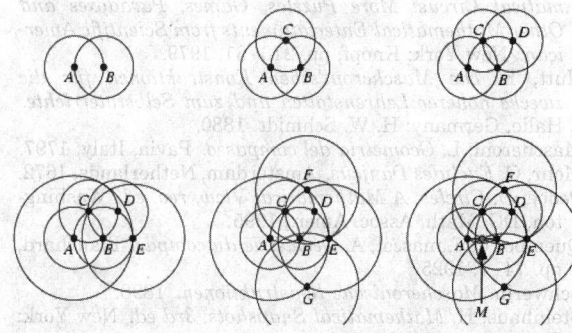

An example of a Mascheroni construction of the midpoint M of a LINE SEGMENT specified by two points A and B illustrated above (Steinhaus 1983, Wells 1991). Without loss of generality, take $AB = 1$.

1. Construct circles centered at A and B passing through B and A. These are unit circles centered at $(0, 0)$ and $(1, 0)$.
2. Locate C, the indicated intersection of circles A and B, and draw a circle centered on C passing through points A and B. This circle has center $(1/2, \sqrt{3}/2)$ and radius 1.
3. Locate D, the indicated intersection of circles B and C, and draw a circle centered on C passing

through points B and C. This circle has center $(3/2, \sqrt{3}/2)$ and radius 1.

4. Locate E, the indicated intersection of circles B and D, and draw a circle centers on E passing through point C. This circle has center $(2, 0)$ and radius $\sqrt{3}$.

5. Locate F and G, the intersections of circles AE and EC. These points are located at positions $(5/4, \pm\sqrt{39}/4)$.

6. Locate M, the intersection of circles F and G. This point has position $(1/2, 0)$, and is therefore the desired MIDPOINT of $\overline{AB}$.

Pedoe (1995, pp. xviii-xix) also gives a Mascheroni solution.

See also COMPASS, GEOMETRIC CONSTRUCTION, NEUSIS CONSTRUCTION, STEINER CONSTRUCTION, STRAIGHTEDGE

References

Ball, W. W. R. and Coxeter, H. S. M. *Mathematical Recreations and Essays, 13th ed.* New York: Dover, pp. 96–7, 1987.

Bogomolny, A. "Geometric Constructions with the Compass Alone." http://www.cut-the-knot.com/do_you_know/compass.html.

Courant, R. and Robbins, H. "Constructions with Other Tools. Mascheroni Constructions with Compass Alone." §3.5 in *What is Mathematics?: An Elementary Approach to Ideas and Methods, 2nd ed.* Oxford, England: Oxford University Press, pp. 146–58, 1996.

Dörrie, H. "Mascheroni's Compass Problem." §33 in *100 Great Problems of Elementary Mathematics: Their History and Solutions.* New York: Dover, pp. 160–64, 1965.

Gardner, M. "Mascheroni Constructions." Ch. 17 in *Mathematical Circus: More Puzzles, Games, Paradoxes and Other Mathematical Entertainments from Scientific American.* New York: Knopf, pp. 216–31, 1979.

Hutt, E. *Die Mascheroni'schen Konstruktionen für die zwecke höherer Lehrenstalten und zum Selbstuterrichte.* Halle, Germany: H. W. Schmidt, 1880.

Mascheroni, L. *Geometria del compasso.* Pavia, Italy, 1797.

Mohr, G. *Euclides Danicus.* Amsterdam, Netherlands, 1672.

Pedoe, D. *Circles: A Mathematical View, rev. ed.* Washington, DC: Math. Assoc. Amer., 1995.

Quemper de Lanascol, A. *Géométrie du compas.* Blanchard, pp. 74–7, 1925.

Schwerin. *Mascheronische Konstruktionen.* 1898.

Steinhaus, H. *Mathematical Snapshots, 3rd ed.* New York: Dover, pp. 141–42, 1999.

Wells, D. *The Penguin Dictionary of Curious and Interesting Geometry.* London: Penguin, pp. 148–49, 1991.

Maschke's Theorem

If a MATRIX GROUP is reducible, then it is completely reducible, i.e., if the MATRIX GROUP is equivalent to the MATRIX GROUP in which every MATRIX has the reduced form

$$\begin{bmatrix} D_i^{(1)} & X_i \\ 0 & D_i^{(2)} \end{bmatrix},$$

then it is equivalent to the MATRIX GROUP obtained by putting $X_i = 0$.

See also MATRIX GROUP

References

Lomont, J. S. *Applications of Finite Groups.* New York: Dover, p. 49, 1987.

Mason's abc Theorem

MASON'S THEOREM

Mason's Theorem

Let there be three POLYNOMIALS $a(x)$, $b(x)$, and $c(x)$ with no common factors such that

$$a(x) + b(x) = c(x).$$

Then the number of distinct ROOTS of the three POLYNOMIALS is one or more greater than their largest degree. The theorem was first proved by Stothers (1981).

Mason's theorem may be viewed as a very special case of a Wronskian estimate (Chudnovsky and Chudnovsky 1984). The corresponding Wronskian identity in the proof by Lang (1993) is

$$c^3 * W(a, b, c) = W(W(a, c), W(b, c)),$$

so if a, b, and c are linearly dependent, then so are $W(a, c)$ and $W(b, c)$. More powerful Wronskian estimates with applications toward Diophantine approximation of solutions of linear differential equations may be found in Chudnovsky and Chudnovsky (1984) and Osgood (1985).

The RATIONAL FUNCTION case of FERMAT'S LAST THEOREM follows trivially from Mason's theorem (Lang 1993, p. 195).

See also ABC CONJECTURE

References

Chudnovsky, D. V. and Chudnovsky, G. V. "The Wronskian Formalism for Linear Differential Equations and Padé Approximations." *Adv. Math.* **53**, 28–4, 1984.

Lang, S. "Old and New Conjectured Diophantine Inequalities." *Bull. Amer. Math. Soc.* **23**, 37–5, 1990.

Lang, S. *Algebra, 3rd ed.* Reading, MA: Addison-Wesley, 1993.

Mason, R. C. *Diophantine Equations over Functions Fields.* Cambridge, England: Cambridge University Press, 1984.

Osgood, C. F. "Sometimes Effective Thue-Siegel-Roth-Schmidt-Nevanlinna Bounds, or Better." *J. Number Th.* **21**, 347–89, 1985.

Stothers, W. W. "Polynomial Identities and Hauptmodulen." *Quart. J. Math. Oxford Ser. II* **32**, 349–70, 1981.

Masser-Gramain Constant

N.B. A detailed online essay by S. Finch was the starting point for this entry.

Let $f(z)$ be an ENTIRE FUNCTION such that $f(n)$ is an INTEGER for each POSITIVE INTEGER n. Then Pólya (1915) showed that if

$$\limsup_{r \to \infty} \frac{\ln M_r}{r} < \ln 2 = 0.693\ldots, \tag{1}$$

where

$$M_r = \sup_{|z| \le r} |f(x)| \tag{2}$$

is the SUPREMUM, then f is a POLYNOMIAL. Furthermore, $\ln 2$ is the best constant (i.e., counterexamples exist for every smaller value).

If $f(z)$ is an ENTIRE FUNCTION with $f(n)$ a GAUSSIAN INTEGER for each GAUSSIAN INTEGER n, then Gelfond (1929) proved that there exists a constant α such that

$$\limsup_{r \to \infty} \frac{\ln M_r}{r^2} < \alpha \tag{3}$$

implies that f is a POLYNOMIAL. Gramain (1981, 1982) showed that the best such constant is

$$\alpha = \frac{\pi}{2e} = 0.578\ldots \tag{4}$$

Maser (1980) proved the weaker result that f must be a POLYNOMIAL if

$$\limsup_{r \to \infty} \frac{\ln M_r}{r^2} < \alpha_0 = \frac{1}{2} \exp\left(-\delta + \frac{4c}{\pi}\right), \tag{5}$$

where

$$c = \gamma\beta(1) + \beta'(1) = 0.642454398948114\ldots, \tag{6}$$

γ is the EULER-MASCHERONI CONSTANT, $\beta(z)$ is the DIRICHLET BETA FUNCTION,

$$\delta \equiv \lim_{n \to \infty}\left(\sum_{k=2}^{n} \frac{1}{\pi r k^2} - \ln n\right), \tag{7}$$

and r_k is the minimum NONNEGATIVE r for which there exists a COMPLEX NUMBER z for which the CLOSED DISK with center z and radius r contains at least k distinct GAUSSIAN INTEGERS. Gosper gave

$$c = \pi\left\{-\ln[\Gamma(\tfrac{1}{4})] + \tfrac{3}{4}\pi + \tfrac{1}{2}\ln 2 + \tfrac{1}{2}\gamma\right\}. \tag{8}$$

Gramain and Weber (1985, 1987) have obtained

$$1.811447299 < \delta < 1.897327177, \tag{9}$$

which implies

$$0.1707339 < \alpha_0 < 0.1860446. \tag{10}$$

Gramain (1981, 1982) conjectured that

$$\alpha_0 = \frac{1}{2e}, \tag{11}$$

which would imply

$$\delta = 1 + \frac{4c}{\pi} = 1.822825249\ldots. \tag{12}$$

References

Finch, S. "Favorite Mathematical Constants." http://www.mathsoft.com/asolve/constant/masser/masser.html.

Gramain, F. "Sur le théorème de Fukagawa-Gel'fond." *Invent. Math.* **63**, 495–06, 1981.

Gramain, F. "Sur le théorème de Fukagawa-Gel'fond-Gruman-Masser." *Séminaire Delange-Pisot-Poitou (Théorie des Nombres), 1980–981.* Boston, MA: Birkhäuser, 1982.

Gramain, F. and Weber, M. "Computing and Arithmetic Constant Related to the Ring of Gaussian Integers." *Math. Comput.* **44**, 241–45, 1985.

Gramain, F. and Weber, M. "Computing and Arithmetic Constant Related to the Ring of Gaussian Integers." *Math. Comput.* **48**, 854, 1987.

Masser, D. W. "Sur les fonctions entières à valeurs entières." *C. R. Acad. Sci. Paris Sér. A-B* **291**, A1-A4, 1980.

Mastermind

References

Bewersdorff, J. *Glück, Logik and Bluff: Mathematik im Spiel: Methoden, Ergebnisse und Grenzen.* Wiesbaden, Germany: Vieweg, 1998.

Bogomolny, A. and Greenwell, D. "Cut the Knot: Invitation to Mastermind." http://www.maa.org/editorial/knot/Mastermind.html.

Chvatal, V. "Mastermind." *Combinatorica* **3**, 325–29, 1983.

Erdos, P. and C. Rényi, C. "On Two Problems in Information Theory." *Magyar Tud. Akad. Mat. Kut. Int. Közl.* **8**, 229–42, 1963.

Greenwell, D. L. "Mastermind." Submitted to *J. Recr. Math.*

Guy, R. "The Strong Law of Small Numbers." In *The Lighter Side of Mathematics* (Ed. R. K. Guy and R. E. Woodrow). Washington, DC: Math. Assoc. Amer., 1994.

Knuth, D. E. "The Computer as a Master Mind." *J. Recr. Math.* **9**, 1–, 1976–7.

Koyama, K. and Lai, T. W. "An Optimal Mastermind Strategy." *J. Recr. Math.* **25**, 251–56, 1993.

Mitchell, M. "MasterMind ® Mathematics." Key Curriculum Press, 1999.

Neuwirth, E. "Some Strategies for Mastermind." *Z. für Operations Research* **26**, B257-B278, 1982.

Matching

A matching on a GRAPH G is a set of edges of G such that no two of them share a vertex in common. The largest possible matching consists of $n/2$ edges, and such a matching is called a perfect matching. Although not all graphs have perfect matchings, a maximum matching exists for each graph.

The maximum matching in a BIPARTITE GRAPH can be found using BipartiteMatching[g] in the *Mathematica* add-on package DiscreteMath`Combinatorica` (which can be loaded with the command <<DiscreteMath`). The maximum matching on a general graph can be found using MaximalMatching[g] in the same package.

See also BERGE'S THEOREM, MARRIAGE THEOREM, PERFECT MATCHING, STABLE MARRIAGE PROBLEM

References

Hopcroft, J. and Karp, R. "An $n^{5/2}$ Algorithm for Maximum Matching in Bipartite Graphs." *SIAM J. Comput.*, 225–31, 1975.

Lovász, L. and Plummer, M. D. *Matching Theory.* Amsterdam, Netherlands: North-Holland, 1986.

Skiena, S. "Matching." §6.4 in *Implementing Discrete Mathematics: Combinatorics and Graph Theory with Mathematica.* Reading, MA: Addison-Wesley, pp. 240–46, 1990.

Match Problem

Given n matches (i.e., rigid unit line segments), find the number of topologically distinct planar arrangements which can be made (Gardner 1991). In this problem, two matches laid end-to-end with no third match at their meeting point are considered equivalent to a single match, so triangles are equivalent to squares, n-match tails are equivalent to 1-match tails, etc.

Solutions to the match problem are PLANAR TOPOLOGICAL GRAPHS on e edges, and the first few values for $e = 1, 1, 3, 5, 10, 19, 39, \ldots$ (Sloane's A003055).

See also CIGARETTES, MATCHSTICK GRAPH, PLANAR GRAPH, POLYNEMA, TOPOLOGICAL GRAPH

References

Gardner, M. "The Problem of the Six Matches." In *The Unexpected Hanging and Other Mathematical Diversions.* Chicago, IL: Chicago University Press, pp. 79–1, 1991.

Sloane, N. J. A. Sequences A003055/M2464 in "An On-Line Version of the Encyclopedia of Integer Sequences." http://www.research.att.com/~njas/sequences/eisonline.html.

Matchstick Construction

Every point which can be constructed with a STRAIGHTEDGE and COMPASS, and no other points, can be constructed using identical matchsticks (i.e., identical movable line segments). Wells (1991) gives matchstick constructions which bisect a line segment and construct a SQUARE.

See also GEOMETRIC CONSTRUCTION, MASCHERONI CONSTRUCTION, NEUSIS CONSTRUCTION, STEINER CONSTRUCTION

References

Dawson, T. R. "'Match-Stick' Geometry." *Math. Gaz.* **23**, 161–68, 1939.

Wells, D. *The Penguin Dictionary of Curious and Interesting Geometry.* London: Penguin, p. 149, 1991.

Matchstick Graph

A PLANAR GRAPH whose EDGES are all unit line segments. The minimal number of EDGES for matchstick graphs of various degrees are given in the table below. The minimal degree 1 matchstick graph is a single EDGE, and the minimal degree 2 graph is an EQUILATERAL TRIANGLE.

n	e	v
1	1	2
2	3	3
3	12	8
4	≤ 42	

Mathematical Induction

INDUCTION

Mathematics

Mathematics is a broad-ranging field of study in which the properties and interactions of idealized objects are examined. Whereas mathematics began merely as a calculational tool for computation and tabulation of quantities, it has blossomed into an extremely rich and diverse set of tools, terminologies, and approaches which range from the purely abstract to the utilitarian.

Bertrand Russell once whimsically defined mathematics as "The subject in which we never know what we are talking about nor whether what we are saying is true" (Bergamini 1969).

The term "mathematics" is often shortened to "math" in informal American speech and, consistent with the British penchant for adding superfluous letters, "maths" in British English.

See also METAMATHEMATICS

References

Bergamini, D. *Mathematics.* New York: Time-Life Books, p. 9, 1969.

Mathematics Contests

There are several regular mathematics competitions available to students. The International Mathematical Olympiad is perhaps the largest, while the William Lowell Putnam Competition is another important contest.

The International Mathematical Olympiad (IMO) is the yearly world championship of mathematics for high school students and is held in a different country each year. The first IMO was held in 1959 in Romania, but the contest has gradually expanded to include students from more than 80 different countries.

The William Lowell Putnam Mathematics Competition is a North American math contest for college students. Each year, on the first Saturday in December, more than 2000 students spend six hours in two sittings trying to solve 12 problems. The majority of the problems are very difficult, in the sense that their solution may require a nonstandard and creative approach. It is very rare for students to be able to solve all the problems, let alone the majority of them. The test can be taken both by individual and by teams, and the winners or their schools receive a small monetary compensation. Results for a given exam usually become available in early April of the following year.

The International Mathematical Contest in Modeling (MCM) is a competition that challenges teams of undergraduate students to clarify, analyze, and propose solutions to open-ended problems. Problems are chosen with the advice of experts in industry and government, and the best papers are submitted to be published in professional journals.

See also MATHEMATICS PRIZES, UNSOLVED PROBLEMS

References

COMAP: The Consortium for Mathematics and Its Applications. "Abut MCM." http://www.comap.com/undergraduate/contests/mcm/about.html.
"International Mathematics Olympiad." http://imo.math.ca/ and http://olympiads.win.tue.nl/imo/.
"William Lowell Putnam Competition." http://www.unl.edu/amc/putnam/.

Mathematics Prizes

Several prizes are awarded periodically for outstanding mathematical achievement. There is no Nobel Prize in mathematics, and the most prestigious mathematical award is known as the FIELDS MEDAL. In rough order of importance, other awards are the $100,000 Wolf Prize of the Wolf Foundation of Israel, the Leroy P. Steele Prize of the American Mathematical Society, followed by the Bôcher Memorial Prize, Frank Nelson Cole Prizes in Algebra and Number Theory, and the Delbert Ray Fulkerson Prize, all presented by the American Mathematical Society.

The Clay Mathematics Institute of Cambridge, Massachusetts (CMI) has named seven "Millennium Prize Problems," selected by focusing on important classic questions in mathematics that have resisted solution over the years. A $7 million prize fund has been established for the solution to these problems, with $1 million allocated to each. The problems consist of the RIEMANN HYPOTHESIS, POINCARÉ CONJECTURE, HODGE CONJECTURE, SWINNERTON-DYER CONJEC-

TURE, solution of the Navier-Stokes equation, formulation of Yang-Mills theory , and determination of whether NP-PROBLEMS are actually P-PROBLEMS.

See also FIELDS MEDAL, MATHEMATICS CONTESTS, UNSOLVED PROBLEMS, WOLFSKEHL PRIZE

References

American Mathematical Society. "AMS Funds and Prizes." http://www.ams.org/secretary/prizes.html.
Clay Mathematics Institute. "Millennium Prize Problems." http://www.claymath.org/prize_problems/.
MacTutor History of Mathematics Archives. "The Fields Medal." http://www-groups.dcs.st-and.ac.uk/~history/Societies/FieldsMedal.html. "Winners of the Bôcher Prize of the AMS." http://www-groups.dcs.st-and.ac.uk/~history/Societies/AMSBocherPrize.html. "Winners of the Frank Nelson Cole Prize of the AMS." http://www-groups.dcs.st-and.ac.uk/~history/Societies/AMSColePrize.html.
MacTutor History of Mathematics Archives. "Mathematical Societies, Medals, Prizes, and Other Honours." http://www-groups.dcs.st-and.ac.uk/~history/Societies/.
Monastyrsky, M. *Modern Mathematics in the Light of the Fields Medals.* Wellesley, MA: A. K. Peters, 1997.
"Wolf Prize Recipients in Mathematics." http://www.aquanet.co.il/wolf/wolf5.html.

Mathematics Problems

HILBERT'S PROBLEMS, LANDAU'S PROBLEMS, PROBLEM

MathieuC

MATHIEU FUNCTION

MathieuCharacteristicA

MATHIEU CHARACTERISTIC EXPONENT

MathieuCharacteristicB

MATHIEU CHARACTERISTIC EXPONENT

Mathieu Characteristic Exponent

MATHIEU CHARACTERISTIC EXPONENT

MathieuCPrime

MATHIEU FUNCTION

Mathieu Differential Equation

$$\frac{d^2V}{dv^2} + [a - 2q\cos(2v)]V = 0 \qquad (1)$$

(Abramowitz and Stegun 1972; Zwillinger 1997, p. 125), having solution

$$y = C_1 C(a, q, v) + C_2 S(a, q, v), \qquad (2)$$

where $C(a, q, v)$ and $S(a, q, v)$ are MATHIEU FUNCTIONS. The equation arises in separation of variables of the HELMHOLTZ DIFFERENTIAL EQUATION in ELLIPTIC CYLINDRICAL COORDINATES. Whittaker and Watson (1990) use a slightly different form to define the MATHIEU FUNCTIONS.

The modified Mathieu differential equation

$$\frac{d^2U}{du^2} - [a - 2q\cosh(2u)]U = 0 \qquad (3)$$

(Iyanaga and Kawada 1980, p. 847; Zwillinger 1997, p. 125) arises in SEPARATION OF VARIABLES of the HELMHOLTZ DIFFERENTIAL EQUATION in ELLIPTIC CYLINDRICAL COORDINATES, and has solutions

$$y = C_1 C(a, q, -iu) + C_2 S(a, q - iu). \qquad (4)$$

The associated Mathieu differential equation is given by

$$y'' + [(1 - 2r)\cot x]y' + (a + k^2 \cos^2 x)y = 0 \qquad (5)$$

(Ince 1956, p. 403; Zwillinger 1997, p. 125).

See also HILL'S DIFFERENTIAL EQUATION, MATHIEU FUNCTION, WHITTAKER-HILL DIFFERENTIAL EQUATION

References

Abramowitz, M. and Stegun, C. A. (Eds.). *Handbook of Mathematical Functions with Formulas, Graphs, and Mathematical Tables, 9th printing.* New York: Dover, p. 722, 1972.

Campbell, R. *Théorie générale de l'équation de Mathieu et de quelques autres équations différentielles de la mécanique.* Paris: Masson, 1955.

Ince, E. L. *Ordinary Differential Equations.* New York: Dover, 1956.

Iyanaga, S. and Kawada, Y. (Eds.). *Encyclopedic Dictionary of Mathematics.* Cambridge, MA: MIT Press, p. 847, 1980.

Morse, P. M. and Feshbach, H. *Methods of Theoretical Physics, Part I.* New York: McGraw-Hill, pp. 556–57, 1953.

Whittaker, E. T. and Watson, G. N. *A Course in Modern Analysis, 4th ed.* Cambridge, England: Cambridge University Press, 1990.

Zwillinger, D. (Ed.). *CRC Standard Mathematical Tables and Formulae.* Boca Raton, FL: CRC Press, 1995.

Zwillinger, D. *Handbook of Differential Equations, 3rd ed.* Boston, MA: Academic Press, p. 125, 1997.

Mathieu Function

The Mathieu functions are the solutions to the MATHIEU DIFFERENTIAL EQUATION

$$\frac{d^2V}{dv^2} - [a - 2q\cos(2v)]V = 0. \qquad (1)$$

Even solutions are denoted $C(a, q, z)$ and odd solutions by $S(a, q, z)$. These are returned by the *Mathematica* functions MathieuC[a, q, z] and MathieuS[a, q, z], respectively. These functions appear in physical problems involving elliptical shapes or periodic potentials. The Mathieu functions have the special values

$$C(a, 0, z) = \cos(\sqrt{a}z) \qquad (2)$$

$$S(a, 0, z) = \sin(\sqrt{a}z). \qquad (3)$$

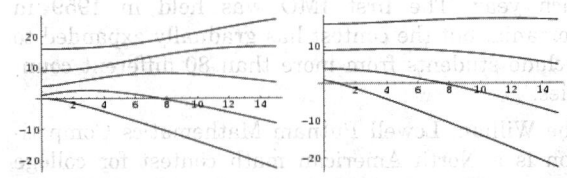

For nonzero q, the Mathieu functions are only periodic in z for certain values of a. Such characteristic values are given by the *Mathematica* functions MathieuCharacteristicA[r, q] and MathieuCharacteristicB[r, q] with r an integer or rational number. These values are often denoted a_r and b_r. For integer r, the even and odd Mathieu functions with characteristic values a_r and b_r are often denoted $ce_r(z, q)$ and $se_r(z, q)$, respectively (Abramowitz and Stegun 1972, p. 725). The left plot above shows a_r for $r = 0, 1, ..., 4$ and the right plot shows b_r for $r = 1, ..., 4$.

Whittaker and Watson (1990, p. 405) define the Mathieu function based on the equation

$$\frac{d^2u}{dz^2} + [a + 16q\cos(2z)]u = 0. \qquad (4)$$

This equation is closely related to HILL'S DIFFERENTIAL EQUATION. For an EVEN Mathieu function,

$$G(\eta) = \lambda \int_{-\pi}^{\pi} e^{k\cos\eta\cos\theta} G(\theta)\,d\theta, \qquad (5)$$

where $k \equiv \sqrt{32q}$. For an ODD Mathieu function,

$$G(\eta) = \lambda \int_{-\pi}^{\pi} \sin(k\sin\eta\sin\theta)G(\theta)\,d\theta. \qquad (6)$$

Both EVEN and ODD functions satisfy

$$G(\eta) = \lambda \int_{-\pi}^{\pi} e^{ik\sin\eta\sin\theta}G(\theta)\,d\theta. \qquad (7)$$

Letting $\zeta \equiv \cos^2 z$ transforms the MATHIEU DIFFERENTIAL EQUATION to

$$4\zeta(1 - \zeta)\frac{d^2u}{d\zeta^2} + 2(1 - 2\zeta)\frac{du}{d\zeta} + (a - 16q + 32q\zeta)u = 0. \qquad (8)$$

See also MATHIEU CHARACTERISTIC EXPONENT, MATHIEU DIFFERENTIAL EQUATION

References

Abramowitz, M. and Stegun, C. A. (Eds.). "Mathieu Functions." Ch. 20 in *Handbook of Mathematical Functions with Formulas, Graphs, and Mathematical Tables, 9th printing.* New York: Dover, pp. 721–46, 1972.

Gradshteyn, I. S. and Ryzhik, I. M. "Mathieu Functions." §6.9 and 8.6 in *Tables of Integrals, Series, and Products, 6th ed.* San Diego, CA: Academic Press, pp. 800–04 and 1006–013, 2000.

Humbert, P. *Fonctions de Lamé et Fonctions de Mathieu.* Paris: Gauthier-Villars, 1926.

Mechel, F. P. *Mathieu Functions: Formulas, Generation, Use.* Stuttgart, Germany: Hirzel, 1997.

Morse, P. M. and Feshbach, H. *Methods of Theoretical Physics, Part I.* New York: McGraw-Hill, pp. 562–68 and 633–42, 1953.

Whittaker, E. T. and Watson, G. N. *A Course in Modern Analysis, 4th ed.* Cambridge, England: Cambridge University Press, 1990.

Mathieu Groups

The first SIMPLE SPORADIC GROUPS discovered. M_{11}, M_{12}, M_{22}, M_{23}, M_{24} were discovered in 1861 and 1873 by Mathieu. Frobenius showed that all the Mathieu groups are SUBGROUPS of M_{24}.

The Mathieu groups are most simply defined as AUTOMORPHISM GROUPS of STEINER SYSTEMS, as summarized in the following table.

Mathieu group	Steiner system
M_{11}	$S(4, 5, 11)$
M_{12}	$S(5, 6, 12)$
M_{22}	$S(3, 6, 22)$
M_{23}	$S(4, 7, 23)$
M_{24}	$S(5, 8, 24)$

M_{11} and M_{23} are TRANSITIVE PERMUTATION GROUPS of 11 and 23 elements. The ORDERS of the Mathieu groups are

$$|M_{11}| = 2^4 \cdot 3^2 \cdot 5 \cdot 11$$

$$|M_{12}| = 2^6 \cdot 3^3 \cdot 5 \cdot 11$$

$$|M_{22}| = 2^7 \cdot 3^2 \cdot 5 \cdot 7 \cdot 11$$

$$|M_{23}| = 2^7 \cdot 3^2 \cdot 5 \cdot 7 \cdot 11 \cdot 23$$

$$|M_{24}| = 2^{10} \cdot 3^3 \cdot 5 \cdot 7 \cdot 11 \cdot 23.$$

See also AUTOMORPHISM GROUP, SIMPLE GROUP, SPORADIC GROUP, STEINER SYSTEM, TRANSITIVE GROUP, WITT GEOMETRY

References

Conway, J. H. and Sloane, N. J. A. "The Golay Codes and the Mathieu Groups." Ch. 11 in *Sphere Packings, Lattices, and Groups, 2nd ed.* New York: Springer-Verlag, pp. 299–30, 1993.

Dixon, J. and Mortimer, B. *Permutation Groups.* New York: Springer-Verlag, 1996.

Rotman, J. J. Ch. 9 in *An Introduction to the Theory of Groups, 4th ed.* New York: Springer-Verlag, 1995.

Wilson, R. A. "ATLAS of Finite Group Representation." http://for.mat.bham.ac.uk/atlas/html/contents.html#spo.

MathieuS

MATHIEU FUNCTION

MathieuSPrime

MATHIEU FUNCTION

Matrix

The TRANSFORMATION given by the system of equations

$$x'_1 = a_{11}x_1 + a_{12}x_2 + \ldots + a_{1n}x_n$$

$$x'_2 = a_{21}x_1 + a_{22}x_2 + \ldots + a_{2n}x_n$$

$$\vdots$$

$$x'_m = a_{m1}x_1 + a_{m2}x_2 + \ldots + a_{mn}x_n$$

is denoted by the MATRIX EQUATION

$$\begin{bmatrix} x'_1 \\ x'_2 \\ \vdots \\ x'_m \end{bmatrix} = \begin{bmatrix} a_{11} & a_{12} & \cdots & a_{1n} \\ a_{21} & a_{22} & \cdots & a_{2n} \\ \vdots & \vdots & \ddots & \vdots \\ a_{m1} & a_{m2} & \cdots & a_{mn} \end{bmatrix} \begin{bmatrix} x_1 \\ x_2 \\ \vdots \\ x_n \end{bmatrix}.$$

In concise notation, this could be written

$$\mathbf{x}' = \mathsf{A}\mathbf{x},$$

where $\mathbf{x}'$ and $\mathbf{x}$ are VECTORS and A is called an $m \times n$ matrix. An $m \times n$ matrix consists of m rows and n columns, and the set of $m \times n$ matrices with real coefficients is sometimes denoted $\mathbb{R}^{m \times n}$. To remember which index refers to which direction, identify the indices of the last (i.e., lower right) term, so the indices m, n of the last element in the above matrix identifies it as an $m \times n$ matrix.

A matrix is said to be SQUARE if $m = n$, and RECTANGULAR if $m \neq n$. An $m \times 1$ matrix is called a COLUMN VECTOR, and a $1 \times n$ matrix is called a ROW VECTOR. Special types of SQUARE MATRICES include the IDENTITY MATRIX I, with $A_2 A_3$ (where δ_{ij} is the KRONECKER DELTA) and the DIAGONAL MATRIX $a_{ij} = c_i \delta_{ij}$ (where c_i are a set of constants).

For every linear transformation there exists one and only one corresponding matrix. Conversely, every matrix corresponds to a unique linear transformation. The matrix is an important concept in mathematics, and was first formulated by Sylvester and Cayley.

Two matrices may be added (MATRIX ADDITION) or multiplied (MATRIX MULTIPLICATION) together to yield a new matrix. Other common operations on a single matrix are diagonalization, inversion (MATRIX INVERSE), and transposition (matrix TRANSPOSE). The DETERMINANT det(A) or |A| of a matrix A is a very important quantity which appears in many diverse applications. Matrices provide a concise notation which is extremely useful in a wide range of problems involving linear equations (e.g., LEAST SQUARES FITTING).

See also ADJACENCY MATRIX, ADJUGATE MATRIX, ALTERNATING SIGN MATRIX, ANTISYMMETRIC MATRIX, BLOCK MATRIX, BOHR MATRIX, BOURQUE-LIGH CONJECTURE, CARTAN MATRIX, CIRCULANT MATRIX, CONDITION NUMBER, CRAMER'S RULE, DETERMINANT, DIAGONAL MATRIX, DIRAC MATRICES, EIGENVECTOR, ELEMENTARY MATRIX, ELEMENTARY ROW AND COLUMN OPERATIONS, EQUIVALENT MATRIX, FOURIER MATRIX, GRAM MATRIX, HILBERT MATRIX, HYPERMATRIX, IDENTITY MATRIX, ILL-CONDITIONED MATRIX, INCIDENCE MATRIX, IRREDUCIBLE MATRIX, KAC MATRIX, LEAST COMMON MULTIPLE MATRIX, LU DECOMPOSITION, MARKOV MATRIX, MATRIX ADDITION, MATRIX DECOMPOSITION THEOREM, MATRIX INVERSE, MATRIX MULTIPLICATION, MCCOY'S THEOREM, MINIMAL MATRIX, NORMAL MATRIX, PAULI MATRICES, PERMUTATION MATRIX, POSITIVE DEFINITE MATRIX, RANDOM MATRIX, RATIONAL CANONICAL FORM, REDUCIBLE MATRIX, ROTH'S REMOVAL RULE, SHEAR MATRIX, SINGULAR MATRIX, SKEW SYMMETRIC MATRIX, SMITH NORMAL FORM, SPARSE MATRIX, SPECIAL MATRIX, SQUARE MATRIX, STOCHASTIC MATRIX, SUBMATRIX, SYMMETRIC MATRIX, TOURNAMENT MATRIX

References

Arfken, G. "Matrices." §4.2 in *Mathematical Methods for Physicists, 3rd ed.* Orlando, FL: Academic Press, pp. 176–91, 1985.

Bapat, R. B. *Linear Algebra and Linear Models, 2nd ed.* New York: Springer-Verlag, 2000.

Frazer, R. A.; Duncan, W. J.; and Collar, A. R. *Elementary Matrices and Some Applications to Dynamics and Differential Equations.* Cambridge, England: Cambridge University Press, 1955.

Lütkepohl, H. *Handbook of Matrices.* New York: Wiley, 1996.

Meyer, C. D. *Matrix Analysis and Applied Linear Algebra.* Philadelphia, PA: SIAM, 2000.

Zhang, F. *Matrix Theory: Basic Results and Techniques.* New York: Springer-Verlag, 1999.

Matrix Addition

Denote the sum of two MATRICES A and B (of the same dimensions) by $C = A + B$. The sum is defined by adding entries with the same indices

$$c_{ij} \equiv a_{ij} + b_{ij}$$

over all i and j. For example,

$$\begin{bmatrix} a_{11} & a_{12} \\ a_{21} & a_{22} \end{bmatrix} + \begin{bmatrix} b_{11} & b_{12} \\ b_{21} & b_{22} \end{bmatrix} = \begin{bmatrix} a_{11} + b_{11} & a_{12} + b_{12} \\ a_{21} + b_{21} & a_{22} + b_{22} \end{bmatrix}.$$

Matrix addition is therefore both COMMUTATIVE and ASSOCIATIVE.

See also MATRIX, MATRIX MULTIPLICATION

Matrix Decomposition

Matrix decomposition refers to the transformation of a given matrix (often assumed to be a SQUARE MATRIX) into a given canonical form.

See also CHOLESKY DECOMPOSITION, JORDAN MATRIX DECOMPOSITION, MATRIX DECOMPOSITION THEOREM, LQ DECOMPOSITION, LU DECOMPOSITION, ORTHOGONAL DECOMPOSITION, QR DECOMPOSITION, SCHUR DECOMPOSITION, SINGULAR VALUE DECOMPOSITION

Matrix Decomposition Theorem

Let P be a MATRIX of EIGENVECTORS of a given MATRIX A and D a MATRIX of the corresponding EIGENVALUES. Then A can be written

$$A = PDP^{-1}, \tag{1}$$

where D is a DIAGONAL MATRIX and the columns of P are ORTHOGONAL VECTORS. If P is not a SQUARE MATRIX, then it cannot have a MATRIX INVERSE. However, if P is $m \times n$ (with $m > n$), then A can be written using a so-called SINGULAR VALUE DECOMPOSITION OF THE FORM

$$A = UDV^T, \tag{2}$$

where U and V are $n \times n$ SQUARE MATRICES with ORTHOGONAL columns so that

$$U^T U = V^T V = 1. \tag{3}$$

See also SINGULAR VALUE DECOMPOSITION

References

Press, W. H.; Flannery, B. P.; Teukolsky, S. A.; and Vetterling, W. T. "Singular Value Decomposition." §2.6 in *Numerical Recipes in FORTRAN: The Art of Scientific Computing, 2nd ed.* Cambridge, England: Cambridge University Press, pp. 51–3, 1992.

Matrix Diagonalization

Diagonalizing a MATRIX is equivalent to finding the EIGENVECTORS and EIGENVALUES. The EIGENVALUES make up the entries of the diagonalized MATRIX, and the EIGENVECTORS make up the new set of axes corresponding to the DIAGONAL MATRIX.

See also DIAGONAL MATRIX, EIGENVALUE, EIGENVECTOR

References

Arfken, G. "Diagonalization of Matrices." §4.6 in *Mathematical Methods for Physicists, 3rd ed.* Orlando, FL: Academic Press, pp. 217–29, 1985.

Matrix Direct Product

The matrix direct product gives the MATRIX of the LINEAR TRANSFORMATION induced by the TENSOR PRODUCT of the original VECTOR SPACES. More precisely, suppose that

$$S : V_1 \rightarrow W_1 \tag{1}$$

and

$$T : V_2 \rightarrow W_2 \tag{2}$$

are given by $S(x) = Ax$ and $T(y) = By$. Then

$$S \otimes T : V_1 \otimes V_2 \rightarrow W_1 \otimes W_2 \tag{3}$$

is determined by

$$S \otimes T(x \otimes y) = (Ax) \otimes (By) = (A \otimes B)(x \otimes y). \tag{4}$$

Given an $m \times n$ MATRIX A and a $p \times q$ MATRIX B, their direct product $C = A \otimes B$ is an $(mp) \times (nq)$ MATRIX with elements defined by

$$c_{\alpha\beta} = a_{ij} b_{kl}, \tag{5}$$

where

$$\alpha \equiv p(i-1) + k \tag{6}$$

$$\beta \equiv q(j-1) + l. \tag{7}$$

In *Mathematica*, the matrix direct product can be formed using the following code.

```
< <LinearAlgebra`MatrixManipulation';
        MatrixDirectProduct[a_List?MatrixQ,
b_List?MatrixQ] :=
  BlockMatrix[Outer[Times, a, b]]
]
```

For example, the matrix direct product of the 2×2 MATRIX A and the 3×2 MATRIX B is given by the following 6×4 MATRIX,

$$A \otimes B = \begin{bmatrix} a_{11}B & a_{12}B \\ a_{21}B & a_{22}B \end{bmatrix} \tag{8}$$

$$= \begin{bmatrix} a_{11}b_{11} & a_{11}b_{12} & a_{12}b_{11} & a_{12}b_{12} \\ a_{11}b_{21} & a_{11}b_{22} & a_{12}b_{21} & a_{12}b_{22} \\ a_{11}b_{31} & a_{11}b_{32} & a_{12}b_{31} & a_{12}b_{32} \\ a_{21}b_{11} & a_{21}b_{12} & a_{22}b_{11} & a_{22}b_{12} \\ a_{21}b_{21} & a_{21}b_{22} & a_{22}b_{21} & a_{22}b_{22} \\ a_{21}b_{31} & a_{21}b_{32} & a_{22}b_{31} & a_{22}b_{32} \end{bmatrix}. \tag{9}$$

See also DIRECT PRODUCT, MATRIX MULTIPLICATION, TENSOR DIRECT PRODUCT

References

Schafer, R. D. *An Introduction to Nonassociative Algebras.* New York: Dover, p. 12, 1996.

Matrix Direct Sum

The construction of a BLOCK MATRIX from a set of SQUARE MATRICES, i.e.,

$$\otimes_{i=1}^{n} A_i = \mathrm{diag}(A_1, A_2, \ldots, A_n) = \begin{bmatrix} A_1 & & & \\ & A_2 & & \\ & & \ddots & \\ & & & A_n \end{bmatrix}.$$

See also BLOCK MATRIX

References

Ayres, F. Jr. *Theory and Problems of Matrices.* New York: Schaum, pp. 13–4, 1962.

Matrix Equality

Two MATRICES A and B are said to be equal IFF

$$a_{ij} \equiv b_{ij}$$

for all i, j. Therefore,

$$\begin{bmatrix} 1 & 2 \\ 3 & 4 \end{bmatrix} = \begin{bmatrix} 1 & 2 \\ 3 & 4 \end{bmatrix},$$

while

$$\begin{bmatrix} 1 & 2 \\ 3 & 4 \end{bmatrix} \neq \begin{bmatrix} 0 & 2 \\ 3 & 4 \end{bmatrix}.$$

See also EQUIVALENT MATRIX

Matrix Equation

Nonhomogeneous matrix equations OF THE FORM

$$A\mathbf{x} = \mathbf{b} \tag{1}$$

can be solved by taking the MATRIX INVERSE to obtain

$$\mathbf{x} = A^{-1}\mathbf{b}. \tag{2}$$

This equation will have a nontrivial solution IFF the DETERMINANT $\det(A) \neq 0$. In general, more numerically stable techniques of solving the equation include GAUSSIAN ELIMINATION, LU DECOMPOSITION, or the SQUARE ROOT METHOD.

For a homogeneous $n \times n$ MATRIX equation

$$\begin{bmatrix} a_{11} & a_{12} & \cdots & a_{1n} \\ a_{21} & a_{22} & \cdots & a_{2n} \\ \vdots & \vdots & \ddots & \vdots \\ a_{n1} & a_{n2} & \cdots & a_{nn} \end{bmatrix} \begin{bmatrix} x_1 \\ x_2 \\ \vdots \\ x_n \end{bmatrix} = \begin{bmatrix} 0 \\ 0 \\ \vdots \\ 0 \end{bmatrix} \tag{3}$$

to be solved for the x_is, consider the DETERMINANT

$$\begin{vmatrix} a_{11} & a_{12} & \cdots & a_{1n} \\ a_{21} & a_{22} & \cdots & a_{2n} \\ \vdots & \vdots & \ddots & \vdots \\ a_{n1} & a_{n2} & \cdots & a_{nn} \end{vmatrix}. \qquad (4)$$

Now multiply by x_1, which is equivalent to multiplying the first column (or any column) by x_1,

$$x_1 \begin{vmatrix} a_{11} & a_{12} & \cdots & a_{1n} \\ a_{21} & a_{22} & \cdots & a_{2n} \\ \vdots & \vdots & \ddots & \vdots \\ a_{n1} & a_{n2} & \cdots & a_{nn} \end{vmatrix} = \begin{vmatrix} a_{11}x_1 & a_{12} & \cdots & a_{1n} \\ a_{21}x_1 & a_{22} & \cdots & a_{2n} \\ \vdots & \vdots & \ddots & \vdots \\ a_{n1}x_1 & a_{n2} & \cdots & a_{nn} \end{vmatrix}. \qquad (5)$$

The value of the DETERMINANT is unchanged if multiples of columns are added to other columns. So add x_2 times column 2, ..., and x_n times column n to the first column to obtain

$$x_1 \begin{vmatrix} a_{11} & a_{12} & \cdots & a_{1n} \\ a_{21} & a_{22} & \cdots & a_{2n} \\ \vdots & \vdots & \ddots & \vdots \\ a_{n1} & a_{n2} & \cdots & a_{nn} \end{vmatrix}$$

$$= \begin{vmatrix} a_{11}x_1 + a_{12}x_2 + \ldots + a_{1n}x_n & a_{12} & \cdots & a_{1n} \\ a_{21}x_1 + a_{22}x_2 + \ldots + a_{2n}x_n & a_{22} & \cdots & a_{2n} \\ \vdots & \vdots & \ddots & \vdots \\ a_{n1}x_1 + a_{n2}x_2 + \ldots + a_{nn}x_n & a_{n2} & \cdots & a_{nn} \end{vmatrix}. \qquad (6)$$

But from the original MATRIX, each of the entries in the first columns is zero since

$$a_{i1}x_1 + a_{i2}x_2 + \ldots + a_{in}x_n = 0, \qquad (7)$$

so

$$\begin{vmatrix} 0 & a_{12} & \cdots & a_{1n} \\ 0 & a_{22} & \cdots & a_{2n} \\ \vdots & \vdots & \ddots & \vdots \\ 0 & a_{n2} & \cdots & a_{nn} \end{vmatrix} = 0. \qquad (8)$$

Therefore, if there is an $x_1 \neq 0$ which is a solution, the DETERMINANT is zero. This is also true for $x_2, \ldots, x_n$, so the original homogeneous system has a nontrivial solution for all x_is only if the DETERMINANT is 0. This approach is the basis for CRAMER'S RULE.

Given a numerical solution to a matrix equation, the solution can be iteratively improved using the following technique. Assume that the numerically obtained solution to

$$A\mathbf{x} = \mathbf{b} \qquad (9)$$

is $\mathbf{x}_1 = \mathbf{x} + \delta\mathbf{x}_1$, where $\delta\mathbf{x}_1$ is an error term. The first solution therefore gives

$$A\mathbf{x}_1 = A(\mathbf{x} + \delta\mathbf{x}_1) = \mathbf{b} + \delta\mathbf{b} \qquad (10)$$

$$A\delta\mathbf{x}_1 = \delta\mathbf{b}, \qquad (11)$$

where $\delta\mathbf{b}$ is found by solving (10)

$$\delta\mathbf{b} = A\mathbf{x}_1 - \mathbf{b}. \qquad (12)$$

Combining (11) and (12) then gives

$$\delta\mathbf{x}_1 = A^{-1}\delta\mathbf{b} = A^{-1}(A\mathbf{x}_1 - \mathbf{b}) = \mathbf{x}_1 - A^{-1}\mathbf{b}. \qquad (13)$$

See also CRAMER'S RULE, GAUSSIAN ELIMINATION, LU DECOMPOSITION, MATRIX, MATRIX ADDITION, MATRIX INVERSE, MATRIX MULTIPLICATION, NORMAL EQUATION, SQUARE ROOT METHOD

MatrixExp
MATRIX EXPONENTIAL

Matrix Exponential

The POWER SERIES that defines the EXPONENTIAL MAP e^x also defines a map between MATRICES. In particular,

$$\exp(A) \equiv e^A = \sum_{n=0}^{\infty} \frac{A^n}{n!} \qquad (1)$$

$$= I + A + \frac{AA}{2!} + \frac{AAA}{3!} + \ldots, \qquad (2)$$

converges for any SQUARE MATRIX A, where I is the IDENTITY MATRIX. The matrix exponential is implemented in *Mathematica* as `MatrixExp[m]`.

In some cases, it is a simple matter to express the exponent. For example, when A is a DIAGONAL MATRIX, exponentiation can be performed simply by exponentiating each of the diagonal elements. For example, given a diagonal matrix

$$A = \begin{bmatrix} a_1 & 0 & \cdots & 0 \\ 0 & a_2 & \cdots & 0 \\ \vdots & \vdots & \ddots & \vdots \\ 0 & 0 & \cdots & a_k \end{bmatrix}, \qquad (3)$$

The matrix exponential is given by

$$\exp(A) = \begin{bmatrix} e^{a_1} & 0 & \cdots & 0 \\ 0 & e^{a_2} & \cdots & 0 \\ \vdots & \vdots & \ddots & \vdots \\ 0 & 0 & \cdots & e^{a_k} \end{bmatrix}. \qquad (4)$$

Since most matrices are DIAGONALIZABLE, it is easiest to diagonalize the matrix before exponentiating it.

When A is a NILPOTENT MATRIX, the exponential is given by a MATRIX POLYNOMIAL because some power of A vanishes. For example, when

$$A = \begin{bmatrix} 0 & x & z \\ 0 & 0 & y \\ 0 & 0 & 0 \end{bmatrix}, \qquad (5)$$

then

$$\exp(A) = \begin{bmatrix} 1 & x & z + \frac{1}{2}xy \\ 0 & 1 & y \\ 0 & 0 & 1 \end{bmatrix} \qquad (6)$$

and $A^3 = 0$.

For the ZERO MATRIX $A = 0$,

$$e^0 = I, \tag{7}$$

i.e., the IDENTITY MATRIX. In general,

$$e^A e^{-A} = e^0 = I, \tag{8}$$

so the exponential of a matrix is always invertible, with inverse the exponent of the negative of the matrix. However, in general, the formula

$$e^A e^B = e^{A+B} \tag{9}$$

holds only when A and B COMMUTE, i.e.,

$$[A,\ B] = AB - BA = 0. \tag{10}$$

For example,

$$\exp\left(\begin{bmatrix} 0 & -x \\ 0 & 0 \end{bmatrix} + \begin{bmatrix} 0 & 0 \\ x & 0 \end{bmatrix}\right) = \begin{bmatrix} \cos x & -\sin x \\ \sin x & \cos x \end{bmatrix}, \tag{11}$$

while

$$\exp\left(\begin{bmatrix} 0 & -x \\ 0 & 0 \end{bmatrix}\right)\exp\left(\begin{bmatrix} 0 & 0 \\ x & 0 \end{bmatrix}\right) = \begin{bmatrix} 1 & -x \\ 0 & 1 \end{bmatrix}\begin{bmatrix} 1 & 0 \\ x & 1 \end{bmatrix}$$

$$= \begin{bmatrix} 1 - x^2 & -x \\ x & 1 \end{bmatrix}. \tag{12}$$

See also EXPONENTIAL FUNCTION, EXPONENTIAL MAP, MATRIX, MATRIX POWER

Matrix Fraction

A pair of matrices ND^{-1} or $D^{-1}N$, where N is the matrix NUMERATOR and D is the DENOMINATOR.

See also FRACTION

Matrix Group

A GROUP in which the elements are SQUARE MATRICES, the group multiplication law is MATRIX MULTIPLICATION, and the group inverse is simply the MATRIX INVERSE. Every matrix group is equivalent to a unitary matrix group (Lomont 1987, pp. 47–8).

See also MASCHKE'S THEOREM

References

Lomont, J. S. "Matrix Groups." §3.1 in *Applications of Finite Groups.* New York: Dover, pp. 46–2, 1987.

Matrix Inverse

The inverse of a SQUARE MATRIX A, sometimes called a reciprocal matrix, is a matrix A^{-1} such that

$$AA^{-1} = I, \tag{1}$$

where I is the IDENTITY MATRIX. Courant and Hilbert (1989, p. 10) use the notation $\breve{A}$ to denote the inverse matrix.

A SQUARE MATRIX A has an inverse IFF the DETERMINANT $|A| \neq 0$ (Lipschutz 1991, p. 45) A matrix possessing an inverse is called NONSINGULAR, or invertible. The matrix inverse of a SQUARE MATRIX m may be taken in *Mathematica* using the function `Inverse[m]`.

For a 2×2 MATRIX

$$A \equiv \begin{bmatrix} a & b \\ c & d \end{bmatrix} \tag{2}$$

the inverse is

$$A^{-1} = \frac{1}{|A|}\begin{bmatrix} d & -b \\ -c & a \end{bmatrix} = \frac{1}{ad-bc}\begin{bmatrix} d & -b \\ -c & a \end{bmatrix}. \tag{3}$$

For a 3×3 MATRIX,

$$A^{-1} = \frac{1}{|A|}\begin{bmatrix} \begin{vmatrix} a_{22} & a_{23} \\ a_{32} & a_{33} \end{vmatrix} & \begin{vmatrix} a_{13} & a_{12} \\ a_{33} & a_{32} \end{vmatrix} & \begin{vmatrix} a_{12} & a_{13} \\ a_{22} & a_{23} \end{vmatrix} \\ \begin{vmatrix} a_{23} & a_{21} \\ a_{33} & a_{31} \end{vmatrix} & \begin{vmatrix} a_{11} & a_{13} \\ a_{31} & a_{33} \end{vmatrix} & \begin{vmatrix} a_{13} & a_{11} \\ a_{23} & a_{21} \end{vmatrix} \\ \begin{vmatrix} a_{21} & a_{22} \\ a_{31} & a_{32} \end{vmatrix} & \begin{vmatrix} a_{12} & a_{11} \\ a_{32} & a_{31} \end{vmatrix} & \begin{vmatrix} a_{11} & a_{12} \\ a_{21} & a_{22} \end{vmatrix} \end{bmatrix}. \tag{4}$$

A general $n \times n$ matrix can be inverted using methods such as the GAUSS-JORDAN ELIMINATION, GAUSSIAN ELIMINATION, or LU DECOMPOSITION.

The inverse of a PRODUCT AB of MATRICES A and B can be expressed in terms of A^{-1} and B^{-1}. Let

$$C \equiv AB. \tag{5}$$

Then

$$B = A^{-1}AB = A^{-1}C \tag{6}$$

and

$$A = ABB^{-1} = CB^{-1}. \tag{7}$$

Therefore,

$$C = AB = (CB^{-1})(A^{-1}C) = CB^{-1}A^{-1}C, \tag{8}$$

so

$$CB^{-1}A^{-1} = I, \tag{9}$$

where I is the IDENTITY MATRIX, and

$$B^{-1}A^{-1} = C^{-1} = (AB)^{-1}. \tag{10}$$

See also GAUSS-JORDAN ELIMINATION, GAUSSIAN ELIMINATION, LU DECOMPOSITION, MATRIX, MATRIX ADDITION, MATRIX MULTIPLICATION, MOORE-PENROSE GENERALIZED MATRIX INVERSE, NONSINGULAR MATRIX, SINGULAR MATRIX, STRASSEN FORMULAS

References

Ayres, F. Jr. *Theory and Problems of Matrices.* New York: Schaum, p. 11, 1962.

Ben-Israel, A. and Greville, T. N. E. *Generalized Inverses: Theory and Applications.* New York: Wiley, 1977.

Courant, R. and Hilbert, D. *Methods of Mathematical Physics, Vol. 1.* New York: Wiley, 1989.

Lipschutz, S. "Invertible Matrices." *Schaum's Outline of Theory and Problems of Linear Algebra, 2nd ed.* New York: McGraw-Hill, pp. 44–5, 1991.

Nash, J. C. *Compact Numerical Methods for Computers: Linear Algebra and Function Minimisation, 2nd ed.* Bristol, England: Adam Hilger, pp. 24–6, 1990.

Press, W. H.; Flannery, B. P.; Teukolsky, S. A.; and Vetterling, W. T. "Is Matrix Inversion an N^3 Process?" §2.11 in *Numerical Recipes in FORTRAN: The Art of Scientific Computing, 2nd ed.* Cambridge, England: Cambridge University Press, pp. 95–8, 1992.

Rosser, J. B. "A Method of Computing Exact Inverses of Matrices with Integer Coefficients." *J. Res. Nat. Bur. Standards Sect. B.* **49**, 349–58, 1952.

Matrix Multiplication

The product C of two MATRICES A and B is defined by

$$c_{ik} = a_{ij}b_{jk}, \qquad (1)$$

where j is summed over for all possible values of i and k. Therefore, in order for multiplication to be defined, the dimensions of the MATRICES must satisfy

$$(n \times m)(m \times p) = (n \times p), \qquad (2)$$

where $(a \times b)$ denotes a MATRIX with a rows and b columns. Writing out the product explicitly,

$$
\begin{bmatrix} c_{11} & c_{12} & \cdots & c_{1p} \\ c_{21} & c_{22} & \cdots & c_{2p} \\ \vdots & \vdots & \ddots & \vdots \\ c_{n1} & c_{n2} & \cdots & c_{np} \end{bmatrix}
$$
$$
= \begin{bmatrix} a_{11} & a_{12} & \cdots & a_{1m} \\ a_{21} & a_{22} & \cdots & a_{2m} \\ \vdots & \vdots & \ddots & \vdots \\ a_{n1} & a_{n2} & \cdots & a_{nm} \end{bmatrix} \begin{bmatrix} b_{11} & b_{12} & \cdots & b_{1p} \\ b_{21} & b_{22} & \cdots & b_{2p} \\ \vdots & \vdots & \ddots & \vdots \\ b_{m1} & b_{m2} & \cdots & b_{mp} \end{bmatrix},
$$
$$(3)$$

where

$$
\begin{aligned}
c_{11} &= a_{11}b_{11} + a_{12}b_{21} + \ldots + a_{1m}b_{m1} \\
c_{12} &= a_{11}b_{12} + a_{12}b_{22} + \ldots + a_{1m}b_{m2} \\
c_{1p} &= a_{11}b_{1p} + a_{12}b_{2p} + \ldots + a_{1m}b_{mp} \\
c_{21} &= a_{21}b_{11} + a_{22}b_{21} + \ldots + a_{2m}b_{m1} \\
c_{22} &= a_{21}b_{12} + a_{22}b_{22} + \ldots + a_{2m}b_{m2} \\
c_{2p} &= a_{21}b_{1p} + a_{22}b_{2p} + \ldots + a_{2m}b_{mp} \\
c_{n1} &= a_{n1}b_{11} + a_{n2}b_{21} + \ldots + a_{nm}b_{m1} \\
c_{n2} &= a_{n1}b_{12} + a_{n2}b_{22} + \ldots + a_{nm}b_{m2} \\
c_{np} &= a_{n1}b_{1p} + a_{n2}b_{2p} + \ldots + a_{nm}b_{mp}.
\end{aligned}
$$

Matrix multiplication is ASSOCIATIVE, as can be seen by taking

$$[(ab)c]_{ij} = (ab)_{ik}c_{kj} = (a_{il}b_{lk})c_{kj}. \qquad (4)$$

Now, since a_{il}, b_{lk}, and c_{kj} are SCALARS, use the ASSOCIATIVITY of SCALAR MULTIPLICATION to write

$$(a_{il}b_{lk})c_{kj} = a_{il}(b_{lk}c_{kj}) = a_{il}(bc)_{lj} = [a(bc)]_{ij}. \qquad (5)$$

Since this is true for all i and j, it must be true that

$$(ab)c = a(bc). \qquad (6)$$

That is, matrix multiplication is ASSOCIATIVE. However, matrix multiplication is *not*, in general, COMMUTATIVE (although it is COMMUTATIVE if A and B are DIAGONAL and of the same dimension).

The product of two BLOCK MATRICES is given by multiplying each block

$$
\begin{bmatrix}
o & o & & & & \\
o & o & & & & \\
& & o & & & \\
& & & o & o & o \\
& & & o & o & o \\
& & & o & o & o
\end{bmatrix}
\begin{bmatrix}
x & x & & & & \\
x & x & & & & \\
& & x & & & \\
& & & x & x & x \\
& & & x & x & x \\
& & & x & x & x
\end{bmatrix}
$$
$$
= \begin{bmatrix}
\begin{bmatrix} o & o \\ o & o \end{bmatrix}\begin{bmatrix} x & x \\ x & x \end{bmatrix} & & \\
& [o][x] & \\
& & \begin{bmatrix} o & o & o \\ o & o & o \\ o & o & o \end{bmatrix}\begin{bmatrix} x & x & x \\ x & x & x \\ x & x & x \end{bmatrix}
\end{bmatrix}.
$$
$$(7)$$

See also LINEAR TRANSFORMATION, MATRIX, MATRIX ADDITION, MATRIX INVERSE, STRASSEN FORMULAS

References

Arfken, G. *Mathematical Methods for Physicists, 3rd ed.* Orlando, FL: Academic Press, pp. 178–79, 1985.

Higham, N. "Exploiting Fast Matrix Multiplication within the Level 3 BLAS." *ACM Trans. Math. Soft.* **16**, 352–68, 1990.

Matrix Norm

Given a SQUARE MATRIX A with COMPLEX (or REAL) entries, a MATRIX NORM |A| is a NONNEGATIVE number associated with A having the properties

1. $\|A\| > 0$ when $A \neq 0$ and $\|A\| = 0$ IFF $A = 0$,
2. $\|kA\| = |k|\|A\|$ for any SCALAR k,
3. $\|A + B\| \leq \|A\| + \|B\|$,
4. $\|AB\| \leq \|A\|\|B\|$

For an $n \times n$ MATRIX A and an $n \times n$ UNITARY MATRIX U,

$$\|AU\| = \|UA\| = \|A\|.$$

Let $\lambda_1, \ldots, \lambda_n$ be the EIGENVALUES of A, then

$$\frac{1}{\|A^{-1}\|} \leq |\lambda| \leq \|A\|.$$

The MAXIMUM ABSOLUTE COLUMN SUM NORM $\|A\|_1$, SPECTRAL NORM $\|A\|_2$, and MAXIMUM ABSOLUTE ROW SUM NORM $\|A\|_\infty$ satisfy

$$\|A\|_2^2 \leq \|A\|_1 \leq \|A\|_\infty.$$

Matrix norms are implemented as `MatrixNorm[m, p]` in the *Mathematica* add-on package `LinearAlgebra`MatrixMultiplication`` (which can be loaded with the command `<<LinearAlgebra`)`, where $p = 1$, 2, or ∞.

For a SQUARE MATRIX, the SPECTRAL NORM, which is the SQUARE ROOT of the maximum EIGENVALUE of A*A (where A* is the ADJOINT MATRIX), is often referred to as "the" matrix norm.

See also COMPATIBLE, HILBERT-SCHMIDT NORM, MAXIMUM ABSOLUTE COLUMN SUM NORM, MAXIMUM ABSOLUTE ROW SUM NORM, NATURAL NORM, NORM, POLYNOMIAL NORM, SPECTRAL NORM, SPECTRAL RADIUS, VECTOR NORM

References

Gradshteyn, I. S. and Ryzhik, I. M. *Tables of Integrals, Series, and Products, 6th ed.* San Diego, CA: Academic Press, pp. 1114–125, 2000.

Matrix p-Norm
MATRIX NORM

Matrix Polynomial

A polynomial with matrix coefficients. An nth order matrix polynomial in a variable t is given by

$$P(t) = A_0 + A_1 t + A_2 t^2 + \ldots + A_n t^n, \qquad (1)$$

where A_k are $p \times p$ square matrices.

If the entries of the matrices are real independent variates with a standard normal distribution, then the expected number of real solutions is given by

$$E_{n,\,p} = \sqrt{\pi}\, E\, \frac{\Gamma(\frac{1}{2}(p+1))}{\Gamma(\frac{1}{2}p)}, \qquad (2)$$

where

$$E_n = \begin{cases} \sqrt{2} \sum_{k=0}^{n/2-1} \dfrac{(4k-1)!!}{(4k)!!} & \text{for } n \text{ even} \\ 1 + \sqrt{2} \sum_{k=1}^{(n-1)/2} \dfrac{(4k-3)!!}{(4k-2)!!} & \text{for } n \text{ odd} \end{cases} \qquad (3)$$

(Edelman and Kostlan 1995).

See also CAYLEY-HAMILTON THEOREM, MATRIX POWER, NILPOTENT MATRIX, POLYNOMIAL MATRIX

References

Edelman, A. and Kostlan, E. "How Many Zeros of a Random Polynomial are Real?" *Bull. Amer. Math. Soc.* **32**, 1–7, 1995.
Faddeeva, V. N. *Computational Methods of Linear Algebra.* New York: Dover, p. 13, 1958.

Matrix Polynomial Identity
CAYLEY-HAMILTON THEOREM

Matrix Power

The power A^n of a MATRIX A for n a nonnegative integer is defined as the MATRIX PRODUCT of n copies of A,

$$A^n = \underbrace{A \cdots A}_{n}.$$

A matrix to the zeroth power is defined to be the IDENTITY MATRIX of the same dimensions, $A^0 = I$. The MATRIX INVERSE is commonly denoted A^{-1}, which should not be interpreted to mean $1/A$.

See also MATRIX EXPONENTIAL, MATRIX MULTIPLICATION, MATRIX POLYNOMIAL, NILPOTENT MATRIX, PERIODIC MATRIX

Matrix Product
The result of a MATRIX MULTIPLICATION.

See also PRODUCT

Matrix Transpose
TRANSPOSE

Matrix Tree Theorem

The number of nonidentical SPANNING TREES of a GRAPH G is equal to any COFACTOR of the DEGREE MATRIX of G minus the ADJACENCY MATRIX of G (Skiena 1990, p. 235).

See also SPANNING TREE

References

Chaiken, S. "A Combinatorial Proof of the All-Minors Matrix Tree Theorem." *SIAM J. Alg. Disc. Methods* **3**, 319–29, 1982.
Kirchhoff, G. "Über die Auflösung der Gleichungen, auf welche man bei der untersuchung der linearen verteilung galvanischer Ströme geführt wird." *Ann. Phys. Chem.* **72**, 497–08, 1847.
Skiena, S. *Implementing Discrete Mathematics: Combinatorics and Graph Theory with Mathematica.* Reading, MA: Addison-Wesley, p. 235, 1990.

Matroid

Roughly speaking, a matroid is a finite set together with a generalization of a concept from linear algebra that satisfies a natural set of properties for that concept. For example, the finite set could be the rows of a MATRIX, and the generalizing concept could be linear dependence and independence of any subset of rows of the MATRIX.

Formally, a matroid consists of a finite set M of elements together with a family $\mathfrak{C} = \{C_1, C_1, \ldots\}$ of nonempty subsets of M, called circuits, which satisfy the axioms

1. No PROPER SUBSET of a circuit is a circuit,
2. If $x \in C_1 \cap C_2$ and $C_1 \neq C_2$, then $C_1 \cup C_2 - \{x\}$ contains a circuit.

(Harary 1994, p. 40).

An equivalent definition considers a matroid as a finite set M of elements together with a family of subsets of M, called independent sets, such that

1. The EMPTY SET is independent,
2. Every SUBSET of an independent set is independent,
3. For every subset A of M, all maximal independent sets contained in A have the same number of elements.

(Harary 1994, pp. 40–1).

The number of simple matroids (or COMBINATORIAL GEOMETRIES) with $n = 0, 1, \ldots$ points are 1, 1, 2, 4, 9, 26, 101, 950, ... (Sloane's A002773), and the number of matroids on $n = 0, 1, \ldots$ points are 1, 2, 4, 8, 17, 38, 98, 306, 1724, ... (Sloane's A055545; Oxley 1993, p. 473). (The value for $n = 5$ given by Oxley 1993, p. 42, is incorrect.)

See also COMBINATORIAL GEOMETRY, GRAPHOID, ORIENTED MATROID

References

Björner, A.; Las Vergnas, M.; Sturmfels, B.; White, N.; and Ziegler, G. *Oriented Matroids, 2nd ed.* Cambridge, England: Cambridge University Press, 1999.
Blackburn, J. E.; Crapo, H. H.; and Higgs, D. A. "A Catalogue of Combinatorial Geometries." *Math. Comput.* **27**, 155–66, 1973.
Crapo, H. H. and Rota, G.-C. "On the Foundations of Combinatorial Theory. II. Combinatorial Geometries." Cambridge, MA: MIT Press, 109–33, 1970.
Harary, F. "Matroids." *Graph Theory.* Reading, MA: Addison-Wesley, pp. 40–1, 1994.
Minty, G. "On the Axiomatic Foundations of the Theories of Directed Linear Graphs, Electric Networks, and Network-Programming." *J. Math. Mech.* **15**, 485–20, 1966.
Oxley, J. G. *Matroid Theory.* Oxford, England: Oxford University Press, 1993.
Papadimitriou, C. H. and Steiglitz, K. *Combinatorial Optimization: Algorithms and Complexity.* Englewood Cliffs, NJ: Prentice-Hall, 1982.
Richter-Gebert, J. and Ziegler, G. M. In *Handbook of Discrete and Computational Geometry* (Ed. J. E. Goodman and J. O'Rourke). Boca Raton, FL: CRC Press, pp. 111–12, 1997.
Sloane, N. J. A. Sequences A002773/M1197 and A055545 in "An On-Line Version of the Encyclopedia of Integer Sequences." http://www.research.att.com/~njas/sequences/eisonline.html.
Sloane, N. J. A. and Plouffe, S. Figure M1197 in *The Encyclopedia of Integer Sequences.* San Diego: Academic Press, 1995.
Tutte, W. T. "Lectures on Matroids." *J. Res. Nat. Bur. Stand. Sect. B* **69**, 1–7, 1965.
Whitely, W. "Matroids and Rigid Structures." In *Matroid Applications, Encyclopedia of Mathematics and Its Applications* (Ed. N. White), Vol. 40. New York: Cambridge University Press, pp. 1–3, 1992.
Whitney, H. "On the Abstract Properties of Linear Dependence." *Amer. J. Math.* **57**, 509–33, 1935.

Maurer Rose

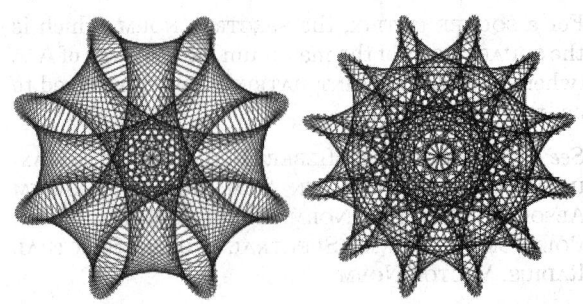

$n = 4$, $d = 120$, $n = 6$, $d = 72$. A Maurer rose is a plot of a "walk" along an n- (or $2n$-) leafed ROSE in steps of a fixed number d degrees, including all cosets.

See also STARR ROSE

References

Maurer, P. "A Rose is a Rose..." *Amer. Math. Monthly* **94**, 631–45, 1987.
Wagon, S. *Mathematica in Action.* New York: W. H. Freeman, pp. 96–02, 1991.

Max
MAXIMUM

Maximal Ideal

A maximal ideal of a RING R is an IDEAL I, not equal to R, such that there are no IDEALS "in between" I and R. In other words, if J is an IDEAL which contains I as a SUBSET, then either $J = I$ or $J = R$. For example, $n\mathbb{Z}$ is a maximal ideal of $\mathbb{Z}$ IFF n is PRIME, where $\mathbb{Z}$ is the RING of INTEGERS.

Only in a LOCAL RING is there just one maximal ideal. For instance, in the integers, $\mathfrak{a} = \langle p \rangle$ is a maximal ideal whenever p is prime.

A maximal ideal $\mathfrak{m}$ is always a PRIME IDEAL, and the QUOTIENT RING $A/\mathfrak{m}$ is always a FIELD. In general, not all prime ideals are maximal.

See also IDEAL, MAXIMAL IDEAL THEOREM, PRIME IDEAL, QUOTIENT RING, REGULAR LOCAL RING, RING

Maximal Ideal Theorem

The proposition that every PROPER IDEAL of a BOOLEAN ALGEBRA can be extended to a MAXIMAL IDEAL. It is equivalent to the BOOLEAN REPRESENTATION THEOREM, which can be proved without using the AXIOM OF CHOICE (Mendelson 1997, p. 121).

See also BOOLEAN REPRESENTATION THEOREM

References

Lós, J. "Sur la théorème de Gödel sur les theories indénombrables." *Bull. de l'Acad. Polon. des Sci.* **3**, 319–20, 1954.

Mendelson, E. *Introduction to Mathematical Logic, 4th ed.* London: Chapman & Hall, p. 121, 1997.

Rasiowa, H. and Sikorski, R. "A Proof of the Completeness Theorem of Gödel." *Fund. Math.* **37**, 193–00, 1951.

Rasiowa, H. and Sikorski, R. "A Proof of the Skolem-Löwenheim Theorem." *Fund. Math.* **38**, 230–32, 1952.

Maximally Linearly Independent

A set of VECTORS is maximally linearly independent if including any other VECTOR in the VECTOR SPACE would make it LINEARLY DEPENDENT (i.e., if any other VECTOR in the SPACE can be expressed as a LINEAR COMBINATION of elements of a maximal set–the BASIS).

See also BASIS, LINEARLY DEPENDENT VECTORS, VECTOR, VECTOR SPACE

Maximal Sum-Free Set

A maximal sum-free set is a set $\{a_1, a_2, \ldots, a_n\}$ of distinct NATURAL NUMBERS such that a maximum l of them satisfy $a_{i_j} + a_{i_k} \neq a_m$ for $1 \leq j < k \leq l$, $1 \leq m \leq n$.

See also MAXIMAL ZERO-SUM-FREE SET

References

Guy, R. K. "Maximal Sum-Free Sets." §C14 in *Unsolved Problems in Number Theory, 2nd ed.* New York: Springer-Verlag, pp. 128–29, 1994.

Maximal Tori Theorem

Let T be a maximal torus of a group G, then T intersects every CONJUGACY CLASS of G, i.e., every element $g \in G$ is conjugate to a suitable element in T. The theorem is due to É. Cartan.

References

Hsiang, W. Y. *Lectures on Lie Groups.* Singapore: World Scientific, p. 42, 2000.

Maximal Zero-Sum-Free Set

A set having the largest number k of distinct residue classes modulo m so that no SUBSET has zero sum.

See also MAXIMAL SUM-FREE SET

References

Guy, R. K. "Maximal Zero-Sum-Free Sets." §C15 in *Unsolved Problems in Number Theory, 2nd ed.* New York: Springer-Verlag, pp. 129–31, 1994.

Maximum

The largest value of a set, function, etc. The maximum value of a set of elements $A = \{a_i\}_{i=1}^{N}$ is denoted $\max A$ or $\max_i a_i$, and is equal to the last element of a sorted (i.e., ordered) version of A. For example, given the set $\{3, 5, 4, 1\}$, the sorted version is $\{1, 3, 4, 5\}$, so the maximum is 5. The maximum and MINIMUM are the simplest ORDER STATISTICS.

A continuous FUNCTION may assume a maximum at a single point or may have maxima at a number of points. A GLOBAL MAXIMUM of a FUNCTION is the largest value in the entire RANGE of the FUNCTION, and a LOCAL MAXIMUM is the largest value in some local neighborhood.

For a function $f(x)$ which is CONTINUOUS at a point x_0, a NECESSARY but not SUFFICIENT condition for $f(x)$ to have a RELATIVE MAXIMUM at $x = x_0$ is that x_0 be a CRITICAL POINT (i.e., $f(x)$ is either not DIFFERENTIABLE at x_0 or x_0 is a STATIONARY POINT, in which case $f'(x_0) = 0$).

The FIRST DERIVATIVE TEST can be applied to CONTINUOUS FUNCTIONS to distinguish maxima from MINIMA. For twice differentiable functions of one variable, $f(x)$, or of two variables, $f(x, y)$, the SECOND DERIVATIVE TEST can sometimes also identify the nature of an EXTREMUM. For a function $f(x)$, the EXTREMUM TEST succeeds under more general conditions than the SECOND DERIVATIVE TEST.

See also CRITICAL POINT, EXTREMUM, EXTREMUM TEST, FIRST DERIVATIVE TEST, GLOBAL MAXIMUM, INFLECTION POINT, LOCAL MAXIMUM, MIDRANGE, MINIMUM, ORDER STATISTIC, SADDLE POINT (FUNCTION), SECOND DERIVATIVE TEST, STATIONARY POINT

References

Abramowitz, M. and Stegun, C. A. (Eds.). *Handbook of Mathematical Functions with Formulas, Graphs, and Mathematical Tables, 9th printing.* New York: Dover, p. 14, 1972.

Niven, I. *Maxima and Minima without Calculus.* Washington, DC: Math. Assoc. Amer., 1982.

Press, W. H.; Flannery, B. P.; Teukolsky, S. A.; and Vetterling, W. T. "Minimization or Maximization of Functions." Ch. 10 in *Numerical Recipes in FORTRAN: The Art of Scientific Computing, 2nd ed.* Cambridge, England: Cambridge University Press, pp. 387–48, 1992.

Tikhomirov, V. M. *Stories About Maxima and Minima.* Providence, RI: Amer. Math. Soc., 1991.

Maximum Absolute Column Sum Norm

The NATURAL NORM induced by the $L1$-NORM is called the maximum absolute column sum norm and is defined by

$$\|A\|_1 = \max_j \sum_{i=1}^{n} |a_{ij}|$$

for a MATRIX A. This MATRIX NORM is implemented as `MatrixNorm[m, 1]` in the *Mathematica* add-on package `LinearAlgebra`MatrixMultiplication`` (which can be loaded with the command `<<LinearAlgebra`).

See also L_1-Norm, Matrix Norm, Maximum Absolute Row Sum Norm, Spectral Norm

Maximum Absolute Row Sum Norm

The NATURAL NORM induced by the L-INFINITY-NORM is called the maximum absolute row sum norm and is defined by

$$\|A\|_\infty = \max_i \sum_{j=1}^n |a_{ij}|$$

for a MATRIX A. This MATRIX NORM is implemented as `MatrixNorm[m, Infinity]` in the *Mathematica* add-on package `LinearAlgebra`MatrixMultiplication` (which can be loaded with the command `<<LinearAlgebra`).

See also L-Infinity-Norm, Matrix Norm, Maximum Absolute Column Sum Norm, Spectral Norm

Maximum Clique Problem

Party Problem

Maximum Entropy Method

A DECONVOLUTION ALGORITHM (sometimes abbreviated MEM) which functions by minimizing a smoothness function ("ENTROPY") in an image. Maximum entropy is also called the ALL-POLES MODEL or AUTOREGRESSIVE MODEL. For images with more than a million pixels, maximum entropy is faster than the CLEAN algorithm.

MEM is commonly employed in astronomical synthesis imaging. In this application, the resolution depends on the signal-to-noise ratio, which must be specified. Therefore, resolution is image dependent and varies across the map. MEM is also biased, since the ensemble average of the estimated noise is NONZERO. However, this bias is much smaller than the NOISE for pixels with a SNR $\gg 1$. It can yield super-resolution, which can usually be trusted to an order of magnitude in SOLID ANGLE.

Two definitions of "ENTROPY" normalized to the flux in the image are

$$H_1 \equiv \sum_k \ln\left(\frac{I_k}{M_k}\right) \tag{1}$$

$$H_2 \equiv -\sum_k I_k \ln\left(\frac{I_k}{M_k e}\right), \tag{2}$$

where M_k is a "default image" and I_k is the smoothed image. Several unnormalized entropy measures (Cornwell 1982, p. 3) are given by

$$H_3 \equiv -\sum f_i \ln(f_i) \tag{3}$$

$$H_4 \equiv \sum \ln(f_i) \tag{4}$$

$$H_5 \equiv -\sum \frac{1}{\ln(f_i)} \tag{5}$$

$$H_6 \equiv -\sum \frac{1}{[\ln(f_i)]^2} \tag{6}$$

$$H_7 \equiv \sum \sqrt{\ln(f_i)}. \tag{7}$$

See also Deconvolution, LUCY

References

Cornwell, T. J. "Can CLEAN be Improved?" VLA Scientific Memorandum No. 141, March 1982.
Cornwell, T. and Braun, R. "Deconvolution." Ch. 8 in *Synthesis Imaging in Radio Astronomy: Third NRAO Summer School, 1988* (Ed. R. A. Perley, F. R. Schwab, and A. H. Bridle). San Francisco, CA: Astronomical Society of the Pacific, pp. 167–83, 1989.
Christiansen, W. N. and Högbom, J. A. *Radiotelescopes, 2nd ed.* Cambridge, England: Cambridge University Press, pp. 217–18, 1985.
Narayan, R. and Nityananda, R. "Maximum Entropy Restoration in Astronomy." *Ann. Rev. Astron. Astrophys.* **24**, 127–70, 1986.
Press, W. H.; Flannery, B. P.; Teukolsky, S. A.; and Vetterling, W. T. "Power Spectrum Estimation by the Maximum Entropy (All Poles) Method" and "Maximum Entropy Image Restoration." §13.7 and 18.7 in *Numerical Recipes in FORTRAN: The Art of Scientific Computing, 2nd ed.* Cambridge, England: Cambridge University Press, pp. 565–69 and 809–17, 1992.
Thompson, A. R.; Moran, J. M.; and Swenson, G. W. Jr. §3.2 in *Interferometry and Synthesis in Radio Astronomy.* New York: Wiley, pp. 349–52, 1986.

Maximum Flow, Minimum Cut Theorem

The maximum flow between vertices v_i and v_j in a GRAPH G is exactly the weight of the smallest set of edges to disconnect G with v_i and v_j in different components (Ford and Fulkerson 1962; Skiena 1990, p. 178).

See also Network Flow

References

Ford, L. R. and Fulkerson, D. R. *Flows in Networks.* Princeton, NJ: Princeton University Press, 1962.
Skiena, S. *Implementing Discrete Mathematics: Combinatorics and Graph Theory with Mathematica.* Reading, MA: Addison-Wesley, 1990.

Maximum Independent Set Problem

This problem is NP-COMPLETE (Garey and Johnson 1983).

References

Garey, M. R. and Johnson, D. S. *Computers and Intractability: A Guide to the Theory of NP-Completeness.* New York: W. H. Freeman, 1983.

Skiena, S. "Maximum Independent Set." §5.6.3. in *Implementing Discrete Mathematics: Combinatorics and Graph Theory with Mathematica.* Reading, MA: Addison-Wesley, pp. 218–19, 1990.

Maximum Likelihood

The procedure of finding the value of one or more parameters for a given statistic which makes the *known* LIKELIHOOD distribution a MAXIMUM. The maximum likelihood estimate for a parameter μ is denoted $\hat{\mu}$.

For a BERNOULLI DISTRIBUTION,

$$\frac{d}{d\theta}\left[\binom{N}{N_p}\theta^{N_p}(1-\theta)^{N_q}\right] = N_p(1-\theta) - \theta N_q = 0, \quad (1)$$

so maximum likelihood occurs for $\theta = p$. If p is not known ahead of time, the likelihood function is

$$f(x_1, \ldots, x_n|p) = P(X_1 = x_1, \ldots, X_n = x_n|p)$$
$$= p^{x_1}(1-p)^{1-x_1} \cdots p^{x_n}(1-p)^{1-x_1 n} = p^{\sum x_i}(1-p)^{\sum(1-x_i)}$$
$$= p^{\sum x_i}(1-p)^{n-\sum x_i}, \quad (2)$$

where $x = 0$ or 1, and $i = 1, ..., n$.

$$\ln f = \sum x_i \ln p + \left(n - \sum x_i\right)\ln(1-p) \quad (3)$$

$$\frac{d(\ln f)}{dp} = \frac{\sum x_i}{p} - \frac{n - \sum x_i}{1-p} = 0 \quad (4)$$

$$\sum x_i - p\sum x_i = np - p\sum x_i \quad (5)$$

$$\hat{p} = \frac{\sum x_i}{n}. \quad (6)$$

For a GAUSSIAN DISTRIBUTION,

$$f(x_1, \ldots, x_n|\mu, \sigma) = \prod \frac{1}{\sigma\sqrt{2\pi}} e^{-(x_i-\mu)^2/2\sigma^2}$$

$$= \frac{(2\pi)^{-n/2}}{\sigma^n}\exp\left[-\frac{\sum (x_i - \mu)^2}{2\sigma^2}\right] \quad (7)$$

$$\ln f = -\tfrac{1}{2}n\ln(2\pi) - n\ln\sigma - \frac{\sum (x_i - \mu)^2}{2\sigma^2} \quad (8)$$

$$\frac{\partial(\ln f)}{\partial\mu} = \frac{\sum (x_i - \mu)}{\sigma^2} = 0 \quad (9)$$

gives

$$\hat{\mu} = \frac{\sum x_i}{n}. \quad (10)$$

$$\frac{\partial(\ln f)}{\partial\sigma} = -\frac{n}{\sigma} + \frac{\sum (x_i - \mu)^2}{\sigma^3} \quad (11)$$

gives

$$\hat{\sigma} = \sqrt{\frac{\sum (x_i - \hat{\mu})^2}{n}}. \quad (12)$$

Note that in this case, the maximum likelihood STANDARD DEVIATION is the sample STANDARD DEVIATION, which is a BIASED ESTIMATOR for the population STANDARD DEVIATION.

For a weighted GAUSSIAN DISTRIBUTION,

$$f(x_1, \ldots, x_n|\mu, \sigma) = \prod \frac{1}{\sigma_i\sqrt{2\pi}} e^{-(x_i-\mu)^2/2\sigma_i^2}$$

$$= \frac{(2\pi)^{-n/2}}{\sigma^n}\exp\left[-\frac{\sum (x_i - \mu)^2}{2\sigma^2}\right] \quad (13)$$

$$\ln f = -\tfrac{1}{2}n\ln(2\pi) - n\sum \ln\sigma_i - \sum \frac{(x_i - \mu)^2}{2\sigma_i^2} \quad (14)$$

$$\frac{\partial(\ln f)}{\partial\mu} = \sum \frac{(x_i - \mu)}{\sigma_i^2} = \sum \frac{x_i}{\sigma_i^2} - \mu\sum \frac{1}{\sigma_i^2} = 0 \quad (15)$$

gives

$$\hat{\mu} = \frac{\sum \dfrac{x_i}{\sigma_i^2}}{\sum \dfrac{1}{\sigma_i^2}}. \quad (16)$$

The VARIANCE of the MEAN is then

$$\sigma_\mu^2 = \sum \sigma_i^2\left(\frac{\partial\mu}{\partial x_i}\right)^2. \quad (17)$$

But

$$\frac{\partial\mu}{\partial x_i} = \frac{\partial}{\partial x_i}\frac{\sum (x_i/\sigma_i^2)}{\sum (1/\sigma_i^2)} = \frac{1/\sigma_i^2}{\sum (1/\sigma_i^2)}. \quad (18)$$

so

$$\sigma_\mu^2 = \sum \sigma_i^2\left(\frac{1/\sigma_i^2}{\sum (1/\sigma_i^2)}\right)$$

$$\sum \frac{1/\sigma_i^2}{\left[\sum (1/\sigma_i^2)\right]^2} = \frac{1}{\sum (1/\sigma_i^2)}. \quad (19)$$

For a POISSON DISTRIBUTION,

$$f(x_1, \ldots, x_n|\lambda) = \frac{e^{-\lambda}\lambda^{x_1}}{x_1!} \cdots \frac{e^{-\lambda}\lambda^{x_n}}{x_n!} = \frac{e^{-n\lambda}\lambda^{\sum x_i}}{x_1!\cdots x_n!} \quad (20)$$

$$\ln f = -n\lambda + (\ln\lambda)\sum x_i - \ln\left(\prod x_i!\right) \quad (21)$$

$$\frac{d(\ln f)}{\lambda} = -n + \frac{\sum x_i}{\lambda} = 0 \quad (22)$$

$$\hat{\lambda} = \frac{\sum x_i}{n}. \tag{23}$$

See also BAYESIAN ANALYSIS

References

Press, W. H.; Flannery, B. P.; Teukolsky, S. A.; and Vetterling, W. T. "Least Squares as a Maximum Likelihood Estimator." §15.1 in *Numerical Recipes in FORTRAN: The Art of Scientific Computing, 2nd ed.* Cambridge, England: Cambridge University Press, pp. 651–55, 1992.

Maximum Modulus Principle

Let $U \subseteq \mathbb{C}$ be a DOMAIN, and let f be an ANALYTIC FUNCTION on U. Then if there is a point $z_0 \in U$ such that $|f(z_0)| \geq |f(z)|$ for all $z \in U$, then f is constant. The following slightly sharper version can also be formulated. Let $U \subseteq \mathbb{C}$ be a DOMAIN, and let f be an ANALYTIC FUNCTION on U. Then if there is a point $z_0 \in U$ at which $|f|$ has a LOCAL MAXIMUM, then f is constant.

Furthermore, let $U \subseteq \mathbb{C}$ be a bounded domain, and let f be a continuous function on the CLOSED SET $\bar{U}$ that is analytic on U. Then the maximum value of $|f|$ on $\bar{U}$ (which always exists) occurs on the boundary ∂U. In other words,

$$\max_{\bar{U}} |f| = \max_{\partial U} |f|.$$

The maximum modulus theorem is not always true on an unbounded domain.

See also MINIMUM MODULUS PRINCIPLE, MODULUS (COMPLEX NUMBER)

References

Krantz, S. G. "The Maximum Modulus Principle" and "Boundary Maximum Modulus Theorem." §5.4.1 and 5.4.2 in *Handbook of Complex Analysis.* Boston, MA: Birkhäuser, pp. 76–7, 1999.

Max Sequence

A sequence defined from a FINITE sequence $a_0, a_1, \ldots, a_n$ by defining $a_{n+1} = \max_i(a_i + a_{n-i})$.

See also MEX SEQUENCE

References

Guy, R. K. "Max and Mex Sequences." §E27 in *Unsolved Problems in Number Theory, 2nd ed.* New York: Springer-Verlag, pp. 227–28, 1994.

Maxwell Distribution

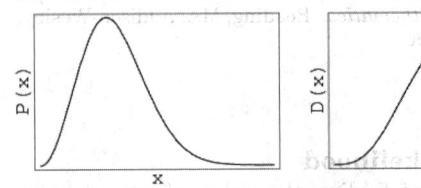

The distribution of speeds of molecules in thermal equilibrium as given by statistical mechanics. The probability and cumulative distributions over the range $x \in [0, \infty)$ are

$$P(x) = \sqrt{\frac{2}{\pi}} a^{3/2} x^2 e^{-ax^2/2} \tag{1}$$

$$D(x) = \frac{2\gamma(\frac{3}{2}, \frac{1}{2}ax^2)}{\sqrt{\pi}} \tag{2}$$

$$= \mathrm{erf}\left(x\sqrt{\frac{a}{2}}\right) - e^{-ax^2/2}\sqrt{\frac{2a}{\pi}}, \tag{3}$$

where $\gamma(a, x)$ is an incomplete GAMMA FUNCTION and $\mathrm{erf}(x)$ is ERF. The RAW MOMENTS are

$$\mu'_n = \frac{2^{1+n/2} a^{-n/2} \Gamma(\frac{1}{2}(3+n))}{\sqrt{\pi}}. \tag{4}$$

$$\mu' = 2\sqrt{\frac{2}{\pi a}} \tag{5}$$

$$\mu'_2 = \frac{3}{a} \tag{6}$$

$$\mu'_3 = 8\sqrt{\frac{2}{a^3 \pi}} \tag{7}$$

$$\mu'_4 = \frac{15}{2} \tag{8}$$

(Papoulis 1984, p. 149), and the MEAN, VARIANCE, SKEWNESS, and KURTOSIS are given by

$$\mu = 2\sqrt{\frac{2}{\pi a}} \tag{9}$$

$$\sigma^2 = \frac{3\pi - 8}{\pi a} \tag{10}$$

$$\gamma_1 = \frac{8}{3}\sqrt{\frac{2}{3\pi}} \tag{11}$$

$$\gamma_2 = -\frac{4}{3}. \tag{12}$$

See also EXPONENTIAL DISTRIBUTION, GAUSSIAN DISTRIBUTION, RAYLEIGH DISTRIBUTION

References

Papoulis, A. *Probability, Random Variables, and Stochastic Processes, 2nd ed.* New York: McGraw-Hill, pp. 104 and 149, 1984.

Spiegel, M. R. *Theory and Problems of Probability and Statistics.* New York: McGraw-Hill, p. 119, 1992.

von Seggern, D. *CRC Standard Curves and Surfaces.* Boca Raton, FL: CRC Press, p. 252, 1993.

Maxwell Equations

The system of PARTIAL DIFFERENTIAL EQUATIONS describing classical electromagnetism and therefore of central importance in physics. In the so-called cgs system of units, the Maxwell equations are given by

$$\nabla \cdot \mathbf{D} = 4\pi\rho \tag{1}$$

$$\nabla \times \mathbf{E} = -\frac{1}{c}\frac{\partial \mathbf{B}}{\partial t} \tag{2}$$

$$\nabla \cdot \mathbf{B} = 0 \tag{3}$$

$$\nabla \times \mathbf{H} = \frac{4\pi}{c}\mathbf{J} + \frac{1}{c}\frac{\partial \mathbf{D}}{\partial t}, \tag{4}$$

where **D** is the effective electric field in a dielectric , ρ is the charge density, **E** is the electric field, c is the speed of light, **B** is the imposed magnetic field, **H** is the effective magnetic field in a dielectric, and **J** is the current density. As usual, $\nabla \cdot \mathbf{V}$ is the DIVERGENCE and $\nabla \times \mathbf{V}$ is the CURL.

In the MKS system of units, the equations are written

$$\nabla \cdot \mathbf{D} = \frac{\rho}{\epsilon_0} \tag{5}$$

$$\nabla \times \mathbf{E} = -\frac{\partial \mathbf{B}}{\partial t} \tag{6}$$

$$\nabla \cdot \mathbf{B} = 0 \tag{7}$$

$$\nabla \times \mathbf{H} = \mu_0 \mathbf{J} + \epsilon_0 \mu_0 \frac{\partial \mathbf{D}}{\partial t}, \tag{8}$$

where ϵ_0 is the permittivity of free space and μ_0 is the permeability of free space.

See also DIRAC EQUATION

References

Jackson, J. D. *Classical Electrodynamics, 3rd ed.* New York: Wiley, p. 177, 1998.

Zwillinger, D. *Handbook of Differential Equations, 3rd ed.* Boston, MA: Academic Press, p. 138, 1997.

Maxwell's Equations

The system of PARTIAL DIFFERENTIAL EQUATIONS describing electromagnetism. In the so-called cgs system of units, they are given by

$$\nabla \cdot \mathbf{D} \tag{1}$$

$$4\pi\rho \tag{2}$$

$$\nabla \times \mathbf{E} \tag{3}$$

$$-\frac{1}{c}\frac{\partial \mathbf{B}}{\partial t} \tag{4}$$

where **D** is the electric induction, ρ is the charge density, **B** is the magnetic field, **H** is the magnetic induction, c is the speed of light, **J** is the current density, and **E** is the electric field.

References

Jackson, J. D. *Classical Electrodynamics, 3rd ed.* New York: Wiley, p. 177, 1998.

Zwillinger, D. *Handbook of Differential Equations, 3rd ed.* Boston, MA: Academic Press, p. 138, 1997.

May's Theorem

Simple majority vote is the only procedure which is ANONYMOUS, DUAL, and MONOTONIC.

References

May, K. "A Set of Independent Necessary and Sufficient Conditions for Simple Majority Decision." *Econometrica* **20**, 680–84, 1952.

May-Thomason Uniqueness Theorem

For every infinite LOOP SPACE MACHINE E, there is a natural equivalence of spectra between EX and Segal's spectrum $\mathbf{B}X$.

References

May, J. P. and Thomason, R. W. "The Uniqueness of Infinite Loop Space Machines." *Topology* **17**, 205–24, 1978.

Weibel, C. A. "The Mathematical Enterprises of Robert Thomason." *Bull. Amer. Math. Soc.* **34**, 1–3, 1996.

Maze

A maze is a drawing of impenetrable line segments (or curves) with "paths" between them. The goal of the maze is to start at one given point and find a path which reaches a second given point.

References

Bellman, R.; Cooke, K. L.; and Lockett, J. A. *Algorithms, Graphs, and Computers.* New York: Academic Press, pp. 94–00, 1970.

Dantzig, G. B. "All Shortest Routes in a Graph." *Operations Res. Techn. Rep. 66–.* Stanford, CA: Stanford University, pp. 346–65, Sept. 1961.

Gardner, M. "Mazes." Ch. 10 in *The Second Scientific American Book of Mathematical Puzzles & Diversions: A New Selection.* New York: Simon and Schuster, pp. 112–18, 1961.

Gardner, M. "Three-Dimensional Maze." §6.3 in *The Sixth Book of Mathematical Games from Scientific American.* Chicago, IL: University of Chicago Press, pp. 49–0, 1984.

Hu, T. C. and Torres, W. T. "Shortcut in the Decomposition Algorithm for Shortest Paths in a Network." *IBM J. Res. Devel.* **13**, 387–90, Jul. 1969.

Jablan, S. "Roman Mazes." http://members.tripod.com/~modularity/mazes.htm.

Lee, C. Y. "An Algorithm for Path Connections and Its Applications." *IRE Trans. Elec. Comput.* **EC-10**, 346–65, 1961.

Matthews, W. H. *Mazes and Labyrinths: Their History and Development.* New York: Dover, 1970.

Moore, E. F. "The Shortest Path through a Maze." *Ann. Comput. Lab. Harvard University* **30**, 285–92, 1959.

Pappas, T. "Mazes." *The Joy of Mathematics.* San Carlos, CA: Wide World Publ./Tetra, pp. 192–94, 1989.

Phillips, A. "The Topology of Roman Mazes." *Leonardo* **25**, 321–29, 1992.

Shepard, W. *Mazes and Labyrinths: A Book of Puzzles.* New York: Dover, 1961.

Weisstein, E. W. "Books about Mazes." http://www.treasure-troves.com/books/Mazes.html.

Mazur's Theorem

The generalization of the SCHÖNFLIES THEOREM to n-D. A smoothly embedded n-HYPERSPHERE in an $(n+1)$-HYPERSPHERE separates the $(n+1)$-HYPERSPHERE into two components, each HOMEOMORPHIC to $(n+1)$-BALLS. It can be proved using MORSE THEORY.

See also BALL, HYPERSPHERE, MORSE THEORY

M'Cay Circle

McCay Circle

McCay Circle

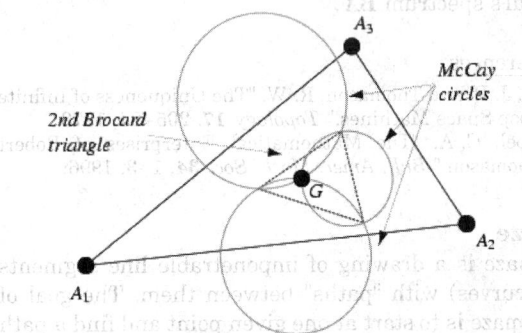

The three circumcircles through the CENTROID G of a given triangle $\Delta A_1 A_2 A_3$ and the pairs of the vertices of the second BROCARD TRIANGLE are called the McCay circles (Johnson 1929, p. 306).

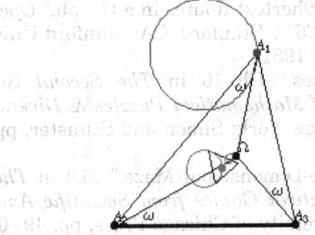

If the VERTEX A_1 of a TRIANGLE describes a NEUBERG CIRCLE N_1, then its CENTROID G describes one of the McCay circles (Johnson 1929, p. 290), which has RADIUS,

$$r = \tfrac{1}{6} a_1 \sqrt{\cot^2 \omega - 3},$$

1/3 that of the NEUBERG CIRCLE, where a_1 is the length of the edge $A_2 A_3$ and ω is the BROCARD ANGLE (Johnson 1929, p. 307). In the above figure, the inner triangle is the second BROCARD TRIANGLE of $\Delta A_1 A_2 A_3$, whose two indicated edges are concyclic with G on the McCay circle.

See also BROCARD TRIANGLES, CIRCLE, CONCURRENT, MEDIAN POINT, NEUBERG CIRCLE

References

Coolidge, J. L. *A Treatise on the Geometry of the Circle and Sphere.* New York: Chelsea, pp. 83–4 and 128–29, 1971.

Johnson, R. A. *Modern Geometry: An Elementary Treatise on the Geometry of the Triangle and the Circle.* Boston, MA: Houghton Mifflin, pp. 290 and 306–07, 1929.

Lachlan, R. *An Elementary Treatise on Modern Pure Geometry.* London: Macmillian, pp. 145 and 222, 1893.

M'Cay, W. S. "On Three Circles Related to a Triangle." *Trans. Roy. Irish Acad.* **28**, 453–70, 1885.

McCoy's Theorem

If two SQUARE $n \times n$ MATRICES A and B are simultaneously upper triangularizable by similarity transforms, then there is an ordering $a_1, ..., a_n$ of the EIGENVALUES of A and $b_1, ..., b_n$ of the EIGENVALUES of B so that, given any POLYNOMIAL $p(x, y)$ in noncommuting variables, the EIGENVALUES of $p(A, B)$ are the numbers $p(a_i, b_i)$ with $i = 1, ..., n$. McCoy's theorem states the converse: If every POLYNOMIAL exhibits the correct EIGENVALUES in a consistent ordering, then A and B are simultaneously triangularizable.

References

Luchins, E. H. and McLoughlin, M. A. "In Memoriam: Olga Taussky-Todd." *Not. Amer. Math. Soc.* **43**, 838–47, 1996.

McGee Graph

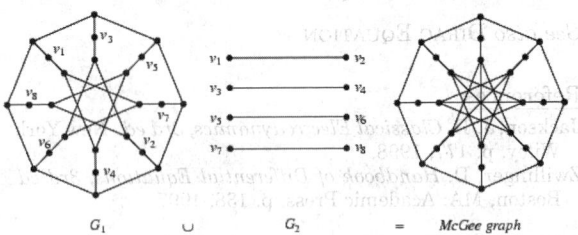

The unique 7-CAGE GRAPH (right figure) consisting of the union of the two leftmost subgraphs illustrated above. It has 24 nodes, 36 edges, and all nodes have degree 3. Its AUTOMORPHISM GROUP is of size 32. The graph is not vertex-transitive, having orbits of length 8 and 16. It was discovered by McGee (1960) and proven unique by Tutte (1966) (Wong 1982).

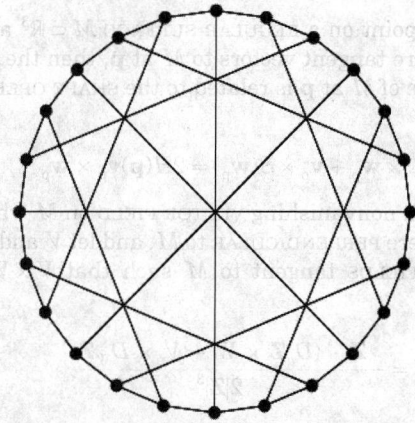

An alternative embedding is illustrated above.

See also CAGE GRAPH

References

Bondy, J. A. and Murty, U. S. R. *Graph Theory with Applications.* New York: North Holland, p. 237, 1976.

Harary, F. *Graph Theory.* Reading, MA: Addison-Wesley, pp. 174–75, 1994.

McGee, W. F. "A Minimal Cubic Graph of Girth Seven." *Canad. Math. Bull.* **3**, 149–52, 1960.

Royle, G. "Cubic Cages." http://www.cs.uwa.edu.au/~gordon/cages/.

Tutte, W. T. *Connectivity in Graphs.* Toronto, Ontario: University of Toronto Press, 1966.

Weisstein, E. W. "Graphs." MATHEMATICA NOTEBOOK GRAPHS.M.

Wong, P. K. "Cages--A Survey." *J. Graph Th.* **6**, 1–2, 1982.

McLaughlin Group

The SPORADIC GROUP *McL*.

References

Wilson, R. A. "ATLAS of Finite Group Representation." http://for.mat.bham.ac.uk/atlas/html/McL.html.

McMahon's Theorem

PRICE'S THEOREM

McNugget Number

A number which can be obtained by adding together orders of McDonald's® Chicken McNuggets™ (prior to consuming any), which originally came in boxes of 6, 9, and 20. All integers are McNugget numbers except 1, 2, 3, 4, 5, 7, 8, 10, 11, 13, 14, 16, 17, 19, 22, 23, 25, 28, 31, 34, 37, and 43. Since the Happy Meal™-sized nugget box (4 to a box) can now be purchased separately, the modern McNugget numbers are LINEAR COMBINATIONS of 4, 6, 9, and 20. These new-fangled numbers are much less interesting than before, with only 1, 2, 3, 5, 7, and 11 remaining as non-McNugget numbers.

The GREEDY ALGORITHM can be used to find a McNugget expansion of a given INTEGER.

See also COMPLETE SEQUENCE, GREEDY ALGORITHM

References

Vardi, I. *Computational Recreations in Mathematica.* Reading, MA: Addison-Wesley, pp. 19–0 and 233–34, 1991.

Wilson, D. `rec.puzzles` newsgroup posting, March 20, 1990.

Mean

A mean is HOMOGENEOUS and has the property that a mean μ of a set of numbers x_i satisfies

$$\min(x_1, \ldots, x_n) \leq \mu \leq \max(x_1, \ldots, x_n).$$

There are several statistical quantities called means, e.g., ARITHMETIC-GEOMETRIC MEAN, GEOMETRIC MEAN, HARMONIC MEAN, QUADRATIC MEAN, ROOT-MEAN-SQUARE. However, the quantity referred to as "the" mean is the ARITHMETIC MEAN, also called the AVERAGE.

An interesting empirical relationship between the mean, median, and mode which appears to hold for unimodal curves of moderate asymmetry is given by

$$\text{mean} - \text{mode} \approx 3(\text{mean} - \text{median})$$

(Kenney and Keeping 1962, p. 53), which is the basis for the definition of the PEARSON MODE SKEWNESS.

See also ARITHMETIC-GEOMETRIC MEAN, AVERAGE, GENERALIZED MEAN, GEOMETRIC MEAN, HARMONIC MEAN, PEARSON MODE SKEWNESS, QUADRATIC MEAN, REVERSION TO THE MEAN, ROOT-MEAN-SQUARE

References

Kenney, J. F. and Keeping, E. S. "Averages," "Relation Between Mean, Median, and Mode," and "Relative Merits of Mean, Median, and Mode." §3.1 and §4.8–.9 in *Mathematics of Statistics, Pt. 1, 3rd ed.* Princeton, NJ: Van Nostrand, pp. 32 and 52–4, 1962.

Mean Absolute Deviation

The mean absolute deviation (often inaccurately called the MEAN DEVIATION), is defined by

$$\text{M.A.D} = \frac{1}{N} \sum_{i=1}^{N} f_i |x_i - \bar{x}|,$$

where the SAMPLE SIZE is N, the samples have values x_i, the MEAN is $\bar{x}$, and f_i is an ABSOLUTE FREQUENCY.

See also MEAN DEVIATION

References

Kenney, J. F. and Keeping, E. S. "Mean Absolute Deviation." §6.4 in *Mathematics of Statistics, Pt. 1, 3rd ed.* Princeton, NJ: Van Nostrand, pp. 76–7 1962.

Mean Caliper Diameter

MEAN TANGENT DIAMETER

Mean Cluster Count Per Site

s-CLUSTER

Mean Cluster Density

s-CLUSTER

Mean Curvature

Let κ_1 and κ_2 be the PRINCIPAL CURVATURES, then their MEAN

$$H = \tfrac{1}{2}(\kappa_1 + \kappa_2) \tag{1}$$

is called the mean curvature. Let R_1 and R_2 be the radii corresponding to the PRINCIPAL CURVATURES, then the MULTIPLICATIVE INVERSE of the mean curvature H is given by the MULTIPLICATIVE INVERSE of the HARMONIC MEAN,

$$H \equiv \frac{1}{2}\left(\frac{1}{R_1} + \frac{1}{R_2}\right) = \frac{R_1 + R_2}{2R_1 R_2}. \tag{2}$$

In terms of the GAUSSIAN CURVATURE K,

$$H = \tfrac{1}{2}(R_1 + R_2)K. \tag{3}$$

The mean curvature of a REGULAR SURFACE in $\mathbb{R}^3$ at a point **p** is formally defined as

$$H(p) = \tfrac{1}{2}\operatorname{Tr}(S(\mathbf{p})) \tag{4}$$

where S is the SHAPE OPERATOR and $\operatorname{Tr}(S)$ denotes the TRACE. For a MONGE PATCH with $z = h(x, y)$,

$$H = \frac{(1 + h_v^2)h_{uu} - 2h_u h_v h_{uv} + (1 + h_u^2)h_{vv}}{2(1 + h_u^2 + h_v^2)^{3/2}} \tag{5}$$

(Gray 1997, p. 399).

If $\mathbf{x} : U \to \mathbb{R}^3$ is a REGULAR PATCH, then the mean curvature is given by

$$H = \frac{eG - 2fF + gE}{2(EG - F^2)}, \tag{6}$$

where E, F, and G are coefficients of the first FUNDAMENTAL FORM and e, f, and g are coefficients of the second FUNDAMENTAL FORM (Gray 1997, p. 377). It can also be written

$$
\begin{aligned}
H = & \frac{\det(\mathbf{x}_{uu}\mathbf{x}_u\mathbf{x}_v)|\mathbf{x}_u|^2 - 2\det(\mathbf{x}_{uv}\mathbf{x}_u\mathbf{x}_v)(\mathbf{x}_u \cdot \mathbf{x}_v)}{2[|\mathbf{x}_u|^2|\mathbf{x}_v| - (\mathbf{x}_u \cdot \mathbf{x}_v)^2]^{3/2}} \\
& + \frac{\det(\mathbf{x}_{vv}\mathbf{x}_u\mathbf{x}_v)|\mathbf{x}_u|^2}{2[|\mathbf{x}_u|^2|\mathbf{x}_v|^2 - (\mathbf{x}_u \cdot \mathbf{x}_v)^2]^{3/2}}
\end{aligned}
\tag{7}
$$

Gray (1997, p. 380).

The GAUSSIAN and mean curvature satisfy

$$H^2 \geq K, \tag{8}$$

with equality only at UMBILIC POINTS, since

$$H^2 - K = \tfrac{1}{4}(\kappa_1 - \kappa_2)^2. \tag{9}$$

If **p** is a point on a REGULAR SURFACE $M \subset \mathbb{R}^3$ and $\mathbf{v}_\mathrm{p}$ and $\mathbf{w}_\mathrm{p}$ are tangent vectors to M at **p**, then the mean curvature of M at **p** is related to the SHAPE OPERATOR S by

$$S(\mathbf{v}_\mathrm{p}) \times \mathbf{w}_\mathrm{p} + \mathbf{v}_\mathrm{p} \times S(\mathbf{w}_\mathrm{p}) = 2H(\mathbf{p})\mathbf{v}_\mathrm{p} \times \mathbf{w}_\mathrm{p} \tag{10}$$

Let **Z** be a nonvanishing VECTOR FIELD on M which is everywhere PERPENDICULAR to M, and let V and W be VECTOR FIELDS tangent to M such that $V \times W = \mathbf{Z}$, then

$$H = -\frac{\mathbf{Z} \cdot (D_v\mathbf{Z} \times W + V \times D_W\mathbf{Z})}{2|\mathbf{Z}|^3} \tag{11}$$

(Gray 1997, p. 410).

Wente (1985, 1986, 1987) found a nonspherical finite surface with constant mean curvature, consisting of a self-intersecting three-lobed toroidal surface. A family of such surfaces exists.

See also GAUSSIAN CURVATURE, LAGRANGE'S EQUATION, MINIMAL SURFACE, PRINCIPAL CURVATURES, SHAPE OPERATOR

References

Gray, A. "The Gaussian and Mean Curvatures." §16.5 in *Modern Differential Geometry of Curves and Surfaces with Mathematica, 2nd ed.* Boca Raton, FL: CRC Press, pp. 373–80, 1997.

Isenberg, C. *The Science of Soap Films and Soap Bubbles.* New York: Dover, p. 108, 1992.

Peterson, I. *The Mathematical Tourist: Snapshots of Modern Mathematics.* New York: W. H. Freeman, pp. 69–0, 1988.

Wente, H. C. "A Counterexample in 3-Space to a Conjecture of H. Hopf." In *Workshop Bonn 1984, Proceedings of the 25th Mathematical Workshop Held at the Max-Planck Institut für Mathematik, Bonn, June 15–2, 1984* (Ed. F. Hirzebruch, J. Schwermer, and S. Suter). New York: Springer-Verlag, pp. 421–29, 1985.

Wente, H. C. "Counterexample to a Conjecture of H. Hopf." *Pac. J. Math.* **121**, 193–43, 1986.

Wente, H. C. "Immersed Tori of Constant Mean Curvature in $\mathbb{R}^3$." In *Variational Methods for Free Surface Interfaces, Proceedings of a Conference Held in Menlo Park, CA, Sept. 7–2, 1985* (Ed. P. Concus and R. Finn). New York: Springer-Verlag, pp. 13–4, 1987.

Mean Deviation

The MEAN of the ABSOLUTE DEVIATIONS,

$$\mathrm{MD} \equiv \frac{1}{N}\sum_{i=1}^{N}|x_i - \bar{x}|,$$

where $\bar{x}$ is the MEAN of the distribution.

See also ABSOLUTE DEVIATION

Mean Distribution

For an infinite population with MEAN μ, VARIANCE σ^2, SKEWNESS γ_1, and KURTOSIS γ_2, the corresponding quantities for the *distribution of means* are

$$\mu_{\bar{x}} = \mu \tag{1}$$

$$\sigma_{\bar{x}}^2 = \frac{\sigma^2}{N} \tag{2}$$

$$\gamma_{1,\bar{x}} = \frac{\gamma_1}{\sqrt{N}} \tag{3}$$

$$\gamma_{2,\bar{x}} = \frac{\gamma_2}{N}. \tag{4}$$

For a population of M (Kenney and Keeping 1962, p. 181),

$$\mu_{\bar{x}}^{(M)} = \mu \tag{5}$$

$$\sigma^{2(M)} = \frac{\sigma^2}{N} \frac{M-N}{M-1}. \tag{6}$$

References

Kenney, J. F. and Keeping, E. S. *Mathematics of Statistics, Pt. 1, 3rd ed.* Princeton, NJ: Van Nostrand, 1962.

Mean Run Count Per Site
s-RUN

Mean Run Density
s-RUN

Mean Square Error
ROOT-MEAN-SQUARE

Mean Tangent Diameter
This entry contributed by ROD MACKERT

The mean tangent diameter of a solid, also known as the mean caliper diameter, is the caliper dimension obtained by averaging over all orientations.

See also INNER QUERMASS, STEREOLOGY

References

Hilliard J. E. "The Calculation of the Mean Caliper Diameter of a Body for Use in the Analysis of the Number of Particles per Unit Volume." In *Stereology* (Ed. H. Elias). New York: Springer-Verlag, pp. 211–15, 1967.
Russ, J. C. "Size Distributions." In *Practical Stereology.* New York: Plenum, pp. 53–2, 1986.

Mean-Value Property

Let a function $h : U \to \mathbb{R}$ be continuous on an OPEN SET $U \subseteq \mathbb{C}$. Then h is said to have the ϵ_{z_0}-property if, for each $z_0 \in U$, there exists an $\epsilon_{z_0} > 0$ such that $\bar{D}(z_0, \epsilon_{z_0}) \subseteq U$, where $\bar{D}$ is a closed disk, and for every $0 < \epsilon < \epsilon_{z_0}$,

$$h(z_0) = \frac{1}{2\pi} \int_0^{2\pi} h(z_0 + \epsilon e^{i\theta}) \, d\theta.$$

If h has the mean-value property, then h is harmonic.
See also HARMONIC FUNCTION

References

Krantz, S. G. "The Mean Value Property on Circles." §7.4.1 in *Handbook of Complex Analysis.* Boston, MA: Birkhäuser, p. 94, 1999.

Mean-Value Theorem
Let $f(x)$ be DIFFERENTIABLE on the OPEN INTERVAL (a, b) and CONTINUOUS on the CLOSED INTERVAL $[a, b]$. Then there is at least one point c in (a, b) such that

$$f'(c) = \frac{f(b) - f(a)}{b - a}.$$

See also EXTENDED MEAN-VALUE THEOREM, GAUSS'S MEAN-VALUE THEOREM

References

Gradshteyn, I. S. and Ryzhik, I. M. *Tables of Integrals, Series, and Products, 6th ed.* San Diego, CA: Academic Press, pp. 1097–098, 2000.
Jeffreys, H. and Jeffreys, B. S. "Mean-Value Theorems." §1.13 in *Methods of Mathematical Physics, 3rd ed.* Cambridge, England: Cambridge University Press, pp. 49–0, 1988.

Measurable Function

A function $f : X \to \mathbb{R}$ is measurable if, for every real number a, the set

$$\{x \in X \text{ such that } f(x) > a\}$$

is MEASURABLE. When $X = \mathbb{R}$ with LEBESGUE MEASURE, or more generally any BOREL MEASURE, then all CONTINUOUS functions are measurable. In fact, practically any function that can be described is measurable. Measurable functions are CLOSED under addition and multiplication, but not composition.

The measurable functions form one of the most general classes of REAL FUNCTIONS. They are one of the basic objects of study in ANALYSIS, both because of their wide practical applicability and the aesthetic appeal of their generality. Whether a function $f : X \to \mathbb{R}$ is measurable depends on the MEASURE μ on X, and, in particular, it only depends on the SIGMA ALGEBRA of MEASURABLE SETS in X. Sometimes, the MEASURE on X may be assumed to be a standard measure. For instance, a measurable function on $\mathbb{R}$ is usually measurable with respect to LEBESGUE MEASURE.

From the point of view of MEASURE THEORY, subsets with measure zero do not matter. Often, instead of actual real-valued functions, EQUIVALENCE CLASSES of functions are used. Two functions are equivalent if the subset of the domain X where they differ has MEASURE ZERO.

See also BOREL MEASURE, LEBESGUE MEASURE, MEASURE, MEASURE SPACE, MEASURE THEORY, REAL FUNCTION, SIGMA ALGEBRA

Measurable Set

If F is a SIGMA ALGEBRA and A is a SUBSET of X, then A is called measurable if A is a member of F. X need not have, a priori, a topological structure. Even if it does, there may be no connection between the open sets in the topology and the given SIGMA ALGEBRA.

See also MEASURABLE SPACE, SIGMA ALGEBRA

Measurable Space

A SET considered together with the SIGMA ALGEBRA on the SET.

See also MEASURABLE SET, MEASURE SPACE, SIGMA ALGEBRA

Measure

The terms "measure," "measurable," etc., have very precise technical definitions (usually involving SIGMA ALGEBRAS) which makes them a little difficult to understand. However, the technical nature of the definitions is extremely important, since it gives a firm footing to concepts which are the basis for much of ANALYSIS (including some of the slippery underpinnings of CALCULUS).

For example, every definition of an INTEGRAL is based on a particular measure: the RIEMANN INTEGRAL is based on JORDAN MEASURE, and the LEBESGUE INTEGRAL is based on LEBESGUE MEASURE. The study of measures and their application to INTEGRATION is known as MEASURE THEORY.

A measure is formally defined as a NONNEGATIVE MAP $m : F \to \mathbb{R}$ (the reals) such that $m(\varnothing) = 0$ and, if A_n is a COUNTABLE SEQUENCE in F and the A_n are pairwise DISJOINT, then

$$m\left(\bigcup_n A_n\right) = \sum_n m(A_n)$$

If, in addition, $m(X) = 1$ for X a MEASURE SPACE, then m is said to be a PROBABILITY MEASURE.

A measure m may also be defined on SETS other than those in the SIGMA ALGEBRA F. By adding to F all sets to which m assigns measure zero, we again obtain a SIGMA ALGEBRA and call this the "completion" of F with respect to m. Thus, the completion of a SIGMA ALGEBRA is the smallest SIGMA ALGEBRA containing F and all sets of measure zero.

See also ALMOST EVERYWHERE, BOREL MEASURE, ERGODIC MEASURE, EULER MEASURE, GAUSS MEASURE, HAAR MEASURE, HAUSDORFF MEASURE, HEL-

SON-SZEGO MEASURE, INTEGRAL, JORDAN MEASURE, LEBESGUE MEASURE, LIOUVILLE MEASURE, MAHLER MEASURE, MEASURABLE SPACE, MEASURE ALGEBRA, MEASURE SPACE, MINKOWSKI MEASURE, NATURAL MEASURE, PROBABILITY MEASURE, RADON MEASURE, WIENER MEASURE

References

Czyz, J. *Paradoxes of Measures and Dimensions Originating in Felix Hausdorff's Ideas.* Singapore: World Scientific, 1994.

Measure Algebra

A Boolean SIGMA ALGEBRA which possesses a MEASURE.

Measure Polytope

HYPERCUBE

Measure-Preserving Transformation

ENDOMORPHISM

Measure Space

A measure space is a MEASURABLE SPACE possessing a NONNEGATIVE MEASURE. Examples of measure spaces include n-D EUCLIDEAN SPACE with LEBESGUE MEASURE and the unit interval with LEBESGUE MEASURE (i.e., probability).

See also LEBESGUE MEASURE, MEASURABLE SPACE

Measure Theory

The mathematical theory of how to perform INTEGRATION in arbitrary MEASURE SPACES.

See also ALMOST EVERYWHERE CONVERGENCE, CANTOR SET, FATOU'S LEMMA, FRACTAL, INTEGRAL, INTEGRATION, LEBESGUE'S DOMINATED CONVERGENCE THEOREM, MEASURABLE FUNCTION, MEASURABLE SET, MEASURABLE SPACE, MEASURE, MEASURE SPACE, MONOTONE CONVERGENCE THEOREM, POINTWISE CONVERGENCE

References

Doob, J. L. *Measure Theory.* New York: Springer-Verlag, 1994.
Evans, L. C. and Gariepy, R. F. *Measure Theory and Finite Properties of Functions.* Boca Raton, FL: CRC Press, 1992.
Gordon, R. A. *The Integrals of Lebesgue, Denjoy, Perron, and Henstock.* Providence, RI: Amer. Math. Soc., 1994.
Halmos, P. R. *Measure Theory.* New York: Springer-Verlag, 1974.
Henstock, R. *The General Theory of Integration.* Oxford, England: Clarendon Press, 1991.
Kestelman, H. *Modern Theories of Integration, 2nd rev. ed.* New York: Dover, 1960.
Kingman, J. F. C. and Taylor, S. J. *Introduction to Measure and Probability.* Cambridge, England: Cambridge University Press, 1966.
Rao, M. M. *Measure Theory And Integration.* New York: Wiley, 1987.

Strook, D. W. *A Concise Introduction to the Theory of Integration, 2nd ed.* Boston, MA: Birkhäuser, 1994.
Weisstein, E. W. "Books about Measure Theory." http://www.treasure-troves.com/books/MeasureTheory.html.

Measure Zero

A set of points capable of being enclosed in intervals whose total length is arbitrarily small.

See also ALMOST EVERYWHERE

References

Jeffreys, H. and Jeffreys, B. S. " "Measure Zero": "Almost Everywhere"." §1.1013 in *Methods of Mathematical Physics, 3rd ed.* Cambridge, England: Cambridge University Press, pp. 29–0, 1988.

Mechanical Quadrature

GAUSSIAN QUADRATURE

Mecon

Buckminster Fuller's term for the TRUNCATED OCTAHEDRON.

See also DYMAXION

Medial Axis

The boundaries of the cells of a VORONOI DIAGRAM.

Medial Circle

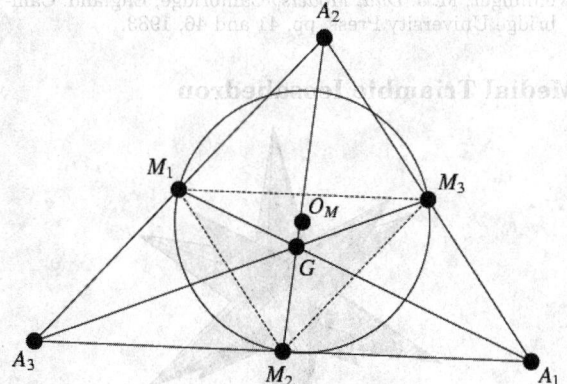

The CIRCUMCIRCLE of the MEDIAL TRIANGLE $\Delta M_1 M_2 M_3$ of a given triangle $\Delta A_1 A_2 A_3$.

See also CIRCUMCIRCLE, MEDIAL TRIANGLE, MEDIAN (TRIANGLE), SPIEKER CIRCLE

Medial Deltoidal Hexecontahedron

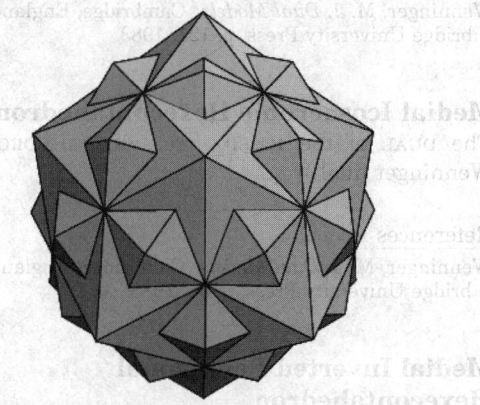

The DUAL of the RHOMBIDODECADODECAHEDRON U_{38} and Wenninger dual W_{76}.

See also DUAL POLYHEDRON, RHOMBIDODECADODECA-HEDRON

References

Wenninger, M. J. *Dual Models.* Cambridge, England: Cambridge University Press, p. 84, 1983.

Medial Disdyakis Triacontahedron

The 30-faced DUAL of the TRUNCATED DODECADODE-CAHEDRON and Wenninger dual W_{98}.

See also ARCHIMEDEAN SOLID, ICOSIDODECAHEDRON, TRUNCATED DODECADODECAHEDRON

References

Wenninger, M. J. *Dual Models.* Cambridge, England: Cambridge University Press, p. 96, 1983.

Medial Hexagonal Hexecontahedron

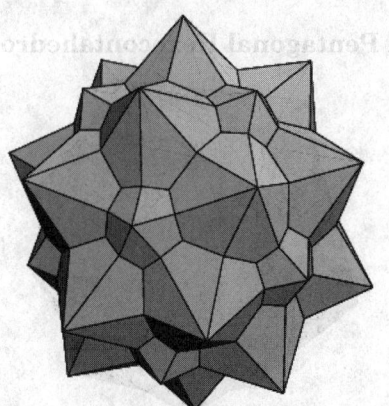

The DUAL of the SNUB ICOSIDODECADODECAHEDRON U_{44} and Wenninger dual W_{112}.

See also DUAL POLYHEDRON, SNUB ICOSIDODECADO-DECAHEDRON

References

Wenninger, M. J. *Dual Models.* Cambridge, England: Cambridge University Press, p. 121, 1983.

Medial Icosacronic Hexecontahedron

The DUAL of the ICOSIDODECADODECAHEDRON and Wenninger dual W_{83}.

References

Wenninger, M. J. *Dual Models.* Cambridge, England: Cambridge University Press, p. 85, 1983.

Medial Inverted Pentagonal Hexecontahedron

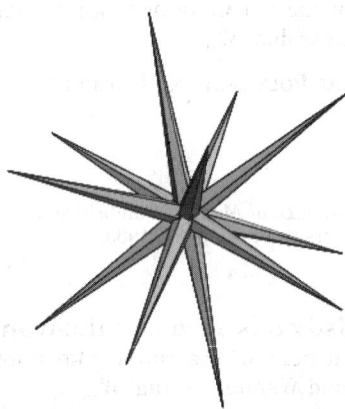

The DUAL of the INVERTED SNUB DODECADODECAHEDRON U_{60} and Wenninger dual W_{114}.

See also DUAL POLYHEDRON, INVERTED SNUB DODECADODECAHEDRON

References

Wenninger, M. J. *Dual Models.* Cambridge, England: Cambridge University Press, p. 124, 1983.

Medial Pentagonal Hexecontahedron

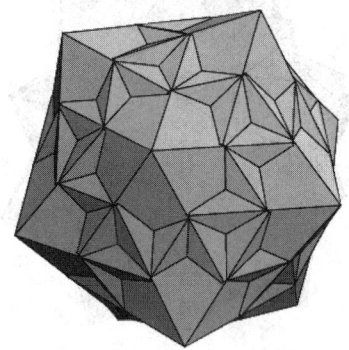

The DUAL of the SNUB DODECADODECAHEDRON U_{40} and Wenninger dual W_{111}.

See also DUAL POLYHEDRON, SNUB DODECADODECAHEDRON

References

Wenninger, M. J. *Dual Models.* Cambridge, England: Cambridge University Press, p. 120, 1983.

Medial Rhombic Triacontahedron

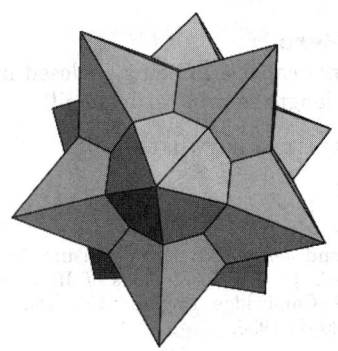

A ZONOHEDRON which is the DUAL of the DODECADODECAHEDRON U_{36} and Wenninger dual W_{73}. The medial rhombic triacontahedron contains interior pentagrammic vertices which are, however, hidden from view (Wenninger 1983, p. 41). The solid is also called the SMALL STELLATED TRIACONTAHEDRON. The CONVEX HULL of the DODECADODECAHEDRON is an ICOSIDODECAHEDRON and the dual of the ICOSIDODECAHEDRON is the RHOMBIC TRIACONTAHEDRON, so the dual of the DODECADODECAHEDRON (i.e., the medial rhombic triacontahedron) is one of the RHOMBIC TRIACONTAHEDRON STELLATIONS (Wenninger 1983, p. 41).

See also DUAL POLYHEDRON, DODECADODECAHEDRON, RHOMBIC TRIACONTAHEDRON STELLATIONS

References

Coxeter, H. S. M. *Regular Polytopes, 3rd ed.* New York: Dover, 1973.
Cundy, H. and Rollett, A. "Small Stellated Triacontahedron. V $(5 \cdot \frac{5}{2})^2$." §3.9.3 in *Mathematical Models, 3rd ed.* Stradbroke, England: Tarquin Pub., p. 125, 1989.
Wenninger, M. J. *Dual Models.* Cambridge, England: Cambridge University Press, pp. 41 and 46, 1983.

Medial Triambic Icosahedron

The DUAL of the DITRIGONAL DODECADODECAHEDRON U_{41} and Wenninger dual W_{80}, whose outward appearance is the same as the GREAT TRIAMBIC ICOSAHEDRON (the dual of the GREAT DITRIGONAL ICOSIDODECAHEDRON), since the internal vertices are hidden from view. The medial triambic icosahedron has hidden pentagrammic faces, while the GREAT TRIAMBIC ICOSAHEDRON has hidden triangular faces (Wenninger 1983, pp. 45 and 47–0).

The CONVEX HULL of the SMALL DITRIGONAL ICOSIDODECAHEDRON is a regular DODECAHEDRON, whose dual is the ICOSAHEDRON, so the dual of the SMALL DITRIGONAL ICOSIDODECAHEDRON (i.e., the medial triambic icosahedron) is one of the ICOSAHEDRON STELLATIONS (Wenninger 1983, p. 42).

See also DUAL POLYHEDRON, DITRIGONAL DODECADODECAHEDRON, GREAT TRIAMBIC ICOSAHEDRON, ICOSAHEDRON STELLATIONS, UNIFORM POLYHEDRON

References

Wenninger, M. J. *Dual Models.* Cambridge, England: Cambridge University Press, pp. 41 and 46, 1983.
Wenninger, M. J. "Ninth Stellation of the Icosahedron." §34 in *Polyhedron Models.* New York: Cambridge University Press, p. 55, 1989.

Medial Triangle

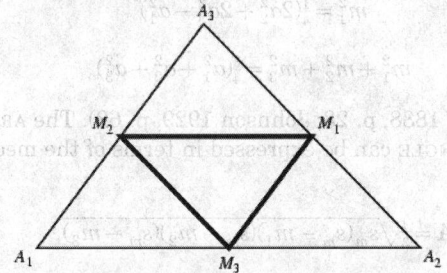

The TRIANGLE $\Delta M_1 M_2 M_3$ formed by joining the MIDPOINTS of the sides of a TRIANGLE $\Delta A_1 A_2 A_3$. The medial triangle is sometimes also called the AUXILIARY TRIANGLE (Dixon 1991). The medial triangle has TRILINEAR COORDINATES

$$A' = 0 : b^{-1} : c^{-1}$$

$$B' = a^{-1} : 0 : c^{-1}$$

$$C' = a^{-1} : b^{-1} : 0.$$

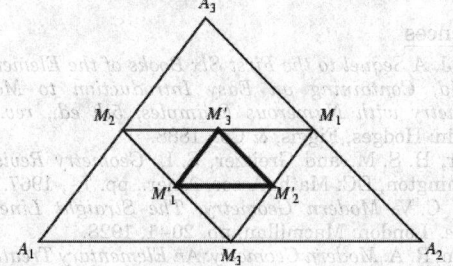

The medial triangle $\Delta M_1' M_2' M_3'$ of the medial triangle

$\Delta M_1 M_2 M_3$ of a TRIANGLE $\Delta A_1 A_2 A_3$ is similar to $\Delta A_1 A_2 A_3$.

The INCIRCLE of the medial triangle is called the SPIEKER CIRCLE, and its INCENTER is called the SPIEKER CENTER. The CIRCUMCIRCLE of the medial triangle is called the MEDIAL CIRCLE.

See also ANTICOMPLEMENTARY TRIANGLE, CLEAVANCE CENTER, CLEAVER, SPIEKER CENTER, SPIEKER CIRCLE

References

Coxeter, H. S. M. and Greitzer, S. L. "The Medial Triangle and Euler Line." §1.7 in *Geometry Revisited.* Washington, DC: Math. Assoc. Amer., pp. 18–0, 1967.
Dixon, R. *Mathographics.* New York: Dover, p. 56, 1991.

Medial Triangle Locus Theorem

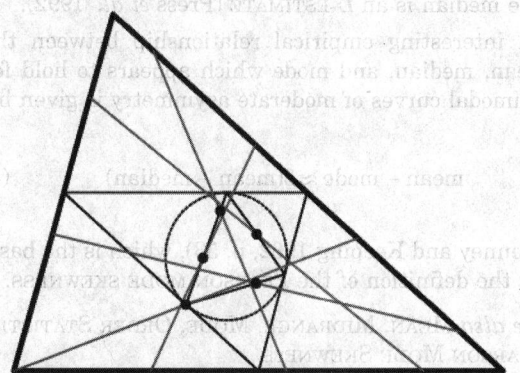

Given an original triangle (thick line), find the MEDIAL TRIANGLE (outer thin line) and its INCIRCLE. Take the PEDAL TRIANGLE (inner thin line) of the MEDIAL TRIANGLE with the INCENTER as the PEDAL POINT. Now pick any point on the original triangle, and connect it to the point located a half-PERIMETER away (gray lines). Then the locus of the MIDPOINTS of these lines (the ●s in the above diagram) is the PEDAL TRIANGLE.

References

Honsberger, R. *More Mathematical Morsels.* Washington, DC: Math. Assoc. Amer., pp. 261–67, 1991.
Tsintsifas, G. "Solution to Problem 674." *Crux Math.* **8,** 256–57, 1982.

Median (Statistics)

The middle value of a distribution (if the sample size N is odd) or average of the two middle items (if N is even), denoted $\mu_{1/2}$ or $\tilde{x}$. For a normal population, the mean μ is the most efficient (in the sense that no other unbiased statistic for estimating μ can have smaller VARIANCE) estimate (Kenney and Keeping 1962, p. 211). The efficiency of the median, measured as the ratio of the variance of the mean to the variance of the median, depends on the sample size $N \equiv 2n + 1$ as

$$\frac{4n}{\pi(2n+)}, \tag{1}$$

which tends to the value $2/\pi \approx 0.637$ as N becomes large (Kenney and Keeping 1962, p. 211). Although, the median is less efficient than the MEAN, it is less sensitive to outliers than the MEAN

For large N samples with population median $\tilde{x}_0$,

$$\mu_{\bar{x}} = \tilde{x}_0 \tag{2}$$

$$\sigma_{\bar{x}}^2 = \frac{1}{8Nf^2(\tilde{x}_0)}. \tag{3}$$

The median is an L-ESTIMATE (Press *et al.* 1992).

An interesting empirical relationship between the mean, median, and mode which appears to hold for unimodal curves of moderate asymmetry is given by

$$\text{mean} - \text{mode} \approx 3(\text{mean} - \text{median}) \tag{4}$$

(Kenney and Keeping 1962, p. 53), which is the basis for the definition of the PEARSON MODE SKEWNESS.

See also MEAN, MIDRANGE, MODE, ORDER STATISTIC, PEARSON MODE SKEWNESS

References

Huang, J. S. "Third-Order Expansion of Mean Squared Error of Medians." *Stat. Prob. Let.* **42**, 185–92, 1999.
Kenney, J. F. and Keeping, E. S. "The Median," "Relation Between Mean, Median, and Mode," "Relative Merits of Mean, Median, and Mode," and "The Median." §3.2, 4.8–.9, and 13.13 in *Mathematics of Statistics, Pt. 1, 3rd ed.* Princeton, NJ: Van Nostrand, pp. 32–5, 52–4, 211–12, 1962.
Press, W. H.; Flannery, B. P.; Teukolsky, S. A.; and Vetterling, W. T. *Numerical Recipes in FORTRAN: The Art of Scientific Computing, 2nd ed.* Cambridge, England: Cambridge University Press, p. 694, 1992.
Zwillinger, D. (Ed.). *CRC Standard Mathematical Tables and Formulae.* Boca Raton, FL: CRC Press, p. 602, 1995.

Median (Tetrahedron)

The lines joining the vertices of a TETRAHEDRON to the centroids of the opposite faces are called medians.

See also COMMANDINO'S THEOREM, TETRAHEDRON

References

Altshiller-Court, N. *Modern Pure Solid Geometry.* New York: Chelsea, p. 51, 1979.

Median (Triangle)

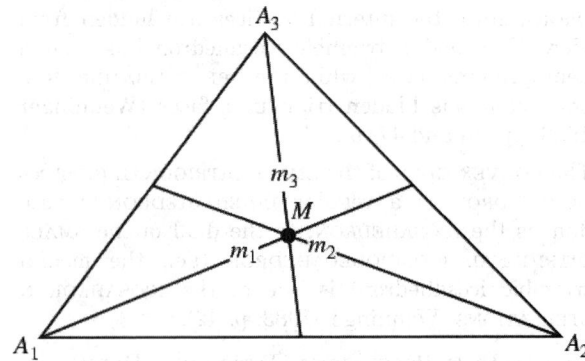

The median of a triangle is the CEVIAN from one of its VERTICES to the MIDPOINT of the opposite side. The three medians of any TRIANGLE are CONCURRENT (Casey 1888, p. 3), meeting in the TRIANGLE'S CENTROID (Durell 1928), which has TRILINEAR COORDINATES $1/a : 1/b : 1/c$. In addition, the medians of a TRIANGLE divide one another in the ratio 2:1 (Casey 1888, p. 3). A median also bisects the AREA of a TRIANGLE.

Let m_i denote the length of the median of the ith side a_i. Then

$$m_1^2 = \tfrac{1}{4}(2a_2^2 + 2a_3^2 - a_1^2) \tag{1}$$

$$m_1^2 + m_2^2 + m_3^2 = \tfrac{3}{4}(a_1^2 + a_2^2 + a_3^2) \tag{2}$$

(Casey 1888, p. 23; Johnson 1929, p. 68). The AREA of a TRIANGLE can be expressed in terms of the medians by

$$A = \tfrac{4}{3}\sqrt{s_m(s_m - m_1)(s_m - m_2)(s_m - m_3)}, \tag{3}$$

where

$$s_m \equiv \tfrac{1}{2}(m_1 + m_2 + m_3). \tag{4}$$

A median triangle is a TRIANGLE whose sides are equal and PARALLEL to the medians of a given TRIANGLE. The median triangle of the median triangle is similar to the given TRIANGLE in the ratio 3/4.

See also BIMEDIAN, COMEDIAN TRIANGLES, COMMANDINO'S THEOREM, EXMEDIAN, EXMEDIAN POINT, HERONIAN TRIANGLE, MEDIAL TRIANGLE

References

Casey, J. *A Sequel to the First Six Books of the Elements of Euclid, Containing an Easy Introduction to Modern Geometry with Numerous Examples, 5th ed., rev. enl.* Dublin: Hodges, Figgis, & Co., 1888.
Coxeter, H. S. M. and Greitzer, S. L. *Geometry Revisited.* Washington, DC: Math. Assoc. Amer., pp. 7–, 1967.
Durell, C. V. *Modern Geometry: The Straight Line and Circle.* London: Macmillan, pp. 20–1, 1928.
Johnson, R. A. *Modern Geometry: An Elementary Treatise on the Geometry of the Triangle and the Circle.* Boston, MA: Houghton Mifflin, pp. 68, 173–75, 282–83, 1929.

Lachlan, R. *An Elementary Treatise on Modern Pure Geometry.* London: Macmillian, p. 62, 1893.

Median Point

CENTROID (TRIANGLE)

Mediant

Given a FAREY SEQUENCE with consecutive terms h/k and h'/k', then the mediant is defined as the reduced form of the fraction $(h+h')/(k+k')$.

See also FAREY SEQUENCE

References

Conway, J. H. and Guy, R. K. "Farey Fractions and Ford Circles." *The Book of Numbers.* New York: Springer-Verlag, pp. 152–54, 1996.

Mediating Plane

MEDIATOR

Mediator

The PLANE through the MIDPOINT of a LINE SEGMENT and perpendicular to that segment, also called a mediating plane. The term "mediator" was introduced by J. Neuberg (Altshiller-Court 1979, p. 298).

See also MIDPOINT, PLANE

References

Altshiller-Court, N. *Modern Pure Solid Geometry.* New York: Chelsea, p. 1, 1979.

Meeussen Sequence

A Meeussen sequence is an increasing sequence of positive integers $(m_1, m_2, ...)$ such that $m_1 = 1$, every nonnegative integer is the sum of a subset of the $\{m_i\}$, and each integer $m_i - 1$ is the sum of a unique such subset. Cook and Kleber (2000) show that Meeussen sequences are isomorphic to TOURNAMENT SEQUENCES.

See also TOURNAMENT SEQUENCE

References

Cook, M. and Kleber, M. "Tournament Sequences and Meeussen Sequences." *Electronic J. Combinatorics* **7**, No. 1, R44, 1–6, 2000. http://www.combinatorics.org/Volume_7/v7i1toc.html#R44.

Mega

A LARGE NUMBER defined as

$$\textcircled{2} = \boxed{2} = \boxed{\triangle 2} = \boxed{\triangle 2^2} = \boxed{4^4} = \boxed{256}$$

where the CIRCLE NOTATION $\textcircled{n}$ denotes "n in n squares," and triangles and squares are expanded in terms of STEINHAUS-MOSER NOTATION (Steinhaus

1983, pp. 28–9). Here, the typographical error of Steinhaus has been corrected.

See also CIRCLE NOTATION, LARGE NUMBER, MEGISTRON, MOSER, STEINHAUS-MOSER NOTATION

References

Steinhaus, H. *Mathematical Snapshots, 3rd ed.* New York: Dover, pp. 28–9, 1999.

Megistron

A very LARGE NUMBER defined in terms of CIRCLE NOTATION by Steinhaus (1983) as $\textcircled{10}$.

See also MEGA, MOSER

References

Steinhaus, H. *Mathematical Snapshots, 3rd ed.* New York: Dover, pp. 28–9, 1999.

Mehler-Dirichlet Integral

$$P_n(\cos \alpha) = \frac{\sqrt{2}}{\pi} \int_0^\alpha \frac{\cos[(n + \tfrac{1}{2})\phi]}{\sqrt{\cos \phi - \cos \alpha}} \, d\phi,$$

where $P_n(x)$ is a LEGENDRE POLYNOMIAL.

Mehler-Fock Transform

The integral transform defined by

$$g(x) = \int_1^\infty t^{1/4 - \nu/2}(t-1)^{1/4 - \nu/2} P_{-1/2 + ix}^{\nu - 1/2}(2t-1) f(t) \, dt$$

(Samko *et al.* 1993, p. 761) or

$$g(x) = \int_1^\infty P_{-1/2 + ix}^k(t) f(t) \, dt$$

(Samko *et al.* 1993, p. 24), where $P_n(z)$ is a LEGENDRE POLYNOMIAL.

References

Marichev, O. I. Eqn. 8.42 in *Handbook of Integral Transforms of Higher Transcendental Functions: Theory and Algorithmic Tables.* Chichester, England: Ellis Horwood, 1982.

Samko, S. G.; Kilbas, A. A.; and Marichev, O. I. *Fractional Integrals and Derivatives.* Yverdon, Switzerland: Gordon and Breach, pp. 24 and 761, 1993.

Mehler Quadrature

JACOBI-GAUSS QUADRATURE

Mehler's Bessel Function Formula

$$J_0(x) = \frac{2}{\pi} \int_0^\infty \sin(x \cosh t) \, dt,$$

where $J_0(x)$ is a zeroth order BESSEL FUNCTION OF THE FIRST KIND.

References

Iyanaga, S. and Kawada, Y. (Eds.). *Encyclopedic Dictionary of Mathematics.* Cambridge, MA: MIT Press, p. 1472, 1980.

Mehler's Hermite Polynomial Formula

$$\sum_{n=0}^{\infty} \frac{H_n(x)H_n(y)}{n!}\left(\frac{1}{2}w\right)^n$$
$$= (1+4w^2)^{-1/2}\exp\left[\frac{2xyw-(x^2+y^2)w^2}{1-w^2}\right],$$

where $H_n(x)$ is a HERMITE POLYNOMIAL.

References

Almqvist, G. and Zeilberger, D. "The Method of Differentiating Under the Integral Sign." *J. Symb. Comput.* **10**, 571–91, 1990.
Foata, D. "A Combinatorial Proof of the Mehler Formula." *J. Comb. Th. Ser. A* **24**, 250–59, 1978.
Petkovsek, M.; Wilf, H. S.; and Zeilberger, D. *A = B.* Wellesley, MA: A. K. Peters, pp. 194–95, 1996.
Rainville, E. D. *Special Functions.* New York: Chelsea, p. 198, 1971.
Szego, G. *Orthogonal Polynomials, 4th ed.* Providence, RI: Amer. Math. Soc., p. 380, 1975.

Meijer's G-Function

A very general function which reduces to simpler special functions in many common cases. Meijer's G-function is defined by

$$G_{p,q}^{m,n}\left(x\left|\begin{matrix}a_1, \ldots, a_p\\b_1, \ldots, b_p\end{matrix}\right.\right) \equiv$$

$$\frac{1}{2\pi i}\int_{\gamma_L}\frac{\prod_{j=1}^m \Gamma(b_j-z)\prod_{j=1}^n \Gamma(1-a_j+z)}{\prod_{j=m+1}^q \Gamma(1-b_j+z)\prod_{j=n+1}^q \Gamma(q_j-z)}\,x^z\,dz,$$

(1)

where $\Gamma(z)$ is the GAMMA FUNCTION. The CONTOUR γ_L lies between the POLES of $\Gamma(1-a_i-z)$ and the POLES of $\Gamma(b_i+z)$ (Wolfram 1999, p. 772; Gradshteyn and Ryzhik 2000, pp. 896–03 and 1068–071). Prudnikov *et al.* (1990) contains an extensive nearly 200-page listing of formulas for the Meijer G-function. The function is built into *Mathematica 4.0* as MeijerG[{{a1, ..., an}, {a(n+1), ..., ap}}, {{b1, ..., bm}, {b(m+1), ..., bq}}, z].

Special cases include

$$G_{12}^{21}\left(z\left|\begin{matrix}1,1\\1,0\end{matrix}\right.\right) = \ln(z+1)$$

(2)

$$G_{12}^{21}\left(z\left|\begin{matrix}1,1\\1,1\end{matrix}\right.\right) = \frac{z}{z+1}$$

(3)

$$G_{10}^{02}\left(\frac{1}{2}z\left|0\frac{1}{2}\right.\right) = \frac{\cos(\sqrt{2z})}{\sqrt{\pi}}$$

(4)

$$G_{01}^{10}(z|1-a) = e^{-1/z}z^{-a}.$$

(5)

See also BARNES' G-FUNCTION, FOX'S H-FUNCTION, G-TRANSFORM, KAMPE DE FERIET FUNCTION, MACROBERT'S E-FUNCTION, RAMANUJAN G- AND G-FUNCTIONS

References

Adamchik, V. "The Evaluation of Integrals of Bessel Functions via G-Function Identities." *J. Comput. Appl. Math.* **64**, 283–90, 1995.
Erdélyi, A.; Magnus, W.; Oberhettinger, F.; and Tricomi, F. G. "Definition of the G-Function" et seq. §5.3–.6 in *Higher Transcendental Functions, Vol. 1.* New York: Krieger, pp. 206–22, 1981.
Gradshteyn, I. S. and Ryzhik, I. M. *Tables of Integrals, Series, and Products, 6th ed.* San Diego, CA: Academic Press, 2000.
Luke, Y. L. *The Special Functions and Their Approximations, 2 vols.* New York: Academic Press, 1969.
Mathai, A. M. *A Handbook of Generalized Special Functions for Statistical and Physical Sciences.* New York: Oxford University Press, 1993.
Meijer, C. S. "Multiplikationstheoreme für di Funktion $G_{p,q}^{m,n}(z)$." *Proc. Nederl. Akad. Wetensch.* **44**, 1062–070, 1941.
Meijer, C. S. "On the G-Function. II." *Proc. Nederl. Akad. Wetensch.* **49**, 344–56, 1946.
Meijer, C. S. "On the G-Function. III." *Proc. Nederl. Akad. Wetensch.* **49**, 457–69, 1946.
Meijer, C. S. "On the G-Function. IV." *Proc. Nederl. Akad. Wetensch.* **49**, 632–41, 1946.
Meijer, C. S. "On the G-Function. V." *Proc. Nederl. Akad. Wetensch.* **49**, 765–72, 1946.
Meijer, C. S. "On the G-Function. VI." *Proc. Nederl. Akad. Wetensch.* **49**, 936–43, 1946.
Meijer, C. S. "On the G-Function. VII." *Proc. Nederl. Akad. Wetensch.* **49**, 1063–072, 1946.
Meijer, C. S. "On the G-Function. VIII." *Proc. Nederl. Akad. Wetensch.* **49**, 1165–175, 1946.
Prudnikov, A. P.; Brychkov, Yu. A.; and Marichev, O. I. "Evaluation of Integrals and the Mellin Transform." *Itogi Nauki i Tekhniki, Seriya Matemat. Analiz* **27**, 3–46, 1989.
Prudnikov, A. P.; Marichev, O. I.; and Brychkov, Yu. A. *Integrals and Series, Vol. 3: More Special Functions.* Newark, NJ: Gordon and Breach, 1990.
Wolfram, S. *The Mathematica Book, 4th ed.* Cambridge, England: Cambridge University Press, 1999.

Meijer Transform

The INTEGRAL TRANSFORM

$$(Kf)(x) = \int_{-\infty}^{\infty}\sqrt{xt}K_\nu(xt)f(t)\,dt$$

where $K_\nu(x)$ is a MODIFIED BESSEL FUNCTION OF THE SECOND KIND.

References

Samko, S. G.; Kilbas, A. A.; and Marichev, O. I. *Fractional Integrals and Derivatives.* Yverdon, Switzerland: Gordon and Breach, p. 23, 1993.

Meissel's Formula

A modification of LEGENDRE'S FORMULA for the PRIME COUNTING FUNCTION $\pi(x)$. It starts with

$$\lfloor x \rfloor = 1 + \sum_{1 \le i \le a} \left\lfloor \frac{x}{p_i} \right\rfloor - \sum_{1 \le i \le j \le a} \left\lfloor \frac{x}{p_i p_j} \right\rfloor$$
$$+ \sum_{1 \le i \le j \le k \le a} \left\lfloor \frac{x}{p_i p_j p_k} \right\rfloor - \ldots + \pi(x) - a + P_2(x, a)$$
$$+ P_3(x, a) + \ldots, \qquad (1)$$

where $\lfloor x \rfloor$ is the FLOOR FUNCTION, $P_2(x, a)$ is the number of INTEGERS $p_i p_j \le x$ with $a + 1 \le j \le j$, and $P_3(x, a)$ is the number of INTEGERS $p_i p_j p_k < x$ with $a + 1 \le i \le j \le k$. Identities satisfied by the Ps include

$$P_2(x, a) = \sum \left[\pi\left(\frac{x}{p_i}\right) - (i - 1) \right] \qquad (2)$$

for $p_a < p_i \le \sqrt{x}$ and

$$P_3(x, a) = \sum_{i > a} P_2\left(\frac{x}{p_i}, a\right)$$

$$= \sum_{i=a+1}^{c} \sum_{j=i}^{\pi(\sqrt{x/p_i})} \left[\pi\left(\frac{x}{p_i p_j}\right) - (j - 1) \right]. \qquad (3)$$

Meissel's formula is

$$\pi(x) = \lfloor x \rfloor - \sum_{i=1}^{c} \left\lfloor \frac{x}{p_i} \right\rfloor + \sum_{1 \le i \le j \le c} \left\lfloor \frac{x}{p_i p_j} \right\rfloor - \ldots$$
$$+ \tfrac{1}{2}(b + c - 2)(b - c + 1) - \sum_{c \le i \le b} \pi\left(\frac{x}{p_i}\right), \qquad (4)$$

where

$$b \equiv \pi(x^{1/2}) \qquad (5)$$
$$c \equiv \pi(x^{1/3}). \qquad (6)$$

Taking the derivation one step further yields LEHMER'S FORMULA.

See also LEGENDRE'S FORMULA, LEHMER'S FORMULA, PRIME COUNTING FUNCTION

References
Gram. *Acta Math.* **17**, 301–14, 1893.
Hardy, G. H. *Ramanujan: Twelve Lectures on Subjects Suggested by His Life and Work, 3rd ed.* New York: Chelsea, p. 46, 1999.
Mathews, G. B. Ch. 10 in *Theory of Numbers.* New York: Chelsea, 1961.
Meissel. *Math. Ann.* **25**, 251–57, 1885.
Riesel, H. "Meissel's Formula." *Prime Numbers and Computer Methods for Factorization, 2nd ed.* Boston, MA: Birkhäuser, p. 12, 1994.
Séroul, R. "Meissel's Formula." §8.7.3 in *Programming for Mathematicians.* Berlin: Springer-Verlag, pp. 179–81, 2000.

Meixner-Pollaczek Polynomial

The hypergeometric orthogonal polynomial defined by

$$P_n^{(\lambda)}(x;\ \phi) = \frac{(2\lambda)_n}{n!} e^{in\phi} {}_2F_1(-n,\ \lambda + ix;\ 2\lambda;\ 1 - e^{-2i\phi}),$$

where $(x)_n$ is the POCHHAMMER SYMBOL. The first few are given by

$$P_0^{(\lambda)}(x;\ \phi) = 1$$
$$P_1^{(\lambda)}(x;\ \phi) = 2(\lambda \cos \phi + x \sin \phi)$$
$$P_2^{(\lambda)}(x;\ \phi) = x^2 + \lambda^2 + (\lambda^2 + \lambda - x^2) \cos(2\phi)$$
$$+ (1 + 2\lambda)x \sin(2\phi).$$

References
Koekoek, R. and Swarttouw, R. F. "Meixner-Pollaczek." §1.7 in *The Askey-Scheme of Hypergeometric Orthogonal Polynomials and its q-Analogue.* Delft, Netherlands: Technische Universiteit Delft, Faculty of Technical Mathematics and Informatics Report 98–7, pp. 37–8, 1998. ftp://www.twi.tudelft.nl/publications/tech-reports/1998/DUT-TWI-98-7.ps.gz.

Meixner Polynomial of the First Kind

Polynomials $m_k(x;\ \beta,\ c)$ which form the SHEFFER SEQUENCE for

$$g(t) = \left(\frac{1 - c}{1 - ce^t} \right)^{\beta} \qquad (1)$$

$$f(t) = \frac{1 - e^t}{c^{-1} - e^t} \qquad (2)$$

and have GENERATING FUNCTION

$$\sum \frac{m_k(x;\ \beta,\ c)}{k!} t^k = \left(1 - \frac{t}{c} \right)(1 - t)^{-x-\beta}. \qquad (3)$$

The are given in terms of the HYPERGEOMETRIC SERIES by

$$m_n^{(\gamma,\mu)}(x) = (\gamma)_n {}_2F_1(-n,\ -x;\ \gamma;\ 1 - \mu^{-1}), \qquad (4)$$

where $(x)_n$ is the POCHHAMMER SYMBOL (Koepf 1998, p. 115). The first few are

$$m_0(x;\ \beta,\ c) = 1$$
$$m_1(x;\ \beta,\ c) = b + x\left(1 - \frac{1}{c} \right)$$
$$m_2(x;\ \beta,\ c)$$
$$= \frac{b(b + 1)c^2 + (c - 1)(2bc + c + 1)x + (c - 1)^2 x^2}{c^2}$$

Koekoek and Swarttouw (1998) defined the Meixner polynomials without the POCHHAMMER SYMBOL as

$$M'_n(x;\ \beta,\ c) = {}_2F_1(-n,\ -x;\ \beta;\ 1-1/c). \tag{5}$$

The KRAWTCHOUK POLYNOMIALS are a special case of the Meixner polynomials of the first kind.

See also KRAWTCHOUK POLYNOMIAL, MEIXNER POLYNOMIAL OF THE SECOND KIND, SHEFFER SEQUENCE

References

Chihara, T. S. *An Introduction to Orthogonal Polynomials.* New York: Gordon and Breach, p. 175, 1978.

Erdélyi, A.; Magnus, W.; Oberhettinger, F.; and Tricomi, F. G. *Higher Transcendental Functions, Vol. 2.* New York: Krieger, pp. 224–25, 1981.

Koekoek, R. and Swarttouw, R. F. "Meixner." §1.9 in *The Askey-Scheme of Hypergeometric Orthogonal Polynomials and its q-Analogue.* Delft, Netherlands: Technische Universiteit Delft, Faculty of Technical Mathematics and Informatics Report 98–7, pp. 45–6, 1998. ftp://www.twi.tudelft.nl/publications/tech-reports/1998/DUT-TWI-98–7.ps.gz.

Koepf, W. *Hypergeometric Summation: An Algorithmic Approach to Summation and Special Function Identities.* Braunschweig, Germany: Vieweg, p. 115, 1998.

Roman, S. *The Umbral Calculus.* New York: Academic Press, 1984.

Szego, G. *Orthogonal Polynomials, 4th ed.* Providence, RI: Amer. Math. Soc., p. 35, 1975.

Meixner Polynomial of the Second Kind

The polynomials $M_k(x;\ \delta,\ \eta)$ which form the SHEFFER SEQUENCE for

$$g(t) = \{[1 + \delta f(t)]^2 + [f(t)]^2\}^{\eta/2} \tag{1}$$

$$f(t) = \tan\left(\frac{t}{1 + \delta t}\right) \tag{2}$$

which have GENERATING FUNCTION

$$\sum_{k=0}^{\infty} \frac{M_k(x;\ \delta,\ \eta)}{k!} t^k$$

$$= [(1 + \delta t)^2]^{-\eta/2} \exp\left(\frac{x \tan^{-1} t}{1 - \delta \tan^{-1} t}\right). \tag{3}$$

The first few are

$$M_0(x;\ \delta,\ \eta) = 1$$
$$M_1(x;\ \delta,\ \eta) = x - \delta\eta$$
$$M_2(x;\ \delta,\ \eta) = x^2 + 2\delta(1 - \eta)x + \eta[(\eta + 1)\delta^2 - 1].$$

See also MEIXNER POLYNOMIAL OF THE FIRST KIND, SHEFFER SEQUENCE

References

Chihara, T. S. *An Introduction to Orthogonal Polynomials.* New York: Gordon and Breach, p. 179, 1978.

Roman, S. *The Umbral Calculus.* New York: Academic Press, 1984.

Mellin-Barnes Integral

A type of integral containing gamma functions in their integrands. A typical such integral is given by

$$f(z) = \frac{1}{2\pi i} \int_{\gamma - i\infty}^{\gamma + i\infty} \frac{\Gamma(a_1 + A_1 s)\ldots\Gamma(a_n + A_n s)}{\Gamma(c_1 + C_1 s)\ldots\Gamma(c_p + C_p s)}$$

$$\times \frac{\Gamma(b_1 - B_1 s)\ldots\Gamma(b_n - B_n s)}{\Gamma(d_1 - D_1 s)\ldots\Gamma(d_q - D_q s)} z^s\, ds,$$

where γ is real, A_j, B_j, C_j, and D_j are positive, and the CONTOUR is a straight line parallel to the IMAGINARY AXIS with indentations if necessary to avoid poles of the integrand.

References

Barnes, E. W. "A New Development in the Theory of the Hypergeometric Functions." *Proc. London Math. Soc.* **6**, 141–77, 1908.

Dixon, A. L. and Ferrar, W. L. "A Class of Discontinuous Integrals." *Quart. J. Math. (Oxford Ser.)* **7**, 81–6, 1936.

Erdélyi, A.; Magnus, W.; Oberhettinger, F.; and Tricomi, F. G. "Mellin-Barnes Integrals." §1.19 in *Higher Transcendental Functions, Vol. 1.* New York: Krieger, pp. 49–0, 1981.

Mellin, H. "Om Definita Integraler." *Acta Societatis Scientiarum Fennicae* **20**, No. 7, 1–9, 1895.

Mellin, H. "Abrißeiner einheitlichen Theorie der Gamma- und der hypergeometrischen Funktionen." *Math. Ann.* **68**, 305–37, 1909.

Pincherle, S. *Atti d. R. Academia dei Lincei, Ser. 4, Rendiconti* **4**, 694–00 and 792–99, 1888.

Ramanujan, S. *Collected Papers.* New York: Chelsea, p. 216, 1962.

Whittaker, E. T. and Watson, G. N. *A Course in Modern Analysis, 4th ed.* Cambridge, England: Cambridge University Press, p. 289, 1990.

Mellin's Formula

$$\frac{e^{\gamma\psi_0(x)}\Gamma(x)}{\Gamma(x + \gamma)} = \prod_{n=0}^{\infty}\left(1 + \frac{\gamma}{n + x}\right)e^{-\gamma/(n+x)}, \tag{1}$$

where $\psi_0(x)$ is the DIGAMMA FUNCTION, $\Gamma(x)$ is the GAMMA FUNCTION, and γ is the EULER-MASCHERONI CONSTANT.

See also DIGAMMA FUNCTION, GAMMA FUNCTION

References

Erdélyi, A.; Magnus, W.; Oberhettinger, F.; and Tricomi, F. G. *Higher Transcendental Functions, Vol. 1.* New York: Krieger, p. 6, 1981.

Mellin Transform

The INTEGRAL TRANSFORM defined by

$$\phi(z) = \int_0^{\infty} t^{z-1} f(t)\, dt \tag{1}$$

$$f(t) = \frac{1}{2\pi i} \int_{c-i\infty}^{c+i\infty} t^{-z} \phi(z)\, dz. \tag{2}$$

The transform $\phi(z)$ exists if the integral

$$\int_0^\infty |f(x)|x^{k-1}\, dx \qquad (3)$$

is bounded for some $k > 0$, in which case the inverse $f(t)$ exists with $c > k$. The functions $\phi(z)$ and $f(t)$ are called a Mellin transform pair, and either can be computed if the other is known.

The following table gives Mellin transforms of common functions (Bracewell 1999, p. 255). Here, δ is the DELTA FUNCTION, $H(x)$ is the HEAVISIDE STEP FUNCTION, $\Gamma(z)$ is the GAMMA FUNCTION, $B(z; a, b)$ is the INCOMPLETE BETA FUNCTION, erfc z is the complementary error function ERFC, and Si(z) is the SINE INTEGRAL.

$f(t)$	$\phi(z)$	convergence
$\delta(t-a)$	a^{z-1}	
$H(t-a)$	$-\dfrac{a^z}{z}$	$a>0,\ z<0$
$H(a-t)$	$\dfrac{a^z}{z}$	$a>0,\ z>0$
$t^n H(t-a)$	$-\dfrac{a^{n+z}}{n+z}$	$a>0,$
		$\Re[z+n]<0$
$t^n H(a-t)$	$\dfrac{a^{n+z}}{n+z}$	$a>0,$
		$\Re[n+z]>0$
e^{-at}	$a^{-z}\Gamma(z)$	$\Re[a],\ \Re[z]>0$
e^{-t^2}	$\frac{1}{2}\Gamma\left(\frac{1}{2}z\right)$	$\Re[z]>0$
$\sin t$	$\Gamma(z)\sin\left(\frac{1}{2}\pi z\right)$	$-1<\Re[z]<1$
$\cos t$	$\Gamma(z)\cos\left(\frac{1}{2}\pi z\right)$	$0<\Re[z]<1$
$\dfrac{1}{1+t}$	$\pi\csc(\pi z)$	$0<\Re[z]<1$
$\dfrac{1}{(1+t)^a}$	$\dfrac{\Gamma(a-z)\Gamma(z)}{\Gamma(a)}$	$\Re[a-z]>0,$
		$\Re[z]>0$
$\dfrac{1}{1+t^2}$	$\frac{1}{2}\pi\csc\left(\frac{1}{2}\pi z\right)$	$0<\Re[z]<2$
$(1-t)^{a-1}H(1-t)$	$\dfrac{\Gamma(a)\Gamma(z)}{\Gamma(a+z)}$	$\Re[a],\ \Re[z]>0$

$(t-1)^{-a}H(t-1)$	$\dfrac{\Gamma(1-a)\Gamma(a-z)}{\Gamma(1-x)}$	$\Re[a-z]>0,$
		$\Re[a]<1$
$\ln(1+t)$	$\dfrac{\pi\csc(\pi z)}{z}$	$-1<\Re[z]<0$
$\frac{1}{2}\pi-\tan^{-1}t$	$\dfrac{\pi\sec(\frac{1}{2}\pi z)}{2z}$	$0<\Re[z]<1$
erfc t	$\dfrac{\Gamma(\frac{1}{2}(1+z))}{\sqrt{\pi}z}$	$\Re[z]>0$
Si(t)	$-\dfrac{1}{z}\Gamma(z)\sin(\frac{1}{2}\pi z)$	$\Re[z]>-1$
$\dfrac{t^a}{1-t}H(t-a)$	$-B(a^{-1};1-a-z;0)$	$a>1,\ \Re[a+z]<1$

See also FOURIER TRANSFORM, INTEGRAL TRANSFORM, STRASSEN FORMULAS

References

Arfken, G. *Mathematical Methods for Physicists, 3rd ed.* Orlando, FL: Academic Press, p. 795, 1985.

Bracewell, R. *The Fourier Transform and Its Applications, 3rd ed.* New York: McGraw-Hill, pp. 254–57, 1999.

Gradshteyn, I. S. and Ryzhik, I. M. "Mellin Transform." §17.41 in *Tables of Integrals, Series, and Products, 6th ed.* San Diego, CA: Academic Press, pp. 1193–197, 2000.

Morse, P. M. and Feshbach, H. *Methods of Theoretical Physics, Part I.* New York: McGraw-Hill, pp. 469–71, 1953.

Oberhettinger, F. *Tables of Mellin Transforms.* New York: Springer-Verlag, 1974.

Prudnikov, A. P.; Brychkov, Yu. A.; and Marichev, O. I. "Evaluation of Integrals and the Mellin Transform." *Itogi Nauki i Tekhniki, Seriya Matemat. Analiz* **27**, 3–46, 1989.

Zwillinger, D. (Ed.). *CRC Standard Mathematical Tables and Formulae.* Boca Raton, FL: CRC Press, p. 567, 1995.

Melnikov-Arnold Integral

$$A_m(\lambda) = \int_{-\infty}^\infty \cos\left[\frac{1}{2}\,m\,\phi(t) - \lambda t\right]dt,$$

where the function

$$\phi(t) \equiv 4\tan^{-1}(e^t) - \pi$$

describes the motion along the pendulum SEPARATRIX. Chirikov (1979) has shown that this integral has the approximate value

$$A_m(\lambda) \approx \begin{cases} \dfrac{4\pi(2\lambda)^{m-1}}{\Gamma(m)}\,e^{-\pi\lambda/2} & \text{for } \lambda>0 \\[2ex] \dfrac{4e^{-\pi|\lambda|/2}}{(2|l|)^{m+1}}\,\Gamma(m+1)\sin(\pi m) & \text{for } \lambda<0. \end{cases}$$

References

Chirikov, B. V. "A Universal Instability of Many-Dimensional Oscillator Systems." *Phys. Rep.* **52**, 264–79, 1979.

Melodic Sequence

If a_1, a_2, a_3, ... is an ARTISTIC SEQUENCE, then $1/a_1$, $1/a_2$, $1/a_3$, ... is a melodic sequence. The RECURRENCE RELATION obeyed by melodic series is

$$b_{i+3} = \frac{b_i b_{i+2}^2}{b_{i+1}^2} + \frac{b_{i+2}^2}{b_{i+1}} - b_{i+2}.$$

See also ARTISTIC SEQUENCE

References

Duffin, R. J. "On Seeing Progressions of Constant Cross Ratio." *Amer. Math. Monthly* **100**, 38–7, 1993.

MEM

MAXIMUM ENTROPY METHOD

Memoryless

A variable x is memoryless with respect to t if, for all s with $t \neq 0$,

$$P(x > s + t | x > t) = P(x > s). \tag{1}$$

Equivalently,

$$\frac{P(x > s + t, \, x > t)}{P(x > t)} = P(x > s) \tag{2}$$

$$P(x > s + t) = P(x > s)P(x > t). \tag{3}$$

The EXPONENTIAL DISTRIBUTION, which satisfies

$$P(x > t) = e^{-\lambda t} \tag{4}$$

$$P(x > s + t) = e^{-\lambda(s+t)}, \tag{5}$$

and therefore

$$P(x > s + t) = P(x > s)P(x > t) = e^{-\lambda s}e^{-\lambda t} = e^{-\lambda(s+t)}, \tag{6}$$

is the only memoryless random distribution.

See also EXPONENTIAL DISTRIBUTION

Ménage Number

MARRIED COUPLES PROBLEM

Ménage Problem

MARRIED COUPLES PROBLEM

Menasco's Theorem

For a BRAID with M strands, R components, P positive crossings, and N negative crossings,

$$\begin{cases} P - N \leq U_+ + M - R & \text{if } P \geq N \\ P - N \leq U_- + M - R & \text{if } P \leq N, \end{cases}$$

where $U_\pm$ are the smallest number of positive and negative crossings which must be changed to crossings of the opposite sign. These inequalities imply

BENNEQUIN'S CONJECTURE. Menasco's theorem can be extended to arbitrary knot diagrams.

See also BENNEQUIN'S CONJECTURE, BRAID, UNKNOTTING NUMBER

References

Cipra, B. "From Knot to Unknot." *What's Happening in the Mathematical Sciences, Vol. 2.* Providence, RI: Amer. Math. Soc., pp. 8–3, 1994.
Menasco, W. W. "The Bennequin-Milnor Unknotting Conjectures." *C. R. Acad. Sci. Paris Sér. I Math.* **318**, 831–36, 1994.

Menelaus' Theorem

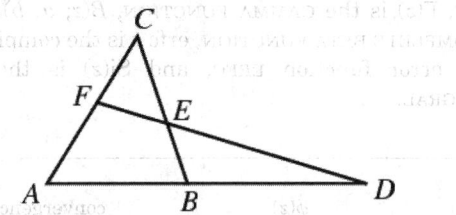

For TRIANGLES in the PLANE,

$$AD \cdot BE \cdot CF = BD \cdot CE \cdot AF. \tag{1}$$

For SPHERICAL TRIANGLES,

$$\sin AD \cdot \sin BE \cdot \sin CF = \sin BD \cdot \sin CF \cdot \sin AF \tag{2}$$

This can be generalized to n-gons $P = [V_1, \ldots, V_n]$, where a transversal cuts the side $V_i V_{i+1}$ in W_i for $i = 1, \ldots, n$, by

$$\prod_{i=1}^{n} \left[\frac{V_i W_i}{W_i V_{i+1}} \right] = (-1)^n. \tag{3}$$

Here, $AD \| CD$ and

$$\left[\frac{AB}{CD} \right] \tag{4}$$

is the ratio of the lengths $[A, B]$ and $[C, D]$ with a PLUS or MINUS SIGN depending if these segments have the same or opposite directions (Grünbaum and Shepard 1995). The case $n = 3$ is PASCH'S AXIOM.

See also CEVA'S THEOREM, HOEHN'S THEOREM, PASCH'S AXIOM

References

Beyer, W. H. (Ed.). *CRC Standard Mathematical Tables, 28th ed.* Boca Raton, FL: CRC Press, p. 122, 1987.
Coxeter, H. S. M. and Greitzer, S. L. "Menelaus's Theorem." §3.4 in *Geometry Revisited.* Washington, DC: Math. Assoc. Amer., pp. 66–7, 1967.
Durell, C. V. *Modern Geometry: The Straight Line and Circle.* London: Macmillan, pp. 42–4, 1928.
Graustein, W. C. *Introduction to Higher Geometry.* New York: Macmillan, p. 81, 1930.
Grünbaum, B. and Shepard, G. C. "Ceva, Menelaus, and the Area Principle." *Math. Mag.* **68**, 254–68, 1995.

Honsberger, R. "The Theorem of Menelaus." Ch. 13 in *Episodes in Nineteenth and Twentieth Century Euclidean Geometry*. Washington, DC: Math. Assoc. Amer., pp. 147–54, 1995.

Pedoe, D. *Circles: A Mathematical View, rev. ed.* Washington, DC: Math. Assoc. Amer., p. xxi, 1995.

Wells, D. *The Penguin Dictionary of Curious and Interesting Geometry*. London: Penguin, p. 150, 1991.

Menger's n-Arc Theorem

Let G be a GRAPH with A and B two disjoint n-tuples of VERTICES. Then either G contains n pairwise disjoint AB-paths, each connecting a point of A and a point of B, or there exists a set of fewer than n VERTICES that separate A and B.

Harary (1994, pp. 47) states the theorem as "the minimum number of points separating two nonadjacent points s and t is the maximum number of disjoint $s-t$ paths." Skiena (1990, p. 178) states the theorem as "a graph is K-CONNECTED GRAPH IFF every pair of vertices is joined by at least k vertex-disjoint paths" (Menger 1927, Whitney 1932).

See also K-CONNECTED GRAPH

References

Harary, F. *Graph Theory*. Reading, MA: Addison-Wesley, 1994.

Menger, K. "Zur allgemeinen Kurventheorie." *Fund. Math.* **10**, 95–15, 1927.

Menger, K. *Kurventheorie*. Leipzig, Germany: Teubner, 1932.

Whitney, H. "Congruent Graphs and the Connectivity of Graphs." *Amer. J. Math.* **54**, 150–68, 1932.

Menger Sponge

A FRACTAL which is the 3-D analog of the SIERPINSKI CARPET. Let N_n be the number of filled boxes, L_n the length of a side of a hole, and V_n the fractional VOLUME after the nth iteration.

$$N_n = 20^n \tag{1}$$

$$L_n = \left(\frac{1}{3}\right)^n = 3^{-n} \tag{2}$$

$$V_n = L_n^3 N_n = \left(\frac{20}{27}\right)^n. \tag{3}$$

The CAPACITY DIMENSION is therefore

$$d_{\mathrm{cap}} = -\lim_{n\to\infty}\frac{\ln N_n}{\ln L_n} = -\lim_{n\to\infty}\frac{\ln(20^n)}{\ln(3^{-n})} = \frac{\ln 20}{\ln 3}$$

$$= \frac{\ln(2^5 \cdot 5)}{\ln 3} = \frac{2\ln 2 + \ln 5}{\ln 3} = 2.726833028\ldots \tag{4}$$

J. Mosely is leading an effort to construct a large Menger sponge out of old business cards.

See also SIERPINSKI CARPET, TETRIX

References

Dickau, R. "Sierpinski-Menger Sponge Code and Graphic." http://www.mathsource.com/cgi-bin/msitem22?0206–10.

Dickau, R. M. "Menger (Sierpinski) Sponge." http://forum.swarthmore.edu/advanced/robertd/sponge.html.

Mosely, J. "Menger's Sponge (Depth 3)." http://world.std.com/~j9/sponge/.

Weisstein, E. W. "Fractals." MATHEMATICA NOTEBOOK FRACTAL.M.

Werbeck, S. "A Journey into Menger's Sponge." http://pages.hotbot.com/arts/werbeck/.

Menger's Theorem

MENGER'S N-ARC THEOREM

Menn's Surface

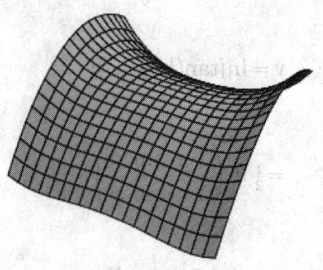

A surface given by the PARAMETRIC EQUATIONS

$$x(u,\ v) = u$$

$$y(u,\ v) = v$$

$$x(u,\ v) = au^4 + u^2 v - v^2.$$

References

Gray, A. *Modern Differential Geometry of Curves and Surfaces with Mathematica, 2nd ed.* Boca Raton, FL: CRC Press, p. 956, 1997.

Mensuration Formula

A mensuration formula is simply a formula for computing the length-related properties of an object (such as AREA, CIRCUMRADIUS, etc., of a POLYGON) based on other known lengths, areas, etc. Beyer (1987) gives a collection of such formulas for various plane and solid geometric figures.

References

Beyer, W. H. *CRC Standard Mathematical Tables, 28th ed.* Boca Raton, FL: CRC Press, pp. 121–33, 1987.

Mercator Projection

The following equations place the X-AXIS of the projection on the equator and the Y-AXIS at LONGITUDE λ_0, where λ is the LONGITUDE and ϕ is the LATITUDE.

$$x = \lambda - \lambda_0 \qquad (1)$$

$$y = \ln[\tan(\tfrac{1}{4}\pi + \tfrac{1}{2}\phi)] \qquad (2)$$

$$= \tfrac{1}{2} \ln\left(\frac{1 + \sin \phi}{1 - \sin \phi}\right) \qquad (3)$$

$$= \sinh^{-1}(\tan \phi) \qquad (4)$$

$$= \tanh^{-1}(\sin \phi) \qquad (5)$$

$$= \ln(\tan \phi + \sec \phi). \qquad (6)$$

The inverse FORMULAS are

$$\phi = 2 \tan^{-1}(e^y) - \tfrac{1}{2}\pi \qquad (7)$$

$$= \tan^{-1}(\sinh y) \qquad (8)$$

$$= \text{gd } y \qquad (9)$$

$$\lambda = x + \lambda_0, \qquad (10)$$

where gd y is the GUDERMANNIAN FUNCTION. LOXODROMES are straight lines and GREAT CIRCLES are curved.

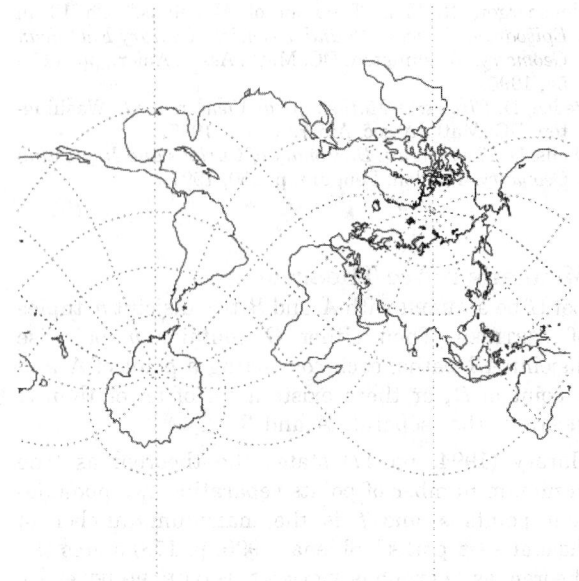

An oblique form of the Mercator projection is illustrated above. It has equations

$$x = \frac{\tan^{-1}[\tan \phi \cos \phi_p + \sin \phi_p \sin(\lambda - \lambda_0)]}{\cos(\lambda - \lambda_0)} \qquad (11)$$

$$y = \tfrac{1}{2} \ln\left(\frac{1 + A}{1 - A}\right) = \tanh^{-1} A, \qquad (12)$$

where

$$\lambda_p =$$

$$\tan^{-1}\left(\frac{\cos \phi_1 \sin \phi_2 \cos \lambda_1 - \sin \phi_1 \cos \phi_2 \cos \lambda_2}{\sin \phi_1 \cos \phi_2 \sin \lambda_2 - \cos \phi_1 \sin \phi_2 \sin \lambda_1}\right) \qquad (13)$$

$$\phi_p = \tan^{-1}\left(-\frac{\cos(\lambda_p - \lambda_1)}{\tan \phi_1}\right) \qquad (14)$$

$$A = \sin \phi_p \sin \phi - \cos \phi_p \cos \phi \sin(\lambda - \lambda_0). \qquad (15)$$

The inverse FORMULAS are

$$\phi = \sin^{-1}\left(\sin \phi_p \tanh y + \frac{\cos \phi_p \sin x}{\cosh y}\right) \qquad (16)$$

$$\lambda = \lambda_0 + \tan^{-1}\left(\frac{\sin \phi_p \sin x - \cos \phi_p \sinh y}{\cos x}\right). \qquad (17)$$

There is also a transverse form of the Mercator projection, illustrated above (Deetz and Adams 1934, Snyder 1987). It is given by the equations

$$x = \tfrac{1}{2} \ln\left(\frac{1+B}{1-B}\right) = \tanh^{-1} B \tag{18}$$

$$y = \tan^{-1}\left[\frac{\tan\phi}{\cos(\lambda-\lambda_0)}\right] - \phi_0 \tag{19}$$

$$\phi = \sin^{-1}\left(\frac{\sin D}{\cosh x}\right) \tag{20}$$

$$\lambda = \lambda_0 + \tan^{-1}\left(\frac{\sinh x}{\cos D}\right), \tag{21}$$

where

$$B \equiv \cos\phi\,\sin(\lambda-\lambda_0) \tag{22}$$

$$D \equiv y + \phi_0. \tag{23}$$

Finally, the "universal transverse Mercator projection" is a MAP PROJECTION which maps the SPHERE into 60 zones of 6° each, with each zone mapped by a transverse Mercator projection with central MERIDIAN in the center of the zone. The zones extend from 80° S to 84° N (Dana).

See also GUDERMANNIAN FUNCTION, SPHERICAL SPIRAL

References

Dana, P. H. "Map Projections." http://www.colorado.edu/geography/gcraft/notes/mapproj/mapproj_f.html.

Deetz, C. H. and Adams, O. S. *Elements of Map Projection with Applications to Map and Chart Construction, 4th ed.* Washington, DC: U. S. Coast and Geodetic Survey Special Pub. 68, 1934.

Snyder, J. P. *Map Projections--A Working Manual.* U. S. Geological Survey Professional Paper 1395. Washington, DC: U. S. Government Printing Office, pp. 38–5, 1987.

Mercator Series

The TAYLOR SERIES for the NATURAL LOGARITHM

$$\ln(1+x) = x - \tfrac{1}{2}x^2 + \tfrac{1}{3}x^3 - \cdots$$

which was found by Newton, but independently discovered and first published by Mercator in 1668.

See also LOGARITHMIC NUMBER, NATURAL LOGARITHM

Mercer's Theorem

RIEMANN-LEBESGUE LEMMA

Meredith Graph

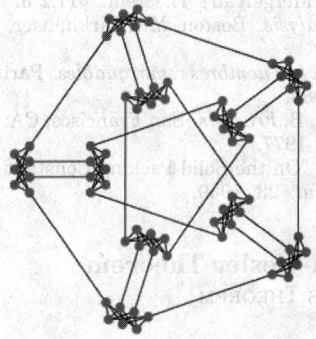

A counterexample to the conjecture that every 4-regular 4-connected graph is HAMILTONIAN.

See also HAMILTONIAN GRAPH

References

Bondy, J. A. and Murty, U. S. R. *Graph Theory with Applications.* New York: North Holland, pp. 236–39, 1976.

Meredith, G. H. J. "Regular *n*-valent *n*-connected nonhamiltonian non-*n*-edge-colorable Graphs." *J. Combin. Th. B* **14**, 55–0, 1973.

Mergelyan's Theorem

Mergelyan's theorem can be stated as follows (Krantz 1999). Let $K \subseteq \mathbb{C}$ be compact and suppose $\mathbb{C}^* \backslash K$ has only finitely many connected components. If $f \in C(K)$ is holomorphic on the interior of K and if $\epsilon > 0$, then there is a RATIONAL FUNCTION $r(z)$ with poles in $\mathbb{C}^* \backslash K$ such that

$$\max_{z\in K} |f(z) - r(z)| < \epsilon. \tag{1}$$

A consequence is that if $P = \{D_1, D_2, \ldots\}$ is an infinite set of disjoint OPEN DISKS D_n of radius r_n such that the union is almost the unit DISK. Then

$$\sum_{n=1}^{\infty} r_n = \infty. \tag{2}$$

Define

$$M_x(P) \equiv \sum_{n=1}^{\infty} r_n^x. \tag{3}$$

Then there is a number $e(P)$ such that $M_x(P)$ diverges for $x < e(P)$ and converges for $x > e(P)$. The above theorem gives

$$1 < e(P) < 2. \tag{4}$$

There exists a constant which improves the inequality, and the best value known is

$$S = 1.306951\ldots. \tag{5}$$

See also RUNGE'S THEOREM

References

Krantz, S. G. "Mergelyan's Theorem." §11.2 in *Handbook of Complex Analysis*. Boston, MA: Birkhäuser, pp. 146–47, 1999.
Le Lionnais, F. *Les nombres remarquables*. Paris: Hermann, pp. 36–7, 1983.
Mandelbrot, B. B. *Fractals*. San Francisco, CA: W. H. Freeman, p. 187, 1977.
Melzack, Z. A. "On the Solid Packing Constant for Circles." *Math. Comput.* **23**, 1969.

Mergelyan-Wesler Theorem

MERGELYAN'S THEOREM

Meridian

A line of constant LONGITUDE on a SPHEROID (or SPHERE). More generally, a meridian of a SURFACE OF REVOLUTION is the intersection of the surface with a PLANE containing the axis of revolution.

See also LATITUDE, LONGITUDE, PARALLEL (SURFACE OF REVOLUTION), SURFACE OF REVOLUTION

References

Gray, A. *Modern Differential Geometry of Curves and Surfaces with Mathematica, 2nd ed.* Boca Raton, FL: CRC Press, p. 238, 1997.

Meromorphic Function

A meromorphic function is a single-valued function that is ANALYTIC in all but possibly a discrete subset of its domain, and at those singularities it must go to infinity like a POLYNOMIAL (i.e., these exceptional points must be POLES and not ESSENTIAL SINGULARITIES). A simpler definition states that a meromorphic function is a function $f(z)$ OF THE FORM

$$f(z) = \frac{g(z)}{h(z)}$$

where $g(z)$ and $h(z)$ are ENTIRE FUNCTIONS with $h(z) \neq 0$ (Krantz 1999, p. 64).

A meromorphic function therefore has only possibly finite, isolated POLES and zeros and no ESSENTIAL SINGULARITIES in its domain. A meromorphic function with an infinite number of poles is exemplified by $\csc(1/z)$ on the PUNCTURED $U = D \setminus \{0\}$, where D is the open unit disk.

An equivalent definition of a meromorphic function is a complex analytic MAP to the RIEMANN SPHERE.

The word derives from the Greek $\mu\varepsilon\rho\sigma\varsigma$ (*meros*), meaning "part," and $\mu\sigma\rho\varphi\eta$ (*morphe*), meaning "form" or "appearance."

See also ANALYTIC FUNCTION, ENTIRE FUNCTION, ESSENTIAL SINGULARITY, HOLOMORPHIC FUNCTION, POLE, REAL ANALYTIC FUNCTION, RIEMANN SPHERE

References

Knopp, K. "Meromorphic Functions." Ch. 2 in *Theory of Functions Parts I and II, Two Volumes Bound as One, Part II*. New York: Dover, pp. 34–7, 1996.
Krantz, S. G. "Meromorphic Functions and Singularities at Infinity." §4.6 in *Handbook of Complex Analysis*. Boston, MA: Birkhäuser, pp. 63–8, 1999.
Morse, P. M. and Feshbach, H. *Methods of Theoretical Physics, Part I*. New York: McGraw-Hill, pp. 382–83, 1953.

Mersenne Number

A number OF THE FORM

$$M_n \equiv 2^n - 1 \tag{1}$$

for n an INTEGER is known as a Mersenne number. The Mersenne numbers are therefore 2-REPDIGITS, and also the numbers obtained by setting $x = 1$ in a FERMAT POLYNOMIAL. The first few are 1, 3, 7, 15, 31, 63, 127, 255, ... (Sloane's A000225).

The number of digits D in the Mersenne number M_n is

$$D = \lfloor \log(2^n - 1) + 1 \rfloor, \tag{2}$$

where $\lfloor x \rfloor$ is the FLOOR FUNCTION, which, for large n, gives

$$D \approx \lfloor n \log 2 + 1 \rfloor \approx \lfloor 0.301029n + 1 \rfloor$$
$$= \lfloor 0.301029n \rfloor + 1. \tag{3}$$

In order for the Mersenne number M_n to be PRIME, n must be PRIME. This is true since for COMPOSITE n with factors r and s, $n = rs$. Therefore, $2^n - 1$ can be written as $2^{rs} - 1$, which is a BINOMIAL NUMBER and can be factored. Since the most interest in Mersenne numbers arises from attempts to factor them, many authors prefer to define a Mersenne number as a number of the above form

$$M_p = 2^p - 1 \tag{4}$$

but with p restricted to PRIME values.

The search for MERSENNE PRIMES is one of the most computationally intensive and actively pursued areas of advanced and distributed computing.

See also CUNNINGHAM NUMBER, DOUBLE MERSENNE NUMBER, EBERHART'S CONJECTURE, FERMAT NUMBER, LUCAS-LEHMER TEST, MERSENNE PRIME, PERFECT NUMBER, REPUNIT, RIESEL NUMBER, SIERPINSKI NUMBER OF THE SECOND KIND, SOPHIE GERMAIN PRIME, SUPERPERFECT NUMBER, WHEAT AND CHESSBOARD PROBLEM, WIEFERICH PRIME

References

Dickson, L. E. *History of the Theory of Numbers, Vol. 1: Divisibility and Primality*. New York: Chelsea, p. 13, 1952.

Gardner, M. "Mathematical Games: About the Remarkable Similarity between the Icosian Game and the Towers of Hanoi." *Sci. Amer.* **196**, 150–56, May 1957.

Hardy, G. H. and Wright, E. M. *An Introduction to the Theory of Numbers, 5th ed.* Oxford, England: Clarendon Press, pp. 15–6 and 22, 1979.

Pappas, T. "Mersenne's Number." *The Joy of Mathematics*. San Carlos, CA: Wide World Publ./Tetra, p. 211, 1989.

Shanks, D. *Solved and Unsolved Problems in Number Theory, 4th ed.* New York: Chelsea, pp. 14, 18–9, 22, and 29–0, 1993.

Sloane, N. J. A. Sequences A000225/M2655 in "An On-Line Version of the Encyclopedia of Integer Sequences." http://www.research.att.com/~njas/sequences/eisonline.html.

Steinhaus, H. *Mathematical Snapshots, 3rd ed.* New York: Dover, pp. 23–4, 1999.

Mersenne Prime

A MERSENNE NUMBER which is PRIME is called a Mersenne prime. In order for the Mersenne number M_n defined by

$$M_n \equiv 2^n - 1$$

for n an INTEGER to be PRIME, n must be PRIME. This is true since for COMPOSITE n with factors r and s, $n = rs$. Therefore, $2^n - 1$ can be written as $2^{rs} - 1$, which is a BINOMIAL NUMBER and can be factored. Every MERSENNE PRIME gives rise to a PERFECT NUMBER. The first few Mersenne primes are 3, 7, 31, 127, 8191, 131071, 524287, 2147483647, ... (Sloane's A000668) corresponding to $n = 2, 3, 5, 7, 13, 17, 19, 31, 61, 89, ...$ (Sloane's A000043).

If $n \equiv 3 \pmod 4$ is a PRIME, then $2n + 1$ DIVIDES M_n IFF $2n + 1$ is PRIME. It is also true that PRIME divisors of $2^p - 1$ must have the form $2kp + 1$ where k is a POSITIVE INTEGER and simultaneously of either the form $8n + 1$ or $8n - 1$ (Uspensky and Heaslet). A PRIME factor p of a Mersenne number $M_q = 2^q - 1$ is a WIEFERICH PRIME IFF $p^2 | 2^q - 1$, Therefore, MERSENNE PRIMES are *not* WIEFERICH PRIMES. All known Mersenne numbers M_p with p PRIME are SQUARE-FREE. However, Guy (1994) believes that there are M_p which are not SQUAREFREE.

TRIAL DIVISION is often used to establish the COMPOSITENESS of a potential Mersenne prime. This test immediately shows M_p to be COMPOSITE for $p = 11, 23, 83, 131, 179, 191, 239,$ and 251 (with small factors 23, 47, 167, 263, 359, 383, 479, and 503, respectively). A much more powerful primality test for M_p is the LUCAS-LEHMER TEST.

It has been conjectured that there exist an infinite number of Mersenne primes, although finding them is computationally very challenging. The table below gives the index p of known Mersenne primes (Sloane's A000043) M_p, together with the number of digits, discovery years, and discoverer. A similar table has

been compiled by C. Caldwell. Note that the region after the 35th known Mersenne prime has not been completely searched, so identification of "the" 36th and larger Mersenne primes are tentative. L. Welsh maintains an extensive bibliography and history of Mersenne numbers. G. Woltman has organized a distributed search program via the Internet in which hundreds of volunteers use their personal computers to perform pieces of the search.

#	p	Digits	Year	Discoverer (Reference)
1	2	1	Antiquity	
2	3	1	Antiquity	
3	5	2	Antiquity	
4	7	3	Antiquity	
5	13	4	1461	Reguis 1536, Cataldi 1603
6	17	6	1588	Cataldi 1603
7	19	6	1588	Cataldi 1603
8	31	10	1750	Euler 1772
9	61	19	1883	Pervouchine 1883, Seelhoff 1886
10	89	27	1911	Powers 1911
11	107	33	1913	Powers 1914
12	127	39	1876	Lucas 1876
13	521	157	1952	Lehmer 1952–, Robinson 1952
14	607	183	1952	Lehmer 1952–, Robinson 1952
15	1279	386	1952	Lehmer 1952–, Robinson 1952
16	2203	664	1952	Lehmer 1952–, Robinson 1952
17	2281	687	1952	Lehmer 1952–, Robinson 1952
18	3217	969	1957	Riesel 1957
19	4253	1281	1961	Hurwitz 1961
20	4423	1332	1961	Hurwitz 1961
21	9689	2917	1963	Gillies 1964
22	9941	2993	1963	Gillies 1964
23	11213	3376	1963	Gillies 1964
24	19937	6002	1971	Tuckerman 1971
25	21701	6533	1978	Noll and Nickel 1980
26	23209	6987	1979	Noll 1980
27	44497	13395	1979	Nelson and Slowinski 1979
28	86243	25962	1982	Slowinski 1982
29	110503	33265	1988	Colquitt and Welsh 1991
30	132049	39751	1983	Slowinski 1988
31	216091	65050	1985	Slowinski 1989
32	756839	227832	1992	Gage and Slowinski 1992
33	859433	258716	1994	Gage and Slowinski 1994
34	1257787	378632	1996	Slowinski and Gage
35	1398269	420921	1996	Armengaud, Woltman, *et al.*
36?	2976221	895832	1997	Spence (Devlin 1997)

| 37? | 3021377 | 909526 | 1998 | Clarkson, Woltman, *et al.* |
| 38? | 6972593 | 2098960 | 1999 | Hajratwala 1999 |

See also CUNNINGHAM NUMBER, DOUBLE MERSENNE NUMBER, FERMAT-LUCAS NUMBER, FERMAT NUMBER, FERMAT NUMBER (LUCAS), FERMAT POLYNOMIAL, LUCAS-LEHMER TEST, MERSENNE NUMBER, PERFECT NUMBER, REPUNIT, SUPERPERFECT NUMBER

References

Bateman, P. T.; Selfridge, J. L.; and Wagstaff, S. S. "The New Mersenne Conjecture." *Amer. Math. Monthly* **96**, 125–28, 1989.

Ball, W. W. R. and Coxeter, H. S. M. *Mathematical Recreations and Essays, 13th ed.* New York: Dover, p. 66, 1987.

Beiler, A. H. Ch. 3 in *Recreations in the Theory of Numbers: The Queen of Mathematics Entertains.* New York: Dover, 1966.

Bell, E. T. *Mathematics: Queen and Servant of Science.* Washington, DC: Math. Assoc. Amer., 1987.

Caldwell, C. "Mersenne Primes: History, Theorems and Lists." http://www.utm.edu/research/primes/mersenne.shtml.

Caldwell, C. K. "The Top Twenty: Mersenne Primes." http://www.utm.edu/research/primes/lists/top20/Mersenne.html.

Caldwell, C. "GIMPS Finds a Prime! $2^{1398269} - 1$ is Prime." http://www.utm.edu/research/primes/notes/1398269/.

Caldwell, C. "GIMPS Finds a Multi-Million Digit Prime!." http://www.utm.edu/research/primes/notes/6972593/.

Colquitt, W. N. and Welsh, L. Jr. "A New Mersenne Prime." *Math. Comput.* **56**, 867–70, 1991.

Conway, J. H. and Guy, R. K. "Mersenne's Numbers." In *The Book of Numbers.* New York: Springer-Verlag, pp. 135–37, 1996.

Devlin, K. "World's Largest Prime." *FOCUS: Newsletter Math. Assoc. Amer.* **17**, 1, Dec. 1997.

Dickson, L. E. *History of the Theory of Numbers, Vol. 1: Divisibility and Primality.* New York: Chelsea, p. 13, 1952.

Gardner, M. *The Sixth Book of Mathematical Games from Scientific American.* Chicago, IL: University of Chicago Press, p. 85, 1984.

Gardner, M. "Patterns in Primes are a Clue to the Strong Law of Small Numbers." *Sci. Amer.* **243**, 18–8, Dec. 1980.

Gillies, D. B. "Three New Mersenne Primes and a Statistical Theory." *Math Comput.* **18**, 93–7, 1964.

Guy, R. K. "Mersenne Primes. Repunits. Fermat Numbers. Primes of Shape $k \cdot 2^n + 2$ [sic]." §A3 in *Unsolved Problems in Number Theory, 2nd ed.* New York: Springer-Verlag, pp. 8–3, 1994.

Haghighi, M. "Computation of Mersenne Primes Using a Cray X-MP." *Intl. J. Comput. Math.* **41**, 251–59, 1992.

Hardy, G. H. and Wright, E. M. *An Introduction to the Theory of Numbers, 5th ed.* Oxford, England: Clarendon Press, pp. 14–6, 1979.

Kraitchik, M. "Mersenne Numbers and Perfect Numbers." §3.5 in *Mathematical Recreations.* New York: W. W. Norton, pp. 70–3, 1942.

Kravitz, S. and Berg, M. "Lucas' Test for Mersenne Numbers $6000 < p < 7000$." *Math. Comput.* **18**, 148–49, 1964.

Lehmer, D. H. "On Lucas's Test for the Primality of Mersenne's Numbers." *J. London Math. Soc.* **10**, 162–65, 1935.

Leyland, P. ftp://sable.ox.ac.uk/pub/math/factors/mersenne.

Mersenne, M. *Cogitata Physico-Mathematica.* 1644.

Mersenne Organization. "GIMPS Discovers 36th Known Mersenne Prime, $2^{2976221} - 1$ is Now the Largest Known Prime." http://www.mersenne.org/2976221.htm.

Mersenne Organization. "GIMPS Discovers 37th Known Mersenne Prime, $2^{3021377} - 1$ is Now the Largest Known Prime." http://www.mersenne.org/3021377.htm.

Mersenne Organization. "GIMPS Finds First Million-Digit Prime, Stakes Claim to $50,000 EFF Award. $2^{6,972,593} - 1$ is Now the Largest Known Prime." http://www.mersenne.org/6972593.htm.

Noll, C. and Nickel, L. "The 25th and 26th Mersenne Primes." *Math. Comput.* **35**, 1387–390, 1980.

Powers, R. E. "The Tenth Perfect Number." *Amer. Math. Monthly* **18**, 195–96, 1911.

Powers, R. E. "Note on a Mersenne Number." *Bull. Amer. Math. Soc.* **40**, 883, 1934.

Sloane, N. J. A. Sequences A000043/M0672 and A000668/M2696 in "An On-Line Version of the Encyclopedia of Integer Sequences." http://www.research.att.com/~njas/sequences/eisonline.html.

Slowinski, D. "Searching for the 27th Mersenne Prime." *J. Recreat. Math.* **11**, 258–61, 1978–979.

Slowinski, D. *Sci. News* **139**, 191, 9/16/1989.

Tuckerman, B. "The 24th Mersenne Prime." *Proc. Nat. Acad. Sci. USA* **68**, 2319–320, 1971.

Uhler, H. S. "A Brief History of the Investigations on Mersenne Numbers and the Latest Immense Primes." *Scripta Math.* **18**, 122–31, 1952.

Uspensky, J. V. and Heaslet, M. A. *Elementary Number Theory.* New York: McGraw-Hill, 1939.

Weisstein, E. W. "Mersenne Numbers." MATHEMATICA NOTEBOOK MERSENNE.M.

Welsh, L. "Marin Mersenne." http://www.scruznet.com/~luke/mersenne.htm.

Welsh, L. "Mersenne Numbers & Mersenne Primes Bibliography." http://www.scruznet.com/~luke/biblio.htm.

Woltman, G. "The GREAT Internet Mersenne Prime Search." http://www.mersenne.org/prime.htm.

Mertens Conjecture

Given MERTENS FUNCTION defined by

$$M(n) \equiv \sum_{k=1}^{n} \mu(k), \qquad (1)$$

where $\mu(n)$ is the MÖBIUS FUNCTION, Mertens (1897) conjecture states that

$$|M(x)| < x^{1/2} \qquad (2)$$

for $x > 1$. The conjecture has important implications, since the truth of any equality OF THE FORM

$$|M(x)| \leq cx^{1/2} \qquad (3)$$

for any fixed c (the form of Mertens conjecture with $c = 1$) would imply the RIEMANN HYPOTHESIS. In 1885, Stieltjes claimed that he had a proof that $M(x)x^{-1/2}$ always stayed between two fixed bounds. However, it seems likely that Stieltjes was mistaken.

Mertens conjecture was proved false by Odlyzko and te Riele (1985). Their proof is indirect and does not produce a specific counterexample, but it does show that

$$\lim_{x \to \infty} \sup M(x)x^{-1/2} > 1.06 \qquad (4)$$

$$\liminf_{x \to \infty} M(x)x^{-1/2} < -1.009. \qquad (5)$$

Odlyzko and te Riele (1985) believe that there are no counterexamples to Mertens conjecture for $x \le 10^{20}$, or even 10^{30}. Pintz (1987) subsequently showed that at least one counterexample to the conjecture occurs for $x \le 10^{65}$, using a weighted integral average of $M(x)/x$ and a discrete sum involving nontrivial zeros of the RIEMANN ZETA FUNCTION.

It is still not known if

$$\limsup_{x \to \infty} |M(x)| |x|^{-1/2} = \infty, \qquad (6)$$

although it seems very probable (Odlyzko and te Riele 1985).

See also MERTENS FUNCTION, MÖBIUS FUNCTION, RIEMANN HYPOTHESIS

References

Anderson, R. J. "On the Mertens Conjecture for Cusp Forms." *Mathematika* **26**, 236–49, 1979.

Anderson, R. J. "Corrigendum: 'On the Mertens Conjecture for Cusp Forms.'" *Mathematika* **27**, 261, 1980.

Devlin, K. "The Mertens Conjecture." *Irish Math. Soc. Bull.* **17**, 29–3, 1986.

Grupp, F. "On the Mertens Conjecture for Cusp Forms." *Mathematika* **29**, 213–26, 1982.

Hardy, G. H. *Ramanujan: Twelve Lectures on Subjects Suggested by His Life and Work, 3rd ed.* New York: Chelsea, p. 64, 1999.

Jurkat, W. and Peyerimhoff, A. "A Constructive Approach to Kronecker Approximation and Its Application to the Mertens Conjecture." *J. reine angew. Math.* **286/287**, 322–40, 1976.

Mertens, F. "Uuml;ber eine zahlentheoretische Funktion." *Sitzungsber. Akad. Wiss. Wien IIa* **106**, 761–30, 1897.

Odlyzko, A. M. and te Riele, H. J. J. "Disproof of the Mertens Conjecture." *J. reine angew. Math.* **357**, 138–60, 1985.

Pintz, J. "An Effective Disproof of the Mertens Conjecture." *Astérique* **147–48**, 325–33 and 346, 1987.

te Riele, H. J. J. "Some Historical and Other Notes About the Mertens Conjecture and Its Recent Disproof." *Nieuw Arch. Wisk.* **3**, 237–43, 1985.

Mertens Constant

N.B. Portions of this entry based on a detailed online essay by S. Finch.

A constant related to the TWIN PRIMES CONSTANT which appears in HARMONIC SERIES for the SUM of reciprocal PRIMES

$$\sum_{p \text{ prime}}^{x} \frac{1}{p} = \ln \ln x + B_1 + o(1), \qquad (1)$$

which is given by

$$B_1 = \gamma + \sum_{p \text{ prime}} \left[\ln(1 - p^{-1}) + \frac{1}{p} \right] \approx 0.2614972128, \qquad (2)$$

where γ is the EULER-MASCHERONI CONSTANT (Rosser and Schoenfeld 1962; Le Lionnais 1983; Ellison and Ellison 1985; Hardy and Wright 1985). According to Lindqvist and Peetre (1997), this was shown independently by Meissel in 1866 and Mertens (1874). (2) is equivalent to

$$\prod_{p \le x} \left(1 - \frac{1}{p} \right) \sim \frac{e^{-\gamma}}{\ln x}, \qquad (3)$$

where γ is the EULER-MASCHERONI CONSTANT (Hardy 1999, p. 57). Knuth (1998) gives 40 digits of B_1, and Gourdon and Sebah give 100 digits. The constant is sometimes known as Kronecker's constant (Schroeder 1997).

A rapidly converging series for B_1 is given by

$$B_1 = \gamma + \sum_{m=2}^{\infty} \frac{\mu(m)}{m} \ln[\zeta(m)], \qquad (4)$$

where γ is the EULER-MASCHERONI CONSTANT, $\zeta(n)$ is the RIEMANN ZETA FUNCTION, and $\mu(n)$ is the MÖBIUS FUNCTION (Flajolet and Vardi 1996, Schroeder 1997, Knuth 1998).

The constant B_1 also occurs in the SUMMATORY FUNCTION of the number of DISTINCT PRIME FACTORS $\omega(k)$,

$$\sum_{k=2}^{n} \omega(k) = n \ln \ln n + B_1 n + o(n) \qquad (5)$$

(Hardy and Wright 1979, p. 355).

The related constant

$$B_2 = \gamma + \sum_{p \text{ prime}} \left[\ln(1 - p^{-1}) + \frac{1}{p-1} \right] \approx 1.034653 \qquad (6)$$

appears in the SUMMATORY FUNCTION of the DIVISOR FUNCTION $\sigma_0(n) = \Omega(n)$,

$$\sum_{k=2}^{n} \Omega(k) = n \ln \ln n + B_2 + o(n) \qquad (7)$$

(Hardy and Wright 1979, p. 355).

Another related series is

$$\lim_{n \to \infty} \left(\sum_{k=1}^{\pi(n)} \frac{\ln p_k}{p_k} - \ln n \right) = -\gamma - \sum_{j=2}^{\infty} \sum_{k=1}^{\infty} \frac{\ln p_k}{p_k^j}$$

$$\equiv -C_2 = -1.3325822757\ldots \qquad (8)$$

(Rosser and Schoenfeld 1962, Montgomery 1971, Finch).

See also BRUN'S CONSTANT, HARMONIC SERIES, PRIME FACTORS, PRIME NUMBER, TWIN PRIMES CONSTANT

References

Ellison, W. J. and Ellison, F. *Prime Numbers.* New York: Wiley, 1985.

Finch, S. "Favorite Mathematical Constants." http://www.mathsoft.com/asolve/constant/hdmrd/hdmrd.html.

Flajolet, P. and Vardi, I. "Zeta Function Expansions of Classical Constants." Unpublished manuscript. 1996. http://pauillac.inria.fr/algo/flajolet/Publications/landau.ps.

Gourdon, X. and Sebah, P. "Some Constants from Number Theory." http://xavier.gourdon.free.fr/Constants/Miscellaneous/constantsNumTheory.html.

Hardy, G. H. *Ramanujan: Twelve Lectures on Subjects Suggested by His Life and Work, 3rd ed.* New York: Chelsea, 1999.

Hardy, G. H. and Wright, E. M. "Mertens's Theorem." §22.8 in *An Introduction to the Theory of Numbers, 5th ed.* Oxford, England: Oxford University Press, pp. 351–53 and 355, 1979.

Ingham, A. E. *The Distribution of Prime Numbers.* London: Cambridge University Press, pp. 22–4, 1990.

Knuth, D. E. *The Art of Computer Programming, Vol. 2: Seminumerical Algorithms, 3rd ed.* Reading, MA: Addison-Wesley, 1998.

Landau, E. *Handbuch der Lehre von der Verteilung der Primzahlen, 3rd ed.* New York: Chelsea, pp. 100–02, 1974.

Le Lionnais, F. *Les nombres remarquables.* Paris: Hermann, p. 24, 1983.

Lindqvist, P. and Peetre, J. "On the Remainder in a Series of Mertens." *Expos. Math.* **15**, 467–78, 1997.

Mertens, F. *J. für Math.* **78**, 46–2, 1874.

Montgomery, H. L. *Topics in Multiplicative Number Theory.* New York: Springer-Verlag, 1971.

Rosser, J. B. and Schoenfeld, L. "Approximate Formulas for Some Functions of Prime Numbers." *Ill. J. Math.* **6**, 64–4, 1962.

Schroeder, M. R. *Number Theory in Science and Communication, with Applications in Cryptography, Physics, Digital Information, Computing, and Self-Similarity, 3rd ed.* New York: Springer-Verlag, 1997.

Mertens Function

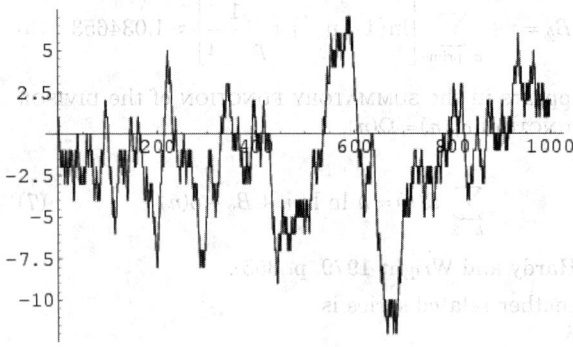

The summary function

$$M(n) \equiv \sum_{k=1}^{n} \mu(k), \qquad (1)$$

where $\mu(n)$ is the MÖBIUS FUNCTION. The first few values are 1, 0, -1, -1, -2, -1, -2, -2, -2, -1, -2, -2, ... (Sloane's A002321). The first few values of n at which $M(n) = 0$ are 2, 39, 40, 58, 65, 93, 101, 145, 149, 150, ... (Sloane's A028442).

The Mertens function is related to the number of SQUAREFREE integers up to n, which is the sum from 1 to n of the absolute value of $\mu(k)$,

$$\sum_{k=1}^{n} |\mu(k)| \sim \frac{6}{\pi^2} n + \mathcal{O}(\sqrt{n}). \qquad (2)$$

The Mertens function obeys

$$\sum_{n=1}^{x} M\left(\frac{x}{n}\right) = 1 \qquad (3)$$

(Lehman 1960). The analytic form is unsolved, although MERTENS CONJECTURE that

$$|M(x)| < x^{1/2} \qquad (4)$$

has been disproved.

Lehman (1960) gives an algorithm for computing $M(x)$ with $\mathcal{O}(x^{2/3+\epsilon})$ operations, while the Lagarias-Odlyzko (1987) algorithm for computing the PRIME COUNTING FUNCTION $\pi(x)$ can be modified to give $M(x)$ in $\mathcal{O}(x^{3/5+\epsilon})$ operations.

See also MERTENS CONJECTURE, MÖBIUS FUNCTION, SQUAREFREE

References

Lagarias, J. and Odlyzko, A. "Computing $\pi(x)$: An Analytic Method." *J. Algorithms* **8**, 173–91, 1987.

Lehman, R. S. "On Liouville's Function." *Math. Comput.* **14**, 311–20, 1960.

Lehmer, D. H. *Guide to Tables in the Theory of Numbers.* Bulletin No. 105. Washington, DC: National Research Council, pp. 7–0, 1941.

Odlyzko, A. M. and te Riele, H. J. J. "Disproof of the Mertens Conjecture." *J. reine angew. Math.* **357**, 138–60, 1985.

Sloane, N. J. A. Sequences A002321/M0102 and A028442 in "An On-Line Version of the Encyclopedia of Integer Sequences." http://www.research.att.com/~njas/sequences/eisonline.html.

Sterneck, R. D. von. "Empirische Untersuchung über den Verlauf der zahlentheoretischer Funktion $\sigma(n) = \sum_{x=1}^{n} \mu(x)$ im Intervalle von 0 bis 150 000." *Sitzungsber. der Kaiserlichen Akademie der Wissenschaften Wien, Math.-Naturwiss. Klasse 2a* **106**, 835–024, 1897.

Mertens Theorem

$$\lim_{x \to \infty} \frac{\prod_{\substack{2 \leq p \leq x \\ p \text{ prime}}} \left(1 - \frac{1}{p}\right)}{\frac{e^{-\gamma}}{\ln x}} = 1,$$

where γ is the EULER-MASCHERONI CONSTANT and $e^{-\gamma} = 0.56145\ldots$.

See also EULER PRODUCT

References

Hardy, G. H. and Wright, E. M. *An Introduction to the Theory of Numbers, 5th ed.* Oxford, England: Oxford University Press, p. 351, 1979.

Riesel, H. *Prime Numbers and Computer Methods for Factorization, 2nd ed.* Boston, MA: Birkhäuser, pp. 66–7, 1994.

Mertz Apodization Function

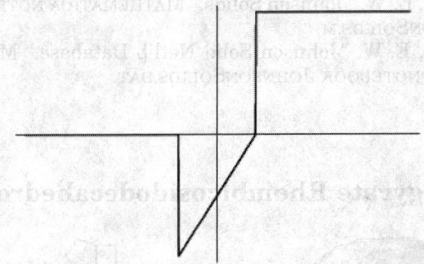

An asymmetrical APODIZATION FUNCTION defined by

$$M(x, b, d) = \begin{cases} 0 & \text{for } x < -b \\ (x-b)/(2b) & \text{for } -b < x < b \\ 1 & \text{for } b < x < b+2d \\ 0 & \text{for } x < b+2d, \end{cases}$$

where the two-sided portion is $2b$ long (total) and the one-sided portion is $b+2d$ long (Schnopper and Thompson 1974, p. 508). The APPARATUS FUNCTION is

$$M_A(k, b, d) = \frac{\sin[2\pi k(b + 2d)]}{2\pi k}$$

$$+ i \left\{ \frac{\cos[2\pi k(b + 2d)]}{2\pi k} - \frac{\sin(2b)}{4\pi^2 k^2 b} \right\}.$$

References

Schnopper, H. W. and Thompson, R. I. "Fourier Spectrometers." In *Methods of Experimental Physics* **12A**. New York: Academic Press, pp. 491–29, 1974.

Mesh

See also FINITE ELEMENT METHOD, LATTICE POINT, MESH SIZE

References

Bern, M. and Plassmann, P. "Mesh Generation." Ch. 6 in *Handbook of Computational Geometry* (Ed. J.-R. Sack and J. Urrutia). Amsterdam, Netherlands: North-Holland, pp. 291–32, 2000.

Mesh Size

When a CLOSED INTERVAL $[a, b]$ is partitioned by points $a < x_1 < x_2 < \ldots < x_{n-1} < b$, the lengths of the resulting intervals between the points are denoted $\Delta x_1, \Delta x_2, \ldots, \Delta x_n$, and the value max Δx_k is called the mesh size of the partition.

See also INTEGRAL, LOWER SUM, RIEMANN INTEGRAL, UPPER SUM

Mesokurtic

A distribution with zero KURTOSIS ($\gamma_2 = 0$).

See also KURTOSIS, LEPTOKURTIC

M-Estimate

A ROBUST ESTIMATION based on maximum likelihood argument.

See also L-ESTIMATE, R-ESTIMATE

References

Press, W. H.; Flannery, B. P.; Teukolsky, S. A.; and Vetterling, W. T. "Robust Estimation." §15.7 in *Numerical Recipes in FORTRAN: The Art of Scientific Computing*, 2nd ed. Cambridge, England: Cambridge University Press, pp. 694–00, 1992.

Metabiaugmented Dodecahedron

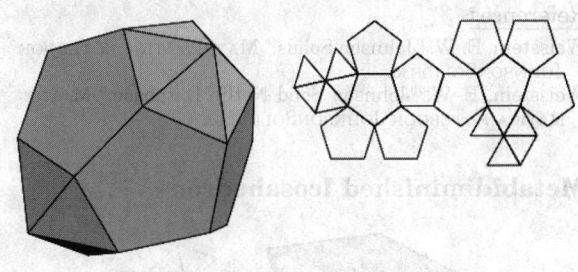

JOHNSON SOLID J_{60}.

References

Weisstein, E. W. "Johnson Solids." MATHEMATICA NOTEBOOK JOHNSONSOLIDS.M.
Weisstein, E. W. "Johnson Solid Netlib Database." MATHEMATICA NOTEBOOK JOHNSONSOLIDS.DAT.

Metabiaugmented Hexagonal Prism

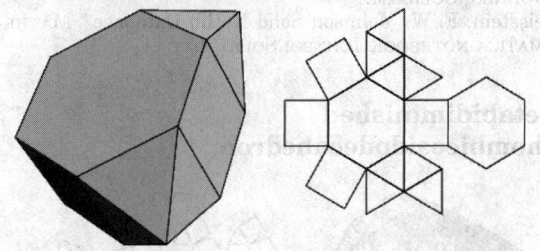

JOHNSON SOLID J_{56}.

References

Weisstein, E. W. "Johnson Solids." MATHEMATICA NOTEBOOK JOHNSONSOLIDS.M.
Weisstein, E. W. "Johnson Solid Netlib Database." MATHEMATICA NOTEBOOK JOHNSONSOLIDS.DAT.

Metabiaugmented Truncated Dodecahedron

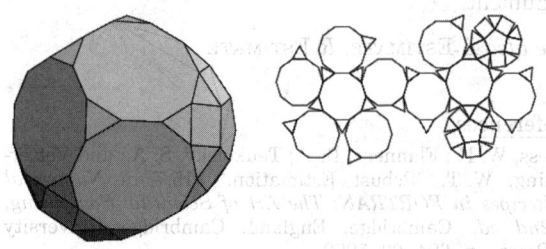

Johnson Solid J_{70}.

References

Weisstein, E. W. "Johnson Solids." Mathematica notebook JohnsonSolids.m.
Weisstein, E. W. "Johnson Solid Netlib Database." Mathematica notebook JohnsonSolids.dat.

Metabidiminished Icosahedron

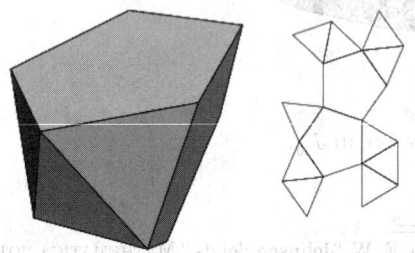

Johnson Solid J_{62}.

References

Weisstein, E. W. "Johnson Solids." Mathematica notebook JohnsonSolids.m.
Weisstein, E. W. "Johnson Solid Netlib Database." Mathematica notebook JohnsonSolids.dat.

Metabidiminished Rhombicosidodecahedron

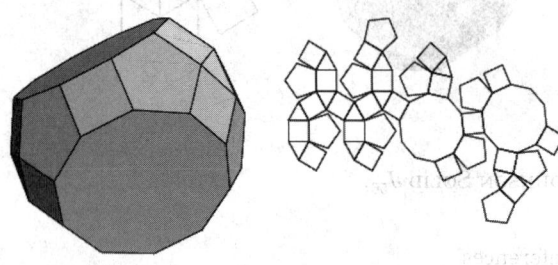

Johnson Solid J_{81}.

References

Weisstein, E. W. "Johnson Solids." Mathematica notebook JohnsonSolids.m.
Weisstein, E. W. "Johnson Solid Netlib Database." Mathematica notebook JohnsonSolids.dat.

References

Weisstein, E. W. "Johnson Solids." Mathematica notebook JohnsonSolids.m.
Weisstein, E. W. "Johnson Solid Netlib Database." Mathematica notebook JohnsonSolids.dat.

Metabigyrate Rhombicosidodecahedron

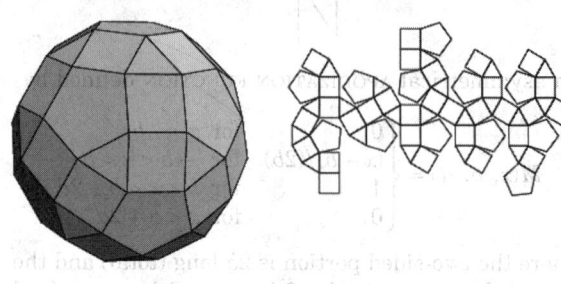

Johnson Solid J_{74}.

References

Weisstein, E. W. "Johnson Solids." Mathematica notebook JohnsonSolids.m.
Weisstein, E. W. "Johnson Solid Netlib Database." Mathematica notebook JohnsonSolids.dat.

Metacyclic Group

See also Cyclic Group

References

Mac Lane, S. and Birkhoff, G. *Algebra.* New York: Macmillan, p. 462, 1967.

Metadrome

A metadrome is a number whose hexadecimal digits are in strict ascending order. The first few are 0, 1, 2, 3, 4, 5, 6, 7, 8, 9, 10, 11, 12, 13, 14, 15, 17, 18, 19, 20, ... (Sloane's A023784). The first few numbers which are *not* metadromes are 16, 17, 32, 33, 34, ..., corresponding to 10_{16}, 11_{16}, 20_{16}, 21_{16}, 22_{16}, ...

See also Digit, Hexadecimal, Katadrome, Nialpdrome, Plaindrome

References

Sloane, N. J. A. Sequences A023784 in "An On-Line Version of the Encyclopedia of Integer Sequences." http://www.research.att.com/~njas/sequences/eisonline.html.
Weisstein, E. W. "Integer Sequences." Mathematica notebook IntegerSequences.m.

Metagyrate Diminished Rhombicosidodecahedron

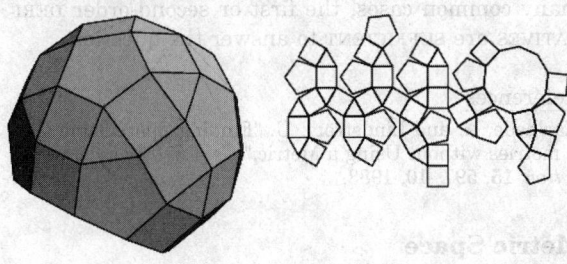

JOHNSON SOLID J_{78}.

References

Weisstein, E. W. "Johnson Solids." MATHEMATICA NOTEBOOK JOHNSONSOLIDS.M.
Weisstein, E. W. "Johnson Solid Netlib Database." MATHEMATICA NOTEBOOK JOHNSONSOLIDS.DAT.

Metalogic
METAMATHEMATICS

Metamathematics

The branch of LOGIC dealing with the study of the combination and application of mathematical symbols, sometimes called METALOGIC. Metamathematics is the study of MATHEMATICS itself, and one of its primary goals is to determine the nature of mathematical reasoning (Hofstadter 1989).

See also LOGIC, MATHEMATICS

References

Birkhoff, G. and Mac Lane, S. *A Survey of Modern Algebra,* *5th ed.* New York: Macmillan, p. 326, 1996.
Chaitin, G. J. *The Unknowable.* New York: Springer-Verlag, 1999.
Hofstadter, D. R. *Gödel, Escher, Bach: An Eternal Golden* *Braid.* New York: Vintage Books, p. 23, 1989.

Meteorology Theorem

Somewhere on the Earth, there is a pair of ANTIPODAL POINTS having simultaneously the same temperature and pressure.

References

Dodson, C. T. J. and Parker, P. E. *A User's Guide to* *Algebraic Topology.* Dordrecht, Netherlands: Kluwer, pp. 121 and 284, 1997.

Method

A particular way of doing something, sometimes also called an ALGORITHM or PROCEDURE. (According to Petkovsek *et al.* (1996), "a method is a trick that has worked at least twice.")

References

Petkovsek, M.; Wilf, H. S.; and Zeilberger, D. *A = B.* Wellesley, MA: A. K. Peters, p. 117, 1996.

Method of Exclusions

A method used by Gauss to solve the quadratic DIOPHANTINE EQUATION OF THE FORM

$$mx^2 + ny^2 = A$$

(Dickson 1992, pp. 391 and 407).

References

Dickson, L. E. *History of the Theory of Numbers, Vol. 2:* *Diophantine Analysis.* New York: Chelsea, p. 407, 1992.

Method of False Position

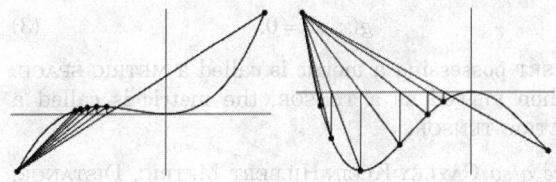

An ALGORITHM for finding ROOTS which retains that prior estimate for which the function value has opposite sign from the function value at the current best estimate of the root. In this way, the method of false position keeps the root bracketed (Press *et al.* 1992).

Using the two-point form of the line

$$y - y_1 = \frac{f(x_{n-1}) - f(x_1)}{x_{n-1} - x_1}(x_n - x_1)$$

with $y = 0$, using $y_1 = f(x_1)$, and solving for x_n therefore gives the iteration

$$x_n = x_1 - \frac{x_{n-1} - x_1}{f(x_{n-1}) - f(x_1)} f(x_1).$$

See also BRENT'S METHOD, RIDDERS' METHOD, SECANT METHOD

References

Abramowitz, M. and Stegun, C. A. (Eds.). *Handbook of* *Mathematical Functions with Formulas, Graphs, and* *Mathematical Tables, 9th printing.* New York: Dover, p. 18, 1972.
Chabert, J.-L. (Ed.). "Methods of False Position." Ch. 3 in *A* *History of Algorithms: From the Pebble to the Microchip.* New York: Springer-Verlag, pp. 83–12, 1999.
Press, W. H.; Flannery, B. P.; Teukolsky, S. A.; and Vetterling, W. T. "Secant Method, False Position Method, and Ridders' Method." §9.2 in *Numerical Recipes in FORTRAN: The Art of Scientific Computing, 2nd ed.* Cambridge, England: Cambridge University Press, pp. 347–52, 1992.
Whittaker, E. T. and Robinson, G. "The Rule of False Position." §49 in *The Calculus of Observations: A Treatise*

on Numerical Mathematics, 4th ed. New York: Dover, pp. 92–4, 1967.

Method of Reduction

METHOD OF EXCLUSIONS

Metric

A NONNEGATIVE function $g(x, y)$ describing the "DISTANCE" between neighboring points for a given SET. A metric satisfies the TRIANGLE INEQUALITY

$$g(x, y) + g(y, z) \geq g(x, z) \qquad (1)$$

and is SYMMETRIC, so

$$g(x, y) = g(y, x). \qquad (2)$$

A metric also satisfies

$$g(x, x) = 0. \qquad (3)$$

A SET possessing a metric is called a METRIC SPACE. When viewed as a TENSOR, the metric is called a METRIC TENSOR.

See also CAYLEY-KLEIN-HILBERT METRIC, DISTANCE, FRENCH METRO METRIC, FUNDAMENTAL FORMS, HYPERBOLIC METRIC, METRIC ENTROPY, METRIC EQUIVALENCE PROBLEM, METRIC SPACE, METRIC TENSOR, PART METRIC, RIEMANNIAN METRIC, ULTRAMETRIC

References
Gray, A. "Metrics on Surfaces." Ch. 15 in *Modern Differential Geometry of Curves and Surfaces with Mathematica, 2nd ed.* Boca Raton, FL: CRC Press, pp. 341–58, 1997.

Metric Entropy

Also known as KOLMOGOROV ENTROPY, KOLMOGOROV-SINAI ENTROPY, or KS Entropy. The metric entropy is 0 for nonchaotic motion and > 0 for CHAOTIC motion.

References
Ott, E. *Chaos in Dynamical Systems.* New York: Cambridge University Press, p. 138, 1993.

Metric Equivalence Problem

1. Find a complete system of invariants, or
2. decide when two METRICS differ only by a coordinate transformation.

The most common statement of the problem is, "Given METRICS g and g', does there exist a coordinate transformation from one to the other?" Christoffel and Lipschitz (1870) showed how to decide this question for two RIEMANNIAN METRICS. The solution by É. Cartan requires computation of the 10th order COVARIANT DERIVATIVES. The demonstration was simplified by A. Karlhede using the

TETRAD formalism so that only seventh order COVARIANT DERIVATIVES need be computed. however, in many common cases, the first or second-order DERIVATIVES are SUFFICIENT to answer the question.

References
Karlhede, A. and Lindström, U. "Finding Space-Time Geometries without Using a Metric." *Gen. Relativity Gravitation* **15**, 597–10, 1983.

Metric Space

A SET S with a global distance FUNCTION (the METRIC g) which, for every two points x, y in S, gives the DISTANCE between them as a NONNEGATIVE REAL NUMBER $g(x, y)$. A metric space must also satisfy

1. $g(x, y) = 0$ IFF $x = y$,
2. $g(x, y) = g(y, x)$,
3. The TRIANGLE INEQUALITY $g(x, y) + g(y, z) \geq g(x, z)$.

See also UNIVERSAL METRIC SPACE

References
Munkres, J. R. *Topology: A First Course.* Englewood Cliffs, NJ: Prentice-Hall, 1975.
Rudin, W. *Principles of Mathematical Analysis.* New York: McGraw-Hill, 1976.

Metric Tensor

A TENSOR, also called a RIEMANNIAN METRIC, which is symmetric and POSITIVE DEFINITE. Very roughly, the metric tensor g_{ij} is a function which tells how to compute the distance between any two points in a given SPACE. Its components can be viewed as multiplication factors which must be placed in front of the differential displacements dx_i in a generalized PYTHAGOREAN THEOREM

$$ds^2 = g_{11}\, dx_1^2 + g_{12}\, dx_1\, dx_2 + g_{22}\, dx_2^2 + \ldots. \qquad (1)$$

In EUCLIDEAN SPACE, $g_{ij} = \delta_{ij}$ where δ is the KRONECKER DELTA (which is 0 for $i \neq j$ and 1 for $i = j$), reproducing the usual form of the PYTHAGOREAN THEOREM

$$ds^2 = dx_1^2 + dx_2^2 + \ldots. \qquad (2)$$

The metric tensor is defined abstractly as an INNER PRODUCT of every TANGENT SPACE of a MANIFOLD such that the INNER PRODUCT is a symmetric, nondegenerate, BILINEAR FORM on a VECTOR SPACE. This means that it takes two VECTORS $\mathbf{v}$, $\mathbf{w}$ as arguments and produces a REAL NUMBER $\langle \mathbf{v}, \mathbf{w} \rangle$ such that

$$\langle k\mathbf{v}, \mathbf{w} \rangle = k\langle \mathbf{v}, \mathbf{w} \rangle = \langle \mathbf{v}, k\mathbf{w} \rangle \qquad (3)$$

$$\langle \mathbf{v} + \mathbf{w}, \mathbf{x} \rangle = \langle \mathbf{v}, \mathbf{x} \rangle + \langle \mathbf{w}, \mathbf{x} \rangle \qquad (4)$$

$$\langle \mathbf{v}, \mathbf{w} + \mathbf{x} \rangle = \langle \mathbf{v}, \mathbf{w} \rangle + \langle \mathbf{v}, \mathbf{x} \rangle \qquad (5)$$

$$\langle \mathbf{v}, \mathbf{w} \rangle = \langle \mathbf{w}, \mathbf{v} \rangle \qquad (6)$$

$$\langle \mathbf{v}, \mathbf{v} \rangle \geq 0, \tag{7}$$

with equality IFF $\mathbf{v} = 0$.

In coordinate NOTATION (with respect to the basis),

$$g^{\alpha\beta} = \vec{e}_\alpha \cdot \vec{e}_\beta \tag{8}$$

$$g_{\alpha\beta} = \vec{e}_\alpha \cdot \vec{e}_\beta. \tag{9}$$

$$g_{\mu\nu} \equiv \frac{\partial \xi^\zeta}{\partial x^\mu} \frac{\partial \xi^\beta}{\partial x^\nu} \eta_{\alpha\beta}, \tag{10}$$

where $\eta_{\alpha\beta}$ is the MINKOWSKI METRIC. This can also be written

$$g = D^{\mathrm{T}} \eta D, \tag{11}$$

where

$$D_{\alpha\mu} \equiv \frac{\partial \xi^\alpha}{\partial x^\mu} \tag{12}$$

$$D_{\alpha\mu}^{\mathrm{T}} \equiv D_{\mu\alpha} \tag{13}$$

$$\frac{\partial}{\partial x^m} g_{il} g^{lk} = \frac{\partial}{\partial x^m} \delta_i^k \tag{14}$$

gives

$$g_{il} \frac{\partial g^{lk}}{\partial x^m} = -g^{lk} \frac{\partial g_{il}}{\partial x^m}. \tag{15}$$

The metric is POSITIVE DEFINITE, so a metric's DISCRIMINANT is POSITIVE. For a metric in 2-space,

$$g \equiv g_{11} g_{22} - g_{12}^2 > 0. \tag{16}$$

The ORTHOGONALITY of CONTRAVARIANT and COVARIANT metrics stipulated by

$$g_{ik} g^{ij} = \delta_k^j \tag{17}$$

for $i = 1, \dots, n$ gives n linear equations relating the $2n$ quantities g_{ij} and g^{ij}. therefore, if n metrics are known, the others can be determined.

in 2-space,

$$g^{11} = \frac{g_{22}}{g} \tag{18}$$

$$g^{12} = g^{21} = -\frac{g_{12}}{g} \tag{19}$$

$$g^{22} = \frac{g_{11}}{g}. \tag{20}$$

if g is symmetric, then

$$g_{\alpha\beta} = g_{\beta\alpha} \tag{21}$$

$$g^{\alpha\beta} = g^{\beta\alpha} \tag{22}$$

in EUCLIDEAN SPACE (and all other symmetric

SPACES),

$$g_\alpha^\beta = g_\alpha^\beta = \delta_\alpha^\beta, \tag{23}$$

so

$$g_{\alpha\alpha} = \frac{1}{g^{\alpha\alpha}}. \tag{24}$$

The ANGLE ϕ between two parametric curves is given by

$$\cos \phi = \hat{\mathbf{r}}_1 \cdot \hat{\mathbf{r}}_2 = \frac{\mathbf{r}_1}{g_1} \cdot \frac{\mathbf{r}_2}{g_2} = \frac{g_{12}}{g_1 g_2}, \tag{25}$$

so

$$\sin \phi = \frac{\sqrt{g}}{g_1 g_2} \tag{26}$$

and

$$|\mathbf{r}_1 \times \mathbf{r}_2| = g_1 g_2 \sin \phi = \sqrt{g}. \tag{27}$$

The LINE ELEMENT can be written

$$ds^2 = dx_i \, dx_i = g_{ij} \, dq_i \, dq_j \tag{28}$$

where EINSTEIN SUMMATION has been used. But

$$dx_i = \frac{\partial x_i}{\partial q_1} \, dq_1 + \frac{\partial x_i}{\partial q_2} \, dq_2 + \frac{\partial x_i}{\partial q_3} \, dq_3 = \frac{\partial x_i}{\partial q_j} \, dq_j, \tag{29}$$

so

$$g_{ij} = \sum_k \frac{\partial^2 x_k}{\partial q_i \, \partial q_j}. \tag{30}$$

For ORTHOGONAL coordinate systems, $g_{ij} = 0$ for $i \neq j$, and the LINE ELEMENT becomes (for 3-space)

$$\begin{aligned}
ds^2 &= g_{11} \, dq_1^2 + g_{22} \, dq_2^2 + g_{33} \, dq_3^2 \\
&= (h_1 \, dq_1)^2 + (h_2 \, dq_2)^2 + (h_3 \, dq_3)^2,
\end{aligned} \tag{31}$$

where $h_i \equiv \sqrt{g_{ii}}$ are called the SCALE FACTORS.

See also CURVILINEAR COORDINATES, DISCRIMINANT (METRIC), LICHNEROWICZ CONDITIONS, LINE ELEMENT, METRIC, METRIC EQUIVALENCE PROBLEM, MINKOWSKI SPACE, SCALE FACTOR, SPACE

Metropolis Algorithm
SIMULATED ANNEALING

Mex
The MINIMUM excluded value. The mex of a SET S of NONNEGATIVE INTEGERS is the least NONNEGATIVE INTEGER *not* in the set.

See also MEX SEQUENCE

References

Guy, R. K. "Max and Mex Sequences." §E27 in *Unsolved Problems in Number Theory, 2nd ed.* New York: Springer-Verlag, pp. 227–28, 1994.

Mex Sequence

A sequence defined from a FINITE sequence $a_0, a_1, ...,$ a_n by defining $a_{n+1} = \text{mex}_i(a_i + a_{n-i})$, where mex is the MEX (minimum excluded value).

See also MAX SEQUENCE, MEX

References

Guy, R. K. "Max and Mex Sequences." §E27 in *Unsolved Problems in Number Theory, 2nd ed.* New York: Springer-Verlag, pp. 227–28, 1994.

Mian-Chowla Sequence

The sequence produced by starting with $a_1 = 1$ and applying the GREEDY ALGORITHM in the following way: for each $k \geq 2$, let a_k be the least INTEGER exceeding a_{k-1} for which $a_j + a_k$ are all distinct, with $1 \leq j \leq k$. This procedure generates the sequence 1, 2, 4, 8, 13, 21, 31, 45, 66, 81, 97, 123, 148, 182, 204, 252, 290, ... (Sloane's A005282). The RECIPROCAL sum of the sequence,

$$S \equiv \sum_{i=1}^{\infty} \frac{1}{a_i}$$

satisfies

$$2.158435 \leq S \leq 2.158677$$

(R. Lewis).

See also A-SEQUENCE, B_2-SEQUENCE

References

Mian, A. M. and Chowla, S. D. "On the B_2-Sequences of Sidon." *Proc. Nat. Acad. Sci. India* **A14**, 3–, 1944.
Guy, R. K. "B_2-Sequences." §E28 in *Unsolved Problems in Number Theory, 2nd ed.* New York: Springer-Verlag, pp. 228–29, 1994.
Sloane, N. J. A. Sequences A005282/M1094 in "An On-Line Version of the Encyclopedia of Integer Sequences." http://www.research.att.com/~njas/sequences/eisonline.html.

Mice Problem

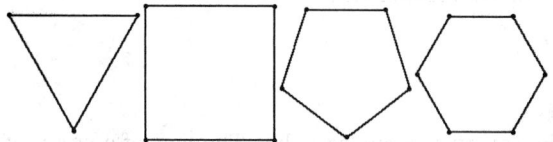

n mice start at the corners of a regular n-gon of unit side length, each heading towards its closest neighboring mouse in a counterclockwise direction at constant speed. The mice each trace out a LOGARITHMIC SPIRAL, meet in the center of the POLYGON, and travel a distance

$$d_n = \frac{1}{1 - \cos\left(\dfrac{2\pi}{n}\right)}.$$

The first few values for $n = 2, 3, ...,$ are

$$\frac{1}{2}, \ \frac{2}{3}, \ 1, \ \frac{1}{5}\left(5 + \sqrt{5}\right), \ 2, \ \frac{1}{1 - \cos\left(\dfrac{2\pi}{7}\right)},$$

$$2 + \sqrt{2}, \ \frac{1}{1 - \cos\left(\dfrac{2\pi}{9}\right)}, \ 3 + \sqrt{5}, \ ...,$$

giving the numerical values 0.5, 0.666667, 1, 1.44721, 2, 2.65597, 3.41421, 4.27432, 5.23607, ... The curve formed by connecting the mice at regular intervals of time is an attractive figure called a WHIRL.

The problem is also variously known as the (three, four, etc.) (bug, dog, etc.) problem. It can be generalized to irregular polygons and mice traveling at differing speeds (Bernhart 1959). Miller (1871) considered three mice in general positions with speeds adjusted to keep paths similar and the triangle similar to the original.

See also APOLLONIUS PURSUIT PROBLEM, PURSUIT CURVE, SPIRAL, TRACTRIX, WHIRL

References

Bernhart, A. "Polygons of Pursuit." *Scripta Math.* **24**, 23–0, 1959.
Brocard, H. "Solution of Lucas's Problem." *Nouv. Corresp. Math.* **3**, 280, 1877.
Clapham, A. J. *Rec. Math. Mag.*, Aug. 1962.
Gardner, M. *The Scientific American Book of Mathematical Puzzles and Diversions.* New York: NY: Simon and Schuster, 1959.
Gardner, M. *The Sixth Book of Mathematical Games from Scientific American.* Chicago, IL: University of Chicago Press, pp. 240–43, 1984.
Good, I. J. "Pursuit Curves and Mathematical Art." *Math. Gaz.* **43**, 34–5, 1959.
Lucas, E. "Problem of the Three Dogs." *Nouv. Corresp. Math.* **3**, 175–76, 1877.
Madachy, J. S. *Madachy's Mathematical Recreations.* New York: Dover, pp. 201–04, 1979.
Miller, R. K. Problem 16. *Cambridge Math. Tripos Exam.* January 5, 1871.
Steinhaus, H. *Mathematical Snapshots, 3rd ed.* New York: Dover, p. 136, 1999.
Weisstein, E. W. "Mice Problem." MATHEMATICA NOTEBOOK MICEPROBLEM.M.
Wells, D. *The Penguin Dictionary of Curious and Interesting Geometry.* London: Penguin, pp. 201–02, 1991.
Wilson, J. "Problem: Four Dogs." http://jwilson.coe.uga.edu/emt725/Four.Dogs/four.dogs.html.

Microlocal Analysis

References

Demuth, M.; Schrohe, E.; Schulze, B.-E.; and Sjöstrand, J. (Eds.). *Spectral Theory, Microlocal Analysis, Singular Manifolds.* Berlin: Akademie Verlag, 1997.

Grigis, A. and Sjöstrand, J. *Microlocal Analysis for Differential Operators: An Introduction.* Cambridge, England: Cambridge University Press, 1994.

Sjöstrand, J. "Singularités analytiques microlocales." *Astérisque* **95**, 1–66, 1982.

Mid-Arc Points

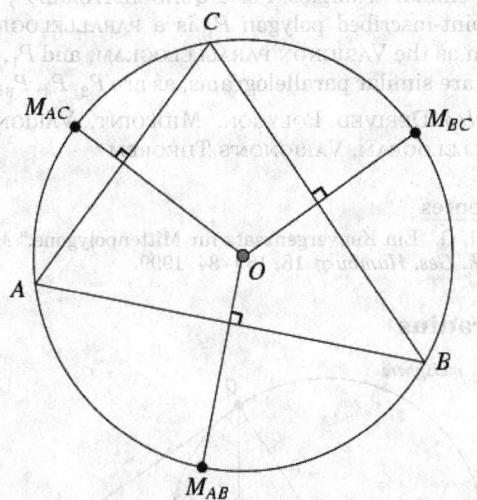

The mid-arc points M_{AB}, M_{AC}, and M_{BC} of a TRIANGLE ΔABC are the points on the CIRCUMCIRCLE of the triangle which lie half-way along each of the three ARCS determined by the vertices (Johnson 1929). These points arise in the definition of the FUHRMANN CIRCLE and FUHRMANN TRIANGLE, and lie on the extensions of the PERPENDICULAR BISECTORS of the triangle sides drawn from the CIRCUMCENTER O.

Kimberling (1988, 1994) and Kimberling and Veldkamp (1987) define the mid-arc points as the POINTS which have TRIANGLE CENTER FUNCTIONS

$$\alpha_1 = \left[\cos\left(\tfrac{1}{2}B\right) + \cos\left(\tfrac{1}{2}C\right)\right]\sec\left(\tfrac{1}{2}A\right)$$

$$\alpha_2 = \left[\cos\left(\tfrac{1}{2}B\right) + \cos\left(\tfrac{1}{2}C\right)\right]\csc\left(\tfrac{1}{2}A\right).$$

See also ARC, CYCLIC QUADRILATERAL, FUHRMANN CIRCLE, FUHRMANN TRIANGLE

References

Johnson, R. A. *Modern Geometry: An Elementary Treatise on the Geometry of the Triangle and the Circle.* Boston, MA: Houghton Mifflin, pp. 228–29, 1929.

Kimberling, C. "Problem 804." *Nieuw Archief voor Wiskunde* **6**, 170, 1988.

Kimberling, C. "Central Points and Central Lines in the Plane of a Triangle." *Math. Mag.* **67**, 163–87, 1994.

Kimberling, C. and Veldkamp, G. R. "Problem 1160 and Solution." *Crux Math.* **13**, 298–99, 1987.

Midcircle

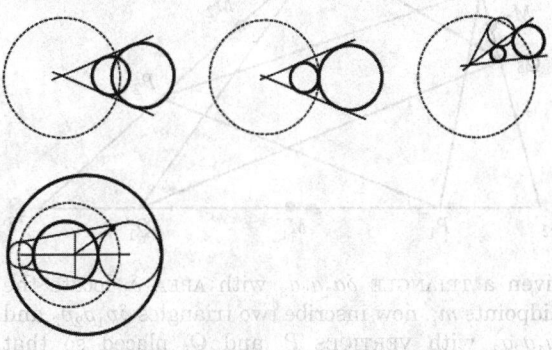

The midcircle of two given CIRCLES is the CIRCLE which would INVERT the circles into each other. Dixon (1991) gives constructions for the midcircle for four of the five possible configurations. In the case of the two given CIRCLES tangent to each other, there are two midcircles.

See also INVERSION, INVERSION CIRCLE

References

Dixon, R. *Mathographics.* New York: Dover, pp. 66–8, 1991.

Middlespoint

MITTENPUNKT

Midpoint

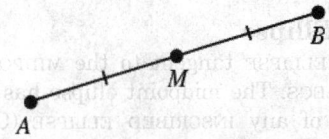

The point on a LINE SEGMENT dividing it into two segments of equal length. The midpoint of a line segment is easy to locate by first constructing a LENS using circular arcs, then connecting the cusps of the LENS. The point where the cusp-connecting line intersects the segment is then the midpoint (Pedoe 1995, p. xii). It is more challenging to locate the midpoint using only a COMPASS (i.e., a MASCHERONI CONSTRUCTION).

In a RIGHT TRIANGLE, the midpoint of the HYPOTENUSE is equidistant from the three VERTICES (Dunham 1990).

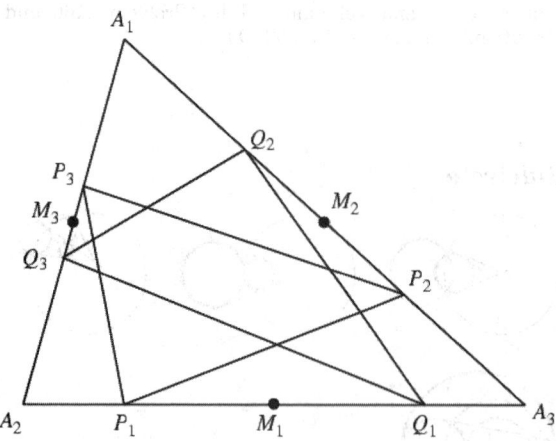

Given a TRIANGLE $\delta a_1 a_2 a_3$ with AREA δ, locate the midpoints m_i. now inscribe two triangles $\delta p_1 p_2 p_3$ and $\delta q_1 q_2 q_3$ with VERTICES P_i and Q_i placed so that $\overline{P_i M_i} = \overline{Q_i M_i}$. Then $\Delta P_1 P_2 P_3$ and $\Delta Q_1 Q_2 Q_3$ have equal areas

$$\Delta_P = \Delta_Q$$
$$= \Delta \left[1 - \left(\frac{m_1}{a_1} + \frac{m_2}{a_2} + \frac{m_3}{a_3} \right) + \frac{m_2 m_2}{a_2 a_3} + \frac{m_3 m_1}{a_3 a_1} + \frac{m_1 m_2}{a_1 a_2} \right],$$

where a_i are the sides of the original triangle and m_i are the lengths of the MEDIANS (Johnson 1929).

See also ANTICENTER, ARCHIMEDES' MIDPOINT THEOREM, BIMEDIAN, BRAHMAGUPTA'S THEOREM, BROCARD MIDPOINT, CIRCLE-POINT MIDPOINT THEOREM, CLEAVER, DROZ-FARNY THEOREM, LINE SEGMENT, MALTITUDE, MASCHERONI CONSTRUCTION, MEDIAN (TRIANGLE), MEDIATOR, MIDPOINT ELLIPSE

References

Dunham, W. *Journey through Genius: The Great Theorems of Mathematics.* New York: Wiley, pp. 120–21, 1990.
Johnson, R. A. *Modern Geometry: An Elementary Treatise on the Geometry of the Triangle and the Circle.* Boston, MA: Houghton Mifflin, p. 80, 1929.

Midpoint Ellipse

The unique ELLIPSE tangent to the MIDPOINTS of a TRIANGLE'S LEGS. The midpoint ellipse has the maximum AREA of any INSCRIBED ELLIPSE (Chakerian 1979). Under an AFFINE TRANSFORMATION, the midpoint ellipse can be transformed into the INCIRCLE of an EQUILATERAL TRIANGLE.

See also AFFINE TRANSFORMATION, ELLIPSE, INCIRCLE, MIDPOINT, TRIANGLE

References

Central Similarities. University of Minnesota College Geometry Project. Distributed by International Film Bureau, Inc.
Chakerian, G. D. "A Distorted View of Geometry." Ch. 7 in *Mathematical Plums* (Ed. R. Honsberger). Washington, DC: Math. Assoc. Amer., pp. 135–36 and 145–46, 1979.

Pedoe, D. "Thinking Geometrically." *Amer. Math. Monthly* **77**, 711–21, 1970.

Midpoint Polygon

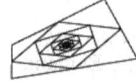 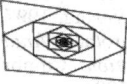  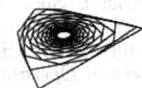

A DERIVED POLYGON with side ratios chosen as $r = 1/2$ so that inscribed polygons are constructed by connecting the midpoints of the base polygon. For a TRIANGLE P, the midpoint-inscribed polygons P_1, P_2, ... are similar triangles. For a QUADRILATERAL P, the midpoint-inscribed polygon P_1 is a PARALLELOGRAM known as the VARIGNON PARALLELOGRAM, and P_1, P_3, P_5, ... are similar parallelograms, as are P_2, P_4, P_6, ...

See also DERIVED POLYGON, MIDPOINT, VARIGNON PARALLELOGRAM, VARIGNON'S THEOREM

References

Tischel, G. "Ein Konvergenzsatz für Mittenpolygone." *Mitt. Math. Ges. Hamburg* **18**, 169–84, 1999.

Midradius

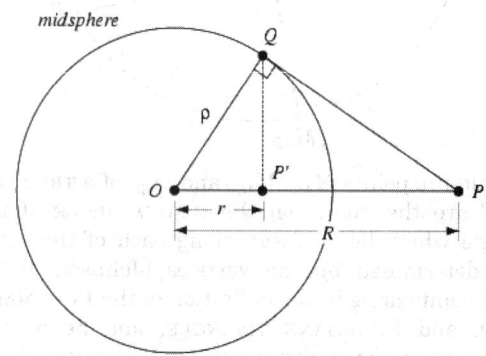

The RADIUS ρ of the MIDSPHERE of a POLYHEDRON, also called the interradius. Let P be a point on the original polyhedron and P' the corresponding point P on the dual. Then because P and P' are INVERSE POINTS, the radii $r = OP'$, $R = OP$, and $\rho = OQ$ satisfy

$$rR = \rho^2.$$

The above figure shows a plane section of a midsphere.

Let r be the INRADIUS the dual polyhedron, R CIRCUMRADIUS of the original polyhedron, and a the side length of the original polyhedron. (For a PLATONIC SOLID or ARCHIMEDEAN SOLID, r is not only the INRADIUS of the dual polyhedron, but also the INRADIUS of the original polyhedron.) For a REGULAR POLYHEDRON with SCHLÄFLI SYMBOL $\{q, p\}$, the DUAL POLYHEDRON is $\{p, q\}$. Then

$$r^2 = \left[a \csc \left(\frac{\pi}{p} \right) \right]^2 + R^2 = a^2 + \rho^2 \qquad (1)$$

$$\rho^2 = \left[a \cot\left(\frac{\pi}{p}\right) \right]^2 + R^2. \qquad (2)$$

Furthermore, let θ be the ANGLE subtended by the EDGE of an ARCHIMEDEAN SOLID. Then

$$r = \tfrac{1}{2} a \, \cos\left(\tfrac{1}{2}\theta\right) \cot\left(\tfrac{1}{2}\theta\right) \qquad (3)$$

$$\rho = \tfrac{1}{2} a \cot\left(\tfrac{1}{2}\theta\right) \qquad (4)$$

$$R = \tfrac{1}{2} a \csc\left(\tfrac{1}{2}\theta\right), \qquad (5)$$

so

$$r : \rho : R = \cos\left(\tfrac{1}{2}\theta\right) : 1 : \sec\left(\tfrac{1}{2}\theta\right) \qquad (6)$$

(Cundy and Rollett 1989). Expressing the midradius in terms of the INRADIUS r and CIRCUMRADIUS R gives

$$\rho = \tfrac{1}{2}\sqrt{2}\sqrt{r^2 + r\sqrt{r^2 + a^2}}$$

$$= \sqrt{R^2 - \tfrac{1}{4}a^2} \qquad (7)$$

for an ARCHIMEDEAN SOLID.

References

Cundy, H. and Rollett, A. *Mathematical Models, 3rd ed.* Stradbroke, England: Tarquin Pub., pp. 126–27, 1989.

Midrange

$$\text{midrange}[f(x)] \equiv \tfrac{1}{2}\{\max[f(x)] + \min[f(x)]\}.$$

See also MAXIMUM, MEAN, MEDIAN (STATISTICS), MINIMUM

References

Zwillinger, D. (Ed.). *CRC Standard Mathematical Tables and Formulae.* Boca Raton, FL: CRC Press, p. 602, 1995.

Midsphere

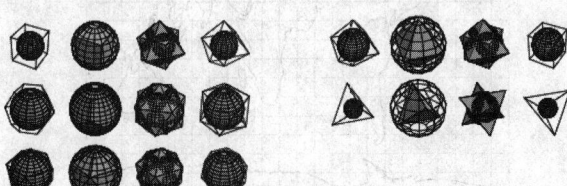

The SPHERE with respect to which the VERTICES of a POLYHEDRON are the POLES of the planes of the faces of the DUAL POLYHEDRON (and vice versa), also called the intersphere, reciprocating sphere, or INVERSION SPHERE. The midsphere touches all EDGES of a SEMIREGULAR or REGULAR POLYHEDRON, as well as the edges of the dual of that solid (Cundy and Rollett 1989, p. 117). The radius ρ of the midsphere is called

the MIDRADIUS. The figure above shows the Platonic solids and their duals, with the CIRCUMSPHERE of the solid, MIDSPHERE, and INSPHERE of the dual superposed.

See also CIRCUMSPHERE, DUAL POLYHEDRON, INSPHERE, MIDRADIUS, POLE (INVERSION)

References

Coxeter, H. S. M. *Regular Polytopes, 3rd ed.* New York: Dover, p. 16, 1973.

Cundy, H. and Rollett, A. *Mathematical Models, 3rd ed.* Stradbroke, England: Tarquin Pub., 1989.

Midvalue

CLASS MARK

Midy's Theorem

If the period of a REPEATING DECIMAL for a/p has an EVEN number of digits, the sum of the two halves is a string of 9s, where p is PRIME and a/p is a REDUCED FRACTION.

See also DECIMAL EXPANSION, REPEATING DECIMAL

References

Rademacher, H. and Toeplitz, O. *The Enjoyment of Mathematics: Selections from Mathematics for the Amateur.* Princeton, NJ: Princeton University Press, pp. 158–60, 1957.

Mikusinski's Problem

Is it possible to cover completely the surface of a SPHERE with congruent, nonoverlapping arcs of GREAT CIRCLES? Conway and Croft (1964) proved that it can be covered with half-open arcs, but not with open arcs. They also showed that the PLANE can be covered with congruent closed and half-open segments, but not with open ones.

References

Conway, J. H. and Croft, H. T. "Covering a Sphere with Great-Circle Arcs." *Proc. Cambridge Phil. Soc.* **60**, 787–00, 1964.

Gardner, M. "Point Sets on the Sphere." Ch. 12 in *Knotted Doughnuts and Other Mathematical Entertainments.* New York: W. H. Freeman, pp. 145–54, 1986.

Milin Conjecture

An INEQUALITY which IMPLIES the correctness of the ROBERTSON CONJECTURE (Milin 1971). de Branges (1985) proved this conjecture, which led to the proof of the full BIEBERBACH CONJECTURE.

See also BIEBERBACH CONJECTURE, ROBERTSON CONJECTURE

References

de Branges, L. "A Proof of the Bieberbach Conjecture." *Acta Math.* **154**, 137–52, 1985.

Milin, I. M. "The Area Method in the Theory of Univalent Functions." *Dokl. Acad. Nauk SSSR* **154**, 264–67, 1964.
Milin, I. M. *Univalent Functions and Orthonormal Systems.* Providence, RI: Amer. Math. Soc., 1977.
Stewart, I. *From Here to Infinity: A Guide to Today's Mathematics.* Oxford, England: Oxford University Press, p. 165, 1996.

Mill Curve

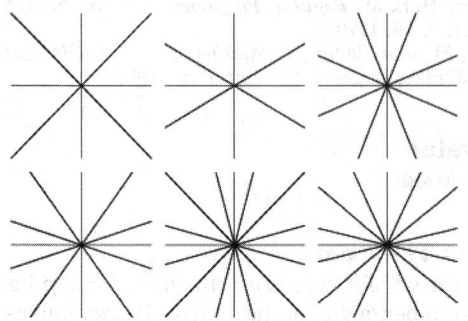

The n-roll mill curve is given by the equation

$$x^n - \binom{n}{2}x^{n-2}y^2 + \binom{n}{4}x^{n-4}y^4 - \cdots = a^n,$$

where $\binom{n}{k}$ is a BINOMIAL COEFFICIENT.

References

von Seggern, D. *CRC Standard Curves and Surfaces.* Boca Raton, FL: CRC Press, p. 86, 1993.

Miller-Aškinuze Solid

ELONGATED SQUARE GYROBICUPOLA

Miller Cylindrical Projection

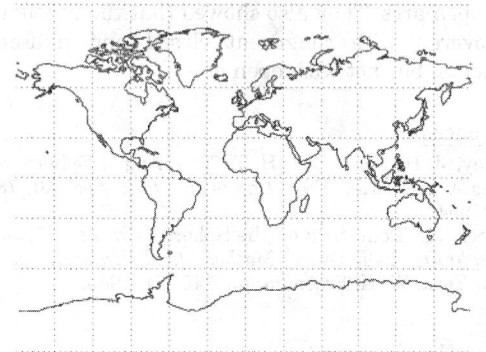

A MAP PROJECTION given by the following transformation,

$$x = \lambda - \lambda_0 \qquad (1)$$

$$y = \tfrac{5}{4} \ln\left[\tan\left(\tfrac{1}{4}\pi + \tfrac{2}{5}\phi\right)\right] \qquad (2)$$

$$= \tfrac{5}{4}\sinh^{-1}\left[\tan\left(\tfrac{4}{5}\phi\right)\right]. \qquad (3)$$

Here x and y are the plane coordinates of a projected point, λ is the longitude of a point on the globe, λ_0 is central longitude used for the projection, and ϕ is the latitude of the point on the globe. The inverse FORMULAS are

$$\phi = \tfrac{5}{2}\tan^{-1}\left(e^{4y/5}\right) - \tfrac{5}{8}\pi = \tfrac{5}{4}\tan^{-1}\left[\sinh\left(\tfrac{4}{5}y\right)\right] \qquad (4)$$

$$\lambda = \lambda_0 + x. \qquad (5)$$

See also EQUIDISTANT PROJECTION, MILLER EQUIDISTANT PROJECTION

References

Miller, O. M. "Notes on a Cylindrical World Map Projection." *Geograph. Rev.* **32**, 424–30, 1942.
Snyder, J. P. *Map Projections--A Working Manual.* U. S. Geological Survey Professional Paper 1395. Washington, DC: U. S. Government Printing Office, pp. 86–9, 1987.
United States Geological Survey. *National Atlas of the United States.* Washington, DC: USGS, pp. 330–31, 1970.

Miller Equidistant Projection

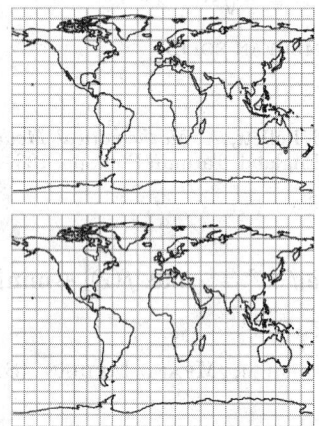

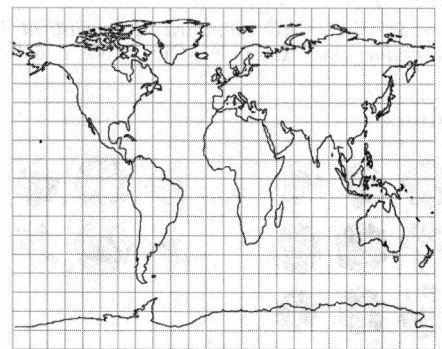

Several CYLINDRICAL EQUIDISTANT PROJECTIONS were devised by R. Miller. Miller's projections have standard parallels of $\phi_1 = 37°30'$ (giving minimal overall scale distortion), $\phi_1 = 43°$ (giving minimal scale distortion over continents), and $\phi_1 = 50°28'$ (Miller 1949).

See also CYLINDRICAL EQUIDISTANT PROJECTION, MILLER CYLINDRICAL PROJECTION

References

Miller, R. "An Equi-Rectangular Map Projection." *Geography Rev.* **34**, 196–01, 1949.
Miller, R. "Correction to: An Equi-Rectangular Map Projection." *Geography* **36**, 270, 1951.
Snyder, J. P. *Flattening the Earth: Two Thousand Years of Map Projections.* Chicago, IL: University of Chicago Press, 1993.

Miller's Algorithm

For a catastrophically unstable recurrence in one direction, any seed values for consecutive x_j and x_{j+1} will converge to the desired sequence of functions in the opposite direction times an unknown normalization factor.

Miller's Primality Test

If a number fails this test, it is not a PRIME. If the number passes, it *may* be a PRIME. A number passing Miller's test is called a STRONG PSEUDOPRIME to base a. If a number n does not pass the test, then it is called a WITNESS for the COMPOSITENESS of n. If n is an ODD, POSITIVE COMPOSITE NUMBER, then n passes Miller's test for at most $(n-1)/4$ bases with $1 \le a \le -1$ (Long 1995). There is no analog of CARMICHAEL NUMBERS for STRONG PSEUDOPRIMES.

The only COMPOSITE NUMBER less than 2.5×10^{13} which does not have 2, 3, 5, or 7 as a WITNESS is 3215031751. Miller showed that any composite n has a WITNESS less than $70(\ln n)^2$ if the RIEMANN HYPOTHESIS is true.

See also ADLEMAN-POMERANCE-RUMELY PRIMALITY TEST, STRONG PSEUDOPRIME

References

Long, C. T. Th. 4.21 in *Elementary Introduction to Number Theory, 3rd ed.* Prospect Heights, IL: Waveland Press, 1995.

Miller's Solid

ELONGATED SQUARE GYROBICUPOLA

Milliard

In British, French, and German usage, one milliard equals 10^9. American usage does not have a number called the milliard, instead using the term BILLION to denote 10^9.

See also BILLION, LARGE NUMBER, MILLION, TRILLION

Millin Series

The series with sum

$$S' \equiv \sum_{n=0}^{\infty} \frac{1}{F_{2^n}} = \frac{1}{2}\left(7 - \sqrt{5}\right),$$

where F_k is a FIBONACCI NUMBER (Honsberger 1985).

See also FIBONACCI NUMBER

References

Honsberger, R. *Mathematical Gems III.* Washington, DC: Math. Assoc. Amer., pp. 135–37, 1985.

Million

The number $1,000,000 = 10^6$. While one million in the "American" system of numbers means the same thing as one million in the "British" system, the words BILLION, TRILLION, etc., refer to *different numbers* in the two naming systems. Fortunately, in recent years, the "American" system has become common in both the United States and Britain.

While Americans may say "Thanks a million" to express gratitude, Norwegians offer "Thanks a thousand" ("tusen takk").

See also BILLION, LARGE NUMBER, MILLIARD, THOUSAND, TRILLION

Mills' Constant

N.B. A detailed online essay by S. Finch was the starting point for this entry.

Mills (1947) proved the existence of a constant

$$\theta = 1.306377883863080690\ldots \qquad (1)$$

(Sloane's A051021) such that

$$f(n) = \lfloor \theta^{3^n} \rfloor \qquad (2)$$

is PRIME for all $n \ge 1$, where $\lfloor x \rfloor$ is the FLOOR FUNCTION. It is not, however, known if θ is IRRATIONAL. The first few values of $f(n)$ are 2, 11, 1361, 2521008887, ... (Sloane's A051254).

Mills' proof was based on the following theorem by Hoheisel (1930) and Ingham (1937). Let p_n be the nth PRIME, then there exists a constant K such that

$$p_{n+1} - p_n < K p_n^{5/8} \qquad (3)$$

for all n. This has more recently been strengthened to

$$p_{n+1} - p_n < K p_n^{1051/1920} \qquad (4)$$

(Mozzochi 1986). If the RIEMANN HYPOTHESIS is true, then Cramér (1937) showed that

$$p_{n+1} - p_n = \mathcal{O}\left(\ln p_n \sqrt{p_n}\right) \qquad (5)$$

(Finch).

Hardy and Wright (1979) and Ribenboim (1996) point out that, despite the beauty of such PRIME FORMULAS, they do not have any practical consequences. In fact, unless the exact value of θ is known, the PRIMES themselves must be known in advance to determine θ. The numbers generated by $f(n)$ grow very rapidly, with the first few being 2, 11, 1361, ...

A generalization of Mills' theorem to an arbitrary sequence of POSITIVE INTEGERS is given as an exercise by Ellison and Ellison (1985). Consequently, infinitely many values for θ other than the number $1.3063\ldots$ are possible.

See also CEILING FUNCTION, PRIME FORMULAS, PRIME NUMBER

References

Caldwell, C. "Mills' Theorem--A Generalization." http://www.utm.edu/research/primes/notes/proofs/A3n.html.

Ellison, W. and Ellison, F. *Prime Numbers.* New York: Wiley, pp. 31–2, 1985.

Finch, S. "Favorite Mathematical Constants." http://www.mathsoft.com/asolve/constant/mills/mills.html.

Hardy, G. H. and Wright, E. M. *An Introduction to the Theory of Numbers, 5th ed.* Oxford, England: Clarendon Press, 1979.

Mills, W. H. "A Prime-Representing Function." *Bull. Amer. Math. Soc.* **53**, 604, 1947.

Mozzochi, C. J. "On the Difference Between Consecutive Primes." *J. Number Th.* **24**, 181–87, 1986.

Nagell, T. *Introduction to Number Theory.* New York: Wiley, p. 65, 1951.

Ribenboim, P. *The New Book of Prime Number Records.* New York: Springer-Verlag, pp. 186–87, 1996.

Ribenboim, P. *The Little Book of Big Primes.* New York: Springer-Verlag, pp. 109–10, 1991.

Sloane, N. J. A. Sequences A051021 and A051254 in "An On-Line Version of the Encyclopedia of Integer Sequences." http://www.research.att.com/~njas/sequences/eisonline.html.

Mills-Robbins-Rumsey Determinant Formula

$$\det\begin{pmatrix} i+j+\mu \\ 2i-j \end{pmatrix}_{i,\,j=0}^{n-1} = 2^{-n}\prod_{k=0}^{n-1}\Delta_{2k}(2\mu),$$

where μ is an indeterminate, $\Delta_0(\mu)=2$,

$$\Delta_{2j}(\mu) = \frac{(\mu+2j+2)_j\left(\tfrac{1}{2}\mu_2 j + \tfrac{3}{2}\right)_{j-1}}{(j)_j\left(\tfrac{1}{2}\mu+j+\tfrac{3}{2}\right)_{j-1}},$$

for $j=1,\,2,\,\ldots$, and $(x)_j = x(x+1)\cdots(x+j-1)$ is the RISING FACTORIAL (Mills *et al.* 1987, Andrews and Burge 1993).

References

Andrews, G. E. and Burge, W. H. "Determinant Identities." *Pacific J. Math.* **158**, 1–4, 1993.

Mills, W. H.; Robbins, D. P.; and Rumsey, H. Jr. "Enumeration of a Symmetry Class of Plane Partitions." *Discrete Math.* **67**, 43–5, 1987.

Petkovsek, M. and Wilf, H. S. "A High-Tech Proof of the Mills-Robbins-Rumsey Determinant Formula." *Electronic J. Combinatorics* **3**, No. 2, R19, 1–, 1996. http://www.combinatorics.org/Volume_3/volume3_2.html.

Milne's Method

A PREDICTOR-CORRECTOR METHOD for solution of ORDINARY DIFFERENTIAL EQUATIONS. The third-order equations for predictor and corrector are

$$y_{n+1} = y_{n-3} + \tfrac{4}{3}h(2y_n' - y_{n-1}' + 2y_{n-2}') + \mathcal{O}(h^5)$$
$$y_{n+1} = y_{n-1} + \tfrac{1}{3}h(y_{n-1}' - 4y_n' + y_{n+1}') + \mathcal{O}(h^5).$$

Abramowitz and Stegun (1972) also give the fifth order equations and formulas involving higher derivatives.

See also ADAMS' METHOD, GILL'S METHOD, PREDICTOR-CORRECTOR METHODS, RUNGE-KUTTA METHOD

References

Abramowitz, M. and Stegun, C. A. (Eds.). *Handbook of Mathematical Functions with Formulas, Graphs, and Mathematical Tables, 9th printing.* New York: Dover, pp. 896–97, 1972.

Milnor's Conjecture

The UNKNOTTING NUMBER for a TORUS KNOT $(p,\,q)$ is $(p-1)(q-1)/2$. This 40-year-old CONJECTURE was proved (Adams 1994) in Kronheimer and Mrowka (1993, 1995).

See also TORUS KNOT, UNKNOTTING NUMBER

References

Adams, C. C. *The Knot Book: An Elementary Introduction to the Mathematical Theory of Knots.* New York: W. H. Freeman, p. 113, 1994.

Kronheimer, P. B. and Mrowka, T. S. "Gauge Theory for Embedded Surfaces. I." *Topology* **32**, 773–26, 1993.

Kronheimer, P. B. and Mrowka, T. S. "Gauge Theory for Embedded Surfaces. II." *Topology* **34**, 37–7, 1995.

Milnor's Theorem

If a COMPACT MANIFOLD M has NONNEGATIVE RICCI CURVATURE, then its FUNDAMENTAL GROUP has at most POLYNOMIAL growth. On the other hand, if M has NEGATIVE curvature, then its FUNDAMENTAL GROUP has exponential growth in the sense that $n(\lambda)$ grows exponentially, where $n(\lambda)$ is (essentially) the number of different "words" of length λ which can be made in the FUNDAMENTAL GROUP.

References

Chavel, I. *Riemannian Geometry: A Modern Introduction.* New York: Cambridge University Press, 1994.

Min

MINIMUM

Mincut

Let $G=(V,\,E)$ be a (not necessarily simple) UNDIRECTED edge-weighted graph with nonnegative weights. A cut C of G is any nontrivial subset of V, and the weight of the cut is the sum of weights of edges crossing the cut. A mincut is then defined as a cut of G of minimum weight. The problem is NP-complete for general graphs, but polynomial-time solvable for trees.

See also BOOLEAN FUNCTION, WEIGHTED GRAPH

References

Stoer, M. and Wagner, F. "A Simple Min Cut Algorithm." *Algorithms--ESA '94*, LNCS 855, 141–47, 1994.

Minimal Cover

A minimal cover is a COVER for which removal of any single member destroys the covering property. For example, of the five COVERS of $\{1, 2\}$, namely $\{\{1\}, \{2\}\}$, $\{\{1, 2\}\}$, $\{\{1\}, \{1, 2\}\}$, $\{\{2\}, \{1, 2\}\}$, and $\{\{1\}, \{2\}, \{1, 2\}\}$, only $\{\{1\}, \{2\}\}$ and $\{\{1, 2\}\}$ are minimal covers. Similarly, the minimal covers of $\{1, 2, 3\}$ are given by $\{\{1\}, \{2\}, \{3\}\}$, $\{\{1, 2\}, \{3\}\}$, $\{\{1, 3\}, \{2\}\}$, $\{\{1, 2\}, \{2, 3\}\}$, $\{\{1, 2\}, \{2, 3\}\}$, $\{\{1, 2, 3\}\}$, $\{\{1, 2\}, \{1, 3\}\}$, $\{\{1, 2\}, \{2, 3\}\}$. The number of minimal covers of n members for $n = 1$, 2, ..., are 1, 2, 8, 49, 462, 6424, 129425, ... (Sloane's A046165).

Let $\mu(n, k)$ be the number of minimal covers of $\{1, \ldots, n\}$ with k members. Then

$$\mu(n, k) = \frac{1}{k!} \sum_{m=k}^{\alpha_k} \binom{2^k - k - 1}{m - k} m! s(n, m),$$

where $\binom{n}{k}$ is a BINOMIAL COEFFICIENT, $s(n, m)$ is a STIRLING NUMBER OF THE SECOND KIND, and

$$\alpha_k = \min(n, 2^k - 1).$$

Special cases include $\mu(n, 1) = 1$ and $\mu(n, 2) = s(n + 1, 3)$. The table below gives the a triangle of $\mu(n, k)$ (Sloane's A035348).

n	$k = 1$	$k = 2$	$k = 3$	$k = 4$	$k = 5$	$k = 6$	$k = 7$
Sloane	Sloane	Sloane's A003468	Sloane's A016111	Sloane's A046166	Sloane's A046167	Sloane's A057668
	A000392						
1	1						
2	1						
3	1	6	1				
4	1	25	22	1			
5	1	90	305	65	1		
6	1	301	3410	2540	171	1	
7	1	966	33621	77350	17066	420	1
8	1	3925	305382	2022951	1298346	100814	988

See also COVER, LEW K-GRAM, STIRLING NUMBER OF THE SECOND KIND

References

Hearne, T. and Wagner, C. "Minimal Covers of Finite Sets." *Disc. Math.* **5**, 247–51, 1973.
Macula, A. J. "Covers of a Finite Set." *Math. Mag.* **67**, 141–44, 1994.
Macula, A. J. "Lewis Carroll and the Enumeration of Minimal Covers." *Math. Mag.* **68**, 269–74, 1995.
Sloane, N. J. A. Sequences A000392, A003468, A016111, A035348, A046165, A046166, A046167, A046168, and
A057668 in "An On-Line Version of the Encyclopedia of Integer Sequences." http://www.research.att.com/~njas/sequences/eisonline.html.

Minimal Discriminant

FREY CURVE

Minimal Matrix

A MATRIX with 0 DETERMINANT whose DETERMINANT becomes NONZERO when any element on or below the diagonal is changed from 0 to 1. An example is

$$\mathsf{M} = \begin{bmatrix} 1 & -1 & 0 & 0 \\ 0 & 0 & -1 & 0 \\ 1 & 1 & 1 & -1 \\ 0 & 0 & 1 & 0 \end{bmatrix}.$$

There are $2^n - 1$ minimal SPECIAL MATRICES of size $n \times n$.

See also SPECIAL MATRIX

References

Knuth, D. E. "Problem 10470." *Amer. Math. Monthly* **102**, 655, 1995.

Minimal Polynomial (Matrix)

The minimal polynomial of a matrix A is the polynomial in A of smallest degree n such that

$$p(\mathsf{A}) = \sum_{i=0}^{n} c_i \mathsf{A}^i = 0. \tag{1}$$

The minimal polynomial divides any polynomial q with $q(\mathsf{A}) = 0$ and, in particular, it divides the CHARACTERISTIC POLYNOMIAL. If the CHARACTERISTIC POLYNOMIAL factors as

$$\text{char}(\mathsf{A})(x) = (x - \lambda_1)^{n_1} \ldots (x - \lambda_k)^{n_k}, \tag{2}$$

then its minimal polynomial is

$$p(x) = (x - \lambda_1)^{m_1} \ldots (x - \lambda_k)^{m_k} \tag{3}$$

with $1 \le m_i \le n_i$.

For example, the CHARACTERISTIC POLYNOMIAL of the $n \times n$ ZERO MATRIX is $(-1)^n x^n$, and its minimal polynomial is x. The CHARACTERISTIC POLYNOMIAL and minimal polynomial of

$$\begin{bmatrix} 0 & 1 \\ 0 & 0 \end{bmatrix} \tag{4}$$

are the same (up to scalar multiple), x^2.

The following *Mathematica* command will find the minimal polynomial for the SQUARE MATRIX a in the variable x.

```
MinPolyMatrix[a_List,x_]:=
```

```
                                    Modu-
le[{i,n=1,qu={},mnm={Flatten[IdentityMatr-
{Flatten[IdentityMatrix[Length[a]]]}},
   While[Length[qu]==0,
    AppendTo[mnm,Flatten[MatrixPower[a,n]]];
    qu=NullSpace[Transpose[mnm]];
    n++
   ];
   First[qu].Table[x^i,{i,0,n-1}]
  ]
```

See also CAYLEY-HAMILTON THEOREM, CHARACTERIS-
TIC POLYNOMIAL, MINIMAL POLYNOMIAL (ALGEBRAIC
NUMBER), RATIONAL CANONICAL FORM

References

Dummit, D. and Foote, R. *Abstract Algebra.* Englewood
    Cliffs, NJ: Prentice-Hall, 1991.
Herstein, I. §6.7 in *Topics in Algebra, 2nd ed.* New York:
    Wiley, 1975.
Jacobson, N. §3.10 in *Basic Algebra I.* New York:
    W. H. Freeman, 1985.

Minimal Residue

The value b or $b - m$, whichever is smaller in
ABSOLUTE VALUE, where $a \equiv b \pmod{m}$.

See also RESIDUE (CONGRUENCE)

Minimal Set

A SET for which the dynamics can be generated by the
dynamics on any SUBSET.

Minimal Surface

Minimal surfaces are defined as surfaces with zero
MEAN CURVATURE. A minimal surface parametrized
as $\mathbf{x} = (u, v, h(u, v))$ therefore satisfies LAGRANGE'S
EQUATION,

$$\left(1 + f_v^2\right)f_{uu} + 2f_u f_v f_{uv} + \left(1 + f_u^2\right)f_{vv} = 0.$$

Finding a minimal surface of a boundary with
specified constraints is a problem in the CALCULUS
OF VARIATIONS and is sometimes known at PLATEAU'S
PROBLEM. Minimal surfaces may also be characterized as surfaces of minimal SURFACE AREA for given
boundary conditions. A PLANE is a trivial MINIMAL
SURFACE, and the first nontrivial examples (the
CATENOID and HELICOID) were found by Meusnier in
1776 (Meusnier 1785). The problem of finding the
minimum bounding surface of a SKEW QUADRILAT-
ERAL was solved by Schwarz (1890).

Note that while a SPHERE is a "minimal surface" in
the sense that it minimizes the surface area-to-
volume ratio, it does not qualify as a minimal surface
in the sense used by mathematicians.

Euler proved that a minimal surface is planar IFF its
GAUSSIAN CURVATURE is zero at every point so that it
is locally SADDLE-shaped. The EXISTENCE of a solution

to the general case was independently proven by
Douglas (1931) and Radó (1933), although their
analysis could not exclude the possibility of singularities. Osserman (1970) and Gulliver (1973) showed
that a minimizing solution *cannot* have singularities.

The only known complete (boundaryless), embedded
(no self-intersections) minimal surfaces of finite
topology known for 200 years were the CATENOID,
HELICOID, and PLANE. Hoffman discovered a three-
ended GENUS 1 minimal embedded surface, and
demonstrated the existence of an infinite number of
such surfaces. A four-ended embedded minimal surface has also been found. L. Bers proved that any
finite isolated SINGULARITY of a single-valued parameterized minimal surface is removable.

A surface can be parameterized using a ISOTHERMAL
PARAMETERIZATION. Such a parameterization is minimal if the coordinate functions x_k are HARMONIC, i.e.,
$\phi_k(\zeta)$ are ANALYTIC. A minimal surface can therefore
be defined by a triple of ANALYTIC FUNCTIONS such
that $\phi_k \phi_k = 0$. The REAL parameterization is then
obtained as

$$x_k = \Re \int \phi_k(\zeta) \, d\zeta. \tag{1}$$

But, for an ANALYTIC FUNCTION f and a MEROMORPHIC
FUNCTION g, the triple of functions

$$\phi_1(\zeta) = f(1 - g^2) \tag{2}$$

$$\phi_2(\zeta) = if(1 + g^2) \tag{3}$$

$$\phi_3(\zeta) = 2fg \tag{4}$$

are ANALYTIC as long as f has a zero of order $\geq m$ at
every POLE of g of order m. This gives a minimal
surface in terms of the ENNEPER-WEIERSTRASS PARA-
METERIZATION

$$\Re \int \begin{bmatrix} f(1 - g^2) \\ if(1 + g^2) \\ 2fg \end{bmatrix} d\zeta. \tag{5}$$

See also BERNSTEIN MINIMAL SURFACE THEOREM,
BOUR'S MINIMAL SURFACE, BUBBLE, CALCULUS OF
VARIATIONS, CATALAN'S SURFACE, CATENOID, COM-
PLETE MINIMAL SURFACE, COSTA MINIMAL SURFACE,
DOUBLE BUBBLE, ENNEPER'S MINIMAL SURFACE, EN-
NEPER-WEIERSTRASS PARAMETERIZATION, FLAT SUR-
FACE, GYROID, HELICOID, HENNEBERG'S MINIMAL
SURFACE, HOFFMAN'S MINIMAL SURFACE, IMMERSED
MINIMAL SURFACE, LICHTENFELS MINIMAL SURFACE,
LOPEZ MINIMAL SURFACE, MEAN CURVATURE, NIR-
ENBERG'S CONJECTURE, OLIVEIRA'S MINIMAL SUR-
FACE, PARAMETERIZATION, PLANE, PLATEAU'S LAWS,
PLATEAU'S PROBLEM, SCHERK'S MINIMAL SURFACES,
SCHWARZ'S MINIMAL SURFACE, SURFACE AREA, TRI-
NOID.

References

Darboux, G. *Leçons sur la théorie générale des surfaces.* Paris: Gauthier-Villars, 1941.

Dickson, S. "Minimal Surfaces." *Mathematica J.* **1**, 38–0, 1990.

Dierkes, U.; Hildebrandt, S.; Küster, A.; and Wohlraub, O. *Minimal Surfaces, Vol. 1: Boundary Value Problems.* New York: Springer-Verlag, 1992.

Dierkes, U.; Hildebrandt, S.; Küster, A.; and Wohlraub, O. *Minimal Surfaces, Vol. 2: Boundary Regularity.* New York: Springer-Verlag, 1992.

do Carmo, M. P. "Minimal Surfaces." §3.5 in *Mathematical Models from the Collections of Universities and Museums* (Ed. G. Fischer). Braunschweig, Germany: Vieweg, pp. 41–3, 1986.

Douglas, J. "Solution of the Problem of Plateau." *Trans. Amer. Math. Soc.* **33**, 263–21, 1931.

Fischer, G. (Ed.). Plates 93 and 96 in *Mathematische Modelle/Mathematical Models, Bildband/Photograph Volume.* Braunschweig, Germany: Vieweg, pp. 89 and 96, 1986.

Gray, A. "Minimal Surfaces" and "Minimal Surfaces and Complex Variables." Ch. 30 and 31 in *Modern Differential Geometry of Curves and Surfaces with Mathematica, 2nd ed.* Boca Raton, FL: CRC Press, pp. 681–34, 1997.

Gulliver, R. "Regularity of Minimizing Surfaces of Prescribed Mean Curvature." *Ann. Math.* **97**, 275–05, 1973.

Hoffman, D. "The Computer-Aided Discovery of New Embedded Minimal Surfaces." *Math. Intell.* **9**, 8–1, 1987.

Hoffman, D. and Meeks, W. H. III. *The Global Theory of Properly Embedded Minimal Surfaces.* Amherst, MA: University of Massachusetts, 1987.

Isenberg, C. *The Science of Soap Films and Soap Bubbles.* New York: Dover, 1992.

Lagrange. "Essai d'une nouvelle méthode pour déterminer les maxima et les minima des formules intégrales indéfinies." 1776.

Meusnier, J. B. "Mémoire sur la courbure des surfaces." *Mém. des savans étrangers* **10** (lu 1776), 477–10, 1785.

Nitsche, J. C. C. *Introduction to Minimal Surfaces.* Cambridge, England: Cambridge University Press, 1989.

Osserman, R. *A Survey of Minimal Surfaces.* New York: Dover, 1986.

Osserman, R. "A Proof of the Regularity Everywhere of the Classical Solution to Plateau's Problem." *Ann. Math.* **91**, 550–69, 1970.

Osserman, R. (Ed.). *Minimal Surfaces.* Berlin: Springer-Verlag, 1997.

Radó, T. "On the Problem of Plateau." *Ergeben. d. Math. u. ihrer Grenzgebiete.* Berlin: Springer-Verlag, 1933.

Schwarz, H. A. *Gesammelte Mathematische Abhandlungen, 2nd ed.* New York: Chelsea.

Weisstein, E. W. "Books about Minimal Surfaces." http://www.treasure-troves.com/books/MinimalSurfaces.html.

Wells, D. *The Penguin Dictionary of Curious and Interesting Geometry.* London: Penguin, pp. 185–87, 1991.

Minimax Approximation

A minimization of the MAXIMUM error for a fixed number of terms.

See also REMEZ ALGORITHM

Minimax Polynomial

The approximating POLYNOMIAL which has the smallest maximum deviation from the true function. It is closely approximated by the CHEBYSHEV POLYNOMIALS OF THE FIRST KIND.

Minimax Theorem

The fundamental theorem of GAME THEORY which states that every FINITE, ZERO-SUM, two-person GAME has optimal MIXED STRATEGIES. It was proved by John von Neumann in 1928.

Formally, let **X** and **Y** be MIXED STRATEGIES for players A and B. Let A be the PAYOFF MATRIX. Then

$$\max_X \min_Y \mathbf{X}^T A \mathbf{Y} = \min_Y \max_X \mathbf{X}^T A \mathbf{Y} = v,$$

where v is called the VALUE of the GAME and **X** and **Y** are called the solutions. It also turns out that if there is more than one optimal MIXED STRATEGY, there are infinitely many.

See also GAME, GAME THEORY, MIXED STRATEGY

References

Willem, M. *Minimax Theorem.* Boston, MA: Birkhäuser, 1996.

Minimize

INFIMUM

Minimum

The smallest value of a set, function, etc. The minimum value of a set of elements $A = \{a_i\}_{i=1}^N$ is denoted $\min A$ or $\min_i a_i$, and is equal to the first element of a sorted (i.e., ordered) version of A. For example, given the set $\{3, 5, 4, 1\}$, the sorted version is $\{1, 3, 4, 5\}$, so the minimum is 1. The MAXIMUM and minimum are the simplest ORDER STATISTICS.

A continuous FUNCTION may assume a minimum at a single point or may have minima at a number of points. A GLOBAL MINIMUM of a FUNCTION is the smallest value in the entire RANGE of the FUNCTION, while a LOCAL MINIMUM is the smallest value in some local neighborhood.

For a function $f(x)$ which is CONTINUOUS at a point x_0, a NECESSARY but not SUFFICIENT condition for $f(x)$ to have a RELATIVE MINIMUM at $x = x_0$ is that x_0 be a CRITICAL POINT (i.e., $f(x)$ is either not DIFFERENTIABLE at x_0 or x_0 is a STATIONARY POINT, in which case $f'(x_0) = 0$).

The FIRST DERIVATIVE TEST can be applied to CONTINUOUS FUNCTIONS to distinguish minima from MAXIMA. For twice differentiable functions of one variable, $f(x)$, or of two variables, $f(x, y)$, the SECOND DERIVATIVE TEST can sometimes also identify the nature of an EXTREMUM. For a function $f(x)$, the EXTREMUM TEST succeeds under more general conditions than the SECOND DERIVATIVE TEST.

See also CONJUGATE GRADIENT METHOD, CRITICAL POINT, EXTREMUM, FIRST DERIVATIVE TEST, GLOBAL MAXIMUM, INFLECTION POINT, LOCAL MAXIMUM, MAXIMUM, MIDRANGE, ORDER STATISTIC, SADDLE POINT (FUNCTION), SECOND DERIVATIVE TEST, STATIONARY POINT, STEEPEST DESCENT METHOD

References

Abramowitz, M. and Stegun, C. A. (Eds.). *Handbook of Mathematical Functions with Formulas, Graphs, and Mathematical Tables, 9th printing.* New York: Dover, p. 14, 1972.

Brent, R. P. *Algorithms for Minimization Without Derivatives.* Englewood Cliffs, NJ: Prentice-Hall, 1973.

Nash, J. C. "Descent to a Minimum I-II: Variable Metric Algorithms." Chs. 15–6 in *Compact Numerical Methods for Computers: Linear Algebra and Function Minimisation, 2nd ed.* Bristol, England: Adam Hilger, pp. 186–06, 1990.

Niven, I. *Maxima and Minima without Calculus.* Washington, DC: Math. Assoc. Amer., 1982.

Press, W. H.; Flannery, B. P.; Teukolsky, S. A.; and Vetterling, W. T. "Minimization or Maximization of Functions." Ch. 10 in *Numerical Recipes in FORTRAN: The Art of Scientific Computing, 2nd ed.* Cambridge, England: Cambridge University Press, pp. 387–48, 1992.

Tikhomirov, V. M. *Stories About Maxima and Minima.* Providence, RI: Amer. Math. Soc., 1991.

Minimum Clique

CLIQUE

Minimum Gossip Graph

GOSSIPING

Minimum Modulus Principle

Let f be ANALYTIC on a DOMAIN $U \subseteq \mathbb{C}$, and assume that f never vanishes. Then if there is a point $z_0 \in U$ such that $|f(z_0)| \leq |f(z)|$ for all $z \in U$, then f is constant.

Let $U \subseteq \mathbb{C}$ be a bounded domain, let f be a continuous function on the closed set $\bar{U}$ that is analytic on U, and assume that f never vanishes on $\bar{U}$. Then the minimum value of $|f|$ on $\bar{U}$ (which always exists) must occur on ∂U. In other words,

$$\min_{\bar{U}} |f| = \min_{\partial U} |f|.$$

See also MAXIMUM MODULUS PRINCIPLE, MODULUS (COMPLEX NUMBER)

References

Krantz, S. G. "The Minimum Principle." §5.4.3 in *Handbook of Complex Analysis.* Boston, MA: Birkhäuser, p. 77, 1999.

Minimum Spanning Tree

The minimum spanning tree of a WEIGHTED GRAPH is a set of $n - 1$ edges of minimum total weight which form a SPANNING TREE of the graph. When a graph is unweighted, any SPANNING TREE is a minimum spanning tree.

The minimum spanning tree can be found in polynomial time. Common algorithms include those due to Prinn (1957) and Kruskal (1956). The problem can also be formulated using MATROIDS (Papadimitriou and Steiglitz 1982). The minimum spanning tree can be found using the command `MinimumSpanningTree[g]` in the *Mathematica* add-on package `DiscreteMath`Combinatorica`` (which can be loaded with the command `<<DiscreteMath`).

See also SPANNING TREE

References

Fredman, M. L. and Tarjan, R. E. "Fibonacci Heaps and Their Uses in Network Optimization." *J. ACM* **34**, 596–15, 1987.

Graham, R. L. and Hell, P. "On the History of the Minimum Spanning Tree Problem." *Ann. History Comput.* **7**, 43–7, 1985.

Kruskal, J. B. "On the Shortest Spanning Subtree of a Graph and the Traveling Salesman Problem." *Proc. Amer. Math. Soc.* **7**, 48–0, 1956.

Papadimitriou, C. H. and Steiglitz, K. *Combinatorial Optimization: Algorithms and Complexity.* Englewood Cliffs, NJ: Prentice-Hall, 1982.

Prinn, R. C. "Shortest Connection Networks and Some Generalizations." *Bell System Tech. J.* **36**, 1389–401, 1957.

Skiena, S. "Minimum Spanning Tree." §6.2 in *Implementing Discrete Mathematics: Combinatorics and Graph Theory with Mathematica.* Reading, MA: Addison-Wesley, pp. 232–36, 1990.

Minimum Vertex Cover

VERTEX COVER

Minkowski-Bouligand Dimension

In many cases, the HAUSDORFF DIMENSION correctly describes the correction term for a resonator with FRACTAL PERIMETER in Lorentz's conjecture. However, in general, the proper dimension to use turns out to be the Minkowski-Bouligand dimension (Schroeder 1991).

Let $F(r)$ be the AREA traced out by a small CIRCLE with RADIUS r following a fractal curve. Then, providing the LIMIT exists,

$$D_M \equiv \lim_{r \to 0} \frac{\ln F(r)}{-\ln r} + 2$$

(Schroeder 1991). It is conjectured that for all strictly self-similar fractals, the Minkowski-Bouligand dimension is equal to the HAUSDORFF DIMENSION D; otherwise $D_M > D$.

See also HAUSDORFF DIMENSION, MINKOWSKI COVER, MINKOWSKI SAUSAGE

References

Berry, M. V. "Diffractals." *J. Phys.* **A12**, 781–97, 1979.

Hunt, F. V.; Beranek, L. L.; and Maa, D. Y. "Analysis of Sound Decay in Rectangular Rooms." *J. Acoust. Soc. Amer.* **11**, 80–4, 1939.

Lapidus, M. L. and Fleckinger-Pellé, J. "Tambour fractal: vers une résolution de la conjecture de Weyl-Berry pour les valeurs propres du laplacien." *Compt. Rend. Acad. Sci. Paris Math. Sér 1* **306**, 171–75, 1988.

Schroeder, M. *Fractals, Chaos, Power Laws: Minutes from an Infinite Paradise.* New York: W. H. Freeman, pp. 41–5, 1991.

Minkowski Convex Body Theorem

A bounded plane convex region symmetric about a LATTICE POINT and with AREA > 4 must contain at least three LATTICE POINTS in the interior. In n-D, the theorem can be generalized to a region with AREA $> 2^n$, which must contain at least three LATTICE POINTS. The theorem can be derived from BLICHFELDT'S THEOREM.

See also BLICHFELDT'S THEOREM

References

Hilbert, D. and Cohn-Vossen, S. "Minkowski's Theorem." §6.3 in *Geometry and the Imagination.* New York: Chelsea, pp. 41–4, 1999.

Minkowski, H. *Geometrie der Zahlen.* Leipzig, Germany: Teubner, 1912.

Steinhaus, H. *Mathematical Snapshots, 3rd ed.* New York: Dover, p. 99, 1999.

Warmus, W. *Colloq. Math. I* **1**, 45–6, 1947.

Minkowski Cover

The covering of a PLANE CURVE with disks of radius ϵ whose centers lie on the curve.

See also MINKOWSKI-BOULIGAND DIMENSION, MINKOWSKI SAUSAGE

Minkowski Geometry

MINKOWSKI SPACE

Minkowski-Hlawka Theorem

There exist lattices in n-D having HYPERSPHERE PACKING densities satisfying

$$\eta \geq \frac{\zeta(n)}{2^{n-1}},$$

where $\zeta(n)$ is the RIEMANN ZETA FUNCTION. However, the proof of this theorem is nonconstructive and it is still not known how to actually construct packings that are this dense.

See also HERMITE CONSTANTS, HYPERSPHERE PACKING

References

Conway, J. H. and Sloane, N. J. A. *Sphere Packings, Lattices, and Groups, 2nd ed.* New York: Springer-Verlag, pp. 14–6, 1993.

Pach, J. and Agarwal, P. K. *Combinatorial Geometry.* New York: Wiley, 1995.

Minkowski Integral Inequality

If $p > 1$, then

$$\left[\int_a^b |f(x) + g(x)|^p \, dx \right]^{1/p}$$

$$\leq \left[\int_a^b |f(x)|^p \, dx \right]^{1/p} + \left[\int_a^b |g(x)|^p \, dx \right]^{1/p}.$$

See also MINKOWSKI SUM INEQUALITY

References

Abramowitz, M. and Stegun, C. A. (Eds.). *Handbook of Mathematical Functions with Formulas, Graphs, and Mathematical Tables, 9th printing.* New York: Dover, p. 11, 1972.

Gradshteyn, I. S. and Ryzhik, I. M. *Tables of Integrals, Series, and Products, 6th ed.* San Diego, CA: Academic Press, p. 1099, 2000.

Hardy, G. H.; Littlewood, J. E.; and Pólya, G. *Inequalities, 2nd ed.* Cambridge, England: Cambridge University Press, pp. 146–50, 1988.

Minkowski, H. *Geometrie der Zahlen, Vol. 1.* Leipzig, Germany: pp. 115–17, 1896.

Sansone, G. *Orthogonal Functions, rev. English ed.* New York: Dover, p. 33, 1991.

Minkowski Measure

The Minkowski measure of a bounded, CLOSED SET is the same as its LEBESGUE MEASURE.

References

Ko, K.-I. "A Polynomial-Time Computable Curve whose Interior has a Nonrecursive Measure." *Theoret. Comput. Sci.* **145**, 241–70, 1995.

Minkowski Metric

In CARTESIAN COORDINATES,

$$ds^2 = dx^2 + dy^2 + dz^2 \tag{1}$$

$$dr^2 = -c^2 \, dt^2 + dx^2 + dy^2 + dz^2, \tag{2}$$

and

$$g_{\alpha\beta} \equiv \eta_{\alpha\beta} = \begin{bmatrix} -1 & 0 & 0 & 0 \\ 0 & 1 & 0 & 0 \\ 0 & 0 & 1 & 0 \\ 0 & 0 & 0 & 1 \end{bmatrix}. \tag{3}$$

In SPHERICAL COORDINATES,

$$ds^2 = dr^2 + r^2 \, d\theta + r^2 \sin^2 \theta \, d\phi^2 \tag{4}$$

$$dr^2 = -c^2 \, dt^2 + dr^2 + r^2 \, d\theta + r^2 \sin^2 \theta \, d\phi^2, \tag{5}$$

and

$$g = \begin{bmatrix} -1 & 0 & 0 & 0 \\ 0 & 1 & 0 & 0 \\ 0 & 0 & r^2 & 0 \\ 0 & 0 & 0 & r^2 \sin^2 \theta \end{bmatrix}. \tag{6}$$

See also LORENTZ TRANSFORMATION, MINKOWSKI SPACE

Minkowski Sausage

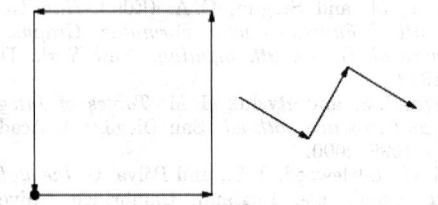

A FRACTAL curve created from the base curve and motif illustrated above (Lauwerier 1991, p. 37). The number of segments after the nth iteration is

$$N_n = 8^n, \tag{1}$$

and

$$\epsilon_n = \left(\frac{1}{4} \right)^n, \tag{2}$$

so the CAPACITY DIMENSION is

$$D \equiv -\lim_{n \to \infty} \frac{\ln N_n}{\ln \epsilon_n} = -\lim_{n \to \infty} \frac{\ln 8^n}{\ln 4^n} = \frac{\ln 8}{\ln 4} = \frac{3 \ln 2}{2 \ln 2} = \frac{3}{2}. \tag{3}$$

The term Minkowski sausage is also used to refer to the MINKOWSKI COVER of a curve.

See also MINKOWSKI-BOULIGAND DIMENSION, MINKOWSKI COVER

References
Lauwerier, H. *Fractals: Endlessly Repeated Geometric Figures.* Princeton, NJ: Princeton University Press, pp. 37–8 and 42, 1991.
Peitgen, H.-O. and Saupe, D. (Eds.). *The Science of Fractal Images.* New York: Springer-Verlag, p. 283, 1988.
Weisstein, E. W. "Fractals." MATHEMATICA NOTEBOOK FRACTAL.M.

Minkowski's Inequalities
If $p > 1$, then Minkowski's integral inequality states that

$$\left[\int_a^b |f(x) + g(x)|^p \, dx \right]^{1/p}$$
$$\leq \left[\int_a^b |f(x)|^p \, dx \right]^{1/p} + \left[\int_a^b |g(x)|^p \, dx \right]^{1/p}.$$

Similarly, if $p > 1$ and a_k, $b_k > 0$, then Minkowski's sum inequality states that

$$\left[\sum_{k=1}^n (a_k + b_k)^p \right]^{1/p} \leq \left(\sum_{k=1}^n a_k^p \right)^{1/p} + \left(\sum_{k=1}^n b_k^p \right)^{1/p}.$$

Equality holds IFF the sequences a_1, a_2, ... and b_1, b_2, ... are proportional.

References
Abramowitz, M. and Stegun, C. A. (Eds.). *Handbook of Mathematical Functions with Formulas, Graphs, and Mathematical Tables, 9th printing.* New York: Dover, p. 11, 1972.
Gradshteyn, I. S. and Ryzhik, I. M. *Tables of Integrals, Series, and Products, 6th ed.* San Diego, CA: Academic Press, pp. 1092 and 1099, 2000.
Hardy, G. H.; Littlewood, J. E.; and Pólya, G. 'Minkowski's Inequality" and "Minkowski's Inequality for Integrals." §2.11, 5.7, and 6.13 in *Inequalities, 2nd ed.* Cambridge, England: Cambridge University Press, pp. 30–2, 123, and 146–50, 1988.
Minkowski, H. *Geometrie der Zahlen, Vol. 1.* Leipzig, Germany: pp. 115–17, 1896.
Sansone, G. *Orthogonal Functions, rev. English ed.* New York: Dover, p. 33, 1991.

Minkowski Space
A 4-D space with the MINKOWSKI METRIC. Alternatively, it can be considered to have a EUCLIDEAN METRIC, but with its VECTORS defined by

$$\begin{bmatrix} x_0 \\ x_1 \\ x_2 \\ x_3 \end{bmatrix} = \begin{bmatrix} ict \\ x \\ y \\ z \end{bmatrix}, \tag{1}$$

where c is the speed of light and i is the IMAGINARY NUMBER $\sqrt{-1}$. Minkowski space unifies Euclidean 3-space plus time (the "fourth dimension") in Einstein's theory of special relativity.

The METRIC of Minkowski space is DIAGONAL with

$$g_{\alpha\alpha} = \frac{1}{g_{\alpha\alpha}}, \tag{2}$$

so

$$\eta^{\beta\delta} = \eta_{\beta\delta}. \tag{3}$$

Let Λ be the TENSOR for a LORENTZ TRANSFORMATION. Then

$$\eta^{\beta\delta} \Lambda^\gamma_{\ \delta} = \Lambda^{\beta\gamma} \tag{4}$$

$$\eta_{\alpha\gamma} \Lambda^{\beta\gamma} = \Lambda^\beta_\alpha \tag{5}$$

$$\Lambda^\beta_\alpha = \eta_{\alpha\gamma}\Lambda^{\beta\gamma} = \eta_{\alpha\gamma}\eta^{\beta\delta}\Lambda^\gamma{}_\delta. \qquad (6)$$

The NECESSARY and SUFFICIENT conditions for a metric $g_{\mu\nu}$ to be equivalent to the Minkowski metric $\eta_{\alpha\beta}$ are that the RIEMANN TENSOR vanishes everywhere ($R^\lambda{}_{\mu\nu\kappa} = 0$) and that at some point $g^{\mu\nu}$ has three POSITIVE and one NEGATIVE EIGENVALUES.

See also LORENTZ TRANSFORMATION, MINKOWSKI METRIC, TWISTOR, TWISTOR SPACE

References

Thompson, A. C. *Minkowski Geometry.* New York: Cambridge University Press, 1996.

Minkowski's Question Mark Function

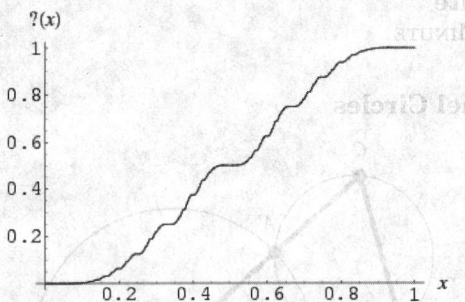

The function $y = ?(x)$ defined by Minkowski for the purpose of mapping the rational numbers in the OPEN INTERVAL $(0, 1)$ into the QUADRATIC IRRATIONAL NUMBERS of $(0, 1)$ in a continuous, order-preserving manner. $?(x)$ takes a number having BINARY expansion $x = 0.a_1 a_2 a_3 \ldots_2$ to the number

$$?(x) = \sum_k \frac{(-1)^{k-1}}{2^{(a_1 + \ldots + a_k)-1}}. \qquad (1)$$

The function satisfies the following properties (Salem 1943).

1. $?(x)$ is strictly increasing.
2. If x is rational, then $?(x)$ is of the form $k/2^s$, with k and s integers.
3. If x is a QUADRATIC IRRATIONAL NUMBER, then the continued fraction is periodic, and hence $?(x)$ is rational.
4. The function is purely singular (Denjoy 1938).

$?(x)$ can also be constructed as

$$?\left(\frac{p + p'}{q + q'}\right) = \frac{?(p/q) + ?(p'/q')}{2}, \qquad (2)$$

where p/q and p'/q' are two consecutive irreducible fractions from the FAREY SEQUENCE. At the nth stage of this definition, $?(x)$ is defined for $2^n + 1$ values of x, and the ordinates corresponding to these values are $x = k/2^n$ for $k = 0, 1, \ldots, 2^n$ (Salem 1943).

The function satisfies the identity

$$?\left(\frac{1}{k^n}\right) = \frac{1}{2^{k^n-1}}. \qquad (3)$$

A few special values include

$$?(0) = 0$$

$$?\left(\frac{1}{3}\right) = \frac{1}{4}$$

$$?\left(\frac{1}{2}\right) = \frac{1}{2}$$

$$?(\phi - 1) = \frac{2}{3}$$

$$?\left(\frac{2}{3}\right) = \frac{3}{4}$$

$$?\left(\frac{1}{2}\sqrt{2}\right) = \frac{4}{5}$$

$$?\left(\frac{1}{2}\sqrt{3}\right) = \frac{84}{85}$$

$$?(1) = 1,$$

where ϕ is the GOLDEN RATIO.

See also DEVIL'S STAIRCASE, FAREY SEQUENCE

References

Conway, J. H. "Contorted Fractions." *On Numbers and Games.* New York: Academic Press, pp. 82–6, 1976.
Denjoy, A. "Sur une fonction réelle de Minkowski." *J. Math. Pures Appl.* **17**, 105–55, 1938.
Girgensohn, R. "Constructing Singular Functions via Farey Fractions." *J. Math. Anal. Appl.* **203**, 127–41, 1996.
Kinney, J. R. "Note on a Singular Function of Minkowski." *Proc. Amer. Math. Soc.* **11**, 788–94, 1960.
Minkowski, H. "Zur Geometrie der Zahlen." In *Gesammelte Abhandlungen, Vol. 2.* New York: Chelsea, pp. 50–1, 1991.
Salem, R. "On Some Singular Monotone Functions which Are Strictly Increasing." *Trans. Amer. Math. Soc.* **53**, 427–39, 1943.
Tichy, R. and Uitz, J. "An Extension of Minkowski's Singular Functions." *Appl. Math. Lett.* **8**, 39–6, 1995.
Viader, P.; Paradis, J.; and Bibiloni, L. "A New Light on Minkowski's ?(x) Function." *J. Number Th.* **73**, 212–27, 1998.

Minkowski Sum

The sum of sets A and B in a VECTOR SPACE, equal to $\{a + b : a \in A,\ b \in B\}$.

References

Skiena, S. S. "Minkowski Sum." §8.6.16 in *The Algorithm Design Manual.* New York: Springer-Verlag, pp. 395–96, 1997.

Minkowski Sum Inequality

If $p > 1$ and $a_k, b_k > 0$, then

$$\left[\sum_{k=1}^{n}(a_k+b_k)^p\right]^{1/p} \le \left(\sum_{k=1}^{n}a_k^p\right)^{1/p} + \left(\sum_{k=1}^{n}b_k^p\right)^{1/p}.$$

Equality holds IFF the sequences $a_1, a_2, \ldots$ and $b_1, b_2,$ $\ldots$ are proportional.

See also MINKOWSKI INTEGRAL INEQUALITY

References

Abramowitz, M. and Stegun, C. A. (Eds.). *Handbook of Mathematical Functions with Formulas, Graphs, and Mathematical Tables, 9th printing.* New York: Dover, p. 11, 1972.
Gradshteyn, I. S. and Ryzhik, I. M. *Tables of Integrals, Series, and Products, 6th ed.* San Diego, CA: Academic Press, p. 1092, 2000.
Hardy, G. H.; Littlewood, J. E.; and Pólya, G. *Inequalities, 2nd ed.* Cambridge, England: Cambridge University Press, pp. 24–6, 1988.

Minor

The reduced DETERMINANT of a DETERMINANT EXPANSION, denoted M_{ij}, which is formed by omitting the ith row and jth column. The minor can be computed in *Mathematica* using

```
Minor[m_List,{i_Integer,j_Integer}] :=
    Drop[Transpose[Drop[Transpose[m],{j}]],{i}]
```

Minors[m] gives the minors of a matrix m, while Minors[m, k] gives the kth minors of m.

See also COFACTOR, DETERMINANT, DETERMINANT EXPANSION BY MINORS

References

Arfken, G. *Mathematical Methods for Physicists, 3rd ed.* Orlando, FL: Academic Press, pp. 169–70, 1985.
Muir, T. "Minors and Expansion." Ch. 4 in *A Treatise on the Theory of Determinants.* New York: Dover, pp. 53–37, 1960.
Skiena, S. *Implementing Discrete Mathematics: Combinatorics and Graph Theory with Mathematica.* Reading, MA: Addison-Wesley, p. 235, 1990.

Minor Axis

SEMIMINOR AXIS

Minor Graph

A "minor" is a sort of SUBGRAPH and is what Kuratowski means when he says "contain." It is roughly a small graph which can be mapped into the big one without merging VERTICES.

Minuend

A quantity from which another (the SUBTRAHEND) is subtracted.

See also MINUS, SUBTRACTION, SUBTRAHEND

Minus

The operation of SUBTRACTION, i.e., a minus b. The operation is denoted $a - b$. The MINUS SIGN "−" is also used to denote a NEGATIVE number, i.e., $-x$.

See also MINUS SIGN, NEGATIVE, PLUS, PLUS OR MINUS, TIMES

Minus or Plus

PLUS OR MINUS

Minus Sign

The symbol "−" which is used to denote a NEGATIVE number or SUBTRACTION.

See also MINUS, PLUS SIGN, SIGN, SUBTRACTION

Minute

ARC MINUTE

Miquel Circles

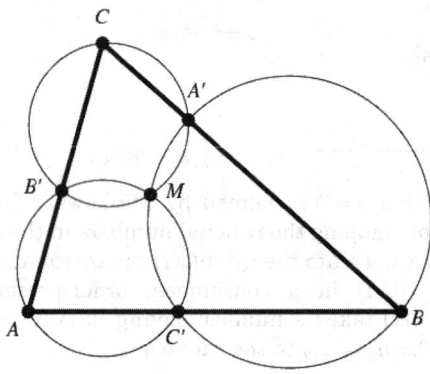

For a TRIANGLE ΔABC and three points $F(r)$, B', and C', one on each of its sides, the three Miquel circles are the circles passing through each VERTEX and its neighboring side points (i.e., $AC'B'$, $BA'C'$, and $CB'A'$). According to MIQUEL'S THEOREM, the Miquel circles are CONCURRENT in a point M known as the MIQUEL POINT. Similarly, there are n Miquel circles for n lines taken $(n-1)$ at a time.

See also CLIFFORD'S CIRCLE THEOREM, MIQUEL POINT, MIQUEL'S THEOREM, MIQUEL TRIANGLE

References

Honsberger, R. *Episodes in Nineteenth and Twentieth Century Euclidean Geometry.* Washington, DC: Math. Assoc. Amer., p. 81, 1995.

Miquel Equation

$$\angle A_2 M A_3 = \angle A_2 A_1 A_3 + \angle P_2 P_1 P_3,$$

where $\angle$ is a DIRECTED ANGLE.

See also DIRECTED ANGLE, MIQUEL'S THEOREM, PIVOT THEOREM

Miquel Five Circles Theorem

References

Johnson, R. A. *Modern Geometry: An Elementary Treatise on the Geometry of the Triangle and the Circle.* Boston, MA: Houghton Mifflin, pp. 131–44, 1929.

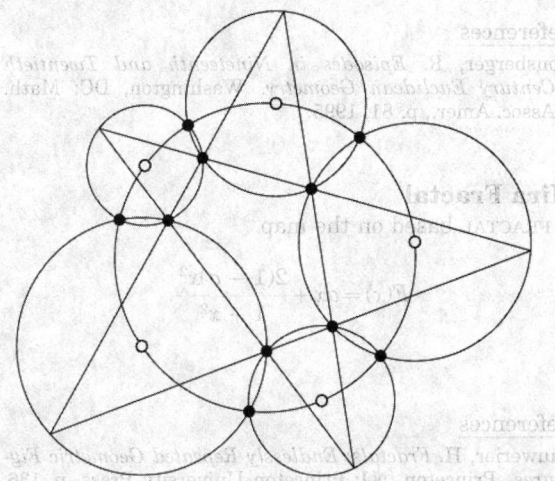

Let five circles with CONCYCLIC centers be drawn such that each intersects its neighbors in two points, with one of these intersections lying itself on the circle of centers. By joining adjacent pairs of the intersection points which do not lie on the circle of center, an (irregular) PENTAGRAM is obtained whose five vertices lie on the circle of centers.

Let the circle of centers have radius r and let the five circles be centered and angular positions θ_i along this circle. The radii r_i of the circles and their angular positions ϕ_i along the circle of centers can then be determined by solving the ten simultaneous equations

$$(\cos\phi_i - \cos\theta_i)^2 + (\sin\phi_i - \sin\theta_i)^2 = \frac{r_i^2}{r^2}$$

$$(\cos\phi_{i-1} - \cos\theta_i)^2 + (\sin\phi_{i-1} - \sin\theta_i)^2 = \frac{r_i^2}{r^2}$$

for $i = 1, ..., 5$, where $\phi_0 \equiv \phi_5$ and $r_0 \equiv r_5$.

See also FIVE DISKS PROBLEM, PENTAGRAM

References

Casey, J. *A Sequel to the First Six Books of the Elements of Euclid, Containing an Easy Introduction to Modern Geometry with Numerous Examples,* 5th ed., rev. enl. Dublin: Hodges, Figgis, & Co., pp. 151–52, 1888.

Weisstein, E. W. "Plane Geometry." MATHEMATICA NOTE-BOOK PlaneGeometry.m.

Wells, D. *The Penguin Dictionary of Curious and Interesting Geometry.* Middlesex, England: Penguin Books, p. 79, 1991.

Miquel Point

The point of CONCURRENCE of the MIQUEL CIRCLES.

See also MIQUEL CIRCLES, MIQUEL'S THEOREM, MIQUEL TRIANGLE

References

Coolidge, J. L. *A Treatise on the Geometry of the Circle and Sphere.* New York: Chelsea, pp. 87–0, 1971.

Honsberger, R. *Episodes in Nineteenth and Twentieth Century Euclidean Geometry.* Washington, DC: Math. Assoc. Amer., p. 81, 1995.

Wells, D. *The Penguin Dictionary of Curious and Interesting Geometry.* London: Penguin, p. 151, 1991.

A generalized version of Miquel's theorem states that given four lines $l_1, ..., l_4$, each intersecting the other three, the four Miquel circles passing through each subset of three intersection points meet in a point known as the 4-Miquel point M. Furthermore, the centers of these four Miquel circles lie on a circle C (Johnson 1929, p. 139). The 4-Miquel point M is given points of the sides of a 4-gon, and locates with respect to the sides.

Moreover, given n lines taken by $(n-1)$ yield n Miquel circles like C, passing through a point P and their centers lie on a circle C.

See also CLIFFORD'S CIRCLE THEOREM, MIQUEL CIRCLES, MIQUEL FIVE CIRCLES THEOREM, MIQUEL EQUATION, MIQUEL TRIANGLE, NINE-POINT CIRCLE, PIVOT THEOREM

References

Honsberger, R. "The Miquel Theorem." Ch. 8 in *Episodes in Nineteenth and Twentieth Century Euclidean Geometry.* Washington, DC: Math. Assoc. Amer., p. 1995.

Miquel's Theorem

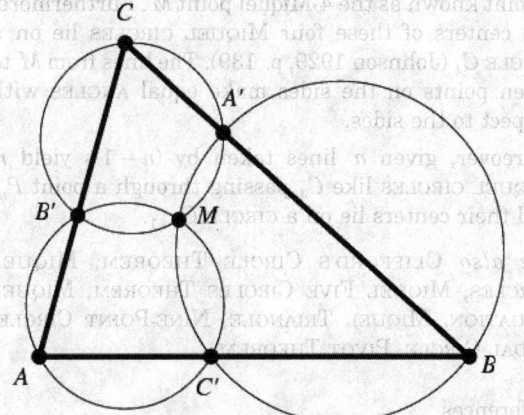

If points A', B', and C' are marked on each side of a TRIANGLE $\triangle ABC$, one on each side (or on a side's extension), then the three MIQUEL CIRCLES (each through a VERTEX and the two marked points on the adjacent sides) are CONCURRENT at a point M called the MIQUEL POINT. This result is a slight generalization of the so-called PIVOT THEOREM.

If M lies in the interior of the triangle, then it satisfies

$$\angle P_2 M P_3 = 180° - \alpha_1$$

$$\angle P_3 M P_1 = 180° - \alpha_2$$

$$\angle P_1 M P_2 = 180° - \alpha_3.$$

The lines from the MIQUEL POINT to the marked points make equal angles with the respective sides. (This is a by-product of the MIQUEL EQUATION.)

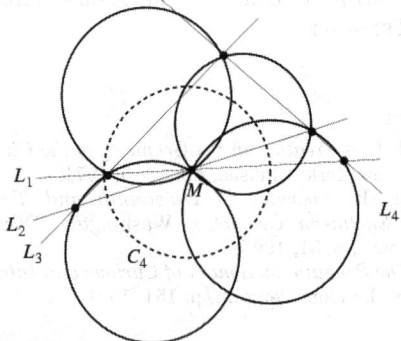

A generalized version of Miquel's theorem states that given four lines $L_1, ..., L_4$ each intersecting the other three, the four MIQUEL CIRCLES passing through each subset of three intersection points of the lines meet in a point known as the 4-Miquel point M. Furthermore, the centers of these four MIQUEL CIRCLES lie on a CIRCLE C_4 (Johnson 1929, p. 139). The lines from M to given points on the sides make equal ANGLES with respect to the sides.

Moreover, given n lines taken by $(n-1)$s yield n MIQUEL CIRCLES like C_4 passing through a point P_n, and their centers lie on a CIRCLE C_{n+1}.

See also CLIFFORD'S CIRCLE THEOREM, MIQUEL CIRCLES, MIQUEL FIVE CIRCLES THEOREM, MIQUEL EQUATION, MIQUEL TRIANGLE, NINE-POINT CIRCLE, PEDAL CIRCLE, PIVOT THEOREM

References

Honsberger, R. "The Miquel Theorem." Ch. 8 in *Episodes in Nineteenth and Twentieth Century Euclidean Geometry*. Washington, DC: Math. Assoc. Amer., pp. 79–6, 1995.
Johnson, R. A. *Modern Geometry: An Elementary Treatise on the Geometry of the Triangle and the Circle*. Boston, MA: Houghton Mifflin, pp. 131–44, 1929.
Wells, D. *The Penguin Dictionary of Curious and Interesting Geometry*. London: Penguin, pp. 151–52, 1991.

Miquel Triangle

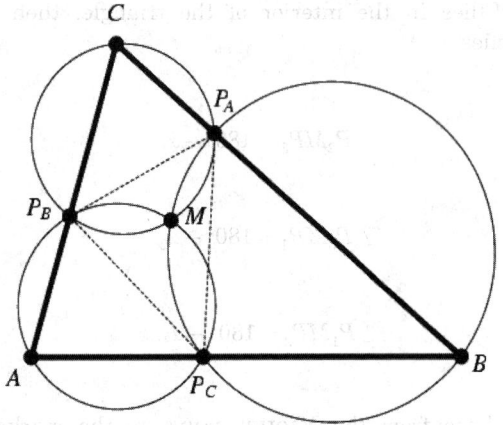

Given a point P and a triangle $\triangle ABC$, the Miquel triangle is the triangle $\triangle P_A P_B P_C$ connecting the side

points P_A, P_B, and P_C of $\triangle ABC$ with respect to which M is the MIQUEL POINT. All Miquel triangles of a given point M are directly similar, and M is the SIMILITUDE CENTER in every case.

See also MIQUEL CIRCLES, MIQUEL POINT, MIQUEL'S THEOREM

References

Honsberger, R. *Episodes in Nineteenth and Twentieth Century Euclidean Geometry*. Washington, DC: Math. Assoc. Amer., p. 81, 1995.

Mira Fractal

A FRACTAL based on the map

$$F(x) = ax + \frac{2(1-a)x^2}{1+x^2}.$$

References

Lauwerier, H. *Fractals: Endlessly Repeated Geometric Figures*. Princeton, NJ: Princeton University Press, p. 136, 1991.

Mirimanoff's Congruence

If the first case of FERMAT'S LAST THEOREM is false for the PRIME exponent p, then $3^{p-1} \equiv 1 \pmod{p^2}$.

See also FERMAT'S LAST THEOREM

Mirror Image

An image of an object obtained by reflecting it in a mirror so that the signs of one of its coordinates are reversed.

AMPHICHIRAL, CHIRAL, ENANTIOMER, HANDEDNESS, REFLECTION, SYMMETRY

References

Coxeter, H. S. M. and Greitzer, S. L. *Geometry Revisited*. Washington, DC: Math. Assoc. Amer., p. 87, 1967.

Mirror Plane

The SYMMETRY OPERATION $(x, y, z) \to (x, y, -z)$, etc., which is equivalent to $\bar{2}$, where the bar denotes an IMPROPER ROTATION.

See also MIRROR IMAGE

Misère Form

A version of NIM-like GAMES in which the player taking the last piece is the loser. For most IMPARTIAL GAMES, this form is much harder to analyze, but it requires only a trivial modification for the game of NIM.

Mitchell Index

The statistical INDEX

$$P_M \equiv \frac{\sum p_n q_a}{\sum p_0 q_a},$$

where p_n is the price per unit in period n and q_n is the quantity produced in period n.

See also INDEX

References

Kenney, J. F. and Keeping, E. S. *Mathematics of Statistics, Pt. 1, 3rd ed.* Princeton, NJ: Van Nostrand, pp. 66–7, 1962.

Miter Surface

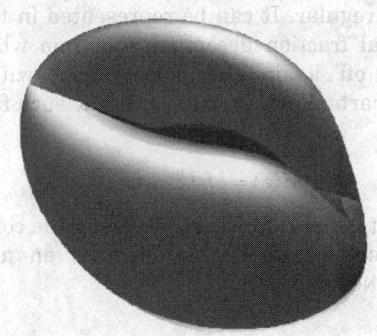

A QUARTIC SURFACE named after its resemblance to the liturgical headdress worn by bishops and given by the equation

$$4x^2(x^2 + y^2 + z^2) - y^2(1 - y^2 - z^2) = 0.$$

See also QUARTIC SURFACE

References

Nordstrand, T. "Surfaces." http://www.uib.no/people/nfytn/surfaces.htm.

Mittag-Leffler Function

$$E_n(x) \equiv \sum_{k=0}^{\infty} \frac{x^k}{\Gamma(nk + 1)}. \tag{1}$$

It is related to the GENERALIZED HYPERBOLIC FUNCTIONS $F_{n,r}^{\alpha}(x)$ by

$$F_{n,0}^1(x) = E_n(x^n). \tag{2}$$

Special values for integer n are

$$E_0(x) = \frac{1}{1-x} \tag{3}$$

$$E_1(x) = e^x \tag{4}$$

$$E_2(x) = \cosh(\sqrt{x}) \tag{5}$$

$$E_3(x) = \tfrac{1}{3}\left[e^{x^{1/3}} + 2e^{-x^{1/3}/2}\cos\left(\tfrac{1}{2}\sqrt{3}x^{1/3}\right)\right] \tag{6}$$

$$E_4(x) = \tfrac{1}{2}\left[\cos(x^{1/4}) + \cosh(x^{1/4})\right], \tag{7}$$

and special values of half-integer n are

$$E_{1/2}(x) = e^{x^2}(1 + \operatorname{erf} x) \tag{8}$$

$$E_{3/2}(x) = \tfrac{1}{3}\left[e^{x^{2/3}} + 2e^{-x^{2/3}/2}\cos\left(\tfrac{1}{2}\sqrt{3}x^{2/3}\right)\right.$$
$$\left. + \frac{4x\,_1F_3\left(1;\ \tfrac{5}{6}, \tfrac{7}{6}, \tfrac{3}{2};\ \tfrac{1}{27}x^2\right)}{\sqrt{\pi}}\right] \tag{9}$$

$$E_{5/2}(x) = \,_0F_4\left(;\ \tfrac{1}{5}, \tfrac{2}{5}, \tfrac{3}{5}, \tfrac{4}{5};\ \tfrac{1}{3125}x^2\right)$$
$$+ \frac{8x\,_1F_5\left(1;\ \tfrac{7}{10}, \tfrac{9}{10}, \tfrac{11}{10}, \tfrac{13}{10}, \tfrac{3}{2};\ \tfrac{1}{3125}x^2\right)}{15\sqrt{\pi}}, \tag{10}$$

where $_pF_q$ are generalized hypergeometric functions, and $_0F_q$ is a generalized confluent hypergeometric function. As can be seen, $E_{1/2}(x)$ is closely related to DAWSON'S INTEGRAL $D_-(x)$.

The more general Mittag-Leffler function

$$E_{m,n} = \sum_{k=0}^{\infty} \frac{x^k}{\Gamma(mk + n)} \tag{11}$$

can also be defined (Wiman 1905, Agarwal 1953, Gorenflo 1987, Miller 1993, Mainardi and Gorenflo 1995, Gorenflo 1998, Sixdeniers *et al.*).

See also DAWSON'S INTEGRAL, GENERALIZED HYPERBOLIC FUNCTIONS

References

Agarwal, R. P. "A propos d'une note de M. Pierre Humbert." *C. R. Acad. Sci. Paris* **236**, 2031–032, 1953.
Gorenflo, R. "Newtonsche Aufheizung, Abelsche Integralgleichungen zweiter Art und Mittag-Leffler-Funktionen." *Z. Naturforsch. A* **42**, 1141–146, 1987.
Gorenflo, R.; Kilbas, A. A.; and Rogosin, S. V. "On the Generalized Mittag-Leffler Type Functions." *Integral Transform. Spec. Funct.* **7**, 215–24, 1998.
Humbert, P. "Quelques résultats relatifs à la fonction de Mittag-Leffler." *C. R. Acad. Sci. Paris* **236**, 1467–468, 1953.
Humbert, P. and Agarwal, R. P. "Sur la fonction de Mittag-Leffler et quelques-unes de ses généralisations." *Bull. Sci. Math. Ser. 2* **77**, 180–85, 1953.
Humbert, P. and Delerue, P. "Sur une extension à deux variables de la fonction de Mittag-Leffler." *C. R. Acad. Sci. Paris* **237**, 1059–060, 1953.

Mainardi, F. and Gorenflo, R. "The Mittag-Leffler Function in the Riemann-Liouville Fractional Calculus." In *Proceedings of the International Conference Dedicated to the Memory of Academician F. D. Gakhov; Held in Minsk, February 16–0, 1996* (Ed. A. A. Kilbas). Minsk, Beloruss: Beloruss. Gos. Univ., Minsk, pp. 215–25, 1996.

Miller, K. S. "The Mittag-Leffler and Related Functions." *Integral Transform. Spec. Funct.* **1**, 41–9, 1993.

Mittag-Leffler, M. G. *C. R. Acad. Sci. Paris Ser. 2* **137**, 554, 1903.

Muldoon, M. E. and Ungar, A. A. "Beyond Sin and Cos." *Math. Mag.* **69**, 3–4, 1996.

Sixdeniers, J.-M.; Penson, K. A.; and Solomon, A. I. "Mittag-Leffler Coherent States." *J. Phys. A: Math. Gen.* **32**, 7543–563, 1999.

Wiman, A. "Uuml;ber den Fundamentalsatz in der Teorie der Funktionen $E_a(x)$." *Acta Math.* **29**, 191–01, 1905.

Mittag-Leffler Polynomial

Polynomials $M_k(x)$ which form the associated SHEFFER SEQUENCE for

$$f(t) = \frac{e^t - 1}{e^t + 1} \tag{1}$$

and have the GENERATING FUNCTION

$$\sum_{k=0}^{\infty} \frac{M_k(x)}{k!} t^k = \left(\frac{1+t}{1-t}\right)^x. \tag{2}$$

An explicit formula is given by

$$M_n(x) = \sum_{k=0}^{n} \binom{n}{k} (n-1)_{n-k} 2^k (x)_k, \tag{3}$$

where $(x)_n$ is a FALLING FACTORIAL, which can be summed in closed form in terms of the HYPERGEOMETRIC FUNCTION, GAMMA FUNCTION, and POLYGAMMA FUNCTION. The binomial identity associated with the SHEFFER SEQUENCE is

$$M_n(x+y) = \sum_{k=0}^{n} \binom{n}{k} M_k(x) M_{n-k}(y). \tag{4}$$

The Mittag-Leffler polynomials satisfy the recurrence formula

$$M_{n+1}(x) = \tfrac{1}{2} x [M_n(x+1) + 2M_n(x) + M_n(x-1)]. \tag{5}$$

The first few Mittag-Leffler polynomials are

$$M_0(x) = 1$$
$$M_1(x) = 2x$$
$$M_2(x) = 4x^2$$
$$M_3(x) = 8x^3 + 4x$$
$$M_4(x) = 16x^4 + 32x^2.$$

The Mittag-Leffler polynomials $M_n(x)$ are related to the PIDDUCK POLYNOMIALS by

$$P_n(x) = \tfrac{1}{2}(e^t + 1) M_n(x) \tag{6}$$

(Roman 1984, p. 127).

See also PIDDUCK POLYNOMIAL

References

Bateman, H. "The Polynomial of Mittag-Leffler." *Proc. Nat. Acad. Sci. USA* **26**, 491–96, 1940.

Roman, S. "The Mittag-Leffler Polynomials." §4.1.6 in *The Umbral Calculus.* New York: Academic Press, pp. 75–8 and 127, 1984.

Mittag-Leffler's Partial Fractions Theorem

Let any finite or infinite set of points having no finite LIMIT POINT be prescribed and associate with each of its points a principal part, i.e., a RATIONAL FUNCTION of the special form

$$h_v(z) = \frac{a_{-1}^{(v)}}{z - z_v} + \frac{a_{-2}^{(v)}}{(z - z_v)^2} + \ldots + \frac{a_{-\alpha_n}^{(v)} u}{(z - z_v)^{\alpha_v}}$$

for $v = 1, 2, \ldots, k$. Then there exists a MEROMORPHIC FUNCTION which has poles with the prescribed principal parts at precisely the prescribed points, and is otherwise regular. It can be represented in the form of a partial fraction decomposition from which one can read off again the poles, along with their principal parts. Further, if $M_0(z)$ is one such function, then

$$M(z) = M_0(z) + G(z)$$

is the most general function satisfying the conditions of the problem, where $G(z)$ denotes an arbitrary ENTIRE FUNCTION.

References

Knopp, K. *Theory of Functions Parts I and II, Two Volumes Bound as One, Part II.* New York: Dover, pp. 37–9, 1996.

Krantz, S. G. "The Mittag-Leffler Theorem." §8.3.6 in *Handbook of Complex Analysis.* Boston, MA: Birkhäuser, pp. 112–13, 1999.

Mittag-Leffler's Theorem

If a function analytic at the origin has no SINGULARITIES other than POLES for finite x, and if we can choose a sequence of contours C_m about $z = 0$ tending to infinity such that $|f(z)|$ never exceeds a given quantity M on any of these contours and $\int |dz/z|$ is uniformly bounded on them, then

$$f(z) = f(0) + \lim[P_m(z) - P_m(0)],$$

where $P_m(z)$ is the sum of the principal parts of $f(z)$ at all POLES α within C_m. If there is a POLE at $z = 0$, then we can replace $f(0)$ by the negative powers and the constant term in the LAURENT SERIES of $f(z)$ about $z = 0$.

References

Jeffreys, H. and Jeffreys, B. S. "Mittag-Leffler's Theorem." §12.006 in *Methods of Mathematical Physics, 3rd ed.* Cambridge, England: Cambridge University Press, pp. 383–86, 1988.

Mittenpunkt

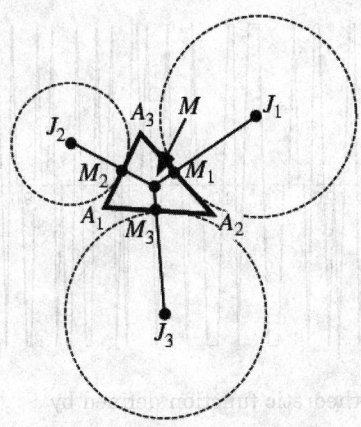

The SYMMEDIAN POINT of the EXCENTRAL TRIANGLE, i.e., the point of concurrence M of the lines from the EXCENTERS J_i through the corresponding TRIANGLE side MIDPOINT M_i. It is also called the MIDDLESPOINT and has TRIANGLE CENTER FUNCTION

$$\alpha = b + c - a = \tfrac{1}{2}\cot A.$$

See also EXCENTER, EXCENTRAL TRIANGLE, NAGEL POINT

References
Baptist, P. *Die Entwicklung der Neueren Dreiecksgeometrie.* Mannheim: Wissenschaftsverlag, p. 72, 1992.
Eddy, R. H. "A Generalization of Nagel's Middlespoint." *Elem. Math.* **45**, 14–8, 1990.
Kimberling, C. "Central Points and Central Lines in the Plane of a Triangle." *Math. Mag.* **67**, 163–87, 1994.
Kimberling, C. "Mittenpunkt." http://cedar.evansville.edu/~ck6/tcenters/class/mitten.html.

Mixed Fraction

An IMPROPER FRACTION $p/q > 1$ written in the form $n + r/s$. In common usage such as cooking recipes, $n + r/s$ is often written as $n\,\tfrac{r}{s}$ (e.g., $1\tfrac{1}{2}$), much to the chagrin of mathematicians, to whom $n\,\tfrac{r}{s}$ means nr/s, *not* $n + r/s$. (The author of this work discovered this fact early in his mathematical career after having points marked off a CALCULUS exam for using the recipe-like notation. Future mathematicians are therefore encouraged to avoid mixed fractions, except perhaps in the kitchen.)

See also FRACTION, IMPROPER FRACTION, PROPER FRACTION

Mixed Indices
MIXED TENSOR

Mixed Partial Derivative

A PARTIAL DERIVATIVE of second or greater order with respect to two or more different variables, for example

$$f_{xy} = \frac{\partial^2 f}{\partial x\, \partial y}.$$

If the mixed partial derivatives exist and are continuous at a point $\mathbf{x}_0$, then they are equal at $\mathbf{x}_0$ regardless of the order in which they are taken.

See also PARTIAL DERIVATIVE

Mixed Strategy

A collection of moves together with a corresponding set of weights which are followed probabilistically in the playing of a GAME. The MINIMAX THEOREM of GAME THEORY states that every finite, zero-sum, two-person game has optimal mixed strategies.

See also GAME THEORY, MINIMAX THEOREM, STRATEGY

Mixed Tensor

A TENSOR having CONTRAVARIANT and COVARIANT indices.

See also CONTRAVARIANT TENSOR, COVARIANT TENSOR, TENSOR

Mnemonic

A mental device used to aid memorization. Common mnemonics for mathematical constants such as E and PI consist of sentences in which the number of letters in each word give successive digits.

See also E, JOSEPHUS PROBLEM, PI

References
Luria, A. R. *The Mind of a Mnemonist: A Little Book about a Vast Memory.* Cambridge, MA: Harvard University Press, 1987.
Weisstein, E. W. "Books about Calculating Prodigies." http://www.treasure-troves.com/books/CalculatingProdigies.html.

Moat-Crossing Problem

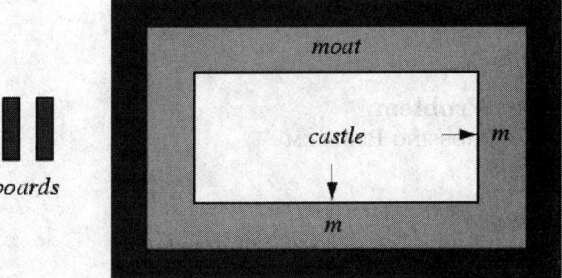

There are two versions of the moat-crossing problem, one geometric and one algebraic. The geometric moat problems asks for the widest moat Rapunzel can cross to escape if she has only two unit-length boards (and no means to nail or otherwise attach them together)? More generally, what is the widest moat which can be crossed using n boards? Matthew Cook has conjectured that the asymptotic solution to this problem is $\mathcal{O}(n^{1/3})$ (Finch).

The algebraic moat-crossing problem asks if it is possible to walk to infinity on the REAL LINE using only steps of bounded lengths and steps on the prime numbers. The answer is negative (Gethner *et al.* 1998). However, the Gaussian moat problem that asks whether it is possible to walk to infinity in the GAUSSIAN INTEGERS using the GAUSSIAN PRIMES as stepping stones and taking steps of bounded length is unresolved. Gethner *et al.* (1998) show that a moat of width $\sqrt{26}$ exists.

References

Finch, S. "Unsolved Mathematics Problems: Moat Crossing Optimization Problem." http://www.mathsoft.com/asolve/moat/moat.html.

Gethner, E. and Stark, H. M. "Periodic Gaussian Moats." *Experiment. Math.* **6**, 251–54, 1997.

Gethner, E.; Wagon, S.; and Wick, B. "A Stroll Through the Gaussian Primes." *Amer. Math. Monthly* **105**, 327–37, 1998.

Guy, R. K. *Unsolved Problems in Number Theory, 2nd ed.* New York: Springer-Verlag, 1994.

Haugland, J. K. "A Walk on Complex Primes." [Norwegian.] *Normat* **43**, 168–70, 1995.

Jordan, J. H. and Rabung, J. R. "A Conjecture of Paul Erdos Concerning Gaussian Primes." *Math. Comput.* **24**, 221–23, 1970.

Montgomery, H. *Ten Lectures on the Interface Between Analytic Number Theory and Harmonic Analysis.* Providence, RI: Amer. Math. Soc., 1994.

Vardi, I. "Prime Percolation." *Experiment. Math.* **7**, 275–89, 1998.

Wagon, S. *Mathematica in Action, 2nd ed.* New York: Springer-Verlag, 1999.

Moat Problem
MOAT-CROSSING PROBLEM

Möbius Band
MÖBIUS STRIP

Möbius Function

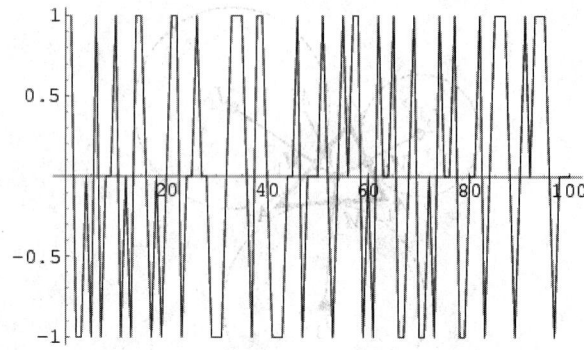

A number theoretic function defined by

$$\mu(n) \equiv$$

$$\begin{cases} 0 & \text{if } n \text{ has one or repeated prime factors} \\ 1 & \text{if } n = 1 \\ (-1)^k & \text{if } n \text{ is a product of } k \text{ distinct primes,} \end{cases} \tag{1}$$

so $\mu(n) \neq 0$ indicates that n is SQUAREFREE. The first few values are 1, -1, -1, 0, -1, 1, -1, 0, 0, 1, -1, 0, ... (Sloane's A008683). The SUMMATORY FUNCTION of the Möbius function is called MERTENS FUNCTION.

The Möbius function has GENERATING FUNCTIONS

$$\sum_{n=1}^{\infty} \frac{\mu(n)}{n^s} = \frac{1}{\zeta(s)} \tag{2}$$

for $\Re[s] > 1$ (Nagell 1951, p. 130), and

$$\sum_{n=1}^{\infty} \frac{\mu(n)x^n}{1-x^n} = x \tag{3}$$

for $|x| < 1$. It also obeys the infinite sums

$$\sum_{n=1}^{\infty} \frac{\mu(n)}{n} = 0 \tag{4}$$

$$\sum_{n=1}^{\infty} \frac{\mu(n)\ln n}{n} = -1 \tag{5}$$

and the INFINITE PRODUCT

$$\prod_{n=1}^{\infty} (1-x^n)^{\mu(n)/n} = e^{-x} \tag{6}$$

for $|x| < 1$ (Bellman 1943; Buck 1944;, Pólya and Szego 1976, p. 126; Robbins 1999). (2) is as "deep" as the PRIME NUMBER THEOREM (Landau 1909, pp. 567–74; Landau 1911; Hardy 1999, p. 24), and behaves asymptotically as

$$\sum_{n \leq x} \mu(n) = \mathcal{O}(xe^{-c\sqrt{\ln x}}) \tag{7}$$

The Möbius function is MULTIPLICATIVE,

$$\mu(mn) = \begin{cases} \mu(m)\mu(n) & \text{if } (m, n) = 1 \\ 0 & \text{if } (m, n) > 1, \end{cases} \qquad (8)$$

and satisfies

$$\sum_{d|n} \mu(d) = \delta_{n1}, \qquad (9)$$

where δ_{ij} is the KRONECKER DELTA, as well as

$$\sum_d \mu(d)\sigma_0\left(\frac{n}{d}\right) = 1, \qquad (10)$$

where $\sigma_0(n)$ is the number of divisors (i.e., DIVISOR FUNCTION of order zero; Nagell 1951, p. 281).

See also BRAUN'S CONJECTURE, MERTENS FUNCTION, MÖBIUS INVERSION FORMULA, MÖBIUS PERIODIC FUNCTION, PRIME ZETA FUNCTION, RIEMANN FUNCTION, SQUAREFREE

References

Abramowitz, M. and Stegun, C. A. (Eds.). "The Möbius Function." §24.3.1 in *Handbook of Mathematical Functions with Formulas, Graphs, and Mathematical Tables, 9th printing.* New York: Dover, p. 826, 1972.

Bellman, R. "Problem 4072." *Amer. Math. Monthly* **50**, 124–25, 1943.

Buck, R. C. "Solution to Problem 4072." *Amer. Math. Monthly* **51**, 410, 1944.

Deléglise, M. and Rivat, J. "Computing the Summation of the Möbius Function." *Experiment. Math.* **5**, 291–95, 1996.

Hardy, G. H. "A Note on the Möbius Function." §4.9 in *Ramanujan: Twelve Lectures on Subjects Suggested by His Life and Work, 3rd ed.* New York: Chelsea, pp. 64–5, 1999.

Hardy, G. H. and Wright, E. M. *An Introduction to the Theory of Numbers, 5th ed.* Oxford: Clarendon Press, p. 236, 1979.

Landau, E. *Handbuch der Lehre von der Verteilung der Primzahlen.* Leipzig, Germany: Teubner, 1909.

Landau, E. *Prac. Matematyczno-Fizycznych* **21**, 97–77, 1910.

Landau, E. *Wiener Sitzungsber.* **120**, 973–88, 1911.

Nagell, T. *Introduction to Number Theory.* New York: Wiley, p. 27, 1951.

Pólya, G. and Szego, G. *Problems and Theorems in Analysis, Vol. 2.* New York: Springer-Verlag, 1976.

Robbins, N. "Some Identities Connecting Partition Functions to Other Number Theoretic Functions." *Rocky Mtn. J. Math.* **29**, 335–45, 1999.

Rota, G.-C. "On the Foundations of Combinatorial Theory I. Theory of Möbius Functions." *Z. für Wahrscheinlichkeitsth.* **2**, 340–68, 1964.

Séroul, R. "The Moebius Function." §2.12 and 8.5 in *Programming for Mathematicians.* Berlin: Springer-Verlag, pp. 19–1 and 167–69, 2000.

Sloane, N. J. A. Sequences A008683 in "An On-Line Version of the Encyclopedia of Integer Sequences." http://www.research.att.com/~njas/sequences/eisonline.html.

Vardi, I. *Computational Recreations in Mathematica.* Redwood City, CA: Addison-Wesley, pp. 7– and 223–25, 1991.

Möbius Group

The equation

$$x_1^2 + x_2^2 + \ldots + x_n^2 - 2x_0 x_\infty = 0$$

represents an n-D HYPERSPHERE $\mathbb{S}^n$ as a quadratic hypersurface in an $(n+1)$-D real projective space $\mathbb{P}^{n+1}$, where x_a are homogeneous coordinates in $\mathbb{P}^{n+1}$. Then the GROUP $M(n)$ of projective transformations which leave $\mathbb{S}^n$ invariant is called the Möbius group.

See also MODULAR GROUP GAMMA

References

Iyanaga, S. and Kawada, Y. (Eds.). "Möbius Geometry." §78A in *Encyclopedic Dictionary of Mathematics.* Cambridge, MA: MIT Press, pp. 265–66, 1980.

Möbius Inversion Formula

The transform inverting the sequence

$$g(n) \equiv \sum_{d|n} f(d) \qquad (1)$$

into

$$f(n) = \sum_{d|n} \mu(d)g\left(\frac{n}{d}\right), \qquad (2)$$

where the sums are over all possible INTEGERS d that DIVIDE n and $\mu(d)$ is the MÖBIUS FUNCTION.

The LOGARITHM of the CYCLOTOMIC POLYNOMIAL

$$\Phi_n(x) = \prod_{d|n}(1 - x^{n/d})^{\mu(d)} \qquad (3)$$

is closely related to the Möbius inversion formula.

See also CYCLOTOMIC POLYNOMIAL, MÖBIUS FUNCTION, MÖBIUS TRANSFORM

References

Hardy, G. H. and Wright, W. M. *An Introduction to the Theory of Numbers, 5th ed.* Oxford, England: Oxford University Press, pp. 91–3, 1979.

Hunter, J. *Number Theory.* London: Oliver and Boyd, 1964.

Landau, E. *Handbuch der Lehre von der Verteilung der Primzahlen, 3rd ed.* New York: Chelsea, pp. 577–80, 1974.

Nagell, T. *Introduction to Number Theory.* New York: Wiley, pp. 28–9, 1951.

Schroeder, M. R. *Number Theory in Science and Communication, 3rd ed.* New York: Springer-Verlag, 1997.

Séroul, R. *Programming for Mathematicians.* Berlin: Springer-Verlag, pp. 19–0, 2000.

Vardi, I. *Computational Recreations in Mathematica.* Redwood City, CA: Addison-Wesley, pp. 7– and 223–25, 1991.

Möbius Periodic Function

A function periodic with period 2π such that

$$p(\theta + \pi) = -p(\theta)$$

for all θ is said to be Möbius periodic.

See also PERIODIC FUNCTION

Möbius Problem

Let $A = \{a_1, a_2, \ldots\}$ be a free Abelian SEMIGROUP, where a_1 is the IDENTITY ELEMENT, and let $\mu(n)$ be the MÖBIUS FUNCTION. Define $\mu(a_n)$ on the elements of the semigroup analogously to the definition of $\mu(n)$ (as $(-1)^r$ if n is the product of r distinct primes) by regarding generators of the semigroup as primes. Then the Möbius problem asks if the properties

1. $a < b$ IMPLIES $ac < bc$ for a, b, $c \in A$, where A has the linear order $a_1 < a_2 < \ldots$,
2. $\mu(a_n) = \mu(n)$ for all n,

imply that

$$a_{m,n} = a_m a_n$$

for all m, $n \geq 1$. Informally, the problem asks "Is the multiplication law on the positive integers uniquely determined by the values of the Möbius function and the property that multiplication respects order?"

The problem is known to be true for all $mn \leq 74$ if $\mu(a_n) = \mu(n)$ for all $n \leq 240$ (Flath and Zulauf 1995).

See also BRAUN'S CONJECTURE, MÖBIUS FUNCTION

References

Flath, A. and Zulauf, A. "Does the Möbius Function Determine Multiplicative Arithmetic?" *Amer. Math. Monthly* **102**, 354–56, 1995.

Möbius Shorts

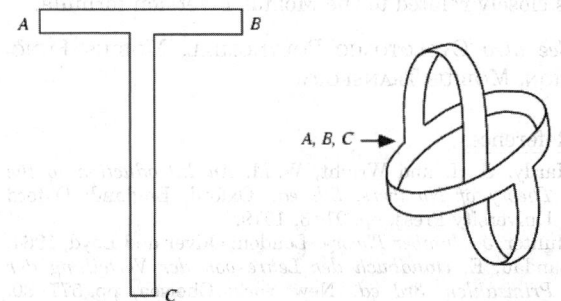

A one-sided surface reminiscent of the MÖBIUS STRIP, attributed to Gourmalin (Bouvier and George 1979, p. 477; Boas 1995). This surface is topologically equivalent to a KLEIN BOTTLE with a hole in it, and is topologically distinct from the MÖBIUS STRIP (Gramain 1984, Stewart 2000b).

See also KLEIN BOTTLE, MÖBIUS STRIP

References

Boas, R. P. Jr. "Möbius Shorts." *Math. Mag.* **68**, 127, 1995.
Bouvier, A. and George, M. *Dictionaire des mathématiques.* Paris: Presses Universitaires de France, 1979.
Gramain, A. *Topology of Surfaces.* Moscow, ID: BCS Associates, 1984.

Stewart, I. "Mathematical Recreations: Reader Feedback." *Sci. Amer.* **282**, 111, May 2000a.
Stewart, I. "Mathematical Recreations: Reader Feedback." *Sci. Amer.* **283**, 101, Sep. 2000b.

Möbius Strip

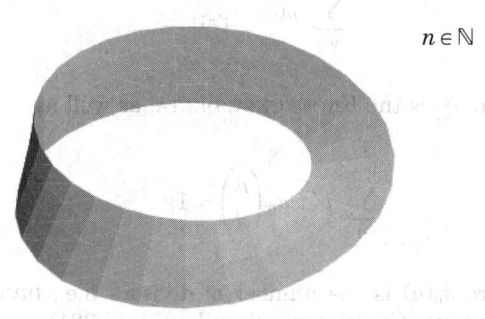

$n \in \mathbb{N}$

A one-sided NONORIENTABLE SURFACE obtained by cutting a closed band into a single strip, giving one of the two ends thus produced a half twist, and then re-attaching the two ends. According to Madachy (1979), the B. F. Goodrich Company patented a conveyor belt in the form of a Möbius strip which lasts twice as long as conventional belts.

A Möbius strip of half-width w with midcircle of radius R and at height $z = 0$ can be represented parametrically by

$$x = \left[R + s \cos\left(\tfrac{1}{2} t\right)\right] \cos t \tag{1}$$

$$y = \left[R + s \cos\left(\tfrac{1}{2} t\right)\right] \sin t \tag{2}$$

$$z = s \sin\left(\tfrac{1}{2} t\right), \tag{3}$$

for $s \in [-w, w]$ and $t \in [0, 2\pi]$.

The coefficients of the FIRST FUNDAMENTAL FORM for this surface are

$$E = 1 \tag{4}$$

$$F = 0 \tag{5}$$

$$G = R^2 + 2Rs \cos\left(\tfrac{1}{2} t\right) + \tfrac{1}{2} s^2 (3 + 2 \cos t), \tag{6}$$

the SECOND FUNDAMENTAL FORM coefficients are

$$e = 0 \tag{7}$$

$$f = \frac{R}{\sqrt{4R^2 + 3s^2 + 2s\left[4 R \cos\left(\tfrac{1}{2} t\right) + s \cos t\right]}} \tag{8}$$

$$g = \frac{\left[2(R^2 + s^2) + 4 Rs \cos\left(\tfrac{1}{2} t\right) + s^2 \cos t\right] \sin\left(\tfrac{1}{2} t\right)}{\sqrt{4R^2 + 3s^2 + 2s\left[4 R \cos\left(\tfrac{1}{2} t\right) + s \cos t\right]}}, \tag{9}$$

the AREA ELEMENT is

$$dS = \sqrt{R^2 + 2Rs\cos\left(\tfrac{1}{2}t\right) + s^2\left(\tfrac{3}{4} + \tfrac{1}{2}\cos t\right)}\, ds \wedge dt,$$

(10)

and the GAUSSIAN and MEAN CURVATURES are

$$K = -\frac{4R^2}{\left\{4R^2 + 3s^2 + 2s\left[4R\cos\left(\tfrac{1}{2}t\right) + s\cos t\right]\right\}^2}$$

(11)

$$H = \frac{2\left[2(R^2 + s^2) + 4Rs\cos\left(\tfrac{1}{2}t\right) + s^2\cos t\right]\sin\left(\tfrac{1}{2}t\right)}{\left\{4R^2 + 3s^2 + 2s\left[4R\cos\left(\tfrac{1}{2}t\right) + s\cos t\right]\right\}^2}.$$

(12)

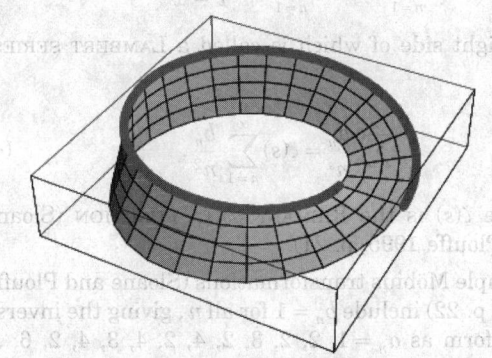

The perimeter of the Möbius strip is given by integrating the complicated function

$$ds = \sqrt{x'^2 + y'^2} = \left[\tfrac{1}{16}w^4\cos^4\left(\tfrac{1}{2}t\right)\right.$$
$$+ \left\{\left[R + w\cos(\tfrac{1}{2}t)\right]\cos t - \tfrac{1}{2}w\sin\left(\tfrac{1}{2}t\right)\sin t\right\}^4$$
$$\left. + \left\{R\sin t + \tfrac{1}{4}w\left[\sin\left(\tfrac{1}{2}t\right) + 3\sin\left(\tfrac{3}{2}t\right)\right]\right\}^4\right]^{1/2} \quad (13)$$

from 0 to 4π, which can unfortunately not be done in closed form. Note that although the *surface* closes at $t = 2\pi$, this corresponds to the bottom edge connecting with the top edge, as illustrated above, so an additional 2π must be traversed to comprise the entire arc length of the bounding edge.

Cutting a Möbius strip, giving it extra twists, and reconnecting the ends produces unexpected figures called PARADROMIC RINGS (Listing and Tait 1847, Ball and Coxeter 1987) which are summarized in the table below.

half-twists	cuts	divs.	result
1	1	2	1 band, length 2
1	1	3	1 band, length 2
			1 Möbius strip, length 1
1	2	4	2 bands, length 2
1	2	5	2 bands, length 2
			1 Möbius strip, length 1
1	3	6	3 bands, length 2
1	3	7	3 bands, length 2
			1 Möbius strip, length 1
2	1	2	2 bands, length 1
2	2	3	3 bands, length 1
2	3	4	4 bands, length 1

A TORUS can be cut into a Möbius strip with an EVEN number of half-twists, and a KLEIN BOTTLE can be cut in half along its length to make two Möbius strips. In addition, two strips on top of each other, each with a half-twist, give a single strip with four twists when disentangled.

There are three possible SURFACES which can be obtained by sewing a Möbius strip to the edge of a DISK: the BOY SURFACE, CROSS-CAP, and ROMAN SURFACE.

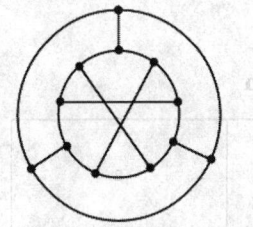

Tietze graph　　　　*embedding on the Möbius band*

The Möbius strip has EULER CHARACTERISTIC $\chi = 1$ (or genus $g = 1/2$), so the HEAWOOD CONJECTURE shows that any set of regions on it can be colored using only six colors, as illustrated above.

See also BOY SURFACE, CROSS-CAP, MAP COLORING, MÖBIUS STRIP DISSECTION, NONORIENTABLE SURFACE, PARADROMIC RINGS, PRISMATIC RING, ROMAN SURFACE, TIETZE GRAPH

References

Ball, W. W. R. and Coxeter, H. S. M. *Mathematical Recreations and Essays, 13th ed.* New York: Dover, pp. 127–28, 1987.

Bondy, J. A. and Murty, U. S. R. *Graph Theory with Applications.* New York: North Holland, p. 243, 1976.

Bogomolny, A. "Möbius Strip." http://www.cut-the-knot.com/do_you_know/moebius.html.

Gardner, M. "Möbius Bands." Ch. 9 in *Mathematical Magic Show: More Puzzles, Games, Diversions, Illusions and Other Mathematical Sleight-of-Mind from Scientific American.* New York: Vintage, pp. 123–36, 1978.

Gardner, M. *The Sixth Book of Mathematical Games from Scientific American.* Chicago, IL: University of Chicago Press, p. 10, 1984.

Gray, A. "The Möbius Strip." §14.3 in *Modern Differential Geometry of Curves and Surfaces with Mathematica, 2nd ed.* Boca Raton, FL: CRC Press, pp. 325–26, 1997.

Hunter, J. A. H. and Madachy, J. S. *Mathematical Diversions.* New York: Dover, pp. 41–5, 1975.

JavaView. "Classic Surfaces from Differential Geometry: Moebius Strip." http://www-sfb288.math.tu-berlin.de/vgp/javaview/demo/surface/common/PaSurface_Moebius-Strip.html.

Kraitchik, M. §8.4.3 in *Mathematical Recreations.* New York: W. W. Norton, pp. 212–13, 1942.

Listing and Tait. *Vorstudien zur Topologie, Göttinger Studien,* Pt. 10, 1847.

Madachy, J. S. *Madachy's Mathematical Recreations.* New York: Dover, p. 7, 1979.

Möbius, A. F. *Werke, Vol. 2.* p. 519, 1858.

Nordstrand, T. "Moebiusband." http://www.uib.no/people/nfytn/moebtxt.htm.

Pappas, T. "The Moebius Strip & the Klein Bottle," "A Twist to the Moebius Strip," "The 'Double' Moebius Strip." *The Joy of Mathematics.* San Carlos, CA: Wide World Publ./Tetra, p. 207, 1989.

Steinhaus, H. *Mathematical Snapshots, 3rd ed.* New York: Dover, pp. 269–74, 1999.

Wagon, S. "Rotating Circles to Produce a Torus or Möbius Strip." §7.4 in *Mathematica in Action.* New York: W. H. Freeman, pp. 229–32, 1991.

Wells, D. *The Penguin Dictionary of Curious and Interesting Geometry.* London: Penguin, pp. 152–53 and 164, 1991.

Möbius Strip Dissection

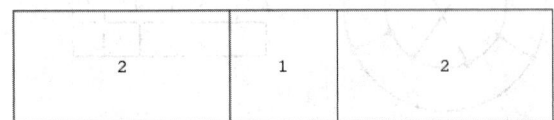

Tiling of a Möbius strip can be performed immediately by carrying over a tiling of a rectangle with the same two-sided SURFACE AREA. However, additional tilings are possible by cutting tiles across glued edges. An example of such a tiling is the strip constructed from a 5×1 RECTANGLE consisting of two halves of a width 2 square (which are rejoined when edges are connected) separated by a 1×1 square (Stewart 1997). Unfortunately, since the long top and bottom edges must be glued together, this example is not constructible out of paper. It also suffers from having the unit square share a boundary with itself. In 1993, S. J. Chapman found a tiling free of the latter defect (although still suffering from the former) which can be constructed using five squares. No similar tiling is possible using fewer tiles (Stewart 1997).

See also CYLINDER DISSECTION, MÖBIUS STRIP, PERFECT SQUARE DISSECTION, TORUS DISSECTION

References

Stewart, I. "Squaring the Square." *Sci. Amer.* **277**, 94–6, July 1997.

Möbius Transform

The transformation of a sequence $a_1, a_2, \ldots$ with

$$a_n = \sum_{d|n} b_d \tag{1}$$

into the sequence $b_1, b_2, \ldots$ via the MÖBIUS INVERSION FORMULA,

$$b_n = \sum_{d|n} \mu\left(\frac{n}{d}\right) a_d. \tag{2}$$

The transformation of b_n to a_n is sometimes called the sum-of-divisors transform. Two other equivalent formulations are given by

$$\sum_{n=1}^{\infty} a_n x^n = \sum_{n=1}^{\infty} b_n \frac{x^n}{1 - x^n}, \tag{3}$$

the right side of which is called a LAMBERT SERIES, and

$$\sum_{n=1}^{\infty} \frac{a_n}{n^s} = \zeta(s) \sum_{n=1}^{\infty} \frac{b_n}{n^2}, \tag{4}$$

where $\zeta(s)$ is the RIEMANN ZETA FUNCTION (Sloane and Plouffe 1995, p. 21).

Example Möbius transformations (Sloane and Plouffe 1995, p. 22) include $b_n = 1$ for all n, giving the inverse transform as $a_n = 1, 2, 2, 3, 2, 4, 2, 4, 3, 4, 2, 6, \ldots$ (Sloane's A000005), the DIVISOR FUNCTION $\sigma_0(n)$ of n. The Möbius transform of $a_n = n$ gives $b_n = 1, 1, 2, 2, 4, 2, 6, 4, 6, 4, 10, 4, 12, \ldots$ (Sloane's A000010), the TOTIENT FUNCTION of n. The inverse Möbius transform of the sequence $b_{2n} = 0$ and $b_{2n+1} = 4(-1)^n$ gives $a_n = 4, 4, 0, 4, 8, 0, 0, 4, 4, \ldots$ (Sloane's A004018), the number of ways $r(n)$ of writing n as a sum of two squares. The inverse Möbius transform of $b_n = 1$ for n prime and $b_n = 0$ for n composite gives the sequence $a_n = 0, 1, 1, 1, 1, 2, 1, 1, 1, \ldots$ (Sloane's A001221), the number of DISTINCT PRIME FACTORS of n.

See also BINOMIAL TRANSFORM, DIVISOR FUNCTION, EULER TRANSFORM, LAMBERT SERIES, MÖBIUS INVERSION FORMULA, MÖBIUS TRANSFORMATION, STIRLING TRANSFORM

References

Bender, E. A. and Goldman, J. R. "On the Applications of Möbius Inversion in Combinatorial Analysis." *Amer. Math. Monthly* **82**, 789–03, 1975.

Bernstein, M. and Sloane, N. J. A. "Some Canonical Sequences of Integers." *Linear Algebra Appl.* **226//228**, 57–2, 1995.

Gessel, I. and Rota, C.-G. (Eds.). *Classic Papers in Combinatorics.* Boston, MA: Birkhäuser, 1987.

Hardy, G. H. and Wright, E. M. §17.10 in *An Introduction to the Theory of Numbers, 5th ed.* Oxford, England: Clarendon Press, 1979.

Rota, G.-C. "On the Foundations of Combinatorial Theory I. Theory of Möbius Functions." *Z. für Wahrscheinlichkeitsth.* **2**, 340–68, 1964.

Sloane, N. J. A. Sequences A000005/M0246, A000010/M0299, A001221/M0056, and A004018/M3218 in "An On-Line Version of the Encyclopedia of Integer Sequences." http://www.research.att.com/~njas/sequences/eisonline.html.

Sloane, N. J. A. and Plouffe, S. *The Encyclopedia of Integer Sequences.* San Diego, CA: Academic Press, 1995.

Stanley, R. P. *Enumerative Combinatorics, Vol. 1.* Cambridge, England: Cambridge University Press, p. 259, 1999.

Möbius Transformation

Let $a \in \mathbb{C}$ and $|a| < 1$, then

$$\varphi_a(z) = \frac{z - a}{1 - \bar{a}z}$$

is a Möbius transformation, where $\bar{a}$ is the COMPLEX CONJUGATE of a. φ_a is a CONFORMAL TRANSFORMATION SELF-MAP of the UNIT DISK D for each a, and specifically of the boundary of the unit disk to itself. The same holds for $(\varphi_a)^{-1} = \varphi_{-a}$.

Any conformal self-map of the UNIT DISK to itself is a composition of a Möbius transformation with a ROTATION, and any conformal self-map f of the unit disk can be written in the form

$$f(z) = \varphi_b(wz)$$

for some Möbius transformation φ_b and some complex number w with $|w| = 1$ (Krantz 1999, p. 81).

See also LINEAR FRACTIONAL TRANSFORMATION

References

Krantz, S. G. "Möbius Transformations." §6.2.2 in *Handbook of Complex Analysis.* Boston, MA: Birkhäuser, p. 81, 1999.

Möbius Triangles

SPHERICAL TRIANGLES into which a SPHERE is divided by the planes of symmetry of a UNIFORM POLYHEDRON.

See also SPHERICAL TRIANGLE, UNIFORM POLYHEDRON

Mock Theta Function

In his last letter to Hardy, Ramanujan defined 17 JACOBI THETA FUNCTION-like functions $F(q)$ with $|q| < 1$ which he called "mock theta functions" (Watson 1936, Ramanujan 1988, pp. 127–31; Ramanujan 2000, pp. 354–55). These functions are Q-SERIES with exponential singularities such that the arguments terminate for some power t^N. In particular, if $f(q)$ is *not* a JACOBI THETA FUNCTION, then it is a mock theta function if, for each ROOT OF UNITY ρ, there is an approximation OF THE FORM

$$f(q) = \sum_{\mu=1}^{M} t^{k_\mu} \exp\left(\sum_{\nu=-1}^{N} c_{\mu\nu} t^\nu\right) + \mathcal{O}(1) \qquad (1)$$

as $t \to 0^+$ with $q = \rho e^{-t}$ (Gordon and McIntosh 2000b).

If, in addition, for every ROOT OF UNITY ρ there are modular forms $h_j^{(\rho)}(q)$ and real numbers α_j and $1 \le j \le J(\rho)$ such that

$$f(q) - \sum_{j=1}^{J(\rho)} q^{\alpha_j} h_j^{(\rho)}(q) \qquad (2)$$

is bounded as q radially approaches ρ, then $f(q)$ is said to be a strong mock theta function (Gordon and McIntosh 2000b).

Ramanujan found an additional three mock theta functions in his "lost notebook" which were subsequently rediscovered by Watson (1936). The first formula on page 15 of Ramanujan's lost notebook relates the functions which Watson calls $\rho(-q)$ and $\omega(-q)$ (equivalent to the third equation on page 63 of Watson's 1936 paper), and the last formula on page 31 of the lost notebook relates what Watson calls $\nu(-q)$ and $\omega(q^2)$ (equivalent to the fourth equation on page 63 of Watson's paper). The orders of these and Ramanujan's original 17 functions were all 3, 5, or 7.

Ramanujan's "lost notebook" also contained several mock theta functions of orders 6 and 10, which, however, were not explicitly identified as mock theta functions by Ramanujan. Their properties have now been investigated in detail (Andrews and Hickerson 1991, Choi 1999).

Examples of the mock theta functions found by Ramanujan include

$$F_0(q) = \sum_{n=0}^{\infty} \frac{q^{2n^2}}{(q; q^2)_n} \qquad (3)$$

$$F_1(q) = \sum_{n=1}^{\infty} \frac{q^{2n(n-1)}}{(q; q^2)_n}. \qquad (4)$$

(Gordon and McIntosh 2000b).

Gordon and McIntosh (2000b) found eight mock theta functions of order 8,

$$S_0(q) = \sum_{n=0}^{\infty} \frac{q^{n^2}(-q; q^2)_n}{(-q^2; q^2)_n} \qquad (5)$$

$$S_1(q) = \sum_{n=0}^{\infty} \frac{q^{n(n+2)}(-q; q^2)_n}{(-q^2; q^2)_n} \qquad (6)$$

$$T_0(q) = \sum_{n=0}^{\infty} \frac{q^{(n+1)(n+2)}(-q^2; q^2)_n}{(-q; q^2)_{n+1}} \qquad (7)$$

$$T_1(q) = \sum_{n=0}^{\infty} \frac{q^{n(n+1)}(-q^2; q^2)_n}{(-q; q^2)_{n+1}} \qquad (8)$$

$$U_0(q) = \sum_{n=0}^{\infty} \frac{q^{n^2}(-q; q^2)_n}{(-q^4; q^4)_n} \tag{9}$$

$$U_1(q) = \sum_{n=0}^{\infty} \frac{q^{(n+1)^2}(-q; q^2)_n}{(-q^2; q^4)_{n+1}} \tag{10}$$

$$V_0(q) = -1 + 2\sum_{n=0}^{\infty} \frac{q^{n^2}(-q; q^2)_n}{(q; q^2)_n} \tag{11}$$

$$= -1 + 2\sum_{n=0}^{\infty} \frac{q^{2n^2}(-q^2; q^4)_n}{(q; q^2)_{2n+1}} \tag{12}$$

$$V_1(q) = \sum_{n=0}^{\infty} \frac{q^{(n+1)^2}(-q; q^2)_n}{(q; q^4)_{n+1}} \tag{13}$$

$$= \sum_{n=0}^{\infty} \frac{q^{2n^2+2n+1}(-q^4; q^4)_n}{(q; q^2)_{2n+2}}. \tag{14}$$

See also JACOBI THETA FUNCTIONS, MORDELL INTEGRAL, Q-SERIES

References

Andrews, G. E. "The Fifth and Seventh Order Mock Theta Functions." *Trans. Amer. Soc.* **293**, 113–34, 1986.

Andrews, G. E. "Mock Theta Functions." *Proc. Sympos. Pure Math.* **49**, 283–98, 1989.

Andrews, G. E. and Hickerson, D. "Ramanujan's "Lost" Notebook VII: The Sixth Order Mock Theta Functions." *Adv. Math.* **89**, 60–05, 1991.

Bellman, R. E. *A Brief Introduction to Theta Functions.* New York: Holt, Rinehart, and Winston, p. 51, 1961.

Choi, Y.-S. "Tenth Order Mock Theta Functions in Ramanujan's Lost Notebook." *Invent. Math.* **136**, 497–69, 1999.

Gordon, B. and McIntosh, R. J. "Modular Transformations of Ramanujan's Fifth and Seventh Order Mock Theta Functions." Submitted to *Invent. Math.* 2000a.

Gordon, B. and McIntosh, R. J. "Some Eighth Order Mock Theta Functions." To appear in *J. London Math. Soc.* 2000b.

Ramanujan, S. *The Lost Notebook and Other Unpublished Manuscripts.* New Delhi, India: Narosa, 1988.

Ramanujan, S. *Collected Papers of Srinivasa Ramanujan* (Ed. G. H. Hardy, S. Aiyar, P. Venkatesvara, and B. M. Wilson). Providence, RI: Amer. Math. Soc., 2000.

Selberg, A. "Über die Mock-Thetafunktionen siebenter Ordnung." *Arch. Math. og Naturvidenskab* **41**, 3–5, 1938.

Watson, G. N. "The Final Problem: An Account of the Mock Theta Functions." *J. London Math. Soc.* **11**, 55–0, 1936.

Watson, G. N. "The Mock Theta Function (2)." *Proc. London Math. Soc.* **42**, 274–04, 1937.

Mod

CONGRUENCE

Mode

The most common value obtained in a set of observations. An interesting empirical relationship between the mean, median, and mode which appears to hold for unimodal curves of moderate asymmetry is given by

$$\text{mean} - \text{mode} \approx 3(\text{mean} - \text{median})$$

(Kenney and Keeping 1962, p. 53), which is the basis for the definition of the PEARSON MODE SKEWNESS.

See also MEAN, MEDIAN (STATISTICS), ORDER STATISTIC, PEARSON MODE SKEWNESS

References

Kenney, J. F. and Keeping, E. S. "The Mode," "Relation Between Mean, Median, and Mode," and "Relative Merits of Mean, Median, and Mode." §4.7–.9 in *Mathematics of Statistics, Pt. 1, 3rd ed.* Princeton, NJ: Van Nostrand, pp. 50–4, 1962.

Zwillinger, D. (Ed.). *CRC Standard Mathematical Tables and Formulae.* Boca Raton, FL: CRC Press, p. 602, 1995.

Model

A well-formed formula B is said to be true for the interpretation M (written $\vDash_M B$) IFF every sequence in Σ (the set of all denumerable sequences of elements of the domain of D), satisfies B. B is said to be false for M IFF no sequence in Σ satisfies B.

Then an interpretation M is said to be a model for a set Γ of well-formed formulas IFF every well-formed formula in Γ is true for M (Mendelson 1997, pp. 59–0).

See also GENERALIZED COMPLETENESS THEOREM

References

Mendelson, E. *Introduction to Mathematical Logic, 4th ed.* London: Chapman & Hall, pp. 59–0, 1997.

Model Completion

Model completion is a term employed when EXISTENTIAL CLOSURE is successful. The formation of the COMPLEX NUMBERS, and the move from affine to projective geometry, are successes of this kind. The theory of existential closure gives a theoretical basis of Hilbert's "method of ideal elements."

References

Manders, K. L. "Interpretations and the Model Theory of the Classical Geometries." In *Models and Sets*. Berlin: Springer-Verlag, pp. 297–30, 1984.

Manders, K. L. "Domain Extension and the Philosophy of Mathematics." *J. Philos.* **86**, 553–62, 1989.

Mode Locking

A phenomenon in which a system being forced at an IRRATIONAL period undergoes rational, periodic motion which persists for a finite range of forcing values. It may occur for strong couplings between natural and forcing oscillation frequencies.

The phenomenon can be exemplified in the CIRCLE MAP when, after q iterations of the map, the new angle differs from the initial value by a RATIONAL NUMBER

$$\theta_{n+q} = \theta_n + \frac{p}{q}.$$

This is the form of the unperturbed CIRCLE MAP with the WINDING NUMBER

$$\Omega = \frac{p}{q}.$$

For Ω not a RATIONAL NUMBER, the trajectory is QUASIPERIODIC.

See also CHAOS, QUASIPERIODIC FUNCTION

Model Theory

Model theory is a general theory of interpretations of AXIOMATIC SET THEORY. It is the branch of LOGIC studying mathematical structures by considering first-order sentences which are true of those structures and the sets which are definable in those structures by first-order FORMULAS (Marker 1996).

Mathematical structures obeying axioms in a system are called "models" of the system. The usual axioms of ANALYSIS are second order and are known to have the REAL NUMBERS as their unique model. Weakening the axioms to include only the first-order ones leads to a new type of model in what is called NONSTANDARD ANALYSIS.

See also KHOVANSKI'S THEOREM, NONSTANDARD ANALYSIS, WILKIE'S THEOREM

References

Doets, K. *Basic Model Theory*. New York: Cambridge University Press, 1996.
Hodges, W. *A Shorter Model Theory*. New York: Cambridge University Press, 1997.
Manzano, M. *Model Theory*. Oxford, England: Oxford University Press, 1999.
Marker, D. "Model Theory and Exponentiation." *Not. Amer. Math. Soc.* **43**, 753–59, 1996.
Stewart, I. "Non-Standard Analysis." In *From Here to Infinity: A Guide to Today's Mathematics*. Oxford, England: Oxford University Press, pp. 80–1, 1996.

Modified Bernoulli Number

The numbers b_{2n} having GENERATING FUNCTION

$$\sum_{n=0}^{\infty} b_{2n} x^{2n} = \frac{1}{2} \ln\left(\frac{e^{x/2} - e^{-x/2}}{\frac{1}{2}x}\right)$$

$$= \frac{1}{2}\ln 2 + \frac{1}{48}x^2 - \frac{1}{5760}x^4 + \frac{1}{362880}x^6 - \dots.$$

For $n = 1$, 2, ..., the denominators are 48, 5760, 362880, 19353600, ... (Sloane's A057868).

See also BERNOULLI NUMBER, KONTSEVICH INTEGRAL

References

Sloane, N. J. A. Sequences A057868 in "An On-Line Version of the Encyclopedia of Integer Sequences." http://www.research.att.com/~njas/sequences/eisonline.html.

Modified Bessel Differential Equation

The second-order ordinary differential equation

$$x^2 \frac{d^2y}{dx^2} + x\frac{dy}{dx} - (x^2 + n^2)y = 0. \qquad (1)$$

The solutions are the MODIFIED BESSEL FUNCTIONS OF THE FIRST and SECOND KINDS, and can be written

$$y = a_1 J_n(-ix) + a_2 Y_n(-ix) \qquad (2)$$

$$= c_1 I_n(x) + c_2 K_n(x), \qquad (3)$$

where $J_n(x)$ is a BESSEL FUNCTION OF THE FIRST KIND, $Y_n(x)$ is a BESSEL FUNCTION OF THE SECOND KIND, $I_n(x)$ is a MODIFIED BESSEL FUNCTION OF THE FIRST KIND, and $K_n(x)$ is MODIFIED BESSEL FUNCTION OF THE SECOND KIND.

If $n = 0$, the modified Bessel differential equation becomes

$$x^2 \frac{d^2y}{dx^2} + x\frac{dy}{dx} - x^2y = 0, \qquad (4)$$

which can also be written

$$\frac{d}{dx}\left(x\frac{dy}{dx}\right) = xy. \qquad (5)$$

References

Abramowitz, M. and Stegun, C. A. (Eds.). §9.6.1 in *Handbook of Mathematical Functions with Formulas, Graphs, and Mathematical Tables, 9th printing*. New York: Dover, 1972.
Zwillinger, D. *Handbook of Differential Equations, 3rd ed.* Boston, MA: Academic Press, p. 121, 1997.

Modified Bessel Function of the First Kind

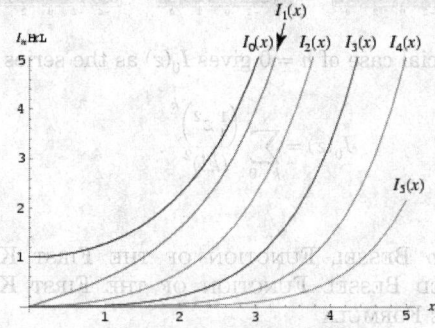

A function $I_n(x)$ which is one of the solutions to the MODIFIED BESSEL DIFFERENTIAL EQUATION and is closely related to the BESSEL FUNCTION OF THE FIRST KIND $J_n(x)$. The above plot shows $I_n(x)$ for $n = 1$, 2, ..., 5. In terms of $J_n(x)$,

$$I_n(x) \equiv i^{-n} J_n(ix) = e^{-n\pi i/2} J_n\left(xe^{i\pi/2}\right). \qquad (1)$$

For a REAL NUMBER v, the function can be computed

using

$$I_\nu(z) = (\tfrac{1}{2} z)^\nu \sum_{k=0}^{\infty} \frac{\left(\tfrac{1}{4} z^2\right)^k}{k!\,\Gamma(\nu + k + 1)}, \qquad (2)$$

where $\Gamma(z)$ is the GAMMA FUNCTION. An integral formula is

$$I_\nu(z) = \frac{1}{\pi} \int_0^\pi e^{z \cos\theta} \cos(\nu\theta)\, d\theta$$

$$- \frac{\sin(\nu\pi)}{\pi} \int_0^\infty e^{-z \cosh t - \nu t}\, dt, \qquad (3)$$

which simplifies for ν an INTEGER n to

$$I_n(z) = \frac{1}{\pi} \int_0^\pi e^{z \cos\theta} \cos(n\theta)\, d\theta \qquad (4)$$

(Abramowitz and Stegun 1972, p. 376).
A derivative identity for expressing higher order modified Bessel functions in terms of $I_0(x)$ is

$$I_n(x) = T_n\left(\frac{d}{dx}\right) I_0(x), \qquad (5)$$

where $T_n(x)$ is a CHEBYSHEV POLYNOMIAL OF THE FIRST KIND.

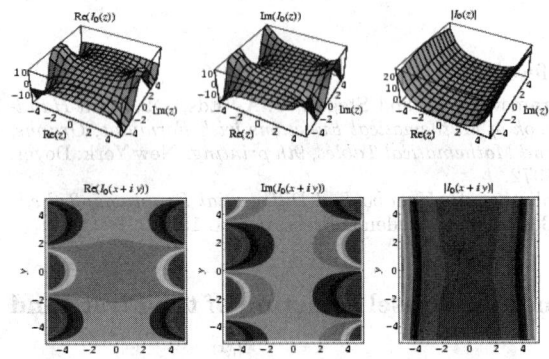

The special case of $n = 0$ gives $I_0(z)$ as the series

$$J_0(z) = \sum_{k=0}^{\infty} \frac{\left(\tfrac{1}{4} z^2\right)^k}{(k!)^2}. \qquad (6)$$

See also BESSEL FUNCTION OF THE FIRST KIND, MODIFIED BESSEL FUNCTION OF THE FIRST KIND, WEBER'S FORMULA

References

Abramowitz, M. and Stegun, C. A. (Eds.). "Modified Bessel Functions I and K." §9.6 in *Handbook of Mathematical Functions with Formulas, Graphs, and Mathematical Tables, 9th printing.* New York: Dover, pp. 374–77, 1972.
Arfken, G. "Modified Bessel Functions, $I_\nu(x)$ and $K_\nu(x)$." §11.5 in *Mathematical Methods for Physicists, 3rd ed.* Orlando, FL: Academic Press, pp. 610–16, 1985.

Finch, S. "Favorite Mathematical Constants." http://www.mathsoft.com/asolve/constant/cntfrc/cntfrc.html.
Press, W. H.; Flannery, B. P.; Teukolsky, S. A.; and Vetterling, W. T. "Bessel Functions of Fractional Order, Airy Functions, Spherical Bessel Functions." §6.7 in *Numerical Recipes in FORTRAN: The Art of Scientific Computing, 2nd ed.* Cambridge, England: Cambridge University Press, pp. 234–45, 1992.
Spanier, J. and Oldham, K. B. "The Hyperbolic Bessel Functions $I_0(x)$ and $I_1(x)$" and "The General Hyperbolic Bessel Function $I_\nu(x)$." Chs. 49–0 in *An Atlas of Functions.* Washington, DC: Hemisphere, pp. 479–87 and 489–97, 1987.

Modified Bessel Function of the Second Kind

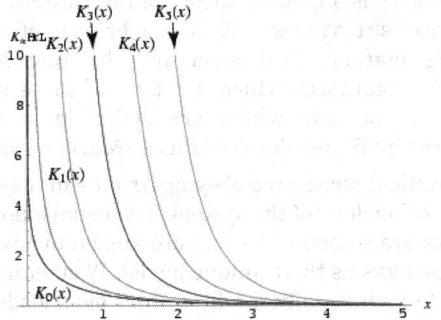

The function $K_n(x)$ which is one of the solutions to the MODIFIED BESSEL DIFFERENTIAL EQUATION. The modified Bessel functions of the second kind are sometimes called the Basset functions (Spanier and Oldham 1987, p. 499) or Macdonald functions (Spanier and Oldham 1987, p. 499; Samko *et al.* 1993, p. 20). $K_n(x)$ is closely related to the MODIFIED BESSEL FUNCTION OF THE FIRST KIND $I_n(x)$ and HANKEL FUNCTION $H_n(x)$,

$$K_n(x) \equiv \tfrac{1}{2} \pi i^{n+1} H_n^{(1)}(ix) \qquad (1)$$

$$= \tfrac{1}{2} \pi i^{n+1} [J_n(ix) + i N_n(ix)] \qquad (2)$$

$$= \frac{\pi}{2} \frac{I_{-n}(x) - I_n(x)}{\sin(n\pi)} \qquad (3)$$

(Watson 1966, p. 185). A sum formula for $K_n(x)$ is

$$K_n(z) = \tfrac{1}{2}(\tfrac{1}{2} z)^{-n} \sum_{k=0}^{n-1} \frac{(n-k-1)!}{k!} (-\tfrac{1}{4} z^2)^k$$

$$+ (-1)^{n+1} \ln(\tfrac{1}{2} z) I_n(z) + (-1)^n \tfrac{1}{2}(\tfrac{1}{2} z)^n$$

$$\times \sum_{k=0}^{\infty} [\psi(k+1) + \psi(n+k+1)] \frac{(\tfrac{1}{4} z^2)^k}{k!(n+k)!}, \quad (4)$$

where ψ is the DIGAMMA FUNCTION (Abramowitz and Stegun 1972). An integral formula is

$$K_\nu(z) = \frac{\Gamma(\nu + \tfrac{1}{2})(2z)^\nu}{\sqrt{\pi}} \int_0^\infty \frac{\cos t\, dt}{(t^2 + z^2)^{\nu + 1/2}} \qquad (5)$$

which, for $\nu = 0$, simplifies to

$$K_0(x) = \int_0^\infty \cos(x \sinh t) \, dt = \int_0^\infty \frac{\cos(xt) \, dt}{\sqrt{t^2 + 1}}. \quad (6)$$

Other identities are

$$K_n(z) = \frac{\sqrt{\pi}}{(n - \frac{1}{2})!} (\tfrac{1}{2} z)^n \int_1^\infty e^{-zx} (x^2 - 1)^{n - 1/2} \, dx \quad (7)$$

for $n > -1/2$ and

$$K_n(z) = \sqrt{\frac{\pi}{2z}} \frac{e^{-z}}{(n - \frac{1}{2})!} \int_0^\infty e^{-t} t^{n - 1/2} \left(1 - \frac{t}{2z}\right)^{n - 1/2} dt \quad (8)$$

$$= \sqrt{\frac{\pi}{2z}} \frac{e^{-z}}{(n - \frac{1}{2})!} \sum_{r=0}^\infty \frac{(n - \frac{1}{2})!}{r! (n - r - \frac{1}{2})!} (2z)^{-r}$$

$$\times \int_0^\infty e^{-t} t^{n + r - 1/2} \, dt. \quad (9)$$

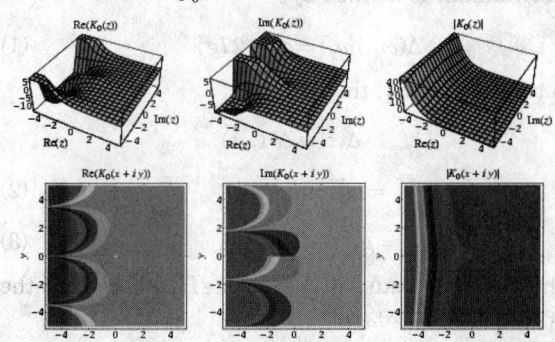

The special case of $n = 0$ gives $K_0(z)$ as the integrals

$$K_0(z) = \int_0^\infty \cos(x \sinh t) \, dt \quad (10)$$

$$= \int_0^\infty \frac{\cos(xt)}{\sqrt{t^2 + 1}} \, dt \quad (11)$$

(Abramowitz and Stegun 1972, p. 376).

References

Abramowitz, M. and Stegun, C. A. (Eds.). "Modified Bessel Functions I and K." §9.6 in *Handbook of Mathematical Functions with Formulas, Graphs, and Mathematical Tables, 9th printing.* New York: Dover, pp. 374–77, 1972.
Arfken, G. "Modified Bessel Functions, $I_\nu(x)$ and $K_\nu(x)$." §11.5 in *Mathematical Methods for Physicists, 3rd ed.* Orlando, FL: Academic Press, pp. 610–16, 1985.
Press, W. H.; Flannery, B. P.; Teukolsky, S. A.; and Vetterling, W. T. "Modified Bessel Functions of Integral Order" and "Bessel Functions of Fractional Order, Airy Functions, Spherical Bessel Functions." §6.6 and 6.7 in *Numerical Recipes in FORTRAN: The Art of Scientific Computing, 2nd ed.* Cambridge, England: Cambridge University Press, pp. 229–45, 1992.
Samko, S. G.; Kilbas, A. A.; and Marichev, O. I. *Fractional Integrals and Derivatives.* Yverdon, Switzerland: Gordon and Breach, p. 20, 1993.
Spanier, J. and Oldham, K. B. "The Basset $K_\nu(x)$." Ch. 51 in *An Atlas of Functions.* Washington, DC: Hemisphere, pp. 499–07, 1987.

Watson, G. N. *A Treatise on the Theory of Bessel Functions, 2nd ed.* Cambridge, England: Cambridge University Press, 1966.

Modified Emden Differential Equation

The second-order ORDINARY DIFFERENTIAL EQUATION

$$y'' + \alpha(x) y' + x^2 y^n = 0.$$

See also EMDEN DIFFERENTIAL EQUATION

References

Leach, P. G. L. "First Integrals for the Modified Emden Equation $\ddot{q} + \alpha(t)\dot{q} + q^n = 0$." *J. Math. Phys.* **26**, 2510–514, 1985.
Zwillinger, D. *Handbook of Differential Equations, 3rd ed.* Boston, MA: Academic Press, p. 122, 1997.

Modified Spherical Bessel Differential Equation

The modified spherical Bessel differential equation is given by the SPHERICAL BESSEL DIFFERENTIAL EQUATION with a NEGATIVE separation constant,

$$r^2 \frac{d^2 R}{dr^2} + 2r \frac{dR}{dr} - r[r^2 + n(n+1)]R = 0.$$

The solutions are called MODIFIED SPHERICAL BESSEL FUNCTIONS.

See also MODIFIED SPHERICAL BESSEL FUNCTION, SPHERICAL BESSEL DIFFERENTIAL EQUATION

References

Abramowitz, M. and Stegun, C. A. (Eds.). §10.2.1 in *Handbook of Mathematical Functions with Formulas, Graphs, and Mathematical Tables, 9th printing.* New York: Dover, pp. 374–77, 1972.
Zwillinger, D. *Handbook of Differential Equations, 3rd ed.* Boston, MA: Academic Press, p. 121, 1997.

Modified Spherical Bessel Function

Solutions to the MODIFIED SPHERICAL BESSEL DIFFERENTIAL EQUATION, given by

$$i_n(x) \equiv \sqrt{\frac{\pi}{2x}} I_{n + 1/2}(x) \quad (1)$$

$$i_0(x) = \frac{\sinh x}{x} \quad (2)$$

$$k_n(x) \equiv \sqrt{\frac{2\pi}{x}} K_{n + 1/2}(x) \quad (3)$$

$$k_0(x) = \frac{e^{-x}}{x}, \quad (4)$$

where $I_n(x)$ is a MODIFIED BESSEL FUNCTION OF THE

FIRST KIND and $K_n(x)$ is a MODIFIED BESSEL FUNCTION OF THE SECOND KIND.

See also MODIFIED BESSEL FUNCTION OF THE FIRST KIND, MODIFIED BESSEL FUNCTION OF THE SECOND KIND

References

Abramowitz, M. and Stegun, C. A. (Eds.). "Modified Spherical Bessel Functions." §10.2 in *Handbook of Mathematical Functions with Formulas, Graphs, and Mathematical Tables, 9th printing.* New York: Dover, pp. 443–45, 1972.

Modified Struve Function

$$\mathscr{L}_v(z) = \left(\tfrac{1}{2}z\right)^{v+1} \sum_{k=0}^{\infty} \frac{\left(\tfrac{1}{2}z\right)^{2k}}{\Gamma\left(k + \tfrac{3}{2}\right)\Gamma\left(k + v + \tfrac{3}{2}\right)}$$

$$= \frac{2\left(\tfrac{1}{2}z\right)^v}{\sqrt{\pi}\,\Gamma\left(v + \tfrac{1}{2}\right)} \int_0^{\pi/2} \sinh(z\cos\theta)\sin^{2v}\theta\,d\theta,$$

where $\Gamma(z)$ is the GAMMA FUNCTION. For integer n, the function is related to the ordinary STRUVE FUNCTION $\mathscr{H}_n(z)$ by

$$\mathscr{L}_n(iz) = -ie^{-n\pi i/2}\mathscr{H}_n(z).$$

The Struve function $\mathscr{L}_v(z)$ is built into *Mathematica* 4.0 as StruveL[n, z].

See also ANGER FUNCTION, STRUVE FUNCTION, WEBER FUNCTIONS

References

Abramowitz, M. and Stegun, C. A. (Eds.). "Modified Struve Function $L_v(x)$." §12.2 in *Handbook of Mathematical Functions with Formulas, Graphs, and Mathematical Tables, 9th printing.* New York: Dover, p. 498, 1972.
Apelblat, A. "Derivatives and Integrals with Respect to the Order of the Struve Functions $H_v(x)$ and $L_v(x)$." *J. Math. Anal. Appl.* **137**, 17–6, 1999.

Modul

MODULE

Modular Angle

Given a MODULUS k in an ELLIPTIC INTEGRAL, the modular angle is defined by $k \equiv \sin\alpha$. An ELLIPTIC INTEGRAL is written $I(\phi|m)$ when the PARAMETER is used, $I(\phi, k)$ when the MODULUS is used, and $I(\phi\backslash\alpha)$ when the modular angle is used.

See also AMPLITUDE, CHARACTERISTIC (ELLIPTIC INTEGRAL), ELLIPTIC INTEGRAL, HALF-PERIOD RATIO, MODULUS (ELLIPTIC INTEGRAL), NOME, PARAMETER

References

Abramowitz, M. and Stegun, C. A. (Eds.). *Handbook of Mathematical Functions with Formulas, Graphs, and*

Mathematical Tables, 9th printing. New York: Dover, p. 590, 1972.

Modular Discriminant

Define $q \equiv e^{2\pi i\tau}$ (cf. the usual NOME), where τ is in the UPPER HALF-PLANE. Then the modular discriminant is defined by

$$\Delta(\tau) \equiv q \prod_{r=1}^{\infty}(1 - q^r)^{24}$$

(Rankin 1977, p. 196; Berndt 1988, p. 326; Milne 2000).

If $g_2(\omega_1, \omega_2)$ and $g_3(\omega_1, \omega_2)$ are the INVARIANTS of a WEIERSTRASS ELLIPTIC FUNCTION $\wp(z|\omega_1, \omega_2) = \wp(z; g_2, g_3)$ with periods ω_1 and ω_2, then the discriminant is defined by

$$\Delta(\omega_1, \omega_2) = g_2^3 - 27g_3^2. \tag{1}$$

Letting $\tau \equiv \omega_2/\omega_1$, then

$$\Delta(\tau) \equiv \Delta(1, \tau)$$

$$= \omega_1^{12}\Delta(\omega_1, \omega_2) \tag{2}$$

$$= g_2^3(\tau) - 27g_3^2(\tau). \tag{3}$$

The FOURIER SERIES of $\Delta(\tau)$ for $\tau \in H$, where H is the UPPER HALF-PLANE, is

$$\Delta(\tau) = (2\pi)^{12}\sum_{n=1}^{\infty}\tau(n)e^{2\pi in\tau}, \tag{4}$$

where $\tau(n)$ is the TAU FUNCTION, and $\tau(n)$ are integers (Apostol 1997, p. 20). The discriminant can also be expressed in terms of the DEDEKIND ETA FUNCTION $\eta(\tau)$ by

$$\Delta(\tau) = (2\pi)^{12}[\eta(\tau)]^24 \tag{5}$$

(Apostol 1997, p. 51).

See also DEDEKIND ETA FUNCTION, INVARIANT (ELLIPTIC FUNCTION), KLEIN'S ABSOLUTE INVARIANT, NOME, TAU FUNCTION, WEIERSTRASS ELLIPTIC FUNCTION

References

Apostol, T. M. "The Discriminant Δ" and "The Fourier Expansions of $\Delta(\tau)$ and $J(\tau)$." §1.11 and 1.15 in *Modular Functions and Dirichlet Series in Number Theory, 2nd ed.* New York: Springer-Verlag, pp. 14 and 20–2, 1997.
Berndt, B. C. *Ramanujan's Notebooks, Part II.* New York: Springer-Verlag, p. 326, 1988.
Milne, S. C. Hankel Determinants of Eisenstein Series. 13 Sep 2000. http://xxx.lanl.gov/abs/math.NT/0009130/.
Nesterenko, Yu. V. §1.2 in *A Course on Algebraic Independence: Lectures at IHP 1999.* http://www.math.jussieu.fr/~nesteren/.
Rankin, R. A. *Modular Forms and Functions.* Cambridge, England: Cambridge University Press, p. 196, 1977.

Modular Equation

The modular equation of degree n gives an algebraic connection OF THE FORM

$$\frac{K'(l)}{K(l)} = n \frac{K'(k)}{K(k)} \tag{1}$$

between the TRANSCENDENTAL COMPLETE ELLIPTIC INTEGRALS OF THE FIRST KIND with moduli k and l. When k and l satisfy a modular equation, a relationship OF THE FORM

$$\frac{M(l, k)\,dy}{\sqrt{(1-y^2)(1-l^2y^2)}} = \frac{dx}{\sqrt{(1-x^2)(1-k^2x^2)}} \tag{2}$$

exists, and M is called the multiplier. In general, if p is an ODD PRIME, then the modular equation is given by

$$\Omega_p(u, v) = (v - u_0)(v - u_1) \cdots (v - u_p), \tag{3}$$

where

$$u_p \equiv (-1)^{(p^2-1)/8}[\lambda(q^p)]^{1/8} \equiv (-1)^{(p^2-1)/8}u(q^p), \tag{4}$$

λ is a ELLIPTIC LAMBDA FUNCTION, and

$$q \equiv e^{i\pi t} \tag{5}$$

(Borwein and Borwein 1987, p. 126). An ELLIPTIC INTEGRAL identity gives

$$\frac{K'(k)}{K(k)} = 2 \frac{K'\left(\dfrac{2\sqrt{k}}{1+k}\right)}{K\left(\dfrac{2\sqrt{k}}{1+k}\right)}, \tag{6}$$

so the modular equation of degree 2 is

$$l = \frac{2\sqrt{k}}{1+k} \tag{7}$$

which can be written as

$$l^2(1+k^2) = 4k. \tag{8}$$

A few low order modular equations written in terms of k and l are

$$\Omega_2 = l^2(1+k)^2 - 4k = 0 \tag{9}$$

$$\Omega_7 = (kl)^{1/4} + (k'l')^{1/4} - 1 = 0 \tag{10}$$

$$\Omega_{23} = (kl)^{1/4} + (k'l')^{1/4} + 2^{2/3}(klk'l')^{1/12} - 1 = 0. \tag{11}$$

In terms of u and v,

$$\Omega_3(u, v) = u^4 - v^4 + 2uv(1 - u^2v^2) = 0 \tag{12}$$

$$\Omega_5(u, v) = v^6 - u^6 + 5u^2v^2(v^2 - u^2) + 4uv(u^4v^4 - 1)$$
$$= \left(\frac{u}{v}\right)^3 + \left(\frac{v}{u}\right)^3 = 2\left(u^2v^2 - \frac{1}{u^2v^2}\right) = 0 \tag{13}$$

$$\Omega_7(u, v) = (1 - u^8)(1 - v^8) - (1 - uv)^8 = 0, \tag{14}$$

where

$$u^2 \equiv \sqrt{k} = \frac{\vartheta_2(q)}{\vartheta_3(q)} \tag{15}$$

and

$$v^2 \equiv \sqrt{l} = \frac{\vartheta_2(q^p)}{\vartheta_3(q^p)}. \tag{16}$$

Here, ϑ_i are JACOBI THETA FUNCTIONS.

A modular equation of degree 2^r for $r \geq 2$ can be obtained by iterating the equation for 2^{r-1}. Modular equations for PRIME p from 3 to 23 are given in Borwein and Borwein (1987).

Quadratic modular identities include

$$\frac{\vartheta_3(q)}{\vartheta_3(q^4)} - 1 = \left[\frac{\vartheta_3^2(q^2)}{\vartheta_3^2(q^4)} - 1\right]^{1/2}. \tag{17}$$

Cubic identities include

$$\left[3\frac{\vartheta_2(q^9)}{\vartheta_2(q)} - 1\right]^3 = 9\frac{\vartheta_2^4(q^3)}{\vartheta_2^4(q)} - 1 \tag{18}$$

$$\left[3\frac{\vartheta_3(q^9)}{\vartheta_3(q)} - 1\right]^3 = 9\frac{\vartheta_3^4(q^3)}{\vartheta_3^4(q)} - 1 \tag{19}$$

$$\left[3\frac{\vartheta_4(q^9)}{\vartheta_4(q)} - 1\right]^3 = 9\frac{\vartheta_4^4(q^3)}{\vartheta_4^4(q)} - 1. \tag{20}$$

A seventh-order identity is

$$\sqrt{\vartheta_3(q)\vartheta_3(q^7)} - \sqrt{\vartheta_4(q)\vartheta_4(q^7)} = \sqrt{\vartheta_2(q)\vartheta_2(q^7)}. \tag{21}$$

From Ramanujan (1913–914),

$$(1+q)(1+q^3)(1+q^5)\cdots = 2^{1/6}q^{1/24}(kk')^{-1/12} \tag{22}$$

$$(1-q)(1-q^3)(1-q^5)\cdots = 2^{1/6}q^{1/24}k^{-1/12}k'^{1/6}. \tag{23}$$

When k and l satisfy a MODULAR EQUATION, a relationship OF THE FORM

$$\frac{M(l, k)\,dy}{\sqrt{(1-y^2)(1-l^2y^2)}} = \frac{dx}{\sqrt{(1-x^2)(1-k^2x^2)}} \tag{24}$$

exists, and M is called the multiplier. The multiplier of degree n can be given by

$$M_n(l, k) \equiv \frac{\vartheta_3^2(q)}{\vartheta_3^2(q^{1/p})} = \frac{K(k)}{K(l)}, \tag{25}$$

where ϑ_i is a JACOBI THETA FUNCTION and $K(k)$ is a complete ELLIPTIC INTEGRAL OF THE FIRST KIND.

The first few multipliers in terms of l and k are

$$M_2(l, k) = \frac{1}{1 + k} = \frac{1 + l'}{2} \tag{26}$$

$$M_3(l, k) = \frac{1 - \sqrt{\dfrac{l^3}{k}}}{1 - \sqrt{\dfrac{k^3}{l}}}. \tag{27}$$

In terms of the u and v defined for MODULAR EQUATIONS,

$$M_3 = \frac{v}{v + 2u^3} = \frac{2v^3 - u}{3u} \tag{28}$$

$$M_5 = \frac{v(1 - uv^3)}{v - u^5} = \frac{u + v^5}{5u(1 + u^3v)} \tag{29}$$

$$M_7 = \frac{v(1 - uv)(1 - uv + (uv)^2)]}{v - u^2}$$

$$= \frac{v^7 - u}{7u(1 - uv)(1 - uv + (uv)^2)]}. \tag{30}$$

See also MODULAR FORM, MODULAR FUNCTION, SCHLÄFLI'S MODULAR FORM

References

Borwein, J. M. and Borwein, P. B. *Pi & the AGM: A Study in Analytic Number Theory and Computational Complexity.* New York: Wiley, pp. 127–32, 1987.
Hanna, M. "The Modular Equations." *Proc. London Math. Soc.* **28**, 46–2, 1928.
Ramanujan, S. "Modular Equations and Approximations to π." *Quart. J. Pure. Appl. Math.* **45**, 350–72, 1913–914.

Modular Form

A function f is said to be an entire modular form of weight k if it satisfies

1. f is analytic in the UPPER HALF-PLANE H,
2. $f\left(\frac{a\tau+b}{c\tau+d}\right) = (c\tau+d)^k f(\tau)$ whenever $\begin{bmatrix} a & b \\ c & d \end{bmatrix}$ is a member of the MODULAR GROUP GAMMA,
3. The FOURIER SERIES of f has the form

$$f(\tau) = \sum_{n=0}^{\infty} c(n)e^{2\pi i n\tau} \tag{1}$$

Care must be taken when consulting the literature because some authors use the term "dimension $-k$" or "degree $-k$" instead of "weight k," and others write k instead of k (Apostol 1997, pp. 114–15). More general types of modular forms (which are not "entire"rpar;

can also be defined which allow poles in H or at $i\infty$. Since KLEIN'S ABSOLUTE INVARIANT J, which is a MODULAR FUNCTION, has a pole at $i\infty$, it is a nonentire modular form of weight 0.

The set of all entire forms of weight k is denoted M_k, which is a linear space over the complex field. The dimension of M_k is 1 for $k = 4, 6, 8, 10,$ and 14 (Apostol 1997, p. 119).

$c(0)$ is the value of f at $i\infty$, and if $c(0) = 0$, the function is called a CUSP FORM. The smallest r such that $c(r) \neq 0$ is called the order of the zero of f at $i\infty$. An estimate for $c(n)$ states that

$$c(n) = \mathcal{O}(n^{2k-1}) \tag{2}$$

if $f \in M_{2k}$ and is not a CUSP FORM (Apostol 1997, p. 135).

If $f \neq 0$ is an entire modular form of weight k, let f have N zeros in the closure of the FUNDAMENTAL REGION R_Γ (omitting the vertices). Then

$$k = 12N + 6N(i) + 4N(\rho) + 12N(i\infty), \tag{3}$$

where $N(p)$ is the order of the zero at a point p (Apostol 1997, p. 115). In addition,

1. The only entire modular forms of weight $k = 0$ are the constant functions.
2. If k is ODD, $k < 0$, or $k = 2$, then the only entire modular form of weight k is the zero function.
3. Every nonconstant entire modular form for weight $k \geq 4$, where k is EVEN.
4. The only entire CUSP FORM of weight $k < 12$ is the zero function.

(Apostol 1997, p. 116).

For f an entire modular form of EVEN weight $k \geq 0$, define $E_0(\tau) = 1$ for all τ. Then f can be expressed in exactly one way as a sum

$$f = \sum_{\substack{r=0 \\ k=12r \neq 2}}^{\lfloor k/12 \rfloor} a_r E_{k-12r} \Delta^r, \tag{4}$$

where a_r are complex numbers, E_n is an EISENSTEIN SERIES, and Δ is the MODULAR DISCRIMINANT of the WEIERSTRASS ELLIPTIC FUNCTION. CUSP FORMS of EVEN weight k are then those sums for which $a_0 = 0$ (Apostol 1997, pp. 117–18). Even more amazingly, every entire modular form f of weight k is a POLYNOMIAL in E_4 and E_6 given by

$$f = \sum_{a, b} c_{a, b} E_4^a E_6^a, \tag{5}$$

where the $c_{a, b}$ are complex numbers and the sum is extended over all integers $a, b \geq 0$ such that $4a + 6b = k$ (Apostol 1998, p. 118).

Modular forms satisfy rather spectacular and special properties resulting from their surprising array of internal symmetries. Hecke discovered an amazing connection between each modular form and a corresponding DIRICHLET L-SERIES. A remarkable connection between rational ELLIPTIC CURVES and modular forms is given by the TANIYAMA-SHIMURA CONJECTURE, which states that any rational ELLIPTIC CURVE is a modular form in disguise. This result was the one proved by Andrew Wiles in his celebrated proof of FERMAT'S LAST THEOREM.

See also CUSP FORM, DIRICHLET SERIES, ELLIPTIC CURVE, ELLIPTIC FUNCTION, FERMAT'S LAST THEOREM, HECKE ALGEBRA, HECKE OPERATOR, MODULAR FUNCTION, SCHLÄFLI'S MODULAR FORM, TANIYAMA-SHIMURA CONJECTURE

References

Apostol, T. M. "Modular Forms with Multiplicative Coefficients." Ch. 6 in *Modular Functions and Dirichlet Series in Number Theory, 2nd ed.* New York: Springer-Verlag, pp. 113–41, 1997.
Hecke, E. "Über Modulfunktionen und die Dirichlet Reihen mit Eulerscher Produktentwicklungen. I." *Math. Ann.* **114**, 1–8, 1937.
Knopp, M. I. *Modular Functions in Analytic Number Theory.* New York: Chelsea, 1993.
Koblitz, N. *Introduction to Elliptic Curves and Modular Forms.* New York: Springer-Verlag, 1993.
Rankin, R. A. *Modular Forms and Functions.* Cambridge, England: Cambridge University Press, 1977.
Sarnack, P. *Some Applications of Modular Forms.* Cambridge, England: Cambridge University Press, 1993.

Modular Function

A function is said to be modular (or "elliptic modular") if it satisfies:

1. f is MEROMORPHIC in the UPPER HALF-PLANE H,
2. $f(A\tau) = f(\tau)$ for every MATRIX A in the MODULAR GROUP GAMMA,
3. The LAURENT SERIES of f has the form

$$f(\tau) = \sum_{n=-m}^{m} a(n)e^{2\pi i n \tau}$$

(Apostol 1997, p. 34). Every RATIONAL FUNCTION of KLEIN'S ABSOLUTE INVARIANT J is a modular function, and every modular function can be expressed as a RATIONAL FUNCTION of J (Apostol 1997, p. 40).

An important property of modular functions is that if f is modular and not identically 0, then the number of zeros of f is equal to the number of poles of f in the closure of the FUNDAMENTAL REGION R_Γ (Apostol 1997, p. 34).

See also DIRICHLET SERIES, ELLIPTIC FUNCTION, ELLIPTIC LAMBDA FUNCTION, ELLIPTIC MODULAR FUNCTION, KLEIN'S ABSOLUTE INVARIANT, MODULAR

EQUATION, MODULAR FORM, MODULAR GROUP GAMMA, MODULAR GROUP GAMMA0, MODULAR GROUP LAMBDA

References

Apostol, T. M. *Modular Functions and Dirichlet Series in Number Theory, 2nd ed.* New York: Springer-Verlag, 1997.
Askey, R. In *Ramanujan International Symposium* (Ed. N. K Thakare). pp. 1–3.
Borwein, J. M. and Borwein, P. B. "Elliptic Modular Functions." §4.3 in *Pi & the AGM: A Study in Analytic Number Theory and Computational Complexity.* New York: Wiley, pp. 112–16, 1987.
Rademacher, H. "Zur Theorie der Modulfunktionen." *J. reine angew. Math.* **167**, 312–36, 1932.
Rankin, R. A. *Modular Forms and Functions.* Cambridge, England: Cambridge University Press, 1977.
Schoeneberg, B. *Elliptic Modular Functions: An Introduction.* Berlin: New York: Springer-Verlag, 1974.
Weisstein, E. W. "Books about Modular Functions." http://www.treasure-troves.com/books/ModularFunctions.html.

Modular Group

MODULAR GROUP GAMMA, MODULAR GROUP GAMMA0, MODULAR GROUP LAMBDA

Modular Group Gamma

The GROUP Γ of all MÖBIUS TRANSFORMATIONS OF THE FORM

$$\tau' = \frac{a\tau + b}{c\tau + d}, \tag{1}$$

where a, b, c, and d are integers with $ab - bc = 1$. The group can be represented by the 2×2 matrix

$$A = \begin{bmatrix} a & b \\ c & d \end{bmatrix}, \tag{2}$$

where $\det(A) = 1$. Every $A \in \Gamma$ can be expressed in the form

$$A = T^{n_1}ST^{n_2}S \cdots ST^{n_k}, \tag{3}$$

where

$$S = \begin{bmatrix} 0 & -1 \\ 1 & 0 \end{bmatrix} \tag{4}$$

$$T = \begin{bmatrix} 1 & 1 \\ 0 & 1 \end{bmatrix}, \tag{5}$$

although the representation is not unique (Apostol 1997, pp. 28–9).

See also KLEIN'S ABSOLUTE INVARIANT, MÖBIUS TRANSFORMATION, MODULAR GROUP GAMMA0, MODULAR GROUP LAMBDA, THETA FUNCTIONS, UNIMODULAR TRANSFORMATION

References

Apostol, T. M. "The Modular Group and Modular Functions." Ch. 2 in *Modular Functions and Dirichlet Series in

Number Theory, 2nd ed. New York: Springer-Verlag, pp. 17 and 26–6, 1997.

Borwein, J. M. and Borwein, P. B. *Pi & the AGM: A Study in Analytic Number Theory and Computational Complexity.* New York: Wiley, p. 113, 1987.

Modular Group Gamma0

Let q be a POSITIVE INTEGER, then $\Gamma_0(q)$ is defined as the set of all matrices $\begin{bmatrix} a & b \\ c & d \end{bmatrix}$ in the MODULAR GROUP GAMMA Γ with $c \equiv 0 \pmod q$. $\Gamma_0(q)$ is a SUBGROUP of Γ. For any PRIME p, the set

$$R_\Gamma \cup \bigcup_{k=0}^{p-1} ST^k(R_\Gamma)$$

is a FUNDAMENTAL REGION of the subgroup $\Gamma_0(q)$, where $S\tau = -1/\tau$ and $T\tau = \tau + 1$ (Apostol 1997).

See also MODULAR GROUP GAMMA0, MODULAR GROUP LAMBDA

References

Apostol, T. M. "The Subgroup $\Gamma_0(q)$" and "Fundamental Region $\Gamma_0(q)$." §4.2–.3 in *Modular Functions and Dirichlet Series in Number Theory, 2nd ed.* New York: Springer-Verlag, pp. 75–8, 1997.

Modular Group Lambda

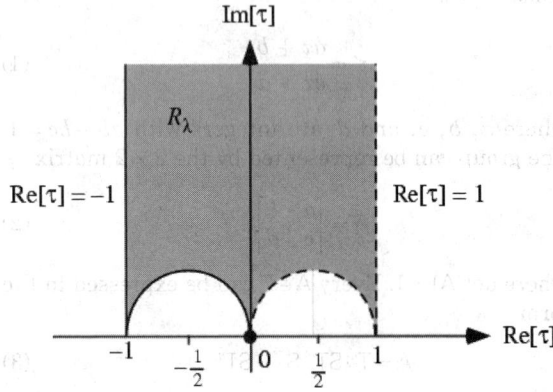

The set λ of linear MÖBIUS TRANSFORMATIONS w which satisfy

$$w(t) = \frac{at+b}{ct+d},$$

where a and d are ODD and b and c are EVEN. λ is a SUBGROUP of the MODULAR GROUP GAMMA, and is also called the THETA SUBGROUP. The FUNDAMENTAL REGION of the modular lambda group is illustrated above.

See also MODULAR GROUP GAMMA

References

Borwein, J. M. and Borwein, P. B. *Pi & the AGM: A Study in Analytic Number Theory and Computational Complexity.* New York: Wiley, pp. 113–14, 1987.

ModularLambda

ELLIPTIC LAMBDA FUNCTION

Modular Lattice

A LATTICE which satisfies the identity

$$(x \wedge y) \vee (x \wedge z) = x \wedge (y \vee (x \wedge z))$$

is said to be modular.

See also DISTRIBUTIVE LATTICE

References

Grätzer, G. *Lattice Theory: First Concepts and Distributive Lattices.* San Francisco, CA: W. H. Freeman, pp. 35–6, 1971.

Modular System

A set M of all POLYNOMIALS in s variables, $x_1, ..., x_s$ such that if P, P_1, and P_2 are members, then so are $P_1 + P_2$ and QP, where Q is any POLYNOMIAL in $x_1, ..., x_s$.

See also HILBERT'S THEOREM, MODULE, MODULAR SYSTEM BASIS

Modular System Basis

A basis of a MODULAR SYSTEM M is any set of POLYNOMIALS B_1, B_2, ...of M such that every POLYNOMIAL of M is expressible in the form

$$R_1 B_1 + R_2 B_2 + ...,$$

where R_1, R_2, ...are POLYNOMIALS.

Modular Transformation

MODULAR EQUATION

Modulation Theorem

The important property of FOURIER TRANSFORMS that $\mathcal{F}[\cos(2\pi k_0 x)f(x)]$ can be expressed in terms of $\mathcal{F}[f(x)] = F(k)$ as follows,

$$\mathcal{F}[\cos(2\pi k_0 x)f(x)] = \tfrac{1}{2}[F(k-k_0) + F(k+k_0)].$$

See also FOURIER TRANSFORM

References

Bracewell, R. "Modulation Theorem." *The Fourier Transform and Its Applications, 3rd ed.* New York: McGraw-Hill, p. 108, 1999.

Module

A mathematical object in which things can be added together COMMUTATIVELY by multiplying COEFFICIENTS and in which most of the rules of manipulating VECTORS hold. A module is abstractly very similar to a VECTOR SPACE, although in modules, COEFFICIENTS are taken in RINGS which are much more general algebraic objects than the FIELDS used in VECTOR SPACES. A module taking its coefficients in a RING R is called a module over R, or a R-MODULE.

Modules are the basic tool of HOMOLOGICAL ALGEBRA. Examples of modules include the set of INTEGERS $\mathbb{Z}$, the cubic lattice in d dimensions $\mathbb{Z}^d$, and the GROUP RING of a GROUP.

$\mathbb{Z}$ is a module over itself. It is CLOSED under ADDITION and SUBTRACTION (although it is SUFFICIENT to require closure under SUBTRACTION). Numbers OF THE FORM $n\alpha$ for $n \in \mathbb{Z}$ and α a fixed integer form a submodule since, for all $(n, m) \in \mathbb{Z}$,

$$n\alpha \pm m\alpha = (n \pm m)\alpha$$

and $(n \pm m)$ is still in $\mathbb{Z}$.

Given two INTEGERS a and b, the smallest module containing a and b is the module for their GREATEST COMMON DIVISOR, $\alpha = \mathrm{GCD}(a, b)$.

See also DIFFERENT, DIRECT SUM, DISCRIMINANT (MODULE), FIELD, GRADED MODULE, GROUP RING, HOMOLOGICAL ALGEBRA, MODULAR SYSTEM, R-MODULE, RING, SUBMODULE, VERMA MODULE, VECTOR SPACE

References
Beachy, J. A. *Introductory Lectures on Rings and Modules.* Cambridge, England: Cambridge University Press, 1999.

Berrick, A. J. and Keating, M.E *An Introduction to Rings and Modules with K-Theory in View.* Cambridge, England: Cambridge University Press, 2000.

Birkhoff, G. and Mac Lane, S. *A Survey of Modern Algebra, 3rd ed.* New York: Macmillian, p. 390, 1996.

Dummit, D. S. and Foote, R. M. *Abstract Algebra, 2nd ed.* Englewood Cliffs, NJ: Prentice-Hall, 1998.

Herstein, I. N. "Modules." §1.1 in *Noncommutative Rings.* Washington, DC: Math. Assoc. Amer., pp. 1–, 1968.

Nagell, T. "Moduls, Rings, and Fields." §6 in *Introduction to Number Theory.* New York: Wiley, pp. 19–1, 1951.

Riesel, H. "Modules." *Prime Numbers and Computer Methods for Factorization, 2nd ed.* Boston, MA: Birkhäuser, pp. 239–40, 1994.

Module Direct Sum

The direct sum of modules A and B is the module

$$A \oplus B = \{a \oplus b \mid a \in A, \ b \in B\}, \tag{1}$$

where all algebraic operations are defined componentwise. In particular, suppose that A and B are left R-modules, then

$$a_1 \oplus b_1 + a_2 \oplus b_2 = (a_1 + a_2) \oplus (b_1 + b_2) \tag{2}$$

and

$$r(a \oplus b) = (ra \oplus rb), \tag{3}$$

where r is an element of the RING R. The direct sum of an arbitrary family of MODULES over the same RING is also defined. If J is the indexing set for the family of MODULES, then the direct sum is represented by the collection of functions with finite support from J to

the union of all these MODULES such that the function sends $j \in J$ to an element in the MODULE indexed by j.

The dimension of a direct sum is the sum of the dimensions of the quantities summed. The significant property of the direct sum is that it is the COPRODUCT in the CATEGORY of MODULES. This general definition gives as a consequence the definition of the direct sum $A \oplus B$ of ABELIAN GROUPS A and B (since they are $\mathbb{Z}$-modules, i.e., MODULES over the INTEGERS) and the direct sum of VECTOR SPACES (since they are MODULES over a FIELD). Note that the direct sum of Abelian groups is the same as the GROUP DIRECT PRODUCT, but that the term direct sum is not used for groups which are NON-ABELIAN.

Whenever C is a MODULE, with module homomorphisms $f_A : A \to C$ and $f_B : B \to C$, then there is a module homomorphism $f_A : A \oplus B \to C$, given by $f(a \oplus b) = f_A(a) + f_B(b)$. Note that this map is well-defined because addition in modules is commutative. Sometimes direct sum is preferred over direct product when the coproduct property is emphasized.

See also COPRODUCT, DIRECT SUM, GROUP DIRECT PRODUCT, MODULE

References
Beachy, J. A. *Introductory Lectures on Rings and Modules.* Cambridge, England: Cambridge University Press, pp. 11 and 80, 1999.

Moduli Space

This entry contributed by EDGAR VAN TUYLL

In ALGEBRAIC GEOMETRY classification problems, an ALGEBRAIC VARIETY (or other appropriate space in other parts of geometry) whose points correspond to the equivalence classes of the objects to be classified in some natural way. Moduli space can be thought of as the space of EQUIVALENCE CLASSES of COMPLEX STRUCTURES on a fixed surface of GENUS g, where two COMPLEX STRUCTURES are deemed "the same" if they are equivalent by CONFORMAL MAPPING.

See also ALGEBRAIC VARIETY, COMPLEX STRUCTURE

References
Kirwan, F. "Introduction to Moduli Spaces." In *Proceedings of the EWM Workshop on Moduli Spaces, Oxford, EWM.* 1999.

Naber, G. L. *Topology, Geometry and Gauge Fields: Foundations.* New York: Springer-Verlag, 1997.

Polchinski, J. G. *String Theory: An Introduction to the Bosonic String.* Cambridge, England: Cambridge University Press, 1998.

Modulo

CONGRUENCE

Modulo Multiplication Group

A FINITE GROUP M_m of RESIDUE CLASSES prime to m under multiplication mod m. M_m is ABELIAN of ORDER $\phi(m)$, where $\phi(m)$ is the TOTIENT FUNCTION. The following table gives the modulo multiplication groups of small orders, where $\mathbb{Z}_n$ denotes the CYCLIC GROUP of order n.

M_m	Group	$\phi(m)$	Elements
M_2	$\langle e \rangle$	1	1
M_3	$\mathbb{Z}_2$	2	1, 2
M_4	$\mathbb{Z}_2$	2	1, 3
M_5	$\mathbb{Z}_4$	4	1, 2, 3, 4
M_6	$\mathbb{Z}_2$	2	1, 5
M_7	$\mathbb{Z}_6$	6	1, 2, 3, 4, 5, 6
M_8	$\mathbb{Z}_2 \times \mathbb{Z}_2$	4	1, 3, 5, 7
M_9	$\mathbb{Z}_6$	6	1, 2, 4, 5, 7, 8
M_{10}	$\mathbb{Z}_4$	4	1, 3, 7, 9
M_{11}	$\mathbb{Z}_{10}$	10	1, 2, 3, 4, 5, 6, 7, 8, 9, 10
M_{12}	$\mathbb{Z}_2 \times \mathbb{Z}_2$	4	1, 5, 7, 11
M_{13}	$\mathbb{Z}_{12}$	12	1, 2, 3, 4, 5, 6, 7, 8, 9, 10, 11, 12
M_{14}	$\mathbb{Z}_6$	6	1, 3, 5, 9, 11, 13
M_{15}	$\mathbb{Z}_2 \times \mathbb{Z}_4$	8	1, 2, 4, 7, 8, 11, 13, 14
M_{16}	$\mathbb{Z}_2 \times \mathbb{Z}_4$	8	1, 3, 5, 7, 9, 11, 13, 15
M_{17}	$\mathbb{Z}_{16}$	16	1, 2, 3, ..., 16
M_{18}	$\mathbb{Z}_6$	6	1, 5, 7, 11, 13, 17
M_{19}	$\mathbb{Z}_{18}$	18	1, 2, 3, ..., 18
M_{20}	$\mathbb{Z}_2 \times \mathbb{Z}_4$	8	1, 3, 7, 9, 11, 13, 17, 19
M_{21}	$\mathbb{Z}_2 \times \mathbb{Z}_6$	12	1, 2, 4, 5, 7, 8, 10, 11, 13, 16, 17, 19
M_{22}	$\mathbb{Z}_{10}$	10	1, 3, 5, 7, 9, 13, 15, 17, 19, 21
M_{23}	$\mathbb{Z}_{22}$	22	1, 2, 3, ..., 22
M_{24}	$\mathbb{Z}_2 \times \mathbb{Z}_2 \times \mathbb{Z}_2$	8	1, 5, 7, 11, 13, 17, 19, 23

M_m is a CYCLIC GROUP (which occurs exactly when m has a PRIMITIVE ROOT) IFF m is of one of the forms $m = 2$, 4, p^n, or $2p^n$, where p is an ODD PRIME and $n \geq 1$ (Shanks 1993, p. 92).

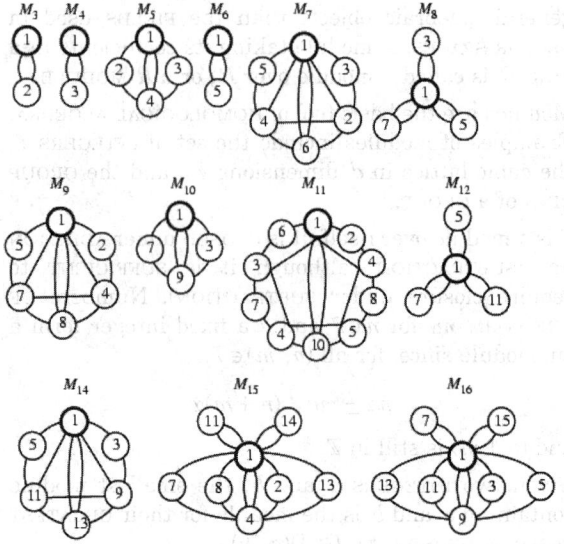

ISOMORPHIC modulo multiplication groups can be determined using a particular type of factorization of $\phi(m)$ as described by Shanks (1993, pp. 92–3). To perform this factorization (denoted ϕ_m), factor m in the standard form

$$m = p_1^{a_1} p_2^{a_2} \cdots p_n^{a_n}. \qquad (1)$$

Now write the factorization of the TOTIENT FUNCTION involving each power of an ODD PRIME

$$\phi(p_i^{a_i}) = (p_i - 1)p_i^{a_i - 1} \qquad (2)$$

as

$$\phi(p_i^{a_i}) = \left\langle q_1^{b_1} \right\rangle \left\langle q_2^{b_2} \right\rangle \cdots \left\langle q_s^{b_s} \right\rangle \left\langle p_i^{a_i - 1} \right\rangle, \qquad (3)$$

where

$$p_i - 1 = q_1^{b_1} q_2^{b_2} \cdots q_s^{b_s}, \qquad (4)$$

$\langle q^b \rangle$ denotes the explicit expansion of q^b (i.e., $5^2 = 25$), and the last term is omitted if $a_i = 1$. If $p_1 = 2$, write

$$\phi(2^{a_1}) = \begin{cases} 2 & \text{for } a_1 = 2 \\ 2\langle 2^{a_1 - 2} \rangle & \text{for } a_1 > 2. \end{cases} \qquad (5)$$

Now combine terms from the odd and even primes. For example, consider $m = 104 = 2^3 \cdot 13$. The only odd prime factor is 13, so factoring gives $13 - 1 = 12 = \langle 2^2 \rangle \langle 3 \rangle = 3 \cdot 4$. The rule for the powers of 2 gives $2^3 = 2\langle 2^{3-2} \rangle = 2\langle 2 \rangle = 2 \cdot 2$. Combining these two gives $\phi_{104} = 2 \cdot 2 \cdot 3 \cdot 4$. Other explicit values of ϕ_m are given below.

$$\phi_3 = 2$$

$$\phi_4 = 2$$

$$\phi_5 = 4$$

$$\phi_6 = 2$$

$$\phi_{15} = 2 \cdot 4$$

$$\phi_{16} = 2 \cdot 4$$

$$\phi_{17} = 16$$

$$\phi_{104} = 2 \cdot 2 \cdot 3 \cdot 4$$

$$\phi_{105} = 2 \cdot 2 \cdot 3 \cdot 4.$$

M_m and M_n are isomorphic IFF ϕ_m and ϕ_n are identical. More specifically, the abstract GROUP corresponding to a given M_m can be determined explicitly in terms of a GROUP DIRECT PRODUCT of CYCLIC GROUPS of the so-called CHARACTERISTIC FACTORS, whose product is denoted Φ_n. This representation is obtained from ϕ_m as the set of products of largest powers of each factor of ϕ_m. For example, for ϕ_{104}, the largest power of 2 is $4 = 2^2$ and the largest power of 3 is $3 = 3^1$, so the first characteristic factor is $4 \times 3 = 12$, leaving $2 \cdot 2$ (i.e., only powers of two). The largest power remaining is $2 = 2^1$, so the second CHARACTERISTIC FACTOR is 2, leaving 2, which is the third and last CHARACTERISTIC FACTOR. Therefore, $\Phi_{104} = 2 \cdot 2 \cdot 4$, and the group M_m is isomorphic to $\mathbb{Z}_2 \times \mathbb{Z}_2 \times \mathbb{Z}_4$.

The following table summarizes the isomorphic modulo multiplication groups M_n for the first few n and identifies the corresponding abstract GROUP. No M_m is ISOMORPHIC to $\mathbb{Z}_8$, Q_8, or D_4. However, every finite ABELIAN GROUP is isomorphic to a SUBGROUP of M_m for infinitely many different values of m (Shanks 1993, p. 96). CYCLE GRAPHS corresponding to M_n for small n are illustrated above, and more complicated CYCLE GRAPHS are illustrated by Shanks (1993, pp. 87–2).

Group	Isomorphic M_m
$\langle e \rangle$	M_2
$\mathbb{Z}_2$	M_3, M_4, M_6
$\mathbb{Z}_4$	M_5, M_{10}
$\mathbb{Z}_2 \times \mathbb{Z}_2$	M_8, M_{12}
$\mathbb{Z}_6$	M_7, M_9, M_{14}, M_{18}
$\mathbb{Z}_2 \times \mathbb{Z}_4$	$M_{15}, M_{16}, M_{20}, M_{30}$
$\mathbb{Z}_2 \times \mathbb{Z}_2 \times \mathbb{Z}_2$	M_{24}
$\mathbb{Z}_{10}$	M_{11}, M_{22}
$\mathbb{Z}_{12}$	M_{13}, M_{26}
$\mathbb{Z}_2 \times \mathbb{Z}_6$	$M_{21}, M_{28}, M_{36}, M_{42}$
$\mathbb{Z}_{16}$	M_{17}, M_{34}
$\mathbb{Z}_2 \times \mathbb{Z}_8$	M_{32}
$\mathbb{Z}_2 \times \mathbb{Z}_2 \times \mathbb{Z}_4$	M_{40}, M_{48}, M_{60}
$\mathbb{Z}_{18}$	$M_{19}, M_{27}, M_{38}, M_{54}$
$\mathbb{Z}_{20}$	M_{25}, M_{50}
$\mathbb{Z}_2 \times \mathbb{Z}_{10}$	M_{33}, M_{44}, M_{66}
$\mathbb{Z}_{22}$	M_{23}, M_{46}
$\mathbb{Z}_2 \times \mathbb{Z}_{12}$	$M_{35}, M_{39}, M_{45}, M_{52}, M_{70}, M_{78}, M_{90}$
$\mathbb{Z}_{28}$	M_{29}, M_{58}
$\mathbb{Z}_{30}$	M_{31}, M_{62}
$\mathbb{Z}_{36}$	M_{37}, M_{74}

See also CHARACTERISTIC FACTOR, TOTIENT FUNCTION, RIGHT CARD, RESIDUE CLASS

References

Riesel, H. *The Structure of the Group M_n.* *Prime Numbers and Computer Methods for Factorization, 2nd ed.* Boston, MA: Birkhäuser, pp. 270–72, 1994.

The number of CHARACTERISTIC FACTORS r of M_m for $m = 1, 2, \ldots$ are 1, 1, 1, 1, 1, 1, 1, 2, 1, 1, 1, 2, ... (Sloane's A046072). The number of QUADRATIC RESIDUES in M_m for $m > 2$ are given by $\phi(m)/2^r$ (Shanks 1993, p. 95). The first few for $m = 1, 2, \ldots$ are 0, 1, 1, 1, 2, 1, 3, 1, 3, 2, 5, 1, 6, ... (Sloane's A046073).

In the table below, $\phi(n)$ is the TOTIENT FUNCTION (Sloane's A000010) factored into CHARACTERISTIC FACTORS, $\lambda(n)$ is the CARMICHAEL FUNCTION (Sloane's A011773), and g_i are the smallest generators of the group M_n (of which there is a number equal to the number of CHARACTERISTIC FACTORS).

n	$\phi(n)$	$\lambda(n)$	g_i	n	$\phi(n)$	$\lambda(n)$	g_i
3	2	2	2	27	18	18	2
4	2	2	3	28	$2 \cdot 6$	6	13, 3
5	4	2	2	29	28	28	2
6	2	2	2	30	$2 \cdot 4$	4	11, 7
7	6	6	3	31	30	30	3
8	$2 \cdot 2$	2	7, 3	32	$2 \cdot 8$	8	31, 3
9	6	6	2	33	$2 \cdot 10$	10	10, 2
10	4	4	3	34	16	16	3
11	10	10	2	35	$2 \cdot 12$	12	6, 2
12	$2 \cdot 2$	2	5, 7	36	$2 \cdot 6$	6	19, 5
13	12	12	2	37	36	36	2
14	6	6	3	38	18	18	3
15	$2 \cdot 4$	4	14, 2	39	$2 \cdot 12$	12	38, 2
16	$2 \cdot 4$	4	15, 3	40	$2 \cdot 2 \cdot 4$	4	39, 11, 3
17	16	16	3	41	40	40	6
18	6	6	5	42	$2 \cdot 6$	6	13, 5
19	18	18	2	43	42	42	3
20	$2 \cdot 4$	4	19, 3	44	$2 \cdot 10$	10	43, 3
21	$2 \cdot 6$	6	20, 2	45	$2 \cdot 12$	12	44, 2
22	10	10	7	46	22	22	5

23	22	22	5	47	46	46	5
24	$2 \cdot 2 \cdot 2$	2	5, 7, 13	48	$2 \cdot 2 \cdot 4$	4	47, 7, 5
25	20	20	2	49	42	42	3
26	12	12	7	50	20	20	3

See also CHARACTERISTIC FACTOR, CYCLE GRAPH, FINITE GROUP, RESIDUE CLASS

References

Riesel, H. "The Structure of the Group M_n." *Prime Numbers and Computer Methods for Factorization, 2nd ed.* Boston, MA: Birkhäuser, pp. 270–72, 1994.

Shanks, D. *Solved and Unsolved Problems in Number Theory, 4th ed.* New York: Chelsea, pp. 61–2 and 92, 1993.

Sloane, N. J. A. Sequences A000010/M0299, A011773, A046072, and A046073 in "An On-Line Version of the Encyclopedia of Integer Sequences." http://www.research.-att.com/~njas/sequences/eisonline.html.

Weisstein, E. W. "Groups." MATHEMATICA NOTEBOOK GROUPS.M.

Modulus

The word modulus has several different meanings in mathematics with respect to complex numbers, congruences, elliptic integrals, quadratic invariants, sets, etc.

See also MODULUS (COMPLEX NUMBER), MODULUS (CONGRUENCE), MODULUS (ELLIPTIC INTEGRAL), MODULUS (QUADRATIC INVARIANTS), MODULUS (SET)

Modulus (Complex Number)

The modulus of a COMPLEX NUMBER z is denoted $|z|$.

$$|x + iy| \equiv \sqrt{x^2 + y^2} \tag{1}$$

$$\left| re^{i\phi} \right| = |r|. \tag{2}$$

Let $c_1 \equiv Ae^{i\phi_1}$ and $c_2 \equiv Be^{i\phi_2}$ be two COMPLEX NUMBERS. Then

$$\left| \frac{c_1}{c_2} \right| = \left| \frac{Ae^{i\phi_1}}{Be^{i\phi_2}} \right| = \frac{A}{B} \left| e^{i(\phi_1 - \phi_2)} \right| = \frac{A}{B} \tag{3}$$

$$\frac{|c_1|}{|c_2|} = \frac{|Ae^{i\phi_1}|}{|Be^{i\phi_2}|} = \frac{A}{B} \frac{|e^{i\phi_1}|}{|e^{i\phi_2}|} = \frac{A}{B}, \tag{4}$$

so

$$\left| \frac{c_1}{c_2} \right| = \frac{|c_1|}{|c_2|}. \tag{5}$$

Also,

$$|c_1 c_2| = |(Ae^{i\phi_1})(Be^{i\phi_2})| = AB|e^{i(\phi_1 + \phi_2)}| = AB \tag{6}$$

$$|c_1||c_2| = |Ae^{i\phi_1}||Be^{i\phi_2}| = AB|e^{i\phi_1}||e^{i\phi_2}| = AB, \tag{7}$$

so

$$|c_1 c_2| = |c_1||c_2| \tag{8}$$

and, by extension,

$$|z^n| = |z|^n. \tag{9}$$

The only functions satisfying identities OF THE FORM

$$|f(x + iy)| = |f(x) + f(iy)| \tag{10}$$

are $f(z) = Az$, $f(z) = A \sin(bz)$, and $f(z) = A \sinh(bz)$ (Robinson 1957).

See also ABSOLUTE SQUARE, ARGUMENT (COMPLEX NUMBER), COMPLEX NUMBER, IMAGINARY PART, MAXIMUM MODULUS PRINCIPLE, MINIMUM MODULUS PRINCIPLE, REAL PART

References

Abramowitz, M. and Stegun, C. A. (Eds.). *Handbook of Mathematical Functions with Formulas, Graphs, and Mathematical Tables, 9th printing.* New York: Dover, p. 16, 1972.

Krantz, S. G. "Modulus of a Complex Number." §1.1.4 n *Handbook of Complex Analysis.* Boston, MA: Birkhäuser, pp. 2–, 1999.

Robinson, R. M. "A Curious Mathematical Identity." *Amer. Math. Monthly* **64**, 83–5, 1957.

Modulus (Congruence)

The modulus of a CONGRUENCE $a \equiv b \pmod{m}$ is the number m. It is the "base" with respect to which a CONGRUENCE is computed (i.e., m gives the number of multiples of a that are "thrown out"). For example, when computing the time of day using a 12-hour clock obtained by adding four hours to 9:00, the answer, 1:00, is obtained by taking $9 + 4 \equiv 1 \pmod{12}$ (i.e., adding the hours with modulus 12).

In many computer languages (such as FORTRAN or *Mathematica*), the COMMON RESIDUE of $b \pmod{m}$ is written `mod(b,m)` (FORTRAN) or `Mod[b,m]` (*Mathematica*).

See also CONGRUENCE

Modulus (Elliptic Integral)

A parameter k used in ELLIPTIC INTEGRALS and ELLIPTIC FUNCTIONS defined to be $k \equiv \sqrt{m}$, where m is the PARAMETER. An ELLIPTIC INTEGRAL is written $I(\phi, k)$ when the modulus is used. It can be computed explicitly in terms of JACOBI THETA FUNCTIONS of zero argument:

$$k = \frac{\vartheta_2^2(0, q)}{\vartheta_3^2(0, q)}. \tag{1}$$

The REAL period $K(k)$ and IMAGINARY period $K'(k) = K(k') = K(\sqrt{1 - k^2})$ are given by

$$4K(k) = 2\pi \vartheta_3^2(0|\tau) \tag{2}$$

$$2iK'(k) = \pi\tau\vartheta_3^2(0|\tau), \qquad (3)$$

where $K(k)$ is a complete ELLIPTIC INTEGRAL OF THE FIRST KIND and the complementary modulus is defined by

$$k'^2 \equiv 1 - k^2, \qquad (4)$$

with k the modulus.

See also AMPLITUDE, CHARACTERISTIC (ELLIPTIC INTEGRAL), COMPLEMENTARY MODULUS, ELLIPTIC FUNCTION, ELLIPTIC INTEGRAL, ELLIPTIC INTEGRAL SINGULAR VALUE, HALF-PERIOD RATIO, JACOBI THETA FUNCTIONS, MODULAR ANGLE, NOME, PARAMETER

References

Abramowitz, M. and Stegun, C. A. (Eds.). *Handbook of Mathematical Functions with Formulas, Graphs, and Mathematical Tables, 9th printing.* New York: Dover, p. 590, 1972.

Borwein, J. M. and Borwein, P. B. *Pi & the AGM: A Study in Analytic Number Theory and Computational Complexity.* New York: Wiley, p. 35, 1987.

Tölke, F. "Parameterfunktionen." Ch. 3 in *Praktische Funktionenlehre, zweiter Band: Theta-Funktionen und spezielle Weierstraßsche Funktionen.* Berlin: Springer-Verlag, pp. 83–15, 1966.

Modulus (Quadratic Invariants)

The quantity $ps - rq$ obtained by letting

$$x = pX + qY \qquad (1)$$

$$y = rX + sY \qquad (2)$$

in

$$ax^2 + 2bxy + cy^2 \qquad (3)$$

so that

$$A = ap^2 + 2bpr + cr^2 \qquad (4)$$

$$B = apq + b(ps + qr) + crs \qquad (5)$$

$$C = aq^2 + 2bqs + cs^2 \qquad (6)$$

and

$$B^2 - AC = (ps - rq)^2(b^2 - ac), \qquad (7)$$

is called the modulus.

Modulus (Set)

The name for the SET of INTEGERS modulo m, denoted $\mathbb{Z}\backslash m\mathbb{Z}$. If m is a PRIME p, then the modulus is a FINITE FIELD $\mathbb{F}_p = \mathbb{Z}\backslash p\mathbb{Z}$.

Moebius

MÖBIUS FUNCTION, MÖBIUS GROUP, MÖBIUS INVERSION FORMULA, MÖBIUS PERIODIC FUNCTION, MÖBIUS PROBLEM, MÖBIUS SHORTS, MÖBIUS STRIP, MÖBIUS STRIP DISSECTION, MÖBIUS TRANSFORMATION, MÖBIUS TRIANGLES

MoebiusMu

MÖBIUS FUNCTION

Moessner's Theorem

Write down the POSITIVE INTEGERS in row one, cross out every k_1th number, and write the partial sums of the remaining numbers in the row below. Now cross off every k_2th number and write the partial sums of the remaining numbers in the row below. Continue. For every POSITIVE INTEGER $k > 1$, if every kth number is ignored in row 1, every $(k-1)$th number in row 2, and every $(k + 1 - i)$th number in row i, then the kth row of partial sums will be the kth POWERS 1^k, 2^k, 3^k, ...

References

Conway, J. H. and Guy, R. K. "Moessner's Magic." In *The Book of Numbers.* New York: Springer-Verlag, pp. 63–5, 1996.

Honsberger, R. *More Mathematical Morsels.* Washington, DC: Math. Assoc. Amer., pp. 268–77, 1991.

Long, C. T. "On the Moessner Theorem on Integral Powers." *Amer. Math. Monthly* **73**, 846–51, 1966.

Long, C. T. "Strike it Out--Add it Up." *Math. Mag.* **66**, 273–77, 1982.

Moessner, A. "Eine Bemerkung über die Potenzen der natürlichen Zahlen." *S.-B. Math.-Nat. Kl. Bayer. Akad. Wiss.* **29**, 1952.

Paasche, I. "Ein neuer Beweis des moessnerischen Satzes." *S.-B. Math.-Nat. Kl. Bayer. Akad. Wiss.* **1952**, 1–, 1953.

Paasche, I. "Ein zahlentheoretische-logarithmischer 'Rechenstab'." *Math. Naturwiss. Unterr.* **6**, 26–8, 1953–4.

Paasche, I. "Eine Verallgemeinerung des moessnerschen Satzes." *Compositio Math.* **12**, 263–70, 1956.

Mohammed Sign

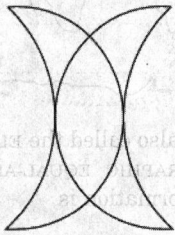

A curve consisting of two mirror-reversed intersecting crescents. This curve can be traced UNICURSALLY.

See also UNICURSAL CIRCUIT

Moiré Pattern

An interference pattern produced by overlaying similar but slightly offset templates. Møiré patterns can also be created by plotting series of curves on a computer screen. Here, the interference is provided by the discretization of the finite-sized pixels.

See also CIRCLES-AND-SQUARES FRACTAL

References

Amidror, I. *The Theory of the Møiré Phenomenon.* Dordrecht, Netherlands: Kluwer, 1999.

Cassin, C. *Visual Illusions in Motion with Møiré Screens: 60 Designs and 3 Plastic Screens.* New York: Dover, 1997.

Gardner, M. *The Sixth Book of Mathematical Games from Scientific American.* Chicago, IL: University of Chicago Press, pp. 229–30, 1984.

Grafton, C. B. *Optical Designs in Motion with Møiré Overlays.* New York: Dover, 1976.

Oster, G. and Nishijima, Y. "Møiré Patterns." *Sci. Amer.*, May 1963.

Strong, C. L. "The Amateur Scientist." *Sci. Amer.*, Nov. 1964.

Molenbroek's Equation

The PARTIAL DIFFERENTIAL EQUATION

$$\nabla^2 \phi = M_\infty^2 \left\{ \phi_x^2 \phi_{xx} + 2\phi_x \phi_y \phi_{xy} + \phi_y^2 \phi_{yy} \right.$$

$$\left. + \tfrac{1}{2}(\gamma - 1)(\phi_x^2 + \phi_y^2 - 1)\left(\phi_{xx} + \phi_{yy} + \epsilon \, \frac{\phi_y}{y} \right) \right\}$$

(Cole and Cook 1986, p. 34; Zwillinger 1997, p. 134).

References

Cole, J. D. and Cook, P. *Transonic Aerodynamics.* New York: North-Holland, p. 34, 1986.

Zwillinger, D. *Handbook of Differential Equations, 3rd ed.* Boston, MA: Academic Press, p. 134, 1997.

Mollweide Projection

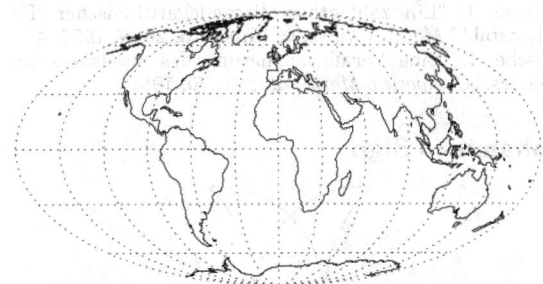

A MAP PROJECTION also called the ELLIPTICAL PROJECTION or HOMOLOGRAPHIC EQUAL-AREA PROJECTION. The forward transformation is

$$x = \frac{2\sqrt{2}(\lambda - \lambda_0)\cos\theta}{\pi} \tag{1}$$

$$y = 2^{1/2} \sin\theta, \tag{2}$$

where θ is given by

$$2\theta + \sin(2\theta) = \pi \sin\phi. \tag{3}$$

NEWTON'S METHOD can then be used to compute θ' iteratively from

$$\Delta\theta' = -\frac{\theta' + \sin\theta' - \pi\sin\phi}{1 + \cos\theta'}, \tag{4}$$

where

$$\theta' = \tfrac{1}{2}\theta' \tag{5}$$

or, better yet,

$$\theta' = 2\sin^{-1}\left(\frac{2\phi}{\pi}\right) \tag{6}$$

can be used as a first guess. The inverse FORMULAS are

$$\phi = \sin^{-1}\left[\frac{2\theta + \sin(2\theta)}{\pi}\right] \tag{7}$$

$$\lambda = \lambda_0 + \frac{\pi x}{2\sqrt{2}\cos\theta} \tag{8}$$

where

$$\theta = \sin^{-1}\left(\frac{y}{\sqrt{2}}\right). \tag{9}$$

References

Snyder, J. P. *Map Projections--A Working Manual.* U. S. Geological Survey Professional Paper 1395. Washington, DC: U. S. Government Printing Office, pp. 249–52, 1987.

Mollweide's Formulas

$$\frac{b - c}{a} = \frac{\sin[\tfrac{1}{2}(B - C)]}{\cos(\tfrac{1}{2}A)}$$

$$\frac{c - a}{b} = \frac{\sin[\tfrac{1}{2}(C - A)]}{\cos(\tfrac{1}{2}B)}$$

$$\frac{a - b}{c} = \frac{\sin[\tfrac{1}{2}(A - B)]}{\cos(\tfrac{1}{2}C)}.$$

See also NEWTON'S FORMULAS, TRIANGLE, TRIGONOMETRY

References

Beyer, W. H. *CRC Standard Mathematical Tables, 28th ed.* Boca Raton, FL: CRC Press, p. 146, 1987.

Moment

The nth RAW MOMENT μ_n' (i.e., moment about zero) of a distribution $P(x)$ is defined by

$$\mu_n' = \langle x^n \rangle, \tag{1}$$

where

$$\langle f(x) \rangle = \begin{cases} \sum f(x)P(x) & \text{discrete distribution} \\ \int f(x)P(x)\,dx & \text{continuous distribution} \end{cases} \tag{2}$$

μ_1', the MEAN, is usually simply denoted $\mu = \mu_1$. If the moment is instead taken about a point a,

$$\mu_n(a) = \langle (x-a)^n \rangle = \sum (x-a)^n P(x). \qquad (3)$$

A STATISTICAL DISTRIBUTION is *not* uniquely specified by its moments, although it is by its CHARACTERISTIC FUNCTION.

The moments are most commonly taken about the MEAN. These so-called CENTRAL MOMENTS are denoted μ_n and are defined by

$$\mu_n \equiv \langle (x-\mu)^n \rangle, \qquad (4)$$

$$= \int (x-\mu)^n P(x)\, dx, \qquad (5)$$

with $\mu_1 = 0$. The second moment about the MEAN is equal to the VARIANCE

$$\mu_2 = \sigma^2, \qquad (6)$$

where $\sigma = \sqrt{\mu_2}$ is called the STANDARD DEVIATION. The related CHARACTERISTIC FUNCTION is defined by

$$\phi^{(n)}(0) \equiv \left[\frac{d^n \phi}{dt^n}\right]_{t=0} = i^n \mu(0). \qquad (7)$$

The moments may be simply computed using the MOMENT-GENERATING FUNCTION,

$$\mu'_n = M^{(n)}(0). \qquad (8)$$

See also ABSOLUTE MOMENT, CHARACTERISTIC FUNCTION, CHARLIER'S CHECK, CUMULANT-GENERATING FUNCTION, FACTORIAL MOMENT, KURTOSIS, MEAN, MOMENT-GENERATING FUNCTION, MOMENT PROBLEM, MOMENT SEQUENCE, SKEWNESS, STANDARD DEVIATION, STANDARDIZED MOMENT, VARIANCE

References

Papoulis, A. *Probability, Random Variables, and Stochastic Processes, 2nd ed.* New York: McGraw-Hill, pp. 145–49, 1984.

Press, W. H.; Flannery, B. P.; Teukolsky, S. A.; and Vetterling, W. T. "Moments of a Distribution: Mean, Variance, Skewness, and So Forth." §14.1 in *Numerical Recipes in FORTRAN: The Art of Scientific Computing, 2nd ed.* Cambridge, England: Cambridge University Press, pp. 604–09, 1992.

Momental Skewness

$$\alpha^{(m)} \equiv \tfrac{1}{2}\gamma_1 = \frac{\mu_3}{2\sigma^3},$$

where γ_1 is the FISHER SKEWNESS.
See also FISHER SKEWNESS, SKEWNESS

Moment-Generating Function

Given a RANDOM VARIABLE $x \in R$, if there exists an $h > 0$ such that for $|t| < h$, then

$$M(t) \equiv \langle e^{tx} \rangle$$

$$= \begin{cases} \sum_R e^{tx} P(x) & \text{for a discrete distribution} \\ \int_{-\infty}^{\infty} e^{tx} P(x)\, dx & \text{for a continuous distribution} \end{cases}$$

$$(1)$$

is the moment-generating function.

$$M(t) = \int_{-\infty}^{\infty} \left(1 + tx + \frac{1}{2!}t^2x^2 + \dots\right) P(x)\, dx$$

$$= +tm_1 + \tfrac{1}{2!}t^2m_2 + \cdots, \qquad (3)$$

where m_r is the rth MOMENT about zero. The moment-generating function satisfies

$$M_{x+y}(t) = \langle e^{t(x+y)} \rangle = \langle e^{tx}e^{ty} \rangle = \langle e^{tx} \rangle \langle e^{ty} \rangle = M_x(t)M_y(t). \qquad (4)$$

If $M(t)$ is differentiable at zero, then the nth MOMENTS about the ORIGIN are given by $M^{(n)}(0)$

$$M(t) = \langle e^{tx} \rangle \quad M(0) = 1 \qquad (5)$$

$$M'(t) = \langle xe^{tx} \rangle \quad M'(0) = \langle x \rangle \qquad (6)$$

$$M''(t) = \langle x^2 e^{tx} \rangle \quad M''(0) = \langle x^2 \rangle \qquad (7)$$

$$M^{(n)}(t) = \langle x^n e^{tx} \rangle \quad M^{(n)}(0) = \langle x^n \rangle. \qquad (8)$$

The MEAN and VARIANCE are therefore

$$\mu \equiv \langle x \rangle = M'(0) \qquad (9)$$

$$\sigma^2 \equiv \langle x^2 \rangle - \langle x \rangle^2 = M''(0) - [M'(0)]^2. \qquad (10)$$

It is also true that

$$\mu_n = \sum_{j=0}^{n} \binom{n}{j} (-1)^{n-j} \mu'_j (\mu'_1)^{n-j}, \qquad (11)$$

where $\mu'_0 = 1$ and μ'_j is the jth moment about the origin.

It is sometimes simpler to work with the LOGARITHM of the moment-generating function, which is also called the CUMULANT-GENERATING FUNCTION, and is defined by

$$R(t) \equiv \ln[M(t)] \qquad (12)$$

$$R'(t) = \frac{M'(t)}{M(t)} \qquad (13)$$

$$R''(t) = \frac{M(t)M''(t) - [M'(t)]^2}{[M(t)]^2} \qquad (14)$$

But $M(0) = \langle 1 \rangle = 1$, so

$$\mu = M'(0) = R'(0) \qquad (15)$$

$$\sigma^2 = M''(0) - [M'(0)]^2 = R''(0) \qquad (16)$$

See also CHARACTERISTIC FUNCTION (PROBABILITY), CUMULANT, CUMULANT-GENERATING FUNCTION, MOMENT

References
Kenney, J. F. and Keeping, E. S. "Moment-Generating and Characteristic Functions," "Some Examples of Moment-Generating Functions," and "Uniqueness Theorem for Characteristic Functions." §4.6–.8 in *Mathematics of Statistics, Pt. 2, 2nd ed.* Princeton, NJ: Van Nostrand, pp. 72–7, 1951.

Moment Problem
The moment problem, also called "Hausdorff's moment problem "or the "little moment problem," may be stated as follows. Given a sequence of numbers $\{\mu_n\}_{n=0}^{\infty}$, under what conditions is it possible to determine a function $\alpha(t)$ of bounded variation in the interval $(0, 1)$ such that

$$\mu_n = \int_0^1 t^n \, d\alpha(t)$$

for $n = 0, 1, \ldots$. Such a sequence is called a MOMENT SEQUENCE, and Hausdorff (1921) was the first to obtain necessary and sufficient conditions for a sequence to be a MOMENT SEQUENCE.

See also MOMENT, MOMENT SEQUENCE

References
Hausdorff, F. "Summationsmethoden und Momentfolgen. I." *Math. Z.* **9**, 74–09, 1921.
Hausdorff, F. "Summationsmethoden und Momentfolgen. II." *Math. Z.* **9**, 280–99, 1921.
Leviatan, D. "A Generalized Moment Problem." *Israel J. Math.* **5**, 97–03, 1967.
Widder, D. V. "The Moment Problem." Ch. 3 in *The Laplace Transform*. Princeton, NJ: Princeton University Press, pp. 100–01, 1941.

Moment Sequence
A moment sequence is a sequence $\{\mu_n\}_{n=0}^{\infty}$ defined for $n = 0, 1, \ldots$ by

$$\mu_n = \int_0^1 t^n \, d\alpha(t),$$

where $\alpha(t)$ is a function of bounded variation in the interval $(0, 1)$.

See also MOMENT, MOMENT PROBLEM

Monad
A mathematical object which consists of a set of a single element. The YIN-YANG is also known as the monad.

See also HEXAD, QUARTET, QUINTET, TETRAD, TRIAD, YIN-YANG

Money-Changing Problem
COIN PROBLEM

Monge-Ampère Differential Equation
A second-order PARTIAL DIFFERENTIAL EQUATION OF THE FORM

$$Hr + 2Ks + Lt + M + N(rt - s^2) = 0, \tag{1}$$

where $H, K, L, M,$ and N are functions of $x, y, z, p,$ and q, and $r, s, t, p,$ and q are defined by

$$r = \frac{\partial^2 z}{\partial x^2} \tag{2}$$

$$s = \frac{\partial^2 z}{\partial x \, \partial y} \tag{3}$$

$$t = \frac{\partial^2 z}{\partial y^2} \tag{4}$$

$$p = \frac{\partial z}{\partial x} \tag{5}$$

$$q = \frac{\partial z}{\partial y}. \tag{6}$$

The solutions are given by a system of differential equations given by Iyanaga and Kawada (1980).

Other equations called the Monge-Ampère equation are

$$u_{xy}^2 - u_x u_y = f(x, y, u, u_x, u_y) \tag{7}$$

(Moon and Spencer 1969, p. 171; Zwillinger 1997, p. 134) and

$$\begin{vmatrix} u_{x_1 x_1} & u_{x_1 x_2} & \cdots & u_{x_1 x_n} \\ u_{x_2 x_1} & u_{x_2 x_2} & \cdots & u_{x_2 x_n} \\ \cdots & \cdots & \ddots & \cdots \\ u_{x_n x_1} & u_{x_n x_2} & \cdots & u_{x_n x_n} \end{vmatrix} = f(u, \mathbf{x}, \nabla u) \tag{8}$$

(Gilberg and Trudinger 1983, p. 441; Zwillinger 1997, p. 134).

References
Caffarelli, L. A. and Milman, M. *Monge Ampère Equation: Applications to Geometry and Optimization*. Providence, RI: Amer. Math. Soc., 1999.
Fairlie, D. B. and Leznov, A. N. The General Solution of the Complex Monge-Ampère Equation in a Space of Arbitrary Dimension. 16 Sep 1999. http://xxx.lanl.gov/abs/solv-int/9909014/.
Gilberg, D. and Trudinger, N. S. *Elliptic Partial Differential Equations of Second Order*. Berlin: Springer-Verlag, p. 441, 1983.
Iyanaga, S. and Kawada, Y. (Eds.). "Monge-Ampère Equations." §276 in *Encyclopedic Dictionary of Mathematics*. Cambridge, MA: MIT Press, pp. 879–80, 1980.
Moon, P. and Spencer, D. E. *Partial Differential Equations*. Lexington, MA: Heath, p. 171, 1969.

Monge Patch

A Monge patch is a PATCH $\mathbf{x} : U \to \mathbb{R}^3$ OF THE FORM

$$\mathbf{x}(u, v) = (u, v, h(u, v)), \tag{1}$$

where U is an OPEN SET in $\mathbb{R}^2$ and $h : U \to \mathbb{R}$ is a differentiable function. The coefficients of the first FUNDAMENTAL FORM are given by

$$E = 1 + h_u^2 \tag{2}$$

$$F = h_u h_v \tag{3}$$

$$G = 1 + h_v^2 \tag{4}$$

and the second FUNDAMENTAL FORM by

$$e = \frac{h_{uu}}{\sqrt{1 + h_u^2 + h_v^2}} \tag{5}$$

$$f = \frac{h_{uv}}{\sqrt{1 + h_u^2 + h_v^2}} \tag{6}$$

$$g = \frac{g_{vv}}{\sqrt{1 + h_u^2 + h_v^2}}. \tag{7}$$

For a Monge patch, the GAUSSIAN CURVATURE and MEAN CURVATURE are

$$K = \frac{h_{uu} h_{vv} - h_{uv}^2}{\left(1 + h_u^2 + h_v^2\right)^2} \tag{8}$$

$$H = \frac{(1 - h_v^2) h_{uu} - 2 h_u h_v h_{uv} + (1 - h_u^2) h_{vv}}{2(1 + h_u^2 + h_v^2)^{3/2}}. \tag{9}$$

See also MONGE'S FORM, PATCH

References

Gray, A. "A Monge Patch." *Modern Differential Geometry of Curves and Surfaces with Mathematica, 2nd ed.* Boca Raton, FL: CRC Press, pp. 398–01, 1997.

Monge Point

The point of concurrence of the six PLANES in MONGE'S TETRAHEDRON THEOREM.

See also MANNHEIM'S THEOREM, MONGE'S TETRAHEDRON THEOREM, PLANE, TETRAHEDRON

References

Altshiller-Court, N. "The Monge Point." §4.2c in *Modern Pure Solid Geometry.* New York: Chelsea, pp. 69–1, 1979.
Forder, H. G. "Article 1006. A Theorem in Coolidge's 'Circle and Sphere.'" *Math. Gaz.* **15**, pp. 470–71, 1930–931.
Lez, H. and Dugrais, M. "Solution des questions proposées dans les Nouvelles Annales: Question 906." *Nouvelles ann. de math.* **8**, 173, 1869.
Monge, G. *Corresp. sur l'École Polytech.* **2**, 266, 1795.
Thompson, H. F. "A Geometrical Proof of a Theorem Connected with the Tetrahedron." *Proc. Edinburgh Math. Soc.* **17**, 51–3, 1908–909.

Monge's Chordal Theorem

RADICAL CENTER

Monge's Circle Theorem

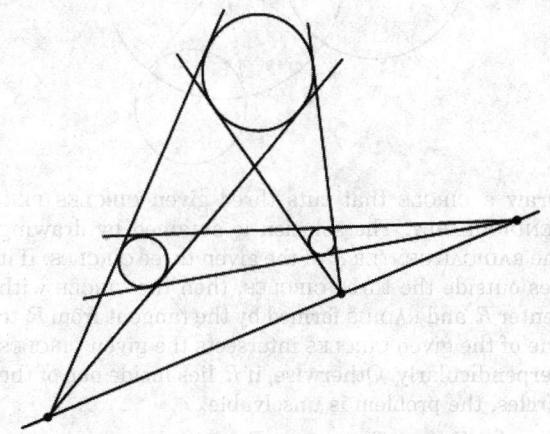

Draw three nonintersecting CIRCLES in the plane, and the common tangent line for each pair of two. The points of intersection of the three pairs of tangent lines lie on a straight line.

Monge's theorem has a 3-D analog which states that the apexes of the CONES defined by four SPHERES, taken two at a time, lie in a PLANE (when the CONES are drawn with the SPHERES on the same side of the apex; Wells 1991).

See also CIRCLE TANGENTS

References

Coxeter, H. S. M. "The Problem of Apollonius." *Amer. Math. Monthly* **75**, 5–5, 1968.
Graham, L. A. Problem 62 in *Ingenious Mathematical Problems and Methods.* New York: Dover, 1959. Ogilvy, C. S. *Excursions in Geometry.* New York: Dover, pp. 115–17, 1990.
Petersen, J. *Methods and Theories for the Solution of Problems of Geometrical Constructions, Applied to 410 Problems.* London: Sampson Low, Marston, Searle & Rivington, pp. 92–3, 1879.
Walker, W. "Monge's Theorem in Many Dimensions." *Math. Gaz.* **60**, 185–88, 1976.
Wells, D. *The Penguin Dictionary of Curious and Interesting Geometry.* London: Penguin, pp. 153–54, 1991.

Monge's Form

A SURFACE given by the form $z = F(x, y)$.

See also MONGE PATCH

Monge's Problem

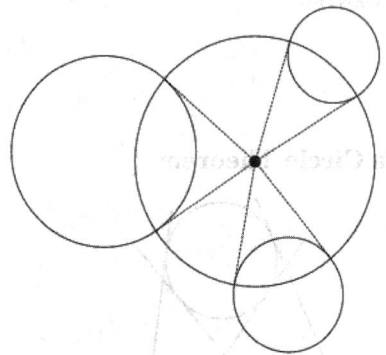

Draw a CIRCLE that cuts three given CIRCLES PERPENDICULARLY. The solution is obtained by drawing the RADICAL CENTER R of the given three CIRCLES. If it lies outside the three CIRCLES, then the CIRCLE with center R and RADIUS formed by the tangent from R to one of the given CIRCLES intersects the given CIRCLES perpendicularly. Otherwise, if R lies inside one of the circles, the problem is unsolvable.

See also CIRCLE TANGENTS, RADICAL CENTER

References
Dörrie, H. "Monge's Problem." §31 in *100 Great Problems of Elementary Mathematics: Their History and Solutions.* New York: Dover, pp. 151–54, 1965.

Monge's Shuffle

A SHUFFLE in which CARDS from the top of the deck in the left hand are alternatively moved to the bottom and top of the deck in the right hand. If the deck is shuffled m times, the final position x_m and initial position x_0 of a card are related by

$$2^{m+1}x_m = (4p+1)\left[2^{m-1} + (-1)^{m-1}\left(2^{m-2} + \cdots + 2 + 1\right)\right]$$
$$+ (-1)^{m-1}2x_0 + 2^m + (-1)^{m-1}$$

for a deck of $2p$ cards (Kraitchik 1942).

See also CARDS, SHUFFLE

References
Conway, J. H. and Guy, R. K. "Fractions Cycle into Decimals." In *The Book of Numbers.* New York: Springer-Verlag, pp. 157–63, 1996.
Kraitchik, M. "Monge's Shuffle." §12.2.14 in *Mathematical Recreations.* New York: W. W. Norton, pp. 321–23, 1942.

Monge's Tetrahedron Theorem

The six PLANES through the midpoints of the edges of a TETRAHEDRON and perpendicular to the opposite edges CONCUR in a point known as the MONGE POINT.

See also MONGE POINT, PLANE, TETRAHEDRON

References
Altshiller-Court, N. "The Monge Theorem." §228 in *Modern Pure Solid Geometry.* New York: Chelsea, p. 69, 1979.
Forder, H. G. *Math. Gaz.* **15**, p. 470, 1930–931.
Lez, H. and Dugrais, M. "Solution des questions proposées dans les Nouvelles Annales: Question 906." *Nouvelles ann. de math.* **8**, 173, 1869.
Monge, G. *Corresp. sur l'École Polytech.* **2**, 266, 1795.
Thompson, H. F. "A Geometrical Proof of a Theorem Connected with the Tetrahedron." *Proc. Edinburgh Math. Soc.* **17**, 51–3, 1908–909.

Monge's Theorem

MONGE'S CIRCLE THEOREM, MONGE'S TETRAHEDRON THEOREM

Monica Set

The nth Monica set M_n is defined as the set of COMPOSITE NUMBERS x for which $n|S(x) - S_p(x)$, where

$$x = a_0 + a_1(10^1) + \cdots + a_d(10^d) = p_1 p_2 \cdots p_n, \qquad (1)$$

and

$$S(x) = \sum_{j=0}^{d} a_j \qquad (2)$$

$$S_p(x) = \sum_{i=1}^{m} S(p_i) \qquad (3)$$

Every Monica set has an infinite number of elements. The Monica set M_n is a subset of the SUZANNE SET S_n. If x is a SMITH NUMBER, then it is a member of the Monica set M_n for all $n \in \mathbb{N}$. For any INTEGER $k > 1$, if x is a k-SMITH NUMBER, then $x \in M_{k-1}$.

See also SUZANNE SET

References
Smith, M. "Cousins of Smith Numbers: Monica and Suzanne Sets." *Fib. Quart.* **34**, 102–04, 1996.

Monic Polynomial

A POLYNOMIAL $x^n + a_{n-1}x^{n-1} + \cdots + a_1 x + a_0$ in which the COEFFICIENT of the highest ORDER term is 1.

See also MONOMIAL

Monkey and Coconut Problem

A DIOPHANTINE problem (i.e., one whose solution must be given in terms of INTEGERS) which seeks a solution to the following problem. Given n men and a pile of coconuts, each man in sequence takes $(1/n)$th of the coconuts left after the previous man removed his (i.e., a_1 for the first man, a_2, for the second, ..., a_n for the last) and gives m coconuts (specified in the problem to be the *same* number for each man) which do not divide equally to a monkey. When all n men have so divided, they divide the remaining coconuts n ways (i.e., taking an additional a coconuts each), and give the m coconuts which are left over to the monkey. If m is the same at each division, then how many coconuts N were there originally? The solution is equivalent to solving the $n+1$ DIOPHANTINE EQUATIONS

$$N = na_1 + m$$
$$N - a_1 - m = na_2 + m$$
$$N - a_1 - a_2 - 2m = na_3 + m \quad (1)$$
$$\vdots$$
$$N - a_1 - a_2 - a_3 - \cdots - a_n - nm = na + m,$$

which can be rewritten as

$$N = na_1 + m$$
$$(n-1)a_1 = na_2 + m$$
$$(n-1)a_1 = na_3 + m \quad (2)$$
$$\vdots$$
$$(n-1)a_{n-1} = na_n + m$$
$$(n-1)a_a = na + m.$$

Since there are $n+1$ equations in the $n+2$ unknowns a_1, a_2, ..., a_n, a, and N, the solutions span a 1-dimensional space (i.e., there is an infinite family of solution parameterized by a single value). The solution to these equations can be given by

$$N = kn^{n+1} - m(n-1), \quad (3)$$

where k is an arbitrary INTEGER (Gardner 1961).

For the particular case of $n=5$ men and $m=1$ left over coconuts, the 6 equations can be combined into the single DIOPHANTINE EQUATION

$$1,024N = 15,625a + 11,529, \quad (4)$$

where a is the number given to each man in the last division. The smallest POSITIVE solution in this case is $N = 15,621$ coconuts, corresponding to $k=1$ and $a = 1,023$; Gardner 1961). The following table shows how this rather large number of coconuts is divided under the scheme described above.

Removed	Given to Monkey	Left
		15,621
3,124	1	12,496
2,499	1	9,996
1,999	1	7,996
1,599	1	6,396
1,279	1	5,116
5 × 1,023	1	0

If no coconuts are left for the monkey after the final n-way division (Williams 1926), then the original number of coconuts is

$$\begin{cases} (1+nk)n^n - (n-1) & n \text{ odd} \\ (n-1+nk)n^n - (n-1) & n \text{ even}. \end{cases} \quad (5)$$

The smallest POSITIVE solution for case $n=5$ and $m=1$ is $N = 3,121$ coconuts, corresponding to $k=1$ and 1,020 coconuts in the final division (Gardner 1961). The following table shows how these coconuts are divided.

Removed	Given to Monkey	Left
		3,121
624	1	2,496
499	1	1,996
399	1	1,596
319	1	1,276
255	1	1,020
5 × 204	0	0

A different version of the problem having a solution of 79 coconuts is considered by Pappas (1989).

See also DIOPHANTINE EQUATION, PELL EQUATION

References

Anning, N. "Monkeys and Coconuts." *Math. Teacher* **54**, 560–62, 1951.
Bowden, J. "The Problem of the Dishonest Men, the Monkeys, and the Coconuts." In *Special Topics in Theoretical Arithmetic*. Lancaster, PA: Lancaster Press, pp. 203–12, 1936.
Gardner, M. "The Monkey and the Coconuts." Ch. 9 in *The Second Scientific American Book of Puzzles & Diversions: A New Selection*. New York: Simon and Schuster, pp. 104–11, 1961.
Kirchner, R. B. "The Generalized Coconut Problem." *Amer. Math. Monthly* **67**, 516–19, 1960.
Moritz, R. E. "Solution to Problem 3,242." *Amer. Math. Monthly* **35**, 47–8, 1928.
Ogilvy, C. S. and Anderson, J. T. *Excursions in Number Theory*. New York: Dover, pp. 52–4, 1988.
Olds, C. D. *Continued Fractions*. New York: Random House, pp. 48–0, 1963.
Pappas, T. "The Monkey and the Coconuts." *The Joy of Mathematics*. San Carlos, CA: Wide World Publ./Tetra, pp. 226–27 and 234, 1989.
Williams, B. A. "Coconuts." *The Saturday Evening Post*, Oct. 9, 1926.

Monkey Saddle

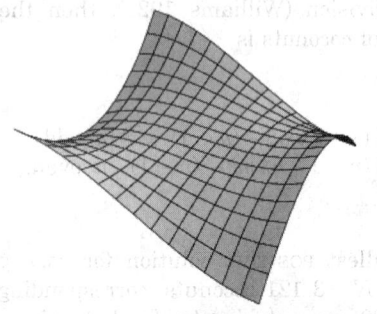

A SURFACE which a monkey can straddle with both his two legs and his tail. A simple Cartesian equation for such a surface is

$$z = x(x^2 - 3y^2), \tag{1}$$

which can also be given by the PARAMETRIC EQUATIONS

$$x(u, v) = u \tag{2}$$

$$y(u, v) = v \tag{3}$$

$$z(u, v) = u^3 - 3uv^2. \tag{4}$$

The coefficients of the coefficients of the FIRST FUNDAMENTAL FORM of the monkey saddle are

$$E = 1 + 9(u^2 - v^2)^2 \tag{5}$$

$$F = -18uv(u^2 - v^2) \tag{6}$$

$$G = 1 + 36u^2v^2 \tag{7}$$

and the SECOND FUNDAMENTAL FORM coefficients are

$$e = \frac{6u}{\sqrt{1 + 9(u^2 + v^2)^2}} \tag{8}$$

$$f = -\frac{6v}{\sqrt{1 + 9(u^2 + v^2)^2}} \tag{9}$$

$$g = -\frac{6u}{\sqrt{1 + 9(u^2 + v^2)^2}}, \tag{10}$$

giving RIEMANNIAN METRIC

$$ds^2 = [1 + (3u^2 - 3v^2)^2] \, du^2 - 2[18uv(u^2 - v^2)] \, du \, dv + (1 + 36u^2v^2) \, dv^2, \tag{11}$$

AREA ELEMENT

$$dA = \sqrt{1 + 9(u^2 + v^2)^2} \, du \wedge dv, \tag{12}$$

and GAUSSIAN and MEAN CURVATURES

$$K = -\frac{36(u^2 + v^2)}{[1 + 9(u^2 + v^2)^2]^2} \tag{13}$$

$$H = \frac{27u(-u^4 + 2u^2v^2 + 3v^4)}{[1 + 9(u^2 + v^2)^2]^{3/2}} \tag{14}$$

(Gray 1997). Every point of the monkey saddle except the origin has NEGATIVE GAUSSIAN CURVATURE.

See also CROSSED TROUGH, PARTIAL DERIVATIVE

References

Coxeter, H. S. M. *Introduction to Geometry, 2nd ed.* New York: Wiley, p. 365, 1969.

Gray, A. "Monkey Saddle." *Modern Differential Geometry of Curves and Surfaces with Mathematica, 2nd ed.* Boca Raton, FL: CRC Press, pp. 299–01, 382–83, and 408, 1997.

Hilbert, D. and Cohn-Vossen, S. *Geometry and the Imagination.* New York: Chelsea, p. 202, 1999.

Monochromatic Forced Triangle

Given a COMPLETE GRAPH K_n which is two-colored, the number of forced monochromatic TRIANGLES is at least

$$\begin{cases} \frac{1}{3}u(u-1)(u-2) & \text{for } n = 2u \\ \frac{2}{3}(u-1)(4u+1) & \text{for } n = 4u+1 \\ \frac{2}{3}u(u+1)(4u-1) & \text{for } n = 4u+3. \end{cases}$$

The first few numbers of monochromatic forced triangles are 0, 0, 0, 0, 0, 2, 4, 8, 12, 20, 28, 40, ... (Sloane's A014557).

See also COMPLETE GRAPH, EXTREMAL GRAPH

References

Goodman, A. W. "On Sets of Acquaintances and Strangers at Any Party." *Amer. Math. Monthly* **66**, 778–83, 1959.

Sloane, N. J. A. Sequences A014557 in "An On-Line Version of the Encyclopedia of Integer Sequences." http://www.research.att.com/~njas/sequences/eisonline.html.

Monodromy

A general concept in CATEGORY THEORY involving the globalization of local MORPHISMS.

See also CATEGORY THEORY, HOLONOMY, MORPHISM

Monodromy Group

A technically defined GROUP characterizing a system of linear differential equations

$$y_j' = \sum_{k=1}^{n} a_{jk}(x)y_k$$

for $j = 1, ..., n$, where a_{jk} are COMPLEX ANALYTIC FUNCTIONS of x in a given COMPLEX DOMAIN.

See also HILBERT'S 21ST PROBLEM, RIEMANN P-SERIES

References

Iyanaga, S. and Kawada, Y. (Eds.). "Monodromy Groups." §253B in *Encyclopedic Dictionary of Mathematics.* Cambridge, MA: MIT Press, p. 793, 1980.

Monodromy Theorem

If a COMPLEX FUNCTION f is ANALYTIC in a DISK contained in a simply connected DOMAIN D and f can be ANALYTICALLY CONTINUED along every polygonal arc in D, then f can be ANALYTICALLY CONTINUED to a single-valued ANALYTIC FUNCTION on all of D!

See also ANALYTIC CONTINUATION

References

Flanigan, F. J. *Complex Variables: Harmonic and Analytic Functions.* New York: Dover, p. 234, 1983.
Knopp, K. "The Monodromy Theorem." §25 in *Theory of Functions Parts I and II, Two Volumes Bound as One, Part I.* New York: Dover, pp. 105–11, 1996.
Krantz, S. G. "The Monodromy Theorem." §10.3.5 in *Handbook of Complex Analysis.* Boston, MA: Birkhäuser, p. 134, 1999.

Monogenic Function

If

$$\lim_{z \to z_0} \frac{f(z) - f(z_0)}{z - z_0}$$

is the same for all paths in the COMPLEX PLANE, then $f(z)$ is said to be monogenic at z_0. Monogenic therefore essentially means having a single DERIVATIVE at a point. Functions are either monogenic or have infinitely many DERIVATIVES (in which case they are called POLYGENIC); intermediate cases are not possible.

See also POLYGENIC FUNCTION

References

Newman, J. R. *The World of Mathematics, Vol. 3.* New York: Simon & Schuster, p. 2003, 1956.

Monohedral Tiling

A TILING in which all tiles are congruent.

See also ANISOHEDRAL TILING, ISOHEDRAL TILING, TILING

References

Berglund, J. "Is There a k-Anisohedral Tile for $k \geq 5$?" *Amer. Math. Monthly* **100**, 585–88, 1993.
Grünbaum, B. and Shephard, G. C. "The 81 Types of Isohedral Tilings of the Plane." *Math. Proc. Cambridge Philos. Soc.* **82**, 177–96, 1977.

Monoid

A GROUP-like object which fails to be a GROUP because elements need not have an inverse within the object. A monoid S must also be ASSOCIATIVE and have an IDENTITY ELEMENT $I \in S$ such that for all $a \in S$, $1a = a1 = a$. A monoid is therefore a SEMIGROUP with an IDENTITY ELEMENT. A monoid must contain at least one element.

The numbers of free idempotent monoids on n letters are 1, 2, 7, 160, 332381, ... (Sloane's A005345).

See also BINARY OPERATOR, GROUP, SEMIGROUP

References

Rosenfeld, A. *An Introduction to Algebraic Structures.* New York: Holden-Day, 1968.
Sloane, N. J. A. Sequences A005345/M1820 in "An On-Line Version of the Encyclopedia of Integer Sequences." http://www.research.att.com/~njas/sequences/eisonline.html.

Monomial

A POLYNOMIAL consisting of a product of powers of variables, e.g., x, xy^2, x^2y^3z, etc. Constant coefficients are sometimes also allowed in front of a monomial.

One monomial is said to divide another if the powers of its variables are no greater than the corresponding powers in the second monomial. For example, x^2y divides x^3y but does not divide xy^3. A monomial m is said to reduce with respect to a polynomial if the leading monomial of that polynomial divides m. For example, x^2y reduces with respect to $2xy + x + 3$ because xy divides x^2y, and te result of this reduction is $x^2y - x(2xy + x + 3)/2$, or $-x^2/2 - 3x/2$. A polynomial can therefore be reduced by reducing its monomials beginning with the greatest and proceeding downward. Similarly, a polynomial can be reduced with respect to a set of polynomials by reducing in turn with respect to each element in that set. A polynomial is fully reduced if none of its monomials can be reduced (Lichtblau 1996).

See also BINOMIAL, GRÖBNER BASIS, MONIC POLYNOMIAL, POLYNOMIAL, TRINOMIAL

References

Lichtblau, D. "Gröbner Bases in *Mathematica* 3.0." *Mathematica J.* **6**, 81–8, 1996.

Monomial Order

"$u < v$ implies $uw < vw$" for all monomials u, v, and w. Examples of monomial orders are the LEXICOGRAPHIC ORDER and the total degree order.

See also WELL ORDERED SET

Monomino

The unique 1-POLYOMINO, consisting of a single SQUARE.

See also DOMINO, TRIOMINO

References

Gardner, M. "Polyominoes." Ch. 13 in *The Scientific American Book of Mathematical Puzzles & Diversions.* New York: Simon and Schuster, pp. 124–40, 1959.

Monomorph

An INTEGER which is expressible in only one way in the form $x^2 + Dy^2$ or $x^2 - Dy^2$ where x^2 is RELATIVELY PRIME to Dy^2. If the INTEGER is expressible in more than one way, it is called a POLYMORPH.

See also ANTIMORPH, IDONEAL NUMBER, PELL EQUATION, POLYMORPH

Monomorphism

A MORPHISM $f : Y \to X$ in a CATEGORY is a monomorphism if, for any two MORPHISMS $u, v : Z \to Y$, $fu = fv$ implies that $u = v$.

See also CATEGORY, MORPHISM

Monotone

Another word for monotonic.

See also MONOTONIC FUNCTION, MONOTONIC SEQUENCE, MONOTONIC VOTING

Monotone Convergence Theorem

If $\{f_n\}$ is a sequence of MEASURABLE FUNCTIONS, with $0 \le f_n \le f_{n+1}$ for every n, then

$$\int \lim_{n \to \infty} f_n \, d\mu = \lim_{n \to \infty} \int f_n \, d\mu$$

Monotone Decreasing

Always decreasing; never remaining constant or increasing. Also called strictly decreasing.

Monotone Increasing

Always increasing; never remaining constant or decreasing. Also called strictly increasing.

Monotone Triangle

A monotone triangle (also called a strict Gelfand pattern or a gog triangle) of order n is a NUMBER TRIANGLE with n numbers along each side and the base containing entries between 1 and n such that there is strict increase across rows and weak increase diagonally up or down to the right. There is a bijection between monotone triangles of order n and ALTERNATING SIGN MATRICES of order n obtained by letting the kth row of the triangle equal the positions of 1s in the sum of the first k rows of an ALTERNATING SIGN MATRIX, as illustrated below.

$$\begin{bmatrix} 0 & 0 & 0 & 1 & 0 \\ 0 & 1 & 0 & -1 & 1 \\ 1 & -1 & 0 & 1 & 0 \\ 0 & 0 & 1 & 0 & 0 \\ 0 & 1 & 0 & 0 & 0 \end{bmatrix} \leftrightarrow \begin{matrix} 4 \\ 2 \quad 5 \\ 1 \quad 4 \quad 5 \\ 1 \quad 3 \quad 4 \quad 5 \\ 1 \quad 2 \quad 3 \quad 4 \quad 5 \end{matrix}$$

$$(0, 0, 0, 1, 0) \to 4$$

$$(0, 0, 0, 1, 0) + (0, 1, 0, -1, 1)$$
$$= (0, 1, 0, 0, 1) \to 2 \quad 5$$

$$(0, 1, 0, 0, 1) + (1, -1, 0, 1, 0,)$$
$$= (1, 0, 0, 1, 1) \to 1 \quad 4 \quad 5$$

$$(1, 0, 0, 1, 1) + (0, 0, 1, 0, 0)$$
$$= (1, 0, 1, 1, 1) \to 1 \quad 3 \quad 4 \quad 5$$

$$(1, 0, 1, 1, 1) + (0, 1, 0, 0, 0)$$
$$= (1, 1, 1, 1, 1) \to 1 \quad 2 \quad 3 \quad 4 \quad 5$$

References

Bressoud, D. and Propp, J. "How the Alternating Sign Matrix Conjecture was Solved." *Not. Amer. Math. Soc.* **46**, 637–46.

Monotonic Function

A function which is either entirely NONINCREASING or NONDECREASING. A function is monotonic if its first DERIVATIVE (which need not be continuous) does not change sign.

See also COMPLETELY MONOTONIC FUNCTION, MONOTONE, MONOTONE DECREASING, MONOTONE INCREASING, NONDECREASING FUNCTION, NONINCREASING FUNCTION

Monotonic Sequence

A SEQUENCE $\{a_n\}$ such that either (1) $a_{i+1} \ge a_i$ for every $i \ge 1$, or (2) $a_{i+1} \le a_i$ for every $i \ge 1$.

Monotonic Voting

A term in SOCIAL CHOICE THEORY meaning a change favorable for X does not hurt X.

See also ANONYMOUS, DUAL VOTING, VOTING

Monster Group

The highest order SPORADIC GROUP M. It has ORDER

$$2^{46} \cdot 3^{20} \cdot 5^9 \cdot 7^6 \cdot 11^2 \cdot 13^3 \cdot 17 \cdot 19 \cdot 23 \cdot 29 \cdot 31$$
$$\cdot \, 41 \cdot 47 \cdot 59 \cdot 71,$$

and is also called the FRIENDLY GIANT GROUP. It was constructed in 1982 by Robert Griess as a GROUP of ROTATIONS in 196,883-D space.

See also BABY MONSTER GROUP, BIMONSTER, LEECH LATTICE

References

Conway, J. H.; Curtis, R. T.; Norton, S. P.; Parker, R. A.; and Wilson, R. A. *Atlas of Finite Groups: Maximal Subgroups and Ordinary Characters for Simple Groups.* Oxford, England: Clarendon Press, p. viii, 1985.
Conway, J. H. and Norton, S. P. "Monstrous Moonshine." *Bull. London Math. Soc.* **11**, 308–39, 1979.

Conway, J. H. and Sloane, N. J. A. "The Monster Group and its 196884-Dimensional Space" and "A Monster Lie Algebra?" Chs. 29–0 in *Sphere Packings, Lattices, and Groups, 2nd ed.* New York: Springer-Verlag, pp. 554–71, 1993.

Wilson, R. A. "ATLAS of Finite Group Representation." http://for.mat.bham.ac.uk/atlas/html/M.html.

Monte Carlo Integration

In order to integrate a function over a complicated DOMAIN D, Monte Carlo integration picks random points over some simple DOMAIN D' which is a superset of D, checks whether each point is within D, and estimates the AREA of D (VOLUME, n-D CONTENT, etc.) as the AREA of D' multiplied by the fraction of points falling within D'. Monte Carlo integration is implemented in *Mathematica* as NIntegrate[f, ..., Method->MonteCarlo].

An estimate of the uncertainty produced by this technique is given by

$$\int f \, dV \approx V\langle f \rangle \pm \sqrt{\frac{\langle f^2 \rangle - \langle f \rangle^2}{N}}.$$

See also MONTE CARLO METHOD, NUMERICAL INTEGRATION, QUASI-MONTE CARLO INTEGRATION

References

Hammersley, J. M. "Monte Carlo Methods for Solving Multivariable Problems." *Ann. New York Acad. Sci.* **86**, 844–74, 1960.

Press, W. H.; Flannery, B. P.; Teukolsky, S. A.; and Vetterling, W. T. "Simple Monte Carlo Integration" and "Adaptive and Recursive Monte Carlo Methods." §7.6 and 7.8 in *Numerical Recipes in FORTRAN: The Art of Scientific Computing, 2nd ed.* Cambridge, England: Cambridge University Press, pp. 295–99 and 306–19, 1992.

Ueberhuber, C. W. "Monte Carlo Techniques." §12.4.4 in *Numerical Computation 2: Methods, Software, and Analysis.* Berlin: Springer-Verlag, pp. 124–25 and 132–38, 1997.

Weinzierl, S. Introduction to Monte Carlo Methods. 23 Jun 200. http://xxx.lanl.gov/abs/hep-ph/0006269/.

Monte Carlo Method

Any method which solves a problem by generating suitable random numbers and observing that fraction of the numbers obeying some property or properties. The method is useful for obtaining numerical solutions to problems which are too complicated to solve analytically. It is named by S. Ulam, who in 1946 became the first mathematician to dignify this approach with a name, in honor of a relative having a propensity to gamble (Hoffman 1998, p. 239).

The most common application of the Monte Carlo method is MONTE CARLO INTEGRATION.

See also MARKOV CHAIN, MONTE CARLO INTEGRATION, STOCHASTIC GEOMETRY

References

Gamerman, D. *Markov Chain Monte Carlo: Stochastic Simulation for Bayesian Inference.* Boca Raton, FL: CRC Press, 1997.

Gilks, W. R.; Richardson, S.; and Spiegelhalter, D. J. (Eds.). *Markov Chain Monte Carlo in Practice.* Boca Raton, FL: Chapman & Hall, 1996.

Hoffman, P. *The Man Who Loved Only Numbers: The Story of Paul Erdos and the Search for Mathematical Truth.* New York: Hyperion, pp. 238–39, 1998.

Manno, I. *Introduction to the Monte Carlo Method.* Budapest, Hungary: Akadémiai Kiadó, 1999.

Mikhailov, G. A. *Parametric Estimates by the Monte Carlo Method.* Utrecht, Netherlands: VSP, 1999.

Niederreiter, H. and Spanier, J. (Eds.). *Monte Carlo and Quasi-Monte Carlo Methods 1998, Proceedings of a Conference held at the Claremont Graduate University, Claremont, California, USA, June 22–6, 1998.* Berlin: Springer-Verlag, 2000. Sobol, I. M. *A Primer for the Monte Carlo Method.* Boca Raton, FL: CRC Press, 1994.

Montel's Theorem

Let $f(z)$ be an ANALYTIC FUNCTION of z, regular in the half-strip S defined by $a < x < b$ and $y > 0$. If $f(z)$ is bounded in S and tends to a limit l as $y \to \infty$ for a certain fixed value ξ of x between a and b, then $f(z)$ tends to this limit l on every line $x = x_0$ in S, and $f(z) \to l$ uniformly for $a + \delta \le x_0 \le b - \delta$.

See also VITALI'S CONVERGENCE THEOREM

References

Krantz, S. G. "Montel's Theorem, First Version and Montel's Theorem, Second Version." §8.4.3 and 8.4.4 in *Handbook of Complex Analysis.* Boston, MA: Birkhäuser, p. 114, 1999.

Titchmarsh, E. C. *The Theory of Functions, 2nd ed.* Oxford, England: Oxford University Press, p. 170, 1960.

Monty Hall Dilemma

MONTY HALL PROBLEM

Monty Hall Problem

The Monty Hall problem is named for its similarity to the *Let's Make a Deal* television game show hosted by Monty Hall. The problem is stated as follows. Assume that a room is equipped with three doors. Behind two are goats, and behind the third is a shiny new car. You are asked to pick a door, and will win whatever is behind it. Let's say you pick door 1. Before the door is opened, however, someone who knows what's behind the doors (Monty Hall) opens *one of the other* two doors, revealing a goat, and asks you if you wish to change your selection to the third door (i.e., the door which neither you picked nor he opened). The Monty Hall problem is deciding whether you do.

The correct answer is that you *do* want to switch. If you do not switch, you have the expected 1/3 chance of winning the car, since no matter whether you initially picked the correct door, Monty will show you a door with a goat. But after Monty has eliminated one of the doors for you, you obviously do not improve your chances of winning to better than 1/3 by sticking with your original choice. If you now switch doors, however, there is a 2/3 chance you will win the car (counterintuitive though it seems).

d_1	d_2	Winning Probability
pick	stick	1/3
pick	switch	2/3

The problem can be generalized to four doors as follows. Let one door conceal the car, with goats behind the other three. Pick a door d_1. Then the host will open one of the nonwinners and give you the option of switching. Call your new choice (which could be the same as d_1 if you don't switch) d_2. The host will then open a second nonwinner, and you must decide for choice d_3 if you want to stick to d_2 or switch to the remaining door. The probabilities of winning are shown below for the four possible strategies.

d_1	d_2	d_3	Winning Probability
pick	stick	stick	2/8
pick	switch	stick	3/8
pick	stick	switch	6/8
pick	switch	switch	5/8

The above results are characteristic of the best strategy for the n-stage Monty Hall problem: stick until the last choice, then switch.

See also ALLAIS PARADOX

References

Barbeau, E. "The Problem of the Car and Goats." *CMJ* **24**, 149, 1993.
Bogomolny, A. "Monty Hall Dilemma." http://www.cut-the-knot.com/hall.html.
Dewdney, A. K. *200% of Nothing*. New York: Wiley, 1993.
Donovan, D. "The WWW Tackles the Monty Hall Problem." http://math.rice.edu/~ddonovan/montyurl.html.
Ellis, K. M. "The Monty Hall Problem." http://www.io.com/~kmellis/monty.html.
Gardner, M. *Aha! Gotcha: Paradoxes to Puzzle and Delight*. New York: W. H. Freeman, 1982.
Gillman, L. "The Car and the Goats." *Amer. Math. Monthly* **99**, 3, 1992.
Hoffman, P. *The Man Who Loved Only Numbers: The Story of Paul Erdos and the Search for Mathematical Truth*. New York: Hyperion, pp. 233–40, 1998.
Selvin, S. "A Problem in Probability." *Amer. Stat.* **29**, 67, 1975.
vos Savant, M. *The Power of Logical Thinking*. New York: St. Martin's Press, 1996.

Moore Graph

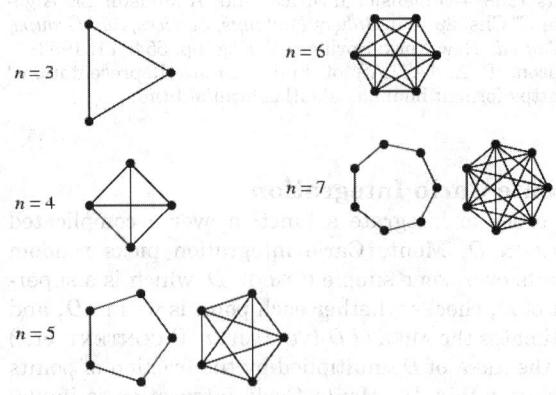

A GRAPH of type (d, k) is a REGULAR GRAPH of vertex degree $d > 2$ and GRAPH DIAMETER k which contains the maximum possible number of nodes,

$$n(d, k) = 1 + d \sum_{r=1}^{k} (d-1)^{r-1} = \frac{d(d-1)^k - 2}{d - 2}$$

(Bannai and Ito 1973). Equivalently, it is a (d, g)-CAGE GRAPH, where d is the vertex degree and g is the GIRTH, with an EXCESS of zero (Wong 1982). Moore graphs are also called minimal (v, g)-graphs (Wong 1982), and are DISTANCE-REGULAR.

Hoffman and Singleton (1960) first used the term "Moore graph," and showed that there is a unique Moore graph for types (3, 2) and (7, 2), but no other $(d, 2)$ Moore graphs with the possible exception of (57, 2) (Bannai and Ito 1973). Bannai and Ito (1973) subsequently showed that there exist no Moore graphs of type (d, k) with GRAPH DIAMETER $k \geq 4$ and valence $d > 2$. Equivalently, a (v, g)-Moore graph exists only if (1) $g = 5$ and $v = 3, 7$, or (possibly) 57, or (2) $g = 6$, 8, or 12 (Wong 1982). This settled the existence and uniqueness problem from finite Moore graphs with the exception of the case (57, 2), which is still open. A proof of this theorem, sometimes called the HOFFMAN-SINGLETON THEOREM, is difficult (Hoffman and Singleton 1960, Feit and Higman 1964, Damerell 1973, Bannai and Ito 1973), but can be found in Biggs (1993).

The (3, 5)-Moore graph is the PETERSEN GRAPH, and the (7, 5)-Moore graph is the HOFFMAN-SINGLETON GRAPH. The existence of a (57, 5)-graph remains an open question.

See also CAGE GRAPH, DISTANCE-REGULAR GRAPH, GENERALIZED POLYGON, GIRTH, GRAPH DIAMETER, HOFFMAN-SINGLETON GRAPH, HOFFMAN-SINGLETON THEOREM, PETERSEN GRAPH, REGULAR GRAPH

References

Aschbacher, M. "The Non-Existence of Rank Three Permutation Group of Degree 3250 and Subdegree 57." *J. Algebra* **19**, 538–40, 1971.

Bannai, E. and Ito, T. "On Moore Graphs." *J. Fac. Sci. Univ. Tokyo Ser. A* **20**, 191–08, 1973.

Biggs, N. L. Ch. 23 in *Algebraic Graph Theory, 2nd ed.* Cambridge, England: Cambridge University Press, 1993.

Bosák, J. "Cubic Moore Graphs." *Mat. Casopis Sloven. Akad. Vied* **20**, 72–0, 1970.

Bosák, J. "Partially Directed Moore Graphs." *Math. Slovaca* **29**, 181–96, 1979.

Damerell, R. M. "On Moore Graphs." *Proc. Cambridge Philos. Soc.* **74**, 227–36, 1973.

Feit, W. and Higman, G. "The Non-Existence of Certain Generalized Polygons." *J. Algebra* **1**, 114–31, 1964.

Friedman, H. D. "On the Impossibility of Certain Moore graphs." *J. Combin. Th. B* **10**, 245–52, 1971.

Godsil, C. D. "Problems in Algebraic Combinatorics." *Electronic J. Combinatorics* **2**, F1 1–0, 1995. http://www.combinatorics.org/Volume_2/volume2.html#F1.

Hoffman, A. J. and Singleton, R. R. "On Moore Graphs of Diameter 2 and 3." *IBM J. Res. Develop.* **4**, 497–04, 1960.

McKay, B. D. and Stanton, R. G. "The Current Status of the Generalised Moore Graph Problem." In *Combinatorial Mathematics VI (Armidale 1978)*. New York: Springer-Verlag, pp. 21–1, 1979.

Wong, P. K. "Cages--A Survey." *J. Graph Th.* **6**, 1–2, 1982.

Moore-Penrose Generalized Matrix Inverse

Given an $m \times n$ MATRIX B, the Moore-Penrose generalized MATRIX INVERSE (sometimes called the pseudoinverse) is a unique $n \times m$ MATRIX B^+ which satisfies

$$BB^+B = B \tag{1}$$

$$B^+BB^+ = B^+ \tag{2}$$

$$(BB^+)^T = BB^+ \tag{3}$$

$$(B^+B)^T = B^+B. \tag{4}$$

It is also true that

$$z = B^+c \tag{5}$$

is the shortest length LEAST SQUARES solution to the problem

$$B = c. \tag{6}$$

If the inverse of (B^TB) exists, then

$$B^+ = (B^TB)^{-1}B^T, \tag{7}$$

where B^T is the matrix TRANSPOSE, as can be seen by premultiplying both sides of (7) by B^T to create a SQUARE MATRIX which can then be inverted,

$$B^TB_z = B^Tc, \tag{8}$$

giving

$$z = (B^TB)^{-1}B^Tc \equiv B^+c. \tag{9}$$

See also LEAST SQUARES FITTING, MATRIX INVERSE

References

Ben-Israel, A. and Greville, T. N. E. *Generalized Inverses: Theory and Applications.* New York: Wiley, 1977.

Lawson, C. and Hanson, R. *Solving Least Squares Problems.* Englewood Cliffs, NJ: Prentice-Hall, 1974.

Penrose, R. "A Generalized Inverse for Matrices." *Proc. Cambridge Phil. Soc.* **51**, 406–13, 1955.

Mordell Conjecture

DIOPHANTINE EQUATIONS that give rise to surfaces with two or more holes have only finite many solutions in GAUSSIAN INTEGERS with no common factors. Fermat's equation has $(n-1)(n-2)/2$ HOLES, so the Mordell conjecture implies that for each INTEGER $n \geq 3$, the FERMAT EQUATION has at most a finite number of solutions. This conjecture was proved by Faltings (1984).

See also ABC CONJECTURE, FERMAT EQUATION, FERMAT'S LAST THEOREM, SAFAREVICH CONJECTURE, SHIMURA-TANIYAMA CONJECTURE

References

Elkies, N. D. "ABC Implies Mordell." *Internat. Math. Res. Not.* **7**, 99–09, 1991.

Faltings, G. "Die Vermutungen von Tate und Mordell." *Jahresber. Deutsch. Math.-Verein* **86**, 1–3, 1984.

Ireland, K. and Rosen, M. "The Mordell Conjecture." §20.3 in *A Classical Introduction to Modern Number Theory, 2nd ed.* New York: Springer-Verlag, pp. 340–42, 1990.

van Frankenhuysen, M. "The ABC Conjecture Implies Roth's Theorem and Mordell's Conjecture." *Mat. Contemp.* **16**, 45–2, 1999.

Mordell Integral

The integral

$$\phi(t, u) = \int \frac{e^{\pi itx^2 + 2\pi iux}}{e^{2\pi ix} - 1} \, dx$$

which is related to the JACOBI THETA FUNCTIONS, MOCK THETA FUNCTIONS, RIEMANN ZETA FUNCTION, and SIEGEL THETA FUNCTION.

See also JACOBI THETA FUNCTIONS, MOCK THETA FUNCTION, RIEMANN ZETA FUNCTION, SIEGEL THETA FUNCTION

Mordell-Weil Theorem

For ELLIPTIC CURVES over the RATIONALS $\mathbb{Q}$, the GROUP of RATIONAL POINTS is always FINITELY GENERATED (i.e., there always exists a finite set of generators for the GROUP). This theorem was proved by Mordell in 1921 and extended by Weil in 1928 to ABELIAN VARIETIES over NUMBER FIELDS.

See also ELLIPTIC CURVE

References

Ireland, K. and Rosen, M. "The Mordell-Weil Theorem." Ch. 19 in *A Classical Introduction to Modern Number*

Theory, 2nd ed. New York: Springer-Verlag, pp. 319–38, 1990.

Nagell, T. "Rational Points on Plane Algebraic Curves. Mordell's Theorem." §69 in *Introduction to Number Theory.* New York: Wiley, pp. 253–60, 1951.

Morera's Theorem

If $f(z)$ is continuous in a region D and satisfies

$$\oint_\gamma f \, dz = 0$$

for all closed CONTOURS γ in D, then $f(z)$ is ANALYTIC in D.

See also CAUCHY INTEGRAL THEOREM, CONTOUR INTEGRATION

References

Arfken, G. *Mathematical Methods for Physicists, 3rd ed.* Orlando, FL: Academic Press, pp. 373–74, 1985.

Krantz, S. G. *Handbook of Complex Analysis.* Boston, MA: Birkhäuser, p. 26, 1999.

Morgado Identity

There are several results known as the Morgado identity. The first is

$$F_n F_{n+1} F_{n+2} F_{n+4} F_{n+5} F_{n+6} + L_{n+3}^2$$
$$= [F_{n+3}(2F_{n+2}F_{n+4} - F_{n+3}^2)]^2, \quad (1)$$

where F_n is a FIBONACCI NUMBER and L_n is a LUCAS NUMBER (Morgado 1987, Dujella 1995).

An second Morgado identity is satisfied by GENERALIZED FIBONACCI NUMBERS w_n,

$$4w_n w_{n+1} w_{n+2} w_{n+4} w_{n+5} w_{n+6}$$
$$+ e^2 q^{2n}(w_n U_4 U_5 - w_{n+1} U_2 U_6 - w_n U_1 U_8)^2$$
$$= (w_{n+1} w_{n+2} w_{n+6} + w_n w_{n+4} w_{n+5})^2, \quad (2)$$

where

$$e \equiv pab - qa^2 - b^2 \quad (3)$$

$$U_n \equiv w_n(0, 1; \; p, \; q) \quad (4)$$

(Morgado 1987, Dujella 1996).

See also FIBONACCI NUMBER, GENERALIZED FIBONACCI NUMBER

References

Dujella, A. "Diophantine Quadruples for Squares of Fibonacci and Lucas Numbers." *Portugaliae Math.* **52**, 305–18, 1995.

Dujella, A. "Generalized Fibonacci Numbers and the Problem of Diophantus." *Fib. Quart.* **34**, 164–75, 1996.

Morgado, J. "Note on Some Results of A. F. Horadam and A. G. Shannon Concerning a Catalan's Identity on Fibonacci Numbers." *Portugaliae Math.* **44**, 243–52, 1987.

Morgan-Voyce Polynomial

Polynomials related to the BRAHMAGUPTA POLYNOMIALS. They are defined by the RECURRENCE RELATIONS

$$b_n(x) = xB_{n-1}(x) + b_{n-1}(x) \quad (1)$$

$$B_n(x) = (x+1)B_{n-1}(x) + b_{n-1}(x) \quad (2)$$

for $n \geq 1$, with

$$b_0(x) = B_0(x) = 1. \quad (3)$$

Alternative recurrences are

$$b_n(x) = (x+2)b_{n-1}(x) - b_{n-2}(x) \quad (4)$$

$$B_n(x) = (x+2)B_{n-1}(x) - B_{n-2}(x) \quad (5)$$

with $b_1(x) = 1 + x$ and $B_1(x) = 2 + x$, and

$$b_{n+1}b_{n-1} - b_n^2 = x. \quad (6)$$

$$B_{n+1}B_{n-1} - B_n^2 = -1 \quad (7)$$

The polynomials can be given explicitly by the sums

$$B_n(x) = \sum_{k=0}^n \binom{n+k-1}{n-k} x^k \quad (8)$$

$$b_n(x) = \sum_{k=0}^n \binom{n+k}{n-k} x^k. \quad (9)$$

Defining the MATRIX

$$Q = \begin{bmatrix} x+2 & -1 \\ 1 & 0 \end{bmatrix} \quad (10)$$

gives the identities

$$Q^n = \begin{bmatrix} B_n & -B_{n-1} \\ B_{n-1} & -B_{n-2} \end{bmatrix} \quad (11)$$

$$Q^n - Q^{n-1} = \begin{bmatrix} b_n & -b_{n-1} \\ b_{n-1} & -b_{n-2} \end{bmatrix}. \quad (12)$$

Defining

$$\cos \theta = \tfrac{1}{2}(x+2) \quad (13)$$

$$\cosh \phi = \tfrac{1}{2}(x+2) \quad (14)$$

gives

$$B_n(x) = \frac{\sin[(n+1)\theta]}{\sin \theta} \quad (15)$$

$$B_n(x) = \frac{\sinh[(n+1)\phi]}{\sinh \phi} \quad (16)$$

and

$$b_n(x) = \frac{\cos\left[\tfrac{1}{2}(2n+1)\theta\right]}{\cos\left(\tfrac{1}{2}\theta\right)} \quad (17)$$

$$b_n(x) = \frac{\cosh\left[\frac{1}{2}(2n+1)\phi\right]}{\cosh\left(\frac{1}{2}\theta\right)}. \tag{18}$$

The Morgan-Voyce polynomials are related to the FIBONACCI POLYNOMIALS $F_n(x)$ by

$$b_n(x^2) = F_{2n+1}(x) \tag{19}$$

$$B_n(x^2) = \frac{1}{x}\, F_{2n+2}(x) \tag{20}$$

(Swamy 1968).

$B_n(x)$ satisfies the ORDINARY DIFFERENTIAL EQUATION

$$x(x+4)y'' + 3(x+2)y' - n(n+2)y = 0, \tag{21}$$

and $b_n(x)$ the equation

$$x(x+4)y'' + 2(x+1)y' - n(n+1)y = 0. \tag{22}$$

These and several other identities involving derivatives and integrals of the polynomials are given by Swamy (1968).

See also BRAHMAGUPTA POLYNOMIAL, FIBONACCI POLYNOMIAL

References

Lahr, J. "Fibonacci and Lucas Numbers and the Morgan-Voyce Polynomials in Ladder Networks and in Electric Line Theory." In *Fibonacci Numbers and Their Applications* (Ed. G. E. Bergum, A. N. Philippou, and A. F. Horadam). Dordrecht, Netherlands: Reidel, 1986.
Morgan-Voyce, A. M. "Ladder Network Analysis Using Fibonacci Numbers." *IRE Trans. Circuit Th.* **CT-6**, 321–22, Sep. 1959.
Swamy, M. N. S. "Properties of the Polynomials Defined by Morgan-Voyce." *Fib. Quart.* **4**, 73–1, 1966.
Swamy, M. N. S. "More Fibonacci Identities." *Fib. Quart.* **4**, 369–72, 1966.
Swamy, M. N. S. "Further Properties of Morgan-Voyce Polynomials." *Fib. Quart.* **6**, 167–75, 1968.

Morley Centers

The CENTROID of MORLEY'S TRIANGLE is called Morley's first center. It has TRIANGLE CENTER FUNCTION

$$\alpha = \cos\left(\tfrac{1}{3}A\right) + 2\cos\left(\tfrac{1}{3}B\right)\cos\left(\tfrac{1}{3}C\right).$$

The PERSPECTIVE CENTER of MORLEY'S TRIANGLE with reference TRIANGLE ABC is called Morley's second center. The TRIANGLE CENTER FUNCTION is

$$\alpha = \sec\left(\tfrac{1}{3}A\right).$$

See also CENTROID (GEOMETRIC), MORLEY'S THEOREM, PERSPECTIVE CENTER

References

Kimberling, C. "Central Points and Central Lines in the Plane of a Triangle." *Math. Mag.* **67**, 163–87, 1994.

Kimberling, C. "1st and 2nd Morley Centers." http://cedar.evansville.edu/~ck6/tcenters/recent/morley.html.
Oakley, C. O. and Baker, J. C. "The Morley Trisector Theorem." *Amer. Math. Monthly* **85**, 737–45, 1978.

Morley's Formula

$$\sum_{k=0}^{\infty}\left[\frac{(m)_k}{k!}\right]^3 = 1 + \left(\frac{m}{1}\right)^3 + \left[\frac{m(m+1)}{1\cdot 2}\right]^3 + \cdots$$

$$= \frac{\Gamma\left(1 - \frac{3}{2}m\right)}{\left[\Gamma\left(1 - \frac{1}{2}m\right)\right]^3}\,\cos\left(\tfrac{1}{2}m\pi\right),$$

where $(m)_k$ is a POCHHAMMER SYMBOL and $\Gamma(z)$ is the GAMMA FUNCTION. This is a special case of the identity

$$\sum_{k=0}^{\infty}\left[\frac{(m)_k}{k!}\right]^n = {}_nF_{n-1}(\underbrace{m,\ldots,m}_{n};\ \underbrace{1,\ldots,1}_{n-1};\ 1).$$

See also GAMMA FUNCTION

References

Hardy, G. H. *Ramanujan: Twelve Lectures on Subjects Suggested by His Life and Work,* 3rd ed. New York: Chelsea, pp. 104 and 111, 1999.

Morley's Theorem

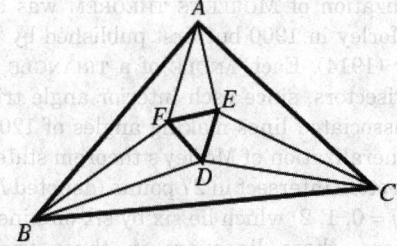

The points of intersection of the adjacent TRISECTORS of the ANGLES of any TRIANGLE $\triangle ABC$ are the VERTICES of an EQUILATERAL TRIANGLE $\triangle DEF$ known as MORLEY'S TRIANGLE. Taylor and Marr (1914) give two geometric proofs and one trigonometric proof.

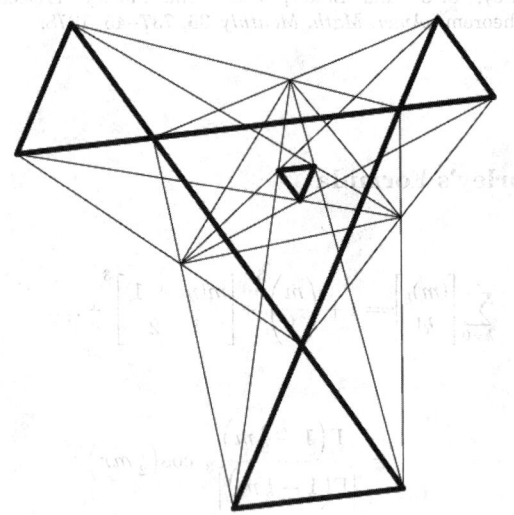

An even more beautiful result is obtained by taking the intersections of the exterior, as well as interior, angle trisectors, as shown above. In addition to the interior EQUILATERAL TRIANGLE formed by the interior trisectors, four additional equilateral triangles are obtained, three of which have sides which are extensions of a central triangle (Wells 1991).

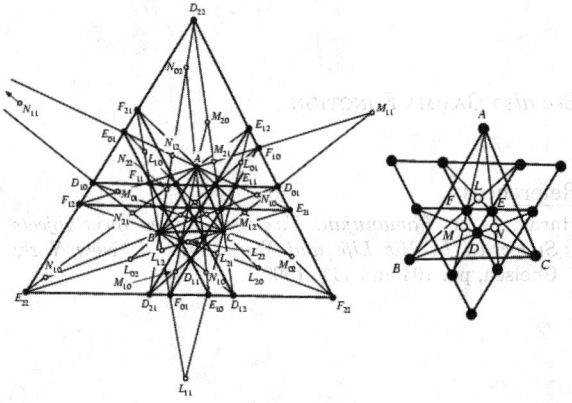

A generalization of MORLEY'S THEOREM was discovered by Morley in 1900 but first published by Taylor and Marr (1914). Each ANGLE of a TRIANGLE $\triangle ABC$ has six trisectors, since each interior angle trisector has two associated lines making angles of 120° with it. The generalization of Morley's theorem states that these trisectors intersect in 27 points (denoted D_{ij}, E_{ij}, F_{ij}, for $i, j = 0, 1, 2$) which lie six by six on nine lines. Furthermore, these lines are in three triples of PARALLEL lines, $(D_{22}E_{22}, E_{12}D_{21}, F_{10}F_{01})$, $(D_{22}F_{22}, F_{21}D_{12}, E_{01}E_{10})$, and $(E_{22}F_{22}, F_{12}E_{21}, D_{10}D_{01})$, making ANGLES of 60° with one another (Taylor and Marr 1914, Johnson 1929, p. 254).

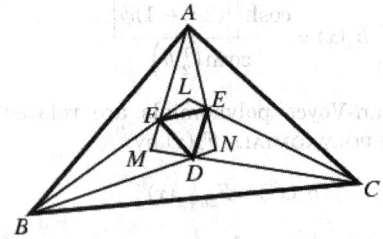

Let L, M, and N be the other trisector-trisector intersections, and let the 27 points L_{ij}, M_{ij}, N_{ij} for $i, j = 0, 1, 2$ be the ISOGONAL CONJUGATES of D, E, and F. Then these points lie 6 by 6 on 9 CONICS through $\triangle ABC$. In addition, these CONICS meet 3 by 3 on the CIRCUMCIRCLE, and the three meeting points form an EQUILATERAL TRIANGLE whose sides are PARALLEL to those of $\triangle DEF$.

See also CONIC SECTION, MORLEY CENTERS, TRISECTION

References

Child, J. M. "Proof of Morley's Theorem." *Math. Gaz.* **11**, 171, 1923.

Coxeter, H. S. M. and Greitzer, S. L. "Morley's Theorem." §2.9 in *Geometry Revisited.* Washington, DC: Math. Assoc. Amer., pp. 47–0, 1967.

Gardner, M. *Martin Gardner's New Mathematical Diversions from Scientific American.* New York: Simon and Schuster, pp. 198 and 206, 1966.

Honsberger, R. "Morley's Theorem." Ch. 8 in *Mathematical Gems I.* Washington, DC: Math. Assoc. Amer., pp. 92–8, 1973.

Johnson, R. A. *Modern Geometry: An Elementary Treatise on the Geometry of the Triangle and the Circle.* Boston, MA: Houghton Mifflin, pp. 253–56, 1929.

Kimberling, C. "Hofstadter Points." *Nieuw Arch. Wiskunder* **12**, 109–14, 1994.

Lebesgue, H. "Sur les *n*-sectrices d'un triangle." *L'enseign. math.* **38**, 39–8, 1939.

Marr, W. L. "Morley's Trisection Theorem: An Extension and Its Relation to the Circles of Apollonius." *Proc. Edinburgh Math. Soc.* **32**, 136–50, 1914.

Morley, F. "On Reflexive Geometry." *Trans. Amer. Math. Soc.* **8**, 14–4, 1907.

Naraniengar, M. T. *Mathematical Questions and Their Solutions from the Educational Times* **15**, 47, 1909.

Oakley, C. O. and Baker, J. C. "The Morley Trisector Theorem." *Amer. Math. Monthly* **85**, 737–45, 1978.

Pappas, T. "Trisecting & the Equilateral Triangle." *The Joy of Mathematics.* San Carlos, CA: Wide World Publ./Tetra, p. 174, 1989.

Steinhaus, H. *Mathematical Snapshots, 3rd ed.* New York: Dover, p. 6, 1999.

Taylor, F. G. "The Relation of Morley's Theorem to the Hessian Axis and Circumcentre." *Proc. Edinburgh Math. Soc.* **32**, 132–35, 1914.

Taylor, F. G. and Marr, W. L. "The Six Trisectors of Each of the Angles of a Triangle." *Proc. Edinburgh Math. Soc.* **32**, 119–31, 1914.

Weisstein, E. W. "Plane Geometry." MATHEMATICA NOTEBOOK PLANEGEOMETRY.M.

Wells, D. *The Penguin Dictionary of Curious and Interesting Geometry.* London: Penguin, pp. 154–55, 1991.

Morley's Triangle

An EQUILATERAL TRIANGLE considered by MORLEY'S THEOREM with side lengths

$$8R \sin\left(\tfrac{1}{3}A\right) \sin\left(\tfrac{1}{3}B\right) \sin\left(\tfrac{1}{3}C\right),$$

where R is the CIRCUMRADIUS of the original TRIANGLE.

See also MORLEY'S THEOREM

Morphism

A morphism is a map between two objects in an abstract CATEGORY.

1. A general morphism is called a HOMOMORPHISM,
2. A morphism $f : Y \to X$ in a CATEGORY is a MONOMORPHISM if, for any two morphisms $u, v : Z \to Y$, $fu = fv$ implies that $u = v$,
3. A morphism $f : Y \to X$ in a CATEGORY is an EPIMORPHISM if, for any two morphisms $u, v : X \to Z$, $uf = vf$ implies $u = v$,
4. A bijective morphism is called an ISOMORPHISM (if there is an isomorphism between two objects, then we say they are isomorphic),
5. A surjective morphism from an object to itself is called an ENDOMORPHISM, and
6. An ISOMORPHISM between an object and itself is called an AUTOMORPHISM.

See also AUTOMORPHISM, CATEGORY, CATEGORY THEORY, EPIMORPHISM, HOMEOMORPHISM, HOMOMORPHISM, ISOMORPHISM, MONOMORPHISM, OBJECT

Morrie's Law

$$\cos(20°) \cos(40°) \cos(80°) = \tfrac{1}{8}.$$

An identity communicated to Feynman as a child by a boy named Morrie Jacobs (Gleick 1992, p. 47). Feynman remembered this fact all his life and referred to it in a letter to Jacobs in 1987 (Gleick 1992, p. 450). It is a special case of the general identity

$$2^k \prod_{j=0}^{k-1} \cos(2^j a) = \frac{\sin(2^k a)}{\sin a},$$

with $k = 3$ and $a = 20°$ (Beyer *et al.* 1996).

See also TRIGONOMETRY VALUES PI/9

References
Anderson, E. C. "Morrie's Law and Experimental Mathematics." To appear in *J. Recr. Math.*
Beyer, W. A.; Louck, J. D.; Zeilberger, D. "A Generalization of a Curiosity that Feynman Remembered All His Life." *Math. Mag.* **69**, 43–4, 1996.
Gleick, J. *Genius: The Life and Science of Richard Feynman.* New York: Pantheon Books, pp. 47 and 450, 1992.

Morse Function

This entry contributed by SERGEI DUZHIN AND S. CHMUTOV

A function for which all CRITICAL POINTS are non-degenerate and all CRITICAL LEVELS are different.

See also KONTSEVICH INTEGRAL, MORSE KNOT

Morse Inequalities

Topological lower bounds in terms of BETTI NUMBERS for the number of critical points form a smooth function on a smooth MANIFOLD.

Morse Knot

This entry contributed by SERGEI DUZHIN AND S. CHMUTOV

A KNOT K embedded in $\mathbb{R}^3 = \mathbb{C}_z \times \mathbb{R}_t$, where the three-dimensional space $\mathbb{R}^3$ is represented as a direct product of a complex line $\mathbb{C}$ with coordinate z and a real line $\mathbb{R}$ with coordinate t, in such a way that the coordinate t is a MORSE FUNCTION on K.

See also KNOT, KONTSEVICH INTEGRAL, MORSE FUNCTION

Morse-Rosen Differential Equation

The second-order ORDINARY DIFFERENTIAL EQUATION

$$y'' + \left[\frac{\alpha}{\cosh^2(ax)} + \beta \tanh(ax) + \gamma \right] y = 0.$$

References
Barut, A. O.; Inomata, A.; and Wilson, R. "Algebraic Treatment of Second Pöschl-Teller, Morse-Rosen, and Eckart Equations." *J. Phys. A: Math. Gen.* **20**, 4083–096, 1987.
Zwillinger, D. *Handbook of Differential Equations, 3rd ed.* Boston, MA: Academic Press, p. 125, 1997.

Morse Theory

A generalization of CALCULUS OF VARIATIONS which draws the relationship between the stationary points of a smooth real-valued function on a MANIFOLD and the global topology of the MANIFOLD. For example, if a COMPACT MANIFOLD admits a function whose only stationary points are a maximum and a minimum, then the manifold is a SPHERE. Technically speaking, Morse theory applied to a FUNCTION g on a MANIFOLD W with $g(M) = 0$ and $g(M') = 1$ shows that every COBORDISM can be realized as a finite sequence of SURGERIES. Conversely, a sequence of SURGERIES gives a COBORDISM.

There are a number of classical applications of Morse theory, including counting geodesics on a RIEMANN SURFACE and determination of the topology of a LIE GROUP (Bott 1960, Milnor 1963). Morse theory has received much attention in the last two decades as a result of the paper by Witten (1982) which relates Morse theory to quantum field theory and also directly connects the stationary points of a smooth function to differential forms on the manifold.

See also CALCULUS OF VARIATIONS, COBORDISM, MAZUR'S THEOREM, SURGERY

References

Bott, R. *Morse Theory and Its Applications to Homotopy Theory.* Bonn, Germany: Universität Bonn, 1960.
Chang, K. C. *Infinite Dimensional Morse Theory and Multiple Solution Problems.* Boston, MA: Birkhäuser, 1993.
Goresky, M. and MacPherson, R. *Stratified Morse Theory.* New York: Springer-Verlag, 1988.
Milnor, J. W. *Morse Theory.* Princeton, NJ: Princeton University Press, 1963.
Rassias, G. (Ed.). *Morse Theory and Its Applications.*
Veverka, J. F. *The Morse Theory and Its Application to Solid State Physics.* Kingston, Ontario, Canada: Queen's University, 1966.
Witten, E. "Supersymmetry and Morse Theory." *J. Diff. Geom.* **17**, 661–92, 1982.

Morse-Thue Sequence
THUE-MORSE SEQUENCE

Mortal
A nonempty finite set of $n \times n$ INTEGER MATRICES for which there exists some product of the MATRICES in the set which is equal to the zero MATRIX.

See also INTEGER MATRIX, MORTALITY PROBLEM

Mortality Problem
For a given n, is the problem of determining if a set is MORTAL solvable? $n = 1$ is solvable, $n = 2$ is unknown, and $n \geq 3$ is unsolvable.

See also MORTAL

Morton-Franks-Williams Inequality
Let E be the largest and e the smallest POWER of ℓ in the HOMFLY POLYNOMIAL of an oriented LINK, and i be the BRAID INDEX. Then the MORTON-FRANKS-WILLIAMS INEQUALITY holds,

$$i \geq \tfrac{1}{2}(E - e) + 1$$

(Franks and Williams 1985, Morton 1985). The inequality is sharp for all PRIME KNOTS up to 10 crossings with the exceptions of 09–42, 09–49, 10–32, 10–50, and 10–56.

See also BRAID INDEX

References

Franks, J. and Williams, R. F. "Braids and the Jones Polynomial." *Trans. Amer. Math. Soc.* **303**, 97–08, 1987.

Mosaic
TESSELLATION

Moser
The very LARGE NUMBER consisting of the number 2 inside a MEGA-gon.

See also MEGA, MEGISTRON

Moser-de Bruijn Sequence
The sequence of numbers which are sums of distinct powers of 4. The first few are 0, 1, 4, 5, 16, 17, 20, 21, 64, 65, 68, 69, 80, 81, 84, ... (Sloane's A000695). These numbers also satisfy the interesting properties that the sum of their BINARY digits equals the sum of their QUATERNARY digits, and that they have identical representations in BINARY and NEGABINARY.

See also BINARY, NEGABINARY, QUATERNARY

References

Allouche, J.-P. and Shallit, J. "The Ring of k-Regular Sequences." *Theor. Comput. Sci.* **98**, 163–97, 1992.
de Bruijn, N. G. "Some Direct Decompositions of the Set of Integers." *Math. Comput.* **18**, 537–46, 1964.
Moser, L. "An Application of Generating Series." *Math. Mag.* **35**, 37–8, 1962.
Sloane, N. J. A. Sequences A000695/M3259 in "An On-Line Version of the Encyclopedia of Integer Sequences." http://www.research.att.com/~njas/sequences/eisonline.html.

Moser's Circle Problem
CIRCLE DIVISION BY CHORDS

Moss's Egg

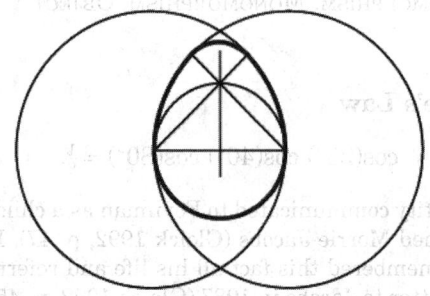

An OVAL whose construction is illustrated in the above diagram.

See also EGG, OVAL

References

Dixon, R. *Mathographics.* New York: Dover, p. 5, 1991.

Mott Polynomial
Polynomials $s_k(x)$ which form the SHEFFER SEQUENCE for

$$f(t) = -\frac{2t}{1 - t^2}$$

and have GENERATING FUNCTION

$$\sum_{k=0}^{\infty} \frac{s_k(x)}{k!} t^k = \exp\left[\frac{x\left(1-\sqrt{1+t^2}\right)}{t}\right].$$

The first few are

$$s_0(x) = 1$$

$$s_1(x) = -\tfrac{1}{2}x$$

$$s_2(x) = \tfrac{1}{4}x^2$$

$$s_3(x) = \tfrac{1}{8}(-x^3 + 6x)$$

$$s_4(x) = \tfrac{1}{16}(x^4 - 24x^2)$$

$$s_5(x) = \tfrac{1}{32}(-x^5 + 60x^3 - 240x).$$

References

Erdélyi, A.; Magnus, W.; Oberhettinger, F.; and Tricomi, F. G. *Higher Transcendental Functions, Vol. 3.* New York: Krieger, p. 251, 1981.

Roman, S. *The Umbral Calculus.* New York: Academic Press, 1984.

Motzkin Number

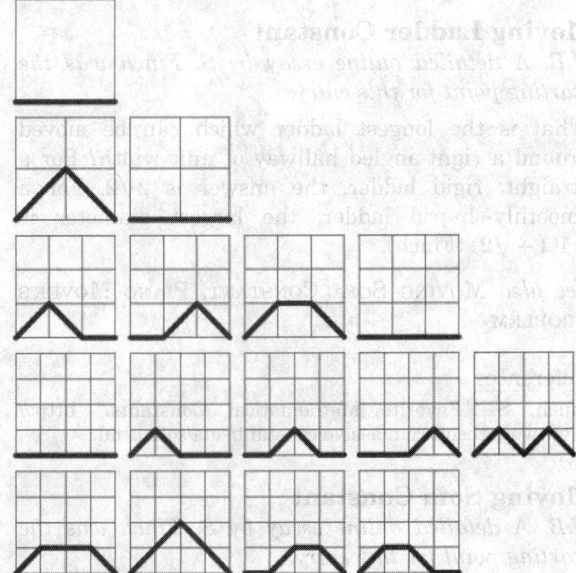

The Motzkin numbers enumerate various combinatorial objects. Donaghey and Shapiro (1977) give 14 different manifestations of these numbers. In particular, they give the number of paths from $(0, 0)$ to $(n, 0)$ which never dip below $y = 0$ and are made up only

of the steps $(1, 0)$, $(1, 1)$, and $(1, -1)$, i.e., $\rightarrow$, $\nearrow$, and $\searrow$. The first are 1, 2, 4, 9, 21, 51, ... (Sloane's A001006). The Motzkin number GENERATING FUNCTION $M(z)$ satisfies

$$M = 1 + xM + x^2 M^2 \tag{1}$$

and is given by

$$M(x) = \frac{1 - x - \sqrt{1 - 2x - 3x^2}}{2x^2}$$

$$= 1 + x + 2x^2 + 4x^3 + 9x^4 + 21x^5 + \ldots, \tag{2}$$

or by the RECURRENCE RELATION

$$M_n = M_{n-1} + \sum_{k=0}^{n-2} M_k M_{n-2-k} \tag{3}$$

with $M_0 = 1$. The Motzkin number M_n is also given by

$$M_n = -\frac{1}{2} \sum_{\substack{a+b=n+2 \\ a \geq 0,\, b \geq 0}} (-3)^a \binom{\frac{1}{2}}{a} \binom{\frac{1}{2}}{b} \tag{4}$$

$$= \frac{(-1)^{n+1}}{2^{2n+5}} \sum_{\substack{a+b=n+2 \\ a \geq 0,\, b \geq 0}} \frac{(-3)^a}{(2a-1)(2b-1)} \binom{2a}{a} \binom{2b}{b}, \tag{5}$$

where $\binom{n}{k}$ is a BINOMIAL COEFFICIENT.

See also CATALAN NUMBER, KING WALK, SCHRÖDER NUMBER

References

Barcucci, E.; Pinzani, R.; and Sprugnoli, R. "The Motzkin Family." *Pure Math. Appl. Ser. A* **2**, 249–79, 1991.

Dickau, R. M. "Delannoy and Motzkin Numbers." http://www.prairienet.org/~pops/delannoy.html.

Donaghey, R. "Restricted Plane Tree Representations of Four Motzkin-Catalan Equations." *J. Combin. Th. Ser. B* **22**, 114–21, 1977.

Donaghey, R. and Shapiro, L. W. "Motzkin Numbers." *J. Combin. Th. Ser. A* **23**, 291–01, 1977.

Kuznetsov, A.; Pak, I.; and Postnikov, A. "Trees Associated with the Motzkin Numbers." *J. Combin. Th. Ser. A* **76**, 145–47, 1996.

Motzkin, T. "Relations Between Hypersurface Cross Ratios, and a Combinatorial Formula for Partitions of a Polygon, for Permanent Preponderance, and for Nonassociative Products." *Bull. Amer. Math. Soc.* **54**, 352–60, 1948.

Sloane, N. J. A. Sequences A001006/M1184 in "An On-Line Version of the Encyclopedia of Integer Sequences." http://www.research.att.com/~njas/sequences/eisonline.html.

Moufang Identities

For all x, y, a in an ALTERNATIVE ALGEBRA $\mathfrak{A}$,

$$(xax)y = x[a(xy)] \tag{1}$$

$$y(xax) = [(yx)a]x \tag{2}$$

$$(xy)(ax) = x(ya)x \qquad (3)$$

(Schafer 1996, p. 28).

References

Schafer, R. D. *An Introduction to Nonassociative Algebras.* New York: Dover, 1996.

Moufang Plane

A PROJECTIVE PLANE in which every line is a translation line is called a Moufang plane.

References

Colbourn, C. J. and Dinitz, J. H. (Eds.). *CRC Handbook of Combinatorial Designs.* Boca Raton, FL: CRC Press, p. 710, 1996.

Mousetrap

A PERMUTATION problem invented by Cayley. Let the numbers 1, 2, ..., n be written on a set of cards, and shuffle this deck of cards. Now, start counting using the top card. If the card chosen does not equal the count, move it to the bottom of the deck and continue counting forward. If the card chosen *does* equal the count, discard the chosen card and begin counting again at 1. The game is won if all cards are discarded, and lost if the count reaches $n + 1$.

The number of ways the cards can be arranged such that at least one card is in the proper place for $n = 1$, 2, ... are 1, 1, 4, 15, 76, 455, ... (Sloane's A002467).

References

Cayley, A. "A Problem in Permutations." *Quart. Math. J.* **1**, 79, 1857.

Cayley, A. "On the Game of Mousetrap." *Quart. J. Pure Appl. Math.* **15**, 8–0, 1877.

Cayley, A. "A Problem on Arrangements." *Proc. Roy. Soc. Edinburgh* **9**, 338–42, 1878.

Cayley, A. "Note on Mr. Muir's Solution of a Problem of Arrangement." *Proc. Roy. Soc. Edinburgh* **9**, 388–91, 1878.

Guy, R. K. "Mousetrap." §E37 in *Unsolved Problems in Number Theory, 2nd ed.* New York: Springer-Verlag, pp. 237–38, 1994.

Guy, R. K. and Nowakowski, R. J. "Mousetrap." In *Combinatorics, Paul Erdos is Eighty, Vol. 1* (Ed. D. Miklós, V. T. Sós, and T. Szonyi). Budapest: János Bolyai Mathematical Society, pp. 193–06, 1993.

Muir, T. "On Professor Tait's Problem of Arrangement." *Proc. Roy. Soc. Edinburgh* **9**, 382–87, 1878.

Muir, T. "Additional Note on a Problem of Arrangement." *Proc. Roy. Soc. Edinburgh* **11**, 187–90, 1882.

Mundfrom, D. J. "A Problem in Permutations: The Game of 'Mousetrap'." *European J. Combin.* **15**, 555–60, 1994.

Sloane, N. J. A. Sequences A002467/M3507, A002468/M2945, and A002469/M3962 in "An On-Line Version of the Encyclopedia of Integer Sequences." http://www.research.att.com/~njas/sequences/eisonline.html.

Steen, A. "Some Formulae Respecting the Game of Mousetrap." *Quart. J. Pure Appl. Math.* **15**, 230–41, 1878.

Tait, P. G. *Scientific Papers, Vol. 1.* Cambridge, England: University Press, p. 287, 1898.

Mouth

A PRINCIPAL VERTEX x_i of a SIMPLE POLYGON P is called a mouth if the diagonal $[x_{i-1}, x_{i+1}]$ is an extremal diagonal (i.e., the interior of $[x_{i-1}, x_{i+1}]$ lies in the exterior of P).

See also ANTHROPOMORPHIC POLYGON, EAR, ONE-MOUTH THEOREM

References

Toussaint, G. "Anthropomorphic Polygons." *Amer. Math. Monthly* **122**, 31–5, 1991.

Moving Average

Given a SEQUENCE $\{a_i\}_{i=1}^N$ an n-moving average is a new sequence $\{s_i\}_{i=1}^{N-n+1}$ defined from the a_i by taking the AVERAGE of subsequences of n terms,

$$s_i = \frac{1}{n} \sum_{j=1}^{i+n-1} a_j.$$

See also MEAN, SPENCER'S 15-POINT MOVING AVERAGE, SPENCER'S FORMULA

References

Kenney, J. F. and Keeping, E. S. "Moving Averages." §14.2 in *Mathematics of Statistics, Pt. 1, 3rd ed.* Princeton, NJ: Van Nostrand, pp. 221–23, 1962.

Whittaker, E. T. and Robinson, G. "Graduation, or the Smoothing of Data." Ch. 11 in *The Calculus of Observations: A Treatise on Numerical Mathematics, 4th ed.* New York: Dover, pp. 285–16, 1967.

Moving Ladder Constant

N.B. A detailed online essay by S. Finch was the starting point for this entry.

What is the longest ladder which can be moved around a right-angled hallway of unit width? For a straight, rigid ladder, the answer is $2\sqrt{2}$. For a smoothly-shaped ladder, the largest diameter is $\geq 1(1 + \sqrt{2})$ (Finch).

See also MOVING SOFA CONSTANT, PIANO MOVER'S PROBLEM

References

Finch, S. "Favorite Mathematical Constants." http://www.mathsoft.com/asolve/constant/sofa/sofa.html.

Moving Sofa Constant

N.B. A detailed online essay by S. Finch was the starting point for this entry.

What is the sofa of greatest AREA S which can be moved around a right-angled hallway of unit width? Hammersley (Croft *et al.* 1994) showed that

$$S \geq \frac{\pi}{2} + \frac{2}{\pi} = 2.2074\ldots \qquad (1)$$

Gerver (1992) found a sofa with larger AREA and provided arguments indicating that it is either optimal or close to it. The boundary of Gerver's sofa is a complicated shape composed of 18 ARCS. Its AREA can be given by defining the constants A, B, ϕ, and θ by solving

$$A(\cos\theta - \cos\phi) - 2B\sin\phi + (\theta - \phi - 1)\cos\theta$$
$$-\sin\theta + \cos\phi + \sin\phi = 0 \qquad (2)$$

$$A(3\sin\theta + \sin\phi) - 2B\cos\phi + 3(\theta - \phi - 1)\sin\theta$$
$$+3\cos\theta - \sin\phi + \cos\phi = 0 \qquad (3)$$

$$A\cos\phi - (\sin\phi + \tfrac{1}{2} - \tfrac{1}{2}\cos\phi + B\sin\phi) = 0 \qquad (4)$$

$$(A + \tfrac{1}{2}\pi - \phi - \theta) - [B - \tfrac{1}{2}(\theta - \phi)(1 + A) - \tfrac{1}{4}(\theta - \phi)^2] = 0. \qquad (5)$$

This gives

$$A = 0.094426560843653\ldots \qquad (6)$$
$$B = 1.399203727333547\ldots \qquad (7)$$
$$\phi = 0.039177364790084\ldots \qquad (8)$$
$$\theta = 0.681301509382725\ldots \qquad (9)$$

Now define

$$r(\alpha) \equiv \begin{cases} \tfrac{1}{2} \\ \quad \text{for } 0 \le \alpha < \phi \\ \tfrac{1}{2}(1 + A + \alpha - \phi) \\ \quad \text{for } \phi \le \alpha < \theta \\ A + \alpha - \phi \\ \quad \text{for } \theta \le \alpha < \tfrac{1}{2}\pi - \theta \\ B - \tfrac{1}{2}\left(\tfrac{1}{2}\pi - \alpha - \phi\right)(1 + A) - \tfrac{1}{4}\left(\tfrac{1}{2}\pi - \alpha - \phi\right)^2 \\ \quad \text{for } \tfrac{1}{2}\pi - \theta \le \alpha < \tfrac{1}{2}\pi - \phi, \end{cases} \qquad (10)$$

where

$$s(\alpha) \equiv 1 - r(\alpha) \qquad (11)$$

$$u(\alpha) \equiv \begin{cases} B - \tfrac{1}{2}(\alpha - \phi)(1 + A) & \text{for } \phi \le \alpha < \theta \\ \quad -\tfrac{1}{4}(\alpha - \phi)^2 \\ A + \tfrac{1}{2}\pi - \phi - \alpha & \text{for } \theta \le \alpha < \tfrac{1}{4}\pi \end{cases} \qquad (12)$$

$$D_u(\alpha) = \frac{du}{d\alpha}$$
$$= \begin{cases} -\tfrac{1}{2}(1 + A) - \tfrac{1}{2}(\alpha - \phi) & \text{for } \phi \le \alpha < \theta \\ -1 & \text{if } \theta \le \alpha < \tfrac{1}{4}\pi. \end{cases} \qquad (13)$$

Finally, define the functions

$$y_1(\alpha) \equiv 1 - \int_0^\alpha r(t)\sin t\, dt \qquad (14)$$

$$y_2(\alpha) \equiv 1 - \int_0^\alpha s(t)\sin t\, dt \qquad (15)$$

$$y_3(\alpha) \equiv 1 - \int_0^\alpha s(t)\sin t\, dt - u(\alpha)\sin\alpha. \qquad (16)$$

The AREA of the optimal sofa is given by

$$A = 2\int_0^{\pi/2 - \phi} y_1(\alpha)r(\alpha)\cos\alpha\, d\alpha$$
$$+ 2\int_0^\theta y_2(\alpha)s(\alpha)\cos\alpha\, d\alpha$$
$$+ 2\int_\phi^{\pi/4} y_3(\alpha)[u(\alpha)\sin\alpha - D_u(\alpha)\cos\alpha - s(\alpha)\cos\alpha]\, d\alpha$$
$$= 2.21953166887197\ldots \qquad (17)$$

(Finch).

See also PIANO MOVER'S PROBLEM

References

Croft, H. T.; Falconer, K. J.; and Guy, R. K. *Unsolved Problems in Geometry.* New York: Springer-Verlag, 1994.
Finch, S. "Favorite Mathematical Constants." http://www.mathsoft.com/asolve/constant/sofa/sofa.html.
Gerver, J. L. "On Moving a Sofa Around a Corner." *Geometriae Dedicata* **42**, 267–83, 1992.
Stewart, I. *Another Fine Math You've Got Me Into....* New York: W. H. Freeman, 1992.

Mrs. Perkins' Quilt

The DISSECTION of a SQUARE of side n into a number S_n of smaller squares. Unlike a PERFECT SQUARE DISSECTION, however, the smaller SQUARES need not be all different sizes. In addition, only prime dissections are considered so that patterns which can be dissected on lower order SQUARES are not permitted. The smallest numbers of RELATIVELY PRIME dissections of an $n \times n$ quilt for $n = 1, 2, \ldots$, are 1, 4, 6, 7, 8, 9, 9, 10, 10, 11, 11, 11, 11, 12, … (Sloane's A005670).

See also PERFECT SQUARE DISSECTION

References

Conway, J. H. "Mrs. Perkins's Quilt." *Proc. Cambridge Phil. Soc.* **60**, 363–68, 1964.
Croft, H. T.; Falconer, K. J.; and Guy, R. K. §C3 in *Unsolved Problems in Geometry.* New York: Springer-Verlag, 1991.
Dudeney, H. E. Problem 173 in *Amusements in Mathematics.* New York: Dover, 1917.
Dudeney, H. E. Problem 177 in *536 Puzzles & Curious Problems.* New York: Scribner, 1967.
Gardner, M. "Mrs. Perkins' Quilt and Other Square-Packing Problems." Ch. 11 in *Mathematical Carnival: A New Round-Up of Tantalizers and Puzzles from Scientific American.* New York: Vintage, 1977.
Sloane, N. J. A. Sequences A005670/M3267 in "An On-Line Version of the Encyclopedia of Integer Sequences." http://www.research.att.com/~njas/sequences/eisonline.html.
Trustrum, G. B. "Mrs. Perkins's Quilt." *Proc. Cambridge Phil. Soc.* **61**, 7–1, 1965.

M-Tree

A TREE not having the COMPLETE BIPARTITE GRAPH $K_{1,2}$ with base at the vertex of degree two as a limb (Lu *et al.* 1993, Lu 1996).

See also TREE

References

Lu, T. "The Enumeration of Trees with and without Given Limbs." *Disc. Math.* **154**, 153–65, 1996.
Lu, T. J.; Read, R. C.; and Palmer, E. M. "On the Enumeration of Trees with Certain Local Restrictions." *Congr. Numer.* **95**, 183–02, 1993.

Much Greater

A strong INEQUALITY in which a is not only GREATER than b, but *much* greater (by some convention), is denoted $a \gg b$. For an astronomer, "much" may mean by a factor of 100 (or even 10), while for a mathematician, it might mean by a factor of 10^4 (or even much more).

See also GREATER, MUCH LESS

Much Less

A strong INEQUALITY in which a is not only LESS than b, but *much* less (by some convention) is denoted $a \ll b$.

See also LESS, MUCH GREATER

Mud Cracks

RIGHT ANGLE

Mu Function

The 2-argument μ-function is defined by

$$\mu(x, \beta) \equiv \int_0^\infty \frac{x^t t^\beta \, dt}{\Gamma(\beta + 1)\Gamma(t + 1)},$$

where $\Gamma(z)$ is the GAMMA FUNCTION (Erdélyi *et al.* 1981, p. 388; Prudnikov *et al.* 1990, p. 798; Gradshteyn and Ryzhik 2000, p. 1109), while the 3-argument function is defined by

$$\mu(x, \beta, \alpha) \equiv \int_0^\infty \frac{x^{\alpha+t} t^\beta \, dt}{\Gamma(\beta + 1)\Gamma(\alpha + t + 1)}$$

(Prudnikov *et al.* 1990, p. 798; Gradshteyn and Ryzhik 2000, p. 1109).

See also LAMBDA FUNCTION, NU FUNCTION

References

Erdélyi, A.; Magnus, W.; Oberhettinger, F.; and Tricomi, F. G. *Higher Transcendental Functions, Vol. 1.* New York: Krieger, p. 388, 1981.
Erdélyi, A.; Magnus, W.; Oberhettinger, F.; and Tricomi, F. G. Ch. 18 in *Higher Transcendental Functions, Vol. 3.* New York: Krieger, p. 217, 1981.
Gradshteyn, I. S. and Ryzhik, I. M. "The Functions $\nu(x)$, $\nu(x, a)$, $\mu(x, \beta)$, $\mu(x, \beta, \alpha)$, $\lambda(x, y)$." §9.64 in *Tables of Integrals, Series, and Products, 6th ed.* San Diego, CA: Academic Press, p. 1109, 2000.
Prudnikov, A. P.; Marichev, O. I.; and Brychkov, Yu. A. *Integrals and Series, Vol. 3: More Special Functions.* Newark, NJ: Gordon and Breach, 1990.

μ Molecule

MANDELBROT SET

Muirhead's Theorem

A NECESSARY and SUFFICIENT condition that $[\alpha']$ should be comparable with $[\alpha]$ for all POSITIVE values of the a is that one of (α') and (α) should be majorized by the other. If $(\alpha') \prec (\alpha)$, then

$$[\alpha'] \leq [\alpha],$$

with equality only when $((\alpha'))$ and (α) are identical or when all the a are equal. See Hardy *et al.* (1988) for a definition of notation.

References

Hardy, G. H.; Littlewood, J. E.; and Pólya, G. "Muirhead's Theorem" and "Proof of Muirhead's Theorem." §2.18 and 2.19 in *Inequalities, 2nd ed.* Cambridge, England: Cambridge University Press, pp. 44–8, 1988.
Muirhead, R. F. "Some Methods Applicable to Identities and Inequalities of Symmetric Algebraic Functions of n Letters." *Proc. Edinburgh Math. Soc.* **21**, 144–57, 1903.

Müller-Lyer Illusion

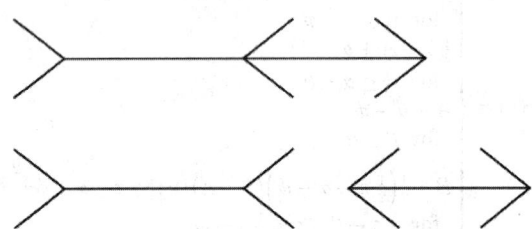

An optical ILLUSION in which the orientation of arrowheads makes one LINE SEGMENT look longer than another. In the above figure, the LINE SEGMENTS on the left and right are of equal length in both cases.

See also ILLUSION, POGGENDORFF ILLUSION, PONZO'S ILLUSION, VERTICAL-HORIZONTAL ILLUSION

References

Fineman, M. *The Nature of Visual Illusion.* New York: Dover, p. 153, 1996.
Luckiesh, M. *Visual Illusions: Their Causes, Characteristics & Applications.* New York: Dover, p. 93, 1965.

Muller's Method

Generalizes the SECANT METHOD of root finding by using quadratic 3-point interpolation

$$q \equiv \frac{x_n - x_{n-1}}{x_{n-1} - x_{n-2}}. \tag{1}$$

Then define

$$A \equiv qP(x_n) - q(1+q)P(x_{n-1}) + q^2 P(x_{n-2}) \quad (2)$$

$$B \equiv (2q+1)P(x_n) - (1+q)^2 P(x_{n-1}) + q^2 P(x_{n-2}) \quad (3)$$

$$C \equiv (1+q)P(x_n), \quad (4)$$

and the next iteration is

$$x_{n+1} = x_n - (x_n - x_{n-1}) \frac{2C}{\max\left(B \pm \sqrt{B^2 - 4AC}\right)}. \quad (5)$$

This method can also be used to find COMPLEX zeros of ANALYTIC FUNCTIONS.

References

Press, W. H.; Flannery, B. P.; Teukolsky, S. A.; and Vetterling, W. T. *Numerical Recipes in FORTRAN: The Art of Scientific Computing, 2nd ed.* Cambridge, England: Cambridge University Press, p. 364, 1992.

Mulliken Symbols

Symbols used to identify irreducible representations of GROUPS:

A = singly degenerate state which is symmetric with respect to ROTATION about the principal C_n axis,

B = singly DEGENERATE state which is antisymmetric with respect to ROTATION about the principal C_n axis,

E = doubly DEGENERATE,

T = triply DEGENERATE,

X_g = (gerade, symmetric) the sign of the wavefunction does not change on INVERSION through the center of the atom,

X_u = (ungerade, antisymmetric) the sign of the wavefunction changes on INVERSION through the center of the atom,

X_1 = (on a or b) the sign of the wavefunction does not change upon ROTATION about the center of the atom,

X_2 = (on a or b) the sign of the wavefunction changes upon ROTATION about the center of the atom,

$'$ = symmetric with respect to a horizontal symmetry plane σ_h,

$''$ = antisymmetric with respect to a horizontal symmetry plane σ_h.

See also CHARACTER TABLE, GROUP THEORY, IRREDUCIBLE REPRESENTATION

References

Cotton, F. A. *Chemical Applications of Group Theory, 3rd ed.* New York: Wiley, pp. 90–1, 1990.

Multiamicable Numbers

Two integers n and $m < n$ are (α, β)-multiamicable if

$$\sigma(m) - m = \alpha n$$

and

$$\sigma(n) - n = \beta m,$$

where $\sigma(n)$ is the DIVISOR FUNCTION and α, β are POSITIVE INTEGERS. If $\alpha = \beta = 1$, (m, n) is an AMICABLE PAIR.

m cannot have just one distinct prime factor, and if it has precisely two prime factors, then $\alpha = 1$ and m is EVEN. Small multiamicable numbers for small α, β are given by Cohen *et al.* (1995). Several of these numbers are reproduced in the table below.

α	β	m	n
1	6	76455288	183102192
1	7	52920	152280
1	7	16225560	40580280
1	7	90863136	227249568
1	7	16225560	40580280
1	7	70821324288	177124806144
1	7	199615613902848	499240550375424

See also AMICABLE PAIR, DIVISOR FUNCTION

References

Cohen, G. L; Gretton, S.; and Hagis, P. Jr. "Multiamicable Numbers." *Math. Comput.* **64**, 1743–753, 1995.

Multichoose

The number of MULTISETS of length k on n symbols is sometimes termed "n multichoose k," denoted $\left(\!\binom{n}{k}\!\right)$ by analogy with the BINOMIAL COEFFICIENT. n multichoose k is given by the simple formula

$$\left(\!\binom{n}{k}\!\right) = n^k,$$

giving the following array of numbers.

$k \backslash n$	1	2	3	4
1	1	1	1	1
2	2	4	8	16
3	3	9	27	81
4	4	16	64	256

See also BINOMIAL COEFFICIENT, CHOOSE, MULTINOMIAL COEFFICIENT, MULTISET

References

Schneiderman, E. R. *Mathematics: A Discrete Introduction.* Pacific Grove, CA: Brooks/Cole, 2000.

Multidigital Number

HARSHAD NUMBER

Multidimensional Continued Fraction Algorithm

INTEGER RELATION

Multifactorial

A generalization of the FACTORIAL and DOUBLE FACTORIAL,

$$n! = n(n-1)(n-2) \cdots 2 \cdot 1 \qquad (1)$$

$$n!! = n(n-2)(n-4) \cdots \qquad (2)$$

$$n!!! = n(n-3)(n-6) \cdots, \qquad (3)$$

etc., where the products run through positive integers.

The FACTORIALS $n!$ for $n = 1, 2, ...,$ are 1, 2, 6, 24, 120, 720, ... (Sloane's A000142); the DOUBLE FACTORIALS $n!!$ are 1, 2, 3, 8, 15, 48, 105, ... (Sloane's A006882); the triple factorials $n!!!$ are 1, 2, 3, 4, 10, 18, 28, 80, 162, 280, ... (Sloane's A007661); and the quadruple factorials $n!!!!$ are 1, 2, 3, 4, 5, 12, 21, 32, 45, 120, ... (Sloane's A007662).

Letting $\mathrm{fac}_k(n)$ denote the k-multifactorial of n,

$$\mathrm{fac}_k(n) = \begin{cases} \prod_{i=1}^{n/k} ik & \text{for } (k, n) \neq 1 \\ \prod_{i=0}^{\lfloor n/k \rfloor} n - ik & \text{for } (k, n) = 1, \end{cases} \qquad (4)$$

Define $r \equiv n/k$ then gives

$$\mathrm{fac}_k(n) = \begin{cases} k^r r! & \text{for } (k, n) \neq 1 \\ (-k)^{1+\lfloor r \rfloor}(-r)_{1+r} & \text{for } (k, n) = 1, \end{cases} \qquad (5)$$

where $(x)_n$ is the POCHHAMMER SYMBOL.

See also DOUBLE FACTORIAL, FACTORIAL, GAMMA FUNCTION, POCHHAMMER SYMBOL

References

Sloane, N. J. A. Sequences A000142/M1675, A006882/M0876, A007661/M0596, and A007662/M0534 in "An On-Line Version of the Encyclopedia of Integer Sequences." http://www.research.att.com/~njas/sequences/eisonline.html.

Multifractal

References

Mandelbrot, B. B. *Multifractals and 1/f Noise: Wild Self-Affinity in Physics (1963–976).* New York: Springer-Verlag, 1998.

Multifractal Measure

A MEASURE for which the Q-DIMENSION D_q varies with q.

References

Ott, E. *Chaos in Dynamical Systems.* New York: Cambridge University Press, 1993.

Multigrade Equation

A (k, l)-multigrade equation is a DIOPHANTINE EQUATION OF THE FORM

$$\sum_{i=1}^{l} n_i^j = \sum_{i=1}^{l} m_i^j$$

for $j = 1, ..., k$, where $\mathbf{m}$ and $\mathbf{n}$ are l-VECTORS. Multigrade identities remain valid if a constant is added to each element of $\mathbf{m}$ and $\mathbf{n}$ (Madachy 1979), so multigrades can always be put in a form where the minimum component of one of the vectors is 1.

Moessner and Gloden (1944) give a bevy of multigrade equations. Small-order examples are the (2, 3)-multigrade with $\mathbf{m} = \{1, 6, 8\}$ and $\mathbf{n} = \{2, 4, 9\}$:

$$\sum_{i=1}^{3} m_i^1 = \sum_{i=1}^{3} n_i^1 = 15$$

$$\sum_{i=1}^{3} m_i^2 = \sum_{i=1}^{3} n_i^2 = 101,$$

the (3, 4)-multigrade with $\mathbf{m} = \{1, 5, 8, 12\}$ and $\mathbf{n} = \{2, 3, 10, 11\}$:

$$\sum_{i=1}^{4} m_i^1 = \sum_{i=1}^{4} n_i^1 = 26$$

$$\sum_{i=1}^{4} m_i^2 = \sum_{i=1}^{4} n_i^2 = 234$$

$$\sum_{i=1}^{4} m_i^3 = \sum_{i=1}^{4} n_i^3 = 2366,$$

and the (4, 6)-multigrade with $\mathbf{m} = \{1, 5, 8, 12, 18, 19\}$ and $\mathbf{n} = \{2, 3, 9, 13, 16, 20\}$:

$$\sum_{i=1}^{6} m_i^1 = \sum_{i=1}^{6} n_i^1 = 63$$

$$\sum_{i=1}^{6} m_i^2 = \sum_{i=1}^{6} n_i^2 = 919$$

$$\sum_{i=1}^{6} m_i^3 = \sum_{i=1}^{6} n_i^3 = 15057$$

$$\sum_{i=1}^{6} m_i^3 = \sum_{i=1}^{6} n_i^4 = 260755$$

(Madachy 1979).

A spectacular example with $k = 9$ and $l = 10$ is given by $\mathbf{n} = \{\pm 12, \pm 11881, \pm 20231, \pm 20885, \pm 23738\}$ and $\mathbf{m} = \{\pm 436, \pm 11857, \pm 20499, \pm 20667, \pm 23750\}$ (Guy 1994), which has sums

$$\sum_{i=1}^{9} m_i^1 = \sum_{i=1}^{9} n_i^1 = 0$$

$$\sum_{i=1}^{9} m_i^2 = \sum_{i=1}^{9} n_i^2 = 3100255070$$

$$\sum_{i=1}^{9} m_i^3 = \sum_{i=1}^{9} n_i^3 = 0$$

$$\sum_{i=1}^{9} m_i^4 = \sum_{i=1}^{9} n_i^4 = 1390452894778220678$$

$$\sum_{i=1}^{9} m_i^5 = \sum_{i=1}^{9} n_i^5 = 0$$

$$\sum_{i=1}^{9} m_i^6 = \sum_{i=1}^{9} n_i^6 = 666573454337853049941719510$$

$$\sum_{i=1}^{9} m_i^7 = \sum_{i=1}^{9} n_i^7 = 0$$

$$\sum_{i=1}^{9} m_i^8 = \sum_{i=1}^{9} n_i^8$$
$$= 330958142560259813821203262692838598$$

$$\sum_{i=1}^{9} m_i^9 = \sum_{i=1}^{9} n_i^9 = 0.$$

Rivera considers multigrade equations involving primes, consecutive primes, etc.

See also DIOPHANTINE EQUATION, PROUHET-TARRY-ESCOTT PROBLEM

References

Chen, S. "Equal Sums of Like Powers: On the Integer Solution of the Diophantine System." http://www.nease.net/~chin/eslp/

Gloden, A. *Mehrgeradige Gleichungen.* Groningen, Netherlands: Noordhoff, 1944.

Gloden, A. "Sur la multigrade $A_1, A_2, A_3, A_4, A_5 =^k B_1, B_2, B_3, B_4, B_5$ ($k = 1, 3, 5, 7$)." *Revista Euclides* **8**, 383–84, 1948.

Guy, R. K. *Unsolved Problems in Number Theory, 2nd ed.* New York: Springer-Verlag, p. 143, 1994.

Kraitchik, M. "Multigrade." §3.10 in *Mathematical Recreations.* New York: W. W. Norton, p. 79, 1942.

Madachy, J. S. *Madachy's Mathematical Recreations.* New York: Dover, pp. 171–73, 1979.

Moessner, A. and Gloden, A. "Einige Zahlentheoretische Untersuchungen und Resultate." *Bull. Sci. École Polytech. de Timisoara* **11**, 196–19, 1944.

Rivera, C. "Problems & Puzzles: Puzzle Multigrade Relations.-065." http://www.primepuzzles.net/puzzles/puzz_065.htm.

Weisstein, E. W. "Like Powers." MATHEMATICA NOTEBOOK LIKEPOWERS.M.

Multigraph

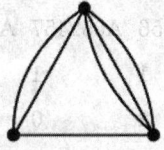

multigraph

A non-SIMPLE GRAPH in which no LOOPS are permitted, but multiple edges between any two nodes are.

See also HYPERGRAPH, PSEUDOGRAPH, SIMPLE GRAPH

References

Harary, F. *Graph Theory.* Reading, MA: Addison-Wesley, p. 10, 1994.

Skiena, S. *Implementing Discrete Mathematics: Combinatorics and Graph Theory with Mathematica.* Reading, MA: Addison-Wesley, p. 89, 1990.

Multilinear

A basis, form, function, etc., in two or more variables is said to be multilinear if it is linear in each variable separately.

See also BILINEAR FUNCTION, LINEAR OPERATOR, MULTILINEAR BASIS, MULTILINEAR FORM

Multilinear Basis

See also BILINEAR BASIS

Multimagic Series

A set n distinct numbers taken from the interval $[1, n^2]$ form a MAGIC SERIES if their sum is the nth MAGIC CONSTANT

$$M_n = \tfrac{1}{2} n(n^2 + 1)$$

(Kraitchik 1942, p. 143). If the sum of the kth powers of these numbers is the MAGIC CONSTANT of degree k for all $k \in [1, p]$, then they are said to form a pth order MULTIMAGIC SERIES. Here, the magic constant $M_n^{(j)}$ of degree k is defined as $1/n$ times the sum of the first n^2 kth powers,

$$M_n^{(k)} = \frac{1}{n} \sum_{i=1}^{n^2} i^k = \frac{H_{n^2}^{(-p)}}{n},$$

where $H_n^{(k)}$ is a HARMONIC NUMBER of order k.

For example $\{2, 8, 9, 15\}$ is bimagic since $2 + 8 + 9 + 15 = 34$ and $2^2 + 8^2 + 9^2 + 15^2 = 374$.

The numbers of magic series of various lengths n are gives in the following table for small orders k (Kraitchik 1942, p. 76).

n	$k = 1$	$k = 2$	$k = 3$	$k = 4$
Sloane	A052456	A052457	A052458	
1	1	1	1	1
2	2	0	0	0
3	8	0	0	0
4	86	2	2	0
5	1,394	8	2	0
6	32,134	98	0	0
7	957,332	1,844	0	0
8		38,039	115	
9			41	
10				
11			961	

See also Magic Series

References
Kraitchik, M. "Multimagic Squares." §7.10 in *Mathematical Recreations.* New York: W. W. Norton, pp. 176–78, 1942.
Sloane, N. J. A. Sequences A052456, A052457, and A052458 in "An On-Line Version of the Encyclopedia of Integer Sequences." http://www.research.att.com/~njas/sequences/eisonline.html.

Multimagic Square
A magic square is p-multimagic if the square formed by replacing each element by its kth power for $k = 1$, 2, ..., p is also magic. A 2-multimagic square is called a bimagic square, and a 3-multimagic square is called a trimagic square.

See also Bimagic Square, Magic Square, Trimagic Square

References
Kraitchik, M. "Multimagic Squares." §7.10 in *Mathematical Recreations.* New York: W. W. Norton, pp. 176–78, 1942.

Multinomial
An algebraic expression containing more than one term (cf., binomial). The term is also used to refer to a polynomial.

See also Binomial, Multinomial Coefficient, Multinomial Series, Polynomial

Multinomial Coefficient
The multinomial coefficients

$$(n_1, n_2, \ldots, n_k)! = \frac{(n_1 + n_2 + \cdots + n_k)!}{n_1! n_2! \cdots n_3!}$$

are the terms in the multinomial series expansion. The multinomial coefficient is returned by the *Mathematica* function `Multinomial[n1, n2, ...]`. The number of distinct permutations in a multiset of k distinct elements of multiplicity n_i $(1 \le i \le k)$ is $(n_1, \ldots, n_k)$ (Skiena 1990, p. 12). The multinomial coefficients satisfy

$$(n_1, n_2, n_3, \ldots) = (n_1 + n_2, n_3, \ldots)(n_1, n_2)$$

$$= (n_1 + n_2 + n_3, \ldots)(n_1, n_2, n_3) = \ldots$$

(Gosper 1972).

The content V of the d-dimensional region $\Sigma_{k=1}^{d} |x_k|^{p_k} < 1$ is given by

$$V = 2^d \left(\sum_{k=1}^{d} p_k^{-1}, p_1^{-1}, p_2^{-1}, \ldots, p_d^{-1} \right).$$

See also Binomial Coefficient, Choose, Dyson's Conjecture, Multichoose, Multinomial Series, Q-Multinomial Coefficient, Zeilberger-Bressoud Theorem

References
Abramowitz, M. and Stegun, C. A. (Eds.). "Multinomial Coefficients." §24.1.2 in *Handbook of Mathematical Functions with Formulas, Graphs, and Mathematical Tables, 9th printing.* New York: Dover, pp. 823–24, 1972.
Gosper, R. W. Item 42 in Beeler, M.; Gosper, R. W.; and Schroeppel, R. *HAKMEM.* Cambridge, MA: MIT Artificial Intelligence Laboratory, Memo AIM-239, p. 16, Feb. 1972.
Skiena, S. *Implementing Discrete Mathematics: Combinatorics and Graph Theory with Mathematica.* Reading, MA: Addison-Wesley, 1990.
Spiegel, M. R. *Theory and Problems of Probability and Statistics.* New York: McGraw-Hill, p. 113, 1992.

Multinomial Distribution
Let a set of random variates $X_1, X_2, \ldots, X_n$ have a probability function

$$P(X_1 = x_1, \ldots, X_n = x_n) = \frac{N!}{\prod_{i=1}^{n} x_i!} \prod_{i=1}^{n} \theta_i^{x_i} \quad (1)$$

where x_i are positive integers such that

$$\sum_{i=1}^{n} x_i = N, \quad (2)$$

and θ_i are constants with $\theta_i > 0$ and

$$\sum_{i=1}^{n} \theta_i = 1. \qquad (3)$$

Then the joint distribution of $X_1, ..., X_n$ is a multinomial distribution and $P(X_1 = x_1, ..., X_n = x_n)$ is given by the corresponding coefficient of the MULTINOMIAL SERIES

$$(\theta_1 + \theta_2 + ... + \theta_n)^N. \qquad (4)$$

In the words, if $X_1, X_2, ..., X_n$ are mutually independent events with $P(X_1) = \theta_1, ..., P(x_n) = \theta_n$. Then the probability that X_1 occurs x_1 times, ..., X_n occurs x_n times is given by

$$P_N(x_1, x_2, ..., x_n) = \frac{N!}{x_1! \cdots x_n!} \theta_1^{x_1} \cdots \theta_n^{x_n}. \qquad (5)$$

(Papoulis 1984, p. 75).

The MEAN and VARIANCE of X_i are

$$\mu_i = N\theta_i \qquad (6)$$

$$\sigma_i^2 = N\theta_i(1 - \theta_i). \qquad (7)$$

The COVARIANCE of X_i and X_j is

$$\sigma_{ij}^2 = -N\theta_i\theta_j. \qquad (8)$$

See also BINOMIAL DISTRIBUTION, MULTINOMIAL COEFFICIENT

References

Beyer, W. H. *CRC Standard Mathematical Tables, 28th ed.* Boca Raton, FL: CRC Press, p. 532, 1987.

Papoulis, A. *Probability, Random Variables, and Stochastic Processes, 2nd ed.* New York: McGraw-Hill, 1984.

Multinomial Series

A generalization of the BINOMIAL SERIES discovered by Johann Bernoulli and Leibniz.

$$(a_1 + a_2 + ... + a_k)^n$$
$$= \sum_{n_1, n_2, ..., n_k} \frac{n!}{n_1! n_2! ... n_k!} a_1^{n_1} a_2^{n_2} ... a_k^{n_k},$$

where $n \equiv n_1 + n_2 + ... + n_k$. The multinomial series arises in a generalization of the BINOMIAL DISTRIBUTION called the MULTINOMIAL DISTRIBUTION.

See also BINOMIAL SERIES, MULTINOMIAL DISTRIBUTION

Multinomial Theorem

MULTINOMIAL SERIES

Multinormal Distribution

GAUSSIAN MULTIVARIATE DISTRIBUTION

Multiperfect Number

A number n is k-multiperfect (also called a k-MULTIPLY PERFECT NUMBER or k-PLUPERFECT NUMBER) if

$$\sigma(n) = kn$$

for some INTEGER $k > 2$, where $\sigma(n)$ is the DIVISOR FUNCTION. The value of k is called the CLASS. The special case $k = 2$ corresponds to PERFECT NUMBERS P_2, which are intimately connected with MERSENNE PRIMES (Sloane's A000396). The number 120 was long known to be 3-multiply perfect (P_3) since

$$\sigma(120) = 3 \cdot 120.$$

The following table gives the first few P_n for $n = 2, 3, ..., 6$.

2	A000396	6, 28, 496, 8128, ...,
3	A005820	120, 672, 523776, 459818240, 1476304896, 51001180160
4	A027687	30240, 32760, 2178540, 23569920, ...
5	A046060	14182439040, 31998395520, 518666803200, ...
6	A046061	154345556085770649600, 9186050031556349952000, ...

In 1900-901, Lehmer proved that P_3 has at least three distinct PRIME FACTORS, P_4 has at least four, P_5 at least six, P_6 at least nine, and P_7 at least 14.

As of 1911, 251 pluperfect numbers were known (Carmichael and Mason 1911). As of 1929, 334 pluperfect numbers were known, many of them found by Poulet. Franqui and García (1953) found 63 additional ones (five P_5s, 29 P_6s, and 29 P_7s), several of which were known to Poulet but had not been published, bringing the total to 397. Brown (1954) discovered 110 pluperfects, including 31 discovered but not published by Poulet and 25 previously published by Franqui and García (1953), for a total of 482. Franqui and García (1954) subsequently discovered 57 additional pluperfects (3 P_6s, 52 P_7s, and 2 P_8s), increasing the total known to 539.

An outdated database is maintained by R. Schroeppel, who lists 2,094 multiperfects, and up-to-date lists by J. L. Moxham (2000b) and A. Flammenkamp. It is believed that all multiperfect numbers of index 3, 4, 5, 6, and 7 are known. The number of known n-multiperfect numbers are 1, 37, 6, 36, 65, 245, 516, 1134, 1982, 183, 0, 0, ... (Moxham 2000b, Flammenkamp, Woltman 2000). Moxham (2000a) found the largest known multiperfect number, approximately equal to 7.3×10^{1345}, on Feb. 13, 2000.

If n is a P_5 number such that $3 \nmid n$, then $3n$ is a P_4 number. If $3n$ is a P_{4k} number such that $3 \nmid n$, then n is a P_{3k} number. If n is a P_3 number such that 3 (but not 5 and 9) DIVIDES n, then $45n$ is a P_4 number.

See also E-MULTIPERFECT NUMBER, FRIENDLY PAIR, HYPERPERFECT NUMBER, INFINARY MULTIPERFECT NUMBER, MERSENNE PRIME, PERFECT NUMBER, UNITARY MULTIPERFECT NUMBER

References

Beck, W. and Najar, R. "A Lower Bound for Odd Triperfects." *Math. Comput.* **38**, 249–51, 1982.
Brown, A. L. "Multiperfect Numbers." *Scripta Math.* **20**, 103–06, 1954.
Cohen, G. L. and Hagis, P. Jr. "Results Concerning Odd Multiperfect Numbers." *Bull. Malaysian Math. Soc.* **8**, 23–6, 1985.
Dickson, L. E. *History of the Theory of Numbers, Vol. 1: Divisibility and Primality.* New York: Chelsea, pp. 33–8, 1952.
Flammenkamp, A. "Multiply Perfect Numbers." http://www.uni-bielefeld.de/~achim/mpn.html.
Franqui, B. and García, M. "Some New Multiply Perfect Numbers." *Amer. Math. Monthly* **60**, 459–62, 1953.
Franqui, B. and García, M. "57 New Multiply Perfect Numbers." *Scripta Math.* **20**, 169–71, 1954.
Guy, R. K. "Almost Perfect, Quasi-Perfect, Pseudoperfect, Harmonic, Weird, Multiperfect and Hyperperfect Numbers." §B2 in *Unsolved Problems in Number Theory, 2nd ed.* New York: Springer-Verlag, pp. 45–3, 1994.
Helenius, F. W. "Multiperfect Numbers (MPFNs)." http://home.netcom.com/~fredh/mpfn/.
Madachy, J. S. *Madachy's Mathematical Recreations.* New York: Dover, pp. 149–51, 1979.
Moxham, J. L. "New Largest MPFN." mpfn@cs.arizona.edu posting, 13 Feb. 2000a.
Moxham, J. L. "New MPFNs for per3.6 server." mpfn@cs.arizona.edu posting, 19 Sep 2000b.
Poulet, P. *La Chasse aux nombres, Vol. 1.* Brussels, pp. 9–7, 1929.
Schroeppel, R. "Multiperfect Numbers-Multiply Perfect Numbers-Pluperfect Numbers-MPFNs." Rev. Dec. 13, 1995. ftp://ftp.cs.arizona.edu/xkernel/rcs/mpfn.html.
Schroeppel, R. (moderator). mpfn mailing list. e-mail rcs@cs.arizona.edu to subscribe.
Sloane, N. J. A. Sequences A000396/M4186, A005820/M5376, A027687, A046060, and A046061 in "An On-Line Version of the Encyclopedia of Integer Sequences." http://www.research.att.com/~njas/sequences/eisonline.html.
Woltman, G. "5 new MPFNs." mpfn@cs.arizona.edu posting, 23 Sep 2000.

Multiple

A multiple of a number x is any quantity $y = nx$ with n an integer. If x and y are integers, then x is called a FACTOR y.

Multiple Analysis of Variance

MANOVA

Multiple-Angle Formulas

Expressions OF THE FORM $\sin(nx)$, $\cos(nx)$, and $\tan(nx)$ can be expressed in terms of $\sin x$ and $\cos x$ only using the EULER FORMULA and BINOMIAL THEOREM. For $\sin(nx)$,

$$\sin(nx) = \frac{e^{inx} - e^{-inx}}{2i} = \frac{(e^{ix})^n - (e^{-ix})^n}{2i}$$

$$= \frac{(\cos x + i \sin x)^n - (\cos x - i \sin x)^n}{2i}$$

$$= \sum_{k=0}^{n} \binom{n}{k} \frac{\cos^k x (i \sin x)^{n-k} - \cos^k x(-i \sin x)^{n-k}}{2i}$$

$$= \sum_{k=0}^{n} \binom{n}{k} \cos^k x \sin^{n-k} x \, \frac{i^{n-k} - (-i)^{n-k}}{2i}$$

$$= \sum_{k=0}^{n} \binom{n}{k} \cos^k x \sin^{n-k} x \sin[\tfrac{1}{2}(n-k)\pi]. \quad (1)$$

Particular cases for multiple angle formulas for $\sin x$ are given by

$$\sin(2x) = 2 \sin x \cos x \quad (2)$$

$$\sin(3x) = 3 \sin x - 4 \sin^3 x \quad (3)$$

$$\sin(4x) = 4 \sin x \cos x - 8 \sin^3 x \cos x \quad (4)$$

$$\sin(5x) = 5 \cos^4 \sin x - 10 \cos^2 x \sin^3 x + \sin^5 x. \quad (5)$$

The function $\sin(nx)$ can also be expressed as a polynomial in $\sin x$ (for n odd) or $\cos x$ times a polynomial in $\sin x$ as

$$\sin(nx) = \begin{cases} (-1)^{(n-1)/2} T_n(\sin x) & \text{for } n \text{ odd} \\ (-1)^{n/2-1} \cos x U_n(\sin x) & \text{for } n \text{ even,} \end{cases} \quad (6)$$

where T_n is a CHEBYSHEV POLYNOMIAL OF THE FIRST KIND and U_n is a CHEBYSHEV POLYNOMIAL OF THE SECOND KIND. The first few cases are

$$\sin(2x) = 2 \cos x \sin x \quad (7)$$

$$\sin(3x) = 3 \sin x - 4 \sin^3 x \quad (8)$$

$$\sin(4x) = \cos x(4 \sin x - 8 \sin^3 x) \quad (9)$$

$$\sin(5x) = 5 \sin x - 20 \sin^3 x + 16 \sin^5 x. \quad (10)$$

Similarly, $\sin(nx)$ can be expressed as $\sin x$ times a polynomial in $\cos x$ as

$$\sin(nx) = \sin x \, U_{n-1}(\cos x). \quad (11)$$

The first few cases are

$$\sin(2x) = 2 \cos x \sin x \quad (12)$$

$$\sin(3x) = \sin x(-1 + 4 \cos^2 x) \quad (13)$$

$$\sin(4x) = \sin x(-4 \cos x + 8 \cos^3 x) \quad (14)$$

$$\sin(5x) = \sin x(1 - 12 \cos^2 x + 16 \cos^4 x). \quad (15)$$

Bromwich (1991) gave the formula

$$\sin(na) =$$

$$\begin{cases} nx - \dfrac{n(n^2 - 1^2)x^3}{3!} + \dfrac{n(n^2 - 1^2)(n^2 - 3^2)x^5}{5!} - \cdots \\ \quad \text{for } n \text{ odd} \\ n\cos a\left[x - \dfrac{(n^2 - 2^2)x^3}{3!} + \dfrac{(n^2 - 2^2)(n^2 - 4^2)x^5}{5!} - \cdots\right] \\ \quad \text{for } n \text{ even}, \end{cases} \tag{16}$$

where $x = \sin a$.

For $\cos(nx)$, the multiple-angle formula can be derived as

$$\cos(nx) = \frac{e^{inx} + e^{-inx}}{2i} = \frac{(e^{ix})^n + (e^{-ix})^n}{2}$$

$$= \frac{(\cos x + i\sin x)^n + (\cos x - i\sin x)^n}{2}$$

$$= \sum_{k=0}^{n} \binom{n}{k} \frac{\cos^k x(i\sin x)^{n-k} + \cos^k x(-i\sin x)^{n-k}}{2}$$

$$= \sum_{k=0}^{n} \binom{n}{k} \cos^k x \sin^{n-k} x \, \frac{i^{n-k} + (-i)^{n-k}}{2}$$

$$= \sum_{k=0}^{n} \binom{n}{k} \cos^k x \sin^{n-k} x \cos\left[\tfrac{1}{2}(n-k)\pi\right]. \tag{17}$$

The first few values are

$$\cos(2x) = \cos^2 x - \sin^2 x \tag{18}$$

$$\cos(3x) = 4\cos^3 x - 3\cos x \sin x \tag{19}$$

$$\cos(4x) = \cos^4 x - 6\cos^2 x \sin^2 x + \sin^4 x \tag{20}$$

$$\cos(5x) = \cos^5 x - 10\cos^3 x \sin^2 x + 5\cos x \sin^4 x. \tag{21}$$

The function $\cos(nx)$ can also be expressed as a polynomial in $\sin x$ (for n even) or $\cos x$ times a polynomial in $\sin x$ as

$$\cos(nx) = \begin{cases} (-1)^{n-1/2}\cos x \, U_{n-1}(\sin x) & \text{for } n \text{ odd} \\ (-1)^{n/2}T_n(\sin x) & \text{for } n \text{ even}. \end{cases} \tag{22}$$

The first few cases are

$$\cos(2x) = 1 - 2\sin^2 x \tag{23}$$

$$\cos(3x) = \cos x(1 - 4\sin^2 x) \tag{24}$$

$$\cos(4x) = \cos x(1 - 12\sin^2 x + 16\sin^4 x) \tag{25}$$

$$\cos(5x) = 1 - 8\sin^2 x + 8\sin^4 x. \tag{26}$$

Similarly, $\cos(nx)$ can be expressed as a polynomial in $\cos x$ as

$$\cos(nx) = T_n(\cos x) \tag{27}$$

The first few cases are

$$\cos(2x) = -1 + 2\cos^2 x \tag{28}$$

$$\cos(3x) = -3\cos x + 4\cos^3 x \tag{29}$$

$$\cos(4x) = 1 - 8\cos^2 x + 8\cos^4 x \tag{30}$$

$$\cos(5x) = 5\cos x - 20\cos^3 x + 16\cos^5 x. \tag{31}$$

Bromwich (1991) gave the formula

$$\cos(na) =$$

$$\begin{cases} \cos a\left[1 - \dfrac{(n^2 - 1^2)x^2}{2!} + \dfrac{(n^2 - 1^2)(n^2 - 3^2)x^4}{4!} - \cdots\right] \\ \quad n \text{ odd} \\ 1 - \dfrac{n^2 x^2}{2!} + \dfrac{n^2(n^2 - 2^2)x^4}{4!} - \cdots \quad n \text{ even}, \end{cases} \tag{32}$$

where $x = \sin a$.

The first few multiple-angle formulas for $\tan(nx)$ are

$$\tan(2x) = \frac{2\tan x}{1 - \tan^2 x} \tag{33}$$

$$\tan(3x) = \frac{3\tan x - \tan^3 x}{1 - 3\tan^2 x} \tag{34}$$

$$\tan(4x) = \frac{4\tan x - 4\tan^3 x}{1 - 6\tan^2 x + \tan^4 x} \tag{35}$$

are given by Beyer (1987, p. 139) for up to $n = 6$.

Multiple angle formulas can also be written using the RECURRENCE RELATIONS

$$\sin(nx) = 2\sin[(n-1)x]\cos x - \sin[(n-2)x] \tag{36}$$

$$\cos(nx) = 2\cos[(n-1)x]\cos x - \cos[(n-2)x] \tag{37}$$

$$\tan(nx) = \frac{\tan[(n-1)x] + \tan x}{1 - \tan[(n-1)x]\tan x}. \tag{38}$$

See also DOUBLE-ANGLE FORMULAS, HALF-ANGLE FORMULAS, HYPERBOLIC FUNCTIONS, PROSTHAPHAERESIS FORMULAS, TRIGONOMETRIC ADDITION FORMULAS, TRIGONOMETRIC FUNCTIONS, TRIGONOMETRY

References

Beyer, W. H. *CRC Standard Mathematical Tables, 28th ed.* Boca Raton, FL: CRC Press, 1987.

Bromwich, T. J. I'a. and MacRobert, T. M. *An Introduction to the Theory of Infinite Series, 3rd ed.* New York: Chelsea, pp. 202–07, 1991.

Multiple-Free Set

DOUBLE-FREE SET, SUM-FREE SET, TRIPLE-FREE SET

Multiple Integral

A set of integrals taken over $n > 1$ variables

$$\underbrace{\int \ldots \int}_{n} f(x_1, \ldots, x_n) \, dx_1 \ldots dx_n \qquad (1)$$

is called a multiple integral. An nth order integral corresponds, in general, to an n-D VOLUME (CONTENT), with $n = 2$ corresponding to an AREA. In an indefinite multiple integral, the order in which the integrals are carried out can be varied at will; for definite multiple integrals, care must be taken to correctly transform the limits if the order is changed.

See also FUBINI THEOREM, INTEGRAL, MONTE CARLO INTEGRATION, REPEATED INTEGRAL

References

Kaplan, W. "Double Integrals" and "Triple Integrals and Multiple Integrals in General." §4.3–.4 in *Advanced Calculus, 4th ed.* Reading, MA: Addison-Wesley, pp. 228–35, 1991.
Press, W. H.; Flannery, B. P.; Teukolsky, S. A.; and Vetterling, W. T. "Multidimensional Integrals." §4.6 in *Numerical Recipes in FORTRAN: The Art of Scientific Computing, 2nd ed.* Cambridge, England: Cambridge University Press, pp. 155–58, 1992.

Multiple Point

MULTIPLE ROOT

Multiple Regression

A REGRESSION giving conditional expectation values of a given variable in terms of two or more other variables.

See also LEAST SQUARES FITTING, MULTIVARIATE ANALYSIS, NONLINEAR LEAST SQUARES FITTING

References

Chatterjee, S.; Hadi, A.; and Price, B. "Multiple Linear Regression." Ch. 3 in *Regression Analysis by Example, 3rd ed.* New York: Wiley, pp. 51–4, 2000.
Edwards, A. L. *Multiple Regression and the Analysis of Variance and Covariance.* San Francisco, CA: W. H. Freeman, 1979.

Multiple Root

A ROOT with MULTIPLICITY $n \geq 2$, also called a multiple point.

See also MULTIPLICITY, ROOT, SIMPLE ROOT

References

Krantz, S. G. "Zero of Order n." §5.1.3 in *Handbook of Complex Analysis.* Boston, MA: Birkhäuser, p. 70, 1999.

Multiple-Valued Function

A function for which several distinct functional values correspond (as a result of different continuations) to one and the same point (Knopp 1996, p. 94).

See also BRANCH CUT, RIEMANN SURFACE, SINGLE-VALUED FUNCTION

References

Knopp, K. "Multiple-Valued Functions." Section II in *Theory of Functions Parts I and II, Two Volumes Bound as One, Part II.* New York: Dover, pp. 93–46, 1996.

Multiplicand

A quantity that is multiplied by another (the MULTIPLIER). For example, in the expression $a \times b$, b is the multiplicand.

See also MULTIPLICATION, MULTIPLIER

Multiplication

In simple algebra, multiplication is the process of calculating the result when a number a is taken b times. The result of a multiplication is called the PRODUCT of a and b, and each of the numbers a and b is called a FACTOR of the PRODUCT ab. Multiplication is denoted $a \times b$, $a \cdot b$, $(a)(b)$, or simply ab. The symbol $\times$ is known as the MULTIPLICATION SIGN. Normal multiplication is ASSOCIATIVE, COMMUTATIVE, and DISTRIBUTIVE.

More generally, multiplication can also be defined for other mathematical objects such as GROUPS, MATRICES, SETS, and TENSORS.

Karatsuba and Ofman (1962) discovered that multiplication of two n digit numbers can be done with a BIT COMPLEXITY of less than n^2 using an algorithm now known as KARATSUBA MULTIPLICATION.

Multiplication of numbers x and y carried out in base b can be implemented in *Mathematica* as

```
Multiply[{x_,y_},b_]:=FromDigits[
      ListConvolve[IntegerDigits[x,    b],
IntegerDigits[y, b],
   {1, -1}, 0], b]
```

See also ADDITION, BIT COMPLEXITY, COMPLEX MULTIPLICATION, DIVISION, FACTOR, KARATSUBA MULTIPLICATION, MATRIX MULTIPLICATION, MULTIPLICAND, MULTIPLIER, PRODUCT, RUSSIAN MULTIPLICATION, SCALAR MULTIPLICATION, SUBTRACTION, TIMES

References

Beck, G. "Long Multiplication and Division." MATHEMATICA NOTEBOOK LONGDIVISION.NB.
Cundy, H. M. "What Is $\times$?" *Math. Gaz.* **43**, 101, 1959.
Karatsuba, A. and Ofman, Yu. "Multiplication of Many-Digital Numbers by Automatic Computers." *Doklady Akad. Nauk SSSR* **145**, 293–94, 1962. Translation in *Physics-Doklady* **7**, 595–96, 1963.

Multiplication Magic Square

128	1	32
4	16	64
8	256	2

A square which is magic under multiplication instead of addition (the operation used to define a conventional MAGIC SQUARE) is called a multiplication magic square. Unlike (normal) MAGIC SQUARES, the n^2 entries for an nth order multiplicative magic square are not required to be consecutive. The above multiplication magic square has a multiplicative magic constant of 4,096.

See also ADDITION-MULTIPLICATION MAGIC SQUARE, MAGIC SQUARE

References
Hunter, J. A. H. and Madachy, J. S. "Mystic Arrays." Ch. 3 in *Mathematical Diversions*. New York: Dover, pp. 30–1, 1975.
Madachy, J. S. *Madachy's Mathematical Recreations*. New York: Dover, pp. 89–1, 1979.

Multiplication Principle

If one event can occur in m ways and a second can occur independently of the first in n ways, then the two events can occur in mn ways.

Multiplication Sign

The symbol $\times$ used to denote MULTIPLICATION, i.e., $a \times b$ denotes a times b.

The symbol $\times$ is also used to denote a GROUP DIRECT PRODUCT, a CARTESIAN PRODUCT, or a direct product in the appropriate category (such as a Cartesian product of manifolds when it is implied that the smooth structure is the natural product structure.) The similar symbol $\otimes$ is reserved for a tensor product, which may rear its head in several guises, representations, bundles, modules.

Multiplication Table

A multiplication table is an array showing the result of applying a BINARY OPERATOR to elements of a given set S.

	1	2	3	4	5	6	7	8	9	10
1	1	2	3	4	5	6	7	8	9	10
2	2	4	6	8	10	12	14	16	18	20
3	3	6	9	12	15	18	21	24	27	30
4	4	8	12	16	20	24	28	32	36	40
5	5	10	15	20	25	30	35	40	45	50
6	6	12	18	24	30	36	42	48	54	60
7	7	14	21	28	35	42	49	56	63	70
8	8	16	24	32	40	48	56	64	72	80
9	9	18	27	36	45	54	63	72	81	90
10	10	20	30	40	50	60	70	80	90	100

See also BINARY OPERATOR, TRUTH TABLE

Multiplicative Character

A continuous HOMEOMORPHISM of a GROUP into the NONZERO COMPLEX NUMBERS. A multiplicative character ω gives a REPRESENTATION on the 1-D SPACE $\mathbb{C}$ of COMPLEX NUMBERS, where the REPRESENTATION action by $g \in G$ is multiplication by $\omega(g)$. A multiplicative character is UNITARY if it has ABSOLUTE VALUE 1 everywhere.

See also GRÖSSENCHARAKTER, UNITARY MULTIPLICATIVE CHARACTER

References
Knapp, A. W. "Group Representations and Harmonic Analysis, Part II." *Not. Amer. Math. Soc.* **43**, 537–49, 1996.

Multiplicative Digital Root

Consider the process of taking a number, multiplying its DIGITS, then multiplying the DIGITS of numbers derived from it, etc., until the remaining number has only one DIGIT. The number of multiplications required to obtain a single DIGIT from a number n is called the MULTIPLICATIVE PERSISTENCE of n, and the DIGIT obtained is called the multiplicative digital root of n.

For example, the sequence obtained from the starting number 9876 is (9876, 3024, 0), so 9876 has a MULTIPLICATIVE PERSISTENCE of two and a multiplicative digital root of 0. The multiplicative digital roots of the first few positive integers are 1, 2, 3, 4, 5, 6, 7, 8, 9, 0, 1, 2, 3, 4, 5, 6, 7, 8, 9, 0, 2, 4, 6, 8, 0, 2, 4, 6, 8, 0, 3, 6, 9, 2, 5, 8, 2, ... (Sloane's A031347).

n	Sloane	numbers having multiplicative digital root n
0	A034048	0, 10, 20, 25, 30, 40, 45, 50, 52, 54, 55, 56, 58, ...
1	A002275	1, 11, 111, 1111, 11111, 111111, 1111111, 11111111, ...
2	A034049	2, 12, 21, 26, 34, 37, 43, 62, 73, 112, 121, 126, ...
3	A034050	3, 13, 31, 113, 131, 311, 1113, 1131, 1311, 3111, ...
4	A034051	4, 14, 22, 27, 39, 41, 72, 89, 93, 98, 114, 122, ...

5	A034052	5, 15, 35, 51, 53, 57, 75, 115, 135, 151, 153, 157, ...
6	A034053	6, 16, 23, 28, 32, 44, 47, 48, 61, 68, 74, 82, 84, ...
7	A034054	7, 17, 71, 117, 171, 711, 1117, 1171, 1711, 7111, ...
8	A034055	8, 18, 24, 29, 36, 38, 42, 46, 49, 63, 64, 66, 67, ...
9	A034056	9, 19, 33, 91, 119, 133, 191, 313, 331, 911, 1119, ...

See also ADDITIVE PERSISTENCE, DIGITADDITION, DIGITAL ROOT, MULTIPLICATIVE PERSISTENCE

References

Sloane, N. J. A. Sequences A002275, A031347, A034048, A034049, A034050, A034051, A034052, A034053, A034054, A034055, and A034056 in "An On-Line Version of the Encyclopedia of Integer Sequences." http://www.research.att.com/~njas/sequences/eisonline.html.

Multiplicative Function

A function $f(m)$ is called multiplicative if $(m, m') = 1$ (i.e., the statement that m and m' are RELATIVELY PRIME) implies

$$f(mm') = f(m)f(m').$$

Examples of multiplicative functions are the MÖBIUS FUNCTION and TOTIENT FUNCTION.

See also COMPLETELY MULTIPLICATIVE FUNCTION, MÖBIUS FUNCTION, QUADRATIC RESIDUE, TOTIENT FUNCTION

Multiplicative Inverse

The multiplicative inverse of a REAL or COMPLEX NUMBER z is its RECIPROCAL $1/z$. For complex $z = x + iy$,

$$\frac{1}{z} = \frac{1}{x + iy} = \frac{x}{x^2 + y^2} - i\frac{y}{x^2 + y^2}.$$

Multiplicative Number Theory

See also ADDITIVE NUMBER THEORY, NUMBER THEORY

References

Davenport, H. *Multiplicative Number Theory*, 2nd ed. New York: Springer-Verlag, p. 110, 1980.
Montgomery, H. L. *Topics in Multiplicative Number Theory.* New York: Springer-Verlag, 1971.

Multiplicative Order

Let n be a positive number having PRIMITIVE ROOTS. If g is a PRIMITIVE ROOT of n, then the numbers $1, g, g^2, ..., g^{\phi(n)-1}$ form a REDUCED RESIDUE SYSTEM modulo n, where $\phi(n)$ is the TOTIENT FUNCTION. In this set, there are $\phi(\phi(n))$ PRIMITIVE ROOTS, and these are the numbers g^c, where c is RELATIVELY PRIME to $\phi(n)$. If a is an arbitrary integer RELATIVELY PRIME to n, then there exists among the numbers $0, 1, 2, ..., \phi(n-1)$ exactly one number μ such that

$$a \equiv g^\mu \pmod{n}. \tag{1}$$

The number μ is then called the generalized multiplicative order of a with respect to the base g modulo n. Note that Nagell (1951, p. 112) instead uses the term "index" and writes

$$\mu = \text{ind}_g\, a \pmod{n}. \tag{2}$$

For example, the number 7 in the least positive PRIMITIVE ROOT of $n = 41$, and since $15 \equiv 7^3 \pmod{41}$, the number 15 has multiplicative order 3 with respect to base 7 (modulo 41) (Nagell 1951, p. 112). The generalized multiplicative order is implemented in *Mathematica* as MultiplicativeOrder[a, n, {g1}], or more generally as MultiplicativeOrder[a, n, {g1, g2, ...}].

If the PRIMITIVE ROOTS $g_1 = -1$ and $g_2 = 1$ are chosen, the resulting function is called the SUBORDER FUNCTION and is denoted $\text{sord}_n(a)$. If the single PRIMITIVE ROOT $g_1 = 1$ is chosen, then the function reduces to "the" (i.e., ungeneralized) multiplicative order, denoted $\text{ord}_n(a)$, implemented in *Mathematica* as MultiplicativeOrder[a, n]. This function is sometimes also known as the discrete logarithm (or, more confusingly, as the "index," a term which Nagell applied to the case of general g).

See also CONGRUENCE, HAUPT-EXPONENT, ORDER (MODULO), PRIMITIVE ROOT, SUBORDER FUNCTION

References

Nagell, T. "The Index Calculus." §33 in *Introduction to Number Theory.* New York: Wiley, pp. 111–15, 1951.
Odlyzko, A. "Discrete Logarithms: The Past and the Future." http://www.research.att.com/~amo/doc/discrete.logs.future.ps.

Multiplicative Perfect Number

A number n for which the PRODUCT of DIVISORS is equal to n^2. The first few are 1, 6, 8, 10, 14, 15, 21, 22, ... (Sloane's A007422).

See also PERFECT NUMBER

References

Sloane, N. J. A. Sequences A007422/M4068 in "An On-Line Version of the Encyclopedia of Integer Sequences." http://www.research.att.com/~njas/sequences/eisonline.html.

Multiplicative Persistence

Multiply all the digits of a number n by each other, repeating with the product until a single DIGIT is obtained. The number of steps required is known as the multiplicative persistence, and the final DIGIT obtained is called the MULTIPLICATIVE DIGITAL ROOT of n.

For example, the sequence obtained from the starting number 9876 is (9876, 3024, 0), so 9876 has an multiplicative persistence of two and a MULTIPLICATIVE DIGITAL ROOT of 0. The multiplicative persistences of the first few positive integers are 0, 0, 0, 0, 0, 0, 0, 0, 0, 1, 1, 1, 1, 1, 1, 1, 1, 1, 1, 1, 1, 1, 1, 1, 2, 2, 2, 2, 2, 1, 1, 1, 1, 2, 2, 2, 2, 2, 2, 3, 1, 1, ... (Sloane's A031346). The smallest numbers having multiplicative persistences of 1, 2, ... are 10, 25, 39, 77, 679, 6788, 68889, 2677889, 26888999, 3778888999, 277777788888899, ... (Sloane's A003001; Wells 1986, p. 78). There is no number $< 10^{50}$ with multiplicative persistence > 11 (Wells 1986, p. 78). It is conjectured that the maximum number lacking the DIGIT 1 with persistence 11 is

$$77777733332222222222222222222$$

There is a stronger conjecture that there is a maximum number lacking the DIGIT 1 for each persistence ≥ 2.

The maximum multiplicative persistence in base 2 is 1. It is conjectured that all powers of $2 > 2^{15}$ contain a 0 in base 3, which would imply that the maximum persistence in base 3 is 3 (Guy 1994).

The multiplicative persistence of an n-DIGIT number is also called its LENGTH. The maximum lengths for $n = 1$-, 2-, 3-, ..., digit numbers are 0, 4, 5, 6, 7, 7, 8, 9, 9, 10, 10, 10, ... (Sloane's A014553; Beeler 1972, Gottlieb 1969–970). The numbers of n-digit numbers having maximal multiplicative persistence for $n = 1$, 2, ..., are 10 (which includes the number 0), 1, 9, 12, 20, 2430, ... (Sloane's A046148). The smallest n-digit numbers with maximal multiplicative persistence are 0, 77, 679, 6788, 68889, 168889, ... (Sloane's A046149). The largest n-digit numbers with maximal multiplicative persistence are 9, 77, 976, 8876, 98886, 997762, ... (Sloane's A046150). The number of distinct n-digit numbers (except for 0s) are given by $\binom{10+n-1}{n} - 1$ which, for $n = 1$, 2, 3, ..., gives 54, 219, 714, 2001, 5004, 11439, ... (Sloane's A035927).

The concept of multiplicative persistence can be generalized to multiplying the kth powers of the digits of a number and iterating until the result remains constant. All numbers other than REPUNITS, which converge to 1, converge to 0. The number of iterations required for the kth powers of a number's digits to converge to 0 is called its k-multiplicative persistence. The following table gives the n-multiplicative persistences for the first few positive integers.

n	Sloane	n-Persistences
2	Sloane's A031348	0, 7, 6, 6, 3, 5, 5, 4, 5, 1, ...
3	Sloane's A031349	0, 4, 5, 4, 3, 4, 4, 3, 3, 1, ...,
4	Sloane's A031350	0, 4, 3, 3, 3, 3, 2, 2, 3, 1, ...
5	Sloane's A031351	0, 4, 4, 2, 3, 3, 2, 3, 2, 1, ...
6	Sloane's A031352	0, 3, 3, 2, 3, 3, 3, 3, 3, 1, ...
7	Sloane's A031353	0, 4, 3, 3, 3, 3, 3, 2, 3, 1, ...
8	Sloane's A031354	0, 3, 3, 3, 2, 4, 2, 3, 2, 1, ...
9	Sloane's A031355	0, 3, 3, 3, 3, 2, 3, 2, 3, 2, 1, ...
10	Sloane's A031356	0, 2, 2, 2, 3, 2, 3, 2, 2, 1, ...

Erdos suggested ignoring all zeros and showed that at most $c \ln \ln n$ steps are needed to reduce n to a single digit, where c depends on the base.

The smallest primes with multiplicative persistences $n = 1$, 2, 3, ... are 2, 29, 47, 277, 769, 8867, 186889, 2678789, 26899889, 3778888999, 277777788888989, ... (Sloane's A046500).

See also 196-ALGORITHM, ADDITIVE PERSISTENCE, DIGITADDITION, DIGITAL ROOT, KAPREKAR NUMBER, LENGTH (NUMBER), MULTIPLICATIVE DIGITAL ROOT, NARCISSISTIC NUMBER, RECURRING DIGITAL INVARIANT

References

Beeler, M. Item 56 in Beeler, M.; Gosper, R. W.; and Schroeppel, R. *HAKMEM.* Cambridge, MA: MIT Artificial Intelligence Laboratory, Memo AIM-239, p. 22, Feb. 1972.

Gottlieb, A. J. Problems 28–9 in "Bridge, Group Theory, and a Jigsaw Puzzle." *Techn. Rev.* **72**, unpaginated, Dec. 1969.

Gottlieb, A. J. Problem 29 in "Integral Solutions, Ladders, and Pentagons." *Techn. Rev.* **72**, unpaginated, Apr. 1970.

Guy, R. K. "The Persistence of a Number." §F25 in *Unsolved Problems in Number Theory, 2nd ed.* New York: Springer-Verlag, pp. 262–63, 1994.

Rivera, C. "Problems & Puzzles: Puzzle Primes & Persistence.-022." http://www.primepuzzles.net/puzzles/puzz_022.htm.

Sloane, N. J. A. "The Persistence of a Number." *J. Recr. Math.* **6**, 97–8, 1973.

Sloane, N. J. A. Sequences A003001/M4687, A014553, A031346, and A046500 in "An On-Line Version of the Encyclopedia of Integer Sequences." http://www.research.att.com/~njas/sequences/eisonline.html.

Wells, D. *The Penguin Dictionary of Curious and Interesting Numbers.* Middlesex, England: Penguin Books, p. 78, 1986.

Multiplicative Primitive Residue Class Group
MODULO MULTIPLICATION GROUP

Multiplicity
The word multiplicity is a general term meaning "the number of values for which a given condition holds." For example, the term is used to refer to the value of the TOTIENT VALENCE FUNCTION or the number of times a given polynomial equation has a ROOT at a given point.

Let z_0 be a ROOT of a function f, and let n be the least positive integer n such that $f^{(n)}(z_0) \neq 0$. Then the POWER SERIES of f about z_0 begins with the nth term,

$$f(z) = \sum_{j=n}^{\infty} \frac{1}{j!} \left.\frac{\partial^j f}{\partial z^j}\right|_{z=z_0} (z - z_0)^j,$$

and f is said to have a ROOT of multiplicity (or "order") n. If $n = 1$, the ROOT is called a SIMPLE ROOT (Krantz 1999, p. 70).

See also DEGENERATE, MULTIPLE ROOT, NOETHER'S FUNDAMENTAL THEOREM, ROOT, SIMPLE ROOT, TOTIENT VALENCE FUNCTION

References
Krantz, S. G. "Zero of Order n." §5.1.3 in *Handbook of Complex Analysis*. Boston, MA: Birkhäuser, p. 70, 1999.

Multiplier
A quantity by which another (the MULTIPLICAND) is multiplied. For example, in the expression $a \times b$, a is the multiplier.

The term "multiplier" also has a special meaning in the theory of MODULAR FUNCTION.

See also MODULAR FUNCTION, MULTIPLICAND, MULTIPLICATION

Multiply Connected

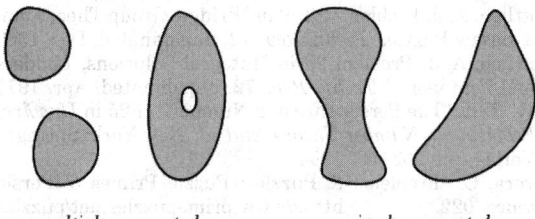

multiply connected   *simply connected*

A set which is CONNECTED but not SIMPLY CONNECTED is called multiply connected. A SPACE is n-MULTIPLY CONNECTED if it is $(n-1)$-connected and if every MAP from the n-SPHERE into it extends continuously over the $(n+1)$-DISK

A theorem of Whitehead says that a SPACE is infinitely connected IFF it is contractible.

See also CONNECTIVITY, LOCALLY PATHWISE-CONNECTED PATHWISE-CONNECTED, SIMPLY CONNECTED

Multiply Perfect Number
MULTIPERFECT NUMBER

Multipolynomial Quadratic Sieve
QUADRATIC SIEVE

Multisection
SERIES MULTISECTION

Multiset
A SET-like object in which order is ignored, but multiplicity is explicitly significant. Therefore, multisets $\{1, 2, 3\}$ and $\{2, 1, 3\}$ are equivalent, but $\{1, 1, 2, 3\}$ and $\{1, 2, 3\}$ differ.

See also LIST, MULTICHOOSE, MULTINOMIAL COEFFICIENT, SET

References
Skiena, S. *Implementing Discrete Mathematics: Combinatorics and Graph Theory with Mathematica*. Reading, MA: Addison-Wesley, p. 12, 1990.

Multistable
A structure such as a polyhedron which can change form from one stable configuration to another with only a slight transient nondestructive elastic stretch (Goldberg 1978). The simplest example of a polyhedron having multistable forms is Wunderlich's bistable JUMPING OCTAHEDRON (Cromwell 1991, pp. 222–23).

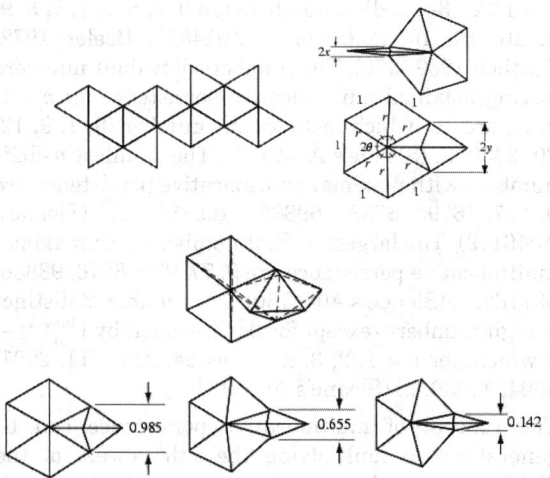

Goldberg (1978) give two tristable polyhedra: one having 12 faces and one having 20. Goldberg's bistable icosahedron, illustrated above, consists of two adjoined PENTAGONAL DIPYRAMIDS, each with two adjacent triangles (one on top and one on bottom) omitted (Goldberg 1978; Wells 1991; Cromwell 1997, pp. 222 and 224). The variables in the schematic above are connected by the equations

$$\sin\theta = \frac{1}{2r}$$

$$x^2 = 1 - r^2$$

$$y = r\sin(5\theta) = r(5\sin\theta - 20\sin^3\theta + 15\sin^5\theta)$$

$$= r\sin\theta(5 - 20\sin^2\theta + 16\sin^4\theta)$$

$$= \frac{1}{2}\left(5 - \frac{5}{r^2} + \frac{1}{r^4}\right).$$

Plugging in $r^2 = 1 - x^2$ and setting $y = x$ gives the QUINTIC EQUATION

$$2x^5 - 4x^2 - 4x^3 + 5x^2 + 2x - 1 = 0,$$

which has smallest positive solution $x \approx 0.327267$. Goldberg gives $(x, y) = (0.071, 0.49)$ and $(0.49, 0.071)$ as other solutions, although it's not clear where these come from.

See also JUMPING OCTAHEDRON

References

Efimow, N. W. "Flachenverbiegung im Grossen." Berlin: Akademie-Verlag, p. 130, 1957.
Goldberg, M. "Unstable Polyhedral Structures." *Math. Mag.* **51**, 165–70, 1978.
Wunderlich, W. "Starre, kippende, wackelige und bewegliche Achtflache." *Elem. Math.* **20**, 25–2, 1965.

Multivalued Function

A FUNCTION which assumes two or more distinct values at one or more points in its DOMAIN.

See also BRANCH CUT, BRANCH POINT

References

Morse, P. M. and Feshbach, H. "Multivalued Functions." §4.4 in *Methods of Theoretical Physics, Part I*. New York: McGraw-Hill, pp. 398–08, 1953.

Multivariate Analysis

The study of random distributions involving more than one variable.

See also GAUSSIAN JOINT VARIABLE THEOREM, MULTIPLE REGRESSION, MULTIVARIATE FUNCTION

References

Abramowitz, M. and Stegun, C. A. (Eds.). *Handbook of Mathematical Functions with Formulas, Graphs, and Mathematical Tables, 9th printing*. New York: Dover, pp. 927–28, 1972.
Feinstein, A. R. *Multivariable Analysis*. New Haven, CT: Yale University Press, 1996.
Hair, J. F. Jr. *Multivariate Data Analysis with Readings, 4th ed.* Englewood Cliffs, NJ: Prentice-Hall, 1995.
Schafer, J. L. *Analysis of Incomplete Multivariate Data*. Boca Raton, FL: CRC Press, 1997.

Sharma, S. *Applied Multivariate Techniques*. New York: Wiley, 1996.

Multivariate Distribution

GAUSSIAN MULTIVARIATE DISTRIBUTION

Multivariate Function

A FUNCTION of more than one variable.

See also MULTIVARIATE ANALYSIS, UNIVARIATE FUNCTION

Multivariate Polynomial

A POLYNOMIAL in more than one variable, e.g.,

$$P(x, y) = a_{22}x^2y^2 + a_{21}x^2y + a_{12}xy^2 + a_{11}xy + a_{10}x + a_{01}y + a_{00}.$$

See also POLYNOMIAL, UNIVARIATE POLYNOMIAL

Multivariate Theorem

GAUSSIAN JOINT VARIABLE THEOREM

Mu Molecule

MANDELBROT SET

Müntz Space

A Müntz space is a technically defined SPACE

$$M(\Lambda) \equiv \text{span}\{x^{\lambda_0}, x^{\lambda_1}, \ldots\}$$

which arises in the study of function approximations.

Müntz's Theorem

Müntz's theorem is a generalization of the WEIERSTRASS APPROXIMATION THEOREM, which states that any continuous function on a closed and bounded interval can be uniformly approximated by POLYNOMIALS involving constants and any INFINITE SEQUENCE of POWERS whose RECIPROCALS diverge.

In technical language, Müntz's theorem states that the MÜNTZ SPACE $M(\Lambda)$ is dense in $C[0, 1]$ IFF

$$\sum_{i=1}^{\infty} \frac{1}{\lambda_i} = \infty.$$

See also WEIERSTRASS APPROXIMATION THEOREM

References

Borwein, P. and Erdélyi, T. "Müntz's Theorem." §4.2 in *Polynomials and Polynomial Inequalities*. New York: Springer-Verlag, pp. 171–05, 1995.

Mutant Knot

Given an original KNOT K, the knots produced by MUTATIONS together with K itself are called mutant knots. Mutant knots are often difficult to distinguish. For instance, mutants have the same HOMFLY POLYNOMIALS and HYPERBOLIC KNOT volume. Many but not all mutants also have the same GENUS (KNOT).

See also KNOT, MUTATION

Mutation

Consider a KNOT as being formed from two TANGLES. The following three operations are called mutations.

1. Cut the knot open along four points on each of the four strings coming out of T_2, flipping T_2 over, and gluing the strings back together.
2. Cut the knot open along four points on each of the four strings coming out of T_2, flipping T_2 to the right, and gluing the strings back together.
3. Cut the knot, rotate it by $180°$, and reglue. This is equivalent to performing (1), then (2).

Mutations applied to an alternating KNOT projection always yield an ALTERNATING KNOT. The mutation of a KNOT is always another KNOT (a opposed to a LINK).

See also KNOT, MUTANT KNOT, TANGLE

References

Adams, C. C. *The Knot Book: An Elementary Introduction to the Mathematical Theory of Knots.* New York: W. H. Freeman, p. 49, 1994.

Mutual Energy

Let Ω be a SPACE with MEASURE $\mu \geq 0$, and let $\Phi(P, Q)$ be a real function on the PRODUCT SPACE $\Omega \times \Omega$. When

$$(\mu, v) = \int \int \Phi(P, Q) \, d\mu(Q) \, dv(P)$$

$$= \int \Phi(P, \mu) \, dv(P)$$

exists for measures $\mu, v \geq 0$, (μ, v) is called the mutual energy. (μ, μ) is then called the ENERGY.

See also ENERGY

References

Iyanaga, S. and Kawada, Y. (Eds.). "General Potential." §335.B in *Encyclopedic Dictionary of Mathematics.* Cambridge, MA: MIT Press, p. 1038, 1980.

Mutual Information

This entry contributed by ERIK G. MILLER

The mutual information between two discrete RANDOM VARIABLES X and Y is defined to be

$$I(X; Y) = \sum_{x \in \chi} \sum_{y \in Y} p(x, y) \ln\left(\frac{p(x, y)}{p(x)p(y)}\right). \quad (1)$$

bits. Additional properties are

$$I(X; Y) = I(Y; X), \quad (2)$$

$$I(X; Y) \geq 0, \quad (3)$$

and

$$I(X; Y) = H(X) + H(Y) - H(X, Y), \quad (4)$$

where $H(X)$ is the ENTROPY of the RANDOM VARIABLE X and $H(X, Y)$ is the joint entropy of these variables.

See also ENTROPY

References

Cover, T. M. and Thomas, J. A. *Elements of Information Theory.* New York: Wiley, pp. 18–6, 1991.

Mutually Exclusive Events

n events are said to be mutually exclusive if the occurrence of any one of them precludes any of the others. Therefore, for events $X_1, ..., X_n$, the CONDITIONAL PROBABILITY is $P(X_i|X_j) = 0$ for all $j \neq i$.

Mutually Exclusive Sets

DISJOINT SETS

Mutually Singular

Let M be a SIGMA ALGEBRA M, and let λ_1 and λ_2 be MEASURES on M. If there EXISTS a pair of disjoint SETS A and B such that λ_1 is CONCENTRATED on A and λ_2 is CONCENTRATED on B, then λ_1 and λ_2 are said to be mutually singular, written $\lambda_1 \perp \lambda_2$.

See also ABSOLUTELY CONTINUOUS, CONCENTRATED, SIGMA ALGEBRA

References

Rudin, W. *Functional Analysis, 2nd ed.* New York: McGraw-Hill, p. 121, 1991.

Myriad

The Greek word for 10,000.

Myriagon

A 10,000-sided POLYGON.

Mystic Pentagram

PENTAGRAM

N

N

The SET of NATURAL NUMBERS (the POSITIVE INTEGERS Z_+ 1, 2, 3, ...; Sloane's A000027), denoted $\mathbb{N}$, also called the WHOLE NUMBERS. Like whole numbers, there is no general agreement on whether 0 should be included in the list of natural numbers.

Due to lack of standard terminology, the following terms are recommended in preference to "COUNTING NUMBER," "natural number," and "WHOLE NUMBER."

set	name	symbol
..., −2, −1, 0, 1, 2, ...	INTEGERS	Z
1, 2, 3, 4, ...	POSITIVE INTEGERS	Z+
0, 1, 2, 3, 4, ...	NONNEGATIVE INTE-GERS	Z*
0, −1, −2, −3, −4, ...	NONPOSITIVE INTE-GERS	
−1, −2, −3, −4, ...	NEGATIVE INTEGERS	Z_

See also C, CARDINAL NUMBER, COUNTING NUMBER, I, INTEGER, Q, R, WHOLE NUMBER, Z, Z+

References

Sloane, N. J. A. Sequences A000027/M0472 in "An On-Line Version of the Encyclopedia of Integer Sequences." http://www.research.att.com/~njas/sequences/eisonline.html.

Nabla

DEL, LAPLACIAN

Nagel Line

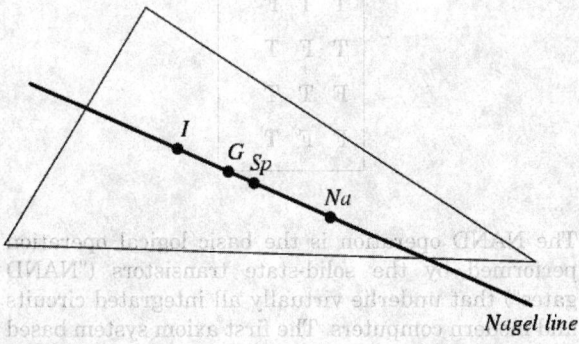

Nagel line

The Nagel line is the term proposed for the first time in this work for the line on which the INCENTER I, CENTROID G, SPIEKER CENTER Sp, and NAGEL POINT Na lie. The points satisfy

$$ISp = SpNa$$

$$IG = \tfrac{1}{2}GNa.$$

See also CENTROID (TRIANGLE), INCENTER, NAGEL POINT, SPIEKER CENTER

References

Honsberger, R. "The Nagel Point M and the Spieker Circle." §1.4 in *Episodes in Nineteenth and Twentieth Century Euclidean Geometry*. Washington, DC: Math. Assoc. Amer., pp. 5–13, 1995.

Nagel Point

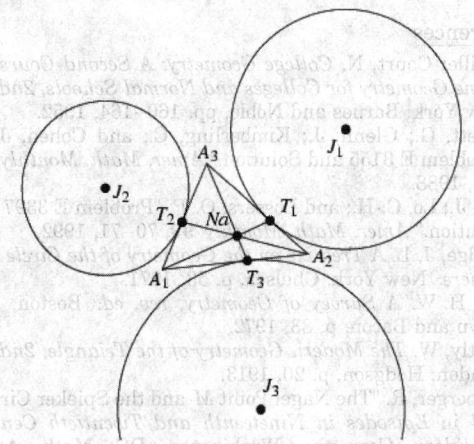

Let T_1 be the point at which the J_1-EXCIRCLE meets the side A_2A_3 of a TRIANGLE $\Delta A_1A_2A_3$, and define T_2 and T_3 similarly. Then the lines T_1, T_2, and T_3 CONCUR in the NAGEL POINT Na (sometimes denoted M)

The points T_1, T_2, and T_3 can also be constructed as the points which bisect the PERIMETER of $\Delta A_1A_2A_3$ starting at A_1, A_2, and A_3. Then the lines A_1T_1, A_2T_2, and A_3T_3 (sometimes called SPLITTERS) concur in the Nagel point Na. For this reason, the Nagel point is sometimes known as the BISECTED PERIMETER POINT (Bennett *et al.* 1988, Chen *et al.* 1992, Kimberling 1994), although the CLEAVANCE CENTER is also a bisected perimeter point.

The Nagel point has TRIANGLE CENTER FUNCTION

$$\alpha = \frac{b + c - a}{a}.$$

The Nagel point lies on the NAGEL LINE. The ORTHOCENTER and Nagel point form a DIAMETER of the FUHRMANN CIRCLE.

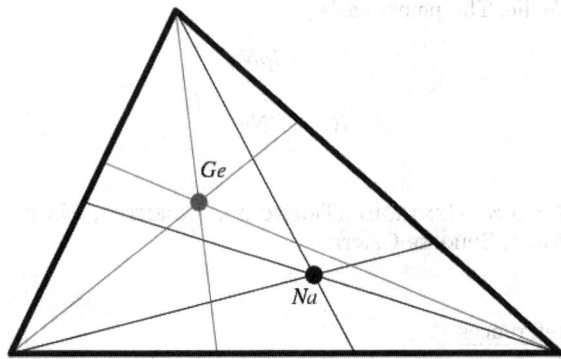

The Nagel point Na is also the ISOTOMIC CONJUGATE POINT of the GERGONNE POINT Ge.

See also CLEAVANCE CENTER, EXCENTER, EXCENTRAL TRIANGLE, EXCIRCLE, FUHRMANN CIRCLE, GERGONNE POINT, MITTENPUNKT, NAGEL LINE, SPLITTER, TRISECTED PERIMETER POINT

References

Altshiller-Court, N. *College Geometry: A Second Course in Plane Geometry for Colleges and Normal Schools, 2nd ed.* New York: Barnes and Noble, pp. 160–164, 1952.

Bennett, G.; Glenn, J.; Kimberling, C.; and Cohen, J. M. "Problem E 3155 and Solution." *Amer. Math. Monthly* **95**, 874, 1988.

Chen, J.; Lo, C.-H.; and Lossers, O. P. "Problem E 3397 and Solution." *Amer. Math. Monthly* **99**, 70–71, 1992.

Coolidge, J. L. *A Treatise on the Geometry of the Circle and Sphere.* New York: Chelsea, p. 53, 1971.

Eves, H. W. *A Survey of Geometry, rev. ed.* Boston, MA: Allyn and Bacon, p. 83, 1972.

Gallatly, W. *The Modern Geometry of the Triangle, 2nd ed.* London: Hodgson, p. 20, 1913.

Honsberger, R. "The Nagel Point M and the Spieker Circle." §1.4 in *Episodes in Nineteenth and Twentieth Century Euclidean Geometry.* Washington, DC: Math. Assoc. Amer., pp. 5–13, 1995.

Johnson, R. A. *Modern Geometry: An Elementary Treatise on the Geometry of the Triangle and the Circle.* Boston, MA: Houghton Mifflin, pp. 184 and 225–226, 1929.

Kimberling, C. "Central Points and Central Lines in the Plane of a Triangle." *Math. Mag.* **67**, 163–187, 1994.

Kimberling, C. "Nagel Point." http://cedar.evansville.edu/ ~ck6/tcenters/class/nagel.html.

Nagel, C. H. *Untersuchungen über die wichtigsten zum Dreiecke gehörigen Kreise. Eine Abhandlung aus dem Gebiete der reinen Geometrie.* Leipzig, Germany, 1836.

Nahm's Equation

The system of PARTIAL DIFFERENTIAL EQUATIONS

$$U_t = [V, W] \tag{1}$$

$$V_t = [W, U] \tag{2}$$

$$W_t = [U, V], \tag{3}$$

where $[A, B]$ denotes the COMMUTATOR.

References

Steeb, W.-H. and Louw, J. A. "Nahm's Equations, Singular Point Analysis, and Integrability." *J. Math. Phys.* **27**, 2458–2460, 1986.

Zwillinger, D. *Handbook of Differential Equations, 3rd ed.* Boston, MA: Academic Press, p. 139, 1997.

Naive Set Theory

A branch of mathematics which attempts to formalize the nature of the SET using a minimal collection of independent axioms. Unfortunately, as discovered by its earliest proponents, naive set theory quickly runs into a number of PARADOXES (such as RUSSELL'S PARADOX), so a less sweeping and more formal theory known as AXIOMATIC SET THEORY must be used.

See also AXIOMATIC SET THEORY, RUSSELL'S PARADOX, SET THEORY

NAND

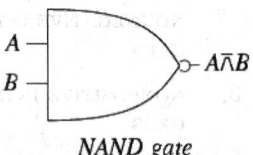

NAND gate

A CONNECTIVE in LOGIC equivalent to the composition NOT AND that yields TRUE if any condition is TRUE, and FALSE if all conditions are TRUE. A NAND B is equivalent to $!(A \wedge B)$, where $!A$ denotes NOT and $\wedge$ denotes AND. In PROPOSITIONAL CALCULUS, the term ALTERNATIVE DENIAL is used to refer to the NAND connective. Notations for NAND include $A \bar{\wedge} B$ and $A|B$ (Mendelson 1997, p. 26). The NAND operation is implemented in *Mathematica* 4.1 as Nand[A, B, ...]. The circuit diagram symbol for an NAND gate is illustrated above.

The BINARY NAND operator has the following TRUTH TABLE (Mendelson 1997, p. 27).

A	B	$A \bar{\wedge} B$
T	T	F
T	F	T
F	T	T
F	F	T

The NAND operation is the basic logical operation performed by the solid-state transistors ("NAND gates") that underlie virtually all integrated circuits and modern computers. The first axiom system based on NAND was given by Henry Sheffer in 1913. In their landmark tome, Whitehead and Russell (1927) promoted NAND as the appropriate foundation for axiomatic logic.

The AND function $A \wedge B$ can be written in terms of NANDs as

$$A \wedge B = (A \overline{\wedge} B) \overline{\wedge} (A \overline{\wedge} B).$$

See also AND, BINARY OPERATOR, CONNECTIVE, INTERSECTION, NOR, NOT, OR, TRUTH TABLE, XNOR, XOR

References

Mendelson, E. *Introduction to Mathematical Logic, 4th ed.* London: Chapman & Hall, 1997.

Simpson, R. E. "The NAND Gate." §12.5.5 in *Introductory Electronics for Scientists and Engineers, 2nd ed.* Boston, MA: Allyn and Bacon, pp. 548–550, 1987.

Whitehead, A. N. and Russell, B. *Principia Mathematica.* New York: Cambridge University Press, 1927.

Napierian Logarithm

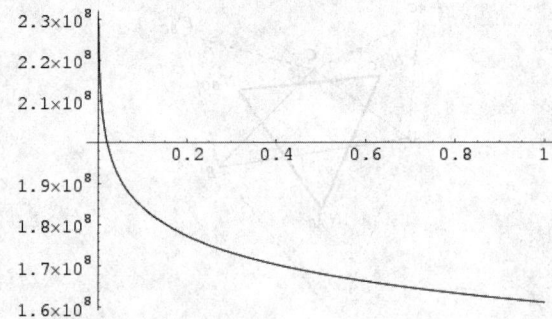

Write a number N as

$$N = 10^7 \left(1 - 10^{-7}\right)^L,$$

then L is the Napierian logarithm of N. This was the original definition of a LOGARITHM, and can be given in terms of the modern LOGARITHM as

$$L(N) = -\frac{\log\left(\frac{n}{10^7}\right)}{\log\left(\frac{10^7}{10^7 - 1}\right)}.$$

The Napierian logarithm decreases with increasing numbers and does not satisfy many of the fundamental properties of the modern LOGARITHM, e.g.,

$$N \log(xy) \neq N \log x + N \log y.$$

Napier's Analogies

Let a SPHERICAL TRIANGLE have sides a, b, and c with A, B, and C the corresponding opposite angles. Then

$$\frac{\sin\left[\frac{1}{2}(A - B)\right]}{\sin\left[\frac{1}{2}(A + B)\right]} = \frac{\tan\left[\frac{1}{2}(a - b)\right]}{\tan\left(\frac{1}{2}c\right)} \tag{1}$$

$$\frac{\cos\left[\frac{1}{2}(A - B)\right]}{\cos\left[\frac{1}{2}(A + B)\right]} = \frac{\tan\left[\frac{1}{2}(a + b)\right]}{\tan\left(\frac{1}{2}c\right)} \tag{2}$$

$$\frac{\sin\left[\frac{1}{2}(a - b)\right]}{\sin\left[\frac{1}{2}(a + b)\right]} = \frac{\tan\left[\frac{1}{2}(A - B)\right]}{\cot\left(\frac{1}{2}C\right)} \tag{3}$$

$$\frac{\cos\left[\frac{1}{2}(a - b)\right]}{\cos\left[\frac{1}{2}(a + b)\right]} = \frac{\tan\left[\frac{1}{2}(A + B)\right]}{\cot\left(\frac{1}{2}C\right)} \tag{4}$$

(Smart 1960, p. 23).

See also SPHERICAL TRIANGLE, SPHERICAL TRIGONOMETRY

References

Beyer, W. H. *CRC Standard Mathematical Tables, 28th ed.* Boca Raton, FL: CRC Press, pp. 131 and 147–150, 1987.

Harris, J. W. and Stocker, H. *Handbook of Mathematics and Computational Science.* New York: Springer-Verlag, pp. 109–110, 1998.

Smart, W. M. *Text-Book on Spherical Astronomy, 6th ed.* Cambridge, England: Cambridge University Press, 1960.

Zwillinger, D. (Ed.). "Spherical Geometry and Trigonometry." §6.4 in *CRC Standard Mathematical Tables and Formulae.* Boca Raton, FL: CRC Press, pp. 468–471, 1995.

Napier's Bones

Numbered rods which can be used to perform MULTIPLICATION. This process is also called RABDOLOGY.

See also GENAILLE RODS

References

Gardner, M. "Napier's Bones." Ch. 7 in *Knotted Doughnuts and Other Mathematical Entertainments.* New York: W. H. Freeman, pp. 85–93, 1986.

Pappas, T. "Napier's Bones." *The Joy of Mathematics.* San Carlos, CA: Wide World Publ./Tetra, pp. 64–65, 1989.

Napier's Constant

E

Napier's Inequality

For $b > a > 0$,

$$\frac{1}{b} < \frac{\ln b - \ln a}{b - a} < \frac{1}{a}.$$

References

Nelsen, R. B. "Napier's Inequality (Two Proofs)." *College Math. J.* **24**, 165, 1993.

Napier's Rules

NAPIER'S ANALOGIES

Napkin Ring

SPHERICAL RING

Napoleon Points

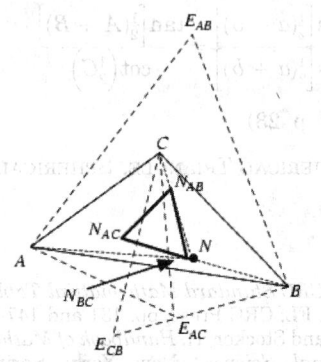

The inner Napoleon point N is the CONCURRENCE of lines drawn between VERTICES of a given TRIANGLE ΔABC and the opposite VERTICES of the corresponding inner NAPOLEON TRIANGLE $\Delta N_{AB}N_{AC}N_{BC}$. The TRIANGLE CENTER FUNCTION of the inner Napoleon point is

$$\alpha = \csc\left(A - \tfrac{1}{6}\pi\right).$$

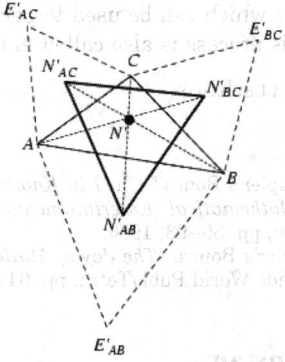

The outer Napoleon point N' is the CONCURRENCE of lines drawn between VERTICES of a given TRIANGLE ΔABC and the opposite VERTICES of the corresponding outer NAPOLEON TRIANGLE $\Delta N'_{AB}N'_{AC}N'_{BC}$. The TRIANGLE CENTER FUNCTION of the point is

$$\alpha = \csc\left(A + \tfrac{1}{6}\pi\right).$$

See also FERMAT POINTS, NAPOLEON'S THEOREM, NAPOLEON TRIANGLES

References

Casey, J. *Analytic Geometry, 2nd ed.* Dublin: Hodges, Figgis, & Co., pp. 442–444, 1893.
Kimberling, C. "Central Points and Central Lines in the Plane of a Triangle." *Math. Mag.* **67**, 163–187, 1994.

Napoleon Triangles

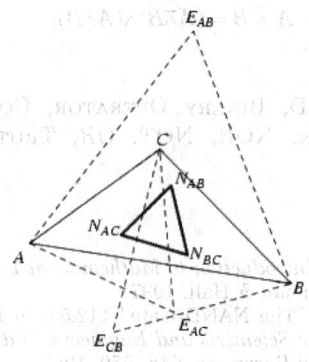

The inner Napoleon triangle is the TRIANGLE $\Delta N_{AB}N_{AC}N_{BC}$ formed by the centers of internally erected EQUILATERAL TRIANGLES ΔABE_{AB}, ΔACE_{AC}, and ΔBCE_{BC} on the sides of a given TRIANGLE ΔABC. It is an EQUILATERAL TRIANGLE.

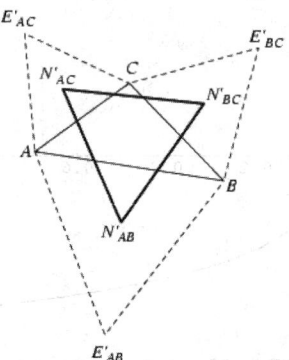

The outer Napoleon triangle is the TRIANGLE $\Delta N'_{AB}N'_{AC}N'_{BC}$ formed by the centers of externally erected EQUILATERAL TRIANGLES $\Delta ABE'_{AB}$, $\Delta ACE'_{AC}$, and $\Delta BCE'_{BC}$ on the sides of a given TRIANGLE ΔABC. It is also an EQUILATERAL TRIANGLE.

See also EQUILATERAL TRIANGLE, NAPOLEON POINTS, NAPOLEON'S THEOREM

References

Belenkiy, I. "New Features of Napoleon's Triangles." *J. Geom.* **66**, 17–26, 1999.
Coxeter, H. S. M. and Greitzer, S. L. "Napoleon Triangles." §3.3 in *Geometry Revisited.* Washington, DC: Math. Assoc. Amer., pp. 60–65, 1967.
Rigby, J. F. "Napoleon Revisited." *J. Geom.* **33**, 129–146, 1988.
Yaglom, I. M. *Geometric Transformations I.* New York: Random House, pp. 38 and 93, 1962.

Napoleon's Problem

Given the center of a CIRCLE, divide the CIRCLE into four equal arcs using a COMPASS alone (a MASCHERONI CONSTRUCTION).

See also CIRCLE, COMPASS, MASCHERONI CONSTRUCTION

Napoleon's Theorem

References

Mascheroni, L. *Geometria del compasso.* 1797.
Quemper de Lanascol, A. *Géométrie du compas.* Blanchard, pp. 74–77, 1925.
Schwerin. *Mascheronische Konstruktionen.* 1898.

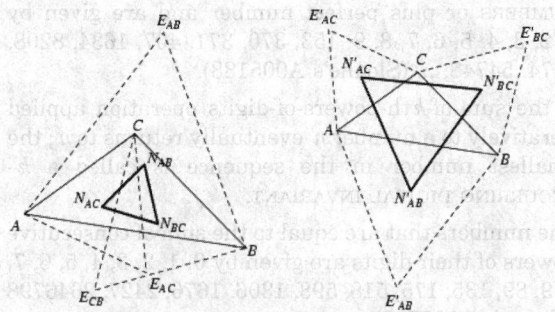

If EQUILATERAL TRIANGLES are erected externally on the sides of any TRIANGLE, then the centers form an EQUILATERAL TRIANGLE (the outer NAPOLEON TRIANGLE). Furthermore, the inner NAPOLEON TRIANGLE is also EQUILATERAL, and the difference between the areas of the outer and inner Napoleon triangles equals the AREA of the original TRIANGLE (Wells 1991, p. 156).

Drawing the centers of one EQUILATERAL TRIANGLE inwards and two outwards gives a 30°-30°-120° TRIANGLE (Wells 1991, p. 156).

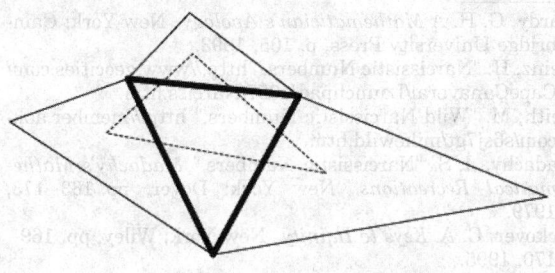

Napoleon's theorem has a very beautiful generalization in the case of externally constructed triangles: If SIMILAR triangles *of any shape* are constructed externally on a triangle such that each is rotated relative to its neighbors and *any* three corresponding points of these triangles are connected, the result is a triangle which is SIMILAR to the external triangles (Wells 1991, pp. 156–157).

See also EQUILATERAL TRIANGLE, FERMAT POINTS, NAPOLEON POINTS, NAPOLEON TRIANGLES, SIMILAR

References

Coxeter, H. S. M. and Greitzer, S. L. *Geometry Revisited.* Washington, DC: Math. Assoc. Amer., pp. 60–65, 1967.
Pappas, T. "Napoleon's Theorem." *The Joy of Mathematics.* San Carlos, CA: Wide World Publ./Tetra, p. 57, 1989.

Schmidt, F. "200 Jahre französische Revolution--Problem und Satz von Napoleon." *Didaktik der Mathematik* **19**, 15–29, 1990.
Wells, D. *The Penguin Dictionary of Curious and Interesting Geometry.* London: Penguin, pp. 74–75 and 156–158, 1991.
Wentzel, J. E. "Converses of Napoleon's Theorem." *Amer. Math. Monthly* **99**, 339–351, 1992.

Nappe

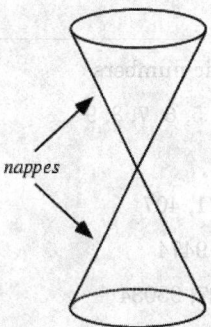

nappes

One of the two pieces of a DOUBLE CONE (i.e., two CONES placed apex to apex).

See also BICONE, CONE, DOUBLE CONE

Narain G-Transform

The INTEGRAL TRANSFORM defined by

$$(K\phi)(x) = \int_{-\infty}^{\infty} G_{pq}^{mn}\left(xt\left|\begin{matrix}(a_p)\\(b_q)\end{matrix}\right.\right)\phi(t)dt,$$

where G_{pq}^{mn} is MEIJER'S G-FUNCTION.

References

Samko, S. G.; Kilbas, A. A.; and Marichev, O. I. *Fractional Integrals and Derivatives.* Yverdon, Switzerland: Gordon and Breach, p. 23, 1993.

Narayana Polynomial

References

Sulanke, R. A. "Counting Lattice Paths by Narayana Polynomials." *Electronic J. Combinatorics* **7**, No. 1, R40, 1–9, 2000. http://www.combinatorics.org/Volume_7/v7i1toc.html.

Narcissistic Number

An n-DIGIT number which is the SUM of the nth POWERS of its DIGITS is called an n-narcissistic number, or sometimes an ARMSTRONG NUMBER or PERFECT DIGITAL INVARIANT (Madachy 1979). The smallest example other than the trivial 1-DIGIT numbers is

$$153 = 1^3 + 5^3 + 3^3. \tag{1}$$

The series of smallest narcissistic numbers of n digits are 0, (none), 153, 1634, 54748, 548834, ... (Sloane's

A014576). Hardy (1993) wrote, "There are just four numbers, after unity, which are the sums of the cubes of their digits: $153 = 1^3 + 5^3 + 3^3$, $370 = 3^3 + 7^3 + 0^3$, $371 = 3^3 + 7^3 + 1^3$, and $407 = 4^3 + 0^3 + 7^3$. These are odd facts, very suitable for puzzle columns and likely to amuse amateurs, but there is nothing in them which appeals to the mathematician." The following table gives the generalization of these "unappealing" numbers to other POWERS (Madachy 1979, p. 164).

n	n-narcissistic numbers
1	0, 1, 2, 3, 4, 5, 6, 7, 8, 9
2	none
3	153, 370, 371, 407
4	1634, 8208, 9474
5	54748, 92727, 93084
6	548834
7	1741725, 4210818, 9800817, 9926315
8	24678050, 24678051, 88593477
9	146511208, 472335975, 534494836, 912985153
10	4679307774

A total of 88 NARCISSISTIC NUMBERS exist in base 10, as proved by D. Winter in 1985 and verified by D. Hoey. These numbers exist for only 1, 3, 4, 5, 6, 7, 8, 9, 10, 11, 14, 16, 17, 19, 20, 21, 23, 24, 25, 27, 29, 31, 32, 33, 34, 35, 37, 38, and 39 digits. It can easily be shown that base-10 n-narcissistic numbers can exist only for $n \leq 60$, since

$$n \cdot 9^n < 10^{n-1} \tag{2}$$

for $n > 60$. The largest base-10 narcissistic number is the 39-narcissistic

$$115132219018763992565095597973971522401. \tag{3}$$

A table of the largest known narcissistic numbers in various BASES is given by Pickover (1995). A tabulation of narcissistic numbers in various bases is given by (Corning).

A closely related set of numbers generalize the narcissistic number to n-DIGIT numbers which are the sums of *any* single POWER of their DIGITS. For example, 4150 is a 4-DIGIT number which is the sum of fifth POWERS of its DIGITS. Since the number of digits is not equal to the power to which they are taken for such numbers, they are *not* narcissistic numbers. The smallest numbers which are sums of *any* single positive power of their digits are 1, 2, 3, 4, 5, 6, 7, 8, 9, 153, 370, 371, 407, 1634, 4150, 4151,

8208, 9474, ... (Sloane's A023052), with powers 1, 1, 1, 1, 1, 1, 1, 1, 1, 3, 3, 3, 3, 4, 5, 5, 4, 4, ... (Sloane's A046074).

The smallest numbers which are equal to the nth powers of their digits for $n = 3$, 4, ..., are 153, 1634, 4150, 548834, 1741725, ... (Sloane's A003321). The n-digit numbers equal to the sum of nth powers of their digits (a finite sequence) are called ARMSTRONG NUMBERS or plus perfect number and are given by 1, 2, 3, 4, 5, 6, 7, 8, 9, 153, 370, 371, 407, 1634, 8208, 9474, 54748, ... (Sloane's A005188).

If the sum-of-kth-powers-of-digits operation applied iteratively to a number n eventually returns to n, the smallest number in the sequence is called a k-RECURRING DIGITAL INVARIANT.

The numbers that are equal to the sum of consecutive powers of their digits are given by 0, 1, 2, 3, 4, 5, 6, 7, 8, 9, 89, 135, 175, 518, 598, 1306, 1676, 2427, 2646798 (Sloane's A032799), e.g.,

$$2646798 = 2^1 + 6^2 + 4^3 + 6^4 + 7^5 + 9^6 + 8^7. \tag{4}$$

See also ADDITIVE PERSISTENCE, DIGITAL ROOT, DIGITADDITION, HARSHAD NUMBER, KAPREKAR NUMBER, MULTIPLICATIVE DIGITAL ROOT, MULTIPLICATIVE PERSISTENCE, POWERFUL NUMBER, RECURRING DIGITAL INVARIANT, VAMPIRE NUMBER

References

Hardy, G. H. *A Mathematician's Apology.* New York: Cambridge University Press, p. 105, 1993.
Heinz, H. "Narcissistic Numbers." http://www.geocities.com/CapeCanaveral/Launchpad/4057/Narciss.htm.
Keith, M. "Wild Narcissistic Numbers." http://member.aol.com/s6sj7gt/mikewild.htm.
Madachy, J. S. "Narcissistic Numbers." *Madachy's Mathematical Recreations.* New York: Dover, pp. 163–173, 1979.
Pickover, C. A. *Keys to Infinity.* New York: Wiley, pp. 169–170, 1995.
Rivera, C. "Problems & Puzzles: Puzzle Narcissistic and Handsome Primes.-015." http://www.primepuzzles.net/puzzles/puzz_015.htm.
Rumney, M. "Digital Invariants." *Recr. Math. Mag.* No. 12, 6–8, Dec. 1962.
Sloane, N. J. A. Sequences A005188/M0488, A003321/M5403, A014576, A023052, A032799, and A046074 in "An On-Line Version of the Encyclopedia of Integer Sequences." http://www.research.att.com/~njas/sequences/eisonline.html.
Weisstein, E. W. "Narcissistic Numbers." MATHEMATICA NOTEBOOK NARCISSISTIC.DAT.

Narumi Polynomial

Polynomials $s_k(x; a)$ which form the SHEFFER SEQUENCE for

$$g(t) = \left(\frac{e^t - 1}{t} \right)^{-a} \tag{1}$$

$$f(t) = e^t - 1 \qquad (2)$$

which have GENERATING FUNCTION

$$\sum_{k=0}^{\infty} \frac{s_k(x)}{k!} t^k = \left[\frac{t}{\ln(1+t)}\right]^a (1+t)^x. \qquad (3)$$

The first few are

$$s_0(x;a) = 1$$
$$s_1(x;a) = \tfrac{1}{2}(2x+a)$$
$$s_2(x;a) = \tfrac{1}{12}[12x^2 + 12(a-1)x + a(3a-5)].$$

References

Boas, R. P. and Buck, R. C. *Polynomial Expansions of Analytic Functions, 2nd print., corr.* New York: Academic Press, p. 37, 1964.

Erdélyi, A.; Magnus, W.; Oberhettinger, F.; and Tricomi, F. G. *Higher Transcendental Functions, Vol. 3.* New York: Krieger, p. 258, 1981.

Roman, S. *The Umbral Calculus.* New York: Academic Press, 1984.

Nash Equilibrium

A set of MIXED STRATEGIES for finite, noncooperative GAMES of two or more players in which no player can improve his payoff by unilaterally changing strategy.

See also FIXED POINT, GAME, MIXED STRATEGY, NASH'S THEOREM

Nash's Embedding Theorem

Two real algebraic manifolds are equivalent IFF they are analytically homeomorphic (Nash 1952).

See also EMBEDDING

References

Kowalczyk, A. "Whitney's and Nash's Embedding Theorems for Differential Spaces." *Bull. Acad. Polon. Sci. Sér. Sci. Math.* **28**, 385–390, 1981.

Masahiro, S. *Nash Manifolds.* Berlin: Springer-Verlag, 1987.

Nash, J. "Real Algebraic Manifolds." *Ann. Math.* **56**, 405–421, 1952.

Nash's Theorem

A theorem in GAME THEORY which guarantees the existence of a NASH EQUILIBRIUM for MIXED STRATEGIES in finite, noncooperative GAMES of two or more players.

See also MIXED STRATEGY, NASH EQUILIBRIUM

Nasik Square

PANMAGIC SQUARE

Nasty Knot

An UNKNOT which can only be unknotted by first increasing the number of crossings.

Natural Boundary

This entry contributed by JONATHAN DEANE

Consider a POWER SERIES in a complex variable z

$$g(z) = \sum_{n=0}^{\infty} a_n z^n \qquad (1)$$

that is convergent within the OPEN DISK $C : |z| < R$. Convergence is limited to within C by the presence of at least one SINGULARITY on the BOUNDARY ∂C of C. If the singularities on C are so densely packed that ANALYTIC CONTINUATION cannot be carried out on a path that crosses C, then C is said to form a natural boundary for the function $g(z)$.

As an example, consider the function

$$f(z) = \sum_{n=0}^{\infty} z^{2^n} = z + z^2 + z^4 + \ldots \qquad (2)$$

Then $f(z)$ formally satisfies the FUNCTIONAL EQUATION

$$f(z) = z + f(z^2). \qquad (3)$$

The series (2) clearly converges within $C_1 : |z| < 1$. Now consider $z = 1$. Equation (3) tells us that $f(1) = 1 + f(1)$ which can only be satisfied if $f(1) = \infty$. Considering now $z = -1$, equation (3) becomes $f(-1) = -1 + \infty$ and hence $f(-1) = \infty$. Substituting z^2 for z in equation (3) then gives

$$f(z^2) = z^2 + f(z^4) = f(z) - z. \qquad (4)$$

from which it follows that

$$f(z) = z + z^2 + f(z^4). \qquad (5)$$

Now consider z equal to any of the fourth roots of unity, ± 1, $\pm i$, for example $z = -i$. Then $f(-i) = -i - 1 + f(1) = \infty$. Applying this procedure recursively shows that $f(z)$ is infinite for any z such that $z^{2^n} = 1$ with $n = 0, 1, 2, \ldots$. In any arc of the circle ∂C_1 of finite length there will therefore be an infinite number of points for which $f(z)$ is infinite and so C_1 constitutes a natural boundary for $f(z)$.

A function that has a natural boundary is said to be a LACUNARY FUNCTION.

See also BOUNDARY, LACUNARY FUNCTION

References

Ash, R. B. Ch. 3 in *Complex Variables.* New York: Academic Press, 1971.

Natural Density

NATURAL INVARIANT

Natural Equation

A natural equation is an equation which specifies a curve independent of any choice of coordinates or parameterization. The study of natural equations began with the following problem: given two functions of one parameter, find the SPACE CURVE for which the functions are the CURVATURE and TORSION.

Euler gave an integral solution for plane curves (which always have TORSION $\tau = 0$). Call the ANGLE between the TANGENT line to the curve and the X-AXIS ϕ the TANGENTIAL ANGLE, then

$$\phi = \int \kappa(s)\,ds, \qquad (1)$$

where κ is the CURVATURE. Then the equations

$$\kappa = \kappa(s) \qquad (2)$$
$$\tau = 0, \qquad (3)$$

where τ is the TORSION, are solved by the curve with PARAMETRIC EQUATIONS

$$x = \int \cos\phi \, ds \qquad (4)$$

$$y = \int \sin\phi \, ds. \qquad (5)$$

The equations $\kappa = \kappa(s)$ and $\tau = \tau(s)$ are called the natural (or INTRINSIC) equations of the space curve. An equation expressing a plane curve in terms of s and RADIUS OF CURVATURE R (or κ) is called a CESÀRO EQUATION, and an equation expressing a plane curve in terms of s and ϕ is called a WHEWELL EQUATION.

Among the special planar cases which can be solved in terms of elementary functions are the CIRCLE, LOGARITHMIC SPIRAL, CIRCLE INVOLUTE, and EPICYCLOID. Enneper showed that each of these is the projection of a HELIX on a CONIC surface of revolution along the axis of symmetry. The above cases correspond to the CYLINDER, CONE, PARABOLOID, and SPHERE.

See also CESÀRO EQUATION, INTRINSIC EQUATION, WHEWELL EQUATION

References

Cesàro, E. *Lezioni di Geometria Intrinseca.* Napoli, Italy, 1896.
Euler, L. *Comment. Acad. Petropolit.* **8**, 66–85, 1736.
Gray, A. *Modern Differential Geometry of Curves and Surfaces with Mathematica, 2nd ed.* Boca Raton, FL: CRC Press, pp. 138–139, 1997.
Melzak, Z. A. *Companion to Concrete Mathematics, Vol. 2.* New York: Wiley, 1976.
Struik, D. J. *Lectures on Classical Differential Geometry.* New York: Dover, pp. 26–28, 1988.

Natural Independence Phenomenon

A type of mathematical result which is considered by most logicians as more natural than the METAMATHEMATICAL incompleteness results first discovered by Gödel. Finite combinatorial examples include GOOD-STEIN'S THEOREM, a finite form of RAMSEY'S THEOREM, and a finite form of KRUSKAL'S TREE THEOREM (Kirby and Paris 1982; Smorynski 1980, 1982, 1983; Gallier 1991).

See also GÖDEL'S INCOMPLETENESS THEOREM, GOODSTEIN'S THEOREM, KRUSKAL'S TREE THEOREM, RAMSEY'S THEOREM

References

Gallier, J. "What's so Special about Kruskal's Theorem and the Ordinal Gamma[0]? A Survey of Some Results in Proof Theory." *Ann. Pure and Appl. Logic* **53**, 199–260, 1991.
Kirby, L. and Paris, J. "Accessible Independence Results for Peano Arithmetic." *Bull. London Math. Soc.* **14**, 285–293, 1982.
Smorynski, C. "Some Rapidly Growing Functions." *Math. Intell.* **2**, 149–154, 1980.
Smorynski, C. "The Varieties of Arboreal Experience." *Math. Intell.* **4**, 182–188, 1982.
Smorynski, C. "'Big' News from Archimedes to Friedman." *Not. Amer. Math. Soc.* **30**, 251–256, 1983.

Natural Invariant

Let $\rho(x)dx$ be the fraction of time a typical dynamical ORBIT spends in the interval $[x, x+dx]$, and let $\rho(x)$ be normalized such that

$$\int_0^\infty \rho(x)dx = 1$$

over the entire interval of the map. Then the fraction the time an ORBIT spends in a finite interval $[a, b]$, is given by

$$\int_a^b \rho(x)dx.$$

The natural invariant is also called the INVARIANT DENSITY or NATURAL DENSITY.

Natural Logarithm

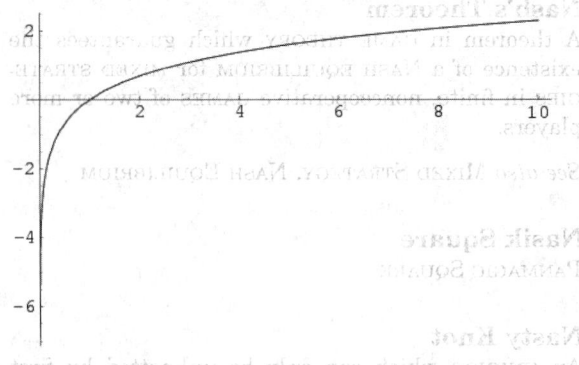

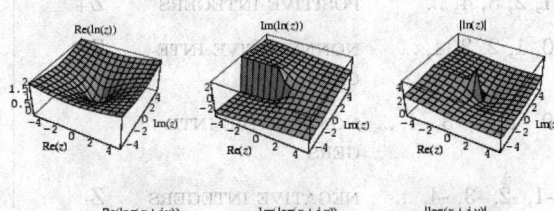

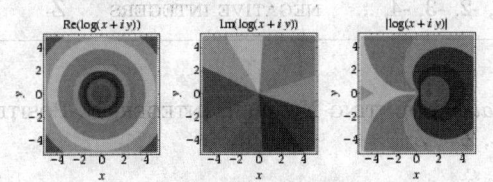

The LOGARITHM having base E, where

$$e = 2.718281828\ldots, \tag{1}$$

which can be defined

$$\ln x \equiv \int_1^x \frac{dt}{t} \tag{2}$$

for $x > 0$. The natural logarithm can also be defined by

$$\ln x = \lim_{x \to \infty} \left(x^{1/n} - 1\right)^n. \tag{3}$$

The symbol $\ln x$ is used in physics and engineering to denote the natural logarithm, while mathematicians commonly use the notation $\log x$. In this work, $\ln x = \log_e x$ denotes a natural logarithm, whereas $\log x = \log_{10} x$ denotes the COMMON LOGARITHM. Common and natural logarithms can be expressed in terms of each other as

$$\ln x = \frac{\log_{10} x}{\log_{10} e} \tag{4}$$

$$\log_{10} x = \frac{\ln x}{\ln 10}. \tag{5}$$

The natural logarithm is especially useful in CALCULUS because its DERIVATIVE is given by the simple equation

$$\frac{d}{dx} \ln x = \frac{1}{x}, \tag{6}$$

whereas logarithms in other bases have the more complicated DERIVATIVE

$$\frac{d}{dx} \log_b x = \frac{1}{x \ln b}. \tag{7}$$

The natural logarithm can be analytically continued to COMPLEX NUMBERS as

$$\ln z \equiv \ln|z| + i \arg(z), \tag{8}$$

where $|z|$ is the MODULUS and $\arg(z)$ is the ARGUMENT

The MERCATOR SERIES

$$\ln(1+x) = x - \tfrac{1}{2}x^2 + \tfrac{1}{3}x^3 - \ldots \tag{9}$$

gives a TAYLOR SERIES for the natural logarithm.

CONTINUED FRACTION representations of logarithmic functions include

$$\ln(1+x) = \cfrac{x}{1 + \cfrac{1^2 x}{2 + \cfrac{1^2 x}{3 + \cfrac{2^2 x}{4 + \cfrac{2^2 x}{5 + \cfrac{3^2 x}{6 + \cfrac{3^2 x}{7 + \cdots}}}}}}} \tag{10}$$

$$\ln\left(\frac{1+x}{1-x}\right) = \cfrac{2x}{1 - \cfrac{x^2}{3 - \cfrac{4x^2}{5 - \cfrac{9x^2}{7 - \cfrac{16x^2}{9 - \cdots}}}}} \tag{11}$$

For a COMPLEX NUMBER z, the natural logarithm satisfies

$$\ln z = \ln\left[re^{i(\theta + 2n\pi)}\right] = \ln r + i(\theta + 2n\pi) \tag{12}$$

$$PV(\ln z) = \ln r + i\theta, \tag{13}$$

where PV is the PRINCIPAL VALUE.

Some special values of the natural logarithm are

$$\ln 1 = 0 \tag{14}$$

$$\ln 0 = -\infty \tag{15}$$

$$\ln(-1) = \pi i \tag{16}$$

$$\ln(\pm i) = \pm \tfrac{1}{2}\pi i. \tag{17}$$

An identity for the natural logarithm of 2 discovered using the PSLQ ALGORITHM is

$$(\ln 2)^2 = 2 \sum_{i=1}^{\infty} \frac{p_i}{2^i i^2} \{p_i\} = \{2, -10, -7, -10, 2, -1\}, \tag{18}$$

where $\{p_i\}$ is given by the periodic sequence obtained by appending copies of $\{2, -10, -7, -10, 2, -1\}$ (in other words, $p_i \equiv p_{[(i-1)(\bmod 6)]+1}$ for $i > 6$) (Bailey *et al.* 1995, Bailey and Plouffe).

See also COMMON LOGARITHM, E, LG, LOGARITHM

References

Bailey, D.; Borwein, P.; and Plouffe, S. "On the Rapid Computation of Various Polylogarithmic Constants." http://www.cecm.sfu.ca/~pborwein/PAPERS/P123.ps.

Bailey, D. and Plouffe, S. "Recognizing Numerical Constants." http://www.cecm.sfu.ca/organics/papers/bailey/.
Gourdon, X. and Sebah, P. "The Constant ln2." http://xavier.gourdon.free.fr/Constants/Log2/log2.html.

Natural Measure

$\mu_i(\epsilon)$, sometimes denoted $P_i(\epsilon)$, is the probability that element i is populated, normalized such that

$$\sum_{i=1}^{N} \mu_i(\epsilon) = 1.$$

See also INFORMATION DIMENSION, Q-DIMENSION

Natural Norm

Let $\|z\|$ be a VECTOR NORM of a VECTOR $\mathbf{z}$ such that

$$\|\mathsf{A}\| = \max_{\|\mathbf{z}\|=1} \|\mathsf{A}\mathbf{z}\|.$$

Then $\|\mathsf{A}\|$ is a MATRIX NORM which is said to be the natural norm INDUCED (or SUBORDINATE) to the VECTOR NORM $\|\mathbf{z}\|$. For any natural norm,

$$\|\mathsf{I}\| = 1,$$

where I is the IDENTITY MATRIX. The natural matrix norms induced by the $L1$-NORM, $L2$-NORM, and L-INFINITY-NORM are called the MAXIMUM ABSOLUTE COLUMN SUM NORM, SPECTRAL NORM, and MAXIMUM ABSOLUTE ROW SUM NORM, respectively.

See also $L1$-NORM, $L2$-NORM, MATRIX NORM, MAXIMUM ABSOLUTE COLUMN SUM NORM, SPECTRAL NORM, VECTOR NORM

References
Gradshteyn, I. S. and Ryzhik, I. M. *Tables of Integrals, Series, and Products, 6th ed.* San Diego, CA: Academic Press, p. 1115, 2000.

Natural Number

A POSITIVE INTEGER 1, 2, 3, ... (Sloane's A000027). The set of natural numbers is denoted N or Z+. Unfortunately, 0 is sometimes also included in the list of "natural" numbers (Bourbaki 1968, Halmos 1974), and there seems to be no general agreement about whether to include it. In fact, Ribenboim (1996) states "Let P be a set of natural numbers; whenever convenient, it may be assumed that $0 \in P$."

Due to lack of standard terminology, the following terms are recommended in preference to "COUNTING NUMBER," "natural number," and "WHOLE NUMBER."

set	name	symbol
..., -2, -1, 0, 1, 2, ...	INTEGERS	Z
1, 2, 3, 4, ...	POSITIVE INTEGERS	Z+
0, 1, 2, 3, 4, ...	NONNEGATIVE INTEGERS	Z*
0, -1, -2, -3, -4, ...	NONPOSITIVE INTEGERS	
-1, -2, -3, -4, ...	NEGATIVE INTEGERS	Z-

See also COUNTING NUMBER, INTEGER, N, POSITIVE, Z, Z-, Z+, Z*

References
Bourbaki, N. *Elements of Mathematics: Theory of Sets.* Paris, France: Hermann, 1968.
Courant, R. and Robbins, H. "The Natural Numbers." Ch. 1 in *What is Mathematics?: An Elementary Approach to Ideas and Methods, 2nd ed.* Oxford, England: Oxford University Press, pp. 1–20, 1996.
Halmos, P. R. *Naive Set Theory.* New York: Springer-Verlag, 1974.
Ribenboim, P. "Catalan's Conjecture." *Amer. Math. Monthly* **103**, 529–538, 1996.
Sloane, N. J. A. Sequences A000027/M0472 in "An On-Line Version of the Encyclopedia of Integer Sequences." http://www.research.att.com/~njas/sequences/eisonline.html.
Welbourne, E. "The Natural Numbers." http://www.chaos.org.uk/~eddy/math/found/natural.html.

Natural Perspective

PERSPECTIVE

Naught

The British word for "ZERO." It is often used to indicate 0 subscripts, so a_0 would be spoken as "a naught."

See also ZERO

Navier's Equation

The general equation of fluid flow

$$(\lambda + 2\mu)\nabla(\nabla \cdot \mathbf{u}) - \mu\nabla \times (\nabla \times \mathbf{u}) = \rho \frac{\partial^2 \mathbf{u}}{\partial t^2},$$

where μ and λ are coefficients of viscosity, $\mathbf{u}$ is the velocity of the fluid parcel, and ρ is the fluid density.

See also NAVIER-STOKES EQUATION

References
Eringen, A. C. and Suhubi, E. S. Ch. 5 in *Elastodynamics, Vol. 2.* New York: Academic Press, 1975.
Zwillinger, D. *Handbook of Differential Equations, 3rd ed.* Boston, MA: Academic Press, p. 139, 1997.

Navier-Stokes Equation

The equation of incompressible fluid flow,

$$\frac{\partial \mathbf{u}}{\partial t} + \mathbf{u} \cdot \nabla \mathbf{u} = \frac{\nabla P}{\rho} + \nu \nabla^2 \mathbf{u},$$

where ν is the kinematic viscosity, $\mathbf{u}$ is the velocity of the fluid parcel, P is the pressure, and ρ is the fluid density.

See also NAVIER'S EQUATION

References

Landau, L. D. and Lifschitz, E. M. *Fluid Mechanics, 2nd ed.* Oxford, England: Pergamon Press, p. 15, 1982.
Zwillinger, D. *Handbook of Differential Equations, 3rd ed.* Boston, MA: Academic Press, p. 139, 1997.

Navigation Problem

A problem in the CALCULUS OF VARIATIONS. Let a vessel traveling at constant speed c navigate on a body of water having surface velocity

$$u = u(x, y)$$

$$v = v(x, y).$$

The navigation problem asks for the course which travels between two points in minimal time.

References

Sagan, H. *Introduction to the Calculus of Variations.* New York: Dover, pp. 226–228, 1992.

nc

JACOBI ELLIPTIC FUNCTIONS

N-Cluster

A LATTICE POINT configuration with no three points COLLINEAR and no four CONCYCLIC. An example is the 6-cluster $(0, 0)$, $(132, -720)$, $(546, -272)$, $(960, -720)$, $(1155, 540)$, $(546, 1120)$. Call the RADIUS of the smallest CIRCLE centered at one of the points of an N-cluster which contains all the points in the N-cluster the EXTENT. Noll and Bell (1989) found 91 nonequivalent prime 6-clusters of EXTENT less than 20937, but found no 7-clusters.

References

Guy, R. K. *Unsolved Problems in Number Theory, 2nd ed.* New York: Springer-Verlag, p. 187, 1994.
Noll, L. C. and Bell, D. I. "n-clusters for $1 < n < 7$." *Math. Comput.* **53**, 439–444, 1989.

n-Cube

HYPERCUBE, POLYCUBE

nd

JACOBI ELLIPTIC FUNCTIONS

Near Noble Number

A REAL NUMBER $0 < \nu < 1$ whose CONTINUED FRACTION is periodic, and the periodic sequence of terms is composed of a string of 1s followed by an INTEGER $n > 1$,

$$\nu = [\underbrace{1, 1, \ldots, 1}_{p}, n]. \tag{1}$$

This can be written in the form

$$\nu = [\underbrace{1, 1, \ldots, 1}_{p}, n, \nu^{-1}], \tag{2}$$

which can be solved to give

$$\nu = \frac{1}{2} n \left(\sqrt{1 + 4 \frac{nF_{p-1} + F_{p-2}}{n^2 F_p}} - 1 \right), \tag{3}$$

where F_n is a FIBONACCI NUMBER. The special case $n = 2$ gives

$$\nu = \sqrt{\frac{F_{p+2}}{F_p}} - 1. \tag{4}$$

See also NOBLE NUMBER

References

Schroeder, M. R. *Number Theory in Science and Communication: With Applications in Cryptography, Physics, Digital Information, Computing, and Self-Similarity, 2nd enl. ed., corr. printing.* Berlin: Springer-Verlag, 1990.
Schroeder, M. "Noble and Near Noble Numbers." In *Fractals, Chaos, Power Laws: Minutes from an Infinite Paradise.* New York: W. H. Freeman, pp. 392–394, 1991.

Nearest Integer Function

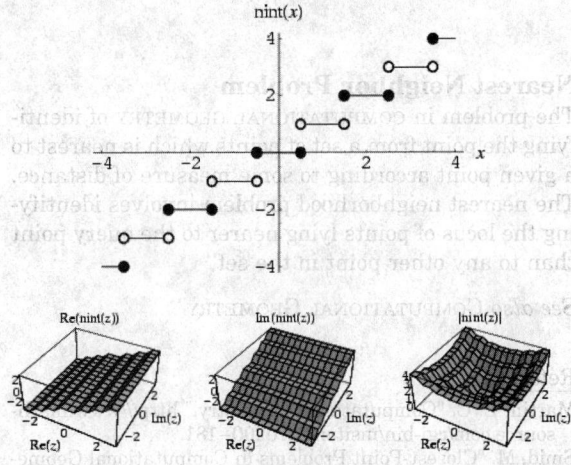

The nearest integer function nint(x) of x, illustrated above and also called nint or the round function, is defined such that $[x]$ is the INTEGER closest to x. Since this definition is ambiguous for half-integers, the additional rule that half-integers are always rounded to even numbers is usually added in order to avoid statistical biasing. For example, $[1.5] = 2$, $[2.5] = 2$, $[3.5] = 4$, $[4.4] = 4$, etc. This convention is followed in the C math.h library function rint, as well as in *Mathematica*, where the nearest integer function is implemented as Round[x].

Although the notation $\lceil x \rfloor$ is sometimes used to denote the nearest integer function (Hastad *et al.* 1989), this notation is rather cumbersome and is not recommended. Also note that while $[x]$ is used to denote the nearest integer function in this work, $[x]$ is also commonly used to denote the FLOOR FUNCTION $\lfloor x \rfloor$.

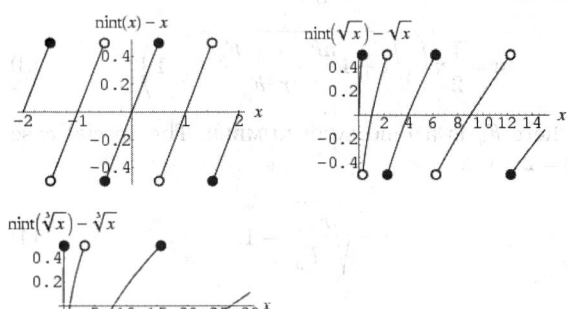

The plots above illustrate $x^{1/n} - [x^{1/n}]$ for small n.

See also CEILING FUNCTION, FLOOR FUNCTION, NINT ZETA FUNCTION, STAIRCASE FUNCTION

References

Hastad, J.; Just, B.; Lagarias, J. C.; and Schnorr, C. P. "Polynomial Time Algorithms for Finding Integer Relations Among Real Numbers." *SIAM J. Comput.* **18**, 859–881, 1988.

Nearest Neighbor Problem

The problem in COMPUTATIONAL GEOMETRY of identifying the point from a set of points which is nearest to a given point according to some measure of distance. The nearest neighborhood problem involves identifying the locus of points lying nearer to the query point than to any other point in the set.

See also COMPUTATIONAL GEOMETRY

References

Martin, E. C. "Computational Geometry." http://www.mathsource.com/cgi-bin/msitem22?0200-181.
Smid, M. "Closest-Point Problems in Computational Geometry." Ch. 20 in *Handbook of Computational Geometry* (Ed. J.-R. Sack and J. Urrutia). Amsterdam, Netherlands: North-Holland, pp. 877–935, 2000.

Skiena, S. S. "Nearest Neighbor Search." §8.6.5 in *The Algorithm Design Manual*. New York: Springer-Verlag, pp. 361–363, 1997.

Near-Integer

ALMOST INTEGER

Nearly-Poised

Let GENERALIZED HYPERGEOMETRIC FUNCTION

$$_pF_q\begin{bmatrix} \alpha_1, \alpha_2, \ldots, \alpha_p \\ \beta_1, \beta_2, \ldots, \beta_q \end{bmatrix};z \tag{1}$$

have $p = q + 1$. Then the generalized hypergeometric function is said to be nearly-poised of the first kind if

$$\beta_1 + \alpha_2 = \ldots = \beta_q + \alpha_{q+1}. \tag{2}$$

(omitting the initial equality in the definition for WELL-POISED), and nearly-poised of the second kind if

$$1 + \alpha_1 = \beta_1 + \alpha_2 = \ldots = \beta_{q-1} + \alpha_q. \tag{3}$$

See also GENERALIZED HYPERGEOMETRIC FUNCTION, K-BALANCED, NEARLY-POISED, SAALSCHÜTZIAN

References

Bailey, W. N. *Generalised Hypergeometric Series.* Cambridge, England: Cambridge University Press, pp. 11–12, 1935.
Koepf, W. *Hypergeometric Summation: An Algorithmic Approach to Summation and Special Function Identities.* Braunschweig, Germany: Vieweg, p. 43, 1998.
Whipple, F. J. W. "On Well-Poised Series, Generalized Hypergeometric Series Having Parameters in Pairs, Each Pair with the Same Sum." *Proc. London Math. Soc.* **24**, 247–263, 1926.

Near-Pencil

An arrangement of $n \geq 3$ points such that $n-1$ of them are COLLINEAR.

See also GENERAL POSITION, ORDINARY LINE, PENCIL

References

Guy, R. K. "Unsolved Problems Come of Age." *Amer. Math. Monthly* **96**, 903–909, 1989.
Kelly, L. M. and Moser, W. O. J. "On the Number of Ordinary Lines Determined by n Points." *Canad. J. Math.* **1**, 210–219, 1958.

Necessary

A CONDITION which must hold for a result to be true, but which does not guarantee it to be true. If a CONDITION is both NECESSARY and SUFFICIENT, then the result is said to be true IFF the CONDITION holds.

See also SUFFICIENT

References

Jeffreys, H. and Jeffreys, B. S. "Necessary: Sufficient." §1.036 in *Methods of Mathematical Physics, 3rd ed.*

Cambridge, England: Cambridge University Press, pp. 10–11, 1988.

Necker Cube

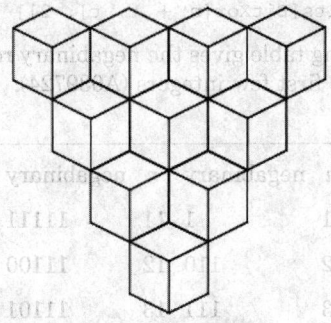

An ILLUSION in which a 2-D drawing of an array of CUBES appears to simultaneously protrude from and intrude into the page.

References

Fineman, M. *The Nature of Visual Illusion.* New York: Dover, pp. 25 and 118, 1996.
Jablan, S. "Impossible Figures." http://members.tripod.com/ ~modularity/impos.htm.
Newbold, M. "Animated Necker Cube." http://dogfeathers.-com/java/necker.html.

Necklace

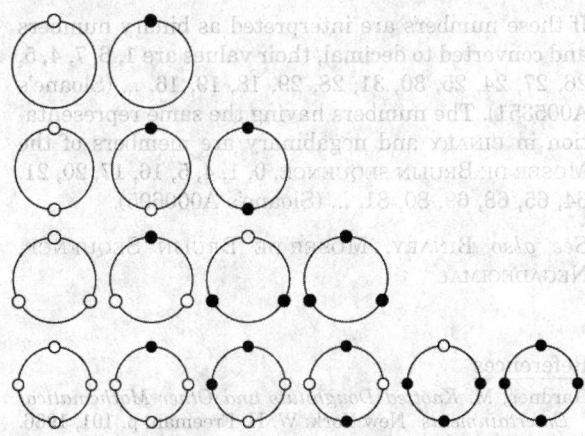

In the technical COMBINATORIAL sense, an a-ary necklace of length n is a string of n characters, each of a possible types. Rotation is ignored, in the sense that $b_1 b_2 \ldots b_n$ is equivalent to $b_k b_{k+1} \ldots b_n b_1 b_2 \ldots b_{k-1}$ for any k.

In FIXED necklaces, reversal of strings is respected, so they represent circular collections of beads in which the necklace may not be picked up out of the PLANE (i.e., opposite orientations are not considered equivalent). The number of fixed necklaces of length n composed of a types of beads $N(n, a)$ is given by

$$N(n, a) = \frac{1}{n} \sum_{i=1}^{v(n)} \phi(d_i) a^{n/d_i}, \quad (1)$$

where d_i are the DIVISORS of n with $d_1 \equiv 1$, d_2, ..., $d_{v(n)} \equiv n$, $v(n)$ is the number of DIVISORS of n, and $\phi(x)$ is the TOTIENT FUNCTION.

For FREE necklaces, opposite orientations (MIRROR IMAGES) are regarded as equivalent, so the necklace *can* be picked up out of the PLANE and flipped over. The number $N'(n, a)$ of such necklaces composed of n beads, each of a possible colors, is given by

$$N'(n, a) = \frac{1}{2n}$$

$$\times \begin{cases} \sum_{i=1}^{v(n)} \phi(d_i) a^{n/d_i} + n a^{(n+1)/2} & \text{for } n \text{ odd} \\ \sum_{i=1}^{v(n)} \phi(di) a^{n/d_i} + \frac{1}{2} n (1 + a) a^{n/2} & \text{for } n \text{ even.} \end{cases}$$

For $a = 2$ and $n = p$ an ODD PRIME, this simplifies to

$$N'(p, 2) = \frac{2^{p-1} - 1}{p} + 2^{(p-1)/2} + 1.$$

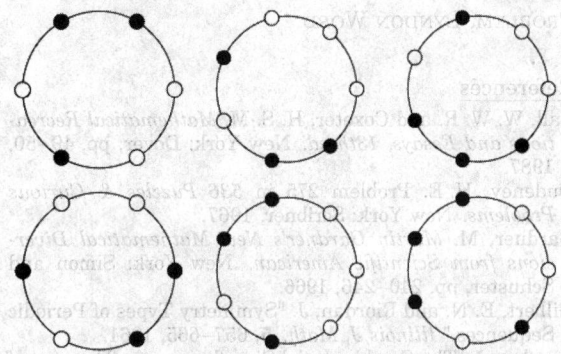

A table of the first few numbers of necklaces for $a = 2$ and $a = 3$ follows. Note that $N(n, 2)$ is larger than $N'(n, 2)$ for $n \geq 6$. For $n = 6$, the necklace 110100 is inequivalent to its MIRROR IMAGE 0110100, accounting for the difference of 1 between $N(6, 2)$ and $N'(6, 2)$. Similarly, the two necklaces 0010110 and 0101110 are inequivalent to their reversals, accounting for the difference of 2 between $N(7, 2)$ and $N'(7, 2)$.

n	$N(n, 2)$	$N'(n, 2)$	$N'(n, 3)$
Sloane	Sloane's A000031	Sloane's A000029	Sloane's A027671
1	2	2	3
2	3	3	6
3	4	4	10
4	6	6	21
5	8	8	39
6	14	13	92
7	20	18	198
8	36	30	498

9	60	46	1219
10	108	78	3210
11	188	126	8418
12	352	224	22913
13	632	380	62415
14	1182	687	173088
15	2192	1224	481598

Ball and Coxeter (1987) consider the problem of finding the number of distinct arrangements of n people in a ring such that no person has the same two neighbors two or more times. For 8 people, there are 21 such arrangements.

See also ANTOINE'S NECKLACE, DE BRUIJN SEQUENCE, FIXED, FREE, IRREDUCIBLE POLYNOMIAL, JOSEPHUS PROBLEM, LYNDON WORD

References

Ball, W. W. R. and Coxeter, H. S. M. *Mathematical Recreations and Essays, 13th ed.* New York: Dover, pp. 49–50, 1987.
Dudeney, H. E. Problem 275 in *536 Puzzles & Curious Problems.* New York: Scribner, 1967.
Gardner, M. *Martin Gardner's New Mathematical Diversions from Scientific American.* New York: Simon and Schuster, pp. 240–246, 1966.
Gilbert, E. N. and Riordan, J. "Symmetry Types of Periodic Sequences." *Illinois J. Math.* **5**, 657–665, 1961.
Riordan, J. "The Combinatorial Significance of a Theorem of Pólya." *J. SIAM* **4**, 232–234, 1957.
Riordan, J. *An Introduction to Combinatorial Analysis.* New York: Wiley, p. 162, 1980.
Ruskey, F. "Information on Necklaces, Lyndon Words, de Bruijn Sequences." http://www.theory.csc.uvic.ca/~cos/inf/neck/NecklaceInfo.html.
Skiena, S. "Polya's Theory of Counting." §1.2.6 in *Implementing Discrete Mathematics: Combinatorics and Graph Theory with Mathematica.* Reading, MA: Addison-Wesley, pp. 25–26, 1990.
Sloane, N. J. A. Sequences A000029/M0563, A000031/M0564, A001869/M3860, and A027671 in "An On-Line Version of the Encyclopedia of Integer Sequences." http://www.research.att.com/~njas/sequences/eisonline.html.
Weisstein, E. W. "Integer Sequences." MATHEMATICA NOTEBOOK INTEGERSEQUENCES.M.

Needle

BUFFON-LAPLACE NEEDLE PROBLEM, BUFFON'S NEEDLE PROBLEM, KAKEYA NEEDLE PROBLEM

Negabinary

The negabinary representation of a number n is given by the coefficients $a_n a_{n-1} \ldots a_1 a_0$ in

$$n = \sum_{i=0} a_i(-2)^i = \ldots + a_2(-2)^2 + a_1(-2)^1 + a_0(-2)^0,$$

where $a_i = 0, 1$. Conversion of n to negabinary can be done using the *Mathematica* code

```
Negabinary[n_Integer] := Module[{t = (2/
3)(4^Floor[Log[4, Abs[n] + 1] + 2] - 1)},
IntegerDigits[BitXor[n + t, t], 2]]
```

The following table gives the negabinary representations for the first few integers (A039724).

n	negabinary	n	negabinary
1	1	11	11111
2	110	12	11100
3	111	13	11101
4	100	14	10010
5	101	15	10011
6	11010	16	10000
7	11011	17	10001
8	11000	18	10110
9	11001	19	10111
10	11110	20	10100

If these numbers are interpreted as binary numbers and converted to decimal, their values are 1, 6, 7, 4, 5, 26, 27, 24, 25, 30, 31, 28, 29, 18, 19, 16, ... (Sloane's A005351). The numbers having the same representation in BINARY and negabinary are members of the MOSER-DE BRUIJN SEQUENCE, 0, 1, 4, 5, 16, 17, 20, 21, 64, 65, 68, 69, 80, 81, ... (Sloane's A000695).

See also BINARY, MOSER-DE BRUIJN SEQUENCE, NEGADECIMAL

References

Gardner, M. *Knotted Doughnuts and Other Mathematical Entertainments.* New York: W. H. Freeman, p. 101, 1986.
Sloane, N. J. A. Sequences A000695/M3259, A005351/M4059, and A039724 in "An On-Line Version of the Encyclopedia of Integer Sequences." http://www.research.att.com/~njas/sequences/eisonline.html.

Negadecimal

The negadecimal representation of a number n is given by the coefficients $a_n a_{n-1} \ldots a_1 a_0$ in

$$n = \sum_{i=0} a_i(-10)^i = \ldots a_2(-10)^2 + a_1(-10)^1 + a_0(-10)^0,$$

where $a_i = 0, 1, \ldots, 9$. The following table gives the negabinary representations for the first few integers (A039723).

n	negadecimal	n	negadecimal	n	negadecimal
1	1	11	191	21	181
2	2	12	192	22	182
3	3	13	193	23	183
4	4	14	194	24	184
5	5	15	195	25	185
6	6	16	196	26	186
7	7	17	197	27	187
8	8	18	198	28	188
9	9	19	199	29	189
10	190	20	180	30	170

The numbers having the same DECIMAL and negadecimal representations are those which are sums of distinct powers of 100: 1, 2, 3, 4, 5, 6, 7, 8, 9, 100, 101, 102, 103, 104, 105, 106, 107, 108, 109, 200, ... (Sloane's A051022).

See also DECIMAL, NEGABINARY

References

Sloane, N. J. A. Sequences A039723 and A051022 in "An On-Line Version of the Encyclopedia of Integer Sequences." http://www.research.att.com/~njas/sequences/eisonline.html.

Negation

The operation of interchanging true and false in a logical statement. The negation of A is often called "NOT-A," and can be denoted !A, or with the NEGATION SIGN $\neg$, so not-A is written $\neg A$.

Note that in computer languages such as C, perl, and *Mathematica*, not-A is denoted !A. In FORTRAN, not-A is written .not.A, where A is a variable of logical type.

See also NEGATION SIGN, NOT

Negation Sign

The symbol $\neg$ used to denote the NEGATION operation ("NOT") in symbolic logic, also called "logical not."

See also NOT

References

Bringhurst, R. *The Elements of Typographic Style, 2nd ed.* Point Roberts, WA: Hartley and Marks, p. 281, 1997.

Negative

A quantity less than ZERO (< 0), denoted with a MINUS SIGN, i.e., $-x$.

See also NONNEGATIVE, NONPOSITIVE, NONZERO, POSITIVE, ZERO

References

Wells, D. *The Penguin Dictionary of Curious and Interesting Numbers.* Middlesex, England: Penguin Books, pp. 20–21, 1986.

Negative Binomial Distribution

Also known as the PASCAL DISTRIBUTION and PÓLYA DISTRIBUTION. The probability of $r-1$ successes and x failures in $x+r-1$ trials, and success on the $(x+r)$th trial is

$$p\left[\binom{x+r-1}{r-1}p^{r-1}(1-p)^{[(x+r-1)-(r-1)]}\right]$$
$$= \left[\binom{x+r-1}{r-1}p^{r-1}(1-p)^x\right]p$$
$$= \binom{x+r-1}{r-1}p^r(1-p)^x, \tag{1}$$

where $\binom{n}{k}$ is a BINOMIAL COEFFICIENT. Let

$$P = \frac{1-p}{p} \tag{2}$$

$$Q = \frac{1}{p}. \tag{3}$$

The CHARACTERISTIC FUNCTION is given by

$$\phi(t) = \left(Q - Pe^{it}\right)^{-r}, \tag{4}$$

and the MOMENT-GENERATING FUNCTION by

$$M(t) = \langle e^{tz}\rangle = \sum_{x=0}^{\infty} e^{tx}\binom{x+r-1}{r-1}p^r(1-p)^x, \tag{5}$$

but, since $\binom{N}{n} = \binom{N}{N-m}$,

$$M(t) = p^r \sum_{x=0}^{\infty}\binom{x+r-1}{x}[(1-p)e^t]^x$$
$$= p^r[1-(1-p)e^t]^{-r} \tag{6}$$

$$M'(t) = p^r(-r)[1-(1-p)e^t]^{-r-1}(p-1)e^t$$
$$= p^r(1-p)r[1-(1-p)e^t]^{-r-1}e^t \tag{7}$$

$$M''(t) = (1-p)rp^r(1-e^t+pe^t)^{-r-2}$$
$$\times(-1-e^tr+e^tpr)e^t \tag{8}$$

$$M'''(t) = (1-p)rp^r(1-e^t+e^tp)^{-r-3}$$
$$\times[1+e^t(1-p+3r-3pr)+r^2e^{2t}(1-p)^2]e^t. \tag{9}$$

The MOMENTS about zero $K(u)$ are therefore

$$\mu_1' = \mu = \frac{r(1-p)}{p} = \frac{rq}{p} \tag{10}$$

$$\mu_2' = \frac{r(1-p)[1-r(p-1)]}{p^2} = \frac{rq(1-rq)}{p^2} \tag{11}$$

$$\mu_3' = \frac{(1-p)r(2-p+3r-3pr+r^2-2pr^2+p^2r^2)}{p^3} \tag{12}$$

$$\mu_4' = \frac{(-1+p)r(-6+6p-p^2-11r+15pr-4p^2r-6r^2}{p^4}$$
$$+ \frac{12pr^2-6p^2r^2-r^3+3pr^3-3p^2r^3+p^3r^3)}{p^4}. \tag{13}$$

(Beyer 1987, p. 487, apparently gives the MEAN incorrectly.) The MOMENTS about the mean are

$$\mu_2 = \sigma^2 = \frac{r(1-p)}{p^2} \tag{14}$$

$$\mu_3 = \frac{r(2-3p+p^2)}{p^3} = \frac{r(p-1)(p-2)}{p^3} \tag{15}$$

$$\mu_4 = \frac{r(1-p)(6-6p+p^2+3r-3pr)}{p^4}. \tag{16}$$

The MEAN, VARIANCE, SKEWNESS and KURTOSIS are then

$$\mu = \frac{r(1-p)}{p} \tag{17}$$

$$\gamma_1 = \frac{\mu_3}{\sigma^3} = \frac{r(p-1)(p-2)}{p^3}\left[\frac{p^2}{r(1-p)}\right]^{3/2}$$
$$= \frac{r(2-p)(1-p)}{p^3}\frac{p^3}{r(1-p)\sqrt{1-p}}$$
$$= \frac{2-p}{\sqrt{r(1-p)}} \tag{18}$$

$$\gamma_2 = \frac{\mu_4}{\sigma^4} - 3$$
$$= \frac{-6+6p-p^2-3r+3pr}{(p-1)r}, \tag{19}$$

which can also be written

$$\mu = nP \tag{20}$$

$$\mu_2 = nPQ \tag{21}$$

$$\gamma_1 = \frac{Q+P}{\sqrt{rPQ}} \tag{22}$$

$$\gamma_2 = \frac{1+6PQ}{rPQ} - 3. \tag{23}$$

The first CUMULANT is

$$\kappa_1 = nP, \tag{24}$$

and subsequent CUMULANTS are given by the RECURRENCE RELATION

$$\kappa_{r+1} = PQ\frac{d\kappa_r}{dQ}. \tag{25}$$

References
Beyer, W. H. *CRC Standard Mathematical Tables*, 28th ed. Boca Raton, FL: CRC Press, p. 533, 1987.
Spiegel, M. R. *Theory and Problems of Probability and Statistics.* New York: McGraw-Hill, p. 118, 1992.

Negative Binomial Series
The SERIES which arises in the BINOMIAL THEOREM for NEGATIVE integer n,

$$(x+a)^{-n} = \sum_{k=0}^{\infty}\binom{-n}{k}x^k a^{-n-k}$$
$$= \sum_{k=0}^{\infty}(-1)^k\binom{n+k-1}{k}x^k a^{-n-k}.$$

For $a=1$, the negative binomial series simplifies to

$$(x+1)^{-n} = 1-nx+\tfrac{1}{2}n(n+1)x^2-\tfrac{1}{6}n(n+1)(n+2)$$
$$+\dots.$$

See also BINOMIAL SERIES, BINOMIAL THEOREM

Negative Definite Matrix
A negative definite matrix is a HERMITIAN MATRIX all of whose EIGENVALUES are negative.

See also NEGATIVE SEMIDEFINITE MATRIX, POSITIVE DEFINITE MATRIX, POSITIVE SEMIDEFINITE MATRIX

References
Marcus, M. and Minc, H. *A Survey of Matrix Theory and Matrix Inequalities.* New York: Dover, p. 69, 1992.

Negative Integer
Z_-

Negative Likelihood Ratio
The term negative likelihood ratio is also used (especially in medicine) to test nonnested complementary hypotheses as follows,

$$\text{NLR} = \frac{[\text{true negative rate}]}{[\text{false negative rate}]} = \frac{[\text{specificity}]}{1-[\text{sensitivity}]}.$$

See also LIKELIHOOD RATIO, SENSITIVITY, SPECIFICITY

Negative Pedal Curve

Given a curve C and O a fixed point called the PEDAL POINT, then for a point P on C, draw a LINE PERPENDICULAR to OP. The ENVELOPE of these LINES as P describes the curve C is the negative pedal of C.

See also PEDAL CURVE

References

Lawrence, J. D. *A Catalog of Special Plane Curves.* New York: Dover, pp. 46–49, 1972.
Lockwood, E. H. "Negative Pedals." Ch. 19 in *A Book of Curves.* Cambridge, England: Cambridge University Press, pp. 156–159, 1967.

Negative Semidefinite Matrix

A negative semidefinite matrix is a HERMITIAN MATRIX all of whose EIGENVALUES are nonpositive.

See also NEGATIVE DEFINITE MATRIX, POSITIVE DEFINITE MATRIX, POSITIVE SEMIDEFINITE MATRIX

References

Marcus, M. and Minc, H. *A Survey of Matrix Theory and Matrix Inequalities.* New York: Dover, p. 69, 1992.

Neighborhood

The word neighborhood is a word with many different levels of meaning in mathematics. One of the most general concepts of a neighborhood of a point $\mathbf{x} \in \mathbb{R}^n$ (also called an epsilon-neighborhood or infinitesimal OPEN SET) is the set of points inside an n-BALL with center $\mathbf{x}$ and RADIUS $\epsilon > 0$.

See also BALL, OPEN SET

Neile's Parabola

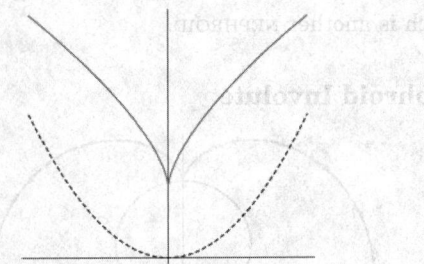

The solid curve in the above figure which is the EVOLUTE of the PARABOLA (dashed curve). In CARTESIAN COORDINATES,

$$y = \tfrac{3}{4}(2x)^{2/3} + \tfrac{1}{2}$$

Neile's parabola is also called the SEMICUBICAL PARABOLA, and was discovered by William Neile in 1657. It was the first nontrivial ALGEBRAIC CURVE to have its ARC LENGTH computed. Wallis published the method in 1659, giving Neile the credit (MacTutor Archive).

See also PARABOLA EVOLUTE

References

MacTutor History of Mathematics Archive. "Neile's Semi-Cubical Parabola." http://www-groups.dcs.st-and.ac.uk/~history/Curves/Neiles.html.

Nelder-Mead Method

A direct search method of optimization that works moderately well for stochastic problems. It is based on evaluating a function at the vertices of a SIMPLEX, then iteratively shrinking the simplex as better points are found until some desired bound is obtained (Nelder and Mead 1965).

See also STOCHASTIC OPTIMIZATION

References

Lagarias, J. C.; Reeds, J. A.; Wright, M. H.; and Wright, P. E. "Convergence Properties of the Nelder-Mead Algorithm in Low Dimensions." AT&T Bell Laboratories Tech. Rep. Murray Hill, NJ, 1995.
Nelder, J. A. and Mead, R. "A Simplex Method for Function Minimization." *Comput. J.* **7**, 308–313, 1965.
Press, W. H.; Flannery, B. P.; Teukolsky, S. A.; and Vetterling, W. T. *Numerical Recipes in C: The Art of Scientific Computing.* Cambridge, England: Cambridge University Press, 1989.
Walters, F. H.; Parker, L. R. Jr.; Morgan, S. L.; and Deming, S. N. *Sequential Simplex Optimization: A Technique for Improving Quality and Productivity in Research, Development, and Manufacturing.* Boca Raton, FL: CRC Press, 1991.
Woods, D. J. *An Interactive Approach for Solving Multi-Objective Optimization Problems.* Ph.D. thesis. Houston, TX: Rice University, 1985.
Wright, M. H. "The Nelder-Mean Method: Numerical Experimentation and Algorithmic Improvements." AT&T Bell Laboratories Techn. Rep. Murray Hill, NJ.
Wright, M. H. "Direct Search Methods: Once Scorned, Now Respectable." In *Numerical Analysis 1995. Papers from the Sixteenth Dundee Biennial Conference held at the University of Dundee, Dundee, June 27–30, 1995* (Ed. D. F. Griffiths and G. A. Watson). London: Longman, Harlow, pp. 191–208, 1996.

Nephroid

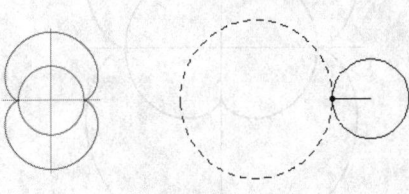

The 2-CUSPED EPICYCLOID is called a nephroid. Since $n = 2$, $a = b/2$, and the equation for r^2 in terms of the parameter ϕ is given by EPICYCLOID equation

$$r^2 = \frac{a^2}{n^2} \left[(n^2 + 2n + 2) - 2(n+1)\cos(n\phi)\right] \qquad (1)$$

with $n = 2$,

$$r^2 = \frac{a^2}{2^2}\left[(2^2 + 2 \cdot 2 + 2) - 2(2+1)\cos(2\phi)\right]$$

$$= \tfrac{1}{4}a^2[10 - 6\cos(2\phi)] = \tfrac{1}{2}a^2[5 - 3\cos(2\phi)], \qquad (2)$$

where

$$\tan\theta = \frac{3\sin\phi - \sin(3\phi)}{3\cos\phi - \cos(3\phi)}. \qquad (3)$$

This can be written

$$\left(\frac{r}{2a}\right)^{2/3} = \left[\sin\left(\tfrac{1}{2}\theta\right)\right]^{2/3} + \left[\cos\left(\tfrac{1}{2}\theta\right)\right]^{2/3}. \qquad (4)$$

The PARAMETRIC EQUATIONS are

$$x = a[3\cos t - \cos(3t)] \qquad (5)$$

$$y = a[3\sin t - \sin(3t)]. \qquad (6)$$

The Cartesian equation is

$$\left(x^2 + y^2 - 4a^2\right)^3 = 108a^4y^2. \qquad (7)$$

The name nephroid means "kidney shaped" and was first used for the two-cusped EPICYCLOID by Proctor in 1878 (MacTutor Archive). The nephroid has ARC LENGTH $24a$ and AREA $12\pi^2 a^2$. The CATACAUSTIC for rays originating at the CUSP of a CARDIOID and reflected by it is a nephroid. Huygens showed in 1678 that the nephroid is the CATACAUSTIC of a CIRCLE when the light source is at infinity. He published this fact in *Traité de la lumière* in 1690 (MacTutor Archive).

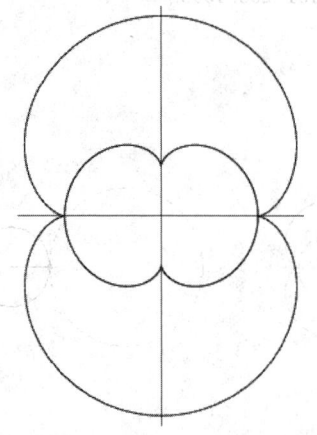

The nephroid can be generated as the ENVELOPE of circles centered on a given circle and tangent to one of the circle's diameters (Wells 1991).

See also ASTROID, DELTOID, FREETH'S NEPHROID

References

Beyer, W. H. *CRC Standard Mathematical Tables, 28th ed.* Boca Raton, FL: CRC Press, p. 221, 1987.

Lawrence, J. D. *A Catalog of Special Plane Curves.* New York: Dover, pp. 169–173, 1972.

Lockwood, E. H. "The Nephroid." Ch. 7 in *A Book of Curves.* Cambridge, England: Cambridge University Press, pp. 62–71, 1967.

MacTutor History of Mathematics Archive. "Nephroid." http://www-groups.dcs.st-and.ac.uk/~history/Curves/Nephroid.html.

Wells, D. *The Penguin Dictionary of Curious and Interesting Geometry.* London: Penguin, p. 158, 1991.

Yates, R. C. "Nephroid." *A Handbook on Curves and Their Properties.* Ann Arbor, MI: J. W. Edwards, pp. 152–154, 1952.

Nephroid Evolute

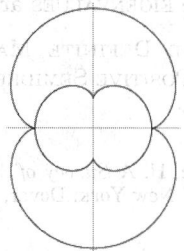

The EVOLUTE of the NEPHROID given by

$$x = \tfrac{1}{2}[3\cos t - \cos(3t)]$$

$$y = \tfrac{1}{2}[3\sin t - \sin(3t)]$$

is given by

$$x = \cos^3 t$$

$$y = \tfrac{1}{4}[3\sin t + \sin(3t)],$$

which is another NEPHROID.

Nephroid Involute

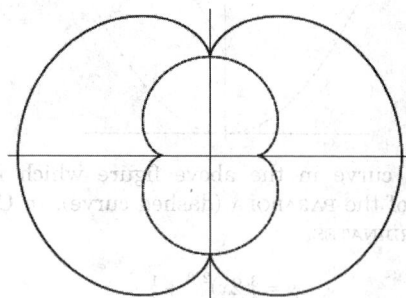

The INVOLUTE of the NEPHROID given by

$$x = \tfrac{1}{2}[3\cos t - \cos(3t)]$$

$$y = \tfrac{1}{2}[3\sin t - \sin(3t)]$$

beginning at the point where the nephroid cuts the y-

AXIS is given by

$$x = 4\cos^3 t$$

$$y = 3\sin t + \sin(3t),$$

another NEPHROID. If the INVOLUTE is begun instead at the CUSP, the result is CAYLEY'S SEXTIC.

Néron-Severi Group

Let V be a complete normal VARIETY, and write $G(V)$ for the group of divisors, $G_n(V)$ for the group of divisors numerically equal to 0, and $G_a(V)$ the group of divisors algebraically equal to 0. Then the finitely generated QUOTIENT GROUP $NS(V) = G(V)/G_a(V)$ is called the Néron-Severi group.

References

Iyanaga, S. and Kawada, Y. (Eds.). *Encyclopedic Dictionary of Mathematics.* Cambridge, MA: MIT Press, p. 75, 1980.

Nerve

The SIMPLICIAL COMPLEX formed from a family of objects by taking sets that have nonempty intersections.

See also DELAUNAY TRIANGULATION, SIMPLICIAL COMPLEX

Nested Hypothesis

Let S be the set of all possibilities that satisfy HYPOTHESIS H, and let S' be the set of all possibilities that satisfy HYPOTHESIS H'. Then H' is a nested hypothesis within H IFF $S' \subset S$, where $\subset$ denotes the PROPER SUBSET.

See also LOG LIKELIHOOD PROCEDURE

Nested Radical

Expressions OF THE FORM

$$\lim_{k \to \infty} x_0 + \sqrt{x_1 + \sqrt{x_2 + \sqrt{\cdots + x_k}}}.$$

Herschfeld (1935) proved that a nested radical of REAL NONNEGATIVE terms converges IFF $(x_n)^{2^{-n}}$ is bounded. He also extended this result to arbitrary POWERS (which include continued square roots and CONTINUED FRACTIONS as well), a result is known as HERSCHFELD'S CONVERGENCE THEOREM.

Nested radicals appear in the computation of PI,

$$\frac{2}{\pi} = \sqrt{\tfrac{1}{2}}\sqrt{\tfrac{1}{2}+\tfrac{1}{2}\sqrt{\tfrac{1}{2}}}\sqrt{\tfrac{1}{2}+\tfrac{1}{2}\sqrt{\tfrac{1}{2}+\tfrac{1}{2}\sqrt{\tfrac{1}{2}}}}\cdots \quad (1)$$

in TRIGONOMETRICAL values of COSINE and SINE for arguments OF THE FORM $\pi/2^n$, e.g.,

$$\sin\left(\frac{\pi}{8}\right) = \tfrac{1}{2}\sqrt{2-\sqrt{2}} \quad (2)$$

$$\cos\left(\frac{\pi}{8}\right) = \tfrac{1}{2}\sqrt{2+\sqrt{2}} \quad (3)$$

$$\sin\left(\frac{\pi}{16}\right) = \tfrac{1}{2}\sqrt{2-\sqrt{2+\sqrt{2}}} \quad (4)$$

$$\cos\left(\frac{\pi}{16}\right) = \tfrac{1}{2}\sqrt{2+\sqrt{2+\sqrt{2}}}, \quad (5)$$

and in the computation of the GOLDEN RATIO,

$$\phi = \sqrt{1+\sqrt{1+\sqrt{1+\sqrt{1+\cdots}}}}. \quad (6)$$

There are a number of general formula for nested radicals (Wong and McGuffin). For example,

$$x = \sqrt{(1-q)x^n + qx^{n-1}\sqrt{(1-q)x^n + qx^{n-1}\sqrt{\cdots}}} \quad (7)$$

which gives as special cases

$$\frac{b+\sqrt{b^2+4a}}{2} = \sqrt{a + b\sqrt{a + b\sqrt{a + b\sqrt{\cdots}}}} \quad (8)$$

$(n = 2, q = 1 - a/x^2, x = b/q)$,

$$x = \sqrt{x^{n-1}\sqrt{x^{n-1}\sqrt{x^{n-1}\sqrt{\cdots}}}} \quad (9)$$

$(q = 1)$, and

$$x = \sqrt{x\sqrt{x\sqrt{x\sqrt{x\sqrt{\cdots}}}}} \quad (10)$$

$(q = 1, n = 2)$. Equation (7) gives rise to

$$q^{(n^k-1/(n-1))}x^{n^j}$$

$$= \sqrt{q^{(n^{k+1}-n)/(n-1)}(1-q)x^{n^{j+1}} + \cdots}$$

$$\cdots + \sqrt{q^{(n^{k+2}-n)/(n-1)}(1-q)x^{n^{j+2}} + \sqrt{\cdots}}, \quad (11)$$

which gives the special case for $q = 1/2$, $n = 2$, $x = 1$, and $k = -1$,

$$\sqrt{2} = \sqrt{\frac{2}{2^{2^0}} + \sqrt{\frac{2}{2^{2^1}} + \sqrt{\frac{2}{2^{2^2}} + \sqrt{\frac{2}{2^{2^3}} + \sqrt{\frac{2}{2^{2^4}} + \cdots}}}}}. \quad (12)$$

Ramanujan discovered

$$x + n + a$$

$$= \sqrt{ax + (n+a)^2 + x\sqrt{a(x+n) + (n+a)^2 + \cdots}}$$

$$\ldots + (x+n)\sqrt{a(x+2n)+(n+a)^2+(x+2n)\sqrt{\ldots}},$$

which gives the special cases

$$x+1 = \sqrt{1+x\sqrt{1+(x+1)\sqrt{1+(x+2)\sqrt{1+\ldots}}}}, \quad (13)$$

for $a = 0$, $n = 1$, and

$$3 = \sqrt{1+2\sqrt{1+3\sqrt{1+4\sqrt{1+5\sqrt{\ldots}}}}} \quad (14)$$

for $a = 0$, $n = 1$, and $x = 2$.

For a nested radical OF THE FORM

$$x = \sqrt{n+\sqrt{n+\sqrt{n+\ldots}}} \quad (15)$$

to be equal a given REAL NUMBER x, it must be true that

$$x = \sqrt{n+\sqrt{n+\sqrt{n+\ldots}}} = \sqrt{n+x}, \quad (16)$$

so

$$x^2 = n+x \quad (17)$$

and

$$x = \tfrac{1}{2}\left(1+\sqrt{4n+1}\right). \quad (18)$$

See also CONTINUED FRACTION, GOLDEN RATIO, HERSCHFELD'S CONVERGENCE THEOREM, PI, SQUARE ROOT

References

Berndt, B. C. *Ramanujan's Notebooks, Part IV.* New York: Springer-Verlag, pp. 14–20, 1994.
Herschfeld, A. "On Infinite Radicals." *Amer. Math. Monthly* **42**, 419–429, 1935.
Landau, S. "A Note on 'Zippel Denesting.'" *J. Symb. Comput.* **13**, 31–45, 1992.
Landau, S. "Simplification of Nested Radicals." *SIAM J. Comput.* **21**, 85–110, 1992.
Landau, S. "How to Tangle with a Nested Radical." *Math. Intell.* **16**, 49–55, 1994.
Landau, S. "$\sqrt{2}+\sqrt{3}$: Four Different Views." *Math. Intell.* **20**, 55–60, 1998.
Pólya, G. and Szego, G. *Problems and Theorems in Analysis, Vol. 1.* New York: Springer-Verlag, 1997.
Sizer, W. S. "Continued Roots." *Math. Mag.* **59**, 23–27, 1986.
Wong, B. and McGuffin, M. "The Museum of Infinite Nested Radicals." http://www.csclub.uwaterloo.ca/~mjmcguff/math/nestedRadicals.html.

Nested Square

The black region in the nested square illustrated above, where the outer boundary is a unit square, has AREA 2.

References

Gardner, M. *The Sixth Book of Mathematical Games from Scientific American.* Chicago, IL: University of Chicago Press, pp. 165–166, 1984.

Net

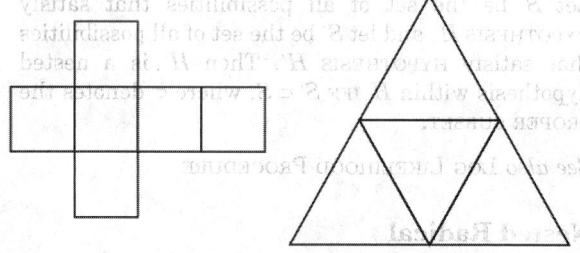

The word net has several meanings in mathematics. It refers to a plane diagram in which the EDGES of a POLYHEDRON are shown. All convex POLYHEDRA have nets, but not all concave polyhedra do (the constituent POLYGONS can overlap one another when a concave POLYHEDRON is flattened out). The GREAT DODECAHEDRON and STELLA OCTANGULA are examples of a concave polyhedron which have nonself-intersecting nets.

A corrected and concatenated version of the Bell Laboratories netlib polyhedron database has been prepared by Weisstein, together with *Mathematica* code to access analytic vertex coordinates and plot nets for all Platonic and Archimedean solids and their duals, as well as the Johnson solids. K. Fukuda has written routines which can unfold convex polyhedra into a planar net.

The term net also has a technical meaning as a generalization of a SEQUENCE, in which context it is also known as a Moore-Smith sequence. In this context, nets is used in general topology and ANALYSIS to imbue non-metrizable topological spaces with convergence properties. This artifice is needed only in spaces which are not FIRST-COUNTABLE, since sequences alone provide an adequate way of dealing with CONTINUITY for FIRST-COUNTABLE SPACES. Nets are used in the study of the RIEMANN INTEGRAL. Formally, a net of a set S is a mapping from a DIRECTED SET D into S.

See also DIRECTED SET, FIBER BUNDLE, FIBER SPACE, FIBRATION, UNFOLDING

References

Bell Laboratories. http://netlib.bell-labs.com/netlib/polyhedra/.
Weisstein, E. W. "Johnson Solids." MATHEMATICA NOTEBOOK JOHNSONSOLIDS.M.
Weisstein, E. W. "Johnson Solid Netlib Database." MATHEMATICA NOTEBOOK JOHNSONSOLIDS.DAT.

Netto's Conjecture

The probability that two elements P_1 and P_2 of a SYMMETRIC GROUP generate the entire GROUP tends to $3/4$ as $n \to \infty$ (Netto 1964, p. 90). The conjecture was proven by Dixon (1969).

See also PERMUTATION GROUP, SYMMETRIC GROUP

References

Dixon, J. D. "The Probability of Generating the Symmetric Group." *Math. Z.* **110**, 199–205, 1969.
Le Lionnais, F. *Les nombres remarquables.* Paris: Hermann, p. 31, 1983.
Netto, E. *The Theory of Substitutions.* New York: Chelsea, p. 90, 1964.

Network

A GRAPH or DIRECTED GRAPH together with a function which assigns a positive real number to each edge (Harary 1994, p. 52).

See also GRAPH, NETWORK FLOW, SINK (DIRECTED GRAPH), SMITH'S NETWORK THEOREM, SOURCE

References

Harary, F. *Graph Theory.* Reading, MA: Addison-Wesley, 1994.

Network Flow

The network flow problem considers a graph G with a set of sources S and sinks T and for which each edge has an assigned capacity (weight), and then asks to find the maximum flow that can be routed from S to T while respecting the given edge capacities. The network flow problem can be solved in time $\mathcal{O}(n^3)$ (Edmonds and Karp 1972; Skiena 1990, p. 237). It has been implemented as `NetworkFlow[g, source, sink]` in the *Mathematica* add-on package `DiscreteMath`Combinatorica`` (which can be loaded with the command `< <DiscreteMath``) and Net-

workFlowEdges[g, *source*, *sink*] in the *Mathematica* add-on package `DiscreteMath`Combinatorica`` (which can be loaded with the command `< <DiscreteMath`).

See also AUGMENTING PATH, MAXIMUM FLOW, MINIMUM CUT THEOREM, NETWORK

References

Edmonds, J. and Karp, R. M. "Theoretical Improvements in Algorithmic Efficiency for Network Flow Problems." *J. ACM* **19**, 248–264, 1972.
Even, S. and Tarjan, R. E. "Network Flow and Testing Graph Connectivity." *SIAM J. Comput.* **4**, 507–518, 1975.
Ford, L. R. and Fulkerson, D. R. *Flows in Networks.* Princeton, NJ: Princeton University Press, 1962.
Gonery, R. E. and Hu, T. C. "Multiterminal Network Flows." *J. SIAM* **9**, 551–570, 1961.
Orlin, J. B. "A Faster Strongly Polynomial Minimum Cost Flow Algorithm." *Proc. 20th ACM Symposium Theorem of Computing.* pp. 377–387, 1988.
Skiena, S. "Network Flow." §6.3 in *Implementing Discrete Mathematics: Combinatorics and Graph Theory with Mathematica.* Reading, MA: Addison-Wesley, pp. 237–239, 1990.
Skiena, S. S. "Network Flow." §8.4.9 in *The Algorithm Design Manual.* New York: Springer-Verlag, pp. 297–300, 1997.
Tarjan, R. E. *Data Structures and Network Algorithms.* Philadelphia, PA: SIAM Press, 1983.

Neuberg Center

The center of a NEUBERG CIRCLE.

See also NEUBERG CIRCLE

Neuberg Circle

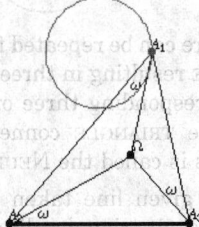

The LOCUS of the VERTEX A_1 of a TRIANGLE on a given base A_2A_3 and with a given BROCARD ANGLE ω is a CIRCLE (actually two circles, one on either side of A_2A_3) known as the Neuberg circle. From the center N_1, the base A_2A_3 subtends the ANGLE 2ω. The equation of the circle can be found by taking the base as $(0, 0)$, $(0, a_1)$ and solving

$$x^2 + y^2 = a_3^2 \tag{1}$$

$$(x - a_1)^2 + y^2 = a_2^2 \tag{2}$$

while eliminating a_2 and a_3 using

$$\cos \omega = \frac{a_1^2 + a_2^2 + a_3^3}{4\Delta} \tag{3}$$

where Δ is the area of the triangle $\Delta A_1A_2A_3$. Solving for x gives

$$x = \frac{1}{2}\left(a_1 \pm \sqrt{\pm 4a_1 y \cot \omega - 4y^2 - 3a_1^2}\right), \quad (4)$$

and squaring and completing the square results in

$$\left(x - \frac{1}{2}a_1\right)^2 + \left(y \pm \frac{1}{2}a_1 \cot \omega\right)^2 = \frac{1}{4}a_1(\cot^2 \omega - 3) \quad (5)$$

Therefore, the Neuberg circle N_1 on this edge has center

$$N_1 = \left(\frac{1}{2}a_1, \pm \frac{1}{2}a_1 \cot \omega\right) \quad (6)$$

(sometimes called the NEUBERG CENTER), and RADIUS

$$r = \frac{1}{2}a_1\sqrt{\cot^2 \omega - 3}.$$

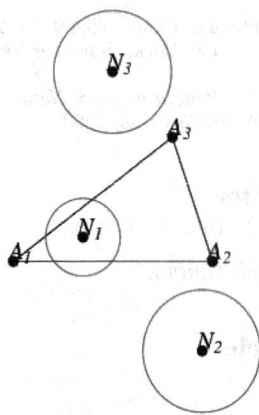

The same procedure can be repeated for the other two sides of a TRIANGLE resulting in three Neuberg circles (with another corresponding three on opposite sides of the edges). The TRIANGLE connecting the three NEUBERG CENTERS is called the NEUBERG TRIANGLE.

On one side of a given line taken as a base, it is possible to construct six triangles directly or inversely similar to a given SCALENE TRIANGLE, and the vertices of these triangles lie on their common Neuberg circles (Johnson 1929, p. 289).

See also BROCARD ANGLE, MCCAY CIRCLE, NEUBERG TRIANGLE

References

Coolidge, J. L. *A Treatise on the Geometry of the Circle and Sphere.* New York: Chelsea, pp. 79–80, 1971.
Emmerich, A. *Die Brocardschen Gebilde und ihre Beziehungen zu den verwandten merkwürdigen Punkten und Kreisen des Dreiecks.* Berlin: Georg Reimer, 1891.
Johnson, R. A. *Modern Geometry: An Elementary Treatise on the Geometry of the Triangle and the Circle.* Boston, MA: Houghton Mifflin, pp. 287–290, 1929.

Neuberg Triangle

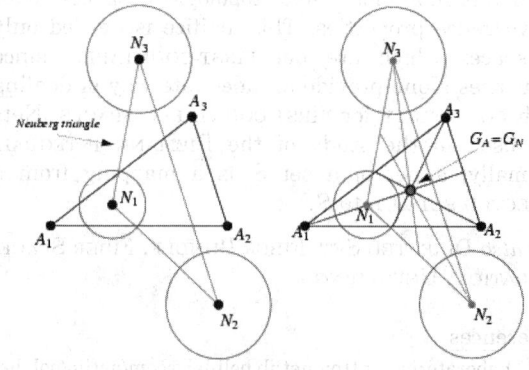

The TRIANGLE $\Delta N_1N_2N_3$ formed by joining a set of three NEUBERG CENTERS (i.e., centers of the NEUBERG CIRCLES) obtained from the edges of a given triangle $\Delta A_1A_2A_3$ (left figure). The CENTROID G_N of $\Delta N_1N_2N_3$ is coincident with the CENTROID G_A of $\Delta A_1A_2A_3$ (Johnson 1929, p. 288; right figure).

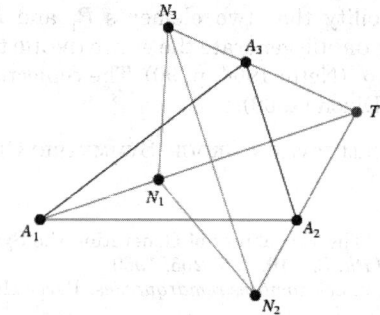

The lines A_1N_1, A_2N_2, and A_3N_3 are concurrent at a point T which Johnson (1929, p. 288) claims (apparently incorrectly) is the TARRY POINT.

See also NEUBERG CIRCLE, TARRY POINT

References

Johnson, R. A. *Modern Geometry: An Elementary Treatise on the Geometry of the Triangle and the Circle.* Boston, MA: Houghton Mifflin, 1929.

Neumann Algebra

VON NEUMANN ALGEBRA

Neumann Boundary Conditions

PARTIAL DIFFERENTIAL EQUATION BOUNDARY CONDITIONS which give the normal derivative on a surface.

See also BOUNDARY CONDITIONS, CAUCHY BOUNDARY CONDITIONS

References

Morse, P. M. and Feshbach, H. *Methods of Theoretical Physics, Part I.* New York: McGraw-Hill, p. 679, 1953.

Neumann Differential Equation

The second-order ORDINARY DIFFERENTIAL EQUATION

$$x^2 y'' + 3xy' + (x^2 + 1 - n^2)y$$
$$= x \cos^2\left(\tfrac{1}{2}n\pi\right) + n \sin^2\left(\tfrac{1}{2}n\pi\right)$$

satisfied by the NEUMANN POLYNOMIALS $O_n(x)$.

See also NEUMANN POLYNOMIAL

References

Gradshteyn, I. S. and Ryzhik, I. M. "Neumann's and Schläfli Polynomials: $O_n(z)$ and $S_n(x)$." §8.59 in *Tables of Integrals, Series, and Products, 6th ed.* San Diego, CA: Academic Press, pp. 989–991, 2000.
Zwillinger, D. *Handbook of Differential Equations, 3rd ed.* Boston, MA: Academic Press, p. 125, 1997.

Neumann Function

BESSEL FUNCTION OF THE SECOND KIND

Neumann Polynomial

Polynomials $O_n(x)$ that can be defined by the sum

$$O_n(x) = \frac{1}{4} \sum_{k=0}^{\lfloor n/2 \rfloor} \frac{n(n-k-1)!}{k!} \left(\tfrac{1}{2}x\right)^{2k-n-1} \quad (1)$$

for $n \geq 1$, where $\lfloor x \rfloor$ is the FLOOR FUNCTION. They obey the RECURRENCE RELATION

$$O_n(x) = -\frac{n}{n-2} O_{n-2}(x) + \frac{2n}{x} O_{n-1}(x)$$
$$+ \frac{2(n-1)}{(n-2)x} \sin^2\left[\tfrac{1}{2}(n-1)\pi\right] \quad (2)$$

for $n \geq 3$. They have the integral representation

$$O_n(x) = \int_0^\infty$$
$$\times \frac{\left(u + \sqrt{u^2 + x^2}\right)^n + \left(u - \sqrt{u^2 + x^2}\right)^n}{2x^{n+1}} e^{-u} du, \quad (3)$$

and the generating function

$$\frac{1}{x - \zeta} = J_0(\zeta)x^{-1} + 2 \sum_{n=1}^\infty J_n(\zeta)O_n(x) \quad (4)$$

(Gradshteyn and Ryzhik 2000, p. 990), and obey the NEUMANN DIFFERENTIAL EQUATION.

The first few Neumann polynomials are given by

$$O_0(x) = \frac{1}{x}$$

$$O_1(x) = \frac{1}{x^2}$$

$$O_2(x) = \frac{x^2 + 4}{x^3}$$

$$O_3(x) = \frac{3x^2 + 24}{x^4}$$

$$O_4(x) = \frac{x^4 + 16x^2 + 192}{x^5}$$

(A057869).

See also NEUMANN DIFFERENTIAL EQUATION, SCHLÄFLI POLYNOMIAL

References

Erdelyi, A.; Magnus, W.; Oberhettinger, F.; and Tricomi, F. G. *Higher Transcendental Functions, Vol. 2.* Krieger, pp. 32–33, 1981.
Gradshteyn, I. S. and Ryzhik, I. M. "Neumann's and Schläfli Polynomials: $O_n(z)$ and $S_n(z)$." §8.59 in *Tables of Integrals, Series, and Products, 6th ed.* San Diego, CA: Academic Press, pp. 989–991, 2000.
Sloane, N. J. A. Sequences A057869 in "An On-Line Version of the Encyclopedia of Integer Sequences." http://www.re-search.att.com/~njas/sequences/eisonline.html.
von Seggern, D. *CRC Standard Curves and Surfaces.* Boca Raton, FL: CRC Press, p. 196, 1993.
Watson, G. N. *A Treatise on the Theory of Bessel Functions, 2nd ed.* Cambridge, England: Cambridge University Press, pp. 298–305, 1966.

Neumann Series (Bessel Function)

A series OF THE FORM

$$\sum_{n=0}^\infty a_n J_{\nu+n}(z), \quad (1)$$

where ν is a REAL and $J_{\nu+n}(z)$ is a BESSEL FUNCTION OF THE FIRST KIND. Special cases are

$$z^\nu = 2^\nu \Gamma\left(\tfrac{1}{2}\nu + 1\right) \sum_{n=0}^\infty \frac{\left(\tfrac{1}{2}z\right)^{\nu/2+n}}{n!} J_{\nu/2+n}(z), \quad (2)$$

where $\Gamma(z)$ is the GAMMA FUNCTION, and

$$\sum_{n=0}^\infty b_n z^{\nu+n} = \sum_{n=0}^\infty a_n \left(\tfrac{1}{2}z\right)^{(\nu+n)/2} J_{(\nu+n)/2}(z), \quad (3)$$

where

$$a_n \equiv \sum_{m=0}^{\lfloor n/2 \rfloor} \frac{2^{\nu+n-2m} \Gamma\left(\tfrac{1}{2}\nu + \tfrac{1}{2}n - m + 1\right)}{m!} b_{n-2m}, \quad (4)$$

and $\lfloor x \rfloor$ is the FLOOR FUNCTION.

See also KAPTEYN SERIES

References

Watson, G. N. *A Treatise on the Theory of Bessel Functions, 2nd ed.* Cambridge, England: Cambridge University Press, 1966.

Neumann Series (Integral Equation)

A FREDHOLM INTEGRAL EQUATION OF THE SECOND KIND

$$\phi(x) = f(x) + \int_a^b K(x,t)\phi(t)dt \qquad (1)$$

may be solved as follows. Take

$$\phi_0(x) \equiv f(x) \qquad (2)$$

$$\phi_1(x) = f(x) + \lambda \int_a^b K(x,t)f(t)dt \qquad (3)$$

$$\phi_2(x) = f(x) + \lambda \int_a^b K(x,t_1)f(t_1)dt_1$$
$$+ \lambda^2 \int_a^b \int_a^b K(x,t_1)K(t_1,t_2)f(t_2)dt_2dt_1 \qquad (4)$$

$$\phi_n(x) = \sum_{i=0}^n \lambda^i u_i(x), \qquad (5)$$

where

$$u_0(x) = f(x) \qquad (6)$$

$$u_1(x) = \int_a^b K(x,t)f(t_1)dt_1 \qquad (7)$$

$$u_2(x) = \int_a^b \int_a^b K(x,t_1)K(t_1,t_2)f(t_2)dt_2dt_1. \qquad (8)$$

$$u_n(x) = \int_a^b \int_a^b \int_a^b K(x,t_1)K(t_1,t_2)\cdots$$
$$\times K(t_{n-1},t_n)f(t_n)dt_n\cdots dt_1. \qquad (9)$$

The Neumann series solution is then

$$\phi(x) = \lim_{n\to\infty} \phi_n(x) = \lim_{n\to\infty} \sum_{i=0}^n \lambda^i u_i(x). \qquad (10)$$

References

Arfken, G. "Neumann Series, Separable (Degenerate) Kernels." §16.3 in *Mathematical Methods for Physicists, 3rd ed.* Orlando, FL: Academic Press, pp. 879–890, 1985.

Neusis Construction

A geometric construction, also called a VERGING CONSTRUCTION, which allows the classical GEOMETRIC CONSTRUCTION rules to be bent in order to permit sliding of a *marked* RULER. Using a Neusis construction, CUBE DUPLICATION, angle TRISECTION, and construction of the regular HEPTAGON are soluble. The CONCHOID OF NICOMEDES can also be used to perform many Neusis constructions (Johnson 1975). Conway and Guy (1996) give Neusis constructions for the 7-, 9-, and 13-gons which are based on angle TRISECTION.

See also CONCHOID OF NICOMEDES, CUBE DUPLICATION, GEOMETRIC CONSTRUCTION, HEPTAGON, MASCHERONI CONSTRUCTION, MATCHSTICK CONSTRUCTION, RULER, STEINER CONSTRUCTION, TRISECTION

References

Conway, J. H. and Guy, R. K. *The Book of Numbers.* New York: Springer-Verlag, pp. 194–200, 1996.
Johnson, C. "A Construction for a Regular Heptagon." *Math. Gaz.* **59**, 17–21, 1975.

Nevanlinna Theory

An analytic refinement of results from COMPLEX analysis such as those codified by PICARD'S LITTLE THEOREM, PICARD'S GREAT THEOREM, and the WEIERSTRASS-CASORATI THEOREM.

See also PICARD'S GREAT THEOREM, PICARD'S LITTLE THEOREM, WEIERSTRASS-CASORATI THEOREM

References

Krantz, S. G. *Handbook of Complex Analysis.* Boston, MA: Birkhäuser, p. 141, 1999.

Neville Theta Function

The functions

$$O_2(x) = \frac{x^2 + 4}{x^3} \qquad (1)$$

$$O_3(x) = \frac{3x^2 + 24}{x^4} \qquad (2)$$

$$O_4(x) = \frac{x^4 + 16x^2 + 192}{x^5} \qquad (3)$$

$$\sum_{n=0}^\infty a_n J_{\nu+n}(z)_1 = J_{\nu+n}(z) \qquad (4)$$

where $z^\nu = 2^\nu \Gamma(\frac{1}{2}\nu + 1)\Sigma_{n=0}^\infty \frac{\left(\frac{1}{2}z\right)^{\nu/2+n}}{n!} J_{\nu/2+n}(z)$ and $\Gamma(z)$ are the JACOBI THETA FUNCTIONS and $\Sigma_{n=0}^\infty b_n z^{\nu+n} = \Sigma_{n=0}^\infty a_n(\frac{1}{2}z)^{(\nu+n)/2} J_{(\nu+n)/2}(z)$ is the complete ELLIPTIC INTEGRAL OF THE FIRST KIND.

See also JACOBI THETA FUNCTION, THETA FUNCTIONS

References

Abramowitz, M. and Stegun, C. A. (Eds.). "Neville's Notation for Theta Functions." §16.36 in *Handbook of Mathematical Functions with Formulas, Graphs, and Mathematical Tables, 9th printing.* New York: Dover, pp. 578–579, 1972.

Neville Theta Functions

The functions

$$\vartheta_s(u) = \frac{H(u)}{H'(0)} \tag{1}$$

$$\vartheta_d(u) = \frac{\Theta(u + K)}{\Theta(k)} \tag{2}$$

$$\vartheta_s(u) = \frac{H(u)}{H(K)} \tag{3}$$

$$\vartheta_n(u) = \frac{\Theta(u)}{\Theta(0)}, \tag{4}$$

where $H(u)$ and $\Theta(u)$ are the JACOBI THETA FUNCTIONS and $K(u)$ is the complete ELLIPTIC INTEGRAL OF THE FIRST KIND.

See also JACOBI THETA FUNCTIONS, THETA FUNCTIONS

References

Abramowitz, M. and Stegun, C. A. (Eds.). "Neville's Notation for Theta Functions." §16.36 in *Handbook of Mathematical Functions with Formulas, Graphs, and Mathematical Tables, 9th printing.* New York: Dover, pp. 578–579, 1972.

Neville's Algorithm

An interpolation ALGORITHM which proceeds by first fitting a POLYNOMIAL P_k of degree 0 through the points (x_k, y_k) for $k = 0 \ldots, n$, i.e., $P_k = y_k$. A second iteration is then performed in which P_{12} is fit through pairs of points, yielding $P_{12}, P_{23}, \ldots$. The procedure is repeated, generating a "pyramid" of approximations until the final result is reached

$$
\begin{array}{c}
P_1 \\
P_2 \quad P_{12} \\
P_3 \quad P_{23} \quad P_{123} \quad P_{1234}. \\
P_4 \quad P_{34} \quad P_{234}
\end{array}
$$

The final result is

$$
P_{i(i+1)\cdots(i+m)} = \frac{(x - x_{i+m})P_{i(i+1)\cdots(i+m-1)}}{x_i - x_{i+m}}
$$
$$
+ \frac{(x_i - x)P_{(i+1)(i+2)\cdots(i+m)}}{x_i - x_{i+m}}.
$$

See also BULIRSCH-STOER ALGORITHM

NevilleThetaC

NEVILLE THETA FUNCTIONS

NevilleThetaD

NEVILLE THETA FUNCTIONS

NevilleThetaN

NEVILLE THETA FUNCTIONS

NevilleThetaS

NEVILLE THETA FUNCTIONS

Newcomb's Paradox

A paradox in DECISION THEORY. Given two boxes, B1 which contains $1000 and B2 which contains either nothing or a million dollars, you may pick either B2 or both. However, at some time before the choice is made, an omniscient Being has predicted what your decision will be and filled B2 with a million dollars if he expects you to take it, or with nothing if he expects you to take both.

See also ALLAIS PARADOX

References

Erickson, G. W. and Fossa, J. A. *Dictionary of Paradox.* Lanham, MD: University Press of America, pp. 137–139, 1998.
Gardner, M. *The Unexpected Hanging and Other Mathematical Diversions.* Chicago, IL: Chicago University Press, 1991.
Gardner, M. "Newcomb's Paradox." Ch. 13 in *Knotted Doughnuts and Other Mathematical Entertainments.* New York: W. H. Freeman, pp. 155–161, 1986.
Nozick, R. "Reflections on Newcomb's Paradox." Ch. 14 in Gardner, M. *Knotted Doughnuts and Other Mathematical Entertainments.* New York: W. H. Freeman, 1986.

Newman-Conway Sequence

The sequence 1, 1, 2, 2, 3, 4, 4, 4, 5, 6, 7, 7, ... (Sloane's A004001) defined by $P(1) = P(2) = 1$ and the RECURRENCE RELATION

$$P(n) = P(P(n - 1)) + P(n - P(n - 1)). \tag{1}$$

It satisfies

$$P(2^k) = 2^{k-1} \tag{2}$$

and

$$P(2n) \le 2P(n). \tag{3}$$

References

Bloom, D. M. "Newman-Conway Sequence." Solution to Problem 1459. *Math. Mag.* **68**, 400–401, 1995.
Sloane, N. J. A. Sequences A004001/M0276 in "An On-Line Version of the Encyclopedia of Integer Sequences." http://www.research.att.com/~njas/sequences/eisonline.html.

Newman's Conjecture

If m is an integer, then for every residue class r (mod m), there are infinitely many nonnegative integers n for which $P(n) \equiv r$ (mod m), where $P(n)$ is the PARTITION FUNCTION P.

See also ERDOS-IVIC CONJECTURE, PARTITION FUNCTION P

References

Newman, M. "Periodicity Modulo m and Divisibility Properties of the Partition Function." *Trans. Amer. Math. Soc.* **97**, 225–236, 1960.

Ono, K. "Distribution of the Partition Functions Modulo m." *Ann. Math.* **151**, 293–307, 2000.

Newton Number

KISSING NUMBER

Newton-Bessel Formula

BESSEL'S FINITE DIFFERENCE FORMULA

Newton-Cotes Formulas

The Newton-Cotes formulas are an extremely useful and straightforward family of NUMERICAL INTEGRATION techniques.

To integrate a function $f(x)$ over some interval $[a, b]$, divide it into n equal parts such that $f_n = f(x_n)$ and $h \equiv (b-a)/n$. Then find POLYNOMIALS which approximate the tabulated function, and integrate them to approximate the AREA under the curve. To find the fitting POLYNOMIALS, use LAGRANGE INTERPOLATING POLYNOMIALS. The resulting formulas are called Newton-Cotes formulas, or QUADRATURE FORMULAS.

Newton-Cotes formulas may be "closed" if the interval $[x_1, x_n]$ is included in the fit, "open" if the points $[x_2, x_{n-1}]$ are used, or a variation of these two. If the formula uses n points (closed or open), the COEFFICIENTS of terms sum to $n-1$.

If the function $f(x)$ is given explicitly instead of simply being tabulated at the values x_i, the best numerical method of integration is called GAUSSIAN QUADRATURE. By picking the intervals at which to sample the function, this procedure produces more accurate approximations (but is significantly more complicated to implement).

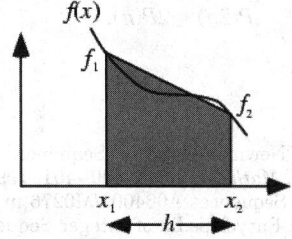

The 2-point closed Newton-Cotes formula is called the TRAPEZOIDAL RULE because it approximates the area under a curve by a TRAPEZOID with horizontal base and sloped top (connecting the endpoints x_1 and x_2). If the first point is x_1, then the other endpoint will be located at

$$x_2 = x_1 + h, \tag{1}$$

and the LAGRANGE INTERPOLATING POLYNOMIAL through the points (x_1, f_1) and (x_2, f_2) is

$$P_2(x) = \frac{x - x_2}{x_1 - x_2} f_1 + \frac{x - x_1}{x_2 - x_1} f_2$$

$$= \frac{x - x_1 - h}{-h} f_1 + \frac{x - x_1}{h} f_2$$

$$\frac{x}{h}(f_2 - f_1) + \left(f_1 + \frac{x_1}{h} f_1 - \frac{x_1}{h} f_2 \right). \tag{2}$$

Integrating over the interval (i.e., finding the area of the trapezoid) then gives

$$\int_{x_1}^{x_2} f(x)dx = \int_{x_1}^{x_1+h} P_2(x)dx$$

$$= \frac{1}{2h}(f_2 - f_1)[x^2]_{x_1}^{x_2} + \left(f_1 + \frac{x_1}{h} f_1 - \frac{x_1}{h} f_2 \right)[x]_{x_1}^{x_2}$$

$$= \frac{1}{2h}(f_2 - f_1)(x_2 + x_1)(x_2 - x_1) + (x_2 - x_1)$$

$$\times \left(f_1 + \frac{x_1}{h} f_1 - \frac{x_1}{h} f_2 \right)$$

$$= \frac{1}{2}(f_2 - f_1)(2x_1 + h) + f_1 h + x_1(f_1 - f_2)$$

$$= x_1(f_2 - f_1) + \frac{1}{2}h(f_2 - f_1) + hf_1 - x_1(f_2 - f_1)$$

$$= \frac{1}{2}h(f_1 + f_2) - \frac{1}{12}h^3 f''(\xi). \tag{3}$$

This is the trapezoidal rule (Ueberhuber 1997, p. 100), with the final term giving the amount of error (which, since $x_1 \le \xi \le x_2$, is no worse than the maximum value of $f''(\xi)$ in this range).

The 3-point rule is known as SIMPSON'S RULE. The ABSCISSAS are

$$x_2 = x_1 + h \tag{4}$$

$$x_3 = x_1 + 2h \tag{5}$$

and the LAGRANGE INTERPOLATING POLYNOMIAL is

$$P_3(x) = \frac{(x - x_2)(x - x_3)}{(x_1 - x_2)(x_1 - x_3)} f_1 + \frac{(x - x_1)(x - x_3)}{(x_2 - x_1)(x_2 - x_3)} f_2$$

$$+ \frac{(x - x_1)(x - x_2)}{(x_3 - x_1)(x_3 - x_2)} f_3$$

$$= \frac{x^2 - x(x_2 + x_3) + x_2 x_3}{h(2h)} f_1$$

$$+ \frac{x^2 - x(x_1 + x_3) + x_1 x_3}{h(-h)} f_2$$

$$+\frac{x^2 - x(x_1 + x_2) + x_1 x_2}{2h(h)}f_3$$

$$=\frac{1}{h^2}\left\{x^2\left(\tfrac{1}{2}f_1 - f_2 - \tfrac{1}{2}f_3\right) + x\left[-\tfrac{1}{2}(2x_1 + 3h)f_1\right.\right.$$

$$\left.+(2x_1 + 2h)f_2 - \tfrac{1}{2}(2x_1 + h)\right] + \left[\tfrac{1}{2}(x_1 + h)(x_1 + 2h)f_1\right.$$

$$\left.\left.-x_1(x_1 + 2h)f_2 + \tfrac{1}{2}x_1(x_1 + h)f_3\right]\right\}. \tag{6}$$

Integrating and simplifying gives

$$\int_{x_1}^{x_2} f(x)dx = \int_{x_1}^{x_1 + 2h} P_3(x)dx$$

$$= \tfrac{1}{3}h(f_1 + 4f_2 + f_3) - \tfrac{1}{90}h^5 f^{(4)}(\xi) \tag{7}$$

(Ueberhuber 1997, p. 100).

The 4-point closed rule is SIMPSON'S 3/8 RULE,

$$\int_{x_1}^{x_4} f(x)dx = \tfrac{3}{8}h(f_1 + 3f_2 + 3f_3 + f_4) - \tfrac{3}{80}h^5 f^{(4)}(\xi) \tag{8}$$

(Ueberhuber 1997, p. 100). The 5-point closed rule is BODE'S RULE,

$$\int_{x_1}^{x_5} f(x)dx = \tfrac{2}{45}h(7f_1 + 32f_2 + 12f_3 + 32f_4 + 7f_5)$$

$$-\tfrac{8}{945}h^7 f^{(6)}(\xi) \tag{9}$$

(Abramowitz and Stegun 1972, p. 886). Higher order rules include the 6-point

$$\int_{x_1}^{x_6} f(x)dx = \tfrac{5}{288}h(19f_1 + 75f_2 + 50f_3 + 50f_4 + 75f_5$$

$$+19f_6) - \tfrac{275}{12096}h^7 f^{(6)}(\xi), \tag{10}$$

7-point

$$\int_{x_1}^{x_7} f(x)dx = \tfrac{1}{140}h(41f_1 + 216f_2 + 27f_3 + 272f_4$$

$$+27f_5 + 216f_6 + 41f_7) - \tfrac{9}{1400}h^9 f^{(8)}(\xi), \tag{11}$$

8-point

$$\int_{x_1}^{x_8} f(x)dx = \tfrac{7}{17280}h(751f_1 + 3577f_2 + 1323f_2 + 2989f_3$$

$$+2989f_5 + 1323f_6 + 3577f_7 + 751f_8)$$

$$-\tfrac{8183}{518400}h^9 f^{(8)}(\xi), \tag{12}$$

9-point

$$\int_{x_1}^{x_9} f(x)dx = \tfrac{4}{14175}h(989f_1 + 5888f_2 - 928f_3$$

$$+10496f_4 + 4540f_5 + 10496f_6 - 928f_7 + 5888f_8 + 989f_9)$$

$$-\tfrac{2368}{467775}h^{11}f^{(10)}(\xi) \tag{13}$$

(Ueberhuber 1997, p. 100), 10-point

$$\int_{x_1}^{x_{10}} f(x)dx = \tfrac{9}{89600}h[2857(f_1 + f_{10})$$

$$+15741(f_2 + f_9) + 1080(f_3 + f_8) + 19344(f_4 + f_7)$$

$$+5788(f_5 + f_6)] - \tfrac{173}{14620}h^{11}f^{(10)}(\xi), \tag{14}$$

and 11-point

$$\int_{x_1}^{x_{11}} f(x)dx = \tfrac{5}{299376}h[16067(f_1 + f_{11})$$

$$+106300(f_2 + f_{10})] - 48525(f_3 + f_9) + 272400(f_4 + f_8)$$

$$-260550(f_5 + f_7) + 427368f_6] - \tfrac{1346350}{326918592}h^{13}f^{(12)}(\xi) \tag{15}$$

rules.

In general, the n-point rule is given by the analytic expression

$$\int_{x_1}^{x_n} f(x)dx = h\sum_{i=1}^{n} H_{n,i}f_i, \tag{16}$$

where

$$H_{n,r+1} = \frac{(-1)^{n-r}}{r!(n-r)!}\int_0^n t(t-1)\cdots(t-r+1)$$

$$\times (t-r-1)\cdots(t-n)dt \tag{17}$$

(Whittaker and Robinson 1967, p. 154).

Closed "extended" rules use multiple copies of lower order closed rules to build up higher order rules. By appropriately tailoring this process, rules with particularly nice properties can be constructed. For n tabulated points, using the TRAPEZOIDAL RULE $(n-1)$ times and adding the results gives

$$\int_{x_1}^{x_n} f(x)dx = \left(\int_{x_1}^{x_2} + \int_{x_2}^{x_3} + \cdots \int_{x_{n-1}}^{x_n}\right)f(x)\,dx$$

$$= \tfrac{1}{2}h[(f_1 + f_2) + (f_2 + f_3) + \cdots + (f_{n-2} + f_{n-1})$$

$$+(f_{n-1} + f_n)]$$

$$= h\left(\tfrac{1}{2}f_1 + f_2 + f_3 + \cdots + f_{n-2} + f_{n-1} + \tfrac{1}{2}f_n\right)$$

$$-\tfrac{1}{12}nh^3 f''(\xi) \tag{18}$$

(Ueberhuber 1997, p. 107). Using a series of refinements on the extended TRAPEZOIDAL RULE gives the method known as ROMBERG INTEGRATION. A 3-point extended rule for ODD n is

$$\int_{x_1}^{x_n} f(x)\,dx = h\left[\left(\tfrac{1}{3}f_1 + \tfrac{4}{3}f_2 + \tfrac{1}{3}f_3\right) + \left(\tfrac{1}{3}f_3 + \tfrac{4}{3}f_4 + \tfrac{1}{3}f_5\right)\right.$$

$$+ \cdots + \left(\tfrac{1}{3}f_{n-4} + \tfrac{4}{3}f_{n-3} + \tfrac{1}{3}f_{n-2}\right) + \left.\left(\tfrac{1}{3}f_{n-2} + \tfrac{4}{3}f_{n-1} + \tfrac{1}{3}f_n\right)\right]$$

$$= \tfrac{1}{3}h(f_1 + 4f_2 + 2f_3 + 4f_4 + 2f_5 + \cdots + 4f_{n-1} + f_n)$$

$$- \tfrac{n-1}{2}\tfrac{1}{90}h^5 f^{(4)}(\xi). \qquad (19)$$

Applying SIMPSON'S 3/8 RULE, then SIMPSON'S RULE (3-point) twice, and adding gives

$$\left[\int_{x_1}^{x_4} + \int_{x_4}^{x_6} + \int_{x_1}^{x_4}\right] f(x)\,dx$$

$$= h\left[\left(\tfrac{3}{8}f_1 + \tfrac{9}{8}f_2 + \tfrac{9}{8}f_3 + \tfrac{3}{8}f_4\right) + \left(\tfrac{1}{3}f_4 + \tfrac{4}{3}f_5 + \tfrac{1}{3}f_6\right)\right.$$

$$+ \left.\left(\tfrac{1}{3}f_6 + \tfrac{4}{3}f_7 + \tfrac{1}{3}f_8\right)\right]$$

$$= h\left[\tfrac{3}{8}f_1 + \tfrac{9}{8}f_2 + \tfrac{9}{8}f_3 + \left(\tfrac{3}{8}+\tfrac{1}{3}\right)f_4 + \tfrac{4}{3}f_5\right.$$

$$+ \left.\left(\tfrac{1}{3}+\tfrac{1}{3}\right)f_6 + \tfrac{4}{3}f_7 + \tfrac{1}{3}f_8\right]$$

$$= h\left(\tfrac{3}{8}f_1 + \tfrac{9}{8}f_2 + \tfrac{9}{8}f_3 + \tfrac{17}{24}f_4 + \tfrac{4}{3}f_5 + \tfrac{2}{3}f_6 + \tfrac{4}{3}f_7 + \tfrac{1}{3}f_8\right). \qquad (20)$$

Taking the next Simpson's 3/8 step then gives

$$\int_{x_8}^{x_{11}} f(x)\,dx = h\left(\tfrac{3}{8}f_8 + \tfrac{9}{8}f_9 + \tfrac{9}{8}f_{10} + \tfrac{3}{8}f_{11}\right). \qquad (21)$$

Combining with the previous result gives

$$\int_{x_1}^{x_{11}} f(x)\,dx = h\left[\tfrac{3}{8}f_1 + \tfrac{9}{8}f_2 + \tfrac{9}{8}f_3 + \tfrac{17}{24}f_4 + \tfrac{4}{3}f_5\right.$$

$$+ \tfrac{2}{3}f_6 + \tfrac{4}{3}f_7 + \left(\tfrac{1}{3}+\tfrac{3}{8}\right)f_8 + \tfrac{9}{8}f_9 + \tfrac{9}{8}f_{10} + \tfrac{3}{8}f_{11}\right]$$

$$= h\left(\tfrac{3}{8}f_1 + \tfrac{9}{8}f_2 + \tfrac{9}{8}f_3 + \tfrac{17}{24}f_4 + \tfrac{4}{3}f_5 + \tfrac{2}{3}f_6 + \tfrac{4}{3}f_7\right.$$

$$+ \left.\tfrac{17}{24}f_8 + \tfrac{9}{8}f_9 + \tfrac{9}{8}f_{10} + \tfrac{3}{8}f_{11}\right), \qquad (22)$$

where terms up to f_{10} have now been completely determined. Continuing gives

$$h\left(\tfrac{3}{8}f_1 + \tfrac{9}{8}f_2 + \tfrac{9}{8}f_3 + \tfrac{17}{24}f_4 + \tfrac{4}{3}f_5 + \tfrac{2}{3}f_6 + \cdots\right.$$

$$+ \left.\tfrac{2}{3}f_{n-5} + \tfrac{4}{3}f_{n-4} + \tfrac{17}{24}f_{n-3} + \tfrac{9}{8}f_{n-2} + \tfrac{9}{8}f_{n-1} + \tfrac{3}{8}f_n\right). \qquad (23)$$

Now average with the 3-point result

$$h\left(\tfrac{1}{3}f_1 + \tfrac{4}{3}f_2 + \tfrac{2}{3}f_3 + \tfrac{4}{3}f_4 + \tfrac{2}{3}f_5 + \tfrac{4}{3}f_{n-1} + \tfrac{1}{3}f_n\right) \qquad (24)$$

to obtain

$$h\left[\tfrac{17}{48}f_1 + \tfrac{59}{48}f_2 + \tfrac{43}{48}f_4 + \tfrac{49}{48}f_4 + (f_5 + f_6 + \cdots + f_{n-5} + f_{n-4})\right.$$

$$+ \left.\tfrac{49}{48}f_{n-3} + \tfrac{43}{38}f_{n-2} + \tfrac{59}{48}f_{n-1} + \tfrac{17}{48}f_n\right] + \mathcal{O}\left(n^{-4}\right). \qquad (25)$$

Note that all the middle terms now have unity

COEFFICIENTS. Similarly, combining a 4-point with the (2+4)-point rule gives

$$h\left(\tfrac{5}{12}f_1 + \tfrac{13}{12}f_2 + f_3 + f_4 + \cdots + f_{n-3} + f_{n-2} + \tfrac{13}{12}f_{n-1} + \tfrac{5}{12}\right)$$

$$+ \mathcal{O}\left(n^{-3}\right). \qquad (26)$$

Other Newton-Cotes rules occasionally encountered include DURAND'S RULE

$$\int_{x_1}^{x_n} f(x)\,dx = h\left(\tfrac{2}{5}f_1 + \tfrac{11}{10}f_2 + f_3 + \cdots + f_{n-2} + \tfrac{11}{10}f_{n-1} + \tfrac{2}{5}f_n\right) \qquad (27)$$

(Beyer 1987), HARDY'S RULE

$$\int_{x_0-3h}^{x_0+3h} f(x)\,dx$$

$$= \tfrac{1}{100}h(28f_{-3} + 162f_{-2} + 22f_0 + 162f_2 + 28f_3)$$

$$+ \tfrac{9}{1400}h^7\left[2f^{(4)}(\xi_2) - h^2 f^{(8)}(\xi_1)\right], \qquad (28)$$

and WEDDLE'S RULE

$$\int_{x_1}^{x_{6n}} f(x)\,dx = \tfrac{3}{10}h(f_1 + 5f_2 + f_3 + 6f_4 + 5f_5 + f_6$$

$$+ \cdots + 5f_{6n-1} + f_{6n}) \qquad (29)$$

(Beyer 1987).

The open Newton-Cotes rules use points outside the integration interval, yielding the 1-point

$$\int_{x_0}^{x_2} f(x)\,dx = 2hf_1, \qquad (30)$$

2-point

$$\int_{x_0}^{x_3} f(x)\,dx = \int_{x_1-h}^{x_1+2h} P_2(x)\,dx$$

$$= \tfrac{1}{2h}(f_2 - f_1)\left[x^2\right]_{x_1-h}^{x_1+2h} + \left(f_1 + \tfrac{x_1}{h}f_1 - \tfrac{x_1}{h}f_2\right)[x]_{x_1-h}^{x_1+2h}$$

$$= \tfrac{3}{2}h(f_1 + f_2) + \tfrac{1}{4}h^3 f''(\xi), \qquad (31)$$

3-point

$$\int_{x_0}^{x_4} f(x)\,dx = \tfrac{4}{3}h(2f_1 - f_2 + 2f_3) + \tfrac{28}{90}h^5 f^{(4)}(\xi), \qquad (32)$$

4-point

$$\int_{x_0}^{x_5} f(x)\,dx = \tfrac{5}{24}h(11f_1 + f_2 + f_3 + 11f_4)$$

$$+ \tfrac{95}{144}h^5 f^{(4)}(\xi), \qquad (33)$$

5-point

$$\int_{x_0}^{x_6} f(x)\,dx = \tfrac{6}{20}h(11f_1 - 14f_2 + 26f_3 - 14f_4 + 11f_5)$$

$$- \tfrac{41}{140}h^7 f^{(6)}(\xi), \tag{34}$$

6-point

$$\int_{x_0}^{x_7} f(x)dx = \tfrac{7}{1440}h(611f_1 - 453f_2 + 562f_3 + 562f_4$$

$$- 453f_5 + 611f_6) - \tfrac{5257}{8640}h^7 f^{(6)}(\xi), \tag{35}$$

and 7-point

$$\int_{x_0}^{x_8} f(x)dx = \tfrac{8}{945}h(460f_1 - 954f_2 + 2196f_3 - 2459f_4$$

$$+ 2196f_5 - 954f_6 + 460f_7) - \tfrac{3956}{14175}h^9 f^{(8)}(\xi) \tag{36}$$

rules.

A 2-point open extended formula is

$$\int_{x_1}^{x_n} f(x)dx = h\left[\left(\tfrac{1}{2}f_1 + f_2 + \ldots + f_{n-1} + \tfrac{1}{2}f_n\right)\right.$$

$$\left. + \tfrac{1}{24}(-f_0 + f_2 + f_{n-1} - f_{n+1})\right] + \tfrac{11(n+1)}{720}h^5 f^{(4)}(\xi). \tag{37}$$

Single interval extrapolative rules estimate the integral in an interval based on the points around it. An example of such a rule is

$$hf_1 + \mathcal{O}\left(h^2 f'\right) \tag{38}$$

$$\tfrac{1}{2}h(3f_1 - f_2) + \mathcal{O}\left(h^3 f''\right) \tag{39}$$

$$\tfrac{1}{12}h(23f_1 - 16f_2 + 5f_3) + \mathcal{O}\left(h^4 f^{(3)}\right) \tag{40}$$

$$\tfrac{1}{24}h(55f_1 - 59f_2 + 37f_3 - 9f_4) + \mathcal{O}\left(h^5 f^{(4)}\right). \tag{41}$$

See also Bode's Rule, Difference Equation, Durand's Rule, Finite Difference, Gaussian Quadrature, Hardy's Rule, Lagrange Interpolating Polynomial, Numerical Integration, Shovelton's Rule, Simpson's Rule, Simpson's 3/8 Rule, Trapezoidal Rule, Weddle's Rule, Woolhouse's Formulas

References

Abramowitz, M. and Stegun, C. A. (Eds.). "Integration." §25.4 in *Handbook of Mathematical Functions with Formulas, Graphs, and Mathematical Tables, 9th printing.* New York: Dover, pp. 885–887, 1972.

Beyer, W. H. (Ed.). *CRC Standard Mathematical Tables, 28th ed.* Boca Raton, FL: CRC Press, p. 127, 1987.

Corbit, D. "Numerical Integration: From Trapezoids to RMS: Object-Oriented Numerical Integration." *Dr. Dobb's J.,* No. 252, 117–120, Oct. 1996.

Daniell, P. J. "Remainders in Interpolation and Quadrature Formulae." *Math. Gaz.* **24**, 238, 1940.

Hildebrand, F. B. *Introduction to Numerical Analysis.* New York: McGraw-Hill, pp. 160–161, 1956.

Press, W. H.; Flannery, B. P.; Teukolsky, S. A.; and Vetterling, W. T. "Classical Formulas for Equally Spaced Abscissas." §4.1 in *Numerical Recipes in FORTRAN: The Art of Scientific Computing, 2nd ed.* Cambridge, England: Cambridge University Press, pp. 124–130, 1992.

Ueberhuber, C. W. *Numerical Computation 2: Methods, Software, and Analysis.* Berlin: Springer-Verlag, 1997.

Whittaker, E. T. and Robinson, G. "The Newton-Cotes Formulae of Integration." §76 in *The Calculus of Observations: A Treatise on Numerical Mathematics, 4th ed.* New York: Dover, pp. 152–156, 1967.

Newton-Gauss Backward Formula

Gauss's Backward Formula

Newton-Gauss Forward Formula

Gauss's Forward Formula

Newton-Girard Formulas

The identities between the elementary symmetric functions $\prod_k(x_1,\ldots,x_n)$ and the sums of nth powers of their variables $S_k = \sum_{k=1}^n x_k$. For $1 \le k \le n$, the identity is

$$(-1)^n n \prod_n (x_1,\ldots,x_k)$$

$$+ \sum_{k=0}^{n-1} (-1)^k S_k(x_1,\ldots,x_k) \prod_k (x_1,\ldots,x_k) = 0, \tag{1}$$

the first few of which are

$$S_1 - \prod_n = 0 \tag{2}$$

$$S_2 - S_1 \prod_1 + 2\prod_2 = 0 \tag{3}$$

$$S_3 - S_2 \prod_1 + S_1 \prod_2 - 3\prod_3 = 0. \tag{4}$$

See also Symmetric Polynomial

References

Séroul, R. "Newton-Girard Formulas." §10.12 in *Programming for Mathematicians.* Berlin: Springer-Verlag, pp. 278–279, 2000.

Newtonian Form

Newton's Divided Difference Interpolation Formula

Newton-Raphson Fractal

Newton's Method

Newton-Raphson Method

Newton's Method

Newton's Backward Difference Formula

$$f_p = f_0 + p\nabla_0 + \tfrac{1}{2!}p(p+1)\nabla_0^2 + \tfrac{1}{3!}p(p+1)(p+2)\nabla_0^3 + \cdots,$$

for $p \in [0,1]$, where ∇ is the BACKWARD DIFFERENCE.

See also NEWTON'S FORWARD DIFFERENCE FORMULA

References

Beyer, W. H. *CRC Standard Mathematical Tables, 28th ed.* Boca Raton, FL: CRC Press, p. 433, 1987.

Newton's Diverging Parabolas

Curves with CARTESIAN equation

$$ay^2 = x(x^2 - 2bx + c)$$

with $a > 0$. The above equation represents the third class of Newton's classification of CUBIC CURVES, which Newton divided into five species depending on the ROOTS of the cubic in x on the right-hand side of the equation. Newton described these cases as having the following characteristics:

1. "All the ROOTS are REAL and unequal. Then the Figure is a diverging Parabola OF THE FORM of a Bell, with an Oval at its Vertex.
2. Two of the ROOTS are equal. A PARABOLA will be formed, either Nodated by touching an Oval, or Punctate, by having the Oval infinitely small.
3. The three ROOTS are equal. This is the NEILIAN PARABOLA, commonly called SEMI-CUBICAL.
4. Only one REAL ROOT. If two of the ROOTS are impossible, there will be a Pure PARABOLA of a Bell-like Form"

(MacTutor Archive).

References

MacTutor History of Mathematics Archive. "Newton's Diverging Parabolas." http://www-groups.dcs.st-and.ac.uk/~history/Curves/Newtons.html.

Newton's Divided Difference Interpolation Formula

Let

$$\pi_n(x) \equiv \prod_{i=1}^{n}(x - x_n), \qquad (1)$$

then

$$f(x) = f_0 + \sum_{k=1}^{n} x_{k-1}(x)[x_0, x_1 \ldots, x_k] + R_n, \qquad (2)$$

where $[x_1, \ldots]$ is a DIVIDED DIFFERENCE, and the remainder is

$$R_n(x) = \pi_n(x)[x_0, \ldots, x_n, x] = \pi_n(x)\frac{f^{(n+1)}(\xi)}{(n+1)} \qquad (3)$$

for $x_0 < \xi < x_n$.

See also DIVIDED DIFFERENCE, FINITE DIFFERENCE

References

Abramowitz, M. and Stegun, C. A. (Eds.). *Handbook of Mathematical Functions with Formulas, Graphs, and Mathematical Tables, 9th printing.* New York: Dover, p. 880, 1972.

Hildebrand, F. B. *Introduction to Numerical Analysis.* New York: McGraw-Hill, pp. 43–44 and 62–63, 1956.

Whittaker, E. T. and Robinson, G. "Newton's Formula for Unequal Intervals." §13 in *The Calculus of Observations: A Treatise on Numerical Mathematics, 4th ed.* New York: Dover, pp. 24–26, 1967.

Newton's Formulas

Let a TRIANGLE have side lengths a, b, and c with opposite angles A, B, and C. Then

$$\frac{b+c}{a} = \frac{\cos\left[\tfrac{1}{2}(B-C)\right]}{\sin\left(\tfrac{1}{2}A\right)}$$

$$\frac{c+a}{b} = \frac{\cos\left[\tfrac{1}{2}(C-A)\right]}{\sin\left(\tfrac{1}{2}B\right)}$$

$$\frac{a+b}{c} = \frac{\cos\left[\tfrac{1}{2}(A-B)\right]}{\sin\left(\tfrac{1}{2}C\right)}$$

See also MOLLWEIDE'S FORMULAS, TRIANGLE

References

Beyer, W. H. *CRC Standard Mathematical Tables, 28th ed.* Boca Raton, FL: CRC Press, p. 146, 1987.

Newton's Forward Difference Formula

A FINITE DIFFERENCE identity giving an interpolated value between tabulated points $\{f_p\}$ in terms of the first value f_0 and the POWERS of the FORWARD DIFFERENCE Δ. For $a \in [0,1]$, the formula states

$$f_a = f_0 + a\Delta + \tfrac{1}{2!}a(a-1)\Delta^2 + \tfrac{1}{3!}a(a-1)(a-2)\Delta^3 + \ldots$$

When written in the form

$$f(x+a) = \sum_{n=0}^{\infty} \frac{(a)_n \Delta^n f(x)}{n!}$$

with $(a)_n$ the POCHHAMMER SYMBOL, the formula looks suspiciously like a finite analog of a TAYLOR SERIES expansion. This correspondence was one of the motivating forces for the development of UMBRAL CALCULUS.

The DERIVATIVE of Newton's forward difference formula gives MARKOFF'S FORMULAS.

See also FINITE DIFFERENCE, MARKOFF'S FORMULAS, NEWTON'S BACKWARD DIFFERENCE FORMULA, NEWTON'S DIVIDED DIFFERENCE INTERPOLATION FORMULA

References

Abramowitz, M. and Stegun, C. A. (Eds.). *Handbook of Mathematical Functions with Formulas, Graphs, and Mathematical Tables, 9th printing.* New York: Dover, p. 880, 1972.

Beyer, W. H. *CRC Standard Mathematical Tables, 28th ed.* Boca Raton, FL: CRC Press, p. 432, 1987.

Whittaker, E. T. and Robinson, G. "The Gregory-Newton Formula of Interpolation" and "An Alternative Form of the Gregory-Newton Formula." §8–9 in *The Calculus of Observations: A Treatise on Numerical Mathematics, 4th ed.* New York: Dover, pp. 10–15, 1967.

Newton's Identities

NEWTON'S RELATIONS

Newton's Iteration

An algorithm for computing the SQUARE ROOT of a number n quadratically as $\lim_{k \to \infty} x_k$,

$$x_{k+1} = \frac{1}{2}\left(x_k + \frac{n}{x_k}\right),$$

where $x_0 = 1$. The first few approximants to $\sqrt{n}$ are given by

$$1, \tfrac{1}{2}(1+n), \frac{1 + 6n + n^2}{4(n+1)},$$

$$\frac{1 + 28n + 70n^2 + 28n^3 + n^4}{8(1+n)(1 + 6n + n^2)}, \dots$$

For $\sqrt{2}$, this gives the convergents as 1, 3/2, 17/12, 577/408, 665857/470832, ... (Sloane's A051008 and A051009).

See also SQUARE ROOT

References

Sloane, N. J. A. Sequences A051008 and A051008 in "An On-Line Version of the Encyclopedia of Integer Sequences." http://www.research.att.com/~njas/sequences/eisonline.html.

Newton's Method

A ROOT-finding ALGORITHM which uses the first few terms of the TAYLOR SERIES of a function $f(x)$ in the vicinity of a suspected ROOT to zero in on the root. It is also called the Newton-Raphson method. For $f(x)$ a POLYNOMIAL, Newton's method is essentially the same as HORNER'S METHOD. The TAYLOR SERIES of $f(x)$ about the point $x + \varepsilon$ is given by

$$f(x + \varepsilon) = f(x) + f'(x)\varepsilon + \tfrac{1}{2}f''(x)\varepsilon^2 + \dots. \tag{1}$$

Keeping terms only to first order,

$$f(x + \varepsilon) \approx f(x) + f'(x)\varepsilon. \tag{2}$$

This expression can be used to estimate the amount of offset ε needed to land closer to the root starting from an initial guess x_0. Setting $f(x_0 + \varepsilon) = 0$ and solving (2) for ε gives

$$\varepsilon_0 = -\frac{f(x_0)}{f'(x_0)}, \tag{3}$$

which is the first-order adjustment to the ROOT's position. By letting $x_1 = x_0 + \varepsilon_0$, calculating a new ε_1, and so on, the process can be repeated until it converges to a root.

Unfortunately, this procedure can be unstable near a horizontal ASYMPTOTE or a LOCAL MINIMUM. However, with a good initial choice of the ROOT's position, the algorithm can by applied iteratively to obtain

$$x_{n+1} = x_n - \frac{f(x_n)}{f'(x_n)} \tag{4}$$

for $n = 1, 2, 3, \dots$. An initial point x_0 that provides safe convergence of Newton's method is called an APPROXIMATE ZERO.

The error ε_{n+1} after the $(n+1)$st iteration is given by

$$\varepsilon_{n+1} = \varepsilon_n + (x_{n+1} - x_n)$$

$$= \varepsilon_n - \frac{f(x_n)}{f'(x_n)}. \tag{5}$$

But

$$f(x_n) = f(x) + f'(x)\varepsilon_n + \tfrac{1}{2}f''(x)\varepsilon_n^2 + \dots$$

$$= f'(x)\varepsilon_n + \tfrac{1}{2}f''(x)\varepsilon_n^2 + \dots \tag{6}$$

$$f'(x_n) = f'(x) + f''(x)\varepsilon_n + \dots, \tag{7}$$

so

$$\frac{f(x_n)}{f'(x_x)} = \frac{f'(x)\varepsilon_n + \tfrac{1}{2}f''(x)\varepsilon_n^2 + \dots}{f'(x)f''(x)\varepsilon_n + \dots}$$

$$\approx \frac{f'(x)\varepsilon + \tfrac{1}{2}f''(x)\varepsilon_n^2}{f'(x) + f''(x)\varepsilon_n} = \varepsilon_n + \frac{f''(x)}{2f'(x)}\varepsilon_n^2, \tag{8}$$

and (5) becomes

$$\varepsilon_{n+1} = \varepsilon_n - \left[\varepsilon_n + \frac{f''(x)}{2f'(x)}\varepsilon_n^2\right] = -\frac{f''(x)}{2f'(x)}\varepsilon_n^2. \tag{9}$$

Therefore, when the method converges, it does so quadratically.

A FRACTAL is obtained by applying Newton's method to finding a ROOT of $z^n - 1 = 0$ (Mandelbrot 1983, Gleick 1988, Peitgen and Saupe 1988, Press *et al.* 1992, Dickau 1997). Iterating for a starting point z_0

gives

$$z_{i+1} = z_i - \frac{z_i^n - 1}{n z_i^{n-1}}. \tag{10}$$

Since this is an nth order POLYNOMIAL, there are n ROOTS to which the algorithm can converge.

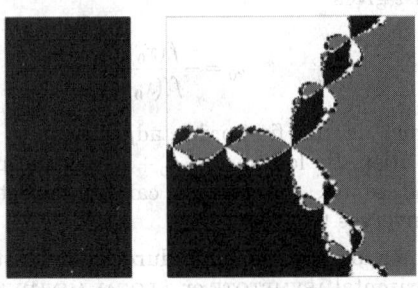

Coloring the BASIN OF ATTRACTION (the set of initial points z_0 which converge to the same ROOT) for each ROOT a different color then gives the above plots, corresponding to $n = 2, 3, 4$, and 5.

See also ALPHA-TEST, APPROXIMATE ZERO, HALLEY'S IRRATIONAL FORMULA, HALLEY'S METHOD, HORNER'S METHOD, HOUSEHOLDER'S METHOD, LAGUERRE'S METHOD

References

Abramowitz, M. and Stegun, C. A. (Eds.). *Handbook of Mathematical Functions with Formulas, Graphs, and Mathematical Tables, 9th printing.* New York: Dover, p. 18, 1972.

Acton, F. S. Ch. 2 in *Numerical Methods That Work.* Washington, DC: Math. Assoc. Amer., 1990.

Arfken, G. *Mathematical Methods for Physicists, 3rd ed.* Orlando, FL: Academic Press, pp. 963–964, 1985.

Boyer, C. B. and Merzbacher, U. C. *A History of Mathematics, 2nd ed.* New York: Wiley, 1991.

Dickau, R. M. "Basins of Attraction for $z^5 = 1$ Using Newton's Method in the Complex Plane." http://forum.swarthmore.edu/advanced/robertd/newtons.html.

Dickau, R. M. "Variations on Newton's Method." http://forum.swarthmore.edu/advanced/robertd/newnewton.html.

Dickau, R. M. "Compilation of Iterative and List Operations." *Mathematica J.* **7**, 14–15, 1997.

Gleick, J. *Chaos: Making a New Science.* New York: Penguin Books, plate 6 (following pp. 114) and p. 220, 1988.

Gourdon, X. and Sebah, P. "Newton's Iteration." http://xavier.gourdon.free.fr/Constants/Algorithms/newton.html.

Householder, A. S. *Principles of Numerical Analysis.* New York: McGraw-Hill, pp. 135–138, 1953.

Mandelbrot, B. B. *The Fractal Geometry of Nature.* San Francisco, CA: W. H. Freeman, 1983.

Newton, I. *Methodus fluxionum et serierum infinitarum.* 1664–1671.

Ortega, J. M. and Rheinboldt, W. C. *Iterative Solution of Nonlinear Equations in Several Variables.* Philadelphia, PA: SIAM, 2000.

Peitgen, H.-O. and Saupe, D. *The Science of Fractal Images.* New York: Springer-Verlag, 1988.

Press, W. H.; Flannery, B. P.; Teukolsky, S. A.; and Vetterling, W. T. "Newton-Raphson Method Using Derivatives" and "Newton-Raphson Methods for Nonlinear Systems of Equations." §9.4 and 9.6 in *Numerical Recipes in FORTRAN: The Art of Scientific Computing, 2nd ed.* Cambridge, England: Cambridge University Press, pp. 355–362 and 372–375, 1992.

Ralston, A. and Rabinowitz, P. §8.4 in *A First Course in Numerical Analysis, 2nd ed.* New York: McGraw-Hill, 1978.

Raphson, J. *Analysis aequationum universalis.* London, 1690.

Whittaker, E. T. and Robinson, G. "The Newton-Raphson Method." §44 in *The Calculus of Observations: A Treatise on Numerical Mathematics, 4th ed.* New York: Dover, pp. 84–87, 1967.

Newton's Parallelogram

Approximates the possible values of y in terms of x if

$$\sum_{i,j=0}^{n} a_{ij} x^i y^j = 0.$$

Newton's Relations

Let s_i be the sum of the products of distinct ROOTS r_j of the POLYNOMIAL equation of degree n

$$a_n x^n + a_{n-1} x^{n-1} + \ldots + a_1 x + a_0 = 0, \tag{1}$$

where the roots are taken i at a time (i.e., s_i is defined as the SYMMETRIC POLYNOMIAL $\prod_i (r_1, \ldots, r_n)$) s_i is defined for $i = 1, \ldots, n$. For example, the first few values of s_i are

$$s_1 = r_1 + r_2 + r_3 + r_4 + \ldots \tag{2}$$

$$s_2 = r_1 r_2 + r_1 r_3 + r_1 r_4 + r_2 r_3 + \ldots \tag{3}$$

$$s_3 = r_1 r_2 r_3 + r_1 r_2 r_4 + r_2 r_3 r_4 + \ldots, \tag{4}$$

and so on. Then

$$s_i = (-1)^i \frac{a_{n-i}}{a_n}. \tag{5}$$

This can be seen for a second DEGREE POLYNOMIAL by multiplying out,

$$\begin{aligned} a_2 x^2 + a_1 x + a_0 &= a_2 (x - r_1)(x - r_2) \\ &= a_2 [x^2 - (r_1 + r_2)x + r_1 r_2], \end{aligned} \tag{6}$$

so

$$s_1 = \sum_{i=1}^{2} r_i = r_1 + r_2 = -\frac{a_1}{a_2} \qquad (7)$$

$$s_2 = \sum_{\substack{i,j=1 \\ i \neq j}}^{2} r_i r_j = r_1 r_2 = \frac{a_0}{a_2}, \qquad (8)$$

and for a third DEGREE POLYNOMIAL,

$$a_3 x^3 + a_2 x^2 + a_1 x + a_0 = a_3 (x - r_1)(x - r_2)(x - r_3)$$

$$= a_3 \left[x^3 - (r_1 + r_2 + r_3) x^2 + (r_1 r_2 + r_1 r_3 + r_2 r_3) x - r_1 r_2 r_3 \right],$$
$$(9)$$

so

$$s_1 = \sum_{i=1}^{3} r_i = -\frac{a_2}{a_3} \qquad (10)$$

$$s_2 = \sum_{\substack{i,j \\ i \neq j}}^{3} r_i r_j = r_1 r_2 + r_1 r_3 + r_2 r_3 = \frac{a_1}{a_3} \qquad (11)$$

$$s_3 = \sum_{\substack{i,j,k \\ i \neq j \neq k}}^{3} r_i r_j r_k = r_1 r_2 r_3 = -\frac{a_0}{a_3}. \qquad (12)$$

See also DISCRIMINANT (POLYNOMIAL), SYMMETRIC POLYNOMIAL

References

Bold, B. *Famous Problems of Geometry and How to Solve Them.* New York: Dover, p. 56, 1982.
Borwein, P. and Erdélyi, T. "Newton's Identities." §1.1.E.2 in *Polynomials and Polynomial Inequalities.* New York: Springer-Verlag, pp. 5–6, 1995.
Coolidge, J. L. *A Treatise on Algebraic Plane Curves.* New York: Dover, pp. 1–2, 1959.

Newton's Theorem

If each of two nonparallel transversals with nonminimal directions meets a given curve in finite points only, then the ratio of products of the distances from the two sets of intersections to the intersection of the lines is independent of the position of the latter point.

References

Coolidge, J. L. *A Treatise on Algebraic Plane Curves.* New York: Dover, p. 189, 1959.

Newton-Stirling Formula

STIRLING'S FINITE DIFFERENCE FORMULA

Next Prime

The next prime function $NP(n)$ gives the smallest PRIME larger than n. The function can be given explicitly as

$$NP(n) = p_{1 + \pi(n)},$$

where p_i is the ith PRIME and $\pi(n)$ is the PRIME COUNTING FUNCTION. For $n = 1, 2, \ldots$ the values are 2, 3, 5, 5, 7, 7, 11, 11, 11, 11, 13, 13, 17, 17, 17, 17, 19, ... (Sloane's A007918).

See also FORTUNATE PRIME, PRIME COUNTING FUNCTION, PRIME NUMBER

References

Sloane, N. J. A. Sequences A007918 in "An On-Line Version of the Encyclopedia of Integer Sequences." http://www.research.att.com/~njas/sequences/eisonline.html.

Nexus Number

A FIGURATE NUMBER built up of the nexus of cells less than n steps away from a given cell. In k-D, the $(n+1)$th nexus number is given by

$$N_{n+1}(k) = \sum_{i=0}^{k} \binom{k}{i} n^i,$$

where $\binom{n}{i}$ is a BINOMIAL COEFFICIENT. The first few k-dimensional nexus numbers are given in the table below.

k	N_{n+1}	name
0	1	unit
1	$1 + 2n$	ODD NUMBER
2	$1 + 3n + 3n^2$	HEX NUMBER
3	$1 + 4n + 6n^2 + 4n^3$	RHOMBIC DODECAHEDRAL NUMBER

See also BINOMIAL SUMS, HEX NUMBER, ODD NUMBER, RHOMBIC DODECAHEDRAL NUMBER

References

Conway, J. H. and Guy, R. K. *The Book of Numbers.* New York: Springer-Verlag, pp. 53–54, 1996.

Neyman-Pearson Lemma

If there exists a critical region C of size α and a NONNEGATIVE constant k such that

$$\frac{\prod_{i=1}^{n} f(x_i | \theta_1)}{\prod_{i=1}^{n} f(x_i | \theta_0)} \geq k$$

for points in C and

$$\frac{\prod_{i=1}^{n} f(x_i|\theta_1)}{\prod_{i=1}^{n} f(x_i|\theta_0)} \leq k$$

for points not in C, then C is a best critical region of size α.

References

Hoel, P. G.; Port, S. C.; and Stone, C. J. "Testing Hypotheses." Ch. 3 in *Introduction to Statistical Theory.* New York: Houghton Mifflin, pp. 56–67, 1971.

Nialpdrome

A nialpdrome is a number whose HEXADECIMAL digits are in nonincreasing order. The first few are 1, 2, 3, 4, 5, 6, 7, 8, 9, 10, 11, 12, 13, 14, 15, 16, 17, 32, 33, 34, 48, 49, 50, ... (Sloane's A023771), corresponding to 1, 2, 3, 4, 5, 6, 7, 8, 9, A, B, C, D, E, F, 10, 11, 20, 21, 22, 30, 31, 32, ...

See also DIGIT, HEXADECIMAL, KATADROME, METADROME, PLAINDROME

References

Sloane, N. J. A. Sequences A023771 in "An On-Line Version of the Encyclopedia of Integer Sequences." http://www.research.att.com/~njas/sequences/eisonline.html.

Nicholson's Formula

Let $J_\nu(z)$ be a BESSEL FUNCTION OF THE FIRST KIND, $Y_\nu(z)$ a BESSEL FUNCTION OF THE SECOND KIND, and $K_\nu(z)$ a MODIFIED BESSEL FUNCTION OF THE FIRST KIND. Also let $\Re[z] > 0$. Then

$$J_\nu^2(z) + Y_\nu^2(z) = \frac{8}{\pi^2} \int_0^\infty K_0(2z \sinh t) \cos(2\nu t) dt.$$

See also DIXON-FERRAR FORMULA, WATSON'S FORMULA

References

Gradshteyn, I. S. and Ryzhik, I. M. Eqn. 6.664.4 in *Tables of Integrals, Series, and Products, 6th ed.* San Diego, CA: Academic Press, p. 727, 2000.
Iyanaga, S. and Kawada, Y. (Eds.). *Encyclopedic Dictionary of Mathematics.* Cambridge, MA: MIT Press, p. 1476, 1980.

Nicomachus's Theorem

The nth CUBIC NUMBER n^3 is a sum of n consecutive ODD NUMBERS, for example

$$1^3 = 1$$
$$2^3 = 3 + 5$$
$$3^3 = 7 + 9 + 11$$
$$4^3 = 13 + 15 + 17 + 19,$$

etc. This identity follows from

$$\sum_{i=1}^{n} [n(n-1) - 1 + 2i] = n^3.$$

It also follows from this fact that

$$\sum_{k=1}^{n} k^3 = \left(\sum_{k=1}^{n} k\right)^2.$$

See also CUBIC NUMBER, ODD NUMBER, ODD NUMBER THEOREM

Nicomedes' Conchoid

CONCHOID OF NICOMEDES

Nielsen Generalized Polylogarithm

A generalization of the POLYLOGARITHM function defined by

$$S_{n,p}(z) = \frac{(-1)^{n+p-1}}{(n-1)!p!} \int_0^1 \frac{(\ln t)^{n-1}[\ln(1-zt)]^p}{t} dt.$$

The function reduces to the usual POLYLOGARITHM for the case

$$S_{n-1,1}(z) = \mathrm{Li}_n(z).$$

The function is implemented in *Mathematica* 4.0 as `PolyLog[n, p, z]`.

See also POLYLOGARITHM

Nielsen-Ramanujan Constants

N.B. A detailed online essay by S. Finch was the starting point for this entry.

N. Nielsen (1909) and Ramanujan (Berndt 1985) considered the integrals

$$a_k = \int_1^2 \frac{(\ln x)^k}{x-1} dx. \tag{1}$$

They found the values for $k = 1$ and 2. The general constants for $k > 3$ were found by Levin (1950) and, much later, independently by V. Adamchik (Finch),

$$a_p = p!\zeta(p+1) - \frac{p(\ln 2)^{p+1}}{p+1} - p! \sum_{k=0}^{p-1}$$
$$\times \frac{\mathrm{Li}_{p+1-k}\left(\frac{1}{2}\right)(\ln 2)^k}{k!}, \tag{2}$$

where $\zeta(z)$ is the RIEMANN ZETA FUNCTION and $\mathrm{Li}_n(x)$ is the POLYLOGARITHM. The first few values are

$$a_1 = \tfrac{1}{2}\zeta(2) = \tfrac{1}{12}\pi^2 \tag{3}$$

$$a_2 = \tfrac{1}{4}\zeta(3) \tag{4}$$

$$a_3 = \tfrac{1}{15}\pi^4$$

$$+ \tfrac{1}{4}\pi^2(\ln 2)^2 - \tfrac{1}{4}(\ln 2)^4 - 6\mathrm{Li}_4\!\left(\tfrac{1}{2}\right) - \tfrac{21}{4}(\ln 2)\zeta(3) \quad (5)$$

$$a_4 = \tfrac{2}{3}\pi^2(\ln 2)^3 - \tfrac{4}{5}(\ln 2)^5 - 24(\ln 2)\mathrm{Li}_4\!\left(\tfrac{1}{2}\right) - 24\mathrm{Li}_5\!\left(\tfrac{1}{2}\right)$$

$$- \tfrac{21}{2}(\ln 2)^2\zeta(3) + 24\zeta(5). \quad (6)$$

See also POLYLOGARITHM, RIEMANN ZETA FUNCTION

References

Berndt, B. C. *Ramanujan's Notebooks, Part I.* New York: Springer-Verlag, 1985.

Borwein, J. M.; Bradley, D. M.; Broadhurst, D. J.; and Losinek, P. "Special Values of Multidimensional Polylogarithms." CECM-98:106, 14 May 1998. http://www.cecm.s-fu.ca/preprints/1998pp.html#98:106.

Finch, S. "Favorite Mathematical Constants." http://www.mathsoft.com/asolve/constant/nielram/nielram.html.

Flajolet, P. and Salvy, B. "Euler Sums and Contour Integral Representation." *Experim. Math.* **7**, 15–35, 1998.

Levin, V. I. "About a Problem of S. Ramanujan" [Russian]. *Uspekhi Mat. Nauk* **5**, 161–166, 1950.

Nielsen's Spiral

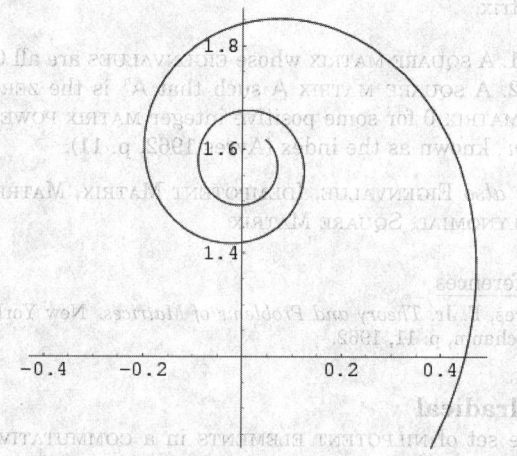

The SPIRAL with PARAMETRIC EQUATIONS

$$x(t) = a\,\mathrm{ci}(t) \quad (1)$$
$$y(t) = a\,\mathrm{si}(t), \quad (2)$$

where $\mathrm{ci}(t)$ is the COSINE INTEGRAL and $\mathrm{si}(t)$ is the SINE INTEGRAL. The CESÀRO equation is

$$k = \frac{e^{s/a}}{a}. \quad (3)$$

See also CORNU SPIRAL, COSINE INTEGRAL, SINE INTEGRAL

References

Gray, A. *Modern Differential Geometry of Curves and Surfaces with Mathematica, 2nd ed.* Boca Raton, FL: CRC Press, pp. 146–147, 1997.

Nil Geometry

The GEOMETRY of the LIE GROUP consisting of REAL MATRICES OF THE FORM

$$\begin{bmatrix} 1 & x & y \\ 0 & 1 & z \\ 0 & 0 & 1 \end{bmatrix},$$

i.e., the HEISENBERG GROUP.

See also HEISENBERG GROUP, LIE GROUP, THURSTON'S GEOMETRIZATION CONJECTURE

Nilalgebra

NILPOTENT ALGEBRA

Nilmanifold

Let N be a NILPOTENT, connected, SIMPLY CONNECTED LIE GROUP, and let D be a discrete SUBGROUP of N with compact right QUOTIENT SPACE. Then N/D is called a nilmanifold.

Nilpotent Algebra

An algebra, also called a nilalgebra, consisting only of NILPOTENT ELEMENTS.

See also NILPOTENT ELEMENT

References

Schafer, R. D. "Nilpotent Algebras." §3.1 in *An Introduction to Nonassociative Algebras.* New York: Dover, pp. 27–32, 1996.

Nilpotent Element

An element B of a RING is nilpotent if there exists a POSITIVE INTEGER k for which $B^k = 0$.

See also ENGEL'S THEOREM

Nilpotent Group

A GROUP G for which the chain of groups

$$I = \mathbb{Z}_0 \subseteq \mathbb{Z}_1 \subseteq \ldots \subseteq \mathbb{Z}_n$$

with $\mathbb{Z}_{k+1}/\mathbb{Z}_k$ (equal to the CENTER of $G/\mathbb{Z}_k$) terminates finitely with 0 is called a nilpotent group. Here, $\mathbb{Z}_n$ denotes a CYCLIC GROUP of order n.

See also CENTER (GROUP), NILPOTENT LIE GROUP

Nilpotent Lie Algebra

A LIE ALGEBRA is nilpotent when its LOWER CENTRAL SERIES $\mathfrak{g}_k$ vanishes for some k. Any nilpotent Lie algebra is also SOLVABLE. The basic example of a nilpotent Lie algebra is the VECTOR SPACE of strictly

UPPER TRIANGULAR MATRICES, such as the Lie algebra of the HEISENBERG GROUP.

The following *Mathematica* function tests whether a Lie algebra $\mathfrak{g}$ is nilpotent, given a list of matrices which is a basis for $\mathfrak{g}$.

```
MatrixBasis[a_-
List]:=Partition[#1,Length[a[[1]]]]&/@
   LatticeReduce[Flatten/@a]
                LieCommutator[a_,b_]:=a.b-b.a
NextLCS[gold_List, {}]={};
         NextLCS[gold_List,      g_List]:=
MatrixBasis[Flatten[Outer[LieCommutator,gold,-
g,1],1]]            NilpotentLieQ[g_List]:=
FixedPoint[NextLCS[g,#1]&,g]=={}
```

For example,

```
borel5 = Flatten[Table[ReplacePart[
                                  Ta-
ble[0,{i,5},{j,5}],1,{k,1}],{k,5},{l,k,5}],1];
   NilpotentLieQ[borel5]
```

yields `False`, while

```
uni5 = Flatten[Table[ReplacePart[
                                  Ta-
ble[0,{i,5},{j,5}],1,{k,1}],{k,5},{l,k-
+1,5}],1];
   NilpotentLieQ[uni5]
```

yields `True`.

See also COMMUTATOR SERIES (LIE ALGEBRA), LIE ALGEBRA, LIE GROUP, LOWER CENTRAL SERIES (LIE ALGEBRA), NILPOTENT LIE GROUP, REPRESENTATION (LIE ALGEBRA), REPRESENTATION (NILPOTENT LIE GROUP), SOLVABLE LIE GROUP, UNIPOTENT

Nilpotent Lie Group

A nilpotent Lie group is a LIE GROUP G which is CONNECTED and whose LIE ALGEBRA is a NILPOTENT LIE ALGEBRA $\mathfrak{g}$. That is, its LOWER CENTRAL SERIES

$$\mathfrak{g}_1[\mathfrak{g},\mathfrak{g}], \mathfrak{g}_2 = [\mathfrak{g},\mathfrak{g}_1], \dots \tag{1}$$

eventually vanishes, $\mathfrak{g}_k = 0$ for some k. So a nilpotent Lie group is a special case of a SOLVABLE LIE GROUP.

The basic example is the GROUP of UPPER TRIANGULAR MATRICES with 1s on their diagonals, e.g.,

$$\begin{bmatrix} 1 & a_{12} & a_{13} \\ 0 & 1 & a_{23} \\ 0 & 0 & 1 \end{bmatrix}, \tag{2}$$

which is called the HEISENBERG GROUP. Its LOWER CENTRAL SERIES is given by

$$\mathfrak{g}_0 = \begin{bmatrix} 0 & b_{12} & b_{13} \\ 0 & 0 & b_{23} \\ 0 & 0 & 0 \end{bmatrix} \tag{3}$$

$$\mathfrak{g}_1 = \begin{bmatrix} 0 & 0 & c_{13} \\ 0 & 0 & 0 \\ 0 & 0 & 0 \end{bmatrix} \tag{4}$$

$$\mathfrak{g}_2 = \begin{bmatrix} 0 & 0 & 0 \\ 0 & 0 & 0 \\ 0 & 0 & 0 \end{bmatrix}. \tag{5}$$

Any real nilpotent Lie group is DIFFEOMORPHIC to EUCLIDEAN SPACE. For instance, the group of matrices in the example above is diffeomorphic to $\mathbb{R}^3$, via the EXPONENTIAL MAPExponential Map (Lie Group). In general, the exponential map of a NILPOTENT LIE ALGEBRA is SURJECTIVE, in contrast to the more general SOLVABLE LIE GROUP.

See also BOREL GROUP, COMMUTATOR SERIES (LIE ALGEBRA), FLAG (VECTOR SPACE), LIE ALGEBRA, LIE GROUP, LOWER CENTRAL SERIES (LIE ALGEBRA), MATRIX, REPRESENTATION, REPRESENTATION (NILPOTENT LIE GROUP), SOLVABLE LIE ALGEBRA, SOLVABLE LIE GROUP, SPLIT SOLVABLE LIE ALGEBRA, UNIPOTENT

References

Knapp, A. W. "Group Representations and Harmonic Analysis, Part II." *Not. Amer. Math. Soc.* **43**, 537–549, 1996.

Nilpotent Matrix

There are two common definitions for a nilpotent matrix.

1. A SQUARE MATRIX whose EIGENVALUES are all 0.
2. A SQUARE MATRIX A such that A^n is the ZERO MATRIX 0 for some positive integer MATRIX POWER n, known as the index (Ayres 1962, p. 11).

See also EIGENVALUE, IDEMPOTENT MATRIX, MATRIX POLYNOMIAL, SQUARE MATRIX

References

Ayres, F. Jr. *Theory and Problems of Matrices.* New York: Schaum, p. 11, 1962.

Nilradical

The set of NILPOTENT ELEMENTS in a COMMUTATIVE RING is an ideal, and it is called the nilradical. Another equivalent description is that it is the intersection of the prime ideals. It could be the zero ideal, as in the case of the integers.

See also ALGEBRAIC GEOMETRY, ALGEBRAIC NUMBER THEORY, IDEAL, JACOBSON RADICAL, RADICAL (IDEAL)

Nim

A game, also called TACTIX, which is played by the following rules. Given one or more piles (NIM-HEAPS), players alternate by taking all or some of the counters in a single heap. The player taking the last counter or stack of counters is the winner. Nim-like games are also called TAKE-AWAY GAMES and DISJUNCTIVE GAMES. If optimal strategies are used, the winner can be determined from any intermediate position by its associated NIM-VALUE.

See also MISÈRE FORM, NIM-VALUE, WYTHOFF'S GAME

References

Ball, W. W. R. and Coxeter, H. S. M. *Mathematical Recreations and Essays, 13th ed.* New York: Dover, pp. 36–38, 1987.

Bogomolny, A. "The Game of Nim." http://www.cut-the-knot.com/bottom_nim.html.

Bouton, C. L. "Nim, A Game with a Complete Mathematical Theory." *Ann. Math. Princeton* **3**, 35–39, 1901–1902.

Gardner, M. "Mathematical Games: Concerning the Game of Nim and Its Mathematical Analysis." *Sci. Amer.* **198**, 104–111, Feb. 1958.

Gardner, M. "Nim and Hackenbush." Ch. 14 in *Wheels, Life, and other Mathematical Amusements.* New York: W. H. Freeman, pp. 142–151, 1983.

Hardy, G. H. and Wright, E. M. *An Introduction to the Theory of Numbers, 5th ed.* Oxford, England: Oxford University Press, pp. 117–120, 1990.

Kraitchik, M. "Nim." §3.12.2 in *Mathematical Recreations.* New York: W. W. Norton, pp. 86–88, 1942.

Nim-Heap

A pile of counters in a game of NIM.

Nim-Sum

NIM-VALUE

Nim-Value

Every position of every IMPARTIAL GAME has a nim-value, making it equivalent to a NIM-HEAP. To find the nim-value (also called the SPRAGUE-GRUNDY NUMBER), take the MEX of the nim-values of the possible moves. The nim-value can also be found by writing the number of counters in each heap in binary, adding without carrying, and replacing the digits with their values mod 2. If the nim-value is 0, the position is SAFE; otherwise, it is UNSAFE. With two heaps, safe positions are (x, x) where $x \in [1, 7]$. With three heaps, (1, 2, 3), (1, 4, 5), (1, 6, 7), (2, 4, 6), (2, 5, 7), and (3, 4, 7).

See also GRUNDY'S GAME, IMPARTIAL GAME, MEX, NIM, SAFE, UNSAFE

References

Ball, W. W. R. and Coxeter, H. S. M. *Mathematical Recreations and Essays, 13th ed.* New York: Dover, pp. 36–38, 1987.

Grundy, P. M. "Mathematics and Games." *Eureka* **2**, 6–8, 1939.

Sprague, R. "Uuml;ber mathematische Kampfspiele." *Tôhoku J. Math.* **41**, 438–444, 1936.

n-in-a-Row

TIC-TAC-TOE

Nine Associated Points Theorem

Any CUBIC CURVE that passes through eight of the nine intersections of two given cubic curves automatically passes through the ninth.

References

Evelyn, C. J. A.; Money-Coutts, G. B.; and Tyrrell, J. A. *The Seven Circles Theorem and Other New Theorems.* London: Stacey International, p. 15, 1974.

Nine Circles Theorem

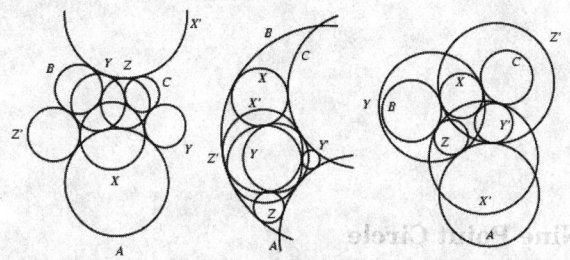

Let A, B, and C be three circles in the plane, and let X be any circle touching B and C. Then build up a chain of circles such that $Y : CAX, Z : ABY, X' : BCZ, Y' : CAX', Z' : ABY', X'' : ABZ'$, where $C : C_1 C_2 C_3$ denotes a circle C tangent to circles C_1, C_2, and C_3. Although there are a number of choices for each successive tangent circle in the chain, if the choice at each stage is made appropriately, then the ninth and final circle X'' coincides with the first circle X (Evelyn *et al.* 1971, p. 58).

See also CIRCLE, SIX CIRCLES THEOREM, SEVEN CIRCLES THEOREM

References

Evelyn, C. J. A.; Money-Coutts, G. B.; and Tyrrell, J. A. "The Nine Circles Theorem." §3.4 in *The Seven Circles Theorem and Other New Theorems.* London: Stacey International, pp. 58–68, 1974.

Tyrrell, J. A. and Powell, M. T. "A Theorem in Circle Geometry." *Bull. London Math. Soc.* **3**, 70–74, 1971.

Nine-j Symbol

WIGNER 9*j*-SYMBOL

Nine-Point Center

The center F (or N) of the NINE-POINT CIRCLE. It has TRIANGLE CENTER FUNCTION

$$\alpha = \cos(B - C) = \cos A + 2 \cos B \cos C$$

$$= bc \left[a^2 b^2 + a^2 c^2 + (b^2 - c^2)^2 \right],$$

and is the MIDPOINT of the line between the CIRCUMCENTER C and ORTHOCENTER H. It lies on the EULER LINE.

See also EULER LINE, LESTER CIRCLE, NINE-POINT CIRCLE, NINE-POINT CONIC

References

Carr, G. S. *Formulas and Theorems in Pure Mathematics,*
    2nd ed. New York: Chelsea, p. 624, 1970.
Coxeter, H. S. M. and Greitzer, S. L. *Geometry Revisited.*
    New York: Random House, p. 21, 1967.
Dixon, R. *Mathographics.* New York: Dover, pp. 57–58,
    1991.
Durell, C. V. *Modern Geometry: The Straight Line and*
    Circle. London: Macmillan, pp. 27–29, 1928.
Kimberling, C. "Central Points and Central Lines in the
    Plane of a Triangle." *Math. Mag.* **67**, 163–187, 1994.
Kimberling, C. "Nine-Point Center." http://cedar.evansvil-
    le.edu/~ck6/tcenters/class/npcenter.html.

Nine-Point Circle

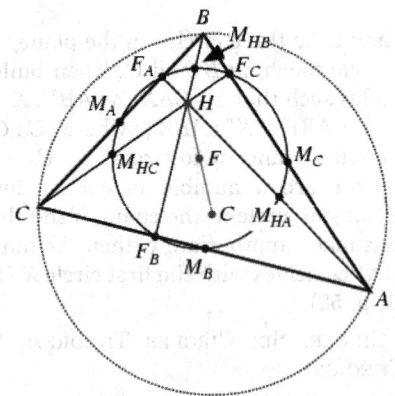

The CIRCLE, also called EULER'S CIRCLE and the
FEUERBACH CIRCLE, which passes through the feet
of the PERPENDICULAR F_A, F_B, and F_C dropped from
the VERTICES of any TRIANGLE ΔABC on the sides
opposite them. Euler showed in 1765 that it also
passes through the MIDPOINTS M_A, M_B, M_C of the
sides of ΔABC.

By FEUERBACH'S THEOREM, the nine-point circle also
passes through the MIDPOINTS M_{HA}, M_{HB}, M_{HC} (now
called the EULER POINTS) of the segments which join
the VERTICES and the ORTHOCENTER H. These three
triples of points make nine in all, giving the circle its
name. The center F of the nine-point circle is called
the NINE-POINT CENTER.

The RADIUS of the nine-point circle is $R/2$, where R is
the CIRCUMRADIUS. The center of KIEPERT'S HYPER-
BOLA lies on the nine-point circle. The nine-point
circle bisects any line from the ORTHOCENTER to a
point on the CIRCUMCIRCLE. The nine-point circle of
the INCENTER and EXCENTERS of a TRIANGLE is the
CIRCUMCIRCLE.

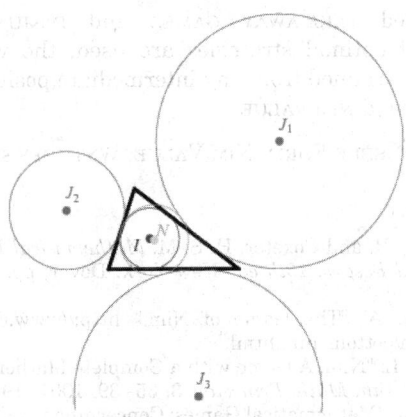

There are four CIRCLES that are tangent all three
sides (or their extensions) of a given TRIANGLE: the
INCIRCLE I and three EXCIRCLES J_1, J_2, and J_3. These
four circles are, in turn, all touched by the nine-point
circle N.

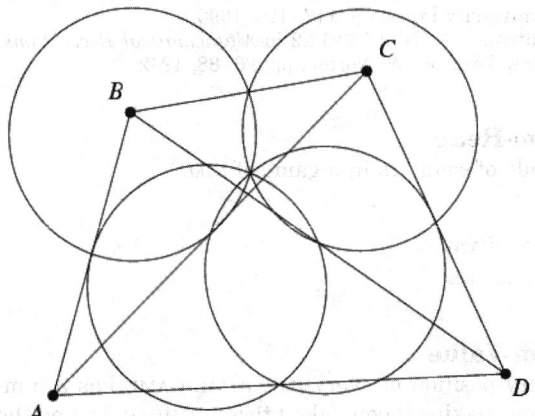

Given four arbitrary points, the four nine-points
circles of the triangles formed by taking three points
at a times are CONCURRENT (Lemoine 1904; Wells
1991, p. 209; Schröder 1999). Moreover, if four points
do not form an ORTHOCENTRIC SYSTEM, then there is a
unique RECTANGULAR HYPERBOLA passing through
them, and its center is given by the intersection of
the nine-point circles of the points taken three at a
time (Wells 1991, p. 209). Finally, the point of concurrence of the four nine-points circles is also the
point of concurrence of the four circles determined by
the feet of the perpendiculars (Schröder 1999).

The sum of the powers of the VERTICES with regard to
the nine-point circle is

$$\tfrac{1}{4}(a_1^2 + a_2^2 + a_3^2).$$

Also,

$$\overline{FA_1}^2 + \overline{FA_2}^2 + \overline{FA_3}^2 + \overline{FH}^2 = 3R^2,$$

where F is the NINE-POINT CENTER, A_i are the
VERTICES, H is the ORTHOCENTER, and R is the
CIRCUMRADIUS. All triangles inscribed in a given

CIRCLE and having the same ORTHOCENTER have the same nine-point circle.

See also COMPLETE QUADRILATERAL, EIGHT-POINT CIRCLE THEOREM, EULER POINT, FEUERBACH'S THEOREM, FONTENÉ THEOREMS, GRIFFITHS' THEOREM, HART CIRCLE, NINE-POINT CENTER, NINE-POINT CONIC, ORTHOCENTRIC SYSTEM, RECTANGULAR HYPERBOLA

References

Altshiller-Court, N. *College Geometry: A Second Course in Plane Geometry for Colleges and Normal Schools, 2nd ed., rev. enl.* New York: Barnes and Noble, pp. 93–97, 1952.

Brand, L. "The Eight-Point Circle and the Nine-Point Circle." *Amer. Math. Monthly* **51**, 84–85, 1944.

Casey, J. *A Sequel to the First Six Books of the Elements of Euclid, Containing an Easy Introduction to Modern Geometry with Numerous Examples, 5th ed., rev. enl.* Dublin: Hodges, Figgis, & Co., pp. 58–61, 1888.

Coolidge, J. L. *A Treatise on the Geometry of the Circle and Sphere.* New York: Chelsea, pp. 40–41, 1971.

Coxeter, H. S. M. and Greitzer, S. L. "The Nine-Point Circle." §1.8 in *Geometry Revisited.* New York: Random House, pp. 20–22, 1967.

Dörrie, H. "The Feuerbach Circle." §28 in *100 Great Problems of Elementary Mathematics: Their History and Solutions.* New York: Dover, pp. 142–144, 1965.

Durell, C. V. *Modern Geometry: The Straight Line and Circle.* London: Macmillan, pp. 27–29, 1928.

F. Gabriel-Marie. *Exercices de géométrie.* Tours, France: Maison Mame, pp. 306–314, 1912.

Gardner, M. *Mathematical Carnival: A New Round-Up of Tantalizers and Puzzles from Scientific American.* New York: Vintage Books, p. 59, 1977.

Guggenbuhl, L. "Karl Wilhelm Feuerbach, Mathematician." Appendix to *Circles: A Mathematical View, rev. ed.* Washington, DC: Math. Assoc. Amer., pp. 89–100, 1995.

Honsberger, R. "The Nine-Point Circle." §1.3 in *Episodes in Nineteenth and Twentieth Century Euclidean Geometry.* Washington, DC: Math. Assoc. Amer., pp. 6–7, 1995.

Johnson, R. A. *Modern Geometry: An Elementary Treatise on the Geometry of the Triangle and the Circle.* Boston, MA: Houghton Mifflin, pp. 165 and 195–212, 1929.

Lachlan, R. "The Nine-Point Circle." §123–125 in *An Elementary Treatise on Modern Pure Geometry.* London: Macmillian, pp. 70–71, 1893.

Lange, J. *Geschichte des Feuerbach'schen Kreises.* Berlin, 1894.

Lemoine, M. T. "Note de géométrie." *Nouv. Ann. Math.* **4**, 400–402, 1904.

Mackay, J. S. "History of the Nine-Point Circle." *Proc. Edinburgh Math. Soc.* **11**, 19–61, 1892.

Ogilvy, C. S. *Excursions in Geometry.* New York: Dover, pp. 119–120, 1990.

Pedoe, D. *Circles: A Mathematical View, rev. ed.* Washington, DC: Math. Assoc. Amer., pp. 1–4, 1995.

Rouché, E. and de Comberousse, C. *Traité de géométrie plane.* Paris: Gauthier-Villars, pp. 306–307, 1900.

Schröder, E. M. "Zwei 8-Kreise-Sätze für Vierecke." *Mitt. Math. Ges. Hamburg* **18**, 105–117, 1999.

Wells, D. *The Penguin Dictionary of Curious and Interesting Numbers.* Middlesex, England: Penguin Books, pp. 73–74, 1986.

Wells, D. *The Penguin Dictionary of Curious and Interesting Geometry.* London: Penguin, pp. 158–159, 1991.

Nine-Point Conic

A CONIC SECTION on which the MIDPOINTS of the sides of any COMPLETE QUADRANGLE lie. The three diagonal points also lie on this conic.

See also COMPLETE QUADRANGLE, CONIC SECTION, NINE-POINT CIRCLE

Nint

NEAREST INTEGER FUNCTION

Nint Zeta Function

Let

$$S_N(s) = \sum_{n=1}^{\infty} \left[\left(n^{1/N} \right) \right]^{-s}, \qquad (1)$$

where $[x]$ denotes NEAREST INTEGER FUNCTION, i.e, the INTEGER closest to x. For $s > 3$,

$$S_2(s) = 2\zeta(s-1) \qquad (2)$$

$$S_3(s) = 3\zeta(s-2) + 4^{-s}\zeta(s) \qquad (3)$$

$$S_4(s) = 4\zeta(s-3) + \zeta(s-1). \qquad (4)$$

$S_N(n)$ is a POLYNOMIAL in π whose COEFFICIENTS are ALGEBRAIC NUMBERS whenever $n - N$ is ODD. The first few values are given explicitly by

$$S_3(4) = \frac{\pi^2}{2} + \frac{\pi^4}{23046} \qquad (5)$$

$$S_5(6) = \frac{5\pi^2}{6} + \frac{\pi^4}{36} + \frac{\pi^6}{4^{12}}$$
$$\times \left(\frac{1}{945} - \frac{170912 + 49928\sqrt{2}}{25} \sqrt{1 - \sqrt{\frac{1}{2}}} \right) \qquad (6)$$

$$S_6(7) = \pi^2 + \frac{\pi^4}{18} + \frac{\pi^6}{2520} + \frac{246013 + 353664\sqrt{2}}{45} \frac{\pi^7}{2^{27}}. \qquad (7)$$

References

Borwein, J. M.; Hsu, L. C.; Mabry, R.; Neu, K.; Roppert, J.; Tyler, D. B.; and de Weger, B. M. M. "Nearest Integer Zeta-Functions." *Amer. Math. Monthly* **101**, 579–580, 1994.

Nirenberg's Conjecture

If the GAUSS MAP of a COMPLETE MINIMAL SURFACE omits a NEIGHBORHOOD of the SPHERE, then the surface is a PLANE. This was proven by Osserman (1959). Xavier (1981) subsequently generalized the result as follows. If the GAUSS MAP of a complete MINIMAL SURFACE omits ≥ 7 points, then the surface is a PLANE.

See also COMPLETE MINIMAL SURFACE, GAUSS MAP, MINIMAL SURFACE, NEIGHBORHOOD

References

do Carmo, M. P. *Mathematical Models from the Collections of Universities and Museums* (Ed. G. Fischer). Braunschweig, Germany: Vieweg, p. 42, 1986.
Osserman, R. "Proof of a Conjecture of Nirenberg." *Comm. Pure Appl. Math.* **12**, 229–232, 1959.
Xavier, F. "The Gauss Map of a Complete Nonflat Minimal Surface Cannot Omit 7 Points on the Sphere." *Ann. Math.* **113**, 211–214, 1981.

Niven Number

HARSHAD NUMBER

Niven's Constant

N.B. A detailed online essay by S. Finch was the starting point for this entry.

Given a POSITIVE INTEGER $m > 1$, let its PRIME FACTORIZATION be written

$$m = p_1^{a_1} p_2^{a_2} p_3^{a_3} \cdots p_k^{a_k}. \tag{1}$$

Define the functions $h(n)$ and $H(n)$ by $h(1) = 1$, $H(1) = 1$, and

$$h(m) = \min(a_1, a_2 \ldots, a_k) \tag{2}$$

$$H(m) = \max(a_1, a_2 \ldots, a_k) \tag{3}$$

Then

$$\lim_{n \to \infty} \frac{1}{n} \sum_{m=1}^{n} h(m) = 1 \tag{4}$$

$$\lim_{n \to \infty} \sqrt{\frac{\sum_{m=1}^{n} h(m) - n}{\sqrt{n}}} = \frac{\zeta\left(\frac{3}{2}\right)}{\zeta(3)}, \tag{5}$$

where $\zeta(z)$ is the RIEMANN ZETA FUNCTION (Niven 1969). Niven (1969) also proved that

$$\lim_{n \to \infty} \frac{1}{n} \sum_{m=1}^{n} H(m) = C, \tag{6}$$

where

$$C = 1 + \left\{ \sum_{j=2}^{\infty} \left[1 - \frac{1}{\zeta(j)} \right] \right\} = 1.705221\ldots \tag{7}$$

(Sloane's A033150).

The CONTINUED FRACTION of Niven's constant is 1, 1, 2, 2, 1, 1, 4, 1, 1, 3, 4, 4, 8, 4, 1, ... (Sloane's A033151). The positions at which the digits 1, 2, ... first occur in the CONTINUED FRACTION are 1, 3, 10, 7, 47, 41, 34, 13, 140, 252, 20, ... (Sloane's A033152). The sequence of largest terms in the CONTINUED FRACTION is 1, 2, 4, 8, 11, 14, 29, 372, 559, ... (Sloane's A033153), which occur at positions 1, 3, 7, 13, 20, 35, 51, 68, 96, ... (Sloane's A033154).

References

Finch, S. "Favorite Mathematical Constants." http://www.mathsoft.com/asolve/constant/niven/niven.html.
Le Lionnais, F. *Les nombres remarquables.* Paris: Hermann, p. 41, 1983.
Niven, I. "Averages of Exponents in Factoring Integers." *Proc. Amer. Math. Soc.* **22**, 356–360, 1969.
Plouffe, S. "The Niven Constant." http://www.lacim.u-qam.ca/piDATA/niven.txt.
Sloane, N. J. A. Sequences A033150, A033151, A033152, A033153, and A033154 in "An On-Line Version of the Encyclopedia of Integer Sequences." http://www.research.att.com/~njas/sequences/eisonline.html.

n-Minex

n-minex is defined as 10^{-n}.

See also N-PLEX

References

Conway, J. H. and Guy, R. K. *The Book of Numbers.* New York: Springer-Verlag, p. 16, 1996.

Nobbs Points

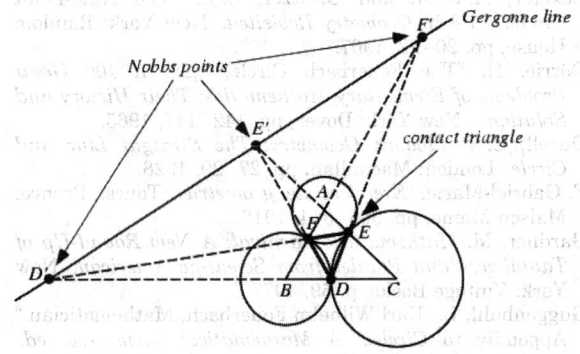

Given a TRIANGLE $\triangle ABC$, construct the CONTACT TRIANGLE $\triangle DEF$. Then the Nobbs points are the three points D', E', and F' from which $\triangle ABC$ and $\triangle DEF$ are PERSPECTIVE, as illustrated above. The Nobbs points are COLLINEAR and fall along the GERGONNE LINE.

See also COLLINEAR, CONTACT TRIANGLE, EVANS POINT, FLETCHER POINT, GERGONNE LINE, PERSPECTIVE TRIANGLES

References

Oldknow, A. "The Euler-Gergonne-Soddy Triangle of a Triangle." *Amer. Math. Monthly* **103**, 319–329, 1996.

Noble Number

A noble number is defined as an IRRATIONAL NUMBER which has a CONTINUED FRACTION which becomes an infinite sequence of 1s at some point,

$$\nu \equiv [a_1, a_2, \ldots, a_n, \bar{1}].$$

The prototype is the GOLDEN RATIO ϕ whose CONTINUED FRACTION is composed *entirely* of 1s, $[\bar{1}]$. Any noble number can be written as

$$v = \frac{A_n + \phi A_{n-1}}{B_n + \phi B_{n+1}},$$

where A_k and B_k are the NUMERATOR and DENOMINATOR of the kth CONVERGENT of $[a_1, a_2, \ldots, a_n]$. The noble numbers are a SUBFIELD of $\mathbb{Q}(\sqrt{5})$.

See also NEAR NOBLE NUMBER

References

Hardy, G. H. and Wright, E. M. *An Introduction to the Theory of Numbers, 5th ed.* Oxford, England: Clarendon Press, p. 236, 1979.
Schroeder, M. "Noble and Near Noble Numbers." In *Fractals, Chaos, Power Laws: Minutes from an Infinite Paradise.* New York: W. H. Freeman, pp. 392–394, 1991.

Node (Algebraic Curve)
ORDINARY DOUBLE POINT

Node (Fixed Point)
A FIXED POINT for which the STABILITY MATRIX has both EIGENVALUES of the same sign (i.e., both are POSITIVE or both are NEGATIVE). If $\lambda_1 < \lambda_2 < 0$, then the node is called STABLE; if $\lambda_1 < \lambda_2 < 0$, then the node is called an UNSTABLE NODE.

See also STABLE NODE, UNSTABLE NODE

Node (Graph)

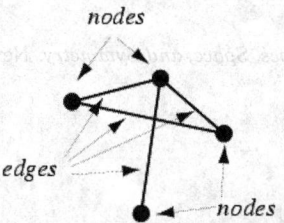

A synonym for a VERTEX of a GRAPH, i.e., one of the points on which the graph may is defined and which may be connected by EDGES. The terms "point," "junction," and 0-simplex are also used (Harary 1994; Skiena 1990, p. 80).
The following tables gives the total numbers of nodes for various classes of graphs on $n = 1, 2, \ldots$ nodes.

graph type	Sloane	total node count for $n = 1, 2, \ldots$ nodes
GRAPH	A055543	1, 4, 12, 44, 170, 936, ...
TREE	A055544	1, 2, 3, 8, 15, 36, 77, 184 ...
LABELED TREE	A000169	1, 2, 9, 64, 625, ...
ROOTED TREE	A055545	1, 2, 6, 16, 45, 120, ...

See also EDGE (GRAPH), GRAPH

References

Harary, F. *Graph Theory.* Reading, MA: Addison-Wesley, 1994.
Skiena, S. *Implementing Discrete Mathematics: Combinatorics and Graph Theory with Mathematica.* Reading, MA: Addison-Wesley, 1990.
Sloane, N. J. A. Sequences A000169/M1946, A055543, A055544, and A055545 in "An On-Line Version of the Encyclopedia of Integer Sequences." http://www.research.att.com/~njas/sequences/eisonline.html.

Noetherian Module
A MODULE M is Noetherian if every submodule is finitely generated.

See also NOETHERIAN RING

Noetherian Ring
An abstract commutative RING satisfying the abstract chain condition.

See also LOCAL RING, NOETHER-LASKER THEOREM

Noether-Lasker Theorem
Let M be a finitely generated MODULE over a commutative NOETHERIAN RING R. Then there exists a finite set $\{N_i | 1 \leq i \leq l\}$ of submodules of M such that

1. $\cap_{i=1}^{l} N_i = 0$ and $\cap_{i \neq i_0} N_i$ is not contained in N_{i_0} for all $1 \leq i_0 \leq l$.
2. Each quotient M/N_i is primary for some prime P_i.
3. The P_i are all distinct for $1 \leq i \leq l$.
4. Uniqueness of the primary component N_i is equivalent to the statement that P_i does not contain P_j for any $j \neq i$.

Noether's Fundamental Theorem
If two curves ϕ and ψ of MULTIPLICITIES $r_i \neq 0$ and $s_i \neq 0$ have only ordinary points or ordinary singular points and CUSPS in common, then every curve which has at least MULTIPLICITY

$$r_i + s_i - 1$$

at every point (distinct or infinitely near) can be written

$$f \equiv \phi \psi' + \psi \phi' = 0,$$

where the curves ϕ' and ψ' have MULTIPLICITIES at least $r_i - 1$ and $s_i - 1$.

References

Coolidge, J. L. *A Treatise on Algebraic Plane Curves.* New York: Dover, pp. 29–30, 1959.

Noether's Symmetry Theorem

An extremely powerful theorem in physics which states that each SYMMETRY of a system leads to a physically conserved quantity. SYMMETRY under TRANSLATION corresponds to momentum conservation, SYMMETRY under ROTATION to angular momentum conservation, SYMMETRY in time to energy conservation, etc.

See also SYMMETRY

Noether's Transformation Theorem

Any irreducible curve may be carried by a factorable CREMONA TRANSFORMATION into one with none but ordinary singular points.

References

Coolidge, J. L. *A Treatise on Algebraic Plane Curves.* New York: Dover, p. 207, 1959.

Noise

An error which is superimposed on top of a true signal. Noise may be random or systematic. Noise can be greatly reduced by transmitting signals digitally instead of in analog form because each piece of information is allowed only discrete values which are spaced farther apart than the contribution due to noise.

CODING THEORY studies how to encode information efficiently, and ERROR-CORRECTING CODES devise methods for transmitting and reconstructing information in the presence of noise.

See also ERROR, STOCHASTIC FUNCTION

References

Abbott, D. and Kiss, L. B. (Eds.). *Proc. 2nd Internat. Conf. Unsolved Problems of Noise and Fluctuations, 11–15 July, Adelaide* Melville, NY: Amer. Inst. Physics Press, 2000.
Davenport, W. B. and Root, W. L. *An Introduction to the Theory of Random Signals and Noise.* New York: IEEE Press, 1987.
McDonough, R. N. and Whalen, A. D. *Detection of Signals in Noise, 2nd ed.* Orlando, FL: Academic Press, 1995.
Pierce, J. R. *Symbols, Signals and Noise: The Nature and Process of Communication.* New York: Harper & Row, 1961.
Vainshtein, L. A. and Zubakov, V. D. *Extraction of Signals from Noise.* New York: Dover, 1970.
van der Ziel, A. *Noise: Sources, Characterization, Measurement.* New York: Prentice-Hall, 1954.
van der Ziel, A. *Noise in Measurement.* New York: Wiley, 1976.
Wax, N. *Selected Papers on Noise and Stochastic Processes.* New York: Dover, 1954.
Weisstein, E. W. "Books about Noise." http://www.treasure-troves.com/books/Noise.html.

Noise Sphere

A mapping of RANDOM NUMBER TRIPLES to points in SPHERICAL COORDINATES according to

$$\theta = 2\pi X_n$$

$$\phi = \pi X_{n+1}$$

$$r = \sqrt{X_{n+2}}$$

in order to detect unexpected structure indicating correlations between triples. When such structure is present (note that this does *not* include the expected bunching of points along the z-axis according to the factor $\sin\phi$ in the spherical volume element), numbers may not be truly RANDOM.

See also BALL POINT PICKING, RANDOM NUMBER, SPHERE POINT PICKING

References

Pickover, C. A. *Computers and the Imagination.* New York: St. Martin's Press, 1991.
Pickover, C. A. "Computers, Randomness, Mind, and Infinity." Ch. 31 in *Keys to Infinity.* New York: W. H. Freeman, pp. 233–247, 1995.
Richards, T. "Graphical Representation of Pseudorandom Sequences." *Computers and Graphics* **13**, 261–262, 1989.

Nolid

An assemblage of faces forming a POLYHEDRON of zero VOLUME (Holden 1991, p. 124).

See also ACOPTIC POLYHEDRON

References

Holden, A. *Shapes, Space, and Symmetry.* New York: Dover, 1991.

Nome

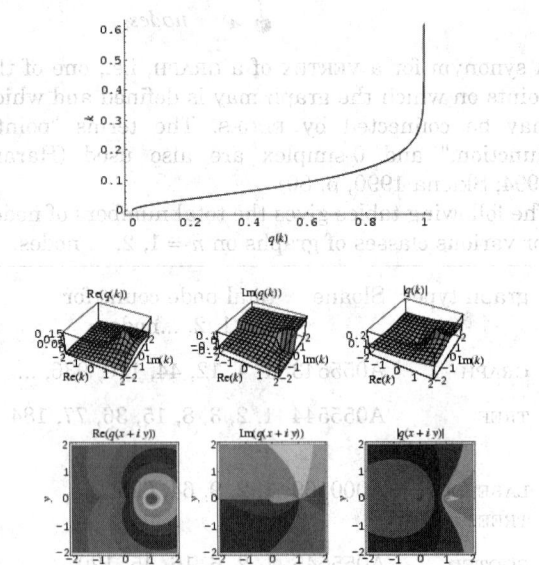

Given a JACOBI THETA FUNCTION, the nome is defined

as

$$q(k) \equiv e^{\pi i \tau} = e^{-\pi K'(k)/K(k)} = e^{-\pi K\left(\sqrt{1-k^2}\right)/K(k)} \quad (1)$$

(Borwein and Borwein 1987, pp. 41, 109 and 114), where τ is the HALF-PERIOD RATIO, $K(k)$ is the complete ELLIPTIC INTEGRAL OF THE FIRST KIND, $m = k^2$ is the PARAMETER, and k is the MODULUS. The nome is implemented in *Mathematica* as Elliptic-NomeQ[m].

Various notations for JACOBI THETA FUNCTIONS involving the nome include

$$\vartheta_i(z, q) \equiv \vartheta(z|\tau), \quad (2)$$

where τ is the HALF-PERIOD RATIO (Whittaker and Watson 1972, p. 464) and

$$\vartheta_i \equiv \vartheta(0, q). \quad (3)$$

See also AMPLITUDE, CHARACTERISTIC (ELLIPTIC INTEGRAL), ELLIPTIC INTEGRAL, HALF-PERIOD RATIO, INVERSE NOME, JACOBI THETA FUNCTIONS, MODULAR ANGLE, MODULAR DISCRIMINANT, MODULUS (ELLIPTIC INTEGRAL), PARAMETER

References

Abramowitz, M. and Stegun, C. A. (Eds.). *Handbook of Mathematical Functions with Formulas, Graphs, and Mathematical Tables, 9th printing.* New York: Dover, p. 591, 1972.
Borwein, J. M. and Borwein, P. B. *Pi & the AGM: A Study in Analytic Number Theory and Computational Complexity.* New York: Wiley, 1987.
Whittaker, E. T. and Watson, G. N. *A Course in Modern Analysis, 4th ed.* Cambridge, England: Cambridge University Press, 1990.

n-Omino

POLYOMINO

Nomogram

A graphical plot which can be used for solving certain types of equations. According to Steinhaus (1983, p. 301), the Nomogram was invented by the French mathematicians Massau and M. P. Ocagne in 1889.

References

Iyanaga, S. and Kawada, Y. (Eds.). "Nomograms." §282 in *Encyclopedic Dictionary of Mathematics.* Cambridge, MA: MIT Press, pp. 891–893, 1980.
Menzel, D. (Ed.). *Fundamental Formulas of Physics, Vol. 1.* New York: Dover, p. 141, 1960.
Steinhaus, H. *Mathematical Snapshots, 3rd ed.* New York: Dover, pp. 92–95 and 301, 1999.
Whittaker, E. T. and Robinson, G. "Nomography." §128 in *The Calculus of Observations: A Treatise on Numerical Mathematics, 4th ed.* New York: Dover, pp. 128–130, 1967.

Nomograph

NOMOGRAM

Non-Abelian

A GROUP or other algebraic object is called non-Abelian is the law of commutativity does not always hold, i.e., if the object is not ABELIAN. For example, the group of INVERTIBLE MATRICES is non-Abelian, as can be seen by comparing

$$\begin{bmatrix} 1 & 0 \\ 0 & -1 \end{bmatrix} \begin{bmatrix} 0 & 1 \\ -1 & 0 \end{bmatrix} = \begin{bmatrix} 0 & 1 \\ 1 & 0 \end{bmatrix} \quad (1)$$

and

$$\begin{bmatrix} 0 & 1 \\ -1 & 0 \end{bmatrix} \begin{bmatrix} 1 & 0 \\ 0 & -1 \end{bmatrix} = \begin{bmatrix} 0 & -1 \\ -1 & 0 \end{bmatrix}. \quad (2)$$

See also ABELIAN, ABELIANIZATION, GROUP, RING

Nonadjacent Vertex Pairs

The following table gives the number of nonadjacent vertex pairs k on graphs of $n = 1, 2, \ldots$ vertices.

k	counts
1	0, 1, 1, 1, 1, 1, 1, ...
2	0, 0, 1, 2, 2, 2, 2, ...
3	0, 0, 1, 3, 4, 5, 5, ...
4	0, 0, 0, 2, 6, 9, 10, ...
5	0, 0, 0, 1, 6, 15, 21, ...

See also ORE GRAPH

Nonagon

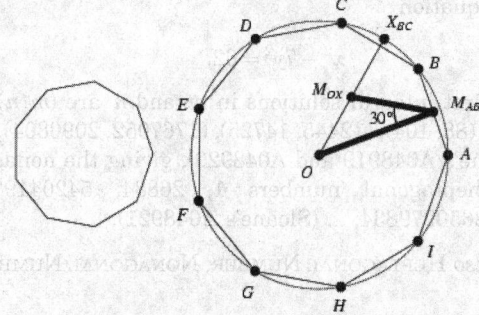

A 9-sided polygon, also known as an enneagon. Although the term "enneagon" is perhaps preferable (since it uses the Greek prefix and suffix instead of the mixed Roman/Greek nonagon), the term "nonagon," which is simpler to spell and pronounce, is used in this work. The REGULAR POLYGON with nine sides and SCHLÄFLI SYMBOL $\{9\}$.

The nonagon cannot be constructed using the classical Greek rules of GEOMETRIC CONSTRUCTION, but

Conway and Guy (1996) give a NEUSIS CONSTRUCTION based on TRISECTION. Madachy (1979) illustrates how to construct a nonagon by folding and knotting a strip of paper. Although the regular nonagon is not a CONSTRUCTIBLE POLYGON, Dixon (1991) gives constructions for several angles which are close approximations to the nonagonal angle $360°/9 = 2\pi/9$, including angles of $\tan^{-1}(5/6) \approx 39.805571°$ and $2\tan^{-1}((\sqrt{3}-1)/2) \approx 40.207819°$.

Given a regular nonagon, let M_{AB} be the MIDPOINT of one side, X_{BC} be the MID-ARC POINT of the arc connecting an adjacent side, and M_{OX} the MIDPOINT of OX_{BC}. Then, amazingly, $\angle OM_{AB}M_{OX} = 30°$ (Karst, quoted in Bankoff and Garfunkel 1973).

See also NONAGRAM, TRIGONOMETRY VALUES PI/9

References

Bankoff, L. and Garfunkel, J. "The Heptagonal Triangle." *Math. Mag.* **46**, 7–19, 1973.
Bold, B. *Famous Problems of Geometry and How to Solve Them.* New York: Dover, pp. 60–61, 1982.
Conway, J. H. and Guy, R. K. *The Book of Numbers.* New York: Springer-Verlag, pp. 194–200, 1996.
Dixon, R. *Mathographics.* New York: Dover, pp. 40–44, 1991.
Madachy, J. S. *Madachy's Mathematical Recreations.* New York: Dover, pp. 60–61, 1979.

Nonagonal Heptagonal Number

A number which is simultaneously a NONAGONAL NUMBER N_m and HEPTAGONAL NUMBER Hep_n and therefore satisfies the DIOPHANTINE EQUATION

$$\tfrac{1}{2}m(7m-5) = \tfrac{1}{2}n(5n-4). \qquad (1)$$

COMPLETING THE SQUARE and rearranging gives

$$(14n-5)^2 - 7(10m-3)^2 = 62. \qquad (2)$$

Defining $x = 14n-5$ and $y = 10m-3$ gives the Pell-like equation

$$x^2 - 7y^2 = 62. \qquad (3)$$

The first integral solutions in m and n are $(m,n) = (1,1), (88, 104), (12445, 14725), (1767052, 2090804), \dots$ (Sloane's A048919 and A048920), giving the nonagonal heptagonal numbers 1, 26884, 542041975, 10928650279834, ... (Sloane's A048921).

See also HEPTAGONAL NUMBER, NONAGONAL NUMBER

References

Sloane, N. J. A. Sequences A048919, A048920, and A048921 in "An On-Line Version of the Encyclopedia of Integer Sequences." http://www.research.att.com/~njas/sequences/eisonline.html.

Nonagonal Hexagonal Number

A number which is simultaneously a NONAGONAL NUMBER N_m and HEXAGONAL NUMBER Hex_n and therefore satisfies the DIOPHANTINE EQUATION

$$\tfrac{1}{2}m(7m-5) = n(2n-1). \qquad (1)$$

COMPLETING THE SQUARE and rearranging gives

$$(14n-5)^2 - 7(4m-1)^2 = 18. \qquad (2)$$

Defining $x = 14n-5$ and $y = 4m-1$ gives the Pell-like equation

$$x^2 - 7y^2 = 18. \qquad (3)$$

This has fundamental solutions $(x,y) = (5,1), (9, 3),$ and $(19, 17)$, giving the family of solutions $(5, 1), (9, 3), (19, 17), (61, 23), (135, 51), (509, 193), \dots$. These give solutions which are integers in m and n of $(m,n) = (1,1), (10, 13), (39025, 51625), \dots$ (Sloane's A048916 and A048917), giving the nonagonal hexagonal numbers 1, 325, 5330229625, 1353857339341, 22184715227362706161, ... (Sloane's A048918).

See also HEXAGONAL NUMBER, NONAGONAL NUMBER

References

Sloane, N. J. A. Sequences A048916, A048917, and A048918 in "An On-Line Version of the Encyclopedia of Integer Sequences." http://www.research.att.com/~njas/sequences/eisonline.html.

See also NONAGONAL NUMBER

Nonagonal Number

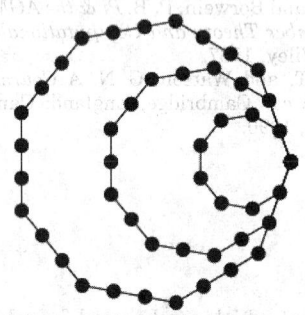

A FIGURATE NUMBER OF THE FORM $n(7n-5)/2$, also called an ENNEAGONAL NUMBER. The first few are 1, 9, 24, 46, 75, 111, 154, 204, ... (Sloane's A001106).
The first few odd nonagonal numbers are 1, 9, 75, 11, 261, 325, ... (Sloane's A028991), and the first few even nonagonal numbers are 24, 46, 154, 204, 396, ... (Sloane's A028992).

See also FIGURATE NUMBER, NONAGONAL HEPTAGONAL NUMBER, NONAGONAL HEXAGONAL NUMBER, NONAGONAL OCTAGONAL NUMBER, NONAGONAL PENTAGONAL NUMBER, NONAGONAL SQUARE NUMBER, NONAGONAL TRIANGULAR NUMBER, POLYGONAL NUMBER

References

Sloane, N. J. A. Sequences A001106/M4604, A028991, and A028992 in "An On-Line Version of the Encyclopedia of

Integer Sequences." http://www.research.att.com/~njas/sequences/eisonline.html.

Nonagonal Octagonal Number

A number which is simultaneously a NONAGONAL NUMBER N_m and OCTAGONAL NUMBER O_n and therefore satisfies the DIOPHANTINE EQUATION

$$\tfrac{1}{2}m(7m-5)=n(3n-2).\qquad(1)$$

COMPLETING THE SQUARE and rearranging gives

$$(14n-5)^2-56(3m-1)^2=19.\qquad(2)$$

Defining $x=14n-5$ and $y=3m-1$ gives the Pell-like equation

$$3x^2-56y^2=19.\qquad(3)$$

The first integral solutions in m and n are $(m,n)=$ (1, 1), (425, 459), (286209, 309141), (192904201, 208360351), ... (Sloane's A048922 and A048923), giving the nonagonal octagonal numbers 1, 631125, 286703855361, 130242107189808901, ... (Sloane's A048924).

See also NONAGONAL NUMBER, OCTAGONAL NUMBER

References

Sloane, N. J. A. Sequences A048922, A048923, and A048924 in "An On-Line Version of the Encyclopedia of Integer Sequences." http://www.research.att.com/~njas/sequences/eisonline.html.

Nonagonal Pentagonal Number

A number which is simultaneously a NONAGONAL NUMBER N_m and PENTAGONAL NUMBER P_n and therefore satisfies the DIOPHANTINE EQUATION

$$\tfrac{1}{2}m(7m-5)=\tfrac{1}{2}n(3n-1).\qquad(1)$$

COMPLETING THE SQUARE and rearranging gives

$$3(14n-5)^2-7(6m-1)^2=68.\qquad(2)$$

Defining $x=14n-5$ and $y=6m+1$ gives the Pell-like equation

$$3x^2-7y^2=68.\qquad(3)$$

This has solutions in (x,y) corresponding to solutions which are integral in m and n of $(m,n)=$ (1, 1), (14, 21), (7189, 10981), (165026, 252081), (86968201, 132846121), ... (Sloane's A048913 and A048914), ... giving the nonagonal pentagonal numbers 1, 651, 180868051, 95317119801, 26472137730696901, ... (Sloane's A048915).

See also NONAGONAL NUMBER, PENTAGONAL NUMBER

References

Sloane, N. J. A. Sequences A048913, A048914, and A048915 in "An On-Line Version of the Encyclopedia of Integer

Sequences." http://www.research.att.com/~njas/sequences/eisonline.html.

Nonagonal Square Number

A number which is simultaneously a NONAGONAL NUMBER N_m and a SQUARE NUMBER S_n and therefore satisfies the DIOPHANTINE EQUATION

$$\tfrac{1}{2}m(7m-5)=n^2.\qquad(1)$$

COMPLETING THE SQUARE and rearranging gives

$$(14n-5)^2-56m^2=25.\qquad(2)$$

Defining $x=14n-5$ and $y=2m^2$ gives the Pell-like equation

$$x^2-14y^2=25.\qquad(3)$$

This has unit solutions $(x,y)=(9,2)$, (23, 6), and (75, 20), which lead to the family of solutions (9, 2), (23, 6), (75, 20), (247, 66), (681, 182), (2245, 600), ... The corresponding integer solutions in n and m are $(n,m)=$ (1, 1), (2, 3), (18, 33), (49, 91), (529, 989), ... (Sloane's A048910 and A048911), giving the nonagonal square numbers 1, 9, 1089, 8281, 978121, 7436529, ... (Sloane's A048912).

See also NONAGONAL NUMBER, SQUARE NUMBER

References

Sloane, N. J. A. Sequences A048910, A048911, and A048912 in "An On-Line Version of the Encyclopedia of Integer Sequences." http://www.research.att.com/~njas/sequences/eisonline.html.

Nonagonal Triangular Number

A number which is simultaneously a NONAGONAL NUMBER N_m and a TRIANGULAR NUMBER T_n and therefore satisfies the DIOPHANTINE EQUATION.

$$\tfrac{1}{2}m(7m-5)=\tfrac{1}{2}n(1+n).\qquad(1)$$

COMPLETING THE SQUARE and rearranging gives

$$(14n-5)^2-7(2m+1)^2=18.\qquad(2)$$

Defining $x=14n-5$ and $y=2m+1$ gives the Pell-like equation

$$x^2-7y^2=18.\qquad(3)$$

This has unit solutions $(x,y)=(5,1)$, (9, 3), and (19, 7), which lead to the family of solutions (5, 1), (9, 3), (19, 7), (61, 23), (135, 51), (299, 113), (971, 367), ... The corresponding integer solutions in n and m are $(n,m)=$ (1, 1), (10, 25), (154, 406), (2449, 6478), ... (Sloane's A048907 and A048908), giving the nonagonal triangular numbers 1, 325, 82621, 20985481, 5330229625, 1353857339341, ... (Sloane's A048909).

See also NONAGONAL NUMBER, TRIANGULAR NUMBER

References

Sloane, N. J. A. Sequences A048907, A048908, and A048909 in "An On-Line Version of the Encyclopedia of Integer Sequences." http://www.research.att.com/~njas/sequences/eisonline.html.

Nonagram

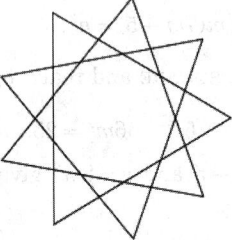

The STAR FIGURE {9/3} composed of three EQUILATERAL TRIANGLES rotated at angles 0°, 40°, and 80°. It has been called the STAR OF GOLIATH by analogy with the STAR OF DAVID (HEXAGRAM).

See also HEXAGRAM, NONAGON, STAR FIGURE, TRIGONOMETRY VALUES PI/9

Nonahedral Graph

A POLYHEDRAL GRAPH having nine vertices. There are 2606 nonisomorphic nonahedral graphs, as first enumerated by Federico (1969; Duijvestijn and Federico 1981).

See also NONAHEDRON, POLYHEDRAL GRAPH

References

Duijvestijn, A. J. W. and Federico, P. J. "The Number of Polyhedral (3-Connected Planar) Graphs." *Math. Comput.* **37**, 523–532, 1981.
Federico, P. J. "Enumeration of Polyhedra: The Number of 9-hedra." *J. Combin. Th.* **7**, 155–161, 1969.

Nonahedron

A nine-faced POLYHEDRON. There are 2606 topologically distinct convex nonahedra, corresponding to the 2606 nonisomorphic NONAHEDRAL GRAPHS.

See also NONAHEDRAL GRAPH

Nonalternating Knot

A KNOT which is not ALTERNATING. Unlike alternating knots, FLYPE moves are *not* sufficient to pass between all minimal diagrams of a given nonalternating knot (Hoste *et al.* 1998). In fact, Thistlethwaite used 13 different moves in generating a list of 16-crossing alternating knots (Hoste *et al.* 1998), and still had 9,868 duplicates out of a list of 1,018,774 knots (Hoste *et al.* 1998).

See also ALTERNATING KNOT, KNOT

References

Hoste, J.; Thistlethwaite, M.; and Weeks, J. "The First 1,701,936 Knots." *Math. Intell.* **20**, 33–48, Fall 1998.

Non-Archimedean Field

See also HENSEL'S LEMMA, NON-ARCHIMEDEAN GEOMETRY, NON-ARCHIMEDEAN VALUATION, VALUATION

Non-Archimedean Geometry

A geometry in which ARCHIMEDES' AXIOM does not hold.

See also ARCHIMEDES' AXIOM, HORN ANGLE, NON-ARCHIMEDEAN FIELD, NON-ARCHIMEDEAN VALUATION

References

Itô, K. (Ed.). §155D in *Encyclopedic Dictionary of Mathematics, 2nd ed., Vol. 2.* Cambridge, MA: MIT Press, p. 611, 1986.

Non-Archimedean Valuation

See also NON-ARCHIMEDEAN FIELD, NON-ARCHIMEDEAN GEOMETRY

Nonarithmetic Progression Sequence

Given two starting numbers (a_1, a_2), the following table gives the unique sequences $\{a_i\}$ that contain no three-term arithmetic progressions.

Sloane	sequence
A003278	1, 2, 4, 5, 10, 11, 13, 14, 28, 29, 31, 32, ...
A033156	1, 3, 4, 6, 10, 12, 13, 15, 28, 30, 31, 33, ...
A033157	1, 4, 5, 8, 10, 13, 14, 17, 28, 31, 32, 35, ...
A033158	1, 5, 6, 8, 12, 13, 17, 24, 27, 32, 34, 38, ...
A033159	2, 3, 5, 6, 11, 12, 14, 15, 29, 30, 32, 33, ...
A033160	2, 4, 5, 7, 11, 13, 14, 16, 29, 31, 32, 34, ...
A033161	2, 5, 6, 9, 11, 14, 15, 18, 29, 32, 33, 36, ...
A033162	3, 4, 6, 7, 12, 13, 15, 16, 30, 31, 33, 34, ...
A033163	3, 5, 6, 8, 12, 14, 15, 17, 30, 32, 33, 35, ...

A033164　4, 5, 7, 8, 13, 14, 16, 17, 31, 32, 34, 35, ...

See also ARITHMETIC SEQUENCE

References

Allouche, J.-P. and Shallit, J. "The Ring of k-Regular Sequences." *Theor. Comput. Sci.* **98**, 163–197, 1992.

Erdos, P. and Turán, P. "On Some Sequences of Integers." *J. London Math. Soc.* **11**, 261–264, 1936.

Gerver, J.; Propp, J.; and Simpson, J. "Greedily Partitioning the Natural Numbers into Sets Free of Arithmetic Progressions." *Proc. Amer. Math. Soc.* **102**, 765–772, 1988.

Guy, R. K. "Theorem of van der Waerden, Szemerédi's Theorem. Partitioning the Integers into Classes; at Least One Contains an A.P." §E10 in *Unsolved Problems in Number Theory, 2nd ed.* New York: Springer-Verlag, pp. 204–209, 1994.

Iacobescu, F. "Smarandache Partition Type and Other Sequences." *Bull. Pure Appl. Sci.* **16E**, 237–240, 1997.

Ibstedt, H. "A Few Smarandache Sequences." *Smarandache Notions J.* **8**, 170–183, 1997.

Sloane, N. J. A. Sequences A003278/M0975, A033156, A033157, A033158, A033159, A033160, A033161, A033162, A033163, and A033164 in "An On-Line Version of the Encyclopedia of Integer Sequences." http://www.research.att.com/~njas/sequences/eisonline.html.

Nonassociative Algebra

An ALGEBRA which does not satisfy

$$a(bc) = (ab)c$$

is called a nonassociative algebra.

See also ALGEBRA, CAYLEY NUMBER, COMPLEX NUMBER, DIVISION ALGEBRA, QUATERNION, REAL NUMBER

References

Kuz'min, E. N. and Shestakov, I. P. "Non-Associative Structures." In *Algebra VI. Combinatorial and Asymptotic Methods of Algebra: Nonassociative Structures* (Ed. A. I. Kostrikin and I. R. Shafarevich). New York: Springer-Verlag, 1995.

Schafer, R. D. *An Introduction to Nonassociative Algebras.* New York: Dover, 1996.

Nonassociative Product

The number of nonassociative n-products with k elements preceding the rightmost left parameter is

$$F(n,k) = F(n-1,k) + F(n-1,k-1)$$

$$= \binom{n+k-2}{k} - \binom{n+k-1}{k-1}$$

where $\binom{n}{k}$ is a BINOMIAL COEFFICIENT. The number of n-products in a nonassociative algebra is

$$F(n) = C_n = \sum_{j=0}^{n-2} F(n,j) = \frac{(2n-2)!}{n!(n-1)!},$$

where C_n is a CATALAN NUMBER, 1, 1, 2, 5, 14, 42, 132, ... (Sloane's A000108).

References

Niven, I. M. *Mathematics of Choice: Or, How to Count Without Counting.* Washington, DC: Math. Assoc. Amer., pp. 140–152, 1965.

Sloane, N. J. A. Sequences A000108/M1459 in "An On-Line Version of the Encyclopedia of Integer Sequences." http://www.research.att.com/~njas/sequences/eisonline.html.

Nonaveraging Sequence

N.B. A detailed online essay by S. Finch was the starting point for this entry.

A sequence of POSITIVE INTEGERS

$$1 \le a_1 < a_2 < a_3 \ldots$$

is a nonaveraging sequence if it contains no three terms which are in an ARITHMETIC PROGRESSION, i.e., terms such that

$$\tfrac{1}{2}(a_i + a_j) = a_k$$

for distinct a_i, a_j, a_k. The EMPTY SET and sets of length one are therefore trivially nonaveraging.

Consider all possible subsets on the integers $S_n = \{1, 2, \ldots, n\}$. There is one nonaveraging sequence on S_0 ($\varnothing$), two on S_1 ($\varnothing$ and $\{1\}$), four on S_2, and so on. For example, 13 of the 16 subjects of S_4 are nonaveraging, with $\{1,2,3\}$, $\{2,3,\}$, and $\{1,2,3,4\}$ excluded. The numbers of nonaveraging subsets on S_0, S_1, ... are 1, 2, 4, 7, 13, 23, 40, ... (Sloane's A051013).

Wróblewski (1984) showed that for infinite nonaveraging sequences,

$$S(A) \equiv \sup_{\text{all nonaveraging sequences}} \sum_{k=1}^{\infty} \frac{1}{a^k} > 3.00849.$$

See also NONDIVIDING SET

References

Abbott, H. L. "On a Conjecture of Erdos and Straus on Non-Averaging Sets of Integers." In *Proceedings of the Fifth British Combinatorial Conference* (Es. C. St. J. A. Nash-Williams and J. Sheehan). Winnipeg, Manitoba, Canada: Utilitas Math. Pub., pp. 1–4, 1976.

Abbott, H. L. "Extremal Problems on Non-Averaging and Non-Dividing Sets." *Pacific J. Math.* **91**, 1–12, 1980.

Abbott, H. L. "On the Erdos-Straus Non-Averaging Set Problem." *Acta Math. Hungar.* **47**, 117–119, 1986.

Behrend, F. "On Sets of Integers which Contain no Three Terms in an Arithmetic Progression." *Proc. Nat. Acad. Sci. USA* **32**, 331–332, 1946.

Finch, S. "Favorite Mathematical Constants." http://www.mathsoft.com/asolve/constant/erdos/erdos.html.

Gerver, J. L. "The Sum of the Reciprocals of a Set of Integers with No Arithmetic Progression of k Terms." *Proc. Amer. Math. Soc.* **62**, 211–214, 1977.

Gerver, J. L. and Ramsey, L. "Sets of Integers with no Long Arithmetic Progressions Generated by the Greedy Algorithm." *Math. Comput.* **33**, 1353–1360, 1979.

Guy, R. K. "Nonaveraging Sets. Nondividing Sets." §C16 in *Unsolved Problems in Number Theory, 2nd ed.* New York: Springer-Verlag, pp. 131–132, 1994.

Sloane, N. J. A. Sequences A051013 in "An On-Line Version of the Encyclopedia of Integer Sequences." http://www.research.att.com/~njas/sequences/eisonline.html.

Straus, E. G. "Non-Averaging Sets." *Proc. Symp. Pure Math* **19**, 215–222, 1971.

Weisstein, E. W. "Integer Sequences." MATHEMATICA NOTEBOOK IntegerSequences.M.

Wróblewski, J. "A Nonaveraging Set of Integers with a Large Sum of Reciprocals." *Math. Comput.* **43**, 261–262, 1984.

Noncentral Distribution

CHI-SQUARED DISTRIBUTION, F-DISTRIBUTION, STUDENT'S T-DISTRIBUTION

Noncommutative Group

A group whose elements do not commute. The simplest noncommutative GROUP is the DIHEDRAL GROUP $D3$ of ORDER six.

See also COMMUTATIVE, FINITE GROUP $D3$, GROUP

Noncommutative Ring

This entry contributed by VIKTOR BENGTSSON

A noncommutative ring R is a RING in which the law of multiplicative commutativity is not satisfied, i.e.,

$$a \cdot b \neq b \cdot a$$

for any two elements $a, b \in R$. In such a case, the elements a and b of the ring R are said not to commute. An important example of a noncommutative ring is the ring $M_n(K)$ consisting of all $n \times n$ matrices whose elements are members of the FIELD K.

See also RING

Nonconformal Map

Let γ be a path in $\mathbb{C}$, $w = f(z)$, and θ and ϕ be the tangents to the curves γ and $f(\gamma)$ at z_0 and w_0. If there is an N such that

$$f^{(N)}(z_0) \neq 0 \tag{1}$$

$$f^{(N)}(z_0) = 0 \tag{2}$$

for all $n < N$ (or, equivalently, if $f'(z)$ has a zero of order $N-1$), then

$$f(z) = f(z_0) + \frac{f^{(N)}(z_0)}{N!}$$
$$\times (z - z_0)^N + \frac{f^{(N+1)}(z_0)}{(N+1)!}(z - z_0)^{N+1} + \cdots \tag{3}$$

$$f(z) - f(z_0)$$
$$= (z - z_0)^N \left[\frac{f^{(N)}(z_0)}{N!} + \frac{f^{(N+1)}(z_0)}{(N+1)!}(z - z_0) + \cdots \right], \tag{4}$$

so the ARGUMENT is

$$\arg[f(z) - f(z_0)] = N \arg(z - z_0) + \arg\left[\frac{f^{(N)}(z_0)}{N!} \right.$$
$$\left. + \frac{f^{(N+1)}(z_0)}{(N+1)!}(z - z_0) + \cdots \right]. \tag{5}$$

As $z \to z_0$, $\arg(z - z_0) \to \theta$ and $|\arg[f(z) - f(z_0)]| \to \phi$,

$$\phi = N\theta + \arg\left[\frac{f^{(N)}(z_0)}{N!} \right] = N\theta + \arg[f^{(N)}(z_0)]. \tag{6}$$

See also CONFORMAL MAPPING

Nonconstructive Proof

A PROOF which indirectly shows a mathematical object exists without providing a specific example or algorithm for producing an example. Nonconstructive proofs are also called existence proofs.

See also EXISTENCE PROBLEM, PROOF

References

Courant, R. and Robbins, H. "The Indirect Method of Proof." §2.4.4 in *What is Mathematics?: An Elementary Approach to Ideas and Methods, 2nd ed.* Oxford, England: Oxford University Press, pp. 86–87, 1996.

Hoffman, P. *The Man Who Loved Only Numbers: The Story of Paul Erdos and the Search for Mathematical Truth.* New York: Hyperion, p. 229, 1998.

Noncototient

A POSITIVE value of n for which $x - \phi(x) = n$ has no solution, where $\phi(x)$ is the TOTIENT FUNCTION. The first few are 10, 26, 34, 50, 52, ... (Sloane's A005278).

See also NONTOTIENT, TOTIENT FUNCTION

References

Guy, R. K. *Unsolved Problems in Number Theory, 2nd ed.* New York: Springer-Verlag, p. 91, 1994.

Sloane, N. J. A. Sequences A005278/M4688 in "An On-Line Version of the Encyclopedia of Integer Sequences." http://www.research.att.com/~njas/sequences/eisonline.html.

Noncylindrical Ruled Surface

A RULED SURFACE parameterization $\mathbf{x}(u,v) = \mathbf{b}(u) + v\mathbf{g}(u)$ is called noncylindrical if $\mathbf{g} \times \mathbf{g}'$ is nowhere 0. A noncylindrical ruled surface always has a parameterization OF THE FORM

$$\mathbf{x}(u,v) = \boldsymbol{\sigma}(u) + v\boldsymbol{\delta}(u),$$

where $|\boldsymbol{\delta}| = 1$ and $\boldsymbol{\sigma}' \cdot \boldsymbol{\delta}' = 0$, where $\boldsymbol{\sigma}$ is called the STRICTION CURVE of $\mathbf{x}$ and $\boldsymbol{\delta}$ the DIRECTOR CURVE.

See also DISTRIBUTION PARAMETER, RULED SURFACE, STRICTION CURVE

References

Gray, A. "Noncylindrical Ruled Surfaces." §19.4 in *Modern Differential Geometry of Curves and Surfaces with Mathematica, 2nd ed.* Boca Raton, FL: CRC Press, pp. 445–448, 1997.

Nondecreasing Function

A function $f(x)$ is said to be nondecreasing on an INTERVAL I if $f(b) \geq f(a)$ for all $b > a$, where $a, b \in I$. Conversely, a function $f(x)$ is said to be nonincreasing on an INTERVAL I if $f(b) \leq f(a)$ for all $b > a$ with $a, b \in I$.

See also DECREASING FUNCTION, MONOTONE DECREASING, MONOTONE INCREASING, NONINCREASING FUNCTION

References

Jeffreys, H. and Jeffreys, B. S. "Increasing and Decreasing Functions." §1.065 in *Methods of Mathematical Physics, 3rd ed.* Cambridge, England: Cambridge University Press, p. 22, 1988.

Nondividing Set

A SET in which no element divides the SUM of any nonempty subset of the other elements. The EMPTY SET and sets of length one are therefore trivially nondividing. Also, any set other than $\{1\}$ which contains 1 is dividing. For example, $\{2, 3, 5\}$ is dividing, since $2|(3 + 5)$ (and $5|(2 + 3)$), but $\{4, 6, 7\}$ is nondividing since 4 divides none of $\{6, 7, (6 + 7)\}$, and similarly for 6 and 7.

Consider all possible subsets on the integers $S_n = \{1, 2, \ldots, n\}$. Then the numbers of nondividing subsets on $S_0, S_1, \ldots$ are 1, 2, 3, 5, 7, 12, 16, 28, 38, 60, ... (Sloane's A051014). For example, the 12 nondividing sets in S_6 are $\emptyset$, $\{1\}$, $\{2\}$, $\{3\}$, $\{4\}$, $\{5\}$, $\{6\}$, $\{2, 3\}$, $\{2, 5\}$, $\{3, 4\}$, $\{3, 5\}$, $\{4, 5\}$, $\{4, 6\}$, $\{5, 6\}$, $\{3, 4, 5\}$, and $\{4, 5, 6\}$.

See also NONAVERAGING SEQUENCE, PRIMITIVE SEQUENCE

References

Abbott, H. L. "Extremal Problems on Non-Averaging and Non-Dividing Sets." *Pacific J. Math.* **91**, 1–12, 1980.
Guy, R. K. "Nonaveraging Sets. Nondividing Sets." §C16 in *Unsolved Problems in Number Theory, 2nd ed.* New York: Springer-Verlag, pp. 131–132, 1994.
Sloane, N. J. A. Sequences A051014 in "An On-Line Version of the Encyclopedia of Integer Sequences." http://www.research.att.com/~njas/sequences/eisonline.html.
Straus, E. G. "Non-Averaging Sets." *Proc. Symp. Pure Math* **19**, 215–222, 1971.
Weisstein, E. W. "Integer Sequences." MATHEMATICA NOTEBOOK INTEGERSEQUENCES.M.

Nonequivalent

If $A \Rightarrow !B$ and $B \Rightarrow !A$ (i.e., $(A \Rightarrow !B) \wedge (B \Rightarrow !A)$, where $!A$ denotes NOT, $\Rightarrow$ denotes IMPLIES, and $\wedge$ denotes AND), then A and B are said to be inequivalent, a relationship which is written symbolically as $A \not\equiv B$, $A \not\Leftrightarrow B$, $A \not= B$ Nonequivalence is implemented in *Mathematica* as Unequal[A, B, ...]. Binary nonequivalence has the same TRUTH TABLE as XOR (i.e., EXCLUSIVE DISJUNCTION), reproduced below.

A	B	$A \not\equiv B$
T	T	F
T	F	T
F	T	T
F	F	F

See also CONNECTIVE, EQUIVALENT, EXCLUSIVE DISJUNCTION, XOR

Nonessential Singularity

REGULAR SINGULAR POINT

Non-Euclidean Geometry

In three dimensions, there are three classes of constant curvature GEOMETRIES. All are based on the first four of EUCLID'S POSTULATES, but each uses its own version of the PARALLEL POSTULATE. The "flat" geometry of everyday intuition is called EUCLIDEAN GEOMETRY (or PARABOLIC GEOMETRY), and the non-Euclidean geometries are called HYPERBOLIC GEOMETRY (or LOBACHEVSKY-BOLYAI-GAUSS GEOMETRY) and ELLIPTIC GEOMETRY (or RIEMANNIAN GEOMETRY). SPHERICAL GEOMETRY is a non-Euclidean 2-D geometry. It was not until 1868 that Beltrami proved that non-Euclidean geometries were as logically consistent as EUCLIDEAN GEOMETRY.

See also ABSOLUTE GEOMETRY, ELLIPTIC GEOMETRY, EUCLID'S POSTULATES, EUCLIDEAN GEOMETRY, HYPERBOLIC GEOMETRY, PARALLEL POSTULATE, SPHERICAL GEOMETRY

References

--. "Welcome to the Non-Euclidean Geometry Homepage." http://members.tripod.com/~noneuclidean/.
Bolyai, J. "Scientiam spatii absolute veritam exhibens: a veritate aut falsitate Axiomatis XI Euclidei (a priori haud unquam decidenda) indepentem: adjecta ad casum falsitatis, quadratura circuli geometrica." Reprinted as "The Science of Absolute Space" in Bonola, R. *Non-Euclidean Geometry, and The Theory of Parallels by Nikolas Lobachevski, with a Supplement Containing The Science of Absolute Space by John Bolyai.* New York: Dover, 1955.
Bonola, R. *Non-Euclidean Geometry, and The Theory of Parallels by Nikolas Lobachevski, with a Supplement Containing The Science of Absolute Space by John Bolyai.* New York: Dover, 1955.
Borsuk, K. *Foundations of Geometry: Euclidean and Bolyai-Lobachevskian Geometry. Projective Geometry.* Amsterdam, Netherlands: North-Holland, 1960.

Carslaw, H. S. *The Elements of Non-Euclidean Plane Geometry and Trigonometry.* London: Longmans, 1916.

Coxeter, H. S. M. *Non-Euclidean Geometry, 6th ed.* Washington, DC: Math. Assoc. Amer., 1988.

Dunham, W. *Journey through Genius: The Great Theorems of Mathematics.* New York: Wiley, pp. 53–60, 1990.

Greenberg, M. J. *Euclidean and Non-Euclidean Geometries: Development and History, 3rd ed.* San Francisco, CA: W. H. Freeman, 1994.

Iversen, B. *An Invitation to Hyperbolic Geometry.* Cambridge, England: Cambridge University Press, 1993.

Iyanaga, S. and Kawada, Y. (Eds.). "Non-Euclidean Geometry." §283 in *Encyclopedic Dictionary of Mathematics.* Cambridge, MA: MIT Press, pp. 893–896, 1980.

Lobachevski, N. Reprinted as "Theory of Parallels" in Bonola, R. *Non-Euclidean Geometry, and The Theory of Parallels by Nikolas Lobachevski, with a Supplement Containing The Science of Absolute Space by John Bolyai.* New York: Dover, 1955.

Martin, G. E. *The Foundations of Geometry and the Non-Euclidean Plane.* New York: Springer-Verlag, 1975.

Pappas, T. "A Non-Euclidean World." *The Joy of Mathematics.* San Carlos, CA: Wide World Publ./Tetra, pp. 90–92, 1989.

Ramsay, A. and Richtmyer, R. D. *Introduction to Hyperbolic Geometry.* New York: Springer-Verlag, 1995.

Sommerville, D. Y. *The Elements of Non-Euclidean Geometry.* London: Bell, 1914.

Sommerville, D. Y. *Bibliography of Non-Euclidean Geometry, 2nd ed.* New York: Chelsea, 1960.

Sved, M. *Journey into Geometries.* Washington, DC: Math. Assoc. Amer., 1991.

Trudeau, R. J. *The Non-Euclidean Revolution.* Boston, MA: Birkhäuser, 1987.

Weisstein, E. W. "Books about Non-Euclidean Geometry." http://www.treasure-troves.com/books/Non-EuclideanGeometry.html.

Woods, F. S. "Non-Euclidean Geometry." Ch. 3 in *Monographs on Topics of Modern Mathematics Relevant to the Elementary Field* (Ed. J. W. A. Young). New York: Dover, pp. 93–147, 1955.

Nonhyperbolic Knot

HYPERBOLIC KNOT, SATELLITE KNOT, TORUS KNOT

Nonic Surface

An ALGEBRAIC SURFACE of degree 9.

See also ALGEBRAIC SURFACE

Nonillion

In the American system, 10^{30}.

See also LARGE NUMBER

Nonincreasing Function

A function $f(x)$ is said to be nonincreasing on an INTERVAL I if $f(b) \leq f(a)$ for all $b > a$, where $a, b \in I$. Conversely, a function $f(x)$ is said to be nondecreasing on an INTERVAL I if $f(b) \geq f(a)$ for all $b > a$ with $a, b \in I$.

See also INCREASING FUNCTION, MONOTONE DECREASING, MONOTONE INCREASING, NONDECREASING FUNCTION

References
Jeffreys, H. and Jeffreys, B. S. "Increasing and Decreasing Functions." §1.065 in *Methods of Mathematical Physics, 3rd ed.* Cambridge, England: Cambridge University Press, p. 22, 1988.

Noninvertible Knot

INVERTIBLE KNOT

Nonlinear Least Squares Fitting

Given a function $f(x)$ of a variable x tabulated at m values $y_1 = f(x_1), \ldots, y_m = f(x_m)$, assume the function is of known analytic form depending on n parameters $f(x; \lambda_1, \ldots, \lambda_n)$, and consider the overdetermined set of m equations

$$y_1 = f(x_1; \lambda_1, \lambda_2, \ldots, \lambda_n) \qquad (1)$$

$$y_m = f(x_m; \lambda_1, \lambda_2, \ldots, \lambda_n). \qquad (2)$$

We desire to solve these equations to obtain the values $\lambda_1, \ldots, \lambda_n$ which best satisfy this system of equations. Pick an initial guess for the λ_i and then define

$$d\beta_i = y_i - f(x_i; \lambda_1, \ldots, \lambda_n). \qquad (3)$$

Now obtain a linearized estimate for the changes $d\lambda_i$ needed to reduce $d\beta_i$ to 0,

$$d\beta_i = \sum_{j=1}^{n} \frac{\partial f}{\partial \lambda_j} d\lambda_j \bigg|_{x_j, \lambda} \qquad (4)$$

for $i = 1, \ldots, n$. This can be written in component form as

$$d\beta_i = A_{ij} d\lambda_i, \qquad (5)$$

where A is the $m \times n$ MATRIX

$$A_{ij} = \begin{pmatrix} \dfrac{\partial f}{d\lambda_1}\bigg|_{x_1, \lambda} & \dfrac{\partial f}{d\lambda_1}\bigg|_{x_1, \lambda} & \cdots \\[2ex] \dfrac{\partial f}{d\lambda_2}\bigg|_{x_2, \lambda} & \dfrac{\partial f}{d\lambda_2}\bigg|_{x_2, \lambda} & \cdots \\[2ex] \vdots & \vdots & \ddots \\[2ex] \dfrac{\partial f}{d\lambda_1}\bigg|_{x_m, \lambda} & \dfrac{\partial f}{d\lambda_n}\bigg|_{x_m, \lambda} & \cdots \end{pmatrix}. \qquad (6)$$

In more concise MATRIX form,

$$d\beta = A d\lambda, \qquad (7)$$

where $d\beta$ and $d\lambda$ are m-VECTORS. Applying the MATRIX TRANSPOSE of A to both sides gives

$$A^T d\beta = (A^T A) d\lambda. \qquad (8)$$

Defining

$$\mathbf{a} \equiv A^T A \qquad (9)$$

$$\mathbf{b} \equiv \mathsf{A}^{\mathrm{T}} d\beta \qquad (10)$$

in terms of the known quantities A and $d\beta$ then gives the MATRIX EQUATION

$$\mathbf{a} d\lambda = \mathbf{b}, \qquad (11)$$

which can be solved for $d\lambda$ using standard matrix techniques such as GAUSSIAN ELIMINATION. This offset is then applied to λ and a new $d\beta$ is calculated. By iteratively applying this procedure until the elements of $d\lambda$ become smaller than some prescribed limit, a solution is obtained. Note that the procedure may not converge very well for some functions and also that convergence is often greatly improved by picking initial values close to the best-fit value. The sum of square residuals is given by $R^2 = d\beta \cdot d\beta$ after the final iteration.

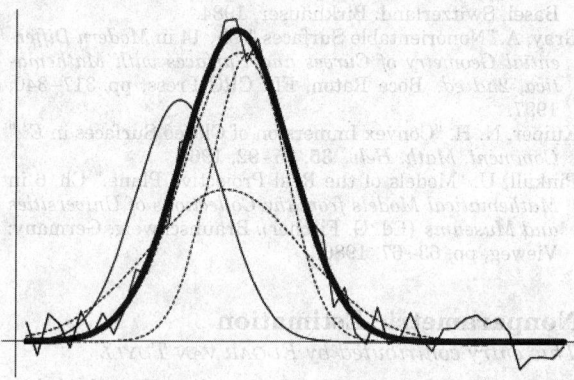

An example of a nonlinear least squares fit to a noisy GAUSSIAN FUNCTION

$$f(x; A, x_0, \sigma) = A e^{-(x-x_0)^2/(2\sigma^2)} \qquad (12)$$

is shown above, where the thin solid curve is the initial guess, the dotted curves are intermediate iterations, and the heavy solid curve is the fit to which the solution converges. The actual parameters are $(A, x_0, \sigma) = (1, 20, 5)$, the initial guess was $(0.8, 15, 4)$, and the converged values are $(1.03105, 20.1369, 4.86022)$, with $R^2 = 0.148461$. The PARTIAL DERIVATIVES used to construct the matrix A are

$$\frac{\partial f}{\partial A} = e^{-(x-x_0)^2/(2\sigma^2)} \qquad (13)$$

$$\frac{\partial f}{\partial x_0} = \frac{A(x-x_0)}{\sigma^2} e^{-(x-x_0)^2/(2\sigma^2)} \qquad (14)$$

$$\frac{\partial f}{\partial \sigma_0} = \frac{A(x-x_0)}{\sigma^3} e^{-(x-x_0)^2/(2\sigma^2)}. \qquad (15)$$

The technique could obviously be generalized to multiple Gaussians, to include slopes, etc., although the convergence properties generally worsen as the number of free parameters is increased.

An analogous technique can be used to solve an overdetermined set of equations. This problem might, for example, arise when solving for the best-fit EULER ANGLES corresponding to a noisy ROTATION MATRIX, in which case there are three unknown angles, but nine correlated matrix elements. In such a case, write the n *different* functions as $f_i(\lambda_1, \dots, \lambda_n)$ for $i = 1, \dots, n$, call their actual values y_i, and define

$$\mathsf{A} = \begin{pmatrix} \left.\dfrac{\partial f_1}{\partial \lambda_1}\right|_{\lambda i} & \left.\dfrac{\partial f_1}{\partial \lambda_2}\right|_{\lambda i} & \cdots & \left.\dfrac{\partial f_1}{\partial \lambda_n}\right|_{\lambda i} \\ \vdots & \vdots & \ddots & \vdots \\ \left.\dfrac{\partial f_m}{\partial \lambda_1}\right|_{\lambda i} & \left.\dfrac{\partial f_m}{\partial \lambda_2}\right|_{\lambda i} & \cdots & \left.\dfrac{\partial f_m}{\partial \lambda_n}\right|_{\lambda i} \end{pmatrix}, \qquad (16)$$

and

$$d\beta = \mathbf{y} - f_i(\lambda_1, \dots, \lambda_n), \qquad (17)$$

where λ_i are the numerical values obtained after the ith iteration. Again, set up the equations as

$$\mathsf{A} d\lambda = d\beta, \qquad (18)$$

and proceed exactly as before.

See also LEAST SQUARES FITTING, LINEAR REGRESSION, MOORE-PENROSE GENERALIZED MATRIX INVERSE

Nonlinear Stability

See also LINEAR STABILITY, LYAPUNOV FUNCTION

Nonnegative

A quantity which is either 0 (ZERO) or POSITIVE, i.e., ≥ 0.

See also NEGATIVE, NONNEGATIVE INTEGER, NONPOSITIVE, NONZERO, POSITIVE, ZERO

Nonnegative Integer

An INTEGER that is either 0 or positive, i.e., a member of the set $\mathbb{Z}^* = \{0\} \cup \mathbb{Z}^+$, where $\mathbf{Z}_+$ denotes the POSITIVE INTEGERS.

See also NEGATIVE INTEGER, NONPOSITIVE INTEGER, POSITIVE INTEGER, $\mathbb{Z}^*$

Nonnegative Partial Sum

The number of sequences with NONNEGATIVE partial sums which can be formed from n 1s and n -1s (Bailey 1996, Brualdi 1992) is given by the CATALAN NUMBERS. Bailey (1996) gives the number of NONNEGATIVE partial sums of n 1s and k -1s $a_1, a_2, \dots, a_{n+k}$, so that

$$a_1 + a_2 + \dots + a_i \geq 0 \qquad (1)$$

for all $1 \leq i \leq n + k$. The closed form expression is

$$\left\{ {n \atop 0} \right\} = 1 \qquad (2)$$

for $n \geq 0$,

$$\left\{ {n \atop 1} \right\} = n \qquad (3)$$

for $n \geq 1$, and

$$\left\{ {n \atop k} \right\} = \frac{(n + 1 - k)(n + 2)(n + 3) \cdots (n + k)}{k!}, \qquad (4)$$

for $n \geq k \geq 2$. Setting $k = n$ then recovers the CATALAN NUMBERS

$$C_n = \left\{ {n \atop n} \right\} = \frac{1}{n + 1} \binom{2n}{n}. \qquad (5)$$

See also CATALAN NUMBER

References

Bailey, D. F. "Counting Arrangements of 1's and -1's." *Math. Mag.* **69**, 128–131, 1996.
Brualdi, R. A. *Introductory Combinatorics, 2nd ed.* New York: Elsevier, 1992.

Nonorientable Surface

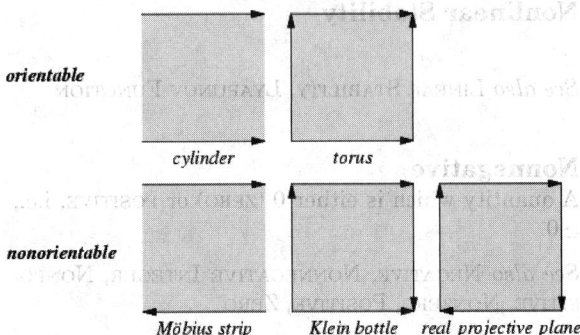

orientable

cylinder   *torus*

nonorientable

Möbius strip   *Klein bottle*   *real projective plane*

A surface such as the MÖBIUS STRIP or KLEIN BOTTLE (Gray 1997, pp. 322–323) on which there exists a closed path such that the directrix is reversed when moved around this path. The REAL PROJECTIVE PLANE is also a nonorientable surface, as are the BOY SURFACE, CROSS-CAP, and ROMAN SURFACE, all of which are homeomorphic to the REAL PROJECTIVE PLANE (Pinkall 1986).

There is a general method for constructing nonorientable surfaces which proceeds as follows (Banchoff 1984, Pinkall 1986). Choose three HOMOGENEOUS POLYNOMIALS of POSITIVE EVEN degree and consider the MAP

$$\mathbf{f} = (f_1(x,y,z), f_2(x,y,z), f_3(x,y,z)) : \mathbb{R}^3 \to \mathbb{R}^3. \qquad (1)$$

Then restricting x, y, and z to the surface of a sphere by writing

$$x = \cos \theta \sin \phi \qquad (2)$$

$$y = \sin \theta \sin \phi \qquad (3)$$

$$z = \cos \phi \qquad (4)$$

and restricting θ to $[0, 2\pi)$ and ϕ to $[0, \pi/2]$ defines a map of the REAL PROJECTIVE PLANE to $\mathbb{R}^3$.

In 3-D, there is no unbounded nonorientable surface which does not intersect itself (Kuiper 1961, Pinkall 1986).

See also BOY SURFACE, CROSS-CAP, KLEIN BOTTLE, MÖBIUS STRIP, ORIENTABLE SURFACE, REAL PROJECTIVE PLANE, ROMAN SURFACE

References

Banchoff, T. "Differential Geometry and Computer Graphics." In *Perspectives of Mathematics: Anniversary of Oberwolfach* (Ed. W. Jager, R. Remmert, and J. Moser). Basel, Switzerland: Birkhäuser, 1984.
Gray, A. "Nonorientable Surfaces." Ch. 14 in *Modern Differential Geometry of Curves and Surfaces with Mathematica, 2nd ed.* Boca Raton, FL: CRC Press, pp. 317–340, 1997.
Kuiper, N. H. "Convex Immersion of Closed Surfaces in E^3." *Comment. Math. Helv.* **35**, 85–92, 1961.
Pinkall, U. "Models of the Real Projective Plane." Ch. 6 in *Mathematical Models from the Collections of Universities and Museums* (Ed. G. Fischer). Braunschweig, Germany: Vieweg, pp. 63–67, 1986.

Nonparametric Estimation

This entry contributed by EDGAR VAN TUYLL

Nonparametric estimation is a statistical method that allows the functional form of a fit to data to be obtained in the absence of any guidance or constraints from theory. As a result, the procedures of nonparametric estimation have no meaningful associated parameters. Two types of nonparametric techniques are artificial neural networks and kernel estimation.

Artificial neural networks model an unknown function by expressing it as a weighted sum of several sigmoids, usually chosen to be logit curves, each of which is a function of all the relevant explanatory variables. This amounts to an extremely flexible functional form for which estimation requires a nonlinear least-squares iterative search algorithm based on gradients.

Kernel estimation specifies $y = m(x) + e$, where $m(x)$ is the conditional expectation of y with no parametric form whatsoever, and the density of the error e is completely unspecified. The N observations y_i and x_i are used to estimate a joint density function for y and x. The density at a point (y_0, x_0) is estimated by seeing what proportion of the N observations are "close to" (y_0, x_0). This procedure involves the use of a function called a kernel to assign weights to nearby observations.

See also NONPARAMETRIC STATISTICS

Nonparametric Statistics

References
Kennedy, P. *A Guide to Econometrics.* Cambridge, MA: MIT Press, 1998.
Pagan, A. R. and Ullah, A. *Non-Parametric Econometrics.* Cambridge, England: Cambridge University Press, 1997.

Nonparametric Statistics

See also NONPARAMETRIC ESTIMATION, PARAMETRIC STATISTICS

References
Brodsky, B. E. and Darkhovsky, B. S. *Non-Parametric Statistical Diagnosis: Problems and Methods.* Dordrecht, Netherlands: Kluwer, 2000.
Sheskin, D. J. *Handbook of Parametric and Nonparametric Statistical Procedures, 2nd ed.* Boca Raton, FL: Chapman & Hall/CRC, 2000.

Nonpositive

A quantity which is either 0 (ZERO) or NEGATIVE, i.e., ≤ 0.

See also NEGATIVE, NONNEGATIVE, NONZERO, POSITIVE, ZERO

Nonpositive Integer

An INTEGER that is either 0 or negative, i.e., a member of the set $\{0\} \cup \mathbb{Z}^-$, where $\mathbb{Z}_-$ denotes the NEGATIVE INTEGERS.

See also NEGATIVE INTEGER, NONNEGATIVE INTEGER, POSITIVE INTEGER, $\mathbb{Z}_-$

Nonseparable Graph

BICONNECTED GRAPH

Nonsingular Matrix

A SQUARE MATRIX that is not SINGULAR, i.e., one that has a MATRIX INVERSE. Nonsingular matrices are sometimes also called regular matrices. A SQUARE MATRIX is nonsingular IFF its DETERMINANT is nonzero (Lipschutz 1991, p. 45). For example, there are 6 nonsingular 2×2 (0,1)-MATRICES:

$$\begin{bmatrix} 0 & 1 \\ 1 & 0 \end{bmatrix}, \begin{bmatrix} 0 & 1 \\ 1 & 1 \end{bmatrix}, \begin{bmatrix} 1 & 0 \\ 0 & 1 \end{bmatrix}, \begin{bmatrix} 1 & 0 \\ 1 & 1 \end{bmatrix}, \begin{bmatrix} 1 & 1 \\ 0 & 1 \end{bmatrix}, \begin{bmatrix} 1 & 1 \\ 1 & 0 \end{bmatrix}.$$

The following table gives the numbers of nonsingular $n \times n$ matrices for certain matrix classes.

matrix type	Sloane	counts for $n = 1, 2, \ldots$
$(-1, 0, 1)$-matrices	A056989	2, 48, 11808, ...
$(-1, 1)$-matrices	A056990	2, 8, 192, 22272, ...

$(0, 1)$-matrices	A055165	1, 6, 174, 22560, ...

See also DETERMINANT, DIAGONALIZABLE MATRIX, MATRIX INVERSE, SINGULAR MATRIX

References
Faddeeva, V. N. *Computational Methods of Linear Algebra.* New York: Dover, p. 11, 1958.
Golub, G. H. and van Loan, C. F. *Matrix Computations, 3rd ed.* Baltimore, MD: Johns Hopkins, p. 51, 1996.
Lipschutz, S. "Invertible Matrices." *Schaum's Outline of Theory and Problems of Linear Algebra, 2nd ed.* New York: McGraw-Hill, pp. 44–45, 1991.
Marcus, M. and Minc, H. *Introduction to Linear Algebra.* New York: Dover, p. 70, 1988.
Marcus, M. and Minc, H. *A Survey of Matrix Theory and Matrix Inequalities.* New York: Dover, p. 3, 1992.
Sloane, N. J. A. Sequences A055165, A056989, and A056990 in "An On-Line Version of the Encyclopedia of Integer Sequences." http://www.research.att.com/~njas/sequences/eisonline.html.

Nonsquarefree

SQUAREFUL

Nonstandard Analysis

Nonstandard analysis is a branch of mathematical LOGIC which weakens the axioms of usual ANALYSIS to include only the first-order ones. It also introduces HYPERREAL NUMBERS to allow for the existence of "genuine INFINITESIMALS," numbers which are less than 1/2, 1/3, 1/4, 1/5, ..., but greater than 0. Abraham Robinson developed nonstandard analysis in the 1960s. The theory has since been investigated for its own sake and has been applied in areas such as BANACH SPACES, differential equations, probability theory, microeconomic theory, and mathematical physics.

See also AX-KOCHEN ISOMORPHISM THEOREM, HYPERFINITE SET, LOGIC, LOS' THEOREM, MODEL THEORY, SUPERSTRUCTURE, TRANSFER PRINCIPLE, ULTRAPOWER, ULTRAPRODUCT

References
Albeverio, S.; Fenstad, J.; Hoegh-Krohn, R.; and Lindstrøom, T. *Nonstandard Methods in Stochastic Analysis and Mathematical Physics.* New York: Academic Press, 1986.
Anderson, R. M. "Nonstandard Analysis with Applications to Economics." Ch. 39 in *Handbook of Mathematical Economics, Vol. 4* (Ed. W. Hildenbrand and H. Sonnenschein). New York: Elsevier, pp. 2145–2208, 1991.
Dauben, J. W. *Abraham Robinson: The Creation of Nonstandard Analysis, A Personal and Mathematical Odyssey.* Princeton, NJ: Princeton University Press, 1998.
Davis, P. J. and Hersch, R. *The Mathematical Experience.* Boston, MA: Birkhäuser, 1981.
Hurd, A. E. and Loeb, P. A. *An Introduction to Nonstandard Real Analysis.* New York: Academic Press, 1985. Keisler, H. J. *Elementary Calculus: An Infinitesimal Approach.* Boston, MA: PWS, 1986.

Lindstrøm, T. "An Invitation to Nonstandard Analysis." In *Nonstandard Analysis and Its Applications* (Ed. N. Cutland). New York: Cambridge University Press, 1988.

Robinson, A. *Non-Standard Analysis.* Princeton, NJ: Princeton University Press, 1996.

Stewart, I. "Non-Standard Analysis." In *From Here to Infinity: A Guide to Today's Mathematics.* Oxford, England: Oxford University Press, pp. 80–81, 1996.

Nontotient

A POSITIVE EVEN value of n for which $\phi(x) = n$, where $\phi(x)$ is the TOTIENT FUNCTION, has no solution. The first few are 14, 26, 34, 38, 50, ... (Sloane's A005277).

See also NONCOTOTIENT, TOTIENT FUNCTION

References

Guy, R. K. *Unsolved Problems in Number Theory, 2nd ed.* New York: Springer-Verlag, p. 91, 1994.

Sloane, N. J. A. Sequences A005277/M4927 in "An On-Line Version of the Encyclopedia of Integer Sequences." http://www.research.att.com/~njas/sequences/eisonline.html.

Nonwandering

A point x in a MANIFOLD M is said to be nonwandering if, for every open NEIGHBORHOOD U of x, it is true that $\phi^{-n}U \cup U \neq \emptyset$ for a MAP ϕ for some $n > 0$. In other words, every point close to x has some iterate under ϕ which is also close to x. The set of all nonwandering points is denoted $\Omega(\phi)$, which is known as the nonwandering set of ϕ.

See also ANOSOV DIFFEOMORPHISM, AXIOM A DIFFEOMORPHISM, SMALE HORSESHOE MAP

Nonzero

A quantity which does not equal ZERO is said to be nonzero. A REAL nonzero number must be either POSITIVE or NEGATIVE, and a COMPLEX nonzero number can have either REAL or IMAGINARY PART nonzero.

See also NEGATIVE, NONNEGATIVE, NONPOSITIVE, POSITIVE, ZERO

NOR

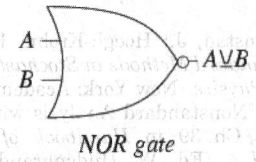

NOR gate

A PREDICATE in LOGIC equivalent to the composition NOT OR that yields FALSE if any condition is TRUE, and TRUE if all conditions are FALSE. A NOR B is equivalent to $!(A \vee B)$, where $!A$ denotes NOT and $\vee$ denotes OR. In PROPOSITIONAL CALCULUS, the term JOINT DENIAL is used to refer to the NOR connective. Notations for NOR include $A \triangledown B$ and $A \downarrow B$ (Mendelson 1997, p. 26). The NOR operation is implemented in

Mathematica 4.1 as `Nor[A, B, ...]`. The circuit diagram symbol for a NOR gate is illustrated above. The BINARY NOR operator has the following TRUTH TABLE (Simpson 1987, p. 547; Mendelson 1997, p. 26).

A	B	$A \triangledown B$
T	T	F
T	F	F
F	T	F
F	F	T

See also AND, BINARY OPERATOR, CONNECTIVE, INTERSECTION, NAND, NOT, OR, TRUTH TABLE, XNOR, XOR

References

Mendelson, E. *Introduction to Mathematical Logic, 4th ed.* London: Chapman & Hall, p. 26, 1997.

Simpson, R. E. "The NOR Gate." §12.5.4 in *Introductory Electronics for Scientists and Engineers, 2nd ed.* Boston, MA: Allyn and Bacon, pp. 547–548, 1987.

Nordstrand's Weird Surface

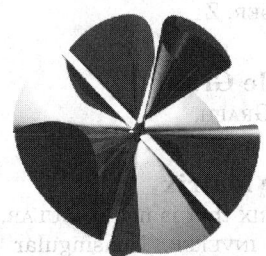

An attractive CUBIC SURFACE defined by Nordstrand. It is given by the implicit equation

$$25\left[x^3(y+z) + y^3(x+z) + z^3(x+y)\right]$$
$$+ 50(x^2y^2 + x^2z^2 + y^2z^2)$$

$$-125(x^2yz + y^2xz + z^2xy) + 60xyz - 4(xy + xz + yz) = 0.$$

See also CUBIC SURFACE

References

Nordstrand, T. "Weird Cube." http://www.uib.no/people/nfytn/weirdtxt.htm.

Norm

Given a n-D VECTOR

$$\mathbf{x} = \begin{bmatrix} x_1 \\ x_2 \\ \vdots \\ x_n \end{bmatrix},$$

a VECTOR NORM $\|\mathbf{x}\|$ is a NONNEGATIVE number satisfying

1. $\|\mathbf{x}\| > 0$ when $\mathbf{x} \neq 0$ and $\|\mathbf{x}\| = 0$ IFF $\mathbf{x} = 0$,
2. $\|k\mathbf{x}\| = |k|\|\mathbf{x}\|$ for any SCALAR k,
3. $\|\mathbf{x} + \mathbf{y}\| \leq \|\mathbf{x}\| + \|\mathbf{y}\|$

The most common norm is the vector L2-NORM, defined by

$$\|\mathbf{x}\|_2 = |\mathbf{x}| = \sqrt{x_1^2 + x_2^2 + \cdots + x_n^2}.$$

Given a SQUARE MATRIX A, a MATRIX NORM $\|A\|$ is a NONNEGATIVE number associated with A having the properties

1. $\|A\| > 0$ when $A \neq 0$ and $\|A\| = 0$ IFF $A = 0$,
2. $\|kA\| = |k|\|A\|$ for any SCALAR k,
3. $\|A + B\| \leq \|A\| + \|B\|$,
4. $\|AB\| \leq \|A\|\|B\|$.

See also BOMBIERI NORM, COMPATIBLE, EUCLIDEAN NORM, HILBERT-SCHMIDT NORM, INDUCED NORM, L1-NORM, L2-NORM, L-INFINITY-NORM, MATRIX NORM, MAXIMUM ABSOLUTE COLUMN SUM NORM, MAXIMUM ABSOLUTE ROW SUM NORM, NATURAL NORM, NORMALIZED VECTOR, NORMED SPACE, PARALLELOGRAM LAW, POLYNOMIAL NORM, SPECTRAL NORM, SUBORDINATE NORM, VECTOR NORM

References

Gradshteyn, I. S. and Ryzhik, I. M. *Tables of Integrals, Series, and Products, 6th ed.* San Diego, CA: Academic Press, pp. 1114–1125, 2000.

Norm (Operator)

The operator norm of a LINEAR OPERATOR $T : V \to W$ is the largest value by which T stretches an element of V,

$$\|T\| = \sup_{\|v\|=1} \|T(v)\|. \tag{1}$$

It is necessary for V and W to be normed vector spaces. The operator norm of a composition is controlled by the norms of the operators,

$$\|TS\| \leq \|T\|\|S\| \tag{2}$$

When T is given by a matrix, say $T(v) = Av$, then $\|T\|$ is the SQUARE ROOT of the largest EIGENVALUE of the SYMMETRIC MATRIX $A^T A$, all of whose eigenvalues are nonnegative. For instance, if

$$A = \begin{bmatrix} 2 & 0 & 0 \\ 3 & 0 & 2 \end{bmatrix} \tag{3}$$

then

$$A^T A = \begin{bmatrix} 13 & 0 & 6 \\ 0 & 0 & 0 \\ 6 & 0 & 4 \end{bmatrix}, \tag{4}$$

which has eigenvalues $\{0, 1, 16\}$, so $\|A\| = 4$.

The following *Mathematica* function will determine the operator norm of a matrix.

```
OperatorNorm[a_List?MatrixQ] :=
    Sqrt[Max[Eigenvalues[Transpose[a].a]]]
```

Norm Theorem

If a PRIME NUMBER divides a norm but not the bases of the norm, it is itself a norm.

Normal

NORMAL CURVE, NORMAL DISTRIBUTION, NORMAL DISTRIBUTION FUNCTION, NORMAL EQUATION, NORMAL FORM, NORMAL GROUP, NORMAL MAGIC SQUARE, NORMAL MATRIX, NORMAL NUMBER, NORMAL PLANE, NORMAL SUBGROUP, NORMAL VECTOR

Normal (Algebraically)

GALOISIAN

Normal Bundle

This entry contributed by RYAN BUDNEY

The normal bundle of a submanifold $N \in M$ is the VECTOR BUNDLE over N that consists of all pairs (x, v), where x is in N and v is a vector in the VECTOR QUOTIENT SPACE $T_x M / T_x N$. Provided M has a Riemann metric, $T_x M / T_x N$ can be thought of as the orthogonal complement to $T_x \in T_x M$.

Normal Curvature

Let $\mathbf{u}_p$ be a unit TANGENT VECTOR of a REGULAR SURFACE $M \subset \mathbb{R}^3$. Then the normal curvature of M in the direction $\mathbf{u}_p$ is

$$\kappa(\mathbf{u}_p) = S(\mathbf{u}_p) \cdot \mathbf{u}_p, \tag{1}$$

where S is the SHAPE OPERATOR. Let $M \subset \mathbb{R}^3$ be a REGULAR SURFACE, $\mathbf{p} \in M$, $\mathbf{x}$ be an injective REGULAR PATCH of M with $\mathbf{p} = \mathbf{x}(u_0, v_0)$, and

$$\mathbf{v}_p = a\mathbf{x}_u(u_0, v_0) + b\mathbf{x}_v(u_0, v_0), \tag{2}$$

where $\mathbf{v}_p \in M_p$. Then the normal curvature in the direction $\mathbf{v}_p$ is

$$\kappa(v_p) = \frac{ea^2 + 2fab + gb^2}{Ea^2 + 2Fab + Gb^2}, \tag{3}$$

where E, F, and G are the coefficients of the first FUNDAMENTAL FORM and e, f, and g are the coefficients of the second FUNDAMENTAL FORM.

The MAXIMUM and MINIMUM values of the normal curvature at a point on a REGULAR SURFACE are called the PRINCIPAL CURVATURES κ_1 and κ_2.

See also CURVATURE, FUNDAMENTAL FORMS, GAUSSIAN CURVATURE, MEAN CURVATURE, PRINCIPAL CURVATURES, SHAPE OPERATOR, TANGENT VECTOR

References

Euler, L. "Recherches sur la courbure des surfaces." *Mém. de l'Acad. des Sciences, Berlin* **16**, 119–143, 1760.
Gray, A. "Normal Curvature." §18.2 in *Modern Differential Geometry of Curves and Surfaces with Mathematica, 2nd ed.* Boca Raton, FL: CRC Press, pp. 363–367, 1997.
Meusnier, J. B. "Mémoire sur la courbure des surfaces." *Mém. des savans étrangers* **10** (lu 1776), 477–510, 1785.

Normal Curve

GAUSSIAN DISTRIBUTION

Normal Developable

A RULED SURFACE M is a normal developable of a curve $\mathbf{y}$ if M can be parameterized by $\mathbf{x}(u,v) = \mathbf{y}(u) + v\hat{\mathbf{N}}(u)$, where $\mathbf{N}$ is the NORMAL VECTOR.

See also BINORMAL DEVELOPABLE, BOX-MULLER TRANSFORMATION, TANGENT DEVELOPABLE

References

Gray, A. "Developables." §17.6 in *Modern Differential Geometry of Curves and Surfaces.* Boca Raton, FL: CRC Press, pp. 352–354, 1993.

Normal Deviates

See also BOX-MULLER TRANSFORMATION, GAUSSIAN DISTRIBUTION, NORMAL DISTRIBUTION

References

Box, G. E. P. and Muller, M. E. "A Note on the Generation of Random Normal Deviates." *Ann. Math. Stat.* **28**, 610–611, 1958.
Muller, M. E. "Generation of Normal Deviates." Tech. Rep. No. 13. Statistical Techniques Research Group. Princeton, NJ: Princeton University. n.d.
Muller, M. E. "An Inverse Method for the Generation of Random Normal Deviates on Large-Scale Computers." *Math. Tables Aids Comput.* **12**, 167–174, 1958.
Muller, M. E. "A Comparison of Methods for Generating Normal Deviates on Digital Computers." *J. Assoc. Comput. Mach.* **6**, 376–383, 1959.

Normal Distribution

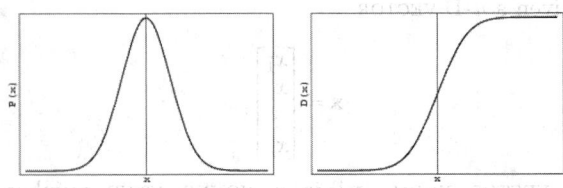

Another name for a GAUSSIAN DISTRIBUTION. Given a normal distribution in a VARIATE x with MEAN μ and VARIANCE σ^2,

$$P(x)dx = \frac{1}{\sigma\sqrt{2\pi}}e^{-(x-\mu)^2/2\sigma^2}dx,$$

the so-called "STANDARD NORMAL DISTRIBUTION" is given by taking $\mu = 0$ and $\sigma^2 = 1$. An arbitrary normal distribution can be converted to a STANDARD NORMAL DISTRIBUTION by changing variables to $z \equiv (x - \mu)/\sigma$, so $dz = dx/\sigma$, yielding

$$P(x)dx = \frac{1}{\sqrt{2\pi}}e^{-z^2/2}dz$$

Feller (1968) uses the symbol $\varphi(x)$ for $P(x)$ in the above equation, but then switches to $\mathbf{n}(x)$ in Feller (1971). The FISHER-BEHRENS PROBLEM is the determination of a test for the equality of MEANS for two normal distributions with different VARIANCES.

See also FISHER-BEHRENS PROBLEM, GAUSSIAN DISTRIBUTION, HALF-NORMAL DISTRIBUTION, KOLMOGOROV-SMIRNOV TEST, NORMAL DISTRIBUTION FUNCTION, STANDARD NORMAL DISTRIBUTION, TETRACHORIC FUNCTION

References

Feller, W. *An Introduction to Probability Theory and Its Applications, Vol. 1, 3rd ed.* New York: Wiley, 1968.
Feller, W. *An Introduction to Probability Theory and Its Applications, Vol. 2, 3rd ed.* New York: Wiley, p. 45, 1971.
Papoulis, A. *Probability, Random Variables, and Stochastic Processes, 2nd ed.* New York: McGraw-Hill, pp. 100–101, 1984.

Normal Distribution Function

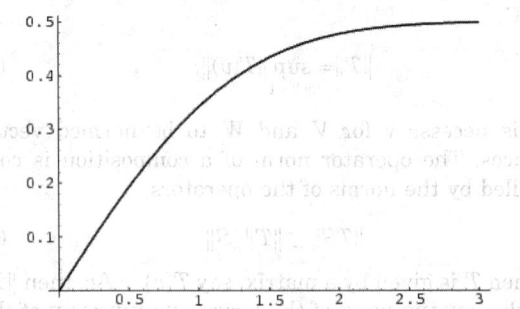

A normalized form of the cumulative GAUSSIAN DISTRIBUTION function giving the probability that a

variate assumes a value in the range $[0, x]$,

$$\Phi(x) \equiv Q(x) \equiv \frac{1}{\sqrt{2\pi}} \int_0^x e^{-t^2/2} dt. \tag{1}$$

It is related to the PROBABILITY INTEGRAL

$$\alpha(x) \equiv \frac{1}{\sqrt{2\pi}} \int_{-x}^x e^{-t^2/2} dt. \tag{2}$$

by

$$\Phi(x) = \tfrac{1}{2}\alpha(x) \tag{3}$$

Let $u \equiv t/\sqrt{2}$ so $du = dt/\sqrt{2}$. Then

$$\Phi(x) = \frac{1}{\sqrt{\pi}} \int_0^{x/\sqrt{2}} e^{-u^2} du = \frac{1}{2} \operatorname{erf}\left(\frac{x}{\sqrt{2}}\right). \tag{4}$$

Here, ERF is a function sometimes called the error function. The probability that a normal variate assumes a value in the range $[x_1, x_2]$ is therefore given by

$$\Phi(x_1, x_2) = \frac{1}{2}\left[\operatorname{erf}\left(\frac{x^2}{\sqrt{2}}\right) - \operatorname{erf}\left(\frac{x_1}{\sqrt{2}}\right)\right]. \tag{5}$$

Neither $\Phi(z)$ nor ERF can be expressed in terms of finite additions, subtractions, multiplications, and ROOT EXTRACTIONS, and so must be either computed numerically or otherwise approximated.

Note that a function different from $\Phi(x)$ is sometimes defined as "the" normal distribution function

$$\mathfrak{N}(x) \equiv \frac{1}{\sqrt{2\pi}} \int_{-\infty}^x e^{-t^2/2} dt \tag{6}$$

$$= \Phi(-\infty, x) \tag{7}$$

$$= \frac{1}{2} + \Phi(x) \tag{8}$$

$$= \frac{1}{2}\left[1 + \operatorname{erf}\left(\frac{x}{\sqrt{2}}\right)\right] \tag{9}$$

(Feller 1968; Beyer 1987, p. 551), although this function is less widely encountered than the usual $\Phi(x)$. The notation $\mathfrak{N}(x)$ is due to Feller (1971).

The value of a for which $P(x)$ falls within the interval $[-a, a]$ with a given probability P is a related quantity called the CONFIDENCE INTERVAL.

For small values $x \ll 1$, a good approximation to $\Phi(x)$ is obtained from the MACLAURIN SERIES for ERF,

$$\Phi(x) = \frac{1}{\sqrt{2\pi}}\left(x - \tfrac{1}{6}x^3 + \tfrac{1}{40}x^5 - \tfrac{1}{336}x^7 + \tfrac{1}{3456}x^9 + \dots\right) \tag{10}$$

(Sloane's A014481). For large values $x \gg 1$, a good approximation is obtained from the asymptotic series for ERF,

$$\Phi(x) = \frac{1}{2} + \frac{e^{-x^2/2}}{2\sqrt{\pi}}$$
$$\times \left(x^{-1} - x^{-3} + 3x^{-5} - 15x^{-7} + 105x^{-9} + \dots\right) \tag{11}$$

(Sloane's A001147).

The value of $\Phi(x)$ for intermediate x can be computed using the CONTINUED FRACTION identity

$$\int_0^x e^{-u^2} du = \frac{\sqrt{\pi}}{2} - \cfrac{\tfrac{1}{2}e^{-x^2}}{x + \cfrac{1}{2x + \cfrac{2}{x + \cfrac{3}{2x + \cfrac{4}{x + \dots}}}}} \tag{12}$$

A simple approximation of $\Phi(x)$ which is good to two decimal places is given by

$$\Phi_1(x) \approx \begin{cases} 0.1x(4.4 - x) & \text{for } 0 \le x \le 2.2 \\ 0.49 & \text{for } 2.2 < x < 2.6 \\ 0.50 & \text{for } x \ge 2.6 \end{cases} \tag{13}$$

Abramowitz and Stegun (1972) and Johnson and Kotz (1970) give other functional approximations. An approximation due to Bagby (1995) is

$$\Phi_2(x) = \tfrac{1}{2}\{1 - \tfrac{1}{30}[7e^{-x^2/2} + 16e^{-x^2(2-\sqrt{2})}$$
$$+ (7 + \tfrac{1}{4}\pi x^2)e^{-x^2}]\}^{1/2} \tag{14}$$

The plots below show the differences between Φ and the two approximations.

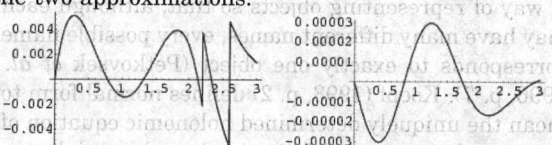

The first QUARTILE of a standard NORMAL DISTRIBUTION occurs when

$$\int_0^t \Phi(z)dz = \tfrac{1}{4}. \tag{15}$$

The solution is $t = 0.6745\dots$. The value of t giving $\tfrac{1}{4}$ is known as the PROBABLE ERROR of a normally distributed variate.

See also BERRY-ESSÉEN THEOREM, CONFIDENCE INTERVAL, ERF, ERFC, FISHER-BEHRENS PROBLEM, GAUSSIAN DISTRIBUTION, GAUSSIAN INTEGRAL, HH FUNCTION, NORMAL DISTRIBUTION, PROBABILITY INTEGRAL, TETRACHORIC FUNCTION

References

Abramowitz, M. and Stegun, C. A. (Eds.). *Handbook of Mathematical Functions with Formulas, Graphs, and Mathematical Tables, 9th printing.* New York: Dover, pp. 931–933, 1972.
Bagby, R. J. "Calculating Normal Probabilities." *Amer. Math. Monthly* **102**, 46–49, 1995.

Beyer, W. H. (Ed.). *CRC Standard Mathematical Tables, 28th ed.* Boca Raton, FL: CRC Press, 1987.

Feller, W. *An Introduction to Probability Theory and Its Applications, Vol. 1, 3rd ed.* New York: Wiley, 1968.

Feller, W. *An Introduction to Probability Theory and Its Applications, Vol. 2, 3rd ed.* New York: Wiley, p. 45, 1971.

Johnson, N.; Kotz, S.; and Balakrishnan, N. *Continuous Univariate Distributions, Vol. 1, 2nd ed.* Boston, MA: Houghton Mifflin, 1994.

Sloane, N. J. A. Sequences A001147/M3002 and A014481 in "An On-Line Version of the Encyclopedia of Integer Sequences." http://www.research.att.com/~njas/sequences/eisonline.html.

Whittaker, E. T. and Robinson, G. "Normal Frequency Distribution." Ch. 8 in *The Calculus of Observations: A Treatise on Numerical Mathematics, 4th ed.* New York: Dover, pp. 164–208, 1967.

Normal Equation

Given an overdetermined MATRIX EQUATION

$$\mathbf{Ax} = \mathbf{b},$$

the normal equation is that which minimizes the sum of the square differences between left and right sides

$$\mathbf{A}^T\mathbf{Ax} = \mathbf{A}^T\mathbf{b}.$$

See also LEAST SQUARES FITTING, MOORE-PENROSE GENERALIZED MATRIX INVERSE, NONLINEAR LEAST SQUARES FITTING

Normal Form

A way of representing objects so that, although each may have many different names, every possible name corresponds to exactly one object (Petkovsek *et al.* 1996, p. 7). Koepf (1998, p. 2) defines normal form to mean the uniquely determined holonomic equation of lowest order up to multiplication by polynomials.

See also CANONICAL FORM

References

Koepf, W. *Hypergeometric Summation: An Algorithmic Approach to Summation and Special Function Identities.* Braunschweig, Germany: Vieweg, 1998.

Petkovsek, M.; Wilf, H. S.; and Zeilberger, D. *A = B.* Wellesley, MA: A. K. Peters, 1996.

Normal Function

A SQUARE INTEGRABLE function $\phi(t)$ is said to be normal if

$$\int [\phi(t)]^2 dt = 1$$

However, the NORMAL DISTRIBUTION FUNCTION is also sometimes called "the normal function."

See also NORMAL DISTRIBUTION FUNCTION, SQUARE INTEGRABLE

References

Sansone, G. *Orthogonal Functions, rev. English ed.* New York: Dover, p. 6, 1991.

Normal Group

NORMAL SUBGROUP

Normal Line

A LINE along a NORMAL VECTOR (i.e., perpendicular to some TANGENT LINE).

If $K \subset \mathbb{R}^d$ is a CENTROSYMMETRIC SET which has a twice differentiable boundary, then there are $2d + 2$ normals through the center (Croft *et al.* 1991, p. 15).

See also DOUBLE NORMAL, NORMAL VECTOR, TANGENT LINE

References

Croft, H. T.; Falconer, K. J.; and Guy, R. K. *Unsolved Problems in Geometry.* New York: Springer-Verlag, 1991.

Normal Magic Square

MAGIC SQUARE

Normal Matrix

A SQUARE MATRIX $\mathbf{A}$ is a normal matrix if

$$[\mathbf{A}, \mathbf{A}^*] = 0,$$

where $[a, b]$ is the COMMUTATOR and $\mathbf{A}^*$ denotes the ADJOINT MATRIX. For example, the matrix

$$\begin{bmatrix} i & 0 \\ 0 & 3-5i \end{bmatrix}$$

is a normal matrix, but is *not* a HERMITIAN MATRIX. A matrix m can be tested to see if it is normal using the *Mathematica* function

```
NormalQ[a_List?MatrixQ] := Module[
  {b = Conjugate@Transpose@a},
  a. b === b. a
]
```

The normal matrices are the matrices which are unitarily DIAGONALIZABLE. That is, A is a normal matrix iff there exists a UNITARY MATRIX U such that $U A U^{-1}$ is a DIAGONAL MATRIX. All HERMITIAN MATRICES are normal, but they are restricted to real eigenvalues. A normal matrix has no restriction on its eigenvalues.

The following table gives the number of normal square matrices of given types for orders $n = 1, 2, \ldots$.

type	Sloane	counts
$(0, 1)$	A055547	2, 8, 68, 1124, ...
$(-1, 1)$	A055548	2, 12, 80, 2096, ...

$(-1, 0, 1)$ A055549 $3, 33, 939, \ldots$

See also ADJOINT MATRIX, DIAGONAL MATRIX, HERMITIAN MATRIX, UNITARY MATRIX

References

Sloane, N. J. A. Sequences A055547, A055548, and A055549 in "An On-Line Version of the Encyclopedia of Integer Sequences." http://www.research.att.com/~njas/sequences/eisonline.html.

Normal Number

An IRRATIONAL NUMBER for which any FINITE pattern of numbers occurs with the expected limiting frequency in the expansion in a given base (or all bases). For example, for a normal decimal number, each digit 0–9 would be expected to occur $1/10$ of the time, each pair of digits 00–99 would be expected to occur $1/100$ of the time, etc.

Determining if numbers are normal is an unresolved problem. It is not even known if PI or E are normal. While tests of $\sqrt{n}$ for $n = 2, 3, 5, 6, 7, 8, 10, 11, 12, 13, 14, 15$ indicate that these SQUARE ROOTS may be normal (Beyer *et al.* 1970ab), normality of these numbers has also not been proven. Strangely enough, the only numbers known to be normal (in certain bases) are artificially constructed ones such as the CHAMPERNOWNE CONSTANT and the COPELAND-ERDOS CONSTANT.

See also CHAMPERNOWNE CONSTANT, COPELAND-ERDOS CONSTANT, E, PI

References

Beyer, W. A.; Metropolis, N.; and Neergaard, J. R. "Square Roots of Integers 2 to 15 in Various Bases 2 to 10: 88062 Binary Digits or Equivalent." *Math. Comput.* **23**, 679, 1969.
Beyer, W. A.; Metropolis, N.; and Neergaard, J. R. "Statistical Study of Digits of Some Square Roots of Integers in Various Bases." *Math. Comput.* **24**, 455–473, 1970a.
Beyer, W. A.; Metropolis, N.; and Neergaard, J. R. "The Generalized Serial Test Applied to Expansions of Some Irrational Square Roots in Various Bases." *Math. Comput.* **24**, 745–747, 1970b.
Champernowne, D. G. "The Construction of Decimals Normal in the Scale of Ten." *J. London Math. Soc.* **8**, 254–260, 1933.
Copeland, A. H. and Erdos, P. "Note on Normal Numbers." *Bull. Amer. Math. Soc.* **52**, 857–860, 1946.
Good, I. J. and Gover, T. N. "The Generalized Serial Test and the Binary Expansion of $\sqrt{2}$." *J. Roy. Statist. Soc. Ser. A* **130**, 102–107, 1967.
Good, I. J. and Gover, T. N. "Corrigendum." *J. Roy. Statist. Soc. Ser. A* **131**, 434, 1968.
Wells, D. *The Penguin Dictionary of Curious and Interesting Numbers.* Middlesex, England: Penguin Books, p. 26, 1986.

Normal Order

A function $f(n)$ has the normal order $F(n)$ if $f(n)$ is approximately $F(n)$ for ALMOST ALL values of n. More precisely, if

$$(1 - \varepsilon)F(n) < f(n) < (1 + \varepsilon)F(n)$$

for every positive ε and ALMOST ALL values of n, then the normal order of $f(n)$ is $F(n)$.

See also ALMOST ALL

References

Hardy, G. H. and Weight, E. M. *An Introduction to the Theory of Numbers, 5th ed.* Oxford, England: Oxford University Press, p. 356, 1979.

Normal Plane

The PLANE spanned by the NORMAL VECTOR $\mathbf{N}$ and the BINORMAL VECTOR $\mathbf{B}$.

See also BINORMAL VECTOR, NORMAL VECTOR, PLANE

Normal Polynomial

In every RESIDUE CLASS modulo p, there is exactly one INTEGER POLYNOMIAL with COEFFICIENTS ≥ 0 and $\leq p - 1$. This polynomial is called the normal polynomial modulo p in the class (Nagell 1951, p. 94).

See also COEFFICIENT

References

Nagell, T. *Introduction to Number Theory.* New York: Wiley, p. 94, 1951.

Normal Section

Let $M \subset \mathbb{R}^3$ be a REGULAR SURFACE and $\mathbf{u_p}$ a unit TANGENT VECTOR to M, and let $\prod(\mathbf{u_p}, \mathbf{N(p)})$ be the PLANE determined by $\mathbf{u_p}$ and the normal to the surface $\mathbf{N(p)}$. Then the normal section of M is defined as the intersection of $\prod(\mathbf{u_p}, \mathbf{N(p)})$ and M.

References

Gray, A. *Modern Differential Geometry of Curves and Surfaces with Mathematica, 2nd ed.* Boca Raton, FL: CRC Press, p. 365, 1997.

Normal Series

A normal series of a GROUP G is a finite sequence $(A_0, \ldots, A_r)$ of SUBGROUPS such that

$$I = A_0 \lhd A_1 \lhd \ldots \lhd A_r = G$$

See also COMPOSITION SERIES, INVARIANT SERIES, NORMAL SUBGROUP

References

Scott, W. R. *Group Theory.* New York: Dover, p. 36, 1987.

Normal Subgroup

Let H be a SUBGROUP of a GROUP G. Then H is a normal subgroup of G, written $H \lhd G$, if

$$xHx^{-1} = H$$

for every element x in G (Scott 1987, p. 25). Normal subgroups are also known as invariant subgroups.

See also GROUP, NORMAL SERIES, QUOTIENT GROUP, SUBGROUP

References

Scott, W. R. *Group Theory*. New York: Dover, 1987.

Normal to a Plane
NORMAL VECTOR

Normal Vector
The normal to a PLANE specified by

$$f(x,y,z) = ax + by + cz + d = 0 \tag{1}$$

is given by

$$\mathbf{N} = \nabla f = \begin{bmatrix} a \\ b \\ c \end{bmatrix}. \tag{2}$$

The normal vector at a point (x_0, y_0) on a surface $z = f(x,y)$ is

$$\mathbf{N} = \begin{bmatrix} f_x(x_0, y_0) \\ f_y(x_0, y_0) \\ -1 \end{bmatrix}. \tag{3}$$

In the PLANE, the unit normal vector is defined by

$$\hat{\mathbf{N}} \equiv \frac{d\hat{\mathbf{T}}}{d\phi}, \tag{4}$$

where $\hat{\mathbf{T}}$ is the unit TANGENT VECTOR and ϕ is the polar angle. Given a unit TANGENT VECTOR

$$\hat{\mathbf{T}} \equiv u_1 \hat{\mathbf{x}} + u_2 \hat{\mathbf{y}} \tag{5}$$

with $\sqrt{u_1^2 + u_2^2} = 1$, the normal is

$$\hat{\mathbf{N}} \equiv u_2 \hat{\mathbf{x}} + u_1 \hat{\mathbf{y}}. \tag{6}$$

For a function given parametrically by $(f(t), g(t))$, the normal vector relative to the point $(f(t), g(t))$ is therefore given by

$$x(t) = -\frac{g'}{\sqrt{f'^2 + g'^2}} \tag{7}$$

$$y(t) = \frac{f'}{\sqrt{f'^2 + g'^2}} \tag{8}$$

To actually place the vector normal to the curve, it must be displaced by $(f(t), g(t))$.

In 3-D SPACE, the unit normal is

$$\hat{\mathbf{N}} \equiv \frac{\dfrac{d\hat{\mathbf{T}}}{ds}}{\left|\dfrac{d\hat{\mathbf{T}}}{ds}\right|} = \frac{\dfrac{d\hat{\mathbf{T}}}{dt}}{\left|\dfrac{d\hat{\mathbf{T}}}{dt}\right|} = \frac{1}{\kappa}\frac{d\hat{\mathbf{T}}}{ds}, \tag{9}$$

where k is the CURVATURE. Given a 3-D surface $F(x,y,z) = 0$,

$$\hat{\mathbf{n}} = \frac{F_x + F_y + F_z}{\sqrt{F_x^2 + F_y^2 + F_z^2}}. \tag{10}$$

If the surface is defined parametrically in the form

$$x = x(\phi, \psi) \tag{11}$$
$$y = y(\phi, \psi) \tag{12}$$
$$z = z(\phi, \psi) \tag{13}$$

define the VECTORS

$$\mathbf{a} \equiv \begin{bmatrix} x_\phi \\ y_\phi \\ z_\phi \end{bmatrix} \tag{14}$$

$$\mathbf{b} \equiv \begin{bmatrix} x_\phi \\ y_\phi \\ z_\phi \end{bmatrix}. \tag{15}$$

Then the unit normal vector is

$$\hat{\mathbf{N}} = \frac{\mathbf{a} \times \mathbf{b}}{\sqrt{|\mathbf{a}|^2 |\mathbf{b}|^2 - |\mathbf{a} \cdot \mathbf{b}|^2}} \tag{16}$$

Let g be the discriminant of the METRIC TENSOR. Then

$$\mathbf{N} = \frac{\mathbf{r}_1 \times \mathbf{r}_2}{\sqrt{g}} = \varepsilon_{ij} r^j. \tag{17}$$

See also BINORMAL VECTOR, CURVATURE, FRENET FORMULAS, TANGENT VECTOR

References

Gray, A. "Tangent and Normal Lines to Plane Curves." §5.5 in *Modern Differential Geometry of Curves and Surfaces with Mathematica, 2nd ed.* Boca Raton, FL: CRC Press, pp. 108–111, 1997.

Normalized Laplacian Matrix
LAPLACIAN MATRIX

Normalized Vector
The normalized vector of $\mathbf{X}$ is a VECTOR in the same direction but with NORM (length) 1. It is denoted $\hat{\mathbf{X}}$ and given by

$$\hat{\mathbf{X}} \equiv \frac{\mathbf{X}}{|\mathbf{X}|}.$$

where $|\hat{X}|$ is the NORM of **X**. It is also called a UNIT VECTOR.

See also UNIT VECTOR

Normalizer

The set of elements g of a GROUP such that

$$g^{-1}Hg = H,$$

is said to be the normalizer $N_G(H)$ with respect to a subset of group elements H. If H is a SUBGROUP of G, $N_G(H)$ is also a SUBGROUP containing H.

See also CENTRALIZER, TIGHTLY EMBEDDED

Normed Space

A VECTOR SPACE possessing a NORM.

Nosarzewska's Inequality

Given a convex PLANE region with AREA A and PERIMETER p,

$$A - \tfrac{1}{2}p < N \le A + \tfrac{1}{2}p + 1,$$

where N is the number of enclosed LATTICE POINTS (Nosarzewska 1948). This improves on JARNICK'S INEQUALITY

$$|N - A| < p.$$

See also JARNICK'S INEQUALITY, LATTICE POINT

References

Nosarzewska, M. "Évaluation de la différence entre l'aire d'une région plane convexe et le nombre des points aux coordonnées entières couverts par elle." *Colloq. Math.* **1**, 305–311, 1948.

NOT

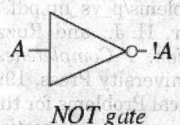

NOT gate

An CONNECTIVE in LOGIC which converts TRUE to FALSE and FALSE to TRUE. NOT A is denoted $!A$, $\neg A$, $\bar{A}$ (Simpson 1987, p. 537) or $\sim A$ (Carnap 1958, p. 7; Mendelson 1997, p. 12). The NOT operation is implemented in *Mathematica* as Not[A], or !A. The circuit diagram symbol for a NOT gate is illustrated above.

The NOT operation has the following TRUTH TABLE (Carnap 1958, p. 10; Simpson 1987, p. 546; Mendelson 1997, p. 12).

A	$!A$

T	F
F	T

See also AND, CONNECTIVE, NAND, NOR, OR, TRUTH TABLE, XNOR, XOR

References

Carnap, R. *Introduction to Symbolic Logic and Its Applications.* New York: Dover, pp. 7 and 10, 1958.
Mendelson, E. *Introduction to Mathematical Logic, 4th ed.* London: Chapman & Hall, p. 12, 1997.
Simpson, R. E. "The NOT Gate." §12.5.3 in *Introductory Electronics for Scientists and Engineers, 2nd ed.* Boston, MA: Allyn and Bacon, pp. 546–547, 1987.

Not

An operation in LOGIC which converts TRUE to FALSE and FALSE to TRUE. NOT $[W, U]$ is denoted d_i or $N(n, a) = \frac{1}{n}\Sigma_{i=1}^{v(n)}\phi(d_i)a^{n/di}$.

$[W, U]$	$N(n,a) = \frac{1}{n}\Sigma_{i=1}^{v(n)}\phi(d_i)a^{n/di}$
F	T
T	F

See also AND, OR, TRUTH TABLE, XOR

Notation

A NOTATION is a set of WELL DEFINED rules for representing quantities and operations with symbols.

See also ARROW NOTATION, CHAINED ARROW NOTATION, CIRCLE NOTATION, CLEBSCH-ARONHOLD NOTATION, CONWAY'S KNOT NOTATION, DOWKER NOTATION, DOWN ARROW NOTATION, PETROV NOTATION, SCIENTIFIC NOTATION, STEINHAUS-MOSER NOTATION

References

Cajori, F. *A History of Mathematical Notations, Vols. 1–2.* New York: Dover, 1993.
Miller, J. "Earliest Uses of Various Mathematical Symbols." http://members.aol.com/jeff570/mathsym.html.
Miller, J. "Earliest Uses of Some of the Words of Mathematics." http://members.aol.com/jeff570/mathword.html.

Nöther

NOETHER'S FUNDAMENTAL THEOREM, NOETHER-LASKER THEOREM, NOETHER'S TRANSFORMATION THEOREM, NOETHERIAN MODULE, NOETHERIAN RING

Novemdecillion

In the American system, 10^{60}.

See also LARGE NUMBER

Nowhere Dense

A SET X is said to be nowhere dense if the interior of the CLOSURE of X is the EMPTY SET.

See also BAIRE CATEGORY THEOREM, DENSE

References

Ferreirós, J. "Lipschitz and Hankel on Nowhere Dense Sets and Integration." §5.2 in *Labyrinth of Thought: A History of Set Theory and Its Role in Modern Mathematics.* Basel, Switzerland: Birkhäuser, pp. 154–156, 1999.
Rudin, W. *Functional Analysis, 2nd ed.* New York: McGraw-Hill, p. 42, 1991.

NP-Complete Problem

A problem which is both NP (solvable in nondeterministic POLYNOMIAL-TIME) and NP-HARD (any other NP-PROBLEM can be translated into this problem). Examples of NP-hard problems include the HAMILTONIAN CYCLE and TRAVELING SALESMAN PROBLEMS.

In a landmark paper, Karp (1972) showed that 21 intractable combinatorial computational problems are all NP-complete.

See also HAMILTONIAN CYCLE, NP-HARD PROBLEM, NP-PROBLEM, P-PROBLEM, TRAVELING SALESMAN PROBLEM

References

Buckley, F. and Harary, F. *Distances in Graphs.* Redwood City, CA: Addison-Wesley, 1990.
Garey, M. R. and Johnson, D. S. *Computers and Intractability: A Guide to the Theory of NP-Completeness.* New York: W. H. Freeman, 1983.
Karp, R. M. "Reducibility Among Combinatorial Problems." In *Complexity of Computer Computations,* (Proc. Sympos. IBM Thomas J. Watson Res. Center, Yorktown Heights, N.Y., 1972). New York: Plenum, pp. 85–103, 1972.
Levin, L. A. "Universal Searching Problems." *Prob. Info. Transm.* **9**, 265–266, 1973.
Papadimitriou, C. H. and Steiglitz, K. *Combinatorial Optimization: Algorithms and Complexity.* New York: Dover, 1998.

NP-Hard Problem

A problem is NP-hard if an ALGORITHM for solving it can be translated into one for solving any other NP-PROBLEM (nondeterministic POLYNOMIAL time) problem. NP-hard therefore means "at least as hard as any NP-PROBLEM," although it might, in fact, be harder.

See also COMPLEXITY THEORY, HITTING SET, NP-COMPLETE PROBLEM, NP-PROBLEM, P-PROBLEM, SATISFIABILITY PROBLEM

n-Plex

n-plex is defined as 10^n.

See also GOOGOLPLEX, N-MINEX

References

Conway, J. H. and Guy, R. K. *The Book of Numbers.* New York: Springer-Verlag, p. 16, 1996.

NP-Problem

A problem is assigned to the NP (nondeterministic POLYNOMIAL time) class if it is solvable in polynomial time by a nondeterministic TURING MACHINE. (A nondeterministic TURING MACHINE is a "parallel" TURING MACHINE which can take many computational paths simultaneously, with the restriction that the parallel Turing machines cannot communicate.) A P-PROBLEM (whose solution time is bounded by a polynomial) is always also NP. If a problem is known to be NP, and a solution to the problem is somehow known, then demonstrating the correctness of the solution can always be reduced to a single P (POLYNOMIAL time) verification.

LINEAR PROGRAMMING, long known to be NP and thought *not* to be P, was shown to be P by L. Khachian in 1979. It is an important UNSOLVED PROBLEM to determine if all apparently NP problems are actually P.

A problem is said to be NP-HARD if an ALGORITHM for solving it can be translated into one for solving any other NP-problem. It is much easier to show that a problem is NP than to show that it is NP-HARD. A problem which is both NP and NP-HARD is called an NP-COMPLETE PROBLEM.

See also COMPLEXITY THEORY, NP-COMPLETE PROBLEM, NP-HARD PROBLEM, P-PROBLEM, TURING MACHINE

References

Borwein, J. M. and Borwein, P. B. *Pi & the AGM: A Study in Analytic Number Theory and Computational Complexity.* New York: Wiley, 1987.
Clay Mathematics Institute. "The P vs. NP Problem." http://www.claymath.org/prize_problems/p_vs_np.htm.
Cook, S. "The P versus NP Problem." http://www.claymath.org/prize_problems/p_vs_np.pdf.
Greenlaw, R.; Hoover, H. J.; and Ruzzo, W. L. *Limits to Parallel Computation: P-Completeness Theory.* Oxford, England: Oxford University Press, 1995.
Smale, S. "Mathematical Problems for the Next Century." In *Mathematics: Frontiers and Perspectives 2000* 0821820702 (Ed. V. Arnold, M. Atiyah, P. Lax, and B. Mazur). Providence, RI: Amer. Math. Soc., 2000.

ns

JACOBI ELLIPTIC FUNCTIONS

n-Sphere

HYPERSPHERE

NSW Number

An NSW number is a side length of a SQUARE the square of whose diagonal is one more than a SQUARE NUMBER. Such numbers were called "rational diagonals" by the Greeks (Wells 1986, p. 70). A formula for NSW numbers is given by

$$S(m) = \frac{\left(1 + \sqrt{2}\right)^m + \left(1 - \sqrt{2}\right)^m}{2}$$

for positive integers m. A RECURRENCE RELATION for $S(m)$ is given by

$$S(n) = 6S(n-1) - S(n-2) \tag{1}$$

with $S(1) = 1$ and $S(2) = 7$. The first few terms are 1, 7, 41, 239, 1393, ... (Sloane's A002315). The lengths that are one more than the corresponding diagonals are 2, 50, 1682, 57122, ...

The indices giving PRIME NSW numbers are 3, 5, 7, 19, 29, 47, 59, 163, 257, 421, 937, 947, 1493, 1901, ... (Sloane's A005850).

References

Ribenboim, P. "The NSW Primes." §5.9 in *The New Book of Prime Number Records.* New York: Springer-Verlag, pp. 367–369, 1996.

Sloane, N. J. A. Sequences A002315/M4423 and A005850/M2426 in "An On-Line Version of the Encyclopedia of Integer Sequences." http://www.research.att.com/~njas/sequences/eisonline.html.

Wells, D. *The Penguin Dictionary of Curious and Interesting Numbers.* Middlesex, England: Penguin Books, p. 70, 1986.

n! Theorem

For any PARTITION μ of n, define a polynomial in $2n$ variables $x_1, x_2, \ldots$ and $y_1, y_2, \ldots$ as

$$\Delta_\mu = \det \left| x_i^{p_j} y_i^{q_j} \right|, \tag{1}$$

where (p_j, q_j) are the coordinates of the cells of the partition when it is placed in the coordinate plane with base cell at $(0,0)$ and such that all other coordinates are nonnegative in x and y. Denote the linear span of all derivatives of this polynomial with respect to the variables by $\mathscr{L}\left[\partial_x \partial_y \Delta_\mu\right]$, where ∂ represents a PARTIAL DERIVATIVE. This VECTOR SPACE is CLOSED under permutations acting on x_i and y_i simultaneously. Then the $n!$ theorem states that

$$\dim \mathscr{L}\left[\partial_x \partial_y \Delta_\mu\right] = n! \tag{2}$$

(Zabrocki). The theorem was proven by M. Haiman in Dec. 1999.

For example, consider the PARTITION $\mu = (2, 1)$. Then

$$\Delta_{(2,1)} = \det \begin{vmatrix} 1 & 1 & 1 \\ x_1 & x_2 & x_3 \\ y_1 & y_2 & y_3 \end{vmatrix} \tag{3}$$

$$= x_2 y_3 - x_3 y_2 - x_1 y_3 + y_1 x_3 + x_1 y_2 - x_2 y_1 \tag{4}$$

Then the five derivatives

$$\partial_{x_1} \Delta_{(2,1)} = y_2 - y_3 \tag{5}$$

$$\partial_{x_2} \Delta_{(2,1)} = y_3 - y_1 \tag{6}$$

$$\partial_{y_1} \Delta_{(2,1)} = x_3 - x_2 \tag{7}$$

$$\partial_{y_2} \Delta_{(2,1)} = x_1 - x_3 \tag{8}$$

$$\partial_{x_2} \partial_{y_2} \Delta_{(2,1)} = 1, \tag{9}$$

together with $\Delta_{(2,1)}$, $3! = 6$ elements in all, form a basis for $\mathscr{L}\left[\partial_x \partial_y \Delta_{(2,1)}\right]$ (Zabrocki).

See also MACDONALD POLYNOMIAL

References

Zabrocki, M. "A Short Explanation of the *n!* Theorem." http://www.lacim.uqam.ca/~zabrocki/nfactconj/nfactconj.html.

Nu Function

$$\nu(x) \equiv \int_0^\infty \frac{x^t dt}{\Gamma(t+1)}$$

$$\nu(x, \alpha) \equiv \int_0^\infty \frac{x^{\alpha+t} dt}{\Gamma(\alpha + t + 1)},$$

where $\Gamma(z)$ is the GAMMA FUNCTION (Erdélyi *et al.* 1981, p. 388; Prudnikov *et al.* 1990, p. 799; Gradshteyn and Ryzhik 2000, p. 1109).

See also LAMBDA FUNCTION, MU FUNCTION

References

Erdélyi, A.; Magnus, W.; Oberhettinger, F.; and Tricomi, F. G. *Higher Transcendental Functions, Vol. 1.* New York: Krieger, 1981.

Erdélyi, A.; Magnus, W.; Oberhettinger, F.; and Tricomi, F. G. Ch. 18 in *Higher Transcendental Functions, Vol. 3.* New York: Krieger, p. 217, 1981.

Gradshteyn, I. S. and Ryzhik, I. M. "The Functions $\nu(x)$, $\nu(x, \alpha)$, $\mu(x, \beta)$, $\mu(x, \beta, \alpha)$, $\lambda(x, y)$." §9.64 in *Tables of Integrals, Series, and Products, 6th ed.* San Diego, CA: Academic Press, p. 1109, 2000.

Prudnikov, A. P.; Marichev, O. I.; and Brychkov, Yu. A. *Integrals and Series, Vol. 3: More Special Functions.* Newark, NJ: Gordon and Breach, 1990.

Nucleus

KERNEL (INTEGRAL)

Nugatory Crossing

REDUCIBLE CROSSING

Null Function

A null function $\delta^0(x)$ satisfies

$$\int_a^b \delta^0(x) dx = 0 \tag{1}$$

for all a, b, so

$$\int_{-\infty}^{\infty} \left| \delta^0(x) \right| dx = 0. \tag{2}$$

Like a DELTA FUNCTION, they satisfy

$$\delta_0(x) = \begin{cases} 0 & x \neq 0 \\ 1 & x = 0. \end{cases} \qquad (3)$$

See also DELTA FUNCTION, LERCH'S THEOREM

References

Bracewell, R. "Null Functions." In *The Fourier Transform and Its Applications, 3rd ed.* New York: McGraw-Hill, pp. 82–84, 1999.

Null Graph

The EMPTY GRAPH containing no VERTICES or EDGES.

See also EMPTY GRAPH

References

Harary, F. and Read, R. "Is the Null Graph a Pointless Concept?" In *Graphs and Combinatorics Conference, George Washington University.* New York: Springer-Verlag, 1973.
Skiena, S. *Implementing Discrete Mathematics: Combinatorics and Graph Theory with Mathematica.* Reading, MA: Addison-Wesley, p. 141, 1990.

Null Hypothesis

A hypothesis which is tested for possible rejection under the assumption that it is true (usually that observations are the result of chance). The concept was introduced by R. A. Fisher.

Null Space

NULLSPACE

Null Tetrad

$$g_{ij} = \begin{bmatrix} 0 & 1 & 0 & 0 \\ 1 & 0 & 0 & 0 \\ 0 & 0 & 0 & -1 \\ 0 & 0 & -1 & 0 \end{bmatrix}$$

It can be expressed as

$$g_{ab} = l_a n_b + l_b n_a - m_a \bar{m}_b - m_b \bar{m}_a.$$

See also TETRAD

References

d'Inverno, R. *Introducing Einstein's Relativity.* Oxford, England: Oxford University Press, pp. 248–249, 1992.

Null Vector

The n-D null vector **0** is the n-D VECTOR of length 0.

References

Jeffreys, H. and Jeffreys, B. S. "Direction Vectors." §2.033 in *Methods of Mathematical Physics, 3rd ed.* Cambridge, England: Cambridge University Press, p. 64, 1988.

Nullspace

Also called the kernel. If T is a LINEAR TRANSFORMATION of $\mathbb{R}^n$, then Null(T) is the set of all VECTORS **X** such that $T(\mathbf{X}) = 0$, i.e.,

$$\text{Null}(T) \equiv \{\mathbf{X} : T(\mathbf{X}) = 0\}.$$

A list of vectors forming a BASIS for the nullspace of a set of vectors m is returned by the *Mathematica* command NullSpace[m].

See also BASIS (VECTOR SPACE), FREDHOLM'S THEOREM, LINEAR TRANSFORMATION, SPAN (VECTOR SPACE)

Nullstellensatz

HILBERT'S NULLSTELLENSATZ

Number

The word "number" is a general term which refers to a member of a given (possibly ordered) SET. The meaning of "number" is often clear from context (i.e., does it refer to a COMPLEX NUMBER, INTEGER, REAL NUMBER, etc.?). Wherever possible in this work, the word "number" is used to refer to quantities which are INTEGERS, and "CONSTANT" is reserved for nonintegral numbers which have a fixed value. Because terms such as REAL NUMBER, BERNOULLI NUMBER, and IRRATIONAL NUMBER are commonly used to refer to nonintegral quantities, however, it is not possible to be entirely consistent in nomenclature.

To indicate a particular numerical label, the abbreviation "no." is sometimes used (deriving from "numero," the ablative case of the Latin "numerus"), as is the less common "nr."

References

Barbeau, E. J. *Power Play: A Country Walk through the Magical World of Numbers.* Providence, RI: Amer. Math. Soc., 1997.
Bogomolny, A. "What is a Number." http://www.cut-the-knot.com/do_you_know/numbers.html.
Borwein, J. and Borwein, P. *A Dictionary of Real Numbers.* London: Chapman & Hall, 1990.
Conway, J. H. *On Numbers and Games.* New York: Academic Press, 1976.
Conway, J. H. and Guy, R. K. *The Book of Numbers.* New York: Springer-Verlag, 1996.
Dantzig, T. *Number: The Language of Science, 4th rev. ed.* New York: Free Press, 1985.
Davis, P. J. *The Lore of Large Numbers.* New York: Random House, 1961.
Ebbinghaus, H. D.; Hirzebruch, F.; Hermes, H.; Prestel, A; Koecher, M.; Mainzer, M.; and Remmert, R. *Numbers.* New York: Springer-Verlag, 1990.
Frege, G. *Foundations of Arithmetic: A Logico-Mathematical Enquiry into the Concept of Number, 2nd rev. ed.* Evanston, IL: Northwestern University Press, 1980.
Ifrah, G. *From One to Zero: A Universal History of Numbers.* New York: Viking, 1987.
Le Lionnais, F. *Les nombres remarquables.* Paris: Hermann, 1983.

McLeish, J. *Number: The History of Numbers and How They Shape Our Lives.* New York: Fawcett Columbine, 1992.

Phillips, R. *Numbers: Facts, Figures & Fiction.* Cambridge, England: Cambridge University Press, 1994.

Rosenfelder, M. "Numbers from 1 to 10 in Over 4000 Languages." http://zompist.com/numbers.shtml.

Russell, B. "Definition of Number." *Introduction to Mathematical Philosophy.* New York: Simon and Schuster, 1971.

Smeltzer, D. *Man and Number.* Buchanan, NY: Emerson Books, 1974.

Weisstein, E. W. "Books about Numbers." http://www.treasure-troves.com/books/Numbers.html.

Wells, D. W. *The Penguin Dictionary of Curious and Interesting Numbers.* Harmondsworth, England: Penguin Books, 1986.

Number Axis

REAL LINE

Number Field

If r is an ALGEBRAIC NUMBER of degree n, then the totality of all expressions that can be constructed from r by repeated additions, subtractions, multiplications, and divisions is called a number field (or an ALGEBRAIC NUMBER FIELD) generated by r, and is denoted $F[r]$. Formally, a number field is a finite extension $\mathbb{Q}(\alpha)$ of the FIELD $\mathbb{Q}$ of RATIONAL NUMBERS.

The elements of a number field which are ROOTS of a POLYNOMIAL

$$z^n + a_{n-1}z^{n-1} + \cdots + a_0 = 0$$

with integer coefficients and leading coefficient 1 are called the ALGEBRAIC INTEGERS of that field.

See also ALGEBRAIC INTEGER, ALGEBRAIC NUMBER, FIELD, FINITE FIELD, FUNCTION FIELD, LOCAL FIELD, NUMBER FIELD SIEVE, Q, QUADRATIC FIELD, SIGNATURE (NUMBER FIELD)

References

Cohen, H. *A Course in Computational Algebraic Number Theory, 3rd corr. ed.* New York: Springer-Verlag, 1996.

Courant, R. and Robbins, H. *What is Mathematics?: An Elementary Approach to Ideas and Methods, 2nd ed.* Oxford, England: Oxford University Press, p. 127, 1996.

Shanks, D. *Solved and Unsolved Problems in Number Theory, 4th ed.* New York: Chelsea, pp. 151–152, 1993.

Number Field Sieve

An extremely fast factorization method developed by Pollard which was used to factor the RSA-130 NUMBER. This method is the most powerful known for factoring general numbers, and has complexity

$$\mathcal{O}\left\{\exp\left[c(\log n)^{1/3}(\log \log n)^{2/3}\right]\right\}, \quad (1)$$

reducing the *exponent* over the CONTINUED FRACTION FACTORIZATION ALGORITHM and QUADRATIC SIEVE. There are three values of c relevant to different flavors of the method (Pomerance 1996). For the "special" case of the algorithm applied to numbers near a large POWER,

$$c = \left(\tfrac{32}{9}\right)^{1/3} = 1.526285\ldots, \quad (2)$$

for the "general" case applicable to any ODD POSITIVE number which is not a POWER,

$$c = \left(\tfrac{64}{9}\right)^{1/3} = 1.922999\ldots, \quad (3)$$

and for a version using many POLYNOMIALS (Coppersmith 1993),

$$c = \tfrac{1}{3}\left(92 + 26\sqrt{13}\right)^{1/3} = 1.901883\ldots \quad (4)$$

See also QUADRATIC SIEVE, RSA NUMBER

References

Coppersmith, D. "Modifications to the Number Field Sieve." *J. Cryptology* **6**, 169–180, 1993.

Coppersmith, D.; Odlyzko, A. M.; and Schroeppel, R. "Discrete Logarithms in GF(p)." *Algorithmics* **1**, 1–15, 1986.

Cowie, J.; Dodson, B.; Elkenbracht-Huizing, R. M.; Lenstra, A. K.; Montgomery, P. L.; Zayer, J. A. "World Wide Number Field Sieve Factoring Record: On to 512 Bits." In *Advances in Cryptology--ASIACRYPT '96 (Kyongju)* (Ed. K. Kim and T. Matsumoto.) New York: Springer-Verlag, pp. 382–394, 1996.

Elkenbracht-Huizing, R.-M. "A Multiple Polynomial General Number Field Sieve." *Algorithmic Number Theory (Talence, 1996).* New York: Springer-Verlag, pp. 99–114, 1996.

Elkenbracht-Huizing, R.-M. "An Implementation of the Number Field Sieve." *Experiment. Math.* **5**, 231–253, 1996.

Elkenbracht-Huizing, R.-M. "Historical Background of the Number Field Sieve Factoring Method." *Nieuw Arch. Wisk.* **14**, 375–389, 1996.

Elkenbracht-Huizing, R.-M. *Factoring Integers with the Number Field Sieve.* Doctor's Thesis, Leiden University, 1997.

Lenstra, A. K. and Lenstra, H. W. Jr. "Algorithms in Number Theory." In *Handbook of Theoretical Computer Science, Volume A: Algorithms and Complexity* (Ed. J. van Leeuwen). New York: Elsevier, pp. 673–715, 1990.

Lenstra, A. K. and Lenstra, H. W. Jr. *The Development of the Number Field Sieve.* Berlin: Springer-Verlag, 1993.

Pomerance, C. "A Tale of Two Sieves." *Not. Amer. Math. Soc.* **43**, 1473–1485, 1996.

Number Field Sieve Factorization Method

An extremely fast factorization method developed by Pollard which was used to factor the RSA-130 NUMBER. This method is the most powerful known for factoring general numbers, and has complexity

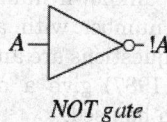

NOT gate

reducing the *exponent* over the CONTINUED FRACTION FACTORIZATION ALGORITHM and QUADRATIC SIEVE FACTORIZATION METHOD. There are three values of c

relevant to different flavors of the method (Pomerance 1996). For the "special" case of the algorithm applied to numbers near a large POWER,

$$\tilde{A}$$

for the "general" case applicable to any ODD POSITIVE number which is not a POWER,

$$\sim A$$

and for a version using many POLYNOMIALS (Coppersmith 1993),

$$10^{60}$$

See also RSA NUMBER

References

Coppersmith, D. "Modifications to the Number Field Sieve." *J. Cryptology* **6**, 169–180, 1993.

Coppersmith, D.; Odlyzko, A. M.; and Schroeppel, R. "Discrete Logarithms in GF(p)." *Algorithmics* **1**, 1–15, 1986.

Cowie, J.; Dodson, B.; Elkenbracht-Huizing, R. M.; Lenstra, A. K.; Montgomery, P. L.; Zayer, J. A. "World Wide Number Field Sieve Factoring Record: On to $S(m) = \frac{(1+\sqrt{2})^m + (1-\sqrt{2})^m}{2}$ Bits." In *Advances in Cryptology--ASIA-CRYPT '96 (Kyongju)* (Ed. K. Kim and T. Matsumoto.) New York: Springer-Verlag, pp. 382–394, 1996.

Elkenbracht-Huizing, R.-M. "A Multiple Polynomial General Number Field Sieve." *Algorithmic Number Theory (Talence, 1996)*. New York: Springer-Verlag, pp. 99–114, 1996.

Elkenbracht-Huizing, R.-M. "An Implementation of the Number Field Sieve." *Experiment. Math.* **5**, 231–253, 1996.

Elkenbracht-Huizing, R.-M. "Historical Background of the Number Field Sieve Factoring Method." *Nieuw Arch. Wisk.* **14**, 375–389, 1996.

Elkenbracht-Huizing, R.-M. *Factoring Integers with the Number Field Sieve.* Doctor's Thesis, Leiden University, 1997.

Lenstra, A. K. and Lenstra, H. W. Jr. "Algorithms in Number Theory." In *Handbook of Theoretical Computer Science, Volume A: Algorithms and Complexity* (Ed. J. van Leeuwen). New York: Elsevier, pp. 673–715, 1990.

Pomerance, C. "A Tale of Two Sieves." *Not. Amer. Math. Soc.* **43**, 1473–1485, 1996.

Number Group

FIELD

Number Guessing

By asking a small number of innocent-sounding questions about an unknown number, it is possible to reconstruct the number with absolute certainty (assuming that the questions are answered correctly). Ball and Coxeter (1987) give a number of sets of questions which can be used.

One of the simplest algorithms uses only three queries that can be used to determine an unknown number n from an audience member.

1. Ask the person to compute $n' = 3n$ (i.e., three times the secret number n) and announce if the result is EVEN or ODD.
2. If you were told that n' is EVEN, ask the person to reveal the number n'' which is half of n'. If you were told that n' is ODD, ask the person to reveal the number n'' which is half of $n' + 1$.
3. Ask the person to reveal the number of times k which 9 divides evenly into $n''' = 3n''$.

The original number n is then given by $2k$ if n' was EVEN, or $2k + 1$ if n' was ODD. For $n = 2m$ even, $n' = 6m$, $n'' = 3m$, $n''' = 9m$, $k = m$, so $2k = 2m = n$. For $n = 2m + 1$ odd, $n' = 6m + 3$, $n'' = 3m + 2$, $n''' = 9m + 6$, $k = m$, so $2k + 1 = 2m + 1 = n$.

Another method asks:

1. Multiply the number n by 5.
2. Add 6 to the product.
3. Multiply the sum by 4.
4. Add 9 to the product.
5. Multiply the sum by 5 and reveal the result n'.

The original number is then given by $n = (n' - 165)/100$, since the above steps give $n' = 5(4(5n + 6) + 9) = 100n + 165$.

See also NUMBER PICKING

References

Bachet, C. G. *Problèmes plaisans et délectables*, 2nd ed. 1624.

Ball, W. W. R. and Coxeter, H. S. M. *Mathematical Recreations and Essays, 13th ed.* New York: Dover, pp. 5–20, 1987.

Kraitchik, M. "To Guess a Selected Number." §3.3 in *Mathematical Recreations.* New York: W. W. Norton, pp. 58–66, 1942.

Number Pattern

It is possible to construct simple functions which produce growing patterns. For example, the BAXTER-HICKERSON FUNCTION

$$f(n) = \tfrac{1}{3}(2 \cdot 10^{5n} - 10^{4n} + 2 \cdot 10^{3n} + 10^{2n} + 10^n + 1)$$

produces the sequence 64037, 6634003367, 666334000333667, ...

See also BAXTER-HICKERSON FUNCTION, NUMBER PYRAMID

Number Picking

Place $2n$ balls in a bag and number them 1 to $2n$, then pick half of them at random. The number of different possible sums for $n = 1, 2, 3, \ldots$ are then 2, 5, 10, 17, 26, ... (Sloane's A002522), or $n^2 + 1$.

See also NUMBER GUESSING

References

Sloane, N. J. A. Sequences A002522 in "An On-Line Version of the Encyclopedia of Integer Sequences." http://www.research.att.com/~njas/sequences/eisonline.html.

Number Pyramid

A set of numbers obeying a pattern like the following,

$$91 \cdot 37 = 3367$$
$$9901 \cdot 3367 = 33336667$$
$$999001 \cdot 333667 = 333333666667$$
$$99990001 \cdot 33336667 = 3333333366666667$$

$$4^2 = 16$$
$$34^2 = 1156$$
$$334^2 = 111556$$

$$7^2 = 49$$
$$67^2 = 4489$$
$$667^2 = 444889.$$

See also AUTOMORPHIC NUMBER, NUMBER PATTERN

References

Heinz, H. "Miscellaneous Number Patterns." http://www.geocities.com/CapeCanaveral/Launchpad/4057/miscnum.htm.

Number Shape

FIGURATE NUMBER

Number Sign

OCTOTHORPE

Number System

BASE (NUMBER)

Number Theoretic Transform

Simplemindedly, a number theoretic transform is a generalization of a FAST FOURIER TRANSFORM obtained by replacing $e^{-2\pi i k/N}$ with an nth PRIMITIVE ROOT OF UNITY. This effectively means doing a transform over the QUOTIENT RING $\mathbb{Z}/p\mathbb{Z}$ instead of the COMPLEX NUMBERS $\mathbb{C}$. The theory is rather elegant and uses the language of FINITE FIELDS and NUMBER THEORY.

See also FAST FOURIER TRANSFORM, FINITE FIELD

References

Arndt, J. "Numbertheoretic Transforms (NTTs)." Ch. 4 in "Remarks on FFT Algorithms." http://www.jjj.de/fxt/.
Cohen, H. *A Course in Computational Algebraic Number Theory.* New York: Springer-Verlag, 1993.

Number Theory

A vast and fascinating field of mathematics, sometimes called "higher arithmetic," consisting of the study of the properties of whole numbers. PRIMES and PRIME FACTORIZATION are especially important in number theory, as are a number of functions such as the DIVISOR FUNCTION, RIEMANN ZETA FUNCTION, and TOTIENT FUNCTION. Excellent introductions to number theory may be found in Ore (1988) and Beiler (1966). The classic history on the subject (now slightly dated) is that of Dickson (1952).

The great difficulty required to prove relatively simple results in number theory prompted no less an authority than Gauss to remark that "it is just this which gives the higher arithmetic that magical charm which has made it the favorite science of the greatest mathematicians, not to mention its inexhaustible wealth, wherein it so greatly surpasses other parts of mathematics." Gauss, often known as the "prince of mathematics," called mathematics the "queen of the sciences,'" and considered number theory the "queen of mathematics" (Beiler 1966, Goldman 1997).

See also ADDITIVE NUMBER THEORY, ARITHMETIC, CONGRUENCE, DIOPHANTINE EQUATION, DIVISOR FUNCTION, GÖDEL'S INCOMPLETENESS THEOREM, MULTIPLICATIVE NUMBER THEORY, PEANO'S AXIOMS, PRIME COUNTING FUNCTION, PRIME FACTORIZATION, PRIME NUMBER, QUADRATIC RECIPROCITY THEOREM, RIEMANN ZETA FUNCTION, TOTIENT FUNCTION

References

Andrews, G. E. *Number Theory.* New York: Dover, 1994.
Andrews, G. E.; Berndt, B. C.; and Rankin, R. A. (Ed.). *Ramanujan Revisited: Proceedings of the Centenary Conference, University of Illinois at Urbana-Champaign, June 1–5, 1987.* Boston, MA: Academic Press, 1988.
Apostol, T. M. *Introduction to Analytic Number Theory.* New York: Springer-Verlag, 1976.
Ayoub, R. G. *An Introduction to the Analytic Theory of Numbers.* Providence, RI: Amer. Math. Soc., 1963.
Beiler, A. H. *Recreations in the Theory of Numbers: The Queen of Mathematics Entertains, 2nd ed.* New York: Dover, 1966.
Bellman, R. E. *Analytic Number Theory: An Introduction.* Reading, MA: Benjamin/Cummings, 1980.
Berndt, B. C. *Ramanujan's Notebooks, Part I.* New York: Springer-Verlag, 1985.
Berndt, B. C. *Ramanujan's Notebooks, Part II.* New York: Springer-Verlag, 1988.
Berndt, B. C. *Ramanujan's Notebooks, Part III.* New York: Springer-Verlag, 1997.
Berndt, B. C. *Ramanujan's Notebooks, Part IV.* New York: Springer-Verlag, 1993.
Berndt, B. C. *Ramanujan's Notebooks, Part V.* New York: Springer-Verlag, 1997.
Berndt, B. C. and Rankin, R. A. *Ramanujan: Letters and Commentary.* Providence, RI: Amer. Math. Soc, 1995.
Borwein, J. M. and Borwein, P. B. *Pi & the AGM: A Study in Analytic Number Theory and Computational Complexity.* New York: Wiley, 1987.
Brown, K. S. "Number Theory." http://www.seanet.com/~ksbrown/inumber.htm.
Burr, S. A. *The Unreasonable Effectiveness of Number Theory.* Providence, RI: Amer. Math. Soc., 1992.
Burton, D. M. *Elementary Number Theory, 4th ed.* Boston, MA: Allyn and Bacon, 1989.
Carmichael, R. D. *The Theory of Numbers, and Diophantine Analysis.* New York: Dover, 1959.

Cohen, H. *Advanced Topics in Computational Number Theory.* New York: Springer-Verlag, 2000.

Cohn, H. *Advanced Number Theory.* New York: Dover, 1980.

Courant, R. and Robbins, H. "The Theory of Numbers." Supplement to Ch. 1 in *What is Mathematics?: An Elementary Approach to Ideas and Methods, 2nd ed.* Oxford, England: Oxford University Press, pp. 21–51, 1996.

Davenport, H. *The Higher Arithmetic: An Introduction to the Theory of Numbers, 6th ed.* Cambridge, England: Cambridge University Press, 1992.

Davenport, H. and Montgomery, H. L. *Multiplicative Number Theory, 2nd ed.* New York: Springer-Verlag, 1980.

Dickson, L. E. *History of the Theory of Numbers, 3 vols.* New York: Chelsea, 1952.

Dudley, U. *Elementary Number Theory.* San Francisco, CA: W. H. Freeman, 1978.

Friedberg, R. *An Adventurer's Guide to Number Theory.* New York: Dover, 1994.

Gauss, C. F. *Disquisitiones Arithmeticae.* New Haven, CT: Yale University Press, 1966.

Goldman, J. R. *The Queen of Mathematics: An Historically Motivated Guide to Number Theory.* Natick, MA: A. K. Peters, 1997.

Guy, R. K. *Unsolved Problems in Number Theory, 2nd ed.* New York: Springer-Verlag, 1994.

Hardy, G. H. and Wright, E. M. *An Introduction to the Theory of Numbers, 5th ed.* Oxford, England: Clarendon Press, 1979.

Hardy, G. H. *Ramanujan: Twelve Lectures on Subjects Suggested by His Life and Work, 3rd ed.* New York: Chelsea, 1959.

Hasse, H. *Number Theory.* Berlin: Springer-Verlag, 1980.

Herkommer, M. A. *Number Theory: A Programmer's Guide.* New York: McGraw-Hill, 1999.

Ireland, K. F. and Rosen, M. I. *A Classical Introduction to Modern Number Theory, 2nd ed.* New York: Springer-Verlag, 1995.

Kato, K.; Kurokawa, N.; and Saito, T. *Number Theory 1: Fermat's Dream.* Providence, RI: Amer. Math. Soc., 2000.

Klee, V. and Wagon, S. *Old and New Unsolved Problems in Plane Geometry and Number Theory.* Washington, DC: Math. Assoc. Amer., 1991.

Koblitz, N. *A Course in Number Theory and Cryptography.* New York: Springer-Verlag, 1987.

Landau, E. *Elementary Number Theory, 2nd ed.* New York: Chelsea, 1999.

Lang, S. *Algebraic Number Theory, 2nd ed.* New York: Springer-Verlag, 1994.

Lenstra, H. W. and Tijdeman, R. (Eds.). *Computational Methods in Number Theory, 2 vols.* Amsterdam: Mathematisch Centrum, 1982.

LeVeque, W. J. *Fundamentals of Number Theory.* New York: Dover, 1996.

Mitrinovic, D. S. and Sandor, J. *Handbook of Number Theory.* Dordrecht, Netherlands: Kluwer, 1995.

Mollin, R. A. *Algebraic Number Theory.* Boca Raton, FL: CRC Press, 1999.

Mollin, R. A. *Fundamental Number Theory with Applications.* Boca Raton, FL: CRC Press, 1998.

Niven, I. M.; Zuckerman, H. S.; and Montgomery, H. L. *An Introduction to the Theory of Numbers, 5th ed.* New York: Wiley, 1991.

Ogilvy, C. S. and Anderson, J. T. *Excursions in Number Theory.* New York: Dover, 1988.

Ore, Ø. *Invitation to Number Theory.* Washington, DC: Math. Assoc. Amer., 1967.

Ore, Ø. *Number Theory and Its History.* New York: Dover, 1988.

Rose, H. E. *A Course in Number Theory, 2nd ed.* Oxford, England: Clarendon Press, 1995.

Rosen, K. H. *Elementary Number Theory and Its Applications, 3rd ed.* Reading, MA: Addison-Wesley, 1993.

Schroeder, M. R. *Number Theory in Science and Communication: With Applications in Cryptography, Physics, Digital Information, Computing, and Self-Similarity, 3rd ed.* New York: Springer-Verlag, 1997.

Shanks, D. *Solved and Unsolved Problems in Number Theory, 4th ed.* New York: Chelsea, 1993.

Sierpinski, W. *250 Problems in Elementary Number Theory.* New York: American Elsevier, 1970.

Uspensky, J. V. and Heaslet, M. A. *Elementary Number Theory.* New York: McGraw-Hill, 1939.

Vinogradov, I. M. *Elements of Number Theory, 5th rev. ed.* New York: Dover, 1954.

Weil, A. *Basic Number Theory, 3rd ed.* Berlin: Springer-Verlag, 1995.

Weil, A. *Number Theory: An Approach Through History From Hammurapi to Legendre.* Boston, MA: Birkhäuser, 1984.

Weisstein, E. W. "Books about Number Theory." http://www.treasure-troves.com/books/NumberTheory.html.

Weyl, H. *Algebraic Theory of Numbers.* Princeton, NJ: Princeton University Press, 1998.

Yildirim, C. Y. and Stepanov, S. A. (Eds.). *Number Theory and Its Applications.* New York: Dekker, 1998.

Young, J. W. A. "The Theory of Numbers." Ch. 7 in *Monographs on Topics of Modern Mathematics Relevant to the Elementary Field* (Ed. J. W. A. Young). New York: Dover, pp. 306–349, 1955.

Number Triangle

Bell Triangle, Clark's Triangle, Euler's Triangle, Leibniz Harmonic Triangle, Lossnitsch's Triangle, Magog Triangle, Monotone Triangle, Pascal's Triangle, Seidel-Entringer-Arnold Triangle, Trinomial Triangle

Number Wall

Quotient-Difference Table

Numerator

The number p in a Fraction p/q.

See also Denominator, Fraction, Rational Number

Numeric Function

A Function $f : A \to B$ such that B is a Set of numbers.

Numerical Derivative

While it is usually much easier to compute a Derivative instead of an Integral (which is a little strange, considering that "more" functions have integrals than derivatives), there are still many applications where derivatives need to be computed numerically. The simplest approach simply uses the definition of the Derivative

$$f'(x) \equiv \lim_{h \to 0} \frac{f(x+h) - f(x)}{h}$$

for some small numerical value of $h \ll 1$.

See also NUMERICAL INTEGRATION

References

Press, W. H.; Flannery, B. P.; Teukolsky, S. A.; and Vetterling, W. T. "Numerical Derivatives." §5.7 in *Numerical Recipes in FORTRAN: The Art of Scientific Computing, 2nd ed.* Cambridge, England: Cambridge University Press, pp. 180–184, 1992.

Weisstein, E. W. "Books about Numerical Methods." http://www.treasure-troves.com/books/NumericalMethods.html.

Numerical Integration

The approximate computation of an INTEGRAL using numerical techniques. The numerical computation of an INTEGRAL is sometimes called QUADRATURE. Ueberhuber (1997, p. 71) uses the word "QUADRATURE" to mean numerical computation of a univariate INTEGRAL, and "CUBATURE" to mean numerical computation of a MULTIPLE INTEGRAL.

There are a wide range of methods available for numerical integration. A good source for such techniques is Press *et al.* (1992).

The most straightforward numerical integration technique uses the NEWTON-COTES FORMULAS (also called QUADRATURE FORMULAS), which approximate a function tabulated at a sequence of regularly spaced INTERVALS by various degree POLYNOMIALS. If the endpoints are tabulated, then the 2- and 3-point formulas are called the TRAPEZOIDAL RULE and SIMPSON'S RULE, respectively. The 5-point formula is called BODE'S RULE. A generalization of the TRAPEZOIDAL RULE is ROMBERG INTEGRATION, which can yield accurate results for many fewer function evaluations.

If the functions are known analytically instead of being tabulated at equally spaced intervals, the best numerical method of integration is called GAUSSIAN QUADRATURE. By picking the abscissas at which to evaluate the function, Gaussianquadrature produces the most accurate approximations possible. However, given the speed of modern computers, the additional complication of the GAUSSIAN QUADRATURE formalism often makes it less desirable than simply brute-force calculating twice as many points on a regular grid (which also permits the already computed values of the function to be re-used). An excellent reference for GAUSSIAN QUADRATURE is Hildebrand (1956).

See also CUBATURE, DOUBLE EXPONENTIAL INTEGRATION, FILON'S INTEGRATION FORMULA, GAUSS-KRONROD QUADRATURE, GREGORY'S FORMULA, INTEGRAL, INTEGRATION, MONTE CARLO INTEGRATION, NUMERICAL DERIVATIVE, QUADRATURE, QUASI-MONTE CARLO INTEGRATION, T-INTEGRATION

References

Corbit, D. "Numerical Integration: From Trapezoids to RMS: Object-Oriented Numerical Integration." *Dr. Dobb's J.*, No. 252, 117–120, Oct. 1996.

Davis, P. J. and Rabinowitz, P. *Methods of Numerical Integration, 2nd ed.* New York: Academic Press, 1984.

Hildebrand, F. B. *Introduction to Numerical Analysis.* New York: McGraw-Hill, pp. 319–323, 1956.

Milne, W. E. *Numerical Calculus: Approximations, Interpolation, Finite Differences, Numerical Integration and Curve Fitting.* Princeton, NJ: Princeton University Press, 1949.

Press, W. H.; Flannery, B. P.; Teukolsky, S. A.; and Vetterling, W. T. *Numerical Recipes in FORTRAN: The Art of Scientific Computing, 2nd ed.* Cambridge, England: Cambridge University Press, 1992.

Ueberhuber, C. W. "Numerical Integration." Ch. 12 in *Numerical Computation 2: Methods, Software, and Analysis.* Berlin: Springer-Verlag, pp. 65–169, 1997.

Weisstein, E. W. "Books about Numerical Methods." http://www.treasure-troves.com/books/NumericalMethods.html.

Whittaker, E. T. and Robinson, G. "Numerical Integration and Summation." Ch. 7 in *The Calculus of Observations: A Treatise on Numerical Mathematics, 4th ed.* New York: Dover, pp. 132–163, 1967.

Numerology

The study of numbers for the supposed purpose of predicting future events or seeking connections with the occult.

See also BEAST NUMBER, NUMBER THEORY

References

Dudley, U. *Numerology, or, What Pythagoras Wrought.* Washington, DC: Math. Assoc. Amer., 1997.

NURBS Curve

A nonuniform rational B-SPLINE curve defined by

$$\mathbf{C}(t) = \frac{\sum_{i=0}^{n} N_{i,p}(t) w_i \mathbf{P}_i}{\sum_{i=0}^{n} N_{i,p}(t) w_i},$$

where p is the order, $N_{i,p}$ are the B-SPLINE basis functions, $\mathbf{P}_i$ are control points, and the weight w_i of $\mathbf{P}_i$ is the last ordinate of the homogeneous point $\mathbf{P}_i^w$. These curves are CLOSED under perspective transformations and can represent CONIC SECTIONS exactly.

See also B-SPLINE, BÉZIER CURVE, NURBS SURFACE

References

Piegl, L. and Tiller, W. *The NURBS Book, 2nd ed.* New York: Springer-Verlag, 1997.

NURBS Surface

A nonuniform rational B-SPLINE surface of degree (p, q) is defined by

$$\mathbf{S}_{(u,v)} = \frac{\sum_{i=0}^{m} \sum_{j=0}^{n} N_{i,p}(u) N_{j,q}(v) w_{i,j} \mathbf{P}_{i,j}}{\sum_{i=0}^{m} \sum_{j=0}^{n} N_{i,p}(u) N_{j,q}(v) w_{i,j}},$$

where $N_{i,p}$ and $N_{j,q}$ are the B-SPLINE basis functions, $\mathbf{P}_{i,j}$ are control points, and the weight $w_{i,j}$ of $\mathbf{P}_{i,j}$ is the last ordinate of the homogeneous point $\mathbf{P}_{i,j}^w$.

See also B-SPLINE, BÉZIER CURVE, NURBS CURVE

Nyquist Frequency

In order to recover all Fourier components of a periodic waveform, it is necessary to sample more than twice as fast as the *highest* waveform frequency v, i.e.,

$$f_{\text{Nyquist}} = 2v.$$

This cutoff frequency f_{Nyquist} above which a signal must be sampled in order to be able to fully reconstruct it is called the Nyquist frequency.

See also Fourier Series, Fourier Transform, Nyquist Sampling, Oversampling, Sampling Theorem

Nyquist Sampling

Sampling at the Nyquist frequency.

See also Sampling Theorem

O

O

The symbol $\mathbb{O}$ is sometimes used to represent CAYLEY NUMBERS (also commonly known as octonions).

See also CAYLEY NUMBER

Obelisk

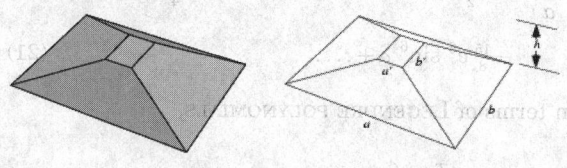

A polyhedron formed by two parallel rectangles, not congruent to each other, whose side faces are trapezoids. The VOLUME is given by

$$V = \tfrac{1}{6} h[(2a + a')b + (2a' + a)b']$$
$$= \tfrac{1}{6} h[(ab + (a + a')(b + b') + a'b'].$$

The distance from the bottom base to the CENTROID is

$$\bar{z} = \frac{h(ab + ab' + a'b + 3a'b')}{2(ab + ab' + a'b + 2a'b')}.$$

The term obelisk is sometimes also used to refer to the DAGGER symbol (Bringhurst 1997, p. 275).

See also DAGGER

References

Bringhurst, R. *The Elements of Typographic Style, 2nd ed.* Point Roberts, WA: Hartley and Marks, 1997.
Harris, J. W. and Stocker, H. "Obelisk." §4.5.3 in *Handbook of Mathematics and Computational Science.* New York: Springer-Verlag, p. 102, 1998.

Obelus

The symbol $\div$ used to indicate DIVISION. In typography, an obelus has a more general definition as any symbol, such as the DAGGER (†), used to indicate a footnote (Bringhurst 1997, p. 225).

See also DIVISION, SOLIDUS

References

Bringhurst, R. *The Elements of Typographic Style, 2nd ed.* Point Roberts, WA: Hartley and Marks, 1997.

Object

A mathematical structure (e.g., a GROUP, VECTOR SPACE, or DIFFERENTIABLE MANIFOLD) in a CATEGORY.

See also MORPHISM

Oblate Ellipsoid

OBLATE SPHEROID

Oblate Spheroid

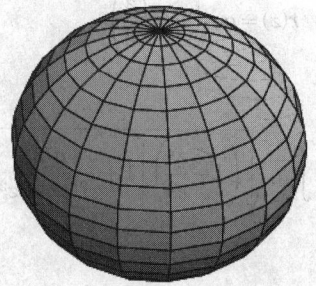

A "squashed" SPHEROID for which the equatorial radius a is greater than the polar radius c, so $a > c$ (called an oblate ellipsoid by Tietze 1965, p. 27). An oblate spheroid is a SURFACE OF REVOLUTION obtained by rotating an ELLIPSE about its minor axis (Hilbert and Cohn-Vossen 1999, p. 10). To first approximation, the shape assumed by a rotating fluid (including the Earth, which is "fluid" over astronomical time scales) is an oblate spheroid. The oblate spheroid can be specified parametrically by the usual SPHEROID equations (for a SPHEROID with z-AXIS as the symmetry axis),

$$x = a \sin v \cos u \tag{1}$$
$$y = a \sin v \sin u \tag{2}$$
$$z = c \cos v, \tag{3}$$

with $a > c$, $u \in [0, 2\pi)$, and $v \in [0, \pi]$. Its Cartesian equation is

$$\frac{x^2 + y^2}{a^2} + \frac{z^2}{c^2} = 1. \tag{4}$$

The ELLIPTICITY of an oblate spheroid is defined by

$$e \equiv \sqrt{1 - \frac{c^2}{a^2}}, \tag{5}$$

so that

$$1 - e^2 = \frac{c^2}{a^2}. \tag{6}$$

The radial distance from center of the spheroid as a function of latitude δ is given by

$$r(\delta) = \sqrt{\frac{a^2 + c^2 + (a - c)(a + c)\cos(2\delta)}{2}} \tag{7}$$

$$= a \sqrt{1 - e^2 \sin^2 \delta}. \tag{8}$$

The SURFACE AREA of an oblate spheroid can be computed as a SURFACE OF REVOLUTION about the z-AXIS,

$$S = 2\pi \int r(z) \sqrt{1 + [r'(z)]^2} \, dz \tag{9}$$

with radius as a function of z given by

$$r(z) = a \sqrt{1 - \left(\frac{z}{c}\right)^2}. \tag{10}$$

Therefore

$$S = 2\pi a \int_{-c}^{c} \sqrt{\left(1 - \frac{z^2}{c^2}\right) \left[1 + \frac{a^2 z^2}{c^2(c^2 - z^2)}\right]} \, dz$$

$$= \frac{\pi}{\sqrt{a^2 - c^2}}$$

$$\times \left[2a^2 \sqrt{a^2 - c^2} + c^2 a \ln\left(\frac{a + \sqrt{a^2 - c^2}}{a - \sqrt{a^2 - c^2}}\right)\right]. \tag{11}$$

Using the identity

$$\sqrt{a^2 - c^2} = ae \tag{12}$$

gives

$$S = 2\pi a^2 + \pi \frac{c^2}{e} \ln\left(\frac{1+e}{1-e}\right) \tag{13}$$

(Beyer 1987, p. 131). Note that this is the conventional form in which the surface area of an oblate spheroid is written, although it is formally equivalent to the conventional form for the PROLATE SPHEROID via the identity

$$\frac{c^2 \pi}{e(a, c)} \ln\left[\frac{1 + e(a, c)}{1 - e(a, c)}\right] = \frac{2\pi ac}{e(c, a)} \sin^{-1}[e(c, a)], \tag{14}$$

where $e(x, y)$ is defined by

$$e(x, y) \equiv \sqrt{1 - \frac{x^2}{y^2}}. \tag{15}$$

The VOLUME of an oblate spheroid can be computed from the formula for a general ELLIPSOID with $b = a$,

$$V = \frac{4}{3}\pi a^2 c \tag{16}$$

(Beyer 1987, p. 131).

An oblate spheroid with its origin at a FOCUS has equation

$$r = \frac{a(1 - e^2)}{1 + e \cos\phi}. \tag{17}$$

Define k and expand up to POWERS of e^6,

$$k \equiv e^2(1 - e^2)^{-1} = e^2(1 + e^2 - 2e^4 + 6e^6 + \ldots)$$

$$= e^2 + e^4 - 2e^6 + \ldots \tag{18}$$

$$k^2 = e^4 + e^6 + \ldots \tag{19}$$

$$k^3 = e^6 + \ldots \tag{20}$$

Expanding r in POWERS of ELLIPTICITY to e^6 therefore yields

$$\frac{r}{a} = 1 - \frac{1}{2}(e^2 + e^4 - 2e^4 + 6e^6)\sin^2 \delta + \frac{3}{4}(e^4 + e^6)\sin^4 \delta$$

$$- \frac{15}{8} e^6 \sin^6 \delta + \ldots. \tag{21}$$

In terms of LEGENDRE POLYNOMIALS,

$$\frac{r}{a} = \left(1 - \frac{1}{6} e^2 - \frac{11}{20} e^4 - \frac{103}{1680} e^6\right)$$

$$+ \left(-\frac{1}{3} e^2 - \frac{5}{42} e^4 - \frac{3}{56} e^6\right) P_2$$

$$+ \left(\frac{3}{35} e^4 + \frac{57}{770} e^6\right) P_4 - \frac{5}{231} e^6 P_6 + \ldots. \tag{22}$$

The ELLIPTICITY may also be expressed in terms of the OBLATENESS (also called FLATTENING), denoted ϵ or f.

$$\epsilon \equiv \frac{a - c}{a} \tag{23}$$

$$c = a(1 - \epsilon) \tag{24}$$

$$c^2 = a^2(1 - \epsilon)^2 \tag{25}$$

$$(1 - \epsilon)^2 = 1 - e^2, \tag{26}$$

so

$$\epsilon = 1 - \sqrt{1 - e^2} \tag{27}$$

and

$$e^2 = 1 - (1 - \epsilon)^2 = 1 - (1 - 2\epsilon + \epsilon^2) = 2\epsilon - \epsilon^2 \tag{28}$$

$$r = a\left[1 + \frac{2\epsilon - \epsilon^2}{(1 - \epsilon)^2} \sin^2 \delta\right]^{-1/2}. \tag{29}$$

Define k and expand up to POWERS of ϵ^6

$$k \equiv (2\epsilon - \epsilon)(1 - \epsilon)^{-2} = (2\epsilon - \epsilon^2)(1 + 2\epsilon - 6\epsilon^2 + \ldots)$$

$$= 2\epsilon + 4\epsilon^4 - 12\epsilon^3 - \epsilon^2 - 2\epsilon^3 + \ldots$$

$$= 2\epsilon + 3\epsilon^2 - 14\epsilon^3 + \ldots \tag{30}$$

$$k^2 = 4\epsilon^2 + 6\epsilon^3 + \ldots \tag{31}$$

$$k^3 = 8\epsilon^3 + \ldots \tag{32}$$

Expanding r in POWERS of the OBLATENESS to ϵ^3 yields

$$\frac{r}{a} = 1 - \tfrac{1}{2}(2\epsilon + 3\epsilon^2 - 14\epsilon^3)\sin^2\delta + \tfrac{3}{4}(4\epsilon^2 + 6\epsilon^3)\sin^4\delta$$

$$+ 8\epsilon^3\sin^6\delta + \dots . \tag{33}$$

In terms of LEGENDRE POLYNOMIALS,

$$\frac{r}{a} = \left(1 - \tfrac{1}{3}\epsilon - \tfrac{2}{5}\epsilon^2 - \tfrac{13}{105}\epsilon^3\right) + \left(-\tfrac{2}{3}\epsilon - \tfrac{1}{7}\epsilon^2 - \tfrac{1}{21}\epsilon^3\right)P_2$$

$$+ \left(\tfrac{12}{35}\epsilon^2 - \tfrac{96}{385}\epsilon^3\right)P_4 - \tfrac{40}{231}\epsilon^3 P_6 + \dots . \tag{34}$$

To find the projection of an oblate spheroid onto a PLANE, set up a coordinate system such that the z-AXIS is towards the observer, and the x-AXIS is in the PLANE of the page. The equation for an oblate spheroid is

$$r(\theta) = a\left[1 + \frac{2\epsilon - \epsilon^2}{(1-\epsilon)^2}\cos^2\theta\right]^{-1/2} . \tag{35}$$

Define

$$k \equiv \frac{2\epsilon - \epsilon^2}{(1-\epsilon)^2}, \tag{36}$$

and $x \equiv \sin\theta$. Then

$$r(\theta) = a[1 + k(1-x^2)]^{-1/2} = a(1 + k - kx^2)^{-1/2} . \tag{37}$$

Now rotate that spheroid about the x-AXIS by an ANGLE B so that the new symmetry axes for the spheroid are $x' \equiv x, y'$, and z'. The projected height of a point in the $x = 0$ PLANE on the y-AXIS is

$$y = r(\theta)\cos(\theta - B) = r(\theta)(\cos\theta\cos B - \sin\theta\sin B)$$

$$= r(\theta)\left(\sqrt{1-x^2}\cos B + x\sin B\right). \tag{38}$$

To find the highest projected point,

$$\frac{dy}{d\theta} = \frac{a\sin(B-\theta)}{(1 + k\cos^2\theta)^{1/2}} + ak\frac{\cos(B-\theta)\cos\theta\sin\theta}{(1 + k\cos^2\theta)^{3/2}}$$

$$= 0 . \tag{39}$$

Simplifying,

$$\tan(B-\theta)(1 + k\cos^2\theta) + k\cos\theta\sin\theta = 0 . \tag{40}$$

But

$$\tan(B - \theta)$$

$$= \frac{\tan B - \tan\theta}{1 + \tan B\tan\theta} = \frac{\tan B - \dfrac{\sin\theta}{\sqrt{1-\sin^2\theta}}}{1 + \tan B\dfrac{\sin\theta}{\sqrt{1-\sin^2\theta}}}$$

$$= \frac{\sqrt{1-\sin^2\theta}\,\tan B - \sin\theta}{\sqrt{1-\sin^2\theta} + \tan B\sin\theta} \tag{41}$$

Plugging (41) into (40),

$$\frac{\sqrt{1-x^2}\,\tan B - x}{\sqrt{1-x^2} + x\tan B}[1 + k(1-x^2)] + kx\sqrt{1-x^2}$$

$$= 0 \tag{42}$$

and performing a number of algebraic simplifications

$$\left(\sqrt{1-x^2}\,\tan B - x\right)(1 + k - kx^2) + kx\sqrt{1-x^2}$$

$$\times\left(\sqrt{1-x^2} + x\tan B\right) = 0 \tag{43}$$

$$\left[(1+k)\sqrt{1-x^2}\,\tan B - kx^2\sqrt{1-x^2}\,\tan B - x - kx + kx^3\right]$$

$$+ \left[kx(1-x^2) + kx^2\sqrt{1-x^2}\,\tan B\right] \tag{44}$$

$$(1+k)\tan B\sqrt{1-x^2} - kx(1-x^2) - x + kx(1-x^2) = 0 \tag{45}$$

$$(1+k)\tan B\sqrt{1-x^2} = x \tag{46}$$

$$(1+k)^2\tan^2 B(1-x^2) = x^2 \tag{47}$$

$$x^2\left[1 + (1+k)^2\tan^2 B\right] = (1+k)^2\tan^2 B \tag{48}$$

finally gives the expression for x in terms of B and k,

$$x^2 = \frac{\tan^2 B(1+k)^2}{1 + (1+k)^2\tan^2 B} . \tag{49}$$

Combine (37) and (38) and plug in for x,

$$y = a\frac{\sqrt{1-x^2}\cos B + x\sin B}{\sqrt{1 + k - kx^2}}$$

$$= a\frac{\cos B + (1+k)\dfrac{\sin^2 B}{\cos B}}{\sqrt{(1+k)[1 + (1+k)\tan^2 B]}}$$

$$= a\frac{\cos^2 B + (1+k)\sin^2 B}{\cos B\sqrt{(1+k)[1 + (1+k)\tan^2 B]}} . \tag{50}$$

Now re-express k in terms of a and c, using $\epsilon \equiv 1 - c/a$,

$$k \equiv \frac{(2-\epsilon)\epsilon}{(1-\epsilon)^2} = \frac{\left(1 + \dfrac{c}{a}\right)\left(1 - \dfrac{c}{a}\right)}{\left(\dfrac{c}{a}\right)^2} = \frac{1 - \left(\dfrac{c}{a}\right)^2}{\left(\dfrac{c}{a}\right)^2}$$

$$= \left(\frac{a}{c}\right)^2 - 1, \tag{51}$$

so

$$1 + k = \left(\frac{a}{c}\right)^2 \tag{52}$$

Plug (51) and (52) into (50) to obtain the SEMIMINOR AXIS of the projected oblate spheroid,

$$c' = a \frac{\cos^2 B + \left(\dfrac{a}{c}\right)^2 \sin^2 B}{\cos B \sqrt{\left(\dfrac{a}{c}\right)^2 \left[1 + \left(\dfrac{a}{c}\right)^2 \tan^2 B\right]}}$$

$$= a \frac{\cos^2 B + \left(\dfrac{a}{c}\right)^2 \sin^2 B}{\dfrac{a}{c}\sqrt{\cos^2 B + \left(\dfrac{a}{c}\right)^2 \sin^2 B}}$$

$$= c \sqrt{\cos^2 B + \left(\dfrac{a}{c}\right)^2 \sin^2 B} = \sqrt{c^2 \cos^2 B + a^2 \sin^2 B}$$

$$= a \sqrt{(1-\epsilon)^2 \cos^2 B + \sin^2 B}. \tag{53}$$

We wish to find the equation for a spheroid which has been rotated about the $x \equiv x'$-axis by ANGLE B, then the z-AXIS by ANGLE P

$$\begin{bmatrix} x' \\ y' \\ z' \end{bmatrix} = \begin{bmatrix} 1 & 0 & 0 \\ 0 & \cos B & \sin B \\ 0 & -\sin B & \cos B \end{bmatrix} \begin{bmatrix} \cos P & 0 & \sin P \\ 0 & 1 & 0 \\ -\sin P & 0 & \cos P \end{bmatrix} \begin{bmatrix} x \\ y \\ z \end{bmatrix}$$

$$= \begin{bmatrix} \cos P & 0 & \sin P \\ -\sin B \sin P & \cos B & \sin B \cos P \\ -\cos B \sin P & -\sin B & \cos B \cos P \end{bmatrix} \begin{bmatrix} x \\ y \\ z \end{bmatrix}. \tag{54}$$

Now, in the original coordinates (x', y', z'), the spheroid is given by the equation

$$\frac{x'^2}{a^2} + \frac{y'^2}{c^2} + \frac{z'^2}{a^2} = 1, \tag{55}$$

which becomes in the new coordinates,

$$\frac{(x \cos P + y \sin P)^2}{a^2}$$

$$+ \frac{(-x \sin B \sin P + z \cos B + y \sin B \cos P)^2}{a^2}$$

$$+ \frac{(-x \cos B \sin P - z \sin B + y \cos B \cos P)^2}{c^2} = 1. \tag{56}$$

Collecting COEFFICIENTS,

$$Ax^2 + By^2 + Cz^2 + Dxy + Exz + Fyz = 1, \tag{57}$$

where

$$A \equiv \frac{\cos^2 P + \sin^2 B \sin^2 P}{a^2} + \frac{\cos^2 B \sin^2 P}{c^2} \tag{58}$$

$$B \equiv \frac{\sin^2 P + \sin^2 B \cos^2 P}{a^2} + \frac{\cos^2 B \cos^2 P}{c^2} \tag{59}$$

$$C \equiv \frac{\cos^2 B}{a^2} + \frac{\sin^2 B}{c^2} \tag{60}$$

$$D \equiv 2 \cos P \sin P \left(\frac{1 - \sin^2 B}{a^2} - \frac{\cos^2 B}{c^2}\right)$$

$$= 2 \cos P \sin P \cos^2 B \left(\frac{1}{a^2} - \frac{1}{c^2}\right) \tag{61}$$

$$E \equiv 2 \sin B \cos B \sin P \left(\frac{1}{b^2} - \frac{1}{a^2}\right) \tag{62}$$

$$F \equiv 2 \sin B \cos B \cos P \left(\frac{1}{a^2} - \frac{1}{b^2}\right). \tag{63}$$

If we are interested in computing z, the radial distance from the symmetry axis of the spheroid (y) corresponding to a point

$$Cz^2 + (Ex + Fy)z + (Ax^2 + By^2 + Dxy - 1)$$

$$= Cz^2 + G(x,y)z + H(x,y) = 0, \tag{64}$$

where

$$G(x,y) \equiv Ex + Fy \tag{65}$$

$$H(x,y) \equiv Ax^2 + By^2 + Dxy - 1. \tag{66}$$

z can now be computed using the quadratic equation when (x, y) is given,

$$z = \frac{-G(x,y) \pm \sqrt{G^2(x,y) - 4CG(x,y)}}{2C}. \tag{67}$$

If $P = 0$, then we have $\sin P = 0$ and $\cos P = 1$, so (58) to (63) and (65) to (66) become

$$A \equiv \frac{1}{a^2} \tag{68}$$

$$B \equiv \frac{\sin^2 B}{a^2} + \frac{\cos^2 B}{b^2} \tag{69}$$

$$C \equiv \frac{\cos^2 B}{a^2} + \frac{\sin^2 B}{b^2} \tag{70}$$

$$D \equiv 0 \tag{71}$$

$$E \equiv 0 \tag{72}$$

$$F \equiv 2 \sin B \cos B \left(\frac{1}{a^2} - \frac{1}{b^2}\right) \tag{73}$$

$$G(x,y) \equiv Fy = 2y \sin B \cos B \left(\frac{1}{a^2} - \frac{1}{b^2}\right) \tag{74}$$

$$H(x,y) \equiv Ax^2 + By^2 - 1$$

$$= \frac{x^2}{a^2} + y^2 \left(\frac{\sin^2 B}{a^2} + \frac{\cos^2 B}{b^2} \right) - 1. \tag{75}$$

See also APPLE, DARWIN-DE SITTER SPHEROID, ELLIPSOID, OBLATE SPHEROIDAL COORDINATES, PROLATE SPHEROID, SPHERE, SPHEROID

References

Beyer, W. H. *CRC Standard Mathematical Tables, 28th ed.* Boca Raton, FL: CRC Press, 1987.
Hilbert, D. and Cohn-Vossen, S. *Geometry and the Imagination.* New York: Chelsea, p. 10, 1999.
Tietze, H. *Famous Problems of Mathematics: Solved and Unsolved Mathematics Problems from Antiquity to Modern Times.* New York: Graylock Press, p. 27, 1965.

Oblate Spheroid Geodesic

The GEODESIC on an OBLATE SPHEROID can be computed analytically, although the resulting expression is much more unwieldy than for a simple SPHERE. A spheroid with equatorial radius a and polar radius c can be specified parametrically by

$$x = a \sin v \cos u \tag{1}$$

$$y = a \sin v \sin u \tag{2}$$

$$z = c \cos v, \tag{3}$$

where $a > c$. Using the first PARTIAL DERIVATIVES

$$\frac{\partial x}{\partial u} = -a \sin v \sin u \quad \frac{\partial x}{\partial v} = a \cos v \cos u \tag{4}$$

$$\frac{\partial y}{\partial u} = a \sin v \cos u \quad \frac{\partial y}{\partial v} = a \cos v \sin u \tag{5}$$

$$\frac{\partial z}{\partial u} = 0 \quad \frac{\partial z}{\partial v} = -c \sin v, \tag{6}$$

and second PARTIAL DERIVATIVES

$$\frac{\partial^2 x}{\partial u^2} = -a \sin v \cos u \quad \frac{\partial^2 x}{\partial v^2} = -a \sin v \cos u \tag{7}$$

$$\frac{\partial^2 y}{\partial u^2} = -a \sin v \sin u \quad \frac{\partial^2 y}{\partial v^2} = -a \sin v \sin u \tag{8}$$

$$\frac{\partial^2 z}{\partial u^2} = 0 \quad \frac{\partial^2 z}{\partial v^2} = -z \cos v, \tag{9}$$

gives the GEODESICS functions as

$$P \equiv \left(\frac{\partial x}{\partial u} \right)^2 + \left(\frac{\partial y}{\partial u} \right)^2 + \left(\frac{\partial z}{\partial u} \right)^2$$

$$= a^2 (\sin^2 v \cos^2 u + \sin^2 v \sin^2 u)$$

$$= a^2 \sin^2 v \tag{10}$$

$$Q \equiv \frac{\partial x}{\partial u} \frac{\partial x}{\partial v} + \frac{\partial y}{\partial u} \frac{\partial y}{\partial v} + \frac{\partial z}{\partial u} \frac{\partial z}{\partial v} = 0 \tag{11}$$

$$R \equiv \left(\frac{\partial x}{\partial v} \right)^2 + \left(\frac{\partial y}{\partial v} \right)^2 + \left(\frac{\partial z}{\partial v} \right)^2$$

$$= a^2 + (c^2 - a^2) \sin^2 v = a^2 (1 - e^2 \sin^2 v), \tag{12}$$

where

$$e = \sqrt{\frac{a^2 - c^2}{a^2}} \tag{13}$$

is the ELLIPTICITY.

Since $Q = 0$ and P and R are explicit functions of v only, we can use the special form of the GEODESIC equation

$$u = \int \sqrt{\frac{R}{P^2 - c_1^2 P}} \, dv = \int \sqrt{\frac{a^2 (1 - e^2 \sin^2 v)}{a^4 \sin^4 v - c_1^2 a^2 \sin^2 v}} \, dv$$

$$= \frac{1}{c_1} \int \sqrt{\frac{1 - e^2 \sin^2 v}{\left(\frac{a}{c_1} \right)^2 \sin^2 v - 1}} \frac{dv}{\sin v}. \tag{14}$$

Integrating gives

$$u =$$

$$-\frac{e^2 F\left(\phi \Big| \frac{(d^2 - 1)e^2}{d^2 - e^2} \right) - d^2 \Pi\left(d^2 - 1, \phi \Big| \frac{(d^2 - 1)e^2}{d^2 - e^2} \right)}{c_1 \sqrt{d^2 - e^2}}, \tag{15}$$

where

$$d \equiv \frac{a}{c_1} \tag{16}$$

$$\cos \phi \equiv \frac{d \cos v}{\sqrt{d^2 - 1}}, \tag{17}$$

$F(\phi|m)$ is an ELLIPTIC INTEGRAL OF THE FIRST KIND with PARAMETER m, and $\Pi(\phi|m, k)$ is an ELLIPTIC INTEGRAL OF THE THIRD KIND.

GEODESICS other than MERIDIANS of an OBLATE SPHEROID undulate between two parallels with latitudes equidistant from the equator. Using the WEIERSTRASS SIGMA FUNCTION and WEIERSTRASS ZETA FUNCTION, the GEODESIC on the OBLATE SPHEROID can be written as

$$x + iy = \kappa \frac{\sigma(a + u)}{\sigma(u)\sigma(a)} e^{u[\eta - \zeta(\omega + a)]} \tag{18}$$

$$x - iy = \kappa \frac{\sigma(a - u)}{\sigma(u)\sigma(a)} e^{-u[\eta - \zeta(\omega + a)]} \tag{19}$$

$$z^2 = \lambda^2 \; \frac{\sigma(\omega'' + u)(\omega'' - u)}{\sigma^2(u)\sigma^2(a)} \qquad (20)$$

(Forsyth 1960, pp. 108–109; Halphen 1886–1891).

The equation of the GEODESIC can be put in the form

$$d\phi = \frac{\sqrt{1 - e^2 \sin^2 v} \, \sin a}{\sqrt{\sin^2 v - \sin^2 a} \, \sin v} \, dv, \qquad (21)$$

where a is the smallest value of v on the curve. Furthermore, the difference in longitude between points of highest and next lowest latitude on the curve is

$$\pi - 2 \; \frac{\sqrt{1 - e^2 \sin^2 a}}{\sin a} \int_0^\kappa \frac{\mathrm{dn}\, u - \mathrm{dn}^2 u}{1 + \cot^2 a \, \mathrm{sn}^2 u} \, du, \quad (22)$$

where the MODULUS of the ELLIPTIC FUNCTION is

$$k = \frac{e \cos a}{\sqrt{1 - e^2 \sin^2 a}} \qquad (23)$$

(Forsyth 1960, p. 446).

See also ELLIPSOID GEODESIC, OBLATE SPHEROID, SPHERE GEODESIC

References

Forsyth, A. R. *Calculus of Variations.* New York: Dover, 1960.

Halphen, G. H. *Traité des fonctions elliptiques et de leurs applications fonctions elliptiques,* Vol. 2. Paris: Gauthier-Villars, pp. 238–243, 1886–1891.

Tietze, H. *Famous Problems of Mathematics: Solved and Unsolved Mathematics Problems from Antiquity to Modern Times.* New York: Graylock Press, pp. 28–29 and 40–41, 1965.

Oblate Spheroidal Coordinates

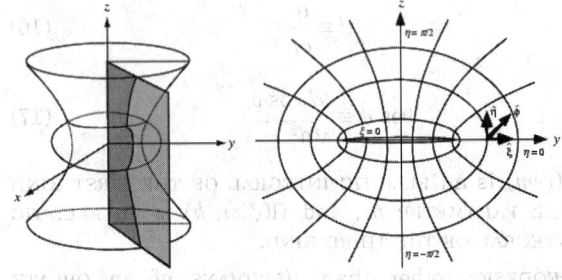

A system of CURVILINEAR COORDINATES in which two sets of coordinate surfaces are obtained by revolving the curves of the ELLIPTIC CYLINDRICAL COORDINATES about the Y-AXIS which is relabeled the Z-AXIS. The third set of coordinates consists of planes passing through this axis.

$$x = a \cosh \xi \cos \eta \cos \phi \qquad (1)$$

$$y = a \cosh \xi \cos \eta \sin \phi \qquad (2)$$

$$z = a \sinh \xi \sin \eta, \qquad (3)$$

where $\xi \in [0, \infty)$, $\eta \in [-\pi/2, \pi/2]$, and $\phi \in [0, 2\pi)$. Arfken (1970) uses (u, v, φ) instead of (ξ, η, ϕ). The SCALE FACTORS are

$$h_\xi = a \sqrt{\sinh^2 \xi + \sin^2 \eta} \qquad (4)$$

$$h_\eta = a \sqrt{\sinh^2 \xi + \sin^2 \eta} \qquad (5)$$

$$h_\phi = a \cosh \xi \cos \eta. \qquad (6)$$

The LAPLACIAN is

$$\nabla^2 f = \frac{1}{a^3 (\sinh^2 \xi + \sinh^2 \eta) \cosh \xi \cos \eta}$$

$$\times \left[\frac{\partial f}{\partial \xi} \left(a \cosh \xi \cos \eta \, \frac{\partial f}{\partial \eta} \right) + \frac{\partial f}{\partial \eta} \left(a \cosh \xi \cos \eta \, \frac{\partial f}{\partial \eta} \right) \right.$$

$$\left. + \frac{a^2 (\sinh^2 \xi + \sinh^2 \eta)}{a \cosh \xi \cos \eta} \frac{\partial^2 f}{\partial \phi^2} \right]$$

$$= \frac{1}{a^3 (\sinh^2 \xi + \sinh^2 \eta) \cosh \xi \cos \eta}$$

$$\times \left[a \sinh \xi \cos \eta \, \frac{\partial f}{\partial \xi} + a \cosh \xi \cos \eta \, \frac{\partial^2 f}{\partial \xi^2} \right.$$

$$\left. + a \sinh \xi \cos \eta \, \frac{\partial f}{\partial \eta} + a \cosh \xi \cos \eta \, \frac{\partial^2 f}{\partial \eta^2} \right]$$

$$+ \frac{1}{a^2 (\sinh^2 \xi + \sinh^2 \eta)} \frac{\partial^2 f}{\partial \phi^2} = \frac{1}{a^2 (\sinh^2 \xi + \sinh^2 \eta)}$$

$$\times \left[\frac{1}{\cosh \xi} \frac{\partial}{\partial \xi} \left(\cosh \xi \, \frac{\partial f}{\partial \xi} \right) + \frac{1}{\cosh \eta} \frac{\partial}{\partial \eta} \left(\cosh \eta \, \frac{\partial f}{\partial \eta} \right) \right]$$

$$+ \frac{1}{a^2 (\cosh^2 \xi + \cos^2 \eta)} \frac{\partial^2 f}{\partial \phi^2} \qquad (7)$$

$$= \frac{1}{\sinh^2 \eta + \sinh^2 \xi}$$

$$\times \left[(\mathrm{sech}^2 \, \xi \, \tan^2 \eta + \sec^2 \tanh^2 \xi) \, \frac{\partial^2}{\partial \phi^2} + \tanh \xi \, \frac{\partial}{\partial \xi} \right.$$

$$\left. + \frac{\partial^2}{\partial \xi^2} - \tan \eta \, \frac{\partial}{\partial \eta} + \frac{\partial^2}{\partial \eta^2} \right]. \qquad (8)$$

An alternate form useful for "two-center" problems is defined by

$$\xi_1 = \sinh \xi \qquad (9)$$

$$\xi_1' = \cosh \xi \qquad (10)$$

$$\xi_2 = \cos \eta \qquad (11)$$

$$\xi_3 = \phi, \qquad (12)$$

where $\xi_1 \in [1, \infty]$, $\xi_2 \in [-1, 1]$, and $\xi_3 \in [0, 2\pi)$. In these coordinates,

$$y = a\xi_1' \xi_2 \sin\xi_3 \tag{13}$$

$$z = a\sqrt{(\xi_1'^2 - 1)(1 - \xi_2^2)} \tag{14}$$

$$x = a\xi_1' \xi_2 \cos\xi_3 \tag{15}$$

(Abramowitz and Stegun 1972). The SCALE FACTORS are

$$h_{\xi_1} = a\sqrt{\frac{\xi_1^2 - \xi_2^2}{\xi_1^2 - 1}} \tag{16}$$

$$h_{\xi_2} = a\sqrt{\frac{\xi_1^2 - \xi_2^2}{1 - \xi_2^2}} \tag{17}$$

$$h_{\xi_3} = a\xi\eta, \tag{18}$$

and the LAPLACIAN is

$$\nabla^2 f = \frac{1}{a^2}\left\{ \frac{1}{\xi_1^2 + \xi_2^2}\frac{\partial}{\partial\xi_1}\left[(\xi_1^2 + 1)\frac{\partial f}{\partial\xi_1}\right] \right.$$
$$+ \frac{1}{\xi_1^2 + \xi_2^2}\frac{\partial}{\partial\xi_2}\left[(1 - \xi_2^2)\frac{\partial f}{\partial\xi_2}\right]$$
$$\left. + \frac{1}{(\xi_1^2 - 1)(1 - \xi_2^2)}\frac{\partial^2 f}{\partial\xi_3^2} \right\}. \tag{19}$$

The HELMHOLTZ DIFFERENTIAL EQUATION is separable.

See also HELMHOLTZ DIFFERENTIAL EQUATION–OBLATE SPHEROIDAL COORDINATES, LATITUDE, LONGITUDE, PROLATE SPHEROIDAL COORDINATES, SPHERICAL COORDINATES

References

Abramowitz, M. and Stegun, C. A. (Eds.). "Definition of Oblate Spheroidal Coordinates." §21.2 in *Handbook of Mathematical Functions with Formulas, Graphs, and Mathematical Tables, 9th printing.* New York: Dover, p. 752, 1972.

Arfken, G. "Prolate Spheroidal Coordinates (u, v, ϕ)." §2.11 in *Mathematical Methods for Physicists, 2nd ed.* Orlando, FL: Academic Press, pp. 107–109, 1970.

Byerly, W. E. *An Elementary Treatise on Fourier's Series, and Spherical, Cylindrical, and Ellipsoidal Harmonics, with Applications to Problems in Mathematical Physics.* New York: Dover, p. 242, 1959.

Moon, P. and Spencer, D. E. "Oblate Spheroidal Coordinates (η, θ, ψ)." Table 1.07 in *Field Theory Handbook, Including Coordinate Systems, Differential Equations, and Their Solutions, 2nd ed.* New York: Springer-Verlag, pp. 31–34, 1988.

Morse, P. M. and Feshbach, H. *Methods of Theoretical Physics, Part I.* New York: McGraw-Hill, p. 663, 1953.

Oblate Spheroidal Wave Function

The wave equation in OBLATE SPHEROIDAL COORDINATES is

$$\nabla^2\Phi + k^2\Phi = \frac{\partial}{\partial\xi_1}\left[(\xi_1^2 + 1)\frac{\partial\Phi}{\partial\xi_1}\right]$$
$$+ \frac{\partial}{\partial\xi_2}\left[(1 - \xi_2^2)\frac{\partial\Phi}{\partial\xi_2}\right] + \frac{\xi_1^2 + \xi_2^2}{(\xi_1^2 + 1)(1 - x_2^2)}\frac{\partial^2\Phi}{\partial\phi^2}$$
$$+ c(\xi_1^2 + \xi_2^2)\Phi = 0, \tag{1}$$

where

$$c \equiv \tfrac{1}{2}ak. \tag{2}$$

Substitute in a trial solution

$$\Phi = R_{mn}(c, \xi_1)S_{mn}(c, \xi_2){}^{\cos}_{\sin}(m\phi). \tag{3}$$

The radial differential equation is

$$\frac{d}{d\xi_2}\left[(1 + \xi_2^2)\frac{d}{d\xi_2} S_{mn}(c, \xi_2)\right]$$
$$- \left(\lambda_{mn} - c^2\xi_2^2 + \frac{m^2}{1 + \xi_2^2}\right)R_{mn}(c, \xi_2) = 0, \tag{4}$$

and the angular differential equation is

$$\frac{d}{d\xi_2}\left[(1 - \xi_2^2)\frac{d}{d\xi_2} S_{mn}(c, \xi_2)\right]$$
$$- \left(\lambda_{mn} - c^2\xi_2^2 + \frac{m^2}{1 - \xi_2^2}\right)R_{mn}(c, \xi_2) = 0 \tag{5}$$

(Abramowitz and Stegun 1972, pp. 753–755; Zwillinger 1997, p. 127).

See also PROLATE SPHEROIDAL WAVE FUNCTION, SPHEROIDAL WAVE FUNCTION

References

Abramowitz, M. and Stegun, C. A. (Eds.). "Spheroidal Wave Functions." Ch. 21 in *Handbook of Mathematical Functions with Formulas, Graphs, and Mathematical Tables, 9th printing.* New York: Dover, pp. 751–759, 1972.

Zwillinger, D. *Handbook of Differential Equations, 3rd ed.* Boston, MA: Academic Press, p. 127, 1997.

Oblateness

FLATTENING

Oblique Angle

An ANGLE which is not a RIGHT ANGLE.

Oblique Cylinder

CYLINDER

Oblique Prism

PRISM

Oblique Triangle

A TRIANGLE that is not a RIGHT TRIANGLE.

See also RIGHT TRIANGLE, TRIANGLE

References
Kern, W. F. and Bland, J. R. *Solid Mensuration with Proofs,*
2nd ed. New York: Wiley, p. 3, 1948.

Oblong Number

PRONIC NUMBER

Obstruction

Obstruction theory studies the extensibility of MAPS using algebraic GADGETS. While the terminology rapidly becomes technical and convoluted (as Iyanaga and Kawada note, "It is extremely difficult to discuss higher obstructions in general since they involve many complexities"), the ideas associated with obstructions are very important in modern ALGEBRAIC TOPOLOGY.

See also ALGEBRAIC TOPOLOGY, CHERN CLASS, EILEN-BERG-MAC LANE SPACE, STIEFEL-WHITNEY CLASS

References
Iyanaga, S. and Kawada, Y. (Eds.). "Obstructions." §300 in
Encyclopedic Dictionary of Mathematics. Cambridge, MA:
MIT Press, pp. 948–950, 1980.

Obtuse Angle

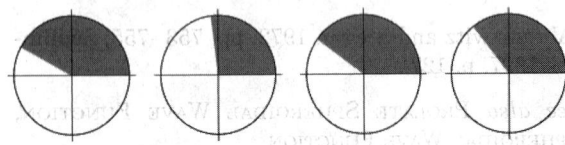

An ANGLE greater than $\pi/2$ RADIANS (90°) and less than π RADIANS (180°).
See also ACUTE ANGLE, FULL ANGLE, OBTUSE TRIANGLE, REFLEX ANGLE, RIGHT ANGLE, STRAIGHT ANGLE

Obtuse Triangle

An obtuse triangle is a TRIANGLE in which one of the ANGLES is an OBTUSE ANGLE. (Obviously, only a single ANGLE in a TRIANGLE can be OBTUSE or it wouldn't be a TRIANGLE.) A triangle must be either obtuse, ACUTE, or RIGHT.
From the LAW OF COSINES, for a triangle with side lengths a, b, and c,

$$\cos C = \frac{a^2 + b^2 - c^2}{2ab},$$

with C the angle opposite side C. For an angle to be

obtuse, $\cos C < 0$. Therefore, an obtuse triangle satisfies one of $a^2 + b^2 < c^2$, $b^2 + c^2 < a^2$, or $c^2 + a^2 < b^2$.

An obtuse triangle can be dissected into no fewer than seven ACUTE TRIANGLES (Wells 1986, p. 71).

A famous problem is to find the chance that three points picked randomly in a PLANE are the VERTICES of an obtuse triangle (Eisenberg and Sullivan 1996). Unfortunately, the solution of the problem depends on the procedure used to pick the "random" points (Portnoy 1994). In fact, it is impossible to pick random variables which are uniformly distributed in the plane (Eisenberg and Sullivan 1996). Guy (1993) gives a variety of solutions to the problem. Woolhouse (1886) solved the problem by picking uniformly distributed points in the unit DISK, and obtained

$$P_2 = 1 - \left(\frac{4}{\pi^2} - \frac{1}{8}\right) = \frac{9}{8} - \frac{4}{\pi^2} = 0.719715\ldots. \quad (1)$$

The problem was generalized by Hall (1982) to n-D BALL TRIANGLE PICKING, and Buchta (1986) gave closed form evaluations for Hall's integrals.

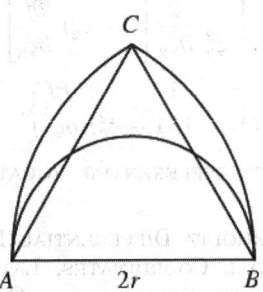

Lewis Carroll (1893) posed and gave another solution to the problem as follows. Call the longest side of a TRIANGLE AB, and call the DIAMETER $2r$. Draw arcs from A and B of RADIUS $2r$. Because the longest side of the TRIANGLE is defined to be AB, the third VERTEX of the TRIANGLE must lie within the region $ABCA$. If the third VERTEX lies within the SEMICIRCLE, the TRIANGLE is an obtuse triangle. If the VERTEX lies *on* the SEMICIRCLE (which will happen with probability 0), the TRIANGLE is a RIGHT TRIANGLE. Otherwise, it is an ACUTE TRIANGLE. The chance of obtaining an obtuse triangle is then the ratio of the AREA of the SEMICIRCLE to that of $ABCA$. The AREA of $ABCA$ is then twice the AREA of a SECTOR minus the AREA of the TRIANGLE.

$$A_{\text{whole figure}} = 2\left(\frac{4\pi r^2}{6}\right) - \sqrt{3}r^2 = r^2\left(\frac{4}{3}\pi - \sqrt{3}\right). \quad (2)$$

Therefore,

$$P = \frac{\frac{1}{2}\pi r^2}{r^2\left(\frac{4}{3}\pi - \sqrt{3}\right)} = \frac{3\pi}{8\pi - 6\sqrt{3}} = 0.63938\ldots. \quad (3)$$

See also ACUTE ANGLE, ACUTE TRIANGLE, BALL TRIANGLE PICKING, OBTUSE ANGLE, RIGHT TRIANGLE, TRIANGLE

References

Buchta, C. "A Note on the Volume of a Random Polytope in a Tetrahedron." *Ill. J. Math.* **30**, 653–659, 1986.

Carroll, L. *Pillow Problems & A Tangled Tale.* New York: Dover, 1976.

Eisenberg, B. and Sullivan, R. "Random Triangles *n* Dimensions." *Amer. Math. Monthly* **103**, 308–318, 1996.

Guy, R. K. "There are Three Times as Many Obtuse-Angled Triangles as There are Acute-Angled Ones." *Math. Mag.* **66**, 175–178, 1993.

Hall, G. R. "Acute Triangles in the *n*-Ball." *J. Appl. Prob.* **19**, 712–715, 1982.

Portnoy, S. "A Lewis Carroll Pillow Problem: Probability on at Obtuse Triangle." *Statist. Sci.* **9**, 279–284, 1994.

Wells, D. *The Penguin Dictionary of Curious and Interesting Numbers.* Middlesex, England: Penguin Books, p. 71, 1986.

Wells, D. G. *The Penguin Book of Interesting Puzzles.* London: Penguin Books, pp. 67 and 248–249, 1992.

Woolhouse, W. S. B. Solution to Problem 1350. *Mathematical Questions, with Their Solutions, from the Educational Times, 1.* London: F. Hodgson and Son, 49–51, 1886.

Ochoa Curve

The ELLIPTIC CURVE

$$3Y^2 = 2X^3 + 386X^2 + 256X - 58195,$$

given in WEIERSTRASS FORM as

$$y^2 = x^3 - 440067x + 106074110.$$

The complete set of solutions to this equation consists of $(x, y) = (-761, 504)$, $(-745, 4520)$, $(-557, 13356)$, $(-446, 14616)$, $(-17, 10656)$, $(91, 8172)$, $(227, 4228)$, $(247, 3528)$, $(271, 2592)$, $(455, 200)$, $(499, 3276)$, $(523, 4356)$, $(530, 4660)$, $(599, 7576)$, $(751, 14112)$, $(1003, 25956)$, $(1862, 75778)$, $(3511, 204552)$, $(5287, 381528)$, $(23527, 3607272)$, $(64507, 16382772)$, $(100102, 31670478)$, and $(1657891, 2134685628)$ (Stroeker and de Weger 1994).

References

Guy, R. K. "The Ochoa Curve." *Crux Math.* **16**, 65–69, 1990.

Ochoa Melida, J. "La ecuacion diofántica $b_0 y^3 - b_1 y^2 + b_2 y - b_3 = z^2$." *Gaceta Math.* 139–141, 1978.

Stroeker, R. J. and de Weger, B. M. M. "On Elliptic Diophantine Equations that Defy Thue's Method: The Case of the Ochoa Curve." *Experiment. Math.* **3**, 209–220, 1994.

Ockham Algebra

References

Blyth, T. S. and Varlet, C. *Ockham Algebras.* Oxford, England: Oxford University Press, 1994.

Octacontagon

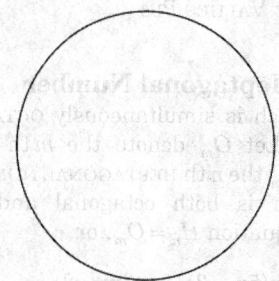

An 80-sided POLYGON.

Octadecagon

An 18-sided POLYGON, sometimes also called an OCTAKAIDECAGON.

See also POLYGON, REGULAR POLYGON, TRIGONOMETRY VALUES PI/18

Octagon

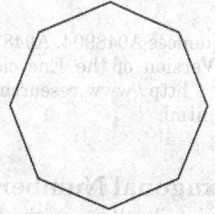

An octagon is an eight-sided POLYGON. The INRADIUS r, CIRCUMRADIUS R, and AREA A of the *regular* octagon can be computed directly from the formulas for a general REGULAR POLYGON with side length s and $n = 8$ sides as

$$r = \tfrac{1}{2} s \cot\left(\frac{\pi}{8}\right) = \tfrac{1}{2}\left(1 + \sqrt{2}\right)s \tag{1}$$

$$R = \tfrac{1}{2}s \csc\left(\frac{\pi}{8}\right) = \tfrac{1}{2}\sqrt{4 + 2\sqrt{2}}s \tag{2}$$

$$A = \tfrac{1}{4}ns^2 \cot\left(\frac{\pi}{8}\right) = 2\left(1 + \sqrt{2}\right)s^2. \tag{3}$$

See also OCTAHEDRON, POLYGON, REGULAR POLYGON, TRIGONOMETRY VALUES PI/8

Octagonal Heptagonal Number

A number which is simultaneously OCTAGONAL and HEPTAGONAL. Let O_m denote the mth OCTAGONAL NUMBER and H_n the nth HEPTAGONAL NUMBER, then a number which is both octagonal and hexagonal satisfies the equation $H_n = O_m$, or

$$\tfrac{1}{2} n(5n - 3) = m(3m - 2). \tag{1}$$

COMPLETING THE SQUARE and rearranging gives

$$3(10n - 3)^2 - 40(3m - 1)^2 = -13. \tag{2}$$

Therefore, defining

$$x \equiv (10n - 3) \tag{3}$$

$$y \equiv 2(3m - 1) \tag{4}$$

gives the second-order Diophantine equation

$$3x^2 - 10y^2 = -13 \tag{5}$$

The first few solutions are $(x, y) = (3, 2)$, $(7, 4)$, $(73, 40)$, $(157, 86)$, ... These give the integer solutions $(1, 1)$, $(345, 315)$, $(166145, 151669)$, ... (Sloane's A048904 and A048905), corresponding to the octagonal heptagonal numbers 1, 297045, 69010153345, ... (Sloane's A048906).

See also HEPTAGONAL NUMBER, OCTAGONAL NUMBER

References

Sloane, N. J. A. Sequences A048904, A048905, and A048906 in "An On-Line Version of the Encyclopedia of Integer Sequences." http://www.research.att.com/~njas/sequences/eisonline.html.

Octagonal Hexagonal Number

A number which is simultaneously OCTAGONAL and HEXAGONAL. Let O_n denote the nth OCTAGONAL NUMBER and H_m the mth HEXAGONAL NUMBER, then a number which is both octagonal and hexagonal satisfies the equation $O_n = H_m$, or

$$n(3n - 2) = m(2m - 1). \tag{1}$$

COMPLETING THE SQUARE and rearranging gives

$$8(3n - 1)^2 - 3(4m - 1)^2 = 5. \tag{2}$$

Therefore, defining

$$x \equiv 2(3n - 1) \tag{3}$$

$$y \equiv 4m - 1 \tag{4}$$

gives the second-order Diophantine equation

$$2x^2 - 3y^2 = 5 \tag{5}$$

The first few solutions are $(x, y) = (2, 1)$, $(4, 3)$, $(16,$

13), (38, 31), (158, 129), (376, 307), ... These give the solutions $(n, m) = (2/3, 1/2)$, $(1, 1)$, $(3, 7/2)$, $(20/3, 8)$, $(80/3, 65/2)$, $(63, 77)$, ..., of which the integer solutions are $(1, 1)$, $(63, 77)$, $(6141, 7521)$, $(601723, 736957)$, ... (Sloane's A046190 and A046191), corresponding to the octagonal hexagonal numbers 1, 11781, 113123361, 1086210502741, ... (Sloane's A046192).

See also HEXAGONAL NUMBER, OCTAGONAL NUMBER, OCTAGONAL PENTAGONAL NUMBER, OCTAGONAL SQUARE NUMBER, OCTAGONAL TRIANGULAR NUMBER

References

Sloane, N. J. A. Sequences A046190, A046191, and A046192 in "An On-Line Version of the Encyclopedia of Integer Sequences." http://www.research.att.com/~njas/sequences/eisonline.html.

Octagonal Number

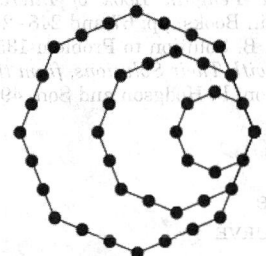

A POLYGONAL NUMBER OF THE FORM $n(3n - 2)$. The first few are 1, 8, 21, 40, 65, 96, 133, 176, ... (Sloane's A000567). The GENERATING FUNCTION for the octagonal numbers is

$$\frac{x(5x + 1)}{(1 - x)^3} = x + 8x^2 + 21x^3 + 40x^4 + \cdots.$$

See also OCTAGONAL HEPTAGONAL NUMBER, OCTAGONAL HEXAGONAL NUMBER, OCTAGONAL PENTAGONAL NUMBER, OCTAGONAL SQUARE NUMBER, OCTAGONAL TRIANGULAR NUMBER

References

Sloane, N. J. A. Sequences A000567/M4493 in "An On-Line Version of the Encyclopedia of Integer Sequences." http://www.research.att.com/~njas/sequences/eisonline.html.

Octagonal Pentagonal Number

A number which is simultaneously OCTAGONAL and PENTAGONAL. Let O_n denote the nth OCTAGONAL NUMBER and P_m the mth PENTAGONAL NUMBER, then a number which is both octagonal and pentagonal satisfies the equation $O_n = P_m$, or

$$n(3n - 2) = \tfrac{1}{2} m(3m - 1). \tag{1}$$

COMPLETING THE SQUARE and rearranging gives

$$(6m - 1)^2 - 8(3n - 1)^2 = -7. \tag{2}$$

Therefore, defining

$$x \equiv (6m - 1) \tag{3}$$

$$y \equiv 2(3n - 1) \tag{4}$$

gives the PELL EQUATION

$$x^2 - 2y^2 = -7. \tag{5}$$

The first few solutions are $(x, y) = (1, 2)$, $(5, 4)$, $(11, 8)$, $(31, 22)$, $(65, 46)$, ... These give the solutions $(n, m) = (1/3, 2/3)$, $(1, 1)$, $(2, 5/3)$, $(16/3, 4)$, $(11, 8)$, ..., of which the integer solutions are $(1, 1)$, $(11, 8)$, $(1025, 725)$, $(12507, 8844)$, ... (Sloane's A046187 and A046188), corresponding to the octagonal pentagonal numbers 1, 176, 1575425, 234631320, 2098015778145, ... (Sloane's A046189).

See also OCTAGONAL NUMBER, PENTAGONAL NUMBER

References

Sloane, N. J. A. Sequences A046187, A046188, and A046188 in "An On-Line Version of the Encyclopedia of Integer Sequences." http://www.research.att.com/~njas/sequences/eisonline.html.

Octagonal Prism

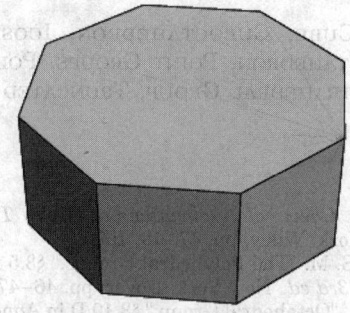

A PRISM composed of octagonal faces. The regular right octagonal prism of unit edge length has SURFACE AREA and VOLUME

$$S = 4\left(3 + \sqrt{2}\right)$$

$$V = 2\left(1 + \sqrt{2}\right).$$

See also PRISM

Octagonal Square Number

A number which is simultaneously OCTAGONAL and SQUARE. Let O_n denote the nth OCTAGONAL NUMBER and T_m the mth SQUARE NUMBER, then a number which is both octagonal and square satisfies the equation $O_n = S_m$, or

$$n(3n - 2) = m^2. \tag{1}$$

COMPLETING THE SQUARE and rearranging gives

$$(3n - 1)^2 - 3m^2 = 1. \tag{2}$$

Therefore, defining

$$x \equiv (3n - 1) \tag{3}$$

$$y \equiv m \tag{4}$$

gives the PELL EQUATION

$$x^2 - 3y^2 = 1 \tag{5}$$

The first few solutions are $(x, y) = (2, 1)$, $(7, 4)$, $(26, 15)$, $(97, 56)$, $(362, 209)$, $(1351, 780)$, ... These give the solutions $(n, m) = (1, 1)$, $(8/3, 4)$, $(9, 15)$, $(98/3, 56)$, $(121, 209)$, ..., of which the integer solutions are $(1, 1)$, $(9, 15)$, $(121, 209)$, $(1681, 2911)$, ... (Sloane's A046184 and A028230), corresponding to the octagonal square numbers 1, 225, 43681, 8473921, 1643897025, ... (Sloane's A036428).

See also OCTAGONAL NUMBER, SQUARE NUMBER

References

Graham, R. L.; Knuth, D. E.; and Patashnik, O. *Concrete Mathematics: A Foundation for Computer Science.* Reading, MA: Addison-Wesley, p. 329, 1990.
Konhauser, J. D. E.; Velleman, D.; and Wagon, S. *Which Way Did the Bicycle Go? And Other Intriguing Mathematical Mysteries.* Washington, DC: Math. Assoc. Amer., p. 104, 1996.
Sloane, N. J. A. Sequences A028230, A036428, and A046184 in "An On-Line Version of the Encyclopedia of Integer Sequences." http://www.research.att.com/~njas/sequences/eisonline.html.

Octagonal Triangular Number

A number which is simultaneously OCTAGONAL and TRIANGULAR. Let O_n denote the nth OCTAGONAL NUMBER and T_m the mth TRIANGULAR NUMBER, then a number which is both octagonal and triangular satisfies the equation $O_n = T_m$, or

$$n(3n - 2) = \tfrac{1}{2}m(m + 1). \tag{1}$$

COMPLETING THE SQUARE and rearranging gives

$$8(3n - 1)^2 - 3(2m + 1)^2 = 5. \tag{2}$$

Therefore, defining

$$x \equiv 2(2n - 1) \tag{3}$$

$$y \equiv 2m + 1 \tag{4}$$

gives the second-order Diophantine equation

$$2x^2 - 3y^2 = 5 \tag{5}$$

The first few solutions are $(x, y) = (2, 1)$, $(4, 3)$, $(16, 13)$, $(38, 31)$, $(158, 129)$, $(376, 307)$, ... These give the solutions $(n, m) = (2/3, 0)$, $(1, 1)$, $(3, 6)$, $(20/3, 15)$, $(80/3, 64)$, $(63, 153)$, ..., of which the integer solutions are $(1, 1)$, $(3, 6)$, $(63, 153)$, $(261, 638)$, $(6141, 15041)$, $(25543, 62566)$, $(601723, 1473913)$, ... (Sloane's

A046181 and A046182), corresponding to the pentagonal hexagonal numbers 1, 21, 11781, 203841, 113123361, ... (Sloane's A046183).

See also HEXAGONAL NUMBER, OCTAGONAL HEXAGONAL NUMBER, PENTAGONAL NUMBER

References

Sloane, N. J. A. Sequences A046181, A046182, and A046183 in "An On-Line Version of the Encyclopedia of Integer Sequences." http://www.research.att.com/~njas/sequences/eisonline.html.

Octagram

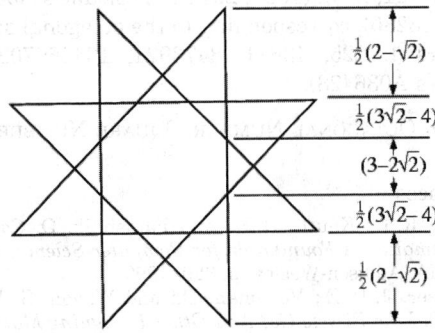

$\frac{1}{2}(2-\sqrt{2})$

$\frac{1}{2}(3\sqrt{2}-4)$

$(3-2\sqrt{2})$

$\frac{1}{2}(3\sqrt{2}-4)$

$\frac{1}{2}(2-\sqrt{2})$

The STAR POLYGON $\{8/3\}$.

Octahedral Graph

A PLATONIC GRAPH on eight nodes. There are 257 topologically distinct octahedral graphs, as first enumerated by Kirkman (1862) and Hermes (1899ab, 1900, 1901; Federico 1969; Duijvestijn and Federico 1981).

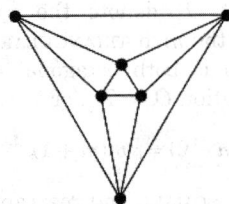

Confusingly, the term "octahedral graph" is also used to refer to the 6-vertex POLYHEDRAL GRAPH having the connectivity of the OCTAHEDRON. It is isomorphic to the CIRCULANT GRAPH $Ci_{1,2}(6)$. Several circular embeddings of this graph are illustrated above. The octahedral graph has 6 nodes, 12 edges, VERTEX CONNECTIVITY 4, EDGE CONNECTIVITY 4, GRAPH DIAMETER 2, GRAPH RADIUS 2, and GIRTH 3. It has CHROMATIC POLYNOMIAL

$$\pi_G(z) = z^6 - 12z^5 + 58z^4 - 137z^3 + 154z^2 - 64z,$$

and CHROMATIC NUMBER 3.

See also CIRCULANT GRAPH, CUBICAL GRAPH, DODECAHEDRAL GRAPH, ICOSAHEDRAL GRAPH, OCTAHEDRON, PLATONIC GRAPH, POLYHEDRAL GRAPH, TETRAHEDRAL GRAPH

References

Bondy, J. A. and Murty, U. S. R. *Graph Theory with Applications.* New York: North Holland, p. 234, 1976.
Duijvestijn, A. J. W. and Federico, P. J. "The Number of Polyhedral (3-Connected Planar) Graphs." *Math. Comput.* **37**, 523–532, 1981.
Federico, P. J. "Enumeration of Polyhedra: The Number of 9-Hedra." *J. Combin. Th.* **7**, 155–161, 1969.
Grünbaum, B. *Convex Polytopes.* New York: Wiley, pp. 288 and 424, 1967.
Hermes, O. "Die Formen der Vielflache. I." *J. reine angew. Math.* **120**, 27–59, 1899a.
Hermes, O. "Die Formen der Vielflache. II." *J. reine angew. Math.* **120**, 305–353, 1899b.
Hermes, O. "Die Formen der Vielflache. III." *J. reine angew. Math.* **122**, 124–154, 1900.
Hermes, O. "Die Formen der Vielflache. IV." *J. reine angew. Math.* **123**, 312–342, 1901.
Kirkman, T. P. "Application of the Theory of the Polyhedra to the Enumeration and Registration of Results." *Proc. Roy. Soc. London* **12**, 341–380, 1862–1863.

Octahedral Group

The POINT GROUP of symmetries of the OCTAHEDRON having order 24 and denoted O_h. It is also the symmetry group of the CUBE, CUBOCTAHEDRON, and TRUNCATED OCTAHEDRON. It has symmetry operations E, $8C_3$, $6C_4$, $6C_2$, $3C_2 = C_4^2$, i, $6S_4$, $8S_6$, $3\sigma_h$, and $6\sigma_4$ (Cotton 1990).

See also CUBE, CUBOCTAHEDRON, ICOSAHEDRAL GROUP, OCTAHEDRON, POINT GROUPS, POLYHEDRAL GROUP, TETRAHEDRAL GROUP, TRUNCATED OCTAHEDRON

References

Cotton, F. A. *Chemical Applications of Group Theory, 3rd ed.* New York: Wiley, pp. 47–49, 1990.
Coxeter, H. S. M. "The Polyhedral Groups." §3.5 in *Regular Polytopes, 3rd ed.* New York: Dover, pp. 46–47, 1973.
Lomont, J. S. "Octahedral Group." §3.10.D in *Applications of Finite Groups.* New York: Dover, p. 81, 1987.

Octahedral Number

A FIGURATE NUMBER which is the sum of two consecutive PYRAMIDAL NUMBERS,

$$O_n = P_{n-1} + P_n = \tfrac{1}{3} n(2n^2 + 1). \tag{1}$$

The first few are 1, 6, 19, 44, 85, 146, 231, 344, 489, 670, 891, 1156, ... (Sloane's A005900). The GENERATING FUNCTION for the octahedral numbers is

$$\frac{x(x+1)^2}{(x-1)^4} = x + 6x^2 + 19x^3 + 44x^4 + \dots. \tag{2}$$

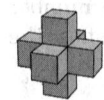

A related set of numbers is the number of cubes in the HAUY CONSTRUCTION of the OCTAHEDRON. Each CROSS SECTION has area

$$S_n = n + 2 \sum_{i=1,3,\ldots,n-2} i = \tfrac{1}{2}(n^2 + 1), \qquad (3)$$

where n is an ODD NUMBER, and adding all CROSS SECTIONS gives

$$HO_k = S_k + 2 \sum_{i=1,3,\ldots,k-2} S_i = \tfrac{1}{6} k = (k^2 + 5), \qquad (4)$$

for k an ODD NUMBER. Re-indexing so that $k = 2n - 1$ gives

$$HO_n = \tfrac{1}{3}(2n-1)(2n^2 - 2n + 3), \qquad (5)$$

the first few values of which are 1, 7, 25, 63, 129, ... (Sloane's A001845). These numbers have the GENERATING FUNCTION

$$f(x) = \frac{(1+x)^3}{(1-x)^4}$$

$$= 1 + 7x + 25x^2 + 63x^3 + 129x^4 + \ldots. \qquad (6)$$

See also HAUY CONSTRUCTION, OCTAHEDRON, TRUNCATED OCTAHEDRAL NUMBER

References

Conway, J. H. and Guy, R. K. *The Book of Numbers.* New York: Springer-Verlag, p. 50, 1996.

Sloane, N. J. A. Sequences A001845/M4384 and A005900/M4128 in "An On-Line Version of the Encyclopedia of Integer Sequences." http://www.research.att.com/~njas/sequences/eisonline.html.

Octahedron

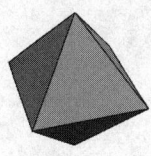 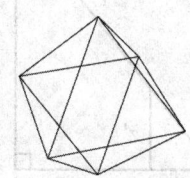 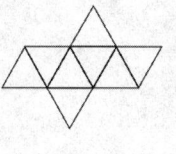

 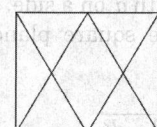 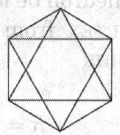 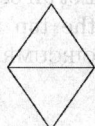

The PLATONIC SOLID P_3 with six VERTICES, 12 EDGES, and eight equivalent EQUILATERAL TRIANGULAR faces, $8\{3\}$. It is also UNIFORM POLYHEDRON U_5 and Wen-

ninger model W_2. It is given by the SCHLÄFLI SYMBOL $\{3, 4\}$ and WYTHOFF SYMBOL $4|23$.

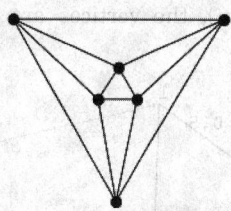

The octahedron of unit side length is the ANTIPRISM of $n = 3$ sides with height $h = \sqrt{6}/3$. The DUAL POLYHEDRON of the octahedron is the CUBE. Like the CUBE, it has the O_h OCTAHEDRAL GROUP of symmetries. The connectivity of the vertices is given by the OCTAHEDRAL GRAPH.

The octahedron has a single STELLATION: the STELLA OCTANGULA. The solid bounded by the two TETRAHEDRA of the STELLA OCTANGULA (left figure) is an octahedron (right figure; Ball and Coxeter 1987).

The following table gives polyhedra which can be constructed by CUMULATION of an octahedron by pyramids of given heights h.

h	$(r+h)/h$	Result
$\sqrt{3} - \tfrac{2}{3}\sqrt{6}$	$5 - 3\sqrt{2}$	SMALL TRIAKIS OCTAHEDRON
$\tfrac{1}{3}\sqrt{6}$	3	STELLA OCTANGULA

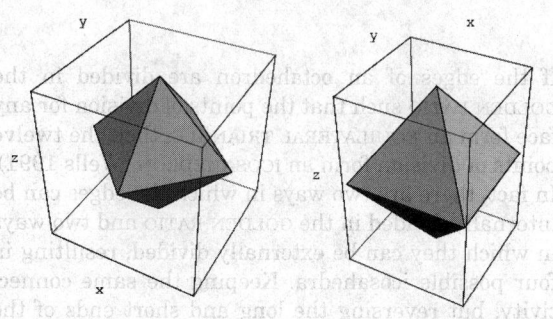

In one orientation (left figure), the VERTICES are given by $(\pm 1, 0, 0)$, $(0, \pm 1, 0)$, $(0, 0, \pm 1)$. In another orientation (right figure), the vertices are $(\pm 1, \pm 1, 0)$ and $(0, 0, \pm \sqrt{2})$.

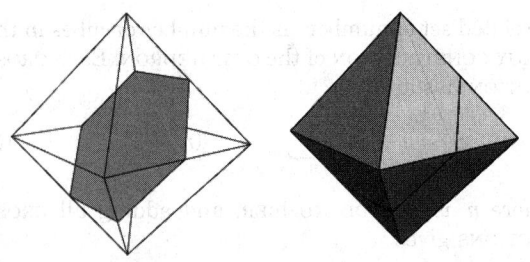

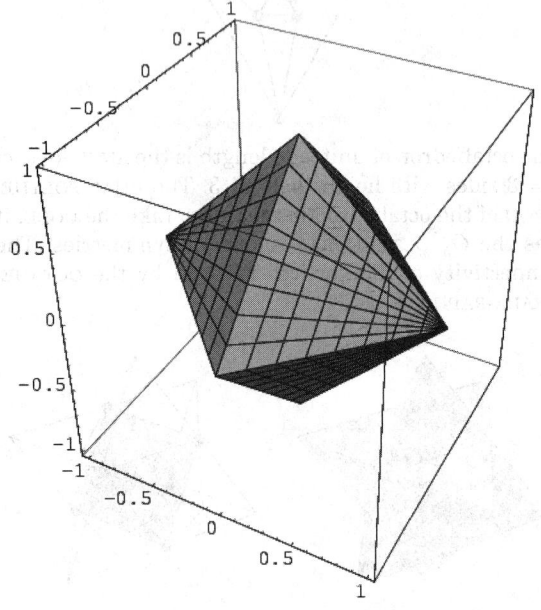

A plane PERPENDICULAR to a C_3 axis of an octahedron cuts the solid in a regular HEXAGONAL CROSS SECTION (Holden 1991, pp. 22–23). Since there are four such axes, there are four possible HEXAGONAL CROSS SECTIONS.

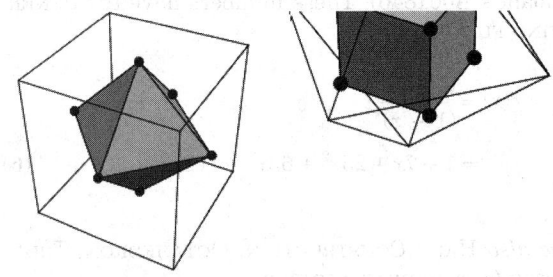

The face planes are $\pm x \pm y \pm z = 1$, so a solid octahedron is given by the equation

$$|x| + |y| + |z| \leq 1. \tag{1}$$

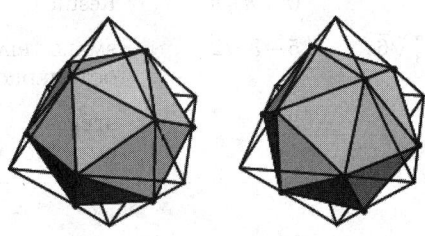

The centers of the faces of an octahedron form a CUBE, and the centers of the faces of a CUBE form an octahedron (Steinhaus 1983, pp. 194–195). Faceted forms of the octahedron include the CUBOCTATRUNCATED CUBOCTAHEDRON and TETRAHEMIHEXAHEDRON.

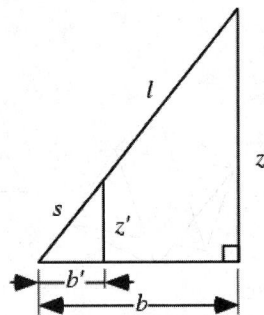

If the edges of an octahedron are divided in the GOLDEN RATIO such that the points of division for any face form an EQUILATERAL TRIANGLE, then the twelve points of division form an ICOSAHEDRON (Wells 1991). In fact, there are two ways in which the edges can be internally divided in the GOLDEN RATIO and two ways in which they can be externally divided, resulting in four possible icosahedra. Keeping the same connectivity, but reversing the long and short ends of the division gives JESSEN'S ORTHOGONAL ICOSAHEDRON.

Let an octahedron be length a on a side. The height of the top VERTEX from the square plane is also the CIRCUMRADIUS

$$R = \sqrt{a^2 - d^2}, \tag{2}$$

where

$$d = \tfrac{1}{2}\sqrt{2}a \tag{3}$$

is the diagonal length, so

$$R = \sqrt{a^2 - \tfrac{1}{2} a^2} = \tfrac{1}{2} \sqrt{2} a \approx 0.70710a. \tag{4}$$

Now compute the INRADIUS.

$$\ell = \tfrac{1}{2} \sqrt{3} a \tag{5}$$

$$b = \tfrac{1}{2} a \tag{6}$$

$$s = \tfrac{1}{2} a \tan 30° = \frac{a}{2\sqrt{3}}, \tag{7}$$

so

$$\frac{s}{\ell} = \frac{1}{2\sqrt{3}} \frac{2}{\sqrt{3}} = \tfrac{1}{3}. \tag{8}$$

Use similar TRIANGLES to obtain

$$b' = \frac{s}{\ell} b = \tfrac{1}{6} a \tag{9}$$

$$z' = \frac{s}{\ell} z = \frac{a}{3\sqrt{2}} \tag{10}$$

$$x = b - b' = \tfrac{1}{2} a - \tfrac{1}{6} a = \tfrac{1}{3} a, \tag{11}$$

so the INRADIUS is

$$r = \sqrt{x^2 - z'^2} = a\sqrt{\tfrac{1}{9} + \tfrac{1}{18}} = \tfrac{1}{6}\sqrt{6} a \approx 0.40824a, \tag{12}$$

and twice the INRADIUS gives the height of the octahedron viewed as a 3-sided ANTIPRISM. The MIDRADIUS of the octahedron is

$$\rho = \tfrac{1}{2} a = 0.5a. \tag{13}$$

The AREA of one face is the AREA of an EQUILATERAL TRIANGLE

$$A = \tfrac{1}{4} \sqrt{3} a^2. \tag{14}$$

The volume is two times the volume of a square-base pyramid,

$$V = 2\left(\tfrac{1}{3} a^2 R\right) = 2\left(\tfrac{1}{3}\right)(a^2)\left(\tfrac{1}{2}\sqrt{2}a\right) = \tfrac{1}{3}\sqrt{2}a^3. \tag{15}$$

The DIHEDRAL ANGLE is

$$\alpha = \cos^{-1}\left(-\tfrac{1}{3}\right) \approx 109.47°. \tag{16}$$

The octahedron can be built using a HAUY CONSTRUCTION. The Hauy octahedral numbers

$$HO_n = \tfrac{1}{3}(2n - 1)(2n^2 - 2n + 3) \tag{17}$$

give another method for calculating the VOLUME of the octahedron,

$$V = \lim_{n\to\infty} HO_n \left(\frac{a}{n\sqrt{2}}\right)^3 = \tfrac{1}{3}\sqrt{2}a^3, \tag{18}$$

in agreement with the result derived above.

See also ANTIPRISM, DÜRER'S SOLID, HAUY CONSTRUCTION, ICOSAHEDRON, JUMPING OCTAHEDRON, OCTAHEDRAL GRAPH, OCTAHEDRAL GROUP, OCTAHEDRON 3-COMPOUND, OCTAHEDRON 5-COMPOUND, PLATONIC SOLID, STELLA OCTANGULA, TRUNCATED OCTAHEDRON

References

Beyer, W. H. *CRC Standard Mathematical Tables, 28th ed.* Boca Raton, FL: CRC Press, p. 228, 1987.

Cundy, H. and Rollett, A. "Octahedron. 3^4." §3.5.3 in *Mathematical Models, 3rd ed.* Stradbroke, England: Tarquin Pub., p. 64, 1989.

Davie, T. "The Octahedron." http://www.dcs.st-and.ac.uk/~ad/mathrecs/polyhedra/octahedron.html.

Harris, J. W. and Stocker, H. "Octahedron." §4.4.4 in *Handbook of Mathematics and Computational Science.* New York: Springer-Verlag, p. 100, 1998.

Holden, A. *Shapes, Space, and Symmetry.* New York: Dover, 1991.

Steinhaus, H. *Mathematical Snapshots, 3rd ed.* New York: Dover, pp. 193–195, 1999.

Wells, D. *The Penguin Dictionary of Curious and Interesting Geometry.* London: Penguin, p. 163, 1991.

Wenninger, M. J. "The Octahedron." Model 2 in *Polyhedron Models.* Cambridge, England: Cambridge University Press, p. 15, 1989.

Octahedron 3-Compound

A POLYHEDRON COMPOUND consisting of three octahedra.

See also OCTAHEDRON, OCTAHEDRON 5-COMPOUND

Octahedron 5-Compound

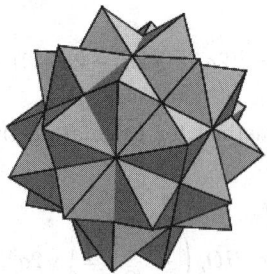

A POLYHEDRON COMPOUND composed of five OCTAHE-DRA occupying the VERTICES of an ICOSAHEDRON. The 30 VERTICES of the compound form an ICOSIDODECA-HEDRON (Ball and Coxeter 1987), and the solid is one of the ICOSAHEDRON STELLATIONS (Wenninger 1983). The octahedron 5-compound is the dual of the CUBE 5-COMPOUND.

Constructing the octahedra as the duals of the CUBE 5-COMPOUND where the cubes have unit edge lengths give a solid with edge lengths

$$s_1 = \sqrt{\tfrac{1}{5}\left(3 - \sqrt{5}\right)} \tag{1}$$

$$s_2 = 2\sqrt{\tfrac{1}{5}\left(7 - 3\sqrt{5}\right)} \tag{2}$$

$$s_3 = \sqrt{7 - 3\sqrt{5}} \tag{3}$$

$$s_4 = 3 - \sqrt{5}. \tag{4}$$

The CIRCUMRADIUS is

$$R = 1, \tag{5}$$

and the SURFACE AREA and VOLUME are

$$S = 20\sqrt{3} \tag{6}$$

$$V = \tfrac{20}{3}. \tag{7}$$

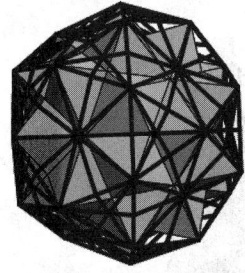 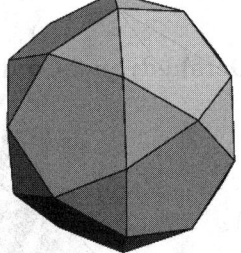

The CONVEX HULL of the octahedron 5-compound is the ICOSIDODECAHEDRON.

See also CUBE 5-COMPOUND, CUBE 5-COMPOUND–OCTAHEDRON 5-COMPOUND, ICOSAHEDRON STELLA-TIONS, ICOSIDODECAHEDRON, OCTAHEDRON, OCTAHE-DRON 3-COMPOUND, OCTAHEDRON 6-COMPOUND, POLYHEDRON COMPOUND, STELLA OCTANGULA

References

Ball, W. W. R. and Coxeter, H. S. M. *Mathematical Recrea-tions and Essays, 13th ed.* New York: Dover, pp. 135 and 137, 1987.
Cundy, H. and Rollett, A. "Five Octahedra About in Icosahedron." §3.10.7 in *Mathematical Models, 3rd ed.* Stradbroke, England: Tarquin Pub., pp. 137–138, 1989.
Wenninger, M. J. *Dual Models.* Cambridge, England: Cam-bridge University Press, p. 55, 1983.
Wenninger, M. J. "Compound of Five Octahedra." §23 in *Polyhedron Models.* New York: Cambridge University Press, p. 43, 1989.

Octahedron 6-Compound

See also OCTAHEDRON, OCTAHEDRON 3-COMPOUND, OCTAHEDRON 5-COMPOUND

Octahedron Stellation

STELLA OCTANGULA

Octahemioctacron

The DUAL POLYHEDRON of the OCTAHEMIOCTAHEDRON U_3 and Wenninger dual W_{68}. When rendered, the octahemioctacron and HEXAHEMIOCTACRON appear the same.

See also DUAL POLYHEDRON, HEXAHEMIOCTACRON, OCTAHEMIOCTAHEDRON, UNIFORM POLYHEDRON

Octahemioctahedron

References

Wenninger, M. J. *Dual Models.* Cambridge, England: Cambridge University Press, p. 104, 1983.

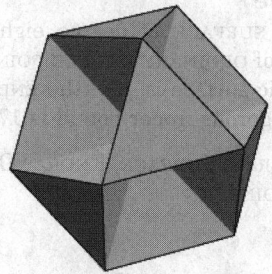

The UNIFORM POLYHEDRON U_3, also called the OCTA-TETRAHEDRON, whose DUAL POLYHEDRON is the OCTA-HEMIOCTACRON. It has WYTHOFF SYMBOL $\frac{3}{2}3|3$. Its faces are $8\{3\} + 4\{6\}$. It is a FACETED CUBOCTAHEDRON. For unit edge length, its CIRCUMRADIUS is

$$R = 1.$$

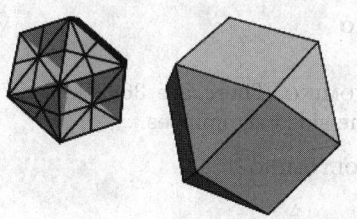

The CONVEX HULL of the octahemioctahedron is the CUBOCTAHEDRON.

References

Wenninger, M. J. *Polyhedron Models.* Cambridge, England: Cambridge University Press, p. 103, 1989.

Octakaidecagon

OCTADECAGON

Octal

The base 8 notational system for representing REAL NUMBERS. The digits used are 0, 1, 2, 3, 4, 5, 6, and 7, so that 8_{10} (8 in base 10) is REPRESENTED AS 10_8 ($10 = 1 \cdot 8^1 + 0 \cdot 8^0$) in base 8. The following table gives the octal equivalents of the first few decimal numbers.

1	1	11	13	21	25
2	2	12	14	22	26
3	3	13	15	23	27
4	4	14	16	24	30
5	5	15	17	25	31
6	6	16	20	26	32
7	7	17	21	27	33
8	10	18	22	28	34
9	11	19	23	29	35
10	12	20	24	30	36

See also BASE (NUMBER), BINARY, DECIMAL, HEXADECIMAL, QUATERNARY, TERNARY

References

Lauwerier, H. *Fractals: Endlessly Repeated Geometric Figures.* Princeton, NJ: Princeton University Press, pp. 9–10, 1991.

Weisstein, E. W. "Bases." MATHEMATICA NOTEBOOK BASES.M.

Wells, D. *The Penguin Dictionary of Curious and Interesting Numbers.* Middlesex, England: Penguin Books, pp. 72–73, 1986.

Octant

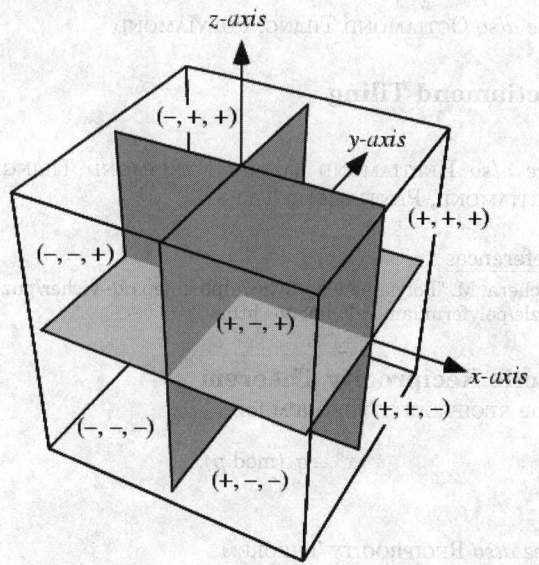

One of the eight regions of SPACE defined by the eight possible combinations of SIGNS $(\pm, \pm, \pm)$ for x, y, and z.

See also QUADRANT

Octatetracontagon

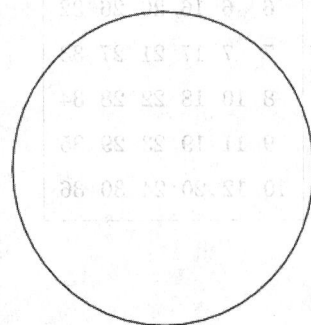

A 48-faced POLYGON.

See also DISDYAKIS DODECAHEDRON, GREAT RHOMBI-CUBOCTAHEDRON (ARCHIMEDEAN)

Octatetrahedron

OCTAHEMIOCTAHEDRON

Octave

A multiple of 2. The word should really be something like "bicade" (by analogy with DECADE) but the "oct" embedded in the stem of the word derives historically to the fact that eight notes correspond to a factor of two in frequency.

See also DECADE

Octiamond

An 8-POLYIAMOND.

See also OCTIAMOND TILING, POLYIAMOND

Octiamond Tiling

See also HEPTIAMOND TILING, HEXIAMOND TILING, OCTIAMOND, PENTIAMOND TILING

References

Vichera, M. "Polyiamonds." http://alpha.ujep.cz/~vicher/puzzle/polyform/iamond/iamonds.htm.

Octic Reciprocity Theorem

The RECIPROCITY THEOREM for

$$x^8 \equiv q \pmod{p}.$$

See also RECIPROCITY THEOREM

References

Aigner, A. "Kriterien zum 8. und 16. Potenzcharakter der Reste 2 und −2." *Deutsche Math.* **4**, 44–52, 1939.

Hasse, H. "Der 2^n-te Potenzcharakter von 2 im Koerper der 2^n-ten Einheitswurzeln." *Rend. Circ. Matem. Palermo* **7**, 185–243, 1958.
Whiteman, A. L. " The Sixteenth Power Residue Character of 2." *Canad. J. Math.* **6**, 364–373, 1954.

Octic Surface

An ALGEBRAIC SURFACE of degree eight. The maximum number of ORDINARY DOUBLE POINTS known to exist on an octic surface is 168 (the ENDRAß OCTICS), although the rigorous upper bound is 174.

See also ALGEBRAIC SURFACE, ENDRAß OCTIC, ORDINARY DOUBLE POINT

Octillion

In the American system, 10^{27}.

See also LARGE NUMBER

Octodecillion

In the American system, 10^{57}.

See also LARGE NUMBER

Octomino

An 8-POLYOMINO. There are 369 FREE, 2725 FIXED, and 704 one-sided octominoes.

See also POLYOMINO

Octonion

CAYLEY NUMBER

Octothorpe

The number sign # sometimes used in mathematics to indicate the number of a quantity satisfying some condition, e.g., $\#\{n : n > 1\}$. The symbol is also used to denote a PRIMORIAL.

References

Bringhurst, R. *The Elements of Typographic Style, 2nd ed.* Point Roberts, WA: Hartley and Marks, p. 282, 1997.

Odd Divisor Function

The sum of powers of ODD DIVISORS of a number. It is the analog of the DIVISOR FUNCTION for odd divisors only and is written $\sigma_k^{(o)}(n)$. For the case $k = 1$,

$$\sigma_1^{(o)}(n) = \sigma_1(n) - 2\sigma_1(n/2),$$

where $\sigma_k(n/2)$ is defined to be 0 if n is ODD. The following table gives the first few $\sigma_k^{(o)}(n)$.

k	Sloane	$\sigma_k^{(o)}(n)$
0	A001227	1, 1, 2, 1, 2, 2, 2, 1, 3, 2, ...
1	A000593	1, 1, 4, 1, 6, 4, 8, 1, 13, 6, ...

2	A050999	1, 1, 10, 1, 26, 10, 50, 1, 91, 26, ...
3	A051000	1, 1, 28, 1, 126, 28, 344, 1, 757, 126, ...
4	A051001	1, 1, 82, 1, 626, 82, 2402, 1, 6643, 626, ...
5	A051002	1, 1, 244, 1, 3126, 244, 16808, 1, 59293, 3126, ...

This function arises in Ramanujan's EISENSTEIN SERIES $L(q)$ and in a RECURRENCE RELATION for the PARTITION FUNCTION P.

See also DIVISOR FUNCTION, EVEN DIVISOR FUNCTION

References

Dickson, L. E. *History of the Theory of Numbers, Vol. 1: Divisibility and Primality.* New York: Chelsea, p. 306, 1952.

Hirzebruch, F. *Manifolds and Modular Forms, 2nd ed.* Braunschweig, Germany: Vieweg, p. 133, 1994.

Riordan, J. *Combinatorial Identities.* New York: Wiley, p. 187, 1979.

Sloane, N. J. A. Sequences A000593/M3197, A001227, A050999, A051000, A051001, and A051002 in "An On-Line Version of the Encyclopedia of Integer Sequences." http://www.research.att.com/~njas/sequences/eisonline.html.

Verhoeff, T. "Rectangular and Trapezoidal Arrangements." *J. Integer Sequences* **2**, #99.1.6, 1999.

Odd Function

An odd function is a function for which $f(x) = -f(-x)$. An EVEN FUNCTION times an odd function is odd.

Odd Graph

An odd graph O_n is a graph having vertices given by the $n-1$-subsets of $\{1, \ldots, 2n-1\}$ such that two vertices are connected by an edge IFF the associated subsets are disjoint (Biggs 1974). The number of nodes in O_n is therefore $\binom{2n-1}{n-1}$, where $\binom{n}{k}$ is a BINOMIAL COEFFICIENT. For $n = 1, 2, \ldots$, the first few values are 1, 3, 10, 35, 126, ... (Sloane's A001700).

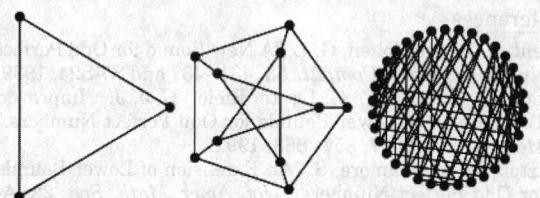

O_2 is isomorphic to the COMPLETE GRAPH K_3, and O_3 is the PETERSEN GRAPH (Skiena 1990, p. 162).

See also COMPLETE GRAPH, ODD NODE, PETERSEN GRAPH

References

Biggs, N. L. *Algebraic Graph Theory, 2nd ed.* Cambridge, England: Cambridge University Press, 1993.

Skiena, S. *Implementing Discrete Mathematics: Combinatorics and Graph Theory with Mathematica.* Reading, MA: Addison-Wesley, 1990.

Sloane, N. J. A. Sequences A001700/M2848 in "An On-Line Version of the Encyclopedia of Integer Sequences." http://www.research.att.com/~njas/sequences/eisonline.html.

Odd Node

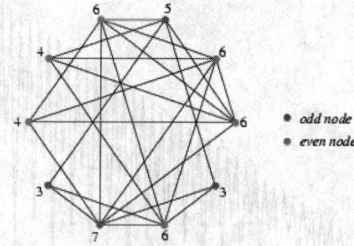

A NODE in a GRAPH is said to be an odd node if its VERTEX DEGREE is ODD.

See also EVEN NODE, GRAPH, NODE (GRAPH), ODD GRAPH, VERTEX DEGREE

Odd Number

An INTEGER OF THE FORM $N = 2n + 1$, where n is an INTEGER. The odd numbers are therefore $\ldots, -3, -1, 1, 3, 5, 7, \ldots$ (Sloane's A005408), which are also the GNOMONIC NUMBERS. The GENERATING FUNCTION for the odd numbers is

$$\frac{x(1+x)}{(x-1)^2} = x + 3x^2 + 5x^3 + 7x^4 + \ldots.$$

Since the odd numbers leave a remainder of 1 when divided by two, $N \equiv 1 \pmod{2}$ for odd N. Integers which are not odd are called EVEN.

See also EVEN NUMBER, GNOMONIC NUMBER, NICOMACHUS'S THEOREM, ODD NUMBER THEOREM, ODD PRIME

References

Commission on Mathematics of the College Entrance Examination Board. *Informal Deduction in Algebra: Properties of Odd and Even Numbers.* Princeton, NJ, 1959.

Sloane, N. J. A. Sequences A005408/M2400 in "An On-Line Version of the Encyclopedia of Integer Sequences." http://www.research.att.com/~njas/sequences/eisonline.html.

Odd Number Theorem

The sum of the first n ODD NUMBERS is a SQUARE NUMBER:

$$\sum_{k=1}^{n} (2k-1) = 2 \sum_{k=1}^{n} k - \sum_{k=1}^{n} 1 = 2 \left[\frac{n(n+1)}{2} \right] - n$$

$$= n(n+1) - n = n^2.$$

See also NICOMACHUS'S THEOREM, ODD NUMBER

Odd Order Theorem

FEIT-THOMPSON THEOREM

Odd Part

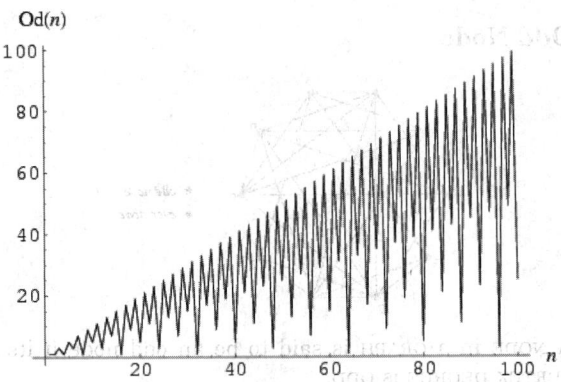

The odd part Od(n) of a positive integer n is defined by

$$\mathrm{Od}(n) = \frac{n}{2^{b(n)}},$$

where $b(n)$ is the exponent of the exact power of 2 dividing n. Od(n) is therefore the product of odd factors of n. The values for $n = 1, 2, ...,$ are 1, 1, 3, 1, 5, 3, 7, 1, 9, 5, 11, ... (Sloane's A000265). The odd part function can be implemented in *Mathematica* as

```
OddPart[n_Integer]                          := n/
2^IntegerExponent[n,2]
```

See also EVEN PART, GREATEST DIVIDING EXPONENT

References

"Problem H-81." *Fib. Quart.* **6**, 52, 1968.

Sloane, N. J. A. Sequences A000265/M2222 in "An On-Line Version of the Encyclopedia of Integer Sequences." http://www.research.att.com/~njas/sequences/eisonline.html.

Odd Perfect Number

In Book IX of *The Elements,* Euclid gave a method for constructing PERFECT NUMBERS (Dickson 1957, p. 3), although this method applies only to *even* perfect numbers. In a 1638 letter to Mersenne, Descartes proposed that every even perfect number is of Euclid's form, and stated that he saw no reason why an odd perfect number could not exist (Dickson 1957, p. 12). Descartes was therefore among the first to consider the existence off odd perfect numbers; prior to Descartes, many authors had implicitly assumed (without proof) that the perfect numbers generated by Euclid's construction comprised all possible perfect numbers (Dickson 1957, pp. 6–12). In 1657, Frenicle repeated Descartes' belief that every even perfect number is of Euclid's form and that there was no reason odd perfect could not exist. Like Frenicle, Euler also considered odd perfect numbers.

To this day, it is not known if any odd perfect numbers exist, although numbers up to 10^{300} have been checked without success, making the existence of odd perfect numbers appear unlikely (Brent *et al.* 1991; Guy 1994, p. 44). The following table summarizes the development of ever-higher bounds for the smallest possible odd perfect number.

author	bound
Kanold (1957)	10^{20}
Tuckerman (1973)	10^{36}
Hagis (1973)	10^{50}
Brent and Cohen (1989)	10^{160}
Brent *et al.* (1991)	10^{300}

Euler showed that an odd perfect number, if it exists, must be OF THE FORM

$$m = p^{4\lambda+1}Q^2, \tag{1}$$

where p is a prime of the form $4n + 1$, a result similar to that derived by Frenicle in 1657 (Dickson 1957, pp. 14 and 19). In 1887, Sylvester conjectured and in 1925, Gradshtein proved that any odd perfect number must have at least six different prime aliquot factors (Ball and Coxeter 1987). If it is not divisible by 3, an odd perfect number must then have at least 11 different prime factors (Hagis 1983). Catalan (1888) proved that if an ODD perfect number is not divisible by 3, 5, or 7, it has at least 26 distinct prime aliquot factors. Stuyvaert (1896) proved that an odd perfect number must be a sum of squares.

See also ODD NUMBER, PERFECT NUMBER

References

Brent, R. P. and Cohen, G. L. "A New Bound for Odd Perfect Numbers." *Math. Comput.* **53**, 431–437 and S7–S24, 1989.

Brent, R. P.; Cohen, G. L.; te Riele, H. J. J. "Improved Techniques for Lower Bounds for Odd Perfect Numbers." *Math. Comput.* **57**, 857–868, 1991.

Buxton, M. and Elmore, S. "An Extension of Lower Bounds for Odd Perfect Numbers." *Not. Amer. Math. Soc.* **22**, A-55, 1976.

Buxton, M. and Stubblefield, B. "On Odd Perfect Numbers." *Not. Amer. Math. Soc.* **22**, A-543, 1975.

Dickson, L. E. *History of the Theory of Numbers, Vol. 1: Divisibility and Primality.* New York: Chelsea, pp. 3–33, 1952.

Guy, R. K. "Perfect Numbers." §B1 in *Unsolved Problems in Number Theory, 2nd ed.* New York: Springer-Verlag, pp. 44–45, 1994.

Hagis, P. Jr. "A Lower Bound for the Set of Odd Perfect Numbers." *Math. Comput.* **27**, 951–953, 1973.

Hagis, P. Jr. "An Outline of a Proof that Every Odd Perfect Number has at Least Eight Prime Factors." *Math. Comput.* **34**, 1027–1032, 1980.

Hagis, P. Jr.; and Cohen, G. L. "Every Odd Perfect Number Has a Prime Factor Which Exceeds 10^6." *Math. Comput.* **67**, 1323–1330, 1998.

Heath-Brown, D. R. "Odd Perfect Numbers." *Math. Proc. Cambridge Philos. Soc.* **115**, 191–196, 1994.

Iannucci, D. E. "The Second Largest Prime Divisor of an Odd Perfect Number Exceeds Ten Thousand." *Math. Comput.* **68**, 1749–1760, 1999.

Iannucci, D. E. "The Third Largest Prime Divisor of an Odd Perfect Number Exceeds One Hundred." *Math. Comput.* **69**, 867–879, 2000.

Kanold, H.-J. "Über mehrfach vollkommene Zahlen. II." *J. reine angew. Math.* **197**, 82–96, 1957.

Subbarao, M. V. "Odd Perfect Numbers: Some New Issues." *Period. Math. Hungar.* **38**, 103–109, 1999.

Tuckerman, B. "Odd Perfect Numbers: A Search Procedure, and a New Lower Bound of 10^{36}." *Not. Amer. Math. Soc.* **15**, 226, 1968.

Tuckerman, B. "A Search Procedure and Lower Bound for Odd Perfect Numbers." *Math. Comp.* **27**, 943–949, 1973.

Odd Prime

Any PRIME NUMBER other than 2 (which is the unique EVEN PRIME).

See also EVEN PRIME, PRIME NUMBER

References

Wells, D. *The Penguin Dictionary of Curious and Interesting Numbers.* Middlesex, England: Penguin Books, p. 44, 1986.

Odd Sequence

A SEQUENCE of n 0s and 1s is called an odd sequence if each of the n SUMS $\sum_{i=1}^{n-k} a_i a_{i+k}$ for $k = 0, 1, ..., n-1$ is odd.

References

Guy, R. K. "Odd Sequences." §E38 in *Unsolved Problems in Number Theory, 2nd ed.* New York: Springer-Verlag, pp. 238–239, 1994.

Odd Triple

TWO-GRAPH

Odds

Betting odds are written in the form $r:s$ (and correspond to the probability of winning $P = s/(r+s)$. Therefore, given a probability P, the odds of winning are $(1/P) - 1 : 1$.

See also FRACTION, RATIO, RATIONAL NUMBER

References

Kraitchik, M. "The Horses." §6.17 in *Mathematical Recreations.* New York: W. W. Norton, pp. 134–135, 1942.

ODE

ORDINARY DIFFERENTIAL EQUATION

Oesterlé-Masser Conjecture

ABC CONJECTURE

Of Order

ASYMPTOTIC NOTATION

Of Shape

OF THE FORM

Of the Form

An expression that is of a given type. For example, all primes $p > 3$ are "of the form" $6n \pm 1$. The term "of shape" is sometimes also used.

See also REPRESENTED AS

References

Wells, D. *The Penguin Dictionary of Curious and Interesting Numbers.* Middlesex, England: Penguin Books, p. 13, 1986.

Offset Curves

PARALLEL CURVES

Offset Rings

SURFACE OF REVOLUTION

Ogive

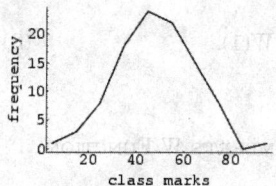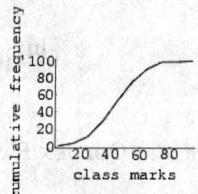

Any continuous cumulative frequency curve, such as the one illustrated above in the right figure.

See also FREQUENCY POLYGON, HISTOGRAM

References

Kenney, J. F. and Keeping, E. S. "Ogive Curves." §2.7 in *Mathematics of Statistics, Pt. 1, 3rd ed.* Princeton, NJ: Van Nostrand, pp. 29–31, 1962.

Oldknow Points

The PERSPECTIVE CENTERS of a triangle and the TANGENTIAL TRIANGLES of its inner and outer SODDY CIRCLES, given by

$$Ol = I + 2Ge$$

$$Ol' = I - 2Ge,$$

where I is the INCENTER and Ge is the GERGONNE POINT.

See also GERGONNE POINT, INCENTER, PERSPECTIVE CENTER, SODDY CIRCLES, TANGENTIAL TRIANGLE

Oliveira's Minimal Surface

References

Oldknow, A. "The Euler-Gergonne-Soddy Triangle of a Triangle." *Amer. Math. Monthly* **103**, 319–329, 1996.

Oliveira's Minimal Surface

See also MINIMAL SURFACE

Oloid

References

Capocasa, C. "Oloid." http://www.blackpoint.net/capocssa/oloid.html.
Schatz, P. "Das Oloid als Wälzkörper." §14 in *Rythmus-forschung und Technik.* Stuttgart: Verlag Freies Geiste-sleben, p. 122, 1975.

Omega Constant

$$W(1) \equiv 0.5671432904\ldots, \tag{1}$$

where $W(x)$ is LAMBERT'S W-FUNCTION. It is available in *Mathematica* using the function ProductLog[1]. $W(1)$ can be considered a sort of "GOLDEN RATIO" for exponentials since

$$\exp[-W(1)] = W(1), \tag{2}$$

giving

$$\ln\left[\frac{1}{W(1)}\right] = W(1). \tag{3}$$

See also GOLDEN RATIO, LAMBERT'S W-FUNCTION

References

Plouffe, S. "The Omega Constant or $W(1)$." http://www.lacim.uqam.ca/piDATA/omega.txt.

Omega Function

LAMBERT'S W-FUNCTION

Omino

POLYOMINO

Omnific Integer

The appropriate notion of INTEGER for SURREAL NUMBERS.

See also SURREAL NUMBER

O'Nan Group

The SPORADIC GROUP $O'N$.

References

Wilson, R. A. "ATLAS of Finite Group Representation." http://for.mat.bham.ac.uk/atlas/html/ON.html.

Onduloid

UNDULOID

One

1

One-Form

A linear real-valued FUNCTION ω^1 of VECTORS $\mathbf{v}$ such that $\omega^1(\mathbf{v}) \mapsto \mathbb{R}$. VECTORS (i.e., CONTRAVARIANT VECTORS or "KETS" $|\psi\rangle$) and one-forms (i.e., COVARIANT VECTORS or "BRAS" $\langle\phi|$) are DUAL to each other. Therefore

$$\omega^1(\mathbf{v}) \equiv \mathbf{v}\big(\omega^1\big) \equiv \big\langle \omega^1, \ \mathbf{v} \big\rangle = \langle\phi|\psi\rangle.$$

The operation of applying the one-form to a VECTOR $\omega^1(\mathbf{v})$ is called CONTRACTION.

See also ANGLE BRACKET, BRA, CONTRAVARIANT VECTOR, COVARIANT VECTOR, DIFFERENTIAL K-FORM, KET, MEROMORPHIC ONE-FORM, TWO-FORM, VECTOR, ZERO-FORM

One-Mouth Theorem

Except for convex polygons, every SIMPLE POLYGON has at least one MOUTH.

See also MOUTH, PRINCIPAL VERTEX, TWO-EARS THEOREM

References

Toussaint, G. "Anthropomorphic Polygons." *Amer. Math. Monthly* **122**, 31–35, 1991.

One-Ninth Constant

N.B. A detailed online essay by S. Finch was the starting point for this entry.

Let $\lambda_{m,\,n}$ be CHEBYSHEV CONSTANTS. Schönhage (1973) proved that

$$\lim_{n\to\infty}\big(\lambda_{0,\,n}\big)^{1/n} = \tfrac{1}{3}. \tag{1}$$

It was conjectured that

$$\Lambda \equiv \lim_{n\to\infty}\big(\lambda_{n,\,n}\big)^{1/n} = \tfrac{1}{9}. \tag{2}$$

Carpenter *et al.* (1984) obtained

$$\Lambda = 0.1076539192\ldots \tag{3}$$

numerically. Gonchar and Rakhmanov (1980) showed that the limit exists and disproved the 1/9 conjecture, showing that Λ is given by

$$\Lambda = \exp\left[-\frac{\pi K\left(\sqrt{1-c^2}\right)}{K(c)}\right], \tag{4}$$

where K is the complete ELLIPTIC INTEGRAL OF THE FIRST KIND, and $c = 0.9089085575485414\ldots$ is the PARAMETER which solves

$$K(k) = 2E(k), \tag{5}$$

and E is the complete ELLIPTIC INTEGRAL OF THE SECOND KIND. This gives the value for Λ computed by Carpenter *et al.* (1984) Λ is also given by the unique POSITIVE ROOT of

$$f(z) = \tfrac{1}{8}, \tag{6}$$

where

$$f(z) \equiv \sum_{j=1}^{\infty} a_j z^j \tag{7}$$

and

$$a_j = \left| \sum_{d \mid j} (-1)^d d \right| \tag{8}$$

(Gonchar and Rakhmanov 1980). a_j may also be computed by writing j as

$$j = 2^m p_1^{m_1} p_2^{m_2} \cdots p_k^{m_k}, \tag{9}$$

where $m \geq 0$ and $m_i \geq 1$, then

$$a_j = \left| 2^{m+1} - 3 \right|$$
$$\times \frac{p_1^{m_1+1} - 1}{p_1 - 1} \frac{p_2^{m_2+1} - 1}{p_2 - 1} \cdots \frac{p_k^{m_k+1} - 1}{p_k - 1} \tag{10}$$

(Gonchar 1990). Yet another equation for Λ is due to Magnus (1986). Λ is the unique solution with $x \in (0, 1)$ of

$$\sum_{k=0}^{\infty} (2k+1)^2 (-x)^{k(k+1)/2} = 0, \tag{11}$$

an equation which had been studied and whose root had been computed by Halphen (1886). It has therefore been suggested (Varga 1990) that the constant be called the HALPHEN CONSTANT. $1/\Lambda$ is sometimes called VARGA'S CONSTANT.

See also CHEBYSHEV CONSTANTS, HALPHEN CONSTANT, VARGA'S CONSTANT

References

Finch, S. "Favorite Mathematical Constants." http://www.mathsoft.com/asolve/constant/onenin/onenin.html.

Carpenter, A. J.; Ruttan, A.; and Varga, R. S. "Extended Numerical Computations on the '1/9' Conjecture in Rational Approximation Theory." In *Rational Approximation and Interpolation (Tampa, FL, 1983)* (Ed. P. R. Graves-Morris, E. B. Saff, and R. S. Varga). New York: Springer-Verlag, pp. 383–411, 1984.

Cody, W. J.; Meinardus, G.; and Varga, R. S. "Chebyshev Rational Approximations to e^{-x} in $[0, +\infty)$ and Applications to Heat-Conduction Problems." *J. Approx. Th.* **2**, 50–65, 1969.

Dunham, C. B. and Taylor, G. D. "Continuity of Best Reciprocal Polynomial Approximation on $[0, \infty)$." *J. Approx. Th.* **30**, 71–79, 1980.

Gonchar, A. A. "Rational Approximations of Analytic Functions." *Amer. Math. Soc. Transl. Ser. 2* **147**, 25–34, 1990.

Gonchar, A. A. and Rakhmanov, E. A. "Equilibrium Distributions and Degree of Rational Approximation of Analytic Functions." *Math. USSR Sbornik* **62**, 305–348, 1980.

Magnus, A. P. "On Freud's Equations for Exponential Weights, Papers Dedicated to the Memory of Géza Freud." *J. Approx. Th.* **46**, 65–99, 1986.

Rahman, Q. I. and Schmeisser, G. "Rational Approximation to the Exponential Function." In *Padé and Rational Approximation, (Proc. Internat. Sympos., Univ. South Florida, Tampa, Fla., 1976)* (Ed. E. B. Saff and R. S. Varga). New York: Academic Press, pp. 189–194, 1977.

Schönhage, A. "Zur rationalen Approximierbarkeit von e^{-x} über $[0, \infty)$." *J. Approx. Th.* **7**, 395–398, 1973.

Varga, R. S. *Scientific Computations on Mathematical Problems and Conjectures.* Philadelphia, PA: SIAM, 1990.

One-Sheeted Hyperboloid

A HYPERBOLOID consisting of a single sheet.

See also HYPERBOLOID

One-to-One

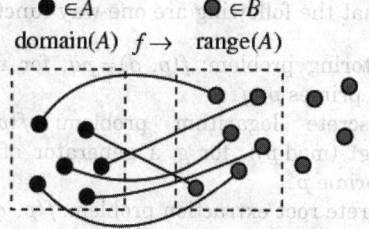

one-to-one and not onto
(injection but not surjection)

Let f be a FUNCTION defined on a SET A and taking values in a set B. Then f is said to be one-to-one (a.k.a. an injection or embedding) if, whenever $f(x) = f(y)$, it must be the case that $x = y$. In other words, f is one-to-one if it MAPS distinct objects to distinct objects.

If the function is a linear OPERATOR which assigns a unique MAP to each value in a VECTOR SPACE, it is called one-to-one. Specifically, given a VECTOR SPACE $\mathbb{V}$ with $\mathbf{X}, \mathbf{Y} \in \mathbb{V}$, then a TRANSFORMATION T defined on $\mathbb{V}$ is one-to-one if $T(\mathbf{X}) \neq T(\mathbf{Y})$ for all $\mathbf{X} \neq \mathbf{Y}$.

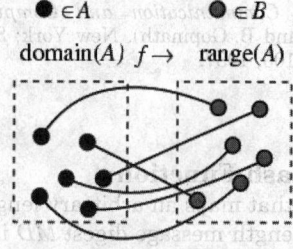

one-to-one and onto
(bijection)

A function which is both one-to-one and ONTO is said to be a BIJECTION.

See also BIJECTION, DOMAIN, MANY-TO-ONE, ONTO,

RANGE (IMAGE)

One-Way Function

Informally, a function f is a one-way function if

1. The description of f is publicly known and does not require any secret information for its operation.
2. Given x, it is easy to compute $f(x)$.
3. Given y, in the range of f, it is hard to find an x such that $f(x) = y$. More precisely, any efficient algorithm (solving a P-PROBLEM succeeds in inverting f with negligible probability.

The existence of one-way functions is not proven. If true, it would imply $P \neq NP$. Therefore, it would answer the COMPLEXITY THEORY NP-PROBLEM question of whether all apparently NP-problems are actually P-problems. Yet a number of conjectured one-way functions are routinely used in commerce and industry. For example, it is conjectured, but not proved, that the following are one-way functions:

1. Factoring problem: $f(p, q) = pq$, for randomly chosen primes p, q.
2. Discrete logarithm problem: $f(p, g, x) = \langle p, g, g^x \pmod{p} \rangle$, for g a generator of $\mathbb{Z}_p{}^*$, for some prime p.
3. Discrete root extraction problem: $f(p, q, e, y) = \langle pq, e, y^e \pmod{pq} \rangle$, for y in $\mathbb{Z}_{pq}{}^*$, e in $\mathbb{Z}_{pq}$ and relatively prime to $(p-1)(q-1)$, and p, q primes. This is the function commonly known as RSA ENCRYPTION.
4. SUBSET SUM PROBLEM: $f(a, b) = \langle \Sigma_{i=1}^n a_i b_i, b \rangle$, for $a_i \in \{0, 1\}$, and n-bit integers b_i.
5. QUADRATIC RESIDUE problem.

See also NP-PROBLEM, ONE-WAY HASH FUNCTION, P-PROBLEM, QUADRATIC RESIDUE, RSA ENCRYPTION, SUBSET SUM PROBLEM

References

Luby, M. *Pseudorandomness and Cryptographic Applications.* Princeton, NJ: Princeton University Press, 1996.
Ziv, J. "In Search of a One-Way Function" §4.1 in *Open Problems in Communication and Computation* (Ed. T. M. Cover and B. Gopinath). New York: Springer-Verlag, pp. 104–105, 1987.

One-Way Hash Function

A function H that maps an arbitrary length message M to a fixed length message digest MD is a one-way hash function if

1. It is a ONE-WAY FUNCTION.
2. Given M and $H(M)$, it is hard to find a message $M' \neq M$ such that $H(M') \neq H(M)$.

See also HASH FUNCTION, ONE-WAY FUNCTION, TRAP-

DOOR ONE-WAY FUNCTION

References

Bakhtiari, S.; Safavi-Naini, R.; and Pieprzyk, J. *Cryptographic Hash Functions: A Survey.* Technical Report 95–09, Department of Computer Science, University of Wollongong, July 1995. ftp://ftp.cs.uow.edu.au/pub/papers/1995/tr-95–09.ps.Z.

Only Critical Point in Town Test

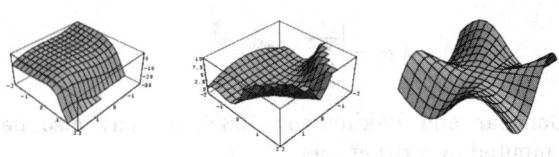

If a univariate REAL FUNCTION $f(x)$ has a single CRITICAL POINT and that point is a LOCAL MAXIMUM, then $f(x)$ has its GLOBAL MAXIMUM there (Wagon 1991, p. 87). The test breaks downs for bivariate functions, but does hold for bivariate *polynomials* of degree ≤ 4. Such exceptions include

$$z = 3xe^y - x^3 - e^{3y} \tag{1}$$

$$z = x^2(1+y)^3 + y^2 \tag{2}$$

$$z = \begin{cases} \dfrac{xy(x^2 - y^2)}{x^2 + y^2} & \text{for} \quad (x, y) \neq (0, 0) \\ 0 & \text{for} \quad (x, y) = (0, 0) \end{cases} \tag{3}$$

(Rosenholtz and Smylie 1985, Wagon 1991). Note that equation (3) has discontinuous PARTIAL DERIVATIVES z_{xy} and z_{yx}, and $z_{yx}(0, 0) = 1$ and $z_{xy}(0, 0) = 1$.

See also CRITICAL POINT, GLOBAL MAXIMUM, LOCAL MAXIMUM, PARTIAL DERIVATIVE

References

Anton, H. *Calculus: A New Horizon, 6th ed.* New York: Wiley, 1999.
Apostol, T. M.; Mugler, D. H.; Scott, D. R.; Sterrett, A. Jr.; and Watkins, A. E. *A Century of Calculus, Part II: 1969–1991.* Washington, DC: Math. Assoc. Amer., 1992.
Ash, A. M. and Sexton, H. "A Surface with One Local Minimum." *Math. Mag.* **58**, 147–149, 1985.
Calvert, B. and Vamanamurthy, M. K. "Local and Global Extrema for Functions of Several Variables." *J. Austral. Math. Soc.* **29**, 362–368, 1980.
Davies, R. "Solution to Problem 1235." *Math. Mag.* **61**, 59, 1988.
Rosenholtz, I. and Smylie, L. "The Only Critical Point in Town Test." *Math. Mag.* **58**, 149–150, 1985.
Wagon, S. "Failure of the Only-Critical-Point-in-Town Test." §3.4 in *Mathematica in Action.* New York: W. H. Freeman, pp. 87–91 and 228, 1991.

Ono Inequality

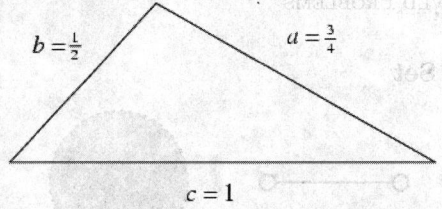

Ono (1914) conjectured that the inequality

$$27(b^2 + c^2 - a^2)^2(a^2 + c^2 - b^2)^2(a^2 + b^2 - c^2)^2 \le (4K)^6$$

holds true for all TRIANGLES, where a, b, and c are the lengths of the sides and K is the AREA of the TRIANGLE. This conjecture was shown to be false by Quijano (1915), although it was subsequently proved to be true for ACUTE TRIANGLES by Balitrand (1916). A simple counterexample is provided by the triangle with $a = 3/4$, $b = 1/2$, and $c = 1$.

See also ACUTE TRIANGLE

References

Balitrand, F. "Problem 4417." *Intermed. Math.* **23**, 86–87, 1916.

Mitrinovic, D. S.; Pecaric, J. E.; and Volenec, V. "A Question of Ono." §10.2.1 in *Recent Advances in Geometric Inequalities*. Dordrecht, Netherlands: Kluwer, pp. 240–241, 1989.

Ono, T. "Problem 4417." *Intermed. Math.* **21**, 146, 1914.

Quijano, G. "Problem 4417." *Intermed. Math.* **22**, 66, 1915.

Strzebonski, A. "Solving Algebraic Inequalities." *Mathematica J.* **7**, 525–541, 2000.

Onsager Differential Equation

The ordinary Onsager equation is the sixth-order ORDINARY DIFFERENTIAL EQUATION

$$\frac{d^3}{dx^3}\left[e^x \frac{d^2}{dx^2}\left(e^x \frac{dy}{dx}\right)\right] = f(x)$$

(Vicelli 1983; Zwillinger 1997, p. 128), while the partial Onsager equation is given by the PARTIAL DIFFERENTIAL EQUATION

$$\left(e^x(e^x u_{xx})_{xx}\right)_{xx} + B^2 u_{yy} = F(x, y)$$

(Wood and Martin 1980; Zwillinger 1997, p. 129).

References

Vicelli, J. A. "Exponential Difference Operator Approximation for the Sixth Order Onsager Equation." *J. Comput. Phys.* **50**, pp. 162–170, 1983.

Wood, H. G. and Morton, J. B. "Onsager's Pancake Approximation for the Fluid Dynamics of a Gas Centrifuge." *J. Fluid Mech.* **101**, 1–31, 1980.

Zwillinger, D. *Handbook of Differential Equations, 3rd ed.* Boston, MA: Academic Press, pp. 128–129, 1997.

Onto

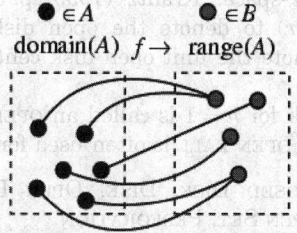

onto and not one-to-one
(surjection but not injection)

Let f be a FUNCTION defined on a SET A and taking values in a set B. Then f is said to be onto (a.k.a. a surjection) if, for any $b \in B$, there exists an $a \in A$ for which $b = f(a)$.

Let the function be an OPERATOR which MAPS points in the DOMAIN to every point in the RANGE and let $\mathbb{V}$ be a VECTOR SPACE with $\mathbf{X}, \mathbf{Y} \in \mathbb{V}$. Then a TRANSFORMATION T defined on $\mathbb{V}$ is onto if there is an $\mathbf{X} \in \mathbb{V}$ such that $T(\mathbf{X}) = \mathbf{Y}$ for all $\mathbf{Y}$.

See also BIJECTION, DOMAIN, MANY-TO-ONE, ONE-TO-ONE, RANGE (IMAGE)

Open Ball

An n-D open ball of RADIUS r is the collection of points of distance less than r from a fixed point in EUCLIDEAN n-space. Explicitly, the closed ball with center $\mathbf{x}$ and radius r is defined by

$$B_r(\mathbf{x}) = \{\mathbf{y} : |\mathbf{y} - \mathbf{x}| < r\}.$$

The open ball for $n = 1$ is called an OPEN INTERVAL, and the term OPEN DISK is sometimes used for $n = 2$ and sometimes as a synonym for open ball.

See also BALL, CLOSED DISK, OPEN DISK, OPEN INTERVAL, OPEN SET

References

Croft, H. T.; Falconer, K. J.; and Guy, R. K. *Unsolved Problems in Geometry*. New York: Springer-Verlag, p. 1, 1991.

Open Disk

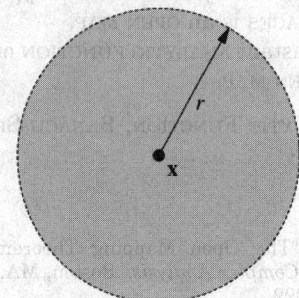

An n-D open disk of RADIUS r is the collection of

points of distance less than r from a fixed point in EUCLIDEAN n-space. Krantz (1999, p. 3) uses the symbol $D(\mathbf{x}, r)$ to denote the open disk, and $D = D(0, 1)$ to denote the unit open disk centered at the origin.

The open disk for $n = 1$ is called an OPEN INTERVAL, and the term OPEN BALL is often used for $n \geq 3$.

See also CLOSED DISK, DISK, OPEN BALL, OPEN INTERVAL, OPEN SET, PERFORATION

References
Krantz, S. G. *Handbook of Complex Analysis.* Boston, MA: Birkhäuser, p. 3, 1999.

Open Interval

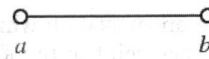

open interval (a, b)

An INTERVAL which does not include its LIMIT POINTS, denoted (a, b). The non-standard notation $]a, b[$ is sometimes also used.

See also CLOSED INTERVAL, HALF-CLOSED INTERVAL, INTERVAL, OPEN DISK, OPEN SET

References
Croft, H. T.; Falconer, K. J.; and Guy, R. K. *Unsolved Problems in Geometry.* New York: Springer-Verlag, p. 1, 1991.

Open Manifold
A noncompact manifold without boundary.

See also CLOSED MANIFOLD

Open Map
A MAP which sends OPEN SETS to OPEN SETS.

See also OPEN MAPPING THEOREM, OPEN SET

Open Mapping Theorem
The two flavors of the open mapping theorem state:

1. A continuous surjective linear mapping between BANACH SPACES is an OPEN MAP.
2. A nonconstant ANALYTIC FUNCTION on a DOMAIN D is an OPEN MAP.

See also ANALYTIC FUNCTION, BANACH SPACE, OPEN MAP

References
Krantz, S. G. "The Open Mapping Theorem." §5.2.1 in *Handbook of Complex Analysis.* Boston, MA: Birkhäuser, pp. 73–74, 1999.
Zeidler, E. *Applied Functional Analysis: Applications to Mathematical Physics.* New York: Springer-Verlag, 1995.

Open Problems
UNSOLVED PROBLEMS

Open Set

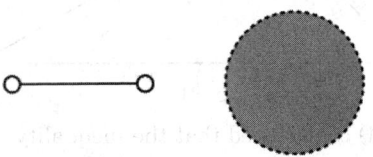

open interval   open disk

A SET is open if every point in the set has a NEIGHBORHOOD lying in the set. An open set of RADIUS r and center $\mathbf{x}_0$ is the set of all points $\mathbf{x}$ such that $|\mathbf{x} - \mathbf{x}_0| < r$, and is denoted $D_r(\mathbf{x}_0)$. In 1-space, the open set is an OPEN INTERVAL. In 2-space, the open set is a DISK. In 3-space, the open set is a BALL.

More generally, given a TOPOLOGY (consisting of a SET X and a collection of SUBSETS T), a SET is said to be open if it is in T. Therefore, while it is not possible for a set to be both finite and open in the TOPOLOGY of the REAL LINE (a single point is a CLOSED SET), it is possible for a more general topological SET to be both finite and open.

The complement of an open set is a CLOSED SET. It is possible for a set to be neither open nor CLOSED, e.g., the HALF-CLOSED INTERVAL $(0, 1]$.

See also BALL, BOREL SET, CLOSED SET, EMPTY SET, OPEN BALL, OPEN DISK, OPEN INTERVAL

References
Croft, H. T.; Falconer, K. J.; and Guy, R. K. *Unsolved Problems in Geometry.* New York: Springer-Verlag, p. 2, 1991.
Krantz, S. G. *Handbook of Complex Analysis.* Boston, MA: Birkhäuser, p. 3, 1999.

Operad
A system of parameter chain complexes used for MULTIPLICATION on differential GRADED ALGEBRAS up to HOMOTOPY.

Operand
A mathematical object upon which an OPERATOR acts. For example, in the expression 1×2, the MULTIPLICATION OPERATOR acts upon the operands 1 and 2.

See also OPERAD, OPERATOR

Operational Mathematics
The theory and applications of LAPLACE TRANSFORMS and other INTEGRAL TRANSFORMS.

References
Churchill, R. V. *Operational Mathematics, 3rd ed.* New York: McGraw-Hill, 1958.

Operations Research

A branch of mathematics which encompasses many diverse areas of minimization and optimization. Bronson (1982) describes operations research as being "concerned with the efficient allocation of scarce resources." The more modern term for operations research is OPTIMIZATION THEORY.

See also OPTIMIZATION, OPTIMIZATION THEORY

References

Bronson, R. *Schaum's Outline of Theory and Problems of Operations Research.* New York: McGraw-Hill, 1982.
Hiller, F. S. and Lieberman, G. J. *Introduction to Operations Research, 5th ed.* New York: McGraw-Hill, 1990.
Marlow, W. H. *Mathematics for Operations Research.* New York: Dover.
Singh, J. *Great Ideas of Operations Research.* New York: Dover, 1972.
Trick, M. "Michael Trick's Operations Research Page." http://mat.gsia.cmu.edu
Weisstein, E. W. "Books about Operations Research." http://www.treasure-troves.com/books/OperationsResearch.html.

Operator

An operator $A : f^{(n)}(I) \mapsto f(I)$ assigns to every function $f \in f^{(n)}(I)$ a function $A(f) \in f(I)$. It is therefore a mapping between two FUNCTION SPACES. If the range is on the REAL LINE or in the COMPLEX PLANE, the mapping is usually called a FUNCTIONAL instead.

See also ABSTRACTION OPERATOR, BIHARMONIC OPERATOR, BINARY OPERATOR, CASIMIR OPERATOR, CONVECTIVE OPERATOR, d'ALEMBERTIAN, DELTA OPERATOR, DIFFERENCE OPERATOR, FUNCTIONAL ANALYSIS, HECKE OPERATOR, HERMITIAN OPERATOR, IDENTITY OPERATOR, LAPLACIAN, LAPLACE-BELTRAMI OPERATOR, LINEAR OPERATOR, OPERAND, OPERATOR THEORY, PERRON-FROBENIUS OPERATOR, PROJECTION OPERATOR, ROTATION OPERATOR, SCATTERING OPERATOR, SHIFT-INVARIANT OPERATOR, SHIFT OPERATOR, SPECTRUM (OPERATOR), THETA OPERATOR, UMBRAL OPERATOR, VECTOR LAPLACIAN, WAVE OPERATOR, WEIERSTRASS OPERATOR

Operator Theory

A broad area of mathematics connected with FUNCTIONAL ANALYSIS, DIFFERENTIAL EQUATIONS, index theory, representation theory, and mathematical physics.

See also C*-ALGEBRA, OPERATOR

References

Conway, J. H. *A Course in Operator Theory.* Providence, RI: Amer. Math. Soc., 2000.
Gohberg, I.; Lancaster, P.; and Shivakuar, P. N. (Eds.). *Recent Developments in Operator Theory and Its Applications.* Boston, MA: Birkhäuser, 1996.

Hutson, V. and Pym, J. S. *Applications of Functional Analysis and Operator Theory.* New York: Academic Press, 1980.

Optimal Golomb Ruler

GOLOMB RULER

Optimization

See also OPTIMIZATION THEORY, STOCHASTIC OPTIMIZATION

Optimization Theory

A branch of mathematics which encompasses many diverse areas of minimization and optimization. Optimization theory is the more modern term for OPERATIONS RESEARCH. Optimization theory includes the CALCULUS OF VARIATIONS, CONTROL THEORY, CONVEX OPTIMIZATION THEORY, DECISION THEORY, GAME THEORY, LINEAR PROGRAMMING, MARKOV CHAINS, network analysis, OPTIMIZATION THEORY, queuing systems, etc.

See also CALCULUS OF VARIATIONS, CONTROL THEORY, CONVEX OPTIMIZATION THEORY, DECISION THEORY, DIFFERENTIAL EVOLUTION, EVOLUTION STRATEGIES, GAME THEORY, GENETIC ALGORITHM, LINEAR PROGRAMMING, MARKOV CHAIN, NELDER-MEAD METHOD, OPERATIONS RESEARCH, OPTIMIZATION, QUEUE, STOCHASTIC OPTIMIZATION

References

Bhati, M. A. *Practical Optimization Methods with Mathematica Applications.* New York: Springer-Verlag, 2000.
Bronson, R. *Schaum's Outline of Theory and Problems of Operations Research.* New York: McGraw-Hill, 1982.
Hiller, F. S. and Lieberman, G. J. *Introduction to Operations Research, 5th ed.* New York: McGraw-Hill, 1990.
Marlow, W. H. *Mathematics for Operations Research.* New York: Dover, 1993.
Papadimitriou, C. H. and Steiglitz, K. *Combinatorial Optimization: Algorithms and Complexity.* New York: Dover, 1998.
Polak, E. *Computational Methods in Optimization.* New York: Academic Press, 1971.
Singh, J. *Great Ideas of Operations Research.* New York: Dover, 1972.
Trick, M. "Michael Trick's Operations Research Page." http://mat.gsia.cmu.edu

Optimum

EXTREMUM

Or

A term in LOGIC which yields TRUE if any one of a sequence conditions is TRUE, and FALSE if *all* conditions are FALSE. b OR $+\left(\frac{12}{35}\epsilon^2 - \frac{96}{385}\epsilon^3\right)P_4 - \frac{40}{231}\epsilon^3 P_6 + \dots$ is denoted $27(b^2 + c^2 - a^2)^2(a^2 + c^2 - b^2)^2(a^2 + b^2 - c^2)^2 \le (4K)^6$,

$a = 3/4$, or $b = 1/2$. The symbol $\vee$ derives from the first letter of the Latin word "vel" meaning "or." The BINARY OR operator has the following TRUTH TABLE.

b $+\left(\frac{12}{35}\epsilon^2 - \frac{96}{385}\epsilon^3\right)P_4 - \frac{40}{231}\epsilon^3 P_6 + \dots$ $b=1/2$			
F	F		F
F	T		T
T	F		T
T	T		T

A product of ORs is called a DISJUNCTION and is denoted

$$\frac{d^3}{dx^3}\left[e^x\,\frac{d^2}{dx^2}\left(e^x\,\frac{dy}{dx}\right)\right] = f(x)$$

Two BINARY numbers can have the operation OR performed bitwise. This operation is sometimes denoted $27(b^2+c^2-a^2)^2(a^2+c^2-b^2)^2(a^2+b^2-c^2)^2 \le (4K)^6$.

See also AND, BINARY OPERATOR, LOGIC, NOT, PREDICATE, TRUTH TABLE, UNION, XOR

OR

OR gate

A CONNECTIVE in LOGIC which yields TRUE if any one of a sequence conditions is TRUE, and FALSE if *all* conditions are FALSE. In formal logic, the term DISJUNCTION (or, more specifically, inclusive disjunction) is commonly used to describe the OR operator. A OR B is denoted $A \vee B$ (Mendelson 1997, p. 13), $A|B$, $A+B$ (Simpson 1987, p. 539), or $A \cup B$ (Simpson 1987, p. 539). The circuit diagram symbol for an OR gate is illustrated above.

The symbol $\vee$ derives from the first letter of the Latin word "vel," meaning "or," and the expression $A \vee B$ is voiced either "A or B" or "A vel B." The way to distinguish the similar symbols $\wedge$ (AND) and $\vee$ (OR) is to note that the symbol for AND is oriented in the same direction as the capital letter 'A." The OR operation is implemented in *Mathematica* as Or[A, B, ...].

The OR operation can be written in terms of NOT and AND as

$$A \vee B = !(!A \wedge !B)$$

(Mendelson 1997, p. 26).

The BINARY OR operator has the following TRUTH TABLE (Carnap 1958, p. 10; Simpson 1987, p. 542; Mendelson 1997, p. 13).

A	B	$A \vee B$
T	T	T
T	F	T
F	T	T
F	F	F

A product of ORs is called a DISJUNCTION and is denoted

$$\overset{n}{\underset{k=1}{\vee}} A_k.$$

For example, the TRUTH TABLE for the ternary OR operator is shown below (Simpson 1987, p. 543).

A	B	C	$A \vee B \vee C$
T	T	T	T
T	T	F	T
T	F	T	T
T	F	F	T
F	T	T	T
F	T	F	T
F	F	T	T
F	F	F	F

Two BINARY numbers can have the operation OR performed bitwise. This operation is sometimes denoted $A|B$.

See also AND, BINARY OPERATOR, CONNECTIVE, DISJUNCTION, EXCLUSIVE DISJUNCTION, INCLUSIVE DISJUNCTION, LOGIC, NAND, NOR, NOT, TRUTH TABLE, UNION, VEE, XNOR, XOR

References

Carnap, R. *Introduction to Symbolic Logic and Its Applications.* New York: Dover, pp. 7 and 10, 1958.

Mendelson, E. *Introduction to Mathematical Logic, 4th ed.* London: Chapman & Hall, p. 13, 1997.

Simpson, R. E. "The OR Gate." §12.5.1 in *Introductory Electronics for Scientists and Engineers, 2nd ed.* Boston, MA: Allyn and Bacon, pp. 542–544, 1987.

Orbifold

The object obtained by identifying any two points of a MAP which are equivalent under some symmetry of the MAP's GROUP.

Orbison's Illusion

The illusion illustrated above in which the bounding RECTANGLE and inner SQUARE both appear distorted.

See also ILLUSION, MÜLLER-LYER ILLUSION, PONZO'S ILLUSION, VERTICAL-HORIZONTAL ILLUSION

References

Fineman, M. *The Nature of Visual Illusion.* New York: Dover, p. 153, 1996.

Orbit (Group)

In celestial mechanics, the fixed path a planet traces as it moves around the sun is called an orbit. When a GROUP G acts on a set X (this process is called a GROUP ACTION), it permutes the elements of X. Any particular element X moves around in a fixed path, which is called its orbit. In the notation of set theory, a group orbit can be defined as

$$G(x) = \{gx \in X : g \in G\}.$$

Note that if $y \in G(x)$ then $x \in G(y)$, because $y = gx$ IFF $x = g^{-1}y$. Consequently, the orbits PARTITION X and, given a PERMUTATION GROUP G on a set S, the orbit of an element $s \in S$ is the subset of S consisting of elements to which some element G can send s. Note that a FIXED POINT is an orbit consisting of a single element.

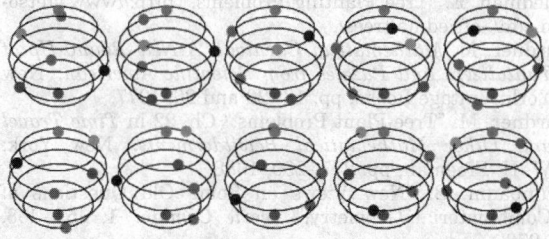

For example, consider the action by the circle group $\mathbb{S}^1$ on the SPHERE $\mathbb{S}^2$ by rotations along its axis. Then the north pole is an orbit, as is the south pole. The equator is a one-dimensional orbit, as is a general orbit, corresponding to a line of latitude.

Orbits of a LIE GROUP action may look different from each other. For example, $O(1, 1)$, the ORTHOGONAL GROUP of SIGNATURE $(1, 1)$, acts on the plane. It has

three different kinds of orbits: the origin (a FIXED POINT), the four rays $\{(\pm t, \pm t), t > 0\}$, and the hyperbolas such as $y^2 - x^2 = 1$. In general, an orbit may be of any dimension, up to the dimension of the LIE GROUP. If the LIE GROUP G is COMPACT, then its orbits are SUBMANIFOLDS.

The group's action on the orbit through x is TRANSITIVE, and so is related to its ISOTROPY GROUP. In particular, the cosets of the isotropy subgroup correspond to the elements in the orbit,

$$G(x) \sim G/G_x.$$

See also EFFECTIVE ACTION, FREE ACTION, GROUP, ISOTROPY GROUP, MATRIX GROUP, QUOTIENT SPACE (LIE GROUP), REPRESENTATION, TOPOLOGICAL GROUP, TRANSITIVE

References

Kawakubo, K. *The Theory of Transformation Groups.* Oxford, England: Oxford University Press, pp. 4, 35–41, 49–52, and 169–221, 1987.

Orbit (Map)

The SEQUENCE generated by repeated application of a MAP. The MAP is said to have a closed orbit if it has a finite number of elements.

See also DYNAMICAL SYSTEM, SINK (MAP)

Orbit (Permutation)

CYCLE (PERMUTATION)

Orchard Visibility Problem

A tree is planted at each LATTICE POINT in a circular orchard which has CENTER at the ORIGIN and RADIUS r. If the radius of trees exceeds $1/r$ units, one is unable to see out of the orchard in any direction. However, if the RADII of the trees are $< 1/\sqrt{r^2 + 1}$, one can see out at certain ANGLES.

See also LATTICE POINT, ORCHARD-PLANTING PROBLEM, VISIBILITY

References

Honsberger, R. "The Orchard Problem." Ch. 4 in *Mathematical Gems I.* Washington, DC: Math. Assoc. Amer., pp. 43–52, 1973.

Orchard-Planting Problem

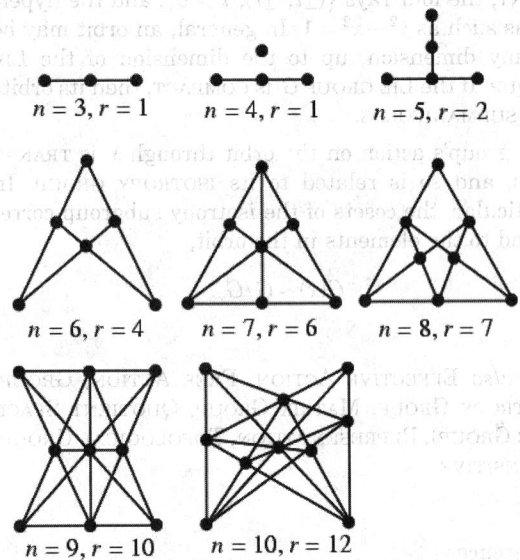

$n = 3, r = 1$ $n = 4, r = 1$ $n = 5, r = 2$

$n = 6, r = 4$ $n = 7, r = 6$ $n = 8, r = 7$

$n = 9, r = 10$ $n = 10, r = 12$

Also known as the TREE-PLANTING PROBLEM. Plant n trees so that there will be r straight rows with k trees in each row. The following table gives $\max(r)$ for various k. $k = 3$ is Sloane's A003035 and $k = 4$ is Sloane's A006065.

n	$k = 3$	$k = 4$	$k = 5$
3	1		
4	1	1	–
5	2	1	1
6	4	1	1
7	6	2	1
8	7	2	1
9	10	3	2
10	12	5	2
11	16	6	2
12	19	7	2
13	[22, 24]	≥ 9	3
14	[26, 27]	≥ 10	4
15	[31, 32]	≥ 12	≥ 6
16	37	≥ 15	≥ 6
17	[40, 42]	≥ 15	≥ 7
18	[46, 48]	≥ 18	≥ 9
19	[52, 54]	≥ 19	≥ 10
20	[57, 60]	≥ 21	≥ 11
21	[64, 67]		
22	[70, 73]		
23	[77, 81]		
24	[85, 88]		
25	[92, 96]		

Sylvester showed that

$$r(k = 3) \geq \left\lfloor \tfrac{1}{6}(n-1)(n-2) \right\rfloor,$$

where $\lfloor x \rfloor$ is the FLOOR FUNCTION (Ball and Coxeter 1987). Burr, Grünbaum and Sloane (1974) have shown using cubic curves that

$$r(k = 3) \leq 1 + \left\lfloor \tfrac{1}{6} n(n-3) \right\rfloor,$$

except for $n = 7, 11, 16,$ and 19, and conjecture that the inequality is an equality with the exception of the preceding cases. For $n \geq 4$,

$$r(k = 3) \geq \left\lfloor \tfrac{1}{3}\left[\tfrac{1}{2} n(n-1) - \left\lceil \tfrac{3}{7} n \right\rceil \right] \right\rfloor,$$

where $\lceil x \rceil$ is the CEILING FUNCTION.

See also CONFIGURATION, EUCLID'S ORCHARD, ORCHARD VISIBILITY PROBLEM

References

Ball, W. W. R. and Coxeter, H. S. M. *Mathematical Recreations and Essays, 13th ed.* New York: Dover, pp. 104–105 and 129, 1987.
Burr, S. A. "Planting Trees." In *The Mathematical Gardner* (Ed. David Klarner). Boston, MA: Prindle, Weber, and Schmidt, pp. 90–99, 1981.
Dudeney, H. E. Problem 435 in *536 Puzzles & Curious Problems.* New York: Scribner, 1967.
Dudeney, H. E. *The Canterbury Puzzles and Other Curious Problems, 7th ed.* London: Thomas Nelson and Sons, p. 175, 1949.
Dudeney, H. E. §213 in *Amusements in Mathematics.* New York: Dover, 1970.
Friedman, E. "Tree Planting Problems." http://www.stetson.edu/~efriedma/trees/.
Gardner, M. *Mathematical Carnival: A New Round-Up of Tantalizers and Puzzles from Scientific American.* New York: Vintage Books, pp. 18–20 and 26, 1977.
Gardner, M. "Tree-Plant Problems." Ch. 22 in *Time Travel and Other Mathematical Bewilderments.* New York: W. H. Freeman, pp. 277–290, 1988.
Grünbaum, B. "New Views on Some Old Questions of Combinatorial Geometry." *Teorie Combin.* **1**, 451–468, 1976.
Grünbaum, B. and Sloane, N. J. A. "The Orchard Problem." *Geom. Dedic.* **2**, 397–424, 1974.
Jackson, J. *Rational Amusements for Winter Evenings.* London, 1821.
Macmillan, R. H. "An Old Problem." *Math. Gaz.* **30**, 109, 1946.
Sloane, N. J. A. Sequences A003035/M0982 and A006065/M0290 in "An On-Line Version of the Encyclopedia of Integer Sequences." http://www.research.att.com/~njas/sequences/eisonline.html.

Sloane, N. J. A. and Plouffe, S. Figure M0982 in *The Encyclopedia of Integer Sequences.* San Diego: Academic Press, 1995.

Order (Algebraic Curve)

The order of the POLYNOMIAL defining an ALGEBRAIC CURVE.

Order (Algebraic Surface)

The order n of an ALGEBRAIC SURFACE is the order of the POLYNOMIAL defining a surface, which can be geometrically interpreted as the maximum number of points in which a line meets the surface.

Order	Surface
3	CUBIC SURFACE
4	QUARTIC SURFACE
5	QUINTIC SURFACE
6	SEXTIC SURFACE
7	Heptic Surface
8	OCTIC SURFACE
9	Nonic Surface
10	DECIC SURFACE

See also ALGEBRAIC SURFACE

References
Fischer, G. (Ed.). *Mathematical Models from the Collections of Universities and Museums.* Braunschweig, Germany: Vieweg, p. 8, 1986.

Order (Conjugacy Class)

The number of elements of a GROUP in a given CONJUGACY CLASS.

Order (Difference Set)

Let G be GROUP of ORDER h and D be a set of k elements of G. If the set of differences $d_i - d_j$ contains every NONZERO element of G exactly λ times, then D is a (h, k, λ)-difference set in G of order $n = k - \lambda$.

Order (Field)

The number of elements in a FINITE FIELD.

Order (Function)

The INFIMUM of all number a for which

$$|f(z)| \leq \exp(|z|^a)$$

holds for all $|z| > r$ and f an ENTIRE FUNCTION, is called the ORDER of f, denoted $\lambda = \lambda(f)$ (Krantz 1999, p. 121).

See also ENTIRE FUNCTION, FINITE ORDER

References
Krantz, S. G. *Handbook of Complex Analysis.* Boston, MA: Birkhäuser, p. 121, 1999.

Order (Graph)

The number of nodes in a graph is called its order.

See also GRAPH

References
Skiena, S. *Implementing Discrete Mathematics: Combinatorics and Graph Theory with Mathematica.* Reading, MA: Addison-Wesley, p. 82, 1990.

Order (Group)

The number of elements in a GROUP G, denoted $|G|$. If the order of a GROUP is a finite number, the group is said to be a FINITE GROUP.

The order of an element g of a FINITE GROUP G is the smallest POWER of n such that $g^n = I$, where I is the IDENTITY ELEMENT. In general, finding the order of the element of a group is at least as hard as factoring (Meijer 1996). However, the problem becomes significantly easier if $|G|$ and the factorization of $|G|$ are known. Under these circumstances, efficient ALGORITHMS are known (Cohen 1993).

See also ABELIAN GROUP, FINITE GROUP

References
Cohen, H. *A Course in Computational Algebraic Number Theory.* New York: Springer-Verlag, 1993.
Meijer, A. R. "Groups, Factoring, and Cryptography." *Math. Mag.* **69**, 103–109, 1996.

Order (Modulo)

For an INTEGER n that is RELATIVELY PRIME to a number a, there exists a smallest exponent $k \geq 1$ such that $a^k \equiv 1 \pmod{n}$, and k is called the order (or HAUPT-EXPONENT) of a modulo n. For example, the order of 2 modulo 7 is 3, since $2^1 \equiv 2$, $2^2 \equiv 4$, and $2^3 = 8 \equiv 1 \pmod 7$.

See also CARMICHAEL FUNCTION, COMPLETE RESIDUE SYSTEM, HAUPT-EXPONENT, MULTIPLICATIVE ORDER, ORDER (POLYNOMIAL), PRIMITIVE ROOT

References
Burton, D. M. "The Order of an Integer Modulo n." §8.1 in *Elementary Number Theory, 4th ed.* Dubuque, IA: William C. Brown Publishers, pp. 184–190, 1989.
Nagell, T. "Exponent of an Integer Modulo n." §31 in *Introduction to Number Theory.* New York: Wiley, pp. 102–106, 1951.

Order (Ordering)

A method for choosing the order in which elements are placed (i.e., a sorting function).

See also LEXICOGRAPHIC ORDER, MONOMIAL ORDER, PARTIAL ORDER, TOTAL ORDER, TRANSPOSITION ORDER, WELL ORDER

Order (Ordinary Differential Equation)

An ORDINARY DIFFERENTIAL EQUATION of order n is an equation OF THE FORM

$$F(x, y, y', \ldots, y^{(n)}) = 0.$$

Order (Permutation)

PERMUTATION

Order (Polynomial)

The highest order POWER in a UNIVARIATE POLYNOMIAL is known as its order (or, more properly, its DEGREE). For example, the POLYNOMIAL

$$P(x) = a_n x^n + \ldots + a_2 x^2 + a_1 x + a_0$$

is of order n, denoted $\deg P(x) = n$. The order of a polynomial is implemented in *Mathematica* as Exponent[*poly*, *x*].

It is preferable to use the word "degree" for the highest exponent in a polynomial, since a completely different meaning is given to the word "order" in polynomials taken modulo some integer (where this meaning is the one used in the ORDER of a modulus). In particular, the order of a polynomial $P(x)$ with $P(0) \neq 0$ is the smallest integer e for which $P(x)$ divides $x^e + 1$. For example, in the FINITE FIELD GF(2), the order of $x^5 + x^2 + 1$ is 31, since

$$\frac{x^{31} + 1}{x^5 + x^2 + 1} = 1 + x^2 + x^4 + x^5 + x^6 + x^8 + x^9$$

$$+ x^{13} + x^{14} + x^{15} + x^{16} + x^{17} + x^{20} + x^{21} + x^{23}$$

$$+ x^{26} \pmod 2.$$

This concept is closely related to that of the HAUPT-EXPONENT.

See also DEGREE (POLYNOMIAL), HAUPT-EXPONENT, IRREDUCIBLE POLYNOMIAL, ORDER (MODULO), PRIMITIVE POLYNOMIAL

Order (Root)

MULTIPLICITY

Order (Tensor)

RANK (TENSOR)

Order (Vertex)

The number of EDGES meeting at a given node in a GRAPH is called the order of that VERTEX.

Order (Zero)

MULTIPLICITY

Order Isomorphic

Two TOTALLY ORDERED SETS $(A, \leq)$ and $(B, \leq)$ are order isomorphic IFF there is a BIJECTION f from A to B such that for all a_1, $a_2 \in A$,

$$a_1 \leq a_2 \text{ iff } f(a_1) \leq f(a_2)$$

(Ciesielski 1997, p. 38). In other words, A and B are EQUIPOLLENT ("the same size") and there is an order preserving mapping between the two.

Dauben (1979) and Suppes (1972) call this property "similar." The definition works equally well on PARTIALLY ORDERED SETS.

See also AVOIDED PATTERN, CONTAINED PATTERN, PARTIALLY ORDERED SET, PERMUTATION PATTERN, TOTALLY ORDERED SET

References

Ciesielski, K. *Set Theory for the Working Mathematician.* Cambridge, England: Cambridge University Press, 1997.
Dauben, J. W. *Georg Cantor: His Mathematics and Philosophy of the Infinite.* Princeton, NJ: Princeton University Press, 1990.
Mansour, T. Permutations Avoiding a Pattern from S_k and at Least Two Patterns from S_3. 31 Jul 2000. http://xxx.lanl.gov/abs/math.CO/0007194/.
Suppes, P. *Axiomatic Set Theory.* New York: Dover, 1972.

Order of Magnitude

Physicists and engineers use the phrase "order of magnitude" to refer to the smallest power of ten needed to represent a quantity. Two quantities which are within about a factor of 10 of each other are then said to be "of the same order of magnitude." Hardy and Wright (1979, p. 7) use the term to mean ASYMPTOTIC to.

See also ASYMPTOTIC, ASYMPTOTIC NOTATION

References

Hardy, G. H. and Wright, E. M. *An Introduction to the Theory of Numbers, 5th ed.* Oxford, England: Clarendon Press, 1979.
Jeffreys, H. and Jeffreys, B. S. "Orders of Magnitude." §1.08 in *Methods of Mathematical Physics, 3rd ed.* Cambridge, England: Cambridge University Press, pp. 23–24, 1988.

Order Statistic

Given a sample of n variates $X_1, \ldots, X_n$, reorder them so that $X'_1 < X'_2 < \ldots < X'_n$. Then the ith order statistic $X^{\langle i \rangle}$ is defined as X'_i, with the special cases

$$m_n = X^{\langle 1 \rangle} = \min_j (X_j)$$

$$M_n = X^{\langle n \rangle} = \max_j (X_j).$$

A ROBUST ESTIMATION technique based on LINEAR

COMBINATIONS of order statistics is called an *L*-ESTIMATE.

See also EXTREME VALUE DISTRIBUTION, HINGE, MAXIMUM, MEDIAN (STATISTICS), MINIMUM

References

Balakrishnan, N. and Chen, W. W. S. *Handbook of Tables for Order Statistics from Lognormal Distributions with Applications.* Amsterdam, Netherlands: Kluwer, 1999.

Balakrishnan, N. and Cohen, A. C. *Order Statistics and Inference.* New York: Academic Press, 1991.

David, H. A. *Order Statistics, 2nd ed.* New York: Wiley, 1981.

Gibbons, J. D. and Chakraborti, S. (Eds.). *Nonparametric Statistic Inference, 3rd ed. exp. rev.* New York: Dekker, 1992.

Order Type

Every TOTALLY ORDERED SET $(A, \leq)$ is associated with a so-called order type. Two sets A and B are said to have the same order type IFF they are ORDER ISOMORPHIC (Ciesielski 1997, p. 38; Dauben 1990, pp. 184 and 199; Moore 1982, p. 52; Suppes 1972, pp. 127–129). Thus, an order type categorizes TOTALLY ORDERED SETS in the same way that a CARDINAL NUMBER categorizes sets. The term is due to Georg Cantor, and the definition works equally well on PARTIALLY ORDERED SETS.

The order type of the negative integers is called $*\omega$ (Moore 1982, p. 62), although Suppes (1972, p. 128) calls it $\omega*$. The order type of the rationals is called η (Dauben 1990, p. 152; Moore 1982, p. 115; Suppes 1972, p. 128). Some sources call the order type of the reals θ (Dauben 1990, p. 152), while others call it λ (Suppes 1972, p. 128).

In general, if α is any order type, then $*\alpha$ is the same type ordered backwards (Dauben 1990, p. 153).

See also CARDINAL NUMBER, ORDER ISOMORPHIC, ORDINAL NUMBER, TOTALLY ORDERED SET

References

Ciesielski, K. *Set Theory for the Working Mathematician.* Cambridge, England: Cambridge University Press, 1997.

Dauben, J. W. *Georg Cantor: His Mathematics and Philosophy of the Infinite.* Princeton, NJ: Princeton University Press, 1990.

Moore, G. H. *Zermelo's Axiom of Choice: Its Origin, Development, and Influence.* New York: Springer-Verlag, 1982.

Suppes, P. *Axiomatic Set Theory.* New York: Dover, 1972.

Ordered Factorization

An ordered factorization is a factorization (not necessarily into prime factors) in which $a \times b$ is considered distinct from $b \times a$. The number of ordered factorizations of n is equal to the number of PERFECT PARTITIONS of $n-1$ (Goulden and Jackson 1983, p. 94).

See also PERFECT PARTITION

References

Goulden, I. P. and Jackson, D. M. Problem 2.5.12 in *Combinatorial Enumeration.* New York: Wiley, p. 94, 1983.

Ordered Geometry

A GEOMETRY constructed without reference to measurement. The only primitive concepts are those of points and intermediacy. There are 10 AXIOMS underlying ordered GEOMETRY.

See also ABSOLUTE GEOMETRY, AFFINE GEOMETRY, GEOMETRY

Ordered List

The number of nondecreasing lists $\{a_1, a_2, \ldots, a_n\}$ consisting of n elements $1 \leq a_i \leq k$ is given by the binomial coefficient

$$N(n, k) = \binom{n+k-1}{n-1}.$$

For example, there are six nondecreasing lists of length 2 for elements chosen from 1 to 3: (1, 1), (1, 2), (1, 3), (2, 2), (2, 3), and (3,3).

Ordered Pair

A PAIR of quantities (a, b) where ordering is significant, so (a, b) is considered distinct from (b, a) for $a \neq b$.

See also LIST, MULTISET, ORDERED PAIRS REPRESENTATION, PAIR, SET, VECTOR

Ordered Pairs Representation

A representation of a GRAPH in which edges are specified as ordered pairs (for a DIRECTED GRAPH), or unordered pairs (for an UNDIRECTED GRAPH). The ordered pairs representation of a graph g may be computed using `ToOrderedPairs[g]` in the *Mathematica* add-on package `DiscreteMath`Combinatorica`` (which can be loaded with the command `<<DiscreteMath``) or `ToUnorderedPairs[g]`. A graph may be constructed from ordered pairs using `FromOrderedPairs[l]`, or from unordered pairs using `FromUnorderedPairs[l]`.

References

Skiena, S. "Ordered Pairs." §3.1.3 in *Implementing Discrete Mathematics: Combinatorics and Graph Theory with Mathematica.* Reading, MA: Addison-Wesley, pp. 87–88, 1990.

Ordered Set

An ambiguous term which is sometimes used to mean a PARTIALLY ORDERED SET and sometimes to mean a TOTALLY ORDERED SET.

Ordered Tree

A ROOTED TREE in which the order of the subtrees is significant. There is a ONE-TO-ONE correspondence between ordered FORESTS with n nodes and BINARY TREES with n nodes.

See also BINARY TREE, FOREST, ROOTED TREE

Ordering

The number of "ARRANGEMENTS" in an ordering of n items is given by either a COMBINATION (order is ignored) or a PERMUTATION (order is significant).

See also ARRANGEMENT, COMBINATION, CUTTING, DERANGEMENT, PARTIAL ORDER, PERMUTATION, SORTING, TOTAL ORDER

Ordering Axioms

The four of HILBERT'S AXIOMS which concern the arrangement of points.

See also CONGRUENCE AXIOMS, CONTINUITY AXIOMS, HILBERT'S AXIOMS, INCIDENCE AXIOMS, PARALLEL POSTULATE

References
Hilbert, D. *The Foundations of Geometry, 2nd ed.* Chicago, IL: Open Court, 1980.
Iyanaga, S. and Kawada, Y. (Eds.). "Hilbert's System of Axioms." §163B in *Encyclopedic Dictionary of Mathematics.* Cambridge, MA: MIT Press, pp. 544–545, 1980.

Ordinal

ORDINAL NUMBER

Ordinal Addition

Let $(A, \leq)$ and $(B, \leq)$ be disjoint TOTALLY ORDERED SETS with ORDER TYPES α and β. Then the ordinal sum is defined at set $(C = A \cup B, \leq)$ where, if c_1 and c_2 are both from the same SUBSET, the order is the same as in the subset, but if c_1 is from A and c_2 is from B, then $c_1 < c_2$ has ORDER TYPE $\alpha + \beta$ (Ciesielski 1997, p. 48; Dauben 1990, p. 104; Moore 1982, p. 40).

One should note that in the infinite case, ORDER TYPE addition is not commutative, although it is associative. For example,

$$1 + \omega = \omega \neq \omega + 1.$$

In addition, $\{a\} \cup \{0, 1, 2, 3, \ldots\}$, with a the least element, is ORDER ISOMORPHIC to $\{0, 1, 2, 3, \ldots\}$, but not to $\{0, 1, 2, 3, \ldots\} \cup \{a\}$, with a the greatest element, since it has a greatest element and the other does not.

An inductive definition for ordinal addition states that for any ORDINAL NUMBER α,

$$\alpha + 0 = \alpha, \qquad (1)$$

and

$$\alpha + (\text{successor to } \beta) = \text{the successor to } (\alpha + \beta). \quad (2)$$

If β is a LIMIT ORDINAL, then $\alpha + \beta$ is the least ordinal greater than any ordinal in the set $\{\alpha + \gamma : \gamma < \beta\}$ (Rubin 1967, p. 188; Suppes 1972, p. 205).

See also ORDINAL EXPONENTIATION, ORDINAL MULTIPLICATION, ORDINAL NUMBER

References
Ciesielski, K. *Set Theory for the Working Mathematician.* Cambridge, England: Cambridge University Press, 1997.
Dauben, J. W. *Georg Cantor: His Mathematics and Philosophy of the Infinite.* Princeton, NJ: Princeton University Press, 1990.
Moore, G. H. *Zermelo's Axiom of Choice: Its Origin, Development, and Influence.* New York: Springer-Verlag, 1982.
Rubin, J. E. *Set Theory for the Mathematician.* New York: Holden-Day, 1967.
Suppes, P. *Axiomatic Set Theory.* New York: Dover, 1972.

Ordinal Comparison

Let $(A, \leq)$ and $(B, \leq)$ be WELL ORDERED SETS with ORDINAL NUMBERS α and β. Then $\alpha < \beta$ IFF A is ORDER ISOMORPHIC to an INITIAL SEGMENT of B (Dauben 1990, p. 199). From this, it can easily be shown that the ORDINAL NUMBERS are TOTALLY ORDERED by the relation. In fact, they are WELL ORDERED by the relation.

See also WELL ORDERED SET

References
Dauben, J. W. *Georg Cantor: His Mathematics and Philosophy of the Infinite.* Princeton, NJ: Princeton University Press, 1990.

Ordinal Exponentiation

Let α and β be any ORDINAL NUMBERS, then ordinal exponentiation is defined so that if $\beta = 0$ then $\alpha^\beta = 1$. If β is not a LIMIT ORDINAL, then choose γ such that $\gamma + 1 = \beta$,

$$alpha^{(\text{successor of } \beta)}(\alpha^\beta) * a.$$

If β is a LIMIT ORDINAL, then if $\alpha = 0$, $\alpha^\beta = 0$. If $\alpha \neq 0$ then, α^β is the least ordinal greater than any ordinal in the set $\{\alpha^\gamma : \gamma < \beta\}$ (Rubin 1967, p. 204; Suppes 1972, p. 215).

Note that this definition is not analogous to the definition for cardinals, since $|\alpha|^{|\beta|}$ may not equal $|\alpha^\beta|$, even though $|\alpha| + |\beta| = |\alpha + \beta|$ and $|\alpha| * |\beta| = |\alpha * \beta|$. Note also that $2^\omega = \omega$.

A familiar example of ordinal exponentiation is the definition of Cantor's first epsilon number. ϵ_0 is the least ordinal such that $\omega^{\epsilon_0} = \epsilon_0$. It can be shown that it is the least ordinal greater than any ordinal in $\{\omega, \omega^\omega, \omega^{\omega^\omega}, \ldots\}$.

References
Rubin, J. E. *Set Theory for the Mathematician.* New York: Holden-Day, 1967.

Suppes, P. *Axiomatic Set Theory.* New York: Dover, 1972.

Ordinal Multiplication

Let $(A, \leq)$ and $(B, \leq)$ be TOTALLY ORDERED SETS. Let $C = A \times B$ be the CARTESIAN PRODUCT and define order as follows. For any a_1, $a_2 \in A$ and b_1, $b_2 \in B$,

1. If $a_1 < a_2$, then $(a_1, b_1) < (a_2, b_2)$,
2. If $a_1 = a_2$, then (a_1, b_1) and (a_2, b_2) compare the same way as b_1, b_2 (i.e., lexicographical order)

(Ciesielski 1997, p. 48; Rubin 1967; Suppes 1972). However, Dauben (1990, p. 104) and Moore (1982, p. 40) define multiplication in the reverse order.

Like addition, multiplication is not commutative, but it is associative,

$$2 * \omega = \omega \neq \omega * 2. \tag{1}$$

An inductive definition for ordinal multiplication states that for any ORDINAL NUMBER α,

$$\alpha * 0 = 0 \tag{2}$$

$$\alpha * (\text{successor to beta}) = \alpha * \beta + \alpha. \tag{3}$$

If β is a LIMIT ORDINAL, then $\alpha + \beta$ is the least ordinal greater than any ordinal in the set $\{\alpha * \gamma : \gamma < \beta\}$ (Suppes 1972, p. 212).

See also ORDINAL ADDITION, ORDINAL EXPONENTIATION, ORDINAL NUMBER, SUCCESSOR

References

Ciesielski, K. *Set Theory for the Working Mathematician.* Cambridge, England: Cambridge University Press, 1997.

Dauben, J. W. *Georg Cantor: His Mathematics and Philosophy of the Infinite.* Princeton, NJ: Princeton University Press, 1990.

Moore, G. H. *Zermelo's Axiom of Choice: Its Origin, Development, and Influence.* New York: Springer-Verlag, 1982.

Rubin, J. E. *Set Theory for the Mathematician.* New York: Holden-Day, 1967.

Suppes, P. *Axiomatic Set Theory.* New York: Dover, 1972.

Ordinal Number

In common usage, an ordinal number is an adjective which describes the numerical position of an object, e.g., first, second, third, etc.

In formal SET THEORY, an ordinal number (sometimes simply called an "ordinal" for short) is one of the numbers in Georg Cantor's extension of the WHOLE NUMBERS. An ordinal number is defined as the ORDER TYPE of a WELL ORDERED SET (Dauben 1990, p. 199; Moore 1982, p. 52; Suppes 1972, p. 129). Finite ordinal numbers are commonly denoted using arabic numerals, while transfinite ordinals as denoted using lower case Greek letters.

It is easy to see that every *finite* TOTALLY ORDERED SET is WELL ORDERED. Any two TOTALLY ORDERED SETS with k elements (for k a nonnegative integer) are ORDER ISOMORPHIC, and therefore have the same

ORDER TYPE (which is also an ordinal number). The ordinals for finite sets are denoted 0, 1, 2, 3, ..., i.e., the integers one less than the corresponding nonnegative integers.

The first transfinite ordinal, denoted ω, is the ORDER TYPE of the set of nonnegative integers (Dauben 1979, p 152; Moore 1982, p. viii; Rubin 1967, pp. 86 and 177; Suppes 1972, p. 128). This is the "smallest" of Cantor's TRANSFINITE NUMBERS, defined to be the smallest ordinal number greater than the ordinal number of the WHOLE NUMBERS. Conway and Guy (1996) denote it with the notation $\omega = \{0, 1, \ldots |\}$.

From the definition of ORDINAL COMPARISON, is follows that the ordinal numbers are a WELL ORDERED SET. In order of increasing size, the ordinal numbers are 0, 1, 2, ..., ω, $\omega + 1$, $\omega + 2$, ..., $\omega + \omega$, $\omega + \omega + 1$, ... The notation of ordinal numbers can be a bit counter-intuitive, e.g., even though $1 + \omega = \omega$, $\omega + 1 > \omega$. The CARDINALITY of the set of countable ordinal numbers is denoted ALEPH-1.

If $(A, \leq)$ is a WELL ORDERED SET with ordinal number α, then the set of all ordinals $< \alpha$ is ORDER ISOMORPHIC to A. This provides the motivation to define an ordinal as the set of all ordinals less that itself. John von Neumann defined a set α to be an ordinal number IFF

1. If β is a member of α, then β is a PROPER SUBSET of α
2. If β and γ are members of α then one of the following is true: $\beta = \gamma$, β is a member of γ, or γ is a member of β.
3. If B is a nonempty PROPER SUBSET of α, then there exists a γ member of B such that the intersection $\gamma \cap B$ is empty.

(Rubin 1967, p. 176; Ciesielski 1997, p. 44). This is the standard representation of ordinals. In this representation,

symbol	elements	description
0	{}	empty set
1	{0}	set of one element
2	{0, 1}	set of two elements
3	{0, 1, 2}	set of three elements
⋮		
ω	{0, 1, 2, ...}	set of all finite ordinals
$\omega + 1$	{0, 1, 2, ..., ω}	
⋮		

ω_1	set of all countable ordinals
$\vdots$	
ω_2	set of all countable and $\aleph_1$ ordinals
$\vdots$	
ω_ω	set all finite ordinals and $\aleph_k$ ordinals for all nonnegative integers k
$\vdots$	

Rubin (1967, p. 272) provides a nice definition of the ω_α ordinals.

Since for any ordinal α, the union $\alpha \cup \alpha$ is a bigger ordinal $\alpha + 1$, there is no largest ordinal, and the class of all ordinals is therefore a PROPER CLASS (as shown by the BURALI-FORTI PARADOX).

Ordinal numbers have some other rather peculiar properties. The sum of two ordinal numbers can take on two different values, the sum of three can take on five values. The first few terms of this sequence are 2, 5, 13, 33, 81, 193, 449, 33^2, $33 \cdot 81$, 81^2, $81 \cdot 193$, 192^2, ... (Conway and Guy 1996, Sloane's A005348). The sum of n ordinals has either $193^a 81^b$ or $33 \cdot 81^a$ possible answers for $n \geq 15$ (Conway and Guy 1996).

$r \times \omega$ is the same as ω, but $\omega \times r$ is equal to $\underbrace{\omega + \ldots + \omega}$. ω^2 is larger than any number OF THE FORM $\omega \times r$, ω^3 is larger than ω^2, and so on.

There exist ordinal numbers which cannot be constructed from smaller ones by finite additions, multiplications, and exponentiations. These ordinals obey CANTOR'S EQUATION. The first such ordinal is

$$\epsilon_0 = \underbrace{\omega^{\omega^{\cdot^{\cdot^{\omega}}}}}_{\omega} = 1 + \omega + \omega^\omega + \omega^{\omega^\omega} + \ldots.$$

The next is

$$\epsilon_1 = (1 + \epsilon_0) + \omega^{\epsilon_0 + 1} + \omega^{\omega^{\epsilon_0 + 1}} + \ldots,$$

then follow ϵ_2, ϵ_3, ..., ϵ_ω, $\epsilon_{\omega+1}$, ..., $\epsilon_{\omega \times 2}$, ..., ϵ_{ω^2}, ϵ_{ω^ω}, ..., ϵ_{ϵ_0}, $\epsilon_{\epsilon_0 + 1}$, ..., $\epsilon_{\epsilon_0 + \omega}$, ..., $\epsilon_{\epsilon_0 + \omega}$, ..., $\epsilon_{\epsilon_0 \times 2}$, ..., ϵ_{ϵ_1}, ..., ϵ_{ϵ_2}, ..., $\epsilon_{\epsilon_\omega}$, ..., $\epsilon_{\epsilon_{\epsilon_1}}$, ..., $\epsilon_{\epsilon_{\epsilon_\omega}}$, ..., $\epsilon_{\epsilon_{\epsilon_0}}$, ... (Conway and Guy 1996).

ORDINAL ADDITION, ORDINAL MULTIPLICATION, and ORDINAL EXPONENTIATION can all be defined. Although these definitions also work perfectly well for ORDER TYPES, this does not seem to be commonly done. There are two methods common used to define operations on the ordinals: one is using sets, and the other is inductively.

See also ALEPH-1, AXIOM OF CHOICE, BURALI-FORTI PARADOX, CANTOR'S EQUATION, CARDINALITY, CARDINAL NUMBER, INITIAL ORDINAL, ORDER STATISTIC, ORDER TYPE, POWER SET, SURREAL NUMBER, WELL

ORDERED SET

References

Cantor, G. *Über unendliche, lineare Punktmannigfältigkeiten, Arbeiten zur Mengenlehre aus dem Jahren 1872–1884.* Leipzig, Germany: Teubner-Archiv zur Mathematik, 1884.

Conway, J. H. and Guy, R. K. "Cantor's Ordinal Numbers." In *The Book of Numbers.* New York: Springer-Verlag, pp. 266–267 and 274, 1996.

Dauben, J. W. *Georg Cantor: His Mathematics and Philosophy of the Infinite.* Princeton, NJ: Princeton University Press, 1990.

Moore, G. H. *Zermelo's Axiom of Choice: Its Origin, Development, and Influence.* New York: Springer-Verlag, 1982.

Suppes, P. *Axiomatic Set Theory.* New York: Dover, 1972.

Sloane, N. J. A. Sequences A005348/M1435 in "An On-Line Version of the Encyclopedia of Integer Sequences." http://www.research.att.com/~njas/sequences/eisonline.html.

Ordinary Differential Equation

An ordinary differential equation (frequently abbreviated ODE) is an equality involving a function and its DERIVATIVES. An ODE of order n is an equation OF THE FORM

$$F(x, y, y', \cdots, y^{(n)}) = 0, \tag{1}$$

where $y' = dy/dx$ is a first DERIVATIVE with respect to x and $y^{(n)} = d^n y/dx^n$ is an nth DERIVATIVE with respect to x. An ODE of order n is said to be linear if it is OF THE FORM

$$a_n(x) y^{(n)} + a_{n-1}(x) y^{(n-1)} + \cdots + a_1(x) y' + a_0(x) y = Q(x). \tag{2}$$

A linear ODE where $Q(x) = 0$ is said to be homogeneous. Confusingly, an ODE OF THE FORM

$$\frac{dy}{dx} = f\left(\frac{y}{x}\right) \tag{3}$$

is also sometimes called "homogeneous."

In general, an nth-order ODE has n linearly independent solutions. Furthermore, any LINEAR COMBINATION of LINEARLY INDEPENDENT FUNCTIONS solutions is also a solution.

Simple theories exist for first-order (INTEGRATING FACTOR) and second-order (STURM-LIOUVILLE THEORY) ordinary differential equations, and arbitrary ODEs with linear constant COEFFICIENTS can be solved when they are of certain factorable forms. Integral transforms such as the LAPLACE TRANSFORM can also be used to solve classes of linear ODEs. Morse and Feshbach (1953, pp. 667–674) give canonical forms and solutions for second-order ODEs.

While there are many general techniques for analytically solving classes of ODEs, the only practical solution technique for complicated equations is to use numerical methods (Milne 1970, Jeffreys and Jeffreys 1988). The most popular of these is the RUNGE-KUTTA

METHOD, but many others have been developed, including the COLLOCATION METHOD and GALERKIN METHOD. A vast amount of research and huge numbers of publications have been devoted to the numerical solution of differential equations, both ordinary and PARTIAL (PDEs) as a result of their importance in fields as diverse as physics, engineering, economics, and electronics.

The solutions to an ODE satisfy EXISTENCE and UNIQUENESS properties. These can be formally established by PICARD'S EXISTENCE THEOREM for certain classes of ODEs. Let a system of first-order ODE be given by

$$\frac{dx_i}{dt} = f_i(x_1, \ldots, x_n, t), \tag{4}$$

for $i = 1, \ldots, n$ and let the functions $f_i(x_1, \ldots, x_n, t)$, where $i = 1, \ldots, n$, all be defined in a DOMAIN D of the $(n + 1)$-D space of the variables $x_1, \ldots, x_n, t$. Let these functions be continuous in D and have continuous first PARTIAL DERIVATIVES $\partial f_i / \partial x_j$ for $i = 1, \ldots, n$ and $j = 1, \ldots, n$ in D. Let $(x_1^0, \ldots, x_n^0)$ be in D. Then there exists a solution of (4) given by

$$x_1 = x_1(t), \ldots, x_n = x_n(t) \tag{5}$$

for $t_0 - \delta < t < t_0 + \delta$ (where $\delta > 0$) satisfying the initial conditions

$$x_1(t_0) = x_1^0, \ldots, x_n(t_0) = x_n^0. \tag{6}$$

Furthermore, the solution is unique, so that if

$$x_1 = x_1{}^*(t), \ldots, x_n = x_n{}^*(t) \tag{7}$$

is a second solution of (4) for $t_0 - \delta < t < t_0 + \delta$ satisfying (6), then $x_i(t) \equiv x_i{}^*(t)$ for $t_0 - \delta < t < t_0 + \delta$. Because every nth-order ODE can be expressed as a system of n first-order differential equations, this theorem also applies to the single nth-order ODE.

An exact FIRST-ORDER ODES is one OF THE FORM

$$p(x, y)\, dx + q(x, y)\, dy = 0, \tag{8}$$

where

$$\frac{\partial p}{\partial y} = \frac{\partial q}{\partial x}. \tag{9}$$

An equation OF THE FORM (8) with

$$\frac{\partial p}{\partial y} \neq \frac{\partial q}{\partial x} \tag{10}$$

is said to be nonexact. If

$$\frac{\dfrac{\partial p}{\partial y} - \dfrac{\partial q}{\partial x}}{q} = f(x) \tag{11}$$

in (8), it has an x-dependent integrating factor. If

$$\frac{\dfrac{\partial q}{\partial x} - \dfrac{\partial p}{\partial y}}{xp - yq} = f(xy) \tag{12}$$

in (8), it has an xy-dependent integrating factor. If

$$\frac{\dfrac{\partial q}{\partial x} - \dfrac{\partial p}{\partial y}}{p} = f(y) \tag{13}$$

in (8), it has a y-dependent integrating factor.

Other special first-order types include cross multiple equations

$$yf(xy)\, dx + xg(xy)\, dy = 0, \tag{14}$$

homogeneous equations

$$\frac{dy}{dx} = f\left(\frac{y}{x}\right), \tag{15}$$

linear equations

$$\frac{dy}{dx} + p(x)y = q(x), \tag{16}$$

and separable equations

$$\frac{dy}{dx} = X(x)Y(y). \tag{17}$$

Special classes of SECOND-ORDER ODES include

$$\frac{d^2y}{dx^2} = f(y, y') \tag{18}$$

(x missing) and

$$\frac{d^2y}{dx^2} = f(x, y') \tag{19}$$

(y missing). A second-order linear homogeneous ODE

$$\frac{d^2y}{dx^2} + P(x)\frac{dy}{dx} + Q(x)y = 0 \tag{20}$$

for which

$$\frac{Q'(x) + 2P(x)Q(x)}{2[Q(x)]^{3/2}} = [\text{constant}] \tag{21}$$

can be transformed to one with constant coefficients. The undamped equation of SIMPLE HARMONIC MOTION is

$$\frac{d^2y}{dx^2} + \omega_0^2 y = 0, \tag{22}$$

which becomes

$$\frac{d^2y}{dx^2} + \beta\,\frac{dy}{dx} + \omega_0^2 y = 0 \tag{23}$$

when damped, and

$$\frac{d^2y}{dx^2} + \beta \frac{dy}{dx} + \omega_0^2 y = A \cos(\omega t) \tag{24}$$

when both forced and damped.

SYSTEMS WITH CONSTANT COEFFICIENTS are of the form

$$\frac{d\mathbf{x}}{dt} = A\mathbf{x}(t) + \mathbf{p}(t). \tag{25}$$

The following are examples of important ordinary differential equations which commonly arise in problems of mathematical physics.

ABEL'S DIFFERENTIAL EQUATION

$$y' = f_0(x) + f_1(x)y + f_2(x)y^2 + f_3(x)y^3 + \dots \tag{26}$$

$$[g_0(x) + g_1(x)y]y' = f_0(x) + f_1(x)y + f_2(x)y^2 + f_3(x)y^3. \tag{27}$$

AIRY DIFFERENTIAL EQUATION

$$\frac{d^2y}{dx^2} - xy = 0. \tag{28}$$

ANGER DIFFERENTIAL EQUATION

$$y'' + \frac{y'}{x} + \left(1 - \frac{v^2}{x^2}\right)y = \frac{x-v}{\pi x^2} \sin(vx). \tag{29}$$

BAER DIFFERENTIAL EQUATIONS

$$(x-a_1)(x-a_2)y'' + \tfrac{1}{2}[2x - (a_1+a_2)]y' - (p^2x + q^2)y = 0, \tag{30}$$

$$(x-a_1)(x-a_2)y'' + \tfrac{1}{2}[2x - (a_1+a_2)]y' - (k^2x^2 - p^2x + q^2)y = 0. \tag{31}$$

BERNOULLI DIFFERENTIAL EQUATION

$$\frac{dy}{dx} + p(x)y = q(x)y^n. \tag{32}$$

BESSEL DIFFERENTIAL EQUATION

$$x^2 \frac{d^2y}{dx^2} + x \frac{dy}{dx} + (\lambda^2 x^2 - n^2)y = 0. \tag{33}$$

BINOMIAL DIFFERENTIAL EQUATION

$$(y')^m = f(x, y). \tag{34}$$

BÔCHER EQUATION

$$y'' + \tfrac{1}{2}\left[\frac{m_1}{x-a_1} + \dots + \frac{m_{n-1}}{x-a_{n-1}}\right]y'$$

$$+ \tfrac{1}{4}\left[\frac{A_0 + A_1x + \dots + A_lx^l}{(x-a_1)^{m_1}(x-a_2)^{m_2}\cdots(x-a_{n-1})^{m_{n-1}}}\right]y = 0. \tag{35}$$

BRIOT-BOUQUET EQUATION

$$x^m \frac{dy}{dx} = f(x, y). \tag{36}$$

CHEBYSHEV DIFFERENTIAL EQUATION

$$(1-x^2)\frac{d^2y}{dx^2} - x \frac{dy}{dx} + \alpha^2 y = 0. \tag{37}$$

CLAIRAUT'S DIFFERENTIAL EQUATION

$$y = x \frac{dy}{dx} + f\left(\frac{dy}{dx}\right). \tag{38}$$

CONFLUENT HYPERGEOMETRIC DIFFERENTIAL EQUATION

$$x \frac{d^2y}{dx^2} + (c-x)\frac{dy}{dx} - ay = 0. \tag{39}$$

D'ALEMBERT'S EQUATION.

$$y = xf(y') + g(y'). \tag{40}$$

DUFFING DIFFERENTIAL EQUATION

$$\ddot{x} + \omega_0^2 x + \beta x^3 = 0. \tag{41}$$

ECKART DIFFERENTIAL EQUATION

$$y'' + \left[\frac{\alpha\eta}{1+\eta} + \frac{\beta\eta}{(1+n)^2} + \gamma\right]y = 0, \tag{42}$$

where $\eta = e^{\delta x}$.

EMDEN-FOWLER DIFFERENTIAL EQUATION

$$(x^p y')' \pm x^\sigma y^n = 0. \tag{43}$$

EULER DIFFERENTIAL EQUATION

$$x^2 \frac{d^2y}{dx^2} + ax \frac{dy}{dx} + by = S(x). \tag{44}$$

HALM'S DIFFERENTIAL EQUATION

$$(1+x^2)^2 y'' + \lambda y = 0. \tag{45}$$

HERMITE DIFFERENTIAL EQUATION

$$\frac{d^2y}{dx^2} - 2x \frac{dy}{dx} + \lambda y = 0. \tag{46}$$

HEUN'S DIFFERENTIAL EQUATION

$$\frac{d^2w}{dx^2} + \left(\frac{\gamma}{x} + \frac{\delta}{x-1} + \frac{\varepsilon}{x-a}\right)\frac{dw}{dx} + \frac{\alpha\beta x - q}{x(x-1)(x-a)} w = 0. \tag{47}$$

HILL'S DIFFERENTIAL EQUATION

$$\frac{d^2y}{dx^2}\left[\theta_0 + 2 \sum_{n=1}^{\infty} \theta_n \cos(2nz)\right] = 0. \tag{48}$$

HYPERGEOMETRIC DIFFERENTIAL EQUATION

$$x(x-1)\frac{d^2y}{dx^2}+[(1+\alpha+\beta)x-\gamma]\frac{dy}{dx}+\alpha\beta y=0. \quad (49)$$

JACOBI DIFFERENTIAL EQUATION

$$(1-x^2)y''+[\beta-\alpha-(\alpha+\beta+2)x]y'+n(n+\alpha+\beta+1)y$$
$$=0. \quad (50)$$

LAGUERRE DIFFERENTIAL EQUATION

$$x\,\frac{d^2y}{dx^2}+(1-x)\frac{dy}{dx}+\lambda y=0. \quad (51)$$

LAMÉ'S DIFFERENTIAL EQUATION

$$(x^2-b^2)(x^2-c^2)\frac{d^2z}{dx^2}+x(x^2-b^2+x^2-c^2)\frac{dz}{dx}$$
$$-[m(m+1)x^2-(b^2+c^2)p]z=0. \quad (52)$$

LANE-EMDEN DIFFERENTIAL EQUATION

$$\frac{1}{\xi^2}\frac{d}{d\xi}\left(\xi^2\frac{d\theta}{d\xi}\right)+\theta^n=0. \quad (53)$$

LEGENDRE DIFFERENTIAL EQUATION

$$(1-x^2)\frac{d^2y}{dx^2}-2x\,\frac{dy}{dx}+\alpha(\alpha+1)y=0. \quad (54)$$

LINEAR CONSTANT COEFFICIENTS

$$a_0\,\frac{d^ny}{dx^n}+\ldots+a_{n-1}\,\frac{dy}{dx}+a_ny=p(x). \quad (55)$$

LOMMEL DIFFERENTIAL EQUATION

$$z^2\,\frac{d^2y}{dz^2}+z\,\frac{dy}{dz}-(z^2+v^2)y=kz^{\mu+1}. \quad (56)$$

LÖWNER'S DIFFERENTIAL EQUATION

$$y'=-y\,\frac{1+\kappa(x)y}{1-\kappa(x)y}.$$

MALMSTÉN'S DIFFERENTIAL EQUATION

$$\frac{d^2y}{dx^2}+\frac{r}{z}\frac{dy}{dx}=\left(Az^m+\frac{s}{z^2}\right)y. \quad (57)$$

MATHIEU DIFFERENTIAL EQUATION

$$\frac{d^2V}{dv^2}+[a-2q\,\cos(2v)]V=0. \quad (58)$$

MODIFIED BESSEL DIFFERENTIAL EQUATION

$$x^2\,\frac{d^2y}{dx^2}+x\,\frac{dy}{dx}-(x^2+n^2)y=0. \quad (59)$$

MODIFIED SPHERICAL BESSEL DIFFERENTIAL EQUATION

$$r^2\,\frac{d^2R}{dr^2}+2r\,\frac{dR}{dr}-[k^2r^2+n(n+1)]R=0. \quad (60)$$

RAYLEIGH DIFFERENTIAL EQUATION

$$y''-\mu\left(1-\tfrac{1}{3}y'^2\right)y'+y=0. \quad (61)$$

RICCATI DIFFERENTIAL EQUATION

$$\frac{dw}{dx}=q_0(x)+q_1(x)w+q_2(x)w^2. \quad (62)$$

RIEMANN P-DIFFERENTIAL EQUATION

$$\frac{d^2u}{dz^2}+\left[\frac{1-\alpha-\alpha'}{z-a}+\frac{1-\beta-\beta'}{z-b}+\frac{1-\gamma-\gamma'}{z-c}\right]\frac{du}{dz}$$
$$+\left[\frac{\alpha\alpha'(a-b)(a-c)}{z-a}+\frac{\beta\beta'(b-c)(b-a)}{z-b}+\frac{\gamma\gamma'(c-a)(c-b)}{z-c}\right]$$
$$\times\frac{u}{(z-a)(z-b)(z-c)}=0. \quad (63)$$

SHARPE'S DIFFERENTIAL EQUATION

$$zy''+y'+(z+A)y=0. \quad (64)$$

SPHERICAL BESSEL DIFFERENTIAL EQUATION

$$r^2\,\frac{d^2R}{dr^2}+2r\,\frac{dR}{dr}+[k^2r^2-n(n+1)]R=0. \quad (65)$$

STRUVE DIFFERENTIAL EQUATION

$$z^2y''+zy'+(z^2-v^2)y=\frac{4\left(\frac{1}{2}z\right)^{v+1}}{\sqrt{\pi}\Gamma\left(v+\frac{1}{2}\right)}. \quad (66)$$

STURM-LIOUVILLE EQUATION

$$\frac{d}{dx}\left[p(x)\frac{dy}{dx}\right]+[\lambda w(x)-q(x)]y=0. \quad (67)$$

ULTRASPHERICAL DIFFERENTIAL EQUATION

$$(1-x^2)y''-(2\alpha+1)xy'+n(n+2\alpha)y=0. \quad (68)$$

VAN DER POL EQUATION

$$y''-\mu\left(1-y^2\right)y'+y=0. \quad (69)$$

WEBER DIFFERENTIAL EQUATION

$$\frac{d^2y}{dz^2}+\left(n+\tfrac{1}{2}-\tfrac{1}{4}z^2\right)y=0. \quad (70)$$

WHITTAKER DIFFERENTIAL EQUATION

$$\frac{d^2u}{dz^2} + \frac{du}{dz} + \left(\frac{k}{z} + \frac{\frac{1}{4} - m^2}{z^2}\right)u = 0. \tag{71}$$

See also ADAMS' METHOD, GREEN'S FUNCTION, ISO-CLINE, LAPLACE TRANSFORM, LEADING ORDER ANALYSIS, MAJORANT, ORDINARY DIFFERENTIAL EQUATION–FIRST-ORDER, ORDINARY DIFFERENTIAL EQUATION–SECOND-ORDER, PARTIAL DIFFERENTIAL EQUATION, RELAXATION METHODS, RUNGE-KUTTA METHOD, SIMPLE HARMONIC MOTION

References
Boyce, W. E. and DiPrima, R. C. *Elementary Differential Equations and Boundary Value Problems, 5th ed.* New York: Wiley, 1992.
Braun, M. *Differential Equations and Their Applications, 4th ed.* New York: Springer-Verlag, 1993.
Carroll, J. "A Composite Integration Scheme for the Numerical Solution of Systems of Ordinary Differential Equations." *J. Comput. Appl. Math.* **25**, 1–13, 1989.
Coddington, E. A. *An Introduction to Ordinary Differential Equations.* New York: Dover, 1989.
Forsyth, A. R. *Theory of Differential Equations, 6 vols.* New York: Dover, 1959.
Forsyth, A. R. *A Treatise on Differential Equations.* New York: Dover, 1997.
Fulford, G.; Forrester, P.; and Jones, A. *Modelling with Differential and Difference Equations.* New York: Cambridge University Press, 1997.
Guterman, M. M. and Nitecki, Z. H. *Differential Equations: A First Course, 3rd ed.* Philadelphia, PA: Saunders, 1992.
Hull, T. E.; Enright, W. H.; Fellen, B. M.; and Sedgwick, A. E. "Comparing Numerical Methods for Ordinary Differential Equations." *SIAM J. Numer. Anal.* **9**, 603–637, 1972.
Hull, T. E.; Enright, W. H.; Fellen, B. M.; and Sedgwick, A. E. "Erratum to 'Comparing Numerical Methods for Ordinary Differential Equations.'" *SIAM J. Numer. Anal.* **11**, 681, 1974.
Ince, E. L. *Ordinary Differential Equations.* New York: Dover, 1956.
Jeffreys, H. and Jeffreys, B. S. "Numerical Solution of Differential Equations." *Methods of Mathematical Physics, 3rd ed.* Cambridge, England: Cambridge University Press, pp. 290–301, 1988.
Kamke, E. *Differentialgleichungen: Lösungsmethoden und Lösungen, Bd. 1: Gewöhnliche Differentialgleichungen, 9. Aufl.* Stuttgart, Germany: Teubner, 1983.
Milne, W. E. *Numerical Solution of Differential Equations.* New York: Dover, 1970.
Morse, P. M. and Feshbach, H. "Ordinary Differential Equations." Ch. 5 in *Methods of Theoretical Physics, Part I.* New York: McGraw-Hill, pp. 492–675, 1953.
Moulton, F. R. *Differential Equations.* New York: Dover, 1958.
Polyanin, A. D. and Zaitsev, V. F. *Handbook of Exact Solutions for Ordinary Differential Equations.* Boca Raton, FL: CRC Press, 1995.
Postel, F. and Zimmermann, P. "A Review of the ODE Solvers of Axiom, Derive, Macsyma, Maple, Mathematica, MuPad, and Reduce." Submitted to *The 5th Rhine Workshop on Computer Algebra.* July 26, 1996. http://www.loria.fr/~zimmerma/ComputerAlgebra/ode_comp.ps.gz.
Press, W. H.; Flannery, B. P.; Teukolsky, S. A.; and Vetterling, W. T. "Integration of Ordinary Differential Equations." Ch. 16 in *Numerical Recipes in FORTRAN: The Art of Scientific Computing, 2nd. ed.* Cambridge, England: Cambridge University Press, pp. 701–744, 1992.
Simmons, G. F. *Differential Equations, with Applications and Historical Notes, 2nd ed.* New York: McGraw-Hill, 1991.
Weisstein, E. W. "Books about Ordinary Differential Equations." http://www.treasure-troves.com/books/OrdinaryDifferentialEquations.html.
Zwillinger, D. *Handbook of Differential Equations, 3rd ed.* Boston, MA: Academic Press, 1997.

Ordinary Differential Equation—First-Order

Given a first-order ORDINARY DIFFERENTIAL EQUATION

$$\frac{dy}{dx} = F(x, y), \tag{1}$$

if $F(x, y)$ can be expressed using SEPARATION OF VARIABLES as

$$F(x, y) = X(x)Y(y), \tag{2}$$

then the equation can be expressed as

$$\frac{dy}{Y(y)} = X(x)\,dx \tag{3}$$

and the equation can be solved by integrating both sides to obtain

$$\int \frac{dy}{Y(y)} = \int X(x)\,dx. \tag{4}$$

Any first-order ODE OF THE FORM

$$\frac{dy}{dx} + p(x)y = q(x) \tag{5}$$

can be solved by finding an INTEGRATING FACTOR $\mu = \mu(x)$ such that

$$\frac{d}{dx}(\mu y) = \mu\frac{dy}{dx} + y\frac{d\mu}{dx} = \mu q(x). \tag{6}$$

Dividing through by μy yields

$$\frac{1}{y}\frac{dy}{dx} + \frac{1}{\mu}\frac{d\mu}{dx} = \frac{q(x)}{y}. \tag{7}$$

However, this condition enables us to explicitly determine the appropriate μ for arbitrary p and q. To accomplish this, take

$$p(x) = \frac{1}{\mu}\frac{d\mu}{dx} \tag{8}$$

in the above equation, from which we recover the original equation (5), as required, in the form

$$\frac{1}{y}\frac{dy}{dx} + p(x) = \frac{q(x)}{y}. \tag{9}$$

But we can integrate both sides of (8) to obtain

$$\int p(x)\,dx = \int \frac{d\mu}{\mu} = \ln \mu + c \qquad (10)$$

$$\mu = e^{\int p(x)\,dx}. \qquad (11)$$

Now integrating both sides of (6) gives

$$\mu y = \int \mu q(x)\,dx + c \qquad (12)$$

(with μ now a known function), which can be solved for y to obtain

$$y = \frac{\int \mu q(x)\,dx + c}{\mu} = \frac{\int e^{\int^x p(x')\,dx'} q(x)\,d(x) + c}{e^{\int^x p(x')\,dx'}}, \qquad (13)$$

where c is an arbitrary constant of integration.

Given an nth-order linear ODE with constant COEF-FICIENTS

$$\frac{d^n y}{dx^n} + a_{n-1}\frac{d^{n-1}y}{dx^{n-1}} + \ldots a_1 \frac{dy}{dx} + a_0 y = Q(x), \qquad (14)$$

first solve the characteristic equation obtained by writing

$$y \equiv e^{rx} \qquad (15)$$

and setting $Q(x) = 0$ to obtain the n COMPLEX ROOTS.

$$r^n e^{rx} + a_{n-1} r^{n-1} e^{rx} + \ldots + a_1 r e^{rx} + a_0 e^{rx} = 0 \qquad (16)$$

$$r^n + a_{n-1} r^{n-1} + \ldots + a_1 r + a_0 = 0. \qquad (17)$$

Factoring gives the ROOTS r_i,

$$(r - r_1)(r - r_2)\cdots(r - r_n) = 0. \qquad (18)$$

For a nonrepeated REAL ROOT r, the corresponding solution is

$$y = e^{rx}. \qquad (19)$$

If a REAL ROOT r is repeated k times, the solutions are degenerate and the linearly independent solutions are

$$y = e^{rx}, y = xe^{rx}, \cdots, y = x^{k-1}e^{rx}. \qquad (20)$$

Complex ROOTS always come in COMPLEX CONJUGATE pairs, $r_\pm = a \pm ib$. For nonrepeated COMPLEX ROOTS, the solutions are

$$y = e^{ax}\cos(bx), y = e^{ax}\sin(bx). \qquad (21)$$

If the COMPLEX ROOTS are repeated k times, the linearly independent solutions are

$$y = e^{ax}\cos(bx), y = e^{ax}\sin(bx), \cdots,$$

$$y = x^{k-1}e^{ax}\cos(bx), y = x^{k-1}e^{ax}\sin(bx). \qquad (22)$$

Linearly combining solutions of the appropriate types with arbitrary multiplicative constants then gives the complete solution. If initial conditions are specified,

the constants can be explicitly determined. For example, consider the sixth-order linear ODE

$$(\tilde{D} - 1)(\tilde{D} - 2)^3(\tilde{D}^2 + \tilde{D} + 1)y = 0, \qquad (23)$$

which has the characteristic equation

$$(r - 1)(r - 2)^3(r^2 + r + 1) = 0. \qquad (24)$$

The roots are 1, 2 (three times), and $(-1 \pm \sqrt{3}i)/2$, so the solution is

$$y = Ae^x + Be^{2x} + Cxe^{2x} + Dx^2 e^{3x} + Ee^{-x/2}\cos\left(\tfrac{1}{2}\sqrt{3}x\right)$$

$$+ Fe^{-x}\sin\left(\tfrac{1}{2}\sqrt{3}x\right). \qquad (25)$$

If the original equation is nonhomogeneous ($Q(x) \neq 0$), now find the particular solution y^* by the method of VARIATION OF PARAMETERS. The general solution is then

$$y(x) = \sum_{i=1}^n c_i y_i(x) + y^*(x), \qquad (26)$$

where the solutions to the linear equations are $y_1(x)$, $y_2(x)$, ..., $y_n(x)$, and $y^*(x)$ is the particular solution.

See also INTEGRATING FACTOR, ORDINARY DIFFEREN-TIAL EQUATION–FIRST-ORDER EXACT, SEPARATION OF VARIABLES, VARIATION OF PARAMETERS

References

Arfken, G. *Mathematical Methods for Physicists, 3rd ed.* Orlando, FL: Academic Press, pp. 440–445, 1985.

Ordinary Differential Equation—First-Order Exact

Consider a first-order ODE in the slightly different form

$$p(x,\, y)\,dx + q(x,\, y)\,dy = 0. \qquad (1)$$

Such an equation is said to be exact if

$$\frac{\partial p}{\partial y} = \frac{\partial q}{\partial x}. \qquad (2)$$

This statement is equivalent to the requirement that a CONSERVATIVE FIELD exists, so that a scalar potential can be defined. For an exact equation, the solution is

$$\int_{(x_0,\, y_0)}^{(x,\, y)} p(x,\, y)\,dx + q(x,\, y)\,dy = c, \qquad (3)$$

where c is a constant.

A first-order ODE (1) is said to be inexact if

$$\frac{\partial p}{\partial y} \neq \frac{\partial q}{\partial x}. \qquad (4)$$

For a nonexact equation, the solution may be obtained by defining an INTEGRATING FACTOR μ of (6) so that the new equation

$$\mu p(x,\, y)\, dx + \mu q(x,\, y)\, dy = 0 \qquad (5)$$

satisfies

$$\frac{\partial}{\partial y}(\mu p) = \frac{\partial}{\partial x}(\mu q), \qquad (6)$$

or, written out explicitly,

$$p\,\frac{\partial \mu}{\partial y} + \mu\,\frac{\partial p}{\partial y} = q\,\frac{\partial \mu}{\partial x} + \mu\,\frac{\partial p}{\partial x}. \qquad (7)$$

This transforms the nonexact equation into an exact one. Solving (7) for μ gives

$$\mu = \frac{q\,\dfrac{\partial \mu}{\partial x} - p\,\dfrac{\partial \mu}{\partial y}}{\dfrac{\partial p}{\partial y} - \dfrac{\partial q}{\partial x}}. \qquad (8)$$

Therefore, if a function μ satisfying (8) can be found, then writing

$$P(x,\, y) = \mu p \qquad (9)$$

$$Q(x,\, y) = \mu q \qquad (10)$$

in equation (5) then gives

$$P(x,\, y)\, dx + Q(x,\, y)\, dy = 0, \qquad (11)$$

which is then an exact ODE. Special cases in which μ can be found include x-dependent, xy-dependent, and y-dependent integrating factors.

Given an inexact first-order ODE, we can also look for an INTEGRATING FACTOR $\mu(x)$ so that

$$\frac{\partial \mu}{\partial y} = 0. \qquad (12)$$

For the equation to be exact in μp and μq, the equation for a first-order nonexact ODE

$$p\,\frac{\partial \mu}{\partial y} + \mu\,\frac{\partial p}{\partial y} = q\,\frac{\partial \mu}{\partial x} + \mu\,\frac{\partial p}{\partial x} \qquad (13)$$

becomes

$$\mu\,\frac{\partial p}{\partial y} = q\,\frac{\partial \mu}{\partial x} + \mu\,\frac{\partial p}{\partial x}. \qquad (14)$$

Solving for $\partial \mu / \partial x$ gives

$$\frac{\partial \mu}{\partial x} = \mu(x)\frac{\dfrac{\partial p}{\partial y} - \dfrac{\partial q}{\partial x}}{q} \equiv f(x,\, y)\mu(x), \qquad (15)$$

which will be integrable if

$$f(x,\, y) \equiv \frac{\dfrac{\partial p}{\partial y} - \dfrac{\partial q}{\partial x}}{q} = f(x), \qquad (16)$$

in which case

$$\frac{d\mu}{\mu} = f(x)\, dx, \qquad (17)$$

so that the equation is integrable

$$\mu(x) = e^{\int f(x)\, dx}, \qquad (18)$$

and the equation

$$[\mu p(x,\, y)]\, dx + [\mu q(x,\, y)]\, dy = 0 \qquad (19)$$

with known $\mu(x)$ is now exact and can be solved as an exact ODE.

Given in an exact first-order ODE, look for an INTEGRATING FACTOR $\mu(x,\, y) = g(xy)$. Then

$$\frac{\partial \mu}{\partial x} = \frac{\partial g}{\partial x}\, y. \qquad (20)$$

$$\frac{\partial \mu}{\partial y} = \frac{\partial g}{\partial y}\, x. \qquad (21)$$

Combining these two,

$$\frac{\partial \mu}{\partial x} = \frac{y}{x}\,\frac{\partial \mu}{\partial y}. \qquad (22)$$

For the equation to be exact in μp and μq, the equation for a first-order nonexact ODE

$$p\,\frac{\partial \mu}{\partial y} + \mu\,\frac{\partial p}{\partial y} = q\,\frac{\partial \mu}{\partial x} + \mu\,\frac{\partial p}{\partial x} \qquad (23)$$

becomes

$$\frac{\partial \mu}{\partial y}\left(p - \frac{y}{x}\, q\right) = \left(\frac{\partial p}{\partial x} - \frac{\partial p}{\partial y}\right)\mu. \qquad (24)$$

Therefore,

$$\frac{1}{x}\,\frac{\partial \mu}{\partial y} = \frac{\dfrac{\partial q}{\partial x} - \dfrac{\partial p}{\partial y}}{xp - yq}\,\mu. \qquad (25)$$

Define a new variable

$$t(x,\, y) \equiv xy, \qquad (26)$$

then $\partial t / \partial y = x$, so

$$\frac{\partial \mu}{\partial t} = \frac{\partial \mu}{\partial y}\frac{\partial y}{\partial t} = \frac{\dfrac{\partial q}{\partial x} - \dfrac{\partial p}{\partial y}}{xp - yq}\,\mu(t) \equiv f(x,\, y)\mu(t), \qquad (27)$$

Now, if

$$f(x,\,y) \equiv \frac{\dfrac{\partial q}{\partial x} - \dfrac{\partial p}{\partial y}}{xp - yq} = f(xy) = f(t), \tag{28}$$

then

$$\frac{\partial \mu}{\partial t} = f(t)\mu(t), \tag{29}$$

so that

$$\mu = e^{\int f(t)\,dt} \tag{30}$$

and the equation

$$[\mu p(x,\,y)]\,dx + [\mu q(x,\,y)]\,dy = 0 \tag{31}$$

is now exact and can be solved as an exact ODE.

Given an inexact first-order ODE, assume there exists an integrating factor

$$\mu = f(y), \tag{32}$$

so $\partial \mu / \partial x = 0$. For the equation to be exact in μp and μq, equation (7) becomes

$$\frac{\partial \mu}{\partial y} = \frac{\dfrac{\partial q}{\partial x} - \dfrac{\partial p}{\partial y}}{p}\,\mu = f(x,\,y)\mu(y). \tag{33}$$

Now, if

$$f(x,\,y) \equiv \frac{\dfrac{\partial q}{\partial x} - \dfrac{\partial p}{\partial y}}{p} = f(y), \tag{34}$$

then

$$\frac{d\mu}{\mu} = f(y)\,dy, \tag{35}$$

so that

$$\mu(y) = e^{\int f(y)\,dy}, \tag{36}$$

and the equation

$$\mu p(x,\,y)\,dx + \mu q(x,\,y)\,dy = 0 \tag{37}$$

is now exact and can be solved as an exact ODE.

Given a first-order ODE OF THE FORM

$$yf(xy)\,dx + xg(xy)\,dy = 0, \tag{38}$$

define

$$v \equiv xy. \tag{39}$$

Then the solution is

$$\begin{cases} \ln x = \displaystyle\int \frac{g(v)\,dv}{c[g(v) - f(v)]} + c & \text{for } g(v) \neq f(v) \\ xy = c & \text{for } g(v) = f(v). \end{cases} \tag{40}$$

If

$$\frac{dy}{dx} = F(x,\,y) = G(v), \tag{41}$$

where

$$v \equiv \frac{y}{x}, \tag{42}$$

then letting

$$y \equiv xv \tag{43}$$

gives

$$\frac{dy}{dx} = x\,dv/dx + v \tag{44}$$

$$x\,\frac{dv}{dx} + v = G(v). \tag{45}$$

This can be integrated by quadratures, so

$$\ln x = \int \frac{dv}{f(v) - v} + c \quad \text{for } f(v) \neq v \tag{46}$$

$$y = cx \quad \text{for } f(v) = v. \tag{47}$$

References

Boyce, W. E. and DiPrima, R. C. *Elementary Differential Equations and Boundary Value Problems,* 4th ed. New York: Wiley, 1986.

Ordinary Differential Equation—Second-Order

An ODE

$$y'' + P(x)y' + Q(x)y = 0 \tag{1}$$

has singularities for finite $x = x_0$ under the following conditions: (a) If either $P(x)$ or $Q(x)$ diverges as $x \to x_0$, but $(x - x_0)P(x)$ and $(x - x_0)^2 Q(x)$ remain finite as $x \to x_0$, then x_0 is called a regular or nonessential singular point. (b) If $P(x)$ diverges faster than $(x - x_0)^{-1}$ so that $(x - x_0)P(x) \to \infty$ as $x \to x_0$, or $Q(x)$ diverges faster than $(x - x_0)^{-2}$ so that $(x - x_0)^2 Q(x) \to \infty$ as $x \to x_0$, then x_0 is called an irregular or essential singularity.

Singularities of equation (1) at infinity are investigated by making the substitution $x \equiv z^{-1}$, so $dx = -z^{-2}dz$, giving

$$\frac{dy}{dx} = -z^2\,\frac{dy}{dz} \tag{2}$$

$$\frac{d^2y}{dx^2} = -z^2\,\frac{d}{dz}\left(-z^2\,\frac{dy}{dz}\right) = -z^2\left(-2z\,\frac{dy}{dz} - z^2\,\frac{d^2y}{dz^2}\right)$$

$$= 2z^3\,\frac{dy}{dz} + z^4\,\frac{d^2y}{dz^2}. \tag{3}$$

Then (1) becomes

$$z^4 \frac{d^2y}{dz^2} + [2z^3 - z^2 P(z)] \frac{dy}{dz} + Q(z)y = 0. \qquad (4)$$

Case (a): If

$$\alpha(z) \equiv \frac{2z - P(z)}{z^2} \qquad (5)$$

$$\beta(z) \equiv \frac{Q(z)}{z^4} \qquad (6)$$

remain finite at $x = \pm\infty$ $(y = 0)$, then the point is ordinary. Case (b): If either $\alpha(z)$ diverges no more rapidly than $1/z$ or $\beta(z)$ diverges no more rapidly than $1/z^2$, then the point is a regular singular point. Case (c): Otherwise, the point is an irregular singular point.

Morse and Feshbach (1953, pp. 667–674) give the canonical forms and solutions for second-order ODEs classified by types of singular points.

For special classes of second-order linear ordinary differential equations, variable COEFFICIENTS can be transformed into constant COEFFICIENTS. Given a second-order linear ODE with variable COEFFICIENTS

$$\frac{d^2y}{dx^2} + p(x) \frac{dy}{dx} = q(x)y = 0. \qquad (7)$$

Define a function $z \equiv y(x)$,

$$\frac{dy}{dx} = \frac{dz}{dx} \frac{dy}{dz} \qquad (8)$$

$$\frac{d^2y}{dx^2} = \left(\frac{dz}{dx}\right)^2 \frac{d^2y}{dz^2} + \frac{d^2z}{dx^2} \frac{dy}{dz} \qquad (9)$$

$$\left(\frac{dz}{dx}\right)^2 \frac{d^2y}{dz^2} + \left[\frac{d^2z}{dx^2} + p(x) \frac{dz}{dx}\right] \frac{dy}{dz} + q(x)y = 0 \qquad (10)$$

$$\frac{d^2y}{dz^2} + \left[\frac{\frac{d^2z}{dx^2} + P(x) \frac{dz}{dx}}{\left(\frac{dz}{dx}\right)^2}\right] \frac{dy}{dz} + \left[\frac{q(x)}{\left(\frac{dz}{dx}\right)^2}\right] y$$

$$\equiv \frac{d^2y}{dz^2} + A \frac{dy}{dz} + By = 0. \qquad (11)$$

This will have constant COEFFICIENTS if A and B are not functions of x. But we are free to set B to an arbitrary POSITIVE constant for $q(x) \geq 0$ by defining z as

$$z \equiv B^{-1/2} \int [q(x)]^{1/2} \, dx. \qquad (12)$$

Then

$$\frac{dz}{dx} = B^{-1/2} [q(x)]^{1/2} \qquad (13)$$

$$\frac{d^2z}{dx^2} = \tfrac{1}{2} B^{-1/2} [q(x)]^{-1/2} q'(x), \qquad (14)$$

and

$$A = \frac{\tfrac{1}{2} B^{-1/2} [q(x)]^{-1/2} q'(x) + B^{-1/2} p(x) [q(x)]^{1/2}}{B^{-1} q(x)}$$

$$= \frac{q'(x) + 2p(x)q(x)}{2[q(x)]^{3/2}} B^{1/2}. \qquad (15)$$

Equation (11) therefore becomes

$$\frac{d^2y}{dz^2} + \frac{q'(x) + 2p(x)q(x)}{2[q(x)]^{3/2}} B^{1/2} \frac{dy}{dz} + By = 0, \qquad (16)$$

which has constant COEFFICIENTS provided that

$$A \equiv \frac{q'(x) + 2p(x)q(x)}{2[q(x)]^{3/2}} B^{1/2} = [\text{constant}]. \qquad (17)$$

Eliminating constants, this gives

$$A' \equiv \frac{q'(x) + 2p(x)q(x)}{2[q(x)]^{3/2}} = [\text{constant}]. \qquad (18)$$

So for an ordinary differential equation in which A' is a constant, the solution is given by solving the second-order linear ODE with constant COEFFICIENTS

$$\frac{d^2y}{dz^2} + A \frac{dy}{dz} + By = 0 \qquad (19)$$

for z, where z is defined as above.

A linear second-order homogeneous differential equation of the general form

$$y'' + P(x)y' + Q(x)y = 0 \qquad (20)$$

can be transformed into standard form

$$z'' + q(x)z = 0 \qquad (21)$$

with the first-order term eliminated using the substitution

$$\ln y \equiv \ln z - \tfrac{1}{2} \int P(x)dx. \qquad (22)$$

Then

$$\frac{y'}{y} = \frac{z'}{z} - \tfrac{1}{2} P(x) \qquad (23)$$

$$\frac{yy'' - y'^2}{y^2} = \frac{zz'' - z'^2}{z^2} - \tfrac{1}{2} P'(x) \qquad (24)$$

$$\frac{y''}{y} - \left(\frac{y'}{y}\right)^2 = \frac{z''}{z} - \frac{z'^2}{z} - \frac{z'^2}{z^2} - \tfrac{1}{2} P'(x) \qquad (25)$$

$$\frac{y''}{y} - \left[\frac{z'}{z} - \tfrac{1}{2}P(x)\right]^2 + \frac{z''}{z} - \frac{z'^2}{z} - \tfrac{1}{2}P'(x)$$

$$= \frac{z'^2}{z^2} - \frac{z'}{z}P(x) + \tfrac{1}{4}P^2(x) + \frac{z''}{z} - \frac{z'^2}{z^2} - \tfrac{1}{2}P'(x), \quad (26)$$

so

$$\frac{y''}{y} + P(x)\frac{y'}{y} + Q(x)$$

$$= -\frac{z'}{z}P(x) + \tfrac{1}{4}P^2(x) + \frac{z''}{z} - \tfrac{1}{2}P'(x) + P(x)\left[\frac{z'}{z} - \tfrac{1}{2}P(x)\right]$$

$$+ Q(x). \quad (27)$$

Therefore,

$$z'' + \left[Q(x) - \tfrac{1}{2}P'(x) - \tfrac{1}{4}P^2(x)\right]z \equiv z''(x) + q(x)z = 0, \quad (28)$$

where

$$q(x) \equiv Q(x) - \tfrac{1}{2}P'(x) - \tfrac{1}{4}P^2(x). \quad (29)$$

If $Q(x) = 0$, then the differential equation becomes

$$y'' + P(x)y' = 0, \quad (30)$$

which can be solved by multiplying by

$$\exp\left[\int^x P(x')\,dx'\right] \quad (31)$$

to obtain

$$0 = \frac{d}{dx}\left\{\exp\left[\int^x P(x')\,dx'\right]\frac{dy}{dx}\right\} \quad (32)$$

$$c_1 = \exp\left[\int^x P(x')\,dx'\right]\frac{dy}{dx} \quad (33)$$

$$y = c_1 \int^x \frac{dx}{\exp\left[\int^x P(x')\,dx'\right]} + c_2. \quad (34)$$

If one solution (y_1) to a second-order ODE is known, the other (y_2) may be found using the REDUCTION OF ORDER method. From ABEL'S DIFFERENTIAL EQUATION IDENTITY

$$\frac{dW}{W} = -P(x)\,dx, \quad (35)$$

where

$$W \equiv y_1 y_2' - y_1' y_2 \quad (36)$$

$$\int_a^x \frac{dW}{W} = \int_a^x P'(x')\,dx' \quad (37)$$

$$\ln\left[\frac{W(x)}{W(a)}\right] = \int_a^x P(x')\,dx' \quad (38)$$

$$W(x) = W(a)\exp\left[-\int_a^x P(x')\,dx'\right]. \quad (39)$$

But

$$W \equiv y_1 y_2' - y_1' y_2 = y_1^2 \frac{d}{dx}\left(\frac{y_2}{y_1}\right). \quad (40)$$

Combining (39) and (40) yields

$$\frac{d}{dx}\left(\frac{y_2}{y_1}\right) = W(a)\frac{\exp\left[-\int_a^x P(x')\,dx'\right]}{y_1^2} \quad (41)$$

$$y_2(x) = y_1(x)W(a)\int_b^x \frac{\exp\left[-\int_a^{x'} P(x'')\,dx''\right]}{\left[y_1(x')\right]^2}\,dx'. \quad (42)$$

Disregarding $W(a)$, since it is simply a multiplicative constant, and the constants a and b, which will contribute a solution which is not linearly independent of (y_1),

$$y_2(x) = y_1(x)\int^x \frac{\exp\left[-\int^{x'} P(x'')\,dx''\right]}{\left[y_1(x')\right]^2}\,dx'. \quad (43)$$

If $P(x) = 0$, this simplifies to

$$y_2(x) = y_1(x)\int^x \frac{dx'}{\left[y_1(x')\right]^2}. \quad (44)$$

For a nonhomogeneous second-order ODE in which the x term does not appear in the function $f(x, y, y')$,

$$\frac{d^2y}{dx^2} = f(y, y') \quad (45)$$

let $v \equiv y'$, then

$$\frac{dv}{dx} = f(v, y) = \frac{dv}{dy}\frac{dy}{dx} = v\frac{dv}{dy}. \quad (46)$$

So the first-order ODE

$$v\frac{dv}{dy} = f(y, v), \quad (47)$$

if linear, can be solved for v as a linear first-order ODE. Once the solution is known,

$$\frac{dy}{dx} = v(y) \quad (48)$$

$$\int \frac{dy}{v(y)} = \int dx. \quad (49)$$

On the other hand, if y is missing from $f(x, y, y')$,

$$\frac{d^2y}{dx^2} = f(x, y'), \quad (50)$$

let $v \equiv y'$, then $v' = y''$, and the equation reduces to

$$v' = f(x, v), \tag{51}$$

which, if linear, can be solved for v as a linear first-order ODE. Once the solution is known,

$$y = \int v(x) \, dx. \tag{52}$$

See also ABEL'S DIFFERENTIAL EQUATION IDENTITY, ADJOINT

References

Arfken, G. "A Second Solution." §8.6 in *Mathematical Methods for Physicists, 3rd ed.* Orlando, FL: Academic Press, pp. 467–480, 1985.

Boyce, W. E. and DiPrima, R. C. *Elementary Differential Equations and Boundary Value Problems, 4th ed.* New York: Wiley, 1986.

Morse, P. M. and Feshbach, H. *Methods of Theoretical Physics, Part I.* New York: McGraw-Hill, pp. 667–674, 1953.

Ordinary Differential Equation—System with Constant Coefficients

To solve the system of differential equations

$$\frac{d\mathbf{x}}{dt} = A\mathbf{x}(t) + \mathbf{p}(t), \tag{1}$$

where A is a MATRIX and $\mathbf{x}$ and $\mathbf{p}$ are VECTORS, first consider the homogeneous case with $\mathbf{p} = 0$. Then the solutions to

$$\frac{d\mathbf{x}}{dt} = a\mathbf{x}(t) \tag{2}$$

are given by

$$\mathbf{x}(t) = e^{at}\mathbf{x}(t). \tag{3}$$

But, by the MATRIX DECOMPOSITION THEOREM, the MATRIX EXPONENTIAL can be written as

$$e^{At} = uDu^{-1}, \tag{4}$$

where the EIGENVECTOR MATRIX is

$$u = [\mathbf{u}_1 \cdots \mathbf{u}_n] \tag{5}$$

and the EIGENVALUE MATRIX is

$$D = \begin{bmatrix} e^{\lambda_1 t} & 0 & \cdots & 0 \\ 0 & e^{\lambda_2 t} & \cdots & 0 \\ \vdots & \vdots & \ddots & 0 \\ 0 & 0 & \cdots & e^{\lambda_n t} \end{bmatrix}. \tag{6}$$

Now consider

$$e^{At}u = uDu^{-1}u = uD$$
$$= \begin{bmatrix} u_{11} & u_{21} & \cdots & u_{n1} \\ u_{12} & u_{22} & \cdots & u_{n2} \\ \vdots & \vdots & \ddots & \vdots \\ u_{1n} & u_{2n} & \cdots & u_{nn} \end{bmatrix} \begin{bmatrix} e^{\lambda_1 t} & 0 & \cdots & 0 \\ 0 & e^{\lambda_2 t} & \cdots & 0 \\ \vdots & \vdots & \ddots & 0 \\ 0 & 0 & \cdots & e^{\lambda_n t} \end{bmatrix}$$
$$= \begin{bmatrix} u_{11}e^{\lambda_1 t} & \cdots & u_{n1}e^{\lambda_n t} \\ u_{11}e^{\lambda_1 t} & \cdots & u_{n2}e^{\lambda_n t} \\ \vdots & \ddots & \vdots \\ u_{n1}e^{\lambda_1 t} & \cdots & u_{n2}e^{\lambda_n t} \end{bmatrix}. \tag{7}$$

The individual solutions are then

$$\mathbf{x}_i = \left(e^{At}u\right) \cdot \hat{\mathbf{e}}_i = \mathbf{u}_i e^{\lambda_i t}, \tag{8}$$

so the homogeneous solution is

$$\mathbf{x} = \sum_{i=1}^{n} c_i \mathbf{u}_i e^{\lambda_i t}, \tag{9}$$

where the c_is are arbitrary constants.

The general procedure is therefore

1. Find the EIGENVALUES of the MATRIX A $(\lambda_1, ..., \lambda_n)$ by solving the CHARACTERISTIC EQUATION.
2. Determine the corresponding EIGENVECTORS $\mathbf{u}_1$, ..., $\mathbf{u}_n$.
3. Compute

$$\mathbf{x}_i \equiv e^{\lambda_i t}\mathbf{u}_i \tag{10}$$

for $i = 1, ..., n$. Then the VECTORS $\mathbf{x}_i$ which are REAL are solutions to the homogeneous equation. If A is a 2×2 matrix, the COMPLEX vectors $\mathbf{x}_j$ correspond to REAL solutions to the homogeneous equation given by $\Re(\mathbf{x}_j)$ and $\Im(\mathbf{x}_j)$.
4. If the equation is nonhomogeneous, find the particular solution given by

$$\mathbf{x}^*(t) = X(t) \int X^{-1}(t)\mathbf{p}(t) \, dt, \tag{11}$$

where the MATRIX X is defined by

$$X(t) \equiv [\mathbf{x}_1 \cdots \mathbf{x}_n]. \tag{12}$$

If the equation is homogeneous so that $\mathbf{p}(t) = \mathbf{0}$, then look for a solution OF THE FORM

$$\mathbf{x} = \xi e^{\lambda t}. \tag{13}$$

This leads to an equation

$$(A - \lambda I)\xi = \mathbf{0}, \qquad (14)$$

so ξ is an EIGENVECTOR and λ an EIGENVALUE.

5. The general solution is

$$\mathbf{x}(t) = \mathbf{x}^*(t) + \sum_{i=1}^{n} c_i \mathbf{x}_i. \qquad (15)$$

Ordinary Double Point

Portions of this entry contributed by SERGEI DUZHIN

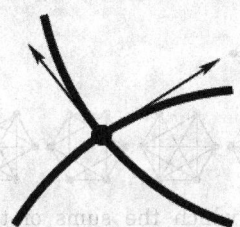

Let $f : \mathbb{R} \to \mathbb{R}^3$ (or $f : \mathbb{S}^1 \to \mathbb{R}^3$) be a SPACE CURVE. Then a point $p \in \operatorname{im}(f) \subset \mathbb{R}^3$ (where $\operatorname{im}(f)$ denotes the IMMERSION of f) is an ordinary double point if its PREIMAGE under f consists of two values t_1 and t_2, and the two TANGENT VECTORS $f'(t_1)$ and $f'(t_2)$ are noncollinear. Geometrically, this means that, in a NEIGHBORHOOD of p, the curve consists of two transverse branches. Ordinary double points are ISOLATED SINGULARITIES having COXETER-DYNKIN DIAGRAM of type A_1, and also called "nodes" or "simple double points."

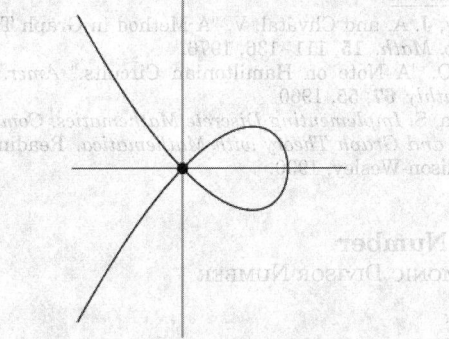

The above plot shows the curve $x^3 - x^2 + y^2 = 0$, which has an ordinary double point at the ORIGIN.

A surface in complex 3-space admits at most finitely many ordinary double points. The maximum possible number of ordinary double points $\mu(d)$ for a surface of degree $d = 1, 2, \ldots$, are $0, 1, 4, 16, 31, 65, 93 \le \mu(7) \le 104$, $168 \le \mu(8) \le 174$, $216 \le \mu(8) \le 246$, $345 \le \mu(10) \le 360$, $425 \le \mu(11) \le 480$, $576 \le \mu(12) \le 645 \ldots$ (Sloane's A046001; Chmutov 1992, Endraß 1995).

$\mu(4) = 16$ was known to Kummer in 1864 (Chmutov 1992), the fact that $\mu(5) = 31$ was proved by Beauville (1980), and $\mu(6) = 65$ was proved by Jaffe and Ruberman (1994). For $d \ge 3$, the following inequality holds:

$$\mu(d) \le \tfrac{1}{2}[d(d-1) - 3]$$

(Endraß 1995). Examples of ALGEBRAIC SURFACES having the maximum (known) number of ordinary double points are given in the following table.

d	$\mu(d)$	Surface
3	4	CAYLEY CUBIC
4	16	KUMMER SURFACE
5	31	DERVISH
6	65	BARTH SEXTIC
7	93	CHMUTOV SURFACE
8	168	ENDRAß OCTIC
9	216	CHMUTOV SURFACE
10	345	BARTH DECIC
11	425	CHMUTOV SURFACE
12	600	SARTI DODECIC

See also ALGEBRAIC SURFACE, BARTH DECIC, BARTH SEXTIC, CAYLEY CUBIC, CHMUTOV SURFACE, CUSP, DERVISH, DOUBLE POINT, ENDRAß OCTIC, ISOLATED SINGULARITY, KUMMER SURFACE, RATIONAL DOUBLE POINT, SARTI DODECIC

References

Basset, A. B. "The Maximum Number of Double Points on a Surface." *Nature* **73**, 246, 1906.

Beauville, A. "Sur le nombre maximum de points doubles d'une surface dans $\mathbb{P}^3$ ($\mu(5) = 31$)." *Journées de géométrie algébrique d'Angers (1979)*. Sijthoff & Noordhoff, pp. 207–215, 1980.

Chmutov, S. V. "Examples of Projective Surfaces with Many Singularities." *J. Algebraic Geom.* **1**, 191–196, 1992.

Endraß, S. "Surfaces with Many Ordinary Nodes." http://enriques.mathematik.uni-mainz.de/kon/docs/Eflaechen.shtml.

Endraß, S. "Flächen mit vielen Doppelpunkten." *DMV-Mitteilungen* **4**, 17–20, Apr. 1995.

Endraß, S. *Symmetrische Fläche mit vielen gewöhnlichen Doppelpunkten.* Ph.D. thesis. Erlangen, Germany, 1996.

Fischer, G. (Ed.). *Mathematical Models from the Collections of Universities and Museums.* Braunschweig, Germany: Vieweg, pp. 12–13, 1986.

Jaffe, D. B. and Ruberman, D. "A Sextic Surface Cannot have 66 Nodes." *J. Algebraic Geom.* **6**, 151–168, 1997.

Kreiss, H. O. "Über syzygetische Flächen." *Ann. Math.* **41**, 105–111, 1955.

Miyaoka, Y. "The Maximal Number of Quotient Singularities on Surfaces with Given Numerical Invariants." *Math. Ann.* **268**, 159–171, 1984.

Sloane, N. J. A. Sequences A046001 in "An On-Line Version of the Encyclopedia of Integer Sequences." http://www.research.att.com/~njas/sequences/eisonline.html.

Togliatti, E. G. "Sulle superficie algebriche col massimo numero di punti doppi." *Rend. Sem. Mat. Torino* **9**, 47–59, 1950.

Varchenko, A. N. "On the Semicontinuity of Spectrum and an Upper Bound for the Number of Singular Points on a Projective Hypersurface." *Dokl. Acad. Nauk SSSR* **270**, 1309–1312, 1983.

Walker, R. J. *Algebraic Curves.* New York: Springer-Verlag, pp. 56–57, 1978.

Ordinary Generating Function

GENERATING FUNCTION

Ordinary Line

Given an arrangement of $n \geq 3$ points, a LINE containing just two of them is called an ordinary line. Kelly and Moser (1958) proved that at least $3n/7$ lines must be ordinary (Guy 1989, p. 903).

See also COLINEAR, GENERAL POSITION, INCIDENT, NEAR-PENCIL, ORDINARY POINT, SPECIAL POINT, SYLVESTER GRAPH

References
Coxeter, H. S. M. "A Problem of Collinear Points." *Amer. Math. Monthly* **55**, 26–28, 1948.

Coxeter, H. S. M. *The Real Projective Plane, 3rd ed.* Cambridge, England: Cambridge University Press, 1993.

de Bruijn, N. G. and Erdos, P. "On a Combinatorial Problem." *Hederl. Adad. Wetenach.* **51**, 1277–1279, 1948.

Dirac, G. A. "Collinearity Properties of Sets of Points." *Quart. J. Math.* **2**, 221–227, 1951.

Erdos, P. "Problem 4065." *Amer. Math. Monthly* **51**, 169, 1944.

Guy, R. K. "Unsolved Problems Come of Age." *Amer. Math. Monthly* **96**, 903–909, 1989.

Kelly, L. M. and Moser, W. O. J. "On the Number of Ordinary Lines Determined by n Points." *Canad. J. Math.* **1**, 210–219, 1958.

Lang, D. W. "The Dual of a Well-Known Theorem." *Math. Gaz.* **39**, 314, 1955.

Motzkin, T. "The Lines and Planes Connecting the Points of a Finite Set." *Trans. Amer. Math. Soc.* **70**, 451–463, 1951.

Sylvester, J. J. "Mathematical Question 11851." *Educational Times* **59**, 98, 1893.

Ordinary Point

A POINT which lies on at least one ORDINARY LINE is called an ordinary point, or sometimes a REGULAR POINT.

See also ORDINARY LINE, REGULAR POINT, SPECIAL POINT, SYLVESTER GRAPH

References
Guy, R. K. "Unsolved Problems Come of Age." *Amer. Math. Monthly* **96**, 903–909, 1989.

Ordinary Surface

A surface which is homeomorphic to a finite collection of spheres, each with a finite number of HANDLES, cross-handles, CROSS-CAPS, and PERFORATIONS. A preliminary version of the CLASSIFICATION THEOREM OF SURFACES states that every surface is ordinary.

References
Francis, G. K. and Weeks, J. R. "Conway's ZIP Proof." *Amer. Math. Monthly* **106**, 393–399, 1999.

Ordinate

The y- (vertical) coordinate of a point in a two dimensional coordinate system. Physicists and astronomers sometimes use the term to refer to the axis itself instead of the distance along it.

See also ABSCISSA, x-AXIS, y-AXIS, z-AXIS

Ore Graph

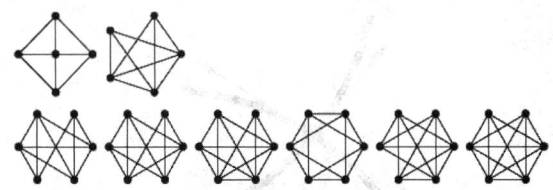

A GRAPH G in which the sums of the degrees of nonadjacent vertices is greater than the number of nodes n for all subsets of nonadjacent vertices (Ore 1960; Skiena 1990, p. 197). Ore graphs are always HAMILTONIAN, and a HAMILTONIAN CIRCUIT in such a graph can be constructed in polynomial time (Bondy and Chvátal 1976; Skiena 1990, p. 197). The numbers of Ore graphs on $n = 5$, 6, ... nodes are 2, 6, 32, ..., the first few of which are illustrated above.

See also HAMILTONIAN CIRCUIT, HAMILTONIAN GRAPH

References
Bondy, J. A. and Chvátal, V. "A Method in Graph Theory." *Disc. Math.* **15**, 111–136, 1976.

Ore, O. "A Note on Hamiltonian Circuits." *Amer. Math. Monthly* **67**, 55, 1960.

Skiena, S. *Implementing Discrete Mathematics: Combinatorics and Graph Theory with Mathematica.* Reading, MA: Addison-Wesley, 1990.

Ore Number

HARMONIC DIVISOR NUMBER

Ore's Conjecture

Define the HARMONIC MEAN of the DIVISORS of n

$$H(n) \equiv \frac{\tau(n)}{\sum_{d|n} \dfrac{1}{d}},$$

where $\tau(n)$ is the TAU FUNCTION (the number of DIVISORS of n). If n is a PERFECT NUMBER, $H(n)$ is an INTEGER. Ore conjectured that if n is ODD, then $H(n)$ is not an INTEGER. This implies that no ODD PERFECT NUMBERS exist.

See also HARMONIC DIVISOR NUMBER, HARMONIC MEAN, PERFECT NUMBER, TAU FUNCTION

Ore's Theorem

If a GRAPH G has $n \geq 3$ VERTICES such that every pair of the n VERTICES which are not joined by an EDGE has a sum of VALENCES which is $\geq n$, then G is HAMILTONIAN.

See also HAMILTONIAN GRAPH

Orientable Surface

A REGULAR SURFACE $M \subset \mathbb{R}^n$ is called orientable if each TANGENT SPACE M_p has a COMPLEX STRUCTURE $J_p : M_p \to M_p$ such that $p \to J_p$ is a continuous function.

See also NONORIENTABLE SURFACE, REGULAR SURFACE

References

Gray, A. *Modern Differential Geometry of Curves and Surfaces with Mathematica, 2nd ed.* Boca Raton, FL: CRC Press, p. 318, 1997.

Orientation (Bundle)

A real VECTOR BUNDLE $\pi : E \to M$ has an orientation if there exists a covering by TRIVIALIZATIONS $U_i \times \mathbb{R}^k$ such that the TRANSITION FUNCTIONS are ORIENTATION preserving. Alternatively, there exists a section of the PROJECTIVIZATION of the top exterior power of the bundle, $P_{\mathbb{R}}(\wedge^k E)$. A bundle is called orientable if there exists an orientation. Hence a bundle E of RANK k is orientable iff $\wedge^k E$ is a TRIVIAL LINE BUNDLE.

An orientation of the TANGENT BUNDLE is equivalent to an orientation on the BASE MANIFOLD. Not all bundles are orientable, as can be seen by the TANGENT BUNDLE of the MÖBIUS STRIP. The nontrivial LINE BUNDLE on the circle is also not orientable.

See also BUNDLE, ORIENTATION (MANIFOLD), ORIENTATION (VECTOR SPACE), VECTOR BUNDLE

References

Spivak, M. *A Comprehensive Introduction to Differential Geometry, Vol. 1, 2nd ed.* Houston, TX: Publish or Perish, pp. 273–383, 1999.

Orientation (Graph)

An orientation of an UNDIRECTED GRAPH G is an assignment of exactly one direction to each of the edges of G. Only connected, bridgeless graphs can have a strong orientation (Robbins 1939; Skiena 1990, p. 174). An oriented COMPLETE GRAPH is called a TOURNAMENT.

See also DIRECTED GRAPH, TOURNAMENT

References

Robbins, H. E. "A Theorem on Graphs with an Application to a Problem of Traffic Control." *Amer. Math. Monthly* **46**, 281–283, 1939.

Skiena, S. *Implementing Discrete Mathematics: Combinatorics and Graph Theory with Mathematica.* Reading, MA: Addison-Wesley, 1990.

Orientation (Manifold)

An orientation on an n-dimensional MANIFOLD is given by a nowhere vanishing DIFFERENTIAL N-FORM. Alternatively, it is an ORIENTATION for the TANGENT BUNDLE. If an orientation exists on M, then M is called orientable.

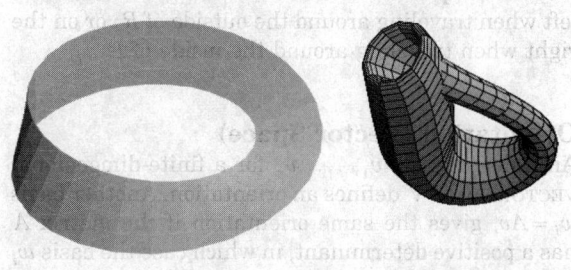

Not all MANIFOLDS are orientable, as exemplified by the MÖBIUS STRIP and the KLEIN BOTTLE, illustrated above.

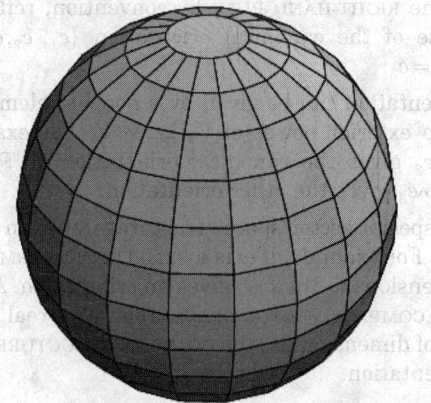

However, an $(n-1)$-dimensional SUBMANIFOLD of $\mathbb{R}^n$ is orientable IFF it has a unit normal vector field. The choice of unit determines the orientation of the submanifold. For example, the SPHERE $\mathbb{S}^2$ is orientable.

Some types of manifolds are always orientable. For instance, COMPLEX MANIFOLDS, including VARIETIES, and also SYMPLECTIC MANIFOLDS are orientable. Also, any unoriented manifold has a double COVER which is oriented.

A map $f : M \to N$ between oriented manifolds of the same dimension is called orientation preserving if the volume form on N pulls back to a positive volume form on M. Equivalently, the differential df maps an ORIENTED BASIS in TM to an ORIENTED BASIS in TN.

See also DIFFERENTIAL FORM, ORIENTATION (BUNDLE), ORIENTATION (VECTOR SPACE), VOLUME FORM

References

Berger, M. *Differential Geometry.* New York: Springer-Verlag, pp. 146–237, 1988.

Spivak, M. *A Comprehensive Introduction to Differential Geometry, Vol. 1, 2nd ed.* Houston, TX: Publish or Perish, pp. 273–383, 1999.

Sternberg, S. *Differential Geometry.* New York: Chelsea, pp. 14–30, 1983.

Orientation (Plane Curve)

A curve has positive orientation if a region R is on the left when traveling around the outside of R, or on the right when traveling around the inside of R.

Orientation (Vector Space)

An ordered BASIS $v_1, \ldots, v_n$ for a finite-dimensional VECTOR SPACE V defines an orientation. Another basis $w_i = A v_i$ gives the same orientation if the matrix A has a positive determinant, in which case the basis w_i is called oriented.

Any VECTOR SPACE has two possible orientations since the DETERMINANT of an INVERTIBLE MATRIX is either positive or negative. For example, in $\mathbb{R}^2$, $\{e_1, e_2\}$ is one orientation and $\{e_2, e_1\} \sim \{e_1, -e_2\}$ is the other orientation. In three dimensions, the CROSS PRODUCT uses the RIGHT-HAND RULE by convention, reflecting the use of the canonical orientation $\{e_1, e_2, e_3\}$ as $e_1 \times e_2 = e_3$.

An orientation can be given by a nonzero element in the top exterior power of V, i.e. $\wedge^n V$. For example, $e_1 \wedge e_2 \wedge e_3$ gives the canonical orientation on $\mathbb{R}^3$ and $-e_1 \wedge e_2 \wedge e_3$ gives the other orientation.

Some special vector space structures imply an orientation. For example, if ω is a SYMPLECTIC FORM on V, of dimension $2n$, then ω^n gives an orientation. Also, if V is a COMPLEX VECTOR SPACE, then as a real vector space of dimension $2n$, the COMPLEX STRUCTURE gives an orientation.

See also ORIENTATION (MANIFOLD), ORIENTATION (VECTOR BUNDLE)

Orientation (Vectors)

Let θ be the ANGLE between two VECTORS. If $0 < \theta < \pi$, the VECTORS are positively oriented. If $\pi < \theta < 2\pi$, the vectors are negatively oriented.

Two vectors in the plane

$$\begin{bmatrix} x_1 \\ x_2 \end{bmatrix} \text{ and } \begin{bmatrix} y_1 \\ y_2 \end{bmatrix}$$

are positively oriented IFF the DETERMINANT

$$D \equiv \begin{vmatrix} x_1 & y_1 \\ x_2 & y_2 \end{vmatrix} > 0,$$

and are negatively oriented IFF the DETERMINANT $D < 0$.

Orientation-Preserving

A nonsingular linear MAP $A : \mathbb{R}^n \to \mathbb{R}^n$ is orientation-preserving if $(A) > 0$.

See also ORIENTATION-REVERSING, ROTATION

Orientation-Reversing

A nonsingular linear MAP $A : \mathbb{R}^n \to \mathbb{R}^n$ is orientation-reversing if $\det(A) < 0$.

See also ORIENTATION-PRESERVING

Oriented Graph

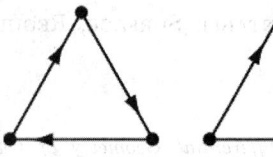

oriented graphs

A DIRECTED GRAPH having no symmetric pair of directed edges.

See also DIRECTED GRAPH

References

Harary, F. *Graph Theory.* Reading, MA: Addison-Wesley, p. 10, 1994.

Oriented Knot

See also KNOT, ORIENTED LINK

References

Cerf, C. "Atlas of Oriented Knots and Links." *Topology Atlas Invited Contributions* **3**, No. 2, 1–32, 1998. http://at.yorku.ca/t/a/i/c/31.htm.

Oriented Link

See also LINK, ORIENTED KNOT

References

Cerf, C. "Atlas of Oriented Knots and Links." *Topology Atlas Invited Contributions* **3**, No. 2, 1–32, 1998. http://at.yorku.ca/t/a/i/c/31.htm.

Oriented Matroid

The oriented matroid of a finite CONFIGURATION of points extracts relative position and orientation information from the CONFIGURATION. An oriented matroid can be described roughly as a MATROID in which every basis is equipped with an orientation (Richter-Gebert and Ziegler 1997, p. 112).

See also CONFIGURATION, MATROID

References

Björner, A.; Las Vergnas, M.; Sturmfels, B.; White, N.; and Ziegler, G. *Oriented Matroids, 2nd ed.* Cambridge, England: Cambridge University Press, 1999.

Richter-Gebert, J. and Ziegler, G. M. "Oriented Matroids." Ch. 6 in *Handbook of Discrete and Computational Geometry* (Ed. J. E. Goodman and J. O'Rourke). Boca Raton, FL: CRC Press, pp. 111–132, 1997.

Origami

The Japanese art of paper folding.

CUBE DUPLICATION and TRISECTION of an ANGLE can be solved using origami, although they cannot be solved using the traditional rules for GEOMETRIC CONSTRUCTIONS.

There are a number of recent very powerful results in origami mathematics. A very general result states that any planar straight-line drawing may be cut out of one sheet of paper by a single straight cut, after appropriate folding (Demaine, Demaine, and Lubiw, 1998, 1999, O'Rourke 1999). Another result is that any polyhedron may be wrapped with a sufficiently large square sheet of paper. This implies that any connected, planar, polygonal region may be covered by a flat origami folded from a single square of paper. Moreover, and 2-coloring of the faces may be realized with paper whose two sides are those colors (Demaine, Demaine, and Mitchell 1999, O'Rourke 1999).

See also FOLDING, GEOMETRIC CONSTRUCTION, MAP FOLDING, STAMP FOLDING, STOMACHION, TANGRAM

References

Andersen, E. "Origami on the Web." http://www.netspace.org/users/ema/oriweb.html.

Biddle, S. and Biddle, M. *The New Origami.* New York: St. Martin's Press, 1993.

Brill, D. *Brilliant Origami: A Collection of Original Designs.* Japan Pub., 1996.

Cerceda, A. and Palacios, V. *Fascinating Origami: 101 Models by Adolfo Cerceda.* New York: Dover, 1997.

Demaine, E. D.; Demaine, M. L.; and Lubiw, A. "Folding and Cutting Paper." In *Proc. Japan Conf. Discrete Comput. Geom.* New York: Springer-Verlag, 1998.

Demaine, E. D.; Demaine, M. L.; and Lubiw, A. "Folding and One Straight Cut Suffice." In *Proc. 10th Ann. ACM-SIAM Sympos. Discrete Alg. (SODA'99).* Baltimore, MD, pp. 891–892, Jan. 1999.

Demaine, E. D.; Demaine, M. L.; and Mitchell, J. S. B. "Folding Flat Silhouettes and Wrapping Polyhedral Packages: New Results in Computation Origami." In *Proc. 15th Ann. ACM Sympos. Comput. Geom.* Miami Beach, FL, pp. 105–114, June 1999.

Eppstein, D. "Origami." http://www.ics.uci.edu/~eppstein/junkyard/origami.html.

Fuse, T. *Unit Origami: Multidimensional Transformations.* Japan Pub., 1990. ISBN: 0870408526.

Geretschläger, R. "Euclidean Constructions and the Geometry of Origami." *Math. Mag.* **68**, 357–371, 1995.

Gurkewitz, R. "Rona's Modular Origami Polyhedra Page." http://www.wcsu.ctstateu.edu/~gurkewitz/homepage.html.

Gurkewitz, R. and Arnstein, B. *3-D Geometric Origami.* New York: Dover, 1996.

Harbin, R. *Origami Step-By-Step.* New York: Dover, 1998.

Harbin, R. *Secrets of Origami: The Japanese Art of Paper Folding.* New York: Dover, 1997.

Kasahara, K. *Origami Omnibus: Paper-Folding for Everyone.* Tokyo: Japan Publications, 1988.

Kasahara, K. and Takahara, T. *Origami for the Connoisseur.* Tokyo: Japan Publications, 1987.

Montroll, J. *Origami Inside-Out.* New York: Dover, 1993.

Montroll, J. *Origami Sculptures, 2nd ed.* Antroll Pub., 1991.

O'Rourke, J. "Computational Geometry Column 36." *SIGACT News* **30**, 35–38, Sep. 1999.

Palacios, V. *Fascinating Origami: 101 Models by Alfredo Cerceda.* New York: Dover, 1997.

Pappas, T. "Mathematics & Paperfolding." *The Joy of Mathematics.* San Carlos, CA: Wide World Publ./Tetra, pp. 48–50, 1989.

Row, T. S. *Geometric Exercises in Paper Folding.* New York: Dover, 1966.

Simon, L.; Arnstein, B.; and Gurkewitz, R. *Modular Origami Polyhedra.* New York: Dover, 1999.

by Takahama, T. *The Complete Origami Collection.* Japan Pub., 1997.

Tomoko, F. *Unit Origami.* Tokyo: Japan Publications, 1990.

Wu, J. "Joseph Wu's Origami Page." http://www.origami.vancouver.bc.ca/.

Origin

The central point ($r = 0$) in POLAR COORDINATES, or the point with all zero coordinates $(0, ..., 0)$ in CARTESIAN COORDINATES. In 3-D, the *x*-AXIS, *y*-AXIS, and *z*-AXIS meet at the origin.

See also OCTANT, QUADRANT, *x*-AXIS, *y*-AXIS, *z*-AXIS

Ornstein's Theorem

An important result in ERGODIC THEORY. It states that any two "Bernoulli schemes" with the same MEASURE-THEORETIC ENTROPY are MEASURE-THEORETICALLY ISOMORPHIC.

See also ERGODIC THEORY, ISOMORPHISM, MEASURE THEORY

Orr's Theorem

If

$$(1-z)^{\alpha+\beta+\gamma-1/2} \, {}_2F_1(2\alpha, \, 2\beta; \, 2\gamma; \, z) = \sum a_n z^n, \qquad (1)$$

where ${}_2F_1(a, \, b; \, c; \, z)$ is a HYPERGEOMETRIC FUNCTION, then

$$
{}_2F_1(\alpha, \, \beta; \, \gamma; \, z) \, {}_2F_1\left(\gamma - \alpha + \tfrac{1}{2}, \, \gamma - \beta + \tfrac{1}{2}; \, \gamma + 1; \, z\right)
$$
$$
= \sum_{(\gamma+1/2)_n/(\gamma+1)_n} a_n z^n. \qquad (2)
$$

Furthermore, if

$$(1-z)^{\alpha+\beta-\gamma-1/2} \, {}_2F_1(2\alpha-1, \, 2\beta; \, 2\gamma-1; \, z)$$
$$= \sum a_n z^n, \qquad (3)$$

then

$$_2F_1(\alpha,\ \beta;\ \gamma;\ z)\Gamma\left(\gamma-\alpha+\tfrac{1}{2},\ \gamma-\beta-\tfrac{1}{2};\ \gamma;\ z\right)$$
$$=\sum_{(\gamma-1/2)_n/(\gamma)_n} a_n z^n, \qquad (4)$$

where $\Gamma(z)$ is the GAMMA FUNCTION (Bailey 1935, p. 84).

References

Bailey, W. N. *Generalised Hypergeometric Series.* Cambridge, England: Cambridge University Press, 1935.

Cayley, A. "On a Theorem Relating to Hypergeometric Series." *Philos. Mag.* **16**, 356–357, 1858. Reprinted in *Collected Papers, Vol. 3*, pp. 268–269.

Edwards, D. "An Expansion in Factorials Similar to Vandermonde's Theorem, and Applications." *Messenger Math.* **52**, 129–136, 1923.

Orr, W. M. "Theorems Relating to the Product of Two Hypergeometric Series." *Trans. Cambridge Philos. Soc.* **17**, 1–15, 1899.

Watson, G. N. "The Theorems of Clausen and Cayley on Products of Hypergeometric Functions." *Proc. London Math. Soc.* **22**, 163–170, 1924.

Whipple, F. J. W. "Algebraic Proofs of the Theorems of Cayley and Orr Concerning the Products of Certain Hypergeometric Series." *J. London Math. Soc.* **2**, 85–90, 1927.

Whipple, F. J. W. "On a Formula Implied in Orr's Theorems Concerning the Products of Hypergeometric Series." *J. London Math. Soc.* **4**, 48–50, 1929.

Orr-Sommerfeld Differential Equation

The ORDINARY DIFFERENTIAL EQUATION

$$\frac{1}{i\alpha R}\left(\frac{d^2}{dx^2}-\alpha^2\right)^2 y$$

$$+\left\{[f(x)-c]\left(\frac{d^2}{dx^2}-\alpha^2\right)-f''(x)\right\}y=0.$$

References

Herron, I. H. "The Orr-Sommerfeld Equations on Infinite Intervals." *SIAM Rev.* **29**, 597–620, 1987.

Zwillinger, D. *Handbook of Differential Equations, 3rd ed.* Boston, MA: Academic Press, p. 120, 1997.

Orthic Axis

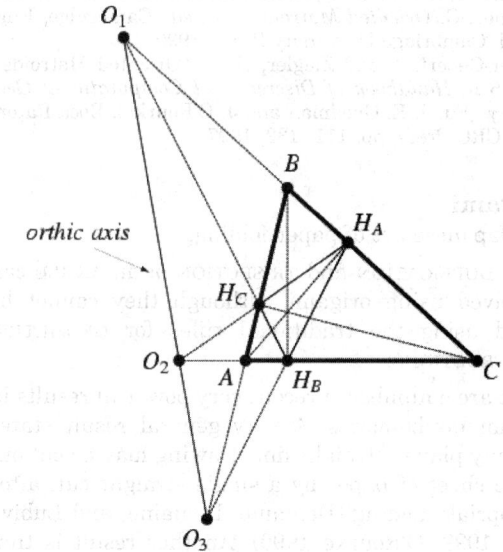

The $\Delta H_A H_B H_C$ be the ORTHIC TRIANGLE of a TRIANGLE $\Delta A_B C$. Then each side of each triangle meets the three sides of the other triangle, and the points of intersection lie on a line $O_1 O_2 O_3$ called the orthic axis.

See also ORTHIC TRIANGLE

References

Honsberger, R. §13.2 (ii) in *Episodes in Nineteenth and Twentieth Century Euclidean Geometry.* Washington, DC: Math. Assoc. Amer., p. 151, 1995.

Orthic Triangle

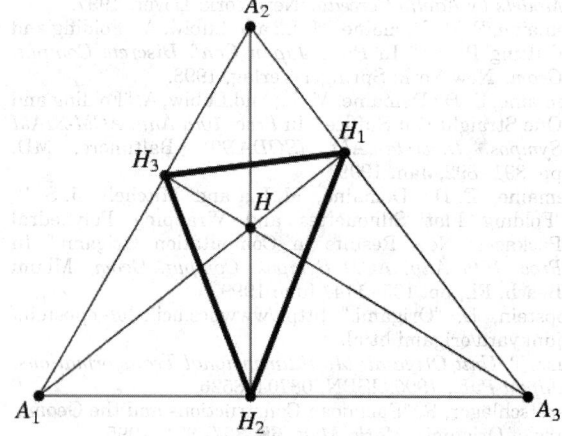

Given a TRIANGLE $\Delta A_1 A_2 A_3$, the TRIANGLE $\Delta H_1 H_2 H_3$ with VERTICES at the feet of the ALTITUDES (perpendiculars from a point to the sides) is called the orthic triangle. The three lines $A_i H_i$ are CONCURRENT at the ORTHOCENTER H of $\Delta A_1 A_2 A_3$. The orthic triangle is therefore the PEDAL TRIANGLE with respect to H.

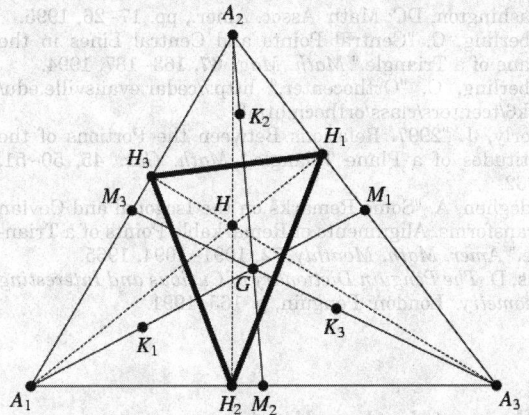

Given a triangle $\Delta A_1A_2A_3$, construct the orthic triangle $\Delta H_1H_2H_3$ and determine the SYMMEDIAN POINTS K_1, K_2, and K_3 of $\Delta A_1H_2H_3$, $\Delta H_1A_2H_3$, and $\Delta H_1H_2A_3$, respectively. Then the SYMMEDIANS K_1, K_2, and K_3 of each corner triangle pass through the MIDPOINTS M_1, M_2, and M_3 of the corresponding sides of the original triangle $\Delta A_1A_2A_3$ (Honsberger 1995, p. 75). Moreover, the lines K_1M_1, K_2M_2, and K_3M_3 CONCUR in the CENTROID of $\Delta A_1A_2A_3$.

The sides of the orthic triangle are parallel to the tangents to the CIRCUMCIRCLE at the vertices (Johnson 1929, p. 172).

The centroid of the orthic triangle has TRIANGLE CENTER FUNCTION

$$\alpha = a^2 \cos(B - C)$$

(Casey 1893, Kimberling 1994). The ORTHOCENTER of the orthic triangle has TRIANGLE CENTER FUNCTION

$$\alpha = \cos(2A)\cos(B - C)$$

(Casey 1893, Kimberling 1994). The SYMMEDIAN POINT of the orthic triangle has TRIANGLE CENTER FUNCTION

$$\alpha = \tan A \cos(B - C)$$

(Casey 1893, Kimberling 1994).

See also ALTITUDE, FAGNANO'S PROBLEM, ORTHOCENTER, PEDAL TRIANGLE, SCHWARZ'S TRIANGLE PROBLEM, SYMMEDIAN POINT

References

Casey, J. *A Treatise on the Analytical Geometry of the Point, Line, Circle, and Conic Sections, Containing an Account of Its Most Recent Extensions, with Numerous Examples, 2nd ed., rev. enl.* Dublin: Hodges, Figgis, & Co., p. 9, 1893.

Coxeter, H. S. M. and Greitzer, S. L. "The Orthic Triangle." §1.6 in *Geometry Revisited.* Washington, DC: Math. Assoc. Amer., pp. 9 and 16–18, 1967.

Honsberger, R. "The Orthic Triangle." §2.3 in *Episodes in Nineteenth and Twentieth Century Euclidean Geometry.* Washington, DC: Math. Assoc. Amer., pp. 21–25, 1995.

Johnson, R. A. *Modern Geometry: An Elementary Treatise on the Geometry of the Triangle and the Circle.* Boston, MA: Houghton Mifflin, 1929.

Kimberling, C. "Central Points and Central Lines in the Plane of a Triangle." *Math. Mag.* **67**, 163–187, 1994.

Orthobicupola

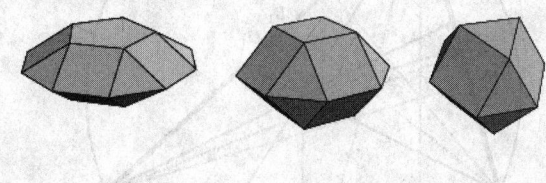

A BICUPOLA in which the bases are in the same orientation.

See also PENTAGONAL ORTHOBICUPOLA, SQUARE ORTHOBICUPOLA, TRIANGULAR ORTHOBICUPOLA

Orthobirotunda

A BIROTUNDA in which the bases are in the same orientation.

Orthocenter

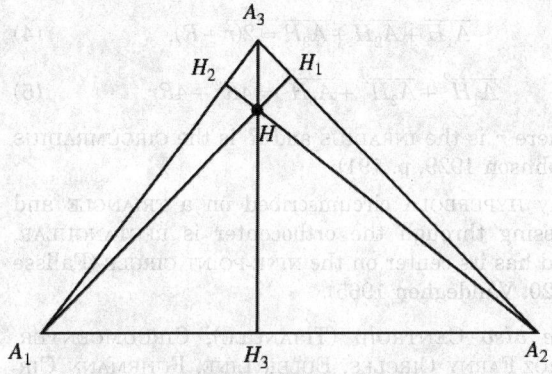

The intersection H of the three ALTITUDES of a TRIANGLE is called the orthocenter. The name was invented by Besant and Ferrers in 1865 while walking on a road leading out of Cambridge, England in the direction of London (Satterly 1962). The TRILINEAR COORDINATES of the orthocenter are

$$\cos B \cos C : \cos C \cos A : \cos A \cos B. \qquad (1)$$

If the TRIANGLE is not a RIGHT TRIANGLE, then (1) can be divided through by $\cos A \cos B \cos C$ to give

$$\sec A : \sec B : \sec C. \qquad (2)$$

If the triangle is ACUTE, the orthocenter is in the interior of the triangle. In a RIGHT TRIANGLE, the orthocenter is the VERTEX of the RIGHT ANGLE.

When the vertices of a triangle are combined with its orthocenter, any one of the points is the orthocenter of the other three, as first noted by Carnot (Wells

1991). These four points therefore form an ORTHO-CENTRIC SYSTEM.

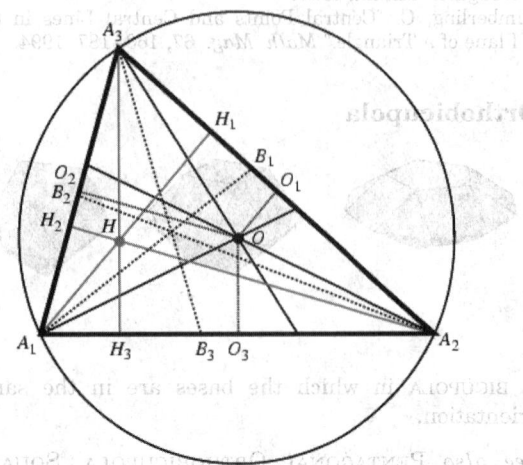

The CIRCUMCENTER O and orthocenter H are ISO-GONAL CONJUGATES. The orthocenter lies on the EULER LINE. The orthocenter and NAGEL POINT form a DIAMETER of the FUHRMANN CIRCLE.

Relationships involving the orthocenter include the following:

$$a_1^2 + a_2^2 + a_3^2 + \overline{A_1H}^2 + \overline{A_2H}^2 + \overline{A_3H}^2 = 12R^2 \quad (3)$$

$$\overline{A_1H} + \overline{A_2H} + \overline{A_3H} = 2(r+R), \quad (4)$$

$$\overline{A_1H}^2 + \overline{A_2H}^2 + \overline{A_3H}^2 = 4R^2 - 4Rr, \quad (5)$$

where r is the INRADIUS and R is the CIRCUMRADIUS (Johnson 1929, p. 191).

Any HYPERBOLA circumscribed on a TRIANGLE and passing through the orthocenter is RECTANGULAR, and has its center on the NINE-POINT CIRCLE (Falisse 1920, Vandeghen 1965).

See also CENTROID (TRIANGLE), CIRCUMCENTER, DROZ-FARNY CIRCLES, EULER LINE, FUHRMANN CIRCLE, INCENTER, ORTHIC TRIANGLE, ORTHOCENTRIC COORDINATES, ORTHOCENTRIC QUADRILATERAL, ORTHOCENTRIC SYSTEM, POLAR CIRCLE

References

Altshiller-Court, N. *College Geometry: A Second Course in Plane Geometry for Colleges and Normal Schools, 2nd ed.* New York: Barnes and Noble, pp. 165–172, 1952.
Carr, G. S. *Formulas and Theorems in Pure Mathematics, 2nd ed.* New York: Chelsea, p. 622, 1970.
Coxeter, H. S. M. and Greitzer, S. L. "More on the Altitudes and Orthocenter of a Triangle." *Geometry Revisited.* Washington, DC: Math. Assoc. Amer., pp. 9 and 36–40, 1967.
Dixon, R. *Mathographics.* New York: Dover, p. 57, 1991.
Falisse, V. *Cours de géométrie analytique plane.* Brussels, Belgium: Office de Publicité, 1920.
Johnson, R. A. *Modern Geometry: An Elementary Treatise on the Geometry of the Triangle and the Circle.* Boston, MA: Houghton Mifflin, pp. 165–172 and 191, 1929.

Honsberger, R. "The Orthocenter." Ch. 2 in *Episodes in Nineteenth and Twentieth Century Euclidean Geometry.* Washington, DC: Math. Assoc. Amer., pp. 17–26, 1995.
Kimberling, C. "Central Points and Central Lines in the Plane of a Triangle." *Math. Mag.* **67**, 163–187, 1994.
Kimberling, C. "Orthocenter." http://cedar.evansville.edu/~ck6/tcenters/class/orthocn.html.
Satterly, J. "2997. Relations Between the Portions of the Altitudes of a Plane Triangle." *Math. Gaz.* **45**, 50–51, 1962.
Vandeghen, A. "Some Remarks on the Isogonal and Cevian Transforms. Alignments of Remarkable Points of a Triangle." *Amer. Math. Monthly* **72**, 1091–1094, 1965.
Wells, D. *The Penguin Dictionary of Curious and Interesting Geometry.* London: Penguin, p. 165, 1991.

Orthocentric Coordinates

Coordinates defined by an ORTHOCENTRIC SYSTEM.

See also TRILINEAR COORDINATES

Orthocentric Line

The common axis of the three altitude planes of a TRIHEDRON.

See also TRIHEDRON

References

Altshiller-Court, N. "The Orthocentric Line." §2.1 in *Modern Pure Solid Geometry.* New York: Chelsea, pp. 27–30, 1979.

Orthocentric Quadrangle

Given four points, A, B, C, and H, let H be the ORTHOCENTER of $\triangle ABC$. Then A is the ORTHOCENTER $\triangle HBC$, B is the ORTHOCENTER of $\triangle HAC$, and C is the ORTHOCENTER of $\triangle HAB$. The configuration $ABCH$ is called an orthocentric quadrangle.

See also ORTHOCENTER, ORTHOCENTRIC QUADRILATERAL, ORTHOCENTRIC SYSTEM

References

Coxeter, H. S. M. and Greitzer, S. L. *Geometry Revisited.* Washington, DC: Math. Assoc. Amer., p. 39, 1967.

Orthocentric Quadrilateral

If two pairs of opposite sides of a COMPLETE QUADRILATERAL are pairs of PERPENDICULAR lines, the QUADRILATERAL is said to be orthocentric. In such a case, the remaining sides are also PERPENDICULAR.

See also ORTHOCENTRIC QUADRANGLE, ORTHOCENTRIC SYSTEM

Orthocentric System

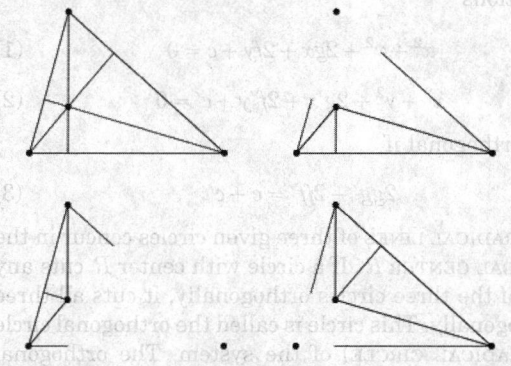

A set of four points, one of which is the ORTHOCENTER of the other three. In an orthocentric system, each point is the ORTHOCENTER of the TRIANGLE of the other three, as illustrated above (Coxeter and Greitzer 1967, p. 39). The INCENTER and EXCENTERS of a TRIANGLE are an orthocentric system.

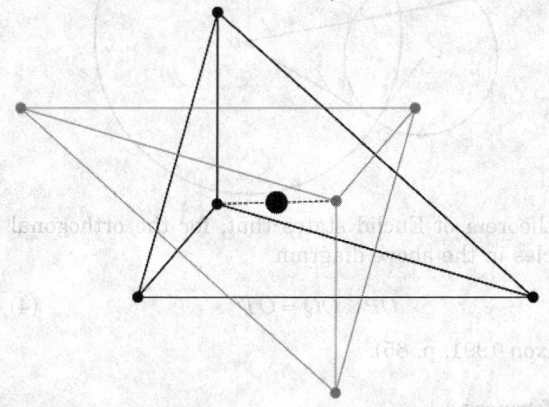

The centers of the CIRCUMCIRCLES of the points in an orthocentric system form another orthocentric system congruent to the first, and are the reflection of the original points in their common NINE-POINT CENTER (Wells 1991).

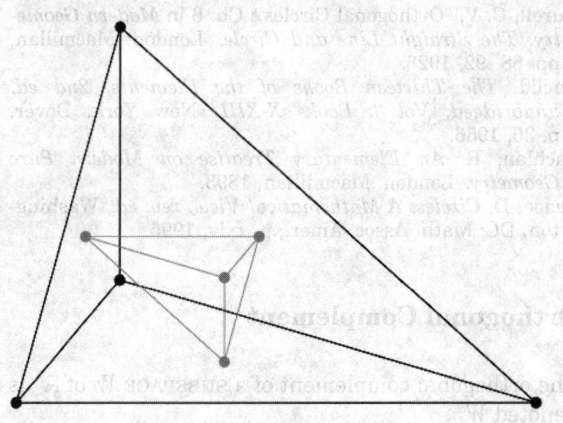

The centroids of the points in an orthocentric system form another orthocentric system similar to the first, but one third the size (Wells 1991).

The sum of the squares of any nonadjacent pair of connectors of an orthocentric system equals the square of the DIAMETER of the CIRCUMCIRCLE. Orthocentric systems are used to define ORTHOCENTRIC COORDINATES.

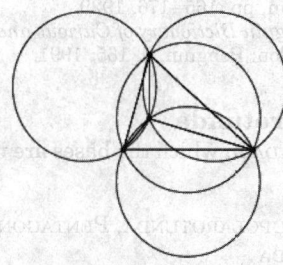

The four CIRCUMCIRCLES of points in an orthocentric system taken three at a time (illustrated above) have equal RADIUS (Wells 1991).

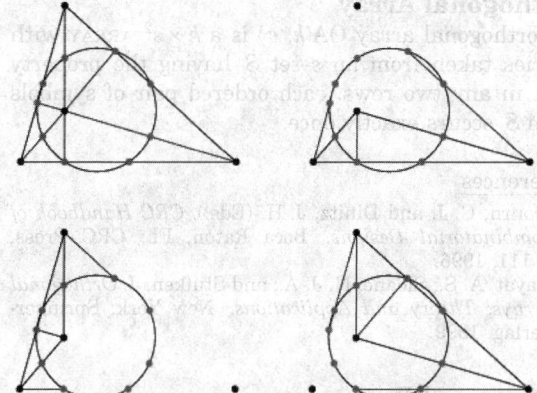

The four triangles of an orthocentric system have a common NINE-POINT CIRCLE, illustrated above. Furthermore, this circle is tangent to the 16 incircles and excircles of the four triangles (Wells 1991).

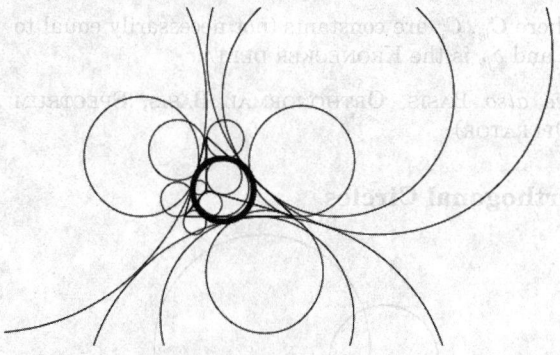

See also ANGLE BISECTOR, CIRCUMCIRCLE, CYCLIC QUADRANGLE, NINE-POINT CIRCLE, ORTHIC TRIANGLE, ORTHOCENTER, ORTHOCENTRIC QUADRANGLE ORTHO-

CENTRIC QUADRILATERAL, POLAR CIRCLE, RIGHT HYPERBOLA

References

Altshiller-Court, N. *College Geometry: A Second Course in Plane Geometry for Colleges and Normal Schools, 2nd ed.* New York: Barnes and Noble, pp. 109–114, 1952.

Johnson, R. A. *Modern Geometry: An Elementary Treatise on the Geometry of the Triangle and the Circle.* Boston, MA: Houghton Mifflin, pp. 165–176, 1929.

Wells, D. *The Penguin Dictionary of Curious and Interesting Geometry.* London: Penguin, p. 165, 1991.

Orthocupolarotunda

A CUPOLAROTUNDA in which the bases are in the same orientation.

See also GYROCUPOLAROTUNDA, PENTAGONAL ORTHOCUPOLAROTUNDA

Orthodrome

GREAT CIRCLE

Orthogonal Array

An orthogonal array OA(k, s) is a $k \times s^2$ ARRAY with entries taken from an s-set S having the property that in any two rows, each ordered pair of symbols from S occurs exactly once.

References

Colbourn, C. J. and Dinitz, J. H. (Eds.). *CRC Handbook of Combinatorial Designs.* Boca Raton, FL: CRC Press, p. 111, 1996.

Hedayat, A. S.; Sloane, N. J. A.; and Stufken, J. *Orthogonal Arrays: Theory and Applications.* New York: Springer-Verlag, 1999.

Orthogonal Basis

A BASIS of vectors **x** which satisfy

$$x_j x_k = C_{jk} \delta_{jk}$$

$$x^\mu x_\nu = C_\nu^\mu \delta_\nu^\mu,$$

where C_{jk}, C_ν^μ are constants (not necessarily equal to 1) and δ_{jk} is the KRONECKER DELTA.

See also BASIS, ORTHONORMAL BASIS, SPECTRUM (OPERATOR)

Orthogonal Circles

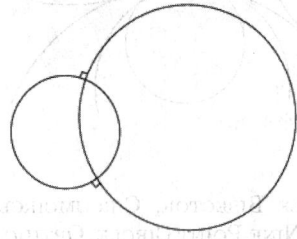

Orthogonal circles are ORTHOGONAL CURVES, i.e., they

cut one another at RIGHT ANGLES. Two CIRCLES with equations

$$x^2 + y^2 + 2gx + 2fy + c = 0 \tag{1}$$

$$x^2 + y^2 + 2g'x + 2f'y + c' = 0 \tag{2}$$

are orthogonal if

$$2gg' + 2ff' = c + c'. \tag{3}$$

The RADICAL LINES of three given circles concur in the RADICAL CENTER R. If a circle with center R cuts any one of the three circles orthogonally, it cuts all three orthogonally. This circle is called the orthogonal circle (or RADICAL CIRCLE) of the system. The orthogonal circle is the LOCUS of a point whose POLARS with respect to the three given circles are concurrent (Lachlan 1893, p. 237).

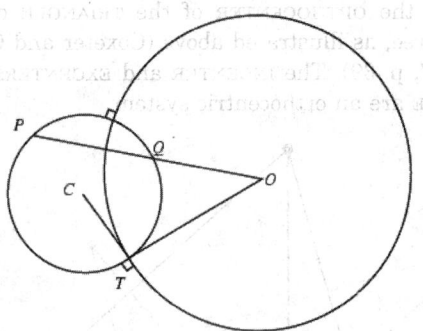

A theorem of Euclid states that, for the orthogonal circles in the above diagram,

$$OP \times OQ = OT^2 \tag{4}$$

(Dixon 1991, p. 65).

References

Casey, J. *A Sequel to the First Six Books of the Elements of Euclid, Containing an Easy Introduction to Modern Geometry with Numerous Examples, 5th ed., rev. enl.* Dublin: Hodges, Figgis, & Co., p. 42, 1888.

Dixon, R. *Mathographics.* New York: Dover, pp. 65–66, 1991.

Durell, C. V. "Orthogonal Circles." Ch. 8 in *Modern Geometry: The Straight Line and Circle.* London: Macmillan, pp. 88–92, 1928.

Euclid. *The Thirteen Books of the Elements, 2nd ed. unabridged, Vol. 3: Books X-XIII.* New York: Dover, p. 36, 1956.

Lachlan, R. *An Elementary Treatise on Modern Pure Geometry.* London: Macmillian, 1893.

Pedoe, D. *Circles: A Mathematical View, rev. ed.* Washington, DC: Math. Assoc. Amer., p. xxiv, 1995.

Orthogonal Complement

The orthogonal complement of a SUBSPACE W of $\mathbb{R}^n$ is denoted $W^\perp$.

See also FREDHOLM'S THEOREM, ORTHOGONAL DECOMPOSITION

Orthogonal Coordinate System

A system of CURVILINEAR COORDINATES in which each family of surfaces intersects the others at right angles.

Orthogonal CURVILINEAR COORDINATES satisfy the additional constraint that

$$\hat{\mathbf{u}}_i \cdot \hat{\mathbf{u}}_j = \delta_{ij}. \tag{1}$$

Therefore, the LINE ELEMENT becomes

$$d\mathbf{s}^2 = d\mathbf{r} \cdot d\mathbf{r} = h_1^2 \, du_1^2 + h_2^2 \, du_2^2 + h_3^2 \, du_3^2 \tag{2}$$

and the VOLUME ELEMENT is

$$dV = |(h_1 \hat{\mathbf{u}}_1 \, du_1) \cdot (h_2 \hat{\mathbf{u}}_2 \, du_2) \times (h_3 \hat{\mathbf{u}}_3 \, du_3)|$$

$$= h_1 h_2 h_3 \, du_1 \, du_2 \, du_3$$

$$= \left| \frac{\partial r}{\partial u_1} \cdot \frac{\partial r}{\partial u_2} \times \frac{\partial r}{\partial u_3} \right| du_1 \, du_2 \, du_3$$

$$= \begin{vmatrix} \dfrac{\partial x}{\partial u_1} & \dfrac{\partial x}{\partial u_2} & \dfrac{\partial x}{\partial u_3} \\ \dfrac{\partial y}{\partial u_1} & \dfrac{\partial y}{\partial u_2} & \dfrac{\partial y}{\partial u_3} \\ \dfrac{\partial z}{\partial u_1} & \dfrac{\partial z}{\partial u_2} & \dfrac{\partial z}{\partial u_3} \end{vmatrix} du_1 \, du_2 \, du_3$$

$$= \left| \frac{\partial(x, y, z)}{\partial(u_1, u_2, u_3)} \right| du_1 \, du_2 \, du_3, \tag{3}$$

where the latter is the JACOBIAN.

For surfaces of first degree, the only 3-D coordinate system of surfaces having orthogonal intersections is CARTESIAN COORDINATES (Moon and Spencer 1988, p. 1). Including degenerate cases, there are 11 sets of quadratic surfaces having orthogonal coordinates. Furthermore, LAPLACE'S EQUATION and the HELMHOLTZ DIFFERENTIAL EQUATION are separable in all of these coordinate systems (Moon and Spencer 1988, p. 1).

Planar orthogonal curvilinear coordinate systems of degree two or less include 2-D CARTESIAN COORDINATES and POLAR COORDINATES.

3-D orthogonal curvilinear coordinate systems of degree two or less include BIPOLAR CYLINDRICAL COORDINATES, BISPHERICAL COORDINATES, 3-D CARTESIAN COORDINATES, CONFOCAL ELLIPSOIDAL COORDINATES, CONFOCAL PARABOLOIDAL COORDINATES, CONICAL COORDINATES, CYCLIDIC COORDINATES, CYLINDRICAL COORDINATES, ELLIPSOIDAL COORDINATES, ELLIPTIC CYLINDRICAL COORDINATES, OBLATE SPHEROIDAL COORDINATES, PARABOLIC COORDINATES, PARABOLIC CYLINDRICAL COORDINATES, PARABOLOIDAL COORDINATES, PROLATE SPHEROIDAL COORDINATES, SPHERICAL COORDINATES, and TOROIDAL COORDINATES. These are degenerate cases of the CONFOCAL ELLIPSOIDAL COORDINATES.

Orthogonal coordinate systems can also be built from fourth-order (in particular, CYCLIDIC COORDINATES) and higher surfaces (Bôcher 1894), but are generally less important in solving physical problems than are quadratic surfaces (Moon and Spencer 1988, p. 1).

See also CHANGE OF VARIABLES THEOREM, CURL, CURVILINEAR COORDINATES, CYCLIDIC COORDINATES, DIVERGENCE, GRADIENT, JACOBIAN, LAPLACIAN, SKEW COORDINATE SYSTEM

References

Arfken, G. "Curvilinear Coordinates" and "Differential Vector Operators." §2.1 and 2.2 in *Mathematical Methods for Physicists, 3rd ed.* Orlando, FL: Academic Press, pp. 86–90 and 90–94, 1985.

Bôcher, M. *Über die Reihenentwicklungen der Potentialtheorie.* Leipzig, Germany: Teubner, 1894.

Darboux, G. *Sur une classe remarquable de courbes et de surfaces algébriques et sur la théorie des imaginaires.* Paris: Hermann, 1896.

Darboux, G. *Leçons sur les systemes orthogonaux et les coordonnées curvilignes.* Paris: Gauthier-Villars, 1910.

Gradshteyn, I. S. and Ryzhik, I. M. *Tables of Integrals, Series, and Products, 6th ed.* San Diego, CA: Academic Press, pp. 1084–1088, 2000.

Lamé, G. *Leçons sur les coordonnées curvilignes et leurs diverses applications.* Paris: Mallet-Bachelier, 1859.

Moon, P. and Spencer, D. E. "Eleven Coordinate Systems." §1 in *Field Theory Handbook, Including Coordinate Systems, Differential Equations, and Their Solutions, 2nd ed.* New York: Springer-Verlag, pp. 1–48, 1988.

Morse, P. M. and Feshbach, H. "Curvilinear Coordinates" and "Table of Properties of Curvilinear Coordinates." §1.3 in *Methods of Theoretical Physics, Part I.* New York: McGraw-Hill, pp. 21–31 and 115–117, 1953.

Müller, E. "Die verschiedenen Koordinatensysteme." S. 596 in *Encyk. Math. Wissensch., Bd. III.1.1.* Leipzig, Germany: Teubner, 1907–1910.

See also CURVILINEAR COORDINATES

Orthogonal Curves

Two intersecting curves which are PERPENDICULAR at their INTERSECTION are said to be orthogonal.

Orthogonal Decomposition

This entry contributed by VIKTOR BENGTSSON

The orthogonal decomposition of a VECTOR $\mathbf{y}$ in $\mathbb{R}^n$ is the sum of a vector in a SUBSPACE W of $\mathbb{R}^n$ and a vector in the ORTHOGONAL COMPLEMENT $W^\perp$ to W.

The orthogonal decomposition theorem states that if W is a SUBSPACE of $\mathbb{R}^n$, then each vector $\mathbf{y}$ in $\mathbb{R}^n$ can be written uniquely in the form

$$\mathbf{y} = \hat{\mathbf{y}} + \mathbf{x},$$

where $\hat{\mathbf{y}}$ is in W and $\mathbf{z}$ is in $W^\perp$. In fact, if $\{\mathbf{u}_1, \mathbf{u}_2, \ldots, \mathbf{u}_p\}$ is any ORTHOGONAL BASIS of W, then

$$\hat{\mathbf{y}} = \frac{\mathbf{y} \cdot \mathbf{u}_1}{\mathbf{u}_1 \cdot \mathbf{u}_1} \mathbf{u}_1 + \frac{\mathbf{y} \cdot \mathbf{u}_2}{\mathbf{u}_2 \cdot \mathbf{u}_2} \mathbf{u}_2 + \ldots + \frac{\mathbf{y} \cdot \mathbf{u}_p}{\mathbf{u}_p \cdot \mathbf{u}_p} \mathbf{u}_p,$$

and $\mathbf{z} = \mathbf{y} - \hat{\mathbf{y}}$.

Geometrically, $\hat{\mathbf{y}}$ is the ORTHOGONAL PROJECTION of $\mathbf{y}$ onto the SUBSPACE W and $\mathbf{z}$ is a vector orthogonal to $\hat{\mathbf{y}}$

See also FREDHOLM'S THEOREM, LU DECOMPOSITION, QR DECOMPOSITION

References

Golub, G. and van Loan, C. *Matrix Computations, 3rd ed.* Baltimore, MD: Johns Hopkins University Press, 1996.

Orthogonal Functions

Two functions $f(x)$ and $g(x)$ are orthogonal on the interval $a \leq x \leq b$ if

$$\langle f(x) | g(x) \rangle \equiv \int_a^b f(x) g(x) \, dx = 0.$$

See also ORTHOGONAL POLYNOMIALS, ORTHONORMAL FUNCTIONS

Orthogonal Group

For every DIMENSION $n > 0$, the orthogonal group $O(n)$ is the GROUP of $n \times n$ ORTHOGONAL MATRICES. These matrices form a GROUP because they are CLOSED under multiplication and taking inverses.

Thinking of a matrix as given by n^2 coordinate functions, the set of matrices is identified with $\mathbb{R}^{n^2}$. The orthogonal matrices are the solutions to the n^2 equations

$$AA^T = I, \tag{1}$$

where I is the IDENTITY MATRIX, which are redundant. Only $n(n+1)/2$ of these are independent, leaving $n(n-1)/2$ "free variables." In fact, the orthogonal group is a smooth $n(n-1)/2$ dimensional SUBMANIFOLD.

Because the orthogonal group is a group and a manifold, it is a LIE GROUP. $O(n)$ has a TANGENT SPACE at the identity that is the LIE ALGEBRA of SKEW SYMMETRIC MATRICES $o(n)$. In fact, the orthogonal group is a COMPACT LIE GROUP.

The DETERMINANT of an ORTHOGONAL MATRIX is either 1 or -1, and so the orthogonal group has two COMPONENTS. The component containing the identity is a the SPECIAL ORTHOGONAL GROUP $SO(n)$. For example, The GROUP $O(2)$ has GROUP ACTION on the plane that is a rotation:

$$O(2) = \left\{ \begin{bmatrix} \cos\theta & -\sin\theta \\ \sin\theta & \cos\theta \end{bmatrix} \right\} \cup \left\{ \begin{bmatrix} -\cos\theta & \sin\theta \\ \sin\theta & \cos\theta \end{bmatrix} \right\}, \tag{2}$$

where θ is any real number in $[0, 2\pi)$. These matrices preserve the QUADRATIC FORM $x^2 + y^2$, and so they also preserve CIRCLES $x^2 + y^2 = r^2$, which are the ORBITS.

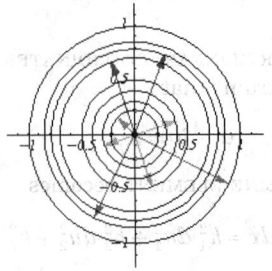

As a manifold, $O(2)$ is a one dimensional, two disjoint copies of the CIRCLE. The SUBGROUP $SO(2)$ is not a NORMAL SUBGROUP, so $O(2)$ is the SEMIDIRECT PRODUCT of the circle $SO(2)$ and $\mathbb{Z}_2$.

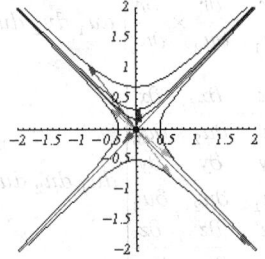

There are several generalizations of the orthogonal group. First, it is possible to define the orthogonal group for any SYMMETRIC QUADRATIC FORM Q with SIGNATURE (p, q). The group of matrices A which preserve Q, that is,

$$Q(v, w) = Q(Av, Aw), \tag{3}$$

is denoted $O(p, q)$. The LORENTZ GROUP is $O(3, 1)$. For example, the matrices

$$A = \begin{bmatrix} \cosh t & \sinh t \\ \sinh t & \cosh t \end{bmatrix} \tag{4}$$

are elements of $O(1, 1)$. They preserve the QUADRATIC FORM $x^2 - y^2$ so they preserve the HYPERBOLAS $x^2 - y^2 = c$.

Instead of using real numbers for the coefficients, it is possible to use coefficients from any FIELD $\mathbb{F}$, in which case it is denoted $O(n, \mathbb{F})$. The orthogonal matrices still satisfy $AA^t = I$. For example, $O(2, \mathbb{F}_{23})$ contains

$$\begin{bmatrix} 11 & 15 \\ 15 & 12 \end{bmatrix}, \tag{5}$$

and has 48 elements in total.

Of course, $O(p, q, \mathbb{F})$ denotes the group of matrices which preserve the SYMMETRIC QUADRATIC FORM of SIGNATURE (p, q), with coefficients in the field $\mathbb{F}$. When $\mathbb{F}$ is not $\mathbb{R}$ or $\mathbb{C}$, these are called LIE-TYPE GROUPS.

When the coefficients are COMPLEX NUMBERS, it is called the complex orthogonal group, which is much different from the UNITARY GROUP. For example, matrices OF THE FORM

$$A = \begin{bmatrix} \cos z & -\sin z \\ \sin z & \cos z \end{bmatrix} \qquad (6)$$

are in $O(2, \mathbb{C})$. In particular, $O(n, \mathbb{C})$ is not COMPACT. The equations defining $O(n)$ in AFFINE SPACE are polynomials of degree two. Consequently, $O(n)$ is a LINEAR ALGEBRAIC GROUP.

The numbers of subgroups $s(n)$ of orders $n = 1, 2, 3,$ ... in the orthogonal group $O(3)$ are 1, 3, 1, 5, 1, 5, 1, 7, 1, 5, 1, 8, ... (Sloane's A001051), i.e., a repeating sequence of copies of $\{1, 5, 1, 7\}$ with the exceptions $s(2) = 3$, $s(4) = 5$, $s(12) = 8$, $s(24) = 10$, and $s(48) = s(60) = s(120) = 8$.

See also DETERMINANT, GENERAL ORTHOGONAL GROUP, GROUP, FIELD, LAPLACIAN, LIE ALGEBRA, LIE GROUP, LIE-TYPE GROUP, LINEAR ALGEBRAIC GROUP, ORTHOGONAL GROUP REPRESENTATIONS, ORTHOGONAL MATRIX, ORTHOGONAL TRANSFORMATION, ORTHONORMAL BASIS, PROJECTIVE GENERAL ORTHOGONAL GROUP, PROJECTIVE SPECIAL ORTHOGONAL GROUP, RIEMANNIAN METRIC, SPECIAL ORTHOGONAL GROUP, SUBMANIFOLD, SYMMETRIC QUADRATIC FORM, UNITARY GROUP, VECTOR SPACE

References

Arfken, G. "Orthogonal Group, O_3^+." *Mathematical Methods for Physicists, 3rd ed.* Orlando, FL: Academic Press, pp. 252–253, 1985.
Wilson, R. A. "ATLAS of Finite Group Representation." http://for.mat.bham.ac.uk/atlas/html/contents.html#orth.

Orthogonal Group Representations

Two representations of a GROUP χ_i and χ_j are said to be orthogonal if

$$\sum_R \chi_i(R)\chi_j(R) = 0$$

for $i \neq j$, where the sum is over all elements R of the representation.

See also GROUP

Orthogonal Lines

Two or more LINES or LINE SEGMENTS which are PERPENDICULAR are said to be orthogonal.

See also ORTHOGONAL CURVES, PERPENDICULAR, RIGHT ANGLE

Orthogonal Matrix

A $n \times n$ matrix A is an orthogonal matrix if

$$AA^T = I, \qquad (1)$$

where A^T is the TRANSPOSE of A and I is the IDENTITY

MATRIX. In particular, an orthogonal matrix is always invertible, and

$$A^{-1} = A^T \qquad (2)$$

(Note that transpose is a much simpler computation than inverse.) For example,

$$A = \frac{1}{\sqrt{2}} \begin{bmatrix} 1 & 1 \\ 1 & -1 \end{bmatrix} \qquad (3)$$

$$A = \frac{1}{3} \begin{bmatrix} 2 & -2 & 1 \\ 1 & 2 & 2 \\ 2 & 1 & -2 \end{bmatrix} \qquad (4)$$

are orthogonal matrices. A matrix m can be tested to see if it is orthogonal using the *Mathematica* function

```
OrthogonalQ[m_List?MatrixQ]  :=
(Transpose[m].m = = IdentityMatrix@Length@m)
```

The rows of an orthogonal matrix are an ORTHONORMAL BASIS. That is, each row has length one, and are mutually perpendicular. Similarly, the columns are also an orthonormal basis. In fact, given any orthonormal basis, the matrix whose rows are that basis is an orthogonal matrix. It is automatically the case that the columns are another orthonormal basis.

The orthogonal matrices are precisely those matrices which preserve the INNER PRODUCT

$$\langle v, w \rangle = \langle A_v, A_w \rangle. \qquad (5)$$

Also, the determinant of A is either 1 or -1. As a subset of $\mathbb{R}^{n^2}$, the orthogonal matrices are not CONNECTED since the determinant is a CONTINUOUS FUNCTION. Instead, there are two COMPONENTS corresponding to whether the determinant is 1 or -1. The orthogonal matrices with $A = 1$ are rotations, and such a matrix is called a SPECIAL ORTHOGONAL MATRIX.

The product of two orthogonal matrices is another orthogonal matrix. In addition, the inverse of an orthogonal matrix is an orthogonal matrix, as is the IDENTITY MATRIX. Hence the set of orthogonal matrices form a GROUP, called the ORTHOGONAL GROUP $O(n)$.

See also EULER'S ROTATION THEOREM, INNER PRODUCT, ORTHOGONAL GROUP, ORTHOGONAL TRANSFORMATION, ORTHOGONALITY CONDITION, ORTHONORMAL BASIS, ROTATION, ROTATION MATRIX, ROTOINVERSION, SKEW SYMMETRIC MATRIX, SPECIAL ORTHOGONAL MATRIX, SPIN GROUP, UNITARY MATRIX

References

Arfken, G. "Orthogonal Matrices." *Mathematical Methods for Physicists, 3rd ed.* Orlando, FL: Academic Press, pp. 191–205, 1985.
Goldstein, H. "Orthogonal Transformations." §4–2 in *Classical Mechanics, 2nd ed.* Reading, MA: Addison-Wesley, 132–137, 1980.

Orthogonal Polynomials

Orthogonal polynomials are classes of POLYNOMIALS $\{p_n(x)\}$ over a range $[a, b]$ which obey an ORTHOGONALITY relation

$$\int_a^b w(x)p_m(x)p_n(x)\,dx = \delta_{mn}c_n, \qquad (1)$$

where $w(x)$ is a WEIGHTING FUNCTION and δ is the KRONECKER DELTA. If $c_n = 1$, then the POLYNOMIALS are not only orthogonal, but orthonormal.

Orthogonal polynomials have very useful properties in the solution of mathematical and physical problems. Just as FOURIER SERIES provide a convenient method of expanding a periodic function in a series of linearly independent terms, orthogonal polynomials provide a natural way to solve, expand, and interpret solutions to many types of important DIFFERENTIAL EQUATIONS. Orthogonal polynomials are especially easy to generate using GRAM-SCHMIDT ORTHONORMALIZATION. Abramowitz and Stegun (1972, pp. 774–775) give a table of common orthogonal polynomials.

Type	Interval	$w(x)$	c_n
CHEBYSHEV POLYNOMIAL OF THE FIRST KIND	$[-1, 1]$	$(1-x^2)^{-1/2}$	$\begin{cases} \frac{1}{2}\pi & \text{for } n=0 \\ \pi & \text{otherwise} \end{cases}$
CHEBYSHEV POLYNOMIAL OF THE SECOND KIND	$[-1, 1]$	$\sqrt{1-x^2}$	$\frac{1}{2}\pi$
HERMITE POLYNOMIAL	$(-\infty, \infty)$	e^{-x^2}	$\sqrt{\pi}\,2^n\,n!$
JACOBI POLYNOMIAL	$(-1, 1)$	$(1-x)^\alpha(1+x)^\beta$	h_n
LAGUERRE POLYNOMIAL	$[0, \infty)$	e^{-x}	1
LAGUERRE POLYNOMIAL (Associated)	$[0, \infty)$	$x^k e^{-x}$	$\frac{(n+k)!}{n!}$
LEGENDRE POLYNOMIAL	$[-1, 1]$	1	$\frac{2}{2n+1}$
ULTRASPHERICAL POLYNOMIAL	$[-1, 1]$	$(1-x^2)^{\alpha-1/2}$	$\begin{cases} \frac{2^{1-2\alpha}\,\pi\Gamma(n+2\alpha)}{n!(n+\alpha)\,[\Gamma(\alpha)]^2} & \text{for } \alpha\neq0 \\ \frac{2\pi}{n^2} & \text{for } \alpha\neq0 \end{cases}$

In the above table, the normalization constant is the value of

$$c_n \equiv \int w(x)[p_n(x)]^2\,dx \qquad (2)$$

and

$$h_n \equiv \frac{2^{\alpha+\beta+1}}{2n+\alpha+\beta+1}\frac{\Gamma(n+\alpha+1)\Gamma(n+\beta+1)}{n!\,\Gamma(n+\alpha+\beta+1)}, \qquad (3)$$

where $\Gamma(z)$ is a GAMMA FUNCTION.

The ROOTS of orthogonal polynomials possess many rather surprising and useful properties. For instance, let $x_1 < x_2 < \dots < x_n$ be the ROOTS of the $p_n(x)$ with $x_0 = a$ and $x_{n+1} = b$. Then each interval $[x_\nu, x_{\nu+1}]$ for $\nu = 0, 1, \dots, n$ contains exactly one ROOT of $p_{n+1}(x)$. Between two ROOTS of $p_n(x)$ there is at least one ROOT of $p_m(x)$ for $m > n$.

Let c be an arbitrary REAL constant, then the POLYNOMIAL

$$p_{n+1}(x) - cp_n(x) \qquad (4)$$

has $n + 1$ distinct REAL ROOTS. If $c > 0\,(c < 0)$, these ROOTS lie in the interior of $[a, b]$, with the exception of the greatest (least) ROOT which lies in $[a, b]$ only for

$$c \le \frac{p_{n+1}(b)}{p_n(b)}\quad\left(c \ge \frac{p_{n+1}(a)}{p_n(a)}\right). \qquad (5)$$

The following decomposition into partial fractions holds

$$\frac{p_n(x)}{p_{n+1}(x)} = \sum_{\nu=0}^{n}\frac{l_\nu}{x-\xi}, \qquad (6)$$

where $\{\xi_\nu\}$ are the ROOTS of $p_{n+1}(x)$ and

$$l_\nu = \frac{p_n(\xi_\nu)}{p'_{n+1}(\xi_\nu)}$$

$$= \frac{p'_{n+1}(\xi_\nu)p_n(\xi_\nu) - p'_n(\xi_\nu)'p_{n+1}(\xi_\nu)}{[p'_{n+1}(\xi_\nu)]^2} > 0. \qquad (7)$$

Another interesting property is obtained by letting $\{p_n(x)\}$ be the orthonormal set of POLYNOMIALS associated with the distribution $d\alpha(x)$ on $[a, b]$. Then the CONVERGENTS R_n/S_n of the CONTINUED FRACTION

$$\frac{1}{A_1x + B_1} - \frac{C_2}{A_2x + B_2} - \frac{C_3}{A_3x + B_3} - \dots - \frac{C_n}{A_nx + B_n}$$
$$+ \dots \qquad (8)$$

are given by

$$R_n = R_n(x)$$
$$= c_0^{-3/2}\sqrt{c_0c_2 - c_1^2}\int_a^b\frac{p_n(x) - p_n(t)}{x - t}\,d\alpha(t) \qquad (9)$$

$$S_n = S_n(x) = \sqrt{c_0}\,p_n(x), \qquad (10)$$

where $n = 0, 1, \dots$ and

$$c_n = \int_a^b x^n\,d\alpha(x). \qquad (11)$$

Furthermore, the ROOTS of the orthogonal polynomials $p_n(x)$ associated with the distribution $d\alpha(x)$ on the interval $[a, b]$ are REAL and distinct and are located in the interior of the interval $[a, b]$.

See also CHEBYSHEV POLYNOMIAL OF THE FIRST KIND, CHEBYSHEV POLYNOMIAL OF THE SECOND KIND, GRAM-SCHMIDT ORTHONORMALIZATION, HERMITE POLYNOMIAL, JACOBI POLYNOMIAL, KRAWTCHOUK POLYNOMIAL, LAGUERRE POLYNOMIAL, LEGENDRE POLYNOMIAL, ORTHOGONAL FUNCTIONS, SPHERICAL HARMONIC, ULTRASPHERICAL POLYNOMIAL, ZERNIKE POLYNOMIAL

References

Abramowitz, M. and Stegun, C. A. (Eds.). "Orthogonal Polynomials." Ch. 22 in *Handbook of Mathematical Functions with Formulas, Graphs, and Mathematical Tables, 9th printing.* New York: Dover, pp. 771–802, 1972.
Arfken, G. "Orthogonal Polynomials." *Mathematical Methods for Physicists, 3rd ed.* Orlando, FL: Academic Press, pp. 520–521, 1985.
Chihara, T. S. *An Introduction to Orthogonal Polynomials.* New York: Gordon and Breach, 1978.
Gautschi, W.; Golub, G. H.; and Opfer, G. (Eds.) *Applications and Computation of Orthogonal Polynomials, Conference at the Mathematical Research Institute Oberwolfach, Germany, March 22–28, 1998.* Basel, Switzerland: Birkhäuser, 1999.
Iyanaga, S. and Kawada, Y. (Eds.). "Systems of Orthogonal Functions." Appendix A, Table 20 in *Encyclopedic Dictionary of Mathematics.* Cambridge, MA: MIT Press, p. 1477, 1980.
Koekoek, R. and Swarttouw, R. F. *The Askey-Scheme of Hypergeometric Orthogonal Polynomials and its q-Analogue.* Delft, Netherlands: Technische Universiteit Delft, Faculty of Technical Mathematics and Informatics Report 98–17, 1–168, 1998. ftp://www.twi.tudelft.nl/publications/tech-reports/1998/DUT-TWI-98–17.ps.gz.
Nikiforov, A. F.; Uvarov, V. B.; and Suslov, S. S. *Classical Orthogonal Polynomials of a Discrete Variable.* New York: Springer-Verlag, 1992.
Sansone, G. *Orthogonal Functions.* New York: Dover, 1991.
Szego, G. *Orthogonal Polynomials, 4th ed.* Providence, RI: Amer. Math. Soc., pp. 44–47 and 54–55, 1975.

Orthogonal Projection

A PROJECTION of a figure by parallel rays. In such a projection, tangencies are preserved. Parallel lines project to parallel lines. The ratio of lengths of parallel segments is preserved, as is the ratio of areas.

Any TRIANGLE can be positioned such that its shadow under an orthogonal projection is EQUILATERAL. Also, the MEDIANS of a TRIANGLE project to the MEDIANS of the image TRIANGLE. ELLIPSES project to ELLIPSES, and any ELLIPSE can be projected to form a CIRCLE. The center of an ELLIPSE projects to the center of the image ELLIPSE. The CENTROID of a TRIANGLE projects to the CENTROID of its image. Under an ORTHOGONAL TRANSFORMATION, the MIDPOINT ELLIPSE can be transformed into a CIRCLE INSCRIBED in an EQUILATERAL TRIANGLE.

SPHEROIDS project to ELLIPSES (or CIRCLE in the DEGENERATE case).

In an orthogonal projection, any vector v can be written $v = v_W + v_{W^\perp}$, so

$$\langle v, Pw \rangle = \langle v_W, Pw \rangle = \langle Pv, w \rangle,$$

and the PROJECTION MATRIX is a SYMMETRIC MATRIX IFF the PROJECTION is orthogonal. The following *Mathematica* function will test whether a PROJECTION MATRIX is an orthogonal projection.

```
OrthogProjectionMatrixQ[a_List?MatrixQ] :=
    (a.a = = a && Transpose[a] = = a)
```

The following *Mathematica* function gives the PROJECTION MATRIX for orthogonal projection onto a subspace spanned by a given basis.

```
< <LinearAlgebra`Orthogonalization`;
  OrthogProjectMatrixOntoBasis[a_List?MatrixQ]
  :=
    Module[{a1 = GramSchmidt[a]},
    Transpose[a1].a1]
  ]
```

For instance, OrthogProjectMatrixOntoBasis[{{1, 2, 3}}] yields {{1/14, 1/7, 3/14}, {1/7, 2/7, 3/7}, {3/14, 3/7, 9/14}}.

See also PROJECTION, PROJECTION MATRIX

Orthogonal Rotation Group

ORTHOGONAL GROUP

Orthogonal Set

A subset $\{v_1, \ldots, v_k\}$ of a VECTOR SPACE V, with the INNER PRODUCT $\langle, \rangle$, is called orthogonal if $\langle v_i, v_j \rangle = 0$ when $i \neq j$. That is, the vectors are mutually PERPENDICULAR.

Note that there is no restriction on the lengths of the vectors. If the vectors in an orthogonal set all have length one, then they are ORTHONORMAL.

The notion of orthogonal makes sense for an abstract VECTOR SPACE over any field as long as there is a SYMMETRIC QUADRATIC FORM. The usual orthogonal sets and groups in EUCLIDEAN SPACE can be generalized, with applications to special relativity, DIFFERENTIAL GEOMETRY, and ABSTRACT ALGEBRA.

See also CLIFFORD ALGEBRA, HOMOGENEOUS SPACE, HYPERBOLIC SPACE, LIE GROUP, LORENTZIAN INNER PRODUCT, ORTHOGONAL GROUP, ORTHOGONAL TRANSFORMATION, ORTHONORMAL BASIS, SYMMETRIC QUADRATIC FORM

Orthogonal Subspaces

Two SUBSPACES S_1 and S_2 of $\mathbb{R}^n$ are said to be orthogonal if $\mathbf{v}_1 \cdot \mathbf{v}_2 = 0$ for all $\mathbf{v}_1 \in S_1$ and all $\mathbf{v}_2 \in S_2$.

Orthogonal Surfaces

Families of surfaces which are mutually orthogonal. Up to three families of surfaces may be orthogonal in 3-D. The simplest example of three orthogonal surfaces in 3-D are orthogonal planes, but three confocal conic surfaces are also mutually orthogonal.

References

Wells, D. *The Penguin Dictionary of Curious and Interesting Geometry.* London: Penguin, p. 166, 1991.

Orthogonal Tensors

Orthogonal CONTRAVARIANT and COVARIANT satisfy

$$g_{ik}g^{ij} = \delta_k^j,$$

where δ_j^k is the KRONECKER DELTA.

See also CONTRAVARIANT TENSOR, COVARIANT TENSOR

Orthogonal Transformation

An orthogonal transformation is a LINEAR TRANSFORMATION $T : V \to V$ which preserves a SYMMETRIC INNER PRODUCT. In particular, an orthogonal transformation (technically, an *orthonormal transformation*) preserves lengths of vectors and angles between vectors,

$$\langle v, w \rangle = \langle Tv, Tw \rangle. \tag{1}$$

In addition, an orthogonal transformation is either a rigid ROTATION or a ROTOINVERSION (a rotation followed by a flip). (Flipping and then rotating can be realized by first rotating in the reverse direction and then flipping). Orthogonal transformations correspond to and may be represented using ORTHOGONAL MATRICES.

The set of orthonormal transformations forms the ORTHOGONAL GROUP, and an orthonormal transformation can be realized by an ORTHOGONAL MATRIX.

Any linear transformation in 3-D

$$x_1' = a_{11}x_1 + a_{12}x_2 + x_{13}x_3 \tag{2}$$

$$x_2' = a_{21}x_1 + a_{22}x_2 + a_{23}x_3 \tag{3}$$

$$x_3' = a_{31}x_1 + a_{32}x_2 + a_{33}x_3 \tag{4}$$

satisfying the ORTHOGONALITY CONDITION

$$a_{ij}a_{ik} = \delta_{jk}, \tag{5}$$

where EINSTEIN SUMMATION has been used and δ_{ij} is the KRONECKER DELTA, is an orthogonal transformation. If $A : \mathbb{R}^n \to \mathbb{R}^n$ is an orthogonal transformation, then $\det(A) = \pm 1$.

See also INNER PRODUCT, LIE GROUP, LINEAR TRANSFORMATION, LORENTZ TRANSFORMATION, MATRIX, ORTHOGONAL MATRIX, ORTHOGONAL GROUP, ORTHOGONALITY CONDITION, SPIN GROUP, ROTATION, ROTOINVERSION, SYMMETRIC QUADRATIC FORM

References

Goldstein, H. "Orthogonal Transformations." §4–2 in *Classical Mechanics, 2nd ed.* Reading, MA: Addison-Wesley, 132–137, 1980.
Gray, A. *Modern Differential Geometry of Curves and Surfaces with Mathematica, 2nd ed.* Boca Raton, FL: CRC Press, pp. 128–129, 1997.

Orthogonal Vectors

Two vectors **u** and **v** whose DOT PRODUCT is $\mathbf{u} \cdot \mathbf{v} = 0$ (i.e., the vectors are PERPENDICULAR) are said to be orthogonal. In 3-space, three vectors can be mutually perpendicular.

See also DOT PRODUCT, ORTHONORMAL VECTORS, PERPENDICULAR

Orthogonality Condition

A linear transformation

$$x_1' = a_{11}x_1 + a_{12}x_2 + x_{13}x_3$$

$$x_2' = a_{21}x_1 + a_{22}x_2 + a_{23}x_3$$

$$x_3' = a_{31}x_1 + a_{32}x_2 + a_{33}x_3,$$

is said to be an ORTHOGONAL TRANSFORMATION if it satisfies the orthogonality condition

$$a_{ij}a_{ik} = \delta_{jk},$$

where EINSTEIN SUMMATION has been used and δ_{ij} is the KRONECKER DELTA.

See also ORTHOGONAL TRANSFORMATION

References

Goldstein, H. "Orthogonal Transformations." §4–2 in *Classical Mechanics, 2nd ed.* Reading, MA: Addison-Wesley, pp. 132–137, 1980.

Orthogonality Theorem

GROUP ORTHOGONALITY THEOREM

Orthographic Projection

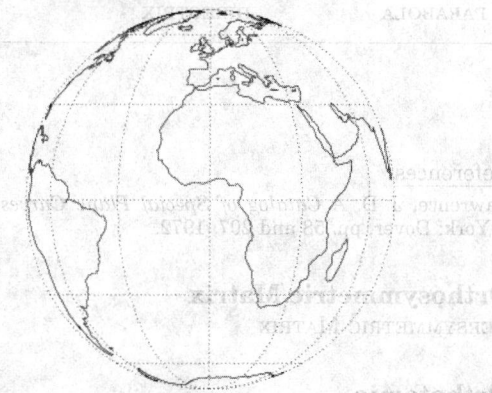

A projection from infinity which preserves neither AREA nor angle.

$$x = \cos \phi \, \sin(\lambda - \lambda_0) \tag{1}$$

$$y = \cos \phi_1 \sin \phi - \sin \phi_1 \cos \phi \cos(\lambda - \lambda_0). \tag{2}$$

The inverse FORMULAS are

$$\phi = \sin^{-1}\left(\cos c \sin \phi_1 + \frac{y \sin c \cos \phi_1 +}{\rho} \right) \tag{3}$$

$$\lambda = \lambda_0 + \tan^{-1}\left(\frac{x \sin c}{\rho \cos \phi_1 \cos c - y \sin \phi_1 \sin c} \right), \tag{4}$$

where

$$\rho = \sqrt{x^2 + y^2} \tag{5}$$

$$c = \sin^{-1} \rho. \tag{6}$$

References
Snyder, J. P. *Map Projections--A Working Manual.* U. S. Geological Survey Professional Paper 1395. Washington, DC: U. S. Government Printing Office, pp. 145–153, 1987.

Orthologic Triangles
Two TRIANGLES $A_1 B_1 C_1$ and $A_2 B_2 C_2$ are orthologic if the perpendiculars from the VERTICES A_1, B_1, C_1 on the sides $B_2 C_2$, $A_2 C_2$, and $A_2 B_2$ pass through one point. This point is known as the orthology center of TRIANGLE 1 with respect to TRIANGLE 2.

Orthomorphic Projection
CONFORMAL PROJECTION

Orthonormal Basis
A subset $\{v_1, \ldots, v_k\}$ of a VECTOR SPACE V, with the INNER PRODUCT $\langle , \rangle$, is called orthonormal if $\langle v_i, v_j \rangle = 0$ when $i \neq j$. That is, the vectors are mutually PERPENDICULAR. Moreover, they are all required to have length one: $\langle v_i, v_i \rangle = 1$.

An orthonormal set must be linearly independent, and so it is a BASIS for the space it SPANS. Such a basis is called an orthonormal basis.

The simplest example of an orthonormal basis is the standard basis e_i for EUCLIDEAN SPACE $\mathbb{R}^n$. The vector e_i is the vector with all 0s except for a 1 in the ith coordinate. For example, $e_1 = (1, 0, \ldots, 0)$. A rotation (or flip) through the origin will send an orthonormal set to another orthonormal set. In fact, given any orthonormal basis, there is a rotation, or rotation combined with a flip, which will send the orthonormal basis to the standard basis. These are precisely the transformations which preserve the inner product, and are called ORTHOGONAL TRANSFORMATIONS.

Usually when one needs a basis to do calculations, it is convenient to use an orthonormal basis. For example, the formula for a PROJECTION is much simpler with an orthonormal basis. The savings in effort make it worthwhile to find an orthonormal basis before doing such a calculation. GRAM-SCHMIDT ORTHONORMALIZATION is a popular way to find an orthonormal basis.

Another instance when orthonormal bases arise is as a set of EIGENVECTORS for a SYMMETRIC MATRIX. For a general matrix, the set of eigenvectors may not be orthonormal, or even be a basis.

See also BASIS (VECTOR SPACE), DOT PRODUCT, INNER PRODUCT, KRONECKER DELTA, LIE GROUP, LORENTZIAN INNER PRODUCT, MATRIX, ORTHOGONAL BASIS ORTHOGONAL MATRIX, ORTHOGONAL GROUP, ORTHOGONAL TRANSFORMATION, PROJECTION (VECTOR SPACE), SYMMETRIC QUADRATIC FORM

Orthonormal Functions
A pair of functions $\phi_i(x)$ and $\phi_j(x)$ are orthonormal if they are ORTHOGONAL and each normalized. These two conditions can be succinctly written as

$$\int_a^b \phi_i(x)\phi_j(x)w(x)\,dx = \delta_{ij},$$

where $w(x)$ is a WEIGHTING FUNCTION and δ_{ij} is the KRONECKER DELTA.

See also ORTHOGONAL POLYNOMIALS

Orthonormal Transformation
ORTHOGONAL TRANSFORMATION

Orthonormal Vectors
UNIT VECTORS which are ORTHOGONAL are said to be orthonormal.

See also ORTHOGONAL VECTORS

Orthoplex
CROSS POLYTOPE

Orthopole

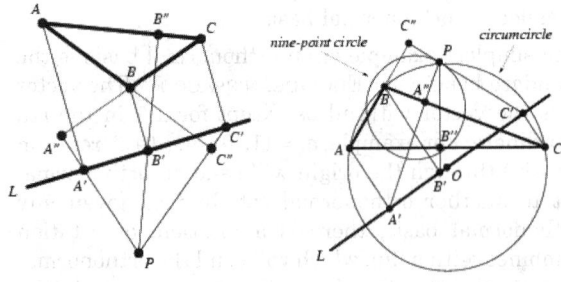

If perpendiculars A', B', and C' are dropped on any line L from the vertices of a TRIANGLE $\triangle ABC$, then the perpendiculars to the opposite sides from their FEET A'', B'', and C'' are CONCURRENT at a point P called the orthopole. The orthopole of a line lies on the SIMSON LINE which is PERPENDICULAR to it (Honsberger 1995, p. 130). If a line crosses the CIRCUMCIRCLE of a triangle, the SIMSON LINES of the points of intersection meet at the orthopole of the line. Also, the orthopole of a line through the CIRCUMCENTER O of a triangle $\triangle ABC$ lies on that triangle's NINE-POINT CIRCLE (Honsberger 1995, p. 127).

If the line L is displaced PARALLEL to itself, the orthopole moves along a line PERPENDICULAR to L a distance equal to the displacement. If L is the SIMSON LINE of a point P, then P is called the POLE of L (Honsberger 1995, p. 128).

See also NINE-POINT CIRCLE, POLE (SIMSON LINE), RIGBY POINTS, SIMSON LINE

References
Honsberger, R. "The Orthopole." Ch. 11 in *Episodes in Nineteenth and Twentieth Century Euclidean Geometry.* Washington, DC: Math. Assoc. Amer., pp. 125–136, 1995.
Johnson, R. A. *Modern Geometry: An Elementary Treatise on the Geometry of the Triangle and the Circle.* Boston, MA: Houghton Mifflin, p. 247, 1929.
Ramler, O. J. "The Orthopole Loci of Some One-Parameter Systems of Lines Referred to a Fixed Triangle." *Amer. Math. Monthly* **37**, 130–136, 1930.

Orthoptic Curve

An ISOPTIC CURVE formed from the locus of TANGENTS meeting at RIGHT ANGLES. The orthoptic of a PARABOLA is its DIRECTRIX. The orthoptic of a central CONIC was investigated by Monge and is a CIRCLE concentric with the CONIC SECTION. The orthoptic of an ASTROID is a CIRCLE.

Curve	Orthoptic
ASTROID	QUADRIFOLIUM
CARDIOID	CIRCLE or LIMAÇON
DELTOID	CIRCLE

LOGARITHMIC SPIRAL	equal LOGARITHMIC SPIRAL
PARABOLA	DIRECTRIX

References
Lawrence, J. D. *A Catalog of Special Plane Curves.* New York: Dover, pp. 58 and 207, 1972.

Orthosymmetric Matrix
PERSYMMETRIC MATRIX

Orthotomic

Given a source S and a curve γ, pick a point on γ and find its tangent T. Then the LOCUS of reflections of S about tangents T is the orthotomic curve (also known as the secondary CAUSTIC). The INVOLUTE of the orthotomic is the CAUSTIC. For a parametric curve $(f(t), g(t))$ with respect to the point (x_0, y_0), the orthotomic is

$$x = x_0 - \frac{2g'[f'(g - y_0) - g'(f - x_0)]}{f'^2 + g'^2}$$

$$y = y_0 + \frac{2f'[f'(g - y_0) - g'(f - x_0)]}{f'^2 + g'^2}$$

See also CAUSTIC, INVOLUTE

References
Lawrence, J. D. *A Catalog of Special Plane Curves.* New York: Dover, p. 60, 1972.

Orthotope

A PARALLELOTOPE whose edges are all mutually PERPENDICULAR. The orthotope is a generalization of the RECTANGLE and RECTANGULAR PARALLELEPIPED.

See also RECTANGLE, RECTANGULAR PARALLELEPIPED

References
Coxeter, H. S. M. *Regular Polytopes, 3rd ed.* New York: Dover, pp. 122–123, 1973.

Osborne's Rule

The prescription that a TRIGONOMETRY identity can be converted to an analogous identity for HYPERBOLIC FUNCTIONS by expanding, exchanging trigonometric functions with their hyperbolic counterparts, and then flipping the sign of each term involving the product of two HYPERBOLIC SINES. For example, given the identity

$$\cos(x - y) = \cos x \cos y + \sin x \sin y,$$

Osborne's rule gives the corresponding identity

$$\cosh(x-y) = \cosh x \cosh y + \sinh x \sinh y.$$

See also HYPERBOLIC FUNCTIONS, TRIGONOMETRIC FUNCTIONS

Oscillation

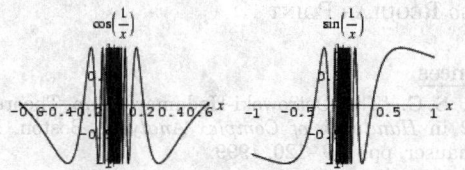

The variation of a FUNCTION which exhibits SLOPE changes, also called the SALTUS of a function. A series may also oscillate, causing it not to converge.

References

Jeffreys, H. and Jeffreys, B. S. "Bounded, Unbounded, Convergent, Oscillatory." §1.041 in *Methods of Mathematical Physics, 3rd ed.* Cambridge, England: Cambridge University Press, pp. 11–12 and 22, 1988.

Oscillation Land

CAROTID-KUNDALINI FUNCTION

Osculating Circle

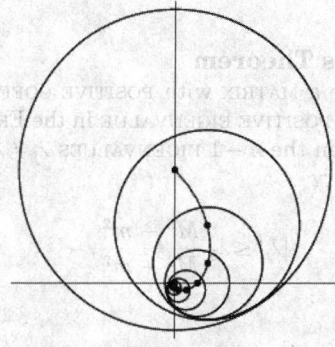

The CIRCLE which shares the same TANGENT as a curve at a given point. Given a plane curve with PARAMETRIC EQUATIONS $(f(t), g(t))$ and parameterized by a variable t, the RADIUS OF CURVATURE of the osculating circle is

$$\rho(t) = \frac{1}{|\kappa(t)|}, \tag{1}$$

where $\kappa(t)$ is the CURVATURE, and the center is

$$x = f - \frac{(f'^2 + g'^2)g'}{f'g'' - f''g'} \tag{2}$$

$$y = g + \frac{(f'^2 + g'^2)f'}{f'g'' - f''g'}. \tag{3}$$

Here, derivatives are taken with respect to the parameter t. Note that the centers of the osculating

circles to a curve form the EVOLUTE to that curve.

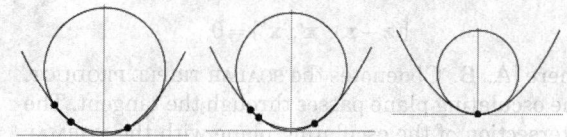

In addition, let $C(t_1, t_2, t_3)$ denote the CIRCLE passing through three points on a curve $(f(t), g(t))$ with $t_1 < t_2 < t_3$. Then the osculating circle C is given by

$$C = \lim_{t_1, t_2, t_3 \to t} C(t_1, t_2, t_3) \tag{4}$$

(Gray 1997).

See also CURVATURE, EVOLUTE, OSCULATING CURVES, RADIUS OF CURVATURE, TANGENT

References

Gardner, M. "The Game of Life, Parts I-III." Chs. 20–22 in *Wheels, Life, and other Mathematical Amusements.* New York: W. H. Freeman, pp. 221, 237, and 243, 1983.
Gray, A. "Osculating Circles to Plane Curves." §5.6 in *Modern Differential Geometry of Curves and Surfaces with Mathematica, 2nd ed.* Boca Raton, FL: CRC Press, pp. 111–115, 1997.

Osculating Curves

An curve $y(x)$ is osculating to $f(x)$ at x_0 if it is TANGENT at x_0 and has the same CURVATURE there. Osculating curves therefore satisfy

$$y^{(k)}(x_0) = f^{(k)}(x_0)$$

for $k = 0$, 1, 2. The point of tangency is called a TACNODE.

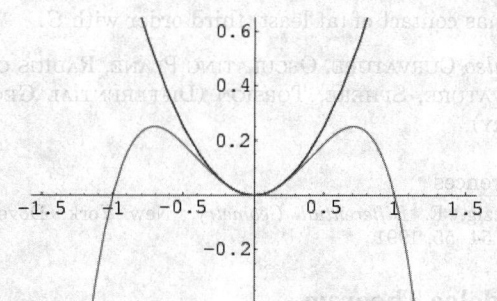

One of simplest examples of a pairs of osculating curves is x^2 and $x^2 - x^4$, which osculate at the point $x_0 = 0$ since for $k = 0$, 1, 2, $y^{(k)}(0) = f^{(k)}(0)$ is equal to 0, 0, and 2.

See also OSCULATING CIRCLE, TACNODE, TANGENT CURVES

Osculating Interpolation

HERMITE'S INTERPOLATING POLYNOMIAL

Osculating Plane

The PLANE spanned by the three points $\mathbf{x}(t)$, $\mathbf{x}(t + h_1)$, and $\mathbf{x}(t + h_2)$ on a curve as h_1, $h_2 \to 0$. Let $\mathbf{z}$ be a point on the osculating plane, then

$$[(\mathbf{z} - \mathbf{x}), \mathbf{x}', \mathbf{x}''] = 0,$$

where $[\mathbf{A}, \mathbf{B}, \mathbf{C}]$ denotes the SCALAR TRIPLE PRODUCT. The osculating plane passes through the tangent. The intersection of the osculating plane with the NORMAL PLANE is known as the PRINCIPAL NORMAL VECTOR. The VECTORS $\mathbf{T}$ and $\mathbf{N}$ (TANGENT VECTOR and NORMAL VECTOR) span the osculating plane.

See also NORMAL VECTOR, OSCULATING SPHERE, SCALAR TRIPLE PRODUCT, TANGENT VECTOR

Osculating Sphere

The center of any SPHERE which has a contact of (at least) first-order with a curve C at a point P lies in the normal plane to C at P. The center of any SPHERE which has a contact of (at least) second-order with C at point P, where the CURVATURE $\kappa > 0$, lies on the polar axis of C corresponding to P. All these SPHERES intersect the OSCULATING PLANE of C at P along a circle of curvature at P. The osculating sphere has center

$$\mathbf{a} = \mathbf{x} + \rho \hat{\mathbf{N}} + \frac{\dot{\rho}}{\tau} \hat{\mathbf{B}}$$

where $\hat{\mathbf{N}}$ is the unit NORMAL VECTOR, $\hat{\mathbf{B}}$ is the unit BINORMAL VECTOR, ρ is the RADIUS OF CURVATURE, and τ is the TORSION, and RADIUS

$$R = \sqrt{\rho^2 + \left(\frac{\dot{\rho}}{\tau}\right)^2},$$

and has contact of (at least) third order with C.

See also CURVATURE, OSCULATING PLANE, RADIUS OF CURVATURE, SPHERE, TORSION (DIFFERENTIAL GEOMETRY)

References

Kreyszig, E. *Differential Geometry.* New York: Dover, pp. 54–55, 1991.

Osedelec Theorem

For an n-D MAP, the LYAPUNOV CHARACTERISTIC EXPONENTS are given by

$$\sigma_i = \lim_{N \to \infty} \ln|\lambda_i(N)|$$

for $i = 1, ..., n$, where λ_i is the LYAPUNOV CHARACTERISTIC NUMBER.

See also LYAPUNOV CHARACTERISTIC EXPONENT, LYAPUNOV CHARACTERISTIC NUMBER

Ostrowski-Hadamard Gap Theorem

Let $0 < p_1 < p_2 < ...$ be integers and suppose that there exists a $\lambda > 1$ such that $p_{j+1}/p_j > \lambda$ for $j = 1, 2,$ .... Suppose that for some sequence of complex

numbers $\{a_j\}$ the POWER SERIES

$$f(z) = \sum_{j=1}^{\infty} a_j z^{p_j}$$

has radius of convergence 1, then no point of ∂D is a REGULAR POINT for f (Krantz 1999, p. 120).

See also REGULAR POINT

References

Krantz, S. G. "The Ostrowski-Hadamard Gap Theorem." §9.2.2 in *Handbook of Complex Analysis.* Boston, MA: Birkhäuser, pp. 119–120, 1999.

Ostrowski's Inequality

If $f(x)$ is a monotonically increasing integrable function on $[a, b]$ with $f(b) \leq 0$, then if g is a REAL function integrable on $[a, b]$,

$$\left| \int_a^b f(x)g(x)\, dx \right| \leq |f(a)| \max_{a \leq \xi \leq b} \left| \int_a^\xi g(x)\, dx \right|.$$

References

Gradshteyn, I. S. and Ryzhik, I. M. *Tables of Integrals, Series, and Products, 6th ed.* San Diego, CA: Academic Press, p. 1100, 2000.

Ostrowski's Theorem

Let $\mathbf{A} = a_{ij}$ be a MATRIX with POSITIVE COEFFICIENTS and λ_0 be the POSITIVE EIGENVALUE in the FROBENIUS THEOREM, then the $n - 1$ EIGENVALUES $\lambda_j \neq \lambda_0$ satisfy the INEQUALITY

$$|\lambda_j| \leq \lambda_0 \frac{M^2 - m^2}{M^2 + m^2},$$

where

$$M = \max_{i, j} a_{ij}$$

$$m = \min_{i, j} a_{ij}$$

and $i, j = 1, 2, ..., n$.

See also FROBENIUS THEOREM

References

Gradshteyn, I. S. and Ryzhik, I. M. *Tables of Integrals, Series, and Products, 6th ed.* San Diego, CA: Academic Press, p. 1121, 2000.

Otter's Theorem

In any TREE, the number of dissimilar points minus the number of dissimilar lines plus the number of symmetry lines equals 1.

See also TREE

References

Harary, F. and Prins, G. "The Number of Homeomorphically Irreducible Trees, and Other Species." *Acta Math.* **101**, 141–162, 1959.

Otter, R. "The Number of Trees." *Ann. Math.* **49**, 583–599, 1948.

Oudor

References

Moon, P. and Spencer, D. E. *Theory of Holors: A Generalization of Tensors.* Cambridge, England: Cambridge University Press, 1986.

Oui-Ja Board Curve

COCHLEOID

Outcome

An outcome is a subset of a PROBABILITY SPACE. Experimental outcomes are not uniquely determined from the description of an experiment, and must be agreed upon to avoid ambiguity (Papoulis 1984, pp. 24–25).

See also EVENT, EXPERIMENT, TRIAL

References

Papoulis, A. *Probability, Random Variables, and Stochastic Processes, 2nd ed.* New York: McGraw-Hill, 1984.

Outdegree

The number of outward directed EDGES from a given VERTEX in a DIRECTED GRAPH.

See also DIRECTED GRAPH, INDEGREE, LOCAL DEGREE

Outer Automorphism Group

A particular type of AUTOMORPHISM GROUP which exists only for GROUPS. For a GROUP G, the outer automorphism group is the QUOTIENT GROUP $\text{Aut}(G)/\text{Inn}(G)$, which is the AUTOMORPHISM GROUP of G modulo its INNER AUTOMORPHISM GROUP.

See also AUTOMORPHISM GROUP, INNER AUTOMORPHISM GROUP, QUOTIENT GROUP

Outer Product

TENSOR DIRECT PRODUCT, TENSOR PRODUCT (VECTOR SPACE)

Outer Quermass

BRIGHTNESS

Outplanar Graph

A graph that can be embedded in the plane such that all vertices lie on the outer face (Skiena 1990, p. 251).

See also PLANAR GRAPH

References

Skiena, S. *Implementing Discrete Mathematics: Combinatorics and Graph Theory with Mathematica.* Reading, MA: Addison-Wesley, 1990.

Oval

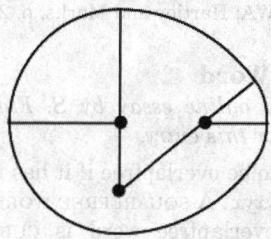

An oval is a curve resembling a squashed CIRCLE but, unlike the ELLIPSE, without a precise mathematical definition. The word oval derived from the Latin word "ovus" for egg. Unlike ellipses, ovals sometimes have only a single axis of reflection symmetry (instead of two).

Ovals can be constructed with a COMPASS by joining together arcs of different radii such that the centers of the arcs lie on a line passing through the join point (Dixon 1991). Albrecht Dürer used this method to design a Roman letter font.

See also CARTESIAN OVALS, CASSINI OVALS, EGG, ELLIPSE, LEMON, OVOID, SUPERELLIPSE

References

Critchlow, K. *Time Stands Still.* London: Gordon Fraser, 1979.

Cundy, H. and Rollett, A. *Mathematical Models, 3rd ed.* Stradbroke, England: Tarquin Pub., 1989.

Dixon, R. *Mathographics.* New York: Dover, pp. 3–11, 1991.

Dixon, R. "The Drawing Out of an Egg." *New Sci.*, July 29, 1982.

Pedoe, D. *Geometry and the Liberal Arts.* London: Peregrine, 1976.

Oval of Descartes

CARTESIAN OVALS

Ovals of Cassini

CASSINI OVALS

Overbar

MACRON

Overdamping

DAMPED SIMPLE HARMONIC MOTION–OVERDAMPING

Overdot

An "overdot" is a raised DOT appearing above a symbol most commonly used in mathematics to indicate a DERIVATIVE taken with respect to time (e.g., $\dot{x} \equiv dx/dt$). The expression $\dot{a}$ is voiced "a dot," and was Newton's notation for derivatives (which he called "FLUXIONS").

See also DERIVATIVE, DOT, DOUBLE DOT

References

Bringhurst, R. *The Elements of Typographic Style, 2nd ed.* Point Roberts, WA: Hartley and Marks, p. 282, 1997.

Overlapfree Word

N.B. A detailed online essay by S. Finch was the starting point for this entry.

A word is said to be overlapfree if it has no subwords OF THE FORM *xyxyx*. A SQUAREFREE WORD is overlapfree, and an overlapfree word is CUBEFREE. The number $t(n)$ of binary overlapfree words of length $n = 1, 2, ...$ are 2, 4, 6, 10, 14, 20, ... (Sloane's A007777), $t(n)$ satisfies

$$p \cdot n^{1.155} \leq t(n) \leq q \cdot n^{1.587} \tag{1}$$

for some constants p and q (Restivo and Selemi 1985, Kobayashi 1988). In addition, while

$$\lim_{n \to \infty} \frac{\ln t(n)}{\ln n} \tag{2}$$

does not exist,

$$1.155 < T_L < 1,276 < 1.332 < T_U < 1.587, \tag{3}$$

where

$$T_L \equiv \liminf_{n \to \infty} \frac{\ln t(n)}{\ln n} \tag{4}$$

$$T_U \equiv \limsup_{n \to \infty} \frac{\ln t(n)}{\ln n} \tag{5}$$

(Cassaigne 1993).

See also CUBEFREE WORD, SQUAREFREE WORD, WORD

References

Cassaigne, J. "Counting Overlap-Free Binary Words." *STACS '93: Tenth Annual Symposium on Theoretical Aspects of Computer Science, Würzburg, Germany, February 25–27, 1993 Proceedings* (Ed. G. Goos, J. Hartmanis, A. Finkel, P. Enjalbert, K. W. Wagner). New York: Springer-Verlag, pp. 216–225, 1993.
Cassaigne, J. *Motifs évitables et régularités dans les mots (Thèse de Doctorat).* Tech. Rep. LITP-TH 94–04. Paris: Institut Blaise Pascal, 1994.
Finch, S. "Favorite Mathematical Constants." http://www.mathsoft.com/asolve/constant/words/words.html.
Kobayashi, Y. "Enumeration of Irreducible Binary Words." *Discrete Appl. Math.* **20**, 221–232, 1988.
Séébold, P. "Overlap-Free Sequences." In *Combinatorics on Words* (Ed. L. J. Cummings). Toronto: Academic Press, pp. 207–215, 1983.

Sloane, N. J. A. Sequences A007777 in "An On-Line Version of the Encyclopedia of Integer Sequences." http://www.research.att.com/~njas/sequences/eisonline.html.

Overlapping Rectangles

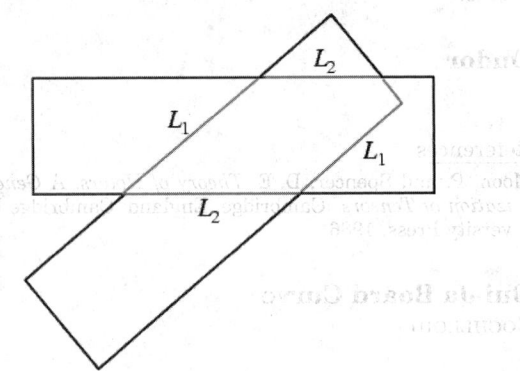

See also RECTANGLE

References

Croft, H. T.; Falconer, K. J.; and Guy, R. K. "Overlapping Convex Bodies." §A12 in *Unsolved Problems in Geometry.* New York: Springer-Verlag, p. 25, 1991.

Overlapping Resonance Method

RESONANCE OVERLAP METHOD

Overline

MACRON, VINCULUM

Oversampling

A signal sampled at a frequency higher than the NYQUIST FREQUENCY is said to be oversampled β times, where the oversampling ratio is defined as

$$\beta \equiv \frac{v_{\text{sampling}}}{v_{\text{Nyquist}}}.$$

See also NYQUIST FREQUENCY, NYQUIST SAMPLING

Ovoid

An egg-shaped curve. Lockwood (1967) calls the NEGATIVE PEDAL CURVE of an ELLIPSE with ECCENTRICITY $e \leq 1/2$ an ovoid.

See also OVAL

References

Lockwood, E. H. *A Book of Curves.* Cambridge, England: Cambridge University Press, p. 157, 1967.

P

p (Prime) Group

X is a p'-group if p does not divide the ORDER of X.

Paasche's Index

The statistical INDEX

$$P_P \equiv \frac{\sum p_n q_n}{\sum p_0 q_n},$$

where p_n is the price per unit in period n and q_n is the quantity produced in period n.

See also INDEX

References

Kenney, J. F. and Keeping, E. S. *Mathematics of Statistics, Pt. 1, 3rd ed.* Princeton, NJ: Van Nostrand, p. 65, 1962.

Packing

The placement of objects so that they touch in some specified manner, often inside a container with specified properties. For example, one could consider a SPHERE PACKING, ELLIPSOID PACKING, POLYHEDRON PACKING, etc.

See also BARLOW PACKING, BOX-PACKING THEOREM, CIRCLE PACKING, COVERING, ELLIPSOID PACKING, GROEMER PACKING, HYPERSPHERE PACKING, KEPLER PROBLEM, KISSING NUMBER PACKING DENSITY, POLYHEDRON PACKING, SPACE-FILLING POLYHEDRON, SPHERE PACKING, SPHERICAL COVERING, SPHERICAL DESIGN, TRIANGLE PACKING

References

Eppstein, D. "Covering and Packing." http://www.ics.uci.edu/~eppstein/junkyard/cover.html.
Friedman, E. "Erich's Packing Center." http://www.stetson.edu/~efriedma/packing.html.

Packing Density

The fraction of a volume filled by a given collection of solids.

See also HYPERSPHERE PACKING, PACKING, SPHERE PACKING

Padé Approximant

Approximants derived by expanding a function as a ratio of two POWER SERIES and determining both the NUMERATOR and DENOMINATOR COEFFICIENTS. Padé approximations are usually superior to TAYLOR EXPANSIONS when functions contain POLES, because the use of RATIONAL FUNCTIONS allows them to be well-represented.

The Padé approximant $R_{L/0}$ corresponds to the MACLAURIN SERIES. When it exists, the $R_{L/M} \equiv$

$[L/M]$ Padé approximant to any POWER SERIES

$$A(x) = \sum_{j=0}^{\infty} a_j x^j \qquad (1)$$

is unique. If $A(x)$ is a TRANSCENDENTAL FUNCTION, then the terms are given by the TAYLOR SERIES about x_0

$$a_n = \frac{1}{n!} A^{(n)}(x_0). \qquad (2)$$

The COEFFICIENTS are found by setting

$$A(x) - \frac{P_L(x)}{Q_M(x)} = 0 \qquad (3)$$

and equating COEFFICIENTS. $Q_M(x)$ can be multiplied by an arbitrary constant which will rescale the other COEFFICIENTS, so an addition constraint can be applied. The conventional normalization is

$$Q_M(0) = 1. \qquad (4)$$

Expanding (3) gives

$$P_L(x) = p_0 + p_1 x + \ldots + p_L x^L \qquad (5)$$

$$Q_M(x) = 1 + q_1 x + \ldots + p_M x^M. \qquad (6)$$

These give the set of equations

$$a_0 = p_0 \qquad (7)$$

$$a_1 + a_0 q_1 = p_1 \qquad (8)$$

$$a_2 + a_1 q_1 + a_0 q_2 = p_2 \qquad (9)$$

$$\vdots$$

$$a_L + a_{L-1} q_1 + \ldots + a_0 q_L = p_L \qquad (10)$$

$$a_{L+1} + a_L q_1 + \ldots + a_{L-M+1} q_M = 0 \qquad (11)$$

$$\vdots$$

$$q_{L+M} + a_{L+M-1} q_1 + \ldots + a_L q_M = 0, \qquad (12)$$

where $a_n = 0$ for $n < 0$ and $q_j = 0$ for $j > M$. Solving these directly gives

$$[L/M] = \frac{\begin{vmatrix} a_{L-m+1} & a_{L-m+2} & \cdots & a_{L+1} \\ \vdots & \vdots & \ddots & \vdots \\ a_L & a_{L+1} & \cdots & a_{L+M} \\ \sum_{j=M}^{L} a_j - M x^j & \sum_{j=M-1}^{L} a_{j-M+1} x^j & \cdots & \sum_{j=0}^{L} a_j x^j \end{vmatrix}}{\begin{vmatrix} a_{L-m+1} & a_{L-m+2} & \cdots & a_{L+1} \\ \vdots & \vdots & \ddots & \vdots \\ a_L & a_{L+1} & \cdots & a_{L+M} \\ x^M & x^{M-1} & \cdots & 1 \end{vmatrix}}, \qquad (13)$$

where sums are replaced by a zero if the lower index exceeds the upper. Alternate forms are

$$[L/M] = \sum_{j=0}^{L-M} a_j x^j + x^{L-M+1} \mathbf{w}_{L/M}^{\mathrm{T}} \mathsf{W}_{L/M}^{-1} \mathbf{w}_{L/M}$$

$$= \sum_{j=0}^{L+n} a_j x^j + x^{L+n+1} \mathbf{w}_{(L+M)/M}^{\mathrm{T}} \mathsf{W}_{L/M}^{-1} \mathbf{w}_{(L+n)/M}$$

for

$$\mathsf{W}_{L/M} = \begin{bmatrix} a_{L-M+1} - xa_{L-M+2} & \cdots & a_L - xa_{L+1} \\ \vdots & \ddots & \vdots \\ a_L - xa_{L+1} & \cdots & a_{L+M-1} - xa_{L+M} \end{bmatrix} \tag{14}$$

$$\mathbf{w}_{L/M} = \begin{bmatrix} a_{L-M+1} \\ a_{L-M+2} \\ \vdots \\ a_L \end{bmatrix}, \tag{15}$$

and $0 \le n \le M$.

For example, the first few Padé approximants for e^x are

$$\exp_{0/0}(x) = 1$$

$$\exp_{0/1}(x) = \frac{1}{1-x}$$

$$\exp_{0/2}(x) = \frac{2}{2 - 2x + x^2}$$

$$\exp_{0/3}(x) = \frac{6}{6 - 6x + 3x^2 - x^3}$$

$$\exp_{1/0}(x) = 1 + x$$

$$\exp_{1/1}(x) = \frac{2+x}{2-x}$$

$$\exp_{1/2}(x) = \frac{6 + 2x}{6 - 4x + x^2}$$

$$\exp_{1/3}(x) = \frac{24 + 6x}{24 - 18x + 6x^2 - x^3}$$

$$\exp_{2/0}(x) = \frac{2 + 2x + x^2}{2}$$

$$\exp_{2/1}(x) = \frac{6 + 4x + x^2}{6 - 2x}$$

$$\exp_{2/2}(x) = \frac{12 + 6x + x^2}{12 - 6x + x^2}$$

$$\exp_{2/3}(x) = \frac{60 + 24x + 3x^2}{60 - 36x + 9x^2 - x^3}$$

$$\exp_{3/0}(x) = \frac{6 + 6x + 3x^2 + x^3}{6}$$

$$\exp_{3/1}(x) = \frac{24 + 18x + 16x^2 + x^3}{24 - 6x}$$

$$\exp_{3/2}(x) = \frac{60 + 36x + 9x^2 + x^3}{60 - 24x + 3x^2}$$

$$\exp_{3/3}(x) = \frac{120 + 60x + 12x^2 + x^3}{120 - 60x + 12x^2 - x^3}.$$

Two-term identities include

$$\frac{P_{L+1}(x)}{Q_{M+1}(x)} - \frac{P'_L(x)}{Q'_M(x)} = \frac{C_{(L+1)/(M+1)}^2 x^{L+M+1}}{Q_{M+1}(x) Q'_M(x)} \tag{16}$$

$$\frac{P_{L+1}(x)}{Q_M(x)} - \frac{P'_L(x)}{Q'_M(x)} = \frac{C_{(L+1)/M} C_{(L+1)/(M+1)} x^{L+M+1}}{Q_M(x) Q'_M(x)} \tag{17}$$

$$\frac{P_L(x)}{Q_{M+1}(x)} - \frac{P'_L(x)}{Q'_M(x)} = \frac{C_{L/(M+1)} C_{(L+1)/(M+1)} x^{L+M+1}}{Q_M(x) Q'_M(x)} \tag{18}$$

$$\frac{P_L(x)}{Q_{M+1}(x)} - \frac{P'_{L+1}(x)}{Q'_M} = \frac{C_{(L+1)/(M+1)}^2 x^{L+M+2}}{Q_{M+1} Q'_M} \tag{19}$$

$$\frac{P_{L+1}}{Q_M(x)} - \frac{P'_{L-1}(x)}{Q'_M(x)} =$$

$$\frac{C_{L/(M+1)} C_{(L+1)/M} x^{L+M} + C_{L/M} C_{(L+1)/(M+1)} x^{L+M+1}}{Q_M(x) Q'_M(x)} \tag{20}$$

$$\frac{P_L(x)}{Q_{M+1}(x)} - \frac{P'_L(x)}{Q'_{M-1}(x)} =$$

$$\frac{C_{L/(M+1)} C_{(L+1)/M} x^{L+M} - C_{L/M} C_{(L+1)/(M+1)} x^{L+M+1}}{Q_{M+1}(x) Q'_{M-1}(x)}, \tag{21}$$

where C is the C-DETERMINANT. Three-term identities can be derived using the FROBENIUS TRIANGLE IDENTITIES (Baker 1975, p. 32).

A five-term identity is

$$S_{(L+1)/M} S_{(L-1)/M} - S_{L/(M+1)} S_{L/(M-1)} = S_{L/M}^2. \tag{22}$$

Cross ratio identities include

$$\frac{\left(R_{L/M} - R_{L/(M+1)}\right)\left(R_{(L+1)/M} - R_{(L+1)/(M+1)}\right)}{\left(R_{L/M} - R_{(L+1)/M}\right)\left(R_{L/(M+1)} - R_{(L+1)/(M+1)}\right)}$$

$$= \frac{C_{L/(M+1)} C_{(L+2)/(M+1)}}{C_{(L+1)/M} C_{(L+1)/(M+2)}} \tag{23}$$

$$\frac{\left(R_{L/M} - R_{(L+1)/(M+1)}\right)\left(R_{(L+1)/M} - R_{L/(M+1)}\right)}{\left(R_{L/M} - R_{L/(M+1)}\right)\left(R_{(L+1)/M} - R_{(L+1)/(M+1)}\right)}$$

$$= \frac{C_{(L+1)/(M+1)}^2 x}{C_{L/(M+1)} C_{(L+2)/(M+1)}} \tag{24}$$

$$\frac{\left(R_{L/M} - R_{(L+1)/(M+1)}\right)\left(R_{(L+1)/M} - R_{L/(M+1)}\right)}{\left(R_{L/M} - R_{(L+1)/M}\right)\left(R_{L/(M+1)} - R_{(L+1)/(M+1)}\right)}$$

$$= \frac{C_{(L+1)/(M+1)}^2 x}{C_{(L+1)/M} C_{(L+1)/(M+2)}} \tag{25}$$

$$\frac{\left(R_{L/M} - R_{(L+1)/(M+1)}\right)\left(R_{L/(M+1)} - R_{(L+1)/M}\right)}{\left(R_{L/M} - R_{(L+1)/(M+1)}\right)\left(R_{(L+1)/(M+1)} - R_{(L+1)/M}\right)}$$

$$= \frac{C_{(L+1)/M}C_{(L+1)/(M+1)}x}{C_{L/(M+1)}C_{(L+2)/M}} \qquad (26)$$

$$\frac{\left(R_{L/M}-R_{(L-1)/(M+1)}\right)\left(R_{(L+1)/M}-R_{L/(M+1)}\right)}{\left(R_{L/M}-R_{(L+1)/M}\right)\left(R_{(L-1)/(M+1)}-R_{L/(M+1)}\right)}$$

$$= \frac{C_{L/(M+1)}C_{(L+1)/(M+1)}x}{C_{(L+1)/M}C_{L/(M+2)}}. \qquad (27)$$

See also C-Determinant, Economized Rational Approximation, Frobenius Triangle Identities

References

Baker, G. A. Jr. "The Theory and Application of The Pade Approximant Method." In *Advances in Theoretical Physics, Vol. 1* (Ed. K. A. Brueckner). New York: Academic Press, pp. 1–58, 1965.

Baker, G. A. Jr. *Essentials of Padé Approximants in Theoretical Physics.* New York: Academic Press, pp. 27–38, 1975.

Baker, G. A. Jr. and Graves-Morris, P. *Padé Approximants.* New York: Cambridge University Press, 1996.

Brent, R. P.; Gustavson, F. G.; and Yun, D. Y. Y. "Fast Solution of Toeplitz Systems of Equations and Computation of Padé Approximants." *J. Algorithms* **1**, 259–295, 1980.

Press, W. H.; Flannery, B. P.; Teukolsky, S. A.; and Vetterling, W. T. "Padé Approximants." §5.12 in *Numerical Recipes in FORTRAN: The Art of Scientific Computing*, 2nd ed. Cambridge, England: Cambridge University Press, pp. 194–197, 1992.

Weisstein, E. W. "Books about Padé Approximants." http://www.treasure-troves.com/books/PadeApproximants.html.

Padé Conjecture

If $P(z)$ is a POWER SERIES which is regular for $|z| \leq 1$ except for m POLES within this CIRCLE and except for $z = +1$, at which points the function is assumed continuous when only points $|z| \leq 1$ are considered, then at least a subsequence of the $[N, N]$ PADÉ APPROXIMANTS are uniformly bounded in the domain formed by removing the interiors of small circles with centers at these POLES and uniformly continuous at $z = +1$ for $|z| \leq 1$.

See also Padé Approximant

References

Baker, G. A. Jr. "The Padé Conjecture and Some Consequences." §II.D in *Advances in Theoretical Physics, Vol. 1* (Ed. K. A. Brueckner). New York: Academic Press, pp. 23–27, 1965.

p-adic Absolute Value

P-ADIC NORM

p-adic Norm

Any NONZERO RATIONAL NUMBER x can be represented by

$$x = \frac{p^a r}{s}, \qquad (1)$$

where p is a PRIME NUMBER, r and s are INTEGERS not DIVISIBLE by p, and a is a unique INTEGER. The p-adic norm of x is then defined by

$$|x|_p = p^{-a}. \qquad (2)$$

Also define the p-adic value

$$|0|_p = 0. \qquad (3)$$

As an example, consider the FRACTION

$$\tfrac{140}{297} = 2^2 \cdot 3^{-3} \cdot 5 \cdot 7 \cdot 11^{-1}. \qquad (4)$$

It has p-adic absolute values given by

$$\left|\tfrac{140}{297}\right|_2 = \tfrac{1}{4} \qquad (5)$$

$$\left|\tfrac{140}{297}\right|_3 = 27 \qquad (6)$$

$$\left|\tfrac{140}{297}\right|_5 = \tfrac{1}{5} \qquad (7)$$

$$\left|\tfrac{140}{297}\right|_7 = \tfrac{1}{7} \qquad (8)$$

$$\left|\tfrac{140}{297}\right|_{11} = 11 \qquad (9)$$

The p-adic norm of a nonzero RATIONAL NUMBER x can be implemented in *Mathematica* as follows.

```
PadicNorm[x_Integer,    p_Integer?PrimeQ]  :=
p^(-IntegerExponent[x, p])
  PadicNorm[x_Rational,  p_Integer?PrimeQ]  :=
PadicNorm[Numerator[x],                    p]/
PadicNorm[Denominator[x], p]
```

The p-adic norm satisfies the relations

1. $|x|_p \geq 0$ for all x,
2. $|x|_p = 0$ IFF $x = 0$,
3. $|xy|_p = |x|_p |y|_p$ for all x and y,
4. $|x+y|_p \leq |x|_p + |y|_p$ for all x and y (the TRIANGLE INEQUALITY), and
5. $|x+y|_p \leq \max(|x|_p, |y|_p)$ for all x and y (the STRONG TRIANGLE INEQUALITY).

In the above, relation 4 follows trivially from relation 5, but relations 4 and 5 are relevant in the more general VALUATION THEORY.

The p-adic norm is the basis for the algebra of P-ADIC NUMBERS.

See also P-ADIC Number

p-adic Number

A p-adic number is an extension of the FIELD of RATIONAL NUMBERS such that CONGRUENCES MODULO POWERS of a fixed PRIME p are related to proximity in the so called "p-adic metric."

Any NONZERO RATIONAL NUMBER x can be represented by

$$x = \frac{p^a r}{s}, \tag{1}$$

where p is a PRIME NUMBER, r and s are INTEGERS not DIVISIBLE by p, and a is a unique INTEGER. Then define the P-ADIC NORM of x by

$$|x|_p = p^{-a}. \tag{2}$$

Also define the p-adic norm

$$|0|_p = 0. \tag{3}$$

The p-adics were probably first introduced by Hensel (1897) in a paper which was concerned with the development of algebraic numbers in POWER SERIES. p-adic numbers were then generalized to VALUATIONS by Kürschák (1913). Hasse (1923) subsequently formulated the LOCAL-GLOBAL PRINCIPLE (now usually called the HASSE PRINCIPLE), which is one of the chief applications of LOCAL FIELD theory. Skolem's p-adic method, which is used in attacking certain DIOPHANTINE EQUATIONS, is another powerful application of p-adic numbers. Another application is the theorem that the HARMONIC NUMBERS H_n are never INTEGERS (except for H_1). A similar application is the proof of the VON STAUDT-CLAUSEN THEOREM using the p-adic valuation, although the technical details are somewhat difficult. Yet another application is provided by the MAHLER-LECH THEOREM.

Every RATIONAL x has an "essentially" unique p-adic expansion ("essentially" since zero terms can always be added at the beginning)

$$x = \sum_{j=m}^{\infty} a_j p^j, \tag{4}$$

with m an INTEGER, a_j the INTEGERS between 0 and $p-1$ inclusive, and where the sum is convergent with respect to p-adic valuation. If $x \neq 0$ and $a_m \neq 0$, then the expansion is unique. Burger and Struppeck (1996) show that for p a PRIME and n a POSITIVE INTEGER,

$$|n!|_p = p^{-(n-A_p(n))/(p-1)}, \tag{5}$$

where the p-adic expansion of n is

$$n = a_0 + a_1 p + a_2 p^2 + \ldots + a_L p^L, \tag{6}$$

and

$$A_p(n) = a_0 + a_1 + a_2 + \ldots + a_L. \tag{7}$$

For sufficiently large n,

$$|n!|_p \leq p^{-n/(2p-2)}. \tag{8}$$

The p-adic valuation on $\mathbb{Q}$ gives rise to the p-adic metric

$$d(x,y) = |x-y|_p, \tag{9}$$

which in turn gives rise to the p-adic topology. It can be shown that the rationals, together with the p-adic metric, do not form a COMPLETE METRIC SPACE. The completion of this space can therefore be constructed, and the set of p-adic numbers $\mathbb{Q}_p$ is defined to be this completed space.

Just as the REAL NUMBERS are the completion of the RATIONALS $\mathbb{Q}$ with respect to the usual absolute valuation $|x-y|$, the p-adic numbers are the completion of $\mathbb{Q}$ with respect to the p-adic valuation $|x-y|_p$. The p-adic numbers are useful in solving DIOPHANTINE EQUATIONS. For example, the equation $X^2 = 2$ can easily be shown to have no solutions in the field of 2-adic numbers (we simply take the valuation of both sides). Because the 2-adic numbers contain the rationals as a subset, we can immediately see that the equation has no solutions in the RATIONALS. So we have an immediate proof of the irrationality of $\sqrt{2}$.

This is a common argument that is used in solving these types of equations: in order to show that an equation has no solutions in $\mathbb{Q}$, we show that it has no solutions in an EXTENSION FIELD. For another example, consider $X^2 + 1 = 0$. This equation has no solutions in $\mathbb{Q}$ because it has no solutions in the reals $\mathbb{R}$, and $\mathbb{Q}$ is a subset of $\mathbb{R}$.

Now consider the converse. Suppose we have an equation that does have solutions in $\mathbb{R}$ and in all the $\mathbb{Q}_p$ for every PRIME p. Can we conclude that the equation has a solution in $\mathbb{Q}$? Unfortunately, in general, the answer is no, but there are classes of equations for which the answer is yes. Such equations are said to satisfy the HASSE PRINCIPLE.

See also AX-KOCHEN ISOMORPHISM THEOREM, DIOPHANTINE EQUATION, HARMONIC NUMBER, HASSE PRINCIPLE, LOCAL FIELD, LOCAL-GLOBAL PRINCIPLE, MAHLER-LECH THEOREM, P-ADIC NORM, PRODUCT FORMULA, VALUATION, VALUATION THEORY, VON STAUDT-CLAUSEN THEOREM

References

Burger, E. B. and Struppeck, T. "Does $\sum_{n=0}^{\infty} \frac{1}{n!}$ Really Converge? Infinite Series and p-adic Analysis." *Amer. Math. Monthly* **103**, 565–577, 1996.

Cassels, J. W. S. and Scott, J. W. *Local Fields.* Cambridge, England: Cambridge University Press, 1986.

Gouvêa, F. Q. *P-adic Numbers: An Introduction,* 2nd ed. New York: Springer-Verlag, 1997.

Hasse, H. "Über die Darstellbarkeit von Zahlen durch quadratische Formen im Körper der rationalen Zahlen." *J. reine angew. Math.* **152**, 129–148, 1923.

Hasses, H. "Die Normenresttheorie relativ-Abelscher Zahlkörper als Klassenkörpertheorie in Kleinen." *J. reine angew. Math.* **162**, 145–154, 1930.

Hensel, K. "Über eine neue Begründung der Theorie der algebraischen Zahlen." *Jahresber. Deutsch. Math. Verein* **6**, 83–88, 1897.

Kakol, J.; De Grande-De Kimpe, N.; and Perez-Garcia, C. (Eds.). *p-adic Functional Analysis.* New York: Dekker, 1999.

Koblitz, N. *P*-adic Numbers, *P*-adic Analysis, and Zeta-Functions, 2nd ed. New York: Springer-Verlag, 1984.

Koch, H. "Valuations." Ch. 4 in *Number Theory: Algebraic Numbers and Functions.* Providence, RI: Amer. Math. Soc., pp. 103–139, 2000.

Mahler, K. *P*-adic Numbers and Their Functions, 2nd ed. Cambridge, England: Cambridge University Press, 1981.

Ostrowski, A. "Über sogennante perfekte Körper." *J. reine angew. Math.* **147**, 191–204, 1917.

Vladimirov, V. S. Tables of Integrals of Complex-Valued Functions of *p.*-adic Arguments 22 Nov 1999. http://xxx.lanl.gov/abs/math-ph/9911027/.

Weisstein, E. W. "Books about P-adic Numbers." http://www.treasure-troves.com/books/P-adicNumbers.html.

Padovan Sequence

The INTEGER SEQUENCE defined by the RECURRENCE RELATION

$$P(n) = P(n-2) + P(n-3)$$

with the initial conditions $P(0) = P(1) = P(2) = 1$. The RECURRENCE RELATION can be solved explicitly, giving

$$P(n) = \frac{1+r_1}{r_1^{n+2}(2+3r_1)} + \frac{1+r_2}{r_2^{n+2}(2+3r_2)} + \frac{1+r_3}{r_3^{n+2}(2+3r_3)},$$

where r_n is the nth root of

$$x^3 + x^2 - 1 = 0.$$

The first few terms are 1, 1, 2, 2, 3, 4, 5, 7, 9, 12, ... (Sloane's A000931).

The ratio $\lim_{n \to \infty} P(n)/P(n-1)$ is called the PLASTIC CONSTANT.

See also PERRIN SEQUENCE, PLASTIC CONSTANT

References

Sloane, N. J. A. Sequences A000931/M0284 in "An On-Line Version of the Encyclopedia of Integer Sequences." http://www.research.att.com/~njas/sequences/eisonline.html.

Stewart, I. "Tales of a Neglected Number." *Sci. Amer.* **274**, 102–103, June 1996.

Painlevé Property

Following the work of Fuchs in classifying first-order ORDINARY DIFFERENTIAL EQUATIONS, Painlevé studied second-order ODEs OF THE FORM

$$\frac{d^2 y}{dx^2} = F(y', y, x),$$

where F is ANALYTIC in x and rational in y and y'. Painlevé found 50 types whose only movable SINGULARITIES are ordinary POLES. This characteristic is known as the Painlevé property. Six of the transcendents define new transcendents known as PAINLEVÉ TRANSCENDENTS, and the remaining 44 can be integrated in terms of classical transcendents, quadratures, or the PAINLEVÉ TRANSCENDENTS.

See also PAINLEVÉ TRANSCENDENTS

Painlevé Transcendents

There are six Painlevé transcendents, corresponding to second-order ordinary differential equations whose only movable singularities are ordinary poles and which cannot be integrated in terms of other known functions or transcendents.

$$y'' = 6y^2 + x \tag{1}$$

$$y = 2y^3 + xy + \alpha \tag{2}$$

$$y'' = \frac{y'^2}{y} - \frac{y'}{x} + \frac{\alpha y^2 + \beta}{x} + \gamma y^3 + \frac{\delta}{y} \tag{3}$$

$$y'' = \frac{y'^2}{2y} + \frac{3}{2} y^3 + 4xy^2 + 2(x^2 - \alpha)y + \frac{\beta}{y} \tag{4}$$

$$y = \left(\frac{1}{2y} + \frac{1}{y-1} \right) y'^2 - \frac{y'}{x} + \frac{(y-1)^2}{x^2} \left(\alpha y + \frac{\beta}{y} \right)$$
$$+ \frac{\gamma y}{x} + \frac{\delta y(y+1)}{y-1} \tag{5}$$

$$y = \frac{1}{2} \left(\frac{1}{y} + \frac{1}{y-1} + \frac{1}{y-x} \right) y'^2 + \left(\frac{1}{x} + \frac{1}{x-1} + \frac{1}{y-x} \right) y'$$
$$+ \frac{y(y-1)(y-x)}{x^2(x-1)^2} \left[\alpha + \frac{\beta x}{y^2} + \frac{\gamma(x-1)}{(y-1)^2} + \frac{\delta x(x-1)}{(y-x)^2} \right] \tag{6}$$

(Painlevé 1906; Ince 1956, p. 345; Zwillinger 1997, pp. 125–126). All Painlevé transcendents have first integrals for special values of their parameters except (2). Five of the transcendents were found by Painlevé and his students; the sixth transcendent was found by Gambier and contains the other five as limiting cases (Garnier 1916ab; Ince 1956, p. 345).

See also PAINLEVÉ PROPERTY, TRANSCENDENTAL FUNCTION

References

Garnier, R. "Étude de l'intégrale générale de l'équation (VI) de M. Painlevé dans le voisinage de ses singularités transcendantes." *C. R. Acad. Sci. Paris* **162**, 939–942, 1916a.

Garnier, R. "Étude de l'intégrale générale de l'équation (VI) de M. Painlevé dans le voisinage de ses singularités transcendantes." *C. R. Acad. Sci. Paris* **163**, 8–10, 1916b.

Garnier, R. "Étude de l'intégrale générale de l'équation (VI) de M. Painlevé dans le voisinage de ses singularités transcendantes." *C. R. Acad. Sci. Paris* **163**, 118, 1916c.

Ince, E. L. "The Painlevé Transcendents" and "The First Painlevé Transcendent: Freedom from Movable Branch Points." §14.4 and 14.41 in *Ordinary Differential Equations.* New York: Dover, pp. 345–347, 1956.

Painlevé, P. "Sur l'irréducibilité des transcendantes uniformes définie par les équations différentielles du second ordre." *C. R. Acad. Sci. Paris* **135**, 411–415, 1902.

Painlevé, P. "Démonstration de l'irréducibilité absolue de l'équation $y = 6y^2 + x$." *C. R. Acad. Sci. Paris* 641–647, 1902.

Painlevé, P. "Sur les transcendantes uniformed définies par l'équation $y = 6y^2 + x$." *C. R. Acad. Sci. Paris* **135**, 757–761, 1902.

Painlevé, P. "Sur l'irréducibilité de l'équation: $y = 6y^2 + x$." *C. R. Acad. Sci. Paris* **135**, 1020–1025, 1902.

Painlevé, P. "Sur les équations différentielles du second ordre à points critiques fixes." *C. R. Acad. Sci. Paris* **143**, 1111–1117, 1906.

Zwillinger, D. (Ed.). *CRC Standard Mathematical Tables and Formulae.* Boca Raton, FL: CRC Press, p. 414, 1995.

Zwillinger, D. *Handbook of Differential Equations, 3rd ed.* Boston, MA: Academic Press, pp. 125–126, 1997.

Pair

A SET of two numbers or objects linked in some way is said to be a pair. The pair a and b is usually denoted (a, b), and is generally considered to be ordered. In certain circumstances, pairs are also called BROTHERS or TWINS.

See also AMICABLE PAIR, AUGMENTED AMICABLE PAIR, BROWN NUMBERS, FRIENDLY PAIR, HEXAD, HOMOGENEOUS NUMBERS, IMPULSE PAIR, IRREGULAR PAIR, LAX PAIR, LONG EXACT SEQUENCE OF A PAIR AXIOM, MONAD, ORDERED PAIR, PERKO PAIR, QUADRUPLET, QUASIAMICABLE PAIR, QUINTUPLET, REDUCED AMICABLE PAIR, SMITH BROTHERS, TRIAD, TRIPLET, TWIN PEAKS, TWIN PRIMES, TWINS, UNITARY AMICABLE PAIR, WILF-ZEILBERGER PAIR, ZIP-PAIR

Pair Sum

Given an AMICABLE PAIR (m, n), the quantity

$$\sigma(m) = \sigma(n) = s(m) + s(n) = m + n$$

is called the pair sum, where $\sigma(n)$ is the DIVISOR FUNCTION and $s(n)$ is the RESTRICTED DIVISOR FUNCTION.

See also AMICABLE PAIR

Paired t-Test

Given two paired sets X_i and Y_i of n measured values, the paired t-test determines if they differ from each other in a significant way. Let

$$\hat{X}_i = (X_i - \bar{X}_i)$$

$$\hat{Y}_i = (Y_i - \bar{Y}_i),$$

then define t by

$$t = (\bar{X} - \bar{Y}) \sqrt{\frac{n(n-1)}{\sum_{i=1}^{n} (\hat{X}_i - \hat{Y}_i)^2}}.$$

This statistic has $n - 1$ DEGREES OF FREEDOM.

A table of STUDENT'S T-DISTRIBUTION confidence intervals can be used to determine the significance level at which two distributions differ.

See also FISHER SIGN TEST, HYPOTHESIS TESTING, STUDENT'S T-DISTRIBUTION, WILCOXON SIGNED RANK TEST

References

Goulden, C. H. *Methods of Statistical Analysis, 2nd ed.* New York: Wiley, pp. 50–55, 1956.

Paley Class

The Paley class of a POSITIVE INTEGER $m \equiv 0 \pmod 4$ is defined as the set of all possible QUADRUPLES (k, e, q, n) where

$$m = 2^e(q^n + 1),$$

q is an ODD PRIME, and

$$k = \begin{cases} 0 & \text{if } q = 0 \\ 1 & \text{if } q^n - 3 \equiv 0 \pmod 4 \\ 2 & \text{if } q^n - 1 \equiv 0 \pmod 4 \\ \text{undefined} & \text{otherwise.} \end{cases}$$

See also HADAMARD MATRIX, PALEY CONSTRUCTION

Paley Construction

HADAMARD MATRICES H_n can be constructed using FINITE FIELD GF(p^m) when $p = 4l - 1$ and m is ODD. Pick a representation r RELATIVELY PRIME to p. Then by coloring white $\lfloor (p-1)/2 \rfloor$ (where $\lfloor x \rfloor$ is the FLOOR FUNCTION) distinct equally spaced RESIDUES mod p (r^0, r, r^2, ...; r^0, r^2, r^4, ...; etc.) *in addition to* 0, a HADAMARD MATRIX is obtained if the POWERS of r (mod p) run through $< \lfloor (p-1)/2 \rfloor$. For example,

$$n = 12 = 11^1 + 1 = 2(5 + 1) = 2^2(2 + 1)$$

is of this form with $p = 11 = 4 \times 3 - 1$ and $m = 1$. Since $m = 1$, we are dealing with GF(11), so pick $p = 2$ and compute its RESIDUES (mod 11), which are

$$p^0 \equiv 1$$
$$p^1 \equiv 2$$
$$p^2 \equiv 4$$
$$p^3 \equiv 8$$
$$p^4 \equiv 16 \equiv 5$$
$$p^5 \equiv 10$$
$$p^6 \equiv 20 \equiv 9$$
$$p^7 \equiv 18 \equiv 7$$
$$p^8 \equiv 14 \equiv 3$$
$$p^9 \equiv 6$$
$$p^{10} \equiv 12 \equiv 1.$$

Picking the first $\lfloor 11/2 \rfloor = 5$ RESIDUES and adding 0 gives: 0, 1, 2, 4, 5, 8, which should then be colored in the MATRIX obtained by writing out the RESIDUES increasing to the left and up along the border (0 through $p-1$, followed by ∞), then adding horizontal and vertical coordinates to get the residue to place in each square.

$$\begin{bmatrix} \infty & \infty & \infty & \infty & \infty & \infty & \infty & \infty & \infty & \infty & \infty & \infty \\ 10 & 0 & 1 & 2 & 3 & 4 & 5 & 6 & 7 & 8 & 9 & \infty \\ 9 & 10 & 0 & 1 & 2 & 3 & 4 & 5 & 6 & 7 & 8 & \infty \\ 8 & 9 & 10 & 0 & 1 & 2 & 3 & 4 & 5 & 6 & 7 & \infty \\ 7 & 8 & 9 & 10 & 0 & 1 & 2 & 3 & 4 & 5 & 6 & \infty \\ 6 & 7 & 8 & 9 & 10 & 0 & 1 & 2 & 3 & 4 & 5 & \infty \\ 5 & 6 & 7 & 8 & 9 & 10 & 0 & 1 & 2 & 3 & 4 & \infty \\ 4 & 5 & 6 & 7 & 8 & 9 & 10 & 0 & 1 & 2 & 3 & \infty \\ 3 & 4 & 5 & 6 & 7 & 8 & 9 & 10 & 0 & 1 & 2 & \infty \\ 2 & 3 & 4 & 5 & 6 & 7 & 8 & 9 & 10 & 0 & 1 & \infty \\ 1 & 2 & 3 & 4 & 5 & 6 & 7 & 8 & 9 & 10 & 0 & \infty \\ 0 & 1 & 2 & 3 & 4 & 5 & 6 & 7 & 8 & 9 & 10 & \infty \end{bmatrix}$$

H_{16} can be trivially constructed from $H_4 \otimes H_4$. H_{20} cannot be built up from smaller MATRICES, so use $n = 20 = 19 + 1 = 2(3^2 + 1) = 2^2(2^2 + 1)$. Only the first form can be used, with $p = 19 = 4 \times 5 - 1$ and $m = 1$. We therefore use GF(19), and color 9 RESIDUES plus 0 white. H_{24} can be constructed from $H_2 \otimes H_{12}$.

Now consider a more complicated case. For $n = 28 = 3^3 + 1 = 2(13 + 1)$, the only form having $p = 4l - 1$ is the first, so use the GF(3^3) field. Take as the modulus the IRREDUCIBLE POLYNOMIAL $x^3 + 2x + 1$, written 1021. A four-digit number can always be written using only three digits, since $1000 - 1021 \equiv 0012$ and $2000 - 2012 \equiv 0021$. Now look at the moduli starting with 10, where each digit is considered separately. Then

$$\begin{aligned} x^0 &\equiv 1 & x^1 &\equiv 10 & x^2 &\equiv 100 \\ x^3 &\equiv 1000 \equiv 12 & x^4 &\equiv 120 & x^5 &\equiv 1200 \equiv 212 \\ x^6 &\equiv 2120 \equiv 111 & x^7 &\equiv 1100 \equiv 122 & x^8 &\equiv 1220 \equiv 202 \\ x^9 &\equiv 2020 \equiv 11 & x^{10} &\equiv 110 & x^{11} &\equiv 1100 \equiv 112 \\ x^{12} &\equiv 1120 \equiv 102 & x^{13} &\equiv 1020 \equiv 2 & x^{14} &\equiv 20 \\ x^{15} &\equiv 200 & x^{16} &\equiv 2000 \equiv 21 & x^{17} &\equiv 210 \\ x^{18} &\equiv 2100 \equiv 121 & x^{19} &\equiv 1210 \equiv 222 & x^{20} &\equiv 2220 \equiv 211 \\ x^{21} &\equiv 2110 \equiv 101 & x^{22} &\equiv 101 \equiv 22 & x^{23} &\equiv 220 \\ x^{24} &\equiv 2200 \equiv 221 & x^{25} &\equiv 2210 \equiv 201 & x^{26} &\equiv 2010 \equiv 1 \end{aligned}$$

Taking the alternate terms gives white squares as 000, 001, 020, 021, 022, 100, 102, 110, 111, 120, 121, 202, 211, and 221.

References

Ball, W. W. R. and Coxeter, H. S. M. *Mathematical Recreations and Essays, 13th ed.* New York: Dover, pp. 107–109 and 274, 1987.

Beth, T.; Jungnickel, D.; and Lenz, H. *Design Theory, 2nd ed. rev.* Cambridge, England: Cambridge University Press, 1998.

Geramita, A. V. *Orthogonal Designs: Quadratic Forms and Hadamard Matrices.* New York: Dekker, 1979.

Kitis, L. "Paley's Construction of Hadamard Matrices." http://www.mathsource.com/cgi-bin/msitem?0205–760.

Paley's Theorem

Proved in 1933. If q is an ODD PRIME or $q = 0$ and n is any POSITIVE INTEGER, then there is a HADAMARD

MATRIX of order

$$m = 2^e(q^n + 1),$$

where e is any POSITIVE INTEGER such that $m \equiv 0 \pmod 4$. If m is of this form, the matrix can be constructed with a PALEY CONSTRUCTION. If m is divisible by 4 but not OF THE FORM (1), the PALEY CLASS is undefined. However, HADAMARD MATRICES have been shown to exist for all $m \equiv 0 \pmod 4$ for $m < 428$.

See also HADAMARD MATRIX, PALEY CLASS, PALEY CONSTRUCTION

Palindrome Number

PALINDROMIC NUMBER

Palindromic Number

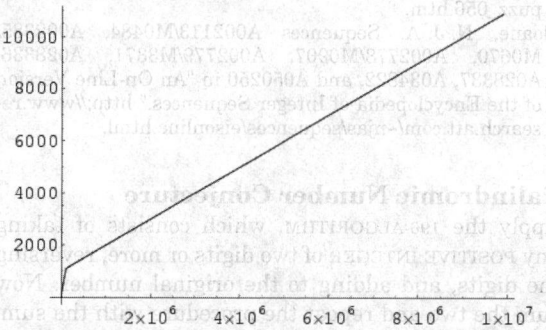

A symmetrical number which is written in some base b as $a_1 a_2 \cdots a_2 a_1$. The first few are 0, 1, 2, 3, 4, 5, 6, 7, 8, 9, 11, 22, 33, 44, 55, 66, 77, 88, 99, 101, 111, 121, ... (Sloane's A002113). The number of palindromic numbers less than a given number are illustrated in the plot above. The number of palindromic numbers less than 10, 10^2, 10^3, ... are 9, 18, 108, 198, 1098, 1998, 10998, ... (Sloane's A050250).

The sum of the reciprocals of the palindromic numbers converges to a constant ≈ 3.36977 (Rivera), where this value has been computed using all palindromic numbers $\leq 10^7$.

The first few n for which the PRONIC NUMBER P_n is palindromic are 1, 2, 16, 77, 538, 1621, ... (Sloane's A028336), and the first few palindromic numbers which are PRONIC are 2, 6, 272, 6006, 289982, ... (Sloane's A028337). The first few numbers whose squares are palindromic are 1, 2, 3, 11, 22, 26, ... (Sloane's A002778), and the first few palindromic squares are 1, 4, 9, 121, 484, 676, ... (Sloane's A002779).

There are no palindromic square n-digit numbers for $n = 2, 4, 8, 10, 14, 18, 20, 24, 30, ...$ (Sloane's A034822).

See also DEMLO NUMBER, PALINDROMIC NUMBER CONJECTURE, PALINDROMIC PRIME, REVERSAL

References

Beiler, A. H. *Recreations in the Theory of Numbers: The Queen of Mathematical Entertains.* New York: Dover, 1964.

De Geest, P. "Palindromic Numbers and Other Recreational Topics." http://www.ping.be/~ping6758/index.shtml.

De Geest, P. "Palindromic Products of Two Consecutive Integers." http://www.ping.be/~ping6758/consec.htm.

De Geest, P. "Palindromic Squares." http://www.ping.be/~ping6758/square.htm.

Dr. Pete. "The Math Forum. Ask Dr. Math: Questions & Answers from Our Archives. Palindromic Numbers." http://forum.swarthmore.edu/dr.math/problems/akyildiz1.4.98.html.

Dr. Rob. "The Math Forum. Ask Dr. Math: Questions & Answers from Our Archives. Palindromic Numbers." http://forum.swarthmore.edu/dr.math/problems/stang4.8.14.97.html.

Keith, M. "On General Palindromic Numbers." http://www.seanet.com/~ksbrown/kmath359.htm

Pappas, T. "Numerical Palindromes." *The Joy of Mathematics.* San Carlos, CA: Wide World Publ./Tetra, p. 146, 1989.

Rivera, C. "Problems & Puzzles: Puzzle The Honaker's Constant.-056." http://www.primepuzzles.net/puzzles/puzz_056.htm.

Sloane, N. J. A. Sequences A002113/M0484, A002385/M0670, A002778/M0907, A002779/M3371, A028336, A028337, A034822, and A050250 in "An On-Line Version of the Encyclopedia of Integer Sequences." http://www.research.att.com/~njas/sequences/eisonline.html.

Palindromic Number Conjecture

Apply the 196-ALGORITHM, which consists of taking any POSITIVE INTEGER of two digits or more, reversing the digits, and adding to the original number. Now sum the two and repeat the procedure with the sum. Of the first 10,000 numbers, only 251 do not produce a PALINDROMIC NUMBER in ≤ 23 steps (Gardner 1979).

It was therefore conjectured that *all* numbers will eventually yield a PALINDROMIC NUMBER. However, the conjecture has been proven false for bases which are a POWER of 2, and seems to be false for base 10 as well. Among the first 100,000 numbers, 5,996 numbers apparently never generate a PALINDROMIC NUMBER (Gruenberger 1984). The first few are 196, 887, 1675, 7436, 13783, 52514, 94039, 187088, 1067869, 10755470, ... (Sloane's A006960).

It is conjectured, but not proven, that there are an infinite number of palindromic PRIMES. With the exception of 11, palindromic PRIMES must have an ODD number of digits.

See also 196-ALGORITHM, DEMLO NUMBER

References

Gardner, M. *Mathematical Circus: More Puzzles, Games, Paradoxes and Other Mathematical Entertainments from Scientific American.* New York: Knopf, pp. 242–245, 1979.

Gruenberger, F. "How to Handle Numbers with Thousands of Digits, and Why One Might Want to." *Sci. Amer.* **250**, 19–26, Apr. 1984.

Sloane, N. J. A. Sequences A006960/M5410 in "An On-Line Version of the Encyclopedia of Integer Sequences." http://www.research.att.com/~njas/sequences/eisonline.html.

Palindromic Prime

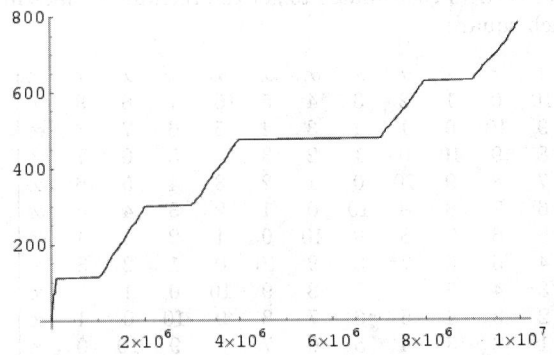

The first few palindromic PRIMES are 2, 3, 5, 7, 11, 101, 131, 151, 181, 191, 313, 353, 373, 383, 727, 757, 787, ... (Sloane's A002385; Beiler 1964, p. 228). The number of palindromic primes less than a given number are illustrated in the plot above. The number of palindromic numbers having $n = 1, 2, 3, ...$ digits are 4, 1, 15, 0, 93, 0, 668, 0, 5172, ... (Sloane's A016115; De Geest) and the total number of palindromic primes less than 10, 10^2, 10^3, ... are 4, 5, 20, 20, 113, 113, 781, ... (Sloane's A050251).

The sum of the reciprocals of the palindromic primes converges to ≈ 1.32398, where this value has been computed using all palindromic primes $\leq 10^{11}$ (M. Keith).

Palindromic primes formed from the reflected decimal expansion of PI include 3, 313,

$$31415926535897932384626433833462648323979853562951413,$$

... (Sloane's A039954).

The first few n such that both n and p_n are palindromic (where p_n is the nth prime) are given by 1, 2, 3, 4, 5, 8114118, ... (Sloane's A046942; Rivera), corresponding to p_n of 2, 3, 5, 7, 11, 143787341 (Sloane's A046941; Rivera).

Palindromic primes OF THE FORM

$$pp_n(x) = x^n + (x+1)^n$$

for $n = 2$ include 5, 181, 313, 3187813, ... (Sloane's A050239; De Geest, Rivera), which occur for $x = 1, 9, 12, 1262, ...$ (Sloane's A050236; De Geest, Rivera), with no others for $n < 10^{20}$ and $x < 2 \times 10^{10}$ (De Geest). Dubner (1999) found

$$P = 10^3 5352 + 2049402 * 10^1 7673 + 1,$$

which, at 35,353 digits is believed to be the largest known prime that is not OF THE FORM, $ab^n \pm 1$.

See also PALINDROMIC NUMBER

References

Beiler, A. H. *Recreations in the Theory of Numbers: The Queen of Mathematical Entertains.* New York: Dover, 1964.

De Geest, P. "Palindromic Numbers and Other Recreational Topics." http://www.ping.be/~ping6758/index.shtml.

De Geest, P. "Palindromic Prime Statistics--The Table." http://www.ping.be/~ping6758/palprim1.htm.

De Geest, P. "Palindromic Prime Page 3." http://www.ping.be/~ping6758/palprim3.htm.

De Geest, P. "Palindromic Sums of Squares of Consecutive Integers." http://www.ping.be/~ping6758/sumsquare.htm.

Dubner, H. "Palindromic prime record: 35353 digits." nmbrthry@listserv.nodak.edu posting, 14 Nov 1999.

Rivera, C. "Problems & Puzzles: Puzzle Pal-Primes and Sum of Powers.-014." http://www.primepuzzles.net/puzzles/puzz_014.htm.

Rivera, C. "Problems & Puzzles: Puzzle Pi Such that Pi is Palprime & i = Palindrome.-051." http://www.primepuzzles.net/puzzles/puzz_051.htm.

Rivera, C. "Problems & Puzzles: Puzzle The Honaker's Constant.-056." http://www.primepuzzles.net/puzzles/puzz_056.htm.

Sloane, N. J. A. Sequences A002385/M0670, A016115, A039954, A046941, A046942, A050251, A050236, and A050239 in "An On-Line Version of the Encyclopedia of Integer Sequences." http://www.research.att.com/~njas/sequences/eisonline.html.

Palprime

PALINDROMIC PRIME

Pancake Cutting

CIRCLE DIVISION BY LINES

Pancake Sorting Problem

Assume that n numbered pancakes are stacked, and that a spatula can be used to reverse the order of the top k pancakes for $2 \leq k \leq n$. Then the pancake sorting problem asks how many such "prefix reversals" are sufficient to sort an arbitrary stack (Skiena 1990, p. 48).

See also PANCAKE THEOREM

References

Gates, W. and Papadimitriou, C. "Bounds for Sorting by Prefix Reversal." *Discr. Math.* **27**, 47–57, 1979.

Skiena, S. *Implementing Discrete Mathematics: Combinatorics and Graph Theory with Mathematica.* Reading, MA: Addison-Wesley, 1990.

Pancake Theorem

The 2-D version of the HAM SANDWICH THEOREM.

See also HAM SANDWICH THEOREM, PANCAKE SORTING PROBLEM

Pancyclic Graph

A simple unlabeled GRAPH on n vertices is called pancyclic if it contains cycles of all lengths, 3, 4, ..., n.

Pandiagonal Square

PANMAGIC SQUARE

Pandigital Fraction

A FRACTION containing the digits 1 through 9 is called a pandigital fraction. The following table gives the number of pandigital fractions which represent simple unit fractions. The numbers of pandigital fractions for 1/1, 1/2, 1/3, ... are 0, 12, 2, 4, 12, 3, 7, 46, 3, ... (Sloane's A054383).

f	#	fractions					
$\frac{1}{2}$	12	$\frac{6729}{13458}$,	$\frac{6792}{13584}$,	$\frac{6927}{13854}$,	$\frac{7269}{14538}$,	$\frac{7293}{14586}$,	$\frac{7329}{14658}$,
		$\frac{7692}{15384}$,	$\frac{7923}{15846}$,	$\frac{7932}{15864}$,	$\frac{9267}{18534}$,	$\frac{9273}{18546}$,	$\frac{9327}{18654}$
$\frac{1}{3}$	2	$\frac{5823}{17469}$,	$\frac{5832}{17496}$				
$\frac{1}{4}$	4	$\frac{3942}{15768}$,	$\frac{4392}{17568}$,	$\frac{5796}{23184}$,	$\frac{7956}{31824}$		
$\frac{1}{5}$	12	$\frac{2697}{13485}$,	$\frac{2769}{13845}$,	$\frac{2937}{14685}$,	$\frac{2967}{14835}$,	$\frac{2973}{14865}$,	$\frac{3297}{16485}$,
		$\frac{3729}{18645}$,	$\frac{6297}{31485}$,	$\frac{7629}{38145}$,	$\frac{9237}{46185}$,	$\frac{9627}{48135}$,	$\frac{9723}{48615}$
$\frac{1}{6}$	3	$\frac{2943}{17658}$,	$\frac{4653}{27918}$,	$\frac{5697}{34182}$			
$\frac{1}{7}$	7	$\frac{2394}{16758}$,	$\frac{2637}{18459}$,	$\frac{4527}{31689}$,	$\frac{5274}{36918}$,	$\frac{5418}{37926}$,	$\frac{5976}{41832}$,
		$\frac{7614}{53298}$					
$\frac{1}{8}$	46	$\frac{3187}{25496}$,	$\frac{4589}{36712}$,	$\frac{4591}{36728}$,	$\frac{4689}{37512}$,	$\frac{4691}{37528}$,	$\frac{4769}{38152}$,
		$\frac{5237}{41896}$,	$\frac{5371}{42968}$,	$\frac{5789}{46312}$,	$\frac{5791}{46328}$,	$\frac{5839}{46712}$,	$\frac{5892}{47136}$,
		$\frac{5916}{47328}$,	$\frac{5921}{47368}$,	$\frac{6479}{51832}$,	$\frac{6741}{53928}$,	$\frac{6789}{54312}$,	$\frac{6791}{54328}$,
		$\frac{6839}{54712}$,	$\frac{7123}{56984}$,	$\frac{7312}{58496}$,	$\frac{7364}{58912}$,	$\frac{7416}{59328}$,	$\frac{7421}{59368}$,
		$\frac{7894}{63152}$,	$\frac{7941}{63528}$,	$\frac{8174}{65392}$,	$\frac{8179}{65432}$,	$\frac{8394}{67152}$,	$\frac{8419}{67352}$,
		$\frac{8439}{67512}$,	$\frac{8932}{71456}$,	$\frac{8942}{71536}$,	$\frac{8953}{71624}$,	$\frac{8954}{71632}$,	$\frac{9156}{73248}$,
		$\frac{9158}{73264}$,	$\frac{9182}{73456}$,	$\frac{9316}{74528}$,	$\frac{9321}{74568}$,	$\frac{9352}{74816}$,	$\frac{9416}{75328}$,
		$\frac{9421}{75368}$,	$\frac{9523}{76184}$,	$\frac{9531}{76248}$,	$\frac{9541}{76328}$		
$\frac{1}{9}$	3	$\frac{6381}{57429}$,	$\frac{6471}{58239}$,	$\frac{8361}{75249}$			
$\frac{1}{10}$	0						
$\frac{1}{11}$	0						
$\frac{1}{12}$	4	$\frac{3816}{45792}$,	$\frac{6129}{73548}$,	$\frac{7461}{89532}$,	$\frac{7632}{91584}$		

See also PANDIGITAL NUMBER

References

Friedman, M. J. *Scripta Math.* **8**.

Sloane, N. J. A. Sequences A054383 in "An On-Line Version of the Encyclopedia of Integer Sequences." http://www.research.att.com/~njas/sequences/eisonline.html.

Wells, D. *The Penguin Dictionary of Curious and Interesting Numbers.* Middlesex, England: Penguin Books, p. 27, 1986.

Pandigital Number

A decimal INTEGER which contains each of the digits from 0 to 9 (and whose leading digit must be nonzero).

The first few pandigital numbers are 1023456789, 1023456798, 1023456879, 1023456897, 1023456978, ... (Sloane's A050278). A 10-digit pandigital number is always divisible by 9 since

$$\sum_{i=0}^{9} i = 45.$$

This passes the DIVISIBILITY TEST for 9 since $4 + 5 = 9$. The smallest pandigital primes must therefore have 11 digits (no two of which can be 0). The first few pandigital primes are therefore 10123457689, 10123465789, 10123465897, 10123485679, ... (Sloane's A050288).

If zeros are excluded, the first few "zeroless" pandigital numbers are 123456789, 123456798, 123456879, 123456897, 123456978, 123456987, ... (Sloane's A050289), and the first few zeroless pandigital primes are 1123465789, 1123465879, 1123468597, 1123469587, 1123478659, ... (Sloane's A050290).

The sum of the first 32423 (a PALINDROMIC NUMBER) consecutive PRIMES is 5897230146, which is pandigital (Honaker). No other PALINDROMIC NUMBER shares this property.

Numbers n that give zeroless pandigital numbers when the Fibonacci recurrence

$$a(n) = a(n-1) + a(n-2)$$

with $a(1) = 1$ and $a(2) = n$ is applied are 718, 1790, 1993, 2061, 2259, 3888, 3960, 4004, 4396, 5093, 5832, 7031, 7310, 7712, 8039, 8955, 9236, ...

See also PANDIGITAL FRACTION, PERSISTENT NUMBER

References

De Geest, P. "The Nine Digits Page." http://www.ping.be/~ping6758/ninedigits.htm.

Sloane, N. J. A. Sequences A050278, A050288, A050289, and A050290 in "An On-Line Version of the Encyclopedia of Integer Sequences." http://www.research.att.com/~njas/sequences/eisonline.html.

Panmagic Square

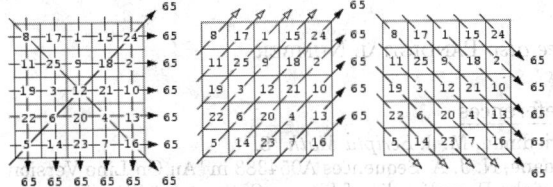

If *all* the DIAGONALS—including those obtained by "wrapping around" the edges—of a MAGIC SQUARE sum to the same MAGIC CONSTANT, the square is said to be a panmagic square (Kraitchik 1942, pp. 143 and 189–191). (Only the rows, columns, and *main diagonals* must sum to the same constant for the usual type of magic square.) The terms DIABOLIC SQUARE (Hunter and Madachy 1975, p. 24; Madachy 1979, p. 87),

PANDIAGONAL SQUARE (Hunter and Madachy 1975, p. 24), and NASIK SQUARE (Madachy 1979, p. 87) are sometimes also used.

No panmagic squares exist of order 3 or any order $4k + 2$ for k an INTEGER. The Siamese method for generating MAGIC SQUARES produces panmagic squares for orders $6k \pm 1$ with ordinary vector $(2, 1)$ and break vector $(1, -1)$.

1	15	24	8	17
23	7	16	5	14
20	4	13	22	6
12	21	10	19	3
9	18	2	11	25

The LO SHU is not panmagic, but it is an ASSOCIATIVE MAGIC SQUARE. Order four squares can be panmagic or ASSOCIATIVE, but not both. Order five squares are the smallest which can be both ASSOCIATIVE and panmagic, and 16 distinct ASSOCIATIVE panmagic squares exist, one of which is illustrated above (Gardner 1988).

The number of distinct panmagic squares of order 1, 2, ... are 1, 0, 0, 384, 3600, 0, ... (Sloane's A027567, Hunter and Madachy 1975). Panmagic squares are related to HYPERCUBES.

See also ASSOCIATIVE MAGIC SQUARE, HYPERCUBE, FRANKLIN MAGIC SQUARE, LO SHU, MAGIC SQUARE

References

Gardner, M. *The Second Scientific American Book of Mathematical Puzzles & Diversions: A New Selection.* New York: Simon and Schuster, pp. 135–137, 1961.

Gardner, M. "Magic Squares and Cubes." Ch. 17 in *Time Travel and Other Mathematical Bewilderments.* New York: W. H. Freeman, pp. 213–225, 1988.

Hunter, J. A. H. and Madachy, J. S. "Mystic Arrays." Ch. 3 in *Mathematical Diversions.* New York: Dover, pp. 24–25, 1975.

Kraitchik, M. "Panmagic Squares." §7.9 in *Mathematical Recreations.* New York: W. W. Norton, pp. 143 and 174–176, 1942.

Madachy, J. S. *Madachy's Mathematical Recreations.* New York: Dover, p. 87, 1979.

Rosser, J. B. and Walker, R. J. "The Algebraic Theory of Diabolical Squares." *Duke Math. J.* **5**, 705–728, 1939.

Sloane, N. J. A. Sequences A027567 in "An On-Line Version of the Encyclopedia of Integer Sequences." http://www.research.att.com/~njas/sequences/eisonline.html.

Pantograph

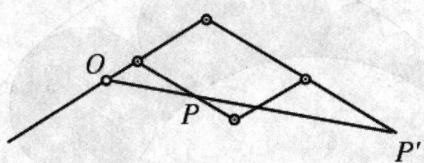

A LINKAGE invented in 1630 by Christoph Scheiner for making a scaled copy of a given figure. The linkage is pivoted at O; hinges are denoted $\odot$. By placing a PENCIL at P (or P'), a DILATED image is obtained at P' (or P).

See also HOMOTHETIC, LINKAGE

References

Cundy, H. and Rollett, A. *Mathematical Models, 3rd ed.* Stradbroke, England: Tarquin Pub., pp. 232–233, 1989.

Coxeter, H. S. M. *Introduction to Geometry, 2nd ed.* New York: Wiley, pp. 69–70, 1969.

Durell, C. V. *Modern Geometry: The Straight Line and Circle.* London: Macmillan, p. 5, 1928.

Wells, D. *The Penguin Dictionary of Curious and Interesting Geometry.* London: Penguin, pp. 167–168, 1991.

Papal Cross

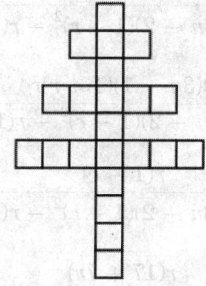

See also CROSS

Paper Folding

FOLDING, ORIGAMI

Pappus Chain

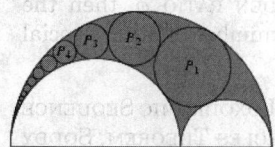 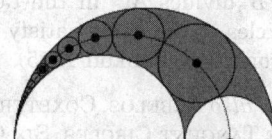

In the ARBELOS, construct a chain of TANGENT CIRCLES starting with the CIRCLE TANGENT to the two small interior semicircles and the large exterior one. This is called a Pappus chain (left figure).

In a Pappus chain, the distance from the center of the first INSCRIBED CIRCLE P_1 to the bottom line is twice the CIRCLE'S RADIUS, from the second CIRCLE P_2 is four times the RADIUS, and for the nth CIRCLE P_n is $2n$

times the RADIUS. Furthermore, the centers of the circles P_i lie on an ELLIPSE (right figure).

If $r \equiv AB/AC$, then the center and radius of the nth circle P_n in the Pappus chain are

$$x_n = \frac{r(1+r)}{2[n^2(1-r)^2+r]} \tag{1}$$

$$y_n = \frac{nr(1-r)}{n^2(1-r)^2+r} \tag{2}$$

$$r_n = \frac{(1-r)r}{2\left[n^2(1-r)^2+r\right]}. \tag{3}$$

This general result simplifies to $r_n = 1/(6+n^2)$ for $r = 2/3$ (Gardner 1979). Further special cases when $AC = 1+AB$ are considered by Gaba (1940).

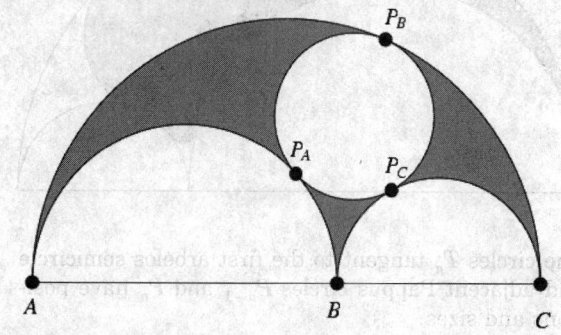

The positions of the points of tangency for the first circle are

$$x_A = \frac{r}{(1-r)^2} \tag{4}$$

$$y_A = \frac{r(1-r)}{(1-r)^2} \tag{5}$$

$$x_B = \frac{r(1+r)}{1+r^2} \tag{6}$$

$$y_B = \frac{r(1-r)}{1+r^2} \tag{7}$$

$$x_C = \frac{r^2}{1-2r+2r^2} \tag{8}$$

$$y_C = \frac{r(1-r)}{1-2r+2r^2}. \tag{9}$$

The centers of the CIRCLES lie on an ELLIPSE, and the DIAMETER of the nth CIRCLE P_n is $(1/n)$th PERPENDICULAR distance to the base of the SEMICIRCLE. This result was known to Pappus, who referred to it as an ancient theorem (Hood 1961, Cadwell 1966, Gardner 1979, Bankoff 1981). The simplest proof is via INVERSIVE GEOMETRY. Eliminating n from the equations for x_n and y_n gives

$$4rx^2 - 2r(1+r)x + (1+r)^2y^2 = 0 \quad (10)$$

$$4r\left[x - \tfrac{1}{4}(1+r)\right]^2 + (1+r^2)y^2 = \tfrac{1}{4}r(1+r)^2 \quad (11)$$

$$\left[\frac{x - \tfrac{1}{4}(1+r)}{\tfrac{1}{4}(1+r)}\right]^2 + \left(\frac{y}{\tfrac{1}{2}\sqrt{r}}\right)^2 = 1, \quad (12)$$

which is the equation of an ellipse with center $((1+r)/4, 0)$ and semimajor and semiminor axes $(1+r)/4$ and $\sqrt{r}/2$ respectively.

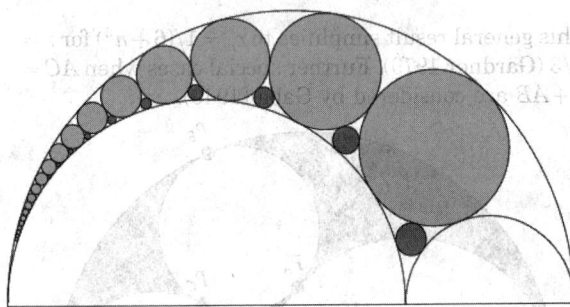

The circles T_n tangent to the first arbelos semicircle and adjacent Pappus circles P_{n-1} and P_n have positions and sizes

$$x'_n = \frac{r(7+r)}{2[4 + 4n(n-1)(1-r)^2 + r(r-1)]} \quad (13)$$

$$y'_n = \frac{2(2n-1)r(1-r)}{4 + 4n(n-1)(1-r)^2 + r(r-1)} \quad (14)$$

$$r'_n = \frac{r(1-r)}{2[4 + 4n(n-1)(1-r)^2 + r(r-1)]}. \quad (15)$$

A special case of this problem with $r = 1/2$ (giving equal circles forming the arbelos) was considered in a Japanese temple tablet (Sangaku) problem from 1788 in the Tokyo prefecture (Rothman 1998). In this case, the solution simplifies to

$$x'_n = \frac{15}{2(15 - 4n + 4n^2)} \quad (16)$$

$$y'_n = \frac{2(2n-1)}{15 - 4n + 4n^2} \quad (17)$$

$$r'_n = \frac{1}{2(15 - rn + 4n^2)}. \quad (18)$$

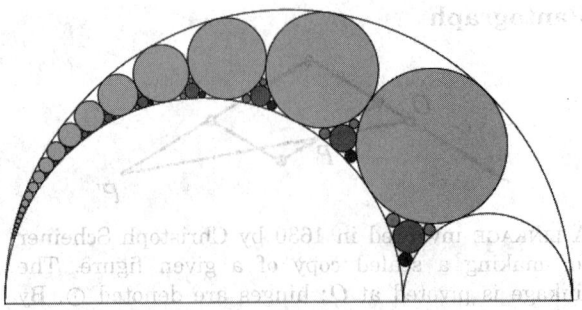

Furthermore, the positions and radii of the three tangent circles surrounding this circle can also be found analytically, and are given by

$$x_n^{(1)} = \frac{r(17+r)}{2\left[12 + 3n(3n-4)(1-r)^2 + r(4r-7)\right]} \quad (19)$$

$$y_n^{(1)} = \frac{3(3n-2)(1-r)r}{12 + 3n(3n-4)(1-r)^2 + r(4r-7)} \quad (20)$$

$$r_n^{(1)} = \frac{r(1-r)}{2\left[12 + 3n(3n-4)(1-r)^2 + r(4r-7)\right]} \quad (21)$$

$$x_n^{(2)} = \frac{r(17+r)}{2\left[9 + 3n(3n-2)(1-r)^2 - r(1-r)\right]} \quad (22)$$

$$y_n^{(2)} = \frac{3(3n-1)(1-r)r}{9 + 3n(3n-2)(1-r)^2 - r(1-r)} \quad (23)$$

$$r_n^{(2)} = \frac{r(1-r)}{2\left[9 + 3n(3n-2)(1-r)^2 - r(1-r)\right]} \quad (24)$$

$$x_n^{(3)} = \frac{r(17+7r)}{2\left[9 + 12n(n-1)(1-r)^2 + r(4r-1)\right]} \quad (25)$$

$$y_n^{(3)} = \frac{6(2n-1)(1-r)r}{9 + 12n(n-1)(1-r)^2 + r(4r-1)} \quad (26)$$

$$r_n^{(3)} = \frac{r(1-r)}{2[9 + 12n(n-1)(1-r)^2 + r(4r-1)]}. \quad (27)$$

If B divides AC in the GOLDEN RATIO ϕ, then the circles in the chain satisfy a number of other special properties (Bankoff 1955).

See also ARBELOS, COXETER'S LOXODROMIC SEQUENCE OF TANGENT CIRCLES, SIX CIRCLES THEOREM, SODDY CIRCLES, STEINER CHAIN

References

Bankoff, L. "The Golden Arbelos." *Scripta Math.* **21**, 70–76, 1955.

Bankoff, L. "Are the Twin Circles of Archimedes Really Twins?" *Math. Mag.* **47**, 214–218, 1974.

Bankoff, L. "How Did Pappus Do It?" In *The Mathematical Gardner* (Ed. D. Klarner). Boston, MA: Prindle, Weber, and Schmidt, pp. 112–118, 1981.

Casey, J. *A Sequel to the First Six Books of the Elements of Euclid, Containing an Easy Introduction to Modern Geometry with Numerous Examples, 5th ed., rev. enl.* Dublin: Hodges, Figgis, & Co., p. 103, 1888.
Gaba, M. G. "On a Generalization of the Arbelos." *Amer. Math. Monthly* **47**, 19–24, 1940.
Gardner, M. "Mathematical Games: The Diverse Pleasures of Circles that Are Tangent to One Another." *Sci. Amer.* **240**, 18–28, Jan. 1979.
Hood, R. T. "A Chain of Circles." *Math. Teacher* **54**, 134–137, 1961.
Johnson, R. A. *Modern Geometry: An Elementary Treatise on the Geometry of the Triangle and the Circle.* Boston, MA: Houghton Mifflin, p. 117, 1929.
Rothman, T. "Japanese Temple Geometry." *Sci. Amer.* **278**, 85–91, May 1998.
Steiner, J. *Jacob Steiner's gesammelte Werke, Band I.* Bronx, NY: Chelsea, p. 47, 1971.

Pappus-Guldinus Theorem

PAPPUS'S CENTROID THEOREM

Pappus's Centroid Theorem

The SURFACE AREA S of a SURFACE OF REVOLUTION generated by the revolution of a curve about an external axis is equal to the product of the arc length s of the generating curve and the distance d_1 traveled by the curve's centroid $\bar{x}_1$,

$$S = sd_1 = 2\pi s\bar{x}_1.$$

Similarly, the VOLUME V of a SOLID OF REVOLUTION generated by the revolution of a lamina about an external axis is equal to the product of the area A of the lamina and the distance d_2 traveled by the lamina's centroid $\bar{x}_2$,

$$V = Ad_2 = 2\pi A\bar{x}_2.$$

The following table summarizes the surface areas and volumes calculated using Pappus's centroid theorem for various solids and surfaces of revolution.

SOLID	SECTION	s	$\bar{x}_1$	S	A	$\bar{x}_2$	V
CONE	RIGHT TRIANGLE	$\sqrt{r^2+h^2}$	$\frac{1}{2}r$	$\pi r\sqrt{r^2+h^2}$	$\frac{1}{2}hr$	$\frac{1}{3}hr$	$\frac{1}{3}\pi r^2$
CYLINDER	CIRCLE	h	$\frac{1}{2}r$	$2\pi rh$	hr	$\frac{1}{2}r$	$\pi r^2 h$
SPHERE	SEMI-CIRCLE	πr	$\frac{2r}{\pi}$	$4\pi r^2$	$\frac{1}{2}\pi r^2$	$\frac{4r}{3\pi}$	$\frac{4}{3}\pi r^3$

See also CENTROID (GEOMETRIC), CROSS SECTION, PERIMETER, SOLID OF REVOLUTION, SURFACE AREA, SURFACE OF REVOLUTION, TOROID, TORUS

References

Beyer, W. H. *CRC Standard Mathematical Tables, 28th ed.* Boca Raton, FL: CRC Press, p. 132, 1987.
Harris, J. W. and Stocker, H. "Guldin's Rules." §4.1.3 in *Handbook of Mathematics and Computational Science.* New York: Springer-Verlag, p. 96, 1998.

Kern, W. F. and Bland, J. R. "Theorem of Pappus." §40 in *Solid Mensuration with Proofs, 2nd ed.* New York: Wiley, pp. 110–115, 1948.

Pappus's Harmonic Theorem

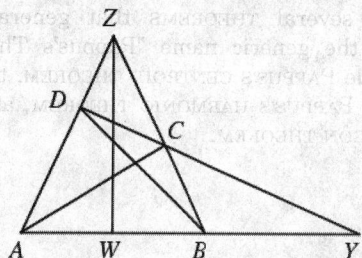

AW, AB, and AY in the above figure are in a HARMONIC RANGE.

See also CEVA'S THEOREM, HARMONIC RANGE, MENELAUS' THEOREM

References

Coxeter, H. S. M. and Greitzer, S. L. *Geometry Revisited.* Washington, DC: Math. Assoc. Amer., pp. 67–68, 1967.

Pappus's Hexagon Theorem

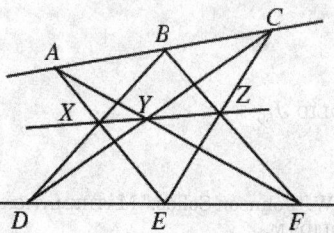

If A, B, and C are three points on one LINE, D, E, and F are three points on another LINE, and AE meets BD at X, AF meets CD at Y, and BF meets CE at Z, then the three points X, Y, and Z are COLLINEAR. Pappus's hexagon theorem is dual to DESARGUES' THEOREM according to the DUALITY PRINCIPLE of PROJECTIVE GEOMETRY.

See also BRIANCHON'S THEOREM, CAYLEY-BACHARACH THEOREM, DESARGUES' THEOREM, DUALITY PRINCIPLE, PASCAL'S THEOREM, PROJECTIVE GEOMETRY

References

Coxeter, H. S. M. and Greitzer, S. L. "Pappus's Theorem." §3.5 in *Geometry Revisited.* Washington, DC: Math. Assoc. Amer., pp. 67–70, 1967.
Eves, H. "Pappus' Theorem." §6.2.6 in *A Survey of Geometry, rev. ed.* Boston, MA: Allyn & Bacon, pp. 79 and 250–251, 1965.
Johnson, R. A. "Theorem of Pappus." §388 in *Modern Geometry: An Elementary Treatise on the Geometry of the Triangle and the Circle.* Boston, MA: Houghton Mifflin, pp. 237–238, 1929.
Ogilvy, C. S. *Excursions in Geometry.* New York: Dover, pp. 92–94, 1990.
Pappas, T. "Pappus' Theorem & the Nine Coin Puzzle." *The Joy of Mathematics.* San Carlos, CA: Wide World Publ./ Tetra, p. 163, 1989.

Wells, D. *The Penguin Dictionary of Curious and Interesting Geometry.* London: Penguin, pp. 168–169, 1991.

Pappus's Theorem

There are several THEOREMS that generally are known by the generic name "Pappus's Theorem." They include PAPPUS'S CENTROID THEOREM, the PAPPUS CHAIN, PAPPUS'S HARMONIC THEOREM, and PAPPUS'S HEXAGON THEOREM.

Parabiaugmented Dodecahedron

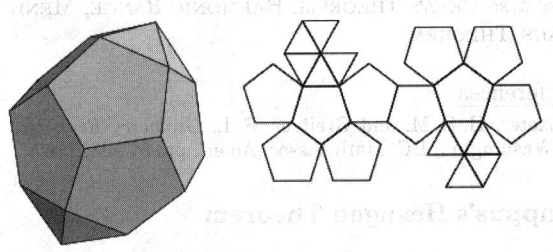

JOHNSON SOLID J_{59}.

References

Weisstein, E. W. "Johnson Solids." MATHEMATICA NOTEBOOK JOHNSONSOLIDS.M.

Weisstein, E. W. "Johnson Solid Netlib Database." MATHEMATICA NOTEBOOK JOHNSONSOLIDS.DAT.

Parabiaugmented Hexagonal Prism

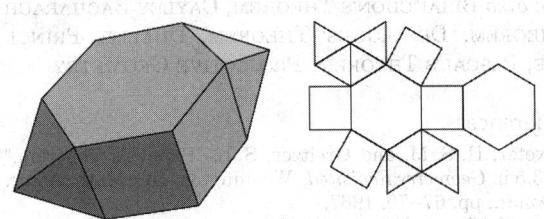

JOHNSON SOLID J_{55}.

References

Weisstein, E. W. "Johnson Solids." MATHEMATICA NOTEBOOK JOHNSONSOLIDS.M.

Weisstein, E. W. "Johnson Solid Netlib Database." MATHEMATICA NOTEBOOK JOHNSONSOLIDS.DAT.

Parabiaugmented Truncated Dodecahedron

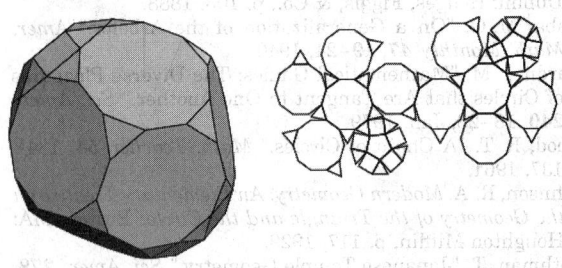

JOHNSON SOLID J_{69}.

References

Weisstein, E. W. "Johnson Solids." MATHEMATICA NOTEBOOK JOHNSONSOLIDS.M.

Weisstein, E. W. "Johnson Solid Netlib Database." MATHEMATICA NOTEBOOK JOHNSONSOLIDS.DAT.

Parabidiminished Rhombicosidodecahedron

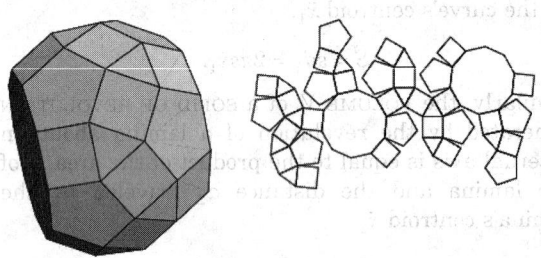

JOHNSON SOLID J_{80}.

References

Weisstein, E. W. "Johnson Solids." MATHEMATICA NOTEBOOK JOHNSONSOLIDS.M.

Weisstein, E. W. "Johnson Solid Netlib Database." MATHEMATICA NOTEBOOK JOHNSONSOLIDS.DAT.

Parabigyrate Rhombicosidodecahedron

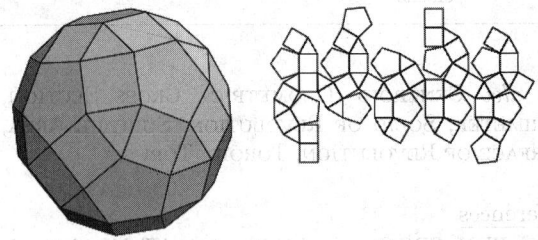

JOHNSON SOLID J_{73}.

References

Weisstein, E. W. "Johnson Solids." MATHEMATICA NOTEBOOK JOHNSONSOLIDS.M.

Weisstein, E. W. "Johnson Solid Netlib Database." MATHEMATICA NOTEBOOK JOHNSONSOLIDS.DAT.

Parabola

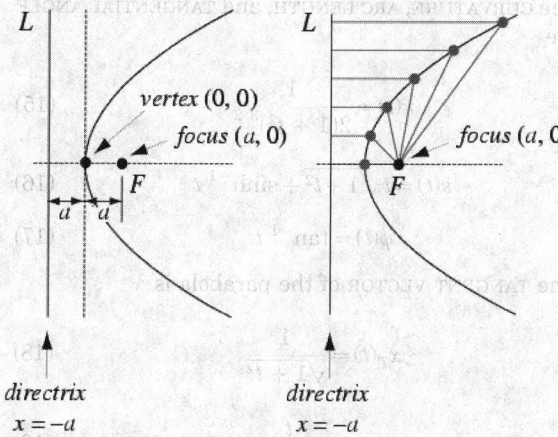

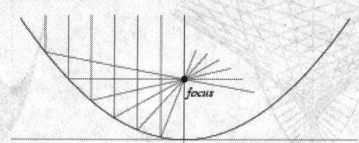

The set of all points in the PLANE equidistant from a given LINE L (the DIRECTRIX) and a given point F not on the line (the FOCUS). The FOCAL PARAMETER (i.e., the distance between the directrix and focus) is therefore given by $p = 2a$, where a is the distance from the vertex to the directrix or focus.

The parabola was studied by Menaechmus in an attempt to achieve CUBE DUPLICATION. Menaechmus solved the problem by finding the intersection of the two parabolas $x^2 = y$ and $y^2 = 2x$. Euclid wrote about the parabola, and it was given its present name by Apollonius. Pascal considered the parabola as a projection of a CIRCLE, and Galileo showed that projectiles falling under uniform gravity follow parabolic paths. Gregory and Newton considered the CATACAUSTIC properties of a parabola which bring parallel rays of light to a focus (MacTutor Archive), as illustrated above.

For a parabola opening to the right with vertex at (0, 0), the equation in CARTESIAN COORDINATES is

$$\sqrt{(x-a)^2 + y^2} = x + a \qquad (1)$$

$$(x-a)^2 + y^2 = (x+a)^2 \qquad (2)$$

$$x^2 - 2ax + a^2 + y^2 = x^2 + 2ax + a^2 \qquad (3)$$

$$y^2 = 4ax. \qquad (4)$$

The quantity $4a$ is known as the LATUS RECTUM. If the vertex is at (x_0, y_0) instead of (0, 0), the equation of the parabola is

$$(y - y_0)^2 = 4a(x - x_0). \qquad (5)$$

If the parabola instead opens upwards, its equation is

$$x^2 = 4ay. \qquad (6)$$

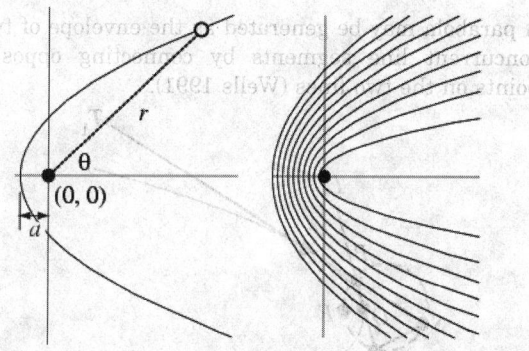

In POLAR COORDINATES, the equation of a parabola with parameter a and center (0, 0) is given by

$$r = -\frac{2a}{1 + \cos\theta} \qquad (7)$$

(left figure). The equivalence with the Cartesian form can be seen by setting up a coordinate system $(x', y') = (x - a, y)$ and plugging in $r = \sqrt{x'^2 + y'^2}$ and $\theta = \tan^{-1}(y'/x')$ to obtain

$$\sqrt{(x-a)^2 + y^2} = \frac{2a}{1 + \dfrac{x-a}{\sqrt{(x-a)^2 + y^2}}} \qquad (8)$$

Expanding and collecting terms,

$$a + x + \sqrt{(a-x)^2 + y^2} = 0, \qquad (9)$$

so solving for y^2 gives (4). A set of confocal parabolas is shown in the figure on the right.

In PEDAL COORDINATES with the PEDAL POINT at the FOCUS, the equation is

$$p^2 = ar. \qquad (10)$$

The parametric equations for the parabola are

$$x = at^2 \qquad (11)$$

$$y = 2at \qquad (12)$$

or

$$x = \frac{t^2}{4a} \qquad (13)$$

$$y = t. \qquad (14)$$

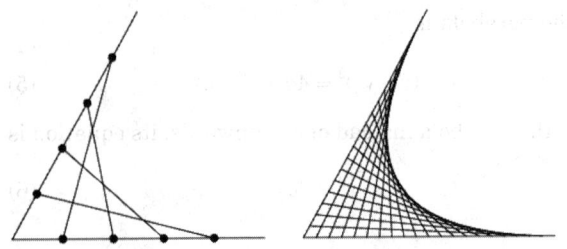

A parabola may be generated as the envelope of two concurrent line segments by connecting opposite points on the two lines (Wells 1991).

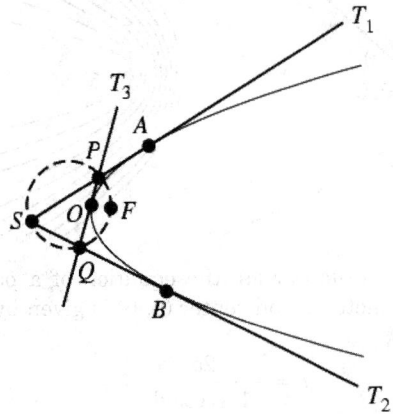

In the above figure, the lines SPA, SQB, and POQ are tangent to the parabola at points A, B, and O, respectively. Then $SP/PA = QO/OP = BQ/QS$ (Wells 1991). Moreover, the CIRCUMCIRCLE of $\triangle PQS$ passes through the FOCUS F (Honsberger 1995, p. 47). In addition, the foot of the perpendicular to a tangent to a parabola from the FOCUS always lies on the tangent at the vertex (Honsberger 1995, p. 48).

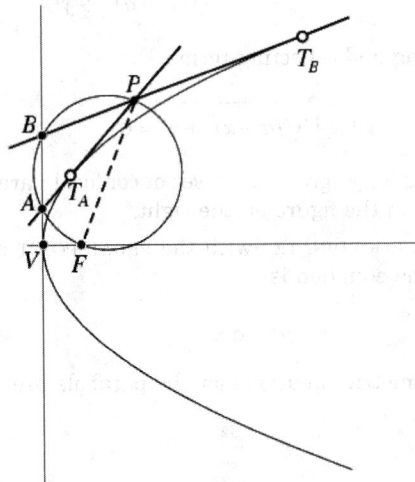

Given an arbitrary point P located "outside" a parabola, the tangent or tangents to the parabola through P can be constructed by drawing the CIRCLE having PF as a DIAMETER, where F is the FOCUS. Then locate the points A and B at which the circle cuts the VERTICAL TANGENT through V. The points T_A and T_B (which can collapse to a single point in the degenerate

case) are then the points of tangency of the lines PA and PB and the parabola (Wells 1991).

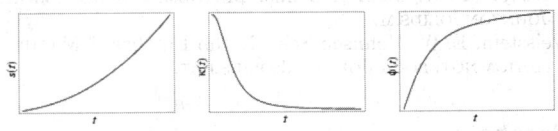
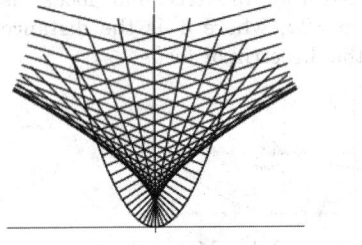
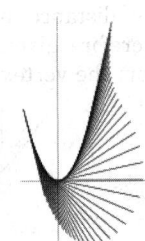

The CURVATURE, ARC LENGTH, and TANGENTIAL ANGLE are

$$\kappa(t) = \frac{1}{2(1 + t^2)^{3/2}} \tag{15}$$

$$s(t) = t\sqrt{1 + t^2} + \sinh^{-1} t \tag{16}$$

$$\phi(t) = \tan^{-1} t. \tag{17}$$

The TANGENT VECTOR of the parabola is

$$x_T(t) = \frac{1}{\sqrt{1 + t^2}} \tag{18}$$

$$y_T(t) = \frac{t}{\sqrt{1 + t^2}}. \tag{19}$$

The plots below show the normal and tangent vectors to a parabola.

See also CONIC SECTION, ELLIPSE, HYPERBOLA, QUADRATIC CURVE, REFLECTION PROPERTY, TSCHIRNHAUSEN CUBIC PEDAL CURVE

References

Beyer, W. H. *CRC Standard Mathematical Tables, 28th ed.* Boca Raton, FL: CRC Press, pp. 198 and 222–223, 1987.

Casey, J. "The Parabola." Ch. 5 in *A Treatise on the Analytical Geometry of the Point, Line, Circle, and Conic Sections, Containing an Account of Its Most Recent Extensions, with Numerous Examples, 2nd ed., rev. enl.* Dublin: Hodges, Figgis, & Co., pp. 173–200, 1893.

Coxeter, H. S. M. "Conics." §8.4 in *Introduction to Geometry, 2nd ed.* New York: Wiley, pp. 115–119, 1969.

Hilbert, D. and Cohn-Vossen, S. *Geometry and the Imagination.* New York: Chelsea, p. 4, 1999.

Honsberger, R. *Episodes in Nineteenth and Twentieth Century Euclidean Geometry.* Washington, DC: Math. Assoc. Amer., p. 47, 1995.

Lawrence, J. D. *A Catalog of Special Plane Curves.* New York: Dover, pp. 67–72, 1972.

Lockwood, E. H. "The Parabola." Ch. 1 in *A Book of Curves.* Cambridge, England: Cambridge University Press, pp. 2–12, 1967.

MacTutor History of Mathematics Archive. "Parabola." http://www-groups.dcs.st-and.ac.uk/~history/Curves/ Parabola.html.

Pappas, T. "The Parabolic Ceiling of the Capitol." *The Joy of Mathematics.* San Carlos, CA: Wide World Publ./Tetra, pp. 22–23, 1989.

Wells, D. *The Penguin Dictionary of Curious and Interesting Geometry.* London: Penguin, pp. 169–172, 1991.

Yates, R. C. "Conics." *A Handbook on Curves and Their Properties.* Ann Arbor, MI: J. W. Edwards, pp. 36–56, 1952.

Parabola Caustic

The CAUSTIC of a PARABOLA with rays PERPENDICULAR to the axis of the PARABOLA is TSCHIRNHAUSEN CUBIC.

Parabola Evolute

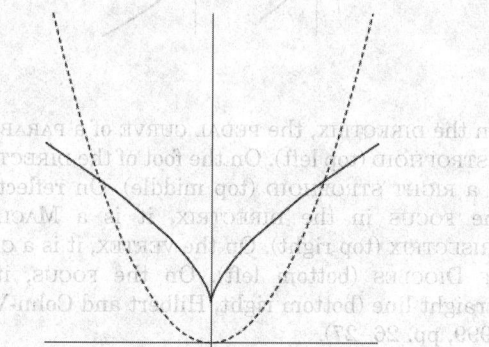

Given a PARABOLA

$$y = x^2, \tag{1}$$

the parametric equations of the parabola are

$$x = t \tag{2}$$

$$y = t^2, \tag{3}$$

and the derivatives are

$$x' = 1 \tag{4}$$

$$x'' = 0 \tag{5}$$

$$y' = 2t \tag{6}$$

$$y'' = 2. \tag{7}$$

The RADIUS OF CURVATURE is therefore given by

$$R = \frac{(x'^2 + y'^2)^{3/2}}{x'y'' - x''y'} = \tfrac{1}{2}(1 + 4t^2)^{3/2}. \tag{8}$$

The TANGENT VECTOR is

$$\hat{\mathbf{T}} = \frac{1}{\sqrt{1 + 4t^2}} \begin{bmatrix} 1 \\ 2t \end{bmatrix}, \tag{9}$$

so the parametric equations of the evolute are

$$\xi = -4t^3 \tag{10}$$

$$\eta = \tfrac{1}{2} + 3t^2, \tag{11}$$

and

$$-\tfrac{1}{4}\,\xi = t^3 \tag{12}$$

$$\tfrac{1}{3}\left(\eta - \tfrac{1}{2}\right) = t^2 \tag{13}$$

$$\tfrac{1}{3}\left(\eta - \tfrac{1}{2}\right) = \left(-\tfrac{1}{4}\,\xi\right)^{2/3} \tag{14}$$

$$\tfrac{1}{3}\left(\eta - \tfrac{1}{2}\right) = \left(-\frac{2\xi}{8}\right)^{2/3} = \tfrac{1}{4}(2\xi)^{2/3}. \tag{15}$$

The EVOLUTE is therefore

$$\eta = \tfrac{3}{4}(2\xi)^{2/3} + \tfrac{1}{2}. \tag{16}$$

This is known as NEILE'S PARABOLA and is a SEMI-CUBICAL PARABOLA. From a point above the evolute three normals can be drawn to the PARABOLA, while only one normal can be drawn to the PARABOLA from a point below the EVOLUTE.

See also NEILE'S PARABOLA, PARABOLA, SEMICUBICAL PARABOLA

Parabola Inverse Curve

The INVERSE CURVE for a PARABOLA given by

$$x = at^2 \tag{1}$$

$$y = 2at \tag{2}$$

with INVERSION CENTER (x_0, y_0) and INVERSION RADIUS k is

$$x = x_0 + \frac{k(at^2 - x_0)}{(at^2 + x_0)^2 + (2at - y_0)^2} \tag{3}$$

$$y = y_0 + \frac{k(2at - y_0)}{(at^2 + x_0)^2 + (2at - y_0)^2}. \tag{4}$$

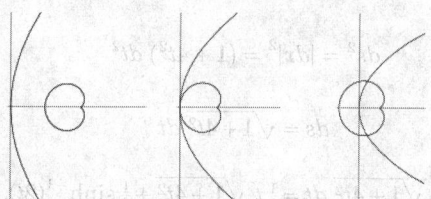

For $(x_0, y_0) = (a, 0)$ at the FOCUS, the INVERSE CURVE is the CARDIOID

$$x = a + \frac{k(t^2 - 1)}{a(1 + t^2)^2} \tag{5}$$

$$y = \frac{2kt}{a(1+t^2)^2}. \qquad (6)$$

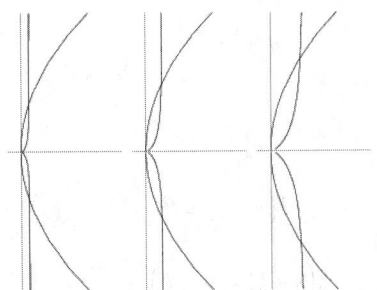

For $(x_0, y_0) = (0, 0)$ at the VERTEX, the INVERSE CURVE is the CISSOID OF DIOCLES

$$x = \frac{k}{a(4+t^2)} \qquad (7)$$

$$y = \frac{2k}{at(4+t^2)}. \qquad (8)$$

Parabola Involute

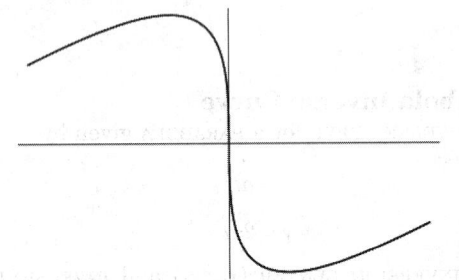

$$\frac{d\mathbf{r}}{dt} = \begin{bmatrix} 1 \\ 2t \end{bmatrix} \qquad (1)$$

$$\hat{\mathbf{T}} = \frac{1}{\sqrt{1+4t^2}} \begin{bmatrix} 1 \\ 2t \end{bmatrix} \qquad (2)$$

$$ds^2 = |d\mathbf{r}|^2 = (1+4t^2)\, dt^2 \qquad (3)$$

$$ds = \sqrt{1+4t^2}\, dt \qquad (4)$$

$$s = \int \sqrt{1+4t^2}\, dt = \tfrac{1}{2} t \sqrt{1+4t^2} + \tfrac{1}{4} \sinh^{-1}(2t), \qquad (5)$$

so the equation of the INVOLUTE is

$$\mathbf{r}_i = \mathbf{r} - s\hat{\mathbf{T}} = \begin{bmatrix} t \\ t^2 \end{bmatrix} - \frac{\tfrac{1}{2} t \sqrt{1+4t^2} + \tfrac{1}{4} \sinh^{-1}(2t)}{\sqrt{1+4t^2}} \begin{bmatrix} 1 \\ 2t \end{bmatrix}$$

$$= \frac{1}{2\sqrt{1+4t^2}} \begin{bmatrix} t - \tfrac{1}{2} \sinh^{-1}(2t) \\ -\sinh^{-1}(2t) \end{bmatrix}. \qquad (6)$$

Parabola Pedal Curve

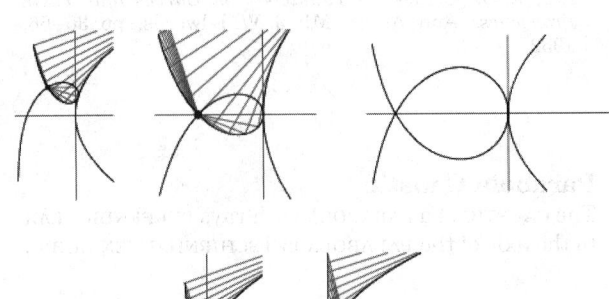

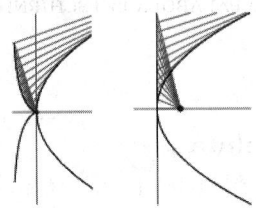

On the DIRECTRIX, the PEDAL CURVE of a PARABOLA is a STROPHOID (top left). On the foot of the DIRECTRIX, it is a RIGHT STROPHOID (top middle). On reflection of the FOCUS in the DIRECTRIX, it is a MACLAURIN TRISECTRIX (top right). On the VERTEX, it is a CISSOID OF DIOCLES (bottom left). On the FOCUS, it is a straight line (bottom right; Hilbert and Cohn-Vossen 1999, pp. 26–27).

References

Hilbert, D. and Cohn-Vossen, S. *Geometry and the Imagination.* New York: Chelsea, 1999.
Lawrence, J. D. *A Catalog of Special Plane Curves.* New York: Dover, pp. 94–97, 1972.

Parabolic Coordinates

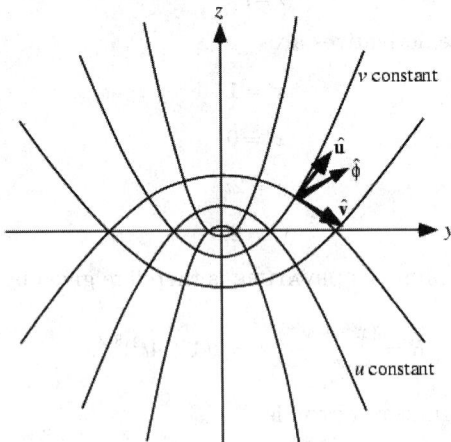

A system of CURVILINEAR COORDINATES in which two sets of coordinate surfaces are obtained by revolving the parabolas of PARABOLIC CYLINDRICAL COORDINATES about the X-AXIS, which is then relabeled the z-AXIS. There are several notational conventions. Whereas (u, v, θ) is used in this work, Arfken (1970) uses (ξ, η, φ).

The equations for the parabolic coordinates are

$$x = uv \cos \theta \tag{1}$$

$$y = uv \sin \theta \tag{2}$$

$$z = \tfrac{1}{2}(u^2 - v^2), \tag{3}$$

where $u \in [0, \infty)$, $v \in [0, \infty)$, and $\theta \in [0, 2\pi)$. To solve for u, v, and θ, examine

$$x^2 + y^2 + z^2 = u^2 v^2 + \tfrac{1}{4}(u^4 - 2u^2 v^2 + v^4)$$

$$= \tfrac{1}{4}(u^4 + 2u^2 v^2 + v^4)$$

$$= \tfrac{1}{4}(u^2 + v^2)^2, \tag{4}$$

so

$$\sqrt{x^2 + y^2 + z^2} = \tfrac{1}{2}(u^2 + v^2) \tag{5}$$

and

$$\sqrt{x^2 + y^2 + z^2} + z = u^2 \tag{6}$$

$$\sqrt{x^2 + y^2 + z^2} - z = v^2. \tag{7}$$

We therefore have

$$u = \sqrt{\sqrt{x^2 + y^2 + z^2} + z} \tag{8}$$

$$v = \sqrt{\sqrt{x^2 + y^2 + z^2} - z} \tag{9}$$

$$\theta = \tan^{-1}\left(\frac{y}{x}\right). \tag{10}$$

The SCALE FACTORS are

$$h_u = \sqrt{u^2 + v^2} \tag{11}$$

$$h_v = \sqrt{u^2 + v^2} \tag{12}$$

$$h_\theta = uv. \tag{13}$$

The LINE ELEMENT is

$$ds^2 = (u^2 + v^2)(du^2 + dv^2) + u^2 v^2 \, d\theta^2, \tag{14}$$

and the VOLUME ELEMENT is

$$dV = uv(u^2 + v^2) \, du \, dv \, d\theta. \tag{15}$$

The LAPLACIAN is

$$\nabla^2 f = \frac{1}{uv(u^2 + v^2)}\left[\frac{\partial}{\partial u}\left(uv \frac{\partial f}{\partial u}\right) + \frac{\partial}{\partial v}\left(uv \frac{\partial f}{\partial v}\right)\right]$$

$$+ \frac{1}{u^2 v^2}\frac{\partial^2 f}{\partial \theta^2}$$

$$= \frac{1}{u^2 + v^2}\left[\frac{1}{u}\frac{\partial}{\partial u}\left(u \frac{\partial f}{\partial u}\right) + \frac{1}{v}\frac{\partial}{\partial v}\left(v \frac{\partial f}{\partial v}\right)\right]$$

$$+ \frac{1}{u^2 v^2}\frac{\partial^2 f}{\partial \theta^2}$$

$$= \frac{1}{u^2 + v^2}\left(\frac{1}{u}\frac{\partial f}{\partial u} + \frac{\partial^2 f}{\partial u^2} + \frac{1}{v}\frac{\partial f}{\partial v} + \frac{\partial^2 f}{\partial v^2}\right)$$

$$+ \frac{1}{u^2 v^2}\frac{\partial^2 f}{\partial \theta^2}. \tag{16}$$

The HELMHOLTZ DIFFERENTIAL EQUATION is SEPARABLE in parabolic coordinates.

See also CONFOCAL PARABOLOIDAL COORDINATES, HELMHOLTZ DIFFERENTIAL EQUATION–PARABOLIC COORDINATES, PARABOLIC CYLINDRICAL COORDINATES

References

Arfken, G. "Parabolic Coordinates (ξ, η, ϕ)." §2.12 in *Mathematical Methods for Physicists, 2nd ed.* Orlando, FL: Academic Press, pp. 109–112, 1970.

Moon, P. and Spencer, D. E. "Parabolic Coordinates (μ, v, ψ)." Table 1.08 in *Field Theory Handbook, Including Coordinate Systems, Differential Equations, and Their Solutions, 2nd ed.* New York: Springer-Verlag, pp. 34–36, 1988.

Morse, P. M. and Feshbach, H. *Methods of Theoretical Physics, Part I.* New York: McGraw-Hill, p. 660, 1953.

Parabolic Cyclide

A CYCLIDE formed by INVERSION of a STANDARD TORUS when INVERSION SPHERE is tangent to the TORUS.

See also CYCLIDE, INVERSION, INVERSION SPHERE, PARABOLIC HORN CYCLIDE, PARABOLIC RING CYCLIDE, PARABOLIC SPINDLE CYCLIDE

Parabolic Cylinder

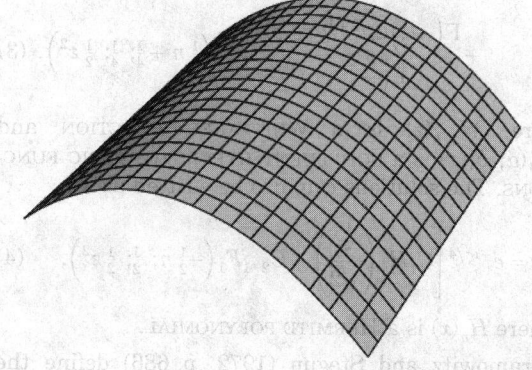

A QUADRATIC SURFACE given by the equation

$$x^2 + 2rz = 0.$$

References

Hilbert, D. and Cohn-Vossen, S. *Geometry and the Imagination.* New York: Chelsea, p. 12, 1999.

Parabolic Cylinder Differential Equation

The second-order ORDINARY DIFFERENTIAL EQUATION

$$y'' + (ax^2 + bx + c) = 0$$

(Abramowitz and Stegun 1972, p. 686; Zwillinger 1995, p. 414; Zwillinger 1997, p. 126) whose solutions are called PARABOLIC CYLINDER FUNCTIONS.

See also PARABOLIC CYLINDER FUNCTION

References

Abramowitz, M. and Stegun, C. A. (Eds.). "Parabolic Cylinder Function." Ch. 19 in *Handbook of Mathematical Functions with Formulas, Graphs, and Mathematical Tables, 9th printing.* New York: Dover, pp. 685–700, 1972.
Zwillinger, D. (Ed.). *CRC Standard Mathematical Tables and Formulae.* Boca Raton, FL: CRC Press, p. 414, 1995.
Zwillinger, D. *Handbook of Differential Equations, 3rd ed.* Boston, MA: Academic Press, p. 126, 1997.

Parabolic Cylinder Function

These functions are sometimes called WEBER FUNCTIONS. Whittaker and Watson (1990, p. 347) define the parabolic cylinder functions as solutions to the WEBER DIFFERENTIAL EQUATION

$$y''(z) + \left(n + \tfrac{1}{2} - \tfrac{1}{4}z^2\right)y(z) = 0. \tag{1}$$

The two independent solutions are given by $y = D_n(z)$ and $D_{-n-1}\left(ze^{i\pi/2}\right)$, where

$$D_n(z) = 2^{n/2+1/4}z^{-1/2}W_{n/2+1/4,\,-1/4}\left(\tfrac{1}{2}z^2\right) \tag{2}$$

$$= \frac{\Gamma\left(\tfrac{1}{2}\right)2^{n/2+1/4}z^{-1/2}}{\Gamma\left(\tfrac{1}{2} - \tfrac{1}{2}n\right)} \,{}_1F_1\left(\tfrac{1}{2}n + \tfrac{1}{4};\ -\tfrac{1}{4};\ \tfrac{1}{2}z^2\right)$$

$$+ \frac{\Gamma\left(-\tfrac{1}{2}\right)2^{n/2+1/4}z^{-1/2}}{\Gamma\left(-\tfrac{1}{2}n\right)} \,{}_1F_1\left(\tfrac{1}{2}n + \tfrac{1}{4};\ \tfrac{1}{4};\ \tfrac{1}{2}z^2\right). \tag{3}$$

Here, $W_{a,\,b}(z)$ is a WHITTAKER FUNCTION and ${}_1F_1(a;\ b;\ z)$ is a CONFLUENT HYPERGEOMETRIC FUNCTIONS. The solutions can also be written as

$$y = e^{-z^2/4}\left[C_1 H_n\left(\frac{z}{\sqrt{2}}\right) + C_2\,{}_1F_1\left(-\tfrac{1}{2}n;\ \tfrac{1}{2};\ \tfrac{1}{2}z^2\right)\right], \tag{4}$$

where $H_n(x)$ is a HERMITE POLYNOMIAL.

Abramowitz and Stegun (1972, p. 686) define the parabolic cylinder functions as solutions to

$$y'' + (ax^2 + bx + c) = 0, \tag{5}$$

sometimes called the PARABOLIC CYLINDER DIFFERENTIAL EQUATION (Zwillinger 1995, p. 414; Zwillinger 1997, p. 126). This can be rewritten by COMPLETING THE SQUARE,

$$y'' + \left[a\left(x + \frac{b}{2a}\right)^2 - \frac{b^2}{4a} + c\right]y = 0. \tag{6}$$

Now letting

$$u = x + \frac{b}{2a} \tag{7}$$

$$du = dx \tag{8}$$

gives

$$\frac{d^2y}{du^2} + (au^2 + d)y = 0 \tag{9}$$

where

$$d \equiv \frac{b^2}{4a} + c. \tag{10}$$

Equation (5) has the two standard forms

$$y'' - \left(\tfrac{1}{4}x^2 + a\right)y = 0 \tag{11}$$

$$y'' + \left(\tfrac{1}{4}x^2 - a\right)y = 0. \tag{12}$$

For a general a, the EVEN and ODD solutions to (11) are

$$y_1(x) = e^{-x^2/4}\,{}_1f_1\left(\tfrac{1}{2}a + \tfrac{1}{4};\ \tfrac{1}{2};\ \tfrac{1}{2}x^2\right) \tag{13}$$

$$y_2(x) = xe^{-x^2/4}\,{}_1f_1\left(\tfrac{1}{2}a + \tfrac{3}{4};\ \tfrac{3}{2};\ \tfrac{1}{2}x^2\right), \tag{14}$$

where ${}_1F_1(a;\ b;\ z)$ is a CONFLUENT HYPERGEOMETRIC FUNCTION. If $y(a,\ x)$ is a solution to (11), then (12) has solutions

$$y(\pm ia,\ xe^{\mp i\pi/4}),\ y(\pm ia,\ -xe^{\mp i\pi/4}). \tag{15}$$

Abramowitz and Stegun (1972, p. 687) define standard solutions to (11) as

$$U(a,\ x) = \cos\left[\pi\left(\tfrac{1}{4} + \tfrac{1}{2}a\right)\right]Y_1 - \sin\left[\pi\left(\tfrac{1}{4} + \tfrac{1}{2}a\right)\right]Y_2 \tag{16}$$

$$V(a,\ x) = \frac{\sin\left[\pi\left(\tfrac{1}{4} + \tfrac{1}{2}a\right)\right]Y_1 + \cos\left[\pi\left(\tfrac{1}{4} + \tfrac{1}{2}a\right)\right]Y_2}{\Gamma\left(\tfrac{1}{2} - a\right)}, \tag{17}$$

where

$$Y_1 \equiv \frac{1}{\sqrt{\pi}}\,\frac{\Gamma\left(\tfrac{1}{4} - \tfrac{1}{2}a\right)}{2^{a/2+1/4}}\,y_1$$

$$= \frac{1}{\sqrt{\pi}} \frac{\Gamma\left(\frac{1}{4} - \frac{1}{2}a\right)}{2^{a/2+1/4}} e^{-x^2/4} \, _1F_1\left(\frac{1}{2}a + \frac{1}{4}; \frac{1}{2}; \frac{1}{2}x^2\right) \quad (18)$$

$$Y_2 \equiv \frac{1}{\sqrt{\pi}} \frac{\Gamma\left(\frac{3}{4} - \frac{1}{2}a\right)}{2^{a/2+1/4}} y_2$$

$$= \frac{1}{\sqrt{\pi}} \frac{\Gamma\left(\frac{3}{4} - \frac{1}{2}a\right)}{2^{a/2+1/4}} xe^{-x^2/4} \, _1F_1\left(\frac{1}{2}a + \frac{3}{4}; \frac{3}{2}; \frac{1}{2}x^2\right) \quad (19)$$

In terms of Whittaker and Watson's functions,

$$U(a, x) = D_{-a-1/2}(x) \quad (20)$$

$$V(a, x) = \frac{\Gamma\left(\frac{1}{2} + a\right)\left[\sin(\pi a)D_{-a-1/2}(x) + D_{-a-1/2}(-x)\right]}{\pi}.$$

$$(21)$$

For NONNEGATIVE INTEGER n, the solution D_n reduces to

$$D_n(x) = 2^{-n/2}e^{-x^2/4}H_n\left(\frac{x}{\sqrt{2}}\right) = e^{-x^2/4}\mathrm{He}_n(x), \quad (22)$$

where $H_n(x)$ is a HERMITE POLYNOMIAL and He^n is a modified HERMITE POLYNOMIAL.

The parabolic cylinder functions D_ν satisfy the RECURRENCE RELATIONS

$$D_{\nu+1}(z) - zD_\nu(z) + \nu D_{\nu-1}(z) = 0 \quad (23)$$

$$D'_\nu(z) + \tfrac{1}{2} zD_\nu(z) - \nu D_{\nu-1}(z) = 0. \quad (24)$$

The parabolic cylinder function for integral n can be defined in terms of an integral by

$$D_n(z) = \frac{1}{\pi}\int_0^\pi \sin(n\theta - z\sin\theta)\, d\theta \quad (25)$$

(Watson 1966, p. 308), which is similar to the ANGER FUNCTION. The result

$$\int_{-\infty}^\infty D_m(x)D_n(x)\, dx = \delta_{mn}n!\sqrt{2\pi}, \quad (26)$$

where δ_{ij} is the KRONECKER DELTA, can also be used to determine the COEFFICIENTS in the expansion

$$f(z) = \sum_{n=0}^\infty a_n D_n \quad (27)$$

as

$$a_n = \frac{1}{n!\sqrt{2\pi}}\int_{-\infty}^\infty D_n(t)f(t)\, dt. \quad (28)$$

For ν real,

$$\int_0^\infty [D_\nu(t)]^2\, dt$$

$$= \pi^{1/2}2^{-3/2}\frac{\phi_0\left(\frac{1}{2} - \frac{1}{2}\nu\right) - \phi_0\left(-\frac{1}{2}\nu\right)}{\Gamma(-\nu)} \quad (29)$$

(Gradshteyn and Ryzhik 2000, p. 885, 7.711.3), where $\Gamma(z)$ is the GAMMA FUNCTION and $\phi_0(z)$ is the POLYGAMMA FUNCTION of order 0.

See also ANGER FUNCTION, BESSEL FUNCTION, DARWIN'S EXPANSIONS, HH FUNCTION, STRUVE FUNCTION

References

Abramowitz, M. and Stegun, C. A. (Eds.). "Parabolic Cylinder Function." Ch. 19 in *Handbook of Mathematical Functions with Formulas, Graphs, and Mathematical Tables, 9th printing*. New York: Dover, pp. 685–700, 1972.

Gradshteyn, I. S. and Ryzhik, I. M. *Tables of Integrals, Series, and Products, 6th ed.* San Diego, CA: Academic Press, 2000.

Iyanaga, S. and Kawada, Y. (Eds.). "Parabolic Cylinder Functions (Weber Functions)." Appendix A, Table 20.III in *Encyclopedic Dictionary of Mathematics*. Cambridge, MA: MIT Press, p. 1479, 1980.

Jeffreys, H. and Jeffreys, B. S. "The Parabolic Cylinder, Hermite, and Hh Functions" et seq. §23.08–23.081 in *Methods of Mathematical Physics, 3rd ed.* Cambridge, England: Cambridge University Press, pp. 620–627, 1988.

Spanier, J. and Oldham, K. B. "The Parabolic Cylinder Function $D_\nu(x)$." Ch. 46 in *An Atlas of Functions*. Washington, DC: Hemisphere, pp. 445–457, 1987.

Watson, G. N. *A Treatise on the Theory of Bessel Functions, 2nd ed.* Cambridge, England: Cambridge University Press, 1966.

Whittaker, E. T. and Watson, G. N. *A Course in Modern Analysis, 4th ed.* Cambridge, England: Cambridge University Press, 1990.

Zwillinger, D. (Ed.). *CRC Standard Mathematical Tables and Formulae*. Boca Raton, FL: CRC Press, p. 414, 1995.

Zwillinger, D. *Handbook of Differential Equations, 3rd ed.* Boston, MA: Academic Press, p. 126, 1997.

Parabolic Cylindrical Coordinates

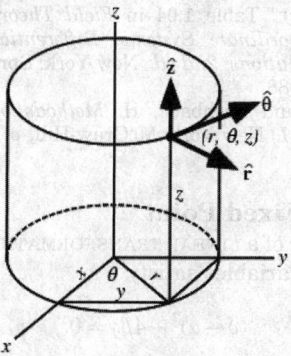

A system of CURVILINEAR COORDINATES. There are several different conventions for the orientation and designation of these coordinates. Arfken (1970) defines coordinates $(\xi,\ \eta,\ z)$ such that

$$x = \xi\eta \tag{1}$$

$$y = \tfrac{1}{2}(\eta^2 - \xi^2) \tag{2}$$

$$z = z. \tag{3}$$

In this work, following Morse and Feshbach (1953), the coordinates $(u,\ v,\ z)$ are used instead. In this convention, the traces of the coordinate surfaces of the xy-PLANE are confocal PARABOLAS with a common axis. The u curves open into the NEGATIVE X-AXIS; the v curves open into the POSITIVE X-AXIS. The u and v curves intersect along the Y-AXIS.

$$x = \tfrac{1}{2}(u^2 - v^2) \tag{4}$$

$$y = uv \tag{5}$$

$$z = z, \tag{6}$$

where $u \in [0,\ \infty)$, $v \in [0,\ \infty)$, and $z \in (-\infty,\ \infty)$. The SCALE FACTORS are

$$h_1 = \sqrt{u^2 + v^2} \tag{7}$$

$$h_2 = \sqrt{u^2 + v^2} \tag{8}$$

$$h_3 = 1. \tag{9}$$

LAPLACE'S EQUATION is

$$\nabla^2 f = \frac{1}{u^2 + v^2}\left(\frac{\partial^2 f}{\partial u^2} + \frac{\partial^2 f}{\partial v^2}\right) + \frac{\partial^2 f}{\partial z^2}. \tag{10}$$

The HELMHOLTZ DIFFERENTIAL EQUATION is SEPARABLE in parabolic cylindrical coordinates.

See also CONFOCAL PARABOLOIDAL COORDINATES, HELMHOLTZ DIFFERENTIAL EQUATION–PARABOLIC CYLINDRICAL COORDINATES, PARABOLIC COORDINATES

References

Arfken, G. "Parabolic Cylinder Coordinates $(\xi,\ \eta,\ z)$." §2.8 in *Mathematical Methods for Physicists, 2nd ed.* Orlando, FL: Academic Press, p. 97, 1970.
Moon, P. and Spencer, D. E. "Parabolic-Cylinder Coordinates $(\mu,\ v,\ z)$." Table 1.04 in *Field Theory Handbook, Including Coordinate Systems, Differential Equations, and Their Solutions, 2nd ed.* New York: Springer-Verlag, pp. 21–24, 1988.
Morse, P. M. and Feshbach, H. *Methods of Theoretical Physics, Part I.* New York: McGraw-Hill, p. 658, 1953.

Parabolic Fixed Point

A FIXED POINT of a LINEAR TRANSFORMATION for which the rescaled variables satisfy

$$(\delta - \alpha)^2 + 4\beta\gamma = 0.$$

See also ELLIPTIC FIXED POINT (MAP), HYPERBOLIC FIXED POINT (MAP), LINEAR TRANSFORMATION

Parabolic Geometry
EUCLIDEAN GEOMETRY

Parabolic Horn Cyclide

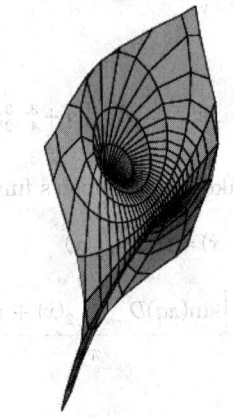

A PARABOLIC CYCLIDE formed by INVERSION of a HORN TORUS when the INVERSION SPHERE is tangent to the TORUS.

See also CYCLIDE, INVERSION, INVERSION SPHERE, PARABOLIC RING CYCLIDE, PARABOLIC SPINDLE CYCLIDE

Parabolic Partial Differential Equation

A PARTIAL DIFFERENTIAL EQUATION of second-order, i.e., one OF THE FORM

$$Au_{xx} + 2Bu_{xy} + Cu_{yy} + Du_x + Eu_y + F = 0, \tag{1}$$

is called parabolic if the MATRIX

$$Z \equiv \begin{bmatrix} A & B \\ B & C \end{bmatrix} \tag{2}$$

satisfies $\det(Z) = 0$. The HEAT CONDUCTION EQUATION and other diffusion equations are examples. Initial-boundary conditions are used to give

$$u(x,\ t) = g(x,\ t) \quad \text{for } x \in \partial\Omega,\ t > 0 \tag{3}$$

$$u(x,\ 0) = v(x) \quad \text{for } x \in \Omega, \tag{4}$$

where

$$u_{xx} = f(u_x,\ u_y,\ u,\ x,\ y) \tag{5}$$

holds in Ω.

See also BOUNDARY CONDITIONS, BOUNDARY VALUE PROBLEM, ELLIPTIC PARTIAL DIFFERENTIAL EQUATION, HYPERBOLIC PARTIAL DIFFERENTIAL EQUATION, INITIAL VALUE PROBLEM, PARTIAL DIFFERENTIAL EQUATION

Parabolic Point

A point $\mathbf{p}$ on a REGULAR SURFACE $M \in \mathbb{R}^3$ is said to be parabolic if the GAUSSIAN CURVATURE $K(\mathbf{p}) = 0$ but $S(\mathbf{p}) \neq 0$ (where S is the SHAPE OPERATOR), or equivalently, exactly one of the PRINCIPAL CURVATURES κ_1 and κ_2 is 0.

See also ANTICLASTIC, ELLIPTIC POINT, GAUSSIAN CURVATURE, HYPERBOLIC POINT, PLANAR POINT, SYNCLASTIC

References

Gray, A. *Modern Differential Geometry of Curves and Surfaces with Mathematica, 2nd ed.* Boca Raton, FL: CRC Press, p. 375, 1997.

Parabolic Ring Cyclide

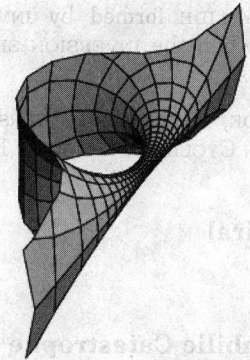

A PARABOLIC CYCLIDE formed by INVERSION of a RING TORUS when the INVERSION SPHERE is tangent to the TORUS.

See also CYCLIDE, INVERSION, INVERSION SPHERE, PARABOLIC HORN CYCLIDE, PARABOLIC SPINDLE CY-CLIDE

Parabolic Rotation

The MAP

$$x' = x + 1 \tag{1}$$

$$y' = 2x + y + 1, \tag{2}$$

which leaves the PARABOLA

$$x'^2 - y' = (x+1)^2 - (2x+y+1) = x^2 - y \tag{3}$$

invariant.

See also PARABOLA, ROTATION

Parabolic Rule

SIMPSON'S RULE

Parabolic Segment

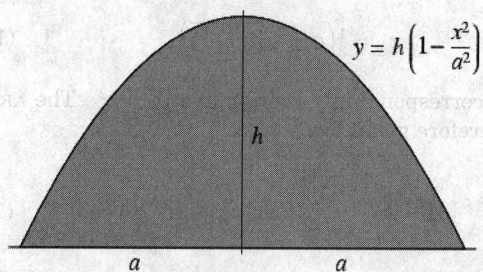

The ARC LENGTH of the parabolic segment

$$y = h\left(1 - \frac{x^2}{a^2}\right) \tag{1}$$

illustrated above is given by

$$s = \int_{-a}^{a} \sqrt{1 + y'^2}\, dx = 2\int_{0}^{a} \sqrt{1 + y'^2}\, dx \tag{2}$$

$$= \frac{1}{2}\sqrt{a^2 + 4h^2} + \frac{a^2}{4h}\ln\left(\frac{2h + \sqrt{a^2 + 4h^2}}{a}\right), \tag{3}$$

and the AREA is given by

$$A = \int_{-a}^{a} = h\left(1 - \frac{x^2}{a^2}\right) dx = \tfrac{4}{3}ah \tag{4}$$

(Kern and Bland 1948, p. 4). The weighted mean of y is

$$\langle y \rangle = \text{int}_{-a}^{a} \int_{0}^{h\left(1 - x^2/a^2\right)} y\, dx\, dy = \tfrac{8}{15}ah^2, \tag{5}$$

so the CENTROID is then given by

$$\bar{y} = \frac{\langle y \rangle}{A} = \tfrac{2}{5}h. \tag{6}$$

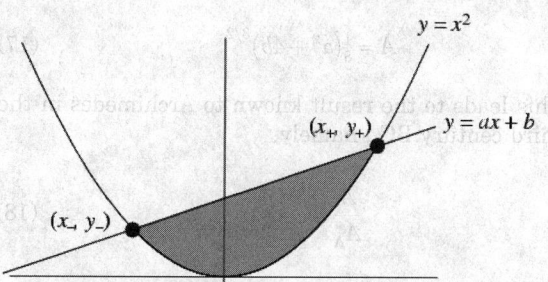

The AREA of the cut-off parabolic segment contained between the curves

$$y = x^2 \tag{7}$$

$$y = ax + b \tag{8}$$

can be found by eliminating y,

$$x^2 - ax - b = 0, \tag{9}$$

so the points of intersection are

$$x_\pm = \tfrac{1}{2}\left(a \pm \sqrt{a^2 + 4b}\right), \qquad (10)$$

with corresponding y-coordinates $y_\pm = x_\pm^2$. The AREA is therefore given by

$$A = \int_{a-\sqrt{a^2+4b}}^{a+\sqrt{a^2+4b}} \left[(ax+b) - x^2\right] dx \qquad (11)$$

$$= \tfrac{1}{6}(a^2 + 4b)\sqrt{a^2 + 4b} = \tfrac{1}{6}(a^2 + 4b)^{3/2}. \qquad (12)$$

The maximum AREA of a TRIANGLE inscribed in this segment will have two of its VERTICES at the intersections (x_-, y_-) and (x_+, y_+), and the third at a point (x^*, y^*) to be determined. From the general equation for a triangle, the AREA of the inscribed triangle is given by the DETERMINANT equation

$$A_\Delta = \begin{vmatrix} x^- & y^- & 1 \\ x^+ & y^+ & 1 \\ x^* & y^* & 1 \end{vmatrix}. \qquad (13)$$

Plugging in and using $y_* = x_*^2$ gives

$$A_\Delta = \tfrac{1}{2}[b + (a - x^*)x^*]\sqrt{a^2 + 4b}. \qquad (14)$$

To find the maximum AREA, differentiable with respect to x^* and set to 0 to obtain

$$\frac{\partial A_\Delta}{\partial x_*} = \tfrac{1}{2}(a - 2x^*)\sqrt{a^2 + 4b} = 0, \qquad (15)$$

so

$$x_* = \tfrac{1}{2}a. \qquad (16)$$

Plugging (16) into (14) then gives

$$A = \tfrac{1}{8}(a^2 + 4b)^{3/2}. \qquad (17)$$

This leads to the result known to Archimedes in the third century BC , namely

$$\frac{A}{A_\Delta} = \frac{\frac{1}{6}}{\frac{1}{8}} = \frac{4}{3}. \qquad (18)$$

See also CENTROID (GEOMETRIC), PARABOLA, SEGMENT

References

Beyer, W. H. (Ed.). *CRC Standard Mathematical Tables,* *28th ed.* Boca Raton, FL: CRC Press, p. 125, 1987.
Kern, W. F. and Bland, J. R. *Solid Mensuration with Proofs,* *2nd ed.* New York: Wiley, p. 4, 1948.

Parabolic Spindle Cyclide

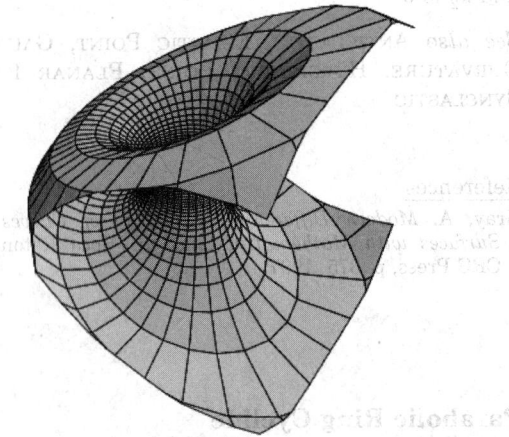

A PARABOLIC CYCLIDE formed by INVERSION of a SPINDLE TORUS when the INVERSION SPHERE is tangent to the TORUS.

See also CYCLIDE, INVERSION, INVERSION SPHERE, PARABOLIC HORN CYCLIDE, PARABOLIC RING CYCLIDE

Parabolic Spiral

FERMAT'S SPIRAL

Parabolic Umbilic Catastrophe

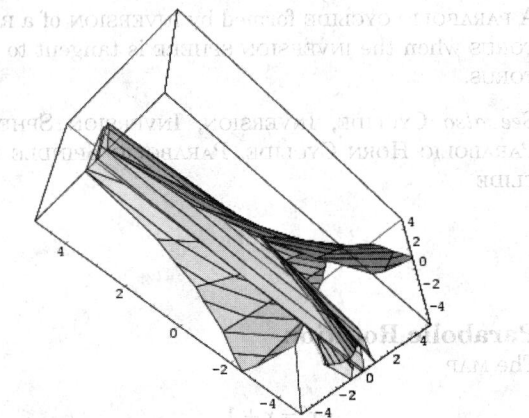

A CATASTROPHE which can occur for four control factors and two behavior axes. The parabolic umbilic catastrophe is given by the unfolding $F(x, y, w, t, u, v) = y^4 + x^2 y + ux^2 + vy^2 + wx + ty$ of $f(x, y) = y^4 + x^2 y$.

See also CATASTROPHE THEORY

References

Sanns, W. *Catastrophe Theory with Mathematica: A Geometric Approach.* Germany: DAV, 2000.

Parabolic-Cylinder Coordinates

PARABOLIC CYLINDRICAL COORDINATES

Paraboloid

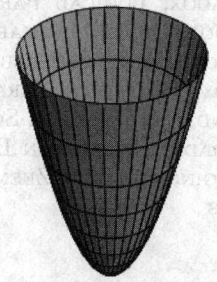

The SURFACE OF REVOLUTION of the PARABOLA which is the shape used in the reflectors of automobile headlights (Steinhaus 1983, p. 242; Hilbert and Cohn-Vossen 1999). It is a QUADRATIC SURFACE which can be specified by the Cartesian equation

$$z = b(x^2 + y^2). \tag{1}$$

The paraboloid which has radius a at height h is then given parametrically by

$$x(u, v) = a\sqrt{u/h}\,\cos v \tag{2}$$

$$y(u, v) = a\sqrt{u/h}\,\sin v \tag{3}$$

$$z(u, v) = u, \tag{4}$$

where $u \geq 0$, $v \in [0,\, 2\pi)$.

The coefficients of the FIRST FUNDAMENTAL FORM are given by

$$E = 1 + \frac{a^2}{4hu} \tag{5}$$

$$F = 0 \tag{6}$$

$$G = \frac{a^2 u}{h} \tag{7}$$

and the SECOND FUNDAMENTAL FORM coefficients are

$$e = \frac{a^2}{2u\sqrt{a^4 + 4a^2 hu}} \tag{8}$$

$$f = 0 \tag{9}$$

$$g = \frac{2a^2 u}{\sqrt{a^4 + 4a^2 hu}} \tag{10}$$

The AREA ELEMENT is then

$$dS = \frac{\sqrt{a^4 + 4a^2 hu}}{2h}\, du \wedge dv, \tag{11}$$

giving SURFACE AREA

$$S = \int_0^{2\pi} \int_0^h dS = \frac{\pi a}{6h^2}\left[\left(a^2 + 4h^2\right)^{3/2} - a^3\right]. \tag{12}$$

The GAUSSIAN CURVATURE is given by

$$K = \frac{4h^2}{\left(a^2 + 4hu\right)^2}, \tag{13}$$

and the MEAN CURVATURE

$$H = \frac{2h(a^2 + 2hu)}{(a^2 + 4hu)\sqrt{a^4 + 4a^2 hu}}. \tag{14}$$

The VOLUME of the paraboloid of height h is then

$$V = \pi \int_0^h \frac{a^2 z}{h}\, dz = \tfrac{1}{2}\pi a^2 h. \tag{15}$$

The weighted mean of z over the paraboloid is

$$\langle z \rangle = \pi \int_0^h \frac{a^2 z}{h} z\, dz = \tfrac{1}{3}\pi a^2 h^2. \tag{16}$$

The CENTROID is then given by

$$\bar{z} = \frac{\langle z \rangle}{V} = \tfrac{2}{3}h \tag{17}$$

(Beyer 1987).

See also ELLIPTIC PARABOLOID, HYPERBOLIC PARABOLOID, PARABOLA

References

Beyer, W. H. (Ed.). *CRC Standard Mathematical Tables, 28th ed.* Boca Raton, FL: CRC Press, p. 133, 1987.

Gray, A. "The Paraboloid." §13.5 in *Modern Differential Geometry of Curves and Surfaces with Mathematica, 2nd ed.* Boca Raton, FL: CRC Press, pp. 307–308, 1997.

Harris, J. W. and Stocker, H. "Paraboloid of Revolution." §4.10.2 in *Handbook of Mathematics and Computational Science.* New York: Springer-Verlag, p. 112, 1998.

Hilbert, D. and Cohn-Vossen, S. *Geometry and the Imagination.* New York: Chelsea, pp. 10–11, 1999.

Steinhaus, H. *Mathematical Snapshots, 3rd ed.* New York: Dover, 1999.

Paraboloid Geodesic

A GEODESIC on a PARABOLOID has differential parameters defined by

$$P \equiv \left(\frac{\partial x}{\partial u}\right)^2 + \left(\frac{\partial y}{\partial u}\right)^2 + \left(\frac{\partial z}{\partial u}\right)^2$$

$$= 1 + \frac{\cos^2 v}{4u} + \frac{\sin^2 v}{4u} = 1 + \frac{1}{4u} \tag{1}$$

$$Q \equiv \frac{\partial^2 x}{\partial u\, \partial v} + \frac{\partial^2 y}{\partial u\, \partial v} + \frac{\partial^2 z}{\partial u\, \partial v}$$

$$= 0 + u\cos^2 v + u\sin^2 v = u \tag{2}$$

$$R \equiv 0 - \frac{\sin v}{2\sqrt{u}} + \frac{\cos v}{2\sqrt{u}} = \frac{1}{2\sqrt{u}}(\cos v - \sin v). \tag{3}$$

The GEODESIC is then given by solving the EULER-LAGRANGE DIFFERENTIAL EQUATION

$$\frac{\frac{\partial P}{\partial v}+2v'\frac{\partial Q}{\partial v}+v'^2\frac{\partial R}{\partial v}}{2\sqrt{P+2Qv'+Rv'^2}}-\frac{d}{du}\left(\frac{Q+Rv'}{\sqrt{P+2Qv'+Rv'^2}}\right)=0.$$

(4)

As given by Weinstock (1974), the solution simplifies to

$$u-c^2=u(1+4c^2)$$
$$\times\sin^2\left\{v-2c\ln\left[k\left(2\sqrt{u-c^2}+\sqrt{4u+1}\right)\right]\right\}.$$

(5)

See also GEODESIC

References

Weinstock, R. *Calculus of Variations, with Applications to Physics and Engineering.* New York: Dover, p. 45, 1974.

Paraboloidal Coordinates

CONFOCAL PARABOLOIDAL COORDINATES

Paracompact Space

A paracompact space is a HAUSDORFF SPACE such that every open COVER has a LOCALLY FINITE open REFINEMENT. Paracompactness is a very common property that TOPOLOGICAL SPACES satisfy. Paracompactness is similar to the compactness property, but generalized for slightly "bigger" SPACES. All MANIFOLDS (e.g, second countable and Hausdorff) are paracompact.

See also HAUSDORFF SPACE, LOCALLY FINITE SPACE, MANIFOLD, TOPOLOGICAL SPACE

Paracycle

ASTROID

Paradox

A statement which appears self-contradictory or contrary to expectations, also known as an ANTINOMY. Curry (1977, p. 5) uses the term PSEUDOPARADOX to describe an apparent paradox for which, however, there is no underlying actual contradiction. Bertrand Russell classified known logical paradoxes into seven categories.

Ball and Coxeter (1987) give several examples of geometrical paradoxes.

See also ALLAIS PARADOX, ARISTOTLE'S WHEEL PARADOX, ARROW'S PARADOX, BANACH-TARSKI PARADOX, BARBER PARADOX, BERNOULLI'S PARADOX, BERRY PARADOX, BERTRAND'S PARADOX, BUCHOWSKI PARADOX, BURALI-FORTI PARADOX, CANTOR'S PARADOX, CATALOGUE PARADOX, COASTLINE PARADOX, COIN PARADOX, ELEVATOR PARADOX, EPIMENIDES PARADOX, EUBULIDES PARADOX, GRELLING'S PARADOX, HAUSDORFF PARADOX, HEMPEL'S PARADOX, HETERO-

LOGICAL PARADOX, HYPERGAME, LEONARDO'S PARADOX, LIAR'S PARADOX, LOGICAL PARADOX, POTATO PARADOX, PSEUDOPARADOX, RICHARD'S PARADOX, RUSSELL'S PARADOX, SAINT PETERSBURG PARADOX, SIEGEL'S PARADOX, SIMPSON'S PARADOX, SKOLEM PARADOX, SMARANDACHE PARADOX, SOCRATES' PARADOX, SORITES PARADOX, THOMPSON LAMP PARADOX, UNEXPECTED HANGING PARADOX, ZEEMAN'S PARADOX, ZENO'S PARADOXES

References

Ball, W. W. R. and Coxeter, H. S. M. *Mathematical Recreations and Essays, 13th ed.* New York: Dover, pp. 84–86, 1987.
Bunch, B. *Mathematical Fallacies and Paradoxes.* New York: Dover, 1982.
Carnap, R. *Introduction to Symbolic Logic and Its Applications.* New York: Dover, 1958.
Church, A. "Paradoxes, Logical." In *The Dictionary of Philosophy, rev. enl. ed.* (Ed. D. D Runes). New York: Rowman and Littlefield, p. 224, 1984.
Curry, H. B. *Foundations of Mathematical Logic.* New York: Dover, 1977.
Czyz, J. *Paradoxes of Measures and Dimensions Originating in Felix Hausdorff's Ideas.* Singapore: World Scientific, 1994.
Erickson, G. W. and Fossa, J. A. *Dictionary of Paradox.* Lanham, MD: University Press of America, 1998.
Kasner, E. and Newman, J. R. "Paradox Lost and Paradox Regained." In *Mathematics and the Imagination.* Redmond, WA: Tempus Books, pp. 193–222, 1989.
Northrop, E. P. *Riddles in Mathematics: A Book of Paradoxes.* Princeton, NJ: Van Nostrand, 1944.
O'Beirne, T. H. *Puzzles and Paradoxes.* New York: Oxford University Press, 1965.
Quine, W. V. "Paradox." *Sci. Amer.* **206**, 84–96, Apr. 1962.

Paradromic Rings

Rings produced by cutting a strip that has been given m half twists and been re-attached into n equal strips (Ball and Coxeter 1987, pp. 127–128).

See also MÖBIUS STRIP

References

Ball, W. W. R. and Coxeter, H. S. M. *Mathematical Recreations and Essays, 13th ed.* New York: Dover, pp. 127–128, 1987.

Paragyrate Diminished Rhombicosidodecahedron

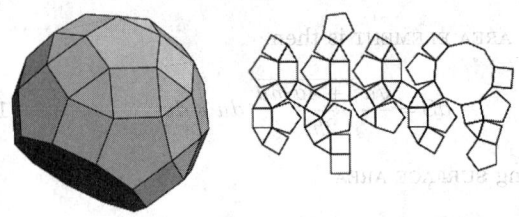

JOHNSON SOLID J_{77}.

Parallel

References

Weisstein, E. W. "Johnson Solids." MATHEMATICA NOTEBOOK JOHNSONSOLIDS.M.
Weisstein, E. W. "Johnson Solid Netlib Database." MATHEMATICA NOTEBOOK JOHNSONSOLIDS.DAT.

Parallel

Two lines in 2-dimensional EUCLIDEAN SPACE are said to be parallel if they do not intersect. In 3-dimensional EUCLIDEAN SPACE, parallel lines not only fail to intersect, but also maintain a constant separation between points closest to each other on the two lines. (Lines in 3-space which are not parallel but do not intersect are called SKEW LINES.)

In a NON-EUCLIDEAN GEOMETRY, the concept of parallelism must be modified from its intuitive meaning. This is accomplished by changing the so-called PARALLEL POSTULATE. While this has counterintuitive results, the geometries so defined are still completely self-consistent.

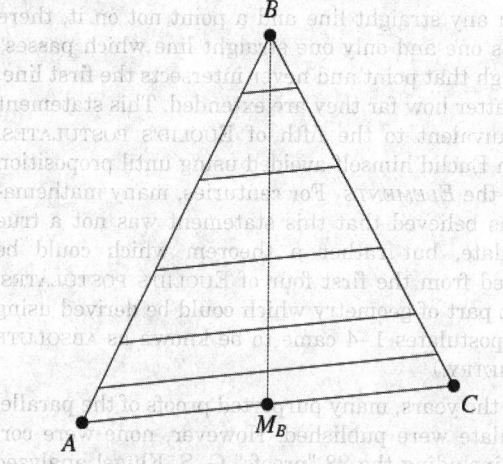

In a TRIANGLE ΔABC, a MEDIAN BM_B bisects all segments parallel to a given side AC (Honsberger 1995, p. 87).

See also ABSOLUTE GEOMETRY, ANTIPARALLEL, HYPERPARALLEL, LINE, NON-EUCLIDEAN GEOMETRY, PARALLEL CURVES, PARALLEL LINE AND PLANE, PARALLEL LINES, PARALLEL PLANES, PARALLEL POSTULATE PERPENDICULAR, SKEW LINES

References

Honsberger, R. "Parallels and Antiparallels." §9.1 in *Episodes in Nineteenth and Twentieth Century Euclidean Geometry*. Washington, DC: Math. Assoc. Amer., pp. 87–88, 1995.
Kern, W. F. and Bland, J. R. *Solid Mensuration with Proofs*, 2nd ed. New York: Wiley, p. 9, 1948.

Parallel (Surface of Revolution)

A parallel of a SURFACE OF REVOLUTION is the intersection of the surface with a PLANE orthogonal to the axis of revolution.

See also MERIDIAN, SURFACE OF REVOLUTION

References

Gray, A. *Modern Differential Geometry of Curves and Surfaces with Mathematica, 2nd ed.* Boca Raton, FL: CRC Press, p. 458, 1997.

Parallel Axiom

PARALLEL POSTULATE

Parallel Class

A set of blocks, also called a RESOLUTION CLASS, that partition the set V, where (V, B) is a balanced incomplete BLOCK DESIGN.

See also BLOCK DESIGN, RESOLVABLE

References

Abel, R. J. R. and Furino, S. C. "Resolvable and Near Resolvable Designs." §I.6 in *The CRC Handbook of Combinatorial Designs* (Ed. C. J. Colbourn and J. H. Dinitz). Boca Raton, FL: CRC Press, pp. 87–94, 1996.

Parallel Curves

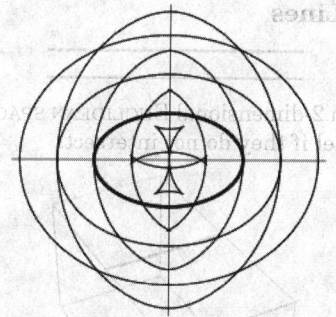

Parallel curves, frequently called "offset curves" in computer graphics applications, are curves which are displaced from a base curve by a constant offset, either positive or negative, in the direction of the curve's normal. The two branches of the parallel curve a distance k away from a parametrically represented base curve $(f(t), g(t))$ are

$$x = f \pm \frac{kg'}{\sqrt{f'^2 + g'^2}}$$

$$y = g \mp \frac{kf'}{\sqrt{f'^2 + g'^2}},$$

where $f' = df/dt$ and $g' = dg/dt$. The above figure shows curves parallel to a CIRCLE, ELLIPSE, and 3-

petalled ROSE, where the base curves are indicated in red.

See also PARALLEL, PARALLEL LINES

References

Gray, A. "Parallel Curves." §5.7 in *Modern Differential Geometry of Curves and Surfaces with Mathematica, 2nd ed.* Boca Raton, FL: CRC Press, pp. 115–117, 1997.
Lawrence, J. D. *A Catalog of Special Plane Curves.* New York: Dover, pp. 42–43, 1972.
Yates, R. C. "Parallel Curves." *A Handbook on Curves and Their Properties.* Ann Arbor, MI: J. W. Edwards, pp. 155–159, 1952.

Parallel Line and Plane

A line and a plane are parallel if they do not intersect.

See also PARALLEL, PARALLEL LINES, PARALLEL PLANES

References

Kern, W. F. and Bland, J. R. *Solid Mensuration with Proofs, 2nd ed.* New York: Wiley, p. 9, 1948.

Parallel Lines

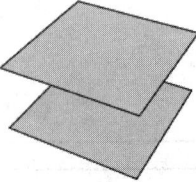

Two lines in 2-dimensional EUCLIDEAN SPACE are said to be parallel if they do not intersect.

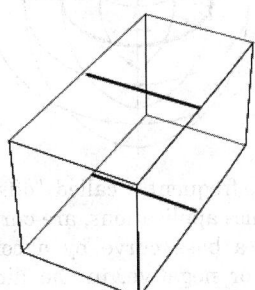

In 3-dimensional EUCLIDEAN SPACE, parallel lines not only fail to intersect, but also maintain a constant separation between points closest to each other on the two lines. Therefore, parallel lines in 3-space lie in a single PLANE (Kern and Blank 1948, p. 9). Lines in 3-space which are not parallel but do not intersect are called SKEW LINES.

See also PARALLEL, PARALLEL CURVES, PARALLEL LINE AND PLANE, PARALLEL PLANES, PARALLEL POSTULATE, SKEW LINES

References

Kern, W. F. and Bland, J. R. *Solid Mensuration with Proofs, 2nd ed.* New York: Wiley, p. 9, 1948.

Parallel Planes

A second image showing two parallel planes appears here.

Two planes that do not intersect are said to be parallel.

See also PARALLEL, PARALLEL LINES, PARALLEL PLANES, PLANE

References

Kern, W. F. and Bland, J. R. *Solid Mensuration with Proofs, 2nd ed.* New York: Wiley, p. 9, 1948.

Parallel Postulate

Portions of this entry contributed by MATTHEW SZUDZIK

Given any straight line and a point not on it, there "exists one and only one straight line which passes" through that point and never intersects the first line, no matter how far they are extended. This statement is equivalent to the fifth of EUCLID'S POSTULATES, which Euclid himself avoided using until proposition 29 in the *ELEMENTS*. For centuries, many mathematicians believed that this statement was not a true postulate, but rather a theorem which could be derived from the first four of EUCLID'S POSTULATES. (That part of geometry which could be derived using only postulates 1–4 came to be known as ABSOLUTE GEOMETRY.)

Over the years, many purported proofs of the parallel postulate were published. However, none were correct, including the 28 "proofs" G. S. Klügel analyzed in his dissertation of 1763 (Hofstadter 1989). The main motivation for all of this effort was that Euclid's parallel postulate did not seem as "intuitive" as the other axioms, but it was needed to prove important results. John Wallis proposed a new axiom that implied the parallel postulate and was also intuitively appealing. His "axiom" states that any triangle can be made bigger or smaller without distorting its proportions or angles (Greenberg 1994, pp. 152–153). However, Wallis's axiom never caught on.

In 1823, Janos Bolyai and Lobachevsky independently realized that entirely self-consistent "NON-EUCLIDEAN GEOMETRIES" could be created in which the parallel postulate *did not hold.* (Gauss had also discovered but suppressed the existence of non-Euclidean geometries.)

As stated above, the parallel postulate describes the type of geometry now known as PARABOLIC GEOMETRY. If, however, the phrase "exists one and only one straight line which passes" is replaced by "exist no line which passes," or "exist at least two lines which pass," the postulate describes equally valid (though less intuitive) types of geometries known as ELLIPTIC and HYPERBOLIC GEOMETRIES, respectively.

The parallel postulate is equivalent to the EQUI-DISTANCE POSTULATE, PLAYFAIR'S AXIOM, PROCLUS' AXIOM, the TRIANGLE POSTULATE, and the PYTHAGOREAN THEOREM. There is also a single parallel axiom in HILBERT'S AXIOMS which is equivalent to Euclid's parallel postulate.

S. Brodie has shown that the parallel postulate is equivalent to the PYTHAGOREAN THEOREM.

See also ABSOLUTE GEOMETRY, EUCLID'S AXIOMS, EUCLIDEAN GEOMETRY, HILBERT'S AXIOMS, NON-EUCLIDEAN GEOMETRY, PLAYFAIR'S AXIOM, PYTHAGOREAN THEOREM, TRIANGLE POSTULATE

References

Brodie, S. E. "The Pythagorean Theorem Is Equivalent to the Parallel Postulate." http://www.cut-the-knot.com/triangle/pythpar/PTimpliesPP.html.
Dixon, R. *Mathographics.* New York: Dover, p. 27, 1991.
Greenberg, M. J. *Euclidean and Non-Euclidean Geometries: Development and History, 3rd ed.* San Francisco, CA: W. H. Freeman, 1994.
Hilbert, D. *The Foundations of Geometry, 2nd ed.* Chicago, IL: Open Court, 1980.
Hofstadter, D. R. *Gödel, Escher, Bach: An Eternal Golden Braid.* New York: Vintage Books, pp. 88–92, 1989.
Iyanaga, S. and Kawada, Y. (Eds.). "Hilbert's System of Axioms." §163B in *Encyclopedic Dictionary of Mathematics.* Cambridge, MA: MIT Press, pp. 544–545, 1980.

Parallelepiped

In 3-D, a parallelepiped is a PRISM whose faces are all PARALLELOGRAMS. The volume of a 3-D parallelepiped is given by the SCALAR TRIPLE PRODUCT

$$V_{\text{parallelepiped}} = |\mathbf{A} \cdot (\mathbf{B} \times \mathbf{C})|$$

$$= |\mathbf{C} \cdot (\mathbf{A} \times \mathbf{B})| = |\mathbf{B} \cdot (\mathbf{C} \times \mathbf{A})|.$$

In n-D, a parallelepiped is the POLYTOPE spanned by n VECTORS $\mathbf{v}_1, ..., \mathbf{v}_n$ in a VECTOR SPACE over the reals,

$$\text{span}(\mathbf{v}_1, ..., \mathbf{v}_n) = t_1 \mathbf{v}_1 + ... + t_n \mathbf{v}_n,$$

where $t_i \in [0, 1]$ for $i = 1, ..., n$. In the usual interpretation, the VECTOR SPACE is taken as EUCLIDEAN SPACE, and the CONTENT of this parallelepiped is given by

$$\text{abs}(\det(\mathbf{v}_1, ..., \mathbf{v}_n)),$$

where the sign of the determinant is taken to be the "orientation" of the "oriented volume" of the parallelepiped.

Given k vectors $v_1, ..., v_k$ in n-dimensional space, their CONVEX HULL (along with the ZERO VECTOR)

$$\left\{ \sum t_i v_i \,\middle|\, 0 \le t_i \le 1 \right\} \tag{1}$$

is called a parallelepiped, generalizing the notion of a parallelogram, or rather its interior, in the plane. If the number of vectors is equal to the dimension, then

$$\mathsf{A} = (v_1 \ldots v_k) \tag{2}$$

is a SQUARE MATRIX, and the volume of the parallelepiped is given by $|\det \mathsf{A}|$, where the columns of A are given by the vectors v. More generally, a parallelepiped has k dimensional volume given by $\left| \det \mathsf{A}^T \mathsf{A} \right|^{1/2}$.

When the vectors are TANGENT VECTORS, then the parallelepiped represents an infinitesimal k-dimensional VOLUME ELEMENT. Integrating this volume can give formulas for the volumes of k-dimensional objects in n-dimensional space. More intrinsically, the parallelepiped corresponds to a DECOMPOSABLE element of the EXTERIOR ALGEBRA $\Lambda^k \mathbb{R}^n$.

See also DETERMINANT, DIFFERENTIAL K-FORM, EXTERIOR ALGEBRA, PARALLELOGRAM, PRISMATOID, RECTANGULAR PARALLELEPIPED, VOLUME ELEMENT, VOLUME INTEGRAL, ZONOHEDRON

References

Phillips, A. W. and Fisher, I. *Elements of Geometry.* New York: Amer. Book Co., 1896.

Parallelism

ANGLE OF PARALLELISM

Parallelizable

A HYPERSPHERE $\mathbb{S}^n$ is parallelizable if there exist n cuts containing linearly independent tangent vectors. There exist only three parallelizable spheres: $\mathbb{S}^1$, $\mathbb{S}^2$, and $\mathbb{S}^7$ (Adams 1962, Le Lionnais 1983).

See also SPHERE

References

Adams, J. F. "On the Non-Existence of Elements of Hopf Invariant One." *Bull. Amer. Math. Soc.* **64**, 279–282, 1958.
Adams, J. F. "On the Non-Existence of Elements of Hopf Invariant One." *Ann. Math.* **72**, 20–104, 1960.
Le Lionnais, F. *Les nombres remarquables.* Paris: Hermann, p. 49, 1983.

Parallelogram

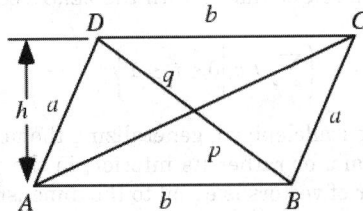

A QUADRILATERAL with opposite sides parallel (and therefore opposite angles equal). A quadrilateral with equal sides is called a RHOMBUS, and a parallelogram whose ANGLES are all RIGHT ANGLES is called a RECTANGLE. The DIAGONALS of a parallelogram bisect each other (Casey 1888, p. 2).

A parallelogram of base b and height h has AREA

$$A = bh = ab \sin A = ab \sin B. \tag{1}$$

The height of a parallelogram is

$$h = a \sin A = a \sin B, \tag{2}$$

and the DIAGONALS p and q are

$$p = \sqrt{a^2 + b^2 - 2ab \cos A} \tag{3}$$

$$q = \sqrt{a^2 + b^2 - 2ab \cos B} \tag{4}$$

$$= \sqrt{a^2 + b^2 + 2ab \cos A} \tag{5}$$

(Beyer 1987).

The sides a, b, c, d and diagonals p, q of a parallelogram satisfy

$$p^2 + q^2 = a^2 + b^2 + c^2 + d^2 \tag{6}$$

(Casey 1888, p. 22).

The AREA of the parallelogram with sides formed by the VECTORS (a, c) and (b, d) is

$$A = \det\left(\begin{bmatrix} a & b \\ c & d \end{bmatrix}\right) = |ad - bc|. \tag{7}$$

Given a parallelogram P with area $A(P)$ and linear transformation T, the AREA of $T(P)$ is

$$A(T(P)) = \begin{vmatrix} a & b \\ c & d \end{vmatrix} A(P). \tag{8}$$

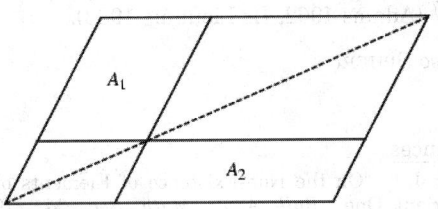

As shown by Euclid, if lines parallel to the sides are drawn through any point on a diagonal of a parallelogram, then the parallelograms not containing segments of that diagonal are equal in AREA (and

conversely), so in the above figure, $A_1 = A_2$ (Johnson 1929).

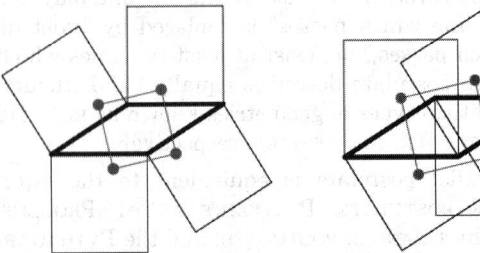

The centers of four SQUARES erected either internally or externally on the sides of a parallelograms are the vertices of a SQUARE (Yaglom 1962, pp. 96–97; Coxeter and Greitzer 1967, p. 84).

See also DIAMOND, LOZENGE, PARALLELOGRAM ILLUSION, PARALLELOGRAM LAW, QUADRILATERAL, RECTANGLE, RHOMBUS, SQUARE, VARIGNON PARALLELOGRAM, WITTENBAUER'S PARALLELOGRAM

References

Beyer, W. H. (Ed.). *CRC Standard Mathematical Tables, 28th ed.* Boca Raton, FL: CRC Press, p. 123, 1987.

Casey, J. *A Sequel to the First Six Books of the Elements of Euclid, Containing an Easy Introduction to Modern Geometry with Numerous Examples, 5th ed., rev. enl.* Dublin: Hodges, Figgis, & Co., 1888.

Coxeter, H. S. M. and Greitzer, S. L. *Geometry Revisited.* Washington, DC: Math. Assoc. Amer., p. 84, 1967.

Harris, J. W. and Stocker, H. "Parallelogram." §3.6.3 in *Handbook of Mathematics and Computational Science.* New York: Springer-Verlag, p. 83, 1998.

Kern, W. F. and Bland, J. R. *Solid Mensuration with Proofs, 2nd ed.* New York: Wiley, p. 3, 1948.

Johnson, R. A. *Modern Geometry: An Elementary Treatise on the Geometry of the Triangle and the Circle.* Boston, MA: Houghton Mifflin, p. 61, 1929.

Yaglom, I. M. *Geometric Transformations I.* New York: Random House, pp. 96–97, 1962.

Parallelogram Illusion

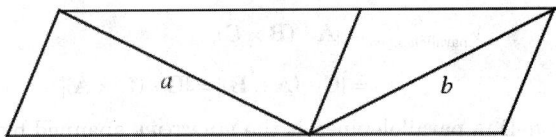

In the above figure, the sides a and b have the same length, appearances to the contrary.

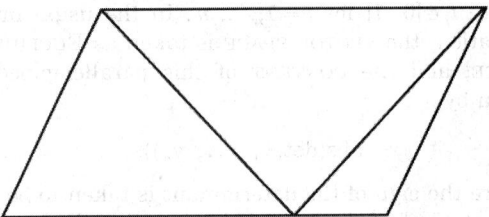

In the related illusion illustrated above, the interior

lines appear to be of different lengths, despite the fact that they are the same (Wells 1991).

References

Wells, D. *The Penguin Dictionary of Curious and Interesting Geometry.* London: Penguin, pp. 86–87, 1991.

Parallelogram Law

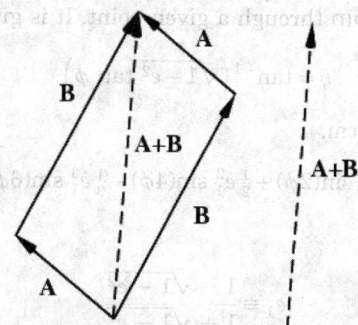

The parallelogram law gives the rule for VECTOR ADDITION of vectors **A** and **B**. The sum **A** + **B** of the vectors is obtained by placing them head to tail and drawing the vector from the free tail to the free head. Let $|\cdot|$ denote the NORM of a quantity. Then the quantities x and y are said to satisfy the parallelogram law if

$$\|x+y\|^2 + \|x-y\|^2 = 2\|x\|^2 + 2\|y\|^2.$$

If the NORM is defined as $|f| = \sqrt{\langle f | f \rangle}$ (the so-called L_2-NORM), then the law will always hold.

See also L_2-NORM, NORM, VECTOR, VECTOR ADDITION

References

Arfken, G. *Mathematical Methods for Physicists, 3rd ed.* Orlando, FL: Academic Press, pp. 1–2, 1985.
Jeffreys, H. and Jeffreys, B. S. *Methods of Mathematical Physics, 3rd ed.* Cambridge, England: Cambridge University Press, p. 58, 1988.

Parallelohedron

A special class of ZONOHEDRON. There are five parallelohedra with an infinity of equal and similarly situated replicas which are SPACE-FILLING POLYHEDRA: the CUBE, ELONGATED DODECAHEDRON, hexagonal PRISM, RHOMBIC DODECAHEDRON, and TRUNCATED OCTAHEDRON.

See also PARALLELOTOPE, SPACE-FILLING POLYHEDRON

References

Coxeter, H. S. M. *Regular Polytopes, 3rd ed.* New York: Dover, p. 29, 1973.

Parallelotope

Move a point Π_0 along a LINE from an initial point to a final point. It traces out a LINE SEGMENT Π_1. When Π_1 is translated from an initial position to a final position, it traces out a PARALLELOGRAM Π_2. When Π_2 is translated, it traces out a PARALLELEPIPED Π_3. The generalization of Π_n to n-D is then called a parallelotope. Π_n has 2^n vertices and

$$N_k = 2^{n-k} \binom{n}{k}$$

Π_ks, where $\binom{n}{k}$ is a BINOMIAL COEFFICIENT and $k = 0$, $1, \ldots, n$ (Coxeter 1973). These are also the coefficients of $(x+2)^n$.

See also HONEYCOMB, HYPERCUBE, ORTHOTOPE, PARALLELOHEDRON

References

Coxeter, H. S. M. *Regular Polytopes, 3rd ed.* New York: Dover, pp. 122–123, 1973.
Klee, V. and Wagon, S. *Old and New Unsolved Problems in Plane Geometry and Number Theory.* Washington, DC: Math. Assoc. Amer., 1991.
Zaks, J. "Neighborly Families of Congruent Convex Polytopes." *Amer. Math. Monthly* **94**, 151–155, 1987.

Paralogic Triangles

At the points where a line cuts the sides of a TRIANGLE $\Delta A_1 A_2 A_3$, perpendiculars to the sides are drawn, forming a TRIANGLE $\Delta B_1 B_2 B_3$ similar to the given TRIANGLE. The two triangles are also in perspective. One point of intersection of their CIRCUMCIRCLES is the SIMILITUDE CENTER, and the other is the PERSPECTIVE CENTER. The CIRCUMCIRCLES meet ORTHOGONALLY.

See also CIRCUMCIRCLE, ORTHOGONAL CIRCLES, PERSPECTIVE CENTER, SIMILITUDE CENTER

References

Johnson, R. A. *Modern Geometry: An Elementary Treatise on the Geometry of the Triangle and the Circle.* Boston, MA: Houghton Mifflin, pp. 258–262, 1929.

Parameter

A parameter m used in ELLIPTIC INTEGRALS defined to be $m \equiv k^2$, where k is the MODULUS. An ELLIPTIC INTEGRAL is written $I(\phi|m)$ when the parameter is used. The complementary parameter is defined by

$$m' \equiv 1 - m, \tag{1}$$

where m is the parameter. Let q be the NOME, k the MODULUS, and $m \equiv k^2$ the PARAMETER. Then

$$q(m) = e^{-\pi K'(m)/K(m)} \tag{2}$$

where $K(m)$ is the complete ELLIPTIC INTEGRAL OF THE FIRST KIND. Then the inverse of $q(m)$ is given by

$$m(q) = \frac{\vartheta_2^4(q)}{\vartheta_3^4(q)}, \tag{3}$$

where ϑ_i is a JACOBI THETA FUNCTION.

See also AMPLITUDE, CHARACTERISTIC (ELLIPTIC INTEGRAL), ELLIPTIC INTEGRAL, ELLIPTIC INTEGRAL OF THE FIRST KIND, HALF-PERIOD RATIO, JACOBI THETA FUNCTIONS, MODULAR ANGLE, MODULUS (ELLIPTIC INTEGRAL), NOME, PARAMETER

References

Abramowitz, M. and Stegun, C. A. (Eds.). *Handbook of Mathematical Functions with Formulas, Graphs, and Mathematical Tables, 9th printing.* New York: Dover, p. 590, 1972.

Parameter (Quadric)

The number θ in the QUADRIC

$$\frac{x^2}{a^2 + \theta} + \frac{y^2}{b^2 + \theta} + \frac{z^2}{c^2 + \theta} = 1$$

is called the parameter.

See also QUADRIC

Parameterization

The specification of a curve, surface, etc., by means of one or more variables which are allowed to take on values in a given specified range.

See also ISOTHERMAL PARAMETERIZATION, PARAMETRIC EQUATIONS, REGULAR PARAMETERIZATION, REPARAMETERIZATION, SURFACE PARAMETERIZATION

Parametric Equations

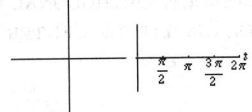

Parametric equations are a set of equations that express a set of quantities as explicit functions of a number of independent variables, known as "parameters." For example, while the equation of a CIRCLE in CARTESIAN COORDINATES can be given by $r^2 = x^2 + y^2$, one set of parametric equations for the circle are given by

$$x = r \cos t$$

$$y = r \sin t,$$

illustrated above. Note that parametric representations are generally nonunique, so the same quantities may be expressed by a number of different parameterizations. A single parameter is usually represented with the parameter t, while the symbols u and v are commonly used for parametric equations in two parameters.

Parametric equations provide a convenient way to represent curves and surfaces, as implemented, for example, in the *Mathematica* commands Parame-

tricPlot[{*x*, *y*}, {*t*, *t1*, *t2*}] and Parametric-Plot3D[{*x*, *y*, *z*}, {*u*, *u1*, *u2*}, {*v*, *v1*, *v2*}].

Parametric Latitude

An AUXILIARY LATITUDE also called the REDUCED LATITUDE and denoted η or θ. It gives the LATITUDE on a SPHERE of RADIUS a for which the parallel has the same radius as the parallel of geodetic latitude ϕ and the ELLIPSOID through a given point. It is given by

$$\eta = \tan^{-1}\left(\sqrt{1 - e^2} \, \tan \phi\right).$$

In series form,

$$\eta = \phi - e_1 \sin(2\phi) + \tfrac{1}{2} e_1^2 \sin(4\phi) - \tfrac{1}{3} e_1^3 \sin(6\phi) + \ldots,$$

where

$$e_1 \equiv \frac{1 - \sqrt{1 - e^2}}{1 + \sqrt{1 - e^2}}.$$

See also AUXILIARY LATITUDE, ELLIPSOID, LATITUDE, SPHERE

References

Adams, O. S. "Latitude Developments Connected with Geodesy and Cartography with Tables, Including a Table for Lambert Equal-Area Meridional Projections." Spec. Pub. No. 67. U. S. Coast and Geodetic Survey, 1921.
Snyder, J. P. *Map Projections--A Working Manual.* U. S. Geological Survey Professional Paper 1395. Washington, DC: U. S. Government Printing Office, p. 18, 1987.

Parametric Statistics

See also NONPARAMETRIC STATISTICS

References

Sheskin, D. J. *Handbook of Parametric and Nonparametric Statistical Procedures, 2nd ed.* Boca Raton, FL: Chapman & Hall/CRC, 2000.

Parametric Test

A STATISTICAL TEST in which assumptions are made about the underlying distribution of observed data.

Parametrization

PARAMETERIZATION

Parenthesis

One of the symbols (or) used to denote grouping. Parentheses have a great many specialized meanings in mathematics. A few of these are described below.

1. Parentheses are used in mathematical expressions to denote modifications to normal order of

operations (precedence rules). In an expression like $(3+5) \times 7$, the part of the expression within the parentheses, $(3+5) = 8$, is evaluated first, and then this result is used in the rest of the expression. Nested parentheses work similarly, since parts of expressions within parentheses are also considered expressions. Parentheses are also used in this manner to clarify order of operations in confusing or abnormally large expressions.

2. A parenthesis can be used to denote an open end of an INTERVAL. For example, $[0, 5)$ denotes the HALF-OPEN INTERVAL which includes all real numbers from 0 to 5 except 5 itself.

3. Parentheses are used to enclose the variables of a FUNCTION in the form $f(x)$, which means that values of the function f are dependent upon the values of x.

4. Large parentheses around two numbers, one above the other, denotes a BINOMIAL COEFFICIENT $\binom{n}{k}$.

5. Parentheses around a set of two or more numbers, as in (a, b, c), denote an n-tuple of numbers that are linked in some special way.

6. Large parentheses around an array of numbers, e.g., $\left(\begin{smallmatrix} a & b \\ c & d \end{smallmatrix}\right)$ indicate a MATRIX. (However, in this work, the symbol $\left[\begin{smallmatrix} a & b \\ c & d \end{smallmatrix}\right]$ is used instead.)

7. Parentheses may also be used to denote the GREATEST COMMON DIVISOR, e.g., $(54, 21) \equiv$ GCD$(54, 21) = 3$.

8. Parenthesis are used to denote a CONGRUENCE, as in $a \equiv d \pmod{m}$.

See also ANGLE BRACKET, BRACE, SQUARE BRACKET

References

Bringhurst, R. *The Elements of Typographic Style, 2nd ed.* Point Roberts, WA: Hartley and Marks, p. 282, 1997.

Pareto Distribution

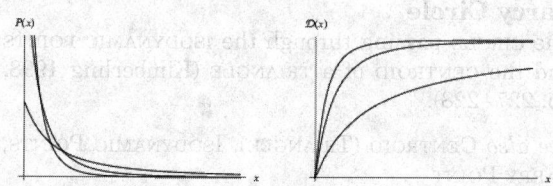

The distribution with probability density function and distribution function

$$P(x) = \frac{ab^a}{x^{a+1}} \tag{1}$$

$$D(x) = 1 - \left(\frac{b}{x}\right)^a \tag{2}$$

defined over the interval $x \geq b$. The RAW MOMENTS are

$$\mu_1' = \frac{ab}{a-1} \tag{3}$$

$$\mu_2' = \frac{ab^2}{a-2} \tag{4}$$

$$\mu_3' = \frac{ab^3}{a-3} \tag{5}$$

$$\mu_4' = \frac{ab^4}{a-4} \tag{6}$$

and the CENTRAL MOMENTS are

$$\mu_2 = \frac{ab^2}{(a-1)^2(a-2)} \tag{7}$$

$$\mu_3 = \frac{2a(a+1)b^3}{(a-1)^3(a-2)(a-3)} \tag{8}$$

$$\mu_4 = \frac{3a(3a^3+a+2)b^4}{(a-1)^4(a-2)(a-3)(a-4)} \tag{9}$$

Giving MEAN, VARIANCE, SKEWNESS, and KURTOSIS

$$\mu = \frac{ab}{a-1} \tag{10}$$

$$\sigma^2 = \frac{ab^2}{(a-1)^2(a-2)} \tag{11}$$

$$\gamma_1 = \sqrt{\frac{a-2}{a}}\frac{2(a+1)}{a-3} \tag{12}$$

$$\gamma_2 = \frac{6(a^3+a^2-6a-2)}{a(a-3)(a-4)} \tag{13}$$

References

von Seggern, D. *CRC Standard Curves and Surfaces.* Boca Raton, FL: CRC Press, p. 252, 1993.

Parity

The parity of an integer is its attribute of being EVEN or ODD. Thus, it can be said that 6 and 14 have the same parity (since both are EVEN), whereas 7 and 12 have opposite parity (since 7 is ODD and 12 is EVEN).

More specifically, the parity of an integer n can be defined as the sum of the bits in BINARY representation, computed modulo 2. The parities of the first few integers (starting with 0) are therefore 0, 1, 1, 0, 1, 0, 0, 1, 1, 0, 0, ... (Sloane's A010060), summarized in the following table.

N	Binary	Parity	N	Binary	Parity
1	1	1	11	1011	1
2	10	1	12	1100	0
3	11	0	13	1101	1
4	100	1	14	1110	1
5	101	0	15	1111	0
6	110	0	16	10000	1
7	111	1	17	10001	0
8	1000	1	18	10010	0
9	1001	0	19	10011	1
10	1010	0	20	10100	0

The parity function obeys the sum identity

$$\sum_{k=0}^{2^{n+1}-1} (-1)^{P(k)} (k+r)^n = 0$$

for any n. For example, for $n=2$ and $r=0$,

$$1 - 4 - 9 + 16 - 25 + 36 + 49 - 64 = 0.$$

The constant generated by the sequence of parity digits $0.011010011\ldots_2$ is called the THUE-MORSE CONSTANT.

See also BINARY, EVEN NUMBER, ODD NUMBER, THUE-MORSE CONSTANT

References

Commission on Mathematics of the College Entrance Examination Board. *Informal Deduction in Algebra: Properties of Odd and Even Numbers.* Princeton, NJ, 1959.

Gardner, M. "Parity Checks." Ch. 8 in *The Sixth Book of Mathematical Games from Scientific American.* Chicago, IL: University of Chicago Press, pp. 71–78, 1984.

Sloane, N. J. A. Sequences A010060 in "An On-Line Version of the Encyclopedia of Integer Sequences." http://www.research.att.com/~njas/sequences/eisonline.html.

Parity Constant

THUE-MORSE CONSTANT

Parking Constant

RÉNYI'S PARKING CONSTANTS

Parodi's Theorem

The EIGENVALUES λ satisfying $P(\lambda) = 0$, where $P(\lambda)$ is the CHARACTERISTIC POLYNOMIAL, lie in the unions of the DISKS

$$|z| \leq 1$$

$$|z + b_1| \leq \sum_{j=1}^{n} |b_j|.$$

References

Gradshteyn, I. S. and Ryzhik, I. M. *Tables of Integrals, Series, and Products, 6th ed.* San Diego, CA: Academic Press, p. 1119, 2000.

Parrondo's Paradox

Two losing gambling games can be set up so that when they are played one after the other, they become winning. There are many ways to construct such scenarios, the simplest of which uses three biased coins (Harmer and Abbott 1999).

References

Doering, C. R. "Randomly Rattled Ratchets." *Il Nuovo Cimento* **17D**, 685–697, 1995.

Harmer, G. P. and Abbott, D. "Losing Strategies Can Win by Parrondo's Paradox." *Nature* **402**, 864, 1999.

Harmer, G. P.; Abbott, D.; Taylor, P. G.; and Parrondo, J. M. R. "Parrondo's Paradoxical Games and the Discrete Brownian Ratchet." In *Proc. 2nd Internat. Conf. Unsolved Problems of Noise and Fluctuations, 11–15 July, Adelaide* (Ed. D. Abbott and L. B. Kiss). Melville, NY: Amer. Inst. Physics Press, pp. 189–200, 2000.

Harmer, G. P.; Abbott, D.; Taylor, P. G.; Pearce, C. E. M.; and Parrondo, J. M. R. "Information Entropy and Parrondo's Discrete-Time Ratchet." In *Proc. Stochastic and Chaotic Dynamics in the Lakes, 16–20 August, Ambleside, UK* (Ed. P. V. E. McClintock). Melville, NY: Amer. Inst. Physics Press, pp. 544–549, 2000.

McClintock, P. V. E. "Unsolved Problems of Noise." *Nature* **401**, 23–25, 1999.

Pearce, C. E. M. "Entropy, Markov Information Sources and Parrondo Games." In *Proc. 2nd Internat. Conf. Unsolved Problems of Noise and Fluctuations, 11–15 July, Adelaide* (Ed. D. Abbott and L. B. Kiss). Melville, NY: Amer. Inst. Physics Press, pp. 207–212, 2000.

Pearce, C. E. M. "On Parrondo's Paradoxical Games." In *Proc. 2nd Internat. Conf. Unsolved Problems of Noise and Fluctuations, 11–15 July, Adelaide* (Ed. D. Abbott and L. B. Kiss). Melville, NY: Amer. Inst. Physics Press, pp. 201–206, 2000.

Parry Circle

The CIRCLE passing through the ISODYNAMIC POINTS and the CENTROID of a TRIANGLE (Kimberling 1998, pp. 227–228).

See also CENTROID (TRIANGLE), ISODYNAMIC POINTS, PARRY POINT

References

Kimberling, C. "Triangle Centers and Central Triangles." *Congr. Numer.* **129**, 1–295, 1998.

Parry Point

The intersection of the PARRY CIRCLE and the CIRCUMCIRCLE of a TRIANGLE. The TRILINEAR COORDINATES of the Parry point are

$$\frac{a}{2a^2 - b^2 - c^2} : \frac{b}{2b^2 - c^2 - a^2} : \frac{c}{2c^2 - a^2 - b^2}$$

(Kimberling 1998, pp. 227–228).

See also PARRY CIRCLE

References

Kimberling, C. "Parry Point." http://cedar.evansville.edu/
~ck6/tcenters/recent/parry.html.
Kimberling, C. "Triangle Centers and Central Triangles."
Congr. Numer. **129**, 1–295, 1998.

Parseval's Integral

The POISSON INTEGRAL with $n = 0$,

$$J_0(z) = \frac{1}{\left[\Gamma\left(n + \frac{1}{2}\right)\right]^2} \int_0^\pi \cos(z \cos\theta)\, d\theta,$$

where $J_0(z)$ is a BESSEL FUNCTION OF THE FIRST KIND
and $\Gamma(x)$ is a GAMMA FUNCTION.

Parseval's Relation

Let $F(v)$ and $G(v)$ be the FOURIER TRANSFORMS of $f(t)$
and $g(t)$, respectively. Then

$$\int_{-\infty}^\infty f(t)\bar{g}(t)\, dt$$

$$= \int_{-\infty}^\infty \left[\int_{-\infty}^\infty F(v)e^{-2\pi i v t}\, dv\right]\left[\int_{-\infty}^\infty \bar{G}(v')e^{2\pi i v' t}\, dv'\right] dt$$

$$= \int_{-\infty}^\infty F(v) \int_{-\infty}^\infty \bar{G}(v')\left[\int_{-\infty}^\infty e^{2\pi i t(v'-v)}\, dt\right] dv'\, dv$$

$$= \int_{-\infty}^\infty F(v)\left[\int_{-\infty}^\infty \bar{G}(v')\delta(v'-v)\, dv'\right] dv$$

$$= \int_{-\infty}^\infty F(v)\bar{G}(v)\, dv,$$

where $\bar{z}$ denotes the COMPLEX CONJUGATE.

See also FOURIER TRANSFORM, PARSEVAL'S THEOREM

References

Arfken, G. *Mathematical Methods for Physicists, 3rd ed.*
Orlando, FL: Academic Press, p. 425, 1985.

Parseval's Theorem

Let $E(t)$ be a continuous function and $E(t)$ and E_v be
FOURIER TRANSFORM pairs so that

$$E(t) \equiv \int_{-\infty}^\infty E_v e^{-2\pi i v t}\, dv \qquad (1)$$

$$\bar{E}(t) \equiv \int_{-\infty}^\infty \bar{E}_v e^{2\pi i v' t}\, dv', \qquad (2)$$

where $\bar{z}$ denotes the COMPLEX CONJUGATE. Then

$$\int_{-\infty}^\infty |E(t)|^2\, dt = \int_{-\infty}^\infty E(t)\bar{E}(t)\, dt$$

$$= \int_{-\infty}^\infty \left[\int_{-\infty}^\infty E_v e^{-2\pi i v t}\, dv \int_{-\infty}^\infty \bar{E}_v e^{2\pi i v' t}\, dv'\right] dt$$

$$= \int_{-\infty}^\infty \int_{-\infty}^\infty \int_{-\infty}^\infty E_v \bar{E}_v e^{2\pi i t(v'-v)}\, dv\, dv'\, dt$$

$$= \int_{-\infty}^\infty \int_{-\infty}^\infty \int_{-\infty}^\infty E_v \bar{E}_v e^{2\pi i t(v'-v)}\, dt\, dv\, dv'$$

$$= \int_{-\infty}^\infty \int_{-\infty}^\infty \delta(v'-v)\, E_v \bar{E}_{v'}\, dv\, dv'$$

$$= \int_{-\infty}^\infty E_v \bar{E}_{v'}\, dv = \int_{-\infty}^\infty |E_v|^2\, dv. \qquad (3)$$

where $\delta(x - x_0)$ is the DELTA FUNCTION.
For finite FOURIER TRANSFORM pairs h_k and H_n,

$$\sum_{k=0}^{N-1} |h_k|^2 = \frac{1}{N} \sum_{n=0}^{N-1} |H_n|^2. \qquad (4)$$

If a function has a FOURIER SERIES given by

$$f(x) = \frac{1}{2} a_0 + \sum_{n=1}^\infty a_n \cos(nx) + \sum_{n=1}^\infty b_n \sin(nx), \qquad (5)$$

then BESSEL'S INEQUALITY becomes an equality
known as Parseval's theorem. From (5),

$$[f(x)]^2 = \frac{1}{4} a_0^2 + a_0 \sum_{n=1}^\infty [a_n \cos(nx) + b_n \sin(nx)]$$

$$+ \sum_{n=1}^\infty \sum_{m=1}^\infty [a_n a_m \cos(nx) \cos(mx)$$

$$+ a_n b_m \cos(nx) \sin(mx)$$

$$+ a_m b_n \sin(nx) \cos(mx)$$

$$+ b_n b_m \sin(nx) \sin(mx)]. \qquad (6)$$

Integrating

$$\int_{-\pi}^\pi [f(x)]^2\, dx$$

$$= \frac{1}{4} a_0^2 \int_{-\pi}^\pi dx$$

$$+ a_0 \int_{-\pi}^\pi \sum_{n=1}^\infty [a_n \cos(nx) + b_n \sin(nx)]\, dx$$

$$+ \int_{-\pi}^\pi \sum_{n=1}^\infty \sum_{m=1}^\infty [a_n a_m \cos(nx) \cos(mx)$$

$$+ a_n b_m \cos(nx) \sin(mx) + a_m b_n \sin(nx) \cos(mx)$$

$$+ b_n b_m \sin(nx) \sin(mx)]\, dx = \frac{1}{4} a_0^2 (2\pi) + 0$$

$$+\sum_{n=1}^{\infty}\sum_{m=1}^{\infty}[a_n a_m \pi \delta_{nm} + 0 + 0 + b_n b_m \pi \delta_{nm}], \qquad (7)$$

so

$$\frac{1}{\pi}\int_{-\pi}^{\pi}[f(x)]^2 \, dx = \frac{1}{2}a_0^2 + \sum_{n=1}^{\infty}(a_n^2 + b_n^2). \qquad (8)$$

For a generalized FOURIER SERIES with a COMPLETE BASIS $\{\phi_i\}_{i=1}^{\infty}$, an analogous relationship holds. For a COMPLEX FOURIER SERIES,

$$\frac{1}{2\pi}\int_{-\pi}^{\pi}|f(x)|^2 \, dx = \sum_{n=-\infty}^{\infty}|a_n|^2. \qquad (9)$$

References

Gradshteyn, I. S. and Ryzhik, I. M. *Tables of Integrals, Series, and Products, 6th ed.* San Diego, CA: Academic Press, p. 1101, 2000.

Part Metric

A METRIC defined by

$$d(z, \, w) = \sup\left\{\ln\left|\left[\frac{u(z)}{u(w)}\right]\right| : u \in H^+\right\},$$

where H^+ denotes the POSITIVE HARMONIC FUNCTIONS on a DOMAIN. The part metric is invariant under CONFORMAL MAPS for any DOMAIN.

References

Bear, H. S. "Part Metric and Hyperbolic Metric." *Amer. Math. Monthly* **98**, 109–123, 1991.

Partial Derivative

Partial derivatives are defined as derivatives of a function of multiple variables when all but the variable of interest are held fixed during the differentiation.

$$\frac{\partial f}{\partial x_m} \equiv$$

$$\lim_{h \to 0} \frac{f(x_1, \, \ldots, \, x_m + h, \, \ldots, \, x_n) - f(x_1, \, \ldots, \, x_m, \, \ldots, \, x_n)}{h}.$$

$$(1)$$

The above partial derivative is sometimes denoted f_{x_m} for brevity. For a "nice" 2-D function $f(x, y)$ (i.e., one for which $f, f_x, f_y, f_{xy}, f_{yx}$ exist and are continuous in a NEIGHBORHOOD (a, b)), then $f_{xy}(a, b) = f_{yx}(a, b)$. Partial derivatives involving more than one variable are called MIXED PARTIAL DERIVATIVES.

For nice functions, mixed partial derivatives must be equal regardless of the order in which the differentiation is performed so, for example,

$$f_{xy} = f_{yx} \qquad (2)$$

$$f_{xxy} = f_{xyx} = f_{yxx}. \qquad (3)$$

For an EXACT DIFFERENTIAL,

$$df = \left(\frac{\partial f}{\partial x}\right)_y \, dx + \left(\frac{\partial f}{\partial y}\right)_x \, dy, \qquad (4)$$

so

$$\left(\frac{\partial y}{\partial x}\right)_f = -\frac{\left(\dfrac{\partial f}{\partial x}\right)_y}{\left(\dfrac{\partial f}{\partial y}\right)_x}. \qquad (5)$$

A differential equation expressing one or more quantities in terms of partial derivatives is called a PARTIAL DIFFERENTIAL EQUATION. Partial differential equations are extremely important in physics and engineering, and are in general difficult to solve.

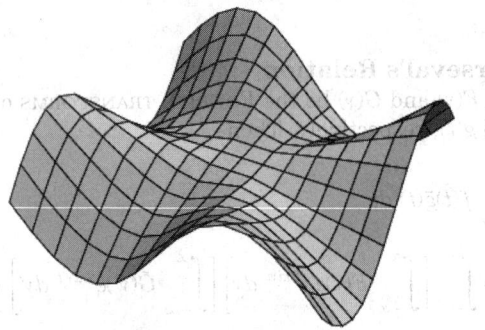

If the continuity requirement for MIXED PARTIALS is dropped, it is possible to construct functions for which MIXED PARTIALS are *not* equal. An example is the function

$$f(x, \, y) = \begin{cases} \dfrac{xy(x^2 - y^2)}{x^2 + y^2} & \text{for}(x, \, y) \neq (0, \, 0) \\ 0 & \text{for}(x, \, y) = (0, \, 0), \end{cases} \qquad (6)$$

which has $f_{xy}(0, \, 0) = -1$ and $f_{yx}(0, \, 0) = 1$ (Wagon 1991). This function is depicted above and by Fischer (1986).

Abramowitz and Stegun (1972) give FINITE DIFFERENCE versions for partial derivatives.

See also ABLOWITZ-RAMANI-SEGUR CONJECTURE, DERIVATIVE, MIXED PARTIAL DERIVATIVE, MONKEY SADDLE, PARTIAL DIFFERENTIAL EQUATION

References

Abramowitz, M. and Stegun, C. A. (Eds.). *Handbook of Mathematical Functions with Formulas, Graphs, and Mathematical Tables, 9th printing.* New York: Dover, pp. 883–885, 1972.
Fischer, G. (Ed.). Plate 121 in *Mathematische Modelle/ Mathematical Models, Bildband/Photograph Volume.* Braunschweig, Germany: Vieweg, p. 118, 1986.

Thomas, G. B. and Finney, R. L. §16.8 in *Calculus and Analytic Geometry, 9th ed.* Reading, MA: Addison-Wesley, 1996.

Wagon, S. *Mathematica in Action.* New York: W. H. Freeman, pp. 83–85, 1991.

Partial Differential Equation

A partial differential equation (PDE) is an equation involving functions and their PARTIAL DERIVATIVES; for example, the WAVE EQUATION

$$\frac{\partial^2 \psi}{\partial x^2} + \frac{\partial^2 \psi}{\partial y^2} + \frac{\partial^2 \psi}{\partial z^2} = \frac{1}{v^2}\frac{\partial^2 \psi}{\partial t^2}. \tag{1}$$

in general, partial differential equations are much more difficult to solve analytically than are ORDINARY DIFFERENTIAL EQUATIONS. They may sometimes be solved using a BÄCKLUND TRANSFORMATION, CHARAC-TERISTIC, GREEN'S FUNCTION, INTEGRAL TRANSFORM, LAX PAIR, SEPARATION OF VARIABLES, or–when all else fails (which it frequently does)–numerical methods.

Fortunately, partial differential equations of second-order are often amenable to analytical solution. Such PDEs are of the form

$$Au_{xx} + 2Bu_{xy} + Cu_{yy} + Du_x + Eu_y + F = 0. \tag{2}$$

Second-order PDEs are then classified according to the properties of the MATRIX

$$Z \equiv \begin{bmatrix} A & B \\ B & C \end{bmatrix} \tag{3}$$

as ELLIPTIC, HYPERBOLIC, or PARABOLIC.

If Z is a POSITIVE DEFINITE MATRIX, i.e., $\det(Z) > 0$, the PDE is said to be ELLIPTIC. LAPLACE'S EQUATION and POISSON'S EQUATION are examples. Boundary conditions are used to give the constraint $u(x, y) = g(x, y)$ on $\partial\Omega$, where

$$u_{xx} + u_{yy} = f(u_x, u_y, u, x, y) \tag{4}$$

holds in Ω.

If $\det(Z) < 0$, the PDE is said to be HYPERBOLIC. The WAVE EQUATION is an example of a hyperbolic partial differential equation. Initial-boundary conditions are used to give

$$u(x, y, t) = g(x, y, t) \quad \text{for } x \in \partial\Omega, \ t > 0 \tag{5}$$

$$u(x, y, 0) = v_0(x, y) \quad \text{in } \Omega \tag{6}$$

$$u_t(x, y, 0) = v_1(x, y) \quad \text{in } \Omega, \tag{7}$$

where

$$u_{xy} = f(u_x, u_t, x, y) \tag{8}$$

holds in Ω.

If $\det(Z) = 0$, the PDE is said to be parabolic. The HEAT CONDUCTION EQUATION equation and other diffusion equations are examples. Initial-boundary conditions are used to give

$$u(x, t) = g(x, t) \quad \text{for } x \in \partial\Omega, \ t > 0 \tag{9}$$

$$u(x, 0) = v(x) \quad \text{for } x \in \Omega, \tag{10}$$

where

$$u_{xx} = f(u_x, u_y, u, x, y) \tag{11}$$

holds in Ω.

See also BÄCKLUND TRANSFORMATION, BOUNDARY CONDITIONS, CHARACTERISTIC (PARTIAL DIFFERENTIAL EQUATION), ELLIPTIC PARTIAL DIFFERENTIAL EQUATION, GREEN'S FUNCTION, HYPERBOLIC PARTIAL DIFFERENTIAL EQUATION, INTEGRAL TRANSFORM, JOHNSON'S EQUATION, LAX PAIR, MONGE-AMPÈRE DIFFERENTIAL EQUATION, PARABOLIC PARTIAL DIF-FERENTIAL EQUATION, SEPARATION OF VARIABLES

References

Arfken, G. "Partial Differential Equations of Theoretical Physics." §8.1 in *Mathematical Methods for Physicists, 3rd ed.* Orlando, FL: Academic Press, pp. 437–440, 1985.

Bateman, H. *Partial Differential Equations of Mathematical Physics.* New York: Dover, 1944.

Conte, R. Exact Solutions of Nonlinear Partial Differential Equations by Singularity Analysis. 13 Sep 2000. http://xxx.lanl.gov/abs/nlin.SI/0009024/.

Folland, G. B. *Introduction to Partial Differential Equations, 2nd ed.* Princeton, NJ: Princeton University Press, 1996.

Kevorkian, J. *Partial Differential Equations: Analytical Solution Techniques, 2nd ed.* New York: Springer-Verlag, 2000.

Morse, P. M. and Feshbach, H. "Standard Forms for Some of the Partial Differential Equations of Theoretical Physics." *Methods of Theoretical Physics, Part I.* New York: McGraw-Hill, pp. 271–272, 1953.

Press, W. H.; Flannery, B. P.; Teukolsky, S. A.; and Vetter-ling, W. T. "Partial Differential Equations." Ch. 19 in *Numerical Recipes in FORTRAN: The Art of Scientific Computing, 2nd ed.* Cambridge, England: Cambridge University Press, pp. 818–880, 1992.

Sobolev, S. L. *Partial Differential Equations of Mathematical Physics.* New York: Dover, 1989.

Sommerfeld, A. *Partial Differential Equations in Physics.* New York: Academic Press, 1964.

Taylor, M. E. *Partial Differential Equations, Vol. 1: Basic Theory.* New York: Springer-Verlag, 1996.

Taylor, M. E. *Partial Differential Equations, Vol. 2: Qualitative Studies of Linear Equations.* New York: Springer-Verlag, 1996.

Taylor, M. E. *Partial Differential Equations, Vol. 3: Nonlinear Equations.* New York: Springer-Verlag, 1996.

Webster, A. G. *Partial Differential Equations of Mathematical Physics, 2nd corr. ed.* New York: Dover, 1955.

Weisstein, E. W. "Books about Partial Differential Equations." http://www.treasure-troves.com/books/PartialDif-ferentialEquations.html.

Zwillinger, D. *Handbook of Differential Equations, 3rd ed.* Boston, MA: Academic Press, 1997.

Partial Fraction Decomposition

A RATIONAL FUNCTION $P(x)/Q(x)$ can be rewritten using what is known as partial fraction decomposition. This procedure often allows integration to be performed on each term separately by inspection. For each factor of $Q(x)$ the form $(ax + b)^m$, introduce terms

$$\frac{A_1}{ax+b}+\frac{A_2}{(ax+b)^2}+\ldots+\frac{A_m}{(ax+b)^m}. \quad (1)$$

For each factor OF THE FORM $(ax^2+bx+c)^m$, introduce terms

$$\frac{A_1x+B_1}{ax^2+bx+c}+\frac{A_2x+B_2}{(ax^2+bx+c)^2}+\ldots$$
$$+\frac{A_mx+B_m}{(ax^2+bx+c)^m}. \quad (2)$$

Then write

$$\frac{P(x)}{Q(x)}=\frac{A_1}{ax+b}+\ldots+\frac{A_2x+B_2}{ax^2+bx+c}+\ldots \quad (3)$$

and solve for the A_is and B_is.

Partial fraction decomposition is implemented in *Mathematica* 4.0 as Apart.

References

Beyer, W. H. *CRC Standard Mathematical Tables, 28th ed.* Boca Raton, FL: CRC Press, pp. 13–15, 1987.

Partial Integration

INTEGRATION BY PARTS

Partial Latin Square

In a normal $n\times n$ LATIN SQUARE, the entries in each row and column are chosen from a "global" set of n objects. Like a Latin square, a partial Latin square has no two rows or columns which contain the same two symbols. However, in a partial Latin square, each cell is assigned one of its own set of n possible "local" (and distinct) symbols, chosen from an overall set of more than three distinct symbols, and these symbols may vary from location to location. For example, given the possible symbols $\{1, 2, \ldots, 6\}$ which must be arranged as

$$\{1, 2, 3\} \quad \{1, 3, 4\} \quad \{2, 5, 6\}$$
$$\{2, 3, 5\} \quad \{1, 2, 3\} \quad \{4, 5, 6\}$$
$$\{4, 3, 6\} \quad \{3, 5, 6\} \quad \{2, 3, 5\},$$

the 3×3 partial Latin square

$$\begin{array}{ccc} 1 & 3 & 2 \\ 2 & 4 & 5 \\ 6 & 5 & 3 \end{array}$$

can be constructed.

See also DINITZ PROBLEM, LATIN SQUARE

References

Cipra, B. "Quite Easily Done." In *What's Happening in the Mathematical Sciences* **2**, pp. 41–46, 1994.

Partial Order

A RELATION "$\leq$" is a partial order on a SET S if it has:

1. Reflexivity: $a\leq a$ for all $a\in S$.
2. Antisymmetry: $a\leq b$ and $b\leq a$ implies $a=b$.
3. Transitivity: $a\leq b$ and $b\leq c$ implies $a\leq c$.

For a partial order, the size of the longest CHAIN (ANTICHAIN) is called the LENGTH (WIDTH). A partially ordered set is also called a poset.

A largest set of unrelated vertices in a PARTIAL ORDER can be found using MaximumAntichain[g] in the *Mathematica* add-on package DiscreteMath`Combinatorica` (which can be loaded with the command < <DiscreteMath`). MinimumChainPartition[g] in the *Mathematica* add-on package DiscreteMath`Combinatorica` (which can be loaded with the command < <DiscreteMath`) partitions a partial order into a minimum number of CHAINS.

See also ANTICHAIN, CHAIN, FENCE POSET, IDEAL (PARTIAL ORDER), LENGTH (PARTIAL ORDER), LINEAR EXTENSION, PARTIALLY ORDERED SET, TOTAL ORDER, WIDTH (PARTIAL ORDER)

References

Ruskey, F. "Information on Linear Extension." http://www.theory.csc.uvic.ca/~cos/inf/pose/LinearExt.html.
Skiena, S. "Partial Orders." §5.4 in *Implementing Discrete Mathematics: Combinatorics and Graph Theory with Mathematica.* Reading, MA: Addison-Wesley, pp. 203–209, 1990.

Partial Quotient

If the SIMPLE CONTINUED FRACTION of a REAL NUMBER x is given by

$$x=a_0+\cfrac{1}{a_1+\cfrac{1}{a_2+\cfrac{1}{a_3+\ldots}}},$$

then the quantities a_i are called partial quotients.

See also CONTINUED FRACTION, CONVERGENT, SIMPLE CONTINUED FRACTION

Partially Ordered Set

A partially ordered set (or poset) is a SET taken together with a PARTIAL ORDER on it. Formally, a partially ordered set is defined as an ordered pair $P=(X,\leq)$, where X is called the GROUND SET of P and $\leq$ is the PARTIAL ORDER of P.

See also CIRCLE ORDER, COVER RELATION, DOMINANCE, GROUND SET, HASSE DIAGRAM, INTERVAL ORDER, ISOMORPHIC POSETS, ORDER ISOMORPHIC, PARTIAL ORDER, POSET DIMENSION, REALIZER, RELATION

References

Dushnik, B. and Miller, E. W. "Partially Ordered Sets." *Amer. J. Math.* **63**, 600–610, 1941.

Fishburn, P. C. *Interval Orders and Interval Sets: A Study of Partially Ordered Sets.* New York: Wiley, 1985.

Skiena, S. "Partial Orders." §5.4 in *Implementing Discrete Mathematics: Combinatorics and Graph Theory with Mathematica.* Reading, MA: Addison-Wesley, pp. 203–209, 1990.

Trotter, W. T. *Combinatorics and Partially Ordered Sets: Dimension Theory.* Baltimore, MD: Johns Hopkins University Press, 1992.

Particularly Well-Behaved Functions

Functions which have DERIVATIVES of all orders at all points and which, together with their DERIVATIVES, fall off at least as rapidly as $|x|^{-n}$ as $|x| \to \infty$, no matter how large n is.

See also REGULAR SEQUENCE

Partisan Game

A GAME for which each player has a different set of moves in any position. Every position in an IMPARTIAL GAME has a NIM-VALUE.

Partition

A partition is a way of writing an INTEGER n as a sum of POSITIVE INTEGERS where the order of the summands is not significant, possibly subject to one or more additional constraints. By convention, partitions are normally written from largest to smallest summands (Skiena 1990, p. 51), e.g., $10 = 3 + 2 + 2 + 2 + 1$. PartitionsQ[p] in the *Mathematica* add-on package DiscreteMath`Combinatorica` (which can be loaded with the command <<DiscreteMath`) tests a list to determine that it consists of positive integers and therefore is a valid partition. Andrews (1998, p. 1) used the notation $\lambda \vdash n$ to indicate "a sequence λ is a partition of n," and the notation $^{a_1}2^{a_2} \cdots)$ to abbreviate the partition $\{\underbrace{1, \ldots, 1}_{a_1}, \underbrace{2, \ldots, 2}_{a_2}, \ldots\}$.

Particular types of partition functions include the PARTITION FUNCTION P, giving the number of partitions of a number as a sum of smaller integers without regard to order, and PARTITION FUNCTION Q, giving the number of ways of writing the INTEGER n as a sum of POSITIVE INTEGERS without regard to order and with the constraint that all INTEGERS in each sum are distinct. The PARTITION FUNCTION B, which gives the number of partitions of n in which no parts are multiples of k is sometimes also used (Gordon and Ono 1997).

The EULER transform b_n gives the number of partitions of n into integer parts of which there are a_1 different types of parts of size 1, a_2 of size 2, etc. For example, if $a_n = 1$ for all n, then b_n is the number of partitions of n into integer parts. Similarly, if $a_n =$ 1 for n prime and $a_n = 0$ for n composite, then b_n is the number of partitions of n into prime parts (Sloane and Plouffe 1995, p. 21).

A partition of a number n into a sum of elements of a list L can be determined using a GREEDY ALGORITHM. The following table gives the number of partitions of n into a sum of positive powers p for multiples of n.

n	$p = 1$	$p = 2$	$p = 3$	$p = 4$
	Sloane's A000041	Sloane's A001156	Sloane's A003108	Sloane's A046042
10	42	4	2	1
50	204226	104	10	4
100	190569292	1116	39	9
150	40853235313	6521	97	15
200	3972999029388	27482	208	24
250	2.307×10^{14}		388	34
300	9.253×10^{15}		683	49

See also AMENABLE NUMBER, CONJUGATE PARTITION, DURFEE SQUARE, ELDER'S THEOREM, FERRERS DIAGRAM, GÖLLNITZ'S THEOREM, GRAPHICAL PARTITION, GREEDY ALGORITHM, PARTITION FUNCTION B, PARTITION FUNCTION P, PARTITION FUNCTION Q, PERFECT PARTITION, PLANE PARTITION, PRIME PARTITION, SELF-CONJUGATE PARTITION, SET PARTITION, SOLID PARTITION, STANLEY'S THEOREM

References

Andrews, G. E. *The Theory of Partitions.* Cambridge, England: Cambridge University Press, 1998.

Dickson, L. E. "Partitions." Ch. 3 in *History of the Theory of Numbers, Vol. 2: Diophantine Analysis.* New York: Chelsea, pp. 101–164, 1952.

Gordon, B. and Ono, K. "Divisibility of Certain Partition Functions by Powers of Primes." *Ramanujan J.* **1**, 25–34, 1997.

Hardy, G. H. and Wright, E. M. "Partitions." Ch. 19 in *An Introduction to the Theory of Numbers, 5th ed.* Oxford, England: Clarendon Press, pp. 273–296, 1979.

Savage, C. "Gray Code Sequences of Partitions." *J. Algorithms* **10**, 577–595, 1989.

Skiena, S. "Partitions." §2.1 in *Implementing Discrete Mathematics: Combinatorics and Graph Theory with Mathematica.* Reading, MA: Addison-Wesley, pp. 51–59, 1990.

Sloane, N. J. A. Sequences A000041/M0663, A001156/M0221, A003108/M0209, and A046042 in "An On-Line Version of the Encyclopedia of Integer Sequences." http://www.research.att.com/~njas/sequences/eisonline.html.

Sloane, N. J. A. and Plouffe, S. *The Encyclopedia of Integer Sequences.* San Diego, CA: Academic Press, 1995.

Partition Function b

The number of partitions of n in which no parts are multiples of k is sometimes denoted $b_k(n)$ (Gordon and Ono 1997). $b_k(n)$ is also the number of partitions of n into at most $k-1$ copies of each part.

$b_2(n) = Q(n)$, where $Q(n)$ is the PARTITION FUNCTION Q, and $b_p(n)$ is the number of irreducible p-modular representations of the SYMMETRIC GROUP S_n. The generating function for $b_k(n)$ is given by

$$\sum_{n=0}^{\infty} b_k(n) x^n = \prod_{n=1}^{\infty} \frac{1 - x^{kn}}{1 - x^n}. \tag{1}$$

The following table gives the first few values of $b_k(n)$ for small k.

k	Sloane	$b_k(n)$
2	A000009	1, 1, 2, 2, 3, 4, 5, 6, 8, 10, 12, 15, 18, 22, ...
3	A000726	1, 2, 2, 4, 5, 7, 9, 13, 16, 22, 27, 36, 44, 57, ...
4	A001935	1, 2, 3, 4, 6, 9, 12, 16, 22, 29, 38, 50, 64, 82, ...
5	A035959	1, 2, 3, 5, 6, 10, 13, 19, 25, 34, 44, 60, 76, 100, ...

Gordon and Ono (1997) show that

$$b_5(5n + 4) \equiv 0 \pmod{5} \tag{2}$$

$$b_7(7n + 5) \equiv 0 \pmod{7} \tag{3}$$

$$b_{11}(11n + 6) \equiv 0 \pmod{11}. \tag{4}$$

Defining $S_k(N; M)$ as the number of positive integers $n \leq N$ for which $b_k(n) \equiv 0 \pmod{M}$, Gordon and Ono (1997) proved that if $p_i^{a_i} \geq \sqrt{k}$, then

$$\lim_{N \to \infty} \frac{S_k(N; p_i^j)}{N} = 1 \tag{5}$$

for all j, where $k = p_1^{a_1} p_2^{a_2} \cdots p_m^{a_m}$.

References

Andrews, G. E. *The Theory of Partitions.* Cambridge, England: Cambridge University Press, p. 109, 1998.

Carlitz, L. "Generating Functions and Partition Problems." In *Theory of Numbers* (Ed. A. L. Whiteman). Providence, RI: Amer. Math. Soc., pp. 144–169, 1965.

Cayley, A. "A Memoir on the Transformation of Elliptic Functions." *Collected Mathematical Papers, Vol. 9.* London: Cambridge University Press, p. 128, 1889–1897.

Honsberger, R. *Mathematical Gems III.* Washington, DC: Math. Assoc. Amer., p. 241, 1985.

Gordon, B. and Ono, K. "Divisibility of Certain Partition Functions By Powers of Primes." *Ramanujan J.* **1**, 25–34, 1997.

Sloane, N. J. A. Sequences A000009/M0281, A000726/M0316, A001935/M0566, and A035959 in "An On-Line Version of the Encyclopedia of Integer Sequences." http://www.research.att.com/~njas/sequences/eisonline.html.

Partition Function P

$P(n)$, denotes also denoted $p(n)$, gives the number of ways of writing the INTEGER n as a sum of POSITIVE INTEGERS, where the order of summands is not considered significant. By convention, partitions are usually ordered from largest to smallest (Skiena 1990, p. 51). For example, since 4 can be written

$$4 = 4$$
$$= 3 + 1$$
$$= 2 + 2$$
$$= 2 + 1 + 1$$
$$= 1 + 1 + 1 + 1 \tag{1}$$

it follows that $P(4) = 5$. The function $P(n)$ is implemented in *Mathematica* as PartitionsP[n]. The values of $P(n)$ for $n = 1, 2, ...,$ are 1, 2, 3, 5, 7, 11, 15, 22, 30, 42, ... (Sloane's A000041). The following table gives the value of $P(n)$ for selected small n.

n	$P(n)$
50	204226
100	190569292
200	3972999029388
300	9253082936723602
400	6727090051741041926
500	2300165032574323995027
600	458004788008144308553622
700	60378285202834474611028659
800	5733052172321422504456911979
900	415873681190459054784114365430
1000	24061467864032622473692149727991

```
     6  · · · · · ·
    +3  · · ·
    +3  · · ·
    +2  · ·
    +1  ·
   = 15
```

When explicitly listing the partitions of a number n, the simplest form is the so-called *natural representation* which simply gives the sequence of numbers in the representation (e.g., (2, 1, 1) for the number $4 = 2 + 1 + 1$). The *multiplicity representation* instead gives the number of times each number occurs together with that number (e.g., (2, 1), (1, 2) for $4 = 2 \cdot 1 + 1 \cdot 2$). The FERRERS DIAGRAM is a pictorial representation of a partition. For example, the diagram above illustrates the FERRERS DIAGRAM of the partition $6 + 3 + 3 + 2 + 1 = 15$.

Euler gave a GENERATING FUNCTION for $P(n)$ using the Q-SERIES

$$(q)_\infty \equiv \prod_{m=1}^\infty (1 - q^m) = \sum_{n=-\infty}^\infty (-1)^n q^{(3n+1)/2} \quad (2)$$

$$= 1 - q - q^2 + q^5 + q^7 - q^{12} - q^{15} + q^{22} + q^{26} + \ldots \quad (3)$$

Here, the exponents are generalized PENTAGONAL NUMBERS 0, 1, 2, 5, 7, 12, 15, 22, 26, 35, ... (Sloane's A001318) and the sign of the kth term (counting 0 as the 0th term) is $(-1)^{\lfloor (k+1)/2 \rfloor}$ (with $\lfloor x \rfloor$ the FLOOR FUNCTION). Then the partition numbers $P(n)$ are given by the GENERATING FUNCTION

$$\frac{1}{(q)_\infty} = \sum_{n=0}^\infty P(n)q^n = 1 + q + 2q^2 + 3q^3 + 5q^4 + \ldots \quad (4)$$

(Hirschhorn 1999). Hirschhorn (1999) gives the additional beautiful identity

$$\frac{1}{(q)_\infty} = \frac{(q)_\infty^9}{(q)_\infty^{10}} = \frac{((q)_\infty^3)^3}{((q)_\infty^5)^2}. \quad (5)$$

Another GENERATING FUNCTION is given by

$$\sum_{n=0}^\infty P(n)t^n = \left(\frac{2t^{1/8}}{\vartheta_1'(0, \sqrt{t})} \right)^{1/3}, \quad (6)$$

where $\vartheta_1'(0, x)$ is the derivative of the JACOBI THETA FUNCTION of the first kind.

The number of partitions of a number n into m parts is equal to the number of partitions into parts of which the largest is m, and the number of partitions into at most m parts is equal to the number of partitions into parts which do not exceed m. Both these results follow immediately from noting that a FERRERS DIAGRAM can be read either row-wise or column-wise (although the default order is row-wise; Hardy 1999, p. 83).

For example, if $a_n = 1$ for all n, then the EULER TRANSFORM b_n is the number of partitions of n into integer parts.

Euler invented a GENERATING FUNCTION which gives rise to a POWER SERIES in $P(n)$,

$$P(n) = \sum_{k=1}^n (-1)^{k+1}$$
$$\times \left[P\left(n - \tfrac{1}{2}k(3k-1)\right) + P\left(n - \tfrac{1}{2}k(3k+1)\right) \right] \quad (7)$$

(Skiena 1990, p. 57). Other recurrence formulas include

$$P(2n+1) = P(n) + \sum_{k=1}^\infty \left[P(n - 4k^2 - 3k) + P(n - 4k^2 + 3k) \right]$$
$$- \sum_{k=1}^\infty (-1)^k \left[P(2n+1 - 3k^2 + k) + P(2n+1 - 3k^2 - k) \right] \quad (8)$$

and

$$P(n) = \frac{1}{n} \sum_{k=0}^{n-1} \sigma(n-k)P(k), \quad (9)$$

where $\sigma(n)$ is the DIVISOR FUNCTION (Skiena 1990, p. 77; Berndt 1994, p. 108), as well as the identity

$$\sum_{k=\lceil -(\sqrt{24n+1}+1)/6 \rceil}^{\lfloor (\sqrt{24n+1}-1)/6 \rfloor} (-1)^k P\left(n - \tfrac{1}{2}k(3k+1)\right) = 0, \quad (10)$$

where $\lfloor x \rfloor$ is the FLOOR FUNCTION and $\lceil x \rceil$ is the CEILING FUNCTION.

A RECURRENCE RELATION involving the PARTITION FUNCTION Q is given by

$$P(n) = \sum_{k=0}^{\lfloor n/2 \rfloor} Q(n-2k)P(k). \quad (11)$$

Atkin and Swinnerton-Dyer (1954) obtained the unexpected identities

$$\sum_{n=0}^\infty P(5n)q^n$$
$$\equiv \prod_{n=1}^\infty \frac{(1 - q^{5n-3})(1 - q^{5n-2})(1 - q^{5n})}{(1 - q^{5n-4})^2(1 - q^{5n-1})^2} \pmod 5 \quad (12)$$

$$\sum_{n=0}^\infty P(5n+1)q^n$$
$$\equiv \prod_{n=1}^\infty \frac{(1 - q^{5n})}{(1 - q^{5n-4})(1 - q^{5n-1})} \pmod 5 \quad (13)$$

$$\sum_{n=0}^\infty P(5n+2)q^n$$
$$\equiv 2 \prod_{n=1}^\infty \frac{(1 - q^{5n})}{(1 - q^{5n-3})(1 - q^{5n-2})} \pmod 5 \quad (14)$$

$$\sum_{n=0}^\infty P(5n+3)q^n$$
$$\equiv 3 \prod_{n=1}^\infty \frac{(1 - q^{5n-4})(1 - q^{5n-1})(1 - q^{5n})}{(1 - q^{5n-3})^2(1 - q^{5n-2})^2} \pmod 5 \quad (15)$$

(Hirschhorn 1999).

MacMahon obtained the beautiful RECURRENCE RELATION

$$P(n) - P(n-1) - P(n-2) + P(n-5) + P(n-7)$$

$$-P(n-12) - P(n-15) + \ldots = 0, \qquad (16)$$

where the sum is over generalized PENTAGONAL NUMBERS $\leq n$ and the sign of the kth term is $(-1)^{\lfloor (k+1)/2 \rfloor}$, as above. Ramanujan stated without proof the remarkable identities

$$P(4) + P(9)x + P(14)x^2 + \ldots$$

$$= 5 \frac{[(1-x^5)(1-x^{10})(1-x^{15}) \cdots]^5}{[(1-x)(1-x^2)(1-x^3) \cdots]^6} \qquad (17)$$

(Darling 1921; Mordell 1922; Hardy 1999, pp. 89–90), and

$$P(5) + P(12)x + P(17)x^2 + \ldots$$

$$= 7 \frac{[(1-x^7)(1-x^{14})(1-x^{21}) \cdots]^3}{[(1-x)(1-x^2)(1-x^3) \cdots]^4}$$

$$+ 49x \frac{[(1-x^7)(1-x^{14})(1-x^{21}) \cdots]^7}{[(1-x)(1-x^2)(1-x^3) \cdots]^8} \qquad (18)$$

(Mordell 1922; Hardy 1999, pp. 89–90).

Hardy and Ramanujan (1918) used the CIRCLE METHOD and MODULAR FUNCTIONS to obtain the asymptotic solution

$$P(n) \sim \frac{1}{4n\sqrt{3}} e^{\pi\sqrt{2n/3}} \qquad (19)$$

(Hardy 1999, p. 116), which was also independently discovered by Uspensky (1920). Rademacher (1937) subsequently obtained an exact convergent series solution which yields the Hardy-Ramanujan formula (19) as the first term:

$$P(n) = \frac{1}{\pi\sqrt{2}} \sum_{k=1}^{\infty} A_k(n)\sqrt{k}$$

$$\times \left\{ \frac{d}{dn'} \left[\frac{\sinh\left(\sqrt{\frac{2}{3}\left(n' - \frac{1}{24}\right)}\right)}{\sqrt{n' - \frac{1}{24}}} \right] \right\}_{n'=n}, \qquad (20)$$

where

$$A_k(n) = \sum_{h=1}^{k} \delta_{\mathrm{GCD}(h,\,k),\,1}$$

$$\times \exp\left[\pi i \sum_{j=1}^{k-1} \frac{i}{k}\frac{hj}{k}\left(\frac{hj}{k} - \left\lfloor \frac{hj}{k} \right\rfloor - \frac{1}{2} \right) - \frac{2\pi ihn}{k} \right], \qquad (21)$$

δ_{mn} is the KRONECKER DELTA, and $\lfloor x \rfloor$ is the FLOOR FUNCTION (Hardy 1999, pp. 120–121). The remainder after N terms is

$$R(N) < CN^{-1/2} + D\sqrt{\frac{N}{n}}\sinh\left(\frac{K\sqrt{n}}{N}\right), \qquad (22)$$

where C and D are fixed constants (Apostol 1997, pp. 104–110; Hardy 1999, pp. 121 and 128). Rather amazingly, the CONTOUR used by Rademacher involves FAREY SEQUENCES and FORD CIRCLES (Apostol 1997, pp. 102–104; Hardy 1999, pp. 121–122). In 1942, Erdos showed that the formula of Hardy and Ramanujan could be derived by elementary means (Hoffman 1998, p. 91).

With $f(x)$ as defined above, Ramanujan also showed that

$$5 \frac{(q^5)_\infty^5 (x^5)}{(q)_\infty^6} = \sum_{m=0}^{\infty} P(5m+4)x^m. \qquad (23)$$

Ramanujan also found numerous PARTITION FUNCTION P CONGRUENCES.

Let $f_O(x)$ be the GENERATING FUNCTION for the number of partitions $P_O(n)$ of n containing ODD numbers only and $f_D(x)$ be the GENERATING FUNCTION for the number of partitions $P_D(n)$ of n without duplication, then

$$f_O(x) = f_D(x) = \prod_{k=1,\,3,\,\ldots}^{\infty} \sum_{i=0}^{\infty} x^{ik}$$

$$= \frac{1}{\prod_{k=1,\,3,\,\ldots}^{\infty} 1 - x^k}$$

$$= \prod_{k=1}^{\infty} (1 + x^k) = 1 + x + x^2$$

$$+ 2x^3 + 2x^4 + 3x^5 + \ldots, \qquad (24)$$

as discovered by Euler (Honsberger 1985; Andrews 1998, p. 5; Hardy 1999, p. 86), giving the first few values of $P_O(n) = P_D(n)$ for $n = 0, 1, \ldots$ as 1, 1, 1, 2, 2, 3, 4, 5, 6, 8, 10, ... (Sloane's A000009). The identity

$$\prod_{k=1}^{\infty} (1 + z^k) = \prod_{k=1}^{\infty} (1 + z^{2k-1})^{-1}, \qquad (25)$$

$$= 1 - x - x^2 + x^5 + x^7 - x^{12} - x^{15} + \ldots \qquad (26)$$

$$= 1 + \sum_{k=1}^{\infty} c_k, \qquad (27)$$

where

$$c_k = \begin{cases} (-1)^n & \text{for } k \text{ of the form } \frac{1}{2}n(3n \pm 1) \\ 0 & \text{otherwise,} \end{cases} \qquad (28)$$

which is the GENERATING FUNCTION for the difference between the number of partitions into an even number of unequal parts and the number of partitions in an odd number of unequal parts, is known as the EULER IDENTITY (Hardy 1999, p. 84).

Let $P_E(n)$ be the number of partitions of EVEN numbers only, and let $P_{EO}(n)$ $(P_{DO}(n))$ be the number of partitions in which the parts are all EVEN (ODD) and all different. Then the GENERATING FUNCTION of $P_{DO}(n)$ is given by

$$f_{DO}(n) = \prod_{k=1, 3, \dots}^{\infty} 1 + x^k \qquad (29)$$

(Hardy 1999, p. 86), and the first few values of are 1, 1, 0, 1, 1, 1, 1, 1, 2, 2, 2, 2, 3, 3, 3, 4, ... (Sloane's A000700). Some additional GENERATING FUNCTIONS are given by Honsberger (1985, pp. 241–242)

$$\sum_{n=1}^{\infty} P_{\text{no even part repeated}}(n)x^n$$
$$= \prod_{k=1}^{\infty} (1 - x^{2k-1})^{-1}(1 + x^{2k}) \qquad (30)$$

$$\sum_{n=1}^{\infty} P_{\text{no part occurs more than 3 times}}(n)x^n$$
$$= \prod_{k=1}^{\infty} (1 + x^k + x^{2k} + x^{3k}) \qquad (31)$$

$$\sum_{n=1}^{\infty} P_{\text{no part divisible by 4}}(n)x^n = \prod_{k=1}^{\infty} \frac{1 - x^{4k}}{1 - x^k} \qquad (32)$$

$$\sum_{n=1}^{\infty} P_{\text{no part occurs more than } d \text{ times}}(n)x^n$$
$$= \prod_{k=1}^{\infty} \sum_{i=0}^{d} x^{ik} = \prod_{k=1}^{\infty} \frac{1 - x^{(d+1)k}}{1 - x^k} \qquad (33)$$

$$\sum_{n=1}^{\infty} P_{\text{every part occurs 2, 3, or 5 times}}(n)x^n$$
$$= \prod_{k=1}^{\infty} (1 + x^{2k} + x^{3k} + x^{5k})$$
$$= \prod_{k=1}^{\infty} (1 + x^{2k})(1 + x^{3k}) = \prod_{k=1}^{\infty} \frac{1 - x^{4k}}{1 - x^{2k}} \frac{1 - x^{6k}}{1 - x^{3k}} \qquad (34)$$

$$\sum_{n=1}^{\infty} P_{\text{no part occurs exactly once}}(n)x^n$$
$$= (1 + x^{2k} + x^{3k} + \dots) = \prod_{k} \frac{1 + x^{6k}}{(1 - x^{2k})(1 - x^{3k})}. \qquad (35)$$

Some additional interesting theorems following from these (Honsberger 1985, pp. 64–68 and 143–146) are:

1. The number of partitions of n in which no EVEN part is repeated is the same as the number of partitions of n in which no part occurs more than three times and also the same as the number of partitions in which no part is divisible by four.
2. The number of partitions of n in which no part occurs more often than d times is the same as the

number of partitions in which no term is a multiple of $d + 1$.
3. The number of partitions of n in which each part appears either 2, 3, or 5 times is the same as the number of partitions in which each part is CONGRUENT mod 12 to either 2, 3, 6, 9, or 10.
4. The number of partitions of n in which no part appears exactly once is the same as the number of partitions of n in which no part is CONGRUENT to 1 or 5 mod 6.
5. The number of partitions in which the parts are all EVEN and different is equal to the absolute difference of the number of partitions with ODD and EVEN parts.

$P(n)$ satisfies the inequality

$$P(n) \le \tfrac{1}{2}[(n+1) + P(n-1)] \qquad (36)$$

(Honsberger 1991).

$P(n, k)$, also written $P_k(n)$, is the number of ways of writing n as a sum of k terms or, equivalently, the number of partitions into parts of which the largest is k. The latter can be enumerated by Partitions[n, k] in the *Mathematica* add-on package Discrete-Math`Combinatorica` (which can be loaded with the command << DiscreteMath`). For example, the $P(5, 3) = 5$ partitions of 5 of which the largest member is ≤ 3 are {3, 2}, {3, 1, 1}, {2, 2, 1}, {2, 1, 1, 1}, and {1, 1, 1, 1, 1}. Similarly, the five partitions of 5 into three or fewer parts are {5}, {4, 1}, {3, 2}, {3, 1, 1}, and {2, 2, 1}.

$P(n, k)$ is implemented as ConstrainedInteger-PartitionsP[n, k] in the *Mathematica* add-on package DiscreteMath`IntegerPartitions` (which can be loaded with the command << DiscreteMath`), and can be computed from the RECURRENCE RELATION

$$P(n, k) = P(n-1, k-1) + P(n-k, k) \qquad (37)$$

(Skiena 1990, p. 58; Ruskey) with $P(n, k) = 0$ for $k > n$, $P(n, n) = 1$, and $P(n, 0) = 0$. The triangle of $P(k, n)$ is given by

1

1 1

1 1 1

1 2 1 1

1 2 2 1 1

1 3 3 2 1 1

(Sloane's A008284). The number of partitions of n with largest part k is the same as $P(n, k)$.

The RECURRENCE RELATION can be solved exactly to give

$$P(n, 1) = 1 \qquad (38)$$

$$P(n, 2) = \tfrac{1}{4}\left[2n - 1 + (-1)^n\right] \qquad (39)$$

$$P(n, 3) = \tfrac{1}{72}\left[6n^2 - 7 - 9(-1)^n + 16\cos\left(\tfrac{2}{3}n\pi\right)\right] \qquad (40)$$

$$P(n, 4) = \tfrac{1}{864}\{3(n+1)[2n(n+2) - 13 + 9(-1)^n]$$

$$-96\cos\left(\tfrac{2}{3}n\pi\right) + 108(-1)^{n/2}\bmod(n+1, 2)$$

$$+32\sqrt{3}\sin(\tfrac{2}{3}n\pi)\}, \qquad (41)$$

where $P(n, k) = 0$ for $n < k$. The functions $P(n, k)$ can also be given explicitly for the first few values of k in the simple forms

$$P(n, 2) = \left\lfloor \tfrac{1}{2}n \right\rfloor \qquad (42)$$

$$P(n, 3) = \left[\tfrac{1}{12}n^2 \right], \qquad (43)$$

where $\lfloor x \rfloor$ is the FLOOR FUNCTION and $[x]$ is the NINT function (Honsberger 1985, pp. 40–45). A similar treatment by B. Schwennicke defines

$$t_k(n) = n + \tfrac{1}{4}k(k - 3) \qquad (44)$$

and then yields

$$P(n, 2) = \left[\tfrac{1}{2}t_2(n) \right] \qquad (45)$$

$$P(n, 3) = \left[\tfrac{1}{12}t_2^3(n) \right] \qquad (46)$$

$$P(n, 4) = \begin{cases} \left[\tfrac{1}{144}t_4^3(n) - \tfrac{1}{48}t_4(n) \right] & \text{for } n \text{ even} \\ \left[\tfrac{1}{144}t_4^3(n) - \tfrac{1}{12}t_4(n) \right] & \text{for } n \text{ odd.} \end{cases} \qquad (47)$$

Hardy and Ramanujan (1918) obtained the exact asymptotic formula

$$P(n) = \sum_{k < \alpha\sqrt{n}} P_k(n) + \mathcal{O}(n^{-1/4}), \qquad (48)$$

where α is a constant. However, the sum

$$\sum_{k=1}^{\infty} P_k(n) \qquad (49)$$

diverges, as first shown by Lehmer (1937).

See also ALCUIN'S SEQUENCE, CONJUGATE PARTITION, ELDER'S THEOREM, EULER IDENTITY, FERRERS DIAGRAM, GÖLLNITZ'S THEOREM, PARTITION FUNCTION P CONGRUENCES, PARTITION FUNCTION Q, PENTAGONAL NUMBER, PENTAGONAL NUMBER THEOREM, PLANE PARTITION, RANDOM PARTITION, ROGERS-RAMANUJAN IDENTITIES, SELF-CONJUGATE PARTITION, STANLEY'S THEOREM, SUM OF SQUARES FUNCTION, TAU FUNCTION

References

Abramowitz, M. and Stegun, C. A. (Eds.). "Unrestricted Partitions." §24.2.1 in *Handbook of Mathematical Functions with Formulas, Graphs, and Mathematical Tables, 9th printing.* New York: Dover, p. 825, 1972.

Adler, H. "Partition Identities--From Euler to the Present." *Amer. Math. Monthly* **76**, 733–746, 1969.

Adler, H. "The Use of Generating Functions to Discover and Prove Partition Identities." *Two-Year College Math. J.* **10**, 318–329, 1979.

Andrews, G. E. *The Theory of Partitions.* Cambridge, England: Cambridge University Press, 1998.

Apostol, T. M. Ch. 4 in *Introduction to Analytic Number Theory.* New York: Springer-Verlag, 1976.

Apostol, T. M. "Rademacher's Series for the Partition Function." Ch. 5 in *Modular Functions and Dirichlet Series in Number Theory, 2nd ed.* New York: Springer-Verlag, pp. 94–112, 1997.

Atkin, A. O. L. and Swinnerton-Dyer, P. "Some Properties of Partitions." *Proc. London Math. Soc.* **4**, 84–106, 1954.

Berndt, B. C. *Ramanujan's Notebooks, Part IV.* New York: Springer-Verlag, 1994.

Comtet, L. *Advanced Combinatorics: The Art of Finite and Infinite Expansions, rev. enl. ed.* Dordrecht, Netherlands: Reidel, p. 307, 1974.

Conway, J. H. and Guy, R. K. *The Book of Numbers.* New York: Springer-Verlag, pp. 94–96, 1996.

David, F. N.; Kendall, M. G.; and Barton, D. E. *Symmetric Function and Allied Tables.* Cambridge, England: Cambridge University Press, p. 219, 1966.

Gupta, H. "A Table of Partitions." *Proc. London Math. Soc.* **39**, 142–149, 1935.

Gupta, H. "A Table of Partitions (II)." *Proc. London Math. Soc.* **42**, 546–549, 1937.

Gupta, H.; Gwyther, A. E.; and Miller, J. C. P. *Tables of Partitions.* London: Royal Society Mathematical Tables, Vol. 4, 1958.

Hardy, G. H. "Ramanujan's Work on Partitions" and "Asymptotic Theory of Partitions." Chs. 6 and 8 in *Ramanujan: Twelve Lectures on Subjects Suggested by His Life and Work, 3rd ed.* New York: Chelsea, pp. 83–100 and 113–131, 1999.

Hardy, G. H. and Ramanujan, S. "Asymptotic Formulae in Combinatory Analysis." *Proc. London Math. Soc.* **17**, 75–115, 1918.

Hardy, G. H. and Wright, E. M. *An Introduction to the Theory of Numbers, 5th ed.* Oxford, England: Clarendon Press, 1979.

Hoffman, P. *The Man Who Loved Only Numbers: The Story of Paul Erdos and the Search for Mathematical Truth.* New York: Hyperion, 1998.

Honsberger, R. *Mathematical Gems III.* Washington, DC: Math. Assoc. Amer., pp. 40–45 and 64–68, 1985.

Honsberger, R. *More Mathematical Morsels.* Washington, DC: Math. Assoc. Amer., pp. 237–239, 1991.

Jackson, D. and Goulden, I. *Combinatorial Enumeration.* New York: Academic Press, 1983.

Lehmer, D. H. "On the Hardy-Ramanujan Series for the Partition Function." *J. London Math. Soc.* **12**, 171–176, 1937.

Lehmer, D. H. "On a Conjecture of Ramanujan." *J. London Math. Soc.* **11**, 114–118, 1936.

Lehmer, D. H. "The Series for the Partition Function." *Trans. Amer. Math. Soc.* **43**, 271–295, 1938.

Lehmer, D. H. "On the Remainders and Convergence of the Series for the Partition Function." *Trans. Amer. Math. Soc.* **46**, 362–373, 1939.

MacMahon, P. A. "Note of the Parity of the Number which Enumerates the Partitions of a Number." *Proc. Cambridge Philos. Soc.* **20**, 281–283, 1921.

MacMahon, P. A. "The Parity of $p(n)$, the Number of Partitions of n, when $n \leq 1000$." *J. London Math. Soc.* **1**, 225–226, 1926.

MacMahon, P. A. *Combinatory Analysis.* New York: Chelsea, 1960.

Rademacher, H. "Zur Theorie der Modulfunktionen." *J. reine angew. Math.* **167**, 312–336, 1932.

Rademacher, H. "On the Partition Function $p(n)$." *Proc. London Math. Soc.* **43**, 241–254, 1937.

Rademacher, H. "On the Expansion of the Partition Function in a Series." *Ann. Math.* **44**, 416–422, 1943.

Ruskey, F. "Information of Numerical Partitions." http://www.theory.csc.uvic.ca/~cos/inf/nump/NumPartition.html.

Sloane, N. J. A. Sequences A000009/M0281, A000041/M0663, A000700/M0217, A001318/M1336, and A008284 in "An On-Line Version of the Encyclopedia of Integer Sequences." http://www.research.att.com/~njas/sequences/eisonline.html.

Sloane, N. J. A. and Plouffe, S. *The Encyclopedia of Integer Sequences.* San Diego, CA: Academic Press, 1995.

Uspensky, J. V. "Asymptotic Formulae for Numerical Functions Which Occur in the Theory of Partitions.' *Bull. Acad. Sci. URSS* **14**, 199–218, 1920.

Partition Function P Congruences

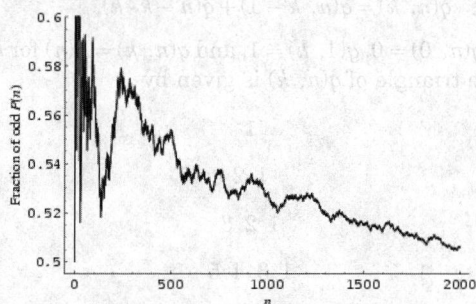

The fraction of odd values of the PARTITION FUNCTION P is roughly 50%, independent of n, whereas odd values of $Q(n)$ occur with ever decreasing frequency as n becomes large. Kolberg (1959) proved that there are infinitely many even and odd values of $P(n)$.

Leibniz noted that $P(n)$ is prime for $n = 2, 3, 4, 5, 6$, but not 7. In fact, values of n for which $P(n)$ is PRIME are 2, 3, 4, 5, 6, 13, 36, 77, 132, 157, 168, 186, ... (Sloane's A046063), corresponding to 2, 3, 5, 7, 11, 101, 17977, 10619863, ... (Sloane's A049575). Numbers which cannot be written as a PRODUCT of $P(n)$ are 13, 17, 19, 23, 26, 29, 31, 34, 37, 38, 39, ... (Sloane's A046064), corresponding to numbers of nonisomorphic ABELIAN GROUPS which are not possible for any group order.

Ramanujan conjectured a number of amazing and unexpected CONGRUENCES involving $P(n)$. In particular, he proved

$$P(5m + 4) \equiv 0 \pmod 5 \tag{1}$$

using RAMANUJAN'S IDENTITY (Darling 1919; Hardy and Wright 1979; Drost 1997; Hardy 1999, pp. 87–88; Hirschhorn 1999). Ramanujan (1919) also showed that

$$P(25m + 24) \equiv 0 \pmod{5^2}, \tag{2}$$

and Krecmar (1933) proved that

$$P(125m + 99) \equiv 0 \pmod{5^3}. \tag{3}$$

Watson (1938) then proved the general congruence

$$P(n) \equiv 0 \pmod{5^a} \quad \text{if } 24n \equiv 1 \pmod{5^a} \tag{4}$$

(Gordon and Hughes 1981; Hardy 1999, p. 89). For $a = 1, 2, ...$, the corresponding minimal values of n are 4, 24, 99, 599, 2474, 14974, 61849, ... (Sloane's A052463). However, the even more general congruences

$$P(125m + 74, 99, 124) \equiv 0 \pmod{5^3} \tag{5}$$

$$P(3125m + 1849, 2474, 3099) \equiv 0 \pmod{5^5} \tag{6}$$

seem also to hold.

Ramanujan showed that

$$P(7m + 5) \equiv 0 \pmod 7 \tag{7}$$

(Darling 1919), which can be derived using the EULER IDENTITY and JACOBI TRIPLE PRODUCT (Hardy 1999, pp. 87–88), and also that

$$P(49m + 47) \equiv 0 \pmod{7^2} \tag{8}$$

(Hardy 1999, p. 90). He conjectured that in general

$$P(n) \equiv 0 \pmod{7^b} \quad \text{if } 24n \equiv 1 \pmod{7^b} \tag{9}$$
[incorrect]

(Gordon and Hughes 1981, Hardy 1999), although Gupta (1936) showed that this is *false* when $b = 3$. Watson (1938) subsequently formulated and proved the modified relation

$$P(n) \equiv 0 \pmod{7^b} \quad \text{if } 24n \equiv 1 \pmod{7^{2b-2}} \tag{10}$$

for $b \geq 2$. For $b = 1, 2, ...$, the corresponding minimal values of n are 0, 47, 2301, 112747, ... (Sloane's A052464). However, the even more general congruences

$$P(49m + 19, 33, 40, 47) \equiv 0 \pmod{7^2} \tag{11}$$

appear to hold.

Ramanujan showed that

$$P(11m + 6) \equiv 0 \pmod{11} \tag{12}$$

holds (Gordon and Hughes 1981; Hardy 1999, pp. 87–88), and conjectured the general relation

$$P(n) \equiv 0 \pmod{11^c} \quad \text{if } 24n \equiv 1 \pmod{11^c}. \tag{13}$$

This was finally proved by Atkin (1967). For $c = 1, 2, ...$, the corresponding minimal values of n are 6, 116, 721, 14031, ... (Sloane's A052465).

Atkin and O'Brien (1967) proved

$$P(169n - 7) \equiv \kappa_d P(n) \pmod{13^d} \tag{14}$$

$$\text{if } 24n \equiv 1 \pmod{13^d},$$

where κ_d is an integer depending only on d (Gordon and Hughes 1981). For $d = 1, 2, \ldots$, the corresponding minimal values of n are 6, 162, 1007, 27371, ... (Sloane's A052466).

Subbarao (1966) conjectured that in every ARITHMETIC PROGRESSION r (mod t), there are infinitely many integers $N \equiv r$ (mod t) for which $P(N)$ is EVEN, and infinitely many integer $M \equiv r$ (mod t) for which $P(M)$ is ODD.

See also CONGRUENCE, ERDOS-IVIC CONJECTURE, NEWMAN'S CONJECTURE, PARTITION FUNCTION P, PARTITION FUNCTION q, PARTITION FUNCTION Q, PARTITION FUNCTION Q CONGRUENCES

References

Atkin, A. O. L. "Proof of a Conjecture of Ramanujan." *Glasgow Math. J.* **8**, 14–32, 1967.

Atkin, A. O. L. and O'Brien, J. N. "Some Properties of $p(n)$ and $c(n)$ Modulo Powers of 13." *Trans. Amer. Math. Soc.* **126**, 442–459, 1967.

Chowla, S. "Congruence Properties of Partitions." *J. London Math. Soc.* **9**, 247, 1934.

Darling, H. B. C. "Proofs of Certain Identities and Congruences Enunciated by S. Ramanujan." *Proc. London Math. Soc.* **19**, 350–372, 1921.

Darling, H. B. C. "On Mr. Ramanujan's Congruence Properties of $p(n)$." *Proc. Cambridge Philos. Soc.* **19**, 217–218, 1919.

Drost, J. L. "A Shorter Proof of the Ramanujan Congruence mod 5." *Amer. Math. Monthly* **104**, 963–964, 1997.

Getz, J. "On Congruence Properties of the Partition Function." *Internat. J. Math. Math. Sci.* **23**, 493–496, 2000.

Gordon, B. and Hughes, K. "Ramanujan Congruences for $q(n)$." In *Analytic Number Theory, Proceedings of the Conference Held at Temple University, Philadelphia, Pa., May 12–15, 1980* (Ed. M. I. Knopp). New York: Springer-Verlag, pp. 333–359, 1981.

Gupta, H. "On a Conjecture of Ramanujan." *Proc. Indian Acad. Sci. (A)* **4**, 625–629, 1936.

Hardy, G. H. *Ramanujan: Twelve Lectures on Subjects Suggested by His Life and Work, 3rd ed.* New York: Chelsea, 1999.

Hardy, G. H. and Wright, E. M. *An Introduction to the Theory of Numbers, 5th ed.* Oxford, England: Clarendon Press, 1979.

Hirschhorn, M. D. "Another Short Proof of Ramanujan's Mod 5 Partition Congruences, and More." *Amer. Math. Monthly* **106**, 580–583, 1999.

Kolberg, O. "Note on the Parity of the Partition Function." *Math. Scand.* **7**, 377–378, 1959.

Krecmar, W. "Sur les propriétés de la divisibilité d'une fonction additive." *Bull. Acad. Sci. URSS* **7**, 763–800, 1933.

Lehmer, D. H. "An Application of Schläfli's Modular Equation to a Conjecture of Ramanujan." *Bull. Amer. Math. Soc.* **44**, 84–90, 1938.

Mordell, L. J. "Note on Certain Modular Relations Considered by Messrs Ramanujan, Darling and Rogers." *Proc. London Math. Soc.* **20**, 408–416, 1922.

Ono, K. "Parity of the Partition Function in Arithmetic Progressions." *J. reine. angew. Math.* **472**, 1–15, 1996.

Ono, K. "The Partition Function in Arithmetic Progressions." *Math. Ann.* **312**, 251–260, 1998.

Ono, K. "Distribution of the Partition Function Modulo m." *Ann. Math.* **151**, 293–307, 2000.

Ramanujan, S. "Some Properties of $p(n)$, the Number of Partitions of n." *Proc. Cambridge Philos. Soc.* **19**, 207–210, 1919.

Ramanujan, S. "Congruence Properties of Partitions." *Math. Z.* **9**, 147–153, 1921.

Sloane, N. J. A. Sequences A046063, A046064, A049575, A052462, A052463, A052464, A052465, and A052466 in "An On-Line Version of the Encyclopedia of Integer Sequences." http://www.research.att.com/~njas/sequences/eisonline.html.

Subbarao, M. V. "Some Remarks on the Partition Function." *Amer. Math. Monthly* **73**, 851–854, 1966.

Watson, G. N. "Ramanujans Vermutung über Zerfällungsanzahlen." *J. für Math.* **179**, 97–128, 1938.

Partition Function q

The number of PARTITIONS of n with $\leq k$ summands is denoted $q(n, k)$ or $q_k(n)$. For example, $q(10, 2) = 6$, since there are six partitions of 10 into two or fewer parts: $\{10\}$, $\{9, 1\}$, $\{8, 2\}$, $\{7, 3\}$, $\{6, 4\}$, and $\{5, 5\}$. The $q(n, k)$ satisfy the RECURRENCE RELATION

$$q(n, k) = q(n, k - 1) + q(n - k, k), \tag{1}$$

with $q(n, 0) = 0$, $q(1, k) = 1$, and $q(n, k) = P(n)$ for $k \geq n$. The triangle of $q(n, k)$ is given by

$$1$$
$$1 \quad 2$$
$$1 \quad 2 \quad 3$$
$$1 \quad 3 \quad 4 \quad 5$$
$$1 \quad 3 \quad 5 \quad 6 \quad 7$$
$$1 \quad 4 \quad 7 \quad 9 \quad 10 \quad 11$$

(Sloane's A026820).

See also PARTITION FUNCTION P, PARTITION FUNCTION Q

References

Sloane, N. J. A. Sequences A026820 in "An On-Line Version of the Encyclopedia of Integer Sequences." http://www.research.att.com/~njas/sequences/eisonline.html.

Partition Function Q

$Q(n)$ gives the number of ways of writing the INTEGER n as a sum of POSITIVE INTEGERS without regard to order with the constraint that all INTEGERS in a given partition are *distinct*. For example, $Q(10) = 10$, since the partitions of 10 into distinct parts are $\{1, 2, 3, 4\}$, $\{2, 3, 5\}$, $\{1, 4, 5\}$, $\{1, 3, 6\}$, $\{4, 6\}$, $\{1, 2, 7\}$, $\{3, 7\}$, $\{2, 8\}$, $\{1, 9\}$, $\{10\}$. The $Q(n)$ function is implemented in *Mathematica* as PartitionsQ[n]. $Q(0)$ is generally defined to be 1. The values for $n = 1, 2, \ldots$ are 1, 1, 2, 2, 3, 4, 5, 6, 8, 10, ... (Sloane's A000009).

The GENERATING FUNCTION for $Q(n)$ is

$$G(x) = \prod_{n=1}^{\infty} (1 + x^n) \tag{1}$$

$$= \frac{1}{\prod_{n=0}^{\infty} (1 - x^{2n+1})} \tag{2}$$

$$= \prod_{n=1}^{\infty} \frac{1 - x^{2n}}{1 - x^n} \tag{3}$$

$$= 1 + x + x^2 + 2x^3 + 2x^4 + 3x^5 + \ldots. \tag{4}$$

This can also be interpreted as another form of the JACOBI TRIPLE PRODUCT, written in terms of the Q-FUNCTIONS as

$$Q_1 Q_2 Q_3 = 1 \tag{5}$$

(Borwein and Borwein 1987, p. 64).

A RECURRENCE RELATION is given by $Q(0) = Q(1) = 1$ and

$$Q(n) = \frac{1}{n} \sum_{k=1}^{n} [s(k) - 2s(k/2)] Q(n-k), \tag{6}$$

where

$$s(n) = \begin{cases} \sigma_1(n) & \text{for } n \text{ an integer} \\ 0 & \text{otherwise,} \end{cases} \tag{7}$$

and

$$\sigma_1(n) \equiv s(n) - 2s(n/2) \tag{8}$$

is the ODD DIVISOR FUNCTION giving the sum of odd divisors of n: 1, 1, 4, 1, 6, 4, 8, ... (Sloane's A000593; Abramowitz and Stegun 1972, p. 826).

$Q(n)$ satisfies the inequality

$$Q(n) \le \frac{1}{2}[Q(n+1) + Q(n-1)] \tag{9}$$

for $n \ge 4$. $Q(n)$ has the ASYMPTOTIC SERIES

$$Q(n) \sim \frac{e^{\pi\sqrt{n/3}}}{4 \cdot 3^{1/4} n^{3/4}} \tag{10}$$

(Abramowitz and Stegun 1972, p. 826).

A Rademacher-like convergent series for $Q(n)$ is given by

$$Q(n) = \frac{1}{2}\sqrt{2} \sum_{k=1}^{\infty} A_{2k-1}(n)$$

$$\times \left\{ \frac{d}{dn'} \left[J_0\left(\frac{\pi i}{2k-1}, \sqrt{\frac{1}{3}\left(n' + \frac{1}{24}\right)} \right) \right] \right\}_{n'=n}, \tag{11}$$

where

$$A_k(n) = \sum_{h=1}^{k} \delta_{\text{GCD}(h, k), 1}$$

$$\times \exp\left[\pi i \sum_{j=1}^{k-1} \frac{i}{k} \frac{j}{k} \left(\frac{hj}{k} - \left\lfloor \frac{hj}{k} \right\rfloor - \frac{1}{2} \right) - \frac{2\pi i hn}{k} \right], \tag{12}$$

where δ_{mn} is the KRONECKER DELTA, $\lfloor x \rfloor$ is the FLOOR FUNCTION, and $J_0(x)$ is the zeroth order BESSEL FUNCTION OF THE FIRST KIND (Abramowitz and Stegun 1972, p. 825). (11) can also be written explicitly as

$$Q(n) = \frac{\pi^2\sqrt{2}}{24} \sum_{k=1}^{\infty} \frac{A_{2k-1}(n)}{(1-2k)^2} {}_0F_1\left(; 2; \frac{(1+24n)\pi^2}{288(1-2k)^2} \right), \tag{13}$$

where ${}_0F_1(; a; b; z)$ is a GENERALIZED HYPERGEOMETRIC FUNCTION.

Let $Q(n, k)$ denote the number of ways of partitioning n into exactly k *distinct* parts. For example, $Q(10, 3) = 4$ since there are four partitions of 10 into three distinct parts: $\{1, 2, 7\}$, $\{1, 3, 6\}$, $\{1, 4, 5\}$, and $\{2, 3, 5\}$. $Q(n, k)$ is given by

$$Q(n, k) = P\left(n - \binom{k}{2}, k \right), \tag{14}$$

where $P(n)$ is the PARTITION FUNCTION P and $\binom{n}{k}$ is a BINOMIAL COEFFICIENT (Comtet 1974, p. 116). The following table gives the first few values of $Q(n, k)$ (Sloane's A008289; Comtet 1974, pp. 115–116).

$n\backslash k$	1	2	3	4
1	1			
2	1			
3	1	1		
4	1	1		
5	1	2		
6	1	2	1	
7	1	3	1	
8	1	3	2	
9	1	4	3	
10	1	4	4	1

See also ODD DIVISOR FUNCTION, PARTITION FUNCTION P, PARTITION FUNCTION q, PARTITION FUNCTION Q CONGRUENCES

References

Abramowitz, M. and Stegun, C. A. (Eds.). "Partitions into Distinct Parts." §24.2.2 in *Handbook of Mathematical Functions with Formulas, Graphs, and Mathematical Tables, 9th printing.* New York: Dover, pp. 825–826, 1972.

Borwein, J. M. and Borwein, P. B. *Pi & the AGM: A Study in Analytic Number Theory and Computational Complexity.* New York: Wiley, 1987.

Comtet, L. *Advanced Combinatorics: The Art of Finite and Infinite Expansions, rev. enl. ed.* Dordrecht, Netherlands: Reidel, p. 114–115, 1974.

Skiena, S. *Implementing Discrete Mathematics: Combinatorics and Graph Theory with Mathematica.* Reading, MA: Addison-Wesley, p. 58, 1990.

Sloane, N. J. A. Sequences A000009/M0281, A000593/M3197, and A008289 in "An On-Line Version of the Encyclopedia of Integer Sequences." http://www.research.att.com/~njas/sequences/eisonline.html.

Partition Function Q Congruences

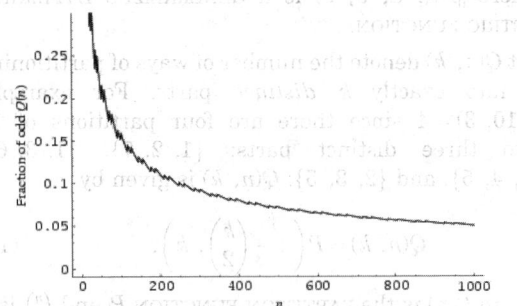

Odd values of $Q(n)$ are 1, 1, 3, 5, 27, 89, 165, 585, ... (Sloane's A051044), and occur with ever decreasing frequency as n becomes large (unlike $P(n)$, for which the fraction of odd values remains roughly 50%). This follows from the PENTAGONAL NUMBER THEOREM which gives

$$G(x) = \prod_{n=1}^{\infty}(1+x^n) \equiv \prod_{n=1}^{\infty}(1-x^n)$$

$$\equiv \sum_{n=-\infty}^{\infty} x^{(3n^2+n)/2} \pmod 2 \qquad (1)$$

(Gordon and Ono 1997), so $Q(n)$ is ODD IFF n is OF THE FORM $k(3k \pm 1)/2$, i.e., 1, 5, 12, 22, 35, ... or 2, 7, 15, 26, 40, ...

The values of n for which $Q(n)$ is PRIME are 3, 4, 5, 7, 22, 70, 100, 495, 1247, 2072, 320397, ... (Sloane's A035359), with no others for $n \leq 3,015,000$ (Weisstein, May 6, 2000). These values correspond to 2, 2, 3, 5, 89, 29927, 444793, 602644050950309, ... (Sloane's A051005). It is not known if $Q(n)$ is infinitely often prime, but Gordon and Ono (1997) proved that it is "almost always" divisible by any given power of 2 (1997).

Gordon and Hughes (1981) showed that

$$Q(n) \equiv 0 \pmod{5^a} \quad \text{if } 24n \equiv -1 \pmod{5^{2a+1}} \qquad (2)$$

and

$$Q(n) \equiv 49n + 2 \pmod{\lambda_b Q(n)} 7^b \qquad (3)$$
$$\text{if } 24n \equiv -1 \pmod{7^b},$$

where λ_b is an integer depending only on b.

See also PARTITION FUNCTION P, PARTITION FUNCTION P CONGRUENCES, PARTITION FUNCTION Q

References

Gordon, B. and Hughes, K. "Ramanujan Congruences for $q(n)$." In *Analytic Number Theory, Proceedings of the Conference Held at Temple University, Philadelphia, Pa., May 12–15, 1980* (Ed. M. I. Knopp). New York: Springer-Verlag, pp. 333–359, 1981.

Gordon, B. and Ono, K. "Divisibility of Certain Partition Functions by Powers of Primes." *Ramanujan J.* **1**, 25–34, 1997.

Sloane, N. J. A. Sequences A035359, A051005, and A051044 in "An On-Line Version of the Encyclopedia of Integer Sequences." http://www.research.att.com/~njas/sequences/eisonline.html.

Partition of Unity

Given a SMOOTH MANIFOLD M with an OPEN COVER U_i, a partition of unity is a collection of smooth, nonnegative functions ψ_i, such that the support of ψ_i is contained in U_i and $\Sigma_i \psi_i = 1$ everywhere. Often one requires that the U_i have COMPACT CLOSURE, which can be interpreted as finite, or bounded, open sets. In the case that the U_i is a LOCALLY finite cover, any point $x \in M$ has only finitely many i with $\psi_i(x) \neq 0$.

A partition of unity can be used to patch together objects defined locally. For instance, there always exist smooth GLOBAL VECTOR FIELDS, possibly vanishing somewhere, but not identically zero. Cover M with coordinate charts U_i such that only finitely many overlap at any point. On each coordinate chart U_i, there are the local vector fields $\partial/\partial x_j$. Label these $v_{i,j}$ and, for each chart, pick the vector field $v_{i,1} = \partial/\partial x_1$. Then $\Sigma_i \psi_i v_{i,1}$ is a global vector field. The sum converges because at any x, only finitely many $\psi_i(x) \neq 0$.

Other applications require the objects to be interpreted as functions, or a generalization of functions called SECTIONS, such as a RIEMANNIAN METRIC. By viewing such a metric as a section of a bundle, it is easy to show the existence of a smooth metric on any smooth manifold. The proof uses a partition of unity and is similar to the one used above.

Strictly speaking, the sum $\Sigma_i \psi_i$ doesn't have to be identically UNITY for the arguments to work. It goes with the name, because at every point the functions partition the value 1. Also, it is convenient when considered from the point of view of CONVEXITY.

See also CONVEX SET, OPEN COVER RIEMANNIAN METRIC, SECTION, SMOOTH MANIFOLD, VECTOR FIELD

PartitionsP

PARTITION FUNCTION P

PartitionsQ

PARTITION FUNCTION Q

Party Problem

Also known as the MAXIMUM CLIQUE PROBLEM. Find the minimum number of guests that must be invited so that at least m will know each other or at least n will not know each other. The solutions are known as RAMSEY NUMBERS.

See also CLIQUE, RAMSEY NUMBER

References

Hoffman, P. *The Man Who Loved Only Numbers: The Story of Paul Erdos and the Search for Mathematical Truth.* New York: Hyperion, p. 52, 1998.

Parzen Apodization Function

An APODIZATION FUNCTION similar to the BARTLETT FUNCTION.

See also APODIZATION FUNCTION, BARTLETT FUNCTION

References

Press, W. H.; Flannery, B. P.; Teukolsky, S. A.; and Vetterling, W. T. *Numerical Recipes in FORTRAN: The Art of Scientific Computing, 2nd ed.* Cambridge, England: Cambridge University Press, p. 547, 1992.

Pascal Distribution

NEGATIVE BINOMIAL DISTRIBUTION

Pascal Lines

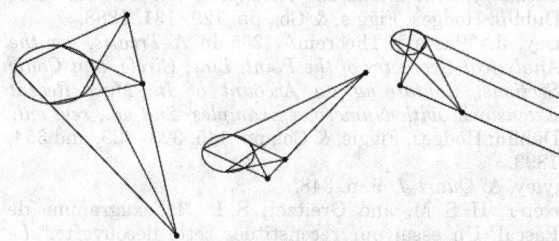

The lines containing the three points of the intersection of the three pairs of opposite sides of a (not necessarily regular) HEXAGON.

There are 6! (i.e., 6 FACTORIAL) possible ways of taking all VERTICES in any order, but among these are six equivalent CYCLIC PERMUTATIONS and two possible orderings, so the total number of different

hexagons (not all simple) is

$$\frac{6!}{2 \cdot 6} = \frac{720}{12} = 60.$$

There are therefore a total of 60 Pascal lines created by connecting VERTICES in any order.

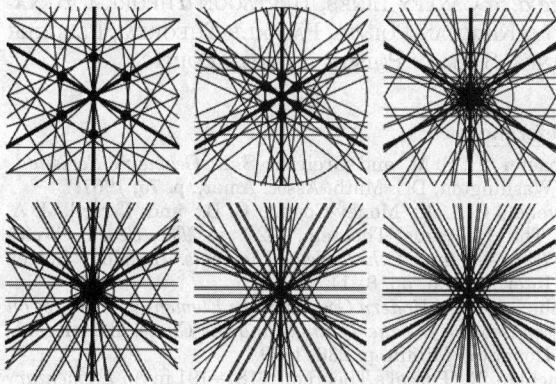

The 60 Pascal lines form a very complicated pattern which can be visualized most easily in the degenerate case of a regular hexagon inscribed in a circle, as illustrated above for magnifications ranging over five powers of 2. Only 45 lines are visible in this figure since each of the three thick lines (located at $60°$ angles to each other) represents a degenerate group of four Pascal lines, and six of the Pascal lines are LINES AT INFINITY (Wells 1991).

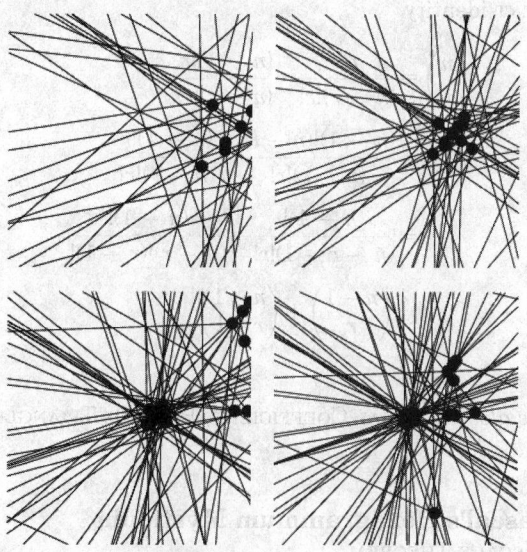

The pattern for a general ellipse and hexagon (illustrated above) is much more complicated, and is difficult to distinguish from a clutter of lines.

The 60 Pascal lines intersect three at a time through 20 STEINER POINTS (some of which are shown as the filled circles in the above figures). In the symmetrical case of the regular hexagon inscribed in a CIRCLE, the 20 Steiner points degenerate into seven distinct points arranged at the vertices and center of a regular hexagon centered at the origin of the circle. The 60 Pascal line also intersect three at a time in 60 KIRKMAN POINTS. Each Steiner point lines together with three Kirkman points on a total of 20 CAYLEY LINES. There is a dual relationship between the 60 Pascal lines and the 60 KIRKMAN POINTS.

See also CAYLEY LINES, HEPTAGON THEOREM, HEXA-GON, KIRKMAN POINTS, PASCAL'S THEOREM, PLÜCKER LINES, SALMON POINTS, STEINER POINTS

References

Coxeter, H. S. M. and Greitzer, S. L. *Geometry Revisited.* Washington, DC: Math. Assoc. Amer., p. 75, 1967.

Evelyn, C. J. A.; Money-Coutts, G. B.; and Tyrrell, J. A. "The Heptagon Theorem." §2.1 in *The Seven Circles Theorem and Other New Theorems.* London: Stacey International, pp. 8–11, 1974.

Johnson, R. A. *Modern Geometry: An Elementary Treatise on the Geometry of the Triangle and the Circle.* Boston, MA: Houghton Mifflin, p. 236, 1929.

Lachlan, R. "Pascal's Theorem." §181–191 in *An Elementary Treatise on Modern Pure Geometry.* London: Macmillian, pp. 113–119, 1893.

Wells, D. *The Penguin Dictionary of Curious and Interesting Geometry.* London: Penguin, pp. 172–173, 1991.

Pascal's Formula

Each subsequent row of PASCAL'S TRIANGLE is obtained by adding the two entries diagonally above. This follows immediately from the BINOMIAL COEFFICIENT identity

$$\binom{n}{r} \equiv \frac{n!}{(n-r)!\,r!} = \frac{(n-1)!\,n}{(n-r)!\,r!}$$

$$= \frac{(n-1)!(n-r)}{(n-r)!\,r!} + \frac{(n-1)!\,r}{(n-r)!\,r!}$$

$$= \frac{(n-1)!}{(n-r-1)!\,r!} + \frac{(n-1)!}{(n-r)!(r-1)!}$$

$$= \binom{n-1}{r} + \binom{n-1}{r-1}.$$

See also BINOMIAL COEFFICIENT, PASCAL'S TRIANGLE

Pascal's Hexagrammum Mysticum

PASCAL'S THEOREM

Pascal's Limaçon

LIMAÇON

Pascal's Rule

PASCAL'S FORMULA

Pascal's Theorem

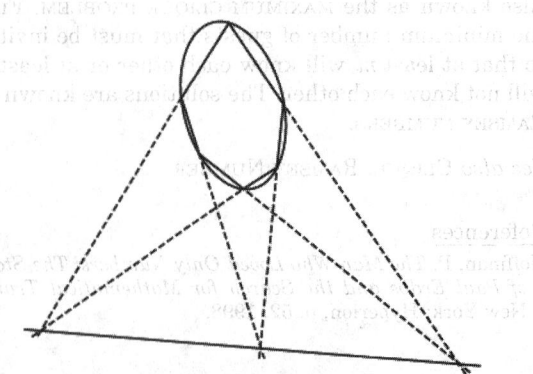

The dual of BRIANCHON'S THEOREM (Casey 1888, p. 146), discovered by B. Pascal in 1640 when he was just 16 years old (Leibniz 1640; Wells 1986, p. 69). It states that, given a (not necessarily REG-ULAR, or even CONVEX) HEXAGON inscribed in a CONIC SECTION, the three pairs of the continuations of opposite sides meet on a straight LINE, called the PASCAL LINE.

See also BRAIKENRIDGE-MACLAURIN CONSTRUCTION, BRIANCHON'S THEOREM, CAYLEY-BACHARACH THEO-REM, CONIC SECTION, DUALITY PRINCIPLE, HEXAGON, PAPPUS'S HEXAGON THEOREM, PASCAL LINES, STEI-NER POINTS, STEINER'S THEOREM

References

Casey, J. *A Sequel to the First Six Books of the Elements of Euclid, Containing an Easy Introduction to Modern Geometry with Numerous Examples,* 5th ed., rev. enl. Dublin: Hodges, Figgis, & Co., pp. 129–131, 1888.

Casey, J. "Pascal's Theorem." §255 in *A Treatise on the Analytical Geometry of the Point, Line, Circle, and Conic Sections, Containing an Account of Its Most Recent Extensions, with Numerous Examples,* 2nd ed., rev. enl. Dublin: Hodges, Figgis, & Co., pp. 145, 328–329, and 354, 1893.

Cayley, A. *Quart J.* **9**, p. 348.

Coxeter, H. S. M. and Greitzer, S. L. "L'hexagramme de Pascal. Un essai pur reconstituer cette découverte." *Le Jeune Scientifique (Joliette, Quebec)* **2**, 70–72, 1963.

Coxeter, H. S. M. and Greitzer, S. L. "Pascal's Theorem." §3.8 in *Geometry Revisited.* Washington, DC: Math. Assoc. Amer., pp. 74–76, 1967.

Durell, C. V. *Modern Geometry: The Straight Line and Circle.* London: Macmillan, p. 44, 1928.

Evelyn, C. J. A.; Money-Coutts, G. B.; and Tyrrell, J. A. "Extensions of Pascal's and Brianchon's Theorems." Ch. 2 in *The Seven Circles Theorem and Other New Theorems.* London: Stacey International, pp. 8–30, 1974.

Forder, H. G. *Higher Course Geometry.* Cambridge, Eng-land: Cambridge University Press, p. 13, 1931.

Graustein, W. C. *Introduction to Higher Geometry.* New York: Macmillan, pp. 260–261, 1930.

Johnson, R. A. §386 in *Modern Geometry: An Elementary Treatise on the Geometry of the Triangle and the Circle.* Boston, MA: Houghton Mifflin, pp. 236–237, 1929.

Lachlan, R. "Pascal's Theorem." §181–191 in *An Elementary Treatise on Modern Pure Geometry.* London: Macmillian, pp. 113–119, 1893.

Leibniz, G. Letter to M. Périer. In *Œuvres de B. Pascal,* Vol. 5 (Ed. Bossut). p. 459.

Ogilvy, C. S. *Excursions in Geometry.* New York: Dover, pp. 105–106, 1990.

Pappas, T. "The Mystic Hexagram." *The Joy of Mathematics.* San Carlos, CA: Wide World Publ./Tetra, p. 118, 1989.

Perfect, H. *Topics in Geometry.* London: Pergamon, p. 26, 1963.

Salmon, G. §267 and "Notes: Pascal's Theorem, Art. 267" in *A Treatise on Conic Sections, 6th ed.* New York: Chelsea, pp. 245–246 and 379–382, 1960.

Spieker, T. *Lehrbuch der ebene Geometrie.* Potsdam, Germany, 1888.

Veronese. "Nuovi Teremi sull' Hexagrammum Mysticum." *Real. Accad. dei Lincei.* 1877.

Wells, D. *The Penguin Dictionary of Curious and Interesting Numbers.* Middlesex, England: Penguin Books, p. 69, 1986.

Wells, D. *The Penguin Dictionary of Curious and Interesting Geometry.* London: Penguin, p. 173, 1991.

Pascal's Triangle

A TRIANGLE of numbers arranged in staggered rows such that

$$a_{nr} \equiv \frac{n!}{r!(n-r)!} \equiv \binom{n}{r}, \tag{1}$$

where $\binom{n}{r}$ is a BINOMIAL COEFFICIENT. The triangle was studied by B. Pascal, although it had been described centuries earlier by Chinese mathematician Yanghui (about 500 years earlier, in fact) and the Persian astronomer-poet Omar Khayyám. It is therefore known as the Yanghui triangle in China. Starting with $n = 0$, the TRIANGLE is

$$1$$

$$1 \quad 1$$

$$1 \quad 2 \quad 1$$

$$1 \quad 3 \quad 3 \quad 1$$

$$1 \quad 4 \quad 6 \quad 4 \quad 1$$

$$1 \quad 5 \quad 10 \quad 10 \quad 5 \quad 1$$

$$1 \quad 6 \quad 15 \quad 20 \quad 15 \quad 6 \quad 1$$

(Sloane's A007318). PASCAL'S FORMULA shows that each subsequent row is obtained by adding the two entries diagonally above,

$$\binom{n}{r} = \frac{n!}{(n-r)!r!} = \binom{n-1}{r} + \binom{n-1}{r-1}. \tag{2}$$

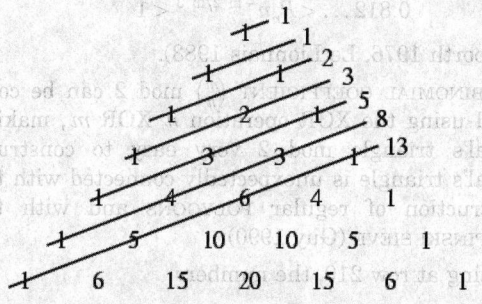

In addition, the "SHALLOW DIAGONALS" of Pascal's triangle sum to FIBONACCI NUMBERS,

$$\sum_{k=1}^{n} \binom{k}{n-k}$$

$$= \frac{(-1)^n \, {}_3F_2\left(1, \, 2, \, 1-n; \, \tfrac{1}{2}(3-n), \, 2-\tfrac{1}{2}n; \, -4\right)}{\pi(2 - 3n + n^2)}$$

$$= F_{n+1}, \tag{3}$$

where ${}_3F_2(a, b, c; \, d, e; \, z)$ is a GENERALIZED HYPERGEOMETRIC FUNCTION.

Pascal's triangle contains the FIGURATE NUMBERS along its diagonals. It can be shown that

$$\sum_{i=1}^{n} a_{ij} = \frac{n+1}{j+1} \, a_{nj} = a_{(n+1),(j+1)} \tag{4}$$

and

$$\binom{m+1}{1} \sum k^m + \binom{m+1}{2} \sum k^{m-1}$$

$$+ \dots + \binom{m+1}{m} \sum k = (n+1)[(n+1)^m - 1]. \tag{5}$$

The "shallow diagonals" sum to the FIBONACCI SEQUENCE, i.e.,

$$1 = 1$$

$$1 = 1$$

$$2 = 1 + 1$$

$$3 = 2 + 1$$

$$5 = 1 + 3 + 1$$

$$8 = 3 + 4 + 1. \tag{6}$$

In addition,

$$\sum_{j=1}^{i} a_{ij} = 2^i - 1. \tag{7}$$

It is also true that the first number after the 1 in each row divides all other numbers in that row IFF it is a PRIME. If P_n is the number of ODD terms in the first n rows of the Pascal triangle, then

$$0.812\ldots < P_n n^{-\ln 2/\ln 3} < 1 \qquad (8)$$

(Harborth 1976, Le Lionnais 1983).

The BINOMIAL COEFFICIENT $\binom{m}{n}$ mod 2 can be computed using the XOR operation n XOR m, making Pascal's triangle mod 2 very easy to construct. Pascal's triangle is unexpectedly connected with the construction of regular POLYGONS and with the SIERPINSKI SIEVE (Guy 1990).

Starting at row 210, the numbers

$$120 = \binom{10}{3} = \binom{10}{7} = \binom{16}{2} = \binom{16}{14} = \binom{120}{1}$$
$$= \binom{120}{119} \qquad (9)$$

$$210 = \binom{10}{4} = \binom{10}{6} = \binom{21}{2} = \binom{21}{19} = \binom{210}{1}$$
$$= \binom{210}{209} \qquad (10)$$

$$3003 = \binom{14}{6} = \binom{14}{8} = \binom{15}{5} = \binom{15}{10} = \binom{78}{2}$$
$$= \binom{78}{76} \qquad (11)$$

have appeared six times, more than any other number (excluding 1), and remain the most common numbers in the triangle up to at least row 1436.

Guy (1990) gives another several unexpected properties of Pascal's triangle.

See also BELL TRIANGLE, BINOMIAL COEFFICIENT, BINOMIAL THEOREM, BRIANCHON'S THEOREM, CATALAN'S TRIANGLE, CLARK'S TRIANGLE, EULER'S TRIANGLE, FIBONACCI NUMBER, FIGURATE NUMBER TRIANGLE, LEIBNIZ HARMONIC TRIANGLE, LOSSNITSCH'S TRIANGLE, NUMBER TRIANGLE, PASCAL'S FORMULA, POLYGON, SEIDEL-ENTRINGER-ARNOLD TRIANGLE, SIERPINSKI SIEVE, TRINOMIAL TRIANGLE

References

Conway, J. H. and Guy, R. K. "Pascal's Triangle." In *The Book of Numbers.* New York: Springer-Verlag, pp. 68–70, 1996.
Courant, R. and Robbins, H. *What is Mathematics?: An Elementary Approach to Ideas and Methods, 2nd ed.* Oxford, England: Oxford University Press, p. 17, 1996.
Guy, R. K. "The Second Strong Law of Small Numbers." *Math. Mag.* **63**, 3–20, 1990.
Harborth, H. "Number of Odd Binomial Coefficients." *Not. Amer. Math. Soc.* **23**, 4, 1976.
Le Lionnais, F. *Les nombres remarquables.* Paris: Hermann, p. 31, 1983.
Pappas, T. "Pascal's Triangle, the Fibonacci Sequence & Binomial Formula," "Chinese Triangle," and "Probability and Pascal's Triangle." *The Joy of Mathematics.* San Carlos, CA: Wide World Publ./Tetra, pp. 40–41 88, and 184–186, 1989.
Sloane, N. J. A. Sequences A007318/M0082 in "An On-Line Version of the Encyclopedia of Integer Sequences." http://www.research.att.com/~njas/sequences/eisonline.html.

Smith, D. E. *A Source Book in Mathematics.* New York: Dover, p. 86, 1984.
Steinhaus, H. *Mathematical Snapshots, 3rd ed.* New York: Dover, pp. 284–285, 1999.
Wells, D. *The Penguin Dictionary of Curious and Interesting Geometry.* London: Penguin, pp. 174–175, 1991.

Pascal's Wager

"God is or He is not...Let us weigh the gain and the loss in choosing...'God is.' If you gain, you gain all, if you lose, you lose nothing. Wager, then, unhesitatingly, that He is."

References

Erickson, G. W. and Fossa, J. A. *Dictionary of Paradox.* Lanham, MD: University Press of America, pp. 150–151, 1998.

Pasch's Axiom

In the plane, if a line intersects one side of a TRIANGLE and misses the three VERTICES, then it must intersect one of the other two sides. This is a special case of the generalized MENELAUS' THEOREM with $n = 3$.

See also HELLY'S THEOREM, MENELAUS' THEOREM, PASCH'S THEOREM

Pasch's Theorem

A theorem stated in 1882 which cannot be derived from EUCLID'S POSTULATES. Given points a, b, c, and d on a LINE, if it is known that the points are ordered as (a, b, c) and (b, c, d), then it is also true that (a, b, d).

See also EUCLID'S POSTULATES, LINE, PASCH'S AXIOM

Pass Equivalent

Two KNOTS are pass equivalent if there exists a sequence of pass moves taking one to the other. Every KNOT is either pass equivalent to the UNKNOT or TREFOIL KNOT. These two knots are not pass equivalent to each other, but the ENANTIOMERS of the TREFOIL KNOT are pass equivalent. A KNOT has ARF INVARIANT 0 if the KNOT is pass equivalent to the UNKNOT and 1 if it is pass equivalent to the TREFOIL KNOT.

See also ARF INVARIANT, KNOT, KNOT MOVE, PASS MOVE, TREFOIL KNOT, UNKNOT

References

Adams, C. C. *The Knot Book: An Elementary Introduction to the Mathematical Theory of Knots.* New York: W. H. Freeman, pp. 223–228, 1994.

Pass Move

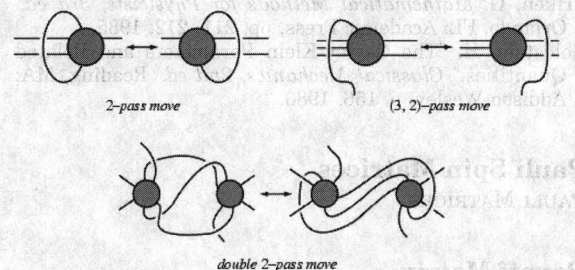

2–pass move   *(3, 2)–pass move*

double 2–pass move

A change in a knot projection such that a pair of oppositely oriented strands are passed through another pair of oppositely oriented strands.

See also KNOT MOVE, PASS EQUIVALENT

References

Adams, C. C. *The Knot Book: An Elementary Introduction to the Mathematical Theory of Knots.* New York: W. H. Freeman, pp. 223–228, 1994.

Hoste, J.; Thistlethwaite, M.; and Weeks, J. "The First 1,701,936 Knots." *Math. Intell.* **20**, 33–48, Fall 1998.

Patch

A patch (also called a LOCAL SURFACE) is a differentiable mapping $\mathbf{x}: U \to \mathbb{R}^n$, where U is an open subset of $\mathbb{R}^2$. More generally, if A is any SUBSET of $\mathbb{R}^2$, then a map $\mathbf{x}: A \to \mathbb{R}^n$ is a patch provided that $\mathbf{x}$ can be extended to a differentiable map from U into $\mathbb{R}^n$, where U is an open set containing A. Here, $\mathbf{x}(U)$ (or more generally, $\mathbf{x}(A)$) is called the TRACE of $\mathbf{x}$.

See also GAUSS MAP, INJECTIVE PATCH, MONGE PATCH, REGULAR PATCH, TRACE (MAP)

References

Gray, A. "Patches in $\mathbb{R}^n$" and "Patches in $\mathbb{R}^3$." §12.1 and 12.2 in *Modern Differential Geometry of Curves and Surfaces with Mathematica, 2nd ed.* Boca Raton, FL: CRC Press, pp. 269–278, 1997.

Path

A path γ is a continuous mapping $\gamma: [a, b] \mapsto \mathbb{C}$, where $\gamma(a)$ is the initial point and $\gamma(b)$ is the final point. It is often written parametrically as $\sigma(t)$.

See also CHAIN (GRAPH), CONTOUR, CURVE, EULERIAN CIRCUIT, GRAPH CYCLE, HAMILTONIAN CIRCUIT, UNICURSAL CIRCUIT

Path Graph

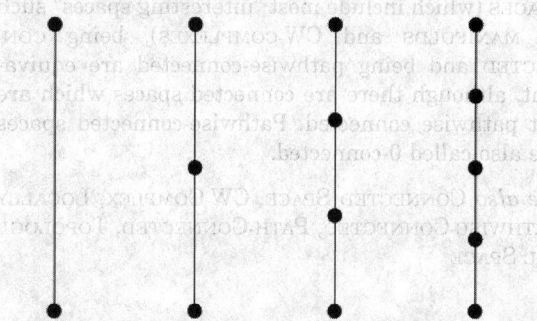

The path P_n is a TREE with two nodes of VERTEX DEGREE 1, and the other $n-2$ nodes of VERTEX DEGREE 2. Path graphs P_n are always GRACEFUL for $n > 4$.

See also CHAIN (GRAPH), GRACEFUL GRAPH, HAMILTONIAN PATH, TREE

Path Integral

Let γ be a PATH given parametrically by $\sigma(t)$. Let s denote ARC LENGTH from the initial point. Then

$$\int_\gamma f(s) \, ds = \int_\gamma f(\sigma(t)) |\sigma'(t)| \, dt$$

$$= \int_\gamma f(x(t), y(t), z(t)) |\sigma'(t)| \, dt.$$

See also LINE INTEGRAL

References

Press, W. H.; Flannery, B. P.; Teukolsky, S. A.; and Vetterling, W. T. "Evaluation of Functions by Path Integration." §5.14 in *Numerical Recipes in FORTRAN: The Art of Scientific Computing, 2nd ed.* Cambridge, England: Cambridge University Press, pp. 201–204, 1992.

Path Length

EXTERNAL PATH LENGTH, INTERNAL PATH LENGTH

Path-Connected

See also ARCWISE-CONNECTED, CONNECTED SET, LOCALLY PATHWISE-CONNECTED, PATHWISE-CONNECTED

Path-Connected Set

See also ARCWISE-CONNECTED SET, CONNECTED SET

Pathwise-Connected

A TOPOLOGICAL SPACE X is pathwise-connected IFF for every two points $x, y \in X$, there is a CONTINUOUS FUNCTION f from [0,1] to X such that $f(0) = x$ and $f(1) = y$. Roughly speaking, a SPACE X is pathwise-connected if, for every two points in X, there is a path connecting them. For LOCALLY PATHWISE-CONNECTED SPACES (which include most "interesting spaces" such as MANIFOLDS and CW-COMPLEXES), being CONNECTED and being pathwise-connected are equivalent, although there are connected spaces which are not pathwise connected. Pathwise-connected spaces are also called 0-connected.

See also CONNECTED SPACE, CW-COMPLEX, LOCALLY PATHWISE-CONNECTED, PATH-CONNECTED, TOPOLOGICAL SPACE

Patriarchal Cross
GAULLIST CROSS

Patterson Quadrature
GAUSS-KRONROD QUADRATURE

Pauli Matrices
Matrices which arise in Pauli's treatment of spin in quantum mechanics. They are defined by

$$\sigma_1 = \sigma_x \equiv \mathsf{P}_1 \equiv \begin{bmatrix} 0 & 1 \\ 1 & 0 \end{bmatrix} \tag{1}$$

$$\sigma_2 = \sigma_y \equiv \mathsf{P}_2 \equiv \begin{bmatrix} 0 & i \\ -i & 0 \end{bmatrix} \tag{2}$$

$$\sigma_3 = \sigma_z \equiv \mathsf{P}_3 \equiv \begin{bmatrix} 1 & 0 \\ 0 & -1 \end{bmatrix}. \tag{3}$$

The Pauli matrices plus the 2×2 IDENTITY MATRIX I form a complete set, so any 2×2 matrix A can be expressed as

$$\mathsf{A} = c_0 \mathsf{I} + c_1 \sigma_1 + c_2 \sigma_2 + c_3 \sigma_3. \tag{4}$$

The associated matrices

$$\sigma_+ \equiv 2 \begin{bmatrix} 0 & 1 \\ 0 & 0 \end{bmatrix} \tag{5}$$

$$\sigma_- \equiv 2 \begin{bmatrix} 0 & 0 \\ 1 & 0 \end{bmatrix} \tag{6}$$

$$\sigma^2 \equiv 3 \begin{bmatrix} 1 & 0 \\ 0 & 1 \end{bmatrix} \tag{7}$$

can also be defined. The Pauli spin matrices satisfy the identities

$$\sigma_i \sigma_j = \mathsf{I} \delta_{ij} + \epsilon_{ijk} i \sigma_k \tag{8}$$

$$\sigma_i \sigma_j = \sigma_j \sigma_i = 2 \sigma_{ij} \tag{9}$$

$$\sigma_x p_x + \sigma_y p_y + \sigma_z p_z = \sqrt{p_x^2 + p_y^2 + p_z^2}. \tag{10}$$

See also DIRAC MATRICES, QUATERNION

References
Arfken, G. *Mathematical Methods for Physicists, 3rd ed.* Orlando, FL: Academic Press, pp. 211–212, 1985.
Goldstein, H. "The Cayley-Klein Parameters and Related Quantities." *Classical Mechanics, 2nd ed.* Reading, MA: Addison-Wesley, p. 156, 1980.

Pauli Spin Matrices
PAULI MATRICES

Payoff Matrix
An $m \times n$ MATRIX which gives the possible outcome of a two-person ZERO-SUM GAME when player A has m possible moves and player B n moves. The analysis of the MATRIX in order to determine optimal strategies is the aim of GAME THEORY. The so-called "augmented" payoff matrix is defined as follows:

$$\mathsf{G} = \begin{bmatrix} P_0 & P_1 & P_2 & \cdots & P_n & P_{n+1} & P_{n+2} & \cdots & P_{n+m} \\ 0 & 1 & 1 & \cdots & 0 & 0 & 0 & \cdots & 0 \\ -1 & a_{11} & a_{12} & \cdots & a_{1n} & 1 & 0 & \cdots & 0 \\ -1 & a_{21} & a_{22} & \cdots & a_{2n} & 0 & 1 & \cdots & 0 \\ \vdots & \vdots & \vdots & \ddots & \vdots & \vdots & \vdots & \ddots & \vdots \\ -1 & a_{m1} & a_{m2} & \cdots & a_{mn} & 0 & 0 & \cdots & 1 \end{bmatrix}.$$

See also GAME THEORY, ZERO-SUM GAME

P-Circle
SPIEKER CIRCLE

PC-Point
PEDAL-CEVIAN POINT

Peacock's Tail
One name for the figure used by Euclid to prove the PYTHAGOREAN THEOREM.

See also BRIDE'S CHAIR, WINDMILL

Peano Arithmetic
The theory of NATURAL NUMBERS defined by the five PEANO'S AXIOMS. Paris and Harrington (1977) gave the first "natural" example of a statement which is true for the integers but unprovable in Peano arithmetic (Spencer 1983).

See also KREISEL CONJECTURE, NATURAL INDEPENDENCE PHENOMENON, NUMBER THEORY, PEANO'S AXIOMS

References
Kirby, L. and Paris, J. "Accessible Independence Results for Peano Arithmetic." *Bull. London Math. Soc.* **14**, 285–293, 1982.
Paris, J. and Harrington, L. "A Mathematical Incompleteness in Peano Arithmetic." In *Handbook of Mathematical Logic* (Ed. J. Barwise). Amsterdam, Netherlands: North-Holland, pp. 1133–1142, 1977.

Spencer, J. "Large Numbers and Unprovable Theorems." *Amer. Math. Monthly* **90**, 669–675, 1983.

Peano Curve

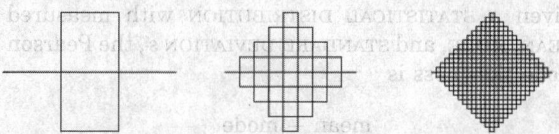

A FRACTAL curve which can be written as a LINDENMAYER SYSTEM.

See also DRAGON CURVE, HILBERT CURVE, LINDENMAYER SYSTEM, SIERPINSKI CURVE

References

Dickau, R. M. "Two-Dimensional L-Systems." http://forum.swarthmore.edu/advanced/robertd/lsys2d.html.
Hilbert, D. "Uuml;ber die stetige Abbildung einer Linie auf ein Flachenstück." *Math. Ann.* **38**, 459–460, 1891.
Peano, G. "Sur une courbe, qui remplit une aire plane." *Math. Ann.* **36**, 157–160, 1890.
Wagon, S. *Mathematica in Action.* New York: W. H. Freeman, p. 207, 1991.
Weisstein, E. W. "Fractals." MATHEMATICA NOTEBOOK FRACTAL.M.

Peano Surface

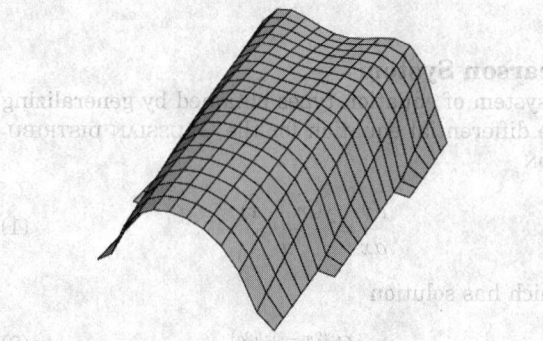

The function

$$f(x, y) = (2x^2 - y)(y - x^2)$$

which does *not* have a LOCAL MAXIMUM at (0, 0), despite criteria commonly touted in the second half of the 1800s which indicated the contrary.

See also LOCAL MAXIMUM

References

Fischer, G. (Ed.). Plate 122 in *Mathematische Modelle/ Mathematical Models, Bildband/Photograph Volume.* Braunschweig, Germany: Vieweg, p. 119, 1986.
Leitere, J. "Functions." §7.1.2 in *Mathematical Models from the Collections of Universities and Museums* (Ed. G. Fischer). Braunschweig, Germany: Vieweg, pp. 70–71, 1986.

Peano-Gosper Curve

A PLANE-FILLING CURVE originally called a FLOWSNAKE by R. W. Gosper and M. Gardner. Mandelbrot (1977) subsequently coined the name Peano-Gosper curve. The GOSPER ISLAND bounds the space that the Peano-Gosper curve fills.

See also DRAGON CURVE, EXTERIOR SNOWFLAKE, GOSPER ISLAND, HILBERT CURVE, KOCH SNOWFLAKE, PEANO CURVE, SIERPINSKI ARROWHEAD CURVE, SIERPINSKI CURVE

References

Dickau, R. M. "Two-Dimensional L-Systems." http://forum.swarthmore.edu/advanced/robertd/lsys2d.html.
Mandelbrot, B. B. *Fractals: Form, Chance, & Dimension.* San Francisco, CA: W. H. Freeman, 1977.
Weisstein, E. W. "Fractals." MATHEMATICA NOTEBOOK FRACTAL.M.

Peano's Axioms

1. Zero is a number.
2. If a is a number, the successor of a is a number.
3. ZERO is not the successor of a number.
4. Two numbers of which the successors are equal are themselves equal.
5. (INDUCTION AXIOM.) If a set S of numbers contains ZERO and also the successor of every number in S, then every number is in S.

Peano's axioms are the basis for the version of NUMBER THEORY known as PEANO ARITHMETIC.

See also INDUCTION AXIOM, PEANO ARITHMETIC

Pear Curve

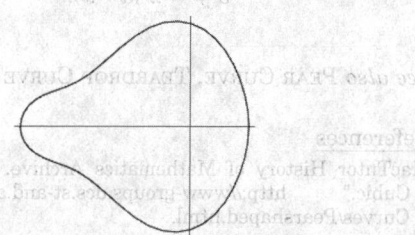

The LEMNISCATE L_3 in the iteration towards the MANDELBROT SET. In CARTESIAN COORDINATES with a constant r, the equation is given by

$$r^2 = (x^2 + y^2)(1 + 2x + 5x^2 + 6x^3 + 6x^4 + 4x^5 + x^6$$
$$- 3y^2 - 2xy^2 + 8x^2y^2 + 8x^3y^2 + 3x^4y^2 + 2y^4$$
$$+ 4xy^4 + 3x^2y^4 + y^6).$$

See also PEAR-SHAPED CURVE

Pearls of Sluze

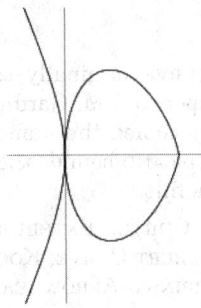

$$y^m = kx^n(a-x)^b.$$

The curves with integer n, b, and m were studied by de Sluze between 1657 and 1698. The name "Pearls of Sluze" was given to these curves by Blaise Pascal (MacTutor Archive).

References
MacTutor History of Mathematics Archive. "Pearls of Sluze." http://www-groups.dcs.st-and.ac.uk/~history/Curves/Pearls.html.

Pear-Shaped Curve

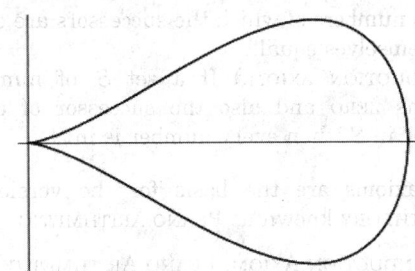

A curve given by the Cartesian equation

$$b^2y^2 = x^3(a-x).$$

See also PEAR CURVE, TEARDROP CURVE

References
MacTutor History of Mathematics Archive. "Pear-Shaped Cubic." http://www-groups.dcs.st-and.ac.uk/~history/Curves/Pearshaped.html.

Pearson Kurtosis
Let μ_4 be the fourth CENTRAL MOMENT of random variable and μ_2 its second CENTRAL MOMENT (i.e., the VARIANCE). Then the Pearson kurtosis is defined by

$$\beta_2 \equiv \frac{\mu_4}{\mu_2^2}.$$

See also CENTRAL MOMENT, FISHER KURTOSIS, KURTOSIS

Pearson Mode Skewness
Given a STATISTICAL DISTRIBUTION with measured MEAN, MODE, and STANDARD DEVIATION s, the Pearson mode skewness is

$$\frac{\text{mean} - \text{mode}}{s}.$$

See also MEAN, MODE, PEARSON SKEWNESS, PEARSON'S SKEWNESS COEFFICIENTS, SKEWNESS

Pearson Skewness
Let a STATISTICAL DISTRIBUTION have third MOMENT μ_3 and STANDARD DEVIATION σ, then the Pearson skewness is defined by

$$\beta_1 = \left(\frac{\mu_3}{\sigma^3}\right)^2.$$

See also FISHER SKEWNESS, PEARSON'S SKEWNESS COEFFICIENTS, SKEWNESS

Pearson System
A system of equation types obtained by generalizing the differential equation for the GAUSSIAN DISTRIBUTION

$$\frac{dy}{dx} = \frac{y(m-x)}{a}, \tag{1}$$

which has solution

$$y = Ce^{(2m-x)x/(2a)}, \tag{2}$$

to

$$\frac{dy}{dx} = \frac{y(m-x)}{a + bx + cx^2}, \tag{3}$$

which has solution

$$y = C\left(a + bx + cx^2\right)^{-1/(2c)}$$

$$\times \exp\left[\frac{(b + 2cm)\tan^{-1}\left(\dfrac{b + 2cx}{\sqrt{4ac - b^2}}\right)}{c\sqrt{4ac - b^2}}\right]. \tag{4}$$

Let c_1, c_2 be the roots of $a + bx + cx^2$. Then the possible types of curves are

0. $b = c = 0$, $a > 0$. E.g., NORMAL DISTRIBUTION.

I. $b^2/4ac < 0$, $c_1 \le x \le c_2$. E.g., BETA DISTRIBUTION.

II. $b^2/4ac = 0$, $c < 0$, $-c_1 \le x \le c_1$ where $c_1 \equiv \sqrt{-c/a}$.

III. $b^2/4ac = \infty$, $c = 0$, $c_1 \le x < \infty$ where $c_1 \equiv -a/b$. E.g., GAMMA DISTRIBUTION. This case is intermediate to cases I and VI.

IV. $0 < b^2/4ac < 1$, $-\infty < x < \infty$.

V. $b^2/4ac = 1$, $c_1 \le x < \infty$ where $c_1 \equiv -b/2a$. Intermediate to cases IV and VI.

VI. $b^2/4ac > 1$, $c_1 \le x < \infty$ where c_1 is the larger root. E.g., BETA PRIME DISTRIBUTION.

VII. $b^2/4ac = 0$, $c > 0$, $-\infty < x < \infty$. E.g., STUDENT'S T-DISTRIBUTION.

Classes IX-XII are discussed in Pearson (1916). See also Craig (in Kenney and Keeping 1951).

If a Pearson curve possesses a MODE, it will be at $x = m$. Let $y(x) = 0$ at c_1 and c_2, where these may be $-\infty$ or ∞. If yx^{r+2} also vanishes at c_1, c_2, then the rth MOMENT and $(r+1)$th MOMENTS exist.

$$\int_{c_1}^{c_2} \frac{dy}{dx}(ax^r + bx^{r+1} + cx^{r+2})\, dx$$

$$= \int_{c_1}^{c_2} y(mx^r - x^{r+1})\, dx, \tag{5}$$

giving

$$\left[y(ax^r + bx^{r+1} + cx^{r+2}) \right]_{c_1}^{c_2} - \int_{c_1}^{c_2} y\left[arx^{r-1} + b(r+1)x^r \right.$$

$$\left. + c(r+2)x^{r+1} \right] dx$$

$$= \int_{c_1}^{c_2} y(mx^r - x^{r+1})\, dx \tag{6}$$

$$0 - \int_{c_1}^{c_2} y\left[arx^{r-1} + b(r+1)x^r + c(r+2)x^{r+1} \right] dx$$

$$= \int_{c_1}^{c_2} y(mx^r - x^{r+1})\, dx. \tag{7}$$

Now define the raw rth moment by

$$v_r = \int_{c_1}^{c_2} yx^r\, dx, \tag{8}$$

so combining (7) with (8) gives

$$arv_{r-1} + b(r+1)v_r + c(r+2)v_{r+1} = -mv_r + v_{r+1}. \tag{9}$$

For $r = 0$,

$$b + 2cv_1 = -m + v_1, \tag{10}$$

so

$$v_1 = \frac{m + b}{1 - 2c}, \tag{11}$$

and for $r = 1$,

$$a + 2bv_1 + 3cv_2 = -mv_1 + v_2, \tag{12}$$

so

$$v_2 = \frac{a + (m + 2b)v_1}{1 - 3c}. \tag{13}$$

Combining (11), (13), and the definitions

$$v_1 = 0 \tag{14}$$

$$v_2 = \mu_2 = 1 \tag{15}$$

obtained by letting $t \equiv (x - v_1)/\sigma$ and solving simultaneously gives $b = -m$ and $a = 1 - 3c$. Writing

$$\alpha_r = \mu_r = v_r \tag{16}$$

then allows the general recurrence to be written

$$(1 - 3c)r\alpha_{r-1} - mr\alpha_r + [c(r+2) - 1]\alpha_{r+1} = 0. \tag{17}$$

For the special cases $r = 2$ and $r = 3$, this gives

$$2m + (1 - 4c)\alpha_3 = 0. \tag{18}$$

$$3(1 - 3c) - 3m\alpha_3 - (1 - 5c)\alpha_4 = 0, \tag{19}$$

so the SKEWNESS and KURTOSIS are

$$\gamma_1 = \alpha_3 = \frac{2m}{4c - 1} \tag{20}$$

$$\gamma_2 = \alpha_4 - 3 = \frac{6(m^2 - 4c^2 + c)}{(4c - 1)(5c - 1)}. \tag{21}$$

The parameters a, b, and c can therefore be written

$$a = 1 - 3c \tag{22}$$

$$b = -m = \frac{\gamma_1}{2(1 + 2\delta)} \tag{23}$$

$$c = \frac{\delta}{2(1 + 2\delta)}, \tag{24}$$

where

$$\delta \equiv \frac{2\gamma_2 - 3\gamma_1^2}{\gamma_2 + 6}. \tag{25}$$

References

Craig, C. C. "A New Exposition and Chart for the Pearson System of Frequency Curves." *Ann. Math. Stat.* **7**, 16–28, 1936.

Kenney, J. F. and Keeping, E. S. *Mathematics of Statistics, Pt. 2, 2nd ed.* Princeton, NJ: Van Nostrand, p. 107, 1951.

Pearson, K. "Second Supplement to a Memoir on Skew Variation." *Phil. Trans. A* **216**, 429–457, 1916.

Pearson Type III Distribution

A skewed distribution which is similar to the BINOMIAL DISTRIBUTION when $p \ne q$ (Abramowitz and Stegun 1972, p. 930).

$$y = k(t+A)^{A^2-1}e^{-At}, \qquad (1)$$

for $t \in [0, \infty)$ where

$$A \equiv 2/\gamma \qquad (2)$$

$$K \equiv \frac{A^{A^2}e^{-A^2}}{\Gamma(A^2)}, \qquad (3)$$

$\Gamma(z)$ is the GAMMA FUNCTION, and t is a standardized variate. Another form is

$$P(x) = \frac{1}{\beta\Gamma(p)}\left(\frac{x-\alpha}{\beta}\right)^{p-1}\exp\left(-\frac{x-\alpha}{\beta}\right). \qquad (4)$$

For this distribution, the CHARACTERISTIC FUNCTION is

$$\phi(t) = e^{i\alpha t}(1-i\beta t)^{-p}, \qquad (5)$$

and the MEAN, VARIANCE, SKEWNESS, and KURTOSIS are

$$\mu = \alpha + p\beta \qquad (6)$$

$$\sigma^2 = p\beta^2 \qquad (7)$$

$$\gamma_1 = \frac{2}{\sqrt{p}} \qquad (8)$$

$$\gamma_2 = \frac{6}{p}. \qquad (9)$$

See also PEARSON TYPE IV DISTRIBUTION

References
Abramowitz, M. and Stegun, C. A. (Eds.). *Handbook of Mathematical Functions with Formulas, Graphs, and Mathematical Tables, 9th printing.* New York: Dover, 1972.

Pearson Type IV Distribution

See also PEARSON TYPE III DISTRIBUTION

References
Nagahara, Y. "The PDF and CF of Pearson Type IV Distributions and the ML Estimation of the Parameters." *Stat. Prob. Let.* **43**, 251–264, 1999.

Pearson-Cunningham Function
CUNNINGHAM FUNCTION

Pearson's Correlation
CORRELATION COEFFICIENT

Pearson's Function

$$I\left(\frac{X_s^2}{\sqrt{2(k-1)}}, \frac{k-3}{2}\right) \equiv \frac{\Gamma\left(\frac{1}{2}\chi_s^2, \frac{k-1}{2}\right)}{\Gamma\left(\frac{k-1}{2}\right)},$$

where $\Gamma(x)$ is the GAMMA FUNCTION.

See also CHI-SQUARED TEST, GAMMA FUNCTION

Pearson's Skewness Coefficients
Given a STATISTICAL DISTRIBUTION with measured MEAN, MEDIAN, MODE, and STANDARD DEVIATION s, Pearson's first skewness coefficient is

$$\frac{3[\text{mean}] - [\text{mode}]}{s},$$

and the second coefficient is

$$\frac{3[\text{mean}] - [\text{median}]}{s}.$$

See also FISHER SKEWNESS, PEARSON SKEWNESS, SKEWNESS

References
Kenney, J. F. and Keeping, E. S. *Mathematics of Statistics, Pt. 1, 3rd ed.* Princeton, NJ: Van Nostrand, pp. 101–102, 1962.

Peaucellier Cell
PEAUCELLIER INVERSOR

Peaucellier Inversor

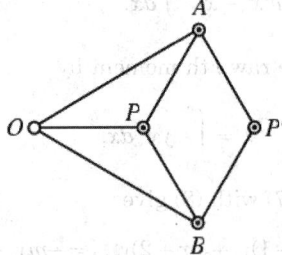

A LINKAGE with six rods which draws the inverse of a given curve. When a pencil is placed at P, the inverse is drawn at P' (or vice versa). If a seventh rod (dashed) is added (with an additional pivot), P is kept on a circle and the locus traced out by P' is a straight line. It therefore converts circular motion to linear motion without sliding, and was discovered in 1864. Another LINKAGE which performs this feat using hinged squares had been published by Sarrus in 1853 but ignored. Coxeter (1969, p. 428) shows that

$$OP \times OP' = OA^2 - PA^2.$$

See also HART'S INVERSOR, KEMPE LINKAGE, LINKAGE

References

Bogomolny, A. "Peaucellier Linkage." http://www.cut-the-knot.com/pythagoras/invert.html.
Courant, R. and Robbins, H. *What is Mathematics?: An Elementary Approach to Ideas and Methods.* Oxford, England: Oxford University Press, p. 156, 1978.
Coxeter, H. S. M. *Introduction to Geometry, 2nd ed.* New York: Wiley, pp. 82–83, 1969.
Durell, C. V. *Modern Geometry: The Straight Line and Circle.* London: Macmillan, p. 117, 1928.
Ogilvy, C. S. *Excursions in Geometry.* New York: Dover, pp. 46–48, 1990.
Rademacher, H. and Toeplitz, O. *The Enjoyment of Mathematics: Selections from Mathematics for the Amateur.* Princeton, NJ: Princeton University Press, pp. 121–126, 1957.
Sarrus. *Comptes Rendus de l'Académie de Paris* **36**, 1036, 1853.
Smith, D. E. *A Source Book in Mathematics.* New York: Dover, p. 324, 1994.
Steinhaus, H. *Mathematical Snapshots, 3rd ed.* New York: Dover, p. 139, 1999.
Wells, D. *The Penguin Dictionary of Curious and Interesting Geometry.* London: Penguin, pp. 120 and 181–182, 1991.

Peaucellier's Linkage

PEAUCELLIER INVERSOR

Pedal

PEDAL CURVE

Pedal-Cevian Point

If the PEDAL TRIANGLE of a point P in a TRIANGLE ΔABC is a CEVIAN TRIANGLE, then the point P is called the pedal-cevian point of ΔABC with respect to the PEDAL TRIANGLE.

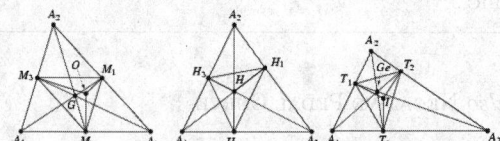

The CIRCUMCENTER O, ORTHOCENTER H, and INCENTER I of a triangle $\Delta A_1 A_2 A_3$ are always pedal-Cevian points, with corresponding pedal triangles given by the MEDIAL TRIANGLE $\Delta M_1 M_2 M_3$, ORTHIC TRIANGLE $\Delta H_1 H_2 H_3$, and CONTACT TRIANGLE $\Delta T_1 T_2 T_3$, respectively, and PEDAL POINTS the CENTROID G, ORTHOCENTER H, and GERGONNE POINT Ge, respectively (Honsberger 1995, p. 142). If P is a pedal-Cevian point of a triangle, then so is its ISOTOMIC CONJUGATE POINT Q, as is its reflection P' in the CIRCUMCENTER (Honsberger 1995, p. 143).

See also CEVIAN, CEVIAN TRIANGLE, PEDAL POINT, PEDAL TRIANGLE

References

Honsberger, R. *Episodes in Nineteenth and Twentieth Century Euclidean Geometry.* Washington, DC: Math. Assoc. Amer., pp. 142–143, 1995.

Pedal Circle

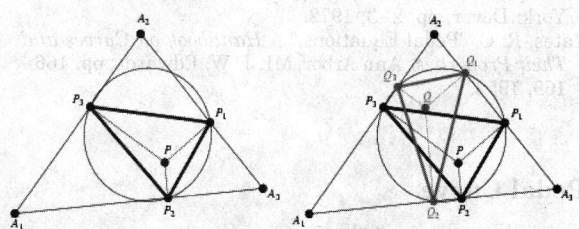

The pedal circle with respect to a PEDAL POINT P of a TRIANGLE $\Delta A_1 A_2 A_3$ is the CIRCUMCIRCLE of the PEDAL TRIANGLE $\Delta P_1 P_2 P_3$ with respect to P. Amazingly, the vertices of the PEDAL TRIANGLE $\Delta Q_1 Q_2 Q_3$ of the ISOGONAL CONJUGATE point Q of P also lie on the same circle (Honsberger 1995). If the PEDAL POINT is taken as the INCENTER, the pedal circle is given by the INCIRCLE.

The radius of the pedal circle of a point P is

$$r = \frac{\overline{A_1 P} \cdot \overline{A_2 P} \cdot \overline{A_3 P}}{2\left(R^2 - \overline{OP}^2\right)}$$

(Johnson 1929, p. 141).

When P is on a side of the TRIANGLE, the line between the two perpendiculars is called the PEDAL LINE. Given four points, no three of which are COLLINEAR, then the four PEDAL CIRCLES of each point for the TRIANGLE formed by the other three have a common point through which the NINE-POINT CIRCLES of the four TRIANGLES pass.

See also FONTENÉ THEOREMS, GRIFFITHS' THEOREM, MIQUEL POINT, NINE-POINT CIRCLE, PEDAL LINE, PEDAL TRIANGLE

References

Coolidge, J. L. *A Treatise on the Geometry of the Circle and Sphere.* New York: Chelsea, p. 50, 1971.
Fontené, G. "Sur le cercle pédal." *Nouv. Ann. Math.* **65**, 55–58, 1906.
Honsberger, R. *More Mathematical Morsels.* Washington, DC: Math. Assoc. Amer., p. 54, 1991.
Honsberger, R. "The Pedal Circle." §7.4 (viii) in *Episodes in Nineteenth and Twentieth Century Euclidean Geometry.* Washington, DC: Math. Assoc. Amer., pp. 67–69, 1995.
Johnson, R. A. *Modern Geometry: An Elementary Treatise on the Geometry of the Triangle and the Circle.* Boston, MA: Houghton Mifflin, 1929.

Pedal Coordinates

The pedal coordinates of a point P with respect to the curve C and the PEDAL POINT O are the radial distance r from O to P and the PERPENDICULAR distance p from O to the line L tangent to C at P.

See also PEDAL CURVE, PEDAL POINT

References

Lawrence, J. D. *A Catalog of Special Plane Curves*. New York: Dover, pp. 2–3, 1972.

Yates, R. C. "Pedal Equations." *A Handbook on Curves and Their Properties*. Ann Arbor, MI: J. W. Edwards, pp. 166–169, 1952.

Pedal Curve

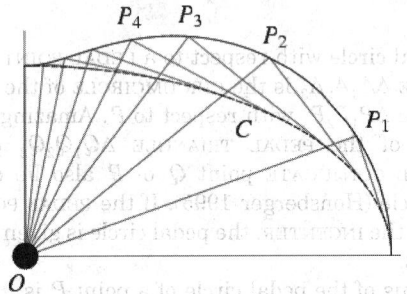

The pedal of a curve C with respect to a point O is the LOCUS of the foot of the PERPENDICULAR from P to the TANGENT to the curve. More precisely, given a curve C, the pedal curve P of C with respect to a fixed point O (called the PEDAL POINT) is the locus of the point P of intersection of the PERPENDICULAR from O to a TANGENT to C. The parametric equations for a curve $(f(t), g(t))$ relative to the PEDAL POINT (x_0, y_0) are given by

$$x = \frac{x_0 f'^2 + fg'^2 + (y_0 - g)f'g'}{f'^2 + g'^2}$$

$$y = \frac{gf'^2 + y_0 g'^2 + (x_0 - f)f'g'}{f'^2 + g'2^2}$$

When a CLOSED CURVE rolls on a straight line, the AREA between the line and ROULETTE after a complete revolution by any point on the curve is twice the AREA of the pedal curve (taken with respect to the generating point) of the rolling curve.

The following table gives the pedal curves for a number of common special curves.

Curve	PEDAL POINT	Pedal Curve
ASTROID	center	QUADRIFOLIUM
CARDIOID	cusp	CAYLEY'S SEXTIC
CIRCLE	any point	LIMAÇON
CIRCLE	on CIRCUMFERENCE	CARDIOID
CIRCLE	center of CIRCLE	ARCHIMEDEAN SPIRAL
CISSOID OF DIOCLES	FOCUS	CARDIOID
DELTOID	center	TRIFOLIUM
DELTOID	cusp	simple FOLIUM
DELTOID	on curve	unsymmetric double folium
DELTOID	vertex	double folium
ELLIPSE	FOCUS	CIRCLE
EPICYCLOID	center	ROSE
HYPERBOLA	center	LEMNISCATE
HYPERBOLA	FOCUS	CIRCLE
HYPOCYCLOID	center	ROSE
LINE	any point	point
LOGARITHMIC SPIRAL	pole	LOGARITHMIC SPIRAL
PARABOLA	FOCUS	LINE
PARABOLA	foot of DIRECTRIX	RIGHT STROPHOID
PARABOLA	on DIRECTRIX	STROPHOID
PARABOLA	reflection of FOCUS by DIRECTRIX	MACLAURIN TRISECTRIX
PARABOLA	vertex	CISSOID OF DIOCLES
SINUSOIDAL SPIRAL	pole	SINUSOIDAL SPIRAL
TSCHIRNHAUSEN CUBIC	center	PARABOLA

See also NEGATIVE PEDAL CURVE

References

Hilbert, D. and Cohn-Vossen, S. *Geometry and the Imagination*. New York: Chelsea, p. 25, 1999.

Lawrence, J. D. *A Catalog of Special Plane Curves*. New York: Dover, pp. 46–49 and 204, 1972.

Lockwood, E. H. "Pedal Curves." Ch. 18 in *A Book of Curves*. Cambridge, England: Cambridge University Press, pp. 152–155, 1967.

Yates, R. C. "Pedal Curves." *A Handbook on Curves and Their Properties*. Ann Arbor, MI: J. W. Edwards, pp. 160–165, 1952.

Pedal Line

Mark a point P on a side of a TRIANGLE and draw the perpendiculars from the point to the two other sides.

The line between the feet of these two perpendiculars is called the pedal line.

See also PEDAL TRIANGLE, SIMSON LINE

Pedal Point

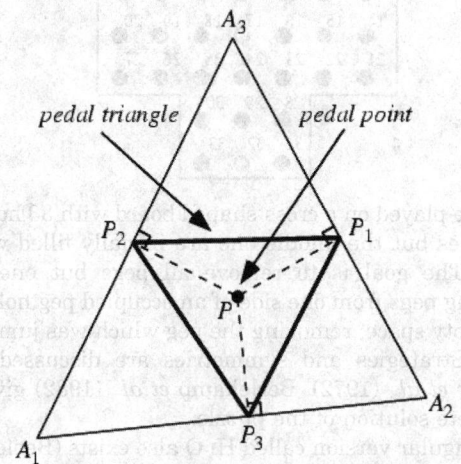

pedal triangle   *pedal point*

The fixed point with respect to which a PEDAL CURVE or PEDAL TRIANGLE is drawn.

See also PEDAL-CEVIAN POINT, PEDAL CURVE, PEDAL TRIANGLE

References

Coxeter, H. S. M. and Greitzer, S. L. *Geometry Revisited.* New York: Random House, p. 22, 1967.

Pedal Triangle

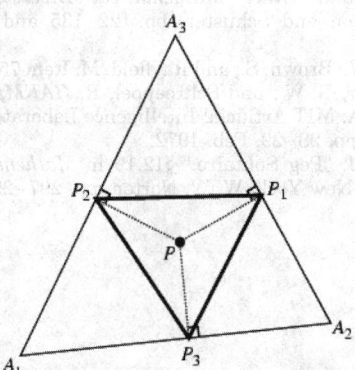

Given a point P, the pedal triangle of P is the TRIANGLE whose VERTICES are the feet of the perpendiculars from P to the side lines. The pedal triangle of a TRIANGLE with TRILINEAR COORDINATES $\alpha : \beta : \gamma$ and angles A, B, and C has VERTICES with TRILINEAR COORDINATES

$$0 : \beta + \alpha \cos C : \gamma + \alpha \cos B \qquad (1)$$

$$\alpha + \beta \cos C : 0 : \gamma + \beta \cos A \qquad (2)$$

$$\alpha + \gamma \cos B : \beta + \gamma \cos A : 0. \qquad (3)$$

The SYMMEDIAN POINT of a triangle is the CENTROID of its pedal triangle (Honsberger 1995, pp. 72–74).

The third pedal triangle is similar to the original one. This theorem can be generalized to: the nth pedal n-gon of any n-gon is similar to the original one. It is also true that

$$P_2 P_3 = A_1 P \sin \alpha_1 \qquad (4)$$

(Johnson 1929, pp. 135–136; Stewart 1940; Coxeter and Greitzer 1967, p. 25). The AREA A of the pedal triangle of a point P is proportional to the POWER of P with respect to the CIRCUMCIRCLE,

$$A = \tfrac{1}{2}\left(R^2 - \overline{OP}^2\right) \sin \alpha_1 \sin \alpha_2 \sin \alpha_3$$

$$= \frac{R^2 - \overline{OP}^2}{4R^2} \Delta \qquad (5)$$

(Johnson 1929, pp. 139–141).

The only closed BILLIARDS path of a single circuit in an ACUTE TRIANGLE is the pedal triangle. There are an infinite number of multiple-circuit paths, but all segments are parallel to the sides of the pedal triangle (Wells 1991).

See also ANTIPEDAL TRIANGLE, FAGNANO'S PROBLEM, ORTHIC TRIANGLE, PEDAL CIRCLE, PEDAL LINE

References

Coxeter, H. S. M. and Greitzer, S. L. "Pedal Triangles." §1.9 in *Geometry Revisited.* Washington, DC: Math. Assoc. Amer., pp. 22–26, 1967.
Honsberger, R. *Episodes in Nineteenth and Twentieth Century Euclidean Geometry.* Washington, DC: Math. Assoc. Amer., pp. 67–74, 1995.
Johnson, R. A. *Modern Geometry: An Elementary Treatise on the Geometry of the Triangle and the Circle.* Boston, MA: Houghton Mifflin, 1929.
Stewart, B. M. "Cyclic Properties of Miquel Polygons." *Amer. Math. Monthly* **47**, 462–466, 1940.

Peg

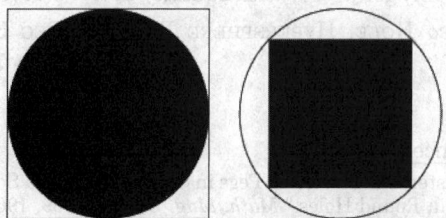

The answer to the question "which fits better, a round peg in a square hole, or a square peg in a round hole?" can be interpreted as asking which is larger, the ratio of the AREA of a CIRCLE to its circumscribed SQUARE, or the AREA of the SQUARE to its circumscribed CIRCLE? In 2-D, the ratios are $\pi/4$ and $2/\pi$, respectively. Therefore, a round peg fits better into a square hole than a square peg fits into a round hole (Wells

1986, p. 74).

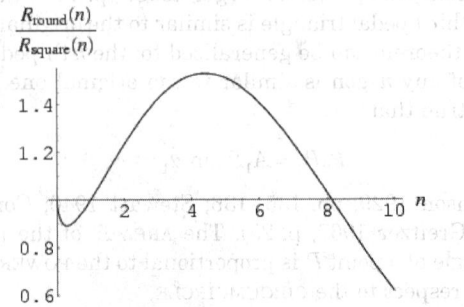

However, this result is true only in dimensions $n < 9$, and for $n \geq 9$, the unit n-hypersphere fits more closely into the 9-hypercube than vice versa (Singmaster; Wells 1986, p. 74). This can be demonstrated by noting that the formulas for the content $V(n)$ of the unit n-ball, the content $V_c(n)$ of its circumscribed HYPERCUBE, and the content $V_i(n)$ of its inscribed HYPERCUBE are given by

$$V(n) = \frac{\pi^{n/2}}{\Gamma\left(\frac{1}{2}n + 1\right)} \qquad (1)$$

$$V_c(n) = 2^n \qquad (2)$$

$$V_i(n) = \frac{2^n}{n^{n/2}}. \qquad (3)$$

The ratios in question are then

$$R_{\text{round peg}} = \frac{V(n)}{V_c(n)} = \frac{\pi^{n/2}}{2^n \Gamma\left(\frac{1}{2}n + 1\right)} \qquad (4)$$

$$R_{\text{square peg}} = \frac{V_i(n)}{V_c(n)} = \frac{2^{\Gamma\left(\frac{1}{2}n + 1\right)}}{n^{n/2} n^{n/2}} \qquad (5)$$

(Singmaster 1964). As illustrated above, $R_{\text{round}} < R_{\text{square}}$ only for $n < 9$, with equality at $n \approx 8.13785$.

See also HOLE, HYPERSPHERE PACKING, PEG SOLITAIRE

References

Singmaster, D. "On Round Pegs in Square Holes and Square Pegs in Round Holes." *Math. Mag.* **37**, 335–339, 1964.
Wells, D. *The Penguin Dictionary of Curious and Interesting Numbers.* Middlesex, England: Penguin Books, p. 74, 1986.

Peg Knot
CLOVE HITCH

Peg Solitaire

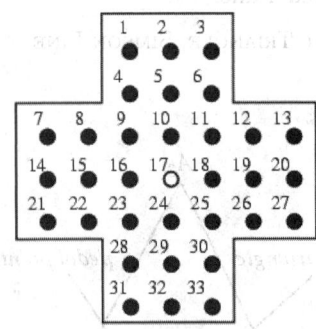

A game played on a cross-shaped board with 33 holes. All holes but the middle one are initially filled with pegs. The goal is to remove all pegs but one by jumping pegs from one side of an occupied peg hole to an empty space, removing the peg which was jumped over. Strategies and symmetries are discussed by Gosper *et al.* (1972). Berlekamp *et al.* (1982) give a complete solution of the puzzle.

A triangular version called HI-Q also exists (Beeler *et al.* 1972, Item 76). Kraitchik (1942) considers a board with one additional hole placed at the vertices of the central right angles.

See also HI-Q

References

Beasley, J. D. *The Ins and Outs of Peg Solitaire.*
Berlekamp, E. R.; Conway, J. H; and Guy, R. K. Ch. 23 in *Winning Ways for Your Mathematical Plays, Vol. 2: Games in Particular.* London: Academic Press, 1982.
Gardner, M. "Peg Solitaire." Ch. 11 in *The Unexpected Hanging and Other Mathematical Diversions.* New York: Simon and Schuster, pp. 122–135 and 250–251, 1969.
Gosper, R. W.; Brown, S.; and Rayfield, M. Item 75 in Beeler, M.; Gosper, R. W.; and Schroeppel, R. *HAKMEM.* Cambridge, MA: MIT Artificial Intelligence Laboratory, Memo AIM-239, pp. 28–29, Feb. 1972.
Kraitchik, M. "Peg Solitaire." §12.19 in *Mathematical Recreations.* New York: W. W. Norton, pp. 297–298, 1942.

Peg Top
PIRIFORM

Peirce Decomposition

Let $\mathfrak{A}$ be a finite-dimensional power-associative algebra, then $\mathfrak{A}$ is the vector space DIRECT SUM

$$\mathfrak{A} = \mathfrak{A}_{11} + \mathfrak{A}_{10} + \mathfrak{A}_{01} + \mathfrak{A}_{00},$$

where $\mathfrak{A}_{ij}$, with $i, j = 0, 1$ is the subspace of $\mathfrak{A}$ defined by

$$\mathfrak{A}_{ij} = \{x_{ij} : ex_{ij} = ix_{ij}, \ x_{ij}e = jx_{ij}\}$$

for $i, j = 0, 1$, where e is an idempotent.

References

Schafer, R. D. "The Peirce Decomposition." §3.2 in *An Introduction to Nonassociative Algebras*. New York: Dover, pp. 32–37, 1996.

Peirce's Theorem

The only linear associative algebra in which the coordinates are REAL NUMBERS and products vanish only if one factor is zero are the FIELD OF REAL NUMBERS, the FIELD OF COMPLEX NUMBERS, and the algebra of QUATERNIONS with REAL COEFFICIENTS.

See also COMPLEX NUMBER, QUATERNION, REAL NUMBER, WEIERSTRASS'S THEOREM

References

Schafer, R. D. "The Peirce Decomposition." §3.2 in *An Introduction to Nonassociative Algebras*. New York: Dover, pp. 32–37, 1996.

p-Element

SEMISIMPLE ELEMENT

p-Elementary Subgroup

A p-elementary subgroup of a FINITE GROUP G is a SUBGROUP H which is the GROUP DIRECT PRODUCT

$$H = C_n \times P,$$

where P is a P-GROUP, C_n is a cyclic group, and p does not divide n.

See also GROUP, GROUP DIRECT PRODUCT, INDUCED REPRESENTATION, P-GROUP

Pell Equation

A special case of the quadratic DIOPHANTINE EQUATION having the form

$$x^2 - Dy^2 = 1, \tag{1}$$

where $D > 0$ is a nonsquare NATURAL NUMBER (Dickson 1952). The equation

$$x^2 - Dy^2 = \pm 4 \tag{2}$$

arising in the computation of FUNDAMENTAL UNITS is sometimes also called the Pell equation (Dörrie 1965, Itô 1987), and Dörrie calls the positive form of (2) the FERMAT DIFFERENCE EQUATION. While Fermat deserves credit for being the first to extensively study the equation, the erroneous attribution to Pell was perpetrated by none other than Euler himself (Nagell 1951, p. 197; Dickson 1957, p. 341; Burton 1989). The Pell equation was also solved by the Indian mathematician Bhaskara. Pell equations are extremely important in NUMBER THEORY, and arise in the investigation of numbers which are FIGURATE in more than one way, for example, simultaneously square and triangular.

The equation has an obvious generalization to the Pell-like equation

$$ax^2 \pm by^2 = c, \tag{3}$$

as well as the general second-order bivariate Diophantine equation

$$ax^2 + bxy + cy^2 + dx + ey + f = 0. \tag{4}$$

However, several different technique are required to solve this equation for arbitrary values of a, b, and c. In a future release of *Mathematica*, the command Reduce will find solutions to the general equation (4), when they exist.

Pell equations OF THE FORM (1), as well as certain cases of the analogous equation with a minus sign on the right,

$$x^2 - Dy^2 = -1, \tag{5}$$

can be solved by finding the CONTINUED FRACTION $[a_0, a_1, \ldots]$ of $\sqrt{D}$. Note that although the equation (5) is solvable for only certain values of D, the continued fraction technique provides solutions when they exist, and always in the case of (1), for which a solution always exists. A necessary condition that (5) be solvable is that all odd prime factors of D be OF THE FORM $4n + 1$, and that D cannot be DOUBLY EVEN (i.e., divisible by 4). However, these conditions are not SUFFICIENT for a solution to exist, as demonstrated by the equation $x^2 - 34y^2 = -1$, which has no solutions in integers (Nagell 1951, pp. 201 and 204).

In all subsequent discussion, ignore the trivial solution $x = 1$, $y = 0$. Let p_n/q_n denote the nth CONVERGENT $[a_0, a_1, \ldots, a_n]$, then we will have solved (1) or (5) if we can find a CONVERGENT which obeys the identity

$$p_n^2 - Dq_n^2 = (-1)^{n+1}. \tag{6}$$

Amazingly, this turns out to *always* be possible as a result of the fact that the CONTINUED FRACTION of a QUADRATIC SURD always becomes periodic at some term a_{r+1}, where $a_{r+1} = 2a_0$, i.e.,

$$\sqrt{D} = [a_0, \overline{a_1, \ldots, a_r, 2a_0}]. \tag{7}$$

To compute the CONTINUED FRACTION convergents to $\sqrt{D}$, use the usual RECURRENCE RELATIONS

$$a_0 = \left\lfloor \sqrt{D} \right\rfloor$$

$$p_0 = a_0 \tag{8}$$

$$p_1 = a_0 a_1 + 1 \tag{9}$$

$$p_n = a_n p_{n-1} + p_{n-2} \tag{10}$$

$$q_0 = 1 \tag{11}$$

$$q_1 = a_1 \tag{12}$$

$$q_n = a_n q_{n-1} + q_{n-2}, \tag{13}$$

where $\lfloor x \rfloor$ is the FLOOR FUNCTION. For reasons to be explained shortly, also compute the two additional quantities P_n and Q_n defined by

$$P_0 = 0 \tag{14}$$

$$P_1 = a_0 \tag{15}$$

$$P_n = a_{n-1} Q_{n-1} - P_{n-1} \tag{16}$$

$$Q_0 = 1 \tag{17}$$

$$Q_1 = D - a_0^2 \tag{18}$$

$$Q_n = \frac{D - P_n^2}{Q_{n-1}} \tag{19}$$

$$a_n = \left\lfloor \frac{a_0 + P_n}{Q_n} \right\rfloor. \tag{20}$$

Now, two important identities satisfied by CONTINUED FRACTION convergents are

$$p_n q_{n-1} - p_{n-1} q_n = (-1)^{n+1} \tag{21}$$

$$p_n^2 - D q_n^2 = (-1)^{n+1} Q_{n+1} \tag{22}$$

(Beiler 1966, p. 262), so both linear

$$ax - by = \pm 1 \tag{23}$$

and quadratic

$$x^2 - D y^2 = \pm c \tag{24}$$

equations are solved simply by finding an appropriate continued fraction.

Let $a_{r+1} = 2a_0$ be the term at which the continued fraction becomes periodic (which will always happen for a quadratic surd). For the Pell equation

$$x^2 - D y^2 = 1 \tag{25}$$

with r ODD, $(-1)^{r+1}$ is POSITIVE and the solution in terms of smallest INTEGERS is $x = p_r$ and $y = q_r$, where p_r / q_r is the rth CONVERGENT. If r is EVEN, then $(-1)^{r+1}$ is NEGATIVE, but

$$p_{2r+1}^2 - D q_{2r+1}^2 = 1, \tag{26}$$

so the solution in smallest INTEGERS is $x = p_{2r+1}$, $y = q_{2r+1}$. Summarizing,

$$(x, y) = \begin{cases} (p_r, q_r) & \text{for } r \text{ odd} \\ (p_{2r+1}, p_{2r+1}) & \text{for } r \text{ even.} \end{cases} \tag{27}$$

The equation

$$x^2 - D y^2 = -1 \tag{28}$$

can be solved analogously to the equation with $+1$ on the right side IFF r is EVEN, but has no solution if r is odd,

$$(x, y) = \begin{cases} (p_r, q_r) & \text{for } r \text{ even} \\ \text{no solution} & \text{for } r \text{ odd.} \end{cases} \tag{29}$$

Given one solution $(x, y) = (p, q)$ (which can be found as above), a whole family of solutions can be found by taking each side to the nth POWER,

$$x^2 - D y^2 = \left(p^2 - D q^2 \right)^n = 1. \tag{30}$$

Factoring gives

$$\left(x + \sqrt{D} y \right) \left(x - \sqrt{D} y \right) = \left(p + \sqrt{D} q \right)^n \left(p - \sqrt{D} q \right)^n \tag{31}$$

and

$$x + \sqrt{D} y = \left(p + \sqrt{D} q \right)^n \tag{32}$$

$$x - \sqrt{D} y = \left(p - \sqrt{D} q \right)^n, \tag{33}$$

which gives the family of solutions

$$x = \frac{\left(p + q\sqrt{D} \right)^n + \left(p - q\sqrt{D} \right)^n}{2} \tag{34}$$

$$y = \frac{\left(p + q\sqrt{D} \right)^n - \left(p - q\sqrt{D} \right)^n}{2\sqrt{D}}. \tag{35}$$

These solutions also hold for

$$x^2 - D y^2 = -1, \tag{36}$$

except that n can take on only ODD values.

The following table gives the smallest integer solutions (x, y) to the Pell equation with constant $D \leq 102$ (Beiler 1966, p. 254). SQUARE $D = d^2$ are not included, since they would result in an equation OF THE FORM

$$x^2 - d^2 y^2 = x^2 - (dy)^2 = x^2 - y'^2 = 1, \tag{37}$$

which has no solutions (since the difference of two SQUARES cannot be 1).

D	x	y	D	x	y
2	3	2	54	485	66
3	2	1	55	89	12
5	9	4	56	15	2
6	5	2	57	151	20
7	8	3	58	19603	2574
8	3	1	59	530	69
10	19	6	60	31	4
11	10	3	61	1766319049	226153980

D	x	y	D	x	y
12	7	2	62	63	8
13	649	180	63	8	1
14	15	4	65	129	16
15	4	1	66	65	8
17	33	8	67	48842	5967
18	17	4	68	33	4
19	170	39	69	7775	936
20	9	2	70	251	30
21	55	12	71	3480	413
22	197	42	72	17	2
23	24	5	73	2281249	267000
24	5	1	74	3699	430
26	51	10	75	26	3
27	26	5	76	57799	6630
28	127	24	77	351	40
29	9801	1820	78	53	6
30	11	2	79	80	9
31	1520	273	80	9	1
32	17	3	82	163	18
33	23	4	83	82	9
34	35	6	84	55	6
35	6	1	85	285769	30996
37	73	12	86	10405	1122
38	37	6	87	28	3
39	25	4	88	197	21
40	19	3	89	500001	53000
41	2049	320	90	19	2
42	13	2	91	1574	165
43	3482	531	92	1151	120
44	199	30	93	12151	1260
45	161	24	94	2143295	221064
46	24335	3588	95	39	4
47	48	7	96	49	5
48	7	1	97	62809633	6377352
50	99	14	98	99	10
51	50	7	99	10	1
52	649	90	101	201	20
53	66249	9100	102	101	10

The first few minimal values of x and y for nonsquare D are 3, 2, 9, 5, 8, 3, 19, 10, 7, 649, ... (Sloane's A033313) and 2, 1, 4, 2, 3, 1, 6, 3, 2, 180, ... (Sloane's A033317), respectively. The values of D having $x = 2$, 3, ... are 3, 2, 15, 6, 35, 12, 7, 5, 11, 30, ... (Sloane's A033314) and the values of D having $y = 1, 2, ...$ are 3, 2, 7, 5, 23, 10, 47, 17, 79, 26, ... (Sloane's A033318). Values of the incrementally largest minimal x are 3, 9, 19, 649, 9801, 24335, 66249, ... (Sloane's A033315) which occur at $D = 2, 5, 10, 13, 29, 46, 53, 61, 109, 181, ...$ (Sloane's A033316). Values of the incrementally largest minimal y are 2, 4, 6, 180, 1820, 3588, 9100, 226153980, ... (Sloane's A033319), which occur at $D = 2, 5, 10, 13, 29, 46, 53, 61, ...$ (Sloane's A033320).

The more complicated Pell-like equation

$$x^2 - Dy^2 = c \tag{38}$$

with $|c| < \sqrt{D}$ has solution IFF c is one of the values $(-1)^k Q_k$ for $k = 1, 2, ..., r$ computed in the process of finding the convergents to $\sqrt{D}$ (where, as above, $a_{r+1} = 2a_0$ is the term at which the continued fraction becomes periodic). If $|c| > \sqrt{D}$, the procedure is significantly more complicated (Beiler 1966, p. 265; Dickson 1992, pp. 387–388) and is discussed by Gérardin (1910) and Chrystal (1961).

Regardless of how it is found, if a single solution $x = p$, $y = q$ to (38) is known, other solutions can be found. Let p and q be solutions to (38), and r and s solutions to the "unit" form

$$x^2 - Dy^2 = 1. \tag{39}$$

Then the identity

$$\left(p^2 - Dq^2\right)\left(r^2 - Ds^2\right) = (pr \pm Dqs)^2 - D(ps \pm qr)^2$$

$$= c \tag{40}$$

allows larger solutions $(x, y) = (pr \pm Dqs, ps \pm qr)$ to the c equation to be found by using incrementally larger values of the (r, s), which can be easily computed using the standard technique for the Pell equation. Such a family of solutions does not necessarily generate *all* solutions, however. For example, the equation

$$x^2 - 10y^2 = 9 \tag{41}$$

has *three* distinct sets of fundamental solutions, $(x, y) = (7, 2)$, $(13, 4)$, and $(57, 18)$. Using (40), these generate the solutions shown in the following table, from which the set of all solutions $(7, 2)$, $(13, 4)$, $(57, 18)$, $(253, 80)$, $(487, 154)$, $(2163, 684)$, $(9607, 3038)$, ... can be generated.

fundamental	generated solutions
(7, 2)	(253, 80), (9607, 3038), (364813, 115364), (13853287, 4380794), ...
(13, 4)	(487, 154), (18493, 5848), (702247, 222070), (26666893, 8432812), ...
(57, 18)	(2163, 684), (82137, 25974), (3119043, 986328), (118441497, 37454490), ...

The case

$$ax^2 - by^2 = c \qquad (42)$$

can be reduced to the one above by multiplying through by a,

$$(ax)^2 - (ab)y^2 = ac, \qquad (43)$$

finding solutions in $(x' \equiv ax, y)$, and then selecting those for which x'/a is an integer.

According to Dickson (1992, pp. 408 and 411), the equation

$$ax^2 + by^2 = c \qquad (44)$$

with a, b, $c > 0$, which has either no solutions or a finite number of solutions, was solved by Gauss (1863) using the METHOD OF EXCLUSIONS and considered by Euler (1773) and Nasimoff (1885), although Euler's methods were incomplete (Dickson 1992, p. 378; Smith 1965). According to Itô (1987), this equation can be solved completely using solutions to Pell's equation. Nasimoff (1885) applied Jacobi elliptic functions to express the number of solutions of this equation for a, c ODD (Dickson 1992, p. 411). Additional discussion including the connection with elliptic functions is given in Dickson (1992, pp. 387–391).

The special case of $a = 1$ and c prime was solved by Cornacchia (Cornacchia 1908, Cox 1989, Wagon 1990). Solution for $a = 1$, $b \geq 1$, and odd c is implemented in *Mathematica* as QuadraticRepresentation[b, c] in the *Mathematica* add-on package NumberTheory`NumberTheoryFunctions` (which can be loaded with the command << NumberTheory`). A deterministic algorithm for finding all primitive solutions to (44) for a, b, $c > 0$ fixed relatively prime integers, $c \geq a + b + 1$, and $(c, ab) = 1$ was given by Hardy *et al.* (1990). This algorithm generalizes those of Hermite (1848), Serret (1848), Brillhart (1972), Cornacchia (1908), and Wilker (1980). It requires factorization of c, and has worst case running time of $\mathcal{O}\left(c^{1/4}(\ln c)^3(\ln \ln c)\right)(\ln \ln \ln c)$, independent of a and b.

See also BINARY QUADRATIC FORM, DIOPHANTINE EQUATION, DIOPHANTINE EQUATION–2ND POWERS, FUNDAMENTAL UNIT, HILBERT SYMBOL, LAGRANGE NUMBER (DIOPHANTINE EQUATION), MONOMORPH, POLYMORPH

References

Beiler, A. H. "The Pellian." Ch. 22 in *Recreations in the Theory of Numbers: The Queen of Mathematics Entertains.* New York: Dover, pp. 248–268, 1966.

Brillhart, J. "Note on Representing a Prime as a Sum of Two Squares." *Math. Comput.* **26**, 1011–1013, 1972.

Burton, D. M. *Elementary Number Theory, 4th ed.* Boston, MA: Allyn and Bacon, pp. 379–382 and 392, 1989.

Chrystal, G. *Textbook of Algebra, 2nd ed., Vol. 2.* New York: Chelsea, pp. 478–486, 1961.

Cipolla, M. "Un metodo per la risoluzione della congruenza di secondo grado." *Rend. Accad. Sci. Fis. Mat. Napoli* **9**, 154–163, 1903.

Cohn, H. "Pell's Equation." §6.9 in *Advanced Number Theory.* New York: Dover, pp. 110–111, 1980.

Cornacchia, G. "Su di un metodo per la risoluzione in numeri unteri dell' equazione $\Sigma_{h=0}^n c_h x^{n-h} y^h = P$." *Giornale di Matematiche di Battaglini* **46**, 33–90, 1908.

Cox, D. A. *Primes OF THE FORM $x^2 + ny^2$.* New York: Wiley, 1989.

Degan, C. F. *Canon Pellianus.* Copenhagen, Denmark, 1817.

Dickson, L. E. "Pell Equation: $ax^2 + bx + c$ Made Square." Ch. 12 in *History of the Theory of Numbers, Vol. 2: Diophantine Analysis.* New York: Chelsea, pp. 341–400, 1952.

Dörrie, H. *100 Great Problems of Elementary Mathematics: Their History and Solutions.* New York: Dover, 1965.

Euler, L. *Novi Comm. Acad. Petrop.* **18**, 218, 1773. Reprinted in *Opera Omnia, Vol. 3*, p. 310.

Euler, L. *Comm. Arith.* 570. Reprinted in *Opera Omnia, Vol. 3*, p. 310.

Gérardin, A. "Formules de récurrence." *Sphinx-Oedipe* **5**, 17–29, 1910.

Hardy, K.; Muskat, J. B.; and Williams, K. S. "A Deterministic Algorithm for Solving $n = fu^2 + gv^2$ in Coprime Integers u and v." *Math. Comput.* **55**, 327–343, 1990.

Hermite, C. "Note au sujet de l'article précédent." *J. Math. Pures Appl.* **13**, 15, 1848.

Itô, K. (Ed.). *Encyclopedic Dictionary of Mathematics, 2nd ed, Vol. 1.* Cambridge, MA: MIT Press, p. 450, 1987.

Lagarias, J. C. "On the Computational Complexity of Determining the Solvability or Unsolvability of the Equation $X^2 - Dy^2 = -1$." *Trans. Amer. Math. Soc.* **260**, 485–508, 1980.

Nagell, T. "The Diophantine Equation $x^2 - Dy^2 = 1$," "The Diophantine Equation $x^2 - Dy^2 = -1$," and "The Diophantine Equation $u^2 - Dv^2 = C$." §56–58 in *Introduction to Number Theory.* New York: Wiley, pp. 195–212, 1951.

Nasimoff, P. S. Ch. 1 in *Application of Elliptic Functions to the Theory of Numbers.* Moscow, 1885. French summary in *Ann. sci. de l'École normale supér.* **5**, 23–31, 1888.

Serret, J. A. "Sur un théorème rélatif aux nombres enti'eres." *J. Math. Pures Appl.* **13**, 12–14, 1848.

Sloane, N. J. A. Sequences A033313, A033314, A033315, A033316, A033317, A033318, A033319, and A033320 in "An On-Line Version of the Encyclopedia of Integer Sequences." http://www.research.att.com/~njas/sequences/eisonline.html.

Smith, H. J. S. *Collected Mathematical Papers I.* New York: Chelsea, pp. 195–202, 1965.

Smarandache, F. "Un metodo de resolucion de la ecuacion diofantica." *Gaz. Math.* **1**, 151–157, 1988.

Smarandache, F. " Method to Solve the Diophantine Equation $ax^2 - by^2 + c = 0$." In *Collected Papers, Vol. 1.* Lupton, AZ: Erhus University Press, 1996.

Stillwell, J. C. *Mathematics and Its History.* New York: Springer-Verlag, 1989.

Wagon, S. "The Euclidean Algorithm Strikes Again." *Amer. Math. Monthly* **97**, 124–125, 1990.

Weisstein, E. W. "Integer Sequences." MATHEMATICA NOTEBOOK IntegerSequences.m.

Whitford, E. E. *Pell Equation.* New York: Columbia University Press, 1912.

Wilker, P. "An efficient Algorithmic Solution of the Diophantine Equation $u^2 + 5v^2 = m$." *Math. Comput.* **35**, 1347–1352, 1980.

Pell-Lucas Number

Pell Number

Pell-Lucas Polynomial

Pell Polynomial

Pell Number

The numbers obtained by the U_ns in the Lucas sequence with $P = 2$ and $Q = -1$. They and the Pell-Lucas numbers (the V_ns in the Lucas sequence) satisfy the recurrence relation

$$P_n = 2P_{n-1} + P_{n-2}. \tag{1}$$

Using P_i to denote a Pell number and Q_i to denote a Pell-Lucas number,

$$P_{m+n} = P_m P_{n+1} + P_{m-1} P_n \tag{2}$$

$$P_{m+n} = 2P_m Q_n - (-1)^n P_{m-n}, \tag{3}$$

$$P_{2^t m} = P_m (2Q_m)(2Q_{2m})(2Q_{4m}) \cdots (2Q_{2^{t-1}m}) \tag{4}$$

$$Q_m^2 = 2P_m^2 + (-1)^m \tag{5}$$

$$Q_{2m} = 2Q_m^2 - (-1)^m. \tag{6}$$

The Pell numbers have $P_0 = 0$ and $P_1 = 1$ and are 0, 1, 2, 5, 12, 29, 70, 169, 408, 985, 2378, ... (Sloane's A000129). The Pell-Lucas numbers have $Q_0 = 2$ and $Q_1 = 2$ and are 2, 2, 6, 14, 34, 82, 198, 478, 1154, 2786, 6726, ... (Sloane's A002203).

The only TRIANGULAR Pell number is 1 (McDaniel 1996).

See also Brahmagupta Polynomial, Pell Polynomial

References

McDaniel, W. L. "Triangular Numbers in the Pell Sequence." *Fib. Quart.* **34**, 105–107, 1996.

Ram, R. "Pell Numbers Formulae." http://users.tellurian.net/hsejar/maths/pell/.

Sloane, N. J. A. Sequences A000129/M1413 and A002203/M0360 in "An On-Line Version of the Encyclopedia of Integer Sequences." http://www.research.att.com/~njas/sequences/eisonline.html.

Pell Polynomial

The Pell polynomials $P(x)$ and Lucas-Pell polynomials $Q(x)$ are generated by a Lucas polynomial sequence using generator $(2x, 1)$. This gives recursive equations for $P(x)$ from $P_0(x) = P_1(x) = 1$ and

$$P_{n+2}(x) = 2xP_{n+1}(x) + P_n(x). \tag{1}$$

The first few are

$$P_1 = 1$$

$$P_2 = 2x$$

$$P_3 = 4x^2 - 1$$

$$P_4 = 8x^3 - 4x$$

$$P_5 = 16x^4 - 12x^2 + 1.$$

The Pell-Lucas numbers are defined recursively by $q_0(x) = 1$, $q_1(x) = x$ and

$$q_{n+2}(x) = 2xq_{n+1}(x) + q_n(x), \tag{2}$$

together with

$$Q_n(x) \equiv 2q_n(x). \tag{3}$$

The first few are

$$Q_1 = 2x$$

$$Q_2 = 4x^2 - 2$$

$$Q_3 = 8x^3 - 6x$$

$$Q_4 = 16x^4 - 16x^2 + 2$$

$$Q_5 = 32x^5 - 40x^3 + 10x.$$

See also Lucas Polynomial Sequence

References

Horadam, A. F. and Mahon, J. M. "Pell and Pell-Lucas Polynomials." *Fib. Quart.* **23**, 7–20, 1985.

Mahon, J. M. M. A. (Honors) thesis, The University of New England. Armidale, Australia, 1984.

Sloane, N. J. A. Sequences A000129/M1413 in "An On-Line Version of the Encyclopedia of Integer Sequences." http://www.research.att.com/~njas/sequences/eisonline.html.

Pell Sequence

Pell Number

Pencil

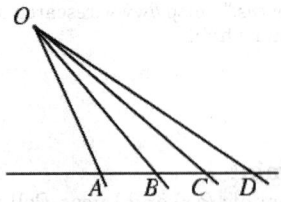

The set of all LINES through a point. The term was first used by Desargues (Cremona 1960, p. x). The six angles of any pencils of four rays $O\{ABCD\}$ are connected by the relation

$$\sin BOC \sin AOD + \sin COA \sin BOD$$

$$+\sin AOB \sin COD = 0$$

and the lengths satisfy

$$BC \cdot AD + CA \cdot BD + AB \cdot CD = 0$$

(Lachlan 1893).

Woods (1961) uses the term pencil as a synonym for RANGE, and Altshiller-Court (1979, p. 12) uses the term to mean SHEAF OF PLANES.

See also NEAR-PENCIL, PERSPECTIVITY, RANGE (LINE SEGMENT), SECTION (PENCIL), SHEAF OF PLANES

References

Altshiller-Court, N. *Modern Pure Solid Geometry.* New York: Chelsea, 1979.
Cremona, L. *Elements of Projective Geometry, 3rd ed.* New York: Dover, 1960.
Lachlan, R. "Relations Connecting the Angles of a Pencil." §29 in *An Elementary Treatise on Modern Pure Geometry.* London: Macmillian, pp. 16–18, 1893.
Graustein, W. C. *Introduction to Higher Geometry.* New York: Macmillan, p. 36, 1930.
Woods, F. S. *Higher Geometry: An Introduction to Advanced Methods in Analytic Geometry.* New York: Dover, pp. 8 and 11–12, 1961.

Pencil of Coaxal Circles

COAXAL CIRCLES

Pencil of Planes

SHEAF OF PLANES

Peninsula Surface

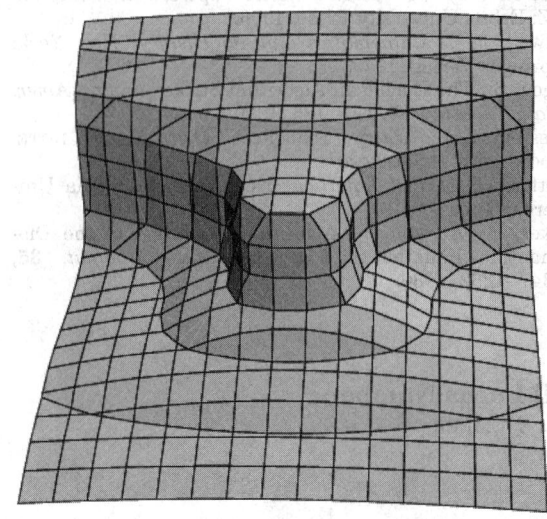

A QUINTIC SURFACE given by the equation

$$x^2 + y^3 + z^5 = 1.$$

See also QUINTIC SURFACE

Penrose Stairway

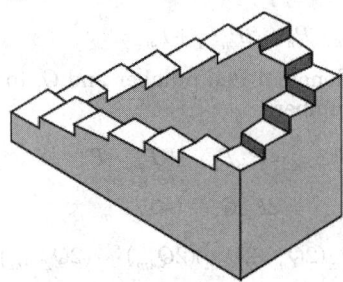

An IMPOSSIBLE FIGURE (also called the SCHROEDER STAIRS) in which a stairway in the shape of a square appears to circulate indefinitely while still possessing normal steps. The Dutch artist M. C. Escher included Penrose stairways in many of his mind-bending illustrations.

See also IMPOSSIBLE FIGURE

References

Hofstadter, D. R. *Gödel, Escher, Bach: An Eternal Golden Braid.* New York: Vintage Books, p. 15, 1989.
Jablan, S. "Impossible Figures." http://members.tripod.com/~modularity/impos.htm.
Pappas, T. "Optical Illusions and Computer Graphics." *The Joy of Mathematics.* San Carlos, CA: Wide World Publ./Tetra, p. 5, 1989.
Robinson, J. O. and Wilson, J. A. "The Impossible Colonnade and Other Variations of a Well-Known Figure." *Brit. J. Psych.* **64**, 363–365, 1973.

Penrose Tiles

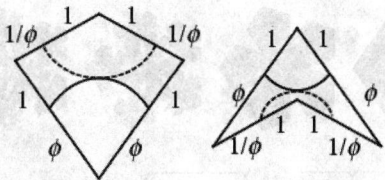

A pair of shapes which tile the plane only aperiodically (when the markings are constrained to match at borders). The two tiles, illustrated above, are called the "KITE" and "DART."

To see how the plane may be tiled aperiodically using the kite and dart, divide the kite into acute and obtuse tiles, shown above. Now define "deflation" and "inflation" operations. The deflation operator takes an acute TRIANGLE to the union of two ACUTE TRIANGLES and one OBTUSE, and the OBTUSE TRIANGLE goes to an ACUTE and an OBTUSE TRIANGLE. These operations are illustrated below.

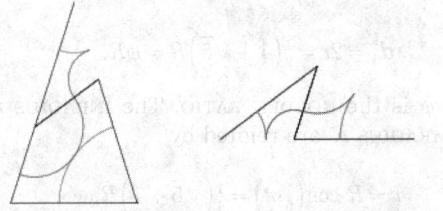

When applied to a collection of tiles, the deflation operator leads to a more refined collection. The operators do not respect tile boundaries, but do respect the half tiles defined above. There are two ways to obtain aperiodic TILINGS with 5-fold symmetry about a single point. they are known an the "star" and "sun" configurations, and are show below.

Higher order versions can then be obtained by deflation. For example, the following are third-order deflations:

References

Gardner, M. "Extraordinary Nonperiodic Tiling that Enriches the Theory of Tiles." *Sci. Amer.* 110–119, Dec. 1977.

Gardner, M. "Penrose Tiling" and "Penrose Tiling II." Chs. 1–2 in *Penrose Tiles and Trapdoor Ciphers... and the Return of Dr. Matrix, reissue ed.* New York: W. H. Freeman, pp. 1–29, 1989.

Hurd, L. P. "Penrose Tiles." http://www.mathsource.com/cgi-bin/msitem?0206–772.

Peterson, I. *The Mathematical Tourist: Snapshots of Modern Mathematics.* New York: W. H. Freeman, pp. 86–95, 1988.

Radin, C. *Miles of Tiles.* Providence, RI: Amer. Math. Soc., pp. 2 and 34–36, 1999.

Smith, T. "Penrose Tilings and Wang Tilings." http://www.innerx.net/personal/tsmith/pwtile.html.

Vichera, M. "Penrose Tiling." http://alpha.ujep.cz/~vicher/puzzle/penrose/penr.htm.

Wagon, S. "Penrose Tiles." §4.3 in *Mathematica in Action.* New York: W. H. Freeman, pp. 108–117, 1991.

Wells, D. *The Penguin Dictionary of Curious and Interesting Geometry.* London: Penguin, pp. 175–177, 1991.

Penrose Triangle

TRIBAR

Penrose Tribar

TRIBAR

Pentabolo

A 5-POLYABOLO.

Pentacle

PENTAGRAM

Pentacontagon

A 50-sided POLYGON.

Pentacube

This entry contributed by RONALD M. AARTS

A POLYCUBE composed of 5 cubes. There are 29 distinct three-dimensional pentacubes (Bouwkamp 1981). Of these, the 12 planar pentacubes (corresponding to solid pentominoes), are well known. Among the nonplanar pentacubes, there are five that have at least one plane of symmetry; each of them is its own mirror image. The remaining 12 pentacubes come in mirror image pairs.

See also POLYCUBE

References

Bouwkamp, C. J. "Packing Handed Pentacubes." In *The Mathematical Gardner* (Ed. D. Klarner). Boston, MA: Prindle, Weber, 1981.

Pentad

A group of five elements.

See also MONAD, PAIR, QUADRUPLET, QUINTUPLET, TETRAD, TRIAD, TRIPLET, TWINS

Pentadecagon

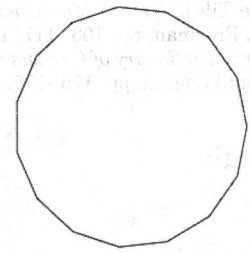

A 15-sided POLYGON, sometimes also called the PENTAKAIDECAGON. For a regular pentadecagon with side length 1, the INRADIUS r, CIRCUMRADIUS R, and AREA A are

$$r = \frac{1}{2}\left(\sqrt{3} + \sqrt{5 + 2\sqrt{5}}\right)$$

$$R = \frac{1}{4}\left(\sqrt{3} + \sqrt{15} + \sqrt{2}\sqrt{5 + \sqrt{5}}\right)$$

$$A = \frac{15}{8}\left(\sqrt{3} + \sqrt{15} + \sqrt{2}\sqrt{5 + \sqrt{5}}\right)$$

See also POLYGON, REGULAR POLYGON, TRIGONOMETRY VALUES PI/15

Pentaflake

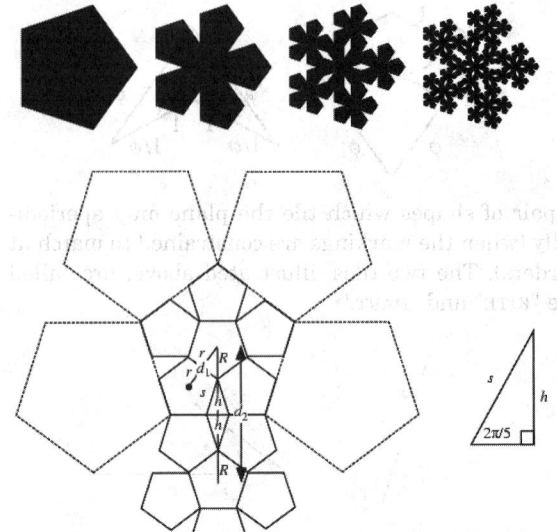

A FRACTAL with 5-fold symmetry. As illustrated above, five PENTAGONS can be arranged around an identical PENTAGON to form the first iteration of the pentaflake. This cluster of six pentagons has the shape of a pentagon with five triangular wedges removed. This construction was first noticed by Albrecht Dürer (Dixon 1991).

For a pentagon of side length 1, the first ring of pentagons has centers at RADIUS

$$d_1 = 2r = \frac{1}{2}\left(1 + \sqrt{5}\right)R = \phi R, \tag{1}$$

where ϕ is the GOLDEN RATIO. The INRADIUS r and CIRCUMRADIUS R are related by

$$r = R\cos\left(\frac{1}{5}\pi\right) = \frac{1}{4}\left(\sqrt{5} + 1\right)R, \tag{2}$$

and these are related to the side length s by

$$s = 2\sqrt{R^2 - r^2} = \frac{1}{2}R\sqrt{10 - 2\sqrt{5}}. \tag{3}$$

The height h is

$$h = s\sin\left(\frac{2}{5}\pi\right) = \frac{1}{4}s\sqrt{10 + 2\sqrt{5}} = \frac{1}{2}\sqrt{5}R, \tag{4}$$

giving a RADIUS of the second ring as

$$d_2 = 2(R + h) = \left(2 + \sqrt{5}\right)R = \phi^3 R. \tag{5}$$

Continuing, the nth pentagon ring is located at

$$d_n = \phi^{2n-1}. \tag{6}$$

Now, the length of the side of the first pentagon compound is given by

$$s_2 = 2\sqrt{(2r + R)^2 - (h + R)^2} = R\sqrt{5 + 2\sqrt{5}}, \tag{7}$$

so the ratio of side lengths of the original pentagon to that of the compound is

$$\frac{s_2}{s} = \frac{R\sqrt{5 + 2\sqrt{5}}}{\frac{1}{2}R\sqrt{10 - 2\sqrt{5}}} = 1 + \phi. \tag{8}$$

We can now calculate the dimension of the pentaflake fractal. Let N_n be the number of black pentagons and L_n the length of side of a pentagon after the n iteration,

$$N_n = 6^n \tag{9}$$

$$L_n = (1 + \phi)^{-n}. \tag{10}$$

The CAPACITY DIMENSION is therefore

$$d_{\text{cap}} = -\lim_{n \to \infty} \frac{\ln N_n}{\ln L_n} = \frac{\ln 6}{\ln(1 + \phi)} = \frac{\ln 2 + \ln 3}{\ln(1 + \phi)} \tag{11}$$

See also PENTAGON

References

Dixon, R. *Mathographics.* New York: Dover, pp. 186–188, 1991.

Weisstein, E. W. "Fractals." MATHEMATICA NOTEBOOK FRACTAL.M.

Wells, D. *The Penguin Dictionary of Curious and Interesting Geometry.* London: Penguin, p. 104, 1991.

Pentagon

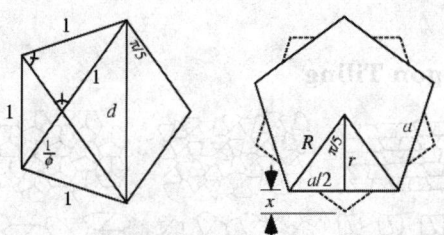

The regular convex 5-gon is called the pentagon. By SIMILAR TRIANGLES in the figure on the left,

$$\frac{d}{1} = \frac{1}{\frac{1}{\phi}} = \phi, \tag{1}$$

where d is the diagonal distance. But the dashed vertical line connecting two nonadjacent VERTICES is the same length as the diagonal one, so

$$\phi = 1 + \frac{1}{\phi} \tag{2}$$

$$\phi^2 - \phi - 1. \tag{3}$$

Solving the QUADRATIC EQUATION gives

$$\frac{1 + \sqrt{1 + 4}}{2}, \tag{4}$$

and taking the plus sign gives the GOLDEN RATIO

$$\phi = \frac{1}{2}\left(1 + \sqrt{5}\right). \tag{5}$$

(Taking the minus sign instead gives $1/\phi$.)

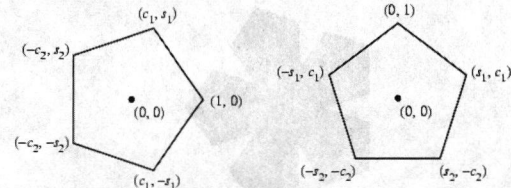

The coordinates of the VERTICES relative to the center of the pentagon with unit sides are given as shown in the above figure, with

$$c_1 = \cos\left(\frac{2\pi}{5}\right) = \frac{1}{4}\left(\sqrt{5} - 1\right) \tag{6}$$

$$c_2 = \cos\left(\frac{4\pi}{5}\right) = \frac{1}{4}\left(\sqrt{5} + 1\right) \tag{7}$$

$$s_1 = \sin\left(\frac{2\pi}{5}\right) = \frac{1}{4}\sqrt{10 + 2\sqrt{5}} \tag{8}$$

$$s_2 = \sin\left(\frac{4\pi}{5}\right) = \frac{1}{4}\sqrt{10 - 2\sqrt{5}}. \tag{9}$$

For a REGULAR POLYGON, the CIRCUMRADIUS, INRADIUS, SAGITTA, and AREA are given by

$$R_n = \frac{1}{2} a \csc\left(\frac{\pi}{n}\right) \tag{10}$$

$$r_n = \frac{1}{2} a \cot\left(\frac{\pi}{n}\right) \tag{11}$$

$$x_n = R_n - r_n = \frac{1}{2} a \tan\left(\frac{\pi}{2n}\right) \tag{12}$$

$$A_n = \frac{1}{4} n a^2 \cot\left(\frac{\pi}{n}\right). \tag{13}$$

Plugging in $n = 5$ gives

$$R = \frac{1}{2} a \csc\left(\frac{1}{5}\pi\right) = \frac{1}{10} a\sqrt{50 + 10\sqrt{5}} \tag{14}$$

$$r = \frac{1}{2} a \cot\left(\frac{1}{5}\pi\right) = \frac{1}{10} a\sqrt{25 + 10\sqrt{5}} \tag{15}$$

$$x = \frac{1}{2} a \frac{1}{10}\sqrt{25 - 10\sqrt{5}} \tag{16}$$

$$A = \frac{1}{4} a^2 \sqrt{25 + 10\sqrt{5}}. \tag{17}$$

Five pentagons can be arranged around an identical pentagon to form the first iteration of the "PENTA-

FLAKE," which itself has the shape of a pentagon with five triangular wedges removed. For a pentagon of side length 1, the first ring of pentagons has centers at radius ϕ, the second ring at ϕ^3, and the nth at ϕ^{2n-1}.

In proposition IV.11, Euclid showed how to inscribe a regular pentagon in a CIRCLE. Ptolemy also gave a RULER and COMPASS construction for the pentagon in his epoch-making work *The Almagest*. While Ptolemy's construction has a SIMPLICITY of 16, a GEOMETRIC CONSTRUCTION using CARLYLE CIRCLES can be made with GEOMETROGRAPHY symbol $2S_1 + S_2 + 8C_1 + 0C_2 + 4C_3$, which has SIMPLICITY 15 (De Temple 1991).

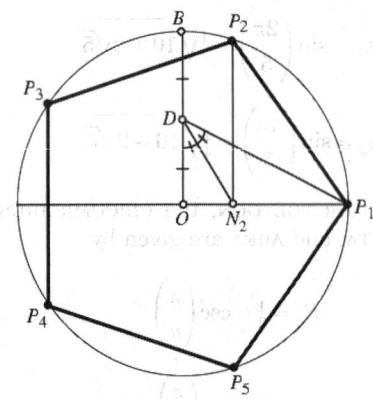

pentagon construction

The following elegant construction for the pentagon is due to Richmond (1893). Given a point, a CIRCLE may be constructed of any desired RADIUS, and a DIAMETER drawn through the center. Call the center O, and the right end of the DIAMETER P_1. The DIAMETER PERPENDICULAR to the original DIAMETER may be constructed by finding the PERPENDICULAR BISECTOR. Call the upper endpoint of this PERPENDICULAR DIAMETER B. For the pentagon, find the MIDPOINT of OB and call it D. Draw DP_1, and BISECT $\angle ODP_1$, calling the intersection point with OP_1 N_2. Draw N_2P_2 PARALLEL to OB, and the first two points of the pentagon are P_1 and P_2, and copying the angle $\angle P_1OP_2$ then gives the remaining points P_3, P_4, and P_5 (Coxeter 1969, Wells 1991).

Madachy (1979) illustrates how to construct a pentagon by folding and knotting a strip of paper.

See also CYCLIC PENTAGON, DECAGON, DISSECTION, FIVE DISKS PROBLEM, HOME PLATE, PENTAFLAKE, PENTAGRAM, POLYGON, TRIGONOMETRY VALUES PI/5

References

Ball, W. W. R. and Coxeter, H. S. M. *Mathematical Recreations and Essays, 13th ed.* New York: Dover, pp. 95–96, 1987.

Coxeter, H. S. M. *Introduction to Geometry, 2nd ed.* New York: Wiley, pp. 26–28, 1969.

De Temple, D. W. "Carlyle Circles and the Lemoine Simplicity of Polygonal Constructions." *Amer. Math. Monthly* **98**, 97–108, 1991.

Dickson, L. E. "Regular Pentagon and Decagon." §8.17 in *Monographs on Topics of Modern Mathematics Relevant to the Elementary Field* (Ed. J. W. A. Young). New York: Dover, pp. 368–370, 1955.

Dixon, R. *Mathographics.* New York: Dover, p. 17, 1991.

Dudeney, H. E. *Amusements in Mathematics.* New York: Dover, p. 38, 1970.

Fukagawa, H. and Pedoe, D. "Pentagons." §4.3 in *Japanese Temple Geometry Problems.* Winnipeg, Manitoba, Canada: Charles Babbage Research Foundation, pp. 49 and 132–134, 1989.

Madachy, J. S. *Madachy's Mathematical Recreations.* New York: Dover, p. 59, 1979.

Pappas, T. "The Pentagon, the Pentagram & the Golden Triangle." *The Joy of Mathematics.* San Carlos, CA: Wide World Publ./Tetra, pp. 188–189, 1989.

Richmond, H. W. "A Construction for a Regular Polygon of Seventeen Sides." *Quart. J. Pure Appl. Math.* **26**, 206–207, 1893.

Wantzel, M. L. "Recherches sur les moyens de reconnaître si un Problème de Géométrie peut se résoudre avec la règle et le compas." *J. Math. pures appliq.* **1**, 366–372, 1836.

Wells, D. *The Penguin Dictionary of Curious and Interesting Geometry.* London: Penguin, p. 211, 1991.

Pentagon Tiling

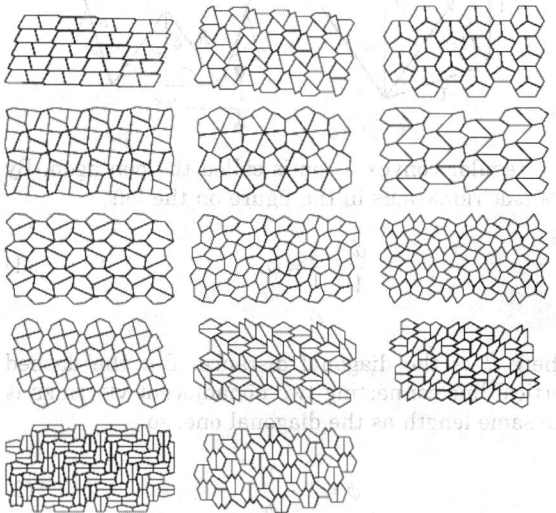

There are at least 14 classes of convex PENTAGONAL tilings (Steinhaus 1983, p. 75; Wells 1991, pp. 177–179; Pegg), as illustrated above. It has not been proven whether these 14 cases exhaust all possible tilings, but no others are known.

See also TILING

References

Bowers, P. L. and Stephenson, K. "A 'Regular' Pentagonal Tiling of the Plane." Submitted to *Conformal Geom. Dynamics*.

Pegg, E. Jr. "The 14 Different Types of Pentagons that Tile the Plane." http://www.mathpuzzle.com/tilepent.html.

Steinhaus, H. *Mathematical Snapshots, 3rd ed.* New York: Dover, 1999.

Wells, D. *The Penguin Dictionary of Curious and Interesting Geometry.* London: Penguin, pp. 177–179, 208, and 211, 1991.

Pentagonal Antiprism

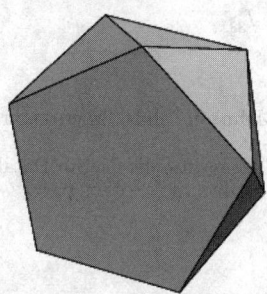

An ANTIPRISM and UNIFORM POLYHEDRON U_{77} whose DUAL POLYHEDRON is the PENTAGONAL DELTAHEDRON.

Pentagonal Cupola

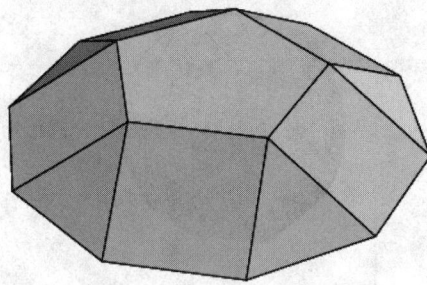

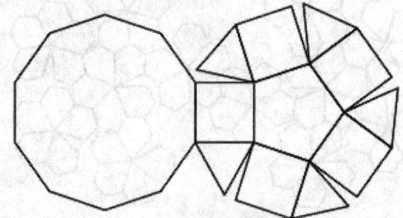

JOHNSON SOLID J_5. The bottom 10 VERTICES are

$$\left(\pm\frac{\left(1+\sqrt{5}\right)\sqrt{5+\sqrt{5}}}{4\sqrt{2}},\ \pm\tfrac{1}{2},\ 0\right),$$

$$\left(\pm\frac{\left(1+\sqrt{5}\right)\sqrt{5-\sqrt{5}}}{4\sqrt{2}},\ \pm\frac{3+\sqrt{5}}{2},\ 0\right),$$

$$\left(0,\ \pm\tfrac{1}{2}\left(1+\sqrt{5}\right),0\right)$$

and the top five vertices are

$$\left(\frac{\sqrt{5+\sqrt{5}}}{\sqrt{10}},\ 0,\ \frac{\sqrt{5-\sqrt{5}}}{\sqrt{10}}\right),$$

$$\left(\frac{\left(\sqrt{5}-1\right)\sqrt{5+\sqrt{5}}}{4\sqrt{10}},\ \pm\tfrac{1}{4}\left(1+\sqrt{5}\right),\ \frac{\sqrt{5-\sqrt{5}}}{\sqrt{10}}\right),$$

$$\left(\frac{-\left(\sqrt{5}+1\right)\sqrt{5+\sqrt{5}}}{4\sqrt{10}},\ \pm\tfrac{1}{2},\ \frac{\sqrt{5-\sqrt{5}}}{\sqrt{10}}\right).$$

Pentagonal Deltahedron

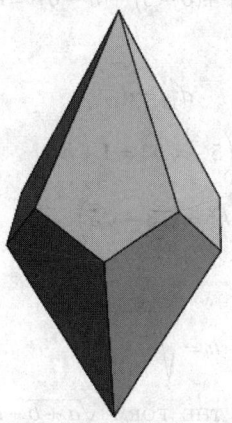

A TRAPEZOHEDRON which is the DUAL POLYHEDRON of the PENTAGONAL ANTIPRISM U_{77}.

See also DUAL POLYHEDRON, PENTAGONAL ANTIPRISM, TRAPEZOHEDRON

Pentagonal Dipyramid

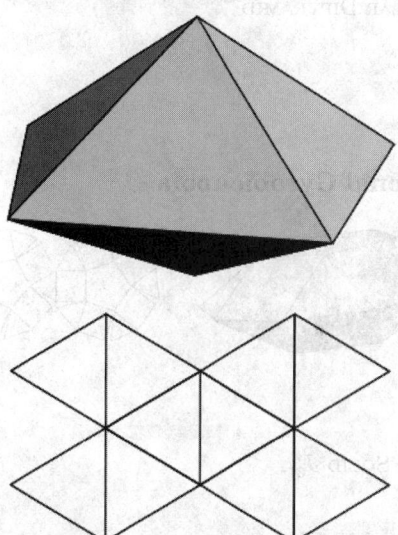

The pentagonal dipyramid is one of the convex DELTAHEDRA, and JOHNSON SOLID J_{13}. It is also the DUAL POLYHEDRON of the PENTAGONAL PRISM U_{76}. The

distance between two adjacent VERTICES on the base of the PENTAGON is

$$d_{12}^2 = \left[1 - \cos\left(\tfrac{2}{5}\pi\right)\right]^2 + \sin^2\left(\tfrac{2}{5}\pi\right)$$

$$= \left[1 - \tfrac{1}{4}\left(\sqrt{5} - 1\right)\right]^2 + \left[\frac{(1+\sqrt{5})\sqrt{5 - \sqrt{5}}}{4\sqrt{2}}\right]^2$$

$$= \tfrac{1}{2}\left(5 - \sqrt{5}\right), \tag{1}$$

and the distance between the apex and one of the base points is

$$d_{1h}^2 = (0 - 1)^2 + (0 - 0)^2 + (h - 0)^2 = 1 + h^2. \tag{2}$$

But

$$d_{12}^2 = d_{12}^2 \tag{3}$$

$$\tfrac{1}{2}\left(5 - \sqrt{5}\right) = 1 + h^2 \tag{4}$$

$$h^2 = \tfrac{1}{2}\left(3 - \sqrt{5}\right), \tag{5}$$

and

$$h = \sqrt{\frac{3 - \sqrt{5}}{2}}. \tag{6}$$

This root is OF THE FORM $\sqrt{a + b + c}$, so applying SQUARE ROOT simplification gives

$$h = \tfrac{1}{2}\left(\sqrt{5} - 1\right) \equiv \phi - 1, \tag{7}$$

where ϕ is the GOLDEN MEAN.

See also DELTAHEDRON, DIPYRAMID, GOLDEN MEAN, ICOSAHEDRON, JOHNSON SOLID, RIGIDITY THEOREM, TRIANGULAR DIPYRAMID

Pentagonal Gyrobicupola

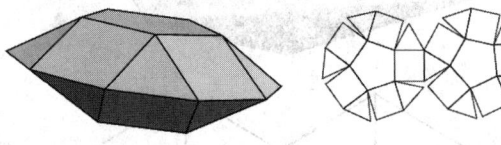

JOHNSON SOLID J_{31}.

References

Weisstein, E. W. "Johnson Solids." MATHEMATICA NOTEBOOK JOHNSONSOLIDS.M.
Weisstein, E. W. "Johnson Solid Netlib Database." MATHEMATICA NOTEBOOK JOHNSONSOLIDS.DAT.

Pentagonal Gyrocupolarotunda

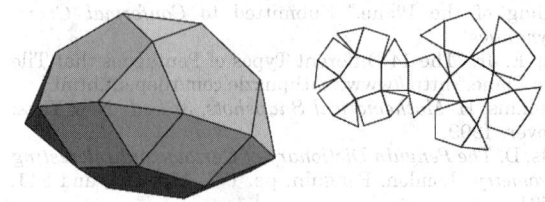

JOHNSON SOLID J_{33}.

References

Weisstein, E. W. "Johnson Solids." MATHEMATICA NOTEBOOK JOHNSONSOLIDS.M.
Weisstein, E. W. "Johnson Solid Netlib Database." MATHEMATICA NOTEBOOK JOHNSONSOLIDS.DAT.

Pentagonal Hexecontahedron

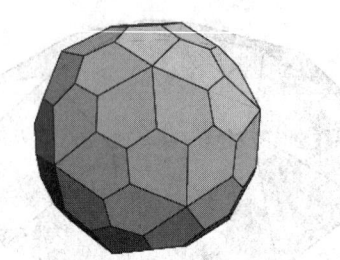

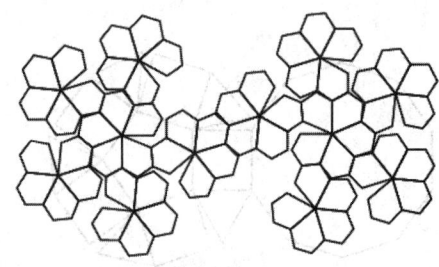

The 60-faced DUAL POLYHEDRON of the SNUB DODECAHEDRON A_8 and Wenninger dual W_{18}.

See also ARCHIMEDEAN DUAL, ARCHIMEDEAN SOLID, HEXECONTAHEDRON, SNUB DODECAHEDRON

References

Wenninger, M. J. *Dual Models.* Cambridge, England: Cambridge University Press, p. 29, 1983.

Pentagonal Icositetrahedron

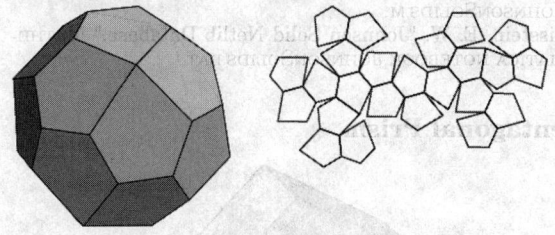

The 24-faced DUAL POLYHEDRON of the SNUB CUBE A_7 and Wenninger dual W_{17}. The mineral cuprite (Cu_2O) forms in pentagonal icositetrahedral crystals (Steinhaus 1983, pp. 207 and 209). The dual formed from a SNUB CUBE with unit edge length has side lengths given by the unique positive real roots of

$$2s_1^6 - 4s_1^4 + 4s_1^2 - 1 = 0 \tag{1}$$

$$32s_1^6 - 32s_1^4 + 8s_1^2 - 1 = 0. \tag{2}$$

The CIRCUMRADIUS R is given by the unique positive real root of

$$128r^6 - 224r^4 - 24r^2 - 1 = 0. \tag{3}$$

The SURFACE AREA S given by the positive real root of

$$S^6 - 684S^4 + 142560S^2 - 9879408 = 0, \tag{4}$$

and VOLUME V given by the positive real root of

$$8V^6 - 452V^4 + 462V^2 - 121 = 0. \tag{5}$$

See also ARCHIMEDEAN DUAL, ARCHIMEDEAN SOLID, ICOSITETRAHEDRON, SNUB CUBE, SNUB CUBE-PENTAGONAL ICOSITETRAHEDRON COMPOUND

References

Steinhaus, H. *Mathematical Snapshots, 3rd ed.* New York: Dover, 1999.

Wenninger, M. J. *Dual Models.* Cambridge, England: Cambridge University Press, p. 28, 1983.

Pentagonal Number

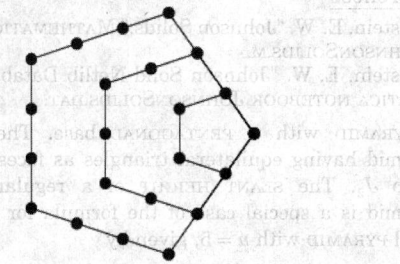

A POLYGONAL NUMBER OF THE FORM $n(3n-1)/2$. The first few are 1, 5, 12, 22, 35, 51, 70, ... (Sloane's A000326). The GENERATING FUNCTION for the pentagonal numbers is

$$\frac{x(2x+1)}{(1-x)^3} = x + 5x^2 + 12x^3 + 22x^4 + \dots.$$

Every pentagonal number is 1/3 of a TRIANGULAR NUMBER.

The so-called generalized pentagonal numbers are given by $n(3n-1)/2$ with $n = 0, \pm 1, \pm 2, \dots$, the first few of which are 0, 1, 2, 5, 7, 12, 15, 22, 26, 35, ... (Sloane's A001318).

See also HEPTAGONAL PENTAGONAL NUMBER, HEXAGONAL PENTAGONAL NUMBER, OCTAGONAL PENTAGONAL NUMBER, PARTITION FUNCTION P, PENTAGONAL NUMBER THEOREM, PENTAGONAL SQUARE NUMBER, PENTAGONAL TRIANGULAR NUMBER, POLYGONAL NUMBER, TRIANGULAR NUMBER

References

Guy, R. K. "Sums of Squares." §C20 in *Unsolved Problems in Number Theory, 2nd ed.* New York: Springer-Verlag, pp. 136–138, 1994.

Pappas, T. "Triangular, Square & Pentagonal Numbers." *The Joy of Mathematics.* San Carlos, CA: Wide World Publ./Tetra, p. 214, 1989.

Silverman, J. H. *A Friendly Introduction to Number Theory.* Englewood Cliffs, NJ: Prentice Hall, 1996.

Sloane, N. J. A. Sequences A000326/M3818 and A001318/M1336 in "An On-Line Version of the Encyclopedia of Integer Sequences." http://www.research.att.com/~njas/sequences/eisonline.html.

Pentagonal Number Theorem

$$\prod_{k=1}^{\infty}\left(1-x^k\right) = \sum_{k=-\infty}^{\infty}(-1)^k x^{k(3k+1)/2} \tag{1}$$

$$= 1 + \sum_{k=-1}^{\infty}(-1)^k\left[x^{k(3k-1)/2} + x^{k(3k+1)/2}\right], \tag{2}$$

where $n(3n+1)/2$ are generalized PENTAGONAL NUMBERS. Related equalities are

$$\prod_{k=1}^{\infty}\left(1-x^k t\right) = \sum_{n=0}^{\infty}\frac{(-1)^n x^{n(n+1)/2} t^n}{\prod_{k=1}^{n}(1-x^k)} \tag{3}$$

$$\prod_{k=1}^{\infty}\left(1-x^k t\right)^{-1} = \sum_{n=0}^{\infty}\frac{t^n}{\prod_{k=1}^{n}(1-x^k)}. \tag{4}$$

See also PARTITION FUNCTION P, PARTITION FUNCTION Q, PENTAGONAL NUMBER, RAMANUJAN THETA FUNCTIONS

References

Bailey, W. N. *Generalised Hypergeometric Series.* Cambridge, England: Cambridge University Press, p. 72, 1935.

Borwein, J. M. and Borwein, P. B. *Pi & the AGM: A Study in Analytic Number Theory and Computational Complexity.* New York: Wiley, p. 64, 1987.

Hardy, G. H. *Ramanujan: Twelve Lectures on Subjects Suggested by His Life and Work,* 3rd ed. New York: Chelsea, pp. 83–85, 1999.

Pentagonal Orthobicupola

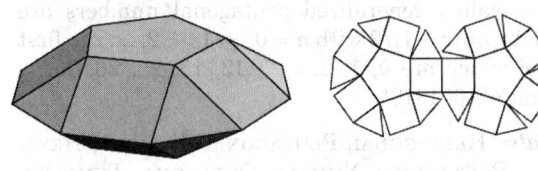

JOHNSON SOLID J_{30}.

References

Weisstein, E. W. "Johnson Solids." MATHEMATICA NOTEBOOK JOHNSONSOLIDS.M.

Weisstein, E. W. "Johnson Solid Netlib Database." MATHEMATICA NOTEBOOK JOHNSONSOLIDS.DAT.

Pentagonal Orthobirotunda

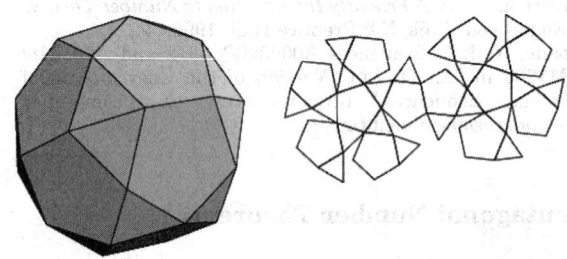

JOHNSON SOLID J_{34}.

References

Weisstein, E. W. "Johnson Solids." MATHEMATICA NOTEBOOK JOHNSONSOLIDS.M.

Weisstein, E. W. "Johnson Solid Netlib Database." MATHEMATICA NOTEBOOK JOHNSONSOLIDS.DAT.

Pentagonal Orthocupolarontunda

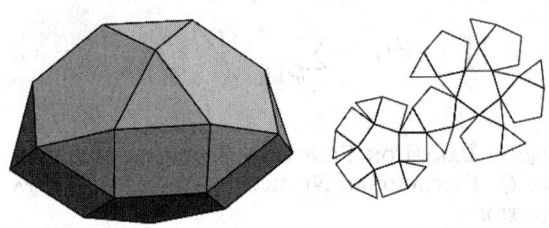

JOHNSON SOLID J_{32}.

References

Weisstein, E. W. "Johnson Solids." MATHEMATICA NOTEBOOK JOHNSONSOLIDS.M.

Weisstein, E. W. "Johnson Solid Netlib Database." MATHEMATICA NOTEBOOK JOHNSONSOLIDS.DAT.

Pentagonal Prism

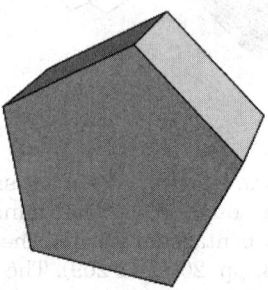

A PRISM, HEPTAHEDRON, and UNIFORM POLYHEDRON U_{76} whose DUAL POLYHEDRON is the PENTAGONAL DIPYRAMID. The SURFACE AREA and VOLUME for the pentagonal prism of unit edge length are

$$S = \tfrac{1}{2}\left(10 + \sqrt{5\left(5 + 2\sqrt{5}\right)}\right)$$

$$V = \tfrac{1}{4}\sqrt{5\left(5 + 2\sqrt{5}\right)}.$$

See also HEPTAHEDRON, PENTAGRAMMIC PRISM

Pentagonal Pyramid

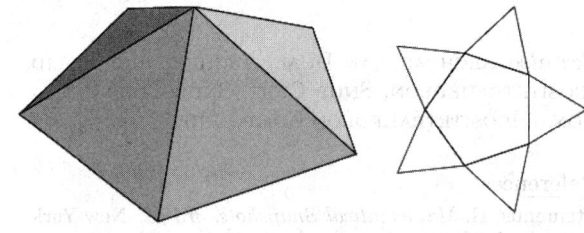

JOHNSON SOLID J_2.

References

Weisstein, E. W. "Johnson Solids." MATHEMATICA NOTEBOOK JOHNSONSOLIDS.M.

Weisstein, E. W. "Johnson Solid Netlib Database." MATHEMATICA NOTEBOOK JOHNSONSOLIDS.DAT.

A PYRAMID with a PENTAGONAL base. The pentagonal pyramid having equilateral triangles as faces is JOHNSON SOLID J_2. The SLANT HEIGHT of a regular pentagonal pyramid is a special case of the formula for a regular n-gonal PYRAMID with $n = 5$, given by

$$s = \sqrt{h^2 + \tfrac{1}{10}\left(5 + \sqrt{5}\right)a^2}, \tag{1}$$

where h is the height and a is the length of a side of the base.

See also PENTAGON, PYRAMID

Pentagonal Pyramidal Number

A FIGURATE NUMBER corresponding to a PENTAGONAL PYRAMID. The first few are 1, 6, 18, 40, 75, ... (Sloane's A002411). The GENERATING FUNCTION for the pentagonal pyramidal numbers is

$$\frac{x(2x+1)}{(x-1)^4} = x + 6x^2 + 18x^3 + 40x^4 + \cdots.$$

The odd pentagonal pyramidal numbers are given by 1, 75, 405, 1183, 2601, ... (Sloane's A015223), having squares 1, 5625, 164025, ... (Sloane's A014799), while the even pentagonal pyramidal numbers are given by 6, 18, 40, 126, 196, 288, ... (Sloane's A015224), having squares 36, 324, 1600, 15876, ... (Sloane's A014800).

See also PENTAGONAL NUMBER, PYRAMIDAL NUMBER

References

Sloane, N. J. A. Sequences A002411/M4116, A014799, A014800, A015223, and A015224 in "An On-Line Version of the Encyclopedia of Integer Sequences." http://www.research.att.com/~njas/sequences/eisonline.html.

Pentagonal Rotunda

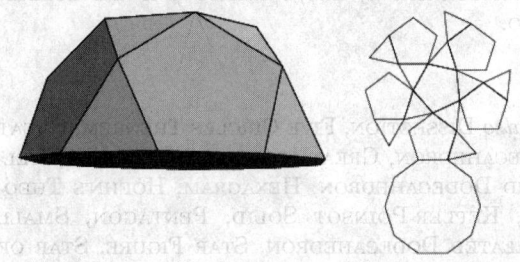

Half of an ICOSIDODECAHEDRON, denoted R_5. It has 10 triangular and five pentagonal faces separating a PENTAGONAL ceiling and a DODECAHEDRAL floor. It is JOHNSON SOLID J_6, and the only true ROTUNDA.

See also ICOSIDODECAHEDRON, JOHNSON SOLID, ROTUNDA

Pentagonal Square Number

A number which is simultaneously a PENTAGONAL NUMBER P_n and a SQUARE NUMBER S_m. Such numbers exist when

$$\tfrac{1}{2}n(3n-1) = m^2. \tag{1}$$

COMPLETING THE SQUARE gives

$$\tfrac{1}{2}n(3n-1) = \tfrac{3}{2}\left(n^2 - \tfrac{1}{3}n\right) = \tfrac{3}{2}\left(n - \tfrac{1}{6}\right)^2 - \tfrac{3}{72} = m^2 \tag{2}$$

$$\tfrac{3}{6}(6n-1)^2 - \tfrac{3}{2} = 36m^2 \tag{3}$$

$$(6n-1)^2 - 24m^2 = 1. \tag{4}$$

Substituting $x = 6n-1$ and $y = 2m$ gives the PELL EQUATION

$$x^2 - 6y^2 = 1, \tag{5}$$

which has solutions $(x, y) = (5, 2)$, $(49, 20)$, $(495, 198)$, ... In terms of (n, m), these give $(1,1)$, $(25/3, 10)$, $(81, 99)$, $(2401/3, 980)$, $(7921, 9701)$, ..., of which the whole number solutions are $(n, m) = (1, 1)$, $(81, 99)$, $(7921, 9701)$, $(776161, 950599)$, ... (Sloane's A046172 and A046173), corresponding to the pentagonal square numbers 1, 9801, 94109401, 903638458801, 8676736387298001, ... (Sloane's A036353).

Rathbun has searched for pentagonal square triangular numbers up to index 2000, but found none other than the trivial number 1.

See also PENTAGONAL NUMBER, SQUARE NUMBER

References

Silverman, J. H. *A Friendly Introduction to Number Theory.* Englewood Cliffs, NJ: Prentice Hall, 1996.
Sloane, N. J. A. Sequences A036353, A046172, and A046173 in "An On-Line Version of the Encyclopedia of Integer Sequences." http://www.research.att.com/~njas/sequences/eisonline.html.

Pentagonal Triangular Number

A number which is simultaneously a PENTAGONAL NUMBER P_n and TRIANGULAR NUMBER T_m. Such numbers exist when

$$\tfrac{1}{2}n(3n-1) = \tfrac{1}{2}m(m+1). \tag{1}$$

COMPLETING THE SQUARE gives

$$(6n-1)^2 - 3(2m+)^2 = -2. \tag{2}$$

Substituting $x = 6n-1$ and $y = 2m+1$ gives the Pell-like quadratic Diophantine equation

$$x^2 - 3y^2 = -2, \tag{3}$$

which has solutions $(x, y) = (5, 3)$, $(19, 11)$, $(71, 41)$, $(265, 153)$, ... In terms of (n, m), these give $(1, 1)$, $(10/3, 5)$, $(12, 20)$, $(133/3, 76)$, $(165, 285)$, ..., of which the whole number solutions are $(n, m) = (1, 1)$, $(12, 20)$, $(165, 285)$, $(2296, 3976)$, ... (Sloane's A046174 and A046175), corresponding to the pentagonal triangular numbers 1, 210, 40755, 7906276, 1533776805, ... (Sloane's A014979).

Rathbun has searched for pentagonal square triangular numbers up to index 2000, but found none other than the trivial number 1.

See also PENTAGONAL NUMBER, TRIANGULAR NUMBER

References

Silverman, J. H. *A Friendly Introduction to Number Theory.* Englewood Cliffs, NJ: Prentice Hall, 1996.

Sloane, N. J. A. Sequences A014979, A046174, and A046175 in "An On-Line Version of the Encyclopedia of Integer Sequences." http://www.research.att.com/~njas/sequences/eisonline.html.

Pentagram

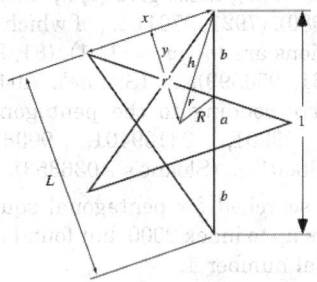

The STAR POLYGON {5/2}, also called the PENTACLE, PENTALPHA, or PENTANGLE. In the above figure, the pentagram has side length 1, and the indicated lengths are given by

$$a = \sqrt{5} - 2 \tag{1}$$

$$b = \tfrac{1}{2}\left(3 - \sqrt{5}\right) \tag{2}$$

$$r = \tfrac{1}{2}\, a \cot\left(\frac{\pi}{5}\right) = \tfrac{1}{2}\sqrt{\tfrac{1}{5}\left(5 - 2\sqrt{5}\right)} \tag{3}$$

$$R = \tfrac{1}{2}\, a \csc\left(\frac{\pi}{5}\right) = \sqrt{\tfrac{1}{10}\left(25 - 11\sqrt{5}\right)} \tag{4}$$

$$h = \sqrt{b^2 - \left(\tfrac{1}{2}a\right)^2} = \tfrac{1}{2}\sqrt{5 - 2\sqrt{5}} \tag{5}$$

$$x = 2(r + h)\sin\left(\frac{\pi}{5}\right) = \tfrac{1}{2}\left(\sqrt{5} - 1\right) \tag{6}$$

$$r' = \sqrt{(h + r)^2 + \left(\tfrac{1}{2}x\right)^2} = \tfrac{1}{2}\sqrt{\tfrac{1}{10}\left(5 + \sqrt{5}\right)} \tag{7}$$

$$y = r' - R = \tfrac{1}{2}\sqrt{\tfrac{1}{2}\left(25 + 11\sqrt{5}\right)} \tag{8}$$

$$L = \sqrt{1 - \tfrac{1}{4}x^2} = \tfrac{1}{2}\sqrt{\tfrac{1}{2}\left(5 + \sqrt{5}\right)}. \tag{9}$$

This gives the ratio

$$\frac{b}{a} = \phi, \tag{10}$$

where ϕ is the GOLDEN RATIO (Wells 1986, p. 36).

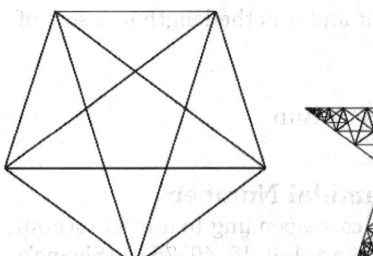
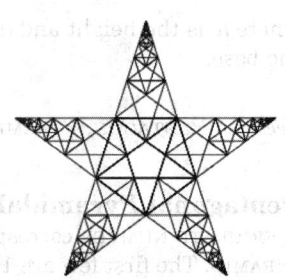

A series of embedded pentagrams can be constructed to form a larger pentagram, as illustrated above (Williams 1979, p. 53). If the central pentagram has center (0, 0) and CIRCUMRADIUS 1, then the subsequent pentagrams have radii

$$r_n = \phi^{-n}$$

and centers

$$x_n = -\tfrac{1}{4}\left(1 - \phi^{-n}\right)\sqrt{50 + 22\sqrt{5}}$$

$$y_n = \tfrac{1}{2}\,\phi\left(1 - \phi^{-n}\right)$$

modulo rotation by $2\pi k/5$, where ϕ is the GOLDEN RATIO.

See also DISSECTION, FIVE CIRCLES THEOREM, GREAT DODECAHEDRON, GREAT ICOSAHEDRON, GREAT STELLATED DODECAHEDRON, HEXAGRAM, HOEHN'S THEOREM, KEPLER-POINSOT SOLID, PENTAGON, SMALL STELLATED DODECAHEDRON, STAR FIGURE, STAR OF LAKSHMI

References

Ogilvy, C. S. *Excursions in Geometry.* New York: Dover, pp. 122–125, 1990.

Pappas, T. "The Pentagon, the Pentagram & the Golden Triangle." *The Joy of Mathematics.* San Carlos, CA: Wide World Publ./Tetra, pp. 188–189, 1989.

Schwartzman, S. *The Words of Mathematics: An Etymological Dictionary of Mathematical Terms Used in English.* Washington, DC: Math. Assoc. Amer., 1994.

Steinhaus, H. *Mathematical Snapshots, 3rd ed.* New York: Dover, p. 211, 1999.

Wells, D. *The Penguin Dictionary of Curious and Interesting Numbers.* Middlesex, England: Penguin Books, p. 36, 1986.

Williams, R. *The Geometrical Foundation of Natural Structure: A Source Book of Design.* New York: Dover, 1979.

Pentagrammic Antiprism

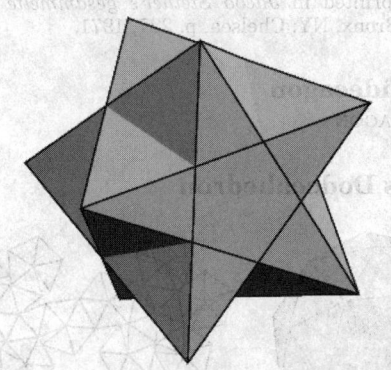

An ANTIPRISM and UNIFORM POLYHEDRON U_{79} whose DUAL POLYHEDRON is the PENTAGRAMMIC DELTAHEDRON.

Pentagrammic Concave Deltahedron

The DUAL POLYHEDRON of the PENTAGRAMMIC CROSSED ANTIPRISM U_{80}.

See also DUAL POLYHEDRON, PENTAGRAMMIC CROSSED ANTIPRISM

Pentagrammic Crossed Antiprism

An ANTIPRISM and UNIFORM POLYHEDRON U_{80} whose

DUAL POLYHEDRON is the PENTAGRAMMIC CONCAVE DELTAHEDRON.

Pentagrammic Deltahedron

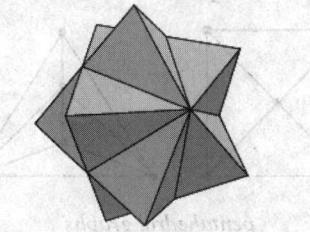

The DUAL POLYHEDRON of the PENTAGRAMMIC ANTIPRISM U_{79}.

See also DUAL POLYHEDRON, PENTAGRAMMIC ANTIPRISM

Pentagrammic Dipyramid

The DUAL POLYHEDRON of the PENTAGRAMMIC PRISM U_{78}.

See also DUAL POLYHEDRON, PENTAGRAMMIC PRISM

Pentagrammic Prism

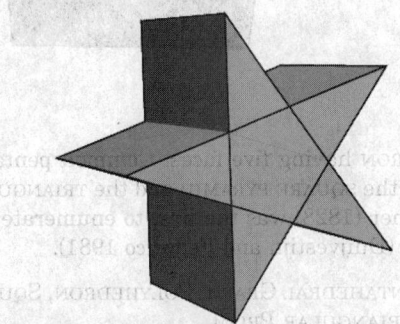

A PRISM, self-intersecting HEPTAHEDRON, and UNIFORM POLYHEDRON U_{78} whose DUAL POLYHEDRON is the PENTAGRAMMIC DIPYRAMID.

See also HEPTAHEDRON, PENTAGONAL PRISM

Pentagrammic Pyramid

See also PYRAMID

Pentahedral Graph

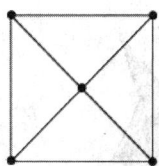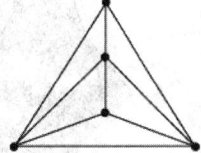

pentahedral graphs

A POLYHEDRAL GRAPH on five nodes. There are two topologically distinct pentahedral graphs, corresponding to the skeletons of the SQUARE PYRAMID (left figure) and TRIANGULAR DIPYRAMID (right figure). The pentahedral graphs were first enumerated by Steiner (1828; Duijvestijn and Federico 1981).

See also POLYHEDRAL GRAPH, SQUARE PYRAMID, TRIANGULAR DIPYRAMID.

References

Duijvestijn, A. J. W. and Federico, P. J. "The Number of Polyhedral (3-Connected Planar) Graphs." *Math. Comput.* **37**, 523–532, 1981.
Steiner, J. "Problème de situation." *Ann. de Math* **19**, 36, 1828. Reprinted in *Jacob Steiner's gesammelte Werke, Band I.* Bronx, NY: Chelsea, p. 227, 1971.

Pentahedron

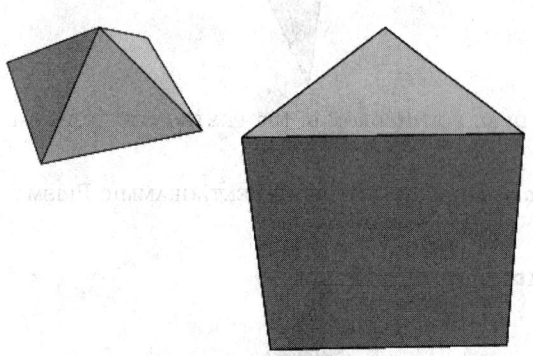

A POLYHEDRON having five faces. Common pentahedra include the SQUARE PYRAMID and the TRIANGULAR PRISM. Steiner (1828) was the first to enumerate the pentahedra (Duijvestijn and Federico 1981).

See also PENTAHEDRAL GRAPH, POLYHEDRON, SQUARE PYRAMID, TRIANGULAR PRISM

References

Duijvestijn, A. J. W. and Federico, P. J. "The Number of Polyhedral (3-Connected Planar) Graphs." *Math. Comput.* **37**, 523–532, 1981.

Steiner, J. "Problème de situation." *Ann. de Math.* **19**, 36, 1828. Reprinted in *Jacob Steiner's gesammelte Werke, Band I.* Bronx, NY: Chelsea, p. 227, 1971.

Pentakaidecagon

PENTADECAGON

Pentakis Dodecahedron

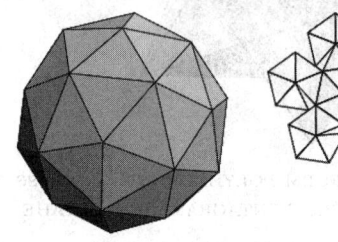

The 60-faced DUAL POLYHEDRON of the TRUNCATED ICOSAHEDRON A_{11} and Wenninger dual W_9. It can be constructed by CUMULATION of a unit edge-length DODECAHEDRON by a pyramid with height $\frac{1}{19}\sqrt{\frac{1}{5}(65+22\sqrt{5})}$. Taking the dual of a TRUNCATED ICOSAHEDRON with unit edge lengths gives a pentakis dodecahedron with edge lengths

$$s_1 = \tfrac{1}{19}\left(18\sqrt{5}-9\right) \tag{1}$$

$$s_2 = \tfrac{3}{2}\left(\sqrt{5}-1\right). \tag{2}$$

Normalizing so that $s_1 = 1$, the SURFACE AREA and VOLUME are

$$S = \tfrac{5}{3}\sqrt{\tfrac{1}{2}\left(421+63\sqrt{5}\right)} \tag{3}$$

$$V = \tfrac{5}{36}\left(41+25\sqrt{5}\right). \tag{4}$$

See also ARCHIMEDEAN DUAL, ARCHIMEDEAN SOLID, DUAL POLYHEDRON, HEXECONTAHEDRON, TRUNCATED ICOSAHEDRON

References

Wenninger, M. J. *Dual Models.* Cambridge, England: Cambridge University Press, p. 18, 1983.

Pentalpha

PENTAGRAM

Pentangle

PENTAGRAM

Pentaspherical Space

The set of all points **x** that can be put into one-to-one correspondence with sets of essentially distinct values of five homogeneous coordinates $x_0 : x_1 : x_2 : x_3 : x_4$, not all simultaneously zero, which are connected by the relation

$$\mathbf{x} \cdot \mathbf{x} = x_0^2 + x_1^2 + x_2^2 + x_3^2 + x_4^2 = 0. \qquad (1)$$

See also TETRACYCLIC PLANE

References

Coolidge, J. L. "Pentaspherical Space." Ch. 7 in *A Treatise on the Geometry of the Circle and Sphere*. New York: Chelsea, pp. 282–305, 1971.

Pentatope

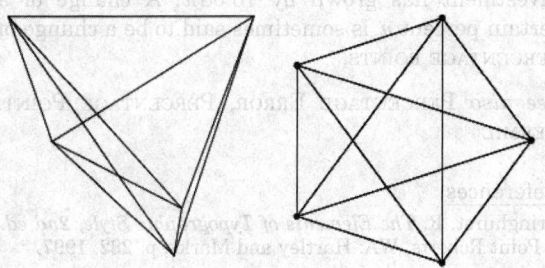

The simplest regular figure in 4-D, representing the 4-D analog of the solid TETRAHEDRON. It is also called the 5-cell, since it consists of five vertices. The pentatope is the 4-D SIMPLEX, and can be viewed as a regular TETRAHEDRON $ABCD$ in which a point E along the fourth dimension through the center of $ABCD$ is chosen so that $EA = EB = EC = ED = AB$. The pentatope has SCHLÄFLI SYMBOL $\{3, 3, 3\}$. The pentatope is self-dual, has 5 3-D facets (each the shape of a TETRAHEDRON), 10 ridges (faces), 10 edges, and 5 vertices. In the above figure, the pentatope is shown projected onto one of the four mutually perpendicular 3-spaces within the 4-space obtained by dropping one of the four vertex components (R. Towle).

See also 16-CELL, 24-CELL, 120-CELL, 600-CELL, HYPERCUBE, POLYTOPE, SIMPLEX, TETRAHEDRON

References

Gardner, M. *The Sixth Book of Mathematical Games from Scientific American.* Chicago, IL: University of Chicago Press, pp. 187–188, 1984.
Wells, D. *The Penguin Dictionary of Curious and Interesting Geometry.* London: Penguin, pp. 179–180 and 210, 1991.

Pentatope Number

A FIGURATE NUMBER which is given by

$$Ptop_n = \tfrac{1}{4} Te_n(n+3) = \tfrac{1}{24} n(n+1)(n+2)(n+3),$$

where Te_n is the nth TETRAHEDRAL NUMBER. The first few pentatope numbers are 1, 5, 15, 35, 70, 126, ... (Sloane's A000332). The GENERATING FUNCTION for the pentatope numbers is

$$\frac{x}{(1-x)^5} = x + 5x^2 + 15x^3 + 35x^4 + \ldots.$$

See also FIGURATE NUMBER, TETRAHEDRAL NUMBER

References

Conway, J. H. and Guy, R. K. *The Book of Numbers.* New York: Springer-Verlag, pp. 55–57, 1996.
Sloane, N. J. A. Sequences A000332/M3853 in "An On-Line Version of the Encyclopedia of Integer Sequences." http://www.research.att.com/~njas/sequences/eisonline.html.

Pentiamond

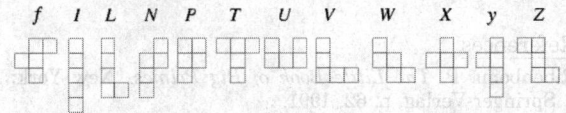
pentiamonds

One of the four 5-polyiamonds are called pentiamonds.

See also PENTIAMOND TILING, POLYIAMOND

Pentiamond Tiling

See also HEPTIAMOND TILING, HEXIAMOND TILING, OCTIAMOND TILING, PENTIAMOND

References

Vichera, M. "Polyiamonds." http://alpha.ujep.cz/~vicher/puzzle/polyform/iamond/iamonds.htm.

Pentomino

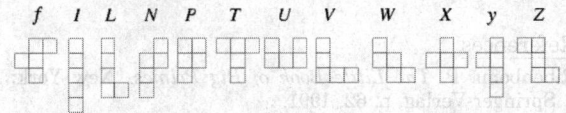

The twelve 5-POLYOMINOES illustrated above and known by the letters of the alphabet they most closely resemble: $f, I, L, N, P, T, U, V, W, X, y, Z$ (Gardner 1960, Golomb 1995). Another common naming convention replaces f, I, L, and N with R, O, Q, and S so that all letters from O to Z are used (Berlekamp *et al.* 1982). In particular, in the LIFE CELLULAR AUTOMATON, the f-pentomino is always known as the r-pentomino. The I, L, and T pentominoes can also be called the 5-STRAIGHT POLYOMINO, L-POLYOMINO, and T-POLYOMINO, respectively.

See also DOMINO, HEXOMINO, HEPTOMINO, OCTOMINO, POLYOMINO, TETROMINO, TRIOMINO

References

Ball, W. W. R. and Coxeter, H. S. M. *Mathematical Recreations and Essays, 13th ed.* New York: Dover, pp. 110–111, 1987.

Berlekamp, E. R.; Conway, J. H; and Guy, R. K. *Winning Ways for Your Mathematical Plays, Vol. 1: Games in General.* London: Academic Press, 1982.

Berlekamp, E. R.; Conway, J. H; and Guy, R. K. *Winning Ways for Your Mathematical Plays, Vol. 2: Games in Particular.* London: Academic Press, 1982.

Dudeney, H. E. "The Broken Chessboard." Problem 74 in *The Canterbury Puzzles and Other Curious Problems, 7th ed.* London: Thomas Nelson and Sons, pp. 119–120, 1949.

Gardner, M. "Mathematical Games: About the Remarkable Similarity between the Icosian Game and the Towers of Hanoi." *Sci. Amer.* **196**, 150–156, May 1957.

Gardner, M. "Mathematical Games: More About the Shapes that Can Be Made with Complex Dominoes." *Sci. Amer.* **203**, 186–198, Nov. 1960.

Golomb, S. W. *Polyominoes: Puzzles, Patterns, Problems, and Packings, 2nd ed.* Princeton, NJ: Princeton University Press, 1995.

Hunter, J. A. H. and Madachy, J. S. *Mathematical Diversions.* New York: Dover, pp. 80–86, 1975.

Lei, A. "Pentominoes." http://www.cs.ust.hk/~philipl/omino/pento.html.

Madachy, J. S. "Pentominoes: Some Solved and Unsolved Problems." *J. Rec. Math.* **2**, 181–188, 1969.

O'Beirne, T. H. "Pentominoes and Hexiamonds." *New Scientist* **12**, 379–380, 1961.

Ruskey, F. "Information on Pentomino Puzzles." http://www.theory.csc.uvic.ca/~cos/inf/misc/PentInfo.html.

Smith, A. "Pentomino Relationships." http://www.snaffles.demon.co.uk/pentanomes/.

Pépin's Test

A test for the PRIMALITY of FERMAT NUMBERS $F_n = 2^{2^n} + 1$, with $n \geq 2$ and $k \geq 2$. Then the two following conditions are equivalent:

1. F_n is PRIME and $(k/F_n) = -1$, where (n/k) is the JACOBI SYMBOL,
2. $k^{(F_n-1)/2} \equiv -1 \pmod{F_n}$.

k is usually taken as 3 as a first test.

See also FERMAT NUMBER, PÉPIN'S THEOREM

References

Ribenboim, P. *The Little Book of Big Primes.* New York: Springer-Verlag, p. 62, 1991.

Shanks, D. *Solved and Unsolved Problems in Number Theory, 4th ed.* New York: Chelsea, pp. 119–120, 1993.

Pépin's Theorem

The FERMAT NUMBER F_n is PRIME IFF

$$3^{2^{2^n-1}} \equiv -1 \pmod{F_n}.$$

See also FERMAT NUMBER, PÉPIN'S TEST, SELFRIDGE-HURWITZ RESIDUE

Per Cent

PERCENT

Per Mil

PERMIL

Per Mille

PERMIL

Percent

The use of percentages is a way of expressing RATIOS in terms of whole numbers. Given a RATIO or FRACTION, it is converted to a percentage by multiplying by 100 and appending a "percentage sign" %. For example, if an investment grows from a number $P = 13.00$ to a number $A = 22.50$, then A is $22.50/13.00 = 1.7308$ times as much as P, or 173.08%, and the investment has grown by 73.08%. A change of a certain percent n is sometimes said to be a change of PERCENTAGE POINTS.

See also PERCENTAGE ERROR, PERCENTAGE POINT, PERMIL

References

Bringhurst, R. *The Elements of Typographic Style, 2nd ed.* Point Roberts, WA: Hartley and Marks, p. 282, 1997.

Percent Sign

The symbol % used to indicate PERCENT.

References

Bringhurst, R. *The Elements of Typographic Style, 2nd ed.* Point Roberts, WA: Hartley and Marks, p. 282, 1997.

Percentage

PERCENT, PERCENTAGE ERROR, PERCENTAGE POINT

Percentage Error

The percentage error is 100% times the RELATIVE ERROR.

See also ABSOLUTE ERROR, ERROR PROPAGATION, PERCENT, RELATIVE ERROR

References

Abramowitz, M. and Stegun, C. A. (Eds.). *Handbook of Mathematical Functions with Formulas, Graphs, and Mathematical Tables, 9th printing.* New York: Dover, p. 14, 1972.

Percentage Point

1%.

See also BASIS POINT, PERCENT

Percentile

The kth percentile P_k is that value of x, say x_k, which corresponds to a CUMULATIVE FREQUENCY of $Nk/100$.

See also QUANTILE, QUARTILE

Percolation Theory

References
Kenney, J. F. and Keeping, E. S. "Percentile Ranks." §3.6 in *Mathematics of Statistics, Pt. 1, 3rd ed.* Princeton, NJ: Van Nostrand, pp. 38–39, 1962.

Percolation Theory

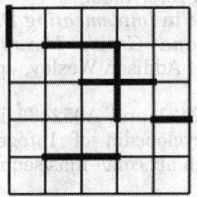 

bond percolation   *site percolation*

Percolation theory deals with fluid flow (or any other similar process) in random media. If the medium is a set of regular LATTICE POINTS, then there are two types of percolation. A SITE PERCOLATION considers the lattice vertices as the relevant entities; a BOND PERCOLATION considers the lattice edges as the relevant entities.

See also BOND PERCOLATION, CAYLEY TREE, CLUSTER, CLUSTER PERIMETER, LATTICE ANIMAL, PERCOLATION THRESHOLD, POLYOMINO, RANDOM WALK, *s*-CLUSTER, *s*-RUN, SITE PERCOLATION

References
Deutscher, G.; Zallen, R.; and Adler, J. (Eds.). *Percolation Structures and Processes.* Bristol: Adam Hilger, 1983.
Finch, S. "Favorite Mathematical Constants." http://www.mathsoft.com/asolve/constant/rndprc/rndprc.html.
Grimmett, G. *Percolation.* New York: Springer-Verlag, 1989.
Grimmett, G. *Percolation and Disordered Systems.* Berlin: Springer-Verlag, 1997.
Kesten, H. *Percolation Theory for Mathematicians.* Boston, MA: Birkhäuser, 1982.
Stauffer, D. and Aharony, A. *Introduction to Percolation Theory, 2nd ed.* London: Taylor & Francis, 1992.
Weisstein, E. W. "Books about Percolation Theory." http://www.treasure-troves.com/books/PercolationTheory.html.

Percolation Threshold

The critical fraction of lattice points which must be filled to create a continuous path of nearest neighbors from one side to another. The following table is from Stauffer and Aharony (1992, p. 17).

Lattice	Site	Bond
Cubic (Body-Centered)	0.246	0.1803
Cubic (Face-Centered)	0.198	0.119
Cubic (Simple)	0.3116	0.2488
Diamond	0.43	0.388
Honeycomb	0.6962	0.65271
4-Hypercubic	0.197	0.1601
5-Hypercubic	0.141	0.1182
6-Hypercubic	0.107	0.0942
7-Hypercubic	0.089	0.0787
Square	0.592746	0.50000
Triangular	0.50000	0.34729

The square bond value is 1/2 exactly, as is the triangular site. $p_c = 2 \sin(\pi/18)$ for the triangular bond and $p_c = 1 - 2 \sin(\pi/18)$ for the honeycomb bond. An exact answer for the square site percolation threshold is not known.

See also PERCOLATION THEORY

References
Essam, J. W.; Gaunt, D. S.; and Guttmann, A. J. "Percolation Theory at the Critical Dimension." *J. Phys. A* **11**, 1983–1990, 1978.
Finch, S. "Favorite Mathematical Constants." http://www.mathsoft.com/asolve/constant/rndprc/rndprc.html.
Kesten, H. *Percolation Theory for Mathematicians.* Boston, MA: Birkhäuser, 1982.
Stauffer, D. and Aharony, A. *Introduction to Percolation Theory, 2nd ed.* London: Taylor & Francis, 1992.

Perfect Box

EULER BRICK

Perfect Code

See also ERROR-CORRECTING CODE, HAMMING CODE

References
MacWilliams, F. J. and Sloane, N. J. A. *The Theory of Error-Correcting Codes.* Amsterdam, Netherlands: North-Holland, 1977.

Perfect Cubic Polynomial

A perfect cubic POLYNOMIAL can be factored into a linear and a quadratic term,

$$x^3 + y^3 = (x+y)(x^2 - xy + y^2)$$

$$x^3 - y^3 = (x-y)(x^2 + xy + y^2).$$

See also CUBIC EQUATION, PERFECT SQUARE, POLYNOMIAL

Perfect Cuboid

EULER BRICK

Perfect Difference Set

A SET of RESIDUES $\{a_1, a_2, \ldots, a_{k+1}\}$ (mod n) such that every NONZERO RESIDUE can be uniquely expressed in the form $a_i - a_j$. Examples include $\{1, 2, 4\}$ (mod 7) and $\{1, 2, 5, 7\}$ (mod 13). A

NECESSARY condition for a difference set to exist is that n be OF THE FORM $k^2 + k + 1$. A SUFFICIENT condition is that k be a PRIME POWER. Perfect sets can be used in the construction of PERFECT RULERS.

See also PERFECT RULER

References

Guy, R. K. "Modular Difference Sets and Error Correcting Codes." §C10 in *Unsolved Problems in Number Theory, 2nd ed.* New York: Springer-Verlag, pp. 118–121, 1994.

Perfect Digital Invariant

NARCISSISTIC NUMBER

Perfect Graph

A GRAPH G such that for every INDUCED SUBGRAPH of G, the size of the largest CLIQUE equals the CHROMATIC NUMBER. A graph can be tested to see if it is perfect using PerfectQ[g] in the *Mathematica* add-on package DiscreteMath`Combinatorica` (which can be loaded with the command <<DiscreteMath`). Determining if a graph is perfect requires solving two NP-COMPLETE PROBLEMS (Skiena 1990, p. 219).

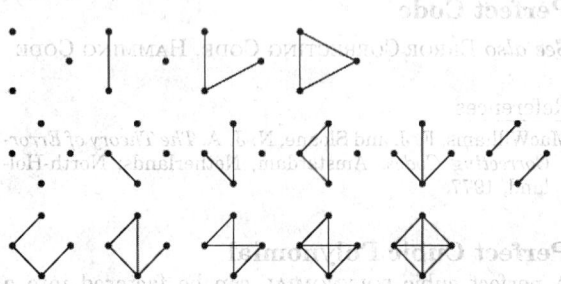

The numbers of perfect graphs on $n = 1, 2, \ldots$ nodes are 1, 2, 4, 11, 33, 148, 906, ... (Sloane's A052431).

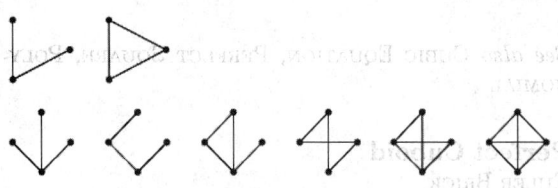

The numbers of perfect CONNECTED GRAPHS on $n = 1, 2, \ldots$ nodes are 1, 1, 2, 6, 20, 105, 724, ... (Sloane's A052433).

See also CHROMATIC NUMBER, CLIQUE, INDUCED

SUBGRAPH, PERFECT GRAPH THEOREM, STRONG PERFECT GRAPH CONJECTURE

References

Golumbic, M. C. *Algorithmic Graph Theory and Perfect Graphs.* New York: Academic Press, 1980.
Skiena, S. "Perfect Graphs." §5.6.4 in *Implementing Discrete Mathematics: Combinatorics and Graph Theory with Mathematica.* Reading, MA: Addison-Wesley, p. 219, 1990.
Sloane, N. J. A. Sequences A052431 and A052433 in "An On-Line Version of the Encyclopedia of Integer Sequences." http://www.research.att.com/~njas/sequences/eisonline.html.

Perfect Graph Theorem

The GRAPH COMPLEMENT of a PERFECT GRAPH is itself perfect (Fulkerson 1971; Lovász 1972; Skiena 1990, p. 219).

See also PERFECT GRAPH, STRONG PERFECT GRAPH CONJECTURE

References

Fulkerson, D. R. "Blocking and Anti-Blocking Pairs of Polyhedra." *Math. Program.* **1**, 168–194, 1971.
Lovász, L. "Normal Hypergraphs and the Perfect Graph Conjecture." *Disc. Math.* **2**, 253–267, 1972.
Skiena, S. *Implementing Discrete Mathematics: Combinatorics and Graph Theory with Mathematica.* Reading, MA: Addison-Wesley, 1990.

Perfect Group

References

Holt, D. G. and Plesken, W. *Perfect Groups.* Oxford, England: Clarendon Press, 1989.

Perfect Information

A class of GAME in which players move alternately and each player is completely informed of previous moves. FINITE, ZERO-SUM, two-player GAMES with perfect information (including checkers and chess) have a SADDLE POINT, and therefore one or more optimal strategies. However, the optimal strategy may be so difficult to compute as to be effectively impossible to determine (as in the game of CHESS).

See also FINITE GAME, GAME, ZERO-SUM GAME

Perfect Magic Cube

19	497	255	285	432	78	324	162
303	205	451	33	148	370	128	414
336	174	420	66	243	273	31	509
116	402	160	382	463	45	291	193
486	8	266	236	89	443	181	343
218	316	54	472	357	135	393	107
185	347	85	439	262	232	490	12
389	103	361	139	58	476	214	312

134	360	106	396	313	219	469	55
442	92	342	184	5	487	233	267
473	59	309	215	102	392	138	364
229	263	9	491	346	188	438	88
371	145	415	125	208	302	36	450
429	163	321	500	18	288	254	
48	462	196	290	403	113	383	157
276	242	512	30	175	333	67	417

306	212	478	64	141	367	97	387
14	496	226	260	433	83	349	191
109	399	129	355	466	52	318	224
337	179	445	95	238	272	2	484
199	293	43	457	380	154	408	118
507	25	279	245	72	422	172	330
412	122	376	150	39	453	203	297
168	326	76	426	283	249	503	21

423	69	331	169	28	506	248	278
155	377	119	405	296	198	460	42
252	292	24	502	327	165	427	73
456	38	300	202	123	409	151	373
82	436	190	352	493	15	257	227
366	144	386	100	209	307	61	479
269	239	481	3	178	340	94	448
49	467	221	319	398	112	354	132

381	159	401	115	194	292	46	464
65	419	173	335	510	32	274	244
34	452	206	304	413	127	369	147
286	256	498	20	161	323	77	431
140	362	104	390	311	213	475	57
440	86	348	166	11	489	231	261
471	53	315	217	108	394	136	358
235	265	7	485	344	182	444	90

492	10	264	230	87	437	187	345
216	310	60	474	363	137	391	101
183	341	91	441	260	234	488	6
395	105	359	133	56	470	220	314
29	511	241	275	418	68	334	176
289	195	461	47	158	384	114	404
322	164	430	80	253	287	17	499
416	146	372	449	35	301	207	

96	446	180	338	483	1	271	237
356	130	400	110	223	317	51	465
259	225	13	192	350	84	434	
63	477	211	305	388	98	368	142
425	75	325	167	22	504	250	284
149	375	121	411	298	204	454	40
246	280	26	508	329	171	421	71
458	44	294	200	117	407	153	379

201	299	37	455	374	152	410	124
501	23	281	251	74	428	166	328
406	120	378	156	41	459	197	295
170	332	70	424	277	247	505	27
320	222	466	50	131	353	111	397
4	482	240	270	447	93	339	177
99	385	143	365	480	62	308	210
351	189	435	81	228	258	16	494

A perfect magic cube is a MAGIC CUBE for which the CROSS SECTION diagonals, as well as the space diagonals, sum to the MAGIC CONSTANT. Perfect magic cubes are impossible for orders 3 and 4 (Schroeppel 1972, Gardner 1988), but it is not known if such cubes can exist for order 5 or 6 (Wells 1986, p. 72). Although no perfect magic cubes of order five are known, any such cube must have a central value of 63 (Schroeppel 1972; Gardner 1988).

Langman (1962) constructed a perfect magic cube of order seven, and others were found by R. Schroeppel and Ernst Straus (Wells 1986, p. 72). An order-eight perfect magic cube was published anonymously in 1875 (Barnard 1888, Gardner 1976, Benson and Jacoby 1981, Gardner 1988). The construction of such a cube is discussed in Ball and Coxeter (1987). Rosser and Walker rediscovered the order-eight cube in the late 1930s (but did not publish it), and Myers independently discovered the cube illustrated above in 1970 (Wells 1986, p. 72; Gardner 1988). Order 9 and 11 magic cubes have also been discovered, but none of order 10 (Gardner 1988).

See also MAGIC CUBE, SEMIPERFECT MAGIC CUBE

References

Ball, W. W. R. and Coxeter, H. S. M. *Mathematical Recreations and Essays, 13th ed.* New York: Dover, pp. 216–224, 1987.

Barnard, F. A. P. "Theory of Magic Squares and Cubes." *Mem. Nat. Acad. Sci.* **4**, 209–270, 1888.

Benson, W. H. and Jacoby, O. *Magic Cubes: New Recreations.* New York: Dover, 1981.

Gardner, M. *Sci. Amer.*, Jan. 1976.

Gardner, M. "Magic Squares and Cubes." Ch. 17 in *Time Travel and Other Mathematical Bewilderments.* New York: W. H. Freeman, pp. 213–225, 1988.

Langman, H. *Play Mathematics.* New York: Hafner, pp. 75–76, 1962.

Schroeppel, R. Item 50 in Beeler, M.; Gosper, R. W.; and Schroeppel, R. *HAKMEM.* Cambridge, MA: MIT Artificial Intelligence Laboratory, Memo AIM-239, p. 18, Feb. 1972.

Wells, D. *The Penguin Dictionary of Curious and Interesting Numbers.* Middlesex, England: Penguin Books, p. 72, 1986.

Perfect Matching

A MATCHING of a GRAPH containing $n/2$ edges, the largest possible. Not all graphs have a perfect matching, although all graphs do have a *maximal* matching (Skiena 1990, p. 240). Every CUBIC GRAPH without BRIDGES has a perfect matching (Skiena 1990, p. 244).

See also K-FACTOR, MATCHING

References

Skiena, S. *Implementing Discrete Mathematics: Combinatorics and Graph Theory with Mathematica.* Reading, MA: Addison-Wesley, 1990.

Perfect Number

Perfect numbers are INTEGERS n such that

$$n = s(n), \qquad (1)$$

where $s(n)$ is the RESTRICTED DIVISOR FUNCTION (i.e., the SUM of PROPER DIVISORS of n), or equivalently

$$\sigma(n) = 2n, \qquad (2)$$

where $\sigma(n)$ is the DIVISOR FUNCTION (i.e., the SUM of DIVISORS of n including n itself). The first few perfect numbers are 6, 28, 496, 8128, ... (Sloane's A000396). This follows from the fact that

$$6 = 1 + 2 + 3$$

$$28 = 1 + 2 + 4 + 7 + 14$$

$$496 = 1 + 2 + 4 + 8 + 16 + 31 + 62 + 124 + 248,$$

etc. Perfect numbers were deemed to have important numerological properties by the ancients, and were extensively studied by the Greeks, including Euclid.

Perfect numbers are intimately connected with a class of numbers known as MERSENNE PRIMES. This can be demonstrated by considering a perfect number P OF THE FORM $P = q2^{p-1}$ where q is PRIME. Then

$$\sigma(P) = 2P, \qquad (3)$$

and using

$$\sigma(q) = q + 1 \qquad (4)$$

for q prime, and

$$\sigma(2^x) = 2^{x+1} - 1 \qquad (5)$$

gives

$$\sigma(q2^{p-1}) = \sigma(q)\sigma(2^{p-1}) = (q+1)(2^p - 1)$$

$$= 2q2^{p-1} = q2^p \qquad (6)$$

$$q(2^p - 1) + 2^p - 1 = q2^p \qquad (7)$$

$$q = 2^p - 1. \qquad (8)$$

Therefore, if $M_p \equiv q = 2^p - 1$ is PRIME, then

$$P = \tfrac{1}{2}(M_p + 1)M_p = 2^{p-1}(2^p - 1) \qquad (9)$$

is a perfect number, as was stated in Proposition IX.36 of Euclid's *ELEMENTS* (Dickson 1957, p. 3; Dunham 1990). The first few perfect numbers are summarized in the following table.

#	p	P
1	2	6
2	3	28
3	5	496
4	7	8128
5	13	33550336
6	17	8589869056
7	19	137438691328
8	31	2305843008139952128

While many of Euclid's successors implicitly assumed that *all* perfect numbers were of the form (9) (Dickson 1952, pp. 3–33), the precise statement that all *even* perfect numbers are of this form. This was considered in a 1638 letter from Descartes to Mersenne (Dickson 1957, p. 12), and proving or disproving that Euclid's construction gives all possible even perfect numbers was prosed to Fermat in a 1658 letter from Frans van Schooten (Dickson 1957, p. 14). In a posthumous paper, Euler (Euler 1849) provided the first proof that Euclid's construction gives all possible even perfect numbers (Dickson 1957, p. 19).

It is known that all EVEN perfect numbers (except 6) end in 16, 28, 36, 56, 76, or 96 (Lucas 1891) and have DIGITAL ROOT 1. Every perfect number OF THE FORM $2^p(2^{p+1} - 1)$ can be written

$$2^p(2^{p+1} - 1) = \sum_{k=1}^{p/2} (2k-1)^3. \qquad (10)$$

All EVEN perfect numbers $P > 6$ are OF THE FORM

$$P = 1 + 9T_n, \qquad (11)$$

where T_n is a TRIANGULAR NUMBER

$$T_n = \tfrac{1}{2}n(n+1) \qquad (12)$$

such that $n = 8j + 2$ (Eaton 1995, 1996). In addition, all even perfect numbers are HEXAGONAL NUMBERS, so it follows that perfect numbers are always the sum of consecutive POSITIVE INTEGERS starting at 1, for example,

$$6 = \sum_{n=1}^{3} n \qquad (13)$$

$$28 = \sum_{n=1}^{7} n \qquad (14)$$

$$496 = \sum_{n=1}^{31} n \qquad (15)$$

(Singh 1997).

It is not known if any ODD PERFECT NUMBERS exist, although numbers up to 10^{300} have been checked (Brent *et al.* 1991; Guy 1994, p. 44) without success.

The sum of reciprocals of all the divisors of a perfect number is 2, since

$$\underbrace{n + \ldots + c + b + a}_{n} = 2n \qquad (16)$$

$$\frac{n}{a} + \frac{n}{b} + \ldots = 2n \qquad (17)$$

$$\frac{1}{a} + \frac{1}{b} + \ldots = 2. \qquad (18)$$

If $s(n) > n$, n is said to be an ABUNDANT NUMBER. If $s(n) < n$, n is said to be a DEFICIENT NUMBER. And if $s(n) = kn$ for a POSITIVE INTEGER $k > 1$, n is said to be a MULTIPERFECT NUMBER of order k.

The only even perfect number OF THE FORM $x^3 + 1$ is 28 (Makowski 1962).

See also ABUNDANT NUMBER, ALIQUOT SEQUENCE, AMICABLE NUMBERS, DEFICIENT NUMBER, DIVISOR FUNCTION, *e*-PERFECT NUMBER, HARMONIC NUMBER, HYPERPERFECT NUMBER, INFINARY PERFECT NUMBER, MERSENNE NUMBER, MERSENNE PRIME, MULTIPERFECT NUMBER, MULTIPLICATIVE PERFECT NUMBER, ODD PERFECT NUMBER, PLUPERFECT NUMBER, PSEUDOPERFECT NUMBER, QUASIPERFECT NUMBER, SEMIPERFECT NUMBER, SMITH NUMBER, SOCIABLE NUMBERS, SUBLIME NUMBER, SUPER UNITARY PERFECT NUMBER, SUPERPERFECT NUMBER, UNITARY PERFECT NUMBER, WEIRD NUMBER

References

Ball, W. W. R. and Coxeter, H. S. M. *Mathematical Recreations and Essays, 13th ed.* New York: Dover, pp. 66–67, 1987.

Brent, R. P.; Cohen, G. L. L.; and te Riele, H. J. J. "Improved Techniques for Lower Bounds for Odd Perfect Numbers." *Math. Comput.* **57**, 857–868, 1991.

Conway, J. H. and Guy, R. K. "Perfect Numbers." In *The Book of Numbers*. New York: Springer-Verlag, pp. 136–137, 1996.

Dickson, L. E. "Notes on the Theory of Numbers." *Amer. Math. Monthly* **18**, 109–111, 1911.

Dickson, L. E. *History of the Theory of Numbers, Vol. 1: Divisibility and Primality*. New York: Chelsea, pp. 3–33, 1952.

Dunham, W. *Journey through Genius: The Great Theorems of Mathematics*. New York: Wiley, p. 75, 1990.

Eaton, C. F. "Problem 1482." *Math. Mag.* **68**, 307, 1995.

Eaton, C. F. "Perfect Number in Terms of Triangular Numbers." Solution to Problem 1482. *Math. Mag.* **69**, 308–309, 1996.

Gardner, M. "Perfect, Amicable, Sociable." Ch. 12 in *Mathematical Magic Show: More Puzzles, Games, Diversions, Illusions and Other Mathematical Sleight-of-Mind from Scientific American*. New York: Vintage, pp. 160–171, 1978.

Guy, R. K. "Perfect Numbers." §B1 in *Unsolved Problems in Number Theory, 2nd ed.* New York: Springer-Verlag, pp. 44–45, 1994.

Iannucci, D. E. "The Second Largest Prime Divisor of an Odd Perfect Number Exceeds Ten Thousand." *Math. Comput.* **68**, 1749–1760, 1999.

Kraitchik, M. "Mersenne Numbers and Perfect Numbers." §3.5 in *Mathematical Recreations*. New York: W. W. Norton, pp. 70–73, 1942.

Madachy, J. S. *Madachy's Mathematical Recreations*. New York: Dover, pp. 145 and 147–151, 1979.

Makowski, A. "Remark on Perfect Numbers." *Elemente Math.* **17**, 109, 1962.

Powers, R. E. "The Tenth Perfect Number." *Amer. Math. Monthly* **18**, 195–196, 1911.

Séroul, R. "Perfect Numbers." §8.3 in *Programming for Mathematicians*. Berlin: Springer-Verlag, pp. 163–165, 2000.

Shanks, D. *Solved and Unsolved Problems in Number Theory, 4th ed.* New York: Chelsea, pp. 1–13 and 25–29, 1993.

Singh, S. *Fermat's Enigma: The Epic Quest to Solve the World's Greatest Mathematical Problem*. New York: Walker, pp. 11–13, 1997.

Sloane, N. J. A. Sequences A000396/M4186 in "An On-Line Version of the Encyclopedia of Integer Sequences." http://www.research.att.com/~njas/sequences/eisonline.html.

Smith, H. J. "Perfect Numbers." http://pweb.netcom.com/~hjsmith/Perfect.html.

Souissi, M. *Un Texte Manuscrit d'Ibn Al-Banna' Al-Marrakusi sur les Nombres Parfaits, Abondants, Deficients, et Amiables*. Karachi, Pakistan: Hamdard Nat. Found., 1975.

Wagon, S. "Perfect Numbers." *Math. Intell.* **7**, 66–68, 1985.

Zachariou, A. and Zachariou, E. "Perfect, Semi-Perfect and Ore Numbers." *Bull. Soc. Math. Grèce (New Ser.)* **13**, 12–22, 1972.

Perfect Partition

A PARTITION of n whose elements *uniquely* generate any number $1, 2, ..., n$. The following table gives the first several perfect partitions for small n.

n	perfect partitions
1	$\{1\}$
2	$\{1, 1\}$
3	$\{2, 1\}, \{1, 1, 1\}$
4	$\{1, 1, 1, 1\}$
5	$\{3, 1, 1\}, \{2, 2, 1\}, \{1, 1, 1, 1, 1\}$
6	$\{1, 1, 1, 1, 1, 1\}$

The numbers of perfect partitions of n for $n = 1, 2, ...$ are given by 1, 1, 2, 1, 3, 1, 4, 2, 3, ... (Sloane's A002033). For p^k a PRIME POWER, the number of perfect partitions $a(p^k)$ is given by

$$a(p^k) = 2^{k-1}.$$

Let $b(n) = a(n+1)$, then $b(n)$ is given by the RECURRENCE RELATION

$$b(n) = \sum_{\substack{d \mid n \\ d \neq n}} b(d).$$

The number of perfect partitions of n is equal to the number of ordered factorizations of $n+1$ (Goulden and Jackson 1983, p. 94).

See also PARTITION

References

Cohen, D. I. A. *Basic Techniques of Combinatorial Theory*. New York: Wiley and Sons, p. 97, 1978.

Goulden, I. P. and Jackson, D. M. Problem 2.5.12 in *Combinatorial Enumeration*. New York: Wiley, 1983.

Honsberger, R. *Mathematical Gems III*. Washington, DC: Math. Assoc. Amer., pp. 140–143, 1985.

Riordan, J. "An Introduction to Combinatorial Analysis." In (Ed.). , pp. , .

Sloane, N. J. A. Sequences A002033/M0131 and A035341 in "An On-Line Version of the Encyclopedia of Integer Sequences." http://www.research.att.com/~njas/sequences/eisonline.html.

Perfect Proportion

Since

$$\frac{2a}{a+b} = \frac{2ab}{(a+b)b}, \tag{1}$$

it follows that

$$\frac{a}{\frac{a+b}{2}} = \frac{\frac{2ab}{a+b}}{b}, \tag{2}$$

so

$$\frac{a}{A} = \frac{H}{b}, \tag{3}$$

where A and H are the ARITHMETIC MEAN and HARMONIC MEAN of a and b. This relationship was purportedly discovered by Pythagoras.

See also ARITHMETIC MEAN, HARMONIC MEAN

Perfect Rectangle

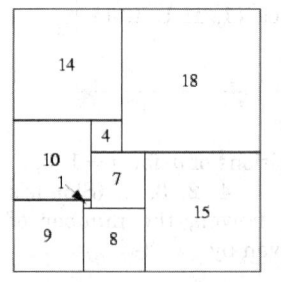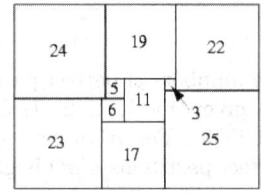

9-*square rectangle*　　10-*square rectangle*

A RECTANGLE which cannot be built up of SQUARES all of different sizes is called an imperfect rectangle. A RECTANGLE which can be built up of SQUARES all of different sizes is called perfect. The number of perfect rectangles of orders 8, 9, 10, ... are 0, 2, 6, 22, 67, 213, 744, 2609, ... (Sloane's A002839) and the corresponding numbers of imperfect rectangles are 0, 1, 0, 0, 9, 34, 103, 283, ... (Sloane's A002882).

See also PERFECT SQUARE DISSECTION, RECTANGLE TILING

References

Bouwkamp, C. J. "On the Dissection of Rectangles into Squares. I." *Indag. Math.* **8**, 724–736, 1946.
Bouwkamp, C. J. "On the Dissection of Rectangles into Squares. II." *Indag. Math.* **9**, 43–56, 1947.
Bouwkamp, C. J. "On the Dissection of Rectangles into Squares. III." *Indag. Math.* **9**, 57–63, 1947.
Brooks, R. L.; Smith, C. A. B.; Stone, A. H.; and Tutte, W. T. "The Dissection of Rectangles into Squares." *Duke Math. J.* **7**, 312–340, 1940.
Croft, H. T.; Falconer, K. J.; and Guy, R. K. "Squaring the Square." §C2 in *Unsolved Problems in Geometry.* New York: Springer-Verlag, pp. 81–83, 1991.
Descartes, B. "Division of a Square into Rectangles." *Eureka,* No. 34, 31–35, 1971.
Duijvestijn, A. J. W. *Electronic Computation of Squared Rectangles.* Thesis. Eindhoven, Netherlands: Technische Hogeschool, 1962.
Moron, Z. "O rozkładach prostokatów na kwadraty." *Przeglad matematyczno-fizyczny* **3**, 152–153, 1925.
Sloane, N. J. A. Sequences A002839/M1658 and A002882/M4614 in "An On-Line Version of the Encyclopedia of Integer Sequences." http://www.research.att.com/~njas/sequences/eisonline.html.
Stewart, I. "Squaring the Square." *Sci. Amer.* **277**, 94–96, July 1997.
Wells, D. *The Penguin Dictionary of Curious and Interesting Numbers.* Middlesex, England: Penguin Books, p. 73, 1986.

Perfect Ruler

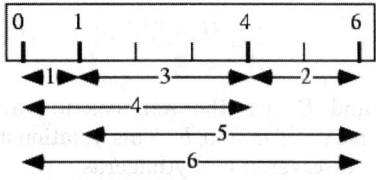

A type of RULER considered by Guy (1994) which has

k distinct marks spaced such that the distances between marks can be used to measure all the distances 1, 2, 3, 4, ... up to some maximum distance $n > k$. Such a ruler can be constructed from a PERFECT DIFFERENCE SET by subtracting one from each element. For example, the PERFECT DIFFERENCE SET $\{1, 2, 5, 7\}$ gives 0, 1, 4, 6, which can be used to measure $1-0 = 1$, $6-4 = 2$, $4-1 = 3$, $4-0 = 4$, $6-1 = 5$, $6-0 = 6$ (so we get 6 distances with only four marks).

See also GOLOMB RULER, PERFECT DIFFERENCE SET, RULER

References

Guy, R. K. "Modular Difference Sets and Error Correcting Codes." §C10 in *Unsolved Problems in Number Theory,* 2nd ed. New York: Springer-Verlag, pp. 118–121, 1994.

Perfect Set

A SET P is called perfect if $P = P'$, where P' is the DERIVED SET of P.

See also DERIVED SET, SET

Perfect Shuffle

Gale (1992) considered the following problem. Take an infinite deck of cards labeled 1, 2, 3, 4, 5, 6, ... At step n, pick up the top n cards and interlace them with the next n cards. This is called a perfect n-shuffle. For example, after step two, we have 3, 2, 4, 1, 5, 6, 7, ... For step there, pick up 3, 2, 4 and shuffle them in, giving 1, 3, 5, 2, 6, 4, 7, 8, 9, ... Iterate this process. It is conjectured that eventually every number appears on top of the deck.

The cards on top of deck at the nth step are 1, 2, 3, 1, 6, 5, 9, 1, 4, 2, 16, 10, 12, ... (Sloane's A035485). The step at which card n first appears on top the deck is given by 0, 1, 2, 8, 5, 4, 78, 37, ... (Sloane's A035490). The position of the first card after the nth shuffle is 1, 2, 4, 1, 2, 4, 8, 1, 2, 4, 8, 16, 7, 14, 28, ... (Sloane's A035492). The order in which new cards appear on top for the first time is 1, 2, 3, 6, 5, 9, 4, 16, 10, ... (Sloane's A035493). The order in which record new high cards appear on top for the first time is 1, 2, 3, 6, 9, 16, ... (Sloane's A035494).

See also KIMBERLING SHUFFLE, SHUFFLE

References

Gale, D. "Mathematical Entertainments: Careful Card-Shuffling and Cutting Can Create Chaos." *Math. Intell.* **14**, 54–56, 1992.
Gale, D. *Tracking the Automatic Ant and Other Mathematical Explorations, A Collection of Mathematical Entertainments Columns from The Mathematical Intelligencer.* New York: Springer-Verlag, 1998.
Sloane, N. J. A. Sequences A035485, A035490, A035492, A035493, and A035494 in "An On-Line Version of the Encyclopedia of Integer Sequences." http://www.research.att.com/~njas/sequences/eisonline.html.

Perfect Square

The term perfect square is used to refer to a SQUARE NUMBER, a PERFECT SQUARE DISSECTION, or a factorable quadratic polynomial OF THE FORM $a^2 \pm 2ab + b^2 = (a \pm b)^2$.

See also PERFECT SQUARE DISSECTION, QUADRATIC EQUATION, SQUARE NUMBER, SQUAREFREE

Perfect Square Dissection

A SQUARE which can be DISSECTED into a number of smaller SQUARES with no two equal is called a PERFECT SQUARE DISSECTION (or a SQUARED SQUARE). Square dissections in which the squares need not be different sizes are called MRS. PERKINS' QUILTS. If no subset of the SQUARES forms a RECTANGLE, then the perfect square is called "simple."

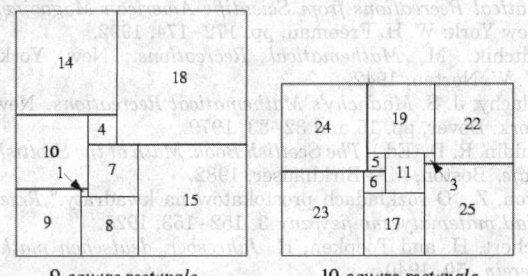

9-square rectangle   *10-square rectangle*

Moroz (1925) constructed a 33×32 PERFECT RECTANGLE composed of nine squares of different sizes (Descartes 1971), but Lusin claimed that perfect squares were impossible to construct. This assertion was proved erroneous when a 55-SQUARE perfect square was published by R. Sprague in 1939 (Wells 1991). Reichert and Toepkin (1940) proved that a RECTANGLE cannot be dissected into fewer than nine different SQUARES (Steinhaus 1983, p. 297).

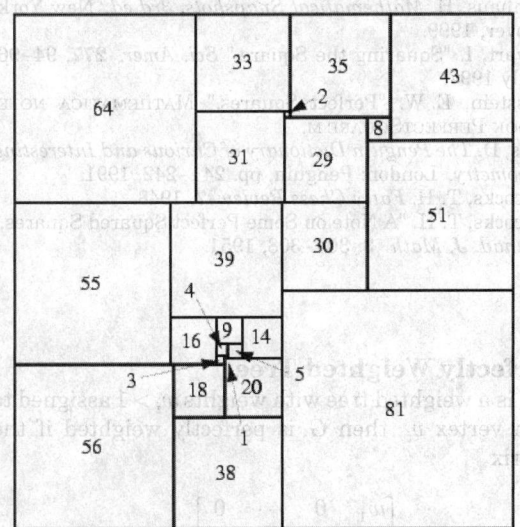

24-square perfect square

A 24-SQUARE perfect square was subsequently found

by Willcocks (Willcocks 1948, 1951; Steinhaus 1983, pp. 8–9).

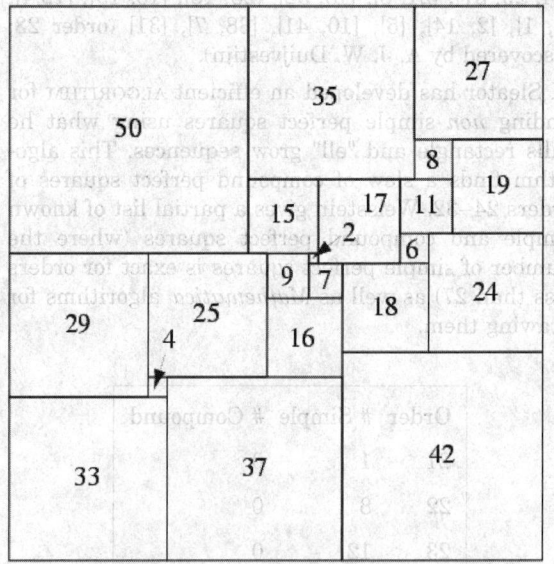

21-square perfect square

There is a unique simple perfect square of order 21 (the lowest possible order), discovered in 1978 by A. J. W. Duijvestijn (Bouwkamp and Duijvestijn 1992). It is composed of 21 squares with total side length 112, and is illustrated above. There is a simple notation (sometimes called Bouwkamp code) used to describe perfect squares. In this notation, brackets are used to group adjacent squares with flush tops, and then the groups are sequentially placed in the highest (and leftmost) possible slots. For example, the 21-square illustrated above is denoted [50, 35, 27], [8, 19], [15, 17, 11], [6, 24], [29, 25, 9, 2], [7, 18], [16], [42], [4, 37], [33].

A compound 26-perfect square having side length 608 was discovered in 1940 (Brooks *et al.* 1940; Kraitchik 1942, p. 198). Beiler (1966) illustrates a compound 28-square and a simple 38-square. Gardner (1961, pp. 203 and 206) illustrates compound 39- and 24-squares.

The number of simple perfect squares of order n for $n \geq 21$ are 1, 8, 12, 26, 160, 441, ... (Sloane's A006983). Duijvestijn's Table I gives a list of the 441 simple perfect squares of order 26, the smallest with side length 212 and the largest with side length 825. Skinner (1993) gives the smallest possible side length (and smallest order for each) as 110 (22), 112 (21), 120 (24), 139 (22), 140 (23), ... for simple perfect squared squares, and 175 (24), 235 (25), 288 (26), 324 (27), 325 (27), ... for compound perfect squared squares.

There are actually three simple perfect squares having side length 110. They are [60, 50], [23, 27], [24, 22, 14], [7, 16], [8, 6], [12, 15], [13], [2, 28], [26], [4, 21, 3], [18], [17] (order 22; discovered by A. J. W. Duijvestijn); [60, 50], [27, 23], [24, 22, 14],

[4, 19], [8, 6], [3, 12, 16], [9], [2, 28], [26], [21], [1, 18], [17] (order 22; discovered by T. H. Willcocks); and [44, 29, 37], [21, 8], [13, 32], [28, 16], [15, 19], [12,4], [3, 1], [2, 14], [5], [10, 41], [38, 7], [31] (order 23; discovered by A. J. W. Duijvestijn).

D. Sleator has developed an efficient ALGORITHM for finding *non*-simple perfect squares using what he calls rectangle and "ell" grow sequences. This algorithm finds a slew of compound perfect squares of orders 24–32. Weisstein gives a partial list of known simple and compound perfect squares (where the number of simple perfect squares is exact for orders less than 27) as well as *Mathematica* algorithms for drawing them.

Order	# Simple	# Compound
21	1	0
22	8	0
23	12	0
24	26	1
25	160	1
26	441	2
27	?	2
28	?	4
29	?	2
30	?	3
31	?	2
32	?	2
38	1	0
39	?	1
69	1	0

See also BLANCHE'S DISSECTION, CYLINDER DISSECTION, DISSECTION, EQUILATERAL TRIANGLE PACKING, FAULT-FREE RECTANGLE, KLEIN BOTTLE DISSECTION, MÖBIUS STRIP DISSECTION, MRS. PERKINS' QUILT, PERFECT RECTANGLE, PROJECTIVE PLANE DISSECTION, TORUS DISSECTION

References

Ball, W. W. R. and Coxeter, H. S. M. *Mathematical Recreations and Essays, 13th ed.* New York: Dover, pp. 115–116, 1987.

Beiler, A. H. *Recreations in the Theory of Numbers: The Queen of Mathematics Entertains.* New York: Dover, pp. 157–161, 1966.

Bouwkamp, C. J. and Duijvestijn, A. J. W. "Catalogue of Simple Perfect Squared Squares of Orders 21 Through 25." Eindhoven Univ. Technology, Dept. Math, Report 92-WSK-03, Nov. 1992.

Brooks, R. L.; Smith, C. A. B.; Stone, A. H.; and Tutte, W. T. "The Dissection of Rectangles into Squares." *Duke Math. J.* **7**, 312–340, 1940.

Croft, H. T.; Falconer, K. J.; and Guy, R. K. "Squaring the Square." §C2 in *Unsolved Problems in Geometry.* New York: Springer-Verlag, pp. 81–83, 1991.

Descartes, B. "Division of a Square into Rectangles." *Eureka,* No. 34, 31–35, 1971.

Duijvestijn, A. J. W. "A Simple Perfect Square of Lowest Order." *J. Combin. Th. Ser. B* **25**, 240–243, 1978.

Duijvestijn, A. J. W. "A Lowest Order Simple Perfect 2×1 Squared Rectangle." *J. Combin. Th. Ser. B* **26**, 372–374, 1979.

Duijvestijn, A. J. W. ftp://ftp.cs.utwente.nl/pub/doc/dvs/TableI.

Gardner, M. "Squaring the Square." Ch. 17 in *The Second Scientific American Book of Mathematical Puzzles & Diversions: A New Selection.* New York: Simon and Schuster, pp. 186–209, 1961.

Gardner, M. *Fractal Music, Hypercards, and More: Mathematical Recreations from Scientific American Magazine.* New York: W. H. Freeman, pp. 172–174, 1992.

Kraitchik, M. *Mathematical Recreations.* New York: W. W. Norton, 1942.

Madachy, J. S. *Madachy's Mathematical Recreations.* New York: Dover, pp. 15 and 32–33, 1979.

Mauldin, R. D. (Ed.). *The Scottish Book: Math at the Scottish Cafe.* Boston, MA: Birkhäuser, 1982.

Moron, Z. "O rozkładach prostokatów na kwadraty." *Przeglad matematyczno-fizyczny* **3**, 152–153, 1925.

Reichert, H. and Toepken, H. *Jahresber. deutschen math. Verein.* **50**, 1940.

Skinner, J. D. II. *Squared Squares: Who's Who & What's What.* Published by the author, 1993.

Sloane, N. J. A. Sequences A006983/M4482 in "An On-Line Version of the Encyclopedia of Integer Sequences." http://www.research.att.com/~njas/sequences/eisonline.html.

Sloane, N. J. A. and Plouffe, S. Figure M4482 in *The Encyclopedia of Integer Sequences.* San Diego: Academic Press, 1995.

Smith, C. A. B. and Tutte, W. T. "A Class of Self-Dual Maps." *Canad. J. Math.* **2**, 179–196, 1950.

Sprague, R. "Beispiel einer Zerlegung des Quadrats in lauter verschiedene Quadrate." *Math. Z.* **45**, 607–608, 1939.

Steinhaus, H. *Mathematical Snapshots, 3rd ed.* New York: Dover, 1999.

Stewart, I. "Squaring the Square." *Sci. Amer.* **277**, 94–96, July 1997.

Weisstein, E. W. "Perfect Squares." MATHEMATICA NOTEBOOK PERFECTSQUARE.M.

Wells, D. *The Penguin Dictionary of Curious and Interesting Geometry.* London: Penguin, pp. 241–242, 1991.

Willcocks, T. H. *Fairy Chess Review* **7**, 1948.

Willcocks, T. H. "A Note on Some Perfect Squared Squares." *Canad. J. Math.* **3**, 304–308, 1951.

Perfectly Weighted Tree

If G is a weighted tree with weights $w_i > 1$ assigned to each vertex v_i, then G is perfectly weighted if the matrix

$$M_G = \begin{bmatrix} w_1 & 0 & \cdots & 0 \\ 0 & w_2 & \cdots & 0 \\ \vdots & \ddots & \ddots & \vdots \\ 0 & 0 & \ddots & w_n \end{bmatrix} - \mathrm{adj}(G),$$

where akj(G) is the ADJACENCY MATRIX of G (Butske *et al.* 1999).

See also ADJACENCY MATRIX

References

Brenton, L. and Drucker, D. "Perfect Graphs and Complex Surface Singularities with Perfect Local Fundamental Group." *Tôhoku Math. J.* **41**, 507–525, 1989.

Butske, W.; Jaje, L. M.; and Mayernik, D. R. "The Equation $\Sigma_{p/N} \, 1/p + 1/N = 1$, Pseudoperfect Numbers, and Partially Weighted Graphs." *Math. Comput.* **69**, 407–420, 1999.

Perforation

The portion of a SURFACE left when an OPEN DISK is removed from it.

See also OPEN DISK

References

Francis, G. K. and Weeks, J. R. "Conway's ZIP Proof." *Amer. Math. Monthly* **106**, 393–399, 1999.

Periapsis

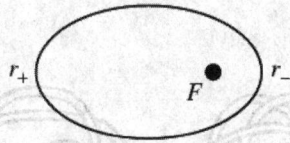

The smallest radial distance of an ELLIPSE as measured from a FOCUS. Taking $v = 0$ in the equation of an ELLIPSE

$$r = \frac{a(1 - e^2)}{1 + e \cos v}$$

gives the periapsis distance

$$r_- = a(1 - e).$$

Periapsis for an orbit around the Earth is called perigee, and periapsis for an orbit around the Sun is called perihelion.

See also APOAPSIS, ECCENTRICITY, ELLIPSE, FOCUS

Perigon

An ANGLE of 2π radians $= 360°$ corresponding to the CENTRAL ANGLE of an entire CIRCLE.

Perimeter

The ARC LENGTH along the boundary of a closed 2-D region. The perimeter of a CIRCLE is called the CIRCUMFERENCE.

See also CIRCUMFERENCE, CLUSTER PERIMETER, HONEYCOMB CONJECTURE, SEMIPERIMETER

Perimeter Polynomial

A sum over all CLUSTER PERIMETERS.

Period Doubling

A characteristic of some systems making a transition to CHAOS. Doubling is followed by quadrupling, etc. An example of a map displaying period doubling is the LOGISTIC MAP.

See also CHAOS, LOGISTIC MAP

Period Ratio

HALF-PERIOD RATIO

Period Three Theorem

Li and Yorke (1975) proved that *any* 1-D system which exhibits a regular CYCLE of period three will also display regular CYCLES of every other length as well as completely CHAOTIC CYCLES.

See also CHAOS, CYCLE (MAP)

References

Li, T. Y. and Yorke, J. A. "Period Three Implies Chaos." *Amer. Math. Monthly* **82**, 985–92, 1975.

Periodic Function

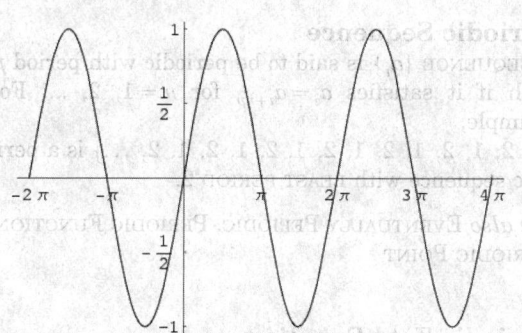

A FUNCTION $f(x)$ is said to be periodic with period p if

$$f(x) = f(x + np)$$

for $n = 1, 2, \ldots$. For example, the SINE function $\sin x$, illustrated above, is periodic with period 2π (as well as with period -2π, 4π, 6π, etc.).

The CONSTANT FUNCTION $f(x) = 0$ is periodic with any period R for all NONZERO REAL NUMBERS R, so there is no concept analogous to the LEAST PERIOD for constant functions.

See also ALMOST PERIODIC FUNCTION, DOUBLY PERIODIC FUNCTION, LEAST PERIOD, PERIODIC POINT, PERIODIC SEQUENCE

References

Knopp, K. "Periodic Functions." Ch. 3 in *Theory of Functions Parts I and II, Two Volumes Bound as One, Part II.* New York: Dover, pp. 58–92, 1996.

Morse, P. M. and Feshbach, H. *Methods of Theoretical Physics, Part I.* New York: McGraw-Hill, pp. 425–427, 1953.

Spanier, J. and Oldham, K. B. "Periodic Functions." Ch. 36 in *An Atlas of Functions*. Washington, DC: Hemisphere, pp. 343–349, 1987.

Periodic Matrix

A SQUARE MATRIX A such that the MATRIX POWER $A^{k+1} = A$ for k a positive integer is called a periodic matrix. If k is the least such integer, then the matrix is said to have period k. If $k = 1$, then $A^2 = A$ and A is called IDEMPOTENT.

See also MATRIX POWER

References

Ayres, F. Jr. *Theory and Problems of Matrices*. New York: Schaum, p. 11, 1962.

Periodic Point

A point x_0 is said to be a periodic point of a FUNCTION f of period n if $f^n(x_0) = x_0$, where $f_0(x) = x$ and $f^n(x)$ is defined recursively by $f^n(x) = f(f^{n-1}(x))$.

See also LEAST PERIOD, PERIODIC FUNCTION, PERIODIC SEQUENCE

Periodic Sequence

A SEQUENCE $\{a_i\}$ is said to be periodic with period p with if it satisfies $a_i = a_{i+np}$ for $n = 1$, 2, ... For example,
$\{1, 2, 1, 2, 1, 2, 1, 2, 1, 2, 1, 2, 1, 2, \ldots\}$ is a periodic sequence with LEAST PERIOD 2.

See also EVENTUALLY PERIODIC, PERIODIC FUNCTION, PERIODIC POINT

Periodic Zeta Function

$$F(x, s) = \sum_{m=1}^{\infty} \frac{e^{2\pi i m x}}{m^s}$$
$$= \psi_s\left(e^{2\pi i x}\right),$$

where $\psi_s(x)$ is the POLYGAMMA FUNCTION.

See also POLYGAMMA FUNCTION, RIEMANN ZETA FUNCTION, ZETA FUNCTION

References

Apostol, T. M. *Modular Functions and Dirichlet Series in Number Theory, 2nd ed.* New York: Springer-Verlag, p. 55, 1997.

Periodogram

A graphical plot with ABSCISSA given by the number p of consecutive numbers constituting a single period and ORDINATE given by the correlation ratio η. The equation of the periodogram is

$$\eta^2 = \frac{\frac{a^2}{2m^2} \sin^2\left(\frac{m\pi p}{T}\right) + \frac{\sigma_b^2}{m}}{\frac{1}{2} a^2 + \sigma_b^2},$$

where each of the terms of the sequence u_x consists of a simple periodic part of period T, together with a part which does not involve this periodicity b_x, so

$$u_x = a \sin\left(\frac{2\pi x}{T}\right) + b_x,$$

σ_b is the standard deviation of the bs, σ is the standard deviation of the us, and m is the number of periods covered by the observations.

See also TIME SERIES ANALYSIS

References

Schuster. *Terrestrial Magnetism* **3**, 24, 1898.
Whittaker, E. T. and Robinson, G. "The Periodogram in the Neighbourhood of a True Period" and "An Example of Periodogram Analysis." §174–175 in *The Calculus of Observations: A Treatise on Numerical Mathematics, 4th ed.* New York: Dover, pp. 346–362, 1967.

Perko Pair

The KNOTS 10-161 and 10-162 illustrated above. For many years, they were listed as separate knots in Little (1885) and all similar tables, including the pictorial enumeration of Rolfsen (1976, Appendix C). They were identified as identical by Perko (1974), who found that they are related to one another by the so-called PERKO MOVE (Perko 1974, Hoste *et al.* 1998). Although these knots are equivalent, their diagrams have different WRITHES (Hoste *et al.* 1998).

See also PERKO MOVE

References

Hoste, J.; Thistlethwaite, M.; and Weeks, J. "The First 1,701,936 Knots." *Math. Intell.* **20**, 33–48, Fall 1998.
Little, C. N. "On Knots, with a Census of Order Ten." *Trans. Connecticut Acad. Sci.* **18**, 374–378, 1885.
Perko, K. A. Jr. "On the Classification of Knots." *Proc. Amer. Math. Soc.* **45**, 262–266, 1974.
Rolfsen, D. "Table of Knots and Links." Appendix C in *Knots and Links*. Wilmington, DE: Publish or Perish Press, pp. 280–287, 1976.

Permanence of Algebraic Form

All ELEMENTARY FUNCTIONS can be extended to the COMPLEX PLANE. Such definitions agree with the real definitions on the x-AXIS and constitute an ANALYTIC CONTINUATION.

See also ANALYTIC CONTINUATION, ELEMENTARY FUNCTION, PERMANENCE OF MATHEMATICAL RELATIONS PRINCIPLE

References

Arfken, G. *Mathematical Methods for Physicists, 3rd ed.* Orlando, FL: Academic Press, p. 380, 1985.

Permanence of Mathematical Relations Principle

CONTINUITY PRINCIPLE

Permanent

An analog of a DETERMINANT where all the signs in the expansion by MINORS are taken as POSITIVE. The permanent of a MATRIX $\mathbf{A}$ is the coefficient of $x_1 \ldots x_n$ in

$$\prod_{i=1}^{n} (a_{i1}x_1 + a_{i2}x_2 + \ldots + a_{in}x_n)$$

(Vardi 1991). Another equation is the RYSER FORMULA

$$\operatorname{perm}(a_{ij}) = (-1)^n \sum_{a \subseteq \{1, \ldots, n\}} (-1)^{|s|} \prod_{i=1}^{n} \sum_{j \in s} a_{ij},$$

where the SUM is over all SUBSETS of $\{1, \ldots, n\}$, and $|s|$ is the number of elements in s (Vardi 1991). Muir (1960, p. 19) uses the notation $|_+|_+$ to denote a permanent. The permanent can be implemented in *Mathematica* as

```
Permanent[m_List]  :=  With[{v = Array[x,
Length[m]]},
    Coefficient[Times @@ (m.v), Times @@ v] ]
```

The computation of permanents has been studied fairly extensively in algebraic complexity theory. The complexity of the best-known algorithms grows as the exponent of the matrix size (Knuth 1998, p. 499), which would appear to be very surprising, given the permanent's similarity to the tractable DETERMINANT.

If $\mathbf{M}$ is a UNITARY MATRIX, then

$$|\operatorname{perm}(\mathbf{M})| \leq 1$$

(Minc 1978, p. 25; Vardi 1991). The maximum permanent for an $n \times n$ BINARY MATRIX is $n!$, corresponding to all elements 1.

See also DETERMINANT, FROBENIUS-KÖNIG THEOREM, IMMANANT, RYSER FORMULA, SCHUR MATRIX

References

Borovskikh, Y. V. and Korolyuk, V. S. *Random Permanents.* Philadelphia, PA: Coronet Books, 1994.

Comtet, L. "Permanents." §4.9 in *Advanced Combinatorics: The Art of Finite and Infinite Expansions, rev. enl. ed.* Dordrecht, Netherlands: Reidel, pp. 197–198, 1974.

Knuth, D. E. *The Art of Computer Programming, Vol. 1: Fundamental Algorithms, 3rd ed.* Reading, MA: Addison-Wesley, p. 51, 1997.

Knuth, D. E. *The Art of Computer Programming, Vol. 2: Seminumerical Algorithms, 3rd ed.* Reading, MA: Addison-Wesley, pp. 499 and 515–516, 1998.

Minc, H. *Permanents.* Reading, MA: Addison-Wesley, 1978.

Muir, T. §27 in *A Treatise on the Theory of Determinants.* New York: Dover, p. 19 1960.

Valiant, L. G. *Theoret. Comp. Sci.* **8**, 189–201, 1979.

Vardi, I. "Permanents." §6.1 in *Computational Recreations in Mathematica.* Reading, MA: Addison-Wesley, pp. 108 and 110–112, 1991.

Permil

The use of permil (a.k.a. parts per thousand) is a way of expressing RATIOS in terms of whole numbers. Given a RATIO or FRACTION, it is converted to a permil-age by multiplying by 1000 and appending a "mil sign" %0. For example, if an investment grows from a number $P = 13.00$ to a number $A = 22.50$, then A is $22.50/13.00 = 1.7308$ times as much as P, or 1730.8%0.

See also PERCENT

References

Bringhurst, R. *The Elements of Typographic Style, 2nd ed.* Point Roberts, WA: Hartley and Marks, p. 283, 1997.

Permutation

The rearrangement of elements in an ordered list S into a ONE-TO-ONE correspondence with S itself, also called an "arrangement number" or "order." The number of permutations on a set of n elements is given by $n!$ (n FACTORIAL; Uspensky 1937, p. 18). For example, there are $2! = 2 \cdot 1 = 2$ permutations of $\{1, 2\}$, namely $\{1, 2\}$ and $\{2, 1\}$, and $3! = 3 \cdot 2 \cdot 1 = 6$ permutations of $\{1, 2, 3\}$, namely $\{1, 2, 3\}$, $\{1, 3, 2\}$, $\{2, 1, 3\}$, $\{2, 3, 1\}$, $\{3, 1, 2\}$, and $\{3, 2, 1\}$. The permutations of a list can be found in *Mathematica* using the command `Permutations[list]`. A list of length n can be tested to see if it is a permutation of 1, ..., n with the command `PermutationQ[list]` in the *Mathematica* add-on package `DiscreteMath`Combinatorica`` (which can be loaded with the command `<<DiscreteMath`).

Sedgewick (1977) summarized a number of algorithms for generating permutations, and identifies the minimum change permutation algorithm of Heap (1963) to be generally the fastest (Skiena 1990, p. 10). Another method of enumerating permutations was given by Johnson (1963; Séroul 2000, pp. 213–218).

The number of ways of obtaining an *ordered* subset of k elements from a set of n elements is given by

$$_nP_k \equiv \frac{n!}{(n-k)!} \qquad (1)$$

(Uspensky 1937, p. 18). For example, there are $4!/2! = 12$ 2-subsets of $\{1, 2, 3, 4\}$, namely $\{1, 2\}$, $\{1, 3\}$, $\{1, 4\}$, $\{2, 1\}$, $\{2, 3\}$, $\{2, 4\}$, $\{3, 1\}$, $\{3, 2\}$, $\{3, 4\}$, $\{4, 1\}$, $\{4, 2\}$, and $\{4, 3\}$. The *unordered* subsets containing k elements are known as the K-SUBSETS of a given set.

A representation of a permutation as a product of CYCLES is unique (up to the ordering of the cycles). An example of a cyclic decomposition is $(\{1, 3, 4\}, \{2\})$, corresponding to the permutations $(1 \rightarrow 3, 3 \rightarrow 4, 4 \rightarrow 1)$ and $(2 \rightarrow 2)$, which combine to give $\{4, 2, 1, 3\}$. Muir (1960, p. 8) uses the notation (1237)(4568) to denote the ordered permutation (12345678), and (1237)(4568) to denote (12374568).

Any permutation is also a product of TRANSPOSITIONS. Permutations are commonly denoted in LEXICO-GRAPHIC or TRANSPOSITION ORDER. There is a correspondence between a PERMUTATION and a pair of YOUNG TABLEAUX known as the SCHENSTED CORRESPONDENCE.

The number of wrong permutations of n objects is $[n!/e]$ where $[x]$ is the NINT function. A permutation of n ordered objects in which no object is in its natural place is called a DERANGEMENT (or sometimes, a COMPLETE PERMUTATION) and the number of such permutations is given by the SUBFACTORIAL $!n$.

Using

$$(x+y)^n = \sum_{r=0}^{n} \binom{n}{r} x^{n-r} y^r \qquad (2)$$

with $x = y = 1$ gives

$$2^n = \sum_{r=0}^{n} \binom{n}{r}, \qquad (3)$$

so the number of ways of choosing $0, 1, \ldots,$ or n at a time is 2^n.

The set of all permutations of a set of elements $1, \ldots, n$ can be obtained using the following recursive procedure

```
1   2
 \ /
 2   1
```
(4)

```
  1     2   3
       /
  1   3     2
     /
  3   1     2
  |
  3   2     1
       \
  2   3     1
       \ /
  2     1   3
```
(5)

Let the set of INTEGERS $1, 2, \ldots, n$ be permuted and the resulting sequence be divided into increasing RUNS. As n approaches INFINITY, the average length of the nth RUN is denoted L_n. The first few values are

$$L_1 = e - 1 = 1.71828818\ldots \qquad (6)$$

$$L_2 = e^2 - 2e = 1.9524\ldots \qquad (7)$$

$$L_3 = e^3 - 3e^2 + \tfrac{3}{2}e = 1.9957\ldots, \qquad (8)$$

where E is the base of the NATURAL LOGARITHM (Knuth 1973, Le Lionnais 1983).

See also ALTERNATING PERMUTATION, BINOMIAL COEFFICIENT, CIRCULAR PERMUTATION, COMBINATION, COMPLETE PERMUTATION, CYCLE (PERMUTATION), DERANGEMENT, DISCORDANT PERMUTATION, EULERIAN NUMBER, K-SUBSET, LINEAR EXTENSION, PERMUTATION INVERSION, PERMUTATION MATRIX, PERMUTATION PATTERN, PERMUTATION SYMBOL, RANDOM PERMUTATION, SUBFACTORIAL, TRANSPOSITION

References

Bogomolny, A. "Graphs." http://www.cut-the-knot.com/do_you_know/permutation.html.

Conway, J. H. and Guy, R. K. "Arrangement Numbers." In *The Book of Numbers.* New York: Springer-Verlag, p. 66, 1996.

Dickau, R. M. "Permutation Diagrams." http://forum.swarthmore.edu/advanced/robertd/permutations.html.

Heap, B. R. "Permutations by Interchanges." *Computer J.* **6**, 293–294, 1963.

Johnson, S. M. "Generation of Permutations by Adjacent Transpositions." *Math. Comput.* **17**, 282–285, 1963.

Knuth, D. E. *The Art of Computer Programming, Vol. 1: Fundamental Algorithms, 3rd ed.* Reading, MA: Addison-Wesley, 1998.

Kraitchik, M. "The Linear Permutations of n Different Things." §10.1 in *Mathematical Recreations.* New York: W. W. Norton, pp. 239–240, 1942.

Le Lionnais, F. *Les nombres remarquables.* Paris: Hermann, pp. 41–42, 1983.

Muir, T. *A Treatise on the Theory of Determinants.* New York: Dover, 1960.

Ruskey, F. "Information on Permutations." http://www.theory.csc.uvic.ca/~cos/inf/perm/PermInfo.html.

Sedgewick, R. "Permutation Generation Methods." *Comput. Surveys* **9**, 137–164, 1977.

Séroul, R. "Permutations: Johnson's [sic] Algorithm." §8.15 in *Programming for Mathematicians.* Berlin: Springer-Verlag, pp. 213–218, 2000.

Skiena, S. "Permutations." §1.1 in *Implementing Discrete Mathematics: Combinatorics and Graph Theory with*

Mathematica. Reading, MA: Addison-Wesley, pp. 3–16, 1990.

Sloane, N. J. A. Sequences A000142/M1675 in "An On-Line Version of the Encyclopedia of Integer Sequences." http://www.research.att.com/~njas/sequences/eisonline.html.

Trotter, H. F. "Perm (Algorithm 115)." *Comm. ACM* **5**, 434–435, 1962.

Uspensky, J. V. *Introduction to Mathematical Probability.* New York: McGraw-Hill, p. 18, 1937.

Permutation Ascent

An ascent is a pair of adjacent positions in a PERMUTATION which are out of order. k ascents imply $k + 1$ PERMUTATION RUNS (Skiena 1990, p. 31).

See also PERMUTATION, PERMUTATION RUN

References

Graham, R. L.; Knuth, D. E.; and Patashnik, O. *Concrete Mathematics: A Foundation for Computer Science, 2nd ed.* Reading, MA: Addison-Wesley, 1994.

Knuth, D. E. *The Art of Computer Programming, Vol. 3: Sorting and Searching, 2nd ed.* Reading, MA: Addison-Wesley, 1998.

Mannila, H. "Measures of Presortedness and Optimal Sorting Algorithms." *IEE Trans. Comput.* **34**, 318–325, 1985.

Skiena, S. "Runs and Eulerian Numbers." §1.3.4 in *Implementing Discrete Mathematics: Combinatorics and Graph Theory with Mathematica.* Reading, MA: Addison-Wesley, pp. 30–31, 1990.

Permutation Graph

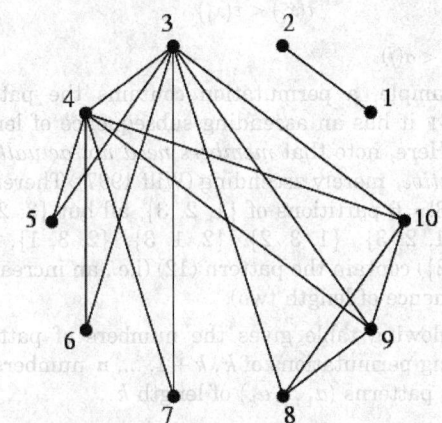

For a PERMUTATION α in the SYMMETRIC GROUP S_p, the α-permutation graph of a LABELED GRAPH G is the GRAPH UNION of two disjoint copies of G (say, G_1 and G_2), together with the lines joining point v_i of G_i with $v_{\alpha(i)}$ of G_2 (Harary 1994, p. 175). Skiena (1990, p. 28) defined a permutation graph G_p as a GRAPH whose edges $\{v_i, v_j\}$ correspond exactly to (i, j) being a PERMUTATION INVERSION is some PERMUTATION p, i.e., $i < j$ but j occurs before i in p.

The above graph corresponds to the permutation $\{2, 1, 5, 6, 7, 10, 9, 4, 8, 3\}$, which has PERMUTATION INVERSION $\{2, 1, 10, 8, 3, 4, 5, 9, 7, 6\}$.

See also PERMUTATION, PERMUTATION INVERSION

References

Atallah, M. J.; Manacher, G. K.; and Urrutia, J. "Finding a Minimum Independent Dominating Set in a Permutation Graph." *Discr. Appl. Math.* **21**, 177–183, 1988.

Brandstadt, A. and Kratsch, D. "On Domination Problems for Permutation and Other Graphs." *Theoret. Comput. Sci.* **54**, 181–198, 1987.

Harary, F. *Graph Theory.* Reading, MA: Addison-Wesley, 1994.

Skiena, S. *Implementing Discrete Mathematics: Combinatorics and Graph Theory with Mathematica.* Reading, MA: Addison-Wesley, 1990.

Permutation Group

A FINITE GROUP of order $n!$ consisting of substitutions of n elements for each other. For instance, the 24 PERMUTATIONS on four elements form a permutation group, and one the operations in this group is the permutation $\{4, 2, 1, 3\}$, which rearranges the elements $\{A, B, C, D\}$ in the order $\{D, B, A, C\}$. A permutation group of two elements is called a TRANSPOSITION.

Every SUBSTITUTION GROUP with > 2 elements can be written as a product of transpositions. For example,

$$(abc) = (ab)(ac)$$

$$(abcde) = (ab)(ac)(ad)(ae).$$

CONJUGACY CLASSES of elements which are interchanged are called CYCLES (in the above example, the CYCLES are $\{\{1, 3, 4\}, \{2\}\}$).

Two PERMUTATIONS form a group only if one is the identity element and the other is an INVOLUTION, i.e., a PERMUTATION which is its own inverse (Skiena 1990, p. 20).

See also CAYLEY'S GROUP THEOREM, CYCLE (PERMUTATION), GROUP, INVOLUTION (PERMUTATION), NETTO'S CONJECTURE, PERMUTATION, SUBSTITUTION GROUP, TRANSPOSITION

References

Cameron, P. *Permutation Groups.* New York: Cambridge University Press, 1999.

Furst, M.; Hopcroft, J.; and Luks, E. "Polynomial Time Algorithms for Permutation Groups." In *Proc. Symp. Foundations Computer Sci.* IEEE, pp. 36–41, 1980.

Roberts, F. S. *Applied Combinatorics.* Englewood Cliffs, NJ: Prentice-Hall, 1984.

Skiena, S. "Permutation Groups." §1.2 in *Implementing Discrete Mathematics: Combinatorics and Graph Theory with Mathematica.* Reading, MA: Addison-Wesley, pp. 17–26, 1990.

Wielandt, H. *Finite Permutation Groups.* New York: Academic Press, 1964.

Permutation Index

The index of a PERMUTATION p is defined as the sum of all subscripts j such that $p_j > p_{j+1}$, for $1 \le j \le n$. MacMahon (1960) proved that the number of permutations of size n having index k is the same as the number having exactly k inversions (Skiena 1990,

p. 29). The permutation index can be computed as Index[*p*] in the *Mathematica* add-on package DiscreteMath`Combinatorica` (which can be loaded with the command < <DiscreteMath`).

See also PERMUTATION

References

Knuth, D. E. *The Art of Computer Programming, Vol. 3: Sorting and Searching, 2nd ed.* Reading, MA: Addison-Wesley, 1998.

MacMahon, P. A. *Combinatory Analysis, 2 vols.* New York: Chelsea, 1960.

Skiena, S. *Implementing Discrete Mathematics: Combinatorics and Graph Theory with Mathematica.* Reading, MA: Addison-Wesley, 1990.

Permutation Inversion

A pair of elements (p_i, p_j) is called an inversion in a permutation p if $i > j$ and $p_i < p_j$. For example, in the permutation $a_6 a_5 a_7 a_3 a_8$ contains the four inversions $a_7 a_3$, $a_5 a_3$, $a_6 a_3$, and $a_6 a_5$. Inversions are pairs which are out of order, and are important in sorting algorithms (Skiena 1990, p. 27).

The total number of inversions can be obtained by summing the elements of the INVERSION VECTOR, and is implemented as Inversions[*p*] in the *Mathematica* add-on package DiscreteMath`Combinatorica` (which can be loaded with the command < <DiscreteMath`). The number of inversions in any PERMUTATION is the same as the number of interchanges of consecutive elements *necessary* to arrange them in their natural order (Muir 1960, p. 1). The value $(-1)^{i(p)}$ can be found in *Mathematica* using Signature[*p*].

The number of inversions in a PERMUTATION is equal to that of its inverse permutation (Skiena 1990, p. 29; Knuth 1998). If, from any permutation, another is formed by interchanging two elements, then the difference between the number of inversions in the two is always an ODD NUMBER.

See also INVERSE PERMUTATION, INVERSION VECTOR, PERMUTATION, PERMUTATION SYMBOL

References

Knuth, D. E. *The Art of Computer Programming, Vol. 3: Sorting and Searching, 2nd ed.* Reading, MA: Addison-Wesley, 1998.

Mannila, H. "Measures of Presortedness and Optimal Sorting Algorithms." *IEEE Trans. Comput.* **34**, 318–325, 1985.

Muir, T. *A Treatise on the Theory of Determinants.* New York: Dover, 1960.

Skiena, S. "Encroaching Lists as a Measure of Presortedness." *BIT* **28**, 775–784, 1988.

Skiena, S. "Inversions and Inversion Vectors." §1.3 in *Implementing Discrete Mathematics: Combinatorics and Graph Theory with Mathematica.* Reading, MA: Addison-Wesley, pp. 27–31, 1990.

Permutation Matrix

A MATRIX p_{ij} obtained by permuting the ith and jth rows of the IDENTITY MATRIX with $i < j$. Every row and column therefore contain precisely a single 1, and every permutation corresponds to a unique permutation matrix. A permutation matrix is nonsingular, so the DETERMINANT is always NONZERO.

In addition, a permutation matrix satisfies

$$p_{ij}^2 = I,$$

where I is the IDENTITY MATRIX. Applying to another MATRIX, $p_{ij}A$ gives A with the ith and jth rows interchanged, and Ap_{ij} gives A with the ith and jth columns interchanged.

Interpreting the 1s in an $n \times n$ permutation matrix as ROOKS gives an allowable configuration of nonattacking ROOKS on an $n \times n$ CHESSBOARD.

See also ALTERNATING SIGN MATRIX, ELEMENTARY MATRIX, IDENTITY, PERMUTATION, ROOK NUMBER

Permutation Pattern

Let $F(n, \sigma)$ denote the number of permutations on the SYMMETRIC GROUP S_n which avoid $\sigma \in S_n$ as a sub-pattern, where "τ contains σ as a subpattern" is interpreted to mean that there exist $1 \le x_1 \le x_2 \le \ldots \le x_k \le n$ such that for $1 \le i, j \le k$,

$$\tau(x_i) < \tau(x_j) \qquad (1)$$

IFF $\sigma(i) < \sigma(j)$.

For example, a permutation contains the pattern (123) IFF it has an ascending subsequence of length three. Here, note that *members need not actually be consecutive,* merely ascending (Wilf 1997). Therefore, of the $3! = 6$ partitions of $\{1, 2, 3\}$, all but $\{3, 2, 1\}$ (i.e., $\{1, 2, 3\}$, $\{1, 3, 2\}$, $\{2, 1, 3\}$, $\{2, 3, 1\}$, and $\{3, 1, 2\}$) contain the pattern (12) (i.e., an increasing subsequence of length two).

The following table gives the numbers of pattern-matching permutations of k, $k + 1$, ..., n numbers for various patterns $(a_1 \ldots a_k)$ of length k.

pattern	Sloane	number of pattern-matching permutations
1	A000142	1, 2, 6, 24, 120, 720, 5040, ...
12	A033312	1, 5, 23, 119, 719, 5039, 40319, ...
α_3	A056986	1, 10, 78, 588, 4611, 38890, ...
1234	A000000	1, 17, 207, ...
1342	A000000	1, 17, 208, ...

The following table gives the numbers of pattern-avoiding permutations of $\{1, \ldots, n\}$ for various sets of patterns.

Wilf class	Sloane	number of pattern-avoiding permutations
α_3	A000108	1, 2, 5, 14, 42, 132, ...
123, 132, 213	A000027	1, 2, 3, 4, 5, 6, 7, 8, 9, 10, ...
132, 231, 321	A000027	1, 2, 3, 4, 5, 6, 7, 8, 9, 10, ...
123, 132, 3214	A000073	1, 2, 4, 7, 13, 24, 44, 81, 149, ...
123, 132, 3241	A000071	1, 2, 7, 12, 20, 33, 54, 88, 143, ...
123, 132, 3412	A000124	1, 2, 4, 7, 11, 16, 22, 29, 37, 46, ...
123, 231, $\alpha_4^{(1)}$	A004275	1, 2, 4, 6, 8, 10, 12, 14, 16, 18, ...
123, 231, $\alpha_4^{(2)}$	A000124	1, 2, 4, 7, 11, 16, 22, 29, 37, 46, ...
123, 231, 4321		1, 2, 4, 6, 3, 1, 0, ...
132, 213, 1234	A000073	1, 2, 4, 7, 13, 24, 44, 81, 149, ...
213, 231, $\alpha_4^{(3)}$	A000124	1, 2, 4, 7, 11, 16, 22, 29, 37, 46, ...

Abbreviations used in the above table are summarized below.

abbreviation	patterns in class
α_3	123, 132, 213, 232, 312, 321
$\alpha_4^{(1)}$	1432, 2143, 3214, 4132, 4213, 4312
$\alpha_4^{(2)}$	1234, 1243, 1324, 1342, 1423, 2134, 2314, 2341, 2413, 2431, 3124, 3142, 3241, 3412, 3421, 4123, 4231
$\alpha_4^{(3)}$	1234, 1243, 1423, 1432

See also CONTAINED PATTERN, ORDER ISOMORPHIC, PERMUTATION, PERMUTATION PATTERN, STANLEY-WILF CONJECTURE, WILF CLASS, WILF EQUIVALENT

References

Arratia, R. "On the Stanley-Wilf Conjecture for the Number of Permutations Avoiding a Given Patter." *Electronic J. Combinatorics* **6**, No. 1, N1, 1–4, 1999. http://www.combinatorics.org/Volume_6/v6i1toc.html.

Billey, S.; Jockusch, W.; and Stanley, R. P. "Some Combinatorial Properties of Schubert Polynomials." *J. Alg. Combin.* **2**, 345–374, 1993.

Guibert, O. "Permutations sans sous séquence interdite." *Mémoire de Diplôme d'Etudes Approfondies de L'Université Bordeaux I.* 1992.

Mansour, T. Permutations Avoiding a Pattern from S_k and at Least Two Patterns from S_3. 31 Jul 2000. http://xxx.lanl.gov/abs/math.CO/0007194/.

Simon, R. and Schmidt, F. W. "Restricted Permutations." *Europ. J. Combin.* **6**, 383–406, 1985.

Sloane, N. J. A. Sequences A000027/M0472, A000071/M1056, A000073/M1074, A000108/M1459, A000124/M1041, A000142/M1675, A004275, A033312, and A056986 in "An On-Line Version of the Encyclopedia of Integer Sequences." http://www.research.att.com/~njas/sequences/eisonline.html.

Stankova, Z. E. "Forbidden Subsequences." *Disc. Math.* **132**, 291–316, 1994.

West, J. "Generating Trees and Forbidden Subsequences." *Disc. Math.* **157**, 363–372, 1996.

Wilf, H. "On Crossing Numbers, and Some Unsolved Problems." In *Combinatorics, Geometry, and Probability: A Tribute to Paul Erdos. Papers from the Conference in Honor of Erdos' 80th Birthday Held at Trinity College, Cambridge, March 1993* (Ed. B. Bollobás and A. Thomason). Cambridge, England: Cambridge University Press, pp. 557–562, 1997.

Permutation Pseudotensor

PERMUTATION TENSOR

Permutation Run

A set of ascending sequences in a PERMUTATION is called a run. A sorted permutation consists of a single run, whereas a reverse permutation consists of n runs, each of length 1. Runs are closely related to PERMUTATION ASCENTS, with n runs implying $n - 1$ ascents (Skiena 1990, p. 31). The number of runs in a permutation can be computed using Runs[p] in the *Mathematica* add-on package DiscreteMath`Combinatorica` (which can be loaded with the command < <DiscreteMath`). The number of permutations of length n with exactly k runs is given by the EULERIAN NUMBER $\left\langle {n \atop k} \right\rangle$.

Surprisingly, the expected length of the first run is shorter than the expected length of the second run (Gassner 1967; Skiena 1990, p. 30; Knuth 1998).

See also EULERIAN NUMBER, PERMUTATION, PERMUTATION ASCENT, RUN

References

Gassner, B. J. "Sorting by Replacement Selection." *Comm. ACM* **10**, 89–93, 1967.

Graham, R. L.; Knuth, D. E.; and Patashnik, O. *Concrete Mathematics: A Foundation for Computer Science, 2nd ed.* Reading, MA: Addison-Wesley, 1994.

Knuth, D. E. *The Art of Computer Programming, Vol. 3: Sorting and Searching, 2nd ed.* Reading, MA: Addison-Wesley, 1998.

Mannila, H. "Measures of Presortedness and Optimal Sorting Algorithms." *IEE Trans. Comput.* **34**, 318–325, 1985.

Skiena, S. "Runs and Eulerian Numbers." §1.3.4 in *Implementing Discrete Mathematics: Combinatorics and Graph Theory with Mathematica.* Reading, MA: Addison-Wesley, pp. 30–31, 1990.

Permutation Symbol

A three-index object sometimes called the Levi-Civita symbol or signature, and defined by

$$\epsilon_{ijk} = \begin{cases} 0 & \text{for } i=j,\ j=k,\ \text{or } k=i \\ +1 & \text{for } (i,\ j,\ k) \in \{(1,\ 2,\ 3),\ (2,\ 3,\ 1),\ (3,\ 1,\ 2)\} \\ -1 & \text{for } (i,\ j,\ k) \in \{(1,\ 3,\ 2),\ (3,\ 2,\ 1),\ (2,\ 1,\ 3)\}. \end{cases}$$

$$(1)$$

The permutation symbol is implemented in *Mathematica* as Signature[*list*]. The permutation symbol satisfies

$$\delta_{ij}\epsilon_{ijk} = 0 \qquad (2)$$

$$\epsilon_{ipq}\epsilon_{jpq} = 2\delta_{ij} \qquad (3)$$

$$\epsilon_{ijk}\epsilon_{ijk} = 6 \qquad (4)$$

$$\epsilon_{ijk}\epsilon_{pqk} = \delta_{ip}\delta_{jq} - \delta_{iq}\delta_{jp}, \qquad (5)$$

where δ_{ij} is the KRONECKER DELTA. The symbol can be defined as the SCALAR TRIPLE PRODUCT of unit vectors in a right-handed coordinate system,

$$\epsilon_{ijk} \equiv \hat{\mathbf{x}}_i \cdot (\hat{\mathbf{x}}_j \times \hat{\mathbf{x}}_k). \qquad (6)$$

The symbol can also be interpreted as a TENSOR, in which case it is called the PERMUTATION TENSOR.

The symbol can be generalized to an arbitrary number of elements, in which case the permutation symbol is $(-1)^{i(p)}$, where $i(p)$ is the number of transpositions of pairs of elements (i.e., PERMUTATION INVERSIONS) that must be composed to build up the permutation p (Skiena 1990). This type of symbol arises in computation of determinants of $n \times n$ matrices. The number of permutations on n symbols having signature -1 is $n!/2$, which is also the number of permutations having signature $+1$.

See also CYCLE (PERMUTATION), PERMUTATION, PERMUTATION INVERSION, PERMUTATION TENSOR, TRANSPOSITION

References

Arfken, G. *Mathematical Methods for Physicists, 3rd ed.* Orlando, FL: Academic Press, pp. 132–133, 1985.

Jeffreys, H. and Jeffreys, B. S. *Methods of Mathematical Physics, 3rd ed.* Cambridge, England: Cambridge University Press, pp. 69–74, 1988.

Skiena, S. "Signature." §1.2.5 in *Implementing Discrete Mathematics: Combinatorics and Graph Theory with*

Mathematica. Reading, MA: Addison-Wesley, pp. 24–25, 1990.

Permutation Tensor

A PSEUDOTENSOR which is ANTISYMMETRIC under the interchange of any two slots. Recalling the definition of the PERMUTATION SYMBOL in terms of a SCALAR TRIPLE PRODUCT of the Cartesian unit vectors,

$$\epsilon_{ijk} \equiv \hat{\mathbf{x}}_i \cdot (\hat{\mathbf{x}}_j \times \hat{\mathbf{x}}_k) = [\hat{\mathbf{x}}_i,\ \hat{\mathbf{x}}_j,\ \hat{\mathbf{x}}_k], \qquad (1)$$

the pseudotensor is a generalization to an arbitrary BASIS defined by

$$\epsilon_{\alpha\beta\cdots\mu} = \sqrt{|g|}[\alpha,\ \beta,\ \ldots,\ \mu] \qquad (2)$$

$$\epsilon^{\alpha\beta\cdots\mu} = \frac{[\alpha,\ \beta,\ \ldots,\ \mu]}{\sqrt{|g|}}, \qquad (3)$$

where

$$[\alpha,\ \beta,\ \ldots,\ \mu]$$
$$= \begin{cases} 1 & \text{the arguments are an even permutation} \\ -1 & \text{the arguments are an odd permutation} \\ 0 & \text{two or more arguments are equal,} \end{cases}$$

$$(4)$$

and $g \equiv \det(g_{\alpha\beta})$, where $g_{\alpha\beta}$ is the METRIC TENSOR. $\epsilon(\mathbf{x}_1,\ \ldots,\ \mathbf{x}_n)$ is NONZERO IFF the VECTORS are LINEARLY INDEPENDENT.

See also KRONECKER DELTA, PERMUTATION SYMBOL, SCALAR TRIPLE PRODUCT

Permutation Tests

See also BOOTSTRAP METHODS, JACKKNIFE, HYPOTHESIS TESTING, RESAMPLING STATISTICS

References

Good, P. I. *Permutation Tests: A Practical Guide to Resampling Methods for Testing Hypotheses, 2nd ed.* New York: Springer-Verlag, 2000.

Perpendicular

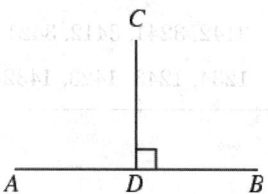

Two lines, vectors, planes, etc., are said to be perpendicular if they meet at a RIGHT ANGLE. In $\mathbb{R}^n$, two VECTORS **A** and **B** are PERPENDICULAR if their DOT

PRODUCT

$$\mathbf{A} \cdot \mathbf{B} = 0.$$

In $\mathbb{R}^2$, a LINE with SLOPE $m_2 = -1/m_1$ is PERPENDICULAR to a LINE with SLOPE m_1. Perpendicular objects are sometimes said to be "orthogonal."

In the above figure, the LINE SEGMENT AB is perpendicular to the LINE SEGMENT CD. This relationship is commonly denoted with a small SQUARE at the vertex where perpendicular objects meet, as shown above, and is denoted $AB \perp CD$.

See also ORTHOGONAL LINES, ORTHOGONAL VECTORS, PARALLEL, PERPENDICULAR BISECTOR, PERPENDICULAR FOOT, RIGHT ANGLE

References

Kern, W. F. and Bland, J. R. *Solid Mensuration with Proofs,* 2nd ed. New York: Wiley, p. 10, 1948.

Perpendicular Bisector

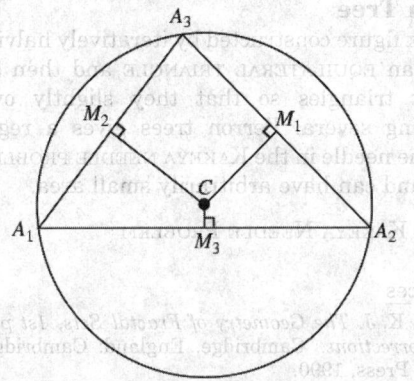

The perpendicular bisectors of a TRIANGLE $\Delta A_1 A_2 A_3$ are lines passing through the MIDPOINT M_i of each side which are PERPENDICULAR to the given side. A TRIANGLE'S three perpendicular bisectors meet (Casey 1888, p. 9) at a point C known as the CIRCUMCENTER (Durell 1928), which is also the center of the TRIANGLE'S CIRCUMCIRCLE.

See also CIRCUMCENTER, MIDPOINT, PERPENDICULAR, PERPENDICULAR FOOT

References

Casey, J. *A Sequel to the First Six Books of the Elements of Euclid, Containing an Easy Introduction to Modern Geometry with Numerous Examples,* 5th ed., rev. enl. Dublin: Hodges, Figgis, & Co., 1888.
Durell, C. V. *Modern Geometry: The Straight Line and Circle.* London: Macmillan, pp. 19–20, 1928.

Perpendicular Foot

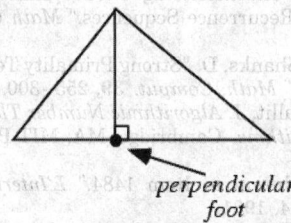

perpendicular foot

The FOOT of the PERPENDICULAR is the point on the leg opposite a given vertex of a TRIANGLE at which the PERPENDICULAR passing through that vertex intersects the side. The length of the LINE SEGMENT from the vertex to the perpendicular foot is called the ALTITUDE of the TRIANGLE.

When a line is drawn from a POINT to a PLANE, its intersection with the PLANE is known as the foot.

See also ALTITUDE, FOOT, PERPENDICULAR, PERPENDICULAR BISECTOR, TAYLOR CIRCLE

References

Coxeter, H. S. M. and Greitzer, S. L. *Geometry Revisited.* Washington, DC: Math. Assoc. Amer., p. 9, 1967.
Kern, W. F. and Bland, J. R. *Solid Mensuration with Proofs,* 2nd ed. New York: Wiley, p. 9, 1948.

Perrin Pseudoprime

If p is PRIME, then $p|P(p)$, where $P(p)$ is a member of the PERRIN SEQUENCE 3, 0, 2, 3, 2, 5, 5, 7, 10, 12, 17, ... (Sloane's A001608). A Perrin pseudoprime is a COMPOSITE NUMBER n such that $n|P(n)$. Several "unrestricted" Perrin pseudoprimes are known, the smallest of which are 271441, 904631, 16532714, 24658561, ... (Sloane's A013998).

Adams and Shanks (1982) discovered the smallest unrestricted Perrin pseudoprime after unsuccessful searches by Perrin (1899), Malo (1900), Escot (1901), and Jarden (1966). (A 1996 article by Stewart's stating that no Perrin pseudoprimes were then known was incorrect.)

Grantham (1996) generalized the definition of Perrin pseudoprime with parameters (r, s) to be an ODD COMPOSITE NUMBER n for which either

1. $(\Delta/n) = 1$ and n has an S-SIGNATURE, or
2. $(\Delta/n) = -1$ and n has a Q-SIGNATURE,

where (a/b) is the JACOBI SYMBOL. All the 55 Perrin pseudoprimes less than 50×10^9 have been computed by Kurtz *et al.* (1986). All have S-SIGNATURE, and form the sequence Sloane calls "restricted" Perrin pseudoprimes: 27664033, 46672291, 102690901, ... (Sloane's A018187).

See also PERRIN SEQUENCE, PSEUDOPRIME

References

Adams, W. W. "Characterizing Pseudoprimes for Third-Order Linear Recurrence Sequences." *Math Comput.* **48**, 1–15, 1987.

Adams, W. and Shanks, D. "Strong Primality Tests that Are Not Sufficient." *Math. Comput.* **39**, 255–300, 1982.

Bach, E. and Shallit, J. *Algorithmic Number Theory, Vol. 1: Efficient Algorithms.* Cambridge, MA: MIT Press, p. 305, 1996.

Escot, E.-B. "Solution to Item 1484." *L'Intermédiare des Math.* **8**, 63–64, 1901.

Grantham, J. "Frobenius Pseudoprimes." http://www.clark.net/pub/grantham/pseudo/pseudo1.ps

Holzbaur, C. "Perrin Pseudoprimes." http://ftp.ai.univie.ac.at/perrin.html.

Jarden, D. *Recurring Sequences.* Jerusalem: Riveon Lematematika, 1966.

Kurtz, G. C.; Shanks, D.; and Williams, H. C. "Fast Primality Tests for Numbers Less than $50 \cdot 10^9$." *Math. Comput.* **46**, 691–701, 1986.

Perrin, R. "Item 1484." *L'Intermédiare des Math.* **6**, 76–77, 1899.

Ribenboim, P. *The New Book of Prime Number Records, 3rd ed.* New York: Springer-Verlag, p. 135, 1996.

Sloane, N. J. A. Sequences A001608/M0429, A013998, and A018187 in "An On-Line Version of the Encyclopedia of Integer Sequences." http://www.research.att.com/~njas/sequences/eisonline.html.

Stewart, I. "Tales of a Neglected Number." *Sci. Amer.* **274**, 102–103, June 1996.

Perrin Sequence

The INTEGER SEQUENCE defined by the recurrence

$$P(n) = P(n-2) + P(n-3) \qquad (1)$$

with the initial conditions $P(0) = 3$, $P(1) = 0$, $P(2) = 2$. This RECURRENCE RELATION is the same as that for the PADOVAN SEQUENCE but with different initial conditions. The first few terms for $n = 0, 1, \ldots$, are 3, 0, 2, 3, 2, 5, 5, 7, 10, 12, 17, ... (Sloane's A001608). $P(n)$ is the solution of a third-order linear homogeneous DIFFERENCE EQUATION having characteristic equation

$$x^3 - x - 1 = 0, \qquad (2)$$

discriminant -23, and ROOTS

$$\alpha \approx 1.324717957 \qquad (3)$$

$$\beta \approx -0.6623589786 + 0.5622795121i \qquad (4)$$

$$\gamma \approx -0.6623589786 - 0.5622795121i. \qquad (5)$$

The solution is then

$$P(n) = \alpha^n + \beta^n + \gamma^n, \qquad (6)$$

where

$$P(n) \sim \alpha^n. \qquad (7)$$

Perrin (1899) investigated the sequence and noticed that if n is PRIME, then $n|P(n)$. The first statement of this fact is attributed to É. Lucas in 1876 by Stewart (1996). Perrin also searched for but did not find any COMPOSITE NUMBER n in the sequence such that $n|P(n)$. Such numbers are now known as PERRIN

PSEUDOPRIMES. Malo (1900), Escot (1901), and Jarden (1966) subsequently investigated the series and also found no PERRIN PSEUDOPRIMES. Adams and Shanks (1982) subsequently found that 271,441 is such a number.

See also PADOVAN SEQUENCE, PERRIN PSEUDOPRIME, SIGNATURE (RECURRENCE RELATION)

References

Adams, W. and Shanks, D. "Strong Primality Tests that Are Not Sufficient." *Math. Comput.* **39**, 255–300, 1982.

Escot, E.-B. "Solution to Item 1484." *L'Intermédiare des Math.* **8**, 63–64, 1901.

Jarden, D. *Recurring Sequences.* Jerusalem: Riveon Lematematika, 1966.

Perrin, R. "Item 1484." *L'Intermédiare des Math.* **6**, 76–77, 1899.

Stewart, I. "Tales of a Neglected Number." *Sci. Amer.* **274**, 102–103, June 1996.

Sloane, N. J. A. Sequences A001608/M0429 in "An On-Line Version of the Encyclopedia of Integer Sequences." http://www.research.att.com/~njas/sequences/eisonline.html.

Perron Integral

An integral which is equivalent to the DENJOY INTEGRAL "in the restricted sense."

See also DENJOY INTEGRAL

Perron Tree

A convex figure constructed by iteratively halving the base of an EQUILATERAL TRIANGLE and then sliding adjacent triangles so that they slightly overlap. Combining several Perron trees gives a region in which the needle in the KAKEYA NEEDLE PROBLEM can rotate, and can have arbitrarily small area.

See also KAKEYA NEEDLE PROBLEM

References

Falconer, K. J. *The Geometry of Fractal Sets, 1st pbk. ed., with corrections.* Cambridge, England: Cambridge University Press, 1990.

Wells, D. *The Penguin Dictionary of Curious and Interesting Geometry.* London: Penguin, pp. 128–129, 1991.

Perron-Frobenius Operator

An OPERATOR which describes the time evolution of densities in PHASE SPACE. The OPERATOR can be defined by

$$\rho_{n+1} = \tilde{L}\rho_n,$$

where ρ_n are the NATURAL DENSITIES after the nth iteration of a map f. This can be explicitly written as

$$\tilde{L}\rho(y) = \sum_{x \in f^{-1}(y)} \frac{\rho(x)}{|f'(x)|}.$$

See also FROBENIUS-PERRON EQUATION

Perron-Frobenius Theorem

References
Berman, A. and Plemmons, R. *Nonnegative Matrices in the Mathematical Sciences*. New York: Academic Press, 1979.
Beck, C. and Schlögl, F. "Transfer Operator Methods." Ch. 17 in *Thermodynamics of Chaotic Systems*. Cambridge, England: Cambridge University Press, pp. 190–203, 1995.

Perron-Frobenius Theorem

If all elements a_{ij} of an IRREDUCIBLE MATRIX A are NONNEGATIVE, then $R = \min M_\lambda$ is an EIGENVALUE of A and all the EIGENVALUES of A lie on the DISK

$$|z| \le R,$$

where, if $\lambda = (\lambda_1, \ldots, \lambda_2, \ldots, \lambda_n)$ is a set of NONNEGATIVE numbers (which are not all zero),

$$M_\lambda = \inf\left\{\mu : \mu\lambda_i > \sum_{j=1}^{n} |a_{ij}|\lambda_j, \ 1 \le i \le n\right\}$$

and $R = \min M_\lambda$. Furthermore, if A has exactly p EIGENVALUES ($p \le n$) on the CIRCLE $|z| = R$, then the set of all its EIGENVALUES is invariant under rotations by $2\pi/p$ about the ORIGIN.

See also WIELANDT'S THEOREM

References
Gradshteyn, I. S. and Ryzhik, I. M. *Tables of Integrals, Series, and Products, 6th ed.* San Diego, CA: Academic Press, p. 1121, 2000.

Perron's Formula

$$A^*(x) = \sum_{\lambda_n \le x}' a_n = \frac{1}{2\pi i}\int_{c-i\infty}^{c+i\infty} f(s)\frac{e^{sx}}{s}\,ds,$$

where

$$f(s) = \sum a_n e^{-\lambda_n s}.$$

References
Hardy, G. H. *Ramanujan: Twelve Lectures on Subjects Suggested by His Life and Work, 3rd ed.* New York: Chelsea, 1999.
Hardy, G. H. and Riesz. *The General Theory of Dirichlet's Series.* p. 12.

Perron's Theorem

If $\mu = (\mu_1, \mu_2, \ldots, \mu_n)$ is an arbitrary set of POSITIVE numbers, then all EIGENVALUES λ of the $n \times n$ MATRIX $a = a_{ij}$ lie on the DISK $|z| \le m_\mu$, where

$$m_\mu = \max_{1 \le i \le n} \sum_{j=1}^{n} \frac{\mu_j}{\mu_i}|a_{ij}|.$$

References
Gradshteyn, I. S. and Ryzhik, I. M. *Tables of Integrals, Series, and Products, 6th ed.* San Diego, CA: Academic Press, p. 1121, 2000.
MacCluer, C. R. "The Many Proofs and Applications of Perron's Theorem." *SIAM Rev.* **42**, 487–498, 2000.
Perron, O. "Grundlagen für eine Theorie des Jacobischen Kettenbruchalgorithmus." *Math. Ann.* **64**, 11–76, 1907.

Persistence

ADDITIVE PERSISTENCE, MULTIPLICATIVE PERSISTENCE, PERSISTENT NUMBER, PERSISTENT PROCESS

Persistent Number

An n-persistent number is a POSITIVE INTEGER k which contains the digits 0, 1, ..., 9 (i.e., is a PANDIGITAL NUMBER), and for which $2k, \ldots, nk$ also share this property. No ∞-persistent numbers exist. However, the number $k = 1234567890$ is 2-persistent, since $2k = 2469135780$ but $3k = 3703703670$, and the number $k = 526315789473684210$ is 18-persistent. There exists at least one k-persistent number for each POSITIVE INTEGER k.

n	Sloane	n-persistent
1	A051264	1023456798, 1023456897, 1023456978, 1023456987, ...
2	A051018	1023456789, 1023456879, 1023457689, 1023457869, ...
3	A051019	1052674893, 1052687493, 1052746893, 1052748693, ...
4	A051020	1053274689, 1089467253, 1253094867, 1267085493, ...

See also ADDITIVE PERSISTENCE, MULTIPLICATIVE PERSISTENCE, PANDIGITAL NUMBER

References
Honsberger, R. *More Mathematical Morsels.* Washington, DC: Math. Assoc. Amer., pp. 15–18, 1991.
Sloane, N. J. A. Sequences A051018, A051019, A051020, and A051264 in "An On-Line Version of the Encyclopedia of Integer Sequences." http://www.research.att.com/~njas/sequences/eisonline.html.

Persistent Process

A FRACTAL PROCESS for which $H > 1/2$, so $r > 0$.

See also ANTIPERSISTENT PROCESS, FRACTAL PROCESS

Perspective

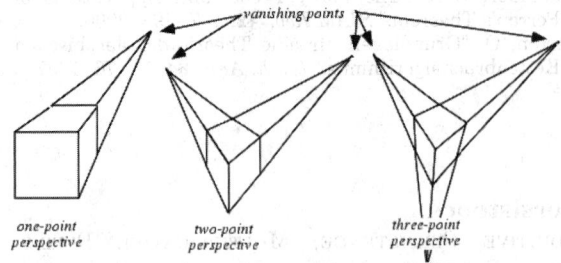

one-point perspective *two-point perspective* *three-point perspective*

Perspective is the art and mathematics of realistically depicting 3-D objects in a 2-D plane, sometimes called CENTRIC or NATURAL PERSPECTIVE to distinguish it from BICENTRIC PERSPECTIVE. The study of the projection of objects in a plane is called PROJECTIVE GEOMETRY. The principles of perspective drawing were elucidated by the Florentine architect F. Brunelleschi (1377–1446). These rules are summarized by Dixon (1991):

1. The horizon appears as a line.
2. Straight lines in space appear as straight lines in the image.
3. Sets of PARALLEL lines meet at a VANISHING POINT.
4. Lines PARALLEL to the picture plane appear PARALLEL and therefore have no VANISHING POINT.

There is a graphical method for selecting vanishing points so that a CUBE or box appears to have the correct dimensions (Dixon 1991).

See also BICENTRIC PERSPECTIVE, LEONARDO'S PARADOX, PERSPECTIVE AXIS, PERSPECTIVE CENTER, PERSPECTIVE COLLINEATION, PERSPECTIVE TRIANGLES, PERSPECTIVITY, PROJECTION, PROJECTIVE GEOMETRY, VANISHING POINT, ZEEMAN'S PARADOX

References

de Vries, V. *Perspective.* New York: Dover, 1968.
Dixon, R. "Perspective Drawings." Ch. 3 in *Mathographics.* New York: Dover, pp. 79–88, 1991.
Lambert, J. H. *Freie Perspective, 2nd ed.* Zürich, 1774.
Parramon, J. M. *Perspective--How to Draw.* Barcelona, Spain: Parramon Editions, 1984.
Steinhaus, H. *Mathematical Snapshots, 3rd ed.* New York: Dover, pp. 157–159, 1999.

Perspective Axis

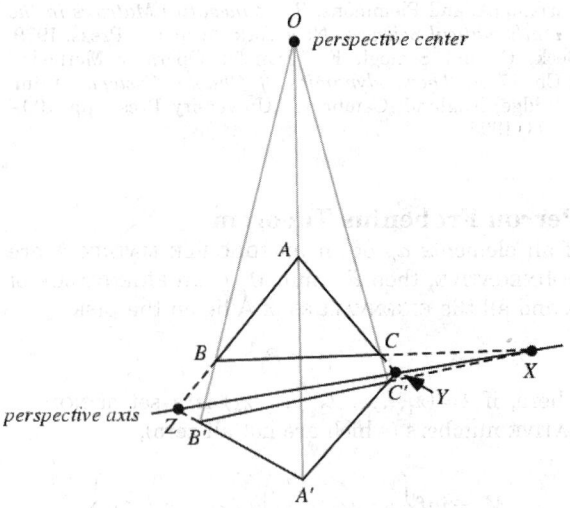

The line joining the three collinear points of intersection of the extensions of corresponding sides in PERSPECTIVE TRIANGLES, sometimes also called the homology axis.

See also PERSPECTIVE CENTER, PERSPECTIVE TRIANGLES, SONDAT'S THEOREM

Perspective Center

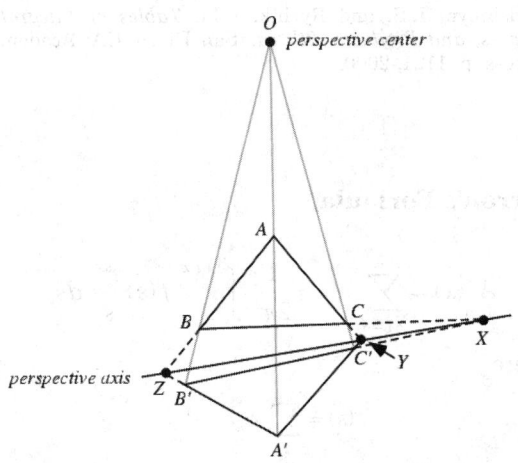

The point at which the three LINES connecting the VERTICES of PERSPECTIVE TRIANGLES (from a point) CONCUR, sometimes also called the homology center or pole.

See also PERSPECTIVE AXIS, PERSPECTIVE TRIANGLES

Perspective Collineation

A perspective collineation with center O and axis o is a COLLINEATION which leaves all lines through O and points of o invariant. Every perspective collineation is a PROJECTIVE COLLINEATION.

See also COLLINEATION, ELATION, HOMOLOGY (GEOMETRY), PROJECTIVE COLLINEATION

References

Coxeter, H. S. M. *Introduction to Geometry, 2nd ed.* New York: Wiley, pp. 247–248, 1969.

Perspective Triangles

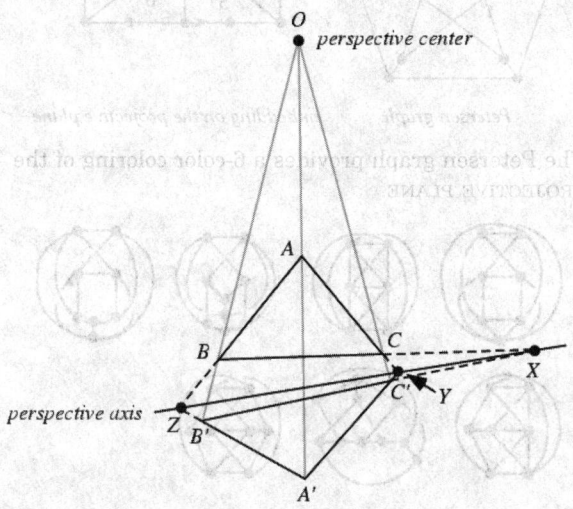

perspective center

perspective axis

Two TRIANGLES ΔABC and $\Delta A'B'C'$ are perspective from a line if the extensions of their three pairs of corresponding sides meet in COLLINEAR points X, Y, and Z. The line joining these points is called the PERSPECTIVE AXIS.

Two TRIANGLES are perspective from a point if their three pairs of corresponding VERTICES are joined by lines which meet in a point of CONCURRENCE O. This point is called the PERSPECTIVE CENTER, or sometimes the homology center or pole.

DESARGUES' THEOREM guarantees that if two TRIANGLES are perspective from a point, they are perspective from a line (called the PERSPECTIVE AXIS). Triangles in perspective are sometimes said to be homologous or copolar.

See also DESARGUES' THEOREM, DILATION, HOMOTHETIC TRIANGLES, PARALOGIC TRIANGLES, PERSPECTIVE AXIS, PERSPECTIVE CENTER

References

Coxeter, H. S. M. and Greitzer, S. L. "Perspective Triangles; Desargues's Theorem." §3.6 in *Geometry Revisited*. Washington, DC: Math. Assoc. Amer., pp. 70–72, 1967.
Lachlan, R. "Triangles in Perspective" and "Relations Between Two Triangles in Perspective." §160–180 in *An Elementary Treatise on Modern Pure Geometry*. London: Macmillian, pp. 100–113, 1893.

Perspectivity

A correspondence between two RANGES that are sections of one PENCIL by two distinct lines.

See also PENCIL, PROJECTIVITY, RANGE (LINE SEGMENT)

Persymmetric Matrix

A SQUARE MATRIX with constant SKEW DIAGONALS. Such matrices are sometimes known as orthosymmetric in older literature.

See also DIAGONAL MATRIX, SKEW DIAGONAL, SKEW SYMMETRIC MATRIX, SYMMETRIC MATRIX

References

Mays, M. E. and Wojciechowski, J. "A Determinant Property of Catalan Numbers." *Disc. Math.* **211**, 125–133, 2000.

Pesin Theory

The theory of non-uniformly hyperbolic DIFFEOMORPHISMS.

See also DIFFEOMORPHISM

References

Katok, A. "Lyapunov Exponents, Entropy, and Periodic Orbits for Diffeomorphisms." *Pub. Math. (IHS)* **51**, 137–173, 1980.
Katok, A. and Strelcyn, J.-M. *Invariant Manifolds, Entropy and Billiards, Smooth Maps with Singularities.* Berlin: Springer-Verlag, 1988.
Newhouse, S. "Continuity Properties of Entropy." *Ann. Math.* **129**, 215–237, 1989.
Newhouse, S. "Entropy and Volume." *Ergodic Th. Dynam. Sys.* **8**, 283–299, 1989.
Pollicott, M. *Lectures on Ergodic Theory and Pesin Theory on Compact Manifolds.* Cambridge, England: Cambridge University Press, 1993.

Peters Polynomial

Polynomials $s_k(x; \lambda, \mu)$ which are a generalization of the BOOLE POLYNOMIALS, form the SHEFFER SEQUENCE for

$$g(t) = (1 + e^{\lambda t})^{\mu} \qquad (1)$$

$$f(t) = e^t - 1 \qquad (2)$$

and have GENERATING FUNCTION

$$\sum_{k=0}^{\infty} \frac{s_k(x; \lambda, \mu)}{k!} t^k = [1 + (1+t)^{\lambda}]^{-\mu} (1+t)^x. \qquad (3)$$

The first few are

$$s_0(x; \lambda, \mu) = 2^{-\mu}$$

$$s_1(x; \lambda, \mu) = 2^{-(\mu+1)}(2x - \lambda\mu)$$

$$s_2(x; \lambda, \mu) = 2^{-(\mu+2)}[4x(x-1) + (2 - 4x)\lambda\mu + \mu(\mu-1)\lambda^2].$$

References

Boas, R. P. and Buck, R. C. *Polynomial Expansions of Analytic Functions, 2nd print., corr.* New York: Academic Press, p. 37, 1964.
Roman, S. "The Peters Polynomial." §4.6 in *The Umbral Calculus.* New York: Academic Press, p. 128, 1984.

Rota, G.-C.; Kahaner, D.; Odlyzko, A. "On the Foundations of Combinatorial Theory. VIII: Finite Operator Calculus." *J. Math. Anal. Appl.* **42**, 684–760, 1973.

Peters Projection

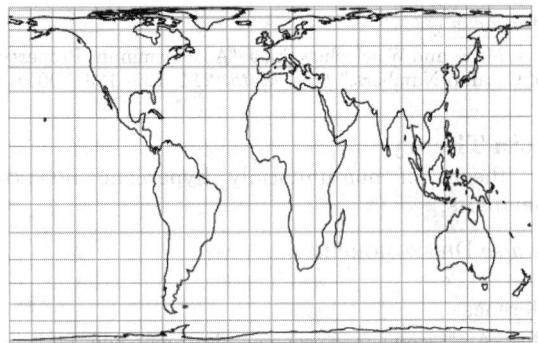

A CYLINDRICAL EQUAL-AREA PROJECTION that de-emphasizes the exaggeration of areas at high latitudes by shifting the standard LATITUDE to $\phi_s = 44.138°$ (or sometimes 45° or 47°; Dana).

See also BALTHASART PROJECTION, BEHRMANN CYLINDRICAL EQUAL-AREA PROJECTION, CYLINDRICAL EQUAL-AREA PROJECTION, CYLINDRICAL PROJECTION, EQUAL-AREA PROJECTION, GALL ORTHOGRAPHIC PROJECTION, LAMBERT AZIMUTHAL EQUAL-AREA PROJECTION, PETERS PROJECTION

References

Dana, P. H. "Map Projections." http://www.colorado.edu/geography/gcraft/notes/mapproj/mapproj_f.html.

Petersen Graph

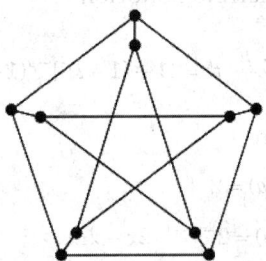

"The" Petersen graph is the GRAPH illustrated above possessing ten nodes, all of whose nodes have DEGREE 3 (Saaty and Kainen 1986, Harary 1994, p. 89). The Petersen graph is the only smallest-GIRTH graph which has no Tait coloring, and is the unique 5-CAGE GRAPH (Harary 1994, p. 175). It is the complement of the LINE GRAPH of the COMPLETE GRAPH K_5 (Skiena 1990, p. 139), and the ODD GRAPH O_3 (Skiena 1990, p. 162). It is depicted on the cover of the journal *Discrete Mathematics*. The Petersen graph is the

smallest HYPOHAMILTONIAN GRAPH

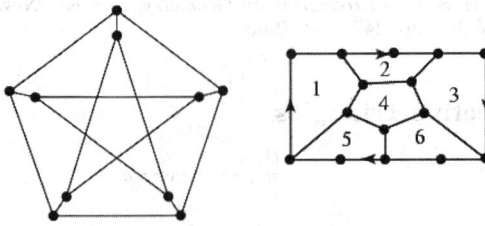

Petersen graph   *embedding on the projective plane*

The Petersen graph provides a 6-color coloring of the PROJECTIVE PLANE.

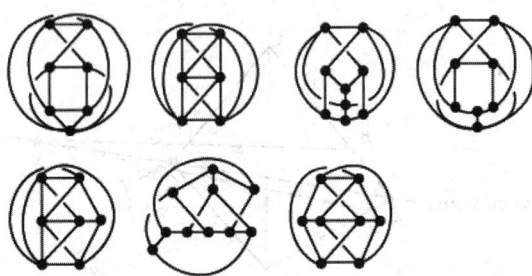

The seven graphs obtainable from the COMPLETE GRAPH K_6 by repeated triangle-Y exchanges are also called Petersen graphs, where the three EDGES forming the TRIANGLE are replaced by three EDGES and a new VERTEX that form a Y, and the reverse operation is also permitted. A GRAPH is intrinsically linked IFF it contains one of the seven Petersen graphs (Robertson *et al.* 1993).

See also CAGE GRAPH, GIRTH, HOFFMAN-SINGLETON GRAPH, HYPOHAMILTONIAN GRAPH, ODD GRAPH

References

Adams, C. C. *The Knot Book: An Elementary Introduction to the Mathematical Theory of Knots.* New York: W. H. Freeman, pp. 221–222, 1994.

Bondy, J. A. and Murty, U. S. R. *Graph Theory with Applications.* New York: North Holland, pp. 236 and 243, 1976.

Harary, F. *Graph Theory.* Reading, MA: Addison-Wesley, pp. 89 and 112, 1994.

Hoffman, A. J. and Singleton, R. R. "On Moore Graphs of Diameter Two and Three." *IBM J. Res. Develop.* **4**, 497–504, 1960.

Holton, D A. and Sheehan, J. (Eds.). *The Petersen Graph.* Cambridge, England: Cambridge University Press, 1993.

Robertson, N.; Seymour, P. D.; and Thomas, R. "Linkless Embeddings of Graphs in 3-Space." *Bull. Amer. Math. Soc.* **28**, 84–89, 1993.

Saaty, T. L. and Kainen, P. C. *The Four-Color Problem: Assaults and Conquest.* New York: Dover, p. 102, 1986.

Skiena, S. *Implementing Discrete Mathematics: Combinatorics and Graph Theory with Mathematica.* Reading, MA: Addison-Wesley, pp. 139 and 191, 1990.

Weisstein, E. W. "Graphs." MATHEMATICA NOTEBOOK GRAPHS.M.

Wong, P. K. "Cages--A Survey." *J. Graph Th.* **6**, 1–22, 1982.

Petersen-Shoute Theorem

A beautiful general theory of which the following two statements are special cases.

1. If ΔABC and $\Delta A'B'C'$ are two DIRECTLY SIMILAR triangles, while $\Delta AA'A''$, $\Delta BB'B''$, and $\Delta CC'C''$ are three DIRECTLY SIMILAR triangles, then $\Delta A''B''C''$ is directly similar to ΔABC.

2. When all the points P on AB are related by a SIMILARITY TRANSFORMATION to all the points P' on $A'B'$, the points dividing the segment PP' in a given ratio are distant and collinear, or else they coincide.

See also DIRECTLY SIMILAR, SIMILARITY TRANSFORMATION

References

Coxeter, H. S. M. and Greitzer, S. L. *Geometry Revisited.* Washington, DC: Math. Assoc. Amer., pp. 95–100, 1967.
Forder, H. G. *Higher Course Geometry.* Cambridge, England: Cambridge University Press, p. 53, 1931.
Petersen, J. *Methods and Theories for the Solution of Problems of Geometrical Constructions Applied to 410 Problems.* New York: Stechert, p. 74, 1923. Reprinted in *String Figures and Other Monographs.* New York: Chelsea, 1960.

Peterson-Mainardi-Codazzi Equations

$$\frac{\partial e}{\partial v} - \frac{\partial f}{\partial u} = e\Gamma_{12}^1 + f(\Gamma_{12}^2 - \Gamma_{11}^1) - g\Gamma_{11}^2 \quad (1)$$

$$\frac{\partial f}{\partial v} - \frac{\partial g}{\partial u} = e\Gamma_{22}^1 + f(\Gamma_{22}^2 - \Gamma_{12}^1) - g\Gamma_{12}^2, \quad (2)$$

where e, f, and g are coefficients of the second FUNDAMENTAL FORM and Γ_{ij}^k are CHRISTOFFEL SYMBOLS OF THE SECOND KIND. Therefore,

$$\frac{\partial e}{\partial v} = \frac{1}{2}E_v\left(\frac{e}{E} + \frac{g}{G}\right) \quad (3)$$

$$\frac{\partial g}{\partial u} = \frac{1}{2}G_u\left(\frac{e}{E} + \frac{g}{G}\right) \quad (4)$$

$$\frac{\partial(\ln f)}{\partial u} = \Gamma_{11}^1 - \Gamma_{12}^2 \quad (5)$$

$$\frac{\partial(\ln f)}{\partial v} = \Gamma_{22}^2 - \Gamma_{12}^1 \quad (6)$$

$$\frac{\partial}{\partial u}\left(\frac{\ln f}{\sqrt{EG - F^2}}\right) = -2\Gamma_{12}^2 \quad (7)$$

$$\frac{\partial}{\partial v}\left(\frac{\ln f}{\sqrt{EG - F^2}}\right) = -2\Gamma_{12}^1, \quad (8)$$

where E, F, and G are coefficients of the first FUNDAMENTAL FORM.

References

Gray, A. "The Peterson-Mainardi-Codazzi Equations." §28.3 in *Modern Differential Geometry of Curves and Surfaces with Mathematica, 2nd ed.* Boca Raton, FL: CRC Press, pp. 649–652, 1997.
Green, A. E. and Zerna, W. *Theoretical Elasticity, 2nd ed.* New York: Dover, p. 37, 1992.

Petersson Conjecture

Petersson considered the absolutely converging DIRICHLET L-SERIES

$$\phi(s) = \prod_p \frac{1}{1 - c(p)p^{-s} + p^{2k-1}p^{-2s}}.$$

Writing the DENOMINATOR as

$$1 - c(p)x + p^{2k-1}x^2 = (1 - r_1 x)(1 - r_2 x),$$

where

$$r_1 + r_2 = c(p)$$

and

$$r_1 r_2 = p^{2k-1},$$

Petersson conjectured that r_1 and r_2 are always COMPLEX CONJUGATE, which implies

$$|r_1| = |r_2| = p^{k-1/2}$$

and

$$|c(p)| \leq 2p^{k-1/2}.$$

This conjecture was proven by Deligne (1974), which also proved the TAU CONJECTURE as a special case. Deligne was awarded the FIELDS MEDAL for his proof.

See also DIRICHLET L-SERIES, TAU CONJECTURE

References

Apostol, T. M. *Modular Functions and Dirichlet Series in Number Theory, 2nd ed.* New York: Springer-Verlag, p. 140, 1997.
Deligne, P. "La conjecture de Weil. I." *Inst. Hautes Études Sci. Publ. Math.* **43**, 273–307, 1974.
Deligne, P. "La conjecture de Weil. II." *Inst. Hautes Études Sci. Publ. Math.* **52**, 137–252, 1980.

Peter-Weyl Theorem

Establishes completeness for a group REPRESENTATION.

References

Huang, J.-S. "The Peter-Weyl Theorem." §8.5 in *Lectures on Representation Theory.* Singapore: World Scientific, pp. 99–103, 1999.
Knapp, A. W. "Group Representations and Harmonic Analysis, Part II." *Not. Amer. Math. Soc.* **43**, 537–549, 1996.

Petrie Polygon

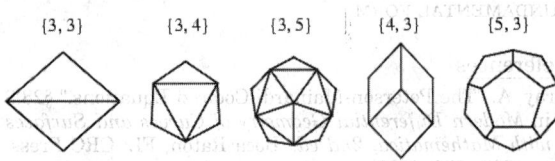

{3, 3}   {3, 4}   {3, 5}   {4, 3}   {5, 3}

A SKEW POLYGON such that every two consecutive sides (but no three) belong to a face of a regular POLYHEDRON. Every REGULAR POLYHEDRON can be orthogonally projected onto a plane in such a way that one Petrie polygon becomes a REGULAR POLYGON with the remainder of the projection interior to it. The Petrie polygon of the POLYHEDRON $\{p, q\}$ has h sides, where

$$\cos^2\left(\frac{\pi}{h}\right) = \cos^2\left(\frac{\pi}{p}\right) + \cos^2\left(\frac{\pi}{q}\right).$$

The Petrie polygons shown above correspond to the PLATONIC SOLIDS.

See also PLATONIC SOLID, REGULAR POLYGON, REGULAR POLYHEDRON, SKEW POLYGON

References
Ball, W. W. R. and Coxeter, H. S. M. *Mathematical Recreations and Essays, 13th ed.* New York: Dover, p. 135, 1987.
Coxeter, H. S. M. "Petrie Polygons." §2.6 in *Regular Polytopes, 3rd ed.* New York: Dover, pp. 24–25, 1973.

Petrov Notation

A TENSOR notation which considers the RIEMANN TENSOR $R_{\lambda\mu\nu\kappa}$ as a matrix $R_{(\lambda\mu)(\nu\kappa)}$ with indices $\lambda\mu$ and $\nu\kappa$.

References
Weinberg, S. *Gravitation and Cosmology: Principles and Applications of the General Theory of Relativity.* New York: Wiley, p. 142, 1972.

Petty Projection Inequality

An affine isoperimetric inequality.

References
Lutwak, E. "Selected Affine Isoperimetric Inequalities." In *Handbook of Convex Geometry* (Ed. P. M. Gruber and J. M. Wills). Amsterdam, Netherlands: North-Holland, pp. 151–176, 1993.

Pfaff Transformation

When $|x| < 1/2$,

$$(1-x)^{-a} {}_2F_1(a, b; c; -x/(1-x)) = {}_2F_1(a, c-b; c; x).$$

References
Koepf, W. *Hypergeometric Summation: An Algorithmic Approach to Summation and Special Function Identities.* Braunschweig, Germany: Vieweg, pp. 39–40, 1998.

Pfaffian

An analog of the determinant for NUMBER TRIANGLES defined as a signed sum indexed by set partitions of $\{1, \ldots, n\}$ into pairs of elements. The Pfaffian is the square root of the determinant of the corresponding skew symmetric matrix.

References
Bressoud, D. and Propp, J. "How the Alternating Sign Matrix Conjecture was Solved." *Not. Amer. Math. Soc.* **46**, 637–646.

Pfaffian Form

A 1-FORM

$$\omega = \sum_{i=1}^{n} a_i(x)\, dx_i$$

such that

$$\omega = 0.$$

References
Knuth, D. E. "Overlapping Pfaffians." *Electronic J. Combinatorics* **3**, No. 2, R5, 1–13, 1996. http://www.combinatorics.org/Volume_3/volume3_2.html#R5.

p-Form

DIFFERENTIAL K-FORM

p-Good Path

A LATTICE PATH from one point to another is p-good if it lies completely below the line

$$y = (p-1)x.$$

Hilton and Pederson (1991) show that the number of p-good paths from $(1, q-1)$ to $(k, n-k)$ under the condition $2 \leq k \leq n-p+1 \leq p(k-1)$ is

$$\binom{n-q}{k-1} - \sum_{j=1}^{\ell} {}_p d_{qj} \binom{n-pj}{k-j},$$

where $\binom{a}{b}$ is a BINOMIAL COEFFICIENT, and

$$\ell \equiv \left\lfloor \frac{n-k}{p-1} \right\rfloor,$$

where $\lfloor x \rfloor$ is the FLOOR FUNCTION.

See also CATALAN NUMBER, LATTICE PATH, SCHRÖDER NUMBER

References

Hilton, P. and Pederson, J. "Catalan Numbers, Their Generalization, and Their Uses." *Math. Intel.* **13**, 64–75, 1991.

p-Group

When p is a PRIME NUMBER, then a p-group is a GROUP, all of whose elements have order some power of p. For a FINITE GROUP, the equivalent definition is that the number of elements in G is a power of p. In fact, every FINITE GROUP has subgroups which are p-groups by the SYLOW THEOREMS, in which case they are called SYLOW P-SUBGROUPS.

Sylow proved that every GROUP of this form has a power-commutator representation on n generators defined by

$$a_i^p = \prod_{k=i+1}^{n} a_k^{\beta(i,\,k)} \tag{1}$$

for $0 \le \beta(i,\,k) < p$, $1 \le i \le n$ and

$$[a_j,\,a_i] = \prod_{k=j+1}^{n} a_k^{\beta(i,\,j,\,k)} \tag{2}$$

for $0 \le \beta(i,\,j,\,k) < p$, $1 \le i < j \le n$. If (p^m) is a PRIME POWER and $f(p^m)$ is the number of GROUPS of order (p^m), then

$$f(p^m) = p^{Am^3}, \tag{3}$$

where

$$\lim_{m \to \infty} A = \tfrac{2}{27} \tag{4}$$

(Higman 1960ab).

See also GROUP, GROUP DIRECT PRODUCT, ORDER (GROUP), SYLOW P-SUBGROUP, SYLOW THEOREMS

References

Higman, G. "Enumerating p-Groups. I. Inequalities." *Proc. London Math. Soc.* **10**, 24–30, 1960a.
Higman, G. "Enumerating p-Groups. II. Problems Whose Solution is PORC." *Proc. London Math. Soc.* **10**, 566–582, 1960b.

Phase

The angular position of a quantity. For example, the phase of a function $\cos(\omega t + \phi_0)$ as a function of time is

$$\phi(t) = \omega t + \phi_0.$$

The ARGUMENT of a COMPLEX NUMBER is sometimes also called the phase.

See also ARGUMENT (COMPLEX NUMBER), COMPLEX NUMBER, PHASOR, RETARDANCE

Phase Space

For a function or object with n DEGREES OF FREEDOM, the n-D SPACE which is accessible to the function or object is called its phase space.

See also WORLD LINE

Phase Transition

Erdos and Rényi (1960) showed that for many monotone-increasing properties of RANDOM GRAPHS, graphs of a size slightly less than a certain threshold are very unlikely to have the property, whereas graphs with a few more EDGES are almost certain to have it. This is known as a PHASE TRANSITION (Janson *et al.* 2000, p. 103).

See also RANDOM GRAPH

References

Erdos, P. and Rényi, A. "On the Evolution of Random Graphs." *Publ. Math. Inst. Hungar. Acad. Sci.* **5**, 17–61, 1960.
Janson, S.; uczak, T.; and Rucinski, A. "The Phase Transition." Ch. 5 in *Random Graphs*. New York: Wiley, pp. 103–138, 2000.

Phasor

The representation, beloved of engineers and physicists, of a COMPLEX NUMBER in terms of a COMPLEX exponential

$$x + iy = |z|e^{i\phi}, \tag{1}$$

where I (called J by engineers) is the IMAGINARY NUMBER and the MODULUS and ARGUMENT (also called PHASE) are

$$|z| = \sqrt{x^2 + y^2} \tag{2}$$

$$\phi = \tan^{-1}\left(\frac{y}{x}\right). \tag{3}$$

Here, ϕ (sometimes also denoted θ) is called the ARGUMENT or the PHASE. It corresponds to the counterclockwise ANGLE from the POSITIVE REAL AXIS, i.e., the value of ϕ such that $x = \cos\phi$ and $y = \sin\phi$. The special kind of INVERSE TANGENT used here takes into account the quadrant in which z lies and is returned by the FORTRAN command ATAN2(X,Y) and the *Mathematica* command ArcTan[x, y], and is often restricted to the range $-\pi < \theta \le \pi$. In the degenerate case when $x = 0$,

$$\phi = \begin{cases} -\tfrac{1}{2}\pi & \text{if } y < 0 \\ \text{undefined} & \text{if } y = 0 \\ \tfrac{1}{2}\pi & \text{if } y > 0 \end{cases} \tag{4}$$

It is trivially true that

$$\sum_i \Re[\psi_i] = \Re\left[\sum_i \psi_i\right]. \tag{5}$$

Now consider a SCALAR FUNCTION $\psi \equiv \psi_0 e^{i\phi}$. Then

$$I \equiv [\Re(\psi)]^2 = \left[\tfrac{1}{2}(\psi + \bar{\psi})\right]^2 = \tfrac{1}{4}(\psi + \bar{\psi})^2$$

$$= \tfrac{1}{4}(\psi^2 + 2\psi\bar{\psi} + \bar{\psi}^2), \tag{6}$$

where $\bar{\psi}$ is the COMPLEX CONJUGATE. Look at the time averages of each term,

$$\langle \psi^2 \rangle = \langle \psi_0^2 e^{2i\phi} \rangle = \psi_0^2 \langle e^{2i\phi} \rangle = 0 \tag{7}$$

$$\langle \psi\bar{\psi} \rangle = \langle \psi_0^2 e^{i\phi} \psi_0 e^{-i\phi} \rangle = \psi_0^2 = |\psi|^2 \tag{8}$$

$$\langle \bar{\psi}^2 \rangle = \langle \psi_0^2 e^{-2i\phi} \rangle = \psi_0^2 \langle e^{-2i\phi} \rangle = 0. \tag{9}$$

Therefore,

$$\langle I \rangle = \tfrac{1}{2}|\psi|^2. \tag{10}$$

Consider now two scalar functions

$$\psi_1 \equiv \psi_{1,0} e^{i(kr_1 + \phi_1)} \tag{11}$$

$$\psi_2 \equiv \psi_{2,0} e^{i(kr_2 + \phi_2)}. \tag{12}$$

Then

$$I \equiv [\Re(\psi_1) + \Re(\psi_2)]^2 = \tfrac{1}{4}[(\psi_1 + \bar{\psi}_1) + (\psi_2 + \bar{\psi}_2)]^2$$

$$= \tfrac{1}{4}[(\psi_1 + \bar{\psi}_1)^2 + (\psi_2 + \bar{\psi}_2)^2$$

$$+ 2(\psi_1\psi_2 + \psi_1\bar{\psi}_2 + \bar{\psi}_1\psi_2 + \bar{\psi}_1\bar{\psi}_2)] \tag{13}$$

$$\langle I \rangle = \tfrac{1}{4}[2\psi_1\bar{\psi}_1 + 2\psi_2\bar{\psi}_2 + 2\psi_1\bar{\psi}_2 + 2\bar{\psi}_1\psi_2]$$

$$= \tfrac{1}{2}[\psi_1(\bar{\psi}_1 + \bar{\psi}_2) + \psi_2(\bar{\psi}_1 + \bar{\psi}_2)]$$

$$= \tfrac{1}{2}(\psi_1 + \psi_2)(\bar{\psi}_1 + \bar{\psi}_2) = \tfrac{1}{2}|\psi_1 + \psi_2|^2. \tag{14}$$

In general,

$$\langle I \rangle = \frac{1}{2}\left|\sum_{i=1}^n \psi_i\right|^2. \tag{15}$$

See also AFFIX, ARGUMENT (COMPLEX NUMBER), CIS, COMPLEX MULTIPLICATION, COMPLEX NUMBER, EXPONENTIAL FUNCTION, INVERSE TANGENT, MODULUS (COMPLEX NUMBER), PHASE

References

Krantz, S. G. "Polar Form of a Complex Number," §1.2.4 in *Handbook of Complex Analysis*. Boston, MA: Birkhäuser, pp. 8–10, 1999.

Phi Curve

An ADJOINT CURVE which bears a special relation to the base curve.

References

Coolidge, J. L. *A Treatise on Algebraic Plane Curves*. New York: Dover, p. 310, 1959.

Phi Number System

For every POSITIVE INTEGER n, there is a corresponding finite sequence of distinct INTEGERS $k_1, ..., k_m$ such that

$$n = \phi^{k_1} + \ldots + \phi^{k_m},$$

where ϕ is the GOLDEN RATIO.

See also GOLDEN RATIO

References

Bergman, G. "A Number System with an Irrational Base." *Math. Mag.* **31**, 98–110, 1957.
Knuth, D. *The Art of Computer Programming, Vol. 1: Fundamental Algorithms, 3rd ed.* Reading, MA: Addison-Wesley, 1997.
Rousseau, C. "The Phi Number System Revisited." *Math. Mag.* **68**, 283–284, 1995.

Phi-Four Equation

The PARTIAL DIFFERENTIAL EQUATION

$$u_H - u_{xx} - u + u^3 = 0.$$

References

Calogero, F. and Degasperis, A. *Spectral Transform and Solitons: Tools to Solve and Investigate Nonlinear Evolution Equations.* New York: North-Holland, p. 60, 1982.
Zwillinger, D. *Handbook of Differential Equations, 3rd ed.* Boston, MA: Academic Press, p. 134, 1997.

Philo Line

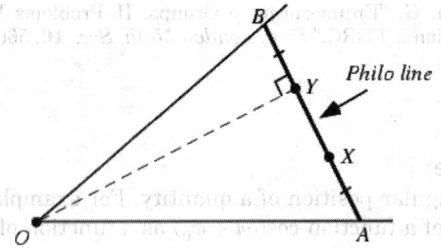

Given two intersecting lines OA and AB forming an angle with vertex at O and a point X inside the angle $\angle AOB$, the Philo line (or Philon line) is the shortest LINE SEGMENT AB touching both lines and passing through X. The line is named for Philo of Byzantium who considered the line while attempting to duplicate the cube. The line can be constructed by finding $OY \perp$

AB such that $AX = BY$ (Wells 1991).

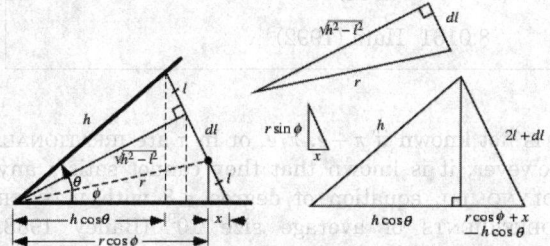

The distances along the angle edges x and h and the lengths along the Philo line l and dl can be computed by solving the simultaneous equations

$$r^2 \sin^2 \phi + x^2 = l^2$$

$$h^2 - l^2 = (r \cos \phi + x)^2 - (l + dl)^2$$

$$(2l + dl)^2 = h^2 \sin^2 \theta + (r \cos \theta + x - h \cos \theta)^2$$

$$(h^2 - l^2) + dl^2 = r^2,$$

where θ is the VERTEX ANGLE and the point X has POLAR COORDINATES (r, ϕ).

References

Eves, H. "Philo's Line." *Scripta Math.* **24**, 141–148, 1959.
Eves, H. W. *A Survey of Geometry, Vol. 2.* Boston, MA: Allyn and Bacon, pp. 39 and 234–238, 1965.
Wells, D. *The Penguin Dictionary of Curious and Interesting Geometry.* London: Penguin, pp. 182–183, 1991.
Wells, D. G. *You Are a Mathematician: A Wise and Witty Introduction to the Joy of Numbers.* New York: Wiley, 1997.

Philon Line

PHILO LINE

Phragmén-Lindelöf Theorem

Let $f(z)$ be an ANALYTIC FUNCTION in an angular domain $W : |\arg z| < \alpha\pi/2$. Suppose there is a constant M such that for each $\epsilon > 0$, each finite boundary point has a NEIGHBORHOOD such that $|f(z)| < M + \epsilon$ on the intersection of D with this NEIGHBORHOOD, and that for some POSITIVE number $\beta > \alpha$ for sufficiently large $|z|$, the INEQUALITY $|f(z)| < \exp(|z|^{1/\beta})$ holds. Then $|f(z)| \le M$ in D.

References

Iyanaga, S. and Kawada, Y. (Eds.). *Encyclopedic Dictionary of Mathematics.* Cambridge, MA: MIT Press, p. 160, 1980.

Phyllotaxis

The beautiful arrangement of leaves in some plants, called phyllotaxis, obeys a number of subtle mathematical relationships. For instance, the florets in the head of a sunflower form two oppositely directed spirals: 55 of them clockwise and 34 counterclockwise. Surprisingly, these numbers are consecutive FIBONACCI NUMBERS. The ratios of alternate FIBO-

NACCI NUMBERS are given by the convergents to ϕ^{-2}, where ϕ is the GOLDEN RATIO, and are said to measure the fraction of a turn between successive leaves on the stalk of a plant: 1/2 for elm and linden, 1/3 for beech and hazel, 2/5 for oak and apple, 3/8 for poplar and rose, 5/13 for willow and almond, etc. (Coxeter 1969, Ball and Coxeter 1987). A similar phenomenon occurs for DAISIES, pineapples, pinecones, cauliflowers, and so on.

Lilies, irises, and the trillium have three petals; columbines, buttercups, larkspur, and wild rose have five petals; delphiniums, bloodroot, and cosmos have eight petals; corn marigolds have 13 petals; asters have 21 petals; and daisies have 34, 55, or 89 petals–all FIBONACCI NUMBERS.

See also DAISY, FIBONACCI NUMBER, SPIRAL

References

Ball, W. W. R. and Coxeter, H. S. M. *Mathematical Recreations and Essays, 13th ed.* New York: Dover, pp. 56–57, 1987.
Church, A. H. *The Relation of Phyllotaxis to Mechanical Laws.* London: Williams and Norgate, 1904.
Church, A. H. *On the Interpretation of Phenomena of Phyllotaxis.* Riverside, NJ: Hafner, 1968.
Conway, J. H. and Guy, R. K. "Phyllotaxis." In *The Book of Numbers.* New York: Springer-Verlag, pp. 113–125, 1995.
Cook, T. A. *The Curves of Life, Being an Account of Spiral Formations and Their Application to Growth in Nature, To Science and to Art.* New York: Dover, 1979.
Coxeter, H. S. M. "The Golden Section and Phyllotaxis." Ch. 11 in *Introduction to Geometry, 2nd ed.* New York: Wiley, 1969.
Coxeter, H. S. M. "The Role of Intermediate Convergents in Tait's Explanation for Phyllotaxis." *J. Algebra* **10**, 167–175, 1972.
Coxeter, H. S. M. "The Golden Section, Phyllotaxis, and Wythoff's Game." *Scripta Mathematica* **19**, 135–143, 1953.
Dixon, R. "The Mathematics and Computer Graphics of Spirals in Plants." *Leonardo* **16**, 86–90, 1983.
Dixon, R. *Mathographics.* New York: Dover, 1991.
Douady, S. and Couder, Y. "Phyllotaxis as a Self-Organized Growth Process." In *Growth Patterns in Physical Sciences and Biology* (Ed. J. M. Garcia-Ruiz *et al.*). New York: Plenum, 1993.
Hargittai, I. and Pickover, C. A. (Eds.). *Spiral Symmetry.* New York: World Scientific, 1992.
Hunter, J. A. H. and Madachy, J. S. *Mathematical Diversions.* New York: Dover, pp. 20–22, 1975.
Jean, R. V. "Number-Theoretic Properties of Two-Dimensional Lattices." *J. Number Th.* **29**, 206–223, 1988.
Jean, R. V. "On the Origins of Spiral Symmetry in Plants." In *Spiral Symmetry.* (Ed. I. Hargittai and C. A. Pickover). New York: World Scientific, pp. 323–351, 1992.
Jean, R. V. *Phyllotaxis: A Systematic Study in Plant Morphogenesis.* New York: Cambridge University Press, 1994.
Pappas, T. "The Fibonacci Sequence & Nature." *The Joy of Mathematics.* San Carlos, CA: Wide World Publ./Tetra, pp. 222–225, 1989.
Prusinkiewicz, P. and Lindenmayer, A. *The Algorithmic Beauty of Plants.* New York: Springer-Verlag, 1990.
Steinhaus, H. *Mathematical Snapshots, 3rd ed.* New York: Dover, p. 138, 1999.
Stevens, P. S. *Patterns in Nature.* London: Peregrine, 1977.

Stewart, I. "Daisy, Daisy, Give Me Your Answer, Do." *Sci. Amer.* **200**, 96–99, Jan. 1995.

Thompson, D. W. *On Growth and Form.* Cambridge, England: Cambridge University Press, 1952.

Vogel, H. "A Better Way to Construct the Sunflower Head." *Math. Biosci.* **44**, 179–189, 1979.

Wells, D. *The Penguin Dictionary of Curious and Interesting Numbers.* Middlesex, England: Penguin Books, pp. 65–66, 1986.

Pi

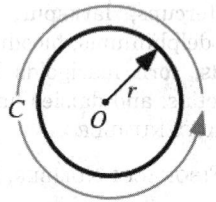

A REAL NUMBER denoted π which is defined as the ratio of a CIRCLE's CIRCUMFERENCE C to its DIAMETER $\pi = 2r$,

$$\pi \equiv \frac{C}{d} = \frac{C}{2r} \qquad (1)$$

It is equal to

$$\pi = $$

$$3.1415926535897932384626433832795028841971\ldots \qquad (2)$$

(Sloane's A000796). PI's DIGITS have many interesting properties, although not very much is known about their analytic properties. PI's CONTINUED FRACTION is given by [3, 7, 15, 1, 292, 1, 1, 1, ...] (Sloane's A001203).

π is known to be IRRATIONAL (Lambert 1761, Legendre 1794, Hermite 1873, Nagell 1951, Niven 1956, Struik 1969, Königsberger 1990, Schröder 1993, Stevens 1999). In 1794, Legendre also proved that π^2 is IRRATIONAL (Wells 1986, p. 76). π is also TRANSCENDENTAL (Lindemann 1882). An immediate consequence of Lindemann's proof of the transcendence of π also proved that the GEOMETRIC PROBLEM OF ANTIQUITY known as CIRCLE SQUARING is impossible. A simplified, but still difficult, version of Lindemann's proof is given by Klein (1955).

It is also known that π is not a LIOUVILLE NUMBER (Mahler 1953). The following table summarizes progress in computing upper bounds on the IRRATIONALITY MEASURE for π. It is likely that the exponent can be reduced to $2 + \epsilon$, where ϵ is an infinitesimally small number (Borwein *et al.* 1989).

upper bound	reference
20	Mahler (1953), Le Lionnais (1983, p. 50)
14.65	Chudnovsky and Chudnovsky (1984)
8.0161	Hata (1992)

It is not known if $\pi + e$, π/e, or $\ln \pi$ are IRRATIONAL. However, it is known that they cannot satisfy any POLYNOMIAL equation of degree ≤ 8 with INTEGER COEFFICIENTS of average size 10^9 (Bailey 1988, Borwein *et al.* 1989).

J. H. Conway has shown that there is a sequence of fewer than 40 FRACTIONS $F_1, F_2, \ldots$ with the property that if you start with 2^n and repeatedly multiply by the first of the F_i that gives an integer result until a POWER of 2 (say, 2^k) occurs, then k is the nth decimal digit of π.

π crops up in all sorts of unexpected places in mathematics besides CIRCLES and SPHERES. For example, it occurs in the normalization of the GAUSSIAN DISTRIBUTION, in the distribution of PRIMES, in the construction of numbers which are very close to INTEGERS (the RAMANUJAN CONSTANT), and in the probability that a pin dropped on a set of PARALLEL lines intersects a line (BUFFON'S NEEDLE PROBLEM). Pi also appears as the average ratio of the actual length and the direct distance between source and mouth in a meandering river (Støllum 1996, Singh 1997).

A brief history of NOTATION for pi is given by Castellanos (1988). π is sometimes known as LUDOLPH'S CONSTANT after Ludolph van Ceulen (1539–1610), a Dutch π calculator. The symbol π was first used by English mathematician William Jones in 1706, and subsequently adopted by Euler. In *Measurement of a Circle,* Archimedes (ca. 225 BC) obtained the first rigorous approximation by INSCRIBING and CIRCUMSCRIBING $6 \cdot 2^n$-gons on a CIRCLE using the ARCHIMEDES ALGORITHM. Using $n = 4$ (a 96-gon), Archimedes obtained

$$3 + \tfrac{10}{71} < \pi < 3 + \tfrac{1}{7} \qquad (3)$$

(Wells 1986, p. 49; Shanks 1993, p. 140).

The Bible contains two references (I Kings 7:23 and Chronicles 4:2) which give a value of 3 for π (Wells 1986, p. 48). It should be mentioned, however, that both instances refer to a value obtained from physical measurements and, as such, are probably well within the bounds of experimental uncertainty. I Kings 7:23 states, "Also he made a molten sea of ten cubits from brim to brim, round in compass, and five cubits in height thereof; and a line thirty cubits did compass it round about." This implies $\pi = C/d = 30/10 = 3$. The Babylonians gave an estimate of π as $3 + 1/8 = 3.125$. The Egyptians did better still, obtaining $2^8/3^4 = 3.1605\ldots$ in the Rhind papyrus, and 22/7 elsewhere. The Chinese geometers, however, did best of all, rigorously deriving π to 6 decimal places.

There are many, many FORMULAS FOR PI, from the simple to the very complicated.

Ramanujan (1913–14) and Olds (1963) give geometric constructions for 355/113. Gardner (1966, pp. 92–93) gives a geometric construction for $3 + 16/113 = 3.1415929\ldots$. Dixon (1991) gives constructions for $6/5(1 + \phi) = 3.141640\ldots$ and $\sqrt{4 + [3 - \tan(30°)]^2} = 3.141533\ldots$. Constructions for approximations of π are approximations to CIRCLE SQUARING (which is itself impossible).

See also ALMOST INTEGER, ARCHIMEDES ALGORITHM, BRENT-SALAMIN FORMULA, BUFFON-LAPLACE NEEDLE PROBLEM, BUFFON'S NEEDLE PROBLEM, CIRCLE, CIRCUMFERENCE, DIAMETER, DIRICHLET BETA FUNCTION, DIRICHLET ETA FUNCTION, DIRICHLET LAMBDA FUNCTION, e, EULER-MASCHERONI CONSTANT, GAUSSIAN DISTRIBUTION, MACLAURIN SERIES, MACHIN'S FORMULA, MACHIN-LIKE FORMULAS, PI APPROXIMATIONS, PI CONTINUED FRACTION, PI DIGITS, PI FORMULAS, PI WORDPLAY, RADIUS, RELATIVELY PRIME, RIEMANN ZETA FUNCTION, SPHERE, TRIGONOMETRY

References

Almkvist, G. and Berndt, B. "Gauss, Landen, Ramanujan, and Arithmetic-Geometric Mean, Ellipses, π, and the Ladies Diary." *Amer. Math. Monthly* **95**, 585–608, 1988.

Almkvist, G. "Many Correct Digits of π, Revisited." *Amer. Math. Monthly* **104**, 351–353, 1997.

Arndt, J. "Cryptic Pi Related Formulas." http://www.jjj.de/hfloat/pise.dvi.

Arndt, J. and Haenel, C. *Pi: Algorithmen, Computer, Arithmetik.* Berlin: Springer-Verlag, 1998.

Assmus, E. F. "Pi." *Amer. Math. Monthly* **92**, 213–214, 1985.

Bailey, D. H. "Numerical Results on the Transcendence of Constants Involving π, e, and Euler's Constant." *Math. Comput.* **50**, 275–281, 1988a.

Bailey, D. H. "The Computation of π to 29,360,000 Decimal Digit using Borwein's' Quartically Convergent Algorithm." *Math. Comput.* **50**, 283–296, 1988b.

Bailey, D.; Borwein, P.; and Plouffe, S. "On the Rapid Computation of Various Polylogarithmic Constants." http://www.cecm.sfu.ca/~pborwein/PAPERS/P123.ps.

Ball, W. W. R. and Coxeter, H. S. M. *Mathematical Recreations and Essays, 13th ed.* New York: Dover, p. 55 and 274, 1987.

Beck, G. and Trott, M. "Calculating Pi from Antiquity to 1996." http://library.wolfram.com/demos/v4/Calculating-Pi.nb.

Beckmann, P. *A History of Pi, 3rd ed.* New York: Dorset Press, 1989.

Beeler, M. *et al.* Item 140 in Beeler, M.; Gosper, R. W.; and Schroeppel, R. *HAKMEM.* Cambridge, MA: MIT Artificial Intelligence Laboratory, Memo AIM-239, p. 69, Feb. 1972.

Berggren, L.; Borwein, J.; and Borwein, P. *Pi: A Source Book.* New York: Springer-Verlag, 1997.

Bellard, F. "Fabrice Bellard's Pi Page." http://www-stud.enst.fr/~bellard/pi/.

Berndt, B. C. *Ramanujan's Notebooks, Part IV.* New York: Springer-Verlag, 1994.

Blatner, D. *The Joy of Pi.* New York: Walker, 1997.

Blatner, D. "The Joy of Pi." http://www.joyofpi.com/.

Borwein, P. B. "Pi and Other Constants." http://www.cecm.sfu.ca/~pborwein/PISTUFF/Apistuff.html.

Borwein, J. M. "Ramanujan Type Series." http://www.cecm.sfu.ca/organics/papers/borwein/paper/html/local/omlink9/html/node1.html.

Borwein, J. M. and Borwein, P. B. *Pi & the AGM: A Study in Analytic Number Theory and Computational Complexity.* New York: Wiley, 1987a.

Borwein, J. M. and Borwein, P. B. "Ramanujan's Rational and Algebraic Series for $1/\pi$." *Indian J. Math.* **51**, 147–160, 1987b.

Borwein, J. M. and Borwein, P. B. "More Ramanujan-Type Series for $1/\pi$." In *Ramanujan Revisited.* Boston, MA: Academic Press, pp. 359–374, 1988.

Borwein, J. M. and Borwein, P. B. "Class Number Three Ramanujan Type Series for $1/\pi$." *J. Comput. Appl. Math.* **46**, 281–290, 1993.

Borwein, J. M.; Borwein, P. B.; and Bailey, D. H. "Ramanujan, Modular Equations, and Approximations to Pi, or How to Compute One Billion Digits of Pi." *Amer. Math. Monthly* **96**, 201–219, 1989.

Brown, K. S. "Rounding Up to Pi." http://www.seanet.com/~ksbrown/kmath001.htm.

Calvet, C. "First Communication. A) Secrets of Pi: Strange Things in a Mathematical Train." http://www.terravista.pt/guincho/1219/1a_index_uk.html.

Castellanos, D. "The Ubiquitous Pi. Part I." *Math. Mag.* **61**, 67–98, 1988.

Castellanos, D. "The Ubiquitous Pi. Part II." *Math. Mag.* **61**, 148–163, 1988.

Chan, J. "As Easy as Pi." *Math Horizons,* Winter 1993, pp. 18–19, 1993.

Choong, Daykin, and Rathbone. *Math. Comput.* **25**, 387, 1971.

Chudnovsky, D. V. and Chudnovsky, G. V. *Padé and Rational Approximations to Systems of Functions and Their Arithmetic Applications.* Berlin: Springer-Verlag, 1984.

Chudnovsky, D. V. and Chudnovsky, G. V. "Approximations and Complex Multiplication According to Ramanujan." In *Ramanujan Revisited: Proceedings of the Centenary Conference* (Ed. G. E. Andrews, B. C. Berndt, and R. A. Rankin). Boston, MA: Academic Press, pp. 375–472, 1987.

Conway, J. H. and Guy, R. K. "The Number π." In *The Book of Numbers.* New York: Springer-Verlag, pp. 237–239, 1996.

David, Y. "On a Sequence Generated by a Sieving Process." *Riveon Lematematika* **11**, 26–31, 1957.

Dixon, R. "The Story of Pi (π)." §4.3 in *Mathographics.* New York: Dover, pp. 44–49 and 98–101, 1991.

Dunham, W. "A Gem from Isaac Newton." Ch. 7 in *Journey through Genius: The Great Theorems of Mathematics.* New York: Wiley, pp. 106–112 and 155–183, 1990.

Exploratorium. "π Page." http://www.exploratorium.edu/learning_studio/pi/.

Finch, S. "Favorite Mathematical Constants." http://www.mathsoft.com/asolve/constant/pi/pi.html.

Flajolet, P. and Vardi, I. "Zeta Function Expansions of Classical Constants." Unpublished manuscript. 1996. http://pauillac.inria.fr/algo/flajolet/Publications/landau.ps.

Gardner, M. "Memorizing Numbers." Ch. 11 in *The Scientific American Book of Mathematical Puzzles and Diversions.* New York: Simon and Schuster, p. 103, 1959.

Gardner, M. "The Transcendental Number Pi." Ch. 8 in *Martin Gardner's New Mathematical Diversions from Scientific American.* New York: Simon and Schuster, pp. 91–102, 1966.

Gosper, R. W. *Table of Simple Continued Fraction for π and the Derived Decimal Approximation.* Stanford, CA: Artificial Intelligence Laboratory, Stanford University, Oct. 1975. Reviewed in *Math. Comput.* **31**, 1044, 1977.

Gourdon, X. and Sebah, P. "The Constant π." http://xavier.gourdon.free.fr/Constants/Pi/pi.html.

Hardy, G. H. *A Course of Pure Mathematics, 10th ed.* Cambridge, England: Cambridge University Press, 1952.

Hata, M. "Improvement in the Irrationality Measures of π and π^2." *Proc. Japan. Acad. Ser. A Math. Sci.* **68**, 283–286, 1992.

Havermann, H. "Continued Fraction expansion of Pi: 20,000,000 terms." http://www.lacim.uqam.ca/piDATA/.

Hermite, C. "Sur quelques approximations algébriques." *J. reine angew. Math.* **76**, 342–344, 1873. Reprinted in *Oeuvres complètes, Tome III.* Paris: Hermann, pp. 146–149, 1912.

Hobsen, E. W. *Squaring the Circle.* New York: Chelsea, 1988.

Kanada, Y. "New World Record of Pi: 51.5 Billion Decimal Digits." http://www.cecm.sfu.ca/personal/jborwein/Kanada_50b.html.

Klein, F. *Famous Problems.* New York: Chelsea, 1955.

Knopp, K. §32, 136, and 138 in *Theory and Application of Infinite Series.* New York: Dover, p. 238, 1990.

Königsberger, K. *Analysis 1.* Berlin: Springer-Verlag, 1990.

Laczkovich, M. "On Lambert's Proof of the Irrationality of π." *Amer. Math. Monthly* **104**, 439–443, 1997.

Lambert, J. H. "Mémoire sur quelques propriétés remarquables des quantités transcendantes circulaires et logarithmiques." *Mémoires de l'Academie des sciences de Berlin* **17**, 265–322, 1761.

Le Lionnais, F. *Les nombres remarquables.* Paris: Hermann, pp. 22 and 50, 1983.

Lindemann, F. "Uuml;ber die Zahl π." *Math. Ann.* **20**, 213–225, 1882.

Lopez, A. "Indiana Bill Sets the Value of π to 3." http://www.cs.unb.ca/~alopez-o/math-faq/mathtext/node18.html.

MacTutor Archive. "Pi Through the Ages." http://www-groups.dcs.st-and.ac.uk/~history/HistToPi_through_the_ages.html.

Mahler, K. "On the Approximation of π." *Nederl. Akad. Wetensch. Proc. Ser. A.* **56**/*Indagationes Math.* **15**, 30–42, 1953.

Nagell, T. "Irrationality of the numbers e and π." §13 in *Introduction to Number Theory.* New York: Wiley, pp. 38–40, 1951.

Niven, I. M. *Irrational Numbers.* New York: Wiley, 1956.

Ogilvy, C. S. "Pi and Pi-Makers." Ch. 10 in *Excursions in Mathematics.* New York: Dover, pp. 108–120, 1994.

Olds, C. D. *Continued Fractions.* New York: Random House, pp. 59–60, 1963.

Pappas, T. "Probability and π." *The Joy of Mathematics.* San Carlos, CA: Wide World Publ./Tetra, pp. 18–19, 1989.

Peterson, I. *Islands of Truth: A Mathematical Mystery Cruise.* New York: W. H. Freeman, pp. 178–186, 1990.

Pickover, C. A. *Keys to Infinity.* New York: Wiley, p. 62, 1995.

Plouffe, S. "Plouffe's Inverter: Table of Current Records for the Computation of Constants." http://www.lacim.uqam.ca/pi/records.html.

Plouffe, S. "1 Billion Digits of Pi." http://www.lacim.uqam.ca/piDATA/PI/.

Plouffe, S. "PiHex: A Distributed Effort to Calculate Pi." http://www.cecm.sfu.ca/projects/pihex/.

Plouffe, S. "Plouffe's Inverter: A Few Approximations of Pi." http://www.lacim.uqam.ca/pi/approxpi.html.

Plouffe, S. "The π Page." http://www.cecm.sfu.ca/pi/.

Plouffe, S. "Plouffe's Inverter: Table of Current Records for the Computation of Constants." http://www.lacim.uqam.ca/pi/records.html.

Plouffe, S. "Table of Computation of Pi from 2000 BC to Now." http://www.cecm.sfu.ca/projects/ISC/Pihistory.html.

Preston, R. "Mountains of Pi." *New Yorker* **68**, 36–67, Mar. 2, 1992. http://www.lacim.uqam.ca/plouffe/Chudnovsky.html.

Project Mathematics. "The Story of Pi." Videotape. http://www.projmath.caltech.edu/storypi.htm.

Rabinowitz, S. and Wagon, S. "A Spigot Algorithm for the Digits of π." *Amer. Math. Monthly* **102**, 195–203, 1995.

Ramanujan, S. "Modular Equations and Approximations to π." *Quart. J. Pure. Appl. Math.* **45**, 350–372, 1913–1914.

Rivera, C. "Problems & Puzzles: Puzzle The Best Approximation to Pi with Primes.-050." http://www.primepuzzles.net/puzzles/puzz_050.htm.

Rudio, F. "Archimedes, Huygens, Lambert, Legendre." In *Vier Abhandlungen über die Kreismessung.* Leipzig, Germany, 1892.

Schröder, E. M. "Zur Irrationalität von π^2 und π." *Mitt. Math. Ges. Hamburg* **13**, 249, 1993.

Shanks, D. "Dihedral Quartic Approximations and Series for π." *J. Number. Th.* **14**, 397–423, 1982.

Shanks, D. *Solved and Unsolved Problems in Number Theory, 4th ed.* New York: Chelsea, 1993.

Singh, S. *Fermat's Enigma: The Epic Quest to Solve the World's Greatest Mathematical Problem.* New York: Walker, pp. 17–18, 1997.

Sloane, N. J. A. Sequences A000796/M2218, A001203/M2646, A001901, A002485/M3097, A002486/M4456, A002491/M1009, A007509/M2061, A025547, A032510, A032523 A033089, A033090, A036903, and A046126 in in "An On-Line Version of the Encyclopedia of Integer Sequences." http://www.research.att.com/~njas/sequences/eisonline.html.

Smith, D. E. "The History and Transcendence of π." Ch. 9 in *Monographs on Topics of Modern Mathematics Relevant to the Elementary Field* (Ed. J. W. A. Young). New York: Dover, pp. 388–416, 1955.

Stevens, J. "Zur Irrationalität von π." *Mitt. Math. Ges. Hamburg* **18**, 151–158, 1999.

Støllum, H.-H. "River Meandering as a Self-Organization Process." *Science* **271**, 1710–1713, 1996.

Stoschek, E. "Modul 33: Algames with Numbers" http://marvin.sn.schule.de/~inftreff/modul33/task33.htm.

Struik, D. *A Source Book in Mathematics, 1200–1800.* Cambridge, MA: Harvard University Press, 1969.

Vardi, I. *Computational Recreations in Mathematica.* Reading, MA: Addison-Wesley, p. 159, 1991.

Viète, F. *Uriorum de rebus mathematicis responsorum,* liber VIII, 1593.

Wagon, S. "Is π Normal?" *Math. Intel.* **7**, 65–67, 1985.

Wells, D. *The Penguin Dictionary of Curious and Interesting Numbers.* Middlesex, England: Penguin Books, pp. 48–55 and 76, 1986.

Whitcomb, C. "Notes on Pi (π)." http://witcombe.sbc.edu/earthmysteries/EMPi.html.

Woon, S. C. "Problem 1441." *Math. Mag.* **68**, 72–73, 1995.

Pi Approximations

KOCHANSKY'S APPROXIMATION is the ROOT of

$$9x^4 - 240x^2 + 1492. \tag{1}$$

given by

$$\pi \approx \sqrt{\tfrac{40}{3} - \sqrt{12}} \approx 3.141533. \tag{2}$$

An approximation involving the GOLDEN MEAN is

$$\pi \approx \tfrac{6}{5}\phi^2 = \frac{6}{5}\left(\frac{\sqrt{5}+1}{2}\right)^2 = \tfrac{3}{5}\left(3+\sqrt{5}\right) = 3.14164\ldots. \tag{3}$$

Some approximations due to Ramanujan include

$$\pi \approx \frac{19\sqrt{7}}{16} \tag{4}$$

$$\approx \frac{7}{3}\left(1 + \frac{1}{5}\sqrt{3}\right) \tag{5}$$

$$\approx \frac{9}{5} + \sqrt{\frac{9}{5}} \tag{6}$$

$$\approx \left(\frac{2143}{22}\right)^{1/4}$$

$$= \left(9^2 + \frac{19^2}{22}\right)^{1/4} \tag{7}$$

$$= \left(102 - \frac{2222}{22^{22}}\right)^{1/4} \tag{8}$$

$$= \left(97 + \frac{1}{2} - \frac{1}{11}\right)^{1/4} \tag{9}$$

$$= \left(97 + \frac{9}{22}\right)^{1/4} \tag{10}$$

$$\approx \frac{63}{25}\left(\frac{17 + 15\sqrt{5}}{7 + 15\sqrt{5}}\right) \tag{11}$$

$$\approx \frac{355}{113}\left(1 - \frac{0.003}{3533}\right) \tag{12}$$

$$\approx \frac{12}{\sqrt{130}}\ln\left[\frac{(3 + \sqrt{13})(\sqrt{8} + \sqrt{10})}{2}\right] \tag{13}$$

$$\approx \frac{24}{\sqrt{142}}\ln\left[\frac{\sqrt{10 + 11\sqrt{2}} + \sqrt{10 + 7\sqrt{2}}}{2}\right] \tag{14}$$

$$\approx \frac{12}{\sqrt{190}}\ln\left[\left(3 + \sqrt{10}\right)\left(\sqrt{8} + \sqrt{10}\right)\right] \tag{15}$$

$$\approx \frac{12}{\sqrt{310}}\ln\left[\frac{1}{4}\left(3 + \sqrt{5}\right)\left(2 + \sqrt{2}\right)\right.$$
$$\left. \times \left(5 + 2\sqrt{10} + \sqrt{61 + 20\sqrt{10}}\right)\right] \tag{16}$$

$$\approx \frac{4}{\sqrt{522}}\ln\left[\left(\frac{5 + \sqrt{29}}{\sqrt{2}}\right)^3\left(5\sqrt{29} + 11\sqrt{6}\right)\right.$$
$$\left. \times \left(\sqrt{\frac{9 + 3\sqrt{6}}{4}} + \sqrt{\frac{5 + 3\sqrt{6}}{4}}\right)^6\right] \tag{17}$$

which are accurate to 3, 4, 4, 8, 8, 9, 14, 15, 15, 18, 23, 31 digits, respectively (Ramanujan 1913–1914; Hardy 1952, p. 70; Wells 1986, p. 54; Berndt 1994, pp. 48–49 and 88–89).

S. Irvine noted that (0), giving an approximation to π good to 8 digits, can be written in a form using all digits 0–9,

$$\pi \approx \left(\frac{2143}{22}\right)^{1/4} = 0 + \sqrt{\sqrt{\sqrt{3^3 + \frac{19^2}{78 - 56}}}} \tag{18}$$

(S. Plouffe). E. Pegg notes that

$$0 + 3 + \frac{1 - (9 - 8^{-5}) - 6}{7 + 2^{-4}}$$

$$= \frac{2335469214202557776949708833181535571}{743402939681157856549274558663885931} \tag{19}$$

approximates π to 9 digits.

Castellanos (1988) gives a slew of curious formulas:

$$\pi \approx (2e^3 + e^8)^{1/7} \tag{20}$$

$$\approx \left(\frac{553}{311 + 1}\right)^2 \tag{21}$$

$$\approx \left(\frac{3}{14}\right)^4\left(\frac{193}{5}\right)^2 \tag{22}$$

$$\approx \left(\frac{296}{167}\right)^2 \tag{23}$$

$$\approx \left(\frac{66^3 + 86^2}{55^3}\right)^2 \tag{24}$$

$$\approx 1.09999901 \cdot 1.19999911 \cdot 1.39999931$$
$$\cdot 1.69999961 \tag{25}$$

$$\approx \frac{47^3 + 20^3}{30^3} - 1 \tag{26}$$

$$\approx 2 + \sqrt{1 + \left(\frac{413}{750}\right)^2} \tag{27}$$

$$\approx \left(\frac{77729}{254}\right)^{1/5} \tag{28}$$

$$\approx \left(31 + \frac{62^2 + 14}{28^4}\right)^{1/3} \tag{29}$$

$$\approx \frac{1700^3 + 82^3 - 10^3 - 9^3 - 6^3 - 3^3}{69^5} \tag{30}$$

$$\approx \left(95 + \frac{93^4 + 34^4 + 17^4 + 88}{75^4}\right)^{1/4} \tag{31}$$

$$\approx \left(100 - \frac{2125^3 + 214^3 + 30^3 + 37^2}{82^5}\right)^{1/4}, \tag{32}$$

which are accurate to 3, 4, 4, 5, 6, 7, 7, 8, 9, 10, 11, 12, and 13 digits, respectively. An extremely accurate approximation due to Shanks (1982) is

$$\pi \approx \frac{6}{\sqrt{3502}} \ln(2u) + 7.37 \times 10^{-82}, \qquad (33)$$

where u is the product of four simple quartic units. A sequence of approximations due to Plouffe includes

$$\pi \approx 43^{7/23} \qquad (34)$$

$$\approx \frac{\ln 2198}{\sqrt{6}} \qquad (35)$$

$$\approx \left(\frac{13}{4}\right)^{1181/1216} \qquad (36)$$

$$\approx \frac{689}{396 \ln\left(\frac{689}{396}\right)} \qquad (37)$$

$$\approx \left(\frac{2143}{22}\right)^{1/4} \qquad (38)$$

$$\approx \sqrt{\frac{9}{67}} \ln 5280 \qquad (39)$$

$$\approx \left(\frac{63023}{30510}\right)^{1/3} + \frac{1}{4} + \frac{1}{2}\left(\sqrt{5}+1\right) \qquad (40)$$

$$\approx \frac{48}{23} \ln\left(\frac{60318}{13387}\right) \qquad (41)$$

$$\approx \left(228 + \frac{16}{1329}\right)^{1/41} + 2 \qquad (42)$$

$$\approx \frac{125}{123} \ln\left(\frac{28102}{1277}\right) \qquad (43)$$

$$\approx 276694819753963^{1/158} + 2 \qquad (44)$$

$$\approx \frac{\ln 262537412640768744}{\sqrt{163}}, \qquad (45)$$

which are accurate to 4, 5, 7, 7, 8, 9, 10, 11, 11, 11, 23, and 30 digits, respectively.

An approximation due to Stoschek using powers of two and the special number 163 (the largest HEEGNER NUMBER) is given by

$$\pi \approx \frac{2^9}{163} = \frac{512}{163} \approx 3.1411043, \qquad (46)$$

which is good to 3 digits. A fraction with small numerator and denominator which gives is close approximation to π is

$$\frac{311}{99} = 3.14141414\ldots. \qquad (47)$$

Some approximations involving the ninth roots of rational numbers include

$$\pi \approx \left(\frac{4297607660}{144171}\right)^{1/9} \qquad (48)$$

$$\pi \approx \left(\frac{4297607660}{144171}\right)^{1/9}, \qquad (49)$$

which are good to 12 and 15 digits, respectively (P. Galliani).

J. Iuliano found

$$\pi \approx \left(\frac{19}{60} + \frac{1}{\sqrt{3 \cdot 123449}}\right)^{-1}, \qquad (50)$$

which is good to 11 digits. Rivera gives other approximation formulas.

See also PI

References

Berndt, B. C. *Ramanujan's Notebooks, Part IV.* New York: Springer-Verlag, 1994.
Castellanos, D. "The Ubiquitous Pi. Part I." *Math. Mag.* **61**, 67–98, 1988.
Castellanos, D. "The Ubiquitous Pi. Part II." *Math. Mag.* **61**, 148–163, 1988.
Hardy, G. H. *A Course of Pure Mathematics, 10th ed.* Cambridge, England: Cambridge University Press, 1952.
Ramanujan, S. "Modular Equations and Approximations to π." *Quart. J. Pure. Appl. Math.* **45**, 350–372, 1913–1914.
Rivera, C. "Problems & Puzzles: Puzzle The Best Approximation to Pi with Primes.-050." http://www.primepuzzles.net/puzzles/puzz_050.htm.
Shanks, D. "Dihedral Quartic Approximations and Series for π." *J. Number. Th.* **14**, 397–423, 1982.

Pi Continued Fraction

The SIMPLE CONTINUED FRACTION for PI, which gives the "best" approximation of a given order, is [3, 7, 15, 1, 292, 1, 1, 1, 2, 1, 3, 1, 14, 2, 1, 1, 2, 2, 2, 2, ...] (Sloane's A001203; Havermann). The very large term 292 means that the CONVERGENT

$$[3, 7, 15, 1] = [3, 7, 16] = \tfrac{355}{113} = 3.1415929\ldots \qquad (1)$$

is an extremely good approximation. The first few CONVERGENTS are 22/7, 333/106, 355/113, 103993/33102, 104348/33215, ... (Sloane's A002485 and A002486). A nice expression for the third convergent of π is given by

$$\pi \approx 2[1, 1, 1, 3, 32] = \tfrac{355}{113} \approx 3.14159292\ldots \qquad (2)$$

(Stoschek).

Gosper has computed 17,001,303 terms of π's CONTINUED FRACTION (Gosper 1977, Ball and Coxeter 1987), a record which was recently upped to 20,000,000 by H. Havermann in June 1999 (Plouffe). The first occurrences of n in the CONTINUED FRACTION are 4, 9, 1, 30, 40, 32, 2, 44, 130, 100, ... (Sloane's A032523). The smallest integer which does not occur in the first 20,000,000 terms is 2297. The sequence of increasing terms in the CONTINUED FRACTION is 3, 7, 15, 292, 436, 20776, 78629, 179136, 528210, 12996958, 878783625, ... (Sloane's A033089), occurring at positions 1, 2, 3, 5, 308, 432, 28422, 156382, 267314, 453294, 11504931 ... (Sloane's A033090).

The following table gives the first few occurrences of d-digit terms in the CONTINUED FRACTION of π, counting 3 as the 0th (e.g., Choong *et al.* 1971, Beeler *et al.* 1972).

d	Sloane	Terms/Positions
1	Sloane's A048292	3, 7, 1, 1, 1, 1, 2, 1, 3, 1, 2, 1, 1, 2, ...
	Sloane's A048293	0, 1, 3, 5, 6, 7, 8, 9, 10, 11, 13, 14, ...
2	Sloane's A048294	15, 14, 84, 15, 13, 99, 12, 16, 45, 22, ...
	Sloane's A048955	2, 12, 21, 25, 27, 33, 54, 77, 80, 82, ...
3	Sloane's A048956	292, 161, 120, 127, 436, 106, 141, ...
	Sloane's A048957	4, 79, 196, 222, 307, 601, 669, 725, ...
4	Sloane's A048958	1722, 2159, 8277, 1431, 1282, 2050, ...
	Sloane's A048959	3273, 3777, 3811, 4019, 4700, 6209, ...
5	Sloane's A048960	20776, 19055, 19308, 78629, 17538, ...
	Sloane's A048961	431, 15543, 23398, 28421, 51839, ...
6	Sloane's A048962	179136, 528210, 104293, 196030, ...
	Sloane's A048963	156381, 267313, 294467, 513205, ...
7	Sloane's A048964	8093211, 1811791, 3578547, ...
	Sloane's A048965	1118727, 2782369, 2899883, ...
8	Sloane's A048966	12996958, ...
	Sloane's A048967	453293, ...
9	Sloane's A048968	878783625, ...
	Sloane's A048969	11504930, ...

The SIMPLE CONTINUED FRACTION for π does not show any obvious patterns, but clear patterns *do* emerge in the beautiful non-simple CONTINUED FRACTIONS

$$\frac{4}{\pi} = 1 + \cfrac{1^2}{2 + \cfrac{3^2}{2 + \cfrac{5^2}{2 + \cfrac{7^2}{2 + \dotsb}}}} \tag{3}$$

(Brouckner), giving convergents 1, 3/2, 15/13, 105/76, 315/263, ... (Sloane's A025547 and A007509) and

$$\frac{\pi}{2} = 1 - \cfrac{1}{3 - \cfrac{2 \cdot 3}{1 - \cfrac{1 \cdot 2}{3 - \cfrac{4 \cdot 5}{1 - \cfrac{3 \cdot 4}{3 - \cfrac{6 \cdot 7}{1 - \cfrac{5 \cdot 6}{3 - \dotsb}}}}}}} \tag{4}$$

(Stern 1833), giving convergents 1, 2/3, 4/3, 16/15, 64/45, 128/105, ... (Sloane's A001901 and A046126).

See also PI

References

Ball, W. W. R. and Coxeter, H. S. M. *Mathematical Recreations and Essays, 13th ed.* New York: Dover, p. 55 and 274, 1987.

Beeler, M. *et al.* Item 140 in Beeler, M.; Gosper, R. W.; and Schroeppel, R. *HAKMEM.* Cambridge, MA: MIT Artificial Intelligence Laboratory, Memo AIM-239, p. 69, Feb. 1972.

Choong, Daykin, and Rathbone. *Math. Comput.* **25**, 387, 1971.

Gosper, R. W. *Table of Simple Continued Fraction for π and the Derived Decimal Approximation.* Stanford, CA: Artificial Intelligence Laboratory, Stanford University, Oct. 1975. Reviewed in *Math. Comput.* **31**, 1044, 1977.

Havermann, H. "Simple Continued Fraction Expansion of Pi." http://members.home.net/hahaj/cfpi.html.

Lochs, G. "Die ersten 968 Kettenbruchnenner von π." *Monatsh. für Math.* **67**, 311–316, 1963.

Stoschek, E. "Modul 33: Algames with Numbers." http://marvin.sn.schule.de/~inftreff/modul33/task33.htm.

Pi Digits

The calculation of the π's digits has occupied mathematicians since the day of the Rhind papyrus (1500 BC). Ludolph van Ceulen spent much of his life calculating π to 35 places. Although he did not live to publish his result, it was inscribed on his gravestone. Wells (1986, p. 48) discusses a number of other calculations. The calculation of π also figures in the *Star Trek* episode "Wolf in the Fold," in which Captain Kirk and Mr. Spock force an evil entity (composed of pure energy and which feeds on fear) out of the starship *Enterprise*'s computer by commanding the computer to "compute to the last digit the value of pi," thus sending the computer into an infinite loop.

π has recently (Sep. 20, 1999) been computed to a world record $206,158,430,208 \approx 3 \cdot 2^{36}$ DECIMAL DIGITS by Y. Kanada (Kanada, Plouffe). This calculation was done using Borwein's fourth-order convergent algorithm and required 46 hours on a massively parallel 1024-processor Hitachi SR8000 supercomputer. The largest number of digits of π computing using a PC is $6,442,450,944 \approx 3 \cdot 21^{31}$ DECIMAL DIGITS by S. Kondo on Jan. 13, 2000 (Gourdon). One billion digits of π are accessible from Plouffe's web site.

Between April 19, 1998, and Feb. 9, 1999, 126 computers from eighteen different countries set a new record for calculating specific bits of π using a program written by C. Percival. The calculation took a total of about 84,500 CPU hours and was done using idle CPU cycles under Windows 95 and Windows NT. The answer, starting at the 39,999,999,999,997th bit of π is

$$1010000011111001111111110011011100011101$$

$$00010111010110010011111100000, \qquad (1)$$

so the 40 trillionth bit of π is 0 (Plouffe).

In the following, the word "digit" refers to decimal digit after the decimal point. The following table gives the starting positions for strings of n copies of the digit d.

d	n	Sloane	Positions
0	1	Sloane's A050200	32, 50, 54, 65, 71, 77, 85, 97, ...
0	2	Sloane's A050201	307, 360, 601, 602, 855, 856, 973, ...
0	3	Sloane's A050202	601, 855, 1598, 4255, 4793, 7832, ...
0	4	Sloane's A050203	13390, 17534, 17535, 37322, ...
0	5		17534, 211058, 215287, 652115, ...
0	6		1699927, 2328783, 2609392, ...
0	7		3794572, 13310436, 28970114, ...
1	1	Sloane's A050207	1, 3, 37, 40, 49, 68, 94, 95, ...
1	2	Sloane's A050208	94, 153, 154, 174, 362, 395, 427, ...
1	3	Sloane's A050209	153, 983, 3503, 3992, 4508, 6116, ...
1	4		12700, 16732, 32788, 32789, ...
1	5		32788, 120459, 141899, 255945, ...
1	6		255945, 2645268, 3218870, ...
1	7		4657555, 42408103, 70787432, ...
2	1	Sloane's A050214	6, 16, 21, 28, 33, 53, 63, 73, 76, ...
2	2	Sloane's A050215	135, 185, 484, 535, 661, 687, 824, ...
2	3		1735, 1889, 2278, 2376, 3434, ...
2	4		4902, 7964, 12486, 43405, 50271, ...
2	5		65260, 327074, 580735, 619398, ...
2	6		963024, 1637080, 1795773, ...
2	7		82599811, 88301507, ...
3	1	Sloane's A050221	9, 15, 17, 24, 25, 27, 43, 46, 64, ...
3	2	Sloane's A050222	24, 215, 230, 282, 364, 401, 503, ...
3	3		1698, 4928, 6917, 7651, 8413, ...
3	4		28467, 28468, 66846, 79979, ...
3	5		28467, 89085, 146043, 335792, ...
3	6		710100, 710101, 1129019, ...
3	7		710100, 3204765, 12469058, ...
3	8		36488176, ...
4	1	Sloane's A050229	2, 19, 23, 36, 57, 59, 60, 70, 87, ...
4	2	Sloane's A050230	59, 125, 182, 201, 217, 453, 511, ...
4	3		2707, 2928, 3476, 3809, 3866, ...
4	4		54525, 57609, 74544, 75558, ...

4	5		808650, 828499, 828500, ...
4	6		828499, 1264270, 1691163, ...
4	7		17893953, 22931745, 22931746, ...
4	8		22931745, 65122865, ...
5	1	Sloane's A050237	4, 8, 10, 31, 48, 51, 61, 90, ...
5	2	Sloane's A050238	130, 177, 178, 315, 809, 914, ...
5	3		177, 1232, 1450, 2359, 2674, 7245, ...
5	4		24466, 24467, 33172, 39861, ...
5	5		24466, 39861, 205034, 205193, ...
5	6		244453, 253209, 419997, 3517236, ...
5	7		3517236, 9325203, 10519242, ...
6	1	Sloane's A050244	7, 20, 22, 41, 69, 72, 75, 82, ...
6	2	Sloane's A050245	117, 211, 257, 276, 309, 377, 516, ...
6	3		2440, 3151, 4000, 4435, 5403, 6840, ...
6	4		21880, 29868, 32427, 43523, 48439, ...
6	5		48439, 102387, 140744, 250129, ...
6	6		252499, 3813777, 4213896, ...
6	7		8209165, 18696860, 19715001, ...
6	8		45681781, 45681782, 55616210, ...
6	9		45681781, ...
7	1	Sloane's A050253	13, 29, 39, 47, 56, 66, 96, 99, 120, ...
7	2	Sloane's A050254	559, 621, 625, 633, 739, 742, 890, ...
7	3		1589, 1590, 4575, 5241, 5242, 5322, ...
7	4		1589, 5241, 5322, 5863, 29504, ...
7	5		162248, 283693, 322347, 399579, ...
7	6		399579, 452071, 1006927, 2309218, ...
7	7		3346228, 3775287, 14233532, ...
7	8		24658601, 24658602, 82144203, ...
7	9		24658601, ...
8	1	Sloane's A050262	11, 18, 26, 34, 35, 52, 67, 74, 78, ...
8	2	Sloane's A050263	34, 204, 317, 322, 372, 472, 848, ...
8	3		4751, 4752, 4985, 5871, 6070, 6850, ...
8	4		4751, 30796, 59550, 60822, 62383, ...
8	5		213245, 222299, 222300, 493647, ...
8	6		222299, 2418533, 3019042, ...
8	7		4722613, 7820866, 19921876, ...
8	8		46663520, 46663521, ...
8	9		46663520, ...
9	1	Sloane's A050271	5, 12, 14, 30, 38, 42, 44, 45, 55, ...
9	2	Sloane's A050272	44, 79, 459, 705, 747, 762, 763, ...
9	3		762, 763, 764, 765, 2949, 7759, ...
9	4		762, 763, 764, 17988, 19437, 19446, ...
9	5		762, 763, 19446, 56988, 161862, ...
9	6		762, 193034, 1722776, 1722777, ...
9	7		1722776, 3389380, 4313727, ...
9	8		36356642, 66780105, ...

The following table gives the first few positions at which a digit d occurs n times. Note that the sequence 9999998 occurs at decimal 762 (which is sometimes called the FEYNMAN POINT; Wells 1986,

p. 51). This is the largest value of any seven digits in the first million decimals.

d	Sloane	strings of 1, 2, ... ds first occur at
0	Sloane's A050279	32, 307, 601, 13390, 17534, 1699927, ...
1	Sloane's A050280	1, 94, 153, 12700, 32788, 255945, ...
2	Sloane's A050281	6, 135, 1735, 4902, 65260, 963024, ...
3	Sloane's A050282	9, 24, 1698, 28467, 28467, 710100, ...
4	Sloane's A050283	2, 59, 2707, 54525, 808650, 828499, ...
5	Sloane's A050284	4, 130, 177, 24466, 24466, 244453, ...
6	Sloane's A050285	7, 117, 2440, 21880, 48439, 252499, ...
7	Sloane's A050286	13, 559, 1589, 1589, 162248, 399579, ...
8	Sloane's A050287	11, 34, 4751, 4751, 213245, 222299, ...
9	Sloane's A050288	5, 44, 762, 762, 762, 762, 1722776, ...

The first time the BEAST NUMBER 666 appears is decimal 2440. The digits 314159 appear at least six times in the first 10 million decimal places of π (Pickover 1995). The sequence 0123456789 occurs beginning at digits 17,387,594,880, 26,852,899,245, 30,243,957,439, 34,549,153,953, 41,952,536,161, and 43,289,964,000 (cf. Wells 1986, p. 51). The sequence 9876543210 occurs beginning at digits 21,981,157,633, 29,832,636,867, 39,232,573,648, 42,140,457,481, and 43,065,796,214. The sequence 27182818284 (the first few digits of e) occur beginning at digit 45,111,908,393. There are also interesting patterns for $1/\pi$. 0123456789 occurs at 6,214,876,462, 9876543210 occurs at 15,603,388,145 and 51,507,034,812, and 999999999999 occurs at 12,479,021,132 of $1/\pi$.

Scanning the decimal expansion of π until all n-digit numbers have occurred, the last 1-, 2-, ... digit numbers appearing are 0, 68, 483, 6716, 33394, 569540, ... (Sloane's A032510). These end at digits 32, 606, 8555, 99849, 1369564, 14118312, ... (Sloane's A036903).

The last n-digit number seen in the decimal expansion of π for $n = 1, 2, ...$ are 0, 68, 483, 6716, 33394,

569540, 1075656, ... (Sloane's A032150). The last digits of these numbers occur at positions 32, 606, 8555, 99849, ... (Sloane's A036903).

It is *not* known if π is NORMAL (Wagon 1985, Bailey and Crandall 2000), although the first 30 million DIGITS are very UNIFORMLY DISTRIBUTED (Bailey 1988). The following distribution is found for the first n DIGITS of $\pi - 3$. It shows no statistically SIGNIFICANT departure from a UNIFORM DISTRIBUTION (technically, in the CHI-SQUARED TEST, it has a value of $\chi_s^2 = 5.60$ for the first 5×10^{10} terms).

digit	1×10^5	1×10^6	6×10^9	5×10^{10}
0	9,999	99,959	599,963,005	5,000,012,647
1	10,137	99,758	600,033,260	4,999,986,263
2	9,908	100,026	599,999,169	5,000,020,237
3	10,025	100,229	600,000,243	4,999,914,405
4	9,971	100,230	599,957,439	5,000,023,598
5	10,026	100,359	600,017,176	4,999,991,499
6	10,029	99,548	600,016,588	4,999,928,368
7	10,025	99,800	600,009,044	5,000,014,860
8	9,978	99,985	599,987,038	5,000,117,637
9	9,902	100,106	600,017,038	4,999,990,486

The digits of $1/\pi$ are also very uniformly distributed ($\chi_s^2 = 7.04$), as shown in the following table.

digit	5×10^{10}
0	4,999,969,955
1	5,000,113,699
2	4,999,987,893
3	5,000,040,906
4	4,999,985,863
5	4,999,977,583
6	4,999,990,916
7	4,999,985,552
8	4,999,881,183
9	5,000,066,450

See also PI, PI FORMULAS

References

Plouffe, S. "Plouffe's Inverter: Table of Current Records for the Computation of Constants." http://www.lacim.u-qam.ca/pi/records.html.

Adamchik, V. and Wagon, S. "A Simple Formula for π." *Amer. Math. Monthly* **104**, 852–855, 1997.

Bailey, D. H. "The Computation of π to 29,360,000 Decimal Digit using Borwein's' Quartically Convergent Algorithm." *Math. Comput.* **50**, 283–296, 1988.

Bailey, D.; Borwein, P.; and Plouffe, S. "On the Rapid Computation of Various Polylogarithmic Constants." http://www.cecm.sfu.ca/~pborwein/PAPERS/P123.ps.

Bailey, D. H. and Crandall, R. E. "On the Random Character of Fundamental Constant Expansions." Manuscript, Mar. 2000.

Caldwell, C. K. and Dubner, H. "Primes in Pi." *J. Recr. Math.* **29**, 282–289, 1998.

Gourdon, X. and Sebah, P. "PiFast: The Fastest Program to Compute Pi." http://xavier.gourdon.free.fr/Constants/Pi-Program/pifast.html.

Kanada, Y. "Our Latest Record." Sep. 20, 1999. ftp://www.cc.u-tokyo.ac.jp/README.our_latest_record.

Le Lionnais, F. *Les nombres remarquables.* Paris: Hermann, pp. 22 and 50, 1983.

Pickover, C. A. *Keys to Infinity.* New York: Wiley, p. 62, 1995.

Plouffe, S. "1 Billion Digits of Pi." http://www.lacim.uqam.ca/piDATA/PI/.

Rabinowitz, S. and Wagon, S. "A Spigot Algorithm for the Digits of π." *Amer. Math. Monthly* **102**, 195–203, 1995.

Sloane, N. J. A. Sequences A032150 and A036903 in "An On-Line Version of the Encyclopedia of Integer Sequences." http://www.research.att.com/~njas/sequences/eisonline.html.

Smith, H. J. "Computing Pi." http://pweb.netcom.com/~hjsmith/Pi.html.

Wagon, S. "Is π Normal?" *Math. Intel.* **7**, 65–67, 1985.

Wells, D. *The Penguin Dictionary of Curious and Interesting Numbers.* Middlesex, England: Penguin Books, p. 46, 1986.

Wrench, J. W. Jr. "The Evolution of Extended Decimal Approximations to π." *Math. Teacher* **53**, 644–650, 1960.

Pi Formulas

A method similar to Archimedes' can be used to estimate π by starting with an n-gon and then relating the AREA of subsequent $2n$-gons. Let β be the ANGLE from the center of one of the POLYGON's segments,

$$\beta = \tfrac{1}{4}(n-3)\pi, \tag{1}$$

then

$$\pi = \frac{2\sin(2\beta)}{(n-3)\prod_{k=0}^{\infty}\cos(2^{-k}\beta)} \tag{2}$$

(Beckmann 1989, pp. 92–94). Viète (1593) was the first to give an exact expression for π by taking $n = 4$ in the above expression, giving

$$\cos\beta = \sin\beta = \frac{1}{\sqrt{2}} = \tfrac{1}{2}\sqrt{2}, \tag{3}$$

which leads to an INFINITE PRODUCT of NESTED RADICALS,

$$\frac{2}{\pi} = \sqrt{\tfrac{1}{2}}\sqrt{\tfrac{1}{2}+\tfrac{1}{2}\sqrt{\tfrac{1}{2}}}\sqrt{\tfrac{1}{2}+\tfrac{1}{2}\sqrt{\tfrac{1}{2}+\tfrac{1}{2}\sqrt{\tfrac{1}{2}}}}\cdots \tag{4}$$

(Wells 1986, p. 50; Beckmann 1989, p. 95). However, this expression was not rigorously proved to converge until Rudio (1892). A related formula is given by

$$\pi = \lim_{n\to\infty} 2^n \underbrace{\sqrt{2 - \sqrt{2 + \sqrt{2 + \sqrt{2 + \ldots + \sqrt{2}}}}}}_{n}, \tag{5}$$

where the square root term can be written using the iteration

$$\pi_n = \sqrt{\left(\tfrac{1}{2}\pi_{n-1}\right)^2 + \left[1 - \sqrt{1 - \left(\tfrac{1}{2}\pi_{n-1}\right)^2}\right]^2}, \tag{6}$$

where $\pi_0 = \sqrt{2}$ (J. Munkhammer). The formula

$$\pi = 2 \lim_{m\to\infty}$$

$$\times \sum_{n=1}^{m} \sqrt{\left[\sqrt{1 - \left(\frac{n-1}{m}\right)^2} - \sqrt{1 - \left(\frac{n}{m}\right)^2}\right]^2 + \frac{1}{m^2}} \tag{7}$$

is also closely related.

Another exact FORMULA is MACHIN'S FORMULA, which is

$$\frac{\pi}{4} = 4\tan^{-1}\left(\tfrac{1}{5}\right) - \tan^{-1}\left(\tfrac{1}{239}\right). \tag{8}$$

There are three other MACHIN-LIKE FORMULAS, as well as other FORMULAS with more terms. An interesting INFINITE PRODUCT formula due to Euler which relates π and the nth PRIME p_n is

$$\pi = \frac{2}{\prod_{i=n}^{\infty}\left[1 + \dfrac{\sin\left(\tfrac{1}{2}\pi p_n\right)}{p_n}\right]} \tag{9}$$

$$= \frac{2}{\prod_{i=n}^{\infty}\left[1 + \dfrac{(-1)^{(p_n-1)/2}}{p_n}\right]} \tag{10}$$

(Blatner 1997, p. 119), plotted below as a function of the number of terms in the product.

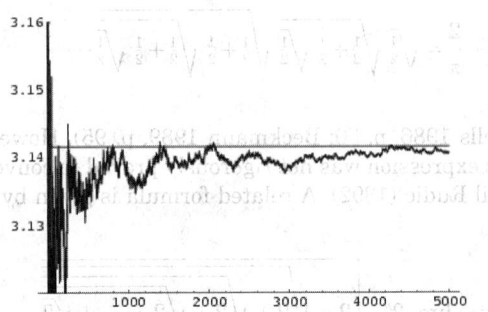

The AREA and CIRCUMFERENCE of the UNIT CIRCLE are given by

$$A = \pi = 4 \int_0^1 \sqrt{1 - x^2} \, dx \tag{11}$$

$$= \lim_{n \to \infty} \frac{4}{n^2} \sum_{k=0}^n \sqrt{n^2 - k^2} \tag{12}$$

and

$$C = 2\pi = 4 \int_0^1 \frac{dx}{\sqrt{1 - x^2}} \tag{13}$$

$$= 4 \int_0^1 \sqrt{1 + \left(\frac{d}{x} \sqrt{1 - x^2} \right)^2} \, dx. \tag{14}$$

The SURFACE AREA and VOLUME of the unit SPHERE are

$$S = 4\pi \tag{15}$$

$$V = \tfrac{4}{3}\pi. \tag{16}$$

Beginning with any POSITIVE INTEGER n, round up to the nearest multiple of $n-1$, then up to the nearest multiple of $n-2$, and so on, up to the nearest multiple of 1. Let $f(n)$ denote the result. Then the ratio

$$\lim_{n \to \infty} \frac{n^2}{f(n)} = \pi \tag{17}$$

(Brown). David (1957) credits this result to Jabotinski and Erdos and gives the more precise asymptotic result

$$f(n) = \frac{n^2}{\pi} + \mathcal{O}\left(n^{4/3}\right). \tag{18}$$

The first few numbers in the sequence $\{f(n)\}$ are 1, 2, 4, 6, 10, 12, 18, 22, 30, 34, ... (Sloane's A002491). A particular case of the WALLIS FORMULA gives

$$\frac{\pi}{2} = \prod_{n=1}^\infty \left[\frac{(2n)^2}{(2n-1)(2n+1)} \right]$$

$$= \frac{2 \cdot 2 \, 4 \cdot 4 \, 6 \cdot 6}{1 \cdot 3 \, 3 \cdot 5 \, 5 \cdot 7} \cdots \tag{19}$$

(Wells 1986, p. 50). This formula can also be written

$$\lim_{n \to \infty} \frac{2^{4n}}{n \binom{2n}{n}^2} = \pi \quad \lim_{n \to \infty} \frac{n[\Gamma(n)]^2}{\left[\Gamma\left(\tfrac{1}{2} + n \right) \right]^2} = \pi, \tag{20}$$

where $\binom{n}{k}$ denotes a BINOMIAL COEFFICIENT and $\Gamma(x)$ is the GAMMA FUNCTION (Knopp 1990). Euler obtained

$$\pi = \sqrt{6 \left(1 + \frac{1}{2^2} + \frac{1}{3^2} + \frac{1}{4^2} + \cdots \right)}, \tag{21}$$

which follows from the special value of the RIEMANN ZETA FUNCTION $\zeta(2) = \pi^2/6$. Similar FORMULAS follow from $\zeta(2n)$ for all POSITIVE INTEGERS n. Gregory and Leibniz found

$$\frac{\pi}{4} = 1 - \frac{1}{3} + \frac{1}{5} + \cdots \tag{22}$$

(Wells 1986, p. 50), which is sometimes known as GREGORY'S FORMULA or the LEIBNIZ SERIES. The error after the nth term of this series in GREGORY'S FORMULA is larger than $(2n)^{-1}$ so this sum converges so slowly that 300 terms are not sufficient to calculate π correctly to two decimal places! However, it can be transformed to

$$\pi = \sum_{k=1}^\infty \frac{3^k - 1}{4^k} \zeta(k+1), \tag{23}$$

where $\zeta(z)$ is the RIEMANN ZETA FUNCTION (Vardi 1991, pp. 157–158; Flajolet and Vardi 1996), so that the error after k terms is $\approx (3/4)^k$. In 1666, Newton used

$$\pi = \tfrac{3}{4}\sqrt{3} + 24 \int_0^{1/4} \sqrt{x - x^2} \, dx \tag{24}$$

$$= \frac{3\sqrt{3}}{4} + 24 \left(\frac{1}{12} - \frac{1}{5 \cdot 2^5} - \frac{1}{28 \cdot 2^7} - \frac{1}{72 \cdot 2^9} - \cdots \right) \tag{25}$$

(Wells 1986, p. 50; Borwein *et al.* 1989). The coefficients can be found from the integral

$$I(x) = \int \sqrt{x - x^2} \, dx$$

$$= \tfrac{1}{4}(2x - 1)\sqrt{x - x^2} - \tfrac{1}{8} \sin^{-1}(1 - 2x) \tag{26}$$

by taking the series expansion of $I(x) - I(0)$ about 0, obtaining

$$I(x) = \tfrac{2}{3} x^{3/2} - \tfrac{1}{5} x^{5/2} - \tfrac{1}{28} x^{7/2} - \tfrac{1}{72} x^{9/2} - \tfrac{5}{704} x^{11/2} + \cdots \tag{27}$$

(Sloane's A054387 and A054388). Using Euler's CON-VERGENCE IMPROVEMENT transformation gives

$$\frac{\pi}{2} = \frac{1}{2} \sum_{n=0}^{\infty} \frac{(n!)^2 2^{n+1}}{(2n+1)!} = \sum_{n=0}^{\infty} \frac{n!}{(2n+1)!!}$$

$$= 1 + \frac{1}{3} + \frac{1 \cdot 2}{3 \cdot 5} + \frac{1 \cdot 2 \cdot 3}{3 \cdot 5 \cdot 7} + \cdots \tag{28}$$

$$= 1 + \frac{1}{3}\left(1 + \frac{2}{5}\left(1 + \frac{3}{7}\left(1 + \frac{4}{9}(1 + \ldots)\right)\right)\right) \tag{29}$$

(Beeler *et al.* 1972, Item 120). This corresponds to plugging $x = 1/\sqrt{2}$ into the POWER SERIES for the HYPERGEOMETRIC FUNCTION $_2F_1(a, b; c; x)$,

$$\frac{\sin^{-1} x}{\sqrt{1-x^2}} = \sum_{i=0}^{\infty} \frac{(2x)^{2i+1}(i!)^2}{2(2i+1)!} = {}_2F_1\left(1, 1; \tfrac{3}{2}; x^2\right)x. \tag{30}$$

Despite the convergence improvement, series (29) converges at only one bit/term. At the cost of a SQUARE ROOT, Gosper has noted that $x = 1/2$ gives 2 bits/term,

$$\tfrac{1}{9}\sqrt{3}\pi = \frac{1}{2} \sum_{i=0}^{\infty} \frac{(i!)^2}{(2i+1)!}, \tag{31}$$

and $x = \sin(\pi/10)$ gives almost 3.39 bits/term,

$$\frac{\pi}{5\sqrt{\phi+2}} = \frac{1}{2} \sum_{i=0}^{\infty} \frac{(i!)^2}{\phi^{2i+1}(2i+1)!}, \tag{32}$$

where ϕ is the GOLDEN RATIO. Gosper also obtained

$$\pi = 3 + \frac{1}{60}\left(8 + \frac{2 \cdot 3}{7 \cdot 8 \cdot 3}\left(13 + \frac{3 \cdot 5}{10 \cdot 11 \cdot 3}\right.\right.$$
$$\left.\left. \times \left(18 + \frac{4 \cdot 7}{13 \cdot 14 \cdot 3}(23 + \ldots)\right)\right)\right). \tag{33}$$

An infinite sum due to Ramanujan is

$$\frac{1}{\pi} = \sum_{n=0}^{\infty} \binom{2n}{n}^3 \frac{42n+5}{2^{12n+4}} \tag{34}$$

(Borwein *et al.* 1989). Further sums are given in Ramanujan (1913–14),

$$\frac{4}{\pi} = \sum_{n=0}^{\infty} \frac{(-1)^n(1123 + 21460n)(2n-1)!!(4n-1)!!}{882^{2n+1}32^n(n!)^3} \tag{35}$$

and

$$\frac{1}{\pi} = \sqrt{8} \sum_{n=0}^{\infty} \frac{(1103 + 26390n)(2n-1)!!(4n-1)!!}{99^{4n+2}32^n(n!)^3}$$

$$= \frac{\sqrt{8}}{9801} \sum_{n=0}^{\infty} \frac{(4n)!(1103 + 26390n)}{(n!)^4 396^{4n}} \tag{36}$$

(Beeler *et al.* 1972, Item 139; Borwein *et al.* 1989). Equation (36) is derived from a modular identity of order 58, although a first derivation was not pre-

sented prior to Borwein and Borwein (1987). The above series both give

$$\pi \approx \frac{9801}{2206\sqrt{2}} = 3.14159273001\ldots \tag{37}$$

(Wells 1986, p. 54) as the first approximation and provide, respectively, about 6 and 8 decimal places per term. Such series exist because of the rationality of various modular invariants. The general form of the series is

$$\sum_{n=0}^{\infty} [a(t) + nb(t)] \frac{(6n)!}{(3n)!(n!)^3} \frac{1}{[j(t)]^n} = \frac{\sqrt{-j(t)}}{\pi}, \tag{38}$$

where t is a QUADRATIC FORM DISCRIMINANT, $j(t)$ is the J-FUNCTION,

$$b(t) = \sqrt{t[1728 - j(t)]} \tag{39}$$

$$a(t) = \frac{b(t)}{6}\left\{1 - \frac{E_4(t)}{E_6(t)}\left[E_2(t) - \frac{6}{\pi\sqrt{t}}\right]\right\}, \tag{40}$$

and the E_i are RAMANUJAN-EISENSTEIN SERIES. A CLASS NUMBER p field involves pth degree ALGEBRAIC INTEGERS of the constants $A = a(t)$, $B = b(t)$, and $C = c(t)$. The fastest converging series that uses only INTEGER terms corresponds to the largest CLASS NUMBER 1 discriminant of $d = -163$ and was formulated by the Chudnovsky brothers (1987). The 163 appearing here is the same one appearing in the fact that $e^{\pi\sqrt{163}}$ (the RAMANUJAN CONSTANT) is very nearly an INTEGER. The series is given by

$$\frac{1}{\pi} = 12 \sum_{n=0}^{\infty} \frac{(-1)^n(6n)!(13591409 + 545140134n)}{(n!)^3(3n)!(640320^3)^{n+1/2}}$$

$$= \frac{163 \cdot 8 \cdot 27 \cdot 7 \cdot 11 \cdot 19 \cdot 127}{640320^{3/2}}$$

$$\times \sum_{n=0}^{\infty}\left(\frac{13591409}{163 \cdot 2 \cdot 9 \cdot 7 \cdot 11 \cdot 19 \cdot 127} + n\right)$$

$$\times \frac{(6n)!}{(3n)!(n!)^3}\frac{(-1)^n}{640320^{3n}} \tag{41}$$

(Borwein and Borwein 1993). This series gives 14 digits accurately per term. The same equation in another form was given by the Chudnovsky brothers (1987) and is used by *Mathematica* to calculate π (Vardi 1991),

$$\pi =$$

$$\frac{426880\sqrt{10005}}{A\left[{}_3F_2\left(\tfrac{1}{6}, \tfrac{1}{2}, \tfrac{5}{6}; 1, 1; B\right) - C\,{}_3F_2\left(\tfrac{7}{6}, \tfrac{3}{2}, \tfrac{11}{6}; 2, 2; B\right)\right]}, \tag{42}$$

where

$$A \equiv 13591409 \tag{43}$$

$$B \equiv -\frac{1}{151931373056000} \tag{44}$$

$$C \equiv \frac{30285563}{1651969144908540723200}. \tag{45}$$

The best formula for CLASS NUMBER 2 (largest discriminant -427) is

$$\frac{1}{\pi} = 12 \sum_{n=0}^{\infty} \frac{(-1)^n (6n)!(A + Bn)}{(n!)^3 (3n)! C^{n+1/2}}, \tag{46}$$

where

$$A \equiv 212175710912\sqrt{61} + 1657145277365 \tag{47}$$

$$B \equiv 13773980892672\sqrt{61} + 107578229802750 \tag{48}$$

$$C \equiv \left[5280\left(236674 + 30303\sqrt{61}\right)\right]^3 \tag{49}$$

(Borwein and Borwein 1993). This series adds about 25 digits for each additional term. The fastest converging series for CLASS NUMBER 3 corresponds to $d = -907$ and gives 37–38 digits per term. The fastest converging CLASS NUMBER 4 series corresponds to $d = -1555$ and is

$$\frac{\sqrt{-C^3}}{\pi} = \sum_{n=0}^{\infty} \frac{(6n)!}{(3n)!(n!)^3} \frac{A + nB}{C^{3n}}, \tag{50}$$

where

$$A = 63365028312971999585426220$$
$$+ 28337702140800842046825600\sqrt{5}$$
$$+ 384\sqrt{5}(10891728551171782004674\ldots$$
$$\ldots 362123952091603856560 17$$
$$+ 487902908657881022\ldots$$
$$\ldots 5077338534541688721351255040\sqrt{4})^{1/2} \tag{51}$$

$$B = 7849910453496627210289749000$$
$$+ 3510586678260932028965606400\sqrt{5}$$
$$+ 2515968\sqrt{3110}(62602083237890016\ldots$$
$$\ldots 3699332265444 4020882161$$
$$+ 2799650273060444296\ldots$$
$$\ldots 5772068907188251 90235\sqrt{5})^{1/2} \tag{52}$$

$$C = -214772995063512240$$
$$- 96049403338648032\sqrt{5}$$

$$- 1296\sqrt{5}(109852345794635503237133 18473$$
$$+ 491274625369236275460739591 2\sqrt{5})^{1/2}, \tag{53}$$

This gives 50 digits per term. Borwein and Borwein (1993) have developed a general ALGORITHM for generating such series for arbitrary CLASS NUMBER. Bellard gives the exotic formula

$$\pi = \frac{1}{740025} \left[\sum_{n=1}^{\infty} \frac{3P(n)}{\binom{7n}{2n} 2^{n-1}} - 20379280 \right], \tag{54}$$

where

$$P(n) \equiv -885673181 n^5 + 3125347237 n^4$$
$$- 2942969225 n^3 + 1031962795 n^2$$
$$- 196882274 n + 10996648. \tag{55}$$

A complete listing of Ramanujan's series for $1/\pi$ found in his second and third notebooks is given by Berndt (1994, pp. 352–354),

$$\frac{4}{\pi} = \sum_{n=0}^{\infty} \frac{(6n + 1)\left(\frac{1}{2}\right)_n^3}{4^n (n!)^3} \tag{56}$$

$$\frac{16}{\pi} = \sum_{n=0}^{\infty} \frac{(42n + 5)\left(\frac{1}{2}\right)_n^3}{(64)^n (n!)^3} \tag{57}$$

$$\frac{32}{\pi} = \sum_{n=0}^{\infty} \frac{(42\sqrt{5}n + 5\sqrt{5} + 30n - 1)\left(\frac{1}{2}\right)_n^3}{(64)^n (n!)^3}$$
$$\times \left(\frac{\sqrt{5} - 1}{2}\right)^{8n} \tag{58}$$

$$\frac{27}{4\pi} = \sum_{n=0}^{\infty} \frac{(15n + 2)\left(\frac{1}{2}\right)_n \left(\frac{1}{3}\right)_n \left(\frac{2}{3}\right)_n}{(n!)^3} \left(\frac{2}{27}\right)^n \tag{59}$$

$$\frac{15\sqrt{3}}{2\pi} = \sum_{n=0}^{\infty} \frac{(33n + 4)\left(\frac{1}{2}\right)_n \left(\frac{1}{3}\right)_n \left(\frac{2}{3}\right)_n}{(n!)^3} \left(\frac{4}{125}\right)^n \tag{60}$$

$$\frac{5\sqrt{5}}{2\pi\sqrt{3}} = \sum_{n=0}^{\infty} \frac{(11n + 1)\left(\frac{1}{2}\right)_n \left(\frac{1}{6}\right)_n \left(\frac{5}{6}\right)_n}{(n!)^3} \left(\frac{4}{125}\right)^n \tag{61}$$

$$\frac{85\sqrt{85}}{18\pi\sqrt{3}} = \sum_{n=0}^{\infty} \frac{(133n + 8)\left(\frac{1}{2}\right)_n \left(\frac{1}{6}\right)_n \left(\frac{5}{6}\right)_n}{(n!)^3} \left(\frac{4}{85}\right)^n \tag{62}$$

$$\frac{4}{\pi} = \sum_{n=0}^{\infty} \frac{(-1)^n (20n + 3)\left(\frac{1}{2}\right)_n \left(\frac{1}{4}\right)_n \left(\frac{3}{4}\right)_n}{(n!)^3 2^{2n+1}} \tag{63}$$

$$\frac{4}{\pi\sqrt{3}} = \sum_{n=0}^{\infty} \frac{(-1)^n (28n + 3)\left(\frac{1}{2}\right)_n \left(\frac{1}{4}\right)_n \left(\frac{3}{4}\right)_n}{(n!)^3 3^n 4^{n+1}} \tag{64}$$

$$\frac{4}{\pi} = \sum_{n=0}^{\infty} \frac{(-1)^n (260n + 23)\left(\frac{1}{2}\right)_n \left(\frac{1}{4}\right)_n \left(\frac{3}{4}\right)_n}{(n!)^3 (18)^{2n+1}} \tag{65}$$

$$\frac{4}{\pi\sqrt{5}} = \sum_{n=0}^{\infty} \frac{(-1)^n (644n + 41)\left(\frac{1}{2}\right)_n \left(\frac{1}{4}\right)_n \left(\frac{3}{4}\right)_n}{(n!)^3 5^n (72)^{2n+1}} \tag{66}$$

$$\frac{4}{\pi} = \sum_{n=0}^{\infty} \frac{(-1)^n (21460n + 1123)\left(\frac{1}{2}\right)_n \left(\frac{1}{4}\right)_n \left(\frac{3}{4}\right)_n}{(n!)^3 (882)^{2n+1}} \tag{67}$$

$$\frac{2\sqrt{3}}{\pi} = \sum_{n=0}^{\infty} \frac{(8n + 1)^n \left(\frac{1}{2}\right)_n \left(\frac{1}{4}\right)_n \left(\frac{3}{4}\right)_n}{(n!)^3 9^n} \tag{68}$$

$$\frac{1}{2\pi\sqrt{2}} = \sum_{n=0}^{\infty} \frac{(10n + 1)^n \left(\frac{1}{2}\right)_n \left(\frac{1}{4}\right)_n \left(\frac{3}{4}\right)_n}{(n!)^3 9^{2n+1}} \tag{69}$$

$$\frac{1}{3\pi\sqrt{3}} = \sum_{n=0}^{\infty} \frac{(40n + 3)\left(\frac{1}{2}\right)_n \left(\frac{1}{4}\right)_n \left(\frac{3}{4}\right)_n}{(n!)^3 (49)^{2n+1}} \tag{70}$$

$$\frac{2}{\pi\sqrt{11}} = \sum_{n=0}^{\infty} \frac{(280n + 19)\left(\frac{1}{2}\right)_n \left(\frac{1}{4}\right)_n \left(\frac{3}{4}\right)_n}{(n!)^3 (99)^{2n+1}} \tag{71}$$

$$\frac{1}{2\pi\sqrt{2}} = \sum_{n=0}^{\infty} \frac{(26390n + 1103)\left(\frac{1}{2}\right)_n \left(\frac{1}{4}\right)_n \left(\frac{3}{4}\right)_n}{(n!)^3 (99)^{4n+2}}. \tag{72}$$

These equations were first proved by Borwein and Borwein (1987, pp. 177–187). Borwein and Borwein (1987b, 1988, 1993) proved other equations of this type, and Chudnovsky and Chudnovsky (1987) found similar equations for other transcendental constants.

Another identity is

$$\pi^2 = 36 \operatorname{Li}_2\left(\frac{1}{2}\right) - 36 \operatorname{Li}_2\left(\frac{1}{4}\right) - 12 \operatorname{Li}_2\left(\frac{1}{8}\right) + 6 \operatorname{Li}_2\left(\frac{1}{64}\right), \tag{73}$$

where L_n is the POLYLOGARITHM. (73) is equivalent to

$$\frac{\pi^2}{36} = \sum_{i=1}^{\infty} \frac{a_i}{2^i i^2} \quad \{a_i\} = \overline{[1, -3, -2, -3, 1, 0]} \tag{74}$$

and

$$\pi^2 = 12 L_2\left(\frac{1}{2}\right) + 6(\ln 2)^2 \tag{75}$$

(Bailey *et al.* 1995).

A SPIGOT ALGORITHM for π is given by Rabinowitz and Wagon (1995). More amazingly still, a closed form expression giving a DIGIT-EXTRACTION ALGORITHM which produces digits of π (or π^2) in base-16 was recently discovered by Bailey *et al.* (Bailey *et al.* 1995, Adamchik and Wagon 1997),

$$\pi = \sum_{n=0}^{\infty} \left(\frac{4}{8n + 1} - \frac{2}{8n + 4} - \frac{1}{8n + 5} - \frac{1}{8n + 6} \right) \left(\frac{1}{16} \right)^n. \tag{76}$$

This formula, sometimes called the BAILEY-BORWEIN-PLOUFFE ALGORITHM can also be written using the shorthand notation

$$\pi = \sum_{i=1}^{\infty} \frac{p_i}{16^{\lfloor i/8 \rfloor} i} \tag{77}$$

$$\{p_i\} = \{4, 0, 0, -2, -1, -1, 0, 0\},$$

where $\{p_i\}$ is given by the periodic sequence obtained by appending copies of $\{4, 0, 0, -2, -1, -1, 0, 0\}$ (in other words, $p_i \equiv p_{[(i-1) \pmod 8]+1}$ for $i > 8$) and $\lfloor x \rfloor$ is the FLOOR FUNCTION. This expression was discovered using the PSLQ ALGORITHM (Ferguson *et al.* 1999) and is equivalent to

$$\pi = \int_0^1 \frac{16y - 16}{y^4 - 2y^3 + 4y - 4} \, dy. \tag{78}$$

A similar formula was subsequently discovered by Ferguson, leading to a 2-D lattice of such formulas which can be generated by these two formulas. A related integral is

$$\pi = \frac{22}{7} - \int_0^1 \frac{x^4(1 - x)^4}{1 + x^2} \, dx \tag{79}$$

(Le Lionnais 1983, p. 22). F. Bellard found the more rapidly converging DIGIT-EXTRACTION ALGORITHM (in HEXADECIMAL)

$$\pi = \frac{1}{2^6} \sum_{n=0}^{\infty} \frac{(-1)^n}{2^{10n}} \left(-\frac{2^5}{4n + 1} - \frac{1}{4n + 3} + \frac{2^8}{10n + 1} \right.$$
$$\left. - \frac{2^6}{10n + 3} - \frac{2}{10n + 5} - \frac{2^2}{10n + 7} + \frac{1}{10n + 9} \right). \tag{80}$$

This formula can be generalized to

$$\pi = \sum_{k=0}^{\infty} \left(\frac{4 + 8r}{8k + 1} - \frac{8r}{8k + 2} - \frac{4r}{8k + 3} - \frac{2 + 8r}{8k + 4} \right.$$
$$\left. - \frac{1 + 2r}{8k + 5} - \frac{1 + 2r}{8k + 6} + \frac{r}{8k + 7} \right) \left(\frac{1}{16} \right)^k \tag{81}$$

for any complex value of r (Adamchik and Wagon), giving the Bailey-Borwein-Plouffe algorithm as the special case $r = 0$.

Related formulas are

$$\pi^2 = \frac{1}{8} \sum_{k=0}^{\infty} \frac{1}{64^k} \left[\frac{144}{(6k + 1)^2} - \frac{216}{(6k + 2)^2} - \frac{72}{(6k + 3)^2} \right.$$
$$\left. - \frac{54}{(6k + 4)^2} + \frac{9}{(6k + 5)^2} \right] \tag{82}$$

and

$$\pi^2 = \sum_{k=0}^{\infty} \frac{1}{16^k} \left[\frac{16}{(8k+1)^2} - \frac{16}{(8k+2)^2} - \frac{8}{(8k+3)^2} \right.$$
$$\left. - \frac{16}{(8k+4)^2} - \frac{4}{(8k+5)^2} - \frac{4}{(8k+6)^2} + \frac{2}{(8k+7)^2} \right]$$
(83)

(Bailey *et al.* 1995, Bailey and Plouffe). More amazingly still, S. Plouffe has devised an algorithm to compute the nth DIGIT of π in any base in $\mathcal{O}(n^3(\log n)^3)$ steps.

A slew of additional identities due to Ramanujan , Catalan, and Newton are given by Castellanos (1988, pp. 86–88), including several involving sums of FIBONACCI NUMBERS. Ramanujan found

$$\sum_{k=0}^{\infty} \frac{(-1)^k(4k+1)[(2k-1)!!]^3}{[(2k)!!]^3}$$
$$= \sum_{k=0}^{\infty} \frac{(-1)^k(4k+1)\left[\Gamma\left(k+\frac{1}{2}\right)\right]^3}{\pi^{3/2}[\Gamma(k+1)]^3} = \frac{2}{\pi}$$
(84)

(Hardy 1923; Hardy 1924; Hardy 1999, p. 7). Gasper quotes the result

$$\pi = \frac{16}{3} \left[\lim_{x \to \infty} x \,_1F_2\left(\tfrac{1}{2};\ 2,\ 3;\ -x^2\right) \right]^{-1},$$
(85)

where $_1F_2$ is a GENERALIZED HYPERGEOMETRIC FUNCTION, and transforms it to

$$\pi = \lim_{x \to \infty} 4x \,_1F_2\left(\tfrac{1}{2};\ \tfrac{3}{2},\ \tfrac{3}{2};\ -x^2\right),$$
(86)

Fascinating results due to Gosper include

$$\lim_{n \to \infty} \prod_{i=n}^{2n} \frac{\pi}{2\tan^{-1} i} = 4^{1/\pi} = 1.554682275\ldots$$
(87)

and

$$\sum_{n=1}^{\infty} \frac{1}{n^2} \cos\left(\frac{9}{n\pi + \sqrt{n^2\pi^2 - 9}}\right) = -\frac{\pi^2}{12e^3}$$
$$= -0.040948222\ldots$$
(88)

Gosper also gives the curious identity

$$\frac{1}{e} \prod_{n=1}^{\infty} \left(\frac{1}{3n} + 1\right)^{3n+1/2}$$
$$= \frac{3 \cdot 3^{1/24} \sqrt{\left(\tfrac{1}{3}\right)!}}{2^{5/6} \exp\left[\frac{\gamma}{3} - \frac{\pi\sqrt{3}}{18} + \frac{\sqrt{3}_1\left(\tfrac{1}{3}\right)}{12\pi} - \frac{2\zeta'(2)}{\pi^2}\right] \pi^{5/6}}$$
$$= 1.01237855722912\ldots$$
(89)

Another curious fact is the ALMOST INTEGER

$$e^\pi - \pi = 19.999099979\ldots,$$
(90)

which can also be written as

$$(\pi + 20)^i = -0.9999999992 - 0.0000388927i \approx -1 \quad (91)$$

$$\cos(\ln(\pi + 20)) \approx -0.9999999992. \quad (92)$$

Applying COSINE a few more times gives

$$\cos(\pi \cos(\pi \cos(\ln(\pi + 20))))$$
$$\approx -1 + 3.9321609261 \times 10^{-35}. \quad (93)$$

π may also be computed using iterative ALGORITHMS. A quadratically converging ALGORITHM due to Borwein is

$$x_0 = \sqrt{2}$$
(94)

$$\pi_0 = 2 + \sqrt{2}$$
(95)

$$y_1 = 2^{1/4}$$
(96)

and

$$x_{n+1} = \frac{1}{2}\left(\sqrt{x_n} + \frac{1}{\sqrt{x_n}}\right)$$
(97)

$$y_{n+1} = \frac{y_n\sqrt{x_n} + \dfrac{1}{\sqrt{x_n}}}{y_n + 1}$$
(98)

$$\pi_n = \pi_{n-1}\frac{x_n + 1}{y_n + 1}.$$
(99)

π_n decreases monotonically to π with

$$\pi_n - \pi < 10^{-2+1}$$
(100)

for $n \geq 2$. The BRENT-SALAMIN FORMULA is another quadratically converging algorithm which can be used to calculate π. A quadratically convergent algorithm for $\pi/\ln 2$ based on an observation by Salamin is given by defining

$$f(k) = k2^{-k/4}\left[\sum_{n=1}^{\infty} 2^{-k\binom{n}{2}}\right]^2,$$
(101)

then writing

$$g_0 \equiv \frac{f(n)}{f(2n)}.$$
(102)

Now iterate

$$g_k = \sqrt{\frac{1}{2}\left(g_{k-1} + \frac{1}{g_{k+1}}\right)}$$
(103)

to obtain

$$\pi = 2(\ln 2)f(n) \prod_{k=1}^{\infty} g_k. \tag{104}$$

A cubically converging ALGORITHM which converges to the nearest multiple of π to f_0 is the simple iteration

$$f_n = f_{n-1} + \sin(f_{n-1}) \tag{105}$$

(Beeler *et al.* 1972). For example, applying to 23 gives the sequence

$$\{23, 22.1537796, 21.99186453, 21.99114858, \ldots\}, \tag{106}$$

which converges to $7\pi \approx 21.99114858$.

A quartically converging ALGORITHM is obtained by letting

$$y_0 = \sqrt{2} - 1 \tag{107}$$

$$\alpha = 6 - 4\sqrt{2}, \tag{108}$$

then defining

$$y_{n+1} = \frac{1 - (1 - y_n^4)^{1/4}}{1 + (1 - y_n^4)^{1/4}} \tag{109}$$

$$\alpha_{n+1} = (1 + y_{n+1})^4 \alpha_n - 2^{2n+3} y_{n+1} (1 + y_{n+1} + y_{n+1}^2). \tag{110}$$

Then

$$\pi = \lim_{n \to \infty} \frac{1}{\alpha_n} \tag{111}$$

and α_n converges to $1/\pi$ quartically with

$$\alpha_n - \frac{1}{\pi} < 16 \cdot 4^n e^{-2\pi \cdot 4^n} \tag{112}$$

(Borwein and Borwein 1987, Bailey 1988, Borwein *et al.* 1989). This ALGORITHM rests on a MODULAR EQUATION identity of order 4.

A quintically converging ALGORITHM is obtained by letting

$$s_0 = 5\left(\sqrt{5} - 2\right) \tag{113}$$

$$\alpha_0 = \frac{1}{2}. \tag{114}$$

Then let

$$s_{n+1} = \frac{25}{\left(z + \dfrac{x}{z} + 1\right)^2 s_n}, \tag{115}$$

where

$$x = \frac{5}{s_n} - 1 \tag{116}$$

$$y = (x - 1)^2 + 7 \tag{117}$$

$$z = \left[\frac{1}{2} x \left(y + \sqrt{y^2 - 4x^3}\right)\right]^{1/5}. \tag{118}$$

Finally, let

$$\alpha_{n+1} = s_n^2 \alpha_n - 5^n \left[\frac{1}{2}(s_n^2 - 5) + \sqrt{s_n(s_n^2 - 2s_n + 5)}\right], \tag{119}$$

then

$$0 < \alpha_n - \frac{1}{\pi} < 16 \cdot 5^n e^{-\pi 5^n} \tag{120}$$

(Borwein *et al.* 1989). This ALGORITHM rests on a MODULAR EQUATION identity of order 5.

Another ALGORITHM is due to Woon (1995). Define $a(0) \equiv 1$ and

$$a(n) = \sqrt{1 + \left[\sum_{k=0}^{n-1} a(k)\right]^2}. \tag{121}$$

It can be proved by induction that

$$a(n) = \csc\left(\frac{\pi}{2^{n+1}}\right). \tag{122}$$

For $n = 0$, the identity holds. If it holds for $n \le t$, then

$$a(t + 1) = \sqrt{1 + \left[\sum_{k=0}^{t} \csc\left(\frac{\pi}{2^{k+1}}\right)\right]^2}, \tag{123}$$

but

$$\csc\left(\frac{\pi}{2^{k+1}}\right) = \cot\left(\frac{\pi}{2^{k+2}}\right) - \cot\left(\frac{\pi}{2^{k+1}}\right), \tag{124}$$

so

$$\sum_{k=0}^{t} \csc\left(\frac{\pi}{2^{k+1}}\right) = \cot\left(\frac{\pi}{2^{t+2}}\right). \tag{125}$$

Therefore,

$$a(t + 1) = \csc\left(\frac{\pi}{2^{t+2}}\right), \tag{126}$$

so the identity holds for $n = t + 1$ and, by induction, for all NONNEGATIVE n, and

$$\lim_{n \to \infty} \frac{2^{n+1}}{a(n)} = \lim_{n \to \infty} 2^{n+1} \sin\left(\frac{\pi}{2^{n+1}}\right)$$

$$= \lim_{n \to \infty} 2^{n+1} \frac{\pi}{2^{n+1}} \frac{\sin\left(\dfrac{\pi}{2^{n+1}}\right)}{\dfrac{\pi}{2^{n+1}}}.$$

$$= \pi \lim_{\theta \to 0} \frac{\sin \theta}{\theta} = \pi. \tag{127}$$

Additional series in which π appears are

$$\tfrac{1}{4}\pi\sqrt{2} = 1 + \tfrac{1}{3} - \tfrac{1}{5} - \tfrac{1}{7} + \tfrac{1}{9} + \tfrac{1}{11} - \cdots \tag{128}$$

$$\tfrac{1}{4}(\pi - 3) = \frac{1}{2 \cdot 3 \cdot 4} - \frac{1}{4 \cdot 5 \cdot 6} + \frac{1}{6 \cdot 7 \cdot 8} - \cdots \tag{129}$$

$$\frac{\pi^2}{8} = 1 + \frac{1}{3^2} + \frac{1}{5^2} + \frac{1}{7^2} + \cdots \tag{130}$$

(Wells 1986, p. 53).

Other iterative ALGORITHMS are the ARCHIMEDES ALGORITHM, which was derived by Pfaff in 1800, and the BRENT-SALAMIN FORMULA. Borwein *et al.* (1989) discuss pth order iterative algorithms.

π satisfies the INEQUALITY

$$\left(1 + \frac{1}{\pi}\right)^{\pi+1} \approx 3.14097 < \pi. \tag{131}$$

See also PI

References

Adamchik, V. and Wagon, S. "A Simple Formula for π." *Amer. Math. Monthly* **104**, 852–855, 1997.

Adamchik, V. and Wagon, S. "Pi: A 2000-Year Search Changes Direction." http://members.wri.com/victor/articles/pi.html.

Bailey, D. H. "Numerical Results on the Transcendence of Constants Involving π, e, and Euler's Constant." *Math. Comput.* **50**, 275–281, 1988a.

Bailey, D. H. "The Computation of π to 29,360,000 Decimal Digit using Borwein's Quartically Convergent Algorithm." *Math. Comput.* **50**, 283–296, 1988b.

Bailey, D. H.; Borwein, P.; and Plouffe, S. "On the Rapid Computation of Various Polylogarithmic Constants." *Math. Comput.* **66**, 903–913, 1997.

Beckmann, P. *A History of Pi, 3rd ed.* New York: Dorset Press, 1989.

Beeler, M. *et al.* Item 140 in Beeler, M.; Gosper, R. W.; and Schroeppel, R. *HAKMEM.* Cambridge, MA: MIT Artificial Intelligence Laboratory, Memo AIM-239, p. 69, Feb. 1972.

Berndt, B. C. *Ramanujan's Notebooks, Part IV.* New York: Springer-Verlag, 1994.

Blatner, D. *The Joy of Pi.* New York: Walker, 1997.

Borwein, J. M. and Borwein, P. B. *Pi & the AGM: A Study in Analytic Number Theory and Computational Complexity.* New York: Wiley, 1987.

Borwein, J. M. and Borwein, P. B. "Ramanujan's Rational and Algebraic Series for $1/\pi$." *Indian J. Math.* **51**, 147–160, 1987b.

Borwein, J. M. and Borwein, P. B. "More Ramanujan-Type Series for $1/\pi$." In *Ramanujan Revisited.* Boston, MA: Academic Press, pp. 359–374, 1988.

Borwein, J. M.; Borwein, P. B.; and Bailey, D. H. "Ramanujan, Modular Equations, and Approximations to Pi, or How to Compute One Billion Digits of Pi." *Amer. Math. Monthly* **96**, 201–219, 1989.

Borwein, J. M. and Borwein, P. B. "Class Number Three Ramanujan Type Series for $1/\pi$." *J. Comput. Appl. Math.* **46**, 281–290, 1993.

Brown, K. S. "Rounding Up to Pi." http://www.seanet.com/~ksbrown/kmath001.htm.

Castellanos, D. "The Ubiquitous Pi. Part I." *Math. Mag.* **61**, 67–98, 1988.

Castellanos, D. "The Ubiquitous Pi. Part II." *Math. Mag.* **61**, 148–163, 1988.

Chudnovsky, D. V. and Chudnovsky, G. V. "Approximations and Complex Multiplication According to Ramanujan." In *Ramanujan Revisited: Proceedings of the Centenary Conference* (Ed. G. E. Andrews, B. C. Berndt, and R. A. Rankin). Boston, MA: Academic Press, pp. 375–472, 1987.

David, Y. "On a Sequence Generated by a Sieving Process." *Riveon Lematematika* **11**, 26–31, 1957.

Ferguson, H. R. P.; Bailey, D. H.; and Arno, S. "Analysis of PSLQ, An Integer Relation Finding Algorithm." *Math. Comput.* **68**, 351–369, 1999.

Finch, S. "Unsolved Mathematics Problems: The Miraculous Bailey-Borwein-Plouffe Pi Algorithm." http://www.mathsoft.com/asolve/plouffe/plouffe.html.

Flajolet, P. and Vardi, I. "Zeta Function Expansions of Classical Constants." Unpublished manuscript. 1996. http://pauillac.inria.fr/algo/flajolet/Publications/landau.ps.

Hardy, G. H. "Some Formulae of Ramanujan." *Proc. London Math. Soc.* (Records of Proceedings at Meetings) **22**, xii–xiii, 1924.

Hardy, G. H. "A Chapter from Ramanujan's Note-Book." *Proc. Cambridge Philos. Soc.* **21**, 492–503, 1923.

Hardy, G. H. *Ramanujan: Twelve Lectures on Subjects Suggested by His Life and Work, 3rd ed.* New York: Chelsea, 1999.

Ramanujan, S. "Modular Equations and Approximations to π." *Quart. J. Pure. Appl. Math.* **45**, 350–372, 1913–1914.

Sloane, N. J. A. Sequences A054387 and A054388 in "An On-Line Version of the Encyclopedia of Integer Sequences." http://www.research.att.com/~njas/sequences/eisonline.html.

Vardi, I. *Computational Recreations in Mathematica.* Reading, MA: Addison-Wesley, p. 159, 1991.

Viète, F. *Uriorum de rebus mathematicis responsorum,* liber VIII, 1593.

Wells, D. *The Penguin Dictionary of Curious and Interesting Numbers.* Middlesex, England: Penguin Books, 1986.

Woon, S. C. "Problem 1441." *Math. Mag.* **68**, 72–73, 1995.

Pi Heptomino

A HEPTOMINO in the shape of the Greek character PI.

Pi Wordplay

A short mnemonic for remembering the first eight DECIMAL DIGITS of π is "May I have a large container of coffee?" giving 3.1415926 (Gardner 1959; Gardner 1966, p. 92; Eves 1990, p. 122, Davis 1993, p. 9). "But

I must a while endeavour to reckon right" gives nine correct digits (3.1.4159265). A more substantial mnemonic giving 15 digits (3.14159265358979) is "How I want a drink, alcoholic of course, after the heavy lectures involving quantum mechanics," originally due to Sir James Jeans (Gardner 1966, p. 92; Castellanos 1988, p. 152; Eves 1990, p. 122; Davis 1993, p. 9; Blatner 1997, p. 112). A slight extension of this adds the phrase "All of thy geometry, Herr Planck, is fairly hard," giving 24 digits in all (3.14159265358979323846264).

An even more extensive rhyming mnemonic giving 31 digits is "Now I will a rhyme construct, By chosen words the young instruct. Cunningly devised endeavour, Con it and remember ever. Widths in circle here you see, Sketched out in strange obscurity." (Note that the British spelling of "endeavour" is required here.)

The following stanzas are the first part of a poem written by M. Keith based on Edgar Allen Poe's "The Raven." The entire poem gives 740 digits; the fragment below gives only the first 80 (Blatner 1997, p. 113). Words with ten letters represent the digit 0, and those with 11 or more digits are taken to represent two digits.

Poe, E.: Near a Raven.

Midnights so dreary, tired and weary.

Silently pondering volumes extolling all by-now obsolete lore.

During my rather long nap-the weirdest tap!

An ominous vibrating sound disturbing my chamber's antedoor.

'This,' I whispered quietly, 'I ignore.' Perfectly, the intellect remembers: the ghostly fires, a glittering ember.

Inflamed by lightning's outbursts, windows cast penumbras upon this floor. Sorrowful, as one mistreated, unhappy thoughts I heeded:

That inimitable lesson in elegance–Lenore–

Is delighting, exciting... nevermore.

An extensive collection of π mnemonics in many languages is maintained by A. P. Hatzipolakis. Other mnemonics in various languages are given by Castellanos (1988) and Blatner (1997, pp. 112–118).

Keith (1999) considered the set of letters obtained by writing π to base 26 with digits $0 = A$, $1 = B$, ..., $25 = Z$, so that

$$\pi = D.DRSQLOLYRTRODNLHNQTG \ldots.$$

Then the sequence of the first Webster-sanctioned n-letter words in this expression is given by o, lo, rod, trod, steel, oxygen, subplot, ... Additional 6-letter words are: prinky, Libyan, and thingy. The positions of the starting letter of the first n-letter words are 6, 5, 11, 10, 6570, 11582, 115042, ...

See also PI

References

Blatner, D. *The Joy of Pi.* New York: Walker, 1997.
Castellanos, D. "The Ubiquitous Pi. Part II." *Math. Mag.* **61**, 148–163, 1988.
Davis, D. M. *The Nature and Power of Mathematics.* Princeton, NJ: Princeton University Press, 1993.
Eves, H. *An Introduction to the History of Mathematics, 6th ed.* Philadelphia, PA: Saunders, 1990.
Gardner, M. "Memorizing Numbers." Ch. 11 in *The Scientific American Book of Mathematical Puzzles and Diversions.* New York: Simon and Schuster, p. 103, 1959.
Gardner, M. "The Transcendental Number Pi." Ch. 8 in *Martin Gardner's New Mathematical Diversions from Scientific American.* New York: Simon and Schuster, pp. 91–102, 1966.
Hatzipolakis, A. P. "PiPhilology." http://users.hol.gr/~xpolakis/piphil.html.
Keith, M. "The Pi Code." *Word Ways* **32**, Nov. 1999.
Sallows, L. "Base 27: The Key to a New Gematria." *Word Ways* **26**, 67–77, May 1993.

Piano Mover's Problem

N.B. A detailed online essay by S. Finch was the starting point for this entry.

Given an open subset U in n-D space and two compact subsets C_0 and C_1 of U, where C_1 is derived from C_0 by a continuous motion, is it possible to move C_0 to C_1 while remaining entirely inside U?

See also MOVING LADDER CONSTANT, MOVING SOFA CONSTANT

References

Buchberger, B.; Collins, G. E.; and Kutzler, B. "Algebraic Methods in Geometry." *Annual Rev. Comput. Sci.* **3**, 85–119, 1988.
Feinberg, E. B. and Papadimitriou, C. H. "Finding Feasible Points for a Two-point Body." *J. Algorithms* **10**, 109–119, 1989.
Finch, S. "Favorite Mathematical Constants." http://www.mathsoft.com/asolve/constant/sofa/sofa.html.
Leven, D. and Sharir, M. "An Efficient and Simple Motion Planning Algorithm for a Ladder Moving in Two-Dimensional Space Amidst Polygonal Barriers." *J. Algorithms* **8**, 192–215, 1987.

Picard Variety

Let V be a VARIETY, and write $G(V)$ for the set of divisors, $G_l(V)$ for the set of divisors linearly equivalent to 0, and $G_a(V)$ for the group of divisors algebraically equal to 0. Then $G_a(V)/G_l(V)$ is called the Picard variety. The ALBANESE VARIETY is dual to the Picard variety.

See also ALBANESE VARIETY

References

Iyanaga, S. and Kawada, Y. (Eds.). *Encyclopedic Dictionary of Mathematics.* Cambridge, MA: MIT Press, p. 75, 1980.

Picard's Existence Theorem

If f is a continuous function that satisfies the LIPSCHITZ CONDITION

$$|f(x,\,t)-f(y,\,t)|\le L|x-y|$$

in a surrounding of $(x_0,\,t_0)\in\Omega\subset\mathbb{R}\times\mathbb{R}^n=\{(x,\,t):|x-x_0|<b,\,|t-t_0|<a\}$, then the differential equation

$$\frac{df}{dx}=f(x,\,t)$$

$$x(t_0)=x_0$$

has a unique solution $x(t)$ in the interval $|t-t_0|<d$, where $d=\min(a,\,b/B)$, min denotes the MINIMUM, $B=\sup|f(t,\,x)|$, and sup denotes the SUPREMUM.

See also LIPSCHITZ CONDITION, ORDINARY DIFFERENTIAL EQUATION

Picard's Great Theorem

Every nonconstant ENTIRE FUNCTION attains every complex value with at most one exception (Apostol 1997). Furthermore, every ANALYTIC FUNCTION assumes every complex value, with possibly one exception, infinitely often in any NEIGHBORHOOD of an ESSENTIAL SINGULARITY.

See also ANALYTIC FUNCTION, ESSENTIAL SINGULARITY, NEIGHBORHOOD, PICARD'S LITTLE THEOREM

References

Apostol, T. M. "Application to Picard's Theorem." §2.9 in *Modular Functions and Dirichlet Series in Number Theory, 2nd ed.* New York: Springer-Verlag, pp. 43–44, 1997.
Krantz, S. G. "Picard's Great Theorem." §10.5.3 in *Handbook of Complex Analysis.* Boston, MA: Birkhäuser, p. 140, 1999.

Picard's Little Theorem

Any ENTIRE ANALYTIC FUNCTION whose RANGE omits two points must be a CONSTANT FUNCTION.

Of course, an ENTIRE FUNCTION that omits a single point from its range need not be a constant, as illustrated by the function e^z, which is entire but omits the point $z=0$ from its range.

See also ENTIRE FUNCTION, PICARD'S GREAT THEOREM

References

Krantz, S. G. "Picard's Little Theorem." §10.5.2 in *Handbook of Complex Analysis.* Boston, MA: Birkhäuser, p. 140, 1999.

Picard's Theorem

PICARD'S GREAT THEOREM

Pick's Formula

PICK'S THEOREM

Pick's Theorem

Let A be the AREA of a simply closed LATTICE POLYGON. Let B denote the number of LATTICE POINTS on the EDGES and I the number of points in the interior of the POLYGON. Then

$$A=I+\tfrac{1}{2}B-1.$$

The FORMULA has been generalized to 3-D and higher dimensions using EHRHART POLYNOMIALS.

See also BLICHFELDT'S THEOREM, EHRHART POLYNOMIAL, LATTICE POINT, MINKOWSKI CONVEX BODY THEOREM

References

Coxeter, H. S. M. *Introduction to Geometry, 2nd ed.* New York: Wiley, p. 209, 1969.
DeTemple, D. "Pick's Formula: A Retrospective." *Math. Notes Washington State Univ.* **32**, Nov. 1989.
Diaz, R. and Robins, S. "Pick's Formula via the Weierstrass $\wp$-Function." *Amer. Math. Monthly* **102**, 431–437, 1995.
Ewald, G. *Combinatorial Convexity and Algebraic Geometry.* New York: Springer-Verlag, 1996.
Gardner, M. *The Sixth Book of Mathematical Games from Scientific American.* Chicago, IL: University of Chicago Press, p. 215, 1984.
Grünbaum, B. and Shephard, G. C. "Pick's Theorem." *Amer. Math. Monthly* **100**, 150–161, 1993.
Haigh, G. "A 'Natural' Approach to Pick's Theorem." *Math. Gaz.* **64**, 173–, 1980.
Hammer, J. *Unsolved Problems Concerning Lattice Points.* London: Pitman, 1977.
Kelley, D. A. "Areas of Simple Polygons." *Pentagon* **20**, 3–11, 1960.
Khan, M. R. "A Counting Formula for Primitive Tetrahedra in Z^3." *Amer. Math. Monthly* **106**, 525–533, 1999.
Morelli, R. "Pick's Theorem and the Todd Class of a Toric Variety." *Adv. Math.* **100**, 183–231, 1993.
Niven, I. and Zuckerman, H. S. "Lattice Points and Polygonal Area." *Amer. Math. Monthly* **74**, 1195, 1967.
Pick, G. "Geometrisches zur Zahlentheorie." *Sitzenber. Lotos (Prague)* **19**, 311–319, 1899.
Steinhaus, H. "O polu figur płaskich." *Przeglad Mat.-Fiz.*, 1924.
Steinhaus, H. *Mathematical Snapshots, 3rd ed.* New York: Dover, pp. 96–98, 1999.
Wells, D. *The Penguin Dictionary of Curious and Interesting Geometry.* London: Penguin, pp. 183–184, 1991.

Picone's Theorem

Let $f(x)$ be integrable in $[-1,\,1]$, let $(1-x^2)f(x)$ be of bounded variation in $[-1,\,1]$, let M' denote the least upper bound of $|f(x)(1-x^2)|$ in $[-1,\,1]$, and let V' denote the total variation of $f(x)(1-x^2)$ in $[-1,\,1]$. Given the function

$$F(x)=F(-1)+\int_1^x f(x)\,dx,$$

then the terms of its LEGENDRE SERIES

$$F(x)\sim\sum_{n=0}^{\infty}a_nP_n(x)$$

$$a_n = \tfrac{1}{2}(2n+1)\int_{-1}^{1} F(x)P_n(x)\,dx,$$

where $P_n(x)$ is a LEGENDRE POLYNOMIAL, satisfy the inequalities

$$|a_n P_n(x)| < \begin{cases} 8\sqrt{\dfrac{2}{\pi}}\,\dfrac{M'+V'}{(1-\delta^2)^{1/4}}\,n^{-3/2} & \text{for } |x| \le \delta < 1 \\[2mm] 2(M'+V')n^{-1} & \text{for } |x| \le 1 \end{cases}$$

for $n \ge 1$ (Sansone 1991).

See also JACKSON'S THEOREM, LEGENDRE SERIES

References

Picone, M. *Appunti di Analise Superiore.* Naples, Italy, p. 260, 1940.
Sansone, G. *Orthogonal Functions, rev. English ed.* New York: Dover, pp. 203–205, 1991.

PID

A popular acronym for "PRINCIPAL IDEAL DOMAIN." In engineering circles, the acronym PID refers to the "PROPORTIONAL-INTEGRAL-DERIVATIVE METHOD" algorithm for controlling systems.

See also PRINCIPAL IDEAL DOMAIN, PRINCIPAL IDEAL RING, PROPORTIONAL-INTEGRAL-DERIVATIVE METHOD

Pidduck Polynomial

Polynomials $P_k(x)$ which form the SHEFFER SEQUENCE for

$$g(t) = \frac{2t}{e^t - 1} \qquad (1)$$

$$f(t) = \frac{e^t - 1}{e^t + 1} \qquad (2)$$

and have GENERATING FUNCTION

$$\sum_{k=0}^{\infty} \frac{P_k(x)}{k!}\, t^k = \frac{t}{1-t}\left(\frac{1+t}{1-t}\right)^x. \qquad (3)$$

The first few are

$$\begin{aligned} P_0(x) &= 1 \\ P_1(x) &= 2x+1 \\ P_2(x) &= 4x^2\,4x+2 \\ P_3(x) &= 8x^3 + 12x^2 + 16x + 6. \end{aligned}$$

The Pidduck polynomials are related to the MITTAG-LEFFLER POLYNOMIALS $M_n(x)$ by

$$P_n(x) = \tfrac{1}{2}(e^t+1)M_n(x) \qquad (4)$$

(Roman 1984, p. 127).

See also MITTAG-LEFFLER POLYNOMIAL, SHEFFER SEQUENCE

References

Bateman, H. "The Polynomial of Mittag-Leffler." *Proc. Nat. Acad. Sci. USA* **26**, 491–496, 1940.
Boas, R. P. and Buck, R. C. *Polynomial Expansions of Analytic Functions, 2nd print., corr.* New York: Academic Press, p. 38, 1964.
Erdélyi, A.; Magnus, W.; Oberhettinger, F.; and Tricomi, F. G. *Higher Transcendental Functions, Vol. 3.* New York: Krieger, p. 248, 1981.
Roman, S. *The Umbral Calculus.* New York: Academic Press, 1984.

Pie Chart

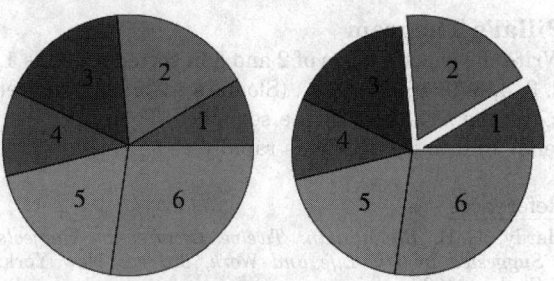

A chart made by plotting the numeric values of a set of quantities as a set of adjacent circular wedges with arc lengths proportional to the total amount. All wedges taken together comprise an entire disk. One or more segments are slightly separated from the disk center for emphasis in a so-called "exploded" pie chart.

See also BAR CHART, HISTOGRAM

References

Kenney, J. F. and Keeping, E. S. *Mathematics of Statistics, Pt. 1, 3rd ed.* Princeton, NJ: Van Nostrand, p. 23, 1962.

Pie Cutting

CIRCLE DIVISION BY LINES, CYLINDER CUTTING, PANCAKE THEOREM, PIZZA THEOREM

Piecewise Circular Curve

A curve composed exclusively of circular ARCS.

See also ARC, FLOWER OF LIFE, LENS, REULEAUX POLYGON, REULEAUX TRIANGLE, SALINON, SEED OF LIFE, TRIANGLE ARCS, YIN-YANG

References

Banchoff, T. and Giblin, P. "On The Geometry Of Piecewise Circular Curves." *Amer. Math. Monthly* **101**, 403–416, 1994.

Piecewise Continuous

A function or curve is piecewise continuous if it is CONTINUOUS on all but a finite number of points at which certain matching conditions are sometimes required.

See also CONTINUOUS, CONTINUOUS FUNCTION

Pigeonhole Principle

DIRICHLET'S BOX PRINCIPLE

Pillai's Conjecture

For every $k > 1$, there exist only finite many pairs of POWERS (p, p') with p and p' NATURAL NUMBERS and $k = p' - p$.

References

Ribenboim, P. "Catalan's Conjecture." *Amer. Math. Monthly* **103**, 529–538, 1996.

Pillai's Theorem

Write the exact powers of 2 and 3 in sorted order as 1, 2, 3, 4, 8, 9, 16, 27, 32, ... (Sloane's A006899), and let u_n be the nth term in the sequence. Then $u_{n+1} - u_n$ tends to infinity nearly as rapidly as u_n.

References

Hardy, G. H. *Ramanujan: Twelve Lectures on Subjects Suggested by His Life and Work, 3rd ed.* New York: Chelsea, 1999.
Pillai. *J. Indian Math. Soc.* **19**, 1–11, 1931.
Sloane, N. J. A. Sequences A006899/M0588 in "An On-Line Version of the Encyclopedia of Integer Sequences." http://www.research.att.com/~njas/sequences/eisonline.html.

Pilot Vector

VECTOR SPHERICAL HARMONIC

Pinch Point

A singular point such that every NEIGHBORHOOD of the point intersects itself. Pinch points are also called Whitney singularities or branch points.

Pincherle Derivative

Let $\mathbf{x} : p(x) \to xp(x)$, then for any operator T,

$$T' = T\mathbf{x} - \mathbf{x}T$$

is called the Pincherle derivative of T. If T is a SHIFT-INVARIANT OPERATOR, then its Pincherle derivative is also a SHIFT-INVARIANT OPERATOR.

References

Pincherle, S. "Operatori lineari e coefficienti di fattoriali." *Alti Accad. Naz. Lincei, Rend. Cl. Fis. Mat. Nat. (6)* **18**, 417–519, 1933.
Rota, G.-C.; Kahaner, D.; Odlyzko, A. "On the Foundations of Combinatorial Theory. VIII: Finite Operator Calculus." *J. Math. Anal. Appl.* **42**, 684–760, 1973.

Pinching Theorem

Let $g(x) \le f(x) \le h(x)$ for all x in some OPEN INTERVAL containing a. If

$$\lim_{x \to a} g(x) = \lim_{x \to a} h(x) = L,$$

then $\lim_{x \to a} f(x) = L$.

See also LIMIT, SQUEEZING THEOREM

Pine Cone Number

FIBONACCI NUMBER

Piriform

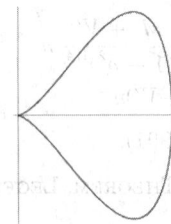

A plane curve also called the PEG TOP and given by the CARTESIAN equation

$$a^4 y^2 = b^2 x^3 (2a - x) \tag{1}$$

and the parametric curves

$$x = a(1 + \sin t) \tag{2}$$

$$y = b \cos t (1 + \sin t) \tag{3}$$

for $t \in [-\pi/2, \pi/2]$. It was studied by G. de Long-champs in 1886. The generalization to a QUARTIC 3-D surface

$$(x^4 - x^3) + y^2 + z^2 = 0, \tag{4}$$

is shown below (Nordstrand).

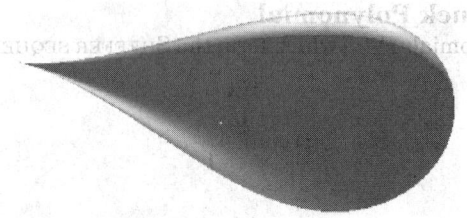

See also BUTTERFLY CURVE, DUMBBELL CURVE, EIGHT CURVE, HEART SURFACE, PEAR CURVE

References

Cundy, H. and Rollett, A. *Mathematical Models, 3rd ed.* Stradbroke, England: Tarquin Pub. p. 71, 1989.
Lawrence, J. D. *A Catalog of Special Plane Curves.* New York: Dover, pp. 148–150, 1972.
Nordstrand, T. "Surfaces." http://www.uib.no/people/nfytn/surfaces.htm.

Pisot Constant

PISOT-VIJAYARAGHAVAN CONSTANT

Pisot-Vijayaraghavan Constant

Let θ be a number greater than 1, λ a POSITIVE number, and

$$\text{frac}(x) \equiv x - \lfloor x \rfloor \tag{1}$$

denote the FRACTIONAL PART of x, where $\lfloor x \rfloor$ is the

FLOOR FUNCTION. Then for a given λ, the sequence of numbers $\mathrm{frac}(\lambda\theta^n)$ for $n = 1, 2, \ldots$ is an EQUIDISTRIBUTED SEQUENCE in the interval $(0, 1)$ when θ does not belong to a λ-dependent exceptional set S of MEASURE ZERO (Koksma 1935). Pisot (1938) and Vijayaraghavan (1941) independently studied the exceptional values of θ, and Salem (1943) proposed calling such values Pisot-Vijayaraghavan numbers.

Pisot (1938) proved that if θ is chosen such that there exists a $\lambda \neq 0$ for which the series

$$\sum_{n=0}^{\infty} \sin^2(\pi\lambda\theta)^n \qquad (2)$$

converges, then θ is an ALGEBRAIC INTEGER whose conjugates all (except for itself) have modulus < 1, and λ is an ALGEBRAIC INTEGER of the FIELD $K(\theta)$. Vijayaraghavan (1940) proved that the set of Pisot-Vijayaraghavan numbers has infinitely many LIMIT POINTS.

Salem (1944) proved that the set of Pisot-Vijayaraghavan constants is closed. The proof of this theorem is based on the LEMMA that for a Pisot-Vijayaraghavan constant θ, there always exists a number λ such that $1 \leq \lambda < \theta$ and the following inequality is satisfied,

$$\sum_{n=0}^{\infty} \sin^2(\pi\lambda\theta^n) \leq \frac{\pi^2(2\theta + 1)^2}{(\theta - 1)^2}. \qquad (3)$$

The smallest Pisot-Vijayaraghavan constant is given by the POSITIVE ROOT $\theta_0 \approx 1.32372$ of

$$x^3 - x - 1 = 0. \qquad (4)$$

This number was identified as the smallest known by Salem (1944), and proved to be the smallest possible by Siegel (1944). Siegel also identified the next smallest Pisot-Vijayaraghavan constant θ_1 as the root of

$$x^4 - x^3 - 1 = 0. \qquad (5)$$

showed that θ_1 and θ_2 are isolated in S, and showed that the roots of each POLYNOMIAL

$$x^n(x^2 - x - 1) + x^2 - 1 \quad n = 1, 2, 3, \ldots \qquad (6)$$

$$x^n - \frac{x^{n+1} - 1}{x^2 - 1} \quad n = 3, 5, 7, \ldots \qquad (7)$$

$$x^n - \frac{x^{n-1} - 1}{x - 1} \quad n = 3, 5, 7, \ldots \qquad (8)$$

belong to S, where $\theta_0 = \phi$ (the GOLDEN MEAN) is the accumulation point of the set (in fact, the smallest; Le Lionnais 1983, p. 40).

Some small Pisot-Vijayaraghavan constants and their POLYNOMIALS are given in the following table. The latter two entries are from Boyd (1977).

k	number	order	POLYNOMIAL
0	1.3247179572	3	1 0 -1 -1
1	1.3802775691	4	1 -1 0 0 -1
	1.6216584885	16	1 -2 2 -3 2 -2 1 0 0 1 -1 2 - 2 2 -2 1 -1
	1.8374664495	20	1 -2 0 1 -1 0 1 -1 0 1 0 -1 0 1 -1 0 1 -1 0 1 -1

All the points in S less than ϕ are known (Dufresnoy and Pisot 1955). Each point of S is a limit point from both sides of the set T of SALEM CONSTANTS (Salem 1945).

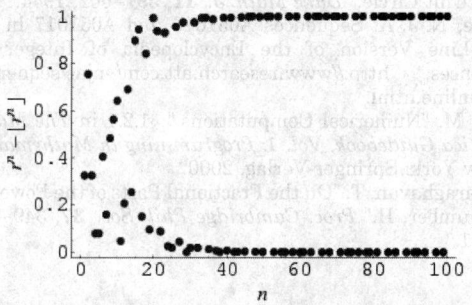

Pisot-Vijayaraghavan constants give rise to ALMOST INTEGERS. For example, the larger the power to which θ_0 is taken, the closer $\theta_0^n - \lfloor\theta_0^n\rfloor$, where $\lfloor x \rfloor$ is the FLOOR FUNCTION, is to either 0 or 1 (Trott 2000). The powers of θ_0 for which this quantity is closer to 0 are 1, 3, 4, 5, 6, 7, 8, 11, 12, 14, 17, ... (Sloane's A051016), and those for which it is closer to 1 are 2, 9, 10, 13, 15, 16, 18, 20, 21, 23, ... (Sloane's A051017).

See also ALMOST INTEGER, EQUIDISTRIBUTED SEQUENCE, SALEM CONSTANTS, WEYL'S CRITERION

References

Bertin, M. J. and Pathiaux-Delefosse, A. *Conjecture de Lehmer et petits nombres de Salem.* Kingston: Queen's Papers in Pure and Applied Mathematics, 1989.

Bertin, M. J.; Decomps-Guilloux, A.; Grandet-Hugot, M.; Pathiaux-Delefosse, M.; and Schreiber, J. P. *Pisot and Salem Numbers.* Basel: Birkhäuser, 1992.

Borwein, P. and Hare, K. G. "Some Computations on Pisot and Salem Numbers." CECM-00:148, 18 May 2000. http://www.cecm.sfu.ca/preprints/2000pp.html#00:148.

Boyd, D. W. "Small Salem Numbers." *Duke Math. J.* **44**, 315–328, 1977.

Boyd, D. W. "Pisot and Salem Numbers in Intervals of the Real Line." *Math. Comput.* **32**, 1244–1260, 1978.

Boyd, D. W. "Pisot Numbers in the Neighbourhood of a Limit Point. II." *Math. Comput.* **43**, 593–602, 1984.

Boyd, D. W. "Pisot Numbers in the Neighbourhood of a Limit Point. I." *J. Number Theory* **21**, 17–43, 1985.

Dufresnoy, J. and Pisot, C. "Étude de certaines fonctions méromorphes bornées sur le cercle unité, application à un ensemble fermé d'entiers algébriques." *Ann. Sci. École Norm. Sup.* **72**, 69–92, 1955.

Erdos, P.; Joó, M.; and Schnitzer, F. J. "On Pisot Numbers." *Ann. Univ. Sci. Budapest, Eotvos Sect. Math.* **39**, 95–99, 1997.

Katai, I. and Kovacs, B. "Multiplicative Functions with Nearly Integer Values." *Acta Sci. Math.* **48**, 221–225, 1985.

Le Lionnais, F. *Les nombres remarquables.* Paris: Hermann, pp. 38 and 148, 1983.

Koksma, J. F. "Ein mengentheoretischer Satz über die Gleichverteilung modulo Eins." *Comp. Math.* **2**, 250–258, 1935.

Pisot, C. "La répartition modulo 1 et les nombres algébriques." *Annali di Pisa* **7**, 205–248, 1938.

Salem, R. "Sets of Uniqueness and Sets of Multiplicity." *Trans. Amer. Math. Soc.* **54**, 218–228, 1943.

Salem, R. "A Remarkable Class of Algebraic Numbers. Proof of a Conjecture of Vijayaraghavan." *Duke Math. J.* **11**, 103–108, 1944.

Salem, R. "Power Series with Integral Coefficients." *Duke Math. J.* **12**, 153–172, 1945.

Siegel, C. L. "Algebraic Numbers whose Conjugates Lie in the Unit Circle." *Duke Math. J.* **11**, 597–602, 1944.

Sloane, N. J. A. Sequences A051016 and A051017 in "An On-Line Version of the Encyclopedia of Integer Sequences." http://www.research.att.com/~njas/sequences/eisonline.html.

Trott, M. "Numerical Computations." §1.2.1 in *The Mathematica Guidebook, Vol. 1: Programming in Mathematica.* New York: Springer-Verlag, 2000.

Vijayaraghavan, T. "On the Fractional Parts of the Powers of a Number, II." *Proc. Cambridge Phil. Soc.* **37**, 349–357, 1941.

Pistol

A 4-POLYHEX.

References

Gardner, M. *Mathematical Magic Show: More Puzzles, Games, Diversions, Illusions and Other Mathematical Sleight-of-Mind from Scientific American.* New York: Vintage, p. 147, 1978.

Pitchfork Bifurcation

Let $f : \mathbb{R} \times \mathbb{R} \to \mathbb{R}$ be a one-parameter family of C^3 maps satisfying

$$f(-x, \mu) = -f(x, \mu) \tag{1}$$

$$\left[\frac{\partial f}{\partial x}\right]_{\mu=0, \, x=0} = 0 \tag{2}$$

$$\left[\frac{\partial^2 f}{\partial x \, \partial \mu}\right]_{0, \, 0} > 0 \tag{3}$$

$$\left[\frac{\partial^3 f}{\partial \mu^3}\right]_{\mu=0, \, x=0} < 0. \tag{4}$$

(Actually, condition (1) can be relaxed slightly.) Then there are intervals having a single stable fixed point and three fixed points (two of which are stable and one of which is unstable). This BIFURCATION is called a pitchfork bifurcation. An example of an equation displaying a pitchfork bifurcation is

$$\dot{x} = \mu x - x^3 \tag{5}$$

(Guckenheimer and Holmes 1997, p. 145).

See also BIFURCATION, TRANSCRITICAL BIFURCATION

References

Guckenheimer, J. and Holmes, P. *Nonlinear Oscillations, Dynamical Systems, and Bifurcations of Vector Fields, 3rd ed.* New York: Springer-Verlag, pp. 145 and 149–150, 1997.

Rasband, S. N. *Chaotic Dynamics of Nonlinear Systems.* New York: Wiley, p. 31, 1990.

Pivot Theorem

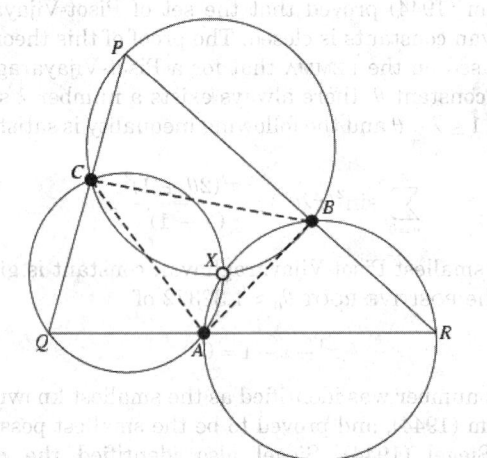

If the VERTICES A, B, and C of TRIANGLE $\triangle ABC$ lie on sides QR, RP, and PQ of the TRIANGLE $\triangle PQR$, then the three CIRCUMCIRCLES CBP, ACQ, and BAR have a common point X. In extended form, this theorem becomes MIQUEL'S THEOREM.

See also CIRCUMCIRCLE, CLIFFORD'S CIRCLE THEOREM, MIQUEL'S THEOREM

References

Coxeter, H. S. M. and Greitzer, S. L. *Geometry Revisited.* New York: Random House, pp. 61–62, 1967.

Forder, H. G. *Geometry.* London: Hutchinson, p. 17, 1960.

Wells, D. *The Penguin Dictionary of Curious and Interesting Geometry.* London: Penguin, p. 184, 1991.

Pivoting

The element in the diagonal of a matrix by which other elements are divided in an algorithm such as GAUSS-JORDAN ELIMINATION is called the pivot element. Partial pivoting is the interchanging of rows and full pivoting is the interchanging of both rows and columns in order to place a particularly "good" element in the diagonal position prior to a particular operation.

See also Gauss-Jordan Elimination

References

Press, W. H.; Flannery, B. P.; Teukolsky, S. A.; and Vetterling, W. T. *Numerical Recipes in FORTRAN: The Art of Scientific Computing, 2nd ed.* Cambridge, England: Cambridge University Press, pp. 29–30, 1992.

Pizza Theorem

If a circular pizza is divided into 8, 12, 16, ...slices by making cuts at equal angles from an arbitrary point, then the sums of the areas of alternate slices are equal.

There is also a second pizza theorem. This one gives the VOLUME of a pizza of thickness a and RADIUS z,

$$pi zz a.$$

Place (Digit)

DIGIT

Place (Field)

A place v of a NUMBER FIELD k is an ISOMORPHISM class of field maps k onto a dense subfield of a nondiscrete locally compact FIELD k_v.

In the function field case, let F be a function field of algebraic functions of one variable over a FIELD K. Then by a place in F, we mean a subset p of F which is the IDEAL of nonunits of some VALUATION RING O over K.

References

Chevalley, C. *Introduction to the Theory of Algebraic Functions of One Variable.* Providence, RI: Amer. Math. Soc., p. 2, 1951.
Knapp, A. W. "Group Representations and Harmonic Analysis, Part II." *Not. Amer. Math. Soc.* **43**, 537–549, 1996.
van der Waerden, B. L. *Algebra, 2 vols.* New York: Springer-Verlag, 1991.

Place (Game)

For n players, $n-1$ games are needed to fairly determine first place, and $n-1+\lg(n-1)$ are needed to fairly determine first and second place.

Place (Riemann Sphere)

The word "place" has a special meaning in complex variables, where it roughly corresponds to a point in the COMPLEX PLANE (except that it reflects the Riemann sheet structure imposed by whatever function is under discussion). For example, if the function in question is $\ln z$, then 1 and $e^{2\pi i}$ are different places.

Plaindrome

A plaindrome is a number whose HEXADECIMAL digits are in nondecreasing order. The first few are 1, 2, 3, 4, 5, 6, 7, 8, 9, 10, 11, 12, 13, 14, 15, 17, 18, 19, 20, 21, 22, 23, 24, ... (Sloane's A023757). The first few which are not plaindromes are 16, 32, 33, 48, 49, 50, 64, ..., corresponding to 10_{16}, 20_{16}, 21_{16}, 30_{16}, 31_{16}, 32_{16}, 64_{16}, ...

See also DIGIT, HEXADECIMAL, KATADROME, METADROME, NIALPDROME

References

Sloane, N. J. A. Sequences A023757 in "An On-Line Version of the Encyclopedia of Integer Sequences." http://www.research.att.com/~njas/sequences/eisonline.html.
Weisstein, E. W. "Integer Sequences." MATHEMATICA NOTEBOOK INTEGERSEQUENCES.M.

Plaited Polyhedron

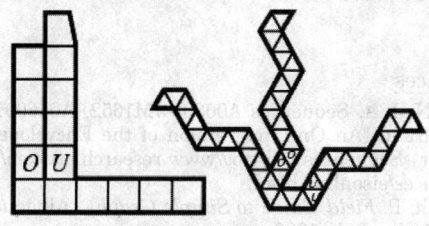

plaited cube   *plaited icosahedron*

There exist POLYHEDRA which can be plaited (braided). Examples include a plaited CUBE and plaited ICOSAHEDRON illustrated above (Pargetter 1959, Wells 1991). In the above figures, heavy lines indicate cuts, thin lines indicate folds, and polygons labeled "O" are placed over polygons labeled "U."

References

Gorham, J. *Plaited Crystal Models.* 1888.
Pargetter, A. R. "Plaited Polyhedra." *Math. Gaz.* **43**, 88–101, 1959.
Wells, D. *The Penguin Dictionary of Curious and Interesting Geometry.* London: Penguin, p. 160, 1991.

Planar Bubble Problem

BUBBLE

Planar Connected Graph

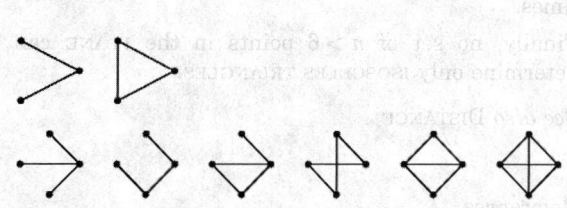

A planar connected graph is a GRAPH which is both planar and connected. The numbers of planar connected graphs with $n = 1, 2, \ldots$ nodes are $1, 1, 1, 2, 6,$ $20, 99, \ldots$ (Sloane's A003094; Steinbach 1990, p. 131). A subset of planar 3-connected graphs are called POLYHEDRAL GRAPHS.

The following table gives the numbers of planar connected graphs having minimal degrees of at least k.

k	Sloane	$n = 1, 2, 3, \ldots$
2	A054381	0, 0, 1, 3, 10, 49, 332, ...

The numbers of planar connected graphs with $n = 1,$ $2, \ldots$ edges are $1, 1, 3, 5, 12, 30, 79, 227, 709, 2318, \ldots$ (Sloane's A046091).

See also CONNECTED GRAPH, PLANAR GRAPH, POLY-HEDRAL GRAPH, POLYNEMA

References

Sloane, N. J. A. Sequences A003094/M1652, A046091, and A054381 in "An On-Line Version of the Encyclopedia of Integer Sequences." http://www.research.att.com/~njas/sequences/eisonline.html.
Steinbach, P. *Field Guide to Simple Graphs.* Albuquerque, NM: Design Lab, 1990.

Planar Distance

For n points in the PLANE, there are at least

$$N_1 = \sqrt{n - \tfrac{3}{4}} - \tfrac{1}{2}$$

different DISTANCES. The minimum DISTANCE can occur only $\leq 3n - 6$ times, and the MAXIMUM DISTANCE can occur $\leq n$ times. Furthermore, no DISTANCE can occur as often as

$$N_2 = \tfrac{1}{4} n \left(1 + \sqrt{8n - 7}\right) < \frac{n^{3/2}}{\sqrt{2}} - \frac{n}{4}$$

times.

Finally, no set of $n > 6$ points in the PLANE can determine only ISOSCELES TRIANGLES.

See also DISTANCE

References

Honsberger, R. "The Set of Distances Determined by n Points in the Plane." Ch. 12 in *Mathematical Gems II*. Washington, DC: Math. Assoc. Amer., pp. 111–135, 1976.

Planar Graph

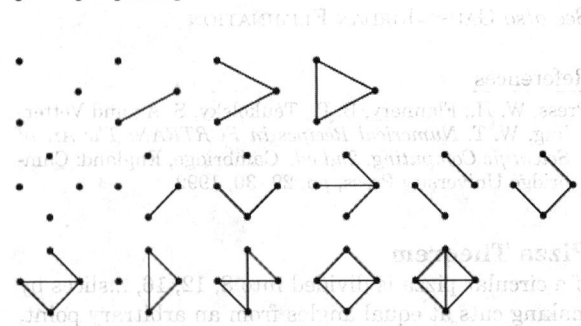

A GRAPH is planar if it can be drawn in a PLANE without EDGES crossing (i.e., it has CROSSING NUMBER 0). The number of planar graphs with $n = 1, 2, \ldots$ nodes are $1, 2, 4, 11, 33, 142, \ldots$ (Sloane's A005470; Wilson 1975, p. 162).

There are a number of efficient algorithms for planarity testing, which are unfortunately all difficult to implement. Most are based on the $o(n^3)$ algorithm of Auslander and Parter (1961; Skiena 1990, p. 247). One implementation is given by PlanarQ[g] in the *Mathematica* add-on package DiscreteMath`Combinatorica` (which can be loaded with the command < <DiscreteMath`), which however should be trusted for only versions 4.1 and higher.

Only planar graphs have DUALS and if G is planar, then G has VERTEX DEGREE ≤ 5. A graph is planar IFF it has a COMBINATORIAL DUAL GRAPH (Harary 1994, p. 115). Any planar graph has a GRAPH EMBEDDING as a PLANAR STRAIGHT LINE GRAPH where edges do not intersect (Fáry 1948; Bryant 1989; Skiena 1990, pp. 100 and 251; Scheinerman and Wilf 1994).

COMPLETE GRAPHS are planar only for $n \leq 4$. The complete BIPARTITE GRAPH $K(3, 3)$ is nonplanar. More generally, Kuratowski proved in 1930 that a graph is planar IFF it does not contain within it any graph which can be CONTRACTED to the pentagonal graph $K(5)$ or the hexagonal graph $K(3, 3)$. K_5 can be decomposed into a union of two planar graphs, giving it a "DEPTH" of $E(K_5) = 2$. Simple CRITERIA for determining the depth of graphs are not known. Beineke and Harary (1964, 1965) have shown that if $n \not\equiv 4$ (mod 6), then

$$E(K_n) = \left\lfloor \tfrac{1}{6}(n + 7) \right\rfloor.$$

The DEPTHS of the graphs K_n for $n = 4, 10, 22, 28, 34,$ and 40 are $1, 3, 4, 5, 6,$ and 7 (Meyer 1970).

All TREES are planar, as is a CYCLE GRAPH, GRID GRAPH, or WHEEL GRAPH. Every planar graph on nine vertices has a nonplanar complement (Battle *et al.* 1962; Skiena 1990, p. 250).

The following table gives the numbers of planar graphs having minimal degrees of at least k.

k	Sloane	$n = 1, 2, 3, \ldots$
2	A049370	0, 0, 1, 3, 10, 50, 335, ...
3	A049371	0, 0, 0, 1, 2, 9, 46, 386, ...
4	A049372	0, 0, 0, 0, 0, 1, 1, 4, 14, 69, ...
5	A049373	0, 0, 0, 0, 0, 0, 0, 0, 0, 0, 0, 1, 0, 1, 1, 5, ...

See also BARNETTE'S CONJECTURE, COMPLETE GRAPH, DUAL GRAPH, FABRY IMBEDDING, INTEGRAL DRAWING, KURATOWSKI REDUCTION THEOREM, OUTPLANAR GRAPH, PLANAR CONNECTED GRAPH, PLANAR STRAIGHT LINE GRAPH, POLYHEDRAL GRAPH, STEINITZ'S THEOREM, UTILITY GRAPH

References

Auslander, L. and Parter, S. "On Imbedding Graphs in the Sphere." *J. Math. Mechanics* **10**, 517–523, 1961.

Battle, J.; Harary, F.; and Kodama, Y. "Every Planar Graph with Nine Points has a Nonplanar Complement." *Bull. Amer. Math. Soc.* **68**, 569–571, 1962.

Beineke, L. W. and Harary, F. "On the Thickness of the Complete Graph." *Bull. Amer. Math. Soc.* **70**, 618–620, 1964.

Beineke, L. W. and Harary, F. "The Thickness of the Complete Graph." *Canad. J. Math.* **17**, 850–859, 1965.

Booth, K. S. and Lueker, G. S. "Testing for the Consecutive Ones Property, Interval Graphs, and Graph Planarity using PQ-Tree Algorithms." *J. Comput. System Sci.* **13**, 335–379, 1976.

Bryant, V. W. "Straight Line Representation of Planar Graphs." *Elem. Math.* **44**, 64–66, 1989.

Cai, J.; Han, X.; and Tarjan, R. "New Solutions to Four Planar Graph Problems." Technical Report. New York University, 1990.

Di Battista, G.; Eades, P.; Tamassia, R.; and Tollis, I. G. *Graph Drawing: Algorithms for the Visualization of Graphs.* Englewood Cliffs, NJ: Prentice-Hall, 1998.

Eades, P. and Tamassia, R. "Algorithms for Drawing Graphs: An Annotated Bibliography." Technical Report CS-89–09. Department of Computer Science. Providence, RI: Brown University, Feb. 1989.

Even, S. *Graph Algorithms.* Rockville, MD: Computer Science Press, 1979.

Fáry, I. "On Straight Line Representations of Planar Graphs." *Acta Sci. Math. (Szeged)* **11**, 229–233, 1948.

Friedman, E. "Large Regular Graphs with Small Diameter." http://www.stetson.edu/~efriedma/planar/.

Gardner, M. *The Sixth Book of Mathematical Games from Scientific American.* Chicago, IL: University of Chicago Press, pp. 91–94, 1984.

Harary, F. "Planarity." Ch. 11 in *Graph Theory.* Reading, MA: Addison-Wesley, pp. 102–125, 1994.

Hopcroft, J. and Tarjan, R. "Efficiency Planarity Testing." *J. ACM* **21**, 549–568, 1974.

Le Lionnais, F. *Les nombres remarquables.* Paris: Hermann, p. 56, 1983.

Meyer, J. "L'épaisseur des graphes completes K_{34} et K_{40}." *J. Comp. Th.* **9**, 1970.

Schneinerman, E. and Wilf, H. S. "The Rectilinear Crossing Number of a Complete Graph and Sylvester's 'Four Point'

Problem of Geometric Probability." *Amer. Math. Monthly* **101**, 939–943, 1994.

Skiena, S. "Planar Graphs." §6.5 in *Implementing Discrete Mathematics: Combinatorics and Graph Theory with Mathematica.* Reading, MA: Addison-Wesley, pp. 247–253, 1990.

Sloane, N. J. A. Sequences A005470/M1252, A049370, A049371, A049372, and A049373 in "An On-Line Version of the Encyclopedia of Integer Sequences." http://www.research.att.com/~njas/sequences/eisonline.html.

Steinbach, P. *Field Guide to Simple Graphs.* Albuquerque, NM: Design Lab, 1990.

Stony Brook Algorithm Repository. §.4.12. "Detection and Embedding." http://www.cs.sunysb.edu/~algorith/files/planar-drawing.shtml.

Wagon, S. "Coloring Planar Maps and Graphs." Ch. 24 in *Mathematica in Action, 2nd ed.* New York: Springer-Verlag, pp. 507–537, 1999.

Whitney, H. "Non-Separable and Planar Graphs." *Trans. Amer. Math. Soc.* **34**, 339–362, 1932.

Whitney, H. "Planar Graphs." *Fund. Math.* **21**, 73–84, 1933.

Wilson, R. J. *Introduction to Graph Theory.* London: Longman, 1975.

Planar Point

A point **p** on a REGULAR SURFACE $M \in \mathbb{R}^3$ is said to be planar if the GAUSSIAN CURVATURE $K(\mathbf{p}) = 0$ and $S(\mathbf{p}) = 0$ (where S is the SHAPE OPERATOR), or equivalently, both of the PRINCIPAL CURVATURES κ_1 and κ_2 are 0.

See also ANTICLASTIC, ELLIPTIC POINT, GAUSSIAN CURVATURE, HYPERBOLIC POINT, PARABOLIC POINT, SYNCLASTIC

References

Gray, A. *Modern Differential Geometry of Curves and Surfaces with Mathematica, 2nd ed.* Boca Raton, FL: CRC Press, p. 375, 1997.

Planar Polygon

Flat polygons embedded in 3-D space can be transformed into a congruent planar polygon as follows. First, translate the starting vertex to (0, 0, 0) by subtracting it from each vertex of the polygon. Then find the normal **n** to the polygon by taking the CROSS PRODUCT of the first and last vertices. Now, let **A** be the rotation matrix for EULER ANGLES ψ, θ, and ϕ, and solve

$$\mathbf{A} \begin{bmatrix} n_x \\ n_y \\ \sqrt{1 - n_x^2 - n_y^2} \end{bmatrix} = \begin{bmatrix} 0 \\ 0 \\ 1 \end{bmatrix} \tag{1}$$

for $\cos \psi$ and $\cos \theta$ (after first expressing sines in terms of cosines using $\cos x = \sqrt{1 - \sin^2 x}$. The result is

$$\phi = \pm \left(\frac{n_y}{\sqrt{n_x^2 + n_y^2}} \right) \tag{2}$$

$$\theta = \pm\sqrt{1 - n_x^2 - n_y^2}. \tag{3}$$

The signs are chosen as follows:

$$\psi = \cos^{-1}\left[-\text{sgn}(n_x)\frac{n_y}{\sqrt{n_x^2 + n_y^2}}\right] \tag{4}$$

$$\theta = \cos^{-1}\left[-\text{sgn}(n_x n_z)\sqrt{1 - n_x^2 - n_y^2}\right]. \tag{5}$$

Plugging these back in and applying to the original polygon then gives a polygon whose vertices all have one component zero. This component can then be dropped. The only special cases which need to be taken into account are $|n_z| = 1$, in which case the polygon is parallel to the xy-plane and the third components can be immediately dropped. The second occurs when $n_x = 0$, in which case there is no component of the normal vector along the x-AXIS, so the Euler rotation will not work. However, simply picking a different starting vertex from which to calculate the normal resolves this degenerate case.

See also POLYGON

Planar Space

Let (ξ_1, ξ_2) be a locally EUCLIDEAN coordinate system. Then

$$ds^2 = d\xi_1^2 + d\xi_2^2. \tag{1}$$

Now plug in

$$d\xi_1 = \frac{\partial\xi_1}{\partial x_1}dx_1 + \frac{\partial\xi_1}{\partial x_2}dx_2 \tag{2}$$

$$d\xi_2 = \frac{\partial\xi_2}{\partial x_1}dx_1 + \frac{\partial\xi_2}{\partial x_2}dx_2 \tag{3}$$

to obtain

$$ds^2 = \left[\left(\frac{\partial\xi_1}{\partial x_1}\right)^2 + \left(\frac{\partial\xi_2}{\partial x_1}\right)^2\right]dx_1^2$$
$$+ 2\left[\frac{\partial\xi_1}{\partial x_1}\frac{\partial\xi_1}{\partial x_2} + \frac{\partial\xi_2}{\partial x_1}\frac{\partial\xi_2}{\partial x_2}\right]dx_1\,dx_2$$
$$+ \left[\left(\frac{\partial\xi_1}{\partial x_2}\right)^2 + \left(\frac{\partial\xi_2}{\partial x_2}\right)^2\right]dx_2^2. \tag{4}$$

Reading off the COEFFICIENTS from

$$ds^2 = g_{11}\,dx_1^2 + 2g_{12}\,dx_1\,dx_2 + g_{22}(dx_2)^2 \tag{5}$$

gives

$$g_{11} = \left(\frac{\partial\xi_1}{\partial x_1}\right)^2 + \left(\frac{\partial\xi_2}{\partial x_1}\right)^2 \tag{6}$$

$$g_{12} = \frac{\partial\xi_1}{\partial x_1}\frac{\partial\xi_1}{\partial x_2} + \frac{\partial\xi_2}{\partial x_1}\frac{\partial\xi_2}{\partial x_2} \tag{7}$$

$$g_{22} = \left(\frac{\partial\xi_1}{\partial x_2}\right)^2 + \left(\frac{\partial\xi_2}{\partial x_2}\right)^2. \tag{8}$$

Making a change of coordinates $(x_1, x_2) \to (x'_1, x'_2)$ gives

$$g'_{11} = \left(\frac{\partial\xi_1}{\partial x'_1}\right)^2 + \left(\frac{\partial\xi_2}{\partial x'_1}\right)^2$$
$$= \left(\frac{\partial\xi_1}{\partial x_1}\frac{\partial x_1}{\partial x'_1} + \frac{\partial\xi_1}{\partial x_2}\frac{\partial x_2}{\partial x'_1}\right)^2 + \left(\frac{\partial\xi_2}{\partial x_1}\frac{\partial x_1}{\partial x'_1} + \frac{\partial\xi_2}{\partial x_2}\frac{\partial x_2}{\partial x'_1}\right)^2$$
$$= g_{11}\left(\frac{\partial x_1}{\partial x'_1}\right)^2 + 2g_{12}\frac{\partial x_1}{\partial x'_1}\frac{\partial x_2}{\partial x'_1} + g_{22}\left(\frac{\partial x_2}{\partial x'_1}\right)^2 \tag{9}$$

$$g'_{12} = \frac{\partial\xi_1}{\partial x_1}\frac{\partial x_1}{\partial x'_1}\frac{\partial\xi_1}{\partial x_2}\frac{\partial x_2}{\partial x'_2} + \frac{\partial\xi_2}{\partial x_1}\frac{\partial x_1}{\partial x'_1}\frac{\partial\xi_2}{\partial x_2}\frac{\partial x_2}{\partial x'_2}$$
$$= g_{12}\frac{\partial x_1}{\partial x'_1}\frac{\partial x_2}{\partial x'_2} \tag{10}$$

$$g'_{22} = g_{11}\left(\frac{\partial x_1}{\partial x'_1}\right)^2 + 2g_{12}\frac{\partial x_1}{\partial x'_2}\frac{\partial x_2}{\partial x'_2} + g_{22}\left(\frac{\partial x_2}{\partial x'_2}\right)^2. \tag{11}$$

Planar Straight Line Graph

A GRAPH EMBEDDING of a PLANAR GRAPH in which only straight line segments are used to connect the VERTICES. Fáry (1948) showed that every PLANAR GRAPH has an EMBEDDING which is a planar straight line graph with noncrossing edges (Bryant 1989; Skiena 1990, pp. 100 and 251; Schneinerman and Wilf 1994). de Fraysseix *et al.* (1988) give an algorithm for constructing a planar straight line for a graph of order n by placing the vertices on a $(2n - 4) \times (n - 2)$ grid (Skiena 1990, p. 251).

See also PLANAR GRAPH, RECTILINEAR CROSSING NUMBER

References

Bryant, V. W. "Straight Line Representation of Planar Graphs." *Elem. Math.* **44**, 64–66, 1989.

de Fraysseix, H.; Pach, J; and Pollack, R. "Small Sets Supporting Fáry Embeddings of Planar Graphs." *Proc. of the 20th Symposium on the Theory of Computing.* ACM, pp. 426–433, 1988.

Fáry, I. "On Straight Line Representations of Planar Graphs." *Acta Sci. Math. (Szeged(* **11**, 229–233, 1948.

Schneinerman, E. and Wilf, H. S. "The Rectilinear Crossing Number of a Complete Graph and Sylvester's 'Four Point' Problem of Geometric Probability." *Amer. Math. Monthly* **101**, 939–943, 1994.

Skiena, S. *Implementing Discrete Mathematics: Combinatorics and Graph Theory with Mathematica.* Reading, MA: Addison-Wesley, 1990.

Plancherel's Theorem

$$\int_{-\infty}^{\infty} f(x)\bar{g}(x)\,dx = \int_{-\infty}^{\infty} F(s)\bar{G}(s)\,ds,$$

where $F(s) \equiv \mathcal{F}[f(x)]$ and $\mathcal{F}$ denotes a FOURIER TRANSFORM and $\bar{z}$ is the COMPLEX CONJUGATE. If f and g are real

$$\int_{-\infty}^{\infty} f(x)g(-x)\,dx = \int_{-\infty}^{\infty} F(s)G(s)\,ds.$$

See also FOURIER TRANSFORM, PARSEVAL'S THEOREM

Planck's Radiation Function

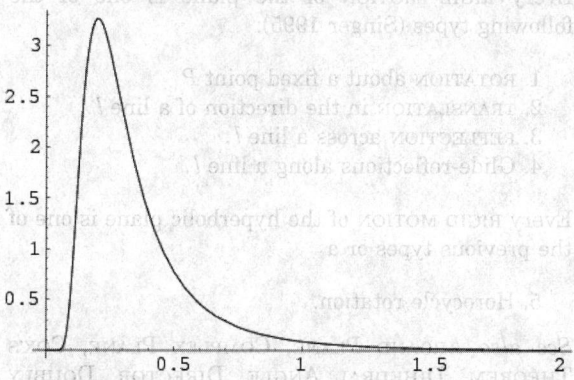

The function

$$f(x) = \frac{15}{\pi^4}\frac{1}{x^5(e^{1/x}-1)}, \tag{1}$$

which is normalized so that

$$\int_0^{\infty} f(x)\,dx = 1. \tag{2}$$

The first and second RAW MOMENTS are

$$\mu_1' = \frac{30\zeta(3)}{\pi^4} \tag{3}$$

$$\mu_2' = \frac{5}{2\pi^2}, \tag{4}$$

but higher order raw moments do not exist since the corresponding integrals do not converge.
It has a MAXIMUM at $x \approx 0.201405$, where

$$f'(x) = \frac{5x - e^{1/x}(5x-1)}{x^7(e^{1/x}-1)^2} = 0, \tag{5}$$

and inflection points at $x \approx 0.11842$ and $x \approx 0.283757$, where

$$f''(x) = \frac{e^{1/x}(1+e^{1/x}) + 6x(e^{1/x}-1)[e^{1/x}(5x-2)-5x]}{(e^{1/x}-1)^3 x^9}$$

$$= 0. \tag{6}$$

References

Abramowitz, M. and Stegun, C. A. (Eds.). "Planck's Radiation Function." §27.2 in *Handbook of Mathematical Functions with Formulas, Graphs, and Mathematical Tables, 9th printing.* New York: Dover, p. 999, 1972.

Plane

A plane is a 2-D DOUBLY RULED SURFACE spanned by two linearly independent vectors. The generalization of the plane to higher DIMENSIONS is called a HYPERPLANE. The angle between two intersecting planes is known as the DIHEDRAL ANGLE.

In intercept form, a plane passing through the points $(a, 0, 0)$, $(0, b, 0)$ and $(0, 0, c)$ is given by

$$\frac{x}{a} + \frac{y}{b} + \frac{z}{c} = 1. \tag{1}$$

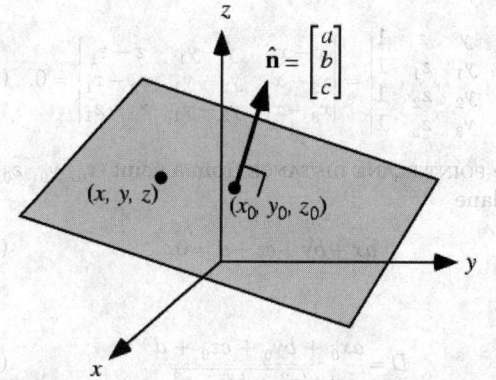

The equation of a plane PERPENDICULAR to the NONZERO VECTOR $\hat{\mathbf{n}} = (a, b, c)$ through the point (x_0, y_0, z_0) is

$$\begin{bmatrix} a \\ b \\ c \end{bmatrix} \cdot \begin{bmatrix} x-x_0 \\ y-y_0 \\ z-z_0 \end{bmatrix} = a(x-x_0) + b(y-y_0) + c(z-z_0) = 0, \tag{2}$$

so

$$ax + by + cz + d = 0. \tag{3}$$

where

$$d \equiv -ax_0 - by_0 - cz_0. \tag{4}$$

A plane specified in this form therefore has x-, y-, and z-intercepts at

$$x = -\frac{d}{a} \tag{5}$$

$$y = -\frac{d}{b} \tag{6}$$

$$z = -\frac{d}{c}, \tag{7}$$

and lies at a DISTANCE

$$h = \frac{|d|}{\sqrt{a^2 + b^2 + c^2}} \tag{8}$$

from the ORIGIN.

The plane through P_1 and parallel to (a_1, b_1, c_1) and (a_2, b_2, c_2) is

$$\begin{vmatrix} x - x_1 & y - y_1 & z - z_1 \\ a_1 & b_1 & c_1 \\ a_2 & b_2 & c_2 \end{vmatrix} = 0. \tag{9}$$

The plane through points P_1 and P_2 parallel to direction (a, b, c) is

$$\begin{vmatrix} x - x_1 & y - y_1 & z - z_1 \\ x_2 - x_1 & y_2 - y_1 & z_2 - z_1 \\ a & b & c \end{vmatrix} = 0. \tag{10}$$

The three-point form is

$$\begin{vmatrix} x & y & z & 1 \\ x_1 & y_1 & z_1 & 1 \\ x_2 & y_2 & z_2 & 1 \\ x_3 & y_3 & z_3 & 1 \end{vmatrix} = \begin{vmatrix} x - x_1 & y - y_1 & z - z_1 \\ x_2 - x_1 & y_2 - y_1 & z_2 - z_1 \\ x_3 - x_1 & y_3 - y_1 & z_3 - z_1 \end{vmatrix} = 0. \tag{11}$$

The POINT-PLANE DISTANCE from a point (x_0, y_0, z_0) to a plane

$$ax + by + cz + d = 0 \tag{12}$$

is

$$D = \frac{ax_0 + by_0 + cz_0 + d}{\pm \sqrt{a^2 + b^2 + c^2}}. \tag{13}$$

The DIHEDRAL ANGLE between the planes

$$A_1 x + B_1 y + C_1 z + D_1 = 0 \tag{14}$$

$$A_2 x + B_2 y + C_2 z + D_2 = 0 \tag{15}$$

which have normal vectors $\mathbf{N}_1 = (A_1, B_1, C_1)$ and $\mathbf{N}_2 = (A_2, B_2, C_2)$ is simply given via the DOT PRODUCT of the normals,

$$\cos \theta = \mathbf{N}_1 \cdot \mathbf{N}_2$$

$$= \frac{A_1 A_2 + B_1 B_2 + C_1 C_2}{\sqrt{A_1^2 + B_1^2 + C_1^2} \sqrt{A_2^2 + B_2^2 + C_2^2}}. \tag{16}$$

In order to specify the relative distances of $n > 1$ points in the plane, $1 + 2(n-2) = 2n - 3$ coordinates are needed, since the first can always be placed at $(0, 0)$ and the second at $(x, 0)$, where it defines the x-AXIS. The remaining $n - 2$ points need two coordinates each. However, the total number of distances is

$$_n C_2 = \binom{n}{2} = \frac{n!}{2!(n-2)!} = \tfrac{1}{2} n(n-1), \tag{17}$$

where $\binom{n}{k}$ is a BINOMIAL COEFFICIENT, so the distances between points are subject to m relationships, where

$$m \equiv \tfrac{1}{2} n(n-1) - (2n-3) = \tfrac{1}{2}(n-2)(n-3). \tag{18}$$

For $n = 2$ and $n = 3$, there are no relationships. However, for a QUADRILATERAL (with $n = 4$), there is one (Weinberg 1972).

It is impossible to pick random variables which are uniformly distributed in the plane (Eisenberg and Sullivan 1996). In 4-D, it is possible for four planes to intersect in exactly one point. For every set of n points in the plane, there exists a point O in the plane having the property such that *every* straight line through O has at least 1/3 of the points on each side of it (Honsberger 1985).

Every RIGID MOTION of the plane is one of the following types (Singer 1995):

1. ROTATION about a fixed point P.
2. TRANSLATION in the direction of a line l.
3. REFLECTION across a line l.
4. Glide-reflections along a line l.

Every RIGID MOTION of the hyperbolic plane is one of the previous types or a

5. Horocycle rotation.

See also ARGAND PLANE, COMPLEX PLANE, COX'S THEOREM, DIHEDRAL ANGLE, DIRECTOR, DOUBLY RULED SURFACE, ELLIPTIC PLANE, FANO PLANE, HYPERPLANE, ISOCLINAL PLANE, LINE-PLANE INTERSECTION, MEDIATOR, MOUFANG PLANE, NIRENBERG'S CONJECTURE, NORMAL SECTION, POINT-PLANE DISTANCE, PROJECTIVE PLANE

References

Beyer, W. H. *CRC Standard Mathematical Tables, 28th ed.* Boca Raton, FL: CRC Press, pp. 208–209, 1987.

Eisenberg, B. and Sullivan, R. "Random Triangles n Dimensions." *Amer. Math. Monthly* **103**, 308–318, 1996.

Honsberger, R. *Mathematical Gems III.* Washington, DC: Math. Assoc. Amer., pp. 189–191, 1985.

Kern, W. F. and Bland, J. R. "Lines and Planes in Space." §4 in *Solid Mensuration with Proofs, 2nd ed.* New York: Wiley, pp. 9–12, 1948.

Singer, D. A. "Isometries of the Plane." *Amer. Math. Monthly* **102**, 628–631, 1995.

Weinberg, S. *Gravitation and Cosmology: Principles and Applications of the General Theory of Relativity.* New York: Wiley, p. 7, 1972.

Plane Chart

EQUIRECTANGULAR PROJECTION

Plane Curve

A CURVE which lies in a single PLANE. A plane curve may be closed or open. Curves which are interesting for some reason and whose properties have therefore been investigates are called "special" curves (Lawrence 1972). Some of the most common open curves are the LINE, PARABOLA, and HYPERBOLA, and some of the most common closed curves are the CIRCLE and ELLIPSE.

See also ALGEBRAIC CURVE, CURVE, SPACE CURVE, SPHERICAL CURVE

References

Coolidge, J. L. *A Treatise on Algebraic Plane Curves.* New York: Dover, p. 30, 1959.

Gray, A. "Famous Plane Curves." Ch. 3 in *Modern Differential Geometry of Curves and Surfaces with Mathematica, 2nd ed.* Boca Raton, FL: CRC Press, pp. 49–74, 1997.

Hilbert, D. and Cohn-Vossen, S. "Plane Curves." §1 in *Geometry and the Imagination.* New York: Chelsea, pp. 1–7, 1999.

Lawrence, J. D. *A Catalog of Special Plane Curves.* New York: Dover, 1972.

Lockwood, E. H. *A Book of Curves.* Cambridge, England: Cambridge University Press, 1961.

MacTutor History of Mathematics Archive. http://www-groups.dcs.st-and.ac.uk/~history/Curves/Curves.html.

Weisstein, E. W. "Plane Curves." MATHEMATICA NOTEBOOK CURVES.M.

Yates, R. C. *A Handbook on Curves and Their Properties.* Ann Arbor, MI: J. W. Edwards, 1947.

Plane Cutting

PLANE DIVISION BY CIRCLES, PLANE DIVISION BY ELLIPSES, PLANE DIVISION BY LINES

Plane Division by Circles

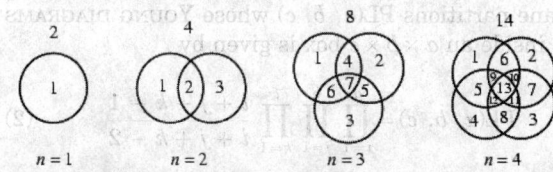

$n = 1$ $n = 2$ $n = 3$ $n = 4$

Consider n intersecting CIRCLES. The maximal number of regions into which these divide the PLANE are

$$N(n) = n^2 - n + 2,$$

giving values for $n = 1, 2, \ldots$ of 2, 4, 8, 14, 22, 32, 44, 58, ... (Sloane's A014206).

See also ARRANGEMENT, CIRCLE, CIRCLE DIVISION BY LINES, PLANE DIVISION BY ELLIPSES, PLANE DIVISION BY LINES, SPACE DIVISION BY SPHERES

References

Indiana School Mathematics J. **14**, No. 4, p. 4, 1979.

Konhauser, J. D. E.; Velleman, D.; and Wagon, S. *Which Way Did the Bicycle Go? And Other Intriguing Mathematical Mysteries.* Washington, DC: Math. Assoc. Amer., p. 177, 1996.

Problem Q736. *Parabola* **24**, 22, 1988.

Sloane, N. J. A. Sequences A014206 in "An On-Line Version of the Encyclopedia of Integer Sequences." http://www.research.att.com/~njas/sequences/eisonline.html.

Yaglom, A. M. and Yaglom, I. M. *Challenging Mathematical Problems with Elementary Solutions, Vol. 1.* New York: Dover, pp. 102–106, 1987.

Plane Division by Ellipses

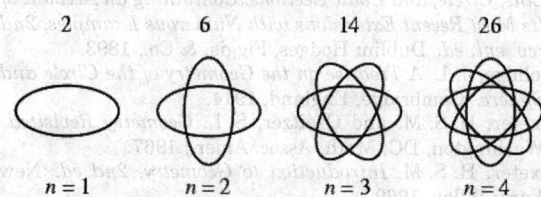

$n = 1$ $n = 2$ $n = 3$ $n = 4$

Consider n intersecting ELLIPSES. The maximal number of regions into which these divide the PLANE are

$$N(n) = 2n^2 - 2n + 2 = 2(n^2 - n + 1),$$

giving values for $n = 1, 2, \ldots$ of 2, 6, 14, 26, 42, 62, 86, 114, ...

See also ARRANGEMENT, CIRCLE DIVISION BY LINES, ELLIPSE, PLANE DIVISION BY CIRCLES, PLANE DIVISION BY LINES

References

Problem Q607. *Parabola* **20**, 27, 1984.

Plane Division by Lines

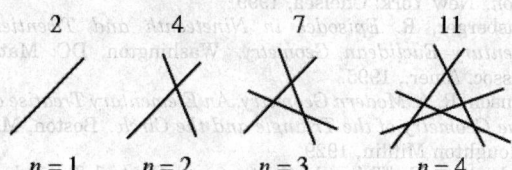

$n = 1$ $n = 2$ $n = 3$ $n = 4$

The maximal number of regions into which n lines divide a PLANE are

$$N(n) = \tfrac{1}{2}(n^2 + n + 2)$$

which, for $n = 1, 2, \ldots$ gives 2, 4, 7, 11, 16, 22, ... (Sloane's A000124), the same maximal number of regions into which a circle can be divided by n lines.

See also ARRANGEMENT, CIRCLE DIVISION BY LINES, LINE, PLANE DIVISION BY CIRCLES, PLANE DIVISION BY ELLIPSES

References

Sloane, N. J. A. Sequences A000124/M1041 in "An On-Line Version of the Encyclopedia of Integer Sequences." http://www.research.att.com/~njas/sequences/eisonline.html.

Plane Geometry

That portion of GEOMETRY dealing with figures in a PLANE, as opposed to SOLID GEOMETRY. Plane geometry deals with the CIRCLE, LINE, POLYGON, etc.

See also CONSTRUCTIBLE POLYGON, GEOMETRIC CONSTRUCTION, GEOMETRY, SOLID GEOMETRY, SPHERICAL GEOMETRY

References

Altshiller-Court, N. *College Geometry: A Second Course in Plane Geometry for Colleges and Normal Schools, 2nd ed., rev. enl.* New York: Barnes and Noble, 1952.

Casey, J. *A Treatise on the Analytical Geometry of the Point, Line, Circle, and Conic Sections, Containing an Account of Its Most Recent Extensions with Numerous Examples, 2nd rev. enl. ed.* Dublin: Hodges, Figgis, & Co., 1893.

Coolidge, J. L. *A Treatise on the Geometry of the Circle and Sphere.* Cambridge, England, 1914.

Coxeter, H. S. M. and Greitzer, S. L. *Geometry Revisited.* Washington, DC: Math. Assoc. Amer., 1967.

Coxeter, H. S. M. *Introduction to Geometry, 2nd ed.* New York: Wiley, 1969.

Dixon, R. *Mathographics.* New York: Dover, 1991.

Durell, C. V. *Modern Geometry: The Straight Line and Circle.* London: Macmillan, 1928.

Fuhrmann, W. *Synthetische Beweise Planimetrische Sätze.* Berlin, 1890.

Gallatly, W. *The Modern Geometry of the Triangle, 2nd ed.* London: Hodgson, 1913.

Heath, T. L. *The Thirteen Books of the Elements, 2nd ed., Vol. 1: Books I and II.* New York: Dover, 1956.

Heath, T. L. *The Thirteen Books of the Elements, 2nd ed., Vol. 2: Books III-IX.* New York: Dover, 1956.

Heath, T. L. *The Thirteen Books of the Elements, 2nd ed., Vol. 3: Books X-XIII.* New York: Dover, 1956.

Henderson, D. W. *Experiencing Geometry: On Plane and Sphere.* Englewood Cliffs, NJ: Prentice-Hall, 1995.

Hilbert, D. *The Foundations of Geometry.* Chicago, IL: Open Court, 1980.

Hilbert, D. and Cohn-Vossen, S. *Geometry and the Imagination.* New York: Chelsea, 1999.

Honsberger, R. *Episodes in Nineteenth and Twentieth Century Euclidean Geometry.* Washington, DC: Math. Assoc. Amer., 1995.

Johnson, R. A. *Modern Geometry: An Elementary Treatise on the Geometry of the Triangle and the Circle.* Boston, MA: Houghton Mifflin, 1929.

Kimberling, C. "Triangle Centers and Central Triangles." *Congr. Numer.* **129**, 1-295, 1998.

Klee, V. "Some Unsolved Problems in Plane Geometry." *Math. Mag.* **52**, 131-145, 1979.

Klee, V. and Wagon, S. *Old and New Unsolved Problems in Plane Geometry and Number Theory, rev. ed.* Washington, DC: Math. Assoc. Amer., 1991.

Lachlan, R. *An Elementary Treatise on Modern Pure Geometry.* London: Macmillan, 1893.

McClelland, W. J. *Geometry of the Circle.* London, 1891.

Pedoe, D. *Circles: A Mathematical View, rev. ed.* Washington, DC: Math. Assoc. Amer., 1995.

Rouché, E. and de Comberousse, C. *Traité de Géométrie, nouv. éd., vol. 1: Géométrie plane.* Paris: Gauthier-Villars, 1922.

Russell, J. W. *Elementary Pure Geometry.* Oxford, 1893.

Simon, M. *Über die Entwicklung der Elementargeometrie im XIX Jahrhundert.* Berlin, 1906.

Weisstein, E. W. "Plane Geometry." Mathematica NOTE-BOOK PLANEGEOMETRY.M.

Weisstein, E. W. "Books about Plane Geometry." http://www.treasure-troves.com/books/PlaneGeometry.html.

Plane Graph

Planar Graph

Plane Partition

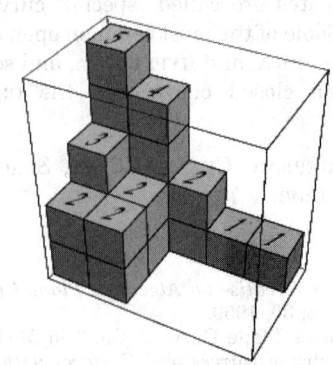

$$5 \quad 4 \quad 2 \quad 1 \quad 1$$
$$3 \quad 2$$
$$2 \quad 2$$

A two-dimensional array of INTEGERS nonincreasing both left to right and top to bottom which add up to a given number, i.e., $n_{ij} \geq n_{i(j+1)}$ and $n_{ij} \geq n_{(i+1)j}$. For example, a planar partition of 22 is illustrated above. The GENERATING FUNCTION for the number PL(n) of planar partitions of n is

$$\sum_{n=0}^{\infty} \mathrm{PL}(n)x^n = \frac{1}{\prod_{k=1}^{\infty}(1-x^k)^k}$$

$$= 1 + x + 3x^2 + 6x^3 + 13x^4 + 24x^5 + \ldots \quad (1)$$

(Sloane's A000219, MacMahon 1912b, Speciner 1972, Bender and Knuth 1972, Bressoud and Propp 1999). MacMahon (1960) also showed that the number of plane partitions PL(a, b, c) whose YOUNG DIAGRAMS fit inside an $a \times b \times c$ box is given by

$$\mathrm{PL}(a, b, c) = \prod_{i=1}^{a} \prod_{j=1}^{b} \prod_{k=1}^{c} \frac{i+j+k-1}{i+j+k-2} \quad (2)$$

(Bressoud and Propp 1999, Fulmek and Krattenthaler 2000). Expanding out the products gives

$$\mathrm{PL}(a, b, c) = \prod_{i=1}^{a} \frac{\Gamma(i)\Gamma(b+c+i)}{\Gamma(b+i)\Gamma(c+i)} \quad (3)$$

$$= \frac{G(a+1)G(b+1)G(c+1)G(a+b+c+1)}{G(a+b+1)G(a+c+1)G(b+c+1)}, \quad (4)$$

where $G(n)$ is BARNES' G-FUNCTION. Taking $n = a = b = c$ gives

$$\mathrm{PL}(n, n, n) = \prod_{i=1}^{n} \frac{\Gamma(i)\Gamma(i+2n)}{[\Gamma(i+n)]^2} \quad (5)$$

$$= \frac{[G(n+1)]^3 G(3n+1)}{[G(2n+1)]^3}, \quad (6)$$

the first few terms of which are 2, 20, 980, 232848, 267227532, 1478619421136, ... (Sloane's A008793).

Amazingly, PL(a, b, c) also gives the number of HEXAGON TILINGS by RHOMBI for a hexagon of side lengths a, b, c, a, b, c (David and Tomei 1989, Fulmek and Krattenthaler 2000).

The concept of planar partitions can also be generalized to cubic partitions.

See also CYCLICALLY SYMMETRIC PLANE PARTITION, DESCENDING PLANE PARTITION, HEXAGON TILING, PARTITION, MACDONALD'S PLANE PARTITION CONJECTURE, SOLID PARTITION, TOTALLY SYMMETRIC SELF-COMPLEMENTARY PLANE PARTITION, YOUNG DIAGRAM

References

Bender, E. A. and Knuth, D. E. "Enumeration of Plane Partitions." *J. Combin. Theory Ser. A.* **13**, 40–54, 1972.

Bressoud, D. *Proofs and Confirmations: The Story of the Alternating Sign Matrix Conjecture.* Cambridge, England: Cambridge University Press, 1999.

Bressoud, D. and Propp, J. "How the Alternating Sign Matrix Conjecture was Solved." *Not. Amer. Math. Soc.* **46**, 637–646.

Cohn, H.; Larsen, M.; and Propp, J. "The Shape of a Typical Boxed Plane Partition." *New York J. Math.* **4**, 137–166, 1998.

David, G. and Tomei, C. "The Problem of the Calissons." *Amer. Math. Monthly* **96**, 429–431, 1989.

Fulmek, M. and Krattenthaler, C. "The Number of Rhombus Tilings of a Symmetric Hexagon which Contains a Fixed Rhombus on the Symmetry Axes, II." *Europ. J. Combin.* **21**, 601–640, 2000.

Knuth, D. E. "A Note on Solid Partitions." *Math. Comput.* **24**, 955–961, 1970.

MacMahon, P. A. "Memoir on the Theory of the Partitions of Numbers. V: Partitions in Two-Dimensional Space." *Phil. Trans. Roy. Soc. London Ser. A* **211**, 75–110, 1912a.

MacMahon, P. A. "Memoir on the Theory of the Partitions of Numbers. VI: Partitions in Two-Dimensional Space, to which is Added an Adumbration of the Theory of Partitions in Three-Dimensional Space." *Phil. Trans. Roy. Soc. London Ser. A* **211**, 345–373, 1912b.

MacMahon, P. A. §429 and 494 in *Combinatory Analysis, Vol. 2.* New York: Chelsea, 1960.

Mills, W. H.; Robbins, D. P.; and Rumsey, H. Jr. "Proof of the Macdonald Conjecture." *Invent. Math.* **66**, 73–87, 1982.

Sloane, N. J. A. Sequences A000219/M2566 and A008793 in "An On-Line Version of the Encyclopedia of Integer Sequences." http://www.research.att.com/~njas/sequences/eisonline.html.

Speciner, M. Item 18 in Beeler, M.; Gosper, R. W.; and Schroeppel, R. *HAKMEM.* Cambridge, MA: MIT Artificial Intelligence Laboratory, Memo AIM-239, p. 10, Feb. 1972.

Stanley, R. P. "Symmetry of Plane Partitions." *J. Combin. Th. Ser. A* **3**, 103–113, 1986.

Stanley, R. P. "A Baker's Dozen of Conjectures Concerning Plane Partitions." In *Combinatoire Énumérative* (Ed. G. Labelle and P. Leroux). New York: Springer-Verlag, 285–293, 1986.

Plane Symmetry Groups
WALLPAPER GROUPS

Plane-Filling Curve
PLANE-FILLING FUNCTION

Plane-Filling Function

A SPACE-FILLING FUNCTION which maps a 1-D INTERVAL into a 2-D area. Plane-filling functions were thought to be impossible until Hilbert discovered the HILBERT CURVE in 1891.

Plane-filling functions are often (imprecisely) defined to be the "limit" of an infinite sequence of specified curves which "fill" the PLANE without "HOLES," hence the more popular term PLANE-FILLING CURVE. The term "plane-filling function" is preferable to "PLANE-FILLING CURVE" because "curve" informally connotes "GRAPH" (i.e., range) of some continuous function, but the GRAPH of a plane-filling function is a solid patch of 2-space with no evidence of the order in which it was traced (and, for a dense set, retraced). Actually, all that is needed to rigorously define a plane-filling function is an arbitrarily refinable correspondence between contiguous subintervals of the domain and contiguous subareas of the range.

True plane-filling functions are not ONE-TO-ONE. In fact, because they map closed intervals onto closed areas, they cannot help but overfill, revisiting at least twice a dense subset of the filled area. Thus, every point in the filled area has *at least* one inverse image.

See also HILBERT CURVE, PEANO CURVE, PEANO-GOSPER CURVE, SCHOENBERG CURVE, SIERPINSKI CURVE, SPACE-FILLING FUNCTION, SPACE-FILLING POLYHEDRON

References

Bogomolny, A. "Plane Filling Curves." http://www.cut-the-knot.com/do_you_know/hilbert.html.

Wagon, S. "A Space-Filling Curve." §6.3 in *Mathematica in Action.* New York: W. H. Freeman, pp. 196–209, 1991.

Plane-Line Intersection
LINE-PLANE INTERSECTION

Planted Planar Tree
A planted plane tree (V, E, v, α) is defined as a vertex set V, edges set E, ROOT v, and order relation α on V which satisfies

1. For x, $y \in V$ if $\rho(x) < \rho(y)$, then $x \alpha y$, where $\rho(x)$ is the length of the path from v to x,
2. If $\{r, s\}$, $\{x, y\} \in E$, $\rho(r) = \rho(x) = \rho(s) - 1 = \rho(y) - 1$ and $r \alpha x$, then $s \alpha y$

(Klarner 1969, Chorneyko and Mohanty 1975). The CATALAN NUMBERS give the number of planar trivalent planted trees.

See also CATALAN NUMBER, PLANTED TREE, TREE

References

Chorneyko, I. Z. and Mohanty, S. G. "On the Enumeration of Certain Sets of Planted Plane Trees." *J. Combin. Th. Ser. B* **18**, 209–221, 1975.
Harary, F.; Prins, G.; and Tutte, W. T. "The Number of Plane Trees." *Indag. Math.* **26**, 319–327, 1964.
Klarner, D. A. "A Correspondence Between Sets of Trees." *Indag. Math.* **31**, 292–296, 1969.

Planted Tree

A planted tree is a ROOTED TREE whose ROOT NODE has VERTEX DEGREE 1. The number of planted trees of n nodes is T_{n-1}, where T_{n-1} is the number of ROOTED TREES of $n - 1$ vertices (Harary 1994, pp. 188–190), so there are 1, 1, 1, 2, 4, 9, 20, ... (Sloane's A000081) planted trees of $n = 1, 2, 3, ...$ vertices.

See also ROOTED TREE, TREE

References

Harary, F. *Graph Theory.* Reading, MA: Addison-Wesley, 1994.
Sloane, N. J. A. Sequences A000081/M1180 in "An On-Line Version of the Encyclopedia of Integer Sequences." http://www.research.att.com/~njas/sequences/eisonline.html.

Plastic Constant

The limiting ratio of the successive terms of the PADOVAN SEQUENCE, $P = 1.32471795\ldots$. It is given exactly by the unique real root of $x^3 - x - 1 = 0$.

See also PADOVAN SEQUENCE

References

Stewart, I. "Tales of a Neglected Number." *Sci. Amer.* **274**, 102–103, Jun. 1996.

Plat

A BRAID in which strands are intertwined in the center and are free in "handles" on either side of the diagram.

Plate Carre

EQUIRECTANGULAR PROJECTION

Plateau Curves

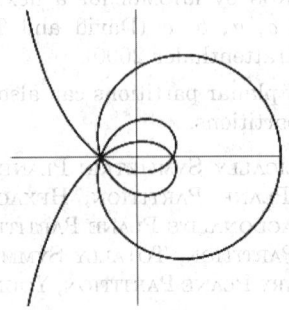

A curve studied by the Belgian physicist and mathematician Joseph Plateau. It has Cartesian equation

$$x = \frac{a \, \sin[(m + n)t]}{\sin[(m - n)t]}$$

$$y = \frac{2a \, \sin(mt) \sin(nt)}{\sin[(m - n)t]}.$$

If $m = 2n$, the Plateau curve degenerates to a CIRCLE with center $(1, 0)$ and radius 2.

References

MacTutor History of Mathematics Archive. "Plateau Curves." http://www.groups.dcs.st-and.ac.uk/~history/Curves/Plateau.html.

Plateau's Equation

The PARTIAL DIFFERENTIAL EQUATION

$$(1 + u_x^2)u_{xx} - 2u_x u_y u_{xy} + (1 + u_y^2)u_{yy} = 0.$$

References

Bateman, H. *Partial Differential Equations of Mathematical Physics.* New York: Dover, p. 501, 1944.
Zwillinger, D. *Handbook of Differential Equations, 3rd ed.* Boston, MA: Academic Press, p. 134, 1997.

Plateau's Laws

BUBBLES can meet only at ANGLES of 120° (for two BUBBLES) and 109°28′16″ (for three BUBBLES), where the exact value of 109.5° is the TETRAHEDRAL DIHEDRAL ANGLE. This was proved by Jean Taylor using MEASURE THEORY to study AREA minimization. The DOUBLE BUBBLE is AREA minimizing, but it is not known if the triple BUBBLE is also AREA minimizing. It is also unknown if empty chambers trapped inside can minimize AREA for $n \geq 3$ BUBBLES.

See also BUBBLE, CALCULUS OF VARIATIONS, DOUBLE BUBBLE, MINIMAL SURFACE, PLATEAU'S PROBLEM

References

Morgan, F. "Mathematicians, including Undergraduates, Look at Soap Bubbles." *Amer. Math. Monthly* **101**, 343–351, 1994.

Taylor, J. E. "The Structure of Singularities in Soap-Bubble-Like and Soap-Film-Like Minimal Surfaces." *Ann. Math.* **103**, 489–539, 1976.

Plateau's Problem

The problem in CALCULUS OF VARIATIONS to find the MINIMAL SURFACE of a boundary with specified constraints (usually having no singularities on the surface). In general, there may be one, multiple, or no MINIMAL SURFACES spanning a given closed curve in space. The EXISTENCE of a solution to the general case was independently proven by Douglas (1931) and Radó (1933), although their analysis could not exclude the possibility of singularities. Osserman (1970) and Gulliver (1973) showed that a minimizing solution *cannot* have singularities.

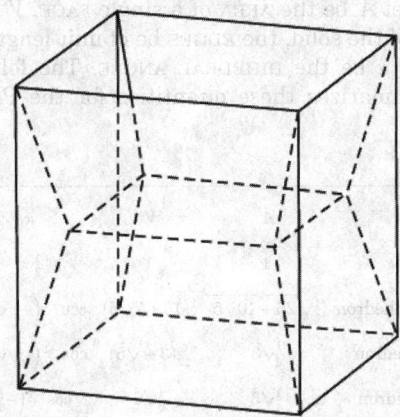

The problem is named for the Belgian physicist who solved some special cases experimentally using soap films and wire frames (Isenberg 1992, Wells 1991). The illustration above shows the 13-polygon surface obtained for a cubical wire frame.

See also BUBBLE, CALCULUS OF VARIATIONS, DOUBLE BUBBLE, MINIMAL SURFACE, PLATEAU'S LAWS, STEINER TREE, TRAVELING SALESMAN PROBLEM

References
Cundy, H. and Rollett, A. *Mathematical Models, 3rd ed.* Stradbroke, England: Tarquin Pub., pp. 48–49, 1989.
Douglas, J. "Solution of the Problem of Plateau." *Trans. Amer. Math. Soc.* **33**, 263–321, 1931.
Gulliver, R. "Regularity of Minimizing Surfaces of Prescribed Mean Curvature." *Ann. Math.* **97**, 275–305, 1973.
Isenberg, C. *The Science of Soap Films and Soap Bubbles.* New York: Dover, 1992.
Osserman, R. "A Proof of the Regularity Everywhere of the Classical Solution to Plateau's Problem." *Ann. Math.* **91**, 550–569, 1970.
Osserman, R. "Plateau's Problem." §1, Appendix in *A Survey of Minimal Surfaces.* New York: Dover, pp. 143–145, 1986.
Radó, T. "On the Problem of Plateau." *Ergeben. d. Math. u. ihrer Grenzgebiete.* Berlin: Springer-Verlag, 1933.
Steinhaus, H. *Mathematical Snapshots, 3rd ed.* New York: Dover, pp. 119–121, 1999.
Stuwe, M. *Plateau's Problem and the Calculus of Variations.* Princeton, NJ: Princeton University Press, 1989.

Wells, D. *The Penguin Dictionary of Curious and Interesting Geometry.* London: Penguin, pp. 185–187, 1991.

Platonic Graph

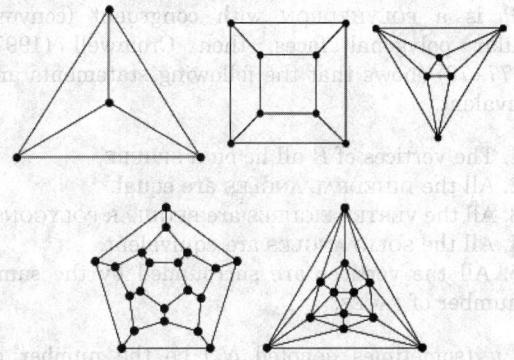

A POLYHEDRAL GRAPH corresponding to the SKELETON of a PLATONIC SOLID. The five platonic graphs, the TETRAHEDRAL GRAPH, CUBICAL GRAPH, OCTAHEDRAL GRAPH, DODECAHEDRAL GRAPH, and ICOSAHEDRAL GRAPH, are illustrated above. They are special cases of SCHLEGEL GRAPHS.

See also PLATONIC SOLID, POLYHEDRAL GRAPH, SCHLEGEL GRAPH

References
Bondy, J. A. and Murty, U. S. R. *Graph Theory with Applications.* New York: North Holland, p. 234, 1976.

Platonic Solid

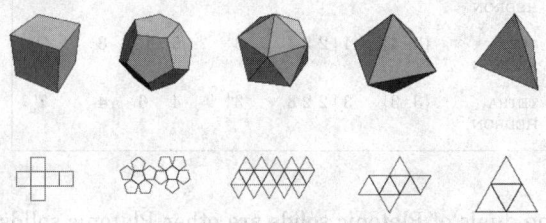

The Platonic solids, also called the regular solids or regular polyhedra, are CONVEX POLYHEDRA with equivalent faces composed of congruent CONVEX REGULAR POLYGONS. There are exactly five such solids (Steinhaus 1983, pp. 252–256): the CUBE, DODECAHEDRON, ICOSAHEDRON, OCTAHEDRON, and TETRAHEDRON, as was proved by Euclid in the last proposition of the *ELEMENTS*. The Platonic solids are sometimes also called "cosmic figures" (Cromwell 1997), although this term is sometimes used to refer collectively to both the Platonic solids *and* KEPLER-POINSOT SOLIDS (Coxeter 1973).

The Platonic solids were known to the ancient Greeks, and were described by Plato in his *Timaeus* ca. 350 BC. In this work, Plato equated the TETRAHEDRON with the "element" fire, the CUBE with earth, the ICOSAHEDRON with water, the OCTAHEDRON with air, and the DODECAHEDRON with the stuff of which the constellations and heavens were made (Cromwell 1997).

If P is a POLYHEDRON with congruent (convex) regular polygonal faces, then Cromwell (1997, pp. 77–78) shows that the following statements are equivalent.

1. The vertices of P all lie on a SPHERE.
2. All the DIHEDRAL ANGLES are equal.
3. All the VERTEX FIGURES are REGULAR POLYGONS.
4. All the SOLID ANGLES are equivalent.
5. All the vertices are surrounded by the same number of FACES.

Let v (sometimes denoted N_0) be the number of VERTICES, e (or N_1) the number of EDGES, and f (or N_2) the number of FACES. The following table gives the SCHLÄFLI SYMBOL, WYTHOFF SYMBOL, and C&R symbol, the number of vertices v, edges e, and faces f, and the POINT GROUPS for the Platonic solids (Wenninger 1989).

Solid	Schläfli symbol	Wythoff Symbol	C&R Symbol	v	e	f	Group
CUBE	$\{4, 3\}$	3 | 2 2 4	4^3	8	12	6	O_h
DODECA-HEDRON	$\{5, 3\}$	3 | 2 2 5	5^3	20	30	12	I_h
ICOSA-HEDRON	$\{3, 5\}$	5 | 2 2 3	3^5	12	30	20	I_h
OCTA-HEDRON	$\{3, 4\}$	4 | 2 2 3	3^4	6	12	8	O_h
TETRA-HEDRON	$\{3, 3\}$	3 | 2 2 3	3^3	4	6	4	T_d

The duals of Platonic solids are other Platonic solids and, in fact, the dual of the TETRAHEDRON is another TETRAHEDRON. Let r be the INRADIUS, ρ the MIDRADIUS, and R the CIRCUMRADIUS of a given Platonic solid. Then

$$rR = \rho^2.$$

The following two tables give the analytic and numerical values of these distances for Platonic solids with unit side length.

Solid	r	ρ	R
CUBE	$\frac{1}{2}$	$\frac{1}{2}\sqrt{2}$	$\frac{1}{2}\sqrt{3}$
DODECAHEDRON	$\frac{1}{20}\sqrt{250+110\sqrt{5}}$	$\frac{1}{4}(3+\sqrt{5})$	$\frac{1}{4}(\sqrt{15}+\sqrt{3})$
ICOSAHEDRON	$\frac{1}{12}(3\sqrt{3}+\sqrt{15})$	$\frac{1}{4}(1+\sqrt{5})$	$\frac{1}{4}\sqrt{10+2\sqrt{5}}$
OCTAHEDRON	$\frac{1}{6}\sqrt{6}$	$\frac{1}{2}$	$\frac{1}{2}\sqrt{2}$
TETRAHEDRON	$\frac{1}{12}\sqrt{6}$	$\frac{1}{4}\sqrt{2}$	$\frac{1}{4}\sqrt{6}$

Solid	r	ρ	R
CUBE	0.5	0.70711	0.86603
DODECAHEDRON	1.11352	1.30902	1.40126
ICOSAHEDRON	0.75576	0.80902	0.95106
OCTAHEDRON	0.40825	0.5	0.70711
TETRAHEDRON	0.20412	0.35355	0.61237

Finally, let A be the AREA of a single FACE, V be the VOLUME of the solid, the EDGES be of unit length on a side, and α be the DIHEDRAL ANGLE. The following table summarizes these quantities for the Platonic solids.

Solid	A	V	α
Cube	1	1	$\frac{1}{2}\pi$
Dodecahedron	$\frac{1}{4}\sqrt{25+10\sqrt{5}}$	$\frac{1}{4}(15+7\sqrt{5})$	$\cos^{-1}\left(-\frac{1}{5}\sqrt{5}\right)$
Icosahedron	$\frac{1}{4}\sqrt{3}$	$\frac{5}{12}(3+\sqrt{5})$	$\cos^{-1}\left(-\frac{1}{3}\sqrt{5}\right)$
Octahedron	$\frac{1}{4}\sqrt{3}$	$\frac{1}{3}\sqrt{2}$	$\cos^{-1}\left(-\frac{1}{3}\right)$
Tetrahedron	$\frac{1}{4}\sqrt{3}$	$\frac{1}{12}\sqrt{2}$	$\cos^{-1}\left(\frac{1}{3}\right)$

The number of EDGES meeting at a VERTEX is $2e/v$. The SCHLÄFLI SYMBOL can be used to specify a Platonic solid. For the solid whose faces are p-gons (denoted $\{p\}$), with q touching at each VERTEX, the symbol is $\{p, q\}$. Given p and q, the number of VERTICES, EDGES, and faces are given by

$$N_0 = \frac{4p}{4 - (p - 2)(q - 2)}$$

$$N_1 = \frac{2pq}{4 - (p - 2)(q - 2)}$$

$$N_2 = \frac{4q}{4 - (p - 2)(q - 2)}.$$

The plots above show scaled duals of the Platonic solid embedded in a CUMULATED form of the original solid, where the scaling is chosen so that the dual edges lie at the incenters of the original faces (Wenninger 1983, pp. 8–9).

Since the Platonic solids are convex, the CONVEX HULL of each Platonic solid is the solid itself. MINIMAL SURFACES for Platonic solid frames are illustrated in Isenberg (1992, pp. 82–83).

See also ARCHIMEDEAN SOLID, CATALAN SOLID, JOHNSON SOLID, KEPLER-POINSOT SOLID, QUASIREGULAR POLYHEDRON, UNIFORM POLYHEDRON

References

Artmann, B. "Symmetry Through the Ages: Highlights from the History of Regular Polyhedra." In *In Eves' Circles* (Ed. J. M. Anthony). Washington, DC: Math. Assoc. Amer., pp. 139–148, 1994.

Ball, W. W. R. and Coxeter, H. S. M. "Polyhedra." Ch. 5 in *Mathematical Recreations and Essays, 13th ed.* New York: Dover, pp. 131–136, 1987.

Behnke, H.; Bachman, F.; Fladt, K.; and Kunle, H. (Eds.). *Fundamentals of Mathematics, Vol. 2: Geometry.* Cambridge, MA: MIT Press, p. 272, 1974.

Beyer, W. H. (Ed.). *CRC Standard Mathematical Tables, 28th ed.* Boca Raton, FL: CRC Press, pp. 128–129, 1987.

Bogomolny, A. "Regular Polyhedra." http://www.cut-the-knot.com/do_you_know/polyhedra.html.

Bourke, P. "Platonic Solids (Regular Polytopes in 3D)." http://www.swin.edu.au/astronomy/pbourke/geometry/platonic/.

Coxeter, H. S. M. *Regular Polytopes, 3rd ed.* New York: Dover, pp. 1–17, 93, and 107–112, 1973.

Critchlow, K. *Order in Space: A Design Source Book.* New York: Viking Press, 1970.

Cromwell, P. R. *Polyhedra.* New York: Cambridge University Press, pp. 51–57, 66–70, and 77–78, 1997.

Dunham, W. *Journey through Genius: The Great Theorems of Mathematics.* New York: Wiley, pp. 78–81, 1990.

Gardner, M. "The Five Platonic Solids." Ch. 1 in *The Second Scientific American Book of Mathematical Puzzles & Diversions: A New Selection.* New York: Simon and Schuster, pp. 13–23, 1961.

Harris, J. W. and Stocker, H. "Regular Polyhedron." §4.4 in *Handbook of Mathematics and Computational Science.* New York: Springer-Verlag, pp. 99–101, 1998.

Heath, T. *A History of Greek Mathematics, Vol. 1: From Thales to Euclid.* New York: Dover, p. 162, 1981.

Hume, A. "Exact Descriptions of Regular and Semi-Regular Polyhedra and Their Duals." *Computing Science Tech. Rep.*, No. 130. Murray Hill, NJ: AT&T Bell Laboratories, 1986.

Isenberg, C. *The Science of Soap Films and Soap Bubbles.* New York: Dover, 1992.

Kepler, J. *Opera Omnia, Vol. 5.* Frankfort, p. 121, 1864.

Kern, W. F. and Bland, J. R. "Regular Polyhedrons." In *Solid Mensuration with Proofs, 2nd ed.* New York: Wiley, pp. 116–119, 1948.

Meserve, B. E. *Fundamental Concepts of Geometry.* New York: Dover, 1983.

Nooshin, H.; Disney, P. L.; and Champion, O. C. "Properties of Platonic and Archimedean Polyhedra." Table 12.1 in "Computer-Aided Processing of Polyhedric Configurations." Ch. 12 in *Beyond the Cube: The Architecture of Space Frames and Polyhedra* (Ed. J. F. Gabriel). New York: Wiley, pp. 360–361, 1997.

Ogilvy, C. S. *Excursions in Geometry.* New York: Dover, pp. 129–131, 1990.

Pappas, T. "The Five Platonic Solids." *The Joy of Mathematics.* San Carlos, CA: Wide World Publ./Tetra, pp. 39 and 110–111, 1989.

Pedagoguery Software. Poly. http://www.peda.com/poly/.

Steinhaus, H. *Mathematical Snapshots, 3rd ed.* New York: Dover, pp. 191–201, 1999.

Rawles, B. A. "Platonic and Archimedean Solids--Faces, Edges, Areas, Vertices, Angles, Volumes, Sphere Ratios." http://www.intent.com/sg/polyhedra.html.

Robertson, S. A. and Carter, S. "On the Platonic and Archimedean Solids." *J. London Math. Soc.* **2**, 125–132, 1970.

Sharp, A. *Geometry Improv'd: 1. By a Large and Accurate Table of Segments of Circles, with Compendious Tables for Finding a True Proportional Part, Exemplify'd in Making out Logarithms from them, there Being a Table of them for all Primes to 1100, True to 61 Figures. 2. A Concise Treatise of Polyhedra, or Solid Bodies, of Many Bases.* London: R. Mount, p. 87, 1717.

Steinhaus, H. "Platonic Solids, Crystals, Bees' Heads, and Soap." Ch. 8 in *Mathematical Snapshots, 3rd ed.* New York: Dover, pp. 199–201 and 252–256, 1983.

Waterhouse, W. "The Discovery of the Regular Solids." *Arch. Hist. Exact Sci.* **9**, 212–221, 1972–1973.

Wells, D. *The Penguin Dictionary of Curious and Interesting Numbers.* Middlesex, England: Penguin Books, pp. 60–61, 1986.

Wells, D. *The Penguin Dictionary of Curious and Interesting Geometry.* London: Penguin, pp. 187–188, 1991.

Wenninger, M. "The Five Regular Convex Polyhedra and Their Duals." Ch. 1 in *Dual Models.* Cambridge, England: Cambridge University Press, pp. 7–13, 1983.

Wenninger, M. J. *Polyhedron Models.* Cambridge, England: Cambridge University Press, 1971.

Platonic Solid Stellations

The only STELLATIONS of PLATONIC SOLIDS which are UNIFORM POLYHEDRA are the three DODECAHEDRON STELLATIONS and the GREAT ICOSAHEDRON.

See also DODECAHEDRON STELLATIONS, ICOSAHEDRON STELLATIONS, STELLA OCTANGULA

Plato's Number

A vaguely specified number appearing in *The Republic* which involves 216 and 12,960,000.

References

Heath, T. L. *Aristarchus of Samos: The Ancient Copernicus.* New York: Dover, pp. 171–172, 1981.

Plato. *The Republic.* New York: Oxford University Press, 1994.

Wells, D. G. *The Penguin Dictionary of Curious and Interesting Numbers.* London: Penguin, p. 144, 1986.

Platykurtic

A distribution with FISHER KURTOSIS $\gamma_2 < 0$ (and therefore having a flattened shape).

See also FISHER KURTOSIS

p-Layer

The p-layer of H, $L_{p'}(H)$ is the unique minimal NORMAL SUBGROUP of H which maps onto $E(H/O_{p'}(H))$.

See also Bp-THEOREM, Lp'-BALANCE THEOREM, SIGNALIZER FUNCTOR THEOREM

Playfair's Axiom

Through any point in space, there is exactly one straight line PARALLEL to a given straight line. This AXIOM is equivalent to the PARALLEL POSTULATE.

See also PARALLEL POSTULATE

References

Dunham, W. "Hippocrates' Quadrature of the Lune." Ch. 1 in *Journey through Genius: The Great Theorems of Mathematics.* New York: Wiley, p. 54, 1990.

Henderson, D. W. *Experiencing Geometry: On Plane and Sphere.* Englewood Cliffs, NJ: Prentice-Hall, 1995.

Playfair, J. *Elements of Geometry: Containing the First Six Books of Euclid, with a Supplement on the Circle and the Geometry of Solids to which are added Elements of Plane and Spherical Trigonometry.* New York: W. E. Dean.

Plethysm

A group theoretic operation which is useful in the study of complex atomic spectra. A plethysm takes a set of functions of a given symmetry type $\{\mu\}$ and forms from them symmetrized products of a given degree r and other symmetry type $\{v\}$. A plethysm

$$\{\mu\} \otimes \{v\} = \sum \{\lambda\}$$

satisfies the rules

$$A \otimes (BC) = (A \otimes B)(A \otimes C) = A \otimes BA \otimes C,$$

$$A \otimes (B \pm C) = A \otimes B \pm A \otimes C,$$

$$(A \otimes B) \otimes C = A \otimes (B \otimes C)$$

$$(A + B) \otimes \{\lambda\} = \sum \Gamma_{\mu v\lambda}(A \otimes \{\mu\})(B \otimes \{v\}),$$

where $\Gamma_{\mu v \lambda}$ is the coefficient of $\{\lambda\}$ in $\{\mu\}\{v\}$,

$$(A - B) \otimes \{\lambda\} = \sum (-1)^r \Gamma_{\mu v \lambda}(A \otimes \{\mu\})(B \otimes \{\tilde{v}\}),$$

where $\{\tilde{v}\}$ is the partition of r conjugate to $\{v\}$, and

$$(AB) \otimes \{\lambda\} = \sum g_{\mu v \lambda}(A \otimes \{\mu\})(B \otimes \{v\}),$$

where $g_{\mu v \lambda}$ is the coefficient of $\{\lambda\}$ in the inner product $\{\mu\} \circ (v)$ (Wybourne 1970).

References

Littlewood, D. E. "Polynomial Concomitants and Invariant Matrices." *J. London Math. Soc.* **11**, 49–55, 1936.

Wybourne, B. G. "The Plethysm of *S*-Functions" and "Plethysm and Restricted Groups." Chs. 6–7 in *Symmetry Principles and Atomic Spectroscopy.* New York: Wiley, pp. 49–68, 1970.

Plot

GRAPH (FUNCTION)

Plot3D

GRAPH (FUNCTION)

Plouffe's Constant

N.B. A detailed online essay by S. Finch was the starting point for this entry.

Define the function

$$\rho(x) \equiv \begin{cases} 1 & \text{for } x < 0 \\ 0 & \text{for } x \geq 0. \end{cases} \tag{1}$$

Let

$$a_n = \sin(2^n) = \begin{cases} \sin 1 & \text{for } n = 0 \\ 2a_0\sqrt{1 - a_0^2} & \text{for } n = 1 \\ 2a_{n-1}(1 - 2a_{n-2}^2) & \text{for } n \geq 2, \end{cases} \tag{2}$$

then

$$\sum_{n=0}^{\infty} \frac{\rho(a_n)}{2^{n+1}} = \frac{1}{2\pi}. \tag{3}$$

For

$$b_n = \cos(2^n) = \begin{cases} \cos 1 & \text{for } n = 0 \\ 2b_{n-1}^2 - 1 & \text{for } n \geq 1, \end{cases} \tag{4}$$

and

$$\sum_{n=0}^{\infty} \frac{\rho(b_n)}{2^{n+1}} = 0.4756260767\ldots. \tag{5}$$

Letting

$$c_n = \tan(2^n) = \begin{cases} \tan 1 & \text{for } n = 0 \\ \dfrac{2c_{n-1}}{1 - c_{n-1}^2} & \text{for } n \geq 1, \end{cases} \tag{6}$$

then

$$\sum_{n=0}^{\infty} \frac{\rho(c_n)}{2^{n+1}} = \frac{1}{\pi}. \tag{7}$$

Plouffe asked if the above processes could be "inverted." He considered

$$\alpha_n = \sin\left(2^n \sin^{-1} \cdot \frac{1}{2}\right)$$

$$= \begin{cases} \frac{1}{2} & \text{for } n = 0 \\ \frac{1}{2}\sqrt{3} & \text{for } n = 1 \\ 2\alpha_{n-1}(1 - 2\alpha_{n-2}^2) & \text{for } n \geq 2, \end{cases} \tag{8}$$

giving

$$\sum_{n=0}^{\infty} \frac{\rho(\alpha_n)}{2^{n+1}} = \frac{1}{12}, \tag{9}$$

and

$$\beta_n = \cos\left(2^n \cos^{-1} \frac{1}{2}\right) = \begin{cases} \frac{1}{2} & \text{for } n = 0 \\ 2\beta_{n-1}^2 - 1 & \text{for } n \geq 1, \end{cases} \tag{10}$$

giving

$$\sum_{n=0}^{\infty} \frac{\rho(\beta_n)}{2^{n+1}} = \frac{1}{2},\qquad (11)$$

and

$$\gamma_n = \tan\left(2^n \tan^{-1} \tfrac{1}{2}\right) = \begin{cases} \frac{1}{2} & \text{for } n=0 \\ \frac{2\gamma_{n-1}}{1-\gamma_{n-1}^2} & \text{for } n\geq 1, \end{cases}\qquad (12)$$

giving

$$\sum_{n=0}^{\infty} \frac{\rho(\alpha_n)}{2^{n+1}} = \frac{1}{\pi}\tan^{-1}\left(\tfrac{1}{2}\right).\qquad (13)$$

The latter is known as Plouffe's constant (Plouffe 1997). The positions of the 1s in the BINARY expansion of this constant are 3, 6, 8, 9, 10, 13, 21, 23, ... (Sloane's A004715).

Borwein and Girgensohn (1995) extended Plouffe's γ_n to *arbitrary* REAL x, showing that if

$$\xi_n = \tan(2^n \tan^{-1} x)$$

$$= \begin{cases} x & \text{for } n=0 \\ \frac{2\xi_{n-1}}{1-\xi_{n-1}^2} & \text{for } n\geq 1 \text{ and } |\xi_{n-1}|\neq 1 \\ -\infty & \text{for } n\geq 1 \text{ and } |\xi_{n-1}|=1, \end{cases}\qquad (14)$$

then

$$\sum_{n=0}^{\infty} \frac{\rho(\xi_n)}{2^{n+1}} = \begin{cases} \frac{\tan^{-1} x}{\pi} & \text{for } x\geq 0 \\ 1+\frac{\tan^{-1} x}{\pi} & \text{for } x<0. \end{cases}\qquad (15)$$

Borwein and Girgensohn (1995) also give much more general recurrences and formulas.

References

Borwein, J. M. and Girgensohn, R. "Addition Theorems and Binary Expansions." *Canad. J. Math.* **47**, 262–273, 1995.
Finch, S. "Favorite Mathematical Constants." http://www.mathsoft.com/asolve/constant/plff/plff.html.
Plouffe, S.. "The Computation of Certain Numbers Using a Ruler and Compass." *J. Integer Sequences* **1**, No. 98.1.3, 1998. http://www.research.att.com/~njas/sequences/JIS/compass.html.
Sloane, N. J. A. Sequences A004715 in "An On-Line Version of the Encyclopedia of Integer Sequences." http://www.research.att.com/~njas/sequences/eisonline.html.

Plücker Characteristics

The CLASS m, ORDER n, number of NODES δ, number of CUSPS κ, number of STATIONARY TANGENTS (INFLECTION POINTS) ι, number of BITANGENTS τ, and GENUS p.

See also ALGEBRAIC CURVE, BITANGENT, CUSP, GENUS (SURFACE), INFLECTION POINT, NODE (ALGEBRAIC CURVE), STATIONARY TANGENT

Plücker Coordinates

GRASSMANN COORDINATES

Plücker Lines

The 60 PASCAL LINES of a HEXAGON inscribed in a CONIC SECTION intersect three at a time through 20 STEINER POINTS. There is a dual relationship between the 15 Plücker lines and the 15 SALMON POINTS.

See also KIRKMAN POINTS, PASCAL LINES, PASCAL'S THEOREM, SALMON POINTS, STEINER POINTS

References

Johnson, R. A. *Modern Geometry: An Elementary Treatise on the Geometry of the Triangle and the Circle.* Boston, MA: Houghton Mifflin, pp. 236–237, 1929.
Plücker, M. *J. reine angew. Math.* **5**, p. 274.
Salmon, G. "Notes: Pascal's Theorem, Art. 267" in *A Treatise on Conic Sections, 6th ed.* New York: Chelsea, pp. 379–382, 1960.
Wells, D. *The Penguin Dictionary of Curious and Interesting Geometry.* London: Penguin, p. 172, 1991.

Plücker Relations

PLÜCKER'S EQUATIONS

Plücker's Conoid

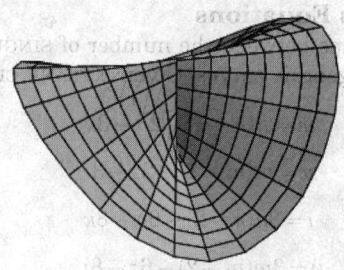

A RULED SURFACE sometimes also called the CYLINDROID. von Seggern (1993) gives the general functional form as

$$ax^2 + by^2 - zx^2 - zy^2 = 0,\qquad (1)$$

whereas Fischer (1986) and Gray (1997) give

$$z = \frac{2xy}{(x^2+y^2)}.\qquad (2)$$

A polar parameterization therefore gives

$$x(r,\ \theta) = r\cos\theta\qquad (3)$$
$$y(r,\ \theta) = r\sin\theta\qquad (4)$$
$$z(r,\ \theta) = 2\cos\theta\sin\theta.\qquad (5)$$

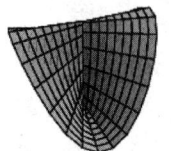

A generalization of Plücker's conoid to n folds is given by

$$x(r, \theta) = r \cos \theta \qquad (6)$$

$$y(r, \theta) = r \sin \theta \qquad (7)$$

$$z(r, \theta) = \sin(n\theta) \qquad (8)$$

(Gray 1997). The cylindroid is the inversion of the CROSS-CAP (Pinkall 1986).

See also CROSS-CAP, RIGHT CONOID, RULED SURFACE

References

Fischer, G. (Ed.). *Mathematical Models from the Collections of Universities and Museums.* Braunschweig, Germany: Vieweg, pp. 4–5, 1986.
Gray, A. "Plücker's Conoid." *Modern Differential Geometry of Curves and Surfaces with Mathematica, 2nd ed.* Boca Raton, FL: CRC Press, pp. 435–437, 1997.
Pinkall, U. *Mathematical Models from the Collections of Universities and Museums* (Ed. G. Fischer). Braunschweig, Germany: Vieweg, p. 64, 1986.
von Seggern, D. *CRC Standard Curves and Surfaces.* Boca Raton, FL: CRC Press, p. 288, 1993.

Plücker's Equations

Relationships between the number of SINGULARITIES of plane algebraic curves. Given a PLANE CURVE,

$$m = n(n-1) - 2\delta - 3\kappa \qquad (1)$$

$$n = m(m-1) - 2\tau - 3\iota \qquad (2)$$

$$\iota = 3n(n-2) - 6\delta - 8\kappa \qquad (3)$$

$$\kappa = 3m(m-2) - 6\tau - 8\iota, \qquad (4)$$

where m is the CLASS, n the ORDER, δ the number of NODES, κ the number of CUSPS, ι the number of STATIONARY TANGENTS (INFLECTION POINTS), and τ the number of BITANGENTS. Only three of these equations are LINEARLY INDEPENDENT.

See also ALGEBRAIC CURVE, BIOCHE'S THEOREM, BITANGENT, CUSP, GENUS (SURFACE), INFLECTION POINT, KLEIN'S EQUATION, NODE (ALGEBRAIC CURVE), STATIONARY TANGENT

References

Boyer, C. B. *A History of Mathematics.* New York: Wiley, pp. 581–582, 1968.
Coolidge, J. L. *A Treatise on Algebraic Plane Curves.* New York: Dover, pp. 99–118, 1959.
Graustein, W. C. *Introduction to Higher Geometry.* New York: Macmillan, pp. 220–222, 1930.

Plumbing

The plumbing of a p-sphere and a q-sphere is defined as the disjoint union of $\mathbb{S}^p \times \mathbb{S}^q$ and $\mathbb{D}^p \times \mathbb{S}^q$ with their common $\mathbb{D}^p \times \mathbb{D}^q$, identified via the identity homeomorphism.

See also HYPERSPHERE

References

Rolfsen, D. *Knots and Links.* Wilmington, DE: Publish or Perish Press, p. 180, 1976.

Pluperfect Number

MULTIPLY PERFECT NUMBER

Plurisubharmonic Function

An upper semicontinuous function whose restrictions to all complex lines are subharmonic (where defined). These functions were introduced by P. Lelong and Oka in the early 1940s. Examples of such a function are the logarithms of moduli of holomorphic functions.

References

Range, R. M. and Anderson, R. W. "Hans-Joachim Bremmermann, 1926–1996." *Not. Amer. Math. Soc.* **43**, 972–976, 1996.

Plus

The ADDITION of two quantities, i.e., a plus b. The operation is denoted $a + b$, and the symbol $+$ is called the PLUS SIGN. Floating point ADDITION is sometimes denoted $\oplus$.

See also ADDITION, MINUS, PLUS OR MINUS, TIMES

Plus or Minus

The symbol $\pm$ is used to denote a quantity which should be both added and subtracted, as in $a \pm b$. The symbol can be used to denote a range of uncertainty, or to denote a pair of quantities, such as the roots given by the QUADRATIC FORMULA

$$x\pm = \frac{-b \pm \sqrt{b^2 - 4ac}}{2a}.$$

When order is relevant, the symbol $a \mp b$ is also used, so an expression OF THE FORM $x \pm y \mp z$ is interpreted as $x + y - z$ or $x - y + z$. In contrast, the expression $x \pm y \pm z$ is interpreted to mean the set of four quantities $x + y + z$, $x - y + z$, $x + y - z$, and $x - y - z$.

See also MINUS, MINUS SIGN, PLUS, PLUS SIGN, SIGN

Plus Perfect Number

ARMSTRONG NUMBER

Plus Sign

The symbol "+" which is used to denote a POSITIVE number or to indicate ADDITION.

See also ADDITION, MINUS SIGN, SIGN

Plutarch Numbers

In *Moralia,* the Greek biographer and philosopher Plutarch states "Chrysippus says that the number of compound propositions that can be made from only ten simple propositions exceeds a million. (Hipparchus, to be sure, refuted this by showing that on the affirmative side there are 103,049 compound statements, and on the negative side 310,952.)" These numbers are known as the Plutarch numbers.

103,049 can be interpreted as the number s_{10} of BRACKETINGS on ten letters (Stanley 1997, Habsieger *et al.* 1998). Similarly, Plutarch's second number is given by $(s_{10} + s_{11})/2 = 310,954$ (Habsieger *et al.* 1998).

References

Biermann, K.-R. and Mau, J. "Überprüfung einer frühen Anwendung der Kombinatorik in der Logik." *J. Symbolic Logic* **23**, 129–132, 1958.

Biggs, N. L. "The Roots of Combinatorics." *Historia Mathematica* **6**, 109–136, 1979.

Habsieger, L.; Kazarian, M.; and Lando, S. "On the Second Number of Plutarch." *Amer. Math. Monthly* **105**, 446, 1998.

Heath, T. L. *A History of Greek Mathematics, Vol. 2: From Aristarchus to Diophantus.* New York: Dover, p. 256, 1981.

Kneale, W. and Kneale, M. *The Development of Logic.* Oxford, England: Oxford University Press, p. 162, 1971.

Neugebauer, O. *A History of Ancient Mathematical Astronomy.* New York: Springer-Verlag, p. 338, 1975.

Plutarch. §VIII.9 in *Moralia, Vol. 9.* Cambridge, MA: Harvard University Press, p. 732, 1961.

Stanley, R. P. *Enumerative Combinatorics, Vol. 1.* Cambridge, England: Cambridge University Press, p. 63, 1996.

Stanley, R. P. "Hipparchus, Plutarch, Schröder, and Hough." *Amer. Math. Monthly* **104**, 344–350, 1997.

Pochhammer Symbol

The Pochhammer symbol

$$(x)_n \equiv \frac{\Gamma(x+n)}{\Gamma(x)} = x(x+1)\cdots(x+n-1) = \frac{\Gamma(x+n)}{\Gamma(x)} \quad (1)$$

(Abramowitz and Stegun 1972, p. 256; Spanier 1987; Koepf 1998, p. 5) for $n \geq 0$ is an unfortunate notation used in the theory of special functions for the RISING FACTORIAL, which is denoted $x^{(n)}$ (Roman 1984, p. 5) or $\langle x \rangle_n$ (Comtet 1974, p. 6) in combinatorics. In combinatorial usage, $(x)_n$ denotes the FALLING FACTORIAL. Extreme caution is therefore needed in interpreting the notations $(x)_n$ and $x^{(n)}$.

The Pochhammer symbol $(x)_n$ obeys the transformation due to Euler

$$\sum_{n=0}^{\infty} \frac{(a)_n}{n!} a_n z^n = (1-z)^{-a} \sum_{n=0}^{\infty} \frac{(a)_n}{n!} \Delta^n a_0 \left(\frac{z}{1-z}\right)^n, \quad (2)$$

where Δ is the FORWARD DIFFERENCE and

$$\Delta^k a_0 = \sum_{m=0}^{k} (-1)^m \binom{k}{m} a_{k-m} \quad (3)$$

(Nørlund 1955).

The sum of $1/(k)_p$ can be done in closed form as

$$\sum_{k=1}^{n} \frac{1}{(k)_p} = \frac{1}{(p-1)\Gamma(p)} - \frac{n\Gamma(n)}{(p-1)\Gamma(n+p)} \quad (4)$$

for $p > 1$.

See also FACTORIAL, FALLING FACTORIAL, GENERALIZED HYPERGEOMETRIC FUNCTION, HANKEL'S SYMBOL, HARMONIC LOGARITHM, HYPERGEOMETRIC FUNCTION, KRAMP'S SYMBOL

References

Abramowitz, M. and Stegun, C. A. (Eds.). *Handbook of Mathematical Functions with Formulas, Graphs, and Mathematical Tables, 9th printing.* New York: Dover, 1972.

Comtet, L. *Advanced Combinatorics: The Art of Finite and Infinite Expansions, rev. enl. ed.* Dordrecht, Netherlands: Reidel, 1974.

Erdélyi, A.; Magnus, W.; Oberhettinger, F.; and Tricomi, F. G. *Higher Transcendental Functions, Vol. 1.* New York: Krieger, p. 52, 1981.

Koepf, W. *Hypergeometric Summation: An Algorithmic Approach to Summation and Special Function Identities.* Braunschweig, Germany: Vieweg, 1998.

Nørlund, N. E. "Hypergeometric Functions." *Acta Math.* **94**, 289–349, 1955.

Roman, S. *The Umbral Calculus.* New York: Academic Press, p. 5, 1984.

Spanier, J. and Oldham, K. B. "The Pochhammer Polynomials $(x)_n$." Ch. 18 in *An Atlas of Functions.* Washington, DC: Hemisphere, pp. 149–165, 1987.

Pocklington-Lehmer Test

POCKLINGTON'S THEOREM

Pocklington's Criterion

Let p be an ODD PRIME, k be an INTEGER such that $p \nmid k$ and $1 \leq k \leq 2(p+1)$, and

$$N \equiv 2kp + 1.$$

Then the following are equivalent

1. N is PRIME.
2. $\mathrm{GCD}(a^k + 1, N) = 1$,

where GCD is the GREATEST COMMON DENOMINATOR. This is a modified version of the original theorem due to Lehmer.

References

Pocklington, H. C. "The Determination of the Prime or Composite Nature of Large Numbers by Fermat's Theorem." *Proc. Cambridge Phil. Soc.* **18**, 29–30, 1914/16.

Pocklington's Theorem

Let $n-1 = FR$ where F is the factored part of a number

$$F = p_1^{a_1} \cdots p_r^{a_r}, \qquad (1)$$

where $(R, F) = 1$, and $R < \sqrt{n}$. If there exists a b_i for $i = 1, \ldots, r$ such that

$$b_i^{n-1} \equiv 1 \ (\mathrm{mod}\ n) \qquad (2)$$

$$\mathrm{GCD}\left(b_i^{(n-1)/p_i} - 1, \, n\right) = 1, \qquad (3)$$

then n is a PRIME.

Poggendorff Illusion

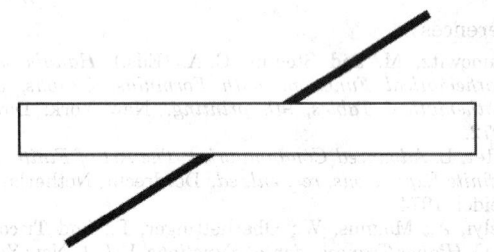

The illusion that the two ends of a straight LINE SEGMENT passing behind an obscuring RECTANGLE are offset when, in fact, they are aligned. The Poggendorff illusion was discovered in 1860 by physicist and scholar J. C. Poggendorff, editor of *Annalen der Physik und Chemie*, after receiving a letter from astronomer F. Zöllner. In his letter, Zöllner described an illusion he noticed on a fabric design in which parallel lines intersected by a pattern of short diagonal lines appear to diverge (ZÖLLNER'S ILLUSION). Pondering this illusion, Poggendorff noticed and described another illusion resulting from the apparent misalignment of a diagonal line; an illusion which today bears his name (IllusionWorks).

See also ILLUSION, MÜLLER-LYER ILLUSION, PONZO'S ILLUSION, VERTICAL-HORIZONTAL ILLUSION, ZÖLLNER'S ILLUSION

References

Burmester, E. "Beiträge zu experimentellen Bestimmung geometrisch-optischer Täuschungen." *Z. Psychologie* **12**, 355–394, 1896.
Day, R. H. and Dickenson, R. G. "The Components of the Poggendorff Illusion." *Brit. J. Psychology* **67**, 537–552, 1976.
Fineman, M. "Poggendorff's Illusion." Ch. 19 in *The Nature of Visual Illusion.* New York: Dover, pp. 151–159, 1996.
Gilliam, B. "A Depth Processing Theory of the Poggendorff Illusion." *Perception & Psychophys.* **10**, 211–216, 1971.
Gillam, B. "Geometrical Illusions." *Sci. Amer.* **242**, 102–111, 1980.

Greene, E. "The Corner Poggendorff." *Perception* **17**, 65–70, 1988.
IllusionWorks. "Poggendorf [sic]." http://www.illusion-works.com/html/poggendorf.html.
Lucas, A. and Fisher, G. H. "Illusions in concrete situations: II. Experimental Studies of the Poggendorff Illusion." *Ergonomics* **12**, 395–402, 1969.
Robinson, J. O. *The Psychology of Visual Illusion.* London: Hutchinson, 1972.
Rock, I. *Perception.* New York: W. H. Freeman, 1984.
Schiffman, H. *Sensation and Perception.* New York: Wiley, 1995.
Spivey-Knowlton, M. J. and Bridgeman, B. "Spatial Context Affects the Poggendorff Illusion." *Perception & Psychophys.* **53**, 467–474, 1993.

Pohlke's Theorem

The principal theorem of AXONOMETRY, first published without proof by Pohlke in 1860. It states that three segments of arbitrary length $a'x'$, $a'y'$, and $a'z'$ which are drawn in a PLANE from a point a' under arbitrary ANGLES form a parallel projection of three equal segments ax, ay, and az from the ORIGIN of three PERPENDICULAR coordinate axes. However, only one of the segments or one of the ANGLES may vanish.

See also AXONOMETRY

References

Schwarz, H. A. *J. reine angew. Math.* **63**, 309–314, 1864.
Steinhaus, H. *Mathematical Snapshots, 3rd ed.* New York: Dover, pp. 170–171, 1999.

Pohlmeyer-Lund-Regge Equation

The system of PARTIAL DIFFERENTIAL EQUATIONS

$$u_{xx} - u_{yy} \pm \sin u \cos u + \frac{\cos u}{\sin^3 u}(v_x^2 - v_y^2) = 0 \qquad (1)$$

$$(v_x \cot^2 u)_x = (v_y \cot^2 u)_y. \qquad (2)$$

References

Calogero, F. and Degasperis, A. *Spectral Transform and Solitons: Tools to Solve and Investigate Nonlinear Evolution Equations.* New York: North-Holland, p. 61, 1982.
Zwillinger, D. *Handbook of Differential Equations, 3rd ed.* Boston, MA: Academic Press, p. 139, 1997.

Poincaré Conjecture

The conjecture that every SIMPLY CONNECTED 3-MANIFOLD is HOMEOMORPHIC to the 3-SPHERE. This conjecture was first proposed in 1904 by H. Poincaré (Poincaré 1953, pp. 486 and 498), and subsequently generalized to the conjecture that every COMPACT n-MANIFOLD is HOMOTOPY-equivalent to the n-sphere IFF it is HOMEOMORPHIC to the n-SPHERE. The generalized statement reduces to the original conjecture for $n = 3$.

The $n = 1$ case of the generalized conjecture is trivial, the $n = 2$ case is classical, $n = 3$ remains open, $n = 4$

was proved by Freedman (1982) (for which he was awarded the 1986 FIELDS MEDAL), $n = 5$ by Zeeman (1961), $n = 6$ by Stallings (1962), and $n \geq 7$ by Smale in 1961. Smale subsequently extended his proof to include $n \geq 5$.

See also COMPACT MANIFOLD, HOMEOMORPHIC, HOMOTOPY, MANIFOLD, PROPERTY P, SIMPLY CONNECTED, SPHERE, THURSTON'S GEOMETRIZATION CONJECTURE

References

Adams, C. C. "The Poincaré Conjecture, Dehn Surgery, and the Gordon-Luecke Theorem." §9.3 in *The Knot Book: An Elementary Introduction to the Mathematical Theory of Knots.* New York: W. H. Freeman, pp. 257–263, 1994.

Batterson, S. *Stephen Smale: The Mathematician Who Broke the Dimension Barrier.* Providence, RI: Amer. Math. Soc., 2000. Bing, R. H. "Some Aspects of the Topology of 3-Manifolds Related to the Poincaré Conjecture." In *Lectures on Modern Mathematics, Vol. II* (Ed. T. L. Saaty). New York: Wiley, pp. 93–128, 1964.

Birman, J. "Poincaré's Conjecture and the Homotopy Group of a Closed, Orientable 2-Manifold." *J. Austral. Math. Soc.* **17**, 214–221, 1974.

Clay Mathematics Institute. "The Poincaré Conjecture." http://www.claymath.org/prize_problems/poincare.htm.

Freedman, M. H. "The Topology of Four-Differentiable Manifolds." *J. Diff. Geom.* **17**, 357–453, 1982.

Gabai, D. "Valentin Poenaru's Program for the Poincaré Conjecture." In *Geometry, Topology, & Physics, Conf. Proc. Lecture Notes Geom. Topol., VI* (Ed. S.-T. Yau). Cambridge, MA: International Press, pp. 139–166, 1995.

Gillman, D. and Rolfsen, D. "The Zeeman Conjecture for Standard Spines is Equivalent to the Poincaré Conjecture." *Topology* **22**, 315–323, 1983.

Jakobsche, W. "The Bing-Borsuk Conjecture is Stronger than the Poincaré Conjecture." *Fund. Math.* **106**, 127–134, 1980.

Milnor, J. "The Poincaré Conjecture." http://www.claymath.org/prize_problems/poincare.pdf.

Papakyriakopoulos, C. "A Reduction of the Poincaré Conjecture to Group Theoretic Conjectures." *Ann. Math.* **77**, 250–205, 1963.

Poincaré, H. *Œuvres de Henri Poincaré, tome VI.* Paris: Gauthier-Villars, pp. 486 and 498, 1953.

Rourke, C. "Algorithms to Disprove the Poincaré Conjecture." *Turkish J. Math.* **21**, 99–110, 1997.

Stallings, J. "The Piecewise-Linear Structure of Euclidean Space." *Proc. Cambridge Philos. Soc.* **58**, 481–488, 1962.

Smale, S. "Generalized Poincaré's Conjecture in Dimensions Greater than Four." *Ann. Math.* **74**, 391–406, 1961.

Smale, S. "The Story of the Higher Dimensional Poincaré Conjecture (What Actually Happened on the Beaches of Rio)." *Math. Intell.* **12**, 44–51, 1990.

Smale, S. "Mathematical Problems for the Next Century." In *Mathematics: Frontiers and Perspectives 2000* 0821820702 (Ed. V. Arnold, M. Atiyah, P. Lax, and B. Mazur). Providence, RI: Amer. Math. Soc., 2000.

Thickstun, T. L. "Open Acyclic 3-Manifolds, a Loop Theorem, and the Poincaré Conjecture." *Bull. Amer. Math. Soc.* **4**, 192–194, 1981.

Zeeman, E. C. "The Generalised Poincaré Conjecture." *Bull. Amer. Math. Soc.* **67**, 270, 1961.

Zeeman, E. C. "The Poincaré Conjecture for $n \geq 5$." In *Topology of 3-Manifolds and Related Topics, Proceedings of the University of Georgia Institute, 1961.* Englewood Cliffs, NJ: Prentice-Hall, pp. 198–204, 1961.

Poincaré Disk

POINCARÉ HYPERBOLIC DISK

Poincaré Duality

The BETTI NUMBERS of a compact orientable n-MANIFOLD satisfy the relation

$$b_i = b_{n-i}.$$

See also BETTI NUMBER, INTERSECTION (HOMOLOGY)

Poincaré Formula

The POLYHEDRAL FORMULA generalized to a surface of GENUS g,

$$V - E + F = \chi(g)$$

where V is the number of VERTICES, E is the number of EDGES, F is the number of faces, and

$$\chi(g) \equiv 2 - 2g$$

is called the EULER CHARACTERISTIC.

See also EULER CHARACTERISTIC, GENUS (SURFACE), POLYHEDRAL FORMULA

References

Coxeter, H. S. M. "Poincaré's Proof of Euler's Formula." Ch. 9 in *Regular Polytopes, 3rd ed.* New York: Dover, pp. 165–172, 1973.

Eppstein, D. "Fourteen Proofs of Euler's Formula: $V - E + F = 2$." http://www.ics.uci.edu/~eppstein/junkyard/euler/.

Poincaré Group

LORENTZ GROUP

Poincaré Hyperbolic Disk

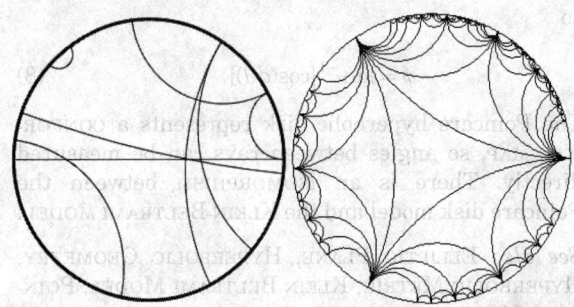

A 2-D space having HYPERBOLIC GEOMETRY defined as the DISK $\{x \in \mathbb{R}^2 : |x| < 1\}$, with HYPERBOLIC METRIC

$$ds^2 = \frac{dx^2 + dy^2}{(1 - r^2)^2}. \tag{1}$$

The Poincaré disk is a model for HYPERBOLIC GEOMETRY in which a line is REPRESENTED AS an ARC of a CIRCLE whose ends are PERPENDICULAR to the DISK's boundary (and DIAMETERS are also permitted). Two arcs which do not meet correspond to parallel rays, arcs which meet orthogonally correspond to PERPENDICULAR lines, and arcs which meet on the boundary are a pair of limits rays.

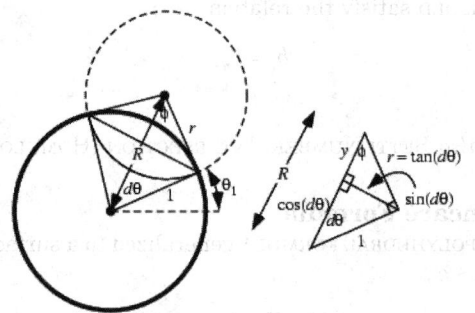

The endpoints of any arc can be specified by two angles around the disk θ_1 and θ_2. Define

$$\theta \equiv \tfrac{1}{2}(\theta_1 + \theta_2) \tag{2}$$

$$d\theta \equiv \tfrac{1}{2}|\theta_1 - \theta_2| \tag{3}$$

Then trigonometry shows that in the above diagram,

$$r = \tan(d\theta) \tag{4}$$

$$y = \sin(d\theta)\tan(d\theta), \tag{5}$$

so the radius of the circle forming the arc is

$$R = \cos(d\theta) + y = \sec(d\theta) \tag{6}$$

and its center is located at $R(\cos\theta, \sin\theta)$. The half-angle subtended by the arc is then

$$\sin\phi = \frac{\sin(d\theta)}{\tan(d\theta)} = \cos(d\theta), \tag{7}$$

so

$$\phi = \sin^{-1}[\cos(d\theta)]. \tag{8}$$

The Poincaré hyperbolic disk represents a CONFORMAL MAP, so angles between rays can be measured directly. There is an ISOMORPHISM between the Poincaré disk model and the KLEIN-BELTRAMI MODEL.

See also ELLIPTIC PLANE, HYPERBOLIC GEOMETRY, HYPERBOLIC METRIC, KLEIN-BELTRAMI MODEL, POINCARÉ METRIC

References

Anderson, J. W. "The Poincaré Disc Model." §4.1 in *Hyperbolic Geometry*. New York: Springer-Verlag, pp. 95–104, 1999.
Goodman-Strauss, C. "Compass and Straightedge in the Poincaré Disk." To be submitted.
Wells, D. *The Penguin Dictionary of Curious and Interesting Geometry*. London: Penguin, pp. 188–189, 1991.

Poincaré Manifold

A nonsimply connected 3-manifold also called a DODECAHEDRAL SPACE.

References

Rolfsen, D. *Knots and Links*. Wilmington, DE: Publish or Perish Press, pp. 245, 290, and 308, 1976.

Poincaré Metric

The METRIC

$$ds^2 = \frac{dx^2 + dy^2}{\left(1 - |z|^2\right)^2}$$

of the POINCARÉ HYPERBOLIC DISK.

See also POINCARÉ HYPERBOLIC DISK

Poincaré Separation Theorem

Let $\{\mathbf{y}^k\}$ be a set of orthonormal vectors with $k = 1, 2, \ldots, K$, such that the INNER PRODUCT $(\mathbf{y}^k, \mathbf{y}^k) = 1$. Then set

$$\mathbf{x} = \sum_{k=1}^{K} u_k \mathbf{y}^k \tag{1}$$

so that for any SQUARE MATRIX $\mathbf{A}$ for which the product $\mathbf{Ax}$ is defined, the corresponding QUADRATIC FORM is

$$(\mathbf{x}, \mathbf{Ax}) = \sum_{k,l=1}^{K} u_k u_l \left(\mathbf{y}^k, \mathbf{Ay}^l\right) \tag{2}$$

Then if

$$\mathbf{B}_k = \left(\mathbf{y}^k, \mathbf{Ay}^l\right) \tag{3}$$

for $k, l = 1, 2, \ldots, K$, it follows that

$$\lambda_i(\mathbf{B}_K) \leq \lambda_1(\mathbf{A}) \tag{4}$$

$$\lambda_{K-j}(\mathbf{B}_K) \geq \lambda_{N-j}(\mathbf{A}) \tag{5}$$

for $i = 1, 2, \ldots, K$ and $j = 0, 1, \ldots, K-1$.

References

Gradshteyn, I. S. and Ryzhik, I. M. *Tables of Integrals, Series, and Products*, 6th ed. San Diego, CA: Academic Press, p. 1120, 2000.

Poincaré-Bertrand Theorem

For $s_1, s_2 = \pm 1$,

$$\lim_{\substack{\epsilon_1 \to 0 \\ \epsilon_2 \to 0}} \frac{1}{x_1 - is_1\epsilon_1} \frac{1}{x_2 - is_2\epsilon_2}$$

$$= \left[PV\left(\frac{1}{x_1}\right) + i\pi s_1 \delta(x_1) \right]\left[PV\left(\frac{1}{x_2}\right) + i\pi s_2 \delta(x_2) \right]$$

$$+\pi^2 \delta(x_1)\delta(x_2), \tag{1}$$

where $\delta(x)$ is the DELTA FUNCTION and PV denotes the CAUCHY PRINCIPAL VALUE.

See also DELTA FUNCTION

Poincaré-Birkhoff Fixed Point Theorem

For the rational curve of an unperturbed system with ROTATION NUMBER r/s under a map T (for which every point is a FIXED POINT of T^s), only an even number of FIXED POINTS $2ks$ ($k = 1, 2, ...$) will remain under perturbation. These FIXED POINTS are alternately stable (ELLIPTIC) and unstable (HYPERBOLIC). Around each elliptic fixed point there is a simultaneous application of the Poincaré-Birkhoff fixed point theorem and the KAM THEOREM, which leads to a self-similar structure on all scales.

The original formulation was: Given a CONFORMAL ONE-TO-ONE transformation from an ANNULUS to itself that advances points on the outer edge positively and on the inner edge negatively, then there are at least two fixed points.

It was conjectured by Poincaré from a consideration of the three-body problem in celestial mechanics and proved by Birkhoff.

Poincaré-Birkhoff-Witt Theorem

Every LIE ALGEBRA L is isomorphic to a SUBALGEBRA of some LIE ALGEBRA A^-, where the ASSOCIATIVE ALGEBRA A may be taken to be the linear operators over a VECTOR SPACE V.

See also ASSOCIATIVE, LIE ALGEBRA, VECTOR SPACE

References

Jacobson, N. *Lie Algebras.* New York: Dover, pp. 159–160, 1979.
Schafer, R. D. *An Introduction to Nonassociative Algebras.* New York: Dover, p. 3, 1996.

Poincaré-Fuchs-Klein Automorphic Function

$$f(z) = \frac{k}{(cz+d)^r} f\left(\frac{az+b}{cz+d}\right)$$

where $\Im[z] > 0$.

See also AUTOMORPHIC FUNCTION

Poincaré-Hopf Index Theorem

The index of a VECTOR FIELD with finitely many zeros on a compact, oriented MANIFOLD is the same as the EULER CHARACTERISTIC of the MANIFOLD.

See also GAUSS-BONNET FORMULA

Poincaré's Holomorphic Lemma

Solutions to HOLOMORPHIC differential equations are themselves HOLOMORPHIC FUNCTIONS of time, initial conditions, and parameters.

See also POINCARÉ'S LEMMA

Poincaré's Lemma

Poincaré's lemma says that on a CONTRACTIBLE MANIFOLD, all CLOSED FORMS are EXACT. While $d^2 = 0$ implies that all exact forms are closed, it is not always true that all closed forms are exact. The Poincaré lemma is used to show that closed forms represent COHOMOLOGY CLASSES.

See also COHOMOLOGY, COHOMOLOGY CLASS, CLOSED FORM, DE RHAM COHOMOLOGY, DIFFERENTIAL FORM, EXACT FORM, EXTERIOR DERIVATIVE, MANIFOLD, POINCARÉ'S HOLOMORPHIC LEMMA, STOKES' THEOREM, WEDGE PRODUCT

Poincaré's Theorem

If $\nabla \times \mathbf{F} = 0$ (i.e., $\mathbf{F}(\mathbf{x})$ is an IRROTATIONAL FIELD) in a simply connected neighborhood $U(\mathbf{x})$ of a point $\mathbf{x}$, then in this neighborhood, $\mathbf{F}$ is the GRADIENT of a SCALAR FIELD $\phi(\mathbf{x})$,

$$\mathbf{F}(\mathbf{x}) = -\nabla\phi(\mathbf{x}) \tag{1}$$

for $\mathbf{x} \in U(\mathbf{x})$, where ∇ is the gradient operator. Consequently, the GRADIENT THEOREM gives

$$\int_\sigma \mathbf{F} \cdot ds = \phi(\mathbf{x}_1) - \phi(\mathbf{x}_2) \tag{2}$$

for any path σ located completely within $U(\mathbf{x})$, starting at $\mathbf{x}_1$ and ending at $\mathbf{x}_2$.

This means that if $\nabla \times \mathbf{F} = 0$, the LINE INTEGRAL of $\mathbf{F}$ is path-independent.

See also CONSERVATIVE FIELD, GRADIENT THEOREM, IRROTATIONAL FIELD, LINE INTEGRAL

Poinsot Solid

KEPLER-POINSOT SOLID

Poinsot's Spirals

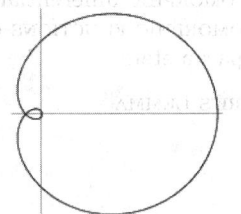

$$r \sinh(n\theta) = a.$$

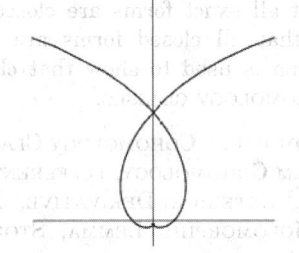

$$r \operatorname{csch}(n\theta) = a.$$

References
Lawrence, J. D. *A Catalog of Special Plane Curves.* New York: Dover, pp. 192 and 194, 1972.

Point

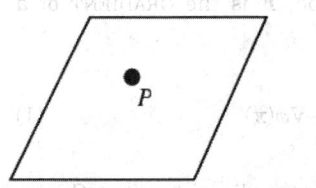 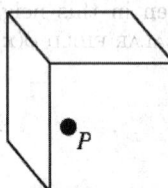

A 0-DIMENSIONAL mathematical object which can be specified in n-D space using n coordinates. Although the notion of a point is intuitively rather clear, the mathematical machinery used to deal with points and point-like objects can be surprisingly slippery. This difficulty was encountered by none other than Euclid himself who, in his *ELEMENTS*, gave the vague definition of a point as "that which has no part."

The basic geometric structures of higher DIMENSIONAL geometry–the LINE, PLANE, SPACE, and HYPERSPACE–are all built up of infinite numbers of points arranged in particular ways.

The DECIMAL POINT in a DECIMAL EXPANSION is voiced as "point" in the United States, e.g., 3.1415 is voiced "three point one four one five," whereas a COMMA is used for this purpose in continental Europe.

See also ACCUMULATION POINT, BOUNDARY POINT, BRANCH POINT, COMMA, CONCUR, CONCURRENT, CRITICAL POINT, DOUBLE POINT, ENDPOINT, FIXED POINT, ISOLATED POINT, LIMIT POINT, MIDPOINT, ORDINARY POINT, SINGULAR POINT (ALGEBRAIC CURVE), SINGULAR POINT (FUNCTION)

References
Casey, J. "The Point." Ch. 1 in *A Treatise on the Analytical Geometry of the Point, Line, Circle, and Conic Sections, Containing an Account of Its Most Recent Extensions, with Numerous Examples,* 2nd ed., rev. enl. Dublin: Hodges, Figgis, & Co., pp. 1–29, 1893.
Lachlan, R. "Special Points Connected with a Triangle." §112–117 in *An Elementary Treatise on Modern Pure Geometry.* London: Macmillian, pp. 62–66, 1893.

Point at Infinity

P is the point on the line AB such that $\overline{PA}/\overline{PB} = 1$. It can also be thought of as the point of intersection of two PARALLEL lines. In 1639, Desargues (1864) became the first to consider the point at infinity (Cremona 1960, p. ix), although Poncelet was the first to systematically employ the point at infinity (Graustein 1930).

The term point at infinity is also used for COMPLEX INFINITY (Krantz 1999, p. 82).

See also COMPLEX INFINITY, LINE AT INFINITY

References
Behnke, H.; Bachmann, F.; Fladt, K.; and Suss, W. (Eds.). Ch. 7 in *Fundamentals of Mathematics, Vol. 3: Points at Infinity.* Cambridge, MA: MIT Press, 1974.
Cremona, L. *Elements of Projective Geometry, 3rd ed.* New York: Dover, 1960.
Desargues, G. "Brouillon-projet d'une atteinte aux événements des recontres d'un cône avec un plan." *Œuvres de Desargues, réunies et analysées par M. Pudra, tome 1.* Paris, pp. 104, 105, and 205, 1864.
Durell, C. V. *Modern Geometry: The Straight Line and Circle.* London: Macmillan, p. 38, 1928.
Graustein, W. C. *Introduction to Higher Geometry.* New York: Macmillan, p. 30, 1930.
Krantz, S. G. *Handbook of Complex Analysis.* Boston, MA: Birkhäuser, p. 82, 1999.
Lachlan, R. "Point at Infinity." §9 in *An Elementary Treatise on Modern Pure Geometry.* London: Macmillian, pp. 5–6, 1893.

Point Circle

Members of a COAXAL SYSTEM satisfy

$$x^2 + y^2 + 2\lambda x + c = (x + \lambda)^2 + y^2 + c - \lambda^2 = 0$$

for values of λ. Picking $\lambda^2 = c$ then gives the two circles

$$\left(x \pm \sqrt{c}\right)^2 + y^2 = 0$$

of zero RADIUS, known as point circles. The two point circles $(\pm\sqrt{c}, 0)$, real or imaginary, are called the LIMITING POINTS of the COAXAL SYSTEM.

See also COAXAL SYSTEM, LIMITING POINT

References

Durell, C. V. *Modern Geometry: The Straight Line and Circle*. London: Macmillan, p. 123, 1928.

Point Connectivity

VERTEX CONNECTIVITY

Point Distances

The maximum distance between n points in 3-D can occur no more than $2n - 2$ times. Also, there exists a fixed number c such that no distance determined by a set of n points in 3-D space occurs more than $cn^{5/3}$ times. The maximum distance can occur no more than $\left\lfloor \frac{1}{4} n^2 \right\rfloor$ times in 4-D, where $\lfloor x \rfloor$ is the FLOOR FUNCTION.

See also POINT-LINE DISTANCE–2-D, POINT-LINE DISTANCE–3-D, POINT-POINT DISTANCE–2-D, POINT-POINT DISTANCE–3-D, SPAN (GEOMETRY)

References

Honsberger, R. *Mathematical Gems II*. Washington, DC: Math. Assoc. Amer., pp. 122–123, 1976.

Point Estimation Theory

A theory of constructing initial conditions that provides safe convergence of a numerical root-finding algorithm for an equation $f(z) = 0$. Point estimation theory treats convergence conditions and the domain of convergence using only information about f at the initial point z_0 (Petkovic *et al.* 1997, p. 1). An initial point that provides safe convergence of NEWTON'S METHOD is called an APPROXIMATE ZERO.

Point estimation theory should not be confusion with POINT ESTIMATORS of probability theory.

See also ALPHA-TEST, APPROXIMATE ZERO, NEWTON'S METHOD, POINT ESTIMATOR

References

Lehmann, E. L. and Casella, G. *Theory of Point Estimation*. New York: Springer-Verlag, 1998.
Petkovic, M. S.; Herceg, D. D.; and Ilic, S. M. *Point Estimation Theory and Its Applications*. Novi Sad, Yugoslavia: Institute of Mathematics, 1997.

Point Estimator

An ESTIMATOR of the actual values of population.

See also POINT ESTIMATION THEORY

Point Groups

A point group is a group of symmetry operations which all leave at least one point unmoved. Although an isolated object may have an arbitrary SCHÖNFLIES SYMBOL, the requirement that symmetry be present in a lattice requires that only 1, 2, 3, and 6-fold symmetry axes are possible (the CRYSTALLOGRAPHY RESTRICTION), which restricts the number of possible so-called CRYSTALLOGRAPHIC POINT GROUPS to 32.

See also CRYSTALLOGRAPHIC POINT GROUPS, CRYSTALLOGRAPHY RESTRICTION, SCHÖNFLIES SYMBOL, SPACE GROUPS

References

Hahn, T. (Ed.). *International Tables for Crystallography, vol. A, 4th ed.* Dordrecht, Netherlands: Kluwer, p. 752, 1995.

Point Lattice

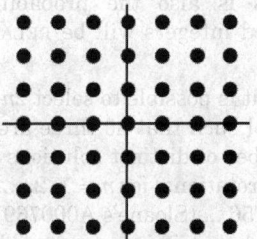

A regularly spaced array of points falling along regularly spaced line. The grid lines can be oriented to form unit cells in the shape of a square, rectangle, hexagon, etc. However, unless otherwise specified, point lattices are generally taken to refer to points in a square array, i.e., points with coordinates $(m, n, \cdots)$, where m, n, ... are INTEGERS. Such an array is often called a GRID or a MESH. Point lattices are frequently simply called "lattices," which unfortunately conflicts with the same term applied to ordered sets treated in LATTICE THEORY.

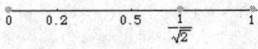

Formally, a lattice is a DISCRETE SUBGROUP of EUCLIDEAN SPACE, assuming it contains the origin. That is, a lattice is closed under addition and inverses, and every point has a neighborhood in which it is the only lattice point. The common examples are $\mathbb{Z} \subset \mathbb{R}$ and $\mathbb{Z}^2 \subset \mathbb{R}^2$. Usually, a lattice is defined to have full rank, i.e., a lattice in $\mathbb{R}^n$ is the SUBGROUP

$$\{a_1 v_1 + \cdots a_n v_n\}, \tag{1}$$

where the a_i are integers and v_i are LINEARLY INDEPENDENT vectors. Note that a lattice needs at most n elements to generate it. For example, the subgroup $\{a_1 + a_2 \sqrt{2}\} \subset \mathbb{R}$ requires two generators but is not DISCRETE, and is not a lattice. The above illustration shows that the subgroup generated by 1 and $1/\sqrt{2}$ is not a lattice by showing $a + b/\sqrt{2}$ for successive $b \in [0, 1]$.

The FRACTION of lattice points VISIBLE from the ORIGIN, as derived in Castellanos (1988, pp. 155–156), is

$$\frac{N'(r)}{N(r)} = \frac{\frac{24}{\pi^2}r^2 + \mathcal{O}(r\ln r)}{4r^2 + \mathcal{O}(r)}$$

$$= \frac{\frac{6}{\pi^2} + \mathcal{O}\left(\frac{\ln r}{r}\right)}{1 + \mathcal{O}\left(\frac{1}{r}\right)}$$

$$= \frac{6}{\pi^2}. \tag{2}$$

Therefore, this is also the probability that two randomly picked integers will be RELATIVELY PRIME to one another.

For $2 \le n \le 32$, it is possible to select $2n$ lattice points with $x, y \in [1, n]$ such that no three are in a straight LINE. The number of distinct solutions (not counting reflections and rotations) for $n = 1, 2, \ldots$, are 1, 1, 4, 5, 11, 22, 57, 51, 156 ... (Sloane's A000769). For large n, it is conjectured that it is only possible to select at most $(c + \epsilon)n$ lattice points with no three COLLINEAR, where

$$c = \left(2\pi^2/3\right)^{1/3} \approx 1.87 \tag{3}$$

(Guy and Kelly 1968; Guy 1994, p. 242). The number of the n^2 lattice points $x, y \in [1, n]$ which can be picked with no four CONCYCLIC is $\mathcal{O}(n^{2/3} - \epsilon)$ (Guy 1994, p. 241).

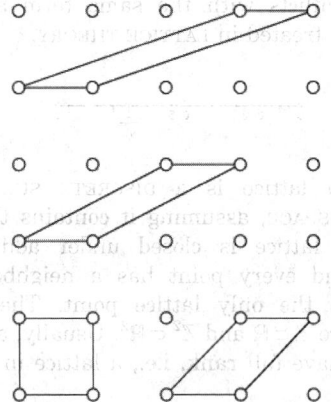

Any PARALLELOGRAM on the lattice in which two opposite sides each have length 1 has unit area (Hilbert and Cohn-Vossen 1999, pp. 33–34).

A special set of POLYGONS defined on the regular lattice are the GOLYGONS. A NECESSARY and SUFFICIENT condition that a linear transformation transforms a lattice to itself is that it be UNIMODULAR. M. Ajtai has shown that there is no efficient ALGORITHM for finding any fraction of a set of spanning vectors in a lattice having the shortest lengths unless there is an efficient algorithm for all of them (of which none is known). This result has potential applications to cryptography and authentication (Cipra 1996).

See also BARNES-WALL LATTICE, BLICHFELDT'S THEOREM, BROWKIN'S THEOREM, CIRCLE LATTICE POINTS, COXETER-TODD LATTICE, EHRHART POLYNOMIAL, ELLIPTIC CURVE, GAUSS'S CIRCLE PROBLEM, GOLYGON, INTEGRATION LATTICE, JARNICK'S INEQUALITY, LATTICE PATH, LATTICE SUM, LEECH LATTICE, MINKOWSKI CONVEX BODY THEOREM, MODULAR LATTICE, N-CLUSTER, NOSARZEWSKA'S INEQUALITY, PICK'S THEOREM, RANDOM WALK, SCHINZEL'S THEOREM, SCHRÖDER NUMBER, TORUS, UNIT LATTICE, VISIBLE POINT, VORONOI POLYGON

References

Apostol, T. *Introduction to Analytic Number Theory.* New York: Springer-Verlag, 1995.

Castellanos, D. "The Ubiquitous Pi." *Math. Mag.* **61**, 67–98, 1988.

Cipra, B. "Lattices May Put Security Codes on a Firmer Footing." *Science* **273**, 1047–1048, 1996.

Eppstein, D. "Lattice Theory and Geometry of Numbers." http://www.ics.uci.edu/~eppstein/junkyard/lattice.html.

Gardner, M. "The Lattice of Integer." Ch. 21 in *The Sixth Book of Mathematical Games from Scientific American.* Chicago, IL: University of Chicago Press, pp. 208–219, 1984.

Guy, R. K. "Gauss's Lattice Point Problem," "Lattice Points with Distinct Distances," "Lattice Points, No Four on a Circle," and "The No-Three-in-a-Line Problem." §F1, F2, F3, and F4 in *Unsolved Problems in Number Theory, 2nd ed.* New York: Springer-Verlag, pp. 240–244, 1994.

Guy, R. K. and Kelly, P. A. "The No-Three-in-Line-Problem." *Canad. Math. Bull.* **11**, 527–531, 1968.

Hammer, J. *Unsolved Problems Concerning Lattice Points.* London: Pitman, 1977.

Hilbert, D. and Cohn-Vossen, S. "Regular Systems of Points." Ch. 2 in *Geometry and the Imagination.* New York: Chelsea, pp. 32–93, 1999.

Knupp, P. and Steinberg, S. *Fundamentals of Grid Generation.* Boca Raton, FL: CRC Press, 1994.

Nagell, T. "Lattice Points and Point Lattices." §11 in *Introduction to Number Theory.* New York: Wiley, pp. 32–34, 1951.

Sloane, N. J. A. Sequences A000769/M3252 in "An On-Line Version of the Encyclopedia of Integer Sequences." http://www.research.att.com/~njas/sequences/eisonline.html.

Thompson, J. F.; Soni, B.; and Weatherill, N. *Handbook of Grid Generation.* Boca Raton, FL: CRC Press, 1998.

Point Picking

In finding the average area $\bar{A}_R$ of a triangle chosen from a closed, bounded, convex region R of the plane, then $\bar{A}_{T(R)} = \bar{A}_R$, for T any nonsingular affine transformation of the plane.

See also 18-POINT PROBLEM, BALL LINE PICKING, BALL TRIANGLE PICKING, CUBE LINE PICKING, CUBE POINT PICKING, CUBE TETRAHEDRON PICKING, CUBE TRIANGLE PICKING, DISCREPANCY THEOREM, DISK LINE PICKING, DISK POINT PICKING, DISK TRIANGLE PICKING, HAPPY END PROBLEM, PLANAR DISTANCE, POINT-POINT DISTANCE–1-D, POINT-POINT DISTANCE–2-D, POINT-POINT DISTANCE–3-D, SIMPLEX POINT PICKING, SPHERE LINE PICKING, SPHERE POINT PICKING, SPHERE TETRAHEDRON PICKING, SYLVESTER'S FOUR-POINT PROBLEM, TRIANGLE POINT PICKING

References

Pfiefer, R. E. "The Historical Development of J. J. Sylvester's Four Point Problem." *Math. Mag.* **62**, 309–17, 1989.

Point Probability

The portion of the probability distribution which has a *P*-VALUE *equal to* the observed *P*-VALUE.

See also TAIL PROBABILITY

Point-Line Distance—2-D

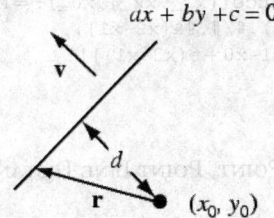

Given a line $ax + by + c = 0$ and a point (x_0, y_0), in slope-intercept form, the equation of the line is

$$y = -\frac{a}{b}x - \frac{c}{b}, \tag{1}$$

so the line has SLOPE $-a/b$. Points on the line have the vector coordinates

$$\begin{bmatrix} x \\ -\frac{a}{b}x - \frac{c}{d} \end{bmatrix} = \begin{bmatrix} 0 \\ -\frac{c}{d} \end{bmatrix} - \frac{1}{b}\begin{bmatrix} -b \\ a \end{bmatrix}x. \tag{2}$$

Therefore, the VECTOR

$$\begin{bmatrix} -b \\ a \end{bmatrix} \tag{3}$$

is PARALLEL to the line, and the VECTOR

$$\mathbf{v} = \begin{bmatrix} a \\ b \end{bmatrix} \tag{4}$$

is PERPENDICULAR to it. Now, a VECTOR from the point to the line is given by

$$\mathbf{r} = \begin{bmatrix} x - x_0 \\ y - y_0 \end{bmatrix} \tag{5}$$

Projecting $\mathbf{r}$ onto $\mathbf{v}$,

$$d = |\mathrm{proj}_\mathbf{v}\mathbf{r}| = \frac{|\mathbf{v} \cdot \mathbf{r}|}{\mathbf{v}} = |\hat{\mathbf{v}} \cdot \mathbf{r}| = \frac{|a(x - x_0) + b(y - y_0)|}{\sqrt{a^2 + b^2}}$$

$$= \frac{|ax + by - ax_0 - by_0|}{\sqrt{a^2 + b^2}}$$

$$= \frac{|ax_0 + by_0 + c|}{\sqrt{a^2 + b^2}}. \tag{6}$$

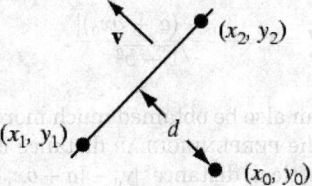

If the line is represented by the endpoints of a VECTOR (x_1, y_1) and (x_2, y_2), then the PERPENDICULAR VECTOR is

$$\mathbf{v} = \begin{bmatrix} y_2 - y_1 \\ -(x_2 - x_1) \end{bmatrix} \tag{7}$$

$$\hat{\mathbf{v}} = \frac{1}{s}\begin{bmatrix} y_2 - y_1 \\ -(x_2 - x_1) \end{bmatrix}, \tag{8}$$

where

$$s = |\mathbf{v}| = \sqrt{(x_2 - x_1)^2 + (y_2 - y_1)^2}, \tag{9}$$

so the distance is

$$d = |\hat{\mathbf{v}} \cdot \mathbf{r}| = \frac{|(y_2 - y_1)(x_0 - x_1) - (x_2 - x_1)(y_0 - y_1)|}{\sqrt{(x_2 - x_1)^2 + (y_2 - y_1)^2}}. \tag{10}$$

The distance from a point (x_0, y_0) to the line $y = a + bx$ can also be computed using simple VECTOR algebra. Let $\mathbf{L}$ be a VECTOR in the same direction as the line

$$\mathbf{L} = \begin{bmatrix} x \\ a + bx \end{bmatrix} - \begin{bmatrix} 0 \\ a \end{bmatrix} = \begin{bmatrix} x \\ bx \end{bmatrix} \tag{11}$$

$$\hat{\mathbf{L}} = \frac{1}{\sqrt{b^2 + 1}} \begin{bmatrix} 1 \\ b \end{bmatrix}. \tag{12}$$

A given point on the line is

$$\mathbf{x} = \begin{bmatrix} x_0 \\ y_0 \end{bmatrix} - \begin{bmatrix} 0 \\ -a \end{bmatrix} = \begin{bmatrix} x_0 \\ y_0 - a \end{bmatrix}, \tag{13}$$

so the point-line distance is

$$\mathbf{r} = (\mathbf{x} \cdot \hat{\mathbf{L}})\hat{\mathbf{L}} - \mathbf{x}$$

$$= \frac{1}{1+b^2}\left(\begin{bmatrix} x_0 \\ y_0 - a \end{bmatrix} \cdot \begin{bmatrix} 1 \\ v \end{bmatrix}\right)\begin{bmatrix} 1 \\ b \end{bmatrix} - \begin{bmatrix} x_0 \\ y_0 - a \end{bmatrix}$$

$$= \frac{y_0 - (a + bx_0)}{1+b^2}\begin{bmatrix} b \\ -1 \end{bmatrix}. \tag{14}$$

Therefore,

$$d = |\mathbf{r}| = \frac{|y_0 - (a + bx_0)|}{\sqrt{1+b^2}}. \tag{15}$$

This result can also be obtained much more simply by noting that the PERPENDICULAR distance is just $\cos\theta$ times the vertical distance $|y_0 - (a + bx_1)|$. But the SLOPE b is just $\tan\theta$, so

$$\sin^2\theta + \cos^2\theta = 1 \Rightarrow \tan^2\theta + 1 = \frac{1}{\cos^2\theta}, \tag{16}$$

and

$$\cos\theta = \frac{1}{\sqrt{1 + \tan^2\theta}} = \frac{1}{\sqrt{1+b^2}}. \tag{17}$$

The PERPENDICULAR distance is then

$$d = \frac{|y_0 - (a + bx_1)|}{\sqrt{1+b^2}}, \tag{18}$$

the same result as before.

See also LINE, POINT, POINT-LINE DISTANCE–3-D

Point-Line Distance—3-D

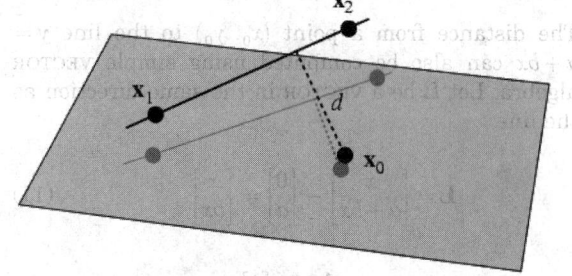

Let a line in 3-D be specified by two points $\mathbf{x}_1$ and $\mathbf{x}_2$ lying on it, so a vector along the line is given by

$$\mathbf{v} = \begin{bmatrix} x_1 + (x_2 - x_1)t \\ y_1 + (y_2 - y_1)t \\ z_1 + (z_2 - z_1)t \end{bmatrix}. \tag{1}$$

The distance between a point on the line with parameter t and a point (x_0, y_0, z_0) is therefore

$$r^2 = [x_1 - x_0 + (x_2 - x_1)t]^2 + [y_1 - y_0$$

$$+ (y_2 - y_1)t]^2 + [z_1 - z_0 + (z_2 - z_1)t]^2. \tag{2}$$

To minimize the distance, set $d(r^2)/dt = 0$ and solve for t to obtain $t = f/g$, where

$$f = (x_1 - x_0)(x_2 - x_1) + (y_1 - y_0)(y_2 - y_1)$$

$$+ (z_1 - z_0)(z_2 - z_1) \tag{3}$$

$$g = (x_2 - x_1)^2 + (y_2 - y_1)^2 + (z_2 - z_1)^2, \tag{4}$$

and the minimum distance can then be found by plugging t into (2) and taking the SQUARE ROOT. This can be implemented in *Mathematica* as

```
PointLineDistance[{x1_,x2_},x0_]:=Module[
  {t=-(x1-x0).#/#.#&[x2-x1]},
  Sqrt[#.#&[x1-x0+t(x2-x1)]]
]
```

See also LINE, POINT, POINT-LINE DISTANCE–2-D

Point-Plane Distance

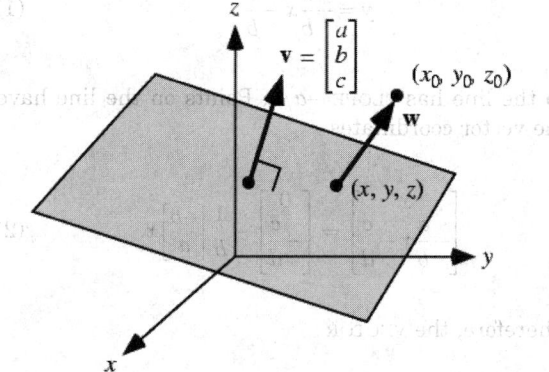

Given a PLANE

$$ax + by + cz + d = 0 \tag{1}$$

and a point (x_0, y_0, z_0), the NORMAL to the PLANE is given by

$$\mathbf{v} = \begin{bmatrix} a \\ b \\ c \end{bmatrix}, \tag{2}$$

and a VECTOR from the plane to the point is given by

$$\mathbf{w} = -\begin{bmatrix} x - x_0 \\ y - y_0 \\ z - z_0 \end{bmatrix}. \tag{3}$$

Projecting **w** onto **v**,

$$D = |\text{proj}_\mathbf{v}\, \mathbf{w}| = \frac{|\mathbf{v} \cdot \mathbf{w}|}{|\mathbf{v}|}$$

$$= \frac{|a(x - x_0) + b(y - y_0) + c(z - z_0)|}{\sqrt{a^2 + b^2 + c^2}}$$

$$= \frac{|ax + by + cz - ax_0 - by_0 - cz_0|}{\sqrt{a^2 + b^2 + c^2}}$$

$$= \frac{|ax_0 + by_0 + cz_0 + d|}{\sqrt{a^2 + b^2 + c^2}}. \tag{4}$$

Given three points $\mathbf{x}_i$ for $i = 1, 2, 3$, compute the unit normal

$$\hat{\mathbf{n}} = \frac{(\mathbf{x}_2 - \mathbf{x}_1) \times (\mathbf{x}_3 - \mathbf{x}_1)}{|(\mathbf{x}_2 - \mathbf{x}_1) \times (\mathbf{x}_3 - \mathbf{x}_1)|}. \tag{5}$$

Then the distance from a point $\mathbf{x}_0$ to the plane containing the three points is given by

$$D_i = \hat{\mathbf{n}} \cdot (\mathbf{x}_i - \mathbf{x}_0), \tag{6}$$

where $\mathbf{x}_i$ is any of the three points. Expanding out the coordinates shows that

$$D \equiv D_1 = D_2 = D_3, \tag{7}$$

as it must since all points are in the same plane, although this is far from obvious based on the above vector equation.

See also PROJECTION THEOREM

Point-Point Distance—1-D

Given a unit LINE SEGMENT $[0, 1]$, pick two points at random on it. Call the first point x_1 and the second point x_2. Find the distribution of distances d between points. The probability of the points being a (POSITIVE) distance d apart (i.e., without regard to ordering) is given by

$$P(d) = \frac{\int_0^1 \int_0^1 \delta(d - |x_2 - x_1|)\, dx_1\, dx_2}{\int_0^1 \int_0^1 dx_1\, dx_2}$$

$$= (1 - d)[H(1 - d) - H(d - 1) + H(d) - H(-d)]$$

$$= \begin{cases} 2(1 - d) & \text{for } 0 \le d \le 1 \\ 0 & \text{otherwise,} \end{cases} \tag{1}$$

where δ is the DIRAC DELTA FUNCTION and H is the HEAVISIDE STEP FUNCTION. The MOMENTS are then

$$\mu'_m = \int_0^1 d^m p(d)\, dd = 2 \int_0^1 d^m (1 - d)\, dd$$

$$= 2 \left[\frac{d^{m+1}}{m + 1} - \frac{d^{m+2}}{m + 2} \right]_0^1$$

$$= 2 \left(\frac{1}{m + 1} - \frac{1}{m + 2} \right) = 2 \left[\frac{(m + 2) - (m + 1)}{(m + 1)(m + 2)} \right]$$

$$= \frac{2}{(m + 1)(m + 2)}$$

$$= \begin{cases} \dfrac{1}{(n + 1)(2n + 1)} & \text{for } m = 2n \\[2mm] \dfrac{1}{(n + 1)(2n + 3)} & \text{for } m = 2n + 1 \end{cases} \tag{2}$$

(Uspensky 1934, p. 257), giving RAW MOMENTS

$$\mu'_1 = \tfrac{1}{3} \tag{3}$$

$$\mu'_2 = \tfrac{1}{6} \tag{4}$$

$$\mu'_3 = \tfrac{1}{10} \tag{5}$$

$$\mu'_4 = \tfrac{1}{15}. \tag{6}$$

The MOMENTS can also be computed directly without explicit knowledge of the distribution

$$\mu'_1 = \frac{\int_0^1 \int_0^1 |x_2 - x_1|\, dx_1\, dx_2}{\int_0^1 \int_0^1 dx_1\, dx_2}$$

$$= \int_0^1 \int_0^1 |x_2 - x_1|\, dx_1\, dx_2$$

$$= \int\!\!\!\int_{x_2 - x_1 > 0} (x_2 - x_1)\, dx_1\, dx_2 + \int\!\!\!\int_{x_2 - x_1 < 0} (x_1 - x_2)\, dx_1\, dx_2$$

$$= \int_0^1 \int_{x_1}^1 (x_2 - x_1)\, dx_1\, dx_2 + \int_0^1 \int_0^{x_1} (x_2 - x_1)\, dx_1\, dx_2$$

$$= \int_0^1 \left[\tfrac{1}{2} x_2^2 - x_1 x_2 \right]_{x_1}^1 dx_1 + \int_0^1 \left[x_1 x_2 - \tfrac{1}{2} x_2^2 \right]_0^{x_1} dx_1$$

$$= \int_0^1 \left[\left(\tfrac{1}{2} - x_1 \right) - \left(\tfrac{1}{2} x_1^2 - x_1^2 \right) \right] dx_1$$

$$\quad + \int_0^1 \left[\left(x_1^2 - \tfrac{1}{2} x_1^2 \right) - (0 - 0) \right] dx_1$$

$$= \int_0^1 \left(\tfrac{1}{2} - x_1 + x_1^2 \right) dx_1 = \left[\tfrac{1}{2} x_1 - \tfrac{1}{2} x_1^2 + \tfrac{1}{3} x_1^3 \right]_0^1$$

$$= \left(\tfrac{1}{2} - \tfrac{1}{2} + \tfrac{1}{3} \right) - (0 - 0 + 0) = \tfrac{1}{3} \tag{7}$$

$$\mu'_2 = \int_0^1 \int_0^1 (|x_2 - x_1|)^2\, dx_2\, dx_1$$

$$= \int_0^1 \int_0^1 (x_2 - x_1)^2 \, dx_1 \, dx_2$$

$$= \int_0^1 \int_0^1 (x_2^2 - 2x_1 x_2 + x_1^2) \, dx_1 \, dx_2$$

$$= \int_0^1 \left[\tfrac{1}{3} x_2^3 - x_1 x_2^2 + x_1^2 x_2 \right]_0^1 \, dx_1$$

$$= \int_0^1 \left(\tfrac{1}{3} - x_1 + x_1^2 \right) dx_1 = \left[\tfrac{1}{3} x_1^3 - \tfrac{1}{2} x_1^2 + \tfrac{1}{3} x_1 \right]_0^1$$

$$= \tfrac{1}{3} - \tfrac{1}{2} + \tfrac{1}{3} = \tfrac{1}{6}. \tag{8}$$

The CENTRAL MOMENTS are therefore

$$\mu_2 = \mu_2' - {\mu_1'}^2 = \tfrac{1}{6} - \left(\tfrac{1}{3} \right)^2 = \tfrac{1}{18} \tag{9}$$

$$\mu_3 = \mu_3' - 3\mu_2'\mu_1' + 2(\mu_1')^3 = \tfrac{1}{135} \tag{10}$$

$$\mu_4 = \mu_4' - 4\mu_3'\mu_1' + 6\mu_2'(\mu_1')^2 - 3(\mu_1')^4 = \tfrac{1}{135}, \tag{11}$$

so the MEAN, VARIANCE, SKEWNESS, and KURTOSIS are

$$\mu = \mu_1' = \tfrac{1}{3} \tag{12}$$

$$\sigma^2 = \mu_2 = \tfrac{1}{18} \tag{13}$$

$$\gamma_1 = \frac{\mu_3}{\sigma^3} = \tfrac{2}{5}\sqrt{2} \tag{14}$$

$$\gamma_2 = \frac{\mu_4}{\sigma^4} - 3 = -\tfrac{3}{5}. \tag{15}$$

The probability distribution of the distance between two points randomly picked on a LINE SEGMENT is germane to the problem of determining the access time of computer hard drives. In fact, the average access time for a hard drive is precisely the time required to seek across 1/3 of the tracks (Benedict 1995).

See also POINT-POINT DISTANCE–2-D, POINT-POINT DISTANCE–3-D, POINT-QUADRATIC DISTANCE, SPHERE POINT PICKING

References

Arfken, G. *Mathematical Methods for Physicists, 3rd ed.* Orlando, FL: Academic Press, pp. 930–31, 1985.
Benedict, B. *Using Norton Utilities for the Macintosh.* Indianapolis, IN: Que, pp. B-8-B-9, 1995.
Uspensky, J. V. *Introduction to Mathematical Probability.* New York: McGraw-Hill, p. 257, 1937.

Point-Point Distance—2-D

Given two points in the PLANE, find the curve which minimizes the distance between them. The LINE ELEMENT is given by

$$ds = \sqrt{dx^2 + dy^2}, \tag{1}$$

so the ARC LENGTH between the points x_1 and x_2 is

$$L = \int ds = \int_{x_1}^{x_2} \sqrt{1 + y'^2} \, dx, \tag{2}$$

where $y' \equiv dy/dx$ and the quantity we are minimizing is

$$f = \sqrt{1 + y'^2}. \tag{3}$$

Finding the derivatives gives

$$\frac{\partial f}{\partial y} = 0 \tag{4}$$

$$\frac{d}{dx} \frac{\partial f}{\partial y'} = \frac{d}{dx} \left[(1 + y'^2)^{-1/2} y' \right], \tag{5}$$

so the EULER-LAGRANGE DIFFERENTIAL EQUATION becomes

$$\frac{\partial f}{\partial y} - \frac{d}{dx} \frac{\partial f}{\partial y'} = \frac{d}{dx} \left(\frac{y'}{\sqrt{1 + y'^2}} \right) = 0. \tag{6}$$

Integrating and rearranging,

$$\frac{y'}{\sqrt{1 + y'^2}} = c \tag{7}$$

$$y'^2 = c^2 (1 + y'^2) \tag{8}$$

$$y'^2 (1 - c^2) = c^2 \tag{9}$$

$$y' = \frac{c}{\sqrt{1 - c^2}} \equiv a. \tag{10}$$

The solution is therefore

$$y = ax + b, \tag{11}$$

which is a straight LINE. Now verify that the ARC LENGTH is indeed the straight-line distance between the points. a and b are determined from

$$y_1 = ax_1 + b. \tag{12}$$

$$y_2 = ax_2 + b. \tag{13}$$

Writing (12) and (13) as a MATRIX EQUATION gives

$$\begin{bmatrix} y_1 \\ y_2 \end{bmatrix} = \begin{bmatrix} x_1 & 1 \\ x_2 & 1 \end{bmatrix} \begin{bmatrix} a \\ b \end{bmatrix} \tag{14}$$

$$\begin{bmatrix} a \\ b \end{bmatrix} = \begin{bmatrix} x_1 & 1 \\ x_2 & 1 \end{bmatrix}^{-1} \begin{bmatrix} y_1 \\ y_2 \end{bmatrix}$$

$$= \frac{1}{x_1 - x_2} \begin{bmatrix} 1 & -1 \\ -x_2 & x_1 \end{bmatrix}^{-1} \begin{bmatrix} y_1 \\ y_2 \end{bmatrix}, \tag{15}$$

so

$$a = \frac{y_1 - y_2}{x_1 - x_2} = \frac{y_2 - y_1}{x_2 - x_1} \tag{16}$$

$$b = \frac{x_1 y_2 - x_2 y_1}{x_1 - x_2} \tag{17}$$

$$L = \int_{x_1}^{x_2} \sqrt{1 + y'^2}\, dy = (x_2 - x_1)\sqrt{1 + a^2}$$

$$= (x_2 - x_1)\sqrt{1 + \left(\frac{y_2 - y_1}{x_2 - x_1}\right)^2}$$

$$= \sqrt{(x_2 - x_1)^2 + (y_2 - y_1)^2}, \tag{18}$$

as expected.

The shortest distance between two points on a SPHERE is the so-called GREAT CIRCLE distance.

See also CALCULUS OF VARIATIONS, CIRCLE TRIANGLE PICKING, GREAT CIRCLE, POINT-POINT DISTANCE–1-D, POINT-POINT DISTANCE–3-D, POINT-QUADRATIC DISTANCE, SPHERE POINT PICKING

References

Arfken, G. *Mathematical Methods for Physicists, 3rd ed.* Orlando, FL: Academic Press, pp. 930–31, 1985.

Point-Point Distance—3-D

The LINE ELEMENT is

$$ds = \sqrt{dx^2 + dy^2 + dz^2}, \tag{1}$$

so the ARC LENGTH between the points x_1 and x_2 is

$$L = \int ds = \int_{x_1}^{x_2} \sqrt{1 + y'^2 + z'^2}\, dx \tag{2}$$

and the quantity we are minimizing is

$$f = \sqrt{1 + y'^2 + z'^2}. \tag{3}$$

Finding the derivatives gives

$$\frac{\partial f}{\partial y} = 0 \tag{4}$$

$$\frac{\partial f}{\partial z} = 0 \tag{5}$$

and

$$\frac{\partial f}{\partial y'} = \frac{y'}{\sqrt{1 + y'^2 + z'^2}} \tag{6}$$

$$\frac{\partial f}{\partial z'} = \frac{z'}{\sqrt{1 + y'^2 + z'^2}}, \tag{7}$$

so the EULER-LAGRANGE DIFFERENTIAL EQUATIONS become

$$\frac{d}{dx}\left(\frac{y'}{\sqrt{1 + y'^2 + z'^2}}\right) = 0 \tag{8}$$

$$\frac{d}{dx}\left(\frac{z'}{\sqrt{1 + y'^2 + z'^2}}\right) = 0. \tag{9}$$

These give

$$\frac{y'}{\sqrt{1 + y'^2 + z'^2}} = c_1 \tag{10}$$

$$\frac{z'}{\sqrt{1 + y'^2 + z'^2}} = c_2. \tag{11}$$

Taking the ratio,

$$z' = \frac{c_2}{c_1}\, y' \tag{12}$$

$$\frac{y'}{\sqrt{1 + y'^2 + \left(\frac{c_2}{c_1}\right)^2 y'^2}} = c_1 \tag{13}$$

$$y'^2 = c_1^2\left[1 + y'^2 + \left(\frac{c_2}{c_1}\right)^2 y'^2\right] = c_1^2 + y'^2(c_1^2 + c_2^2), \tag{14}$$

which gives

$$y'^2 = \frac{c_1^2}{1 - c_1^2 - c_2^2} \equiv a_1^2 \tag{15}$$

$$z'^2 = \left(\frac{c_2}{c_1}\right)^2 y'^2 = \frac{c_2^2}{1 - c_1^2 - c_2^2} \equiv b_1^2. \tag{16}$$

Therefore, $y' = a_1$ and $z' = b_1$, so the solution is

$$\begin{bmatrix} x \\ y \\ z \end{bmatrix} = \begin{bmatrix} x \\ a_1 x + a_0 \\ b_1 x + b_0 \end{bmatrix}, \tag{17}$$

which is the parametric representation of a straight line with parameter $x \in [x_1, x_2]$. Verifying the ARC LENGTH gives

$$L = \sqrt{1 + a_1^2 + b_1^2}(x_2 - x_1) \tag{18}$$

where

$$\begin{bmatrix} y_1 \\ y_2 \end{bmatrix} = \begin{bmatrix} x_1 & 1 \\ x_2 & 1 \end{bmatrix}\begin{bmatrix} a_1 \\ a_0 \end{bmatrix} \tag{19}$$

$$\begin{bmatrix} z_1 \\ z_2 \end{bmatrix} = \begin{bmatrix} x_1 & 1 \\ x_2 & 1 \end{bmatrix}\begin{bmatrix} b_1 \\ b_0 \end{bmatrix}. \tag{20}$$

See also POINT-POINT DISTANCE–1-D, POINT-POINT DISTANCE–2-D, POINT-QUADRATIC DISTANCE

Point-Quadratic Distance

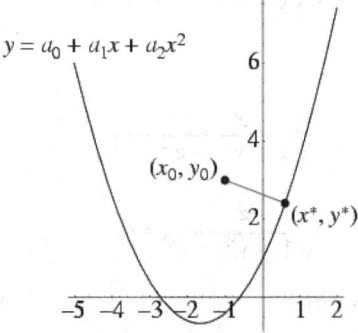

To find the minimum distance between a point in the plane (x_0, y_0) and a quadratic PLANE CURVE

$$y = a_0 + a_1 x + a_2 x^2, \tag{1}$$

note that the square of the distance is

$$r^2 = (x - x_0)^2 + (y - y_0)^2$$
$$= (x - x_0)^2 + (a_0 + a_1 x + a_2 x^2 - y_0)^2. \tag{2}$$

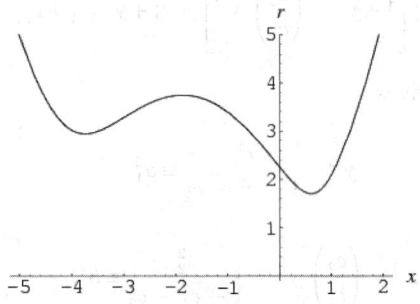

Minimizing the distance squared is equivalent to minimizing the distance (since r^2 and $|r|$ have minima at the same point), so take

$$\frac{\partial (r^2)}{\partial x} = 2(x - x_0) + 2(a_0 + a_1 x + a_2 x^2 - y_0)(a_1 + 2a_2 x)$$

$$= 0 \tag{3}$$

$$x - x_0 + a_0 a_1 + a_1^2 + a_1 a_2 x^2 - a_1 y_0 + 2a_0 a_2 x$$
$$+ 2a_1 a_2 x^2 + 2a_2^2 x^3 - 2a_2 y_0 x = 0 \tag{4}$$

$$2a_2^2 x^3 + 3a_1 a_2 x^2 + (a_1^2 + 2a_0 a_2 - 2a_2 y_0 + 1)x$$
$$+ (a_0 a_1 - a_1 y_0 - x_0) = 0. \tag{5}$$

Minimizing the distance to find the closest point (x^*, y^*) therefore requires solution of a CUBIC EQUATION.

See also POINT-POINT DISTANCE–1-D, POINT-POINT DISTANCE–2-D, POINT-POINT DISTANCE–3-D

Points Problem

SHARING PROBLEM

Point-Set Topology

The low-level language of TOPOLOGY, which is not really considered a separate "branch" of TOPOLOGY. Point-set topology, also called set-theoretic topology or general topology, is the study of the general abstract nature of continuity or "closeness" on SPACES. Basic point-set topological notions are ones like CONTINUITY, DIMENSION, COMPACTNESS, and CONNECTEDNESS. The INTERMEDIATE VALUE THEOREM (which states that if a path in the real line connects two numbers, then it passes over every point between the two) is a basic topological result. Others are that EUCLIDEAN n-space is HOMEOMORPHIC to EUCLIDEAN m-space IFF $m = n$, and that REAL valued functions achieve maxima and minima on COMPACT SETS.

Foundational point-set topological questions are ones like "when can a topology on a space be derived from a metric?" Point-set topology deals with differing notions of continuity and compares them, as well as dealing with their properties. Point-set topology is also the ground-level of inquiry into the geometrical properties of spaces and continuous functions between them, and in that sense, it is the foundation on which the remainder of topology (ALGEBRAIC, DIFFERENTIAL, and LOW-DIMENSIONAL) stands.

See also ALGEBRAIC TOPOLOGY, DIFFERENTIAL TOPOLOGY, LOW-DIMENSIONAL TOPOLOGY, TOPOLOGY

References
Bing, R. H. "Elementary Point Set Topology." *Amer. Math. Monthly* **67**, 1960.
Ferreirós, J. "Origins of the Theory of Point-Sets." Ch. 5 in *Labyrinth of Thought: A History of Set Theory and Its Role in Modern Mathematics.* Basel, Switzerland: Birkhäuser, pp. 95–7, 1999.
Sutherland, W. A. *An Introduction to Metric & Topological Spaces.* New York: Oxford University Press, 1975.
Vaidyanathaswamy, R. *Set Topology.* New York: Dover, 1999.

Pointwise Convergence

The hypothesis is that, for X is a MEASURE SPACE, $f_n(x) \to f(x)$ for each $x \in X$, as $n \to \infty$. The hypothesis may be weakened to ALMOST EVERYWHERE CONVERGENCE.

See also ALMOST EVERYWHERE CONVERGENCE

References
Browder, A. *Mathematical Analysis: An Introduction.* New York: Springer-Verlag, 1996.

Pointwise Dimension

$$D_p(\mathbf{x}) \equiv \lim_{\epsilon \to 0} \frac{\ln \mu(B_\epsilon(\mathbf{x}))}{\ln \epsilon},$$

where $B_\epsilon(\mathbf{x})$ is an n-D BALL of RADIUS ϵ centered at $\mathbf{x}$ and μ is the PROBABILITY MEASURE.

See also BALL, PROBABILITY MEASURE

References

Nayfeh, A. H. and Balachandran, B. *Applied Nonlinear Dynamics: Analytical, Computational, and Experimental Methods.* New York: Wiley, pp. 541–45, 1995.

Poised

NEARLY-POISED, WELL-POISED

Poisson Bracket

Let F and G be infinitely differentiable functions of x and p. Then the Poisson bracket is defined by

$$(F,\ G) = \sum_{v=1}^{n} \left(\frac{\partial F}{\partial p_v} \frac{\partial G}{\partial x_v} - \frac{\partial G}{\partial p_v} \frac{\partial F}{\partial x_v} \right).$$

If F and G are functions of x and p only, then the LAGRANGE BRACKET $[F,\ G]$ collapses the Poisson bracket $(F,\ G)$.

See also LAGRANGE BRACKET, LIE BRACKET

References

Iyanaga, S. and Kawada, Y. (Eds.). *Encyclopedic Dictionary of Mathematics.* Cambridge, MA: MIT Press, p. 1004, 1980.

Poisson Distribution

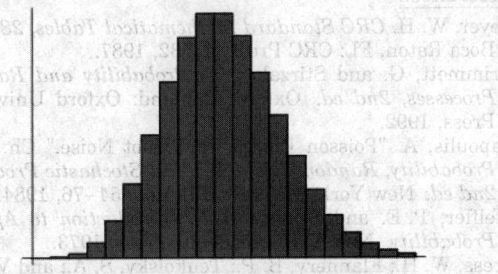

Given a POISSON PROCESS, the probability of k changes occurring in a given interval is given by the limit of the BINOMIAL DISTRIBUTION

$$P_B(k) = \frac{n!}{k!(n-k)!} \left(\frac{v}{n} \right)^k \left(1 - \frac{v}{n} \right)^{n-k}. \tag{1}$$

As the number of trials becomes very large, (1)

approaches the distribution

$$P(k) = \lim_{n \to \infty} P_B(k)$$

$$= \lim_{n \to \infty} \frac{n(n-1)\cdots(n-k+1)}{n^k} \frac{v^k}{k!}$$

$$\times \left(1 - \frac{v}{n} \right)^n \left(1 - \frac{v}{n} \right)^{-k}$$

$$= 1 \cdot \frac{v^k}{k!} \cdot e^{-v} \cdot 1 = \frac{v^k e^{-v}}{k!}, \tag{2}$$

which is called the Poisson distribution (Papoulis 1984, pp. 101 and 554; Pfeiffer and Schum 1973, p. 200).

The Poisson distribution is normalized so that the sum of probabilities equals 1, since

$$\sum_{k=0}^{\infty} P(k) = e^{-v} \sum_{k=0}^{\infty} \frac{v^k}{k!} = e^{-v} e^v = 1. \tag{3}$$

The ratio of probabilities is given by

$$\frac{P(k=i+1)}{P(k=i)} = \frac{\dfrac{v^{i+1}e^{-v}}{(i+1)!}}{\dfrac{v^i}{e^{-v} v^i}} = \frac{v}{i+1}. \tag{4}$$

The MOMENT-GENERATING FUNCTION of the Poisson distribution is given by

$$M(t) = \sum_{k=0}^{\infty} e^{tk} \frac{v^k e^{-v}}{k!} = e^{-v} \sum_{k=0}^{\infty} \frac{(ve^t)^k}{k!}$$

$$= e^{-v} e^{ve^t} = e^{v(e^t - 1)} \tag{5}$$

$$M'(t) = ve^t e^{v(e^t - 1)} \tag{6}$$

$$M''(t) = (ve^t)^2 e^{v(e^t - 1)} + ve^t e^{v(e^t - 1)} \tag{7}$$

$$R(t) \equiv \ln M(t) = v(e^t - 1) \tag{8}$$

$$R'(t) = ve^t \tag{9}$$

$$R''(t) = ve^t, \tag{10}$$

so

$$\mu = R'(0) = v \tag{11}$$

$$\sigma^2 = R''(0) = v \tag{12}$$

(Papoulis 1984, p. 554).

The RAW MOMENTS can also be computed directly by summation, which yields an unexpected connection with STIRLING NUMBERS OF THE SECOND KIND,

$$\sum_{k=0}^{\infty} \frac{e^{-x} x^k}{k!} k^n = \sum_{k=1}^{n} x^k S(n,\ k), \tag{13}$$

so

$$\mu_2' = v(1 + v) \tag{14}$$

$$\mu_3' = v\left(1 + 3v + v^2\right) \tag{15}$$

$$\mu_4' = v\left(1 + 7v + 6v^2 + v^3\right). \tag{16}$$

The CENTRAL MOMENTS can then be computed as

$$\mu_2 = v \tag{17}$$

$$\mu_3 = v \tag{18}$$

$$\mu_4 = v(1 + 3v), \tag{19}$$

so the MEAN, VARIANCE, SKEWNESS, and KURTOSIS are

$$\mu = v \tag{20}$$

$$\sigma^2 = v \tag{21}$$

$$\gamma_1 \equiv \frac{\mu_3}{\sigma^3} = \frac{v}{v^{3/2}} = v^{-1/2} \tag{22}$$

$$\gamma_2 \equiv \frac{\mu_4}{\sigma^4} - 3 = \frac{v(1 + 3v)}{v} - 3$$

$$= \frac{v + 3v^2 - 3v^2}{v^2} = v^{-1}. \tag{23}$$

The CHARACTERISTIC FUNCTION for the Poisson distribution is

$$\phi(t) = e^{v\left(e^{it} - 1\right)} \tag{24}$$

(Papoulis 1984, pp. 154 and 554), and the CUMULANT-GENERATING FUNCTION is

$$K(h) = v\left(e^h - 1\right) = v\left(h + \frac{1}{2!}\, h^2 + \frac{1}{3!}\, h^3 + \ldots\right), \tag{25}$$

so

$$\kappa_r = v. \tag{26}$$

The Poisson distribution can also be expressed in terms of

$$\lambda \equiv \frac{v}{x}, \tag{27}$$

the rate of changes, so that

$$P(k) = \frac{(\lambda x)^k e^{-\lambda x}}{k!}. \tag{28}$$

The MOMENT-GENERATING FUNCTION of a Poisson distribution in two variables is given by

$$M(t) = e^{(v_1 + v_2)(e^t - 1)}. \tag{29}$$

If the independent variables $x_1, x_2, \ldots, x_N$ have Poisson distributions with parameters $\mu_1, \mu_2, \ldots, \mu_N$, then

$$X = \sum_{j=1}^{N} x_j \tag{30}$$

has a Poisson distribution with parameter

$$\mu = \sum_{j=1}^{N} \mu_j. \tag{31}$$

This can be seen since the CUMULANT-GENERATING FUNCTION is

$$K_j(h) = \mu_j\left(e^h - 1\right), \tag{32}$$

$$K \equiv \sum_j K_j(h) = \left(e^h - 1\right) \sum_j \mu_j = \mu\left(e^h - 1\right). \tag{33}$$

A generalization of the Poisson distribution has been used by Saslaw (1989) to model the observed clustering of galaxies in the universe. The form of this distribution is given by

$$f_b(N) = \frac{\bar{N}(1 - b)}{N!}\left[\bar{N}(1 - b) + Nb\right]^{N-1} e^{\bar{N}(1-b) - Nb}, \tag{34}$$

where N is the number of galaxies in a volume V, $\bar{N} = \bar{n}V$, $\bar{n}$ is the average density of galaxies, and $b = -W/(2K) \approx 0.70 \pm 0.05$, with $0 \leq b < 1$ is the ratio of gravitational energy to the kinetic energy of peculiar motions, Letting $b = 0$ gives

$$f_0(N) = \frac{e^{-\bar{N}} \bar{N}^N}{N!}, \tag{35}$$

which is indeed a Poisson distribution with $v = \bar{N}$. Similarly, letting $b = 1$ gives $f_1(N) = 0$.

See also BINOMIAL DISTRIBUTION, POISSON PROCESS, POISSON THEOREM

References

Beyer, W. H. *CRC Standard Mathematical Tables, 28th ed.* Boca Raton, FL: CRC Press, p. 532, 1987.

Grimmett, G. and Stirzaker, D. *Probability and Random Processes, 2nd ed.* Oxford, England: Oxford University Press, 1992.

Papoulis, A. "Poisson Process and Shot Noise." Ch. 16 in *Probability, Random Variables, and Stochastic Processes, 2nd ed.* New York: McGraw-Hill, pp. 554–76, 1984.

Pfeiffer, P. E. and Schum, D. A. *Introduction to Applied Probability.* New York: Academic Press, 1973.

Press, W. H.; Flannery, B. P.; Teukolsky, S. A.; and Vetterling, W. T. "Incomplete Gamma Function, Error Function, Chi-Square Probability Function, Cumulative Poisson Function." §6.2 in *Numerical Recipes in FORTRAN: The Art of Scientific Computing, 2nd ed.* Cambridge, England: Cambridge University Press, pp. 209–14, 1992.

Saslaw, W. C. "Some Properties of a Statistical Distribution Function for Galaxy Clustering." *Astrophys. J.* **341**, 588–98, 1989.

Spiegel, M. R. *Theory and Problems of Probability and Statistics.* New York: McGraw-Hill, pp. 111–12, 1992.

Poisson Integral

There are at least two integrals called the Poisson integral. The first is also known as BESSEL'S SECOND INTEGRAL,

$$J_n(z) = \frac{\left(\frac{1}{2}\right)^n}{\Gamma\left(n + \frac{1}{2}\right)\Gamma\left(\frac{1}{2}\right)} \int_0^\pi \cos(z \cos \theta) \sin^{2n} \theta \, d\theta,$$

where $J_n(z)$ is a BESSEL FUNCTION OF THE FIRST KIND and $\Gamma(x)$ is a GAMMA FUNCTION. It can be derived from SONINE'S INTEGRAL. With $n = 0$, the integral becomes PARSEVAL'S INTEGRAL.

In complex analysis, let $u : U \to \mathbb{R}$ be a HARMONIC FUNCTION on a NEIGHBORHOOD of the CLOSED DISK $\bar{D}(0, 1)$, then for any point z_0 in the OPEN DISK $D(0, 1)$,

$$u(z_0) = \frac{1}{2\pi} \int_0^{2\pi} u(e^{i\psi}) \frac{1 - |z_0|^2}{|z_0 - e^{i\psi}|^2} \, d\psi.$$

In polar coordinates on $\bar{D}(0, R)$,

$$u(z_0) = \frac{1}{2\pi} \int_0^{2\pi} K(r, \theta)\phi(z_0 + re^{i\theta}) \, d\theta, \qquad (1)$$

where $R = |z_0|$ and $K(r, \theta)$ is the POISSON KERNEL. For a CIRCLE,

$$u(x, y) = \frac{1}{2\pi} \int_0^{2\pi} u(a \cos \phi, a \sin \phi)$$

$$\times \frac{a^2 - R^2}{a^2 + R^2 - 2ar \cos(\theta - \phi)} \, d\phi. \qquad (2)$$

For a SPHERE,

$$u(x, y, z) = \frac{1}{4\pi a} \int \int_S u \frac{a^2 - R^2}{\left(a^2 + R^2 - 2aR \cos \theta\right)^{3/2}} \, dS, \qquad (3)$$

where

$$\cos \theta \equiv \mathbf{x} \cdot \boldsymbol{\xi}. \qquad (4)$$

See also BESSEL FUNCTION OF THE FIRST KIND, CIRCLE, HARMONIC FUNCTION, PARSEVAL'S INTEGRAL, POISSON KERNEL, SONINE'S INTEGRAL, SPHERE

References

Krantz, S. G. "The Poisson Integral." §7.3.1 in *Handbook of Complex Analysis*. Boston, MA: Birkhäuser, pp. 92–3, 1999.
Morse, P. M. and Feshbach, H. *Methods of Theoretical Physics, Part I*. New York: McGraw-Hill, pp. 373–74, 1953.

Poisson Integral Representation

$$j_n(z) = \frac{z^n}{2^{n+1}n!} \int_0^\pi \cos(z \cos \theta) \sin^{2n+1} \theta \, d\theta,$$

where $j_n(z)$ is a SPHERICAL BESSEL FUNCTION OF THE FIRST KIND.

Poisson Kernel

The KERNEL in the POISSON INTEGRAL, given by

$$K(\psi) = \frac{1}{2\pi} \frac{1 - |z_0|^2}{|z_0 - e^{i\psi}|^2} \qquad (1)$$

for the open UNIT DISK $D(0, 1)$. Writing $z_0 = re^{i\theta}$ and taking $D(0, R)$ gives

$$K(r, \theta) \equiv \frac{1}{2\pi} \, \Re\left[\frac{R + re^{i\theta}}{R - re^{i\theta}}\right]$$

$$= \frac{1}{2\pi} \, \Re\left[\frac{(R + re^{i\theta})(R - re^{-i\theta})}{(R - re^{i\theta})(R - re^{-i\theta})}\right]$$

$$= \frac{1}{2\pi} \, \Re\left[\frac{R^2 - rR(e^{i\theta} - e^{-i\theta}) - r^2}{R^2 - rR(e^{i\theta} + e^{-i\theta}) + r^2}\right]$$

$$= \frac{1}{2\pi} \, \Re\left[\frac{R^2 + 2irR \sin \theta - r^2}{R^2 - 2Rr \cos \theta + r^2}\right]$$

$$= \frac{1}{2\pi} \frac{R^2 - r^2}{R^2 - 2Rr \cos \theta + r^2} \qquad (2)$$

(Krantz 1999, p. 93).
In 3-D,

$$u(\mathbf{y}) = \frac{R(R^2 - a^2)}{4\pi} \int_0^{2\pi} \int_0^\pi \frac{f(\theta, \phi) \sin \theta \, d\theta \, d\phi}{(R^2 + a^2 - 2aR \cos \gamma)^{3/2}}, \qquad (3)$$

where $a = |\mathbf{y}|$ and

$$\cos \gamma = \mathbf{y} \cdot \begin{bmatrix} R \cos \theta \sin \phi \\ R \sin \theta \sin \phi \\ R \cos \phi \end{bmatrix}. \qquad (4)$$

The Poisson kernel for the n-BALL is

$$P(\mathbf{x}, \mathbf{z}) = \frac{1}{2 - n}(D_{\mathbf{n}}\mathbf{v})(\mathbf{z}), \qquad (5)$$

where $D_{\mathbf{n}}$ is the outward normal derivative at point $\mathbf{z}$ on a unit n-sphere and

$$\mathbf{v}(\mathbf{z}) = |\mathbf{z} - \mathbf{x}|^{2-n} - |\mathbf{x}|^{2-n}\left|\frac{\mathbf{x}}{|\mathbf{x}|^2}\right|^{2-n}. \qquad (6)$$

Let u be harmonic on a neighborhood of the closed UNIT DISK $\bar{D}(0, 1)$, then the reproducing property of the Poisson kernal states that for $z \in D(0, 1)$,

$$u(z) = \frac{1}{2\pi} \int_0^{2\pi} u(e^{i\psi}) \frac{1 - |z|^2}{|z - e^{i\psi}|^2} \, d\psi \qquad (7)$$

(Krantz 1999, p. 94).

See also DIRICHLET PROBLEM, HARMONIC FUNCTION, MEAN-VALUE PROPERTY, POISSON INTEGRAL, POISSON KERNEL

References

Gradshteyn, I. S. and Ryzhik, I. M. *Tables of Integrals, Series, and Products, 6th ed.* San Diego, CA: Academic Press, p. 1090, 2000.
Krantz, S. G. "The Poisson Kernel." §7.3.2 in *Handbook of Complex Analysis.* Boston, MA: Birkhäuser, p. 93, 1999.

Poisson Manifold

A smooth MANIFOLD with a POISSON BRACKET defined on its FUNCTION SPACE.

Poisson Process

A Poisson is a process satisfying the following properties.

1. The numbers of changes in nonoverlapping intervals are independent for all intervals.
2. The probability of exactly one change in a sufficiently small interval $h \equiv 1/n$ is $P = vh \equiv v/n$, where v is the probability of one change and n is the number of TRIALS.
3. The probability of two or more changes in a sufficiently small interval h is essentially 0.

In the limit of the number of trials becoming large, the resulting distribution is called a POISSON DISTRIBUTION.

See also POISSON DISTRIBUTION

References

Grimmett, G. and Stirzaker, D. *Probability and Random Processes, 2nd ed.* Oxford, England: Oxford University Press, 1992.
Papoulis, A. *Probability, Random Variables, and Stochastic Processes, 2nd ed.* New York: McGraw-Hill, pp. 548–49, 1984.

Poisson Sum Formula

A special case of the general result

$$\sum_{n=-\infty}^{\infty} f(x+n) = \sum_{k=-\infty}^{\infty} e^{2\pi i k x} \int_{-\infty}^{\infty} f(x') e^{-2\pi i k x'} \, dx' \qquad (1)$$

with $x = 0$, yielding

$$\sum_{n=-\infty}^{\infty} f(n) = \sum_{k=-\infty}^{\infty} \int_{-\infty}^{\infty} f(x') e^{-2\pi i k x'} \, dx'. \qquad (2)$$

Given f a nonnegative, continuous, decreasing, and Riemann integrable function of $[0, \infty)$, define

$$\psi(x) = \sqrt{\frac{2}{\pi}} \int_0^{\infty} \phi(t) \cos(xt) \, dt. \qquad (3)$$

Then

$$\sqrt{\alpha} \left[\frac{1}{2} f(0) + \sum_{n=1}^{\infty} f(n\alpha) \right] = \sqrt{\beta} \left[\frac{1}{2} g(0) + \sum_{n=1}^{\infty} g(n\beta) \right] \qquad (4)$$

whenever $\alpha\beta = 2\pi$, from which it follows that

$$\sqrt{\alpha} \left[\frac{1}{2} + \sum_{n=1}^{\infty} e^{-\alpha^2 n^2/2} \right] = \sqrt{\beta} \left[\frac{1}{2} + \sum_{n=1}^{\infty} e^{-\beta^2 n^2/2} \right] \qquad (5)$$

(Apostol 1974, Borwein 1987).

References

Apostol, T. M. *Mathematical Analysis.* Reading, MA: Addison-Wesley, pp. 332–33, 1974.
Borwein, J. M. and Borwein, P. B. "Poisson Summation." §2.2 in *Pi & the AGM: A Study in Analytic Number Theory and Computational Complexity.* New York: Wiley, pp. 36–0, 1987.
Hardy, G. H. *Ramanujan: Twelve Lectures on Subjects Suggested by His Life and Work, 3rd ed.* New York: Chelsea, p. 14, 1999.
Morse, P. M. and Feshbach, H. *Methods of Theoretical Physics, Part I.* New York: McGraw-Hill, pp. 466–67, 1953.

Poisson Theorem

Poisson's theorem give the estimate

$$\frac{n!}{k!(n-k)!} p^k q^{n-k} \sim e^{-np} \frac{(np)^k}{k!}$$

for the probability of an event occurring k times in n trials with $n \gg 1$, $p \ll 1$, and $np \approx npq \gg 1$.

See also POISSON DISTRIBUTION

References

Papoulis, A. *Probability, Random Variables, and Stochastic Processes, 2nd ed.* New York: McGraw-Hill, p. 71, 1984.

Poisson Trials

A number s of TRIALS in which the probability of success p_i varies from trial to trial. Let x be the number of successes, then

$$\text{var}(x) = spq - s\sigma_p^2, \qquad (1)$$

where σ_p^2 is the VARIANCE of p_i and $q \equiv (1-p)$. Uspensky has shown that

$$P(s, x) = \beta \frac{m^x e^{-m}}{x!}, \qquad (2)$$

where

$$\beta = [1 - \theta g(x)] e^{h(x)} \qquad (3)$$

$$g(x) = \frac{(s-x)m^3}{3(s-m)^3} + \frac{x^3}{2s(s-x)} \tag{4}$$

$$h(x) = \frac{mx}{s} - \frac{m^2}{2s^2}(s-x) - \frac{x(x-1)}{2s}$$

$$= p\left[\frac{x}{2}\left(1+\frac{1}{m}\right) - \frac{(x-m)^2}{2m}\right] \tag{5}$$

and $\theta \in (0, 1)$. The probability that the number of successes is at least x is given by

$$Q_m(x) = \sum_{r=x}^{\infty} \frac{m^r e^{-m}}{r!}. \tag{6}$$

Uspensky gives the true probability that there are at least x successes in s trials as

$$P_{ms}(x) = Q_m(x) + \Delta, \tag{7}$$

where

$$|\Delta| < \begin{cases} (e^\chi - 1)Q_m(x+1) & \text{for } Q_m(x+1) \geq \frac{1}{2} \\ (e^\chi - 1)[1 - Q_m(x+1)] & \text{for } Q_m(x+1) \leq \frac{1}{2} \end{cases} \tag{8}$$

$$\chi = \frac{m + \frac{1}{4} + \frac{m^3}{s}}{2(s-m)}. \tag{9}$$

See also TRIAL

Poisson-Boltzmann Differential Equation

The ORDINARY DIFFERENTIAL EQUATION

$$y'' + \frac{k}{x}y' + \delta e^y = 0.$$

References

Chambré, P. L. "On the Solution of the Poisson-Boltzmann Equation with Application to the Theory of Thermal Explosions." *J. Chem. Phys.* **20**, 1795–797, 1952.

Zwillinger, D. *Handbook of Differential Equations, 3rd ed.* Boston, MA: Academic Press, p. 126, 1997.

Poisson-Charlier Function

$$\rho_n(v, x) \equiv \frac{(1 + v - n)}{\sqrt{n!x^n}} \, {}_1F_1(-n; \, 1 + v - n; \, x),$$

where $(a)_n$ is a POCHHAMMER SYMBOL and ${}_1F_1(a; \, b; \, z)$ is a CONFLUENT HYPERGEOMETRIC FUNCTION.

See also POISSON-CHARLIER POLYNOMIAL

Poisson-Charlier Polynomial

The Poisson-Charlier polynomials $c_k(x; a)$ form a SHEFFER SEQUENCE with

$$g(t) = e^{a(e^t - 1)} \tag{1}$$

$$f(t) = a(e^t - 1), \tag{2}$$

giving the GENERATING FUNCTION

$$\sum_{k=0}^{\infty} \frac{c_k(x; a)}{k!} \, t^k = e^{-t}\left(\frac{a+t}{a}\right)^x. \tag{3}$$

The Sheffer identity is

$$c_n(x + y; \, a) = \sum_{k=0}^{n} \binom{n}{k} a^{k-n} c_k(y; \, a)(x)_{n-k}, \tag{4}$$

where $(x)_n$ is a FALLING FACTORIAL (Roman 1984, p. 121). The polynomials satisfy the RECURRENCE RELATION

$$c_{n+1}(x; \, a) = a^{-1}xc_n(x-1; \, a) - c_n(x; \, a). \tag{5}$$

These polynomials belong to the distribution $d\alpha(x)$ where $\alpha(x)$ is a STEP FUNCTION with JUMP

$$j(x) = e^{-a}a^x(x!)^{-1} \tag{6}$$

at $x = 0, 1, \ldots$ for $a > 0$. They are given by the formulas

$$c_n(x; \, a) = \sum_{v=0}^{n} (-1)^{n-v}\binom{n}{v}v!a^{-v}\binom{x}{v} \tag{7}$$

$$= \sum_{k=0}^{n} \binom{n}{k}(-1)^{n-k}a^{-k}(x)_k \tag{8}$$

$$= a^n(-1)^n[j(x)]^{-1}\Delta^n j(x-n) \tag{9}$$

$$= a^{-n}n!L_n^{x-n}(a) \tag{10}$$

$$= \sum_{j=0}^{n} x^j \sum_{k=0}^{n}\binom{n}{k}(-1)^{n-k}a^{-k}s(k, j) \tag{11}$$

where $\binom{n}{k}$ is a BINOMIAL COEFFICIENT, $(x)_n$ is a FALLING FACTORIAL, $L_n^k(x)$ is an associated LAGUERRE POLYNOMIAL, $s(n, m)$ is a STIRLING NUMBER OF THE FIRST KIND, and

$$\Delta f(x) = f(x+1) - f(x) \tag{12}$$

$$\Delta^n f(x) = \Delta\left[\Delta^{n-1}f(x)\right]$$

$$= f(x+n) - \binom{n}{1}f(x+n-1) + \ldots + (-1)^n f(x). \tag{13}$$

They are normalized so that

$$\sum_{k=0}^{\infty} j(k)c_n(k; \, a)c_m(k; \, a) = a^{-n}n!\delta_{nm}, \tag{14}$$

where δ_{mn} is the DELTA FUNCTION.

The first few polynomials are

$$c_0(x; a) = 1$$

$$c_1(x; a) = -\frac{a - x}{a}$$

$$c_2(x; a) = \frac{a^2 - x - 2ax + x^2}{a^2}$$

$$c_3(x; a) = -\frac{a^3 - 2x - 3ax - 3a^2x + 3x^2 + 3ax^2 - x^3}{a^3}.$$

See also LAGUERRE POLYNOMIAL, POISSON-CHARLIER FUNCTION, SHEFFER SEQUENCE

References

Erdélyi, A.; Magnus, W.; Oberhettinger, F.; and Tricomi, F. G. *Higher Transcendental Functions, Vol. 2.* New York: Krieger, p. 226, 1981.

Jordan, C. *Calculus of Finite Differences, 3rd ed.* New York: Chelsea, p. 473, 1965.

Roman, S. "The Poisson-Charlier Polynomials." §4.3.3 in *The Umbral Calculus.* New York: Academic Press, pp. 119–22, 1984.

Szego, G. *Orthogonal Polynomials, 4th ed.* Providence, RI: Amer. Math. Soc., pp. 34–5, 1975.

Poisson's Bessel Function Formula

For $\Re[\nu] > -1/2$,

$$J_\nu(z) = \left(\frac{z}{2}\right)^\nu \frac{2}{\sqrt{\pi}\Gamma\left(\nu + \frac{1}{2}\right)} \int_0^{\pi/2} \cos(z \cos t) \sin^{2\nu} t \, dt,$$

where $J_\nu(z)$ is a BESSEL FUNCTION OF THE FIRST KIND, and $\Gamma(z)$ is the GAMMA FUNCTION.

References

Iyanaga, S. and Kawada, Y. (Eds.). *Encyclopedic Dictionary of Mathematics.* Cambridge, MA: MIT Press, p. 1472, 1980.

Poisson's Equation

A second-order PARTIAL DIFFERENTIAL EQUATION arising in physics,

$$\nabla^2 \psi = -4\pi\rho.$$

If $\rho = 0$, it reduces LAPLACE'S EQUATION. It is also related to the HELMHOLTZ DIFFERENTIAL EQUATION

$$\nabla^2 \psi + k^2 \psi = 0.$$

See also HELMHOLTZ DIFFERENTIAL EQUATION, LAPLACE'S EQUATION, VECTOR POISSON EQUATION

References

Arfken, G. "Gauss's Law, Poisson's Equation." §1.14 in *Mathematical Methods for Physicists, 3rd ed.* Orlando, FL: Academic Press, pp. 74–8, 1985.

Morse, P. M. and Feshbach, H. *Methods of Theoretical Physics, Part I.* New York: McGraw-Hill, p. 271, 1953.

Zwillinger, D. (Ed.). *CRC Standard Mathematical Tables and Formulae.* Boca Raton, FL: CRC Press, p. 417, 1995.

Zwillinger, D. *Handbook of Differential Equations, 3rd ed.* Boston, MA: Academic Press, p. 129, 1997.

Poke Move

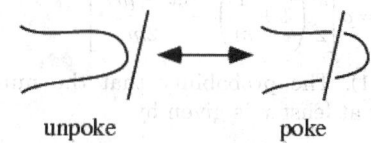

unpoke   poke

The REIDEMEISTER MOVE of type II.

See also KNOT MOVE, REIDEMEISTER MOVES

Poker

Poker is a CARD game played with a normal deck of 52 CARDS. Sometimes, additional cards called "jokers" are also used. In straight or draw poker, each player is normally dealt a hand of five cards. Depending on the variant, players then discard and redraw CARDS, trying to improve their hands. Bets are placed at each discard step. The number of possible distinct five-card hands is

$$N = \binom{52}{5} = 2,598,960,$$

where $\binom{n}{k}$ is a BINOMIAL COEFFICIENT.

There are special names for specific types of hands. A royal flush is an ace, king, queen, jack, and 10, all of one suit. A straight flush is five consecutive cards all of the same suit (but not a royal flush), where an ace may count as either high or low. A full house is three-of-a-kind and a pair. A flush is five cards of the same suit (but not a royal flush or straight flush). A straight is five consecutive cards (but not a royal flush or straight flush), where an ace may again count as either high or low.

The probabilities of being dealt five-card poker hands of a given type (before discarding and with no jokers) on the initial deal are given below (Packel 1981). As usual, for a hand with probability P, the ODDS against being dealt it are $(1/r) - 1 : 1$.

Hand	Exact Probability	Probability	ODDS
royal flush	$\frac{4}{N} = \frac{1}{649,740}$	1.54×10^{-6}	649,739.0:1
straight flush	$\frac{4(10) - 4}{N} = \frac{3}{216,580}$	1.39×10^{-5}	72,192.3:1
four of a kind	$\frac{13(48)}{N} = \frac{1}{4,165}$	2.40×10^{-4}	4,164.0:1
full house	$\frac{13\binom{4}{3}12\binom{4}{2}}{N} = \frac{6}{4,165}$	1.44×10^{-3}	693.2:1
flush	$\frac{4\binom{13}{5} - 36 - 4}{N} = \frac{1,277}{649,740}$	1.97×10^{-3}	507.8:1

straight	$\dfrac{10(4^5) - 36 - 4}{N}$	$= \frac{5}{1,274}$	3.92×10^{-3}	$253.8{:}1$
three of a kind	$\dfrac{13\binom{4}{3}\dfrac{(48)(44)}{2!}}{N}$	$= \frac{88}{4,165}$	0.0211	$46.3{:}1$
two pair	$\dfrac{13\binom{4}{2}12\binom{4}{2}}{2!}\dfrac{44}{N}$	$= \frac{198}{4,165}$	0.0475	$20.0{:}1$
one pair	$\dfrac{13\binom{4}{2}\dfrac{(48)(44)(40)}{3!}}{N}$	$= \frac{352}{833}$	0.423	$1.366{:}1$

Gadbois (1996) gives probabilities for hands if two jokers are included, and points out that it is *impossible* to rank hands in any single way which is consistent with the relative frequency of the hands.

See also BRIDGE CARD GAME, CARDS

References
Cheung, Y. L. "Why Poker is Played with Five Cards." *Math. Gaz.* **73**, 313–15, 1989.
Conway, J. H. and Guy, R. K. "Choice Numbers with Repetitions." In *The Book of Numbers.* New York: Springer-Verlag, pp. 70–1, 1996.
Friedman, E. "Erich's Poker Page." http://www.stetson.edu/~efriedma/poker/.
Gadbois, S. "Poker with Wild Cards--A Paradox?" *Math. Mag.* **69**, 283–85, 1996.
Jacoby, O. *Oswald Jacoby on Poker.* New York: Doubleday, 1981.
Packel, E. W. *The Mathematics of Games and Gambling.* Washington, DC: Math. Assoc. Amer., 1981.
Rubens, J. *Win at Poker.* New York: Dover.
Sarrett, P. "Poker Game Variants." http://gamereport.com/poker/.

Polar

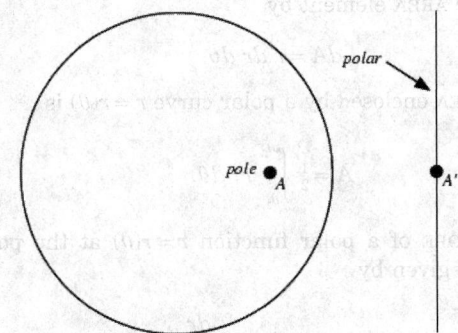

If two points A and A' are INVERSE (sometimes called conjugate) with respect to a CIRCLE (the INVERSION CIRCLE), then the straight LINE through A' which is PERPENDICULAR to the line of the points AA' is called the polar of A with respect to the CIRCLE, and A is called the POLE of the polar.

An incidence-preserving transformation in which points and lines are transformed into their POLES and polars is called RECIPROCATION (a.k.a. constructing the dual).

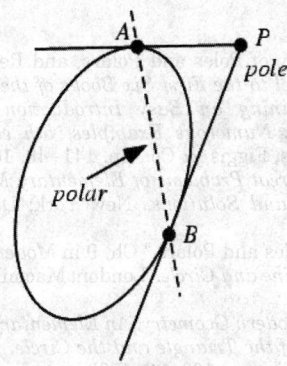

The concept of poles and polars can also be generalized to arbitrary CONIC SECTIONS. If two tangents to a CONIC SECTION at points A and B meet at P, then P is called the POLE of the line AB with respect to the conic and AB is said to be the polar of the point P with respect to the conic (Wells 1991).

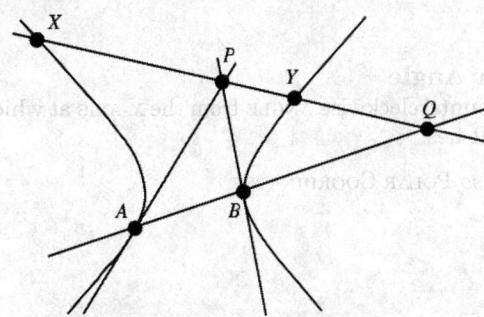

In the above figure, let a line through the polar P meet a conic section at point X and Y, and let the line XY intersect the polar line AB and Q. Then $\{XPYQ\}$ form a HARMONIC RANGE (Wells 1991).

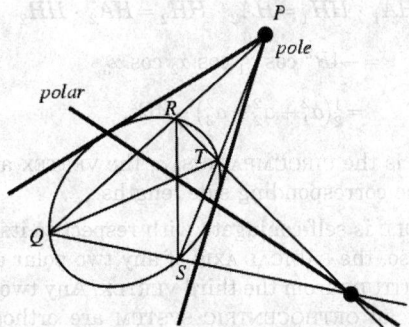

In the above figure, let two lines through the polar P meet a conic at points P and Q and S and T. Then QT and RS are concurrent on the polar (Wells 1991).

The concept can be generalized even further to an arbitrary ALGEBRAIC CURVE so that every point has a polar with respect to the curve and every line has a pole (Wells 1991).

See also APOLLONIUS' PROBLEM, DUAL POLYHEDRON, INVERSE POINTS, INVERSION CIRCLE, POLARITY, POLE (INVERSION), RECIPROCAL, RECIPROCATION, SALMON'S THEOREM, TRILINEAR POLAR

References

Casey, J. "Theory of Poles and Polars, and Reciprocation." §6.7 in *A Sequel to the First Six Books of the Elements of Euclid, Containing an Easy Introduction to Modern Geometry with Numerous Examples, 5th ed., rev. enl.* Dublin: Hodges, Figgis, & Co., pp. 141–48, 1888.

Dörrie, H. *100 Great Problems of Elementary Mathematics: Their History and Solutions.* New York: Dover, p. 157, 1965.

Durell, C. V. "Poles and Polars." Ch. 9 in *Modern Geometry: The Straight Line and Circle.* London: Macmillan, pp. 93–7, 1928.

Johnson, R. A. *Modern Geometry: An Elementary Treatise on the Geometry of the Triangle and the Circle.* Boston, MA: Houghton Mifflin, pp. 100–06, 1929.

Lachlan, R. "Poles and Polars." §243–57 in *An Elementary Treatise on Modern Pure Geometry.* London: Macmillian, pp. 151–57, 1893.

Wells, D. *The Penguin Dictionary of Curious and Interesting Geometry.* London: Penguin, pp. 190–91, 1991.

Polar Angle

The counterclockwise ANGLE from the x-AXIS at which a point lies.

See also POLAR COORDINATES

Polar Circle

Given a TRIANGLE, the polar circle has center at the ORTHOCENTER H. Call H_i the FEET of the ALTITUDE. Then the RADIUS is

$$r^2 = \overline{HA_1} \cdot \overline{HH_1} = \overline{HA_2} \cdot \overline{HH_2} = \overline{HA_2} \cdot \overline{HH_2} \quad (1)$$

$$= -4R^2 \cos \alpha_1 \cos \alpha_2 \cos \alpha_3 \quad (2)$$

$$= \tfrac{1}{2}(a_1^2 + a_2^2 + a_3^2) - 4R^2, \quad (3)$$

where R is the CIRCUMRADIUS, α_i the VERTEX angles, and a_i the corresponding side lengths.

A TRIANGLE is self-conjugate with respect to its polar circle. Also, the RADICAL AXIS of any two polar circles is the ALTITUDE from the third VERTEX. Any two polar circles of an ORTHOCENTRIC SYSTEM are orthogonal. The polar circles of the triangles of a COMPLETE QUADRILATERAL constitute a COAXAL SYSTEM conjugate to that of the circles on the diagonals.

See also COAXAL SYSTEM, ORTHOCENTRIC SYSTEM, POLAR, POLE (INVERSION), RADICAL AXIS

References

Coxeter, H. S. M. and Greitzer, S. L. *Geometry Revisited.* Washington, DC: Math. Assoc. Amer., pp. 136–38, 1967.

Johnson, R. A. *Modern Geometry: An Elementary Treatise on the Geometry of the Triangle and the Circle.* Boston, MA: Houghton Mifflin, pp. 176–81, 1929.

Polar Coordinates

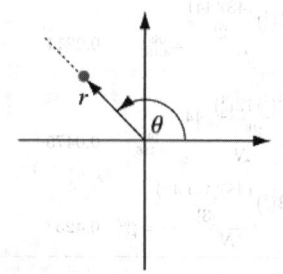

The polar coordinates r (the radial coordinate) and θ (the angular coordinate) are defined in terms of CARTESIAN COORDINATES by

$$x = r \cos \theta \quad (1)$$

$$y = r \sin \theta, \quad (2)$$

where r is the radial distance from the ORIGIN, and θ is the counterclockwise angle from the x-AXIS. In terms of x and y,

$$r = \sqrt{x^2 + y^2} \quad (3)$$

$$\theta = \tan^{-1}\left(\frac{y}{x}\right). \quad (4)$$

The ARC LENGTH of a polar curve given by $r = r(\theta)$ is

$$s = \int_{\theta_1}^{\theta_2} \sqrt{r^2 + \left(\frac{dr}{d\theta}\right)^2}\, d\theta. \quad (5)$$

The LINE ELEMENT is given by

$$ds^2 = r^2\, d\theta^2, \quad (6)$$

and the AREA element by

$$dA = r\, dr\, d\theta. \quad (7)$$

The AREA enclosed by a polar curve $r = r(\theta)$ is

$$A = \frac{1}{2}\int_{\theta_1}^{\theta_2} r^2\, d\theta. \quad (8)$$

The SLOPE of a polar function $r = r(\theta)$ at the point (r,θ) is given by

$$m = \frac{r + \tan\theta\, \dfrac{dr}{d\theta}}{-r\tan\theta + \dfrac{dr}{d\theta}}. \quad (9)$$

The ANGLE between the tangent and radial line at the point (r,θ) is

$$\psi = \tan^{-1}\left(\frac{r}{\dfrac{dr}{d\theta}}\right). \quad (10)$$

Polar Coordinates

A polar curve is symmetric about the x-AXIS if replacing θ by $-\theta$ in its equation produces an equivalent equation, symmetric about the y-AXIS if replacing θ by $\pi - \theta$ in its equation produces an equivalent equation, and symmetric about the origin if replacing r by $-r$ in its equation produces an equivalent equation.

In Cartesian coordinates, the POSITION VECTOR and its derivatives are

$$\mathbf{r} = \sqrt{x^2 + y^2}\,\hat{\mathbf{r}} \tag{11}$$

$$\dot{\mathbf{r}} = \dot{\hat{\mathbf{r}}}\sqrt{x^2 + y^2} + \hat{\mathbf{r}}(x^2 + y^2)^{-1/2}(x\dot{x} + y\dot{y}) \tag{12}$$

$$\hat{\mathbf{r}} = \frac{x\hat{\mathbf{x}} + y\hat{\mathbf{y}}}{\sqrt{x^2 + y^2}} \tag{13}$$

$$\dot{\hat{\mathbf{r}}} = \frac{\dot{x}\hat{\mathbf{x}} + \dot{y}\hat{\mathbf{y}}}{\sqrt{x^2 + y^2}} - \tfrac{1}{2}(x^2 + y^2)^{-3/2}(2)(x\dot{x} + y\dot{y})(x\hat{\mathbf{x}} + y\hat{\mathbf{y}})$$

$$= \frac{(x\dot{y} + y\dot{x})(x\hat{\mathbf{y}} - y\hat{\mathbf{x}})}{(x^2 + y^2)^{3/2}}. \tag{14}$$

In polar coordinates, the UNIT VECTORS and their derivatives are

$$\mathbf{r} \equiv \begin{bmatrix} r\cos\theta \\ r\sin\theta \end{bmatrix} \tag{15}$$

$$\hat{\mathbf{r}} \equiv \frac{\dfrac{d\mathbf{r}}{dr}}{\left|\dfrac{d\mathbf{r}}{dr}\right|} = \begin{bmatrix} \cos\theta \\ \sin\theta \end{bmatrix} \tag{16}$$

$$\hat{\boldsymbol{\theta}} \equiv \frac{\dfrac{d\theta}{d\theta}}{\left|\dfrac{d\theta}{d\theta}\right|} = \begin{bmatrix} -\sin\theta \\ \cos\theta \end{bmatrix} \tag{17}$$

$$\dot{\hat{\mathbf{r}}} = \begin{bmatrix} -\sin\theta\,\dot{\theta} \\ \cos\theta\,\dot{\theta} \end{bmatrix} = \dot{\theta}\hat{\boldsymbol{\theta}} \tag{18}$$

$$\dot{\hat{\boldsymbol{\theta}}} = \begin{bmatrix} -\cos\theta\,\dot{\theta} \\ \sin\theta\,\dot{\theta} \end{bmatrix} = -\dot{\theta}\hat{\mathbf{r}} \tag{19}$$

$$\dot{\mathbf{r}} = \begin{bmatrix} -r\sin\theta\,\dot{\theta} + \cos\theta\,\dot{r} \\ r\cos\theta\,\dot{\theta} + \sin\theta\,\dot{r} \end{bmatrix} = r\dot{\theta}\hat{\boldsymbol{\theta}} + \dot{r}\hat{\mathbf{r}} \tag{20}$$

$$\ddot{\mathbf{r}} = \dot{r}\dot{\theta}\hat{\boldsymbol{\theta}} + r\ddot{\theta}\hat{\boldsymbol{\theta}} + r\dot{\theta}\dot{\hat{\boldsymbol{\theta}}} + \ddot{r}\hat{\mathbf{r}} + \dot{r}\dot{\hat{\mathbf{r}}}$$

$$= \dot{r}\dot{\theta}\hat{\boldsymbol{\theta}} + r\ddot{\theta}\hat{\boldsymbol{\theta}} + r\dot{\theta}(-\dot{\theta}\hat{\mathbf{r}}) + \ddot{r}\hat{\mathbf{r}} + \dot{r}\dot{\theta}\hat{\boldsymbol{\theta}}$$

$$= (\ddot{r} - r\dot{\theta}^2)\hat{\mathbf{r}} + (2\dot{r}\dot{\theta} + r\ddot{\theta})\hat{\boldsymbol{\theta}}$$

$$= (\ddot{r} - r\dot{\theta}^2)\hat{\mathbf{r}} + \frac{1}{r}\frac{d}{dt}\left(r^2\dot{\theta}\right)\hat{\boldsymbol{\theta}}. \tag{21}$$

See also CARDIOID, CIRCLE, CISSOID, CONCHOID, CURVILINEAR COORDINATES, CYLINDRICAL COORDINATES, EQUIANGULAR SPIRAL, LEMNISCATE, LIMAÇON, ROSE

Polar Line

POLAR

Polar Reciprocals

INVERSE POINTS

Polar Reciprocation

INVERSE POINTS, RECIPROCATION

Polar Representation (Complex Number)

PHASOR

Polar Representation (Measure)

A polar representation of a COMPLEX MEASURE μ is analogous to the polar representation of a COMPLEX NUMBER as $z = re^{i\theta}$, where $r = |z|$,

$$d\mu = e^{i\theta}d|\mu|. \tag{1}$$

The analog of absolute value is the TOTAL VARIATION MEASURE $|\mu|$, and θ is replaced by a MEASURABLE real-valued function θ. Or sometimes one writes h with $|h| = 1$ instead of $e^{i\theta}$.

More precisely, for any measurable set E,

$$\mu(E) = \int_E e^{i\theta}\,d|\mu|, \tag{2}$$

where the integral is the LEBESGUE INTEGRAL. It is natural to extend the definition of the Lebesgue integral to complex measures using the polar representation

$$\int f\,d\mu = \int e^{i\theta}f\,d|\mu|. \tag{3}$$

See also ABSOLUTELY CONTINUOUS, COMPLEX MEASURE, FUNDAMENTAL THEOREMS OF CALCULUS, LEBESGUE MEASURE, POLAR REPRESENTATION (MEASURE), RADON-NIKODYM THEOREM

References

Rudin, W. *Real and Complex Analysis.* New York: McGraw-Hill, pp. 124–25, 1987.

Polarity

A PROJECTIVE CORRELATION of period two. In a polarity, a is called the POLAR of A, and A the POLE a.

See also CHASLES'S THEOREM, CORRELATION (GEOMETRIC), POLAR, POLE (INVERSION), PROJECTIVE CORRELATION

References

Coxeter, H. S. M. *Introduction to Geometry, 2nd ed.* New York: Wiley, p. 248, 1969.

Polarized Telephone
GOSSIPING

Pole

A HOLOMORPHIC FUNCTION f has a pole of order m at a point $z = z_0$ if, in the LAURENT SERIES, $a_n = 0$ for $n < -m$ and $a_m \neq 0$. Equivalently, f has a pole of order n at z_0 if n is the smallest POSITIVE INTEGER for which $(z - z_0)^n f(z)$ is holomorphic at z_0. A holomorphic function f has a pole at infinity if

$$\lim_{z \to \infty} f(z) = \infty.$$

A nonconstant polynomial $P(z)$ has a pole at infinity of order $\deg P$, i.e., the DEGREE of P.

The basic example of a pole is $f = 1/z^n$, which has a single pole of order n at $z = 0$. A simple *Mathematica* function which finds the poles of a RATIONAL FUNCTION is given by

```
Poles[f_,    z_]    :=       Union[z    /.
{ToRules[Roots[Denominator[Together[D[f,   z]]]
== 0, z]]}]
```

A HOLOMORPHIC FUNCTION whose only singularities are poles is called a MEROMORPHIC FUNCTION.

See also ARGUMENT PRINCIPLE, ESSENTIAL SINGULARITY, HOLOMORPHIC FUNCTION, LAURENT SERIES, MEROMORPHIC FUNCTION, POLE (INVERSION), REMOVABLE SINGULARITY, RESIDUE (COMPLEX ANALYSIS), SIMPLE POLE, SINGULAR POINT (FUNCTION)

References

Arfken, G. *Mathematical Methods for Physicists, 3rd ed.* Orlando, FL: Academic Press, pp. 396–97, 1985.
Knopp, K. "Essential and Non-Essential Singularities or Poles." §31 in *Theory of Functions Parts I and II, Two Volumes Bound as One, Part I.* New York: Dover, pp. 123–26, 1996.
Krantz, S. G. "Removable Singularities, Poles, and Essential Singularities." §4.1.4 in *Handbook of Complex Analysis.* Boston, MA: Birkhäuser, p. 42, 1999.

Pole (Inversion)

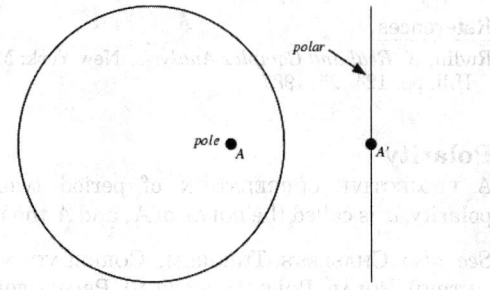

If two points A and A' are INVERSE with respect to a

CIRCLE (the INVERSION CIRCLE), then the straight line through A' which is PERPENDICULAR to the line of the points AA' is called the POLAR of the POINT A with respect to the CIRCLE, and A is called the pole of the POLAR.

An incidence-preserving transformation in which points and lines are transformed into their poles and POLARS is called a RECIPROCATION.

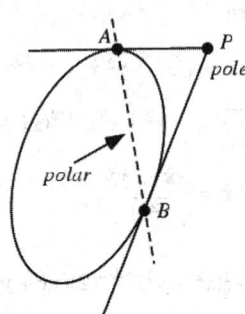

The concept of poles and polars can also be generalized to arbitrary CONIC SECTIONS. If two tangents to a CONIC SECTION at points A and B meet at P, then P is called the pole of the line AB with respect to the conic and AB is said to be the POLAR of the point P with respect to the conic (Wells 1991). Let a line through P meet a conic at points X and Y and its polar AB and Q. Then X, Y, P, and Q are a HARMONIC RANGE (Wells 1991). Furthermore, if two lines through a pole P meet a conic at points Q and R and points S and T, then the lines QT and SR meet on the polar, as do the lines QS and RT.

The concept can be generalized even further to an arbitrary ALGEBRAIC CURVE so that every point has a polar with respect to the curve and every line has a pole (Wells 1991).

See also DIAGONAL TRIANGLE, INVERSE POINTS, INVERSION CIRCLE, POLAR, POLARITY, RECIPROCAL, RECIPROCATION, TRILINEAR POLAR

References

Casey, J. "Theory of Poles and Polars, and Reciprocation." §6.7 in *A Sequel to the First Six Books of the Elements of Euclid, Containing an Easy Introduction to Modern Geometry with Numerous Examples, 5th ed., rev. enl.* Dublin: Hodges, Figgis, & Co., pp. 141–48, 1888.
Dörrie, H. *100 Great Problems of Elementary Mathematics: Their History and Solutions.* New York: Dover, p. 157, 1965.
Durell, C. V. "Poles and Polars." Ch. 9 in *Modern Geometry: The Straight Line and Circle.* London: Macmillan, pp. 93–7, 1928.
Johnson, R. A. *Modern Geometry: An Elementary Treatise on the Geometry of the Triangle and the Circle.* Boston, MA: Houghton Mifflin, pp. 100–06, 1929.
Lachlan, R. "Poles and Polars." §243–57 in *An Elementary Treatise on Modern Pure Geometry.* London: Macmillian, pp. 151–57, 1893.
Wells, D. *The Penguin Dictionary of Curious and Interesting Geometry.* London: Penguin, pp. 190–91, 1991.

Pole (Origin)

ORIGIN

Pole (Perspective)

PERSPECTIVE CENTER

Pole (Simson Line)

If a line L is the SIMSON LINE of a point P on the CIRCUMCIRCLE of a TRIANGLE, then P is called the pole of L (Honsberger 1995, p. 128).

See also SIMSON LINE

References

Honsberger, R. *Episodes in Nineteenth and Twentieth Century Euclidean Geometry.* Washington, DC: Math. Assoc. Amer., p. 128, 1995.

Policeman on Point Duty Curve

CRUCIFORM

Polignac's Conjecture

DE POLIGNAC'S CONJECTURE

Polish Notation

REVERSE POLISH NOTATION

Polish Space

The HOMEOMORPHIC image of a so-called "complete separable" METRIC SPACE. The continuous image of a Polish space is called a SOUSLIN SET.

See also DESCRIPTIVE SET THEORY, STANDARD SPACE

Pollaczek Polynomial

Let $a > |b|$, and write

$$h(\theta) = \frac{a \cos \theta + b}{2 \sin \theta}. \tag{1}$$

Then define $P_n(x; a, b)$ by the GENERATING FUNCTION

$$f(x, w) = f(\cos \theta, w) = \sum_{n=0}^{\infty} P_n(x; a, b)w^n$$

$$= (1 - we^{i\theta})^{-1/2 + ih(\theta)}(1 - we^{i\theta})^{-1/2 - ih(\theta)}. \tag{2}$$

The GENERATING FUNCTION may also be written

$$f(x, w) = \left(1 - 2xw + w^2\right)^{-1/2}$$

$$\times \exp\left[(ax + b) \sum_{m=1}^{\infty} \frac{w^m}{m} U_{m-1}(x)\right], \tag{3}$$

where $U_m(x)$ is a CHEBYSHEV POLYNOMIAL OF THE SECOND KIND.

Pollaczek polynomials satisfy the RECURRENCE RELATION

$$nP_n(x; a, b) = [(2n - 1 + 2a)x + 2b]P_{n-1}(x; a, b)$$

$$-(n-1)P_{n-2}(x; a, b) \tag{4}$$

for $n = 2, 3, ...$ with

$$P_0 = 1 \tag{5}$$

$$P_1 = (2a + 1)x + 2b. \tag{6}$$

In terms of the HYPERGEOMETRIC FUNCTION $_2F_1(a, b; c; x)$,

$$P_n(\cos \theta; a; b)$$

$$= e^{in\theta} \, _2F_1\left(-n, \tfrac{1}{2} + ih(\theta); 1; 1 - e^{-2i\theta}\right). \tag{7}$$

They obey the orthogonality relation

$$\int_{-1}^{1} P_n(x; a, b)P_m(x; a, b)w(x; a, b)\,dx$$

$$= \left[n + \tfrac{1}{2}(a + 1)\right]^{-1} \delta_{nm}, \tag{8}$$

where δ_{mn} is the KRONECKER DELTA, for $n, m = 0, 1, ...$, with the WEIGHT FUNCTION

$$w(\cos \theta; a, b) = e^{(2\theta - \pi)h(\theta)} \{\cosh[\pi h(\theta)]\}^{-1}. \tag{9}$$

References

Szego, G. *Orthogonal Polynomials, 4th ed.* Providence, RI: Amer. Math. Soc., pp. 393–00, 1975.

Pollard Monte Carlo Factorization Method

POLLARD RHO FACTORIZATION METHOD

Pollard p-1 Factorization Method

A PRIME FACTORIZATION ALGORITHM which can be implemented in a single-step or double-step form. In the single-step version, PRIMES p are found if $p - 1$ is a product of small PRIMES by finding an m such that

$$m \equiv c^q \pmod{n},$$

where $p - 1 | q$, with q a large number and $(c, n) = 1$. Then since $p - 1 | q$, $m \equiv 1 \pmod{p}$, so $p | m - 1$. There is therefore a good chance that $n \nmid m - 1$, in which case $\text{GCD}(m - 1, n)$ (where GCD is the GREATEST COMMON DIVISOR) will be a nontrivial divisor of n.

In the double-step version, a PRIMES p can be factored if $p - 1$ is a product of small PRIMES and a single larger PRIME.

See also PRIME FACTORIZATION ALGORITHMS, WILLIAMS $P+1$ FACTORIZATION METHOD

References

Bressoud, D. M. *Factorization and Prime Testing.* New York: Springer-Verlag, pp. 67–9, 1989.
Pollard, J. M. "Theorems on Factorization and Primality Testing." *Proc. Cambridge Phil. Soc.* **76**, 521–28, 1974.

Pollard Rho Factorization Method

A PRIME FACTORIZATION ALGORITHM also known as POLLARD MONTE CARLO FACTORIZATION METHOD. Let $x_0 = 2$, then compute

$$x_{i+1} = x_i^2 - x_i + 1 \pmod{n}.$$

If $\text{GCD}(x_{2i} - x_i, n) > 1$, then n is COMPOSITE and its factors are found. In modified form, it becomes BRENT'S FACTORIZATION METHOD. In practice, almost any unfactorable POLYNOMIAL can be used for the iteration ($x^2 - 2$, however, cannot). Under worst conditions, the ALGORITHM can be very slow.

See also BRENT'S FACTORIZATION METHOD, PRIME FACTORIZATION ALGORITHMS

References

Brent, R. P. "Some Integer Factorization Algorithms Using Elliptic Curves." *Austral. Comp. Sci. Comm.* **8**, 149–63, 1986.
Bressoud, D. M. *Factorization and Prime Testing.* New York: Springer-Verlag, pp. 61–7, 1989.
Eldershaw, C. and Brent, R. P. "Factorization of Large Integers on Some Vector and Parallel Computers."
Montgomery, P. L. "Speeding the Pollard and Elliptic Curve Methods of Factorization." *Math. Comput.* **48**, 243–64, 1987.
Pollard, J. M. "A Monte Carlo Method for Factorization." *Nordisk Tidskrift for Informationsbehandlung (BIT)* **15**, 331–34, 1975.
Vardi, I. *Computational Recreations in Mathematica.* Reading, MA: Addison-Wesley, pp. 83 and 102–03, 1991.

Poloidal Field

A VECTOR FIELD resembling a magnetic multipole which has a component along the z-AXIS of a SPHERE and continues along lines of LONGITUDE.

See also DIVERGENCELESS FIELD, TOROIDAL FIELD

References

Stacey, F. D. *Physics of the Earth, 2nd ed.* New York: Wiley, p. 239, 1977.

Pólya Conjecture

Let n be a POSITIVE INTEGER and $r(n)$ the number of (not necessarily distinct) PRIME FACTORS of n (with $r(1) = 0$). Let $O(m)$ be the number of POSITIVE INTEGERS $\leq m$ with an ODD number of PRIME FACTORS, and $E(m)$ the number of POSITIVE INTEGERS $\leq m$ with an EVEN number of PRIME FACTORS. Pólya conjectured that

$$L(m) \equiv E(m) - O(m) = \sum_{n=1}^{m} \lambda(n)$$

is ≤ 0, where $\lambda(n)$ is the LIOUVILLE FUNCTION.

The conjecture was made in 1919, and disproven by Haselgrove (1958) using a method due to Ingham (1942). Lehman (1960) found the first explicit counterexample, $L(906, 180, 359) = 1$, and the smallest counterexample $m = 906,150,257$ was found by Tanaka (1980). The first n for which $L(n) = 0$ are $n = 2$, 4, 6, 10, 16, 26, 40, 96, 586, 906150256, ... (Tanaka 1980, Sloane's A028488). It is unknown if $L(x)$ changes sign infinitely often (Tanaka 1980).

See also ANDRICA'S CONJECTURE, LIOUVILLE FUNCTION, PRIME FACTORS

References

Haselgrove, C. B. "A Disproof of a Conjecture of Pólya." *Mathematika* **5**, 141–45, 1958.
Ingham, A. E. "On Two Conjectures in the Theory of Numbers." *Amer. J. Math.* **64**, 313–19, 1942.
Lehman, R. S. "On Liouville's Function." *Math. Comput.* **14**, 311–20, 1960.
Sloane, N. J. A. Sequences A028488 in "An On-Line Version of the Encyclopedia of Integer Sequences." http://www.research.att.com/~njas/sequences/eisonline.html.
Tanaka, M. "A Numerical Investigation on Cumulative Sum of the Liouville Function" [sic]. *Tokyo J. Math.* **3**, 187–89, 1980.

Pólya Distribution

NEGATIVE BINOMIAL DISTRIBUTION

Pólya Enumeration Theorem

A very general theorem which allows the number of discrete combinatorial objects of a given type to be enumerated (counted) as a function of their "order." The most common application is in the counting of the number of GRAPHS of n nodes, TREES and ROOTED TREES with n branches, GROUPS of order n, etc. The theorem is an extension of the CAUCHY-FROBENIUS LEMMA, which is sometimes also called BURNSIDE'S LEMMA, the PÓLYA-BURNSIDE LEMMA, the CAUCHY-FROBENIUS LEMMA, or even "the LEMMA THAT IS NOT BURNSIDE'S!"

Pólya enumeration is implemented as [g, m], in the *Mathematica* add-on package DiscreteMath`Combinatorica` (which can be loaded with the command < <DiscreteMath`) which returns the polynomial giving the number of colorings with M colors of a structure defined by a PERMUTATION GROUP g.

See also CAUCHY-FROBENIUS LEMMA, GRAPH, GROUP, ROOTED TREE, TREE

References

Harary, F. "The Number of Linear, Directed, Rooted, and Connected Graphs." *Trans. Amer. Math. Soc.* **78**, 445–63, 1955.
Harary, F. "Pólya's Enumeration Theorem." *Graph Theory.* Reading, MA: Addison-Wesley, pp. 180–84, 1994.
Pólya, G. "Kombinatorische Anzahlbestimmungen für Gruppen, Graphen, und chemische Verbindungen." *Acta Math.* **68**, 145–54, 1937.
Roberts, F. S. *Applied Combinatorics.* Englewood Cliffs, NJ: Prentice-Hall, 1984.
Skiena, S. "Pólya's Theory of Counting." §1.2.6 in *Implementing Discrete Mathematics: Combinatorics and Graph*

Theory with Mathematica. Reading, MA: Addison-Wesley, pp. 25–6, 1990.

Tucker, A. *Applied Combinatorics, 3rd ed.* New York: Wiley, 1995.

Pólya Polynomial

The POLYNOMIAL giving the number of colorings with m colors of a structure defined by a PERMUTATION GROUP.

See also PERMUTATION GROUP, PÓLYA ENUMERATION THEOREM

Polyabolo

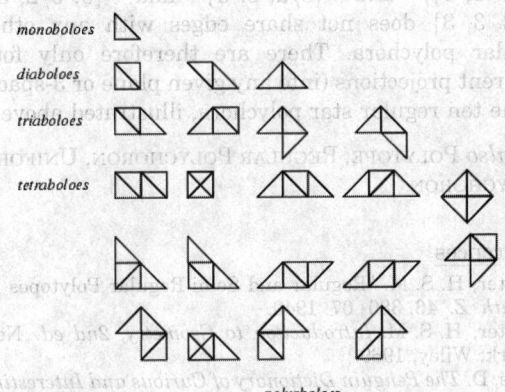

monoboloes

diaboloes

triaboloes

tetraboloes

polyaboloes

An analog of the POLYOMINO composed of n ISOSCELES RIGHT TRIANGLES joined along edges of the same length. Polyaboloes are sometimes also called polytans. The number of fixed polyaboloes composed of n triangles are 1, 3, 4, 14, 30, 107, 318, 1106, 3671, ... (Sloane's A006074).

See also DIABOLO, HEXABOLO, PENTABOLO, POLYABOLO TILING, POLYIAMOND, TETRABOLO, TRIABOLO

References

Sloane, N. J. A. Sequences A006074/M2379 in "An On-Line Version of the Encyclopedia of Integer Sequences." http://www.research.att.com/~njas/sequences/eisonline.html.

Vichera, M. "Polyforms." http://alpha.ujep.cz/~vicher/puzzle/polyforms.htm.

Polyabolo Tiling

See also POLYABOLO

References

Vichera, M. "Polytans." http://alpha.ujep.cz/~vicher/puzzle/polyform/tan/tan.htm.

Pólya-Burnside Lemma

CAUCHY-FROBENIUS LEMMA, PÓLYA ENUMERATION THEOREM

Pólya's Random Walk Constants

N.B. A detailed online essay by S. Finch was the starting point for this entry.

Let $p(d)$ be the probability that a RANDOM WALK on a d-D lattice returns to the origin. Pólya (1921) proved that

$$p(1) = p(2) = 1, \qquad (1)$$

but

$$p(d) < 1 \qquad (2)$$

for $d > 2$. Watson (1939), McCrea and Whipple (1940), Domb (1954), and Glasser and Zucker (1977) showed that

$$p(3) = 1 - \frac{1}{u(3)} = 0.3405373296\ldots, \qquad (3)$$

where

$$u(3) = \frac{3}{(2\pi)^3} \int_{-\pi}^{\pi} \int_{-\pi}^{\pi} \int_{-\pi}^{\pi} \frac{dx\,dy\,dz}{3 - \cos x - \cos y - \cos z} \qquad (4)$$

$$= \frac{12}{\pi^2} \left(18 + 12\sqrt{2} - 10\sqrt{3} - 7\sqrt{6} \right)$$
$$\times \left\{ K\left[\left(2 - \sqrt{3}\right)\left(\sqrt{3} - \sqrt{2}\right) \right] \right\}^2 \qquad (5)$$

$$= 3\left(18 + 12\sqrt{2} - 10\sqrt{3} - 7\sqrt{6} \right)$$
$$\times \left[1 + 2\sum_{k=1}^{\infty} \exp\left(-k^2 \pi \sqrt{6}\right) \right]^4 \qquad (6)$$

$$= \frac{\sqrt{6}}{32\pi^3} \Gamma\left(\tfrac{1}{24}\right) \Gamma\left(\tfrac{5}{24}\right) \Gamma\left(\tfrac{7}{24}\right) \Gamma\left(\tfrac{11}{24}\right) \qquad (7)$$

$$= 1.5163860592\ldots. \qquad (8)$$

Here, $K(k)$ is a complete ELLIPTIC INTEGRAL OF THE FIRST KIND and $\Gamma(z)$ is the GAMMA FUNCTION. Closed forms for $d > 3$ are not known, but Montroll (1956) showed that

$$p(d) = 1 - [u(d)]^{-1}, \qquad (9)$$

where

$$u(d) = \frac{d}{(2\pi)^d} \underbrace{\int_{-\pi}^{\pi} \int_{-\pi}^{\pi} \cdots \int_{-\pi}^{\pi}}_{d}$$
$$\times \left(d - \sum_{k=1}^{d} \cos x_k \right)^{-1} dx_1\,dx_2 \cdots dx_d$$
$$= \int_0^{\infty} \left[I_0\left(\frac{t}{d}\right) \right]^d e^{-t}\,dt, \qquad (10)$$

and $I_0(z)$ is a MODIFIED BESSEL FUNCTION OF THE FIRST KIND. Numerical values of $p(d)$ from Montroll (1956) and Flajolet (Finch) are given in the following table.

d	p(d)
3	0.3405086322
4	0.1932016706
5	0.1351786098
6	0.1047154956
7	0.0858449341
8	0.0729126492

See also RANDOM WALK

References

Finch, S. "Favorite Mathematical Constants." http://www.mathsoft.com/asolve/constant/polya/polya.html.

Domb, C. "On Multiple Returns in the Random-Walk Problem." *Proc. Cambridge Philos. Soc.* **50**, 586–91, 1954.

Glasser, M. L. and Zucker, I. J. "Extended Watson Integrals for the Cubic Lattices." *Proc. Nat. Acad. Sci. U.S.A.* **74**, 1800–801, 1977.

McCrea, W. H. and Whipple, F. J. W. "Random Paths in Two and Three Dimensions." *Proc. Roy. Soc. Edinburgh* **60**, 281–98, 1940.

Montroll, E. W. "Random Walks in Multidimensional Spaces, Especially on Periodic Lattices." *J. SIAM* **4**, 241–60, 1956.

Watson, G. N. "Three Triple Integrals." *Quart. J. Math., Oxford Ser. 2* **10**, 266–76, 1939.

Pólya-Vinogradov Inequality

Let χ be a nonprincipal character (mod q). Then

$$\sum_{n=M+1}^{M+N} \chi(n) \ll \sqrt{q} \ln q,$$

where $\ll$ indicates MUCH LESS THAN.

References

Davenport, H. "The Pólya-Vinogradov Inequality." Ch. 23 in *Multiplicative Number Theory, 2nd ed.* New York: Springer-Verlag, pp. 135–38, 1980.

Pólya, G. "Uuml;ber die Verteilung der quadratischen Reste und Nichtreste." *Nachr. Königl. Gesell. Wissensch. Göttingen, Math.-phys. Klasse,* 21–9, 1918.

Vinogradov. *Perm. Univ. Fiz.-Mat. ob.-vo Zh.* **1**, 18–4 and 94–8, 1918.

Polychoron

A POLYTOPE in 4-D. Polychora are bounded by polyhedra.

The NECESSARY condition for the polychoron with SCHLÄFLI SYMBOL $\{p, q, r\}$ to be a finite polytope is

$$\cos\left(\frac{\pi}{q}\right) < \sin\left(\frac{\pi}{p}\right)\sin\left(\frac{\pi}{r}\right).$$

SUFFICIENCY can be established by consideration of

the six figures satisfying this condition.

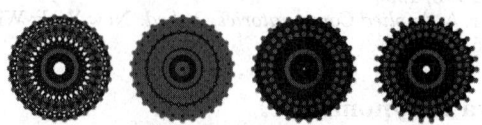

Nine of the ten star polychora can be obtained by faceting $\{3, 3, 5\}$; in other words, they have the same vertices as $\{3, 3, 5\}$. The tenth, $\{5/2, 3, 3\}$, can be obtained by faceting $\{5, 3, 3\}$. In addition, of the ten regular star polychora, several share the same edges: $\{3, 3, 5\}$, $\{3, 5, 5/2\}$, $\{5, 5/2, 5\}$, and $\{5, 3, 5/2\}$; $\{3, 3, 5/2\}$, $\{3, 5/2, 5\}$, $\{5/2, 5, 5/2\}$, and $\{5/2, 3, 5\}$; and $\{5/2, 5, 3\}$ and $\{5, 5/2, 3\}$. $\{5/2, 3, 3\}$ does not share edges with any other regular polychora. There are therefore only four different projections (into any given plane or 3-space) of the ten regular star polychora, illustrated above.

See also POLYTOPE, REGULAR POLYCHORON, UNIFORM POLYCHORON

References

Coxeter, H. S. M. "Regular and Semi-Regular Polytopes I." *Math. Z.* **46**, 380–07, 1940.

Coxeter, H. S. M. *Introduction to Geometry, 2nd ed.* New York: Wiley, 1969.

Wells, D. *The Penguin Dictionary of Curious and Interesting Geometry.* London: Penguin, 1991.

Polyconic Projection

A class of map projections in which the parallels are represented by a system of non-concentric circular arcs with centers lying on the straight line representing the central meridian (Lee 1944). The term was first applied by Hunt, and later extended by Tissot (1881).

$$x = \cot\phi \sin E \tag{1}$$

$$y = (\phi - \phi_0) + \cot\phi(1 - \cos E), \tag{2}$$

where

$$E = (\lambda - \lambda_0)\sin\phi. \tag{3}$$

The inverse FORMULAS are

$$\lambda = \frac{\sin^{-1}(x \tan\phi)}{\sin\phi} + \lambda_0, \tag{4}$$

and ϕ is determined from

$$\Delta\phi = -\frac{A(\phi\tan\phi + 1) - \phi - \frac{1}{2}(\phi^2 + B)\tan\phi}{\frac{\phi - A}{\tan\phi} - 1}, \quad (5)$$

where $\phi_0 = A$ and

$$A = \phi_0 + y \quad (6)$$

$$B = x^2 + A^2. \quad (7)$$

References

Beaman, W. M. *Topographic Mapping.* Washington, DC: U. S. Geol. Survey Bull. 788-E, p. 167, 1928.

Birdseye, C. H. *Formulas and Tables for the Construction of Polyconic Projections.* U. S. Geological Survey, Bulletin 809, 1929.

Hunt. Appendix 39 in *Report for the U.S. Coast and Geodetic Survey.* 1853.

Lee, L. P. "The Nomenclature and Classification of Map Projections." *Empire Survey Rev.* **7**, 190–00, 1944.

Snyder, J. P. *Map Projections--A Working Manual.* U. S. Geological Survey Professional Paper 1395. Washington, DC: U. S. Government Printing Office, pp. 124–37, 1987.

Tissot, A. *Mémoir sur la représentation des surfaces et les projections des cartes géographiques.* Paris: Gauthier-Villars, 1881.

Polycube

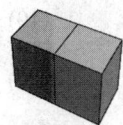

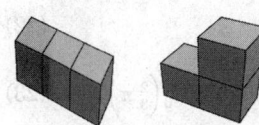

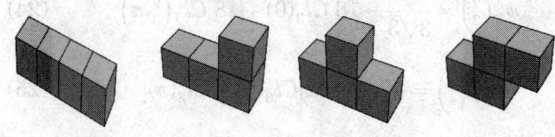

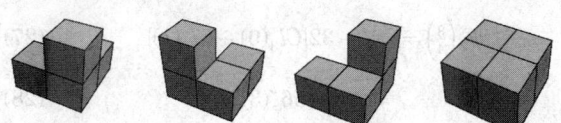

3-D generalization of the POLYOMINOES to n-D. The number of polycubes $N(n)$ composed of n CUBES are 1, 1, 2, 8, 29, 166, 1023, ... (Sloane's A000162, Ball and Coxeter 1987).

There are 1390 distinct ways to pack the eight polycubes of order $n = 4$ into a $2 \times 4 \times 4$ box (Beeler 1972).

See also CONWAY PUZZLE, CUBE DISSECTION, DIABOLICAL CUBE, PENTACUBE, SLOTHOUBER-GRAATSMA PUZZLE, SOMA CUBE

References

Ball, W. W. R. and Coxeter, H. S. M. *Mathematical Recreations and Essays, 13th ed.* New York: Dover, pp. 112–13, 1987.

Beeler, M. Item 112 in Beeler, M.; Gosper, R. W.; and Schroeppel, R. *HAKMEM.* Cambridge, MA: MIT Artificial Intelligence Laboratory, Memo AIM-239, pp. 48–0, Feb. 1972.

Bouwkamp, C. J. "Packing Handed Pentacubes." In *The Mathematical Gardner* (Ed. D. Klarner). Boston, MA: Prindle, Weber, 1981.

Gardner, M. *The Second Scientific American Book of Mathematical Puzzles & Diversions: A New Selection.* New York: Simon and Schuster, pp. 76–7, 1961.

Gardner, M. "Polycubes." Ch. 3 in *Knotted Doughnuts and Other Mathematical Entertainments.* New York: W. H. Freeman, pp. 28–3, 1986.

Keller, M. "Counting Polyforms." http://members.aol.com/wgreview/polyenum.html.

Sloane, N. J. A. Sequences A000162/M1845 in "An On-Line Version of the Encyclopedia of Integer Sequences." http://www.research.att.com/~njas/sequences/eisonline.html.

Polycyclic Group

See also SOLVABLE GROUP

References

Roseblade, J. E.; Goldie, A. W.; and Wehrfritz, B. A. F. *Three Lectures on Polycyclic Groups.* London: Queen Mary College, 1973.

Segal, D. *Polycyclic Groups.* Cambridge, England: Cambridge University Press, 1983.

Polydisk

Let $\mathbf{c} = (c_1, \ldots, c_n)$ be a point in $\mathbb{C}^n$, then the open polydisk is defined by

$$S = \{z : |z_j - c_j| < |z_j^0 - c_j|\}$$

for $j = 1, \ldots, n$.

See also DISK, OPEN DISK

References

Iyanaga, S. and Kawada, Y. (Eds.). *Encyclopedic Dictionary of Mathematics.* Cambridge, MA: MIT Press, p. 100, 1980.

Polyfrob

POLYHEX

Polygamma Function

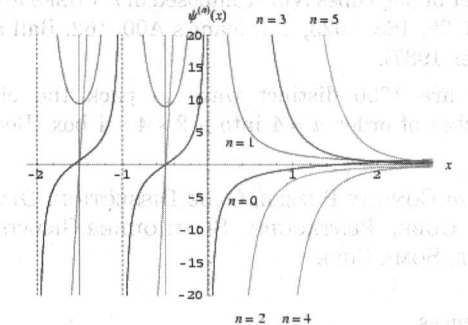

A SPECIAL FUNCTION which is given by the $(n+1)$st DERIVATIVE of the LOGARITHM of the GAMMA FUNCTION $\Gamma(z)$ (or, depending on the definition, of the FACTORIAL $z!$). This is equivalent to the nth normal derivative of the LOGARITHMIC DERIVATIVE of $\Gamma(z)$ (or $z!$) and, in the former case, to the nth normal derivative of the DIGAMMA FUNCTION $\psi_0(z)$. Because of this ambiguity in definition, two different notations are sometimes (but not always) used, namely

$$\psi_n(z) = \frac{d^{n+1}}{dz^{n+1}} \ln[\Gamma(z)]$$

$$= \frac{d^n}{dz^n} \frac{\Gamma'(z)}{\Gamma(z)} = \frac{d^n}{dz^n} \psi_0(z) \qquad (1)$$

$$= (-1)^{n+1} n! \sum_{k=0}^{\infty} \frac{1}{(z+k)^{n+1}} \qquad (2)$$

$$= (-1)^{n+1} n! \zeta(n+1, z), \qquad (3)$$

where $\zeta(a, z)$ is the HURWITZ ZETA FUNCTION, and

$$F_n(z) \equiv \frac{d^{n+1}}{dz^{n+1}} \ln z!. \qquad (4)$$

The two notations are connected by

$$\psi_n(z) = F_n(z-1). \qquad (5)$$

Unfortunately, Morse and Feshbach (1953) adopt a notation no longer in standard use in which Morse and Feshbach's "$\psi_n(z)$" is equal to $\psi_{n-1}(z)$ in the usual notation. Also note that the function $\psi_0(z)$ is equivalent to the DIGAMMA FUNCTION $\Psi(z)$. $\psi_n(z)$ is implemented in *Mathematica* as `PolyGamma[n, z]`.
The polygamma function obeys the RECURRENCE RELATION

$$\psi_n(z+1) = \psi_n(z) + (-1)^n n! z^{-n-1}, \qquad (6)$$

the reflection FORMULA

$$\psi_n(1-z) + (-1)^{n+1} \psi_n(z) = (-1)^n \pi \frac{d^n}{dz^n} \cot(\pi z), \qquad (7)$$

and the multiplication FORMULA,

$$\psi_n(mz) = \delta_{n0} \ln m + \frac{1}{m^{n+1}} \sum_{k=1}^{m-1} \psi_n\left(z + \frac{k}{m}\right), \qquad (8)$$

where δ_{mn} is the KRONECKER DELTA.
In general, special values for integral indices are given by

$$\psi_n(1) = (-1)^{n+1} n! \zeta(n+1) \qquad (9)$$

$$\psi_n\left(\tfrac{1}{2}\right) = (-1)^{n+1} n! \left(2^{n+1} - 1\right) \zeta(n+1), \qquad (10)$$

giving

$$\psi_1\left(\tfrac{1}{2}\right) = \tfrac{1}{2}\pi^2 \qquad (11)$$

$$\psi_1(1) = \zeta(2) = \tfrac{1}{6}\pi^2 \qquad (12)$$

$$\psi_2(1) = -2\zeta(3), \qquad (13)$$

$$\psi_3\left(\tfrac{1}{2}\right) = \pi^4 \qquad (14)$$

and so on.
The polygamma function can be expressed in terms of CLAUSEN FUNCTIONS for RATIONAL arguments and integer indices. Special cases are given by

$$\psi_1\left(\tfrac{1}{3}\right) = \tfrac{2}{3}\pi^2 + 3\sqrt{3}\, Cl_2\left(\tfrac{2}{3}\pi\right) \qquad (15)$$

$$\psi_1\left(\tfrac{2}{3}\right) = \tfrac{2}{3}\pi^2 - 3\sqrt{3}\, Cl_2\left(\tfrac{2}{3}\pi\right) \qquad (16)$$

$$\psi_1\left(\tfrac{1}{4}\right) = \pi^2 + 8\, Cl_2\left(\tfrac{1}{2}\pi\right) \qquad (17)$$

$$= \pi^2 + 8K \qquad (18)$$

$$\psi_1\left(\tfrac{3}{4}\right) = \pi^2 - 8\, Cl_2\left(\tfrac{1}{2}\pi\right) \qquad (19)$$

$$= \pi^2 - 8K \qquad (20)$$

$$\psi_2\left(\tfrac{1}{2}\right) = -8[Cl_3(0) - Cl_3(\pi)] \qquad (21)$$

$$= 14\zeta(3) \qquad (22)$$

$$\psi_2\left(\tfrac{1}{3}\right) = -\frac{4\pi^3}{3\sqrt{3}} - 18\, Cl_3(0) + 18\, Cl_3\left(\tfrac{2}{3}\pi\right) \qquad (23)$$

$$\psi_2\left(\tfrac{2}{3}\right) = \frac{4\pi^3}{3\sqrt{3}} - 18\, Cl_3(0) + 18\, Cl_3\left(\tfrac{2}{3}\pi\right) \qquad (24)$$

$$\psi_2\left(\tfrac{1}{4}\right) = -2\pi^3 - 32[Cl_3(0) - Cl_3(\pi)] \qquad (25)$$

$$= -2\pi^3 - 56\zeta(3) \qquad (26)$$

$$\psi_2\left(\tfrac{3}{4}\right) = 2\pi^3 - 32[Cl_3(0) - Cl_3(\pi)] \qquad (27)$$

$$= 2\pi^3 - 56\zeta(3) \qquad (28)$$

$$\psi_2\left(\tfrac{1}{6}\right) = -182\zeta(3) - 4\sqrt{3}\,\pi^3 \qquad (29)$$

$$\psi_2\!\left(\tfrac{5}{6}\right) = -182\zeta(3) + 4\sqrt{3}\,\pi^3 \tag{30}$$

$$\psi_3\!\left(\tfrac{1}{3}\right) = \tfrac{8}{3}\pi^4 + 162\sqrt{3}\,Cl_4\!\left(\tfrac{2}{3}\pi\right) \tag{31}$$

$$\psi_3\!\left(\tfrac{2}{3}\right) = \tfrac{8}{3}\pi^4 - 162\sqrt{3}\,Cl_4\!\left(\tfrac{2}{3}\pi\right) \tag{32}$$

$$\psi_3\!\left(\tfrac{1}{4}\right) = 8\pi^4 + 768\,Cl_4\!\left(\tfrac{1}{2}\pi\right) \tag{33}$$

$$= 8\pi^4 + 768\beta(4) \tag{34}$$

$$\psi_3\!\left(\tfrac{3}{4}\right) = 8\pi^4 - 768\,Cl_4\!\left(\tfrac{1}{2}\pi\right) \tag{35}$$

$$= 8\pi^4 - 768\beta(4), \tag{36}$$

where K is CATALAN'S CONSTANT, $\zeta(z)$ is the RIEMANN ZETA FUNCTION, and $\beta(z)$ is the DIRICHLET BETA FUNCTION.

See also CATALAN'S CONSTANT, CLAUSEN FUNCTION, DIGAMMA FUNCTION, DIRICHLET BETA FUNCTION, GAMMA FUNCTION, PERIODIC ZETA FUNCTION, RIEMANN ZETA FUNCTION, STIRLING'S SERIES

References

Abramowitz, M. and Stegun, C. A. (Eds.). "Polygamma Functions." §6.4 in *Handbook of Mathematical Functions with Formulas, Graphs, and Mathematical Tables, 9th printing.* New York: Dover, p. 260, 1972.

Adamchik, V. S. "Polygamma Functions of Negative Order." *J. Comput. Appl. Math.* **100**, 191–99, 1999.

Arfken, G. "Digamma and Polygamma Functions." §10.2 in *Mathematical Methods for Physicists, 3rd ed.* Orlando, FL: Academic Press, pp. 549–55, 1985.

Davis, H. T. *Tables of the Higher Mathematical Functions.* Bloomington, IN: Principia Press, 1933.

Kolbig, V. "The Polygamma Function $\psi_k(x)$ for $x = 1/4$ and $x = 3/4$." *J. Comput. Appl. Math.* **75**, 43–6, 1996.

Morse, P. M. and Feshbach, H. *Methods of Theoretical Physics, Part I.* New York: McGraw-Hill, pp. 422–24, 1953.

Polygenic Function

A function which has infinitely many DERIVATIVES at a point. If a function is not polygenic, it is MONOGENIC.

See also MONOGENIC FUNCTION

References

Newman, J. R. *The World of Mathematics, Vol. 3.* New York: Simon & Schuster, p. 2003, 1956.

Polygon

A closed plane figure with n sides. If all sides and angles are equivalent, the polygon is called REGULAR. Polygons can be CONVEX, concave, or STAR. The word "polygon" derives from the Greek $\pi o \lambda \upsilon$ (*poly*) meaning "many" and $\gamma \omega \nu \iota \alpha$ (*gonia*) meaning "angle."

The AREA of a planar CONVEX POLYGON with VERTICES $(x_1, y_1), ..., (x_n, y_n)$ is

$$A = \frac{1}{2}\left(\begin{vmatrix} x_1 & x_2 \\ y_1 & y_2 \end{vmatrix} + \begin{vmatrix} x_2 & x_3 \\ y_2 & y_3 \end{vmatrix} + \cdots + \begin{vmatrix} x_n & x_1 \\ y_n & y_1 \end{vmatrix}\right), \tag{1}$$

which can be written

$$A = \tfrac{1}{2}(x_1 y_2 - x_2 y_1 + x_2 y_3 - x_3 y_2 + \cdots + x_{n-1} y_n$$
$$- x_n y_{n-1} + x_n y_1 - x_1 y_n), \tag{2}$$

where the signs can be found from the following diagram.

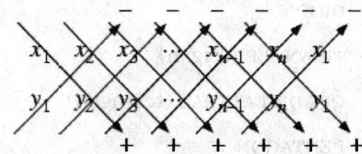

The AREA of a polygon is defined to be POSITIVE if the points are arranged in a counterclockwise order, and NEGATIVE if they are in clockwise order (Beyer 1987).

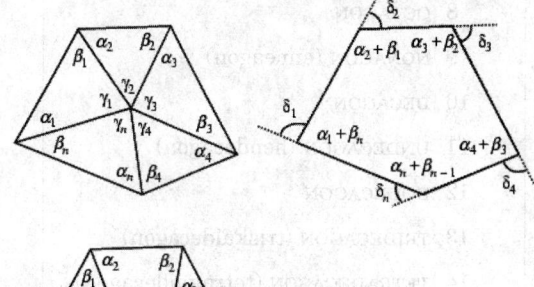

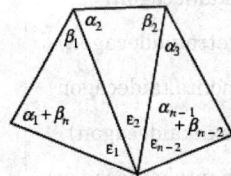

The sum I of interior angles in the top left diagram of a dissected polygon is

$$I \equiv \sum_{i=1}^{n}(\alpha_i + \beta_i) = \sum_{i=1}^{n}(\alpha_i + \beta_i + \gamma_i) - \sum_{i=1}^{n}\gamma_i. \tag{3}$$

But

$$\sum_{i=1}^{n}\gamma_i = 360° \tag{4}$$

and the sum of ANGLES of the n TRIANGLES is

$$\sum_{i=1}^{n}(\alpha_i + \beta_i + \gamma_i) = \sum_{i=1}^{n}(180°) = n(180°). \tag{5}$$

Therefore,

$$I = n(180°) - 360° = (n-2)180°. \tag{6}$$

The same equation can be derived using EXTERIOR ANGLES (top right figure) or a triangulation from a single vertex (bottom figure).

The following table gives the names for polygons with n sides. The words for polygons with $n \geq 5$ sides (e.g., PENTAGON, HEXAGON, HEPTAGON, etc.) can refer to either REGULAR or non-regular polygons, depending on context. It is therefore always best to specify "regular n-gon" explicitly. For some polygons, several different terms are used interchangeably, e.g., nonagon and enneagon both refer to the polygon with $n = 9$ sides.

n	polygon
2	DIGON
3	TRIANGLE (trigon)
4	QUADRILATERAL (tetragon)
5	PENTAGON
6	HEXAGON
7	HEPTAGON
8	OCTAGON
9	NONAGON (enneagon)
10	DECAGON
11	UNDECAGON (hendecagon)
12	DODECAGON
13	TRIDECAGON (triskaidecagon)
14	TETRADECAGON (tetrakaidecagon)
15	PENTADECAGON (pentakaidecagon)
16	HEXADECAGON (hexakaidecagon)
17	HEPTADECAGON (heptakaidecagon)
18	OCTADECAGON (octakaidecagon)
19	ENNEADECAGON (enneakaidecagon)
20	ICOSAGON
30	TRIACONTAGON
40	TETRACONTAGON
50	PENTACONTAGON
60	HEXACONTAGON
70	HEPTACONTAGON
80	OCTACONTAGON
90	ENNEACONTAGON
100	HECTOGON
10000	MYRIAGON

See also 257-GON, 65537-GON, ANTHROPOMORPHIC POLYGON, BICENTRIC POLYGON, CARNOT'S POLYGON THEOREM, CHAOS GAME, CONVEX POLYGON, CYCLIC POLYGON, DE MOIVRE NUMBER, DERIVED POLYGON, DIAGONAL (POLYGON), EQUIANGULAR POLYGON, EQUILATERAL POLYGON, EQUILATERAL TRIANGLE, EULER'S POLYGON DIVISION PROBLEM, HEPTADECAGON, HEXAGON, HEXAGRAM, ILLUMINATION PROBLEM, JORDAN POLYGON, LOZENGE, OCTAGON, PARALLELOGRAM, PASCAL'S THEOREM, PENTAGON, PENTAGRAM, PETRIE POLYGON, PLANAR POLYGON, POLYGON CIRCUMSCRIBING CONSTANT, POLYGON INSCRIBING CONSTANT, POLYGONAL KNOT, POLYGONAL NUMBER, POLYGONAL SPIRAL, POLYGON TRIANGULATION, POLYGRAM, POLYHEDRAL FORMULA, POLYHEDRON, POLYTOPE, QUADRANGLE, QUADRILATERAL, REGULAR POLYGON, REULEAUX POLYGON, RHOMBUS, ROTOR, ROULETTE, SIMPLE POLYGON, SIMPLICITY, SQUARE, STAR POLYGON, TRAPEZIUM, TRAPEZOID, TRIANGLE, VISIBILITY, VORONOI POLYGON, WALLACE-BOLYAI-GERWEIN THEOREM

References

Beyer, W. H. *CRC Standard Mathematical Tables, 28th ed.* Boca Raton, FL: CRC Press, pp. 124-25 and 196, 1987.

Polygon Circumscribing Constant

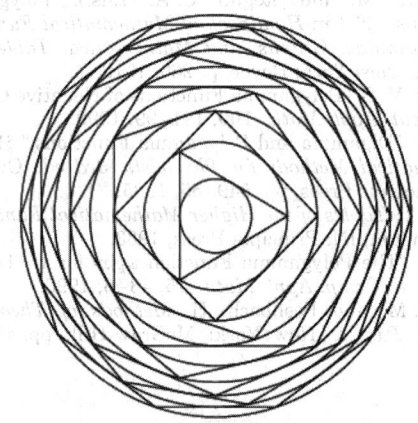

If a TRIANGLE is CIRCUMSCRIBED about a CIRCLE, another CIRCLE around the TRIANGLE, a SQUARE outside the CIRCLE, another CIRCLE outside the SQUARE, and so on. From POLYGONS, the CIRCUMRADIUS and INRADIUS for an n-gon are

$$R = \tfrac{1}{2} s \csc\left(\frac{\pi}{n}\right) \tag{1}$$

$$r = \tfrac{1}{2} s \cot\left(\frac{\pi}{n}\right), \tag{2}$$

where s is the side length. Therefore,

$$\frac{R}{r} = \frac{1}{\cos\left(\dfrac{\pi}{n}\right)} = \sec\left(\frac{\pi}{n}\right), \tag{3}$$

and an infinitely nested set of circumscribed polygons

and circles has

$$K \equiv \frac{r_{\text{final circle}}}{r_{\text{initial circle}}} = \sec\left(\frac{\pi}{3}\right) \sec\left(\frac{\pi}{4}\right) \sec\left(\frac{\pi}{5}\right) \dots \quad (4)$$

Kasner and Newman (1989) and Haber (1964) state that $K = 12$, but this is incorrect. Write

$$K = \prod_{n=3}^{\infty} \frac{1}{\cos\left(\frac{\pi}{n}\right)} \quad (5)$$

$$\ln K = -\sum_{n=3}^{\infty} \ln(\cos x). \quad (6)$$

Define

$$y_0(x) \equiv -\ln(\cos x) = \frac{1}{2}x^2 + \frac{1}{12}x^4 + \frac{1}{45}x^6 + \frac{17}{2520}x^8 + \dots \quad (7)$$

Now define

$$y_1(x) = \frac{1}{2}ax^2, \quad (8)$$

with

$$y_1\left(\frac{\pi}{3}\right) = y_0\left(\frac{\pi}{3}\right) \quad (9)$$

$$\frac{1}{2}a\left(\frac{\pi}{3}\right)^2 = \ln 2, \quad (10)$$

so

$$a = 2\left(\frac{3}{\pi}\right)^2 \ln 2, \quad (11)$$

and

$$y_2(x) = \frac{9 \ln 2}{\pi^2} x^2. \quad (12)$$

But $y_2(x) > y_1(x)$ for $x \in (0, \pi/3)$, so

$$\sum_{n=3}^{\infty} y_2\left(\frac{\pi}{n}\right) > -\sum_{n=3}^{\infty} \ln\left[\cos\left(\frac{\pi}{n}\right)\right] \quad (13)$$

$$\ln K < \sum_{n=3}^{\infty} y_2\left(\frac{\pi}{n}\right) \frac{9 \ln 2}{\pi^2} \sum_{n=3}^{\infty} \left(\frac{\pi}{n}\right)^2 = 9 \ln 2 \sum_{n=3}^{\infty} \frac{1}{n^2}$$

$$= 9 \ln 2 \left(\sum_{n=1}^{\infty} \frac{1}{n^2} - \sum_{n=1}^{2} \frac{1}{n^2}\right) = 9 \ln 2 \left[\zeta(2) - \frac{5}{4}\right]$$

$$= 9 \ln 2 \left(\frac{\pi^2}{6} - \frac{5}{4}\right) = 2.4637 \quad (14)$$

$$K < e^{2.4637} = 11.75. \quad (15)$$

If the next term is included,

$$y_2(x) = a\left(\frac{1}{2}x^2 + \frac{1}{12}x^4\right). \quad (16)$$

As before,

$$y_2\left(\frac{\pi}{3}\right) = y_0\left(\frac{\pi}{3}\right) \quad (17)$$

$$a = \frac{972 \ln 2}{\pi^2(54 + \pi^2)}, \quad (18)$$

so

$$y_2(x) = \frac{972 \ln 2}{\pi^2(54 + \pi^2)}\left(\frac{1}{2}x^2 + \frac{1}{12}x^4\right) \quad (19)$$

$$\ln K < \frac{972 \ln 2}{\pi^2(54 + \pi^2)} \sum_{n=3}^{\infty} \left[\frac{1}{2}\left(\frac{\pi}{n}\right)^2 + \frac{1}{12}\left(\frac{\pi}{n}\right)^4\right]$$

$$= \frac{972 \ln 2}{\pi^2(54 + \pi^2)} \left\{\frac{1}{2}\left[\zeta(2) - \frac{5}{4}\right] + \frac{\pi^2}{12}\left[\zeta(4) - 1 - \frac{1}{2^4}\right]\right\}$$

$$= \frac{972 \ln 2}{\pi^2(54 + \pi^2)} \left[\frac{1}{2}\left(\frac{\pi^2}{6} - \frac{5}{4}\right) + \frac{\pi^2}{12}\left(\frac{\pi^2}{90} - 1 - \frac{1}{2^4}\right)\right]$$

$$= \frac{9(8\pi^6 - 45\pi^2 - 5400) \ln 2}{80(\pi^2 + 54)} = 2.255, \quad (20)$$

and

$$K < e^{2.255} = 9.535. \quad (21)$$

The process can be automated using computer algebra, and the first few bounds are 11.7485, 9.53528, 8.98034, 8.8016, 8.73832, 8.71483, 8.70585, 8.70235, 8.70097, and 8.70042. In order to obtain this accuracy by direct multiplication of the terms, more than 10,000 terms are needed. The limit is

$$K = 8.700036625\dots. \quad (22)$$

Bouwkamp (1965) produced the following INFINITE PRODUCT formulas

$$K = \frac{2}{\pi} \prod_{m=1}^{\infty} \prod_{n=1}^{\infty} \left[1 - \frac{1}{m^2\left(n + \frac{1}{2}\right)^2}\right] \quad (23)$$

$$= 6 \exp\left\{\sum_{k=1}^{\infty} \frac{[\lambda(2k) - 1]2^{2k}[\zeta(2k) - 1 - 2^{-2k}]}{k}\right\}, \quad (24)$$

where $\zeta(x)$ is the RIEMANN ZETA FUNCTION and $\lambda(x)$ is the DIRICHLET LAMBDA FUNCTION. Bouwkamp (1965) also produced the formula with accelerated convergence

$$K = \frac{1}{12}\sqrt{6}\,\pi^4 \left(1 - \frac{1}{2}\pi^2 + \frac{1}{24}\pi^4\right)\left(1 - \frac{1}{8}\pi^2 + \frac{1}{384}\pi^4\right)$$

$$\times \csc\left(\frac{\pi^2}{\sqrt{6 + 2\sqrt{3}}}\right)\csc\left(\frac{\pi^2}{\sqrt{6 - 2\sqrt{3}}}\right)B, \quad (25)$$

where

$$B \equiv \prod_{n=3}^{\infty}\left(1 - \frac{\pi^2}{2n^2} + \frac{\pi^4}{24n^4}\right)\sec\left(\frac{\pi}{n}\right) \quad (26)$$

(cited in Pickover 1995).

See also POLYGON INSCRIBING CONSTANT

References

Bouwkamp, C. "An Infinite Product." *Indag. Math.* **27**, 40–6, 1965.
Finch, S. "Favorite Mathematical Constants." http://www.mathsoft.com/asolve/constant/infprd/infprd.html.
Haber, H. "Das Mathematische Kabinett." *Bild der Wissenschaft* **2**, 73, Apr. 1964.
Kasner, E. and Newman, J. R. *Mathematics and the Imagination.* Redmond, WA: Microsoft Press, pp. 311–12, 1989.
Pappas, T. "Infinity & Limits." *The Joy of Mathematics.* San Carlos, CA: Wide World Publ./Tetra, p. 180, 1989.
Pickover, C. A. "Infinitely Exploding Circles." Ch. 18 in *Keys to Infinity.* New York: W. H. Freeman, pp. 147–51, 1995.
Pinkham, R. S. "Mathematics and Modern Technology." *Amer. Math. Monthly* **103**, 539–45, 1996.
Plouffe, S. "Product(cos(Pi/n),n = 3..infinity)." http://www.la-cim.uqam.ca/piDATA/productcos.txt.

Polygon Construction

GEOMETRIC CONSTRUCTION, GEOMETROGRAPHY, POLYGON, SIMPLICITY

Polygon Division Problem

EULER'S POLYGON DIVISION PROBLEM

Polygon Fractal

CHAOS GAME

Polygon Inscribing Constant

If a TRIANGLE is inscribed in a CIRCLE, another CIRCLE inside the TRIANGLE, a SQUARE inside the CIRCLE, another CIRCLE inside the SQUARE, and so on,

$$K' \equiv \frac{r_{\text{final circle}}}{r_{\text{initial circle}}} = \cos\left(\frac{\pi}{3}\right)\cos\left(\frac{\pi}{4}\right)\cos\left(\frac{\pi}{5}\right)\ldots.$$

Numerically,

$$K' = \frac{1}{K} = \frac{1}{8.7000366252\ldots} = 0.1149420448\ldots,$$

where K is the POLYGON CIRCUMSCRIBING CONSTANT. Kasner and Newman's (1989) assertion that $K = 1/12$ is incorrect.

Let a convex POLYGON be inscribed in a CIRCLE and divided into TRIANGLES from diagonals from one VERTEX. The sum of the RADII of the CIRCLES inscribed in these TRIANGLES is the same independent of the VERTEX chosen (Johnson 1929, p. 193).

See also POLYGON CIRCUMSCRIBING CONSTANT

References

Finch, S. "Favorite Mathematical Constants." http://www.mathsoft.com/asolve/constant/infprd/infprd.html.
Johnson, R. A. *Modern Geometry: An Elementary Treatise on the Geometry of the Triangle and the Circle.* Boston, MA: Houghton Mifflin, 1929.
Kasner, E. and Newman, J. R. *Mathematics and the Imagination.* Redmond, WA: Microsoft Press, pp. 311–12, 1989.
Pappas, T. "Infinity & Limits." *The Joy of Mathematics.* San Carlos, CA: Wide World Publ./Tetra, p. 180, 1989.
Plouffe, S. "Product(cos(Pi/n),n = 3..infinity)." http://www.la-cim.uqam.ca/piDATA/productcos.txt.

Polygon Tiling

See also HEXAGON TILING, PENTAGON TILING, QUADRILATERAL TILING, SQUARE TILING, TILING, TRIANGLE TILING

References

Laczkovich, M. "Tilings of Polygons with Similar Triangles." *Combinatorica* **10**, 281–06, 1990.

Polygon Triangle Picking

The mean area of a TRIANGLE picked inside a regular n-gon of unit area is

$$\bar{A} = \frac{9\cos^2\omega + 52\cos\omega + 44}{36n^2\sin^2\omega}, \quad (1)$$

where $\omega \equiv 2\pi/n$ (Alikoski 1939; Solomon 1978; Croft *et al.* 1991, p. 54). Prior to Alikoski's work, only the special cases $n = 3$, 4, 6, 8, and ∞ had been determined. The first few cases are summarized in the following table, where $\bar{A}_7$ is the largest root of

$$784147392x^3 - 84015792x^2 + 2125620x - 15289 = 0, \quad (2)$$

and $\bar{A}_9$ is the largest root of

$$24794911296x^3 - 2525407632x^2 + 55366092x$$
$$-312427 = 0. \quad (3)$$

n	$\bar{A}_n$	problem
3	$\frac{1}{12}$	TRIANGLE TRIANGLE PICKING
4	$\frac{11}{144}$	SQUARE TRIANGLE PICKING
5	$\frac{1}{180}\left(9 + 2\sqrt{5}\right)$	

6	$\frac{289}{3888}$	HEXAGON TRIANGLE PICKING
7	$\bar{A}_7$	
8	$\frac{1}{2304}(97 + 52\sqrt{2})$	
9	$\bar{A}_9$	
10	$\frac{1}{18000}(745 + 262\sqrt{5})$	

See also HEXAGON TRIANGLE PICKING, SQUARE TRIANGLE PICKING, SYLVESTER'S FOUR-POINT PROBLEM, TRIANGLE TRIANGLE PICKING

References

Alikoski, H. A. "Uuml;ber das Sylvestersche Vierpunktproblem." *Ann. Acad. Sci. Fenn.* **51**, No. 7, 1–0, 1939.

Croft, H. T.; Falconer, K. J.; and Guy, R. K. *Unsolved Problems in Geometry.* New York: Springer-Verlag, 1991.

Kendall, M. G. "Exact Distribution for the Shape of Random Triangles in Convex Sets." *Adv. Appl. Prob.* **17**, 308–29, 1985.

Kendall, M. G. and Le, H.-L. "Exact Shape Densities for Random Triangles in Convex Polygons." *Adv. Appl. Prob.* **1986 Suppl.**, 59–2, 1986.

Solomon, H. *Geometric Probability.* Philadelphia, PA: SIAM, pp. 109–14, 1978.

Polygon Triangulation

EULER'S POLYGON DIVISION PROBLEM, TESSELLATION, TRIANGULATION

Polygonal Knot

A KNOT equivalent to a POLYGON in $\mathbb{R}^3$, also called a TAME KNOT. For a polygonal knot K, there exists a PLANE such that the orthogonal projection π on it satisfies the following conditions:

1. The image $\pi(K)$ has no multiple points other than a FINITE number of double points.
2. The projections of the vertices of K are not double points of $\pi(K)$.

Such a projection $\pi(K)$ is called a regular knot projection.

References

Iyanaga, S. and Kawada, Y. (Eds.). *Encyclopedic Dictionary of Mathematics.* Cambridge, MA: MIT Press, p. 735, 1980.

Polygonal Number

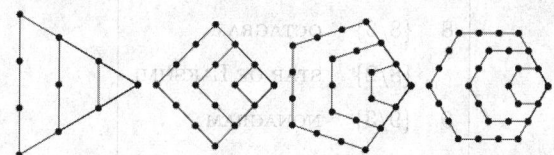

A type of FIGURATE NUMBER which is a generalization of TRIANGULAR, SQUARE, etc., numbers to an arbitrary

n-gonal number. The above diagrams graphically illustrate the process by which the polygonal numbers are built up. Starting with the nth TRIANGULAR NUMBER T_n, then

$$n + T_{n-1} = T_n. \tag{1}$$

Now note that

$$n + 2T_{n-1} = n^2 = S_n \tag{2}$$

gives the nth SQUARE NUMBER,

$$n + 3T_{n-1} = \tfrac{1}{2}n(3n-1) = P_n, \tag{3}$$

gives the nth PENTAGONAL NUMBER, and so on. The general polygonal number can be written in the form

$$p_r^n = \tfrac{1}{2}r[(r-1)n - 2(r-2)] = \tfrac{1}{2}r[(n-2)r - (n-4)], \tag{4}$$

where p_r^n is the rth n-gonal number (Savin 2000). For example, taking $n = 3$ in (4) gives a TRIANGULAR NUMBER, $n = 4$ gives a SQUARE NUMBER, etc.

Fermat proposed that every number is expressible as *at most k k-gonal* numbers (FERMAT'S POLYGONAL NUMBER THEOREM). Fermat claimed to have a proof of this result, although this proof has never been found. Jacobi, Lagrange (1772), and Euler all proved the square case, and Gauss proved the triangular case in 1796. In 1813, Cauchy proved the proposition in its entirety.

An arbitrary number N can be checked to see if it is a n-gonal number as follows. Note the identity

$$8(n-2)p_n^r + (n-4)^2 = (2rn - 4r - n + 4)^2, \tag{5}$$

so $8(n-2)N + (n-4)^2 = S^2$ must be a PERFECT SQUARE. Therefore, if it is not, the number cannot be n-gonal. If it is a PERFECT SQUARE, then solving

$$S = 2rn - 4r - n + 4 \tag{6}$$

for the rank r gives

$$r = \frac{S + n - 4}{2(n-2)}. \tag{7}$$

An n-gonal number is equal to the sum of the $(n-1)$-gonal number of the same RANK and the TRIANGULAR NUMBER of the previous RANK.

See also CENTERED POLYGONAL NUMBER, DECAGONAL NUMBER, FERMAT'S POLYGONAL NUMBER THEOREM, FIGURATE NUMBER, HEPTAGONAL NUMBER, HEXAGONAL NUMBER, NONAGONAL NUMBER, OCTAGONAL NUMBER, PENTAGONAL NUMBER, PYRAMIDAL NUMBER, SQUARE NUMBER, TRIANGULAR NUMBER

References

Abramovich, S.; Fujii, T.; and Wilson, J. W. "Multiple-Application Medium for the Study of Polygonal Numbers." http://jwilson.coe.uga.edu/Texts.Folder/AFW/AFWarticle.html.

Beiler, A. H. "Ball Games." Ch. 18 in *Recreations in the Theory of Numbers: The Queen of Mathematics Entertains.* New York: Dover, pp. 184–99, 1966.

Cauchy, A. "Démonstration du théorème général de Fermat sur les nombres polygones." *Oeuvres, 2e. serie, Vol. 6.* pp. 320–53.

Dickson, L. E. *History of the Theory of Numbers, Vol. 1: Divisibility and Primality.* New York: Chelsea, pp. 3–3, 1952.

Guy, K. "Every Number is Expressible as a Sum of How Many Polygonal Numbers?" *Amer. Math. Monthly* **101**, 169–72, 1994.

Nathanson, M. B. "Sums of Polygonal Numbers." In *Analytic Number Theory and Diophantine Problems: Proceedings of a Conference at Oklahoma State University, 1984* (Ed. A. Adolphson *et al.*). Boston, MA: Birkhäuser, pp. 305–16, 1987.

Pappas, T. "Triangular, Square & Pentagonal Numbers." *The Joy of Mathematics.* San Carlos, CA: Wide World Publ./Tetra, p. 214, 1989.

Savin, A. "Shape Numbers." *Quantum* **11**, 14–8, 2000.

Sloane, N. J. A. Sequences A000217/M2535 in "An On-Line Version of the Encyclopedia of Integer Sequences." http://www.research.att.com/~njas/sequences/eisonline.html.

Sloane, N. J. A. and Plouffe, S. Figure M2535 in *The Encyclopedia of Integer Sequences.* San Diego: Academic Press, 1995.

Polygonal Spiral

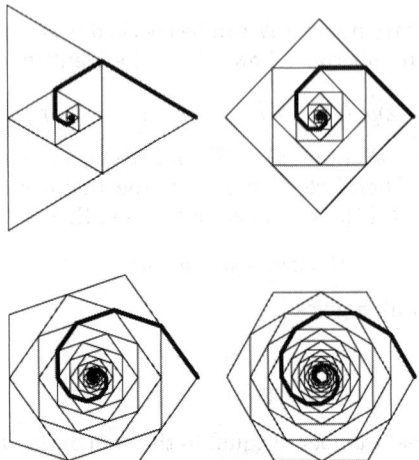

The length of the polygonal spiral is found by noting that the ratio of INRADIUS to CIRCUMRADIUS of a REGULAR POLYGON of n sides is

$$\frac{r}{R} = \frac{\cot\left(\dfrac{\pi}{n}\right)}{\csc\left(\dfrac{\pi}{n}\right)} = \cos\left(\frac{\pi}{n}\right). \tag{1}$$

The total length of the spiral for an n-gon with side

length s is therefore

$$L = \tfrac{1}{2}s \sum_{k=0}^{\infty} \cos^k\left(\frac{\pi}{n}\right) = \frac{s}{2\left[1 - \cos\left(\dfrac{\pi}{n}\right)\right]}. \tag{2}$$

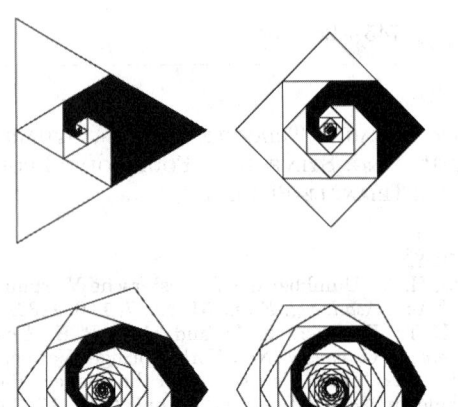

Consider the solid region obtained by filling in subsequent triangles which the spiral encloses. The AREA of this region, illustrated above for n-gons of side length s, is

$$A = \tfrac{1}{4}s^2 \cot\left(\frac{\pi}{n}\right). \tag{3}$$

The shaded triangular polygonal spiral is a REP-4-TILE.

See also REP-TILE

References

Sandefur, J. T. "Using Self-Similarity to Find Length, Area, and Dimension." *Amer. Math. Monthly* **103**, 107–20, 1996.

Polygram

A self-intersecting STAR POLYGON such as the PENTAGRAM or HEXAGRAM.

n	symbol	polygram
5	$\{5/2\}$	PENTAGRAM
6	$\{6/2\}$	HEXAGRAM
7	$\{7/2\}$	Heptagram
8	$\{8/3\}$	OCTAGRAM
	$\{8/2\}$	STAR OF LAKSHMI
9	$\{9/3\}$	NONAGRAM
10	$\{10/3\}$	DECAGRAM

Lachlan (1893) defines polygram to be a figure consisting of n straight lines.

See also DECAGRAM, HEXAGRAM, OCTAGRAM, PENTAGRAM, STAR FIGURE, STAR OF LAKSHMI, STAR POLYGON

References

Lachlan, R. *An Elementary Treatise on Modern Pure Geometry.* London: Macmillian, p. 83, 1893.

Polyhedral Formula

A formula relating the number of VERTICES V, FACES F, and EDGES E of a simply connected (i.e., GENUS 0) POLYHEDRON (or POLYGON). It was discovered independently by Euler (1752) and Descartes, so it is also known as the Descartes-Euler polyhedral formula. Although the formula holds for some non-CONVEX POLYHEDRA, it does not hold for STELLATED POLYHEDRA.

The polyhedral formula states

$$V + F - E = 2, \tag{1}$$

where $V = N_0$ is the number of VERTICES, $E = N_1$ is the number of EDGES, and $F = N_2$ is the number of FACES. For a proof, see Courant and Robbins (1978, pp. 239–40).

The FORMULA was generalized to n-D POLYTOPES by Schläfli (Coxeter 1968, p. 233),

$$\Pi_1 : N_0 = 2 \tag{2}$$

$$\Pi_2 : N_0 - N_1 = 0 \tag{3}$$

$$\Pi_3 : N_0 - N_1 + N_2 = 2 \tag{4}$$

$$\Pi_4 : N_0 - N_1 + N_2 - N_3 = 0 \tag{5}$$

$$\Pi_n : N_0 - N_1 + N_2 - \ldots + (-1)^{n-1} N_{n-1} = 1 - (-1)^n. \tag{6}$$

and proved by Poincaré (Poincaré 1893; Coxeter 1973, pp. 166–71; Williams 1979, pp. 24–5).

For GENUS g surfaces, the formula can be generalized to the POINCARÉ FORMULA

$$\chi \equiv V - E + F = \chi(g), \tag{7}$$

where

$$\chi(g) = 2 - 2g, \tag{8}$$

is the EULER CHARACTERISTIC, sometimes also known as the EULER-POINCARÉ CHARACTERISTIC. The polyhedral formula corresponds to the special case $g = 0$.

There exist polytopes which do not satisfy the polyhedral formula, the most prominent of which are the GREAT DODECAHEDRON $\{5, \frac{5}{2}\}$ and SMALL STELLATED DODECAHEDRON $\{\frac{5}{2}, 5\}$, which no less than Schläfli himself refused to recognize (Schläfli 1901, p. 134) since for these solids,

$$N_0 - N_1 + N_2 = 12 - 30 + 12 = -6 \tag{9}$$

(Coxeter 1973, p. 172).

See also DEHN INVARIANT, EULER CHARACTERISTIC, DESCARTES TOTAL ANGULAR DEFECT, GENUS (SURFACE), POINCARÉ FORMULA, POLYHEDRAL GRAPH, POLYTOPE

References

Aigner, M. and Ziegler, G. M. "Three Applications of Euler's Formula." Ch. 10 in *Proofs from the Book.* Berlin: Springer-Verlag, 1998.
Beyer, W. H. (Ed.). *CRC Standard Mathematical Tables, 28th ed.* Boca Raton, FL: CRC Press, p. 128, 1987.
Courant, R. and Robbins, H. *What is Mathematics?: An Elementary Approach to Ideas and Methods.* Oxford, England: Oxford University Press, 1978.
Coxeter, H. S. M. *The Beauty of Geometry: Twelve Essays.* New York: Dover, 1999.
Coxeter, H. S. M. "Euler's Formula." and "Poincaré's Proof of Euler's Formula." §1.6 and Ch. 9 in *Regular Polytopes, 3rd ed.* New York: Dover, pp. 9–1 and 165–72, 1973.
Euler, L. "Elementa doctrine solidorum." *Novi comm. acad. scientiarum imperialis petropolitanae* **4**, 109–60, 1752–753. Reprinted in *Opera, Vol. 26*, pp. 71–2.
Poincaré, H. "Sur la généralisation d'un théorème d'Euler relatif aux polyèdres." *Comptes rendus hebdomadaires des séances de l'Académie des Sciences* **117**, 144–45, 1893.
Schläfli, L. "Theorie der vielfachen Kontinuität." *Denkschriften der Schweizerischen naturforschenden Gessel.* **38**, 1–37, 1901.
Steinhaus, H. *Mathematical Snapshots, 3rd ed.* New York: Dover, pp. 252–53, 1999.
Williams, R. *The Geometrical Foundation of Natural Structure: A Source Book of Design.* New York: Dover, 1979.

Polyhedral Graph

An n-polyhedral graph (sometimes called a c-net) is a 3-CONNECTED SIMPLE PLANAR GRAPH on n nodes. Every CONVEX POLYHEDRON can be represented in the plane or on the surface of a sphere by a 3-connected PLANAR GRAPH. Conversely, by a theorem of Steinitz as restated by Grünbaum (1967, p. 235), every 3-connected planar graph can be realized as a CONVEX POLYHEDRON (Duijvestijn and Federico 1981). Polyhedral graphs are sometimes simply known as "polyhedra" (which is rather confusing since the term "polyhedron" more commonly refers to a solid with n faces, not n vertices).

The number of distinct polyhedral graphs having $V = 1, 2, \ldots$ vertices (or equivalently $F = 1, 2, \ldots$ faces) are 0, 0, 0, 1, 2, 7, 34, 257, 2606, ... (Sloane's A000944; Grünbaum 1967, p. 424; Duijvestijn and Federico 1981; Dillencourt 1992; Croft *et al.* 1994). There is therefore a single TETRAHEDRAL GRAPH, two PENTAHEDRAL GRAPHS, etc. There is no known formula for enumerating the number of nonisomorphic polyhedral graphs by numbers of edges E, vertices V, or faces F (Harary and Palmer 1973, p. 224; Duijvestijn and Federico 1981).

V	#	graph name
4	1	TETRAHEDRAL GRAPH
5	2	PENTAHEDRAL GRAPH
6	7	HEXAHEDRAL GRAPH
7	34	HEPTAHEDRAL GRAPH
8	257	OCTAHEDRAL GRAPH
9	2606	NONAHEDRAL GRAPH
10	32300	DECAHEDRAL GRAPH

Duijvestijn and Federico (1981) enumerated the polyhedral graphs on E edges, obtaining 1, 0, 1, 2, 2, 4, 12, 22, 58, 158, 448, ... (Sloane's A002840) for $E = 6, 7, 8, \ldots$.

See also CUBICAL GRAPH, DODECAHEDRAL GRAPH, ICOSAHEDRAL GRAPH, K-CONNECTED GRAPH, OCTAHEDRAL GRAPH, PLANAR CONNECTED GRAPH, PLANAR GRAPH, PLATONIC GRAPH, POLYHEDRAL FORMULA, POLYHEDRAL GROUP, POLYTOPAL GRAPH, SCHLEGEL GRAPH, SIMPLE GRAPH, SKELETON, TETRAHEDRAL GRAPH

References

Bouwkamp, C. J.; Duijvestijn, A. J. W.; and Medema, P. *Table of c*-Nets of Orders 8 to 19, Inclusive, 2 vols. Unpublished manuscript. Eindhoven, Netherlands: Philips Research Laboratories, 1960.
Croft, H. T.; Falconer, K. J.; and Guy, R. K. §B15 in *Unsolved Problems in Geometry.* New York: Springer-Verlag, 1991.
Dillencourt, M. B. "Polyhedra of Small Orders and Their Hamiltonian Properties." Tech. Rep. 92–1, Info. and Comput. Sci. Dept. Irvine, CA: Univ. Calif. Irvine, 1992.
Duijvestijn, A. J. W. "List of 3-Connected Planar Graphs with 6 to 22 Edges." Unpublished computer tape. Enschede, Netherlands: Twente Univ. Technology, 1979.
Duijvestijn, A. J. W. and Federico, P. J. "The Number of Polyhedral (3-Connected Planar) Graphs." *Math. Comput.* **37**, 523–32, 1981.
Federico, P. J. "Enumeration of Polyhedra: The Number of 9-Hedra." *J. Combin. Th.* **7**, 155–61, 1969.
Federico, P. J. "The Number of Polyhedra." *Philips Res. Rep.* **30**, 220–31, 1975.
Grünbaum, B. *Convex Polytopes.* New York: Wiley, 1967.
Grünbaum, B. "Polytopal Graphs." In *Studies in Graph Theory, Part II* (Ed. D. R. Fulkerson). Washington, DC: Math. Assoc. Amer., pp. 201–24, 1975.
Harary, F. and Palmer, E. M. *Graphical Enumeration.* New York: Academic Press, 1973.
Sloane, N. J. A. Sequences A000944/M1796 and A002840/M0339 in "An On-Line Version of the Encyclopedia of Integer Sequences." http://www.research.att.com/~njas/sequences/eisonline.html.
Tutte, W. T. "A Theory of 3-Connected Graphs." *Indag. Math.* **23**, 451–55, 1961.
Tutte, W. T. "On the Enumeration of Convex Polyhedra." *J. Combin. Th. Ser. B* **28**, 105–26, 1980.

Polyhedral Group

One of the symmetry groups of the PLATONIC SOLIDS. There are three polyhedral groups: the TETRAHEDRAL GROUP of order 12, the OCTAHEDRAL GROUP of order 24, and the ICOSAHEDRAL GROUP of order 60.

See also ICOSAHEDRAL GROUP, OCTAHEDRAL GROUP, PLATONIC SOLID, POLYHEDRAL GRAPH, TETRAHEDRAL GROUP

References

Coxeter, H. S. M. "The Polyhedral Groups." §3.5 in *Regular Polytopes, 3rd ed.* New York: Dover, pp. 46–7, 1973.

Polyhedron

The word polyhedron has slightly different meanings in geometry and ALGEBRAIC GEOMETRY. In geometry, a polyhedron is simply a 3-D solid which consists of a collection of POLYGONS, usually joined at their EDGES. The word derives from the Greek *poly* (many) plus the Indo-European *hedron* (seat). A polyhedron is the 3-D version of the more general POLYTOPE (in the geometric sense), which can be defined in arbitrary dimension. The plural of polyhedron is "polyhedra" (or sometimes "polyhedrons").

The term "polyhedron" is used somewhat differently in ALGEBRAIC TOPOLOGY, where it is defined as a space that can be built from such "building blocks" as line segments, triangles, tetrahedra, and their higher dimensional analogs by "gluing them together" along their faces (Munkres 1993, p. 2). More specifically, it can be defined as the UNDERLYING SPACE of a SIMPLICIAL COMPLEX (with the additional constraint sometimes imposed that the complex be finite; Munkres 1993, p. 9). In the usual definition, a polyhedron can be viewed as an intersection of half-spaces, while a POLYTOPE is a *bounded* polyhedron.

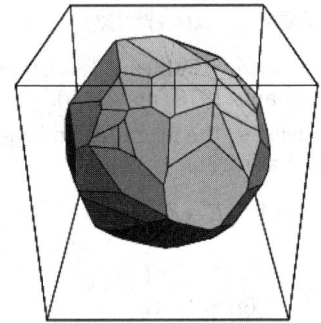

A CONVEX POLYHEDRON can be formally defined as the set of solutions to a system of linear inequalities

$$m x \leq \mathbf{b},$$

where m is a real $s \times 3$ MATRIX and $\mathbf{b}$ is a real s-VECTOR. Although usage varies, most authors additional require that a solution be bounded for it to define a CONVEX POLYHEDRON. An example of a convex polyhedron is illustrated above.

A polyhedron is said to be regular if its FACES and VERTEX FIGURES are REGULAR (not necessarily CONVEX) polygons (Coxeter 1973, p. 16). Using this definition, there are a total of nine REGULAR POLYHEDRA, five being the CONVEX Platonic solids and four being the CONCAVE (stellated) KEPLER-POINSOT SOLIDS. However, the term "regular polyhedra" is sometimes used to refer exclusively to the PLATONIC SOLIDS (Cromwell 1997, p. 53). The DUAL POLYHEDRA of the PLATONIC SOLIDS are not new polyhedra, but are themselves PLATONIC SOLIDS.

A CONVEX polyhedron is called SEMIREGULAR if its FACES have a similar arrangement of nonintersecting regular plane CONVEX polygons of two or more different types about each VERTEX (Holden 1991, p. 41). These solids are more commonly called the ARCHIMEDEAN SOLIDS, and there are 13 of them. The DUAL POLYHEDRA of the ARCHIMEDEAN SOLIDS are 13 new (and beautiful) solids, sometimes called the CATALAN SOLIDS.

A QUASIREGULAR POLYHEDRON is the solid region interior to two DUAL REGULAR POLYHEDRA (Coxeter 1973, pp. 17–0). There are only two CONVEX QUASIREGULAR POLYHEDRA: the CUBOCTAHEDRON and ICOSIDODECAHEDRON. There are also infinite families of PRISMS and ANTIPRISMS.

There exist exactly 92 CONVEX POLYHEDRA with REGULAR POLYGONAL faces (and not necessarily equivalent vertices). They are known as the JOHNSON SOLIDS. Polyhedra with identical VERTICES related by a symmetry operation are known as UNIFORM POLYHEDRA. There are 75 such polyhedra in which only two faces may meet at an EDGE, and 76 in which any EVEN number of faces may meet. Of these, 37 were discovered by Badoureau in 1881 and 12 by Coxeter and Miller ca. 1930.

Polyhedra can be superposed on each other (with the sides allowed to pass through each other) to yield additional POLYHEDRON COMPOUNDS. Those made from REGULAR POLYHEDRA have symmetries which are especially aesthetically pleasing. The graphs corresponding to polyhedra skeletons are called SCHLEGEL GRAPHS.

Behnke *et al.* (1974) have determined the symmetry groups of all polyhedra symmetric with respect to their VERTICES.

See also ACOPTIC POLYHEDRON, APEIROGON, ARCHIMEDEAN SOLID, CANONICAL POLYHEDRON, CATALAN SOLID, CONVEX POLYHEDRON, CUBE, CUMULATION, DICE, DIGON, DODECAHEDRON, DUAL POLYHEDRON, ECHIDNAHEDRON, FLEXIBLE POLYHEDRON, HAUY CONSTRUCTION, HEXAHEDRON, HOLYHEDRON, HYPERBOLIC POLYHEDRON, ICOSAHEDRON, ISOHEDRON, JESSEN'S ORTHOGONAL ICOSAHEDRON JOHNSON SOLID, KEPLER-POINSOT SOLID, NOLID, OCTAHEDRON, PETRIE POLYGON, PLAITED POLYHEDRON, PLATONIC SOLID, POLYCHORON, POLYHEDRON COLORING, POLYHEDRON COMPOUND, POLYTOPE, PRISMATOID, QUADRICORN, QUASIREGULAR POLYHEDRON, RIGID POLYHEDRON, RIGIDITY THEOREM, SCHWARZ'S POLYHEDRON, SHAKY POLYHEDRON, SEMIREGULAR POLYHEDRON, SKELETON, STELLATION, TETRAHEDRON, TRUNCATION, UNIFORM POLYHEDRON, ZONOHEDRON

References

Ball, W. W. R. and Coxeter, H. S. M. "Polyhedra." Ch. 5 in *Mathematical Recreations and Essays, 13th ed.* New York: Dover, pp. 130–61, 1987.

Behnke, H.; Bachman, F.; Fladt, K.; and Kunle, H. (Eds.). *Fundamentals of Mathematics, Vol. 2: Geometry.* Cambridge, MA: MIT Press, 1974.

Bulatov, V. "Polyhedra Collection." http://www.physics.orst.edu/~bulatov/polyhedra/.

Coxeter, H. S. M. *Regular Polytopes, 3rd ed.* New York: Dover, 1973.

Critchlow, K. *Order in Space: A Design Source Book.* New York: Viking Press, 1970.

Cromwell, P. R. *Polyhedra.* New York: Cambridge University Press, 1997.

Cundy, H. and Rollett, A. *Mathematical Models, 3rd ed.* Stradbroke, England: Tarquin Pub., 1989.

Davie, T. "Books and Articles about Polyhedra and Polytopes." http://www.dcs.st-andrews.ac.uk/~ad/mathrecs/polyhedra/polyhedrabooks.html.

Davie, T. "The Regular (Platonic) and Semi-Regular (Archimedean) Solids." http://www.dcs.st-andrews.ac.uk/~ad/mathrecs/polyhedra/polyhedratopic.html.

Eppstein, D. "Geometric Models." http://www.ics.uci.edu/~eppstein/junkyard/model.html.

Eppstein, D. "Polyhedra and Polytopes." http://www.ics.uci.edu/~eppstein/junkyard/polytope.html.

Gabriel, J. F. (Ed.). *Beyond the Cube: The Architecture of Space Frames and Polyhedra.* New York: Wiley, 1997.

Hart, G. "Annotated Bibliography." http://www.georgehart.com/virtual-polyhedra/references.html.

Hart, G. "Virtual Polyhedra." http://www.georgehart.com/virtual-polyhedra/vp.html.

Hilton, P. and Pedersen, J. *Build Your Own Polyhedra.* Reading, MA: Addison-Wesley, 1994.

Holden, A. *Shapes, Space, and Symmetry.* New York: Dover, 1991.

Kern, W. F. and Bland, J. R. "Polyhedrons." §41 in *Solid Mensuration with Proofs, 2nd ed.* New York: Wiley, pp. 115–19, 1948.

Lyusternik, L. A. *Convex Figures and Polyhedra.* New York: Dover, 1963.

Malkevitch, J. "Milestones in the History of Polyhedra." In *Shaping Space: A Polyhedral Approach* (Ed. M. Senechal and G. Fleck). Boston, MA: Birkhäuser, pp. 80–2, 1988.

Miyazaki, K. *An Adventure in Multidimensional Space: The Art and Geometry of Polygons, Polyhedra, and Polytopes.* New York: Wiley, 1983.

Munkres, J. R. *Elements of Algebraic Topology.* Perseus Press, 1993.

Paeth, A. W. "Exact Dihedral Metrics for Common Polyhedra." In *Graphic Gems II* (Ed. J. Arvo). New York: Academic Press, 1991.

Pappas, T. "Crystals-Nature's Polyhedra." *The Joy of Mathematics.* San Carlos, CA: Wide World Publ./Tetra, pp. 38–9, 1989.

Pearce, P. *Structure in Nature Is a Strategy for Design.* Cambridge, MA: MIT Press, 1990.

Pedagoguery Software. `Poly.` http://www.peda.com/poly/.

Pugh, A. *Polyhedra: A Visual Approach.* Berkeley: University of California Press, 1976.

Schaaf, W. L. "Regular Polygons and Polyhedra." Ch. 3, §4 in *A Bibliography of Recreational Mathematics.* Washington, DC: National Council of Teachers of Math., pp. 57–0, 1978.

Virtual Image. "Polytopia I" and "Polytopia II" CD-ROMs. http://ourworld.compuserve.com/homepages/vir_image/html/polytopiai.html and http://ourworld.compuserve.com/homepages/vir_image/html/polytopiaii.html.

Weisstein, E. W. "Books about Solid Geometry." http://www.treasure-troves.com/books/SolidGeometry.html.

Williams, R. *The Geometrical Foundation of Natural Structure: A Source Book of Design.* New York: Dover, 1979.

Polyhedron Coloring

Define a valid "coloring" to occur when no two faces with a common EDGE share the same color. Given two colors, there is a single way to color an OCTAHEDRON (Ball and Coxeter 1987, pp. 238–39). Given three colors, there is one way to color a CUBE (Ball and Coxeter 1987, pp. 238–39) and 144 ways to color an ICOSAHEDRON (Ball and Coxeter 1987, pp. 239–42). Given four colors, there are two distinct ways to color a TETRAHEDRON (Ball and Coxeter 1987, p. 238) and four ways to color a DODECAHEDRON, consisting of two enantiomorphous ways (Steinhaus 1983, pp. 196–98; Ball and Coxeter 1987, p. 238). Given five colors, there are four ways to color an ICOSAHEDRON. Given six colors, there are 30 ways to color a CUBE (Steinhaus 1983, p. 167).

See also COLORING, CUBE, DODECAHEDRON, ICOSAHEDRON, OCTAHEDRON, PLATONIC SOLID, POLYHEDRON, TETRAHEDRON

References

Ball, W. W. R. and Coxeter, H. S. M. *Mathematical Recreations and Essays, 13th ed.* New York: Dover, 238–42, 1987.

Cundy, H. and Rollett, A. *Mathematical Models, 3rd ed.* Stradbroke, England: Tarquin Pub., pp. 82–3, 1989.

Steinhaus, H. *Mathematical Snapshots, 3rd ed.* New York: Dover, 1999.

Polyhedron Compound

A polyhedron compound is an arrangement of a number of interpenetrating polyhedra, either all the same or of several distinct types, usually having visually attractive symmetric properties. The following table gives some common polyhedron compounds.

solid	vertices
CUBE 2-COMPOUND	
CUBE 3-COMPOUND	
CUBE 4-COMPOUND	
CUBE 5-COMPOUND	DODECAHEDRON
CUBE-OCTAHEDRON COMPOUND	both
DODECAHEDRON 2-COMPOUND	
DODECAHEDRON 3-COMPOUND	
DODECAHEDRON 5-COMPOUND	
DODECAHEDRON-ICOSAHEDRON COMPOUND	both
DODECAHEDRON-SMALL TRIAMBIC ICOSAHEDRON COMPOUND	both
GREAT DODECAHEDRON-SMALL STELLATED DODECAHEDRON COMPOUND	both
GREAT ICOSAHEDRON-GREAT STELLATED DODECAHEDRON COMPOUND	both
OCTAHEDRON 3-COMPOUND	
OCTAHEDRON 5-COMPOUND	ICOSIDODECAHEDRON
STELLA OCTANGULA	CUBE
TETRAHEDRON 4-COMPOUND	
TETRAHEDRON 5-COMPOUND	DODECAHEDRON
TETRAHEDRON 10-COMPOUND	DODECAHEDRON

In Coxeter's NOTATION, d distinct VERTICES of $\{m, n\}$ taken c times are denoted

$$c\{m, n\}[d\{p, q\}], \tag{1}$$

or faces of $\{s, t\}$ e times

$$[d\{p, q\}]e\{s, t\}, \tag{2}$$

or both

$$c\{m, n\}[d\{p, q\}]e\{s, t\}. \tag{3}$$

See also CUBE 2-COMPOUND, CUBE 3-COMPOUND, CUBE 4-COMPOUND, CUBE 5-COMPOUND, CUBE 20-COMPOUND, CUBE-OCTAHEDRON COMPOUND, DODECAHEDRON 2-COMPOUND, DODECAHEDRON 3-COMPOUND, DODECAHEDRON 5-COMPOUND, DODECAHEDRON-ICOSAHEDRON COMPOUND, DODECAHEDRON-SMALL TRIAMBIC ICOSAHEDRON COMPOUND, OCTAHEDRON 3-COMPOUND, OCTAHEDRON 5-COMPOUND, STELLA OCTANGULA, TETRAHEDRON 4-COMPOUND, TETRAHEDRON 5-COMPOUND, TETRAHEDRON 10-COMPOUND

References

Cundy, H. and Rollett, A. "Regular Compounds." §3.10 in *Mathematical Models, 3rd ed.* Stradbroke, England: Tarquin Pub., pp. 129–42, 1989.

Hart, G. "Compounds of Cubes." http://www.georgehart.com/virtual-polyhedra/compound-cubes-info.html.

Wells, D. *The Penguin Dictionary of Curious and Interesting Geometry.* London: Penguin, pp. 37–8, 1991.

Wenninger, M. J. "Some Interesting Polyhedral Compounds." Ch. 5 in *Dual Models.* Cambridge, England: Cambridge University Press, pp. 143–48, 1983.

Polyhedron Dissection

A DISSECTION of one or more polyhedra into other shapes.

See also CUBE DISSECTION, DIABOLICAL CUBE, POLYCUBE, SOMA CUBE, WALLACE-BOLYAI-GERWEIN THEOREM

References

Bulatov, V. "Compounds of Uniform Polyhedra." http://www.physics.orst.edu/~bulatov/polyhedra/uniform_compounds/.

Coffin, S. T. *The Puzzling World of Polyhedral Dissections.* New York: Oxford University Press, 1990.

Coffin, S. T. and Rausch, J. R. *The Puzzling World of Polyhedral Dissections CD-ROM.* Puzzle World Productions, 1998.

Polyhedron Dual

DUAL POLYHEDRON

Polyhedron Hinging

RIGIDITY THEOREM

Polyhedron Packing

A packing of polyhedron in 3-D space. A polyhedron which can pack with no holes or gaps is said to be a SPACE-FILLING POLYHEDRON. Betke and Henk (1999) present an efficient algorithm for computing the density of a densest lattice packing of an arbitrary polyhedron, and explicitly calculate the densities for the PLATONIC and ARCHIMEDEAN SOLIDS.

See also KELVIN'S CONJECTURE, PACKING, SPACE-FILLING POLYHEDRON

References

Betke, U. and Henk, M. "Densest Lattice Packings of 3-Polytopes." Preprint. Erwin Schrödinger Institute for Mathematical Physics. Vienna, Austria, Sep. 7, 1999. ftp://ftp.esi.ac.at/pub/Preprints/esi747.ps.

Polyhex

planar polyhexes

$n = 1$

$n = 2$

$n = 3$

$n = 4$

bee   bar   pistol   propeller   worm   arch   wave

An analog of the POLYOMINOES and POLYIAMONDS in which collections of regular hexagons are arranged with adjacent sides. They are also called HEXES, HEXAS, or POLYFROBS (Beeler 1972). For the 4-hexes (tetrahexes), the possible arrangements are known as the BEE, BAR, PISTOL, PROPELLER, WORM, ARCH, and WAVE.

	catafusenes	perifusenes	multiple internal nodes
$n = 1$			
$n = 2$			
$n = 3$			
$n = 4$			

A simple connected polyhex is called a fusene. Let the number of internal vertices of a polyhex be denoted n_i. Then catafusenes (or catacondensed fusenes) have $n_i = 0$ (and are therefore also called "tree-like"), and perifusenes (or pericondensed fusenes) have $n_i = 1$. The numbers of catafusenes composed of n polyhexes are sometimes called Harary-Read numbers, and have the impressive GENERATING FUNCTION

$$H(x) = \tfrac{1}{24} x^{-2} \{12 + 24x + 48x^2 - 24x^3$$
$$+ [(1-x)(1-5x)]^{3/2} - 3(5x+3)$$
$$\times \sqrt{(1-x^2)(1-5x^2)} - 4\sqrt{(1-x^3)(1-5x^3)}\}$$

$$= x + x^2 + 2x^3 + 5x^4 + 12x^5 + 37x^6 + \ldots$$

(Harary and Read 1970, Cyvin *et al.* 1993). Polyhexes may also be classified on the basis of being geometrically planar (called nonhelicenic) or geometrically nonplanar (called helicenic). Fusenes include the helicenes.

one-sided 4-polyhexes

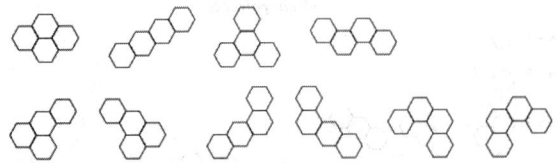

"One-sided" polyhexes are considered to be FIXED in the plane, and so mirror images are counted separately.

The following table gives the numbers of n-polyhexes that are geometrically planar (Klarner 1967, Balaban and Harary 1968, Harary and Read 1970, Lunnon 1972, Gardner 1978, Knop *et al.* 1984, Gardner 1988), catafusenes (Harary and Read 1970, Beinecke and Pippert 1974, Knop *et al.* 1984, Cyvin *et al.* 1993), cata- and planar, cata- and simply connected, and one-sided.

n	planar	cata-	cata- planar	cata- simpl.	one-sided
Sloane	A000228	A002216	A038142	A018190	A006535
1	1	1	1	1	1
2	1	1	1	1	1
3	3	2	2	3	3
4	7	5	5	7	10
5	22	12	12	22	33
6	82	37	36	81	147
7	333	123	118	331	620
8	1448	446	411	1435	2821
9	6572	1689	1489	6505	12942
10	30490	6693	5572	30086	60639
11	143552	27034		141229	286190
12	683101	111630		669584	1364621
13	3274826	467262		3198256	6545430
14		1981353		15367577	
15		8487400		74207910	
16		36695369		359863778	
17		159918120		1751594643	
18		701957539			
19		3101072051			
20		13779935438			
21		61557789660			
22		276327463180			
23		1245935891922			
24		5640868033058			

See also POLYHEX TILING, POLYIAMOND, POLYKING, POLYOMINO

References

Balaban, A. T. "Enumeration of Cyclic Graphs." In *Chemical Applications of Graph Theory* (Ed. A. T. Balaban). London: Academic Press, pp. 63–05, 1976.

Balaban, A. T. and Harary, F. "Chemical Graphs V: Enumeration and Proposed Nomenclature of Benzenoid Cata-Condensed Polycyclic Aromatic Hydrocarbons." *Tetrahedron* **24**, 2505–506, 1968.

Balasubramanian, K.; Kauffman, J. J.; Koski, W. S.; and Balaban, A. T. "Graph Theoretical Characterization and Computer Generation of Certain Carcinogenic Benzenoid Hydrocarbons and Identification." *J. Comput. Chem.* **1**, 149–57, 1980.

Beeler, M. Item 112 in Beeler, M.; Gosper, R. W.; and Schroeppel, R. *HAKMEM.* Cambridge, MA: MIT Artificial Intelligence Laboratory, Memo AIM-239, pp. 48–0, Feb. 1972.

Beineke, L. W. and Pippert, R. E. "On the Enumeration of Planar Trees of Hexagons." *Glasgow Math. J.* **15**, 131–47.

Cyvin, S. J.; Brunvoll, J.; Xiaofeng, G.; and Fuji, Z. "Number of Perifusenes with One Internal Vertex." *Rev. Roumaine Chem.* **38**, 65–7, 1993.

Dias, J. R. "A Periodic Table for Polycyclic Aromatic Hydrocarbons. 1. Isomer Enumeration of Fused Polycyclic Aromatic Hydrocarbon." *J. Chem. Inf. Comput. Sci.* **22**, 15–2, 1982.

Dias, J. R. "A Periodic Table for Polycyclic Aromatic Hydrocarbons. 2. Polycyclic Aromatic Hydrocarbons Containing Tetragonal, Pentagonal, Heptagonal, and Octagonal Rings." *J. Chem. Inf. Comput. Sci.* **22**, 139–52, 1982.

Dias, J. R. "A Periodic Table for Polycyclic Aromatic Hydrocarbons. 3. Enumeration of All the Polycyclic Conjugated Isomers of Pyrene Having Ring Sizes Ranging from 3 to 9." *Math. Chem (Mülheim/Ruhr)* **14**, 83–38, 1983.

Gardner, M. "Polyhexes and Polyaboloes." Ch. 11 in *Mathematical Magic Show: More Puzzles, Games, Diversions, Illusions and Other Mathematical Sleight-of-Mind from Scientific American.* New York: Vintage, pp. 146–59, 1978.

Gardner, M. "Tiling with Polyominoes, Polyiamonds, and Polyhexes." Ch. 14 in *Time Travel and Other Mathematical Bewilderments.* New York: W. H. Freeman, pp. 175–87, 1988.

Golomb, S. W. *Polyominoes: Puzzles, Patterns, Problems, and Packings, 2nd ed.* Princeton, NJ: Princeton University Press, pp. 92–3, 1994.

Harary, F. "Graphical Enumeration Problems." In *Graph Theory and Theoretical Physics* (Ed. F. Harary). London: Academic Press, pp. 1–1, 1967.

Harary, F. *Graph Theory.* Reading, MA: Addison-Wesley, pp. 178–97, 1994.

Harary, F. and Palmer, E. M. *Graphical Enumeration.* New York: Academic Press, 1973.

Harary, F. and Read, R. C. "The Enumeration of Tree-Like Polyhexes." *Proc. Edinburgh Math. Soc.* **17**, 1–3, 1970.

Keller, M. "Counting Polyforms." http://members.aol.com/wgreview/polyenum.html.

Klarner, D. A. "Cell Growth Problems." In *Canad. J. Math* **19**, 851–63, 1967.

Knop, J. V.; Szymanski, K.; Jericevic, Z.; and Trinajstic, N. "On the Total Number of Polyhexes." *Match: Commun. Math. Chem.*, No. 16, 119–34, Aug. 1984.

Lunnon, W. F. "Counting Hexagonal and Triangular Polyominoes." In *Graph Theory and Computing* (Ed. R. C. Read). New York: Academic Press, pp. 87–00, 1972.

Palmer, E. M. "Variations of the Cell Growth Problem." In *Graph Theory and Applications: Proceedings of the Conference at Western Michigan University, Kalamazoo, Mich., May 10–3, 1972* (Ed. Y. Alavi, D. R. Lick, and A. T. White). New York: Springer-Verlag, pp. 214–23, 1972.

Sloane, N. J. A. Sequences A000228/M2682, A002216/M1426, A006535/M2846, A018190, and A038142 in "An On-Line Version of the Encyclopedia of Integer Sequences." http://www.research.att.com/~njas/sequences/eisonline.html.

Vichera, M. "Polyforms." http://alpha.ujep.cz/~vicher/puzzle/polyforms.htm.

von Seggern, D. *CRC Standard Curves and Surfaces.* Boca Raton, FL: CRC Press, pp. 342–43, 1993.

Weisstein, E. W. "Polyominoes." MATHEMATICA NOTEBOOK POLYOMINO.M.

Weisstein, E. W. "Books about Polyominoes." http://www.treasure-troves.com/books/Polyominoes.html.

Polyhex Tiling

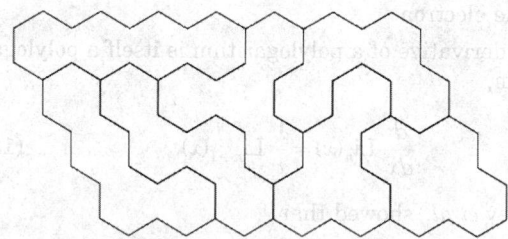

There are no tilings of the EQUILATERAL TRIANGLE of side length 7 by *all* the polyhexes of order $n = 4$. There are nine distinct solutions of *all* the polyhexes of order $n = 4$ which tile a PARALLELOGRAM of base length 7 and side length 4, one of which is illustrated above (Beeler 1972).

See also POLYHEX, POLYIAMOND TILING, POLYOMINO TILING

References

Beeler, M. Item 112 in Beeler, M.; Gosper, R. W.; and Schroeppel, R. *HAKMEM.* Cambridge, MA: MIT Artificial Intelligence Laboratory, Memo AIM-239, pp. 48–0, Feb. 1972.

Polyiamond

1   △

2   ◁▷

3   ◁▷△

4   polyiamond shapes

5   polyiamond shapes

6   polyiamond shapes

A generalization of the POLYOMINOES using a collection of equal-sized EQUILATERAL TRIANGLES (instead of SQUARES) arranged with coincident sides. Polyiamonds are sometimes simply known as IAMONDS.

The number of two-sided (i.e., can be picked up and flipped, so MIRROR IMAGE pieces are considered

identical) polyiamonds made up of n triangles are 1, 1, 1, 3, 4, 12, 24, 66, 160, 448, ... (Sloane's A000577). The number of one-sided polyiamonds composed of n triangles are 1, 1, 1, 4, 6, 19, 43, 121, ... (Sloane's A006534). One of the 160 9-polyiamonds has a hole (Gardner 1984, p. 174).

The top row of HEXIAMONDS in the above figure are known as the BAR, CROOK, CROWN, SPHINX, SNAKE, and YACHT. The bottom row of 6-polyiamonds are known as the CHEVRON, SIGNPOST, LOBSTER, HOOK, HEXAGON, and BUTTERFLY.

See also POLYABOLO, POLYHEX, POLYIAMOND TILING, POLYOMINO

References

Beeler, M. Item 112 in Beeler, M.; Gosper, R. W.; and Schroeppel, R. *HAKMEM.* Cambridge, MA: MIT Artificial Intelligence Laboratory, Memo AIM-239, pp. 48–0, Feb. 1972.

Gardner, M. "Mathematical Games." *Sci. Amer.* **211**, Dec. 1964.

Gardner, M. "Polyiamond." Ch. 18 in *The Sixth Book of Mathematical Games from Scientific American.* Chicago, IL: University of Chicago Press, pp. 173–82, 1984.

Golomb, S. W. *Polyominoes: Puzzles, Patterns, Problems, and Packings, 2nd ed.* Princeton, NJ: Princeton University Press, pp. 90–2, 1994.

Keller, M. "Counting Polyforms." http://members.aol.com/wgreview/polyenum.html.

O'Beirne, T. H. "Pentominoes and Hexiamonds." *New Scientist* **12**, 379–80, 1961.

Pegg, E. Jr. "Iamonds." http://www.mathpuzzle.com/iamond.htm.

Reeve, J. E. and Tyrrell, J. A. "Maestro Puzzles." *Math. Gaz.* **45**, 97–9, 1961.

Sloane, N. J. A. Sequences A000577/M2374 and A006534/M3287 in "An On-Line Version of the Encyclopedia of Integer Sequences." http://www.research.att.com/~njas/sequences/eisonline.html.

Torbijn, I. P. J. "Polyiamonds." *J. Recr. Math.* **2**, 216–27, 1969.

Vichera, M. "Polyforms." http://alpha.ujep.cz/~vicher/puzzle/polyforms.htm.

von Seggern, D. *CRC Standard Curves and Surfaces.* Boca Raton, FL: CRC Press, pp. 342–43, 1993.

Weisstein, E. W. "Polyominoes." MATHEMATICA NOTEBOOK POLYOMINO.M.

Polyiamond Tiling

HEPTIAMOND TILING, HEXIAMOND TILING, OCTIAMOND TILING, PENTIAMOND TILING

Polyking

POLYPLET

PolyLog

POLYLOGARITHM

Polylogarithm

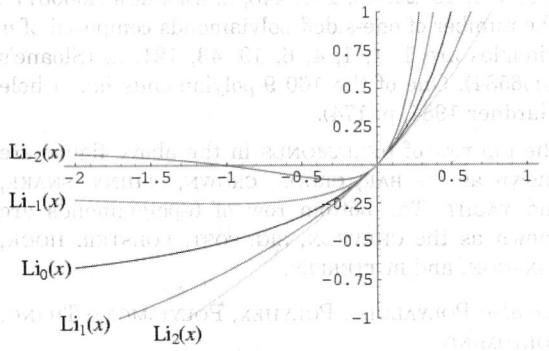

The function

$$\mathrm{Li}_n(z) \equiv \sum_{k=1}^{\infty} \frac{z^k}{k^n}, \tag{1}$$

Also known as Jonquière's function. (Note that the similar NOTATION Li(z) is used for the LOGARITHMIC INTEGRAL.) The polylogarithm is also denoted $F(z, n)$ and equal to

$$\mathrm{Li}_n(z) = z\Phi(z, n, 1), \tag{2}$$

where $\Phi(z, n, a)$ is the LERCH TRANSCENDENT (Erdélyi *et al.* 1981, p. 30). The polylogarithm arises in Feynman diagram integrals (and, in particular, in the computation of quantum electrodynamics corrections to the electrons gyromagnetic ratio), and the special cases $n = 2$ and $n = 3$ are called the DILOGARITHM and TRILOGARITHM, respectively.

The polylogarithm of NEGATIVE INTEGER order arises in sums OF THE FORM

$$\sum_{k=1}^{\infty} k^n r^k = \mathrm{Li}_{-n}(r) = \frac{r}{(1-r)^{n+1}} \sum_{i=1}^{n} \left\langle \begin{matrix} n \\ i \end{matrix} \right\rangle r^{n-i}, \tag{3}$$

where $\left\langle \begin{matrix} n \\ i \end{matrix} \right\rangle$ is an EULERIAN NUMBER.

Special forms of low-order polylogarithms include

$$\mathrm{Li}_{-2}(x) = \frac{x(x+1)}{(1-x)^3} \tag{4}$$

$$\mathrm{Li}_{-1}(x) = \frac{x}{(1-x)^2} \tag{5}$$

$$\mathrm{Li}_0(x) = \frac{x}{1-x} \tag{6}$$

$$\mathrm{Li}(x) = -\ln(1-x). \tag{7}$$

At arguments -1 and 1, the general polylogarithms become

$$\mathrm{Li}_n(-1) = -\eta(n) \tag{8}$$

$$\mathrm{Li}_n(1) = \zeta(n), \tag{9}$$

where $\eta(x)$ is the DIRICHLET ETA FUNCTION and $\zeta(x)$ is the RIEMANN ZETA FUNCTION. The polylogarithm for argument $1/2$ can also be evaluated analytically for small n,

$$\mathrm{Li}_1\left(\tfrac{1}{2}\right) = \ln 2 \tag{10}$$

$$\mathrm{Li}_2\left(\tfrac{1}{2}\right) = \tfrac{1}{12}[\pi^2 - 6(\ln 2)^2] \tag{11}$$

$$\mathrm{Li}_3\left(\tfrac{1}{2}\right) = \tfrac{1}{24}[4(\ln 2)^3 - 2\pi^2 \ln 2 + 21\zeta(3)]. \tag{12}$$

No similar formulas of this type are known for higher orders (Lewin 1991, p. 2). $\mathrm{Li}_4(1/2)$ appears in the third-order correction term in the gyromagnetic ratio of the electron.

The derivative of a polylogarithm is itself a polylogarithm,

$$\frac{d}{dx} \mathrm{Li}_n(x) = \frac{1}{x} \mathrm{Li}_{n-1}(x). \tag{13}$$

Bailey *et al.* showed that

$$\frac{\mathrm{Li}_m\left(\frac{1}{64}\right)}{6^{m-1}} - \frac{\mathrm{Li}_m\left(\frac{1}{8}\right)}{3^{m-1}} - \frac{2 \mathrm{Li}_m\left(\frac{1}{4}\right)}{2^{m-1}} + \frac{4 \mathrm{Li}_m\left(\frac{1}{2}\right)}{9} - \frac{5(-\ln 2)^m}{9m!}$$

$$+ \frac{\pi^2(-\ln 2)^{m-2}}{54(m-2)!} - \frac{\pi^4(-\ln 2)^{m-4}}{486(m-4)!} - \frac{403\zeta(5)(-\ln 2)^{m-5}}{1296(m-5)!}$$

$$= 0. \tag{14}$$

No general ALGORITHM is know for the integration of polylogarithms of functions.

See also DILOGARITHM, EULERIAN NUMBER, LEGENDRE'S CHI-FUNCTION, LOGARITHMIC INTEGRAL, NIELSEN GENERALIZED POLYLOGARITHM, NIELSEN-RAMANUJAN CONSTANTS, TRILOGARITHM

References

Bailey, D.; Borwein, P.; and Plouffe, S. "On the Rapid Computation of Various Polylogarithmic Constants." http://www.cecm.sfu.ca/~pborwein/PAPERS/P123.ps.

Bailey, D. H. and Broadhurst, D. J. A Seventeenth-Order Polylogarithm Ladder. 20 Jun 1999. http://xxx.lanl.gov/abs/math.CA/9906134/.

Borwein, J. M.; Bradley, D. M.; Broadhurst, D. J.; and Losinek, P. "Special Values of Multidimensional Polylogarithms." CECM-98:106, 14 May 1998. http://www.cecm.sfu.ca/preprints/1998pp.html#98:106.

Borwein, J. M.; Bradley, D. M.; Broadhurst, D. J.; and Losinek, P. Special Values of Multidimensional Polylogarithms. 8 Oct 1999. http://xxx.lanl.gov/abs/math.CA/9910045/.

Berndt, B. C. *Ramanujan's Notebooks, Part IV.* New York: Springer-Verlag, pp. 323–26, 1994.

Erdélyi, A.; Magnus, W.; Oberhettinger, F.; and Tricomi, F. G. *Higher Transcendental Functions, Vol. 1.* New York: Krieger, pp. 30–1, 1981.

Lewin, L. *Dilogarithms and Associated Functions.* London: Macdonald, 1958.

Lewin, L. *Polylogarithms and Associated Functions.* New York: North-Holland, 1981.

Lewin, L. (Ed.). *Structural Properties of Polylogarithms.* Providence, RI: Amer. Math. Soc., 1991.

Nielsen, N. *Der Euler'sche Dilogarithms.* Leipzig, Germany: Halle, 1909.

Prudnikov, A. P.; Marichev, O. I.; and Brychkov, Yu. A. "The Generalized Zeta Function $\zeta(s, x)$, Bernoulli Polynomials $B_n(x)$, Euler Polynomials $E_n(x)$, and Polylogarithms $\text{Li}_n(x)$." §1.2 in *Integrals and Series, Vol. 3: More Special Functions.* Newark, NJ: Gordon and Breach, pp. 23-4, 1990.

Truesdell, C. A. *Ann. Math.* **46**, 114-57, 1945.

Zagier, D. "Special Values and Functional Equations of Polylogarithms." Appendix A in *Structural Properties of Polylogarithms* (Ed. L. Lewin). Providence, RI: Amer. Math. Soc., 1991.

Polymorph

An INTEGER which is expressible in more than one way in the form $x^2 + Dy^2$ or $x^2 - Dy^2$ where x^2 is RELATIVELY PRIME to Dy^2. If the INTEGER is expressible in only one way, it is called a MONOMORPH.

See also ANTIMORPH, IDONEAL NUMBER, MONOMORPH, PELL EQUATION

Polymorph Tessellation

TESSELLATION

Polynema

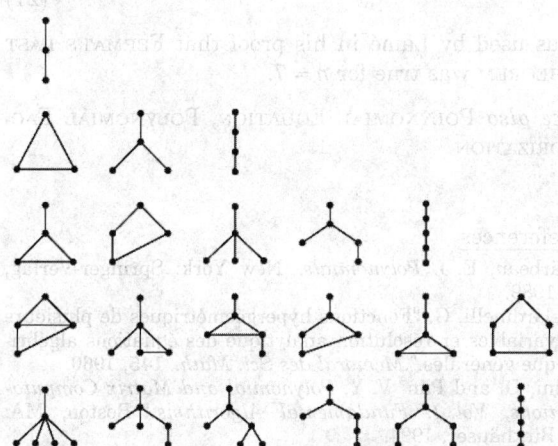

A polynema of order n is Kyrmse's term for a CONNECTED GRAPH having n edges. An n-polynema must therefore have either n or $n+1$ nodes. The numbers of n-polynemas for $n = 1, 2 \ldots$ are 1, 1, 3, 5, 12, 30, 79, 227, ... (Sloane's A002905). Polynemas are related to a graphical construction problem called the MATCH PROBLEM (Gardner 1991).

See also CONNECTED GRAPH, MATCH PROBLEM, PLANAR CONNECTED GRAPH, TREE

References

Gardner, M. "The Problem of the Six Matches." In *The Unexpected Hanging and Other Mathematical Diversions.* Chicago, IL: Chicago University Press, pp. 79-1, 1991.

Kyrmse, R. http://users.sti.com.br/rkyrmse/POLIN-E.htm.

Sloane, N. J. A. Sequences A002905/M2486 in "An On-Line Version of the Encyclopedia of Integer Sequences." http://www.research.att.com/~njas/sequences/eisonline.html.

Polynomial

A POLYNOMIAL is a mathematical expression involving a series of POWERS in one or more variables multiplied by COEFFICIENTS. A POLYNOMIAL in one variable (i.e., a univariate polynomial) with constant COEFFICIENTS is given by

$$a_n x^n + \ldots + a_2 x^2 + a_1 x + a_0. \tag{1}$$

The highest POWER in a univariate polynomial is called its ORDER. A POLYNOMIAL in two variables (i.e., a bivariate polynomial) with constant COEFFICIENTS is given by

$$a_{nm} x^n y^m + \ldots + a_{22} x^2 y^2 + a_{21} x^2 y + a_{12} xy^2 + a_{11} xy + a_{10} x + a_{01} y + a_{00} \tag{2}$$

Exchanging the COEFFICIENTS of a univariate polynomial end-to-end produces a polynomial

$$a_0 x^n + a_1 x^{n-1} + \ldots + a_{n-1} x + a_n = 0 \tag{3}$$

whose ROOTS are RECIPROCALS $1/x_i$ of the original ROOTS x_i.

HORNER'S RULE provides a computationally efficient method of forming a polynomial from a list of its coefficients, and can be implemented in *Mathematica* as follows.

```
PolynomialFromCoefs[l_List, x_] := Fold[x#1 +
#2 &, 0, l]
```

The following table gives special names given to polynomials of low orders.

ORDER	Polynomial Type
1	LINEAR EQUATION
2	QUADRATIC EQUATION
3	CUBIC EQUATION
4	QUARTIC EQUATION
5	QUINTIC EQUATION
6	SEXTIC EQUATION

Polynomials of fourth degree may be computed using three multiplications and five additions if a few quantities are calculated first (Press *et al.* 1989):

$$a_0 + a_1 x + a_2 x^2 + a_3 x^3 + a_4 x^4$$

$$= [(Ax + B)^2 + Ax + C][(Ax + B)^2 + D] + E, \tag{4}$$

where

$$A \equiv (a_4)^{1/4} \tag{5}$$

$$B \equiv \frac{a_3 - A^3}{4A^3} \tag{6}$$

$$D \equiv 3B^2 + 8B^3 + \frac{a_1 A - 2a_2 B}{A^2} \tag{7}$$

$$C \equiv \frac{a_2}{A^2} - 2B - 6B^2 - D \tag{8}$$

$$E \equiv a_0 - B^4 - B^2(C + D) - CD. \tag{9}$$

Similarly, a POLYNOMIAL of fifth degree may be computed with four multiplications and five additions, and a POLYNOMIAL of sixth degree may be computed with four multiplications and seven additions.

Polynomials of orders one to four are solvable using only rational operations and finite ROOT EXTRACTIONS. A first-order equation is trivially solvable. A second-order equation is soluble using the QUADRATIC EQUATION. A third-order equation is solvable using the CUBIC EQUATION. A fourth-order equation is solvable using the QUARTIC EQUATION. It was proved by Abel and Galois using GROUP THEORY that general equations of fifth and higher order cannot be solved rationally with finite ROOT EXTRACTIONS (ABEL'S IMPOSSIBILITY THEOREM).

However, the general QUINTIC EQUATION may be given in terms of the JACOBI THETA FUNCTIONS, or HYPERGEOMETRIC FUNCTIONS in one variable. Hermite and Kronecker proved that higher order POLYNOMIALS are not soluble in the same manner. Klein showed that the work of Hermite was implicit in the GROUP properties of the ICOSAHEDRON. Klein's method of solving the quintic in terms of HYPERGEOMETRIC FUNCTIONS in one variable can be extended to the sextic, but for higher order POLYNOMIALS, either HYPERGEOMETRIC FUNCTIONS in several variables or "Siegel functions" must be used (Belardinelli 1960, King 1996, Chow 1999). In the 1880s, Poincaré created functions which give the solution to the nth order POLYNOMIAL equation in finite form. These functions turned out to be "natural" generalizations of the ELLIPTIC FUNCTIONS.

Given an nth degree polynomial, the ROOTS can be found by finding the EIGENVALUES of the MATRIX

$$\begin{bmatrix} -a_0/a_n & -a_1/a_n & -a_2/a_n & \cdots & -1 \\ 1 & 0 & 0 & \cdots & 0 \\ 0 & 1 & 0 & \cdots & 0 \\ \vdots & \vdots & 1 & \ddots & 0 \\ 0 & 0 & 0 & \cdots & 0 \end{bmatrix}. \tag{10}$$

This method can be computationally expensive, but is fairly robust at finding close and multiple roots.

Polynomial identities involving sums and differences of like POWERS include

$$x^2 - y^2 = (x - y)(x + y) \tag{11}$$

$$x^3 - y^3 = (x - y)(x^2 + xy + y^2) \tag{12}$$

$$x^3 + y^3 = (x + y)(x^2 - xy + y^2) \tag{13}$$

$$x^4 - y^4 = (x - y)(x + y)(x^2 + y^2) \tag{14}$$

$$x^5 - y^5 = (x - y)(x^4 + x^3 y + x^2 y^2 + xy^3 + y^4) \tag{15}$$

$$x^5 + y^5 = (x + y)(x^4 - x^3 y + x^2 y^2 - xy^3 + y^4) \tag{16}$$

$$x^6 - y^6 = (x - y)(x + y)(x^2 + xy + y^2)(x^2 - xy + y^2) \tag{17}$$

$$x^6 + y^6 = (x^2 + y^2)(x^4 - x^2 y^2 + y^4). \tag{18}$$

Further identities include

$$\left(x_1^2 - Dy_1^2\right)\left(x_2^2 - Dy_2^2\right)$$
$$= (x_1 x_2 + Dy_1 y_2)^2 - D(x_1 y_2 + x_2 y_1)^2 \tag{19}$$

$$\left(x_1^2 + Dy_1^2\right)\left(x_2^2 + Dy_2^2\right)$$
$$= (x_1 x_2 \pm Dy_1 y_2)^2 + D(x_1 y_2 \mp x_2 y_1)^2. \tag{20}$$

The identity

$$(X + Y + Z)^7 - (X^7 + Y^7 + Z^7) = 7(X + Y)(X + Z)(Y + Z)$$
$$\times [(X^2 + Y^2 + Z^2 + XY + XZ + YZ)^2 + XYZ(X + Y + Z)] \tag{21}$$

was used by Lamé in his proof that FERMAT'S LAST THEOREM was true for $n = 7$.

See also POLYNOMIAL EQUATION, POLYNOMIAL FACTORIZATION

References

Barbeau, E. J. *Polynomials.* New York: Springer-Verlag, 1989.

Belardinelli, G. "Fonctions hypergéométriques de plusieurs variables er résolution analytique des équations algébrique générales." *Mémoral des Sci. Math.* **145**, 1960.

Bini, D. and Pan, V. Y. *Polynomial and Matrix Computations, Vol. 1: Fundamental Algorithms.* Boston, MA: Birkhäuser, 1994.

Borwein, P. and Erdélyi, T. *Polynomials and Polynomial Inequalities.* New York: Springer-Verlag, 1995.

Chow, T. Y. "What is a Closed-Form Number." *Amer. Math. Monthly* **106**, 440–48, 1999.

Cockle, J. "Notes on the Higher Algebra." *Quart. J. Pure Applied Math.* **4**, 49–7, 1861.

Cockle, J. "Notes on the Higher Algebra (Continued)." *Quart. J. Pure Applied Math.* **5**, 1–7, 1862.

King, R. B. *Beyond the Quartic Equation.* Boston, MA: Birkhäuser, 1996.

Mignotte, M. and Stefanescu, D. *Polynomials: An Algorithmic Approach.* Singapore: Springer-Verlag, 1999.

Press, W. H.; Flannery, B. P.; Teukolsky, S. A.; and Vetterling, W. T. *Numerical Recipes in C: The Art of Scientific Computing.* Cambridge, England: Cambridge University Press, 1989.

Project Mathematics. "Polynomials." Videotape. http://www.projmath.caltech.edu/polynom.htm.

Ram, R. "Sums of Powers." http://users.tellurian.net/hsejar/maths/sumsofpowers/.

Weisstein, E. W. "Books about Polynomials." http://www.treasure-troves.com/books/Polynomials.html.

Polynomial Bar Norm
POLYNOMIAL NORM

Polynomial Bracket Norm
BOMBIERI NORM

Polynomial Curve

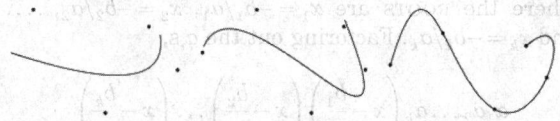

A curve obtained by fitting POLYNOMIALS to each ordinate of an ordered sequence of points. The above plots show POLYNOMIAL curves where the order of the fitting POLYNOMIAL varies from $p-3$ to $p-1$, where p is the number of points.

Polynomial curves have several undesirable features, including a nonintuitive variation of fitting curve with varying COEFFICIENTS, and numerical instability for high orders. SPLINES such as the BÉZIER CURVE are therefore used more commonly.

See also BÉZIER CURVE, POLYNOMIAL, SPLINE

Polynomial Equation
An EQUATION of the form

$$P(x) = 0,$$

where $P(x)$ is a POLYNOMIAL.

See also POLYNOMIAL

Polynomial Factorization
A FACTOR of a POLYNOMIAL $P(x)$ of degree n is a POLYNOMIAL $Q(x)$ of degree less than n which can be multiplied by another POLYNOMIAL $R(x)$ of degree less than n to yield $P(x)$, i.e., a POLYNOMIAL $Q(x)$ such that

$$P(x) = Q(x)R(x).$$

For example, since

$$x^2 - 1 = (x+1)(x-1),$$

both $x-1$ and $x+1$ are FACTORS of $x^2 - 1$. Polynomial factorization can be performed in *Mathematica* using `Factor[poly]`.

The COEFFICIENTS of factor POLYNOMIALS are often required to be REAL NUMBERS or INTEGERS but could, in general, be COMPLEX NUMBERS. The FUNDAMENTAL THEOREM OF ALGEBRA states that a POLYNOMIAL $P(z)$ of degree n has n values z_i (some of which are possibly degenerate) for which $P(z_i) = 0$. Such values are called POLYNOMIAL ROOTS.

See also FACTOR, FACTORIZATION, FUNDAMENTAL THEOREM OF ALGEBRA, KRONECKER'S ALGORITHM, POLYNOMIAL ROOTS, PRIME FACTORIZATION

References
Abbott, J.; Shoup, V.; and Zimmerman, P. "Factorization in $\mathbb{Z}[x]$: The Searching Phase." To appear in ISSAC'2000 Proceedings.
Kaltofen, E. "Polynomial Factorization." In *Computer Algebra: Symbolic and Algebraic Computation, 2nd ed.* (Ed. B. Buchberger, G. E. Collins, R. Loos, and R. Albrecht). Vienna: Springer-Verlag, pp. 95–13, 1983.
Lenstra, A. K.; Lenstra, H. W.; and Lovász, L. "Factoring Polynomials with Rational Coefficients." *Math. Ann.* **261**, 515–34, 1982.
Séroul, R. "Factoring a Polynomial with Integral Coefficients." §10.14 in *Programming for Mathematicians.* Berlin: Springer-Verlag, pp. 286–95, 2000.
van Hoeij, M. "Factoring Polynomials and the Knapsack Problem." Preprint. http://www.math.fsu.edu/~aluffi/archive/paper124.ps.gz.

Polynomial Height
The l_∞-POLYNOMIAL NORM defined for a polynomial $P = a_k x^k + \ldots + a_1 x + a_0$ by

$$\|P\|_\infty = \max_k |a_k|.$$

Note that some authors (especially in the area of Diophantine analysis) use $|P|$ as a shorthand for $\|P\|_\infty$, while others (especially in the area of computational complexity) used $|P|$ to denote the l_2-norm $\|P\|_2$ (Zippel 1993, p. 174).

See also POLYNOMIAL NORM

References
Zippel, R. "Heights of Polynomials." §11.1 in *Effective Polynomial Computation.* Boston, MA: Kluwer, pp. 174–75, 1993.

Polynomial Map
A map OF THE FORM

$$\phi_{\mathbf{f}} : K^n \to K^n$$

$$\phi_{\mathbf{f}} : (a_1, \ldots, a_n) \mapsto (f_1(\mathbf{a}), \ldots, f_1(\mathbf{a})),$$

where $\mathbf{f} = (f_1, \ldots, f_n) \in (K[X_1, \ldots, X_n])^m$ in a FIELD K, and $\mathbf{a} = (a_1, \ldots, a_n)$.

See also INVERTIBLE POLYNOMIAL MAP, JACOBIAN CONJECTURE

References
Becker, T. and Weispfenning, V. *Gröbner Bases: A Computational Approach to Commutative Algebra.* New York: Springer-Verlag, p. 330, 1993.

Polynomial Matrix
A MATRIX whose entries are POLYNOMIALS.

See also MATRIX POLYNOMIAL

References
Pascoletti, A. "Polynomial Matrix Utilities." http://www.mathsource.com/cgi-bin/msitem?0207-51.

Polynomial Norm

For a POLYNOMIAL

$$P = \sum_{k=0}^{n} a_k z^k, \qquad (1)$$

several classes of norms are commonly defined. The l_p-norm is defined as

$$\|P\|_p \equiv \left(\sum_{k=0}^{n} |a_k|^p \right) \qquad (2)$$

for $p \geq 1$, giving the special cases

$$\|P\|_1 \equiv \sum_j |a_k| \qquad (3)$$

$$\|P\|_2 = \sqrt{\sum_k |a_k|^2} \qquad (4)$$

$$\|P\|_\infty = \max_k |a_k|. \qquad (5)$$

Here, $\|P\|_\infty$ is called the POLYNOMIAL HEIGHT. Note that some authors (especially in the area of Diophantine analysis) use $|P|$ as a shorthand for $\|P\|_\infty$ and $|P|$ as a shorthand for $\|P_2\|$, while others (especially in the area of computational complexity) used $|P|$ to denote the l_2-norm $\|P\|_2$ and (Zippel 1993, p. 174).

Another class of norms is the L_p-norms, defined by

$$\|P\|_{L_p} = \left(\int_0^{2\pi} |P(e^{i\theta})| \, \frac{d\theta}{2\pi} \right)^{1/p} \qquad (6)$$

for $p \geq 1$, giving the special cases

$$\|P\|_{L_1} = \int_0^{2\pi} |P(e^{i\theta})| \, \frac{d\theta}{2\pi}$$

$$\|P\|_{L_2} = \sqrt{\int_0^{2\pi} |P(e^{i\theta})|^2 \, \frac{d\theta}{2\pi}}$$

$$\|P\|_{L_\infty} = \sup_{|z|=1} |P(z)|$$

(Borwein and Erdélyi 1995, p. 6).

See also BOMBIERI NORM, MATRIX NORM, NORM, UNIT CIRCLE, VECTOR NORM

References

Borwein, P. and Erdélyi, T. "Norms on $\mathscr{P}_n$." §1.1.E.3 in *Polynomials and Polynomial Inequalities*. New York: Springer-Verlag, pp. 6–, 1995.
Zippel, R. *Effective Polynomial Computation*. Boston, MA: Kluwer, 1993.

Polynomial Remainder Theorem

If the COEFFICIENTS of the POLYNOMIAL

$$d_n x^n + d_{n-1} x^{n-1} + \ldots + d_0 = 0 \qquad (1)$$

are specified to be INTEGERS, then integral ROOTS must have a NUMERATOR which is a factor of d_0 and a DENOMINATOR which is a factor of d_n (with either sign possible). This follows since a POLYNOMIAL of ORDER n with k integral ROOTS can be expressed as

$$(a_1 x + b_1)(a_2 x + b_2) \cdots (a_k x + b_k)(c_{n-k} x^{n-k} + \ldots + c_0)$$
$$= 0, \qquad (2)$$

where the ROOTS are $x_1 = -b_1/a_1$, $x_2 = -b_2/a_2$, ..., and $x_k = -b_k/a_k$. Factoring out the a_is,

$$a_1 a_2 \ldots a_k \left(x - \frac{b_1}{a_1} \right) \left(x - \frac{b_2}{a_2} \right) \ldots \left(x - \frac{b_k}{a_k} \right)$$
$$\times (c_{n-k} x^{n-k} + \ldots + c_0) = 0. \qquad (3)$$

Now, multiplying through,

$$a_1 a_2 \ldots a_k c_{n-k} x^n + \ldots + b_1 b_2 \ldots b_k c_0 = 0, \qquad (4)$$

where we have not bothered with the other terms. Since the first and last COEFFICIENTS are d_n and d_0, all the integral roots of (1) are OF THE FORM [factors of d_0]/[factors of d_n].

References

Bold, B. *Famous Problems of Geometry and How to Solve Them.* New York: Dover, p. 34, 1982.
Niven, I. M. *Numbers: Rational and Irrational.* New York: Random House, 1961.

Polynomial Ring

The RING $R[x]$ of POLYNOMIALS in a variable x.

See also MODULE, POLYNOMIAL, RING

Polynomial Roots

A root of a polynomial $P(z)$ is a number z_i such that $P(z_i) = 0$. The FUNDAMENTAL THEOREM OF ALGEBRA states that a POLYNOMIAL $P(z)$ of degree n has n roots, some of which may be degenerate. For example, the roots of the polynomial

$$x^3 - 2x^2 - x + 2 = (x - 2)(x - 1)(x + 1) \qquad (1)$$

are -1, 1, and 2. Finding roots of a polynomial is therefore equivalent to POLYNOMIAL FACTORIZATION into factors of degree 1. The roots of a polynomial equation may be found in *Mathematica* using `Roots[lhs == rhs, var]`.

Let the ROOTS of the polynomial

$$P(x) \equiv a_n x^n + a_{n-1} x^{n-1} + \ldots a_1 x + a_0 \qquad (2)$$

be denoted r_1, r_2, ..., r_n. Then NEWTON'S RELATIONS are

$$\sum r_i = -\frac{a_{n-1}}{a_n} \qquad (3)$$

$$\sum r_i r_j = \frac{a_{n-2}}{a_n} \tag{4}$$

$$\sum r_1 r_2 \cdots r_k = (-1)^k \frac{a_{n-k}}{a_n}. \tag{5}$$

These can be derived by writing

$$P(x) = a_n (x - r_1)(x - r_2) \cdots (x - r_n), \tag{6}$$

expanding, and then comparing the coefficients with (2).

Any POLYNOMIAL can be numerically factored, although different ALGORITHMS have different strengths and weaknesses.

If the COEFFICIENTS of the POLYNOMIAL

$$d_n x^n + d_{n-1} x^{n-1} + \ldots + d_0 = 0 \tag{7}$$

are specified to be INTEGERS, then integral roots must have a NUMERATOR which is a factor of d_0 and a DENOMINATOR which is a factor of d_n (with either sign possible). This is known as the POLYNOMIAL REMAINDER THEOREM.

If there are no NEGATIVE ROOTS of a POLYNOMIAL (as can be determined by DESCARTES' SIGN RULE), then the GREATEST LOWER BOUND is 0. Otherwise, write out the COEFFICIENTS, let $n = -1$, and compute the next line. Now, if any COEFFICIENTS are 0, set them to minus the sign of the next higher COEFFICIENT, starting with the second highest order COEFFICIENT. If all the signs alternate, n is the greatest lower bound. If not, then subtract 1 from n, and compute another line. For example, consider the POLYNOMIAL

$$y = 2x^4 + 2x^3 - 7x^2 + x - 7. \tag{8}$$

Performing the above ALGORITHM then gives

0	2	2	−7	1	−7
−1	2	0	−7	8	−15
−	2	−1	−7	8	−15
−2	2	−2	−3	7	−21
−3	2	−4	5	−14	35

so the greatest lower bound is −3.

If there are no POSITIVE ROOTS of a POLYNOMIAL (as can be determined by DESCARTES' SIGN RULE), the LEAST UPPER BOUND is 0. Otherwise, write out the COEFFICIENTS of the POLYNOMIALS, including zeros as necessary. Let $n = 1$. On the line below, write the highest order COEFFICIENT. Starting with the second-highest COEFFICIENT, add n times the number just written to the original second COEFFICIENT, and write it below the second COEFFICIENT. Continue through

order zero. If all the COEFFICIENTS are NONNEGATIVE, the least upper bound is n. If not, add one to x and repeat the process again. For example, take the POLYNOMIAL

$$y = 2x^4 - x^3 - 7x^2 + x - 7. \tag{9}$$

Performing the above ALGORITHM gives

0	2	−1	−7	1	−7
1	2	1	−6	−5	−12
2	2	3	−1	−1	−9
3	2	5	8	25	68

so the LEAST UPPER BOUND is 3.

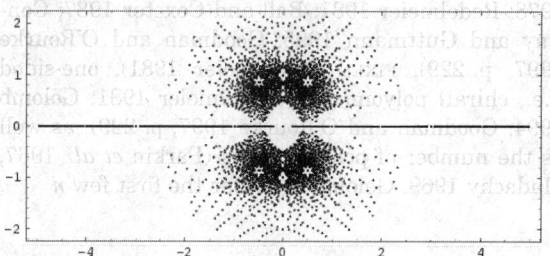

Plotting the roots in the complex plane of all polynomials up to some degree with integer coefficients less than some cutoff integer in absolute value shows the beautiful structure illustrated above (Trott 2000).

See also BAIRSTOW'S METHOD, DESCARTES' SIGN RULE, GRAEFFE'S METHOD, JENKINS-TRAUB METHOD, LAGUERRE'S METHOD, LEHMER-SCHUR METHOD, MAEHLY'S PROCEDURE, MULLER'S METHOD, POLYNOMIAL FACTORIZATION, ROOT, ZASSENHAUS-BERLEKAMP ALGORITHM

References

Bharucha-Reid, A. T. and Sambandham, M. *Random Polynomials.* New York: Academic Press, 1986.
Odlyzko, A. M.; and Poonen, B. *L'Enseignement Math.* **39**, 317, 1993.
Pan, V. Y. "Solving a Polynomial Equation: Some History and Recent Progress." *SIAM Rev.* **39**, 187–20, 1997.
Trott, M. "Numerical Computations." §1.2.1 in *The Mathematica Guidebook, Vol. 1: Programming in Mathematica.* New York: Springer-Verlag, 2000.

Polynomial Sequence

A SEQUENCE of POLYNOMIALS $p_i(x)$, for $i = 0, 1, 2, \ldots$, where $p_i(x)$ is exactly of degree i for all i.

See also BASIC POLYNOMIAL SEQUENCE, POLYNOMIAL

Polynomial Series

MULTINOMIAL SERIES

Polynomial-Time

See also NP-PROBLEM, P-PROBLEM

Polyomino

A generalization of the DOMINO, originally called "super-dominoes" by Gardner (1957). An n-polyomino (or "n-omino"rpar; is defined as a collection of n squares of equal size arranged with coincident sides. FREE polyominoes can be picked up and flipped, so mirror image pieces are considered identical, whereas FIXED polyominoes are distinct if they have different chirality or orientation. FIXED polyominoes are also called LATTICE ANIMALS.

Redelmeier (1981) computed the number of FREE and FIXED polyominoes for $n \leq 24$, and Mertens (1990) gives a simple computer program. The following table gives the number of FREE (Lunnon 1971, 1972; Read 1978; Redelmeier 1981; Ball and Coxeter 1987; Conway and Guttmann 1995; Goodman and O'Rourke 1997, p. 229), FIXED (Redelmeier 1981), one-sided (i.e., chiral) polyominoes (Redelmeier 1981; Golomb 1994; Goodman and O'Rourke 1997, p. 229), as well as the number of possible holes (Parkin *et al.* 1967, Madachy 1969, Golomb 1994) for the first few n

n	FREE	FIXED	one-sided	poss. holes
Sloane	A000105	A014559	A000988	A001419
1	1	1	1	0
2	1	2	1	0
3	2	6	2	0
4	5	19	7	0
5	12	53	18	0
6	35	216	60	0
7	108	760	196	1
8	369	2725	704	6
9	1285	9910	2500	37
10	4655	39446	9189	195
11	17073	125268	33896	979
12	63600	505861	126759	4663
13	238591	1903890	476270	21474
14	901971	7204874	1802312	96496
15	3426576	27394666	6849777	425365
16	13079255	104592937	26152418	
17	50107909	400795844	100203194	
18	192622052	1540820542	385221143	
19	742624232	5940738676	1485200848	
20	2870671950	22964779660	5741256764	
21	11123060678	88983512783	22245940545	
22	43191857688	345532572678	86383382827	
23	168047007728	1344372335524	336093325058	
24	654999700403	5239988770268	1309998125640	

The best currently known bounds on the number of n-polyominoes are

$$3.72^n < P(n) < 4.65^n$$

(Eden 1961, Klarner 1967, Klarner and Rivest 1973, Ball and Coxeter 1987).

There is a single unique 2-omino (the DOMINO), and two distinct 3-ominoes (the straight- and L-TRIOMINOES). The 4-ominoes (TETROMINOES) are known as the STRAIGHT, L, T, SQUARE, and SKEW TETROMINOES. The 5-ominoes (PENTOMINOES) are called f, I, L, N, P, T, U, V, W, X, y, and Z (Golomb 1995). Another common naming scheme replaces f, I, L, and N with R, O, Q, and S so that all letters from O to Z are used (Berlekamp *et al.* 1982).

See also COLUMN-CONVEX POLYOMINO, CONVEX POLYOMINO, DOMINO, HEXOMINO, LATTICE POLYGON, MONOMINO, PENTOMINO, POLYABOLO, POLYCUBE, POLYHEX, POLYIAMOND, POLYKING, POLYPLET, ROW-CONVEX POLYOMINO, SELF-AVOIDING POLYGON, TETROMINO, TRIOMINO

References

Atkin, A. O. L. and Birch, B. J. (Eds.). *Computers in Number Theory: Proc. Sci. Research Council Atlas Symposium No. 2 Held at Oxford from 18–3 Aug., 1969.* New York: Academic Press, 1971.

Ball, W. W. R. and Coxeter, H. S. M. *Mathematical Recreations and Essays, 13th ed.* New York: Dover, pp. 109–13, 1987.

Beeler, M. Item 112 in Beeler, M.; Gosper, R. W.; and Schroeppel, R. *HAKMEM.* Cambridge, MA: MIT Artificial Intelligence Laboratory, Memo AIM-239, pp. 48–0, Feb. 1972.

Beineke, L. W. and Wilson, R. J. (Eds.). *Selected Topics in Graph Theory.* New York: Academic Press, pp. 417–44, 1978.

Berlekamp, E. R.; Conway, J. H; and Guy, R. K. *Winning Ways for Your Mathematical Plays, Vol. 1: Games in General.* London: Academic Press, 1982.

Berlekamp, E. R.; Conway, J. H; and Guy, R. K. *Winning Ways for Your Mathematical Plays, Vol. 2: Games in Particular.* London: Academic Press, 1982.

Bousquet-Mélou, M.; Guttmann, A. J.; Orrick, W. P.; and Rechnitzer, A. Inversion Relations, Reciprocity and Polyominoes. 23 Aug 1999. http://xxx.lanl.gov/abs/math.CO/9908123/.

Conway, A. R. and Guttmann, A. J. "On Two-Dimensional Percolation." *J. Phys. A: Math. Gen.* **28**, 891–04, 1995.

Eden, M. "A Two-Dimensional Growth Process." *Proc. Fourth Berkeley Symposium Math. Statistics and Probability, Held at the Statistical Laboratory, University of*

California, June 30-July 30, 1960. Berkeley, CA: University of California Press, pp. 223–39, 1961.

Finch, S. "Favorite Mathematical Constants." http://www.mathsoft.com/asolve/constant/rndprc/rndprc.html.

Gardner, M. "Mathematical Games: About the Remarkable Similarity between the Icosian Game and the Towers of Hanoi." *Sci. Amer.* **196**, 150–56, May 1957.

Gardner, M. "Polyominoes and Fault-Free Rectangles." Ch. 13 in *Martin Gardner's New Mathematical Diversions from Scientific American.* New York: Simon and Schuster, pp. 150–61, 1966.

Gardner, M. "Polyominoes and Rectification." Ch. 13 in *Mathematical Magic Show: More Puzzles, Games, Diversions, Illusions and Other Mathematical Sleight-of-Mind from Scientific American.* New York: Vintage, pp. 172–87, 1978.

Golomb, S. W. "Checker Boards and Polyominoes." *Amer. Math. Monthly* **61**, 675–82, 1954.

Golomb, S. W. *Polyominoes: Puzzles, Patterns, Problems, and Packings, 2nd ed.* Princeton, NJ: Princeton University Press, 1995.

Goodman, J. E. and O'Rourke, J. (Eds.). *Handbook of Discrete & Computational Geometry.* Boca Raton, FL: CRC Press, 1997.

Keller, M. "Counting Polyforms." http://members.aol.com/wgreview/polyenum.html.

Klarner, D. A. "Cell Growth Problems." *Can. J. Math.* **19**, 851–63, 1967.

Klarner, D. A. and Riverst, R. "A Procedure for Improving the Upper Bound for the Number of *n*-ominoes." *Can. J. Math.* **25**, 585–02, 1973.

Lei, A. "Bigger Polyominoes." http://www.cs.ust.hk/~philipl/omino/bigpolyo.html.

Lei, A. "Polyominoes." http://www.cs.ust.hk/~philipl/omino/omino.html.

Lunnon, W. F. "Counting Polyominoes." In *Computers in Number Theory* (Ed. A. O. L. Atkin and B. J. Brich). London: Academic Press, pp. 347–72, 1971.

Lunnon, W. F. "Counting Hexagonal and Triangular Polyominoes." In *Graph Theory and Computing* (Ed. R. C. Read). New York: Academic Press, 1972.

Madachy, J. S. "Pentominoes: Some Solved and Unsolved Problems." *J. Rec. Math.* **2**, 181–88, 1969.

Martin, G. *Polyominoes: A Guide to Puzzles and Problems in Tiling.* Washington, DC: Math. Assoc. Amer., 1991.

Marzetta, A. "List of Polyominoes of order 4..7." http://wwwjn.inf.ethz.ch/ambros/polyo-list.html.

Mertens, S. "Lattice Animals--A Fast Enumeration Algorithm and New Perimeter Polynomials." *J. Stat. Phys.* **58**, 1095–108, 1990.

Parkin, T. R.; Lander, L. J.; and Parkin, D. R. "Polyomino Enumeration Results." *SIAM Fall Meeting.* Santa Barbara, CA, 1967.

Read, R. C. "Contributions to the Cell Growth Problem." *Canad. J. Math.* **14**, 1–0, 1962.

Read, R. C. "Some Applications of Computers in Graph Theory." In *Selected Topics in Graph Theory* (Ed. L. W. Beineke and R. J. Wilson). New York: Academic Press, pp. 417–44, 1978.

Redelmeier, D. H. "Counting Polyominoes: Yet Another Attack." *Discrete Math.* **36**, 191–03, 1981.

Ruskey, F. "Information on Polyominoes." http://www.theory.csc.uvic.ca/~cos/inf/misc/PolyominoInfo.html.

Schroeppel, R. Item 77 in Beeler, M.; Gosper, R. W.; and Schroeppel, R. *HAKMEM.* Cambridge, MA: MIT Artificial Intelligence Laboratory, Memo AIM-239, p. 30, Feb. 1972.

Sloane, N. J. A. Sequences A000105/M1425, A001419/M4226, and A014559 in "An On-Line Version of the Encyclopedia of Integer Sequences." http://www.research.att.com/~njas/sequences/eisonline.html.

Vichera, M. "Polyforms." http://alpha.ujep.cz/~vicher/puzzle/polyforms.htm.

von Seggern, D. *CRC Standard Curves and Surfaces.* Boca Raton, FL: CRC Press, pp. 342–43, 1993.

Weisstein, E. W. "Polyominoes." MATHEMATICA NOTEBOOK POLYOMINO.M.

Weisstein, E. W. "Books about Polyominoes." http://www.treasure-troves.com/books/Polyominoes.html.

Wells, D. *The Penguin Dictionary of Curious and Interesting Geometry.* London: Penguin, p. 117, 1991.

Wells, D. *Recreations in Logic.* New York: Dover, 1979.

Polyomino Tiling

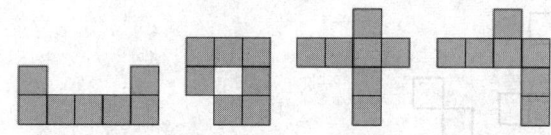

A TILING of the PLANE by specified types of POLYOMINOES. Interestingly, the FIBONACCI NUMBER F_{n+1} gives the number of ways for 2×1 DOMINOES to cover a $2 \times n$ checkerboard. Each MONOMINO, DOMINO, TRIOMINO, TETROMINO, PENTOMINO, and HEXOMINO tiles the plane, with requiring flipping. In addition, each heptomino, with the exception of the four illustrated above, can tile the plane, also without flipping (Schroeppel 1972).

Consider now those collections of *all* n-ominoes which form a RECTANGLE. The polynomials of orders $n = 1$ and $n = 2$ form only a SQUARE and RECTANGLE, respectively. The two polyominoes of order $n = 3$ cannot form a rectangle, nor can the five polyominoes of order $n = 4$ or the 35 polyominoes of order $n = 6$ (Beeler 1972). There are several rectangles formed by the 12 polyominoes of order $n = 5$, as summarized in the following table (Beeler 1972).

Size	Solutions
3×20	2
4×15	368
5×12	1010
6×10	2339
$2\ 5 \times 6$	2
8×8 with 2×2 hole	65

See also DOMINO, FIBONACCI NUMBER, POLYHEX TILING, POLYIAMOND TILING, POLYOMINO

References

Beeler, M. Item 112 in Beeler, M.; Gosper, R. W.; and Schroeppel, R. *HAKMEM.* Cambridge, MA: MIT Artificial Intelligence Laboratory, Memo AIM-239, pp. 48–0, Feb. 1972.

Friedman, E. "Puzzle of the Month (February 1999)." http://www.stetson.edu/~efriedma/mathmagic/0299.html.

Gardner, M. "Tiling with Polyominoes, Polyiamonds, and Polyhexes." Ch. 14 in *Time Travel and Other Mathematical Bewilderments*. New York: W. H. Freeman, pp. 177–87, 1988.

Schroeppel, R. Item 109 in Beeler, M.; Gosper, R. W.; and Schroeppel, R. *HAKMEM*. Cambridge, MA: MIT Artificial Intelligence Laboratory, Memo AIM-239, p. 48, Feb. 1972.

Vichera, M. "Polyominoes." http://alpha.ujep.cz/~vicher/puzzle/polyform/minio/polynom.htm.

Weisstein, E. W. "Books about Polyominoes." http://www.treasure-troves.com/books/Polyominoes.html.

Polyplet

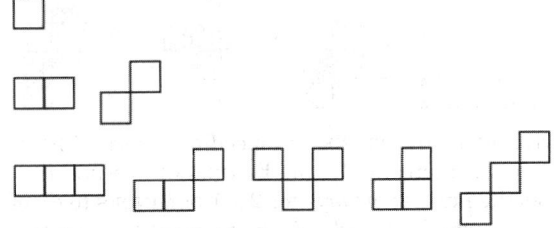

A POLYOMINO-like object made by attaching squares joined either at sides or corners. Because neighboring squares can be in relation to one another as KINGS may move on a CHESSBOARD, polyplets are sometimes also called POLYKINGS. The number of n-polyplets (with holes allowed) are 1, 2, 5, 22, 94, 524, 3031, ... (Sloane's A030222). The number of n-polyplets having bilateral symmetry are 1, 2, 4, 10, 22, 57, 131, ... (Sloane's A030234). The number of n-polyplets not having bilateral symmetry are 0, 0, 1, 12, 72, 467, 2900, ... (Sloane's A030235). The number of fixed n-polyplets are 1, 4, 20, 110, 638, 3832, ... (Sloane's A030232). The number of one-sided n-polyplets are 1, 2, 6, 34, 166, 991, ... (Sloane's A030233).

See also POLYIAMOND, POLYOMINO

References

Sloane, N. J. A. Sequences A030222, A030232, A030233, A030234, and A030235 in "An On-Line Version of the Encyclopedia of Integer Sequences." http://www.research.att.com/~njas/sequences/eisonline.html.

Polystigm

Lachlan's terms for a collection of n points.

See also POLYGRAM, TETRASTIGM

References

Lachlan, R. *An Elementary Treatise on Modern Pure Geometry*. London: Macmillian, p. 83, 1893.

Polytan

POLYABOLO

Polytopal Graph

A GRAPH G is called d-polytopal if there exists a d-dimensional CONVEX POLYTOPE P such that the vertices and edges of G are in a one-to-one inci-

dence-preserving correspondence with those of P. In other words G is d-polytopal IFF it is isomorphic to the 1-SKELETON of some convex d-polytopes P. If $d = 3$, the graph is called a POLYHEDRAL GRAPH.

See also POLYHEDRAL GRAPH

References

Grünbaum, B. "Polytopal Graphs." In *Studies in Graph Theory, Part II* (Ed. D. R. Fulkerson). Washington, DC: Math. Assoc. Amer., pp. 201–24, 1975.

Polytope

The word polytope is used to mean a number of related, but slightly different mathematica objects. A convex polytope may be defined as the CONVEX HULL of a finite set of points (which are always bounded), or as a *bounded* intersection of a finite set of half-spaces. Coxeter (1973, p. 118) defines polytope as the general term of the sequence "POINT, LINE SEGMENT, POLYGON, POLYHEDRON, ...," or more specifically as a finite region of n-dimensional space enclosed by a finite number of hyperplanes. The special name POLYCHORON is sometimes given to a 4-D polytope. However, in ALGEBRAIC TOPOLOGY, the UNDERLYING SPACE of a SIMPLICIAL COMPLEX is sometimes called a polytope (Munkres 1993, p. 8). The word "polytope" was introduced by Alicia Boole, the somewhat colorful daughter of logician George Boole (MacHale 1985).

The part of the polytope that lies in one of the bounding hyperplanes is called a cell. A 4-D polytope is sometimes called a POLYCHORON. Explicitly, a d-dimensional polytope may be specified as the set of solutions to a system of linear inequalities

$$\mathsf{m}\mathbf{x} \le \mathbf{b},$$

where m is a real $s \times d$ MATRIX and $\mathbf{b}$ is a real s-VECTOR. The positions of the vertices given by the above equations may be found using a process called VERTEX ENUMERATION.

A regular polytope is a generalization of the PLATONIC SOLIDS to an arbitrary DIMENSION. The regular polytopes were discovered before 1852 by the Swiss mathematician Ludwig Schläfli. For n-D with $n \ge 5$, there are only three regular convex polytopes: the HYPERCUBE, CROSS POLYTOPE, and regular SIMPLEX, which are analogs of the CUBE, OCTAHEDRON, and TETRAHEDRON (Coxeter 1969; Wells 1991, p. 210).

See also 16-CELL, 24-CELL, 120-CELL, 600-CELL, CROSS POLYTOPE, EDGE (POLYTOPE), FACE, FACET, HYPERCUBE, INCIDENCE MATRIX, LINE SEGMENT, PENTATOPE, POINT, POLYCHORON, POLYGON, POLYHEDRON, POLYTOPE STELLATIONS, PRIMITIVE POLYTOPE, RIDGE, SIMPLEX, TESSERACT, UNIFORM POLYCHORON, VERTEX (POLYHEDRON)

References

Bisztriczky, T.; McMullen, P., Schneider, R.; and Weiss, A. W. (Eds.). *Polytopes: Abstract, Convex, and Computational.* Dordrecht, Netherlands: Kluwer, 1994.

Coxeter, H. S. M. "Regular and Semi-Regular Polytopes I." *Math. Z.* **46**, 380–07, 1940.

Coxeter, H. S. M. *Introduction to Geometry, 2nd ed.* New York: Wiley, 1969.

Eppstein, D. "Polyhedra and Polytopes." http://www.ics.uci.edu/~eppstein/junkyard/polytope.html.

Fukuda, K. "Polytope Movie Page." http://www.ifor.math.ethz.ch/~fukuda/polymovie/polymovie.html.

MacHale, D. *George Boole: His Life and Work.* Dublin, Ireland: Boole, 1985.

Munkres, J. R. *Analysis on Manifolds.* Reading, MA: Addison-Wesley, 1991.

Sullivan, J. "Generating and Rendering Four-Dimensional Polytopes." *Mathematica J.* **1**, 76–5, 1991.

Weisstein, E. W. "Books about Polyhedra." http://www.treasure-troves.com/books/Polyhedra.html.

Wells, D. *The Penguin Dictionary of Curious and Interesting Geometry.* London: Penguin, 1991.

Polytope Stellations

There are 10 stellated regular 4-polytopes (Wells 1991, p. 209).

See also POLYTOPE, STELLATION

References

Wells, D. *The Penguin Dictionary of Curious and Interesting Geometry.* London: Penguin, 1991.

Polytropic Differential Equation

LANE-EMDEN DIFFERENTIAL EQUATION

Poncelet Transform

PONCELET TRANSVERSE

Poncelet Transverse

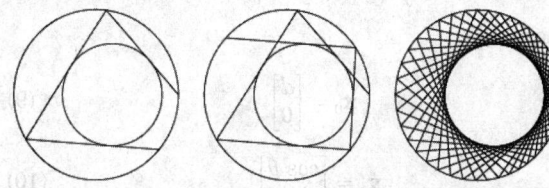

Let a CIRCLE C_1 lie inside another CIRCLE C_2. From any point on C_2, draw a tangent to C_1 and extend it to C_2. From the point, draw another tangent, etc. For n tangents, the result is called an n-sided Poncelet transverse.

If, on the circle of circumscription there is one point of origin for which a four-sided Poncelet transverse is closed, then the four-sided transverse will also close for any other point of origin on the circle (Dörrie 1965).

See also BICENTRIC POLYGON, BICENTRIC QUADRILATERAL, PONCELET'S PORISM

References

Dörrie, H. *100 Great Problems of Elementary Mathematics: Their History and Solutions.* New York: Dover, p. 192, 1965.

Poncelet's Closure Theorem

PONCELET'S PORISM

Poncelet's Coaxal Theorem

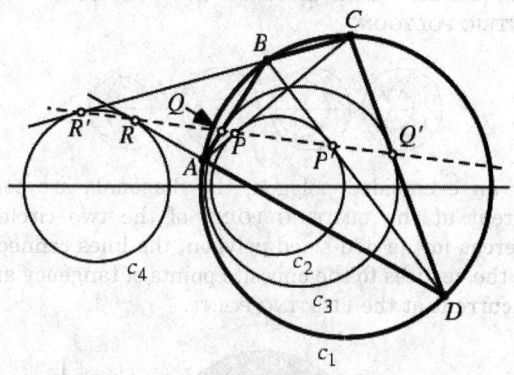

If a CYCLIC QUADRILATERAL $ABCD$ is inscribed in a circle c_1 of a COAXAL SYSTEM such that one pair AC of connectors touches another circle c_2 of the system at P, then each pair of opposite connectors will touch a circle of the system (BD at P' on c_2, AB at Q on c_3, CD at Q' on c_3, DA at R on c_4, and CB at R' on c_4), and the six points of contact P, P', Q, Q', R, and R' will be COLLINEAR.

The general theorem states that if $A_1, A_2, ..., A_n$ are any number of points taken in order on a CIRCLE of a give COAXAL SYSTEM so that $A_1A_2, A_2A_3, ..., A_{n-1}A_n$ touch respectively $n - 1$ fixed circles $X_1, X_2, ..., X_{n-1}$ of the system, then A_nA_1 must touch a fixed circle X_n of the system. Further, if $A_1A_2, A_2A_3, ..., A_{n-1}A_n$ touch respectively any $n - 1$ of the circles $X_1, X_2, ..., X_n$, then A_nA_1 must touch the remaining CIRCLE.

See also COAXAL SYSTEM

References

Lachlan, R. "Poncelet's Theorem." §334–42 in *An Elementary Treatise on Modern Pure Geometry.* London: Macmillian, pp. 209–17, 1893.

Poncelet's Continuity Principle

PERMANENCE OF MATHEMATICAL RELATIONS PRINCIPLE

Poncelet's Porism

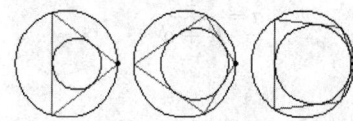

If an n-sided PONCELET TRANSVERSE constructed for two given CONIC SECTIONS is closed for one point of

origin, it is closed for any position of the point of origin. Specifically, given one ELLIPSE inside another, if there exists one CIRCUMINSCRIBED (simultaneously inscribed in the outer and circumscribed on the inner) n-gon, then any point on the boundary of the outer ELLIPSE is the vertex of some CIRCUMINSCRIBED n-gon. If the conic is taken as a circle (Casey 1888, pp. 124–26) , then a polygon which has both an incenter and a circumcenter (and for which the transveRsals would therefore close) is called a BI-CENTRIC POLYGON.

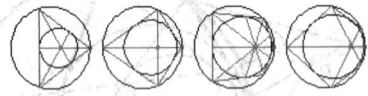

For an even-sided polygon, the diagonals are concurrent at the LIMITING POINT of the two circles, whereas for an odd-sided polygon, the lines connecting the vertices to the opposite points of tangency are concurrent at the LIMITING POINT.

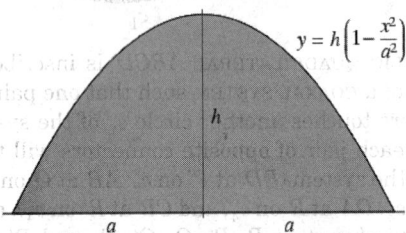

$$y = h\left(1 - \frac{x^2}{a^2}\right)$$

Inverting about either of the two LIMIT POINTS gives two concentric circles. However, the n-gonal sides become arcs of circles in the process, so this sort of simple INVERSION does not provide an automatic proof of the theorem (as happens in STEINER'S PORISM, for example).

Fuss (1792) derived formulas not only for the BI-CENTRIC QUADRILATERAL, but also the bicentric PEN-TAGON, HEXAGON, HEPTAGON, and OCTAGON, as did Steiner (Fuss 1792; Steiner 1827; Jacobi 1881; Dörrie 1965, p. 192). Chaundy (1923) exhibited porisms for $n = 3, 4, 5, 6, 7, 8, 9, 10, 12, 14, 16, 18, 20$, as well as erroneous expressions for several other values (Kerawala 1947). Richelot derived the expression for $n = 11$. In fact, there is a general analytic expression relating the CIRCUMRADIUS R, INRADIUS r, and offset between the CIRCUMCENTER and INCENTER d for a bicentric polygon. Given R, r, and d, define

$$a = \frac{1}{R + d} \tag{1}$$

$$b = \frac{1}{R - d} \tag{2}$$

$$c = \frac{1}{r}. \tag{3}$$

Now let

$$\lambda = 1 + \frac{2c^2(a^2 - b^2)}{a^2(b^2 - c^2)} \tag{4}$$

$$\omega = \cosh^{-1}\lambda, \tag{5}$$

and define the MODULUS as

$$k^2 = 1 - e^{-2\omega}. \tag{6}$$

Then the condition for an n-gon to be bicentric is

$$\mathrm{sc}\left(\frac{K(k)}{n}, k\right) = \frac{c\sqrt{b^2 - a^2} + b\sqrt{c^2 - a^2}}{a(b + c)}, \tag{7}$$

where $\mathrm{sc}(x, k)$ is a JACOBI ELLIPTIC FUNCTION and $K(k)$ is a complete ELLIPTIC INTEGRAL OF THE FIRST KIND (Richelot 1830, Kerawala 1947). Kerawala (1947) was able to establish many porisms in simple explicit form without resorting to the use of elliptic functions.

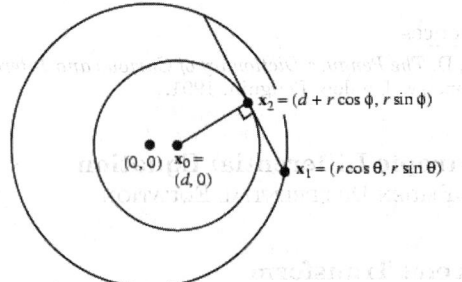

For the two circles illustrated above, the tangent on the inner circle can be determined by solving

$$(\mathbf{x}_2 - \mathbf{x}_1) \cdot (\mathbf{x}_2 - \mathbf{x}_0) = 0, \tag{8}$$

where

$$\mathbf{x}_0 = \begin{bmatrix} d \\ 0 \end{bmatrix} \tag{9}$$

$$\mathbf{x}_1 = \begin{bmatrix} \cos\theta \\ \sin\theta \end{bmatrix} \tag{10}$$

$$\mathbf{x}_2 = \begin{bmatrix} d + r\cos\phi \\ r\sin\phi \end{bmatrix}, \tag{11}$$

r is the radius of the inner circle, x is the offset of the inner circle, θ is the given position on the outer circle, and ϕ is the angle around the inner circle at which the tangent occurs. Taking the DOT PRODUCT and simplifying gives

$$r + d\cos\phi - \cos(\phi - \theta) = 0. \tag{12}$$

When this is solved for ϕ, the point at which the extension of this line intersects the outer circle again can be found using the standard equation of a CIRCLE-LINE INTERSECTION.

The degrees d_n of the algebraic equations relating a, b, and c for $n = 3, 4, ...,$ are 1, 2, 3, 4, 6, 8, 9, 12, 15, 16, 21, 24, 24, 32, 36, ... (Sloane's A002348; Kerawala 1947). Let the PRIME FACTORIZATION of n be written as

$$n = 2^{z_0} \prod_i p_i^{z_i}, \tag{13}$$

then d_n in general is given by

$$d_n = \frac{4^{z_0}}{8} \prod_i p_i^{2(\alpha_i - 1)}(p_i^2 - 1). \tag{14}$$

In the following expressions, write

$$e_0 \equiv a + b + c \tag{15}$$

$$e_1 \equiv -a + b + c \tag{16}$$

$$e_2 \equiv a - b + c \tag{17}$$

$$e_3 \equiv a + b - c \tag{18}$$

$$E_1 \equiv -a^2 + b^2 + c^2 \tag{19}$$

$$E_2 \equiv a^2 - b^2 + c^2 \tag{20}$$

$$E_3 \equiv a^2 + b^2 - c^2 \tag{21}$$

$$F_1 \equiv -E_2 E_3 + E_3 E_1 + E_1 E_2 \tag{22}$$

$$F_2 \equiv E_2 E_3 - E_3 E_1 + E_1 E_2 \tag{23}$$

$$F_3 \equiv E_2 E_3 + E_3 E_1 - E_1 E_2 \tag{24}$$

$$F_0 \equiv E_2 E_3 + E_3 E_1 + E_1 E_2 \equiv e_0 e_1 e_2 e_3 \tag{25}$$

$$g_0 \equiv E_1 E_2 E_3 + 2ab E_1 E_2 + 2bc E_2 E_3 + 2ca E_3 E_1 \tag{26}$$

$$g_1 \equiv E_1 E_2 E_3 - 2ab E_1 E_2 + 2bc E_1 E_2 - 2ca E_3 E_1 \tag{27}$$

$$g_2 \equiv E_1 E_2 E_3 - 2ab E_1 E_2 - 2bc E_2 E_3 + 2ca E_3 E_1 \tag{28}$$

$$g_3 \equiv E_1 E_2 E_3 + 2ab E_1 E_2 - 2bc E_2 E_3 - 2ca E_3 E_1 \tag{29}$$

following Kerawala (1947), and

$$p = \frac{R + d}{r} \tag{30}$$

$$q = \frac{R - d}{r} \tag{31}$$

following Richelot (1830).

The equation for a bicentric triangle ($n = 3$), i.e., any triangle, may be variously written as

$$a + b = c \tag{32}$$

$$(R + d)^{-1} + (R - d)^{-1} = r^{-1} \tag{33}$$

$$\sqrt{R - d - r} + \sqrt{R + d - r} = \sqrt{2R} \tag{34}$$

$$(p - 1)(q - 1) = 1 \tag{35}$$

(Richelot 1830),

$$R^2 - 2Rr - d^2 = 0 \tag{36}$$

(Steiner 1827; F. Gabriel-Marie 1912, pp. 497–01; Kerawala 1947; Altshiller-Court 1957, pp. 85–7; Wells 1991). The latter is sometimes known as the EULER TRIANGLE FORMULA.

For a BICENTRIC QUADRILATERAL ($n = 4$), the radii and offset are connected by the equation

$$a^2 + b^2 = c^2, \tag{37}$$

(Kerawala 1947), which expands to

$$\frac{1}{(R - d)^2} + \frac{1}{(R + d)^2} = \frac{1}{r^2} \tag{38}$$

(Davis; Durége; Casey 1888, pp. 109–10; F. Gabriel-Marie 1912, pp. 321 and 814–16; Johnson 1929; Dörie 1965). This can also be written

$$(R^2 - d^2)^2 = 2r^2(R^2 + d^2), \tag{39}$$

$$(R + r + d)(R + r - d)(R - r + d)(R - r - d) = r^4 \tag{40}$$

(Steiner 1827), or

$$(p^2 - 1)(q^2 - 1) = 1 \tag{41}$$

(Richelot 1830).

The relationship for a bicentric PENTAGON ($n = 5$) is

$$r(R - d) = (R + d)\sqrt{(R - r + d)(R - r - d)}$$

$$+ (R + d)\sqrt{2R(R - r - d)} \tag{42}$$

(Steiner 1827) or

$$4p^2 q^2(p - 1)(q - 1) = (p^2 + q^2 - p^2 q^2)^2 \tag{43}$$

(Richelot 1830). A number of alternative forms are given by

$$(a + b)(b + c)(c + a) = a^3 + b^3 + c^3 \tag{44}$$

$$(a + b + c)^3 = 4(a^3 + b^3 + c^3) \tag{45}$$

$$(-a + b + c)(a - b + c)(a + b - c) + 4abc = 0 \tag{46}$$

$$\begin{vmatrix} e_0 & e_3 & e_2 \\ e_3 & e_0 & e_1 \\ e_2 & e_1 & e_0 \end{vmatrix} = 0 \tag{47}$$

$$a(-a^2 + b^2 + c^2) + b(a^2 - b^2 + c^2) + c(a^2 + b^2 - c^2)$$

$$+ 2abc = 0, \tag{48}$$

and

$$e_0^{-1} + e_1^{-1} + e_2^{-1} + e_3^{-1} = 0 \qquad (49)$$

(Kerawala 1947).

For $n = 6$,

$$3(R^2 - d^2)^4 = 4r^2(R^2 + d^2)(R^2 - d^2) + 16r^4 d^2 R^2 \qquad (50)$$

(Steiner 1827),

$$4p^2 q^2 (p^2 - 1)(q^2 - 1) = (p^2 + q^2 - p^2 q^2)^2 \qquad (51)$$

(Richelot 1830),

$$F_3 = 0, \qquad (52)$$

or

$$E_1^{-1} + E_2^{-1} + E_3^{-1} \qquad (53)$$

(Kerawala 1947).

For $n = 7$,

$$g_3 = 0 \qquad (54)$$

(Jacobi 1881, Kerawala 1947)

For $n = 8$,

$$E_1^{-2} + E_2^{-2} = E_3^{-2} \qquad (55)$$

(Kerawala 1947), which can also be written in the form

$$16p^4 q^4 (p^2 - 1)(q^2 - 1) = (p^2 + q^2 - p^2 q^2)^4, \qquad (56)$$

(Richelot 1830, Jacobi 1881). The equation given by Steiner (1827) contains (at least one) typographical error.

For $n = 9$,

$$aF_2 F_3 + bF_3 F_1 - cF_1 F_2 = 0. \qquad (57)$$

for $n = 10$,

$$16p^2 q^2 (p^2 - 1)(q^2 - 1)[p^4 q^4 - (p^2 - q^2)^2]^2$$
$$= \{[p^4 - (p^2 q^2 - q^2)^2] + [q^4 - (p^2 q^2 - p^2)^2]^2$$
$$+ [p^4 q^4 - (p^2 - q^2)^2]\}^2 \qquad (58)$$

(Richelot 1989).

For $n = 12$,

$$64p^4 q^4 (p^2 - 1)(q^2 - 1)[p^4 q^4 - (p^2 - q^2)^2]^2$$
$$= \{[p^4 - (p^2 q^2 - q^2)^2] + [q^4 - (p^2 q^2 - p^2)^2]^2$$
$$+ [p^4 q^4 - (p^2 - q^2)^2]\}^2 \qquad (59)$$

(Richelot 1989).

For $n = 14$,

$$g_1 = 0. \qquad (60)$$

For $n = 16$,

$$E_2^{-2} + E_3^{-2} = E_1^{-2}, \qquad (61)$$

(Kerawala 1947) or

$$64p^4 q^4 (p^2 - 1)(q^2 - 1)\{p^4 q^4 - (p^2 - q^2)^2]$$
$$\times (p^2 + q^2 - p^2 q^2)\}^4$$
$$= \{[p^4 - (p^2 q^2 - q^2)^2] + [q^4 - (p^2 q^2 - p^2)^2]^2$$
$$+ [p^4 q^4 - (p^2 - q^2)^2]^2\}^4 \qquad (62)$$

(Richelot 1830).

Weill (1878) gives an algorithm for finding approximate solutions (d, r, R) for porisms with even n. The following table gives the approximate relations for fixed $R \ll 1$.

n	d/R	r/R	error
6	$\frac{1}{2}$	$\frac{3}{4}$	$\frac{243}{128} R^8$
8	$\frac{1}{4}$	$\frac{15}{4} r$	$\frac{2955538440751415296}{6568408355712890625} R^{16}$
10	$\frac{1}{10}\sqrt{10}$	$\frac{9}{40}\sqrt{10}$	

See also Bicentric Polygon, Bicentric Quadrilateral, Billiards, Circle-Line Intersection, Collinear, Cyclic Quadrilateral, Euler Triangle Formula, Poncelet Transverse, Triquetra, Weill's Theorem

References

Allanson, B. "Bicentric Polygons" java applet. http://www.a-delaide.net.au/~allanson/bimovie.html.

Appell, P. and Lacour, E. *Principes de la théorie des fonctions elliptiques et applications.* Paris: Gauthier-Villars, pp. 138–39 and 227–43, 1922.

Barth, W. and Bauer, T. "Poncelet Theorems." *Expos. Math.* **14**, 125–44, 1996.

Barth, W. and Michel, J. "Modular Curves and Poncelet Polygons." *Math. Ann.* **295**, 25–9, 1993.

Bos, H. J. M.; Kers, C.; Oort, F.; and Raven, D. W. "Poncelet's Closure Theorem, Its History, Its Modern Formulation, a Comparison of Its Modern Proof with Those by Poncelet and Jacobi, and Some Mathematical Remarks Inspired by These Early Proofs." *Expos. Math.* **5**, 289–64, 1987.

Casey, J. *A Sequel to the First Six Books of the Elements of Euclid, Containing an Easy Introduction to Modern Geometry with Numerous Examples,* 5th ed., rev. enl. Dublin: Hodges, Figgis, & Co., 1888.

Cayley, A. *Philos. Mag.* **5**, 281–84, 1853.

Cayley, A. *Philos. Mag.* **6**, 99–02, 1853.

Cayley, A. "Developments on the Porism of the In-and-Circumscribed Polygon." *Philos. Mag.* **7**, 339–45, 1854.

Cayley, A. *Phil. Trans. Roy. Soc. London* **151**, 225–39, 1861.

Chaundy, T. W. *Proc. London Math. Soc.* **22**, 104–23, 1923.

Chaundy, T. W. *Proc. London Math. Soc.* **25**, 17–4, 1926.

Clifford, W. K. *Proc. London Math. Soc.* **7**, 29–8.

Clifford, W. K. *Proc. London Math. Soc.* **7**, 225–33.

Clifford, W. K. *Proc. Cambridge Phil. Soc.*, 120–23, 1868.

Darboux, G. *Comte Rendus de l'Acadamie de Sciences* **90**, 1880.

Darboux, G. *Principles de géométrie analytique, Vol. 3.* Paris, pp. 250–87, 1917.

Davis, M. A. *Educ. Times* **32**.

Dörrie, H. *100 Great Problems of Elementary Mathematics: Their History and Solutions.* New York: Dover, pp. 192–93, 1965.

Durége. *Theorie der Elliptischen Functionen.* p. 185.

Fuss, N. *Nova Acta Petropol.* **10**, 1792.

Fuss, N. "De Polygonis symmetrice irregularibus circulo simul inscriptis et circumscriptis." *Nova Acta Petropol.* **13**, 166–89, 1798.

F. Gabriel-Marie. *Exercices de Géométrie.* Tours, France: Maison Mame, 1912.

Griffiths, P. and Harris, J. "A Poncelet Theorem in Space." *Comment. Math. Helv.* **52**, 145–60, 1977.

Griffiths, P. and Harris, J. "On Cayley's Explicit Solution to Poncelet's Porism." *Enseign. Math.* **24**, 31–0, 1978.

Hart. *Quart. J. Math.*, 1857.

Jacobi, C. G. J. "Ueber die Anwendung der elliptischen Transcendenten auf ein bekanntes Problem der Elementargeometrie." *J. reine angew. Math.* **3**, 376–87, 1823. Reprinted in *Gesammelte Werke, Vol. 1.* Providence, RI: Amer. Math. Soc., pp. 278–93, 1969.

Johnson, R. A. *Modern Geometry: An Elementary Treatise on the Geometry of the Triangle and the Circle.* Boston, MA: Houghton Mifflin, pp. 91–6, 1929.

Kerawala, S. M. "Poncelet Porism in Two Circles." *Bull. Calcutta Math. Soc.* **39**, 85–05, 1947.

Lebesgue, H. "Polygones de Poncelet." Ch. 4 in *Les Coniques.* Paris: Gauthier-Villars, pp. 115–49, 1955. Reprint of "Exposé gémoétrique d'un mémoire de Cayley sur les Polygones de Poncelet." *Ann. de la Faculté des Sci. de l'Université de Toulouse* **14**, 1922.

Lelieuvre, A. "Sur les polygones de Poncelet." *L'enseign. math.* **2**, 410–23, 1900.

Lelieuvre, A. "Sur les polygones de Poncelet." *L'enseign. math.* **3**, 115–17, 1901.

Moutard, M. "Recherches analytiques sur les polygones simultanément inscrits et circonscrits à deux coniques." Appendix to Poncelet, J. V. *Traité des propriétés projectives des figures: ouvrage utile à qui s'occupent des applications de la géométrie descriptive et d'opérations géométriques sur le terrain, Vol. 1, 2nd ed.* Paris: Gauthier-Villars, pp. 535–60, 1865–6.

Poncelet, J. V. *Traité des propriétés projectives des figures: ouvrage utile à qui s'occupent des applications de la géométrie descriptive et d'opérations géométriques sur le terrain, Vols. 1–, 2nd ed.* Paris: Gauthier-Villars, 1865–6.

Previato, E. "Poncelet's Theorem in Space." *Proc. Amer. Math. Soc.* **127**, 2547–556, 1999.

Richelot, F. J. "Anwendung der elliptischen Transcendenten auf die sphärischen Polygone; welche zugleich einem kleinen Kreise der Kugel eingeschrieben und einem andern umgeschrieben sind." *J. reine angew. Math.* **5**, 250–67, 1830.

Richelot. *J. reine angew. Math.* **38**, p. 353.

Rosanes, J. and Pasch, M. "Über das einem Kegelschnitte umbeschriebene und einem andern einbeschriebene Polygon." *J. reine angew. Math.* **64**, 126–66, 1865.

Rosanes, J. and Pasch, M. "Über eine algebraische Aufgabe, welche einer Gattung geometrischer Probleme zu Grunde liegt." *J. reine angew. Math.* **70**, 169–73, 1869.

Sloane, N. J. A. Sequences A002348/M0549 in "An On-Line Version of the Encyclopedia of Integer Sequences." http://www.research.att.com/~njas/sequences/eisonline.html.

Steiner, J. §26.57 in "Aufgaben und Lehrsätze, erstere aufzulösen, leztere zu beweisen." *J. reine angew. Math.* **2**, 289, 1827.

Titchmarsh, E. C. *Messenger Math.* **52**, 42, 1922.

Weill, M. and Bützberger. "Sur les polygones inscrits et circonscrits à la fois à deux cercle." *Journal de Liouville,* 3me série **4**, 7–2, 1878.

Weill, M. "Sur une classe de polygones de Poncelet." *Bull. de la Soc. Math. France* **29**, 199–08, 1901.

Wells, D. *The Penguin Dictionary of Curious and Interesting Geometry.* New York: Viking Penguin, pp. 192–93, 1992.

Poncelet-Steiner Theorem

All Euclidean GEOMETRIC CONSTRUCTIONS can be carried out with a STRAIGHTEDGE alone if, in addition, one is given the RADIUS of a single CIRCLE and its center. The theorem was suggested by Poncelet in 1822 and proved by Steiner in 1833. A construction using STRAIGHTEDGE alone is called a STEINER CONSTRUCTION.

See also GEOMETRIC CONSTRUCTION, STEINER CONSTRUCTION

References

Dörrie, H. "Steiner's Straight-Edge Problem." §34 in *100 Great Problems of Elementary Mathematics: Their History and Solutions.* New York: Dover, pp. 165–70, 1965.

Steiner, J. *Geometric Constructions with a Ruler, Given a Fixed Circle with Its Center.* New York: Scripta Mathematica, 1950.

Pong Hau K'i

A Chinese TIC-TAC-TOE-like game.

See also TIC-TAC-TOE

References

Evans, R. "Pong Hau K'i." *Games and Puzzles* **53**, 19, 1976.

Straffin, P. D. Jr. "Position Graphs for Pong Hau K'i and Mu Torere." *Math. Mag.* **68**, 382–86, 1995.

Pons Asinorum

An elementary theorem in geometry whose name means "asses' bridge," perhaps in reference to the fact that fools would be unable to pass this point in their geometric studies. The theorem states that the ANGLES at the base of an ISOSCELES TRIANGLE (defined as a TRIANGLE with two legs of equal length) are equal and appears as the fifth proposition in Book I of Euclid's *ELEMENTS*.

See also ISOSCELES TRIANGLE, PYTHAGOREAN THEOREM

References

Dunham, W. *Journey through Genius: The Great Theorems of Mathematics.* New York: Wiley, p. 38, 1990.

Wells, D. *The Penguin Dictionary of Curious and Interesting Geometry.* London: Penguin, pp. 193–94, 1991.

Pontryagin Class

The ith Pontryagin class of a VECTOR BUNDLE is $(-1)^i$ times the ith CHERN CLASS of the complexification of the VECTOR BUNDLE. It is also in the $4i$th cohomology group of the base SPACE involved.

See also CHERN CLASS, STIEFEL-WHITNEY CLASS

Pontryagin Duality

Let G be a locally compact ABELIAN GROUP. Let G^* be the group of all homeomorphisms $G \to R/Z$, in the compact open topology. Then G^* is also a locally compact ABELIAN GROUP, where the asterisk defines a contravariant equivalence of the category of locally compact Abelian groups with itself. The natural mapping $G \to (G^*)^*$, sending g to G, where $G(f) = f(g)$, is an isomorphism and a HOMEOMORPHISM. Under this equivalence, compact groups are sent to discrete groups and vice versa.

See also ABELIAN GROUP, HOMEOMORPHISM

Pontryagin Maximum Principle

A result in CONTROL THEORY. Define

$$H(\psi, x, u) \equiv (\psi, f(x, u)) \equiv \sum_{a=0}^{n} \psi_a f^a(x, u).$$

Then in order for a control $u(t)$ and a trajectory $x(t)$ to be optimal, it is NECESSARY that there exist NONZERO absolutely continuous vector function $\psi(t) = (\psi_0(t), \psi_1(t), \ldots, \psi_n(t))$ corresponding to the functions $u(t)$ and $x(t)$ such that

1. The function $H(\psi(t), x(t), u)$ attains its maximum at the point $u = u(t)$ almost everywhere in the interval $t_0 \leq t \leq t_1$,

$$H(\psi(t), x(t), u(t)) = \max_{u \in U} H(\psi(t), x(t), u).$$

2. At the terminal time t_1, the relations $\psi_0(t_1) \leq 0$ and $H(\psi(t_1), x(t_1), u(t_1)) = 0$ are satisfied.

See also CONTROL THEORY

References
Iyanaga, S. and Kawada, Y. (Eds.). "Pontrjagin's [sic] Maximum Principle." §88C in *Encyclopedic Dictionary of Mathematics*. Cambridge, MA: MIT Press, pp. 295–96, 1980.

Pontryagin Number

The Pontryagin number is defined in terms of the PONTRYAGIN CLASS of a MANIFOLD as follows. For any collection of PONTRYAGIN CLASSES such that their cup product has the same DIMENSION as the MANIFOLD, this cup product can be evaluated on the MANIFOLD's FUNDAMENTAL CLASS. The resulting number is called the Pontryagin number for that combination of Pontryagin classes. The most important aspect of Pontryagin numbers is that they are COBORDISM invariant. Together, Pontryagin and STIEFEL-WHITNEY NUMBERS determine an oriented manifold's oriented COBORDISM class.

See also CHERN NUMBER, STIEFEL-WHITNEY NUMBER

Ponzo's Illusion

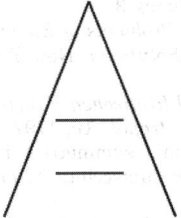

The upper HORIZONTAL line segment in the above figure appears to be longer than the lower line segment despite the fact that both are the same length.

See also ILLUSION, MÜLLER-LYER ILLUSION, POGGENDORFF ILLUSION, VERTICAL-HORIZONTAL ILLUSION

References
Fineman, M. *The Nature of Visual Illusion.* New York: Dover, p. 153, 1996.

Pop

An action which removes a single element from the top of a QUEUE or STACK, turning the LIST $(a_1, a_2, \ldots, a_n)$ into $(a_2, \ldots, a_n)$ and yielding the element a_1.

See also PUSH, STACK

Population

The word population has a number of distinct but closely related meanings in statistics.

1. A finite and actually existing group of objects which, although possibly large, can be enumerated in theory (e.g., people living in the United States).
2. A generalization from experience which is indefinitely large (e.g., the total number of throws that might conceivably by made in unlimited time with a particular pair of dice). Any actual set of throws can then be regarded as a SAMPLE drawn from this practically infinite population.
3. A purely hypothetically population which can be completely described mathematically.

See also SAMPLE

References
Kenney, J. F. and Keeping, E. S. "Populations and Samples." §7.1 in *Mathematics of Statistics, Pt. 1, 3rd ed.* Princeton, NJ: Van Nostrand, pp. 90–1, 1962.

Population Comparison

Let x_1 and x_2 be the number of successes in variates taken from two populations. Define

$$\hat{p}_1 \equiv \frac{x_1}{n_1} \tag{1}$$

$$\hat{p}_2 \equiv \frac{x_2}{n_2} \tag{2}$$

The ESTIMATOR of the difference is then $\hat{p}_1 - \hat{p}_2$. Doing a Z-TRANSFORM,

$$z = \frac{(\hat{p}_1 - \hat{p}_2) - (p_1 - p_2)}{\sigma_{\hat{p}_1 - \hat{p}_2}}, \tag{3}$$

where

$$\sigma_{\hat{p}_1 - \hat{p}_2} \equiv \sqrt{\sigma_{\hat{p}_1}^2 - \sigma_{\hat{p}_2}^2}. \tag{4}$$

The STANDARD ERROR is

$$SE_{\hat{p}_1 - \hat{p}_2} = \sqrt{\frac{\hat{p}_1(1 - \hat{p}_1)}{n_1} + \frac{\hat{p}_2(1 - \hat{p}_2)}{n_2}} \tag{5}$$

$$SE_{\bar{x}_1 - \bar{x}_2} = \sqrt{\frac{s_1^2}{n_1} + \frac{s_2^2}{n_2}} \tag{6}$$

$$s_{pool}^2 = \frac{(n_1 - 1)s_1^2 + (n_2 - 1)s_2^2}{n_1 + n_2 - 2}. \tag{7}$$

See also Z-TRANSFORM (POPULATION)

References
Gonick, L. and Smith, W. *The Cartoon Guide to Statistics*. New York: Harper Perennial, pp. 162–71, 1993.

Population Growth
The differential equation describing exponential growth is

$$\frac{dN}{dt} = \frac{N}{\tau}. \tag{1}$$

This can be integrated directly

$$\int_{N_0}^{N} \frac{dN}{N} = \int_0^t \frac{dt}{\tau} \tag{2}$$

$$\ln\left(\frac{N}{N_0}\right) = \frac{t}{\tau}. \tag{3}$$

Exponentiating,

$$N(t) = N_0 e^{t/\tau} \tag{4}$$

Defining $N(t = 1) = N_0 e^\alpha$ gives $\tau = 1/\alpha$ in (4), so

$$N(t) = N_0 e^{\alpha t}. \tag{5}$$

This equation is called the LAW OF GROWTH, and the quantity α in this equation is sometimes known as the MALTHUSIAN PARAMETER.

Consider a more complicated growth law

$$\frac{dN}{dt} = \left(\frac{\alpha t - 1}{t}\right) N, \tag{6}$$

where $\alpha > 1$ is a constant. This can also be integrated directly

$$\frac{dN}{N} = \left(\alpha - \frac{1}{t}\right) dt \tag{7}$$

$$\ln N = \alpha t - \ln t + C \tag{8}$$

$$N(t) = \frac{C e^{\alpha t}}{t}. \tag{9}$$

Note that this expression blows up at $t = 0$. We are given the INITIAL CONDITION that $N(t = 1) = N_0 e^\alpha$, so $C = N_0$.

$$N(t) = N_0 \frac{e^{\alpha t}}{t}. \tag{10}$$

The t in the DENOMINATOR of (10) greatly suppresses the growth in the long run compared to the simple growth law.

The LOGISTIC GROWTH CURVE, defined by

$$\frac{dN}{dt} = \frac{r(K - N)}{N} \tag{11}$$

is another growth law which frequently arises in biology. It has a rather complicated solution for $N(t)$.

See also GOMPERTZ CURVE, GROWTH, LAW OF GROWTH, LIFE EXPECTANCY, LOGISTIC GROWTH CURVE, LOTKA-VOLTERRA EQUATIONS, MAKEHAM CURVE, MALTHUSIAN PARAMETER, SURVIVORSHIP CURVE

References
Steinhaus, H. *Mathematical Snapshots, 3rd ed.* New York: Dover, pp. 290–95, 1999.

Porism
An archaic type of mathematical proposition whose historical purpose is not entirely known. In modern usage, the term "porism" is used instead of "theorem" for a small number of results for historical reasons.

See also AXIOM, LEMMA, POSTULATE, PONCELET'S PORISM, PRINCIPLE, STEINER'S PORISM, THEOREM

Porous Medium Equation
The PARTIAL DIFFERENTIAL EQUATION

$$u_t = \nabla \cdot (u^m \nabla u).$$

Porter's Constant

References

Elliott, C. M.; Herrero, M. A.; King, J. R.; and Ockendon, J. R. "The Mesa Problem: Diffusion Patterns for $u_t = \nabla \cdot (u^m \nabla u)$ as $m \to +\infty$." *IMA J. Appl. Math.* **7**, 147–54, 1986.
Zwillinger, D. *Handbook of Differential Equations, 3rd ed.* Boston, MA: Academic Press, p. 134, 1997.

Porter's Constant

N.B. A detailed online essay by S. Finch was the starting point for this entry. The constant appearing in FORMULAS for the efficiency of the EUCLIDEAN ALGORITHM,

$$C = \frac{6 \ln 2}{\pi^2} \left[3 \ln 2 + 4\gamma - \frac{24}{\pi^2} \zeta'(2) - 2 \right] - \frac{1}{2}$$

$$= 1.4670780794 \ldots,$$

where γ is the EULER-MASCHERONI CONSTANT and $\zeta(z)$ is the RIEMANN ZETA FUNCTION.

See also EUCLIDEAN ALGORITHM

References

Finch, S. "Favorite Mathematical Constants." http://www.mathsoft.com/asolve/constant/porter/porter.html.
Porter, J. W. "On a Theorem of Heilbronn." *Mathematika* **22**, 20–8, 1975.

Pósa's Conjecture

Dirac (1952) proved that if the minimum VERTEX DEGREE $\delta(G) \geq n/2$ for a graph G on $n \geq 3$ nodes, then G contains a HAMILTONIAN CIRCUIT (Bollobás 1978, Komlós *et al.* 1998).

In 1962, Pósa conjectured that $G(V, E)$ contains a square of a HAMILTONIAN CIRCUIT if $\delta(G) \geq 2n/3$ (Erdos 1964, p. 159; Komlós *et al.* 1998), where a graph $G(V, E)$ contains the SQUARE of a HAMILTONIAN CIRCUIT if there is a HAMILTONIAN CIRCUIT $H = (x_1, x_2, \ldots, x_n, x_{n+1} = x_1)$ such that $(x_i, x_{i+2}) \in E(G)$, for $i = 1, 2, \ldots, n$.

Komlós *et al.* (1996) proved that there exists a natural number n_0 such that if a graph G has order $n \geq n_0$ and minimum degree at least $2n/3$, then G contains the square of a Hamiltonian circuit. This proved Pósa's conjecture (Erdos 1964) for sufficiently large n. Kierstead and Quintana (1998) proved Pósa's conjecture for graphs G containing a 4-clique K_4.

The conjecture was generalized by Seymour (1974) to state that if $\delta(G) \geq kn/(k+1)$, then G contains the kth power of a HAMILTONIAN CIRCUIT (Komlós *et al.* 1998).

See also HAMILTONIAN CIRCUIT, PÓSA'S CONJECTURE, SEYMOUR CONJECTURE

References

Dirac, G. A. "Some Theorems on Abstract Graphs." *Proc. London Math. Soc.* **2**, 69–1, 1952.
Erdos, P. "Problem 9." In *Theory of Graphs and Its Applications, Proceedings of the Symposium held in Smolenice in June 1963* (Ed. M. Fiedler). Prague, Czechoslovakia: Publishing House of the Czechoslovak Academy of Sciences, p. 159, 1964.
Fan, G. and Kierstead, H. A. "Hamiltonian Square-Paths." *J. Combin. Theory Ser. B* **67**, 167–82, 1996.
Kierstead, H. A. and Quintana, J. "Square Hamiltonian Cycles in Graphs with Maximal 4-Cliques." *Disc. Math.* **178**, 81–2, 1998.
Komlós, J.; Sárkozy, G. N.; and Szemerédi, E. "On the Square of a Hamiltonian Cycle in Dense Graphs." In *Random Structures Algorithms* **9**, 193–11, 1996.
Seymour, P. Problem Section in *Combinatorics: Proceedings of the British Combinatorial Conference, 1973* (Ed. T. P. McDonough and V. C. Mavron). Cambridge, England: Cambridge University Press, pp. 201–02, 1974.

Pósa's Theorem

There are several related theorems involving HAMILTONIAN CIRCUITS of graphs that are associated with Pósa.

Let G be a SIMPLE GRAPH with n VERTICES.

1. If, for every k in $1 \leq k < (n-1)/2$, the number of VERTICES of VERTEX DEGREE not exceeding k is less than k, and
2. If, for n ODD, the number of VERTICES with VERTEX DEGREE not exceeding $(n-1)/2$ is less than or equal to $(n-1)/2$,

then G contains a HAMILTONIAN CIRCUIT.

Kronk (1969) generalized this result as follows. Let G be a SIMPLE GRAPH with n VERTICES, and let $0 \leq k \leq n - 2$. Then the following conditions are SUFFICIENT for G to be k-line Hamiltonian:

1. For all integers j with $k + 1 \leq j < (n + k - 1)/2$, the number of VERTICES of VERTEX DEGREE not exceeding j is less than $j - k$,
2. The number of points of degree not exceeding $(n + k - 1)/2$ does not exceed $(n - k - 1)/2$.

Pósa (1963) generalized a result of Dirac by proving that every FINITE SIMPLE GRAPH G with a sufficiently large valencies of all (or, in some cases, of ALMOST ALL) vertices and with a sufficiently large number of vertices satisfies one of the following conditions.

1. G has a Hamiltonian line containing all edges of given disjoint paths (Theorem 1),
2. G has a circuit with a "large" number of vertices (Theorems 2 and 3), or
3. G has a "small" number of disjoint circuits containing all vertices of the graph (Theorems 4 and 5).

References

Bollobás, B. *Extremal Graph Theory.* New York: Academic Press, 1978.

Bondy, J. A. "Cycles in Graphs." In *Combinatorial Structures and their Applications (Proc. Calgary Internat. Conf., Calgary, Alta., 1969)*. New York: Gordon and Breach, pp. 15–8, 1970.

Dirac, G. A. "Some Theorems on Abstract Graphs." *Proc. London Math. Soc.* **2**, 69–1, 1952.

Komlós, J.; Sárkozy, G. N.; and Szemerédi, E. "Proof of the Seymour Conjecture for Large Graphs." *Ann. Comb.* **2**, 43–0, 1998.

Kronk, H. V. "Variations on a Theorem of Pósa." In *The Many Facets of Graph Theory (Proc. Conf., Western Mich. Univ., Kalamazoo, Mich., 1968)*. Berlin: Springer-Verlag, pp. 193–97, 1969.

Lick, D. R. "*n*-Hamiltonian Connected Graphs." *Duke Math. J.* **37**, 387–92, 1970.

Marshall, C. W. *Applied Graph Theory*. New York: Wiley, 1971.

Nash-Williams, C. St. J. A. "Hamiltonian Lines in Graphs Whose Vertices Have Sufficiently Large Valencies." In *Combinatorial Theory and Its Applications, III (Proc. Colloq., Balatonfüred, 1969)*. Amsterdam, Netherlands: North-Holland, pp. 813–19, 1970.

Nash-Williams, C. St. J. A. "Hamiltonian Lines in Infinite Graphs with Few Vertices of Small Valency." *Aequationes Math.* **7**, 59–1, 1971.

Pósa, L. "On the Circuits of Finite Graphs." *Magyar Tud. Akad. Mat. Kutató Int. Kozl.* **8**, 355–61, 1963.

Pöschl-Teller Differential Equations

The first and second Pöschl-Teller differential equations are given by

$$y'' - \left\{ a^2 \left[\frac{\kappa(\kappa - 1)}{\sin^2(ax)} + \frac{\lambda(\lambda - 1)}{\cos^2(ax)} \right] - b^2 \right\} y = 0$$

and

$$y'' - \left\{ a^2 \left[\frac{\kappa(\kappa - 1)}{\sinh^2(ax)} + \frac{\lambda(\lambda - 1)}{\cosh^2(ax)} \right] - b^2 \right\} y = 0$$

respectively.

References

Barut, A. O.; Inomata, A.; and Wilson, R. "Algebraic Treatment of Second Pöschl-Teller, Morse-Rosen, and Eckart Equations." *J. Phys. A: Math. Gen.* **20**, 4083–4096, 1987.

Zwillinger, D. *Handbook of Differential Equations, 3rd ed.* Boston, MA: Academic Press, p. 126, 1997.

Poset

PARTIALLY ORDERED SET

Poset Dimension

The DIMENSION of a POSET $P = (X, \leq)$ is the size of the smallest REALIZER of P. Equivalently, it is the smallest INTEGER d such that P is ISOMORPHIC to a DOMINANCE order in $\mathbb{R}^d$.

See also DIMENSION, DOMINANCE, ISOMORPHIC POSETS, REALIZER

References

Dushnik, B. and Miller, E. W. "Partially Ordered Sets." *Amer. J. Math.* **63**, 600–10, 1941.

Trotter, W. T. *Combinatorics and Partially Ordered Sets: Dimension Theory*. Baltimore, MD: Johns Hopkins University Press, 1992.

Position Four-Vector

The CONTRAVARIANT FOUR-VECTOR arising in special and general relativity,

$$x^\mu = \begin{bmatrix} x^0 \\ x^1 \\ x^2 \\ x^3 \end{bmatrix} \equiv \begin{bmatrix} ct \\ x \\ y \\ z \end{bmatrix},$$

where c is the speed of light and t is time. Multiplication of two four-vectors gives the spacetime interval

$$I = g_{\mu\nu} x^\mu x^\nu = (x^0)^2 - (x^1)^2 - (x^2)^2 - (x^3)^2$$

$$= (ct)^2 - (x^1)^2 - (x^2)^2 - (x^3)^2$$

See also FOUR-VECTOR, LORENTZ TRANSFORMATION, QUATERNION

Position Vector

RADIUS VECTOR

Positive

A quantity $x > 0$, which may be written with an explicit PLUS SIGN for emphasis, $+x$.

See also NEGATIVE, NONNEGATIVE, PLUS SIGN, ZERO

Positive Definite Function

A positive definite FUNCTION f on a GROUP G is a FUNCTION for which the MATRIX $\{f(x_i x_j^{-1})\}$ is always POSITIVE SEMIDEFINITE Hermitian.

References

Knapp, A. W. "Group Representations and Harmonic Analysis, Part II." *Not. Amer. Math. Soc.* **43**, 537–49, 1996.

Positive Definite Matrix

A HERMITIAN MATRIX $\mathbf{A}$ is called positive definite if

$$(\mathbf{Av}) \cdot \mathbf{v} > 0 \qquad (1)$$

for all VECTORS $\mathbf{v} \neq 0$. This is equivalent to the requirement that all EIGENVALUES be POSITIVE, and to the requirement that the DETERMINANTS associated with *all* upper-left SUBMATRICES are POSITIVE.

The DETERMINANT of a positive definite matrix is POSITIVE, but the converse is not necessarily true (i.e., a matrix with a POSITIVE DETERMINANT is not necessarily positive definite).

The numbers of positive definite $n \times n$ matrices of given types are summarized in the following table. For example, the three positive definite 2×2 (0,1)-

MATRICES are

$$\begin{bmatrix} 1 & 0 \\ 0 & 1 \end{bmatrix}, \begin{bmatrix} 1 & 0 \\ 1 & 1 \end{bmatrix}, \begin{bmatrix} 1 & 1 \\ 0 & 1 \end{bmatrix}, \quad (2)$$

all of which have eigenvalue 1 with degeneracy of two.

| (0, 1)-matrix | A000000 | 0, 3, 25, 543, ... |
| (−1, 0, 1)-matrix | A000000 | 0, 5, 133, ... |

A REAL SYMMETRIC MATRIX A is positive definite IFF there exists a REAL nonsingular MATRIX M such that

$$\mathsf{A} = \mathsf{MM}^{\mathsf{T}} \quad (3)$$

where M^{T} is the TRANSPOSE. A 2×2 SYMMETRIC MATRIX

$$\begin{bmatrix} a & b \\ b & c \end{bmatrix} \quad (4)$$

is positive definite if

$$a v_1^2 + 2 b v_1 v_2 + c v_2^2 > 0 \quad (5)$$

for all $\mathbf{v} = (v_1, v_2) \neq 0$.

A HERMITIAN MATRIX A is positive definite if

1. $a_{ii} > 0$ for all i,
2. $a_{ii} a_{jj} > |a_{ij}|^2$ for $i \neq j$,
3. The element of largest modulus lies on the leading diagonal,
4. $\det(\mathsf{A}) > 0$.

See also DETERMINANT, EIGENVALUE, HERMITIAN MATRIX, MATRIX, NEGATIVE DEFINITE MATRIX, NEGATIVE SEMIDEFINITE MATRIX, POSITIVE SEMIDEFINITE MATRIX

References

Gradshteyn, I. S. and Ryzhik, I. M. *Tables of Integrals, Series, and Products, 6th ed.* San Diego, CA: Academic Press, p. 1106, 2000.
Marcus, M. and Minc, H. *Introduction to Linear Algebra.* New York: Dover, p. 182, 1988.
Marcus, M. and Minc, H. "Positive Definite Matrices." §4.12 in *A Survey of Matrix Theory and Matrix Inequalities.* New York: Dover, p. 69, 1992.

Positive Definite Quadratic Form

A QUADRATIC FORM $Q(\mathbf{x})$ is said to be positive definite if $Q(\mathbf{x}) > 0$ for $\mathbf{x} \neq 0$. A REAL QUADRATIC FORM in n variables is positive definite IFF its canonical form is

$$Q(\mathbf{z}) = z_1^2 + z_2^2 + \ldots + z_n^2. \quad (1)$$

A BINARY QUADRATIC FORM

$$F(x, y) = a_{11} x^2 + 2 a_{12} xy + a_{22} y^2 \quad (2)$$

of two REAL variables is positive definite if it is > 0 for any $(x, y) \neq (0, 0)$, therefore if $a_{11} > 0$ and the DISCRIMINANT $a \equiv a_{11} a_{22} - a_{12}^2 > 0$. A BINARY QUADRATIC FORM is positive definite if there exist NONZERO x and y such that

$$\left(ax^2 + 2bxy + cy^2 \right)^2 \leq \tfrac{4}{3} \left| ac - b^2 \right| \quad (3)$$

(Le Lionnais 1983).

A QUADRATIC FORM $(\mathbf{x}, \mathsf{A}\mathbf{x})$ is positive definite IFF every EIGENVALUE of A is POSITIVE. A QUADRATIC FORM $Q = (\mathbf{x}, \mathsf{A}\mathbf{x})$ with A a HERMITIAN MATRIX is positive definite if all the principal minors in the top-left corner of A are POSITIVE, in other words

$$a_{11} > 0 \quad (4)$$

$$\begin{vmatrix} a_{11} & a_{12} \\ a_{21} & a_{22} \end{vmatrix} > 0 \quad (5)$$

$$\begin{vmatrix} a_{11} & a_{12} & a_{13} \\ a_{21} & a_{22} & a_{23} \\ a_{31} & a_{32} & a_{33} \end{vmatrix} > 0 \quad (6)$$

See also INDEFINITE QUADRATIC FORM, LYAPUNOV'S FIRST THEOREM, POSITIVE SEMIDEFINITE QUADRATIC FORM, QUADRATIC FORM

References

Gradshteyn, I. S. and Ryzhik, I. M. *Tables of Integrals, Series, and Products, 6th ed.* San Diego, CA: Academic Press, p. 1106, 2000.
Le Lionnais, F. *Les nombres remarquables.* Paris: Hermann, p. 38, 1983.

Positive Definite Sequence

This entry contributed by RONALD M. AARTS

A sequence $\{\mu_n\}_{n=0}^{\infty}$ is positive definite if the moment of every nonnegative polynomial which is not identically zero is greater than zero (Widder 1941, p. 132). Here, the moment of a polynomial

$$P_n(x) = \sum_{m=0}^{n} a_m x^m$$

with respect to the sequence $\{\mu_n\}_{n=0}^{\infty}$ is defined as

$$M(P_n(x)) = \sum_{m=0}^{n} a_m \mu_m$$

(Widder 1941, p. 102).

References

Widder, D. V. *The Laplace Transform.* Princeton, NJ: Princeton University Press, 1941.

Positive Definite Tensor

A TENSOR g whose discriminant satisfies

$$g \equiv g_{11}g_{22} - g_{12}^2 > 0.$$

Positive Integer

The positive integers are the numbers 1, 2, 3, ..., sometimes called the counting numbers or natural numbers.

See also Z_+

Positive Measure

A positive measure is a MEASURE which is a function from the measurable sets of a MEASURE SPACE to the nonnegative real numbers. Sometimes, this is what is meant by MEASURE, while "positive" is used to distinguish it from an arbitrary COMPLEX MEASURE.

See also COMPLEX MEASURE, JORDAN MEASURE DECOMPOSITION, LEBESGUE INTEGRAL, MEASURE, MEASURE SPACE, POLAR REPRESENTATION (MEASURE)

Positive Semidefinite Matrix

A positive semidefinite matrix is a HERMITIAN MATRIX all of whose EIGENVALUES are nonnegative.

See also NEGATIVE DEFINITE MATRIX, NEGATIVE SEMIDEFINITE MATRIX, POSITIVE DEFINITE MATRIX

References

Marcus, M. and Minc, H. *Introduction to Linear Algebra.* New York: Dover, p. 182, 1988.
Marcus, M. and Minc, H. *A Survey of Matrix Theory and Matrix Inequalities.* New York: Dover, p. 69, 1992.

Positive Semidefinite Quadratic Form

A QUADRATIC FORM $Q(\mathbf{x})$ is positive semidefinite if it is never < 0, but is 0 for some $\mathbf{x} \neq 0$. The QUADRATIC FORM, written in the form $(\mathbf{x}, \mathbf{A}\mathbf{x})$, is positive semidefinite IFF every EIGENVALUE of $\mathbf{A}$ is NONNEGATIVE.

See also INDEFINITE QUADRATIC FORM, POSITIVE DEFINITE QUADRATIC FORM

References

Gradshteyn, I. S. and Ryzhik, I. M. *Tables of Integrals, Series, and Products, 6th ed.* San Diego, CA: Academic Press, p. 1106, 2000.

Postage Stamp Problem

Consider a SET $A_k = \{a_1, a_2, \dots a_k\}$ of INTEGER denomination postage stamps with $1 = a_1 < a_2 \dots < a_k$. Suppose they are to be used on an envelope with room for no more than h stamps. The postage stamp problem then consists of determining the smallest INTEGER $N(h, A_k)$ which cannot be represented by a LINEAR COMBINATION $\Sigma_{i=1}^k x_i a_i$ with $x_i \geq 0$ and

$\Sigma_{i=1}^k x_i < h$. Exact solutions exist for arbitrary A_k for $k = 2$ and 3. The $k = 2$ solution is

$$n(h, A_2) = (h + 3 - a_2)a_2 - 2$$

for $h \geq a_2 - 2$. The general problem consists of finding

$$n(h, k) = \max_{A_k} n(h, A_k).$$

It is known that

$$n(h, 2) = \left\lfloor \tfrac{1}{4}(h^2 + 6h + 1) \right\rfloor,$$

(Stöhr 1955, Guy 1994), where $\lfloor x \rfloor$ is the FLOOR FUNCTION, the first few values of which are 2, 4, 7, 10, 14, 18, 23, 28, 34, 40, ... (Sloane's A014616).

See also HARMONIOUS GRAPH, INTEGER RELATION, STAMP FOLDING, STÖHR SEQUENCE, SUBSET SUM PROBLEM

References

Guy, R. K. "The Postage Stamp Problem." §C12 in *Unsolved Problems in Number Theory, 2nd ed.* New York: Springer-Verlag, pp. 123–27, 1994.
Mossige, S. "The Postage Stamp Problem: An Algorithm to Determine the h-Range on the h-Range Formula on the Extremal Basis Problem for $k = 4$." *Math. Comput.* **69**, 325–37, 2000.
Sloane, N. J. A. Sequences A014616 in "An On-Line Version of the Encyclopedia of Integer Sequences." http://www.research.att.com/~njas/sequences/eisonline.html.
Stöhr, A. "Gelöste und ungelöste Fragen über Basen der natürlichen Zahlenreihe I, II." *J. reine angew. Math.* **194**, 111–40, 1955.

Posterior Distribution

BAYESIAN ANALYSIS

Postnikov System

An iterated FIBRATION of EILENBERG-MAC LANE SPACES. Every TOPOLOGICAL SPACE has this HOMOTOPY type.

See also EILENBERG-MAC LANE SPACE, FIBRATION, HOMOTOPY

Postulate

A statement, also known as an AXIOM, which is taken to be true without PROOF. Postulates are the basic structure from which LEMMAS and THEOREMS are derived. The whole of EUCLIDEAN GEOMETRY, for example, is based on five postulates known as EUCLID'S POSTULATES.

See also ARCHIMEDES' POSTULATE, AXIOM, BERTRAND'S POSTULATE, CONJECTURE, EQUIDISTANCE POSTULATE, EUCLID'S FIFTH POSTULATE, EUCLID'S POSTULATES, LEMMA, PARALLEL POSTULATE, PORISM, PROOF, THEOREM, TRIANGLE POSTULATE

Potato Paradox

You buy 100 pounds of potatoes and are told that they are 99% water. After leaving them outside, you discover that they are now 98% water. The weight of the dehydrated potatoes is then a surprising 50 pounds!

References
Paulos, J. A. *A Mathematician Reads the Newspaper.* New York: BasicBooks, p. 81, 1995.

Potential Function

The term used in physics and engineering for a HARMONIC FUNCTION. Potential functions are extremely useful, for example, in electromagnetism, where they reduce the study of a 3-component VECTOR FIELD to a 1-component SCALAR FUNCTION.

See also HARMONIC FUNCTION, LAPLACE'S EQUATION, SCALAR POTENTIAL, VECTOR POTENTIAL

Potential Theory

The study of HARMONIC FUNCTIONS (also called POTENTIAL FUNCTIONS).

See also HARMONIC FUNCTION, SCALAR POTENTIAL, VECTOR POTENTIAL

References
Kellogg, O. D. *Foundations of Potential Theory.* New York: Dover, 1953.
MacMillan, W. D. *The Theory of the Potential.* New York: Dover, 1958.
Weisstein, E. W. "Books about Potential Theory." http://www.treasure-troves.com/books/PotentialTheory.html.

Pothenot Problem

SNELLIUS-POTHENOT PROBLEM

Poulet Number

A FERMAT PSEUDOPRIME to base 2, denoted psp(2), i.e., a COMPOSITE ODD INTEGER n such that

$$2^{n-1} \equiv 1 \pmod{n}.$$

The first few Poulet numbers are 341, 561, 645, 1105, 1387, ... (Sloane's A001567). Pomerance *et al.* (1980) computed all 21,853 Poulet numbers less than 25×10^9. The numbers less than 10^2, 10^3, ..., are 0, 3, 22, 78, 245, ... (Sloane's A055550).

Pomerance has shown that the number of Poulet numbers less than x for sufficiently large x satisfy

$$\exp\left[(\ln x)^{5/14}\right] < P_2(x) < x \exp\left(-\frac{\ln x \ln \ln \ln x}{2 \ln \ln x}\right)$$

(Guy 1994).

A Poulet number all of whose DIVISORS d satisfy $d|2^d - 2$ is called a SUPER-POULET NUMBER. There are an infinite number of Poulet numbers which are not SUPER-POULET NUMBERS. Shanks (1993) calls *any* integer satisfying $2^{n-1} \equiv 1 \pmod{n}$ (i.e., not limited to ODD composite numbers) a FERMATIAN.

See also FERMAT PSEUDOPRIME, PSEUDOPRIME, ROTKIEWICZ THEOREM, SUPER-POULET NUMBER

References
Guy, R. K. *Unsolved Problems in Number Theory, 2nd ed.* New York: Springer-Verlag, pp. 28–9, 1994.
Pinch, R. G. E. "The Pseudoprimes Up to 10^{13}." ftp://ftp.dpmms.cam.ac.uk/pub/PSP/.
Pomerance, C.; Selfridge, J. L.; and Wagstaff, S. S. Jr. "The Pseudoprimes to $25 \cdot 10^9$." *Math. Comput.* **35**, 1003–026, 1980. Available electronically from ftp://sable.ox.ac.uk/pub/math/primes/ps2.Z.
Shanks, D. *Solved and Unsolved Problems in Number Theory, 4th ed.* New York: Chelsea, pp. 115–17, 1993.
Sloane, N. J. A. Sequences A001567/M5441 and A055550 in "An On-Line Version of the Encyclopedia of Integer Sequences." http://www.research.att.com/~njas/sequences/eisonline.html.

Power

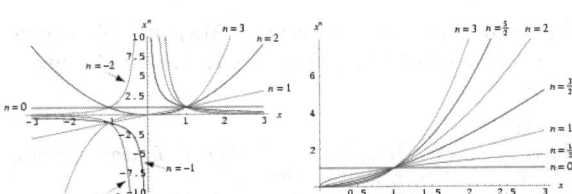

The exponent to which a given quantity is raised is known as its POWER. The expression x^a is therefore known as "x to the ath POWER." The power may be an integer, REAL NUMBER, or COMPLEX NUMBER. However, the power of a real number to a non-integer power is not necessarily itself a real number. For example, $x^{1/2}$ is real only for $x \geq 0$. The rules for combining quantities containing powers are called the EXPONENT LAWS.

While the simple equation

$$a^x = x$$

cannot be solved for x using traditional elementary functions, the solution can be given in terms of LAMBERT'S W-FUNCTION as

$$x = -\frac{W(-\ln a)}{\ln a},$$

where $\ln a$ is the NATURAL LOGARITHM of a.

Special names given to various powers are listed in the following table.

Power	Name
1/2	SQUARE ROOT
1/3	CUBE ROOT
2	SQUARED
3	CUBED

The largest powers p which numbers $n = 1, 2, 3, \ldots$ can be represented in the form $n = a^p$ are 1, 1, 1, 2, 1, 1, 1, 3, 2, 1, ... (Sloane's A052409), with corresponding values of a given by 1, 2, 3, 2, 5, 6, 7, 2, 3, 10, ... (Sloane's A052410).

The POWER SUM of the first n POSITIVE INTEGERS is given by FAULHABER'S FORMULA,

$$\sum_{k=1}^{n} k^p = \frac{1}{p+1} \sum_{k=1}^{p+1} (-1)^{\delta_{kp}} \binom{p+1}{k} B_{p+1-k} n^k,$$

where δ_{kp} is the KRONECKER DELTA, $\binom{n}{k}$ is a BINOMIAL COEFFICIENT, and B_k is a BERNOULLI NUMBER.

Let s_n be the largest INTEGER that is not the SUM of distinct nth powers of POSITIVE INTEGERS (Guy 1994). The first few values for $n = 2, 3, \ldots$ are 128, 12758, 5134240, 67898771, ... (Sloane's A001661).

CATALAN'S CONJECTURE states that 8 and 9 (2^3 and 3^2) are the only consecutive POWERS (excluding 0 and 1), i.e., the only solution to CATALAN'S DIOPHANTINE PROBLEM. This CONJECTURE has not yet been proved or refuted, although R. Tijdeman has proved that there can be only a finite number of exceptions should the CONJECTURE not hold. It is also known that 8 and 9 are the only consecutive CUBIC and SQUARE NUMBERS (in either order). Hyyro and Makowski proved that there do not exist three consecutive POWERS (Ribenboim 1996).

Very few numbers OF THE FORM $n^p \pm 1$ are PRIME (where composite powers $p = kb$ need not be considered, since $n(kb) \pm 1 = \left(n^k\right)^b \pm 1$). The only PRIME NUMBERS of the form $n^p - 1$ for $n \le 100$ and PRIME $2 \le p \le 10$ correspond to $n = 2$, i.e., $2^2 - 1 = 3, 2^3 - 1 = 7, 2^5 - 1 = 31, \ldots$. The only PRIME NUMBERS of the form $n^p + 1$ for $n \le 100$ and PRIME $2 \le p \le 10$ correspond to $p = 2$ with $n = 1, 2, 4, 6, 10, 14, 16, 20, 24, 26, \ldots$ (Sloane's A005574).

There are no nontrivial solutions to the equation

$$1^n + 2^n + \ldots + m^n = (m+1)^n$$

for $m \le 10^{2,000,000}$ (Guy 1994, p. 153).

See also APOCALYPTIC NUMBER, BIQUADRATIC NUMBER, CATALAN'S CONJECTURE, CATALAN'S DIOPHANTINE PROBLEM, CUBE ROOT, CUBED, CUBIC NUMBER, DIGIT-SHIFTING CONSTANTS, EXPONENT, EXPONENT LAWS, FAULHABER'S FORMULA, FIGURATE NUMBER, MOESSNER'S THEOREM, NARCISSISTIC NUMBER, POWER (CIRCLE), POWER RULE, SQUARE NUMBER, SQUARE ROOT, SQUARED, SUM, TRUNCATED POWER FUNCTION, WARING'S PROBLEM

References

Barbeau, E. J. *Power Play: A Country Walk through the Magical World of Numbers.* Washington, DC: Math. Assoc. Amer., 1997.
Beyer, W. H. "Laws of Exponents." *CRC Standard Mathematical Tables, 28th ed.* Boca Raton, FL: CRC Press, pp. 158 and 223, 1987.
Guy, R. K. "Diophantine Equations." Ch. D in *Unsolved Problems in Number Theory, 2nd ed.* New York: Springer-Verlag, pp. 137, 139–98, and 153–54, 1994.
Ribenboim, P. "Catalan's Conjecture." *Amer. Math. Monthly* **103**, 529–38, 1996.
Sloane, N. J. A. Sequences A001661/M5393, A005574/M1010, A052409, and A052410 in "An On-Line Version of the Encyclopedia of Integer Sequences." http://www.research.att.com/~njas/sequences/eisonline.html.
Spanier, J. and Oldham, K. B. "The Integer Powers $(bx+c)^n$ and x^n" and "The Noninteger Powers x^v." Ch. 11 and 13 in *An Atlas of Functions.* Washington, DC: Hemisphere, pp. 83–0 and 99–06, 1987.

Power (Circle)

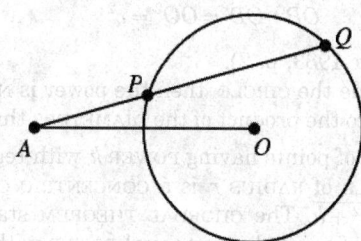

The POWER of a fixed point A with respect to a CIRCLE of RADIUS r and center O is defined by the product

$$p \equiv AP \times AQ, \qquad (1)$$

where P and Q are the intersections of a line through A with the circle. The term "power" was first used in this way by Jacob Steiner (Steiner 1826; Coxeter and Greitzer 1967, p. 30). Amazingly, p (sometimes written k^2) is *independent* of the choice of the line APQ (Coxeter 1969, p. 81).

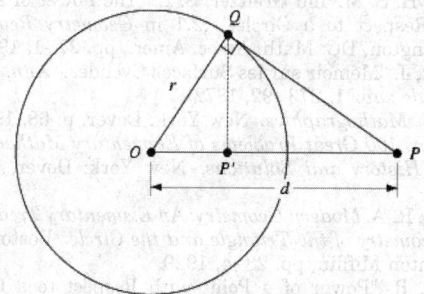

Now consider a point P not necessarily on the circumference of the circle. If $d = OP$ is the distance between P and the circle's center O, then the power of the point P relative to the circle is

$$p = d^2 - r^2. \tag{2}$$

If P is outside the CIRCLE, its power is POSITIVE and equal to the square of the length of the segment PQ from P to the tangent Q to the CIRCLE through P,

$$p = PQ^2 = d^2 - r^2. \tag{3}$$

If OP lies along the X-AXIS, then the angle θ around the circle at which Q lies is given by solving

$$\left[(d - \cos \theta)^2 + \sin^2 \theta \right] + 1 = d^2 \tag{4}$$

for θ, giving

$$\theta = \pm \sec^{-1} d \tag{5}$$

for coordinates

$$(x, y) = r \left(\pm \frac{1}{d}, \sqrt{\frac{d^2 - 1}{d^2}} \right). \tag{6}$$

The points P and P' are INVERSE POINTS, also called polar reciprocals, with respect to the INVERSION CIRCLE if

$$OP \cdot OP' = OQ^2 = r^2 \tag{7}$$

(Wenninger 1983, p. 2).

If P is inside the CIRCLE, then the power is NEGATIVE and equal to the product of the DIAMETERS through P.

The LOCUS of points having POWER k with regard to a fixed CIRCLE of RADIUS r is a CONCENTRIC CIRCLE of RADIUS $\sqrt{r^2 + k}$. The CHORDAL THEOREM states that the LOCUS of points having equal POWER with respect to two given nonconcentric CIRCLES is a line called the RADICAL LINE (or CHORDAL; Dörrie 1965).

See also CHORDAL THEOREM, COAXAL CIRCLES, INVERSE POINTS, INVERSION CIRCLE, INVERSION RADIUS, INVERSIVE DISTANCE, RADICAL LINE

References

Coxeter, H. S. M. *Introduction to Geometry, 2nd ed.* New York: Wiley, 1969.

Coxeter, H. S. M. and Greitzer, S. L. "The Power of a Point with Respect to a Circle." §2.1 in *Geometry Revisited.* Washington, DC: Math. Assoc. Amer., pp. 27–1, 1967.

Darboux, J. "Mémoir sur les Surfaces Cyclides." *Ann. l'École Normale sup.* **1**, 273–92, 1872.

Dixon, R. *Mathographics.* New York: Dover, p. 68, 1991.

Dörrie, H. *100 Great Problems of Elementary Mathematics: Their History and Solutions.* New York: Dover, p. 153, 1965.

Johnson, R. A. *Modern Geometry: An Elementary Treatise on the Geometry of the Triangle and the Circle.* Boston, MA: Houghton Mifflin, pp. 28–4, 1929.

Lachlan, R. "Power of a Point with Respect to a Circle." §300–03 in *An Elementary Treatise on Modern Pure Geometry.* London: Macmillian, pp. 183–85, 1893.

Pedoe, D. *Circles: A Mathematical View, rev. ed.* Washington, DC: Math. Assoc. Amer., pp. xxii–xxiv, 1995.

Steiner, J. "Einige geometrische Betrachtungen." *J. reine angew. Math.* **1**, 161–84, 1826.

Wenninger, M. J. *Dual Models.* Cambridge, England: Cambridge University Press, 1983.

Power (Statistics)

The probability of getting a positive result for a given test which should produce a positive result.

See also PREDICTIVE VALUE, SENSITIVITY, SPECIFICITY, STATISTICAL TEST

Power (Triangle)

The total power of a TRIANGLE is defined by

$$P \equiv \tfrac{1}{2}(a_1^2 + a_2^2 + a_3^2), \tag{1}$$

where a_i are the side lengths, and the "partial power" is defined by

$$p_1 = \tfrac{1}{2}(a_2^2 + a_3^2 - a_1^2). \tag{2}$$

Then

$$p_1 = a_2 a_3 \cos \alpha_1 \tag{3}$$

$$P = p_1 + p_2 + p_3 \tag{4}$$

$$P^2 + p_1^2 + p_2^2 + p_3^2 = a_1^4 + a_2^4 + a_3^4 \tag{5}$$

$$\Delta = \tfrac{1}{2}\sqrt{p_2 p_3 + p_3 p_1 + p_1 p_2} \tag{6}$$

$$p_1 = \overline{A_1 H_2} \cdot \overline{A_1 A_3} \tag{7}$$

$$\frac{a_1 p_1}{\cos \alpha_1} = a_1 a_2 a_3 = 4 \Delta R \tag{8}$$

$$p_1 \tan \alpha_1 = p_2 \tan \alpha_2 = p_3 \tan \alpha_3, \tag{9}$$

where Δ is the AREA of the TRIANGLE and H_i are the FEET of the ALTITUDES. finally, if a side of the TRIANGLE and the value of any partial power are given, then the LOCUS of the third VERTEX is a CIRCLE or straight line.

See also ALTITUDE, FOOT, TRIANGLE

References

Johnson, R. A. *Modern Geometry: An Elementary Treatise on the Geometry of the Triangle and the Circle.* Boston, MA: Houghton Mifflin, pp. 260–61, 1929.

Power Associative Algebra

An ALGEBRA in which the ASSOCIATOR $(x, x, x) = 0$. The SUBALGEBRA generated by one element is associative.

See also ASSOCIATOR

References

Schafer, R. D. *An Introduction to Non-Associative Algebras.* New York: Dover, 1995.

Power Center

RADICAL CENTER

Power Curve

The curve with TRILINEAR COORDINATES $a^t : b^t : c^t$ for a given POWER t.

See also POWER POINT

References

Kimberling, C. "Major Centers of Triangles." *Amer. Math. Monthly* **104**, 431–38, 1997.

Power Line

RADICAL AXIS

Power Point

Triangle centers with TRIANGLE CENTER FUNCTIONS OF THE FORM $\alpha = a^n$ are called nth power points. The 0th power point is the INCENTER, with TRIANGLE CENTER FUNCTION $\alpha = 1$.

See also INCENTER, TRIANGLE CENTER FUNCTION

References

Groenman, J. T. and Eddy, R. H. "Problem 858 and Solution." *Crux Math.* **10**, 306–07, 1984.
Kimberling, C. "Problem 865." *Crux Math.* **10**, 325–27, 1984.
Kimberling, C. "Central Points and Central Lines in the Plane of a Triangle." *Math. Mag.* **67**, 163–87, 1994.

Power Polynomial

The power polynomials x^n are an associated SHEFFER SEQUENCE with

$$f(t) = t, \tag{1}$$

giving GENERATING FUNCTION

$$\sum_{k=0}^{\infty} \frac{x^k}{k!} t^k = e^x t \tag{2}$$

and BINOMIAL IDENTITY

$$(x+y)^n = \sum_{k=0}^{n} \binom{n}{k} x^k y^{n-k}. \tag{3}$$

See also SHEFFER SEQUENCE

References

Roman, S. "The Sequence x^n." §4.1.1 in *The Umbral Calculus.* New York: Academic Press, p. 55, 1984.

Power Rule

The DERIVATIVE of the POWER x^n is given by

$$\frac{d}{dx}(x^n) = nx^{n-1}.$$

See also CHAIN RULE, DERIVATIVE, EXPONENT LAWS, PRODUCT RULE

References

Anton, H. *Calculus: A New Horizon, 6th ed.* New York: Wiley, p. 131, 1999.

Power Series

A power series in a variable z is an infinite SUM OF THE FORM

$$\sum_{n}^{\infty} a_i z^i, \tag{1}$$

where $n \geq 0$ and a_i are INTEGERS, REAL NUMBERS, COMPLEX NUMBERS, or any other quantities of a given type.

A CONJECTURE of Pólya is that if a FUNCTION has a power series with INTEGER COEFFICIENTS and RADIUS OF CONVERGENCE 1, then either the FUNCTION is RATIONAL or the UNIT CIRCLE is a natural boundary.

A generalized POWER sum $a(h)$ for $h = 0, 1, \dots$ is given by

$$a(h) = \sum_{i=1}^{m} A_i(h)\alpha_i^h, \tag{2}$$

with distinct NONZERO ROOTS α_i, COEFFICIENTS $A_i(h)$ which are POLYNOMIALS of degree $n_i - 1$ for POSITIVE INTEGERS n_i, and $i \in [1, m]$. The generalized POWER sum has order

$$n \equiv \sum_{i=m}^{m} n_i. \tag{3}$$

For any power series, one of the following is true:

1. The series converges only for $x = 0$.
2. The series converges absolutely for all x.
3. The series converges absolutely for all x in some finite open interval $(-R, R)$ and diverges if $x < -R$ or $x > R$. At the points $x = R$ and $x = -R$, the series may converge absolutely, converge conditionally, or diverge.

To determine the interval of convergence, apply the RATIO TEST for ABSOLUTE CONVERGENCE and solve for x. A power series may be differentiated or integrated within the interval of convergence. Convergent power series may be multiplied and divided (if there is no division by zero).

$$\sum_{k=1}^{\infty} k^{-p} \qquad (4)$$

CONVERGES if $p > 1$ and DIVERGES if $0 < p \le 1$.

See also BINOMIAL SERIES, CONVERGENCE TESTS, FORMAL POWER SERIES, LAURENT SERIES, MACLAURIN SERIES, MULTINOMIAL SERIES, *P*-SERIES, POLYNOMIAL, POWER SET, QUOTIENT-DIFFERENCE ALGORITHM, RADIUS OF CONVERGENCE, RECURRENCE SEQUENCE, SERIES, SERIES REVERSION, TAYLOR SERIES

References

Arfken, G. "Power Series." §5.7 in *Mathematical Methods for Physicists, 3rd ed.* Orlando, FL: Academic Press, pp. 313–21, 1985.

Hanrot, G.; Quercia, M.; and Zimmerman, P. "Speeding Up the Division and Square Root of Power Series." Report RR-3973. INRIA, Jul 2000. http://www.inria.fr.RRRT/RR-3973.html.

Myerson, G. and van der Poorten, A. J. "Some Problems Concerning Recurrence Sequences." *Amer. Math. Monthly* **102**, 698–05, 1995.

Niven, I. "Formal Power Series." *Amer. Math. Monthly* **76**, 871–89, 1969.

Pólya, G. *Mathematics and Plausible Reasoning, Vol. 2: Patterns of Plausible Inference.* Princeton, NJ: Princeton University Press, p. 46, 1990.

Power Set

Given a SET S, the power set of S is the SET of all SUBSETS of S. The order of a POWER set of a SET of order n is 2^n. Power sets are larger than the SETS associated with them. The power set of S is variously denoted 2^S or $\mathscr{P}(S)$.

The power set of a given set s can be found using `Subsets[s]` in the *Mathematica* add-on package `DiscreteMath`Combinatorica`` (which can be loaded with the command `<<DiscreteMath`). A concise implementation in *Mathematica* is given by

```
PowerSet[s_List] :=    Distribute[Thread[{{},
List /@ s}, List, {2, 2}],
  List, List, List, Join]
```

See also SET, SUBSET

Power Spectrum

For a given signal, the power spectrum gives a plot of the portion of a signal's power (energy per unit time) falling within given frequency bins. The most common way of generating a power spectrum is by using a FOURIER TRANSFORM, but other techniques such as the MAXIMUM ENTROPY METHOD can also be used.

References

Press, W. H.; Flannery, B. P.; Teukolsky, S. A.; and Vetterling, W. T. "Power Spectra Estimation Using the FFT" and "Power Spectrum Estimation by the Maximum Entropy (All Poles) Method." §13.4 and 13.7 in *Numerical Recipes*

in FORTRAN: The Art of Scientific Computing, 2nd ed. Cambridge, England: Cambridge University Press, pp. 542–51 and 565–69, 1992.

Power Sum

An analytic solution for a SUM of POWERS of integers is

$$S_p(n) = \sum_{k=1}^{n} k^p = \zeta(-p) - \zeta(-p,\ 1+n) = H_n^{(-p)}, \qquad (1)$$

where $\zeta(z)$ is the RIEMANN ZETA FUNCTION, $\zeta(z;\ a)$ is the HURWITZ ZETA FUNCTION, and $H_n^{(k)}$ is a generalized HARMONIC NUMBER. For the special case of p a POSITIVE INTEGER, FAULHABER'S FORMULA gives the SUM explicitly as

$$S_p(n) = \frac{1}{p+1} \sum_{k=1}^{p+1} (-1)^{\delta_{kp}} \binom{p+1}{k} B_{p+1-k} n^k, \qquad (2)$$

where δ_{kp} is the KRONECKER DELTA, $\binom{n}{k}$ is a BINOMIAL COEFFICIENT, and B_k is a BERNOULLI NUMBER. Written explicitly in terms of a sum of POWERS,

$$S_p(n) = \frac{B_k p!}{k!(p-k+1)!} n^{p-k+1}. \qquad (3)$$

It is also true that the COEFFICIENTS of the terms in such an expansion sum to 1, as stated by Bernoulli without proof (Boyer 1943).

Computing the sums for $p = 1, \ldots, 10$ gives

$$\sum_{k=1}^{n} k = \tfrac{1}{2}(n^2 + n) \qquad (4)$$

$$\sum_{k=1}^{n} k^2 = \tfrac{1}{6}(2n^3 + 3n^3 + n) \qquad (5)$$

$$\sum_{k=1}^{n} k^3 = \tfrac{1}{4}(n^4 + 2n^3 + n^2) \qquad (6)$$

$$\sum_{k=1}^{n} k^4 = \tfrac{1}{30}(6n^5 + 15n^4 + 10n^3 - n) \qquad (7)$$

$$\sum_{k=1}^{n} k^5 = \tfrac{1}{12}(2n^6 + 6n^5 + 5n^4 - n^2) \qquad (8)$$

$$\sum_{k=1}^{n} k^6 = \tfrac{1}{42}(6n^7 + 21n^6 + 21n^5 - 7n^3 + n) \qquad (9)$$

$$\sum_{k=1}^{n} k^7 = \tfrac{1}{24}(3n^8 + 12n^7 + 14n^6 - 7n^4 + 2n^2) \qquad (10)$$

$$\sum_{k=1}^{n} k^8 = \tfrac{1}{90}(10n^9 + 45n^8 + 60n^7 - 42n^5 + 20n^3 - 3n) \qquad (11)$$

$$\sum_{k=1}^{n} k^9 = \tfrac{1}{20}\big(2n^{10} + 10n^9 + 15n^8 - 14n^6 + 10n^4 - 3n^2\big) \tag{12}$$

$$\sum_{k=1}^{n} k^{10} = \tfrac{1}{66}\big(6n^{11} + 33n^{10} + 55n^9 - 66n^7$$
$$+ 66n^5 - 33n^3 + 5n\big). \tag{13}$$

$$\sum_{k=1}^{n} k = \tfrac{1}{2}\, n(n+1) \tag{14}$$

$$\sum_{k=1}^{n} k^2 = \tfrac{1}{6}\, n(n+1)(2n+1) \tag{15}$$

$$\sum_{k=1}^{n} k^3 = \tfrac{1}{4}\, n^2(n+1)^2 \tag{16}$$

$$\sum_{k=1}^{n} k^4 = \tfrac{1}{30}\, n(n+1)(2n+1)(3n^2+3n-1) \tag{17}$$

$$\sum_{k=1}^{n} k^5 = \tfrac{1}{12}\, n^2(n+1)^2(2n^2+2n-1) \tag{18}$$

$$\sum_{k=1}^{n} k^6 = \tfrac{1}{42}\, n(n+1)(2n+1)(3n^4+6n^3-3n+1) \tag{19}$$

$$\sum_{k=1}^{n} k^7 = \tfrac{1}{24}\, n^2(n+1)^2(3n^4+6n^3-n^2-4n+2) \tag{20}$$

$$\sum_{k=1}^{n} k^8 = \tfrac{1}{90}\, n(n+1)(2n+1)$$
$$\times \big(5n^6 + 15n^5 + 5n^4 - 15n^3 - n^2 + 9n - 3\big) \tag{21}$$

$$\sum_{k=1}^{n} k^9 = \tfrac{1}{20}\, n^2(n+1)^2(n^2+n-1)$$
$$\times \big(2n^4 + 4n^3 - n^2 - 3n + 3\big) \tag{22}$$

$$\sum_{k=1}^{n} k^{10} = \tfrac{1}{60}\, n(n+1)(2n+1)(n^2+n-1)$$
$$\times \big(3n^6 + 9n^5 + 2n^4 - 11n^3 + 10n - 5\big). \tag{23}$$

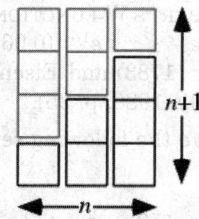

A simple graphical proof of the special case of $S_1(n) = n(n+1)/2$ can also be given by constructing a sequence of stacks of boxes, each 1 unit across and k units high, where $k = 1, 2, ..., n$. Now add a rotated copy on top, as in the above figure. Note that the resulting figure has WIDTH n and HEIGHT $n+1$, and so has AREA $n(n+1)$. The desired sum is half this, so the AREA of the boxes in the sum is $n(n+1)/2$. Since the boxes are of unit width, this is also the value of the sum.

The sum $S_1(n) = n(n+1)/2$ can also be computed using the first EULER-MACLAURIN INTEGRATION FORMULA

$$\sum_{k=1}^{n} f(k) = \int_1^n f(x)\, dx + \tfrac{1}{2} f(1) + \tfrac{1}{2} f(n)$$
$$+ \tfrac{1}{2!} B_2[f'(n) - f'(1)] + \dots \tag{24}$$

with $f(k) = k$. Then

$$\sum_{k=1}^{n} k = \int_1^n x\, dx + \tfrac{1}{2}\cdot 1 + \tfrac{1}{2}\cdot n + \tfrac{1}{6}(1-1) + \dots$$
$$= \tfrac{1}{2}(n^2 - 1) - \tfrac{1}{2} + h + \tfrac{1}{2} n = \tfrac{1}{2}\, n(n+1). \tag{25}$$

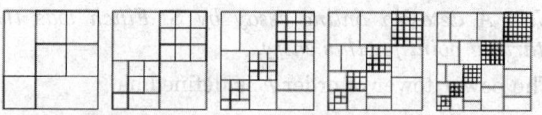

The surprising identity

$$S_3(n) = \sum_{k=1}^{n} k^3 = \left(\sum_{k=1}^{n} k\right)^2, \tag{26}$$

known as NICOMACHUS'S THEOREM, can also be illustrated graphically (Wells 1991, pp. 198–99).

Schultz (1980) showed that the sum $S_k(n)$ can be found by writing

$$S_k(n) = A_{k+1} n^{k+1} + \dots + A_1 n \tag{27}$$

and solving the system of $k+1$ equations

$$\sum_{i=j+1}^{k+1} (-1)^{i-j+1} \binom{i}{j} A_i = 0 \tag{28}$$

for $0 \le j \le k$ (Guo and Qi 1999).

$S_i(n)$ is related to the BINOMIAL THEOREM by

$$(1+n)^{k+1} = 1 + \sum_{i=0}^{k} \binom{k+1}{i} S_i(n) \tag{29}$$

(Guo and Qi 1999).

See also DIOPHANTINE EQUATION, FAULHABER'S FORMULA, MULTIGRADE EQUATION, NICOMACHUS'S THEOREM, SUM

References

Boyer, C. B. "Pascal's Formula for the Sums of Powers of the Integers." *Scripta Math.* **9**, 237–44, 1943.

Brualdi, R. A. *Introductory Combinatorics, 3rd ed.* New York: Elsevier, p. 119, 1997.

Cao, J.-T. "A Method of Summing Series and Some Corollaries" [Chinese]. *Math. Pract. Th.* **20**, 77–4, 1990.

Conway, J. H. and Guy, R. K. *The Book of Numbers.* New York: Springer-Verlag, p. 106, 1996.

Guo, S.-L. and Qi, F. "Recursion Formulae for $\Sigma_{m=1}^n m^k$." *J. Anal. Appl.* **18**, 1123–130, 1999.

Schultz, H. J. "The Sums of the kth Powers of the First n Integers." *Amer. Math. Monthly* **87**, 478–81, 1980.

Struik, D. *A Source Book in Mathematics, 1200–800.* Cambridge, MA: Harvard University Press, 1969.

Wells, D. *The Penguin Dictionary of Curious and Interesting Geometry.* London: Penguin, pp. 198–99, 1991.

Yang, B.-C. "Formulae Related to Bernoulli Number and for Sums of the Same Power of Natural Numbers" [Chinese]. *Math. Pract. Th.* **24**, 52–6 and 74, 1994.

Zhang, N.-Y. "Euler's Number and Some Sums Related to Zeta Function" [Chinese]. *Math. Pract. Th.* **20**, 62–0, 1990.

Power Tower

N.B. A detailed online essay by S. Finch was the starting point for this entry.

The power tower of order k is defined as

$$a \uparrow\uparrow k \equiv \underbrace{a^{a^{\cdot^{\cdot^{a}}}}}_{k}, \qquad (1)$$

where $\uparrow$ is Knuth's (1976) ARROW NOTATION, which in turn is defined by

$$a \uparrow^k n = a \uparrow^{k-1} \left[a \uparrow^k (n-1) \right]. \qquad (2)$$

Rucker (1995, p. 74) uses the notation

$$^k a \equiv \underbrace{a^{a^{\cdot^{\cdot^{a}}}}}_{n}, \qquad (3)$$

and refers to this operation as "tetration." A power tower can be implemented in *Mathematica* as

```
PowerTower[a_, k_] := Fold[Power[a, #] &, 1,
Table[a, {k}]]
```

The following table gives values of $\underbrace{a^{a^{\cdot^{\cdot^{a}}}}}_{n}$ for $a = 1, 2,$ ... for small n.

n	$\underbrace{a^{a^{\cdot^{\cdot^{a}}}}}_{}$
1	1, 2, 3, 4, 5, 6, 7, 8, 9, 10, ...
2	1, 4, 27, 256, 3125, 46656, ...
3	1, 16, 7.63×10^{12}, 1.34×10^{154}, ...
4	1, 65536, ...

The following table gives $\underbrace{a^{a^{\cdot^{\cdot^{a}}}}}_{n}$ for $n = 1, 2,$ ... for small a.

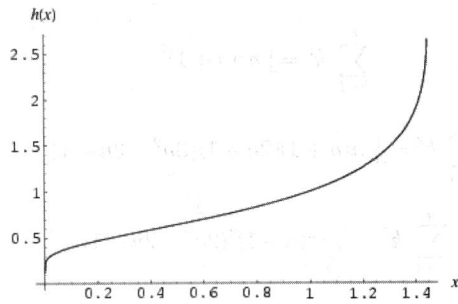

a	$\underbrace{a^{a^{\cdot^{\cdot^{a}}}}}_{n}$
1	1, 1, 1, 1, 1, 1, ...
2	2, 4, 16, 65536, 2.00×10^{19728}, ...
3	3, 27, 7.63×10^{12}, ...
4	4, 256, 1.34×10^{154}, ...

The value of the infinite power tower $h(x) = x \uparrow\uparrow \infty = x^{x^{x^{\cdot^{\cdot}}}}$, where x^{x^x} is an abbreviation for $x^{(x^x)}$, can be computed analytically by writing

$$x^{x^{\cdot^{\cdot}}} = h(x) \qquad (4)$$

taking the logarithm of both sides and plugging back in to obtain

$$x^{x^{\cdot^{\cdot}}} \ln x = h(x) \ln x = \ln[h(x)]. \qquad (5)$$

Solving for $h(x)$ gives

$$h(x) = -\frac{W(-\ln x)}{\ln x}, \qquad (6)$$

where $W(x)$ is LAMBERT'S W-FUNCTION (Corless *et al.*). $h(x)$ converges IFF $e^{-e} \leq x \leq e^{1/e}$ ($0.0659 \leq x \leq 1.4446$), as shown by Euler (1783) and Eisenstein (1844) (Le Lionnais 1983, Wells 1986, p. 35).

Knoebel (1981) gave the following series for $h(z)$

$$h(z) = 1 + \ln x + \frac{3^2 (\ln z)^2}{3!} + \frac{4^3 (\ln z)^3}{4!} + \cdots \qquad (7)$$

(Vardi 1991), and a CONTINUED FRACTION due to Khovanskii (1963) is

$$x^{1/x} = 1 + \cfrac{2(x-1)}{x^2 + 1 - \cfrac{(x^2-1)(x-1)^2}{3x(x+1) - \cfrac{(4x^2-1)(x-1)^2}{5x(x+1) - \cfrac{(9x^2-1)(x-1)^2}{7x(x+1) - \dots}}}}. \tag{8}$$

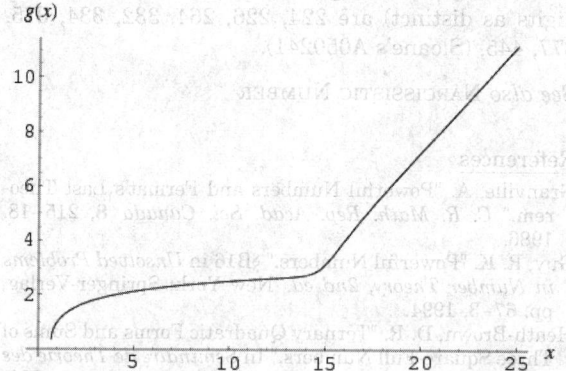

$g(x)$

The related function

$$g(x) = x^{(1/x)^{(1/x)^{\cdot^{\cdot^{\cdot}}}}} \tag{9}$$

converges only for $x \geq e^{-1/e}$, that is, $x \geq 0.692$. The value it converges to is the inverse of x^x which, for $x < e^e$ (i.e., $x < 15.154$), is given by

$$g(x) = \frac{\ln x}{W(\ln x)} \tag{10}$$

for $e^{-1/e} \leq x \leq e^e$.

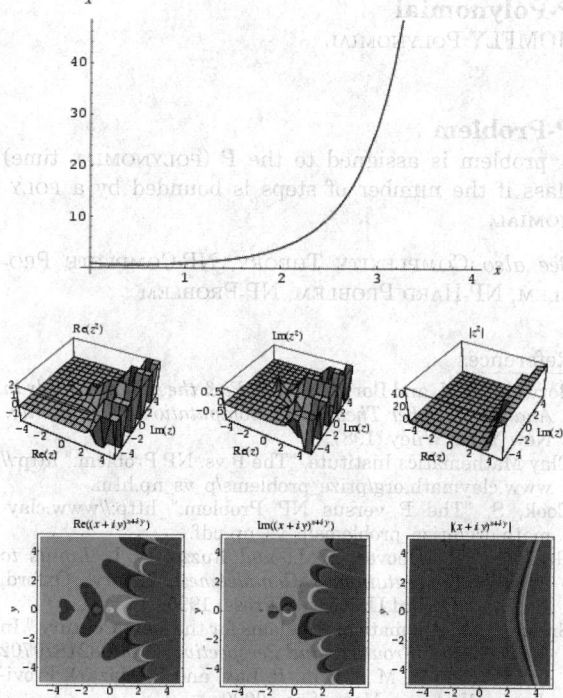

x^x

The function x^x is plotted above along the real line and in the complex plane. It has a minimum where

$$\frac{d}{dx} x^x = x^x (1 + \ln x) = 0, \tag{11}$$

which has solution $x = 1/e$. At this point, the function takes on the value $e^{-1/e}$.

Some interesting related integrals are

$$\int_0^1 x^x \, dx = \sum_{n=1}^{\infty} \frac{(-1)^{n+1}}{n^n} = 0.7834305107\dots \tag{12}$$

$$\int_0^1 x^{-x} \, dx = \sum_{n=1}^{\infty} \frac{1}{n^n} = 1.2912859971\dots \tag{13}$$

(Spiegel 1968, Abramowitz and Stegun 1972).

See also ACKERMANN FUNCTION, ARROW NOTATION, FERMAT NUMBER, LAMBERT'S *W*-FUNCTION, MILLS' CONSTANT, STEINER'S PROBLEM

References

Abramowitz, M. and Stegun, C. A. (Eds.). *Handbook of Mathematical Functions with Formulas, Graphs, and Mathematical Tables, 9th printing.* New York: Dover, 1972.

Ash, J. M. "The Limit of $x^{x^{\cdot^{\cdot^{\cdot x}}}}$ as x Tends to Infinity." *Math. Mag.* **69**, 207–09, 1996.

Baker, I. N. and Rippon, P. J. "Convergence of Infinite Exponentials." *Ann. Acad. Sci. Fennicæ Ser. A. I. Math.* **8**, 179–86, 1983.

Baker, I. N. and Rippon, P. J. "Iteration of Exponential Functions." *Ann. Acad. Sci. Fennicæ Ser. A. I. Math.* **9**, 49–7, 1984.

Baker, I. N. and Rippon, P. J. "A Note on Complex Iteration." *Amer. Math. Monthly* **92**, 501–04, 1985.

Barrow, D. F. "Infinite Exponentials." *Amer. Math. Monthly* **43**, 150–60, 1936.

Corless, R. M.; Gonnet, G. H.; Hare, D. E. G.; Jeffrey, D. J.; and Knuth, D. E. "On the Lambert W Function." *Adv. Comput. Math.* **5**, 329–59, 1996.

Creutz, M. and Sternheimer, R. M. "On the Convergence of Iterated Exponentiation, Part I." *Fib. Quart.* **18**, 341–47, 1980.

Creutz, M. and Sternheimer, R. M. "On the Convergence of Iterated Exponentiation, Part II." *Fib. Quart.* **19**, 326–35, 1981.

de Villiers, J. M. and Robinson, P. N. "The Interval of Convergence and Limiting Functions of a Hyperpower Sequence." *Amer. Math. Monthly* **93**, 13–3, 1986.

Eisenstein, G. "Entwicklung von $\alpha^{\alpha^{\alpha^{\cdot^{\cdot^{\cdot}}}}}$." *J. reine angew. Math.* **28**, 49–2, 1844.

Elstrodt, J. "Iterierte Potenzen." *Math. Semesterber.* **41**, 167–78, 1994.

Euler, L. "De serie Lambertina Plurimisque eius insignibus proprietatibus." *Acta Acad. Scient. Petropol.* **2**, 29–1, 1783. Reprinted in Euler, L. *Opera Omnia I6: Commentationes Algebraicae.* pp. 350–69.

Finch, S. "Favorite Mathematical Constants." http://www.mathsoft.com/asolve/constant/itrexp/itrexp.html.

Ginsburg, J. "Iterated Exponentials." *Scripta Math.* **11**, 340–53, 1945.

Khovanskii, A. N. *The Application of Continued Fractions and Their Generalizations to Problems in Approximation Theory.* Groningen, Netherlands: P. Noordhoff, 1963.

Knoebel, R. A. "Exponentials Reiterated." *Amer. Math. Monthly* **88**, 235-52, 1981.

Knuth, D. E. "Mathematics and Computer Science: Coping with Finiteness. Advances in our Ability to Compute are Bringing us Substantially Closer to Ultimate Limitations." *Science* **194** 1235-242, 1976.

Länger, H. "An Elementary Proof of the Convergence of Iterated Exponentials." *Elem. Math.* **51**, 75-7, 1996.

Le Lionnais, F. *Les nombres remarquables.* Paris: Hermann, pp. 22 and 39, 1983.

Mauerer, H. "Über die Funktion x^x für ganzzahliges Argument (Abundanzen)." *Mitt. Math. Gesell. Hamburg* **4**, 33-0, 1901.

Meyerson, M. D. "The x^x Spindle." *Math. Mag.* **69**, 198-06, 1996.

Rippon, P. J. "Infinite Exponentials." *Math. Gaz.* **67**, 189-96, 1983.

Rucker, R. *Infinity and the Mind: The Science and Philosophy of the Infinite.* Princeton, NJ: Princeton University Press, 1995.

Spiegel, M. R. *Mathematical Handbook of Formulas and Tables.* New York: McGraw-Hill, 1968.

Vardi, I. *Computational Recreations in Mathematica.* Reading, MA: Addison-Wesley, pp. 11-2 and 226-29, 1991.

Weber, R. O. and Roumeliotis, J. "$i \wedge i \wedge i \wedge$...." *Austral. Math. Soc. Gaz.* **22**, 182-84, 1995.

Wells, D. *The Penguin Dictionary of Curious and Interesting Numbers.* Middlesex, England: Penguin Books, p. 35, 1986.

Powerfree

A POSITIVE INTEGER n is kth powerfree if there is no number d such that $d^k | n$ (d^k divides n), i.e., there are no kth powers or higher in the PRIME FACTORIZATION of n. A number which is free of all powers is therefore SQUAREFREE.

See also BIQUADRATEFREE, CUBEFREE, PRIME NUMBER, SQUAREFREE

References

Baake, M.; Moody, R. V.; and Pleasants, P. A. B. Diffraction from Visible Lattice Points and kth Power Free Integers. 19 Jun 1999. http://xxx.lanl.gov/abs/math.MG/9906132/.

Powerful Number

An INTEGER m such that if $p | m$, then $p^2 | m$, is called a powerful number. The first few are 1, 4, 8, 9, 16, 25, 27, 32, 36, 49, ... (Sloane's A001694). Powerful numbers are always OF THE FORM $a^2 b^3$ for a, $b \geq 1$.

Not every NATURAL NUMBER is the sum of two powerful numbers, but Heath-Brown (1988) has shown that every sufficiently large NATURAL NUMBER is the sum of at most three powerful numbers. There are infinitely many pairs of consecutive powerful numbers, but Erdos has conjectured that there do not exist three consecutive powerful numbers. The CONJECTURE that there are no powerful number triples implies that there are infinitely many Wieferich primes (Granville 1986, Vardi 1991).

A separate usage of the term powerful number is for numbers which are the sums of *any* positive powers of their digits (not necessarily the same for each digit).

The first few are 1, 2, 3, 4, 5, 6, 7, 8, 9, 24, 43, 63, 89, ... (Sloane's A007532). These are also called handsome numbers by Rivera, and are a special case of the NARCISSISTIC NUMBERS. Powerful numbers representable in two distinct ways (*not* counting different powers of duplicated digits as distinct) are 264, 373, 375, 2132, 2223, 2241, 2243, 2245, 2263, (Sloane's A050240). Powerful numbers representable in two distinct ways (counting different powers of duplicated digits as distinct) are 224, 226, 264, 332, 334, 375, 377, 445, (Sloane's A050241).

See also NARCISSISTIC NUMBER

References

Granville, A. "Powerful Numbers and Fermat's Last Theorem." *C. R. Math. Rep. Acad. Sci. Canada* **8**, 215-18, 1986.

Guy, R. K. "Powerful Numbers." §B16 in *Unsolved Problems in Number Theory, 2nd ed.* New York: Springer-Verlag, pp. 67-3, 1994.

Heath-Brown, D. R. "Ternary Quadratic Forms and Sums of Three Square-Full Numbers." In *Séminaire de Theorie des Nombres, Paris 1986-7* (Ed. C. Goldstein). Boston, MA: Birkhäuser, pp. 137-63, 1988.

Ribenboim, P. "Catalan's Conjecture." *Amer. Math. Monthly* **103**, 529-38, 1996.

Rivera, C. "Problems & Puzzles: Puzzle Narcissistic and Handsome Primes.-015." http://www.primepuzzles.net/puzzles/puzz_015.htm.

Sloane, N. J. A. Sequences A001694/M3325, A007532/M0487, A050240, and A050241 in "An On-Line Version of the Encyclopedia of Integer Sequences." http://www.research.att.com/~njas/sequences/eisonline.html.

Vardi, I. *Computational Recreations in Mathematica.* Reading, MA: Addison-Wesley, pp. 59-2, 1991.

P-Polynomial

HOMFLY POLYNOMIAL

P-Problem

A problem is assigned to the P (POLYNOMIAL time) class if the number of steps is bounded by a POLYNOMIAL.

See also COMPLEXITY THEORY, NP-COMPLETE PROBLEM, NP-HARD PROBLEM, NP-PROBLEM

References

Borwein, J. M. and Borwein, P. B. *Pi & the AGM: A Study in Analytic Number Theory and Computational Complexity.* New York: Wiley, 1987.

Clay Mathematics Institute. "The P vs. NP Problem." http://www.claymath.org/prize_problems/p_vs_np.htm.

Cook, S. "The P versus NP Problem." http://www.claymath.org/prize_problems/p_vs_np.pdf.

Greenlaw, R.; Hoover, H. J.; and Ruzzo, W. L. *Limits to Parallel Computation: P-Completeness Theory.* Oxford, England: Oxford University Press, 1995.

Smale, S. "Mathematical Problems for the Next Century." In *Mathematics: Frontiers and Perspectives 2000* 0821820702 (Ed. V. Arnold, M. Atiyah, P. Lax, and B. Mazur). Providence, RI: Amer. Math. Soc., 2000.

Practical Number

A number n is practical if for all $k \le n$, k is the sum of distinct proper divisors of n. Defined in 1948 by A. K. Srinivasen. All even PERFECT NUMBERS are practical. The number

$$m = 2^{n-1}(2^{n}-1)$$

is practical for all $n = 2, 3, \ldots$. The first few practical numbers are 1, 2, 4, 6, 8, 12, 16, 18, 20, 24, 28, 30, 32, 36, 40, 42, 48, 54, 56, ... (Sloane's A005153). G. Melfi has computed twins, triplets, and 5-tuples of practical numbers. The first few 5-tuples are 12, 18, 30, 198, 306, 462, 1482, 2550, 4422, ...

References

Melfi, G. "On Two Conjectures About Practical Numbers." *J. Number Th.* **56**, 205–10, 1996.
Sloane, N. J. A. Sequences A005153/M0991 in "An On-Line Version of the Encyclopedia of Integer Sequences." http://www.research.att.com/~njas/sequences/eisonline.html.

Prandtl's Boundary Layer Equations

The system of PARTIAL DIFFERENTIAL EQUATIONS

$$u_t + uu_x + vu_y = U_t + UU_x + \frac{\mu}{\rho} u_{yy}$$

$$u_x + v_y = 0.$$

References

Iyanaga, S. and Kawada, Y. (Eds.). *Encyclopedic Dictionary of Mathematics.* Cambridge, MA: MIT Press, p. 672, 1980.
Zwillinger, D. *Handbook of Differential Equations, 3rd ed.* Boston, MA: Academic Press, p. 139, 1997.

Pratt Certificate

A primality certificate based on FERMAT'S LITTLE THEOREM CONVERSE. Although the general idea had been well-established for some time, Pratt became the first to prove that the certificate tree was of polynomial size and could also be verified in polynomial time. He was also the first to observe that the tree implies that PRIMES are in the complexity class NP.

To generate a Pratt certificate, assume that n is a POSITIVE INTEGER and $\{p_i\}$ is the set of PRIME FACTORS of $n - 1$. Suppose there exists an INTEGER x (called a "WITNESS") such that $x^{n-1} \equiv 1 \pmod{n}$ but $x^e \not\equiv 1 \pmod{n}$ whenever e is one of $(n-1)/p_i$. Then FERMAT'S LITTLE THEOREM CONVERSE states that n is PRIME (Wagon 1991, pp. 278–79).

By applying FERMAT'S LITTLE THEOREM CONVERSE to n and recursively to each purported factor of $n - 1$, a certificate for a given PRIME NUMBER can be generated. Stated another way, the Pratt certificate gives a proof that a number a is a PRIMITIVE ROOT of the multiplicative GROUP (mod p) which, along with the fact that a has order $p - 1$, proves that p is a PRIME.

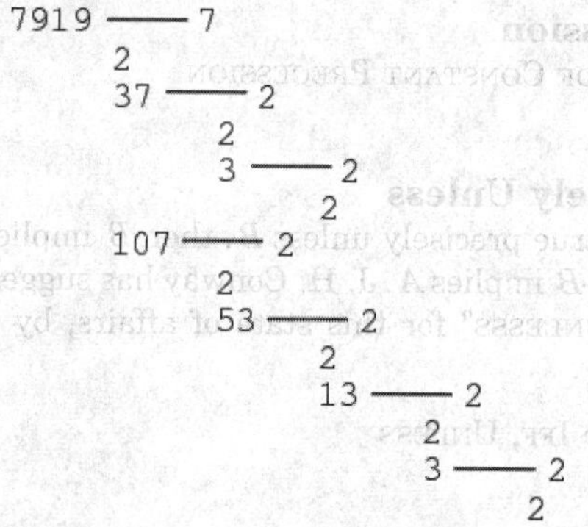

```
7919 ——— 7
  2
  37 ——— 2
     2
     3 ——— 2
        2
107 ——— 2
     2
     53 ——— 2
        2
        13 ——— 2
           2
           3 ——— 2
              2
```

The figure above gives a certificate for the primality of $n = 7919$. The numbers to the right of the dashes are WITNESSES to the numbers to left. The set $\{p_i\}$ for $n - 1 = 7918$ is given by $\{2, 37, 107\}$. Since $7^{7918} \equiv 1 \pmod{7919}$ but $7^{7918/2}$, $7^{7918/37}$, $7^{7918/107} \not\equiv 1 \pmod{7919}$, 7 is a WITNESS for 7919. The PRIME divisors of $7918 = 7919-$ are 2, 37, and 107. 2 is a so-called "self-WITNESS" (i.e., it is recognized as a PRIME without further ado), and the remainder of the witnesses are shown as a nested tree. Together, they certify that 7919 is indeed PRIME. Because it requires the FACTORIZATION of $n - 1$, the METHOD of Pratt certificates is best applied to small numbers (or those numbers n known to have easily factorable $n - 1$).

A Pratt certificate is quicker to generate for small numbers than are other types of primality certificates. The *Mathematica* task `ProvablePrimeQ[n]` in the *Mathematica* add-on package `NumberTheory`-`PrimeQ`` (which can be loaded with the command $<$ $<$ `NumberTheory``)therefore generates an ATKIN-GOLDWASSER-KILIAN-MORAIN CERTIFICATE only for numbers above a certain limit (10^{10} by default), and a Pratt certificate for smaller numbers.

See also ATKIN-GOLDWASSER-KILIAN-MORAIN CERTIFICATE, FERMAT'S LITTLE THEOREM CONVERSE, PRIMALITY CERTIFICATE, WITNESS

References

Pratt, V. "Every Prime Has a Succinct Certificate." *SIAM J. Comput.* **4**, 214–20, 1975.
Wagon, S. *Mathematica in Action.* New York: W. H. Freeman, pp. 278–85, 1991.
Wilf, H. §4.10 in *Algorithms and Complexity.* Englewood Cliffs, NJ: Prentice-Hall, 1986.

Pratt-Kasapi Theorem

HOEHN'S THEOREM

Precedes

The relationship x precedes y is written $x \prec y$. The relation x precedes or is equal to y is written $x \preccurlyeq y$.

See also SUCCEEDS

Precession

CURVE OF CONSTANT PRECESSION

Precisely Unless

If A is true precisely unless B, then B implies not-A and not-B implies A. J. H. Conway has suggested the term "UNLESSS" for this state of affairs, by analogy with IFF.

See also IFF, UNLESS

Predecessor

α is called a predecessor if there is no ORDINAL NUMBER β such that $\beta + 1 = \alpha$.

See also ORDINAL NUMBER, SUCCESSOR

Predicate

An operator in LOGIC which returns either TRUE or FALSE.

See also AND, FALSE, NAND, NOR, NOT, OR, PREDICATE CALCULUS, TRUE, XNOR, XOR

Predicate Calculus

The branch of formal LOGIC, also called functional calculus, that deals with representing the logical connections between statements as well as the statements themselves.

See also GÖDEL'S INCOMPLETENESS THEOREM, LOGIC, PREDICATE, PROPOSITIONAL CALCULUS

Predictability

Predictability at a time τ in the future is defined by

$$\frac{R(x(t),\, x(t+\tau))}{H(x(t))},$$

and linear predictability by

$$\frac{L(x(t),\, x(t+\tau))}{H(x(t))},$$

where R and L are the REDUNDANCY and LINEAR REDUNDANCY, and H is the ENTROPY.

Prediction Paradox

UNEXPECTED HANGING PARADOX

Prediction Theory

The problem of forecasting future values $X_{t+\tau}$ ($\tau > 0$) of a weakly stationary process $\{X_t\}$ from the known values X_s ($s \le t$).

See also TIME SERIES ANALYSIS

References

Itô, K. (Ed.). "Prediction Theory." §395D in *Encyclopedic Dictionary of Mathematics, 2nd ed., Vol. 3.* Cambridge, MA: MIT Press, pp. 1463–465, 1987.

Predictive Value

The positive predictive value is the probability that a test gives a true result for a true statistic. The negative predictive value is the probability that a test gives a false result for a false statistic.

See also POWER (STATISTICS), SENSITIVITY, SPECIFICITY, STATISTICAL TEST

Predictor-Corrector Methods

A general set of methods for integrating ORDINARY DIFFERENTIAL EQUATIONS. Predictor-corrector methods proceed by extrapolating a polynomial fit to the derivative from the previous points to the new point (the predictor step), then using this to interpolate the derivative (the corrector step). Press *et al.* (1992) opine that predictor-corrector methods have been largely supplanted by the BULIRSCH-STOER and RUNGE-KUTTA METHODS, but predictor-corrector schemes are still in common use.

See also ADAMS' METHOD, GILL'S METHOD, MILNE'S METHOD, RUNGE-KUTTA METHOD

References

Abramowitz, M. and Stegun, C. A. (Eds.). *Handbook of Mathematical Functions with Formulas, Graphs, and Mathematical Tables, 9th printing.* New York: Dover, pp. 896–97, 1972.

Arfken, G. *Mathematical Methods for Physicists, 3rd ed.* Orlando, FL: Academic Press, pp. 493–94, 1985.

Press, W. H.; Flannery, B. P.; Teukolsky, S. A.; and Vetterling, W. T. "Multistep, Multivalue, and Predictor-Corrector Methods." §16.7 in *Numerical Recipes in FORTRAN: The Art of Scientific Computing, 2nd ed.* Cambridge, England: Cambridge University Press, pp. 740–44, 1992.

Preimage

Given $f : X \to Y$, the image of x is $f(x)$. The preimage of y is then $f^{-1}(y) = \{x | f(x) = y\}$, or all x whose image is y. Images are in the range, while preimages are in the domain (or they are empty).

Present Value

The present value v_n of a single payment made at n periods in the future is

$$v_n = \frac{p}{(1+r)^n}, \qquad (1)$$

where n is the number of periods until payment, p is the payment amount, and r is the periodic discount rate. The present value v_∞ of equal payments made each successive period in perpetuity (a.k.a. the present value of a perpetuity) is given by

$$v_\infty = \sum_{n=1}^\infty \frac{p}{(1+r)^n} = \frac{p}{r}. \tag{2}$$

The present value v' of equal payments made each successive period for n periods (a.k.a. the present value of an annuity) is given by

$$v' = v_\infty - v_n = \frac{p}{r}\left[1 - \frac{1}{(1+r)^n}\right], \tag{3}$$

where p is the periodic payment amount.

See also INTEREST

Pretzel Curve

KNOT CURVE

Pretzel Knot

A KNOT obtained from a TANGLE which can be represented by a FINITE sequence of INTEGERS.

See also TANGLE

References

Adams, C. C. *The Knot Book: An Elementary Introduction to the Mathematical Theory of Knots.* New York: W. H. Freeman, p. 48, 1994.

Pretzel Transformation

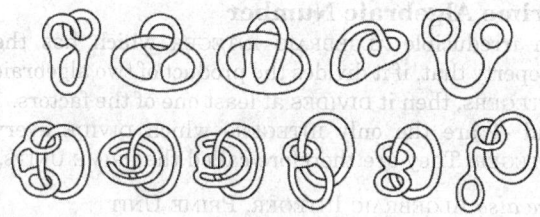

A topological transformation in which a surface is made out of an infinitely elastic material which, however, may not be torn or cut. Using this simple prescription gives the amazing two conversions illustrated above, the first of which untangles two interlocked rings connected by a band, and the second of which unloops one of two rings connected by a band and threaded by a band (Wells 1991).

See also TOPOLOGY

References

Wells, D. *The Penguin Dictionary of Curious and Interesting Geometry.* London: Penguin, p. 194, 1991.

Price's Theorem

Consider a GAUSSIAN BIVARIATE DISTRIBUTION in variables **x** and **y** with COVARIANCE

$$\rho = \rho_{11} = \langle \mathbf{xy} \rangle - \langle \mathbf{x} \rangle \langle \mathbf{y} \rangle$$

and an arbitrary function $g(x, y)$. Then the expected value of the random variable $g(\mathbf{x}, \mathbf{y})$

$$\langle g(\mathbf{x}, \mathbf{y}) \rangle = \int_{-\infty}^\infty \int_{-\infty}^\infty g(x, y) f(x, y)\, dx\, dy$$

satisfies

$$\frac{\partial^n \langle g(\mathbf{x}, \mathbf{y}) \rangle}{\partial \rho^n} = \left\langle \frac{\partial^{2n} g(\mathbf{x}, \mathbf{y})}{\partial \mathbf{x}^n\, \partial \mathbf{y}^n} \right\rangle.$$

See also COVARIANCE, GAUSSIAN BIVARIATE DISTRIBUTION

References

McMahon, E. L. "An Extension of Price's Theorem." *IEEE Trans. Inform. Th.* **10**, 168–71, 1964.
Papoulis, A. "Price's Theorem and Join Moments." *Probability, Random Variables, and Stochastic Processes, 2nd ed.* New York: McGraw-Hill, pp. 226–28, 1984.
Price, R. "A Useful Theorem for Non-Linear Devices Having Gaussian Inputs." *IEEE Trans. Inform. Th.* **4**, 69–2, 1958.

Primality Certificate

A short set of data that proves the primality of a number. A certificate can, in general, be checked much more quickly than the time required to generate the certificate. Varieties of primality certificates include the PRATT CERTIFICATE and ATKIN-GOLDWASSER-KILIAN-MORAIN CERTIFICATE.

See also ATKIN-GOLDWASSER-KILIAN-MORAIN CERTIFICATE, COMPOSITENESS CERTIFICATE, PRATT CERTIFICATE

References

Wagon, S. "Prime Certificates." §8.7 in *Mathematica in Action.* New York: W. H. Freeman, pp. 277–85, 1991.

Primality Test

A test to determine whether or not a given number is PRIME. The RABIN-MILLER STRONG PSEUDOPRIME TEST is a particularly efficient ALGORITHM used by *Mathematica* version 2.2. Like many such algorithms, it is a probabilistic test using PSEUDOPRIMES, and can potentially (although with very small probability) falsely identify a COMPOSITE NUMBER as PRIME (although not vice versa). Unlike PRIME FACTORIZATION, primality testing is believed to be a P-PROBLEM (Wagon 1991). In order to guarantee primality, an almost certainly slower algorithm capable of generating a PRIMALITY CERTIFICATE must be used.

See also ADLEMAN-POMERANCE-RUMELY PRIMALITY

TEST, FERMAT'S LITTLE THEOREM CONVERSE, FERMAT'S PRIMALITY TEST, FERMAT'S THEOREM, LUCAS-LEHMER TEST, MILLER'S PRIMALITY TEST, PÉPIN'S TEST, POCKLINGTON'S THEOREM, PROTH'S THEOREM, PSEUDOPRIME, RABIN-MILLER STRONG PSEUDOPRIME TEST, WARD'S PRIMALITY TEST, WILSON'S THEOREM

References

Beauchemin, P.; Brassard, G.; Crépeau, C.; Goutier, C.; and Pomerance, C. "The Generation of Random Numbers that are Probably Prime." *J. Crypt.* **1**, 53–4, 1988.

Brillhart, J.; Lehmer, D. H.; Selfridge, J.; Wagstaff, S. S. Jr.; and Tuckerman, B. *Factorizations of $b^n \pm 1$, $b = 2$, 3, 5, 6, 7, 10, 11, 12 Up to High Powers, rev. ed.* Providence, RI: Amer. Math. Soc., pp. lviii-lxv, 1988.

Cohen, H. and Lenstra, A. K. "Primality Testing and Jacobi Sums." *Math. Comput.* **42**, 297–30, 1984.

Knuth, D. E. *The Art of Computer Programming, Vol. 2: Seminumerical Algorithms, 3rd ed.* Reading, MA: Addison-Wesley, 1998.

Riesel, H. *Prime Numbers and Computer Methods for Factorization, 2nd ed.* Boston, MA: Birkhäuser, 1994.

Wagon, S. *Mathematica in Action.* New York: W. H. Freeman, pp. 15–7, 1991.

Williams, H. C. *Edouard Lucas and Primality Testing.* New York: Wiley, 1998.

Primary

Each factor $p_i^{\alpha_i}$ in an INTEGER'S PRIME FACTORIZATION is called a primary.

Primary Pseudoperfect Number

An integer N which is a product of distinct primes and which satisfies

$$\frac{1}{N} + \sum_{p|N} \frac{1}{p} = 1$$

(Butske *et al.* 1999). The first few are 2, 6, 42, 1806, 47058, ... (Sloane's A054377).

The similar equation

$$-\frac{1}{N} + \sum_{p|N} \frac{1}{p} = 1$$

arises in the definition of GIUGA NUMBERS.

See also GIUGA NUMBER, SEMIPERFECT NUMBER

References

Borwein, D.; Borwein, J. M.; Borwein, P. B.; and Girgensohn, R. "Giuga's Conjecture on Primality." *Amer. Math. Monthly* **103**, 40–0, 1996.

Butske, W.; Jaje, L. M.; and Mayernik, D. R. "The Equation $\Sigma_{p|N} 1/p + 1/N = 1$, Pseudoperfect Numbers, and Partially Weighted Graphs." *Math. Comput.* **69**, 407–20, 1999.

Cao, Z.; Liu, R.; and Zhang, L. "On the Equation $\Sigma_{j=1}^s (\frac{1}{x_j}) + \frac{1}{(x_j + \cdots + x_n)}$ and Znám's Problem." *J. Number Th.* **27**, 206–11, 1987.

Ke, Z. and Sun, Q. "On the Representation of 1 by Unit Fractions." *Sichuan Daxue Xuebao* **1**, 13–9, 1964.

Sloane, N. J. A. Sequences A054377 in "An On-Line Version of the Encyclopedia of Integer Sequences." http://www.research.att.com/~njas/sequences/eisonline.html.

Primary Representation

Let π be a UNITARY REPRESENTATION of a GROUP G on a separable HILBERT SPACE, and let $R(\pi)$ be the smallest weakly closed algebra of bounded linear operators containing all $\pi(g)$ for $g \in G$. Then π is primary if the center of $R(\pi)$ consists of only scalar operations.

See also REPRESENTATION

References

Knapp, A. W. "Group Representations and Harmonic Analysis, Part II." *Not. Amer. Math. Soc.* **43**, 537–49, 1996.

Prime

A symbol used to distinguish one quantity x' ("x''") from another related x. Primes are most commonly used to denote

1. Transformed coordinates,
2. Conjugate points,
3. DERIVATIVES,
4. The COMPLEMENT F' of a set F,
5. As an alternate notation for TRANSPOSE.

See also DOUBLE PRIME, PRIME ALGEBRAIC NUMBER, PRIME NUMBER

References

Bringhurst, R. *The Elements of Typographic Style, 2nd ed.* Point Roberts, WA: Hartley and Marks, p. 283, 1997.

Prime Algebraic Number

An irreducible ALGEBRAIC INTEGER which has the property that, if it divides the product of two algebraic INTEGERS, then it DIVIDES at least one of the factors. 1 and -1 are the only INTEGERS which DIVIDE every INTEGER. They are therefore called the PRIME UNITS.

See also ALGEBRAIC INTEGER, PRIME UNIT

Prime Arithmetic Progression

An arithmetic progression of primes is a set of primes OF THE FORM $mk + n$ for fixed m and n and consecutive k, i.e., $\{n, m+n, 2m+n, \ldots\}$. For example, 199, 409, 619, 829, 1039, 1249, 1459, 1669, 1879, 2089 is a 10-term arithmetic progression of primes with difference 210. Let P be an increasing arithmetic progression of n PRIMES with minimal difference $d > 0$. If a PRIME $p \leq n$ does not divide d, then the elements of P must assume all residues modulo p, specifically, some element of P must be divisible by p. Since P contains only primes, this element must be equal to p.

Let the number of PRIMES OF THE FORM $mk+n$ less than x be denoted $\pi_{m,\,n}(x)$. Then

$$\lim_{x\to\infty}\frac{\pi_{a,\,b}(x)}{\mathrm{Li}(x)}=\frac{1}{\phi(a)},$$

where $\mathrm{Li}(x)$ is the LOGARITHMIC INTEGRAL and $\phi(x)$ is the TOTIENT FUNCTION.

If $d < n\#$ (where $n\#$ is the PRIMORIAL of n), then some prime $p \le n$ does not divide d, and that prime p is in P. Thus, in order to determine if P has $d < n\#$, we need only check a finite number of possible P (those with $d < n\#$ and containing prime $p \le n$) to see if they contain only primes. If not, then $d \ge n\#$. If $d = n\#$, then the elements of P cannot be made to cover all residues of any prime p. The PRIME PATTERNS CONJECTURE then asserts that there are infinitely many arithmetic progressions of primes with difference d.

A computation shows that the smallest possible common difference for a set of n or more PRIMES in arithmetic progression for $n = 1, 2, 3, \ldots$ is 0, 1, 2, 6, 6, 30, 150, 210, 210, 210, 2310, 2310, 30030, 30030, 30030, 510510, ... (Sloane's A033188, Ribenboim 1989, Dubner and Nelson 1997, Wilson). The values up to $n = 13$ are rigorous, while the remainder are lower bounds which assume the validity of the PRIME PATTERNS CONJECTURE and are simply given by $p_{n-7}\#$, where p_i is the ith PRIME. The smallest first terms of arithmetic progressions of n primes *with minimal differences* are 2, 2, 3, 5, 5, 7, 7, 199, 199, 199, 60858179, 147692845283, 14933623, 856378247603, ... (Sloane's A033189; Wilson).

Smaller first terms are possible for nonminimal n-term progressions. Examples include the 8-term progression $11 + 1210230k$ for $k = 0, 1, \ldots, 7$, the 12-term progression $23143 + 30030k$ for $k = 0, 1, \ldots, 11$ (Golubev 1969, Guy 1994), and the 13-term arithmetic progression $766439 + 510510k$ for $k = 0, 1, \ldots, 12$ (Guy 1994).

The largest known set of primes in ARITHMETIC SEQUENCE is 22,

11, 410, 337, 580, 553 + 4, 609, 098, 694, 200k

for $k = 0, 1, \ldots, 21$ (Pritchard *et al.* 1995, UTS School of Mathematical Sciences).

The largest known sequence of *consecutive* PRIMES in ARITHMETIC PROGRESSION (i.e., all the numbers between the first and last term in the progression, except for the members themselves, are composite) is ten, given by

100, 996, 972, 469, 714, 247, 637, 786, 655, 587, 969,

840, 329 509, 324, 689, 190, 041, 803, 603, 417, 758,

904, 341, 703, 348, 882, 159, 067, 229, 719 + 210k

for $k = 0, 1, \ldots, 9$ (Sloane's A033290), discovered by Harvey Dubner, Tony Forbes, Manfred Toplic, *et al.*

on March 2, 1998. This beats the record of nine consecutive primes set on January 15, 1998 by the same investigators,

99, 679, 432, 066, 701, 086, 484, 490, 653, 695, 853,

561, 638, 982, 364, 080, 991, 618, 395, 774, 048, 585,

529, 071, 475, 461, 114, 799, 677, 694, 651 + 210k

for $k = 0, 1, \ldots, 8$ (two sequences of nine are now known), the progression of eight consecutive primes given by

43, 804, 034, 644, 029, 893, 325, 717, 710, 709, 965,

599, 930, 101, 479, 007, 432, 825, 862, 862, 446, 333,

961, 919, 524, 977, 985, 103, 251, 510, 661 + 210k

for $k = 0, 1, \ldots, 7$, discovered by Harvey Dubner, Tony Forbes, *et al.* on November 7, 1997 (several are now known), and the progression of seven given by

1, 089, 533, 431, 247, 059, 310, 875, 780, 378, 922, 957, 732,

908, 036, 492, 993, 138, 195, 385, 213, 105, 561, 742, 150,

447, 308, 967, 213, 141, 717, 486, 151 + 210k,

for $k = 0, 1, \ldots, 6$, discovered by H. Dubner and H. K. Nelson on Aug. 29, 1995 (Peterson 1995, Dubner and Nelson 1997). The smallest sequence of six consecutive PRIMES in arithmetic progression is

$$121, 174, 811 + 30k$$

for $k = 0, 1, \ldots, 5$ (Lander and Parkin 1967, Dubner and Nelson 1997). According to Dubner *et al.*, a trillion-fold increase in computer speed is needed before the search for a sequence of 11 consecutive primes is practical, so they expect the ten-primes record to stand for a long time to come.

It is conjectured that there are arbitrarily long sequences of PRIMES in ARITHMETIC PROGRESSION (Guy 1994). W. Roonguthai found the largest known arithmetic progression of three primes, $(3, 1593 \cdot 2^{27757} + 1, 1593 \cdot 2^{27758} - 1)$, with common difference $1593 \cdot 2^{27757} - 2$ (Roonguthai 1999).

See also ARITHMETIC PROGRESSION, CUNNINGHAM CHAIN, DIRICHLET'S THEOREM, LINNIK'S THEOREM, PRIME CONSTELLATION, PRIME-GENERATING POLYNOMIAL, PRIME NUMBER THEOREM, PRIME PATTERNS CONJECTURE, PRIME QUADRUPLET

References

Abel, U. and Siebert, H. "Sequences with Large Numbers of Prime Values." *Amer. Math. Monthly* **100**, 167–69, 1993.

Caldwell, C. K. "Cunningham Chain." http://www.utm.edu/research/primes/glossary/CunninghamChain.html.

Courant, R. and Robbins, H. "Primes in Arithmetical Progressions." §1.2b in Supplement to Ch. 1 in *What is Mathematics?: An Elementary Approach to Ideas and Methods, 2nd ed.* Oxford, England: Oxford University Press, pp. 26–7, 1996.

Davenport, H. "Primes in Arithmetic Progression" and "Primes in Arithmetic Progression: The General Modulus." Chs. 1 and 4 in *Multiplicative Number Theory, 2nd ed.* New York: Springer-Verlag, pp. 1–1 and 27–4, 1980.

Dubner, H. *J. Recr. Math.* **20**, 211–13, 1988.

Dubner, H. and Nelson, H. "Seven Consecutive Primes in Arithmetic Progression." *Math. Comput.* **66**, 1743–749, 1997.

Forbes, T. "Searching for 9 Consecutive Primes in Arithmetic Progression." http://www.ltkz.demon.co.uk/ar2/9primes.htm.

Forman, R. "Sequences with Many Primes." *Amer. Math. Monthly* **99**, 548–57, 1992.

Gardner, M. "Primes in Arithmetic Progression." In Press, R. *Mathematical Sciences Calendar 1988.*

Golubev, V. A. "Faktorisation der Zahlen der Form $x^3 \pm 4x^2 + 3x \pm 1$." *Anz. Österreich. Akad. Wiss. Math.-Naturwiss. Kl.* 184–91, 1969.

Guy, R. K. "Arithmetic Progressions of Primes" and "Consecutive Primes in A.P." §A5 and A6 in *Unsolved Problems in Number Theory, 2nd ed.* New York: Springer-Verlag, pp. 15–7 and 18, 1994.

Lander, L. J. and Parkin, T. R. "Consecutive Primes in Arithmetic Progression." *Math. Comput.* **21**, 489, 1967.

Madachy, J. S. *Madachy's Mathematical Recreations.* New York: Dover, pp. 154–55, 1979.

Nelson, H. L. "There Is a Better Sequence." *J. Recr. Math.* **8**, 39–3, 1975.

Peterson, I. "Progressing to a Set of Consecutive Primes." *Sci. News* **148**, 167, Sep. 9, 1995.

Pritchard, P. A.; Moran, A.; and Thyssen, A. "Twenty-Two Primes in Arithmetic Progression." *Math. Comput.* **64**, 1337–339, 1995.

Ramaré, O. and Rumely, R. "Primes in Arithmetic Progressions." *Math. Comput.* **65**, 397–25, 1996.

Ribenboim, P. *The Book of Prime Number Records, 2nd ed.* New York: Springer-Verlag, p. 224, 1989.

Roonguthai, W. "Record Arithmetic Progression of Primes." NMBRTHRY@LISTSERV.NODAK.EDU mailing list posting. Feb. 4, 1999.

Shanks, D. "Primes in Some Arithmetic Progressions and a General Divisibility Theorem." §104 in *Solved and Unsolved Problems in Number Theory, 4th ed.* New York: Chelsea, pp. 104–09, 1993.

Sloane, N. J. A. Sequences A033188, A033189, and A033290 in "An On-Line Version of the Encyclopedia of Integer Sequences." http://www.research.att.com/~njas/sequences/eisonline.html.

UTS School of Mathematical Sciences. "Primes in Arithmetic Progression." http://www.maths.uts.edu.au/numericon/prime2.html.

Weintraub, S. "Consecutive Primes in Arithmetic Progression." *J. Recr. Math.* **25**, 169–71, 1993.

Zimmerman, P. http://www.loria.fr/~zimmerma/records/8primes.announce.

Prime Array

Find the $m \times n$ ARRAY of single digits which contains the maximum possible number of PRIMES, where allowable PRIMES may lie along any horizontal, vertical, or diagonal line. For $m = n = 2$, 11 PRIMES are maximal and are contained in the two distinct arrays

$$A(2,\ 2) = \begin{bmatrix} 1 & 3 \\ 4 & 7 \end{bmatrix},\ \begin{bmatrix} 1 & 3 \\ 7 & 9 \end{bmatrix},$$

giving the PRIMES (3, 7, 13, 17, 31, 37, 41, 43, 47, 71, 73) and (3, 7, 13, 17, 19, 31, 37, 71, 73, 79, 97),

respectively. For the 3×2 array, 18 PRIMES are maximal and are contained in the arrays

$$A(3,\ 2) = \begin{bmatrix} 1 & 1 & 3 \\ 9 & 7 & 4 \end{bmatrix},\ \begin{bmatrix} 1 & 7 & 2 \\ 3 & 5 & 9 \end{bmatrix},\ \begin{bmatrix} 1 & 7 & 2 \\ 4 & 3 & 9 \end{bmatrix},$$

$$\begin{bmatrix} 1 & 7 & 5 \\ 4 & 3 & 9 \end{bmatrix},\ \begin{bmatrix} 1 & 7 & 9 \\ 3 & 2 & 5 \end{bmatrix},\ \begin{bmatrix} 1 & 7 & 9 \\ 4 & 3 & 2 \end{bmatrix},$$

$$\begin{bmatrix} 1 & 7 & 9 \\ 4 & 3 & 4 \end{bmatrix},\ \begin{bmatrix} 3 & 1 & 6 \\ 4 & 7 & 9 \end{bmatrix},\ \begin{bmatrix} 3 & 7 & 6 \\ 4 & 1 & 9 \end{bmatrix}.$$

The best 3×3 array is

$$A(3,\ 3) = \begin{bmatrix} 1 & 1 & 3 \\ 7 & 5 & 4 \\ 9 & 3 & 7 \end{bmatrix},$$

which contains 30 primes: 3, 5, 7, 11, 13, 17, 31, 37, 41, 43, 47, 53, 59, 71, 73, 79, 97, 113, 157, 179, ... (Sloane's A032529). This array was found by Rivera and Ayala and shown by Weisstein in May 1999 to be maximal and unique (modulo reflection and rotation).

The best 4×4 arrays known are

$$\begin{bmatrix} 1 & 1 & 3 & 9 \\ 6 & 4 & 5 & 1 \\ 7 & 3 & 9 & 7 \\ 3 & 9 & 2 & 9 \end{bmatrix},\ \begin{bmatrix} 1 & 1 & 3 & 9 \\ 7 & 6 & 9 & 2 \\ 5 & 4 & 7 & 9 \\ 1 & 7 & 3 & 3 \end{bmatrix},$$

$$\begin{bmatrix} 1 & 7 & 3 & 3 \\ 9 & 4 & 2 & 1 \\ 6 & 5 & 9 & 1 \\ 7 & 7 & 3 & 9 \end{bmatrix},\ \begin{bmatrix} 3 & 1 & 6 & 7 \\ 7 & 5 & 1 & 4 \\ 9 & 2 & 9 & 3 \\ 3 & 3 & 7 & 3 \end{bmatrix},$$

all of which contain 63 PRIMES. The first was found by C. Rivera and J. Ayala in 1998, and the other three by James Bonfield on April 13, 1999.

The best 5×5 prime arrays known are

$$\begin{bmatrix} 1 & 1 & 9 & 3 & 3 \\ 9 & 9 & 5 & 6 & 3 \\ 8 & 9 & 4 & 1 & 7 \\ 3 & 3 & 7 & 3 & 1 \\ 3 & 2 & 9 & 3 & 9 \end{bmatrix},\ \begin{bmatrix} 3 & 3 & 1 & 9 & 9 \\ 8 & 3 & 9 & 1 & 1 \\ 2 & 7 & 4 & 5 & 7 \\ 1 & 9 & 6 & 7 & 3 \\ 9 & 7 & 9 & 1 & 9 \end{bmatrix},$$

each of which contains 116 PRIMES. The first was found by C. Rivera and J. Ayala in 1998, and the second by Wilfred Whiteside on April 17, 1999.

The best 6×6 prime arrays known are

$$\begin{bmatrix} 1 & 3 & 9 & 1 & 9 & 9 \\ 3 & 1 & 7 & 2 & 3 & 4 \\ 9 & 9 & 4 & 7 & 9 & 3 \\ 9 & 1 & 5 & 7 & 1 & 3 \\ 9 & 8 & 3 & 6 & 1 & 7 \\ 9 & 1 & 7 & 3 & 3 & 3 \end{bmatrix},\ \begin{bmatrix} 1 & 3 & 9 & 1 & 9 & 9 \\ 9 & 1 & 7 & 2 & 3 & 4 \\ 6 & 9 & 4 & 7 & 9 & 3 \\ 7 & 1 & 5 & 7 & 1 & 3 \\ 9 & 8 & 3 & 6 & 1 & 7 \\ 9 & 1 & 7 & 3 & 3 & 3 \end{bmatrix},$$

$$\begin{bmatrix} 3&1&7&3&3&3 \\ 9&9&5&6&3&9 \\ 1&1&8&1&4&2 \\ 1&3&6&3&7&3 \\ 3&4&9&1&9&9 \\ 3&7&9&3&7&9 \end{bmatrix}, \begin{bmatrix} 3&1&7&3&3&3 \\ 9&9&5&6&3&9 \\ 1&1&8&1&4&2 \\ 1&3&6&3&7&3 \\ 3&4&9&1&9&9 \\ 3&7&9&3&7&9 \end{bmatrix},$$

$$\begin{bmatrix} 3&1&7&3&3&3 \\ 9&9&5&6&3&9 \\ 1&1&8&1&4&2 \\ 1&3&6&3&7&3 \\ 3&4&9&1&9&9 \\ 9&7&9&3&7&9 \end{bmatrix}, \begin{bmatrix} 3&1&7&3&3&3 \\ 9&9&5&6&3&9 \\ 1&1&8&1&4&5 \\ 1&3&6&3&7&3 \\ 3&4&9&1&9&9 \\ 9&9&9&2&3&3 \end{bmatrix}.$$

each of which contain 187 primes. One was found by S. C. Root, and the others by M. Oswald in 1998.

The best 7×7 prime array known is

$$\begin{bmatrix} 3&1&3&7&3&3&9 \\ 9&9&2&3&3&3&3 \\ 6&9&7&7&8&9&4 \\ 7&6&1&5&9&1&9 \\ 7&7&3&4&2&1&1 \\ 9&9&4&7&9&3&9 \\ 3&3&7&1&9&9&9 \end{bmatrix},$$

which contains 281 primes and was found by Wilfred Whiteside on April 29, 1999.

The best 8×8 prime array known is

$$\begin{bmatrix} 3&3&1&3&9&1&3&3 \\ 6&9&3&3&7&3&9&7 \\ 7&9&9&6&8&5&7&1 \\ 9&7&9&9&1&2&4&9 \\ 1&3&2&1&1&3&9&9 \\ 6&3&9&1&9&4&6&3 \\ 6&3&8&5&3&7&9&3 \\ 9&1&3&1&3&9&3&3 \end{bmatrix}$$

which contains 382 primes and was found by Wilfred Whiteside On Oct. 31, 1999.

Heuristic arguments by Rivera and Ayala suggest that the maximum possible number of primes in 4×4, 5×5, and 6×6 arrays are 58–3, 112–21, and 205–18, respectively.

See also Array, Prime Arithmetic Progression, Prime Constellation, Prime String

References

Dewdney, A. K. "Computer Recreations: How to Pan for Primes in Numerical Gravel." *Sci. Amer.* **259**, 120–23, July 1988.
Lee, G. "Winners and Losers." *Dragon User.* May 1984.
Lee, G. "Gordon's Paradoxically Perplexing Primesearch Puzzle." http://www.geocities.com/MotorCity/7983/prime-search.html.
Rivera, C. "Problems & Puzzles: Puzzle The Gordon Lee Puzzle.-061." http://www.primepuzzles.net/puzzles/puzz_061.htm.

Sloane, N. J. A. Sequences A032529 in "An On-Line Version of the Encyclopedia of Integer Sequences." http://www.research.att.com/~njas/sequences/eisonline.html.
Weisstein, E. W. "Prime Arrays." Mathematica notebook PrimeArray.m.

Prime Circle

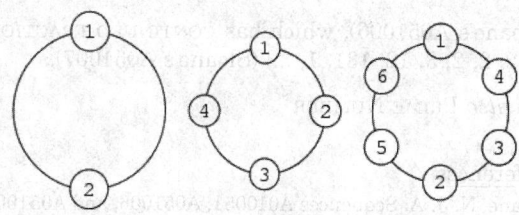

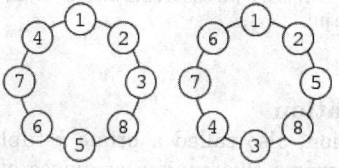

A prime circle of order $2n$ is a free Circular Permutation of the numbers from 1 to $2n$ with adjacent Pairs summing to a Prime. The number of prime circles for $n = 1, 2, \ldots$, are 1, 1, 1, 2, 48, 512, ... (Sloane's A051252). The prime circles for the first few even orders are given in the table below.

$2n$	prime circles
2	$\{1, 2\}$
4	$\{1, 2, 3, 4\}$
6	$\{1, 4, 3, 2, 5, 6\}$
8	$\{1, 2, 3, 8, 5, 6, 7, 4\}$, $\{1, 2, 5, 8, 3, 4, 7, 6\}$

See also Circular Permutation

References

Filz, A. "Problem 1046." *J. Recr. Math.* **14**, 64, 1982.
Filz, A. "Problem 1046." *J. Recr. Math.* **15**, 71, 1983.
Guy, R. K. *Unsolved Problems in Number Theory, 2nd ed.* New York: Springer-Verlag, pp. 105–06, 1994.
Sloane, N. J. A. Sequences A051252 in "An On-Line Version of the Encyclopedia of Integer Sequences." http://www.research.att.com/~njas/sequences/eisonline.html.

Prime Cluster
Prime Constellation

Prime Constant
The characteristic function

$$f(n) = \begin{cases} 1 & n \text{ is prime} \\ 0 & n \text{ otherwise} \end{cases} \tag{1}$$

therefore has first few values 0, 1, 1, 0, 1, 0, 1, 0, 0, 0, 1, 0, 1, 0, 0, 0, 1, 0, 1, ... (Sloane's A010051). The constant obtained by concatenating these digits in binary is therefore

$$P \equiv 0.011010100\ldots_2$$
$$= 0.4146825098511116602481\ldots \quad (2)$$

(Sloane's A051006), which has CONTINUED FRACTION [0, 2, 2, 2, 3, 12, 131, 1, ...] (Sloane's A051007).

See also PRIME NUMBER

References

Sloane, N. J. A. Sequences A010051, A051006, and A051007 in "An On-Line Version of the Encyclopedia of Integer Sequences." http://www.research.att.com/~njas/sequences/eisonline.html.

Prime Constellation

A prime constellation, also called a prime k-tuple, prime k-tuplet, or prime cluster, is a sequence of k consecutive numbers such that the difference between the first and last is, in some sense, the least possible. More precisely, a prime k-tuplet is a sequence of consecutive PRIMES $(p_1, p_2, \ldots, p_k)$ with $p_k - p_1 = s(k)$, where $s(k)$ is the smallest number s for which there exist k integers $b_1 < b_2 < \ldots < b_k$, $b_k - b_1 = s$ and, for every PRIME q, not all the residues modulo q are represented by $b_1, b_2, \ldots, b_k$ (Forbes). For each k, this definition excludes a finite number of clusters at the beginning of the prime number sequence. For example, (97, 101, 103, 107, 109) satisfies the conditions of the definition of a prime 5-tuplet, but (3, 5, 7, 11, 13) does not because all three residues modulo 3 are represented (Forbes).

A prime double with $s(2) = 2$ is OF THE FORM $(p, p+2)$ and is called a pair of TWIN PRIMES. Prime doubles OF THE FORM $(p, p+4)$ are called COUSIN PRIMES, and prime doubles OF THE FORM $(p, p+6)$ are called SEXY PRIMES.

A prime triplet has $s(3) = 6$. The constellation $(p, p+2, p+4)$ cannot exist, except for $p = 3$, since one of p, $p+2$, and $p+4$ must be divisible by three. However, there are several types of prime triplets which can exist: $(p, p+2, p+6)$, $(p, p+4, p+6)$, $(p, p+6, p+12)$.

A PRIME QUADRUPLET is a constellation of four successive PRIMES with minimal distance $s(4) = 8$, and is of the form $(p, p+2, p+6, p+8)$. The sequence $s(n)$ therefore begins 2, 6, 8, and continues 12, 16, 20, 26, 30, ... (Sloane's A008407). Another quadruplet constellation is $(p, p+6, p+12, p+18)$.

Hardy and Wright (1979, p. 5) conjecture, and it seems almost certain to be true, that there are infinitely many TWIN PRIMES $(p, p+2)$ and PRIME TRIPLETS OF THE FORM $(p, p+2, p+6)$ and $(p, p+4, p+6)$.

The first FIRST HARDY-LITTLEWOOD CONJECTURE states that the numbers of constellations $\leq x$ are asymptotically given by

$$P_x(p, p+2) \sim 2 \prod_{p \geq 3} \frac{p(p-2)}{(p-1)^2} \int_2^x \frac{dx'}{(\ln x')^2}$$
$$= 1.320323632 \int_2^x \frac{dx'}{(\ln x')^2} \quad (1)$$

$$P_x(p, p+4) \sim 2 \prod_{p \geq 3} \frac{p(p-2)}{(p-1)^2} \int_2^x \frac{dx'}{(\ln x')^2}$$
$$= 1.320323632 \int_2^x \frac{dx'}{(\ln x')^2} \quad (2)$$

$$P_x(p, p+6) \sim 4 \prod_{p \geq 3} \frac{p(p-2)}{(p-1)^2} \int_2^x \frac{dx'}{(\ln x')^2}$$
$$= 2.640647264 \int_2^x \frac{dx'}{(\ln x')^2} \quad (3)$$

$$P_x(p, p+2, p+6) \sim \frac{9}{2} \prod_{p \geq 5} \frac{p^2(p-3)}{(p-1)^3} \int_2^x \frac{dx'}{(\ln x')^3}$$
$$= 2.858248596 \int_2^x \frac{dx'}{(\ln x')^3} \quad (4)$$

$$P_x(p, p+4, p+6) \sim \frac{9}{2} \prod_{p \geq 5} \frac{p^2(p-3)}{(p-1)^3} \int_2^x \frac{dx'}{(\ln x')^3}$$
$$= 2.858248596 \int_2^x \frac{dx'}{(\ln x')^3} \quad (5)$$

$$P_x(p, p+2, p+6, p+8) \sim \frac{27}{2} \prod_{p \geq 5} \frac{p^3(p-4)}{(p-1)^4} \int_2^x \frac{dx'}{(\ln x')^4}$$
$$= 4.151180864 \int_2^x \frac{dx'}{(\ln x')^4} \quad (6)$$

$$P_x(p, p+4, p+6, p+10)$$
$$\sim 27 \prod_{p \geq 5} \frac{p^3(p-4)}{(p-1)^4} \int_2^x \frac{dx'}{(\ln x')^4}$$
$$= 8.302361728 \int_2^x \frac{dx'}{(\ln x')^4} \quad (7)$$

These numbers are sometimes called the HARDY-LITTLEWOOD CONSTANTS. (1) is sometimes called the extended TWIN PRIME CONJECTURE, and

$$C_{p, p+2} = 2\Pi_2, \quad (8)$$

where Π_2 is the TWIN PRIMES CONSTANT. Riesel (1994) remarks that the HARDY-LITTLEWOOD CONSTANTS can be computed to arbitrary accuracy without needing the infinite sequence of primes.

The integrals above have the analytic forms

$$\int_2^x \frac{dx'}{(\ln x')^2} = \text{Li}(x) + \frac{2}{\ln 2} - \frac{n}{\ln n} \quad (9)$$

$$\int_2^x \frac{dx'}{(\ln x')^4} = \tfrac{1}{2}\text{Li}(x) - \frac{x(1+\ln x)}{(\ln x)^2} + \frac{1}{\ln 2} + \frac{1}{(\ln n)^2} \quad (10)$$

$$\int_2^x \frac{dx'}{(\ln x')^3} = \frac{1}{6}\left\{ \text{Li}(x) + \frac{2\left[2 + \ln 2 + (\ln 2)^2\right]}{(\ln 2)^3} \right.$$

$$\left. - \frac{n[2 + \ln n + (\ln n)^2]}{(\ln n)^3} \right\}, \quad (11)$$

where Li(x) is the LOGARITHMIC INTEGRAL.

The following table gives the number of prime constellations $\leq 10^8$, and the second table gives the values predicted by the Hardy-Littlewood formulas.

Count	10^5	10^6	10^7	10^8
$(p, p+2)$	1224	8169	58980	440312
$(p, p+4)$	1216	8144	58622	440258
$(p, p+6)$	2447	16386	117207	879908
$(p, p+2, p+6)$	259	1393	8543	55600
$(p, p+4, p+6)$	248	1444	8677	55556
$(p, p+2, p+6, p+8)$	38	166	899	4768
$(p, p+6, p+12, p+18)$	75	325	1695	9330

Hardy-Littlewood	10^5	10^6	10^7	10^8
$(p, p+2)$	1249	8248	58754	440368
$(p, p+4)$	1249	8248	58754	440368
$(p, p+6)$	2497	16496	117508	880736
$(p, p+2, p+6)$	279	1446	8591	55491
$(p, p+4, p+6)$	279	1446	8591	55491
$(p, p+2, p+6, p+8)$	53	184	863	4735
$(p, p+6, p+12, p+18)$				

Consider prime constellations in which each term is OF THE FORM n^2+1. Hardy and Littlewood showed that the number of prime constellations of this form $< x$ is given by

$$P(x) \sim C\sqrt{x}(\ln x)^{-1}, \quad (12)$$

where

$$C = \prod_{\substack{p>2 \\ p\ \text{prime}}} \left[1 - \frac{(-1)^{(p-1)/2}}{p-1} \right] = 1.3727\ldots \quad (13)$$

(Le Lionnais 1983).

Forbes gives a list of the "top ten" prime k-tuples for $2 \leq k \leq 17$. The largest known 14-constellations are (11319107721272355839 + 0, 2, 8, 14, 18, 20, 24, 30, 32, 38, 42, 44, 48, 50), (10756418345074847279 + 0, 2, 8, 14, 18, 20, 24, 30, 32, 38, 42, 44, 48, 50), (6808488664768715759 + 0, 2, 8, 14, 18, 20, 24, 30, 32, 38, 42, 44, 48, 50), (6120794469172998449 + 0, 2, 8, 14, 18, 20, 24, 30, 32, 38, 42, 44, 48, 50), (5009128141636113611 + 0, 2, 6, 8, 12, 18, 20, 26, 30, 32, 36, 42, 48, 50).

The largest known prime 15-constellations are (842443436396333356306067 + 0, 2, 6, 12, 14, 20, 24, 26, 30, 36, 42, 44, 50, 54, 56), (898520899795145760433 7+0, 2, 6, 12, 14, 20, 26, 30, 32, 36, 42, 44, 50, 54, 56), (359458541346697269469 7+0, 2, 6, 12, 14, 20, 26, 30, 32, 36, 42, 44, 50, 54, 56), (351438337546154123257 7+0, 2, 6, 12, 14, 20, 26, 30, 32, 36, 42, 44, 50, 54, 56), (349386450998591260948 7+0, 2, 6, 12, 14, 20, 24, 26, 30, 36, 42, 44, 50, 54, 56).

The largest known prime 16-constellations are (3259125690557440336637 + 0, 2, 6, 12, 14, 20, 26, 30, 32, 36, 42, 44, 50, 54, 56, 60), (1522014304823128379267 + 0, 2, 6, 12, 14, 20, 26, 30, 32, 36, 42, 44, 50, 54, 56, 60), (47710850533373130107 + 0, 2, 6, 12, 14, 20, 26, 30, 32, 36, 42, 44, 50, 54, 56, 60), (13, 17, 19, 23, 29, 31, 37, 41, 43, 47, 53, 59, 61, 67, 71, 73).

The largest known prime 17-constellations are (3259125690557440336631 + 0, 6, 8, 12, 18, 20, 26, 32, 36, 38, 42, 48, 50, 56, 60, 62, 66), (17, 19, 23, 29, 31, 37, 41, 43, 47, 53, 59, 61, 67, 71, 73, 79, 83) (13, 17, 19, 23, 29, 31, 37, 41, 43, 47, 53, 59, 61, 67, 71, 73, 79).

Smith (1957) found 8 consecutive primes spaced like the cluster $\{p_n\}_{n=5}^{12}$ (Gardner 1980). K. Conrow and J. J. Devore have found 15 consecutive primes spaced like the cluster $\{p_n\}_{n=5}^{19}$ given by $\{16323737455275581 18190 + p_n\}_{n=5}^{19}$, the first member of which is 16323737455275581 18201.

Rivera tabulates the smallest examples of k consecutive primes ending in a given digit $d = 1, 3, 7,$ or 9 for $k = 5$ to 11. For example, 216401, 216421, 216431, 216451, 216481 is the smallest set of five consecutive primes ending in the digit 1.

See also CLUSTER PRIME, COMPOSITE RUNS, COUSIN PRIMES, PRIME ARITHMETIC PROGRESSION, K-TUPLE CONJECTURE, PRIME K-TUPLES CONJECTURE, PRIME QUADRUPLET, PRIME TRIPLET, SEXY PRIMES, TWIN PRIMES

References

Cohen, H. "High Precision Computation of Hardy-Littlewood Constants." Preprint. http://www.math.u-bordeaux.fr/~cohen/hardylw.dvi.
Forbes, T. "Prime k-tuplets." http://www.ltkz.demon.co.uk/ktuplets.htm.
Forbes, T. "Prime Clusters and Cunningham Chains." *Math. Comput.* **68**, 1739–748, 1999.
Gardner, M. "Mathematical Games." *Sci. Amer.* **243**, Dec. 1980.
Guy, R. K. "Patterns of Primes." §A9 in *Unsolved Problems in Number Theory, 2nd ed.* New York: Springer-Verlag, pp. 23–5, 1994.
Rivera, C. "Problems & Puzzles: Puzzle Consecutive Primes and Ending Digits.-016." http://www.primepuzzles.net/puzzles/puzz_016.htm.
Smith, H. F. "On a Generalization of the Prime Pair Problem." *Math. Tables Aids Comput.* **11**, 249–54, 1957.
Le Lionnais, F. *Les nombres remarquables.* Paris: Hermann, p. 38, 1983.
Riesel, H. *Prime Numbers and Computer Methods for Factorization, 2nd ed.* Boston, MA: Birkhäuser, pp. 60–4, 1994.
Sloane, N. J. A. Sequences A008407 in "An On-Line Version of the Encyclopedia of Integer Sequences." http://www.research.att.com/~njas/sequences/eisonline.html.

Prime Counting Function

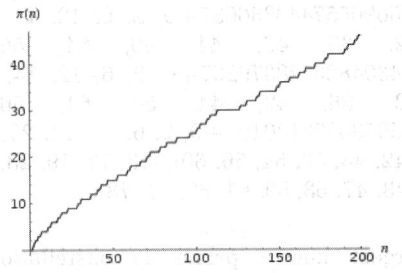

The function $\pi(n)$ giving the number of PRIMES $\leq n$ (Shanks 1993, p. 15). For example, there are no primes ≤ 1, so $\pi(1) = 0$; there is a single prime (2) ≤ 2, so $\pi(2) = 1$; there are two primes (2 and 3) ≤ 3, so $\pi(3) = 2$; and so on. The first few values for $n = 1, 2, \ldots$ are 0, 1, 2, 2, 3, 3, 4, 4, 4, 4, 5, 5, 6, 6, 6, ... (Sloane's A000720).

The following table gives the values of $\pi(n)$ for powers of 10 (Sloane's A006880; Hardy and Wright 1979, p. 4; Shanks 1993, pp. 242–43; Ribenboim 1996, p. 237). The value for $\pi(10^{20})$ comes from Deleglise and Rivat (1996). Note that $\pi(10^9)$ is incorrectly given as 50,847,478 in Hardy and Wright (1979) and Hardy (1999).

n	$\pi(10^n)$
3	168
4	1,229
5	9,592
6	78,498
7	664,579
8	5,761,455
9	50,847,534
10	455,052,511
11	4,118,054,813
12	37,607,912,018
13	346,065,536,839
14	3,204,941,750,802
15	29,844,570,422,669
16	279,238,341,033,925
17	2,623,557,157,654,233
18	24,739,954,287,740,860
19	234,057,667,276,344,607
20	2,220,819,602,560,918,840

One of the most fundamental and important results in NUMBER THEORY is the asymptotic value of $\pi(n)$ as n becomes large. The correct formula is

$$\pi(n) \sim \mathrm{li}(n), \tag{1}$$

where $\mathrm{li}(x)$ is the LOGARITHMIC INTEGRAL, which is known as the PRIME NUMBER THEOREM.

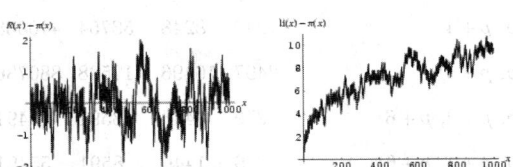

The following table compares the prime counting function $\pi(x)$, LOGARITHMIC INTEGRAL $\mathrm{li}\,x$, and RIEMANN PRIME NUMBER FORMULA $R(x)$ for small x. Note that the values given by Hardy (1999, p. 26) for $x = 10^9$ are incorrect.

x	$\pi(x)$	$\mathrm{li}\,x - \pi(x)$	$R(x) - \pi(x)$
100000	9592	38	−5
1000000	78498	130	29

2000000	148933	122	−9
3000000	216816	155	0
4000000	283146	206	33
5000000	348513	125	−64
6000000	412849	228	24
7000000	476648	179	−38
8000000	539777	223	−6
9000000	602489	187	−53
10000000	664579	339	88
100000000	5761455	754	97
1000000000	50847534	1701	−79

The prime counting function can be expressed by LEGENDRE'S FORMULA, LEHMER'S FORMULA, MAPES' METHOD, or MEISSEL'S FORMULA. A brief history of attempts to calculate $\pi(n)$ is given by Berndt (1994). The following table is taken from Riesel (1994), where $\mathcal{O}(x)$ is ASYMPTOTIC NOTATION.

Method	Time	Storage
Legendre	$\mathcal{O}(x)$	$\mathcal{O}(x^{1/2})$
Meissel	$\mathcal{O}\left(x/(\ln x)^3\right)$	$\mathcal{O}(x^{1/2}/\ln x)$
Lehmer	$\mathcal{O}\left(x/(\ln x)^4\right)$	$\mathcal{O}(x^{1/3}/\ln x)$
Mapes'	$\mathcal{O}(x^{0.7})$	$\mathcal{O}(x^{0.7})$
Lagarias-Miller-Odlyzko	$\mathcal{O}(x^{2/3+\epsilon})$	$\mathcal{O}(x^{1/3+\epsilon})$
Lagarias-Odlyzko 1	$\mathcal{O}(x^{3/5+\epsilon})$	$\mathcal{O}(x^{\epsilon})$
Lagarias-Odlyzko 2	$\mathcal{O}(x^{1/2+\epsilon})$	$\mathcal{O}(x^{1/4+\epsilon})$

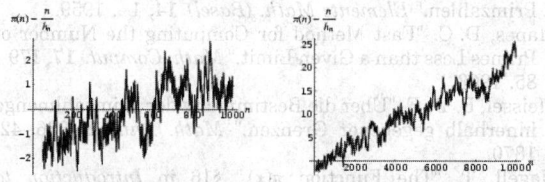

An approximate formula due to Locker-Ernst (Locker-Ernst 1959, Panaitopol 1999), illustrated above, is given by

$$\pi(n) \approx \frac{n}{h_n}, \qquad (2)$$

where h_n is related to the HARMONIC NUMBER H_n by $h_n = H_n - 3/2$. This formula is within ≈ 2 of the actual value for $50 \leq n \leq 1000$. The values for which $\pi - n/h_n > 0$ are 1, 109, 113, 114, 199, 200, 201, ...

(Sloane's A051046). Panaitopol (1999) shows that this quantity is positive for all $n \geq 1429$.

An upper limit for $\pi(n)$ is given by

$$\pi(n) < \frac{2n - 6}{\ln n} \qquad (3)$$

(Rosser and Schoenfeld 1962). Hardy and Wright (1979, p. 414) give the formula

$$\pi(n) = -1 + \sum_{j=3}^{n}\left[(j-2)! - j\left\lfloor\frac{(j-2)!}{j}\right\rfloor\right], \qquad (4)$$

where $\lfloor x \rfloor$ is the FLOOR FUNCTION.

A modified version of the prime counting function is given by

$$\pi_0(p) \equiv \begin{cases} \pi(p) & \text{for } p \text{ composite} \\ \pi(p) - \frac{1}{2} & \text{for } p \text{ prime} \end{cases}$$

$$\pi_0(p) = \sum_{n=1}^{\infty} \frac{\mu(x) f\left(x^{1/n}\right)}{n},$$

where $\mu(n)$ is the MÖBIUS FUNCTION and $f(x)$ is the RIEMANN FUNCTION.

The notation $\pi_{a,b}$ is also used to denote the number of PRIMES OF THE FORM $ak + b$ (Shanks 1993, pp. 21–2). Groups of EQUINUMEROUS values of $\pi_{a,b}$ include $(\pi_{3,1}, \pi_{3,2})$, $(\pi_{4,1}, \pi_{4,3})$, $(\pi_{5,1}, \pi_{5,2}, \pi_{5,3}, \pi_{5,4})$, $(\pi_{6,1}, \pi_{6,5})$, $(\pi_{7,1}, \pi_{7,2}, \pi_{7,3}, \pi_{7,4}, \pi_{7,5}, \pi_{7,6})$, $(\pi_{8,1}, \pi_{8,3}, \pi_{8,5}, \pi_{8,7})$, $(\pi_{9,1}, \pi_{9,2}, \pi_{9,4}, \pi_{9,5}, \pi_{9,7}, \pi_{9,8})$, and so on. The values of $\pi_{n,k}$ for small n are given in the following table for the first few powers of ten (Shanks 1993).

n	$\pi_{3,1}(n)$	$\pi_{3,2}(n)$	$\pi_{4,1}(n)$	$\pi_{4,3}(n)$
10^1	1	2	1	2
10^2	11	13	11	13
10^3	80	87	80	87
10^4	611	617	609	619
10^5	4784	4807	4783	4808
10^6	39231	39266	39175	39322
10^7	332194	332384	332180	332398

n	$\pi_{5,1}(n)$	$\pi_{5,2}(n)$	$\pi_{5,3}(n)$	$\pi_{5,4}(n)$
10^1	0	2	1	0
10^2	5	7	7	5
10^3	40	47	42	38
10^4	306	309	310	303

10^5	2387	2412	2402	2390
10^6	19617	19622	19665	19593
10^7	166104	166212	166230	166032

n	$\pi_{6,1}(n)$	$\pi_{6,5}(n)$
10^1	1	1
10^2	11	12
10^3	80	86
10^4	611	616
10^5	4784	4806
10^6	39231	39265

n	$\pi_{7,1}$	$\pi_{7,2}$	$\pi_{7,3}$	$\pi_{7,4}$	$\pi_{7,5}$	$\pi_{7,6}$
10^1	0	1	1	0	1	0
10^2	3	4	5	3	5	4
10^3	28	27	30	26	29	27
10^4	203	203	209	202	211	200
10^5	1593	1584	1613	1601	1604	1596
10^6	13063	13065	13105	13069	13105	13090

n	$\pi_{8,1}(n)$	$\pi_{8,3}(n)$	$\pi_{8,5}(n)$	$\pi_{8,7}(n)$
10^1	0	1	1	1
10^2	5	7	6	6
10^3	37	44	43	43
10^4	295	311	314	308
10^5	2384	2409	2399	2399
10^6	19552	19653	19623	19669
10^7	165976	166161	166204	166237

Note that since $\pi_{8,1}(n)$, $\pi_{8,3}(n)$, $\pi_{8,5}(n)$, and $\pi_{8,7}(n)$ are EQUINUMEROUS,

$$\pi_{4,1}(n) = \pi_{8,1}(n) + \pi_{8,5}$$

$$\pi_{4,3}(n) = \pi_{8,3}(n) + \pi_{8,7}$$

are also equinumerous.

Erdos proved that there exist at least one PRIME OF THE FORM $4k + 1$ and at least one PRIME of the form $4k + 3$ between n and $2n$ for all $n > 6$.

The smallest x such that $x \geq n\pi(x)$ for $n = 2, 3, \ldots$ are 2, 27, 96, 330, 1008, $\ldots$ (Sloane's A038625), and the corresponding $\pi(x)$ are 1, 9, 24, 66, 168, 437, $\ldots$ (Sloane's A038626). The number of solutions of $x \geq n\pi(x)$ for $n = 2, 3, \ldots$ are 4, 3, 3, 6, 7, 6, $\ldots$ (Sloane's A038627).

See also BERTELSEN'S NUMBER, CHEBYSHEV'S THEOREM, EQUINUMEROUS, LEGENDRE'S CONSTANT, LEGENDRE'S FORMULA, LEHMER-SCHUR METHOD, LOGARITHMIC INTEGRAL, MAPES' METHOD, PRIME ARITHMETIC PROGRESSION, PRIME NUMBER, PRIME NUMBER THEOREM, RIEMANN PRIME NUMBER FORMULA

References

Berndt, B. C. *Ramanujan's Notebooks, Part IV.* New York: Springer-Verlag, pp. 134–35, 1994.

Brent, R. P. "Irregularities in the Distribution of Primes and Twin Primes." *Math. Comput.* **29**, 43–6, 1975.

Deleglise, M. and Rivat, J. "Computing $\pi(x)$: The Meissel, Lehmer, Lagarias, Miller, Odlyzko Method." *Math. Comput.* **65**, 235–45, 1996.

Finch, S. "Favorite Mathematical Constants." http://www.mathsoft.com/asolve/constant/hrdyltl/hrdyltl.html.

Forbes, T. "Prime k-tuplets." http://www.ltkz.demon.co.uk/ktuplets.htm.

Guiasu, S. "Is There Any Regularity in the Distribution of Prime Numbers at the Beginning of the Sequence of Positive Integers?" *Math. Mag.* **68**, 110–21, 1995.

Hardy, G. H. *Ramanujan: Twelve Lectures on Subjects Suggested by His Life and Work, 3rd ed.* New York: Chelsea, 1999.

Hardy, G. H. and Wright, E. M. *An Introduction to the Theory of Numbers, 5th ed.* Oxford, England: Clarendon Press, 1979.

Lagarias, J.; Miller, V. S.; and Odlyzko, A. "Computing $\pi(x)$: The Meissel-Lehmer Method." *Math. Comput.* **44**, 537–60, 1985.

Lagarias, J. and Odlyzko, A. "Computing $\pi(x)$: An Analytic Method." *J. Algorithms* **8**, 173–91, 1987.

Locker-Ernst, L. "Bemerkung über die Verteilung der Primzahlen." *Elemente Math. (Basel)* **14**, 1–, 1959.

Mapes, D. C. "Fast Method for Computing the Number of Primes Less than a Given Limit." *Math. Comput.* **17**, 179–85, 1963.

Meissel, E. D. F. "Über die Bestimmung der Primzahlmenge innerhalb gegebener Grenzen." *Math. Ann.* **2**, 636–42, 1870.

Nagell, T. "The Function $\pi(x)$." §16 in *Introduction to Number Theory.* New York: Wiley, pp. 54–7, 1951.

Panaitopol, L. "Several Approximations of $\pi(x)$." *Math. Ineq. Appl.* **2**, 317–24, 1999.

Ribenboim, P. *The New Book of Prime Number Records, 3rd ed.* New York: Springer-Verlag, 1996.

Riesel, H. "The Number of Primes Below x." *Prime Numbers and Computer Methods for Factorization, 2nd ed.* Boston, MA: Birkhäuser, pp. 10–2, 1994.

Rosser, J. B. and Schoenfeld, L. "Approximate Formulas for Some Functions of Prime Numbers." *Illinois J. Math.* **6**, 64–7, 1962.

Séroul, R. "The Function pi(x)." §8.7 in *Programming for Mathematicians.* Berlin: Springer-Verlag, pp. 175–81, 2000.

Shanks, D. *Solved and Unsolved Problems in Number Theory, 4th ed.* New York: Chelsea, 1993.

Sloane, N. J. A. Sequences A000720/M0256, A006880/M3608, A038625, A038626, A038627, A052434, and A052435 in "An On-Line Version of the Encyclopedia of Integer Sequences." http://www.research.att.com/~njas/sequences/eisonline.html.

Vardi, I. *Computational Recreations in Mathematica.* Reading, MA: Addison-Wesley, pp. 74–6, 1991.

Prime Cut

Find two numbers such that $x^2 \equiv y^2 \pmod{n}$. If you know the GREATEST COMMON DIVISOR of n and $x - y$, there exists a high probability of determining a PRIME factor. Taking small numbers x which additionally give small PRIMES $x^2 \equiv p \pmod{n}$ further increases the chances of finding a PRIME FACTOR.

See also GREATEST COMMON DIVISOR

Prime Decomposition

PRIME FACTORIZATION

Prime Difference Function

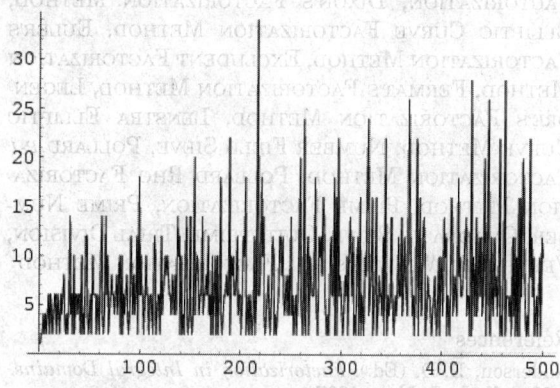

$$d_n \equiv p_{n+1} - p_n.$$

The first few values are 1, 2, 2, 4, 2, 4, 2, 4, 6, 2, 6, 4, 2, 4, 6, 6, ... (Sloane's A001223). Rankin has shown that

$$d_n > \frac{c \ln n \, \ln \, \ln n \, \ln \, \ln \, \ln \, \ln n}{(\ln \, \ln \, \ln n)^2}$$

for infinitely many n and for some constant c (Guy 1994).

An integer n is called a JUMPING CHAMPION if n is the most frequently occurring difference between consecutive primes $n \leq N$ for some N (Odlyzko *et al.*).

See also ANDRICA'S CONJECTURE, GILBREATH'S CONJECTURE, GOOD PRIME, JUMPING CHAMPION, PÓLYA CONJECTURE, PRIME GAPS, SHANKS' CONJECTURE, TWIN PEAKS

References

Bombieri, E. and Davenport, H. "Small Differences Between Prime Numbers." *Proc. Roy. Soc. A* **293**, 1–8, 1966.

Erdos, P.; and Straus, E. G. "Remarks on the Differences Between Consecutive Primes." *Elem. Math.* **35**, 115–18, 1980.

Guy, R. K. "Gaps between Primes. Twin Primes" and "Increasing and Decreasing Gaps." §A8 and A11 in *Unsolved Problems in Number Theory, 2nd ed.* New York: Springer-Verlag, pp. 19–3 and 26–7, 1994.

Odlyzko, A.; Rubinstein, M.; and Wolf, M. "Jumping Champions." http://www.research.att.com/~amo/doc/recent.html.

Riesel, H. "Difference Between Consecutive Primes." *Prime Numbers and Computer Methods for Factorization, 2nd ed.* Boston, MA: Birkhäuser, p. 9, 1994.

Sloane, N. J. A. Sequences A001223/M0296 in "An On-Line Version of the Encyclopedia of Integer Sequences." http://www.research.att.com/~njas/sequences/eisonline.html.

Prime Diophantine Equations

$k + 2$ is PRIME IFF the 14 DIOPHANTINE EQUATIONS in 26 variables

$$wz + h + j - q = 0 \tag{1}$$

$$(gk + 2g + k + 1)(h + j) + h - z = 0 \tag{2}$$

$$16(k + 1)^3(k + 2)(n + 1)^2 + 1 - f^2 = 0 \tag{3}$$

$$2n + p + q + z - e = 0 \tag{4}$$

$$e^3(e + 2)(a + 1)^2 + 1 - o^2 = 0 \tag{5}$$

$$(a^2 - 1)y^2 + 1 - x^2 = 0 \tag{6}$$

$$16r^2y^4(a^2 - 1) + 1 - u^2 = 0 \tag{7}$$

$$n + l + v - y = 0 \tag{8}$$

$$(a^2 - 1)l^2 + 1 - m^2 = 0 \tag{9}$$

$$ai + k + 1 - l - i = 0 \tag{10}$$

$$\left\{ \left[a + u^2(u^2 - a) \right]^2 - 1 \right\}(n + 4 \, dy)^2 + 1 - (x + cu)^2 = 0 \tag{11}$$

$$p + l(a - n - 1) + b(2an + 2a - n^2 - 2n - 2) - m = 0 \tag{12}$$

$$q + y(a - p - 1) + s(2ap + 2a - p^2 - 2p - 2) - x = 0 \tag{13}$$

$$z + pl(a - p) + t(2ap - p^2 - 1) - pm = 0 \tag{14}$$

have a solution in POSITIVE INTEGERS (Riesel 1994, p. 40).

See also PRIME-GENERATING POLYNOMIAL

References

Riesel, H. *Prime Numbers and Computer Methods for Factorization, 2nd ed.* Boston, MA: Birkhäuser, 1994.

Prime Divisor

If $f(x)$ is a nonconstant INTEGER POLYNOMIAL and c is an integer such that $f(c)$ is divisible by the prime p, that p is called a prime divisor of the polynomial $f(x)$

(Nagell 1951, p. 81). Every INTEGER POLYNOMIAL $f(x)$ which is not a constant has an infinite number of prime divisors (Nagell 1951, p. 82).

See also BAUER'S THEOREM, INTEGER POLYNOMIAL

References

Nagell, T. "Prime Divisors of Integral Polynomials." §25 in *Introduction to Number Theory.* New York: Wiley, pp. 81–3, 1951.

Prime Factorization

The FACTORIZATION of a numbers into its constituent PRIMES, also called prime decomposition. Given a POSITIVE INTEGER $n \geq 2$, the prime factorization is written

$$n = p_1^{\alpha_1} p_2^{\alpha_2} \cdots p_k^{\alpha_k},$$

where the p_is are the k PRIME FACTORS, each of order α_i. Each factor $p_i^{\alpha_i}$ is called a PRIMARY. The first few prime factorizations (the number 1, by definition, has a prime factorization of "1") are given in the following table.

1	1	11	11
2	2	12	$2^2 \cdot 3$
3	3	13	13
4	2^2	14	$2 \cdot 7$
5	5	15	$3 \cdot 5$
6	$2 \cdot 3$	16	2^4
7	7	17	17
8	2^3	18	$2 \cdot 9$
9	3^2	19	19
10	$2 \cdot 5$	20	$2^2 \cdot 5$

The number of *digits* in the prime factorization of $n = 1, 2, ...,$ are 1, 1, 1, 2, 1, 2, 1, 2, 2, 2, 2, 3, (Sloane's A050252).

In general, prime factorization is a difficult problem, and many sophisticated PRIME FACTORIZATION ALGORITHMS have been devised for special types of numbers.

See also DISTINCT PRIME FACTORS, ECONOMICAL NUMBER, EQUIDIGITAL NUMBER, FACTORIZATION, PRIMARY, PRIME FACTORIZATION, PRIME FACTORIZATION ALGORITHMS, PRIME FACTORS, PRIME NUMBER, ROUND NUMBER, ROUNDNESS, WASTEFUL NUMBER

Prime Factorization Algorithms

Many ALGORITHMS have been devised for determining the PRIME FACTORS of a given number (a process called PRIME FACTORIZATION). They vary quite a bit in sophistication and complexity. It is very difficult to build a general-purpose algorithm for this computationally "hard" problem, so any additional information which is known about the number in question or its factors can often be used to save a large amount of time.

The simplest method of finding factors is so-called "DIRECT SEARCH FACTORIZATION" (a.k.a. TRIAL DIVISION). In this method, all possible factors are systematically tested using trial division to see if they actually DIVIDE the given number. It is practical only for very small numbers.

The fastest-known fully proven deterministic algorithm is the Pollard-Strassen method (Pomerance 1987; Hardy *et al.* 1990).

See also BRENT'S FACTORIZATION METHOD, CLASS GROUP FACTORIZATION METHOD, CONTINUED FRACTION FACTORIZATION ALGORITHM, DIRECT SEARCH FACTORIZATION, DIXON'S FACTORIZATION METHOD, ELLIPTIC CURVE FACTORIZATION METHOD, EULER'S FACTORIZATION METHOD, EXCLUDENT FACTORIZATION METHOD, FERMAT'S FACTORIZATION METHOD, LEGENDRE'S FACTORIZATION METHOD, LENSTRA ELLIPTIC CURVE METHOD, NUMBER FIELD SIEVE, POLLARD *P-1* FACTORIZATION METHOD, POLLARD RHO FACTORIZATION METHOD, PRIME FACTORIZATION, PRIME NUMBER, QUADRATIC SIEVE, QUITEPRIME, TRIAL DIVISION, VERYPRIME, WILLIAMS *P+1* FACTORIZATION METHOD

References

Anderson, D. D. (Ed.). *Factorization in Integral Domains.* New York: Dekker, 1997.
Bressoud, D. M. *Factorization and Prime Testing.* New York: Springer-Verlag, 1989.
Brillhart, J.; Lehmer, D. H.; Selfridge, J.; Wagstaff, S. S. Jr.; and Tuckerman, B. *Factorizations of* $b^n \pm 1$, $b = 2$, 3, 5, 6, 7, 10, 11, 12 *Up to High Powers*, rev. ed. Providence, RI: Amer. Math. Soc., liv-lviii, 1988.
Dickson, L. E. "Methods of Factoring." Ch. 14 in *History of the Theory of Numbers, Vol. 1: Divisibility and Primality.* New York: Chelsea, pp. 357–74, 1952.
Hardy, K.; Muskat, J. B.; and Williams, K. S. "A Deterministic Algorithm for Solving $n = fu^2 + gv^2$ in Coprime Integers u and v." *Math. Comput.* **55**, 327–43, 1990.
Lenstra, A. K. and Lenstra, H. W. Jr. "Algorithms in Number Theory." In *Handbook of Theoretical Computer Science, Volume A: Algorithms and Complexity* (Ed. J. van Leeuwen). New York: Elsevier, pp. 673–15, 1990.
Odlyzko, A. M. "The Complexity of Computing Discrete Logarithms and Factoring Integers." §4.5 in *Open Problems in Communication and Computation* (Ed. T. M. Cover and B. Gopinath). New York: Springer-Verlag, pp. 113–16, 1987.
Odlyzko, A. M. "The Future of Integer Factorization." *CryptoBytes: The Technical Newsletter of RSA Laboratories* **1**, No. 2, 5–2, 1995.
Pomerance, C. "Fast, Rigorous Factorization and Discrete Logarithm Algorithms." In *Discrete Algorithms and Com-*

plexity (Ed. D. S. Johnson, T. Nishizeki, A. Nozaki, and H. S. Wilf). New York: Academic Press, pp. 119–43, 1987.

Pomerance, C. "Analysis and Comparison of Some Integer Factorization Algorithms." In *Computational Methods in Number Theory, Part 1* (Ed. H. W. Lenstra and R. Tijdeman). Amsterdam, Netherlands: Mathematisch Centrum, pp. 89–39, 1982.

Pomerance, C. "A Tale of Two Sieves." *Not. Amer. Math. Soc.* **43**, 1473–485, 1996.

Riesel, H. "Algebraic Factors." Appendix 6 in *Prime Numbers and Computer Methods for Factorization, 2nd ed.* Boston, MA: Birkhäuser, pp. 304–16, 1994.

Weisstein, E. W. "Books about Prime Numbers." http://www.treasure-troves.com/books/PrimeNumbers.html.

Williams, H. C. and Shallit, J. O. "Factoring Integers Before Computers." In *Mathematics of Computation 1943–993, Fifty Years of Computational Mathematics* (Ed. W. Gautschi). Providence, RI: Amer. Math. Soc., pp. 481–31, 1994.

Prime Factors

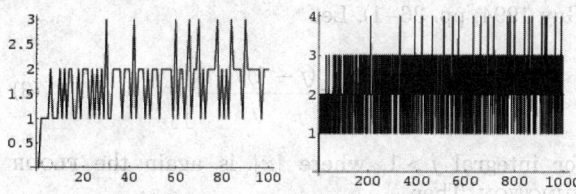

The number of DISTINCT PRIME FACTORS of a number n is denoted $\omega(n)$. $\omega(n)$ therefore corresponds to a prime factorization OF THE FORM

$$n = p_1^{\alpha_1} p_2^{\alpha_2} \cdots p_{\omega(n)}^{\alpha_{\omega(n)}}. \tag{1}$$

The first few values for $n = 1, 2, \ldots$ are 0, 1, 1, 1, 1, 2, 1, 1, 1, 2, 1, 2, 1, 2, 2, 1, 1, 2, 1, 2, ... (Sloane's A001221).

The first few numbers u_n which are products of an odd number of distinct prime factors (Hardy 1999, p. 64; Ramanujan 2000, pp. xxiv and 21) are 2, 3, 5, 7, 11, 13, 17, 19, 23, 29, 30, 31, 37, 41, 42 43, 47, ... (Sloane's A030059). u_n satisfies

$$\sum_{n=1}^{\infty} \frac{1}{u_n^s} = \frac{1}{2} \left[\frac{\zeta(s)}{\zeta(2s)} - \zeta(s) \right] \tag{2}$$

(Hardy 1999, pp. 64–5). In addition, if $U(n)$ is the number of u_k with $k \leq n$, then

$$U(x) \sim \frac{3x}{\pi^2} \tag{3}$$

(Hardy 1999, pp. 64–5).

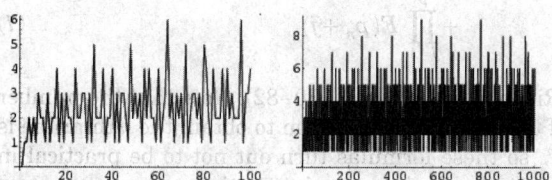

The number of *not necessarily distinct* prime factors of a number n is denoted $r(n)$. The first few values for

$n = 1, 2, \ldots$ are 0, 1, 1, 2, 1, 2, 1, 3, 2, 2, 1, 3, 1, 2, 2, 4, 1, 3, 1, 3, ... (Sloane's A001222). If n is chosen at random between 1 and x, then the probability that $r(n) \leq \ln n \ln n + c\sqrt{\ln \ln x}$ approaches

$$\frac{1}{\sqrt{2\pi}} \int_{-\infty}^{c} e^{-u^2/2} \, du \tag{4}$$

(Knuth 1998, p. 384). In addition, the average value $\bar{t}$ of $r(n) - \ln \ln x$ for $1 \leq n \leq x$ approaches

$$\bar{t} = \gamma + \sum_{p \text{ prime}} \left[\ln \left(1 - \frac{1}{p} \right) + \frac{1}{p-1} \right] \tag{5}$$

$$= \gamma + \sum_{n=2}^{\infty} \frac{\phi(n) \ln[\zeta(n)]}{n} \tag{6}$$

$$\approx 1.0345638819, \tag{7}$$

where γ is the EULER-MASCHERONI CONSTANT, $\phi(n)$ is the TOTIENT FUNCTION, and $\zeta(n)$ is the RIEMANN ZETA FUNCTION.

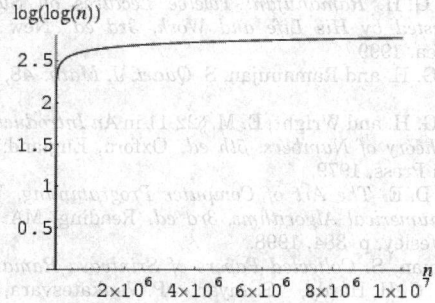

The average orders of both $\omega(n)$ and $r(n)$ are

$$\omega(n) \sim \ln \ln n \tag{8}$$

(Hardy 1999, p. 51). More precisely,

$$\sum_{n \leq x} \omega(n) = x \ln \ln x + Ax + \mathcal{O}\left(\frac{x}{\ln x} \right) \tag{9}$$

$$\sum_{n \leq x} r(n) = x \ln \ln x + Bx + \mathcal{O}\left(\frac{x}{\ln x} \right) \tag{10}$$

for appropriate constants A and B (Hardy and Ramanujan 1917; Hardy and Wright 1979, p. 355; Hardy 1999, p. 57), where $\mathcal{O}(x)$ is ASYMPTOTIC NOTATION.

The following table gives the prime factors for the positive integers ≤ 50.

1	1	11	11	21	$3 \cdot 7$	31	31	41	41
2	2	12	$2^2 \cdot 3$	22	$2 \cdot 11$	32	2^5	42	$2 \cdot 3 \cdot 7$
3	3	13	13	23	23	33	$3 \cdot 11$	43	43
4	2^2	14	$2 \cdot 7$	24	$2^3 \cdot 3$	34	$2 \cdot 17$	44	$2^2 \cdot 11$

5	5	15	$3 \cdot 5$	25	5^2	35	$5 \cdot 7$	45	$3^3 \cdot 5$
6	$2 \cdot 3$	16	2^4	26	$2 \cdot 13$	36	$2^2 \cdot 3^2$	46	$2 \cdot 23$
7	7	17	17	27	3^3	37	37	47	47
8	2^3	18	$2 \cdot 3^2$	28	$2^2 \cdot 7$	38	$2 \cdot 19$	48	$2^4 \cdot 3$
9	3^2	19	19	29	29	39	$3 \cdot 13$	49	7^2
10	$2 \cdot 5$	20	$2^2 \cdot 5$	30	$2 \cdot 3 \cdot 5$	40	$2^3 \cdot 5$	50	$2 \cdot 5^2$

See also DICKMAN FUNCTION, DISTINCT PRIME FACTORS, DIVISOR FUNCTION, GREATEST PRIME FACTOR, LEAST PRIME FACTOR, LIOUVILLE FUNCTION, MERTENS CONSTANT, PÓLYA CONJECTURE, PRIME FACTORIZATION ALGORITHMS, PRIMITIVE PRIME FACTOR, ROUND NUMBER

References

Erdos, P. and Kac, M. "The Gaussian Law of Errors in the Theory of Additive Number Theoretic Functions." *Amer. J. Math.* **26**, 738–42, 1940.
Hardy, G. H. *Ramanujan: Twelve Lectures on Subjects Suggested by His Life and Work, 3rd ed.* New York: Chelsea, 1999.
Hardy, G. H. and Ramanujan, S. *Quart. J. Math.* **48**, 76–2, 1917.
Hardy, G. H. and Wright, E. M. §22.11 in *An Introduction to the Theory of Numbers, 5th ed.* Oxford, England: Clarendon Press, 1979.
Knuth, D. E. *The Art of Computer Programming, Vol. 2: Seminumerical Algorithms, 3rd ed.* Reading, MA: Addison-Wesley, p. 384, 1998.
Ramanujan, S. *Collected Papers of Srinivasa Ramanujan* (Ed. G. H. Hardy, S. Aiyar, P. Venkatesvara, and B. M. Wilson). Providence, RI: Amer. Math. Soc., 2000.
Sloane, N. J. A. Sequences A001222/M0094, A001221/M0056, and A030059 in "An On-Line Version of the Encyclopedia of Integer Sequences." http://www.research.att.com/~njas/sequences/eisonline.html.
Turán, P. "On a Theorem of Hardy and Ramanujan." *J. London Math. Soc.* **9**, 274–76, 1934.
Turán, P. "Über einige Verallgemeinerungen eines Satzes von Hardy und Ramanujan." *J. London Math. Soc.* **11**, 125–33, 1936.

Prime Field

A FINITE FIELD $GF(p)$ where p is PRIME.

Prime Formulas

There exist a variety of formulas for producing either the nth prime as a function of n, or else taking on only prime values. However, all such formula require either extremely accurate knowledge of some unknown constant, or else effectively require knowledge of the primes ahead of time in order to use the formula (Dudley 1969, Ribenboim 1996, p. 186).

For example, there exists a CONSTANT $\theta = 1.3063\ldots$ (Sloane's A051021) known as MILLS' CONSTANT such that

$$f(n) = \left\lfloor \theta^{3^n} \right\rfloor, \tag{1}$$

where $\lfloor x \rfloor$ is the FLOOR FUNCTION, is prime for all $n \geq 1$ (Ribenboim 1996, p. 186). The first few values of $f(n)$ are 2, 11, 1361, 2521008887, ... (Sloane's A051254). It is not known if θ is IRRATIONAL. There also exists a CONSTANT $\omega \approx 1.9287800$ such that

$$g(n) = \left\lfloor 2^{2^{\cdot^{\cdot^{\cdot^{2^\omega}}}}} \right\rfloor \tag{2}$$

(Wright 1951; Ribenboim 1996, p. 186) is prime for every $n \geq 1$. The first few values of $g(n)$ are 3, 13, 16381, ... In the case of both $f(n)$ and $g(n)$, the numbers at $n = 4$ grow so rapidly that an extremely precise value of θ or ω is needed in order to obtain the correct value. Values for $n \geq 5$ are hopeless.

Explicit FORMULAS exist for the nth prime both as a function of n and in terms of the primes 2, ..., p_{n-1} (Hardy and Wright 1979, pp. 5–, 344–45, and 414; Guy 1994, pp. 36–1). Let

$$F(j) = \left\lfloor \cos^2 \left[\pi \frac{(j-1)! + 1}{j} \right] \right\rfloor \tag{3}$$

for integral $j > 1$, where $\lfloor x \rfloor$ is again the FLOOR FUNCTION. Then

$$p_n = 1 + \sum_{m=1}^{2^n} \left\lfloor \left\lfloor \frac{n}{\sum_{j=1}^m F(j)} \right\rfloor^{1/n} \right\rfloor \tag{4}$$

$$= 1 + \sum_{m=1}^{2^n} \left\lfloor \left\lfloor \frac{n}{1 + \pi(m)} \right\rfloor^{1/n} \right\rfloor, \tag{5}$$

where $\pi(m)$ is the PRIME COUNTING FUNCTION. This formula conceals the prime numbers j as those for which $F(j) = 1$, i.e., the values of $F(j)$ are 1, 1, 1, 0, 1, 0, 1, 0, 0, 0, 1, ...

Gandhi gave the formula in which p_{n+1} is the unique integer such that

$$1 < 2^{p_{n+1}} \left(\sum_{d \mid p_n\#} \frac{\mu(d)}{2^d - 1} - \frac{1}{2} \right) < 2, \tag{6}$$

where $p_n\#$ is the PRIMORIAL function (Gandhi 1971, Eynden 1972, Golomb 1974) and $\mu(n)$ is the MÖBIUS FUNCTION. It is also true that

$$p_{n+1} = 1 + p_n + F(p_n + 1) + F(p_n + 1) + F(p_n + 2)$$
$$+ \prod_{j=1}^p F(p_n + j) \tag{7}$$

(Ribenboim 1996, pp. 180–82). Note that the number of terms in the summation to obtain the nth prime is 2^n, so these formulas turn out not to be practical in the study of primes. An interesting INFINITE PRODUCT formula due to Euler which relates π and the nth PRIME p_n is

$$\pi = \frac{2}{\prod_{i=n}^{\infty}\left[1 + \dfrac{\sin\left(\frac{1}{2}\pi p_n\right)}{p_n}\right]} \tag{8}$$

$$= \frac{2}{\prod_{i=n}^{\infty}\left[1 + \dfrac{(-1)^{(p_n-1)/2}}{p_n}\right]} \tag{9}$$

(Blatner 1997). Hardy and Wright (1979, p. 414) give the formula

$$p_n = 1 + \sum_{j=1}^{2^n} f(n,\,\pi(j)), \tag{10}$$

for $n > 3$, where

$$f(x,\,y) = \begin{cases} 0 & \text{for } x = y \\ \frac{1}{2}\left[1 + \dfrac{x-y}{|x-y|}\right] & \text{for } x \ne y \end{cases} \tag{11}$$

and

$$\pi(n) = -1 + \sum_{j=3}^{n}\left[(j-2)! - j\left\lfloor\frac{(j-2)!}{j}\right\rfloor\right] \tag{12}$$

(correcting a sign error), where $\lfloor x \rfloor$ is the FLOOR FUNCTION.

A double sum for the nth prime p_n is

$$p_n = 1 + \sum_{k=1}^{2(\lfloor n \ln n \rfloor + 1)}\left[1 - \left\lfloor\frac{\sum_{j=2}^{k} 1 + \lfloor s(j) \rfloor}{n}\right\rfloor\right], \tag{13}$$

where

$$s(j) \equiv -\frac{\sum_{s=1}^{j}\left(\left\lfloor\frac{j}{s}\right\rfloor - \left\lfloor\frac{j-1}{s}\right\rfloor\right) - 2}{j} \tag{14}$$

(Ruiz 2000).

B. M. Bredihin proved that

$$f(x,\,y) = x^2 + y^2 + 1 \tag{15}$$

takes prime values for infinitely many integral pairs $(x,\,y)$ (Honsberger 1976, p. 30). In addition, the function

$$f(x,y) = \tfrac{1}{2}(y-1)\big\lfloor |B^2(x,\,y)-1| - (B^2(x,\,y)-1) \big\rfloor + 2, \tag{16}$$

where

$$B(x,\,y) = x(y+1) - (y!+1), \tag{17}$$

$y!$ is the FACTORIAL, and $\lfloor x \rfloor$ is the FLOOR FUNCTION, generates only prime numbers for POSITIVE INTEGER arguments. It not only generates every prime number, but generates ODD PRIMES exactly once each, with all other values being 2 (Honsberger 1976,

p. 33). For example,

$$f(1,\,2) = 3 \tag{18}$$

$$f(5,\,4) = 5 \tag{19}$$

$$f(103,\,6) = 7, \tag{20}$$

with no new primes generated for $x,\,y \le 1000$.

Conway (Guy 1983, Conway and Guy 1996, p. 147) gives an algorithm for generating primes based on 14 fractions, but it is actually just a concealed version of a SIEVE.

See also MILLS' CONSTANT, PRIME NUMBER, SIEVE

References

Blatner, D. *The Joy of Pi.* New York: Walker, p. 110, 1997.

Conway, J. H. and Guy, R. K. *The Book of Numbers.* New York: Springer-Verlag, p. 130, 1996.

Dudley, U. "History of Formula for Primes." *Amer. Math. Monthly* **76**, 23–8, 1969.

Finch, S. "Favorite Mathematical Constants." http://www.mathsoft.com/asolve/constant/mills/mills.html.

Gandhi, J. M. "Formulae for the Nth Prime." *Proc. Washington State University Conferences on Number Theory.* pp. 96–07, 1971.

Gardner, M. "Patterns and Primes." Ch. 9 in *The Sixth Book of Mathematical Games from Scientific American.* Chicago, IL: University of Chicago Press, pp. 79–0, 1984.

Guy, R. K. "Conway's Prime Producing Machine." *Math. Mag.* **56**, 26–3, 1983.

Guy, R. K. "Prime Numbers," "Formulas for Primes," and "Products Taken Over Primes." Ch. A, §A17, and §B48 in *Unsolved Problems in Number Theory, 2nd ed.* New York: Springer-Verlag, pp. 3–3, 36–1 and 102–03, 1994.

Hardy, G. H. and Wright, E. M. "Prime Numbers" and "The Sequence of Primes." §1.2 and 1.4 in *An Introduction to the Theory of Numbers, 5th ed.* Oxford, England: Clarendon Press, pp. 1–, 1979.

Honsberger, R. *Mathematical Gems II.* Washington, DC: Math. Assoc. Amer., 1976.

Mills, W. H. "A Prime-Representing Function." *Bull. Amer. Math. Soc.* **53**, 604, 1947.

Ribenboim, P. *The New Book of Prime Number Records.* New York: Springer-Verlag, 1996.

Ruiz, S. M. "The General Term of the Prime Number Sequence and the Smarandache Prime Function." *Smarandache Notions J.* **11**, 59–1, 2000.

Sloane, N. J. A. Sequences A051021 and A051254 in "An On-Line Version of the Encyclopedia of Integer Sequences." http://www.research.att.com/~njas/sequences/eisonline.html.

Wright, E. M. "A Prime-Representing Function." *Amer. Math. Monthly* **58**, 616–18, 1951.

Prime Gaps

Letting

$$d_n \equiv p_{n+1} - p_n \tag{1}$$

be the PRIME DIFFERENCE FUNCTION, Rankin has showed that

$$d_n > \frac{c \ln n \ln \, \ln n \ln \, \ln \, \ln \, \ln n}{(\ln \, \ln \, \ln n)^2} \qquad (2)$$

for infinitely many n and for some constant c (Guy 1994).

Let $p(d)$ be the smallest PRIME following d or more consecutive COMPOSITE NUMBERS. The largest known is

$$p(804) = 90,874,329,412,297. \qquad (3)$$

The largest known prime gap is of length 4247, occurring following $10^{314} - 1929$ (Baugh and O'Hara 1992), although this gap is almost certainly not maximal (i.e., there probably exists a smaller number having a gap of the same length following it). Cramér (1937) and Shanks (1964) conjectured that a maximal gap $p(n)$ of length n first appears at approximately

$$p(n) \sim \exp(\sqrt{n}). \qquad (4)$$

Wolf conjectures a slightly different form

$$p(n) \sim \sqrt{n} \, \exp(\sqrt{n}), \qquad (5)$$

which agrees better with numerical evidence.

Wolf conjectures that the maximal gap $G(n)$ between two consecutive primes less than n appears approximately at

$$G(n) \sim \frac{n}{\pi(n)} [2 \ln \, \pi(n) - \ln n + \ln(2C_2)] \equiv g(n), \qquad (6)$$

where $\pi(n)$ is the PRIME COUNTING FUNCTION and C_2 is the TWIN PRIMES CONSTANT. Setting $\pi(n) \sim n/\ln n$ reduces to Cramer's conjecture for large n,

$$G(n) \sim (\ln n)^2. \qquad (7)$$

Let $c(n)$ be the smallest starting INTEGER $c(n)$ for a run of n consecutive COMPOSITE NUMBERS, also called a COMPOSITE RUN. No general method other than exhaustive searching is known for determining the first occurrence for a maximal gap, although arbitrarily large gaps exist (Nicely 1998). The first few $c(n)$ for $n = 1, 2, \ldots$ are 4, 8, 8, 24, 24, 90, 90, 114, ... (Sloane's A030296).

The following table gives the sequence of maximal prime gaps, omitting degenerate runs which are part of a run with greater n. It is a complete list of smallest maximal runs up to 10^{16} (Nicely, pers. comm., May 30, 2000). $c(n)$ in this table is given by Sloane's A008950, and n by Sloane's A008996. The ending integers for the run corresponding to $c(n)$ are given by Sloane's A008995. Young and Potler (1989) determined the first occurrences of prime gaps up to 72,635,119,999,997, with all first occurrences found between 1 and 673. Nicely (1998) extended the list of maximal prime gaps to a length of 915, denoting gap lengths by the difference of bounding PRIMES, $c(n) - 1$.

n	$c(n)$	n	$c(n)$
1	4	319	2,300,942,550
3	8	335	3,842,610,774
5	24	353	4,302,407,360
7	90	381	10,726,904,660
13	114	383	20,678,048,298
17	524	393	22,367,084,960
19	888	455	25,056,082,088
21	1,130	463	42,652,618,344
33	1,328	467	127,976,334,672
35	9,552	473	182,226,896,240
43	15,684	485	241,160,024,144
51	19,610	489	297,501,075,800
71	31,398	499	303,371,455,242
85	155,922	513	304,599,508,538
95	360,654	515	416,608,695,822
111	370,262	531	461,690,510,012
113	492,114	533	614,487,453,424
117	1,349,534	539	738,832,927,928
131	1,357,202	581	1,346,294,310,750
147	2,010,734	587	1,408,695,493,610
153	4,652,354	601	1,968,188,556,461
179	17,051,708	651	2,614,941,710,599
209	20,831,324	673	7,177,162,611,713
219	47,326,694	715	13,828,048,559,701
221	122,164,748	765	19,581,334,192,423
233	189,695,660	777	42,842,283,925,352
247	191,912,784	803	90,874,329,411,493
249	387,096,134	805	171,231,342,420,521
281	436,273,010	905	218,209,405,436,543
287	1,294,268,492	915	1,189,459,969,825,483
291	1,453,168,142	923	1,686,994,940,955,803
		1131	1,693,182,318,746,371

See also JUMPING CHAMPION, PRIME CONSTELLATION, PRIME DIFFERENCE FUNCTION, SHANKS' CONJECTURE

References

Baugh, D. and O'Hara, F. "Large Prime Gaps." *J. Recr. Math.* **24**, 186–87, 1992.

Berndt, B. C. *Ramanujan's Notebooks, Part IV.* New York: Springer-Verlag, pp. 133–34, 1994.

Bombieri, E. and Davenport, H. "Small Differences Between Prime Numbers." *Proc. Roy. Soc. A* **293**, 1–8, 1966.

Brent, R. P. "The First Occurrence of Large Gaps Between Successive Primes." *Math. Comput.* **27**, 959–63, 1973.

Brent, R. P. "The Distribution of Small Gaps Between Successive Primes." *Math. Comput.* **28**, 315–24, 1974.

Brent, R. P. "The First Occurrence of Certain Large Prime Gaps." *Math. Comput.* **35**, 1435–436, 1980.

Cramér, H. "On the Order of Magnitude of the Difference Between Consecutive Prime Numbers." *Acta Arith.* **2**, 23–6, 1937.

Guy, R. K. "Gaps between Primes. Twin Primes" and "Increasing and Decreasing Gaps." §A8 and A11 in *Unsolved Problems in Number Theory, 2nd ed.* New York: Springer-Verlag, pp. 19–3 and 26–7, 1994.

Lander, L. J. and Parkin, T. R. "On First Appearance of Prime Differences." *Math. Comput.* **21**, 483–88, 1967.

Nicely, T. R. "New Maximal Prime Gaps and First Occurrences." *Math. Comput.* **68**, 1311–315, 1999.

Nicely, T. R. and Nyman, B. "First Occurrence of a Prime Gap of 1000 or Greater." Submitted to *Math. Comput.*

Rivera, C. "Problems & Puzzles: Puzzle Distinct, Increasing & Decreasing Gaps.-011." http://www.primepuzzles.net/puzzles/puzz_011.htm.

Shanks, D. "On Maximal Gaps Between Successive Primes." *Math. Comput.* **18**, 646–51, 1964.

Sloane, N. J. A. Sequences A008950, A008995, A008996, and A030296 in "An On-Line Version of the Encyclopedia of Integer Sequences." http://www.research.att.com/~njas/sequences/eisonline.html.

Young, J. and Potler, A. "First Occurrence Prime Gaps." *Math. Comput.* **52**, 221–24, 1989.

Prime Group

When the ORDER h of a finite GROUP is a PRIME NUMBER, there is only one possible GROUP of ORDER h. Furthermore, the GROUP is CYCLIC.

See also P-GROUP

Prime Ideal

An IDEAL I such that if $ab \in I$, then either $a \in I$ or $b \in I$. For example, in the integers, the IDEAL $\mathfrak{a} = \langle p \rangle$ (i.e., the multiples of p) is prime whenever p is a PRIME NUMBER.

Prime ideals are useful when the ring in question is not necessarily a PRINCIPAL IDEAL DOMAIN, e.g., $\mathfrak{a} = \langle 2, \sqrt{6} \rangle$ in $\mathbb{Z}[\sqrt{6}]$. The general element of $\mathfrak{a}$ can be written as $2a + b\sqrt{6}$ where a and b can be any integers. Suppose that

$$\left(x_1 + x_2\sqrt{6} \right)\left(y_1 + y_2\sqrt{6} \right) = 2a + b\sqrt{6},$$

then $x_1 y_1 + 6x_2 y_2 = 2a$. So either x_1 or y_1 has to be even. The corresponding factor $\left(x_1 + x_2\sqrt{6} \right)$ or $\left(y_1 + y_2\sqrt{6} \right)$ has to be in $\mathfrak{a} = \langle 2, \sqrt{6} \rangle$. Hence, the ideal $\mathfrak{a}$ is prime. Note that this ring does not have UNIQUE FACTORIZATION since $2 \cdot 3 = 6 = \sqrt{6} \cdot \sqrt{6}$.

One consequence of the definition is that the set of elements not in a prime ideal, $R - \mathfrak{p}$, is CLOSED under multiplication. This allows one to LOCALIZE at $\mathfrak{p}$ by considering the RING OF FRACTIONS. This ring is analogous to the construction of the rationals as fractions of integers, except that the denominator must be in $R - \mathfrak{p}$. The only MAXIMAL IDEAL in this ring is the EXTENSION of $\mathfrak{p}$.

From the perspective of ALGEBRAIC GEOMETRY, ideals correspond to VARIETIES. Because multiplication corresponds to union (such as $xy = 0$ implies $x = 0$ or $y = 0$), a prime ideal corresponds to an IRREDUCIBLE VARIETY.

See also DEDEKIND RING, IDEAL, IRREDUCIBLE VARIETY, KRULL DIMENSION, MAXIMAL IDEAL, STICKELBERGER RELATION, STONE SPACE

Prime Knot

A KNOT other than the UNKNOT which cannot be expressed as a sum of two other KNOTS, neither of which is unknotted. A KNOT which is not prime is called a COMPOSITE KNOT. It is often possible to combine two prime knots to create two different COMPOSITE KNOTS, depending on the orientation of the two. Schubert (1949) showed that every knot can be uniquely decomposed (up to the order in which the decomposition is performed) as a KNOT SUM of prime knots.

There is no known FORMULA for giving the number of distinct prime knots as a function of the number of crossings. The numbers of distinct prime knots having $n = 1, 2, \dots$ crossings are 0, 0, 1, 2, 3, 7, 21, 49, 165, 552, 2176, 9988, ... (Sloane's A002863). Hoste *et al.* (1998) computed the number of distinct prime knots of n crossing up to $n = 16$. Let $N(n)$ be the number of distinct PRIME KNOTS of n crossings, *counting CHIRAL versions of the same knot separately*. Then

$$\tfrac{1}{3}(2^{n-2} - 1) \leq N(n) \lesssim e^n$$

(Ernst and Summers 1987). Welsh has shown that the number of knots is bounded by an exponential in n, and it is also known that

$$\lim \sup[N(n)]^{1/n} < 13.5$$

(Welsh 1991, Hoste *et al.* 1998, Thistlethwaite 1998).

Menasco (1984) showed that a reduced alternating diagram represents a prime knot IFF the diagram is itself prime ("an alternating knot is prime IFF it looks prime"; Hoste *et al.* 1998).

See also COMPOSITE KNOT, KNOT

References

Ernst, C. and Sumners, D. W. "The Growth of the Number of Prime Knots." *Math. Proc. Cambridge Philos. Soc.* **102**, 303–15, 1987.

Hoste, J.; Thistlethwaite, M.; and Weeks, J. "The First 1,701,936 Knots." *Math. Intell.* **20**, 33–8, Fall 1998.

Menasco, W. "Closed Incompressible Surfaces in Alternating Knot and Link Complements." *Topology* **23**, 37–4, 1984.

Schubert, H. *Sitzungsber. Heidelberger Akad. Wiss., Math.-Naturwiss. Klasse, 3rd Abhandlung.* 1949.

Sloane, N. J. A. Sequences A002863/M0851 in "An On-Line Version of the Encyclopedia of Integer Sequences." http://www.research.att.com/~njas/sequences/eisonline.html.

Sloane, N. J. A. and Plouffe, S. Figure M0851 in *The Encyclopedia of Integer Sequences.* San Diego: Academic Press, 1995.

Thistlethwaite, M. "On the Structure and Scarcity of Alternating Links and Tangles." *J. Knot Th. Ramifications* **7**, 981–004, 1998.

Welsh, D. J. A. "On the Number of Knots and Links." *Colloq. Math. Soc. J. Bolyai* **60**, 713–18, 1991.

Prime k-Tuple

PRIME CONSTELLATION

Prime k-Tuples Conjecture

K-TUPLE CONJECTURE

Prime k-Tuplet

PRIME CONSTELLATION

Prime Manifold

An *n*-MANIFOLD which cannot be "nontrivially" decomposed into other *n*-MANIFOLDS.

See also MANIFOLD

Prime Number

A prime number (or prime integer, often simply called a "prime" for short) is a POSITIVE INTEGER $p > 1$ that has no positive integer DIVISORS other than 1 and p itself. (More concisely, a prime number p is a POSITIVE INTEGER having exactly one positive divisor other than 1.) For example, the only divisors of 13 are 1 and 13, making 13 a prime number, while the number 24 has divisors 1, 2, 3, 4, 6, 8, 12, and 24 (corresponding to the factorization $24 = 2^3 \cdot 3$), making 24 *not* a prime number. POSITIVE INTEGERS other than 1 which are not prime are called COMPOSITE NUMBERS. The number 1 is a special case which is considered neither prime nor composite (Wells 1986, p. 31).

Although the number 1 used to be considered a prime (Lehmer 1909; Lehmer 1914; Hardy and Wright 1979, p. 11; Sloane and Plouffe 1995, p. 33; Hardy 1999, p. 46), it requires special treatment in so many definitions and applications involving primes greater than or equal to 2 that it is usually placed into a class of its own. As noted by Tietze (1965, p. 2), "Why is the number 1 made an exception? This is a problem that schoolboys often argue about, but since it is a question of definition, it is not arguable." The smallest prime is therefore 2. However, since 2 is the only EVEN PRIME, it is also somewhat special, the set of all primes excluding 2 is called the "ODD PRIMES." Note also that while 2 is considered a prime today, at one time it was not (Tietze 1965, p. 18; Tropfke 1921, p. 96). *Excluding* 1 and *including* 2, the first few primes are 2, 3, 5, 7, 11, 13, 17, 19, 23, 29, 31, 37, ... (Sloane's A000040; Hardy and Wright 1979, p. 3), and the SET of primes is sometimes denoted $\mathbb{P}$.

While the term "prime number" commonly refers to prime positive integers, other types of primes are also defined, such as the GAUSSIAN PRIMES.

The function which gives the number of primes less than a number n is denoted $\pi(n)$ and is called the PRIME COUNTING FUNCTION. The theorem giving an asymptotic form for $\pi(n)$ is called the PRIME NUMBER THEOREM. Prime numbers can be generated by sieving processes (such as the ERATOSTHENES SIEVE), and LUCKY NUMBERS, which are also generated by sieving, appear to share some interesting asymptotic properties with the primes. Prime numbers satisfy many strange and wonderful properties. Although there exist explicit PRIME FORMULAS (i.e., formulas which either generate primes for all values or else the nth prime as a function of n), they are contrived to such an extent that they are of little practical value.

Many PRIME FACTORIZATION ALGORITHMS have been devised for determining the prime factors of a given INTEGER, a process known as factorization or prime factorization. They vary quite a bit in sophistication and complexity. It is *very* difficult to build a general-purpose algorithm for this computationally "hard" problem, so any additional information which is known about the number in question or its factors can often be used to save a large amount of time. It should be emphasized that although no efficient algorithms are known for factoring arbitrary primes, it has not been *proved* that no such algorithm exists. It is therefore conceivable that a suitably clever person could devise a general method of factoring which would render the vast majority of encryption schemes in current widespread use, including those used by banks and governments, easily breakable.

Because of their importance in encryption algorithms such as RSA ENCRYPTION, prime numbers can be important commercial commodities. In fact, Roger Schlafly has obtained U.S. Patent 5,373,560 (12/13/94) on the following two primes (expressed in hexadecimal notation):

```
98A3DF52AEAE9799325CB258D767EBD1F4630E9B

9E21732A4AFB1624BA6DF911466AD8DA960586F4

A0D5E3C36AF099660BDDC1577E54A9F402334433

ACB14BCB
```

and

```
93E8965DAFD9DFECFD00B466B68F90EA68AF5DC9
```

FED915278D1B3A137471E65596C37FED0C7829FF

8F8331F81A2700438ECDCC09447DC397C685F397

294F722BCC484AEDF28BED25AAAB35D35A65DB1F

D62C9D7BA55844FEB1F9401E671340933EE43C54

E4DC459400D7AD61248B83A2624835B31FFF2D95

95A5B90B276E44F9.

The FUNDAMENTAL THEOREM OF ARITHMETIC states that any POSITIVE INTEGER can be represented in exactly one way as a PRODUCT of primes. EUCLID'S SECOND THEOREM demonstrated that there are an infinite number of primes. However, it is not known if there are an infinite number of primes OF THE FORM $n^2 + 1$ (Hardy and Wright 1979, p. 19; Ribenboim 1996, pp. 206–08), whether there are an INFINITE number of TWIN PRIMES (the TWIN PRIME CONJECTURE), or if a prime can always be found between n^2 and $(n + 1)^2$ (Hardy and Wright 1979, p. 415; Ribenboim 1996, pp. 397–98). The latter two of these are two of LANDAU'S PROBLEMS.

The simplest method of finding factors is so-called "DIRECT SEARCH FACTORIZATION" (a.k.a. TRIAL DIVISION). In this method, all possible factors are systematically tested using trial division to see if they actually DIVIDE the given number. It is practical only for very small numbers. More general (and complicated) methods include the ELLIPTIC CURVE FACTORIZATION METHOD and NUMBER FIELD SIEVE factorization method.

It has been proven that the set of prime numbers is a DIOPHANTINE SET (Ribenboim 1991, pp. 106–07). Ramanujan also showed that

$$\frac{d\pi(x)}{dx} \sim \frac{1}{x \ln x} \sum_{n=1}^{\infty} \frac{\mu(n)}{n} x^{1/n}, \qquad (1)$$

where $\pi(x)$ is the PRIME COUNTING FUNCTION and $\mu(n)$ is the MÖBIUS FUNCTION (Berndt 1994, p. 117).

With the exception of 2 and 3, all primes are of the form $p = 6n \pm 1$, i.e., $p \equiv 6 \pmod{1, 5}$. For n an INTEGER ≥ 2, n is prime IFF

$$\binom{n-1}{k} \equiv (-1)^k \pmod{n} \qquad (2)$$

for $k = 0, 1, ..., n - 1$ (Deutsch 1996), where $\binom{n}{k}$ is a BINOMIAL COEFFICIENT. In addition, an integer n is prime IFF

$$\phi(n) + \sigma(n) = 2n. \qquad (3)$$

The first few composite n for which $n|[\phi(n) + \sigma(n)]$ are $n = 312, 560, 588, 1400, 23760, ...$ (Sloane's A011774; Guy 1997), with a total of 18 such numbers less than 2×10^7.

Cheng (1979) showed that for x sufficiently large, there always exist at least two prime factors between

$(x - x^\alpha)$ and x for $\alpha \geq 0.477...$ (Le Lionnais 1983, p. 26). Let $f(n)$ be the number of decompositions of n into two or more consecutive primes. Then

$$\lim_{x \to \infty} \frac{1}{x} \sum_{n=1}^{x} f(n) = \ln 2 \qquad (4)$$

(Moser 1963, Le Lionnais 1983, p. 30).

The probability that the GREATEST PRIME FACTOR of a RANDOM integer n is greater than $\sqrt{n}$ is $\ln 2$ (Schroeppel 1972). The probability that two INTEGERS picked at random are RELATIVELY PRIME is $[\zeta(2)]^{-1} = 6/\pi^2$, where $\zeta(x)$ is the RIEMANN ZETA FUNCTION (Cesaro and Sylvester 1883). Given three INTEGERS chosen at random, the probability that no common factor will divide them all is

$$[\zeta(3)^{-1}] \approx 1.20206^{-1} \approx 0.831907, \qquad (5)$$

where $\zeta(3)$ is APÉRY'S CONSTANT. In general, the probability that n random numbers lack a p th POWER common divisor is $[\zeta(np)]^{-1}$ (Beeler *et al.* 1972, Item 53).

Large primes include the large MERSENNE PRIMES, FERRIER'S PRIME, and $391581 \cdot 2^{216193} - 1$ (Cipra 1989). The largest known prime as of 1999 is the MERSENNE PRIME $2^{6972593} - 1$.

Primes consisting of consecutive DIGITS (counting 0 as coming after 9) include 2, 3, 5, 7, 23, 67, 89, 4567, 78901, ... (Sloane's A006510).

See also ADLEMAN-POMERANCE-RUMELY PRIMALITY TEST, ALMOST PRIME, ANDRICA'S CONJECTURE, BERTRAND'S POSTULATE, BROCARD'S CONJECTURE, BRUN'S CONSTANT, CARMICHAEL'S CONJECTURE, CARMICHAEL FUNCTION, CARMICHAEL NUMBER, CHEBYSHEV FUNCTIONS, CHEBYSHEV-SYLVESTER CONSTANT, CHEN'S THEOREM, CHINESE HYPOTHESIS, COMPOSITE NUMBER, COMPOSITE RUNS, COPELAND-ERDOS CONSTANT, CRAMER CONJECTURE, CUNNINGHAM CHAIN, CYCLOTOMIC POLYNOMIAL, DE POLIGNAC'S CONJECTURE, DIRICHLET'S THEOREM, DIVISOR, ERDOS-KAC THEOREM, EUCLID'S THEOREMS, FEIT-THOMPSON CONJECTURE, FERMAT NUMBER, FERMAT QUOTIENT, FERRIER'S PRIME, FORTUNATE PRIME, FUNDAMENTAL THEOREM OF ARITHMETIC, GIGANTIC PRIME, GIUGA'S CONJECTURE, GOLDBACH CONJECTURE, GOOD PRIME, GRIMM'S CONJECTURE, HARDY-RAMANUJAN THEOREM, HOME PRIME, IRREGULAR PRIME, KUMMER'S CONJECTURE, LANDAU'S PROBLEMS, LEHMER'S PROBLEM, LINNIK'S THEOREM, LONG PRIME, MERSENNE NUMBER, MERTENS FUNCTION, MILLER'S PRIMALITY TEST, MIRIMANOFF'S CONGRUENCE, MÖBIUS FUNCTION, PALINDROMIC NUMBER, PÉPIN'S TEST, PILLAI'S CONJECTURE, POULET NUMBER, PRIMARY, PRIME ARRAY, PRIME CIRCLE, PRIME CONSTANT, PRIME FACTORIZATION ALGORITHMS, PRIME FORMULAS, PRIME NUMBER OF MEASUREMENT, PRIME NUMBER THEOREM, PRIME POWER SYMBOL, PRIME PRODUCTS, PRIME STRING,

PRIME SUMS, PRIME TRIANGLE, PRIME ZETA FUNC-
TION, PRIMITIVE PRIME FACTOR, PRIMORIAL, PROB-
ABLE PRIME, PSEUDOPRIME, REGULAR PRIME,
RIEMANN FUNCTION, ROTKIEWICZ THEOREM, SCHNIR-
ELMANN'S THEOREM, SELFRIDGE'S CONJECTURE, SEMI-
PRIME, SHAH-WILSON CONSTANT, SIERPINSKI'S
COMPOSITE NUMBER THEOREM, SIERPINSKI'S PRIME
SEQUENCE THEOREM, SMOOTH NUMBER, SOLDNER'S
CONSTANT, SOPHIE GERMAIN PRIME, TITANIC PRIME,
TOTIENT FUNCTION, TOTIENT VALENCE FUNCTION,
TWIN PRIMES, TWIN PRIMES CONSTANT, VINOGRA-
DOV'S THEOREM, VON MANGOLDT FUNCTION, WAR-
ING'S CONJECTURE, WEAKLY PRIME, WIEFERICH
PRIME, WILSON PRIME, WILSON QUOTIENT, WILSON'S
THEOREM, WITNESS, WOLSTENHOLME'S THEOREM,
ZSIGMONDY THEOREM

References

Berndt, B. C. "Ramanujan's Theory of Prime Numbers."
Ch. 24 in *Ramanujan's Notebooks, Part IV*. New York:
Springer-Verlag, 1994.
Caldwell, C. "Largest Primes." http://www.utm.edu/re-
search/primes/largest.html.
Caldwell, C. K. "The Top Twenty: Largest Known Primes."
http://www.utm.edu/research/primes/lists/top20/Lar-
gest.html.
Cheng, J. R. "On the Distribution of Almost Primes in an
Interval II." *Sci. Sinica* **22**, 253–75, 1979.
Cipra, B. A. "Math Team Vaults Over Prime Record."
Science **245**, 815, 1989.
Conway, J. H. and Guy, R. K. *The Book of Numbers*. New
York: Springer-Verlag, p. 130, 1996.
Courant, R. and Robbins, H. "The Prime Numbers." §1 in
Supplement to Ch. 1 in *What is Mathematics?: An Ele-
mentary Approach to Ideas and Methods, 2nd ed.* Oxford,
England: Oxford University Press, pp. 21–1, 1996.
Davenport, H. *Multiplicative Number Theory, 2nd ed.* New
York: Springer-Verlag, 1980.
Deutsch, E. "Problem 1494." *Math. Mag.* **69**, 143, 1996.
Dickson, L. E. "Factor Tables, Lists of Primes." Ch. 13 in
*History of the Theory of Numbers, Vol. 1: Divisibility and
Primality*. New York: Chelsea, pp. 347–56, 1952.
Ellison, W. J. and Ellison, F. *Prime Numbers*. New York:
Wiley, 1985.
Eynden, C. V. "A Proof of Gandhi's Formula for the nth
Prime." *Amer. Math. Monthly* **79**, 625, 1972.
Giblin, P. J. *Primes and Programming: Computers and
Number Theory*. New York: Cambridge University Press,
1994.
Glaisher, J. *Factor Tables for the Sixth Million: Containing
the Least Factor of Every Number Not Divisible by 2, 3, or
5 Between 5,000,000 and 6,000,000*. London: Taylor and
Francis, 1883.
Golomb, S W. "A Direct Interpretation of Gandhi's For-
mula." *Amer. Math. Monthly* **81**, 752–54.
Guy, R. K. "Divisors and Desires." *Amer. Math. Monthly*
104, 359–60, 1997.
Guy, R. K. "Prime Numbers," "Formulas for Primes," and
"Products Taken Over Primes." Ch. A, §A17, and §B48 in
Unsolved Problems in Number Theory, 2nd ed. New York:
Springer-Verlag, pp. 3–3, 36–1 and 102–03, 1994.
Hardy, G. H. Ch. 2 in *Ramanujan: Twelve Lectures on
Subjects Suggested by His Life and Work, 3rd ed.* New
York: Chelsea, 1978.
Hardy, G. H. and Wright, E. M. "Prime Numbers" and "The
Sequence of Primes." §1.2 and 1.4 in *An Introduction to the
Theory of Numbers, 5th ed.* Oxford, England: Clarendon
Press, pp. 1–, 1979.
Honaker, G. L. Jr. "Prime Curios!" http://www.utm.edu/
research/primes/curios/.
Honsberger, R. *Mathematical Gems II*. Washington, DC:
Math. Assoc. Amer., p. 30, 1976.
Kraitchik, M. "Prime Numbers." §3.9 in *Mathematical
Recreations*. New York: W. W. Norton, pp. 78–9, 1942.
Le Lionnais, F. *Les nombres remarquables*. Paris: Hermann,
pp. 26, 30, and 46, 1983.
Lehmer, D. N. *Factor Table for the First Ten Millions*.
Washington, DC: Carnegie Institution, 1909.
Lehmer, D. N. *List of Prime Numbers from 1 to 10,006,721*.
Washington, DC: Carnegie Institution, 1914.
Moser, L. "Notes on Number Theory III. On the Sum of
Consecutive Primes." *Can. Math. Bull.* **6**, 159–61, 1963.
Nagell, T. "Primes." §3 in *Introduction to Number Theory*.
New York: Wiley, pp. 13–4, 1951.
Ore, Ø. *Number Theory and Its History*. New York: Dover,
1988.
Pappas, T. "Prime Numbers." *The Joy of Mathematics*. San
Carlos, CA: Wide World Publ./Tetra, pp. 100–01, 1989.
Ramachandra, K. "Many Famous Conjectures on Primes;
Meagre But Precious Progress of a Deep Nature." *Proc.
Indian Nat. Sci. Acad. Part A* **64**, 643–50, 1998.
Ribenboim, P. *The Little Book of Big Primes*. New York:
Springer-Verlag, 1991.
Ribenboim, P. "Prime Number Records." *Coll. Math. J.* **25**,
280–90, 1994.
Ribenboim, P. *The New Book of Prime Number Records*.
New York: Springer-Verlag, 1996.
Riesel, H. *Prime Numbers and Computer Methods for
Factorization, 2nd ed.* Boston, MA: Birkhäuser, 1994.
Schinzel, A. and Sierpinski, W. "Sur certains hypothèses
concernant les nombres premiers." *Acta Arith.* **4**, 185–08,
1958.
Schinzel, A. and Sierpinski, W. Erratum to "Sur certains
hypothèses concernant les nombres premiers." *Acta Arith.*
5, 259, 1959.
Schroeppel, R. Item 29 in Beeler, M.; Gosper, R. W.; and
Schroeppel, R. *HAKMEM*. Cambridge, MA: MIT Artificial
Intelligence Laboratory, Memo AIM-239, p. 13, Feb. 1972.
Sloane, N. J. A. Sequences A000040/M0652, A006510/
M0679, A010051, A011774, and A046024 in "An On-Line
Version of the Encyclopedia of Integer Sequences." http://
www.research.att.com/~njas/sequences/eisonline.html.
Sloane, N. J. A. and Plouffe, S. *The Encyclopedia of Integer
Sequences*. San Diego, CA: Academic Press, 1995.
Tietze, H. "Prime Numbers and Prime Twins." Ch. 1 in
*Famous Problems of Mathematics: Solved and Unsolved
Mathematics Problems from Antiquity to Modern Times*.
New York: Graylock Press, pp. 1–0, 1965.
Torelli, G. *Sulla totalità dei numeri primi fino ad un limite
assegnato*. Naples, Italy: Tip. della Reale accad. della
scienze fisiche e matematiche, 1901.
Tropfke, J. *Geschichte der Elementar-Mathematik, Band 1*.
Berlin, Germany: p. 96, 1921.
Wagon, S. "Primes Numbers." Ch. 1 in *Mathematica in
Action*. New York: W. H. Freeman, pp. 11–7, 1991.
Weisstein, E. W. "Books about Prime Numbers." http://
www.treasure-troves.com/books/PrimeNumbers.html.
Wells, D. *The Penguin Dictionary of Curious and Interesting
Numbers*. Middlesex, England: Penguin Books, p. 31,
1986.
Zaiger, D. "The First 50 Million Prime Numbers." *Math.
Intel.* **0**, 221–24, 1977.

Prime Number of Measurement

The set of numbers generated by excluding the SUMS
of two or more consecutive earlier members is called

the prime numbers of measurement, or sometimes the SEGMENTED NUMBERS. The first few terms are 1, 2, 4, 5, 8, 10, 14, 15, 16, 21, ... (Sloane's A002048). Excluding two *and* three terms gives the sequence 1, 2, 4, 5, 8, 10, 12, 14, 15, 16, 19, 20, 21, ... (Sloane's A005242).

See also SUM-FREE SET

References

Guy, R. K. "MacMahon's Prime Numbers of Measurement." §E30 in *Unsolved Problems in Number Theory, 2nd ed.* New York: Springer-Verlag, pp. 230–31, 1994.

Sloane, N. J. A. Sequences A002048/M0972 and A005242/M0971 in "An On-Line Version of the Encyclopedia of Integer Sequences." http://www.research.att.com/~njas/sequences/eisonline.html.

Prime Number Theorem

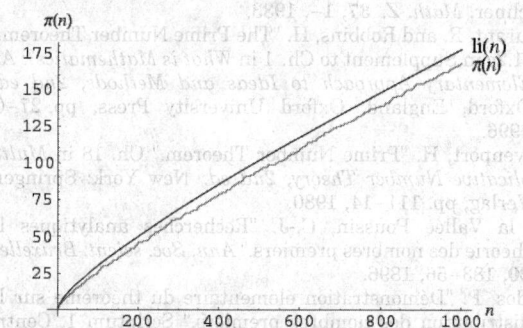

The theorem giving an asymptotic form for the PRIME COUNTING FUNCTION $\pi(n)$, which counts the number of PRIMES less than some INTEGER n. Legendre (1808) suggested that, for large n,

$$\pi(n) \sim \frac{n}{A \ln n + B}, \tag{1}$$

with $A = 1$ and $B = -1.08366$ (where B is sometimes called LEGENDRE'S CONSTANT), a formula which is correct in the leading term only (Nagell 1951, p. 54; Wagon 1991, pp. 28–9). In 1791, Gauss became the first to suggest instead

$$\pi(n) \sim \frac{n}{\ln n}. \tag{2}$$

Gauss later refined his estimate to

$$\pi(n) \sim \text{li}(n), \tag{3}$$

where $\text{li}(n)$ is the LOGARITHMIC INTEGRAL. This function has $n/\ln n$ as the leading term and has been shown to be a better estimate than $n/\ln n$ alone. The statement (3) is often known as "the" prime number theorem and was proved independently by Hadamard (1896) and de la Vallée Poussin (1896). A plot of $\pi(n)$ (lower curve) and $\text{li}(n)$ is shown above for $n \leq 1000$. For small n, it has been checked and always found that $\pi(n) < \text{li}(n)$. However, Skewes proved that the

first crossing of $\pi(n) < \text{li}(n) = 0$ occurs before $10^{10^{10^{34}}}$ (the SKEWES NUMBER). The upper bound for the crossing has subsequently been reduced to 10^{371}. Littlewood (1914) proved that the INEQUALITY reverses infinitely often for sufficiently large n (Ball and Coxeter 1987). Lehman (1966) proved that at least 10^{500} reversals occur for numbers with 1166 or 1167 DECIMAL DIGITS.

Chebyshev put limits on the RATIO

$$\frac{7}{8} < \frac{\pi(n)}{\dfrac{n}{\ln n}} < \frac{9}{8} \tag{4}$$

(Landau 1927; Nagell 1951, p. 55; Landau 1974; Hardy and Wright 1979, Ch. 22; Ingham 1990; Rubinstein and Sarnak 1994; Hardy 1999, p. 27), and showed that if the LIMIT

$$\lim_{n \to \infty} \frac{\pi(n)}{\dfrac{n}{\ln n}} \tag{5}$$

existed, then it would be 1.

Hadamard and Vallée Poussin proved the prime number theorem by showing that the RIEMANN ZETA FUNCTION $\zeta(z)$ has no zeros OF THE FORM $1 + it$, in the sense that no deeper properties of $\zeta(s)$ are required for the proof (Smith 1994, p. 128; Hardy 1999, pp. 58–0). Wiener (1951) allowed this somewhat vague statement to be interpreted literally (Hardy 1999, pp. 34 and 46), and this proof was simplified by Landau (1932) and Bochner (1933).

Hadamard's proof depends on the simple trigonometric inequality

$$3 + 4 \cos \theta + \cos(2\theta) = 2(1 + \cos \theta)^2 \geq 0 \tag{6}$$

(Hardy 1999, p. 58). Vallée Poussin (1899) showed that

$$\pi(x) = \text{li}(x) + \mathcal{O}\left(\frac{x}{\ln x} e^{-a\sqrt{\ln x}}\right) \tag{7}$$

for some constant a (Knuth 1997, p. 381), where $\mathcal{O}(x)$ is ASYMPTOTIC NOTATION. A simplified proof was found by Erdos (1949) and Selberg (1950) (Ball and Coxeter 1987, p. 63), although an unfortunate priority dispute over the joint work marred the otherwise beautiful proof (Hoffman 1998, pp. 39–1). An elementary proof of the prime number theorem, following Selberg, is the final section in Nagell's 1951 textbook.

The error term in (7) has subsequently improved to

$$\pi(x) = \text{li}(x) + \mathcal{O}\left(x \exp\left(-\frac{A(\ln x)^{3/5}}{(\ln \ln x)^{1/5}}\right)\right) \tag{8}$$

(Walfisz 1963; Riesel 1994, p. 56; Knuth 1997, p. 382). Ingham (1930) proved the prime number theorem using the identity of Ramanujan

$$\sum_{n=1}^{\infty} \frac{\sigma_a(n)\sigma_b(n)}{n^s} = \frac{\zeta(s)\zeta(s-a)\zeta(s-b)\zeta(s-a-b)}{\zeta(2s-a-b)}, \quad (9)$$

where $\sigma_a(n)$ is the DIVISOR FUNCTION (Hardy 1999, pp. 59–0).

Riemann estimated the PRIME COUNTING FUNCTION with

$$\pi(n) \sim \ln(n) - \tfrac{1}{2}\operatorname{li}(n^{1/2}), \quad (10)$$

which is a better approximation than li(n) for $n < 10^7$. Riemann (1859) also suggested the RIEMANN FUNCTION

$$R(x) = \sum_{n=1}^{\infty} \frac{\mu(n)}{n}\operatorname{li}(x^{1/n}), \quad (11)$$

where μ is the MÖBIUS FUNCTION (Wagon 1991, p. 29). An even better approximation for small n (by a factor of 10 for $n < 10^9$) is the GRAM SERIES.

The prime number theorem is equivalent to either

$$\lim_{x\to\infty} \frac{\theta(x)}{x} = 1 \quad (12)$$

or

$$\lim_{x\to\infty} \frac{\psi(x)}{x} = 1, \quad (13)$$

where θ and $\psi(x)$ are the CHEBYSHEV FUNCTIONS. Chebyshev showed that the only possible limit of these expressions was 1, but was not able to prove existence of the limit (Hardy 1999, p. 28).

The RIEMANN HYPOTHESIS is equivalent to the assertion that

$$|\operatorname{Li}(x) - \pi(x)| \le c\sqrt{x}\ln x \quad (14)$$

for some value of c (Ingham 1990, p. 83; Landau 1974, pp. 378–88; Ball and Coxeter 1987; Hardy 1999, p. 26). Some limits obtained without assuming the RIEMANN HYPOTHESIS are

$$\pi(x) = \operatorname{Li}(x) + \mathcal{O}[xe^{-\ln\,x^{1/2}/15}] \quad (15)$$

$$\pi(x) = \operatorname{Li}(x) + \mathcal{O}\left[xe^{-0.009(\ln x)^{3/5}/(\ln\ln x)^{1/5}}\right]. \quad (16)$$

Ramanujan showed that for sufficiently large x,

$$\pi^2(x) < \frac{ex}{\ln x}\,\pi\!\left(\frac{x}{e}\right). \quad (17)$$

The largest known PRIME for which the inequality fails is 38,358,837,677 (Berndt 1994, pp. 112–13). The related inequality

$$\operatorname{Li}^2(x) < \frac{ex}{\ln x}\operatorname{Li}\!\left(\frac{x}{e}\right) \quad (18)$$

is true for $x \ge 2418$ (Berndt 1994, p. 114).

See also BERTRAND'S POSTULATE, CHEBYSHEV FUNCTIONS, CHEBYSHEV'S THEOREM, DIRICHLET'S THEOREM, GRAM SERIES, PRIME COUNTING FUNCTION, RIEMANN FUNCTION, SELBERG'S FORMULA, SKEWES NUMBER

References

Apostol, T. M. *Introduction to Analytic Number Theory.* New York: Springer-Verlag, 1976.

Ball, W. W. R. and Coxeter, H. S. M. *Mathematical Recreations and Essays, 13th ed.* New York: Dover, pp. 62–4, 1987.

Berndt, B. C. *Ramanujan's Notebooks, Part IV.* New York: Springer-Verlag, 1994.

Bochner. *Math. Z.* **37**, 1–, 1933.

Courant, R. and Robbins, H. "The Prime Number Theorem." §1.2c in Supplement to Ch. 1 in *What is Mathematics?: An Elementary Approach to Ideas and Methods, 2nd ed.* Oxford, England: Oxford University Press, pp. 27–0, 1996.

Davenport, H. "Prime Number Theorem." Ch. 18 in *Multiplicative Number Theory, 2nd ed.* New York: Springer-Verlag, pp. 111–14, 1980.

de la Vallée Poussin, C.-J. "Recherches analytiques la théorie des nombres premiers." *Ann. Soc. scient. Bruxelles* **20**, 183–56, 1896.

Erdos, P. "Démonstration élémentaire du théorème sur la distribution des nombres premiers." Scriptum 1, Centre Mathématique, Amsterdam, 1949.

Hadamard, J. "Sur la distribution des zéros de la fonction $\zeta(s)$ et ses conséquences arithmétiques (')." *Bull. Soc. math. France* **24**, 199–20, 1896.

Hardy, G. H. "The Proof of the Prime Number Theorem" and "Second Approximation of the Proof." §2.5 and 2.6 in *Ramanujan: Twelve Lectures on Subjects Suggested by His Life and Work, 3rd ed.* New York: Chelsea, pp. 16, 27, and 28–3, 1999.

Hardy, G. H. and Wright, E. M. "Statement of the Prime Number Theorem." §1.8 in *An Introduction to the Theory of Numbers, 5th ed.* Oxford, England: Clarendon Press, pp. 9–0, 1979.

Hoffman, P. *The Man Who Loved Only Numbers: The Story of Paul Erdos and the Search for Mathematical Truth.* New York: Hyperion, 1998.

Ingham, A. E. "Note on Riemann's ζ-Function and Dirichlet's L-Functions." *J. London Math. Soc.* **5**, 107–12, 1930.

Ingham, A. E. *The Distribution of Prime Numbers.* London: Cambridge University Press, p. 83, 1990.

Knuth, D. E. *The Art of Computer Programming, Vol. 2: Seminumerical Algorithms, 3rd ed.* Reading, MA: Addison-Wesley, 1998.

Landau, E. *Vorlesungen über Zahlentheorie, Vol. 1.* New York: Chelsea, pp. 79–6, 1970.

Landau, E. *Berliner Sitzungsber.*, 514–21, 1932.

Landau, E. *Handbuch der Lehre von der Verteilung der Primzahlen, 3rd ed.* New York: Chelsea, 1974.

Legendre, A. M. *Essai sur la Théorie des Nombres.* Paris: Duprat, 1808.

Lehman, R. S. "On the Difference $\pi(x) - \operatorname{li}(x)$." *Acta Arith.* **11**, 397–10, 1966.

Littlewood, J. E. "Sur les distribution des nombres premiers." *C. R. Acad. Sci. Paris* **158**, 1869–872, 1914.

Lu, W. C. "On the Elementary Proof of the Prime Number Theorem with a Remainder Term." *Rocky Mountain J. Math.* **29**, 979, 1999.

Nagell, T. "The Prime Number Theorem." Ch. 8 in *Introduction to Number Theory.* New York: Wiley, pp. 275–99, 1951.

Riemann, G. F. B. "Über die Anzahl der Primzahlen unter einer gegebenen Grösse." *Monatsber. Königl. Preuss. Akad. Wiss. Berlin*, 671, 1859.

Riesel, H. "The Remainder Term in the Prime Number Theorem." *Prime Numbers and Computer Methods for Factorization, 2nd ed.* Boston, MA: Birkhäuser, p. 6, 1994.

Rubinstein, M. and Sarnak, P. "Chebyshev's Bias." *Experimental Math.* **3**, 173–97, 1994.

Selberg, A. "An Elementary Proof of the Prime Number Theorem." *Ann. Math.* **50**, 305–13, 1949.

Shanks, D. "The Prime Number Theorem." §1.6 in *Solved and Unsolved Problems in Number Theory, 4th ed.* New York: Chelsea, pp. 15–7, 1993.

Smith, D. E. *A Source Book in Mathematics.* New York: Dover, 1994.

Vallée Poussin, C. *Mém. Couronnés Acad. Roy. Belgique* **59**, 1–4, 1899.

Wagon, S. *Mathematica in Action.* New York: W. H. Freeman, pp. 25–5, 1991.

Walfisz, A. Ch. 5 in *Weyl'sche Exponentialsummen in der neueren Zahlentheorie.* Berlin: Deutscher Verlag der Wissenschaften, 1963.

Wiener, N. §19 et seq. in *The Fourier Integral and Certain of Its Applications.* New York: Dover, 1951.

Prime Pairs

TWIN PRIMES

Prime Partition

A prime partition of a POSITIVE INTEGER $n \geq 2$ is a set of PRIMES p_i which sum to n. For example, there are three prime partitions of 7 since

$$7 = 7 = 2 + 5 = 2 + 2 + 3.$$

The number of prime partitions of $n = 2, 3, \ldots$ are 1, 1, 1, 2, 2, 3, 3, 4, 5, 6, 7, 9, 10, 12, 14, 17, 19, 23, 26, ... (Sloane's A000607). If $a_n = 1$ for n prime and $a_n = 0$ for n composite, then the EULER TRANSFORM b_n gives the number of partitions of n into prime parts (Sloane and Plouffe 1995, p. 21).

The minimum number of primes needed to sum to $n = 2, 3, \ldots$ are 1, 1, 2, 1, 2, 1, 2, 2, 2, 1, 2, 1, 2, 2, 2, 1, 2, ... (Sloane's A051034). The maximum number of primes needed to sum to n is just $\lfloor n/2 \rfloor$, 0, 0, 1, 1, 2, 2, 3, 3, 4, 4, 5, 5, 6, 6, 7, 7, ... (Sloane's A004526), corresponding to a representation in terms of all 2s for an even number or one 3 and the rest 2s for an odd number.

The numbers which can be represented by a single prime are obviously the primes themselves. Composite numbers which can be REPRESENTED AS the sum of two primes are 4, 6, 8, 9, 10, 12, 14, 15, 16, 18, 20, 21, 22, ... (Sloane's A051035), and composite numbers which are not the sum of fewer than three primes are 27, 35, 51, 57, 65, 77, 87, 93, 95, 117, 119, ..., (Sloane's A025583). The conjecture that *no* numbers require

four or more primes is called the GOLDBACH CONJECTURE.

See also GOLDBACH CONJECTURE, PARTITION, PARTITION FUNCTION P, SCHNIRELMANN'S THEOREM

References

Berndt, B.C. and Wilson, B. M. "Chapter 5 of Ramanujan's Second Notebook." In *Analytic Number Theory: Proceedings of the Conference Held at Temple University, Philadelphia, Pa., May 12–5, 1980* (Ed. M. I. Knopp). Berlin: Springer-Verlag, pp. 49–8, 1981.

Chawla, L. M. and Shad, S. A. "On a Trio-Set of Partition Functions and Their Tables." *J. Natural Sciences and Mathematics* **9**, 87–6, 1969.

Gupta, O. P. and Luthra, S. "Partitions into Primes." *Proc. Nat. Inst. Sci. India. Part A* **21**, 181–84, 1955.

Gupta, H. "Partitions into Distinct Primes." *Proc. Nat. Inst. Sci. India. Part A* **21**, 185–87, 1955.

Guy, R. K. "The Strong Law of Small Numbers." *Amer. Math. Monthly* **95**, 697–12, 1988.

Sloane, N. J. A. Sequences A000607/M0265, A004526, A025583, A051034, and A051035 in "An On-Line Version of the Encyclopedia of Integer Sequences." http://www.research.att.com/~njas/sequences/eisonline.html.

Sloane, N. J. A. and Plouffe, S. *The Encyclopedia of Integer Sequences.* San Diego, CA: Academic Press, 1995.

Prime Patterns Conjecture

K-TUPLE CONJECTURE

Prime Pi

PRIME COUNTING FUNCTION

Prime Polynomial

PRIME-GENERATING POLYNOMIAL

Prime Power

A PRIME or integer power of a PRIME. The first few are 2, 3, 4, 5, 7, 8, 9, 11, 13, 16, 17, 19, 23, 25, ... (Sloane's A000961). The first few prime powers with power ≥ 2 are given by 4, 8, 9, 16, 25, 27, 32, 49, 64, 81, ... (Sloane's A025475). The number of prime powers (≥ 2) up to x does not exceed

$$x^{1/2} + x^{1/3} + x^{1/4} + \ldots = \mathcal{O}\left(x^{1/2} \ln x\right)$$

(Hardy 1999, p. 27).

The following table gives prime kth powers.

k	Sloane	prime kth powers
1	A000040	2, 3, 5, 7, 11, 13, 17, 19, 23, ...
2	A001248	4, 9, 25, 49, 121, 169, 289, 361, ...
3	A030078	8, 27, 125, 343, 1331, 2197, 4913, ...
4	A030514	16, 81, 625, 2401, 14641, 28561, 83521, ...
5	A050997	32, 243, 3125, 16807, 161051, 371293, ...

See also PRIME NUMBER, SOLITARY NUMBER

References

Hardy, G. H. *Ramanujan: Twelve Lectures on Subjects Suggested by His Life and Work,* 3rd ed. New York: Chelsea, 1999.

Sloane, N. J. A. Sequences A000040/M0652, A000961/M0517, A001248, A025475, A030078, A030514, and A050997 in "An On-Line Version of the Encyclopedia of Integer Sequences." http://www.research.att.com/~njas/sequences/eisonline.html.

Prime Power Conjecture

An Abelian planar DIFFERENCE SET of order n exists only for n a PRIME POWER. Gordon (1994) has verified it to be true for $n < 2,000,000$.

See also DIFFERENCE SET

References

Gordon, D. M. "The Prime Power Conjecture is True for $n < 2,000,000$." *Electronic J. Combinatorics* **1**, R6 1–, 1994. http://www.combinatorics.org/Volume_1/volume1.html#R6.

Prime Power Symbol

The symbol $p^e \| n$ means, for p a PRIME, that $p^e \| n$, but $p^{e+1} \nmid n$.

Prime Products

The product of primes

$$p_n \# \equiv \prod_{k=1}^{n} p_k, \qquad (1)$$

with p_n the nth prime, is called the PRIMORIAL function, by analogy with the FACTORIAL function.

The EULER PRODUCT gives

$$e^\gamma = \lim_{n \to \infty} \frac{1}{\ln n} \prod_{k=1}^{n} \frac{1}{1 - \dfrac{1}{p_k}}, \qquad (2)$$

where γ is the EULER-MASCHERONI CONSTANT. There is also an amazing infinite product formula for primes given by

$$\prod_{k=1}^{\infty} \frac{p_k^2 + 1}{p_k^2 - 1} = \frac{5}{2}. \qquad (3)$$

(Ramanujan; Le Lionnais 1983, p. 46).

See also EULER PRODUCT, PRIME NUMBER, PRIME SUMS, PRIMORIAL

References

Grosswald, E. "Some Number Theoretical Products." *Rev. Columbiana Mat.* **21** 231–42, 1987.

Le Lionnais, F. *Les nombres remarquables.* Paris: Hermann, p. 46, 1983.

Uchiyama, S. "On Some Products Involving Primes." *Proc. Amer. Math. Soc.* **28**, 629–30, 1971.

Prime Quadratic Effect

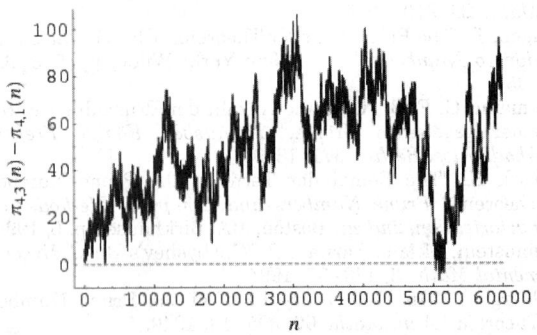

Let $\pi_{m,n}(x)$ denote the number of PRIMES $\leq x$ which are congruent to n modulo m. Then one might expect that

$$\Delta(x) \equiv \pi_{4,3}(x) - \pi_{4,1}(x) \sim \tfrac{1}{2}\,\pi\!\left(x^{1/2}\right) > 0$$

(Berndt 1994). Although this is true for small numbers, Hardy and Littlewood showed that $\Delta(x)$ changes sign infinitely often. The effect was first noted by Chebyshev in 1853, and is sometimes called the CHEBYSHEV PHENOMENON. It was subsequently studied by Shanks (1959), Hudson (1980), and Bays and Hudson (1977, 1978, 1979). The effect was also noted by Ramanujan, who incorrectly claimed that $\lim_{x \to \infty} \Delta(x) = \infty$ (Berndt 1994).

The values at which $\Delta(x) = 0$ are $x = 2946$, 50378, 50380, 50382, 50392, 50414, ... (Sloane's A051024), corresponding to $\pi(x) = 26861$, 616841, 616849, 616877, 617011, ... (Sloane's A051025).

References

Bays, C. and Hudson, R. H. "The Mean Behavior of Primes in Arithmetic Progressions." *J. reine angew. Math.* **296**, 80–9, 1977.

Bays, C. and Hudson, R. H. "On the Fluctuations of Littlewood for Primes of the Form $4n \pm 1$." *Math. Comput.* **32**, 281–86, 1978.

Bays, C. and Hudson, R. H. "Numerical and Graphical Description of All Axis Crossing Regions for the Moduli 4 and 8 which Occur Before 10^{12}." *Internat. J. Math. Math. Sci.* **2**, 111–19, 1979.

Berndt, B. C. *Ramanujan's Notebooks, Part IV.* New York: Springer-Verlag, pp. 135–36, 1994.

Hudson, R. H. "A Common Principle Underlies Riemann's Formula, the Chebyshev Phenomenon, and Other Subtle Effects in Comparative Prime Number Theory. I." *J. reine angew. Math.* **313**, 133–50, 1980.

Shanks, D. "Quadratic Residues and the Distribution of Primes." *Math. Comput.* **13**, 272–84, 1959.

Sloane, N. J. A. Sequences A051024 and A051025 in "An On-Line Version of the Encyclopedia of Integer Sequences." http://www.research.att.com/~njas/sequences/eisonline.html.

Prime Quadruplet

A PRIME CONSTELLATION of four successive PRIMES with minimal distance $(p, p+2, p+6, p+8)$. The term was coined by Paul Stäckel (1892–919; Tietze 1965, p. 19). The quadruplet $(2, 3, 5, 7)$ has smaller minimal distance, but it is an exceptional special case. With the exception of (5, 7, 11, 13), a prime quadruple must be OF THE FORM $(30n + 11, 30n + 13, 30n + 17, 30n + 19)$. The first few values of n which give prime quadruples are $n = 0, 3, 6, 27, 49, 62, 69, 108, 115, \ldots$ (Sloane's A014561), and the first few values of p are 5 (the exceptional case), 11, 101, 191, 821, 1481, 1871, 2081, 3251, 3461, ... (Sloane's A007530). The number of prime quadruplets with largest member less than 10^1, 10^2, ..., are 1, 2, 5, 12, 38, 166, 899, 4768, ... (Sloane's A050258; Nicely 1999).

The asymptotic FORMULA for the frequency of prime quadruples is analogous to that for other PRIME CONSTELLATIONS,

$$P_x(p, p+2, p+6, p+8) \sim \frac{27}{2} \prod_{p \geq 5} \frac{p^3(p-4)}{(p-1)^4} \int_2^x \frac{dx}{(\ln x)^4}$$

$$= 4.151180864 \int_2^x \frac{dx}{(\ln x)^4},$$

where $c = 4.15118\ldots$ is the Hardy-Littlewood constant for prime quadruplets.

Roonguthai found the large prime quadruplets with

$$p = 10^{99} + 349781731$$

$$p = 10^{199} + 21156403891$$

$$p = 10^{299} + 140159459341$$

$$p = 10^{399} + 34993836001$$

$$p = 10^{499} + 883750143961$$

$$p = 10^{599} + 1394283756151$$

$$p = 10^{699} + 547634621251$$

(Roonguthai). Forbes found the large quadruplet with

$$p = 76912895956636885 \left(2^{3279} - 2^{1093}\right) - 6 \cdot 2^{1093} - 7.$$

See also PRIME ARITHMETIC PROGRESSION, PRIME CONSTELLATION, PRIME *K*-TUPLES CONJECTURE, PRIME TRIPLET, SEXY PRIMES, TWIN PRIMES

References

Hardy, G. H. and Wright, E. M. *An Introduction to the Theory of Numbers, 5th ed.* New York: Oxford University Press, 1979.
Forbes, T. "Prime *k*-tuplets." http://www.ltkz.demon.co.uk/ktuplets.htm.
Forbes, T. "Large Prime Quadruplets." nmbrthry@list-serv.nodak.edu. Sep. 17, 1998.
Nicely, T. R. "Enumeration to 1.6×10^{15} of the Prime Quadruplets." Submitted to *Math. Comput.*
Rademacher, H. *Lectures on Elementary Number Theory.* New York: Blaisdell, 1964.
Riesel, H. *Prime Numbers and Computer Methods for Factorization, 2nd ed.* Boston, MA: Birkhäuser, pp. 61–2, 1994.
Roonguthai, W. "Large Prime Quadruplets." http://www.mathsoft.com/asolve/constant/hrdyltl/roonguth.html.
Sloane, N. J. A. Sequences A007530/M3816, A014561, and A050258 in "An On-Line Version of the Encyclopedia of Integer Sequences." http://www.research.att.com/~njas/sequences/eisonline.html.
Tietze, H. *Famous Problems of Mathematics: Solved and Unsolved Mathematics Problems from Antiquity to Modern Times.* New York: Graylock Press, p. 19, 1965.

Prime Representation

Let $a \neq b$, A, and B denote POSITIVE INTEGERS satisfying

$$(a, b) = 1 \quad (A, B) = 1$$

(i.e., both pairs are RELATIVELY PRIME), and suppose every PRIME $p \equiv B \pmod{A}$ with $(p, 2ab) = 1$ is expressible if the form $ax^2 - by^2$ for some INTEGERS x and y. Then every PRIME q such that $q \equiv -B \pmod{A}$ and $(q, 2ab) = 1$ is expressible in the form $bX^2 - aY^2$ for some INTEGERS X and Y (Halter-Koch 1993, Williams 1991).

Prime Form	Representation
$4n + 1$	$x^2 + y^2$
$8n + 1$, $8n + 3$	$x^2 + 2y^2$
$8n \pm 1$	$x^2 - 2y^2$
$6n + 1$	$x^2 + 3y^2$
$12n + 1$	$x^2 - 3y^2$
$20n + 1$, $20n + 9$	$x^2 + 5y^2$
$10n + 1$, $10n + 9$	$x^2 - 5y^2$
$14n + 1$, $14n + 9$, $14n + 25$	$x^2 + 7y^2$
$28n + 1$, $28n + 9$, $28n + 25$	$x^2 - 7y^2$
$30n + 1$, $30n + 49$	$x^2 + 15y^2$
$60n + 1$, $60n + 49$	$x^2 - 15y^2$
$30n - 7$, $30n + 17$	$5x^2 + 3y^2$
$60n - 7$, $60n + 17$	$5x^2 - 3y^2$
$24n + 1$, $24n + 7$	$x^2 + 6y^2$
$24n + 1$, $24n + 19$	$x^2 - 6y^2$
$24n + 5$, $24n + 11$	$2x^2 + 3y^2$
$24n + 5$, $24n - 1$	$2x^2 - 3y^2$

References

Berndt, B. C. *Ramanujan's Notebooks, Part IV.* New York: Springer-Verlag, pp. 70–3, 1994.

Halter-Koch, F. "A Theorem of Ramanujan Concerning Binary Quadratic Forms." *J. Number. Theory* **44**, 209–13, 1993.

Williams, K. S. "On an Assertion of Ramanujan Concerning Binary Quadratic Forms." *J. Number Th.* **38**, 118–33, 1991.

Prime Ring

A RING for which the product of any pair of IDEALS is zero only if one of the two IDEALS is zero. All SIMPLE RINGS are prime.

See also IDEAL, RING, SEMIPRIME RING, SIMPLE RING

Prime Sequence

PRIME ARITHMETIC PROGRESSION, PRIME ARRAY, PRIME-GENERATING POLYNOMIAL, SIERPINSKI'S PRIME SEQUENCE THEOREM

Prime Signature

The prime signature of a positive integer n is a sorted list of exponents a_i in the PRIME FACTORIZATION

$$n = p_1^{a_1} p_2^{a_2} \cdots.$$

The prime signature of n can therefore be computed in *Mathematica* as

```
PrimeSignature[1]              :=              {1}
PrimeSignature[n_Integer?Positive] :=
    Sort[Transpose[FactorInteger[n]][[2]]]
```

See also PRIME FACTORIZATION

Prime Spiral

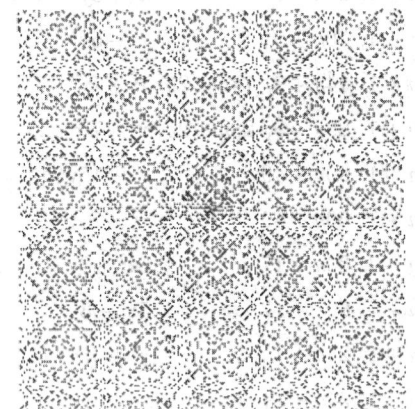

The numbers arranged in a SPIRAL

```
5  4  3
6  1  2
7  8  9
```

with PRIMES indicated in black, as first drawn by

S. Ulam. Unexpected patterns of diagonal lines are apparent in such a plot, as illustrated in the above 199×199 grid. M. Charpentier has written a Post-Script file which can be downloaded to a printer and draws a prime spiral.

See also PRIME-GENERATING POLYNOMIAL

References

Charpentier, M. "Prime Numbers in PostScript." http://www.cs.unh.edu/~charpov/Programming/PostScript-primes/.

Dewdney, A. K. "Computer Recreations: How to Pan for Primes in Numerical Gravel." *Sci. Amer.* **259**, 120–23, July 1988.

Gardner, M. *The Sixth Book of Mathematical Games from Scientific American.* Chicago, IL: University of Chicago Press, pp. 80–3 and 88–9, 1984.

Goddard, T. "Ulam Spiral." http://www.d4maths.co.uk/mirage/ulam.htm.

Hoffman, P. *The Man Who Loved Only Numbers: The Story of Paul Erdos and the Search for Mathematical Truth.* New York: Hyperion, pp. 105–09, 1998.

Lane, C. "Prime Spiral." http://www.best.com/~cdl/Prime-SpiralApplet.html.

Leatherland, A. J. F. "The Mysterious Prime Spiral Phenomenon." http://yoyo.cc.monash.edu.au/~bunyip/primes/#spiral.

Morin, D. "Le Village Premier." http://platon.lacitec.on.ca/~dmorin/applet/village/.

Stein, M. L.; Ulam, S. M.; and Wells, M. B. "A Visual Display of Some Properties of the Distribution of Primes." *Amer. Math. Monthly* **71**, 516–20, 1964.

Weisstein, E. W. "Prime Spiral." MATHEMATICA NOTEBOOK PRIMESPIRAL.M.

Prime String

TRUNCATABLE PRIME

Prime Subfield

The prime subfield of a FIELD F is the SUBFIELD of F generated by the multiplicative identity 1_F of F. It is isomorphic to either $\mathbb{Q}$ (if the CHARACTERISTIC is 0), or the FINITE FIELD $\mathbb{F}_p = \mathbb{Z}/p\mathbb{Z}$ (if the CHARACTERISTIC is p).

See also SUBFIELD

References

Dummit, D. S. and Foote, R. M. *Abstract Algebra, 2nd ed.* Englewood Cliffs, NJ: Prentice-Hall, p. 423, 1998.

Prime Sum

$$60n - 7, \quad 60n + 17$$

Let

$$5x^2 - 3y^2$$

be the sum of the first n PRIMES. The first few terms are 2, 5, 10, 17, 28, 41, 58, 77, ... (Sloane's A007504). Bach and Shallit (1996) show that

$$24n + 1, \quad 24n + 7$$

and provide a general technique for estimating such sums.

The first few values of n such that $x^2 + 6y^2$ are 1, 23, 53, 853, 11869, 117267, 339615, 3600489, 96643287, ... (Sloane's A045345). The corresponding values of $24n + 1$, $24n + 19$ are 2, 874, 5830, 2615298, 712377380, 86810649294, 794712005370, 105784534314378, 92542301212047102, ... (Sloane's A050247; Rivera), and the values of $x^2 - 6y^2$ are 2, 38, 110, 3066, 60020, 740282, 2340038, 29380602, 957565746, ... (Sloane's A050248; Rivera).

See also PRIMORIAL

References

Bach, E. and Shallit, J. §2.7 in *Algorithmic Number Theory, Vol. 1: Efficient Algorithms.* Cambridge, MA: MIT Press, 1996.

Rivera, C. "Problems & Puzzles: Puzzle The Average Prime number, $24n + 5$, $24n + 11$.-031." http://www.primepuzzles.net/puzzles/puzz_031.htm.

Sloane, N. J. A. Sequences A007504/M1370, A045345, A050247, and A050248 in "An On-Line Version of the Encyclopedia of Integer Sequences." http://www.research.att.com/~njas/sequences/eisonline.html.

Prime Sums

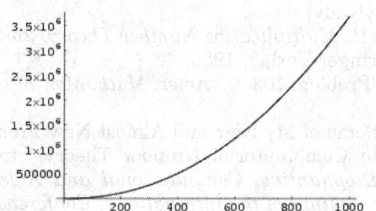

Let

$$\sum(n) \equiv \sum_{i=1}^{n} p_i \tag{1}$$

be the sum of the first n PRIMES (i.e., the sum analog of the PRIMORIAL function). The first few terms are 2, 5, 10, 17, 28, 41, 58, 77, ... (Sloane's A007504). Bach and Shallit (1996) show that

$$\sum(n) \sim \frac{n^2}{2\log n}, \tag{2}$$

and provide a general technique for estimating such sums.

The first few values of n such that $n|\Sigma(n)$ are 1, 23, 53, 853, 11869, 117267, 339615, 3600489, 96643287, ... (Sloane's A045345). The corresponding values of $\Sigma(n)$ are 2, 874, 5830, 2615298, 712377380, 86810649294, 794712005370, 105784534314378, 92542301212047102, ... (Sloane's A050247; Rivera), and the values of $n/\Sigma(n)$ are 2, 38, 110, 3066, 60020, 740282, 2340038, 29380602, 957565746, ... (Sloane's A050248; Rivera).

In 1737, Euler showed that the sum of the reciprocals of the primes diverges

$$\sum_{k=1}^{\infty} \frac{1}{p_k} = \infty \tag{3}$$

(Nagell 1951, p. 59; Hardy and Wright 1979, pp. 17 and 22), although it does so very slowly. The sum exceeds 1, 2, 3, ... after 3, 59, 361139, ... (Sloane's A046024) primes, and its asymptotic equation is

$$\sum_{\substack{p=2 \\ p \text{ prime}}}^{x} \frac{1}{p} = \ln \ln x + B_1 + o(1), \tag{4}$$

where B_1 is MERTENS CONSTANT (Hardy and Wright 1979, p. 351). Dirichlet showed the even stronger result that

$$\sum_{\substack{\text{prime } p \equiv b \pmod{a} \\ (a, b) = 1}} \frac{1}{p} = \infty \tag{5}$$

(Davenport 1980, p. 34). Despite the divergence of the sum of reciprocal primes, the ALTERNATING SERIES

$$\sum_{k=1}^{\infty} \frac{(-1)^k}{p_k} \approx -0.2696065 \tag{6}$$

converges (Robinson and Potter 1971, Finch), but it is not known if the sum

$$\sum_{k=1}^{\infty} (-1)^k \frac{k}{p_k} \tag{7}$$

does (Guy 1994, p. 203; Erdos 1998; Finch).

There are also classes of sums of reciprocal primes with sign determined by congruences on k, for example

$$\sum_{k=2}^{\infty} \frac{c_k}{p_k} \approx 0.3349813253 \tag{8}$$

where

$$c_k = \begin{cases} -1 & \text{for } pk \equiv 1 \pmod{4} \\ 1 & \text{for } pk \equiv 3 \pmod{4} \end{cases} \tag{9}$$

(Glaisher 1891b, Finch) which, is not known to converge, while

$$\sum_{k=2}^{\infty} \frac{c_k}{p_k^2} \approx 0.094619828 \tag{10}$$

does converge (Glaisher 1893, Finch). It is not known if

$$\sum_{k=1}^{\infty} \frac{d_k}{p_k} \approx 0.6419448385 \tag{11}$$

converges, where

$$d_k = \begin{cases} -1 & \text{for } pk \equiv 1 \pmod 3 \\ 1 & \text{for } pk \equiv 2 \pmod 3 \\ 0 & \text{for } pk \equiv 0 \pmod 3 \end{cases} \tag{12}$$

(Glaisher 1891c, Finch).

Although $\Sigma\, 1/p$ diverges, Brun (1919) showed that

$$\sum_{\substack{p \\ p+2 \text{ prime}}} \frac{1}{p} = B < \infty, \tag{13}$$

where B is BRUN'S CONSTANT. The function defined by

$$P(n) \equiv \sum_{p=1}^{\infty} \frac{1}{p_k^n} \tag{14}$$

taken over the primes converges for $n > 1$ and is a generalization of the RIEMANN ZETA FUNCTION known as the PRIME ZETA FUNCTION.

A rapidly converging series for the MERTENS CONSTANT

$$B_1 = \gamma + \sum_{k=1}^{\infty} \left[\ln\left(1 - p_k^{-1}\right) + \frac{1}{p_k} \right] \approx 0.2614972128 \tag{15}$$

is given by

$$B_1 = \gamma + \sum_{m=2}^{\infty} \frac{\mu(m)}{m} \ln[\zeta(m)], \tag{16}$$

where γ is the EULER-MASCHERONI CONSTANT, $\zeta(n)$ is the RIEMANN ZETA FUNCTION, and $\mu(n)$ is the MÖBIUS FUNCTION (Flajolet and Vardi 1996, Schroeder 1997, Knuth 1998). A similar formula gives the sum

$$\sum_{k=1}^{\infty} \frac{1}{p_k^2} \sum_{k=1}^{\infty} \frac{\mu(k)}{k} \ln(\zeta(2k)) \approx 0.45224742 \tag{17}$$

The sum

$$\sum_{k=1}^{\infty} \frac{1}{(p_k - 1)^2} \approx 1.3750649947 \tag{18}$$

is also finite (Glaisher 1891a; Cohen; Finch).

Some curious sums satisfied by primes p include

$$\sum_{k=1}^{p-1} \left\lfloor \frac{k^3}{p} \right\rfloor = \frac{(p-2)(p-1)(p+1)}{4} \tag{19}$$

$$\sum_{k=1}^{(p-1)(p-2)} \left\lfloor (k_p)^{1/3} \right\rfloor = \tfrac{1}{4}(3p-5)(p-2)(p-1) \tag{20}$$

(Doster 1993),

$$\sum_{k=1}^{\infty} x^k \ln k = \sum_{p \text{ prime}} \sum_{k=1}^{\infty} \frac{x^{p^k}}{1 - x^{p^k}}, \tag{21}$$

and

$$\sum_{k=1}^{\infty} (-1)^{k-1} e^{-kx} \ln k = -\ln 2 \sum_{k=1}^{\infty} \frac{1}{e^{2^k x} - 1}$$
$$+ \sum_{\substack{p \text{ an} \\ \text{odd prime}}} \ln p \sum_{k=1}^{\infty} \frac{1}{e^{p^k x} + 1} \tag{22}$$

(Berndt 1994, p. 114).

See also MERTENS CONSTANT, PRIME NUMBER, PRIME PRODUCTS, PRIME ZETA FUNCTION, PRIMORIAL

References

Bach, E. and Shallit, J. §2.7 in *Algorithmic Number Theory, Vol. 1: Efficient Algorithms.* Cambridge, MA: MIT Press, 1996.

Berndt, B. C. "Ramanujan's Theory of Prime Numbers." Ch. 24 in *Ramanujan's Notebooks, Part IV.* New York: Springer-Verlag, 1994.

Brun, V. "La serie $1/5 + 1/7 + \ldots$ est convergente ou finie." *Bull. Sci. Math.* **43**, 124–28, 1919.

Cohen, H. "High Precision Computation of Hardy-Littlewood Constants." Preprint. http://www.math.u-bordeaux.fr/~cohen/hardylw.dvi.

Davenport, H. *Multiplicative Number Theory, 2nd ed.* New York: Springer-Verlag, 1980.

Doster, D. "Problem 10346." *Amer. Math. Monthly* **100**, 951, 1993.

Erdos, P. "Some of My New and Almost New Problems and Results in Combinatorial Number Theory." In *Number Theory: Diophantine, Computational and Algebraic Aspects. Proceedings of the International Conference Held in Eger, July 29-August 2, 1996* (Ed. K. Gyory, A. Petho and V. T. Sós). Berlin: de Gruyter, pp. 169–80, 1998.

Finch, S. "Favorite Mathematical Constants." http://www.mathsoft.com/asolve/constant/hdmrd/hdmrd.html.

Flajolet, P. and Vardi, I. "Zeta Function Expansions of Classical Constants." Unpublished manuscript. 1996. http://pauillac.inria.fr/algo/flajolet/Publications/landau.ps.

Glaisher, J. W. L. "On the Sums of the Inverse Powers of the Prime Numbers." *Quart. J. Pure Appl. Math.* **25**, 347–62, 1891a.

Glaisher, J. W. L. "On the Series $1/3 - 1/5 + 1/7 + 1/11 - 1/13 - \ldots$." *Quart. J. Pure Appl. Math.* **25**, 375–83, 1891b.

Glaisher, J. W. L. "On the Series $1/2 + 1/5 - 1/7 + 1/11 - 1/13 - \ldots$." *Quart. J. Pure Appl. Math.* **25**, 48–5, 1891c.

Glaisher, J. W. L. "On the Series $1/3^2 - 1/5^2 + 1/7^2 + 1/11^2 - 1/13 - \ldots$." *Quart. J. Pure Appl. Math.* **26**, 33–7, 1893.

Guy, R. K. "A Series and a Sequence Involving Primes." §E7 in *Unsolved Problems in Number Theory, 2nd ed.* New York: Springer-Verlag, p. 203, 1994.

Hardy, G. H. and Wright, E. M. "Prime Numbers" and "The Sequence of Primes." §1.2 and 1.4 in *An Introduction to the Theory of Numbers, 5th ed.* Oxford, England: Clarendon Press, pp. 1–, 17, 22, and 251, 1979.

Knuth, D. E. *The Art of Computer Programming, Vol. 2: Seminumerical Algorithms, 3rd ed.* Reading, MA: Addison-Wesley, 1998.

Moree, P. "Approximation of Singular Series and Automata." *Manuscripta Math.* **101**, 385–99, 2000.

Nagell, T. *Introduction to Number Theory.* New York: Wiley, 1951.

Rivera, C. "Problems & Puzzles: Puzzle 031.-The Average Prime Number, $APN(k) = S(p_k)/k$." .htm" target = "extwin">http://www.primepuzzles.net/puzzles/ puzz_The Average Prime Number, $APN(k) = S(p_k)/k$.htm.

Robinson, H. P. and Potter, E. *Mathematical Constants.* Report UCRL-20418. Berkeley, CA: University of California, 1971.

Schroeder, M. R. *Number Theory in Science and Communication, with Applications in Cryptography, Physics, Digital Information, Computing, and Self-Similarity, 3rd ed.* New York: Springer-Verlag, 1997.

Sloane, N. J. A. Sequences A007504/M1370, A045345, A046024, A050247, and A050248 in "An On-Line Version of the Encyclopedia of Integer Sequences." http://www.research.att.com/~njas/sequences/eisonline.html.

Prime Theta Function

CHEBYSHEV FUNCTIONS

Prime Triangle

A triangle with rows containing the numbers $\{1, 2, \ldots, n\}$ that begins with 1, ends with n, and such that the SUM of each two consecutive entries being a PRIME. Rows 2 to 6 are unique,

$$*$$

$$1 \quad 2$$

$$1 \quad 2 \quad 3$$

$$1 \quad 2 \quad 3 \quad 4$$

$$1 \quad 4 \quad 3 \quad 2 \quad 5$$

$$1 \quad 4 \quad 3 \quad 2 \quad 5 \quad 6$$

(Sloane's A051237) but there are multiple possibilities starting with row 7. For example, the two possibilities for row 7 are $\{1, 4, 3, 2, 5, 6, 7, \}$ and $\{1, 6, 5, 2, 3, 4, 7\}$. The number of possible rows ending with $n = 1, 2, \ldots$, are 0, 1, 1, 1, 1, 1, 2, 4, 7, 24, 80, ... (Sloane's A036440).

See also PASCAL'S TRIANGLE

References

Guy, R. K. *Unsolved Problems in Number Theory, 2nd ed.* New York: Springer-Verlag, p. 106, 1994.

Kenney, M. J. "Student Math Notes." *NCTM News Bulletin.* Nov. 1986.

Sloane, N. J. A. Sequences A036440 and A051237 in "An On-Line Version of the Encyclopedia of Integer Sequences." http://www.research.att.com/~njas/sequences/ eisonline.html.

Prime Triplet

A prime triplet is a PRIME CONSTELLATION OF THE FORM $(p, p+2, p+6)$, $(p, p+4, p+6)$, etc. Hardy and Wright (1979, p. 5) conjecture, and it seems almost certain to be true, that there are infinitely many prime triplets OF THE FORM $(p, p+2, p+6)$ and $(p, p+4, p+6)$.

Triplet	Sloane	First Member
$(p, p+2, p+6)$	Sloane's A022004	5, 11, 17, 41, 101, 107, ...
$(p, p+2, p+8)$	Sloane's A046134	3, 5, 11, 29, 59, 71, 101, ...
$(p, p+2, p+12)$	Sloane's A046135	5, 11, 17, 29, 41, 59, 71, ...
$(p, p+4, p+6)$	Sloane's A022005	7, 13, 37, 67, 97, 103, ...
$(p, p+4, p+10)$	Sloane's A046136	3, 7, 13, 19, 37, 43, 79, ...
$(p, p+4, p+12)$	Sloane's A046317	7, 19, 67, 97, 127, 229, ...
$(p, p+6, p+8)$	Sloane's A046138	5, 11, 23, 53, 101, 131, ...
$(p, p+6, p+10)$	Sloane's A046139	7, 13, 31, 37, 61, 73, 97, ...
$(p, p+6, p+12)$	Sloane's A046140	5, 7, 11, 17, 31, 41, 47, ...
$(p, p+8, p+12)$	Sloane's A046141	5, 11, 29, 59, 71, 89, 101, ...

See also PRIME CONSTELLATION, PRIME QUADRUPLET, TWIN PRIMES

References

Hardy, G. H. and Wright, E. M. *An Introduction to the Theory of Numbers, 5th ed.* Oxford, England: Clarendon Press, 1979.

Rivera, C. "Problems & Puzzles: Puzzle Prime Triplets in Arithmetic Progression.-034." http://www.primepuzzles.net/puzzles/puzz_034.htm.

Prime Unit

1 and -1 are the only INTEGERS which divide every INTEGER. They are therefore called the prime units.

See also INTEGER, PRIME NUMBER, UNIT

Prime Zeta Function

The prime zeta function

$$P(n) \equiv \sum_p \frac{1}{p^n}, \tag{1}$$

where the sum is taken over PRIMES is a generalization of the RIEMANN ZETA FUNCTION

$$\zeta(n) \equiv \sum_{k=1} \frac{1}{k^n}, \qquad (2)$$

where the sum is over *all* integers. The prime zeta function can be expressed in terms of the RIEMANN ZETA FUNCTION by

$$\ln \zeta(n) = -\sum_{p \geq 2} \ln(1 - p^{-n}) = \sum_{p \geq 2} \sum_{k=1}^{\infty} \frac{p^{-kn}}{k}$$

$$= \sum_{k=1}^{\infty} \frac{1}{k} \sum_{p \geq 2} p^{-kn} = \sum_{k=1}^{\infty} \frac{P(kn)}{k}. \qquad (3)$$

Inverting then gives

$$P(n) = \sum_{k=1}^{\infty} \frac{\mu(k)}{k} \ln \zeta(kn), \qquad (4)$$

where $\mu(k)$ is the MÖBIUS FUNCTION (Cohen 2000). $P(1)$, The analog of the HARMONIC SERIES, diverges, but convergence of the series for $n > 1$ is quadratic. ARTIN'S CONSTANT C_{Artin} is connected with $P(n)$ by

$$\ln C_{\text{Artin}} = -\sum_{n=2}^{\infty} \frac{(\mu_n - 1)P(n)}{n}, \qquad (5)$$

where

$$u_n = u_{n-1} + u_{n-2} \qquad (6)$$

with $u_1 = 1$, $u_2 = 3$ (Ribenboim 1998, Gourdon and Sebah).

The values of $P(n)$ for the first few integers n starting with two are

$$P(2) \approx 0.452247 \qquad (7)$$

$$P(3) \approx 0.174763 \qquad (8)$$

$$P(4) \approx 0.0769931 \qquad (9)$$

$$P(5) \approx 0.035755. \qquad (10)$$

Merrifield (1881) computed $P(n)$ for n up to 35 to 15 digits, and Liénard (1948) computed $P(n)$ up to $n = 167$ to 50 digits (Ribenboim 1996). Gourdon gives values to 60 digits for $2 \geq n \leq 8$.

See also ARTIN'S CONSTANT, HARMONIC SERIES, MÖBIUS FUNCTION, PRIME SUMS, RIEMANN ZETA FUNCTION, ZETA FUNCTION

References

Cohen, H. "High Precision Computation of Hardy-Littlewood Constants." Preprint. http://www.math.u-bordeaux.fr/~cohen/hardylw.dvi.
Gourdon, X. and Sebah, P. "Some Constants from Number Theory." http://xavier.gourdon.free.fr/Constants/Miscellaneous/constantsNumTheory.html.
Hardy, G. H. and Weight, E. M. *An Introduction to the Theory of Numbers,* 5th ed. Oxford, England: Oxford University Press, pp. 355–56, 1979.
Liénard, R. *Tables fondamentales à 50 décimales des sommes S_n, u_n, Σ_n.* Paris: Centre de Docum. Univ., 1948.
Merrifield, C. W. "The Sums of the Series of Reciprocals of the Prime Numbers and of Their Powers." *Proc. Roy. Soc. London* **33**, 4–0, 1881.
Ribenboim, P. *The New Book of Prime Number Records.* New York: Springer-Verlag, 1996.

Prime-Distance Graph

A DISTANCE GRAPH with distance set given by the set of prime numbers.

See also DISTANCE GRAPH

References

Eggleton, R. B.; Erdos, P.; and Skilton, D. K. "Coloring the Real Line." *J. Combin. Th. B* **39**, 86–00, 1985.
Eggleton, R. B.; Erdos, P.; and Skilton, D. K. "Research Problem 77." *Discr. Math.* **58**, 323, 1986.
Eggleton, R. B.; Erdos, P.; and Skilton, D. K. "Coloring Prime Distance Graphs." *Graphs Combin.* **6**, 17–2, 1990.
Maehara, H. "Distance Graphs in Euclidean Space." *Ryukyu Math. J.* **5**, 33–1, 1992.

Primefree Sequence

A sequence whose terms are never prime. Graham proved that there exist primefree sequences generated by Fibonacci-like recurrences OF THE FORM

$$a_n = a_{n-1} + a_{n-2}$$

for $(a_1, a_2) = 1$, i.e., RELATIVELY PRIME. However, the purported example given by Hoffman (1998, p. 159) in fact contains prime terms for $n = 138$, 163, 190, 523, ...

References

Hoffman, P. *The Man Who Loved Only Numbers: The Story of Paul Erdos and the Search for Mathematical Truth.* New York: Hyperion, 1998.

Prime-Generating Polynomial

Legendre showed that there is no RATIONAL algebraic function which always gives PRIMES. In 1752, Goldbach showed that no POLYNOMIAL with INTEGER COEFFICIENTS can give a PRIME for all integer values (Nagell 1951, p. 65; Hardy and Wright 1979, pp. 18 and 22). However, there exists a POLYNOMIAL in 10 variables with INTEGER COEFFICIENTS such that the set of PRIMES equals the set of POSITIVE values of this POLYNOMIAL obtained as the variables run through all NONNEGATIVE INTEGERS, although it is really a set of DIOPHANTINE EQUATIONS in disguise (Ribenboim 1991).

Polynomial	Range	Sloane	Reference
$36n^2 - 810n + 2753$	$[0, 44]$	A050268	Fung and Ruby
$47n^2 - 1701n + 10181$	$[0, 42]$	A050267	Fung and Ruby

$n^2 + n + 41$	[0, 39]	A005846	Euler
$2n^2 + 29$	[0, 28]	A033542	Legendre
$n^2 + n + 17$	[0, 15]	A033541	Legendre
$4n^2 + 4n + 59$	[0, 13]	A048988	
$2n^2 + 11$	[0, 10]	A050265	
$n^3 + n^2 + 17$	[0, 10]	A050266	

The above table gives some low-order polynomials which generate only PRIMES for the first few NON-NEGATIVE values (Mollin and Williams 1990). The best-known of these formulas is that due to Euler (Euler 1772; Nagell 1951, p. 65; Gardner 1984, p. 83; Ball and Coxeter 1987),

$$n^2 + n + 41. \tag{1}$$

which gives distinct primes for the 40 consecutive integers $n = 0$ to 39. ($n^2 - n + 41$ gives the same 40 primes for $n = 1$ to 40.) By transforming the formula to

$$n^2 - 79n + 1601 = (n - 40)^2 + (n - 40) + 41, \tag{2}$$

primes are obtained for 80 consecutive integers, corresponding to the 40 primes given by the above formula taken twice each (Hardy and Wright 1979, p. 18).

Le Lionnais (1983) has christened numbers p such that the Euler-like polynomial

$$n^2 + n + p \tag{3}$$

is PRIME for $n = 0, 1, ..., p - 2$ as LUCKY NUMBERS OF EULER (where the case $p = 41$ corresponds to Euler's formula). Rabinowitz (1913) showed that for a PRIME $p > 0$, Euler's polynomial represents a PRIME for $n \in [0, p - 2]$ (excluding the trivial case $p = 3$) IFF the FIELD $\mathbb{Q}(\sqrt{1 - 4p})$ has CLASS NUMBER $h = 1$ (Rabinowitz 1913, Le Lionnais 1983, Conway and Guy 1996). As established by Stark (1967), there are only nine numbers $-d$ such that $h(-d) = 1$ (the HEEGNER NUMBERS -2, -3, -7, -11, -19, -43, -67, and -163), and of these, only 7, 11, 19, 43, 67, and 163 are of the required form. Therefore, the only LUCKY NUMBERS OF EULER are 2, 3, 5, 11, 17, and 41 (le Lionnais 1983, Sloane's A014556), and there does not exist a better prime-generating polynomial of Euler's form. The connection between the numbers 163 and 43 and some of the prime-rich polynomials listed above can be seen explicitly by writing

$$x^2 + x + 41 = \left(x + \tfrac{1}{2}\right)^2 + \tfrac{163}{4} \tag{4}$$

$$x^2 + x + 11 = \left(x + \tfrac{1}{2}\right)^2 + \tfrac{43}{4}, \tag{5}$$

etc.

Euler also considered quadratics OF THE FORM

$$2x^2 + p \tag{6}$$

and showed this gives PRIMES for $x \in [0, p - 1]$ for PRIME $p > 0$ IFF $\mathbb{Q}(\sqrt{-2p})$ has CLASS NUMBER 2, which permits only $p = 3, 5, 11,$ and 29. Baker (1971) and Stark (1971) showed that there are no such FIELDS for $p > 29$. Similar results have been found for POLYNOMIALS OF THE FORM

$$px^2 + px + n \tag{7}$$

(Hendy 1974).

See also CLASS NUMBER, HEEGNER NUMBER, LUCKY NUMBER OF EULER, PRIME ARITHMETIC PROGRESSION, PRIME DIOPHANTINE EQUATIONS, SCHINZEL'S HYPOTHESIS

References

Abel, U. and Siebert, H. "Sequences with Large Numbers of Prime Values." *Am. Math. Monthly* **100**, 167–69, 1993.

Baker, A. "Linear Forms in the Logarithms of Algebraic Numbers." *Mathematika* **13**, 204–16, 1966.

Baker, A. "Imaginary Quadratic Fields with Class Number Two." *Ann. Math.* **94**, 139–52, 1971.

Ball, W. W. R. and Coxeter, H. S. M. *Mathematical Recreations and Essays, 13th ed.* New York: Dover, p. 60, 1987.

Boston, N. and Greenwood, M. L. "Quadratics Representing Primes." *Amer. Math. Monthly* **102**, 595–99, 1995.

Conway, J. H. and Guy, R. K. "The Nine Magic Discriminants." In *The Book of Numbers.* New York: Springer-Verlag, pp. 224–26, 1996.

Courant, R. and Robbins, H. *What is Mathematics?: An Elementary Approach to Ideas and Methods, 2nd ed.* Oxford, England: Oxford University Press, p. 26, 1996.

Dudley, U. "History of Formula for Primes." *Amer. Math. Monthly* **76**, 23–8, 1969.

Euler, L. *Nouveaux Mémoires de l'Académie royale des Sciences.* Berlin, p. 36, 1772.

Forman, R. "Sequences with Many Primes." *Amer. Math. Monthly* **99**, 548–57, 1992.

Gardner, M. *The Sixth Book of Mathematical Games from Scientific American.* Chicago, IL: University of Chicago Press, pp. 83–4, 1984.

Garrison, B. "Polynomials with Large Numbers of Prime Values." *Amer. Math. Monthly* **97**, 316–17, 1990.

Hardy, G. H. and Wright, E. M. *An Introduction to the Theory of Numbers, 5th ed.* Oxford, England: Clarendon Press, 1979.

Hendy, M. D. "Prime Quadratics Associated with Complex Quadratic Fields of Class Number 2." *Proc. Amer. Math. Soc.* **43**, 253–60, 1974.

Hoffman, P. *The Man Who Loved Only Numbers: The Story of Paul Erdos and the Search for Mathematical Truth.* New York: Hyperion, pp. 108–09, 1998.

Le Lionnais, F. *Les nombres remarquables.* Paris: Hermann, pp. 88 and 144, 1983.

Mollin, R. A. and Williams, H. C. "Class Number Problems for Real Quadratic Fields." *Number Theory and Cryptology; LMS Lecture Notes Series* **154**, 1990.

Nagell, T. "Primes in Special Arithmetical Progressions." §44 in *Introduction to Number Theory.* New York: Wiley, pp. 60 and 153–55, 1951.

Rabinowitz, G. "Eindeutigkeit der Zerlegung in Primzahlfaktoren in quadratischen Zahlkörpern." *Proc. Fifth Internat. Congress Math.* (Cambridge) **1**, 418–21, 1913.

Ribenboim, P. *The Little Book of Big Primes.* New York: Springer-Verlag, 1991.

Sloane, N. J. A. Sequences A005846/M5273, A014556, A033541, A033542, A048988, A050265, A050266, A050267, and A050268 in "An On-Line Version of the Encyclopedia of Integer Sequences." http://www.research.-att.com/~njas/sequences/eisonline.html.

Stark, H. M. "A Complete Determination of the Complex Quadratic Fields of Class Number One." *Michigan Math. J.* **14**, 1–7, 1967.

Stark, H. M. "An Explanation of Some Exotic Continued Fractions Found by Brillhart." In *Computers in Number Theory, Proc. Science Research Council Atlas Symposium No. 2 held at Oxford, from 18–3 August, 1969* (Ed. A. O. L. Atkin and B. J. Birch). London: Academic Press, 1971.

Stark, H. M. "A Transcendence Theorem for Class Number Problems." *Ann. Math.* **94**, 153–73, 1971.

Primequad

PRIME QUADRUPLET

Primes

The set of PRIME NUMBERS, sometimes denoted $\mathbb{P}$, and implemented in *Mathematica* as Primes. In *Mathematica*, a quantity can be tested to determine if it is in the domain of prime numbers using Element[n, Primes], which is equivalent to PrimeQ[n].

See also PRIME NUMBER

Primitive Abundant Number

An ABUNDANT NUMBER for which all PROPER DIVISORS are DEFICIENT is called a primitive abundant number (Guy 1994, p. 46). The first few ODD primitive abundant numbers are 945, 1575, 2205, 3465, ... (Sloane's A006038).

See also ABUNDANT NUMBER, DEFICIENT NUMBER, HIGHLY ABUNDANT NUMBER, SUPERABUNDANT NUMBER, WEIRD NUMBER

References
Guy, R. K. *Unsolved Problems in Number Theory, 2nd ed.* New York: Springer-Verlag, p. 46, 1994.

Sloane, N. J. A. Sequences A006038/M5486 in "An On-Line Version of the Encyclopedia of Integer Sequences." http://www.research.att.com/~njas/sequences/eisonline.html.

Primitive Character

See also CHARACTER (NUMBER THEORY)

Primitive Element

Given algebraic numbers a_1, ..., a_n it is always possible to find a single ALGEBRAIC NUMBER b such that each of a_1, ..., a_n can be expressed as a polynomial in b with rational coefficients. The number b is then called a primitive element of the EXTENSION FIELD $\mathbb{Q}(a_1, ..., a_n)/\mathbb{Q}$. Stated differently, an ALGEBRAIC NUMBER b is a primitive element of $\mathbb{Q}(a_1, ..., a_n)/\mathbb{Q}$ IFF $\mathbb{Q}(a_1, ..., a_n) = \mathbb{Q}(b)$. Primitive elements are implemented in *Mathematica* as

PrimitiveElement[z, {$a1$, ..., an}] in the *Mathematica* add-on package NumberTheory`PrimitiveElement` (which can be loaded with the command <<NumberTheory`).

For example, a primitive element of $\mathbb{Q}(\sqrt{2}, \sqrt{3})/\mathbb{Q}$ is given by $b = \sqrt{2} + \sqrt{3}$, with

$$\sqrt{2} = \tfrac{1}{2}b(b^2 - 9)$$
$$\sqrt{3} = \tfrac{1}{2}b(11 - b^2).$$

See also EXTENSION FIELD, PRIMITIVE POLYNOMIAL, PRIMITIVE ROOT

References
Loos, R. "Computing in Algebraic Extensions." *Computing*, Suppl. 4, 173–87, 1982.

Primitive Function

INTEGRAL

Primitive Group

A GROUP that has a PRIMITIVE GROUP ACTION.

See also PRIMITIVE GROUP ACTION

Primitive Group Action

A primitive group action is TRANSITIVE and it has no nontrivial BLOCKS. A TRANSITIVE GROUP ACTION that is not primitive is called imprimitive. A group that has a primitive group action is called a PRIMITIVE GROUP.

See also BLOCK (GROUP ACTION), GROUP, PRIMITIVE GROUP, SOCLE, TRANSITIVE GROUP, TRANSITIVE GROUP ACTION

References
Dixon, J. and Mortimer, B. *Permutation Groups.* New York: Springer-Verlag, 1996.

Primitive Polynomial

A polynomial which generates all elements of an EXTENSION FIELD from a base field is called a primitive polynomial. Primitive polynomials are also IRREDUCIBLE POLYNOMIALS. For any PRIME or PRIME POWER q and any POSITIVE INTEGER n, there exists a primitive polynomial of order n over GF(q). There are $\phi(q^n - 1)/n$ primitive polynomials over GF(q), where $\phi(n)$ is the TOTIENT FUNCTION.

Polynomials over the FINITE FIELD GF(2) (i.e., with coefficients either 0 or 1) are primitive if they have ORDER $2^n - 1$, where "order" is used in the specific sense of a HAUPT-EXPONENT or ORDER of a modulo. For example, $x^2 + x + 1 = (x^2 + x + 1)(x + 1) = x^3 + 1$ has order 3, and is therefore primitive (Ruskey). Amazingly, primitive polynomials over GF(2) define a RECURRENCE RELATION which can be used to obtain a new RANDOM bit from the n preceding ones. The numbers of primitive polynomials over GF(2) for $n = 1, 2, ...$ are 1, 1, 2, 2, 6, 6, 18, 16, 48, ... (Sloane's A011260). The following table lists the primitive polynomials (mod 2) of orders 1 through 5.

n	primitive polynomials
1	x
2	$1 + x + x^2$
3	$1 + x + x^3$, $1 + x^2 + x^3$
4	$1 + x + x^4$, $1 + x^3 + x^4$
5	$1 + x^2 + x^5$, $1 + x + x^2 + x^3 + x^5$, $1 + x^3 + x^5$, $1 + x + x^3 + x^4 + x^5$, $1 + x^2 + x^3 + x^4 + x^5$, $1 + x + x^2 + x^4 + x^5$

See also FINITE FIELD, IRREDUCIBLE POLYNOMIAL, ORDER (POLYNOMIAL), POLYNOMIAL, PRIMITIVE ELEMENT, PRIMITIVE ROOT

References

Ruskey, F. "Information on Primitive and Irreducible Polynomials." http://www.theory.csc.uvic.ca/~cos/inf/neck/PolyInfo.html.
Sloane, N. J. A. Sequences A011260/M0107 in "An On-Line Version of the Encyclopedia of Integer Sequences." http://www.research.att.com/~njas/sequences/eisonline.html.
Zierler, N. and Brillhart, J. "On Primitive Trinomials." *Inform. Control* **13**, 541–44, 1968.
Zierler, N. and Brillhart, J. "On Primitive Trinomials (II)." *Inform. Control* **14**, 566–69, 1969.

Primitive Polytope

A POLYTOPE in n-D Euclidean space $\mathbb{R}^n$ whose vertices are integer lattice points but which does not contain any other lattice points in its interior or on its boundary (Khan 1999).

See also HOWE'S THEOREM, POLYTOPE

References

Khan, M. R. "A Counting Formula for Primitive Tetrahedra in Z^3." *Amer. Math. Monthly* **106**, 525–33, 1999.

Primitive Prime Factor

If $n \geq 1$ is the smallest INTEGER such that $P|a^n - b^n$ (or $a^n + b^n$), then p is a primitive prime factor.

See also PRIME FACTORS, PRIMITIVE ROOT

Primitive Pseudoperfect Number

PRIMITIVE SEMIPERFECT NUMBER

Primitive Recursive Function

For-loops (which have a fixed iteration limit) are a special case of while-loops. A function which can be implemented using only for-loops is called primitive recursive. (In contrast, a COMPUTABLE FUNCTION can be coded using a combination of for- and while-loops, or while-loops only.)

The ACKERMANN FUNCTION is the simplest example of a WELL DEFINED TOTAL FUNCTION which is COMPUTABLE but not primitive recursive, providing a counterexample to the belief in the early 1900s that every COMPUTABLE FUNCTION was also primitive recursive (Dötzel 1991).

See also ACKERMANN FUNCTION, COMPUTABLE FUNCTION, TOTAL FUNCTION

References

Dötzel, G. "A Function to End All Functions." *Algorithm: Recreational Programming* **2**, 16–7, 1991.

Primitive Root

A primitive root of a PRIME p is an INTEGER g satisfying $1 \leq g \leq p - 1$ such that the residue classes of $g, g^2, g^3, ..., g^{p-1} = 1$ are all distinct, i.e., $g \pmod{p}$ has ORDER $p - 1$ (Ribenboim 1996, p. 22). If p is a PRIME NUMBER, then there are exactly $\phi(p-1)$ incongruent primitive roots of p (Burton 1989, p. 194).

More generally, if $(g, n) = 1$ (g and n are RELATIVELY PRIME) and g is of ORDER $\phi(n)$ modulo n, where $\phi(n)$ is the TOTIENT FUNCTION, then g is a primitive root of n (Burton 1989, p. 187). In other words, n has g as a primitive root if $g^{\phi(n)} \equiv 1 \pmod{n}$, but $g^k \not\equiv 1 \pmod{n}$ for all positive integers $k < \phi(n)$. A primitive root of a number n (but not necessarily the *smallest* primitive root for composite n) can be computed using the *Mathematica* routine PrimitiveRoot[n] in the *Mathematica* add-on package NumberTheory`NumberTheoryFunctions` (which can be loaded with the command < <NumberTheory`).

If n has a primitive root, then it has exactly $\phi(\phi(n))$ of them (Burton 1989, p. 188). For $n = 1, 2, ...,$ the first few values of $\phi(\phi(n))$ are 1, 1, 1, 1, 2, 1, 2, 2, 2, 2, 4, 2, 4, 2, 4, 4, 8, ... (Sloane's A010554). n has a primitive root if it is OF THE FORM 2, 4, a power p^a, or twice a power $2p^a$, where p is an ODD PRIME and $a \geq 1$ (Burton 1989, p. 204). The first few n for which primitive roots exist are 2, 3, 4, 5, 6, 7, 9, 10, 11, 13, 14, 17, 18, 19, 22, ... (Sloane's A033948), so the number of primitive root of order n for $n = 1, 2, ...$ are 0, 1, 1, 1, 2, 1, 2, 0, 2, 2, 4, 0, 4, ... (Sloane's A046144).

The smallest primitive roots for the first few primes p are 1, 2, 2, 3, 2, 2, 3, 2, 5, 2, 3, 2, 6, 3, 5, 2, 2, 2, ... (Sloane's A001918). Here is table of the primitive roots for the first few n for which a primitive root exists (Sloane's A046147).

n	$g(n)$
2	1
3	2
4	3
5	2, 3
6	5
7	3, 5
9	2, 5
10	3, 7
11	2, 6, 7, 8
13	2, 6, 7, 11

The largest primitive roots for $n = 1, 2, ...,$ are 0, 1, 2, 3, 3, 5, 5, 0, 5, 7, 8, 0, 11, ... (Sloane's A046146). The smallest primitive roots for the first few INTEGERS n are given in the following table (Sloane's A046145), which omits n when $g(n)$ does not exist.

2	1	38	3	94	5	158	3
3	2	41	6	97	5	162	5
4	3	43	3	98	3	163	2
5	2	46	5	101	2	166	5
6	5	47	5	103	5	167	5
7	3	49	3	106	3	169	2
9	2	50	3	107	2	173	2
10	3	53	2	109	6	178	3
11	2	54	5	113	3	179	2
13	2	58	3	118	11	181	2
14	3	59	2	121	2	191	19
17	3	61	2	122	7	193	5
18	5	62	3	125	2	194	5
19	2	67	2	127	3	197	2
22	7	71	7	131	2	199	3
23	5	73	5	134	7	202	3
25	2	74	5	137	3	206	5
26	7	79	3	139	2	211	2
27	2	81	2	142	7	214	5
29	2	82	7	146	5	218	11

31	3	83	2	149	2	223	3
34	3	86	3	151	6	226	3
37	2	89	3	157	5	227	2

Let p be any ODD PRIME $k \geq 1$, and let

$$s \equiv \sum_{j=1}^{p-1} j^k. \tag{1}$$

Then

$$s = \begin{cases} -1 \ (\mathrm{mod}\ p) & \text{for } p-1 | k \\ 0 \ (\mathrm{mod}\ p) & \text{for } p-1 \nmid k \end{cases} \tag{2}$$

(Ribenboim 1996, pp. 22–3). For numbers m with primitive roots, all y satisfying $(p, y) = 1$ are representable as

$$y \equiv g^t \ (\mathrm{mod}\ m), \tag{3}$$

where $t = 0, 1, ..., \phi(m) - 1$, t is known as the index, and y is an INTEGER. Kearnes (1984) showed that for any POSITIVE INTEGER m, there exist infinitely many PRIMES p such that

$$m < g_p < p - m. \tag{4}$$

Call the least primitive root g_p. Burgess (1962) proved that

$$g_p \leq C p^{1/4 + \epsilon} \tag{5}$$

for C and ϵ POSITIVE constants and p sufficiently large (Ribenboim 1996, p. 24).

Matthews (1976) obtained a formula for the "two-dimensional" Artin's constants for the set of primes for which m and n are both primitive roots.

See also ARTIN'S CONJECTURE, ARTIN'S CONSTANT, FULL REPTEND PRIME, MULTIPLICATIVE ORDER, ORDER (MODULO), PRIMITIVE ELEMENT, PRIMITIVE ROOT OF UNITY

References

Abramowitz, M. and Stegun, C. A. (Eds.). "Primitive Roots." §24.3.4 in *Handbook of Mathematical Functions with Formulas, Graphs, and Mathematical Tables, 9th printing.* New York: Dover, p. 827, 1972.
Burgess, D. A. "On Character Sums and L-Series." *Proc. London Math. Soc.* **12**, 193–06, 1962.
Burton, D. M. "The Order of an Integer Modulo n," "Primitive Roots for Primes," and "Composite Numbers Having Primitive Roots." §8.1–.3 in *Elementary Number Theory, 4th ed.* Dubuque, IA: William C. Brown Publishers, pp. 184–05, 1989.
Guy, R. K. "Primitive Roots." §F9 in *Unsolved Problems in Number Theory, 2nd ed.* New York: Springer-Verlag, pp. 248–49, 1994.
Kearnes, K. "Solution of Problem 6420." *Amer. Math. Monthly* **91**, 521, 1984.
Lehmer, D. H. "A Note on Primitive Roots." *Scripta Math.* **26**, 117–19, 1961.

Matthews, K. R. "A Generalization of Artin's Conjecture for Primitive Roots." *Acta Arith.* **29**, 113–46, 1976.

Nagell, T. "Moduli Having Primitive Roots." §32 in *Introduction to Number Theory.* New York: Wiley, pp. 107–11, 1951.

Ribenboim, P. *The New Book of Prime Number Records.* New York: Springer-Verlag, pp. 22–5, 1996.

Riesel, H. *Prime Numbers and Computer Methods for Factorization, 2nd ed.* Boston, MA: Birkhäuser, p. 97, 1994.

Sloane, N. J. A. Sequences A001918/M0242, A010554, and A033948 in "An On-Line Version of the Encyclopedia of Integer Sequences." http://www.research.att.com/~njas/sequences/eisonline.html.

Western, A. E. and Miller, J. C. P. *Tables of Indices and Primitive Roots.* Cambridge, England: Cambridge University Press, pp. xxxvii–xlii, 1968.

Primitive Root of Unity

A number r is an nth ROOT OF UNITY if $r^n = 1$ and a primitive nth root of unity if, in addition, n is the smallest INTEGER of $k = 1, ..., n$ for which $r^k = 1$.

See also PRINCIPAL ROOT OF UNITY, ROOT OF UNITY

References

Nagell, T. *Introduction to Number Theory.* New York: Wiley, p. 157, 1951.

Primitive Semiperfect Number

A SEMIPERFECT NUMBER for which none of its PROPER DIVISORS are pseudoperfect (Guy 1994, p. 46). The first few are 6, 20, 28, 88, 104, 272, ... (Sloane's A006036). Primitive semiperfect numbers are also called primitive pseudoperfect numbers (Guy 1994, p. 46) or irreducible semiperfect numbers. There are infinitely many primitive pseudoperfect numbers which are not HARMONIC DIVISOR NUMBERS, and infinitely many ODD primitive semiperfect numbers.

See also HARMONIC DIVISOR NUMBER, PRIMARY PSEUDOPERFECT NUMBER, SEMIPERFECT NUMBER

References

Guy, R. K. *Unsolved Problems in Number Theory, 2nd ed.* New York: Springer-Verlag, p. 46, 1994.

Sloane, N. J. A. Sequences A006036/M4133 in "An On-Line Version of the Encyclopedia of Integer Sequences." http://www.research.att.com/~njas/sequences/eisonline.html.

Primitive Sequence

A SEQUENCE in which no term DIVIDES any other. Let S_n be the set $\{1, ..., n\}$, then the number of primitive subsets of S_n are 2, 3, 5, 7, 13, 17, 33, 45, 73, 103, 205, 253, ... (Sloane's A051026). For example, the five primitive sequences in S_4 are $\varnothing$, $\{1\}$, $\{2\}$, $\{2, 3\}$, $\{3\}$, $\{3, 4\}$, and $\{4\}$.

See also NONDIVIDING SET

References

Guy, R. K. *Unsolved Problems in Number Theory, 2nd ed.* New York: Springer-Verlag, p. 202, 1994.

Sloane, N. J. A. Sequences A051026 in "An On-Line Version of the Encyclopedia of Integer Sequences." http://www.research.att.com/~njas/sequences/eisonline.html.

Primorial

For the nth PRIME p_n,

$$\text{primorial}(p_n) = p_n\# \equiv \prod_{j=1}^{n} p_j.$$

The values of $p_n\#$ for $n = 1, 2, ...,$ are 2, 6, 30, 210, 2310, 30030, 510510, ... (Sloane's A002110).

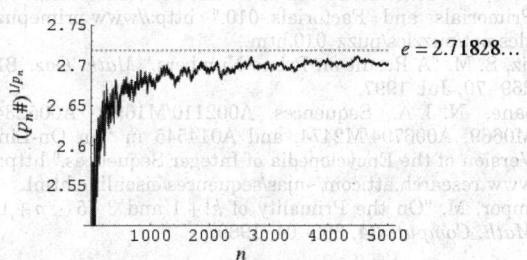

The primorial satisfies the unexpected limit

$$\lim_{n \to \infty} (p_n\#)^{1/p_n} = e$$

(Ruiz 1997), where E is the usual base of the NATURAL LOGARITHM.

$p\# - 1$ is PRIME for PRIMES $p = 3, 5, 11, 41, 89, 317, 337, 991, 1873, 2053, 2377, 4093, 4297, ...$ (Sloane's A006794; Guy 1994), or p_n for $n = 2, 3, 5, 13, 24, 66, 68, 167, 287, 310, 352, 564, 590, ...,$ up to a search limit of $p = 25000$ (Caldwell 1995).

$p\# + 1$ is known to be PRIME for the PRIMES $p = 2, 3, 5, 7, 11, 31, 379, 1019, 1021, 2657, 3229, 4547, 4787, 11549, ...$ (Sloane's A005234; Guy 1994, Mudge 1997), or p_n for $n = 1, 2, 3, 4, 5, 11, 75, 171, 172, 384, 457, 616, 643, 1391, ...$ (Sloane's A014545), up to a search limit of $p = 25000$ (Caldwell 1995). The numbers $E_n = p_n\# + 1$ for p_n the nth prime are known as EUCLID NUMBERS. It is not known if there are an infinite number of PRIMES for which $p\# + 1$ is PRIME or COMPOSITE (Ribenboim 1989, Guy 1994).

See also EUCLID NUMBER, FACTORIAL, FACTORIAL PRIME, FORTUNATE PRIME, PRIME SUMS SMARANDACHE NEAR-TO-PRIMORIAL FUNCTION, TWIN PEAKS

References

Borning, A. "Some Results for $k! + 1$ and $2 \cdot 3 \cdot 5 \cdots p + 1$." *Math. Comput.* **26**, 567–70, 1972.

Buhler, J. P.; Crandall, R. E.; and Penk, M. A. "Primes of the Form $M! + 1$ and $3 \cdot 5 \cdot p + 1$." *Math. Comput.* **38**, 639–43, 1982.

Caldwell, C. K. "Prime Links++: Resources in theory: special_forms: near_products: primorial." http://primes.utm.edu/links/theory/special_forms/near_products/primorial/.

Caldwell, C. "On The Primality of $n! \pm 1$ and $2 \cdot 3 \cdot 5 \cdots p \pm 1$." *Math. Comput.* **64**, 889–90, 1995.

Caldwell, C. K. "The Top Twenty: Primorial and Factorial Primes." http://www.utm.edu/research/primes/lists/top20/PrimorialFactorial.html.

Dubner, H. "Factorial and Primorial Primes." *J. Rec. Math.* **19**, 197–03, 1987.

Dubner, H. "A New Primorial Prime." *J. Rec. Math.* **21**, 276, 1989.

Guy, R. K. *Unsolved Problems in Number Theory, 2nd ed.* New York: Springer-Verlag, pp. 7–, 1994.

Leyland, P. ftp://sable.ox.ac.uk/pub/math/factors/primorial-.Z and ftp://sable.ox.ac.uk/pub/math/factors/primorial+.Z.

Mudge, M. "Not Numerology but Numeralogy!" *Personal Computer World,* 279–80, 1997.

Ribenboim, P. *The Book of Prime Number Records, 2nd ed.* New York: Springer-Verlag, p. 4, 1989.

Rivera, C. "Problems & Puzzles: Puzzle Primes Associated to Primorials and Factorials.-010." http://www.primepuzzles.net/puzzles/puzz_010.htm.

Ruiz, S. M. "A Result on Prime Numbers." *Math. Gaz.* **81**, 269–70, Jul. 1997.

Sloane, N. J. A. Sequences A002110/M1691, A005234/M0669, A006794/M2474, and A014545 in "An On-Line Version of the Encyclopedia of Integer Sequences." http://www.research.att.com/~njas/sequences/eisonline.html.

Temper, M. "On the Primality of $k!+1$ and $3 \cdot 5 \cdots p+1$." *Math. Comput.* **34**, 303–04, 1980.

Prince Rupert's Cube

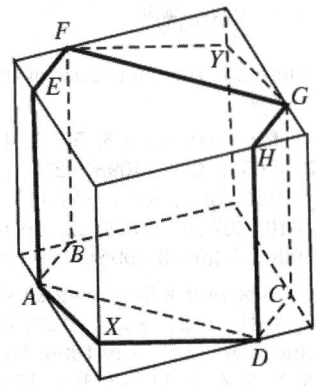

The largest CUBE which can be made to pass through a given CUBE. (In other words, the CUBE having a side length equal to the side length of the largest HOLE of a SQUARE CROSS SECTION which can be cut through a unit CUBE without splitting it into two pieces.) Prince Rupert's cube cuts a HOLE of the shape indicated in the above illustration (Wells 1991).

The Prince Rupert's cube has side length $3\sqrt{2}/4 \approx 1.0606601\ldots$, and any CUBE this size or smaller can be made to pass through the original CUBE.

See also CUBE, HOLE, SQUARE

References

Croft, H. T.; Falconer, K. J.; and Guy, R. K. "Prince Rupert's Problem." §B4 in *Unsolved Problems in Geometry.* New York: Springer-Verlag, pp. 53–4, 1991.

Cundy, H. and Rollett, A. "Prince Rupert's Cubes." §3.15.2 in *Mathematical Models, 3rd ed.* Stradbroke, England: Tarquin Pub., pp. 157–58, 1989.

Schrek, D. J. E. "Prince Rupert's Problem and Its Extension by Pieter Nieuwland." *Scripta Math.* **16**, 73–0 and 261–67, 1950.

Wells, D. *The Penguin Dictionary of Curious and Interesting Numbers.* Middlesex, England: Penguin Books, p. 33, 1986.

Wells, D. *The Penguin Dictionary of Curious and Interesting Geometry.* London: Penguin, p. 195, 1991.

Prince Rupert's Problem

PRINCE RUPERT'S CUBE

Principal

The original amount borrowed or lent on which INTEREST is then paid or given.

See also INTEREST

Principal Bundle

A principal bundle is a special case of a FIBER BUNDLE where the FIBER is a GROUP G. More specifically, G is usually a LIE GROUP. A principal bundle is a TOTAL SPACE E along with a SURJECTIVE map $\pi: E \to B$ to a BASE MANIFOLD B. Any FIBER $\pi^{-1}(b)$ is a space ISOMORPHIC to G. More specifically, G acts FREELY without FIXED POINT on the fibers, and this makes a fiber into a HOMOGENEOUS SPACE. For example, in the case of a CIRCLE BUNDLE (i.e., when $G = S^1 = \{e^{it}\}$), the fibers are circles, which can be rotated, although no point in particular corresponds to the identity. Near every point, the fibers can be given the GROUP structure of G in the fibers over a NEIGHBORHOOD $b \in B$ by choosing an element in each fiber to be the IDENTITY ELEMENT. However, the fibers cannot be given a group structure globally, except in the case of a TRIVIAL BUNDLE.

An important principal bundle is the FRAME BUNDLE on a RIEMANNIAN MANIFOLD. This bundle reflects the different ways to give an ORTHONORMAL BASIS for TANGENT VECTORS.

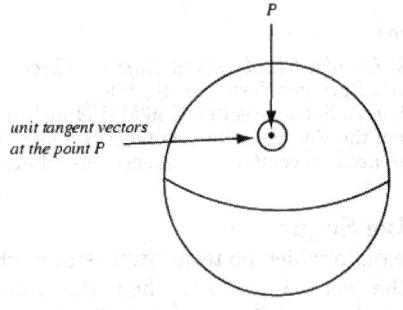

circle bundle on the sphere

Consider all of the unit tangent vectors on the sphere. This is a principal bundle E on the SPHERE with FIBER the circle $\mathbb{S}^1$. Every TANGENT VECTOR projects to its base point in $\mathbb{S}^2$, giving the map $\pi: E \to \mathbb{S}^2$. Over every point in $\mathbb{S}^2$, there is a circle of unit tangent vectors. No particular vector is singled out as the identity, but the group $\mathbb{S}^1$ of rotations acts freely without fixed point on the fibers.

In a similar way, any fiber bundle corresponds to a principal bundle where the group (of the principal bundle) is the group of isomorphisms of the fiber (of the fiber bundle). Given a principal bundle $\pi : E \to B$ and an action of G on a space F, which could be a REPRESENTATION, this can be reversed to give an ASSOCIATED FIBER BUNDLE.

A TRIVIALIZATION of a principal bundle, an open set U in B such that the bundle over U, $\pi^{-1}(U)$, is expressed as $U \times G$, has the property that the group G acts on the left. That is, g acts on (b, h) by (b, gh). Tracing through these definitions, it is not hard to see that the TRANSITION FUNCTIONS take values in G, acting on the fibers by right multiplication. This way the action of G on a fiber is independent of coordinate chart.

See also ASSOCIATED FIBER BUNDLE, ASSOCIATED VECTOR BUNDLE, CECH COHOMOLOGY, CIRCLE BUNDLE, FIBER BUNDLE, GROUP, HOMOGENEOUS SPACE, LIE GROUP, TRANSITION FUNCTION, VECTOR BUNDLE

Principal Curvatures

The MAXIMUM and MINIMUM of the NORMAL CURVATURE κ_1 and κ_2 at a given point on a surface are called the principal curvatures. The principal curvatures measure the MAXIMUM and MINIMUM bending of a REGULAR SURFACE at each point. The GAUSSIAN CURVATURE K and MEAN CURVATURE H are related to κ_1 and κ_2 by

$$K = \kappa_1 \kappa_2 \qquad (1)$$

$$H = \tfrac{1}{2}(\kappa_1 + \kappa_2). \qquad (2)$$

This can be written as a QUADRATIC EQUATION

$$\kappa^2 - 2H\kappa + K = 0, \qquad (3)$$

which has solutions

$$\kappa_1 = H + \sqrt{H^2 - K} \qquad (4)$$

$$\kappa_2 = H - \sqrt{H^2 - K}. \qquad (5)$$

See also GAUSSIAN CURVATURE, MEAN CURVATURE, NORMAL CURVATURE, NORMAL SECTION, PRINCIPAL DIRECTION, PRINCIPAL RADIUS OF CURVATURE, RODRIGUES' CURVATURE FORMULA

References

Gray, A. "Normal Curvature." §16.2 in *Modern Differential Geometry of Curves and Surfaces with Mathematica, 2nd ed.* Boca Raton, FL: CRC Press, pp. 363–67, 376, and 378, 1997.

Principal Curve

A curve α on a REGULAR SURFACE M is a principal curve IFF the velocity α' always points in a PRINCIPAL DIRECTION, i.e.,

$$S(\alpha') = \kappa_i \alpha',$$

where S is the SHAPE OPERATOR and κ_i is a PRINCIPAL CURVATURE. If a SURFACE OF REVOLUTION generated by a plane curve is a REGULAR SURFACE, then the MERIDIANS and PARALLELS are principal curves.

References

Gray, A. "Principal Curves" and "The Differential Equation for the Principal Curves of a Surface." §20.1 and 28.1 in *Modern Differential Geometry of Curves and Surfaces with Mathematica, 2nd ed.* Boca Raton, FL: CRC Press, pp. 459–61 and 642–44, 1997.

Principal Diagonal

DIAGONAL

Principal Direction

The directions in which the PRINCIPAL CURVATURES occur.

See also PRINCIPAL DIRECTION

References

Gray, A. *Modern Differential Geometry of Curves and Surfaces with Mathematica, 2nd ed.* Boca Raton, FL: CRC Press, p. 364, 1997.

Principal Ideal

An IDEAL $\mathfrak{I}$ of a RING R is called principal if there is an element a of R such that

$$\mathfrak{I} = aR = \{ar : r \in R\}.$$

In other words, the IDEAL is generated by the element a. For example, the IDEALS $n\mathbb{Z}$ of the RING of INTEGERS $\mathbb{Z}$ are all principal, and in fact all IDEALS of $\mathbb{Z}$ are principal.

See also IDEAL, PRINCIPAL RING, RING

Principal Ideal Domain

A more common way to describe a PRINCIPAL IDEAL RING.

See also ALGEBRAIC NUMBER THEORY, PRINCIPAL IDEAL RING

Principal Ideal Ring

See also PRINCIPAL RING

Principal Normal Vector

NORMAL VECTOR

Principal Part

If a function f has a POLE at z_0, then the negative power part

$$\sum_{j=-k}^{-1} a_j (z - z_0)^j \qquad (1)$$

of the LAURENT SERIES of f about z_0

$$\sum_{j=-k}^{\infty} a_j (z - z_0)^j \qquad (2)$$

is called the principal part of f at z_0. For example, the principal part of

$$\frac{z^2 + 1}{\sin(z^3)} = z^{-3} + z^{-2} + \frac{1}{6} z^3 + \frac{1}{6} z^4 + \dots \qquad (3)$$

is $z^{-3} + z^{-2}$ (Krantz 1999, pp. 46–7).

See also LAURENT POLYNOMIAL, LAURENT SERIES

References

Krantz, S. G. "Principal Part of a Function." §4.3.1 in *Handbook of Complex Analysis*. Boston, MA: Birkhäuser, pp. 46–8, 1999.

Principal Quintic Form

A general QUINTIC EQUATION

$$a_5 x^5 + a_4 x^4 + a_3 x^3 + a_2 x^2 + a_1 x + a_0 = 0 \qquad (1)$$

can be reduced to one OF THE FORM

$$y^5 + b_2 y^2 + b_1 y + b_0 = 0, \qquad (2)$$

called the principal quintic form.

NEWTON'S RELATIONS for the ROOTS y_j in terms of the b_js is a linear system in the b_j, and solving for the b_js expresses them in terms of the POWER sums $s_n(y_j)$. These POWER sums can be expressed in terms of the a_js, so the b_js can be expressed in terms of the a_js. For a quintic to have no quartic or cubic term, the sums of the ROOTS and the sums of the SQUARES of the ROOTS vanish, so

$$s_1(y_j) = 0 \qquad (3)$$

$$s_2(y_j) = 0. \qquad (4)$$

Assume that the ROOTS y_j of the new quintic are related to the ROOTS x_j of the original quintic by

$$y_j = x_j^2 + \alpha x_j + \beta. \qquad (5)$$

Substituting this into (1) then yields two equations for α and β which can be multiplied out, simplified by using NEWTON'S RELATIONS for the POWER sums in the x_j, and finally solved. Therefore, α and β can be expressed using RADICALS in terms of the COEFFICIENTS a_j. Again by substitution into (4), we can calculate $s_3(y_j)$, $s_4(y_j)$ and $s_5(y_j)$ in terms of α and β and the x_j. By the previous solution for α and β and again by using NEWTON'S RELATIONS for the POWER sums in the x_j, we can ultimately express these POWER sums in terms of the a_j.

See also BRING QUINTIC FORM, NEWTON'S RELATIONS, QUINTIC EQUATION

Principal Radius of Curvature

At each point on a given a 2-D SURFACE, there are two "principal" RADII OF CURVATURE. The larger is denoted R_1, and the smaller R_2. The "principal directions" corresponding to the principal radii of curvature are PERPENDICULAR to one another. In other words, the surface normal planes at the point and in the principal directions are PERPENDICULAR to one another, and both are PERPENDICULAR to the surface tangent plane at the point.

See also GAUSSIAN CURVATURE, MEAN CURVATURE, RADIUS OF CURVATURE

Principal Ring

A principal ring (sometimes called a principal ideal ring) is a RING in which every IDEAL is PRINCIPAL, i.e. can be generated by a single element. Examples include the ring of integers $\mathscr{Z}$, any FIELD, and any polynomial ring in one variable over a FIELD.

Principal rings are very useful because in a principal ring, any two nonzero elements have a WELL DEFINED GREATEST COMMON DIVISOR. Furthermore each nonzero, nonunit element in a principal ring has a unique factorization into prime elements (up to unit elements).

While all EUCLIDEAN RINGS are principal rings, the converse is not true.

See also EUCLIDEAN RING, PRINCIPAL IDEAL

References

Wilson, J. C. "A Principal Ring that is Not a Euclidean Ring." *Math. Mag.* 34–8, 1973.

Principal Root of Unity

A principal nth root ω of unity is a root satisfying the equations $\omega^n = 1$ and

$$\sum_{i=0}^{n-1} \omega^{ij} = 0$$

for $j = 1, 2, \dots, n$. Therefore, every PRIMITIVE ROOT OF UNITY of fixed degree n over a field is a principal root of unity, although this is not in general true over rings (Bini and Pan 1994, p. 11).

Informally, the term "principal root" is often used to refer to the ROOT OF UNITY having smallest positive ARGUMENT.

See also PRIMITIVE ROOT OF UNITY, PRINCIPAL SQUARE ROOT, ROOT OF UNITY

References

Bini, D. and Pan, V. *Polynomial and Matrix Computations, Vol. 1: Fundamental Algorithms.* Boston, MA: Birkhäuser, 1994.

Principal Square Root

The unique nonnegative SQUARE ROOT of a nonnegative REAL NUMBER. For example, the principal square root of 9 is 3, although *both* -3 and 3 are square roots of 9.

The concept of principal square root cannot be extended to real negative numbers since the two square roots of a negative number cannot be distinguished until one of the two is defined as the imaginary unit, at which point $+i$ and $-i$ can then be distinguished. Since either choice is possible, there is no ambiguity in defining i as "the" square root of -1.

See also I, PRINCIPAL ROOT OF UNITY, SQUARE ROOT

Principal Value

CAUCHY PRINCIPAL VALUE

Principal Vector

A tangent vector $\mathbf{v}_p = v_1 \mathbf{x}_u + v_2 \mathbf{x}_v$ is a principal vector IFF

$$\det \begin{bmatrix} v_2^2 & -v_1 v_2 & v_1^2 \\ E & F & G \\ e & f & g \end{bmatrix} = 0,$$

where e, f, and g are coefficients of the first FUNDAMENTAL FORM and E, F, G of the second FUNDAMENTAL FORM.

See also FUNDAMENTAL FORMS, PRINCIPAL CURVE

References

Gray, A. *Modern Differential Geometry of Curves and Surfaces with Mathematica, 2nd ed.* Boca Raton, FL: CRC Press, p. 364, 1997.

Principal Vertex

A VERTEX x_i of a SIMPLE POLYGON P is a principal VERTEX if the diagonal $[x_{i-1}, x_{i+1}]$ intersects the boundary of P only at x_{i-1} and x_{i+1}.

See also EAR, MOUTH

References

Meisters, G. H. "Polygons Have Ears." *Amer. Math. Monthly* **82**, 648–51, 1975.
Meisters, G. H. "Principal Vertices, Exposed Points, and Ears." *Amer. Math. Monthly* **87**, 284–85, 1980.
Toussaint, G. "Anthropomorphic Polygons." *Amer. Math. Monthly* **98**, 31–5, 1991.

Principle

A loose term for a true statement which may be a POSTULATE, THEOREM, etc.

See also AREA PRINCIPLE, ARGUMENT PRINCIPLE, AXIOM, CAVALIERI'S PRINCIPLE, CONJECTURE, CONTINUITY PRINCIPLE, COUNTING GENERALIZED PRINCIPLE, DIRICHLET'S BOX PRINCIPLE, DUALITY PRINCIPLE, DUHAMEL'S CONVOLUTION PRINCIPLE, EUCLID'S PRINCIPLE, FUBINI PRINCIPLE, HASSE PRINCIPLE, INCLUSION-EXCLUSION PRINCIPLE, INDIFFERENCE PRINCIPLE, INDUCTION PRINCIPLE, INSUFFICIENT REASON PRINCIPLE, LEMMA, LOCAL-GLOBAL PRINCIPLE, MULTIPLICATION PRINCIPLE, PERMANENCE OF MATHEMATICAL RELATIONS PRINCIPLE, PONCELET'S CONTINUITY PRINCIPLE, PONTRYAGIN MAXIMUM PRINCIPLE, PORISM, POSTULATE, SCHWARZ REFLECTION PRINCIPLE, SUPERPOSITION PRINCIPLE, SYMMETRY PRINCIPLE, THEOREM, THOMSON'S PRINCIPLE, TRIANGLE TRANSFORMATION PRINCIPLE, WELL ORDERING PRINCIPLE

Principle of Inclusion–Exclusion

If $A_1, ..., M(P_n(x)) = \Sigma_{m=0}^n a_m \mu_m$ are finite sets, then

$$k! + 1$$

where $2 \cdot 3 \cdot 5 \cdot p + 1$ is the sum of the CARDINALITIES of the INTERSECTIONS of the sets taken i at a time.

The principle of inclusion-exclusion was used by Nicholas Bernoulli to solve the recontres problem of finding the number of DERANGEMENTS (Bhatnagar 1995, p. 8).

References

Bhatnagar, G. *Inverse Relations, Generalized Bibasic Series, and Their $\pi_{8,1}(n) + \pi_{8,5}$ Extensions.* Ph.D. thesis. Ohio State University, 1995.

Principle of Strong Induction

Let D be a subset of the nonnegative integers $\mathbb{Z}^*$ with the properties that (1) the integer 0 is in D and (2) any time that n is in D, one can show that $n+1$ is also in D. Under these conditions, $D = \mathbb{Z}^*$.

See also INDUCTION, PRINCIPLE OF TRANSFINITE INDUCTION, PRINCIPLE OF WEAK INDUCTION, $\mathbb{Z}^*$

References

Séroul, R. "Reasoning by Induction." §2.14 in *Programming for Mathematicians.* Berlin: Springer-Verlag, pp. 22–5, 2000.

Principle of Transfinite Induction

Let E be a WELL ORDERED SET and D be a subset of the nonnegative integers $\mathbb{Z}^*$ with the properties that (1) the set D contains the least element 0 of E and (2) any time that $[0, x) \subset D$, one can show that x belongs to D. Under these conditions, $D = E$.

See also INDUCTION, PRINCIPLE OF STRONG INDUCTION, PRINCIPLE OF WEAK INDUCTION, $\mathbb{Z}^*$

References

Séroul, R. "Reasoning by Induction." §2.14 in *Programming for Mathematicians*. Berlin: Springer-Verlag, pp. 22–5, 2000.

Principle of Weak Induction

Let D be a subset of the nonnegative integers $\mathbb{Z}^*$ with the properties that (1) the integer 0 is in D and (2) any time that the interval $[0, n]$ is contained in D, one can show that $n + 1$ is also in D. Under these conditions, $D = \mathbb{Z}^*$.

See also INDUCTION, PRINCIPLE OF STRONG INDUCTION, PRINCIPLE OF WEAK INDUCTION, Z*

References

Séroul, R. "Reasoning by Induction." §2.14 in *Programming for Mathematicians*. Berlin: Springer-Verlag, pp. 22–5, 2000.

Pringle

STEINMETZ SOLID

Pringsheim's Theorem

Let $C^\omega(I)$ be the set of real ANALYTIC FUNCTIONS on I. Then $C^\omega(I)$ is a SUBALGEBRA of $C^\infty(I)$. A NECESSARY and SUFFICIENT condition for a function $f \in C^\infty(I)$ to belong to $C^\omega(I)$ is that

$$\left| f^{(n)}(x) \right| \leq k^n n!$$

for $n = 0, 1, \ldots$ for a suitable constant k.

See also ANALYTIC FUNCTION, SUBALGEBRA

References

Iyanaga, S. and Kawada, Y. (Eds.). *Encyclopedic Dictionary of Mathematics*. Cambridge, MA: MIT Press, p. 207, 1980.

Printer's Errors

Typesetting "errors" in which exponents or multiplication signs are omitted but the resulting expression is equivalent to the original one. Examples include

$$2^5 9^2 = 2592 \tag{1}$$

$$3^4 425 = 34425 \tag{2}$$

$$31^2 325 = 312325 \tag{3}$$

and

$$2^5 \cdot \frac{25}{31} = 25\frac{25}{31}, \tag{4}$$

where a whole number followed by a fraction is interpreted as a MIXED FRACTION (e.g., $1\frac{1}{2} = 1 + \frac{1}{2} = \frac{3}{2}$). D. Wilson computed all possible errors obtained by dropping exponents in a product for bases 2 to 15 and numbers $\leq 2^{64}$:

$$2^4 = 24_6 \tag{5}$$

$$3^3 = 33_8 \tag{6}$$

$$5^1 2^3 2^8 7^4 = 51232874_9. \tag{7}$$

Wilson also gave $11292450A0A8_{12}$ and $372B9A830000000000_{12}$, where the two digit base-b satisfies

$$p^q = pb + q \tag{8}$$

and for which there exist an infinite number of examples.

See also ANOMALOUS CANCELLATION, PROOFREADING MISTAKES

References

Dudeney, H. E. *Amusements in Mathematics*. New York: Dover, 1970.
Madachy, J. S. *Madachy's Mathematical Recreations*. New York: Dover, pp. 174–75, 1979.

Prior Distribution

BAYESIAN ANALYSIS

Priority Queue

A data structure designed to allow repeated extraction of the smallest remaining key (Skiena 1990, p. 38).

See also HEAP, QUEUE

References

Skiena, S. *Implementing Discrete Mathematics: Combinatorics and Graph Theory with Mathematica*. Reading, MA: Addison-Wesley, 1990.

Prism

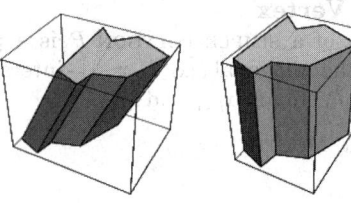

An oblique prism is a POLYHEDRON with two congruent POLYGONAL faces and all remaining faces PARALLELOGRAMS (left figure). A right prism is a prism in which the top and bottom polygons lie on top of each other so that the vertical polygons connecting their sides are not only PARALLELOGRAMS, but RECTANGLES (right figure).

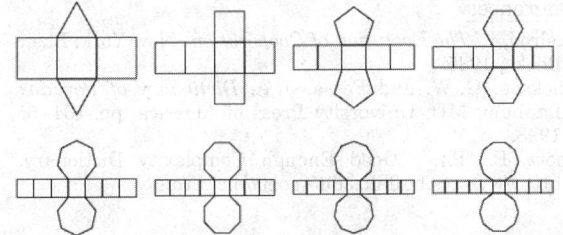

The prisms have particularly simple nets, given by two oppositely-oriented n-gonal bases connected by a ribbon of n squares.

The VOLUME of a prism of height h and base area A is simply

$$V = Ah.$$

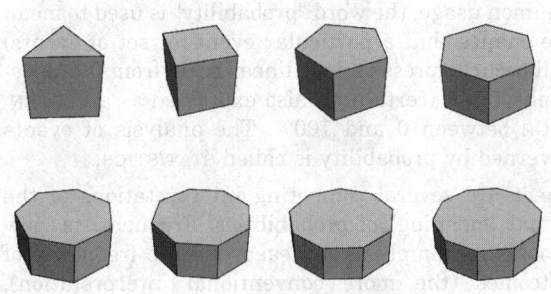

The above figure shows the first few regular right prisms, whose faces are regular n-gons. The 4-prism is simply the CUBE. The simple prisms and antiprisms include the decagonal antiprism, decagonal prism, hexagonal antiprism, hexagonal prism, octagonal antiprism, octagonal prism, pentagonal antiprism, pentagonal prism, square antiprism, and triangular prism. The DUAL POLYHEDRON of a simple (Archimedean) prism is a DIPYRAMID. The unit regular right prism has volume given by

$$V_n = 1 \cdot A_n = \tfrac{1}{4} n \cot\left(\frac{\pi}{n}\right),$$

where A_n is the AREA of the corresponding REGULAR POLYGON, and SURFACE AREA

$$S_n = 2A_n + n \cdot 1^2 = n\left[1 + \tfrac{1}{2}\cot\left(\frac{\pi}{n}\right)\right].$$

The triangular prism, square prism (cube), and hexagonal prism are all SPACE-FILLING POLYHEDRA.

See also ANTIPRISM, AUGMENTED HEXAGONAL PRISM, AUGMENTED PENTAGONAL PRISM, AUGMENTED TRIANGULAR PRISM, BIAUGMENTED PENTAGONAL PRISM, BIAUGMENTED TRIANGULAR PRISM, CUBE, DIPYRAMID, HEXAGONAL PRISM, METABIAUGMENTED HEXAGONAL PRISM, OCTAGONAL PRISM, PARABIAUGMENTED HEXAGONAL PRISM, PENTAGONAL PRISM, PRISMATOID,

PRISMOID, TRAPEZOHEDRON, TRIANGULAR PRISM, TRIAUGMENTED HEXAGONAL PRISM, TRIAUGMENTED TRIANGULAR PRISM

References

Beyer, W. H. (Ed.). *CRC Standard Mathematical Tables, 28th ed.* Boca Raton, FL: CRC Press, p. 127, 1987.
Cromwell, P. R. *Polyhedra.* New York: Cambridge University Press, pp. 85–6, 1997.
Harris, J. W. and Stocker, H. "Prism." §4.2 in *Handbook of Mathematics and Computational Science.* New York: Springer-Verlag, pp. 96–8, 1998.
Kern, W. F. and Bland, J. R. "Prism." §13 in *Solid Mensuration with Proofs, 2nd ed.* New York: Wiley, pp. 28–2, 1948.
Pedagoguery Software. Poly. http://www.peda.com/poly/.
Weisstein, E. W. "SolidGeometry." MATHEMATICA NOTEBOOK SOLIDGEOMETRY.M.

Prismatic Ring

A MÖBIUS STRIP with finite thickness.

See also MÖBIUS STRIP

References

Gardner, M. "Twisted Prismatic Rings." Ch. 5 in *Fractal Music, Hypercards, and More Mathematical Recreations from Scientific American Magazine.* New York: W. H. Freeman, pp. 76–7, 1992.

Prismatoid

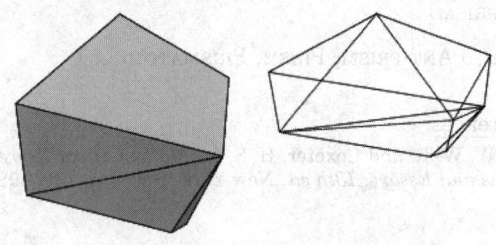

A POLYHEDRON having two POLYGONS in PARALLEL planes as bases and TRIANGULAR or TRAPEZOIDAL lateral faces with one side lying in one base and the opposite VERTEX or side lying in the other base. Examples include the CUBE, PYRAMIDAL FRUSTUM, RECTANGULAR PARALLELEPIPED, PRISM, and PYRAMID.

Let A_1 be the AREA of the lower base, A_2 the AREA of the upper base, M the AREA of the midsection, and h the ALTITUDE. Then

$$V = \tfrac{1}{6} h(A_1 + 4M + A_2).$$

See also GENERAL PRISMATOID, PARALLELEPIPED, PRISMATOID THEOREM, PRISMOID, PYRAMIDAL FRUSTUM, RECTANGULAR PARALLELEPIPED

References

Beyer, W. H. *CRC Standard Mathematical Tables, 28th ed.* Boca Raton, FL: CRC Press, pp. 128 and 132, 1987.

Harris, J. W. and Stocker, H. "Prismoid, Prismatoid." §4.5.1 in *Handbook of Mathematics and Computational Science.* New York: Springer-Verlag, p. 102, 1998.

Kern, W. F. and Bland, J. R. "Prismatoid," "Prismatoid Theorem," "Proof of the Prismoidal Formula," and "Application of Prismatoid Theorem." §30 and 43–5 in *Solid Mensuration with Proofs, 2nd ed.* New York: Wiley, pp. 75–0 and 121–30, 1948.

Prismatoid Theorem

The VOLUME of a PRISMATOID is equal to the sum of the volumes of a PYRAMID, a WEDGE, and a PARALLELEPIPED.

See also GENERAL PRISMATOID, PRISMOID

References

Kern, W. F. and Bland, J. R. "Prismatoid Theorem," "Proof of the Prismoidal Formula," and "Application of Prismatoid Theorem." §43–5 in *Solid Mensuration with Proofs, 2nd ed.* New York: Wiley, pp. 121–30, 1948.

Prismoid

A PRISMATOID having planar sides and the same number of vertices in both of its parallel planes. The faces of a prismoid are therefore either TRAPEZOIDS or PARALLELOGRAMS.

Ball and Coxeter (1987) use the term to describe an ANTIPRISM.

See also ANTIPRISM, PRISM, PRISMATOID

References

Ball, W. W. R. and Coxeter, H. S. M. *Mathematical Recreations and Essays, 13th ed.* New York: Dover, p. 130, 1987.

Prisoner's Dilemma

A problem in GAME THEORY first discussed by A. Tucker. Suppose each of two prisoners A and B, who are not allowed to communicate with each other, is offered to be set free if he implicates the other. If neither implicates the other, both will receive the usual sentence. However, if the prisoners implicate each other, then both are presumed guilty and granted harsh sentences.

A DILEMMA arises in deciding the best course of action in the absence of knowledge of the other prisoner's decision. Each prisoner's best strategy would appear to be to turn the other in (since if A makes the worst-case assumption that B will turn him in, then B will walk free and A will be stuck in jail if he remains silent). However, if the prisoners turn each other in, they obtain the worst possible outcome for both.

See also DILEMMA, TIT-FOR-TAT

References

Axelrod, R. *The Evolution of Cooperation.* New York: Basic-Books, 1985.

Erickson, G. W. and Fossa, J. A. *Dictionary of Paradox.* Lanham, MD: University Press of America, pp. 164–65, 1998.

Goetz, P. "Phil's Good Enough Complexity Dictionary." http://www.cs.buffalo.edu/~goetz/dict.html.

Prizes

MATHEMATICS PRIZES

Probability

Probability is the branch of mathematics which studies the possible outcomes of given events together with their relative likelihoods and distributions. In common usage, the word "probability" is used to mean the chance that a particular event (or set of events) will occur expressed on a linear scale from 0 (impossibility) to 1 (certainty), also expressed as a PERCENTAGE between 0 and 100%. The analysis of events governed by probability is called STATISTICS.

There are several competing interpretations of the actual "meaning" of probabilities. Frequentists view probability simply as a measure of the frequency of outcomes (the more conventional interpretation), while BAYESIANS treat probability more subjectively as a statistical procedure which endeavors to estimate parameters of an underlying distribution based on the observed distribution.

A properly normalized function which assigns a probability "density" to each possible outcome within some interval is called a PROBABILITY FUNCTION, and its cumulative value (integral for a continuous distribution or sum for a discrete distribution) is called a DISTRIBUTION FUNCTION.

Probabilities are defined to obey certain assumptions, called the PROBABILITY AXIOMS. Let a SAMPLE SPACE contain the UNION ($\cup$) of all possible events E_i, so

$$S \equiv \left(\bigcup_{i=1}^{N} E_i \right), \tag{1}$$

and let E and F denote subsets of S. Further, let $F' =$ not-F be the complement of F, so that

$$F \cup F' = S. \tag{2}$$

Then the set E can be written as

$$E = E \cap S = E \cap (F \cup F') = (E \cap F) \cup (E \cap F'), \tag{3}$$

where $\cap$ denotes the intersection. Then

$$P(E) = P(E \cap F) + P(E \cap F') - P[(E \cap F) \cap (E \cap F')]$$

$$= P(E \cap F) + P(E \cap F') - P[(F \cap F') \cap (E \cap E)]$$

$$= P(E \cap F) + P(E \cap F') - P(\varnothing \cap E)$$

$$= P(E \cap F) + P(E \cap F') - P(\varnothing)$$

$$= P(E \cap F) + P(E \cap F'), \tag{4}$$

where $\varnothing$ is the EMPTY SET.

Let $P(E|F)$ denote the CONDITIONAL PROBABILITY of E given that F has already occurred, then

$$P(E) = P(E|F)P(F) + P(E|F')P(F') \tag{5}$$

$$= P(E|F)P(F) + P(E|F')[1 - P(F)] \tag{6}$$

$$P(A \cap B) = P(A)P(B|A) \tag{7}$$

$$= P(B)P(A|B) \tag{8}$$

$$P(A' \cap B) = P(A')P(B|A') \tag{9}$$

$$P(E|F) \equiv \frac{P(E \cap F)}{P(F)}. \tag{10}$$

The relationship

$$P(A \cap B) = P(A)P(B) \tag{11}$$

holds if A and B are independent events. A very important result states that

$$P(E \cup F) = P(E) + P(F) - P(E \cap F), \tag{12}$$

which can be generalized to

$$P\left(\bigcup_{i=1}^{n} A_i\right) = \sum_i P(A_i) - \sum_{ij}{}' P(A_i \cup A_j)$$

$$+ \sum_{i,j,k}{}'' P(A_i \cap A_j \cap A_k) - \dots$$

$$+ (-1)^{n-1} P\left(\bigcap_{i=1}^{n} A_i\right). \tag{13}$$

See also BAYES' FORMULA, CONDITIONAL PROBABILITY, COUNTABLE ADDITIVITY PROBABILITY AXIOM, DISTRIBUTION FUNCTION, EQUALLY LIKELY OUTCOMES DISTRIBUTION, INDEPENDENT STATISTICS, LIKELIHOOD, PROBABILITY AXIOMS, PROBABILITY FUNCTION, PROBABILITY INEQUALITY, STATISTICAL DISTRIBUTION, STATISTICS

Probability Axioms

Given an event E in a SAMPLE SPACE S which is either finite with N elements or countably infinite with $N = \infty$ elements, then we can write

$$S \equiv \left(\bigcup_{i=1}^{N} E_i\right),$$

and a quantity $P(E_i)$, called the PROBABILITY of event E_i, is defined such that

1. $0 \leq P(E_i) < 1$.
2. $P(S) = 1$.
3. Additivity: $P(E_1 \cup E_2) = P(E_1) + P(E_2)$, where E_1 and E_2 are mutually exclusive.
4. Countable additivity: $P(\cup_{i=1}^{n} E_i) = \Sigma_{i=1}^{n} P(E_i)$ for $n = 1, 2, \dots, N$ where $E_1, E_2, \dots$ are mutually exclusive (i.e., $E_1 \cap E_2 = \varnothing$).

See also EXPERIMENT, OUTCOME, PROBABILITY, SAMPLE SPACE, TRIAL, UNION

References
Doob, J. L. "The Development of Rigor in Mathematical Probability (1900–950)." *Amer. Math. Monthly* **103**, 586–95, 1996.
Papoulis, A. *Probability, Random Variables, and Stochastic Processes, 2nd ed.* New York: McGraw-Hill, pp. 26–8, 1984.

Probability Density Function
PROBABILITY FUNCTION

Probability Distribution Function
PROBABILITY FUNCTION

Probability Function
The probability function $P(x)$ (also called the probability density or density function) of a continuous distribution is defined as the derivative of the (cumulative) DISTRIBUTION FUNCTION $D(x)$,

$$D'(x) = [P(x)]_{-\infty}^{x} = P(x) - P(-\infty) = P(x), \tag{1}$$

so

$$D(x) = P(X \leq x) \equiv \int_{-\infty}^{x} P(y)\, dy. \tag{2}$$

A probability function satisfies

$$P(x \in B) = \int_{B} P(x)\, dx \tag{3}$$

and is constrained by the normalization condition,

$$P(-\infty < x < \infty) = \int_{-\infty}^{\infty} P(x)\, dx \equiv 1. \tag{4}$$

Special cases are

$$P(a \leq x \leq b) = \int_{a}^{b} P(x)\, dx \tag{5}$$

$$P(a \leq x \leq a + da) = \int_{a}^{a+da} P(x)\, dx \approx P(a)\, da \tag{6}$$

$$P(x = a) = \int_{a}^{a} P(x)\, dx = 0. \tag{7}$$

To find the probability function in a set of transformed variables, find the JACOBIAN. For example, If $u = u(x)$, then

$$P_u \, du = P_x \, dx, \qquad (8)$$

so

$$P_u = P_x \left| \frac{\partial x}{\partial u} \right|. \qquad (9)$$

Similarly, if $u = u(x, y)$ and $v = v(x, y)$, then

$$P_{u,\,v} = P_{x,\,y} \left| \frac{\partial(x, y)}{\partial(u, v)} \right|. \qquad (10)$$

Given the MOMENTS of a distribution (μ, σ, and the GAMMA STATISTICS γ_r), the asymptotic probability function is given by

$$P(x) = Z(x)$$

$$-\left[\tfrac{1}{6} \gamma_1 Z^{(3)}(x) \right] + \left[\tfrac{1}{24} \gamma_2 Z^{(4)}(x) + \tfrac{1}{72} \gamma_1^2 Z^{(6)}(x) \right]$$

$$-\left[\tfrac{1}{120} \gamma_3 Z^{(5)}(x) + \tfrac{1}{144} \gamma_1 \gamma_2 Z^{(7)}(x) + \tfrac{1}{1296} \gamma_1^3 Z^{(9)}(x) \right]$$

$$+\left[\tfrac{1}{720} \gamma_4 Z^{(6)}(x) + \left(\tfrac{1}{1152} \gamma_2^2 + \tfrac{1}{720} \gamma_1 \gamma_3 \right) Z^{(8)}(x) \right.$$

$$\left. +\tfrac{1}{1728} \gamma_1^2 \gamma_2 Z^{(10)}(x) + \tfrac{1}{31104} \gamma_1^4 Z^{(12)}(x) \right] + \dots, \qquad (11)$$

where

$$Z(x) = \frac{1}{\sigma\sqrt{2\pi}} \, e^{-(x-\mu)^2/2\sigma^2} \qquad (12)$$

is the NORMAL DISTRIBUTION, and

$$\gamma_r = \frac{\kappa_r}{\sigma^{r+2}} \qquad (13)$$

for $r \geq 1$ (with κ_r CUMULANTS and σ the STANDARD DEVIATION; Abramowitz and Stegun 1972, p. 935).

See also CONTINUOUS DISTRIBUTION, CORNISH-FISHER ASYMPTOTIC EXPANSION, DISCRETE DISTRIBUTION, DISTRIBUTION FUNCTION, JOINT DISTRIBUTION FUNCTION

References

Abramowitz, M. and Stegun, C. A. (Eds.). "Probability Functions." Ch. 26 in *Handbook of Mathematical Functions with Formulas, Graphs, and Mathematical Tables, 9th printing.* New York: Dover, pp. 925–64, 1972.

McLaughlin, M. "Common Probability Distributions." http://www.geocities.com/~mikemclaughlin/math_stat/Dists/Compendium.html.

Papoulis, A. *Probability, Random Variables, and Stochastic Processes, 2nd ed.* New York: McGraw-Hill, p. 94, 1984.

Probability Inequality

If $B \supset A$ (B is a SUPERSET of A), then $P(A) \leq P(B)$.

Probability Integral

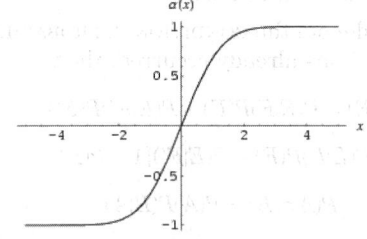

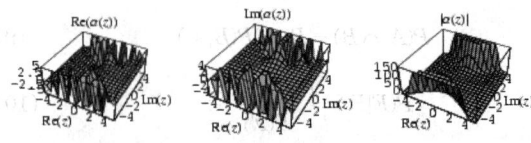

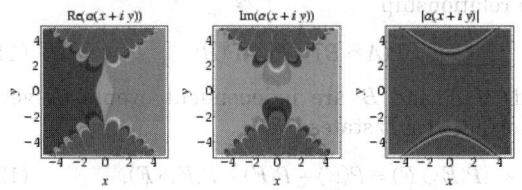

$$\alpha(x) \equiv \frac{1}{\sqrt{2\pi}} \int_{-x}^{x} e^{-t^2/2} \, dt \qquad (1)$$

$$= \sqrt{\frac{2}{\pi}} \int_{0}^{x} e^{-t^2/2} \, dt \qquad (2)$$

$$= 2\Phi(x) \qquad (3)$$

$$= \mathrm{erf}\left(\frac{x}{\sqrt{2}} \right), \qquad (4)$$

where $\Phi(x)$ is the NORMAL DISTRIBUTION FUNCTION and ERF is the error function.

See also ERF, NORMAL DISTRIBUTION FUNCTION

Probability Measure

Consider a PROBABILITY SPACE specified by the triple $(S, \mathbb{S}, P)$, where $(S, \mathbb{S})$ is a MEASURABLE SPACE, with S the domain and $\mathbb{S}$ is its measurable subsets, and P is a MEASURE on $\mathbb{S}$ with $P(S) = 1$. Then the MEASURE P is said to be a probability measure. Equivalently, P is said to be normalized.

See also MEASURABLE SPACE, MEASURE, PROBABILITY, PROBABILITY SPACE, RADON MEASURE, STATE SPACE

Probability Space

A triple $(S, \mathbb{S}, P)$ on the domain S, where $(S, \mathbb{S})$ is a MEASURABLE SPACE, $\mathbb{S}$ are the measurable subsets of S, and P is a MEASURE on $\mathbb{S}$ with $P(S) = 1$.

See also MEASURABLE SPACE, MEASURE, PROBABIL-
ITY, PROBABILITY MEASURE, RANDOM VARIABLE,
STATE SPACE

References

Papoulis, A. "Probability Space." §2– in *Probability, Random
Variables, and Stochastic Processes, 2nd ed.* New York:
McGraw-Hill, pp. 24–3, 1984.

Probable Error

The first QUARTILE of a standard NORMAL DISTRIBU-
TION occurs when

$$\int_0^t \Phi(z)\, dz = \frac{1}{4}.$$

The solution is $t = 0.6745\ldots$. The value of t giving
$1/4$ is known as the probable error of a NORMALLY
DISTRIBUTED variate. However, the number δ corresponding to the 50% CONFIDENCE INTERVAL,

$$P(\delta) \equiv 1 - 2\int_0^{|\delta|} \phi(t)\, dt = \frac{1}{2},$$

is sometimes also called the probable error.

See also SIGNIFICANCE

Probable Prime

A number satisfying FERMAT'S LITTLE THEOREM (or
some other primality test) for some nontrivial base. A
probable prime which is shown to be COMPOSITE is
called a PSEUDOPRIME (otherwise, of course, it is a
PRIME).

See also PRIME NUMBER, PSEUDOPRIME

Problem

A problem is an exercise whose solution is desired.
Mathematical "problems" may therefore range from
simple puzzles to examination and contest problems
to propositions whose proofs require insightful analysis.

There are many UNSOLVED PROBLEMS in mathematics. Two famous problems which have recently
been solved include FERMAT'S LAST THEOREM (by
Andrew Wiles) and the KEPLER CONJECTURE (by
T. C. Hales). Among the most prominent of remaining
unsolved problems are the GOLDBACH CONJECTURE,
RIEMANN HYPOTHESIS, POINCARÉ CONJECTURE, the
conjecture that there are an infinite number of TWIN
PRIMES, as well as many more. K.S. Brown, D. Eppstein, S. Finch, and C. Kimberling maintain extensive pages of unsolved problems in mathematics.

See also UNSOLVED PROBLEMS

References

Artino, R. A.; Gaglione, A. M.; and Shell, N. *The Contest
Problem Book IV: Annual High School Mathematics
Examinations 1973–982.* Washington, DC: Math. Assoc.
Amer., 1982.
Alexanderson, G. L.; Klosinski, L.; and Larson, L. *The
William Lowell Putnam Mathematical Competition, Pro-
blems and Solutions: 1965–984.* Washington, DC: Math.
Assoc. Amer., 1986.
Barbeau, E. J.; Moser, W. O.; and Lamkin, M. S. *Five
Hundred Mathematical Challenges.* Washington, DC:
Math. Assoc. Amer., 1995.
Bold, B. *Famous Problems of Geometry and How to Solve
Them.* New York: Dover, 1964.
Brown, K. S. "Most Wanted List of Elementary Unsolved
Problems." http://www.seanet.com/~ksbrown/mwlist.htm.
Chung, F. and Graham, R. *Erdos on Graphs: His Legacy of
Unsolved Problems.* New York: A. K. Peters, 1998.
Cover, T. M. and Gopinath, B. (Eds.). *Open Problems in
Communication and Computation.* New York: Springer-
Verlag, 1987.
Croft, H. T.; Falconer, K. J.; and Guy, R. K. *Unsolved
Problems in Geometry.* New York: Springer-Verlag, p. 3,
1991.
Dixon, J. D. *Problems in Group Theory.* New York: Dover,
1973.
Dörrie, H. *100 Great Problems of Elementary Mathematics:
Their History and Solutions.* New York: Dover, 1965.
Dudeney, H. E. *Amusements in Mathematics.* New York:
Dover, 1917.
Dudeney, H. E. *The Canterbury Puzzles and Other Curious
Problems, 7th ed.* London: Thomas Nelson and Sons, 1949.
Dudeney, H. E. *536 Puzzles & Curious Problems.* New York:
Scribner, 1967.
Eppstein, D. "Open Problems." http://www.ics.uci.edu/~epp-
stein/junkyard/open.html.
Erdos, P. "Some Combinatorial Problems in Geometry." In
Geometry and Differential Geometry (Ed. R. Artzy and
I. Vaisman). New York: Springer-Verlag, pp. 46–3, 1980.
Fenchel, W. (Ed.). "Problems." In *Proc. Colloquium on
Convexity, 1965.* Københavns Univ. Mat. Inst., pp. 308–
25, 1967.
Finch, S. "Unsolved Mathematical Problems." http://
www.mathsoft.com/asolve/.
Gleason, A. M.; Greenwood, R. E.; and Kelly, L. M. *The
William Lowell Putnam Mathematical Competition, Pro-
blems and Solutions: 1938–964.* Washington, DC: Math.
Assoc. Amer., 1980.
Graham, L. A. *Ingenious Mathematical Problems and Meth-
ods.* New York: Dover, 1959.
Graham, L. A. *The Surprise Attack in Mathematical Pro-
blems.* New York: Dover, 1968.
Greitzer, S. L. *International Mathematical Olympiads,
1959–977.* Providence, RI: Amer. Math. Soc., 1978.
Gruber, P. M. and Schneider, R. "Problems in Geometric
Convexity." In *Contributions to Geometry: Proceedings of
the Geometry-Symposium Held in Siegen, June 28, 1978 to
July 1, 1978* (Ed. J. Tölke and J. M. Wills.) Boston, MA:
Birkhäuser, pp. 255–78, 1979.
Guy, R. K. (Ed.). "Problems." In *The Geometry of Metric and
Linear Spaces.* New York: Springer-Verlag, pp. 233–44,
1974.
Guy, R. K. *Unsolved Problems in Number Theory, 2nd ed.*
New York: Springer-Verlag, p. 21, 1994.
Halmos, P. R. *Problems for Mathematicians Young and Old.*
Washington, DC: Math. Assoc. Amer., 1991.
Hardy, K. and Williams, K. S. *The Green Book of Mathema-
tical Problems.* New York: Dover, 1997.
Hardy, K. and Williams, K. S. *The Red Book of Mathema-
tical Problems.* New York: Dover, 1996.
Herman, J.; Kucera Radan, K.; and Simsa, J. *Equations and
Inequalities: Elementary Problems and Theorems in Alge-
bra and Number Theory.* New York: Springer-Verlag,
2000.

Honsberger, R. *Mathematical Gems I.* Washington, DC: Math. Assoc. Amer., 1973.

Honsberger, R. *Mathematical Gems II.* Washington, DC: Math. Assoc. Amer., 1976.

Honsberger, R. *Mathematical Morsels.* Washington, DC: Math. Assoc. Amer., 1979.

Honsberger, R. *Mathematical Gems III.* Washington, DC: Math. Assoc. Amer., 1985.

Honsberger, R. *More Mathematical Morsels.* Washington, DC: Math. Assoc. Amer., 1991.

Honsberger, R. *From Erdos to Kiev.* Washington, DC: Math. Assoc. Amer., 1995.

Honsberger, R. *In Pólya's Footsteps: Miscellaneous Problems and Essays.* Washington, DC: Math. Assoc. Amer., 1997.

Honsberger, R. (Ed.). *Mathematical Plums.* Washington, DC: Math. Assoc. Amer., 1979.

Inter-IREM Commission. *History of Mathematics: Histories of Problems.* Paris: Ellipses, 1997.

Jacoby, O. and Benson, W. H. *Intriguing Mathematical Problems.* New York: Dover, 1998.

Kimberling, C. "Unsolved Problems and Rewards." http://cedar.evansville.edu/~ck6/integer/unsolved.html.

Klee, V. "Some Unsolved Problems in Plane Geometry." *Math. Mag.* **52**, 131–45, 1979.

Klamkin, M. S. *International Mathematical Olympiads, 1978–985 and Forty Supplementary Problems.* Washington, DC: Math. Assoc. Amer., 1986.

Klamkin, M. S. *U.S.A. Mathematical Olympiads, 1972–986.* Washington, DC: Math. Assoc. Amer., 1988.

Kordemsky, B. A. *The Moscow Puzzles: 359 Mathematical Recreations.* New York: Dover, 1992.

Kurschak, J. and Hajos, G. *Hungarian Problem Book, Based on the Eotvos Competitions, Vol. 1: 1894–905.* New York: Random House, 1963.

Kurschak, J. and Hajos, G. *Hungarian Problem Book, Based on the Eotvos Competitions, Vol. 2: 1906–928.* New York: Random House, 1963.

Larson, L. C. *Problem-Solving Through Problems.* New York: Springer-Verlag, 1983.

Meschkowski, H. *Unsolved and Unsolvable Problems in Geometry.* London: Oliver & Boyd, 1966.

Mott-Smith, G. *Mathematical Puzzles for Beginners and Enthusiasts, 2nd rev. ed.* New York: Dover, 1954.

Ogilvy, C. S. *Tomorrow's Math: Unsolved Problems for the Amateur.* New York: Oxford University Press, 1962.

Ogilvy, C. S. "Some Unsolved Problems of Modern Geometry." Ch. 11 in *Excursions in Geometry.* New York: Dover, pp. 143–53, 1990.

Posamentier, A. S. and Salkind, C. T. *Challenging Problems in Algebra.* New York: Dover, 1997.

Posamentier, A. S. and Salkind, C. T. *Challenging Problems in Geometry.* New York: Dover, 1997.

Rabinowitz, S. (Ed.). *Index to Mathematical Problems 1980–984.* Westford, MA: MathPro Press, 1992.

Reid, L. "Southwest Missouri State University's Problem Corner." http://www.math.smsu.edu/~les/POTW.html.

Salkind, C. T. *The Contest Problem Book I: Problems from the Annual High School Contests 1950–960.* New York: Random House, 1961.

Salkind, C. T. *The Contest Problem Book II: Problems from the Annual High School Contests 1961–965.* Washington, DC: Math. Assoc. Amer., 1966.

Salkind, C. T. and Earl, J. M. *The Contest Problem Book III: Annual High School Contests 1966–972.* Washington, DC: Math. Assoc. Amer., 1973.

Shanks, D. *Solved and Unsolved Problems in Number Theory, 4th ed.* New York: Chelsea, 1993.

Shkliarskii, D. O.; Chentzov, N. N.; and Yaglom, I. M. *The U.S.S.R. Olympiad Problem Book: Selected Problems and Theorems of Elementary Mathematics.* New York: Dover, 1993.

Sierpinski, W. *A Selection of Problems in the Theory of Numbers.* New York: Pergamon Press, 1964. Sierpinski, W. *Problems in Elementary Number Theory.* New York: Elsevier, 1980.

Smarandache, F. *Only Problems, Not Solutions!, 4th ed.* Phoenix, AZ: Xiquan, 1993.

Steinhaus, H. *One Hundred Problems in Elementary Mathematics.* New York: Dover, 1979.

Tietze, H. *Famous Problems of Mathematics.* New York: Graylock Press, 1965.

Trigg, C. W. *Mathematical Quickies: 270 Stimulating Problems with Solutions.* New York: Dover, 1985.

Ulam, S. M. *A Collection of Mathematical Problems.* New York: Interscience Publishers, 1960.

Vakil, R. *A Mathematical Mosaic: Patterns and Problem Solving.* Washington, DC: Math. Assoc. Amer., 1997.

van Mill, J. and Reed, G. M. (Eds.). *Open Problems in Topology.* New York: Elsevier, 1990.

Weisstein, E. W. "Books about Mathematics Problems." http://www.treasure-troves.com/books/MathematicsProblems.html.

Procedure

A specific prescription for carrying out a task or solving a problem. Also called an ALGORITHM, METHOD, or TECHNIQUE

See also BISECTION PROCEDURE, MAEHLY'S PROCEDURE

Proclus' Axiom

If a LINE intersects one of two parallel lines, it must intersect the other also. This AXIOM is equivalent to the PARALLEL AXIOM.

References
Dunham, W. "Hippocrates' Quadrature of the Lune." Ch. 1 in *Journey through Genius: The Great Theorems of Mathematics.* New York: Wiley, p. 54, 1990.

Procrustian Stretch

HYPERBOLIC ROTATION

Product

The term "product" refers to the result of one or more MULTIPLICATIONS. For example, the mathematical statement $a \times b = c$ would be read "a TIMES b EQUALS c," where c is the product.

The product symbol is defined by

$$\prod_{i=1}^{n} f_i \equiv f_1 \cdot f_2 \cdots f_n.$$

Useful product identities include

$$\ln\left(\prod_{i=1}^{\infty} f_i\right) = \sum_{i=1}^{\infty} \ln f_i$$

$$\prod_{i=1}^{\infty} f_i = \exp\left(\sum_{i=1}^{\infty} \ln f_i\right).$$

For $0 \le a_i < 1$, then the products $\prod_{i=1}^{\infty}(1 + a_i)$ and $\prod_{i=1}^{\infty}(1 - a_i)$ converge and diverge as $\prod_{i=1}^{\infty} a_i$.

See also CAUCHY PRODUCT, CROSS PRODUCT, DOT PRODUCT, INNER PRODUCT, JORDAN PRODUCT, MATRIX PRODUCT, MULTIPLICATION, NONASSOCIATIVE PRODUCT, OUTER PRODUCT, SUM, TENSOR PRODUCT, TIMES, VECTOR TRIPLE PRODUCT

References

Guy, R. K. "Products Taken over Primes." §B87 in *Unsolved Problems in Number Theory, 2nd ed.* New York: Springer-Verlag, pp. 102–03, 1994.

Product Formula

Let α be a NONZERO RATIONAL NUMBER $\alpha = \pm p_1^{\alpha_1} p_2^{\alpha_2} \cdots p_L^{\alpha_L}$, where $p_1, \ldots, p_L$ are distinct PRIMES, $\alpha_l \in \mathbb{Z}$ and $\alpha_l \neq 0$. Then

$$|\alpha| \prod_{p \text{ prime}} |\alpha|_p = p_1^{\alpha_1} p_2^{\alpha_2} \cdots p_L^{\alpha_L} p_1^{-\alpha_1} p_2^{-\alpha_2} \cdots p_L^{-\alpha_L} = 1.$$

References

Burger, E. B. and Struppeck, T. "Does $\sum_{n=0}^{\infty} \frac{1}{n}$ Really Converge? Infinite Series and p-adic Analysis." *Amer. Math. Monthly* **103**, 565–77, 1996.

Product Log Function

LAMBERT'S W-FUNCTION

Product Neighborhood

TUBULAR NEIGHBORHOOD

Product Rule

The DERIVATIVE identity

$$\frac{d}{dx}[f(x)g(x)] = \lim_{h \to 0} \frac{f(x+h)g(x+h) - f(x)g(x)}{h}$$

$$= \lim_{h \to 0} \left[\frac{f(x+h)g(x+h) - f(x+h)g(x)}{h} \right.$$

$$\left. + \frac{f(x+h)g(x) - f(x)g(x)}{h} \right]$$

$$= \lim_{h \to 0} \left[f(x+h) \frac{g(x+h) - g(x)}{h} \right.$$

$$\left. + g(x) \frac{f(x+h) - f(x)}{h} \right] = f(x)g'(x) + g(x)f'(x).$$

See also CHAIN RULE, EXPONENT LAWS, QUOTIENT RULE

References

Abramowitz, M. and Stegun, C. A. (Eds.). *Handbook of Mathematical Functions with Formulas, Graphs, and Mathematical Tables, 9th printing.* New York: Dover, p. 11, 1972.

Product Set

CARTESIAN PRODUCT

Product Space

A CARTESIAN PRODUCT equipped with a "product topology" is called a product space (or product topological space, or direct product).

See also CARTESIAN PRODUCT

References

Iyanaga, S. and Kawada, Y. (Eds.). "Product Spaces." §408L *Encyclopedic Dictionary of Mathematics.* Cambridge, MA: MIT Press, pp. 1281–282, 1980.

ProductLog

LAMBERT'S W-FUNCTION

Product-Moment Coefficient of Correlation

CORRELATION COEFFICIENT

Program

A precise sequence of instructions designed to accomplish a given task. The implementation of an ALGORITHM on a computer using a programming language is an example of a program.

See also ALGORITHM

Projection

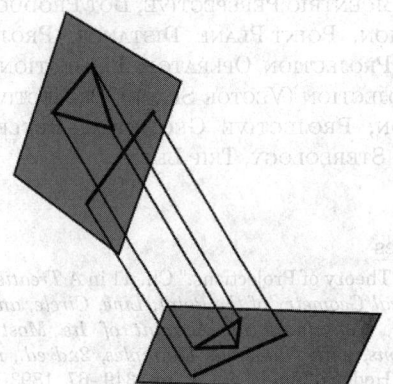

A projection is the transformation of POINTS and LINES in one PLANE onto another PLANE by connecting corresponding points on the two planes with PARAL-LEL lines. This can be visualized as shining a (point) light source (located at infinity) through a translucent sheet of paper and making an image of whatever is drawn on it on a second sheet of paper. The branch of geometry dealing with the properties and invariants of geometric figures under projection is called PROJECTIVE GEOMETRY.

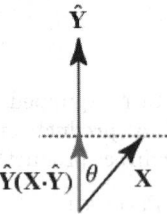

The projection of a VECTOR $\mathbf{a}$ onto a VECTOR $\mathbf{u}$ is given by

$$\mathrm{proj}_{\mathbf{u}}\mathbf{a} = \frac{\mathbf{a} \cdot \mathbf{u}}{|\mathbf{u}|^2}\mathbf{u},$$

where $\mathbf{a} \cdot \mathbf{u}$ is the DOT PRODUCT, and the length of this projection is

$$|\mathrm{proj}_{\mathbf{u}}\mathbf{a}| = \frac{|\mathbf{a} \cdot \mathbf{u}|}{|\mathbf{u}|}.$$

General projections are considered by Foley and VanDam (1983).

The average projected area over all orientations of any ELLIPSOID is 1/4 the total SURFACE AREA. This theorem also holds for any convex solid.

See also BICENTRIC PERSPECTIVE, DOT PRODUCT, MAP PROJECTION, POINT-PLANE DISTANCE, PROJECTION MATRIX, PROJECTION OPERATOR, PROJECTION THEOREM, PROJECTION (VECTOR SPACE), PROJECTIVE COLLINEATION, PROJECTIVE GEOMETRY, REFLECTION, SHADOW, STEREOLOGY, TRIP-LET

References

Casey, J. "Theory of Projections." Ch. 11 in *A Treatise on the Analytical Geometry of the Point, Line, Circle, and Conic Sections, Containing an Account of Its Most Recent Extensions, with Numerous Examples*, 2nd ed., rev. enl. Dublin: Hodges, Figgis, & Co., pp. 349–67, 1893.
Foley, J. D. and VanDam, A. *Fundamentals of Interactive Computer Graphics*, 2nd ed. Reading, MA: Addison-Wesley, 1990.

Projection (Vector Space)

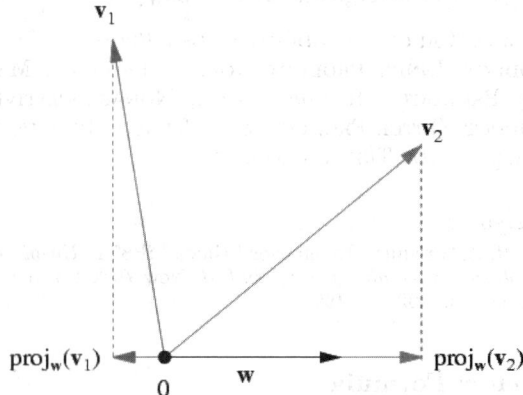

If W is a k-dimensional subspace of a vector space V with inner product $\langle,\rangle$, then it is possible to project vectors from V to W. The most familiar projection is when W is the x-AXIS in the plane. In this case, $P(x, y) = (x, 0)$ is the projection. This projection is an orthogonal projection.

If the SUBSPACE W has an ORTHONORMAL BASIS $\{w_1, \ldots, w_k\}$ then

$$\mathrm{proj}_W(\mathbf{v}) = \sum_{i=1}^{k}\langle \mathbf{v}, w_i\rangle w_i$$

is the orthogonal projection onto W. Any vector $\mathbf{v} \in V$ can be written uniquely as $\mathbf{v} = \mathbf{v}_W + \mathbf{v}_{W^\perp}$, where $\mathbf{v}_W \in W$ and $\mathbf{v}_{W^\perp}$ is in the ORTHOGONAL SUBSPACE $W^\perp$.

A projection is always a LINEAR TRANSFORMATION and can be represented by a PROJECTION MATRIX. In addition, for any projection, there is an inner product for which it is an orthogonal projection.

See also IDEMPOTENT, INNER PRODUCT, PROJECTION MATRIX, ORTHOGONAL SET, PROJECTION, SYMMETRIC MATRIX, VECTOR SPACE

Projection Matrix

A projection matrix P is an $n \times n$ SQUARE MATRIX that gives a PROJECTION from $\mathbb{R}^n$ to a subspace W. The columns of P are the projections of the standard basis vectors, and W is the image of P. A SQUARE MATRIX P is a projection matrix iff $\mathsf{P}^2 = \mathsf{P}$.

The following *Mathematica* function will test if a matrix is a projection matrix.

```
ProjectionMatrixQ[a_List?MatrixQ]  :=  (a.a
= = a)
```

A projection matrix is a SYMMETRIC MATRIX iff the PROJECTION is orthogonal. In an orthogonal projection, any vector v can be written $v = v_W + v_{W^\perp}$, so

$$\langle v, \mathsf{P}w\rangle = \langle v_W, \mathsf{P}w\rangle = \langle \mathsf{P}v, w\rangle. \tag{1}$$

An example of a nonsymmetric projection matrix is

$$P = \begin{bmatrix} 0 & 1 \\ 0 & 1 \end{bmatrix}, \qquad (2)$$

which projects onto the line $y = x$.

The case of a COMPLEX VECTOR SPACE is analogous. A projection matrix is a HERMITIAN MATRIX iff the PROJECTION satisfies

$$\langle v, Pw \rangle = \langle v_W, Pw \rangle = \langle Pv, w \rangle, \qquad (3)$$

where the INNER PRODUCT is the HERMITIAN INNER PRODUCT. Projection operators play a role in quantum mechanics and quantum computing. The following *Mathematica* function gives the Hermitian projection matrix onto a complex subspace, given a basis.

```
< <LinearAlgebra`Orthogonalization';
   HermProjMatrixOntoBasis[a_List?MatrixQ]
:=
   Module[{a1 = GramSchmidt[a, InnerProduct -
> (#1.Conjugate[#2] &) ]},
   Transpose[a1].a1
   ]
```

Any vector in W is fixed by the projection matrix $Pw = w$ for any w in W. Consequently, a projection matrix P has norm equal to one, unless $P = 0$,

$$\|P\| = \sup_{|x|=1} |Px| \le 1. \qquad (4)$$

See also IDEMPOTENT, INNER PRODUCT, PROJECTION (VECTOR SPACE), ORTHOGONAL SET, SYMMETRIC MATRIX

Projection Operator

$$\tilde{p} \equiv |\phi_i(x)\rangle\langle\phi_i(t)|$$

$$\tilde{p} \sum_j c_j |\phi_j(t)\rangle = c_i |\phi_i(x)\rangle$$

$$\sum_i |\phi_i(x)\rangle\langle\phi_i(x)| = 1.$$

See also BRA, KET

Projection Theorem

Let H be a HILBERT SPACE and M a closed subspace of H. Corresponding to any vector $x \in H$, there is a unique vector $m_0 \in M$ such that

$$\|x - m_0\| \le \|x - m\|$$

for all $m \in M$. Furthermore, a necessary and sufficient condition that $m_0 \in M$ be the unique minimizing vector is that $x - m_0$ be orthogonal to M (Luenberger 1997, p. 51).

This theorem can be viewed as a formalization of the result that the closest POINT on a PLANE to a point not

on the PLANE can be found by dropping a perpendicular.

See also POINT-PLANE DISTANCE

References
Luenberger, D. G. *Optimization by Vector Space Methods.* New York: Wiley, 1997.

Projective Algebraic Variety

See also ALGEBRAIC VARIETY, HODGE CONJECTURE

Projective Collineation

A COLLINEATION which transforms every 1-D form projectively. Any COLLINEATION which transforms one range into a projectively related range is a projective collineation. Every PERSPECTIVE COLLINEATION is a projective collineation.

See also COLLINEATION, ELATION, HOMOLOGY (GEOMETRY), PERSPECTIVE COLLINEATION

References
Coxeter, H. S. M. *Introduction to Geometry, 2nd ed.* New York: Wiley, pp. 247–48, 1969.

Projective Correlation

Any CORRELATION which transforms one range into a projectively related PENCIL (or vice versa).

See also CORRELATION (GEOMETRIC), PENCIL

References
Coxeter, H. S. M. *Introduction to Geometry, 2nd ed.* New York: Wiley, p. 248, 1969.

Projective General Linear Group

The projective general linear group $PGL_n(q)$ is the GROUP obtained from the GENERAL LINEAR GROUP $GL_n(q)$ on factoring the scalar MATRICES contained in that group.

See also GENERAL LINEAR GROUP, PROJECTIVE GENERAL ORTHOGONAL GROUP, PROJECTIVE GENERAL UNITARY GROUP

References
Conway, J. H.; Curtis, R. T.; Norton, S. P.; Parker, R. A.; and Wilson, R. A. "The Groups $GL_n(q)$, $SL_n(q)$, $PGL_n(q)$, and $PSL_n(q) = L_n(q)$." §2.1 in *Atlas of Finite Groups: Maximal Subgroups and Ordinary Characters for Simple Groups.* Oxford, England: Clarendon Press, p. x, 1985.

Projective General Orthogonal Group

The projective general orthogonal group $PGO_n(q)$ is the GROUP obtained from the GENERAL ORTHOGONAL GROUP $GO_n(q)$ on factoring the scalar MATRICES contained in that group.

See also GENERAL ORTHOGONAL GROUP, PROJECTIVE

GENERAL LINEAR GROUP, PROJECTIVE GENERAL UNITARY GROUP

References

Conway, J. H.; Curtis, R. T.; Norton, S. P.; Parker, R. A.; and Wilson, R. A. "The Groups $GO_n(q)$, $SO_n(q)$, $PGO_n(q)$, and $PSO_n(q)$, and $O_n(q)$." §2.4 in *Atlas of Finite Groups: Maximal Subgroups and Ordinary Characters for Simple Groups.* Oxford, England: Clarendon Press, pp. xi-xii, 1985.

Projective General Unitary Group

The projective general unitary group $PGU_n(q)$ is the GROUP obtained from the GENERAL UNITARY GROUP $GU_n(q)$ on factoring the scalar MATRICES contained in that group.

See also GENERAL UNITARY GROUP, PROJECTIVE GENERAL LINEAR GROUP, PROJECTIVE GENERAL ORTHOGONAL GROUP, PROJECTIVE GENERAL UNITARY GROUP

References

Conway, J. H.; Curtis, R. T.; Norton, S. P.; Parker, R. A.; and Wilson, R. A. "The Groups $GU_n(q)$, $SU_n(q)$, $PGU_n(q)$, and $PSU_n(q) = U_n(q)$." §2.2 in *Atlas of Finite Groups: Maximal Subgroups and Ordinary Characters for Simple Groups.* Oxford, England: Clarendon Press, p. x, 1985.

Projective Geometry

The branch of GEOMETRY dealing with the properties and invariants of geometric figures under PROJECTION. In older literature, projective geometry is sometimes called "higher geometry," "geometry of position," or "descriptive geometry" (Cremona 1960, pp. v-vi).

The most amazing result arising in projective geometry is the DUALITY PRINCIPLE, which states that a duality exists between theorems such as PASCAL'S THEOREM and BRIANCHON'S THEOREM which allows one to be instantly transformed into the other. More generally, *all* the propositions in projective geometry occur in dual pairs, which have the property that, starting from either proposition of a pair, the other can be immediately inferred by interchanging the parts played by the words "POINT" and "LINE."

The AXIOMS of projective geometry are:

1. If A and B are distinct points on a PLANE, there is at least one LINE containing both A and B.

2. If A and B are distinct points on a PLANE, there is not more than one LINE containing both A and B.

3. Any two LINES in a PLANE have at least one point of the PLANE (which may be the POINT AT INFINITY) in common.

4. There is at least one LINE on a PLANE.

5. Every LINE contains at least three points of the PLANE.

6. All the points of the PLANE do not belong to the same LINE

(Veblen and Young 1910-8, Kasner and Newman 1989).

See also COLLINEATION, DESARGUES' THEOREM, FUNDAMENTAL THEOREM OF PROJECTIVE GEOMETRY, INVOLUTION (LINE), PENCIL, PERSPECTIVITY, PROJECTION, PROJECTIVITY, RANGE (LINE SEGMENT), SECTION (PENCIL)

References

Birkhoff, G. and Mac Lane, S. "Projective Geometry." §9.14 in *A Survey of Modern Algebra, 5th ed.* New York: Macmillan, pp. 275-79, 1996.
Casey, J. "Theory of Projections." Ch. 11 in *A Treatise on the Analytical Geometry of the Point, Line, Circle, and Conic Sections, Containing an Account of Its Most Recent Extensions, with Numerous Examples, 2nd ed., rev. enl.* Dublin: Hodges, Figgis, & Co., pp. 349-67, 1893.
Chasles, M. *Aperçu historique.*
Chasles, M. *Traité de Géométrie supérieure.* Paris, 1852.
Coxeter, H. S. M. *Projective Geometry, 2nd ed.* New York: Springer-Verlag, 1987.
Cremona, L. *Elements of Projective Geometry, 3rd ed.* New York: Dover, 1960.
Kadison, L. and Kromann, M. T. *Projective Geometry and Modern Algebra.* Boston, MA: Birkhäuser, 1996.
Kasner, E. and Newman, J. R. *Mathematics and the Imagination.* Redmond, WA: Microsoft Press, pp. 150-51, 1989.
Lachlan, R. *An Elementary Treatise on Modern Pure Geometry.* London: Macmillian, pp. 119-27, 1893.
Ogilvy, C. S. "Projective Geometry." Ch. 7 in *Excursions in Geometry.* New York: Dover, pp. 86-10, 1990.
Pappas, T. "Art & Projective Geometry." *The Joy of Mathematics.* San Carlos, CA: Wide World Publ./Tetra, pp. 66-7, 1989.
Pedoe, D. and Sneddon, I. A. *An Introduction to Projective Geometry.* New York: Pergamon, 1963.
Poncelet, J.-V. *Traité des Propriétés Projectives.* Paris, 1822.
Reye. *Geometrie der Lage, 2nd ed.* Hannover, Germany, 1877.
Semple, J. G. *Algebraic Projective Geometry.* Oxford, England: Oxford University Press, 1998.
Seidenberg, A. *Lectures in Projective Geometry.* Princeton, NJ: Van Nostrand, 1962.
Staudt, K. G. C. von. *Geometrie der Lage.* Nürnberg, Germany, 1847.
Steiner, J. *Systematische Entwicklung der Abhängigkeit geometrischer Gestalten von einander.* Berlin, 1832.
Struik, D. *Lectures on Projected Geometry.* Reading, MA: Addison-Wesley, 1998.
Veblen, O. and Young, J. W. *Projective Geometry, 2 vols.* Boston, MA: Ginn, 1910-8.

Weisstein, E. W. "Books about Projective Geometry." http://www.treasure-troves.com/books/ProjectiveGeometry.html.

Whitehead, A. N. *The Axioms of Projective Geometry*. New York: Hafner, 1960.

Projective Plane

A projective plane is derived from a usual PLANE by addition of a LINE AT INFINITY. Just as a straight line in projective geometry contains of single POINT AT INFINITY at which the endpoints meet, a plane in projective geometry contains a single LINE AT INFINITY at which the edges of the PLANE meet. A projective plane can be constructed by gluing both pairs of opposite edges of a RECTANGLE together giving both pairs a half-twist. It is a one-sided surface, but cannot be realized in 3-D space without crossing itself.

A finite projective plane of order n is formally defined as a set of $n^2 + n + 1$ POINTS with the properties that:

1. Any two POINTS determine a LINE,
2. Any two LINES determine a POINT,
3. Every POINT has $n + 1$ LINES on it, and
4. Every LINE contains $n + 1$ POINTS.

(Note that some of these properties are redundant.) A projective plane is therefore a SYMMETRIC $(n^2 + n + 1, n + 1, 1)$ BLOCK DESIGN. An AFFINE PLANE of order n exists IFF a projective plane of order n exists.

A finite projective plane exists when the order n is a POWER of a PRIME, i.e., $n = p^a$ for $a \geq 1$. It is conjectured that these are the *only* possible projective planes, but proving this remains one of the most important unsolved problems in COMBINATORICS. The first few orders which *are* powers of primes are 2, 3, 4, 5, 7, 8, 9, 11, 13, 16, ... (Sloane's A000961). The first few orders which are not of this form are 6, 10, 12, 14, 15, ... (Sloane's A024619).

The smallest finite projective plane is of order $n = 2$, and consists of the 7_3 CONFIGURATION known as the FANO PLANE. The remarkable BRUCK-RYSER-CHOWLA THEOREM says that if a projective plane of order n exists, and $n = 1$ or 2 (mod 4), then n is the sum of two SQUARES. This rules out $n = 6$. By answering LAM'S PROBLEM in the negative using massive computer calculations on top of some mathematics, it has been proved that there are no finite projective planes of order 10 (Lam 1991). The status of the order 12 projective plane remains open.

The projective plane of order 2, also known as the FANO PLANE, is denoted PG(2, 2). It has INCIDENCE MATRIX

$$\begin{bmatrix} 1 & 1 & 1 & 0 & 0 & 0 & 0 \\ 1 & 0 & 0 & 1 & 1 & 0 & 0 \\ 1 & 0 & 0 & 0 & 0 & 1 & 1 \\ 0 & 1 & 0 & 1 & 0 & 1 & 0 \\ 0 & 1 & 0 & 0 & 1 & 0 & 1 \\ 0 & 0 & 1 & 1 & 0 & 0 & 1 \\ 0 & 0 & 1 & 0 & 1 & 1 & 0 \end{bmatrix}.$$

Every row and column contains 3 1s, and any pair of rows/columns has a single 1 in common.

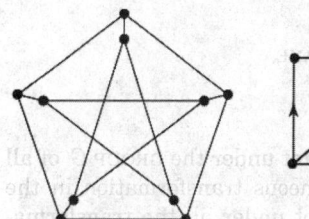 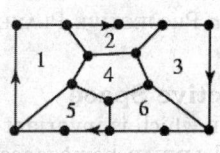

Petersen graph   *embedding on the projective plane*

The projective plane has EULER CHARACTERISTIC 1, and the HEAWOOD CONJECTURE therefore shows that any set of regions on it can be colored using six colors only (Saaty 1986). The Petersen graph provides a 6-color coloring of the PROJECTIVE PLANE.

See also AFFINE PLANE, BLOCK DESIGN, BRUCK-RYSER-CHOWLA THEOREM, CONFIGURATION, FANO PLANE, LAM'S PROBLEM, MAP COLORING, MOUFANG PLANE, PROJECTIVE PLANE PK2, PROJECTIVE SPACE, REAL PROJECTIVE PLANE, SYMMETRIC BLOCK DESIGN

References

Ball, W. W. R. and Coxeter, H. S. M. *Mathematical Recreations and Essays, 13th ed.* New York: Dover, pp. 281–87, 1987.

Bondy, J. A. and Murty, U. S. R. *Graph Theory with Applications*. New York: North Holland, p. 243, 1976.

Bruck, R. H. and Ryser, H. J. "The Nonexistence of Certain Finite Projective Planes." *Canad. J. Math.* **1**, 88–3, 1949.

Lam, C. W. H. "The Search for a Finite Projective Plane of Order 10." *Amer. Math. Monthly* **98**, 305–18, 1991.

Lindner, C. C. and Rodger, C. A. *Design Theory*. Boca Raton, FL: CRC Press, 1997.

Pinkall, U. "Models of the Real Projective Plane." Ch. 6 in *Mathematical Models from the Collections of Universities and Museums* (Ed. G. Fischer). Braunschweig, Germany: Vieweg, pp. 63–7, 1986.

Saaty, T. L. and Kainen, P. C. *The Four-Color Problem: Assaults and Conquest*. New York: Dover, p. 45, 1986.

Sloane, N. J. A. Sequences A000961/M0517 and A024619 in "An On-Line Version of the Encyclopedia of Integer Sequences." http://www.research.att.com/~njas/sequences/eisonline.html.

Wells, D. *The Penguin Dictionary of Curious and Interesting Geometry*. London: Penguin, pp. 72 and 195–97, 1991.

Projective Plane Dissection

Virtually nothing is known about dissection of a PROJECTIVE PLANE using unequal squares.

See also CYLINDER DISSECTION, KLEIN BOTTLE DISSECTION, MÖBIUS STRIP DISSECTION, PERFECT SQUARE DISSECTION, TORUS DISSECTION

References
Stewart, I. "Squaring the Square." *Sci. Amer.* **277**, 94–6, July 1997.

Projective Plane PK²

The 2-D SPACE consisting of the set of TRIPLES

$$\{(a,\ b,\ c): a,\ b,\ c \in K,\ \text{not all zero}\},$$

where triples which are SCALAR multiples of each other are identified.

See also PROJECTIVE PLANE

Projective Space

A SPACE which is invariant under the GROUP G of all general LINEAR homogeneous transformation in the SPACE concerned, but not under all the transformations of any GROUP containing G as a SUBGROUP.

A projective space is the space of 1-D VECTOR SUBSPACES of a given VECTOR SPACE. For REAL VECTOR SPACES, the NOTATION $\mathbb{RP}^n$ or $\mathbb{P}^n$ denotes the REAL projective space of dimension n (i.e., the SPACE of 1-D VECTOR SUBSPACES of $\mathbb{R}^{n+1}$) and $\mathbb{CP}^n$ denotes the COMPLEX projective space of COMPLEX dimension n (i.e., the space of 1-D COMPLEX VECTOR SUBSPACES of $\mathbb{C}^{n+1}$). $\mathbb{P}^n$ can also be viewed as the set consisting of $\mathbb{R}^n$ together with its POINTS AT INFINITY.

See also PROJECTIVE SPACE

Projective Special Linear Group

The projective special linear group $PSL_n(q)$ is the GROUP obtained from the SPECIAL LINEAR GROUP $SL_n(q)$ on factoring by the SCALAR MATRICES contained in that GROUP. It is SIMPLE for $n \geq 2$ except for

$$PSL_2(2) = S_3,$$

$$PSL_3(3) = A_4,$$

and is therefore also denoted $L_n(Q)$.

See also PROJECTIVE SPECIAL ORTHOGONAL GROUP, PROJECTIVE SPECIAL UNITARY GROUP, SPECIAL LINEAR GROUP

References
Conway, J. H.; Curtis, R. T.; Norton, S. P.; Parker, R. A.; and Wilson, R. A. "The Groups $GL_n(q)$, $SL_n(q)$, $PGL_n(q)$, and $PSL_n(q) = L_n(q)$." §2.1 in *Atlas of Finite Groups: Maximal Subgroups and Ordinary Characters for Simple Groups.* Oxford, England: Clarendon Press, p. x, 1985.

Projective Special Orthogonal Group

The projective special orthogonal group $PSO_n(q)$ is the GROUP obtained from the SPECIAL ORTHOGONAL GROUP $SO_n(q)$ on factoring by the SCALAR MATRICES contained in that GROUP. In general, this GROUP is not SIMPLE.

See also PROJECTIVE SPECIAL LINEAR GROUP, PROJECTIVE SPECIAL UNITARY GROUP, SPECIAL ORTHOGONAL GROUP

References
Conway, J. H.; Curtis, R. T.; Norton, S. P.; Parker, R. A.; and Wilson, R. A. "The Groups $GO_n(q)$, $SO_n(q)$, $PGO_n(q)$, and $PSO_n(q)$, and $O_n(q)$." §2.4 in *Atlas of Finite Groups: Maximal Subgroups and Ordinary Characters for Simple Groups.* Oxford, England: Clarendon Press, pp. xi-xii, 1985.

Projective Special Unitary Group

The projective special unitary group $PSU_n(q)$ is the GROUP obtained from the SPECIAL UNITARY GROUP $SU_n(q)$ on factoring by the SCALAR MATRICES contained in that GROUP. $PSU_n(q)$ is SIMPLE except for

$$PSU_2(2) = S_3$$

$$PSU_2(3) = A_4$$

$$PSU_3(2) = 3^2 : Q_8,$$

so it is given the simpler name $U_n(q)$, with $U_2(q) = L_2(q)$.

See also PROJECTIVE SPECIAL LINEAR GROUP, PROJECTIVE SPECIAL ORTHOGONAL GROUP, SPECIAL UNITARY GROUP

References
Conway, J. H.; Curtis, R. T.; Norton, S. P.; Parker, R. A.; and Wilson, R. A. "The Groups $GU_n(q)$, $SU_n(q)$, $PGU_n(q)$, and $PSU_n(q) = U_n(q)$." §2.2 in *Atlas of Finite Groups: Maximal Subgroups and Ordinary Characters for Simple Groups.* Oxford, England: Clarendon Press, p. x, 1985.

Projective Symplectic Group

The projective symplectic group $PSp_n(q)$ is the GROUP obtained from the SYMPLECTIC GROUP $Sp_n(q)$ on factoring by the SCALAR MATRICES contained in that GROUP. $PSp_{2m}(q)$ is SIMPLE except for

$$psp_2(2) = s_3$$

$$psp_2(3) = a_4$$

$$psp_4(2) = s_6,$$

so it is given the simpler name $s_{2m}(q)$, with $s_2(q) = l_2(q)$.

References
Conway, J. H.; Curtis, R. T.; Norton, S. P.; Parker, R. A.; and Wilson, R. A. "The Groups $sp_n(q)$ and $psp_n(q) = s_n q$." §2.3 in *Atlas of Finite Groups: Maximal Subgroups and Ordinary Characters for Simple Groups.* Oxford, England: Clarendon Press, pp. x-xi, 1985.

Projective Variety

PROJECTIVE ALGEBRAIC VARIETY

Projectivity

The product of any number of PERSPECTIVITIES.

See also INVOLUTION (TRANSFORMATION), PERSPECTIVITY

Projectivization

Given a VECTOR SPACE V, its projectivization $P(V)$, sometimes written $P(V-0)$, is the set of EQUIVALENCE CLASSES $x \sim \lambda x$ for any $\lambda \neq 0$ in $V-0$. For example, COMPLEX PROJECTIVE SPACE has HOMOGENEOUS COORDINATES $[x_0, \ldots, x_n]$, with not all $x_i = 0$.

The projectivization is a MANIFOLD with one less dimension than V. In fact, it is covered by the $n+1$ affine COORDINATE CHARTS,

$$U_0 = \{[1, x_1, \ldots, x_n]\}, \ldots, U_n = \{[x_0, \ldots, x_{n-1}, 1]\}.$$

See also COMPLEX PROJECTIVE SPACE, MANIFOLD, VECTOR SPACE

Prolate Cycloid

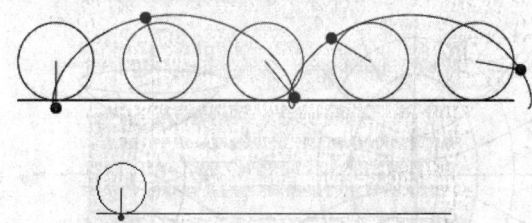

The path traced out by a fixed point at a RADIUS $b > a$, where a is the RADIUS of a rolling CIRCLE, also sometimes called an EXTENDED CYCLOID. The prolate cycloid contains loops, and has PARAMETRIC EQUATIONS

$$x = a\phi - b \sin \phi \tag{1}$$

$$y = a - b \cos \phi. \tag{2}$$

The ARC LENGTH from $\phi = 0$ is

$$s = 2(a+b)E(u), \tag{3}$$

where

$$\sin\left(\tfrac{1}{2}\phi\right) = \operatorname{sn} u \tag{4}$$

$$k^2 = \frac{4ab}{(a+c)^2}. \tag{5}$$

See also CURTATE CYCLOID, CYCLOID, TROCHOID

References

Beyer, W. H. *CRC Standard Mathematical Tables, 28th ed.* Boca Raton, FL: CRC Press, p. 216, 1987.

Harris, J. W. and Stocker, H. *Handbook of Mathematics and Computational Science.* New York: Springer-Verlag, p. 325, 1998.

Lawrence, J. D. *A Catalog of Special Plane Curves.* New York: Dover, pp. 192 and 194–97, 1972.

Lockwood, E. H. *A Book of Curves.* Cambridge, England: Cambridge University Press, p. 146, 1967.

Steinhaus, H. *Mathematical Snapshots, 3rd ed.* New York: Dover, pp. 147–48, 1999.

Zwillinger, D. (Ed.). *CRC Standard Mathematical Tables and Formulae.* Boca Raton, FL: CRC Press, p. 292, 1995.

Prolate Cycloid Evolute

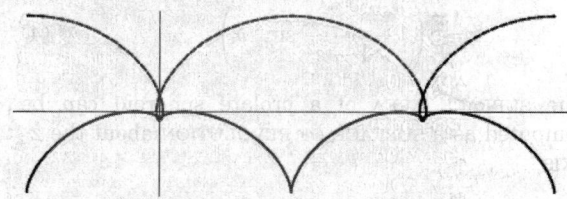

The EVOLUTE of the PROLATE CYCLOID is given by

$$x = \frac{a[-2b\phi + 2a\phi \cos \phi - 2a \sin \phi + b \sin(2\phi)]}{2(a \cos \phi - b)}$$

$$y = \frac{a(a - b \cos \phi)^2}{b(a \cos \phi - b)}.$$

Prolate Spheroid

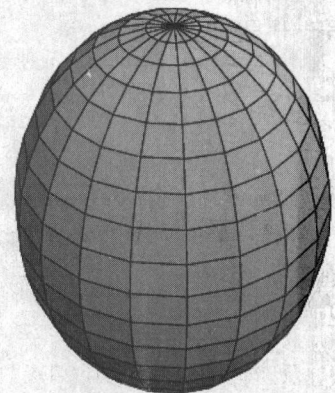

A SPHEROID which is "pointy" instead of "squashed," i.e., one for which the polar radius c is greater than the equatorial radius a, so $c > a$ (called "spindle-shaped ellipsoid" by Tietze 1965, p. 27). A symmetrical egg (i.e., with the same shape at both ends) would approximate a prolate spheroid. A prolate spheroid is a SURFACE OF REVOLUTION obtained by rotating an ELLIPSE about its major axis (Hilbert and Cohn-Vossen 1999, p. 10), and has Cartesian equations

$$\frac{x^2 + y^2}{a^2} + \frac{z^2}{c^2} = 1. \tag{1}$$

The ELLIPTICITY of the prolate spheroid is defined by

$$e \equiv \sqrt{\frac{c^2 - a^2}{c^2}} = \frac{\sqrt{c^2 - a^2}}{c} = \sqrt{1 - \frac{a^2}{c^2}}, \tag{2}$$

so that

$$1 - e^2 = \frac{a^2}{c^2}. \tag{3}$$

Then

$$r = a \left(1 + \frac{e^2}{1 - e^2} \sin^2 \delta \right)^{-1/2}. \tag{4}$$

The SURFACE AREA of a prolate spheroid can be computed as a SURFACE OF REVOLUTION about the z-AXIS,

$$S = 2\pi \int r(z) \sqrt{1 + [r'(z)]^2} \, dz \tag{5}$$

with radius as a function of z given by

$$r(z) = a \sqrt{1 - \left(\frac{z}{c}\right)^2}. \tag{6}$$

The INTEGRAND is then

$$r\sqrt{1 + r'^2} = a \sqrt{1 + \frac{(a-c)(a+c)z^2}{c^4}}, \tag{7}$$

and the integral is given by

$$S = 2\pi a \int_{-c}^{c} \sqrt{1 + \frac{(a-c)(a+c)z^2}{c^4}} \, dz$$

$$= 2\pi a^2 + \frac{2\pi a c^2}{\sqrt{c^2 - a^2}} \sin^{-1}\left(\frac{\sqrt{c^2 - a^2}}{c}\right). \tag{8}$$

Using the identity

$$\sqrt{c^2 - a^2} = ce \tag{9}$$

gives

$$S = 2\pi a^2 + 2\pi \frac{ac}{e} \sin^{-1} e \tag{10}$$

(Beyer 1987, p. 131). Note that this is the conventional form in which the surface area of an prolate spheroid is written, although it is formally equivalent to the conventional form for the OBLATE SPHEROID via the identity

$$\frac{c^2 \pi}{e(a, c)} \ln \left[\frac{1 + e(a, c)}{1 - e(a, c)} \right] = \frac{2\pi ac}{e(c, a)} \sin^{-1}[e(c, a)], \tag{11}$$

where $e(x, y)$ is defined by

$$e(x, y) \equiv \sqrt{1 - \frac{x^2}{y^2}}. \tag{12}$$

The VOLUME of an prolate spheroid can be computed from the formula for a general ELLIPSOID with $b = a$,

$$V = \tfrac{4}{3} \pi a^2 c \tag{13}$$

(Beyer 1987, p. 131).

See also DARWIN-DE SITTER SPHEROID, ELLIPSOID, LEMON, OBLATE SPHEROID, PROLATE SPHEROIDAL COORDINATES, SPHERE, SPHEROID

References

Beyer, W. H. *CRC Standard Mathematical Tables,* 28th ed. Boca Raton, FL: CRC Press, 1987.

Hilbert, D. and Cohn-Vossen, S. *Geometry and the Imagination.* New York: Chelsea, p. 10, 1999.

Tietze, H. *Famous Problems of Mathematics: Solved and Unsolved Mathematics Problems from Antiquity to Modern Times.* New York: Graylock Press, p. 27, 1965.

Wrinch, D. M. "Inverted Prolate Spheroids." *Philos. Mag.* **280**, 1061–070, 1932.

Prolate Spheroidal Coordinates

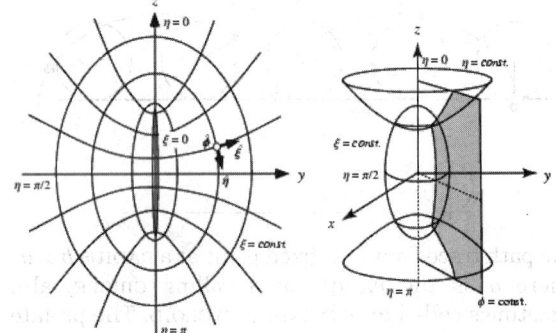

A system of CURVILINEAR COORDINATES in which two sets of coordinate surfaces are obtained by revolving the curves of the ELLIPTIC CYLINDRICAL COORDINATES about the x-AXIS, which is relabeled the z-AXIS. The third set of coordinates consists of planes passing through this axis.

$$x = a \sinh \xi \sin \eta \cos \phi \tag{1}$$

$$y = a \sinh \xi \sin \eta \sin \phi \tag{2}$$

$$z = a \cosh \xi \cos \eta, \tag{3}$$

where $\xi \in [0, \infty)$, $\eta \in [0, \pi]$, and $\phi \in [0, 2\pi)$. Note that several conventions are in common use; Arfken (1970) uses (u, v, φ) instead of (ξ, η, ϕ), and Moon and Spencer (1988, p. 28) use (η, θ, ψ).

In this coordinate system, the SCALE FACTORS are

$$h_\xi = a \sqrt{\sinh^2 \xi + \sin^2 \eta} \tag{4}$$

$$h_\eta = a \sqrt{\sinh^2 \xi + \sin^2 \eta} \tag{5}$$

$$h_\phi = a \sinh \zeta \sin \eta. \qquad (6)$$

The LAPLACIAN is

$$\nabla^2 f = \frac{1}{\sin \eta \sinh \zeta (\sin^2 \eta + \sinh^2 \zeta)}$$

$$\times \left\{ \frac{\partial}{\partial \zeta} \left(\sin \eta \sinh \zeta \, \frac{\partial f}{\partial \zeta} \right) + \frac{\partial}{\partial \eta} \left(\sin \eta \sinh \zeta \, \frac{\partial f}{\partial \eta} \right) \right.$$

$$\left. + \frac{\partial}{\partial \phi} \left[(\operatorname{csch} \zeta \sin \eta + \csc \eta \sinh \zeta) \frac{\partial f}{\partial \phi} \right] \right\}. \qquad (7)$$

$$= \frac{1}{\sin^2 \eta + \sinh^2 \zeta} \left[(\csc^2 \eta + \operatorname{csch}^2 \zeta) \frac{\partial^2 f}{\partial \zeta^2} + \cot \eta \, \frac{\partial f}{\partial \eta} \right.$$

$$\left. + \frac{\partial^2 f}{\partial \eta^2} + \coth \zeta \, \frac{\partial f}{\partial \zeta} + \frac{\partial^2 f}{\partial \zeta^2} \right] \qquad (8)$$

An alternate form useful for "two-center" problems is defined by

$$\xi_1 = \cosh \zeta \qquad (9)$$

$$\xi_2 = \cos \eta \qquad (10)$$

$$\xi_3 = \phi, \qquad (11)$$

where $\xi_1 \in [1, \infty]$, $\xi_2 \in [-1, 1]$, and $\xi_3 \in [0, 2\pi)$ (Abramowitz and Stegun 1972). In these coordinates,

$$z = a \xi_1 \xi_2 \qquad (12)$$

$$x = a \sqrt{(\xi_1^2 - 1)(1 - \xi_2^2)} \cos \xi_3 \qquad (13)$$

$$y = a \sqrt{(\xi_1^2 - 1)(1 - \xi_2^2)} \sin \xi_3. \qquad (14)$$

In terms of the distances from the two FOCI,

$$\xi_1 = \frac{r_1 + r_2}{2a} \qquad (15)$$

$$\xi_2 = \frac{r_1 - r_2}{2a} \qquad (16)$$

$$2a = r_{12} \qquad (17)$$

The SCALE FACTORS are

$$h_{\xi_1} = a \sqrt{\frac{\xi_1^2 - \xi_2^2}{\xi_1^2 - 1}} \qquad (18)$$

$$h_{\xi_2} = a \sqrt{\frac{\xi_1^2 - \xi_2^2}{1 - \xi_2^2}} \qquad (19)$$

$$h_{\xi_3} = a \sqrt{(\xi_1^2 - 1)(1 - \xi_2^2)}, \qquad (20)$$

and the LAPLACIAN is

$$\nabla^2 f = \frac{1}{a^2} \left\{ \frac{1}{\xi_1^2 - \xi_2^2} \frac{\partial}{\partial \xi_1} \left[(\xi_1^2 - 1) \frac{\partial f}{\partial \xi_1} \right] \right.$$

$$+ \frac{1}{\xi_1^2 - \xi_2^2} \frac{\partial}{\partial \xi_2} \left[(1 - \xi_2^2) \frac{\partial f}{\partial} \right]$$

$$\left. + \frac{1}{(\xi_1^2 - 1)(1 - \xi_2^2)} \frac{\partial^2 f}{d \xi_2^2} \right\}. \qquad (21)$$

The HELMHOLTZ DIFFERENTIAL EQUATION is separable in prolate spheroidal coordinates.

See also HELMHOLTZ DIFFERENTIAL EQUATION–PROLATE SPHEROIDAL COORDINATES, LATITUDE, LONGITUDE, OBLATE SPHEROIDAL COORDINATES, SPHERICAL COORDINATES

References

Abramowitz, M. and Stegun, C. A. (Eds.). "Definition of Prolate Spheroidal Coordinates." §21.2 in *Handbook of Mathematical Functions with Formulas, Graphs, and Mathematical Tables, 9th printing.* New York: Dover, p. 752, 1972.

Arfken, G. "Prolate Spheroidal Coordinates (u, v, ϕ)." §2.10 in *Mathematical Methods for Physicists, 2nd ed.* Orlando, FL: Academic Press, pp. 103–07, 1970.

Byerly, W. E. *An Elementary Treatise on Fourier's Series, and Spherical, Cylindrical, and Ellipsoidal Harmonics, with Applications to Problems in Mathematical Physics.* New York: Dover, pp. 243–44, 1959.

Moon, P. and Spencer, D. E. "Prolate Spheroidal Coordinates (η, θ, ψ)." Table 1.06 in *Field Theory Handbook, Including Coordinate Systems, Differential Equations, and Their Solutions, 2nd ed.* New York: Springer-Verlag, pp. 28–0, 1988.

Morse, P. M. and Feshbach, H. *Methods of Theoretical Physics, Part I.* New York: McGraw-Hill, p. 661, 1953.

Wrinch, D. M. "Inverted Prolate Spheroids." *Philos. Mag.* **280**, 1061–070, 1932.

Prolate Spheroidal Wave Function

The WAVE EQUATION in PROLATE SPHEROIDAL COORDINATES is

$$\nabla^2 \Phi + k^2 \Phi = \frac{\partial}{\partial \xi_1} \left[(\xi_1^2 - 1) \frac{\partial \Phi}{\partial \xi_1} \right] + \frac{\partial}{\partial \xi_2} \left[(1 - \xi_2^2) \frac{\partial \Phi}{\partial \xi_2} \right]$$

$$+ \frac{\xi_1^2 - \xi_2^2}{(\xi_1^2 - 1)(1 - x_2^2)} \frac{\partial^2 \Phi}{\partial \phi^2} + c^2(\xi_1^2 - \xi_2^2)\Phi = 0, \quad (1)$$

where

$$c \equiv \tfrac{1}{2} ak. \qquad (2)$$

Substitute in a trial solution

$$\Phi = R_{mn}(c, \xi_1) S_{mn}(c, \xi_2) \frac{\cos}{\sin}(m\phi) \qquad (3)$$

$$\frac{d}{d\xi_1} \left[(\xi_1^2 - 1) \frac{d}{d\xi_1} R_{mn}(c, \xi_1) \right]$$

$$-\left(\lambda_{mn} - c^2\xi_1^2 + \frac{m^2}{\xi_1^2-1}\right)R_{mn}(c,\ \xi_1) = 0. \qquad (4)$$

The radial differential equation is

$$\frac{d}{d\xi_2}\left[(\xi_2^2-1)\frac{d}{d\xi_2}\,S_{mn}(c,\ \xi_2)\right]$$

$$-\left(\lambda_{mn} - c^2\xi_2^2 + \frac{m^2}{\xi_2^2-1}\right)R_{mn}(c,\ \xi_2) = 0. \qquad (5)$$

and the angular differential equation is

$$\frac{d}{d\xi_2}\left[(1-\xi_2^2)\frac{d}{d\xi_2}\,S_{mn}(c,\ \xi_2)\right]$$

$$-\left(\lambda_{mn} - c^2\xi_2^2 + \frac{m^2}{1-\xi_2^2}\right)R_{mn}(c,\ \xi_2) = 0. \qquad (6)$$

Note that these are identical (except for a sign change). The prolate angular function of the first kind is given by

$$S_{mn}^{(1)} = \begin{cases} \sum_{r=1,\,3,\,\dots}^{\infty} d_r(c)P_{m+r}^m(\eta) & \text{for } n-m \text{ odd} \\ \sum_{r=0,\,2,\,\dots}^{\infty} d_r(c)P_{m+r}^m(\eta) & \text{for } n-m \text{ even,} \end{cases} \qquad (7)$$

where $P_k^m(\eta)$ is an associated LEGENDRE POLYNOMIAL. The prolate angular function of the second kind is given by

$$S_{mn}^{(2)} = \begin{cases} \sum_{r=\dots,\,-1,\,1,\,3,\,\dots}^{\infty} d_r(c)Q_{m+r}^m(\eta) & \text{for } n-m \text{ odd} \\ \sum_{r=\dots,\,-2,\,0,\,2,\,\dots}^{\infty} d_r(c)Q_{m+r}^m(\eta) & \text{for } n-m \text{ even,} \end{cases} \qquad (8)$$

where $Q_k^m(\eta)$ is an associated LEGENDRE FUNCTION OF THE SECOND KIND and the COEFFICIENTS d_r satisfy the RECURRENCE RELATION

$$\alpha_k d_{k+2} + (\beta_k - \lambda_{mn})d_k + \gamma_k d_{k-2} = 0, \qquad (9)$$

with

$$\alpha_k = \frac{(2m+k+2)(2m+k+1)c^2}{(2m+2k+3)(2m+2k+3)} \qquad (10)$$

$$\beta_k = (m+k)(m+k+1)$$

$$+\frac{2(m+k)(m+k+1) - 2m^2 - 1}{(2m+2k-1)(2m+2k+3)}c^2 \qquad (11)$$

$$\gamma_k = \frac{k(k-1)c^2}{(2m+2k-3)(2m+2k-1)}. \qquad (12)$$

Various normalization schemes are used for the ds (Abramowitz and Stegun 1972, p. 758). Meixner and Schäfke (1954) use

$$\int_{-1}^{1}[S_{mn}(c,\ \eta)]^2\,d\eta = \frac{2}{2n+1}\frac{(n+m)!}{(n-m)!}. \qquad (13)$$

Stratton *et al.* (1956) use

$$\frac{(n+m)!}{(n-m)!} = \begin{cases} \sum_{r=1,\,3,\,\dots}^{\infty}\dfrac{(r+2m)!}{r!}\,d_r & \text{for } n-m \text{ odd} \\ \sum_{r=0,\,2,\,\dots}^{\infty}\dfrac{(r+2m)!}{r!}\,d_r & \text{for } n-m \text{ even.} \end{cases} \qquad (14)$$

Flammer (1957) uses

$$S_{mn}(c,\ 0) = \begin{cases} P_n^{m+1}(0) & \text{for } n-m \text{ odd} \\ P_n^m(0) & \text{for } n-m \text{ even.} \end{cases} \qquad (15)$$

See also OBLATE SPHEROIDAL WAVE FUNCTION, SPHEROIDAL WAVE FUNCTION

References

Abramowitz, M. and Stegun, C. A. (Eds.). "Spheroidal Wave Functions." Ch. 21 in *Handbook of Mathematical Functions with Formulas, Graphs, and Mathematical Tables, 9th printing.* New York: Dover, pp. 751–59, 1972.

Flammer, C. *Spheroidal Wave Functions.* Stanford, CA: Stanford University Press, 1957.

Meixner, J. and Schäfke, F. W. *Mathieusche Funktionen und Sphäroidfunktionen.* Berlin: Springer-Verlag, 1954.

Rhodes, D. R. "On the Spheroidal Functions." *J. Res. Nat. Bur. Standards--B. Math. Sci.* **74B**, 187–09, Jul.-Sep. 1970.

Stratton, J. A.; Morse, P. M.; Chu, L. J.; Little, J. D. C.; and Corbató, F. J. *Spheroidal Wave Functions.* New York: Wiley, 1956.

Pronic Number

A FIGURATE NUMBER OF THE FORM $P_n = 2T_n = n(n+1)$, where T_n is the nth TRIANGULAR NUMBER. The first few are 2, 6, 12, 20, 30, 42, 56, 72, 90, 110, ... (Sloane's A002378). The GENERATING FUNCTION of the pronic numbers is

$$\frac{2x}{(1-x)^3} = 2x + 6x^2 + 12x^3 + 20x^4 + \dots$$

Kausler (1805) was one of the first to tabulate pronic numbers, creating a list up to $n = 1000$ (Dickson 1952, Vol. 1, p. 357; Vol. 2, p. 233). Pronic numbers are also known as oblong or heteromecic numbers.

McDaniel (1998ab) proved that the only pronic Fibonacci numbers are $F_0 = 0$ and $F_3 = 2$, and the only pronic Lucas number is $L_0 = 2$, rediscovering a result first published by Ming (1995).

The first few n for which P_n are PALINDROMIC are 1, 2, 16, 77, 538, 1621, ... (Sloane's A028336), and the first few PALINDROMIC NUMBERS which are pronic are 2, 6, 272, 6006, 289982, ... (Sloane's A028337).

References

De Geest, P. "Palindromic Products of Two Consecutive Integers." http://www.ping.be/~ping6758/consec.htm.

Dickson, L. E. *History of the Theory of Numbers, Vol. 1: Divisibility and Primality.* New York: Chelsea, p. 357, 1952.

Dickson, L. E. *History of the Theory of Numbers, Vol. 2: Diophantine Analysis.* New York: Chelsea, pp. 6, 232–33, 350, and 407, 1952.

Guy, R. K. "The Second Strong Law of Small Numbers." *Math. Mag* **63**, 3–0, 1990.

McDaniel, W. L. "Pronic Fibonacci Numbers." *Fib. Quart.* **36**, 56–9, 1998.

McDaniel, W. L. "Pronic Lucas Numbers." *Fib. Quart.* **36**, 60–2, 1998.

Ming, L. "Nearly Square Numbers in the Fibonacci and Lucas Sequences" [Chinese]. *J. Chongqing Teachers College*, No. 4, 1–, 1995.

Sloane, N. J. A. Sequences A002378/M1581, A028336, and A028337 in "An On-Line Version of the Encyclopedia of Integer Sequences." http://www.research.att.com/~njas/sequences/eisonline.html.

Kausler, C. F. *Nova Acta Acad. Petrop.* **14**, 268–89, ad annos 1797–, 1805.

Proof

A rigorous mathematical argument which unequivocally demonstrates the truth of a given PROPOSITION. A mathematical statement which has been proven is called a THEOREM.

According to Hardy (1999, pp. 15–6), "all physicists, and a good many quite respectable mathematicians, are contemptuous about proof. I have heard Professor Eddington, for example, maintain that proof, as pure mathematicians understand it, is really quite uninteresting and unimportant, and that no one who is really certain that he has found something good should waste his time looking for proof.... [This opinion], with which I am sure that almost all physicists agree at the bottom of their hearts, is one to which a mathematician ought to have some reply."

There is some debate among mathematicians as to just what constitutes a proof. The FOUR-COLOR THEOREM is an example of this debate, since its "proof" relies on an exhaustive computer testing of many individual cases which cannot be verified "by hand." While many mathematicians regard computer-assisted proofs as valid, some purists do not. There are several computer systems currently under development for automated theorem proving, among them, THƎOREM∀.

See also DEEP THEOREM, PARADOX, PROPOSITION, Q.E.D, REDUCTIO AD ABSURDUM THEOREM, TRIVIAL

References

Aigner, M. and Ziegler, G. M. *Proofs from the Book.* New York: Springer-Verlag, 1999.

Allenby, R. *Numbers and Proofs.* Oxford, England: Oxford University Press, 1997.

Benson, D. C. *The Moment of Proof: Mathematical Epiphanies.* Oxford, England: Oxford University Press, 1999.

Garnier, R. and Taylor, J. *100% Mathematical Proof.* New York: Wiley, 1996.

Hardy, G. H. "Mathematical Proof." *Mind* **38**, 1–5, 1929.

Hardy, G. H. *Ramanujan: Twelve Lectures on Subjects Suggested by His Life and Work, 3rd ed.* New York: Chelsea, 1999.

OMEGA. "Welcome to Omega, the Mathematical Proof Assistant." http://www.ags.uni-sb.de/~omega/primer/.

Pólya, G. *How to Solve It: A New Aspect of Mathematical Method, 2nd ed.* Princeton, NJ: Princeton University Press, 1988.

Pólya, G. *Mathematical Discovery: On Understanding, Learning, and Teaching Problem Solving, 2 vols. in One.* New York: Wiley, 1981.

Pólya, G. *Mathematics and Plausible Reasoning, Vol. 1: Induction and Analogy in Mathematics.* Princeton, NJ: Princeton University Press, 1990.

Pólya, G. *Mathematics and Plausible Reasoning, Vol. 2: Patterns of Plausible Inference.* Princeton, NJ: Princeton University Press, 1990.

Krantz, S. G. *Techniques of Problem Solving.* Providence, RI: Amer. Math. Soc., 1997.

Solow, D. *How to Read and Do Proofs: An Introduction to Mathematical Thought Process, 2nd ed.* New York: Wiley, 1990.

THƎOREM∀ Computer-Supported Mathematical Theorem Proving. http://www.theorema.org.

Vakil, R. *A Mathematical Mosaic: Patterns and Problem Solving.* Washington, DC: Math. Assoc. Amer., 1997.

Wickelgren, W. A. *How to Solve Mathematical Problems: Elements of a Theory of Problems and Problem Solving.* New York: Dover, 1995.

Proofreading Mistakes

If proofreader A finds a mistakes and proofreader B finds b mistakes, c of which were also found by A, how many mistakes were missed by both A and B? Assume there are a total of m mistakes, so proofreader A finds a FRACTION a/m of all mistakes, and also a FRACTION c/b of the mistakes found by B. Assuming these fractions are the same, then solving for m gives

$$m = \frac{ab}{c}.$$

The number of mistakes missed by both is therefore approximately

$$N = m - a - b + c = \frac{(a-c)(b-c)}{c}.$$

See also PRINTER'S ERRORS

References

Pólya, G. "Probabilities in Proofreading." *Amer. Math. Monthly*, **83**, 42, 1976.

Propeller

A 4-POLYHEX.

References

Gardner, M. *Mathematical Magic Show: More Puzzles, Games, Diversions, Illusions and Other Mathematical*

Sleight-of-Mind from Scientific American. New York: Vintage, p. 147, 1978.

Proper Class
A CLASS which is not a SET.

See also CLASS (SET), ORDINAL NUMBER, SET

Proper Cover
Proper covers are defined as COVERS of a set X which do not contain the entire set X itself as a subset (Macula 1994). Of the five covers of $\{1, 2\}$, namely $\{\{1\}, \{2\}\}$, $\{\{1, 2\}\}$, $\{\{1\}, \{1, 2\}\}$, $\{\{2\}, \{1, 2\}\}$, and $\{\{1\}, \{2\}, \{1, 2\}\}$, only $\{\{1\}, \{2\}\}$ does not contain the subset $\{1, 2\}$ and so is the unique proper cover of two elements. In general, the number of proper covers for a set of N elements is

$$|C'(N)| = |C(N)| - \tfrac{1}{4} 2^{2^N}$$

$$= \left[\frac{1}{2} \sum_{k=0}^{N} (-1)^k \binom{N}{k} 2^{2^{N-k}} \right] - \frac{2^{2^N}}{4},$$

the first few of which are 0, 1, 45, 15913, 1073579193, ... (Sloane's A007537).

See also COVER, MINIMAL COVER

References
Macula, A. J. "Covers of a Finite Set." *Math. Mag.* **67**, 141–44, 1994.
Sloane, N. J. A. Sequences A007537/M5287 in "An On-Line Version of the Encyclopedia of Integer Sequences." http://www.research.att.com/~njas/sequences/eisonline.html.

Proper Divisor
A positive proper divisor is a positive DIVISOR of a number n, excluding n itself. For example, 1, 2, and 3 are positive proper divisors of 6, but 6 itself is not. The number of proper divisors of n is therefore given by

$$s_0(n) \equiv \sigma_0(n) - 1,$$

where $\sigma_k(n)$ is the DIVISOR FUNCTION. For $n = 1, 2, ...,$ $s_0(n)$ is therefore given by 0, 1, 1, 2, 1, 3, 1, 3, 2, 3, ... (Sloane's A032741). The largest proper divisors of $n = 2, 3, ...$ are 1, 1, 2, 1, 3, 1, 4, 3, 5, 1, ... (Sloane's A032742).

The term "proper divisor" is sometimes includes negative integer divisors of a number n excluding $-n$. Using this definition, -3, -2, -1, 1, 2, and 3 are the proper divisors of 6, while -6 and 6 are the IMPROPER DIVISORS.

To make matters even more confusing, the proper divisor is often defined so that -1 and 1 are also excluded. Using this alternative definition, the proper divisors of 6 would then be -3, -2, 2, and 3, and the IMPROPER DIVISORS would be -6, -1, 1, and 6.

See also ALIQUANT DIVISOR, ALIQUOT DIVISOR, DIVISOR, IMPROPER DIVISOR

References
Sloane, N. J. A. Sequences A032741 and A032742 in "An On-Line Version of the Encyclopedia of Integer Sequences." http://www.research.att.com/~njas/sequences/eisonline.html.

Proper Fraction
A FRACTION $p/q < 1$. A fraction $p/q > 1$ is called an IMPROPER FRACTION.

See also FRACTION, MIXED FRACTION, IMPROPER FRACTION, REDUCED FRACTION

Proper Integral
An INTEGRAL which has neither limit INFINITE and from which the INTEGRAND does not approach INFINITY at any point in the range of integration.

See also IMPROPER INTEGRAL, INTEGRAL

Proper k-Coloring
K-COLORING

Proper Subfield
See also FIELD, SUBFIELD

Proper Subset
A SUBSET which is not the entire SET. For example, consider a SET $\{1, 2, 3, 4, 5\}$. Then $\{1, 2, 4\}$ and $\{1\}$ are proper subsets, while $\{1, 2, 6\}$ and $\{1, 2, 3, 4, 5\}$ are not.

See also SET, SUBSET

Proper Superset
A SUPERSET which is not the entire SET.

See also SET, SUPERSET

Proper Value
EIGENVALUE

Proper Vector
EIGENVECTOR

Property P
A KNOT having the property that no surgery could possibly yield a counterexample to the POINCARÉ CONJECTURE is said to satisfy Property P (Adams 1994, p. 262).

See also POINCARÉ CONJECTURE

References

Adams, C. C. *The Knot Book: An Elementary Introduction to the Mathematical Theory of Knots.* New York: W. H. Freeman, 1994.

Proportional

If a is (directly) proportional to b, then a/b is a constant. The relationship is written $a \propto b$, which implies

$$a = cb,$$

for some constant c.

See also DIRECTLY PROPORTIONAL, INVERSELY PROPORTIONAL

Proportional-Integral-Derivative Method

A very useful active feedback method for controlling things like temperature control systems, servo motors, and flow control valves.

Proposition

A statement which is to be proved.

Propositional Calculus

The formal basis of LOGIC dealing with the notion and usage of words such as "NOT," "OR," "AND," and "IMPLIES." Many systems of propositional calculus have been devised which attempt to achieve consistency, completeness, and independence of AXIOMS. The term "sentential calculus" is sometimes used as a synonym for propositional calculus.

See also CONNECTIVE, LOGIC, *P*-SYMBOL, PREDICATE CALCULUS

References

Cundy, H. and Rollett, A. *Mathematical Models, 3rd ed.* Stradbroke, England: Tarquin Pub., pp. 254–55, 1989.
Mendelson, E. "The Propositional Calculus." Ch. 1 in *Introduction to Mathematical Logic, 4th ed.* London: Chapman & Hall, pp. 12–4, 1997.
Nidditch, P. H. *Propositional Calculus.* New York: Free Press of Glencoe, 1962.

Propositional Connective

CONNECTIVE

Prosthaphaeresis Formulas

TRIGONOMETRY formulas which convert a product of functions into a sum or difference. The Prosthaphaeresis formulas are

$$\sin \alpha + \sin \beta = 2 \sin\left[\tfrac{1}{2}(\alpha+\beta)\right]\cos\left[\tfrac{1}{2}(\alpha-\beta)\right] \quad (1)$$

$$\sin \alpha - \sin \beta = 2 \cos\left[\tfrac{1}{2}(\alpha+\beta)\right]\sin\left[\tfrac{1}{2}(\alpha-\beta)\right] \quad (2)$$

$$\cos \alpha + \cos \beta = 2 \cos\left[\tfrac{1}{2}(\alpha+\beta)\right]\cos\left[\tfrac{1}{2}(\alpha-\beta)\right] \quad (3)$$

$$\cos \alpha - \cos \beta = -2 \sin\left[\tfrac{1}{2}(\alpha+\beta)\right]\sin\left[\tfrac{1}{2}(\alpha-\beta)\right]. \quad (4)$$

Related formulas are

$$\sin \alpha \sin \beta = \tfrac{1}{2}[\sin(\alpha-\beta)+\sin(\alpha+\beta)] \quad (5)$$

$$\cos \alpha \cos \beta = \tfrac{1}{2}[\cos(\alpha-\beta)+\cos(\alpha+\beta)] \quad (6)$$

$$\cos \alpha \sin \beta = \tfrac{1}{2}[\sin(\alpha+\beta)-\sin(\alpha-\beta)] \quad (7)$$

$$\sin\alpha \sin \beta = \tfrac{1}{2}[\cos(\alpha-\beta)-\cos(\alpha+\beta)]. \quad (8)$$

Multiplying both sides by 2 gives the equations sometimes known as the WERNER FORMULAS.

See also TRIGONOMETRIC ADDITION FORMULAS, TRIGONOMETRIC PRODUCT FORMULAS

Proth's Theorem

For $N = h \cdot 2^n + 1$ with ODD h and $2^n > h$, if there exists an INTEGER a such that

$$a^{(N-1)/2} \equiv -1 \pmod{N},$$

then N is PRIME.

Protractor

A ruled SEMICIRCLE used for measuring and drawing ANGLES.

Prouhet's Problem

PROUHET-TARRY-ESCOTT PROBLEM

Prouhet-Tarry-Escott Problem

Find two *distinct* sets of integers $\{a_1, \ldots, a_n\}$ and $\{b_1, \ldots, b_n\}$, such that for $k = 1, \ldots, m,$

$$\sum_{i=1}^{n} a_i^k = \sum_{i=1}^{n} b_i^k.$$

The Prouhet-Tarry-Escott problem is therefore a special case of a MULTIGRADE EQUATION. A solution with $n = m + 1$ is said to be "ideal," and are of interest because they are minimal solutions of the problem (Borwein and Ingalls 1994).

The smallest symmetric ideal solutions for $m = 9$ was found by Borwein *et al.* (Lisonek 2000),

$$(-313)^k + (-301)^k + (-188)^k + (-100)^k + (-99)^k$$
$$+ 99^k + 100^k + 188^k + 301^k + 313^k$$
$$= (-308)^k + (-307)^k + (-180)^k + (-131)^k + (-71)^k$$
$$+ 71^k + 131^k + 180^k + 307^k + 308^k, \quad (1)$$

as well as the second solution

$$(-515)^k + (-452)^k + (-366)^k + (-189)^k + (-103)^k$$
$$+ 103^k + 189^k + 366^k + 452^k + 515^k$$

$$= (-508)^k + (-417)^k + (-331)^k + (-245)^k + (-18)^k$$

$$+ 18^k + 245^k + 331^k + 471^k + 508^k. \qquad (2)$$

The previous smallest known symmetric ideal solution, found by Letac in the 1940s, is

$$(-23750)^k + (-20667)^k + (-20499)^k + (-11857)^k$$

$$+ (-436)^k + 436^k + 11857^k + 20449^k + 20667^k + 23750^k$$

$$= (-23738)^k + (-20855)^k + (-20231)^k + (-11881)^k$$

$$+ (-12)^k + 12^k + 11881^k + 20231^k + 20885^k + 23738^k. \qquad (3)$$

In 1999, S. Chen found the first ideal solution with $m \geq 10$,

$$0^k + 11^k + 24^k + 65^k + 90^k + 129^k + 173^k + 212^k$$

$$+ 237^k + 278^k + 291^k + 302^k$$

$$= 3^k + 5^k + 30^k + 57^k + 104^k + 116^k + 186^k$$

$$+ 198^k + 245^k + 272^k + 297^k + 299^k, \qquad (4)$$

which is true for $k = 1, 2, \ldots, 11$.

See also MULTIGRADE EQUATION

References

Borwein, P. and Ingalls, C. "The Prouhet-Tarry-Escott Problem Revisited." *Enseign. Math.* **40**, 3–7, 1994. http://www.cecm.sfu.ca/~pborwein/PAPERS/P98.ps.

Chen, S. "The Prouhet-Tarry-Escott Problem." http://member.netease.com/~chin/eslp/TarryPrb.htm.

Dickson, L. E. *History of the Theory of Numbers, Vol. 2: Diophantine Analysis.* New York: Chelsea, pp. 709–10, 1971.

Dorwart, H. L. and Brown, O. E. "The Tarry-Escott Problem." *Amer. Math. Monthly* **44**, 613–26, 1937.

Hahn, L. "The Tarry-Escott Problem." Problem 10284. *Amer. Math. Monthly* **102**, 843–44, 1995.

Hardy, G. H. and Wright, E. M. "The Four-Square Theorem" and "The Problem of Prouhet and Tarry: The Number $P(k, j)$." §20.5 and 21.9 in *An Introduction to the Theory of Numbers, 5th ed.* Oxford, England: Clarendon Press, pp. 302–06 and 328–29, 1979.

Lisonek, P. "New size 10 solutions of the Prouhet-Tarry-Escott Problem." nmbrthry@listserv.nodak.edu posting, 21 Jun 2000.

Wright, E. M. "On Tarry's Problem (I)." *Quart. J. Math. Oxford Ser.* **6**, 216–67, 1935.

Wright, E. M. "The Tarry-Escott and the 'Easier' Waring Problem." *J. reine angew. Math.* **311/312**, 170–73, 1972.

Wright, E. M. "Prouhet's 1851 Solution of the Tarry-Escott Problem of 1910." *Amer. Math. Monthly* **102**, 199–10, 1959.

Prüfer Code

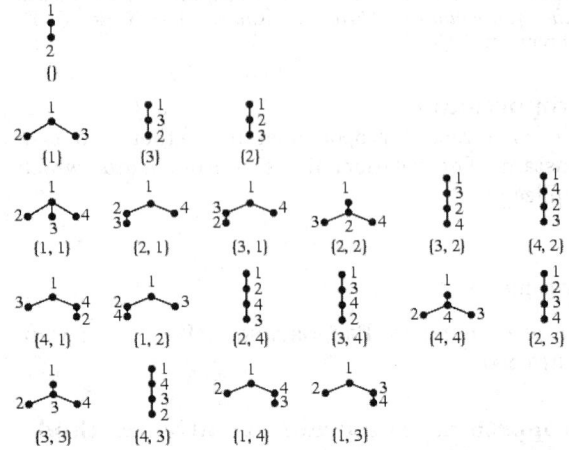

An encoding which provides a bijection between the n^{n-2} LABELED TREES on n nodes and strings of $n-2$ integers chosen from an alphabet of the numbers 1 to n. A LABELED TREE can be converted to a Prüfer code using `LabeledTreeToCode[g]` in the *Mathematica* add-on package `DiscreteMath`Combinatorica`` (which can be loaded with the command `<<DiscreteMath``), and a code can be converted to a LABELED TREE using `CodeToLabeledTree[g]`.

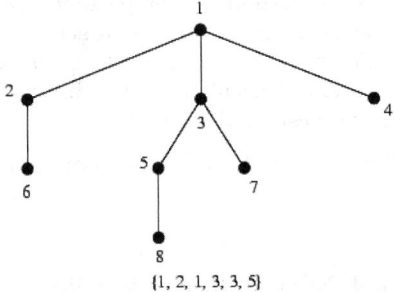

$\{1, 2, 1, 3, 3, 5\}$

Prüfer's bijection is based on the fact that every tree has at least two nodes of degree 1 (i.e., LEAVES). Therefore, the node v which is incident to the lowest labeled leaf is uniquely determined, and v is then taken as the first symbol in the code. This node is then deleted and the procedure is repeated until a single edge is left, giving a total of $n-2$ integers between 1 and n (Skiena 1990). This is demonstrated in the LABELED TREE shown above.

See also LABELED TREE

References

Prüfer, H. "Neuer Beweis eines Satzes über Permutationen." *Arch. Math. Phys.* **27**, 742–44, 1918.

Skiena, S. *Implementing Discrete Mathematics: Combinatorics and Graph Theory with Mathematica.* Reading, MA: Addison-Wesley, 1990.

Prüfer Ring

A metric space $\hat{\mathbb{Z}}$ in which the closure of a congruence class $B(j, m)$ is the corresponding congruence class $\{x \in \hat{\mathbb{Z}} | x \equiv j \pmod{m}\}$.

References

Fontana, M.; Huckaba, J. A.; and Papick, I. J. *Prüfer Domains.* New York: Dekker.
Fried, M. D. and Jarden, M. *Field Arithmetic.* New York: Springer-Verlag, pp. 7–1, 1986.
Postnikov, A. G. *Introduction to Analytic Number Theory.* Providence, RI: Amer. Math. Soc., 1988.

p-Series

A shorthand name for a POWER SERIES with a NEGATIVE exponent, $\Sigma_{k=1}^{\infty} k^{-p}$, where $p > 0$.

See also POWER SERIES, RIEMANN ZETA FUNCTION

Pseudoanalytic Function

A pseudoanalytic function is a function defined using generalized CAUCHY-RIEMANN EQUATIONS. Pseudoanalytic functions come as close as possible to having COMPLEX DERIVATIVES and are nonsingular "quasiregular" functions.

See also ANALYTIC FUNCTION, SEMIANALYTIC, SUBANALYTIC

Pseudocircle

A simple closed curve on a SPHERE that is not necessarily a GREAT CIRCLE but merely intersects as a GREAT CIRCLE would (Billera *et al.* 1999).

See also GREAT CIRCLE

References

Billera, L. J.; Brown, K. S.; and Diaconis, P. "Random Walks and Plane Arrangements in Three Dimensions." *Amer. Math. Monthly* **106**, 497–01, 1999.
Björner, A; Las Vargnas, M.; Sturmfels, B.; White, N.; and Ziegler, G. M. *Oriented Manifolds.* Cambridge, England: Cambridge University Press, 1993.
Grünbaum, B. *Arrangements and Spreads.* Providence, RI: Amer. Math. Soc., 1972.
Ziegler, G. M. *Lectures on Polytopes.* New York: Springer-Verlag, 1995.

Pseudoconic Projection

A MAP PROJECTION in which the parallels are represented by concentric circular arcs and the meridians by concurrent curves.

References

Lee, L. P. "The Nomenclature and Classification of Map Projections." *Empire Survey Rev.* **7**, 190–00, 1944.

Pseudocrosscap

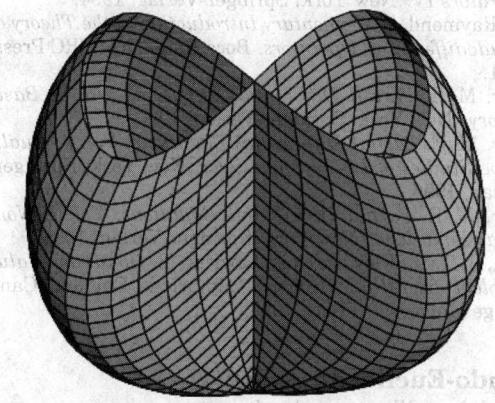

A surface constructed by placing a family of figure-eight curves into $\mathbb{R}^3$ such that the first and last curves reduce to points. The surface has PARAMETRIC EQUATIONS

$$x(u, v) = (1 - u^2) \sin v$$

$$y(u, v) = (1 - u^2) \sin(2v)$$

$$z(u, v) = u.$$

References

Gray, A. *Modern Differential Geometry of Curves and Surfaces with Mathematica,* 2nd ed. Boca Raton, FL: CRC Press, p. 337, 1997.

Pseudocylindrical Projection

A projection in which latitude lines are parallel but meridians are curves.

See also CYLINDRICAL PROJECTION, ECKERT IV PROJECTION, ECKERT VI PROJECTION, MOLLWEIDE PROJECTION, ROBINSON PROJECTION, SINUSOIDAL PROJECTION

References

Dana, P. H. "Map Projections." http://www.colorado.edu/geography/gcraft/notes/mapproj/mapproj_f.html.
Lee, L. P. "The Nomenclature and Classification of Map Projections." *Empire Survey Rev.* **7**, 190–00, 1944.

Pseudodifferential Operator

References

Folland, G. B. *Introduction to Partial Differential Equations,* 2nd ed. Princeton, NJ: Princeton University Press, 1996.
Hormander, L. *The Analysis of Linear Partial Differential Operators I: Distribution Theory and Fourier Analysis,* 2nd ed. New York: Springer-Verlag, 1990.
Hormander, L. *The Analysis of Linear Partial Differential Operators II.* New York: Springer-Verlag, 1983.
Hormander, L. *The Analysis of Linear Partial Differential Operators III.* New York: Springer-Verlag, 1985.

Hormander, L. *The Analysis of Linear Partial Differential Operators IV.* New York: Springer-Verlag, 1994.

Saint Raymond, X. *Elementary Introduction to the Theory of Pseudodifferential Operators.* Boca Raton, FL: CRC Press, 1991.

Taylor, M. E. *Partial Differential Equations, Vol. 1: Basic Theory.* New York: Springer-Verlag, 1996.

Taylor, M. E. *Partial Differential Equations, Vol. 2: Qualitative Studies of Linear Equations.* New York: Springer-Verlag, 1996.

Taylor, M. E. *Partial Differential Equations, Vol. 3: Nonlinear Equations.* New York: Springer-Verlag, 1996.

Wloka, J. T.; Rowley, B.; and Lawruk, B. *Boundary Value Problems for Elliptic Systems.* Cambridge, England: Cambridge University Press, 1995.

Pseudo-Euclidean Space

A Euclidean-like space having LINE ELEMENT

$$ds^2 = (dz^1)^2 + \ldots + (dz^p)^2 - (dz^{p+1})^2 - \ldots - (dz^{p+q})^2,$$

having dimension $m = p + q$ (Rosen 1965). In contrast, the signs would be all be positive for a EUCLIDEAN SPACE.

See also CAMPBELL'S THEOREM, EUCLIDEAN SPACE

References

Rosen, J. "Embedding of Various Relativistic Spaces in Pseudo-Euclidean Spaces." *Rev. Mod. Phys.* **37**, 204–14, 1965.

Pseudograph

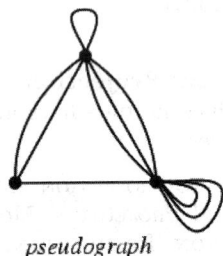

pseudograph

A non-SIMPLE GRAPH in which both LOOPS and multiple edges are permitted.

See also HYPERGRAPH, LOOP (GRAPH), MULTIGRAPH, SIMPLE GRAPH

References

Harary, F. *Graph Theory.* Reading, MA: Addison-Wesley, p. 10, 1994.

Skiena, S. *Implementing Discrete Mathematics: Combinatorics and Graph Theory with Mathematica.* Reading, MA: Addison-Wesley, p. 89, 1990.

Pseudogroup

An algebraic structure whose elements consist of selected HOMEOMORPHISMS between open subsets of a SPACE, with the composition of two transformations defined on the largest possible domain. The "germs" of the elements of a pseudogroup form a GROUPOID (Weinstein 1996).

See also GROUP, GROUPOID, INVERSE SEMIGROUP

References

Weinstein, A. "Groupoids: Unifying Internal and External Symmetry." *Not. Amer. Math. Soc.* **43**, 744–52, 1996.

Pseudoinverse

MOORE-PENROSE GENERALIZED MATRIX INVERSE

Pseudolemniscate Case

The case of the WEIERSTRASS ELLIPTIC FUNCTION with invariants $g_2 = -1$ and $g_3 = 0$.

See also EQUIANHARMONIC CASE, LEMNISCATE CASE, WEIERSTRASS ELLIPTIC FUNCTION

References

Abramowitz, M. and Stegun, C. A. (Eds.). "Pseudo-Lemniscate Case ($g_2 = -1$, $g_3 = 0$)." §18.15 in *Handbook of Mathematical Functions with Formulas, Graphs, and Mathematical Tables, 9th printing.* New York: Dover, pp. 662–63, 1972.

Pseudoparadox

Curry (1977, p. 5) uses the term pseudoparadox to describe an apparent PARADOX, such as the CATALOGUE PARADOX, for which there is no underlying actual contradiction.

See also HYPERGAME, PARADOX

References

Curry, H. B. *Foundations of Mathematical Logic.* New York: Dover, p. 5, 1977.

Pseudoperfect Number

SEMIPERFECT NUMBER

Pseudoprime

A pseudoprime is a COMPOSITE NUMBER which passes a test or sequence of tests which fail for most COMPOSITE NUMBERS. Unfortunately, some authors drop the "COMPOSITE" requirement, calling any number which passes the specified tests a pseudoprime even if it is PRIME. Pomerance, Selfridge, and Wagstaff (1980) restrict their use of "pseudoprime" to ODD COMPOSITE NUMBERS. "Pseudoprime" used without qualification means FERMAT PSEUDOPRIME.

CARMICHAEL NUMBERS are ODD COMPOSITE numbers which are pseudoprimes to every base; they are sometimes called ABSOLUTE PSEUDOPRIMES. The following table gives the number of FERMAT PSEUDOPRIMES psp(2), EULER-JACOBI PSEUDOPRIMES ejpsp(2), and STRONG PSEUDOPRIMES spsp(2) to the base 2, as well as CARMICHAEL NUMBERS CN which are less the first few powers of 10 (Guy 1994).

10^n	psp(2)	ejpsp(2)	spsp(2)	CN
Sloane	A055550	A055551	A055552	A055553
Sloane Counts	A001567	A047713	A001262	A002997
10^1	0	0	0	0
10^2	0	0	0	0
10^3	3	1	0	1
10^4	22	12	5	7
10^5	78	36	16	16
10^6	245	114	46	43
10^7	750	375	162	105
10^8	2057	1071	488	255
10^9	5597	2939	1282	646
10^{10}	14884	7706	3291	1547
10^{11}	38975	20417	8607	3605
10^{12}	101629	53332	22407	8241
10^{13}	264239	124882	58897	19279

See also CARMICHAEL NUMBER, ELLIPTIC PSEUDO-PRIME, EULER PSEUDOPRIME, EULER-JACOBI PSEUDO-PRIME, EXTRA STRONG LUCAS PSEUDOPRIME, FERMAT PSEUDOPRIME, FIBONACCI PSEUDOPRIME, FROBENIUS PSEUDOPRIME, LUCAS PSEUDOPRIME, PERRIN PSEU-DOPRIME, PROBABLE PRIME, SOMER-LUCAS PSEUDO-PRIME, STRONG ELLIPTIC PSEUDOPRIME, STRONG FROBENIUS PSEUDOPRIME, STRONG LUCAS PSEUDO-PRIME, STRONG PSEUDOPRIME

References

Caldwell, C. K. "Prime Links++: Resources in theory: finding_and_proving: probable_primality." http://prime-s.utm.edu/links/theory/finding_and_proving/probable_-primality/.
Grantham, J. "Frobenius Pseudoprimes." http://www.clark.net/pub/grantham/pseudo/pseudo1.ps
Grantham, J. "Pseudoprimes/Probable Primes." http://www.clark.net/pub/grantham/pseudo/.
Guy, R. K. "Pseudoprimes. Euler Pseudoprimes. Strong Pseudoprimes." §A12 in *Unsolved Problems in Number Theory, 2nd ed.* New York: Springer-Verlag, pp. 27–0, 1994.
Pinch, R. G. E. "The Pseudoprimes Up to 10^{13}." ftp://ftp.dpmms.cam.ac.uk/pub/PSP/.
Pomerance, C.; Selfridge, J. L.; and Wagstaff, S. S. "The Pseudoprimes to $25 \cdot 10^9$." *Math. Comput.* **35**, 1003–026, 1980. Available electronically from ftp://sable.ox.ac.uk/pub/math/primes/ps2.Z.
Sloane, N. J. A. Sequences A001262, A001567/M5441, A002997/M5462, A047713, A055550, A055551, A055552, and A055553 in "An On-Line Version of the Encyclopedia of Integer Sequences." http://www.research.att.com/~njas/sequences/eisonline.html.

Pseudorandom Number

A slightly archaic term for a computer-generated RANDOM NUMBER. The prefix pseudo- is used to distinguish this type of number from a "truly" RANDOM NUMBER generated by a random physical process such as radioactive decay.

See also RANDOM NUMBER

References

Luby, M. *Pseudorandomness and Cryptographic Applications.* Princeton, NJ: Princeton University Press, 1996.
Press, W. H.; Flannery, B. P.; Teukolsky, S. A.; and Vetter-ling, W. T. *Numerical Recipes in FORTRAN: The Art of Scientific Computing, 2nd ed.* Cambridge, England: Cambridge University Press, p. 266, 1992.

Pseudorhombicuboctahedron

ELONGATED SQUARE GYROBICUPOLA

Pseudo-Riemannian Manifold

A pseudo-Riemannian manifold is a manifold which has a metric that is of the signature $\mathrm{diag}(-, +, \ldots, +)$, as compared to a RIEMANNIAN MANIFOLD, which has a signature of all positive signs.

See also CAMPBELL'S THEOREM, RIEMANNIAN MANI-FOLD

Pseudoscalar

A SCALAR which reverses sign under inversion is called a pseudoscalar. The SCALAR TRIPLE PRODUCT

$$\mathbf{A} \cdot (\mathbf{B} \times \mathbf{C})$$

is a pseudoscalar. Given a transformation MATRIX A,

$$S' = \det|\mathbf{A}|S,$$

where det is the DETERMINANT.

See also PSEUDOTENSOR, PSEUDOVECTOR, SCALAR

References

Arfken, G. "Pseudotensors, Dual Tensors." §3.4 in *Mathematical Methods for Physicists, 3rd ed.* Orlando, FL: Academic Press, pp. 128–37, 1985.

Pseudosmarandache Function

The pseudosmarandache function $Z(n)$ is the smallest integer such that

$$\sum_{k=1}^{Z(n)} k = \tfrac{1}{2} Z(n)[Z(n)+1]$$

is divisible by n. The values for $n = 1, 2, \ldots$ are 1, 3, 2, 7, 4, 3, 6, 15, 8, 4, ... (Sloane's A011772; Kashihara 1996; Russo 2000, p. 4).

See also SMARANDACHE FUNCTION

References

Ashbacher, C. "Problem 514." *Pentagon* **57**, 36, 1997.
Kashihara, K. "Comments and Topics on Smarandache Notions and Problems." Vail: Erhus University Press, 1996.
Russo, F. *A Set of New Smarandache Functions, Sequences, and Conjectures in Numer Theory.* Lupton, AZ: American Research Press, 2000.
Sloane, N. J. A. Sequences A011772 in "An On-Line Version of the Encyclopedia of Integer Sequences." http://www.re-search.att.com/~njas/sequences/eisonline.html.

Pseudosphere

Half the SURFACE OF REVOLUTION generated by a TRACTRIX about its ASYMPTOTE to form a TRACTROID. The surfaces is sometimes also called the ANTISPHERE or TRACTRISOID (Steinhaus 1983, pp. 251). The Cartesian PARAMETRIC EQUATIONS are

$$x = \operatorname{sech} u \cos v \tag{1}$$

$$y = \operatorname{sech} u \sin v \tag{2}$$

$$z = u - \tanh u \tag{3}$$

for $u \geq 0$ and $v \in [0, 2\pi)$.
The coefficients of the FIRST FUNDAMENTAL FORM are

$$E = \tanh^2 u \tag{4}$$

$$F = 0 \tag{5}$$

$$G = \operatorname{sech}^2 u, \tag{6}$$

the SECOND FUNDAMENTAL FORM coefficients are

$$e = -\operatorname{sech} u \tanh u \tag{7}$$

$$f = 0 \tag{8}$$

$$g = \operatorname{sech} u \tanh u, \tag{9}$$

and the surface area element is

$$dS = \operatorname{sech} u \tanh u. \tag{10}$$

The SURFACE AREA is

$$S = \int_0^{2\pi} \int_0^\infty \operatorname{sech} u \tanh u \, du \, dv = 2\pi. \tag{11}$$

The GAUSSIAN and MEAN CURVATURES are

$$K = -1 \tag{12}$$

$$H = \tfrac{1}{2}(\sinh u - \operatorname{csch} u). \tag{13}$$

The pseudosphere therefore has constant NEGATIVE GAUSSIAN CURVATURE, justifying the name "pseudosphere" (i.e., an analog of the SPHERE, which has constant POSITIVE curvature). Its constant NEGATIVE CURVATURE also makes it a model of HYPERBOLIC GEOMETRY. An equation for the GEODESICS on a pseudosphere is given by

$$\cosh^2 u + (v + c)^2 = k^2. \tag{14}$$

See also FUNNEL, GABRIEL'S HORN, HYPERBOLIC GEOMETRY, TRACTRIX

References

Fischer, G. (Ed.). Plate 82 in *Mathematische Modelle / Mathematical Models, Bildband / Photograph Volume.* Braunschweig, Germany: Vieweg, p. 77, 1986.
Gray, A. *Modern Differential Geometry of Curves and Surfaces with Mathematica, 2nd ed.* Boca Raton, FL: CRC Press, pp. 487 and 489–90, 1997.
JavaView. "Classic Surfaces from Differential Geometry: Pseudo Sphere." http://www-sfb288.math.tu-berlin.de/vgp/javaview/demo/surface/common/PaSurface_Pseudo-Sphere.html.
Steinhaus, H. *Mathematical Snapshots, 3rd ed.* New York: Dover, p. 251, 1999.
Wells, D. *The Penguin Dictionary of Curious and Interesting Geometry.* London: Penguin, pp. 199–00, 1991.

Pseudosquare

Given an ODD PRIME p, a SQUARE NUMBER n satisfies $(n/p) = 0$ or 1 for all $p < n$, where (n/p) is the LEGENDRE SYMBOL. A number $n > 2$ which satisfies this relationship but is not a SQUARE NUMBER is called a pseudosquare. The only pseudosquares less than 10^9 are 3 and 6.

See also LEGENDRE SYMBOL, SQUARE NUMBER

Pseudotensor

A TENSOR-like object which reverses sign under inversion. Given a transformation MATRIX A,

$$A'_{ij} = \det|A| a_{ik} a_{jl} A_{kl},$$

where det is the DETERMINANT. A pseudotensor is sometimes also called a TENSOR DENSITY.

See also PSEUDOSCALAR, PSEUDOVECTOR, SCALAR, TENSOR DENSITY

References

Arfken, G. "Pseudotensors, Dual Tensors." §3.4 in *Mathematical Methods for Physicists, 3rd ed.* Orlando, FL: Academic Press, pp. 128–37, 1985.

Pseudovector

A typical VECTOR is transformed to its NEGATIVE under inversion. A VECTOR which is invariant under inversion is called a pseudovector, also called an AXIAL VECTOR in older literature (Morse and Feshbach 1953). The CROSS PRODUCT

$$\mathbf{A} \times \mathbf{B} \tag{1}$$

is a pseudovector, whereas the VECTOR TRIPLE PRODUCT

$$\mathbf{A} \times (\mathbf{B} \times \mathbf{C}) \tag{2}$$

is a VECTOR.

$$[\text{pseudovector}] \times [\text{pseudovector}] = [\text{pseudovector}] \tag{3}$$

$$[\text{vector}] \times [\text{pseudovector}] = [\text{vector}]. \tag{4}$$

Given a transformation MATRIX A,

$$C_i' = \det|\mathsf{A}|a_{ij}C_j. \tag{5}$$

See also PSEUDOSCALAR, TENSOR, VECTOR

References

Arfken, G. "Pseudotensors, Dual Tensors." §3.4 in *Mathematical Methods for Physicists, 3rd ed.* Orlando, FL: Academic Press, pp. 128–37, 1985.
Morse, P. M. and Feshbach, H. *Methods of Theoretical Physics, Part I.* New York: McGraw-Hill, pp. 46–7, 1953.

Psi Function

$$\Psi(z, s, v) \equiv \sum_{n=0}^{\infty} \frac{z^n}{(v + n)^s}$$

for $|z| < 1$ and $v \neq 0, -1, \ldots$ (Gradshteyn and Ryzhik 2000, pp. 1075–076).

See also HURWITZ ZETA FUNCTION, JACOBI THETA FUNCTIONS, RAMANUJAN PSI SUM

References

Gradshteyn, I. S. and Ryzhik, I. M. *Tables of Integrals, Series, and Products, 6th ed.* San Diego, CA: Academic Press, 2000.

p-Signature

Diagonalize a form over the rationals to

$$\text{diag}[p^a \cdot A, \ p^b \cdot B, \ \ldots],$$

where all the entries are INTEGERS and A, B, ...are RELATIVELY PRIME to p. Then the p-signature OF THE FORM (for $p \neq -1, 2$) is

$$p^a + p^b + \ldots + 4k \pmod 8,$$

where k is the number of ANTISQUARES. For $p = -1$, the p-signature is SYLVESTER'S SIGNATURE.

See also SIGNATURE (QUADRATIC FORM)

PSLQ Algorithm

An algorithm which can be used to find INTEGER RELATIONS between real numbers $x_1, \ldots, x_n$ such that

$$a_1 x_1 + a_2 x_2 + \ldots + a_n x_n = 0,$$

with not all $a_i = 0$. Although the algorithm operates by manipulating a lattice, it does not reduce it to a short vector basis, and is therefore *not* a LATTICE REDUCTION algorithm. PSLQ is based on a partial sum of squares scheme (like the PSOS ALGORITHM) implemented using QR DECOMPOSITION. It was developed by Ferguson and Bailey (1992). A much simplified version of the algorithm was subsequently developed by Ferguson *et al.* (1999), which also extends the algorithm to complex numbers and quaternions. Ferguson *et al.* (1999) also demonstrated that PSLQ is distinct from the HJLS ALGORITHM.

The PSLQ algorithm terminates after a number of iterations bounded by a polynomial in n and uses a numerically stable matrix reduction procedure (Ferguson and Bailey 1992). PSLQ tends to be faster than the FERGUSON-FORCADE ALGORITHM and LLL ALGORITHM because of clever techniques that allow machine arithmetic to be used at many intermediate steps. The LLL ALGORITHM, by comparison, must use moderate precision, although generally not as much as the HJLS ALGORITHM.

While the LLL ALGORITHM is a more general LATTICE REDUCTION algorithm than PSLQ, using LLL to obtain integer relations is in some sense a "trick," whereas with PSLQ one gets either a relation or lower bounds on degrees of polynomials and sizes of coefficients for which such a relation must satisfy.

See also FERGUSON-FORCADE ALGORITHM, INTEGER RELATION, LLL ALGORITHM, PSOS ALGORITHM

References

Bailey, D. H.; Borwein, J. M.; and Girgensohn, R. "Experimental Evaluation of Euler Sums." *Exper. Math.* **3**, 17–0, 1994.
Bailey, D. and Plouffe, S. "Recognizing Numerical Constants." http://www.cecm.sfu.ca/organics/papers/bailey/.
Borwein, J. M. and Corless, R. M. "Emerging Tools for Experimental Mathematics." *Amer. Math. Monthly* **106**, 899–09, 1999.
Crandall, R. E. *Topics in Advanced Scientific Computation.* New York: Springer-Verlag, 1996.
Ferguson, H. R. P. and Bailey, D. H. "A Polynomial Time, Numerically Stable Integer Relation Algorithm." RNR Techn. Rept. RNR-91–32, Jul. 14, 1992.
Ferguson, H. R. P.; Bailey, D. H.; and Arno, S. "Analysis of PSLQ, An Integer Relation Finding Algorithm." *Math. Comput.* **68**, 351–69, 1999.

PSOS Algorithm

An INTEGER-RELATION algorithm which is based on a partial sum of squares approach, from which the algorithm takes its name.

See also FERGUSON-FORCADE ALGORITHM, HJLS ALGORITHM, INTEGER RELATION, LLL ALGORITHM, PSLQ ALGORITHM

References

Bailey, D. H. and Ferguson, H. R. P. "Numerical Results on Relations Between Numerical Constants Using a New Algorithm." *Math. Comput.* **53**, 649–56, 1989.
Ferguson, H. "PSOS: A New Integral Relation Finding Algorithm Involving Partial Sums of Squares and No Square Roots." *Abs. Papers Presented to Amer. Math. Soc.* **9**, No. 56 88T-11-5, 214, Mar. 1988.

P-Symbol

A symbol employed in a formal PROPOSITIONAL CALCULUS.

References

Nidditch, P. H. *Propositional Calculus.* New York: Free Press of Glencoe, p. 1, 1962.

p-System

A p-system of a SET S is a sequence of SUBSETS $A_1, A_2, ..., A_p$ of S, among which some may be empty or coinciding with each other.

See also INCLUSION-EXCLUSION PRINCIPLE, K-SUBSET, SUBSET

References

Comtet, L. *Advanced Combinatorics: The Art of Finite and Infinite Expansions, rev. enl. ed.* Dordrecht, Netherlands: Reidel, pp. 176–77, 1974.

Ptolemy Inequality

For a QUADRILATERAL which is not CYCLIC, PTOLEMY'S THEOREM becomes an INEQUALITY:

$$AB \times CD + BC \times DA > AC \times BD.$$

See also PTOLEMY'S THEOREM, QUADRILATERAL

Ptolemy's Theorem

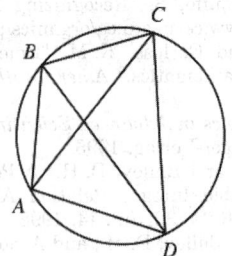

For a CYCLIC QUADRILATERAL, the sum of the products of the two pairs of opposite sides equals the product of the diagonals

$$AB \times CD + BC \times DA = AC \times BD.$$

This fact can be used to derive the TRIGONOMETRY addition formulas.

See also CYCLIC QUADRILATERAL, FUHRMANN'S THEOREM, PTOLEMY INEQUALITY

References

Coolidge, J. L. *A Treatise on the Geometry of the Circle and Sphere.* New York: Chelsea, p. 38, 1971.
Coxeter, H. S. M. and Greitzer, S. L. *Geometry Revisited.* Washington, DC: Math. Assoc. Amer., pp. 42–3, 1967.
Durell, C. V. *Modern Geometry: The Straight Line and Circle.* London: Macmillan, p. 17, 1928.
Wells, D. *The Penguin Dictionary of Curious and Interesting Geometry.* London: Penguin, pp. 200–01, 1991.

Public-Key Cryptography

A type of CRYPTOGRAPHY in which the encoding key is revealed without compromising the encoded message. The two best-known methods are the KNAPSACK PROBLEM and RSA ENCRYPTION.

See also KNAPSACK PROBLEM, RSA ENCRYPTION

References

Diffie, W. and Hellman, M. "New Directions in Cryptography." *IEEE Trans. Info. Th.* **22**, 644–54, 1976.
Flannery, S. and Flannery, D. *In Code: A Mathematical Journey.* Profile Books, 2000.
Hellman, M. E. "The Mathematics of Public-Key Cryptography." *Sci. Amer.* **241**, 130–39, Aug. 1979.
Rivest, R.; Shamir, A.; and Adleman, L. "A Method for Obtaining Digital Signatures and Public-Key Cryptosystems." MIT Memo MIT/LCS/TM-82, 1982.
Wagon, S. "Public-Key Encryption." §1.2 in *Mathematica in Action.* New York: W. H. Freeman, pp. 20–2, 1991.

Puiseux Diagram

A diagram used in the solution of ordinary differential equations OF THE FORM

$$\frac{dw}{dz} = \frac{g(z, w)}{h(z, q)}$$

which vanish when $z = 0$, where

$$g(0, 0) = h(0, 0) = 0$$

(Ince 1956, pp. 298 and 427). The diagram is named in order of French mathematician Vicrot Puiseux.

References

Fine, H. B. "On the Functions Defined by Differential Equations, with an Extension of the Puiseux Polygon Construction to these Equations." *Amer. J. Math.* **11**, 317–28, 1889.
Ince, E. L. *Ordinary Differential Equations.* New York: Dover, 1956.

Puiseux Series

A power series containing fractional exponents (Davenport *et al.* 1993, p. 91).

See also POWER SERIES

References

Davenport, J. H.; Siret, Y.; and Tournier, E. *Computer Algebra: Systems and Algorithms for Algebraic Computation, 2nd ed.* San Diego: Academic Press, pp. 90–2, 1993.
Siegel, C. L. *Topics in Complex Function Theory, Vol. 1: Elliptic Functions and Uniformization Theory.* New York: Wiley, p. 98, 1988.

Puiseux's Theorem

The whole neighborhood of any point y_i of an ALGEBRAIC CURVE may be uniformly represented by a certain finite number of convergent developments in POWER SERIES,

$$x_i = \rho_\nu y_i + a_{vi1}t_v + a_{vi2}t_v^2 + \ldots.$$

References

Coolidge, J. L. *A Treatise on Algebraic Plane Curves.* New York: Dover, p. 207, 1959.
Puiseux, V. "Recherches sur les fonctions algébriques." *J. de math. pures et appl.* **15**, 207, 1850.

Pullback Map

A pullback is a general CATEGORICAL operation appearing in a number of mathematical contexts, sometimes going under a different name. If $T : V \to W$ is a linear transformation between VECTOR SPACES, then $T^* : W^* \to V^*$ (usually called TRANSPOSE MAP or DUAL MAP because its associated matrix is the MATRIX TRANSPOSE of T) is an example of a pullback map.

In the case of a DIFFEOMORPHISM and DIFFERENTIABLE MANIFOLD, a very explicit definition can be formulated. Given an r-form α on a MANIFOLD M_2, define the r-form $T^*(\alpha)$ on M_1 by its action on an r-tuple of tangent vectors $(X_1, \ldots, X_r)$ as the number $T^*(\alpha)(X_1, \ldots, X_r) = \alpha(TX_1, \ldots, TX_r)$. This defines a map on r-forms and is the pullback map.

See also CATEGORY, PUSHFORWARD MAP

Pulse Function

RECTANGLE FUNCTION

Punctured Set

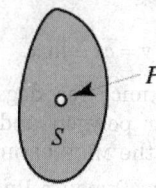

punctured set

A SET S with a single point P removed is called a punctured set, written $S \backslash \{P\}$.

References

Krantz, S. G. *Handbook of Complex Analysis.* Boston, MA: Birkhäuser, pp. 41–2, 1999.

Purser's Theorem

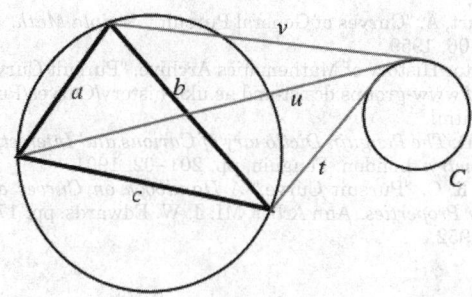

Let t, u, and v be the lengths of the tangents to a CIRCLE C from the vertices of a TRIANGLE with sides of lengths a, b, and c. Then the condition that C is tangent to the CIRCUMCIRCLE of the TRIANGLE is that

$$\pm at \pm bu \pm cv = 0.$$

The theorem was discovered by Casey prior to Purser's independent discovery.

See also CASEY'S THEOREM, CIRCUMCIRCLE

Pursuit Curve

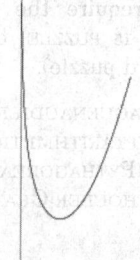

If A moves along a known curve, then P describes a pursuit curve if P is always directed toward A and A and P move with uniform velocities. Pursuit curves were considered in general by the French scientist Pierre Bouguer in 1732, and subsequently by the English mathematician Boole. The case restricting A to a straight line was studied by Arthur Bernhart

(MacTutor Archive). It has CARTESIAN COORDINATES equation

$$y = cx - \ln x.$$

The problem of n mice (or dogs) starting at the corners of a regular polygon and running towards each other is called the MICE PROBLEM.

See also APOLLONIUS PURSUIT PROBLEM, MICE PROBLEM, WHIRL

References

Barton, J. C. and Eliezer, C. J. "On Pursuit Curves." *J. Austral. Math. Soc. Ser. B* **41**, 358–71, 2000.

Bernhart, A. "Curves of Pursuit." *Scripta Math.* **20**, 125–41, 1954.

Bernhart, A. "Curves of Pursuit-II." *Scripta Math.* **23**, 49–5, 1957.

Bernhart, A. "Polygons of Pursuit." *Scripta Math.* **24**, 23–0, 1959.

Bernhart, A. "Curves of General Pursuit." *Scripta Math.* **24**, 189–06, 1959.

MacTutor History of Mathematics Archive. "Pursuit Curve." http://www-groups.dcs.st-and.ac.uk/~history/Curves/Pursuit.html.

Wells, D. *The Penguin Dictionary of Curious and Interesting Geometry.* London: Penguin, pp. 201–02, 1991.

Yates, R. C. "Pursuit Curve." *A Handbook on Curves and Their Properties.* Ann Arbor, MI: J. W. Edwards, pp. 170–71, 1952.

Push

An action which adds a single element to the top of a STACK, turning the STACK $(a_1, a_2, ..., a_n)$ into $(a_0, a_1, a_2, ..., a_n)$.

See also POKE MOVE, POP, STACK

Pushforward Map

See also PULLBACK MAP

Puzzle

A mathematical PROBLEM, usually not requiring advanced mathematics, to which a solution is desired. Puzzles frequently require the rearrangement of existing pieces (e.g., 15 PUZZLE) or the filling in of blanks (e.g., crossword puzzle).

See also 15 PUZZLE, BAGUENAUDIER, CALIBAN PUZZLE, CONWAY PUZZLE, CRYPTARITHMETIC, DISSECTION PUZZLES, ICOSIAN GAME, PYTHAGOREAN SQUARE PUZZLE, RUBIK'S CUBE, SLOTHOUBER-GRAATSMA PUZZLE, T-PUZZLE

References

Bogomolny, A. "Interactive Mathematics Miscellany and Puzzles." http://www.cut-the-knot.com.

Clessa, J. J. *Math and Logic Puzzles for PC Enthusiasts.* New York: Dover.

Costello, M. J. *The Greatest Puzzles of All Time.* New York: Dover.

Dudeney, H. E. *Amusements in Mathematics.* New York: Dover, 1917.

Dudeney, H. E. *The Canterbury Puzzles and Other Curious Problems,* 7th ed. London: Thomas Nelson and Sons, 1949.

Dudeney, H. E. *536 Puzzles & Curious Problems.* New York: Scribner, 1967.

Friedman, E. "Erich's Puzzle Palace." http://www.stetson.edu/~efriedma/puzzle.html.

Fujii, J. N. *Puzzles and Graphs.* Washington, DC: National Council of Teachers, 1966.

Pegg, E. Jr. "Mathpuzzle." http://www.mathpuzzle.com/.

Weisstein, E. W. "Books about Recreational Mathematics." http://www.treasure-troves.com/books/Recreational-Mathematics.html.

Slocum, J. and Botermans, J. *Puzzles Old and New: How to Make and Solve Them.* Seattle, WA: University of Washington Press, 1988.

P-Value

The PROBABILITY that a variate would assume a value greater than or equal to the observed value strictly by chance: $P(z \geq z_{\text{observed}})$.

See also ALPHA VALUE, SIGNIFICANCE

Pyramid

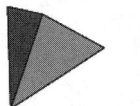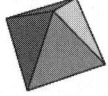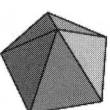

A POLYHEDRON with one face (known as the "base") a POLYGON and all the other faces TRIANGLES meeting at a common VERTEX (known as the "apex"). A right pyramid is a pyramid for which the line joining the centroid of the base and the apex is perpendicular to the base. A regular pyramid is a pyramid whose bases is a REGULAR POLYGON. An n-gonal regular pyramid (denoted Y_n) having EQUILATERAL TRIANGLES as sides is possible only for $n = 3, 4, 5$. These correspond to the TETRAHEDRON, SQUARE PYRAMID, and PENTAGONAL PYRAMID, respectively.

An arbitrary pyramid has a single cross-sectional shape whose lengths scale linearly with height. Therefore, the AREA of a CROSS SECTION scales quadratically with height, decreasing from A_b at the base ($z = 0$) to 0 at the apex (assumed to lie at a height $z = h$). The AREA at a height z above the base is therefore given by

$$A(z) = A_b \, \frac{(h - z)^2}{h^2}. \tag{1}$$

As a result, the VOLUME of a pyramid, regardless of base shape or position of the apex relative to the base, is given by

$$V = \int_0^h A(z) \, dz = A_b \int_0^h \frac{(z - h)^2}{h^2} \, dz = \tfrac{1}{3} A_b h. \tag{2}$$

These results also hold for the CONE, ELLIPTIC CONE, TRIANGULAR PYRAMID, SQUARE PYRAMID, etc.

The CENTROID is the same as for the CONE, given by

$$\bar{z} = \tfrac{1}{4} h. \tag{3}$$

The SURFACE AREA of a pyramid is

$$S = \tfrac{1}{2} ps, \tag{4}$$

where s is the SLANT HEIGHT and p is the base PERIMETER. For a right pyramid with a regular n-gonal base of side length a,

$$s_n = \sqrt{h^2 + R^2} = \sqrt{h^2 + \tfrac{1}{4} a^2 \csc^2\!\left(\frac{\pi}{n}\right)}. \tag{5}$$

This gives the special cases

$$s_3 = \sqrt{h^2 + \tfrac{1}{3} a^2} \tag{6}$$

$$s_4 = \sqrt{h^2 + \tfrac{1}{2} a^2} \tag{7}$$

$$s_5 = \sqrt{h^2 + \tfrac{1}{10}\left(5 + \sqrt{5}\right) a^2} \tag{8}$$

$$s_6 = \sqrt{h^2 + a^2}. \tag{9}$$

Joining two PYRAMIDS together at their bases gives a BIPYRAMID, also called a DIPYRAMID.

See also BIPYRAMID, CUMULATION, ELEVATUM, ELONGATED PYRAMID, GYROELONGATED PYRAMID, HEXAGONAL PYRAMID, INVAGINATUM, PENTAGONAL PYRAMID, PYRAMID, PYRAMIDAL FRUSTUM, SQUARE PYRAMID, TETRAHEDRON, TRIANGULAR PYRAMID, TRUNCATED SQUARE PYRAMID

References

Beyer, W. H. (Ed.). *CRC Standard Mathematical Tables, 28th ed.* Boca Raton, FL: CRC Press, p. 128, 1987.

Harris, J. W. and Stocker, H. "Pyramid." §4.3 in *Handbook of Mathematics and Computational Science.* New York: Springer-Verlag, pp. 98–9, 1998.

Hart, G. "Pyramids, Dipyramids, and Trapezohedra." http://www.georgehart.com/virtual-polyhedra/pyramids-info.html.

Kern, W. F. and Bland, J. R. "Pyramid" and "Regular Pyramid." §20–1 in *Solid Mensuration with Proofs, 2nd ed.* New York: Wiley, pp. 50–3, 1948.

Pyramidal Frustum

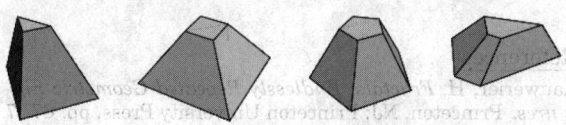

A pyramidal frustum is a FRUSTUM made by chopping the top off a PYRAMID. It is a special case of a PRISMATOID. Let s be the SLANT HEIGHT, p_1 the bottom base PERIMETER, p_2 the top base PERIMETER, A_1 the bottom AREA, and A_2 the top AREA. Then the SURFACE AREA (of the sides) and VOLUME of a pyramidal frustum are given by

$$S = \tfrac{1}{2}(p_1 + p_2)s \tag{1}$$

$$V = \tfrac{1}{3} h \left(A_1 + A_2 + \sqrt{A_1 A_2} \right). \tag{2}$$

The CENTROID of a right pyramidal frustum occurs at a height

$$\bar{z} = \frac{h\left(A_1 + 2\sqrt{A_1 A_2} + 3A_2 \right)}{4\left(A_1 + \sqrt{A_1 A_2} + A_2 \right)} \tag{3}$$

above the bottom base (Harris and Stocker 1998).

The bases of a right n-gonal frustum are regular polygons of side lengths a and b with circumradii

$$R_n = \tfrac{1}{2} c \csc\!\left(\frac{\pi}{n}\right), \tag{4}$$

where c is the side length, so the diagonal connecting corresponding vertices on top and bottom has length

$$x_n = \tfrac{1}{2}(a - b) \csc\!\left(\frac{\pi}{n}\right), \tag{5}$$

and the SLANT HEIGHT is

$$s_n = \sqrt{d^2 + h^2} = \sqrt{\tfrac{1}{4} \csc\!\left(\frac{\pi}{n}\right)(a - b)^2 + h^2}. \tag{6}$$

The triangular ($n = 3$) and square ($n = 4$) right pyramidal frustums therefore have side surface areas

$$S_3 = \tfrac{3}{2}(a + b)\sqrt{\tfrac{1}{3}(a - b)^2 + h^2} \tag{7}$$

$$S_4 = 2(a + b)\sqrt{\tfrac{1}{2}(a - b)^2 + h^2}. \tag{8}$$

The area of a regular n-gon is

$$A_n = \tfrac{1}{4} nc^2 \cot\!\left(\frac{\pi}{n}\right), \tag{9}$$

so the volumes of these frustums are

$$V_3 = \tfrac{1}{12}\sqrt{3}(a^2 + ab + b^2)h \tag{10}$$

$$V_4 = \tfrac{1}{3}(a^2 + ab + b^2)h. \tag{11}$$

See also CONICAL FRUSTUM, FRUSTUM, HERONIAN MEAN, PYRAMID, SPHERICAL SEGMENT, TRUNCATED SQUARE PYRAMID

References

Beyer, W. H. (Ed.). *CRC Standard Mathematical Tables, 28th ed.* Boca Raton, FL: CRC Press, p. 128, 1987.

Dunham, W. *Journey through Genius: The Great Theorems of Mathematics.* New York: Wiley, pp. 3–, 1990.

Eves, H. *A Survey of Geometry, rev. ed.* Boston, MA: Allyn & Bacon, p. 7, 1965.

Harris, J. W. and Stocker, H. "Frustum of a Pyramid." §4.3.2 in *Handbook of Mathematics and Computational Science.* New York: Springer-Verlag, p. 99, 1998.

Kern, W. F. and Bland, J. R. "Frustum of Regular Pyramid." §28 in *Solid Mensuration with Proofs, 2nd ed.* New York: Wiley, pp. 67–1, 1948.

Pyramidal Number

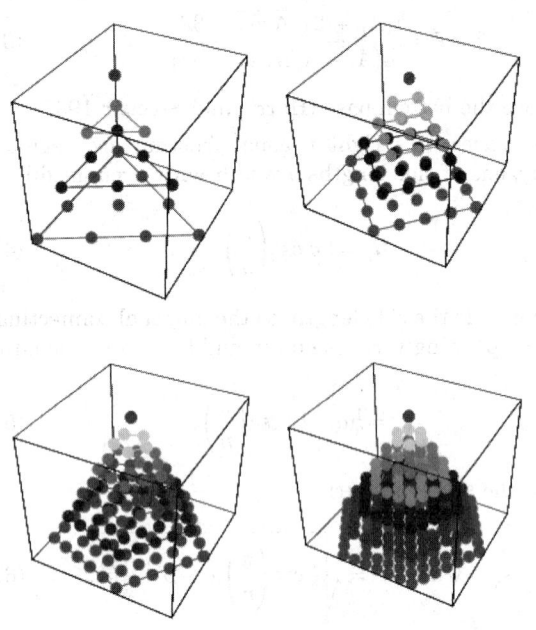

A FIGURATE NUMBER corresponding to a configuration of points which form a pyramid with r-sided REGULAR POLYGON bases can be thought of as a generalized pyramidal number, and has the form

$$P_n^r = \tfrac{1}{6}(n+1)(2p_n^r + n)$$

$$= \tfrac{1}{6} n(n+1)[(r-2)n + (5-r)]. \quad (1)$$

The first few cases are therefore

$$P_n^3 = \tfrac{1}{6} n(n+1)(n+2) \quad (2)$$

$$P_n^4 = \tfrac{1}{6} n(n+1)(2n+1) \quad (3)$$

$$P_n^5 = \tfrac{1}{2} n^2(n+1), \quad (4)$$

so $r = 3$ corresponds to a TETRAHEDRAL NUMBER Te_n, and $r = 4$ to a SQUARE PYRAMIDAL NUMBER P_n.

The pyramidal numbers can also be generalized to 4-D and higher dimensions (Sloane and Plouffe 1995).

See also HEPTAGONAL PYRAMIDAL NUMBER, HEXAGONAL PYRAMIDAL NUMBER, PENTAGONAL PYRAMIDAL NUMBER, SQUARE PYRAMIDAL NUMBER, TETRAHEDRAL NUMBER

References

Conway, J. H. and Guy, R. K. "Tetrahedral Numbers" and "Square Pyramidal Numbers" *The Book of Numbers.* New York: Springer-Verlag, pp. 44–9, 1996.

Sloane, N. J. A. and Plouffe, S. "Pyramidal Numbers." Extended entry for sequence M3382 in *The Encyclopedia of Integer Sequences.* San Diego, CA: Academic Press, 1995.

Pyritohedron

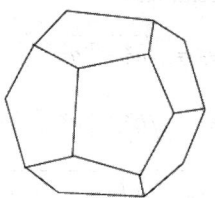

An irregular DODECAHEDRON composed of identical irregular PENTAGONS.

See also DODECAHEDRON, RHOMBIC DODECAHEDRON, TRIGONAL DODECAHEDRON

References

Cotton, F. A. *Chemical Applications of Group Theory, 3rd ed.* New York: Wiley, p. 63, 1990.

Pythagoras Tree

A FRACTAL with symmetric

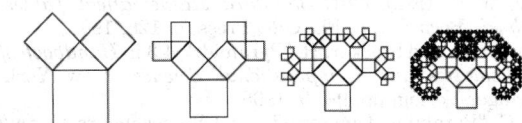

and asymmetric

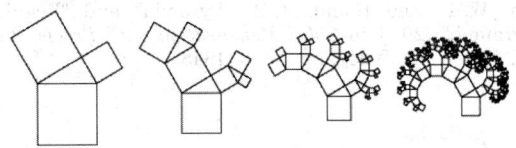

forms.

References

Lauwerier, H. *Fractals: Endlessly Repeated Geometric Figures.* Princeton, NJ: Princeton University Press, pp. 67–7 and 111–13, 1991.

Weisstein, E. W. "Fractals." MATHEMATICA NOTEBOOK FRACTAL.M.

Pythagoras's Constant

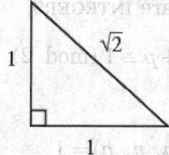

The number

$$\sqrt{2} = 1.4142135623\ldots,$$

which the Pythagoreans proved to be IRRATIONAL. This number is the length of the HYPOTENUSE of an ISOSCELES TRIANGLE with legs of length one, and the statement that it is IRRATIONAL means that it cannot be expressed as a ratio p/q of integers p and q.

Legend has it that the Pythagorean philosopher Hippasus used geometric methods to demonstrate the irrationality of $\sqrt{2}$ while at sea and, upon notifying his comrades of his great discovery, was immediately thrown overboard by the fanatic Pythagoreans.

Theodorus subsequently proved that the square roots of the numbers from 3 to 17 (excluding 4, 9, and 16) are also irrational (Wells 1986, p. 34).

The Babylonians gave the impressive approximation

$$\sqrt{2} \approx 1 + \frac{24}{60} + \frac{51}{60^2} + \frac{10}{60^3} = 1.41421296296296\ldots$$

(Wells 1986, p. 35; Guy 1990; Conway and Guy 1996, pp. 181–82).

See also IRRATIONAL NUMBER, OCTAGON, PYTHAGORAS'S THEOREM, SQUARE

References

Conway, J. H. and Guy, R. K. *The Book of Numbers.* New York: Springer-Verlag, p. 25 and 181–82, 1996.
Finch, S. "Favorite Mathematical Constants." http://www.mathsoft.com/asolve/constant/pythag/pythag.html.
Good, I. J. and Gover, T. N. "The Generalized Serial Test and the Binary Expansion of $\sqrt{2}$." *J. Roy. Statist. Soc. Ser. A* **130**, 102–07, 1967.
Good, I. J. and Gover, T. N. "Corrigendum." *J. Roy. Statist. Soc. Ser. A* **131**, 434, 1968.
Gourdon, X. and Sebah, P. "Pythagore's Constant: $\sqrt{2}$." http://xavier.gourdon.free.fr/Constants/Sqrt2/sqrt2.html.
Guy, R. K. "Review: The Mathematics of Plato's Academy." *Amer. Math. Monthly* **97**, 440–43, 1990.
Nagell, T. *Introduction to Number Theory.* New York: Wiley, p. 34, 1951.
Shanks, D. *Solved and Unsolved Problems in Number Theory, 4th ed.* New York: Chelsea, p. 126, 1993.
Wells, D. *The Penguin Dictionary of Curious and Interesting Numbers.* Middlesex, England: Penguin Books, pp. 34–5, 1986.

Pythagoras's Theorem

Proves that the DIAGONAL d of a SQUARE with sides of integral length s cannot be RATIONAL. Assume d/s is rational and equal to p/q where p and q are INTEGERS

with no common factors. Then

$$d^2 = s^2 + s^2 = 2s^2,$$

so

$$\left(\frac{d}{s}\right)^2 = \left(\frac{p}{q}\right)^2 = 2,$$

and $p^2 = 2q^2$, so p^2 is even. But if p^2 is EVEN, then p is EVEN. Since p/q is defined to be expressed in lowest terms, q must be ODD; otherwise p and q would have the common factor 2. Since p is EVEN, we can let $p \equiv 2r$, then $4r^2 = 2q^2$. Therefore, $q^2 = 2r^2$, and q^2, so q must be EVEN. But q cannot be both EVEN and ODD, so there are no d and s such that d/s is RATIONAL, and d/s must be IRRATIONAL.

In particular, PYTHAGORAS'S CONSTANT $\sqrt{2}$ is IRRATIONAL. Conway and Guy (1996) give a proof of this fact using paper folding, as well as similar proofs for ϕ (the GOLDEN RATIO) and $\sqrt{3}$ using a PENTAGON and HEXAGON.

See also IRRATIONAL NUMBER, PYTHAGORAS'S CONSTANT, PYTHAGOREAN THEOREM

References

Conway, J. H. and Guy, R. K. *The Book of Numbers.* New York: Springer-Verlag, pp. 183–86, 1996.
Gardner, M. *The Sixth Book of Mathematical Games from Scientific American.* Chicago, IL: University of Chicago Press, p. 70, 1984.
Pappas, T. "Irrational Numbers & the Pythagoras Theorem." *The Joy of Mathematics.* San Carlos, CA: Wide World Publ./Tetra, pp. 98–9, 1989.

Pythagorean Extension

An EXTENSION of an arbitrary FIELD F of the form $F\left(\sqrt{1 + \lambda^2}\right)$, where $\lambda \in F$.

See also EXTENSION FIELD, PYTHAGOREAN FIELD

References

Itô, K. (Ed.). §155B in *Encyclopedic Dictionary of Mathematics, 2nd ed., Vol. 2.* Cambridge, MA: MIT Press, p. 611, 1986.

Pythagorean Field

A FIELD F in which any PYTHAGOREAN extension of F coincides with F.

See also PYTHAGOREAN EXTENSION

References

Itô, K. (Ed.). §155B in *Encyclopedic Dictionary of Mathematics, 2nd ed., Vol. 2.* Cambridge, MA: MIT Press, p. 611, 1986.

Pythagorean Fraction

Given a PYTHAGOREAN TRIPLE (a, b, c), the fractions a/b and b/a are called Pythagorean fractions. Diophantus showed that the Pythagorean fractions consist precisely of fractions OF THE FORM $(p^2 - q^2)/(2pq)$.

References

Conway, J. H. and Guy, R. K. "Pythagorean Fractions." In *The Book of Numbers.* New York: Springer-Verlag, pp. 171–73, 1996.

Pythagorean Quadruple

POSITIVE INTEGERS a, b, c, and d which satisfy

$$a^2 + b^2 + c^2 = d^2. \tag{1}$$

For POSITIVE EVEN a and b, there exist such INTEGERS c and d; for POSITIVE ODD a and b, no such INTEGERS exist (Oliverio 1996). Oliverio (1996) gives the following generalization of this result. Let $S = (a_1, \ldots, a_{n-2})$, where a_i are INTEGERS, and let T be the number of ODD INTEGERS in S. Then IFF $T \not\equiv 2$ (mod 4), there exist INTEGERS a_{n-1} and a_n such that

$$a_1^2 + a_2^2 + \ldots + a_{n-1}^2 = a_n^2. \tag{2}$$

A set of Pythagorean quadruples is given by

$$a = 2mp \tag{3}$$

$$b = 2np \tag{4}$$

$$c = p^2 - (m^2 + n^2) \tag{5}$$

$$d = p^2 + (m^2 + n^2), \tag{6}$$

where m, n, and p are INTEGERS,

$$m + n + p \equiv 1 \ (\text{mod } 2), \tag{7}$$

and

$$(m, n, p) = 1 \tag{8}$$

(Mordell 1969). This does not, however, generate all solutions. For instance, it excludes (36, 8, 3, 37). Another set of solutions can be obtained from

$$a = 2mp + 2nq \tag{9}$$

$$b = 2np - 2mq \tag{10}$$

$$c = p^2 + q^2 - (m^2 + n^2) \tag{11}$$

$$d = p^2 + q^2 + (m^2 + n^2) \tag{12}$$

(Carmichael 1915).

See also EULER BRICK, PYTHAGOREAN TRIPLE

References

Carmichael, R. D. *Diophantine Analysis.* New York: Wiley, 1915.
Mordell, L. J. *Diophantine Equations.* London: Academic Press, 1969.
Oliverio, P. "Self-Generating Pythagorean Quadruples and *N*-tuples." *Fib. Quart.* **34**, 98–01, 1996.

Q

q-Abel's Theorem

$$\sum_{y=0}^{m}(-1)^{m-y}q^{\binom{m-y}{2}}\begin{bmatrix}m\\y\end{bmatrix}_q\frac{1-wq^m}{q-wq^y}$$

$$\times(1-wq^y)^m\left(-\frac{1-z}{1-wq^y};q\right)_y$$

$$=(1-z)^m q^{\binom{m}{2}},$$

where $\begin{bmatrix}n\\y\end{bmatrix}_q$ is a Q-BINOMIAL COEFFICIENT.

See also ABEL'S BINOMIAL THEOREM

References

Bhatnagar, G. *Inverse Relations, Generalized Bibasic Series, and their U(n) Extensions.* Ph.D. thesis. Ohio State University, p. 105, 1995.

Chu, W. C. and Hsu, L. C. "Some New Applications of Gould-Hsu Inversions." *J. Combin. Inform. System Sci.* **14**, 1–4, 1990.

q-Analog

A q-analog, also called a Q-EXTENSION or Q-GENERAL-IZATION, is a mathematical expression parameterized by a quantity q which generalizes a known expression and reduces to the known expression in the limit $q \to 1^+$. There are q-analogs of the FACTORIAL, BINOMIAL COEFFICIENT, DERIVATIVE, INTEGRAL, FIBONACCI NUMBERS, and so on. Koornwinder, Suslov, and Bustoz, have even managed some kind of q-Fourier analysis.

q-analogs are based on the observation that

$$\lim_{q\to 1^-}\frac{1-q^\alpha}{1-q}=\alpha,$$

so that the quantity $(1-q^\alpha)/(1-q)$ is sometimes written $[\alpha]$ (Koekoek and Swarttouw 1998, p. 7).

q-analogs also have a combinatorial interpretation based on the fact that one can count the elements of some set S to get the number $\#S$. A so-called "statistic" $f:S\to\mathbb{Z}$ can then be defined which is an integer-valued function on S and separates the elements of S into classes based on what value f takes on the elements. This relationship can be summarized by writing a polynomial in a new variable, usually taken as q, where the coefficient of q^n is $\#\{s\in S:f(s)=n\}$. Evaluating the polynomial at $q=1$ then adds the coefficients together, returning the original S.

The q-analog of a mathematical object is generally called the "q-object", hence Q-BINOMIAL COEFFICIENT, Q-FACTORIAL, etc. There are generally several q-analogs if there is one, and there is sometimes even a multibasic analog with independent $q_1, q_2, \ldots$.

See also D-ANALOG, Q-BETA FUNCTION, Q-BINOMIAL COEFFICIENT, Q-BINOMIAL THEOREM, Q-COSINE, Q-DERIVATIVE, Q-FACTORIAL, Q-GAMMA FUNCTION, Q-POCHHAMMER SYMBOL, Q-SERIES, Q-SINE, Q-VANDERMONDE SUM

References

Exton, H. *q-Hypergeometric Functions and Applications.* New York: Halstead Press, 1983.

Koekoek, R. and Swarttouw, R. F. *The Askey-Scheme of Hypergeometric Orthogonal Polynomials and its q-Analogue.* Delft, Netherlands: Technische Universiteit Delft, Faculty of Technical Mathematics and Informatics Report 98–17, p. 7, 1998. ftp://www.twi.tudelft.nl/publications/tech-reports/1998/DUT-TWI-98–17.ps.gz.

Q-Bar

The algebraic closure of the RATIONAL NUMBERS $\mathbb{Q}$, denoted $\overline{\mathbb{Q}}$. This is equivalent to the set of ALGEBRAIC NUMBERS, sometimes denoted $\mathbb{A}$.

See also ALGEBRAIC NUMBER, ALGEBRAICS, Q

References

Nesterenko, Yu. V. *A Course on Algebraic Independence: Lectures at IHP 1999.* http://www.math.jussieu.fr/~nesteren/.

q-Beta Function

A Q-ANALOG of the BETA FUNCTION

$$B(a,b)=\int_0^1 t^{a-1}(1-t)^{q-1}dt=\frac{\Gamma(a)\Gamma(b)}{\Gamma(a+b)},$$

where $\Gamma(z)$ is a GAMMA FUNCTION, is given by

$$B_q(a,b)\equiv\int_0^1 t^{b-1}(qt;q)_{a-1}d(a,t)=\frac{\Gamma_q(b)\Gamma_q(a)}{\Gamma_q(a+b)},$$

where $\Gamma_q(a)$ is a Q-GAMMA FUNCTION and $(a;q)_n$ is a Q-SERIES coefficient (Andrews 1986, pp. 11–12).

See also Q-FACTORIAL, Q-GAMMA FUNCTION

References

Andrews, G. E. *q-Series: Their Development and Application in Analysis, Number Theory, Combinatorics, Physics, and Computer Algebra.* Providence, RI: Amer. Math. Soc., 1986.

q-Binomial Coefficient

A Q-ANALOG for the BINOMIAL COEFFICIENT, also called a GAUSSIAN COEFFICIENT or a Gaussian polynomial. a q-binomial coefficient is given by

$$\begin{bmatrix}n\\m\end{bmatrix}_q\equiv\frac{(q)_n}{(q)_m(q)_{n-m}}=\prod_{i=0}^{m-1}\frac{1-q^{n-i}}{1-q^{i+1}}, \quad (1)$$

where

$$(q)_k \equiv \prod_{m=1}^{\infty} \frac{1 - q^m}{1 - q^{k+m}} \tag{2}$$

is a Q-SERIES (Koepf 1998, p. 26). For $k, n \in \mathbb{N}$,

$$\begin{bmatrix} n \\ k \end{bmatrix}_q = \frac{[n]_q!}{[k]_q! \, [n-k]_q!}, \tag{3}$$

where $[n]_q!$ is a Q-FACTORIAL (Koepf 1998, p. 30). The q-binomial coefficient can also be defined in terms of the Q-BRACKETS by

$$\begin{bmatrix} n \\ k \end{bmatrix}_q \equiv \begin{cases} \displaystyle\prod_{i=1}^{k} \frac{[n-i+1]_q}{[i]_q} & \text{for } 0 \le k \le n \\ 0 & \text{otherwise.} \end{cases} \tag{4}$$

For $q \to 1^-$, the q-binomial coefficients turn into the usual BINOMIAL COEFFICIENT. The first few q-binomial coefficients are

$$\begin{bmatrix} 2 \\ 1 \end{bmatrix}_q = \frac{1 - q^2}{1 - q} = 1 + q \tag{5}$$

$$\begin{bmatrix} 3 \\ 1 \end{bmatrix}_q = \begin{bmatrix} 3 \\ 2 \end{bmatrix}_q = \frac{1 - q^3}{1 - q} = 1 + q + q^2 \tag{6}$$

$$\begin{bmatrix} 4 \\ 1 \end{bmatrix}_q = \begin{bmatrix} 4 \\ 3 \end{bmatrix}_q = \frac{1 - q^4}{1 - q} = 1 + q + q^2 + q^3 \tag{7}$$

$$\begin{bmatrix} 4 \\ 2 \end{bmatrix}_q = \frac{(1 - q^3)(1 - q^4)}{(1 - q)(1 - q^2)} = 1 + q + 2q^2 + q^3 + q^4. \tag{8}$$

From the definition, it follows that

$$\begin{bmatrix} n \\ 1 \end{bmatrix}_q = \begin{bmatrix} n \\ n-1 \end{bmatrix}_q = \sum_{i=0}^{n-1} q^i \tag{9}$$

Additional identities include

$$\frac{\begin{bmatrix} n+1 \\ k+1 \end{bmatrix}_q}{\begin{bmatrix} n \\ k+1 \end{bmatrix}_q} = \frac{1 - q^{n+1}}{1 - q^{n-k}} \tag{10}$$

$$\frac{\begin{bmatrix} n+1 \\ k+1 \end{bmatrix}_q}{\begin{bmatrix} n+1 \\ k \end{bmatrix}_q} = \frac{1 - q^{n-k+1}}{1 - q^{k+1}}. \tag{11}$$

	{}	{1}	{2}	{1, 1}	{2, 1}	{2, 2}
# boxes:	0	1	2	2	3	4

The q-binomial coefficient $\begin{bmatrix} m+n \\ m \end{bmatrix}_q$ can be interpreted as a polynomial in q whose coefficient q^k counts the number of distinct partitions of k elements which fit inside an $m \times n$ rectangle. For example, the partitions of 1, 2, 3, and 4 are given in the following table.

n	partitions
0	{ }
1	{{1}}
2	{{2}, {1, 1}}
3	{{3}, {2, 1}, {1, 1, 1}}
4	{{4}, {3, 1}, {2, 2}, {2, 1, 1}, {1, 1, 1, 1},}

Of these, { }, {1}, {2}, {1, 1}, {2, 1}, and {2, 2} fit inside a 2×2 box. The counts of these having 0, 1, 2, 3, and 4 elements are 1, 1, 2, 1, and 1, so the (4, 2)-binomial coefficient is given by

$$\begin{bmatrix} 4 \\ 2 \end{bmatrix}_q = 1 + q + 2q^2 + q^3 + q^4, \tag{12}$$

as above.

See also BINOMIAL COEFFICIENT, CAUCHY BINOMIAL THEOREM, Q-SERIES

References

Gasper, G. and Rahman, M. *Basic Hypergeometric Series.* Cambridge, England: Cambridge University Press, 1990.

Koekoek, R. and Swarttouw, R. F. "The q-Gamma Function and the q-Binomial Coefficient." §0.3 in *The Askey-Scheme of Hypergeometric Orthogonal Polynomials and its q-Analogue.* Delft, Netherlands: Technische Universiteit Delft, Faculty of Technical Mathematics and Informatics Report 98–17, pp. 10–11, 1998. ftp://www.twi.tudelft.nl/publications/tech-reports/1998/DUT-TWI-98–17.ps.gz.

Koepf, W. *Hypergeometric Summation: An Algorithmic Approach to Summation and Special Function Identities.* Braunschweig, Germany: Vieweg, p. 26, 1998.

q-Binomial Theorem

The Q-ANALOG of the BINOMIAL THEOREM

$$(1 - z)^n$$

$$= 1 - nz + \frac{n(n-1)}{1 \cdot 2} z^2 - \frac{n(n-1)(n-2)}{1 \cdot 2 \cdot 3} z^3 + \dots$$

is given by

$$\left(1 - \frac{z}{q^n}\right)\left(1 - \frac{z}{q^{n-1}}\right) \cdots \left(1 - \frac{z}{q}\right)$$

$$= 1 - \frac{1 - q^n}{1 - q} \frac{z}{q^n} + \frac{1 - q^n}{1 - q} \frac{1 - q^{n-1}}{1 - q^2} \frac{z^2}{q^{n+(n-1)}}$$

$$- \dots \pm \frac{z^n}{q^{n(n+1)/2}}.$$

Written as a Q-SERIES, the identity becomes

$$\sum_{n=0}^{\infty} \frac{(a;q)_n}{(q;q)_n} z^n = \frac{(az;q)_\infty}{(z;q)_\infty},$$

where

$$(a;q)_n = \prod_{m=0}^{\infty} \frac{(1-aq^m)}{(1-aq^{m+n})}$$

(Heine 1847, p. 303; Andrews 1986). The CAUCHY BINOMIAL THEOREM is a special case of this general theorem.

See also BINOMIAL SERIES, BINOMIAL THEOREM, CAUCHY BINOMIAL THEOREM, RAMANUJAN PSI SUM

References

Andrews, G. E. *q-Series: Their Development and Application in Analysis, Number Theory, Combinatorics, Physics, and Computer Algebra.* Providence, RI: Amer. Math. Soc., p. 10, 1986.

Bhatnagar, G. *Inverse Relations, Generalized Bibasic Series, and their U(n) Extensions.* Ph.D. thesis. Ohio State University, p. 24, 1995.

Gasper, G. "Elementary Derivations of Summation and Transformation Formulas for *q*-Series." In *Fields Inst. Comm.* **14** (Ed. M. E. H. Ismail *et al.*), pp. 55–70, 1997.

Gasper, G. and Rahman, M. *Basic Hypergeometric Series.* Cambridge, England: Cambridge University Press, p. 7, 1990.

Heine, E. "Untersuchungen über die Reihe

$$1 + \frac{(1-q^\alpha)(1-q^\beta)}{(1-q)(1-q^\gamma)} \cdot x + \frac{(1-q^\alpha)(1-q^{\alpha+1})(1-q^\beta)(1-q^{\beta+1})}{(1-q)(1-q^2)(1-q^\gamma)(1-q^{\gamma+1})} \cdot x^2 + \dots".$$

J. reine angew. Math. **34**, 285–328, 1847.

Koepf, W. *Hypergeometric Summation: An Algorithmic Approach to Summation and Special Function Identities.* Braunschweig, Germany: Vieweg, p. 26, 1998.

q-Bracket

The function defined by

$$[k]_q \equiv \frac{1-q^k}{1-q} \tag{1}$$

for integral k. The q-bracket satisfies

$$\lim_{q \to 1^-} [k]_q = k. \tag{2}$$

See also Q-BINOMIAL COEFFICIENT, Q-FACTORIAL

References

Gasper, G. and Rahman, M. *Basic Hypergeometric Series.* Cambridge, England: Cambridge University Press, 1990.

Koepf, W. *Hypergeometric Summation: An Algorithmic Approach to Summation and Special Function Identities.* Braunschweig, Germany: Vieweg, p. 26, 1998.

q-Chu-Vandermonde Identity

A Q-ANALOG of the CHU-VANDERMONDE IDENTITY given by

$$_2\phi_1(q^{-n},b;c;q,cq^n/b) = \frac{(cq^n;q)_\infty (c/b;q)_\infty}{(c;q)_\infty (cq^n/b;q)_\infty} = \frac{(c/b;q)_n}{(c;q)_n},$$

where $_2\phi_1(a,b;c;q,z)$ is the Q-HYPERGEOMETRIC FUNCTION. The identity can also be written as

$$_2\phi_1(q^{-n},b;c;q,q) = \frac{(c/b;q)_n}{(c;q)_n} b^n$$

See also CHU-VANDERMONDE IDENTITY, Q-HYPERGEOMETRIC FUNCTION

References

Bhatnagar, G. *Inverse Relations, Generalized Bibasic Series, and their U(n) Extensions.* Ph.D. thesis. Ohio State University, p. 18, 1995.

Gasper, G. and Rahman, M. *Basic Hypergeometric Series.* Cambridge, England: Cambridge University Press, p. 236, 1990.

Koepf, W. *Hypergeometric Summation: An Algorithmic Approach to Summation and Special Function Identities.* Braunschweig, Germany: Vieweg, p. 43, 1998.

q-Cosine

A Q-ANALOG of the COSINE function, as advocated by R. W. Gosper, is defined by

$$\cos_q(z,q) = \frac{\vartheta_2(z,p)}{\vartheta_2(0,p)}, \tag{1}$$

where $\vartheta_2(z,p)$ is a JACOBI THETA FUNCTION and p is defined via

$$(\ln p)(\ln q) = \pi^2. \tag{2}$$

This is a period 2π, EVEN FUNCTION of unit amplitude with double and triple angle formulas and addition formulas which are analogous to ordinary SINE and COSINE. For example,

$$\cos_q(2z,q) = \cos_q^2(z,q^2) - \sin_q^2(z,q^2), \tag{3}$$

where $\sin_q(z,a)$ is the Q-SINE, and π_q is Q-PI. The q-cosine also satisfies

$$\cos_q(\pi a) = \frac{\sum_{n=-\infty}^{\infty} (-1)^n q^{(n+a)^2}}{\sum_{n=-\infty}^{\infty} (-1)^n q^{n^2}}. \tag{4}$$

See also Q-FACTORIAL, Q-SINE

References

Gosper, R. W. "Experiments and Discoveries in *q*-Trigonometry." Unpublished manuscript.

q-Derivative

The Q-ANALOG of the DERIVATIVE, defined by

$$\left(\frac{d}{dx}\right)_q f(x) = \frac{f(x) - f(qx)}{x - qx}.$$

For example,

$$\left(\frac{d}{dx}\right)_q \sin x = \frac{\sin x - \sin(qx)}{x - qx}$$

$$\left(\frac{d}{dx}\right)_q \ln x = \frac{\ln x - \ln(qx)}{x - qx} = \frac{\ln\left(\frac{1}{q}\right)}{(1-q)x}$$

$$\left(\frac{d}{dx}\right)_q x^2 = \frac{x^2 - q^2 x^2}{x - qx} = (1+q)x$$

$$\left(\frac{d}{dx}\right)_q x^3 = \frac{x^3 - q^3 x^3}{x - qx} = \left(1 + q + q^2\right)x^2.$$

In the LIMIT $q \to 1$, the q-derivative reduces to the usual DERIVATIVE.

See also DERIVATIVE

q-Dimension

$$D_q \equiv \frac{1}{1-q} \lim_{\varepsilon \to 0} \frac{\ln I(q, \varepsilon)}{\ln\left(\frac{1}{\varepsilon}\right)} \tag{1}$$

where

$$I(q, \varepsilon) \equiv \sum_{i=1}^{N} \mu_i^q, \tag{2}$$

ε is the box size, and μ_i is the NATURAL MEASURE. The CAPACITY DIMENSION (a.k.a. box-counting dimension) is given by $q = 0$,

$$D_0 = \frac{1}{1-0} \lim_{\varepsilon \to 0} \frac{\ln\left(\sum_{i=1}^{N(\varepsilon)} 1\right)}{-\ln \varepsilon} = -\lim_{\varepsilon \to 0} \frac{\ln[N(\varepsilon)]}{\ln \varepsilon} \tag{3}$$

If all μ_is are equal, then the CAPACITY DIMENSION is obtained for any q.

The INFORMATION DIMENSION corresponds to $q = 1$ and is given by

$$D_1 = \lim_{q \to 1} D_q = \lim_{q \to 1} \frac{\lim_{\varepsilon \to 0} \frac{\ln\left[\sum_{i=1}^{N(\varepsilon)} \mu_i^q\right]}{-\ln \varepsilon}}{1-q}$$

$$= \lim_{\varepsilon \to 0} \lim_{q \to 1} \frac{\ln\left[\sum_{i=1}^{N(\varepsilon)} \mu_i^q\right]}{(q-1) \ln \varepsilon}. \tag{4}$$

But for the numerator,

$$\lim_{q \to 1} \ln\left(\sum_{i=1}^{N(\varepsilon)} \mu_i^q\right) = \ln\left(\sum_{i=1}^{N(\varepsilon)} \mu_i\right) = \ln 1 = 0, \tag{5}$$

and for the denominator, $\lim_{q \to 1}(q - 1) = 0$, so use L'HOSPITAL'S RULE to obtain

$$D_1 = \lim_{\varepsilon \to 0} \left(\frac{1}{\ln \varepsilon} \lim_{q \to 1} \frac{\sum \mu_i^q \ln \mu_i}{1}\right). \tag{6}$$

Therefore,

$$D_1 = \lim_{\varepsilon \to 0} \left(\frac{\sum_{i=1}^{N(\varepsilon)} \mu_i \ln \mu_i}{\ln \varepsilon}\right) \tag{7}$$

(Ott 1993, p. 79).

D_2 is called the CORRELATION DIMENSION.

If $q_1 > q_2$, then

$$D_{q_1} \le D_{q_2} \tag{8}$$

(Ott 1993, p. 79).

See also CAPACITY DIMENSION, CORRELATION DIMENSION, FRACTAL DIMENSION, INFORMATION DIMENSION

References

Grassberger, P. "Generalized Dimensions of Strange Attractors." *Phys. Lett. A* **97**, 227, 1983.

Hentschel, H. G. E. and Procaccia, I. "The Infinite Number of Generalized Dimensions of Fractals and Strange Attractors." *Physica D* **8**, 435, 1983.

Ott, E. "Measure and the Spectrum of D_q Dimensions." §3.3 in *Chaos in Dynamical Systems*. New York: Cambridge University Press, pp. 78–81, 1993.

Rényi, A. *Probability Theory*. Amsterdam, Netherlands: North-Holland, 1970.

q-Dougall Sum

$$_8\phi_7 \left[\begin{array}{c} a, qa^{1/2}, -qa^{1/2}, b, c, d, e, q^{-N} \\ a^{1/2}, -a^{1/2}, \dfrac{aq}{b}, \dfrac{aq}{c}, \dfrac{aq}{d}, \dfrac{aq}{e}, aq^{N+1}; q, q \end{array}\right]$$

$$= \frac{\left(\dfrac{aq}{bd}; q\right)_N \left(\dfrac{aq}{ed}; q\right)_N (aq; q)_N \left(\dfrac{aq}{be}; q\right)_N}{\left(\dfrac{aq}{bd}; q\right)_N \left(\dfrac{aq}{bed}; q\right)_N \left(\dfrac{aq}{b}; q\right)_N \left(\dfrac{aq}{e}; q\right)_N},$$

where $_8\phi_7$ is a Q-HYPERGEOMETRIC SERIES.

References

Bhatnagar, G. *Inverse Relations, Generalized Bibasic Series, and their U(n) Extensions*. Ph.D. thesis. Ohio State University, p. 36, 1995.

Gasper, G. and Rahman, M. *Basic Hypergeometric Series*. Cambridge, England: Cambridge University Press, p. 35, 1990.

Q.E.D.

An abbreviation for the Latin phrase "quod erat demonstrandum" ("that which was to be demonstrated"), a NOTATION which is often placed at the end of a mathematical PROOF to indicate its completion.

See also PROOF

q-Extension

Q-ANALOG

q-Factorial

The Q-ANALOG of the FACTORIAL (by analogy with the Q-GAMMA FUNCTION). For a an integer, the q-factorial is defined by

$$[k]_q! = \text{faq}(k, q)$$
$$= 1(1+q)\left(1+q+q^2\right)\cdots\left(1+q+\ldots+q^{k-1}\right) \quad (1)$$

$$= \frac{(q;q)_k}{(1-q)^k} \quad (2)$$

(Koepf 1998, p. 26). For $k \in \mathbb{N}$,

$$[k]_q! = \Gamma_q(k+1), \quad (3)$$

where $\Gamma_q(k+1)$ is the Q-GAMMA FUNCTION. The first few values are

$$[1]_q! = 1$$

$$[2]_q! = 1 + q$$

$$[3]_q! = (1+q)\left(1+q+q^2\right)$$

$$= 1 + 2q + 2q^2 + q^3$$

$$[4]_q! = (1+q)\left(1+q+q^2\right)\left(1+q+q^2+q^3\right)$$

$$= 1 + 3q + 5q^2 + 6q^3 + 5q^4 + 3q^5 + q^6.$$

A reflection formula analogous to the GAMMA FUNCTION reflection formula is given by

$$\cos_q(\pi a) = \sin_q\left[\pi\left(\tfrac{1}{2} - a\right)\right]$$
$$= \frac{\pi_q q^{(a-1/2)(a+1/2)}}{\text{faq}\left(a - \tfrac{1}{2}, q^2\right)\text{faq}\left(-\left(a + \tfrac{1}{2}\right), q^2\right)}, \quad (4)$$

where $\cos_q(z)$ is the Q-COSINE, $\sin_q(z)$ is the Q-SINE, and π_q is Q-PI.

See also Q-BETA FUNCTION, Q-BINOMIAL COEFFICIENT, Q-BRACKET, Q-COSINE, Q-GAMMA FUNCTION, Q-PI, Q-SINE

References

Gasper, G. and Rahman, M. *Basic Hypergeometric Series.* Cambridge, England: Cambridge University Press, 1990.
Gosper, R. W. "Experiments and Discoveries in *q*-Trigonometry." Unpublished manuscript.
Koepf, W. *Hypergeometric Summation: An Algorithmic Approach to Summation and Special Function Identities.* Braunschweig, Germany: Vieweg, pp. 26 and 30, 1998.

Q-Function

Let

$$q = e^{-\pi K'/K} = e^{-i\pi\tau}, \quad (1)$$

then

$$Q_0 \equiv \prod_{n=1}^{\infty}\left(1 - q^{2n}\right) \quad (2)$$

$$Q_1 \equiv \prod_{n=1}^{\infty}\left(1 + q^{2n}\right) \quad (3)$$

$$Q_2 \equiv \prod_{n=1}^{\infty}\left(1 + q^{2n-1}\right) \quad (4)$$

$$Q_3 \equiv \prod_{n=1}^{\infty}\left(1 - q^{2n-1}\right). \quad (5)$$

The Q-functions are sometimes written using a lower-case q instead of a capital Q. The Q-functions also satisfy the identities

$$Q_0 Q_1 = Q_0\left(q^2\right) \quad (6)$$

$$Q_0 Q_3 = Q_0\left(q^{1/2}\right) \quad (7)$$

$$Q_2 Q_3 = Q_3\left(q^2\right) \quad (8)$$

$$Q_1 Q_2 = Q_1\left(q^{1/2}\right). \quad (9)$$

The NORMAL DISTRIBUTION FUNCTION $\Phi(x)$ is sometimes also denoted $Q(x)$.

See also HOFSTADTER'S Q-SEQUENCE, JACOBI IDENTITIES, NORMAL DISTRIBUTION FUNCTION, PARTITION FUNCTION Q, Q-SERIES

References

Borwein, J. M. and Borwein, P. B. *Pi & the AGM: A Study in Analytic Number Theory and Computational Complexity.* New York: Wiley, pp. 55 and 63–85, 1987.
Tannery, J. and Molk, J. *Elements de la Théorie des Fonctions Elliptiques,* 4 vols. Paris: Gauthier-Villars et fils, 1893–1902.
Whittaker, E. T. and Watson, G. N. *A Course in Modern Analysis,* 4th ed. Cambridge, England: Cambridge University Press, pp. 469–473 and 488–489, 1990.

q-Gamma Function

A Q-ANALOG of the GAMMA FUNCTION defined by

$$\Gamma_q(x) \equiv \frac{(q;q)_\infty}{(q^x;q)_\infty}(1-q)^{1-x}, \quad (1)$$

where $(x,q)_\infty$ is a Q-SERIES (Koepf 1998, p. 26; Koekoek and Swarttouw 1998). The q-gamma function satisfies

$$\lim_{q \to 1^-} \Gamma_q(x) = \Gamma(x) \quad (2)$$

where $\Gamma(z)$ is the GAMMA FUNCTION, (Andrews 1986). The q-gamma function satisfies the functional equation

$$\Gamma_q(z+1) = \frac{1 - q^z}{1 - q}\Gamma_q(z) \quad (3)$$

with $\Gamma_q(1)$ (Koekoek and Swarttouw 1998), which

simplifies to

$$\Gamma(z+1) = z\Gamma(z) \tag{4}$$

as $q \to 1^-$. A curious identity for the functional equation

$$f(a-b)f(a-c)f(a-d)f(a-e) - f(b)f(c)f(d)f(e)$$
$$= q^b f(a)f(a-b-c)f(a-b-d)f(a-b-e), \tag{5}$$

where

$$b+c+d+e = 2a \tag{6}$$

is given by

$$f(\alpha) = \begin{cases} \sin(k\alpha) & \text{for } q = 1 \\ \dfrac{1}{\Gamma_q(\alpha)\Gamma_q(1-\alpha)} & \text{for } 0 < q < 1, \end{cases} \tag{7}$$

for any k.

See also GAMMA FUNCTION, Q-BETA FUNCTION, Q-FACTORIAL

References

Andrews, G. E. "W. Gosper's Proof that $\lim_{q\to 1^-} \Gamma_q(x) = \Gamma(x)$." Appendix A in *q-Series: Their Development and Application in Analysis, Number Theory, Combinatorics, Physics, and Computer Algebra.* Providence, RI: Amer. Math. Soc., p. 11 and 109, 1986.

Gasper, G. and Rahman, M. *Basic Hypergeometric Series.* Cambridge, England: Cambridge University Press, 1990.

Koekoek, R. and Swarttouw, R. F. "The q-Gamma Function and the q-Binomial Coefficient." §0.3 in *The Askey-Scheme of Hypergeometric Orthogonal Polynomials and its q-Analogue.* Delft, Netherlands: Technische Universiteit Delft, Faculty of Technical Mathematics and Informatics Report 98–17, pp. 10–11, 1998. ftp://www.twi.tudelft.nl/publications/tech-reports/1998/DUT-TWI-98–17.ps.gz.

Koepf, W. *Hypergeometric Summation: An Algorithmic Approach to Summation and Special Function Identities.* Braunschweig, Germany: Vieweg, 1998.

Wenchang, C. Problem 10226 and Solution. "A q-Trigonometric Identity." *Amer. Math. Monthly* **103**, 175–177, 1996.

q-Gauss Identity

A Q-ANALOG of Gauss's theorem due to Jacobi and Heine,

$$_2\phi_1(a,b;c;q,c/(ab)) = \frac{(c/a;q)_\infty (c/b;q)_\infty}{(c;q)_\infty (c/(ab);q)_\infty} \tag{1}$$

for $|c/(ab)| < 1$ (Gordon and McIntosh 1997; Koepf 1998, p. 40), where $_2\phi_1(a,b;c;q,z)$ is a Q-HYPERGEOMETRIC SERIES. A special case for $a = q^{-n}$ is given by

$$\sum_{k=0}^{n} q^{k^2} \begin{bmatrix} n \\ k \end{bmatrix}_q^2 = \frac{(\sqrt{q};q)_n (-\sqrt{q};q)_n (-q;q)_n}{(q;q)_n},$$

where $\begin{bmatrix} n \\ k \end{bmatrix}_q$ is a Q-BRACKET (Koepf 1998, p. 43).

See also Q-CHU-VANDERMONDE IDENTITY, Q-HYPERGEOMETRIC SERIES

References

Bhatnagar, G. *Inverse Relations, Generalized Bibasic Series, and their $U(n)$ Extensions.* Ph.D. thesis. Ohio State University, p. 31, 1995.

Gasper, G. and Rahman, M. *Basic Hypergeometric Series.* Cambridge, England: Cambridge University Press, pp. 10 and 236, 1990.

Gordon, B. and McIntosh, R. J. "Algebraic Dilogarithm Identities." *Ramanujan J.* **1**, 431–448, 1997.

Koepf, W. *Hypergeometric Summation: An Algorithmic Approach to Summation and Special Function Identities.* Braunschweig, Germany: Vieweg, 1998.

q-Generalization

Q-ANALOG

q-Harmonic Series

The series

$$h_q(-r) = \sum_{n=1}^{\infty} \frac{1}{q^n + r} \tag{1}$$

for q an INTEGER other than 0 and ± 1 which is the Q-ANALOG of

$$H_n = \sum_{n=1}^{\infty} \frac{1}{n}. \tag{2}$$

h_q and the related series

$$Ln_q(-r+1) = \sum_{n=1}^{\infty} \frac{(-1)^n}{q^n + r}, \tag{3}$$

which is a q-extension of the NATURAL LOGARITHM $\ln 2$, are irrational for r a RATIONAL NUMBER other than 0 or $-q^n$ (Guy 1994). In fact, Amdeberhan and Zeilberger (1998) showed that the IRRATIONALITY MEASURES of both $h_q(1)$ and $Ln_q(2)$ are 4.80, improving the value of 54.0 implied by Borwein (1991, 1992).

Amdeberhan and Zeilberger (1998) also show that the q-harmonic series and q-extension of $\ln 2$ can be written in the more quickly converging forms

$$h_q(1) = \sum_{n=1}^{\infty} \frac{q^n}{(1-q^n)(q)_n} \tag{4}$$

$$= \sum_{n=1}^{\infty} \frac{1 - q^n - q^{2n}}{(q^n - 1)\binom{2n}{n}_q (q)_n} \tag{5}$$

$$Ln_q(2) = \sum_{n=1}^{\infty} \frac{q^n (q)_n}{(1-q^n)(q^2)_n} \tag{6}$$

$$= \sum_{n=1}^{\infty} \frac{(-1)^{n-1}(q)_n(1-q^{3n})}{(1-q^n)^2 \binom{2n}{n}_q (q^2)_n}, \tag{7}$$

where $\binom{n}{k}_q$ is a Q-BINOMIAL COEFFICIENT and

$$(q)_n = (1-q)(1-q^2)\cdots(1-q^n) \qquad (8)$$

for $n \geq 1$.

See also HARMONIC SERIES, IRRATIONALITY MEASURE

References

Amdeberhan, T. and Zeilberger, D. "q-Apéry Irrationality Proofs by q-WZ Pairs." *Adv. Appl. Math.* **20**, 275–283, 1998.

Borwein, P. B. "On the Irrationality of $\Sigma 1/(q^n + r)$." *J. Number Th.* **37**, 253–259, 1991.

Borwein, P. B. "On the Irrationality of Certain Series." *Math. Proc. Cambridge Philos. Soc.* **112**, 141–146, 1992.

Breusch, R. "Solution to Problem 4518." *Amer. Math. Monthly* **61**, 264–265, 1954.

Erdos, P. "On Arithmetical Properties of Lambert Series." *J. Indian Math. Soc.* **12**, 63–66, 1948.

Erdos, P. "On the Irrationality of Certain Series: Problems and Results." In *New Advances in Transcendence Theory.* Cambridge, England: Cambridge University Press, pp. 102–109, 1988.

Erdos, P. and Kac, M. "Problem 4518." *Amer. Math. Monthly* **60**, 47, 1953.

Guy, R. K. "Some Irrational Series." §B14 in *Unsolved Problems in Number Theory, 2nd ed.* New York: Springer-Verlag, p. 69, 1994.

q-Hypergeometric Function

The modern definition of the q-hypergeometric function is

$$_r\phi_s\begin{bmatrix} \alpha_1, \alpha_2, \ldots, \alpha_r \\ \beta_1, \ldots, \beta_s \end{bmatrix}; q, z$$

$$\equiv \sum_{n=0}^{\infty} \frac{(\alpha_1; q)_n (\alpha_2; q)_n \cdots (\alpha_r; q)_n}{(\beta_1; q)_n \cdots (\beta_s; q)_n} \frac{z^n}{(q; q)_n}$$

$$\times \left[(-1)^n q^{\binom{n}{2}} \right]^{1+s-r}, \qquad (1)$$

where $\binom{n}{2} = \frac{1}{2}n(n-1)$ is a BINOMIAL COEFFICIENT and $(a; q)_n$ is a Q-POCHHAMMER SYMBOL

$$(a; q)_n = (1-a)(1-aq)(1-aq^2)\cdots(1-aq^{n-1}) \qquad (2)$$

$$(a; q)_0 = 1 \qquad (3)$$

(Gasper and Rahman 1990; Bhatnagar 1995, p. 21; Koepf 1998, p. 25).

An old-fashioned definition omits the factor $[(-1)^k q^{\binom{n}{2}}]^{1+s-r}$,

$$_r\phi_s'\begin{bmatrix} \alpha_1, \alpha_2, \ldots, \alpha_r \\ \beta_1, \ldots, \beta_s \end{bmatrix}; q, z$$

$$\equiv \sum_{n=0}^{\infty} \frac{(\alpha_1; q)_n (\alpha_2; q)_n \cdots (\alpha_r; q)_n}{(\beta_1; q)_n \cdots (\beta_s; q)_n} \frac{z^n}{(q; q)_n}, \qquad (4)$$

This is the q-hypergeometric function as defined by Bailey (1935), Slater (1966), Andrews (1986), and Hardy (1999).

A particular case of $_r\phi_s'$ is given by

$$_2\psi_1(a, b; c; q, z) = \sum_{n=0}^{\infty} \frac{(a; q)_n (b; q)_n z^n}{(q; q)_n (c; q)_n} \qquad (5)$$

(Andrews 1986, p. 10). A q-analog of Gauss's theorem (the Q-GAUSS IDENTITY) due to Jacobi and Heine is given by

$$_2\phi_1'(a, b; c; q, c/(ab)) = \frac{(c/a; q)_\infty (c/b; q)_\infty}{(c; q)_\infty (c/(ab); q)_\infty} \qquad (6)$$

for $|c/(ab)| < 1$ (Koepf 1998, p. 40). Heine proved the transformation formula

$$_2\phi_1'(a, b; c; q, z)$$

$$= \frac{(b; q)_\infty (az; q)_\infty}{(c; q)_\infty (z; q)_\infty} {}_2\phi_1(c/b_2 a; az; q, b), \qquad (7)$$

(Andrews 1986, pp. 10–11). Rogers (1893) obtained the formulas

$$_2\phi_1'(a, b; c; q, z)$$

$$= \frac{(c/b; q)_\infty (bz; q)_\infty}{(z; q)_\infty (c; q)_\infty} {}_2\phi_1(b, abz/c; bz; q, c/b) \qquad (8)$$

$$_2\phi_1'(a, b, c; q, z)$$

$$= (abz/c; q)_\infty (z; q)_\infty {}_2\phi 1(c/a, c/b; c; q, abz/c) \qquad (9)$$

(Andrews 1986, pp. 10–11).

The function $_r\phi_s$ has the simple confluent identity

$$\lim_{\alpha_r \to \infty} {}_r\phi_s\begin{bmatrix} \alpha_1, \alpha_2, \ldots, \alpha_r \\ \beta_1, \ldots, \beta_s \end{bmatrix}; q, \frac{z}{\alpha_r}$$

$$= \begin{bmatrix} \alpha_1, \alpha_2, \ldots, \alpha_{r-1} \\ \beta_1, \ldots, \beta_s \end{bmatrix}; q, z. \qquad (10)$$

In the limit $q \to 1^-$,

$$\lim_{q \to 1^-} {}_r\phi_s\begin{bmatrix} q^{\alpha_1} q^{\alpha_2}, \ldots, q^{\alpha_r} \\ q^{\beta_1}, \ldots, q^{\beta_s} \end{bmatrix}; q, (q-1)^{1+s-r} z$$

$$= {}_rF_s\begin{bmatrix} \alpha_1, \alpha_2, \ldots, \alpha_r \\ \beta_2, \ldots, \beta_s \end{bmatrix}; z, \qquad (11)$$

where $_rF_s$ is a GENERALIZED HYPERGEOMETRIC FUNCTION (Koepf 1998, p. 25).

See also GENERALIZED HYPERGEOMETRIC FUNCTION, Q-POCHHAMMER SYMBOL, Q-SAALSCHUETZ SUM, Q-SERIES

References

Andrews, G. E. q-Series: Their Development and Application in Analysis, Number Theory, Combinatorics, Physics, and Computer Algebra. Providence, RI: Amer. Math. Soc., p. 10, 1986.

Bailey, W. N. "Basic Hypergeometric Series." Ch. 8 in *Generalised Hypergeometric Series.* Cambridge, England: Cambridge University Press, pp. 65–72, 1935.

Bhatnagar, G. *Inverse Relations, Generalized Bibasic Series, and their U(n) Extensions.* Ph.D. thesis. Ohio State University, p. 21, 1995.

Gasper, G. and Rahman, M. *Basic Hypergeometric Series.* Cambridge, England: Cambridge University Press, 1990.

Gasper, G. "Elementary Derivations of Summation and Transformation Formulas for *q*-Series." In *Fields Inst. Comm.* **14** (Ed. M. E. H. Ismail *et al.*), pp. 55–70, 1997.

Hardy, G. H. *Ramanujan: Twelve Lectures on Subjects Suggested by His Life and Work, 3rd ed.* New York: Chelsea, pp. 107–111, 1999.

Heine, E. "Über die Reihe

$$1 + \frac{(q^\alpha - 1)(q^\beta - 1)}{(q-1)(q^\gamma - 1)}x + \frac{(q^\alpha - 1)(q^{\alpha+1}-1)(q^\beta - 1)(q^{\beta+1}-1)}{(q-1)(q^2-1)(q^\gamma - 1)(q^{\gamma+1}-1)}x^2 + \dots".$$

J. reine angew. Math. **32**, 210–212, 1846.

Heine, E. "Untersuchungen über die Reihe

$$1 + \frac{(1-q^\alpha)(1-q^\beta)}{(1-q)(1-q^\gamma)} \cdot x + \frac{(1-q^\alpha)(1-q^{\alpha+1})(1-q^\beta)(1-q^{\beta+1})}{(1-q)(1-q^2)(1-q^\gamma)(1-q^{\gamma+1})} \cdot x^2 + \dots".$$

J. reine angew. Math. **34**, 285–328, 1847.

Heine, E. *Theorie der Kugelfunctionen und der verwandten Functionen, Bd. 1.* Berlin: Reimer, pp. 97–125, 1878.

Koepf, W. *Hypergeometric Summation: An Algorithmic Approach to Summation and Special Function Identities.* Braunschweig, Germany: Vieweg, pp. 25–26, 1998.

Krattenthaler, C. "HYP and HYPQ." *J. Symb. Comput.* **20**, 737–744, 1995.

Rogers, L. J. "On a Three-Fold Symmetry in the Elements of Heine's Series." *Proc. London Math. Soc.* **24**, 171–179, 1893.

Slater, L. J. *Generalized Hypergeometric Functions.* Cambridge, England: Cambridge University Press, 1966.

q-Hypergeometric Series

Q-HYPERGEOMETRIC FUNCTION

q-Integral

A *q*-analog of integration

$$\int_q \Phi(x)d(qx)$$

which reduces to

$$\int \Phi(x)dx$$

in the case $q = 1$. A specific case gives

$$\int_{0_t}^\infty \frac{x^{a-1}}{1-x}d(qx) = \frac{\left[\Gamma_q\left(\frac{1}{2}\right)\right]^2}{\sigma_q(a)},$$

where Γ_q is the *q*-Gamma function and σ_q is a doubly periodic sigma function. If $q = 1$, the integral reduces to

$$\int_0^\infty \frac{x^{a-1}}{1-x}dx = \frac{\pi}{\sin(\pi a)}.$$

References

Jackson, F. H. "*q*-Definite Integrals." *Quart. J. Math.* **41**, 163, 1910.

Jackson, F. H. "The *q*-Integral Analogous to Borel's Integral." *Mess. Math.* **47**, 57–64, 1917.

Q-Matrix

FIBONACCI *Q*-MATRIX

q-Multinomial Coefficient

A *Q*-ANALOG of the MULTINOMIAL COEFFICIENT, defined as

$$\frac{[a_1 + \dots + a_n]!}{[a_1]! \dots [a_n]!},$$

where

$$[n]! \equiv (1)(1+q)\cdots\left(1 + q + \dots + q^{n-1}\right).$$

See also MULTINOMIAL COEFFICIENT, ZEILBERGER-BRESSOUD THEOREM

Q-Number

HOFSTADTER'S *Q*-SEQUENCE

q-Pfaff-Saalschuetz Sum

Q-SAALSCHUETZ SUM

q-Pi

The *Q*-ANALOG of PI π_q can be defined by taking $a = 0$ in the *Q*-FACTORIAL

$$\mathrm{faq}(a,q) = 1(1+q)\left(1 + q + q^2\right)\cdots\left(1 + q + \dots + q^{a-1}\right),$$

giving

$$1 = \sin_q\left(\tfrac{1}{2}\pi\right) = \frac{\pi_q}{\mathrm{faq}^2\left(-\frac{1}{2}, q^2\right)q^{1/4}},$$

where $\sin_q(z)$ is the *Q*-SINE. Gosper has developed an iterative algorithm for computing π_q based on the algebraic RECURRENCE RELATION

$$\frac{4\pi_{q^4}}{q^{4+1}}\frac{(q^2 + 1)^2\pi_q^2}{\pi_{q^2}} - \frac{(q^4 + 1)\pi_{q^2}^2}{\pi_{q^4}}$$

q-Pochhammer Symbol

The *Q*-ANALOG of the POCHHAMMER SYMBOL defined by

$$(a;q)k = \begin{cases} \prod_{j=0}^{k-1}(1-aq^j) & \text{if } k > 0 \\ 1 & \text{if } k = 0 \\ \prod_{j=0}^{k}(1-aq^{-j})^{-1} & \text{if } k < 0 \\ \prod_{j=0}^{\infty}(1-aq^j) & \text{if } k = \infty \end{cases} \qquad (1)$$

(Koepf 1998, p. 25). *q*-Pochhammer symbols are frequently called *Q*-SERIES and, for brevity, $(a;q)_k$ is often simply written $(a)_k$.

For $q \to 1^-$,

$$\lim_{q \to 1^-} \frac{(q^{\alpha};q)_k}{(1-q)^k} = (\alpha)_k \tag{2}$$

gives the normal POCHHAMMER SYMBOL $(\alpha)_n$ (Koekoek and Swarttouw 1998, p. 7). The q-Pochhammer symbols are also called q-shifted factorials (Koekoek and Swarttouw 1998, pp. 8–9).

The q-Pochhammer symbol satisfies

$$(a;q)_n = \frac{(a;q)_\infty}{(aq^n;q)_\infty} \tag{3}$$

$$\frac{1-aq^{2n}}{1-a} = \frac{(q\sqrt{a};q)_n(-q\sqrt{a};q)_n}{(\sqrt{a};q)_n(-\sqrt{a};q)_n} \tag{4}$$

$$(a;q)_n(-a;q)_n = (a^2;q^2)_n$$

$$(a;q)_n = (q^{1-n}/a;q)_n(-a)^n q^{\binom{n}{2}} \tag{5}$$

$$(a;q^{-1})_n = (a^{-1};q)_n(-a)^n q^{-\binom{n}{2}} \tag{6}$$

$$(a;q)_{-n} = \frac{1}{(aq^{-n};q)_n} = \frac{(-q/a)^n}{(q/a;q)_n} q^{\binom{n}{2}}, \tag{7}$$

where $\binom{n}{2}$ is a BINOMIAL COEFFICIENT and

$$\binom{n}{2} = \frac{1}{2}n(n-1), \tag{8}$$

as well as many other identities, some of which are given by Koekoek and Swarttouw (1998, p. 9).

A generalized q-Pochhammer symbol can be defined using the concise notation

$$(a_1, a_2, \ldots, a_r; q)_\infty = (a_1;q)_\infty (a_2;q)_\infty \ldots (a_r;q)_\infty \tag{9}$$

(Gordon and McIntosh 2000).

See also POCHHAMMER SYMBOL, Q-SERIES

References

Gordon, B. and McIntosh, R. J. "Some Eighth Order Mock Theta Functions." To appear in *J. London Math. Soc.* 2000.

Koekoek, R. and Swarttouw, R. F. *The Askey-Scheme of Hypergeometric Orthogonal Polynomials and its q-Analogue.* Delft, Netherlands: Technische Universiteit Delft, Faculty of Technical Mathematics and Informatics Report 98–17, p. 7, 1998. ftp://www.twi.tudelft.nl/publications/tech-reports/1998/DUT-TWI-98–17.ps.gz.

Koepf, W. *Hypergeometric Summation: An Algorithmic Approach to Summation and Special Function Identities.* Braunschweig, Germany: Vieweg, pp. 25 and 30, 1998.

Q-Polynomial
BLM/HO POLYNOMIAL

q-Product
Q-FUNCTION

QR Decomposition
Given a MATRIX A, its QR-decomposition is OF THE FORM

$$A = QR,$$

where R is an upper TRIANGULAR MATRIX and Q is an ORTHOGONAL MATRIX, i.e., one satisfying

$$Q^T Q = I$$

where I is the IDENTITY MATRIX. This matrix decomposition can be used to solve linear systems of equations. QR decomposition is implemented in *Mathematica* as QRDecomposition[*m*].

See also CHOLESKY DECOMPOSITION, LU DECOMPOSITION, MATRIX DECOMPOSITION, PSLQ ALGORITHM, SINGULAR VALUE DECOMPOSITION

References

Gentle, J. E. "QR Factorization." §3.2.2 in *Numerical Linear Algebra for Applications in Statistics.* Berlin: Springer-Verlag, pp. 95–97, 1998.

Householder, A. S. *The Numerical Treatment of a Single Non-Linear Equations.* New York: McGraw-Hill, 1970.

Nash, J. C. *Compact Numerical Methods for Computers: Linear Algebra and Function Minimisation,* 2nd ed. Bristol, England: Adam Hilger, pp. 26–28, 1990.

Press, W. H.; Flannery, B. P.; Teukolsky, S. A.; and Vetterling, W. T. "QR Decomposition." §2.10 in *Numerical Recipes in FORTRAN: The Art of Scientific Computing,* 2nd ed. Cambridge, England: Cambridge University Press, pp. 91–95, 1992.

Stewart, G. W. "A Parallel Implementation of the QR Algorithm." *Parallel Comput.* **5**, 187–196, 1987. ftp://thales.cs.umd.edu/pub/reports/piqra.ps.

q-Saalschuetz Sum
A q-analog of the Saalschütz theorem due to Jackson is given by

$$_3\phi_2(q^{-n}, a, b; c, ab/(cq^{n-1}); q, q)$$
$$= \frac{(c/a;q)_n(c/b;q)_n}{(c,q)_n(c/(ab);q)_n} \tag{1}$$

where $_3\phi_2$ is the Q-HYPERGEOMETRIC FUNCTION (Koepf 1998, p. 40; Schilling and Warnaar 1999).

See also Q-HYPERGEOMETRIC FUNCTION

References

Andrews, G. E. *Encyclopedia of Mathematics and Its Applications, Vol. 2: The Theory of Partitions.* Cambridge, England: Cambridge University Press, 1984.

Bailey, W. N. "The Analogue of Saalschütz's Theorem." §8.4 in *Generalised Hypergeometric Series.* Cambridge, England: University Press, p. 68, 1935.

Bhatnagar, G. *Inverse Relations, Generalized Bibasic Series, and their $U(n)$ Extensions.* Ph.D. thesis. Ohio State University, p. 30, 1995.

Carlitz, L. "Remark on a Combinatorial Identity." *J. Combin. Th. Ser. A* **17**, 256–257, 1974.

Gasper, G. and Rahman, M. *Basic Hypergeometric Series.* Cambridge, England: Cambridge University Press, p. 13, 1990.

Gould, H. W. "A New Symmetrical Combinatorial Identity." *J. Combin. Th. Ser. A* **13**, 278–286, 1972.

Koepf, W. *Hypergeometric Summation: An Algorithmic Approach to Summation and Special Function Identities.* Braunschweig, Germany: Vieweg, pp. 25–26, 1998.

Schilling A. and Warnaar, S. O. A Generalization of the *q.*-Saalschütz Sum and the Burge Transform 8 Sep 1999. http://xxx.lanl.gov/abs/math.QA/9909044/.

Watson, G. N. "A New Proof of the Rogers-Ramanujan Identities." *J. London Math. Soc.* **4**, 4–9, 1929.

q-Series

A SERIES involving coefficients OF THE FORM

$$(a;q)_n \equiv (a)_n = \prod_{k=0}^{n-1} (1 - aq^k) \tag{1}$$

$$= \prod_{k=0}^{\infty} \frac{(1 - aq^k)}{(1 - aq^{k+n})} \tag{2}$$

$$= \frac{(a;q)_\infty}{(aq^n;q)_\infty} \tag{3}$$

for $n \geq 1$, also called a Q-POCHHAMMER SYMBOL (Andrews 1986, p. 10). The notation

$$(q)_n \equiv (q;q)_n = \prod_{k=1}^{n-1} (1 - q^k) \tag{4}$$

is also used (Hirschhorn 1999). The symbol for $n \to \infty$ is defined as

$$(a)_\infty \equiv (a;q)_\infty = \prod_{k=0}^{\infty} (1 - aq^k), \tag{5}$$

giving the special case

$$\eta(\tau) = (q;q)_\infty = q^{1/24} \prod_{k=0}^{\infty} (1 - q \cdot q^k)$$

$$= q^{1/24} \prod_{k=1}^{\infty} (1 - q^k), \tag{6}$$

where $q \equiv e^{2\pi i r}$ and $\eta(\tau)$ is called the DEDEKIND ETA FUNCTION.

Identities involving $(q)_\infty$ include

$$(q)_\infty^3 = \sum_{n=0}^{\infty} (-1)^n (2n+1) q^{n(n+1)/2} \tag{7}$$

$$= X + 2qY \tag{8}$$

(Hardy and Wright 1979, Hirschhorn 1999), where

$$X = \prod_{n=1}^{\infty} (1 - q^{25n-15})(1 - q^{25n-10})(1 - q^{25n})$$

$$= \sum_{-\infty}^{\infty} (-1)^n q^{(25n^2-5n)/2} \tag{9}$$

$$Y = \prod_{n=1}^{\infty} (1 - q^{25n-30})(1 - q^{25n-5})(1 - q^{25n})$$

$$= \sum_{-\infty}^{\infty} (-1)^n q^{(25n^2-15n)/2} \tag{10}$$

(Hirschhorn 1999)

The symbols

$$[n] \equiv 1 + q + q^2 + \ldots + q^{n-1} \tag{11}$$

$$[n]! \equiv [n][n-1] \cdots [1] \tag{12}$$

are sometimes also used when discussing q-series.

There are a great many other beautiful identities involving q-series, some of which follow directly by taking the Q-ANALOG of standard combinatorial identities, e.g., the Q-BINOMIAL THEOREM

$$\sum_{n=0}^{\infty} \frac{(a;q)_n z^n}{(q;q)_n} = \frac{(az;q)_\infty}{(z;q)_\infty} \tag{13}$$

($|z| < 1$, $|q| < 1$; Andrews 1986, p. 10), a special case of an identity due to Euler

$$(aq;q)_\infty = \sum_{k=0}^{\infty} \frac{(-1)^k q^{k(k+1)/2} a^k}{(c;q)_k} \tag{14}$$

(Gasper and Rahman 1990, p. 9; Leininger and Milne 1997), and Q-VANDERMONDE SUM

$$_2\phi_1(a, q^{-n}; c; q, q) = \frac{a^n(c/a, q)_n}{(c;q)_n}, \tag{15}$$

where $_2\phi_1(a, b; c; q, z)$ is a Q-HYPERGEOMETRIC SERIES. Other q-series identities, e.g., the JACOBI IDENTITIES, ROGERS-RAMANUJAN IDENTITIES, and Q-HYPERGEOMETRIC identity

$$_2\phi_1(a, b; c; q, z)$$
$$= \frac{(b;q)_\infty (az;q)_\infty}{(c;q)_\infty (z;q)_\infty} \, _2\phi_1(c/b, a; az; q, b), \tag{16}$$

seem to arise out of the blue. Another such example is

$$\sum_{n=0}^{\infty} \frac{(-q;q^2)_n q^{n(n-1)} z^n}{(z;q^2)_n} = \sum_{n=0}^{\infty} \frac{(-zq;q^4)_n q^{n(2n-1)} z^n}{(z;q^2)_{2n+1}} \tag{17}$$

(Gordon and McIntosh 2000).

Asymptotic results for q-series include

$$(q)_\infty = \sqrt{\frac{2\pi}{t}} \exp\left(-\frac{\pi^2}{6t} + \frac{t}{24}\right) + \iota(1) \tag{18}$$

$$(q^2;q^2)_\infty = \sqrt{\frac{\pi}{t}} \exp\left(-\frac{\pi^2}{12t} + \frac{t}{12}\right) + \iota(1) \tag{19}$$

$$(q; q^2)_\infty = \frac{(q)_\infty}{(q^2; q^2)_\infty} = \sqrt{2} \exp\left(-\frac{\pi^2}{12t} - \frac{t}{24}\right) + \imath(1) \quad (20)$$

(Watson 1936, Gordon and McIntosh 2000).

See also BORWEIN CONJECTURES, DEDEKIND ETA FUNCTION, FINE'S EQUATION, GAUSSIAN COEFFICIENT, JACKSON'S IDENTITY, JACOBI IDENTITIES, MOCK THETA FUNCTION, Q-ANALOG, Q-BINOMIAL THEOREM, Q-COSINE, Q-FACTORIAL, Q-FUNCTION, Q-GAMMA FUNCTION, Q-HYPERGEOMETRIC FUNCTION, Q-MULTINOMIAL COEFFICIENT, Q-POCHHAMMER SYMBOL, Q-SINE, RAMANUJAN PSI SUM, RAMANUJAN THETA FUNCTIONS, ROGERS-RAMANUJAN IDENTITIES

References

Andrews, G. E. *q-Series: Their Development and Application in Analysis, Number Theory, Combinatorics, Physics, and Computer Algebra.* Providence, RI: Amer. Math. Soc., 1986.

Berndt, B. C. "*q*-Series." Ch. 27 in *Ramanujan's Notebooks, Part IV.* New York: Springer-Verlag, pp. 261–286, 1994.

Berndt, B. C.; Huang, S.-S.; Sohn, J.; and Son, S. H. "Some Theorems on the Rogers-Ramanujan Continued Fraction in Ramanujan's Lost Notebook." To appears in *Trans. Amer. Math. Soc.*

Bhatnagar, G. "A Multivariable View of One-Variable *q*-Series." In *Special Functions and Differential Equations. Proceedings of the Workshop (WSSF97) held in Madras, January 13–24, 1997)* (Ed. K. S. Rao, R. Jagannathan, G. van den Berghe, and J. Van der Jeugt). New Delhi, India: Allied Pub., pp. 60–72, 1998.

Gasper, G. and Rahman, M. *Basic Hypergeometric Series.* Cambridge, England: Cambridge University Press, 1990.

Gasper, G. "Elementary Derivations of Summation and Transformation Formulas for *q*-Series." In *Fields Inst. Comm.* **14** (Ed. M. E. H. Ismail *et al.*), pp. 55–70, 1997.

Gordon, B. and McIntosh, R. J. "Some Eighth Order Mock Theta Functions." To appear in *J. London Math. Soc.* 2000.

Gosper, R. W. "Experiments and Discoveries in *q*-Trigonometry." Unpublished manuscript.

Hardy, G. H. and Wright, E. M. *An Introduction to the Theory of Numbers, 5th ed.* Oxford, England: Clarendon Press, 1979.

Hirschhorn, M. D. "Another Short Proof of Ramanujan's Mod 5 Partition Congruences, and More." *Amer. Math. Monthly* **106**, 580–583, 1999.

Koekoek, R. and Swarttouw, R. F. *The Askey-Scheme of Hypergeometric Orthogonal Polynomials and its q-Analogue.* Delft, Netherlands: Technische Universiteit Delft, Faculty of Technical Mathematics and Informatics Report 98–17, 1–168, 1998. ftp://www.twi.tudelft.nl/publications/tech-reports/1998/DUT-TWI-98-17.ps.gz.

Leininger, V. E. and Milne, S. C. "Some New Infinite Families of Eta Function Identities." Preprint. http://www.math.ohio-state.edu/~milne/preprints.html.

Watson, G. N. "The Final Problem: An Account of the Mock Theta Functions." *J. London Math. Soc.* **11**, 55–80, 1936.

Weisstein, E. W. "Books about q-Series." http://www.treasure-troves.com/books/q-Series.html.

q-Shifted Factorial

Q-POCHHAMMER SYMBOL

Q-Signature

SIGNATURE (RECURRENCE RELATION)

q-Sine

The Q-ANALOG of the SINE function, as advocated by R. W. Gosper, is defined by

$$\sin_q(z, q) = \frac{\vartheta_1(z, p)}{\vartheta_1\left(\frac{1}{2}\pi, p\right)},$$

where $\vartheta_1(z, p)$ is a JACOBI THETA FUNCTION and p is defined via

$$(\ln p)(\ln q) = \pi^2.$$

This is a period 2π, ODD FUNCTION of unit amplitude with double and triple angle formulas and addition formulas which are analogous to ordinary SINE and COSINE. For example,

$$\sin_q(2z, q) = (q + 1)\frac{\pi_q}{P_{q^2}} \cos_q(z, q^2) \sin_q(z, q^2),$$

where $\cos_q(z, a)$ is the Q-COSINE, and π_q is Q-PI.

See also Q-COSINE, Q-FACTORIAL

References

Gosper, R. W. "Experiments and Discoveries in *q*-Trigonometry." Unpublished manuscript.

Quadrable

A plane figure for which QUADRATURE is possible is said to be quadrable.

Quadrangle

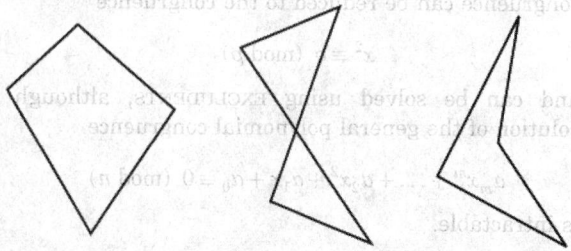

A plane figure consisting of four points, each of which is joined to two other points by a LINE SEGMENT (where the line segments may intersect). A quadrangle may therefore be CONCAVE or CONVEX; if it is CONVEX, it is called a QUADRILATERAL.

See also COMPLETE QUADRANGLE, CYCLIC QUADRANGLE, QUADRILATERAL, TETRASTIGM

References

Coxeter, H. S. M. and Greitzer, S. L. "Collinearity and Concurrence." Ch. 3 in *Geometry Revisited.* Washington, DC: Math. Assoc. Amer., pp. 51–79, 1967.

Durell, C. V. "The Quadrilateral and Quadrangle." Ch. 7 in *Modern Geometry: The Straight Line and Circle.* London: Macmillan, pp. 77–87, 1928.

Quadrant

$x < 0, y > 0$	$x > 0, y > 0$
Quadrant 2	Quadrant 1
Quadrant 3	Quadrant 4
$x < 0, y < 0$	$x > 0, y < 0$

One of the four regions of the PLANE defined by the four possible combinations of SIGNS $(+,+)$, $(+,-)$, $(-,+)$, and $(-,-)$ for (x, y).

See also OCTANT, x-AXIS, y-AXIS

References
Courant, R. and Robbins, H. *What is Mathematics?: An Elementary Approach to Ideas and Methods, 2nd ed.* Oxford, England: Oxford University Press, p. 73, 1996.

Quadratfrei

SQUAREFREE

Quadratic Congruence Equation

A CONGRUENCE OF THE FORM

$$ax^2 + bx + c \equiv 0 \pmod{m},$$

where a, b, and c are INTEGERS. A general quadratic congruence can be reduced to the congruence

$$x^2 \equiv q \pmod{p}$$

and can be solved using EXCLUDENTS, although solution of the general polynomial congruence

$$a_m x^m + \ldots + a_2 x^2 + a_1 x + a_0 \equiv 0 \pmod{n}$$

is intractable.

See also CONGRUENCE, CONGRUENCE EQUATION, EXCLUDENT, LINEAR CONGRUENCE EQUATION

Quadratic Curve

The general bivariate quadratic curve can be written

$$ax^2 + 2bxy + cy^2 + 2dx + 2fy + g = 0. \tag{1}$$

Define the following quantities:

$$\Delta = \begin{vmatrix} a & b & d \\ b & c & f \\ d & f & g \end{vmatrix} \tag{2}$$

$$J = \begin{vmatrix} a & b \\ b & c \end{vmatrix} \tag{3}$$

$$I = a + c \tag{4}$$

$$K = \begin{vmatrix} a & d \\ d & g \end{vmatrix} + \begin{vmatrix} c & f \\ f & g \end{vmatrix}. \tag{5}$$

Then the quadratics are classified into the types summarized in the following table (Beyer 1987). The real (nondegenerate) quadratics (the ELLIPSE, HYPERBOLA, and PARABOLA) correspond to the curves which can be created by the intersection of a PLANE with a (two-NAPPES) CONE, and are therefore known as CONIC SECTIONS.

Curve	Δ	J	Δ/I	K
Coincident Lines	0	0		0
Ellipse (Imaginary)	$\neq 0$	> 0	> 0	
ELLIPSE (Real)	$\neq 0$	> 0	< 0	
HYPERBOLA	$\neq 0$	< 0		
Intersecting Lines (Imaginary)	0	> 0		
Intersecting Lines (Real)	0	< 0		
PARABOLA	$\neq 0$	0		
Parallel Lines (Imaginary)	0	0		> 0
Parallel Lines (Real)	0	0		< 0

It is always possible to eliminate the xy cross term by a suitable ROTATION of the axes. To see this, consider rotation by an arbitrary angle θ. The ROTATION MATRIX is

$$\begin{bmatrix} x \\ y \end{bmatrix} = \begin{bmatrix} \cos\theta & \sin\theta \\ -\sin\theta & \cos\theta \end{bmatrix} \begin{bmatrix} x' \\ y' \end{bmatrix} = \begin{bmatrix} x'\cos\theta + y'\sin\theta \\ -x'\sin\theta + y'\cos\theta \end{bmatrix}, \tag{6}$$

so

$$x = x'\cos\theta + y'\sin\theta \tag{7}$$

$$y = -x'\sin\theta + y'\cos\theta \tag{8}$$

$$xy = -x'^2\cos\theta\sin\theta + x'y'(\cos^2\theta - \sin^2\theta) + y'^2\cos\theta\sin\theta \tag{9}$$

$$x^2 = x'^2\cos^2\theta + 2x'y'\cos\theta\sin\theta + y'^2\sin^2\theta \tag{10}$$

$$y^2 = -x'^2\sin^2\theta - 2x'y'\sin\theta\cos\theta + y'^2\cos^2\theta. \tag{11}$$

Plugging these into (1) gives

$$a\left(x'^2\cos^2\theta + 2x'y'\cos\theta\sin\theta + y'^2\sin^2\theta\right)$$
$$+ 2b(x'\cos\theta + y'\sin\theta) - (-x'\sin\theta + y'\cos\theta)$$

$$+c\left(x'^2 \sin^2 \theta - 2x'y' \cos \theta \sin \theta + y'^2 \cos^2 \theta\right)$$

$$+2d(x' \cos \theta + y' \sin \theta)$$

$$+2f(-x' \sin \theta + y' \cos \theta) + g = 0. \tag{12}$$

$$a\left(x'^2 \cos^2 \theta + 2x'y' \cos \theta + y'^2 \sin^2 \theta\right)$$

$$+2b\left(-x^2 \cos^2 \theta \sin \theta - xy \sin^2 \theta + xy \cos^2 \theta + y^2 \cos \theta \sin \theta\right)$$

$$+c\left(x'^2 \sin^2 \theta - 2x'y' \cos \theta \sin \theta + y'^2 \cos^2 \theta\right)$$

$$+2d(x' \cos \theta + y' \sin \theta)$$

$$+2f(-x' \sin \theta + y' \cos \theta) + g = 0. \tag{13}$$

Grouping terms,

$$x'^2\left(a \cos^2 \theta + c \sin^2 \theta - 2b \cos \theta \sin \theta\right)$$

$$+x'y'\left[2a \cos \theta \sin \theta - 2c \sin \theta \cos \theta + 2b(\cos^2 \theta - \sin^2 \theta)\right]$$

$$+y'^2\left(a \sin^2 \theta + c \cos^2 \theta + 2b \cos \theta \sin \theta\right)$$

$$+x'(2d \cos \theta - 2f \sin \theta) + y'(-2d \sin \theta + 2f \cos \theta)$$

$$+g = 0. \tag{14}$$

Comparing the COEFFICIENTS with (1) gives an equation OF THE FORM

$$a'x'^2 + 2b'x'y' + c'y'^2 + 2d'x' + 2f'y' + g' = 0, \tag{15}$$

where the new COEFFICIENTS are

$$a' = a \cos^2 \theta - 2b \cos \theta \sin \theta + c \sin^2 \theta \tag{16}$$

$$b' = b(\cos^2 \theta - \sin^2 \theta) + (a - c) \sin \theta \cos \theta \tag{17}$$

$$c' = a \sin^2 \theta + 2b \sin \theta \cos \theta + c \cos^2 \theta \tag{18}$$

$$d' = d \cos \theta - f \sin \theta \tag{19}$$

$$f' = -d \sin \theta + f \cos \theta \tag{20}$$

$$g' = g. \tag{21}$$

The cross term $2b'x'y'$ can therefore be made to vanish by setting

$$b' = b(\cos^2 \theta - \sin^2 \theta) - (c - a) \sin \theta \cos \theta$$

$$= b \cos(2\theta) - \tfrac{1}{2}(c - a) \sin(2\theta) = 0. \tag{22}$$

For b' to be zero, it must be true that

$$\cos(2\theta) = \frac{c - a}{2b} \equiv K. \tag{23}$$

The other components are then given with the aid of the identity

$$\cos\left[\cot^{-1}(x)\right] = \frac{x}{\sqrt{1 + x^2}} \tag{24}$$

by defining

$$L \equiv \frac{K}{\sqrt{1 + K^2}}, \tag{25}$$

so

$$\sin \theta = \sqrt{\frac{1 - L}{2}} \tag{26}$$

$$\cos \theta = \sqrt{\frac{1 + L}{2}}. \tag{27}$$

Rotating by an angle

$$\theta = \tfrac{1}{2} \cot^{-1}\left(\frac{c - a}{2b}\right) \tag{28}$$

therefore transforms (1) into

$$a'x'^2 + c'y'^2 + 2d'x' + 2f'y' + g' = 0. \tag{29}$$

COMPLETING THE SQUARE,

$$a'\left(x'^2 + \frac{2d'}{a'}x\right) + c'\left(y'^2 + \frac{2f'}{c'}y'\right) + g' = 0 \tag{30}$$

$$a'\left(x' + \frac{d'}{a'}\right)^2 + c'\left(y' + \frac{f'}{c'}\right)^2 = -g' + \frac{d'^2}{a'} + \frac{f'^2}{c'}. \tag{31}$$

Defining $x'' \equiv x' + d'/a'$, $y'' \equiv y' + f'/c'$, and $g'' \equiv -g' + d'^2/a' + f'^2/c'$ gives

$$a'x'^2 + c'y''^2 = g''. \tag{32}$$

If $g'' \neq 0$, then divide both sides by g''. Defining $a'' \equiv a'/g''$ and $c'' \equiv c'/g''$ then gives

$$a''x''^2 + c''y''^2 = 1. \tag{33}$$

Therefore, in an appropriate coordinate system, the general CONIC SECTION can be written (dropping the primes) as

$$\begin{cases} ax^2 + cy^2 = 1 & a, c, g \neq 0 \\ ax^2 + cy^2 = 0 & a, c \neq 0, g = 0. \end{cases} \tag{34}$$

Consider an equation OF THE FORM $ax^2 + 2bxy + cy^2 = 1$ where $b \neq 0$. Re-express this using t_1 and t_2 in the form

$$ax^2 + 2bxy + cy^2 = t_1 x'^2 + t_2 y'^2. \tag{35}$$

Therefore, rotate the COORDINATE SYSTEM

$$\begin{bmatrix} x' \\ y' \end{bmatrix} = \begin{bmatrix} \cos \theta & \sin \theta \\ -\sin \theta & \cos \theta \end{bmatrix} \begin{bmatrix} x \\ y \end{bmatrix}, \tag{36}$$

so

$$ax^2 + 2bxy + cy^2 = t_1 x'^2 + t_2 y'^2$$

$$= t_1\left(x^2 \cos^2 \theta + 2xy \cos \theta \sin \theta + y^2 \sin^2 \theta\right)$$

$$+ t_2\left(x^2 \sin^2 \theta - 2xy \sin \theta \cos \theta + y^2 \cos^2 \theta\right)$$

$$= x^2 \left(t_1 \cos^2 \theta + t_2 \sin^2 \theta \right) + 2xy \cos \theta \sin \theta (t_1 - t_2)$$

$$+ y^2 \left(t_1 \sin^2 \theta + t_2 \cos^2 \theta \right) \tag{37}$$

and

$$a = t_1 \cos^2 \theta + t_2 \sin^2 \theta \tag{38}$$

$$b = (t_1 - t_2) \cos \theta \sin \theta = \tfrac{1}{2} (t_1 - t_2) \sin(2\theta) \tag{39}$$

$$c = t_1 \sin^2 \theta + t_2 \cos^2 \theta. \tag{40}$$

Therefore,

$$a + c = \left(t_1 \cos^2 \theta + t_2 \sin^2 \theta \right) + \left(t_1 \sin^2 \theta + t_2 \cos^2 \theta \right)$$

$$= t_1 + t_2 \tag{41}$$

$$a - c = t_1 \cos^2 \theta + t_2 \sin^2 \theta - t_1 \sin^2 \theta + t_2 \cos^2 \theta$$

$$= (t_1 - t_2)\left(\cos^2 \theta - \sin^2 \theta \right) = (t_1 - t_2) \cos(2\theta). \tag{42}$$

From (41) and (42),

$$\frac{a - c}{b} = \frac{(t_1 - t_2) \cos(2\theta)}{\tfrac{1}{2}(t_1 - t_2) \sin(2\theta)} = 2 \cot(2\theta), \tag{43}$$

the same angle as before. But

$$\cos(2\theta) = \cos\left[\cot^{-1}\left(\frac{a-c}{2b} \right) \right]$$

$$= \cos\left[\tan^{-1}\left(\frac{2b}{a-c} \right) \right]$$

$$= \frac{1}{\sqrt{1 + \left(\dfrac{2b}{a-c} \right)^2}}, \tag{44}$$

so

$$a - c = \frac{t_1 - t_2}{\sqrt{1 + \left(\dfrac{2b}{a-c} \right)^2}}. \tag{45}$$

Rewriting and copying (41),

$$t_1 - t_2 = (a - c)\sqrt{1 + \left(\frac{2b}{a-c} \right)^2}$$

$$= \sqrt{(a-c)^2 + 4b^2} \tag{46}$$

$$t_1 + t_2 = a + c. \tag{47}$$

Adding (46) and (47) gives

$$t_1 = \tfrac{1}{2}\left[a + c + \sqrt{(a-c)^2 + 4b^2} \right] \tag{48}$$

$$t_2 = a + c - t_1 = \tfrac{1}{2}\left[a + c - \sqrt{(a-c)^2 + 4b^2} \right]. \tag{49}$$

Note that these ROOTS can also be found from

$$(t - t_1)(t - t_2) = t^2 - t(t_1 + t_2) + t_1 t_2 = 0 \tag{50}$$

$$t_2 - t(a + c) + \tfrac{1}{4}\left\{ (a+c)^2 - \left[(a-c)^2 + 4b^2 \right] \right\}$$

$$= t_2 - t(a + c) + \tfrac{1}{4}\left[a^2 + 2ac + c^2 - a^2 + 2ac - c^2 - 4b^2 \right]$$

$$= t^2 - t(a + c) + \left(ac - b^2 \right) = (a - t)(c - t) - b^2$$

$$= \begin{vmatrix} a - t & b \\ b & c - t \end{vmatrix} = (a - t)(c - t) - b^2 = 0. \tag{51}$$

The original problem is therefore equivalent to looking for a solution to

$$\begin{bmatrix} a & b \\ b & c \end{bmatrix} \begin{bmatrix} x \\ y \end{bmatrix} = t \begin{bmatrix} x \\ y \end{bmatrix} \tag{52}$$

$$\begin{bmatrix} ax & bx \\ by & cy \end{bmatrix} \begin{bmatrix} x \\ y \end{bmatrix} = t \begin{bmatrix} x^2 \\ y^2 \end{bmatrix}, \tag{53}$$

which gives the simultaneous equations

$$\begin{cases} ax^2 + bxy = tx^2 \\ bxy + cy^2 = ty^2. \end{cases} \tag{54}$$

Let $\mathbf{X}$ be any point (x, y) with old coordinates and (x', y') be its new coordinates. Then

$$ax^2 + 2bxy + cy^2 = t_+ x'^2 + t_- y'^2 = 1 \tag{55}$$

and

$$x' = \hat{\mathbf{X}}_+ \cdot \begin{bmatrix} x \\ y \end{bmatrix} \tag{56}$$

$$y' = \hat{\mathbf{X}}_- \cdot \begin{bmatrix} x \\ y \end{bmatrix}. \tag{57}$$

If t_+ and t_- are both > 0, the curve is an ELLIPSE. If t_+ and t_- are both < 0, the curve is empty. If t_+ and t_- have opposite SIGNS, the curve is a HYPERBOLA. If either is 0, the curve is a PARABOLA. To find the general form of a quadratic curve in POLAR COORDINATES (as given, for example, in Moulton 1970), plug $x = r \cos \theta$ and $y = r \sin \theta$ into (1) to obtain

$$ar^2 \cos^2 \theta + 2br^2 \cos \theta \sin \theta + cr^2 \sin^2 \theta + 2dr \cos \theta$$

$$+ 2fr \sin \theta + g = 0 \tag{58}$$

$$\left(a \cos^2 \theta + 2b \cos \theta \sin \theta + c \sin^2 \theta \right) + \frac{2}{r}$$

$$\times (d \cos \theta + f \sin \theta) + \frac{g}{r^2} = 0. \tag{59}$$

Define $u \equiv 1/r$. For $g \neq 0$, we can divide through by $2g$,

$$\frac{1}{2}\,u^2+\frac{1}{g}(d\,\cos\,\theta+f\,\sin\,\theta)u+\frac{1}{2g}$$

$$\times(a\,\cos^2\,\theta+2b\,\cos\,\theta\,\sin\,\theta+c\,\sin^2\,\theta)=0. \quad (60)$$

Applying the QUADRATIC FORMULA gives

$$u=-\frac{d}{g}\,\cos\,\theta-\frac{f}{g}\,\sin\,\theta\pm\sqrt{R}, \quad (61)$$

where

$$R\equiv\frac{(d\,\cos\,\theta+f\,\sin\,\theta)^2}{g^2}$$

$$-4\left(\frac{1}{2}\right)\left(\frac{1}{2g}\right)(a\,\cos^2\,\theta+2b\,\cos\,\theta\,\sin\,\theta+c\,\sin^2\,\theta)$$

$$=\frac{d^2}{g^2}\,\cos^2\,\theta+\frac{2df}{g^2}\,\cos\,\theta\,\sin\,\theta+\frac{f^2}{g^2}\,\sin^2\,\theta$$

$$-\frac{1}{g}(a\,\cos^2\,\theta+2b\,\cos\,\theta\,\sin\,\theta+c\,\sin^2\,\theta). \quad (62)$$

Using the trigonometric identities

$$\sin^2\,\theta=1-\cos^2\,\theta \quad (63)$$

$$\sin(2\theta)=2\,\sin\,\theta\,\cos\,\theta, \quad (64)$$

it follows that

$$R=\left(\frac{d^2}{g^2}-\frac{a}{g}-\frac{f^2}{g^2}+\frac{c}{g}\right)\cos^2\,\theta+\left(\frac{df}{g^2}-\frac{b}{g}\right)\sin(2\theta)$$

$$+\left(\frac{f^2}{g^2}-\frac{c}{g}\right)$$

$$=\tfrac{1}{2}[1+\cos(2\theta)]\frac{d^2-ag-f^2+cg}{g^2}+\sin(2\theta)$$

$$\times\left(\frac{df-bg}{g^2}\right)+\frac{f^2-cg}{g^2}\frac{d^2-ag-f^2+cg}{2g^2}\cos(2\theta)$$

$$+\frac{df-db}{g^2}\sin(2\theta)$$

$$+\frac{d^2-ag-f^2+cg+2f^2-2cg}{2g^2}. \quad (65)$$

Defining

$$A\equiv-\frac{f}{g} \quad (66)$$

$$B\equiv-\frac{d}{g} \quad (67)$$

$$C\equiv\frac{df-bg}{g^2} \quad (68)$$

$$D\equiv\frac{d^2-f^2+cg-ag}{2g^2} \quad (69)$$

$$E\equiv\frac{d^2+f^2-ag-cg}{2g^2} \quad (70)$$

then gives the equation

$$u\equiv\frac{1}{r}$$

$$=A\,\sin\,\theta+B\,\cos\,\theta\pm\sqrt{C\,\sin(2\theta)+D\,\cos(2\theta)+E} \quad (71)$$

(Moulton 1970). If $g=0$, then (0) becomes instead

$$u\equiv\frac{1}{r}=-\frac{a\,\cos^2\,\theta+2b\,\cos\,\theta\,\sin\,\theta+c\,\sin^2\,\theta}{2(d\,\cos\,\theta+f\,\sin\,\theta)}. \quad (72)$$

Therefore, the general form of a quadratic curve in polar coordinates is given by

$$u=\begin{cases}A\,\sin\,\theta+B\,\cos\,\theta\\ \pm\sqrt{C\,\sin(2\theta)+D\,\cos(2\theta)+E} & \text{for }g\neq0\\ -\dfrac{a\,\cos^2\,\theta+2b\,\cos\,\theta\,\sin\,\theta+c\,\sin^2\,\theta}{2(d\,\cos\,\theta+f\,\sin\,\theta} & \text{for }g=0.\end{cases}$$

$$(73)$$

See also CONIC SECTION, DISCRIMINANT (QUADRATIC CURVE), ELLIPTIC CURVE

References

Beyer, W. H. *CRC Standard Mathematical Tables*, 28th ed. Boca Raton, FL: CRC Press, pp. 200–201, 1987.

Casey, J. "The General Equation of the Second Degree." Ch. 4 in *A Treatise on the Analytical Geometry of the Point, Line, Circle, and Conic Sections, Containing an Account of Its Most Recent Extensions, with Numerous Examples*, 2nd ed., rev. enl. Dublin: Hodges, Figgis, & Co., pp. 151–172, 1893.

Moulton, F. R. "Law of Force in Binary Stars" and "Geometrical Interpretation of the Second Law." §58 and 59 in *An Introduction to Celestial Mechanics*, 2nd rev. ed. New York: Dover, pp. 86–89, 1970.

Quadratic Effect

PRIME QUADRATIC EFFECT

Quadratic Equation

A quadratic equation is a second-order POLYNOMIAL

$$ax^2+bx+c=0, \quad (1)$$

with $a\neq0$. The roots x can be found by COMPLETING THE SQUARE:

$$x^2+\frac{b}{a}x=-\frac{c}{a} \quad (2)$$

$$\left(x+\frac{b}{2a}\right)^2=-\frac{c}{a}+\frac{b^2}{4a^2}=\frac{b^2-4ac}{4a^2} \quad (3)$$

$$x + \frac{b}{2a} = \frac{\pm \sqrt{b^2 - 4ac}}{2a}. \tag{4}$$

Solving for x then gives

$$x = \frac{-b \pm \sqrt{b^2 - 4ac}}{2a}. \tag{5}$$

This is the QUADRATIC FORMULA.

An alternate form is given by dividing (1) through by x^2:

$$a + \frac{b}{x} + \frac{c}{x^2} = 0 \tag{6}$$

$$c\left(\frac{1}{x^2} + \frac{b}{cx}\right) + a = 0 \tag{7}$$

$$c\left(\frac{1}{x} + \frac{b}{2c}\right)^2 = c\left(\frac{b}{2c}\right)^2 - a = \frac{b^2}{4c} - \frac{4ac}{4c} = \frac{b^2 - 4ac}{4c}. \tag{8}$$

Therefore,

$$\frac{1}{x} + \frac{b}{2c} = \pm \frac{\sqrt{b^2 - 4ac}}{2c} \tag{9}$$

$$\frac{1}{x} = \frac{-b \pm \sqrt{b^2 - 4ac}}{2c} \tag{10}$$

$$x = \frac{2c}{-b \pm \sqrt{b^2 - 4ac}}. \tag{11}$$

This form is helpful if $b^2 \gg 4ac$, in which case the usual form of the QUADRATIC FORMULA can give inaccurate numerical results for one of the ROOTS. This can be avoided by defining

$$q \equiv -\frac{1}{2}\left[b + \operatorname{sgn}(b)\sqrt{b^2 - 4ac}\right] \tag{12}$$

so that b and the term under the SQUARE ROOT sign always have the same sign. Now, if $b > 0$, then

$$q = -\frac{1}{2}\left(b + \sqrt{b^2 - 4ac}\right) \tag{13}$$

$$\frac{1}{q} = \frac{-2}{b + \sqrt{b^2 - 4ac}} \frac{b - \sqrt{b^2 - 4ac}}{b - \sqrt{b^2 - 4ac}}$$

$$= \frac{-2\left(b - \sqrt{b^2 - 4ac}\right)}{b^2 - (b^2 - 4ac)}$$

$$= \frac{-2\left(b - \sqrt{b^2 - 4ac}\right)}{4ac} = \frac{-b + \sqrt{b^2 - 4ac}}{2ac}, \tag{14}$$

so

$$x_1 \equiv \frac{q}{a} = \frac{-b - \sqrt{b^2 - 4ac}}{2a} \tag{15}$$

$$x_2 \equiv \frac{c}{q} = \frac{-b + \sqrt{b^2 - 4ac}}{2a} \tag{16}$$

Similarly, if $b < 0$, then

$$q = -\frac{1}{2}\left(b - \sqrt{b^2 - 4ac}\right) = \frac{1}{2}\left(-b + \sqrt{b^2 - 4ac}\right) \tag{17}$$

$$\frac{1}{q} = \frac{-2}{-b + \sqrt{b^2 - 4ac}} \frac{b + \sqrt{b^2 - 4ac}}{b + \sqrt{b^2 - 4ac}}$$

$$= \frac{2\left(b + \sqrt{b^2 - 4ac}\right)}{-b^2 + (b^2 - 4ac)}$$

$$= \frac{b + \sqrt{b^2 - 4ac}}{-2ac} = \frac{-b - \sqrt{b^2 - 4ac}}{2ac}, \tag{18}$$

so

$$x_1 \equiv \frac{q}{a} = \frac{-b + \sqrt{b^2 - 4ac}}{2a} \tag{19}$$

$$x_2 \equiv \frac{c}{q} = \frac{-b - \sqrt{b^2 - 4ac}}{2a} \tag{20}$$

Therefore, the ROOTS are always given by $x_1 = q/a$ and $x_2 = c/q$.

Now consider the equation expressed in the form

$$a_2 x^2 + a_1 x + a_0 = 0, \tag{21}$$

with solutions z_1 and z_2. These solutions satisfy NEWTON'S RELATIONS

$$z_1 + z_2 = -\frac{a_1}{a_2} \tag{22}$$

$$z_1 z_2 = \frac{a_0}{a_2}. \tag{23}$$

The properties of the SYMMETRIC POLYNOMIALS appearing in NEWTON'S RELATIONS then give

$$z_1^2 + z_2^2 = \frac{a_1^2 - 2a_0 a_2}{a_2^2} \tag{24}$$

$$z_1^3 + z_2^3 = -\frac{a_1^3 - 3a_0 a_1 a_2}{a_2^3} \tag{25}$$

$$z_1^4 + z_2^4 = \frac{a_1^4 - 4a_0 a_1^2 a_2 + 2a_0^2 a_2^2}{a_2^4}. \tag{26}$$

See also CARLYLE CIRCLE, CONIC SECTION, CUBIC EQUATION, DISCRIMINANT (POLYNOMIAL), QUARTIC EQUATION, QUINTIC EQUATION, SEXTIC EQUATION

References

Abramowitz, M. and Stegun, C. A. (Eds.). *Handbook of Mathematical Functions with Formulas, Graphs, and*

Mathematical Tables, 9th printing. New York: Dover, p. 17, 1972.

Beyer, W. H. *CRC Standard Mathematical Tables, 28th ed.* Boca Raton, FL: CRC Press, p. 9, 1987.

Borwein, P. and Erdélyi, T. "Quadratic Equations." §1.1.E.1a in *Polynomials and Polynomial Inequalities.* New York: Springer-Verlag, p. 4, 1995.

Courant, R. and Robbins, H. *What is Mathematics?: An Elementary Approach to Ideas and Methods, 2nd ed.* Oxford, England: Oxford University Press, pp. 91–92, 1996.

King, R. B. *Beyond the Quartic Equation.* Boston, MA: Birkhäuser, 1996.

Press, W. H.; Flannery, B. P.; Teukolsky, S. A.; and Vetterling, W. T. "Quadratic and Cubic Equations." §5.6 in *Numerical Recipes in FORTRAN: The Art of Scientific Computing, 2nd ed.* Cambridge, England: Cambridge University Press, pp. 178–180, 1992.

Spanier, J. and Oldham, K. B. "The Quadratic Function $ax^2 + bx + c$ and Its Reciprocal." Ch. 16 in *An Atlas of Functions.* Washington, DC: Hemisphere, pp. 123–131, 1987.

Quadratic Field

An ALGEBRAIC INTEGER OF THE FORM $a + b\sqrt{D}$ where D is SQUAREFREE forms a quadratic field and is denoted $\mathbb{Q}(\sqrt{D})$. If $D > 0$, the field is called a REAL QUADRATIC FIELD, and if $D < 0$, it is called an IMAGINARY QUADRATIC FIELD. The integers in $\mathbb{Q}(\sqrt{1})$ are simply called "the" INTEGERS. The integers in $\mathbb{Q}(\sqrt{-1})$ are called GAUSSIAN INTEGERS, and the integers in $\mathbb{Q}(\sqrt{-3})$ are called EISENSTEIN INTEGERS. The ALGEBRAIC INTEGERS in an arbitrary quadratic field do not necessarily have unique factorizations. For example, the fields $\mathbb{Q}(\sqrt{-5})$ and $\mathbb{Q}(\sqrt{-6})$ are not uniquely factorable, since

$$21 = 3 \cdot 7 = \left(1 + 2\sqrt{-5}\right)\left(1 - 2\sqrt{-5}\right) \tag{1}$$

$$6 = -\sqrt{6}\left(\sqrt{-6}\right) = 2 \cdot 3, \tag{2}$$

although the above factors are all primes within these fields. All other quadratic fields $\mathbb{Q}\left(\sqrt{D}\right)$ with $|D| \leq 7$ *are* uniquely factorable.

Quadratic fields obey the identities

$$\left(a + b\sqrt{D}\right) \pm \left(c + d\sqrt{D}\right) = (a \pm c) + (b \pm d)\sqrt{D}, \tag{3}$$

$$\left(a + b\sqrt{D}\right)\left(c + d\sqrt{D}\right)$$
$$= (ac + bdD) + (ad + bc)\sqrt{D}, \tag{4}$$

and

$$\frac{a + b\sqrt{D}}{c + d\sqrt{D}} = \frac{ac - bdD}{c^2 - d^2 D} + \frac{(bc - ad)}{c^2 - d^2 D}\sqrt{D} \tag{5}$$

The INTEGERS in the real field $\mathbb{Q}\left(\sqrt{D}\right)$ are of the form $r + sp$, where

$$\rho = \begin{cases} \sqrt{D} & \text{for } D \equiv 2 \text{ or } D \equiv 3 \pmod 4 \\ \dfrac{1}{2}\left(-1 + \sqrt{D}\right) & \text{for } D \equiv 1 \pmod 4. \end{cases} \tag{6}$$

There are exactly 21 quadratic fields in which there is a EUCLIDEAN ALGORITHM, corresponding to $\mathbb{Q}(m)$ for SQUAREFREE integers -11, -7, -3, -2, -1, 2, 3, 5, 6, 7, 11, 13, 17, 19, 21, 29, 33, 37, 41, 57, and 73 (Sloane, N. J. A. Sequences048981). This list was published by Inkeri (1947), but erroneously included the spurious additional term 97 (Barnes and Swinnerton-Dyer 1952; Hardy and Wright 1979, p. 217).

See also ALGEBRAIC INTEGER, EISENSTEIN INTEGER, GAUSSIAN INTEGER, IMAGINARY QUADRATIC FIELD, INTEGER, NUMBER FIELD, REAL QUADRATIC FIELD

References

Barnes, E. S. and Swinnerton-Dyer, H. P. F. "The Inhomogeneous Minima of Binary Quadratic Forms. I." *Acta Math* **87**, 259–323, 1952.

Berg, E. *Fysiogr. Sällsk. Lund. Föhr.* **5**, 1–6, 1935.

Chatland, H. "On the Euclidean Algorithm in Quadratic Number Fields." *Bull. Amer. Math. Soc.* **55**, 948–953, 1949.

Chatland, H. and Davenport, H. "Euclid's Algorithm in Real Quadratic Fields." *Canad. J. Math.* **2**, 289–296, 1950.

Hardy, G. H. and Wright, E. M. "Real Euclidean Fields" and "Real Euclidean Fields (Continued)." §14.8 and 14.9 in *An Introduction to the Theory of Numbers, 5th ed.* Oxford, England: Clarendon Press, pp. 213–217, 1979.

Inkeri, K. "Über den Euklidischen Algorithmus in quadratischen Zahlkörpern." *Ann. Acad. Sci. Fennicae Ser. A. 1. Math.-Phys.*, No. 41, 1–35, 1947.

Koch, H. "Quadratic Number Fields." Ch. 9 in *Number Theory: Algebraic Numbers and Functions.* Providence, RI: Amer. Math. Soc., pp. 275–314, 2000.

LeVeque, W. J. *Topics in Number Theory, Vol. 2.* Reading, MA: Addison-Wesley, p. 57, 1956.

Oppenheim. *Math. Ann.* **109**, 349–352, 1934.

Samuel, P. "Unique Factorization." *Amer. Math. Monthly* **75**, 945–952, 1968.

Stark, H. M. *An Introduction to Number Theory.* Chicago: Markham, p. 294, 1970.

Shanks, D. *Solved and Unsolved Problems in Number Theory, 4th ed.* New York: Chelsea, pp. 153–154, 1993.

Sloane, N. J. A. Sequences A048981 in "An On-Line Version of the Encyclopedia of Integer Sequences." http://www.research.att.com/~njas/sequences/eisonline.html.

Quadratic Form

A quadratic form involving n REAL variables $x_1, x_2, ..., x_n$ associated with the $n \times n$ MATRIX $\mathsf{A} = a_{ij}$ is given by

$$Q(x_1, x_2, \ldots x_n) = a_{ij} x_i x_j, \tag{1}$$

where EINSTEIN SUMMATION has been used. Letting $\mathbf{x}$ be a VECTOR made up of $x_1, ..., x_n$ and $\mathbf{x}^T$ the TRANSPOSE, then

$$Q(\mathbf{x}) = \mathbf{x}^T \mathsf{A} \mathbf{x}, \tag{2}$$

equivalent to

$$Q(\mathbf{x}) = (\mathbf{x}, \mathsf{A}\mathbf{x}) \tag{3}$$

in INNER PRODUCT notation. A BINARY QUADRATIC

FORM is a quadratic form in two variables and has the form

$$Q(x, y) = a_{11}x^2 + 2a_{12}xy + a_{22}y^2. \qquad (4)$$

It is always possible to express an arbitrary quadratic form

$$Q(\mathbf{x}) = a_{ij}x_i x_j, \qquad (5)$$

in the form

$$Q(\mathbf{x}) = (\mathbf{x}, \mathbf{A}\mathbf{x}), \qquad (6)$$

where $\mathbf{A} = a_{ii}$ is a SYMMETRIC MATRIX given by

$$a_{ij} = \begin{cases} a_{ii} & i = j \\ \frac{1}{2}\left(a_{ij} + a_{ji}\right) & i \neq j. \end{cases} \qquad (7)$$

Any REAL quadratic form in n variables may be reduced to the diagonal form

$$Q(x) = \lambda_1 x_1^2 + \lambda_2 x_2^2 + \ldots + \lambda_n x_n^2 \qquad (8)$$

with $\lambda_1 \geq \lambda_2 \geq \cdots \geq \lambda_n$ by a suitable orthogonal point-transformation. Also, two real quadratic forms are equivalent under the group of linear transformations IFF they have the same RANK and SIGNATURE.

See also DISCONNECTED FORM, INDEFINITE QUADRATIC FORM, INNER PRODUCT, INTEGER-MATRIX FORM, POSITIVE DEFINITE QUADRATIC FORM, POSITIVE SEMIDEFINITE QUADRATIC FORM, RANK (QUADRATIC FORM), SIGNATURE (QUADRATIC FORM), SYLVESTER'S INERTIA LAW, SYMMETRIC QUADRATIC FORM

References

Buell, D. A. *Binary Quadratic Forms: Classical Theory and Modern Computations.* New York: Springer-Verlag, 1989.
Conway, J. H. and Fung, F. Y. *The Sensual (Quadratic) Form.* Washington, DC: Math. Assoc. Amer., 1997.
Gradshteyn, I. S. and Ryzhik, I. M. *Tables of Integrals, Series, and Products, 6th ed.* San Diego, CA: Academic Press, pp. 1104–106, 2000.
Kitaoka, Y. *Arithmetic of Quadratic Forms.* Cambridge, England: Cambridge University Press, 1999.
Lam, T. Y. *The Algebraic Theory of Quadratic Forms.* Reading, MA: W. A. Benjamin, 1973.
Weisstein, E. W. "Books about Quadratic Forms." http://www.treasure-troves.com/books/QuadraticForms.html.

Quadratic Formula

The formula giving the ROOTS of a QUADRATIC EQUATION

$$ax^2 + bx + c = 0 \qquad (1)$$

as

$$x = \frac{-b \pm \sqrt{b^2 - 4ac}}{2a}. \qquad (2)$$

An alternate form is given by

$$x = \frac{2c}{-b \pm \sqrt{b^2 - 4ac}}. \qquad (3)$$

See also QUADRATIC EQUATION

Quadratic Integral

To compute an integral OF THE FORM

$$\int \frac{dx}{a + bx + cx^2}, \qquad (1)$$

COMPLETE THE SQUARE in the DENOMINATOR to obtain

$$\int \frac{dx}{a + bx + cx^2} = \frac{1}{c} \int \frac{dx}{\left(x + \dfrac{b}{2c}\right)^2 + \left(\dfrac{a}{c} - \dfrac{b^2}{4c^2}\right)}. \qquad (2)$$

Let $u \equiv x + b/2c$. Then define

$$-A^2 \equiv \frac{a}{c} - \frac{b^2}{4c^2} = \frac{1}{4c^2}\left(4ac - b^2\right) \equiv \frac{1}{4c^2}q, \qquad (3)$$

where

$$q \equiv 4ac - b^2 \qquad (4)$$

is the NEGATIVE of the DISCRIMINANT. If $q < 0$, then

$$A = \frac{1}{2c}\sqrt{-q}. \qquad (5)$$

Now use PARTIAL FRACTION DECOMPOSITION,

$$\frac{1}{c} \int \frac{du}{(u + A)(u - A)} = \frac{1}{c} \int \left(\frac{A_1}{u + A} + \frac{A_2}{u - A}\right) du \qquad (6)$$

$$\left(\frac{A_1}{u + A} + \frac{A_2}{u - A}\right) = \frac{A_1(u - A) + A_2(u + A)}{u^2 - A^2}$$

$$= \frac{(A_1 + A_2)u + A(A_2 - A_1)}{u^2 - A^2}, \qquad (7)$$

so $A_2 + A_1 = 0 \Rightarrow A_2 = -A_1$ and $A(A_2 - A_1) = -2AA_1 = 1 \Rightarrow A_1 = -1/(2A)$. Plugging these in,

$$\frac{1}{c} \int \left(-\frac{1}{2A}\frac{1}{u + A} + \frac{1}{2A}\frac{1}{u - A}\right) du$$

$$= \frac{1}{2Ac}[-\text{In}(u + A) + \text{In}(u - A)]$$

$$= \frac{1}{2Ac}In\left(\frac{u - A}{u + A}\right)$$

$$= \frac{1}{2\left(\dfrac{1}{2c}\right)\sqrt{-q}\,c}\ln\left(\frac{x+\dfrac{b}{2c}-\dfrac{1}{2c}\sqrt{-q}}{x+\dfrac{b}{2c}+\dfrac{1}{2c}\sqrt{-q}}\right)$$

$$= \frac{1}{\sqrt{-q}}\ln\left(\frac{2cx+b-\sqrt{-q}}{2cx+b+\sqrt{-q}}\right) \tag{8}$$

for $q < 0$. Note that this integral is also tabulated in Gradshteyn and Ryzhik (2000, equation 2.172), where it is given with a sign flipped.

References

Gradshteyn, I. S. and Ryzhik, I. M. *Tables of Integrals, Series, and Products, 6th ed.* San Diego, CA: Academic Press, 2000.

Quadratic Invariant

Given the BINARY QUADRATIC FORM

$$ax^2 + 2bxy + cy^2 \tag{1}$$

with DISCRIMINANT $b^2 - ac$, let

$$x = pX + qY \tag{2}$$

$$y = rX + sY. \tag{3}$$

Then

$$a(pX+qY)^2 + 2b(pX+qY)(rX+sY) + c(rX+sY)^2$$

$$= AX^2 + 2BXY + CY^2, \tag{4}$$

where

$$A = ap^2 + 2bpr + cr^2 \tag{5}$$

$$B = apq + b(ps+qr) + crs \tag{6}$$

$$C = aq^2 + 2bqs + cs^2, \tag{7}$$

so

$$B^2 - AC = \left[a^2p^2q^2 + b^2(ps+qr)^2 + c^2r^2s^2\right.$$

$$+2abpq(ps+qr) + 2acpqrs + 2bcrs(ps+qr)\big]$$

$$-(ap^2 + 2bpr + cr^2)(aq^2 + 2bqs + cs^2)$$

$$= a^2p^2q^2 + b^2p^2s^2 + 2b^2pqrs + b^2q^2r^2 + c^2r^2s^2$$

$$+2abp^2qs + 2abpq^2r + 2acpqrs + 2bcprs^2 + 2bcqr^2s$$

$$-a^2p^2q^2 - 2abp^2qs - acp^2s^2 - 2abpq^2r - 4b^2pqrs$$

$$-2bcprs^2 - acq^2r^2 - 2bcqr^2s - c^2r^2s^2$$

$$= b^2p^2s^2 - 2b^2pqrs + b^2q^2r^2 + 2acprs - acp^2s^2$$

$$-acq^2r^2$$

$$= p^2s^2(b^2 - ac) + q^2r^2(b^2 - ac) - 2pqrs(b^2 - ac)$$

$$= (b^2 - ac)(p^2s^2 - 2pqrs + q^2r^2)$$

$$= (ps - rq)^2(b^2 - ac). \tag{8}$$

Surprisingly, this is the same discriminant as before, but multiplied by the factor $(ps-rq)^2$. The quantity $ps - rq$ is called the MODULUS.

See also ALGEBRAIC INVARIANT

Quadratic Irrational Number

An IRRATIONAL NUMBER OF THE FORM

$$\frac{P \pm \sqrt{D}}{Q},$$

where P and Q are INTEGERS and D is a SQUAREFREE INTEGER. Quadratic irrational numbers are sometimes also called quadratic surds. In 1770, Lagrange proved that any quadratic irrational has a CONTINUED FRACTION which is periodic after some point.

See also CONTINUED FRACTION, MINKOWSKI'S QUESTION MARK FUNCTION

Quadratic Map

A 1-D MAP often called "the" quadratic map is defined by

$$x_{n+1} = x_n^2 + c. \tag{1}$$

This is the real version of the complex map defining the MANDELBROT SET. The quadratic map is called attracting if the JACOBIAN $J < 1$, and repelling if $J > 1$. FIXED POINTS occur when

$$x^{(1)} = [x^{(1)}]^2 + c \tag{2}$$

$$\left(x^{(1)}\right)^2 - x^{(1)} + c = 0 \tag{3}$$

$$x_\pm^{(1)} = \tfrac{1}{2}\left(1 \pm \sqrt{1-4c}\right). \tag{4}$$

Period two FIXED POINTS occur when

$$x_{n+2} = x_{n+1}^2 + c = \left(x_n^2 + c\right)^2 + c$$

$$= x_n^4 + 2cx_n^2 + (c^2 + c) = x_n \tag{5}$$

$$x^4 + 2x^2 - x + (cx^2 + c) = (x^2 - x + c)(x^2 + x + 1 + c)$$

$$= 0 \tag{6}$$

$$x_\pm^{(2)} = \tfrac{1}{2}\left[1 \pm \sqrt{1-4(1+c)}\right] = \tfrac{1}{2}\left(1 \pm \sqrt{-3-4c}\right). \tag{7}$$

Period three FIXED POINTS occur when

$$x^6 + x^5 + (3c+1)x^4 + (2c+1)x^3 + (c^2+3c+1)x^2$$

$$+(c+1)^2 x + \left(c^3 + 2c^2 + c + 1\right) = 0. \tag{8}$$

The most general second-order 2-D MAP with an elliptic fixed point at the origin has the form

$$x' = x \cos \alpha - y \sin \alpha + a_{20}x^2 + a_{11}xy + a_{02}y^2 \quad (9)$$

$$y' = x \sin \alpha + y \cos \alpha + b_{20}x^2 + b_{11}xy + b_{02}y^2. \quad (10)$$

The map must have a DETERMINANT of 1 in order to be AREA-preserving, reducing the number of independent parameters from seven to three. The map can then be put in a standard form by scaling and rotating to obtain

$$x' = x \cos \alpha - y \sin \alpha + x^2 \sin \alpha \quad (11)$$

$$y' = x \sin \alpha + y \cos \alpha - x^2 \cos \alpha. \quad (12)$$

The inverse map is

$$x = x' \cos \alpha + y' \sin \alpha \quad (13)$$

$$y = -x' \sin \alpha + y' \cos \alpha + (x' \cos \alpha + y' \sin \alpha)^2. \quad (14)$$

The FIXED POINTS are given by

$$x_i^2 \sin \alpha + 2x_i \cos \alpha - x_{i-1} - x_{i+1} = 0 \quad (15)$$

for $i = 0, ..., n-1$.

See also BOGDANOV MAP, HÉNON MAP, LOGISTIC MAP, LOZI MAP, MANDELBROT SET

Quadratic Mean
ROOT-MEAN-SQUARE

Quadratic Nonresidue
QUADRATIC RESIDUE

Quadratic Phase Array
A method to obtain a signal $C_\ell(z)$ with a flat spectrum $c(\theta; z)$ (such as a pulse), but having a smaller amplitude than the pulse.

$$c(\theta, z) \equiv e^{iz\phi(\theta)} = \sum_{\ell=-\infty}^{\infty} e^{i\ell\theta} C\ell(z), \quad (1)$$

whence

$$C_\ell(z) = 1/(2\pi) \int_{-\pi}^{\pi} e^{i(z\phi(\theta) - \ell\theta)} d\theta, \quad (2)$$

where

$$\phi(\theta) = (1 - |\theta|/\pi)\theta/\pi, \quad (3)$$

with $|\theta| \leq \pi$.

Thus $c(\theta; z)$ and $C_\ell(z)$ are a Fourier pair, and since $|c(\theta, z)| = 1$, it is guaranteed that the sequence C_ℓ has a flat spectrum. The sequence C_ℓ is called the "quadratic phase array."

References
Aarts, R. M. and Janssen, A. J. E. M. "On Analytic Design of Loudspeaker Arrays with Uniform Radiation Characteristics." *J. Acoust. Soc. Amer.* **107**, 287–292, 2000.

Quadratic Reciprocity Law
QUADRATIC RECIPROCITY THEOREM

Quadratic Reciprocity Theorem
Also called the AUREUM THEOREMA (GOLDEN THEOREM) by Gauss. If p and q are distinct ODD PRIMES, then the CONGRUENCES

$$x^2 \equiv q \pmod{p}$$

$$x^2 \equiv p \pmod{q}$$

are both solvable or both unsolvable unless both p and q leave the remainder 3 when divided by 4 (in which case one of the CONGRUENCES is solvable and the other is not). Written symbolically,

$$\left(\frac{p}{q}\right)\left(\frac{q}{p}\right) = (-1)^{(p-1)(q-1)/4},$$

where

$$\left(\frac{p}{q}\right) \equiv \begin{cases} 1 & \text{for } x^2 \equiv p \pmod{q} \text{ solvable for } x \\ -1 & \text{for } x^2 \equiv p \pmod{q} \text{ not solvable for } x \end{cases}$$

is known as a LEGENDRE SYMBOL.

Euler stated the theorem in 1783 without proof. Legendre was the first to publish a proof, but it was fallacious. In 1796, Gauss became the first to publish a correct proof (Nagell 1951, p. 144). The quadratic reciprocity theorem was Gauss's favorite theorem from NUMBER THEORY, and he devised no fewer than eight different proofs of it over his lifetime.

The GENUS THEOREM states that the DIOPHANTINE EQUATION

$$x^2 + y^2 = p$$

can be solved for p a PRIME IFF $p \equiv 1 \pmod{4}$ or $p = 2$.

See also GENUS THEOREM, JACOBI SYMBOL, KRONECKER SYMBOL, LEGENDRE SYMBOL, QUADRATIC RESIDUE, RECIPROCITY THEOREM

References
Courant, R. and Robbins, H. *What is Mathematics?: An Elementary Approach to Ideas and Methods,* 2nd ed. Oxford, England: Oxford University Press, p. 39, 1996.
Ireland, K. and Rosen, M. "Quadratic Reciprocity." Ch. 5 in *A Classical Introduction to Modern Number Theory,* 2nd ed. New York: Springer-Verlag, pp. 50–65, 1990.
Nagell, T. "The Quadratic Reciprocity Law." §41 in *Introduction to Number Theory.* New York: Wiley, pp. 141–145, 1951.
Riesel, H. "The Law of Quadratic Reciprocity." *Prime Numbers and Computer Methods for Factorization,* 2nd ed. Boston, MA: Birkhäuser, pp. 279–281, 1994.
Shanks, D. *Solved and Unsolved Problems in Number Theory,* 4th ed. New York: Chelsea, pp. 42–49, 1993.

Quadratic Recurrence

N.B. A detailed online essay by S. Finch was the starting point for this entry.

A quadratic recurrence is a RECURRENCE RELATION on a SEQUENCE of numbers $\{x_n\}$ expressing x_n as a second degree polynomial in x_k with $k < n$. For example,

$$x_n = x_{n-1}x_{n-2} \tag{1}$$

is a quadratic recurrence. Another simple example is

$$x_n = (x_{n-1})^2 \tag{2}$$

with $x_0 = 2$, which has solution $x_n = 2^{2^n}$. Another example is the number of "strongly" binary trees of height $\leq n$, given by

$$y_n = (y_{n-1})^2 + 1 \tag{3}$$

with $y_0 = 1$. This has solution

$$y_n = \lfloor c^{2^n} \rfloor, \tag{4}$$

where

$$c = \exp\left[\sum_{j=0}^{\infty} 2^{-j-1}\ln\left(1 + y_j^{-2}\right)\right] = 1.502836801\ldots \tag{5}$$

and $\lfloor x \rfloor$ is the FLOOR FUNCTION (Aho and Sloane 1973). A third example is the closest strict under-approximation of the number 1,

$$s_n = \sum_{i=1}^{n} \frac{1}{z_i}, \tag{6}$$

where $1 < z_1 < \ldots < z_n$ are integers. The solution is given by the recurrence

$$z_n = (z_{n-1})^2 - z_{n-1} + 1, \tag{7}$$

with $z_1 = 2$. This has a closed solution as

$$z_n = \left\lfloor d^{2^n} + \tfrac{1}{2} \right\rfloor \tag{8}$$

where

$$d = \tfrac{1}{2}\sqrt{6}\exp\left\{\sum_{j=1}^{\infty} 2^{-j-1}\ln\left[1 + (2z_j - 1)^{-2}\right]\right\}$$
$$= 1.2640847353\ldots \tag{9}$$

(Aho and Sloane 1973). A final example is the well-known recurrence

$$c_n = (c_{n-1})^2 - \mu \tag{10}$$

with $c_0 = 0$ used to generate the MANDELBROT SET.

See also MANDELBROT SET, RECURRENCE RELATION

References

Aho, A. V. and Sloane, N. J. A. "Some Doubly Exponential Sequences." *Fib. Quart.* **11**, 429–437, 1973.

Finch, S. "Favorite Mathematical Constants." http://www.mathsoft.com/asolve/constant/quad/quad.html.

Quadratic Representation

SUM OF SQUARES FUNCTION

Quadratic Residue

If there is an INTEGER x such that

$$x^2 \equiv q \pmod{p}, \tag{1}$$

then q is said to be a quadratic residue (mod p). If not, q is said to be a quadratic nonresidue (mod p). Hardy and Wright (1979, pp. 67–68) use the shorthand notations $q \text{ R } p$ and $q \text{ N } p$, to indicated that q is a quadratic residue or nonresidue, respectively.

For example, $4^2 \equiv 6$, so 6 is a quadratic residue (mod 10). The entire set of quadratic residues (mod 10) are given by 1, 4, 5, 6, and 9, since

$$1^2 \equiv 1 \pmod{10} \quad 2^2 \equiv 4 \pmod{10} \quad 3^2 \equiv 9 \pmod{10}$$

$$4^2 \equiv 6 \pmod{10} \quad 5^2 \equiv 5 \pmod{10} \quad 6^2 \equiv 6 \pmod{10}$$

$$7^2 \equiv 9 \pmod{10} \quad 8^2 \equiv 4 \pmod{10} \quad 9^2 \equiv 1 \pmod{10}$$

making the numbers 2, 3, 7, and 8 the quadratic nonresidues (mod 10).

A list of quadratic residues for $p \leq 29$ is given below (Sloane's A046071), with those numbers $< p$ not in the list being quadratic nonresidues of p.

p	Quadratic Residues
1	(none)
2	1
3	1
4	1
5	1, 4
6	1, 3, 4
7	1, 2, 4
8	1, 4
9	1, 4, 7
10	1, 4, 5, 6, 9
11	1, 3, 4, 5, 9
12	1, 4, 9
13	1, 3, 4, 9, 10, 12
14	1, 2, 4, 7, 8, 9, 11
15	1, 4, 6, 9, 10
16	1, 4, 9

17	1, 2, 4, 8, 9, 13, 15, 16
18	1, 4, 7, 9, 10, 13, 16
19	1, 4, 5, 6, 7, 9, 11, 16, 17
20	1, 4, 5, 9, 16

Given an ODD PRIME p and an INTEGER a, then the LEGENDRE SYMBOL is given by

$$\left(\frac{a}{p}\right) = \begin{cases} 1 & \text{if a is a quadratic residue mod } p \\ -1 & \text{otherwise.} \end{cases} \quad (2)$$

If

$$r^{(p-1)/2} \equiv \pm 1 \pmod{p}, \quad (3)$$

then r is a quadratic residue $(+)$ or nonresidue $(-)$. This can be seen since if r is a quadratic residue of p, then there exists a square x^2 such that $r \equiv x^2 \pmod{p}$, so

$$r^{(p-1)/2} \equiv \left(x^2\right)^{(p-1)/2} \equiv x^{p-1} \pmod{p}, \quad (4)$$

and x^{p-1} is congruent to 1 (mod p) by FERMAT'S LITTLE THEOREM.

Given p and q in the congruence

$$x^2 \equiv q \pmod{p}, \quad (5)$$

x can be explicitly computed for p and q of certain special forms:

$$x = \begin{cases} q^{k+1} \pmod{p} \\ \quad \text{for } p = 4k+3 \\ q^{k+1} \pmod{p} \\ \quad \text{for } p = 8k+5 \text{ and } q^{2k+1} \equiv 1 \pmod{p} \\ \frac{1}{2}(4q)^{k+1}(p+1) \pmod{p} \\ \quad \text{for } p = 8k+5 \text{ and } q^{2k+1} \equiv -1 \pmod{p}. \end{cases} \quad (6)$$

For example, the first form can be used to find x given the quadratic residues $q = 1, 3, 4, 5,$ and 9 (mod $p = 11$, having $k = 2$), whereas the second and third forms determine x given the quadratic residues $q = 1, 3, 4, 9, 10,$ and 12 (mod $p = 13$, having $k = 1$), and $q = 1, 3, 4, 7, 9, 10, 11, 12, 16, 21, 25, 26, 27, 28, 30, 33, 34, 36$ (mod $p = 37$, having $k = 4$).

More generally, let q be a quadratic residue modulo an ODD PRIME p. Choose h such that the LEGENDRE SYMBOL $(h^2 - 4q/p) = -1$. Then defining

$$V_1 = h \quad (7)$$

$$V_2 = h^2 - 2q \quad (8)$$

$$V_i = hV_{i-1} - qV_{i-2} \quad \text{for } i \geq 3, \quad (9)$$

gives

$$V_{2i} = V_i^2 - 2q^i \quad (10)$$

$$V_{2i+1} = V_i V_{i+1} - hn^i, \quad (11)$$

and a solution to the quadratic CONGRUENCE is

$$x = \frac{1}{2}(p+1)V_{(p+1)/2} \pmod{p}. \quad (12)$$

Schoof (1985) gives an algorithm for finding x with running time $\mathcal{O}(\ln n)^{10}$ (Hardy *et al.* 1990). The congruence is solved by the *Mathematica* command `SqrtMod[q, p]` in the *Mathematica* add-on package `NumberTheory`NumberTheoryFunctions`` (which can be loaded with the command `<<NumberTheory`).

The following table gives the PRIMES which have a given number d as a quadratic residue.

d	Primes
-6	$24k + 1, 5, 7, 11$
-5	$20k + 1, 3, 7, 9$
-3	$6k + 1$
-2	$8k + 1, 3$
-1	$4k + 1$
2	$8k \pm 1$
3	$12k \pm 1$
5	$10k \pm 1$
6	$24k \pm 1, 5$

Finding the CONTINUED FRACTION of a SQUARE ROOT $\sqrt{D}$ and using the relationship

$$Q_n = \frac{D - P_n^2}{Q_{n-1}} \quad (13)$$

for the nth CONVERGENT P_n/Q_n gives

$$P_n^2 \equiv -Q_n Q_{n-1} \pmod{D}. \quad (14)$$

Therefore, $-Q_n Q_{n-1}$ is a quadratic residue of D. But since $Q_1 = 1$, $-Q_2$ is a quadratic residue, as must be $-Q_2 Q_3$. But since $-Q_2$ is a quadratic residue, so is Q_3, and we see that $(-1)^{n-1}Q_n$ are all quadratic residues of D. This method is not guaranteed to produce all quadratic residues, but can often produce several small ones in the case of large D, enabling D to be factored.

The number of SQUARES $s(n)$ in $\mathbb{Z}_n$ is related to the number $q(n)$ of quadratic residues in $\mathbb{Z}_n$ by

$$q(p^n) = s(p^n) - s(p^{n-2}) \quad (15)$$

for $n \geq 3$ (Stangl 1996). Both q and s are MULTIPLICATIVE FUNCTIONS.

See also ASSOCIATE, EULER'S CRITERION, JACOBI

SYMBOL, KRONECKER SYMBOL, LEGENDRE SYMBOL, MULTIPLICATIVE FUNCTION, QUADRATIC RECIPROCITY THEOREM, RIEMANN HYPOTHESIS

References

Burgess, D. A. "The Distribution of Quadratic Residues and Non-Residues." *Mathematika* **4**, 106–112, 1975.

Burton, D. M. *Elementary Number Theory, 4th ed.* New York: McGraw-Hill, p. 201, 1997.

Courant, R. and Robbins, H. "Quadratic Residues." §2.3 in Supplement to Ch. 1 in *What is Mathematics?: An Elementary Approach to Ideas and Methods, 2nd ed.* Oxford, England: Oxford University Press, pp. 38–40, 1996.

Guy, R. K. "Quadratic Residues. Schur's Conjecture" and "Patterns of Quadratic Residues." §F5 and F6 in *Unsolved Problems in Number Theory, 2nd ed.* New York: Springer-Verlag, pp. 244–248, 1994.

Hardy, G. H. and Wright, E. M. "Quadratic Residues." §6.5 in *An Introduction to the Theory of Numbers, 5th ed.* Oxford, England: Clarendon Press, pp. 67–68, 1979.

Hilton, P.; Holton, D.; and Pedersen, J. *Mathematical Reflections in a Room with Many Mirrors.* New York: Springer-Verlag, p. 43, 1997.

Nagell, T. "Theory of Quadratic Residues." Ch. 4 in *Introduction to Number Theory.* New York: Wiley, pp. 115 and 132–155, 1951.

Niven, I. and Zuckerman, H. *An Introduction to the Theory of Numbers, 4th ed.* New York: Wiley, p. 84, 1980.

Rosen, K. H. Ch. 9 in *Elementary Number Theory and Its Applications, 3rd ed.* Reading, MA: Addison-Wesley, 1993.

Schoof, R. "Elliptic Curves Over Finite Fields and the Computation of Square Roots mod *p*." *Math. Comput.* **44**, 483–494, 1985.

Séroul, R. "Quadratic Residues." §2.10 in *Programming for Mathematicians.* Berlin: Springer-Verlag, pp. 17–18, 2000.

Shanks, D. *Solved and Unsolved Problems in Number Theory, 4th ed.* New York: Chelsea, pp. 63–66, 1993.

Sloane, N. J. A. Sequences A046071 in "An On-Line Version of the Encyclopedia of Integer Sequences." http://www.research.att.com/~njas/sequences/eisonline.html.

Stangl, W. D. "Counting Squares in $\mathbb{Z}_n$." *Math. Mag.* **69**, 285–289, 1996.

Tonelli, A. "Bemerkung über die Auflösung quadratischer Congruenzen." *Göttingen Nachr.*, 344–346, 1891.

Wagon, S. "Quadratic Residues." §9.2 in *Mathematica in Action.* New York: W. H. Freeman, pp. 292–296, 1991.

Quadratic Sieve

A procedure used in conjunction with DIXON'S FACTORIZATION METHOD to factor large numbers n. Pick values of r given by

$$\lfloor \sqrt{n} \rfloor + k, \tag{1}$$

where $k = 1, 2, \ldots$ and $\lfloor x \rfloor$ is the FLOOR FUNCTION. We are then looking for factors p such that

$$n \equiv r^2 \pmod{p}, \tag{2}$$

which means that only numbers with LEGENDRE SYMBOL $(n/p) = 1$ (less than $N = \pi(d)$ for TRIAL DIVISOR d, where $\pi(d)$ is the PRIME COUNTING FUNCTION) need be considered. The set of PRIMES for which this is true is known as the FACTOR BASE. Next, the CONGRUENCES

$$x^2 \equiv n \pmod{p} \tag{3}$$

must be solved for each p in the FACTOR BASE. Finally, a sieve is applied to find values of $f(r) = r^2 - n$ which can be factored completely using only the FACTOR BASE. GAUSSIAN ELIMINATION is then used as in DIXON'S FACTORIZATION METHOD in order to find a product of the $f(r)$s, yielding a PERFECT SQUARE.

The method requires about $\exp\left(\sqrt{\ln n \, \ln \ln n}\right)$ steps, improving on the CONTINUED FRACTION FACTORIZATION ALGORITHM by removing the 2 under the SQUARE ROOT (Pomerance 1996). The use of multiple POLYNOMIALS gives a better chance of factorization, requires a shorter sieve interval, and is well suited to parallel processing.

See also NUMBER FIELD SIEVE, PRIME FACTORIZATION ALGORITHMS, SMOOTH NUMBER

References

Alford, W. R. and Pomerance, C. "Implementing the Self Initializing Quadratic Sieve on a Distributed Network." In *Number Theoretic and Algebraic Methods in Computer Science, Proc. Internat. Moscow Conf., June-July 1993* (Ed. A. J. van der Poorten, I. Shparlinksi, and H. G. Zimer). Singapore: World Scientific, pp. 163–174, 1995.

Boender, H. and te Riele, H. J. J. "Factoring Integers with Large Prime Variations of the Quadratic Sieve." Preprint. Centrum voor Wiskunde en Informatica, No. NM-R9513, 1995.

Brent, R. P. "Parallel Algorithms for Integer Factorisation." In *Number Theory and Cryptography* (Ed. J. H. Loxton). New York: Cambridge University Press, 26–37, 1990.

Bressoud, D. M. Ch. 8 in *Factorization and Prime Testing.* New York: Springer-Verlag, 1989.

Gerver, J. "Factoring Large Numbers with a Quadratic Sieve." *Math. Comput.* **41**, 287–294, 1983.

Lenstra, A. K. and Manasse, M. S. "Factoring by Electronic Mail." In *Advances in Cryptology--Eurocrypt '89* (Ed. J.-J. Quisquarter and J. Vandewalle). Berlin: Springer-Verlag, pp. 355–371, 1990.

Pomerance, C. "The Quadratic Sieve Factoring Algorithm." In *Advances in Cryptology: Proceedings of EUROCRYPT 84* (Ed. T. Beth, N. Cot, and I. Ingemarsson). New York: Springer-Verlag, pp. 169–182, 1985.

Pomerance, C. "A Tale of Two Sieves." *Not. Amer. Math. Soc.* **43**, 1473–1485, 1996.

Pomerance, C.; Smith, J. W.; and Tuler, R. "A Pipeline Architecture for Factoring Large Integers with the Quadratic Sieve Method." *SIAM J. Comput.* **17**, 387–403, 1988.

Silverman, R. D. "The Multiple Polynomial Quadratic Sieve." *Math. Comput.* **48**, 329–339, 1987.

Quadratic Surd

QUADRATIC IRRATIONAL NUMBER

Quadratic Surface

A second-order ALGEBRAIC SURFACE given by the general equation

$$ax^2 + by^2 + cz^2 + 2fyz + 2gzx + 2hxy + 2px + 2py + 2rz$$
$$+ d = 0. \tag{1}$$

Quadratic surfaces are also called quadrics, and there

are 17 standard-form types. A quadratic surface intersects every plane in a (proper or degenerate) CONIC SECTION. In addition, the CONE consisting of all tangents from a fixed point to a quadratic surface cuts every plane in a CONIC SECTION, and the points of contact of this CONE with the surface form a CONIC SECTION (Hilbert and Cohn-Vossen 1999, p. 12).

Define

$$e = \begin{bmatrix} a & h & g \\ h & b & f \\ g & f & c \end{bmatrix} \qquad (2)$$

$$E = \begin{bmatrix} a & h & g & p \\ h & b & f & q \\ g & f & c & r \\ p & q & r & d \end{bmatrix} \qquad (3)$$

$$\rho_3 = \text{rank } e \qquad (4)$$

$$\rho_4 = \text{rank } E \qquad (5)$$

$$\Delta = \det E, \qquad (6)$$

and k_1, k_2, as k_3 are the roots of

$$\begin{vmatrix} a-x & h & g \\ h & b-x & f \\ g & f & c-x \end{vmatrix} = 0. \qquad (7)$$

Also define

$$k \equiv \begin{cases} 1 & \text{if the signs of nonzero } ks \text{ are the same} \\ 0 & \text{otherwise.} \end{cases} \qquad (8)$$

Then the following table enumerates the 17 quadrics and their properties (Beyer 1987).

Surface	Equation	ρ_3	ρ_4	sgn(Δ)	k
Coincident PLANES	$x^2 = 0$	1	1		
Ellipsoid (Imaginary)	$\frac{x^2}{a^2}+\frac{y^2}{b^2}+\frac{z^2}{c^2}=-1$	3	4	$+$	1
ELLIPSOID (Real)	$\frac{x^2}{a^2}+\frac{y^2}{b^2}+\frac{z^2}{c^2}=1$	3	4	$(-)$	1
Elliptic Cone (Imaginary)	$\frac{x^2}{a^2}+\frac{y^2}{b^2}+\frac{z^2}{c^2}=0$	3	3		1
ELLIPTIC CONE (Real)	$z^2 = \frac{x^2}{a^2}+\frac{y^2}{b^2}$	3	3		0
Elliptic Cylinder (Imaginary)	$\frac{x^2}{a^2}+\frac{y^2}{b^2}=-1$	2	3		1
ELLIPTIC CYLINDER (Real)	$\frac{x^2}{a^2}+\frac{y^2}{b^2}=1$	2	3		1
ELLIPTIC PARABOLOID	$z = \frac{x^2}{a^2}+\frac{y^2}{b^2}$	2	4	$(-)$	1
HYPERBOLIC CYLINDER	$\frac{x^2}{a^2}-\frac{y^2}{b^2}=-1$	2	3		0
HYPERBOLIC PARABOLOID	$z = \frac{y^2}{a^2}-\frac{x^2}{b^2}$	2	4	$+$	0
HYPERBOLOID of one Sheet	$\frac{x^2}{a^2}+\frac{y^2}{b^2}-\frac{z^2}{c^2}=1$	3	4	$+$	0
HYPERBOLOID of two Sheets	$\frac{x^2}{a^2}+\frac{y^2}{b^2}-\frac{z^2}{c^2}=-1$	3	4	$(-)$	0
Intersecting Planes (Imaginary)	$\frac{x^2}{a^2}+\frac{y^2}{b^2}=0$	2	2		1
Intersecting PLANES (Real)	$\frac{x^2}{a^2}-\frac{y^2}{b^2}=0$	2	2		0
PARABOLIC CYLINDER	$x^2 + 2rz = 0$	1	3		
Parallel Planes (Imaginary)	$x^2 = -a^2$	1	2		
Parallel PLANES (Real)	$x^2 = a^2$	1	2		

Of the non-degenerate quadratic surfaces, the ELLIPTIC (and usual) CYLINDER, HYPERBOLIC CYLINDER, ELLIPTIC (and usual) CONE are RULED SURFACES, while the one-sheeted HYPERBOLOID and HYPERBOLIC PARABOLOID are DOUBLY RULED SURFACES.

A curve in which two arbitrary quadratic surfaces in arbitrary positions intersect cannot meet any plane in more than four points (Hilbert and Cohn-Vossen 1999, p. 24).

See also CONE, CONFOCAL QUADRICS, CUBIC SURFACE, CYLINDER, DOUBLY RULED SURFACE, ELLIPSOID, ELLIPTIC CONE, ELLIPTIC CYLINDER, ELLIPTIC PARABOLOID, HYPERBOLIC CYLINDER, HYPERBOLIC PARABOLOID, HYPERBOLOID, PLANE, QUARTIC SURFACE, RULED SURFACE, SURFACE

References

Beyer, W. H. *CRC Standard Mathematical Tables, 28th ed.* Boca Raton, FL: CRC Press, pp. 210–211, 1987.

Hilbert, D. and Cohn-Vossen, S. "The Second-Order Surfaces." §3 in *Geometry and the Imagination.* New York: Chelsea, pp. 12–19, 1999.

Mollin, R. A. *Quadrics.* Boca Raton, FL: CRC Press, 1995.

Quadratrix of Hippias

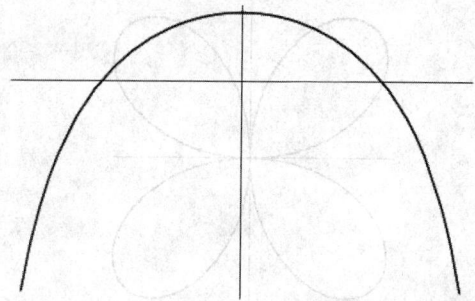

The quadratrix was discovered by Hippias of Elias in 430 BC, and later studied by Dinostratus in 350 BC (MacTutor Archive). It can be used for ANGLE TRISECTION or, more generally, division of an ANGLE into any integral number of equal parts, and CIRCLE SQUARING. In POLAR COORDINATES,

$$\pi\rho = 2r\theta \csc\theta,$$

so

$$r = \frac{\rho\pi\sin\theta}{\theta},$$

which is proportional to the COCHLEOID.

References

Beyer, W. H. *CRC Standard Mathematical Tables, 28th ed.* Boca Raton, FL: CRC Press, p. 223, 1987.

Lawrence, J. D. *A Catalog of Special Plane Curves.* New York: Dover, pp. 195 and 198, 1972.

MacTutor History of Mathematics Archive. "Quadratrix of Hippias." http://www-groups.dcs.st-and.ac.uk/~history/Curves/Quadratrix.html.

Quadrature

The word quadrature has (at least) three incompatible meanings. Integration by quadrature either means solving an INTEGRAL analytically (i.e., symbolically in terms of known functions), or solving of an integral numerically (e.g., GAUSSIAN QUADRATURE, QUADRATURE FORMULAS). Ueberhuber (1997, p. 71) uses the word "quadrature" to mean numerical computation of a univariate INTEGRAL, and "CUBATURE" to mean numerical computation of a MULTIPLE INTEGRAL.

The word quadrature is also used to mean SQUARING: the construction of a square using only COMPASS and STRAIGHTEDGE which has the same AREA as a given geometric figure. If quadrature is possible for a PLANE figure, it is said to be QUADRABLE.

For a function tabulated at given values x_i (so the ABSCISSAS cannot be chosen at will), write the function ϕ as a sum of ORTHONORMAL FUNCTIONS p_j satisfying

$$\int_a^b p_i(x)p_j(x)W(x)dx = \delta_{ij} \tag{1}$$

as

$$\phi(x) = \sum_{j=0}^{\infty} a_j p_j(x), \tag{2}$$

and plug into

$$\int_a^b \phi(x)W(x)\,dx = \int_a^b \sum_{j=1}^m \frac{\pi(x)W(x)}{(x-x_j)\pi'(x_j)}dx\, f(x_j)$$
$$\equiv \sum_{j=1}^m w_j f(x_j), \tag{3}$$

giving

$$\int_a^b \sum_{j=0}^{\infty} a_j p_j(x)W(x)dx = \sum_{i=1}^n w_i \left[\sum_{j=0}^{\infty} a_j p_j(x_j) \right]. \tag{4}$$

But we wish this to hold for all degrees of approximation, so

$$a_j \int_a^b p_j(x)W(x)dx = a_j \sum_{i=1}^n w_i p_j(x_i) \tag{5}$$

$$\int_a^b p_j(x)W(x)dx = \sum_{i=1}^n w_i p_j(x_i). \tag{6}$$

Setting $i = 0$ in (1) gives

$$\int_a^b p_0(x)p_j(x)W(x)dx = \delta_{0j}. \tag{7}$$

The zeroth order orthonormal function can always be taken as $p_0(x) = 1$, so (7) becomes

$$\int_a^b p_j(x)W(x)dx = \delta_{0j} \tag{8}$$

$$= \sum_{i=1}^n w_i p_j(x_i), \tag{9}$$

where (6) has been used in the last step. We therefore have the MATRIX equation

$$\begin{bmatrix} p_0(x_1) & \cdots & p_0(x_n) \\ p_0(x_1) & \cdots & p_1(x_n) \\ \vdots & \ddots & \vdots \\ p_{n-1}(x_1) & \cdots & p_{n-1}(x_n) \end{bmatrix} \begin{bmatrix} w_1 \\ w_2 \\ \vdots \\ w_n \end{bmatrix} = \begin{bmatrix} 1 \\ 0 \\ \vdots \\ 0 \end{bmatrix} \tag{10}$$

which can be inverted to solve for the w_is (Press *et al.* 1992).

See also CALCULUS, CHEBYSHEV-GAUSS QUADRATURE, CHEBYSHEV QUADRATURE, CUBATURE, DERIVATIVE, DOUBLE EXPONENTIAL INTEGRATION, FUNDAMENTAL THEOREM OF GAUSSIAN QUADRATURE, GAUSS-JACOBI MECHANICAL QUADRATURE, GAUSS-KRONROD QUAD-

RATURE, GAUSSIAN QUADRATURE, HERMITE-GAUSS QUADRATURE, HERMITE QUADRATURE, JACOBI-GAUSS QUADRATURE, JACOBI QUADRATURE, LAGUERRE-GAUSS QUADRATURE, LAGUERRE QUADRATURE, LEGENDRE-GAUSS QUADRATURE, LEGENDRE QUADRATURE, LOBATTO QUADRATURE, MECHANICAL QUADRATURE, MEHLER QUADRATURE, NEWTON-COTES FORMULAS, NUMERICAL INTEGRATION, RADAU QUADRATURE, RECURSIVE MONOTONE STABLE QUADRATURE

References

Abramowitz, M. and Stegun, C. A. (Eds.). "Integration." §25.4 in *Handbook of Mathematical Functions with Formulas, Graphs, and Mathematical Tables, 9th printing.* New York: Dover, pp. 885–897, 1972.

Press, W. H.; Flannery, B. P.; Teukolsky, S. A.; and Vetterling, W. T. *Numerical Recipes in FORTRAN: The Art of Scientific Computing, 2nd ed.* Cambridge, England: Cambridge University Press, pp. 365–366, 1992.

Ueberhuber, C. W. *Numerical Computation 2: Methods, Software, and Analysis.* Berlin: Springer-Verlag, p. 71, 1997.

Quadrature Formulas

NEWTON-COTES FORMULAS

Quadri-Amicable Number

AMICABLE QUADRUPLE

Quadric

A quadric is a QUADRATIC SURFACE. A surface OF THE FORM

$$\frac{x^2}{a^2 + \theta} + \frac{y^2}{b^2 + \theta} + \frac{z^2}{c^2 + \theta} = 1$$

is also called a quadric, and θ is said to be the parameter of the quadric.

See also QUADRATIC SURFACE

References

Takahashi, H. "Quadrica Page." http://www2.kawase-h.ed.jp/Teachers/~Takahashi/Quadrica.html.

Quadricorn

A FLEXIBLE POLYHEDRON due to C. Schwabe (with the appearance of having four horns) which flexes from one totally flat configuration to another, passing through intermediate configurations of positive VOLUME.

See also FLEXIBLE POLYHEDRON

Quadrifolium

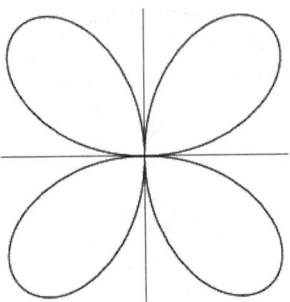

The ROSE with $n = 2$. It has polar equation

$$r = a \sin(2\theta),$$

and Cartesian form

$$\left(x^2 + y^2\right)^3 = 4a^2 x^2 y^2.$$

See also BIFOLIUM, FOLIUM, ROSE, TRIFOLIUM

Quadrilateral

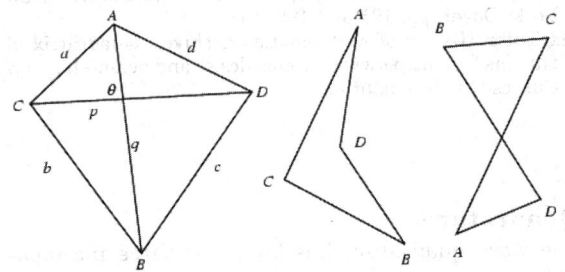

A four-sided POLYGON sometimes (but not very often) also known as a tetragon. If not explicitly stated, all four VERTICES are generally taken to lie in a PLANE. If the points do not lie in a PLANE, the quadrilateral is called a SKEW QUADRILATERAL. There are three topological types of quadrilaterals (Wenninger 1983, p. 50): convex quadrilaterals (left figure), concave quadrilaterals (middle figure), and crossed quadrilaterals (or butterflies, or bow-ties; right figure).

For a planar convex quadrilateral (left figure above), let the lengths of the sides be a, b, c, and d, the SEMIPERIMETER s, and the DIAGONALS p and q. The DIAGONALS are PERPENDICULAR IFF $a^2 + c^2 = b^2 + d^2$. Given any five points in the plane, four will always form a convex quadrilateral. This result is a special case of the so-called HAPPY END PROBLEM (Hoffman 1998, pp. 74–78).

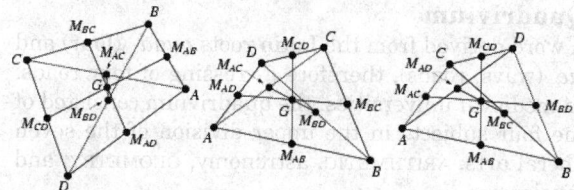

The centroid of the vertices of a quadrilateral occurs at the point of intersection of the BIMEDIANS (i.e., the lines $M_{AB}M_{CD}$ and $M_{AD}M_{BC}$ joining pairs of opposite MIDPOINTS) (Honsberger 1995, pp. 36–37). In addition, it is the MIDPOINT of the line $M_{AC}M_{BD}$ connecting the midpoints of the diagonals AC and BD (Honsberger 1995, pp. 39–40).

An equation for the sum of the squares of side lengths is

$$a^2 + b^2 + c^2 + d^2 = p^2 + q^2 + 4x^2, \tag{1}$$

where x is the length of the line joining the MIDPOINTS of the DIAGONALS (Casey 1888, p. 22). The AREA of a quadrilateral is given by

$$K = \tfrac{1}{2}pq \sin\theta \tag{2}$$

$$= \tfrac{1}{4}(b^2 + d^2 - a^2 - c^2)\tan\theta \tag{3}$$

$$= \tfrac{1}{4}\sqrt{4p^2q^2 - (b^2 + d^2 - a^2 - c^2)^2} \tag{4}$$

$$= \sqrt{(s-a)(s-b)(s-c)(s-d) - abcd\,\cos^2\left[\tfrac{1}{2}(A+B)\right]}, \tag{5}$$

where (4) is known as BRETSCHNEIDER'S FORMULA (Beyer 1987).

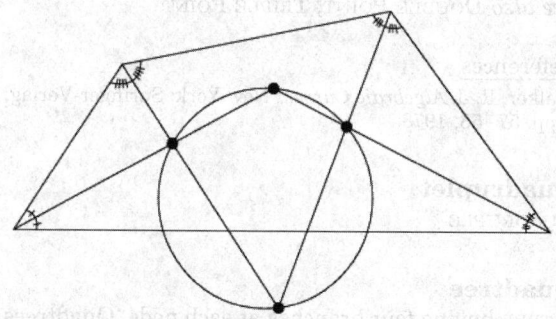

The four ANGLE BISECTORS of a quadrilateral intersect adjacent bisectors in four CONCYCLIC points (Honsberger 1995, p. 35).

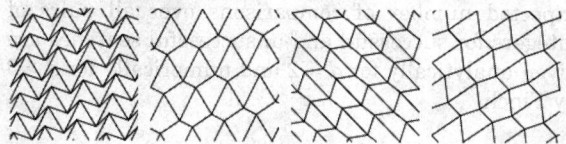

Any non-self-intersecting quadrilateral tiles the plane.

There is a relationship between the six distances d_{12}, d_{13}, d_{14}, d_{23}, d_{24}, and d_{34} between the four points of a quadrilateral (Weinberg 1972):

$$\begin{aligned}
0 = &\; d_{12}^4 d_{34}^2 + d_{13}^4 d_{24}^2 + d_{14}^4 d_{23}^2 + d_{23}^4 d_{14}^2 + d_{24}^4 d_{13}^2 + d_{34}^4 d_{12}^2 \\
&+ d_{12}^2 d_{23}^2 d_{31}^2 + d_{12}^2 d_{24}^2 d_{41}^2 + d_{13}^2 d_{34}^2 d_{41}^2 \\
&+ d_{23}^2 d_{34}^2 d_{42}^2 - d_{12}^2 d_{23}^2 d_{34}^2 - d_{13}^2 d_{32}^2 d_{24}^2 \\
&- d_{12}^2 d_{24}^2 d_{43}^2 - d_{14}^2 d_{42}^2 d_{23}^2 - d_{13}^2 d_{34}^2 d_{42}^2 \\
&- d_{14}^2 d_{43}^2 d_{32}^2 - d_{23}^2 d_{31}^2 d_{14}^2 + d_{21}^2 d_{13}^2 d_{34}^2 \\
&- d_{24}^2 d_{41}^2 d_{13}^2 - d_{21}^2 d_{14}^2 d_{43}^2 - d_{31}^2 d_{12}^2 d_{24}^2 \\
&- d_{32}^2 d_{21}^2 d_{14}^2. \tag{6}
\end{aligned}$$

This can be most simply derived by setting the left side of the CAYLEY-MENGER DETERMINANT

$$288V^2 = \begin{vmatrix} 0 & 1 & 1 & 1 & 1 \\ 1 & 0 & d_{12}^2 & d_{13}^2 & d_{14}^2 \\ 1 & d_{21}^2 & 0 & d_{23}^2 & d_{24}^2 \\ 1 & d_{31}^2 & d_{32}^2 & 0 & d_{34}^2 \\ 1 & d_{41}^2 & d_{42}^2 & d_{43}^2 & 0 \end{vmatrix} \tag{7}$$

equal to 0 (corresponding to a TETRAHEDRON of volume 0), thus giving a relationship between the DISTANCES between vertices of a planar quadrilateral (Uspensky 1948, p. 256).

A special type of quadrilateral is the CYCLIC QUADRILATERAL, for which a CIRCLE can be circumscribed so that it touches each VERTEX. For BICENTRIC QUADRILATERALS, the CIRCUMCIRCLE and INCIRCLE satisfy

$$2r^2(R^2 - s^2) = (R^2 - s^2) - 4r^2s^2, \tag{8}$$

where R is the CIRCUMRADIUS, r in the INRADIUS, and s is the separation of centers. A quadrilateral with two sides PARALLEL is called a TRAPEZOID.

See also ANTICENTER, BICENTRIC QUADRILATERAL, BIMEDIAN, BRAHMAGUPTA'S FORMULA, BRETSCHNEIDER'S FORMULA, BUTTERFLY THEOREM, CAYLEY-MENGER DETERMINANT, COMPLETE QUADRILATERAL, CYCLIC QUADRILATERAL, DIAMOND, EIGHT-POINT CIRCLE THEOREM, EQUILIC QUADRILATERAL, FANO'S AXIOM, LÉON ANNE'S THEOREM, LOZENGE, MALTITUDE, ORTHOCENTRIC QUADRILATERAL, PARALLELOGRAM, PTOLEMY'S THEOREM, RATIONAL QUADRILATERAL, RECTANGLE, RHOMBUS, SKEW QUADRILATERAL, SQUARE, TANGENTIAL QUADRILATERAL, TRAPEZOID, VARIGNON'S THEOREM, VON AUBEL'S THEOREM, WITTENBAUER'S PARALLELOGRAM

References

Beyer, W. H. (Ed.). *CRC Standard Mathematical Tables*, *28th ed.* Boca Raton, FL: CRC Press, p. 123, 1987.

Casey, J. *A Sequel to the First Six Books of the Elements of Euclid, Containing an Easy Introduction to Modern Geometry with Numerous Examples*, *5th ed., rev. enl.* Dublin: Hodges, Figgis, & Co., 1888.

Durell, C. V. "The Quadrilateral and Quadrangle." Ch. 7 in *Modern Geometry: The Straight Line and Circle.* London: Macmillan, pp. 77–87, 1928.

Fukagawa, H. and Pedoe, D. "Circles and Quadrilaterals" and "Quadrilaterals." §3.5 and 4.2 in *Japanese Temple Geometry Problems.* Winnipeg, Manitoba, Canada: Charles Babbage Research Foundation, pp. 43–45, 47–48, and 125–132, 1989.

Harris, J. W. and Stocker, H. "Quadrilaterals." §3.6 in *Handbook of Mathematics and Computational Science.* New York: Springer-Verlag, pp. 82–86, 1998.

Honsberger, R. "On Quadrilaterals." Ch. 4 in *Episodes in Nineteenth and Twentieth Century Euclidean Geometry.* Washington, DC: Math. Assoc. Amer., pp. 35–41, 1995.

Routh, E. J. "Moment of Inertia of a Quadrilateral." *Quart. J. Pure Appl. Math.* **11**, 109–110, 1871.

Uspensky, J. V. *Theory of Equations.* New York: McGraw-Hill, p. 256, 1948.

Weinberg, S. *Gravitation and Cosmology: Principles and Applications of the General Theory of Relativity.* New York: Wiley, p. 7, 1972.

Wenninger, M. J. *Dual Models.* Cambridge, England: Cambridge University Press, 1983.

Quadrilateral of Chords

CYCLIC QUADRILATERAL

Quadrilateral Tiling

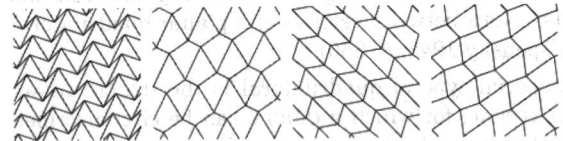

Any nonself-intersecting QUADRILATERAL (Wells 1991, p. 208) tiles the plane, as illustrated above.

References

Wells, D. *The Penguin Dictionary of Curious and Interesting Geometry.* London: Penguin, pp. 177–179, 208, and 211, 1991.

Quadrillion

In the American system, 10^{15}.

See also LARGE NUMBER

Quadriplanar Coordinates

The analog of TRILINEAR COORDINATES for TETRAHEDRA.

See also TETRAHEDRON, TRILINEAR COORDINATES

References

Altshiller-Court, N. *Modern Pure Solid Geometry.* New York: Chelsea, 1979.

Mitrinovic, D. S.; Pecaric, J. E.; and Volenec, V. Ch. 19 in *Recent Advances in Geometric Inequalities.* Dordrecht, Netherlands: Kluwer, 1989.

Woods, F. S. *Higher Geometry: An Introduction to Advanced Methods in Analytic Geometry.* New York: Dover, pp. 193–196, 1961.

Quadrivium

A word derived from the Latin roots *quad-* (four) and *via* (ways, roads), therefore a crossing of four roads. In medieval universities, the quadrivium consisted of the four subjects in the upper division of the seven liberal arts: ARITHMETIC, astronomy, GEOMETRY, and music.

See also TRIVIUM

Quadruple

A group of four elements, also called a QUADRUPLET or TETRAD.

See also AMICABLE QUADRUPLE, DIOPHANTINE QUADRUPLE, MONAD, PAIR, PRIME QUADRUPLET, PYTHAGOREAN QUADRUPLE, QUADRUPLET, QUINTUPLET, TETRAD, TRIAD, TRIPLE, TWINS, VECTOR QUADRUPLE PRODUCT

Quadruple Point

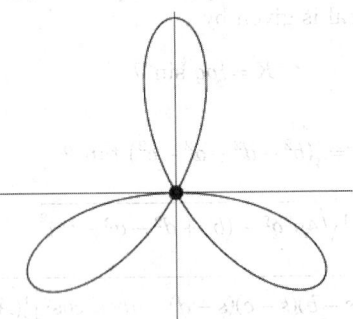

A point where a curve intersects itself along four arcs. The above plot shows the quadruple point at the ORIGIN of the QUADRIFOLIUM $(x^2 + y^2)^3 - 4x^2 y^2 = 0$.

See also DOUBLE POINT, TRIPLE POINT

References

Walker, R. J. *Algebraic Curves.* New York: Springer-Verlag, pp. 57–58, 1978.

Quadruplet

QUADRUPLE

Quadtree

A TREE having four branches at each node. Quadtrees are used in the construction of some multidimensional databases (e.g., cartography, computer graphics, and image processing). For a d-D tree, the expected number of comparisons over all pairs of integers for successful and unsuccessful searches are given analytically for $d = 2$ and numerically for $d \geq 3$ by Finch.

References

de Berg, M.; van Kreveld, M.; Overmans, M.; and Schwarzkopf, O. "Quadtrees: Non-Uniform Mesh Generation."

Ch. 14 in *Computational Geometry: Algorithms and Applications, 2nd rev. ed.* Berlin: Springer-Verlag, pp. 291–306, 2000.

Finch, S. "Favorite Mathematical Constants." http://www.mathsoft.com/asolve/constant/infprd/infprd.html.

Finch, S. "Favorite Mathematical Constants." http://www.mathsoft.com/asolve/constant/qdt/qdt.html.

Finkel, R. A. and Bentley, J. L. "Quad Trees, a Data Structure for Retrieval on Composite Keys." *Acta Informatica* **4**, 1–9, 1974.

Flajolet, P.; Gonnet, G.; Puech, C.; and Robson, J. M. "Analytic Variations on Quadtrees." *Algorithmica* **10**, 473–500, 1993.

Flajolet, P.; Labelle, G.; Laforest, L.; and Salvy, B. "Hypergeometrics and the Cost Structure of Quadtrees." *Random Structure Alg.* **7**, 117–144, 1995. http://pauillac.inria.fr/algo/flajolet/Publications/publist.html.

Gonnet, G. H. and Baeza-Yates, R. Ch. 3 in *Handbook of Algorithms and Data Structures in Pascal and C.* Reading, MA: Addison-Wesley, 1991.

Lauwerier, H. *Fractals: Endlessly Repeated Geometric Figures.* Princeton, NJ: Princeton University Press, pp. 11–13, 1991.

Samet, H. *Applications of Spatial Data Structures: Computer Graphics, Image Processing and GIS.* Reading, MA: Addison-Wesley, 1989.

Samet, H. *The Design and Analysis of Spatial Data Structures.* Reading, MA: Addison-Wesley, 1990.

Quantic

An m-ary n-ic polynomial (i.e., a HOMOGENEOUS POLYNOMIAL with constant COEFFICIENTS of degree n in m independent variables).

See also ALGEBRAIC INVARIANT, FUNDAMENTAL SYSTEM, P-ADIC NUMBER, SYZYGIES PROBLEM

Quantified System

A quantified system of real algebraic equations and inequalities in variables $\{x_1, \ldots, x_n\}$ is an expression

$$QS = Q_1(y_1)(Q_2)(y_2) \cdots Q_m(y_m)S(x_1, \ldots, x_n; y_1, \ldots, y_m),$$

where Q is a QUANTIFIER ($\exists$ or $\forall$) and S is a system of real algebraic equations and inequalities in $\{x_1 \ldots, x_n; y_1, \ldots y_m\}$. By TARSKI'S THEOREM, the solution set of a quantified system of real algebraic equations and inequalities is a SEMIALGEBRAIC SET.

See also QUANTIFIER, SEMIALGEBRAIC SET, TARSKI'S THEOREM

References

Strzebonski, A. "Solving Algebraic Inequalities." *Mathematica J.* **7**, 525–541, 2000.

Quantifier

One of the operations EXISTS $\exists$ or FOR ALL $\forall$. However, there also exist more exotic branches of logic which use quantifiers other than these two.

See also BOUND VARIABLE, EXISTS, FOR ALL, FREE, QUANTIFIED SYSTEM, QUANTIFIER ELIMINATION, UNIVERSAL QUANTIFIER

References

Hall, C. and O'Donnell, J. "Computing with Quantifiers." §3.2 in *Discrete Mathematics Using a Computer.* London: Springer-Verlag, pp. 98–100, 2000.

Quantifier Elimination

Quantifier elimination is the removal of all QUANTIFIERS ($\forall$ and $\exists$) from a quantified system. A first-order theory allows quantifier elimination if, for each quantified formula, there exists an equivalent quantifier-free formula. Examples of such theories include the real numbers with $+$, $*$, $=$, and $>$, and the theory of complex numbers with $+$, $*$, and $=$. Quantifier elimination is implemented in *Mathematica* as Resolve[*expr*].

Unfortunately, it has been proven that the worst-case time complexity for real quantifier elimination is doubly exponential in the number of QUANTIFIER blocks (Weispfenning 1985, Davenport and Heintz 1988, Heintz *et al.* 1989, Caviness and Johnson 1998).

See also CYLINDRICAL ALGEBRAIC DECOMPOSITION, TARSKI'S THEOREM

References

Caviness, B. F. and Johnson, J. R. (Eds.). *Quantifier Elimination and Cylindrical Algebraic Decomposition.* New York: Springer-Verlag, 1998.

Collins, G. E. "Quantifier Elimination for Real Closed Fields by Cylindrical Algebraic Decomposition." In *Proc. 2nd GI Conf. Automata Theory and Formal Languages.* New York: Springer-Verlag, pp. 134–183, 1975.

Collins, G. E. "Quantifier Elimination by Cylindrical Algebraic Decomposition--Twenty Years of Progress." In *Quantifier Elimination and Cylindrical Algebraic Decomposition* (Ed. B. F. Caviness and J. R. Johnson). New York: Springer-Verlag, pp. 8–23, 1998.

Collins, G. E. and Hong, H. "Partial Cylindrical Algebraic Decomposition for Quantifier Elimination." *J. Symb. Comput.* **12**, 299–328, 1991.

Davenport, J. H. "Computer Algebra for Cylindrical Algebraic Decomposition." Report TRITA-NA-8511, NADA, KTH, Stockholm, Sept. 1985.

Davenport, J. and Heintz, J. "Real Quantifier Elimination if Doubly Exponential." *J. Symb. Comput.* **5**, 29–35, 1988.

Dolzmann, A. and Sturm, T. "Simplification of Quantifier-Free Formulae over Ordered Fields." *J. Symb. Comput.* **24**, 209–231, 1997.

Dolzmann, A. and Weispfenning, V. "Local Quantifier Elimination." http://www.fmi.uni-passau.de/~dolzmann/refs/MIP-0003.ps.Z.

Heintz, J.; Roy, R.-F.; and Solerno, P. "Complexité du principe de Tarski-Seidenberg." *C. R. Acad. Sci. Paris Sér. I Math.* **309**, 825–830, 1989.

Loos, R. and Weispfenning, V. "Applying Lattice Quantifier Elimination." *Comput. J.* **36**, 450–461, 1993.

Strzebonski, A. "Solving Algebraic Inequalities." *Mathematica J.* **7**, 525–541, 2000.

Weispfenning, V. "The Complexity of Linear Problems in Fields." *J. Symb. Comput.* **5**, 3–27, 1988.

Quantile

The kth n-tile P_k is that value of x, say x_k, which corresponds to a CUMULATIVE FREQUENCY of Nk/n. If

$n = 4$, the quantity is called a QUARTILE, and if $n = 100$, it is called a PERCENTILE.

See also PERCENTILE, QUARTILE

References
Kenney, J. F. and Keeping, E. S. "Quantiles." §3.5 in *Mathematics of Statistics, Pt. 1, 3rd ed.* Princeton, NJ: Van Nostrand, pp. 37–38, 1962.

Quantity

See also EXPRESSION

Quantization Efficiency

Quantization is a nonlinear process which generates additional frequency components (Thompson *et al.* 1986). This means that the signal is no longer band-limited, so the SAMPLING THEOREM no longer holds. If a signal is sampled at the NYQUIST FREQUENCY, information will be lost. Therefore, sampling faster than the NYQUIST FREQUENCY results in detection of more of the signal and a lower signal-to-noise ratio [SNR]. Let β be the OVERSAMPLING ratio and define

$$\eta_Q \equiv \frac{\text{SNR}_{\text{quant}}}{\text{SNR}_{\text{unquant}}}.$$

Then the following table gives values of η_Q for a number of parameters.

Quantization Levels	$\eta_Q(\beta = 1)$	$\eta_Q(\beta = 2)$
2	0.64	0.74
3	0.81	0.89
4	0.88	0.94

The Very Large Array of 27 radio telescopes in Socorro, New Mexico uses three-level quantization at $\beta = 1$, so $\eta_Q = 0.81$.

See also OVERSAMPLING

References
Thompson, A. R.; Moran, J. M.; and Swenson, G. W. Jr. Fig. 8.3 in *Interferometry and Synthesis in Radio Astronomy.* New York: Wiley, p. 220, 1986.

Quantum Chaos

The study of the implications of CHAOS for a system in the semiclassical (i.e., between classical and quantum mechanical) regime.

References
Ott, E. "Quantum Chaos." Ch. 10 in *Chaos in Dynamical Systems.* New York: Cambridge University Press, pp. 334–362, 1993.

Quarter

The UNIT FRACTION 1/4, also called one-fourth.

See also HALF, KÖBE'S ONE-FOURTH THEOREM, QUARTILE

Quarter Squares Rule

$$\left(\frac{a+b}{2}\right)^2 - \left(\frac{a-b}{2}\right)^2 = ab.$$

Quartet

A SET of four, also called a TETRAD.

See also HEXAD, MONAD, QUINTET, TETRAD, TRIAD

Quartic Curve

A general plane quartic curve is a curve OF THE FORM

$$Ax^4 + By^4 + Cx^3y + Dx^2y^2 + Exy^3 + Fx^3 + Gy^3$$

$$+Hx^2y + Ixy^2 + Jx^2 + Ky^2 + Lxy + Mx + Ny + O = 0. \tag{1}$$

The incidence relations of the 28 bitangents of the general quartic curve can be put into a ONE-TO-ONE correspondence with the vertices of a particular POLYTOPE in 7-D space (Coxeter 1928, Du Val 1931). This fact is essentially similar to the discovery by Schoutte (1910) that the 27 SOLOMON'S SEAL LINES on a CUBIC SURFACE can be connected with a POLYTOPE in 6-D space (Du Val 1931). A similar but less complete relation exists between the tritangent planes of the canonical curve of genus 4 and an 8-D POLYTOPE (Du Val 1931).

The maximum number of DOUBLE POINTS for a nondegenerate quartic curve is three.

A quartic curve OF THE FORM

$$y^2 = (x-a)(x-\xi)(x-\gamma)(x-\delta) \tag{2}$$

can be written

$$\left(\frac{y}{x-\alpha}\right)^2 = \left(1 - \frac{\beta - \alpha}{x - \alpha}\right)\left(1 - \frac{\gamma - \alpha}{x - \alpha}\right)\left(1 - \frac{\delta - \alpha}{x - \alpha}\right), \tag{3}$$

and so is CUBIC in the coordinates

$$X = \frac{1}{x - \alpha} \tag{4}$$

$$Y = \frac{y}{x - \alpha^2}. \tag{5}$$

This transformation is a BIRATIONAL TRANSFORMA-

TION.

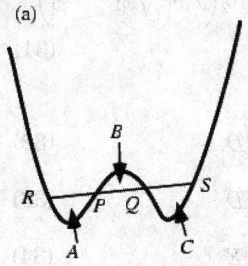
(a)

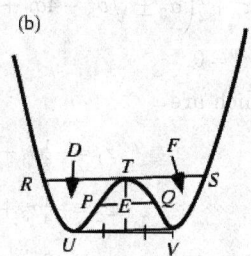
(b)

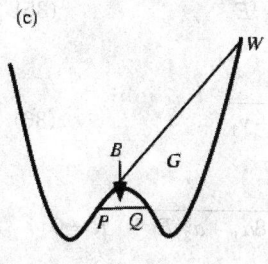
(c)

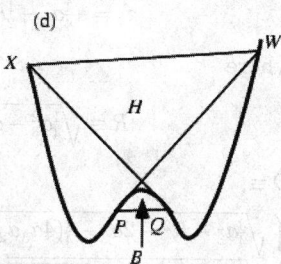
(d)

Let P and Q be the INFLECTION POINTS and R and S the intersections of the line PQ with the curve in Figure (a) above. Then

$$A = C \qquad (6)$$

$$B = 2A. \qquad (7)$$

In Figure (b), let UV be the double tangent, and T the point on the curve whose x coordinate is the average of the x coordinates of U and V. Then $UV \| PQ \| RS$ and

$$D = F \qquad (8)$$

$$E = \sqrt{2}D. \qquad (9)$$

In Figure (c), the tangent at P intersects the curve at W. Then

$$G = 8B. \qquad (10)$$

Finally, in Figure (d), the intersections of the tangents at P and Q are W and X. Then

$$H = 27B \qquad (11)$$

(Honsberger 1991).

See also CUBIC SURFACE, PEAR-SHAPED CURVE, SOLOMON'S SEAL LINES

References

Coxeter, H. S. M. "The Pure Archimedean Polytopes in Six and Seven Dimensions." *Proc. Cambridge Phil. Soc.* **24**, 7–9, 1928.

Du Val, P. "On the Directrices of a Set of Points in a Plane." *Proc. London Math. Soc. Ser. 2* **35**, 23–74, 1933.

Honsberger, R. *More Mathematical Morsels.* Washington, DC: Math. Assoc. Amer., pp. 114–118, 1991.

Schoutte, P. H. "On the Relation Between the Vertices of a Definite Sixdimensional Polytope and the Lines of a Cubic Surface." *Proc. Roy. Akad. Acad. Amsterdam* **13**, 375–383, 1910.

Wells, D. *The Penguin Dictionary of Curious and Interesting Geometry.* London: Penguin, p. 49, 1991.

Quartic Equation

A general quartic equation (also called a BIQUADRATIC EQUATION) is a fourth-order POLYNOMIAL OF THE FORM

$$z^4 + a_3 z^3 + a_2 z^2 + a_1 z + a_0 = 0. \qquad (1)$$

The ROOTS of this equation satisfy NEWTON'S RELATIONS:

$$x_1 + x_2 + x_3 + x_4 = -a_3 \qquad (2)$$

$$x_1 x_2 + x_1 x_3 + x_1 x_4 + x_2 x_3 + x_2 x_4 + x_3 x_4 = a_2 \qquad (3)$$

$$x_1 x_2 x_3 + x_2 x_3 x_4 + x_1 x_2 x_4 + x_1 x_3 x_4 = -a_1 \qquad (4)$$

$$x_1 x_2 x_3 x_4 = a_0, \qquad (5)$$

where the denominators on the right side are all $a_4 \equiv 1$. Writing the quartic in the standard form

$$x^4 + px^2 + qx + r = 0, \qquad (6)$$

the properties of the SYMMETRIC POLYNOMIALS appearing in NEWTON'S RELATIONS then give

$$z_1^2 + z_2^2 + z_3^2 + z_4^2 = -2p \qquad (7)$$

$$z_1^3 + z_2^3 + z_3^3 + z_4^3 = -3p \qquad (8)$$

$$z_1^4 + z_2^4 + z_3^4 + z_4^4 = 2p^2 - 4r \qquad (9)$$

$$z_1^5 + z_2^5 + z_3^5 + z_4^5 = 5pq. \qquad (10)$$

Eliminating p, q, and r, respectively, gives the relations

$$z_1 z_2 (p + z_1^2 + z_1 z_2 + z_2^2) - r = 0 \qquad (11)$$

$$z_1^2 z_2 (z_1 + z_2) - q z_1 - r = 0 \qquad (12)$$

$$q + p z_2 + z_2^3 = 0, \qquad (13)$$

as well as their cyclic permutations.

Ferrari was the first to develop an algebraic technique for solving the general quartic. He applied his technique (which was stolen and published by Cardano) to the equation

$$x^4 + 6x^2 - 60x + 36 = 0 \qquad (14)$$

(Smith 1994, p. 207).

The x^3 term can be eliminated from the general quartic (1) by making a substitution OF THE FORM

$$z \equiv x - \lambda, \qquad (15)$$

so

$$x^4 + (a_3 - 4\lambda)x^3 + (a_2 - 3a_3\lambda + 6\lambda^2)x^2$$

$$+(a_1 - 2a_2\lambda + 3a_3\lambda^2 - 4\lambda^3)x$$

$$+(a_0 - a_1\lambda + a_2\lambda^2 - a_3\lambda^3 + \lambda^4). \tag{16}$$

Letting $\lambda = a_3/4$ so

$$z \equiv x - \tfrac{1}{4}a_3 \tag{17}$$

then gives the standard form

$$x^4 + px^2 + qx + r = 0, \tag{18}$$

where

$$p \equiv a_2 - \tfrac{3}{8}a_3^2 \tag{19}$$

$$q \equiv a_1 - \tfrac{1}{2}a_2 a_3 + \tfrac{1}{8}a_3^3 \tag{20}$$

$$r \equiv a_0 - \tfrac{1}{4}a_1 a_3 + \tfrac{1}{16}a_2 a_3^2 - \tfrac{3}{256}a_3^4. \tag{21}$$

Adding and subtracting $x^2 u + u^2/4$ to (6) gives

$$\left(x^4 + x^2 u + \tfrac{1}{4}u^2\right) - x^2 u - \tfrac{1}{4}u^2 + px^2 + qx + r = 0, \tag{22}$$

which can be rewritten

$$\left(x^2 + \tfrac{1}{2}u\right)^2 - \left[(u - p)x^2 - qx + \left(\tfrac{1}{4}u^2 - r\right)\right] = 0 \tag{23}$$

(Birkhoff and Mac Lane 1965). The first term is a perfect square P^2, and the second term is a perfect square Q^2 for those u such that

$$q^2 = 4(u - p)\left(\tfrac{1}{4}u^2 - r\right). \tag{24}$$

This is the resolvent CUBIC, and plugging a solution u_1 back in gives

$$P^2 - Q^2 = (P + Q)(P - Q), \tag{25}$$

so (23) becomes

$$\left(x^2 + \tfrac{1}{2}u_1 + Q\right)\left(x^2 + \tfrac{1}{2}u_1 - Q\right), \tag{26}$$

where

$$Q \equiv Ax - B \tag{27}$$

$$A \equiv \sqrt{u_1 - p} \tag{28}$$

$$B \equiv -\frac{q}{2A}. \tag{29}$$

Let y_1 be a REAL ROOT of the resolvent CUBIC EQUATION

$$y^3 - a_2 y^2 + (a_1 a_3 - 4a_0)y + (4a_2 a_0 - a_1^2 - a_3^2 a_0)$$
$$= 0. \tag{30}$$

The four ROOTS are then given by the ROOTS of the equation

$$x^2 + \tfrac{1}{2}\left(a_3 \pm \sqrt{a_3^2 - 4a_2 + 4y_1}\right) + \tfrac{1}{2}\left(y_1 \mp \sqrt{y_1^2 - 4a_0}\right)$$
$$= 0, \tag{31}$$

which are

$$z_1 = -\tfrac{1}{4}a_3 + \tfrac{1}{2}R + \tfrac{1}{2}D \tag{32}$$

$$z_2 = -\tfrac{1}{4}a_3 + \tfrac{1}{2}R - \tfrac{1}{2}D \tag{33}$$

$$z_3 = -\tfrac{1}{4}a_3 - \tfrac{1}{2}R + \tfrac{1}{2}E \tag{34}$$

$$z_4 = -\tfrac{1}{4}a_3 - \tfrac{1}{2}R - \tfrac{1}{2}E, \tag{35}$$

where

$$R \equiv \sqrt{\tfrac{1}{4}a_3^2 - a_2 + y_1} \tag{36}$$

$$D \equiv$$

$$\begin{cases} \sqrt{\tfrac{3}{4}a_3^2 - R^2 - 2a_2 - \tfrac{1}{4}(4a_3 a_2 - 8a_1 - a_3^3)R^{-1}} & R \neq 0 \\ \sqrt{\tfrac{3}{4}a_3^2 - 2a_2 + 2\sqrt{y_1^2 - 4a_0}} & R = 0 \end{cases} \tag{37}$$

$$E \equiv$$

$$\begin{cases} \sqrt{\tfrac{3}{4}a_3^2 - R^2 - 2a_2 - \tfrac{1}{4}(4a_3 a_2 - 8a_1 - a_3^3)R^{-1}} & R \neq 0 \\ \sqrt{\tfrac{3}{4}a_3^2 - 2a_2 - 2\sqrt{y_1^2 - 4a_0}} & R = 0 \end{cases} \tag{38}$$

Another approach to solving the quartic (6) defines

$$\alpha \equiv (x_1 + x_2)(x_3 + x_4) = -(x_1 + x_2)^2 \tag{39}$$

$$\beta \equiv (x_1 + x_3)(x_2 + x_4) = -(x_1 + x_3)^2 \tag{40}$$

$$\gamma \equiv (x_1 + x_4)(x_2 + x_3) = -(x_2 + x_3)^2, \tag{41}$$

where the second forms follow from

$$x_1 + x_2 + x_3 + x_4 = -a_3 = 0, \tag{42}$$

and defining

$$h(x) \equiv (x - \alpha)(x - \beta)(x - \gamma) \tag{43}$$

$$= x^3 - (\alpha + \beta + \gamma)x^2 + (\alpha\beta + \alpha\gamma + \beta\gamma)x - \alpha\beta\gamma. \tag{44}$$

This equation can be written in terms of the original coefficients p, q, and r as

$$h(x) = x^3 - 2px^2 + (p^2 - 4r)x + q^2. \tag{45}$$

The roots of this CUBIC EQUATION then give α, β, and γ, and the equations (39) to (41) can be solved for the four roots x_i of the original quartic (Faucette 1996).

See also CUBIC EQUATION, DISCRIMINANT (POLYNOMIAL), QUINTIC EQUATION

References

Abramowitz, M. and Stegun, C. A. (Eds.). *Handbook of Mathematical Functions with Formulas, Graphs, and*

Mathematical Tables, 9th printing. New York: Dover, pp. 17–18, 1972.

Berger, M. §16.4.1–16.4.11.1 in *Geometry I.* New York: Springer-Verlag, 1987.

Beyer, W. H. *CRC Standard Mathematical Tables, 28th ed.* Boca Raton, FL: CRC Press, p. 12, 1987.

Birkhoff, G. and Mac Lane, S. *A Survey of Modern Algebra, 5th ed.* New York: Macmillan, pp. 107–108, 1996.

Borwein, P. and Erdélyi, T. "Quartic Equations." §1.1.E.1e in *Polynomials and Polynomial Inequalities.* New York: Springer-Verlag, p. 4, 1995.

Brown, K. S. "Reducing Quartics to Cubics." http://www.sea-net.com/~ksbrown/kmath296.htm.

Ehrlich, G. §4.16 in *Fundamental Concepts of Abstract Algebra.* Boston, MA: PWS-Kent, 1991.

Faucette, W. M. "A Geometric Interpretation of the Solution of the General Quartic Polynomial." *Amer. Math. Monthly* **103**, 51–57, 1996.

Smith, D. E. *A Source Book in Mathematics.* New York: Dover, 1994.

van der Waerden, B. L. §64 in *Algebra, Vol. 1.* New York: Springer-Verlag, 1993.

Quartic Graph

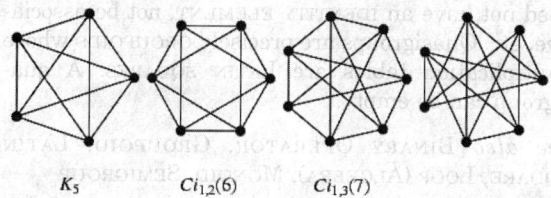

K_5 $Ci_{1,2}(6)$ $Ci_{1,3}(7)$

A quartic graph is a GRAPH which is 4-REGULAR. The unique quartic graph on five nodes is the COMPLETE GRAPH K_5, and the unique quartic graph on six nodes is the CIRCULANT GRAPH $Ci_{1,2}(6)$. There are two quartic graphs on seven nodes, one of which is the CIRCULANT GRAPH $Ci_{1,3}(7)$. The numbers of connected quartic graphs on $n = 1, 2, \ldots$ nodes are 0, 0, 0, 0, 1, 1, 2, 6, 16, 59, ... (Sloane's A006820), the numbers of not necessarily connected quartic graphs are 0, 0, 0, 0, 1, 1, 2, 6, 16, 60, ... (Sloane's A033301), and the numbers of disconnected quartic graphs for $n = 10, 11, \ldots$ are 1, 1, 3, 8, 25, 88, ... (Sloane's A033483; Read and Wilson 1998).

The following tables gives polyhedra whose SKELETONS are quartic.

POLYHEDRON	nodes
OCTAHEDRON	6
CUBOCTAHEDRON	12
SMALL RHOMBICUBOCTAHEDRON	24
ICOSIDODECAHEDRON	30
SMALL RHOMBICOSIDODECAHEDRON	60

See also CUBIC GRAPH, QUINTIC GRAPH, REGULAR GRAPH

References

Colbourn, C. J. and Dinitz, J. H. *CRC Handbook of Combinatorial Designs.* Boca Raton, FL: CRC Press, p. 648, 1996.

Faradzev, I. A. "Constructive Enumeration of Combinatorial Objects." In *Problèmes combinatoires et théorie des graphes (Orsay, 9–13 Juillet 1976).* Paris: Centre Nat. Recherche Scient., pp. 131–135, 1978.

Read, R. C. and Wilson, R. J. *An Atlas of Graphs.* Oxford, England: Oxford University Press, 1998.

Sloane, N. J. A. Sequences A006820/M1617, A033301, and A033483 in "An On-Line Version of the Encyclopedia of Integer Sequences." http://www.research.att.com/~njas/sequences/eisonline.html.

Quartic Reciprocity Theorem

BIQUADRATIC RECIPROCITY THEOREM

Quartic Residue

QUARTIC RECIPROCITY THEOREM

Quartic Surface

An ALGEBRAIC SURFACE of ORDER 4. Unlike CUBIC SURFACES, quartic surfaces have not been fully classified.

See also BOHEMIAN DOME, BURKHARDT QUARTIC, CASSINI SURFACE, CUSHION, CYCLIDE, DESMIC SURFACE, FRESNEL'S ELASTICITY SURFACE, GOURSAT'S SURFACE, KUMMER SURFACE, MITER SURFACE, PIRIFORM, ROMAN SURFACE, SYMMETROID, TETRAHEDROID, TOOTH SURFACE

References

Fischer, G. (Ed.). *Mathematical Models from the Collections of Universities and Museums.* Braunschweig, Germany: Vieweg, p. 9, 1986.

Fischer, G. (Ed.). Plates 40–41, 45–49, and 52–56 in *Mathematische Modelle/Mathematical Models, Bildband/Photograph Volume.* Braunschweig, Germany: Vieweg, pp. 40–41, 45–49, and 52–56, 1986.

Hunt, B. "Some Quartic Surfaces." Appendix B.5 in *The Geometry of Some Special Arithmetic Quotients.* New York: Springer-Verlag, pp. 310–319, 1996.

Jessop, C. *Quartic Surfaces with Singular Points.* Cambridge, England: Cambridge University Press, 1916.

Quartile

One of the four divisions of observations which have been grouped into four equal-sized sets based on their RANK. The quartile including the top RANKED members is called the first quartile and denoted Q_1. The other quartiles are similarly denoted Q_2, Q_3, and Q_4. For N data points with N OF THE FORM $4n + 5$ (for $n = 0, 1, \ldots$), the HINGES are identical to the first and third quartiles.

See also HINGE, INTERQUARTILE RANGE, PERCENTILE, QUANTILE, QUARTILE DEVIATION, QUARTILE VARIATION COEFFICIENT

References

Kenney, J. F. and Keeping, E. S. "Quartiles." §3.3 in *Mathematics of Statistics, Pt. 1, 3rd ed.* Princeton, NJ: Van Nostrand, pp. 35–37, 1962.

Whittaker, E. T. and Robinson, G. *The Calculus of Observations: A Treatise on Numerical Mathematics, 4th ed.* New York: Dover, pp. 184–186, 1967.

Quartile Deviation

$$QD = \tfrac{1}{2}(Q_3 - Q_1),$$

where Q_1 and Q_3 are the first and third QUARTILES and $Q_3 - Q_1$ is the INTERQUARTILE RANGE.

See also INTERQUARTILE RANGE, QUARTILE, QUARTILE VARIATION COEFFICIENT

References

Kenney, J. F. and Keeping, E. S. *Mathematics of Statistics, Pt. 1, 3rd ed.* Princeton, NJ: Van Nostrand, p. 36, 1962.

Quartile Range

INTERQUARTILE RANGE

Quartile Skewness Coefficient

BOWLEY SKEWNESS

Quartile Variation Coefficient

$$V \equiv 100 \frac{Q_3 - Q_1}{Q_3 + Q_1},$$

where Q_1 and Q_3 are the first and third QUARTILES and $Q_3 - Q_1$ is the INTERQUARTILE RANGE.

See also INTERQUARTILE RANGE, QUARTILE, QUARTILE DEVIATION

Quasiamicable Pair

Let $\sigma(m)$ be the DIVISOR FUNCTION of m. Then two numbers m and n are a quasiamicable pair if

$$\sigma(m) = \sigma(n) = m + n + 1.$$

The first few are (48, 75), (140, 195), (1050, 1575), (1648, 1925), ... (Sloane's A005276). Quasiamicable numbers are sometimes called BETROTHED NUMBERS or REDUCED AMICABLE PAIRS.

See also AMICABLE PAIR

References

Beck, W. E. and Najar, R. M. "More Reduced Amicable Pairs." *Fib. Quart.* **15**, 331–332, 1977.

Guy, R. K. "Quasi-Amicable or Betrothed Numbers." §B5 in *Unsolved Problems in Number Theory, 2nd ed.* New York: Springer-Verlag, pp. 59–60, 1994.

Hagis, P. and Lord, G. "Quasi-Amicable Numbers." *Math. Comput.* **31**, 608–611, 1977.

Sloane, N. J. A. Sequences A005276/M5291 in "An On-Line Version of the Encyclopedia of Integer Sequences." http://www.research.att.com/~njas/sequences/eisonline.html.

Quasiconformal Map

A generalized CONFORMAL MAP.

See also BELTRAMI DIFFERENTIAL EQUATION

References

Iyanaga, S. and Kawada, Y. (Eds.). "Quasiconformal Mappings." §347 in *Encyclopedic Dictionary of Mathematics.* Cambridge, MA: MIT Press, pp. 1086–1088, 1980.

Quasigroup

A GROUPOID S such that for all $a, b \in S$, there exist unique $x, y \in S$ such that

$$ax = b$$

$$ya = b.$$

No other restrictions are applied; thus a quasigroup need not have an IDENTITY ELEMENT, not be associative, etc. Quasigroups are precisely GROUPOIDS whose multiplication tables are LATIN SQUARES. A quasigroup can be empty.

See also BINARY OPERATOR, GROUPOID, LATIN SQUARE, LOOP (ALGEBRA), MONOID, SEMIGROUP

References

Albert, A. A. (Ed.). *Studies in Modern Algebra.* Washington, DC: Math. Assoc. Amer., 1963.

van Lint, J. H. and Wilson, R. M. *A Course in Combinatorics.* New York: Cambridge University Press, 1992.

Quasi-Monte Carlo Integration

A method of NUMERICAL INTEGRATION based on equidistributed sequences (Ueberhuber 1997, p. 125). A quasi-Monte Carlo method known as the Halton-Hammersley-Wozniakowski algorithm is implemented in *Mathematica* as NIntegrate[*f*, ...,
Method->QuasiMonteCarlo].

See also CUBATURE, NUMERICAL INTEGRATION, MONTE CARLO INTEGRATION

References

Hammersley, J. M. "Monte Carlo Methods for Solving Multivariable Problems." *Ann. New York Acad. Sci.* **86**, 844–874, 1960.

Ueberhuber, C. W. *Numerical Computation 2: Methods, Software, and Analysis.* Berlin: Springer-Verlag, pp. 124–125, 1997.

Wozniakowski, H. "Average Case Complexity of Multivariate Integration." *Bull. Amer. Math. Soc.* **24**, 185–194, 1991.

Quasiperfect Number

A least ABUNDANT NUMBER, i.e., one such that

$$\sigma(n) = 2n + 1.$$

Quasiperfect numbers are therefore the sum of their nontrivial DIVISORS. No quasiperfect numbers are known, although if any exist, they must be greater than 10^{35} and have seven or more DIVISORS. Singh (1997) called quasiperfect numbers SLIGHTLY EXCESSIVE NUMBERS.

See also ABUNDANT NUMBER, ALMOST PERFECT NUMBER, PERFECT NUMBER

References

Guy, R. K. "Almost Perfect, Quasi-Perfect, Pseudoperfect, Harmonic, Weird, Multiperfect and Hyperperfect Numbers." §B2 in *Unsolved Problems in Number Theory, 2nd ed.* New York: Springer-Verlag, pp. 45–53, 1994.
Singh, S. *Fermat's Enigma: The Epic Quest to Solve the World's Greatest Mathematical Problem.* New York: Walker, p. 13, 1997.

Quasiperiodic Function

WEIERSTRASS SIGMA FUNCTION, WEIERSTRASS ZETA FUNCTION

Quasiperiodic Motion

The type of motion executed by a DYNAMICAL SYSTEM containing two incommensurate frequencies.

Quasirandom Sequence

A sequence of n-tuples that fills n-space more uniformly than uncorrelated random points. Such a sequence is extremely useful in computational problems where numbers are computed on a grid, but it is not known in advance how fine the grid must be to obtain accurate results. Using a quasirandom sequence allows stopping at any point where convergence is observed, whereas the usual approach of halving the interval between subsequent computations requires a huge number of computations between stopping points.

See also PSEUDORANDOM NUMBER, RANDOM NUMBER

References

Press, W. H.; Flannery, B. P.; Teukolsky, S. A.; and Vetterling, W. T. "Quasi- (that is, Sub-) Random Sequences." §7.7 in *Numerical Recipes in FORTRAN: The Art of Scientific Computing, 2nd ed.* Cambridge, England: Cambridge University Press, pp. 299–306, 1992.

Quasiregular Polyhedron

A quasiregular polyhedron is the solid region interior to two DUAL REGULAR POLYHEDRA with SCHLÄFLI SYMBOLS $\{p,q\}$. and $\{q,p\}$. Quasiregular polyhedra are denoted using a SCHLÄFLI SYMBOL OF THE FORM $\begin{Bmatrix} p \\ q \end{Bmatrix}$, with

$$\begin{Bmatrix} p \\ q \end{Bmatrix} = \begin{Bmatrix} q \\ p \end{Bmatrix}. \tag{1}$$

Quasiregular polyhedra have two kinds of regular faces with each entirely surrounded by faces of the other kind, equal sides, and equal dihedral angles. They must satisfy the Diophantine inequality

$$\frac{1}{p} + \frac{1}{q} + \frac{1}{r} > 1. \tag{2}$$

But $p, q \geq 3$, so r must be 2. This means that the possible quasiregular polyhedra have symbols $\begin{Bmatrix} 3 \\ 3 \end{Bmatrix}$, $\begin{Bmatrix} 3 \\ 4 \end{Bmatrix}$, and $\begin{Bmatrix} 3 \\ 5 \end{Bmatrix}$. Now

$$\begin{Bmatrix} 3 \\ 3 \end{Bmatrix} = \{3, 4\} \tag{3}$$

is the OCTAHEDRON, which is a regular PLATONIC SOLID and not considered quasiregular. This leaves only two convex quasiregular polyhedra: the CUBOCTAHEDRON $\begin{Bmatrix} 3 \\ 4 \end{Bmatrix}$ and the ICOSIDODECAHEDRON $\begin{Bmatrix} 3 \\ 5 \end{Bmatrix}$. If nonconvex polyhedra are allowed, then additional quasiregular polyhedra the DODECADODECAHEDRON $\{5, \frac{5}{2}\}$ GREAT ICOSIDODECAHEDRON $\{3, \frac{5}{2}\}$, as well as 12 others (Hart).

For faces to be equatorial $\{h\}$,

$$h = \sqrt{4N_1 + 1} - 1. \tag{4}$$

The EDGES of quasiregular polyhedra form a system of GREAT CIRCLES: the OCTAHEDRON forms three SQUARES, the CUBOCTAHEDRON four HEXAGONS, and the ICOSIDODECAHEDRON six DECAGONS. The VERTEX FIGURES of quasiregular polyhedra are RECTANGLES (Hart). The EDGES are also all equivalent, a property shared only with the completely regular PLATONIC SOLIDS.

See also CUBOCTAHEDRON, DODECADODECAHEDRON, GREAT ICOSIDODECAHEDRON, ICOSIDODECAHEDRON, PLATONIC SOLID

References

Coxeter, H. S. M. "Quasi-Regular Polyhedra." §2–3 in *Regular Polytopes, 3rd ed.* New York: Dover, pp. 17–20, 1973.
Fejes Tóth, L. Ch. 4 in *Regular Figures.* Oxford, England: Pergamon Press, 1964.
Hart, G. "Quasi-Regular Polyhedra." http://www.george-hart.com/virtual-polyhedra/quasi-regular-info.html.
Robertson, S. A. and Carter, S. "On the Platonic and Archimedean Solids." *J. London Math. Soc.* **2**, 125–132, 1970.

Quasirhombicosidodecahedron

GREAT RHOMBICOSIDODECAHEDRON (UNIFORM)

Quasirhombicuboctahedron

GREAT RHOMBICUBOCTAHEDRON (UNIFORM)

Quasisimple Group

A FINITE GROUP L is quasisimple if $L = [L, L]$ and $L/Z(L)$ is a SIMPLE GROUP.

See also COMPONENT, FINITE GROUP, SIMPLE GROUP

Quasithin Theorem

In the classical quasithin case of the QUASI-UNIPOTENT PROBLEM, if a group G does not have a "strongly embedded" SUBGROUP, then G is a GROUP of LIE-TYPE in characteristic 2 of Lie RANK 2 generated by a pair of parabolic SUBGROUPS P_1 and P_2, or G is one of a short list of exceptions.

See also LIE-TYPE GROUP, QUASI-UNIPOTENT PROBLEM

Quasitruncated Cuboctahedron

GREAT TRUNCATED CUBOCTAHEDRON

Quasitruncated Dodecadocahedron

TRUNCATED DODECADODECAHEDRON

Quasitruncated Dodecahedron

TRUNCATED DODECAHEDRON

Quasitruncated Great Stellated Dodecahedron

GREAT STELLATED TRUNCATED DODECAHEDRON

Quasitruncated Hexahedron

STELLATED TRUNCATED HEXAHEDRON

Quasitruncated Small Stellated Dodecahedron

SMALL STELLATED TRUNCATED DODECAHEDRON

Quasi-Unipotent Group

A GROUP G is quasi-unipotent if every element of G of order p is UNIPOTENT for all PRIMES p such that G has p-RANK ≥ 3.

Quasi-Unipotent Problem

QUASITHIN THEOREM

Quaternary

The BASE 4 method of counting in which only the DIGITS 0, 1, 2, and 3 are used. The following table gives the quaternary equivalents of the first few decimal numbers.

1	1	11	23	21	111
2	2	12	30	22	112
3	3	13	31	23	113
4	10	14	32	24	120
5	11	15	33	25	121
6	12	16	100	26	122
7	13	17	101	27	123
8	20	18	102	28	130
9	21	19	103	29	131
10	22	20	110	30	132

These DIGITS have the following MULTIPLICATION TABLE.

×	0	1	2	3
0	0	0	0	0
1	0	1	2	3
2	0	2	10	12
3	0	3	12	21

See also BASE (NUMBER), BINARY, DECIMAL, HEXADECIMAL, MOSER-DE BRUIJN SEQUENCE, OCTAL, TERNARY

References

Lauwerier, H. *Fractals: Endlessly Repeated Geometric Figures.* Princeton, NJ: Princeton University Press, pp. 9–10, 1991.
Weisstein, E. W. "Bases." MATHEMATICA NOTEBOOK BASES.M.

Quaternary Tree

QUADTREE

Quaternion

A member of a *noncommutative* DIVISION ALGEBRA first invented by William Rowan Hamilton. The idea for quaternions occurred to him while be was walking along the Royal Canal on his way to a meeting of the Irish Academy, and Hamilton was so pleased with his discovery that he scratched the fundamental formula of quaternion algebra,

$$i^2 = j^2 = k^2 = ijk = -1, \tag{1}$$

into the stone of the Brougham bridge (Mishchenko and Solovyov 2000). The set of quaternions is denoted $\mathbb{H}$, and the quaternions are a single example of a more general class of HYPERCOMPLEX NUMBERS discovered by Hamilton. While the quaternions are not commutative, they are associative, and they form a GROUP known as the QUATERNION GROUP.

The quaternions can be represented using complex 2×2 MATRICES

$$H = \begin{bmatrix} z & w \\ -\bar{w} & \bar{z} \end{bmatrix} = \begin{bmatrix} a+ib & c+id \\ -c+id & a-ib \end{bmatrix}, \qquad (2)$$

where z and w are COMPLEX NUMBERS, a, b, c, and d are REAL, and $\bar{z}$ is the COMPLEX CONJUGATE of z. A quaternion can be represented using Quaternion[a, b, c, d] in the *Mathematica* add-on package Algebra'Quaternions' (which can be loaded with the command $<\,<$Algebra'), where a, b, c, and d are *explicit real numbers*.

By analogy with the COMPLEX NUMBERS being representable as a sum of REAL and IMAGINARY PARTS, $a \cdot 1 + bi$, a quaternion can also be written as a linear combination

$$H = a\mathsf{U} + b\mathsf{I} + c\mathsf{J} + d\mathsf{K} \qquad (3)$$

of the four matrices

$$\mathsf{U} \equiv \begin{bmatrix} 1 & 0 \\ 0 & 1 \end{bmatrix} \qquad (4)$$

$$\mathsf{I} \equiv \begin{bmatrix} i & 0 \\ 0 & -i \end{bmatrix} \qquad (5)$$

$$\mathsf{J} \equiv \begin{bmatrix} 0 & 1 \\ -1 & 0 \end{bmatrix} \qquad (6)$$

$$\mathsf{K} \equiv \begin{bmatrix} 0 & i \\ i & 0 \end{bmatrix}. \qquad (7)$$

(Note that here, U is used to denote the IDENTITY MATRIX, not I.) The matrices are closely related to the PAULI SPIN MATRICES $\sigma_x, \sigma_y, \sigma_z$, combined with the IDENTITY MATRIX. From the above definitions, it follows that

$$\mathsf{I}^2 = -\mathsf{U} \qquad (8)$$

$$\mathsf{J}^2 = -\mathsf{U} \qquad (9)$$

$$\mathsf{K}^2 = -\mathsf{U}. \qquad (10)$$

Therefore I, J, and K are three essentially different solutions of the matrix equation

$$\mathsf{X}^2 = -\mathsf{U}, \qquad (11)$$

which could be considered the square roots of the negative identity matrix. A LINEAR COMBINATION of basis quaternions with integer coefficients is sometimes called a HAMILTONIAN INTEGER.

In $\mathbb{R}^4$, the basis of the quaternions can be given by

$$i \equiv \begin{bmatrix} 0 & 1 & 0 & 0 \\ -1 & 0 & 0 & 0 \\ 0 & 0 & 0 & 1 \\ 0 & 0 & -1 & 0 \end{bmatrix} \qquad (12)$$

$$j \equiv \begin{bmatrix} 0 & 0 & 0 & -1 \\ 0 & 0 & -1 & 0 \\ 0 & 1 & 0 & 0 \\ 1 & 0 & 0 & 0 \end{bmatrix} \qquad (13)$$

$$k \equiv \begin{bmatrix} 0 & 0 & -1 & 0 \\ 0 & 0 & 0 & 1 \\ 1 & 0 & 0 & 0 \\ 0 & -1 & 0 & 0 \end{bmatrix} \qquad (14)$$

$$1 \equiv \begin{bmatrix} 1 & 0 & 0 & 0 \\ 0 & 1 & 0 & 0 \\ 0 & 0 & 1 & 0 \\ 0 & 0 & 0 & 1 \end{bmatrix}. \qquad (15)$$

The quaternions satisfy the following identities, sometimes known as HAMILTON'S RULES,

$$i^2 = j^2 = k^2 = -1 \qquad (16)$$

$$ij = -ji = k \qquad (17)$$

$$jk = -kj = i \qquad (18)$$

$$ki = -ik = j. \qquad (19)$$

They have the following multiplication table.

	1	i	j	k
1	1	i	j	k
i	i	-1	k	$-j$
j	j	$-k$	-1	i
k	k	j	$-i$	-1

The quaternions ± 1, $\pm i$, $\pm j$, and $\pm k$ form a NON-ABELIAN GROUP of order eight (with multiplication as the group operation) known as Q_8 of $\mathbb{H}$.

The quaternions can be written in the form

$$a = a_1 + a_2 i + a_3 j + a_4 k. \qquad (20)$$

The conjugate quaternion is given by

$$\bar{a} = a_1 - a_2 i - a_3 j - a_4 k. \qquad (21)$$

The sum of two quaternions is then

$$a + b = (a_1 + b_1) + (a_2 + b_2)i + (a_3 + b_3)j \\ + (a_4 + b_4)k, \qquad (22)$$

and the product of two quaternions is

$$ab = (a_1 b_1 - a_2 b_2 - a_3 b_3 - a_4 b_4) \\ + (a_1 b_2 + a_2 b_1 + a_3 b_4 - a_4 b_3)i \\ + (a_1 b_3 - a_2 b_4 + a_3 b_1 + a_4 b_2)j \\ + (a_1 b_4 + a_2 b_3 - a_3 b_2 + a_4 b_1)k, \qquad (23)$$

so the norm is

$$n(a) = \sqrt{a\bar{a}} = \sqrt{\bar{a}a} = \sqrt{a_1^2 + a_2^2 + a_3^2 + a_4^2}. \qquad (24)$$

In this notation, the quaternions are closely related to FOUR-VECTORS.

Quaternions can be interpreted as a SCALAR plus a VECTOR by writing

$$a = a_1 + a_2 i + a_3 j + a_4 k = (a_1, \mathbf{a}), \qquad (25)$$

where $\mathbf{a} \equiv [a_2 a_3 a_4]$.. In this notation, quaternion multiplication has the particularly simple form

$$q_1 q_2 = (s_1, \mathbf{v}_1) \cdot (s_2, \mathbf{v}_2)$$
$$= (s_1 s_2 - \mathbf{v}_1 \cdot \mathbf{v}_2, \ s_1 \mathbf{v}_2 + s_2 \mathbf{v}_1 + \mathbf{v}_1 \times \mathbf{v}_2). \quad (26)$$

Division is uniquely defined (except by zero), so quaternions form a DIVISION ALGEBRA. The inverse of a quaternion is given by

$$a^{-1} = \frac{\bar{a}}{a\bar{a}}, \qquad (27)$$

and the norm is multiplicative

$$n(ab) = n(a)n(b). \qquad (28)$$

In fact, the product of two quaternion norms immediately gives the EULER FOUR-SQUARE IDENTITY.

A rotation about the UNIT VECTOR $\hat{\mathbf{n}}$ by an angle θ can be computed using the quaternion

$$q = (s, \mathbf{v}) = \left(\cos\left(\tfrac{1}{2}\theta\right), \ \hat{\mathbf{n}} \sin\left(\tfrac{1}{2}\theta\right) \right) \qquad (29)$$

(Arvo 1994, Hearn and Baker 1996). The components of this quaternion are called EULER PARAMETERS. After rotation, a point $p = (0, \mathbf{p})$ is then given by

$$p' = qpq^{-1} = qp\bar{q}, \qquad (30)$$

since $n(q) = 1$. A concatenation of two rotations, first q_1 and then q_2, can be computed using the identity

$$q_2(q_1 p \bar{q}_1)\bar{q}_2 = (q_2 q_1)p(\bar{q}_1 \bar{q}_2) = (q_2 q_1)p\overline{q_2 q_1} \qquad (31)$$

(Goldstein 1980).

See also BIQUATERNION, CAYLEY-KLEIN PARAMETERS, COMPLEX NUMBER, DIVISION ALGEBRA, EULER PARAMETERS, FOUR-VECTOR, HAMILTONIAN INTEGER, HYPERCOMPLEX NUMBER, OCTONION, QUATERNION GROUP

References

Altmann, S. L. *Rotations, Quaternions, and Double Groups.* Oxford, England: Clarendon Press, 1986.

Arvo, J. *Graphics Gems II.* New York: Academic Press, pp. 351–354 and 377–380, 1994.

Baker, A. L. *Quaternions as the Result of Algebraic Operations.* New York: Van Nostrand, 1911.

Conway, J. H. and Guy, R. K. *The Book of Numbers.* New York: Springer-Verlag, pp. 230–234, 1996.

Crowe, M. J. *A History of Vector Analysis: The Evolution of the Idea of a Vectorial System.* New York: Dover, 1994.

Dickson, L. E. *Algebras and Their Arithmetics.* New York: Dover, 1960.

Downs, L. "CS184: Using Quaternions to Represent Rotation." http://http.cs.berkeley.edu/~laura/cs184/quat/quaternion.html.

Du Val, P. *Homographies, Quaternions, and Rotations.* Oxford, England: Oxford University Press, 1964.

Ebbinghaus, H. D.; Hirzebruch, F.; Hermes, H.; Prestel, A; Koecher, M.; Mainzer, M.; and Remmert, R. *Numbers.* New York: Springer-Verlag, 1990.

Goldstein, H. *Classical Mechanics, 2nd ed.* Reading, MA: Addison-Wesley, p. 151, 1980.

Hamilton, W. R. *Lectures on Quaternions: Containing a Systematic Statement of a New Mathematical Method.* Dublin: Hodges and Smith, 1853.

Hamilton, W. R. *Elements of Quaternions.* London: Longmans, Green, 1866.

Hamilton, W. R. *The Mathematical Papers of Sir William Rowan Hamilton.* Cambridge, England: Cambridge University Press, 1967.

Hardy, A. S. *Elements of Quaternions.* Boston, MA: Ginn, Heath, & Co., 1881.

Hardy, G. H. and Wright, E. M. "Quaternions." §20.6 in *An Introduction to the Theory of Numbers, 5th ed.* Oxford, England: Clarendon Press, pp. 303–306, 1979.

Hearn, D. and Baker, M. P. *Computer Graphics: C Version, 2nd ed.* Englewood Cliffs, NJ: Prentice-Hall, pp. 419–420 and 617–618, 1996.

Joly, C. J. *A Manual of Quaternions.* London: Macmillan, 1905.

Julstrom, B. A. "Using Real Quaternions to Represent Rotations in Three Dimensions." *UMAP Modules in Undergraduate Mathematics and Its Applications, Module 652.* Lexington, MA: COMAP, Inc., 1992.

Kelland, P. and Tait, P. G. *Introduction to Quaternions, 3rd ed.* London: Macmillan, 1904.

Kuipers, J. B. *Quaternions and Rotation Sequences: A Primer with Applications to Orbits, Aerospace, and Virtual Reality.* Princeton, NJ: Princeton University Press, 1998.

Mishchenko, A. and Solovyov, Y. "Quaternions." *Quantum* **11**, 4–7 and 18, 2000.

Nicholson, W. K. *Introduction to Abstract Algebra, 2nd ed.* New York: Wiley, 1999.

Salamin, G. Item 107 in Beeler, M.; Gosper, R. W.; and Schroeppel, R. *HAKMEM.* Cambridge, MA: MIT Artificial Intelligence Laboratory, Memo AIM-239, pp. 46–47, Feb. 1972.

Shoemake, K. "Animating Rotation with Quaternion Curves." *Computer Graphics* **19**, 245–254, 1985.

Tait, P. G. *An Elementary Treatise on Quaternions, 3rd ed., enl.* Cambridge, England: Cambridge University Press, 1890.

Tait, P. G. "Quaternions." *Encyclopædia Britannica, 9th ed.* ca. 1886. ftp://ftp.netcom.com/pub/hb/hbaker/quaternion/tait/Encyc-Brit.ps.gz.

Weisstein, E. W. "Books about Quaternions." http://www.treasure-troves.com/books/Quaternions.html.

Quaternion Group

The NON-ABELIAN GROUP of order eight formed by the QUATERNIONS ± 1, $\pm i$, $\pm j$, and $\pm k$, denoted Q_8 or $\mathbb{H}$.

See also QUATERNION

Quattuordecillion

In the American system, 10^{45}.

See also LARGE NUMBER

Queens Problem

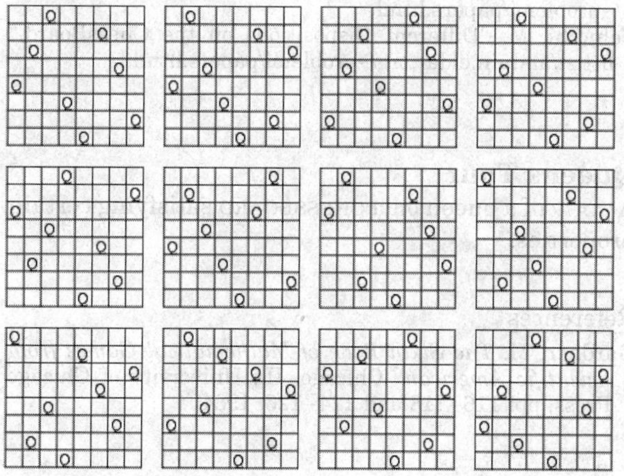

k Queens	n × n	$N_p(k,n)$
2	4	3
3	5	37
3	6	1
4	7	5
5	8	4860

What is the maximum number of queens which can be placed on an $n \times n$ CHESSBOARD such that no two attack one another? The answer is n queens, which gives eight queens for the usual 8×8 board (Madachy 1979; Steinhaus 1983, p. 29). The *number* of different ways the n queens can be placed on an $n \times n$ chessboard so that no two queens may attack each other for the first few n are 1, 0, 0, 2, 10, 4, 40, 92, ... (Sloane's A000170; Madachy 1979; Steinhaus 1983, p. 29). The number of rotationally and reflectively distinct solutions are 1, 0, 0, 1, 2, 1, 6, 12, 46, 92, ... (Sloane's A002562; Dudeney 1970; p. 96). The 12 distinct solutions for $n = 8$ are illustrated above, and the remaining 80 are generated by ROTATION and REFLECTION (Madachy 1979, Steinhaus 1983).

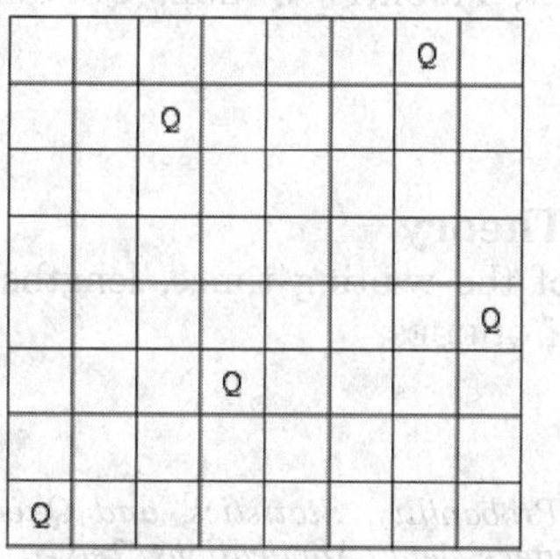

The minimum number of queens needed to occupy or attack all squares of an 8×8 board is 5 (Steinhaus 1983, p. 29). Dudeney (1970, pp. 95–96) gave the following results for the number of distinct arrangements $N_p(k,n)$ of k queens attacking or occupying every square of an $n \times n$ board for which every queen is attacked ("protected") by at least one other, with the $n = 8$ value given by Steinhaus (1983, p. 29). The 4860 solutions in the $n = 5$ case may be obtained from 638 fundamental arrangements by ROTATION and REFLECTION.

Dudeney (1970, pp. 95–96) also gave the following results for the number of distinct arrangements $N_u(k,n)$ of k queens attacking or occupying every square of an $n \times n$ board for which no two queens attack one another (they are "not protected").

k Queens	n × n	$N_u(k,n)$
1	2	1
1	3	1
3	4	2
3	5	2
4	6	17
4	7	1
5	8	91

Vardi (1991) generalizes the problem from a square chessboard to one with the topology of the TORUS. The number of solutions for n queens with n ODD are 1, 0, 10, 28, 0, 88, ... (Sloane's A007705). Vardi (1991) also considers the toroidal "semiqueens" problem, in which a semiqueen can move like a rook or bishop, but only on POSITIVE broken diagonals. The number of solutions to this problem for n queens with n ODD are 1, 3, 15, 133, 2025, 37851, ... (Sloane's A006717), and 0 for EVEN n.

Velucchi gives the solution to the question, "How many different arrangements of k queens are possible on an order n chessboard?" as 1/8th of the COEFFICIENT of $a^k b^{n^2-k}$ in the POLYNOMIAL

$$p(a,b,n) = \begin{cases} (a+b)^{n^2} + 2(a+b)^n (a^2+b^2)^{(n^2-n)/2} \\ \quad + 3(a^2+b^2)^{n^2/2} + 2(a^4+b^4)^{n^2/4} \\ \hfill n \text{ even} \\ (a+b)^{n^2} + 2(a+b)(a^4+b^4)^{(n^2-1)/4} \\ \quad + (a+b)(a^2+b^2)^{(n^2-1)/2} \\ \quad + 4(a+b)^n (a^2+b^2)^{(n^2-n)/2} \\ \hfill n \text{ odd.} \end{cases}$$

Velucchi also considers the nondominating queens problem, which consists of placing n queens on an order n chessboard to leave a maximum number $U(n)$ of unattacked vacant cells. The first few values are 0, 0, 0, 1, 3, 5, 7, 11, 18, 22, 30, 36, 47, 56, 72, 82, ... (Sloane's A001366). The results can be generalized to k queens on an $n \times n$ board.

See also BISHOPS PROBLEM, CHESS, KINGS PROBLEM, KNIGHTS PROBLEM, KNIGHT'S TOUR, ROOKS PROBLEM

References

Ahrens, W. "Das Achtköniginnenproblem." Ch. 9 in *Mathematische Unterhaltungen und Spiele, dritte, verbesserte, anastatisch gedruckte aufl., Bd. 1.* Leipzig, Germany: Teubner, pp. 211–284, 1921.
Ball, W. W. R. and Coxeter, H. S. M. *Mathematical Recreations and Essays, 13th ed.* New York: Dover, pp. 166–169, 1987.
Campbell, P. J. "Gauss and the 8-Queens Problem: A Study in the Propagation of Historical Error." *Historia Math.* **4**, 397–404, 1977.
Dudeney, H. E. "The Eight Queens." §300 in *Amusements in Mathematics.* New York: Dover, p. 89, 1970.
Erbas, C. and Tanik, M. M. "Generating Solutions to the N-Queens Problem Using 2-Circulants." *Math. Mag.* **68**, 343–356, 1995.
Erbas, C.; Tanik, M. M.; and Aliyzaicioglu, Z. "Linear Congruence Equations for the Solutions of the N-Queens Problem." *Inform. Proc. Let.* **41**, 301–306, 1992.
Gardner, M. "Patterns in Primes are a Clue to the Strong Law of Small Numbers." *Sci. Amer.* **243**, 18–28, Dec. 1980.
Garey, M. R. and Johnson, D. S. *Computers and Intractability: A Guide to the Theory of NP-Completeness.* New York: W. H. Freeman, 1983.
Ginsburg, J. "Gauss's Arithmetization of the Problem of n Queens." *Scripta Math.* **5**, 63–66, 1939.
Guy, R. K. "The n Queens Problem." §C18 in *Unsolved Problems in Number Theory, 2nd ed.* New York: Springer-Verlag, pp. 133–135, 1994.
Kraitchik, M. "The Problem of the Queens" and "Domination of the Chessboard." §10.3 and 10.4 in *Mathematical Recreations.* New York: W. W. Norton, pp. 247–256, 1942.
Madachy, J. S. *Madachy's Mathematical Recreations.* New York: Dover, pp. 34–36, 1979.
Riven, I.; Vardi, I.; and Zimmerman, P. "The n-Queens Problem." *Amer. Math. Monthly* **101**, 629–639, 1994.
Riven, I. and Zabih, R. "An Algebraic Approach to Constraint Satisfaction Problems." In *Proc. Eleventh Internat. Joint Conference on Artificial Intelligence, Vol. 1, August 20–25, 1989.* Detroit, MI: IJCAII, pp. 284–289, 1989.
Ruskey, F. "Information on the n Queens Problem." http://www.theory.csc.uvic.ca/~cos/inf/misc/Queen.html.
Sloane, N. J. A. Sequences A000170/M1958, A001366, A002562/M0180, A006717/M3005, and A007705/M4691 in "An On-Line Version of the Encyclopedia of Integer Sequences." http://www.research.att.com/~njas/sequences/eisonline.html.
Sloane, N. J. A. and Plouffe, S. Figure M0180 in *The Encyclopedia of Integer Sequences.* San Diego: Academic Press, 1995.
Steinhaus, H. *Mathematical Snapshots, 3rd ed.* New York: Dover, pp. 29–30, 1999.
Vardi, I. "The n-Queens Problems." Ch. 6 in *Computational Recreations in Mathematica.* Redwood City, CA: Addison-Wesley, pp. 107–125, 1991.

Velucchi, M. "For Me, this Is the Best Chess-Puzzle: Non-Dominating Queens Problem." http://anduin.eldar.org/~problemi/papers.html.
Velucchi, M. "Different Dispositions on the ChessBoard." http://anduin.eldar.org/~problemi/papers.html.

Queens Tour

A TOUR of a queen on a CHESSBOARD satisfying certain properties.

References

Gardner, M. *The Sixth Book of Mathematical Games from Scientific American.* Chicago, IL: University of Chicago Press, pp. 116–118 and 124–126, 1984.

Quermass

BRIGHTNESS, OUTER QUERMASS

Question Mark Function

MINKOWSKI'S QUESTION MARK FUNCTION

Queue

A queue is a special kind of LIST in which elements may only be removed from the bottom by a POP action or added to the top using a PUSH action. Examples of queues include people waiting in line, and submitted jobs waiting to be printed on a printer. The study of queues is called QUEUING THEORY.

See also LIST, PRIORITY QUEUE, QUEUING THEORY, STACK

Queuing Theory

The study of the waiting times, lengths, and other properties of QUEUES.

References

Allen, A. O. *Probability, Statistics, and Queueing Theory with Computer Science Applications, 2nd ed.* Orlando, FL: Academic Press, 1990.
Bunday, B. D. *An Introduction to Queueing Theory.* Oxford, England: Oxford University Press, 1996.
Gross, D. and Harris, C. M. *Fundamentals of Queueing Theory, 3rd ed.* New York: Wiley, 1998.

Quicksort

The fastest known SORTING ALGORITHM (on average, and for a large number of elements), requiring $\mathcal{O}(n \lg n)$ steps. Quicksort is a recursive algorithm which first partitions an array $\{a_i\}_{i=1}^n$ according to several rules (Sedgewick 1978):

1. Some key v is in its final position in the array (i.e., if it is the jth smallest, it is in position a_j).

2. All the elements to the left of a_j are less than or equal to a_j. The elements $a_1, a_2, ..., a_{j-1}$ are called the "left subfile."

3. All the elements to the right of a_j are greater than or equal to a_j. The elements $a_{j+1}, ..., a_n$ are called the "right subfile."

Quicksort was invented by Hoare (1961, 1962), has undergone extensive analysis and scrutiny (Sedgewick 1975, 1977, 1978), and is known to be about twice as fast as the next fastest SORTING algorithm. In the worst case, however, quicksort is a slow n^2 algorithm (and for quicksort, "worst case" corresponds to already sorted).

See also HEAPSORT, SORTING

References

Aho, A. V.; Hopcroft, J. E.; and Ullmann, J. D. *Data Structures and Algorithms.* Reading, MA: Addison-Wesley, pp. 260–270, 1987.

Hoare, C. A. R. "Partition: Algorithm 63," "Quicksort: Algorithm 64," and "Find: Algorithm 65." *Comm. ACM* **4**, 321–322, 1961.

Hoare, C. A. R. "Quicksort." *Computer J.* **5**, 10–15, 1962.

Press, W. H.; Flannery, B. P.; Teukolsky, S. A.; and Vetterling, W. T. "Quicksort." §8.2 in *Numerical Recipes in FORTRAN: The Art of Scientific Computing, 2nd ed.* Cambridge, England: Cambridge University Press, pp. 323–327, 1992.

Sedgewick, R. *Quicksort.* Ph.D. thesis. Stanford Computer Science Report STAN-CS-75–492. Stanford, CA: Stanford University, May 1975.

Sedgewick, R. "The Analysis of Quicksort Programs." *Acta Informatica* **7**, 327–355, 1977.

Sedgewick, R. "Implementing Quicksort Programs." *Comm. ACM* **21**, 847–857, 1978.

Quillen-Lichtenbaum Conjecture

A technical CONJECTURE which connects algebraic K-THEORY to Étale cohomology. The conjecture was made more precise by Dwyer and Friedlander (1982). Thomason (1985) established the first half of this conjecture, but the entire conjecture has not yet been established.

References

Dwyer, W. and Friedlander, E. "Étale K-Theory and Arithmetic." *Bull. Amer. Math. Soc.* **6**, 453–455, 1982.

Thomason, R. W. "Algebraic K-Theory and Étale Cohomology." *Ann. Sci. École Norm. Sup.* **18**, 437–552, 1985.

Weibel, C. A. "The Mathematical Enterprises of Robert Thomason." *Bull. Amer. Math. Soc.* **34**, 1–13, 1996.

Quincunx

The pattern ⁙ of dots on the "5" side of a 6-sided DIE. The word derives from the Latin words for both one and five.

See also DICE

References

Conway, J. H. and Guy, R. K. *The Book of Numbers.* New York: Springer-Verlag, pp. 9 and 22, 1996.

Quindecillion

In the American system, 10^{48}.

See also LARGE NUMBER

Quintet

A SET of five.

See also HEXAD, MONAD, QUARTET, TETRAD, TRIAD

Quintic Equation

Unlike quadratic, cubic, and quartic polynomials, the *general* quintic cannot be solved algebraically in terms of a finite number of ADDITIONS, SUBTRACTIONS, MULTIPLICATIONS, DIVISIONS, and ROOT EXTRACTIONS, as rigorously demonstrated by Abel (ABEL'S IMPOSSIBILITY THEOREM) and Galois. However, certain classes of quintic equations can be solved in this manner.

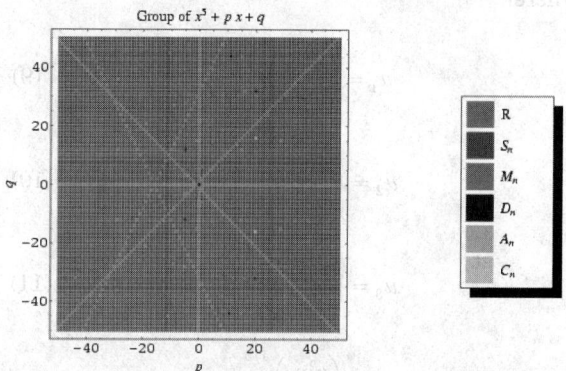

Irreducible quintic equations can be associated with a GALOIS GROUP, which may be a SYMMETRIC GROUP S_n, METACYCLIC GROUP M_n, DIHEDRAL GROUP D_n, ALTERNATING GROUP A_n, or CYCLIC GROUP C_n, as illustrated above.

Euler reduced the general quintic to

$$x^5 - 10qx^2 - p = 0. \qquad (1)$$

A quintic also can be algebraically reduced to PRINCIPAL QUINTIC FORM

$$x^5 + a_2 x^2 + a_1 x + a_0 = 0. \qquad (2)$$

By solving a quartic, a quintic can be algebraically reduced to the BRING QUINTIC FORM

$$x^5 - x - a = 0, \qquad (3)$$

as was first done by Jerrard. Runge (1885) and Cadenhad and Young found a parameterization of solvable quintics in the form

$$x^5 - ax + b = 0, \tag{4}$$

by showing that all irreducible solvable quintics with COEFFICIENTS of x^4, x^3, and x^2 missing have the following form

$$x^5 + \frac{5\mu^4(4v+3)}{v^2+1}x + \frac{5\mu^5(2v+1)(4v+3)}{v^2+1} = 0, \tag{5}$$

where μ and v are RATIONAL. Spearman and Williams (1994) showed that an irreducible quintic OF THE FORM (4) having RATIONAL COEFFICIENTS is solvable by radicals IFF there exist rational numbers $\varepsilon = \pm 1$, $c \geq 0$, and $e \neq 0$ such that

$$a = \frac{5e^4(3 - 4\varepsilon c)}{c^2 + 1} \tag{6}$$

$$b = \frac{-4e^5(11\varepsilon + 2c)}{c^2 + 1} \tag{7}$$

The ROOTS are then

$$x_j = e\left(\omega^j u_1 + \omega^{2j} u_2 + \omega^{3j} u_3 + \omega^{4j} u_4\right), \tag{8}$$

where

$$u_1 = \left(\frac{v_1^2 v_3}{D^2}\right)^{1/5} \tag{9}$$

$$u_2 = \left(\frac{v_3^2 v_4}{D^2}\right)^{1/5} \tag{10}$$

$$u_3 = \left(\frac{v_2^2 v_1}{D^2}\right)^{1/5} \tag{11}$$

$$u_4 = \left(\frac{v_4^2 v_2}{D^2}\right)^{1/5} \tag{12}$$

$$v_1 = \sqrt{D} + \sqrt{D - \varepsilon\sqrt{D}} \tag{13}$$

$$v_2 = -\sqrt{D} - \sqrt{D + \varepsilon\sqrt{D}} \tag{14}$$

$$v_3 = -\sqrt{D} + \sqrt{D + \varepsilon\sqrt{D}} \tag{15}$$

$$v_4 = \sqrt{D} - \sqrt{D - \varepsilon\sqrt{D}} \tag{16}$$

$$D = c^2 + 1. \tag{17}$$

In the case of a solvable quintic, the roots can be found using the formulas of Malfatti (1771), who was the first to "solve" the quintic using a resolvent of sixth degree (Pierpont 1895).

The general quintic can be solved in terms of JACOBI THETA FUNCTIONS, as was first done by Hermite in 1858. Kronecker subsequently obtained the same solution more simply, and Brioshi also derived the equation. To do so, reduce the general quintic

$$a_5 x^5 + a_4 x^4 + a_3 x^3 + a_2 x^2 + a_1 x + a_0 = 0 \tag{18}$$

into BRING QUINTIC FORM

$$x^5 - x + \rho = 0. \tag{19}$$

Then define

$$k \equiv \tan\left[\tfrac{1}{4}\sin^{-1}\left(\frac{16}{25\sqrt{5}\rho^2}\right)\right] \tag{20}$$

$$s \equiv \begin{cases} -\operatorname{sgn}(\Im[\rho]) & \text{for } \Re[\rho] = 0 \\ \operatorname{sgn}(\Re[\rho]) & \text{for } \Re[\rho] \neq 0 \end{cases} \tag{21}$$

$$b = \frac{s(k^2)^{1/8}}{2 \cdot 5^{3/4}\sqrt{k(1-k^2)}} \tag{22}$$

$$q = q(k^2) = e^{i\pi K'(k^2)/K(k^2)}, \tag{23}$$

where k is the MODULUS, $m \equiv k^2$ is the PARAMETER, and q is the NOME. Solving

$$q(m) = e^{i\pi K'(m)/K(m)} \tag{24}$$

for m gives the INVERSE NOME $m(q)$, and the roots of the original quintic are then given by

$$x_1 = (-1)^{3/4} b\left\{\left[m\left(e^{-2\pi i/5}q^{1/5}\right)\right]^{1/8} + i\left[m\left(e^{2\pi i/5}q^{1/5}\right)\right]^{1/8}\right\}$$
$$\times \left\{\left[m\left(e^{-4\pi i/5}q^{1/5}\right)\right]^{1/8} + \left[m\left(e^{4\pi i/5}q^{1/5}\right)\right]^{1/8}\right\}$$
$$\times \left\{\left[m\left(q^{1/5}\right)\right]^{1/8} + q^{5/8}\left(q^5\right)^{-1/8}\left[m\left(q^5\right)\right]^{1/8}\right\} \tag{25}$$

$$x_2 = b\left\{-\left[m\left(q^{1/5}\right)\right]^{1/8} + e^{3\pi i/4}\left[m\left(e^{2\pi i/5}q^{1/5}\right)\right]^{1/8}\right\}$$
$$\times \left\{e^{-3\pi i/4}\left[m\left(e^{-2\pi i/5}q^{1/5}\right)\right]^{1/8} + i\left[m\left(e^{4\pi i/5}q^{1/5}\right)\right]^{1/8}\right\}$$
$$\times \left\{i\left[m\left(e^{-4\pi i/5}q^{1/5}\right)\right]^{1/8} + q^{5/8}\left(q^5\right)^{-1/8}\left[m\left(q^5\right)\right]^{1/8}\right\} \tag{26}$$

$$x_3 = b\left\{e^{-3\pi i/4}\left[m\left(e^{-2\pi i/5}q^{1/5}\right)\right]^{1/8} - i\left[m\left(e^{-4\pi i/5}q^{1/5}\right)\right]^{1/8}\right\}$$
$$\times \left\{-\left[m\left(q^{1/5}\right)\right]^{1/8} - i\left[m\left(e^{4\pi i/5}q^{1/5}\right)\right]^{1/8}\right\}$$
$$\times \left\{e^{-3\pi i/4}\left[m\left(e^{2\pi i/5}q^{1/5}\right)\right]^{1/8} + q^{5/8}\left(q^5\right)^{-1/8}\left[m\left(q^{1/5}\right)\right]^{1/8}\right\} \tag{27}$$

$$x_4 = b\left\{\left[m\left(q^{1/5}\right)\right]^{1/8} - i\left[m\left(e^{-4\pi i/5}q^{1/5}\right)\right]^{1/8}\right\}$$
$$\times \left\{-e^{-3\pi i/4}\left[m\left(e^{2\pi i/5}q^{1/5}\right)\right]^{1/8} - i\left[m\left(e^{4\pi i/5}q^{1/5}\right)\right]^{1/8}\right\}$$

$$\times\left\{e^{-3\pi i/4}\left[m\left(e^{-2\pi i/5}q^{1/5}\right)\right]^{1/8}+q^{5/8}\left(q^5\right)^{-1/8}\left[m\left(q^5\right)\right]^{1/8}\right\} \tag{28}$$

$$x_5 = b\left\{\left[m\left(q^{1/5}\right)\right]^{1/8}-e^{-3\pi i/4}\left[m\left(e^{-2\pi i/5}q^{1/5}\right)\right]^{1/8}\right\}$$

$$\times\left\{-e^{3\pi i/4}\left[m\left(e^{2\pi i/5}q^{1/5}\right)\right]^{1/8}+i\left[m\left(e^{-4\pi i/5}q^{1/5}\right)\right]^{1/8}\right\}$$

$$\times\left\{\left(-i\left[m\left(e^{4\pi i/5}q^{1/5}\right)\right]^{1/8}+q^{5/8}\left(q^5\right)^{-1/8}\left[m\left(q^5\right)\right]^{1/8}\right\}. \tag{29}$$

Felix Klein used a TSCHIRNHAUSEN TRANSFORMATION to reduce the general quintic to the form

$$z^5 + 5az^2 + 5bz + c = 0. \tag{30}$$

He then solved the related ICOSAHEDRAL EQUATION

$$I(z,1,Z) = z^5\left(-1+11z^5+z^{10}\right)^5$$
$$-\left[1+z^{30}-10005\left(z^{10}+z^{20}\right)+522\left(-z^5+z^{25}\right)\right]^2 Z$$
$$= 0, \tag{31}$$

where Z is a function of radicals of a, b, and c. The solution of this equation can be given in terms of HYPERGEOMETRIC FUNCTIONS as

$$\frac{Z^{-1/60} {}_2F_1\left(-\frac{1}{60},\frac{29}{60},\frac{4}{5},1728Z\right)}{Z^{11/60} {}_2F_1\left(\frac{11}{60},\frac{41}{60},\frac{6}{5},1728Z\right)}. \tag{32}$$

Another possible approach uses a series expansion, which gives one root (the first one in the list below) of the BRING QUINTIC FORM

$$t^5 - t - \rho. \tag{33}$$

All five roots can be derived using differential equations (Cockle 1860, Harley 1862). Let

$$F_1(\rho) = F_2(\rho) \tag{34}$$

$$F_2(\rho) = {}_4F_3\left(\frac{1}{5},\frac{2}{5},\frac{3}{5},\frac{4}{5};\frac{1}{2},\frac{3}{4},\frac{5}{4};\frac{3125}{256}\rho^4\right) \tag{35}$$

$$F_3(\rho) = {}_4F_3\left(\frac{9}{20},\frac{13}{20},\frac{17}{20},\frac{21}{20};\frac{3}{4},\frac{5}{4},\frac{3}{2};\frac{3125}{256}\rho^4\right) \tag{36}$$

$$F_4(\rho) = {}_4F_3\left(\frac{7}{10},\frac{9}{10},\frac{11}{10},\frac{13}{10};\frac{5}{4},\frac{3}{2},\frac{7}{4};\frac{3125}{256}\rho^4\right), \tag{37}$$

then the ROOTS are

$$t_1 = -\rho\,{}_4F_3\left(\frac{1}{5},\frac{2}{5},\frac{3}{5},\frac{4}{5};\frac{1}{2},\frac{3}{4},\frac{5}{4};\frac{3125}{256}\rho^4\right) \tag{38}$$

$$t_2 = -F_1(\rho)+\tfrac{1}{4}\rho F_2(\rho)+\tfrac{5}{32}\rho^2 F_3(\rho)+\tfrac{5}{32}\rho^3 F_4(\rho) \tag{39}$$

$$t_3 = -F_1(\rho)+\tfrac{1}{4}\rho F_2(\rho)-\tfrac{5}{32}\rho^2 F_3(\rho)+\tfrac{5}{32}\rho^3 F_4(\rho) \tag{40}$$

$$t_4 = -iF_1(\rho)+\tfrac{1}{4}\rho F_2(\rho)-\tfrac{5}{32}i\rho^2 F_3(\rho)-\tfrac{5}{32}\rho^3 F_4(\rho) \tag{41}$$

$$t_5 = -iF_1(\rho)+\tfrac{1}{4}\rho F_2(\rho)+\tfrac{5}{32}i\rho^2 F_3(\rho)-\tfrac{5}{32}\rho^3 F_4(\rho) \tag{42}$$

This technique gives closed form solutions in terms of HYPERGEOMETRIC FUNCTIONS in one variable for any POLYNOMIAL equation which can be written in the form

$$x^p + bx^q + c. \tag{43}$$

Consider the quintic

$$\prod_{j=0}^{4}\left[x-\left(\omega^j u_1 + \omega^{4j} u_2\right)\right] = 0, \tag{44}$$

where $\omega = e^{2\pi i/5}$ and u_1 and u_2 are COMPLEX NUMBERS. This is called DE MOIVRE'S QUINTIC. Generalize it to

$$\prod_{j=0}^{4}\left[x-\left(\omega^j u_1 + \omega^{2j} u_2 + \omega^{3j} u_3 + \omega^{4j} u_4\right)\right] = 0 \tag{45}$$

Expanding,

$$\left(\omega^j u_1 + \omega^{2j} u_2 + \omega^{3j} u_3 + \omega^{4j} u_4\right)^5$$
$$-5U\left(\omega^j u_1 + \omega^{2j} u_2 + \omega^{3j} u_3 + \omega^{4j} u_4\right)^4$$
$$-5V\left(\omega^j u_1 + \omega^{2j} u_2 + \omega^{3j} u_3 + \omega^{4j} u_4\right)^2$$
$$+5W\left(\omega^j u_1 + \omega^{2j} u_2 + \omega^{3j} u_3 + \omega^{4j} u_4\right)$$
$$+[5(X-Y)-Z] = 0, \tag{46}$$

where

$$U = u_1 u_4 + u2 u_3 \tag{47}$$

$$V = u_1 u_2^2 + u_2 u_4^2 + u_3 u_1^2 + u_4 u_3^2 \tag{48}$$

$$W = u_1^2 u_4^2 + u_2^2 u_3^2 - u_1^3 u_2 - u_2^3 u_4 - u_3^3 u_1 - u_4^3 u_3$$
$$- u_1 u_2 u_3 u_4 \tag{49}$$

$$X = u_1^3 u_3 u_4 + u_2^3 u_1 u_3 + u_3^3 u_2 u_4 + u_4^3 u_1 u_2 \tag{50}$$

$$Y = u_1 u_3^2 u_4^2 + u_2 u_1^2 u_3^2 + u_3 u_2^2 u_4^2 + u_4 u_1^2 u_2^2 \tag{51}$$

$$Z = u_1^5 + u_2^5 + u_3^5 + u_4^5 \tag{52}$$

The u_is satisfy

$$u_1 u_4 + u_2 u_3 = 0 \tag{53}$$

$$u_1 u_2^2 + u_2 u_4^2 + u_3 u_1^2 + u_4 u_3^2 = 0 \tag{54}$$

$$u_1^2 u_4^2 + u_2^2 u_3^2 - u_1^3 u_2 - u_2^3 u_4 - u_3^3 u_1 - u_4^3 u_3 - u_1 u_2 u_3 u_4$$
$$= \tfrac{1}{5}a \tag{55}$$

$$5\left[\left(u_1^3 u_3 u_4 + u_2^3 u_1 u_3 + u_3^3 u_3 u_4 + u_4^3 u_1 u_2\right)\right.$$
$$\left.-\left(u_1 u_3^2 u_4^2 + u_2 u_1^2 u_3^2 + u_3 u_2^2 u_4^2 + u_4 u_1^2 u_2^2\right)\right]$$
$$-\left(u_1^5 + u_2^5 + u_3^5 + u_4^5\right) = b. \tag{56}$$

See also BRING QUINTIC FORM, BRING-JERRARD

QUINTIC FORM, CUBIC EQUATION, DE MOIVRE'S QUINTIC, PRINCIPAL QUINTIC FORM, QUADRATIC EQUATION, QUARTIC EQUATION, SEXTIC EQUATION

References

Birkhoff, G. and Mac Lane, S. "Insolvability of Quintic Equations." §15.8 in *A Survey of Modern Algebra, 5th ed.* New York: Macmillan, pp. 418–421, 1996.

Chowla, S. "On Quintic Equations Soluble by Radicals." *Math. Student* **13**, 84, 1945.

Cockle, J. "Sketch of a Theory of Transcendental Roots." *Phil. Mag.* **20**, 145–148, 1860.

Cockle, J. " On Transcendental and Algebraic Solution--Supplemental Paper." *Phil. Mag.* **13**, 135–139, 1862.

Davis, H. T. *Introduction to Nonlinear Differential and Integral Equations.* New York: Dover, p. 172, 1960.

Drociuk, R. J. On the Complete Solution to the Most General Fifth Degree Polynomial. 3 May 2000. http://xxx.lanl.gov/abs/math.GM/0005026/.

Dummit, D. S. "Solving Solvable Quintics." *Math. Comput.* **57**, 387–401, 1991.

Glashan, J. C. "Notes on the Quintic." *Amer. J. Math.* **8**, 178–179, 1885.

Green, M. L. "On the Analytic Solution of the Equation of Fifth Degree." *Compos. Math.* **37**, 233–241, 1978.

Harley, R. "On the Solution of the Transcendental Solution of Algebraic Equations." *Quart. J. Pure Appl. Math.* **5**, 337–361, 1862.

Harley, R. "A Contribution to the History of the Problem of the Reduction of the General Equation of the Fifth Degree to a Trinomial Form." *Quart. J. Math.* **6**, 38–47, 1864.

Hermite, C. "Sulla risoluzione delle equazioni del quinto grado." *Annali di math. pura ed appl.* **1**, 256–259, 1858.

King, R. B. *Beyond the Quartic Equation.* Boston, MA: Birkhäuser, 1996.

King, R. B. and Cranfield, E. R. "An Algorithm for Calculating the Roots of a General Quintic Equation from Its Coefficients." *J. Math. Phys.* **32**, 823–825, 1991.

Klein, F. "Sull' equazioni dell' Icosaedro nella risoluzione delle equazioni del quinto grado [per funzioni ellittiche]." *Reale Istituto Lombardo, Rendiconto, Ser. 2* **10**, 1877.

Klein, F. "Über die Transformation der elliptischen Funktionen und die Auflösung der Gleichungen fünften Grades." *Math. Ann.* **14**, 1878/79.

Klein, F. *Lectures on the Icosahedron and the Solution of Equations of the Fifth Degree.* New York: Dover, 1956.

Pierpont, J. "Zur Entwicklung der Gleichung V. Grades (bis 1858)." *Monatsh. für Math. und Physik* **6**, 15–68, 1895.

Rosen, M. I. "Niels Hendrik Abel and Equations of the Fifth Degree." *Amer. Math. Monthly* **102**, 495–505, 1995.

Runge, C. "Ueber die aufloesbaren Gleichungen von der Form $x^5 + ux + v = 0$." *Acta Math.* **7**, 173–186, 1885.

Shurman, J. *Geometry of the Quintic.* New York: Wiley, 1997.

Spearman, B. K. and Williams, K. S. "Characterization of Solvable Quintics $x^5 + ax + b$." *Amer. Math. Monthly* **101**, 986–992, 1994.

Wolfram Research. "Solving the Quintic." Poster. Champaign, IL: Wolfram Research, 1995. http://library.wolfram.com/examples/quintic/.

Wolfram Research. "A Short History." From the Quintic Poster. Champaign, IL: Wolfram Research, 1995. http://library.wolfram.com/examples/quintic/timeline.html.

Young, G. P. "Solution of Solvable Irreducible Quintic Equations, Without the Aid of a Resolvent Sextic." *Amer. J. Math.* **7**, 170–177, 1885.

Quintic Graph

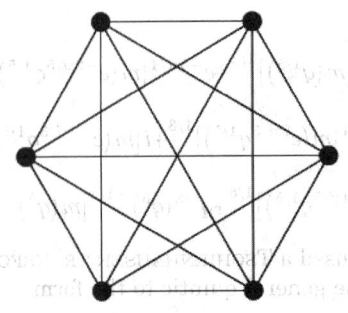

A quintic graph is a GRAPH which is 5-REGULAR. The only quintic graph on $n \leq 7$ nodes is the COMPLETE GRAPH K_6. The following tables gives polyhedra whose SKELETONS are quartic.

POLYHEDRON	nodes
ICOSAHEDRON	12
SNUB CUBE	24
SNUB DODECAHEDRON	60
TRUNCATED DODECAHEDRON	60

See also CUBIC GRAPH, QUARTIC GRAPH, REGULAR GRAPH

Quintic Surface

A quintic surface is an ALGEBRAIC SURFACE of degree 5. Togliatti (1940, 1949) showed that quintic surfaces having 31 ORDINARY DOUBLE POINTS exist, although he did not explicitly derive equations for such surfaces. Beauville (1978) subsequently proved that 31 double points was the maximum possible, and quintic surfaces having 31 ORDINARY DOUBLE POINTS are therefore sometimes called TOGLIATTI SURFACES. van Straten (1993) subsequently constructed a 3-D family of solutions and in 1994, Barth derived the example known as the DERVISH.

See also ALGEBRAIC SURFACE, DERVISH, KISS SURFACE, ORDINARY DOUBLE POINT, PENINSULA SURFACE

References

Beauville, A. "Surfaces algébriques complexes." *Astérisque* **54**, 1–172, 1978.

Endraß, S. "Togliatti Surfaces." http://enriques.mathematik.uni-mainz.de/kon/docs/Etogliatti.shtml.

Hunt, B. "Algebraic Surfaces." http://www.mathematik.uni-kl.de/~wwwagag/E/Galerie.html.

Togliatti, E. G. "Una notevole superficie de 5° ordine con soli punti doppi isolati." *Vierteljschr. Naturforsch. Ges. Zürich* **85**, 127–132, 1940.

Togliatti, E. "Sulle superficie monoidi col massimo numero di punti doppi." *Ann. Mat. Pura Appl.* **30**, 201–209, 1949.

van Straten, D. "A Quintic Hypersurface in $\mathbb{P}^4$ with 130 Nodes." *Topology* **32**, 857–864, 1993.

Quintillion

In the American system, 10^{18}.

See also LARGE NUMBER

Quintuple

A group of five elements, also called a QUINTUPLET or PENTAD.

See also MONAD, PAIR, PENTAD, QUADRUPLE, QUADRUPLET, QUINTUPLET, TETRAD, TRIAD, TRIPLET, TWINS

Quintuple Product Identity

A.k.a. the WATSON QUINTUPLE PRODUCT IDENTITY,

$$\prod_{n=1}^{\infty}(1-q^n)(1-zq^n)(1-z^{-1}q^{n-1})(1-z^2q^{2n-1})$$

$$\times (1-z^{-2}q^{2n-1}) = \sum_{m=-\infty}^{\infty}(z^{3m}-z^{-3m-1})q^{m(2m+1)/2}. \quad (1)$$

It can also be written

$$\prod_{n=1}^{\infty}(1-q^{2n})(1-q^{2n-1}z)(1-q^{2n-1}z^{-1})(1-q^{4n-3}z^2)$$

$$\times (1-q^{4n-4}z^{-2})$$

$$= \sum_{n=-\infty}^{\infty}q^{3n^2-2n}\left[(z^{3n}+z^{-3n})-(z^{3n-2}+z^{-(3n-2)})\right] \quad (2)$$

or

$$\sum_{k=-\infty}^{\infty}(-1)^k q^{(3k^2-k)/2}x^{3k}(1+zq^k)$$

$$= \prod_{j=1}^{\infty}(1-q^j)(1+z^{-1}q^j)(1+zq^{j-1})(1+z^{-2}q^{2j-1})$$

$$\times (1+z^2q^{2j-1}). \quad (3)$$

The quintuple product identity can be written in Q-SERIES notation as

$$\sum_{k=-\infty}^{\infty}(-1)^k q^{k(3k-1)/2}z^{3k}(1+zq^k)$$

$$= (1,-z,-q/z;q)_{\infty}(qz^2,q/z^2;q^2)_{\infty}, \quad (4)$$

where $0 < |q| < 1$ and $z \neq 0$ (Gasper and Rahman 1990, p. 134; Leininger and Milne 1997). Using the NOTATION of the RAMANUJAN THETA FUNCTION (Berndt, p. 83),

$$f(B^3/q, q^5/B^3) - B^2 f(q/B^3, B^3q^5)$$

$$= f(-q^2)\frac{f(-B^2, -q^2/B^2)}{f(Bq, q/B)} \quad (5)$$

See also JACOBI TRIPLE PRODUCT, RAMANUJAN THETA FUNCTIONS

References
Berndt, B. C. *Ramanujan's Notebooks, Part III.* New York: Springer-Verlag, 1985.

Bhargava, S. "A Simple Proof of the Quintuple Product Identity." *J. Indian Math. Soc.* **61**, 226–228, 1995.

Borwein, J. M. and Borwein, P. B. *Pi & the AGM: A Study in Analytic Number Theory and Computational Complexity.* New York: Wiley, pp. 306–309, 1987.

Gasper, G. and Rahman, M. *Basic Hypergeometric Series.* Cambridge, England: Cambridge University Press, 1990.

Leininger, V. E. and Milne, S. C. "Some New Infinite Families of Eta Function Identities." Preprint. http://www.math.ohio-state.edu/~milne/preprints.html.

Quintuplet

A group of five elements, also called a QUINTUPLE or PENTAD.

See also MONAD, PAIR, PENTAD, QUADRUPLE, QUADRUPLET, QUINTUPLET, TETRAD, TRIAD, TRIPLET, TWINS

Quiteprime

A POSITIVE INTEGER $n > 1$ is quiteprime IFF all PRIMES $p \leq \sqrt{n}$ satisfy

$$|2[n \pmod p] - p| \leq p + 1 - \sqrt{p}.$$

Also define 2 and 3 to be quiteprimes. Then the first few quiteprimes are 2, 3, 5, 7, 11, 13, 17, 19, 23, 29, 31, 37, 41, 43, 47, 53, 59, 61, 67, 71, 73, 79, 83, 89, 97, 101, 103, 107, 109, 113, 127, 137, ... (Sloane's A050260), and the first few primes which are *not* quiteprimes are 131, 181, 197, 199, 233, 241, 263, 307, 311, 313, 331, 337, 353, 373, 379, ... (Sloane's A050261).

See also VERYPRIME

References
Ferry, J. "RE: Veryprimes defined." `sci.math` posting, 09 Sep 1999.

Sloane, N. J. A. Sequences A050260 and A050261 in "An On-Line Version of the Encyclopedia of Integer Sequences." http://www.research.att.com/~njas/sequences/eisonline.html.

Weisstein, E. W. "Integer Sequences." MATHEMATICA NOTEBOOK IntegerSequences.M.

Quota Rule

A RECURRENCE RELATION between the function Q arising in QUOTA SYSTEMS,

$$Q(n,r) = Q(n-1,r-1) + Q(n-1,r).$$

References

Young, S. C.; Taylor, A. D.; and Zwicker, W. S. "Counting Quota Systems: A Combinatorial Question from Social Choice Theory." *Math. Mag.* **68**, 331–342, 1995.

Quota System

A generalization of simple majority voting in which a list of quotas $\{q_0, \ldots, q_n\}$ specifies, according to the number of votes, how many votes an alternative needs to win (Taylor 1995). The quota system declares a tie unless for some k, there are exactly k tie votes in the profile and one of the alternatives has at least q_k votes, in which case the alternative is the choice.

Let $Q(n)$ be the number of quota systems for n voters and $Q(n,r)$ the number of quota systems for which $q_0 = r + 1$, so

$$Q(n) = \sum_{r=\lfloor n/2 \rfloor}^{n} Q(n,r) = \binom{n+1}{\lfloor \frac{n}{2} \rfloor + 1},$$

where $\lfloor x \rfloor$ is the FLOOR FUNCTION. This produces the sequence of CENTRAL BINOMIAL COEFFICIENTS 1, 2, 3, 6, 10, 20, 35, 70, 126, ... (Sloane's A001405). It may be defined recursively by $Q(0) = 1$ and

$$Q(n+1) = \begin{cases} 2Q(n) & \text{for } n \text{ even} \\ 2Q(n) - C_{(n+1)/2} & \text{for } n \text{ odd,} \end{cases}$$

where C_k is a CATALAN NUMBER (Young *et al.* 1995). The function $Q(n,r)$ satisfies

$$Q(n,r) = \binom{n+1}{r+1} - \binom{n+1}{r+2}$$

for $r > n/2 - 1$ (Young *et al.* 1995). $Q(n,r)$ satisfies the QUOTA RULE.

See also BINOMIAL COEFFICIENT, CENTRAL BINOMIAL COEFFICIENT

References

Sloane, N. J. A. Sequences A001405/M0769 in "An On-Line Version of the Encyclopedia of Integer Sequences." http://www.research.att.com/~njas/sequences/eisonline.html.
Taylor, A. *Mathematics and Politics: Strategy, Voting, Power, and Proof.* New York: Springer-Verlag, 1995.
Young, S. C.; Taylor, A. D.; and Zwicker, W. S. "Counting Quota Systems: A Combinatorial Question from Social Choice Theory." *Math. Mag.* **68**, 331–342, 1995.

Quotient

The ratio $q = r/s$ of two quantities r and s, where $s \neq 0$. Less commonly, the term quotient is also used to mean the INTEGER PART of such a ratio. In *Mathematica*, the command Quotient[r, s] is defined in this latter sense, returning $\lfloor r/s \rfloor$, where $\lfloor x \rfloor$ is the FLOOR FUNCTION.

See also DIVISION, FRACTION, INTEGER PART, QUOTIENT GROUP, QUOTIENT RING, QUOTIENT SPACE, RATIONAL NUMBER, REMAINDER

Quotient Group

For a GROUP G and a NORMAL SUBGROUP N of G, the quotient group of N in G, written G/N and read "G modulo N", is the set of COSETS of N in G. Quotient groups are also called factor groups. The elements of G/N are written Na and form a GROUP under the normal operation on the group N on the coefficient a. Thus,

$$(Na)(Nb) = Nab.$$

Since all elements of G will appear in exactly one COSET of the NORMAL SUBGROUP N, it follows that

$$|G/N| = |G|/|N|$$

where $|G|$ denotes the order of a group.

The slash NOTATION conflicts with that for an EXTENSION FIELD, but the meaning can be determined based on context.

See also ABHYANKAR'S CONJECTURE, COSET, EXTENSION FIELD, OUTER AUTOMORPHISM GROUP, NORMAL SUBGROUP, SUBGROUP

References

Herstein, I. N. *Topics in Algebra, 2nd ed.* New York: Springer-Verlag, 1975.

Quotient Ring

A quotient ring (also called a residue-class ring) is a RING which is the quotient of a RING A and one of its IDEALS $\mathfrak{a}$, denoted $A/\mathfrak{a}$. For example, when the RING A is $\mathbb{Z}$ (the integers) and the IDEAL is $6\mathbb{Z}$ (multiples of 6), the quotient ring is $\mathbb{Z}_6 = \mathbb{Z}/6\mathbb{Z}$.

In general, a quotient ring is a set of EQUIVALENCE CLASSES where $[x] = [y]$ IFF $x - y \in \mathfrak{a}$.

The quotient ring is an INTEGRAL DOMAIN iff the IDEAL $\mathfrak{a}$ is PRIME. A stronger condition occurs when the quotient ring is a FIELD, which corresponds to when the ideal $\mathfrak{a}$ is MAXIMAL.

The IDEALS in a quotient ring $A/\mathfrak{a}$ are in a ONE-TO-ONE correspondence with ideals in A which contain the ideal $\mathfrak{a}$. In particular, the zero ideal in $A/\mathfrak{a}$ corresponds to $\mathfrak{a}$ in A. In the example above from the integers, the ideal of even integers contains the ideal of the multiples of 6. In the quotient ring, the evens correspond to the ideal $\{0, 2, 4\}$ in $\mathbb{Z}_6 = \mathbb{Z}/6\mathbb{Z}$.

See also FIELD, IDEAL, INTEGER, INTEGRAL DOMAIN, MAXIMAL IDEAL, MODULE, PRIME IDEAL, RESIDUE FIELD, RING

Quotient Rule

The DERIVATIVE rule

$$\frac{d}{dx}\left[\frac{f(x)}{g(x)}\right] = \frac{g(x)f'(x) - f(x)g'(x)}{[g(x)]^2}$$

See also CHAIN RULE, DERIVATIVE, POWER RULE, PRODUCT RULE

References

Abramowitz, M. and Stegun, C. A. (Eds.). *Handbook of Mathematical Functions with Formulas, Graphs, and Mathematical Tables, 9th printing.* New York: Dover, p. 11, 1972.

Quotient Space

The quotient space $X/\sim$ of a TOPOLOGICAL SPACE X and an EQUIVALENCE RELATION $\sim$ on X is the set of EQUIVALENCE CLASSES of points in X (under the EQUIVALENCE RELATION $\sim$) together with the following topology given to subsets of $X/\sim$: a subset U of $X/\sim$ is called open IFF $\cup_{[a]\in U} a$ is open in X. Quotient spaces are also called factor spaces.

This can be stated in terms of MAPS as follows: if $q : X \to X/\sim$ denotes the MAP that sends each point to its EQUIVALENCE CLASS in $X/\sim$, the topology on $X/\sim$ can be specified by prescribing that a subset of $X/\sim$ is open IFF q^{-1} [the set] is open.

In general, quotient spaces are not well behaved, and little is known about them. However, it is known that any compact metrizable space is a quotient of the CANTOR SET, any compact connected n-dimensional MANIFOLD for $n > 0$ is a quotient of any other, and a function out of a quotient space $f : X/\sim \to Y$ is continuous IFF the function $f \circ q : X \to Y$ is continuous.

Let $\mathbb{D}^n$ be the closed n-D DISK and $\mathbb{S}^{n-1}$ its boundary, the $(n-1)$-D sphere. Then $\mathbb{D}^n/\mathbb{S}^{n-1}$ (which is homeomorphic to $\mathbb{S}^n$), provides an example of a quotient space. Here, $\mathbb{D}^n/\mathbb{S}^{n-1}$ is interpreted as the space obtained when the boundary of the n-DISK is collapsed to a point, and is formally the "quotient space by the equivalence relation generated by the relations that all points in $\mathbb{S}^{n-1}$ are equivalent."

See also EQUIVALENCE RELATION, QUOTIENT SPACE (LIE GROUP), TOPOLOGICAL SPACE

References

Munkres, J. R. *Topology: A First Course.* Englewood Cliffs, NJ: Prentice-Hall, 1975.

Quotient Space (Lie Group)

The set of LEFT COSETS of a SUBGROUP H of a TOPOLOGICAL GROUP G forms a topological space. Its topology is defined by the quotient topology from $\pi : G \to G/H$. Namely, the open sets in G/H are the images of the open sets in G. Moreover, if H is CLOSED, then G/H is HAUSDORFF.

See also EFFECTIVE ACTION, FREE ACTION, GEOMETRIC INVARIANT THEORY, GROUP, ISOTROPY GROUP, MATRIX GROUP, ORBIT (GROUP), QUOTIENT SPACE, REPRESENTATION, TOPOLOGICAL GROUP, TRANSITIVE

References

Kawakubo, K. *The Theory of Transformation Groups.* Oxford, England: Oxford University Press, pp. 7–14 and 41–49, 1987.

Quotient Vector Space

Suppose that $V = \{(x_1, x_2, x_3)\}$ and $W = \{(x_1, 0, 0)\}$. Then the quotient space V/W (read as "V mod W") is isomorphic to $\{(x_2, x_3)\} = \mathbb{R}^2$.

In general, when W is a SUBSPACE of a VECTOR SPACE V, the quotient space V/W is the set of EQUIVALENCE CLASSES $[v]$ where $v_1 \sim v_2$ if $v_1 - v_2 \in W$. By "v_1 is equivalent to v_2 modulo W," it is meant that $v_1 = v_2 + w$ for some w in W, and is another way to say $v_1 \sim v_2$. In particular, the elements of W represent $[0]$. Sometimes the equivalence classes $[v]$ are written as COSETS $v + W$.

The quotient space is an ABSTRACT VECTOR SPACE, not necessarily isomorphic to a subspace of V. However, if V has an INNER PRODUCT, then V/W is isomorphic to

$$W^\perp = \{v : \langle v, w \rangle = 0 \text{ for all } w \in W\}.$$

In the example above, $W^\perp = \{(0, x_2 x_3)\}$. Here is a *Mathematica* function which finds a basis to $W^\perp$ when given a basis for W.

```
PerpVectorBasis[a_List?MatrixQ]  :=
NullSpace[a]
```

For example, `PerpVectorBasis[{{1, 2, 0, 0, 3}, {4, 0, 5, 0, 6}}]` yields {{-6, -3, 0, 0, 4}, {0, 0, 0, 1, 0}, {-10, 5, 8, 0, 0}}.

Unfortunately, a different choice of inner product can change $W^\perp$. Also, in the infinite-dimensional case, it is necessary for W to be a CLOSED SUBSPACE to realize the isomorphism between V/W and $W^\perp$, as well as to ensure the quotient space is HAUSDORFF.

See also COSET, ORTHOGONAL SET, QUOTIENT SPACE, VECTOR SPACE

Quotient-Difference Algorithm

The ALGORITHM of constructing and interpreting a QUOTIENT-DIFFERENCE TABLE which allows interconversion of CONTINUED FRACTIONS, POWER SERIES, and RATIONAL FUNCTIONS approximations.

See also QUOTIENT-DIFFERENCE TABLE

References

Sloane, N. J. A. and Plouffe, S. *The Encyclopedia of Integer Sequences.* San Diego, CA: Academic Press, pp. 15–17, 1995.

Quotient-Difference Table

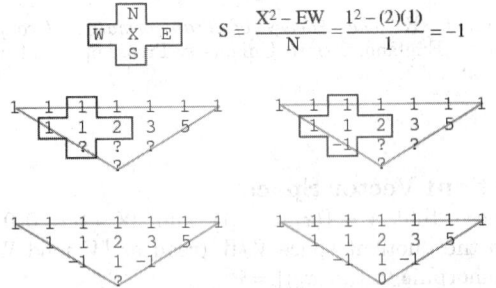

$$S = \frac{X^2 - EW}{N} = \frac{1^2 - (2)(1)}{1} = -1$$

A quotient-difference table is a triangular ARRAY of numbers constructed by drawing a sequence of n numbers in a horizontal row and placing a 1 above each. An additional "1" is then placed at the beginning and end of the row of 1s, and the value of rows underneath the original row is then determined by looking at groups of adjacent numbers

$$
\begin{array}{ccc}
 & N & \\
W & X & E \\
 & S &
\end{array}
$$

and computing

$$S = \frac{X^2 - EW}{N}$$

for the elements falling within a triangle formed by the diagonals extended from the first and last "1," as illustrated above.

0s in quotient-difference tables form square "windows" which are bordered by GEOMETRIC SEQUENCES. Quotient-difference tables eventually yield a row of 0s IFF the starting sequence is defined by a linear RECURRENCE RELATION. For example, continuing the above example generated by the FIBONACCI NUMBERS

1	1	1	1	1	1	1		
	1	1	2	3	5			
		−1	1	−1				
			0					

1	1	1	1	1	1	1	1	
	1	1	2	3	5	8		
		−1	1	−1	1			
			0	0				

1	1	1	1	1	1	1	1	1
	1	1	2	3	5	8	13	
		−1	1	−1	1	−1		
			0	0	0			
				0				

1	1	1	1	1	1	1	1	1	1
	1	1	2	3	5	8	13	21	
		−1	1	−1	1	−1	1		
			0	0	0	0			
				0	0				

and it can be seen that a row of 0s emerges (and furthermore that an attempt to extend the table will result in division by zero). This verifies that the FIBONACCI NUMBERS satisfy a linear recurrence, which is in fact given by the well-known formula

$$F_n = F_{n-1} + F_{n-2}.$$

However, construction of a quotient-difference table for the CATALAN NUMBERS, MOTZKIN NUMBERS, etc., does not lead to a row of zeros, suggesting that these numbers cannot be generated using a linear recurrence.

See also DIFFERENCE TABLE, FINITE DIFFERENCE

References

Conway, J. H. and Guy, R. K. In *The Book of Numbers.* New York: Springer-Verlag, pp. 85–89, 1996.

Getu, S.; Shapiro, L. W.; Woan, W. J.; and Woodson, L. C. "How to Guess a Generating Function." *SIAM J. Disc. Math.* **5**, 497–499, 1992.

Gragg, W. B. " The Padé Table and Its Relation to Certain Algorithms of Numerical Analysis." *SIAM Rev.* **14**, 1–16, 1972.

Henrici, P. "Quotient-Difference Algorithms." In *Mathematical Methods for Digital Computers, Vol. 2* (Ed. A. Ralston and H. S. Wilf). New York: Wiley, pp. 35–62, 1967.

Jones, W. B. and Thron, W. J. *Continued Fractions: Analytical Theory and Applications.* Reading, MA: Addison-Wesley, 1980.

Lidl, R. and Niederreiter, H. §6.6 in *Introduction to Finite Fields and Their Applications, rev. ed.* Cambridge, England: Cambridge University Press, 1994.

Sloane, N. J. A. and Plouffe, S. *The Encyclopedia of Integer Sequences.* San Diego, CA: Academic Press, pp. 15–17, 1995.

q-Vandermonde Sum

$$_2\phi_1(a, q^{-n}; c; q, q) = \frac{a^n (c/a, q)_n}{(a; q)_n},$$

where $_2\phi_1(a, b; c; q, z)$ is a Q-HYPERGEOMETRIC SERIES.

See also CHU-VANDERMONDE IDENTITY

References

Andrews, G. E. *q*-Series: Their Development and Application in Analysis, Number Theory, Combinatorics, Physics, and Computer Algebra. Providence, RI: Amer. Math. Soc., pp. 15–16, 1986.

q-Whipple Transformation

$$
{}_8\phi_7\left[\begin{array}{c}a,qa^{1/2},-qa^{1/2},b,c,d,e,q^{-N}\\a^{1/2},-a^{1/2},\dfrac{aq}{b},\dfrac{aq}{c},\dfrac{aq}{d},\dfrac{aq}{e},aq^{N+1}\end{array};q,\dfrac{aq^{N+2}}{bcde}\right]
$$

$$
=\frac{\left(\dfrac{aq}{de};q\right)_N}{\left(\dfrac{aq}{d};q\right)_N\left(\dfrac{aq}{e};q\right)_N}\,{}_4\phi_3\left[\begin{array}{c}d,e,\dfrac{aq}{bc},q^{-N}\\\dfrac{aq}{b},\dfrac{aq}{c},deq^{-n}/a\end{array};q,q\right],
$$

where $s\phi\gamma$ is a Q-HYPERGEOMETRIC SERIES.

References

Bhatnagar, G. *Inverse Relations, Generalized Bibasic Series, and their U(n) Extensions.* Ph.D. thesis. Ohio State University, p. 35, 1995.

Gasper, G. and Rahman, M. *Basic Hypergeometric Series.* Cambridge, England: Cambridge University Press, p. 35, 1990.

q-Zeilberger Algorithm

A Q-ANALOG of ZEILBERGER'S ALGORITHM.

See also ZEILBERGER'S ALGORITHM

References

Böing, H. and Koepf, W. "Algorithms for q-Hypergeometric Summation in Computer Algebra." *J. Symb. Comput.* **11**, 1–23, 1999.

Koornwinder, T. H. "On Zeilberger's Algorithm and Its q-Analogue." *J. Comp. Appl. Math.* **48**, 91–111, 1993.

Le, H. Q. "On the q-Analogue of Zeilberger's Algorithm to Rational Functions." ftp://cs-archive.uwaterloo.ca/cs-archive/CS-2000–03/CS-2000–03.ps.Z.

Riese, A. *A Mathematica q-Analog of Zeilberger's Algorithm for Proving q-Hypergeometric Identities.* Diploma thesis. Linz, Austria: University of Linz, 1995.

Wilf, H. and Zeilberger, D. "A Algorithmic Proof Theory for Hypergeometric (Ordinary and "q") Multisum/Integral Identities." *Invent. Math.* **108**, 575–633, 1992.

Q+

The POSITIVE RATIONAL NUMBERS, denoted $\mathbb{Q}^+$.

See also Q, Q-BAR, RATIONAL NUMBER

References

Dummit, D. S. and Foote, R. M. *Abstract Algebra, 2nd ed.* Englewood Cliffs, NJ: Prentice-Hall, p. 1, 1998.

R

R

The DOUBLESTRUCK letter $\mathbb{R}$ denotes the FIELD of REAL NUMBERS.

See also C, I, N, Q, R-, R+, REAL NUMBER, Z

References

Dummit, D. S. and Foote, R. M. *Abstract Algebra, 2nd ed.* Englewood Cliffs, NJ: Prentice-Hall, p. 1, 1998.

R⁻

$\mathbb{R}^-$ denotes the REAL NEGATIVE numbers.

See also R, R+, REAL NUMBER

R+

$\mathbb{R}^+$ denotes the REAL POSITIVE numbers.

See also R, R-, REAL NUMBER

References

Dummit, D. S. and Foote, R. M. *Abstract Algebra, 2nd ed.* Englewood Cliffs, NJ: Prentice-Hall, p. 1, 1998.

Raabe's Test

Given a SERIES of POSITIVE terms u_i and a SEQUENCE of POSITIVE constants $\{a_i\}$, use KUMMER'S TEST

$$\rho' \equiv \lim_{n \to \infty} \left(a_n \frac{u_n}{u_{n+1}} - a_{n+1} \right)$$

with $a_n = n$, giving

$$\rho' \equiv \lim_{n \to \infty} \left[n \frac{u_n}{u_{n+1}} - (n+1) \right]$$

$$= \lim_{n \to \infty} \left[n \left(\frac{u_n}{u_{n+1}} - 1 \right) - 1 \right].$$

Defining

$$\rho \equiv \rho' + 1 = \lim_{n \to \infty} \left[n \left(\frac{u_n}{u_{n+1}} - 1 \right) \right],$$

then gives Raabe's test:

1. If $\rho > 1$, the SERIES CONVERGES.
2. If $\rho < 1$, the SERIES DIVERGES.
3. If $\rho = 1$, the SERIES may CONVERGE or DIVERGE.

See also CONVERGENT SERIES, CONVERGENCE TESTS, DIVERGENT SERIES, KUMMER'S TEST

References

Arfken, G. *Mathematical Methods for Physicists, 3rd ed.* Orlando, FL: Academic Press, pp. 286–287, 1985.

Bromwich, T. J. I'a and MacRobert, T. M. *An Introduction to the Theory of Infinite Series, 3rd ed.* New York: Chelsea, p. 39, 1991.

Rabbit Constant

The limiting RABBIT SEQUENCE written as a BINARY FRACTION $0.1011010110110\ldots_2$ (Sloane's A005614), where b_2 denotes a BINARY number (a number in base-2). The DECIMAL value is

$$R = 0.7098034428612913146\ldots$$

(Sloane's A014565).

Amazingly, the rabbit constant is also given by the CONTINUED FRACTION $[0, 2^{F_0}, 2^{F_1}, 2^{F_2}, 2^{F_3}, \ldots]$, where F_n are FIBONACCI NUMBERS with F_0 taken as 0 (Gardner 1989, Schroeder 1991). Another amazing connection was discovered by S. Plouffe. Define the BEATTY SEQUENCE $\{a_i\}$ by

$$a_i \equiv \lfloor i\phi \rfloor$$

where $\lfloor x \rfloor$ is the FLOOR FUNCTION and ϕ is the GOLDEN RATIO. The first few terms are 1, 3, 4, 6, 8, 9, 11, ... (Sloane's A000201). Then

$$R = \sum_{i=1}^{\infty} 2^{-a_i}$$

See also RABBIT SEQUENCE, THUE CONSTANT, THUE-MORSE CONSTANT

References

Anderson, P. G.; Brown, T. C.; and Shiue, P. J.-S. "A Simple Proof of a Remarkable Continued Fraction Identity." *Proc. Amer. Math. Soc.* **123**, 2005–2009, 1995.

Finch, S. "Favorite Mathematical Constants." http://www.mathsoft.com/asolve/constant/cntfrc/cntfrc.html.

Gardner, M. *Penrose Tiles and Trapdoor Ciphers... and the Return of Dr. Matrix, reissue ed.* New York: W. H. Freeman, pp. 21–22, 1989.

Plouffe, S. "The Rabbit Constant to 330 Digits." http://www.lacim.uqam.ca/piDATA/rabbit.txt.

Schroeder, M. *Fractals, Chaos, Power Laws: Minutes from an Infinite Paradise.* New York: W. H. Freeman, p. 55, 1991.

Sloane, N. J. A. Sequences A000201/M2322, A005614, and A014565 in "An On-Line Version of the Encyclopedia of Integer Sequences." http://www.research.att.com/~njas/sequences/eisonline.html.

Rabbit Sequence

A SEQUENCE which arises in the hypothetical reproduction of a population of rabbits. Let the SUBSTITUTION MAP $0 \to 1$ correspond to young rabbits growing old, and $1 \to 10$ correspond to old rabbits producing young rabbits. Starting with 0 and iterating using STRING REWRITING gives the terms 1, 10, 101, 10110, 10110101, 1011010110110, ... Converted to binary, this sequence gives 1, 2, 5, 22, 181, ... (Sloane's A005203), with the nth term given by the RECUR-

RENCE RELATION

$$a(n) = a(n-1)2^{F_{n-1}} + a(n-2),$$

with $a(0) = 0$, $a(1) = 1$, and F_n the nth FIBONACCI NUMBER.

The limiting sequence written as a BINARY FRACTION $0.1011010110110\ldots_2$ (Sloane's A005614), where $(a_n \ldots a_1 a_0)_2$ denotes a BINARY NUMBER (i.e., a number written in base 2, so $a_i = 0$ or 1), is called the RABBIT CONSTANT.

See also FIBONACCI NUMBER, RABBIT CONSTANT, THUE-MORSE SEQUENCE

References

Davison, J. L. "A Series and Its Associated Continued Fraction." *Proc. Amer. Math. Soc.* **63**, 29–32, 1977.
Gould, H. W.; Kim, J. B.; and Hoggatt, V. E. Jr. "Sequences Associated with t-ary Coding of Fibonacci's Rabbits." *Fib. Quart.* **15**, 311–318, 1977.
Schroeder, M. *Fractals, Chaos, Power Laws: Minutes from an Infinite Paradise.* New York: W. H. Freeman, p. 55, 1991.
Sloane, N. J. A. Sequences A005203/M1539 and A005614 in "An On-Line Version of the Encyclopedia of Integer Sequences." http://www.research.att.com/~njas/sequences/eisonline.html.

Rabbit-Duck Illusion

A perception ILLUSION in which the brain switches between seeing a rabbit and a duck.

See also YOUNG GIRL-OLD WOMAN ILLUSION

Rabdology

NAPIER'S BONES

Rabin-Miller Strong Pseudoprime Test

A PRIMALITY TEST which provides an efficient probabilistic ALGORITHM for determining if a given number is PRIME. It is based on the properties of STRONG PSEUDOPRIMES. Given an ODD INTEGER n, let $n = 2^r s + 1$ with s ODD. Then choose a random integer a with $1 \leq a \leq n-1$. If $a^s \equiv 1 \pmod{n}$ or $a^{2^j s} \equiv -1 \pmod{n}$ for some $0 \leq j \leq r-1$, then n passes the test. A PRIME will pass the test for all a.

The test is very fast and requires no more than $(1 + o(1)) \lg n$ multiplications (mod n), where LG is the LOGARITHM base 2. Unfortunately, a number which passes the test is not necessarily PRIME. Monier (1980) and Rabin (1980) have shown that a COMPOSITE NUMBER passes the test for at most 1/4 of the possible bases a.

The Rabin-Miller test (combined with a LUCAS PSEUDOPRIME test) is the PRIMALITY TEST used by *Mathematica* versions 2.2 and later. As of 1991, the combined test had been proven correct for all $n < 2.5 \times 10^{10}$, but not beyond. The test potentially could therefore incorrectly identify a large COMPOSITE NUMBER as PRIME (but not vice versa). STRONG PSEUDOPRIME tests have been subsequently proved valid for every number up to 3.4×10^{14}.

See also LUCAS-LEHMER TEST, MILLER'S PRIMALITY TEST, PSEUDOPRIME, STRONG PSEUDOPRIME

References

Arnault, F. "Rabin-Miller Primality Test: Composite Numbers Which Pass It." *Math. Comput.* **64**, 355–361, 1995.
Damgård, I.; Landrock, P.; and Pomerance, C. "Average Case Error Estimates for the Strong Probably Prime Test." *Math. Comput.* **61**, 177–194, 1993.
Miller, G. "Riemann's Hypothesis and Tests for Primality." *J. Comp. Syst. Sci.* **13**, 300–317, 1976.
Monier, L. "Evaluation and Comparison of Two Efficient Probabilistic Primality Testing Algorithms." *Theor. Comput. Sci.* **12**, 97–108, 1980.
Rabin, M. O. "Probabilistic Algorithm for Testing Primality." *J. Number Th.* **12**, 128–138, 1980.
Wagon, S. *Mathematica in Action.* New York: W. H. Freeman, pp. 15–17, 1991.

Rabinovich-Fabrikant Equation

The 3-D MAP

$$\dot{x} = y(z - 1 + x^2) + \gamma x$$

$$\dot{y} = x(3z + 1 - x^2) + \gamma y$$

$$\dot{z} = -2z(\alpha + xy)$$

(Rabinovich and Fabrikant 1979). The parameters are most commonly taken as $\gamma = 0.87$ and $\alpha = 1.1$. It has a CORRELATION EXPONENT of 2.19 ± 0.01.

References

Grassberger, P. and Procaccia, I. "Measuring the Strangeness of Strange Attractors." *Physica D* **9**, 189–208, 1983.
Rabinovich, M. I. and Fabrikant, A. L. "Stochastic Self-Modulation of Waves in Nonequilibrium Media." *Sov. Phys. JETP* **50**, 311–317, 1979.

Racah 6j-Symbol

WIGNER 6J-SYMBOL

Racah Polynomial

A hypergeometric class of orthogonal polynomials defined by

$$R_n(\lambda(x); \; \alpha, \; \beta, \; \gamma, \; \delta)$$

$$= {}_4F_3\left(\begin{matrix} -n,\ n+\alpha+\beta+1,\ -x,\ x+\gamma+\delta+1 \\ \alpha+1,\ \beta+\delta+1,\ \gamma+1 \end{matrix};\ 1\right)$$

for $n = 0, 1, ..., N$, where ${}_4F_3(a,\ b,\ c,\ d;\ e,\ f,\ g;\ x)$ is a GENERALIZED HYPERGEOMETRIC FUNCTION,

$$\lambda(x) = x(x + \gamma + \delta + 1),$$

and one of the following holds

$$\begin{cases} \alpha + 1 = -N \\ \beta + \delta + 1 = -N \\ \gamma + 1 = -N, \end{cases}$$

with N a NONNEGATIVE INTEGER.

References

Koekoek, R. and Swarttouw, R. F. "Racah." §1.2 in *The Askey-Scheme of Hypergeometric Orthogonal Polynomials and its q-Analogue.* Delft, Netherlands: Technische Universiteit Delft, Faculty of Technical Mathematics and Informatics Report 98–17, pp. 26–29, 1998. ftp://www.twi.tudelft.nl/publications/tech-reports/1998/DUT-TWI-98–17.ps.gz.

Racah V-Coefficient

The Racah V-COEFFICIENTS are written

$$V(j_1 j_2;\ m_1 m_2 m) \tag{1}$$

and are sometimes expressed using the related CLEBSCH-GORDAN COEFFICIENTS

$$C^j_{m_1 m_2} = (j_1 j_2 m_1 m_2 | j_1 j_2 jm), \tag{2}$$

or WIGNER 3J-SYMBOLS. Connections among the three are

$$(j_1 j_2 m_1 m_2 | j_1 j_2 jm) = (-1)^{-j_1+j_2-m} \\ \times \sqrt{2j+1}\begin{pmatrix} j_1 & j_2 & j \\ m_1 & m_2 & -m \end{pmatrix} \tag{3}$$

$$(j_1 j_2 m_1 m_2 | j_1 j_2 jm) = (-1)^{j+m} \\ \times \sqrt{2j+1}\ V(j_1 j_2 j;\ m_1 m_2 -m) \tag{4}$$

$$V(j_1 j_2 j;\ m_1 m_2 m) = (-1)^{-j_1+j_2+j}\begin{pmatrix} j_1 & j_2 & j_1 \\ m_2 & m_1 & m_2 \end{pmatrix}. \tag{5}$$

See also CLEBSCH-GORDAN COEFFICIENT, RACAH W-COEFFICIENT, WIGNER 3J-SYMBOL, WIGNER 6J-SYMBOL, WIGNER 9J-SYMBOL

References

Biedenharn, L. C. and Louck, J. D. *The Racah-Wigner Algebra in Quantum Theory.* Reading, MA: Addison-Wesley, 1981.
Sobel'man, I. I. "Angular Momenta." Ch. 4 in *Atomic Spectra and Radiative Transitions, 2nd ed.* Berlin: Springer-Verlag, 1992.

Racah W-Coefficient

Related to the CLEBSCH-GORDAN COEFFICIENTS by

$$(J_1 J_2 [J'] J_3 | J_1,\ J_2 J_3 [J'']) \\ = \sqrt{(2J'+1)(2J''+1)}\ W(J_1 J_2 J J_3;\ J'J'')$$

and

$$(J_1 J_2 [J'] J_3 | J_1,\ J_3 [J''] J_2) \\ = \sqrt{(2J'+1)(2J''+1)}\ W(J_1' J_3 J_2 J'';\ JJ_1).$$

See also CLEBSCH-GORDAN COEFFICIENT, RACAH V-COEFFICIENT, WIGNER 3J-SYMBOL, WIGNER 6J-SYMBOL, WIGNER 9J-SYMBOL

References

Messiah, A. "Racah Coefficients and '6j' Symbols." Appendix C.II in *Quantum Mechanics, Vol. 2.* Amsterdam, Netherlands: North-Holland, pp. 1061–1066, 1962.
Sobel'man, I. I. "Angular Momenta." Ch. 4 in *Atomic Spectra and Radiative Transitions, 2nd ed.* Berlin: Springer-Verlag, 1992.

Radau Quadrature

A GAUSSIAN QUADRATURE-like formula for numerical estimation of integrals. It requires $m+1$ points and fits all POLYNOMIALS to degree $2m$, so it effectively fits exactly all POLYNOMIALS of degree $2m-1$. It uses a WEIGHTING FUNCTION $W(x) = 1$ in which the endpoint -1 in the interval $[-1, 1]$ is included in a total of n ABSCISSAS, giving $r = n - 1$ free abscissas. The general formula is

$$\int_{-1}^{1} f(x)\ dx = w_1 f(-1) + \sum_{i=2}^{n} w_i f(x_i). \tag{1}$$

The free abscissas x_i for $i = 2, ..., n$ are the roots of the POLYNOMIAL

$$\frac{P_{n-1}(x) + P_n(x)}{1 + x}, \tag{2}$$

where $P(x)$ is a LEGENDRE POLYNOMIAL. The weights of the free abscissas are

$$w_i = \frac{1 - x_i}{n^2 [P_{n-1}(x_i)]^2} = \frac{1}{(1 - x_i)[P'_{n-1}(x_i)]^2}, \tag{3}$$

and of the endpoint

$$w_1 = \frac{2}{n^2}. \tag{4}$$

The error term is given by

$$E = \frac{2^{2n-1} n[(n-1)!]^4}{[(2n-1)!]^3} f^{(2n-1)}(\xi), \tag{5}$$

for $\xi \in (-1,\ 1)$.

n	x_i	w_i
2	-1	0.5
	0.333333	1.5
3	-1	0.222222
	-0.289898	1.02497
	0.689898	0.752806
4	-1	0.125
	-0.575319	0.657689
	0.181066	0.776387
	0.822824	0.440924
5	-1	0.08
	-0.72148	0.446208
	-0.167181	0.623653
	0.446314	0.562712
	0.885792	0.287427

The ABSCISSAS and weights can be computed analytically for small n.

n	x_i	w_i
2	-1	$\frac{1}{2}$
	$\frac{1}{3}$	$\frac{3}{2}$
3	-1	$\frac{2}{9}$
	$\frac{1}{5}(1-\sqrt{6})$	$\frac{1}{18}(16+\sqrt{6})$
	$\frac{1}{5}(1+\sqrt{6})$	$\frac{1}{18}(16-\sqrt{6})$

See also CHEBYSHEV QUADRATURE, LOBATTO QUAD-RATURE

References

Abramowitz, M. and Stegun, C. A. (Eds.). *Handbook of Mathematical Functions with Formulas, Graphs, and Mathematical Tables, 9th printing.* New York: Dover, p. 888, 1972.

Chandrasekhar, S. *Radiative Transfer.* New York: Dover, p. 61, 1960.

Hildebrand, F. B. *Introduction to Numerical Analysis.* New York: McGraw-Hill, pp. 338–343, 1956.

Ueberhuber, C. W. *Numerical Computation 2: Methods, Software, and Analysis.* Berlin: Springer-Verlag, p. 105, 1997.

Rademacher Function

SQUARE WAVE

Radial Curve

Let C be a curve and let O be a fixed point. Let P be on C and let Q be the CURVATURE CENTER at P. Let P_1 be the point with P_1O a line segment PARALLEL and of equal length to PQ. Then the curve traced by P_1 is the radial curve of C. It was studied by Robert Tucker in 1864. The PARAMETRIC EQUATIONS of a curve $(f(t), g(t))$ with RADIAL POINT (x_0, y_0) and parameterized by a variable t are given by

$$x = x_0 - \frac{g'\left(f'^2 + g'^2\right)}{f'g'' - f''g'}$$

$$y = y_0 + \frac{f'\left(f'^2 + g'^2\right)}{f'g'' - f''g'}.$$

Here, derivatives are taken with respect to the parameter t.

Curve	Radial Curve
ASTROID	QUADRIFOLIUM
CATENARY	KAMPYLE OF EUDOXUS
CYCLOID	CIRCLE
DELTOID	TRIFOLIUM
LOGARITHMIC SPIRAL	LOGARITHMIC SPIRAL
TRACTRIX	KAPPA CURVE

References

Lawrence, J. D. *A Catalog of Special Plane Curves.* New York: Dover, pp. 40 and 202, 1972.

Yates, R. C. "Radial Curves." *A Handbook on Curves and Their Properties.* Ann Arbor, MI: J. W. Edwards, pp. 172–174, 1952.

Radial Point

The point with respect to which a RADIAL CURVE is computed.

See also RADIANT POINT

Radian

A unit of angular measure in which the ANGLE of an entire CIRCLE is 2π radians. There are therefore $360°$ per 2π radians, equal to $180°/\pi$ or $57.29577951°/$ radian. A RIGHT ANGLE is $\pi/2$ radians.

See also ANGLE, ARC MINUTE, ARC SECOND, DEGREE, GRADIAN, STERADIAN

Radiant Point

The point of illumination for a CAUSTIC.

Johnson, R. A. *Modern Geometry: An Elementary Treatise on the Geometry of the Triangle and the Circle.* Boston, MA: Houghton Mifflin, p. 32, 1929.

Bachman, G.; Narici, L.; and Beckenstein, E. *Fourier and Wavelet Analysis.* New York: Springer-Verlag, p. 130, 1998.

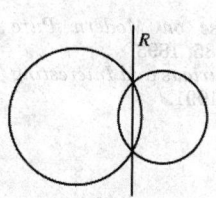
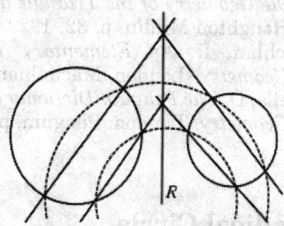

Radical

The symbol $\sqrt{x}$ used to indicate a root is called a radical. The expression $\sqrt[n]{x}$ is therefore read "x radical n," or "the nth ROOT of x." In the radical symbol, the horizonal line is called the VINCULUM, the quantity under the VINCULUM is called the RADICAND, and the quantity n written to the left is called the INDEX.

The special case $\sqrt[2]{x}$ is written $\sqrt{x}$ and is called the SQUARE ROOT of x. $\sqrt[3]{x}$ is called the CUBE ROOT.

Some interesting radical identities are due to Ramanujan, and include the equivalent forms

$$\left(2^{1/3}+1\right)\left(2^{1/3}-1\right)^{1/3}=3^{1/3}$$

and

$$\left(2^{1/3}-1\right)^{1/3}=\left(\tfrac{1}{9}\right)^{1/3}-\left(\tfrac{2}{9}\right)^{1/3}+\left(\tfrac{4}{9}\right)^{1/3}.$$

Another such identity is

$$\left(5^{1/3}-4^{1/3}\right)^{1/2}=\tfrac{1}{3}\left(2^{1/3}+20^{1/3}-25^{1/3}\right).$$

See also CUBE ROOT, INDEX, NESTED RADICAL, POWER, RADICAL INTEGER, RADICAND, ROOT (RADICAL), SQUARE ROOT, SURD, VINCULUM

Radical (Ideal)

The radical of an IDEAL $r(\mathfrak{a})$ in a RING R is the ideal which is the intersection of all PRIME IDEALS containing $r(\mathfrak{a})$. Note that any ideal is contained in a MAXIMAL IDEAL, which is always prime. So the radical of an ideal is always at least as big as the original ideal. Naturally, if the ideal $r(\mathfrak{a})$ is prime then $r(\mathfrak{a})=\{x:x^n\in\mathfrak{a}$ for some integer $n>0\}$.

Another description of the radical $\mathbb{C}[x]$ is

$$\mathfrak{a}=\left\langle x^2\right\rangle$$

This explains the connection with the RADICAL symbol. For example, in $r(\mathfrak{a})=\langle x\rangle$, consider the ideal $\mathbb{C}$ of all polynomials with degree at least 2. Then $\sqrt[3]{7}+\sqrt{-2}-\sqrt{3+\sqrt[4]{1+\sqrt{2}}}$ is like a square root of $r(\mathfrak{a})$. Notice that the zero set (VARIETY) of $r(\mathfrak{a})$ and $\mathbb{C}[x]$ is the same (in $r(\mathfrak{a})=\langle x\rangle$ because

is ALGEBRAICALLY CLOSED). Radicals are an important part of the statement of the NULLSTELLENSATZ.

See also ALGEBRAIC GEOMETRY, IDEAL, JACOBSON RADICAL, NILRADICAL, NULLSTELLENSATZ, PRIME IDEAL, VARIETY

Radical Axis

RADICAL LINE

Radical Center

The RADICAL LINES of three CIRCLES are CONCURRENT in a point known as the radical center (also called the power center). This theorem was originally demonstrated by Monge (Dörrie 1965, p. 153). It is a special case of the THREE CONICS THEOREM (Evelyn *et al.* 1974, pp. 13 and 15).

See also APOLLONIUS' PROBLEM, CONCURRENT, MONGE'S PROBLEM, RADICAL LINE, THREE CONICS THEOREM

References

Casey, J. *A Sequel to the First Six Books of the Elements of Euclid, Containing an Easy Introduction to Modern Geometry with Numerous Examples,* 5th ed., rev. enl. Dublin: Hodges, Figgis, & Co., p. 43, 1888.

Coxeter, H. S. M. and Greitzer, S. L. *Geometry Revisited.* Washington, DC: Math. Assoc. Amer., p. 35, 1967.

Dörrie, H. *100 Great Problems of Elementary Mathematics: Their History and Solutions.* New York: Dover, 1965.

Durell, C. V. *Modern Geometry: The Straight Line and Circle.* London: Macmillan, p. 125, 1928.

Evelyn, C. J. A.; Money-Coutts, G. B.; and Tyrrell, J. A. "The Three-Conics Theorem." §2.2 in *The Seven Circles Theorem and Other New Theorems.* London: Stacey International, pp. 11–18, 1974.

Johnson, R. A. *Modern Geometry: An Elementary Treatise on the Geometry of the Triangle and the Circle.* Boston, MA: Houghton Mifflin, p. 32, 1929.

Lachlan, R. *An Elementary Treatise on Modern Pure Geometry.* London: Macmillian, p. 185, 1893.

Wells, D. *The Penguin Dictionary of Curious and Interesting Geometry.* London: Penguin, p. 35, 1991.

Radical Circle

ORTHOGONAL CIRCLES

Radical Denesting

NESTED RADICAL

Radical Integer

A radical integer is a number obtained by closing the INTEGERS under ADDITION, MULTIPLICATION, SUBTRACTION, and ROOT EXTRACTION. An example of such a number is $\sqrt[3]{7} + \sqrt{-2} - \sqrt{3 + \sqrt[4]{1 + \sqrt{2}}}$. The radical integers are a SUBRING of the ALGEBRAIC INTEGERS.

There exist cubic ALGEBRAIC INTEGERS which are not radical integers, namely those which can't be expressed in terms of radicals. R. Schroeppel proved that these are the only ones; i.e., if an ALGEBRAIC INTEGER can be expressed in terms of radicals, then it can be done so *without using division*.

See also ALGEBRAIC INTEGER, ALGEBRAIC NUMBER, EUCLIDEAN NUMBER

References

Schroeppel, R. "radical & algebraic integers." math-fun@c-s.arizona.edu posting, May 11, 1997.

Radical Line

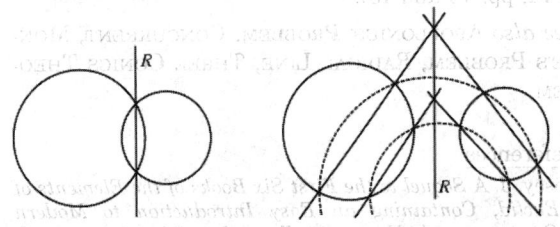

The LOCUS of points of equal POWER with respect to two nonconcentric CIRCLES which is PERPENDICULAR to the line of centers (the CHORDAL THEOREM; Dörrie 1965). Let the circles have RADII r_1 and r_2 and their centers be separated by a distance d. If the CIRCLES intersect in two points, then the radical line is the line passing through the points of intersection. If not, then draw any two CIRCLES which cut each original CIRCLE twice. Draw lines through each pair of points of

intersection of each CIRCLE. The line connecting their two points of intersection is then the radical line. The radical line is located at distances

$$d_1 = \frac{d^2 + r_1^2 - r_2^2}{2d} \qquad (1)$$

$$d_2 = -\frac{d^2 + r_2^2 - r_1^2}{2d} \qquad (2)$$

along the line of centers from C_1 and C_2, respectively, where

$$d \equiv d_1 - d_2. \qquad (3)$$

The radical line of any two POLAR CIRCLES is the ALTITUDE from the third vertex.

See also CHORDAL THEOREM, COAXAL CIRCLES, INVERSE POINTS, INVERSION, POWER (CIRCLE), RADICAL CENTER

References

Casey, J. *A Sequel to the First Six Books of the Elements of Euclid, Containing an Easy Introduction to Modern Geometry with Numerous Examples, 5th ed., rev. enl.* Dublin: Hodges, Figgis, & Co., p. 43, 1888.

Coxeter, H. S. M. *Introduction to Geometry, 2nd ed.* New York: Wiley, p. 86, 1969.

Coxeter, H. S. M. and Greitzer, S. L. "The Radical Axis of Two Circles." §2.2 in *Geometry Revisited.* Washington, DC: Math. Assoc. Amer., pp. 31–34, 1967.

Dixon, R. *Mathographics.* New York: Dover, p. 68, 1991.

Dörrie, H. *100 Great Problems of Elementary Mathematics: Their History and Solutions.* New York: Dover, p. 153, 1965.

Durell, C. V. *Modern Geometry: The Straight Line and Circle.* London: Macmillan, p. 121, 1928.

Johnson, R. A. *Modern Geometry: An Elementary Treatise on the Geometry of the Triangle and the Circle.* Boston, MA: Houghton Mifflin, pp. 28–34 and 176–177, 1929.

Lachlan, R. "The Radical Axis of Two Circles." §304–312 in *An Elementary Treatise on Modern Pure Geometry.* London: Macmillian, pp. 185–189, 1893.

Wells, D. *The Penguin Dictionary of Curious and Interesting Geometry.* London: Penguin, p. 35, 1991.

Radicand

The quantity under a RADICAL sign.

See also CUBE ROOT, RADICAL, ROOT, SQUARE ROOT, VINCULUM

Radius

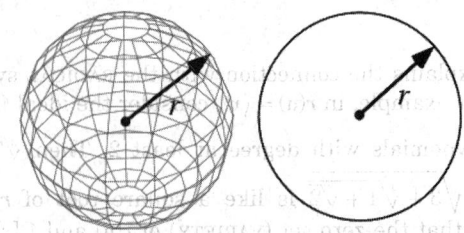

The distance from the center of a CIRCLE to its

PERIMETER, or from the center of a SPHERE to its surface. The radius is equal to half the DIAMETER.

See also BERTRAND'S PROBLEM, CIRCLE, CIRCUMFERENCE, DIAMETER, EXTENT, GRAPH RADIUS, INVERSION RADIUS, KINNEY'S SET, PI, RADIUS OF CONVERGENCE, RADIUS OF CURVATURE, RADIUS OF GYRATION, RADIUS OF TORSION, RADIUS VECTOR, SPHERE

Radius of Convergence

A POWER SERIES $\sum^{\infty} c_k x^k$ will converge only for certain values of x. For instance, $\sum_{k=0}^{\infty} x^k$ converges for $-1 < x < 1$. In general, there is always an interval $(-R, R)$ in which a POWER SERIES converges, and the number R is called the radius of convergence. The quantity R is called the radius of convergence because, in the case of a power series with complex coefficients, the values of x with $|x| < R$ form an OPEN DISK with radius R.

A POWER SERIES always CONVERGES ABSOLUTELY within its radius of convergence. This can be seen by fixing $r = |x|$ and supposing that there exists a SUBSEQUENCE c_{n_i} such that $|c_{n_i}| r^{n_i}$ is UNBOUNDED. Then the POWER SERIES $\sum c_n x^n$ does not CONVERGE (in fact, the terms are unbounded) because it fails the LIMIT TEST. Therefore, for x with $r = |x| > R$, the power series does not converge, where

$$c = \limsup c_n \qquad (1)$$

$$R = \frac{1}{c}, \qquad (2)$$

and lim sup denotes the SUPREMUM LIMIT.

Conversely, suppose that $r < R$. Then for any radius s with $r < s < R$, the terms $c_n x^n$ satisfy

$$|c_n x^n| < \left(\frac{s}{R}\right)^n \qquad (3)$$

for n large enough (depending on s). It is sufficient to fix a value for s in between r and R. Because $s/R < 1$, the power series is dominated by a convergent GEOMETRIC SERIES. Hence, the POWER SERIES converges absolutely by the LIMIT COMPARISON TEST.

See also CONVERGENT SERIES, POWER SERIES, ROOT TEST

References
Levinson, N. and Raymond, R. *Complex Variables.* New York: McGraw-Hill, pp. 349–352, 1970.
Rudin, W. *Principles of Mathematical Analysis.* New York: McGraw-Hill, p. 69, 1976.

Radius of Curvature

The radius of curvature is given by

$$R \equiv \frac{1}{\kappa}, \qquad (1)$$

where κ is the CURVATURE. At a given point on a curve, R is the radius of the OSCULATING CIRCLE. The symbol ρ is sometimes used instead of R to denote the radius of curvature.

Let x and y be given parametrically by

$$x = x(t) \qquad (2)$$

$$y = y(t), \qquad (3)$$

then

$$R = \frac{(x'^2 + y'^2)^{3/2}}{x'y'' - y'x''}, \qquad (4)$$

where $x' = dx/dt$ and $y' = dy/dt$. Similarly, if the curve is written in the form $y = f(x)$, then the radius of curvature is given by

$$R = \frac{\left[1 + \left(\dfrac{dy}{dx}\right)^2\right]^{3/2}}{\dfrac{d^2 y}{dx^2}}. \qquad (5)$$

In POLAR COORDINATES $r = r(\theta)$, the radius of curvature is given by

$$R = \frac{(r^2 + r_\theta^2)^{3/2}}{r^2 + 2r_\theta^2 - rr_{\theta\theta}}, \qquad (6)$$

where $r_\theta = dr/d\theta$ (Gray 1997, p. 89).

See also BEND (CURVATURE), CURVATURE, OSCULATING CIRCLE, RADIUS OF GYRATION, RADIUS OF TORSION, TORSION (DIFFERENTIAL GEOMETRY)

References
Gray, A. *Modern Differential Geometry of Curves and Surfaces with Mathematica,* 2nd ed. Boca Raton, FL: CRC Press, 1997.
Kreyszig, E. *Differential Geometry.* New York: Dover, p. 34, 1991.

Radius of Gyration

A positive number k such that a lamina or solid body with moment of inertia about an axis I and mass m is given by

$$I = mk^2.$$

Pickover (1995) defines a generalization of k as a function R_g quantifying the spatial extent of the structure of a curve and given by

$$R_g = \frac{\sqrt{\int_0^\infty r^2 p(r)\, dr}}{2\int_0^\infty p(r)\, dr},$$

where $p(r)$ is the LENGTH DISTRIBUTION FUNCTION. Small compact patterns have small R_g.

See also RADIUS OF CURVATURE, RADIUS OF TORSION

References

Pickover, C. A. *Keys to Infinity.* New York: Wiley, pp. 204–206, 1995.

Radius of Torsion

$$\sigma \equiv \frac{1}{\tau},$$

where τ is the TORSION. The symbol ϕ is also sometimes used instead of σ.

See also RADIUS OF CURVATURE, TORSION (DIFFERENTIAL GEOMETRY)

References

Kreyszig, E. *Differential Geometry.* New York: Dover, p. 39, 1991.

Radius Vector

The VECTOR $\mathbf{r}$ from the ORIGIN to the current position. It is also called the position vector. The derivative of $\mathbf{r}$ satisfies

$$\mathbf{r} \cdot \frac{d\mathbf{r}}{dt} = \frac{1}{2}\frac{d}{dt}(\mathbf{r} \cdot \mathbf{r}) = \frac{1}{2}\frac{d}{dt}(r^2) = r\frac{dr}{dt} = rv,$$

where v is the magnitude of the VELOCITY (i.e., the SPEED).

See also RADIUS, SPEED, VELOCITY

Radix

The BASE of a number system, i.e., 2 for BINARY, 8 for OCTAL, 10 for DECIMAL, and 16 for HEXADECIMAL. The radix is sometimes called the BASE or SCALE.

See also BASE (NUMBER)

Radon Measure

See also PROBABILITY MEASURE

Radon Transform

An INTEGRAL TRANSFORM whose inverse is used to reconstruct images from medical CT scans. A technique for using Radon transforms to reconstruct a map of a planet's polar regions using a spacecraft in a polar orbit has also been devised (Roulston and Muhleman 1997).

The Radon transform can be defined by

$$R(p, \tau)[f(x, y)] = \int_{-\infty}^{\infty} f(x, \tau + px)\, dx$$

$$= \int_{-\infty}^{\infty}\int_{-\infty}^{\infty} f(x, y)\delta[y - (\tau + px)]\, dy\, dx \equiv U(p, \tau), \quad (1)$$

where p is the SLOPE of a line and τ is its intercept. The inverse Radon transform is

$$f(x, y) = \frac{1}{2\pi}\int_{-\infty}^{\infty}\frac{d}{dy}H[U(p, y - px)]\, dp, \quad (2)$$

where H is a HILBERT TRANSFORM. The transform can also be defined by

$$R'(r, \alpha)[f(x, y)]$$

$$= \int_{-\infty}^{\infty}\int_{-\infty}^{\infty} f(x, y)\delta(r - x\cos\alpha - y\sin\alpha)\, dx\, dy, \quad (3)$$

where r is the PERPENDICULAR distance from a line to the origin and α is the ANGLE formed by the distance VECTOR.

Using the identity

$$\mathscr{F}[R[f(\omega, \alpha)]] = \mathscr{F}^2[f(u, v)], \quad (4)$$

where $\mathscr{F}$ is the FOURIER TRANSFORM, gives the inversion formula

$$f(x, y) = c\int_0^\pi \int_{-\infty}^{\infty} \mathscr{F}[R[f(\omega, \alpha)]]$$

$$\times |\omega|e^{i\omega(x\cos\alpha + y\sin\alpha)}\, d\omega\, d\alpha. \quad (5)$$

The FOURIER TRANSFORM can be eliminated by writing

$$f(x, y) = \int_0^\pi \int_{-\infty}^{\infty} R[f(r, \alpha)]W(r, \alpha, x, y)\, dr\, d\alpha, \quad (6)$$

where W is a WEIGHTING FUNCTION such as

$$W(r, \alpha, x, y) = h(x\cos\alpha + y\sin\alpha - r) = \mathscr{F}^{-1}[|\omega|]. \quad (7)$$

Nievergelt (1986) uses the inverse formula

$$f(x, y) = \frac{1}{\pi}\lim_{c \to 0}\int_0^\pi \int_{-\infty}^{\infty} R[f(r + x\cos\alpha$$

$$+ y\sin\alpha, \alpha)]G_c(r)\, dr\, d\alpha, \quad (8)$$

where

$$G_c(r) = \begin{cases} \dfrac{1}{\pi c^2} & \text{for } |r| \le c \\[2ex] \dfrac{1}{\pi c^2}\left(1 - \dfrac{1}{\sqrt{1 - c^2/r^2}}\right) & \text{for } |r| > c. \end{cases} \quad (9)$$

LUDWIG'S INVERSION FORMULA expresses a function in terms of its Radon transform. $R'(r, \alpha)$ and $R(p, \tau)$ are related by

$$p = \cot \alpha \quad \tau = r \csc \alpha \tag{10}$$

$$r = \frac{\tau}{1 + p^2} \quad \alpha = \cot^{-1} p. \tag{11}$$

The Radon transform satisfies superposition

$$R(p, \tau)[f_1(x, y) + f_2(x, y)] = U_1(p, \tau) + U_2(p, \tau), \tag{12}$$

linearity

$$R(p, \tau)[af(x, y)] = aU(p, \tau), \tag{13}$$

scaling

$$R(p, \tau)\left[f\left(\frac{x}{a}, \frac{y}{b}\right)\right] = |a|U\left(p\frac{a}{b}, \frac{\tau}{b}\right), \tag{14}$$

ROTATION, with R_ϕ ROTATION by ANGLE ϕ

$$R(p, \tau)\left[R_\phi f(x, y)\right] = \frac{1}{|\cos\phi + p\sin\phi|} U$$
$$\times \left(\frac{p - \tan\phi}{1 + p\tan\phi}, \frac{\tau}{\cos\phi + p\sin\phi}\right), \tag{15}$$

and skewing

$$R(p, \tau)[f(ax + by, cx + dy)]$$
$$= \frac{1}{|a + bp|} U\left[\frac{c + dp}{a + bp}, \tau\frac{d - b(c + bd)}{a + bp}\right] \tag{16}$$

(Durrani and Bisset 1984).

The line integral along p, τ is

$$I = \sqrt{1 + p^2} U(p, \tau). \tag{17}$$

The analog of the 1-D CONVOLUTION THEOREM is

$$R(p, \tau)[f(x, y) * g(y)] = U(p, \tau) * g(\tau), \tag{18}$$

the analog of PLANCHEREL'S THEOREM is

$$\int_{-\infty}^{\infty} U(p, \tau)\, d\tau = \int_{-\infty}^{\infty} \int_{-\infty}^{\infty} f(x, y)\, dx\, dy, \tag{19}$$

and the analog of PARSEVAL'S THEOREM is

$$\int_{-\infty}^{\infty} R(p, \tau)[f(x, y)]^2\, d\tau = \int_{-\infty}^{\infty} \int_{-\infty}^{\infty} f^2(x, y)\, dx\, dy. \tag{20}$$

If f is a continuous function on $\mathbb{C}$, integrable with respect to a plane LEBESGUE MEASURE, and

$$\int_l f\, ds = 0 \tag{21}$$

for every (doubly) infinite line l where s is the length measure, then f must be identically zero. However, if the global integrability condition is removed, this result fails (Zalcman 1982, Goldstein 1993).

See also HAMMER'S X-RAY PROBLEMS, TOMOGRAPHY

References

Anger, B. and Portenier, C. *Radon Integrals.* Boston, MA: Birkhäuser, 1992.

Armitage, D. H. and Goldstein, M. "Nonuniqueness for the Radon Transform." *Proc. Amer. Math. Soc.* **117**, 175–178, 1993.

Deans, S. R. *The Radon Transform and Some of Its Applications.* New York: Wiley, 1983.

Durrani, T. S. and Bisset, D. "The Radon Transform and its Properties." *Geophys.* **49**, 1180–1187, 1984.

Esser, P. D. (Ed.). *Emission Computed Tomography: Current Trends.* New York: Society of Nuclear Medicine, 1983.

Gindikin, S. (Ed.). *Applied Problems of Radon Transform.* Providence, RI: Amer. Math. Soc., 1994.

Gradshteyn, I. S. and Ryzhik, I. M. *Tables of Integrals, Series, and Products, 6th ed.* San Diego, CA: Academic Press, 2000.

Helgason, S. *The Radon Transform.* Boston, MA: Birkhäuser, 1980.

Hungerbühler, N. "Singular Filters for the Radon Backprojection." *J. Appl. Analysis* **5**, 17–33, 1998.

Kak, A. C. and Slaney, M. *Principles of Computerized Tomographic Imaging.* IEEE Press, 1988.

Kunyansky, L. A. "Generalized and Attenuated Radon Transforms: Restorative Approach to the Numerical Inversion." *Inverse Problems* **8**, 809–819, 1992.

Nievergelt, Y. "Elementary Inversion of Radon's Transform." *SIAM Rev.* **28**, 79–84, 1986.

Rann, A. G. and Katsevich, A. I. *The Radon Transform and Local Tomography.* Boca Raton, FL: CRC Press, 1996.

Robinson, E. A. "Spectral Approach to Geophysical Inversion Problems by Lorentz, Fourier, and Radon Transforms." *Proc. Inst. Electr. Electron. Eng.* **70**, 1039–1053, 1982.

Roulston, M. S. and Muhleman, D. O. "Synthesizing Radar Maps of Polar Regions with a Doppler-Only Method." *Appl. Opt.* **36**, 3912–3919, 1997.

Shepp, L. A. and Kruskal, J. B. "Computerized Tomography: The New Medical X-Ray Technology." *Amer. Math. Monthly* **85**, 420–439, 1978.

Strichartz, R. S. "Radon Inversion--Variation on a Theme." *Amer. Math. Monthly* **89**, 377–384 and 420–423, 1982.

Weisstein, E. W. "Books about Radon Transforms." http://www.treasure-troves.com/books/RadonTransforms.html.

Zalcman, L. "Uniqueness and Nonuniqueness for the Radon Transform." *Bull. London Math. Soc.* **14**, 241–245, 1982.

Radon Transform—Cylinder

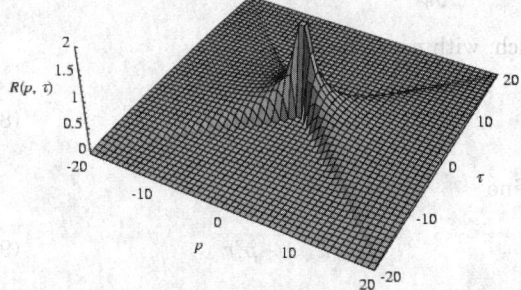

Let the 2-D cylinder function be defined by

$$f(x, y) \equiv \begin{cases} 1 & \text{for } r < R \\ 0 & \text{for } r > R. \end{cases} \tag{1}$$

Then the Radon transform is given by

$$R(p, \, \tau) = \int_{-\infty}^{\infty} \int_{-\infty}^{\infty} f(x, \, y)\delta[y - (\tau + px)] \, dy \, dx, \quad (2)$$

where

$$\delta(x) = \frac{1}{2\pi} \int_{-\infty}^{\infty} e^{-ikx} \quad (3)$$

is the DELTA FUNCTION.

$$R(p, \, \tau) = \frac{1}{2\pi} \int_0^{2\pi} \int_0^R \int_{-\infty}^{\infty} e^{-ik(r \sin \theta - pr \cos \theta)} r \, dr \, d\theta \, dk$$

$$= \frac{1}{2\pi} \int_{-\infty}^{\infty} e^{ikr} \int_0^{2\pi} \int_0^R e^{-ikr(\sin \theta - p \cos \theta)} r \, dr \, d\theta \, dk. \quad (4)$$

Now write

$$\sin \theta - p \cos \theta = \sqrt{1 + p^2} \cos(\theta + \phi) \equiv \sqrt{1 + p^2} \cos \theta', \quad (5)$$

with ϕ a phase shift. Then

$$R(p, \, \tau) = \frac{1}{2\pi} \int_{-\infty}^{\infty} e^{ik\tau}$$

$$\times \int_0^R \left(\int_0^{2\pi} e^{-ik\sqrt{1+p^2}r \cos \theta'} \, d\theta' \right) r \, dr \, dk$$

$$= \frac{1}{2\pi} \int_{-\infty}^{\infty} e^{ik\tau} \int_0^R 2\pi J_0\left(k\sqrt{1+p^2}r\right) r \, dr \, dk$$

$$= \int_{-\infty}^{\infty} e^{ik\tau} \int_0^R J_0\left(k\sqrt{1+p^2}r\right) r \, dr \, dk. \quad (6)$$

Then use

$$\int_0^z t^{n+1} J_n(t) \, dt = z^{n+1} J_{n+1}(z), \quad (7)$$

which, with $n = 0$, becomes

$$\int_0^z t J_0(t) \, dt = z J_1(z). \quad (8)$$

Define

$$t \equiv k\sqrt{1+p^2}r \quad (9)$$

$$dt = k\sqrt{1+p^2} \, dr \quad (10)$$

$$r \, dr = \frac{t \, dt}{k^2(1 + p^2)}, \quad (11)$$

so the inner integral is

$$\int_0^{R\sqrt{1+p^2}} J_0(t) \frac{t \, dt}{k^2(1 + p^2)}$$

$$= \frac{1}{k^2(1 + p^2)} kR\sqrt{1+p^2} J_1\left(kR\sqrt{1+p^2}\right) \quad (12)$$

$$= \frac{J_1\left(kR\sqrt{1+p^2}\right)}{k\sqrt{1+p^2}} R, \quad (13)$$

and the Radon transform becomes

$$R(p, \, \tau) = \frac{R}{\sqrt{1+p^2}} \int_{-\infty}^{\infty} \frac{e^{ik\tau} J_1\left(kR\sqrt{1+p^2}\right)}{k} \, dk$$

$$= \frac{2R}{\sqrt{1+p^2}} \int_0^{\infty} \frac{\cos(kr) J_1\left(kR\sqrt{1+p^2}\right)}{k} \, dk$$

$$= \begin{cases} \dfrac{2}{1+p^2}\sqrt{R^2(1+p^2) - \tau^2} \\ \qquad \text{for } \tau^2 < R^2(1+p^2) \\ 0 \\ \qquad \text{for } \tau^2 \ge R^2(1+p^2). \end{cases} \quad (14)$$

Converting to R' using $p = \cot \alpha$,

$$R'(r, \, \alpha) = \frac{2}{\sqrt{1 + \cot^2 \alpha}} \sqrt{(1 + \cot^2 \alpha)R^2 - r^2 \csc^2 \alpha}$$

$$= \frac{2}{\csc \alpha} \sqrt{\csc^2 \alpha R^2 - r^2 \csc^2 \alpha}$$

$$= 2\sqrt{R^2 - r^2}, \quad (15)$$

which could have been derived more simply by

$$R'(r, \, \alpha) = \int_{-\sqrt{R^2 - r^2}}^{\sqrt{R^2 - r^2}} dy. \quad (16)$$

Radon Transform—Delta Function

For a DELTA FUNCTION at $(x_0, \, y_0)$,

$$R(p, \, \tau) = \int_{-\infty}^{\infty} \int_{-\infty}^{\infty} \delta(x - x_0)\delta(y - y_0)$$

$$\times \delta[y - (\tau + px)] \, dy \, dx$$

$$= \frac{1}{2\pi} \int_{-\infty}^{\infty} \int_{-\infty}^{\infty} \int_{-\infty}^{\infty} e^{-ik[y - (\tau + px)]} \delta(x - x_0)$$

$$\times \delta(y - y_0) \, dk \, dy \, dx$$

$$= \frac{1}{2\pi} \int_{-\infty}^{\infty} e^{ik\tau} \left[\int_{-\infty}^{\infty} e^{-iky} \delta(y - y_0) \, dy \right.$$

$$\left. \times \int_{-\infty}^{\infty} e^{ikpx} \delta(x - x_0) \, dx \right] dk$$

$$= \frac{1}{2\pi} \int_{-\infty}^{\infty} e^{ik\tau} e^{-iky_0} e^{ikpx_0} \, dk.$$

$$= \frac{1}{2\pi} \int_{-\infty}^{\infty} e^{ik(\tau + px_0 - y_0)} \, dk = \delta(\tau + px_0 - y_0).$$

Radon Transform—Gaussian

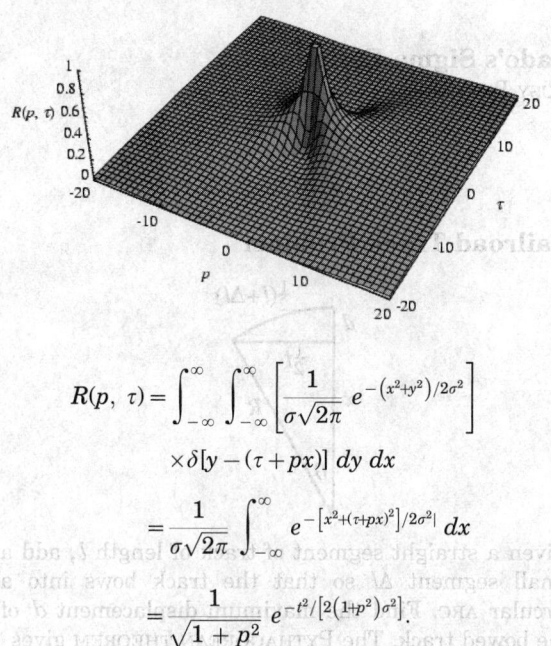

$$R(p, \ \tau) = \int_{-\infty}^{\infty} \int_{-\infty}^{\infty} \left[\frac{1}{\sigma\sqrt{2\pi}} \, e^{-(x^2+y^2)/2\sigma^2} \right]$$
$$\times \delta[y - (\tau + px)] \, dy \, dx$$

$$= \frac{1}{\sigma\sqrt{2\pi}} \int_{-\infty}^{\infty} e^{-[x^2 + (\tau+px)^2]/2\sigma^2|} \, dx$$

$$= \frac{1}{\sqrt{1+p^2}} \, e^{-\tau^2/[2(1+p^2)\sigma^2]}$$

Radon Transform—Square

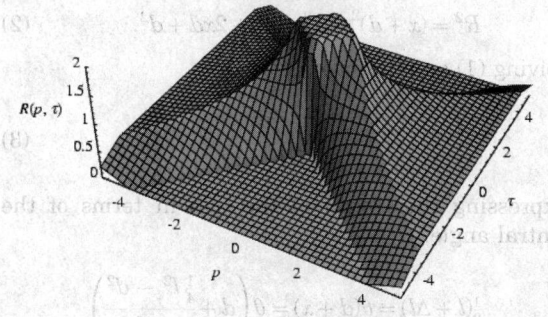

$$R(p, \ \tau) = \int_{-\infty}^{\infty} \int_{-\infty}^{\infty} f(x, y) \delta[y - (\tau + px)] \, dy \, dx, \quad (1)$$

where

$$f(x, y) \equiv \begin{cases} 1 & \text{for } x, \ y \in [-a, \ a] \\ 0 & \text{otherwise} \end{cases} \quad (2)$$

and

$$\delta(x) = \frac{1}{2\pi} \int_{-\infty}^{\infty} e^{-ikx} \quad (3)$$

is the DELTA FUNCTION.

$$R(p, \ r) = \frac{1}{2\pi} \int_{-a}^{a} \int_{-a}^{a} \int_{-\infty}^{\infty} e^{-ik[y-(r+px)]} \, dk \, dy \, dx$$

$$= \frac{1}{2\pi} \int_{-\infty}^{\infty} e^{ikr} \left[\int_{-a}^{a} e^{-ky} \, dy \int_{-a}^{a} e^{ikpx} \, dx \right] dk$$

$$= \frac{1}{2\pi} \int_{-\infty}^{\infty} e^{ikr} \frac{1}{-ik} \left[e^{-iky} \right]_{-a}^{a} \frac{1}{ikp} \left[e^{-ikpx} \right]_{-a}^{a} dk$$

$$= \frac{1}{2\pi} \int_{-\infty}^{\infty} e^{ikr} \frac{1}{k^2 p} [-2i \sin (ka)][2i \sin(kpa)] \, dk$$

$$= \frac{2}{\pi p} \int_{-\infty}^{\infty} \frac{\sin(ka) \sin(kpa) e^{ikr}}{k^2} \, dk$$

$$= \frac{4}{\pi p} \int_{-\infty}^{\infty} \frac{\sin(ka) \sin(kpa) \cos(k\tau)}{k^2} \, dk$$

$$= \frac{2}{\pi p} \int_{-\infty}^{\infty} \frac{\sin[k(\tau + a)] - \sin[k(\tau - a)]}{k^2} \sin(kpa) \, dk$$

$$= \frac{2}{\pi p} \left\{ \int_{0}^{\infty} \frac{\sin[k(\tau + a)] \sin(kpa)}{k^2} \, dk \right.$$

$$\left. - \int_{0}^{\infty} \frac{\sin[k(\tau - a)] \sin(kpa)}{k^2} \, dk \right\}. \quad (4)$$

From Gradshteyn and Ryzhik (2000, equation 3.741.3),

$$\int_{0}^{\infty} \frac{\sin(ax) \sin(bx)}{x^2} \, dx = \tfrac{1}{2} \pi \operatorname{sgn}(ab) \min(|a|, \ |b|), \quad (5)$$

so

$$R(p, \ \tau) = \frac{1}{p} \{ \operatorname{sgn}[(\tau + a)pa] \min(|\tau + a|, \ |pa|)$$

$$- \operatorname{sgn}[(\tau - a)pa] \min(|\tau - a|, \ |pa|) \}. \quad (6)$$

References

Gradshteyn, I. S. and Ryzhik, I. M. *Tables of Integrals, Series, and Products, 6th ed.* San Diego, CA: Academic Press, 2000.

Radon-Nikodym Derivative

When a MEASURE λ is ABSOLUTELY CONTINUOUS with respect to a positive measure μ, then it can be written as

$$\lambda(E) = \int_{E} f \, d\mu.$$

By analogy with the first FUNDAMENTAL THEOREM OF CALCULUS, the function f is called the Radon-Nikodym derivative of λ with respect to μ. Sometimes it is denoted $d\lambda/d\mu$ or $D\lambda/D\mu$.

See also Absolutely Continuous, Complex Measure, Fundamental Theorems of Calculus, Lebesgue Measure, Polar Representation (Measure), Radon-Nikodym Theorem

References

Rudin, W. *Real and Complex Analysis.* New York: McGraw-Hill, p. 122, 1987.

Radon-Nikodym Theorem

The Radon-Nikodym theorem asserts that any ABSOLUTELY CONTINUOUS measure λ with respect to some positive measure μ (which could be LEBESGUE MEASURE or HAAR MEASURE) is given by the integral of some L^1-function f,

$$\lambda(E) = \int_E f \, d\mu. \tag{1}$$

The function f is like a density function for the measure.

A closely related theorem says that any COMPLEX MEASURE λ decomposes into an ABSOLUTELY CONTINUOUS measure λ_a and a singular measure λ_c. This is the LEBESGUE DECOMPOSITION

$$\lambda = \lambda_a + \lambda_c. \tag{2}$$

One consequence of the Radon-Nikodym theorem is that any complex measure has a POLAR REPRESENTATION,

$$d\mu = h \, d|\mu|, \tag{3}$$

with $|h| = 1$.

See also Absolutely Continuous, Complex Measure, Haar Measure, Lebesgue Decomposition (Measure), Lebesgue Measure, Polar Representation (Measure), Singular Measure

References

Doob, J. L. "The Development of Rigor in Mathematical Probability (1900–1950)." *Amer. Math. Monthly* **103**, 586–595, 1996.
Rudin, W. *Real and Complex Analysis.* New York:McGraw-Hill, pp. 121–129, 1987.

Radon's Theorem

Any set of $n + 2$ points in $\mathbb{R}^n$ can always be partitioned in two subsets V_1 and V_2 such that the CONVEX HULLS of V_1 and V_2 intersect.

See also Convex Hull

References

Eckhoff, J. "Helly, Radon, and Carathéodory Type Theorems." Ch. 2.1 in *Handbook of Convex Geometry* (Ed. P. M. Gruber and J. M. Wills). Amsterdam, Netherlands: North-Holland, pp. 389–448, 1993.
McMullen, P. and Shepard, G. C. *Convex Polytopes and the Upper Bound Conjecture.* London: Cambridge University Press, pp. 22–24, 1971.

Peterson, B. B. "The Geometry of Radon's Theorem." *Amer. Math. Monthly* **79**, 949–963, 1972.
Peyerimhoff, N. "Areas and Intersections in Convex Domains." *Amer. Math. Monthly* **104**, 697–704, 1997.
Rado, R. "Theorems on the Intersection of Convex Sets of Points." *J. London Math. Soc.* **27**, 320–328, 1952.
Ziegler, G. M. Ex. 6.0 in *Lectures on Polytopes.* New York: Springer-Verlag, 1994.

Rado's Sigma Function

BUSY BEAVER

Railroad Track Problem

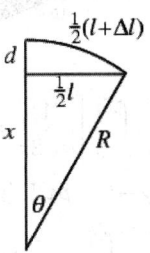

Given a straight segment of track of length l, add a small segment Δl so that the track bows into a circular ARC. Find the maximum displacement d of the bowed track. The PYTHAGOREAN THEOREM gives

$$R^2 = x^2 + (\tfrac{1}{2} l)^2. \tag{1}$$

But R is simply $x + d$, so

$$R^2 = (x + d)^2 = x^2 = x^2 + 2xd + d^2. \tag{2}$$

Solving (1) and (2) for x gives

$$x = \frac{\tfrac{1}{4} l^2 - d^2}{2d}. \tag{3}$$

Expressing the length of the ARC in terms of the central angle,

$$\tfrac{1}{2}(l + \Delta l) = \theta(d + x) = \theta\left(d + \frac{\tfrac{1}{4} l^2 - d^2}{2d}\right)$$

$$= \theta\left(\frac{2d^2 + \tfrac{1}{4} l^2 - d^2}{2d}\right)$$

$$= \theta\left(\frac{d^2 + \tfrac{1}{4} l^2}{2d}\right). \tag{4}$$

But θ is given by

$$\tan \theta = \frac{\tfrac{1}{2} l}{x} = \frac{\tfrac{1}{2} l(2d)}{\tfrac{1}{4} l^2 - d^2} = \frac{dl}{\tfrac{1}{4} l^2 - d^2}, \tag{5}$$

so plugging θ in gives

$$\tfrac{1}{2}(l+\Delta l)=\left(\frac{d^2+\tfrac{1}{4}l^2}{2d}\right)\tan^{-1}\left(\frac{dl}{\tfrac{1}{4}l^2-d^2}\right) \qquad (6)$$

$$d(l+\Delta l)=\left(d^2+\tfrac{1}{4}l^2\right)\tan^{-1}\left(\frac{dl}{\tfrac{1}{4}l^2-d^2}\right). \qquad (7)$$

For $l\gg d$,

$$\frac{dl}{\tfrac{1}{4}l^2\left(1-\dfrac{d^2}{4l^2}\right)}=\frac{4d}{l}\left(1-\frac{4d^2}{l^2}\right)^{-1}\approx\frac{4d}{l}\left(1+\frac{4d}{l^2}\right). \qquad (8)$$

Therefore,

$$d(l+\Delta l)\approx(d^2+\tfrac{1}{4}l^2)$$

$$\times\left\{\frac{4d}{l}\left(1+\frac{4d^2}{l^2}\right)-\frac{1}{3}\left[\frac{4d}{l}\left(1+\frac{4d^2}{l^2}\right)\right]^3\right\}$$

$$\approx\left(d^2+\tfrac{1}{4}l^2\right)\left[\frac{4d}{l}+\frac{16d^3}{l^3}-\frac{1}{3}\left(\frac{4d}{l}\right)^3\right.$$

$$\left.\times\left(1+3\,\frac{4d^2}{l^2}\right)\right]. \qquad (9)$$

Keeping only terms to order $(d/l)^3$,

$$dl+\Delta l\approx\frac{4d^3}{l}+dl+\frac{4d^3}{l}-\frac{16}{3}\frac{d^3}{l} \qquad (10)$$

$$\Delta l\approx\left(8-\tfrac{16}{3}\right)\frac{d^3}{l}=\frac{24-16}{3}\frac{d^3}{l}=\frac{8}{3}\frac{d^3}{l}, \qquad (11)$$

so

$$d^2=\tfrac{3}{8}\,l\Delta l \qquad (12)$$

and

$$d\approx\tfrac{1}{2}\sqrt{\tfrac{3}{2}\,l\Delta l}=\tfrac{1}{4}\sqrt{6l\Delta l}. \qquad (13)$$

If we take $l=1$ mile $=5280$ feet and $\Delta l=1$ foot, then $d\approx44.50$ feet.

References

Abbott, P. "In and Out: Acton's Railroad Problem." *Mathematica J.* **7**, 448–450, 2000.

Acton, F. S. *Numerical Methods That Work, 2nd printing.* Washington, DC: Math. Assoc. Amer., 1990.

Ramanujan 6–10–8 Identity

Let $ad=bc$, then

$$64[(a+b+c)^6+(b+c+d)^6-(c+d+a)^6$$

$$-(d+a+b)^6+(a-d)^6-(b-c)^6]$$

$$\times[(a+b+c)^{10}+(b+c+d)^{10}-(c+d+a)^{10}$$

$$-(d+a+b)^{10}+(a-d)^{10}-(b-c)^{10}]$$

$$=45[(a+b+c)^8+(b+c+d)^8-(c+d+a)^8$$

$$-(d+a+b)^8+(a-d)^8-(b-c)^8]^2. \qquad (1)$$

This can also be expressed by defining

$$F_{2m}(a,\,b,\,c,\,d)=(a+b+c)^{2m}+(b+c+d)^{2m}$$

$$-(c+d+a)^{2m}-(d+a+b)^{2m}+(a-d)^{2m}-(b-c)^{2m} \qquad (2)$$

$$f_{2m}(x,\,y)=(1+x+y)^{2m}+(x+y+xy)^{2m}-(y+xy+1)^{2m}$$

$$-(xy+1+x)^{2m}+(1-xy)^{2m}-(x-y)^{2m}. \qquad (3)$$

Then

$$F_{2m}(a,\,b,\,c,\,d)=a^{2m}f_{2m}(x,\,y), \qquad (4)$$

and identity (1) can then be written

$$64f_6(x,\,y)f_{10}(x,\,y)=45f_8^2(x,\,y). \qquad (5)$$

Incidentally,

$$f_2(x,\,y)=0 \qquad (6)$$

$$f_4(x,\,y)=0. \qquad (7)$$

References

Berndt, B. C. *Ramanujan's Notebooks, Part IV.* New York: Springer-Verlag, pp. 3 and 102–106, 1994.

Berndt, B. C. and Bhargava, S. "A Remarkable Identity Found in Ramanujan's Third Notebook." *Glasgow Math. J.* **34**, 341–345, 1992.

Berndt, B. C. and Bhargava, S. "Ramanujan--For Lowbrows." *Amer. Math. Monthly* **100**, 644–656, 1993.

Bhargava, S. "On a Family of Ramanujan's Formulas for Sums of Fourth Powers." *Ganita* **43**, 63–67, 1992.

Hirschhorn, M. D. "Two or Three Identities of Ramanujan." *Amer. Math. Monthly* **105**, 52–55, 1998.

Nanjundiah, T. S. "A Note on an Identity of Ramanujan." *Amer. Math. Monthly* **100**, 485–487, 1993.

Ramanujan, S. *Notebooks.* New York: Springer-Verlag, pp. 385–386, 1987.

Ramanujan Constant

The IRRATIONAL constant

$$R\equiv e^{\pi\sqrt{163}}$$

$$=262537412640768743.999999999999925\ldots$$

which is very close to an INTEGER. Numbers such as the Ramanujan constant can be found using the theory of MODULAR FUNCTIONS. In fact, the nine HEEGNER NUMBERS (which include 163) share a deep number theoretic property related to some amazing properties of the J-FUNCTION that leads to this sort of near-identity.

Although Ramanujan (1913–14) gave few rather spectacular examples of almost integers (such $e^{\pi\sqrt{58}}$), he did not actually mention particular near-identity give above. In fact, the first to observe this property of 163 was Hermite (1859). The name "Ramanujan's constant" seems to derive from an April Fool's joke played by Martin Gardner (Apr. 1975) on the readers of *Scientific American*. In his column, Gardner claimed that $e^{\pi\sqrt{163}}$ was exactly an INTEGER, and that Ramanujan had conjectured this in his 1914 paper. Gardner admitted his hoax a few months later (Gardner, July 1975).

See also ALMOST INTEGER, CLASS NUMBER, HEEGNER NUMBER, *J*-FUNCTION

References

Ball, W. W. R. and Coxeter, H. S. M. *Mathematical Recreations and Essays, 13th ed.* New York: Dover, p. 387, 1987.

Castellanos, D. "The Ubiquitous Pi. Part I." *Math. Mag.* **61**, 67–98, 1988.

Gardner, M. "Mathematical Games: Six Sensational Discoveries that Somehow or Another have Escaped Public Attention." *Sci. Amer.* **232**, 127–131, Apr. 1975.

Gardner, M. "Mathematical Games: On Tessellating the Plane with Convex Polygons." *Sci. Amer.* **232**, 112–117, Jul. 1975.

Good, I. J. "What is the Most Amazing Approximate Integer in the Universe?" *Pi Mu Epsilon J.* **5**, 314–315, 1972.

Hermite, C. "Sur la théorie des équations modulaires." *C. R. Acad. Sci. (Paris)* **49**, 16–24, 110–118, and 141–144, 1859 *Oeuvres complètes, Tome II*. Paris: Hermann, p. 61, 1912.

Plouffe, S. " $e^{\pi\sqrt{163}}$, the Ramanujan Number." http://www.lacim.uqam.ca/piDATA/ramanujan.txt.

Ramanujan, S. "Modular Equations and Approximations to π." *Quart. J. Pure Appl. Math.* **45**, 350–372, 1913–1914.

Wolfram, S. *The Mathematica Book, 3rd ed.* New York: Cambridge University Press, p. 52, 1996.

Ramanujan Continued Fraction

ROGERS-RAMANUJAN CONTINUED FRACTION

Ramanujan Cos/Cosh Identity

The amazing identity

$$\left[1 + 2\sum_{n=1}^{\infty}\frac{\cos(n\theta)}{\cosh(n\pi)}\right]^{-2} + \left[1 + 2\sum_{n=1}^{\infty}\frac{\cosh(n\theta)}{\cosh(n\pi)}\right]^{-2}$$

$$= \frac{2\Gamma^4\left(\frac{3}{4}\right)}{\pi}$$

for all θ, where $\Gamma(z)$ is the GAMMA FUNCTION. Equating coefficients of θ^0, θ^4, and θ^8 gives some amazing identities for the HYPERBOLIC SECANT.

See also HYPERBOLIC SECANT

Ramanujan Function

The two-argument Ramanujan function is defined by

$$\phi(a, n) \equiv 1 + 2\sum_{k=1}^{n}\frac{1}{(ak)^3 - ak} \qquad (1)$$

$$= 1 - \frac{1}{a}\left(H_{-1/a} + H_{1/a} + 2H_n - H_{n-1/a} - H_{n+1/a}\right). \qquad (2)$$

The one-argument function $\phi(a)$ is then defined as the limiting sum of $\phi(a, n)$ as $n \to \infty$,

$$\phi(a) \equiv \lim_{n\to\infty}\phi(a, n) = 1 + 2\sum_{k=1}^{\infty}\frac{1}{(ak)^3 - ak} \qquad (3)$$

$$= -\frac{1}{a}\left[\psi_0\left(\frac{1}{a}\right) + \psi_0\left(1 - \frac{1}{a}\right) + 2\gamma\right], \qquad (4)$$

$$= 1 - \frac{1}{a}\left(H_{-1/a} + H_{1/a}\right) \qquad (5)$$

where $\psi_0(x)$ is the DIGAMMA FUNCTION, γ is the EULER-MASCHERONI CONSTANT, and H_ν is a HARMONIC NUMBER. The values of $\phi(n)$ for $n = 2, 3, \ldots$ are

$$\phi(2) = 2\ln 2$$

$$\phi(3) = \ln 3$$

$$\phi(4) = \tfrac{3}{2}\ln 2$$

$$\phi(5) = \tfrac{1}{5}\sqrt{5}\ln\phi + \tfrac{1}{2}\ln 5$$

$$\phi(6) = \tfrac{1}{2}\ln 3 + \tfrac{2}{3}\ln 2,$$

where ϕ is the GOLDEN RATIO.

See also HARMONIC NUMBER, RAMANUJAN *G*- AND *G*-FUNCTIONS, TAU FUNCTION

Ramanujan g- and G-Functions

Following Ramanujan (1913–14), write

$$\prod_{k=1, 3, 5, \ldots}^{\infty}\left(1 + e^{-k\pi\sqrt{n}}\right) = 2^{1/4}e^{-\pi\sqrt{n}/24}G_n \qquad (1)$$

$$\prod_{k=1, 3, 5, \ldots}^{\infty}\left(1 + e^{-k\pi\sqrt{n}}\right) = 2^{1/4}e^{-\pi\sqrt{n}/24}g_n. \qquad (2)$$

These satisfy the equalities

$$g_{4n} = 2^{1/4}g_n G_n \qquad (3)$$

$$G_n = G_{1/n} \qquad (4)$$

$$g_n^{-1} = g_{4/n} \qquad (5)$$

$$\tfrac{1}{4} = (g_n G_n)^8(G_n^8 - g_n^8). \qquad (6)$$

G_n and g_n can be derived using the theory of MODULAR FUNCTIONS and can always be expressed as roots of algebraic equations when n is RATIONAL. For simplicity, Ramanujan tabulated g_n for n EVEN and G_n for n ODD. However, (6) allows G_n and g_n to be solved for in terms of g_n and G_n, giving

$$g_n = \tfrac{1}{2}\left(G_n^8 + \sqrt{G_n^{16} - G_n^{-8}}\right)^{1/8} \quad (7)$$

$$G_n = \tfrac{1}{2}\left(g_n^8 + \sqrt{g_n^{16} + G g_n^{-8}}\right)^{1/8}. \quad (8)$$

Using (3) and the above two equations allows g_{4n} to be computed in terms of g_n or G_n

$$g_{4n} = \begin{cases} 2^{1/8} g_n \left(g_n^8 + \sqrt{g_n^{16} + g_n^{-8}}\right)^{1/8} & \text{for } n \text{ even} \\ 2^{1/8} G_n \left(G_n^8 + \sqrt{G_n^{16} + G_n^{-8}}\right)^{1/8} & \text{for } n \text{ odd.} \end{cases} \quad (9)$$

In terms of the PARAMETER k and complementary PARAMETER k',

$$G_n = (2 k_n k_n')^{-1/12} \quad (10)$$

$$g_n = \left(\frac{k_n'^2}{2k}\right)^{1/12}. \quad (11)$$

Here,

$$k_n = \lambda^*(n) \quad (12)$$

is the ELLIPTIC LAMBDA FUNCTION, which gives the value of k for which

$$\frac{K'(k)}{K(k)} = \sqrt{n}. \quad (13)$$

Solving for $\lambda^*(n)$ gives

$$\lambda^*(n) = \tfrac{1}{2}\left[\sqrt{1 + G_n^{-12}} - \sqrt{1 - G_n^{-12}}\right] \quad (14)$$

$$\lambda^*(n) = g_n^6 \left[\sqrt{g_n^{12} + g_n^{-12}} - g_n^6\right]. \quad (15)$$

Analytic values for small values of n can be found in Ramanujan (1913–1914) and Borwein and Borwein (1987), and have been compiled by Weisstein. Ramanujan (1913–1914) contains a typographical error labeling G_{465} as G_{265}.

See also BARNES' G-FUNCTION

References

Borwein, J. M. and Borwein, P. B. *Pi & the AGM: A Study in Analytic Number Theory and Computational Complexity.* New York: Wiley, pp. 139 and 298, 1987.
Ramanujan, S. "Modular Equations and Approximations to π." *Quart. J. Pure. Appl. Math.* **45**, 350–372, 1913–1914.
Weisstein, E. W. "Elliptic Singular Values." MATHEMATICA NOTEBOOK ELLIPTICSINGULAR.M.

Ramanujan Psi Sum

A sum which includes both the JACOBI TRIPLE PRODUCT and the Q-BINOMIAL THEOREM as special cases. Ramanujan's sum is

$$\sum_{n=-\infty}^{\infty} \frac{(a)_n}{(b)_n} x^n = \frac{(ax)_\infty (q/ax)_\infty (q)_\infty (b/a)_\infty}{(x)_\infty (b/ax)_\infty (b)_\infty (q/a)_\infty},$$

where the NOTATION $(q)_k$ denotes Q-SERIES. For $b = q$, this becomes the Q-BINOMIAL THEOREM.

See also JACOBI TRIPLE PRODUCT, Q-BINOMIAL THEOREM, Q-SERIES

Ramanujan Theta Functions

Ramanujan's one-variable theta function is defined by

$$\varphi(q) \equiv \sum_{m=-\infty}^{\infty} q^{m^2}, \quad (1)$$

$$= \vartheta_3(0, q) \quad (2)$$

where $\vartheta_3(0, q)$ is a JACOBI THETA FUNCTION, and is equal to the JACOBI TRIPLE PRODUCT with $z = 1$. Special values include

$$\varphi\left(e^{-\pi\sqrt{2}}\right) = \frac{\Gamma\left(\frac{9}{8}\right)}{\Gamma\left(\frac{5}{4}\right)} \sqrt{\frac{\Gamma\left(\frac{1}{4}\right)}{2^{1/4}\pi}} \quad (3)$$

$$\varphi(e^{-\pi}) = \frac{\pi^{1/4}}{\Gamma\left(\frac{3}{4}\right)}, \quad (4)$$

where $\Gamma(x)$ is a GAMMA FUNCTION.

Another function sometimes given the same symbol is

$$\varphi(q) = \sqrt{\frac{\vartheta_2(0, q)}{\vartheta_3(0, q)}}, \quad (5)$$

where $\vartheta_i(0, q)$ is again a JACOBI THETA FUNCTION, which has special value

$$\varphi\left(-e^{-\pi\sqrt{3}}\right) = \left(4\sqrt{3} - 7\right)^{1/8}. \quad (6)$$

Ramanujan's two-variable theta function is defined by

$$f(a, b) \equiv \sum_{n=-\infty}^{\infty} a^{n(n+1)/2} b^{n(n-1)/2} \quad (7)$$

for $|ab| < 1$ (Berndt *et al.*). It is a generalization of the function $\varphi(x)$

$$f(x, x) = \varphi(x) \quad (8)$$

and satisfies

$$f(-1, a) = 0 \quad (9)$$

$$f(a, b) = f(b, a) = (-a; ab)_\infty (-b; ab)_\infty (ab; ab)_\infty \quad (10)$$

$$f(-q) \equiv f(-q, -q^2) \quad (11)$$

$$= \sum_{k=0}^{\infty} (-1)^k q^{k(2k-1)/2} \sum_{k=1}^{\infty} (-1)^k q^{k(2k+1)/2} \quad (12)$$

$$= (q; q)_\infty \quad (13)$$

(Berndt *et al.*), where $(a; q)_\infty$ is a Q-POCHHAMMER

SYMBOL. (13) is equivalent to EULER'S PENTAGONAL NUMBER THEOREM.

See also EULER'S PENTAGONAL NUMBER THEOREM, JACOBI TRIPLE PRODUCT, Q-SERIES, ROGERS-RAMANUJAN CONTINUED FRACTION, SCHRÖTER'S FORMULA

References

Berndt, B. C.; Huang, S.-S.; Sohn, J.; and Son, S. H. "Some Theorems on the Rogers-Ramanujan Continued Fraction in Ramanujan's Lost Notebook." To appears in *Trans. Amer. Math. Soc.*

Ramanujan-Eisenstein Series
EISENSTEIN SERIES

Ramanujan-Petersson Conjecture

A CONJECTURE for the EIGENVALUES of MODULAR FORMS under HECKE OPERATORS.

Ramanujan's Formula

$$\int_0^\infty \cos(2zt)\,\mathrm{sech}(\pi t)\,dt = \tfrac{1}{2}\,\mathrm{sech}\,z$$

for $|\Im z| < \pi/2$. A related integral is

$$\int_0^\infty \cosh(2zt)\,\mathrm{sech}(\pi t)\,dt = \tfrac{1}{2}\,\mathrm{sech}\,z$$

for $|\Re z| < \pi/2$.

References

Erdélyi, A.; Magnus, W.; Oberhettinger, F.; and Tricomi, F. G. *Higher Transcendental Functions, Vol. 1.* New York: Krieger, p. 11, 1981.

Ramanujan's Hypergeometric Identity

$$1 - \left(\frac{1}{2}\right)^3 + \left(\frac{1\cdot 3}{2\cdot 4}\right)^3 + \ldots = {}_3F_2\left(\begin{matrix}\tfrac{1}{2}, \tfrac{1}{2}, \tfrac{1}{2};\\ 1,\ 1\end{matrix}\ -1\right)$$

$$= \left[{}_2F_1\left(\begin{matrix}\tfrac{1}{4}, \tfrac{1}{4};\\ 1\end{matrix}\ -1\right)\right]^2$$

$$= \frac{\Gamma^2\left(\tfrac{9}{8}\right)}{\Gamma^2\left(\tfrac{5}{4}\right)\Gamma^2\left(\tfrac{7}{8}\right)},$$

where ${}_2F_1(a, b; c; x)$ is a HYPERGEOMETRIC FUNCTION, ${}_3F_2(a, b, c; d; e; x)$ is a GENERALIZED HYPERGEOMETRIC FUNCTION, and $\Gamma(z)$ is a GAMMA FUNCTION.

References

Hardy, G. H. *Ramanujan: Twelve Lectures on Subjects Suggested by His Life and Work, 3rd ed.* New York: Chelsea, p. 106, 1999.

Ramanujan's Hypothesis
TAU CONJECTURE

Ramanujan's Identity

$$5\frac{\phi^5(x^5)}{\phi^6(x)} = \sum_{m=0}^\infty P(5m+4)x^m,$$

where

$$\phi(x) = \prod_{m=1}^\infty (1 - x^m)$$

and $P(n)$ is the PARTITION FUNCTION P.

See also PARTITION FUNCTION P, RAMANUJAN'S SUM IDENTITY

Ramanujan's Integral

$$\int_{-\infty}^\infty \frac{J_{\mu+\xi}(x)}{x^{\mu+\xi}}\,\frac{J_{\nu-\xi}(y)}{y^{\nu-\xi}}\,e^{it\xi}\,d\xi$$

$$= \left[\frac{2\cos\left(\tfrac{1}{2}t\right)}{x^2 e^{-it/2} + y^2 e^{it/2}}\right]^{(\mu+\nu)/2}$$

$$\times J_{\mu+\nu}\left[\sqrt{2\cos\left(\tfrac{1}{2}t\right)(x^2 e^{-it/2} + y^2 e^{it/2})}\right]e^{-it(\nu-\mu)/2},$$

where $J_n(z)$ is a BESSEL FUNCTION OF THE FIRST KIND.

References

Watson, G. N. *A Treatise on the Theory of Bessel Functions, 2nd ed.* Cambridge, England: Cambridge University Press, 1966.

Ramanujan's Interpolation Formula

$$\int_0^\infty x^{s-1}\sum_{k=0}^\infty (-1)^k x^k \phi(k)\,dx = \frac{\pi\phi(-s)}{\sin(s\pi)} \quad (1)$$

$$\int_0^\infty x^{s-1}\sum_{k=0}^\infty (-1)^k \frac{x^k}{k!}\lambda(k)\,dx = \Gamma(s)\lambda(-s), \quad (2)$$

where $\lambda(z)$ is the DIRICHLET LAMBDA FUNCTION and $\Gamma(z)$ is the GAMMA FUNCTION. Equation (2) is obtained from (1) by defining

$$\phi(u) = \frac{\lambda(u)}{\Gamma(1+u)}. \quad (3)$$

These formulas give valid results only for certain classes of functions, and are connected with Mellin transforms (Hardy 1999, p. 15).

References

Hardy, G. H. *Ramanujan: Twelve Lectures on Subjects Suggested by His Life and Work, 3rd ed.* New York: Chelsea, pp. 15 and 186–195, 1999.

Ramanujan's Master Theorem

Suppose that in some NEIGHBORHOOD of $x = 0$,

$$F(x) = \sum_{k=0}^{\infty} \frac{\phi(k)(-x)^k}{k!}.$$

Then

$$\int_0^{\infty} x^{n-1} F(x) \, dx = \Gamma(n)\phi(-n).$$

References

Berndt, B. C. *Ramanujan's Notebooks: Part I.* New York: Springer-Verlag, p. 298, 1985.

Ramanujan's Square Equation

The DIOPHANTINE EQUATION

$$2^n - 7 = x^2.$$

It has been proved that the only solutions to this equation are $n = 3, 4, 5, 7,$ and 15 (Beeler *et al.* 1972, Item 31).

References

Schroeppel, R. C. Item 31 in Beeler, M.; Gosper, R. W.; and Schroeppel, R. *HAKMEM.* Cambridge, MA: MIT Artificial Intelligence Laboratory, Memo AIM-239, p. 14, Feb. 1972.

Ramanujan's Sum

The sum

$$c_q(m) = \sum_{h^*(q)} e^{2\pi i h m/q}, \tag{1}$$

where h runs through the residues RELATIVELY PRIME to q, which is important in the representation of numbers by the sums of squares. If $(q, q') = 1$ (i.e., q and q' are RELATIVELY PRIME), then

$$c_{qq'}(m) = c_q(m)c_{q'}(m). \tag{2}$$

For argument 1,

$$c_b(1) = \mu(b), \tag{3}$$

where μ is the MÖBIUS FUNCTION, and for general m,

$$c_b(m) = \mu\left(\frac{b}{(b, m)}\right) \frac{\phi(b)}{\phi\left(\frac{b}{(b, m)}\right)}. \tag{4}$$

See also MÖBIUS FUNCTION, WEYL'S CRITERION

References

Hardy, G. H. *Ramanujan: Twelve Lectures on Subjects Suggested by His Life and Work, 3rd ed.* New York: Chelsea, pp. 137–143, 1999.

Vardi, I. *Computational Recreations in Mathematica.* Redwood City, CA: Addison-Wesley, p. 254, 1991.

Ramanujan's Sum Identity

If

$$\frac{1 + 53x + 9x^2}{1 - 82x - 82x^2 + x^3} = \sum_{n=1}^{\infty} a_n x^n \tag{1}$$

$$\frac{2 - 26x - 12x^2}{1 - 82x - 82x^2 + x^3} = \sum_{n=0}^{\infty} b_n x^n \tag{2}$$

$$\frac{2 + 8x - 10x^2}{1 - 82x - 82x^2 + x^3} = \sum_{n=0}^{\infty} c_n x^n \tag{3}$$

(Sloane's A051028, A051029, and A051030), then

$$a_n^3 + b_n^3 = c_n^3 + (-1)^n. \tag{4}$$

Hirschhorn (1995) showed that

$$a_n = \tfrac{1}{85}\left[\left(64 + 8\sqrt{85}\right)\alpha^n + \left(64 - 8\sqrt{85}\right)\beta^n - 43(-1)^n\right] \tag{5}$$

$$b_n = \tfrac{1}{85}\left[\left(77 + 7\sqrt{85}\right)\alpha^n + \left(77 - 7\sqrt{85}\right)\beta^n - 16(-1)^n\right] \tag{6}$$

$$c_n = \tfrac{1}{85}\left[\left(93 + 9\sqrt{85}\right)\alpha^n + \left(93 - 9\sqrt{85}\right)\beta^n - 16(-1)^n\right], \tag{7}$$

where

$$\alpha = \tfrac{1}{2}\left(83 + 9\sqrt{85}\right) \tag{8}$$

$$\beta = \tfrac{1}{2}\left(83 - 9\sqrt{85}\right). \tag{9}$$

Hirschhorn (1996) showed that checking the first seven cases $n = 0$ to 6 is sufficient to prove the result.

References

Hirschhorn, M. D. "An Amazing Identity of Ramanujan." *Math. Mag.* **68**, 199–201, 1995.

Hirschhorn, M. D. "A Proof in the Spirit of Zeilberger of an Amazing Identity of Ramanujan." *Math. Mag.* **69**, 267–269, 1996.

Sloane, N. J. A. Sequences A051028, A051029, and A051030 in "An On-Line Version of the Encyclopedia of Integer Sequences." http://www.research.att.com/~njas/sequences/eisonline.html.

Ramanujan's Tau Function

TAU FUNCTION

Ramanujan's Tau-Dirichlet Series

TAU-DIRICHLET SERIES

Ramification Group

References

Koch, H. "Decomposition Group and Ramification Group." §6.1 in *Number Theory: Algebraic Numbers and Functions.* Providence, RI: Amer. Math. Soc., pp. 172–176, 2000.

Ramp Function

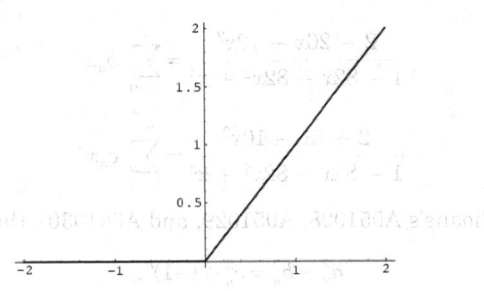

$$R(x) \equiv xH(x) \tag{1}$$

$$= \int_{-\infty}^{x} H(x') \, dx' \tag{2}$$

$$= \int_{-\infty}^{x} H(x')H(x-x') \, dx' \tag{3}$$

$$= H(x) * H(x), \tag{4}$$

where $H(x)$ is the HEAVISIDE STEP FUNCTION and $*$ is the CONVOLUTION. The DERIVATIVE is

$$R'(x) = -H(x). \tag{5}$$

The FOURIER TRANSFORM of the ramp function is given by

$$\mathscr{F}[R(x)] = \int_{-\infty}^{\infty} e^{-2\pi ikx} R(x) \, dx = \pi i\delta'(2\pi k) - \frac{1}{4\pi^2 k^2}, \tag{6}$$

where $\delta(x)$ is the DELTA FUNCTION and $\delta'(x)$ its DERIVATIVE.

See also FOURIER TRANSFORM–RAMP FUNCTION, HEAVISIDE STEP FUNCTION, RECTANGLE FUNCTION, SGN, SQUARE WAVE

Ramphoid Cusp

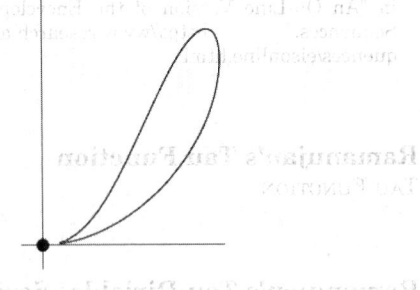

A type of CUSP as illustrated above for the curve $x^4 + x^2y^2 - 2x^2y - xy^2 + y^2 = 0$.

See also CUSP

References

Walker, R. J. *Algebraic Curves.* New York: Springer-Verlag, pp. 57–58, 1978.

Ramsey Number

The Ramsey number $R(m, n)$ gives the solution to the PARTY PROBLEM, which asks the minimum number of guests $R(m, n)$ that must be invited so that at least m will know each other or at least n will not know each other. In the language of GRAPH THEORY, the Ramsey number is the minimum number of vertices $v = R(m, n)$ such that all undirected simple graphs of order v contain a CLIQUE of order m or an INDEPENDENT SET of order n. RAMSEY'S THEOREM states that such a number exists for all m and n.

By symmetry, it is true that

$$R(m, n) = R(n, m). \tag{1}$$

It also must be true that

$$R(m, 2) = m. \tag{2}$$

A generalized Ramsey number is written

$$R(m_1, \ldots, m_k; n) \tag{3}$$

and is the smallest INTEGER r such that, no matter how each n-element SUBSET of an r-element SET is colored with k colors, there exists an i such that there is a SUBSET of size m_i, all of whose n-element SUBSETS are color i. The usual Ramsey numbers are then equivalent to $R(m, n) = R(m, n; 2)$.

Bounds are given by

$$R(k, l) \leq \begin{cases} R(k-1, l) + R(k, l-1) - 1 \\ \quad \text{for } R(k-1, 1) \text{ and } R(k, l-1) \text{ even} \\ R(k-1, l) + R(k, l-1) \\ \quad \text{otherwise} \end{cases} \tag{4}$$

and

$$R(k, k) \leq 4R(k-2, k) + 2 \tag{5}$$

(Chung and Grinstead 1983). Erdos proved that for diagonal Ramsey numbers $R(k, k)$,

$$\frac{k2^{k/2}}{e\sqrt{2}} < R(k, k). \tag{6}$$

This result was subsequently improved by a factor of 2 by Spencer (1975). $R(3, k)$ was known since 1980 to be bounded from above by $c_2 k^2/\ln k$, and Griggs (1983) showed that $c_2 = 5/12$ was an acceptable limit. J.-H. Kim (Cipra 1995) subsequently bounded $R(3, k)$ by a similar expression from below, so

$$c_1 \frac{k^2}{\ln k} \le R(3,\ k) \le c_2 \frac{k^2}{\ln k}. \qquad (7)$$

Burr (1983) gives Ramsey numbers for all 113 graphs with no more than 6 EDGES and no isolated points.

A summary of known results up to 1983 for $R(m,\ n)$ is given in Chung and Grinstead (1983). Radziszowski (1999) maintains an up-to-date list of the best current bounds, reproduced in part in the following table for $R(m,\ n;\ 2)$.

m	n	$R(m, n)$	Reference
3	3	6	Greenwood and Gleason 1955
3	4	9	Greenwood and Gleason 1955
3	5	14	Greenwood and Gleason 1955
3	6	18	Graver and Yackel 1968
3	7	23	Kalbfleisch 1966
3	8	28	McKay and Min 1992
3	9	36	Grinstead and Roberts 1982
3	10	[40, 43]	Exoo 1989, Radziszowski and Kreher 1988
3	11	[46, 51]	Radziszowski and Kreher 1988
3	12	[52, 60]	Exoo 1993, Radziszowski and Kreher 1988, Exoo 1998
3	13	[59, 69]	Piwakowski 1996, Radziszowski and Kreher 1988
3	14	[66, 78]	Exoo (unpub.), Radziszowski and Kreher 1988
3	15	[73, 89]	Wang and Wang 1989, Radziszowski (unpub.)
3	16	≥ 79	Wang and Wang 1989
3	17	≥ 92	WWY
3	18	≥ 98	WWY
3	19	≥ 106	WWY
3	20	≥ 109	WWY
3	21	≥ 122	WWY
3	22	≥ 125	WWY
3	23	≥ 136	WWY
3	26	≥ 150	
4	4	18	Greenwood and Gleason 1955
4	5	25	Mckay and Radziszowski 1995
4	6	[35, 41]	Ex8, MR4
4	7	[49, 61]	
4	8	[55, 84]	Exoo 1998
4	9	[69, 115]	
4	10	[80, 149]	
4	11	[96, 191]	
4	12	[128, 238]	
4	13	[131, 291]	
4	14	[136, 349]	
4	15	[145, 417]	
4	17	≥ 164	
4	18	≥ 182	
4	19	≥ 194	
4	20	≥ 230	
4	21	≥ 242	
4	22	≥ 282	
5	5	[43, 49]	Ex4, MR4
5	6	[58, 87]	Exoo 1993, Walker 1971
5	7	[80, 143]	
5	8	[95, 216]	
5	9	[116, 316]	Exoo 1998
5	10	[141, 442]	
5	11	≥ 153	
5	12	≥ 181	
5	13	≥ 193	
5	14	≥ 221	
5	15	≥ 237	
5	17	≥ 282	
5	19	≥ 338	
5	21	≥ 374	
5	22	≥ 410	
5	23	≥ 432	
5	26	≥ 464	
6	6	[102, 165]	Kalbfleisch 1965, Mac
6	7	[109, 298]	Exoo 1998
6	8	[122, 495]	Exoo 1998
6	9	[153, 780]	
6	10	[167, 1171]	
6	11	≥ 203	
6	12	≥ 224	
6	13	≥ 242	
6	14	≥ 258	
6	15	≥ 338	
6	17	≥ 500	
7	7	[205, 540]	Hill and Irving 1982, Giraud 1973
7	8	[1, 1031]	
7	9	[1, 1713]	
7	10	[1, 2826]	
7	17	≥ 548	
7	19	≥ 618	
7	20	≥ 648	
7	21	≥ 674	
8	8	[282, 1870]	

8	9	[1, 3583]
8	10	[1, 6090]
8	16	≥ 602
8	17	≥ 674
8	20	≥ 752
8	21	≥ 770
9	9	[565, 6588]
9	10	[1, 12677]
10	10	[798, 23581]	Guldan and Tomasta ?
11	11	[522, [522, ∞]]	Guldan and Tomasta ?

Known bounds for generalized Ramsey numbers (multicolor graph numbers) are given in the following table.

$R(\ldots; 2)$	Bounds	Reference
$R(3, 3, 3; 2)$	17	Greenwood and Gleason 1955
$R(3, 3, 3, 3; 2)$	[51, 64]	Chung 1973, Sanchez-Flores 1995
$R(3, 3, 3, 3, 3; 2)$	[162, 317]	
$R(3, 3, 3, 3, 3, 3; 2)$	[500, 1898]	Exoo 1994
$R(3, 3, 3, 4; 2)$	[91, 155]	Robertson 1999, Exoo 1998
$R(3, 3, 3, 5; 2)$	≥ 137	Robertson 1999
$R(3, 3, 3, 6; 2)$	≥ 165	Robertson 1999
$R(3, 3, 3, 7; 2)$	≥ 220	Robertson 1999
$R(3, 3, 3, 9; 2)$	≥ 336	Robertson 1999
$R(3, 3, 3, 11; 2)$	≥ 422	Robertson 1999
$R(3, 3, 4; 2)$	[30, 31]	
$R(3, 3, 4, 4; 2)$	≥ 144	
$R(3, 3, 5; 2)$	[45, 57]	
$R(3, 3, 6; 2)$	≥ 60	
$R(3, 3, 7; 2)$	≥ 72	
$R(3, 3, 9; 2)$	≥ 110	
$R(3, 3, 11; 2)$	≥ 141	
$R(3, 4, 5; 2)$	[80, 161]	Exoo 1998
$R(3, 4, 4; 2)$	[55, 79]	
$R(4, 4, 4; 2)$	[128, 236]	Hill and Irving 1982, Giraud 1973
$R(4, 4, 4, 4; 2)$	≥ 458	
$R(4, 4, 4, 4, 4; 2)$	≥ 942	
$R(5, 5, 5; 2)$	≥ 242	Robertson 1999
$R(6, 6, 6; 2)$	≥ 692	Robertson 1999

Known bounds for hypergraph Ramsey numbers are given in the following table.

$R(\ldots; 3)$	Bounds
$R(4, 4; 3)$	13
$R(4, 4, 4; 3)$	≥ 56
$R(4, 5; 3)$	≥ 33
$R(5, 5; 3)$	≥ 63

See also CLIQUE, CLIQUE NUMBER, COMPLETE GRAPH, EXTREMAL GRAPH, INDEPENDENCE NUMBER, INDEPENDENT SET, IRREDUNDANT RAMSEY NUMBER, RAMSEY'S THEOREM, RAMSEY THEORY, SCHUR NUMBER

References

Burr, S. A. "Generalized Ramsey Theory for Graphs--A Survey." In *Graphs and Combinatorics* (Ed. R. A. Bari and F. Harary). New York: Springer-Verlag, pp. 52–75, 1964.

Burr, S. A. "Diagonal Ramsey Numbers for Small Graphs." *J. Graph Th.* **7**, 57–69, 1983.

Chartrand, G. "The Problem of the Eccentric Hosts: An Introduction to Ramsey Numbers." §5.1 in *Introductory Graph Theory.* New York: Dover, pp. 108–115, 1985.

Chung, F. R. K. "On the Ramsey Numbers $N(3, 3, \ldots, 3; 2)$." *Discrete Math.* **5**, 317–321, 1973.

Chung, F. and Grinstead, C. G. "A Survey of Bounds for Classical Ramsey Numbers." *J. Graph. Th.* **7**, 25–37, 1983.

Cipra, B. "A Visit to Asymptopia Yields Insights into Set Structures." *Science* **267**, 964–965, 1995.

Exoo, G. "On Two Classical Ramsey Numbers of the Form $R(3, n)$." *SIAM J. Discrete Math.* **2**, 488–490, 1989.

Exoo, G. "Announcement: On the Ramsey Numbers $R(4, 6)$, $R(5, 6)$ and $R(3, 12)$." *Ars Combin.* **35**, 85, 1993.

Exoo, G. "A Lower Bound for Schur Numbers and Multicolor Ramsey Numbers of K_3." *Electronic J. Combinatorics* **1**, R8 1–3, 1994. http://www.combinatorics.org/Volume_1/volume1.html#R8.

Exoo, G. "Some New Ramsey Colorings." *Electronic J. Combinatorics* **5**, No. 1, R29, 1–5, 1998. http://www.combinatorics.org/Volume_5/v5i1toc.html.

Folkmann, J. "Notes on the Ramsey Number $N(3, 3, 3, 3)$." *J. Combinat. Theory. Ser. A* **16**, 371–379, 1974.

Fredricksen, H. "Schur Numbers and the Ramsey Numbers $N(3, 3, \ldots, 3; 2)$." *J. Combin. Theory Ser. A* **27**, 376–377, 1979.

Gardner, M. "Mathematical Games: In Which Joining Sets of Points by Lines Leads into Diverse (and Diverting) Paths." *Sci. Amer.* **237**, 18–28, 1977.

Gardner, M. *Penrose Tiles and Trapdoor Ciphers... and the Return of Dr. Matrix, reissue ed.* New York: W. H. Freeman, pp. 240–241, 1989.

Giraud, G. "Une minoration du nombre de quadrangles unicolores et son application a la majoration des nombres de Ramsey binaires bicolors." *C. R. Acad. Sci. Paris A* **276**, 1173–1175, 1973.

Graham, R. L.; Rothschild, B. L.; and Spencer, J. H. *Ramsey Theory, 2nd ed.* New York: Wiley, 1990.

Graver, J. E. and Yackel, J. "Some Graph Theoretic Results Associated with Ramsey's Theorem." *J. Combin. Th.* **4**, 125–175, 1968.

Greenwood, R. E. and Gleason, A. M. "Combinatorial Relations and Chromatic Graphs." *Canad. J. Math.* **7**, 1–7, 1955.

Griggs, J. R. "An Upper Bound on the Ramsey Numbers $R(3, k)$." *J. Comb. Th. A* **35**, 145–153, 1983.

Grinstead, C. M. and Roberts, S. M. "On the Ramsey Numbers $R(3, 8)$ and $R(3, 9)$." *J. Combinat. Th. Ser. B* **33**, 27–51, 1982.

Guldan, F. and Tomasta, P. "New Lower Bounds of Some Diagonal Ramsey Numbers." *J. Graph. Th.* **7**, 149–151, 1983.

Hanson, D. "Sum-Free Sets and Ramsey Numbers." *Discrete Math.* **14**, 57–61, 1976.

Harary, F. "Recent Results on Generalized Ramsey Theory for Graphs." In *Graph Theory and Applications: Proceedings of the Conference at Western Michigan University, Kalamazoo, Mich., May 10–13, 1972* (Ed. Y. Alavi, D. R. Lick, and A. T. White). New York: Springer-Verlag, pp. 125–138, 1972.

Hill, R. and Irving, R. W. "On Group Partitions Associated with Lower Bounds for Symmetric Ramsey Numbers." *European J. Combin.* **3**, 35–50, 1982.

Hoffman, P. *The Man Who Loved Only Numbers: The Story of Paul Erdos and the Search for Mathematical Truth.* New York: Hyperion, pp. 52–53, 1998.

Kalbfleisch, J. G. *Chromatic Graphs and Ramsey's Theorem.* Ph.D. thesis, University of Waterloo, January 1966.

McKay, B. D. and Min, Z. K. "The Value of the Ramsey Number $R(3, 8)$." *J. Graph Th.* **16**, 99–105, 1992.

McKay, B. D. and Radziszowski, S. P. "$R(4, 5) = 25$." *J. Graph. Th* **19**, 309–322, 1995.

Piwakowski, K. "Applying Tabu Search to Determine New Ramsey Numbers." *Electronic J. Combinatorics* **3**, R6 1–4, 1996. http://www.combinatorics.org/Volume_3/volume3.html#R6.

Radziszowski, S. P. "Small Ramsey Numbers." *Electronic J. Combin.* **1**, DS1 1–29, Rev. Jul. 5, 1999. http://www.combinatorics.org/Surveys/.

Radziszowski, S. and Kreher, D. L. "Upper Bounds for Some Ramsey Numbers $R(3, k)$." *J. Combinat. Math. Combin. Comput.* **4**, 207–212, 1988.

Robertson, A. "New Lower Bounds for Some Multicolored Ramsey Numbers." *Electronic J. Combinatorics* **6**, No. 1, R3, 1–6, 1999. http://www.combinatorics.org/Volume_6/v6i1toc.html.

Spencer, J. H. "Ramsey's Theorem--A New Lower Bound." *J. Combinat. Theory Ser. A* **18**, 108–115, 1975.

Wang, Q. and Wang, G. "New Lower Bounds for the Ramsey Numbers $R(3, q)$." *Beijing Daxue Xuebao* **25**, 117–121, 1989.

Whitehead, E. G. "The Ramsey Number $N(3, 3, 3, 3; 2)$." *Discrete Math.* **4**, 389–396, 1973.

Ramsey Theory

The mathematical study of combinatorial objects in which a certain degree of order must occur as the scale of the object becomes large. Ramsey theory is named after Frank Plumpton Ramsey, who did seminal work in this area before his untimely death at age 26 in 1930. The theory was subsequently developed extensively by Erdos.

The classical problem in Ramsey theory is the PARTY PROBLEM, which asks the minimum number of guests $R(m, n)$ that must be invited so that at least m will know each other (i.e., there exists a CLIQUE of order m) or at least n will not know each other (i.e., there exists an INDEPENDENT SET of order n. Here, $R(m, n)$ is called a RAMSEY NUMBER.

A typical result in Ramsey theory states that if some mathematical object is partitioned into finitely many parts, then one of the parts must contain a subobject

of an interesting kind. For example, it is known that if n is large enough and V is an n-dimensional VECTOR SPACE over the FIELD of integers (mod p), then however V is partitioned into r pieces, one of the pieces contains an affine subspace of dimension d.

See also EXTREMAL GRAPH THEORY, GRAHAM'S NUMBER, HAPPY END PROBLEM, PARTY PROBLEM, RAMSEY NUMBER, STRUCTURAL RAMSEY THEORY

References

Burr, S. A. "Generalized Ramsey Theory for Graphs--A Survey." In *Graphs and Combinatorics* (Ed. R. A. Bari and F. Harary). New York: Springer-Verlag, pp. 52–75, 1964.

Erdos, P. and Szekeres, G. "On Some Extremum Problems in Elementary Geometry." *Ann. Univ. Sci. Budapest Eotvos Soc. Math.* **3–4**, 53–62, 1961.

Graham, R. L. and Nesetril, J. "Ramsey Theory in the Work of Paul Erdos." In *The Mathematics of Paul Erdos* (Ed. R. L. Graham and J. Nesetril). Heidelberg, Germany: Springer-Verlag, 1996.

Hoffman, P. *The Man Who Loved Only Numbers: The Story of Paul Erdos and the Search for Mathematical Truth.* New York: Hyperion, pp. 51–57, 1998.

Ramsey's Theorem

A generalization of DILWORTH'S LEMMA. For each m, $n \in \mathbb{N}$ with $m, n \geq 2$, there exists a least INTEGER $R(m, n)$ (the RAMSEY NUMBER) such that no matter how the COMPLETE GRAPH $K_{R(m, n)}$ is two-colored, it will contain a green SUBGRAPH K_m or a red SUBGRAPH K_n. Furthermore,

$$R(m, n) \leq R(m - 1, n) + R(m, n - 1)$$

if $m, n \geq 3$.

The theorem can be equivalently stated that, for all $m \in \mathbb{N}$, there exists an $n \in \mathbb{N}$ such that any COMPLETE DIGRAPH on n VERTICES contains a COMPLETE TRANSITIVE SUBGRAPH of m VERTICES.

Ramsey's theorem is a generalization of the PIGEONHOLE PRINCIPLE since

$$R(\underbrace{2, 2, \ldots, 2}_{t}) = t + 1.$$

See also DILWORTH'S LEMMA, EXTREMAL GRAPH THEORY, GRAPH COLORING, NATURAL INDEPENDENCE PHENOMENON, PARTY PROBLEM, PIGEONHOLE PRINCIPLE, RAMSEY NUMBER, RAMSEY THEORY

References

Graham, R. L.; Rothschild, B. L.; and Spencer, J. H. *Ramsey Theory,* 2nd ed. New York: Wiley, 1990.

Spencer, J. "Large Numbers and Unprovable Theorems." *Amer. Math. Monthly* **90**, 669–675, 1983.

Ramus Tree

A type of BINARY TREE.

See also BINARY TREE

Randelbrot Set

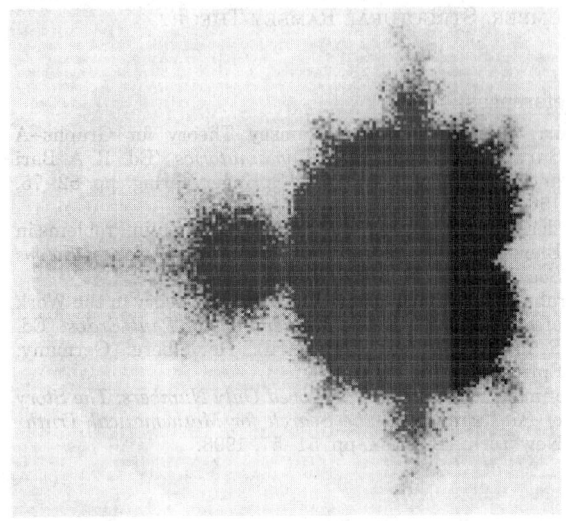

The FRACTAL-like figure obtained by performing the same iteration as for the MANDELBROT SET, but adding a random component R,

$$z_{n+1} = z_n^2 + c + R.$$

In the above plot, $R \equiv R_x + iR_y$, where $R_x, R_y \in [-0.05, 0.05]$.

See also MANDELBROT SET

References
Dickau, R. M. "Randelbrot Set." http://forum.swarthmore.edu/advanced/robertd/randelbrot.html.

Random Close Packing

Random close packing of spheres in three dimensions gives a PACKING DENSITY of only $\eta \approx 0.64$ (Jaeger and Nagel 1992), significantly smaller than the optimal PACKING DENSITY for cubic or hexagonal close packing of 0.74048.

See also SPHERE PACKING

References
--. *Nature* **239**, 488, 1972.
Jaeger, H. M. and Nagel, S. R. "Physics of Granular States." *Science* **255**, 1524, 1992.
Torquato, S.; Truskett, T. M.; and Debenedetti, P. G. "Is Random Close Packing of Spheres Well Defined?" *Phys. Lev. Lett.* **84**, 2064–2067, 2000.

Random Composition

A random composition of a number n in k parts is one of the $\binom{n+k-1}{n}$ possible COMPOSITIONS of n, where $\binom{n}{k}$ is a BINOMIAL COEFFICIENT. A random composition can be given by RandomComposition[n, k] in the *Mathematica* add-on package DiscreteMath`Combinatorica` (which can be loaded with the command << DiscreteMath`).

See also COMPOSITION

References
Nijenhuis, A. and Wilf, H. *Combinatorial Algorithms for Computers and Calculators, 2nd ed.* New York: Academic Press, 1978.
Skiena, S. "Random Partitions." §2.1.5 in *Implementing Discrete Mathematics: Combinatorics and Graph Theory with Mathematica.* Reading, MA: Addison-Wesley, pp. 58–59, 1990.

Random Distribution

A STATISTICAL DISTRIBUTION in which the variates occur with PROBABILITIES asymptotically matching their "true" underlying STATISTICAL DISTRIBUTION is said to be random.

See also RANDOM NUMBER, STATISTICAL DISTRIBUTION

Random Dot Stereogram

STEREOGRAM

Random Fibonacci Sequence

Consider the Fibonacci-like recurrence

$$a_n = \pm a_{n-1} \pm a_{n-2},$$

where $a_0 = 0$, $a_1 = 1$, and each sign is chosen independently and at random with probability 1/2. Surprisingly, Viswanath (2000) showed that

$$\lim_{n \to \infty} |a_n|^{1/n} = 1.13198824\ldots$$

with probability one. This constant can be numerically computed by computing the product of a certain set of RANDOM MATRICES, and taking the SPECTRAL NORM of the result (Viswanath 2000).

See also FIBONACCI NUMBER, RANDOM MATRIX

References
Viswanath, D. "Random Fibonacci Sequences and the Number 1.13198824...." *Math. Comput.* **69**, 1131–1155, 2000.

Random Graph

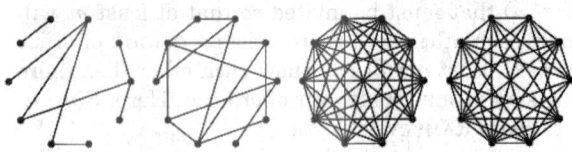

A random graph is a GRAPH in which properties such as the number of NODES, EDGES, and connections between them are determined in some random way. The graphs illustrated above are random graphs on 10 edges with edge probabilities distributed uniformly in [0, 1].

Erdos and Rényi (1960) showed that for many monotone-increasing properties of random graphs, graphs of a size slightly less than a certain threshold are very unlikely to have the property, whereas graphs with a few more EDGES are almost certain to have it. This is known as a PHASE TRANSITION (Janson *et al.* 2000, p. 103). Almost all graphs are connected and nonplanar (Skiena 1990, p. 156).

See also GRAPH, GRAPH THEORY, PHASE TRANSITION

References

Bollobás, B. *Graph Theory: An Introductory Course.* New York: Springer-Verlag, 1979.

Bollobás, B. *Random Graphs.* London: Academic Press, 1985.

Erdos, P. and Rényi, A. "On the Evolution of Random Graphs." *Publ. Math. Inst. Hungar. Acad. Sci.* **5**, 17–61, 1960.

Erdos, P. and Spencer, J. *Probabilistic Methods in Combinatorics.* New York: Academic Press, 1974.

Janson, S.; Luczak, T.; and Rucinski, A. *Random Graphs.* New York: Wiley, 2000.

Kolchin, V. F. *Random Graphs.* New York: Cambridge University Press, 1998.

Palmer, E. M. *Graphical Evolution: An Introduction to the Theory of Random Graphs.* New York: Wiley, 1985.

Skiena, S. "Random Graphs." *Implementing Discrete Mathematics: Combinatorics and Graph Theory with Mathematica.* Reading, MA: Addison-Wesley, pp. 154–160, 1990.

Steele, J. M. "Gibbs' Measures on Combinatorial Objects and the Central Limit Theorem for an Exponential Family of Random Trees." *Prob. Eng. Inform. Sci.* **1**, 47–59, 1987.

Random Matrix

A random matrix is a MATRIX of given type and size whose entries consist of random numbers from some specified distribution.

If n matrices M_i are chosen with probability 1/2 from one of

$$M_+ = \begin{bmatrix} 0 & 1 \\ 1 & 1 \end{bmatrix} \tag{1}$$

$$M_- = \begin{bmatrix} 0 & 1 \\ 1 & -1 \end{bmatrix}, \tag{2}$$

then

$$\lim_{n \to \infty} \frac{\ln\|M_1 \cdots M_n\|}{n} = c, \tag{3}$$

where $e^c = 1.13198824\ldots$ and $\|M\|$ denotes the matrix SPECTRAL NORM (Bougerol and Lacroix 1985, pp. 11 and 157; Viswanath 2000). This is the same constant appearing in the RANDOM FIBONACCI SEQUENCE. The following *Mathematica* code can be used to estimate this constant.

```
n = 100000;
m = Fold[Dot, IdentityMatrix[2],
        {{0,  1}, {1, #}} & /@ ((-
1)^Table[Random[Integer], {n}])
  ]//N;
  Log[Sqrt[Max[Eigenvalues[Transpose[m].m]]]]/
n
```

See also COMPLEX MATRIX, MATRIX, RANDOM FIBONACCI SEQUENCE, REAL MATRIX

References

Bougerol, P. and Lacroix, J. *Random Products of Matrices with Applications to Schrödinger Operators.* Basel, Switzerland: Birkhäuser 1985.

Chassaing, P.; Letac, G.; and Mora, M. "Brocot Sequences and Random Walks on $SL_2(R)$." In *Probability Measures on Groups VII* (Ed. H. Heyer). New York Springer-Verlag, pp. 36–48, 1984.

Furstenberg, H. "Non-Commuting Random Products." *Trans. Amer. Math. Soc.* **108**, 377–428, 1963.

Furstenberg, H. and Kesten, H. "Products of Random Matrices." *Ann. Math. Stat.* **31**, 457–469, 1960.

Katz, M. and Sarnak, P. *Random Matrices, Frobenius Eigenvalues, and Monodromy.* Providence, RI: Amer. Math. Soc., 1999.

Mehta, M. L. *Random Matrices, 2nd rev. enl. ed.* New York: Academic Press, 1991.

Viswanath, D. "Random Fibonacci Sequences and the Number 1.13198824...." *Math. Comput.* **69**, 1131–1155, 2000.

Random Normal Deviates

NORMAL DEVIATES

Random Number

Computer-generated random numbers are sometimes called PSEUDORANDOM NUMBERS, while the term "random" is reserved for the output of unpredictable physical processes. When used without qualification, the word "random" usually means "random with a UNIFORM DISTRIBUTION." Other distributions are, of course possible. For example, the BOX-MULLER TRANSFORMATION allows random numbers with a 2-D uniform distribution to be transformed to corresponding random numbers with a 2-D Gaussian distribution. Similarly, in order to generate a power-law distribution $P(x)$ from a uniform distribution $P(y)$, write $P(x) = Cx^n$ for $x \in [x_0, x_1]$. Then normalization gives

$$\int_{x_0}^{x_1} P(x)\,dx = c\,\frac{[x^{n+1}]_{x_0}^{x_1}}{n+1} = 1, \tag{1}$$

so

$$C = \frac{n+1}{x_1^{n+1} - x_0^{n+1}}. \tag{2}$$

Let y be a uniformly distributed variate on [0, 1]. Then

$$D(x) = \int_{x_0}^{x} P(x') \, dx' = C \int_{x_0}^{x} x'^n \, dx'$$

$$= \frac{C}{n+1}(x^{n+1} - x_0^{n+1}) \equiv y, \qquad (3)$$

and the variate given by

$$x = \left(\frac{n+1}{C} y + x_0^{n+1}\right)^{1/(n+1)}$$

$$= \left[(x_1^{n+1} - x_0^{n+1})y + x_0^{n+1}\right]^{1/(n+1)} \qquad (4)$$

is distributed as $P(x)$.

It is impossible to produce an arbitrarily long string of random digits and prove it is random. Strangely, it is very difficult for humans to produce a string of random digits, and computer programs can be written which, on average, actually predict some of the digits humans will write down based on previous ones.

The LINEAR CONGRUENCE METHOD is one algorithm for generating PSEUDORANDOM NUMBERS. The initial number used as the starting point in a random number generating algorithm is known as the SEED. The goodness of random numbers generated by a given ALGORITHM can be analyzed by examining its NOISE SPHERE.

When generating random numbers over some specified boundary, it is often necessary to normalize the distributions so that each differential area can is equally populated. For example, picking θ and ϕ from uniform distributions *does not* give a uniform distribution for SPHERE POINT PICKING.

See also BAYS' SHUFFLE, BOX-MULLER TRANSFORMATION, CLIFF RANDOM NUMBER GENERATOR, QUASIRANDOM SEQUENCE, RANDOM VARIABLE, SCHRAGE'S ALGORITHM, STOCHASTIC, UNIFORM DISTRIBUTION

References

Bassein, S. "A Sampler of Randomness." *Amer. Math. Monthly* **103**, 483–490, 1996.

Bennett, D. J. *Randomness.* Cambridge, MA: Harvard University Press, 1998.

Bratley, P.; Fox, B. L.; and Schrage, E. L. *A Guide to Simulation, 2nd ed.* New York: Springer-Verlag, 1996.

Dahlquist, G. and Bjorck, A. Ch. 11 in *Numerical Methods.* Englewood Cliffs, NJ: Prentice-Hall, 1974.

Deak, I. *Random Number Generators and Simulation.* New York: State Mutual Book & Periodical Service, 1990.

Forsythe, G. E.; Malcolm, M. A.; and Moler, C. B. Ch. 10 in *Computer Methods for Mathematical Computations.* Englewood Cliffs, NJ: Prentice-Hall, 1977.

Gardner, M. "Random Numbers." Ch. 13 in *Mathematical Carnival: A New Round-Up of Tantalizers and Puzzles from Scientific American.* New York: Vintage, pp. 161–172, 1977.

James, F. "A Review of Pseudorandom Number Generators." *Computer Physics Comm.* **60**, 329–344, 1990.

Kac, M. "What is Random?" *Amer. Sci.* **71**, 405–406, 1983.

Kenney, J. F. and Keeping, E. S. *Mathematics of Statistics, Pt. 1, 3rd ed.* Princeton, NJ: Van Nostrand, pp. 200–201 and 205–207, 1962.

Kenney, J. F. and Keeping, E. S. *Mathematics of Statistics, Pt. 2, 2nd ed.* Princeton, NJ: Van Nostrand, pp. 151–154, 1951.

Knuth, D. E. Ch. 3 in *The Art of Computer Programming, Vol. 2: Seminumerical Algorithms, 3rd ed.* Reading, MA: Addison-Wesley, 1998.

Marsaglia, G. "A Current View of Random Number Generators." In *Computer Science and Statistics: Proceedings of the Symposium on the Interface, 16th, Atlanta, Georgia, March 1984* (Ed. L. Billard). New York: Elsevier, 1985.

Marsaglia, G. "DIEHARD: A Battery of Tests for Random Number Generators." http://stat.fsu.edu/~geo/diehard.html.

Mascagni, M. "Random Numbers on the Web." http://www.ncsa.uiuc.edu/Apps/CMP/RNG/mascagni/www-rng.html.

Nijenhuis, A. and Wilf, H. *Combinatorial Algorithms for Computers and Calculators, 2nd ed.* New York: Academic Press, 1978.

Park, S. and Miller, K. "Random Number Generators: Good Ones are Hard to Find." *Comm. ACM* **31**, 1192–1201, 1988.

Peterson, I. *The Jungles of Randomness: A Mathematical Safari.* New York: Wiley, 1997.

Pickover, C. A. "Computers, Randomness, Mind, and Infinity." Ch. 31 in *Keys to Infinity.* New York: W. H. Freeman, pp. 233–247, 1995.

Press, W. H.; Flannery, B. P.; Teukolsky, S. A.; and Vetterling, W. T. "Random Numbers." Ch. 7 in *Numerical Recipes in FORTRAN: The Art of Scientific Computing, 2nd ed.* Cambridge, England: Cambridge University Press, pp. 266–306, 1992.

Schrage, L. "A More Portable Fortran Random Number Generator." *ACM Trans. Math. Software* **5**, 132–138, 1979.

Schroeder, M. "Random Number Generators." In *Number Theory in Science and Communication, with Applications in Cryptography, Physics, Digital Information, Computing and Self-Similarity, 3rd ed.* New York: Springer-Verlag, pp. 289–295, 1990.

Weisstein, E. W. "Books about Randomness." http://www.treasure-troves.com/books/Randomness.html.

Wilf, H. S. *Combinatorial Algorithms: An Update.* Philadelphia, PA: SIAM, 1989.

Random Partition

A random partition of a number n is one of the $P(n)$ possible PARTITIONS of n, where $P(n)$ is the PARTITION FUNCTION P. A random partition can be given by `RandomPartition[n]` in the *Mathematica* add-on package `DiscreteMath`Combinatorica`` (which can be loaded with the command `<<DiscreteMath``).

See also PARTITION

References

Nijenhuis, A. and Wilf, H. *Combinatorial Algorithms for Computers and Calculators, 2nd ed.* New York: Academic Press, 1978.

Skiena, S. "Random Partitions." §2.1.5 in *Implementing Discrete Mathematics: Combinatorics and Graph Theory with Mathematica.* Reading, MA: Addison-Wesley, pp. 58–59, 1990.

Random Percolation

PERCOLATION THEORY

Random Permutation

A PERMUTATION containing a fixed number n of a random selection from a given set of elements. There are two main algorithms for constructing random permutations. The first constructs a vector of random real numbers and uses them as keys to records containing the integers 1 to n. The second starts with an arbitrary permutation and then exchanges the ith element with a randomly selected one from the first i elements for $i = 1, ..., n$ (Skiena 1990).

There are an average of $n(n-1)/4$ PERMUTATION INVERSIONS in a PERMUTATION on n elements (Skiena 1990, p. 29).

See also PERMUTATION, PERMUTATION INVERSION

References

Moses, L. E. and Oakford, R. V. *Tables of Random Permutations.* Stanford, CA: Stanford University Press, 1963.
Skiena, S. "Random Permutations." §1.1.3 in *Implementing Discrete Mathematics: Combinatorics and Graph Theory with Mathematica.* Reading, MA: Addison-Wesley, pp. 6–9 and 29, 1990.

Random Polygon

A random polygon is a POLYGON generated in some random way. Kendall conjectured that the shape of a random polygon is close to a DISK as the area of the polygon becomes large (Stoyan *et al.* 1987, Kovalenko 1999)

See also CROFTON CELL

References

Kovalenko, I. N. "Proof of David Kendall's Conjecture Concerning the Shape of Large Random Polygons." *Cybern. Sys. Anal.* **33**, 461–467, 1997.
Kovalenko, I. N. "A Simplified Proof of a Conjecture of D. G. Kendall Concerning Shapes of Random Polygons." *J. Appl. Math. Stoch. Anal.* **12**, 301–310, 1999.
Miles, R. E. "A Heuristic Proof of a Long-Standing Conjecture of D. G. Kendall Concerning the Shapes of Certain Large Random Polygons." *Adv. Appl. Prob. (SGSA)* **27**, 397–471, 1997.
Stoyan, D.; Kendall, W. S.; and Mecke, J. *Stochastic Geometry and Its Applications, with a Foreword by D. G. Kendall.* New York: Wiley, 1987.

Random Polynomial

A POLYNOMIAL having random COEFFICIENTS.

See also KAC FORMULA

References

Bharucha-Reid, A. T. and Sambandham, M. *Random Polynomials.* New York: Academic Press, 1986.
Bloch, A. and Pólya, G. "On the Zeros of Certain Algebraic Equations." *Proc. London Math. Soc.* **33**, 102–114, 1932.
Edelman, A. and Kostlan, E. "How Many Zeros of a Random Polynomial are Real?" *Bull. Amer. Math. Soc.* **32**, 1–37, 1995.
Erdos, P. and Turán, P. "On the Distribution of Roots of Polynomials." *Ann. Math.* **51**, 105–119, 1950.
Hammersley, J. "The Zeros of a Random Polynomial." *Proc. Third Berkeley Symp. Math. Stat. Prob.* **2**, 89–111, 1956.
Kac, M. "On the Average Number of Real Roots of a Random Algebraic Equation." *Bull. Amer. Math. Soc.* **49**, 314–320, 1943.
Kac, M. "A Correction to 'On the Average Number of Real Roots of a Random Algebraic Equation'." *Bull. Amer. Math. Soc.* **49**, 938, 1943.
Kostan, E. "On the Distribution of Roots in a Random Polynomial." Ch. 38 in *From Topology to Computation: Proceedings of the Smalefest* (Ed. M. W. Hirsch, J. E. Marsden, and M. Shub). New York: Springer-Verlag, pp. 419–431, 1993.
Littlewood, J. and Offord, A. "On the Number of Real Roots of a Random Algebraic Equation." *J. London Math. Soc.* **13**, 288–295, 1938.
Maslova, N. "On the Distribution of the Number of Reals Roots of a Random Polynomial" [In Russian]. *Teor. Veroyatnost. i Primenen* **19**, 488–500, 1974.
Rice, S. O. "The Distribution of the Maxima of a Random Curve." *Amer. J. Math.* **61**, 409–416, 1939.
Rice, S. O. "Mathematical Analysis of Random Noise." *Bell Syst. Tech. J.* **24**, 45–156, 1945.

Random Tableau

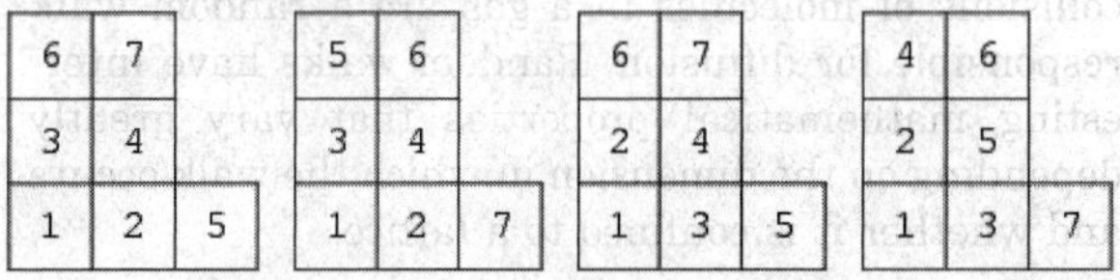

A YOUNG TABLEAU chosen at random from those having a given shape. A random tableau can be generated by RandomTableau[*shape*] in the *Mathematica* add-on package DiscreteMath`Combinatorica` (which can be loaded with the command << DiscreteMath`). The figure above shows four random tableaux of the 21 distinct ones of shape $\{3, 2, 2\}$.

See also YOUNG TABLEAU

References

Nijenhuis, A. and Wilf, H. *Combinatorial Algorithms for Computers and Calculators, 2nd ed.* New York: Academic Press, 1978.
Skiena, S. "Random Tableaux." §2.3.5 in *Implementing Discrete Mathematics: Combinatorics and Graph Theory with Mathematica.* Reading, MA: Addison-Wesley, pp. 72–73, 1990.

Random Variable

A random variable is a measurable function from a PROBABILITY SPACE $(S, \mathbb{S}, P)$ into a MEASURABLE SPACE $(S', \mathbb{S}')$ known as the STATE SPACE (Doob 1996). Papoulis (1984, p. 88) gives the slightly different definition of a random variable X as a REAL FUNCTION whose domain is the PROBABILITY SPACE S and such that:

1. The set $\{X \le x\}$ is an EVENT for any real number x.

2. The probability of the events $\{X = +\infty\}$ and $\{X = -\infty\}$ equals zero.

The abbreviation "r.v." is sometimes used to denote a random variable.

See also PROBABILITY SPACE, RANDOM DISTRIBUTION, RANDOM NUMBER, STATE SPACE, VARIATE

References

Doob, J. L. "The Development of Rigor in Mathematical Probability (1900–1950)." *Amer. Math. Monthly* **103**, 586–595, 1996.
Gikhman, I. I. and Skorokhod, A. V. *Introduction to the Theory of Random Processes.* New York: Dover, 1997.
Papoulis, A. "The Concept of a Ransom Variable." Ch. 4 in *Probability, Random Variables, and Stochastic Processes, 2nd ed.* New York: McGraw-Hill, pp. 83–115, 1984.

Random Walk

A random process consisting of a sequence of discrete steps of fixed length. The random thermal perturbations in a liquid are responsible for a random walk phenomenon known as Brownian motion, and the collisions of molecules in a gas are a random walk responsible for diffusion. Random walks have interesting mathematical properties that vary greatly depending on the dimension in which the walk occurs and whether it is confined to a lattice.

See also MARKOV CHAIN, MARTINGALE, PERCOLATION THEORY, RANDOM WALK–1-D, RANDOM WALK–2-D, RANDOM WALK–3-D, SELF-AVOIDING WALK, SELF-AVOIDING WALK CONNECTIVE CONSTANT

References

Barber, M. N. and Ninham, B. W. *Random and Restricted Walks: Theory and Applications.* New York: Gordon and Breach, 1970.
Chandrasekhar, S. In *Selected Papers on Noise and Stochastic Processes* (Ed. N. Wax). New York: Dover, 1954.
Doyle, P. G. and Snell, J. L. *Random Walks and Electric Networks.* Washington, DC: Math. Assoc. Amer, 1984.
Dykin, E. B. and Uspenskii, V. A. *Random Walks.* New York: Heath, 1963.
Erdos, P. and Révész, P. "Three Problems on the Random Walk in $\mathbb{Z}^d$." *Studia Sci. Math. Hung.* **26**, 309–320, 1991.
Feller, W. *An Introduction to Probability Theory and Its Applications, Vol. 1, 3rd ed.* New York: Wiley, 1968.
Feller, W. *An Introduction to Probability Theory and Its Applications, Vol. 2, 3rd ed.* New York: Wiley, 1971.
Gardner, M. "Random Walks and Gambling" and "Random Walks on the Plane and in Space." Chs. 6–7 in *Mathematical Circus: More Puzzles, Games, Paradoxes, and Other Mathematical Entertainments.* Washington, DC: Math. Assoc. Amer., pp. 66–86, 1992.
Hughes, B. D. *Random Walks and Random Environments, Vol. 1: Random Walks.* New York: Oxford University Press, 1995.
Hughes, B. D. *Random Walks and Random Environments, Vol. 2: Random Environments.* New York: Oxford University Press, 1996.
Lawler, G. F. *Intersections of Random Walks.* Boston, MA: Birkhäuser, 1996.
Révész, P. *Random Walks in Random and Non-Random Environments.* Singapore: World Scientific, 1990.
Spitzer, F. *Principles of Random Walk, 2nd ed.* New York: Springer-Verlag, 1976.
Weiss, G. *Aspects and Applications of the Random Walk.* Amsterdam, Netherlands: North-Holland, 1994.
Weisstein, E. W. "Books about Random Walks." http://www.treasure-troves.com/books/RandomWalks.html.

Random Walk—1-D

Let N steps of equal length be taken along a LINE. Let p be the probability of taking a step to the right, q the probability of taking a step to the left, n_1 the number of steps taken to the right, and n_2 the number of steps taken to the left. The quantities p, q, n_1, n_2, and N are related by

$$p + q = 1 \tag{1}$$

and

$$n_1 + n_2 = N. \tag{2}$$

Now examine the probability of taking exactly n_1 steps out of N to the right. There are $\binom{N}{n_1} = \binom{n_1 + n_2}{n_1}$ ways of taking n_1 steps to the right and n_2 to the left, where $\binom{n}{m}$ is a BINOMIAL COEFFICIENT. The probability of taking a particular ordered sequence of n_1 and n_2 steps is $p^{n_1} q^{n_2}$. Therefore,

$$P(n_1) = \frac{(n_1 + n_2)!}{n_1! n_2!} p^{n_1} q^{n_2} = \frac{N!}{n_1!(N - n_1)!} p^{n_1} q^{N - n_1}, \tag{3}$$

where $n!$ is a FACTORIAL. This is a BINOMIAL DISTRIBUTION and satisfies

$$\sum_{n_1 = 0}^{N} P(n_1) = (p + q)^N = 1^N = 1. \tag{4}$$

The MEAN number of steps n_1 to the right is then

$$\langle n_1 \rangle \equiv \sum_{n_1 = 0}^{N} n_1 P(n_1)$$

$$= \sum_{n_1 = 0}^{N} \frac{N!}{n_1!(N - n_1)!} p^{n_1} q^{N - n_1} n_1, \tag{5}$$

but

$$n_1 p^{n_1} = p \frac{\partial}{\partial p} p^{n_1}, \tag{6}$$

so

$$\langle n_1 \rangle = \sum_{n_1 = 0}^{N} \frac{N!}{n_1!(N - n_1)!} \left(p \frac{\partial}{\partial p} p^{n_1} \right) q^{N - n_1}$$

$$= p \frac{\partial}{\partial p} \sum_{n_1 = 0}^{N} \frac{N!}{n_1!(N - n_1)!} p^{n_1} q^{N - n_1}$$

$$=p\,\frac{\partial}{\partial p}(p+q)^N=pN(p+q)^{N-1}=pN. \quad (7)$$

From the BINOMIAL THEOREM,

$$\langle n_2\rangle=N-\langle n_1\rangle=N(1-p)=qN. \quad (8)$$

The VARIANCE is given by

$$\sigma_{n_1}^2=\langle n_1^2\rangle-\langle n_1\rangle^2. \quad (9)$$

But

$$\langle n_1^2\rangle=\sum_{n_1=0}^{N}\frac{N!}{n_1!(N-n_1)!}\,p^{n_1}q^{N-n_1}n_1^2, \quad (10)$$

so

$$n_1^2 p^{n_1}=n_1\left(p\,\frac{\partial}{\partial p}\right)p^{n_1}=\left(p\,\frac{\partial}{\partial p}\right)^2 p^{n_1}, \quad (11)$$

and

$$\langle n_1^2\rangle=\sum_{n_1=0}^{N}\frac{N!}{n_1!(N-n_1)!}\left(p\,\frac{\partial}{\partial p}\right)^2 p^{n_1}q^{N-n_1}$$

$$=\left(p\,\frac{\partial}{\partial p}\right)^2\sum_{n_1=0}^{N}\frac{N!}{n_1!(N-n_1)!}\,p^{n_1}q^{N-n_1}$$

$$=\left(p\,\frac{\partial}{\partial p}\right)^2(p+q)^N$$

$$=p\,\frac{\partial}{\partial p}[pN(p+q)^{N-1}]$$

$$=p[N(p+q)^{N-1}+pN(N-1)(p+q)^{N-2}]$$

$$=p[N+pN(N-1)]$$

$$=pN[1+pN-p]=(Np)^2+Npq$$

$$=\langle n_1\rangle^2+Npq. \quad (12)$$

Therefore,

$$\sigma_{n_1}^2=\langle n_1^2\rangle-\langle n_1\rangle^2=Npq, \quad (13)$$

and the ROOT-MEAN-SQUARE deviation is

$$\sigma_{n_1}=\sqrt{Npq}. \quad (14)$$

For a large number of total steps N, the BINOMIAL DISTRIBUTION characterizing the distribution approaches a GAUSSIAN DISTRIBUTION.

Consider now the distribution of the distances d_N traveled after a given number of steps,

$$d_N\equiv n_1-n_2=2n_1-N, \quad (15)$$

as opposed to the *number* of steps in a given direction. The above plots show $d_N(p)$ for $N=200$ and three values $p=0.1$, $p=0.5$, and $p=0.9$, respectively. Clearly, weighting the steps toward one direction or the other influences the overall trend, but there is still a great deal of random scatter, as emphasized by the plot below, which shows three random walks all with $p=0.5$.

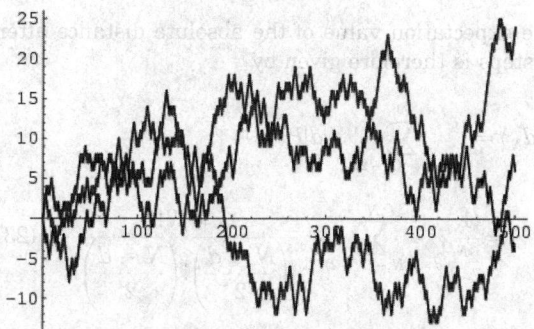

Surprisingly, the most probable number of sign changes in a walk is 0, followed by 1, then 2, etc.

For a random walk with $p=1/2$, the probability $P_N(d)$ of traveling a given distance d after N steps is given in the following table.

steps	−5	−4	−3	−2	−1	0	1	2	3	4	5
0						1					
1					$\frac{1}{2}$	0	$\frac{1}{2}$				
2				$\frac{1}{4}$	0	$\frac{2}{4}$	0	$\frac{1}{4}$			
3			$\frac{1}{8}$	0	$\frac{3}{8}$	0	$\frac{3}{8}$	0	$\frac{1}{8}$		
4		$\frac{1}{16}$	0	$\frac{4}{16}$	0	$\frac{6}{16}$	0	$\frac{4}{16}$	0	$\frac{1}{16}$	
5	$\frac{1}{32}$	0	$\frac{5}{32}$	0	$\frac{10}{32}$	0	$\frac{10}{32}$	0	$\frac{5}{32}$	0	$\frac{1}{32}$

In this table, subsequent rows are found by adding HALF of each cell in a given row to each of the two cells diagonally below it. In fact, it is simply PASCAL'S TRIANGLE padded with intervening zeros and with each row multiplied by an additional factor of 1/2. The COEFFICIENTS in this triangle are given by

$$P_N(d)=\frac{1}{2^N}\binom{N}{\dfrac{d+N}{2}} \quad (16)$$

(Papoulis 1984, p. 291). The moments

$$\mu_p=\sum_{d=-N,\,-(N-2),\,...,\,N}d^p P_N(d) \quad (17)$$

of this distribution of *signed* distances are then given by

$$\mu=0 \quad (18)$$

$$\mu_2 = N \tag{19}$$

$$\mu_3 = 0 \tag{20}$$

$$\mu_4 = N(3N - 2), \tag{21}$$

so the MEAN is $\mu = 0$, the SKEWNESS is $\gamma_1 = 0$, and the KURTOSIS is

$$\gamma_2 = \frac{\mu_4}{\mu_2^2} - 3 = -\frac{2}{N}. \tag{22}$$

The expectation value of the absolute distance after N steps is therefore given by

$$\langle d_N \rangle = \sum_{d = -N, -(N-2), \ldots}^{N} |d| P_N(d)$$

$$= \frac{1}{2^N} \sum_{d = -N, -(N-2), \ldots}^{N} \frac{|d| N!}{\left(\dfrac{N+d}{2}\right)! \left(\dfrac{N-d}{2}\right)!}. \tag{23}$$

This sum can be done symbolically by separately considering the cases N EVEN and N ODD. First, consider EVEN N so that $N \equiv 2J$. Then

$$\langle d_{2,J} \rangle = \frac{N!}{2^N} \Bigg[\sum_{\substack{d = -2J, \\ -2(J-1), \ldots}}^{-2} \frac{|d|}{\left(\dfrac{2J+d}{2}\right)! \left(\dfrac{2J-d}{2}\right)!}$$

$$+ \sum_{d=0} \frac{|d|}{\left(\dfrac{2J+d}{2}\right)! \left(\dfrac{2J-d}{2}\right)!}$$

$$+ \sum_{d=2,4,\ldots}^{2J} \frac{|d|}{\left(\dfrac{2J+d}{2}\right)! \left(\dfrac{2J-d}{2}\right)!} \Bigg]$$

$$= \frac{N!}{2^N} \Bigg[\sum_{\substack{d = -J, \\ -(J-1), \ldots}}^{-1} \frac{|2d|}{\left(\dfrac{2J+2d}{2}\right)! \left(\dfrac{2J-2d}{2}\right)!}$$

$$+ \sum_{d=1,2 \ldots}^{J} \frac{|2d|}{\left(\dfrac{2J+2d}{2}\right)! \left(\dfrac{2J-2d}{2}\right)!} \Bigg]$$

$$= \frac{N!}{2^N} \Bigg[2 \sum_{d=1}^{J} \frac{2d}{(J+d)!(J-d)!} \Bigg]$$

$$= \frac{N!}{2^{N-2}} \sum_{d=1}^{J} \frac{d}{(J+d)!(J-d)!}. \tag{24}$$

But this sum can be evaluated analytically as

$$\sum_{d=1}^{J} \frac{d}{(J+d)!(J-d)!} = \frac{1}{2\Gamma(J)\Gamma(1+J)}. \tag{25}$$

Writing $J = N/2$, plugging back in, and simplifying gives

$$\langle d_{N\ \text{even}} \rangle = \frac{2}{\sqrt{\pi}} \frac{\Gamma\left(\frac{1}{2} + \frac{1}{2}N\right)}{\Gamma\left(\frac{1}{2}N\right)} = \frac{(N-1)!!}{(N-2)!!}, \tag{26}$$

where $N!!$ is the DOUBLE FACTORIAL.

Now consider N ODD, so $N \equiv 2J - 1$. Then

$$\langle d_{N\ \text{odd}} \rangle = \langle d_{2J-1} \rangle$$

$$= \frac{N!}{2^N} \Bigg[\sum_{\substack{d = -(2J-1), \\ -(2J+1), \ldots}}^{-1} \frac{|d|}{\left(\dfrac{2J-1+d}{2}\right)! \left(\dfrac{2J-1-d}{2}\right)!}$$

$$+ \sum_{d=1,3,\ldots}^{2J-1} \frac{|d|}{\left(\dfrac{2J-1+d}{2}\right)! \left(\dfrac{2J-1-d}{2}\right)!} \Bigg]$$

$$= \frac{N!}{2^{N-1}} \Bigg[\sum_{d=1,3,\ldots}^{2J-1} \frac{d}{\left(\dfrac{2J-1+d}{2}\right)! \left(\dfrac{2J-1-d}{2}\right)!} \Bigg]$$

$$= \frac{N!}{2^{N-1}} \Bigg[\sum_{d=2,4,\ldots}^{2J} \frac{d-1}{\left(\dfrac{2J-2+d}{2}\right)! \left(\dfrac{2J-d}{2}\right)!} \Bigg]$$

$$= \frac{N!}{2^{N-1}} \Bigg[\sum_{d=1}^{J} \frac{2d-1}{(J+d-1)!(J-d)!} \Bigg]. \tag{27}$$

But this sum can be evaluated analytically as

$$\sum_{d=1}^{J} \frac{2d-1}{(J+d-1)!(J-d)!} = \frac{1}{[\Gamma(J)]^2}. \tag{28}$$

Writing $J = (N+1)/2$, plugging back in, and simplifying gives

$$\langle d_{N\ \text{odd}} \rangle = \frac{N!}{2^{N-1} \left[\Gamma\left(\frac{1}{2} + \frac{1}{2}N\right) \right]^2}$$

$$= \frac{2}{\sqrt{\pi}} \frac{\Gamma\left(\frac{1}{2}N + 1\right)}{\Gamma\left(\frac{1}{2}N + \frac{1}{2}\right)} = \frac{N!!}{(N-1)!!}. \tag{29}$$

Both the EVEN and ODD solutions can be written in terms of J as

$$\langle d_J \rangle = \frac{2}{\sqrt{\pi}} \frac{\Gamma\left(J + \frac{1}{2}\right)}{\Gamma(J)} = \frac{(2J-1)!!}{(2J-2)!!}, \tag{30}$$

or explicitly in terms of N as

$$\langle d_N \rangle = \begin{cases} \dfrac{(N-1)!!}{(N-2)!!} & \text{for } N \text{ even} \\[2mm] \dfrac{N!!}{(N-1)!!} & \text{for } N \text{ odd.} \end{cases} \qquad (31)$$

The first few values of $\langle d_N \rangle$ are therefore

$$\langle d_0 \rangle = 0$$

$$\langle d_1 \rangle = \langle d_2 \rangle = 1$$

$$\langle d_3 \rangle = \langle d_4 \rangle = \tfrac{3}{2}$$

$$\langle d_5 \rangle = \langle d_6 \rangle = \tfrac{15}{8}$$

$$\langle d_7 \rangle = \langle d_8 \rangle = \tfrac{35}{16}$$

$$\langle d_9 \rangle = \langle d_{10} \rangle = \tfrac{315}{128}$$

$$\langle d_{11} \rangle = \langle d_{12} \rangle = \tfrac{693}{256}$$

$$\langle d_{13} \rangle = \langle d_{14} \rangle = \tfrac{3003}{1024}$$

(Sloane's A001803 and A046161; Abramowitz and Stegun 1972, Prévost 1933, Hughes 1995), which are also given by the GENERATING FUNCTION

$$(1-x)^{-3/2} = 1 + \tfrac{3}{2}x + \tfrac{15}{8}x^2 + \tfrac{35}{16}x^3 + \tfrac{315}{128}x^4 + \dots. \qquad (32)$$

These numbers also arise in the HEADS-MINUS-TAILS DISTRIBUTION.

Now, examine the asymptotic behavior of $\langle d_N \rangle$. The asymptotic expansion of the GAMMA FUNCTION ratio is

$$\frac{\Gamma\left(J + \tfrac{1}{2}\right)}{\Gamma(J)} = \sqrt{J}\left(1 - \frac{1}{8J} + \frac{1}{128J^2} + \dots\right) \qquad (33)$$

(Graham *et al.* 1994), so plugging in the expression for $\langle d_N \rangle$ gives the asymptotic series

$$\langle d_N \rangle = \sqrt{\frac{2N}{\pi}}$$
$$\times \left(1 \mp \frac{1}{4N} + \frac{1}{32N^2} \pm \frac{5}{128N^3} - \frac{21}{2048N^4} \mp \dots\right), \qquad (34)$$

where the top signs are taken for N EVEN and the bottom signs for N ODD. Therefore, for large N,

$$\langle d_N \rangle \sim \sqrt{\frac{2N}{\pi}}, \qquad (35)$$

which is also shown in Mosteller *et al.* (1961, p. 14).

Tóth (2000) has proven that there are no more than three most-visited sites in a simple symmetric random walk in 1-D with unit steps.

See also BINOMIAL DISTRIBUTION, CATALAN NUMBER, HEADS-MINUS-TAILS DISTRIBUTION, P-GOOD PATH,

PÓLYA'S RANDOM WALK CONSTANTS, RANDOM WALK–2-D, RANDOM WALK–3-D, SELF-AVOIDING WALK, WIENER PROCESS

References

Abramowitz, M. and Stegun, C. A. (Eds.). *Handbook of Mathematical Functions with Formulas, Graphs, and Mathematical Tables, 9th printing.* New York: Dover, p. 798, 1972.

Chandrasekhar, S. "Stochastic Problems in Physics and Astronomy." *Rev. Modern Phys.* **15**, 1–89, 1943. Reprinted in *Noise and Stochastic Processes* (Ed. N. Wax). New York: Dover, pp. 3–91, 1954.

Erdos, P. and Révész, P. "On the Favourite Points of Random Walks." *Math. Structures--Comput. Math.--Math. Model. (Sofia)* **2**, 152–157, 1984.

Erdos, P. and Révész, P. "Problems and Results on Random Walks." In *Mathematical Statistics and Probability Theory, Vol. B: Statistical Inference and Methods. Proceedings of the Sixth Pannonian Symposium on Mathematical Statistics Held in Bad Tatzmannsdorf, September 14–20, 1986* (Ed. P. Bauer, F. Koneczny, and W. Wertz). Dordrecht, Netherlands: Reidel, pp. 59–65, 1987.

Feller, W. Ch. 3 in *An Introduction to Probability Theory and Its Applications, Vol. 1, 3rd ed., rev. printing.* New York: Wiley, 1968.

Gardner, M. "Random Walks and Gambling." Ch. 6 in *Mathematical Circus: More Puzzles, Games, Paradoxes, and Other Mathematical Entertainments.* Washington, DC: Math. Assoc. Amer., pp. 66–74, 1992.

Graham, R. L.; Knuth, D. E.; and Patashnik, O. Answer to problem 9.60 in *Concrete Mathematics: A Foundation for Computer Science, 2nd ed.* Reading, MA: Addison-Wesley, 1994.

Hersh, R. and Griego, R. J. "Brownian Motion and Potential Theory." *Sci. Amer.* **220**, 67–74, 1969.

Hughes, B. D. Eq. (7.282) in *Random Walks and Random Environments, Vol. 1: Random Walks.* New York: Oxford University Press, p. 513, 1995.

Kac, M. "Random Walk and the Theory of Brownian Motion." *Amer. Math. Monthly* **54**, 369–391, 1947. Reprinted in *Noise and Stochastic Processes* (Ed. N. Wax). New York: Dover, pp. 295–317, 1954.

Mosteller, F.; Rourke, R. E. K.; and Thomas, G. B. *Probability and Statistics.* Reading, MA: Addison-Wesley, 1961.

Papoulis, A. "Random Walk." *Probability, Random Variables, and Stochastic Processes, 2nd ed.* New York: McGraw-Hill, pp. 290–291, 1984.

Prévost, G. *Tables de Fonctions Sphériques.* Paris: Gauthier-Villars, pp. 156–157, 1933.

Révész, P. *Random Walk in Random and Non-Random Environment.* Singapore: World Scientific, 1990.

Sloane, N. J. A. Sequences A001803/M2986 and A046161 in "An On-Line Version of the Encyclopedia of Integer Sequences." http://www.research.att.com/~njas/sequences/eisonline.html.

Tóth, B. No More than Three Favourite Sites for Simple Random Walk. 26 Apr 2000. http://xxx.lanl.gov/abs/math.PR/0004164/.

Tóth, B. and Werner, W. "Tied Favourite Edges for Simple Random Walk." *Combin., Prob., Comput.* **6**, 359–369, 1997.

Random Walk—2-D

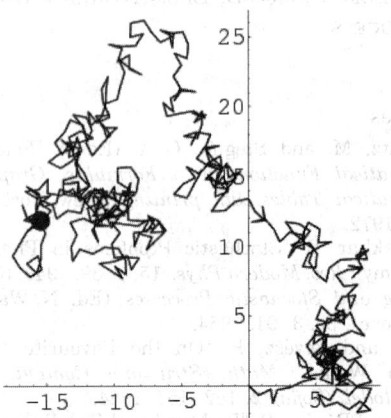

In a PLANE, consider a sum of N 2-D VECTORS with random orientations. Use PHASOR notation, and let the phase of each VECTOR be RANDOM. Assume N unit steps are taken in an arbitrary direction (i.e., with the angle θ uniformly distributed in $[0, 2\pi)$ and *not* on a LATTICE), as illustrated above. The position z in the COMPLEX PLANE after N steps is then given by

$$z = \sum_{j=1}^{N} e^{i\theta_j}, \tag{1}$$

which has ABSOLUTE SQUARE

$$|z|^2 = \sum_{j=1}^{N} e^{i\theta_j} \sum_{k=1}^{N} e^{-i\theta_k} = \sum_{j=1}^{N} \sum_{k=1}^{N} e^{i(\theta_j - \theta_k)}$$

$$= N + \sum_{\substack{j,\,k=1 \\ k \neq j}}^{N} e^{i(\theta_j - \theta_k)}. \tag{2}$$

Therefore,

$$\left\langle |z|^2 \right\rangle = N + \left\langle \sum_{\substack{j,\,k=1 \\ k \neq j}}^{N} e^{i(\theta_j - \theta_k)} \right\rangle. \tag{3}$$

Each step is equally likely to be in any direction, so both θ_j and θ_k are RANDOM VARIABLES with identical MEANS of zero, and their difference is also a random variable. Averaging over this distribution, which has equally likely POSITIVE and NEGATIVE values yields an expectation value of 0, so

$$\left\langle |z|^2 \right\rangle = N. \tag{4}$$

The root-mean-square distance after N unit steps is therefore

$$|z|_{\mathrm{rms}} = \sqrt{N}, \tag{5}$$

so with a step size of l, this becomes

$$d_{\mathrm{rms}} = l\sqrt{N}. \tag{6}$$

In order to travel a distance d

$$N \approx \left(\frac{d}{l}\right)^2 \tag{7}$$

steps are therefore required.

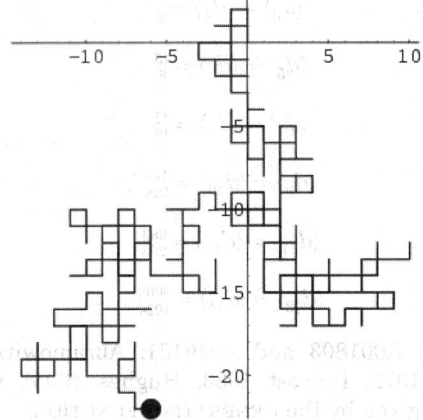

Amazingly, it has been proven that on a 2-D LATTICE, a random walk has unity probability of reaching any point (including the starting point) as the number of steps approaches INFINITY.

See also PÓLYA'S RANDOM WALK CONSTANTS, RANDOM WALK–1-D, RANDOM WALK–3-D

References

McCrea, W. H. and Whipple, F. J. W. "Random Paths in Two and Three Dimensions." *Proc. Roy. Soc. Edinburgh* **60**, 281–298, 1940.

Random Walk—3-D

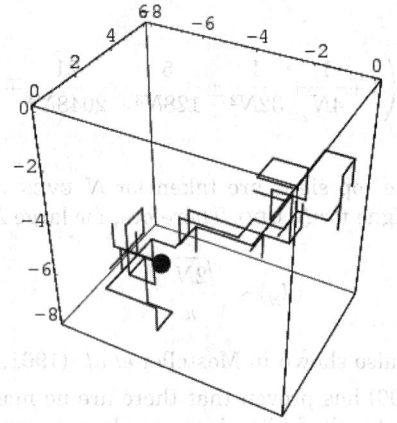

On a 3-D LATTICE, a random walk has *less than* unity probability of reaching any point (including the starting point) as the number of steps approaches infinity. The probability of reaching the starting point

again is 0.3405373296.... This is one of PÓLYA'S RANDOM WALK CONSTANTS.

See also PÓLYA'S RANDOM WALK CONSTANTS, RANDOM WALK–1-D, RANDOM WALK–2-D

References

Glasser, M. L. and Zucker, I. J. "Extended Watson Integrals for the Cubic Lattices." *Proc. Nat. Acad. Sci. U.S.A.* **74**, 1800–1801, 1977.

McCrea, W. H. and Whipple, F. J. W. "Random Paths in Two and Three Dimensions." *Proc. Roy. Soc. Edinburgh* **60**, 281–298, 1940.

Random Young Tableau

RANDOM TABLEAU

Range (Image)

If T is a MAP (a.k.a., FUNCTION, TRANSFORMATION) over a DOMAIN D, then the range of T is defined as

$$\text{Range}(T) = T(D) = \{T(\mathbf{X}) : \mathbf{X} \in D\}.$$

The range $T(D)$ is also called the IMAGE of D under T.

See also DOMAIN, MAP, TRANSFORMATION

Range (Line Segment)

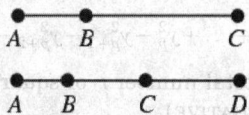

A number of points on a LINE SEGMENT. The term was first used by Desargues (Cremona 1960, p. x). If the points A, B, C, ... lie on a LINE SEGMENT with the coordinates of the points such that $A < B < C$, they are said to form a range, denoted $\{ABC\ldots\}$. Let AB denote the signed distance $B - A$. Then the range $\{ABC\}$ satisfies the relation

$$AB + BC + CA = 0.$$

The range $\{ABCD\}$ satisfies

$$BC \cdot AD + CA \cdot BD + AB \cdot CD = 0$$

and

$$BC \cdot AD^2 + CA \cdot BD^2 + AB \cdot CD^2 + BC \cdot CA \cdot AB = 0,$$

the latter of which holds even when D is not on the line ABC (Lachlan 1893).

Graustein (1930) and Woods (1961) use the term "range" to refer to the totality of points on a straight LINE, making it the dual of a PENCIL.

See also AXIS, HOMOGRAPHIC, LINE, LINE SEGMENT, PENCIL, PERSPECTIVITY, SECTION (PENCIL)

References

Cremona, L. *Elements of Projective Geometry, 3rd ed.* New York: Dover, 1960.

Durell, C. V. "Concurrency and Collinearity." Ch. 4 in *Modern Geometry: The Straight Line and Circle.* London: Macmillan, pp. 37–39, 1928.

Graustein, W. C. *Introduction to Higher Geometry.* New York: Macmillan, p. 40, 1930.

Lachlan, R. *An Elementary Treatise on Modern Pure Geometry.* London: Macmillian, pp. 14–15, 1893.

Woods, F. S. *Higher Geometry: An Introduction to Advanced Methods in Analytic Geometry.* New York: Dover, p. 8, 1961.

Range (Statistics)

$$R \equiv \max(x_i) - \min(x_i). \tag{1}$$

For small samples, the range is a good estimator of the population STANDARD DEVIATION (Kenney and Keeping 1962, pp. 213–214). For a continuous UNIFORM DISTRIBUTION

$$P(x) = \begin{cases} \dfrac{1}{C} & \text{for } 0 < x < C \\ 0 & \text{for } |x| < C, \end{cases} \tag{2}$$

the distribution of the range is given by

$$D(R) = N \left(\frac{R}{C} \right)^{N-1} - (N-1) \left(\frac{R}{C} \right)^N. \tag{3}$$

Given two samples with sizes m and n and ranges R_1 and R_2, let $u \equiv R_1/R_2$. Then

$$D(u) = \begin{cases} \dfrac{m(m-1)n(n-1)}{(m+n)(m+n-1)(m+n-2)} \\ \quad \times [(m+n)u^{m-2} - (m+n-2)u^{m-1}] \\ \quad \text{for } 0 \le u \le 1 \\ \dfrac{m(m-1)n(n-1)}{(m+n)(m+n-1)(m+n-2)} \\ \quad \times [(m+n)u^{-n} - (m+n-2)u^{-n-1}] \\ \quad \text{for } 1 \le u \le \infty. \end{cases} \tag{4}$$

The MEAN is

$$\mu_u = \frac{(m-1)n}{(m+1)(n-2)}, \tag{5}$$

and the MODE is

$$\hat{u} = \begin{cases} \dfrac{(m-2)(m+n)}{(m-1)(m+n-2)} & \text{for } m - n \le 2 \\ \dfrac{(n+1)(m+n-2)}{n(m+n)} & \text{for } m - n \ge 2. \end{cases} \tag{6}$$

References

Kenney, J. F. and Keeping, E. S. "The Range." §6.2 in *Mathematics of Statistics, Pt. 1, 3rd ed.* Princeton, NJ: Van Nostrand, pp. 75–76, 213–214, 1962.

Rank

The word "rank" refers to several unrelated concepts in mathematics involving groups, matrices, quadratic forms, sequences, set theory, statistics, and tensors.

In SET THEORY, rank is a (class) function from SETS to ORDINAL NUMBERS. The rank of a SET is the least ORDINAL NUMBER greater than the rank of any member of the set (Mirimanoff 1917; Moore 1982, pp. 261–262; Rubin 1967, p. 214). The proof that rank is WELL DEFINED uses the AXIOM OF FOUNDATION.

For example, the EMPTY SET {} has rank 0 (since it has no members and 0 is the least ORDINAL NUMBER), {{}} has rank 1 (since {}, its only member, has rank 0), {{{}}} has rank 2, and {{}, {{}}, {{{}}}, ...} has rank ω. Every ORDINAL NUMBER has itself as its rank.

Mirimanoff (1917) showed that, assuming the class of URELEMENTS is a set, for any ORDINAL NUMBER α, the class of all sets having rank α is a SET, i.e., not a PROPER CLASS (Rubin 1967, p. 216) The number of sets having rank k for $k = 0$, 1, ... are 1, 1, 2, 12, 65520, ... (Sloane's A038081), and the number of sets having rank at most k is $\underbrace{2^{2^{\cdot^{\cdot^{2}}}}}_{k}$, 1, 2, 4, 16, 65536, ...

(Sloane's A014221).

The rank of a mathematical object is defined whenever that object is FREE. In general, the rank of a FREE object is the CARDINALITY of the FREE generating SUBSET G.

See also ORDINAL NUMBER, RANK (BUNDLE), RANK (GROUP), RANK (LIE ALGEBRA), RANK (MATRIX), RANK (QUADRATIC FORM), RANK (SEQUENCE), RANK (STATISTICS), RANK (TENSOR)

References

Mirimanoff, D. "Les antinomies de Russell et de Burali-Forti et le problème fondamental de la théorie des ensembles." *Enseign. math.* **19**, 37–52, 1917.
Moore, G. H. *Zermelo's Axiom of Choice: Its Origin, Development, and Influence.* New York: Springer-Verlag, 1982.
Rubin, J. E. *Set Theory for the Mathematician.* New York: Holden-Day, 1967.
Sloane, N. J. A. Sequences A014221 and A038081 in "An On-Line Version of the Encyclopedia of Integer Sequences." http://www.research.att.com/~njas/sequences/eisonline.html.

Rank (Bundle)

The rank of a VECTOR BUNDLE is the DIMENSION of its FIBER. Equivalently, it is the maximum number of linearly independent LOCAL SECTIONS in a TRIVIALIZATION. Naturally, the dimension here is measured in the appropriate CATEGORY. For instance, a real line bundle has fibers isomorphic with $\mathbb{R}$, and a complex line bundle has fibers isomorphic to $\mathbb{C}$, but in both cases their rank is 1.

The rank of the TANGENT BUNDLE of a real MANIFOLD M is equal to the dimension of M. The rank of a trivial bundle $M \times \mathbb{R}^k$ is equal to k. There is no upper bound to the rank of a vector bundle over a fixed manifold M.

See also DIMENSION, FIBER, MANIFOLD, SECTION (BUNDLE), TANGENT BUNDLE, VECTOR BUNDLE

Rank (Group)

For an arbitrary finitely generated ABELIAN GROUP G, the rank of G is defined to be the rank of the FREE generating SUBSET G modulo its TORSION SUBGROUP. For a finitely generated GROUP, the rank is defined to be the rank of its "Abelianization."

See also ABELIAN GROUP, BETTI NUMBER, BURNSIDE PROBLEM, QUASITHIN THEOREM, QUASI-UNIPOTENT GROUP, TORSION (GROUP)

Rank (Matrix)

The rank of a MATRIX or a linear map is the DIMENSION of the range of the matrix or the linear map, corresponding to the number of LINEARLY INDEPENDENT rows or columns of the matrix, or to the number of nonzero singular values of the map.

Rank (Quadratic Form)

For a QUADRATIC FORM Q in the canonical form

$$Q = y_1^2 + y_2^2 + \ldots + y_p^2 - y_{p+1}^2 - y_{p+2}^2 - \ldots - y_r^2,$$

the rank is the total number r of square terms (both POSITIVE and NEGATIVE).

See also SIGNATURE (QUADRATIC FORM)

References

Gradshteyn, I. S. and Ryzhik, I. M. *Tables of Integrals, Series, and Products,* 6th ed. San Diego, CA: Academic Press, p. 1105, 2000.

Rank (Sequence)

The position of a RATIONAL NUMBER in the SEQUENCE $\frac{1}{1}, \frac{1}{2}, \frac{2}{1}, \frac{1}{3}, \frac{2}{1}, \frac{1}{4}, \frac{3}{2}, \frac{2}{1}, \frac{1}{5}$, ..., ordered in terms of increasing NUMERATOR + DENOMINATOR.

See also ENCODING, FAREY SERIES

Rank (Statistics)

The ORDINAL NUMBER of a value in a list arranged in a specified order (usually decreasing).

See also RANK TEST, SPEARMAN RANK CORRELATION COEFFICIENT, WILCOXON RANK SUM TEST, WILCOXON SIGNED RANK TEST, ZIPF'S LAW

Rank (Tensor)

The total number of CONTRAVARIANT and COVARIANT indices of a TENSOR. The rank of a TENSOR is independent of the number of DIMENSIONS of the SPACE.

Rank	Object
0	SCALAR
1	VECTOR
≥ 2	TENSOR

See also CONTRAVARIANT TENSOR, COVARIANT TENSOR, SCALAR, TENSOR, VECTOR

Rank Test

A STATISTICAL TEST making use of the RANKS of data points. Examples include the KOLMOGOROV-SMIRNOV TEST and WILCOXON SIGNED RANK TEST.

See also KOLMOGOROV-SMIRNOV TEST, R-ESTIMATE, RANK (STATISTICS), SPEARMAN RANK CORRELATION COEFFICIENT, STATISTICAL TEST, WILCOXON SIGNED RANK TEST

Ranunculoid

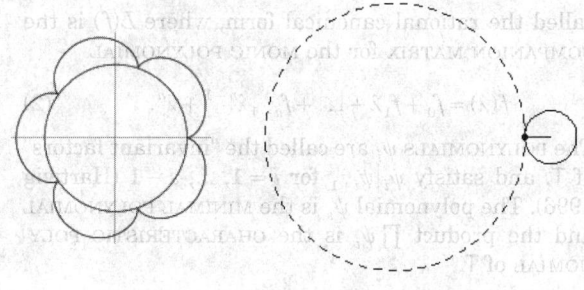

An EPICYCLOID with $n = 5$ cusps, named after the buttercup genus *Ranunculus* (Madachy 1979).

See also CARDIOID, EPICYCLOID, NEPHROID

References

Madachy, J. S. *Madachy's Mathematical Recreations.* New York: Dover, p. 223, 1979.
Pickover, C. A. *Keys to Infinity.* New York: Wiley, pp. 79–80, 1995.

Rapid Rumor Ramification

GOSSIPING

RAT-Free Set

A RAT-free ("right angle triangle-free") set is a set of points, no three of which determine a RIGHT TRIANGLE. Let $f(n)$ be the largest integer such that a RAT-free subset of size $f(n)$ is guaranteed to be contained in any set of n coplanar points. Then the function $f(n)$ is bounded by

$$\sqrt{n} \le f(n) \le 2\sqrt{n}.$$

See also RIGHT TRIANGLE

References

Abbott, H. L. "On a Conjecture of Erdos and Silverman in Combinatorial Geometry." *J. Combin. Th. A* **29**, 380–381, 1980.
Chan, W. K. "On the Largest RAT-FREE Subset of a Finite Set of Points." *Pi Mu Epsilon* **8**, 357–367, 1987.
Honsberger, R. *More Mathematical Morsels.* Washington, DC: Math. Assoc. Amer., pp. 250–251, 1991.
Seidenberg, A. "A Simple Proof of a Theorem of Erdos and Szekeres." *J. London Math. Soc.* **34**, 352, 1959.

Ratio

The ratio of two numbers r and s is written r/s, where r is the NUMERATOR and s is the DENOMINATOR. The ratio of r to s is equivalent to the QUOTIENT r/s. Betting ODDS written as $r : s$ correspond to $s/(r+s)$. A number which can be expressed as a ratio of INTEGERS is called a RATIONAL NUMBER.

See also DENOMINATOR, DIVISION, FRACTION, NUMERATOR, ODDS, QUOTIENT, RATIONAL NUMBER

Ratio Distribution

Given two distributions Y and X with joint probability density function $f(x, y)$, let $U = Y/X$ be the ratio distribution. Then the distribution function of u is

$$D(u) = P(U \le u)$$

$$= P(Y \le uX | X > 0) + P(Y \ge uX | X < 0)$$

$$= \int_0^\infty \int_0^{ux} f(x, y) \, dy \, dx + \int_{-\infty}^0 \int_{ux}^0 f(x, y) \, dy \, dx. \tag{1}$$

The probability function is then

$$P(u) = D'(u) = \int_0^\infty x f(x, ux) \, dx - \int_{-\infty}^0 x f(x, ux) \, dx$$

$$= \int_{-\infty}^\infty |x| f(x, ux) \, dx. \tag{2}$$

For variates with a standard NORMAL DISTRIBUTION, the ratio distribution is a CAUCHY DISTRIBUTION. For a UNIFORM DISTRIBUTION

$$f(x, y) = \begin{cases} 1 & \text{for } x, y \in [0, 1] \\ 0 & \text{otherwise,} \end{cases} \tag{3}$$

$$P(u) = \begin{cases} 0 & u < 0 \\ \int_0^1 x\,dx = \left[\tfrac{1}{2}x^2\right] = \tfrac{1}{2} & \text{for } 0 \le u \le 1 \\ \int_0^{1/u} x\,dx = \left[\tfrac{1}{2}x^2\right]_0^{1/u} = \dfrac{1}{2u^2} & \text{for } u > 1. \end{cases} \quad (4)$$

See also CAUCHY DISTRIBUTION

Ratio Test

Let u_k be a SERIES with POSITIVE terms and suppose

$$\rho \equiv \lim_{k \to \infty} \frac{u_{k+1}}{u_k}.$$

Then

1. If $\rho < 1$, the SERIES CONVERGES.
2. If $\rho > 1$ or $\rho = \infty$, the SERIES DIVERGES.
3. If $\rho = 1$, the SERIES may CONVERGE or DIVERGE.

The test is also called the CAUCHY RATIO TEST or D'ALEMBERT RATIO TEST.

See also CONVERGENCE TESTS

References
Arfken, G. *Mathematical Methods for Physicists, 3rd ed.* Orlando, FL: Academic Press, pp. 282–283, 1985.
Bromwich, T. J. I'a. and MacRobert, T. M. *An Introduction to the Theory of Infinite Series, 3rd ed.* New York: Chelsea, p. 28, 1991.

Rational Approximation

If α is any number and m and n are INTEGERS, then there is a RATIONAL NUMBER m/n for which

$$\left| \alpha - \frac{m}{n} \right| \le \frac{1}{n}. \quad (1)$$

If α is IRRATIONAL and k is any WHOLE NUMBER, there is a FRACTION m/n with $n \le k$ and for which

$$\left| \alpha - \frac{m}{n} \right| \le \frac{1}{nk}. \quad (2)$$

Furthermore, there are an infinite number of FRACTIONS m/n for which

$$\left| \alpha - \frac{m}{n} \right| \le \frac{1}{n^2} \quad (3)$$

(Hilbert and Cohn-Vossen 1999, pp. 40–44).

Hurwitz has shown that for an IRRATIONAL NUMBER ζ

$$\left| \zeta - \frac{h}{k} \right| < \frac{1}{ck^2}, \quad (4)$$

there are infinitely RATIONAL NUMBERS h/k if $0 < c \le$

$\sqrt{5}$, but if $c > \sqrt{5}$, there are some ζ for which this approximation holds for only finitely many h/k.

See also DIRICHLET'S APPROXIMATION THEOREM, HURWITZ'S IRRATIONAL NUMBER THEOREM, IRRATIONALITY MEASURE, KRONECKER'S APPROXIMATION THEOREM, LAGRANGE NUMBER (RATIONAL APPROXIMATION), LIOUVILLE'S APPROXIMATION THEOREM, MARKOV NUMBER, ROTH'S THEOREM, SEGRE'S THEOREM, THUE-SIEGEL-ROTH THEOREM

References
Hilbert, D. and Cohn-Vossen, S. *Geometry and the Imagination.* New York: Chelsea, p. 41, 1999.

Rational Canonical Form

Any SQUARE MATRIX T has a canonical form without any need to EXTEND the FIELD of its coefficients. For instance, if the entries of T are RATIONAL NUMBERS, then so are the entries of its rational canonical form. (The JORDAN CANONICAL FORM may require complex numbers.) There exists an INVERTIBLE MATRIX Q such that

$$Q^{-1}TQ = \text{diag}[L(\psi_1), L(\psi_2), \dots, L(\psi_s)], \quad (1)$$

called the rational canonical form, where $L(f)$ is the COMPANION MATRIX for the MONIC POLYNOMIAL

$$f(\lambda) = f_0 + f_1\lambda + \dots + f_{n-1}\lambda^{n-1} + \lambda^n. \quad (2)$$

The POLYNOMIALS ψ_i are called the "invariant factors" of T, and satisfy $\psi_i | \psi_{i+1}$ for $i = 1, \dots, s-1$ (Hartwig 1996). The polynomial ψ_s is the MINIMAL POLYNOMIAL and the product $\prod \psi_i$ is the CHARACTERISTIC POLYNOMIAL of T.

The rational canonical form is unique, and shows the extent to which the minimal polynomial characterizes a matrix. For example, there is only one 6×6 matrix whose MINIMAL POLYNOMIAL is $(x^2 + 1)^2$, which is

$$\begin{bmatrix} 0 & -1 & 0 & 0 & 0 & 0 \\ 1 & 0 & 0 & 0 & 0 & 0 \\ 0 & 0 & 0 & 0 & 0 & -1 \\ 0 & 0 & 1 & 0 & 0 & 0 \\ 0 & 0 & 0 & 1 & 0 & -2 \\ 0 & 0 & 0 & 0 & 1 & 0 \end{bmatrix} \quad (3)$$

in rational canonical form.

Given a LINEAR TRANSFORMATION $T : V \to V$, the VECTOR SPACE V becomes a $F[x]$-MODULE, that is a MODULE over the RING of polynomials with coefficients in the FIELD F. The VECTOR SPACE determines the field F, which can be taken to be the maximal field containing the entries of a matrix for T. The polynomial x acts on a vector v by $x(v) = T(v)$. The rational canonical form corresponds to writing V as

$$F[x]/(a_1) \oplus \dots \otimes F[x]/(a_s), \quad (4)$$

where (a_i) is the IDEAL generated by the INVARIANT

FACTOR a_i in $F[x]$, the canonical form for any finitely generated module over a PRINCIPAL IDEAL RING such as $F[x]$.

More constructively, given a basis e_i for V, there is a MODULE HOMOMORPHISM

$$t : F[x]^n \to V \tag{5}$$

which is ONTO, given by

$$t\left(\sum p_i(x)e_i\right) = \sum p_i(T)e_i \tag{6}$$

Letting K be the KERNEL,

$$V \cong F[x]^n / K. \tag{7}$$

To construct a basis for the rational canonical form, it is necessary to write K as

$$K \cong \bigoplus_{i-1}^{n-s} F[x] \oplus F[x]/(a_1) \oplus \ldots \oplus F[x]/(a_s), \tag{8}$$

and that is done by finding an appropriate basis for $F[x]^n$ and for K. Such a basis is found by determining matrices P and Q that are invertible $n \times n$ matrices having entries in $F[x]$ (and whose inverses are also in $F[x]$) such that

$$P(x\mathsf{I} - T)Q = \operatorname{diag}(1, \ldots, 1, a_1, \ldots, a_s), \tag{9}$$

where 1 is the IDENTITY MATRIX and $(a_1, \ldots, a_n)$ denotes a DIAGONAL MATRIX. They can be found by using ELEMENTARY MATRIX OPERATIONS.

The above matrix sends a basis for K, written as an n-tuple, to an n-tuple using a new basis f_i for $F[x]^n$, and P gives the linear transformation from the original basis to the one with the f_i. In particular,

$$K =$$
$$\{\beta_1 f_1 + \ldots \beta_{n-s} f_{n-s} + \beta_{n-s+1} a_1 f_{n-s+1} + \ldots + \beta_n a_s f_n\}, \tag{10}$$

where β_i is an arbitrary polynomial in $F[x]$. Setting $z_i = P^{-1}(T)e_{n-s+i}$,

$$V = F[x]z_1 \oplus \ldots \oplus F[x]z_s. \tag{11}$$

In particular, $F[x]z_i$ is the SUBSPACE of V which is generated by $z_i, xz_i, \ldots, x^{n-1}z_i$, where n is the degree of a_i. Therefore, a basis that puts T into rational canonical form is given by

$$\{z_1, Tz_1, \ldots, T^{n_1}z_1, z_2, \ldots, T^{n_2}x_2, \ldots, T^{n_s}z_s\}. \tag{12}$$

See also BLOCK DIAGONAL MATRIX, CHARACTERISTIC POLYNOMIAL, COMPANION MATRIX, FIELD, INVARIANT FACTOR, JORDAN CANONICAL FORM, MATRIX, MINIMAL POLYNOMIAL (MATRIX), PRINCIPAL IDEAL RING, REDUCTION ALGORITHM (PID), SIMILAR MATRICES, SMITH NORMAL FORM

References

Ayres, F. Jr. *Theory and Problems of Matrices*. New York: Schaum, p. 203, 1962.
Dummit, D. and Foote, R. *Abstract Algebra*. Englewood Cliffs, NJ: Prentice-Hall, 1991.
Gantmacher, F. R. *The Theory of Matrices, Vol. 1.* New York: Chelsea, 1960.
Hartwig, R. E. "Roth's Removal Rule and the Rational Canonical Form." *Amer. Math. Monthly* **103**, 332–335, 1996.
Herstein, I. N. *Topics in Algebra, 2nd ed.* New York: Springer-Verlag, p. 162, 1975.
Hoffman, K. and Kunze, K. *Linear Algebra, 3rd ed.* Englewood Cliffs, NJ: Prentice-Hall, 1996.
Jacobson, N. §3.10 in *Basic Algebra I.* New York: W. H. Freeman, 1985.
Lancaster, P. and Tismenetsky, M. *The Theory of Matrices, 2nd ed.* New York: Academic Press, 1985.
Turnbull, H. W. and Aitken, A. C. *An Introduction to the Theory of Canonical Matrices, 2nd impression.* New York: Blackie and Sons, 1945.

Rational Cuboid

EULER BRICK

Rational Diagonal

NSW NUMBER

Rational Distances

It is possible to find six points in the PLANE, no three on a LINE and no four on a CIRCLE (i.e., none of which are COLLINEAR or CONCYCLIC), such that all the mutual distances are RATIONAL. An example is illustrated by Guy (1994, p. 185).

It is not known if a TRIANGLE with INTEGER sides, MEDIANS, and AREA exists (although there are incorrect PROOFS of the impossibility in the literature). However, R. L. Rathbun, A. Kemnitz, and R. H. Buchholz have showed that there are infinitely many triangles with RATIONAL sides (HERONIAN TRIANGLES) with *two* RATIONAL MEDIANS (Guy 1994, p. 188).

See also COLLINEAR, CONCYCLIC, CYCLIC QUADRILATERAL, EQUILATERAL TRIANGLE, EULER BRICK, HERONIAN TRIANGLE, RATIONAL QUADRILATERAL, RATIONAL TRIANGLE, SQUARE, TRIANGLE

References

Guy, R. K. "Six General Points at Rational Distances" and "Triangles with Integer Sides, Medians, and Area." §D20 and D21 in *Unsolved Problems in Number Theory, 2nd ed.* New York: Springer-Verlag, pp. 185–190, 1994.

Rational Domain

FIELD

Rational Double Point

There are nine possible types of ISOLATED SINGULARITIES on a CUBIC SURFACE, eight of them rational double points. Each type of ISOLATED SINGULARITY has an associated normal form and COXETER-DYNKIN DIAGRAM (A_1, A_2, A_3, A_4, A_5, D_4, D_5, E_6 and $\tilde{E}_6$).

The eight types of rational double points (the $\tilde{E}_6$ type being the one excluded) can occur in only 20 combinations on a CUBIC SURFACE (of which Fischer 1986 gives 19): A_1, $2A_1$, $3A_1$, $4A_1$, A_2, (A_2, A_1), $2A_2$, $(2A_2, A_1)$, $3A_2$, A_3, (A_3, A_1), $(A_3, 2A_1)$, A_4, (A_4, A_1), A_5, (A_5, A_1), D_4, D_5, and E_6 (Looijenga 1978, Bruce and Wall 1979, Fischer 1986).

In particular, on a CUBIC SURFACE, precisely those configurations of rational double points occur for which the disjoint union of the COXETER-DYNKIN DIAGRAM is a SUBGRAPH of the COXETER-DYNKIN DIAGRAM $\tilde{E}_6$. Also, a surface specializes to a more complicated one precisely when its graph is contained in the graph of the other one (Fischer 1986).

See also COXETER-DYNKIN DIAGRAM, CUBIC SURFACE, DOUBLE POINT, ISOLATED SINGULARITY, ORDINARY DOUBLE POINT

References

Bruce, J. and Wall, C. T. C. "On the Classification of Cubic Surfaces." *J. London Math. Soc.* **19**, 245–256, 1979.
Fischer, G. (Ed.). *Mathematical Models from the Collections of Universities and Museums.* Braunschweig, Germany: Vieweg, p. 13, 1986.
Fischer, G. (Ed.). Plates 14–31 in *Mathematische Modelle/Mathematical Models, Bildband/Photograph Volume.* Braunschweig, Germany: Vieweg, pp. 17–31, 1986.
Looijenga, E. "On the Semi-Universal Deformation of a Simple Elliptic Hypersurface Singularity. Part II: The Discriminant." *Topology* **17**, 23–40, 1978.
Rodenberg, C. "Modelle von Flächen dritter Ordnung." In *Mathematische Abhandlungen aus dem Verlage Mathematischer Modelle von Martin Schilling.* Halle a. S., 1904.

Rational Function

A QUOTIENT of two polynomials $P(z)$ and $Q(z)$,

$$R(z) \equiv \frac{P(z)}{Q(z)},$$

is called a rational function. More generally, if P and Q are POLYNOMIALS in multiple variables, their quotient is called a (multivariate) rational function.

A rational function has no singularities other than poles in the EXTENDED COMPLEX PLANE. Conversely, if a single-values function has no singularities other than poles in the EXTENDED COMPLEX PLANE, than it is a rational function (Knopp 1996, p. 137). In addition, a rational function can be decomposed into partial fractions (Knopp 1996, p. 139).

See also ABEL'S CURVE THEOREM, CLOSED FORM, FUNDAMENTAL THEOREM OF SYMMETRIC FUNCTIONS, INSIDE-OUTSIDE THEOREM, QUOTIENT-DIFFERENCE ALGORITHM, RATIONAL INTEGER, RATIONAL NUMBER, RIEMANN CURVE THEOREM

References

Knopp, K. "Rational Functions." §35 in *Theory of Functions Parts I and II, Two Volumes Bound as One, Part I.* New York: Dover, pp. 96 and 137–139, 1996.

Rational Integer

A synonym for INTEGER. The word "rational" is sometimes used for emphasis to distinguish it from other types of "integers" such as CYCLOTOMIC INTEGERS, EISENSTEIN INTEGERS, GAUSSIAN INTEGERS, and HAMILTONIAN INTEGERS.

See also CYCLOTOMIC INTEGER, EISENSTEIN INTEGER, GAUSSIAN INTEGER, HAMILTONIAN INTEGER, INTEGER, RATIONAL NUMBER

References

Hardy, G. H. and Wright, E. M. *An Introduction to the Theory of Numbers, 5th ed.* Oxford, England: Clarendon Press, p. 1, 1979.

Rational Number

A number that can be expressed as a FRACTION p/q where p and q are INTEGERS and $q \neq 0$, is called a rational number with NUMERATOR p and DENOMINATOR q. Numbers which are not rational are called IRRATIONAL NUMBERS. The FIELD of rational numbers is denoted $\mathbb{Q}$. Any rational number is trivially also an ALGEBRAIC NUMBER. The set of rational numbers is denoted `Rationals` in *Mathematica*, and a number x can be tested to see if it is rational using the command `Element[x, Rationals]`.

Between any two members of the set of rationals, it is always possible to find another rational number. Therefore, rather counterintuitively, the rational numbers are a continuous set, but at the same time countable.

For a, b, and c any different rational numbers, then

$$\frac{1}{(a-b)^2} + \frac{1}{(b-c)^2} + \frac{1}{(c-a)^2}$$

is the SQUARE of a rational number (Honsberger 1991).

The probability that a random rational number has an EVEN DENOMINATOR is 1/3 (Salamin and Gosper 1972).

It is conjectured that if there exists a REAL NUMBER x for which both 2^x and 3^x are integers, then x is rational. This result would follow from the FOUR EXPONENTIALS CONJECTURE (Finch).

See also ALGEBRAIC INTEGER, ALGEBRAIC NUMBER, ANOMALOUS CANCELLATION, DENOMINATOR, DIRICHLET FUNCTION, FAREY SEQUENCE, FOUR EXPONENTIALS CONJECTURE, FRACTION, INTEGER, IRRATIONAL NUMBER, NUMERATOR, Q, QUOTIENT, TRANSCENDENTAL NUMBER

Rational Point

References
Courant, R. and Robbins, H. "The Rational Numbers." §2.1 in *What is Mathematics?: An Elementary Approach to Ideas and Methods, 2nd ed.* Oxford, England: Oxford University Press, pp. 52–58, 1996.
Finch, S. "Powers of 3/2 Modulo One." http://www.mathsoft.com/asolve/pwrs32/pwrs32.html.
Honsberger, R. *More Mathematical Morsels.* Washington, DC: Math. Assoc. Amer., pp. 52–53, 1991.
Salamin, E. and Gosper, R. W. Item 54 in Beeler, M.; Gosper, R. W.; and Schroeppel, R. *HAKMEM.* Cambridge, MA: MIT Artificial Intelligence Laboratory, Memo AIM-239, p. 18, Feb. 1972.

Rational Point

A K-rational point is a point (X, Y) on an ALGEBRAIC CURVE $f(X, Y) = 0$, where X and Y are in a FIELD K. For example, rational point in the FIELD $\mathbb{Q}$ of ordinary rational numbers is a point (X, Y) satisfying the given equation such that both X and Y are rational numbers.

The rational point may also be a POINT AT INFINITY. For example, take the ELLIPTIC CURVE

$$Y^2 = X^3 + X + 42$$

and homogenize it by introducing a third variable Z so that each term has degree 3 as follows:

$$ZY^2 = X^3 + XZ^2 + 42Z^3.$$

Now, find the points at infinity by setting $Z = 0$, obtaining

$$0 = X^3.$$

Solving gives $X = 0$, Y equal to any value, and (by definition) $Z = 0$. Despite freedom in the choice of Y, there is only a single POINT AT INFINITY because the two triples (X_1, Y_1, Z_1), (X_2, Y_2, Z_2) are considered to be equivalent (or identified) only if one is a scalar multiple of the other. Here, $(0, 0, 0)$ is not considered to be a valid point. The triples $(a, b, 1)$ correspond to the ordinary points (a, b), and the triples $(a, b, 0)$ correspond to the POINTS AT INFINITY, usually called the LINE AT INFINITY.

The rational points on ELLIPTIC CURVES over the FINITE FIELD $\mathrm{GF}(q)$ are 5, 7, 9, 10, 13, 14, 16, ... (Sloane's A005523).

See also ELLIPTIC CURVE, LINE AT INFINITY, POINT AT INFINITY

References
Sloane, N. J. A. Sequences A005523/M3757 in "An On-Line Version of the Encyclopedia of Integer Sequences." http://www.research.att.com/~njas/sequences/eisonline.html.

Rational Quadrilateral

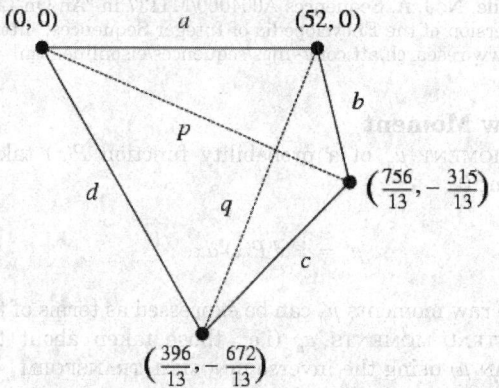

A rational quadrilateral is a QUADRILATERAL for which the sides, DIAGONALS, and AREA are RATIONAL. The simplest case has sides $a = 52$, $b = 25$, $c = 39$, and $d = 60$, DIAGONALS of length $p = 63$ and $q = 56$, and AREA 1764.

See also AREA, DIAGONAL (POLYGON), RATIONAL TRIANGLE

Rational Triangle

A rational triangle is a TRIANGLE all of whose sides are RATIONAL NUMBERS and all of whose ANGLES are RATIONAL numbers of DEGREES. The only such triangle is the EQUILATERAL TRIANGLE (Conway and Guy 1996).

See also EQUILATERAL TRIANGLE, FERMAT'S RIGHT TRIANGLE THEOREM, RATIONAL QUADRILATERAL, RIGHT TRIANGLE

References
Conway, J. H. and Guy, R. K. "The Only Rational Triangle." In *The Book of Numbers.* New York: Springer-Verlag, pp. 201 and 228–239, 1996.

Rationals

RATIONAL NUMBER

RATS Sequence

A sequence produced by the instructions "reverse, add, then sort the digits," where zeros are suppressed. For example, after 668 we get

$$668 + 866 = 1534,$$

so the next term is 1345. Applied to 1, the sequence gives 1, 2, 4, 8, 16, 77, 145, 668, 1345, 6677, 13444, 55778, ... (Sloane's A004000)

See also 196-ALGORITHM, KAPREKAR ROUTINE, REVERSAL, SORT-THEN-ADD SEQUENCE

References

Sloane, N. J. A. Sequences A004000/M1137 in "An On-Line Version of the Encyclopedia of Integer Sequences." http://www.research.att.com/~njas/sequences/eisonline.html.

Raw Moment

A MOMENT μ_n of a probability function $P(x)$ taken about 0,

$$\mu'_n = \int x^n P(x)\,dx. \tag{1}$$

The raw moments μ'_n can be expressed as terms of the CENTRAL MOMENTS μ_n (i.e., those taken about the MEAN μ) using the inverse BINOMIAL TRANSFORM

$$\mu'_n = \sum_{k=0}^{n} \binom{n}{k} \mu_k \mu_1'^{n-k}, \tag{2}$$

with $\mu_0 = 1$ and $\mu_1 = 0$ (Papoulis 1984, p. 146). The first few values are therefore

$$\mu'_2 = \mu_2 + \mu_1'^2 \tag{3}$$

$$\mu'_3 = \mu_3 + 3\mu_2\mu_1'^2 + \mu_1'^4 \tag{4}$$

$$\mu'_4 = \mu_4 + 4\mu_3\mu_1' + 6\mu_2\mu_1'^2 + \mu_1'^4 \tag{5}$$

$$\mu'_5 = \mu_5 + 5\mu_4\mu_1' + 10\mu_3\mu_1'^2 + 10\mu_2\mu_1'^3 + \mu_1'^5. \tag{6}$$

See also ABSOLUTE MOMENT, CENTRAL MOMENT, MEAN, MOMENT

References

Kenney, J. F. and Keeping, E. S. "Moments About the Origin." §7.2 in *Mathematics of Statistics, Pt. 1, 3rd ed.* Princeton, NJ: Van Nostrand, pp. 91–92, 1962.
Papoulis, A. *Probability, Random Variables, and Stochastic Processes, 2nd ed.* New York: McGraw-Hill, 1984.

Ray

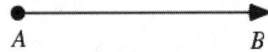

A VECTOR $\overrightarrow{AB}$ from a point A to a point B. In GEOMETRY, a ray is usually taken as a half-infinite LINE with one of the two points A and B taken to be at INFINITY.

See also LINE, VECTOR

Rayleigh Differential Equation

$$y'' - \mu\left(1 - \tfrac{1}{3}y'^2\right)y' + y = 0,$$

where $\mu > 0$. Differentiating and setting $y = y'$ gives the VAN DER POL EQUATION. The equation

$$y'' - \mu(1 - y'^2)y' + y = 0$$

with the 1/3 replaced by 1 is sometimes also called

the Rayleigh differential equation (Birkhoff and Rota 1978, p. 134; Zwillinger 1997, p. 126).

See also RAYLEIGH WAVE EQUATION, VAN DER POL EQUATION

References

Birkhoff, G. and Rota, G.-C. *Ordinary Differential Equations, 3rd ed.* New York: Wiley, p. 134, 1978.
Zwillinger, D. *Handbook of Differential Equations, 3rd ed.* Boston, MA: Academic Press, p. 126, 1997.

Rayleigh Distribution

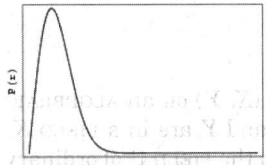 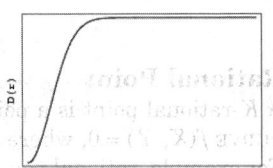

The distribution with PROBABILITY FUNCTION

$$P(r) = \frac{re^{-r^2/2s^2}}{s^2} \tag{1}$$

for $r \in [0, \infty)$. The MOMENTS about 0 are given by

$$\mu'_m \equiv \int_0^\infty r^m P(r)\,dr = s^{-2}\int_0^\infty r^{m+1}e^{-r^2/2s^2}\,dr$$

$$= s^{-2} I_{m+1}\!\left(\frac{1}{2s^2}\right), \tag{2}$$

where $I(x)$ is a GAUSSIAN INTEGRAL (Papoulis 1984, p. 148). The first few of these are

$$I_1(a^{-1}) = \tfrac{1}{2}a \tag{3}$$

$$I_2(a^{-1}) = \tfrac{1}{4}a\sqrt{a\pi} \tag{4}$$

$$I_3(a^{-1}) = \tfrac{1}{2}a^2 \tag{5}$$

$$I_4(a^{-1}) = \tfrac{3}{8}a^2\sqrt{a\pi} \tag{6}$$

$$I_5(a^{-1}) = a^3, \tag{7}$$

so the RAW MOMENTS are

$$\mu'_0 = s^{-2}\tfrac{1}{2}(2s^2) = 1 \tag{8}$$

$$\mu'_1 = s^{-2}\tfrac{1}{4}(2s^2)\sqrt{2s^2\pi} = \tfrac{1}{2}s\sqrt{2\pi} = s\sqrt{\frac{\pi}{2}} \tag{9}$$

$$\mu'_2 = s^{-2}\tfrac{1}{2}(2s^2)^2 = 2s^2 \tag{10}$$

$$\mu'_3 = s^{-2}\tfrac{3}{8}(2s^2)^2\sqrt{2s^2\pi} = \tfrac{3}{2}s^3\sqrt{2\pi} = 3s^3\sqrt{\frac{\pi}{2}} \tag{11}$$

$$\mu'_4 = s^{-2}(2s^2)^3 = 8s^4. \tag{12}$$

The CENTRAL MOMENTS are therefore

$$\mu_2 = \mu_2' - (\mu_1')^2 = \frac{4 - \pi}{2} s^2 \tag{13}$$

$$\mu_3 = \mu_3' - 3\mu_2'\mu_1' + 2(\mu_1')^3 = \sqrt{\frac{\pi}{2}}\,(\pi - 3)s^3 \tag{14}$$

$$\mu_4 = \mu_4' - 4\mu_3'\mu_1' + 6\mu_2'(\mu_1')^2 - 3(\mu - 1')^4$$

$$= \frac{32 - 3\pi^2}{4}\, s^4, \tag{15}$$

so the MEAN, VARIANCE, SKEWNESS, and KURTOSIS are

$$\mu = \mu_1' = s\,\sqrt{\frac{\pi}{2}} \tag{16}$$

$$\sigma^2 = \mu_2 = \frac{4 - \pi}{2}\, s^2 \tag{17}$$

$$\gamma_1 = \frac{\mu_3}{\sigma_3} = \frac{2(\pi - 3)\sqrt{\pi}}{(4 - \pi)^{3/2}} \tag{18}$$

$$\gamma_2 = \frac{\mu_4}{\sigma^4} - 3 = -\frac{6\pi^2 - 24\pi + 16}{(\pi - 4)^2}, \tag{19}$$

The CHARACTERISTIC FUNCTION is

$$\phi(t) = 1 - \sqrt{\frac{\pi}{2}}\, ste^{-s^2 t^2/2}\left[\operatorname{erfi}\left(\frac{st}{\sqrt{2}}\right) - i\right]. \tag{20}$$

See also MAXWELL DISTRIBUTION

References

Papoulis, A. *Probability, Random Variables, and Stochastic Processes, 2nd ed.* New York: McGraw-Hill, pp. 104 and 148, 1984.

Rayleigh Function

The Rayleigh functions $\sigma_n(v)$ for $n = 1, 2, \ldots$, are defined as

$$\sigma_n(v) = \sum_{k=1}^{\infty} j_{vk}^{-2n},$$

where $\pm j_v k$ are the zeros of the BESSEL FUNCTION OF THE FIRST KIND $J_v(z)$ (Watson 1966, p. 502; Gupta and Muldoon 1999). They were used by Euler, Rayleigh, and others to evaluate zeros of Bessel functions.

There is a convolution formula connecting Rayleigh functions of different orders,

$$\sigma_n(v) = \frac{1}{v + n} \sum_{k=1}^{n-1} \sigma_k(v)\sigma_{n-k}(v)$$

(Kishore 1963, Gupta and Muldoon 1999).

See also BESSEL FUNCTION OF THE FIRST KIND

References

Gupta, D. P. and Muldoon, M. E. Riccati Equations and Convolution Formulas for Functions of Rayleigh Type. 24 Oct 1999. http://xxx.lanl.gov/abs/math.CA/9910128/.
Ismail, M. E. H. and Muldoon, M. E. "Bounds for the Small Real and Purely Imaginary Zeros of Bessel and Related Functions." *Meth. Appl. Anal.* **2**, 1–21, 1995.
Kishore, N. "The Rayleigh Function." *Proc. Amer. Math. Soc.* **14**, 527–533, 1963.
Obi, E. C. "The Complete Monotonicity of the Rayleigh Function." *J. Math. Anal. Appl.* **77**, 465–468, 1980.
Watson, G. N. *A Treatise on the Theory of Bessel Functions, 2nd ed.* Cambridge, England: Cambridge University Press, 1966.

Rayleigh Wave Equation

The PARTIAL DIFFERENTIAL EQUATION

$$u_{tt} - u_{xx} = \epsilon\left(u_t - u_t^3\right).$$

See also RAYLEIGH DIFFERENTIAL EQUATION

References

Hall, W. S. "The Rayleigh Wave Equation--An Analysis." *Nonlinear Anal.* **2**, 129–156, 1978.
Zwillinger, D. *Handbook of Differential Equations, 3rd ed.* Boston, MA: Academic Press, p. 134, 1997.

Rayleigh-Ritz Variational Technique

A technique for computing EIGENFUNCTIONS and EIGENVALUES. It proceeds by requiring

$$J = \int_a^b \left[p(x)y_x^2 - q(x)y^2\right] dx \tag{1}$$

to have a STATIONARY VALUE subject to the normalization condition

$$\int_a^b y^2 w(x)\, dx = 1 \tag{2}$$

and the boundary conditions

$$py_x y\big|_a^b = 0. \tag{3}$$

This leads to the STURM-LIOUVILLE EQUATION

$$\frac{d}{dx}\left(p\frac{dy}{dx}\right) + qy + \lambda wy = 0, \tag{4}$$

which gives the stationary values of

$$F[y(x)] = \frac{\displaystyle\int_a^b (py_x^2 - qy^2)\, dx}{\displaystyle\int_a^b y^2 w\, dx} \tag{5}$$

as

$$F[y_n(x)] = \lambda_n, \tag{6}$$

where λ_n are the EIGENVALUES corresponding to the EIGENFUNCTION y_n.

References

Arfken, G. "Rayleigh-Ritz Variational Technique." §17.8 in *Mathematical Methods for Physicists, 3rd ed.* Orlando, FL: Academic Press, pp. 957–961, 1985.

Rayleigh, J. W. "In Finding the Correction for the Open End of an Organ-Pipe." *Phil. Trans.* **161**, 77, 1870.

Ritz, W. "Über eine neue Methode zur Lösung gewisser Variationsprobleme der mathematischen Physik." *J. reine angew. Math.* **135**, 1–61, 1908.

Whittaker, E. T. and Robinson, G. "The Rayleigh-Ritz Method for Minimum Problems." §184 in *The Calculus of Observations: A Treatise on Numerical Mathematics, 4th ed.* New York: Dover, pp. 381–382, 1967.

Rayleigh's Formulas

The formulas

$$j_n(z) = \left(-\frac{1}{z}\frac{d}{dz}\right)^n \frac{\sin z}{z}$$

$$y_n(z) = -z^n \left(-\frac{1}{z}\frac{d}{dz}\right)^n \frac{\cos z}{z}$$

for $n = 0, 1, 2, \ldots$, where $j_n(z)$ is a SPHERICAL BESSEL FUNCTION OF THE FIRST KIND and $y_n(z)$ is a SPHERICAL BESSEL FUNCTION OF THE SECOND KIND.

References

Abramowitz, M. and Stegun, C. A. (Eds.). *Handbook of Mathematical Functions with Formulas, Graphs, and Mathematical Tables, 9th printing.* New York: Dover, p. 439, 1972.

Rayleigh's Theorem

PARSEVAL'S THEOREM

R-Bar

The set of affine EXTENDED REAL NUMBERS.

See also EXTENDED REAL NUMBER (AFFINE)

Re

REAL PART

Real Analysis

That portion of mathematics dealing with functions of real variables. While this includes some portions of TOPOLOGY, it is most commonly used to distinguish that portion of CALCULUS dealing with real as opposed to COMPLEX NUMBERS.

Real Analytic Function

A REAL FUNCTION is said to be analytic if it possesses derivatives of all orders and agrees with its TAYLOR SERIES in the neighborhood of every point.

See also ANALYTIC FUNCTION

Real Axis

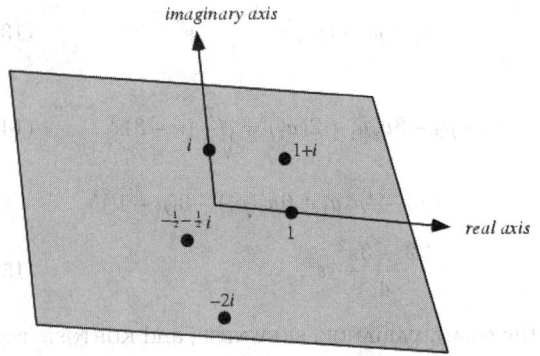

The axis in the COMPLEX PLANE corresponding to zero IMAGINARY PART, $\Im[z] = 0$.

See also COMPLEX PLANE, IMAGINARY AXIS, REAL LINE

Real Function

A FUNCTION whose RANGE is in the REAL NUMBERS is said to be a real function, also called a real-valued function.

See also COMPLEX FUNCTION, SCALAR FUNCTION, VECTOR FUNCTION

Real Line

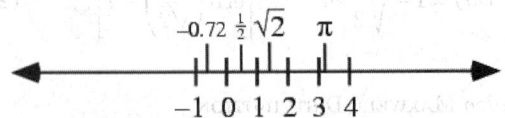

A LINE with a fixed scale so that every REAL NUMBER corresponds to a unique POINT on the LINE. The generalization of the real line to 2-D is called the COMPLEX PLANE.

The term "real line" is also used to distinguish an ordinary LINE from a so-called IMAGINARY LINE which can arise in algebraic geometry.

See also ABSCISSA, COMPLEX PLANE, IMAGINARY AXIS, IMAGINARY LINE, LINE, MOAT-CROSSING PROBLEM, REAL AXIS, REAL SPACE

References

Courant, R. and Robbins, H. *What is Mathematics?: An Elementary Approach to Ideas and Methods, 2nd ed.* Oxford, England: Oxford University Press, p. 57, 1996.

Real Manifold

See also COMPLEX MANIFOLD, MANIFOLD

Real Matrix

A real matrix is a MATRIX whose elements consist entirely of REAL NUMBERS. The set of $m \times n$ real matrices is sometimes denoted $\mathbb{R}^{m \times n}$ (Zwillinger 1995, p. 116).

For a real $n \times n$ matrix, the expected number of real EIGENVALUES is given by

$$E_n = \begin{cases} \sqrt{2} \sum_{k=0}^{n/2-1} \dfrac{(4k-1)!!}{(4k)!!} & \text{for } n \text{ even} \\[2ex] 1 + \sqrt{2} \sum_{k=1}^{(n-1)/2} \dfrac{(4k-3)!!}{(4k-2)!!} & \text{for } n \text{ odd} \end{cases} \tag{1}$$

(Edelman *et al.* 1994, Edelman and Kostlan 1994), which has asymptotic behavior

$$E_n \sim \sqrt{\frac{2n}{\pi}}. \tag{2}$$

GIRKO'S CIRCULAR LAW considers EIGENVALUES λ (possibly complex) of a set of random $n \times n$ REAL MATRICES with entries independent and taken from a standard normal distribution. Then as $n \to \infty$, $\lambda/\sqrt{n}$ is uniformly distributed on the UNIT DISK in the COMPLEX PLANE.

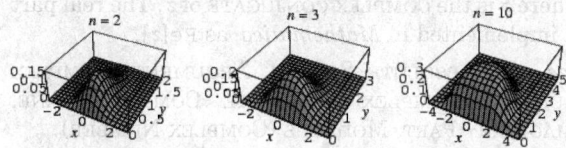

Edelman (1997) proved that the density of a random *complex* pair of eigenvalues $x \pm iy$ of a real $n \times n$ matrix whose elements are taken from a standard normal distribution is

$$\rho_n(x, y) = \sqrt{\frac{2}{\pi}} y e^{y^2 - x^2} \operatorname{erfc}\left(\sqrt{2} y\right) e_{n-2}(x^2 + y^2)$$

$$= \sqrt{\frac{2}{\pi}} e^{2y^2} y \operatorname{erfc}(\sqrt{2} y) \frac{\Gamma(n-1, x^2 + y^2)}{\Gamma(n-1)} \tag{3}$$

for $y \geq 0$, where $\operatorname{erfc}(z)$ is the ERFC (complementary error) function, $e_n(z)$ is the EXPONENTIAL SUM FUNCTION, and $\Gamma(a, x)$ is the upper INCOMPLETE GAMMA FUNCTION. Integrating over the UPPER HALF-PLANE gives half the expected number of complex eigenvalues

$$\int_{-\infty}^{\infty} \int_0^{\infty} \rho_n(x, y) \, dy \, dx = 1 - 2^{n(1-n)/4}. \tag{4}$$

See also COMPLEX MATRIX, GIRKO'S CIRCULAR LAW, INTEGER MATRIX, MATRIX

References

Edelman, A. "The Probability that a Random Real Gaussian Matrix has k Real Eigenvalues, Related Distributions, and the Circular Law." *J. Multivariate Anal.* **60**, 203–232, 1997.

Edelman, A.; Kostlan, E.; and Shub, M. "How Many Eigenvalues of a Random Matrix are Real?" *J. Amer. Math. Soc.* **7**, 247–267, 1994.

Edelman, A. and Kostlan, E. "How Many Zeros of a Random Polynomial are Real?" *Bull. Amer. Math. Soc.* **32**, 1–37, 1995.

Girko, V. L. *Theory of Random Determinants.* Boston, MA: Kluwer, 1990.

Lehmann, N. and Sommers, H.-J. "Eigenvalue Statistics of Random Real Matrices." *Phys. Rev. Let.* **67**, 941–944, 1991.

Mehta, M. L. *Random Matrices, 2nd rev. enl. ed.* New York: Academic Press, 1991.

Zwillinger, D. (Ed.). *CRC Standard Mathematical Tables and Formulae.* Boca Raton, FL: CRC Press, 1995.

Real Measure

A MEASURE that takes on real values.

See also MEASURE

Real Normed Algebra

A finite dimensional ALGEBRA A containing a copy of the reals is a real algebra. Note that this implies that A must be a real VECTOR SPACE. A real normed algebra is a real algebra A with a norm that is preserved by multiplication, i.e., $|a * b| = |a||b|$.

For example, the REAL NUMBERS, the COMPLEX NUMBERS, the QUATERNIONS, and the OCTONIONS are real normed algebras. Multiplication need not be commutative in a real normed algebra (e.g., QUATERNIONS and OCTONIONS are noncommutative), nor does it even need to be associative (e.g., the OCTONIONS).

A real normed algebra A satisfies a number of algebraic restrictions. For example, if the dimension of A is greater than 1, it must contain a copy of the complex numbers. Similarly, if the dimension is greater than 2, it must contain a copy of the QUATERNIONS. And if it is greater than 4, it must contain the OCTONIONS. In fact, these are the only examples, as the OCTONIONS cannot be "doubled" to make a normed algebra.

See also ALGEBRA, COMPLEX NUMBER, OCTONION, QUATERNION, REAL NUMBER, VECTOR SPACE

Real Number

The FIELD of all RATIONAL and IRRATIONAL numbers is called the real numbers, or simply the "reals," and denoted $\mathbb{R}$. The set of real numbers is also called the CONTINUUM, denoted C. The set of reals is called Reals in *Mathematica*, and a number x can be tested to see if it is a member of the reals using the command Element[x, Reals].

The real numbers can be extended with the addition of the IMAGINARY NUMBER I, equal to $\sqrt{-1}$. Numbers OF THE FORM $x + iy$, where x and y are both real, are called COMPLEX NUMBERS, which also form a FIELD. Another extension which includes both the real numbers and the infinite ORDINAL NUMBERS of Georg Cantor is the SURREAL NUMBERS.

Plouffe's "Inverse Symbolic Calculator" includes a huge database of 54 million real numbers which are algebraically related to fundamental mathematical constants and functions.

See also COMPLEX NUMBER, CONTINUUM, EXTENDED REAL NUMBER (AFFINE), EXTENDED REAL NUMBER (PROJECTIVE), i, IMAGINARY NUMBER, INTEGER RELATION, RATIONAL NUMBER, REAL NUMBER PICKING, REAL PART, SURREAL NUMBER

References

Jeffreys, H. and Jeffreys, B. S. "Real Numbers." §1.03 in *Methods of Mathematical Physics, 3rd ed.* Cambridge, England: Cambridge University Press, pp. 5–6, 1988.
Plouffe, S. "Inverse Symbolic Calculator." http://www.cecm.sfu.ca/projects/ISC/.
Plouffe, S. "Plouffe's Inverter." http://www.lacim.uqam.ca/pi/

Real Number Picking

Pick two real numbers x and y at random in $(0, 1)$ with a UNIFORM DISTRIBUTION. What is the PROBABILITY P_{even} that $[x/y]$, where $[r]$ denotes NEAREST INTEGER FUNCTION, is EVEN? The answer may be found as follows.

$$P\left(a < \frac{x}{y} < b\right) = \begin{cases} P(ay < x < by) & \text{for } 0 \le a < b < 1 \\ P\left(\dfrac{x}{b} < y < \dfrac{x}{a}\right) & \text{for } 1 < a < b \end{cases}$$

$$= \begin{cases} \displaystyle\int_0^1 \int_{ay}^{by} dx\,dy = \tfrac{1}{2}(b-a) & \text{for } 0 \le a < b < 1 \\ \displaystyle\int_0^1 \int_{x/b}^{x/a} dy\,dx = \dfrac{1}{2a} - \dfrac{1}{2b} & \text{for } 1 < a < b \end{cases} \quad (1)$$

so

$$P_{even} = P\left(0 < \frac{x}{y} < \tfrac{1}{2}\right) + \sum_{n=1}^{\infty} P\left(2n - \tfrac{1}{2} < \frac{x}{y} < 2n + \tfrac{1}{2}\right)$$

$$= \tfrac{1}{2}\left(\tfrac{1}{2} - 0\right) + \sum_{n=1}^{\infty}\left[\frac{1}{2\left(2n - \tfrac{1}{2}\right)} - \frac{1}{2\left(2n + \tfrac{1}{2}\right)}\right]$$

$$= \tfrac{1}{4} + \sum_{n=1}^{\infty}\left(\frac{1}{4n-1} + \frac{1}{4n-1}\right)$$

$$= \tfrac{1}{4} + \left(\tfrac{1}{3} - \tfrac{1}{5} + \tfrac{1}{7} - \tfrac{1}{9} + \ldots\right) = \tfrac{1}{4} + (1 - \tan^{-1}1)$$

$$= \frac{5}{4} - \frac{\pi}{4} = \tfrac{1}{4}(5 - \pi) \approx 46.460\% \quad (2)$$

(Putnam Exam).

References

Putnam Exam. Problem B-3 in the 54th Putnam Exam.

Real Part

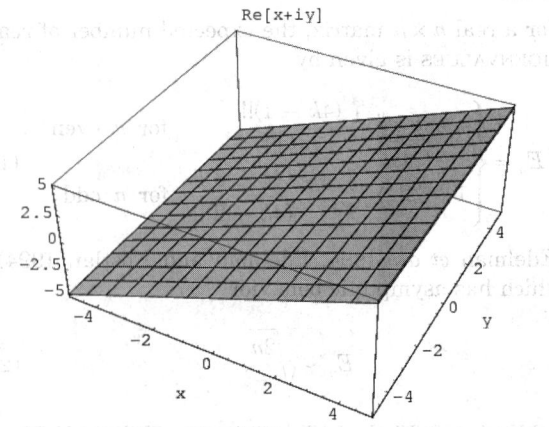

$$\text{Re}[x+iy]$$

The real part $\Re[z]$ of a COMPLEX NUMBER $z = x + iy$ is the REAL NUMBER *not* multiplying i, so $\Re[x + iy] = x$. In terms of z itself,

$$\Re[z] = \tfrac{1}{2}(z + \bar{z}),$$

where $\bar{z}$ is the COMPLEX CONJUGATE of z. The real part is implemented in *Mathematica* as Re[z].

See also ABSOLUTE SQUARE, ARGUMENT (COMPLEX NUMBER), COMPLEX CONJUGATE, COMPLEX PLANE, IMAGINARY PART, MODULUS (COMPLEX NUMBER)

References

Abramowitz, M. and Stegun, C. A. (Eds.). *Handbook of Mathematical Functions with Formulas, Graphs, and Mathematical Tables, 9th printing.* New York: Dover, p. 16, 1972.
Krantz, S. G. *Handbook of Complex Analysis.* Boston, MA: Birkhäuser, p. 2, 1999.

Real Polynomial

A POLYNOMIAL having only REAL NUMBERS as COEFFICIENTS. A polynomial with real coefficients is a product of IRREDUCIBLE POLYNOMIALS of first and second degrees.

See also POLYNOMIAL

Real Projective Plane

real projective plane

The closed topological MANIFOLD, denoted $\mathbb{R}P^2$, which is obtained by projecting the points of a plane E from a fixed point P (not on the plane), with the addition of the LINE AT INFINITY, is called the real projective plane. There is then a one-to-one correspondence between points in E and lines through P. Since each line through P intersects the sphere $\mathbb{S}^2$ centered at P and tangent to E in two ANTIPODAL POINTS, $\mathbb{R}P^2$ can be described as a QUOTIENT SPACE of $\mathbb{S}^2$ by identifying any two such points. The real projective plane is a NONORIENTABLE SURFACE.

The BOY SURFACE, CROSS-CAP, and ROMAN SURFACE are all homeomorphic to the real projective plane and, because $\mathbb{R}P^2$ is nonorientable, these surfaces contain self-intersections (Kuiper 1961, Pinkall 1986).

See also BOY SURFACE, CROSS-CAP, CROSS SURFACE, HENNEBERG'S MINIMAL SURFACE, NONORIENTABLE SURFACE, PROJECTIVE PLANE, REAL PROJECTIVE SPACE, ROMAN SURFACE

References

Apéry, F. *Models of the Real Projective Plane: Computer Graphics of Steiner and Boy Surfaces.* Braunschweig, Germany: Vieweg, 1987.
Coxeter, H. S. M. *The Real Projective Plane, 3rd ed.* Cambridge, England: Cambridge University Press, 1993.
Gray, A. "Realizations of the Real Projective Plane." §14.6 in *Modern Differential Geometry of Curves and Surfaces with Mathematica, 2nd ed.* Boca Raton, FL: CRC Press, pp. 330–335, 1997.
Klein, F. §1.2 in *Vorlesungen über nicht-euklidische Geometrie.* New York: Springer-Verlag, 1968.
Kuiper, N. H. "Convex Immersion of Closed Surfaces in E^3." *Comment. Math. Helv.* **35**, 85–92, 1961.
Pinkall, U. *Mathematical Models from the Collections of Universities and Museums* (Ed. G. Fischer). Braunschweig, Germany: Vieweg, pp. 64–65, 1986.

Real Projective Space

See also COMPLEX PROJECTIVE SPACE, REAL PROJECTIVE PLANE, REAL SPACE

Real Quadratic Field

A QUADRATIC FIELD $\mathbb{Q}(\sqrt{D})$ with $D > 0$.

See also IMAGINARY QUADRATIC FIELD, QUADRATIC FIELD

Real Space

See also COMPLEX SPACE, REAL LINE

Real Vector

A VECTOR whose elements are REAL NUMBERS.

See also COMPLEX VECTOR, REAL NUMBER, VECTOR

Real Vector Bundle

See also VECTOR BUNDLE

Real Vector Space

See also COMPLEX VECTOR SPACE, VECTOR SPACE

Realizer

A SET R of LINEAR EXTENSIONS of a POSET $P = (X, \leq)$ is a realizer of P (and is said to realize P) provided that for all $x, y \in X$, $x \leq y$ IFF x is below y in every member of R.

See also DOMINANCE, LINEAR EXTENSION, PARTIALLY ORDERED SET, POSET DIMENSION

Reals

REAL NUMBER

Real-Valued Function

REAL FUNCTION

Rearrangement Theorem

Each row and each column in the GROUP multiplication table lists each of the GROUP elements once and only once. From this, it follows that no two elements may be in the identical location in two rows or two columns. Thus, each row and each column is a rearranged list of the GROUP elements. Stated otherwise, given a GROUP of n distinct elements $(I, a, b, c, \ldots, n)$, the set of products $(aI, a^2, ab, ac, \ldots, an)$ reproduces the n original distinct elements in a new order.

See also GROUP

Reciprocal

The reciprocal of a REAL or COMPLEX NUMBER $z \neq 0$ is its MULTIPLICATIVE INVERSE $1/z$. The reciprocal of a COMPLEX NUMBER $z = x + iy$ is given by

$$\frac{1}{x + iy} = \frac{x - iy}{x^2 + y^2} = \frac{x}{x^2 + y^2} - \frac{y}{x^2 + y^2}i.$$

Given a geometric figure consisting of an assemblage of points, the POLARS with respect to an INVERSION CIRCLE constitute another figure. These figures are said to be reciprocal with respect to each other. Then there exists a DUALITY PRINCIPLE which states that theorems for the original figure can be immediately applied to the reciprocal figure after suitable modification (Lachlan 1893).

See also INVERSION, POLAR, POLE (INVERSION), RECIPROCAL CURVE, RECIPROCATION

Reciprocal Curve

The reciprocal curve of a given circle is the LOCUS of a point which moves so that its distance from the center of reciprocation varies as its distance from the line which is the reciprocal of the center of the given circle. The reciprocal of a circle is therefore a CONIC SECTION whose FOCUS is the center of reciprocation and whose directrix is the line which corresponds to the center of reciprocation. The conic will be an ELLIPSE, HYPERBOLA, or PARABOLA if the center of reciprocation lies inside, outside, or on the given circle, respectively (Lachlan 1893, p. 181).

See also DUALITY PRINCIPLE, POLAR, POLE (INVERSION), RECIPROCATION

References

Lachlan, R. "Reciprocation." Ch. 11 in *An Elementary Treatise on Modern Pure Geometry.* London: Macmillian, pp. 174–182, 1893.

Reciprocal Difference

The reciprocal differences are closely related to the DIVIDED DIFFERENCE. The first few are explicitly given by

$$\rho(x_0,\ x_1) = \frac{x_0 - x_1}{f_0 - f_1} \tag{1}$$

$$\rho_2(x_0,\ x_1,\ x_2) = \frac{x_0 - x_2}{\rho(x_0,\ x_1) - \rho(x_1,\ x_2)} + f_1 \tag{2}$$

$$\rho_3(x_0,\ x_1,\ x_2,\ x_3)$$

$$= \frac{x_0 - x_3}{\rho_2(x_0,\ x_1,\ x_2) - \rho_2(x_1,\ x_2,\ x_3)} + \rho(x_1,\ x_2) \tag{3}$$

$$\rho_n(x_0,\ x_1,\ \ldots,\ x_n)$$

$$= \frac{x_0 - x_n}{\rho_{n-1}(x_0,\ \ldots,\ x_{n-1}) - \rho_{n-1}(x_1,\ \ldots\ x_n)}$$

$$+ \rho_{n-x}(x_1,\ \ldots,\ x_{n-1}). \tag{4}$$

See also BACKWARD DIFFERENCE, CENTRAL DIFFERENCE, DIVIDED DIFFERENCE, FINITE DIFFERENCE, FORWARD DIFFERENCE

References

Abramowitz, M. and Stegun, C. A. (Eds.). *Handbook of Mathematical Functions with Formulas, Graphs, and Mathematical Tables, 9th printing.* New York: Dover, p. 878, 1972.
Beyer, W. H. (Ed.). *CRC Standard Mathematical Tables, 28th ed.* Boca Raton, FL: CRC Press, p. 443, 1987.

Reciprocal Matrix
MATRIX INVERSE

Reciprocal Permutation
INVERSE PERMUTATION

Reciprocal Polyhedron
DUAL POLYHEDRON

Reciprocal Polynomial

Given a polynomial in a single complex variable with complex coefficients

$$p(z) = a_n z^n + a_{n-1} z^{n-1} + \ldots + a_0,$$

the reciprocal polynomial is defined by

$$p^*(z) \equiv \bar{a}_0 z^n + \bar{a}_1 z^{n-1} + \ldots + \bar{a}_n,$$

where $\bar{a}$ denotes the COMPLEX CONJUGATE.

See also SCHUR TRANSFORM

References

Henrici, P. *Applied and Computational Complex Analysis, Vol. 1: Power Series-Integration-Conformal Mapping-Location of Zeros.* New York: Wiley, p. 492, 1988.

Reciprocating Sphere
MIDSPHERE

Reciprocation

An incidence-preserving transformation in which points are transformed into their POLARS. A PROJECTIVE GEOMETRY-like DUALITY PRINCIPLE holds for reciprocation which states that theorems for the original figure can be immediately applied to the RECIPROCAL figure after suitable modification (Lachlan 1893, pp. 174–182). Reciprocation (or "polar reciprocation") is the strictly proper term for duality. Brückner (1900) gave one the first exact definitions of polar reciprocation for constructing DUAL POLYHEDRA, although the plane geometric version (POLE, POLAR, and POWER of a circle) was considered by none less than Euclid (Wenninger 1983, pp. 1–2).

Lachlan 1893 (pp. 257–265) discusses another type of reciprocation he terms "circular reciprocation." However, the circular reciprocal figure is, in general, more complicated than the original, so the method is not as powerful as the usual polar reciprocation.

See also DUALITY PRINCIPLE, POLAR, POLE (INVERSION), RECIPROCAL

References

Brückner, M. *Vielecke under Vielflache.* Leipzig, Germany: Teubner, 1900.
Casey, J. "Theory of Poles and Polars, and Reciprocation." §6.7 in *A Sequel to the First Six Books of the Elements of Euclid, Containing an Easy Introduction to Modern Geometry with Numerous Examples, 5th ed., rev. enl.* Dublin: Hodges, Figgis, & Co., pp. 141–148, 1888.
Coxeter, H. S. M. and Greitzer, S. L. "Reciprocation." §6.1 in *Geometry Revisited.* Washington, DC: Math. Assoc. Amer., pp. 132–136, 1967.
Lachlan, R. "Reciprocation" and "Circular Reciprocation." Ch. 11 and §405–414 in *An Elementary Treatise on Modern Pure Geometry.* London: Macmillian, pp. 174–182 and 257–265, 1893.
Wenninger, M. J. *Dual Models.* Cambridge, England: Cambridge University Press, pp. 1–6, 1983.

Reciprocity Law
RECIPROCITY THEOREM

Reciprocity Theorem

If there exists a RATIONAL INTEGER x such that, when n, p, and q are POSITIVE INTEGERS,

$$x^n \equiv q \pmod{p},$$

then q is the n-adic residue of p, i.e., q is an n-adic residue of p IFF $x^n \equiv q \pmod{p}$ is solvable for x. Reciprocity theorems relate statements OF THE FORM "p is an n-adic residue of q" with reciprocal statements of the form "q is an n-adic residue of p."

The first case to be considered was $n = 2$ (the QUADRATIC RECIPROCITY THEOREM), of which Gauss gave the first correct proof. Gauss also solved the case $n = 3$ (CUBIC RECIPROCITY THEOREM) using INTEGERS OF THE FORM $a + b\rho$, where ρ is a root of $x^2 + x + 1 = 0$ and a, b are rational INTEGERS. Gauss stated the case $n = 4$ (BIQUADRATIC RECIPROCITY THEOREM) using the GAUSSIAN INTEGERS.

Proof of n-adic reciprocity for PRIME n was given by Eisenstein in 1844–50 and by Kummer in 1850–61. In the 1920s, Artin formulated ARTIN'S RECIPROCITY THEOREM, a general reciprocity law for all orders.

See also ARTIN RECIPROCITY, CLASS FIELD THEORY, CLASS NUMBER, CUBIC RECIPROCITY THEOREM, LANGLANDS PROGRAM, LANGLANDS RECIPROCITY, OCTIC RECIPROCITY THEOREM, QUADRATIC RECIPROCITY THEOREM, QUARTIC RECIPROCITY THEOREM, ROOK RECIPROCITY THEOREM

References

Lemmermeyer, F. *Reciprocity Laws: Their Evolution from Euler to Artin.* Draft. http://www.rzuser.uni-heidelberg.de/~hb3/rec.html.

Lemmermeyer, F. "Bibliography on Reciprocity Laws." http://www.rzuser.uni-heidelberg.de/~hb3/recbib.html.

Nagell, T. "Power Residues. Binomial Congruences." §34 in *Introduction to Number Theory.* New York: Wiley, pp. 115–120, 1951.

Wyman, B. F. "What Is a Reciprocity Law?" *Amer. Math. Monthly* **79**, 571–586, 1972.

Recognize
LATTICE REDUCTION

Recontres Problem
DERANGEMENT

Rectangle

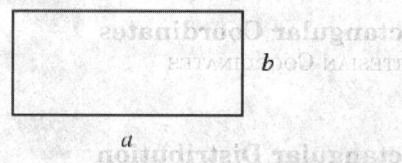

A closed planar QUADRILATERAL with opposite sides of equal lengths a and b, and with four RIGHT ANGLES. The AREA of the rectangle is

$$A = ab,$$

and its DIAGONALS p and q are of length

$$p = q = \sqrt{a^2 + b^2}.$$

A SQUARE is a degenerate rectangle with $a = b$.

A number of important topological surfaces can be constructed from the rectangle. Gluing both pairs of opposite edges together with no twists gives a TORUS, gluing two opposite edges together after giving a half-twist gives a MÖBIUS STRIP, gluing both pairs of opposite edges together giving one pair a half-twist gives a KLEIN BOTTLE, and giving both pairs a half-twist gives a PROJECTIVE PLANE (Stewart 1997).

See also BLANCHE'S DISSECTION, FAULT-FREE RECTANGLE, GOLDEN RECTANGLE, INCOMPARABLE RECTANGLES, KLEIN BOTTLE, MÖBIUS STRIP, OVERLAPPING RECTANGLES, PERFECT RECTANGLE, PROJECTIVE PLANE, RECTANGLE TILING, SQUARE, TORUS

References

Beyer, W. H. (Ed.). *CRC Standard Mathematical Tables,* 28th ed. Boca Raton, FL: CRC Press, p. 122, 1987.

Eppstein, D. "Rectilinear Geometry." http://www.ics.uci.edu/~eppstein/junkyard/rect.html.

Fukagawa, H. and Pedoe, D. "Circle and Rectangles." §3.4 in *Japanese Temple Geometry Problems.* Winnipeg, Manitoba, Canada: Charles Babbage Research Foundation, pp. 43–44 and 125, 1989.

Harris, J. W. and Stocker, H. "Rectangle." §3.6.5 in *Handbook of Mathematics and Computational Science.* New York: Springer-Verlag, p. 84, 1998.

Kern, W. F. and Bland, J. R. *Solid Mensuration with Proofs,* 2nd ed. New York: Wiley, p. 2, 1948.

Rectangle Function

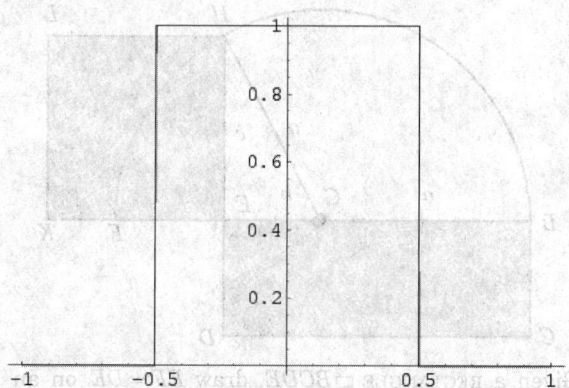

The rectangle function $\Pi(x)$ is a function which is 0 outside the interval $[-1/2, 1/2]$ and unity inside it. It is also called the GATE FUNCTION, PULSE FUNCTION, or

WINDOW FUNCTION, and is defined by

$$\Pi(x) \equiv \begin{cases} 0 & \text{for } |x| > \frac{1}{2} \\ \frac{1}{2} & \text{for } |x| = \frac{1}{2} \\ 1 & \text{for } |x| < \frac{1}{2}. \end{cases} \qquad (1)$$

The function $f(x) = h\Pi((x-c)/b)$ has height h, center c, and full-width b. Identities satisfied by the rectangle function include

$$\Pi(x) = H\left(x + \frac{1}{2}\right) - H\left(x - \frac{1}{2}\right) \qquad (2)$$

$$= H\left(\frac{1}{2} + x\right) + H\left(\frac{1}{2} - x\right) - 1 \qquad (3)$$

$$= H\left(\frac{1}{4} - x^2\right) \qquad (4)$$

$$= \frac{1}{2}\left[\text{sgn}\left(x + \frac{1}{2}\right) - \text{sgn}\left(x - \frac{1}{2}\right)\right], \qquad (5)$$

where $H(x)$ is the HEAVISIDE STEP FUNCTION. The FOURIER TRANSFORM of the rectangle function is given by

$$\mathscr{F}[\Pi(x)] = \int_{-\infty}^{\infty} e^{-2\pi ikx}\Pi(x)\,dx = \text{sinc}(\pi k), \qquad (6)$$

where $\text{sinc}(x)$ is the SINC FUNCTION.

See also ABSOLUTE VALUE, BOXCAR FUNCTION, FOURIER TRANSFORM–RECTANGLE FUNCTION, HEAVISIDE STEP FUNCTION, RAMP FUNCTION, SGN, TRIANGLE FUNCTION, UNIFORM DISTRIBUTION

References

Bracewell, R. "Rectangle Function of Unit Height and Base, $\Pi(x)$." In *The Fourier Transform and Its Applications, 3rd ed.* New York: McGraw-Hill, pp. 52–53, 1999.

Rectangle Squaring

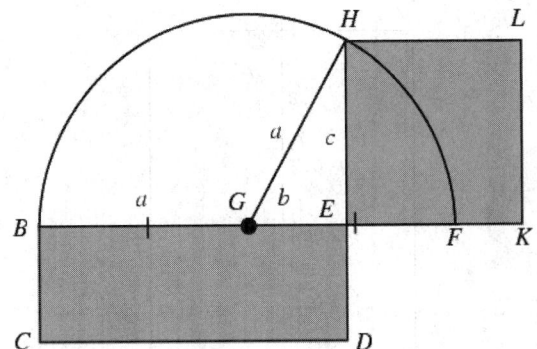

Given a RECTANGLE $\square BCDE$, draw $EF = DE$ on an extension of BE. Bisect BF and call the MIDPOINT G. Now draw a SEMICIRCLE centered at G, and construct the extension of ED which passes through the SEMICIRCLE at H. Then $\square EKLH$ has the same AREA

as $\square BCDE$. This can be shown as follows:

$$A(\square BCDE) = BE \cdot ED = BE \cdot EF$$

$$(a+b)(a-b) = a^2 - b^2 = c^2.$$

References

Dunham, W. "Hippocrates' Quadrature of the Lune." Ch. 1 in *Journey through Genius: The Great Theorems of Mathematics.* New York: Wiley, pp. 13–14, 1990.

Rectangle Tiling

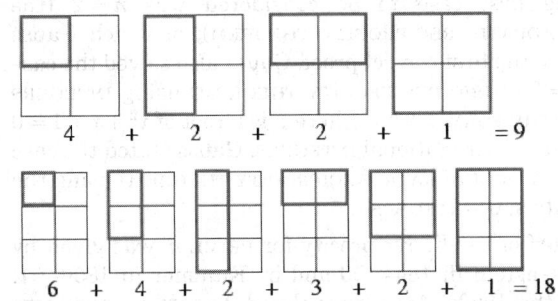

The number of ways $N(m, n)$ in which an $m \times n$ RECTANGLE can be tiled into subrectangles can be computed by counting the number of ways in which the upper right-hand corner can be selected for a given lower left-hand corner. For a lower left-hand corner with coordinates (i, j), there are $(m-i)(n-j)$ possible upper right-hand corners, so

$$N(m, n) = \sum_{i=0}^{m-1} \sum_{j=0}^{n-1} (m-i)(n-j) = \frac{1}{4}m(m+1)n(n+1).$$

Equivalently, $N(m, n)$ is the number of ways of picking two lines out of sets of $m+1$ and $n+1$ lines, giving

$$N(m, n) = \binom{m+1}{2}\binom{n+1}{2} = \frac{1}{4}m(m+1)n(n+1),$$

as before. Particular tilings are shown above for 2×2 and 2×3 rectangles.

See also PERFECT RECTANGLE, RECTANGLE, TRIANGLE TILING

References

Stewart, I. "Squaring the Square." *Sci. Amer.* **277**, 94–96, July 1997.

Rectangular Coordinates
CARTESIAN COORDINATES

Rectangular Distribution
UNIFORM DISTRIBUTION

Rectangular Hyperbola

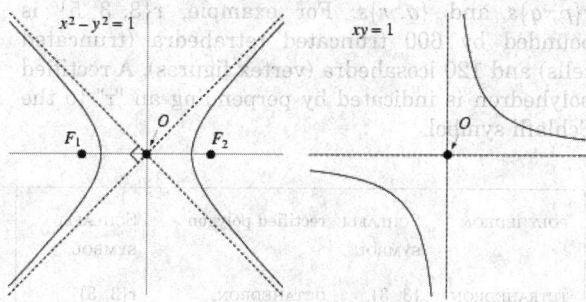

rectangular hyperbola

A HYPERBOLA for which the ASYMPTOTES are PERPEN-DICULAR, also called an EQUILATERAL HYPERBOLA or RIGHT HYPERBOLA. This occurs when the SEMIMAJOR and SEMIMINOR AXES are equal. This corresponds to taking $a = b$, giving eccentricity $e = \sqrt{2}$. Plugging $a = b$ into the general equation of a HYPERBOLA with SEMIMAJOR AXIS parallel to the x-AXIS and SEMIMINOR AXIS parallel to the y-AXIS (i.e., vertical DIRECTRIX),

$$\frac{(x - x_0)^2}{a^2} - \frac{(y - y_0)^2}{b^2} = 1 \quad (1)$$

therefore gives

$$(x - x_0)^2 - (y - y_0)^2 = a^2. \quad (2)$$

The rectangular hyperbola opening to the left and right has polar equation

$$r^2 = a^2 \sec(2\theta), \quad (3)$$

and the rectangular hyperbola opening in the first and third quadrants has the Cartesian equation

$$xy = a^2. \quad (4)$$

The INVERSE CURVE of a rectangular hyperbola with INVERSION CENTER at the center of the hyperbola is a LEMNISCATE (Wells 1991).

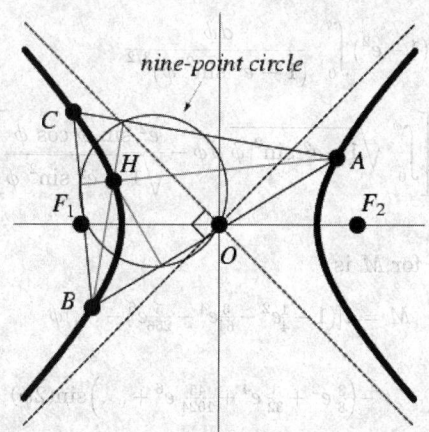

If the three vertices of a TRIANGLE $\triangle ABC$ lie on a

rectangular hyperbola, then so does the ORTHOCEN-TER H (Wells 1991). Equivalently, if four points form an ORTHOCENTRIC SYSTEM, then there is a family of rectangular hyperbolas through the points. Moreover, the LOCUS of centers O of these hyperbolas is the NINE-POINT CIRCLE of the triangle (Wells 1991).

If four points do not form an ORTHOCENTRIC SYSTEM, then there is a unique rectangular hyperbola passing through them, and its center is given by the inter-section of the NINE-POINT CIRCLES of the points taken three at a time (Wells 1991).

See also HYPERBOLA, LEMNISCATE, NINE-POINT CIR-CLE, ORTHOCENTRIC SYSTEM

References

Beyer, W. H. *CRC Standard Mathematical Tables, 28th ed.* Boca Raton, FL: CRC Press, pp. 218–219, 1987.
Courant, R. and Robbins, H. *What is Mathematics?: An Elementary Approach to Ideas and Methods, 2nd ed.* Oxford, England: Oxford University Press, pp. 76–77, 1996.
Coxeter, H. S. M. *Introduction to Geometry, 2nd ed.* New York: Wiley, p. 118, 1969.
Wells, D. *The Penguin Dictionary of Curious and Interesting Geometry.* London: Penguin, p. 209, 1991.

Rectangular Matrix

A MATRIX for which horizontal and vertical dimen-sions are *not* the same (i.e., an $m \times n$ MATRIX with $m \neq n$).

See also MATRIX, SQUARE MATRIX

Rectangular Parallelepiped

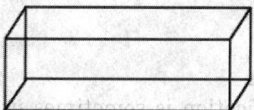

A closed box composed of 3 pairs of rectangular faces placed opposite each other and joined at RIGHT ANGLES to each other. This PARALLELEPIPED therefore corresponds to a rectangular "box." If the lengths of the sides are denoted a, b, and c, then the VOLUME is

$$V = abc, \quad (1)$$

the total SURFACE AREA is

$$S = 2(ab + bc + ca) \quad (2)$$

and the length of the "space" DIAGONAL is

$$d_{abc} = \sqrt{a^2 + b^2 + c^2}. \quad (3)$$

If $a = b = c$, then the rectangular parallelepiped is a CUBE.

See also CUBE, EULER BRICK, PARALLELEPIPED

References

Beyer, W. H. (Ed.). *CRC Standard Mathematical Tables, 28th ed.* Boca Raton, FL: CRC Press, p. 127, 1987.

Kern, W. F. and Bland, J. R. "Rectangular Parallelepiped." §10 in *Solid Mensuration with Proofs, 2nd ed.* New York: Wiley, pp. 21–25, 1948.

Rectangular Projection
EQUIRECTANGULAR PROJECTION

Rectifiable Current
The space of currents arising from rectifiable sets by integrating a differential form is called the space of 2-D rectifiable currents. For C a closed bounded rectifiable curve of a number of components in $\mathbb{R}^3$, C bounds a rectifiable current of least AREA. The theory of rectifiable currents generalizes to m-D surfaces in $\mathbb{R}^n$.

See also INTEGRAL CURRENT, REGULARITY THEOREM

References
Morgan, F. "What is a Surface?" *Amer. Math. Monthly* **103**, 369–376, 1996.

Rectifiable Set
The rectifiable sets include the image of any LIPSCHITZ FUNCTION f from planar domains into $\mathbb{R}^3$. The full set is obtained by allowing arbitrary measurable subsets of countable unions of such images of Lipschitz functions as long as the total AREA remains finite. Rectifiable sets have an "approximate" tangent plane at almost every point.

References
Morgan, F. "What is a Surface?" *Amer. Math. Monthly* **103**, 369–376, 1996.

Rectification
The term rectification is sometimes used to refer to the determination of the length of a curve.

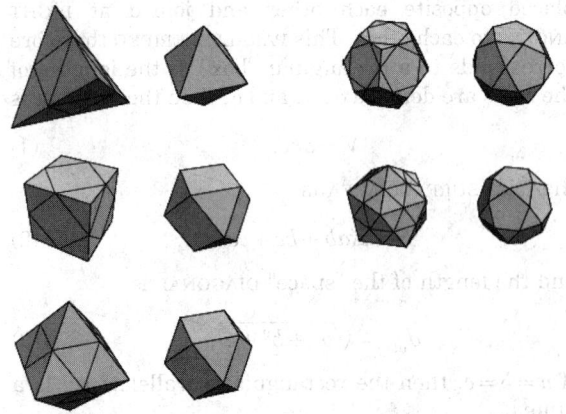

Rectification also refers to the operation which converts the midpoints of the edges of a regular polyhedron to the vertices of the related "rectified" polyhedron. Rectified forms are bounded by a combination of rectified cells and VERTEX FIGURES. There-

fore, a rectified polychoron r$\{p, q, r\}$ is bounded by r$\{p, q\}s$ and $\{q, r\}s$. For example, r$\{3, 3, 5\}$ is bounded by 600 truncated tetrahedra (truncated cells) and 120 icosahedra (vertex figures). A rectified polyhedron is indicated by perpending an "r" to the Schläfli symbol.

POLYHEDRON	SCHLÄFLI SYMBOL	rectified polygon	SCHLÄFLI SYMBOL
TETRAHEDRON	$\{3, 3\}$	OCTAHEDRON	r$\{3, 3\}$ $= \{3, 4\}$
OCTAHEDRON	$\{3, 4\}$	CUBOCTAHEDRON	r$\{3, 4\} = \{{3 \atop 4}\}$
CUBE	$\{4, 3\}$	CUBOCTAHEDRON	r$\{4, 3\} = \{{3 \atop 4}\}$
ICOSAHEDRON	$\{3, 5\}$	ICOSIDODECAHEDRON	r$\{3, 5\} = \{{3 \atop 5}\}$
DODECAHEDRON	$\{5, 3\}$	ICOSIDODECAHEDRON	r$\{5, 3\} = \{{3 \atop 5}\}$
16-CELL	$\{3, 3, 4\}$	24-CELL	r$\{3, 3, 4\}$ $= \{3, 4, 3\}$

Rectification of the six regular POLYCHORA gives five (not six) new POLYCHORA since the rectified 16-CELL r$\{3, 3, 4\}$ is the 24-CELL $\{3, 4, 3\}$.

See also QUADRABLE, SQUARING, STELLATION, TRUNCATION, VERTEX FIGURE

Rectifying Latitude
An AUXILIARY LATITUDE which gives a sphere having correct distances along the meridians. It is denoted μ (or ω) and is given by

$$\mu = \frac{\pi M}{2 M_p}. \tag{1}$$

M_p is evaluated for M at the north pole ($\phi = 90°$), and M is given by

$$M = a\left(1 - e^2\right) \int_0^\phi \frac{d\phi}{\left(1 - e^2 \sin^2 \phi\right)^{3/2}}$$

$$= a\left[\int_0^\phi \sqrt{1 - e^2 \sin^2 \phi}\, d\phi - \frac{e^2 \sin \phi \cos \phi}{\sqrt{1 - e^2 \sin^2 \phi}}\right]. \tag{2}$$

A series for M is

$$M = a[(1 - \tfrac{1}{4}e^2 - \tfrac{3}{64}e^4 - \tfrac{5}{256}e^6 - \ldots)\phi$$

$$- \left(\tfrac{3}{8}e^2 + \tfrac{3}{32}e^4 + \tfrac{45}{1024}e^6 + \ldots\right)\sin(2\phi)$$

$$+ \left(\tfrac{15}{256}e^4 + \tfrac{45}{1024}e^6 + \ldots\right)\sin(4\phi)$$

$$-\left(\tfrac{35}{3072}e^6+\ldots\right)\sin(6\phi)+\ldots], \tag{3}$$

and a series for μ is

$$\mu=\phi-\left(\tfrac{3}{2}e_1-\tfrac{9}{16}e_1^3+\ldots\right)\sin(2\phi)$$

$$+\left(\tfrac{15}{16}e_1^2-\tfrac{15}{32}e_1^4+\ldots\right)\sin(4\phi)$$

$$-\left(\tfrac{35}{48}e_1^3-\ldots\right)\sin(6\phi)+\left(\tfrac{315}{512}e_1^4-\ldots\right)\sin(8\phi)+\ldots, \tag{4}$$

where

$$e_1\equiv\frac{1-\sqrt{1-e^2}}{1+\sqrt{1-e^2}}. \tag{5}$$

The inverse formula is

$$\phi=\mu+\left(\tfrac{3}{2}e_1-\tfrac{27}{32}e_1^3+\ldots\right)\sin(2\mu)$$

$$+\left(\tfrac{21}{16}e_1^2-\tfrac{55}{32}e_1^4+\ldots\right)\sin(4\mu)$$

$$+\left(\tfrac{151}{96}e_1^3-\ldots\right)\sin(6\mu)$$

$$-\left(\tfrac{1097}{512}e^4-\ldots\right)\sin(8\mu)+\ldots \tag{6}$$

See also LATITUDE

References

Adams, O. S. "Latitude Developments Connected with Geodesy and Cartography with Tables, Including a Table for Lambert Equal-Area Meridional Projections." Spec. Pub. No. 67. U. S. Coast and Geodetic Survey, pp. 125–128, 1921.

Snyder, J. P. *Map Projections--A Working Manual.* U. S. Geological Survey Professional Paper 1395. Washington, DC: U. S. Government Printing Office, pp. 16–17, 1987.

Rectifying Plane

The PLANE spanned by the TANGENT VECTOR $\mathbf{T}$ and BINORMAL VECTOR $\mathbf{B}$.

See also BINORMAL VECTOR, TANGENT VECTOR

Rectilinear Crossing Number

The minimum number $\bar{\nu}(G)$ of crossings in a straight line drawing of a graph G in a plane. For a COMPLETE GRAPH of order $n\geq 10$, the rectilinear crossing number is always larger than the general graph crossing number. For the COMPLETE GRAPH K_n with $n=1, 2, \ldots, \bar{\nu}(G)$ is 0, 0, 0, 0, 1, 3, 9, 19, 36, 62, ... (Sloane's A014540; White and Beineke 1978, Schneinerman and Wilf 1994). Although it had long been known that $\bar{\nu}(K_{10})$ was either 61 or 62 (Singer 1971, Gardner 1986), it was finally proven to be 62 by Brodsky *et al.* (2000).

Upper limits have been provided by Singer (1971), who showed that

$$\bar{\nu}(K_n)\leq\tfrac{1}{312}(5n^4-39n^3+91n^2-57n), \tag{1}$$

and Jensen (1971), who showed that

$$\bar{\nu}(K_n)\leq\tfrac{7}{432}n^4+\mathcal{O}(n^3). \tag{2}$$

Bounds for $\bar{\nu}(K_n)$ are given by

$$0.290<\frac{61}{210}\leq\rho=\lim_{n\to\infty}\frac{\bar{\nu}(K_n)}{\binom{n}{4}}\leq\frac{5}{13}<0.385, \tag{3}$$

where $\binom{n}{k}$ is a BINOMIAL COEFFICIENT and the exact value of ρ is not known (Finch).

The rectilinear crossing number has an unexpected connection with SYLVESTER'S FOUR-POINT PROBLEM (Finch).

See also CROSSING NUMBER (GRAPH), PLANAR STRAIGHT LINE GRAPH, SYLVESTER'S FOUR-POINT PROBLEM, TOROIDAL CROSSING NUMBER

References

Brodsky, A.; Durocher, S.; and Gethner, E. "Toward the Rectilinear Crossing Number of K_n: New Drawings, Upper Bounds, and Asymptotics." http://www.cs.ubc.ca/spider/abrodsky/papers/reccr_n.ps.gz.

Brodsky, A.; Durocher, S.; and Gethner, E. The Rectilinear Crossing Number of K_{10} is 62. 22 Sep 2000. http://xxx.lanl.gov/abs/cs.DM/0009023/.

Finch, S. "Favorite Mathematical Constants." http://www.mathsoft.com/asolve/constant/crss/crss.html.

Gardner, M. *Knotted Doughnuts and Other Mathematical Entertainments.* New York: W. H. Freeman, 1986.

Guy, R. K. "Crossing Numbers of Graphs." In *Graph Theory and Applications: Proceedings of the Conference at Western Michigan University, Kalamazoo, Mich., May 10–13, 1972* (Ed. Y. Alavi, D. R. Lick, and A. T. White). New York: Springer-Verlag, pp. 111–124, 1972.

Harary, F. and Hill, A. "On the Number of Crossings in a Complete Graph." *Proc. Edinburgh Math. Soc.* **13**, 333–338, 1962/1963.

Jensen, H. F. "An Upper Bound for the Rectilinear Crossing Number of the Complete Graph." *J. Combin. Th. B* **10**, 212–216, 1971.

Klee, V. "What is the Expected Volume of a Simplex Whose Vertices are Chosen at Random from a Given Convex Body." *Amer. Math. Monthly* **76**, 286–288, 1969.

Schneinerman, E. and Wilf, H. S. "The Rectilinear Crossing Number of a Complete Graph and Sylvester's 'Four Point' Problem of Geometric Probability." *Amer. Math. Monthly* **101**, 939–943, 1994.

Singer, D. "The Rectilinear Crossing Number of Certain Graphs." Unpublished manuscript, 1971. Quoted in Gardner, M. *Knotted Doughnuts and Other Mathematical Entertainments.* New York: W. H. Freeman, 1986.

Sloane, N. J. A. Sequences A014540 in "An On-Line Version of the Encyclopedia of Integer Sequences." http://www.research.att.com/~njas/sequences/eisonline.html.

White, A. T. and Beineke, L. W. "Topological Graph Theory." In *Selected Topics in Graph Theory* (Ed. L. W. Beineke and R. J. Wilson). New York: Academic Press, pp. 15–49, 1978.

Wilf, H. "On Crossing Numbers, and Some Unsolved Problems." In *Combinatorics, Geometry, and Probability:*

A Tribute to Paul Erdos. Papers from the Conference in Honor of Erdos' 80th Birthday Held at Trinity College, Cambridge, March 1993 (Ed. B. Bollobás and A. Thomason). Cambridge, England: Cambridge University Press, pp. 557–562, 1997.

Recurrence Relation

A mathematical relationship expressing f_n as some combination of f_i with $i < n$. The solutions to a linear recurrence can be computed straightforwardly, but QUADRATIC RECURRENCES are not so well understood. The sequence generated by a recurrence relation is called a RECURRENCE SEQUENCE. Perhaps the most famous example of a recurrence relation is the one defining the FIBONACCI NUMBERS,

$$F_n = F_{n-2} + F_{n-1}$$

for $n \geq 3$ and with $F_1 = F_2 = 1$.

See also ARGUMENT ADDITION RELATION, ARGUMENT MULTIPLICATION RELATION, CLENSHAW RECURRENCE FORMULA, QUADRATIC RECURRENCE, RECURRENCE SEQUENCE, REFLECTION RELATION, TRANSLATION RELATION

References

Press, W. H.; Flannery, B. P.; Teukolsky, S. A.; and Vetterling, W. T. "Recurrence Relations and Clenshaw's Recurrence Formula." §5.5 in *Numerical Recipes in FORTRAN: The Art of Scientific Computing, 2nd ed.* Cambridge, England: Cambridge University Press, pp. 172–178, 1992.
Sloane, N. J. A. and Plouffe, S. "Recurrences and Generating Functions" and "Other Methods for Hand Analysis." §2.4 and 2.6 in *The Encyclopedia of Integer Sequences.* San Diego, CA: Academic Press, pp. 9–10 and 13–18, 1995.

Recurrence Sequence

A sequence of numbers generated by a RECURRENCE RELATION is called a recurrence sequence. Perhaps the most famous recurrence sequence is the FIBONACCI NUMBERS.

For a finite linear recurrence sequence of functions

$$s_i(x) = A_i(x) s_{i+1}(x) + B_i(x)$$

where $i = 1, \ldots, r-1$, and $s_r(x) = h(x)$, then

$$s_1(x) = \begin{vmatrix} B_1(x) & -A_1(x) & 0 & \cdots & & 0 \\ B_2(x) & 1 & -A_2(x) & & & 0 \\ B_3(x) & 0 & 1 & \ddots & & \vdots \\ \vdots & \vdots & & \ddots & \ddots & 0 \\ B_{r-1}(x) & 0 & 0 & & & -A_{r-1}(x) \\ h(x) & 0 & 0 & \cdots & & 1 \end{vmatrix} \quad (1)$$

(Mansour 2000).

If a sequence $\{x_n\}$ with $x_1 = x_2 = 1$ is described by a two-term linear RECURRENCE RELATION OF THE FORM

$$x_n = A x_{n-1} + B x_{n-2} \quad (2)$$

for $n \geq 3$ and A and B constants, then the closed form for x_n is given by

$$x_n = \frac{\alpha^n - \beta^n}{\alpha - \beta} \quad (3)$$

where α and β are the ROOTS of the QUADRATIC EQUATION

$$x^2 - Ax - B = 0, \quad (4)$$

$$\alpha = \tfrac{1}{2}\left(A + \sqrt{A^2 + 4B}\right) \quad (5)$$

$$\beta = \tfrac{1}{2}\left(A - \sqrt{A^2 + 4B}\right) \quad (6)$$

For example, the FIBONACCI NUMBERS F_n which are equal to 1, 1, 2, 3, 5, 8, ... for $n = 1, 2, \ldots$, have $A = B = 1$, so $\alpha = (1 + \sqrt{5})/2$ and $\beta = (1 - \sqrt{5})/2$, giving

$$F_n = \frac{\left[\tfrac{1}{2}(1 + \sqrt{5})\right]^n - \left[\tfrac{1}{2}(1 - \sqrt{5})\right]^n}{\sqrt{5}}$$

$$= \frac{(1 + \sqrt{5})^n - (1 - \sqrt{5})^n}{2^n \sqrt{5}}. \quad (7)$$

Grosjean (1993) discusses how to rewrite such "difference of powers of roots" solutions in explicit integer form.

The general second-order linear recurrence

$$x_n = A x_{n-1} + B x_{n-2} \quad (8)$$

for constants A and B with arbitrary x_1 and x_2 has terms

$x_1 = x_1$
$x_2 = x_2$
$x_3 = B x_1 + A x_2$
$x_4 = B x_2 + A B x_1 + A^2 x_2$
$x_5 = B^2 x_1 + 2 A B x_2 + A^2 B x_1 + A^3 x_2$
$x_6 = B^2 x_2 + 2 A B^2 x_1 + 3 A^2 B x_2 + A^3 B x_1 + A^4 x_2$
$x_7 = B^3 x_1 + 4 A^3 B x_2 + 3 A^2 B^2 x_1 + 3 A B^2 x_2 + A^4 B x_1 + A^5 x_2,$

so an arbitrary term can be written as

$$x_n = \sum_{k=0}^{n-2} \binom{\left\lfloor \tfrac{1}{2}(n+k-2) \right\rfloor}{k} A^k B^{\lfloor (n-k-1)/2 \rfloor}$$

$$\times x_1^{[n+k \ (\mathrm{mod}\ 2)]} x_2^{[n+k+1 \ (\mathrm{mod}\ 2)]}. \quad (9)$$

$$= -(A x_1 - x_2) \sum_{k=0}^{n-2} A^{2k-n+2} B^{-k+n-2} \binom{k}{n-k-2}$$

$$+ x_1 \sum_{k=0}^{n-1} A^{2k-n+1} B^{-k+n-1} \binom{k}{n-k-1}. \quad (10)$$

The general linear third-order recurrence

$$x_n = A x_{n-1} + B x_{n-2} + C x_{n-3} \quad (11)$$

has solution

$$x_n = x_1 \left(\frac{\alpha^{-n}}{A + 2\alpha B + 3\alpha^2 C} + \frac{\beta^{-n}}{A + 2\beta B + 3\beta^2 C} \right.$$

$$\left. + \frac{\gamma^{-n}}{A + 2\gamma B + 3\gamma^2 C} \right) - (Ax_1 - x_2)$$

$$\times \left(\frac{\alpha^{1-n}}{A + 2\alpha B + 3\alpha^2 C} + \frac{\beta^{1-n}}{A + 2\beta B + 3\beta^2 B} \right.$$

$$\left. + \frac{\gamma^{1-n}}{A + 2\gamma C + 3\gamma^2 C} \right) - (Bx_1 + Ax_2 - x_3)$$

$$\times \left(\frac{\alpha^{2-n}}{A + 2\alpha B + 3\alpha^2 C} + \frac{\beta^{2-n}}{A + 2\beta B + 3\beta^2 C} \right.$$

$$\left. + \frac{\gamma^{2-n}}{A + 2\gamma B + 3\gamma^2 C} \right), \qquad (12)$$

where α, β, and γ are the roots of the polynomial

$$Cx^3 + Bx^2 + Ax = 1. \qquad (13)$$

A QUOTIENT-DIFFERENCE TABLE eventually yields a line of 0s IFF the starting sequence is defined by a linear RECURRENCE RELATION.

A linear second-order recurrence

$$f_{n+1} = xf_n + yf_{n-1} \qquad (14)$$

can be solved rapidly using a "rate doubling,"

$$f_{n+2} = (x^2 + 2y)f_n - y^2 f_{n-2}, \qquad (15)$$

"rate tripling"

$$f_{n+3} = (x^3 + 3xy)f_n + y^3 f_{n-3}, \qquad (16)$$

or in general, "rate k-tupling" formula

$$f_{n+k} = p_k f_n + q_k f_{n-k}, \qquad (17)$$

where

$$p_0 = 2 \qquad (18)$$

$$p_1 = x \qquad (19)$$

$$p_k = 2(-y)^{k/2} T_k(x/(2i\sqrt{y})) \qquad (20)$$

$$p_{k+1} = xp_k + yp_{k-1} \qquad (21)$$

(here, $T_k(x)$ is a CHEBYSHEV POLYNOMIAL OF THE FIRST KIND) and

$$q_0 = -1 \qquad (22)$$

$$q_1 = y \qquad (23)$$

$$q_k = -(-y)^k \qquad (24)$$

$$q_{k+1} = -yq_k \qquad (25)$$

(Gosper and Salamin 1972).

Let

$$s(X) = \prod_{i=1}^{m} (1 - \alpha_i X)^{n_i} = 1 - s_1 X - \ldots - s_n X^n, \qquad (26)$$

where the generalized POWER sum $a(h)$ for $h = 0, 1, \ldots$ is given by

$$a(h) = \sum_{i=1}^{m} A_i(h)\alpha_i^h, \qquad (27)$$

with distinct NONZERO roots α_i, COEFFICIENTS $A_i(h)$ which are POLYNOMIALS of degree $n_i - 1$ for POSITIVE INTEGERS n_i, and $i \in [1, m]$. Then the sequence $\{a_h\}$ with $a_h = a(h)$ satisfies the RECURRENCE RELATION

$$a_{h+n} = s_i a_{h+n-1} + \ldots + s_n a_h \qquad (28)$$

(Meyerson and van der Poorten 1995).

The terms in a general recurrence sequence belong to a finitely generated RING over the INTEGERS, so it is impossible for every RATIONAL NUMBER to occur in any finitely generated recurrence sequence. If a recurrence sequence vanishes infinitely often, then it vanishes on an arithmetic progression with a common difference 1 that depends only on the roots. The number of values that a recurrence sequence can take on infinitely often is bounded by some INTEGER l that depends only on the roots. There is no recurrence sequence in which each INTEGER occurs infinitely often, or in which every GAUSSIAN INTEGER occurs (Myerson and van der Poorten 1995).

Let $\mu(n)$ be a bound so that a nondegenerate INTEGER recurrence sequence of order n takes the value zero at least $\mu(n)$ times. Then $\mu(2) = 1$, $\mu(3) = 6$, and $\mu(4) \geq 9$ (Myerson and van der Poorten 1995). The maximal case for $\mu(3)$ is

$$a_{n+3} = 2a_{n+2} - 4a_{n+1} + 4a_n \qquad (29)$$

with

$$a_0 = a_1 = 0 \qquad (30)$$

$$a_2 = 1. \qquad (31)$$

The zeros are

$$a_0 = a_1 = a_4 = a_6 = a_{13} = a_{52} = 0 \qquad (32)$$

(Beukers 1991).

See also BINET FORMS, BINET'S FIBONACCI NUMBER FORMULA, FAST FIBONACCI TRANSFORM, FIBONACCI NUMBER, LUCAS SEQUENCE, QUOTIENT-DIFFERENCE TABLE, SKOLEM-MAHLER-LERCH THEOREM

References

Batchelder, P. M. *An Introduction to Linear Difference Equations.* New York: Dover, 1967.
Beukers, F. "The Zero-Multiplicity of Ternary Recurrences." *Composito Math.* **77**, 165–177, 1991.

Gosper, R. W. and Salamin, E. Item 14 in Beeler, M.; Gosper, R. W.; and Schroeppel, R. *HAKMEM.* Cambridge, MA: MIT Artificial Intelligence Laboratory, Memo AIM-239, pp. 8–9, Feb. 1972.

Greene, D. H. and Knuth, D. E. *Mathematics for the Analysis of Algorithms, 3rd ed.* Boston, MA: Birkhäuser, 1990.

Grosjean, C. C. In *Topics in Polynomials of One and Several Variables and Their Applications: Volume Dedicated to the Memory of P.L. Chebyshev (1821–1894)* (Ed. T. M. Rassias, H. M. Srivastava, and A. Yanushauskas). Singapore: World Scientific, 1993.

Levy, H. and Lessman, F. *Finite Difference Equations.* New York: Dover, 1992.

Mansour, T. Permutations Avoiding a Pattern from S_k and at Least Two Patterns from S_3. 31 Jul 2000. http://xxx.lanl.gov/abs/math.CO/0007194/.

Myerson, G. and van der Poorten, A. J. "Some Problems Concerning Recurrence Sequences." *Amer. Math. Monthly* **102**, 698–705, 1995.

Riordan, J. *An Introduction to Combinatorial Analysis.* New York: Wiley, 1980.

Wimp, J. *Computations with Recurrence Relations.* Boston, MA: Pitman, 1984.

Recurring Decimal

REPEATING DECIMAL

Recurring Digital Invariant

To define a recurring digital invariant of order k, compute the sum of the kth powers of the digits of a number n. If this number n' is equal to the original number n, then $n = n'$ is called a k-NARCISSISTIC NUMBER. If not, compute the sums of the kth powers of the digits of n', and so on. If this process eventually leads back to the original number n, the *smallest number* in the sequence $\{n, n', n'', \ldots\}$ is said to be a k-recurring digital invariant. For example,

$$55 : 5^3 + 5^3 = 250$$

$$250 : 2^3 + 5^3 + 0^3 = 133$$

$$133 : 1^3 + 3^3 + 3^3 = 55,$$

so 55 is an order 3 recurring digital invariant. The following table gives recurring digital invariants of orders 2 to 10 (Madachy 1979).

Order	RDI	Cycle Lengths
2	4	8
3	55, 136, 160, 919	3, 2, 3, 2
4	1138, 2178	7, 2
5	244, 8294, 8299, 9044, 9045, 10933,	28, 10, 6, 10, 22, 4, 12, 2, 2
	24584, 58618, 89883	
6	17148, 63804, 93531, 239459, 282595	30, 2, 4, 10, 3
7	80441, 86874, 253074, 376762,	92, 56, 27, 30, 14, 21
	922428, 982108, five more	
8	6822, 7973187, 8616804	
9	322219, 2274831, 20700388, eleven more	
10	20818070, five more	

See also 196-ALGORITHM, ADDITIVE PERSISTENCE, DIGITADDITION, DIGITAL ROOT, HAPPY NUMBER, KAPREKAR NUMBER, NARCISSISTIC NUMBER, VAMPIRE NUMBER

References

Madachy, J. S. *Madachy's Mathematical Recreations.* New York: Dover, pp. 163–165, 1979.

Recursion

A recursive process is one in which objects are defined in terms of other objects of the same type. Using some sort of RECURRENCE RELATION, the entire class of objects can then be built up from a few initial values and a small number of rules. The FIBONACCI NUMBERS are most commonly defined recursively. Care, however, must be taken to avoid SELF-RECURSION, in which an object is defined in terms of itself, leading to an infinite nesting.

See also ACKERMANN FUNCTION, PRIMITIVE RECURSIVE FUNCTION, RECURRENCE RELATION, RECURRENCE SEQUENCE, RECURSIVE FUNCTION, REGRESSION, RICHARDSON'S THEOREM, SELF-RECURSION, SELF-SIMILARITY, TAK FUNCTION

References

Buck, R. C. "Mathematical Induction and Recursive Definitions." *Amer. Math. Monthly* **70**, 128–135, 1963.

Gardner, M. "Infinite Regress." Ch. 22 in *The Sixth Book of Mathematical Games from Scientific American.* Chicago, IL: University of Chicago Press, pp. 220–229, 1984.

Knuth, D. E. "Textbook Examples of Recursion." In *Artificial Intelligence and Mathematical Theory of Computation, Papers in Honor of John McCarthy* (Ed. V. Lifschitz). Boston, MA: Academic Press, pp. 207–229, 1991.

Péter, R. *Rekursive Funktionen.* Budapest: Akad. Kiado, 1951.

Thompson, W. "Recursive Algorithms: A Mixed Blessing." *Computers in Physics* **10**, 25–29, 1996.

Recursive Function

A recursive function is a function generated by (1) ADDITION, (2) MULTIPLICATION, (3) selection of an element from a list, and (4) determination of the truth or falsity of the INEQUALITY $a < b$ according to the technical rules:

1. If F and the sequence of functions $G_1, ..., G_n$ are recursive, then so is $F(G_1, ..., G_n)$.

2. If F is a recursive function such that there is an x for each a with $H(a, x) = 0$, then the smallest x can be obtained recursively.

A TURING MACHINE is capable of computing recursive functions.

See also TURING MACHINE

References

Kleene, S. C. *Introduction to Metamathematics*. Princeton, NJ: Van Nostrand, 1952.

Péter, R. *Rekursive Funktionen*. Budapest: Akad. Kiado, 1951.

Schnorr, C. P. *Rekursive Funktionen und ihre Komplexität*. Stuttgart, Germany: Teubner, 1974.

Recursive Monotone Stable Quadrature

A QUADRATURE (NUMERICAL INTEGRATION) algorithm which has a number of desirable properties.

References

Favati, P.; Lotti, G.; and Romani, F. "Interpolary Integration Formulas for Optimal Composition." *ACM Trans. Math. Software* **17**, 207–217, 1991.

Favati, P.; Lotti, G.; and Romani, F. "Algorithm 691: Improving QUADPACK Automatic Integration Routines." *ACM Trans. Math. Software* **17**, 218–232, 1991.

Red Net

The coloring red of two COMPLETE SUBGRAPHS of $n/2$ points (for EVEN n) in order to generate a BLUE-EMPTY GRAPH.

See also BLUE-EMPTY GRAPH, COMPLETE GRAPH

Red-Black Tree

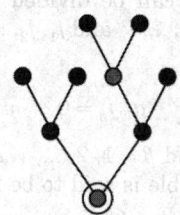

An extended BINARY TREE satisfying the following conditions:

1. Every node has two CHILDREN, each colored either red or black.
2. Every LEAF node is colored black.
3. Every red node has both of its CHILDREN colored black.
4. Every path from the ROOT to a LEAF contains the same number (the "black-height") of black nodes.

Let n be the number of internal nodes of a red-black tree. Then the number of red-black trees for $n = 1, 2,$

... is 2, 2, 3, 8, 14, 20, 35, 64, 122, ... (Sloane's A001131). The number of trees with black roots and red roots are given by Sloane's A001137 and Sloane's A001138, respectively.

Let T_h be the GENERATING FUNCTION for the number of red-black trees of black-height h indexed by the number of LEAVES. Then

$$T_{h+1}(x) = [T_h(x)]^2 + [T_h(x)]^4, \quad (1\mathrm{r_p ar} \qquad (1)$$

where $T_1(x) = x + x^2$. If $T(x)$ is the GENERATING FUNCTION for the number of red-black trees, then

$$T(x) = x + x^2 + T\left(x^2(1+x)^2\right) \qquad (2)$$

(Ruskey). Let $rb(n)$ be the number of red-black trees with n LEAVES, $r(n)$ the number of red-rooted trees, and $b(n)$ the number of black-rooted trees. All three of the quantities satisfy the RECURRENCE RELATION

$$R(n) = \sum_{n/4 \le n \le n/2} \binom{2m}{n-2m} R(m), \qquad (3)$$

where $\binom{n}{k}$ is a BINOMIAL COEFFICIENT, $rb(1) = 1$, $rb(2) = 2$ for $R(n) = rb(n)$, $r(1) = r(3) = 0$, $r(2) = 1$ for $R(n) = r(n)$, and $b(1) = 1$ for $R(n) = b(n)$ (Ruskey).

See also B-TREE

References

Beyer, R. "Symmetric Binary B-Trees: Data Structures and Maintenance Algorithms." *Acta Informat.* **1**, 290–306, 1972.

Binstock, A.; and Rex, J. *Practical Algorithms for Programmers*. Reading, MA: Addison-Wesley, 1995.

Cormen, T.; Leiserson, C.; and Rivest, R. *Introduction to Algorithms*. Cambridge MA: MIT Press, 1990.

Guibas, L. and Sedgewick, R. "A Dichromatic Framework for Balanced Trees." *In Proc. 19th IEEE Symp. Foundations of Computer Science,* pp. 8–21, 1978.

Rivest, R. L.; Leiserson, C. E.; and Cormen, R. H. *Introduction to Algorithms*. New York: McGraw-Hill, 1990.

Ruskey, F. "Information on Red-Black Trees." http://www.theory.csc.uvic.ca/~cos/inf/tree/RedBlackTree.html.

Skiena, S. S. *The Algorithm Design Manual*. New York: Springer-Verlag, pp. 177 and 179, 1997.

Sloane, N. J. A. Sequences A001131, A001137, and A001138 in "An On-Line Version of the Encyclopedia of Integer Sequences." http://www.research.att.com/~njas/sequences/eisonline.html.

Wood, D. *Data Structures, Algorithms, and Performance*. Reading, MA: Addison-Wesley, 1993.

Reduced Amicable Pair

QUASIAMICABLE PAIR

Reduced Fraction

A FRACTION a/b written in lowest terms, i.e., by dividing NUMERATOR and DENOMINATOR through by their GREATEST COMMON DIVISOR (a, b). For example, 2/3 is the reduced fraction of 8/12.

See also FRACTION, IMPROPER FRACTION, MIXED

FRACTION, PROPER FRACTION

Reduced Knot Diagram

A KNOT DIAGRAM in which none of the crossings are REDUCIBLE.

See also KNOT DIAGRAM, REDUCIBLE CROSSING

References

Hoste, J.; Thistlethwaite, M.; and Weeks, J. "The First 1,701,936 Knots." *Math. Intell.* **20**, 33–48, Fall 1998.

Reduced Latitude

PARAMETRIC LATITUDE

Reduced Maxwell-Bloch Equations

The system of PARTIAL DIFFERENTIAL EQUATIONS

$$E_t - v = 0 \qquad (1)$$

$$r_x + \omega v = 0 \qquad (2)$$

$$q_x + Ev = 0 \qquad (3)$$

$$v_x - \omega r - Eq = 0. \qquad (4)$$

References

Calogero, F. and Degasperis, A. *Spectral Transform and Solitons: Tools to Solve and Investigate Nonlinear Evolution Equations.* New York: North-Holland, p. 59, 1982.
Zwillinger, D. *Handbook of Differential Equations, 3rd ed.* Boston, MA: Academic Press, p. 139, 1997.

Reduced Residue System

Any system of $\phi(n)$ integers, where $\phi(n)$ is the TOTIENT FUNCTION, representing all the RESIDUE CLASSES RELATIVELY PRIME to n is called a reduced residue system (Nagell 1951, p. 71).

See also COMPLETE RESIDUE SYSTEM, RESIDUE CLASS

References

Nagell, T. "Residue Classes and Residue Systems." §20 in *Introduction to Number Theory.* New York: Wiley, pp. 69–71, 1951.

Reduced Root System

A ROOT SYSTEM R satisfying the additional property that, if $\alpha \in R$, then the only multiples of α in R are $\pm \alpha$.

See also ROOT SYSTEM

References

Andrews, G. E. *q-Series: Their Development and Application in Analysis, Number Theory, Combinatorics, Physics, and Computer Algebra.* Providence, RI: Amer. Math. Soc., p. 40, 1986.
Humphrey, J. E. *Introduction to Lie Algebras and Representation Theory.* New York: Springer-Verlag, p. 42, 1972.

Reducible Crossing

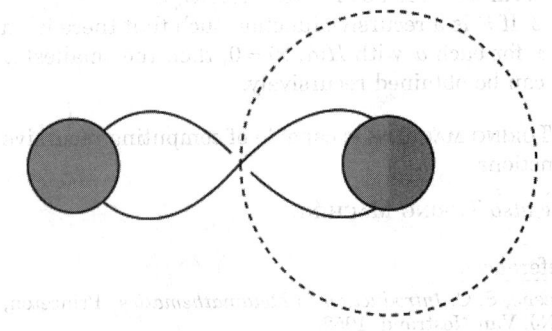

reducible crossing

A crossing in a KNOT DIAGRAM for which there exists a circle in the projection plane meeting the diagram transversely at that crossing, but not meeting the diagram at any other point. Removable crossings can be removed by twisting, and so cannot occur in a KNOT DIAGRAM of minimal CROSSING NUMBER. Reducible crossings are also called nugatory crossings (Tait 1898, Hoste *et al.* 1998) or removable crossings.

See also

See also ALTERNATING KNOT, KNOT DIAGRAM, REDUCED KNOT DIAGRAM

References

Hoste, J.; Thistlethwaite, M.; and Weeks, J. "The First 1,701,936 Knots." *Math. Intell.* **20**, 33–48, Fall 1998.
Tait, P. G. "On Knots I, II, and III." *Scientific Papers, Vol. 1.* Cambridge, England: University Press, pp. 273–347, 1898.

Reducible Matrix

A SQUARE $n \times n$ matrix $\mathsf{A} = a_{ij}$ is called reducible if the indices $1, 2, ..., n$ can be divided into two disjoint nonempty sets $i_1, i_2, ..., i_\mu$ and $j_1, j_2, ..., j_\nu$ (with $\mu + \nu = n$) such that

$$a_{i_\alpha j_\beta} = 0$$

for $\alpha = 1, 2, ..., \mu$ and $\beta = 1, 2, ..., \nu$. A SQUARE MATRIX which is not reducible is said to be IRREDUCIBLE.

See also SQUARE MATRIX

References

Gradshteyn, I. S. and Ryzhik, I. M. *Tables of Integrals, Series, and Products, 6th ed.* San Diego, CA: Academic Press, p. 1103, 2000.

Reducible Representation

IRREDUCIBLE REPRESENTATION

Reductio ad Absurdum

A method of PROOF which proceeds by stating a proposition and then showing that it results in a contradiction, thus demonstrating the proposition to be false. In the words of G. H. Hardy , "*Reductio ad absurdum*, which Euclid loved so much, is one of a mathematician's finest weapons. It is a far finer gambit than any CHESS gambit: a CHESS player may offer the sacrifice of a pawn or even a piece, but a mathematician offers the game" (Coxeter and Greitzer 1967, p. 16; Hardy 1993, p. 34).

See also PROOF

References

Coxeter, H. S. M. and Greitzer, S. L. *Geometry Revisited.* Washington, DC: Math. Assoc. Amer., p. 16, 1967.
Hardy, G. H. *A Mathematician's Apology, reprinted with a foreword by C. P. Snow.* New York: Cambridge University Press, p. 34, 1993.

Reduction of Order

ORDINARY DIFFERENTIAL EQUATION–SECOND-ORDER

Reduction Theorem

If a fixed point is added to each group of a special complete series, then the resulting series is complete.

References

Coolidge, J. L. *A Treatise on Algebraic Plane Curves.* New York: Dover, p. 253, 1959.

Redundancy

$$R(X_1, \ldots X_n) \equiv \sum_{i=1}^{n} H(X_i) - H(X_1, \ldots, X_n),$$

where $H(x_i)$ is the ENTROPY and $H(X_1, \ldots, X_n)$ is the joint ENTROPY. Linear redundancy is defined as

$$L(X_1, \ldots, X_n) \equiv -\tfrac{1}{2} \sum_{i=1}^{n} \ln \sigma_i,$$

where σ_i are EIGENVALUES of the correlation matrix.

See also PREDICTABILITY

References

Fraser, A. M. "Reconstructing Attractors from Scalar Time Series: A Comparison of Singular System and Redundancy Criteria." *Phys. D* **34**, 391–404, 1989.
Palus, M. "Identifying and Quantifying Chaos by Using Information-Theoretic Functionals." In *Time Series Prediction: Forecasting the Future and Understanding the Past* (Ed. A. S. Weigend and N. A. Gerschenfeld). Proc. NATO Advanced Research Workshop on Comparative Time Series Analysis held in Sante Fe, NM, May 14–17, 1992. Reading, MA: Addison-Wesley, pp. 387–413, 1994.

Ree Group

The Ree group $R(q)$ is the AUTOMORPHISM GROUP of a $S(2, q+1, q^3+1)$ STEINER SYSTEM.

See also STEINER SYSTEM

References

Dixon, J. and Mortimer, B. *Permutation Groups.* New York: Springer-Verlag, 1996.

Reeb Foliation

The Reeb foliation of the HYPERSPHERE $\mathbb{S}^3$ is a FOLIATION constructed as the UNION of two solid TORI with common boundary.

See also FOLIATION

References

Rolfsen, D. *Knots and Links.* Wilmington, DE: Publish or Perish Press, pp. 287–288, 1976.

Reed-Sloane Algorithm

An extension to the BERLEKAMP-MASSEY ALGORITHM which applies when the terms of the sequences are integers modulo some given modulus m.

See also BERLEKAMP-MASSEY ALGORITHM

References

Reed and Sloane, N. J. A. *SIAM J. Comput.* **14**, 505, 1985.
Sloane, N. J. A. and Plouffe, S. *The Encyclopedia of Integer Sequences.* San Diego, CA: Academic Press, p. 26, 1995.

Reef Knot

SQUARE KNOT

Re-Entrant Circuit

A GRAPH CYCLE which terminates at the starting point.

See also EULERIAN CIRCUIT, GRAPH CYCLE, HAMILTONIAN CYCLE

Refined Alternating Sign Matrix Conjecture

The fact that the numerators and denominators obtained by taking the ratios of adjacent terms in the triangular array of the number of +1 "bordered" ALTERNATING SIGN MATRICES A_n with a 1 at the top of column k are respectively the numbers in the (2, 1)- and (1, 2)-Pascal triangles which are different from 1. This conjecture was proven by Zeilberger (1996).

See also ALTERNATING SIGN MATRIX, ALTERNATING SIGN MATRIX CONJECTURE

References

Bressoud, D. and Propp, J. "How the Alternating Sign Matrix Conjecture was Solved." *Not. Amer. Math. Soc.* **46**, 637–646.
Zeilberger, D. "Proof of the Refined Alternating Sign Matrix Conjecture." *New York J. Math.* **2**, 59–68, 1996.

Refinement

A refinement X of a COVER Y is a COVER such that every element $x \in X$ is a SUBSET of an element $y \in Y$.

See also COVER

Reflection

The operation of exchanging all points of a mathematical object with their MIRROR IMAGES (i.e., reflections in a mirror). Objects which do not change HANDEDNESS under reflection are said to be AMPHICHIRAL; those that do are said to be CHIRAL.

If the PLANE of reflection is taken as the yz-PLANE, the reflection in 2- or 3-D SPACE consists of making the transformation $x \rightarrow -x$ for each point. Consider an arbitrary point x_0 and a PLANE specified by the equation

$$ax + by + cz + d = 0. \qquad (1)$$

This PLANE has NORMAL VECTOR

$$\mathbf{n} = \begin{bmatrix} a \\ b \\ c \end{bmatrix}, \qquad (2)$$

and the POINT-PLANE DISTANCE is

$$D = \frac{|ax_0 + by_0 + cz_0 + d|}{\sqrt{a^2 + b^2 + c^2}}. \qquad (3)$$

The position of the point reflected in the given plane is therefore given by

$$\mathbf{x}_0' = \mathbf{x}_0 - 2D\hat{\mathbf{n}}$$

$$= \begin{bmatrix} x_0 \\ y_0 \\ z_0 \end{bmatrix} - \frac{2|ax_0 + by_0 + cz_0 + d|}{a^2 + b^2 + c^2} \begin{bmatrix} a \\ b \\ c \end{bmatrix}. \qquad (4)$$

See also AMPHICHIRAL, CHIRAL, DILATION, ENANTIOMER, EXPANSION, GLIDE, HANDEDNESS, IMPROPER ROTATION, INVERSION OPERATION, MIRROR IMAGE, PROJECTION, REFLECTION PROPERTY, REFLECTION RELATION, REFLEXIBLE, ROTATION, ROTOINVERSION, TRANSLATION

References

Addington, S. "The Four Types of Symmetry in the Plane." http://forum.swarthmore.edu/sum95/suzanne/symsusan.html.
Coxeter, H. S. M. and Greitzer, S. L. "Reflection." §4.4 in *Geometry Revisited.* Washington, DC: Math. Assoc. Amer., pp. 86–87, 1967.
Voisin, C. *Mirror Symmetry.* Providence, RI: Amer. Math. Soc., 1999.
Yaglom, I. M. *Geometric Transformations I.* New York: Random House, 1962.

Reflection Formula

REFLECTION RELATION

Reflection Property

In the plane, the reflection property can be stated as three theorems (Ogilvy 1990, pp. 73–77):

1. The LOCUS of the center of a variable CIRCLE, tangent to a fixed CIRCLE and passing through a fixed point inside that CIRCLE, is an ELLIPSE.
2. If a variable CIRCLE is tangent to a fixed CIRCLE and also passes through a fixed point outside the CIRCLE, then the LOCUS of its moving center is a HYPERBOLA.
3. If a variable CIRCLE is tangent to a fixed straight line and also passes through a fixed point not on the line, then the LOCUS of its moving center is a PARABOLA.

Let $\alpha : I \rightarrow \mathbb{R}^2$ be a smooth regular parameterized curve in $\mathbb{R}^2$ defined on an OPEN INTERVAL I, and let F_1 and F_2 be points in $\mathbb{P}^2 \backslash \alpha(I)$, where $\mathbb{P}^n$ is an n-D PROJECTIVE SPACE. Then α has a reflection property with FOCI F_1 and F_2 if, for each point $P \in \alpha(I)$,

1. Any vector normal to the curve α at P lies in the SPAN of the vectors $\overrightarrow{F_1 P}$ and $\overrightarrow{F_2 P}$.
2. The line normal to α at P bisects one of the pairs of opposite ANGLES formed by the intersection of the lines joining F_1 and F_2 to P.

A smooth connected plane curve has a reflection property IFF it is part of an ELLIPSE, HYPERBOLA, PARABOLA, CIRCLE, or straight LINE.

Foci	Sign	Both foci finite	One focus finite	Both foci infinite
distinct	POSITIVE	confocal ellipses	confocal parabolas	parallel lines
distinct	NEGATIVE	confocal hyperbola and perpendicular bisector of interfoci line segment	confocal parabolas	parallel lines
equal		concentric circles		parallel lines

Let $S \in \mathbb{R}^3$ be a smooth CONNECTED SURFACE, and let F_1 and F_2 be points in $\mathbb{P}^3 \backslash S$, where $\mathbb{P}^n$ is an n-D PROJECTIVE SPACE. Then S has a reflection property with FOCI F_1 and F_2 if, for each point $P \in S$,

1. Any vector normal to S at P lies in the SPAN of the vectors $\overrightarrow{F_1 P}$ and $\overrightarrow{F_2 P}$.
2. The line normal to S at P bisects one of the pairs of opposite angles formed by the intersection of the lines joining F_1 and F_2 to P.

A smooth CONNECTED SURFACE has a reflection property IFF it is part of an ELLIPSOID of revolution, a HYPERBOLOID of revolution, a PARABOLOID of revolution, a SPHERE, or a PLANE.

Foci	Sign	Both foci finite	One focus finite	Both foci infinite
distinct	POSITIVE	confocal ellipsoids	confocal paraboloids	parallel planes
distinct	NEGATIVE	confocal hyperboloids and plane perpendicular bisector of interfoci line segment	confocal paraboloids	parallel planes
equal		concentric spheres		parallel planes

See also BILLIARDS

References

Drucker, D. "Euclidean Hypersurfaces with Reflective Properties." *Geometrica Dedicata* **33**, 325–329, 1990.
Drucker, D. "Reflective Euclidean Hypersurfaces." *Geometrica Dedicata* **39**, 361–362, 1991.
Drucker, D. "Reflection Properties of Curves and Surfaces." *Math. Mag.* **65**, 147–157, 1992.
Drucker, D. and Locke, P. "A Natural Classification of Curves and Surfaces with Reflection Properties." *Math. Mag.* **69**, 249–256, 1996.
Ogilvy, C. S. *Excursions in Geometry.* New York: Dover, pp. 73–77, 1990.
Wegner, B. "Comment on 'Euclidean Hypersurfaces with Reflective Properties'." *Geometrica Dedicata* **39**, 357–359, 1991.

Reflection Relation

A mathematical relationship relating $f(-x)$ to $f(x)$, or more generally, $f(a-x)$ to $f(x)$ as in the case of the GAMMA FUNCTION identity

$$\Gamma(z)\Gamma(1-z) = \frac{\pi}{\sin(\pi z)}.$$

See also ARGUMENT ADDITION RELATION, ARGUMENT MULTIPLICATION RELATION, RECURRENCE RELATION, TRANSLATION RELATION

Reflex Angle

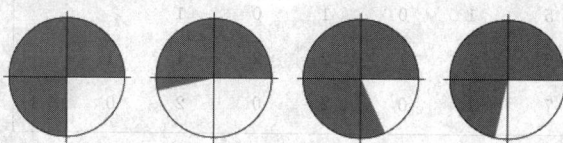

An ANGLE more than $180°$.
See also ACUTE ANGLE, ANGLE, FULL ANGLE, OBTUSE

ANGLE, RIGHT ANGLE, STRAIGHT ANGLE

Reflexible

An object is reflexible if it is superposable with its image in a plane mirror. Also called AMPHICHIRAL.

See also AMPHICHIRAL, CHIRAL, ENANTIOMER, HANDEDNESS, MIRROR IMAGE, REFLECTION

References

Ball, W. W. R. and Coxeter, H. S. M. "Polyhedra." Ch. 5 in *Mathematical Recreations and Essays, 13th ed.* New York: Dover, p. 130, 1987.

Reflexible Map

An AUTOMORPHISM which interchanges the two vertices of a regular map at each edge without interchanging the vertices.

See also EDMONDS' MAP

Reflexive Closure

The reflexive closure of a BINARY RELATION R on a SET X is the minimal REFLEXIVE RELATION R' on X that contains R. Thus $aR'a$ for every element a of X and $aR'b$ for distinct elements a and b, provided that aRb.

See also REFLEXIVE REDUCTION, REFLEXIVE RELATION, RELATION, TRANSITIVE CLOSURE

Reflexive Graph

DIRECTED GRAPH

Reflexive Polyhedron

References

Skarke, H. Reflexive Polyhedra and Their Applications in String and F-Theory. 29 Feb 2000. http://xxx.lanl.gov/abs/hep-th/0002246/.

Reflexive Reduction

The reflexive reduction of a BINARY RELATION R on a SET X is the minimum relation R' on X with the same REFLEXIVE CLOSURE as R. Thus $aR'b$ for any elements a and b of X, provided that a and b are distinct and aRb.

See also REFLEXIVE CLOSURE, RELATION, TRANSITIVE REDUCTION

Reflexive Relation

A RELATION R on a SET S is reflexive provided that xRx for every x in S.

See also RELATION

Reflexivity

A REFLEXIVE RELATION.

Region

An OPEN CONNECTED SET is called a region (sometimes also called a DOMAIN).

Regression

A method for fitting a curve (not necessarily a straight line) through a set of points using some goodness-of-fit criterion. The most common type of regression is LINEAR REGRESSION.

The term regression is sometimes also used to refer to RECURSION.

See also FRACTAL, LEAST SQUARES FITTING, LINEAR REGRESSION, MULTIPLE REGRESSION, NONLINEAR LEAST SQUARES FITTING, RECURSION, REGRESSION COEFFICIENT, SELF-RECURSION

References

Chatterjee, S.; Hadi, A.; and Price, B. *Regression Analysis by Example, 3rd ed.* New York: Wiley, 2000.
Gardner, M. "Infinite Regress." Ch. 22 in *The Sixth Book of Mathematical Games from Scientific American.* Chicago, IL: University of Chicago Press, pp. 220–229, 1984.
Kleinbaum, D. G. and Kupper, L. L. *Applied Regression Analysis and Other Multivariable Methods.* North Scituate, MA: Duxbury Press, 1978.
Passmore, J. "The Infinite Regress." In *Philosophical Reasoning.* New York: Scribner's, 1961.

Regression Coefficient

The slope b of a line obtained using linear LEAST SQUARES FITTING is called the regression coefficient.

See also CORRELATION COEFFICIENT, LEAST SQUARES FITTING

References

Kenney, J. F. and Keeping, E. S. *Mathematics of Statistics, Pt. 2, 2nd ed.* Princeton, NJ: Van Nostrand, p. 254, 1951.

Regula Falsi

FALSE POSITION METHOD

Regular Function

ANALYTIC FUNCTION, HOLOMORPHIC FUNCTION, REGULAR RATIONAL FUNCTION

Regular Graph

A GRAPH is said to be regular of degree r if all LOCAL DEGREES are the same number r. A 0-regular graph is an EMPTY GRAPH, a 1-regular graph consists of disconnected edges, and a 2-regular graph consists of disconnected cycles. The first interesting case is therefore 3-regular graphs, which are called CUBIC GRAPHS (Harary 1994, pp. 14–15). Similarly, 4- and 5-regular graphs are called QUARTIC and QUINTIC GRAPHS, respectively.

For an r-regular graph on n nodes.

$$E = \tfrac{1}{2} nr,$$

where E is the number of EDGES. n-UNITRANSITIVE GRAPHS are sometimes called n-regular (Harary 1994, p. 174).

Let $N(n, r)$ be the number of r-regular graphs with n points. Then $0 \le r \le n - 1$, $N(n, r) = N(n, n - 1, -r)$, and $N(n, r) = 0$ when both n and r are ODD. Zhang and Yang give $N(p, r)$ for $p \le 12$. The numbers of nonisomorphic regular graphs with n nodes are 1, 2, 2, 4, 3, 8, 6, 22, 26, 176, ... (Sloane's A005176; Steinbach 1990). The numbers of nonisomorphic CONNECTED regular graphs of order $n = 1, 2, ...$ are 1, 1, 1, 2, 2, 5, 4, 17, 22, 167, ... (Sloane's A005177; Steinbach 1990)

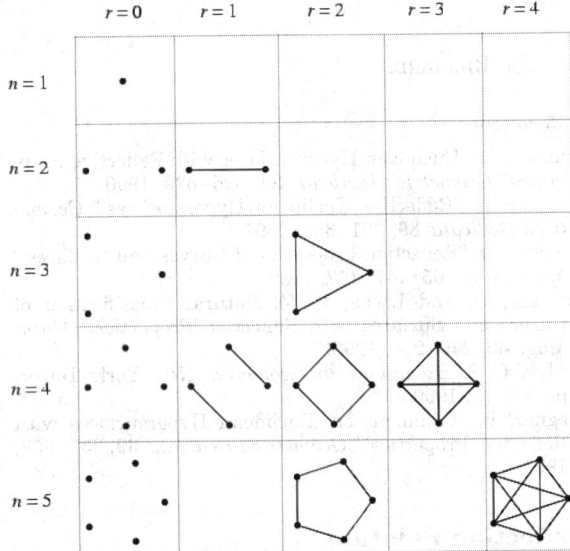

The following table gives the numbers $N(n, r)$ of r-regular graphs for small numbers of nodes n (Sloane's A051031).

n	$N(n, 0)$	$N(n, 1)$	$N(n, 2)$	$N(n, 3)$	$N(n, 4)$	$N(n, 5)$	$N(n, 6)$
1	1						
2	1	1					
3	1	0	1				
4	1	1	1	1			
5	1	0	1	0	1		
6	1	1	2	2	1	1	
7	1	0	2	0	2	0	1

The following table gives the number of connected regular graphs of degree r on $n = r + 1, r + 2, ...$ nodes for n even, and $n = r + 1$, $r + 3$, $r + 5$, ... nodes for n odd.

r	Sloane	Numbers
4	A006820	1, 1, 2, 6, 16, 59, 265, 1544, ...
5	A006821	1, 3, 60, 7848, 3459383, ...
6	A006822	1, 1, 4, 21, 266, 7849, 367860, ...
7	A014377	1, 5, 1547, ...
8	A014378	1, 1, 6, 94, 10786, 3459386, ...
9	A014381	1, 9, 88193, ...
10	A014382	1, 1, 10, 540, 805579, ...
11	A014384	1, 13, 8037796, ...

See also CAGE GRAPH, COMPLETE GRAPH, COMPLETELY REGULAR GRAPH, CONFIGURATION, CUBIC GRAPH, DISTANCE-REGULAR GRAPH, LOCAL DEGREE, MOORE GRAPH, QUARTIC GRAPH, QUINTIC GRAPH, SUPERREGULAR GRAPH

References

Chartrand, G. *Introductory Graph Theory.* New York: Dover, p. 29, 1985.

Colbourn, C. J. and Dinitz, J. H. *CRC Handbook of Combinatorial Designs.* Boca Raton, FL: CRC Press, p. 648, 1996.

Comtet, L. "Asymptotic Study of the Number of Regular Graphs of Order Two on N." §7.3 in *Advanced Combinatorics: The Art of Finite and Infinite Expansions, rev. enl. ed.* Dordrecht, Netherlands: Reidel, pp. 273–279, 1974.

Faradzev, I. A. "Constructive Enumeration of Combinatorial Objects." In *Problèmes combinatoires et théorie des graphes (Orsay, 9–13 Juillet 1976). Colloq. Internat. du C.N.R.S.* Paris: Centre Nat. Recherche Scient., pp. 131–135, 1978.

Gropp, H. "Enumeration of Regular Graphs 100 Years Ago." *Discrete Math.* **101**, 73–85, 1992.

Harary, F. *Graph Theory.* Reading, MA: Addison-Wesley, pp. 14 and 62, 1994.

Petersen, J. "Die Theorie der regulären Graphs." *Acta Math.* **15**, 193–220, 1891.

Read, R. C. and Wilson, R. J. *An Atlas of Graphs.* Oxford, England: Oxford University Press, 1998.

Sachs, H. "On Regular Graphs with Given Girth." In *Theory of Graphs and Its Applications: Proceedings of the Symposium, Smolenice, Czechoslovakia, 1963* (Ed. M. Fiedler). New York: Academic Press, 1964.

Skiena, S. *Implementing Discrete Mathematics: Combinatorics and Graph Theory with Mathematica.* Reading, MA: Addison-Wesley, p. 159, 1990.

Sloane, N. J. A. Sequences A005176/M0303, A005177/M0347, A006820/M1617, A006821/M3168, A006822/M3579, A014377, A014378, A014381, A014382, A014384, A051031 in "An On-Line Version of the Encyclopedia of Integer Sequences." http://www.research.att.com/~njas/sequences/eisonline.html.

Steinbach, P. *Field Guide to Simple Graphs.* Albuquerque, NM: Design Lab, 1990.

Wormald, N. "Generating Random Regular Graphs." *J. Algorithms* **5**, 247–280, 1984.

Zhang, C. X. and Yang, Y. S. "Enumeration of Regular Graphs." *J. Dailan Univ. Tech.* **29**, 389–398, 1989.

Regular Isotopy

The equivalence of MANIFOLDS under continuous deformation within the embedding space. KNOTS of opposite CHIRALITY have AMBIENT ISOTOPY, but not regular isotopy.

See also AMBIENT ISOTOPY

Regular Isotopy Invariant

BRACKET POLYNOMIAL

Regular Local Ring

A regular local ring is a LOCAL RING R with MAXIMAL IDEAL m so that m can be generated with exactly d elements where d is the KRULL DIMENSION of the RING R. Equivalently, R is regular if the VECTOR SPACE m/m^2 has dimension d.

See also KRULL DIMENSION, LOCAL RING, REGULAR RING, RING

References

Eisenbud, D. *Commutative Algebra with a View Toward Algebraic Geometry.* New York: Springer-Verlag, p. 242, 1995.

Regular Matrix

NONSINGULAR MATRIX

Regular Number

A number which has a finite DECIMAL expansion. A number such as $1/3 = 0.33333\ldots$ which is not regular is said to be nonregular.

See also DECIMAL EXPANSION, REPEATING DECIMAL

Regular Parameterization

A parameterization of a SURFACE $\mathbf{x}(u, v)$ in u and v is regular if the TANGENT VECTORS

$$\frac{\partial \mathbf{x}}{\partial u} \quad \text{and} \quad \frac{\partial \mathbf{x}}{\partial v}$$

are always LINEARLY INDEPENDENT.

Regular Patch

A regular patch is a PATCH $\mathbf{x} : U \to \mathbb{R}^n$ for which the JACOBIAN $J(\mathbf{x})(u, v)$ has rank 2 for all $(u, v) \in U$. A PATCH is said to be regular at a point $(u_0, v_0) \in U$ provided that its JACOBIAN has rank 2 at (u_0, v_0). For example, the points at $\phi = \pm\pi/2$ in the standard parameterization of the SPHERE $(\cos\theta \sin\phi, \sin\theta \sin\phi, \cos\phi)$ are not regular.

An example of a PATCH which is regular but not INJECTIVE is the CYLINDER defined parametrically by $(\cos u, \sin u, v)$ with $u \in (-\infty, \infty)$ and $v \in (-2, 2)$. However, if $\mathbf{x} : U \to \mathbb{R}^n$ is an injective regular patch, then $\mathbf{x}$ maps U diffeomorphically onto $\mathbf{x}(U)$.

See also INJECTIVE PATCH, PATCH, REGULAR SURFACE

References

Gray, A. *Modern Differential Geometry of Curves and Surfaces with Mathematica, 2nd ed.* Boca Raton, FL: CRC Press, p. 273, 1997.

Regular Point

If f is ANALYTIC on a DOMAIN U, then a point z_0 on the boundary ∂U is called regular if f extends to be a ANALYTIC FUNCTION on an OPEN SET containing U and also the point z_0 (Krantz 1999, p. 119).

See also ORDINARY POINT

References

Krantz, S. G. *Handbook of Complex Analysis.* Boston, MA: Birkhäuser, p. 119, 1999.

Regular Polychoron

There are sixteen regular polychora, six of which are convex (Wells 1986, p. 68) and ten of which are stellated (Wells 1991, p. 209). The regular convex polychora have four principal types of symmetry axes, and the projections into 3-spaces orthogonal to these may be called the "canonical" projections (R. Towle).

Of the six regular convex polychora, five are typically regarded as being analogous to the Platonic solids: the 4-simplex (a hyper-tetrahedron), the 4-cross polytope (a hyper-octahedron), the 4-cube (a hyper-cube), the 600-cell (a hyper-icosahedron), and the 120-cell (a hyper-dodecahedron). The 24-cell, however, has no perfect analogy in higher or lower spaces (R. Towle). The PENTATOPE and 24-CELL are self-dual, the 16-CELL is the dual of the TESSERACT, and the 600- and 120-CELLS are dual to each other.

The convex regular polychora are listed in the following table (Coxeter 1969, p. 414; Wells 1991, p. 210).

Name	Schläfli Symbol	Class	N_0	N_1	N_2	N_3
PENTATOPE	$\{3, 3, 3\}$	SIMPLEX	5	10	10	5
16-CELL	$\{3, 3, 4\}$	CROSS POLY-TOPE	8	24	32	16
TESSERACT	$\{4, 3, 3\}$	HYPERCUBE	16	32	24	8
24-CELL	$\{3, 4, 3\}$		24	96	96	24
120-CELL	$\{5, 3, 3\}$		600	1200	720	120
600-CELL	$\{3, 3, 5\}$		120	720	1200	600

Here, N_0 is the number of VERTICES, N_1 the number of EDGES, N_2 the number of FACES, and N_3 the number of cells. These quantities satisfy the identity

$$N_0 - N_1 + N_2 - N_3 = 0,$$

which is a version of the POLYHEDRAL FORMULA.

See also POLYCHORON, REGULAR POLYGON, REGULAR POLYHEDRON

References

Coxeter, H. S. M. "Regular and Semi-Regular Polytopes I." *Math. Z.* **46**, 380–407, 1940.
Coxeter, H. S. M. *Introduction to Geometry, 2nd ed.* New York: Wiley, 1969.
Wells, D. *The Penguin Dictionary of Curious and Interesting Numbers.* Middlesex, England: Penguin Books, p. 68, 1986.
Wells, D. *The Penguin Dictionary of Curious and Interesting Geometry.* London: Penguin, 1991.

Regular Polygon

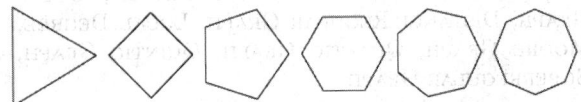

An n-sided POLYGON in which the sides are all the same length and are symmetrically placed about a common center (i.e., the polygon is both EQUIANGULAR and EQUILATERAL). The sum of PERPENDICULARS from any point to the sides of a regular polygon of n sides is n times the APOTHEM. Only certain regular polygons are "CONSTRUCTIBLE" with RULER and STRAIGHTEDGE. The terms EQUILATERAL TRIANGLE and SQUARE refer to the regular 3- and 4-polygons, respectively. The words for POLYGONS with $n \geq 5$ sides (e.g., PENTAGON, HEXAGON, HEPTAGON, etc.) can refer to either regular or non-regular POLYGONS, although the terms generally refer to regular polygons in the absence of specific wording.

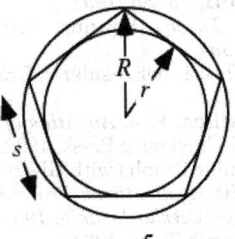

$$n = 5$$

Let s be the side length, r be the INRADIUS, and R the CIRCUMRADIUS of a regular polygon. Then

$$s = 2r \tan\left(\frac{\pi}{n}\right) \tag{1}$$

$$= 2R \sin\left(\frac{\pi}{n}\right) \tag{2}$$

$$r = \tfrac{1}{2} s \cot\left(\frac{\pi}{n}\right) \tag{3}$$

$$= R \cos\left(\frac{\pi}{n}\right) \tag{4}$$

$$R = \tfrac{1}{2} s \csc\left(\frac{\pi}{n}\right) \tag{5}$$

$$= r \sec\left(\frac{\pi}{n}\right) \tag{6}$$

$$A = \tfrac{1}{4} n s^2 \cot\left(\frac{\pi}{n}\right) \tag{7}$$

$$= n r^2 \tan\left(\frac{\pi}{n}\right) \tag{8}$$

$$= \tfrac{1}{2} n R^2 \sin\left(\frac{2\pi}{n}\right). \tag{9}$$

If the number of sides is doubled, then

$$s_{2n} = \sqrt{2R^2 - R\sqrt{4R^2 - s_n^2}} \tag{10}$$

$$A_{2n} = \frac{4rA_n}{2r + \sqrt{4r^2 + s_n^2}}. \tag{11}$$

Furthermore, if p_k and P_k are the PERIMETERS of the regular polygons inscribed in and circumscribed around a given CIRCLE and a_k and A_k their areas, then

$$P_{2n} = \frac{2 p_n P_n}{p_n + P_n} \tag{12}$$

$$p_{2n} = \sqrt{p_n P_{2n}}, \tag{13}$$

and

$$a_{2n} = \sqrt{a_n A_n} \tag{14}$$

$$A_{2n} = \frac{2 a_{2n} A_n}{a_{2n} + A_n} \tag{15}$$

(Beyer 1987, p. 125).

The following table gives parameters for the first few regular polygons, where α is the vertex angle, β is the central angle, r is the INRADIUS, R is the CIRCUMRADIUS, and A is the area (Williams 1979, p. 33).

$\{n\}$	α	β	r	R	A
$\{3\}$	$\tfrac{1}{3}\pi = 60°$	$\tfrac{2}{3}\pi = 120°$	$\tfrac{1}{6}\sqrt{3}$	$\tfrac{1}{3}\sqrt{3}$	$\tfrac{1}{4}\sqrt{3}$
$\{4\}$	$\tfrac{1}{2}\pi = 90°$	$\tfrac{1}{2}\pi = 90°$	$\tfrac{1}{2}$	$\tfrac{1}{2}\sqrt{2}$	1
$\{5\}$	$\tfrac{3}{5}\pi = 108°$	$\tfrac{2}{5}\pi = 72°$	$\tfrac{1}{10}\sqrt{25+10\sqrt{5}}$	$\tfrac{1}{10}\sqrt{50+10\sqrt{5}}$	$\tfrac{1}{4}\sqrt{25+10\sqrt{5}}$
$\{6\}$	$\tfrac{2}{3}\pi = 120°$	$\tfrac{1}{3}\pi = 60°$	$\tfrac{1}{2}\sqrt{3}$	1	$\tfrac{3}{2}\sqrt{3}$
$\{7\}$	$\tfrac{5}{7}\pi = \tfrac{900}{7}°$	$\tfrac{2}{7}\pi = \tfrac{360}{7}°$	$\tfrac{1}{2}\cot\left(\tfrac{1}{7}\pi\right)$	$\tfrac{1}{2}\csc\left(\tfrac{1}{7}\pi\right)$	$\tfrac{7}{4}\cot\left(\tfrac{1}{7}\pi\right)$
$\{8\}$	$\tfrac{3}{4}\pi = 135°$	$\tfrac{1}{4}\pi = 45°$	$\tfrac{1}{2}(1+\sqrt{2})$	$\tfrac{1}{2}\sqrt{4+2\sqrt{2}}$	$2(1+\sqrt{2})$
$\{9\}$	$\tfrac{7}{9}\pi = 140°$	$\tfrac{2}{9}\pi = 40°$	$\tfrac{1}{2}\cot\left(\tfrac{1}{9}\pi\right)$	$\tfrac{1}{2}\csc\left(\tfrac{1}{9}\pi\right)$	$\tfrac{9}{4}\cot\left(\tfrac{1}{9}\pi\right)$
$\{10\}$	$\tfrac{4}{5}\pi = 144°$	$\tfrac{1}{5}\pi = 36°$	$\tfrac{1}{2}\sqrt{5+2\sqrt{5}}$	$\tfrac{1}{2}(1+\sqrt{5})$	$\tfrac{5}{2}\sqrt{5+2\sqrt{5}}$
$\{11\}$	$\tfrac{9}{11}\pi = \tfrac{1620}{11}°$	$\tfrac{2}{11}\pi = \tfrac{360}{11}°$	$\tfrac{1}{2}\cot\left(\tfrac{1}{11}\pi\right)$	$\tfrac{1}{2}\csc\left(\tfrac{1}{11}\pi\right)$	$\tfrac{11}{4}\cot\left(\tfrac{1}{11}\pi\right)$
$\{12\}$	$\tfrac{5}{6}\pi = 150°$	$\tfrac{1}{6}\pi = 30°$	$\tfrac{1}{2}(2+\sqrt{3})$	$\tfrac{1}{2}(\sqrt{2}+\sqrt{6})$	$3(2+\sqrt{3})$

COMPASS and STRAIGHTEDGE constructions dating back to Euclid were capable of inscribing regular polygons of 3, 4, 5, 6, 8, 10, 12, 16, 20, 24, 32, 40, 48, 64, ..., sides. However, this listing is not a complete enumeration of "constructible" polygons. In fact, a regular n-gon is constructible only if $\phi(n)$ is a POWER of 2, where ϕ is the TOTIENT FUNCTION (this is a NECESSARY but not SUFFICIENT condition). More specifically, a regular n-gon ($n \geq 3$) can be constructed by STRAIGHTEDGE and COMPASS (i.e., can have trigonometric functions of its ANGLES expressed in terms of finite SQUARE ROOT extractions) IFF

$$n = 2^k p_1 p_2 \cdots p_s, \tag{16}$$

where k is in INTEGER ≥ 0 and the p_i are distinct FERMAT PRIMES. FERMAT NUMBERS are OF THE FORM

$$F_m = 2^{2^m} + 1, \tag{17}$$

where m is an INTEGER ≥ 0. The only known PRIMES of this form are 3, 5, 17, 257, and 65537.

The fact that this condition was SUFFICIENT was first proved by Gauss in 1796 when he was 19 years old, and it relies on the property of IRREDUCIBLE POLYNOMIALS that ROOTS composed of a finite number of SQUARE ROOT extractions exist only if the order of the equation is OF THE FORM 2^h. That this condition was also NECESSARY was not explicitly proven by Gauss, and the first proof of this fact is credited to Wantzel (1836).

Constructible values of n for $n < 300$ were given by Gauss (Smith 1994), and the first few are 2, 3, 4, 5, 6, 8, 10, 12, 15, 16, 17, 20, 24, 30, 32, 34, 40, 48, 51, 60, 64, 68, 80, 85, 96, 102, 120, 128, 136, 160, 170, 192, ... (Sloane's A003401). Gardner (1977) and independently Watkins (Conway and Guy 1996) noticed that the number of sides for constructible polygons with an ODD number of sides are given by the first 32 rows of PASCAL'S TRIANGLE (mod 2) interpreted as BINARY numbers, giving 1, 3, 5, 15, 17, 51, 85, 255, ... (Sloane's A004729, Conway and Guy 1996, p. 140).

```
        1                    1          1
       1 1                  1 1         3
      1 2 1                1 0 1        5
     1 3 3 1              1 1 1 1      15
    1 4 6 4 1            1 0 0 0 1     17
   1 5 10 10 5 1        1 1 0 0 1 1    51
  1 6 15 20 15 6 1     1 0 1 0 1 0 1   85
 1 7 21 35 35 21 7 1   1 1 1 1 1 1 1 1 255
1 8 28 56 70 56 28 8 1 1 0 0 0 0 0 0 1 257
```

Although constructions for the regular TRIANGLE, SQUARE, PENTAGON, and their derivatives had been given by Euclid, constructions based on the FERMAT PRIMES ≥ 17 were unknown to the ancients. The first explicit construction of a HEPTADECAGON (17-gon) was given by Erchinger in about 1800. Richelot and Schwendenwein found constructions for the 257-GON in 1832, and Hermes spent 10 years on the construction of the 65537-GON at Göttingen around 1900 (Coxeter 1969). Constructions for the EQUILATERAL TRIANGLE and SQUARE are trivial (top figures below). Elegant constructions for the PENTAGON and HEPTADECAGON are due to Richmond (1893) (bottom figures below).

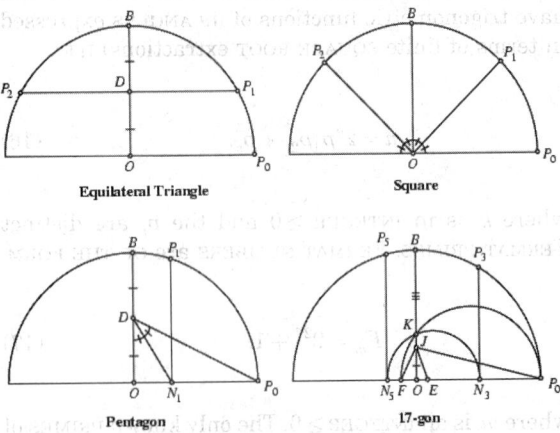

Equilateral Triangle   **Square**

Pentagon   **17-gon**

Given a point, a CIRCLE may be constructed of any desired RADIUS, and a DIAMETER drawn through the center. Call the center O, and the right end of the DIAMETER P_0. The DIAMETER PERPENDICULAR to the original DIAMETER may be constructed by finding the PERPENDICULAR BISECTOR. Call the upper endpoint of this PERPENDICULAR DIAMETER B. For the PENTAGON, find the MIDPOINT of OB and call it D. Draw DP_0, and BISECT $\angle ODP_0$, calling the intersection point with OP_0 N_1. Draw N_1P_1 PARALLEL to OB, and the first two points of the PENTAGON are P_0 and P_1. The construction for the HEPTADECAGON is more complicated, but can be accomplished in 17 relatively simple steps. The construction problem has now been automated (Bishop 1978).

See also 257-GON, 65537-GON, CHAOS GAME, CONSTRUCTIBLE POLYGON, DE MOIVRE NUMBER, EQUILATERAL TRIANGLE, HEPTADECAGON, HEXAGON, HEXAGRAM, OCTAGON, PENTAGON, PENTAGRAM, POLYGON, POLYGON CIRCUMSCRIBING CONSTANT, POLYGON INSCRIBING CONSTANT, SQUARE, STAR POLYGON

References

Bishop, W. "How to Construct a Regular Polygon." *Amer. Math. Monthly* **85**, 186–188, 1978.

Conway, J. H. and Guy, R. K. *The Book of Numbers.* New York: Springer-Verlag, pp. 140 and 197–202, 1996.

Courant, R. and Robbins, H. "Regular Polygons." §3.2 in *What is Mathematics?: An Elementary Approach to Ideas and Methods, 2nd ed.* Oxford, England: Oxford University Press, pp. 122–125, 1996.

Coxeter, H. S.M. *Introduction to Geometry, 2nd ed.* New York: Wiley, 1969.

De Temple, D. W. "Carlyle Circles and the Lemoine Simplicity of Polygonal Constructions." *Amer. Math. Monthly* **98**, 97–108, 1991.

Dickson, L. E. "Constructions with Ruler and Compasses; Regular Polygons." Ch. 8 in *Monographs on Topics of Modern Mathematics Relevant to the Elementary Field* (Ed. J. W. A. Young). New York: Dover, pp. 352–386, 1955.

Gardner, M. *Mathematical Carnival: A New Round-Up of Tantalizers and Puzzles from Scientific American.* New York: Vintage Books, p. 207, 1977.

Gauss, C. F. §365 and 366 in *Disquisitiones Arithmeticae.* Leipzig, Germany, 1801. Translated by A. A Clarke. New Haven, CT: Yale University Press, 1965.

Harris, J. W. and Stocker, H. "Regular n-gons (Polygons)." §3.7 in *Handbook of Mathematics and Computational Science.* New York: Springer-Verlag, pp. 86–89, 1998.

Math Forum. "Naming Polygons and Polyhedra." http://forum.swarthmore.edu/dr.math/faq/faq.polygon.names.html.

Rawles, B. *Sacred Geometry Design Sourcebook: Universal Dimensional Patterns.* Nevada City, CA: Elysian Pub., p. 238, 1997.

Richmond, H. W. "A Construction for a Regular Polygon of Seventeen Sides." *Quart. J. Pure Appl. Math.* **26**, 206–207, 1893.

Sloane, N. J. A. Sequences A003401/M0505 and A004729 in "An On-Line Version of the Encyclopedia of Integer Sequences." http://www.research.att.com/~njas/sequences/eisonline.html.

Smith, D. E. *A Source Book in Mathematics.* New York: Dover, p. 350, 1994.

Tietze, H. Ch. 9 in *Famous Problems of Mathematics.* New York: Graylock Press, 1965.

Wantzel, M. L. "Recherches sur les moyens de reconnaître si un Problème de Géométrie peut se résoudre avec la règle et le compas." *J. Math. pures appliq.* **1**, 366–372, 1836.

Williams, R. "Polygons." §2–1 in *The Geometrical Foundation of Natural Structure: A Source Book of Design.* New York: Dover, pp. 31–33, 1979.

Regular Polyhedron

A polyhedron is said to be regular if its FACES and VERTEX FIGURES are REGULAR (not necessarily CONVEX) polygons (Coxeter 1973, p. 16). Using this definition, there are a total of nine regular polyhedra, five being the CONVEX PLATONIC SOLIDS and four being the CONCAVE (stellated) KEPLER-POINSOT SOLIDS. However, the term "regular polyhedra" is sometimes used to refer exclusively to the CONVEX PLATONIC SOLIDS.

It can be proven that only nine regular solids (in the Coxeter sense) exist by noting that a possible regular polyhedron must satisfy

$$\cos^2\left(\frac{\pi}{p}\right) + \cos^2\left(\frac{\pi}{q}\right) + \cos^2\left(\frac{\pi}{r}\right) = 1.$$

Gordon showed that the only solutions to

$$1 + \cos\phi_1 + \cos\phi_2 + \cos\phi_3 = 0$$

OF THE FORM $\phi_i = \pi m_i / n_i$ are the permutations of $(\frac{2}{3}\pi, \frac{2}{3}\pi, \frac{1}{3}\pi)$ and $(\frac{2}{3}\pi, \frac{2}{5}\pi, \frac{4}{5}\pi)$. This gives three permutations of $(3, 3, 4)$ and six of $(3, 5, \frac{5}{3})$ as possible solutions to the first equation. Plugging back in gives the SCHLÄFLI SYMBOLS of possible regular polyhedra as $\{3, 3\}$, $\{3, 4\}$, $\{4, 3\}$, $\{3, 5\}$, $\{5, 3\}$, $\{3, \frac{5}{2}\}$, $\{\frac{5}{2}, 3\}$, $\{5, \frac{5}{2}\}$, and $\{\frac{5}{2}, 5\}$ (Coxeter 1973, pp. 107–109). The first five of these are the PLATONIC SOLIDS and the remaining four the KEPLER-POINSOT SOLIDS.

Every regular polyhedron has $e + 1$ axes of symmetry, where e is the number of EDGES, and $3h/2$ PLANES of symmetry, where h is the number of sides of the corresponding PETRIE POLYGON.

See also CONVEX POLYHEDRON, KEPLER-POINSOT SOLID, PETRIE POLYGON, PLATONIC SOLID, POLYHEDRON, POLYHEDRON COMPOUND, SPONGE, VERTEX FIGURE

References

Coxeter, H. S. M. "Regular and Semi-Regular Polytopes I." *Math. Z.* **46**, 380–407, 1940.
Coxeter, H. S. M. *Regular Polytopes, 3rd ed.* New York: Dover, pp. 1–17, 93, and 107–112, 1973.
Cromwell, P. R. *Polyhedra.* New York: Cambridge University Press, pp. 85–86, 1997.

Regular Polytope

REGULAR POLYCHORON

Regular Prime

A PRIME which does not DIVIDE the CLASS NUMBER $h(p)$ of the CYCLOTOMIC FIELD obtained by adjoining a PRIMITIVE PTH ROOT OF UNITY to the FIELD of rationals. A PRIME p is regular IFF p does not divide the NUMERATORS of the BERNOULLI NUMBERS B_0, B_2, ..., B_{p-3}. A PRIME which is not regular is said to be an IRREGULAR PRIME.

In 1915, Jensen proved that there are infinitely many IRREGULAR PRIMES. It has not yet been proven that there are an INFINITE number of regular primes (Guy 1994, p. 145). Of the 283,145 PRIMES $< 4 \times 10^6$, 171,548 (or 60.59%) are regular (the conjectured FRACTION is $e^{-1/2} \approx 60.65\%$). The first few are 3, 5, 7, 11, 13, 17, 19, 23, 29, 31, 41, 43, 47, ... (Sloane's A007703).

See also BERNOULLI NUMBER, FERMAT'S THEOREM, IRREGULAR PRIME

References

Buhler, J.; Crandall, R. Ernvall, R.; and Metsankyla, T. "Irregular Primes and Cyclotomic Invariants to Four Million." *Math. Comput.* **61**, 151–153, 1993.
Guy, R. K. *Unsolved Problems in Number Theory, 2nd ed.* New York: Springer-Verlag, p. 145, 1994.
Ribenboim, P. "Regular Primes." §5.1 in *The New Book of Prime Number Records.* New York: Springer-Verlag, pp. 323–329, 1996.
Shanks, D. *Solved and Unsolved Problems in Number Theory, 4th ed.* New York: Chelsea, p. 153, 1993.
Sloane, N. J. A. Sequences A007703/M2411 in "An On-Line Version of the Encyclopedia of Integer Sequences." http://www.research.att.com/~njas/sequences/eisonline.html.

Regular Pyramid

PYRAMID

Regular Ring

In the sense of von Neumann, a regular ring is a RING R such that for all $a \in R$, there exists a $b \in R$ satisfying $a = aba$.

See also REGULAR LOCAL RING, RING

References

Jacobson, N. *Basic Algebra II, 2nd ed.* New York: W. H. Freeman, p. 196, 1989.

Regular Sequence

Let there be two PARTICULARLY WELL-BEHAVED FUNCTIONS $F(x)$ and $p_\tau(x)$. If the limit

$$\lim_{\tau \to 0} \int_{-\infty}^{\infty} p_\tau(x) F(x) \, dx$$

exists, then $p_\tau(x)$ is a regular sequence of PARTICULARLY WELL-BEHAVED FUNCTIONS.

References

Allouche, J.-P. and Shallit, J. "The Ring of k-Regular Sequences." *Theoret. Comput. Sci.* **98**, 16–197, 1992.

Regular Singular Point

Consider a second-order ORDINARY DIFFERENTIAL EQUATION

$$y''P(x)y' + Q(x)y = 0.$$

If $P(x)$ and $Q(x)$ remain FINITE at $x = x_0$, then x_0 is called an ORDINARY POINT. If either $P(x)$ or $Q(x)$ diverges as $x \to x_0$, then x_0 is called a singular point. If either $P(x)$ or $Q(x)$ diverges as $x \to x_0$ but $(x - x_0)P(x)$ and $(x - x_0)^2 Q(x)$ remain FINITE as $x \to x_0$, then $x = x_0$ is called a regular singular point (or NONESSENTIAL SINGULARITY).

See also IRREGULAR SINGULARITY, SINGULAR POINT (DIFFERENTIAL EQUATION)

References

Arfken, G. "Singular Points." §8.4 in *Mathematical Methods for Physicists, 3rd ed.* Orlando, FL: Academic Press, pp. 451–453 and 461–463, 1985.

Regular Singularity

REGULAR SINGULAR POINT

Regular Skew Polyhedron

A regular skew polyhedron is a polyhedron whose faces and VERTEX FIGURES are regular SKEW POLYGONS. There are only three regular skew polyhedra in Euclidean 3-space (Coxeter 1937, Garner 1967), the simplest of which is $\{4, 6|4\}$.

Garner (1967) considered regular skew polyhedra in hyperbolic space $\mathbb{H}^3$, and shows that there are exactly 32 which are derived from honeycombs whose cells and vertex figures are derived from honeycombs whose cells and vertex figures are not inscribed in equidistant surfaces.

See also REGULAR POLYHEDRON

References

Coxeter, H. S. M. "Regular Skew Polyhedra in Three and Four Dimensions." *Proc. London Math. Soc.* **43**, 33–62, 1937.
Garner, C. W. L. "Regular Skew Polyhedra in Hyperbolic Three-Space." *Canad. J. Math.* **19**, 1179–1186, 1967.

Regular Surface

A SUBSET $M \subset \mathbb{R}^n$ is called a regular surface if for each point $p \in M$, there exists a NEIGHBORHOOD V of p in $\mathbb{R}^n$ and a MAP $x : U \to \mathbb{R}^n$ of an OPEN SET $U \subset \mathbb{R}^2$ onto $V \cap M$ such that

1. x is differentiable,
2. $x : U \to V \cap M$ is a HOMEOMORPHISM, and
3. Each map $x : U \to M$ is a REGULAR PATCH.

Any open subset of a regular surface is also a regular surface.

See also REGULAR PATCH

References

Gray, A. "The Definition of a Regular Surface in $\mathbb{R}^n$." §12.4 in *Modern Differential Geometry of Curves and Surfaces with Mathematica, 2nd ed.* Boca Raton, FL: CRC Press, pp. 281–286, 1997.

Regular Triangle Center

A TRIANGLE CENTER is regular IFF there is a TRIANGLE CENTER FUNCTION which is a POLYNOMIAL in Δ, a, b, and c (where Δ is the AREA of the TRIANGLE) such that the TRILINEAR COORDINATES of the center are

$$f(a, b, c) : f(b, c, a) : f(c, a, b).$$

The ISOGONAL CONJUGATE of a regular center is a regular center. Furthermore, given two regular cen-

ters, any two of their HARMONIC CONJUGATE POINTS are also regular centers.

See also ISOGONAL CONJUGATE, TRIANGLE CENTER, TRIANGLE CENTER FUNCTION

Regular Variation

References

Feller, W. *An Introduction to Probability Theory and Its Applications, Vol. 2, 3rd ed.* New York: Wiley, pp. 275–276, 1971.

Regularity Axiom

AXIOM OF FOUNDATION

Regularity Lemma

SZEMERÉDI'S REGULARITY LEMMA

Regularity Theorem

An AREA-minimizing surface (RECTIFIABLE CURRENT) bounded by a smooth curve in $\mathbb{R}^3$ is a smooth submanifold with boundary.

See also MINIMAL SURFACE, RECTIFIABLE CURRENT

References

Morgan, F. "What is a Surface?" *Amer. Math. Monthly* **103**, 369–376, 1996.

Regularized Beta Function

The regularized beta function is defined by

$$I(z; a, b) = \frac{B(z; a, b)}{B(a, b)},$$

where $B(z; a, b)$ is the incomplete BETA FUNCTION and $B(a, b)$ is the complete BETA FUNCTION. The regularized beta function is sometimes also denoted $I_z(a, b)$ and is implemented in *Mathematica* as BetaRegularized[z, a, b]. The four-argument version BetaRegularized[$z1, z2, a, b$] is equivalent to $I(z_2; a, b) - I(z_1; a, b)$.

See also BETA FUNCTION, REGULARIZED GAMMA FUNCTION

Regularized Gamma Function

The regularized gamma functions are defined by

$$P(a, z) = 1 - Q(a, z) \equiv \frac{\gamma(a, z)}{\Gamma(a)} \qquad (1)$$

and

$$Q(a, z) = 1 - P(a, z) \equiv \frac{\Gamma(a, z)}{\Gamma(a)},$$

where $\gamma(a, z)$ and $\Gamma(a, z)$ are INCOMPLETE GAMMA

FUNCTIONS and $\Gamma(a)$ is a complete GAMMA FUNCTION. The function $Q(a, z)$ is implemented in *Mathematica* as GammaRegularized[a, z].

The derivatives of $P(a, z)$ and $Q(a, z)$ are

$$\frac{d}{dz} P(a, z) = \frac{e^{-z} z^{a-1}}{\Gamma(a)} \tag{2}$$

$$\frac{d}{dz} Q(a, z) = \frac{e^{-z} z^{a-1}}{\Gamma(a)}, \tag{3}$$

and the second derivatives are

$$\frac{d^2}{dz^2} P(a, z) = \frac{e^{-z}(a - z - 1) z^{a-2}}{\Gamma(a)} \tag{4}$$

$$\frac{d^2}{dz^2} Q(a, z) = \frac{e^{-z}(1 + z - a) z^{a-2}}{\Gamma(a)} \tag{5}$$

The integrals are

$$\int P(a, z)\, dz = \frac{z\Gamma(a) - z\Gamma(a, z) + \Gamma(a + 1, z)}{\Gamma(a)} \tag{6}$$

$$\int Q(a, z)\, dz = \frac{z\Gamma(a, z) - \Gamma(a + 1, z)}{\Gamma(a)} \tag{7}$$

See also GAMMA FUNCTION, INCOMPLETE GAMMA FUNCTION, REGULARIZED BETA FUNCTION

References

Press, W. H.; Flannery, B. P.; Teukolsky, S. A.; and Vetterling, W. T. *Numerical Recipes in FORTRAN: The Art of Scientific Computing, 2nd ed.* Cambridge, England: Cambridge University Press, pp. 160–161, 1992.

Regularized Long-Wave Equation

The PARTIAL DIFFERENTIAL EQUATION

$$u_t + u_x - 6uu_x - u_{txx} = 0.$$

See also KORTEWEG-DE VRIES EQUATION

References

Calogero, F. and Degasperis, A. *Spectral Transform and Solitons: Tools to Solve and Investigate Nonlinear Evolution Equations.* New York: North-Holland, p. 49, 1982.
Zwillinger, D. *Handbook of Differential Equations, 3rd ed.* Boston, MA: Academic Press, p. 131, 1997.

Regulus

The locus of lines meeting three given SKEW LINES. ("Regulus" is also the name of the brightest star in the constellation Leo.)

Reidemeister Moves

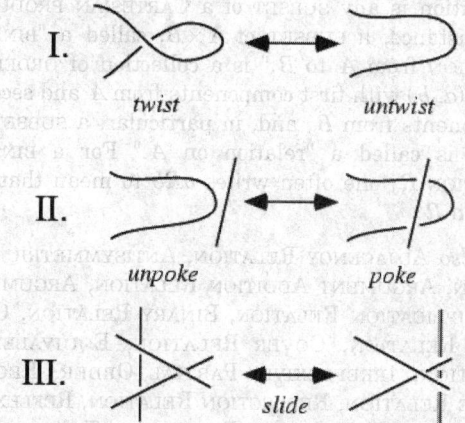

In the 1930s, Reidemeister first rigorously proved that KNOTS exist which are distinct from the UNKNOT. He did this by showing that all KNOT deformations can be reduced to a sequence of three types of "moves," called the (I) TWIST MOVE, (II) POKE MOVE, and (III) SLIDE MOVE. These moves are most commonly called Reidemeister moves, although the term "equivalence moves" is sometimes also used (Aneziris 1999, p. 29).

REIDEMEISTER'S THEOREM guarantees that moves I, II, and III correspond to AMBIENT ISOTOPY (moves II and III alone correspond to REGULAR ISOTOPY). He then defined the concept of COLORABILITY, which is invariant under Reidemeister moves.

See also AMBIENT ISOTOPY, COLORABLE, KNOT MOVE, MARKOV MOVES, REGULAR ISOTOPY, UNKNOT

References

Aneziris, C. N. "The Equivalence Moves." Ch. 4 in *The Mystery of Knots: Computer Programming for Knot Tabulation.* Singapore: World Scientific, pp. 29–33, 1999.
Hoste, J.; Thistlethwaite, M.; and Weeks, J. "The First 1,701,936 Knots." *Math. Intell.* **20**, 33–48, Fall 1998.
Reidemeister, K. "Knotten und Gruppen." *Abh. Math. Sem. Univ. Hamburg* **5**, 7–23, 1927.

Reidemeister's Theorem

Two LINKS can be continuously deformed into each other IFF any diagram of one can be transformed into a diagram of the other by a sequence of REIDEMEISTER MOVES.

See also REIDEMEISTER MOVES

Reinhardt Domain

A Reinhardt domain with center $\mathbf{c}$ is a DOMAIN D in C^n such that whenever D contains z_0, the DOMAIN D also contains the closed POLYDISK.

References

Iyanaga, S. and Kawada, Y. (Eds.). *Encyclopedic Dictionary of Mathematics.* Cambridge, MA: MIT Press, p. 101, 1980.

Relation

A relation is any SUBSET of a CARTESIAN PRODUCT. For instance, a SUBSET of $A \times B$, called a "BINARY RELATION from A to B," is a collection of ORDERED PAIRS (a, b) with first components from A and second components from B, and, in particular, a SUBSET of $A \times A$ is called a "relation on A." For a BINARY RELATION R, one often writes aRb to mean that (a, b) is in R.

See also ADJACENCY RELATION, ANTISYMMETRIC RELATION, ARGUMENT ADDITION RELATION, ARGUMENT MULTIPLICATION RELATION, BINARY RELATION, CLOSURE RELATION, COVER RELATION, EQUIVALENCE RELATION, IRREFLEXIVE, PARTIAL ORDER, RECURRENCE RELATION, REFLECTION RELATION, REFLEXIVE RELATION, SYMMETRIC RELATION, TRANSITIVE, TRANSLATION RELATION

Relational System

This entry contributed by VIKTOR BENGTSSON

A relational system is a structure $\mathcal{R} = (S, \{P_i : i \in I\}, \{f_j : j \in J\})$ consisting of a set S, a collection of relations $P_i (i \in I)$ on S, and a collection of functions $f_j (j \in J)$ on S.

Relative Cumulative Frequency

The CUMULATIVE FREQUENCY in a FREQUENCY DISTRIBUTION divided by the total number of data points.

See also ABSOLUTE FREQUENCY, CUMULATIVE FREQUENCY, FREQUENCY DISTRIBUTION, RELATIVE FREQUENCY

References

Kenney, J. F. and Keeping, E. S. "Frequency Distributions." §1.8 in *Mathematics of Statistics, Pt. 1, 3rd ed.* Princeton, NJ: Van Nostrand, pp. 12–19, 1962.

Relative Degree

DEGREE (EXTENSION FIELD

Relative Entropy

Let a DISCRETE DISTRIBUTION have probability function p_k, and let a second DISCRETE DISTRIBUTION have probability function q_k. Then the relative entropy of p with respect to q, also called the Kullback-Leibler distance, is defined by

$$d = \sum_k p_k \ln\left(\frac{p_k}{q_k}\right).$$

Although relative entropy does not satisfy the triangle inequality and is therefore not a true metric, it satisfies many important mathematical properties. For example, it is a convex function of p_k, is always nonnegative, and equals zero only if $p_k = q_k$.

Relative entropy is a very important concept in quantum information theory, as well as statistical mechanics (Qian 2000).

See also ENTROPY

References

Cover, T. M. and Thomas, J. A. *Elements of Information Theory.* New York: Wiley, 1991.
Qian, H. Relative Entropy: Free Energy Associated with Equilibrium Fluctuations and Nonequilibrium Deviations. 8 Jul 2000. http://xxx.lanl.gov/abs/math-ph/0007010/.

Relative Error

Let the true value of a quantity be x and the measured or inferred value x_0. Then the relative error is defined by

$$\delta x = \frac{\Delta x}{x} = \frac{x_0 - x}{x} = \frac{x_0}{x} - 1,$$

where Δx is the ABSOLUTE ERROR. The relative error of the QUOTIENT or PRODUCT of a number of quantities is less than or equal to the SUM of their relative errors. The PERCENTAGE ERROR is 100% times the relative error.

See also ABSOLUTE ERROR, ERROR PROPAGATION, PERCENTAGE ERROR

References

Abramowitz, M. and Stegun, C. A. (Eds.). *Handbook of Mathematical Functions with Formulas, Graphs, and Mathematical Tables, 9th printing.* New York: Dover, p. 14, 1972.

Relative Extremum

A RELATIVE MAXIMUM or RELATIVE MINIMUM, also called a LOCAL EXTREMUM.

See also EXTREMUM, GLOBAL EXTREMUM, RELATIVE MAXIMUM, RELATIVE MINIMUM

Relative Frequency

The ratio of the ABSOLUTE FREQUENCY to the total number of data points in a FREQUENCY DISTRIBUTION.

See also ABSOLUTE FREQUENCY, CUMULATIVE FREQUENCY, FREQUENCY DISTRIBUTION, RELATIVE CUMULATIVE FREQUENCY

References

Kenney, J. F. and Keeping, E. S. "Frequency Distributions." §1.8 in *Mathematics of Statistics, Pt. 1, 3rd ed.* Princeton, NJ: Van Nostrand, pp. 12–19, 1962.

Relative Maximum

A MAXIMUM within some NEIGHBORHOOD which need not be a GLOBAL MAXIMUM.

See also GLOBAL MAXIMUM, MAXIMUM, RELATIVE MINIMUM

Relative Minimum

A MINIMUM within some NEIGHBORHOOD which need not be a GLOBAL MINIMUM.

See also GLOBAL MINIMUM, MINIMUM, RELATIVE MAXIMUM

Relative Topology

If $A \subset B$ and B has a topology of open sets U_α then the relative topology on A is given by the collection of open sets $U_\alpha \cap A$.

Relatively Prime

Two integers are relatively prime if they share no common positive factors (divisors) except 1. Using the notation (m, n) to denote the GREATEST COMMON DIVISOR, two integers m and n are relatively prime if $(m, n) = 1$. Relatively prime integers are sometimes also called STRANGERS or COPRIME and are denoted $m \perp n$.

The probability that two INTEGERS picked at random are relatively prime is $[\zeta(2)]^{-1} = 6/\pi^2$, where $\zeta(z)$ is the RIEMANN ZETA FUNCTION (Wells 1986, p. 28). This result is related to the fact that the GREATEST COMMON DIVISOR of m and n, $(m, n) = k$, can be interpreted as the number of LATTICE POINTS in the PLANE which lie on the straight LINE connecting the VECTORS $(0, 0)$ and (m, n) (excluding (m, n) itself). In fact, $6/\pi^2$ is the fractional number of LATTICE POINTS VISIBLE from the ORIGIN (Castellanos 1988, pp. 155–156).

Given three INTEGERS chosen at random, the probability that no common factor will divide them all is

$$[\zeta(3)]^{-1} \approx 1.20206^{-1} \approx 0.831907, \qquad (1)$$

where $\zeta(3)$ is APÉRY'S CONSTANT (Wells 1986, p. 29). This generalizes to k random integers (Schoenfeld 1976).

See also DIVISOR, GREATEST COMMON DIVISOR, HAFNER-SARNAK-MCCURLEY CONSTANT, VISIBILITY

References

Castellanos, D. "The Ubiquitous Pi." *Math. Mag.* **61**, 67–98, 1988.
Guy, R. K. *Unsolved Problems in Number Theory, 2nd ed.* New York: Springer-Verlag, pp. 3–4, 1994.
Hoffman, P. *The Man Who Loved Only Numbers: The Story of Paul Erdos and the Search for Mathematical Truth.* New York: Hyperion, pp. 38–39, 1998.
Nagell, T. "Relatively Prime Numbers. Euler's φ-Function." §8 in *Introduction to Number Theory.* New York: Wiley, pp. 23–26, 1951.
Schoenfeld, L. "Sharper Bounds for the Chebyshev Functions $\theta(x)$ and $\psi(x)$, II." *Math. Comput.* **30**, 337–360, 1976.
Wells, D. *The Penguin Dictionary of Curious and Interesting Numbers.* Middlesex, England: Penguin Books, pp. 28–29, 1986.

Relaxation Methods

Methods of solving an ORDINARY DIFFERENTIAL EQUATION by replacing it with a FINITE DIFFERENCE equation on a regular grid spanning the domain of interest. The finite difference equations are then solved using an n-D NEWTON'S METHOD or other similar algorithm.

References

Jeffreys, H. and Jeffreys, B. S. "Relation Methods." §9.18 in *Methods of Mathematical Physics, 3rd ed.* Cambridge, England: Cambridge University Press, pp. 307–312, 1988.
Press, W. H.; Flannery, B. P.; Teukolsky, S. A.; and Vetterling, W. T. "Richardson Extrapolation and the Bulirsch-Stoer Method." §17.3 in *Numerical Recipes in FORTRAN: The Art of Scientific Computing, 2nd ed.* Cambridge, England: Cambridge University Press, pp. 753–763, 1992.

Remainder

In general, a remainder is a quantity "left over" after performing a particular algorithm. The term is most commonly used to refer to the number left over when two integers are divided by each other in INTEGER DIVISION. For example, $55 \backslash 7 = 7$, with a remainder of 6. Of course in real division, there is no such thing as a remainder since, for example, $55/7 = 7 + 6/7$.

The term remainder is also sometimes applied to the RESIDUE of a CONGRUENCE.

See also DIVISION, INTEGER DIVISION, QUOTIENT, RESIDUE (CONGRUENCE)

References

Nagell, T. "Remainders." §2 in *Introduction to Number Theory.* New York: Wiley, pp. 12–13, 1951.

Remainder Theorem

POLYNOMIAL REMAINDER THEOREM

Rembs' Surface

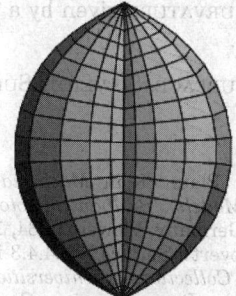

A surface of constant GAUSSIAN CURVATURE that can be given parametrically by

$$x = a(U \cos u - U' \sin u) \qquad (1)$$

$$y = -a(U \sin u - U' \cos u) \qquad (2)$$

$$z = v - aV', \qquad (3)$$

where

$$U \equiv \frac{\cosh\left(u\sqrt{C}\right)}{\sqrt{C}} \qquad (4)$$

$$V \equiv \frac{\cos\left(v\sqrt{C+1}\right)}{\sqrt{C+1}} \qquad (5)$$

$$a \equiv \frac{2V}{(C+1)(U^2 - V^2)}, \qquad (6)$$

and $U' = dU/du$, and $V' = dV/dv$. The value of v is restricted to

$$|v| \leq v_0 \equiv \frac{\pi}{2\sqrt{C+1}} \qquad (7)$$

(Reckziegel 1986), and the values $v = \pm v_0$ correspond to the ends of the cleft in the surface. The surface illustrated above corresponds to $C = 1$.

Rembs' surface has FIRST FUNDAMENTAL FORM coefficients

$$E = \frac{16C(1+C)\cos^2\left(v\sqrt{C+1}\right)\cosh^2\left(u\sqrt{C}\right)}{\left[1 - C\cos\left(2v\sqrt{C+1}\right) + (C+1)\cosh\left(2u\sqrt{C}\right)\right]^2} \qquad (8)$$

$$F = 0 \qquad (9)$$

$$G =$$

$$\frac{\left[1 + 2C + C\cos\left(2v\sqrt{C+1}\right) + (C+1)\cosh\left(2u\sqrt{C}\right)\right]^2}{\left[1 - C\cos\left(2v\sqrt{C+1}\right) + (C+1)\cosh\left(2u\sqrt{C}\right)\right]^2}, \qquad (10)$$

SECOND FUNDAMENTAL FORM coefficients by similar, rather complicated expressions. The GAUSSIAN CURVATURE is

$$K = 1, \qquad (11)$$

with the MEAN CURVATURE given by a rather complicated expression.

See also KUEN SURFACE, SIEVERT'S SURFACE

References

Fischer, G. (Ed.). Plate 88 in *Mathematische Modelle/ Mathematical Models, Bildband/Photograph Volume.* Braunschweig, Germany: Vieweg, p. 84, 1986.
Reckziegel, H. "Sievert's Surface." §3.4.4.3 in *Mathematical Models from the Collections of Universities and Museums* (Ed. G. Fischer). Braunschweig, Germany: Vieweg, pp. 39–40, 1986.
Rembs, E. "Enneper'sche Flächen konstanter positiver Krümmung und Hazzidakissche Transformationen." *Jahrber. DMV* **39**, 278–283, 1930.

Remes Algorithm
REMEZ ALGORITHM

Remez Algorithm
Portions of this entry contributed by CHARLES BOND
Portions of this entry contributed by RONALD M. AARTS

An algorithm for determining optimal coefficients for digital FILTERS. The Remez algorithm in effect goes a step beyond the MINIMAX APPROXIMATION algorithm to give a slightly finer solution to an approximation problem.

The Remez exchange algorithm (Remez 1957) was first studied by Parks and McClellan (1972). The algorithm is an iterative procedure consisting of two steps. One step is the determination of candidate FILTER coefficients $h(n)$ from candidate "alternation frequencies," which involves solving a set of linear equations. The other step is the determination of candidate alternation frequencies from the candidate FILTER coefficients (Lim and Oppenheim 1988). Experience has shown that the algorithm converges very fast, and is widely used in practice to design optimal FILTERS.

A *FORTRAN* implementation is given by Rabiner (1975). A description emphasizing the mathematical foundations rather than digital signal processing applications is given by Cheney (1999), who also spells Remez as Remes (Cheney 1966, p. 96).

See also FILTER, MINIMAX APPROXIMATION

References

Cheney, E. W. *Introduction to Approximation Theory, 2nd ed.* Providence, RI: Amer. Math. Soc., 1999.
Lim, J S. and Oppenheim, A V. (Eds). *Advanced Topics in Signal Processing.* Englewood Cliffs, NJ: Prentice-Hall, 1988.
Parks, T. W. and McClellan, J. J. "Chebyshev Approximation for Nonrecursive Digital Filters with Linear Phase." *IEEE Trans. Circuit Th.* **19**, 189–194, 1972.
Rabiner, L. W. and Gold, B. *Theory and Application of Digital Signal Processing.* Englewood Cliffs, NJ: Prentice-Hall, 1975.
Remez, E. Ya. *General Computational Methods of Chebyshev Approximation.* Atomic Energy Translation 4491. Kiev, 1957.

Removable Crossing
REDUCIBLE CROSSING

Removable Singularity
A SINGULAR POINT z_0 of a FUNCTION $f(z)$ for which it is possible to assign a COMPLEX NUMBER in such a way that $f(z)$ becomes ANALYTIC. A more precise way of defining a removable singularity is as a singularity z_0 of a function $f(z)$ about which the function $f(z)$ is bounded. For example, the point $x_0 = 0$ is a removable singularity in the SINC FUNCTION $\operatorname{sinc} x = \sin x/x$, since this function satisfies $\operatorname{sinc} 0 = 1$.

See also ESSENTIAL SINGULARITY, POLE, RIEMANN REMOVABLE SINGULARITY THEOREM, SINGULAR POINT

(FUNCTION)

References

Krantz, S. G. "Removable Singularities, Poles, and Essential Singularities." §4.1.4 in *Handbook of Complex Analysis*. Boston, MA: Birkhäuser, p. 42, 1999.

Rencontres Number

DERANGEMENT, SUBFACTORIAL

Rendezvous Values

MAGIC GEOMETRIC CONSTANTS

Rényi's Parking Constants

N.B. A detailed online essay by S. Finch was the starting point for this entry.

Given the CLOSED INTERVAL $[0, x]$ with $x > 1$, let 1-D "cars" of unit length be parked randomly on the interval. The MEAN number $M(x)$ of cars which can fit (without overlapping!) satisfies

$$M(x) = \begin{cases} 0 & \text{for } 0 \leq x < 1 \\ 1 + \dfrac{2}{x-1} \displaystyle\int_0^{x-1} M(y)\, dy & \text{for } x \geq 1. \end{cases} \quad (1)$$

The mean density of the cars for large x is

$$m \equiv \lim_{x \to \infty} \frac{M(x)}{x} = \int_0^\infty \exp\left(-2\int_0^x \frac{1-e^{-v}}{y}\, dy\right) dx$$

$$= 0.7475979202\ldots \quad (2)$$

(Sloane's A050996). While the inner integral can be done analytically,

$$f(x) = \gamma + \Gamma(0, x) + \ln x, \quad (3)$$

where γ is the EULER-MASCHERONI CONSTANT and $\Gamma(0, x)$ is the incomplete GAMMA FUNCTION, it is not known how to do the outer one

$$m = \int_0^\infty \exp[-2 f(x)]\, dx \quad (4)$$

$$= e^{-2\gamma} \int_0^\infty \frac{e^{-2\Gamma(0, x)}}{x^2}\, dx \quad (5)$$

$$= 2^{-2\gamma} \int_0^\infty \frac{e^{-2\text{ei}(-x)}}{x^2}, \quad (6)$$

where $\text{ei}(x)$ is the EXPONENTIAL INTEGRAL. The slowly converging series expansion for the integrand is given by

$$\frac{e^{-2\text{ei}(-x)}}{x^2} = 1 - 2x + \tfrac{5}{2}x^2 - \tfrac{22}{9}x^3 + \tfrac{293}{144}x^4 - \tfrac{2711}{1800}x^5 + \ldots \quad (7)$$

(Sloane's A050994 and A050995).

In addition,

$$M(x) = mx + m - 1 + \mathcal{O}(x^{-n}) \quad (8)$$

for all n (Rényi 1958), which was strengthened by Dvoretzky and Robbins (1964) to

$$M(x) = mx + m - 1 + \mathcal{O}\left[\left(\frac{2e}{x}\right)^{x-3/2}\right] \quad (9)$$

Dvoretzky and Robbins (1964) also proved that

$$\inf_{x \leq t \leq x+1} \frac{M(t)+1}{t+1} \leq m \leq \sup_{x \leq t \leq x+1} \frac{M(t)+1}{t+1}. \quad (10)$$

Let $V(x)$ be the variance of the number of cars, then Dvoretzky and Robbins (1964) and Mannion (1964) showed that

$$v \equiv \lim_{z \to \infty} \frac{V(x)}{x}$$

$$= 2\int_0^\infty \left\{ x \int_0^1 e^{-xy} R_2(y)\, dy + x^2 \left[\int_0^\infty e^{-xy} R_1(y)\, dy\right]^2 \right\}$$

$$\times \exp\left(-2\int_0^x \frac{1-e^{-y}}{y}\, dy\right) dx = 0.038156\ldots, \quad (11)$$

where

$$R_1(x) = M(x) - mx - m + 1 \quad (12)$$

$$R_2(x) =$$
$$\begin{cases} (1-m-mx)^2 \\ \quad \text{for } 0 \leq x \leq 1 \\ 4(1-m)^2 \\ \quad \text{for } x = 1 \\ \dfrac{2}{x-1}\left[\displaystyle\int_0^{x-1} R_2(y)\, dy + \int_0^{x-1} R_1(y)R_1(x-y-1)\, dy\right] \\ \quad \text{for } x > 1 \end{cases} \quad (13)$$

and the numerical value is due to Blaisdell and Solomon (1970). Dvoretzky and Robbins (1964) also proved that

$$\inf_{x \leq t \leq x+1} \frac{V(t)}{t+1} \leq v \leq \sup_{x \leq t \leq x+1} \frac{V(t)}{t+1}, \quad (14)$$

and that

$$V(x) = vx + v + \mathcal{O}\left[\left(\frac{4e}{x}\right)^{x-4}\right]. \quad (15)$$

Palasti (1960) conjectured that in 2-D,

$$\lim_{x, y \to \infty} \frac{M(x, y)}{xy} = m^2, \quad (16)$$

but this has not yet been proven or disproven (Finch).

References

Blaisdell, B. E. and Solomon, H. "On Random Sequential Packing in the Plane and a Conjecture of Palasti." *J. Appl. Prob.* **7**, 667–698, 1970.

Dvoretzky, A. and Robbins, H. "On the Parking Problem." *Publ. Math. Inst. Hung. Acad. Sci.* **9**, 209–224, 1964.

Finch, S. "Favorite Mathematical Constants." http://www.mathsoft.com/asolve/constant/renyi/renyi.html.

Mannion, D. "Random Space-Filling in One Dimension." *Publ. Math. Inst. Hung. Acad. Sci.* **9**, 143–154, 1964.

Palasti, I. "On Some Random Space Filling Problems." *Publ. Math. Inst. Hung. Acad. Sci.* **5**, 353–359, 1960.

Rényi, A. "On a One-Dimensional Problem Concerning Random Space-Filling." *Publ. Math. Inst. Hung. Acad. Sci.* **3**, 109–127, 1958.

Sloane, N. J. A. Sequences A050994, A050995, and A050996 in "An On-Line Version of the Encyclopedia of Integer Sequences." http://www.research.att.com/~njas/sequences/eisonline.html.

Solomon, H. and Weiner, H. J. "A Review of the Packing Problem." *Comm. Statist. Th. Meth.* **15**, 2571–2607, 1986.

Repartition

ADÈLE

Repdigit

A number composed of a single digit is called a repdigit. If the digits are all 1s, the repdigit is called a REPUNIT. The BEAST NUMBER 666 is a repdigit.

See also KEITH NUMBER, REPUNIT

Repeated Integral

A repeated integral is an integral taken multiple times over a single variable (as distinguished from a MULTIPLE INTEGRAL, which consists of a number of integrals taken with respect to *different* variables). The first FUNDAMENTAL THEOREM OF CALCULUS states that if $F(x) = D^{-1} f(x)$ is the INTEGRAL of $f(x)$, then

$$\int_0^x f(t)\, dt = F(x) - F(0). \qquad (1)$$

Now, if $F(0) = 0$, then

$$F(x) = \int f(x)\, dx = \int_0^x f(t)\, dt.$$

It follows by induction that if $F(0) = F(F(0)) = \ldots = 0$, then the n-fold integral of $f(x)$ is given by

$$D^{-n} f(x) = \underbrace{\int \cdots \int_0^x}_{n} f(x)\, dx = \int_0^x \frac{f(t)(x-t)^{n-1}}{(n-1)!}\, dt. \qquad (2)$$

Similarly, if $F(x_0) = F(F(x_0)) = \ldots = 0$, then

$$\underbrace{\int \cdots \int_{x_0}^x}_{n} f(x)\, dx = \int_{x_0}^x \frac{f(t)(x-t)^{n-1}}{(n-1)!}\, dt. \qquad (3)$$

See also FRACTIONAL INTEGRAL, FUBINI THEOREM, INTEGRAL, MULTIPLE INTEGRAL

Repeating Decimal

A number whose decimal representation eventually becomes periodic (i.e., the same sequence of digits repeats indefinitely) is called a repeating decimal. Numbers such as 0.5 can be regarded as repeating decimals since $0.5 = 0.5000\ldots = 0.4999\ldots$. All RATIONAL NUMBERS have repeating decimals, e.g., $1/11 = 0.\overline{09}$. However, TRANSCENDENTAL NUMBERS, such as $\pi = 3.141592\ldots$ do not.

If $1/m$ is a repeating decimal and $1/n$ is a terminating decimal, them $1/(mn)$ has a nonperiodic part whose length is that of $1/n$ and a repeating part whose length is that of $1/m$ (Wells 1986, p. 60).

See also CYCLIC NUMBER, DECIMAL EXPANSION, EULER'S TOTIENT RULE, FULL REPTEND PRIME, IRRATIONAL NUMBER, MIDY'S THEOREM, RATIONAL NUMBER, REGULAR NUMBER

References

Ball, W. W. R. and Coxeter, H. S. M. *Mathematical Recreations and Essays, 13th ed.* New York: Dover, pp. 53–54, 1987.

Conway, J. H. and Guy, R. K. *The Book of Numbers.* New York: Springer-Verlag, pp. 167–168, 1996.

Courant, R. and Robbins, H. "Rational Numbers and Periodic Decimals." §2.2.4 in *What is Mathematics?: An Elementary Approach to Ideas and Methods, 2nd ed.* Oxford, England: Oxford University Press, pp. 66–68, 1996.

Wells, D. *The Penguin Dictionary of Curious and Interesting Numbers.* Middlesex, England: Penguin Books, p. 60, 1986.

Repfigit Number

KEITH NUMBER

Replicate

One out of a set of identical observations in a given experiment under identical conditions.

Replicating Symbol

SHAH FUNCTION

Representation

A representation of a GROUP G is a GROUP ACTION of G on a VECTOR SPACE V by INVERTIBLE LINEAR MAPS. For example, the group of two elements $\mathbb{Z}_2 = \{0, 1\}$ has a representation ϕ by $\phi(0)v = v$ and $\phi(1)v = -v$. A representation is a GROUP HOMOMORPHISM $\phi : G \to GL(V)$.

Most groups have many different representations, possibly on different vector spaces. For example, the SYMMETRIC GROUP $S_3 = \{e, (12), (13), (23), (123), (132)\}$ has a representation on $\mathbb{R}$ by

$$\phi_1(\sigma)v = \text{sgn}(\sigma)v, \qquad (1)$$

where sgn(σ) is the SIGNATURE of the PERMUTATION σ. It also has a representation on $\mathbb{R}^3$ by

$$\phi_2(\sigma)(x_1,\ x_2,\ x_3) = \left(x_{\sigma(1)},\ x_{\sigma(2)},\ x_{\sigma(3)}\right). \qquad (2)$$

A representation gives a matrix for each element, and so another representation of S_3 is given by the matrices

$$\begin{bmatrix} 1 & 0 \\ 0 & 1 \end{bmatrix}, \begin{bmatrix} 0 & 1 \\ 1 & 0 \end{bmatrix}, \begin{bmatrix} -1 & 0 \\ -1 & 1 \end{bmatrix},$$

$$\begin{bmatrix} 1 & -1 \\ 0 & -1 \end{bmatrix}, \begin{bmatrix} -1 & 1 \\ -1 & 0 \end{bmatrix}, \begin{bmatrix} 0 & -1 \\ 1 & -1 \end{bmatrix}. \qquad (3)$$

Two representations are considered equivalent if they are conjugates. For example, CONJUGATING the above matrices by

$$\begin{bmatrix} 1 & 19 \\ 0 & 1 \end{bmatrix}$$

gives the following equivalent representation of S_3,

$$\begin{bmatrix} 1 & 0 \\ 0 & 1 \end{bmatrix}, \begin{bmatrix} -19 & -360 \\ 1 & 19 \end{bmatrix}, \begin{bmatrix} 18 & 323 \\ -1 & -18 \end{bmatrix},$$

$$\begin{bmatrix} 1 & 37 \\ 0 & -1 \end{bmatrix}, \begin{bmatrix} 18 & 343 \\ -1 & -19 \end{bmatrix}, \begin{bmatrix} -19 & -343 \\ 1 & 18 \end{bmatrix}. \qquad (4)$$

Any representation V of G can be RESTRICTED to a representation of any subgroup H, in which case, it is denoted Res_H^G. More surprisingly, any representation W on H can be extended to a representation of G, on a larger VECTOR SPACE V, called the INDUCED REPRESENTATION.

Representations have applications to many branches of mathematics, aside from applications to physics and chemistry. The name of the theory depends on the GROUP G and on the VECTOR SPACE V. Different approaches are required depending on whether G is a FINITE GROUP, an infinite DISCRETE GROUP, or a LIE GROUP. Another important ingredient is the field of scalars for V. The vector space V can be infinite dimensional such as a HILBERT SPACE. Also, special kinds of representations may require that a vector space structure is preserved. For instance, a UNITARY REPRESENTATION is a GROUP HOMOMORPHISM $\phi : G \to U(V)$ into the group of UNITARY TRANSFORMATIONS which preserve a HERMITIAN INNER PRODUCT on V.

In favorable situations, such as a finite group, an arbitrary representation will break up into IRREDUCIBLE REPRESENTATIONS, i.e., $V = \oplus V_i$ where the V_i are irreducible. For many groups, the irreducible representations have been classified.

See also GROUP, IRREDUCIBLE REPRESENTATION, MULTIPLICATIVE CHARACTER, ORTHOGONAL GROUP REPRESENTATIONS, PETER-WEYL THEOREM, PRIMARY REPRESENTATION, REPRESENTATION (LIE ALGEBRA), REPRESENTATION RING, REPRESENTATION THEORY, SCHUR'S LEMMA, SEMISIMPLE LIE GROUP, TENSOR PRODUCT (REPRESENTATION), UNITARY REPRESENTATION, VECTOR SPACE

References

Fulton, W. and Harris, J. *Representation Theory.* New York: Springer-Verlag, 1991.
Jacobson, N. *Lie Algebras.* New York: Dover, 1979.
Knapp, A. *Lie Groups: Beyond an Introduction.* Boston, MA: Birkhäuser, 1996.
Knapp, A. W. "Group Representations and Harmonic Analysis, Part II." *Not. Amer. Math. Soc.* **43**, 537–549, 1996.

Representation (Lie Algebra)

A representation of a LIE ALGEBRA $\mathfrak{g}$ is a LINEAR MAP

$$\psi : \mathfrak{g} \to M(V),$$

where $M(V)$ is the set of all linear transformations of a VECTOR SPACE V. In particular, if $V = \mathbb{R}^n$, then $M(V)$ is the set of $n \times n$ square matrices. The map ψ is required to be a map of LIE ALGEBRAS so that

$$\psi([A,\ B]) = \psi(A)\psi(B) - \psi(B)\psi(A)$$

for all $A,\ B \in \mathfrak{g}$. Note that the expression AB only makes sense as a MATRIX PRODUCT in a representation. For example, if A and B are SKEW SYMMETRIC MATRICES, then $AB - BA$ is skew-symmetric, but AB may not be skew symmetric.

The possible IRREDUCIBLE REPRESENTATIONS of complex Lie algebras are determined by the classification of the SEMISIMPLE LIE ALGEBRAS. Any IRREDUCIBLE REPRESENTATION V of a complex LIE ALGEBRA $\mathfrak{g}$ is the TENSOR PRODUCT $V = V_0 \otimes L$, where V_0 is an IRREDUCIBLE REPRESENTATION of the quotient $\mathfrak{g}_{ss}/\mathrm{Rad}(\mathfrak{g})$ of the algebra $\mathfrak{g}$ and its RADICAL, and L is a one-dimensional representation.

A LIE ALGEBRA may be associated with a LIE GROUP, in which case it reflects the local structure of the LIE GROUP. Whenever a LIE GROUP G has a REPRESENTATION on V, its TANGENT SPACE at the identity, which is a LIE ALGEBRA, has a LIE ALGEBRA representation on V given by the differential at the identity. Conversely, if a CONNECTED LIE GROUP G corresponds to the Lie algebra $\mathfrak{g}$, and $\mathfrak{g}$ has a LIE ALGEBRA representation on V, then G has a REPRESENTATION on V given by the MATRIX EXPONENTIAL.

See also IRREDUCIBLE REPRESENTATION, LIE ALGEBRA, LIE GROUP, MATRIX EXPONENTIAL, REPRESENTATION, SIMPLE LIE ALGEBRA, VECTOR SPACE

References

Fulton, W. and Harris, J. *Representation Theory.* New York: Springer-Verlag, 1991.
Jacobson, N. *Lie Algebras.* New York: Dover, 1979.
Knapp, A. *Lie Groups Beyond an Introduction.* Boston, MA: Birkhäuser, 1996.

Representation Theory

See also REPRESENTATION

References

Huang, J.-S. *Lectures on Representation Theory*. Singapore: World Scientific, 1999.

Represented As

An expression describing a form in which a quantity can be written. For example, all primes $p > 3$ can be "represented as" $6n \pm 1$.

See also OF THE FORM

References

Wells, D. *The Penguin Dictionary of Curious and Interesting Numbers*. Middlesex, England: Penguin Books, p. 13, 1986.

Reptend Prime

FULL REPTEND PRIME

Reptile

REP-TILE

Rep-Tile

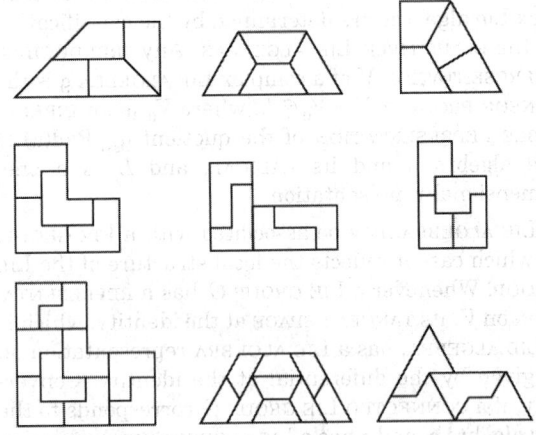

A POLYGON which can be DISSECTED into n smaller copies of itself is called a rep-n-tile. The triangular POLYGONAL SPIRAL is a rep-4-tile.

See also DISSECTION, POLYGONAL SPIRAL

References

Gardner, M. "Rep-Tiles: Replicating Figures on the Plane." Ch. 19 in *The Unexpected Hanging and Other Mathematical Diversions*. Chicago, IL: Chicago University Press, pp. 222–233, 1991.
Langford, C. D. "Uses of a Geometric Puzzle." *Math. Gaz.*, No. 260, 1940.
Wells, D. *The Penguin Dictionary of Curious and Interesting Geometry*. London: Penguin, pp. 213–214, 1991.

Repunit

A (generalized) repunit to the base b is a number OF THE FORM

$$M_n^b = \frac{b^n - 1}{b - 1}.$$

The term "repunit" was coined by Beiler (1966), who also gave the first tabulation of known factors. Repunits $M_n = M_n^2 = 2^n - 1$ with $b = 2$ are called MERSENNE NUMBERS. If $b = 10$, the number is called a repunit (since the digits are all 1s). A number OF THE FORM

$$R_n = \frac{10^n - 1}{10 - 1} = R_n = \frac{10^n - 1}{9}$$

is therefore a (decimal) repunit of order n.

b	Sloane	b-Repunits
2	Sloane's A000225	1, 3, 7, 15, 31, 63, 127, ...
3	Sloane's A003462	1, 4, 13, 40, 121, 364, ...
4	Sloane's A002450	1, 5, 21, 85, 341, 1365, ...
5	Sloane's A003463	1, 6, 31, 156, 781, 3906, ...
6	Sloane's A003464	1, 7, 43, 259, 1555, 9331, ...
7	Sloane's A023000	1, 8, 57, 400, 2801, 19608, ...
8	Sloane's A023001	1, 9, 73, 585, 4681, 37449, ...
9	Sloane's A002452	1, 10, 91, 820, 7381, 66430, ...
10	Sloane's A002275	1, 11, 111, 1111, 11111, ...
11	Sloane's A016123	1, 12, 133, 1464, 16105, 177156, ...
12	Sloane's A016125	1, 13, 157, 1885, 22621, 271453, ...

Williams and Seah (1979) factored generalized repunits for $3 \le b \le 12$ and $2 \le n \le 1000$. A (base-10) repunit can be PRIME only if n is PRIME, since otherwise $10^{ab} - 1$ is a BINOMIAL NUMBER which can be factored algebraically. In fact, if $n = 2a$ is EVEN, then $10^{2a} - 1 = (10^a - 1)(10^a + 1)$.

The number of factors for the base-10 repunits for $n = 1, 2, \ldots$ are 1, 1, 2, 2, 2, 5, 2, 4, 4, 4, 2, 7, 3, ... (Sloane's A046053). The only known base-10 repunit primes R_n are for $n = 2, 19, 23, 317, 1031, 49081$, (Sloane's A004023; Madachy 1979, Williams and Dubner 1986, Ball and Coxeter 1987, Granlund, Dubner 1999). Williams and Dubner (1986) proved R_{1031} to be prime. T. Granlund completed a search up to 45,000 in 1998 using two months of CPU time on a parallel computer. The search was extended by H. Dubner in 1999, culminating in the discovery of the probable prime $R_{49,081}$.

b	Sloane	n of Prime b-Repunits
2	Sloane's A000043	2, 3, 5, 7, 13, 17, 19, 31, 61, 89, 107, 127, 521, 607, ...
3	Sloane's A028491	3, 7, 13, 71, 103, 541, 1091, 1367, 1627, 4177, 9011, 9551, ...
5	Sloane's A004061	3, 7, 11, 13, 47, 127, 149, 181, 619, 929, 3407, 10949, ...
6	Sloane's A004062	2, 3, 7, 29, 71, 127, 271, 509, 1049, 6389, 6883, 10613, ...
7	Sloane's A004063	5, 13, 131, 149, 1699, ...
10	Sloane's A004023	2, 19, 23, 317, 1031, ...
11	Sloane's A005808	17, 19, 73, 139, 907, 1907, 2029, 4801, 5153, 10867, ...
12	Sloane's A004064	2, 3, 5, 19, 97, 109, 317, 353, 701, 9739, ...

Yates (1982) published all the repunit factors for $n \leq 1000$, a portion of which are reproduced in the *Mathematica* notebook by Weisstein. Brillhart *et al.* (1988) gave a table of repunit factors which cannot be obtained algebraically, and a continuously updated version of this table is now maintained on-line. These tables include factors for $10^n - 1$ (with $n \leq 209$ odd) and $10^n + 1$ (for $n \leq 210$ EVEN and ODD) in the files ftp://sable.ox.ac.uk/pub/math/cunningham/10- and ftp://sable.ox.ac.uk/pub/math/cunningham/10+. After algebraically factoring R_n, these types of factors are sufficient for complete factorizations.

A SMITH NUMBER can be constructed from every factored repunit.

See also CUNNINGHAM NUMBER, FERMAT NUMBER, MERSENNE NUMBER, REPDIGIT, SMITH NUMBER

References

Ball, W. W. R. and Coxeter, H. S. M. *Mathematical Recreations and Essays, 13th ed.* New York: Dover, p. 66, 1987.
Beiler, A. H. "11111...111." Ch. 11 in *Recreations in the Theory of Numbers: The Queen of Mathematics Entertains.* New York: Dover, 1966.
Brillhart, J.; Lehmer, D. H.; Selfridge, J.; Wagstaff, S. S. Jr.; and Tuckerman, B. *Factorizations of $b^n \pm 1$, $b = 2$, 3, 5, 6, 7, 10, 11, 12 Up to High Powers, rev. ed.* Providence, RI: Amer. Math. Soc., 1988. Updates are available electronically from ftp://sable.ox.ac.uk/pub/math/cunningham.
Dubner, H. "Generalized Repunit Primes." *Math. Comput.* **61**, 927–930, 1993.
Dudeney, H. E. *The Canterbury Puzzles and Other Curious Problems, 7th ed.* London: Thomas Nelson and Sons, 1949.
Gardner, M. *The Sixth Book of Mathematical Games from Scientific American.* Chicago, IL: University of Chicago Press, pp. 85–86, 1984.
Granlund, T. "Repunits." http://www.swox.com/gmp/repunit.html.
Guy, R. K. "Mersenne Primes. Repunits. Fermat Numbers. Primes of Shape $k \cdot 2^n + 2$." §A3 in *Unsolved Problems in Number Theory, 2nd ed.* New York: Springer-Verlag, pp. 8–13, 1994.
Madachy, J. S. *Madachy's Mathematical Recreations.* New York: Dover, pp. 152–153, 1979.
Ribenboim, P. "Repunits and Similar Numbers." §5.5 in *The New Book of Prime Number Records.* New York: Springer-Verlag, pp. 350–354, 1996.
Sloane, N. J. A. Sequences A000043/M0672, A000225/M2655, A002275, A002450/M3914, A002452/M4733, A003462/M3463, A003463/M4209, A003464/M4425, A004023/M2114, A004023/M2114, A004061/M2620, A004062/M0861, A004063/M3836, A004064/M0744, A005808/M5032, A016123, A016125, A023000, A023001, A028491/M2643, and A046053 in "An On-Line Version of the Encyclopedia of Integer Sequences." http://www.research.att.com/~njas/sequences/eisonline.html.
Snyder, W. M. "Factoring Repunits." *Am. Math. Monthly* **89**, 462–466, 1982.
Weisstein, E. W. "Repunits." MATHEMATICA NOTEBOOK REPUNIT.M.
Williams, H. C. and Dubner, H. "The Primality of $R1031$." *Math. Comput.* **47**, 703–711, 1986.
Williams, H. C. and Seah, E. "Some Primes of the Form $(a^n - 1)/(a - 1)$. *Math. Comput.* **33**, 1337–1342, 1979.
Yates, S. "Peculiar Properties of Repunits." *J. Recr. Math.* **2**, 139–146, 1969.
Yates, S. "Prime Divisors of Repunits." *J. Recr. Math.* **8**, 33–38, 1975.
Yates, S. "The Mystique of Repunits." *Math. Mag.* **51**, 22–28, 1978.
Yates, S. *Repunits and Reptends.* Delray Beach, FL: S. Yates, 1982.

Resampling Statistics

A set of methods that are generally superior to ANOVA for small data sets or where sample distributions are non-normal.

See also BAGGING, BOOSTING, BOOTSTRAP METHODS, HYPOTHESIS TESTING, JACKKNIFE, PERMUTATION TESTS

References

Good, P. I. *Resampling Methods: A Practical Guide to Data Analysis.* New York: Springer-Verlag, 1999.

Good, P. I. *Permutation Tests: A Practical Guide to Resampling Methods for Testing Hypotheses, 2nd ed.* New York: Springer-Verlag, 2000.

Residual

The residual is the sum of deviations from a best-fit curve of arbitrary form.

$$R \equiv \sum [y_i - f(x_i, a_1, \ldots, a_n)]^2.$$

The residual should not be confused with the CORRELATION COEFFICIENT.

Residual vs. Predictor Plot

A plot of y_i versus the ESTIMATOR $e_i \equiv \hat{y}_i - y_i$. Random scatter indicates the model is probably good. A pattern indicates a problem with the model. If the spread in e_i increases as y_i increases, the errors are called HETEROSCEDASTIC.

See also ESTIMATOR

Residue

BIQUADRATIC RESIDUE, COMMON RESIDUE, COMPLETE RESIDUE SYSTEM, CUBIC RESIDUE, MINIMAL RESIDUE, QUADRATIC RESIDUE, RESIDUE CLASS, RESIDUE (COMPLEX ANALYSIS), RESIDUE (CONGRUENCE), RESIDUE INDEX, RESIDUE THEOREM

Residue (Complex Analysis)

The constant a_{-1} in the LAURENT SERIES

$$f(z) = \sum_{n=-\infty}^{\infty} a_n (z - z_0)^n \qquad (1)$$

of $f(z)$ about a point z_0 is called the residue of $f(z)$. Unless z_0 is a POLE of f, its residue is zero. The residue of a function f at a point z_0 may be denoted $\mathrm{Res}_{z=z}(f(z))$. Two basic examples of residues are given by $\mathrm{Res}_{z=0} 1/z = 1$ and $\mathrm{Res}_{z=0} 1/z^n = 0$ for $n > 1$. The residue is implemented in *Mathematica* as Residue[f, {z, z0}].

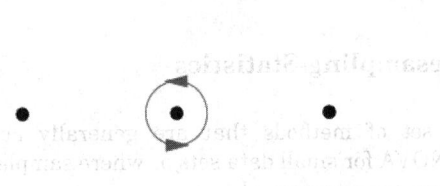

The residue is also defined by

$$\int_\gamma f \, dz, \qquad (2)$$

where γ is clockwise simple closed CONTOUR, small

enough to avoid any other poles of f. In fact, any clockwise path with WINDING NUMBER 1 which does not contain any other poles gives the same result by the CAUCHY INTEGRAL FORMULA. The above diagram shows a suitable CONTOUR for which to define the residue of function, where the poles are indicated as black dots.

It is more natural to consider the residue of a MEROMORPHIC ONE-FORM because it is independent of the choice of coordinate. On a RIEMANN SURFACE, the residue is defined for a MEROMORPHIC ONE-FORM α at a point p by writing $\alpha = f \, dz$ in a coordinate z around p. Then

$$\mathrm{Res}_p \, \alpha = \mathrm{Res}_{z=p} f. \qquad (3)$$

The sum of the residues of $\int f \, dz$ is zero on the RIEMANN SPHERE. More generally, the sum of the residues of a MEROMORPHIC ONE-FORM on a compact RIEMANN SURFACE must be zero.

The residues of a function $f(z)$ may be found without explicitly expanding into a LAURENT SERIES as follows. If $f(z)$ has a POLE of order m at z_0, then $a_n = 0$ for $n < -m$ and $a_{-m} \neq 0$. Therefore,

$$f(z) = \sum_{n=-m}^{\infty} a_n (z - z_0)^n = \sum_{n=0}^{\infty} a_{-m+n} (z - z_0)^{-m+n} \qquad (4)$$

$$(z - z_0)^m f(z) = \sum_{n=0}^{\infty} a_{-m+n} (z - z_0)^n \qquad (5)$$

$$\frac{d}{dz} [(z - z_0)^m f(z)] = \sum_{n=0}^{\infty} n a_{-m+n} (z - z_0)^{n-1}$$

$$= \sum_{n=1}^{\infty} n a_{-m+n} (z - z_0)^{n-1}$$

$$= \sum_{n=0}^{\infty} (n+1) a_{-m+n+1} (z - z_0)^n \qquad (6)$$

$$\frac{d^2}{dz^2} [(z - z_0)^m f(z)] = \sum_{n=0}^{\infty} n(n+1) a_{-m+n+1} (z - z_0)^{n-1}$$

$$= \sum_{n=1}^{\infty} n(n+1) a_{-m+n+1} (z - z_0)^{n-1}$$

$$= \sum_{n=0}^{\infty} (n+1)(n+2) a_{-m+n+2} (z - z_0)^n. \qquad (7)$$

Iterating,

$$\frac{d^{m-1}}{dz^{m-1}}\left[(z-z_0)^m f(z)\right]$$

$$=\sum_{n=0}^{\infty}(n+1)(n+2)(n+m-1)a_{n-1}(z-z_0)^n$$

$$=(m-1)!a_{-1}+\sum_{n=1}^{\infty}(n+1)(n+2)$$

$$\times(n+m-1)a_{n-1}(z-z_0)^{n-1}. \qquad (8)$$

So

$$\lim_{x\to z_0}\frac{d^{m-1}}{dz^{m-1}}\left[(z-z_0)^m f(z)\right]=\lim_{z\to z_0}(m-1)!a_{-1}+0$$

$$=(m-1)!a_{-1}, \qquad (9)$$

and the residue is

$$a_{-1}=\frac{1}{(m-1)!}\frac{d^{m-1}}{dz^{m-1}}\left[(z-z_0)^m f(z)\right]_{z=z_0}. \qquad (10)$$

The residues of a HOLOMORPHIC FUNCTION at its POLES characterize a great deal of the structure of a function, appearing for example in the amazing RESIDUE THEOREM of CONTOUR INTEGRATION.

See also CONTOUR INTEGRATION, LAURENT SERIES, MEROMORPHIC ONE-FORM, POLE, RESIDUE THEOREM, WINDING NUMBER (CONTOUR)

References
Arfken, G. "Calculus of Residues." §7.2 in *Mathematical Methods for Physicists, 3rd ed.* Orlando, FL: Academic Press, pp. 400–421, 1985.
Krantz, S. G. "The Calculus of Residues." §4.4 in *Handbook of Complex Analysis.* Boston, MA: Birkhäuser, pp. 48–51, 1999.

Residue (Congruence)

The number b in the CONGRUENCE $a\equiv b\pmod{m}$ is called the residue of $a\pmod{m}$. The residue of large numbers can be computed quickly using CONGRUENCES. For example, to find $37^{13}\pmod{17}$, note that

$$37\equiv 3$$

$$37^2\equiv 3^2\equiv 9\equiv -8$$

$$37^4\equiv 81\equiv -4$$

$$37^8\equiv 16\equiv -1,$$

so

$$37^{13}\equiv 37^{1+4+8}\equiv 3(-4)(-1)\equiv 12\pmod{17}.$$

See also COMMON RESIDUE, CONGRUENCE, MINIMAL RESIDUE

References
Shanks, D. *Solved and Unsolved Problems in Number Theory, 4th ed.* New York: Chelsea, pp. 55–56, 1993.

Residue Class

The residue classes of a function $f(x)\bmod n$ are all possible values of the RESIDUE $f(x)\pmod{n}$. For example, the residue classes of $x^2\pmod 6$ are $\{0,\ 1,\ 3,\ 4\}$, since

$$0^2\equiv 0\pmod 6$$

$$1^2\equiv 1\pmod 6$$

$$2^2\equiv 4\pmod 6$$

$$3^2\equiv 3\pmod 6$$

$$4^2\equiv 4\pmod 6$$

$$5^2\equiv 1\pmod 6$$

are all the possible residues. A COMPLETE RESIDUE SYSTEM is a set of integers containing one element from each class, so $\{0,\ 1,\ 9,\ 16\}$ would be a COMPLETE RESIDUE SYSTEM for $x^2\pmod 6$, as would $\{0,\ 5,\ 3,\ 4\}$, etc.

The $\phi(m)$ residue classes prime to m form a GROUP under the binary multiplication operation $(\bmod\ m)$, where $\phi(m)$ is the TOTIENT FUNCTION (Shanks 1993) and the GROUP is classed a MODULO MULTIPLICATION GROUP.

See also COMPLETE RESIDUE SYSTEM, CONGRUENCE, CUBIC NUMBER, QUADRATIC RECIPROCITY THEOREM, QUADRATIC RESIDUE, REDUCED RESIDUE SYSTEM, RESIDUE (CONGRUENCE), SQUARE NUMBER

References
Nagell, T. "Residue Classes and Residue Systems." §20 in *Introduction to Number Theory.* New York: Wiley, pp. 69–71, 1951.
Shanks, D. *Solved and Unsolved Problems in Number Theory, 4th ed.* New York: Chelsea, p. 56 and 59–63, 1993.

Residue Field

In a LOCAL RING R, there is only one MAXIMAL IDEAL $\mathfrak{m}$. Hence, R has only one QUOTIENT RING $R/\mathfrak{m}$ which is a FIELD. This field is called the residue field.

See also ALGEBRAIC GEOMETRY, ALGEBRAIC NUMBER THEORY, LOCAL RING

Residue Index

MULTIPLICATIVE ORDER

Residue System

COMPLETE RESIDUE SYSTEM

Residue Theorem

Given an ANALYTIC FUNCTION $f(z)$ whose LAURENT SERIES is given by

$$f(z) = \sum_{n=-\infty}^{\infty} a_n (z - z_0)^n, \quad (1)$$

and integrate term by term using a closed CONTOUR γ encircling z_0,

$$\int_\gamma f(z)\, dz = \sum_{n=-\infty}^{\infty} a_n \int_\gamma (z - z_0)^n\, dz$$

$$= \sum_{n=-\infty}^{-2} a_n \int_\gamma (z - z_0)^n\, dz + a_{-1} \int_\gamma \frac{dz}{z - z_0}$$

$$+ \sum_{n=0}^{\infty} a_n \int_\gamma (z - z_0)^n\, dz. \quad (2)$$

The CAUCHY INTEGRAL THEOREM requires that the first and last terms vanish, so we have

$$\int_\gamma f(x)\, dz = a_{-1} \int_\gamma \frac{dz}{z - z_0}, \quad (3)$$

where a_{-1} is the RESIDUE. Using the CONTOUR $z = \gamma(t) = e^{it} + z_0$ gives

$$\int_\gamma \frac{dz}{z - z_0} = \int_0^{2\pi} \frac{i e^{it}\, dt}{e^{it}} = 2\pi i, \quad (4)$$

so we have

$$\int_\gamma f(z)\, dz = 2\pi i a_{-1}. \quad (5)$$

If the contour γ encloses multiple poles, then the theorem gives the general result

$$\int_\gamma f(z)\, dz = 2\pi i \sum_{a \in A} \operatorname*{Res}_{z = a_i} f(z), \quad (6)$$

where A is the set of poles contained inside the contour. This amazing theorem therefore says that the value of a CONTOUR INTEGRAL for *any* contour in the COMPLEX PLANE depends *only* on the properties of a few very special points *inside* the contour.

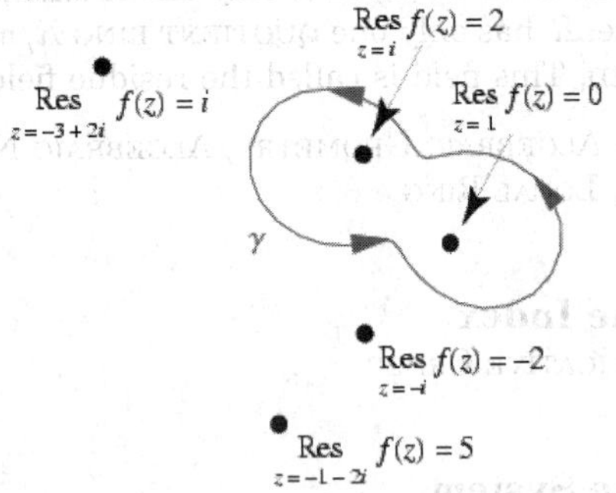

The diagram above shows an example of the residue theorem applied to the illustrated CONTOUR γ and the function

$$g(z) = \frac{3}{(z-1)^2} + \frac{2}{z-i} - \frac{2}{z+i} + \frac{i}{z+3-2i}$$

$$+ \frac{5}{z+1+2i}. \quad (7)$$

Only the poles at 1 and i are contained in the contour, which have residues of 0 and 2, respectively. The values of the CONTOUR INTEGRAL is therefore given by

$$\int_\gamma g(z)\, dz = 2\pi i (0 + 2) = 4\pi i.$$

See also CAUCHY INTEGRAL FORMULA, CAUCHY INTEGRAL THEOREM, CONTOUR, CONTOUR INTEGRAL, CONTOUR INTEGRATION, GROUP RESIDUE THEOREM, LAURENT SERIES, POLE, RESIDUE (COMPLEX ANALYSIS)

References

Knopp, K. "The Residue Theorem." §33 in *Theory of Functions Parts I and II, Two Volumes Bound as One, Part I.* New York: Dover, pp. 129–134, 1996.

Krantz, S. G. "The Residue Theorem." §4.4.2 in *Handbook of Complex Analysis.* Boston, MA: Birkhäuser, pp. 48–49, 1999.

Resistor Network

Consider a network of n resistors R_i so that R_2 may be connected in series or parallel with R_1, R_3 may be connected in series or parallel with the network consisting of R_1 and R_2, and so on. The resistance of two resistors in series is given by

$$R_{net,\ series} = R_1 + R_2,$$

and of two resistors in parallel by

$$R_{net,\ parallel} = \frac{1}{\dfrac{1}{R_1} + \dfrac{1}{R_2}}.$$

The possible values for two resistors with resistances a and b are therefore

$$a + b, \quad \frac{1}{\dfrac{1}{a} + \dfrac{1}{b}},$$

for three resistances a, b, and c are

$$a + b + c, \quad a + \frac{1}{\dfrac{1}{b} + \dfrac{1}{c}}, \quad b + \frac{1}{\dfrac{1}{a} + \dfrac{1}{c}}, \quad c + \frac{1}{\dfrac{1}{a} + \dfrac{1}{b}}$$

$$\frac{1}{\dfrac{1}{a}+\dfrac{1}{b+c}}, \quad \frac{1}{\dfrac{1}{b}+\dfrac{1}{a+c}}, \quad \frac{1}{\dfrac{1}{c}+\dfrac{1}{a+b}}, \quad \frac{1}{\dfrac{1}{a}+\dfrac{1}{b}+\dfrac{1}{c}},$$

and so on. These are obviously all rational numbers, and the numbers of distinct arrangements for $n = 1$, 2, ..., are 1, 2, 8, 46, 332, 2874, ... (Sloane's A005840), which also arises in a completely different context (Stanley 1991).

If the values are restricted to $a = b = \ldots = 1$, then there are 2^{n-1} possible resistances for n 1-Ω resistors, ranging from a minimum of $1/n$ to a maximum of n. Amazingly, the largest denominators for $n = 1$, 2, ... are 1, 2, 3, 5, 8, 13, 21, ..., which are immediately recognizable as the FIBONACCI NUMBERS (Sloane's A000045). The following table gives the values possible for small n.

n	Possible resistances
1	1
2	$\frac{1}{2}, 2$
3	$\frac{1}{3}, \frac{2}{3}, \frac{3}{2}, 3$
4	$\frac{1}{4}, \frac{2}{5}, \frac{3}{5}, \frac{3}{4}, \frac{4}{3}, \frac{5}{3}, \frac{5}{2}, 4$

If the n resistors are given the values 1, 2, ..., n, then the numbers of possible net resistances for 1, 2, ... resistors are 1, 2, 8, 44, 298, 2350, ... (Sloane's A051045). The following table gives the values possible for small n.

n	Possible resistances
1	1
2	$\frac{2}{3}, 3$
3	$\frac{6}{11}, \frac{3}{2}, \frac{11}{3}, 6$
4	$\frac{12}{25}, \frac{12}{11}, \frac{44}{23}, \frac{12}{5}, \frac{50}{11}, \frac{11}{2}, \frac{23}{3}, 10$

See also FIBONACCI NUMBER

References

Amengual, A. "The Intriguing Properties of the Equivalent Resistances of n Equal Resistors Combined in Series and in Parallel." *Amer. J. Phys.* **68**, 175–179, 2000.

Sloane, N. J. A. Sequences A000045/M0692, A005840/M1872, and A051045 in "An On-Line Version of the Encyclopedia of Integer Sequences." http://www.research.att.com/~njas/sequences/eisonline.html.

Stanley, R. P. "A Zonotope Associated with Graphical Degree Sequences." In *Applied Geometry and Discrete Mathematics: The Victor Klee Festschrift* (Ed. P. Gritzmann and B. Sturmfels). Providence, RI: Amer. Math. Soc., pp. 555–570, 1991.

Resolution

Resolution is a widely used word with many different meanings. It can refer to resolution of equations, resolution of singularities (in ALGEBRAIC GEOMETRY), resolution of modules or more sophisticated structures, etc. In a BLOCK DESIGN, a PARTITION R of a BIBD's set of blocks B into PARALLEL CLASSES, each of which in turn partitions the set V, is called a resolution (Abel and Furino 1996).

A resolution of the MODULE M over the RING R is a complex of R-modules C_i and morphisms d_i and a MORPHISM ϵ such that

$$\cdots \to C_i \xrightarrow{d_i} C_{i-1} \to \cdots \to C_0 \xrightarrow{\epsilon} M \to 0$$

satisfying the following conditions:

1. The composition of any two consecutive morphisms is the zero map,
2. For all i, $(\ker d_i)/(\operatorname{im} d_{i+1}) = 0$,
3. $C_0/(\ker \epsilon) \simeq M$,

where ker is the kernel and im is the image. Here, the quotient

$$\frac{(\ker d_i)}{(\operatorname{im} d_{i+1})}$$

is the ith HOMOLOGY GROUP.

If all modules C_i are projective (free), then the resolution is called projective (free). There is a similar concept for resolutions "to the right" of M, which are called injective resolutions.

See also HOMOLOGY GROUP, MODULE, MORPHISM, RING

References

Abel, R. J. R. and Furino, S. C. "Resolvable and Near Resolvable Designs." §I.6 in *The CRC Handbook of Combinatorial Designs* (Ed. C. J. Colbourn and J. H. Dinitz). Boca Raton, FL: CRC Press, pp. 4 and 87–94, 1996.

Jacobson, N. *Basic Algebra II, 2nd ed.* New York: W. H. Freeman, p. 339, 1989.

Resolution Class

PARALLEL CLASS

Resolution Modulus

The least POSITIVE INTEGER m^* with the property that $\chi(y) = 1$ whenever $y \equiv 1 \pmod{m^*}$ and $(y, m) = 1$.

Resolvable

A balanced incomplete BLOCK DESIGN (B, V) is called resolvable if there exists a PARTITION R of its set of blocks B into PARALLEL CLASSES, each of which in turn partitions the set V. The partition R is called a RESOLUTION.

See also BLOCK DESIGN, PARALLEL CLASS

References

Abel, R. J. R. and Furino, S. C. "Resolvable and Near Resolvable Designs." §I.6 in *The CRC Handbook of Combinatorial Designs* (Ed. C. J. Colbourn and J. H. Dinitz). Boca Raton, FL: CRC Press, pp. 4 and 87–94, 1996.

Furino, S.; Miao, Y.; and Yin, J. *Frames and Resolvable Designs: Uses, Constructions, ad Existence.* Boca Raton, FL: CRC Press, 1996.

Resolve
QUANTIFIER ELIMINATION

Resolving Tree

A tree of LINKS obtained by repeatedly choosing a crossing, applying the SKEIN RELATIONSHIP to obtain two simpler LINKS, and repeating the process. The DEPTH of a resolving tree is the number of levels of links, not including the top. The DEPTH of the LINK is the minimal depth for any resolving tree of that LINK.

Resonance Overlap

Isolated resonances in a DYNAMICAL SYSTEM can cause considerable distortion of preserved TORI in their NEIGHBORHOOD, but they do not introduce any CHAOS into a system. However, when two or more resonances are simultaneously present, they will render a system nonintegrable. Furthermore, if they are sufficiently "close" to each other, they will result in the appearance of widespread (large-scale) CHAOS.

To investigate this problem, Walker and Ford (1969) took the integrable Hamiltonian

$$H_0(I_1, I_2) = I_1 + I_2 - I_1^2 - 3I_1I_2 + I_2^2$$

and investigated the effect of adding a 2:2 resonance and a 3:2 resonance

$$H(\mathbf{I}, \theta) = H_0(\mathbf{I}) + \alpha I_1 I_2 \cos(2\theta_1 - 2\theta_2)$$
$$+ \beta I_1^{3/2} I_2 \cos(2\theta_1 - 3\theta_2).$$

At low energies, the resonant zones are well-separated. As the energy increases, the zones overlap and a "macroscopic zone of instability" appears. When the overlap starts, many higher-order resonances are also involved so fairly large areas of PHASE SPACE have their TORI destroyed and the ensuing CHAOS is "widespread" since trajectories are now free to wander between regions that previously were separated by nonresonant TORI.

Walker and Ford (1969) were able to numerically predict the energy at which the overlap of the resonances first occurred. They plotted the θ_2-axis intercepts of the inner 2:2 and the outer 2:3 separatrices as a function of total energy. The energy at which they crossed was found to be identical to that at which 2:2 and 2:3 resonance zones began to overlap.

See also CHAOS, RESONANCE OVERLAP METHOD

References

Walker, G. H. and Ford, J. "Amplitude Instability and Ergodic Behavior for Conservative Nonlinear Oscillator Systems." *Phys. Rev.* **188**, 416–432, 1969.

Resonance Overlap Method

A method for predicting the onset of widespread CHAOS.

See also GREENE'S METHOD

References

Chirikov, B. V. "A Universal Instability of Many-Dimensional Oscillator Systems." *Phys. Rep.* **52**, 264–379, 1979.

Tabor, M. *Chaos and Integrability in Nonlinear Dynamics: An Introduction.* New York: Wiley, pp. 154–163, 1989.

R-Estimate

A ROBUST ESTIMATION based on a RANK TEST.

See also *L*-ESTIMATE, *M*-ESTIMATE, RANK TEST, ROBUST ESTIMATION

References

Press, W. H.; Flannery, B. P.; Teukolsky, S. A.; and Vetterling, W. T. "Robust Estimation." §15.7 in *Numerical Recipes in FORTRAN: The Art of Scientific Computing,* 2nd ed. Cambridge, England: Cambridge University Press, pp. 694–700, 1992.

Restricted Divisor Function

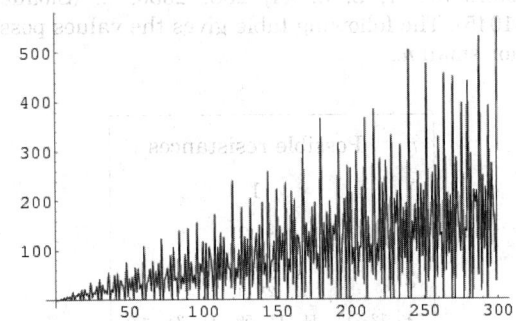

The sum of the ALIQUOT DIVISORS of n, given by

$$s(n) \equiv \sigma(n) - n,$$

where $\sigma(n)$ is the DIVISOR FUNCTION. The first few values are 0, 1, 1, 3, 1, 6, 1, 7, 4, 8, 1, 16, ... (Sloane's A001065).

See also DIVISOR FUNCTION

References

Sloane, N. J. A. Sequences A001065/M2226 in "An On-Line Version of the Encyclopedia of Integer Sequences." http://www.research.att.com/~njas/sequences/eisonline.html.

Restricted Growth Function
RESTRICTED GROWTH STRING

Restricted Growth String

For a SET PARTITION of n elements, the n-character string $a_1 a_2 \ldots a_n$ in which each character gives the BLOCK (**B**$_0$, **B**$_1$, ...) in which the corresponding element belongs is called the restricted growth string (or sometimes the RESTRICTED GROWTH FUNCTION). For example, for the SET PARTITION $\{\{1\}, \{2\}, \{3, 4\}\}$, the restricted growth string would be 0122. If the BLOCKS are "sorted" so that $a_1 = 0$, then the restricted growth string satisfies the INEQUALITY

$$a_{i+1} \leq 1 + \max\{a_1, a_2, \ldots, a_i\}$$

for $i = 1, 2, \ldots, n-1$.

References
Ruskey, F. "Info About Set Partitions." http://www.theory.csc.uvic.ca/~cos/inf/setp/SetPartitions.html.

Restriction (Representation)

A REPRESENTATION of a GROUP G on a VECTOR SPACE V can be restricted to a SUBGROUP H. For example, the SYMMETRIC GROUP on three letters has a representation ϕ on $\mathbb{R}^2$ by

$$\phi(e) = \begin{bmatrix} 1 & 0 \\ 0 & 1 \end{bmatrix} \tag{1}$$

$$\phi(12) = \begin{bmatrix} 0 & 1 \\ 1 & 0 \end{bmatrix} \tag{2}$$

$$\phi(13) = \begin{bmatrix} -1 & 0 \\ -1 & 1 \end{bmatrix} \tag{3}$$

$$\phi(23) = \begin{bmatrix} 1 & -1 \\ 0 & -1 \end{bmatrix} \tag{4}$$

$$\phi(123) = \begin{bmatrix} -1 & 1 \\ -1 & 0 \end{bmatrix} \tag{5}$$

$$\phi(132) = \begin{bmatrix} 0 & -1 \\ 1 & -1 \end{bmatrix} \tag{6}$$

that can be restricted to the subgroup of ORDER 3,

$$\phi(e) = \begin{bmatrix} 1 & 0 \\ 0 & 1 \end{bmatrix} \tag{7}$$

$$\phi(123) = \begin{bmatrix} -1 & 1 \\ -1 & 0 \end{bmatrix} \tag{8}$$

$$\phi(132) = \begin{bmatrix} 0 & -1 \\ 1 & -1 \end{bmatrix} \tag{9}$$

See also FROBENIUS RECIPROCITY, REPRESENTATION, VECTOR SPACE

Resultant

Given a POLYNOMIAL $p(x)$ of degree n with roots α_i, $i = 1, \ldots, n$ and a POLYNOMIAL $q(x)$ of degree m with roots β_j, $j = 1, \ldots, m$, the resultant is defined by

$$\rho(p, q) = \prod_{i=1}^{n} \prod_{j=1}^{m} (\beta_j - \alpha_i).$$

The notation $R(p, q)$ is also used.

There exists an ALGORITHM similar to the EUCLIDEAN ALGORITHM for computing resultants (Pohst and Zassenhaus 1989). The resultant of two polynomials can be computed using the *Mathematica* command `Resultant[poly1, poly2, var]`.

Resultants for a few simple pairs of polynomials include

$$\rho(x - a, \ x - b) = a - b$$

$$\rho((x-a)(x-b), \ x - c) = (a-c)(b-c)$$

$$\rho((x-a)(x-b), \ (x-c)(x-d))$$
$$= (a-c)(b-c)(a-d)(b-d).$$

The resultant is the DETERMINANT of the corresponding SYLVESTER MATRIX. Given p and q, then

$$h(x) = \rho(q(t), \ p(x-t))$$

is a POLYNOMIAL of degree mn, having as its roots all sums OF THE FORM $\alpha_i + \beta_j$.

See also DISCRIMINANT (POLYNOMIAL), SUBRESULTANT, SYLVESTER MATRIX

References
Apostol, T. M. "Resultants of Cyclotomic Polynomials." *Proc. Amer. Math. Soc.* **24**, 457–462, 1970.
Apostol, T. M. "The Resultant of the Cyclotomic Polynomials $F_m(ax)$ and $F_n(bx)$." *Math. Comput.* **29**, 1–6, 1975.
Pohst, M. and Zassenhaus, H. *Algorithmic Algebraic Number Theory.* Cambridge, England: Cambridge University Press, 1989.
Wagon, S. *Mathematica in Action.* New York: W. H. Freeman, p. 348, 1991.

Retardance
A shift in PHASE.

See also PHASE

Reuleaux Polygon

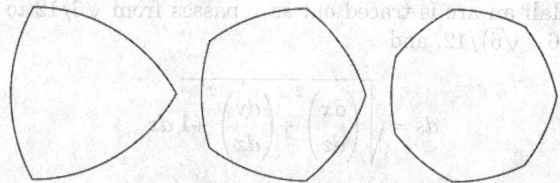

A curvilinear polygon built up of circular ARCS. The Reuleaux polygon is a generalization of the REU-

LEAUX TRIANGLE and, for an ODD NUMBER of sides, is a CURVE OF CONSTANT WIDTH (Gray 1997).

See also CURVE OF CONSTANT WIDTH, DELTA CURVE, REULEAUX TRIANGLE

References

Gray, A. "Reuleaux Polygons." §7.8 in *Modern Differential Geometry of Curves and Surfaces with Mathematica, 2nd ed.* Boca Raton, FL: CRC Press, pp. 176–177, 1997.

Reuleaux, F. *The Kinematics of Machinery.* New York: Dover, 1963.

Wagon, S. *Mathematica in Action.* New York: W. H. Freeman, pp. 52–54, 1991.

Wells, D. *The Penguin Dictionary of Curious and Interesting Geometry.* London: Penguin, pp. 219–220, 1991.

Reuleaux Tetrahedron

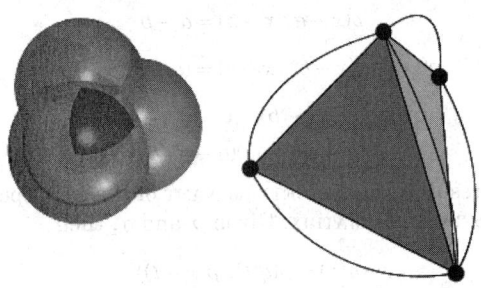

The Reuleaux tetrahedron is the 3-dimensional solid common to four SPHERES of equal radius placed so that the center of each sphere lies on the surface of the other three. The centers of the spheres are therefore located at the vertices of a regular TETRAHEDRON, and the solid consists of an "inflated" tetrahedron with four curved edges.

To analyze the Reuleaux tetrahedron, fix a TETRAHEDRON of unit edge length with its vertices at $(0, 0, -\sqrt{6}/4)$, $(\sqrt{3}/3, 0, \sqrt{6}/12)$, $(-\sqrt{3}/6, 1/2, \sqrt{6}/12)$, and $(-\sqrt{3}/6, -1/2, \sqrt{6}/12)$. Simultaneously solving the equations of three of four spheres for x and y as a function of z then gives

$$x = \tfrac{1}{2}\sqrt{2}z + \tfrac{1}{4}\sqrt{\tfrac{15}{2} - 6z(\sqrt{6}+6z)} \tag{1}$$

$$y = \frac{4\sqrt{3}z + \sqrt{5 - 4z(\sqrt{6}+6z)}}{4\sqrt{2}}. \tag{2}$$

Half an arc is traced out as z passes from $\sqrt{6}/12$ to $(6-\sqrt{6})/12$, and

$$ds = \sqrt{\left(\frac{dx}{dz}\right)^2 + \left(\frac{dy}{dz}\right)^2 + 1}\, dz$$

$$= 3\sqrt{\frac{2}{5 - 4z(\sqrt{6}+6z)}}\, dz, \tag{3}$$

so the ARC LENGTH of the curves connecting the vertices is given by

$$s = \int ds$$

$$= 6\sqrt{2} \int_{\sqrt{6}/12}^{(6-\sqrt{6})/12} \left[5 - 4z\left(\sqrt{6}+6z\right)\right]^{-1/2} dz. \tag{4}$$

Making a change of coordinates,

$$s = \sqrt{3} \int_2^{\sqrt{6}} (6 - u^2)^{-1/2}\, du = \sqrt{3}\cot^{-1}\left(\sqrt{2}\right) \tag{5}$$

$$\approx 1.06604.$$

The VOLUME is significantly trickier to calculate analytically. Set up SPHERICAL COORDINATES from the *centroid* of the TETRAHEDRON, so that the distance from the bottom vertex to the radius vector is 1, i.e.,

$$r^2 \cos^2 \sin^2 \phi + r^2 \sin^2 \theta \sin^2 \phi + \left(r + \tfrac{1}{4}\sqrt{6}\right)^2 = 1, \tag{6}$$

giving

$$r(\theta,\phi) = \tfrac{1}{4}\left[\sqrt{3\cos(2\phi) + 13} - \sqrt{6}\cos\phi\right]. \tag{7}$$

By symmetry, the volume of the Reuleaux tetrahedron is given by

$$V = 24 \int_0^{\pi/3} \int_0^{\phi(\theta)} \int_0^{r(\theta,\phi)} r^2 \sin\phi\, dr\, d\phi\, d\theta. \tag{8}$$

The integral over r can be done immediately,

$$V =$$

$$\tfrac{1}{8} \int_0^{\pi/3} \int_0^{\phi(\theta)} \left[\sqrt{3\cos(2\phi) + 13} - \sqrt{6}\cos\phi\right]^3 \sin\phi\, d\phi\, d\theta. \tag{9}$$

Now parameterize the top right edge as a function of the azimuthal coordinate θ as

$$x = \frac{\cos\theta}{\sqrt{3}\cos\theta + 3\sin\theta} \tag{10}$$

$$y = \frac{\sin\theta}{\sqrt{3}\cos\theta + 3\sin\theta} \tag{11}$$

$$z = \tfrac{1}{12}\sqrt{6}. \tag{12}$$

The polar angle ϕ can then be solved for as a function of θ as

$$\phi(\theta) = \cos^{-1}\left(\frac{z}{\sqrt{x^2+y^2+z^2}}\right)$$

$$= \tan^{-1}\left(\frac{2\sqrt{6}}{\sqrt{3}\cos\theta + 3\sin\theta}\right). \tag{13}$$

The integral over ϕ can be done by making the

change of coordinates

$$u = \frac{2\sqrt{6}}{\sqrt{3}\cos\theta + 3\sin\theta}, \qquad (14)$$

giving

$$V = \int_0^{\pi/3} \frac{1}{32} \Bigg[256 - 45\sqrt{6} + 42\sqrt{6}\cos(2\tan^{-1}u)$$

$$-\frac{58\sqrt{13 + 3\cos(2\tan^{-1}u)}}{\sqrt{1 + u^2}}$$

$$+6\sqrt{13 + 3\cos(2\tan^{-1}u)}\cos(3\tan^{-1}u)$$

$$+3\sqrt{6}\cos(4\tan^{-1}u) \Bigg] d\theta. \qquad (15)$$

Making the change of variables

$$u = \frac{2\sqrt{6(1 + 3t^2)}}{\sqrt{3} + 3\sqrt{3}t} \qquad (16)$$

then gives the volume as

$$V = \int_0^1 \Bigg(\frac{8\sqrt{3}}{1 + 3t^2} - \frac{16\sqrt{2}(3t + 1)(4t^2 + t + 1)^{3/2}}{(3t^2 + 1)(11t^2 + 2t + 3)^2}$$

$$-\frac{\sqrt{2}(249t^2 + 54t + 65)}{(11t^2 + 2t + 3)^2} \Bigg) dt. \qquad (17)$$

This integral can be done analytically, but the analytic form returned by symbolic algebra programs is an extremely complicated expression involving logarithms and inverse tangent functions. After arduous simplification of the expression by hand, the final solution

$$V = \frac{1}{24}\Big[6\sqrt{2} + 16\pi + 57\cos^{-1}\left(\tfrac{17}{81}\right) - 132\tan^{-1}\left(\sqrt{2}\right) \Big] \qquad (18)$$

$$\approx 0.422157733 \qquad (19)$$

is obtained. This solution appears not to have been published previously.

See also HYPERBOLIC TETRAHEDRON, REULEAUX TRIANGLE, SPHERE, SPHERE-SPHERE INTERSECTION, SPHERICAL TRIANGLE, STEINMETZ SOLID, TETRAHEDRON

Reuleaux Triangle

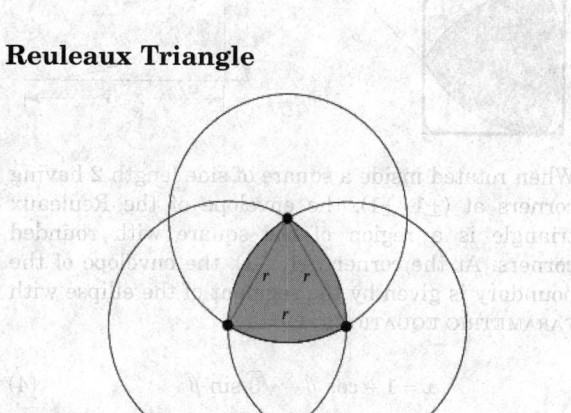

A CURVE OF CONSTANT WIDTH constructed by drawing arcs from each VERTEX of an EQUILATERAL TRIANGLE between the other two VERTICES. The Reuleaux triangle has the smallest AREA for a given width of any CURVE OF CONSTANT WIDTH. Let the arc radius be r. Since the AREA of each meniscus-shaped portion of the Reuleaux triangle is a circular SEGMENT with opening angle $\theta = \pi/3$,

$$A_s = \frac{1}{2}r^2(\theta - \sin\theta) = \left(\frac{\pi}{6} - \frac{\sqrt{3}}{4}\right)r^2. \qquad (1)$$

But the AREA of the central EQUILATERAL TRIANGLE with $a = 1/\sqrt{3}$ is

$$A_t = \frac{1}{4}\sqrt{3}r^2, \qquad (2)$$

so the total AREA is then

$$A = 3A_s + A_t = \frac{1}{2}\left(\pi - \sqrt{3}\right)r^2. \qquad (3)$$

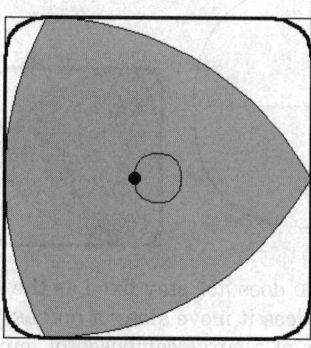

Because it can be rotated inside a SQUARE, as illustrated above, it is the basis for the Harry Watt square drill bit.

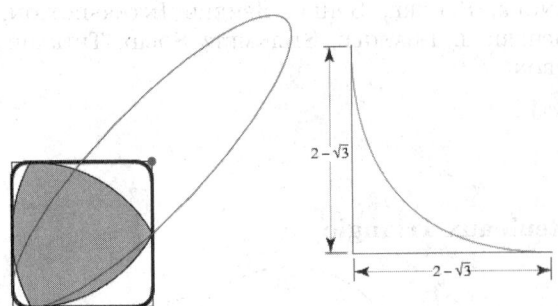

When rotated inside a square of side length 2 having corners at $(\pm 1, \pm 1)$, the envelope of the Reuleaux triangle is a region of the square with rounded corners. At the corner $(-1, -1)$, the envelope of the boundary is given by the segment of the ellipse with PARAMETRIC EQUATIONS

$$x = 1 - \cos \beta - \sqrt{3} \sin \beta \qquad (4)$$

$$y = 1 - \sin \beta - \sqrt{3} \cos \beta \qquad (5)$$

for $\beta \in [\pi/6, \pi/3]$, extending a distance $2 - \sqrt{3}$ from the corner (Gleißner and Zeitler 2000). The ellipse has center $(1, 1)$, semimajor axis $a = 1 + \sqrt{3}$, semiminor axis $b = 1 - \sqrt{3}$, and is rotated by $45°$, which has Cartesian equation

$$x^2 + y^2 - \sqrt{3}xy - \left(2 - \sqrt{3}\right)x - \left(2 - \sqrt{3}\right)y + 1 - \sqrt{3} = 0. \qquad (6)$$

The fractional AREA covered as the Reuleaux triangle rotates is

$$A_{\text{covered}} = 2\sqrt{3} + \tfrac{1}{6}\pi - 3 = 0.9877003907\ldots \qquad (7)$$

Note that Gleißner and Zeitler (2000) fail to simplify their equivalent equation, and then proceed to assert that (7) is erroneous.

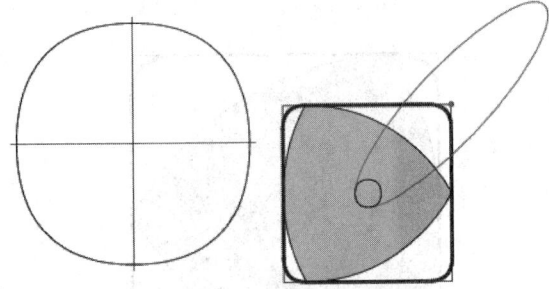

The CENTROID does *not* stay fixed as the TRIANGLE is rotated, nor does it move along a CIRCLE. In fact, the path consists of a curve composed of four arcs of an ELLIPSE (Wagon 1991). For a bounding square of side length 2, the ellipse in the lower-left quadrant has

PARAMETRIC EQUATIONS

$$x = 1 + \cos \beta + \tfrac{1}{3}\sqrt{3}\sin \beta \qquad (8)$$

$$y = 1 + \sin \beta + \tfrac{1}{3}\sqrt{3}\cos \beta \qquad (9)$$

for $\beta \in [\pi/6, \pi/3]$. The ellipse has center $(1, 1)$, semimajor axis $a = 1 + 1/\sqrt{3}$, semiminor axis $b = 1 - 1/\sqrt{3}$, and is rotated by $45°$, which has Cartesian equation

$$3x^2 + 3y^2 - 3\sqrt{3}xy - 3x\left(2 + \sqrt{3}\right) - 3y\left(2 + \sqrt{3}\right)$$
$$+5 - 3\sqrt{3} = 0. \qquad (10)$$

The area enclosed by the locus of the centroid is given by

$$A_{\text{centroid}} = 4 - \tfrac{8}{3}\sqrt{3} + \tfrac{2}{9}\pi \qquad (11)$$

(Gleißner and Zeitler 2000; who again fail to simplify their expression). Note that the CENTROID's path can be closely approximated by a SUPERELLIPSE

$$\left|\frac{x}{a}\right|^r + \left|\frac{y}{a}\right|^r = 1 \qquad (12)$$

with $a = 2\sqrt{3}/3 - 1$ and $r \approx 2.36185$.

See also CURVE OF CONSTANT WIDTH, DELTA CURVE, EQUILATERAL TRIANGLE, FLOWER OF LIFE, PIECEWISE CIRCULAR CURVE, REULEAUX POLYGON, REULEAUX TETRAHEDRON, ROTOR, ROULETTE

References

Blaschke, W. "Konvexe Bereiche gegebener konstanter Breite und kleinsten Inhalts." *Math. Ann.* **76**, 504–513, 1915.

Bogomolny, A. "Shapes of Constant Width." http://www.cut-the-knot.com/do_you_know/cwidth.html.

Croft, H. T.; Falconer, K. J.; and Guy, R. K. *Unsolved Problems in Geometry.* New York: Springer-Verlag, p. 8, 1991.

Dark, H. E. *The Wankel Rotary Engine: Introduction and Guide.* Bloomington, IN: Indiana University Press, 1974.

Eppstein, D. "Reuleaux Triangles." http://www.ics.uci.edu/~eppstein/junkyard/reuleaux.html.

Gardner, M. "Mathematical Games: Curves of Constant Width, One of which Makes it Possible to Drill Square Holes." *Sci. Amer.* **208**, 148–156, Feb. 1963.

Gardner, M. "Curves of Constant Width." Ch. 18 in *The Unexpected Hanging and Other Mathematical Diversions.* Chicago, IL: University of Chicago Press, pp. 212–221, 1991.

Gleißner, W. and Zeitler, H. "The Reuleaux Triangle and Its Center of Mass." *Result. Math.* **37**, 335–344, 2000.

Gray, A. "Reuleaux Polygons." §7.8 in *Modern Differential Geometry of Curves and Surfaces with Mathematica, 2nd ed.* Boca Raton, FL: CRC Press, pp. 176–177, 1997.

Kunkel, P. "Reuleaux Triangle." http://www.nas.com/~kunkel/reuleaux/reuleaux.htm.

Math Forum. "Reuleaux Triangle, Reuleaux Drill." http://mathforum.com/~sarah/HTMLthreads/articletocs/reuleaux.triangle.html.

Peterson, I. "Ivar Peterson's MathLand: Rolling with Reuleaux." Oct. 21, 1996. http://www.maa.org/mathland/mathland_10_21.html.

Rademacher, H. and Toeplitz, O. *The Enjoyment of Mathematics: Selections from Mathematics for the Amateur.* Princeton, NJ: Princeton University Press, 1957.

Reuleaux, F. *The Kinematics of Machinery: Outlines of a Theory of Machines.* London: Macmillan, 1876. Reprinted as *The Kinematics of Machinery.* New York: Dover, 1963.

Smith, S. "Drilling Square Holes." *Math. Teacher* **86**, 579–583, Oct. 1993.

Wagon, S. *Mathematica in Action.* New York: W. H. Freeman, pp. 52–54 and 381–383, 1991.

Yaglom, I. M. and Boltyansky, B. G. *Convex Shapes.* Moscow: Nauka, 1951.

Reversal

The reversal of a decimal number $abc\cdots$ is $\cdots cba$. Ball and Coxeter (1987) consider numbers whose reversals are integral multiples of themselves. PALINDROMIC NUMBERS and numbers ending with a ZERO are trivial examples.

The first few nontrivial examples are 8712, 9801, 87912, 98901, 879912, 989901, 8799912, 9899901, 87128712, 87999912, 98019801, 98999901, ... (Sloane's A031877). The pattern continues for large numbers, with numbers OF THE FORM $87\underbrace{9\cdots9}12$ equal to 4 times their reversals and numbers OF THE FORM $98\underbrace{9\cdots9}01$ equal to 9 times their reversals. In addition, runs of numbers of either of these forms can be concatenated to yield numbers OF THE FORM $87\underbrace{9\cdots9}12\cdots87\underbrace{9\cdots9}12$, equal to 4 times their reversals, and $98\underbrace{9\cdots9}01\cdots98\underbrace{9\cdots9}01$, equal to 9 times their reversals.

The product of a 2-digit number and its reversal is never a SQUARE NUMBER except when the digits are the same (Ogilvy 1988). Numbers whose product is the reversal of the products of their reversals include (221, 312) and (122, 213), since

$$312 \times 221 = 68952$$

$$213 \times 122 = 25986$$

(Ball and Coxeter 1987, p. 14).

See also EMIRP, RATS SEQUENCE

References

Ball, W. W. R. and Coxeter, H. S. M. *Mathematical Recreations and Essays, 13th ed.* New York: Dover, pp. 14–15, 1987.

Edalj, J. Problem 1622. *L'Interméd. Math.* **16**, 34, 1909.

Jonesco, J. Problem 1622. *L'Interméd. Math.* **15**, 128, 1908.

Ogilvy, C. S. and Anderson, J. T. *Excursions in Number Theory.* New York: Dover, pp. 88–89, 1988.

Sloane, N. J. A. Sequences A031877 in "An On-Line Version of the Encyclopedia of Integer Sequences." http://www.research.att.com/~njas/sequences/eisonline.html.

Welsch. Problem 1622. *L'Interméd. Math.* **15**, 278, 1908.

Reverse Greedy Algorithm

An algorithm for computing a UNIT FRACTION.

See also GREEDY ALGORITHM, UNIT FRACTION

References

Eppstein, D. Egypt.ma Mathematica notebook. http://www.ics.uci.edu/~eppstein/numth/egypt/egypt.ma.

Reverse-Then-Add Sequence

An integer sequence produced by the 196-ALGORITHM.

See also 196-ALGORITHM, SORT-THEN-ADD SEQUENCE

Reversible Knot

INVERTIBLE KNOT

Reversible Prime

EMIRP

Reversion of Series

SERIES REVERSION

Reversion to the Mean

This entry contributed by ANTON E. WEISSTEIN

Reversion to the mean is the statistical phenomenon that a random variate which deviates strongly from the mean in a particular direction is likely to be succeeded by an event (independent of the first) that deviates less far in this direction. In other words, an extreme event is likely to be followed by a less extreme event.

Although this phenomenon appears to violate the definition of INDEPENDENT EVENTS, it simply reflects the fact that there are more values from which to choose on the side of the probability distribution closer to the mean than there are on the side corresponding to even more extreme values.

See also MEAN

Reye's Configuration

A configuration of 12 planes and 12 points such that six points lie in every plane and six planes pass through every point. Alternatively, the configuration consists of 16 lines and the same 12 points such that four lines pass through every point and three points lie on every line.

The points consist of the eight vertices of a CUBE together with its center and the three POINTS AT INFINITY where parallel edges of the CUBE meet. The 12 planes are the six faces of the cube and the six planes passing through diagonally opposite edges. The 16 lines consist of the 12 edges and four space diagonals of the cube.

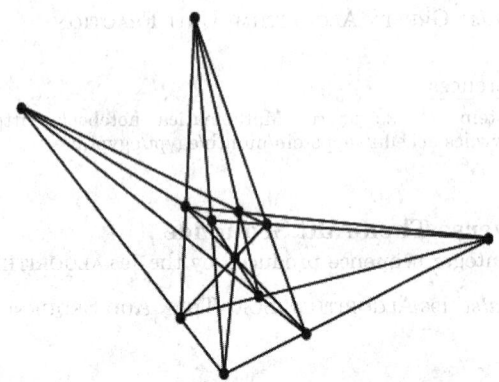

Reye's configuration can be realized without any points at infinity by squashing the cube and bringing the points at infinity to finite positions, as illustrated above.

See also CONFIGURATION

References

Wells, D. *The Penguin Dictionary of Curious and Interesting Geometry.* London: Penguin, pp. 214–215, 1991.

Reznik's Identity

For P and Q POLYNOMIALS in n variables,

$$|P \cdot Q|_2^2 = \sum_{i_1, \ldots, i_n \geq 0}$$
$$\times \frac{|P^{(i_1, \ldots, i_n)}(D_1, \ldots, D_n)Q(x_1, \ldots, x_n)|_2^2}{i_1! \cdots i_n!},$$

where $D_i \equiv \partial/\partial x_i$, $|X|_2$ is the BOMBIERI NORM, and

$$P^{(i_1, \ldots, i_n)} = D_1^{i_1} \cdots D_n^{i_n} P.$$

BOMBIERI'S INEQUALITY follows from this identity.

See also BEAUZAMY AND DÉGOT'S IDENTITY

Rhodonea

ROSE

Rhomb

RHOMBUS

Rhombic Dodecahedral Number

A FIGURATE NUMBER which is constructed as a centered CUBE with a SQUARE PYRAMID appended to each face,

$$RhoDod_n = CCub_n + 6P_{n-1}$$
$$= (2n-1)(2n^2 - 2n + 1), \qquad (1)$$

where $CCub_n$ is a CENTERED CUBE NUMBER and P_n is a PYRAMIDAL NUMBER. The first few are 1, 15, 65, 175, 369, 671, ... (Sloane's A005917). The GENERATING FUNCTION of the rhombic dodecahedral numbers is

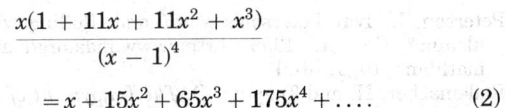

$$\frac{x(1 + 11x + 11x^2 + x^3)}{(x-1)^4}$$
$$= x + 15x^2 + 65x^3 + 175x^4 + \ldots. \qquad (2)$$

A related set of numbers is the number of cubes in the HAUY CONSTRUCTION of the RHOMBIC DODECAHEDRON, given by

$$HauyRhoDod_k = k^3 + 6 \sum_{i=1, 3, \ldots, k-2} i^2, \qquad (3)$$

for k an ODD NUMBER. Re-indexing with $k = 2n - 1$ then gives

$$HauyRhoDod_n = (2n-1)(8n^2 - 14n + 7), \qquad (4)$$

giving the first few values 1, 33, 185, 553, 1233, ... (Sloane's A046142).

See also ESCHER'S SOLID, HAUY CONSTRUCTION, OCTAHEDRAL NUMBER, RHOMBIC DODECAHEDRON

References

Conway, J. H. and Guy, R. K. *The Book of Numbers.* New York: Springer-Verlag, pp. 53–54, 1996.
Sloane, N. J. A. Sequences A005917/M4968 and A046142 in "An On-Line Version of the Encyclopedia of Integer Sequences." http://www.research.att.com/~njas/sequences/eisonline.html.

Rhombic Dodecahedron

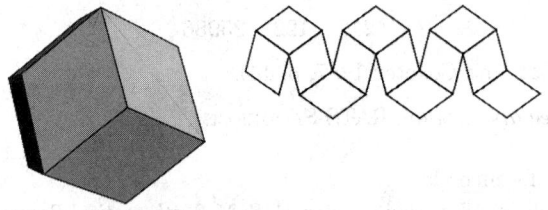

The DUAL POLYHEDRON of the CUBOCTAHEDRON A_1 and Wenninger dual W_{11}. Its sometimes also called the RHOMBOIDAL DODECAHEDRON (Cotton 1990). Its 14 vertices are joined by 12 RHOMBUSES of the dimensions shown in the figure below, where

$$\alpha = 2 \cot^{-1}\sqrt{2} = \cos^{-1}\left(\tfrac{1}{3}\right) \approx 70.53° \qquad (1)$$

$$\beta = 2 \tan^{-1}\sqrt{2} \approx 109.47°. \qquad (2)$$

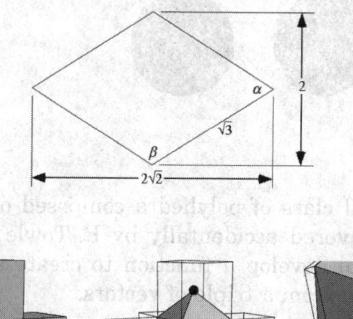

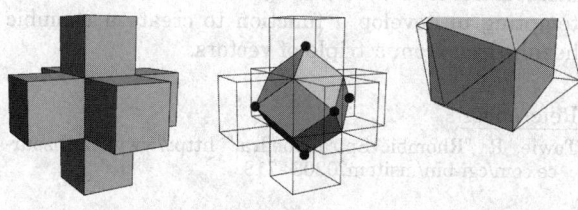

The rhombic dodecahedron can be built up by a placing six cubes on the faces of a seventh, in the configuration of a metal "jack." Joining the centers of the outer cubes with the vertices of the central cube then gives the rhombic dodecahedron. Affixing a SQUARE PYRAMID of height 1/2 on each face of a CUBE having unit edge length results in a rhombic dodecahedron (Brückner 1900, p. 130; Steinhaus 1983, p. 185).

If the rhombic dodecahedron is hinged into six square pyramids along three consecutive face diagonals, the resulting model can be folded into a cube (Wells 1991). One possible construction for the rhombic dodecahedron is known as the BAUSPIEL. It can also be constructed by CUMULATION of a unit edge-length CUBE by a pyramid with height 1/2.

The rhombic dodecahedron is a ZONOHEDRON and a SPACE-FILLING POLYHEDRON (Steinhaus 1983, p. 185). The vertices are given by $(\pm 1, \pm 1, \pm 1)$, $(\pm 2, 0, 0)$, $(0, \pm 2, 0)$, $(0, 0, \pm 2)$.

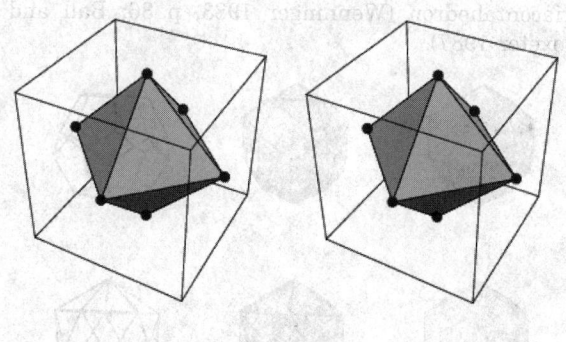

The edges of the CUBE-OCTAHEDRON COMPOUND intersecting in the points plotted above are the diagonals of RHOMBUSES, and the 12 RHOMBUSES form a rhombic dodecahedron (Ball and Coxeter 1987). There are three stellations of the rhombic dodecahedron.

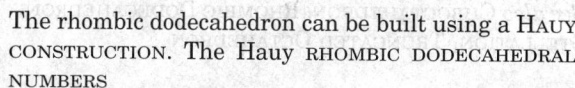

The rhombic dodecahedron can be built using a HAUY CONSTRUCTION. The Hauy RHOMBIC DODECAHEDRAL NUMBERS

$$HRhoDod_n = (2n-1)(8n^2 - 14n + 7) \qquad (3)$$

give a method for calculating the VOLUME of the rhombic dodecahedron,

$$V = \lim_{n \to \infty} HRhoDod_n \left(\frac{a}{n\sqrt{3}}\right)^3 = \frac{16}{9}\sqrt{3}a^3 \qquad (4)$$

(Steinhaus 1983). The SURFACE AREA of a rhombic dodecahedron with unit edge length is

$$S = 8\sqrt{2}. \qquad (5)$$

See also BAUSPIEL, CUBE-OCTAHEDRON COMPOUND, DODECAHEDRON, HAUY CONSTRUCTION, PYRITOHEDRON, RHOMBIC DODECAHEDRON STELLATIONS, RHOMBIC TRIACONTAHEDRON, RHOMBUS, SPHERE PACKING, STEINMETZ SOLID, TRIGONAL DODECAHEDRON, ZONOHEDRON

References

Ball, W. W. R. and Coxeter, H. S. M. *Mathematical Recreations and Essays, 13th ed.* New York: Dover, p. 137, 1987.

Brückner, M. *Vielecke under Vielflache.* Leipzig, Germany, 1900.

Cotton, F. A. *Chemical Applications of Group Theory, 3rd ed.* New York: Wiley, p. 62, 1990.

Cundy, H. and Rollett, A. "Rhombic Dodecahedron. $V(3.4)^2$." §3.8.1 in *Mathematical Models, 3rd ed.* Stradbroke, England: Tarquin Pub., p. 120, 1989.

Steinhaus, H. *Mathematical Snapshots, 3rd ed.* New York: Dover, pp. 185–186, 1999.

Weisstein, E. W. "Polyhedra." MATHEMATICA NOTEBOOK POLYHEDRA.M.

Wells, D. *The Penguin Dictionary of Curious and Interesting Geometry.* London: Penguin, pp. 215–216, 1991.

Wenninger, M. J. *Dual Models.* Cambridge, England: Cambridge University Press, pp. 19, 21, and 34, 1983.

Rhombic Dodecahedron Stellations

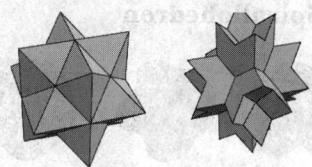

There are three STELLATIONS of the RHOMBIC DODE-CAHEDRON (Wells 1991), two of which are illustrated above. The first stellation can be constructed by drawing diagonals across the square faces of a CUBOCTAHEDRON and connecting centers of these diagonals with the vertices of neighboring squares. The outer edges of the second stellation correspond with those of the TRUNCATED OCTAHEDRON.

See also CUBOCTAHEDRON, RHOMBIC DODECAHEDRON, STELLATION, TRUNCATED OCTAHEDRON

References

Cundy, H. and Rollett, A. "The Stellated Rhombic Dodecahedron." §3.9.5 in *Mathematical Models, 3rd ed.* Stradbroke, England: Tarquin Pub., pp. 127–128, 1989.

Wells, D. *The Penguin Dictionary of Curious and Interesting Geometry.* London: Penguin, pp. 215–216, 1991.

Wenninger, M. J. *Dual Models.* Cambridge, England: Cambridge University Press, p. 36, 1983.

Rhombic Icosahedron

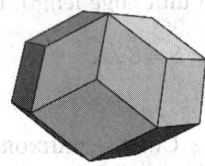

A ZONOHEDRON which can be derived from the RHOMBIC TRIACONTAHEDRON by removing any one of the zones and bringing together the two pieces into which the remainder of the surface is thereby divided.

See also RHOMBIC TRIACONTAHEDRON, ZONOHEDRON

References

Ball, W. W. R. and Coxeter, H. S. M. *Mathematical Recreations and Essays, 13th ed.* New York: Dover, p. 143, 1987.

Bilinski, S. "Über die Rhombenisoeder." *Glasnik Mat.-Fiz. Astron. Drustro Mat. Fiz. Hrvatske Ser. II* **15**, 251–263, 1960.

Weisstein, E. W. "Polyhedra." MATHEMATICA NOTEBOOK POLYHEDRA.M.

Rhombic Polyhedron

A POLYHEDRON with extra square faces, given by the SCHLÄFLI SYMBOL $r\{^p_q\}$.

See also RHOMBIC DODECAHEDRON, RHOMBIC ICOSAHEDRON, RHOMBIC TRIACONTAHEDRON, SNUB POLYHEDRON, TRUNCATED POLYHEDRON

Rhombic Spirallohedron

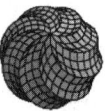

A beautiful class of polyhedra composed of rhombic faces discovered accidentally by R. Towle while attempting to develop a function to create a rhombic hexahedron from a triple of vectors.

References

Towle, R. "Rhombic Spirallohedra." http://www.mathsource.com/cgi-bin/msitem?0208–718.

Rhombic Triacontahedron

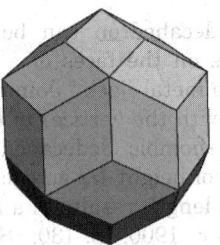

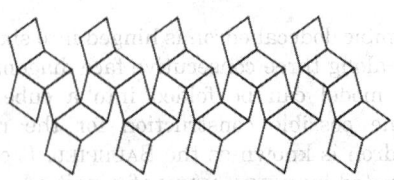

A ZONOHEDRON which is the DUAL POLYHEDRON of the ICOSIDODECAHEDRON A_4 and Wenninger dual W_{12}. It is composed of 30 RHOMBI joined at 32 vertices. The intersecting edges of the DODECAHEDRON-ICOSAHEDRON COMPOUND form the diagonals of 30 RHOMBI which comprise the TRIACONTAHEDRON. The CUBE 5-COMPOUND has the 30 facial planes of the rhombic triacontahedron (Wenninger 1983, p. 36; Ball and Coxeter 1987).

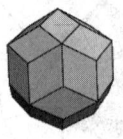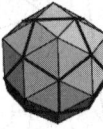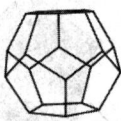

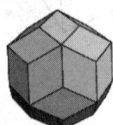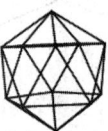

The short diagonals of the faces of the rhombic triacontahedron give the edges of a DODECAHEDRON, while the long diagonals give the edges of the ICOSAHEDRON (Steinhaus 1983, pp. 209–210). Taken

together, the DODECAHEDRON and ICOSAHEDRON give a DODECAHEDRON-ICOSAHEDRON COMPOUND.

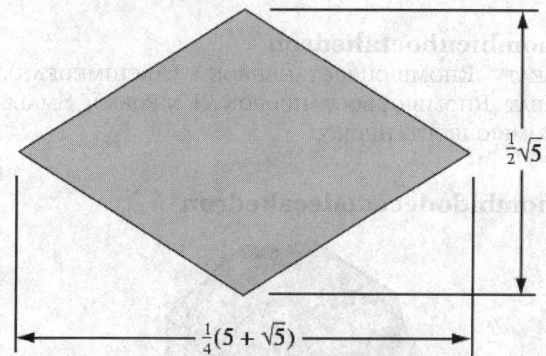

The rhombic triacontahedron generated from an ICOSIDODECAHEDRON of unit edge lengths has edge lengths

$$s = \tfrac{1}{4}\sqrt{\tfrac{5}{2}\left(5+\sqrt{5}\right)}. \qquad (1)$$

and INRADIUS

$$r = \tfrac{1}{8}\left(5+3\sqrt{5}\right). \qquad (2)$$

Normalizing so that $s = 1$, the solid has SURFACE AREA and VOLUME given by

$$S = 12\sqrt{5} \qquad (3)$$

$$V = 4\sqrt{5+2\sqrt{5}}. \qquad (4)$$

See also ARCHIMEDEAN DUAL, ARCHIMEDEAN SOLID, CUBE 5-COMPOUND, DODECAHEDRON, DODECAHEDRON-ICOSAHEDRON COMPOUND, ICOSAHEDRON, ICOSIDODECAHEDRON, RHOMBIC DODECAHEDRON, RHOMBIC TRIACONTAHEDRON STELLATIONS, RHOMBUS, TRIACONTAHEDRON, ZONOHEDRON

References

Ball, W. W. R. and Coxeter, H. S. M. *Mathematical Recreations and Essays, 13th ed.* New York: Dover, p. 137, 1987.
Bulatov, V. "Stellations of Rhombic Triacontahedron." http://www.physics.orst.edu/~bulatov/polyhedra/rtc/.
Cundy, H. and Rollett, A. "Rhombic Triacontahedron." §3.8.2 in *Mathematical Models, 3rd ed.* Stradbroke, England: Tarquin Pub., pp. 121–122 and 127, 1989.
Steinhaus, H. *Mathematical Snapshots, 3rd ed.* New York: Dover, pp. 207 and 209–210, 1999.
Wenninger, M. J. *Dual Models.* Cambridge, England: Cambridge University Press, p. 22, 1983.

Rhombic Triacontahedron Stellations

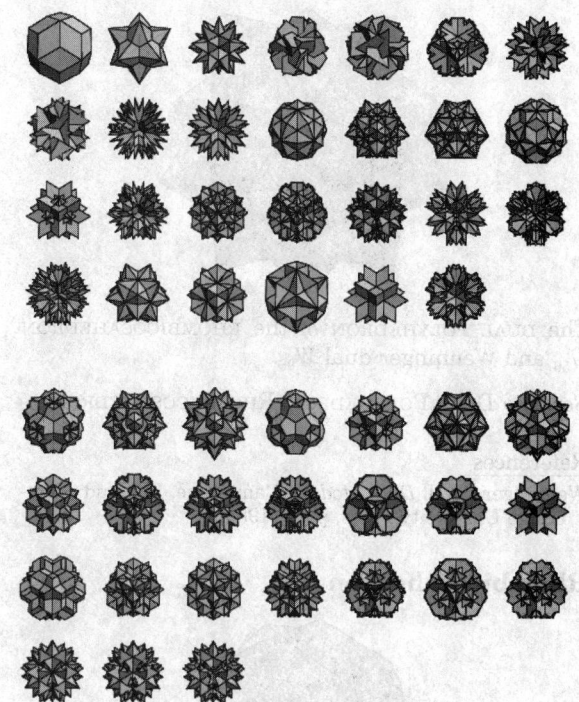

Ede (1958) enumerates 13 basic series of stellations of the rhombic triacontahedron, the total number of which is extremely large. Pawsey (1973) gave a set of restrictions upon which a complete enumeration of stellations can be achieved (Wenninger 1983, p. 36). Messer (1995) describes 226 stellations, some of which are illustrated above.

The CONVEX HULL of the DODECADODECAHEDRON is an ICOSIDODECAHEDRON and the dual of the ICOSIDODECAHEDRON is the RHOMBIC TRIACONTAHEDRON, so the dual of the DODECADODECAHEDRON (the MEDIAL RHOMBIC TRIACONTAHEDRON) is one of the rhombic triacontahedron stellations (Wenninger 1983, p. 41). Another is the GREAT RHOMBIC TRIACONTAHEDRON.

See also GREAT RHOMBIC TRIACONTAHEDRON, MEDIAL RHOMBIC TRIACONTAHEDRON, RHOMBIC TRIACONTAHEDRON, STELLATION

References

Ede, J. D. "Rhombic Triacontahedra." *Math. Gazette* **42**, 98–100, 1958.
Messer, P. W. "Stellations of the Rhombic Triacontahedron and Beyond." *Structural Topology* **21**, 25–46, 1995.
Pawley, G. S "The 227 Triacontahedra." *Geom. Dedicata* **4**, 221–232, 1975.
Wenninger, M. J. *Dual Models.* Cambridge, England: Cambridge University Press, p. 36, 1983.

Rhombicosacron

The DUAL POLYHEDRON of the RHOMBICOSAHEDRON U_{56} and Wenninger dual W_{96}.

See also DUAL POLYHEDRON, RHOMBICOSAHEDRON

References

Wenninger, M. J. *Dual Models.* Cambridge, England: Cambridge University Press, p. 85, 1983.

Rhombicosahedron

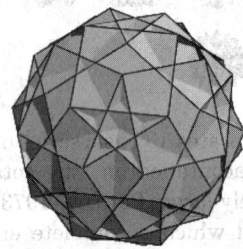

The UNIFORM POLYHEDRON U_{56} and Wenninger model W_{96} whose DUAL POLYHEDRON is the RHOMBICOSACRON. It has WYTHOFF SYMBOL $2\,\frac{5}{2}\,3|$. Its faces are $10\{6\} + 15\{4\} + 15\{\frac{4}{3}\} + 10\{\frac{6}{5}\}$. The CIRCUMRADIUS for unit edge length is

$$R = \tfrac{1}{2}\sqrt{7}.$$

References

Wenninger, M. J. "Rhombicosahedron." Model 96 in *Polyhedron Models.* Cambridge, England: Cambridge University Press, pp. 149–150, 1971.

Rhombicosidodecahedron

BIGYRATE DIMINISHED RHOMBICOSIDODECAHEDRON, DIMINISHED RHOMBICOSIDODECAHEDRON, GREAT RHOMBICOSIDODECAHEDRON (ARCHIMEDEAN), GREAT RHOMBICOSIDODECAHEDRON (UNIFORM), GYRATE BIDIMINISHED RHOMBICOSIDODECAHEDRON, GYRATE RHOMBICOSIDODECAHEDRON, METABIDIMINISHED RHOMBICOSIDODECAHEDRON, METABIGYRATE RHOMBICOSIDODECAHEDRON, METAGYRATE DIMINISHED RHOMBICOSIDODECAHEDRON, PARABIDIMINISHED RHOMBICOSIDODECAHEDRON, PARABIGYRATE RHOMBICOSIDODECAHEDRON, PARAGYRATE DIMINISHED RHOMBICOSIDODECAHEDRON, SMALL RHOMBICOSIDO-

DECAHEDRON, TRIDIMINISHED RHOMBICOSIDODECAHEDRON, TRIGYRATE RHOMBICOSIDODECAHEDRON

Rhombicuboctahedron

GREAT RHOMBICUBOCTAHEDRON (ARCHIMEDEAN), GREAT RHOMBICUBOCTAHEDRON (UNIFORM), SMALL RHOMBICUBOCTAHEDRON

Rhombidodecadodecahedron

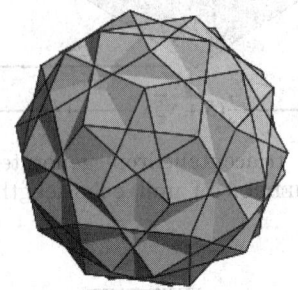

The UNIFORM POLYHEDRON U_{38} whose DUAL POLYHEDRON is the MEDIAL DELTOIDAL HEXECONTAHEDRON. It has SCHLÄFLI SYMBOL r $\{\frac{5}{2}\}$ and WYTHOFF SYMBOL $\frac{5}{2}5|2$. Its faces are $12\{\frac{5}{2}\} + 3\{4\} + 12\{5\}$. The CIRCUMRADIUS for unit edge length is

$$R = \tfrac{1}{2}\sqrt{7}.$$

References

Wenninger, M. J. *Polyhedron Models.* Cambridge, England: Cambridge University Press, pp. 116–117, 1989.

Rhombihexacron

GREAT RHOMBIHEXACRON, SMALL RHOMBIHEXACRON

Rhombihexahedron

GREAT RHOMBIHEXAHEDRON, SMALL RHOMBIHEXAHEDRON

Rhombitruncated Cuboctahedron

GREAT RHOMBICUBOCTAHEDRON (ARCHIMEDEAN)

Rhombitruncated Icosidodecahedron

GREAT RHOMBICOSIDODECAHEDRON (ARCHIMEDEAN)

Rhombohedron

A PARALLELEPIPED bounded by six congruent RHOMBS.

See also PARALLELEPIPED, RHOMB

References

Ball, W. W. R. and Coxeter, H. S. M. *Mathematical Recreations and Essays, 13th ed.* New York: Dover, pp. 142 and 161, 1987.

Rhomboid

A PARALLELOGRAM in which angles are oblique and adjacent sides are of unequal length.

See also BAR (POLYIAMOND), DIAMOND, KITE, LOZENGE, PARALLELOGRAM, QUADRILATERAL, RHOMBUS, SKEW QUADRILATERAL, TRAPEZIUM, TRAPEZOID

References
Gardner, M. *The Sixth Book of Mathematical Games from Scientific American.* Chicago, IL: University of Chicago Press, p. 176, 1984.

Rhomboidal Dodecahedron
RHOMBIC DODECAHEDRON

Rhombus

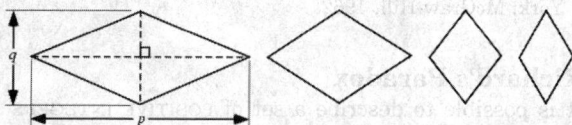

A QUADRILATERAL with both pairs of opposite sides PARALLEL and all sides the same length, i.e., an equilateral PARALLELOGRAM. The word RHOMB is sometimes used instead of rhombus, and a rhombus is sometimes also called a diamond. A rhombus with $2\theta = 45°$ is sometimes called a LOZENGE.

The DIAGONALS p and q of a rhombus are PERPENDICULAR and satisfy

$$p^2 + q^2 = 4a^2.$$

The AREA of a rhombus is given by

$$A = \tfrac{1}{2}pq.$$

See also DIAMOND, HARBORTH'S TILING, KITE, LOZENGE, PARALLELOGRAM, QUADRILATERAL, RHOMBIC DODECAHEDRON, RHOMBIC ICOSAHEDRON, RHOMBIC TRIACONTAHEDRON, RHOMBOID, SKEW QUADRILATERAL, TRAPEZIUM, TRAPEZOID

References
Beyer, W. H. (Ed.). *CRC Standard Mathematical Tables,* 28th ed. Boca Raton, FL: CRC Press, p. 123, 1987.
Harris, J. W. and Stocker, H. "Rhombus." §3.6.4 in *Handbook of Mathematics and Computational Science.* New York: Springer-Verlag, pp. 83–84, 1998.

Rhumb Line
LOXODROME

Ribbon Knot
If the KNOT K is the boundary $K = f(\mathbb{S}^1)$ of a singular disk $f : \mathbb{D} \to \mathbb{S}^3$ which has the property that each self-intersecting component is an arc $A \subset f(\mathbb{D}^2)$ for which $f^{-1}(A)$ consists of two arcs in $\mathbb{D}^2$, one of which is

interior, then K is said to be a ribbon knot. Every ribbon knot is a SLICE KNOT, and it is conjectured that every SLICE KNOT is a ribbon knot.

See also SLICE KNOT

References
Rolfsen, D. *Knots and Links.* Wilmington, DE: Publish or Perish Press, p. 225, 1976.

Ribet's Theorem
If the TANIYAMA-SHIMURA CONJECTURE holds for all semistable ELLIPTIC CURVES, then FERMAT'S LAST THEOREM is true. Before its proof by Ribet in 1986, the theorem had been called the EPSILON CONJECTURE. It had its roots in a surprising result of G. Frey.

See also ELLIPTIC CURVE, EPSILON CONJECTURE, FERMAT'S LAST THEOREM, MODULAR FORM, MODULAR FUNCTION, TANIYAMA-SHIMURA CONJECTURE

Riccati Differential Equation

$$y' = P(z) + Q(z)y + R(z)y^2, \tag{1}$$

where $y' \equiv dy/dz$. The transformation

$$w \equiv -\frac{y'}{yR(z)} \tag{2}$$

leads to the second-order linear homogeneous equation

$$R(z)y'' - [R'(z) + Q(z)R(z)]y' + [R(z)]^2 P(z)y = 0. \tag{3}$$

Another equation sometimes called the Riccati differential equation is

$$z^2 w'' + [z^2 - n(n+1)]w = 0 \tag{4}$$

(Zwillinger 1997, p. 126), which has solutions

$$w = Azj_n(z) + Bzy_n(z), \tag{5}$$

where $j_n(z)$ and $y_n(z)$ are SPHERICAL BESSEL FUNCTIONS OF THE FIRST and SECOND KINDS.

Yet another form of "the" Riccati differential equation is

$$\frac{dy}{dz} = az^n + by^2, \tag{6}$$

which is solvable by algebraic, exponential, and logarithmic functions only when $n = -4m/(2m \pm 1)$, for $m = 0, 1, 2, \ldots.$

References
Abramowitz, M. and Stegun, C. A. (Eds.). "Riccati-Bessel Functions." §10.3 in *Handbook of Mathematical Functions with Formulas, Graphs, and Mathematical Tables, 9th printing.* New York: Dover, p. 445, 1972.
Bender, C. M. and Orszag, S. A. §1.6 in *Advanced Mathematical Methods for Scientists and Engineers.* New York: McGraw-Hill, 1978.

Boyce, W. E. and DiPrima, R. C. *Elementary Differential Equations and Boundary Value Problems, 4th ed.* New York: Wiley, pp. 142–143, 1986.

Glaisher, J. W. L. "On Riccati's Equation." *Quart. J. Pure Appl. Math.* **11**, 267–273, 1871.

Goldstein, M. E. and Braun, W. H. *Advanced Methods for the Solution of Differential Equations.* NASA SP-316. Washington, DC: U.S. Government Printing Office, pp. 45–46, 1973.

Ince, E. L. *Ordinary Differential Equations.* New York: Dover, pp. 23–35 and 295, 1956.

Reid, W. T. *Riccati Differential Equations.* New York: Academic Press, 1972.

Simmons, G. F. *Differential Equations with Applications and Historical Notes.* New York: McGraw-Hill, pp. 62–63, 1972.

Zwillinger, D. (Ed.). *CRC Standard Mathematical Tables and Formulae.* Boca Raton, FL: CRC Press, p. 414, 1995.

Zwillinger, D. "Riccati Equation--1 and Riccati Equation--2." §II.A.75 and II.A.76 in *Handbook of Differential Equations, 3rd ed.* Boston, MA: Academic Press, pp. 121 and 288–291, 1997.

Riccati-Bessel Functions

$$S_n(z) \equiv z j_n(z) = \sqrt{\frac{\pi z}{2}}\, J_{n+1/2}(z)$$

$$C_n(z) \equiv -z n_n(z) = -\sqrt{\frac{\pi z}{2}}\, N_{n+1/2}(z),$$

where $j_n(z)$ and $n_n(z)$ are SPHERICAL BESSEL FUNCTIONS OF THE FIRST and SECOND KIND.

References

Abramowitz, M. and Stegun, C. A. (Eds.). "Riccati-Bessel Functions." §10.3 in *Handbook of Mathematical Functions with Formulas, Graphs, and Mathematical Tables, 9th printing.* New York: Dover, p. 445, 1972.

Ricci Curvature

RICCI CURVATURE TENSOR

Ricci Curvature Tensor

$$R_{\mu\kappa} \equiv R^\lambda{}_{\mu\lambda\kappa},$$

where $R^\lambda{}_{\mu\lambda\kappa}$ is the RIEMANN TENSOR.

Topologically, the Ricci curvature is the mathematical object which controls the growth rate of the volume of metric balls in a MANIFOLD.

See also BISHOP'S INEQUALITY, CAMPBELL'S THEOREM, CURVATURE SCALAR, EINSTEIN TENSOR, MILNOR'S THEOREM, RIEMANN TENSOR

References

Misner, C. W.; Thorne, K. S.; and Wheeler, J. A. *Gravitation.* San Francisco: W. H. Freeman, 1973.

Wald, R. M. *General Relativity.* Chicago, IL: University of Chicago Press, p. 40, 1984.

Weinberg, S. *Gravitation and Cosmology: Principles and Applications of the General Theory of Relativity.* New York: Wiley, pp. 135 and 142, 1972.

Ricci Tensor

RICCI CURVATURE TENSOR

Rice Distribution

$$P(Z) = \frac{Z}{\sigma^2} \exp\left(-\frac{Z^2 + |V|^2}{2\sigma^2}\right) I_0\left(\frac{Z|V|}{\sigma^2}\right),$$

where $I_0(z)$ is a MODIFIED BESSEL FUNCTION OF THE FIRST KIND and $Z > 0$. For a derivation, see Papoulis (1962). For $|V| = 0 = 0$, this reduces to the RAYLEIGH DISTRIBUTION.

See also RAYLEIGH DISTRIBUTION

References

Papoulis, A. *The Fourier Integral and Its Applications.* New York: McGraw-Hill, 1962.

Richard's Paradox

It is possible to describe a set of POSITIVE INTEGERS that cannot be listed in a book containing a set of counting numbers on each consecutively numbered page. Another form of the paradox states that the set of all numerical functions is nondenumerable (Curry 1977).

References

Church, A. "A Bibliography of Symbolic Logic." *J. Symb. Logic* **1**, 121–218, 1936.

Curry, H. B. *Foundations of Mathematical Logic.* New York: Dover, p. 6, 1977.

Erickson, G. W. and Fossa, J. A. *Dictionary of Paradox.* Lanham, MD: University Press of America, pp. 172–173, 1998.

Richardson Extrapolation

The consideration of the result of a numerical calculation as a function of an adjustable parameter (usually the step size). The function can then be fitted and evaluated at $h = 0$ to yield very accurate results. Press *et al.* (1992) describe this process as turning lead into gold. Richardson extrapolation is one of the key ideas used in the popular and robust BULIRSCH-STOER ALGORITHM of solving ORDINARY DIFFERENTIAL EQUATIONS.

See also BULIRSCH-STOER ALGORITHM

References

Acton, F. S. *Numerical Methods That Work, 2nd printing.* Washington, DC: Math. Assoc. Amer., p. 106, 1990.

Jeffreys, H. and Jeffreys, B. S. "L. F. Richardson's Method." §9.091 in *Methods of Mathematical Physics, 3rd ed.* Cambridge, England: Cambridge University Press, p. 288, 1988.

Press, W. H.; Flannery, B. P.; Teukolsky, S. A.; and Vetterling, W. T. "Richardson Extrapolation and the Bulirsch-Stoer Method." §16.4 in *Numerical Recipes in FORTRAN: The Art of Scientific Computing, 2nd ed.* Cambridge, England: Cambridge University Press, pp. 718–725, 1992.

Richardson's Theorem

Let R be the class of expressions generated by

   1. The RATIONAL NUMBERS and the two REAL NUMBERS π and $\ln 2$,
   2. The variable x,
   3. The operations of ADDITION, MULTIPLICATION, and composition, and
   4. The SINE, EXPONENTIAL, and ABSOLUTE VALUE functions.

Then if $E \in R$, the predicate "$E = 0$" is recursively UNDECIDABLE.

See also INTEGER RELATION, RECURSION, UNDECIDABLE

References

Caviness, B. F. "On Canonical Forms and Simplification." *J. Assoc. Comp. Mach.* **17**, 385–396, 1970.
Petkovsek, M.; Wilf, H. S.; and Zeilberger, D. *A = B.* Wellesley, MA: A. K. Peters, 1996.
Richardson, D. "Some Unsolvable Problems Involving Elementary Functions of a Real Variable." *J. Symbolic Logic* **33**, 514–520, 1968.

Riddell's Formula

Riddell's formula for unlabeled graphs is the EULER TRANSFORM relating the number of unlabeled CONNECTED GRAPHS on n nodes satisfying some property with the corresponding total number (not necessarily connected) of GRAPHS on n nodes.

Riddell's formula for labeled graphs is the EXPONENTIAL TRANSFORM relating the number of labeled CONNECTED GRAPHS on n nodes satisfying some property with the corresponding total number (not necessarily connected) of labeled GRAPHS on n nodes.

See also CONNECTED GRAPH, EULER TRANSFORM, EXPONENTIAL TRANSFORM, GRAPH, LABELED GRAPH, UNLABELED GRAPH

References

Cadogan, C. C. "The Möbius Function and Connected Graphs." *J. Combin. Th. B* **11**, 193–200, 1971.
Harary, F. and Palmer, E. M. *Graphical Enumeration.* New York: Academic Press, p. 90, 1973.
Sloane, N. J. A. and Plouffe, S. *The Encyclopedia of Integer Sequences.* San Diego, CA: Academic Press, p. 20, 1995.

Ridders' Method

A variation of the FALSE POSITION METHOD for finding ROOTS which fits the function in question with an exponential.

See also FALSE POSITION METHOD, ROOT

References

Ostrowski, A. M. Ch. 12 in *Solutions of Equations and Systems of Equations, 2nd ed.* New York: Academic Press, 1966.
Press, W. H.; Flannery, B. P.; Teukolsky, S. A.; and Vetterling, W. T. "Secant Method, False Position Method, and Ridders' Method." §9.2 in *Numerical Recipes in FORTRAN: The Art of Scientific Computing, 2nd ed.* Cambridge, England: Cambridge University Press, pp. 347–352, 1992.
Ralston, A. and Rabinowitz, P. §8.3 in *A First Course in Numerical Analysis, 2nd ed.* New York: McGraw-Hill, 1978.
Ridders, C. F. J. "A New Algorithm for Computing a Single Root of a Real Continuous Function." *IEEE Trans. Circuits Systems* **26**, 979–980, 1979.

Ridge

An $(n-2)$-D FACE of an n-D POLYTOPE.

See also POLYTOPE

Riemann Curve Theorem

If two algebraic plane curves with only ordinary singular points and CUSPS are related such that the coordinates of a point on either are RATIONAL FUNCTIONS of a corresponding point on the other, then the curves have the same GENUS (CURVE). This can be stated equivalently as the GENUS of a curve is unaltered by a BIRATIONAL TRANSFORMATION.

References

Coolidge, J. L. *A Treatise on Algebraic Plane Curves.* New York: Dover, p. 120, 1959.

Riemann Differential Equation

RIEMANN P-DIFFERENTIAL EQUATION

Riemann Formula

The solution

$$u(x, y) = \int_0^x d\xi \int_1^y R(\xi, \eta; \ x, y) f(\xi, \eta)\, d\eta, \qquad (1)$$

where $R(x, y; \ \xi, \eta)$ is the RIEMANN FUNCTION of the linear GOURSAT PROBLEM with characteristics $\phi = \psi = 0$ according to the RIEMANN METHOD.

See also GOURSAT PROBLEM, RIEMANN FUNCTION, RIEMANN METHOD

References

Hazewinkel, M. (Managing Ed.). *Encyclopaedia of Mathematics: An Updated and Annotated Translation of the Soviet "Mathematical Encyclopaedia."* Dordrecht, Netherlands: Reidel, p. 289, 1988.

Riemann Function

There are a number of functions in various branches of mathematics known as Riemann functions. Examples include the RIEMANN P-SERIES, RIEMANN-SIEGEL FUNCTIONS, RIEMANN THETA FUNCTION, RIEMANN

ZETA FUNCTION, XI FUNCTION, the function $F(x)$ obtained by Riemann in studying FOURIER SERIES, the function $R(x, y; \xi, \eta)$ appearing in the application of the RIEMANN METHOD for solving the GOURSAT PROBLEM, the function $R(n)$ in the RIEMANN PRIME NUMBER FORMULA, and the function $f(x)$ related to the PRIME COUNTING FUNCTION defined below.

The Riemann function $F(x)$ for a FOURIER SERIES

$$\frac{1}{2} a_0 + \sum_{n=1}^{\infty} [a_n \cos(nx) + b_n \sin(nx)] \tag{1}$$

is obtained by integrating twice term by term to obtain

$$F(x) = \frac{1}{4} a_0 x^2 - \sum_{n=1}^{\infty} \frac{1}{n^2} [a_n \cos(nx) + b_n \sin(nx)]$$
$$+ Cx + D, \tag{2}$$

where C and D are constants (Riemann 1957; Hazewinkel 1988, vol. 8, p. 118).

The Riemann function $R(x, y; \xi, \eta)$ arises in the solution of the linear case of the GOURSAT PROBLEM of solving the HYPERBOLIC PARTIAL DIFFERENTIAL EQUATION

$$\hat{L}u = u_{xy} + au_x + bu_y + cu = f \tag{3}$$

with BOUNDARY CONDITIONS

$$u(0, t) = \phi(t) \tag{4}$$

$$u(t, 1) = \psi(t) \tag{5}$$

$$\phi(1) = \phi(0). \tag{6}$$

Here, $R(x, y; \xi, \eta)$ is defined as the solution of the equation

$$R_{xy} - (aR)_x - (bR)_y + cR = 0 \tag{7}$$

which satisfies the conditions

$$R(\xi, y; \xi, n) = \exp\left[\int_{\eta}^{y} a(\xi, t) dt\right] \tag{8}$$

$$R(x, \eta; \xi, \eta) = \exp\left[\int_{\xi}^{x} b(t, \eta) dt\right] \tag{9}$$

on the characteristics $x = \xi$ and $y = \eta$, where (ξ, η) is a point on the domain Ω on which (8) is defined (Hazewinkel 1988). The solution is then given by the RIEMANN FORMULA

$$u(x, y) = \int_{0}^{x} d\xi \int_{1}^{y} R(\xi, \eta; x, y) f(\xi, \eta) d\eta. \tag{10}$$

This method of solution is called the RIEMANN METHOD.

Riemann defined the function $f(x)$ by

$$f(x) \equiv \sum_{n=1}^{\infty} \frac{\pi(x^{1/n})}{n}$$
$$= \pi(x) + \frac{1}{2} \pi(x^{1/2}) + \frac{1}{3} \pi(x^{1/3}) + \dots \tag{11}$$

(Hardy 1999, p. 30), then the PRIME COUNTING FUNCTION $\pi(x)$ is related to $f(x)$ by

$$\pi(x) = \sum_{n=1}^{\infty} \frac{\mu(n)}{n} f(x^{1/n}), \tag{12}$$

where $\mu(n)$ is the MÖBIUS FUNCTION (Riesel 1994, p. 49). Riemann (1859) proposed that

$$f(x) = \text{li}(x) - \sum_{\rho} \text{li}(x^{\rho}) - \ln 2 + \int_{x}^{\infty} \frac{dt}{t \ln t(t^2 - 1)}, \tag{13}$$

where $\text{li}(x)$ is the LOGARITHMIC INTEGRAL and the sum is over all nontrivial zeros ρ of the RIEMANN ZETA FUNCTION $\zeta(z)$ (Mathews 1892, Ch. 10; Landau 1974, Ch. 19; Ingham 1990, Ch. 4; Hardy 1999, p. 40). This formula was subsequently proved by Mangoldt in 1895 (Riesel 1994, p. 47).

A function related to $f(x)$ is given by

$$J(x) \equiv \begin{cases} \pi(x) + \frac{1}{2} \pi(x^{1/2}) + \frac{1}{3} \pi(x^{1/3}) + \dots - \frac{1}{2m} \\ \quad \text{for } p^m \text{ with } p \text{ prime} \\ \pi(x) + \frac{1}{2} \pi(x^{1/2}) + \frac{1}{3} \pi(x^{1/3}) + \dots \\ \quad \text{otherwise} \end{cases} \tag{14}$$

$$= \lim_{t \to \infty} \frac{1}{2\pi i} \int_{2-iT}^{2+iT} \frac{x^s}{s} \ln \zeta(s) ds, \tag{15}$$

where $\zeta(z)$ is the RIEMANN ZETA FUNCTION. This function satisfies

$$\frac{\ln \zeta(s)}{s} = \int_{1}^{\infty} J(x) x^{-s-1} dx \tag{16}$$

(Riesel 1994, p. 47).

See also CRITICAL STRIP, GOURSAT PROBLEM, LOGARITHMIC INTEGRAL, MANGOLDT FUNCTION, RIEMANN METHOD, PRIME NUMBER THEOREM, RIEMANN PRIME NUMBER FORMULA, RIEMANN ZETA FUNCTION

References

Conway, J. H. and Guy, R. K. *The Book of Numbers.* New York: Springer-Verlag, pp. 144–145, 1996.

Hardy, G. H. *Ramanujan: Twelve Lectures on Subjects Suggested by His Life and Work, 3rd ed.* New York: Chelsea, 1999.

Hazewinkel, M. (Managing Ed.). *Encyclopaedia of Mathematics: An Updated and Annotated Translation of the Soviet "Mathematical Encyclopaedia."* Dordrecht, Netherlands: Reidel, Vol. 4, p. 289 and Vol. 8, p. 125, 1988.

Ingham, A. E. *The Distribution of Prime Numbers.* London: Cambridge University Press, p. 83, 1990.

Knuth, D. E. *The Art of Computer Programming, Vol. 2: Seminumerical Algorithms, 3rd ed.* Reading, MA: Addison-Wesley, 1998.

Landau, E. *Handbuch der Lehre von der Verteilung der Primzahlen, 3rd ed.* New York: Chelsea, 1974.

Mathews, G. B. Ch. 10 in *Theory of Numbers*. New York: Chelsea, 1961.

Ribenboim, P. *The New Book of Prime Number Records*. New York: Springer-Verlag, pp. 224–225, 1996.

Riemann, G. F. B. "Über die Anzahl der Primzahlen unter einer gegebenen Grösse." *Monatsber. Königl. Preuss. Akad. Wiss. Berlin*, 671, 1859.

Riemann, B. "Über die Darstellbarkeit einer Function durch eine trigonometrische Reihe." In *Gesammelte math. Abhandlungen*. New York: Dover, pp. 227–264, 1957.

Riesel, H. "The Riemann Prime Number Formula." *Prime Numbers and Computer Methods for Factorization, 2nd ed.* Boston, MA: Birkhäuser, pp. 50–52, 1994.

Riesel, H. and Göhl, G. "Some Calculations Related to Riemann's Prime Number Formula." *Math. Comput.* **24**, 969–983, 1970.

Wagon, S. *Mathematica in Action*. New York: W. H. Freeman, pp. 28–29 and 362–372, 1991.

Riemann Hypothesis

First published in Riemann (1859), the Riemann hypothesis states that the nontrivial ROOTS of the RIEMANN ZETA FUNCTION

$$\zeta(s) \equiv \sum_{n=1}^{\infty} \frac{1}{n^s}, \tag{1}$$

where $x \in \mathbb{C}$ (the COMPLEX NUMBERS), all lie on the "CRITICAL LINE" $\Re[s] = 1/2$, where $\Re[z]$ denotes the REAL PART of z. The Riemann hypothesis is also known as ARTIN'S CONJECTURE. Wiener showed that the PRIME NUMBER THEOREM is literally equivalent to the assertion that $\zeta(s)$ has no zeros on $\sigma = 1$ (Hardy 1999, pp. 34 and 58–60).

In 1914, Hardy proved that an INFINITE number of values for s can be found for which $\zeta(s) = 0$ and $\Re[s] = 1/2$. However, it is not known if *all* nontrivial roots s satisfy $\Re[s] = 1/2$, so the conjecture remains open. André Weil proved the Riemann hypothesis to be true for field functions (Weil 1948, Eichler 1966, Ball and Coxeter 1987). In 1974, Levinson (1974ab) showed that at least 1/3 of the ROOTS must lie on the CRITICAL LINE (Le Lionnais 1983), a result which has since been sharpened to 40% (Vardi 1991, p. 142). It is known that the zeros are symmetrical placed about the line $\Im[s] = 0$.

The Riemann hypothesis is equivalent to $\Lambda \le 0$, where Λ is the DE BRUIJN-NEWMAN CONSTANT (Csordas *et al.* 1994). It is also equivalent to the assertion that for some constant c,

$$|\mathrm{Li}(x) - \pi(x)| \le c\sqrt{x}\ln x, \tag{2}$$

where $\mathrm{Li}(x)$ is the LOGARITHMIC INTEGRAL and π is the PRIME COUNTING FUNCTION (Wagon 1991). Another equivalent form states that

$$\mathrm{span}_{L^2(0,\,1)}\{\rho_\alpha,\ 0 < \alpha < 1\} = L^2(0,\,1), \tag{3}$$

where

$$\rho_\alpha(t) \equiv \mathrm{frac}\left(\frac{\alpha}{t}\right) - \alpha\,\mathrm{frac}\left(\frac{1}{t}\right), \tag{4}$$

where $\mathrm{frac}(x)$ is the FRACTIONAL PART (Balazard and Saias 2000).

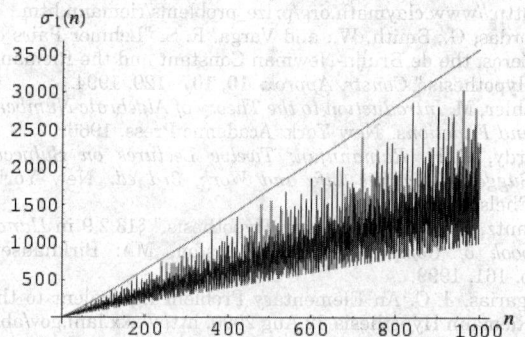

By modifying a criterion of Robin (1984), Lagarias (2000) showed that the Riemann hypothesis is equivalent to the statement that

$$\sigma(n) \le H_n + \exp(H_n)\ln H_n, \tag{5}$$

for all $n \ge 1$, with equality only for $n = 1$, where H_n is a HARMONIC NUMBER and $\sigma(n)$ is the DIVISOR FUNCTION.

There is also a finite analog of the Riemann hypothesis concerning the location of zeros for function fields defined by equations such as

$$ay^l + bz^m + c = 0. \tag{6}$$

This hypothesis, developed by Weil, is analogous to the usual Riemann hypothesis. The number of solutions for the particular cases $(l,\,m) = (2,\,2)$, $(3,\,3)$, $(4,\,4)$, and $(2,\,4)$ were known to Gauss.

The hypothesis has thus far resisted all attempts to prove it, although it has been computationally tested and found to be true for the first $200,000,001$ zeros by Brent *et al.* (1982). Brent's calculation covered zeros $\sigma + it$ in the region $0 < t < 81,702,130.19$. In 2000, Clay Mathematics Institute offered a $1 million prize for proof of the Riemann hypothesis.

See also BERRY CONJECTURE, CRITICAL LINE, CRITICAL STRIP, EXTENDED RIEMANN HYPOTHESIS, GRONWALL'S THEOREM, MERTENS CONJECTURE, MILLS' CONSTANT, PRIME NUMBER THEOREM, RIEMANN ZETA FUNCTION

References

Balazard, M. and Saias, E. "The Nyman-Beurling Equivalent Form for the Riemann Hypothesis." *Expos. Math.* **18**, 131–138, 2000.

Ball, W. W. R. and Coxeter, H. S. M. *Mathematical Recreations and Essays, 13th ed.* New York: Dover, p. 75, 1987.

Bombieri, E. "Problems of the Millennium: The Riemann Hypothesis." http://www.claymath.org/prize_problems/riemann.pdf.

Brent, R. P. "On the Zeros of the Riemann Zeta Function in the Critical Strip." *Math. Comput.* **33**, 1361–1372, 1979.

Brent, R. P.; van de Lune, J.; te Riele, H. J. J.; and Winter, D. T. "On the Zeros of the Riemann Zeta Function in the Critical Strip. II." *Math. Comput.* **39**, 681–688, 1982.

Caldwell, C. K. "Prime Links++: Resources in theory: conjectures: Riemann." http://primes.utm.edu/links/theory/conjectures/Riemann/.

Clay Mathematics Institute. "The Riemann Hypothesis." http://www.claymath.org/prize_problems/riemann.htm.

Csordas, G.; Smith, W.; and Varga, R. S. "Lehmer Pairs of Zeros, the de Bruijn-Newman Constant and the Riemann Hypothesis." *Constr. Approx.* **10**, 107–129, 1994.

Eichler, M. *Introduction to the Theory of Algebraic Numbers and Functions.* New York: Academic Press, 1966.

Hardy, G. H. *Ramanujan: Twelve Lectures on Subjects Suggested by His Life and Work, 3rd ed.* New York: Chelsea, 1999.

Krantz, S. G. "The Riemann Hypothesis." §13.2.9 in *Handbook of Complex Analysis.* Boston, MA: Birkhäuser, p. 161, 1999.

Lagarias, J. C. An Elementary Problem Equivalent to the Riemann Hypothesis 22 Aug 2000. http://xxx.lanl.gov/abs/math.NT/0008177/.

Le Lionnais, F. *Les nombres remarquables.* Paris: Hermann, p. 25, 1983.

Levinson, N. "More than One Third of Zeros of Riemann's Zeta-Function Are on $\sigma = 1/2$." *Adv. Math.* **13**, 383–436, 1974.

Levinson, N. "At Least One Third of Zeros of Riemann's Zeta-Function Are on $\sigma = 1/2$." *Proc. Nat. Acad. Sci. USA* **71**, 1013–1015, 1974.

Odlyzko, A. "The 10^{20}th Zero of the Riemann Zeta Function and 70 Million of Its Neighbors."

Riemann, B. "Über die Anzahl der Primzahlen unter einer gegebenen Grösse," *Mon. Not. Berlin Akad.,* pp. 671–680, Nov. 1859.

Robin, G. "Grandes valeurs de la fonction somme des diviseurs er hypothèse de Riemann." *J. Math. Pures Appl.* **63**, 187–213, 1984.

Sloane, N. J. A. Sequences A002410/M4924 in "An On-Line Version of the Encyclopedia of Integer Sequences." http://www.research.att.com/~njas/sequences/eisonline.html.

Smale, S. "Mathematical Problems for the Next Century." In *Mathematics: Frontiers and Perspectives 2000* 0821820702 (Ed. V. Arnold, M. Atiyah, P. Lax, and B. Mazur). Providence, RI: Amer. Math. Soc., 2000.

te Riele, H. J. J. "Corrigendum to: On the Zeros of the Riemann Zeta Function in the Critical Strip. II." *Math. Comput.* **46**, 771, 1986.

van de Lune, J. and te Riele, H. J. J. "On The Zeros of the Riemann Zeta-Function in the Critical Strip. III." *Math. Comput.* **41**, 759–767, 1983.

van de Lune, J.; te Riele, H. J. J.; and Winter, D. T. "On the Zeros of the Riemann Zeta Function in the Critical Strip. IV." *Math. Comput.* **46**, 667–681, 1986.

Wagon, S. *Mathematica in Action.* New York: W. H. Freeman, p. 33, 1991.

Weil, A. *Sur les courbes algébriques et les variétès qui s'en déduisent.* Paris, 1948.

Wells, D. *The Penguin Dictionary of Curious and Interesting Numbers.* Middlesex, England: Penguin Books, p. 28, 1986.

Riemann Integral

The Riemann integral is the INTEGRAL normally encountered in CALCULUS texts and used by physicists and engineers. Other types of integrals exist (e.g., the LEBESGUE INTEGRAL), but are unlikely to be encountered outside the confines of advanced mathematics texts. In fact, according to Jeffreys and Jeffreys (1988, p. 29), "it appears that cases where these methods [i.e., generalizations of the Riemann integral] are applicable and Riemann's [definition of the integral] is not are too rare in physics to repay the extra difficulty."

The Riemann integral is based on the JORDAN MEASURE, and defined by taking a limit of a RIEMANN SUM,

$$\int_b^a f(x)\, dx \equiv \lim_{\max \Delta x_k \to 0} \sum_{k=1}^{n} f(x_k^*)\Delta x_k \tag{1}$$

$$\iint f(x, y)\, dA \equiv \lim_{\max \Delta A_k \to 0} \sum_{k=1}^{n} f(x_k^*, y_k^*)\Delta A_k \tag{2}$$

$$\iiint f(x, y\, z)\, dV \equiv \lim_{\max \Delta V_k \to 0} \sum_{k=1}^{n} f(x_k^*, y_k^*, z_k^*)\Delta V_k, \tag{3}$$

where $a \le x \le b$ and x_k^*, y_k^*, and z_k^* are arbitrary points in the intervals Δx_k, Δy_k, and Δz_k, respectively. The value $\max \Delta x_k$ is called the MESH SIZE of a partition of the interval $[a, b]$ into subintervals Δx_k.

As an example of the application of the Riemann integral definition, find the AREA under the curve $y = x^r$ from 0 to a. Divide (a, b) into n segments, so $\Delta x_k = \frac{b-a}{n} \equiv h$, then

$$f(x_1) = f(0) = 0 \tag{4}$$

$$f(x_2) = f(\Delta x_k) = h^r \tag{5}$$

$$f(x_3) = f(2\Delta x_k) = (2h)^r. \tag{6}$$

By induction

$$f(x_k) = f([k-1]\Delta x_k) = [(k-1)h]^r = h^r(k-1)^r, \tag{7}$$

so

$$f(x_k)\Delta x_k = h^{r+1}(k-1)^r \tag{8}$$

$$\sum_{k=1}^{n} f(x_k)\Delta x_k = h^{r+1} \sum_{k=1}^{n} (k-1)^r. \tag{9}$$

For example, take $r = 2$.

$$\sum_{k=1}^{n} f(x_k)\Delta x_k = h^3 \sum_{k=1}^{n} (k-1)^2$$

$$= h^3 \left(\sum_{k=1}^{n} k^2 - 2 \sum_{k=1}^{n} k + \sum_{k=1}^{n} 1 \right)$$

$$= h^3 \left[\frac{n(n+1)(2n+1)}{6} - 2\frac{n(n+1)}{2} + n \right], \tag{10}$$

so

$$I \equiv \lim_{n \to \infty} \sum_{k=1}^{n} f(x_k^*) \Delta x_k = \lim_{n \to \infty} \sum_{k=1}^{n} f(x_k) \Delta x_k$$

$$= \lim_{n \to \infty} h^3 \left[\frac{n(n+1)(2n+1)}{6} - 2\frac{n(n+1)}{2} + n \right]$$

$$= a^3 \lim_{n \to \infty} \left[\frac{n(n+1)(2n+1)}{6n^3} - \frac{n(n+1)}{n^3} + \frac{n}{n^3} \right]$$

$$= \tfrac{1}{3} a^3. \tag{11}$$

See also INTEGRAL, RIEMANN SUM

References

Ferreirós, J. "The Riemann Integral." §5.1.2 in *Labyrinth of Thought: A History of Set Theory and Its Role in Modern Mathematics*. Basel, Switzerland: Birkhäuser, pp. 150–153, 1999.

Jeffreys, H. and Jeffreys, B. S. "Integration: Riemann, Stieltjes." §1.10 in *Methods of Mathematical Physics, 3rd ed.* Cambridge, England: Cambridge University Press, pp. 26–36, 1988.

Kestelman, H. "Riemann Integration." Ch. 2 in *Modern Theories of Integration, 2nd rev. ed.* New York: Dover, pp. 33–66, 1960.

Riemann Mapping Theorem

Let z_0 be a point in a simply connected region $R \neq \mathbb{C}$. Then there is a unique ANALYTIC FUNCTION $w = f(z)$ mapping R one-to-one onto the DISK $|w| < 1$ such that $f(z_0) = 0$ and $f'(z_0) = 0$. The COROLLARY guarantees that any two simply connected regions except $\mathbb{R}^2$ can be mapped CONFORMALLY onto each other.

References

Krantz, S. G. "The Riemann Mapping Theorem." §6.4 in *Handbook of Complex Analysis*. Boston, MA: Birkhäuser, pp. 86–87, 1999.

Riemann Method

The method for solving the GOURSAT PROBLEM and CAUCHY PROBLEM for linear HYPERBOLIC PARTIAL DIFFERENTIAL EQUATIONS using a RIEMANN FUNCTION.

See also GREEN'S FUNCTION, RIEMANN FUNCTION

References

Hazewinkel, M. (Managing Ed.). *Encyclopaedia of Mathematics: An Updated and Annotated Translation of the Soviet "Mathematical Encyclopaedia."* Dordrecht, Netherlands: Reidel, Vol. 4, p. 289 and Vol. 8, pp. 125–126, 1988.

Riemann P-Differential Equation

The differential equation

$$\frac{d^2u}{dz^2} + \left[\frac{1 - \alpha - \alpha'}{z-a} + \frac{1 - \beta - \beta'}{z-b} + \frac{1 - \gamma - \gamma'}{z-c} \right] \frac{du}{dz}$$

$$+ \left[\frac{\alpha\alpha'(a-b)(a-c)}{z-a} + \frac{\beta\beta'(b-c)(b-a)}{z-b} \right.$$

$$\left. + \frac{\gamma\gamma'(c-a)(c-b)}{z-c} \right] \frac{u}{(z-a)(z-b)(z-c)} = 0,$$

where

$$\alpha + \alpha' + \beta + \beta' + \gamma + \gamma' = 1,$$

first obtained in the form by Papperitz (1885; Bares 1908). Solutions are RIEMANN *P*-SERIES (Abramowitz and Stegun 1972, pp. 564–565). Zwillinger (1995, p. 414) confusingly calls this equation the "hypergeometric equation."

See also HEUN'S DIFFERENTIAL EQUATION

References

Abramowitz, M. and Stegun, C. A. (Eds.). "Riemann's Differential Equation." §15.6 in *Handbook of Mathematical Functions with Formulas, Graphs, and Mathematical Tables, 9th printing.* New York: Dover, pp. 564–565, 1972.

Barnes, E. W. "A New Development in the Theory of the Hypergeometric Functions." *Proc. London Math. Soc.* **6**, 141–177, 1908.

Morse, P. M. and Feshbach, H. *Methods of Theoretical Physics, Part I.* New York: McGraw-Hill, pp. 541–543, 1953.

Papperitz. *Math. Ann.* **25**, 213, 1885.

Zwillinger, D. (Ed.). *CRC Standard Mathematical Tables and Formulae.* Boca Raton, FL: CRC Press, 1995.

Zwillinger, D. *Handbook of Differential Equations, 3rd ed.* Boston, MA: Academic Press, p. 126, 1997.

Riemann Prime Number Formula

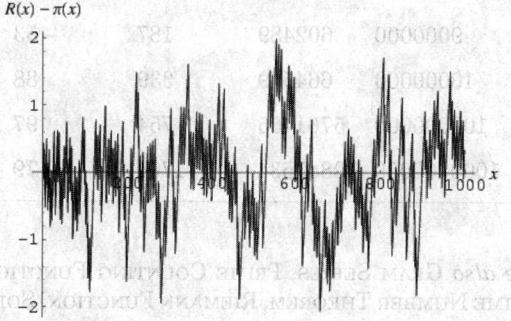

$R(x) - \pi(x)$

Riemann considered

$$R(x) = \sum_{n=1}^{\infty} \frac{\mu(n)}{n} \operatorname{li}(x^{1/n}), \tag{1}$$

obtained by replacing $f(x^{1/n})$ in the RIEMANN FUNCTION with the LOGARITHMIC INTEGRAL $\operatorname{li}(x^{1/n})$., where $\zeta(z)$ is the RIEMANN ZETA FUNCTION and $\mu(n)$ is the MÖBIUS FUNCTION (Hardy 1999, pp. 16 and 23). This series is identical to the GRAM SERIES (Hardy 1999, pp. 24–25). The quantity $R(x) - \pi(x)$ is plotted above. In addition,

$$\pi(x) = R(x) - \sum_\rho R(x^\rho), \qquad (2)$$

where $\pi(x)$ is the PRIME COUNTING FUNCTION and the SUM is over all complex (nontrivial) zeros ρ of $\zeta(s)$, i.e., those in the CRITICAL STRIP so $0 < \Re[\rho] < 1$, interpreted to mean

$$\sum_\rho R(x^\rho) = \lim_{t \to \infty} \sum_{|\Im(\rho)| < t} R(x^\rho). \qquad (3)$$

Riemann conjectured that $R(n) = \pi(n)$ (Knuth 1998, p. 382), but this was disproved by Littlewood in 1914 (Hardy and Littlewood 1918).

Ramanujan independently derived the formula for $R(n)$, but nonrigorously (Berndt 1994, p. 123; Hardy 1999, p. 23). The following table compares $\pi(x)$, $\mathrm{li}\,x$, and $R(x)$ for small x. Note that the values given by Hardy (1999, p. 26) for $x = 10^9$ are incorrect.

x	$\pi(x)$	$\mathrm{li}(x) - \pi(x)$	$R(x) - \pi(x)$
100000	9592	38	−5
1000000	78498	130	29
2000000	148933	122	−9
3000000	216816	155	0
4000000	283146	206	33
5000000	348513	125	−64
6000000	412849	228	24
7000000	476648	179	−38
8000000	539777	223	−6
9000000	602489	187	−53
10000000	664579	339	88
100000000	5761455	754	97
1000000000	50847534	1701	−79

See also GRAM SERIES, PRIME COUNTING FUNCTION, PRIME NUMBER THEOREM, RIEMANN FUNCTION, SOLDNER'S CONSTANT

References

Berndt, B. C. *Ramanujan's Notebooks, Part IV.* New York: Springer-Verlag, 1994.

Hardy, G. H. and Littlewood, J. E. *Acta Math.* **41**, 119–196, 1918.

Hardy, G. H. "The Series $R(x)$." §2.3 in *Ramanujan: Twelve Lectures on Subjects Suggested by His Life and Work, 3rd ed.* New York: Chelsea, 1999.

Knuth, D. E. *The Art of Computer Programming, Vol. 2: Seminumerical Algorithms, 3rd ed.* Reading, MA: Addison-Wesley, 1998.

Riesel, H. "The Riemann Prime Number Formula." *Prime Numbers and Computer Methods for Factorization, 2nd ed.* Boston, MA: Birkhäuser, pp. 50–52, 1994.

Riemann P-Series

The solutions to the RIEMANN P-DIFFERENTIAL EQUATION

$$z \equiv P \left\{ \begin{matrix} a & b & c \\ \alpha & \beta & \gamma; \ z \\ \alpha' & \beta' & \gamma' \end{matrix} \right\}.$$

Solutions are given in terms of the HYPERGEOMETRIC FUNCTION by

$$u_1 = \left(\frac{z-a}{z-b} \right)^\alpha \left(\frac{z-c}{z-b} \right)^\gamma {}_2F_1(\alpha + \beta + \gamma, \ \alpha + \beta' + \gamma;$$
$$1 + \alpha - \alpha'; \ \lambda)$$

$$u_2 = \left(\frac{z-a}{z-b} \right)^{\alpha'} \left(\frac{z-c}{z-b} \right)^\gamma {}_2F_1(\alpha' + \beta + \gamma, \ \alpha' + \beta' + \gamma;$$
$$1 + \alpha' - \alpha; \ \lambda)$$

$$u_3 = \left(\frac{z-a}{z-b} \right)^\alpha \left(\frac{z-c}{z-b} \right)^{\gamma'} {}_2F_1(\alpha + \beta + \gamma', \ \alpha + \beta' + \gamma';$$
$$1 + \alpha - \alpha'; \ \lambda)$$

$$u_4 = \left(\frac{z-a}{z-b} \right)^{\alpha'} \left(\frac{z-c}{z-b} \right)^{\gamma'} {}_2F_1(\alpha' + \beta + \gamma', \ \alpha' + \beta' + \gamma';$$
$$1 + \alpha' - \alpha; \ \lambda)$$

where

$$\lambda = \frac{(z-a)(c-b)}{(z-b)(c-a)}.$$

References

Abramowitz, M. and Stegun, C. A. (Eds.). "Riemann's Differential Equation." §15.6 in *Handbook of Mathematical Functions with Formulas, Graphs, and Mathematical Tables, 9th printing.* New York: Dover, pp. 564–565, 1972.

Morse, P. M. and Feshbach, H. *Methods of Theoretical Physics, Part I.* New York: McGraw-Hill, pp. 541–543, 1953.

Riemann, B. *Abh. d. Ges. d. Wiss. zu Göttingen* **7**, 1857. Reprinted in *Mathematisch Werke*, p. 67, 1892.

Whittaker, E. T. and Watson, G. N. *A Course in Modern Analysis, 4th ed.* Cambridge, England: Cambridge University Press, pp. 283–284, 1990.

Zwillinger, D. (Ed.). *CRC Standard Mathematical Tables and Formulae.* Boca Raton, FL: CRC Press, p. 414, 1995.

Riemann Removable Singularity Theorem

Let $f : D(z_0, r) \backslash \{z_0\} \to \mathbb{C}$ be ANALYTIC and bounded on a PUNCTURED OPEN DISK $D(z_0, r)$, then $\lim_{z \to z_0} f(z)$ exists, and the function defined by $\tilde{f} : D(z_0, r) \to \mathbb{C}$

$$\tilde{f}(z) = \begin{cases} f(z) & \text{for } z \neq z_0 \\ \lim_{z' \to z_0} f(z') & \text{for } z = z_0 \end{cases}$$

is ANALYTIC.

See also REMOVABLE SINGULARITY

References

Krantz, S. G. "The Riemann Removable Singularity Theorem." §4.1.5 in *Handbook of Complex Analysis*. Boston, MA: Birkhäuser, pp. 42–43, 1999.

Riemann Series Theorem

By a suitable rearrangement of terms, a CONDITION-ALLY CONVERGENT SERIES may be made to converge to any desired value, or to DIVERGE.

See also CONDITIONAL CONVERGENCE, DIVERGENT SERIES

References

Bromwich, T. J. I'a. and MacRobert, T. M. *An Introduction to the Theory of Infinite Series, 3rd ed.* New York: Chelsea, p. 74, 1991.
Gardner, M. *Martin Gardner's Sixth Book of Mathematical Games from Scientific American.* New York: Scribner's, p. 171, 1971.

Riemann Space

METRIC SPACE

Riemann Sphere

A 1-D COMPLEX MANIFOLD C*, which is the one-point COMPACTIFICATION of the COMPLEX NUMBERS $\mathbb{C}^* = \mathbb{C} \cup \{\infty\}$, together with two charts. (Here $[522, \infty]$ denoted COMPLEX INFINITY). For all points in the COMPLEX PLANE, the chart is the IDENTITY MAP from the SPHERE (with infinity removed) to the COMPLEX PLANE. For the POINT AT INFINITY, the chart neighborhood is the sphere (with the ORIGIN removed), and the chart is given by sending infinity to 0 and all other points z to $1/z$.

See also C*, COMPLEX INFINITY, COMPLEX PLANE, EXTENDED COMPLEX PLANE

References

Anderson, J. W. "The Riemann Sphere $\bar{\mathbb{C}}$." §1.2 in *Hyperbolic Geometry*. New York: Springer-Verlag, pp. 7–16, 1999.
Knopp, K. *Theory of Functions Parts I and II, Two Volumes Bound as One, Part I.* New York: Dover, p. 4, 1996.
Krantz, S. G. "The Riemann Sphere." §6.3.3 in *Handbook of Complex Analysis*. Boston, MA: Birkhäuser, pp. 83–84, 1999.

Riemann Sum

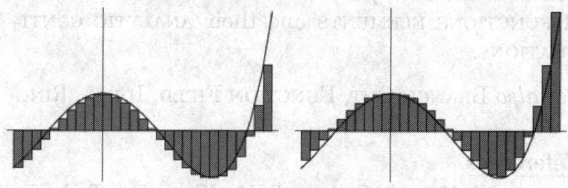

Let a CLOSED INTERVAL $[a, b]$ be partitioned by points $a < x_1 < x_2 < \ldots < x_{n-1} < b$, where the lengths of the resulting intervals between the points are denoted $\Delta x_1, \Delta x_2, \ldots, \Delta x_n$. Let x_k^* be an arbitrary point in the kth subinterval. Then the quantity

$$\sum_{k=1}^{n} f(x_k^*) \Delta x_k$$

is called a Riemann sum for a given function $f(x)$ and partition, and the value $\max \Delta x_k$ is called the MESH SIZE of the partition.

If the LIMIT $\max \Delta x_k \to 0$ exists, this limit is known as the Riemann integral of $f(x)$ over the interval $[a, b]$. The shaded areas in the above plots show the LOWER and UPPER SUMS for a constant MESH SIZE.

See also INTEGRAL, LOWER SUM, MESH SIZE, RIEMANN INTEGRAL, UPPER SUM

References

Anton, H. *Calculus: A New Horizon, 6th ed.* New York: Wiley, pp. 324–327, 1999.

Riemann Surface

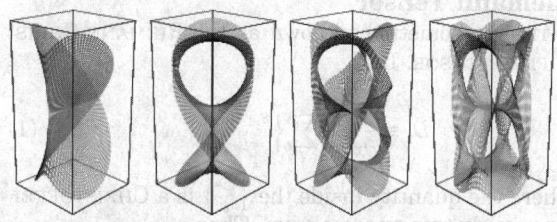

A surface-like configuration which covers the COMPLEX PLANE with several, and in general infinitely many, "sheets." These sheets can have very complicated structures and interconnections (Knopp 1996, pp. 98–99). Riemann surfaces are one way of representing MULTIPLE-VALUED FUNCTIONS; another is BRANCH CUTS. The above plot shows Riemann surfaces for solutions of the equation

$$[w(z)]^d + w(z) + z^{d-1} = 0$$

with $d = 2, 3, 4,$ and 5, where $w(z)$ is LAMBERT'S W-FUNCTION (M. Trott).

The Riemann surface S of the FUNCTION FIELD **K** is the set of nontrivial discrete valuations on **K**. Here, the set S corresponds to the IDEALS of the RING A of INTEGERS of **K** over $\mathbb{C}(z)$. (A consists of the elements of

K that are ROOTS of MONIC POLYNOMIALS over $\mathbb{C}[z]$.) Riemann surfaces provide a geometric visualization of FUNCTIONS ELEMENTS and their ANALYTIC CONTINUATIONS.

See also BRANCH CUT, FUNCTION FIELD, IDEAL, RING

References

Borwein, J. M. and Corless, R. M. "Emerging Tools for Experimental Mathematics." *Amer. Math. Monthly* **106**, 899–909, 1999.

Corless, R. M. and Jeffrey, D. J. "Graphing Elementary Riemann Surfaces." *ACM Sigsam Bulletin: Commun. Comput. Algebra* **32**, 11–17, 1998.

Fischer, G. (Ed.). Plates 123–126 in *Mathematische Modelle/Mathematical Models, Bildband/Photograph Volume.* Braunschweig, Germany: Vieweg, pp. 120–123, 1986.

Knopp, K. *Theory of Functions Parts I and II, Two Volumes Bound as One, Part II.* New York: Dover, pp. 99–118, 1996.

Krantz, S. G. "The Idea of a Riemann Surface." §10.4 in *Handbook of Complex Analysis.* Boston, MA: Birkhäuser, pp. 135–139, 1999.

Mathews, J. H. and Howell, R. W. *Complex Analysis for Mathematics and Engineering, 4th ed.* Boston, MA: Jones and Bartlett, 2000.

Monna, A. F. *Dirichlet's Principle: A Mathematical Comedy of Errors and Its Influence on the Development of Analysis.* Utrecht, Netherlands: Osothoek, Scheltema, and Holkema, 1975.

Trott, M. "Visualization of Riemann Surfaces of Algebraic Functions." *Mathematica J.* **6**, 15–36, 1997.

Trott, M. "Visualization of Riemann Surfaces IIa." *Mathematica J.* **7**, 465–496, 2000.

Trott, M. "Visualization of Riemann Surfaces." http://library.wolfram.com/examples/riemannsurface/.

Riemann Tensor

A TENSOR sometimes known as the RIEMANN-CHRISTOFFEL TENSOR. Let

$$\tilde{D}_s \equiv \frac{\partial}{\partial x^s} - \sum_l \begin{Bmatrix} s & u \\ l \end{Bmatrix}, \tag{1}$$

where the quantity inside the $\begin{Bmatrix} s & u \\ l \end{Bmatrix}$ is a CHRISTOFFEL SYMBOL OF THE SECOND KIND. Then

$$R_{pqrs} \equiv \tilde{D}_q \begin{Bmatrix} p & r \\ s \end{Bmatrix} - \tilde{D}_r \begin{Bmatrix} r & q \\ s \end{Bmatrix}. \tag{2}$$

Broken down into its simplest decomposition in N-D,

$$R_{\lambda\mu\nu\kappa} = \frac{1}{N-2}\left(g_{\lambda\nu}R_{\mu\kappa} - g_{\lambda\kappa}R_{\mu\nu} - g_{\mu\nu}R_{\lambda\kappa} + g_{\mu\kappa}R_{\lambda\nu}\right)$$
$$- \frac{R}{(N-1)(N-2)}\left(g_{\lambda\nu}g_{\mu\kappa} - g_{\lambda\kappa}g_{\mu\nu}\right) + C_{\lambda\mu\nu\kappa}. \tag{3}$$

Here, $R_{\mu\nu}$ is the RICCI TENSOR, R is the CURVATURE SCALAR, and $C_{\lambda\mu\nu\kappa}$ is the WEYL TENSOR.

In terms of the JACOBI TENSOR $J^\mu{}_{\nu\alpha\beta}$,

$$R^\mu{}_{\alpha\nu\beta} = \tfrac{2}{3}\left(J^\mu{}_{\nu\alpha\beta}J^\mu{}_{\beta\alpha\nu}\right). \tag{4}$$

The Riemann tensor is the only tensor that can be constructed from the METRIC TENSOR and its first and second derivatives,

$$R^\alpha{}_{\beta\gamma\delta} = \Gamma^\alpha{}_{\beta\delta,\,\gamma} - \Gamma^\alpha{}_{\beta\gamma,\,\delta} + \Gamma^\mu{}_{\beta\delta}\Gamma^\alpha{}_{\mu\gamma} - \Gamma^\mu{}_{\beta\gamma}\Gamma^\alpha{}_{\mu\delta}, \tag{5}$$

where $\Gamma^\gamma{}_{\alpha\beta}$ are CONNECTION COEFFICIENTS and $A_{,k}$ is a COMMA DERIVATIVE (Schmutzer 1968, p. 108). In 1-D, $R_{1111} = 0$.

The number of independent coordinates in n-D is given by

$$C_n \equiv \tfrac{1}{12}n^2\left(n^2 - 1\right), \tag{6}$$

the "4-D pyramidal numbers," the first few values of which are 0, 1, 6, 20, 50, 105, 196, 336, 540, 825, ... (Sloane's A002415). The number of SCALARS which can be constructed from $R_{\lambda\mu\nu\kappa}$ and $g_{\mu\nu}$ is

$$S_n \equiv \begin{cases} 1 & \text{for } n = 2 \\ \tfrac{1}{12}n(n-1)(n-2)(n+3) & \text{for } n = 1,\ n > 2 \end{cases} \tag{7}$$

(Weinberg 1972). The first few values are then 0, 1, 3, 14, 40, 90, 175, 308, 504, 780, ... (Sloane's A050297).

See also BIANCHI IDENTITIES, CHRISTOFFEL SYMBOL OF THE SECOND KIND, COMMUTATION COEFFICIENT, CONNECTION COEFFICIENT, CURVATURE SCALAR, GAUSSIAN CURVATURE, JACOBI TENSOR, PETROV NOTATION, RICCI TENSOR, WEYL TENSOR

References

Misner, C. W.; Thorne, K. S.; and Wheeler, J. A. *Gravitation.* San Francisco: W. H. Freeman, pp. 220–221, 1973.

Schmutzer, E. *Relativistische Physik (Klassische Theorie).* Leipzig, Germany: Akademische Verlagsgesellschaft, 1968.

Sloane, N. J. A. Sequences A002415/M4135 and A050297 in "An On-Line Version of the Encyclopedia of Integer Sequences." http://www.research.att.com/~njas/sequences/eisonline.html.

Weinberg, S. *Gravitation and Cosmology: Principles and Applications of the General Theory of Relativity.* New York: Wiley, 1972.

Riemann Theta Function

Let the IMAGINARY PART of a $g \times g$ MATRIX F be POSITIVE DEFINITE, and $\mathbf{m} = (m_1, \ldots, m_g)$ be a row VECTOR with coefficients in $\mathbb{Z}$. Then the Riemann theta function is defined by

$$\vartheta(u) = \sum_{\mathbf{m}} \exp\left[2\pi i\left(\mathbf{m}^\mathsf{T} u + \tfrac{1}{2}\,\mathbf{m}\mathsf{F}^\mathsf{T}\mathbf{m}\right)\right].$$

See also JACOBI THETA FUNCTIONS, RAMANUJAN THETA FUNCTIONS, SIEGEL THETA FUNCTION, THETA FUNCTIONS

References

Itô, K. (Ed.). "Abelian Integrals." §3.L in *Encyclopedic Dictionary of Mathematics, 2nd ed., Vol. 1.* Cambridge, MA: MIT Press, p. 9, 1987.

Riemann Xi Function

XI FUNCTION

Riemann Zeta Function

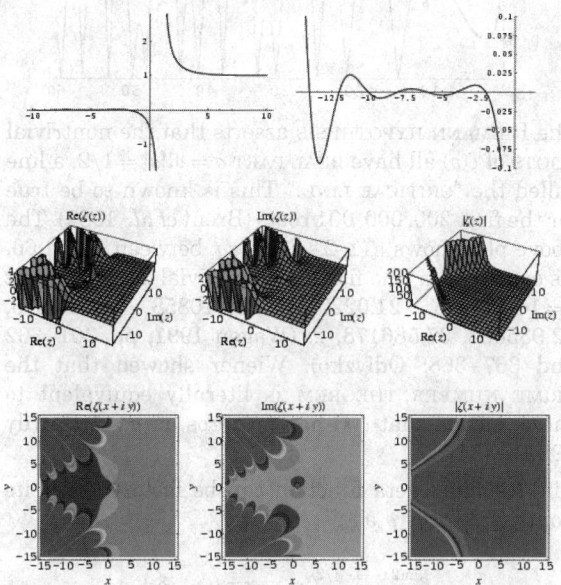

The Riemann zeta function is an extremely important SPECIAL FUNCTION of mathematics and physics which arises in definite integration and is intimately related with very deep results surrounding the PRIME NUMBER THEOREM. While many of the properties of this function have been investigated, there remain important fundamental conjectures (most notably the RIEMANN HYPOTHESIS) which remain unproved to this day.

On the REAL LINE with $x > 1$, the Riemann zeta function can be defined by the integral

$$\zeta(x) \equiv \frac{1}{\Gamma(x)} \int_0^\infty \frac{u^{x-1}}{e^u - 1}\, du, \qquad (1)$$

where $\Gamma(n)$ is the GAMMA FUNCTION. If x is an INTEGER n, then we have the identity

$$\frac{u^{n-1}}{e^u - 1} = \frac{e^{-u} u^{n-1}}{1 - e^{-u}} = e^{-u} u^{n-1} \sum_{k=0}^\infty e^{-ku}$$

$$= \sum_{k=1}^\infty e^{-ku} u^{n-1}, \qquad (2)$$

so

$$\int_0^\infty \frac{u^{n-1}}{e^u - 1}\, du = \sum_{k=1}^\infty \int_0^\infty e^{-ku} u^{n-1}\, du. \qquad (3)$$

To evaluate $\zeta(n)$, let $y \equiv ku$ so that $dy = k\, du$ and plug in the above identity to obtain

$$\zeta(n) = \frac{1}{\Gamma(n)} \sum_{k=1}^\infty \int_0^\infty e^{-ku} u^{n-1}\, du$$

$$= \frac{1}{\Gamma(n)} \sum_{k=1}^\infty \int_0^\infty e^{-y} \left(\frac{y}{k}\right)^{n-1} \frac{dy}{k}$$

$$= \frac{1}{\Gamma(n)} \sum_{k=1}^\infty \frac{1}{k^n} \int_0^\infty e^{-y} y^{n-1}\, dy. \qquad (4)$$

Integrating the final expression in (4) gives $\Gamma(n)$, which cancels the factor $1/\Gamma(n)$ and gives the most common form of the Riemann zeta function,

$$\zeta(n) = \sum_{k=1}^\infty \frac{1}{k^n}. \qquad (5)$$

The Riemann zeta function can also be defined in terms of MULTIPLE INTEGRALS by

$$\zeta(n) = \underbrace{\int_0^1 \cdots \int_0^1}_{n} \frac{\prod_{i=1}^n dx_i}{1 - \prod_{i=1}^n x_i}, \qquad (6)$$

and as a MELLIN TRANSFORM by

$$\int_0^\infty \operatorname{frac}\left(\frac{1}{t}\right) t^{n-1}\, dt = -\frac{\zeta(s)}{s} \qquad (7)$$

for $0 < \Re[s] < 1$, where $\operatorname{frac}(x)$ is the FRACTIONAL PART (Balazard and Saias 2000).

Note that the zeta function has a singularity at $n = 1$, where it reduces to the divergent HARMONIC SERIES.

The Riemann zeta function satisfies the functional equation

$$\zeta(1-s) = 2(2\pi)^{-s} \cos\left(\tfrac{1}{2} s\pi\right) \Gamma(s)\zeta(s) \qquad (8)$$

(Hardy 1999, p. 14; Krantz 1999, p. 160).

As defined above, the zeta function $\zeta(s)$ with $s = \sigma + it$ a COMPLEX NUMBER is defined for $\Re[s] > 1$. However, $\zeta(s)$ has a unique ANALYTIC CONTINUATION to the entire COMPLEX PLANE, excluding the point $s = 1$, which corresponds to a SIMPLE POLE with RESIDUE 1 (Krantz 1999, p. 160). In particular, as $s \to 1$, $\zeta(s)$ obeys

$$\lim_{s \to 1} \zeta(s) - \frac{1}{s-1} = \gamma, \qquad (9)$$

where γ is the EULER-MASCHERONI CONSTANT (Whittaker and Watson 1990, p. 271).

To perform the ANALYTIC CONTINUATION for $\Re[s] > 0$, write

$$\sum_{n=1}^\infty (-1)^n n^{-s} + \sum_{n=1}^\infty n^{-s} = 2 \sum_{n=2,\,4,\,\ldots}^\infty n^{-s}$$

$$= 2 \sum_{k=1}^{\infty} (2k)^{-s} = 2^{1-s} \sum_{n=1}^{\infty} k^{-s} \qquad (10)$$

$$\sum_{n=1}^{\infty} (-1)^n n^{-s} + \zeta(s) = 2^{1-s} \zeta(s). \qquad (11)$$

Therefore,

$$\zeta(s) = \frac{1}{1 - 2^{1-s}} \sum_{n=1}^{\infty} (-1)^{n-1} n^{-s}. \qquad (12)$$

While this form defines $\zeta(s)$ for only the UPPER HALF-PLANE $\Re[s] > 0$, equation (8) can be used to analytically continue it to the rest of the COMPLEX PLANE. Analytic continuation can also be performed using HANKEL FUNCTIONS. A globally convergent series for the Riemann zeta function is given by

$$\zeta(z) = \frac{1}{1 - 2^{1-z}} \sum_{n=0}^{\infty} \frac{1}{2^{n+1}} \sum_{k=0}^{n} (-1)^k \binom{n}{k} (k+1)^{-z}, \qquad (13)$$

where $\binom{n}{k}$ is a BINOMIAL COEFFICIENT.

A generalized Riemann zeta function $\zeta(s, a)$ known as the HURWITZ ZETA FUNCTION can also be defined such that

$$\zeta(s) \equiv \zeta(s, 0). \qquad (14)$$

In the COMPLEX PLANE, trivial zeros of $\zeta(s)$ occur at $s = -2, -4, -6, \ldots$, and nontrivial zeros at

$$s \equiv \sigma + it \qquad (15)$$

for $0 \le \sigma \le 1$. The figures below show the structure of the complex $\zeta(z)$ by plotting $|\zeta(z)|$ and $1/|\zeta(z)|$.

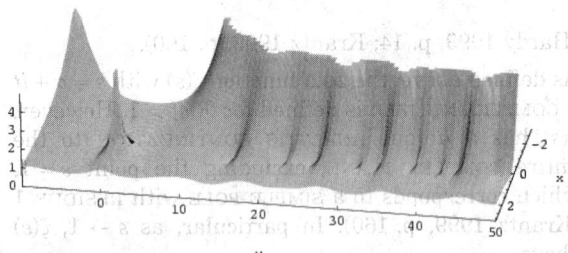

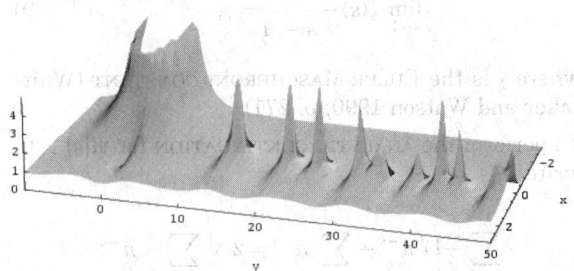

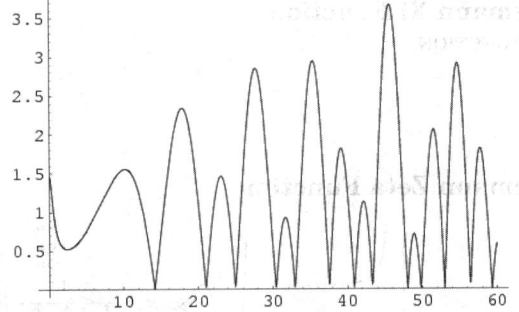

The RIEMANN HYPOTHESIS asserts that the nontrivial ROOTS of $\zeta(s)$ all have REAL PART $\sigma = \Re[s] = 1/2$, a line called the "CRITICAL LINE." This is known to be true for the first $200,000,001$ roots (Brent *et al.* 1982). The above plot shows $|\zeta(1/2 + it)|$ for t between 0 and 60. As can be seen, the first few nontrivial zeros occur at $t = 14.134725$, 21.022040, 25.010858, 30.424876, 32.935062, 37.586178, ... (Wagon 1991, pp. 361–362 and 367–368; Odlyzko). Wiener showed that the PRIME NUMBER THEOREM is literally equivalent to the assertion that $\zeta(s)$ has no zeros on $\sigma = 1$ (Hardy 1999, p. 34).

The Riemann zeta function can be factored over its nontrivial zeros ρ as

$$\zeta(s) = \frac{e^{\ln(2\pi) - 1 - \gamma/2)s}}{2(s-1)\Gamma\left(1 + \frac{1}{2}s\right)} \prod_{\rho} \left(1 - \frac{s}{\rho}\right) e^{s/\rho} \qquad (16)$$

(Voros 1987).

The Riemann zeta function can be split up into

$$\zeta\left(\tfrac{1}{2} + it\right) = z(t) e^{-i\vartheta(t)}, \qquad (17)$$

where $z(t)$ and $\vartheta(t)$ are the RIEMANN-SIEGEL FUNCTIONS. The Riemann zeta function is related to the DIRICHLET LAMBDA FUNCTION $\lambda(v)$ and DIRICHLET ETA FUNCTION $\eta(v)$ by

$$\frac{\zeta(v)}{2^v} = \frac{\lambda(v)}{2^v - 1} = \frac{\eta(v)}{2^v - 2} \qquad (18)$$

and

$$\zeta(v) + \eta(v) = 2\lambda(v) \qquad (19)$$

(Spanier and Oldham 1987). It is related to the LIOUVILLE FUNCTION $\lambda(v)$ by

$$\frac{\zeta(2s)}{\zeta(s)} = \sum_{n=1}^{\infty} \frac{\lambda(n)}{n^s} \qquad (20)$$

(Lehman 1960, Hardy and Wright 1979). Furthermore,

$$\frac{\zeta^2(s)}{\zeta(2s)} = \sum_{n=1}^{\infty} \frac{2^{\omega(n)}}{n^s}, \qquad (21)$$

where $\omega(n)$ is the number of DISTINCT PRIME FACTORS of n (Hardy and Wright 1979, p. 254).

Two sum identities involving $\zeta(n)$ are

$$\sum_{n=2}^{\infty} [\zeta(n) - 1] = 1 \tag{22}$$

$$\sum_{n=2}^{\infty} (-1)^n [\zeta(n) - 1] = \tfrac{1}{2}. \tag{23}$$

The Riemann zeta function is related to the GAMMA FUNCTION $\Gamma(z)$ by

$$\Gamma\left(\frac{s}{2}\right) \pi^{-s/2} \zeta(s) = \Gamma\left(\frac{1-s}{2}\right) \pi^{-(1-s)/2} \zeta(1-s). \tag{24}$$

The DERIVATIVE of the Riemann zeta function is defined by

$$\zeta'(s) = -s \sum_{k=1}^{\infty} k^{-s} \ln k = -\sum_{k=2}^{\infty} \frac{\ln k}{k^s}. \tag{25}$$

As $s \to 0$,

$$\zeta'(0) = -\tfrac{1}{2} \ln(2\pi). \tag{26}$$

$\zeta(n)$ is known to be transcendental for all EVEN n, but the study of the function at ODD n is significantly more difficult. Apéry (1979) finally proved that $\zeta(3)$ to be IRRATIONAL, but no similar results are known for other ODD n. However, Rivoal (2000) recently proved that there are infinitely many integers n such that $\zeta(2n+1)$ is irrational. As a result of Apéry's important discovery, $\zeta(3)$ is sometimes called APÉRY'S CONSTANT. A number of interesting sums for $\zeta(n)$, with n a POSITIVE INTEGER, can be written in terms of binomial coefficients as the BINOMIAL SUMS

$$\zeta(2) = 3 \sum_{k=1}^{\infty} \frac{1}{k^2 \binom{2k}{k}} \tag{27}$$

$$\zeta(3) = \frac{5}{2} \sum_{k=1}^{\infty} \frac{(-1)^{k-1}}{k^3 \binom{2k}{k}} \tag{28}$$

$$\zeta(4) = \frac{36}{17} \sum_{k=1}^{\infty} \frac{1}{k^4 \binom{2k}{k}} \tag{29}$$

(Guy 1994, p. 257). Apéry arrived at his result with the aid of the k^{-3} sum formula above. A relation OF THE FORM

$$\zeta(5) = Z_5 \sum_{k=1}^{\infty} \frac{(-1)^{k-1}}{k^5 \binom{2k}{k}} \tag{30}$$

has been searched for with Z_5 a RATIONAL or ALGE-

BRAIC NUMBER, but if Z_5 is a ROOT of a POLYNOMIAL of degree 25 or less, then the Euclidean norm of the coefficients must be larger than 2×10^{37} (Bailey and Plouffe). Therefore, no such sums for $\zeta(n)$ are known for $n \geq 5$.

The Riemann zeta function may be computed analytically for EVEN n using either CONTOUR INTEGRATION or PARSEVAL'S THEOREM with the appropriate FOURIER SERIES. An unexpected and important formula involving the product of PRIMES was first discovered by Euler in 1737,

$$\zeta(x)(1 - 2^{-x}) = \left(1 + \frac{1}{2^x} + \frac{1}{3^x} + \ldots\right)\left(1 - \frac{1}{2^x}\right)$$

$$= \left(1 + \frac{1}{2^x} + \frac{1}{3^x} + \ldots\right) - \left(\frac{1}{2^x} + \frac{1}{4^x} + \frac{1}{6^x} + \ldots\right) \tag{31}$$

$$\zeta(x)(1 - 2^{-x})(1 - 3^{-x})$$

$$= \left(1 + \frac{1}{3^x} + \frac{1}{5^x} + \frac{1}{7^x} + \ldots\right) - \left(\frac{1}{3^x} + \frac{1}{9^x} + \frac{1}{15^x} + \ldots\right) \tag{32}$$

$$\zeta(x)(1 - 2^{-x})(1 - 3^{-x}) \cdots (1 - p^{-z}) \cdots$$

$$= \zeta(x) \prod_{n=2}^{\infty} (1 - p^{-x}) = 1. \tag{33}$$

Here, each subsequent multiplication by the next PRIME p leaves only terms which are POWERS of p^{-x}. Therefore,

$$\zeta(x) = \left[\prod_{p=2}^{\infty} (1 - p^{-x})\right]^{-1}, \tag{34}$$

where p runs over all PRIMES (Hardy 1999, p. 18; Krantz 1999, p. 159). Euler's product formula can also be written

$$\zeta(s) = (1 - 2^{-s})^{-1} \prod_{\substack{q \equiv 1 \\ (\text{mod } 4)}} (1 - q^{-s})^{-1} \prod_{\substack{r \equiv 3 \\ (\text{mod } 4)}} (1 - r^{-s})^{-1}. \tag{35}$$

For EVEN $n \equiv 2k$,

$$\zeta(n) = \frac{2^{n-1} |B_n| \pi^n}{n!}, \tag{36}$$

where B_n is a BERNOULLI NUMBER. Another intimate connection with the BERNOULLI NUMBERS is provided by

$$B_n = (-1)^{n+1} n \zeta(1-n) \tag{37}$$

for $n \geq 1$, which can be written

$$B_n = -n\zeta(1-n) \qquad (38)$$

for $n \geq 2$. Although no analytic form for $\zeta(n)$ is known for ODD n,

$$\zeta(3) = \frac{1}{2} \sum_{k=1}^{\infty} \frac{1}{k^2} \left(1 + \frac{1}{2} + \ldots + \frac{1}{k} \right) = \frac{1}{2} \sum_{k=1}^{\infty} \frac{h_k}{k^2}, \qquad (39)$$

where h_k is a HARMONIC NUMBER (Stark 1974). In addition, $\zeta(n)$ can be expressed as the sum limit

$$\zeta(n) = \lim_{x \to \infty} \frac{1}{(2x+1)^n} \sum_{k=1}^{x} \left[\cot \left(\frac{k}{2x+1} \right) \right]^n \qquad (40)$$

for $n = 3, 5, \ldots$ (Apostol 1973, given incorrectly in Stark 1974).

For $\mu(n)$ the MÖBIUS FUNCTION,

$$\frac{1}{\zeta(s)} = \sum_{n=1}^{\infty} \frac{\mu(n)}{n^s}. \qquad (41)$$

The values for small integral arguments are

$$\zeta(1) = \infty$$

$$\zeta(2) = \frac{\pi^2}{6}$$

$$\zeta(3) = 1.2020569032\ldots$$

$$\zeta(4) = \frac{\pi^4}{90}$$

$$\zeta(5) = 1.0369277551\ldots$$

$$\zeta(6) = \frac{\pi^6}{945}$$

$$\zeta(7) = 1.0083492774\ldots$$

$$\zeta(8) = \frac{\pi^8}{9450}$$

$$\zeta(9) = 1.0020083928\ldots$$

$$\zeta(10) = \frac{\pi^{10}}{93,555}.$$

Euler gave $\zeta(2)$ to $\zeta(26)$ for EVEN n (Wells 1986, p. 54), and Stieltjes (1993) determined the values of $\zeta(2), \ldots, \zeta(70)$ to 30 digits of accuracy in 1887. The denominators of $\zeta(2n)$ for $n = 1, 2, \ldots$ are 6, 90, 945, 9450, 93555, 638512875, ... (Sloane's A002432).

The value at $n = 0$ is given by

$$\zeta(0) = -\frac{1}{2} \qquad (42)$$

The value $\zeta(-1) = -1/12$ is a deep result of renormalization theory (Elizalde *et al.* 1994, Elizalde 1995). In general,

$$\zeta(-n) = -\frac{B_{n+1}}{n+1} \qquad (43)$$

for $n = 1, 3, \ldots$ where B_n is a BERNOULLI NUMBER, the first few values of which are $-1/12$, $1/120$, $-1/252$, $1/240$, ... (Sloane's A001067 and A006953).

Rapidly converging series for $\zeta(n)$ for n odd were first discovered by Ramanujan (Zucker 1979, Zucker 1984, Berndt 1988, Bailey *et al.* 1997, Cohen 2000). For $n > 1$ and $n \equiv 3 \pmod 4$,

$$\zeta(n) = \frac{2^{n-1}\pi^n}{(n+1)!} \sum_{k=0}^{(n+1)/2} (-1)^{k-1} \binom{n+1}{2k} B_{n+1-2k} B_{2k}$$

$$-2 \sum_{k=1}^{\infty} \frac{1}{k^n(e^{2\pi k} - 1)}, \qquad (44)$$

where B_k is again a BERNOULLI NUMBER and $\binom{n}{k}$ is a BINOMIAL COEFFICIENT. The first few for $n = 3, 7, 11, \ldots$ are 7/180, 19/56700, 1453/425675250, 13687/390769879500, 7708537/21438612514068750, ... (Sloane's A057866 and A057867). For $n \geq 5$ and $n \equiv 1 \pmod 4$, the corresponding formula is slightly messier,

$$\zeta(n) = \frac{(2\pi)^n}{(n+1)!(n-1)}$$

$$\times \sum_{k=0}^{(n+1)/4} (-1)^k (n+1-4k) \binom{n+1}{2k} B_{n+1-2k} B_{2k}$$

$$-2 \sum_{k=1}^{\infty} \frac{e^{2\pi k} \left(1 + \frac{4\pi k}{k-1} \right) - 1}{k^n(e^{2\pi k} - 1)^2}. \qquad (45)$$

Defining

$$S_{\pm}(n) \equiv \sum_{k=1}^{\infty} \frac{1}{k^n(e^{2\pi k} \pm 1)}, \qquad (46)$$

the first few values can then be written

$$\zeta(3) = \frac{7}{180}\pi^3 - 2S_-(3) \qquad (47)$$

$$\zeta(5) = \frac{1}{294}\pi^5 - \frac{72}{35}S_-(5) - \frac{2}{35}S_+(5) \qquad (48)$$

$$\zeta(7) = \frac{19}{56700}\pi^7 - 2S_-(7) \qquad (49)$$

$$\zeta(9) = \frac{125}{3704778}\pi^9 - \frac{992}{495}S_-(9) - \frac{2}{495}S_+(9) \qquad (50)$$

$$\zeta(11) = \frac{1453}{425675250}\pi^{11} - 2S_-(11) \qquad (51)$$

$$\zeta(13) = \frac{89}{257432175}\pi^{13} - \frac{16512}{8255}S_-(13) - \frac{2}{8255}S_+(13) \qquad (52)$$

$$\zeta(15) = \frac{13687}{390769879500}\pi^{15} - 2S_-(15) \qquad (53)$$

$$\zeta(17) = \frac{397549}{112024529867250}\pi^{17} - \frac{261632}{130815}S_-(17)$$

$$- \frac{2}{130815}S_+(17) \qquad (54)$$

$$\zeta(19) = \frac{7708537}{21438612514068750}\pi^{19} - 2S_-(19) \qquad (55)$$

$$\zeta(21) = \frac{68529640373}{1881063815762259253125}\pi^{21} - \frac{4196352}{2098175}S_-(21)$$
$$- \frac{2}{2098175}S_+(21) \qquad (56)$$

(Plouffe).

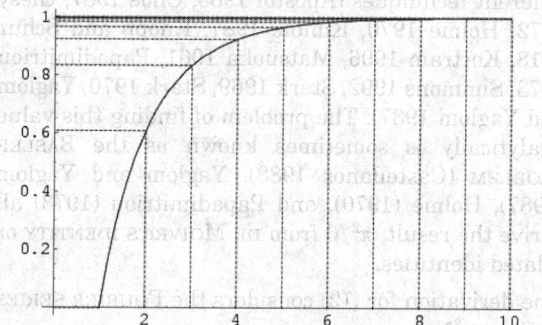

The inverse of the Riemann zeta function $1/\zeta(p)$, plotted above, is the asymptotic density of pth-power-free numbers (i.e., SQUAREFREE numbers, CUBEFREE numbers, etc.). The following table gives the number $Q_p(n)$ of pth-powerfree numbers $\leq n$ for several values of n.

p	$1/\zeta(p)$	$Q_p(10)$	$Q_p(100)$	$Q_p(10^3)$	$Q_p(10^4)$	$Q_p(10^5)$	$Q_p(10^6)$
2	0.607927	7	61	608	6083	60794	607926
3	0.831907	9	85	833	8319	83190	831910
4	0.923938	10	93	925	9240	92395	923939
5	0.964387	10	97	965	9645	96440	964388
6	0.982953	10	99	984	9831	98297	982954

See also ABEL'S FUNCTIONAL EQUATION, BERRY CONJECTURE, CRITICAL LINE, CRITICAL STRIP, DEBYE FUNCTIONS, DIRICHLET BETA FUNCTION, DIRICHLET ETA FUNCTION, DIRICHLET LAMBDA FUNCTION, EULER PRODUCT, HARMONIC SERIES, HURWITZ ZETA FUNCTION, KHINTCHINE'S CONSTANT, LEHMER'S PHENOMENON, PERIODIC ZETA FUNCTION, PRIME NUMBER THEOREM, PSI FUNCTION, RIEMANN HYPOTHESIS, RIEMANN P-SERIES, RIEMANN-SIEGEL FUNCTIONS, RIEMANN ZETA FUNCTION ZETA(2), STIELTJES CONSTANTS, XI FUNCTION

References

Abramowitz, M. and Stegun, C. A. (Eds.). "Riemann Zeta Function and Other Sums of Reciprocal Powers." §23.2 in *Handbook of Mathematical Functions with Formulas, Graphs, and Mathematical Tables, 9th printing.* New York: Dover, pp. 807–808, 1972.

Adamchik, V. S. and Srivastava, H. M. "Some Series of the Zeta and Related Functions." *Analysis* **18**, 131–144, 1998.

Aizenberg, L.; Adamchik, V.; and Levit, V. E. "Approaching the Riemann Hypothesis with *Mathematica*." http://library.wolfram.com/demos/v4/Riemann.nb.

Apéry, R. "Irrationalité de $\zeta(2)$ et $\zeta(3)$." *Astérisque* **61**, 11–13, 1979.

Apostol, T. M. "Another Elementary Proof of Euler's Formula for $\zeta(2n)$." *Amer. Math. Monthly* **80**, 425–431, 1973.

Arfken, G. *Mathematical Methods for Physicists, 3rd ed.* Orlando, FL: Academic Press, pp. 332–335, 1985.

Ayoub, R. "Euler and the Zeta Function." *Amer. Math. Monthly* **81**, 1067–1086, 1974.

Bailey, D. H. "Multiprecision Translation and Execution of Fortran Programs." *ACM Trans. Math. Software.* To appear.

Bailey, D. and Plouffe, S. "Recognizing Numerical Constants." http://www.cecm.sfu.ca/organics/papers/bailey/.

Bailey, D. H.; Borwein, J. M.; and Crandall, R. E. "On the Khintchine Constant." *Math. Comput.* **66**, 417–431, 1997.

Balazard, M. and Saias, E. "The Nyman-Beurling Equivalent Form for the Riemann Hypothesis." *Expos. Math.* **18**, 131–138, 2000.

Balazard, M.; Saias, E.; and Yor, M. "Notes sur la fonction ζ de Riemann, 2." *Adv. Math.* **143**, 284–287, 1999.

Berndt, B. C. Ch. 14 in *Ramanujan's Notebooks, Part II.* New York: Springer-Verlag, 1988.

Borwein, D. and Borwein, J. "On an Intriguing Integral and Some Series Related to $\zeta(4)$." *Proc. Amer. Math. Soc.* **123**, 1191–1198, 1995.

Borwein, J. M.; Bradley, D. M.; and Crandall, R. E. "Computational Strategies for the Riemann Zeta Function." CECM-98:118, 23 Jun 1999. http://www.cecm.sfu.ca/preprints/1999pp.html#98:118.

Brent, R. P. "On the Zeros of the Riemann Zeta Function in the Critical Strip." *Math. Comput.* **33**, 1361–1372, 1979.

Brent, R. P.; van de Lune, J.; te Riele, H. J. J.; and Winter, D. T. "On the Zeros of the Riemann Zeta Function in the Critical Strip. II." *Math. Comput.* **39**, 681–688, 1982.

Castellanos, D. "The Ubiquitous Pi. Part I." *Math. Mag.* **61**, 67–98, 1988.

Cohen, H. "High Precision Computation of Hardy-Littlewood Constants." Preprint. http://www.math.u-bordeaux.fr/~cohen/hardylw.dvi.

Davenport, H. *Multiplicative Number Theory, 2nd ed.* New York: Springer-Verlag, 1980.

Edwards, H. M. *Riemann's Zeta Function.* New York: Academic Press, 1974.

Elizalde, E. *Ten Physical Applications of Spectral Zeta Functions.* Berlin: Springer-Verlag, 1995.

Elizalde, E.; Odintsov, S. D.; Romeo, A.; Bytsenko, A. A.; and Zerbini, S. *Zeta Regularization Techniques With Applications.* River Edge, NJ: World Scientific, 1994.

Farmer, D. W. "Counting Distinct Zeros of the Riemann Zeta-Function." *Electronic J. Combinatorics* **2**, R1 1–5, 1995. http://www.combinatorics.org/Volume_2/volume2.html#R1.

Guy, R. K. "Series Associated with the ζ-Function." §F17 in *Unsolved Problems in Number Theory, 2nd ed.* New York: Springer-Verlag, pp. 257–258, 1994.

Hardy, G. H. *Ramanujan: Twelve Lectures on Subjects Suggested by His Life and Work, 3rd ed.* New York: Chelsea, 1999.

Hardy, G. H. and Wright, E. M. "The Zeta Function." §17.2 in *An Introduction to the Theory of Numbers, 5th ed.* Oxford, England: Clarendon Press, pp. 245–247 and 255, 1979.

Hauss, M. *Verallgemeinerte Stirling, Bernoulli und Euler Zahlen, deren Anwendungen und schnell konvergente Reihen für Zeta Funktionen.* Aachen, Germany: Verlag Shaker, 1995.

Howson, A. G. "Addendum to: 'Euler and the Zeta Function' (Amer. Math. Monthly **81** (1974), 1067–1086) by Raymond Ayoub." *Amer. Math. Monthly* **82**, 737, 1975.

Ivic, A. A. *The Riemann Zeta-Function.* New York: Wiley, 1985.

Ivic, A. A. *Lectures on Mean Values of the Riemann Zeta Function.* Berlin: Springer-Verlag, 1991.

Karatsuba, A. A. and Voronin, S. M. *The Riemann Zeta-Function.* Hawthorne, NY: De Gruyter, 1992.

Katayama, K. "On Ramanujan's Formula for Values of Riemann Zeta-Function at Positive Odd Integers." *Acta Math.* **22**, 149–155, 1973.

Keiper, J. "The Zeta Function of Riemann." *Mathematica Educ. Res.* **4**, 5–7, 1995.

Knopp, K. "4th Example: The Riemann ζ-Function." *Theory of Functions Parts I and II, Two Volumes Bound as One, Part II.* New York: Dover, pp. 51–57, 1996.

Krantz, S. G. "Riemann's Zeta Function." §13.2 in *Handbook of Complex Analysis.* Boston, MA: Birkhäuser, pp. 158–159, 1999.

Le Lionnais, F. *Les nombres remarquables.* Paris: Hermann, p. 35, 1983.

Lehman, R. S. "On Liouville's Function." *Math. Comput.* **14**, 311–320, 1960.

Odlyzko, A. "Andrew Odlyzko: Tables of Zeros of the Riemann Zeta Function." http://www.research.att.com/ ~amo/zeta_tables/.

Odlyzko, A. M. "The 10^{20}th Zero of the Riemann Zeta Function and 70 Million of Its Neighbors." Preprint.

Patterson, S. J. *An Introduction to the Theory of the Riemann Zeta-Function.* New York: Cambridge University Press, 1988.

Plouffe, S. "Identities Inspired from Ramanujan Notebooks." http://www.lacim.uqam.ca/plouffe/identities.html.

Rivoal, T. "La fonction Zeta de Riemann prend une infinité de valeurs irrationnelles aux entiers impairs." *C. R. Acad. Sci.* **331**, 267–270, 2000.

Sloane, N. J. A. Sequences A001067, A002432/M4283, A006953/M2039, A057866, and A057867 in "An On-Line Version of the Encyclopedia of Integer Sequences." http://www.research.att.com/~njas/sequences/eisonline.html.

Spanier, J. and Oldham, K. B. "The Zeta Numbers and Related Functions." Ch. 3 in *An Atlas of Functions.* Washington, DC: Hemisphere, pp. 25–33, 1987.

Stieltjes, T. J. *Oeuvres Complètes, Vol. 2* (Ed. G. van Dijk.) New York: Springer-Verlag, p. 100, 1993.

Titchmarsh, E. C. *The Zeta-Function of Riemann, 2nd ed.* Oxford, England: Oxford University Press, 1987.

Titchmarsh, E. C. and Heath-Brown, D. R. *The Theory of the Riemann Zeta-Function, 2nd ed.* Oxford, England: Oxford University Press, 1986.

Vardi, I. "The Riemann Zeta Function." Ch. 8 in *Computational Recreations in Mathematica.* Reading, MA: Addison-Wesley, pp. 141–174, 1991.

Voros, A. "Spectral Functions, Special Functions and the Selberg Zeta Function." *Commun. Math. Phys.* **110**, 439–465, 1987.

Wagon, S. "The Evidence: Where Are the Zeros of Zeta of *s*?" *Math. Intel.* **8**, 57–62, 1986.

Wagon, S. "The Riemann Zeta Function." §10.6 in *Mathematica in Action.* New York: W. H. Freeman, pp. 353–362, 1991.

Weisstein, E. W. "Books about Riemann Zeta Function." http://www.treasure-troves.com/books/RiemannZetaFunction.html.

Whittaker, E. T. and Watson, G. N. *A Course in Modern Analysis, 4th ed.* Cambridge, England: Cambridge University Press, 1990.

Woon, S C. Generalization of a Relation Between the Riemann Zeta Function and Bernoulli Numbers. 24 Dec 1998. http://xxx.lanl.gov/abs/math.NT/9812143/.

Zucker, I. J. "The Summation of Series of Hyperbolic Functions." *SIAM J. Math. Anal.* **10**, 192–206, 1979.

Zucker, I. J. "Some Infinite Series of Exponential and Hyperbolic Functions." *SIAM J. Math. Anal.* **15**, 406–413, 1984.

Riemann Zeta Function Zeta(2)

The value for $\zeta(2)$ can be found using a number of different techniques (Apostol 1983, Choe 1987, Giesy 1972, Holme 1970, Kimble 1987, Knopp and Schur 1918, Kortram 1996, Matsuoka 1961, Papadimitriou 1973, Simmons 1992, Stark 1969, Stark 1970, Yaglom and Yaglom 1987). The problem of finding this value analytically is sometimes known as the BASLER PROBLEM (Castellanos 1988). Yaglom and Yaglom (1987), Holme (1970), and Papadimitriou (1973) all derive the result, $\pi^2/6$ from DE MOIVRE'S IDENTITY or related identities.

One derivation for $\zeta(2)$ considers the FOURIER SERIES of $f(x) = x^{2n}$

$$f(x) = \tfrac{1}{2} a_0 + \sum_{m=1}^{\infty} a_m \cos(mx) + \sum_{m=1}^{\infty} b_m \sin(mx), \quad (1)$$

which has coefficients given by

$$a_0 = \frac{1}{\pi} \int_{-\pi}^{\pi} f(x)\,dx = \frac{2}{\pi} \int_0^{\pi} x^{2n}\,dx$$

$$= \frac{2}{\pi} \left[\frac{x^{2n+1}}{2n+1} \right]_0^{\pi} = \frac{2\pi^{2n}}{2n+1} \quad (2)$$

$$a_m = \frac{1}{\pi} \int_{-\pi}^{\pi} x^{2n} \cos(mx)\,dx$$

$$= \frac{2}{\pi} \int_0^{\pi} x^{2n} \cos(mx)\,dx \quad (3)$$

$$b_m = \frac{1}{\pi} \int_{-\pi}^{\pi} x^{2n} \sin(mx)\,dx = 0, \quad (4)$$

where the latter is true since the integrand is ODD. Therefore, the FOURIER SERIES is given explicitly by

$$x^{2n} = \frac{\pi^{2n}}{2n+1} + \sum_{m=1}^{\infty} a_m \cos(mx). \quad (5)$$

Now, a_m is given by the COSINE INTEGRAL

$$a_m = \frac{2}{\pi}(-1)^{n+1}(2n)! \left[\sin(mx) \sum_{k=0}^{n} \frac{(-1)^k}{(2k)!\,m^{2n-2k+1}} x^{2k} \right.$$

$$\left. + \cos(mx) \sum_{k=1}^{n} \frac{(-1)^{k+1}}{(2k-3)!\,m^{2n-2k+2}} x^{2k-1} \right]_0^{\pi}. \quad (6)$$

But $\cos(m\pi) = (-1)^m$, and $\sin(m\pi) = \sin 0 = 0$, so

$$a_m = \frac{2}{\pi}(-1)^{n+1}(2n)!(-1)^m \sum_{k=1}^{n} \frac{(-1)^{k+1}}{(2k-3)!\,m^{2n-2k+2}} \pi^{2k-1}$$

$$= (-1)^{m+n} 2(2n)! \sum_{k=1}^{n} \frac{(-1)^k}{(2k-3)! m^{2n-2k+2}} \pi^{2k-2}. \quad (7)$$

Now, if $n = 1$,

$$a_m = (-1)^{m+1} 2(2!) \sum_{k=1}^{1} \frac{(-1)^k}{(2k-3)! m^{4-2k}} \pi^{2k-2}$$

$$= 4(-1)^{m+1} \frac{(-1)}{(-1)! m^2} \pi^0 = \frac{4(-1)^m}{m^2}, \quad (8)$$

so the FOURIER SERIES is

$$x^2 = \frac{\pi^2}{3} + 4 \sum_{m=1}^{\infty} \frac{(-1)^m \cos(mx)}{m^2}. \quad (9)$$

Letting $m \equiv \pi$ gives $\cos(m\pi) = (-1)^m$, so

$$\pi^2 = \frac{\pi^2}{3} + 4 \sum_{m=1}^{\infty} \frac{1}{m^2}, \quad (10)$$

and we have

$$\zeta(2) = \sum_{m=1}^{\infty} \frac{1}{m^2} = \frac{\pi^2}{6}. \quad (11)$$

Higher values of n can be obtained by finding a_m and proceeding as above.

The value $\zeta(2)$ can also be found simply using the ROOT LINEAR COEFFICIENT THEOREM. Consider the equation $\sin z = 0$ and expand $\sin$ in a MACLAURIN SERIES

$$\sin z = z - \frac{z^3}{3!} + \frac{z^5}{5!} + \ldots = 0 \quad (12)$$

$$0 = 1 - \frac{z^2}{3!} + \frac{z^4}{5!} + \ldots = 1 - \frac{w}{3!} + \frac{w^2}{5!} + \ldots, \quad (13)$$

where $w \equiv z^2$. But the zeros of $\sin(z)$ occur at π, 2π, 3π, ..., so the zeros of $\sin w = \sin\sqrt{z}$ occur at π^2, $(2\pi)^2$, ... Therefore, the sum of the roots equals the COEFFICIENT of the leading term

$$\frac{1}{\pi^2} + \frac{1}{(2\pi)^2} + \frac{1}{(3\pi)^2} + \ldots = \frac{1}{3!} = \frac{1}{6}, \quad (14)$$

which can be rearranged to yield

$$\zeta(2) = \frac{\pi^2}{6}. \quad (15)$$

Yet another derivation (Simmons 1992) evaluates the integral using the integral

$$I = \int_0^1 \int_0^1 \frac{dx\, dy}{1 - xy} = \int_0^1 \int_0^1 (1 + xy + x^2 y^2 + \ldots)\, dx\, dy$$

$$= \int_0^1 [(x + \tfrac{1}{2} x^2 y + \tfrac{1}{3} x^3 y^2 + \ldots)]_0^1\, dy$$

$$= \int_0^1 (1 + \tfrac{1}{2} y + \tfrac{1}{3} y^2 + \ldots)\, dy$$

$$= \left[y + \frac{y^2}{2^2} + \frac{y^3}{3^2} + \ldots \right]_0^1 = 1 + \frac{1}{2^2} + \frac{1}{3^2} + \ldots. \quad (16)$$

To evaluate the integral, rotate the coordinate system by $\pi/4$ so

$$x = u \cos\theta - v \sin\theta = \tfrac{1}{2}\sqrt{2}(u - v) \quad (17)$$

$$y = u \sin\theta + v \cos\theta = \tfrac{1}{2}\sqrt{2}(u + v) \quad (18)$$

and

$$xy = \tfrac{1}{2}(u^2 - v^2) \quad (19)$$

$$1 - xy = \tfrac{1}{2}(2 - u^2 + v^2). \quad (20)$$

Then

$$I = 4 \int_0^{\sqrt{2}/2} \int_0^u \frac{du\, dv}{2 - u^2 + v^2}$$

$$+ 4 \int_{\sqrt{2}/2}^{\sqrt{2}} \int_0^{\sqrt{2}-u} \frac{du\, dv}{2 - u^2 + v^2}$$

$$\equiv I_1 + I_2. \quad (21)$$

Now compute the integrals I_1 and I_2.

$$I_1 = 4 \int_0^{\sqrt{2}/2} \left[\int_0^u \frac{dv}{2 - u^2 + v^2} \right] du$$

$$= 4 \int_0^{\sqrt{2}/2} \left[\frac{1}{\sqrt{2 - u^2}} \tan^{-1}\left(\frac{v}{\sqrt{2 - u^2}} \right) \right]_0^u du$$

$$= 4 \int_0^{\sqrt{2}/2} \frac{1}{\sqrt{2 - u^2}} \tan^{-1}\left(\frac{u}{\sqrt{2 - u^2}} \right) du. \quad (22)$$

Make the substitution

$$u = \sqrt{2} \sin\theta \quad (23)$$

$$\sqrt{2 - u^2} = \sqrt{2} \cos\theta \quad (24)$$

$$du = \sqrt{2} \cos\theta\, d\theta, \quad (25)$$

so

$$\tan^{-1}\left(\frac{u}{\sqrt{2 - u^2}} \right) = \tan^{-1}\left(\frac{\sqrt{2} \sin\theta}{\sqrt{2} \cos\theta} \right) = \theta \quad (26)$$

and

$$I_1 = 4 \int_0^{\pi/6} \frac{1}{\sqrt{2} \cos\theta} \theta \sqrt{2} \cos\theta\, d\theta = 2[\theta^2]_0^{\pi/6}$$

$$= \frac{\pi^2}{18}. \quad (27)$$

I_2 can also be computed analytically,

$$I_2 = 4 \int_{\sqrt{2}/2}^{\sqrt{2}} \left[\int_0^{\sqrt{2}-u} \frac{dv}{2 - u^2 + v^2} \right] du$$

$$= 4 \int_{\sqrt{2}/2}^{\sqrt{2}} \left[\frac{1}{\sqrt{2 - u^2}} \tan^{-1}\left(\frac{v}{\sqrt{2 - u^2}} \right) \right]_0^{\sqrt{2}-u} du$$

$$= 4 \int_{\sqrt{2}/2}^{\sqrt{2}} \frac{1}{\sqrt{2 - u^2}} \tan^{-1}\left(\frac{\sqrt{2} - u}{\sqrt{2 - u^2}} \right) du. \qquad (28)$$

But

$$\tan^{-1}\left(\frac{\sqrt{2} - u}{\sqrt{2 - u^2}} \right) = \tan^{-1}\left(\frac{\sqrt{2} - \sqrt{2} \sin \theta}{\sqrt{2} \cos \theta} \right)$$

$$= \tan\left(\frac{1 - \sin \theta}{\cos \theta} \right) = \tan^{-1}\left(\frac{\cos \theta}{1 + \sin \theta} \right)$$

$$= \tan^{-1}\left[\frac{\sin\left(\frac{1}{2} \pi - \theta \right)}{1 + \cos\left(\frac{1}{2} \pi - \theta \right)} \right]$$

$$= \tan^{-1}\left\{ \frac{2 \sin\left[\frac{1}{2}\left(\frac{1}{2} \pi - \theta \right) \right] \cos\left[\frac{1}{2}\left(\frac{1}{2} \pi - \theta \right) \right]}{2 \cos^2\left[\frac{1}{2}\left(\frac{1}{2} \pi - \theta \right) \right]} \right\}$$

$$= \frac{1}{2}\left(\frac{1}{2} \pi - \theta \right), \qquad (29)$$

so

$$I_2 = 4 \int_{\pi/6}^{\pi/2} \frac{1}{\sqrt{2} \cos \theta} \left(\frac{1}{4} \pi - \frac{1}{2} \theta \right) \sqrt{2} \cos \theta \, d\theta$$

$$= 4 \left[\frac{1}{4} \pi \theta - \frac{1}{4} \theta^2 \right]_{\pi/6}^{\pi/2}$$

$$= 4 \left[\left(\frac{\pi^2}{8} - \frac{\pi^2}{16} \right) - \left(\frac{\pi^2}{24} - \frac{\pi^2}{144} \right) \right] = \frac{\pi^2}{9}. \qquad (30)$$

Combining I_1 and I_2 gives

$$\zeta(2) = I_1 + I_2 = \frac{\pi^2}{18} + \frac{\pi^2}{9} = \frac{\pi^2}{6}. \qquad (31)$$

See also RIEMANN ZETA FUNCTION

References

Apostol, T. M. "A Proof That Euler Missed: Evaluating $\zeta(2)$ the Easy Way." *Math. Intel.* **5**, 59–60, 1983.

Choe, B. R. "An Elementary Proof of $\Sigma_{n=1}^{\infty} \frac{1}{n^2} = \frac{\pi^2}{6}$." *Amer. Math. Monthly* **94**, 662–663, 1987.

Giesy, D. P. "Still Another Proof That $\Sigma \, 1/k^2 = \pi^2/6$." *Math. Mag.* **45**, 148–149, 1972.

Holme, F. "Ein enkel beregning av $\Sigma_{k=1}^{\infty} \frac{1}{k^2}$." *Nordisk Mat. Tidskr.* **18**, 91–92 and 120, 1970.

Kimble, G. "Euler's Other Proof." *Math. Mag.* **60**, 282, 1987.

Knopp, K. and Schur, I. "Uuml;ber die Herleitug der Gleichung $\Sigma_{n=1}^{\infty} \frac{1}{n^2} = \frac{\pi^2}{6}$." *Archiv der Mathematik u. Physik* **27**, 174–176, 1918.

Kortram, R. A. "Simple Proofs for $\Sigma_{k=1}^{\infty} \frac{1}{k^2} = \frac{\pi^2}{6}$ and $\sin x = x \prod_{k=1}^{\infty}(1 - \frac{x^2}{k^2 \pi^2})$." *Math. Mag.* **69**, 122–125, 1996.

Matsuoka, Y. "An Elementary Proof of the Formula $\Sigma_{k=1}^{\infty} \frac{1}{k^2} = \frac{\pi^2}{6}$." *Amer. Math. Monthly* **68**, 486–487, 1961.

Papadimitriou, I. "A Simple Proof of the Formula $\Sigma_{k=1}^{\infty} \frac{1}{k^2} = \frac{\pi^2}{6}$." *Amer. Math. Monthly* **80**, 424–425, 1973.

Simmons, G. F. "Euler's Formula $\Sigma_1^{\infty} \, 1/n^2 = \pi^2/6$ by Double Integration." Ch. B. 24 in *Calculus Gems: Brief Lives and Memorable Mathematics.* New York: McGraw-Hill, 1992.

Stark, E. L. "Another Proof of the Formula $\Sigma_{k=1}^{\infty} \frac{1}{k^2} = \frac{\pi^2}{6}$." *Amer. Math. Monthly* **76**, 552–553, 1969.

Stark, E. L. " $1 - \frac{1}{4} + \frac{1}{9} - \frac{1}{16} + \ldots = \frac{\pi^2}{12}$." *Praxis Math.* **12**, 1–3, 1970.

Stark, E. L. "The Series $\Sigma_{k=1}^{\infty} \, k^{-s} \, s = 2, 3, 4, \ldots$, Once More." *Math. Mag.* **47**, 197–202, 1974.

Wells, D. *The Penguin Dictionary of Curious and Interesting Numbers.* Middlesex, England: Penguin Books, p. 40, 1986.

Yaglom, A. M. and Yaglom, I. M. Problem 145 in *Challenging Mathematical Problems with Elementary Solutions, Vol. 2.* New York: Dover, 1987.

Riemann-Christoffel Tensor

RIEMANN TENSOR

Riemann-Finsler Geometry

References

Bao, D.; Chern, S.-S.; and Shen, Z. *An Introduction to Riemann-Finsler Geometry.* New York: Springer-Verlag, 2000.

Riemannian Geometry

The study of MANIFOLDS having a complete RIEMANNIAN METRIC. Riemannian geometry is a general space based on the LINE ELEMENT

$$ds = F\left(x^1, \ldots, x^n; \, dx^1, \ldots, dx^n\right),$$

with $F(x, y) > 0$ for $y \neq 0$ a function on the TANGENT BUNDLE TM. In addition, F is homogeneous of degree 1 in y and OF THE FORM

$$F^2 = g_{ij}(x) \, dx^i \, dx^j$$

(Chern 1996). If this restriction is dropped, the resulting geometry is called FINSLER GEOMETRY.

See also NON-EUCLIDEAN GEOMETRY

References

Besson, G.; Lohkamp, J.; Pansu, P.; and Petersen, P. *Riemannian Geometry.* Providence, RI: Amer. Math. Soc., 1996.

Buser, P. *Geometry and Spectra of Compact Riemann Surfaces.* Boston, MA: Birkhäuser, 1992.

Chavel, I. *Eigenvalues in Riemannian Geometry.* New York: Academic Press, 1984.

Chavel, I. *Riemannian Geometry: A Modern Introduction.* New York: Cambridge University Press, 1994.

Chern, S.-S. "Finsler Geometry is Just Riemannian Geometry without the Quadratic Restriction." *Not. Amer. Math. Soc.* **43**, 959–963, 1996.
do Carmo, M. P. *Riemannian Geometry.* Boston, MA: Birkhäuser, 1992.

Riemannian Geometry (Non-Euclidean)
ELLIPTIC GEOMETRY

Riemannian Manifold
A MANIFOLD possessing a METRIC TENSOR. For a complete Riemannian manifold, the METRIC $d(x, y)$ is defined as the length of the shortest curve (GEODESIC) between x and y.

See also BISHOP'S INEQUALITY, CAMPBELL'S THEOREM, CHEEGER'S FINITENESS THEOREM, PSEUDO-RIEMANNIAN MANIFOLD

Riemannian Metric
Suppose for every point x in a COMPACT MANIFOLD M, an INNER PRODUCT $\langle \, , \, \rangle_x$ is defined on a TANGENT SPACE $T_x M$ of M at x. Then the collection of all these INNER PRODUCTS is called the Riemannian metric. In 1870, Christoffel and Lipschitz showed how to decide when two Riemannian metrics differ by only a coordinate transformation.

See also COMPACT MANIFOLD, LINE ELEMENT, METRIC TENSOR

Riemannian Submersion

See also SUBMERSION

Riemann-Lebesgue Lemma
Sometimes also called MERCER'S THEOREM.

$$\lim_{n \to \infty} \int_a^b K(\lambda, z) C \sin(nz)\, dz = 0$$

for arbitrarily large C and "nice" $K(\lambda, z)$. Gradshteyn and Ryzhik (2000) state the lemma as follows. If $f(x)$ is integrable on $[-\pi, \pi]$, then

$$\lim_{t \to \infty} \int_{-\pi}^{\pi} f(x) \sin(tx)\, dx \to 0$$

and

$$\lim_{t \to \infty} \int_{-\pi}^{\pi} f(x) \cos(tx)\, dx \to 0.$$

References
Gradshteyn, I. S. and Ryzhik, I. M. *Tables of Integrals, Series, and Products,* 6th ed. San Diego, CA: Academic Press, p. 1101, 2000.

Riemann-Roch Theorem
The dimension of a complete series is equal to the sum of the order and index of specialization of any group, less the GENUS of the base curve

$$r = N + i + p.$$

References
Coolidge, J. L. *A Treatise on Algebraic Plane Curves.* New York: Dover, p. 261, 1959.
Koch, H. "The Riemann-Roch Theorem." §5.6 in *Number Theory: Algebraic Numbers and Functions.* Providence, RI: Amer. Math. Soc., pp. 160–164, 2000.
Riemann, B. Grundlagen für eine allgemeine Theorie der Funktionen einer veränlichen komplexen Grösse. Ph.D. dissertation. Göttingen, Germany: University of Göttingen, 1851.

Riemann's Integral Theorem
Associated with an irreducible curve of GENUS (CURVE) p, there are p LINEARLY INDEPENDENT integrals of the first sort. The ROOTS of the integrands are groups of the canonical series, and every such group will give rise to exactly one integral of the first sort.

References
Coolidge, J. L. *A Treatise on Algebraic Plane Curves.* New York: Dover, p. 274, 1959.

Riemann's Moduli Problem
Find an ANALYTIC parameterization of the compact RIEMANN SURFACES in a fixed HOMOMORPHISM class. The AHLFORS-BERS THEOREM proved that RIEMANN'S MODULI SPACE gives the solution.

See also AHLFORS-BERS THEOREM, RIEMANN'S MODULI SPACE

Riemann's Moduli Space
Riemann's moduli space R_p is the space of ANALYTIC EQUIVALENCE CLASSES of RIEMANN SURFACES of fixed GENUS p.

See also AHLFORS-BERS THEOREM, RIEMANN'S MODULI PROBLEM, RIEMANN SURFACE

Riemann-Siegel Functions

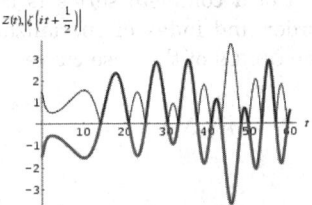

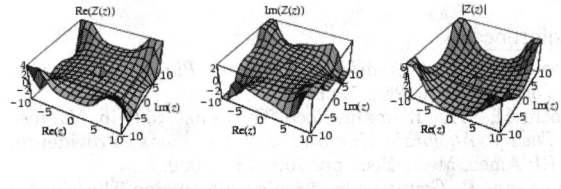

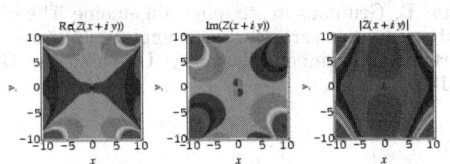

For a REAL POSITIVE t, the Riemann-Siegel Z function is defined by

$$Z(t) \equiv e^{i\theta(t)} \zeta(\tfrac{1}{2} + it).$$

This function is sometimes also called the Hardy function or Hardy Z-function (Karatsuba and Voronin 1992, Borwein *et al.* 1999). The top plot superposes $Z(t)$ (thick line) on $\left|\zeta\left(\tfrac{1}{2} + it\right)\right|$, where $\zeta(z)$ is the RIEMANN ZETA FUNCTION. It has an ASYMPTOTIC SERIES given "approximately" by

$$Z(t) \sim 2 \sum_{k=1}^{\nu(t)} \frac{1}{\sqrt{k}} \cos[\vartheta(t) - t \ln k] + R(t), \quad (1)$$

where

$$\nu(t) = \left\lfloor \sqrt{\frac{t}{2\pi}} \right\rfloor \quad (2)$$

$$R(t) = (-1)^{\nu(t)-1} \left(\frac{t}{2\pi}\right)^{-1/4}$$

$$\times \sum_{k=0}^{\infty} c_k \left(\sqrt{\frac{t}{2\pi}} - \nu(t)\right) \left(\frac{t}{2\pi}\right)^{-k/2} \quad (3)$$

$$c_k(p) = \left[\omega^k\right] \left\{ \exp\left[i\left(\ln\left(\frac{t}{2\pi}\right) - \tfrac{1}{2} t - \tfrac{1}{8} \pi - \vartheta(t)\right)\right]\right.$$

$$\left. \times \left[y^0\right] \left[\left(\sum_{j=0}^{\infty} A_j(y)\omega^j\right) \left(\sum_{j=0}^{\infty} \frac{\psi^{(j)}(p)}{j!} y^j\right)\right]\right\} \quad (4)$$

$$A_0(y) = e^{2\pi i y^2} \quad (5)$$

$$A_j(y) = -\tfrac{1}{2} y A_{j-1}(y) - \frac{1}{32\pi^2} \frac{\partial^2}{\partial y^2} \frac{A_{j-1}(y)}{y} \quad (6)$$

$$\psi(p) = \frac{\cos[2\pi(p^2 - p - \tfrac{1}{16})]}{\cos(2\pi p)} \quad (7)$$

$\lfloor x \rfloor$ is the FLOOR FUNCTION (Edwards 1974), and $[y^k]$ is COEFFICIENT NOTATION. The first few terms $c_k(p)$ are given by

$$c_0(p) = \psi(p) \quad (8)$$

$$c_1(p) = -\frac{\psi^{(3)}(p)}{96\pi^2} \quad (9)$$

$$c_2(p) = \frac{\psi''(p)}{64\pi^2} + \frac{\psi^{(6)}(p)}{18432\pi^4} \quad (10)$$

$$c_3(p) = \frac{\psi'(p)}{64\pi^2} + \frac{\psi^{(5)}(p)}{3840\pi^4} - \frac{\psi^{(9)}(p)}{5308416\pi^6} \quad (11)$$

$$c_4(p) = \frac{\psi(p)}{128\pi^2} + \frac{19\psi^{(4)}(p)}{24576\pi^4} + \frac{11\psi^{(8)}(p)}{5898240\pi^6}$$

$$+ \frac{\psi^{(12)}(p)}{2038431744\pi^8} \quad (12)$$

$$c_5(p) = -\frac{5\psi^{(3)}(p)}{3072\pi^4} - \frac{901\psi^{(7)}(p)}{82575360\pi^6}$$

$$- \frac{7\psi^{(11)}(p)}{849346560\pi^8} - \frac{\psi^{(15)}(p)}{978447237120\pi^{10}}. \quad (13)$$

The numerators and denominators are 1, -1, 1, 1, -1, -1, -1, 1, 19, 11, 1, -5, -901, ... (Sloane's A050276) and 1, 96, 64, 18432, 64, 3840, 5308416, 128, ... (Sloane's A050277), respectively.

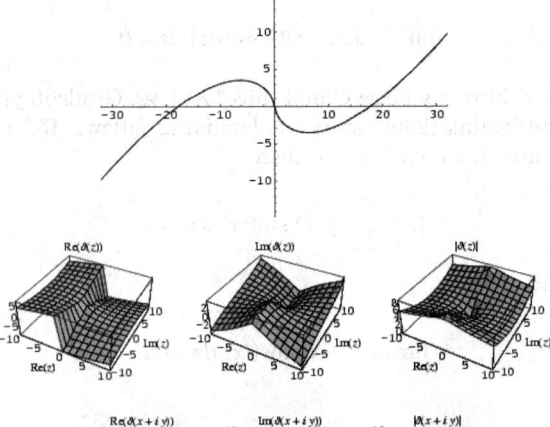

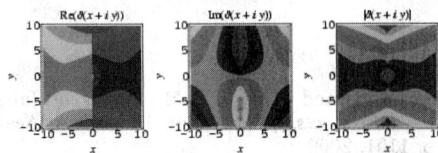

The Riemann-Siegel theta function appearing above is defined by

$$\vartheta(t) \equiv \Im\left[\ln \Gamma\left(\tfrac{1}{4} + \tfrac{1}{2} it\right) - \tfrac{1}{2} t \ln \pi\right]$$

$$= \arg\left[\Gamma\left(\tfrac{1}{4} + \tfrac{1}{2} it\right)\right] - \tfrac{1}{2} t \ln \pi.$$

These functions are implemented in *Mathematica* as `RiemannSiegelZ[z]` and `RiemannSiegelTheta[z]`, illustrated above.

See also RIEMANN ZETA FUNCTION, XI FUNCTION

References

Berry, M. V. "The Riemann-Siegel Expansion for the Zeta Function: High Orders and Remainders." *Proc. Roy. Soc. London A* **450**, 439–462, 1995.

Borwein, J. M.; Bradley, D. M.; and Crandall, R. E. "Computational Strategies for the Riemann Zeta Function." CECM-98:118, 23 Jun 1999. http://www.cecm.sfu.ca/preprints/1999pp.html#98:118.

Brent, R. P. "On the Zeros of the Riemann Zeta Function in the Critical Strip." *Math. Comput.* **33**, 1361–1372, 1979.

Edwards, H. M. *Riemann's Zeta Function.* New York: Academic Press, 1974.

Karatsuba, A. A. and Voronin, S. M. *The Riemann Zeta-Function.* Hawthorn, NY: de Gruyter, 1992.

Odlyzko, A. M. "The 10^{20}th Zero of the Riemann Zeta Function and 70 Million of Its Neighbors." Preprint.

Sloane, N. J. A. Sequences A050276 and A050277 in "An On-Line Version of the Encyclopedia of Integer Sequences." http://www.research.att.com/~njas/sequences/eisonline.html.

Titchmarsh, E. C. *The Theory of the Riemann Zeta Function,* 2nd ed. New York: Clarendon Press, 1987.

van de Lune, J.; te Riele, H. J. J.; and Winter, D. T. "On the Zeros of the Riemann Zeta Function in the Critical Strip. IV." *Math. Comput.* **46**, 667–681, 1986.

Vardi, I. *Computational Recreations in Mathematica.* Reading, MA: Addison-Wesley, p. 143, 1991.

RiemannSiegelTheta

RIEMANN-SIEGEL FUNCTIONS

RiemannSiegelZ

RIEMANN-SIEGEL FUNCTIONS

Riemann-Stieltjes Integral

STIELTJES INTEGRAL

Riemann-Volterra Method

RIEMANN METHOD

Riesel Number

There exist infinitely many ODD INTEGERS k such that $k \cdot 2^n - 1$ is COMPOSITE for every $n \geq 1$. Numbers k with this property are called RIESEL NUMBERS, and analogous numbers with the minus sign replaced by a plus are called SIERPINSKI NUMBERS OF THE SECOND KIND. The smallest known Riesel number is $k = 509, 203$, but there remain 963 smaller candidates (the smallest of which is 659) which generate only composite numbers for all n which have been checked (Ribenboim 1996, p. 358).

Let $a(k)$ be smallest n for which $(2k - 1) \cdot 2^n - 1$ is PRIME, then the first few values are 2, 0, 2, 1, 1, 2, 3, 1, 2, 1, 1, 4, 3, 1, 4, 1, 2, 2, 1, 3, 2, 7, ... (Sloane's A046069), and second smallest n are 3, 1, 4, 5, 3, 26, 7, 2, 4, 3, 2, 6, 9, 2, 16, 5, 3, 6, 2553, ... (Sloane's A046070).

See also CUNNINGHAM NUMBER, MERSENNE NUMBER, SIERPINSKI'S COMPOSITE NUMBER THEOREM, SIERPINSKI NUMBER OF THE SECOND KIND, THÂBIT IBN KURRAH RULE

References

Ribenboim, P. *The New Book of Prime Number Records.* New York: Springer-Verlag, p. 357, 1996.

Riesel, H. "Några stora primtal." *Elementa* **39**, 258–260, 1956.

Riesel, H. *Prime Numbers and Computer Methods for Factorization, 2nd ed.* Basel: Birkhäuser, pp. 394–398, 1994.

Sloane, N. J. A. Sequences A046067, A046068, A046069, and A046070 in "An On-Line Version of the Encyclopedia of Integer Sequences." http://www.research.att.com/~njas/sequences/eisonline.html.

Riesz Representation Theorem

There are a couple of versions of this theorem. Basically, it says that any bounded linear FUNCTIONAL T on the space of compactly supported continuous functions on X is the same as integration against a measure μ,

$$Tf = \int f \, d\mu.$$

Here, the integral is the LEBESGUE INTEGRAL.

Because linear functionals form a VECTOR SPACE, and are not "positive," the measure μ may not be a POSITIVE MEASURE. But if the functional T is positive, in the sense that $f \geq 0$ implies that $Tf \geq 0$, then the measure μ is also positive. In the generality of complex linear functionals, the measure μ is a COMPLEX MEASURE. The measure μ is uniquely determined by T and has the properties of a regular BOREL MEASURE. It must be a finite measure, which corresponds to the boundedness condition on the functional. In fact, the NORM of T, $\|T\|$, is the TOTAL VARIATION MEASURE of X, $|\mu|(X)$.

Naturally, there are some hypotheses necessary for this to make sense. The space X has to be LOCALLY COMPACT and HAUSDORFF, which is not a strong restriction. In fact, for unbounded spaces X, the theorem also applies to functionals on continuous functions which vanish at infinity, in the sense that for any $\epsilon > 0$, there is a compact set K such that for any x not in K, $|f(x)| < \epsilon$ (which is the notion from calculus of $\lim_{x \to \infty} f(x) = 0$).

The Riesz representation theorem is useful in describing the DUAL SPACE to any space which contains the compactly supported continuous functions as a DENSE subspace. Roughly speaking, a linear functional is modified, usually by convolving with a bump function, to a bounded linear functional on the compactly supported continuous functions. Then it can be realized as integration against a measure. Often the measure must be ABSOLUTELY CONTINUOUS, and so the dual is integration against a function.

See also ABSOLUTELY CONTINUOUS, COMPLEX MEASURE, DUAL SPACE, FUNCTIONAL, HILBERT SPACE, LEBESGUE MEASURE, MEASURE SPACE, POLAR REPRESENTATION (MEASURE), RADON-NIKODYM THEOREM, SINGULAR MEASURE

References

Debnath, L. and Mikusinski, P. *Introduction to Hilbert Spaces with Applications.* San Diego, CA: Academic Press, 1990.
Rudin, W. *Real and Complex Analysis.* New York: McGraw-Hill, pp. 40–47 and 129–132, 1987.

Riesz-Fischer Theorem

A function is L_2- (square-) integrable IFF its FOURIER SERIES is L_2-convergent. The application of this theorem requires use of the LEBESGUE INTEGRAL.

See also LEBESGUE INTEGRAL

Riesz's Theorem

Every continuous linear functional $U[f]$ for $f \in C[a, b]$ can be expressed as a STIELTJES INTEGRAL

$$U[f] = \int_a^b f(x) \, dw(x),$$

where $w(x)$ is determined by U and is of bounded variation on $[a, b]$.

See also STIELTJES INTEGRAL

References

Kestelman, H. "Riesz's Theorem." §11.5 in *Modern Theories of Integration, 2nd rev. ed.* New York: Dover, pp. 265–269, 1960.

Riffle Shuffle

A SHUFFLE, also called a FARO SHUFFLE, in which a deck of $2n$ cards is divided into two HALVES which are then alternatively interleaved from the left and right hands (an "in-shuffle") or from the right and left hands (an "out-shuffle"). Using an "in-shuffle," a deck originally arranged as 1 2 3 4 5 6 7 8 would become 5 1 6 2 7 3 8 4. Using an "out-shuffle," the deck order would become 1 5 2 6 3 7 4 8. Riffle shuffles are used in card tricks (Marlo 1958ab, Adler 1973), and also in the theory of parallel processing (Stone 1971, Chen *et al.* 1981).

In general, card k moves to the position originally occupied by the $2k$th card (mod $2n + 1$). Therefore, in-shuffling $2n$ cards $2n$ times (where $2n + 1$ is PRIME) results in the original card order. Similarly, out-shuffling $2n$ cards $2n - 2$ times (where $2n - 1$ is PRIME) results in the original order (Diaconis *et al.* 1983, Conway and Guy 1996). Amazingly, this means that an ordinary deck of 52 cards is returned to its original order after 8 out-shuffles.

Morris (1994) further discusses aspects of the perfect riffle shuffle (in which the deck is cut exactly in half and cards are perfectly interlaced). Ramnath and Scully (1996) give an algorithm for the shortest sequence of in- and out-shuffles to move a card from arbitrary position i to position j. This algorithm works for any deck with an EVEN number of cards and is $\mathcal{O}(\log n)$.

See also CARDS, SHUFFLE

References

Adler, I. "Make Up Your Own Card Tricks." *J. Recr. Math.* **6**, 87–91, 1973.
Ball, W. W. R. and Coxeter, H. S. M. *Mathematical Recreations and Essays, 13th ed.* New York: Dover, pp. 323–325, 1987.
Chen, P. Y.; Lawrie, D. H.; Yew, P.-C.; and Padua, D. A. "Interconnection Networks Using Shuffles." *Computer* **33**, 55–64, Dec. 1981.
Conway, J. H. and Guy, R. K. "Fractions Cycle into Decimals." In *The Book of Numbers.* New York: Springer-Verlag, pp. 163–165, 1996.
Diaconis, P.; Graham, R. L.; and Kantor, W. M. "The Mathematics of Perfect Shuffles." *Adv. Appl. Math.* **4**, 175–196, 1983.
Gardner, M. *Mathematical Carnival: A New Round-Up of Tantalizers and Puzzles from Scientific American.* Washington, DC: Math. Assoc. Amer., 1989.
Herstein, I. N. and Kaplansky, I. *Matters Mathematical.* New York: Harper & Row, 1974.
Mann, B. "How Many Times Should You Shuffle a Deck of Cards." *UMAP J.* **15**, 303–332, 1994.
Marlo, E. *Faro Notes.* Chicago, IL: Ireland Magic Co., 1958a.
Marlo, E. *Faro Shuffle.* Chicago, IL: Ireland Magic Co., 1958b.
Medvedoff, S. and Morrison, K. "Groups of Perfect Shuffles." *Math. Mag.* **60**, 3–14, 1987.
Morris, S. B. and Hartwig, R. E. "The Generalized Faro Shuffle." *Discrete Math.* **15**, 333–346, 1976.
Peterson, I. *Islands of Truth: A Mathematical Mystery Cruise.* New York: W. H. Freeman, pp. 240–244, 1990.
Ramnath, S. and Scully, D. "Moving Card i to Position j with Perfect Shuffles." *Math. Mag.* **69**, 361–365, 1996.
Stone, H. S. "Parallel Processing with the Perfect Shuffle." *IEEE Trans. Comput.* **2**, 153–161, 1971.

Rigby Points

The PERSPECTIVE CENTERS of the TANGENTIAL and CONTACT TRIANGLES of the inner and outer SODDY POINTS. The inner Ri and outer Ri' Rigby points are given by

$$Ri = I + \tfrac{4}{3} Ge$$

$$Ri' = I - \tfrac{4}{3}Ge,$$

where I is the INCENTER and Ge is the GERGONNE POINT.

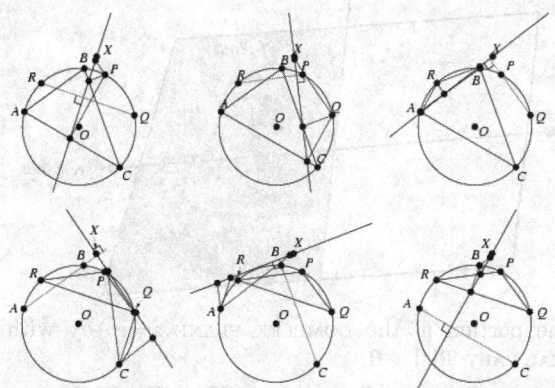

Honsberger (1995) defines a different point which he calls the "Rigby point" X. Let QR be an arbitrary CHORD of the CIRCUMCIRCLE of a given TRIANGLE $\triangle ABC$, and let P be the POLE of the SIMSON LINE S_P with respect to $\triangle ABC$ which is PERPENDICULAR to QR. Then it also turns out that $S_Q \perp PR$ and $S_R \perp PQ$. In addition, $S_A \perp BC$, $S_B \perp AC$, and $S_C \perp AB$ with respect to $\triangle PQR$.

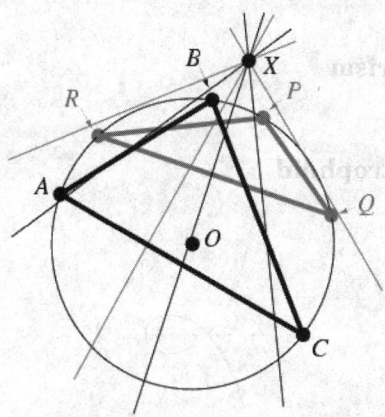

As a result of these remarkable facts, it can be shown that the SIMSON LINES S_P, S_Q, and S_R with respect to $\triangle ABC$ meet in the Rigby point X. Moreover, the SIMSON LINES S_A, S_B, and S_C with respect to $\triangle PQR$ also meet in X, and X is the ORTHOPOLE of AB, BC, and AC with respect to $\triangle PQR$, and of PQ, QR, and PR with respect to $\triangle ABC$. Finally, X is the MIDPOINT of the ORTHOCENTERS of $\triangle ABC$ and $\triangle PQR$ (Honsberger 1996, p. 136).

See also CONTACT TRIANGLE, GERGONNE POINT, GRIFFITHS POINTS, INCENTER, OLDKNOW POINTS, ORTHOPOLE, SIMSON LINE, SODDY POINTS, TANGENTIAL TRIANGLE

References

Honsberger, R. "The Rigby Point." §11.3 in *Episodes in Nineteenth and Twentieth Century Euclidean Geometry.* Washington, DC: Math. Assoc. Amer., pp. 132–136, 1995.

Oldknow, A. "The Euler-Gergonne-Soddy Triangle of a Triangle." *Amer. Math. Monthly* **103**, 319–329, 1996.

Right Angle

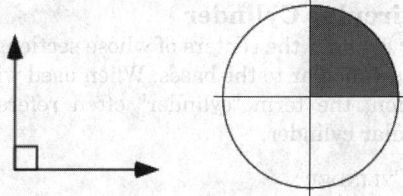

An ANGLE equal to half the ANGLE from one end of a line segment to the other. A right angle is $\pi/2$ radians or $90°$. A TRIANGLE containing a right angle is called a RIGHT TRIANGLE. However, a TRIANGLE cannot contain more than one right angle, since the sum of the two right angles plus the third angle would exceed the $180°$ total possessed by a TRIANGLE.

The patterns of cracks observed in mud which has been dried by the sun form curves which intersect in right angles (Williams 1979, p. 45; Steinhaus 1983, p. 88; Pearce 1990, p. 12).

See also ACUTE ANGLE, FULL ANGLE, OBLIQUE ANGLE, OBTUSE ANGLE, ORTHOGONAL LINES, PERPENDICULAR, RIGHT TRIANGLE, SEMICIRCLE, STRAIGHT ANGLE, THALES' THEOREM

References

Pearce, P. *Structure in Nature Is a Strategy for Design.* Cambridge, MA: MIT Press, 1990.
Steinhaus, H. *Mathematical Snapshots, 3rd ed.* New York: Dover, 1999.
Williams, R. *The Geometrical Foundation of Natural Structure: A Source Book of Design.* New York: Dover, 1979.

Right Circular Cone

A circular cone the centers of whose sections form a line perpendicular to the bases. When used without qualification, the term "cone" often refers to a right circular cone.

See also CONE

References

Kern, W. F. and Bland, J. R. "Right Circular Cone." §25 in *Solid Mensuration with Proofs, 2nd ed.* New York: Wiley, pp. 60–64, 1948.

Right Circular Cylinder

A circular cylinder the centers of whose sections form a line perpendicular to the bases. When used without qualification, the term "cylinder" often refers to a right circular cylinder.

See also CYLINDER

References

Kern, W. F. and Bland, J. R. "Right Circular Cylinder." §17 in *Solid Mensuration with Proofs, 2nd ed.* New York: Wiley, pp. 39–42, 1948.

Right Cone

CONE

Right Conoid

A RULED SURFACE is called a right conoid if it can be generated by moving a straight LINE intersecting a fixed straight LINE such that the LINES are always PERPENDICULAR (Kreyszig 1991, p. 87). Taking the PERPENDICULAR plane as the xy-plane and the line to be the x-AXIS gives the PARAMETRIC EQUATIONS

$$x(u, v) = v \cos \vartheta(u)$$

$$y(u, v) = v \sin \vartheta(u)$$

$$z(u, v) = h(u)$$

(Gray 1997). Taking $h(u) = 2u$ and $\vartheta(u) = u$ gives the HELICOID.

See also HELICOID, PLÜCKER'S CONOID, WALLIS'S CONICAL EDGE

References

Dixon, R. *Mathographics.* New York: Dover, p. 20, 1991.
Gray, A. *Modern Differential Geometry of Curves and Surfaces with Mathematica, 2nd ed.* Boca Raton, FL: CRC Press, pp. 450–452, 1997.
Kreyszig, E. *Differential Geometry.* New York: Dover, 1991.

Right Coset

Consider a countable SUBGROUP H with ELEMENTS h_i and an element x not in H, then $h_i x$ for $i = 1, 2, \ldots$ are the right cosets of the SUBGROUP H with respect to x.

See also COSET, LEFT COSET

Right Cylinder

CYLINDER

Right Half-Plane

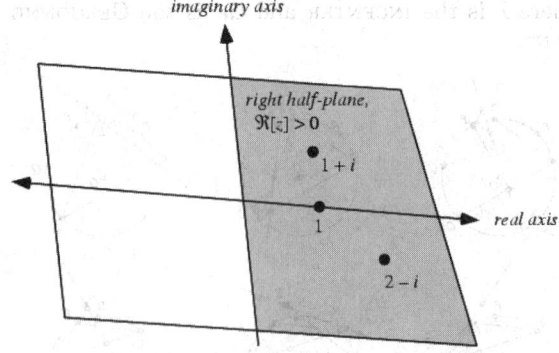

The portion of the COMPLEX PLANE $z = x + iy$ with REAL PART $\Re[z] > 0$.

See also COMPLEX PLANE, LEFT HALF-PLANE, LOWER HALF-PLANE, UPPER HALF-PLANE

Right Hyperbola

RECTANGULAR HYPERBOLA

Right Line

LINE

Right Prism

PRISM

Right Strophoid

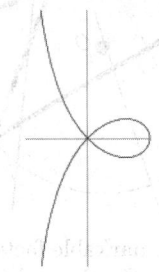

The STROPHOID of a line L with pole O not on L and fixed point O' being the point where the PERPENDICULAR from O to L cuts L is called a right strophoid. It is therefore a general STROPHOID with $a = \pi/2$. The right strophoid is given by the Cartesian equation

$$y^2 = \frac{c - x}{c + x} x^2, \tag{1}$$

or the polar equation

$$r = c \cos(2\theta) \sec \theta. \tag{2}$$

The parametric form of the strophoid is

$$x(t) = \frac{1 - t^2}{t^2 + 1} \tag{3}$$

$$y(t) = \frac{t(t^2 - 1)}{t^2 + 1}. \tag{4}$$

The right strophoid has CURVATURE

$$\kappa(t) = -\frac{4(1 + 3t^2)}{(1 + 6t^2 + t^4)^{3/2}} \tag{5}$$

and TANGENTIAL ANGLE

$$\phi(t) = -2\tan^{-1} t - \tan^{-1}\left(\frac{2t}{1 + t^2}\right). \tag{6}$$

The right strophoid first appears in work by Isaac Barrow in 1670, although Torricelli describes the curve in his letters around 1645 and Roberval found it as the LOCUS of the focus of the conic obtained when the plane cutting the CONE rotates about the tangent at its vertex (MacTutor Archive). The AREA of the loop is

$$A_{\text{loop}} = \tfrac{1}{2}c^2(4 - \pi) \tag{7}$$

(MacTutor Archive).

Let C be the CIRCLE with center at the point where the right strophoid crosses the x-AXIS and radius the distance of that point from the origin. Then the right strophoid is invariant under inversion in the CIRCLE C and is therefore an ANALLAGMATIC CURVE.

See also STROPHOID, TRISECTRIX

References

Gray, A. *Modern Differential Geometry of Curves and Surfaces with Mathematica, 2nd ed.* Boca Raton, FL: CRC Press, p. 92, 1997.

Lawrence, J. D. *A Catalog of Special Plane Curves.* New York: Dover, pp. 100–104, 1972.

Lockwood, E. H. "The Right Strophoid." Ch. 10 in *A Book of Curves.* Cambridge, England: Cambridge University Press, pp. 90–97, 1967.

MacTutor History of Mathematics Archive. "Right Strophoid." http://www-groups.dcs.st-and.ac.uk/~history/Curves/Right.html.

Right Strophoid Inverse Curve

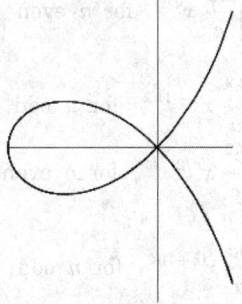

The INVERSE CURVE of a right strophoid is the same strophoid.

Right Triangle

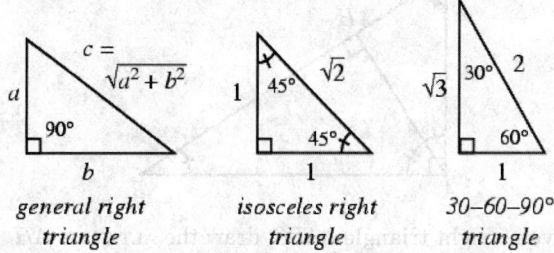

| general right triangle | isosceles right triangle | 30–60–90° triangle |

A TRIANGLE with an ANGLE of 90° ($\pi/2$ radians). The sides a, b, and c of such a TRIANGLE satisfy the PYTHAGOREAN THEOREM. The largest side is conventionally denoted c and is called the HYPOTENUSE. A TRIANGLE that is not a right triangle is sometimes called an OBLIQUE TRIANGLE.

For any three similar shapes on the sides of a right triangle,

$$A_1 + A_2 = A_3, \tag{1}$$

which is equivalent to the PYTHAGOREAN THEOREM.

For a right triangle with sides a, b, and HYPOTENUSE c, let r be the INRADIUS. Then

$$\tfrac{1}{2}ab = \tfrac{1}{2}ra + \tfrac{1}{2}rb + \tfrac{1}{2}rc = \tfrac{1}{2}r(a + b + c). \tag{2}$$

Solving for r gives

$$r = \frac{ab}{a + b + c}. \tag{3}$$

This can also be written in the equivalent forms

$$r = \sqrt{\tfrac{1}{2}(c - a)(c - b)} \tag{4}$$

$$= \tfrac{1}{2}(a + b - c). \tag{5}$$

Now, since any PYTHAGOREAN TRIPLE can be written

$$a = m^2 - n^2 \tag{6}$$

$$b = 2mn \tag{7}$$

$$c = m^2 + n^2, \tag{8}$$

(3) becomes

$$r = \frac{(m^2 - n^2)2mn}{m^2 - n^2 + 2mn + m^2 + n^2} = n(m - n), \tag{9}$$

which is an INTEGER when m and n are integers (Ogilvy and Anderson 1988, p. 68).

The HYPOTENUSE of a right triangle is a DIAMETER of the triangle's CIRCUMCIRCLE, so the CIRCUMRADIUS is given by

$$R = \tfrac{1}{2}c, \tag{10}$$

where c is the HYPOTENUSE.

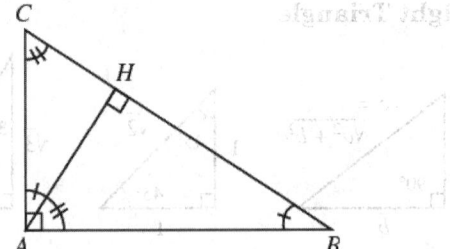

Given a right triangle ΔABC, draw the ALTITUDE AH from the RIGHT ANGLE A. Then the triangles ΔAHC and ΔBHA are similar.

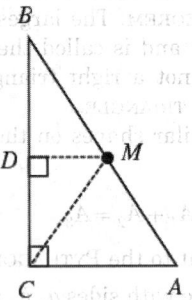

In a right triangle, the MIDPOINT of the HYPOTENUSE is equidistant from the three VERTICES (Dunham 1990). This can be proved as follows. Given ΔABC, let M be the MIDPOINT of AB (so that $AM = BM$). Draw $DM\|CA$, then since ΔBDM is similar to ΔBCA, it follows that $BD = DC$. Since both ΔBDM and ΔCDM are right triangles and the corresponding legs are equal, the HYPOTENUSES are also equal, so we have $AM = BM = CM$ and the theorem is proved.

Fermat showed how to construct an arbitrary number of equiareal nonprimitive right triangles. An analysis of PYTHAGOREAN TRIPLES demonstrates that the right triangle generated by a triple $(m_i^2 - n_i^2,\ 2m_in_i,\ m_i^2 + n_i^2)$ has common AREA

$$A = rs(2r+s)(r+2s)(r+s)(r-s)(r^2+rs+s^2)$$

(Beiler 1966, pp. 126–127). The only EXTREMUM of this function occurs at $(r, s) = (0, 0)$. Since $A(r, s) = 0$ for $r = s$, the smallest AREA shared by three *nonprimitive* right triangles is given by $(r, s) = (1, 2)$, which results in an area of 840 and corresponds to the triplets (24, 70, 74), (40, 42, 58), and (15, 112, 113) (Beiler 1966, p. 126). One can also find quartets of right triangles with the same AREA. The QUARTET having smallest known area is (111, 6160, 6161), (231, 2960, 2969), (518, 1320, 1418), (280, 2442, 2458), with AREA 341,880 (Beiler 1966, p. 127). Guy (1994) gives additional information.

The smallest known AREA shared by three *primitive* right triangles is 13123110, corresponding to the triples (4485, 5852, 7373), (1380, 19019, 19069), and (3059, 8580, 9109) (Beiler 1966, p. 127; Gardner 1984, p. 160).

It is also possible to find sets of three and four Pythagorean triplets having the same PERIMETER (Beiler 1966, pp. 131–132). Lehmer (1900) showed that the number of primitive triples $N(p)$ with PERIMETER less than p is

$$\lim_{p\to\infty} N(p) = \frac{p\ln 2}{\pi^2} = 0.070230\ldots \tag{11}$$

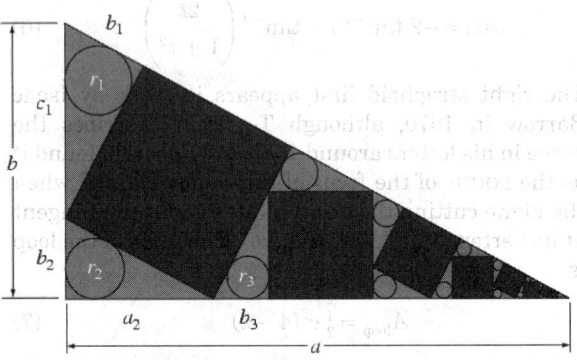

In a given right triangle, an infinite sequence of squares can be nested which alternately lie on the HYPOTENUSE and longest leg. These create a sequence of increasingly smaller similar right triangles. Let the original triangle have legs of lengths a and b and HYPOTENUSE of length $c = \sqrt{a^2 + b^2}$. Also define

$$x \equiv \frac{ac}{ab + c^2} \tag{12}$$

$$y \equiv \tfrac{1}{2}\sqrt{2[c^2 - (a+b)c + ab]}. \tag{13}$$

Then the sides of the n square are of length

$$s_n = bx^n. \tag{14}$$

Number the upper left triangle as 1, and then the remainder by following the "strip" of triangles at adjoining vertices. Then the side lengths of these triangles are

$$a_n = \begin{cases} s_{(n+1)/2} & \text{for } n \text{ odd} \\ \dfrac{ab}{c}\,x^{n/2} & \text{for } n \text{ even} \end{cases} \tag{15}$$

$$b_n = \begin{cases} \dfrac{b^2}{a}\,x^{(n+1)/2} & \text{for } n \text{ odd} \\ \dfrac{b^2}{c}\,x^{n/2} & \text{for } n \text{ even} \end{cases} \tag{16}$$

$$c_n = \begin{cases} \dfrac{bc}{a}\,x^{(n+1)/2} & \text{for } n \text{ odd} \\ s_{n/2} & \text{for } n \text{ even.} \end{cases} \tag{17}$$

The INRADII of the corresponding circles can be found from

$$r_n = \frac{1}{2}\sqrt{\frac{(b_n + c_n - a_n)(c_n + a_n - b_n)(a_n + b_n - c_n)}{a_n + b_n + c_n}},$$
(18)

giving

$$r_n = \begin{cases} \dfrac{b}{a}\, yx^{(n+1)/2} & \text{for } n \text{ odd} \\[2mm] \dfrac{b}{c}\, yx^{n/2} & \text{for } n \text{ even.} \end{cases}$$
(19)

A SANGAKU PROBLEM from 1913 in the Miyagi Prefecture asks for the relationships between the first, third, and fifth inradii (Rothman 1998). This can be solved using elementary TRIGONOMETRY as well as the explicit equations given above, and has solution

$$r_3 = \sqrt{r_1 r_5}.$$
(20)

See also ACUTE TRIANGLE, ARCHIMEDES' MIDPOINT THEOREM, BROCARD MIDPOINT, CIRCLE-POINT MIDPOINT THEOREM, FERMAT'S RIGHT TRIANGLE THEOREM, ISOSCELES TRIANGLE, MALFATTI'S RIGHT TRIANGLE PROBLEM, OBLIQUE TRIANGLE, OBTUSE TRIANGLE, PYTHAGOREAN TRIPLE, QUADRILATERAL, RAT-FREE SET, TRIANGLE, TRIGONOMETRY

References
Beiler, A H. "The Eternal Triangle." Ch. 14 in *Recreations in the Theory of Numbers: The Queen of Mathematics Entertains.* New York: Dover, 1966.
Beyer, W. H. (Ed.). *CRC Standard Mathematical Tables, 28th ed.* Boca Raton, FL: CRC Press, p. 121, 1987.
Dunham, W. *Journey through Genius: The Great Theorems of Mathematics.* New York: Wiley, pp. 120–121, 1990.
Gardner, M. *The Sixth Book of Mathematical Games from Scientific American.* Chicago, IL: University of Chicago Press, pp. 160–161, 1984.
Guy, R. K. "Triangles with Integer Sides, Medians, and Area." §D21 in *Unsolved Problems in Number Theory, 2nd ed.* New York: Springer-Verlag, pp. 188–190, 1994.
Kern, W. F. and Bland, J. R. *Solid Mensuration with Proofs, 2nd ed.* New York: Wiley, p. 2, 1948.
Ogilvy, C. S. and Anderson, J. T. *Excursions in Number Theory.* New York: Dover, p. 68, 1988.
Rothman, T. "Japanese Temple Geometry." *Sci. Amer.* **278**, 85–91, May 1998.
Sierpinski, W. *Pythagorean Triangles.* New York: Academic Press, 1962.
Whitlock, W. P. Jr. "Rational Right Triangles with Equal Areas." *Scripta Math.* **9**, 155–161, 1943.
Whitlock, W. P. Jr. "Rational Right Triangles with Equal Areas." *Scripta Math.* **9**, 265–268, 1943.

Right-Hand Rule

The rule which determines the orientation of the CROSS PRODUCT $\mathbf{u} \times \mathbf{v}$. The right-hand rule states that the orientation of the vectors' cross product is determined by placing $\mathbf{u}$ and $\mathbf{v}$ tail-to-tail, flattening the right hand, extending it in the direction of $\mathbf{u}$, and then curling the fingers in the direction that the angle $\mathbf{v}$ makes with $\mathbf{u}$. The thumb then points in the direction of $\mathbf{u} \times \mathbf{v}$.

A three-dimensional COORDINATE SYSTEM in which the axes satisfy the right-hand rule is called a RIGHT-HANDED COORDINATE SYSTEM, while one that does not is called a LEFT-HANDED COORDINATE SYSTEM.

See also CROSS PRODUCT, LEFT-HANDED COORDINATE SYSTEM, RIGHT-HANDED COORDINATE SYSTEM

Right-Handed Coordinate System

right-handed coordinate system

A three-dimensional COORDINATE SYSTEM in which the axes satisfy the RIGHT-HAND RULE.

See also CROSS PRODUCT, LEFT-HANDED COORDINATE SYSTEM, RIGHT-HAND RULE

Rigid Framework
FRAMEWORK, RIGID GRAPH

Rigid Graph

A FRAMEWORK (or GRAPH) is rigid IFF continuous motion of the points of the configuration maintaining the bar constraints comes from a family of motions of all EUCLIDEAN SPACE which are distance-preserving. A GRAPH that is not rigid is said to be FLEXIBLE (Maehara 1992).

For example, the CYCLE GRAPH C_3 is rigid, while C_4 is flexible. An embedding of the BIPARTITE GRAPH $K_{3,3}$ in the plane is rigid unless its six vertices lie on a CONIC (Bolker and Roth 1980, Maehara 1992).

A GRAPH G is (generically) d-rigid if, for almost all (i.e., an open dense set of) CONFIGURATIONS of p, the FRAMEWORK $G(p)$ is rigid in $\mathbb{R}^d$.

Cauchy (1813) proved the RIGIDITY THEOREM, one of the first results in rigidity theory. Although rigidity problems were of immense interest to engineers, intensive mathematical study of these types of problems has occurred only relatively recently (Connelly 1993, Graver *et al.* 1993).

See also BAR (EDGE), BRACED SQUARE, FLEXIBLE GRAPH, FLEXIBLE POLYHEDRON, FRAMEWORK, JUST RIGID, LAMAN'S THEOREM, LIEBMANN'S THEOREM, RIGID POLYHEDRON, RIGIDITY THEOREM, TENSEGRITY

References

Asimov, L. and Roth, B. "The Rigidity of Graphs." *Trans. Amer. Math. Soc.* **245**, 279–289, 1978.
Bolker, E. D. and Roth, B. "When is a Bipartite Graph a Rigid Framework?" *Pacific J. Math.* **90**, 27–44, 1980.
Cauchy, A. L. "Sur les polygones et les polyèdres." *XVIe Cahier* **IX**, 87–89, 1813.
Connelly, R. "Rigidity." Ch. 1.7 in *Handbook of Convex Geometry, Vol. A* (Ed. P. M. Gruber and J. M. Wills). Amsterdam, Netherlands: North-Holland, pp. 223–271, 1993.
Coxeter, H. S. M. and Greitzer, S. L. *Geometry Revisited.* Washington, DC: Math. Assoc. Amer., p. 56, 1967.
Crapo, H. and Whiteley, W. "Statics of Frameworks and Motions of Panel Structures, A Projective Geometry Introduction." *Structural Topology* **6**, 43–82, 1982.
Croft, H. T.; Falconer, K. J.; and Guy, R. K. "Rigidity of Frameworks." §B14 in *Unsolved Problems in Geometry.* New York: Springer-Verlag, pp. 63–65, 1991.
Dehn, M. "Über die Strakheit knovexer Polyeder." *Math. Ann.* **77**, 466–473, 1916.
Goldberg, M. "Unstable Polyhedral Structures." *Math. Mag.* **51**, 165–170, 1978.
Graver, J.; Servatius, B.; and Servatius, H. *Combinatorial Rigidity.* Providence, RI: Amer. Math. Soc., 1993.
Maehara, H. "Distance Graphs in Euclidean Space." *Ryukyu Math. J.* **5**, 33–51, 1992.
Pegg, E. Jr. "Rigid Nonagon." http://www.mathpuzzle.com/riginona.gif.
Roth, B. "Rigid and Flexible Frameworks." *Amer. Math. Monthly* **88**, 6–21, 1981.

Rigid Motion

A transformation consisting of ROTATIONS and TRANSLATIONS which leaves a given arrangement unchanged.

See also EUCLIDEAN MOTION, PLANE, ROTATION

References

Courant, R. and Robbins, H. *What is Mathematics?: An Elementary Approach to Ideas and Methods, 2nd ed.* Oxford, England: Oxford University Press, p. 141, 1996.
Graustein, W. C. *Introduction to Higher Geometry.* New York: Macmillan, pp. 84–85 and 89–91, 1930.

Rigid Polyhedron

A POLYHEDRON is rigid if it cannot be continuously deformed into another configuration. A rigid polyhedron may have two or more stable forms which cannot be continuously deformed into each other without bending or tearing (Wells 1991).

A structure such as a polyhedron which can change form from one stable configuration to another with only a slight transient nondestructive elastic stretch is called MULTISTABLE (Goldberg 1978).

A non-rigid polyhedron may be "SHAKY" (infinitesimally movable) or FLEXIBLE. An example of a *concave* FLEXIBLE POLYHEDRON with 18 triangular faces was given by Connelly (1978), and a FLEXIBLE POLYHEDRON with only 14 triangular faces was subsequently found by Steffen (Mackenzie 1998).

JESSEN'S ORTHOGONAL ICOSAHEDRON is an example of a SHAKY POLYHEDRON.

See also FLEXIBLE POLYHEDRON, JESSEN'S ORTHOGONAL ICOSAHEDRON, JUMPING OCTAHEDRON, MULTISTABLE, PENTAGONAL DIPYRAMID, RIGID GRAPH, SHAKY POLYHEDRON

References

Cauchy, A. L. "Sur les polygons et le polyhéders." *XVIe Cahier* **IX**, 87–89, 1813.
Connelly, R. "A Flexible Sphere." *Math. Intel.* **1**, 130–131, 1978.
Croft, H. T.; Falconer, K. J.; and Guy, R. K. "Rigidity of Polyhedra." §B13 in *Unsolved Problems in Geometry.* New York: Springer-Verlag, pp. 61–63, 1991.
Cromwell, P. R. "Equality, Rigidity, and Flexibility." Ch. 6 in *Polyhedra.* New York: Cambridge University Press, pp. 219–247, 1997.
Gluck, H. *Almost All Simply Connected Closed Surfaces are Rigid.* Heidelberg, Germany: Springer-Verlag, pp. 225–239, 1975.
Goldberg, M. "Unstable Polyhedral Structures." *Math. Mag.* **51**, 165–170, 1978.
Graver, J.; Servatius, B.; and Servatius, H. *Combinatorial Rigidity.* Providence, RI: Amer. Math. Soc., 1993.
Mackenzie, D. "Polyhedra Can Bend But Not Breathe." *Science* **279**, 1637, 1998.
Wells, D. *The Penguin Dictionary of Curious and Interesting Geometry.* London: Penguin, pp. 161–162, 1991.
Wunderlich, W. "Starre, kippende, wackelige und bewegliche Achtflache." *Elem. Math.* **20**, 25–32, 1965.

Rigidity Theorem

If the faces of a *convex* POLYHEDRON were made of metal plates and the EDGES were replaced by hinges, the POLYHEDRON would be RIGID. The theorem was stated by Cauchy (1813), although a mistake in this paper went unnoticed for more than 50 years.

See also FLEXIBLE POLYHEDRON, RIGID POLYHEDRON, SHAKY POLYHEDRON

References

Cauchy, A. L. "Sur les polygons et le polyhéders." *XVIe Cahier* **IX**, 87–89, 1813.
Cromwell, P. R. "Cauchy's Rigidity Theorem." In *Polyhedra.* New York: Cambridge University Press, pp. 228–233, 1997.
Dehn, M. "Über die Strakheit knovexer Polyeder." *Math. Ann.* **77**, 466–473, 1916.
Goldberg, M. "Unstable Polyhedral Structures." *Math. Mag.* **51**, 165–170, 1978.

Wells, D. *The Penguin Dictionary of Curious and Interesting Geometry*. London: Penguin, pp. 161–162, 1991.

Rigorous

A proof or demonstration is said to be rigorous if the validity of each step and the connections between the steps is explicitly made clear is such a way that the result follows with certainty. "Rigorous" proofs often rely on the postulates and results of formal systems that are themselves considered rigorous under stated conditions.

Ring

A ring (in the mathematical sense) is a SET S together with two BINARY OPERATORS + and * (commonly interpreted as addition and multiplication, respectively) satisfying the following conditions:

1. Additive associativity: For all a, b, $c \in S$, $(a + b) + c = a + (b + c)$,
2. Additive commutativity: For all a, $b \in S$, $a + b = b + a$,
3. Additive identity: There exists an element $0 \in S$ such that for all $a \in S$, $0 + a = a + 0 = a$,
4. Additive inverse: For every $a \in S$ there exists $-a \in S$ such that $a + (-a) = (-a) + a = 0$,
5. Multiplicative associativity: For all a, b, $c \in S$, $a * (b * c) = a * (b * c)$,
6. Left and right distributivity: For all a, b, $c \in S$, $a * (b + c) = (a * b) + (a * c)$ and $(b + c) * a = (b * a) + (c * a)$.

A ring is therefore an ABELIAN GROUP under addition and a SEMIGROUP under multiplication.

The French word for a ring is *anneau*, and the German word is *Ring*, both meaning (not so surprisingly) "ring."

A ring must contain at least one element, but need not contain a multiplicative identity or be commutative. The number of finite rings of n elements for $n = 1, 2, \ldots$, are 1, 2, 2, 11, 2, 4, 2, 52, 11, 4, 2, 22, 2, 4, 4, ... (Sloane's A027623 and A037234; Fletcher 1980). In general, the number of rings of order p^3 for p an ODD PRIME is $3p + 50$ and 52 for $p = 2$ (Ballieu 1947, Gilmer and Mott 1973).

A ring with a multiplicative identity is sometimes called a UNIT RING. Fraenkel (1914) gave the first abstract definition of the ring, although this work did not have much impact.

A ring that is COMMUTATIVE under multiplication, has a unit element, and has no divisors of zero is called an INTEGRAL DOMAIN. A ring which is also a COMMUTATIVE multiplication group is called a FIELD. The simplest rings are the INTEGERS $\mathbb{Z}$, POLYNOMIALS $\mathbb{R}[x]$ and $\mathbb{R}[x, y]$ in one and two variables, and SQUARE $n \times n$ REAL MATRICES.

Rings which have been investigated and found to be of interest are usually named after one or more of their investigators. This practice unfortunately leads to names which give very little insight into the relevant properties of the associated rings.

See also ABELIAN GROUP, ARTINIAN RING, CHOW RING, DEDEKIND RING, DIVISION ALGEBRA, FIELD, GORENSTEIN RING, GROUP, GROUP RING, IDEAL, INTEGRAL DOMAIN, MODULE, NILPOTENT ELEMENT, NOETHERIAN RING, NONCOMMUTATIVE RING, NUMBER FIELD, PRIME RING, PRÜFER RING, QUOTIENT RING, REGULAR RING, RINGOID, SEMIPRIME RING, SEMIRING, SEMISIMPLE RING, SIMPLE RING, UNIT RING, ZERO DIVISOR

References

Allenby, R. B. *Rings, Fields, and Groups: An Introduction to Abstract Algebra*, 2nd ed. Oxford, England: Oxford University Press, 1991.
Ballieu, R. "Anneaux finis; systèmes hypercomplexes de rang trois sur un corps commutatif." *Ann. Soc. Sci. Bruxelles. Sér. I* **61**, 222–227, 1947.
Beachy, J. A. *Introductory Lectures on Rings and Modules*. Cambridge, England: Cambridge University Press, 1999.
Berrick, A. J. and Keating, M.E *An Introduction to Rings and Modules with K-Theory in View*. Cambridge, England: Cambridge University Press, 2000.
Ellis, G. *Rings and Fields*. Oxford, England: Oxford University Press, 1993.
Fletcher, C. R. "Rings of Small Order." *Math. Gaz.* **64**, 9–22, 1980.
Fraenkel, A. "Über die Teiler der Null und die Zerlegung von Ringen." *J. reine angew. Math.* **145**, 139–176, 1914.
Gilmer, R. and Mott, J. "Associative Rings of Order p^3." *Proc. Japan Acad.* **49**, 795–799, 1973.
Kleiner, I. "The Genesis of the Abstract Ring Concept." *Amer. Math. Monthly* **103**, 417–424, 1996.
Nagell, T. "Moduls, Rings, and Fields." §6 in *Introduction to Number Theory*. New York: Wiley, pp. 19–21, 1951.
Sloane, N. J. A. Sequences A027623 and A037234 in "An On-Line Version of the Encyclopedia of Integer Sequences." http://www.research.att.com/~njas/sequences/eisonline.html.
van der Waerden, B. L. *A History of Algebra*. New York: Springer-Verlag, 1985.

Ring Cyclide

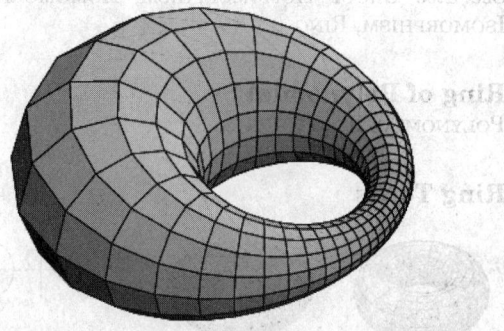

The INVERSION of a RING TORUS. If the INVERSION CENTER lies on the torus, then the ring cyclide degenerates to a PARABOLIC RING CYCLIDE.

See also CYCLIDE, INVERSION, PARABOLIC CYCLIDE, RING CYCLIDE, RING TORUS, SPINDLE CYCLIDE, TORUS

Ring Direct Product

The direct product of the RINGS R_γ, for γ some INDEX SET I, is the set

$$\prod_{\gamma \in I} R_\gamma = \left\{ f : I \to \bigcup_{\gamma \in I} R_\gamma \middle| f(\gamma) \in R_\gamma \text{ all } \gamma \in I \right\}.$$

The ring direct product is confusingly also called the complete direct sum (Herstein 1968).

$$\begin{array}{ccc} X \to G & & X \to G \oplus H \to G \\ \downarrow & \Rightarrow & \downarrow \\ H & & H \end{array}$$

the universal property of a direct product; X factors through G⊕H.

The ring direct product, like the GROUP DIRECT PRODUCT, has the UNIVERSAL PROPERTY that if any ring X has a HOMOMORPHISM to G and a homomorphism to H, then these homomorphisms factor through $G \times H$ in a unique way.

References
Herstein, I. N. *Noncommutative Rings.* Washington, DC: Math. Assoc. Amer., p. 52, 1968.

Ring Function
TOROIDAL FUNCTION

Ring Homomorphism

A ring homomorphism is a map $f : R \to S$ between two RINGS such that

1. Addition is preserved: $f(r_1 + r_2) = f(r_1) + f(r_2)$,
2. The zero element is mapped to zero: $f(0_R) = 0_S$, and
3. Multiplication is preserved: $f(r_1 r_2) = f(r_1) f(r_2)$,

where the operations on the left-hand side is in R and on the right-hand side in S. Note that a homomorphism must preserve the additive inverse map because $f(g) + f(-g) = f(g + -g) = f(0_R) = 0_S$ so $-f(g) = f(-g)$.

See also GROUP HOMOMORPHISM, HOMOMORPHISM, ISOMORPHISM, RING

Ring of Polynomial
POLYNOMIAL RING

Ring Torus

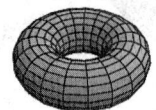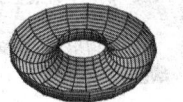

One of the three STANDARD TORI given by the PARAMETRIC EQUATIONS

$$x = (c + a \cos v) \cos u$$

$$y = (c + a \cos v) \sin u$$

$$z = a \sin v$$

with $c > a$. This is the TORUS which is generally meant when the term "torus" is used without qualification. The inversion of a ring torus is a RING CYCLIDE if the INVERSION CENTER does not lie on the torus and a PARABOLIC RING CYCLIDE if it does. The above left figure shows a ring torus, the middle a cutaway, and the right figure shows a CROSS SECTION of the ring torus through the xz-plane.

See also CYCLIDE, HORN TORUS, PARABOLIC RING CYCLIDE, RING CYCLIDE, SPINDLE TORUS, STANDARD TORI, TORUS

References
Gray, A. "Tori." §13.4 in *Modern Differential Geometry of Curves and Surfaces with Mathematica, 2nd ed.* Boca Raton, FL: CRC Press, pp. 304–306, 1997.
Pinkall, U. "Cyclides of Dupin." §3.3 in *Mathematical Models from the Collections of Universities and Museums* (Ed. G. Fischer). Braunschweig, Germany: Vieweg, pp. 28–30, 1986.

Ringoid

A ringoid is a set R with two binary operators, conventionally denoted addition (+) and multiplication (×), where × distributes over + left and right:

$$a(b + c) = ab + ac$$

and

$$(b + c)a = ba + ca.$$

A ringoid can be empty.

See also BINARY OPERATOR, RING, SEMIRING

References
Rosenfeld, A. *An Introduction to Algebraic Structures.* New York: Holden-Day, 1968.

Risch Algorithm

An ALGORITHM for indefinite integration.

See also ELEMENTARY FUNCTION, INDEFINITE INTEGRAL

References
Geddes, K. O.; Czapor, S. R.; and Labahn, G. "The Risch Integration Algorithm." Ch. 12 in *Algorithms for Computer Algebra.* Amsterdam, Netherlands: Kluwer, pp. 511–573, 1992.

Risch, R. "On the Integration of Elementary Functions which are Built Up using Algebraic Operations." Report SP-2801/002/00. Santa Monica, CA: Sys. Dev. Corp., 1968.

Risch, R. "The Problem of Integral in Finite Terms." *Trans. Amer. Math. Soc.* **139**, 167–189, 1969.

Risch, R. "The Solution of the Problem of Integration in Finite Terms." *Bull. Amer. Math. Soc.*, 1–76, 605–608, 1970.

Risch, R. "Algebraic Properties of Elementary Functions of Analysis." *Amer. J. Math.* **101**, 743–759, 1979.

Rising Factorial

There are two notations used for the falling and rising factorials, $(x)_n$ and $x^{(n)}$, which are unfortunately polar opposites of one another. The rising factorial $x^{(n)}$ (sometimes also denoted $\langle x \rangle_n$; Comtet 1974, p. 6), frequently called the POCHHAMMER SYMBOL in the theory of special functions, is defined by

$$x^{(n)} = x(x+1)\cdots(x+n-1). \tag{1}$$

It is related to the GAMMA FUNCTION $\Gamma(z)$ by

$$x^{(n)} = \frac{\Gamma(x+n)}{\Gamma(x)}, \tag{2}$$

where

$$x^{(0)} \equiv 1, \tag{3}$$

and is related to the FALLING FACTORIAL $(x)_n$ by

$$x^{(n)} = (-x)_n (-1)^n. \tag{4}$$

The rising factorial is implemented in *Mathematica* as `Pochhammer[x, n]`.

Note that in combinatorial usage, the FALLING FACTORIAL is denoted $(x)_n$ and the rising factorial is denoted $(x)^{(n)}$ (Comtet 1974, p. 6; Roman 1984, p. 5; Hardy 1999, p. 101), whereas in the calculus of FINITE DIFFERENCES and the theory of special functions, the FALLING FACTORIAL is denoted $x^{(n)}$ and the rising factorial is denoted $(x)_n$ (Roman 1984, p. 5; Abramowitz and Stegun 1972, p. 256; Spanier 1987). Extreme caution is therefore needed in interpreting the meanings of the notations $(x)_n$ and $x^{(n)}$. *In this work, the notation $x^{(n)}$ is used for the rising factorial*, despite the fact that POCHHAMMER SYMBOL, which is another name for the rising factorial, is universally denoted $(x)_n$.

The rising factorial arises in series expansions of HYPERGEOMETRIC FUNCTIONS and GENERALIZED HYPERGEOMETRIC FUNCTIONS. The first few rising factorials are

$$x^{(0)} = 1$$
$$x^{(1)} = x$$
$$x^{(2)} = x(x+1) = x^2 + x$$
$$x^{(3)} = x(x+1)(x+2) = x^3 + 3x^2 + 2x$$
$$x^{(4)} = x(x+1)(x+2)(x+3) = x^4 + 6x^3 + 11x^2 + 6x.$$

Additional identities are

$$\frac{d}{dx} x^{(n)} = x^{(n)}[F(x+n-1) - F(x-1)] \tag{5}$$

$$x^{(n+k)} = (x+n)^k x^{(n)}, \tag{6}$$

where $F(z)$ is the DIGAMMA FUNCTION.

See also CENTRAL FACTORIAL, FACTORIAL, FALLING FACTORIAL, GENERALIZED HYPERGEOMETRIC FUNCTION, HARMONIC LOGARITHM, HYPERGEOMETRIC FUNCTION, POCHHAMMER SYMBOL

References

Abramowitz, M. and Stegun, C. A. (Eds.). *Handbook of Mathematical Functions with Formulas, Graphs, and Mathematical Tables, 9th printing.* New York: Dover, 1972.

Comtet, L. *Advanced Combinatorics: The Art of Finite and Infinite Expansions, rev. enl. ed.* Dordrecht, Netherlands: Reidel, 1974.

Graham, R. L.; Knuth, D. E.; and Patashnik, O. *Concrete Mathematics: A Foundation for Computer Science, 2nd ed.* Reading, MA: Addison-Wesley, 1994.

Hardy, G. H. *Ramanujan: Twelve Lectures on Subjects Suggested by His Life and Work, 3rd ed.* New York: Chelsea, p. 101, 1999.

Roman, S. *The Umbral Calculus.* New York: Academic Press, p. 5, 1984.

Spanier, J. and Oldham, K. B. "The Pochhammer Polynomials $(x)_n$." Ch. 18 in *An Atlas of Functions.* Washington, DC: Hemisphere, pp. 149–165, 1987.

Rivest-Shamir-Adleman Number

RSA NUMBER

R-Module

A MODULE taking its coefficients in a RING R is called a module over R or R-module.

See also MODULE

RMS

ROOT-MEAN-SQUARE

Robbin Constant

$$R = \frac{4}{105} + \frac{17}{105}\sqrt{2} - \frac{2}{35}\sqrt{3} + \frac{1}{5}\ln\left(1+\sqrt{2}\right)$$

$$+ \frac{2}{3}\ln\left(2+\sqrt{3}\right) - \frac{1}{15}\pi = 0.661707182\ldots.$$

See also TRANSFINITE DIAMETER

References

Plouffe, S. "The Robbin Constant." http://www.lacim.u-qam.ca/piDATA/robbin.txt.

Robbins Algebra

Building on work of Huntington (1933), Robbins conjectured that the equations for a Robbins algebra, commutativity, associativity, and the ROBBINS AXIOM

$$!(!(x \vee y) \vee !(x \vee !y)) = x,$$

where $!x$ denotes NOT and $x \vee y$ denotes OR, imply those for a BOOLEAN ALGEBRA. The conjecture was finally proven using a computer (McCune 1997).

See also BOOLEAN ALGEBRA, HUNTINGTON AXIOM, ROBBINS CONJECTURE, ROBBINS AXIOM, WINKLER CONDITIONS

References

Huntington, E. V. "New Sets of Independent Postulates for the Algebra of Logic, with Special Reference to Whitehead and Russell's *Principia Mathematica.*" *Trans. Amer. Math. Soc.* **35**, 274–304, 1933.

Huntington, E. V. "Boolean Algebra. A Correction." *Trans. Amer. Math. Soc.* **35**, 557–558, 1933.

Kolata, G. "Computer Math Proof Shows Reasoning Power." *New York Times*, Dec. 10, 1996.

McCune, W. "Solution of the Robbins Problem." *J. Automat. Reason.* **19**, 263–276, 1997.

McCune, W. "Robbins Algebras are Boolean." http://www-unix.mcs.anl.gov/~mccune/papers/robbins/.

Nelson, E. "Automated Reasoning." http://www.math.princeton.edu/~nelson/ar.html.

Wolfram Research, Inc. "Proof of the Robbins Conjecture." http://library.wolfram.com/demos/v4/Robbins.nb.

Robbins Axiom

The logical axiom

$$R(x, y) \equiv !(!(x \vee y) \vee !(x \vee !y)) = x,$$

where $!x$ denotes NOT and $x \vee y$ denotes OR, that, when taken together with associativity and commutativity, is equivalent to the axioms of BOOLEAN ALGEBRA.

The Robbins operator can be defined in *Mathematica* by

```
Robbins := Function[{x, y}, ! (! (! y \[Or] x)
\[Or] ! (x \[Or] y))]
```

That the Robbins axiom is a true statement in BOOLEAN ALGEBRA can be verified by examining its TRUTH TABLE.

x	y	$R(x, y)$
T	T	T
T	F	T
F	T	F
F	F	F

See also ROBBINS ALGEBRA, ROBBINS CONJECTURE, WOLFRAM AXIOM

Robbins Conjecture

The conjecture that the equations for a Robbins algebra, commutativity, associativity, and the ROBBINS AXIOM

$$!(!(x \vee y) \vee !(x \vee !y)) = x,$$

where $!x$ denotes NOT and $x \vee y$ denotes OR, imply those for a BOOLEAN ALGEBRA. The conjecture was finally proven using a computer (McCune 1997).

See also BOOLEAN ALGEBRA, ROBBINS ALGEBRA, ROBBINS AXIOM

References

Kolata, G. "Computer Math Proof Shows Reasoning Power." *New York Times*, Dec. 10, 1996.

McCune, W. "Solution of the Robbins Problem." *J. Automat. Reason.* **19**, 263–276, 1997.

McCune, W. "Robbins Algebras Are Boolean." http://www-unix.mcs.anl.gov/~mccune/papers/robbins/.

Robbins Equation

$$h(u) = 2u$$

See also ROBBINS ALGEBRA

Robbin's Inequality

If the fourth MOMENT $\mu_4 \neq 0$, then

$$P(|\bar{x} - \mu_4| \geq \lambda) \leq \frac{\mu_4 + 3(N - 1)\sigma^4}{N^3 \lambda^4},$$

where σ^2 is the VARIANCE.

Robbins Number

ALTERNATING SIGN MATRIX

Robbins-Monro Stochastic Approximation

A STOCHASTIC APPROXIMATION method that functions by placing conditions on iterative step sizes and whose convergence is guaranteed under mild conditions. However, the method requires knowledge of the analytical gradient of the function under consideration.

Kiefer and Wolfowitz (1952) developed a finite difference version of the Robbins-Monro method which maintains the nice convergence properties, while obviating the need for knowledge of the analytic form of the gradient.

See also STOCHASTIC APPROXIMATION, STOCHASTIC OPTIMIZATION

References

Kiefer, J. and Wolfowitz, J. "Stochastic Estimation of the Maximum of a Regression Function." *Ann. Math. Stat.* **23**, 462–466, 1952.

Robbins, H. and Munro, S. "A Stochastic Approximation Method." *Ann. Math. Stat.* **22**, 400–407, 1951.

Robertson Condition

For the HELMHOLTZ DIFFERENTIAL EQUATION to be SEPARABLE in a coordinate system, the SCALE FACTORS h_i in the LAPLACIAN

$$\nabla^2 = \sum_{i=1}^{3} \frac{1}{h_1 h_2 h_3} \frac{\partial}{\partial u_i} \left(\frac{h_1 h_2 h_3}{h_i^2} \frac{\partial}{\partial u_i} \right) \quad (1)$$

and the functions $f_i(u_i)$ and Φ_{ij} defined by

$$\frac{1}{f_n} \frac{\partial}{\partial u_n} \left(f_n \frac{\partial X_n}{\partial u_n} \right) + (k_1^2 \Phi_{n1} + k_2^2 \Phi_{n2} + k_3^2 \Phi_{n3}) X_n = 0 \quad (2)$$

must be OF THE FORM of a STÄCKEL DETERMINANT

$$S = |\Phi_{mn}| = \begin{vmatrix} \Phi_{11} & \Phi_{12} & \Phi_{13} \\ \Phi_{21} & \Phi_{22} & \Phi_{23} \\ \Phi_{31} & \Phi_{32} & \Phi_{33} \end{vmatrix} = \frac{h_1 h_2 h_3}{f_1(u_1) f_2(u_2) f_3(u_3)}. \quad (3)$$

See also HELMHOLTZ DIFFERENTIAL EQUATION, LAPLACE'S EQUATION, SEPARATION OF VARIABLES, STÄCKEL DETERMINANT

References

Morse, P. M. and Feshbach, H. *Methods of Theoretical Physics, Part 1.* New York: McGraw-Hill, p. 510, 1953.

Robertson Conjecture

A conjecture due to M. S. Robertson (1936) which treats a UNIVALENT POWER SERIES containing only ODD powers within the UNIT DISK. This conjecture IMPLIES the BIEBERBACH CONJECTURE and follows in turn from the MILIN CONJECTURE. de Branges' proof of the BIEBERBACH CONJECTURE proceeded by proving the MILIN CONJECTURE, thus establishing the Robertson conjecture and hence implying the truth of the BIEBERBACH CONJECTURE.

See also BIEBERBACH CONJECTURE, MILIN CONJECTURE

References

Stewart, I. *From Here to Infinity: A Guide to Today's Mathematics.* Oxford, England: Oxford University Press, p. 165, 1996.

Robertson Graph

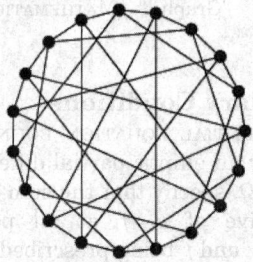

The unique (4, 5)-CAGE GRAPH, which has 19 vertices.

See also CAGE GRAPH

References

Bondy, J. A. and Murty, U. S. R. *Graph Theory with Applications.* New York: North Holland, p. 237, 1976.
Robertson, N. "The Smallest Graph of Girth 5 and Valency 4." *Bull. Amer. Math. Soc.* **70**, 824–825, 1964.
Weisstein, E. W. "Graphs." MATHEMATICA NOTEBOOK GRAPHS.M.
Wong, P. K. "Cages--A Survey." *J. Graph Th.* **6**, 1–22, 1982.

Robertson-Seymour Theorem

A generalization of the KURATOWSKI REDUCTION THEOREM by Robertson and Seymour, which states that the collection of finite GRAPHS is well-quasi-ordered by minor embeddability, from which it follows that Kuratowski's "forbidden minor" embedding obstruction generalizes to higher genus surfaces.

Formally, for a fixed INTEGER $g \geq 0$, there is a finite list of graphs $L(g)$ with the property that a GRAPH C embeds on a surface of genus g IFF it does not contain, as a minor, any of the GRAPHS on the list L.

References

Fellows, M. R. "The Robertson-Seymour Theorems: A Survey of Applications." *Comtemp. Math.* **89**, 1–18, 1987.

Robertson-Wegner Graph

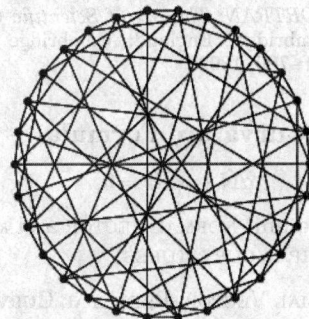

The unique (5, 5)-CAGE GRAPH, which has 30 vertices.

See also CAGE GRAPH

References

Bondy, J. A. and Murty, U. S. R. *Graph Theory with Applications.* New York: North Holland, p. 238, 1976.

Wegner, G. "A Smallest Graph of Girth 5 and Valency 5." *J. Combin. Th. B* **14**, 203–208, 1973.
Weisstein, E. W. "Graphs." Mathematica notebook Graphs.m.

Robin Boundary Conditions

Partial differential equation boundary conditions which, for an elliptic partial differential equation in a region Ω, specify that the sum of αu and the normal derivative of $u = f$ at all points of the boundary of Ω, α and f being prescribed.

Robin's Constant

Transfinite Diameter

Robinson Projection

A pseudocylindrical map projection which distorts shape, area, scale, and distance to create attractive average projection properties.

See also Map Projection, Pseudocylindrical Projection

References

Dana, P. H. "Map Projections." http://www.colorado.edu/geography/gcraft/notes/mapproj/mapproj_f.html.

Robust Estimation

An estimation technique which is insensitive to small departures from the idealized assumptions which have been used to optimize the algorithm. Classes of such techniques include M-estimates (which follow from maximum likelihood considerations), L-estimates (which are linear combinations of order statistics), and R-estimates (based on rank tests).

See also L-Estimate, M-Estimate, R-Estimate

References

Press, W. H.; Flannery, B. P.; Teukolsky, S. A.; and Vetterling, W. T. "Robust Estimation." §15.7 in *Numerical Recipes in FORTRAN: The Art of Scientific Computing*, 2nd ed. Cambridge, England: Cambridge University Press, pp. 694–700, 1992.

Rodrigues' Curvature Formula

$$d\hat{\mathbf{N}} + \kappa_i \, d\mathbf{r} = 0,$$

where $\hat{\mathbf{N}}$ is the unit normal vector and κ_i is one of the two principal curvatures.

See also Normal Vector, Principal Curvatures

Rodrigues Formula

An operator definition of a function. A Rodrigues formula may be converted into a Schläfli integral.

See also Rodrigues' Curvature Formula, Rodrigues' Rotation Formula, Schläfli Integral

Rodrigues' Rotation Formula

This entry contributed by Serge Belongie

Rodrigues' rotation formula gives an efficient method for computing the rotation matrix $\mathsf{R} \in SO(3)$ corresponding to a rotation by an angle $\theta \in \mathbb{R}$ about a fixed axis specified by the unit vector $\omega = (\omega_1, \omega_2, \omega_3) \in \mathbb{R}^3$. R is given by

$$e^{\hat{\omega}\theta} = 1 + \hat{\omega}\sin\theta + \hat{\omega}^2(1 - \cos\theta),$$

where $\hat{\omega}$ denotes the skew symmetric matrix with entries

$$\hat{\omega} = \begin{bmatrix} 0 & -\omega_3 & \omega_2 \\ \omega_3 & 0 & -\omega_1 \\ -\omega_2 & \omega_1 & 0 \end{bmatrix}.$$

See also Rotation Formula, Rotation Matrix

References

Brockett, R. W. "Robotic Manipulators and the Product of Exponentials Formula." In *Mathematical Theory of Networks and Systems. Proceedings of the international symposium held at the Ben Gurion University of the Negev, Beer Sheva, June 20–24, 1983* (Ed. P. A. Fuhrmann). Berlin: Springer-Verlag, pp. 120–127, 1984.
Murray, R. M.; Li, Z.; and Sastry, S. S. *A Mathematical Introduction to Robotic Manipulation.* Boca Raton, FL: CRC Press, 1994.

Rogers L-Function

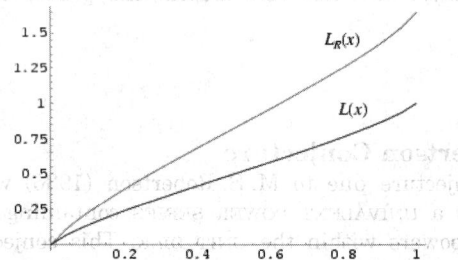

If $\mathrm{Li}_2(x)$ denotes the usual dilogarithm, then there are two variants that are normalized slightly differently, both called the Rogers L-function (Rogers 1907). Bytsko (1999) defines

$$L(x) = \frac{6}{\pi^2}\left[\mathrm{Li}_2(x) + \tfrac{1}{2}\ln x \ln(1-x) \right] \tag{1}$$

$$= \frac{6}{\pi^2}\left[\sum_{n=1}^{\infty} \frac{x^n}{n^2} + \tfrac{1}{2}\ln x \ln(1-x) \right], \tag{2}$$

(which he calls "the" dilogarithm), while Gordon and McIntosh (1997) and Loxton (1991, p. 287) define the

Rogers L-function as

$$L_R(x) = \text{Li}_2(x) + \tfrac{1}{2} \ln x \ln(1-x) \qquad (3)$$

$$= \frac{\pi^2}{6} L(x) \qquad (4)$$

$$= \left[\sum_{n=1}^{\infty} \frac{x^n}{n^2} + \tfrac{1}{2} \ln x \ln(1-x) \right]. \qquad (5)$$

The function $L(x)$ satisfies the concise identity

$$L(x) + L(1-x) = 1 \qquad (6)$$

(Euler 1768), as well as ABEL'S FUNCTIONAL EQUATION

$$L(x) + L(y) = L(xy) + L\left(\frac{x(1-y)}{1-xy} \right) + L\left(\frac{y(1-x)}{1-xy} \right) \qquad (7)$$

(Abel 1988, Bytsko 1999). The duplication formula for $L(x)$ follows from ABEL'S FUNCTIONAL EQUATION and is given by

$$\tfrac{1}{2} L(x^2) = L(x) - L\left(\frac{x}{1+x} \right). \qquad (8)$$

The function has the nice INFINITE SERIES

$$\sum_{k=2}^{\infty} L\left(\frac{1}{k^2} \right) = \tfrac{1}{6} \pi^2 \qquad (9)$$

(Lewin 1982; Loxton 1991, p. 298).
In terms of $L(x)$, the well-known dilogarithm identities become

$$L(0) = 0 \qquad (10)$$

$$L(1-\rho) = \tfrac{2}{5} \qquad (11)$$

$$L\left(\tfrac{1}{2} \right) = \tfrac{1}{2} \qquad (12)$$

$$L(\rho) = \tfrac{3}{5} \qquad (13)$$

$$L(1) = 1 \qquad (14)$$

(Loxton 1991, pp. 287 and 289; Bytsko 1999), where $\rho = (\sqrt{5}-1)/2$.

Numbers $\theta \in (0, 1)$ which satisfy

$$\sum_{k=0}^{n} c_k L(\theta^k) = 0 \qquad (15)$$

for some value of n are called L-ALGEBRAIC NUMBERS. Loxton (1991, p. 289) gives a slew of identities having rational coefficients

$$\sum_{k=0}^{n} \frac{e_k}{k} L(\theta^k) = c \qquad (16)$$

instead of integers, where c is a RATIONAL NUMBER, a corrected and expanded version of which is summar-

ized in the following table. In this table, polynomials $P(x)$ denote the real root of x. Many more similar identities can be found using INTEGER RELATION algorithms.

θ	e_k	c
1	1	1
$\tfrac{1}{2}$	1	$\tfrac{1}{2}$
$\tfrac{1}{2}$	$-1, 6, 3, 0, 0, -3$	$\tfrac{1}{2}$
$\tfrac{1}{3}$	$3, -1$	1
$\tfrac{1}{2}(\sqrt{5}-1)$	1	$\tfrac{3}{5}$
$\tfrac{1}{2}(\sqrt{5}-1)$	$1, -1, -12, 0, 0, 6$	$-\tfrac{3}{5}$
$(\sqrt{5}-2)^{1/3}$	$2, -1$	1
$\sqrt{2}-1$	$2, -1$	$\tfrac{3}{4}$
$\sqrt{2}-1$	$1, 2, 0, -1$	$\tfrac{5}{8}$
$3-2\sqrt{2}$	$5, -2$	1
$\tfrac{1}{2}(\sqrt{3}-1)$	$2, 1, -1$	$\tfrac{5}{6}$
$\sqrt{3}-1$	$2, -3, -1, 0, 0, 1$	$\tfrac{1}{2}$
$2-\sqrt{3}$	$4, 1, 0, -1$	$\tfrac{5}{4}$
$2-\sqrt{3}$	$5, -3, -1, 0, 0, 1$	$\tfrac{4}{3}$
$5-2\sqrt{6}$	$23, -15, -3, 0, 0, 3$	3
$\tfrac{1}{2}(\sqrt{13}-3)$	$4, -2, -2, 0, 0, 1$	$\tfrac{7}{6}$
$\tfrac{1}{6}(\sqrt{13}-1)$	$3, 1, -3, 0, 0, 1$	$\tfrac{4}{3}$
$\tfrac{1}{6}(\sqrt{13}+1)$	$3, -4, -3, 0, 0, 2$	$\tfrac{2}{3}$
$4-\sqrt{15}$	$15, 2, -3, -2$	$\tfrac{5}{2}$
$\tfrac{1}{2}(5-\sqrt{21})$	$7, -1, -3, 0, 0, 1$	$\tfrac{5}{3}$
$\tfrac{1}{2} \sec\left(\tfrac{2}{7} \pi \right),$	$1, -2$	$\tfrac{1}{7}$
$\tfrac{1}{2} \sec\left(\tfrac{1}{7} \pi \right)$	$1, 1$	$\tfrac{5}{7}$
$2 \cos\left(\tfrac{3}{7} \pi \right)$	$1, 1$	$\tfrac{4}{7}$
$\tfrac{1}{2} \sec\left(\tfrac{1}{9} \pi \right)$	$1, 2, -1$	$\tfrac{7}{9}$
$\tfrac{1}{2} \sec\left(\tfrac{2}{9} \pi \right)$	$1, -3, -1, 0, 0, 1$	$-\tfrac{1}{9}$
$2 \cos\left(\tfrac{4}{9} \pi \right)$	$1, -3, -1, 0, 0, 1$	$\tfrac{1}{9}$
$x^3 + 2x - 1$	$1, 5, 0, -4$	1
$x^3 + 2x - 1$	$3, 1, 12, 0, 0, -6$	2
$2x^3 + x - 1$	$2, 1, 3, -2$	$\tfrac{3}{2}$
$x^3 + x - 1$	$2, 6, 3, 0, 0, -3$	3
$x^3 - 3x^2 + 4x - 1$	$5, -9, -6, 0, 0, 6$	1
$x^3 + x^2 - 1$	$1, 6, 6, 0, 0, -6$	2
$x^3 + x^2 + x - 1$	$1, 1, -3$	$\tfrac{1}{2}$
$x^3 + x^2 + x - 1$	$2, 3, 0, -2$	$\tfrac{3}{2}$

Bytsko (1999) gives the additional identities

$$L(\lambda^{-2}) + L((\lambda^2 - 1)^{-2}) = \tfrac{4}{7} \quad (17)$$

$$L(\lambda^{-2}) + L((1 + \lambda)^{-1}) = \tfrac{5}{7} \quad (18)$$

$$L\left(1 - \frac{1}{\sqrt{2}}\right) + L(\sqrt{2} - 1) = \tfrac{3}{4} \quad (19)$$

$$L(\sqrt{\rho}) + L\left(\frac{1}{1 + \sqrt{\rho}}\right) = \tfrac{13}{11} \quad (20)$$

$$L\left(\tfrac{1}{2} - \tfrac{1}{2}\rho\right) + L(2\rho - 1) = \tfrac{1}{2} \quad (21)$$

$$L\left(1 - \tfrac{1}{2}\rho - \tfrac{1}{2}\sqrt{7\rho - 3}\right) + L\left(\tfrac{1}{2}\sqrt{28\rho + 45} - 2\rho - \tfrac{2}{5}\right) = \tfrac{2}{5} \quad (22)$$

$$L(1 - \delta^2) + L((1 + \delta)^{-2}) = \tfrac{2}{5} \quad (23)$$

$$L\left(\tfrac{3}{2} - \tfrac{1}{2}\sqrt{2} - \tfrac{1}{2}\sqrt{2\sqrt{2} - 1}\right)$$
$$+ L\left(\left(\tfrac{3}{2} + \sqrt{2}\right)\sqrt{2\sqrt{2 - 1}} - \tfrac{3}{2} - \tfrac{3}{2}\sqrt{2}\right) = \tfrac{1}{2} \quad (24)$$

$$L(v) - L(\mu^{-1}) = \tfrac{1}{7} \quad (25)$$

where

$$\lambda = 2\cos(\pi/7)$$

$$\rho = \left(\sqrt{5} - 1\right)/2$$

$$\delta = \tfrac{1}{2}\left(\sqrt{3 + 2\sqrt{5}} - 1\right),$$

with δ the positive root of

$$\delta^4 + \delta^3 - \delta - 1 = 0 \quad (26)$$

and $0 < v < 1$ and $\mu > 1$ the real roots of

$$t^6 - 7t^5 + 19t^4 - 28t^3 + 20t^2 - 7t + 1 = 0. \quad (27)$$

Here, (17) and (18) are special cases of the WATSON IDENTITIES and (19) is a special case of ABEL'S DUPLICATION FORMULA with $x = 1/\sqrt{2}$ (Gordon and McIntosh 1997, Bytsko 1999).

Rogers (1907) obtained a dilogarithm identity in m variables with $m^2 + 1$ terms which simplifies to Euler's identity for $m = 1$ and ABEL'S FUNCTIONAL EQUATION for $m = 2$ (Gordon and McIntosh 1997). For $m = 3$, it is equivalent to

$$L(a) + L(b) + L(c) - L(u) - L(v)$$
$$= L(abc) + L(ac/u) + L(bc/v) - L(av/u) - L(bu/v), \quad (28)$$

with

$$av(1 - bc) + bu(1 - ac) = uv(1 - ab) \quad (29)$$

$$v(1 - a) + u(1 - b) = 1 - abc \quad (30)$$

(Gordon and McIntosh 1997).

See also ABEL'S DUPLICATION FORMULA, ABEL'S FUNCTIONAL EQUATION, DILOGARITHM, L-ALGEBRAIC NUMBER, LANDEN'S IDENTITY

References

Abel, N. H. *Oeuvres Completes, Vol. 2* (Ed. L. Sylow and S. Lie). New York: Johnson Reprint Corp., pp. 189–192, 1988.

Bytsko, A. G. *J. Physics A* **32**, 8045, 1999.

Bytsko, A. G. Two-Term Dilogarithm Identities Related to Conformal Field Theory. 9 Nov 1999. http://xxx.lanl.gov/abs/math-ph/9911012/.

Euler, L. *Institutiones calculi integralis, Vol. 1.* pp. 110–113, 1768.

Gordon, B. and McIntosh, R. J. "Algebraic Dilogarithm Identities." *Ramanujan J.* **1**, 431–448, 1997.

Lewin, L. "The Dilogarithm in Algebraic Fields." *J. Austral. Math. Soc. (Ser. A)* **33**, 302–330, 1982.

Lewin, L. (Ed.). *Structural Properties of Polylogarithms.* Providence, RI: Amer. Math. Soc., 1991.

Loxton, J. H. "Partition Identities and the Dilogarithm." Ch. 13 in *Structural Properties of Polylogarithms* (Ed. L. Lewin). Providence, RI: Amer. Math. Soc., pp. 287–299, 1991.

Rogers, L. J. "On Function Sum Theorems Connected with the Series $\Sigma_1^\infty x^n/n^2$." *Proc. London Math. Soc.* **4**, 169–189, 1907.

Watson, G. N. *Quart. J. Math. Oxford Ser.* **8**, 39, 1937.

Rogers-Ramanujan Continued Fraction

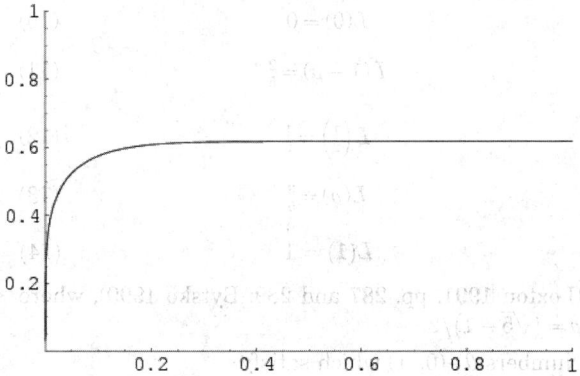

The Rogers-Ramanujan continued fraction is defined by

$$R(q) \equiv \cfrac{q^{1/5}}{1 + \cfrac{q}{1 + \cfrac{q^2}{1 + \cfrac{q^3}{1 + \cdots}}}} \quad (1)$$

(Rogers 1894, Ramanujan 1957, Berndt *et al.*). The coefficients of q^n in the MACLAURIN SERIES of $R(q)/q^{1/5}$ for $n = 0, 1, 2, \ldots$ are 1, -1, 1, 0, -1, 1, -1, 1,

0, -1, 2, -3, ... (Sloane's A007325). The fraction can be given explicitly as

$$R(q) = q^{1/5} \frac{(q;\ q^5)_\infty (q^4;\ q^5)_\infty}{(q^2;\ q^5)_\infty (q^3;\ q^5)_\infty} \tag{2}$$

$$= q^{1/5} \prod_{k=1}^{\infty} \frac{(1 - x^{5k-1})(1 - x^{5k-4})}{(1 - x^{5k-2})(1 - x^{5k-3})} \tag{3}$$

$$= q^{1/5} \frac{f(-q,\ -q^4)}{f(-q^2,\ -q^3)}, \tag{4}$$

where $(a; q)_n$ is a Q-SERIES and $f(a, b)$ is a RAMANU-JAN THETA FUNCTION.
$R(q)$ satisfies the amazing equalities

$$\frac{1}{R(q)} - 1 - R(q) = \frac{f(-q^{1/5})}{q^{1/5} f(-q^5)} \tag{5}$$

$$\frac{1}{[R(q)]^5} - 11 - [R(q)]^5 = \frac{[f(-q)]^6}{q[f(-q^5)]^6} \tag{6}$$

as well as

$$\sum_{n=-\infty}^{\infty} (-1)^n (10n+3) q^{(5n+3)n/2}$$

$$= \left[\frac{3}{[R(q)]^2} + [R(q)]^3 \right] q^{2/5} \left[f(-q^5) \right]^3 \tag{7}$$

$$\sum_{n=-\infty}^{\infty} (-1)^n (10n+1) q^{(5n+1)n/2}$$

$$= \left[\frac{1}{[R(q)]^3} + 3[R(q)]^2 \right] q^{3/5} \left[f(-q^5) \right]^3 \tag{8}$$

(Watson 1929ab; Berndt 1991, pp. 265–267; Berndt *et al.*, Son).
Defining

$$u = R(q) \tag{9}$$

$$u' = -R(-q) \tag{10}$$

$$v = R(q^2) \tag{11}$$

$$w = R(q^4), \tag{12}$$

these quantities satisfy the modular equations

$$uv^2 = \frac{v - u^2}{v + u^2} \tag{13}$$

$$uw = \frac{w^2 - u^2 v}{w + u^2} \tag{14}$$

$$vw^2 = \frac{w - v^2}{w + v^2} \tag{15}$$

$$uu'v^2 = \frac{uu' - v}{u' - u} \tag{16}$$

$$u'w = \frac{u'^2 - w}{v^2 + w} \tag{17}$$

$$-vw = \frac{u'(v^2 - w)}{u'^2 v - w} \tag{18}$$

$$uu'v = \frac{u' - u}{v + uu'} \tag{19}$$

$$vw = \frac{u(v^2 - w)}{u^2 v - w} \tag{20}$$

(Berndt *et al.*).

As discussed by Hardy (1962, pp. xxvii and xxviii), Berndt and Rankin (1995), and Berndt *et al.*, Ramanujan also defined the generalized continued fraction

$$R(a, q) \equiv \cfrac{1}{1 + \cfrac{aq}{1 + \cfrac{aq^2}{1 + \cfrac{aq^3}{1 + \cdots}}}} \tag{21}$$

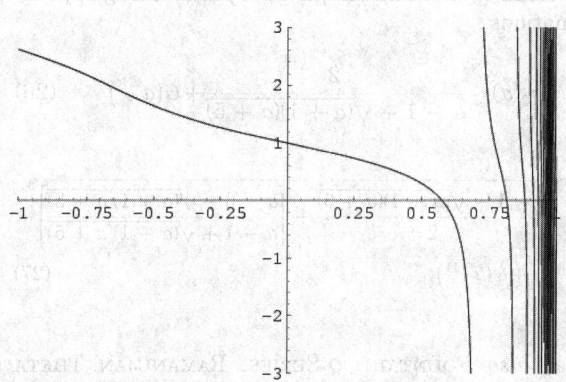

Ramanujan also considered

$$F(a, q) \equiv 1 - \cfrac{aq}{1 - \cfrac{aq^2}{1 - \cfrac{aq^3}{1 - \cdots}}}, \tag{22}$$

$$= \frac{\sum_{k=0}^{\infty} \dfrac{(-a)^k q^{k^2}}{(q)_k}}{\sum_{k=0}^{\infty} \dfrac{(-a)^k q^{k(k+1)}}{(q)_k}}. \tag{23}$$

(Berndt 1991, p. 30; Berndt *et al.*), of which the special case $F(q) = F(1, q)$ is plotted above. Terminating the terms in the continued fraction at a term aq^n gives

$$\frac{\sum_{k=0}^{\lfloor (n+1)/2 \rfloor} \dfrac{(-a)^k q^{k^2} (q)_{n-k+1}}{(q)_k}}{\sum_{k=0}^{\lfloor n/2 \rfloor} \dfrac{(-a)^k q^{(k+1)^2} (q)_{n-k}}{(q)_k (q)_{n-2k}}}$$

$$= 1 - \cfrac{aq}{1 - \cfrac{aq^2}{1 - \cfrac{aq^3}{1 - \cdots - \cfrac{aq^n}{1}}}}, \tag{24}$$

(Berndt *et al.*). The real roots of $F(q)$ are 0.576149, 0.815600, 0.882493, 0.913806, 0.931949, 0.943785, 0.952125, ..., the smallest of which was found by Ramanujan (Berndt *et al.*). $F(q)$ and its smallest positive root are related to the enumeration of coins in a FOUNTAIN (Berndt 1991, Berndt *et al.*) and the study of birth and death processes (Berndt *et al.*, Parthasarathy *et al.* 1998). In general, the least positive root $q_0(a)$ of $F(a, q)$ is given as $a \to \infty$ by

$$q_0(a) \sim \frac{1}{a} - \frac{1}{a^2} + \frac{2}{a^3} - \frac{6}{a^4} + \frac{21}{a^5} - \frac{79}{a^6} + \frac{311}{a^7} - \frac{1266}{a^8}$$

$$+ \frac{5289}{a^9} - \frac{22553}{a^{10}} + \frac{97753}{a^{11}} - \cdots \tag{25}$$

(Berndt *et al.*). Ramanujan gave the amazing approximations

$$q_0(a) \sim \frac{2}{a - 1 + \sqrt{(a+1)(a+5)}} + \mathcal{O}\left(a^{-8}\right) \tag{26}$$

$$\sim \cfrac{1}{\dfrac{a - 1 + \sqrt{(a+1)(a+5)}}{2} + \left[\dfrac{a + 3 - \sqrt{(a+1)(a+5)}}{a - 1 + \sqrt{(a+1)(a+5)}}\right]^3}$$
$$+ \mathcal{O}\left(a^{-11}\right). \tag{27}$$

See also FOUNTAIN, Q-SERIES, RAMANUJAN THETA FUNCTIONS, ROGERS-RAMANUJAN IDENTITIES

References

Andrews, G. E.; Berndt, B. C.; Jacobsen, L.; and Lamphere, R. L. *The Continued Fractions Found in the Unorganized Portion of Ramanujan's Notebooks.* Providence, RI: Amer. Math. Soc., 1992.

Andrews, G. *On the General Rogers-Ramanujan Theorem.* Providence, RI: Amer. Math. Soc., 1974.

Berndt, B. C. *Ramanujan's Notebooks, Part III.* New York: Springer-Verlag, 1991.

Berndt, B. C. "Continued Fractions." Ch. 32 in *Ramanujan's Notebooks, Part V.* New York: Springer-Verlag, pp. 9–88, 1998.

Berndt, B.C. and Chan, H. H. "Some Values for the Rogers-Ramanujan Continued Fraction." *Canad. J. Math.* **47**, 897–914, 1995.

Berndt, B. C.; Chan, H. H.; Huang, S.-S.; Kang, S.-Y.; Sohn, J.; and Son, S. H. "The Rogers-Ramanujan Continued Fraction."

Berndt, B. C.; Chan, H. H.; and Zhang, L.-C. "Explicit Evaluations of the Rogers-Ramanujan Continued Fraction." *J. reine angew. Math.* **480**, 141–159, 1996.

Berndt, B. C.; Huang, S.-S.; Sohn, J.; and Son, S. H. "Some Theorems on the Rogers-Ramanujan Continued Fraction in Ramanujan's Lost Notebook." To appears in *Trans. Amer. Math. Soc.*

Berndt, B. C. and Rankin, R. A. *Ramanujan: Letters and Commentary.* Providence, RI: Amer. Math. Soc, 1995.

Joyce, G. S. "Exact Results for the Activity and Isothermal Compressibility of the Hard-Hexagon Model." *J. Phys. A: Math. Gen.* **21**, L983-L988, 1988.

Parthasarathy, P. R.; Lenin, R. B.; Schoutens, W.; and van Assche, W. "A Birth and Death Process Related to the Rogers-Ramanujan Continued Fraction." *J. Math. Anal. Appl.* **224**, 297–315, 1998.

Ramanathan, K. G. "On Ramanujan's Continued Fraction." *Acta Arith.* **43**, 209–226, 1984.

Ramanathan, K. G. "On the Rogers-Ramanujan Continued Fraction." *Proc. Indian Acad. Sci. (Math. Sci.)* **93**, 67–77, 1984.

Ramanathan, K. G. "Ramanujan's Continued Fraction." *Indian J. Pure Appl. Math.* **16**, 695–724, 1985.

Ramanathan, K. G. "Some Applications of Kronecker's Limit Formula." *J. Indian Math. Soc.* **52**, 71–89, 1987.

Ramanujan, S. *Notebooks (2 Volumes).* Bombay, India: Tata Institute, 1957.

Ramanujan, S. *Collected Papers.* New York: Chelsea, 1962.

Rogers, L. J. "Second Memoir on the Expansion of Certain Infinite Products." *Proc. London Math. Soc.* **25**, 318–343, 1894.

Rogers, L. J. "On a Type of Modular Equations." *Proc. London Math. Soc.* **19**, 387–397, 1920.

Sloane, N. J. A. Sequences A007325/M0415 in "An On-Line Version of the Encyclopedia of Integer Sequences." http://www.research.att.com/~njas/sequences/eisonline.html.

Watson, G. N. "Theorems Stated by Ramanujan (VII): Theorems on Continued Fractions." *J. London Math. Soc.* **4**, 39–48, 1929.

Watson, G. N. "Theorems Stated by Ramanujan (IX): Two Continued Fractions." *J. London Math. Soc.* **4**, 231–237, 1929.

Rogers-Ramanujan Identities

For $|q| < 1$ and using the NOTATION of the RAMANUJAN THETA FUNCTION, the Rogers-Ramanujan identities are

$$\frac{f(-q^5)}{f(-q^2, -q^4)} = \sum_{k=0}^{\infty} \frac{q^{k^2}}{(q)_k} \tag{1}$$

$$\frac{f(-q^5)}{f(-q^2, -q^3)} = \sum_{k=0}^{\infty} \frac{q^{k(k+1)}}{(q)_k}, \tag{2}$$

where $(q)_k$ are Q-SERIES. Written out explicitly (Hardy 1999, pp. 13 and 90),

$$1 + \frac{q}{1-q} + \frac{q^4}{(1-q)(1-q^2)} + \frac{q^9}{(1-q)(1-q^2)(1-q^3)} + \cdots$$

$$= \frac{1}{(1-q)(1-q^6)\ldots(1-q^4)(1-q^9)\ldots}$$

$$= 1 + x + x^2 + x^3 + 2x^4 + 2x^5 + 3x^6 + \cdots \tag{3}$$

(Sloane's A003114), and

$$1 + \frac{q^2}{1-q} + \frac{q^6}{(1-q)(1-q^2)} + \frac{q^{12}}{(1-q)(1-q^2)(1-q^3)} + \ldots$$

$$= \frac{1}{(1-q^2)(1-q^7)\ldots(1-q^3)(1-q^8)\ldots}$$

$$= 1 + x^2 + x^3 + x^4 + x^5 + 2x^6 + \ldots \qquad (4)$$

(Sloane's A003106). These identities can also be written succinctly as

$$1 + \sum_{k=1}^{\infty} \frac{q^{k^2+ak}}{(1-q)(1-q^2)\ldots(1-q^k)}$$

$$= \prod_{j=0}^{\infty} \frac{1}{(1-q^{5j+a+1})(1-q^{5j-a+4})} \qquad (5)$$

where $a = 0, 1$.

Other forms of the Rogers-Ramanujan identities include

$$\sum_k \frac{q^{k^2}}{(q;q)_k(q;q)_{n-k}} = \sum_k \frac{(-1)^k q^{(5k^2-k)/2}}{(q;q)_{n-k}(q;q)_{n+k}} \qquad (6)$$

and

$$\sum_k \frac{2q^{k^2}}{(q;q)_k(q;q)_{n-k}} = \sum_k \frac{(-1)^k(1+q^k)q^{(5k^2-k)/2}}{(q;q)_{n-k}(q;q)_{n+k}} \qquad (7)$$

(Petkovsek *et al.* 1996).

The formulas have a curious history, having been proved by Rogers (1894) in a paper that was completely ignored, then rediscovered (without proof) by Ramanujan sometime before 1913. The formulas were communicated to MacMahon, who published them in his famous text, still without proof. Then, in 1917, Ramanujan accidentally found Roger's 1894 paper while leafing through a journal. In the meantime, Schur (1917) independently rediscovered and published proofs for the identities (Hardy 1999, p. 91). Garsia and Milne (1981ab) gave the first proof of the Rogers-Ramanujan identities to construct a BIJECTION between the relevant classes of partitions (Andrews 1986, p. 59).

Schur showed that (3) has the combinatorial interpretation that the number of partitions of n with minimal difference ≥ 2 is equal to the number of partitions into parts OF THE FORMS $5m + 1$ or $5m + 4$ (Hardy 1999, p. 92). The following table gives the first few values.

n	a_n	min. diff.		$\equiv 1, 4 \pmod 5$	
1	1	1			1
2	1	2			1+1

3	1	3		1+1+1
4	2	4, 3+1		4, 1+1+1+1
5	2	5, 4+1		4+1, 1+1+1+1+1
6	3	6, 5+1, 4+2	5, 4+1+1, 1+1+1+1+1+1

There is a similar combinatorial interpretation for (4). A generalization of the Rogers-Ramanujan identities is given by

$$\sum_{n_1, \ldots, n_{k-1} \geq 0} \frac{x^{N_1^2 + \cdots + N_{k-1}^2 + N_i + \cdots + N_{k-1}}}{(x)_{n_1} \cdots (x)_{n_{k-1}}}$$

$$= \prod_{\substack{r \theta, \pm i \pmod{2k+1}}} \frac{r=1}{1-x^v} \qquad (8)$$

where $1 \leq i \leq k$, $k \geq 2$, x complex with $|x| < 1$, and $N_j = n_j + \cdots n_{k-1}$ (Andrews 1984, p. 111; Fulman 1999). These identities have a number of important applications in mathematical physics (Fulman 1999).

See also ANDREWS-SCHUR IDENTITY, DOUGALL-RAMANUJAN IDENTITY, SLATER'S IDENTITY

References

Andrews, G. E. "The Hard-Hexagon Model and Rogers-Ramanujan Type Identities." *Proc. Nat. Acad. Sci. U.S.A.* **78**, 5290–5292, 1981.

Andrews, G. E. *Encyclopedia of Mathematics and Its Applications, Vol. 2: The Theory of Partitions.* Cambridge, England: Cambridge University Press, pp. 109 and 238, 1984.

Andrews, G. E. *q*-Series: Their Development and Application in Analysis, Number Theory, Combinatorics, Physics, and Computer Algebra. Providence, RI: Amer. Math. Soc., pp. 17–20, 1986.

Andrews, G. E. and Baxter, R. J. "A Motivated Proof of the Rogers-Ramanujan Identities." *Amer. Math. Monthly* **96**, 401–409, 1989.

Andrews, G. E.; Baxter, R. J.; and Forrester, P. J. "Eight-Vertex SOS Model and Generalized Rogers-Ramanujan-Type Identities." *J. Stat. Phys.* **35**, 193–266, 1984.

Bressoud, D. M. *Analytic and Combinatorial Generalizations of the Rogers-Ramanujan Identities.* Providence, RI: Amer. Math. Soc., 1980.

Fulman, J. "The Rogers-Ramanujan Identities, The Finite General Linear Groups, and the Hall-Littlewood Polynomials." *Proc. Amer. Math. Soc.* **128**, 17–25, 1999.

Garsia, A. M. and Milne, S. C. "A Method for Constructing Bijections for Classical Partition Identities." *Proc. Nat. Acad. Sci. USA* **78**, 2026–2028, 1981.

Garsia, A. M. and Milne, S. C. "A Rogers-Ramanujan Bijection." *J. Combin. Th. Ser. A* **31**, 289–339, 1981.

Guy, R. K. "The Strong Law of Small Numbers." *Amer. Math. Monthly* **95**, 697–712, 1988.

Hardy, G. H. *Ramanujan: Twelve Lectures on Subjects Suggested by His Life and Work, 3rd ed.* New York: Chelsea, pp. 13 and 90–99, 1999.

Hardy, G. H. and Wright, E. M. "The Rogers-Ramanujan Identities." §19.13 in *An Introduction to the Theory of Numbers, 5th ed.* Oxford, England: Clarendon Press, pp. 290–294, 1979.

MacMahon, P. A. *Combinatory Analysis, Vol. 2.* New York: Chelsea, pp. 33–36, 1960.

Paule, P. "Short and Easy Computer Proofs of the Rogers-Ramanujan Identities and of Identities of Similar Type." *Electronic J. Combinatorics* **1**, R10 1–9, 1994. http://www.combinatorics.org/Volume_1/volume1.html#R10.

Petkovsek, M.; Wilf, H. S.; and Zeilberger, D. *A = B.* Wellesley, MA: A. K. Peters, p. 117, 1996.

Ramanujan, S. Problem 584. *J. Indian Math. Soc.* **6**, 199–200, 1914.

Robinson, R. M. "Comment to: 'A Motivated Proof of the Rogers-Ramanujan Identities.'" *Amer. Math. Monthly* **97**, 214–215, 1990.

Rogers, L. J. "Second Memoir on the Expansion of Certain Infinite Products." *Proc. London Math. Soc.* **25**, 318–343, 1894.

Rogers, L. J. "On Two Theorems of Combinatory Analysis and Some Allied Identities." *Proc. London Math. Soc.* **16**, 315–336, 1917.

Rogers, L. J. "Proof of Certain Identities in Combinatory Analysis." *Proc. Cambridge Philos. Soc.* **19**, 211–214, 1919.

Schur, I. "Ein Beitrag zur additiven Zahlentheorie und zur Theorie der Kettenbrüche." *Sitzungsber. Preuss. Akad. Wiss. Phys.-Math. Klasse*, pp. 302–321, 1917.

Sloane, N. J. A. Sequences A003106/M0261, A003114/M0266, and A006141/M0260 in "An On-Line Version of the Encyclopedia of Integer Sequences." http://www.research.att.com/~njas/sequences/eisonline.html.

Watson, G. N. "A New Proof of the Rogers-Ramanujan Identities." *J. London Math. Soc.* **4**, 4–9, 1929.

Watson, G. N. "Theorems Stated by Ramanujan (VII): Theorems on Continued Fractions." *J. London Math. Soc.* **4**, 39–48, 1929.

Roller

CURVE OF CONSTANT WIDTH

Rolle's Theorem

Let f be differentiable on (a, b) and continuous on $[a, b]$. If $f(a) = f(b) = 0$, then there is at least one point $c \in (a, b)$ where $f'(c) = 0$.

See also FIXED POINT THEOREM, MEAN-VALUE THEOREM

Rolling Polygon

ROULETTE

Roman Coefficient

A generalization of the BINOMIAL COEFFICIENT whose NOTATION was suggested by Knuth,

$$\left\lfloor \begin{matrix} n \\ k \end{matrix} \right\rfloor = \frac{\lfloor n \rceil !}{\lfloor k \rceil ! \lfloor n - k \rceil !}. \tag{1}$$

The above expression is read "Roman n choose k." Whenever the BINOMIAL COEFFICIENT is defined (i.e., $n \geq k \geq 0$ or $k \geq 0 > n$), the Roman coefficient agrees with it. However, the Roman coefficients are defined for values for which the BINOMIAL COEFFICIENTS are not, e.g.,

$$\left\lfloor \begin{matrix} n \\ -1 \end{matrix} \right\rfloor = \frac{1}{\lfloor n + 1 \rceil} \tag{2}$$

$$\left\lfloor \begin{matrix} 0 \\ k \end{matrix} \right\rfloor = \frac{(-1)^{k + (k > 0)}}{\lfloor k \rceil}, \tag{3}$$

where

$$n < 0 \equiv \begin{cases} 1 & \text{for } n < 0 \\ 0 & \text{for } n \geq 0. \end{cases} \tag{4}$$

The Roman coefficients also satisfy properties like those of the BINOMIAL COEFFICIENT,

$$\left\lfloor \begin{matrix} n \\ k \end{matrix} \right\rfloor = \left\lfloor \begin{matrix} n \\ n - k \end{matrix} \right\rfloor \tag{5}$$

$$\left\lfloor \begin{matrix} n \\ k \end{matrix} \right\rfloor \left\lfloor \begin{matrix} k \\ r \end{matrix} \right\rfloor = \left\lfloor \begin{matrix} n \\ r \end{matrix} \right\rfloor \left\lfloor \begin{matrix} n - r \\ k - r \end{matrix} \right\rfloor \tag{6}$$

an analog of PASCAL'S FORMULA

$$\left\lfloor \begin{matrix} n \\ k \end{matrix} \right\rfloor = \left\lfloor \begin{matrix} n - 1 \\ k \end{matrix} \right\rfloor + \left\lfloor \begin{matrix} n - 1 \\ k - 1 \end{matrix} \right\rfloor, \tag{7}$$

and a curious rotation/reflection law due to Knuth

$$(-1)^{k + (k > 0)} \left\lfloor \begin{matrix} -n \\ k - 1 \end{matrix} \right\rfloor = (-1)^{n + (n > 0)} \left\lfloor \begin{matrix} -k \\ n - 1 \end{matrix} \right\rfloor \tag{8}$$

(Roman 1992).

See also BINOMIAL COEFFICIENT, ROMAN FACTORIAL

References

Roman, S. "The Logarithmic Binomial Formula." *Amer. Math. Monthly* **99**, 641–648, 1992.

Roman Factorial

$$\lfloor n \rceil ! \equiv \begin{cases} n! & \text{for } n \geq 0 \\ \dfrac{(-1)^{-n-1}}{(-n-1)!} & \text{for } n < 0. \end{cases} \tag{1}$$

The Roman factorial arises in the definition of the HARMONIC LOGARITHM and ROMAN COEFFICIENT. It obeys the identities

$$\lfloor n \rceil ! = \lfloor n \rceil \lfloor n - 1 \rceil ! \tag{2}$$

$$\frac{\lfloor n \rceil !}{\lfloor n - k \rceil !} = \lfloor n \rceil \lfloor n - 1 \rceil \cdots \lfloor n - k + 1 \rceil \tag{3}$$

$$\lfloor n \rceil ! \lfloor -n - 1 \rceil ! = (-1)^{n + (n < 0)}, \tag{4}$$

where

$$\lfloor n \rceil \equiv \begin{cases} n & \text{for } n \neq 0 \\ 1 & \text{for } n = 0 \end{cases} \tag{5}$$

and

$$n < 0 \equiv \begin{cases} 1 & \text{for } n < 0 \\ 0 & \text{for } n \geq 0. \end{cases} \quad (6)$$

See also HARMONIC LOGARITHM, HARMONIC NUMBER, ROMAN COEFFICIENT

References

Loeb, D. and Rota, G.-C. "Formal Power Series of Logarithmic Type." *Advances Math.* **75**, 1–118, 1989.

Roman, S. "The Logarithmic Binomial Formula." *Amer. Math. Monthly* **99**, 641–648, 1992.

Roman Numeral

A system of numerical notations used by the Romans. It is an additive (and subtractive) system in which letters are used to denote certain "base" numbers, and arbitrary numbers are then denoted using combinations of symbols. Unfortunately, little is known about the origin of the Roman numeral system (Cajori 1993, p. 30).

Character	Numerical Value
I	1
V	5
X	10
L	50
C	100
D	500
M	1000

For example, the number 1732 would be denoted MDCCXXXII. One additional rule states that, instead of using four symbols to represent a 4, 40, 9, 90, etc., such numbers are instead denoted by preceding the symbol for 5, 50, 10, 100, etc., with a symbol indicating *subtraction*. For example, 4 is denoted IV, 9 as IX, 40 as XL, etc. However, this rule is generally *not* followed on the faces of clocks, where IIII is usually encountered instead of IV. Furthermore, the practice of placing smaller digits before large ones to indicate subtraction of value was hardly ever used by Romans and came into popularity in Europe after the invention of the printing press (Wells 1986, p. 60; Cajori 1993, p. 31).

I̅   II̅   X̅   C̅

100,000   200,000   1,000,000   10,000,000

For large numbers, the Romans placed a partial frame around numbers (open at the bottom), which indicated that the framed number was to be multiplied by 100,000, as illustrated above (Menninger 1992, p. 44; Cajori 1993, p. 32). In more recent practice, the strokes were sometimes written only on the sides, e.g., |X| (Cajori 19993, p. 32). It should also be noted that the Romans themselves never wrote M for 1000, but instead wrote (I) for 1,000, (I)(I) for 2,000, etc., and also occasionally wrote IM, IIM, etc. (Menninger 1992, p. 281; Cajori 1993, p. 32). However, in the Middle Ages, the use of M became quite common. The Romans sometimes used multiple parentheses to denote nested multiplications by 10, so (I) for 1,000, ((I)) for 10,000, (((I))) for 100,000, etc. (Cajori 1993, p. 33).

The Romans also occasionally used a VINCULUM (called a titulus in the Middle Ages) over a Roman numeral to indicate multiplication by 1000, so $\bar{\text{I}} = 1000$, $\bar{\text{II}} = 2000$, etc. (Menninger 1992, p. 281; Cajori 1993, p. 32).

Roman numerals are encountered in the release year for movies and occasionally on the numerals on the faces of watches and clocks, but in few other modern instances. They do have the advantage that ADDITION can be done "symbolically" (and without worrying about the "place" of a given DIGIT) by simply combining all the symbols together, grouping, writing groups of five Is as V, groups of two Vs as X, etc.

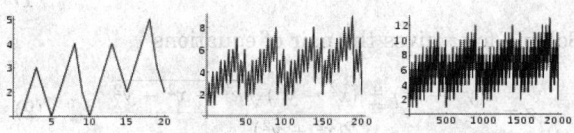

The number of characters in the Roman numerals for 1, 2, 3, 4, 5, 6, 7, 8, 9, 10, ... (i.e, I, II, III, IV, V, VI, VII, VIII, IX, X, ...) are 1, 2, 3, 2, 1, 2, 3, 4, 2, 1, 2, 3, 4, ... (Sloane's A006968). This leads to a scale-invariant FRACTAL-like stairstep pattern which rises in steps then falls abruptly.

References

Cajori, F. *A History of Mathematical Notations, 2 vols. Bound as One, Vol. 1: Notations in Elementary Mathematics.* New York: Dover, pp. 30–37, 1993.

Menninger, K. *Number Words and Number Symbols: A Cultural History of Numbers.* New York: Dover, pp. 44–45 and 281, 1992.

Neugebauer, O. *The Exact Sciences in Antiquity, 2nd ed.* New York: Dover, pp. 4–5, 1969.

Sloane, N. J. A. Sequences A006968/M0417 in "An On-Line Version of the Encyclopedia of Integer Sequences." http://www.research.att.com/~njas/sequences/eisonline.html.

Wells, D. *The Penguin Dictionary of Curious and Interesting Numbers.* Middlesex, England: Penguin Books, pp. 60 and 79, 1986.

Roman Surface

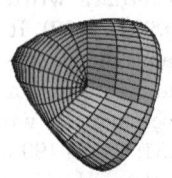 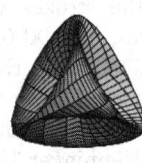

A QUARTIC NONORIENTABLE SURFACE, also known as the STEINER SURFACE. The Roman surface is one of the three possible surfaces obtained by sewing a MÖBIUS STRIP to the edge of a DISK. The other two are the BOY SURFACE and CROSS-CAP, all of which are homeomorphic to the REAL PROJECTIVE PLANE (Pinkall 1986).

The center point of the Roman surface is an ordinary TRIPLE POINT with $(\pm 1, 0, 0) = (0, \pm 1, 0) = (0, 0, \pm 1)$, and the six endpoints of the three lines of self-intersection are singular PINCH POINTS, also known as WHITNEY SINGULARITIES. The Roman surface is essentially six CROSS-CAPS stuck together and contains a double INFINITY of CONICS.

The Roman surface can given by the equation

$$\left(x^2 + y^2 + z^2 - k^2\right)^2 = \left[(z-k)^2 - 2x^2\right]\left[(z+k)^2 - 2y^2\right]. \tag{1}$$

Solving for z gives the pair of equations

$$z = \frac{k(y^2 - x^2) \pm (x^2 - y^2)\sqrt{k^2 - x^2 - y^2}}{2(x^2 + y^2)}. \tag{2}$$

If the surface is rotated by $45°$ about the z-AXIS via the ROTATION MATRIX

$$\mathsf{R}_z(45°) = \frac{1}{\sqrt{2}} \begin{bmatrix} 1 & 1 & 0 \\ -1 & 1 & 0 \\ 0 & 0 & 1 \end{bmatrix} \tag{3}$$

to give

$$\begin{bmatrix} x' \\ y' \\ z' \end{bmatrix} = \mathsf{R}_z(45°) \begin{bmatrix} x \\ y \\ z \end{bmatrix}, \tag{4}$$

then the simple equation

$$x^2 y^2 + x^2 z^2 + y^2 z^2 + 2kxyz = 0 \tag{5}$$

results. The Roman surface can also be generated using the general method for NONORIENTABLE SURFACES using the polynomial function

$$\mathbf{f}(x, y, z) = (xy, yz, zx) \tag{6}$$

(Pinkall 1986). Setting

$$x = \cos u \sin v \tag{7}$$

$$y = \sin u \sin v \tag{8}$$

$$z = \cos v \tag{9}$$

in the former gives

$$x(u, v) = \tfrac{1}{2} \sin(2u) \sin^2 v \tag{10}$$

$$y(u, v) = \tfrac{1}{2} \sin u \cos(2v) \tag{11}$$

$$z(u, v) = \tfrac{1}{2} \cos u \sin(2v) \tag{12}$$

for $u \in [0, 2\pi)$ and $v \in [-\pi/2, \pi/2]$. Flipping $\sin v$ and $\cos v$ and multiplying by 2 gives the form shown by Wang.

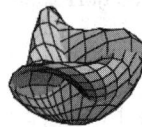 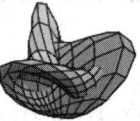 

A HOMOTOPY (smooth deformation) between the Roman surface and BOY SURFACE is given by the equations

$$x(u, v) = \frac{\sqrt{2} \cos(2u) \cos^2 v + \cos u \sin(2v)}{2 - \alpha\sqrt{2} \sin(3u) \sin(2v)} \tag{13}$$

$$y(u, v) = \frac{\sqrt{2} \sin(2u) \cos^2 v - \sin u \sin(2v)}{2 - \alpha\sqrt{2} \sin(3u) \sin(2v)} \tag{14}$$

$$z(u, v) = \frac{3 \cos^2 v}{2 - \alpha\sqrt{2} \sin(3u) \sin(2v)} \tag{15}$$

for $u \in [-\pi/2, \pi/2]$ and $v \in [0, \pi]$ as α varies from 0 to 1. $\alpha = 0$ corresponds to the Roman surface and $\alpha = 1$ to the BOY SURFACE (Wang).

See also BOY SURFACE, CROSS-CAP, HEPTAHEDRON, MÖBIUS STRIP, NONORIENTABLE SURFACE, QUARTIC SURFACE, STEINER SURFACE

References

Fischer, G. (Ed.). *Mathematical Models from the Collections of Universities and Museums.* Braunschweig, Germany: Vieweg, p. 19, 1986.

Fischer, G. (Ed.). Plates 42–44 and 108–114 in *Mathematische Modelle/Mathematical Models, Bildband/Photograph Volume.* Braunschweig, Germany: Vieweg, pp. 42–44 and 108–109, 1986.

Gray, A. "Steiner's Roman Surface." *Modern Differential Geometry of Curves and Surfaces with Mathematica, 2nd ed.* Boca Raton, FL: CRC Press, pp. 331–333, 1997.

Nordstrand, T. "Steiner's Roman Surface." http://www.uib.no/people/nfytn/steintxt.htm.

Pinkall, U. *Mathematical Models from the Collections of Universities and Museums* (Ed. G. Fischer). Braunschweig, Germany: Vieweg, p. 64, 1986.

Roman Symbol

$$\lfloor n \rceil \equiv \begin{cases} n & \text{for } n \neq 0 \\ 1 & \text{for } n = 0. \end{cases}$$

See also ROMAN FACTORIAL, HARMONIC LOGARITHM

References

Roman, S. "The Logarithmic Binomial Formula." *Amer. Math. Monthly* **99**, 641–648, 1992.

Romberg Integration

A powerful NUMERICAL INTEGRATION technique which uses k refinements of the extended TRAPEZOIDAL RULE to remove error terms less than order $\mathcal{O}(N^{-2k})$. The routine advocated by Press *et al.* (1992) makes use of NEVILLE'S ALGORITHM.

References

Acton, F. S. *Numerical Methods That Work, 2nd printing.* Washington, DC: Math. Assoc. Amer., pp. 106–107, 1990.
Dahlquist, G. and Bjorck, A. §7.4.1–7.4.2 in *Numerical Methods.* Englewood Cliffs, NJ: Prentice-Hall, 1974.
Press, W. H.; Flannery, B. P.; Teukolsky, S. A.; and Vetterling, W. T. "Romberg Integration." §4.3 in *Numerical Recipes in FORTRAN: The Art of Scientific Computing, 2nd ed.* Cambridge, England: Cambridge University Press, pp. 134–135, 1992.
Ralston, A. and Rabinowitz, P. §4.10 in *A First Course in Numerical Analysis, 2nd ed.* New York: McGraw-Hill, 1978.
Stoer, J.; and Bulirsch, R. §3.4–3.5 in *Introduction to Numerical Analysis.* New York: Springer-Verlag, 1980.
Ueberhuber, C. W. "Romberg Formulas." §12.3.4 in *Numerical Computation 2: Methods, Software, and Analysis.* Berlin: Springer-Verlag, pp. 110–111, 1997.

Rook Number

The rook numbers r_n^B of an $n \times n$ BOARD B are the number of subsets of size n such that no two elements have the same first or second coordinate. In other word, it is the number of ways of placing n rooks on B such that none attack each other. The rook numbers of a board determine the rook numbers of the complementary board $\bar{B}$, defined to be $\mathbf{d} \times \mathbf{d} \backslash B$. This is known as the ROOK RECIPROCITY THEOREM. The first few rook numbers are 1, 2, 7, 23, 115, 694, 5282, 46066, ... (Sloane's A000903). For an $n \times n$ board, each $n \times n$ PERMUTATION MATRIX corresponds to an allowed configuration of rooks.

See also ROOK RECIPROCITY THEOREM

References

Sloane, N. J. A. Sequences A000903/M1761 in "An On-Line Version of the Encyclopedia of Integer Sequences." http://www.research.att.com/~njas/sequences/eisonline.html.

Rook Reciprocity Theorem

$$\sum_{k=0}^{d} r_k^B (d-k)! x^k = \sum_{k=0}^{d} (-1)^k r_k^B (d-k)! x^k (x+1)^{d-k}.$$

References

Chow, T. Y. "The Path-Cycle Symmetric Function of a Digraph." *Adv. Math.* **118**, 71–98, 1996.
Chow, T. "A Short Proof of the Rook Reciprocity Theorem." *Electronic J. Combinatorics* **3**, R10 1–2, 1996. http://www.combinatorics.org/Volume_3/volume3.html#R10.
Goldman, J. R.; Joichi, J. T.; and White, D. E. "Rook Theory I. Rook Equivalence of Ferrers Boards." *Proc. Amer. Math. Soc.* **52**, 485–492, 1975.
Riordan, J. *An Introduction to Combinatorial Analysis.* New York: Wiley, 1958.

Rooks Problem

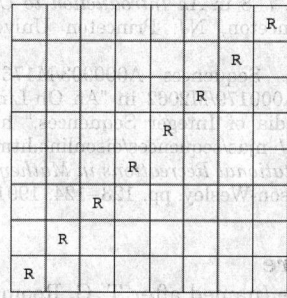

The rook is a CHESS piece which may move any number of spaces either horizontally or vertically per move. The maximum number of nonattacking rooks which may be placed on an $n \times n$ CHESSBOARD is n. This arrangement is achieved by placing the rooks along the diagonal (Madachy 1979). The total number of ways of placing n nonattacking rooks on an $n \times n$ board is $n!$ (Madachy 1979, p. 47). The number of rotationally and reflectively inequivalent ways of placing n nonattacking rooks on an $n \times n$ board are 1, 2, 7, 23, 115, 694, ... (Sloane's A000903; Dudeney 1970, p. 96; Madachy 1979, pp. 46–54).

The minimum number of rooks needed to occupy or attack all spaces on an 8×8 CHESSBOARD is 8 (Madachy 1979), arranged in the same orientation as above.

Consider an $n \times n$ chessboard with the restriction that, for every subset of $\{1, \ldots, n\}$, a rook may not be put in column $s + j \pmod{n}$ when on row j, where the rows are numbered 0, 1, ..., $n - 1$. Vardi (1991) denotes the number of rook solutions so restricted as rook(s, n). rook$(\{1\}, n)$ is simply the number of DERANGEMENTS on n symbols, known as a SUBFACTORIAL. The first few values are 1, 2, 9, 44, 265, 1854, ... (Sloane's A000166). rook$(\{1, 2\}, n)$ is a solution to the MARRIED COUPLES PROBLEM, sometimes known as MÉNAGE NUMBERS. The first few MÉNAGE NUMBERS are -1, 1, 0, 2, 13, 80, 579, ... (Sloane's A000179).

Although simple formulas are not known for general $\{1, \ldots, p\}$, RECURRENCE RELATIONS can be used to compute rook($\{1, \ldots, p\}$, n) in polynomial time for $p = 3, \ldots, 6$ (Metropolis *et al.* 1969, Minc 1978, Vardi 1991).

See also CHESS, MÉNAGE NUMBER, ROOK NUMBER, ROOK RECIPROCITY THEOREM

References

Dudeney, H. E. "The Eight Rooks." §295 in *Amusements in Mathematics*. New York: Dover, p. 88, 1970.

Kraitchik, M. "The Problem of the Rooks" and "Domination of the Chessboard." §10.2 and 10.4 in *Mathematical Recreations*. New York: W. W. Norton, pp. 240–247 and 255–256, 1942.

Madachy, J. S. *Madachy's Mathematical Recreations*. New York: Dover, pp. 36–37, 1979.

Metropolis, M.; Stein, M. L.; and Stein, P. R. "Permanents of Cyclic (0, 1) Matrices." *J. Combin. Th.* **7**, 291–321, 1969.

Minc, H. §3.1 in *Permanents*. Reading, MA: Addison-Wesley, 1978.

Riordan, J. Chs. 7–8 in *An Introduction to Combinatorial Analysis*. Princeton, NJ: Princeton University Press, 1978.

Sloane, N. J. A. Sequences A000903/M1761, A000166/M1937, and A000179/M2062 in "An On-Line Version of the Encyclopedia of Integer Sequences." http://www.research.att.com/~njas/sequences/eisonline.html.

Vardi, I. *Computational Recreations in Mathematica*. Reading, MA: Addison-Wesley, pp. 123–124, 1991.

Room Square

A Room square (named after T. G. Room) of order n (for n EVEN) is an arrangement in an $(n-1) \times (n-1)$ SQUARE MATRIX of n objects such that each cell is either empty or holds exactly two different objects. Furthermore, each object appears once in each row and column and each unordered pair occupies exactly one cell. The Room square of order 2 is shown below.

1,2

The Room square of order 8 is

1,8		5,7		3,4	2,6	
3,7	2,8			6,1		4,5
5,6	4,1	3,8			7,2	
	6,7	5,2	4,8			1,3
2,4		7,1	6,3	5,8		
	3,5		1,2	7,4	6,8	
		4,6		2,3	1,5	7,8

References

Dinitz, J. H. and Stinson, D. R. In *Contemporary Design Theory: A Collection of Surveys* (Ed. J. H. Dinitz and D. R. Stinson). New York: Wiley, 1992.

Gardner, M. "Mathematical Games: On the Remarkable Császár Polyhedron and Its Applications in Problem Solving." *Sci. Amer.* **232**, 102–107, May 1975.

Gardner, M. *Time Travel and Other Mathematical Bewilderments*. New York: W. H. Freeman, pp. 146–147 and 151–152, 1988.

Mullin, R. C. and Nemeth, E. "On Furnishing Room Squares." *J. Combin. Th.* **7**, 266–272, 1969.

Mullin, R. D. and Wallis, W. D. "The Existence of Room Squares." *Aequationes Math.* **13**, 1–7, 1975.

O'Shaughnessy, C. D. "On Room Squares of Order $6m + 2$." *J. Combin. Th.* **13**, 306–314, 1972.

Room, T. G. "A New Type of Magic Square" (Note 2569). *Math. Gaz.* **39**, 307, 1955.

Wallis, W. D. "Solution of the Room Square Existence Problem." *J. Combin. Th.* **17**, 379–383, 1974.

Wallis, W. D.; Street, A. P.; and Wallis, J. S. *Combinatorics: Room Squares, Sum-free Sets, Hadamard Matrices*. New York: Springer-Verlag, 1972.

Root

The roots (sometimes also called "zeros") of an equation

$$f(x) = 0 \tag{1}$$

are the values of x for which the equation is satisfied.

The FUNDAMENTAL THEOREM OF ALGEBRA states that every POLYNOMIAL equation of degree n has exactly n roots, where some roots may have a multiplicity greater than 1 (in which case they are said to be degenerate). In *Mathematica*, the expression Root[f, k] represents the kth root of the POLYNOMIAL $f(x) = 0$.

To find the nth roots of a COMPLEX NUMBER, solve the equation $z^n = w$. Then

$$z^n = |z|^n [\cos(n\theta) + i \sin(n\theta)] = |w|(\cos \phi + i \sin \phi), \tag{2}$$

so

$$|z| = |w|^{1/n} \tag{3}$$

and

$$\arg(z) = \frac{\phi}{n}. \tag{4}$$

Rolle proved that any number has n nth roots (Boyer 1968, p. 476). Householder (1970) gives an algorithm for constructing root-finding algorithms with an arbitrary order of convergence. Special root-finding techniques can often be applied when the function in question is a POLYNOMIAL.

See also BAILEY'S METHOD, BERNOULLI'S METHOD, BISECTION PROCEDURE, BRENT'S METHOD, CROUT'S METHOD, DESCARTES' SIGN RULE, FALSE POSITION METHOD, FUNDAMENTAL THEOREM OF SYMMETRIC FUNCTIONS, GRAEFFE'S METHOD, HALLEY'S IRRATIONAL FORMULA, HALLEY'S METHOD, HALLEY'S RATIONAL FORMULA, HORNER'S METHOD,

HOUSEHOLDER'S METHOD, HUTTON'S METHOD, INSIDE-OUTSIDE THEOREM, ISOGRAPH, JENKINS-TRAUB METHOD, LAGUERRE'S METHOD, LAMBERT'S METHOD, LEHMER-SCHUR METHOD, LIN'S METHOD, MAEHLY'S PROCEDURE, MULLER'S METHOD, MULTIPLICITY, NEWTON'S METHOD, POLYNOMIAL, POLYNOMIAL ROOTS, RIDDERS' METHOD, ROOT DRAGGING THEOREM, ROOT EXTRACTION, ROUCHÉ'S THEOREM, SCHRÖDER'S METHOD, SECANT METHOD, SIMPLE ROOT, STURM FUNCTION, STURM THEOREM, TANGENT HYPERBOLAS METHOD, VANISH, WEIERSTRASS APPROXIMATION THEOREM, ZERO SET

References

Arfken, G. "Appendix 1: Real Zeros of a Function." *Mathematical Methods for Physicists, 3rd ed.* Orlando, FL: Academic Press, pp. 963–967, 1985.

Boyer, C. B. *A History of Mathematics.* New York: Wiley, 1968.

Householder, A. S. *The Numerical Treatment of a Single Nonlinear Equation.* New York: McGraw-Hill, 1970.

Kravanja, P. and van Barel, M. *Computing the Zeros of Analytic Functions.* Berlin: Springer-Verlag, 2000.

McNamee, J. M. "A Bibliography on Roots of Polynomials." *J. Comput. Appl. Math.* **47**, 391–392, 1993.

McNamee, J. M. "A Bibliography on Roots of Polynomials." http://www.elsevier.com/homepage/sac/cam/mcnamee/.

Press, W. H.; Flannery, B. P.; Teukolsky, S. A.; and Vetterling, W. T. "Roots of Polynomials." §9.5 in *Numerical Recipes in FORTRAN: The Art of Scientific Computing, 2nd ed.* Cambridge, England: Cambridge University Press, pp. 362–372, 1992.

Whittaker, E. T. and Robinson, G. "The Numerical Solution of Algebraic and Transcendental Equations." Ch. 6 in *The Calculus of Observations: A Treatise on Numerical Mathematics, 4th ed.* New York: Dover, pp. 78–131, 1967.

Root (Lie Algebra)

The roots of a SEMISIMPLE LIE ALGEBRA $\mathfrak{g}$ are the WEIGHTS occurring in its ADJOINT REPRESENTATION. The set of roots form the ROOT SYSTEM, and are completely determined by $\mathfrak{g}$. It is possible to choose a set of POSITIVE ROOTS, every root α is either positive or $-\alpha$ is positive. The SIMPLE ROOTS are the positive roots which cannot be written as a sum of positive roots.

The simple roots can be considered as a LINEARLY INDEPENDENT finite subset of EUCLIDEAN SPACE, and they generate the ROOT LATTICE. For example, in the SPECIAL LIE ALGEBRA $sl_2\mathbb{C}$ of two by two matrices with zero TRACE, has a basis given by the matrices

$$H = \begin{bmatrix} 1 & 0 \\ 0 & -1 \end{bmatrix}, \, X = \begin{bmatrix} 0 & 1 \\ 0 & 0 \end{bmatrix}, \, Y = \begin{bmatrix} 0 & 0 \\ 1 & 0 \end{bmatrix}.$$

The ADJOINT REPRESENTATION is given by the BRACKETS

$$\mathrm{ad}(H(X)) = [H, \, X] = 2X$$
$$\mathrm{ad}(H(Y)) = [H, \, Y] = -2Y,$$

so there are two roots of $\mathfrak{sl}_2$ given by $\alpha(H) = 2$ and $-\alpha(H) = -2$. The RANK of $\mathfrak{sl}_2\mathbb{C}$ is one, and it has one positive root.

See also CARTAN MATRIX, LIE ALGEBRA, SEMISIMPLE LIE ALGEBRA, WEIGHT (LIE ALGEBRA), WEYL GROUP

References

Fulton, W. and Harris, J. *Representation Theory.* New York:Springer-Verlag, 1991.

Jacobson, N. *Lie Algebras.* New York: Dover, 1979.

Knapp, A. *Lie Groups Beyond an Introduction.* Boston, MA: Birkhäuser, 1996.

Root (Radical)

The nth root (or "nth RADICAL") of a quantity z is a value r such that $z = r^n$, and therefore is the INVERSE FUNCTION to the taking of a POWER. The nth root is denoted $r = \sqrt[n]{z}$ or, using POWER notation, $r = z^{1/n}$. The special case of the SQUARE ROOT is denoted $\sqrt{z}$.

The quantities for which a general FUNCTION equals 0 are also called ROOTS, or sometimes ZEROS.

See also CUBE ROOT, RADICAL, ROOT, SQUARE ROOT, VINCULUM

Root (Tree)

ROOT NODE

Root Dragging Theorem

If any of the ROOTS of a POLYNOMIAL are increased, then *all* of the critical points increase.

References

Anderson, B. "Polynomial Root Dragging." *Amer. Math. Monthly* **100**, 864–866, 1993.

Root Extraction

The operation of taking an nth ROOT of a number.

See also ADDITION, DIVISION, MULTIPLICATION, ROOT (RADICAL), SUBTRACTION

Root Lattice

The root lattice of a SEMISIMPLE LIE ALGEBRA is the DISCRETE LATTICE generated by the ROOTS in $\mathfrak{h}^*$, the DUAL SPACE to the CARTAN SUBALGEBRA.

See also CARTAN MATRIX, LIE ALGEBRA, ROOT (LIE ALGEBRA), ROOT SYSTEM, SEMISIMPLE LIE ALGEBRA, WEIGHT (LIE ALGEBRA), WEIGHT LATTICE, WEYL CHAMBER, WEYL GROUP

References

Fulton, W. and Harris, J. *Representation Theory.* New York: Springer-Verlag, 1991.

Jacobson, N. *Lie Algebras.* New York: Dover, 1979.

Knapp, A. *Lie Groups Beyond an Introduction.* Boston, MA: Birkhäuser, 1996.

Root Linear Coefficient Theorem

The sum of the reciprocals of ROOTS of an equation equals the NEGATIVE COEFFICIENT of the linear term in the MACLAURIN SERIES.

See also NEWTON'S RELATIONS

Root Node

A special node which is designated to turn a TREE into a ROOTED TREE. The root is sometimes also called "EVE" or an "ENDPOINT" (Saaty and Kainen 1986, p. 30) and each of the nodes which is one EDGE further away from a given EDGE is called a CHILD. Nodes connected to the same node are then called SIBLINGS.

See also CHILD, ROOTED TREE, SIBLING, TREE

References

Harary, F. *Graph Theory.* Reading, MA: Addison-Wesley, p. 187, 1994.
Saaty, T. L. and Kainen, P. C. *The Four-Color Problem: Assaults and Conquest.* New York: Dover, 1986.

Root of Unity

The nth ROOTS of UNITY are ROOTS $e^{2\pi i k/n}$ of the CYCLOTOMIC EQUATION

$$x^n = 1,$$

which are known as the DE MOIVRE NUMBERS. The notations ζ_k, ϵ_k, and ϵ_k are variously used to denote the kth nth root of unity.

$+1$ is always an nth root of unity, but -1 is such a root only if n is even.

See also CYCLOTOMIC EQUATION, CYCLOTOMIC POLYNOMIAL, DE MOIVRE'S IDENTITY, DE MOIVRE NUMBER, PRIMITIVE ROOT OF UNITY, PRINCIPAL ROOT OF UNITY, UNITY

References

Courant, R. and Robbins, H. "De Moivre's Formula and the Roots of Unity." §5.3 in *What is Mathematics?: An Elementary Approach to Ideas and Methods, 2nd ed.* Oxford, England: Oxford University Press, pp. 98–100, 1996.
Lam, T. Y. and Leung, K. H. "On Vanishing Sums of Roots of Unity." *J. Algebra* **224**, 91–109, 2000.
Nagell, T. "Arithmetical Properties of the Roots of Unity." Ch. 5 in *Introduction to Number Theory.* New York: Wiley, pp. 156–187, 1951.

Root System

Let E be a Euclidean space, (β, α) be the dot product, and denote the reflection in the hyperplane $P_\alpha = \{\beta \in E | (\beta, \alpha) = 0\}$ by

$$\sigma_\alpha(\beta) = \beta - 2(\beta, \alpha)/(\alpha, \alpha)\alpha = \beta - \langle \beta, \alpha \rangle \alpha,$$

where

$$\langle \beta, \alpha \rangle = \frac{2(\beta, \alpha)}{(\alpha, \alpha)}.$$

Then a subset R of the Euclidean space E is called a root system in E if:

1. R is finite, SPANS E, and does not contain 0,
2. If $\alpha \in R$, the reflection σ_α leaves R invariant, and
3. If α, $\beta \in R$, then $\langle \beta, \alpha \rangle \in \mathbb{Z}$.

The ROOTS of a SEMISIMPLE LIE ALGEBRA are a root system, in a real subspace of the DUAL SPACE to the CARTAN SUBALGEBRA. In this case, the reflections W_α generate the WEYL GROUP, which is the symmetry group of the root system.

See also CARTAN MATRIX, LIE ALGEBRA, MACDONALD'S CONSTANT-TERM CONJECTURE, REDUCED ROOT SYSTEM, ROOT (LIE ALGEBRA), SEMISIMPLE LIE ALGEBRA, WEIGHT (LIE ALGEBRA), WEYL CHAMBER, WEYL'S DENOMINATOR FORMULA, WEYL GROUP

References

Andrews, G. E. *q-Series: Their Development and Application in Analysis, Number Theory, Combinatorics, Physics, and Computer Algebra.* Providence, RI: Amer. Math. Soc., p. 40, 1986.
Fulton, W. and Harris, J. *Representation Theory.* New York: Springer-Verlag, 1991.
Humphrey, J. E. *Introduction to Lie Algebras and Representation Theory.* New York: Springer-Verlag, p. 42, 1972.
Jacobson, N. *Lie Algebras.* New York: Dover, 1979.
Knapp, A. *Lie Groups Beyond an Introduction.* Boston, MA: Birkhäuser, 1996.

Root Test

Let u_k be a SERIES with POSITIVE terms, and let

$$\rho \equiv \lim_{k \to \infty} u_k^{1/k}.$$

1. If $\rho < 1$, the SERIES CONVERGES.
2. If $\rho > 1$ or $\rho = \infty$, the SERIES DIVERGES.
3. If $\rho = 1$, the SERIES may CONVERGE or DIVERGE.

This test is also called the Cauchy root test.

See also CONVERGENCE TESTS

References

Arfken, G. *Mathematical Methods for Physicists, 3rd ed.* Orlando, FL: Academic Press, pp. 281–282, 1985.
Bromwich, T. J. I'a and MacRobert, T. M. *An Introduction to the Theory of Infinite Series, 3rd ed.* New York: Chelsea, pp. 31–39, 1991.

Rooted Tree

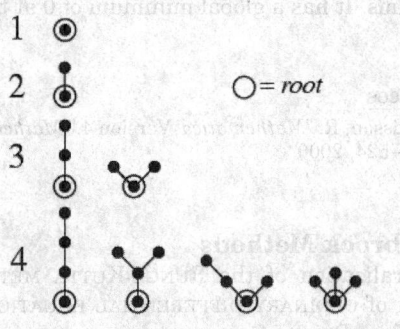

rooted trees

A TREE with a single special ("labeled"rpar; node called the "ROOT" or "eve." A tree which is not rooted is sometimes called a FREE TREE. Denote the number of rooted trees with n nodes by T_n, then the GENERATING FUNCTION is

$$T(x) \equiv \sum_{n=0}^{\infty} T_n x^n = x + x^2 + 2x^3 + 4x^4 + 9x^5 + 20x^6$$

$$+48x^7 + 115x^8 + 286x^9 + 719x^{10} + \ldots \quad (1)$$

(Sloane's A000081). This POWER SERIES satisfies

$$T(x) = x \exp\left[\sum_{r=1}^{\infty} \frac{1}{r} T(x^r)\right] \quad (2)$$

$$t(x) = T(x) - \tfrac{1}{2}\left[T^2(x) - T(x^2)\right], \quad (3)$$

where $t(x)$ is the GENERATING FUNCTION for unrooted TREES. A GENERATING FUNCTION for T_n can be written using a product involving *the sequence itself* as

$$x \prod_{n=1}^{\infty} \frac{1}{(1-x^n)^{T_n}} = \sum_{n=1}^{\infty} T_n x^n. \quad (4)$$

The number of rooted trees can also be calculated from the RECURRENCE RELATION

$$T_{i+1} = \frac{1}{i} \sum_{j=1}^{i} \left(\sum_{d|j} d T_d\right) T_{i-j+1}, \quad (5)$$

with $T_0 = 0$ and $T_1 = 1$, where the second sum is over all d which DIVIDE j (Finch).

See also ORDERED TREE, PLANTED TREE, RED-BLACK TREE, WEAKLY BINARY TREE

References

Finch, S. "Favorite Mathematical Constants." http://www.mathsoft.com/asolve/constant/otter/otter.html.
Harary, F. *Graph Theory.* Reading, MA: Addison-Wesley, pp. 187–190 and 232, 1994.
Nijenhuis, A. and Wilf, H. *Combinatorial Algorithms for Computers and Calculators, 2nd ed.* New York: Academic Press, 1978.
Ruskey, F. "Information on Rooted Trees." http://www.theory.csc.uvic.ca/~cos/inf/tree/RootedTree.html.

Sloane, N. J. A. Sequences A000081/M1180 in "An On-Line Version of the Encyclopedia of Integer Sequences." http://www.research.att.com/~njas/sequences/eisonline.html.
Wilf, H. S. *Combinatorial Algorithms: An Update.* Philadelphia, PA: SIAM, 1989.

Root-Mean-Square

The root-mean-square (RMS) of a variate x, sometimes called the QUADRATIC MEAN, is the SQUARE ROOT of the mean squared value of x:

$$R(x) \equiv \sqrt{\langle x^2 \rangle} \quad (1)$$

$$= \begin{cases} \sqrt{\dfrac{\sum_{i=1}^{n} x_i^2}{n}} & \text{for a discrete distribution} \\[2em] \sqrt{\dfrac{\int P(x)x^2\,dx}{\int P(x)\,dx}} & \text{for a continuous distribution.} \end{cases} \quad (2)$$

Hoehn and Niven (1985) show that

$$R(a_1 + c, \, a_2 + c, \, \ldots, \, a_n + c) < c + R(a_1, \, a_2, \ldots, \, a_n) \quad (3)$$

for any POSITIVE constant c.

Physical scientists often use the term root-mean-square as a synonym for STANDARD DEVIATION when they refer to the SQUARE ROOT of the mean squared deviation of a signal from a given baseline or fit.

See also ARITHMETIC-GEOMETRIC MEAN, ARITHMETIC-HARMONIC MEAN, GENERALIZED MEAN, GEOMETRIC MEAN, HARMONIC MEAN, HARMONIC-GEOMETRIC MEAN, MEAN, MEDIAN (STATISTICS), STANDARD DEVIATION, VARIANCE

References

Hoehn, L. and Niven, I. "Averages on the Move." *Math. Mag.* **58**, 151–156, 1985.
Kenney, J. F. and Keeping, E. S. "Root Mean Square." §4.15 in *Mathematics of Statistics, Pt. 1, 3rd ed.* Princeton, NJ: Van Nostrand, pp. 59–60, 1962.

RootSum

POLYNOMIAL ROOTS

Rosatti's Theorem

There is a one-to-one correspondence between the sets of equivalent correspondences (not of value 0) on an irreducible curve of GENUS (CURVE) p, and the rational COLLINEATIONS of a projective space of $2p-1$ dimensions which leave invariant a space of $p-1$ dimensions. The number of linearly independent correspondences will be that of linearly independent COLLINEATIONS.

References

Coolidge, J. L. *A Treatise on Algebraic Plane Curves.* New York: Dover, p. 339, 1959.

Rose

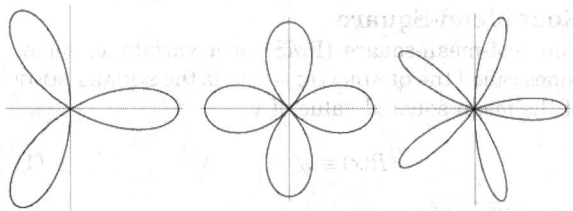

A curve which has the shape of a petalled flower. This curve was named RHODONEA by the Italian mathematician Guido Grandi between 1723 and 1728 because it resembles a rose (MacTutor Archive). The polar equation of the rose is

$$r = a\,\sin(n\theta),$$

or

$$r = a\,\cos(n\theta).$$

If n is ODD, the rose is n-petalled. If n is EVEN, the rose is $2n$-petalled. If n is IRRATIONAL, then there are an infinite number of petals.

The QUADRIFOLIUM is the rose with $n = 2$. The rose is the RADIAL CURVE of the EPICYCLOID.

See also DAISY, MAURER ROSE, STARR ROSE

References

Beyer, W. H. *CRC Standard Mathematical Tables, 28th ed.* Boca Raton, FL: CRC Press, pp. 223–224, 1987.
Hall, L. "Trochoids, Roses, and Thorns--Beyond the Spirograph." *College Math. J.* **23**, 20–35, 1992.
Lawrence, J. D. *A Catalog of Special Plane Curves.* New York: Dover, pp. 175–177, 1972.
MacTutor History of Mathematics Archive. "Rhodonea Curves." http://www-groups.dcs.st-and.ac.uk/~history/Curves/Rhodonea.html.
Wagon, S. "Roses." §4.1 in *Mathematica in Action.* New York: W. H. Freeman, pp. 96–102, 1991.

Rosenbrock Function

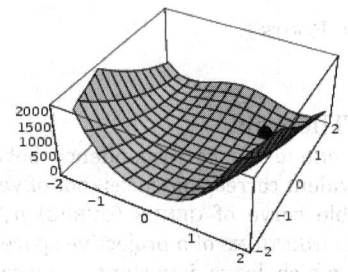

The function

$$f(x,\,y) = (1-x)^2 + 105\left(y - x^2\right)^2$$

that is often used as a test problem for optimization algorithms. It has a global minimum of 0 at the point (1, 1).

References

Germundsson, R. "*Mathematica* Version 4." *Mathematica J.* **7**, 497–524, 2000.

Rosenbrock Methods

A generalization of the RUNGE-KUTTA METHOD for solution of ORDINARY DIFFERENTIAL EQUATIONS, also called KAPS-RENTROP METHODS.

See also RUNGE-KUTTA METHOD

References

Press, W. H.; Flannery, B. P.; Teukolsky, S. A.; and Vetterling, W. T. *Numerical Recipes in FORTRAN: The Art of Scientific Computing, 2nd ed.* Cambridge, England: Cambridge University Press, pp. 730–735, 1992.

Rössler Model

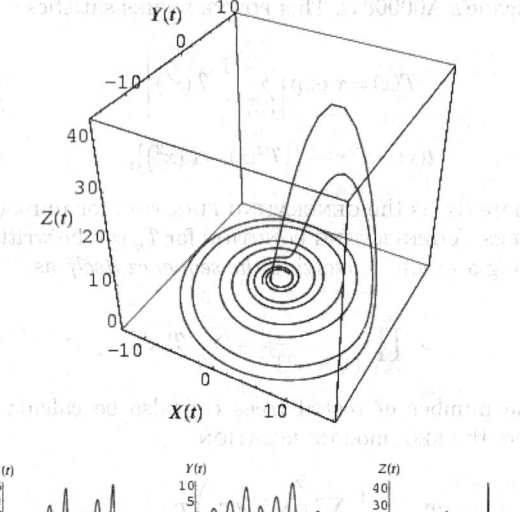

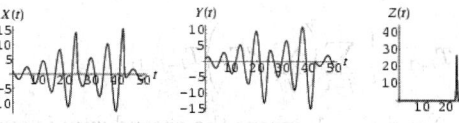

The nonlinear 3-D MAP

$$\dot{X} = (-Y + Z)$$
$$\dot{Y} = X + aY$$
$$\dot{Z} = b + XZ - cZ.$$

See also LORENZ SYSTEM

References

Dickau, R. M. "Rössler Attractor." http://forum.swarthmore.edu/advanced/robertd/rossler.html.
Peitgen, H.-O.; Jürgens, H.; and Saupe, D. §12.3 in *Chaos and Fractals: New Frontiers of Science.* New York: Springer-Verlag, pp. 686–696, 1992.

RotateLeft

CYCLIC PERMUTATION

RotateRight

CYCLIC PERMUTATION

Rotation

The turning of an object or coordinate system by an ANGLE about a fixed point. A rotation is an ORIENTATION-PRESERVING ORTHOGONAL TRANSFORMATION. EULER'S ROTATION THEOREM states that an arbitrary rotation can be parameterized using three parameters. These parameters are commonly taken as the EULER ANGLES. Rotations can be implemented using ROTATION MATRICES.

The rotation SYMMETRY OPERATION for rotation by $360°/n$ is denoted "n." For periodic arrangements of points (, the CRYSTALLOGRAPHY RESTRICTION gives the only allowable rotations as 1, 2, 3, 4, and 6.

See also DILATION, EUCLIDEAN GROUP, EULER ANGLES, EULER PARAMETERS, EULER'S ROTATION THEOREM, EXPANSION, HALF-TURN, IMPROPER ROTATION, INFINITESIMAL ROTATION, INVERSION OPERATION, MIRROR PLANE, ORIENTATION-PRESERVING, ORTHOGONAL TRANSFORMATION, REFLECTION, ROTATION FORMULA, ROTATION GROUP, ROTATION MATRIX, ROTATION OPERATOR, ROTOINVERSION, SHIFT, SPIRAL SIMILARITY, TRANSLATION

References

Addington, S. "The Four Types of Symmetry in the Plane." http://forum.swarthmore.edu/sum95/suzanne/symsusan.html.

Beyer, W. H. (Ed.). *CRC Standard Mathematical Tables, 28th ed.* Boca Raton, FL: CRC Press, p. 211, 1987.

Coxeter, H. S. M. and Greitzer, S. L. "Rotation." §4.2 in *Geometry Revisited.* Washington, DC: Math. Assoc. Amer., pp. 82–85, 1967.

Varshalovich, D. A.; Moskalev, A. N.; and Khersonskii, V. K. "Rotations of Coordinate Systems." §1.4 in *Quantum Theory of Angular Momentum.* Singapore: World Scientific, pp. 21–35, 1988.

Yates, R. C. "Instantaneous Center of Rotation and the Construction of Some Tangents." *A Handbook on Curves and Their Properties.* Ann Arbor, MI: J. W. Edwards, pp. 119–122, 1952.

Rotation Formula

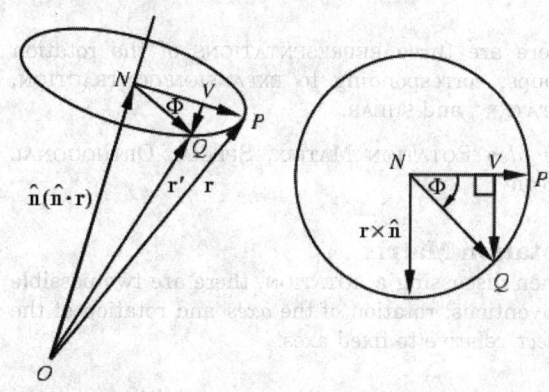

A formula which transforms a given coordinate system by rotating it through a counterclockwise angle Φ about an axis $\hat{\mathbf{n}}$. This formula is used implicitly to transform objects in VRML (virtual reality markup language) using the command Rotation {*angle nx ny nz Phi*}. Referring to the above figure (Goldstein 1980), the equation for the "fixed" vector in the transformed coordinate system (i.e., the above figure corresponds to an ALIAS TRANSFORMATION), is

$$\mathbf{r}' = \overrightarrow{ON} + \overrightarrow{NV} + \overrightarrow{VQ} \tag{1}$$

$$= \hat{\mathbf{n}}(\hat{\mathbf{n}} \cdot \mathbf{r}) + [\mathbf{r} - \hat{\mathbf{n}}(\hat{\mathbf{n}} \cdot \mathbf{r})] \cos \Phi + (\mathbf{r} \times \hat{\mathbf{n}}) \sin \Phi \tag{2}$$

$$= \mathbf{r} \cos \Phi + \hat{\mathbf{n}}(\hat{\mathbf{n}} \cdot \mathbf{r})(1 - \cos \Phi) + (\mathbf{r} \times \hat{\mathbf{n}}) \sin \Phi \tag{3}$$

(Goldstein 1980; Varshalovich *et al.* 1988, p. 24). The ANGLE Φ and unit normal $\hat{\mathbf{n}}$ may also be expressed as EULER ANGLES. In terms of the EULER PARAMETERS,

$$\mathbf{r}' = \mathbf{r}(e_0^2 - e_1^2 - e_2^2 - e_3^2) + 2\mathbf{e}(\mathbf{e} \cdot \mathbf{r}) + 2(\mathbf{r} \times \mathbf{e})e_0. \tag{4}$$

See also ALIAS TRANSFORMATION, ALIBI TRANSFORMATION, EULER ANGLES, EULER PARAMETERS, RODRIGUES' ROTATION FORMULA

References

Gibbs, J. W. and Wilson, E. B. *Vector Analysis: A Text-Book for the use of Students of Mathematics and Physics, Founded Upon the Lectures of J. Willard Gibbs.* New York: Dover, p. 338, 1960.

Goldstein, H. "Finite Rotations." §4–7 in *Classical Mechanics, 2nd ed.* Reading, MA: Addison-Wesley, pp. 164–166, 1980.

Grubin, C. "Derivation of the Quaternion Scheme via the Euler Axis and Angle." *J. Spacecraft* **7**, 1251–1263, 1970.

Hamel, G. *Theoretische Mechanik: Eine Einheitliche Einführung in die Gesamte Mechanik.* Berlin: New York: Springer-Verlag, p. 103, 1949.

Varshalovich, D. A.; Moskalev, A. N.; and Khersonskii, V. K. "Description of Rotations in Terms of Rotation Axis and Rotation Angle." §1.4.2 in *Quantum Theory of Angular Momentum.* Singapore: World Scientific, pp. 23–24, 1988.

Rotation Group

There are three REPRESENTATIONS of the rotation groups, corresponding to EXPANSION/CONTRACTION, ROTATION, and SHEAR.

See also ROTATION MATRIX, SPECIAL ORTHOGONAL GROUP

Rotation Matrix

When discussing a ROTATION, there are two possible conventions: rotation of the *axes* and rotation of the *object* relative to fixed axes.

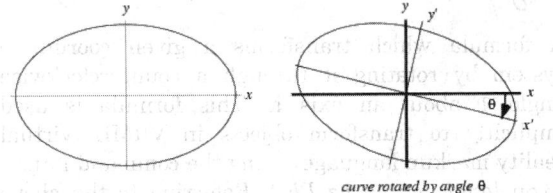

curve rotated by angle θ

In $\mathbb{R}^2$, let a curve be rotated by a clockwise ANGLE θ, so that the original axes of the curve are $\hat{\mathbf{x}}$ and $\hat{\mathbf{y}}$, and the new axes of the curve are $\hat{\mathbf{x}}'$ and $\hat{\mathbf{y}}'$. The MATRIX transforming the original curve to the rotated curve, referred to the original $\hat{\mathbf{x}}$ and $\hat{\mathbf{y}}$ axes, is

$$\mathsf{R}_\theta = \begin{bmatrix} \cos\theta & \sin\theta \\ -\sin\theta & \cos\theta \end{bmatrix}, \tag{1}$$

i.e.,

$$\mathbf{x} = \mathsf{R}_\theta \mathbf{x}'. \tag{2}$$

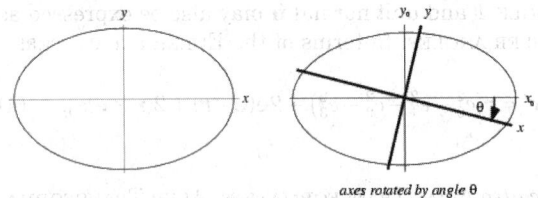

axes rotated by angle θ

On the other hand, let the *axes* with respect to which a curve is measured be rotated by a clockwise ANGLE θ, so that the original axes are $\hat{\mathbf{x}}_0$ and $\hat{\mathbf{y}}_0$, and the new axes are $\hat{\mathbf{x}}$ and $\hat{\mathbf{y}}$. Then the MATRIX transforming the coordinates of the curve with respect to $\hat{\mathbf{x}}$ and $\hat{\mathbf{y}}$ is given by the MATRIX TRANSPOSE of the above matrix:

$$\mathsf{R}'_\theta = \begin{bmatrix} \cos\theta & -\sin\theta \\ \sin\theta & \cos\theta \end{bmatrix}, \tag{3}$$

i.e.,

$$\mathbf{x} = \mathsf{R}'_\theta \mathbf{x}_0. \tag{4}$$

In $\mathbb{R}^3$, rotations of the x-, y-, and z-AXES give the matrices

$$\mathsf{R}_x(\alpha) = \begin{bmatrix} 1 & 0 & 0 \\ 0 & \cos\alpha & \sin\alpha \\ 0 & -\sin\alpha & \cos\alpha \end{bmatrix} \tag{5}$$

$$\mathsf{R}_y(\beta) = \begin{bmatrix} \cos\beta & 0 & -\sin\beta \\ 0 & 1 & 0 \\ \sin\beta & 0 & \cos\beta \end{bmatrix} \tag{6}$$

$$\mathsf{R}_z(\gamma) = \begin{bmatrix} \cos\gamma & \sin\gamma & 0 \\ -\sin\gamma & \cos\gamma & 0 \\ 0 & 0 & 1 \end{bmatrix}. \tag{7}$$

Any ROTATION can be given as a composition of rotations about three axes (EULER'S ROTATION THEOREM), and thus can be represented by a 3×3 MATRIX operating on a VECTOR,

$$\begin{bmatrix} x'_1 \\ x'_2 \\ x'_3 \end{bmatrix} = \begin{bmatrix} a_{11} & a_{12} & a_{13} \\ a_{21} & a_{22} & a_{23} \\ a_{31} & a_{32} & a_{33} \end{bmatrix} \begin{bmatrix} x_1 \\ x_2 \\ x_3 \end{bmatrix}. \tag{8}$$

We wish to place conditions on this matrix so that it is consistent with an ORTHOGONAL TRANSFORMATION (basically, a ROTATION or ROTOINVERSION).

In a ROTATION, a VECTOR must keep its original length, so it must be true that

$$x'_i x'_i = x_i x_i \tag{9}$$

for $i = 1, 2, 3$, where EINSTEIN SUMMATION is being used. Therefore, from the transformation equation,

$$(a_{ij} x_j)(a_{ik} x_k) = x_i x_i. \tag{10}$$

This can be rearranged to

$$a_{ij}(x_j a_{ik}) x_k = a_{ij}(a_{ik} x_j) x_k$$
$$= a_{ij} a_{ik} x_j x_k = x_i x_i. \tag{11}$$

In order for this to hold, it must be true that

$$a_{ij} a_{ik} = \delta_{jk} \tag{12}$$

for $j, k = 1, 2, 3$, where δ_{ij} is the KRONECKER DELTA. This is known as the ORTHOGONALITY CONDITION, and it guarantees that

$$\mathsf{A}^{-1} = \mathsf{A}^\mathrm{T}, \tag{13}$$

and

$$\mathsf{A}^\mathrm{T} \mathsf{A} = \mathsf{I}, \tag{14}$$

where A^T is the MATRIX TRANSPOSE and I is the IDENTITY MATRIX. Equation (14) is the identity which gives the orthogonal matrix its name. Orthogonal matrices have special properties which allow them to be manipulated and identified with particular ease.

Let A and B be two orthogonal matrices. By the ORTHOGONALITY CONDITION, they satisfy

$$a_{ij} a_{ik} = \delta_{jk}, \tag{15}$$

and

$$b_{ij} b_{ik} = \delta_{jk}, \tag{16}$$

where δ_{ij} is the KRONECKER DELTA. Now

$$c_{ij}c_{ik} = (ab)_{ij}(ab)_{jk} = a_{is}b_{sj}a_{it}b_{tk} = a_{is}a_{it}b_{sj}b_{tk}$$
$$= \delta_{st}b_{sj}b_{tk} = b_{tj}b_{tk} = \delta_{jk}, \qquad (17)$$

so the product $C \equiv AB$ of two orthogonal matrices is also orthogonal.

The EIGENVALUES of an orthogonal matrix must satisfy one of the following:

1. All EIGENVALUES are 1.
2. One EIGENVALUE is 1 and the other two are -1.
3. One EIGENVALUE is 1 and the other two are COMPLEX CONJUGATES OF THE FORM $e^{i\theta}$ and $e^{-i\theta}$.

An orthogonal MATRIX A is classified as proper (corresponding to pure ROTATION) if

$$\det(A) = 1, \qquad (18)$$

where $\det(A)$ is the DETERMINANT of A, or improper (corresponding to inversion with possible rotation; ROTOINVERSION) if

$$\det(A) = -1. \qquad (19)$$

See also EULER ANGLES, EULER PARAMETERS, EULER'S ROTATION THEOREM, ROTATION, ROTATION FORMULA

Rotation Number

The period for a QUASIPERIODIC trajectory to pass through the same point in a SURFACE OF SECTION. If the rotation number is IRRATIONAL, the trajectory will densely fill out a curve in the SURFACE OF SECTION. If the rotation number is RATIONAL, it is called the WINDING NUMBER, and only a finite number of points in the SURFACE OF SECTION will be visited by the trajectory.

See also QUASIPERIODIC FUNCTION, SURFACE OF SECTION, WINDING NUMBER (MAP)

Rotation Operator

The rotation operator can be derived from examining an INFINITESIMAL ROTATION

$$\left(\frac{d}{dt}\right)_{\text{space}} = \left(\frac{d}{dt}\right)_{\text{body}} + \boldsymbol{\omega} \times,$$

where d/dt is the time derivative, ω is the ANGULAR VELOCITY, and $\times$ is the CROSS PRODUCT operator.

See also ACCELERATION, ANGULAR ACCELERATION, INFINITESIMAL ROTATION

Roth's Removal Rule

If the matrices A, X, B, and C satisfy

$$AX - XB = C,$$

then

$$\begin{bmatrix} I & X \\ 0 & I \end{bmatrix} \begin{bmatrix} A & C \\ 0 & B \end{bmatrix} \begin{bmatrix} I & -X \\ 0 & I \end{bmatrix} = \begin{bmatrix} A & 0 \\ 0 & B \end{bmatrix},$$

where I is the IDENTITY MATRIX.

References
Roth, W. E. "The Equations $AX - YB = C$ and $AX - XB = C$ in Matrices." *Proc. Amer. Math. Soc.* **3**, 392–396, 1952.
Turnbull, H. W. and Aitken, A. C. *An Introduction to the Theory of Canonical Matrices.* New York: Dover, p. 422, 1961.

Roth's Theorem

For ALGEBRAIC α

$$\left| \alpha - \frac{p}{q} \right| < \frac{1}{q^{2+\epsilon}},$$

with $\epsilon > 0$, has finitely many solutions. Klaus Roth received a FIELDS MEDAL for this result.

See also HURWITZ EQUATION, HURWITZ'S IRRATIONAL NUMBER THEOREM, IRRATIONALITY MEASURE, LAGRANGE NUMBER (RATIONAL APPROXIMATION), LIOUVILLE'S APPROXIMATION THEOREM, MARKOV NUMBER, SEGRE'S THEOREM, SIEGEL'S THEOREM, THUE-SIEGEL-ROTH THEOREM

References
Davenport, H. and Roth, K. F. "Rational Approximations to Algebraic Numbers." *Mathematika* **2**, 160–167, 1955.
Roth, K. F. "Rational Approximations to Algebraic Numbers." *Mathematika* **2**, 1–20, 1955.
Roth, K. F. "Corrigendum to 'Rational Approximations to Algebraic Numbers'." *Mathematika* **2**, 168, 1955.

Rotkiewicz Theorem

If $n > 19$, there exists a POULET NUMBER between n and n^2. The theorem was proved in 1965.

See also POULET NUMBER

References
Rotkiewicz, A. "Les intervalles contenants les nombres pseudopremiers." *Rend. Circ. Mat. Palermo Ser. 2* **14**, 278–280, 1965.
Rotkiewicz, A. "Sur les nombres de Mersenne dépourvus de diviseurs carrés et sur les nombres naturels n, tel que $n^2 - 2^n - 2$." *Mat. Vesnik* **2** (17), 78–80, 1965.
Rotkiewicz, A. "Sur les nombres pseudopremiers carrés." *Elem. Math.* **20**, 39–40, 1965.

Rotoinversion

IMPROPER ROTATION

Rotor

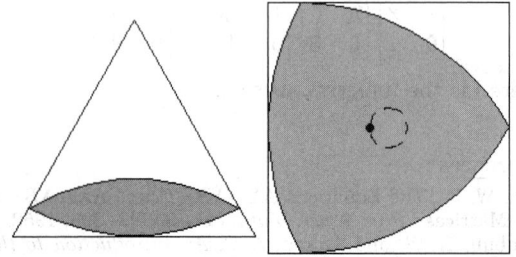

A convex figure that can be rotated inside a POLYGON (or POLYHEDRON) while always touching every side (or face). The least AREA rotor in a SQUARE is the REULEAUX TRIANGLE. The least AREA rotor in an EQUILATERAL TRIANGLE is a LENS with two 60° ARCS of CIRCLES and RADIUS equal to the TRIANGLE ALTITUDE.

There exist nonspherical rotors for the TETRAHEDRON, OCTAHEDRON, and CUBE, but not for the DODECAHEDRON and ICOSAHEDRON.

See also DELTA CURVE, LENS, REULEAUX POLYGON, REULEAUX TRIANGLE, ROULETTE, TRIP-LET

References
Gardner, M. *The Unexpected Hanging and Other Mathematical Diversions.* Chicago, IL: Chicago University Press, p. 219, 1991.
Goldberg, M. "Circular-Arc Rotors in Regular Polygons." *Amer. Math. Monthly* **55**, 392–402, 1948.
Goldberg, M. "Two-Lobed Rotors with Three-Lobed Stators." *J. Mechanisms* **3**, 55–60, 1968.
Steinhaus, H. *Mathematical Snapshots, 3rd ed.* New York: Dover, pp. 151–152, 1999.
Wells, D. *The Penguin Dictionary of Curious and Interesting Geometry.* London: Penguin, pp. 221–222, 1991.

Rotunda

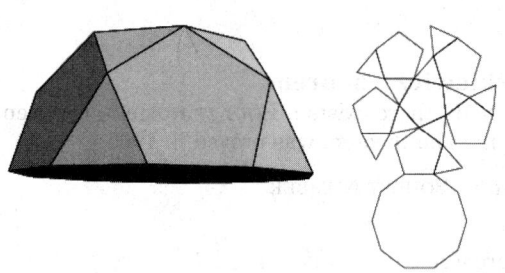

A POLYHEDRON consisting of a n-gon, a parallel $2n$-gon rotated a half-edge turn, and a band of paired triangles separated by pentagons. The only true member giving a polyhedron consisting of all regular polygons with unit edge lengths is the PENTAGONAL ROTUNDA. It corresponds to half of an ICOSIDODECAHEDRON.

See also ELONGATED ROTUNDA, GYROELONGATED ROTUNDA, ICOSIDODECAHEDRON, PENTAGONAL ROTUNDA, TRIANGULAR HEBESPHENOROTUNDA

References
Johnson, N. W. "Convex Polyhedra with Regular Faces." *Canad. J. Math.* **18**, 169–200, 1966.

Rouché's Theorem

Given two functions f and g ANALYTIC in A with γ a simple loop HOMOTOPIC to a point in A, if $|g(z)| < |f(z)|$ for all z on γ, then f and $f + g$ have the same number of ROOTS inside γ.

A stronger version has been proved by Estermann (1962). The strong version also has a converse, as shown by Challener and Rubel (1982).

See also ARGUMENT PRINCIPLE

References
Challener, D. and Rubel, L. "A Converse to Rouché's Theorem." *Amer. Math. Monthly* **89**, 302–305, 1982.
Estermann, T. *Complex Numbers and Functions.* London: Oxford University Press, p. 156, 1962.
Knopp, K. *Theory of Functions Parts I and II, Two Volumes Bound as One, Part II.* New York: Dover, p. 111, 1996.
Krantz, S. G. "Rouché's Theorem." §5.3.1 in *Handbook of Complex Analysis.* Boston, MA: Birkhäuser, p. 74, 1999.
Szego, G. *Orthogonal Polynomials, 4th ed.* Providence, RI: Amer. Math. Soc., p. 22, 1975.

Roulette

The curve traced by a fixed point on a closed convex curve as that curve rolls without slipping along a second curve. The roulettes described by the FOCI of CONICS when rolled upon a line are sections of MINIMAL SURFACES (i.e., they yield MINIMAL SURFACES when revolved about the line) known as UNDULOIDS.

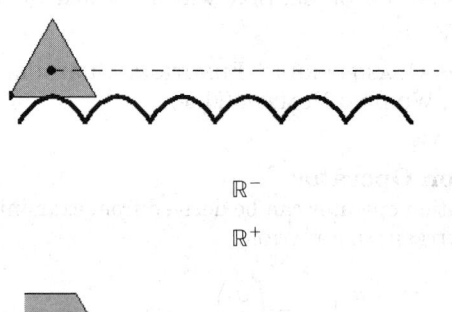

$\mathbb{R}^-$
$\mathbb{R}^+$

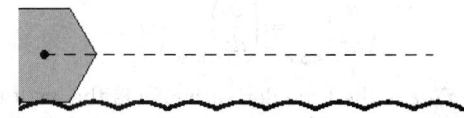

A particularly interesting case of a roulette is a regular n-gon rolling on a "road" composed of a sequence of truncated catenaries, as illustrated above. This motion is smooth in the sense that the CENTROID follows a straight line, although in the case of the rolling EQUILATERAL TRIANGLE, a physical model would be impossible to construct (Wagon 1991). For the rolling SQUARE, the shape of the road is the CATENARY $y = -\cosh x$ truncated at $x = \pm \sinh^{-1} 1$ (Wagon 1991). For a regular n-gon, the Cartesian equation of the corresponding CATENARY is

$$y = -A \cosh\left(\frac{x}{A}\right), \qquad (1)$$

where

$$A \equiv R \cos\left(\frac{\pi}{n}\right). \qquad (2)$$

Curve 1	Curve 2	Pole	Roulette
CIRCLE	exterior CIRCLE	on CIR-CUM-FERENCE	EPICYCLOID
CIRCLE	interior CIRCLE	on CIR-CUM-FERENCE	HYPOCYCLOID
CIRCLE	LINE	on CIR-CUM-FERENCE	CYCLOID
CIRCLE	same CIRCLE	any point	ROSE
CIRCLE INVOLUTE	LINE	CENTER	PARABOLA
CYCLOID	LINE	center	ELLIPSE
ELLIPSE	LINE	FOCUS	elliptic catenary
HYPERBOLA	LINE	FOCUS	hyperbolic catenary
HYPERBOLIC SPIRAL	LINE	ORIGIN	TRACTRIX
LINE	any curve	on LINE	INVOLUTE of the curve
LOGARITHMIC SPIRAL	LINE	any point	LINE
PARABOLA	equal PARABOLA	VERTEX	CISSOID OF DIOCLES
PARABOLA	LINE	FOCUS	CATENARY

See also CATENARY, DELTA CURVE, GLISSETTE, REULEAUX POLYGON, REULEAUX TRIANGLE, ROTOR, UNDULOID

References

Besant, W. H. *Notes on Roulettes and Glissettes, 2nd enl. ed.* Cambridge, England: Deighton, Bell & Co., 1890.
Cundy, H. and Rollett, A. "Roulettes and Involutes." §2.6 in *Mathematical Models, 3rd ed.* Stradbroke, England: Tarquin Pub., pp. 46–55, 1989.
Gardner, M. *The Sixth Book of Mathematical Games from Scientific American.* Chicago, IL: University of Chicago Press, p. 128, 1984.
Hall, L. and Wagon, S. "Mathematical Roads and Wheels." *Math. Mag.* To appear.
Lawrence, J. D. *A Catalog of Special Plane Curves.* New York: Dover, pp. 56–58 and 206, 1972.
Lockwood, E. H. "Roulettes." Ch. 17 in *A Book of Curves.* Cambridge, England: Cambridge University Press, pp. 138–151, 1967.
Wagon, S. *Mathematica in Action.* New York: W. H. Freeman, p. 52, 1991.
Yates, R. C. "Roulettes." *A Handbook on Curves and Their Properties.* Ann Arbor, MI: J. W. Edwards, pp. 175–185, 1952.
Zwillinger, D. (Ed.). "Roulettes (Spirograph Curves)." §8.2 in *CRC Standard Mathematical Tables and Formulae, 3rd ed.* Boca Raton, FL: CRC Press, 1996.

Round

NEAREST INTEGER FUNCTION, ROUND NUMBER, ROUNDNESS

Round Number

A number which is the product of a considerable number of comparatively small factors (Hardy 1999, p. 48). Round numbers are very rare. As Hardy (1999, p. 48) notes, "Half the numbers are divisible by 2, one-third by 3, one-sixth by both 2 and 3, and so on. Surely, then we may expect most numbers to have a *large* number of factors. But the facts seem to show the opposite."

See also HIGHLY COMPOSITE NUMBER, PRIME FACTORS, ROUNDNESS, SMOOTH NUMBER

References

Hardy, G. H. "Round Numbers." Ch. 3 in *Ramanujan: Twelve Lectures on Subjects Suggested by His Life and Work, 3rd ed.* New York: Chelsea, pp. 48–57, 1999.
Hoffman, P. *The Man Who Loved Only Numbers: The Story of Paul Erdos and the Search for Mathematical Truth.* New York: Hyperion, pp. 89–90, 1998.

Rounding

The process of approximating a quantity, be it for convenience or, as in the case of numerical computations, of necessity. If rounding is performed on each of a series of numbers in a long computation, ROUNDING ERROR can become important, especially if division by a small number ever occurs.

See also NEAREST INTEGER FUNCTION, ROUNDING ERROR, SHADOWING THEOREM

References

Mulliss, C. "Significant Figures and Rounding Rules." http://www.angelfire.com/oh/cmulliss/.
Wilkinson, J. H. *Rounding Errors in Algebraic Processes.* New York: Dover, 1994.

Rounding Error

The error produced in a computation by rounding results at one or more intermediate steps, resulting in a result different from that which would be obtained using exact numbers. The most common problems resulting from rounding error occur either when many steps are involved with rounding occurring at each step, when two quantities very close to each other are subtracted, or when a number is divided by a number which is close to zero.

An egregious example of rounding error is provided by a short-lived index devised at the Vancouver stock exchange. At its inception in 1982, the index was given a value of 1000.000. After 22 months of recomputing the index and truncating to three decimal places at each change in market value, the index stood at 524.881, despite the fact that its "true" value should have been 1009.811.

Other sorts of rounding error can also occur. A notorious example is the fate of the Ariane rocket launched on June 4, 1996. In the 37th second of flight, the inertial reference system attempted to convert a 64-bit floating point number to a 16-bit number, but instead triggered an overflow error which was interpreted by the guidance system as flight data, causing the rocket to veer off course and be destroyed. The Patriot missile defense system used during the Gulf War was also rendered ineffective due to roundoff error. The system used an integer timing register which was incremented at intervals of 0.1 s. However, the integers were converted to decimal numbers by multiplying by the BINARY approximation of 0.1,

$$0.00011001100110011001100_2 = \frac{209715}{2097152}.$$

As a result, after 100 hours $(3.6 \times 10^6$ ticks), an error of

$$\left(\frac{1}{10} - \frac{209715}{2097152}\right)(3600 \cdot 100 \cdot 10) = \frac{5625}{16384} \approx 0.3433 \text{ second}$$

had accumulated. This discrepancy caused the Patriot system to continuously recycle itself instead of targeting properly. As a result, an Iraqi Scud missile could not be targeted and was allowed to detonate on a barracks, killing 28 people.

See also ROUNDING

Roundness

Hoffman (1998, p. 90) calls the sum of the exponents in the PRIME FACTORIZATION of a number its roundness. The first few values for $n = 1, 2, \ldots$ are 0, 1, 1, 2, 1, 2, 1, 3, 2, 2, ... (Sloane's A001222).

See also HIGHLY COMPOSITE NUMBER, PRIME FACTORIZATION, ROUND NUMBER

References

Abramowitz, M. and Stegun, C. A. (Eds.). *Handbook of Mathematical Functions with Formulas, Graphs, and Mathematical Tables, 9th printing.* New York: Dover, p. 844, 1972.
Hoffman, P. *The Man Who Loved Only Numbers: The Story of Paul Erdos and the Search for Mathematical Truth.* New York: Hyperion, p. 90, 1998.
Kac, M. *Statistical Independence in Probability, Analysis, and Number Theory.* Buffalo, NY: Math. Assoc. Amer., p. 64, 1959.
Sloane, N. J. A. Sequences A001222/M0094 in "An On-Line Version of the Encyclopedia of Integer Sequences." http://www.research.att.com/~njas/sequences/eisonline.html.

Route

An n-route is defined as a WALK of length n with specified initial point in which no line succeeds itself.

See also TRANSITIVE GRAPH

References

Harary, F. *Graph Theory.* Reading, MA: Addison-Wesley, p. 173, 1994.

Routh-Hurwitz Theorem

Consider the CHARACTERISTIC EQUATION

$$|\lambda I - A| = \lambda^n + b_1 \lambda^{n-1} + \ldots + b_{n-1} \lambda + b_n = 0$$

determining the n EIGENVALUES λ of a REAL $n \times n$ MATRIX A, where I is the IDENTITY MATRIX. Then the EIGENVALUES λ all have NEGATIVE REAL PARTS if

$$\Delta_1 > 0, \Delta_2 > 0, \ldots, \Delta_n > 0,$$

where

$$\Delta_k = \begin{vmatrix} b_1 & 1 & 0 & 0 & 0 & 0 & \cdots & 0 \\ b_3 & b_2 & b_1 & 1 & 0 & 0 & \cdots & 0 \\ b_5 & b_4 & b_3 & b_2 & b_1 & 0 & \cdots & 0 \\ \vdots & \vdots & \vdots & \vdots & \vdots & \vdots & \ddots & \vdots \\ b_{2k-1} & b_{2k-2} & b_{2k-3} & b_{2k-4} & b_{2k-5} & b_{k-6} & \cdots & b_k \end{vmatrix}.$$

See also STABLE POLYNOMIAL

References

Gradshteyn, I. S. and Ryzhik, I. M. *Tables of Integrals, Series, and Products, 6th ed.* San Diego, CA: Academic Press, p. 1119, 2000.
Séroul, R. "Stable Polynomials." §10.13 in *Programming for Mathematicians.* Berlin: Springer-Verlag, pp. 280–286, 2000.

Routh's Theorem

If the sides of a TRIANGLE are divided in the ratios $\lambda : 1$, $\mu : 1$, and $\nu : 1$, the CEVIANS form a central TRIANGLE whose AREA is

$$a = \frac{(\lambda \mu \nu - 1)^2}{(\lambda \mu + \lambda + 1)(\mu \nu + \mu + 1)(\nu \lambda + \nu + 1)} \delta, \qquad (1)$$

where δ is the AREA of the original TRIANGLE. for $\lambda =$

$$\mu = v \equiv n,$$

$$a = \frac{(n-1)^2}{n^2 + n + 1}\delta. \qquad (2)$$

for $n = 1, 2, 3, \ldots$, the areas are 0, 1/7 (Steinhaus 1983, pp. 8–9), 4/13, 3/7, 16/31, 25/43, ... (Sloane's A046162 and A046163). The AREA of the TRIANGLE formed by connecting the division points on each side is

$$A' = \frac{\lambda\mu v + 1}{(\lambda+1)(\mu+1)(v+1)}\Delta. \qquad (3)$$

Routh's theorem gives CEVA'S THEOREM and MENE-LAUS' THEOREM ($\lambda\mu v = -1$) as special cases.

See also CEVA'S THEOREM, CEVIAN, MENELAUS' THE-OREM

References

Bottema, O. "On the Area of a Triangle in Barycentric Coordinates." *Crux. Math.* **8**, 228–231, 1982.

Coxeter, H. S. M. *Introduction to Geometry, 2nd ed.* New York: Wiley, pp. 211–212, 1969.

Dudeney, H. E. *Amusements in Mathematics.* New York: Dover, p. 27, 1970.

Klamkin, M. S. *Crux. Math.* p. 199, 1981.

Mikusinski, J. G. *Ann. Univ. M. Curie-Sklodowska* **1**, 45–50, 1946.

Sloane, N. J. A. Sequences A046162 and A046163 in "An On-Line Version of the Encyclopedia of Integer Sequences." http://www.research.att.com/~njas/sequences/eisonline.html.

Steinhaus, H. *Mathematical Snapshots, 3rd ed.* New York: Dover, 1999.

Row Space

See also COLUMN SPACE

Row Vector

A $1 \times n$ MATRIX

$$[a_{11} \quad a_{12} \quad \cdots \quad a_{1n}].$$

See also COLUMN VECTOR, MATRIX, VECTOR

Row-Convex Polyomino

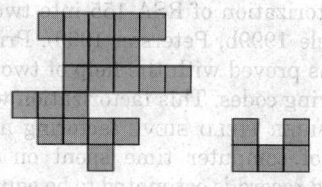

row-convex   *not row-convex*

A row-convex polyomino is a self-avoiding CONVEX POLYOMINO such that the intersection of any horizontal line with the polyomino has at most two connected components. A row-convex polyomino is also called a

horizontally convex polyomino. A COLUMN-CONVEX POLYOMINO is similarly defined.

See also COLUMN-CONVEX POLYOMINO, CONVEX POLY-OMINO, POLYOMINO

RPN

REVERSE POLISH NOTATION

RSA Encryption

A PUBLIC-KEY CRYPTOGRAPHY ALGORITHM which uses PRIME FACTORIZATION as the TRAPDOOR ONE-WAY FUNCTION. Define

$$n \equiv pq \qquad (1)$$

for p and q PRIMES. Also define a private key d and a public key e such that

$$de \equiv 1 \pmod{\phi(n)} \qquad (2)$$

$$(e, \phi(n)) = 1, \qquad (3)$$

where $\phi(n)$ is the TOTIENT FUNCTION, (a, b) denotes the GREATEST COMMON DIVISOR (so $(a, b) = 1$ means that a and b are RELATIVELY PRIME), and $a \equiv b \pmod{m}$ is a CONGRUENCE.

Let the message be converted to a number M. The sender then makes n and e public and sends

$$E = M^e \pmod{n}. \qquad (4)$$

To decode, the receiver (who knows d) computes

$$E^d \equiv (M^e)^d \equiv M^{ed} \equiv M^{N\phi(n)+1} \equiv M \pmod{n}, \qquad (5)$$

since N is an INTEGER. In order to crack the code, d must be found. But this requires factorization of n since

$$\phi(n) = (p-1)(q-1). \qquad (6)$$

Both p and q should be picked so that $p \pm 1$ and $q \pm 1$ are divisible by large PRIMES, since otherwise the POLLARD P-1 FACTORIZATION METHOD or WILLIAMS $P+1$ FACTORIZATION METHOD potentially factor n easily. It is also desirable to have $\phi(\phi(pq))$ large and divisible by large PRIMES.

It is possible to break the cryptosystem by repeated encryption if a unit of $\mathbb{Z}/\phi(n)\mathbb{Z}$ has small ORDER (Simmons and Norris 1977, Meijer 1996), where $\mathbb{Z}/s\mathbb{Z}$ is the RING of INTEGERS between 0 and $s-1$ under addition and multiplication (mod s). Meijer (1996) shows that "almost" every encryption exponent e is safe from breaking using repeated encryption for factors OF THE FORM

$$p = 2p_1 + 1 \qquad (7)$$

$$q = 2q_1 + 1, \qquad (8)$$

where

$$p_1 = 2p_2 + 1 \qquad (9)$$

$$q_1 = 2q_2 + 1, \qquad (10)$$

and $p, p_1, p_2, q, q_1,$ and q_2 are all PRIMES. In this case,

$$\phi(n) = 4p_1q_1 \qquad (11)$$

$$\phi(\phi(n)) = 8p_2q_2. \qquad (12)$$

Meijer (1996) also suggests that p_2 and q_2 should be of order 10^{75}.

Using the RSA system, the identity of the sender can be identified as genuine without revealing his private code.

See also CONGRUENCE, PUBLIC-KEY CRYPTOGRAPHY

References

Coutinho, S. C. *The Mathematics of Ciphers: Number Theory and RSA Cryptography.* Natick, MA: A. K. Peters, 1999.

Flannery, S. and Flannery, D. *In Code: A Mathematical Journey.* Profile Books, 2000.

Honsberger, R. *Mathematical Gems III.* Washington, DC: Math. Assoc. Amer., pp. 166–173, 1985.

Meijer, A. R. "Groups, Factoring, and Cryptography." *Math. Mag.* **69**, 103–109, 1996.

Rivest, R. L. "Remarks on a Proposed Cryptanalytic Attack on the MIT Public-Key Cryptosystem." *Cryptologia* **2**, 62–65, 1978.

Rivest, R.; Shamir, A.; and Adleman, L. "A Method for Obtaining Digital Signatures and Public Key Cryptosystems." *Comm. ACM* **21**, 120–126, 1978.

RSA Laboratories. ® "RSA Factoring Challenge." http://www.rsasecurity.com/rsalabs/challenges/factoring/.

RSA Laboratories. ® "Factoring Challenge: Status." http://www.rsasecurity.com/rsalabs/challenges/factoring/status.html.

Simmons, G. J. and Norris, M. J. "Preliminary Comments on the MIT Public-Key Cryptosystem." *Cryptologia* **1**, 406–414, 1977.

RSA Number

Numbers contained in the "factoring challenge" of RSA Data Security, Inc. An additional number which is not part of the actual challenge is the RSA-129 number. The RSA numbers which have been factored are RSA-100 (Apr. 1991), RSA-110 (Apr. 1992), RSA-120 (Jun. 1993), RSA-129 (Apr. 1994), RSA-130 (Apr. 1996), RSA-140 (Feb. 1999), and RSA-155 (Aug. 1999; Peterson 1999). RSA-150 has not yet been factored.

RSA-129 is a 129-digit number used to encrypt one of the first public-key messages. This message was published by R. Rivest, A. Shamir, and L. Adleman (Gardner 1977), along with the number and a $100 reward for its decryption. Despite belief that the message encoded by RSA-129 "would take millions of years to break," RSA-129 was factored in 1994 using a distributed computation which harnessed networked computers spread around the globe performing a multiple polynomial QUADRATIC SIEVE factorization method. The effort was coordinated by P. Leylad, D. Atkins, and M. Graff. They received 112,011 full factorizations, 1,431,337 single partial factorizations, and 8,881,138 double partial factoriza-

tions out of a factor base of 524,339 PRIMES. The final MATRIX obtained was $188,346 \times 188,346$ square.

The text of the message was "The magic words are squeamish ossifrage" (an ossifrage is a rare, predatory vulture found in the mountains of Europe), and the FACTORIZATION (into a 64-DIGIT number and a 65-DIGIT number) is

$$11438162575788886766923577997614661201021829 6 \cdots$$

$$\cdots 7212423625625618429357069352457338978305971 \cdots$$

$$\cdots 2356395870505898907514759929002687954354 1$$

$$= 3490529510847650949147849619903898133417764 \cdots$$

$$\cdots 6384933878439908205 77 \cdot 3276913299326 \cdots$$

$$\cdots 6709549961988190834461413177642967992 \cdots$$

$$\cdots 942539798288533$$

(Leutwyler 1994, Cipra 1995).

On Feb. 2, 1999, a group led by H. te Riele completed factorization of RSA-140 into two 70-digits primes. Primality of the factors was proved using two different methods. The factorization was found using the NUMBER FIELD SIEVE factorization method, and beat the 130-digit record (for RSA-130) set on April 10, 1996. The amount of computer time spent on this factorization is estimated to be equivalent to 2000 MIPS years. (For the old 130-digit NFS-record, this effort is estimated to be 1000 MIPS years; te Riele 1999.) Sieving was done on about 125 SGI and Sun workstations running at 175 MHz on average, and on about 60 PCs running at 300 MHz on average. The total amount of CPU-time spent on sieving was 8.9 CPU years (te Riele 1999). Sieving started the day before Christmas 1998 and was completed one month later. The relations were collected and required 3.7 GB of memory (te Riele 1999) The filtering of the data and the building of the matrix took one calendar week. The resulting matrix had 4,671,181 rows and 4,704,451 columns, and weight 151,141,999 (32.36 nonzero entries per row). It took almost 100 CPU hours and 810 MB of central memory to find 64 dependencies among the rows of this matrix (te Riele 1999a).

On Aug. 22, 1999, a group led by H. te Riele completed factorization of RSA-155 into two 78-digit primes (te Riele 1999b, Peterson 1999). Primality of the factors was proved with the help of two different primality proving codes. This factorization was found using the NUMBER FIELD SIEVE factoring algorithm. The amount of computer time spent on this new factoring world record is estimated to be equivalent to 8000 MIPS years. Sieving was done on about 160 175–400 MHz SGI and Sun workstations, on 8 300 MHz SGI Origin 2000 processors, on about 120 300–450 MHz Pentium II PCs, and on 4 500 MHz Digital/Compaq boxes. The total amount of CPU-time spent on sieving was 35.7 CPU years estimated to be equivalent to approximately 8000 MIPS years. Calendar time for sieving was 3 1/2 months. The filtering of the data and the building of the matrix were carried out at CWI and took one month. The resulting matrix had 6,699,191 rows, 6,711,336 columns, and weight 417,132,631 (62.27 nonzeros per row). It took 224 CPU hours and 2 GB of central memory on the Cray C916 at the SARA Amsterdam Academic Computer Center to find 64 dependencies among the rows of this matrix (te Riele 1999b).

See also NUMBER FIELD SIEVE

References

Cipra, B. "The Secret Life of Large Numbers." *What's Happening in the Mathematical Sciences, 1995–1996, Vol. 3.* Providence, RI: Amer. Math. Soc., pp. 90–99, 1996.

Cowie, J.; Dodson, B.; Elkenbracht-Huizing, R. M.; Lenstra, A. K.; Montgomery, P. L.; Zayer, J. A. "World Wide Number Field Sieve Factoring Record: On to 512 Bits." In *Advances in Cryptology--ASIACRYPT '96 (Kyongju)* (Ed. K. Kim and T. Matsumoto.) New York: Springer-Verlag, pp. 382–394, 1996.

Gardner, M. "Mathematical Games: A New Kind of Cipher that Would Take Millions of Years to Break." *Sci. Amer.* **237**, 120–124, Aug. 1977.

Klee, V. and Wagon, S. *Old and New Unsolved Problems in Plane Geometry and Number Theory, rev. ed.* Washington, DC: Math. Assoc. Amer., p. 223, 1991.

Leutwyler, K. "Superhack: Forty Quadrillion Years Early, a 129-Digit Code is Broken." *Sci. Amer.* **271**, 17–20, 1994.

Leyland, P. ftp://sable.ox.ac.uk/pub/math/rsa129.

Peterson, I. "Crunching Internet Security Codes." *Sci. News* **156**, 221, Oct. 2, 1999.

RSA Data Security. ® "RSA Factoring Challenge." http://www.rsasecurity.com/rsalabs/challenges/factoring/.

RSA Data Security. ® "What is the RSA Factoring Challenge and What is RSA-129?" http://www.rsasecurity.com/rsa-labs/faq/.

Taubes, G. "Small Army of Code-breakers Conquers a 129-Digit Giant." *Science* **264**, 776–777, 1994.

te Riele, H. "Factorisation of RSA-140." *NMBRTHRY@LISTSERV.NODAK.EDU* mailing list posting, Feb. 4, 1999a.

te Riele, H. "New Factorization Record." *NMBRTHRY@LISTSERV.NODAK.EDU* mailing list posting, Aug. 26, 1999b.

Weisstein, E. W. "RSA Numbers." MATHEMATICA NOTEBOOK RSANUMBERS.M.

Rubber-Sheet Geometry

ALGEBRAIC TOPOLOGY

Rubik's Clock

A puzzle consisting of 18 small clocks. There are 12^{18} possible configurations, although not all are realizable.

See also RUBIK'S CUBE

References

Dénes, J. and Mullen, G. L. "Rubik's Clock and Its Solution." *Math. Mag.* **68**, 378–381, 1995.

Zeilberger, D. "Doron Zeilberger's Maple Packages and Programs: RubikClock." http://www.math.temple.edu/~zeilberg/programs.html.

Rubik's Cube

A $3 \times 3 \times 3$ CUBE in which the 26 subcubes on the outside are internally hinged in such a way that rotation (by a quarter turn in either direction or a half turn) is possible in any plane of cubes. Each of the six sides is painted a distinct color, and the goal of the puzzle is to return the cube to a state in which each side has a single color after it has been randomized by repeated rotations. The PUZZLE was invented in the 1970s by the Hungarian Erno Rubik and sold millions of copies worldwide over the next decade.

The number of possible positions of Rubik's cube is

$$\frac{8!12!3^8 2^{12}}{2 \cdot 3 \cdot 2} = 43,252,003,274,489,856,000$$

(Turner and Gold 1985, Schönert). Hoey showed using the PÓLYA-BURNSIDE LEMMA that there are 901,083,404,981,813,616 positions up to conjugacy by whole-cube symmetries.

Algorithms exist for solving a cube from an arbitrary initial position, but they are not necessarily optimal (i.e., requiring a minimum number of turns). The minimum number of turns required for an arbitrary starting position is still not known, although it is bounded from above. Michael Reid (1995) produced the best proven bound of 29 turns (or 42 "quarter-turns"). The proof involves large tables of "subroutines" generated by computer.

However, Dik Winter has produced a program based on work by Kociemba which has solved each of millions of cubes in at most 21 turns. Recently, Richard Korf (1997) has produced a different algorithm which is practical for cubes up to 18 moves away from solved. Out of 10 randomly generated cubes, one was solved in 16 moves, three required 17 moves, and six required 18 moves.

See also RUBIK'S CLOCK

References

Helms, G. "Rubik's Cube." http://webplaza.pt.lu/public/geo-helm/myweb/cubeold.htm.

Hoey, D. "The Real Size of Cube Space." http://www.math.rwth-aachen.de/~Martin.Schoenert/Cube-Lovers/Dan_Hoey__The_real_size_of_cube_space.html.

Hofstadter, D. R. "Metamagical Themas: The Magic Cube's Cubies are Twiddled by Cubists and Solved by Cubemeisters." *Sci. Amer.* **244**, 20–39, Mar. 1981.

Larson, M. E. "Rubik's Revenge: The Group Theoretical Solution." *Amer. Math. Monthly* **92**, 381–390, 1985.

Longridge, M. "Domain of the Cube." http://web.idirect.com/~cubeman/.

Miller, D. L. W. "Solving Rubik's Cube Using the 'Bestfast' Search Algorithm and 'Profile' Tables." http://www.sunyit.edu/~millerd1/RUBIK.HTM.

Schoenert, M. "Cube Lovers: Index by Date." http://www.math.rwth-aachen.de/~Martin.Schoenert/Cube-Lovers/.

Schönert, M. "Analyzing Rubik's Cube with GAP." http://www-groups.dsc.st-and.ac.uk/~gap/Intro/rubik.html.

Singmaster, D. *Notes on Rubik's 'Magic Cube.'* Hillside, NJ: Enslow Pub., 1981.

Taylor, D. *Mastering Rubik's Cube.* New York: Holt, Rinehart, and Winston, 1981.

Taylor, D. and Rylands, L. *Cube Games: 92 Puzzles & Solutions.* New York: Holt, Rinehart, and Winston, 1981.

Turner, E. C. and Gold, K. F. "Rubik's Groups." *Amer. Math. Monthly* **92**, 617–629, 1985.

Rudin-Shapiro Sequence

Let a number n be written in BINARY as

$$n = (\epsilon_\kappa \epsilon_{k-1} \ldots \epsilon_1 \epsilon_0)_2, \tag{1}$$

and define

$$b_n = \sum_{i=0}^{k-1} \epsilon_i \epsilon_{i+1} \tag{2}$$

as the number of DIGITS BLOCKS of 11s in the BINARY expansion of n. For $n = 0, 1, \ldots, b_n$ is given by 0, 0, 1, 0, 0, 1, 2, 0, 0, 0, 1, 1, 1, 2, 3, ... (Sloane's A014081).

Now define

$$a_n = (-1)^{b_n} \tag{3}$$

as the parity of the number of pairs of consecutive 1s in the BINARY expansion of n. For $n = 0, 1, \ldots$, the first few values are 1, 1, -1, 1, 1, -1, 1, 1, 1, 1, -1, -1, -1, ... (Sloane's A020985).

The SUMMATORY sequence of a_n is the defined by

$$s_n \equiv \sum_{i=0}^{n} a_i, \tag{4}$$

giving the first few terms 2, 3, 2, 3, 4, 3, 4, 5, 6, 7, 6, 5, 4, ... (Sloane's A020986). For the special case $n = 2^{k-1}$, s_n can be computed using the formula

$$s_n = \begin{cases} 2^{k/2} + 1 & \text{if } k \text{ is even} \\ 2^{(k-1)/2} + 1 & \text{if } k \text{ is odd} \end{cases} \tag{5}$$

(Blecksmith and Laud 1995), giving 2, 3, 3, 5, 5, 9, 9, 17, 17, 33, 33, 65, ... (Sloane's A051032).

See also BINARY, DIGIT BLOCK, FOLDING, STOLARSKY-HARBORTH CONSTANT

References

Blecksmith, R. and Laud, P. W. "Some Exact Number Theory Computations via Probability Mechanisms." *Amer. Math. Monthly* **102**, 893–903, 1995.

Brillhart, J.; Erdos, P.; and Morton, P. "On the Sums of the Rudin-Shapiro Coefficients II." *Pac. J. Math.* **107**, 39–69, 1983.

Brillhart, J. and Morton, P. "Über Summen von Rudin-Shapiroschen Koeffizienten." *Ill. J. Math.* **22**, 126–148, 1978.

Mendes France, M. and van der Poorten, A. J. "Arithmetic and Analytic Properties of Paper Folding Sequences." *Bull. Austral. Math. Soc.* **24**, 123–131, 1981.

Sloane, N. J. A. Sequences A014081, A020985, A020986, and A051032 in "An On-Line Version of the Encyclopedia of Integer Sequences." http://www.research.att.com/~njas/sequences/eisonline.html.

Weisstein, E. W. "Integer Sequences." MATHEMATICA NOTEBOOK IntegerSequences.M.

Rudvalis Group

The SPORADIC GROUP Ru.

See also SPORADIC GROUP

References

Wilson, R. A. "ATLAS of Finite Group Representation." http://for.mat.bham.ac.uk/atlas/html/Ru.html.

Ruffini-Horner Method

HORNER'S METHOD

Rule

A usually simple ALGORITHM or IDENTITY. The term is frequently applied to specific orders of NEWTON-COTES FORMULAS.

See also ALGORITHM, BAC-CAB RULE, BODE'S RULE, CHAIN RULE, CRAMER'S RULE, DESCARTES' SIGN RULE, DURAND'S RULE, ESTIMATOR, EULER'S RULE, EULER'S TOTIENT RULE, GOLDEN RULE, HARDY'S RULE, HORNER'S RULE, IDENTITY, L'HOSPITAL'S RULE, LEIBNIZ INTEGRAL RULE, METHOD, OSBORNE'S RULE, PASCAL'S RULE, POWER RULE, PRODUCT RULE, QUARTER SQUARES RULE, QUOTA RULE, QUOTIENT RULE, ROTH'S REMOVAL RULE, RULE OF 72, SIMPSON'S RULE, SLIDE RULE, SUM RULE, TRAPEZOIDAL RULE, WEDDLE'S RULE, ZEUTHEN'S RULE

Rule of 72

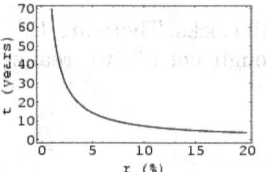 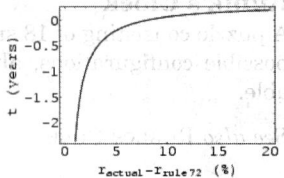

The time required for a given PRINCIPAL to double (assuming $n = 1$ CONVERSION PERIOD) for COMPOUND

INTEREST is given by solving

$$2P = P(1+r)^t, \tag{1}$$

or

$$t = \frac{\ln 2}{\ln(n+r)}, \tag{2}$$

where LN is the NATURAL LOGARITHM. This function can be approximated by the so-called "rule of 72":

$$t \approx \frac{0.72}{r}. \tag{3}$$

The above plots show the actual doubling time t (left plot) and the difference between the actual doubling time and the doubling time calculated using the rule of 72 (right plot) as a function of the interest rate r.

See also COMPOUND INTEREST, INTEREST

References

Avanzini, J. F. *Rapid Debt-Reduction Strategies*. Fort Worth, TX: HIS Pub., 1990.

Ruled Surface

A SURFACE which can be swept out by a moving a LINE in space and therefore has a parameterization OF THE FORM

$$\mathbf{x}(v,\ v) = \mathbf{b}(u) + v\boldsymbol{\delta}(u), \tag{1}$$

where $\mathbf{b}$ is called the DIRECTRIX (also called the BASE CURVE) and δ is the DIRECTOR CURVE. The straight lines themselves are called RULINGS. The rulings of a ruled surface are ASYMPTOTIC CURVES. Furthermore, the GAUSSIAN CURVATURE on a ruled REGULAR SURFACE is everywhere NONPOSITIVE.

Examples of ruled surfaces include the elliptic HYPERBOLOID of one sheet (a DOUBLY RULED SURFACE)

$$\begin{bmatrix} a(\cos u \mp v \sin u) \\ b(\sin u \pm v \cos u) \\ \pm cv \end{bmatrix} = \begin{bmatrix} a\cos u \\ b\sin u \\ 0 \end{bmatrix} \pm v \begin{bmatrix} -a\sin u \\ b\cos u \\ c \end{bmatrix}, \tag{2}$$

the HYPERBOLIC PARABOLOID (a DOUBLY RULED SURFACE)

$$\begin{bmatrix} a(u+v) \\ \pm bv \\ u^2 + 2uv \end{bmatrix} = \begin{bmatrix} au \\ 0 \\ u^2 \end{bmatrix} + v \begin{bmatrix} a \\ \pm b \\ 2u \end{bmatrix}, \tag{3}$$

PLÜCKER'S CONOID

$$\begin{bmatrix} r\cos\theta \\ r\sin\theta \\ 2\cos\theta\sin\theta \end{bmatrix} = \begin{bmatrix} 0 \\ 0 \\ 2\cos\theta\sin\theta \end{bmatrix} + r\begin{bmatrix} \cos\theta \\ \sin\theta \\ 0 \end{bmatrix}, \tag{4}$$

and the MÖBIUS STRIP

$$a\begin{bmatrix} \cos u + v\cos\left(\frac{1}{2}u\right)\cos u \\ \sin u + v\cos\left(\frac{1}{2}u\right)\sin u \\ v\sin\left(\frac{1}{2}u\right) \end{bmatrix}$$

$$= a\begin{bmatrix} \cos u \\ \sin u \\ 0 \end{bmatrix} + au\begin{bmatrix} \cos u\left(\frac{1}{2}u\right)\cos u \\ \cos\left(\frac{1}{2}u\right)\sin u \\ \sin\left(\frac{1}{2}u\right) \end{bmatrix} \tag{5}$$

(Gray 1997).

The only ruled MINIMAL SURFACES are the PLANE and HELICOID (Catalan 1842, do Carmo 1986).

See also ASYMPTOTIC CURVE, CAYLEY'S RULED SURFACE, DEVELOPABLE SURFACE, DIRECTOR CURVE, DIRECTRIX (RULED SURFACE), DOUBLY RULED SURFACE, GENERALIZED CONE, GENERALIZED CYLINDER, HELICOID, NONCYLINDRICAL RULED SURFACE, PLANE, RIGHT CONOID, RULING

References

Catalan E. "Sur les surfaces réglées dont l'aire est un minimum." *J. Math. Pure. Appl.* **7**, 203–211, 1842.
do Carmo, M. P. "The Helicoid." §3.5B in *Mathematical Models from the Collections of Universities and Museums* (Ed. G. Fischer). Braunschweig, Germany: Vieweg, pp. 44–45, 1986.
Fischer, G. (Ed.). Plates 32–33 in *Mathematische Modelle/Mathematical Models, Bildband/Photograph Volume.* Braunschweig, Germany: Vieweg, pp. 32–33, 1986.
Gray, A. "Ruled Surfaces." Ch. 19 in *Modern Differential Geometry of Curves and Surfaces with Mathematica, 2nd ed.* Boca Raton, FL: CRC Press, pp. 431–456, 1993.
Hilbert, D. and Cohn-Vossen, S. *Geometry and the Imagination.* New York: Chelsea, p. 15, 1999.
Steinhaus, H. *Mathematical Snapshots, 3rd ed.* New York: Dover, pp. 242–243, 1999.

Ruler

A STRAIGHTEDGE with markings to indicate distances. Although GEOMETRIC CONSTRUCTIONS are sometimes said to be performed with a ruler and COMPASS, the term STRAIGHTEDGE is preferable to ruler since markings are not allowed by the classical Greek rules.

See also COASTLINE PARADOX, COMPASS, GEOMETRIC CONSTRUCTION, GEOMETROGRAPHY, GOLOMB RULER, PERFECT RULER, SIMPLICITY, SLIDE RULE, STRAIGHTEDGE

References

Smogorzhevskii, A. S. *The Ruler in Geometrical Constructions.* New York: Blaisdell, 1961.

Ruler Function

The exponent of the largest POWER of 2 which DIVIDES a given number $2n$. The values of the ruler function for $n = 1, 2, \ldots$, are 1, 2, 1, 3, 1, 2, 1, 4, 1, 2, ... (Sloane's A001511).

See also 2

References

Guy, R. K. "Cycles and Sequences Containing All Permutations as Subsequences." §E22 in *Unsolved Problems in Number Theory, 2nd ed.* New York: Springer-Verlag, p. 224, 1994.

Sloane, N. J. A. Sequences A001511/M0127 in "An On-Line Version of the Encyclopedia of Integer Sequences." http://www.research.att.com/~njas/sequences/eisonline.html.

Ruling

One of the straight lines sweeping out a RULED SURFACE. The rulings on a ruled surface are ASYMPTOTIC CURVES.

See also ASYMPTOTIC CURVE, DIRECTOR CURVE, DIRECTRIX (RULED SURFACE), RULED SURFACE

Rumors

GOSSIPING

Rumor Spreading

GOSSIPING

Run

A run is a sequence of more than one consecutive identical outcomes, also known as a CLUMP. Given n BERNOULLI TRIALS (say, in the form of COIN TOSSINGS), the probability $P_t(n)$ of a run of t consecutive heads or tails is given by the RECURRENCE RELATION

$$P_t(n) = P_t(n-1) + 2^{-t}[1 - P_t(n-t)], \qquad (1)$$

where $P_t(n) = 0$ for $n < t$ and $P_t(t) = 2^{1-t}$ (Bloom 1996).

Let $R(r, n)$ be the probability that a run of r consecutive heads appears in n independent tosses of a COIN. There is a beautiful formula for $R(r, n)$ given in terms of the coefficients of the GENERATING FUNCTION

$$F_p(r, s) = \frac{p^r s^r (1 - ps)}{1 - s + (1 - p)p^r s^{r+1}} \equiv \sum_{i=r}^{\infty} c_i^p s^i \qquad (2)$$

(Feller 1968, 2nd ed. p. 300), where $0 < p < 1$ is the probability of obtaining a head in a single toss. Then

$$R_p(r, n) = \sum_{i=r}^{n} c_i^p \qquad (3)$$

The following table gives the triangle of numbers $2^n R_{1/2}(r, n)$ for $r = 1, 2, \ldots$ and $n = r, r+1, \ldots, \ldots$ (Sloane's A050227).

$r \backslash n$	1	2	3	4	5	6	7	8
1	1	3	7	15	31	63	127	255
2	0	1	3	8	19	43	94	201
3	0	0	1	3	8	20	47	107
4	0	0	0	1	3	8	20	48
5	0	0	0	0	1	3	8	20
6	0	0	0	0	0	1	3	8
7	0	0	0	0	0	0	1	3
8	0	0	0	0	0	0	0	1

The special case $r = 2$ gives the sequence

$$R_2(n) = 2^{n+1} - F_{n+3}, \qquad (4)$$

where F_n is a FIBONACCI NUMBER, the first few terms of which for $n = 1, 2, \ldots$ are 0, 1, 3, 8, 19, 43, 94, 201, ... (Sloane's A008466). The first few $R_3(n)$ are given by 0, 0, 1, 3, 8, 20, 47, 107, 238, ... Sloane's A050231; the first few $R_4(n)$ are 0, 0, 0, 1, 3, 8, 20, 48, 111, 251, 558, ... (Sloane's A050232); and the first few $R_5(n)$ 0, 0, 0, 0, 1, 3, 8, 20, 48, 112, 255, 571, 1262, ... (Sloane's A050233).

Given n BERNOULLI TRIALS with a probability of success (heads) p, the expected number of tails is $n(1 - p)$, so the expected number of tail runs ≥ 1 is $\approx n(1 - p)p$. Continuing,

$$N_R = n(1 - p)p^R \qquad (5)$$

is the expected number of runs $\geq R$. The longest expected run is therefore given by

$$R = \log_{1/p}[n(1 - p)] \qquad (6)$$

(Gordon *et al.* 1986, Schilling 1990). Given m 0s and n 1s, the number of possible arrangements with u runs is

$$f_u = \begin{cases} 2\binom{m-1}{k-1}\binom{n-1}{k-1} & u \equiv 2k \\ \binom{m-1}{k-1}\binom{n-1}{k-2} + \binom{m-1}{k-2}\binom{n-1}{k-1} & u \equiv 2k+1 \end{cases} \qquad (7)$$

for k an INTEGER, where $\binom{n}{k}$ is a BINOMIAL COEFFICIENT. Then

$$P(u \leq u') = \sum_{u=2}^{u'} \frac{f_u}{\binom{m+n}{m}}. \qquad (8)$$

Feller (1968, pp. 278–279) proved that for $w(n) \equiv 1 - R_{1/2}(3, n)$,

$$\lim_{n \to \infty} w(n)\alpha^{n+1} = \beta, \qquad (9)$$

where

$$\alpha = \tfrac{1}{3}\left[\left(136 + 24\sqrt{33}\right)^{1/3} - 8\left(136 + 24\sqrt{33}\right)^{-1/3} - 2\right]$$

$$= 1.087378025\ldots \tag{10}$$

and

$$\beta = \frac{2 - \alpha}{4 - 3\alpha} = 1.236839845\ldots . \tag{11}$$

The corresponding constants for a RUN of $k > 1$ heads are α_k, the smallest POSITIVE ROOT of

$$1 - x + \left(\tfrac{1}{2}x\right)^{k+1} = 0, \tag{12}$$

and

$$\beta_k = \frac{2 - \alpha}{k + 1 - k\alpha_k}. \tag{13}$$

These are modified for unfair coins with $P(H) = p$ and $P(T) = q = 1 - p$ to α'_k, the smallest POSITIVE ROOT of

$$1 - x + qp^k x^{k+1} = 0, \tag{14}$$

and

$$\beta'_k = \frac{1 - p\alpha'_k}{(k + 1 - k\alpha'_k)p} \tag{15}$$

(Feller 1968, pp. 322–325).

Let $C_t(m, k)$ denote the number of sequences of m indistinguishable objects of type A and k indistinguishable objects of type B in which *no* t-run occurs. The probability that a t-run *does* occur is then given by

$$P_t(m, k) = 1 - \frac{C_t(m, k)}{\binom{m + k}{k}}, \tag{16}$$

where $\binom{a}{b}$ is a BINOMIAL COEFFICIENT. Bloom (1996) gives the following recurrence sequence for $C_t(m, k)$,

$$C_t(m, k) = \sum_{i=0}^{t-1} C_t(m-1, k-i) - \sum_{i=1}^{t-1} C_t(m-t, k-i)$$
$$+ e_t(m, k), \tag{17}$$

where

$$e_t(m, k) \equiv \begin{cases} 1 & \text{if } m = 0 \text{ and } 0 \leq k < t \\ -1 & \text{if } m = t \text{ and } 0 \leq k < t \\ 0 & \text{otherwise.} \end{cases} \tag{18}$$

Another recurrence which has only a fixed number of terms is given by

$$C_t(m, k) = C_t(m-1, k) + C_t(m, k-1)$$
$$- C_t(m-t, k-1)$$

$$- C_t(m-1, k-t) + C_t(m-t, k-t) + e_t^*(m, k), \tag{19}$$

where

$$e_t^*(m, k) \equiv \begin{cases} 1 & \text{if } (m, k) = (0, 0) \text{ or } (t, t) \\ -1 & \text{if } (m, k) = (0, t) \text{ or } (t, 0) \\ 0 & \text{otherwise} \end{cases} \tag{20}$$

(Goulden and Jackson 1983, Bloom 1996). These formulas disprove the assertion of Gardner (1982) that "there will almost always be a clump of six or seven CARDS of the same color" in a normal deck of cards by giving $P_6(26, 26) = 0.46424$.

Bloom (1996) gives the expected number of noncontiguous t-runs in a sequence of m 0s and n 1s as

$$E(n, m, t) = \frac{(m + 1)(n)_t + (n + 1)(m)_t}{(m + n)_t}, \tag{21}$$

where $(a)_n$ is the POCHHAMMER SYMBOL. For $m > 10$, u has an approximately NORMAL DISTRIBUTION with MEAN and VARIANCE

$$\mu_u = 1 + \frac{2mn}{m + n} \tag{22}$$

$$\sigma_u^2 = \frac{2mn(2mn - m - n)}{(m + n)^2(m + n - 1)}. \tag{23}$$

See also COIN TOSSING, EULERIAN NUMBER, PERMUTATION, PERMUTATION RUN, s-RUN

References

Bloom, D. M. "Probabilities of Clumps in a Binary Sequence (and How to Evaluate Them Without Knowing a Lot)." *Math. Mag.* **69**, 366–372, 1996.

Feller, W. *An Introduction to Probability Theory and Its Application, Vol. 1, 3rd ed.* New York: Wiley, 1968.

Finch, S. "Favorite Mathematical Constants." http://www.mathsoft.com/asolve/constant/feller/feller.html.

Gardner, M. *Aha! Gotcha: Paradoxes to Puzzle and Delight.* New York: W. H. Freeman, p. 124, 1982.

Godbole, A. P. "On Hypergeometric and Related Distributions of Order k." *Commun. Stat.: Th. and Meth.* **19**, 1291–1301, 1990.

Godbole, A. P. and Papastavridis, G. (Eds.). *Runs and Patterns in Probability: Selected Papers.* New York: Kluwer, 1994.

Gordon, L.; Schilling, M. F.; and Waterman, M. S. "An Extreme Value Theory for Long Head Runs." *Prob. Th. and Related Fields* **72**, 279–287, 1986.

Goulden, I. P. and Jackson, D. M. *Combinatorial Enumeration.* New York: Wiley, 1983.

Mood, A. M. "The Distribution Theory of Runs." *Ann. Math. Statistics* **11**, 367–392, 1940.

Philippou, A. N. and Makri, F. S. "Successes, Runs, and Longest Runs." *Stat. Prob. Let.* **4**, 211–215, 1986.

Schilling, M. F. "The Longest Run of Heads." *Coll. Math. J.* **21**, 196–207, 1990.

Schuster, E. F. In *Runs and Patterns in Probability: Selected Papers* (Ed. A. P. Godbole and S. Papastavridis). Boston, MA: Kluwer, pp. 91–111, 1994.

Sloane, N. J. A. Sequences A008466, A050227, A050231, A050232, and A050233 in "An On-Line Version of the Encyclopedia of Integer Sequences." http://www.research.att.com/~njas/sequences/eisonline.html.

Runge-Kutta Method

A method of numerically integrating ORDINARY DIF-FERENTIAL EQUATIONS by using a trial step at the midpoint of an interval to cancel out lower-order error terms. The second-order formula is

$$k_1 = hf(s_n, y_n)$$

$$k_2 = hf\left(x_n + \tfrac{1}{2}h, \ y_n + \tfrac{1}{2}k_1\right)$$

$$y_{n+1} = y_n + k_2 + \mathcal{O}(h^3),$$

and the fourth-order formula is

$$k_1 = hf(s_n, y_n)$$

$$k_2 = hf\left(x_n + \tfrac{1}{2}h, \ y_n + \tfrac{1}{2}k_1\right)$$

$$k_3 = hf\left(x_n + \tfrac{1}{2}h, \ y_n + \tfrac{1}{2}k_2\right)$$

$$k_4 = hf(x_n + h, \ y_n + k_3)$$

$$y_{n+1} = y_n + \tfrac{1}{6}k_1 + \tfrac{1}{3}k_2 + \tfrac{1}{3}k_3 + \tfrac{1}{6}k_4 + \mathcal{O}(h^5).$$

(Press *et al.* 1992). This method is reasonably simple and robust and is a good general candidate for numerical solution of differential equations when combined with an intelligent adaptive step-size routine.

See also ADAMS' METHOD, GILL'S METHOD, MILNE'S METHOD, ORDINARY DIFFERENTIAL EQUATION, ROSENBROCK METHODS

References
Abramowitz, M. and Stegun, C. A. (Eds.). *Handbook of Mathematical Functions with Formulas, Graphs, and Mathematical Tables, 9th printing.* New York: Dover, pp. 896–897, 1972.
Arfken, G. *Mathematical Methods for Physicists, 3rd ed.* Orlando, FL: Academic Press, pp. 492–493, 1985.
Cartwright, J. H. E. and Piro, O. "The Dynamics of Runge-Kutta Methods." *Int. J. Bifurcations Chaos* **2**, 427–449, 1992. http://formentor.uib.es/~julyan/TeX/rkpaper/root/root.html.
Kutta, M. W. *Z. für Math. u. Phys.* **46**, 435, 1901.
Lambert, J. D. and Lambert, D. Ch. 5 in *Numerical Methods for Ordinary Differential Systems: The Initial Value Problem.* New York: Wiley, 1991.
Lindelöf, E. *Acta Soc. Sc. Fenn.* **2**, 1938.
Press, W. H.; Flannery, B. P.; Teukolsky, S. A.; and Vetterling, W. T. "Runge-Kutta Method" and "Adaptive Step Size Control for Runge-Kutta." §16.1 and 16.2 in *Numerical Recipes in FORTRAN: The Art of Scientific Computing, 2nd ed.* Cambridge, England: Cambridge University Press, pp. 704–716, 1992.
Runge, C. *Math. Ann.* **46**, 167, 1895.

Runge's Theorem

Let $K \subseteq \mathbb{C}$ be compact, let f be analytic on a neighborhood of K, and let $P \subseteq \mathbb{C}^* \backslash K$ contain at least one point from each connected component of $\mathbb{C}^* \backslash K$. Then for any $\epsilon > 0$, there is a RATIONAL FUNCTION $r(z)$ with poles in P such that

$$\max_{z \in K} |f(z) - r(z)| < \epsilon$$

(Krantz 1999, p. 143).

A polynomial version can be obtained by taking $P = \{\infty\}$. Let $f(x)$ be an ANALYTIC FUNCTION which is REGULAR in the interior of a JORDAN CURVE C and continuous in the closed DOMAIN bounded by C. Then $f(x)$ can be approximated with arbitrary accuracy by POLYNOMIALS (Szego o 1975, p. 5; Krantz 1999, p. 144).

See also ANALYTIC FUNCTION, JORDAN CURVE, MERGELYAN'S THEOREM

References
Krantz, S. G. "Runge's Theorem." §11.1.2 in *Handbook of Complex Analysis.* Boston, MA: Birkhäuser, pp. 143–144, 1999.
Szego, G. *Orthogonal Polynomials, 4th ed.* Providence, RI: Amer. Math. Soc., p. 7, 1975.

Runge-Walsh Theorem

RUNGE'S THEOREM

Run-Length Encoding

A specification of elements in a list as a list of pairs giving the element and number of times it occurs in a run. For example, given the list $\{1, 1, 1, 3, 3, 6, 6, 6, 2, 2, 2, 2, 3, 3, 1, 4, 4\}$, the run-length encoding is $\{\{1, 3\}, \{3, 2\}, \{6, 3\}, \{2, 4\}, \{3, 2, \}, \{1, 1\}, \{4, 2\}\}$. Run-length encoding can be implemented in *Mathematica* as

```
RunLengthEncode[x_List] := (Through[{First,
Length}][#1]] &) /@ Split[x]
```

See also LOOK AND SAY SEQUENCE, RUN

Running Average

MOVING AVERAGE

Running Knot

A KNOT which tightens around an object when strained but slackens when the strain is removed. Running knots are sometimes also known as slip knots or nooses.

References
Owen, P. *Knots.* Philadelphia, PA: Courage, p. 60, 1993.

Russell's Antinomy

Let R be the set of all sets which are not members of themselves. Then R is neither a member of itself nor not a member of itself. Symbolically, let $R = \{x : x \notin x\}$. Then $R \in R$ IFF $R \notin R$.

Bertrand Russell discovered this PARADOX and sent it in a letter to G. Frege just as Frege was completing *Grundlagen der Arithmetik.* This invalidated much of the rigor of the work, and Frege was forced to add a note at the end stating, "A scientist can hardly meet with anything more undesirable than to have the foundation give way just as the work is finished. I was put in this position by a letter from Mr. Bertrand Russell when the work was nearly through the press."

See also BARBER PARADOX, CATALOGUE PARADOX, GRELLING'S PARADOX

References

Courant, R. and Robbins, H. "The Paradoxes of the Infinite." §2.4.5 in *What is Mathematics?: An Elementary Approach to Ideas and Methods, 2nd ed.* Oxford, England: Oxford University Press, p. 78, 1996.
Curry, H. B. *Foundations of Mathematical Logic, 2nd rev. ed.* New York: Dover, p. 4, 1977.
Erickson, G. W. and Fossa, J. A. *Dictionary of Paradox.* Lanham, MD: University Press of America, pp. 175–177, 1998.
Frege, G. *Foundations of Arithmetic: A Logico-Mathematical Enquiry into the Concept of Number, 2nd rev. ed.* Evanston, IL: Northwestern University Press, 1980.
Hoffman, P. *The Man Who Loved Only Numbers: The Story of Paul Erdos and the Search for Mathematical Truth.* New York: Hyperion, p. 116, 1998.
Hofstadter, D. R. *Gödel, Escher, Bach: An Eternal Golden Braid.* New York: Vintage Books, pp. 20–21, 1989.
Mirimanoff, D. "Les antinomies de Russell et de Burali-Forti et le problème fondamental de la théorie des ensembles." *Enseign. math.* **19**, 37–52, 1917.
Whitehead, A. N. and Russell, B. *Principia Mathematica.* New York: Cambridge University Press, pp. 79 and 101, 1927.

Russell's Paradox

RUSSELL'S ANTINOMY

Russian Doll Prime

PRIME STRING

Russian Multiplication

Also called "Ethiopian multiplication." To multiply two numbers a and b, write $a_0 \equiv a$ and $b_0 \equiv b$ in two columns. Under a_0, write $\lfloor a_0/2 \rfloor$, where $\lfloor x \rfloor$ is the FLOOR FUNCTION, and under b_0, write $2b_0$. Continue until $a_i = 1$. Then cross out any entries in the b column which are opposite an EVEN NUMBER in the a column and add the b column. The result is the desired product. For example, for $a = 27$, $b = 35$

27	35
13	70
6	~~140~~
3	280
1	560
	945

See also MULTIPLICATION

References

Wells, D. *The Penguin Dictionary of Curious and Interesting Numbers.* Middlesex, England: Penguin Books, p. 44, 1986.

Russian Roulette

Russian roulette is a GAME of chance in which one or more of the six chambers of a gun are filled with bullets, the magazine is rotated at random, and the gun is fired. The shooter bets on whether the chamber which rotates into place will be loaded. If it is, he loses not only his bet but his life.

A modified version is considered by Blom *et al.* (1996) and Blom (1989). In this variant, the revolver is loaded with a single bullet, and two duelists alternately spin the chamber and fire at themselves until one is killed. The probability that the first duelist is killed is then 6/11.

References

Blom, G. *Probabilities and Statistics: Theory and Applications.* New York: Springer-Verlag, p. 32, 1989.
Blom, G.; Englund, J.-E.; and Sandell, D. "General Russian Roulette." *Math. Mag.* **69**, 293–297, 1996.

Ruth-Aaron Pair

A pair of consecutive numbers $(n, n + 1)$ such that the sums of the prime factors of n and $n + 1$ are equal. They are so named because they were inspired by the pair (714, 715) corresponding to Hank Aaron's record-breaking 715th home run in 1974, breaking Babe Ruth's earlier record of 714 (Hoffman 1998, pp. 179–181). The first few ns giving Ruth-Aaron pairs are 5, 8, 15, 77, 125, 714, 948, ... (Sloane's A039752), corresponding to the sums 5, 6, 8, 18, 15, 29, 86, ... (Sloane's A054378).

Pomerance suspected there were an infinite number of such pairs, and this was almost immediately proved true by P. Erdos (Hoffman 1998, pp. 180–181).

References

Hoffman, P. *The Man Who Loved Only Numbers: The Story of Paul Erdos and the Search for Mathematical Truth.* New York: Hyperion, 1998.
Nelson, C.; Penney, D. E.; and Pomerance, C. "714 and 715." *J. Recr. Math.* **7**, 87–89, 1994.
Peterson, I. "Ivars Peterson's MathLand: Playing with Ruth-Aaron Pairs." http://www.maa.org/mathland/mathland_6_30.html.
Sloane, N. J. A. Sequences A039752 and A054378 in "An On-Line Version of the Encyclopedia of Integer Sequences." http://www.research.att.com/~njas/sequences/eisonline.html.

Rutishauser's Rule

Let m and $m + h$ be two consecutive CRITICAL INDICES of f and let F be $(m + h)$-normal. If the polynomials $\tilde{p}_k^{(n)}$ are defined by

$$\tilde{p}_0^{(n)}(u) \equiv 1 \tag{1}$$

$$\tilde{p}_{k+1}^{(n)}(u) \equiv u\tilde{p}_k^{(n-1)}(u) - q_{m+k+1}^{(n)}\tilde{p}_k^{(n)}(u) \tag{2}$$

for $n = 0, 1, \ldots$ and $k = 0, \ldots, h-1$, then, under the hypothesis below, there exists an infinite set $\mathfrak{N}$ of positive integers such that

$$\lim_{\substack{n \to \infty \\ n \in \mathfrak{N}}} \tilde{p}_h^{(n)}(u) = \tilde{p}_h(u), \tag{3}$$

where

$$\tilde{p}_h(u) \equiv (u - u_{m+1})(u - u_{m+2}) \cdots (u - u_{m+h}). \tag{4}$$

By hypothesis, if $m = 0$, the polynomials $\tilde{p}_k^{(n)}$ are identical to the Hadamard polynomials $p_L^{(n)}$, and if $m > 0$, the algorithm for constructing the $\tilde{p}_k^{(n)}$ is applied to the qd scheme suitably bounded by columns $e_m^{(n)}$ and $e_{m+h}^{(n)}$ (Henrici 1988, pp. 642–643).

See also CRITICAL INDEX

References

Henrici, P. *Applied and Computational Complex Analysis, Vol. 1: Power Series-Integration-Conformal Mapping-Location of Zeros.* New York: Wiley, pp. 642–643, 1988.

Ryser Formula

A formula for the PERMANENT of a MATRIX

$$\mathrm{perm}(a_{ij}) = (-1)^n \sum_{s \subseteq \{1, \ldots, n\}} (-1)^{|s|} \prod_{i=1}^n \sum_{j \in s} a_{ij},$$

where the SUM is over all SUBSETS of $\{1, \ldots, n\}$, and $|s|$ is the number of elements in s. The formula can be optimized by picking the SUBSETS so that only a single element is changed at a time (which is precisely a GRAY CODE), reducing the number of additions from n^2 to n.

It turns out that the number of disks moved after the kth step in the TOWERS OF HANOI is the same as the element which needs to be added or deleted in the kth ADDEND of the Ryser formula (Gardner 1988, Vardi 1991, p. 111).

See also DETERMINANT, GRAY CODE, PERMANENT, TOWERS OF HANOI

References

Gardner, M. "The Icosian Game and the Tower of Hanoi." Ch. 6 in *The Scientific American Book of Mathematical Puzzles & Diversions.* New York: Simon and Schuster, pp. 55–62, 1959.

Knuth, D. E. *The Art of Computer Programming, Vol. 2: Seminumerical Algorithms, 3rd ed.* Reading, MA: Addison-Wesley, p. 515, 1998.

Nijenhuis, A. and Wilf, H. Chs. 7–8 in *Combinatorial Algorithms.* New York: Academic Press, 1975.

Vardi, I. *Computational Recreations in Mathematica.* Reading, MA: Addison-Wesley, p. 111, 1991.

S

Saalschützian

A GENERALIZED HYPERGEOMETRIC FUNCTION

$$
{}_pF_q\left[\begin{matrix}\alpha_1,\ \alpha_2,\ \ldots,\ \alpha_p;\ z\\ \beta_1,\ \beta_2,\ \ldots,\ \beta_q\end{matrix}\right],
$$

is said to be Saalschützian if it is κ-BALANCED with $k = 1$,

$$
\sum_{i=1}^q \beta_i = 1 + \sum_{i=1}^p \alpha_i.
$$

See also GENERALIZED HYPERGEOMETRIC FUNCTION, κ-BALANCED, NEARLY-POISED, WELL-POISED

References

Bailey, W. N. *Generalised Hypergeometric Series.* Cambridge, England: Cambridge University Press, p. 11, 1935.

Koepf, W. *Hypergeometric Summation: An Algorithmic Approach to Summation and Special Function Identities.* Braunschweig, Germany: Vieweg, p. 43, 1998.

Whipple, F. J. W. "Well-Poised Series and Other Generalized Hypergeometric Series." *Proc. London Math. Soc.* **25**, 525–544, 1926.

Saalschütz's Theorem

Mathematics:Calculus and Analysis:Special Functions:Hypergeometric Functions:Generalized Hypergeometric Functions

$$
{}_3F_2\left[\begin{matrix}-x,\ -y,\ -z\\ n+1,\ -x-y-z\end{matrix}\right] = \frac{\Gamma(n+1)\Gamma(x+y+n+1)}{\Gamma(x+n+1)\Gamma(y+n+1)}
$$

$$
\times \frac{\Gamma(y+z+n+1)\Gamma(z+x+n+1)}{\Gamma(z+n+1)(x+y+z+n+1)}, \tag{1}
$$

where ${}_3F_2(a, b, c; d, e; z)$ is a GENERALIZED HYPERGEOMETRIC FUNCTION and $\Gamma(z)$ is the GAMMA FUNCTION. It can be derived from the DOUGALL-RAMANUJAN IDENTITY and written in the symmetric form

$$
{}_3F_2(a, b, c; d, e; 1) = \frac{(d-a)_{|c|}(d-b)_{|c|}}{d_{|c|}(d-a-b)_{|c|}} \tag{2}
$$

for

$$
d + e = a + b + c + 1 \tag{3}
$$

with c a NONPOSITIVE INTEGER and $(a)_n$ the POCHHAMMER SYMBOL (Bailey 1935, p. 9; Petkovsek *et al.* 1996; Koepf 1998, p. 32). If one of a, b, and c is nonpositive but it is not known which, an alternative formulation due to W. Gosper gives the form

$$
{}_3F_2(a, b, c; d, e; 1)
$$

$$
= \frac{\Gamma(d)}{\Gamma(d-a)\Gamma(d-b)\Gamma(d-c)}\frac{\Gamma(e)}{\Gamma(e-a)\Gamma(e-b)(e-c)}
$$

$$
\times \frac{\pi^2}{\cos(\pi d)\cos(\pi e) + \cos(\pi a)\cos(\pi b)\cos(\pi c)}. \tag{4}
$$

which is symmetric in (a, b, c) and (d, e).

If instead

$$
a + b + c + 2 = d + e, \tag{5}
$$

then

$$
{}_3F_2(a, b, c;\ d, e;\ 2)
$$

$$
\pi^2\frac{de-(a+1)(b+1)(c+1)+abc}{\cos(d\pi)\cos(e\pi)-\cos(a\pi)\cos(b\pi)\cos(c\pi)}
$$

$$
\times \frac{\Gamma(d)}{\Gamma(d-a)\Gamma(d-b)\Gamma(d-c)}\frac{\Gamma(e)}{\Gamma(e-a)\Gamma(e-b)\Gamma(e-c)}
$$

$$
\tag{6}
$$

(W. Gosper).

See also DOUGALL-RAMANUJAN IDENTITY, GENERALIZED HYPERGEOMETRIC FUNCTION, KUMMER'S THEOREM

References

Bailey, W. N. "Saalschütz's Theorem." §2.2 in *Generalised Hypergeometric Series.* Cambridge, England: Cambridge University Press, p. 9, 1935.

Dougall, J. "On Vandermonde's Theorem and Some More General Expansions." *Proc. Edinburgh Math. Soc.* **25**, 114–132, 1907.

Hardy, G. H. *Ramanujan: Twelve Lectures on Subjects Suggested by His Life and Work, 3rd ed.* New York: Chelsea, p. 104, 1999.

Koepf, W. *Hypergeometric Summation: An Algorithmic Approach to Summation and Special Function Identities.* Braunschweig, Germany: Vieweg, 1998.

Petkovsek, M.; Wilf, H. S.; and Zeilberger, D. *A = B.* Wellesley, MA: A. K. Peters, pp. 43 and 126, 1996.

Saalschütz, L. "Eine Summationsformel." *Z. für Math. u. Phys.* **35**, 186–188, 1890.

Saalschütz, L. "Über einen Spezialfall der hypergeometrischen Reihe dritter Ordnung." *Z. für Math. u. Phys.* **36**, 278–295 and 321–327, 1891.

Shepard, W. F. "Summation of the Coefficients of Some Terminating Hypergeometric Series." *Proc. London Math. Soc.* **10**, 469–478, 1912.

s-Additive Sequence

A generalization of an ULAM SEQUENCE in which each term is the SUM of two earlier terms in exactly s ways. (s, t)-additive sequences are a further generalization in which each term has exactly s representations as the SUM of t distinct earlier numbers. It is conjectured that 0-additive sequences ultimately have periodic differences of consecutive terms (Guy 1994, p. 233).

See also GREEDY ALGORITHM, STÖHR SEQUENCE, SUM-FREE SET, ULAM SEQUENCE

References

Finch, S. R. "Conjectures about s-Additive Sequences." *Fib. Quart.* **29**, 209–214, 1991.

Finch, S. R. "Are 0-Additive Sequences Always Regular?" *Amer. Math. Monthly* **99**, 671–673, 1992.

Finch, S. R. "On the Regularity of Certain 1-Additive Sequences." *J. Combin. Th. Ser. A.* **60**, 123–130, 1992.

Finch, S. R. "Patterns in 1-Additive Sequences." *Experiment. Math.* **1**, 57–63, 1992.

Finch, S. "Unsolved Mathematics Problems: Ulam s-Additive Sequences." http://www.mathsoft.com/asolve/sadd/sadd.html.

Guy, R. K. *Unsolved Problems in Number Theory, 2nd ed.* New York: Springer-Verlag, pp. 110 and 233, 1994.

Ulam, S. M. *Problems in Modern Mathematics.* New York: Interscience, p. ix, 1964.

Saddle

A SURFACE possessing a SADDLE POINT.

See also HYPERBOLIC PARABOLOID, MONKEY SADDLE, SADDLE POINT (FUNCTION)

Saddle Point (Fixed Point)

HYPERBOLIC FIXED POINT (DIFFERENTIAL EQUATIONS), HYPERBOLIC FIXED POINT (MAP)

Saddle Point (Function)

A POINT of a FUNCTION or SURFACE which is a STATIONARY POINT but not an EXTREMUM. An example of a 1-D FUNCTION with a saddle point is $f(x) = x^3$, which has

$$f'(x) = 3x^2$$

$$f''(x) = 6x$$

$$f'''(x) = 6.$$

This function has a saddle point at $x_0 = 0$ by the EXTREMUM TEST since $f''(x_0) = 0$ and $f'''(x_0) = 6 \neq 0$. An example of a SURFACE with a saddle point is the MONKEY SADDLE.

Saddle Point (Game)

For a general two-player ZERO-SUM GAME,

$$\max_{i \leq m} \min_{j \leq n} a_{ij} \leq \min_{j \leq n} \max_{i \leq m} a_{ij}.$$

If the two are equal, then write

$$\max_{i \leq m} \min_{j \leq n} a_{ij} \leq \min_{j \leq n} \max_{i \leq m} a_{ij} \equiv v,$$

where v is called the VALUE of the GAME. In this case, there exist optimal strategies for the first and second players.

A NECESSARY and SUFFICIENT condition for a saddle point to exist is the presence of a PAYOFF MATRIX element which is both a minimum of its row and a maximum of its column. A GAME may have more than one saddle point, but all must have the same VALUE.

See also GAME, PAYOFF MATRIX, VALUE

References

Dresher, M. "Saddle Points." §1.5 in *The Mathematics of Games of Strategy: Theory and Applications.* New York: Dover, pp. 12–14, 1981.

Llewellyn, D. C.; Tovey, C.; and Trick, M. "Finding Saddlepoints of Two-Person, Zero Sum Games." *Amer. Math. Monthly* **95**, 912–918, 1988.

Saddle Polygon

SKEW POLYGON

Saddle-Node Bifurcation

FOLD BIFURCATION

Safarevich Conjecture

SHAFAREVICH CONJECTURE

Safe

A position in a GAME is safe for a player A if the person who plays next (player B) will lose.

See also GAME, UNSAFE

Sagitta

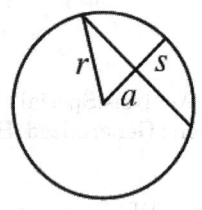 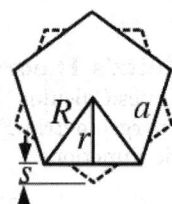

The PERPENDICULAR distance s from an ARC'S MIDPOINT to the CHORD across it, equal to the RADIUS r minus the APOTHEM a,

$$s = r - a. \tag{1}$$

For a REGULAR POLYGON of side length a,

$$s \equiv R - r = \tfrac{1}{2} a \left[\csc\left(\frac{\pi}{n}\right) - \cot\left(\frac{\pi}{n}\right) \right]$$

$$= \tfrac{1}{2} a \tan\left(\frac{\pi}{2n}\right) \tag{2}$$

$$= r \tan\left(\frac{\pi}{n}\right) \tan\left(\frac{\pi}{2n}\right) \tag{3}$$

$$= 2R \sin^2\left(\frac{\pi}{2n}\right). \tag{4}$$

where R is the CIRCUMRADIUS, r the INRADIUS, a is the side length, and n is the number of sides.

See also APOTHEM, CHORD, SECTOR, SEGMENT

Saint Andrew's Cross

A GREEK CROSS rotated by 45°, also called the crux decussata. The MULTIPLICATION SIGN × is based on Saint Andrew's cross (Bergamini 1969).

See also CROSS, GREEK CROSS, MULTIPLICATION SIGN

References

Bergamini, D. *Mathematics*. New York: Time-Life Books, p. 11, 1969.

Saint Anthony's Cross

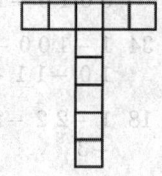

A CROSS also called the tau cross or crux commissa.

See also CROSS

Saint Petersburg Paradox

Consider a game, first proposed by Daniel Bernoulli, in which a player bets on how many TOSSES of a COIN will be needed before it first turns up heads. The player pays a fixed amount initially, and then receives 2^n dollars if the coin comes up heads on the nth toss. The expectation value of the gain is then

$$\tfrac{1}{2}(2) + \tfrac{1}{4}(4) + \tfrac{1}{8}(8) + \ldots = 1 + 1 + 1 + \ldots = \infty$$

dollars, so any finite amount of money can be wagered and the player will still come out ahead on average.

Feller (1968) discusses a modified version of the game in which the player receives nothing if a trial takes more than a fixed number N of tosses. The classical theory of this modified game concluded that ∞ is a fair entrance fee, but Feller notes that "the modern student will hardly understand the mysterious discussions of this 'paradox'."

In another modified version of the game, the player bets $2 that heads will turn up on the first throw, $4 that heads will turn up on the second throw (if it did not turn up on the first), $8 that heads will turn up on the third throw, etc. Then the expected *payoff* is

$$\tfrac{1}{2}(2) + \tfrac{1}{4}(4) + \tfrac{1}{8}(8) + \ldots = 1 + 1 + 1 + \ldots = \infty,$$

so the player can apparently be in the hole by any amount of money and still come out ahead in the end. This paradox can clearly be resolved by making the distinction between the amount of the final payoff

and the net amount won in the game. It is misleading to consider the payoff without taking into account the amount lost on previous bets, as can be shown as follows. At the time the player first wins (say, on the nth toss), he will have lost

$$\sum_{k=1}^{n-1} 2^k = 2^n - 2$$

dollars. In this toss, however, he wins 2^n dollars. This means that the net gain for the player is a whopping $2, no matter how many tosses it takes to finally win. As expected, the large payoff after a long run of tails is exactly balanced by the large amount that the player has to invest. In fact, by noting that the probability of winning on the nth toss is $1/2^n$, it can be seen that the probability distribution for the number of tosses needed to win is simply a GEOMETRIC DISTRIBUTION with $p = 1/2$.

See also COIN TOSSING, GAMBLER'S RUIN, GEOMETRIC DISTRIBUTION, MARTINGALE

References

Ball, W. W. R. and Coxeter, H. S. M. *Mathematical Recreations and Essays, 13th ed.* New York: Dover, pp. 201–202, 1987.
Erickson, G. W. and Fossa, J. A. *Dictionary of Paradox.* Lanham, MD: University Press of America, pp. 13–15, 1998.
Feller, W. "The Petersburg Game." §10.4 in *An Introduction to Probability Theory and Its Applications, Vol. 1, 3rd ed.* New York: Wiley, pp. 235–237, 1968.
Gardner, M. *The Scientific American Book of Mathematical Puzzles & Diversions.* New York: Simon and Schuster, pp. 51–52, 1959.
Kamke, E. *Einführung in die Wahrscheinlichkeitstheorie.* Leipzig, Germany, pp. 82–89, 1932.
Keynes, J. M. K. "The Application of Probability to Conduct." In *The World of Mathematics, Vol. 2* (Ed. K. Newman). Redmond, WA: Microsoft Press, 1988.
Kraitchik, M. "The Saint Petersburg Paradox." §6.18 in *Mathematical Recreations.* New York: W. W. Norton, pp. 138–139, 1942.
Todhunter, I. §391 in *History of the Mathematical Theory of Probability.* New York: Chelsea, p. 221, 1949.

Sal

WALSH FUNCTION

Salamin Formula

BRENT-SALAMIN FORMULA

Salem Constants

Each point of a PISOT-VIJAYARAGHAVAN CONSTANT S is a LIMIT POINT from both sides of a set T known as the Salem constants (Salem 1945). The Salem constants are ALGEBRAIC INTEGERS > 1 in which one or more of the conjugates is on the UNIT CIRCLE with the others inside (Le Lionnais 1983, p. 150). The smallest known Salem number was found by Lehmer (1933) as the largest REAL ROOT of

$$x^{10} + x^9 - x^7 - x^6 - x^5 - x^4 - x^3 + x + 1 = 0,$$

which is

$$\sigma_1 = 1.176280818\ldots$$

(Le Lionnais 1983, p. 35). Boyd (1977) found the following table of small Salem numbers, and suggested that σ_1, σ_2, σ_3, and σ_4 are the smallest Salem numbers. The NOTATION $1\,1\,0\,-1\,-1\,-1$ is short for $1\,1\,0\,-1\,-1\,-1\,-1\,-1\,0\,1\,1$, the coefficients of the above polynomial.

k	σ_k	°	POLYNOMIAL
1	1.1762808183	10	1 1 0 −1 −1 −1
2	1.1883681475	18	1 −1 1 −1 0 0 −1 1 −1 1
3	1.2000265240	14	1 0 0 −1 −1 0 0 1
4	1.2026167437	14	1 0 −1 0 0 0 0 −1
5	1.2163916611	10	1 0 0 0 −1 −1
6	1.2197208590	18	1 −1 0 0 0 0 0 0 −1 1
7	1.2303914344	10	1 0 0 −1 0 −1
8	1.2326135486	20	1 −1 0 0 0 −1 1 0 0 −1 1
9	1.2356645804	22	1 0 −1 −1 0 0 0 1 1 0 −1 −1
10	1.2363179318	16	1 −1 0 0 0 0 0 0 −1
11	1.2375048212	26	1 0 −1 0 0 −1 0 0 −1 0 1 0 0 1
12	1.2407264237	12	1 −1 1 −1 0 0 −1
13	1.2527759374	18	1 0 0 0 0 0 −1 −1 −1 −1
14	1.2533306502	20	1 0 −1 0 0 −1 0 0 0 0 0
15	1.2550935168	14	1 0 −1 −1 0 1 0 −1
16	1.2562211544	18	1 −1 0 0 −1 1 0 0 0 −1
17	1.2601035404	24	1 −1 0 0 −1 1 0 −1 1 −1 0 1 −1
18	1.2602842369	22	1 −1 0 −1 1 0 0 0 −1 1 −1 1
19	1.2612309611	10	1 0 −1 0 0 −1
20	1.2630381399	26	1 −1 0 0 0 0 −1 0 0 0 0 0 0 1
21	1.2672964425	14	1 −1 0 0 0 0 −1 1
22	1.2806381563	8	1 0 0 −1 −1
23	1.2816913715	26	1 0 0 0 0 0 −1 −1 −1 −1 −1 −1 −1 −1
24	1.2824955606	20	1 −2 2 −2 2 −2 1 0 −1 1 −1
25	1.2846165509	18	1 0 0 0 −1 0 −1 −1 0 −1
26	1.2847468215	26	1 −2 1 1 −2 1 0 0 −1 1 0 −1 1 −1
27	1.2850993637	30	1 0 0 0 0 −1 −1 −1 −1 −1 −1 0 0 0 0 1
28	1.2851215202	30	1 −2 2 −2 2 1 0 −1 2 −2 1 0 −1 1 −1 1 −1
29	1.2851856708	30	1 −1 0 0 0 0 0 0 −1 0 0 0 −1 0 0 −1
30	1.2851967268	26	1 0 −1 −1 0 0 0 1 0 −1 −1 0 1 1
31	1.2851991792	44	1 −1 0 0 0 0 0 −1 0 0 0 −1 0 0 0 0 0 0 0 1 0 0 1
32	1.2852354362	30	1 0 −1 0 0 −1 0 −1 0 0 0 1 0 0 1 0 −1
33	1.2854090648	34	1 −1 0 0 −1 1 −1 0 1 −1 1 0 −1 1 −1 0 1 −1
34	1.2863959668	18	1 −2 2 −2 2 −2 2 −3 3 −3
35	1.2867301820	26	1 −1 0 0 −1 1 −1 0 1 −1 1 0 −1 1
36	1.2917414257	24	1 −1 0 0 0 0 −1 0 0 0 0 0 0
37	1.2920391602	20	1 0 −1 0 0 −1 0 0 −1 0 1
38	1.2934859531	10	1 0 −1 −1 0 1
39	1.2956753719	18	1 −1 0 0 −1 1 −1 0 1 −1

See also PISOT-VIJAYARAGHAVAN CONSTANT

References

Boyd, D. W. "Small Salem Numbers." *Duke Math. J.* **44**, 315–328, 1977.

Boyd, D. W. "Pisot and Salem Numbers in Intervals of the Real Line." *Math. Comput.* **32**, 1244–1260, 1978.

Le Lionnais, F. *Les nombres remarquables.* Paris: Hermann, 1983.

Lehmer, D. H. "Factorization of Certain Cyclotomic Functions." *Ann. Math., Ser. 2* **34**, 461–479, 1933.

Salem, R. "Power Series with Integral Coefficients." *Duke Math. J.* **12**, 153–172, 1945.

Stewart, C. L. "Algebraic Integers whose Conjugates Lie Near the Unit Circle." *Bull. Soc. Math. France* **106**, 169–176, 1978.

Salesman Problem

TRAVELING SALESMAN PROBLEM

Salient Point

A point at which two noncrossing branches of a curve meet with different tangents.

See also CUSP

Salinon

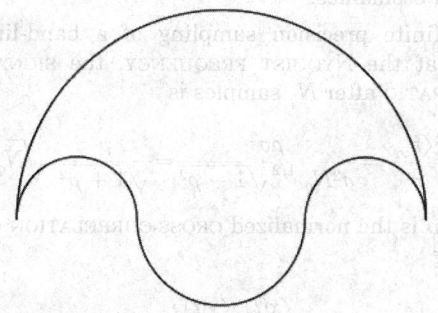

The above figure formed from four connected SEMI-CIRCLES. The word *salinon* is Greek for "salt cellar," which the figure resembles. In his *Book of Lemmas*, Archimedes proved that the salinon has an area equal to the CIRCLE having the line segment joining the top and bottom points as its DIAMETER (Wells 1991).

See also ARBELOS, LUNE, PIECEWISE CIRCULAR CURVE, SEMICIRCLE

References
Schwartzman, S. *The Words of Mathematics: An Etymological Dictionary of Mathematical Terms Used in English.* Washington, DC: Math. Assoc. Amer., p. 192, 1994.
Wells, D. *The Penguin Dictionary of Curious and Interesting Geometry.* London: Penguin, p. 144, 1991.

Sally Sequence
The Sally sequence gives the sequence of lengths of the repetitions which are avoided in the LINUS SEQUENCE. The first few terms are 0, 1, 1, 2, 1, 3, 1, 1, 3, 2, 1, 6, 3, 2, ... (Sloane's A006346).

See also LINUS SEQUENCE

References
Sloane, N. J. A. Sequences A006346/M0126 in "An On-Line Version of the Encyclopedia of Integer Sequences." http://www.research.att.com/~njas/sequences/eisonline.html.
Sloane, N. J. A. and Plouffe, S. Figure M0126 in *The Encyclopedia of Integer Sequences.* San Diego: Academic Press, 1995.

Salmon Points
The 20 CAYLEY LINES generated by a HEXAGON inscribed in a CONIC SECTION pass four at a time though 15 points known as Salmon points (Wells 1991). There is a dual relationship between the 15 Salmon points and the 15 PLÜCKER LINES.

See also CAYLEY LINES, KIRKMAN POINTS, PASCAL LINES, PASCAL'S THEOREM, PLÜCKER LINES, STEINER POINTS

References
Wells, D. *The Penguin Dictionary of Curious and Interesting Geometry.* London: Penguin, p. 172, 1991.

Salmon's Theorem
There are at least two theorems known as Salmon's theorem. This first states that if P and S are two points, PX and SY are the perpendiculars from P and S to the POLARS of S and P, respectively, with respect to a CIRCLE with center O, then $OP/OS = PX/SY$ (Durell 1928).

The second Salmon's theorem states that, given a track bounded by two confocal ELLIPSES, if a ball is rolled so that its trajectory is tangent to the inner ELLIPSE, the ball's trajectory will be tangent to the inner ELLIPSE following all subsequent caroms as well.

See also BILLIARDS, POLAR

References
Durell, C. V. *Modern Geometry: The Straight Line and Circle.* London: Macmillan, p. 95, 1928.
Salmon, G. *A Treatise on Conic Sections.* New York: Chelsea, p. 182, 1960.

Saltus
The word saltus has two different meanings: either a jump or an oscillation of a function.

Sample

See also POPULATION, SAMPLE PROPORTION, SAMPLE SIZE, SAMPLE SPACE, SAMPLE VARIANCE, SAMPLING

References
Kenney, J. F. and Keeping, E. S. "Populations and Samples." §7.1 in *Mathematics of Statistics, Pt. 1, 3rd ed.* Princeton, NJ: Van Nostrand, pp. 90–91, 1962.

Sample Proportion
Let there be x successes out of n BERNOULLI TRIALS. The sample proportion is the fraction of samples which were successes, so

$$\hat{p} = \frac{x}{n}. \qquad (1)$$

For large n, $\hat{p}$ has an approximately NORMAL DISTRIBUTION. Let RE be the RELATIVE ERROR and SE the STANDARD ERROR, then

$$\langle p \rangle = p \qquad (2)$$

$$\mathrm{SE}(\hat{p}) \equiv \sigma(\hat{p}) = \sqrt{\frac{p(1-p)}{n}} \qquad (3)$$

$$\mathrm{RE}(\hat{p}) = \sqrt{\frac{2\hat{p}(1-\hat{p})}{n}}\, \mathrm{erf}^{-1}(\mathrm{CI}), \qquad (4)$$

where CI is the CONFIDENCE INTERVAL and erf x is the ERF function. The number of tries needed to determine p with RELATIVE ERROR RE and CONFIDENCE

INTERVAL CI is

$$n = \frac{2\left[\mathrm{erf}^{-1}(\mathrm{CI})\right]^2 \hat{p}(1-\hat{p})}{(\mathrm{RE})^2}. \quad (5)$$

Sample Size

See also SAMPLE, SAMPLE VARIANCE

Sample Space

Informally, the sample space for a given set of events is the set of all possible values the events may assume. Formally, the set of possible events for a given variate forms a SIGMA ALGEBRA, and sample space is defined as the largest set in the SIGMA ALGEBRA.

See also PROBABILITY SPACE, RANDOM VARIABLE, SAMPLE, SIGMA ALGEBRA, STATE SPACE

Sample Variance

To estimate the population VARIANCE σ^2 from a sample of N elements with a priori *unknown* MEAN (i.e., the MEAN is estimated from the sample itself), we need an unbiased ESTIMATOR for σ^2. This ESTIMATOR is given by K-STATISTIC k_2, where

$$k_2 = \frac{N}{N-1} m_2 \quad (1)$$

and $m_2 \equiv s^2$ is the sample variance

$$s^2 \equiv \frac{1}{N} \sum_{i=1}^{N} (x_i - \bar{x})^2. \quad (2)$$

Note that some authors prefer the definition

$$s'^2 \equiv \frac{1}{N-1} \sum_{i=1}^{N} (x_i - \bar{x})^2, \quad (3)$$

since this makes the sample variance an UNBIASED ESTIMATOR for the population variance.

Also note that, in general, $\sqrt{\hat{\sigma}^2}$ in *not* an UNBIASED ESTIMATOR of σ even if $\hat{\sigma}^2$ is an UNBIASED ESTIMATOR for σ^2).

See also K-STATISTIC, SAMPLE, UNBIASED ESTIMATOR, VARIANCE

Sampling

The selection and implementation of statistical observations in order to estimate properties of an underlying population. Sampling is a vital part of modern polling, market research, and manufactur-

ing, and its proper use is vital in the functioning of modern economies.

For infinite precision sampling of a band-limited signal at the NYQUIST FREQUENCY, the SIGNAL-TO-NOISE RATIO after N_q samples is

$$\mathrm{SNR} = \frac{\langle r_\infty \rangle}{\sigma_\infty} = \frac{\rho \sigma^2}{\sigma^2 N_q^{-1/2} \sqrt{1+\rho^2}} = \frac{\rho}{\sqrt{1+\rho^2}} \sqrt{N_q}, \quad (1)$$

where ρ is the normalized CROSS-CORRELATION COEFFICIENT

$$\rho \equiv \frac{\langle x(t) \rangle \langle y(t) \rangle}{\sqrt{\langle x^2(t) \rangle \langle y^2(t) \rangle}}. \quad (2)$$

For $\rho \ll 1$,

$$\mathrm{SNR} \approx \rho \sqrt{N_q}. \quad (3)$$

The identical result is obtained for oversampling. For undersampling, the SIGNAL-TO-NOISE RATIO decreases (Thompson *et al.* 1986).

See also NYQUIST SAMPLING, OVERSAMPLING, QUANTIZATION EFFICIENCY, SAMPLE, SAMPLING FUNCTION, SHANNON SAMPLING THEOREM, SINC FUNCTION

References

Feuer, A. *Sampling in Digital Signal Processing and Control.* Boston, MA: Birkhäuser, 1996.
Govindarajulu, Z. *Elements of Sampling Theory and Methods.* Upper Saddle River, NJ: Prentice-Hall, 1999.
Thompson, A. R.; Moran, J. M.; and Swenson, G. W. Jr. *Interferometry and Synthesis in Radio Astronomy.* New York: Wiley, pp. 214–216, 1986.

Sampling Function

SHAH FUNCTION

Sampling Theorem

In order for a band-limited (i.e., one with a zero POWER SPECTRUM for frequencies $v > B$) baseband ($v > 0$) signal to be reconstructed fully, it must be sampled at a rate $v \geq 2B$. A signal sampled at $v = 2B$ is said to be NYQUIST SAMPLED, and $v = 2B$ is called the NYQUIST FREQUENCY. No information is lost if a signal is sampled at the NYQUIST FREQUENCY, and no additional information is gained by sampling faster than this rate.

See also ALIASING, NYQUIST FREQUENCY, NYQUIST SAMPLING, OVERSAMPLING

Sampling Theory

The study of SAMPLING

San Marco Fractal

The FRACTAL $J(-3/4, 0)$, where J is the JULIA SET. It slightly resembles the MANDELBROT SET.

See also DENDRITE FRACTAL, DOUADY'S RABBIT FRACTAL, JULIA SET, MANDELBROT SET, SIEGEL DISK FRACTAL

References

Wagon, S. *Mathematica in Action.* New York: W. H. Freeman, p. 173, 1991.

Sandwich Theorem

The LOVÁSZ NUMBER $\vartheta(G)$ of a GRAPH G satisfies

$$\omega(G) \leq \vartheta(\bar{G}) \leq \chi(G).$$

where $\omega(G)$ is the CLIQUE NUMBER and χ is the minimum number of colors needed to color the VERTICES of G. $\vartheta(G)$ can be computed efficiently despite the fact that the computation of the two numbers it lies between is an NP-HARD PROBLEM.

The SQUEEZING THEOREM is also sometimes known as the sandwich theorem.

See also HAM SANDWICH THEOREM, SQUEEZING THEOREM

References

Grötschel, M.; Lovász, L.; and Schrijver, A. "The Ellipsoid Method and Its Consequences in Combinatorial Optimization." *Combinatorica* **1**, 169–197, 1981.
Knuth, D. E. "The Sandwich Theorem." *Electronic J. Combinatorics* **1**, A1 1–48, 1994. http://www.combinatorics.org/Volume_1/volume1.html#A1.

Sangaku Problem

A geometric problem found on a mathematical wooden tablet (in Japan. Such problems typically involve mutually TANGENT CIRCLES or TANGENT SPHERES.

See also CASEY'S THEOREM, CIRCLE INSCRIBING, CYLINDER-SPHERE INTERSECTION, DESCARTES CIRCLE THEOREM, ELLIPSE TANGENT, HEXLET, JAPANESE THEOREM, RIGHT TRIANGLE, TANGENT CIRCLES, TANGENT SPHERES

References

Fukagawa, H. and Sokolowsky, D. *Traditional Japanese Mathematics Problems from the 18th and 19th Centuries.* Singapore: Science Culture Technology Press, in preparation.
Fukagawa, H. and Pedoe, D. *Japanese Temple Geometry Problems.* Winnipeg, Manitoba, Canada: Charles Babbage Research Foundation, 1989.
Mikami, Y. *The Development of Mathematics in China and Japan, 2nd ed.* New York: Chelsea, 1974.
Rothman, T. "Japanese Temple Geometry." *Sci. Amer.* **278**, 85–91, May 1998.
Smith, D. E. and Mikami, Y. *A History of Japanese Mathematics.* Chicago: Open Court, 1914.

Sard's Theorem

The set of "critical values" of a MAP $u : \mathbb{R}^n \to \mathbb{R}^n$ of CLASS C^1 has LEBESGUE MEASURE 0 in $\mathbb{R}^n$.

See also CLASS (MAP), LEBESGUE MEASURE, TRANSVERSAL INTERSECTION

References

Iyanaga, S. and Kawada, Y. (Eds.). *Encyclopedic Dictionary of Mathematics.* Cambridge, MA: MIT Press, p. 682, 1980.

Sarkovskii's Theorem

Order the NATURAL NUMBERS as follows:

$$3 \prec 5 \prec 7 \prec 9 \prec 11 \prec 13 \prec 15 \prec \ldots \prec 2 \cdot 3 \prec 2 \cdot 5 \prec 2 \cdot 7$$

$$\prec 2 \cdot 9 \prec \ldots \prec 2 \cdot 2 \cdot 3 \prec 2 \cdot 2 \cdot 5 \prec 2 \cdot 2 \cdot 7$$

$$\prec 2 \cdot 2 \cdot 9 \prec \ldots \prec 2 \cdot 2 \cdot 2 \cdot 3 \prec \ldots \prec 2^5 \prec 2^4 \prec 2^3 \prec 2^2$$

$$\prec 2 \prec 1.$$

Now let F be a CONTINUOUS FUNCTION from the REALS to the REALS and suppose $p \prec q$ in the above ordering. Then if F has a point of LEAST PERIOD p, then F also has a point of LEAST PERIOD q.

A special case of this general result, also known as Sarkovskii's theorem, states that if a CONTINUOUS REAL function has a PERIODIC POINT with period 3, then there is a PERIODIC POINT of period n for every INTEGER n.

A converse to Sarkovskii's theorem says that if $p \prec q$ in the above ordering, then we can find a CONTINUOUS FUNCTION which has a point of LEAST PERIOD q, but does not have any points of LEAST PERIOD p (Elaydi 1996). For example, there is a CONTINUOUS FUNCTION with no points of LEAST PERIOD 3 but having points of all other LEAST PERIODS.

See also LEAST PERIOD

References

Conway, J. H. and Guy, R. K. "Periodic Points." In *The Book of Numbers.* New York: Springer-Verlag, pp. 207–208, 1996.

Devaney, R. L. *An Introduction to Chaotic Dynamical Systems,* 2nd ed. Reading, MA: Addison-Wesley, 1989.

Elaydi, S. "On a Converse of Sharkovsky's Theorem." *Amer. Math. Monthly* **103**, 386–392, 1996.

Ott, E. *Chaos in Dynamical Systems.* New York: Cambridge University Press, p. 49, 1993.

Sharkovsky, A. N. "Co-Existence of Cycles of a Continuous Mapping of a Line onto Itself." *Ukranian Math. Z.* **16**, 61–71, 1964.

Stefan, P. "A Theorem of Sharkovsky on the Existence of Periodic Orbits of Continuous Endomorphisms of the Real Line." *Comm. Math. Phys.* **54**, 237–248, 1977.

Sárközy's Theorem

A partial solution to the ERDOS SQUAREFREE CONJECTURE which states that the BINOMIAL COEFFICIENT $\binom{2n}{n}$ is never SQUAREFREE for all sufficiently large $n \geq n_0$. Sárközy (1985) showed that if $s(n)$ is the square part of the BINOMIAL COEFFICIENT $\binom{2n}{n}$, then

$$\ln s(n) \sim \left(\sqrt{2} - 2\right)\zeta\left(\tfrac{1}{2}\right)\sqrt{n}.$$

where $\zeta(z)$ is the RIEMANN ZETA FUNCTION. An upper bound on n_0 of $2^{8,000}$ has been obtained.

See also BINOMIAL COEFFICIENT, ERDOS SQUAREFREE CONJECTURE

References

Erdos, P. and Graham, R. L. *Old and New Problems and Results in Combinatorial Number Theory.* Geneva, Switzerland: L'Enseignement Mathématique Université de Genève, Vol. 28, 1980.

Sander, J. W. "A Story of Binomial Coefficients and Primes." *Amer. Math. Monthly* **102**, 802–807, 1995.

Sárközy, A. "On the Divisors of Binomial Coefficients, I." *J. Number Th.* **20**, 70–80, 1985.

Vardi, I. "Applications to Binomial Coefficients." *Computational Recreations in Mathematica.* Reading, MA: Addison-Wesley, pp. 25–28, 1991.

Sarrus Linkage

A LINKAGE which converts circular to linear motion using a hinged square.

See also HART'S INVERSOR, LINKAGE, PEAUCELLIER INVERSOR

Sarrus Number

POULET NUMBER

Sarti Dodecic

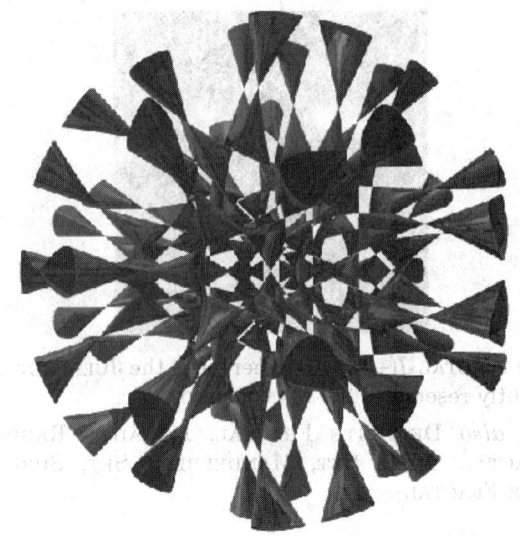

The DODECIC SURFACE defined by

$$X_{12} = 243S_{12} - 22Q_{12} = 0, \tag{1}$$

where

$$Q_{12} = \left(x^2 + y^2 + z^2 + w^2\right)^6 \tag{2}$$

$$S_{12} = 33\sqrt{5}\left(s_{2,\,3}^- + s_{3,\,4}^- + s_{4,\,2}^-\right)$$
$$+ 19\left(s_{2,\,3}^+ + s_{3,\,4}^+ + s_{4,\,2}^+\right)$$
$$+ 10s_{2,\,3,\,4} - 14s_{1,\,0} + 2s_{1,\,1} - 6s_{1,\,2}$$
$$- 352s_{5,\,1} + 336l_5^2 l_1 + 48l_2 l_3 l_4 \tag{3}$$

$$l_1 = x^4 + y^4 + z^4 + w^4 \tag{4}$$

$$l_2 = x^2 y^2 + z^2 w^2 \tag{5}$$

$$l_3 = x^2 z^2 + y^2 w^2 \tag{6}$$

$$l_4 = x^2 w^2 + y^2 z^2 \tag{7}$$

$$l_5 = xyzw \tag{8}$$

$$s_{1,\,0} = l_1(l_2 l_3 + l_2 l_4 + l_3 l_4) \tag{9}$$

$$s_{1,\,1} = l_1^2(l_2 + l_3 + l_4) \tag{10}$$

$$s_{1,\,2} = l_1(l_2^2 + l_3^2 + l_4^2) \tag{11}$$

$$s_{5,\,1} = l_5^2(l_2 + l_3 + l_4) \tag{12}$$

$$s_{2,\,3,\,4} = l_2^3 + l_3^3 + l_4^3 \tag{13}$$

$$s_{2,\,3}^\pm = l_2^2 l_3 \pm l_2 l_3^2 \tag{14}$$

$$s_{3,\,4}^\pm = l_3^2 l_4 \pm l_3 l_4^2 \tag{15}$$

$$s_{4,\,2}^\pm = l_4^2 l_2 \pm l_4 l_2^2. \tag{16}$$

Q_{12} and S_{12} are both invariants of order 12. The Sarti surface is invariant under the BIPOLYHEDRAL GROUP and has exactly 600 ORDINARY POINTS (Endraß). It was discovered by A. Sarti in 1999.

See also ALGEBRAIC SURFACE, BIPOLYHEDRAL GROUP, DODECIC SURFACE

References

Endraß, S. "The Sarti Surface." http://enriques.mathemati-k.uni-mainz.de/kon/docs/Esarti.shtml.

SAS Theorem

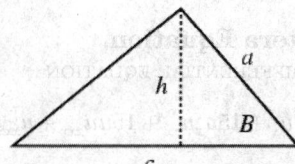

Specifying two sides and the ANGLE between them uniquely determines a TRIANGLE. Let c be the base length and h be the height. Then the AREA is

$$K = \tfrac{1}{2}ch = \tfrac{1}{2}ac\sin B. \qquad (1)$$

The length of the third side is given by the LAW OF COSINES,

$$b^2 = a^2 + c^2 - 2ac\cos B.$$

so

$$b = \sqrt{a^2 + c^2 - 2ac\cos B}. \qquad (2)$$

Using the LAW OF SINES

$$\frac{a}{\sin A} = \frac{b}{\sin B} = \frac{c}{\sin C} \qquad (3)$$

then gives the two other ANGLES as

$$A = \sin^{-1}\left(\frac{a\sin B}{\sqrt{a^2 + c^2 - 2ac\cos B}}\right) \qquad (4)$$

$$C = \sin^{-1}\left(\frac{c\sin B}{\sqrt{a^2 + c^2 - 2ac\cos B}}\right) \qquad (5)$$

See also AAA THEOREM, AAS THEOREM, ASA THEOREM, ASS THEOREM, SSS THEOREM, TRIANGLE

Satellite Knot

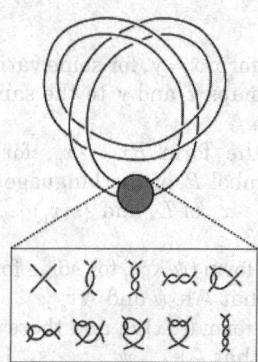

Let K_1 be a knot inside a TORUS, and knot the TORUS in the shape of a second knot (called the COMPANION KNOT) K_2, with certain additional mild restrictions to avoid trivial cases. Then the new knot resulting from K_1 is called the satellite knot K_3. All satellite knots are PRIME (Hoste *et al.* 1998). The illustration above illustrates a satellite knot of the TREFOIL KNOT, which is the form all satellite knots of 16 or fewer crossings take (Hoste *et al.* 1998). Satellites of the trefoil share the trefoil's chirality, and all have wrapping number 2.

Any satellite knot having wrapping number > 2 must have at least 27 crossings, and any satellite of the FIGURE EIGHT KNOT must have at least 17 crossings (Hoste *et al.* 1998). The numbers of satellite knots with n crossings are 0, 0, 0, 0, 0, 0, 0, 0, 0, 0, 0, 0, 2, 2, 6, 10, ... (Sloane's A051765), so the satellite knot of minimal crossing number occurs for 13 crossings.

The only KNOTS which are not HYPERBOLIC KNOTS are TORUS KNOTS and satellite knots (including COMPOSITE KNOTS). No satellite knot is an ALMOST ALTERNATING KNOT. If a COMPANION KNOT has crossing number k and the satellite ravels m times longitudinally around the solid torus, then it is conjectured that the satellite cannot be projected with fewer than km^2 crossings (Hoste *et al.* 1998).

See also ALMOST ALTERNATING KNOT, CABLE KNOT, COMPANION KNOT, COMPOSITE KNOT, HYPERBOLIC KNOT, TORUS KNOT

References

Adams, C. C. *The Knot Book: An Elementary Introduction to the Mathematical Theory of Knots.* New York: W. H. Freeman, pp. 115–118, 1994.

Hoste, J.; Thistlethwaite, M.; and Weeks, J. "The First 1,701,936 Knots." *Math. Intell.* **20**, 33–48, Fall 1998.

Sloane, N. J. A. Sequences A051765 in "An On-Line Version of the Encyclopedia of Integer Sequences." http://www.research.att.com/~njas/sequences/eisonline.html.

Satisfaction

Let $\mathbb{A}$ be a RELATIONAL SYSTEM, and let L be a language which is appropriate for $\mathbb{A}$. Let ϕ be a well-formed formula of L, and let s be a valuation in

$\mathbb{A}$. Then $\mathbb{A} \models_s \phi$ is written provided that one of the following holds:

1. ϕ is of the form $x = y$, for some variables x and y of L, and s maps x and y to the same element of the structure $\mathbb{A}$.
2. ϕ is of the form $R x_1 \cdots x_n$, for some n-ary predicate symbol R of the language L, and some variables $x_1 \cdots x_n$ of L, and $\{s(x_1), \ldots, s(x_n)\}$ is a member of $R^{\mathbb{A}}$.
3. ϕ is of the form $(\psi \wedge \gamma)$, for some formulas ψ and γ of L such that $\mathbb{A} \models_s \psi$ and $\mathbb{A} \models_s \gamma$.
4. ϕ is of the form $((\exists x)\psi)$, and there is an element a of $\mathbb{A}$ such that $\mathbb{A} \models_{s(x|a)} \psi$.

In this case, $\mathbb{A}$ is said to satisfy ϕ with the valuation s.

See also Los' Theorem

References

Bell, J. L. and Slomson, A. B. *Models and Ultraproducts: An Introduction.* Amsterdam, Netherlands: North-Holland, 1969.
Enderton, H. E. *A Mathematical Introduction to Logic.* Boston, MA: Academic Press, 1972.

Satisfiability Problem

Deciding whether a given Boolean formula in conjunctive normal form has an assignment that makes the formula "true." In 1971, Cook showed that the problem is NP-complete.

See also Boolean Algebra

References

Cook, S. A. and Mitchell, D. G. "Finding Hard Instances of the Satisfiability Problem: A Survey." In *Satisfiability Problem: Theory and Applications (Piscataway, NJ, 1996)* (Ed. D. Du, J. Gu, and P. M.Pardalos). Providence, RI: Amer. Math. Soc., pp. 1–17, 1997.

Sausage Conjecture

In n-D for $n \geq 5$ the arrangement of hyperspheres whose convex hull has minimal content is always a "sausage" (a set of hyperspheres arranged with centers along a line), independent of the number of n-spheres. The conjecture was proposed by Fejes Tóth, and solved for dimensions ≥ 42 by Betke *et al.* (1994) and Betke and Henk (1998).

See also Content, Convex Hull, Hypersphere, Hypersphere Packing, Sphere Packing

References

Betke, U.; Henk, M.; and Wills, J. M. "Finite and Infinite Packings." *J. reine angew. Math.* **453**, 165–191, 1994.
Betke, U. and Henk, M. "Finite Packings of Spheres." *Discrete Comput. Geom.* **19**, 197–227, 1998.
Croft, H. T.; Falconer, K. J.; and Guy, R. K. Problem D9 in *Unsolved Problems in Geometry.* New York: Springer-Verlag, 1991.

Fejes Tóth, L. "Research Problems." *Periodica Methematica Hungarica* **6**, 197–199, 1975.

Savitzky-Golay Filter

A low-pass filter which is useful for smoothing data.

See also Filter

References

Press, W. H.; Flannery, B. P.; Teukolsky, S. A.; and Vetterling, W. T. *Numerical Recipes in FORTRAN: The Art of Scientific Computing, 2nd ed.* Cambridge, England: Cambridge University Press, pp. 183 and 644–645, 1992.

Savoy Knot

Figure-of-Eight Knot

Sawada-Kotera Equation

The partial differential equation

$$u_t + 45 u^2 u_x + 15 u_x u_{xx} + 15 u u_{xxx} + u_{xxxxx} = 0.$$

See also Caudrey-Dodd-Gibbon-Sawada-Kotera Equation

References

Matsumo, Y. *Bilinear Transformation Method.* New York: Academic Press, p. 7, 1984.
Zwillinger, D. *Handbook of Differential Equations, 3rd ed.* Boston, MA: Academic Press, p. 134, 1997.

Sawtooth Wave

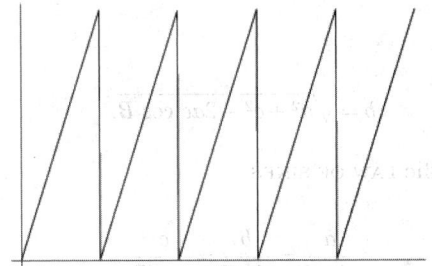

The periodic function given by

$$S(x) = A \, \mathrm{frac}(x/T + \phi). \tag{1}$$

where $\mathrm{frac}(x)$ is the fractional part $\mathrm{frac} \, x \equiv x - \lfloor x \rfloor$, A is the amplitude, T is the period of the wave, and ϕ is its phase. If $\phi = 0$, $A = 1$, and $T = 2L$, then the Fourier series is given by

$$f(x) = \tfrac{1}{2} - \frac{1}{\pi} \sum_{n-1}^{\infty} \frac{1}{n} \sin\left(\frac{n \pi x}{L}\right).$$

See also Fourier Series–Sawtooth Wave, Fractional Part, Staircase Function

References

Spanier, J. and Oldham, K. B. *An Atlas of Functions.* Washington, DC: Hemisphere, p. 74, 1987.

sc

JACOBI ELLIPTIC FUNCTIONS

Scalar

A one-component quantity which is invariant under ROTATIONS of the coordinate system.

See also PSEUDOSCALAR, SCALAR FIELD, SCALAR FUNCTION, SCALAR MULTIPLICATION, SCALAR POTENTIAL, SCALAR TRIPLE PRODUCT, TENSOR, VECTOR

References

Jeffreys, H. and Jeffreys, B. S. "Scalars and Vectors." Ch. 2 in *Methods of Mathematical Physics, 3rd ed.* Cambridge, England: Cambridge University Press, pp. 56–85, 1988.

Scalar Curvature

The scalar curvature (called the "curvature scalar" by Weinberg 1972, p. 135) is given by

$$R \equiv g^{\mu\kappa} R_{\mu\kappa}.$$

where $g^{\mu\kappa}$ is the METRIC TENSOR and $R_{\mu\kappa}$ is the RICCI TENSOR.

See also CURVATURE, EINSTEIN TENSOR, GAUSSIAN CURVATURE, MEAN CURVATURE, METRIC TENSOR, RADIUS OF CURVATURE, RICCI TENSOR, RIEMANN-CHRISTOFFEL TENSOR

References

Misner, C. W.; Thorne, K. S.; and Wheeler, J. A. *Gravitation.* San Francisco: W. H. Freeman, p. 222, 1973.
Wald, R. M. *General Relativity.* Chicago, IL: University of Chicago Press, p. 40, 1984.
Weinberg, S. *Gravitation and Cosmology: Principles and Applications of the General Theory of Relativity.* New York: Wiley, p. 135, 1972.

Scalar Field

A MAP $f : \mathbb{R}^n \mapsto \mathbb{R}$ which assigns each $\mathbf{x}$ a SCALAR FUNCTION $f(\mathbf{x})$.

See also VECTOR FIELD

References

Morse, P. M. and Feshbach, H. "Scalar Fields." §1.1 in *Methods of Theoretical Physics, Part I.* New York: McGraw-Hill, pp. 4–8, 1953.

Scalar Function

A function $f(x_1, \ldots, x_n)$ of one or more variables whose RANGE is one-dimensional, as compared to a VECTOR FUNCTION, whose RANGE is three-dimensional (or, in general, n-dimensional).

See also COMPLEX FUNCTION, REAL FUNCTION, VECTOR FUNCTION

Scalar Multiplication

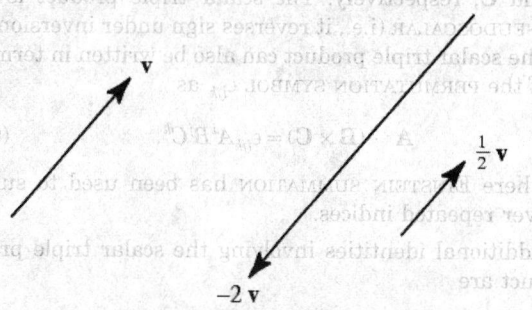

Scalar multiplication refers to the multiplication of a VECTOR by a constant s, producing a vector in the same (for $s > 0$) or opposite (for $s < 0$) direction but of different length. Scalar multiplication is indicated in *Mathematica* by placing a scalar next to a vector (with or without an optional asterisk), $s\{a1, a2, \ldots, an\}$.

See also MULTIPLICATION, VECTOR, VECTOR ADDITION, VECTOR MULTIPLICATION

Scalar Potential

A conservative VECTOR FIELD (for which the CURL $\nabla \times \mathbf{F} = 0$) may be assigned a scalar potential

$$\phi(x, y, z) - \phi(0, 0, 0) \equiv -\int_C \mathbf{F} \cdot d\mathbf{s}$$

$$= -\int_{(0, 0, 0)}^{(x, 0, 0)} F_1(t, 0, 0)\, dt + \int_{(x, 0, 0)}^{(x, y, 0)} F_2(x, t, 0)\, dt$$

$$+ \int_{(x, y, 0)}^{(x, y, z)} F_3(x, y, t)\, dt.$$

where $\int_C \mathbf{F} \cdot d\mathbf{s}$ is a LINE INTEGRAL.

See also LINE INTEGRAL, POTENTIAL FUNCTION, VECTOR POTENTIAL

Scalar Product

DOT PRODUCT

Scalar Triple Product

The scalar triple product of three VECTORS $\mathbf{A}$, $\mathbf{B}$, and $\mathbf{C}$ is denoted $[\mathbf{A}, \mathbf{B}, \mathbf{C}]$ and defined by

$$[\mathbf{A}, \mathbf{B}, \mathbf{C}] \equiv \mathbf{A} \cdot (\mathbf{B} \times \mathbf{C}) \tag{1}$$

$$= \mathbf{B} \cdot (\mathbf{C} \times \mathbf{A}) \tag{2}$$

$$= \mathbf{C} \cdot (\mathbf{A} \times \mathbf{B}) \tag{3}$$

$$= \det(\mathbf{A}(\mathbf{BC})) \tag{4}$$

$$= \begin{vmatrix} A_1 & A_2 & A_3 \\ A_2 & C_2 & B_3 \\ A_3 & C_2 & C_3 \end{vmatrix} \tag{5}$$

where $\mathbf{A} \cdot \mathbf{B}$ denotes a DOT PRODUCT, $\mathbf{A} \times \mathbf{B}$ denotes a CROSS PRODUCT, $\det(A) = |A|$ denotes a DETERMINANT, and A_i, B_i, and C_i are components of the vectors **A**, **B**, and **C**, respectively. The scalar triple product is a PSEUDOSCALAR (i.e., it reverses sign under inversion). The scalar triple product can also be written in terms of the PERMUTATION SYMBOL ϵ_{ijk} as

$$\mathbf{A} \cdot (\mathbf{B} \times \mathbf{C}) = \epsilon_{ijk} A^i B^j C^k. \tag{6}$$

where EINSTEIN SUMMATION has been used to sum over repeated indices.

Additional identities involving the scalar triple product are

$$\mathbf{A} \cdot (\mathbf{B} \times \mathbf{C}) = \mathbf{B} \cdot (\mathbf{C} \times \mathbf{A}) = \mathbf{C} \cdot (\mathbf{A} \times \mathbf{B}) \tag{7}$$

$$\begin{aligned}[\mathbf{A}, \mathbf{B}, \mathbf{C}]\mathbf{D} \\ = [\mathbf{D}, \mathbf{B}, \mathbf{C}]\mathbf{A} + [\mathbf{A}, \mathbf{D}, \mathbf{C}]\mathbf{B} + [\mathbf{A}, \mathbf{B}, \mathbf{D}]\mathbf{C} \end{aligned} \tag{8}$$

$$[\mathbf{q}, \mathbf{q}', \mathbf{q}''][\mathbf{r}, \mathbf{r}', \mathbf{r}''] = \begin{vmatrix} \mathbf{q} \cdot \mathbf{r} & \mathbf{q} \cdot \mathbf{r}' & \mathbf{q} \cdot \mathbf{r}'' \\ \mathbf{q}' \cdot \mathbf{r} & \mathbf{q}' \cdot \mathbf{r}' & \mathbf{q}' \cdot \mathbf{r}'' \\ \mathbf{q}'' \cdot \mathbf{r} & \mathbf{q}'' \cdot \mathbf{r}' & \mathbf{q}'' \cdot \mathbf{r}'' \end{vmatrix}. \tag{9}$$

The VOLUME of a PARALLELEPIPED whose sides are given by the vectors **A**, **B**, and **C** is given by the ABSOLUTE VALUE of the scalar triple product

$$V_{\text{parallelepiped}} = .|\mathbf{A} \cdot (\mathbf{B} \times \mathbf{C})|. \tag{10}$$

See also CROSS PRODUCT, DOT PRODUCT, PARALLELEPIPED, VECTOR MULTIPLICATION, VECTOR TRIPLE PRODUCT

References

Arfken, G. "Triple Scalar Product, Triple Vector Product." §1.5 in *Mathematical Methods for Physicists, 3rd ed.* Orlando, FL: Academic Press, pp. 26–33, 1985.

Jeffreys, H. and Jeffreys, B. S. "The Triple Scalar Product." §2.091 in *Methods of Mathematical Physics, 3rd ed.* Cambridge, England: Cambridge University Press, pp. 74–75, 1988.

Scale

BASE (NUMBER)

Scale Factor

For a diagonal METRIC TENSOR $g_{ij} = g_{ii}\delta_{ij}$, where δ_{ij} is the KRONECKER DELTA, the scale factor is defined by

$$h_i \equiv \sqrt{g_{ii}}. \tag{1}$$

The LINE ELEMENT (first FUNDAMENTAL FORM) is then given by

$$ds^2 = g_{11}\, dx_{11}^2 + g_{22}\, dx_{22}^2 + g_{33}\, dx_{33}^2 \tag{2}$$

$$= h_1^2\, dx_{11}^2 + h_2^2\, dx_{22}^2 + h_3^2\, dx_{33}^2. \tag{3}$$

The scale factor appears in vector derivatives of coordinates in CURVILINEAR COORDINATES.

See also CURVILINEAR COORDINATES, FUNDAMENTAL FORMS, LINE ELEMENT

Scale Invariance

SELF-SIMILARITY

Scalene Triangle

A TRIANGLE with three unequal sides.

See also ACUTE TRIANGLE, EQUILATERAL TRIANGLE, ISOSCELES TRIANGLE, OBTUSE TRIANGLE, TRIANGLE

Scaling

Increasing a plane figure's linear dimensions by a scale factor s increases the PERIMETER $p' \to sp$ and the AREA $A' \to s^2 A$.

See also CONTRACTION (GEOMETRY), EXPANSION, FRACTAL, HOMOTHETIC, SELF-SIMILARITY

Scattering Operator

An OPERATOR relating the past asymptotic state of a DYNAMICAL SYSTEM governed by the Schrödinger equation

$$i \frac{d}{dt} \psi(t) = H\psi(t)$$

to its future asymptotic state.

See also WAVE OPERATOR

Scattering Theory

The mathematical study of the SCATTERING OPERATOR and Schrödinger equation.

See also SCATTERING OPERATOR

References

Yafaev, D. R. *Mathematical Scattering Theory: General Theory.* Providence, RI: Amer. Math. Soc., 1996.

Schaar's Identity

A generalization of the GAUSSIAN SUM. For p and q of opposite PARITY (i.e., one is EVEN and the other is ODD), Schaar's identity states

$$\frac{1}{\sqrt{q}} \sum_{r=0}^{q-1} \epsilon^{-\pi i r^2 p/q} = \frac{\epsilon^{-\pi i/4}}{\sqrt{p}} \sum_{r=0}^{p-1} \epsilon^{\pi i r^2 q/p}.$$

Schaar's identity can also be written so as to be valid for p, q with pq EVEN.

See also GAUSSIAN SUM

References

Borwein, J. M. and Borwein, P. B. *Pi & the AGM: A Study in Analytic Number Theory and Computational Complexity.* New York: Wiley, 1987.

Evans, R. and Berndt, B. "The Determination of Gauss Sums." *Bull. Amer. Math. Soc.* **5**, 107–129, 1981.

Schanuel's Conjecture

Let $\lambda_1, \ldots, \lambda_n \in \mathbb{C}$ be linearly independent over the RATIONALS $\mathbb{Q}$, then

$$\mathbb{Q}\left(\lambda_1, \ldots, \lambda_n, e^{\lambda_1}, \ldots, e^{\lambda_n}\right)$$

has TRANSCENDENCE degree at least n over $\mathbb{Q}$. Schanuel's conjecture implies the LINDEMANN-WEIERSTRASS THEOREM and GELFOND'S THEOREM. If the conjecture is true, then it follows that e and π are ALGEBRAICALLY INDEPENDENT. Mcintyre (1991) proved that the truth of Schanuel's conjecture also guarantees that there are no unexpected exponential-algebraic relations on the INTEGERS $\mathbb{Z}$ (Marker 1996).

At present, a proof of Schanuel's conjecture seems out of reach (Chow 1999).

See also ALGEBRAICALLY INDEPENDENT, CONSTANT PROBLEM, GELFOND'S THEOREM, LINDEMANN-WEIERSTRASS THEOREM

References

Chow, T. Y. "What is a Closed-Form Number." *Amer. Math. Monthly* **106**, 440–448, 1999.
Chudnovsky, G. V. "On the Way to Schanuel's Conjecture." Ch. 3 in *Contributions to the Theory of Transcendental Numbers.* Providence, RI: Amer. Math. Soc., pp. 145–176, 1984.
Lin, F.-C. "Schanuel's Conjecture Implies Ritt's Conjecture." *Chinese J. Math.* **11**, 41–50, 1983.
Macintyre, A. "Schanuel's Conjecture and Free Exponential Rings." *Ann. Pure Appl. Logic* **51**, 241–246, 1991.
Marker, D. "Model Theory and Exponentiation." *Not. Amer. Math. Soc.* **43**, 753–759, 1996.

Schauder Fixed Point Theorem

Let A be a closed convex subset of a BANACH SPACE and assume there exists a continuous MAP T sending A to a countably compact subset $T(A)$ of A. Then T has fixed points.

References

Iyanaga, S. and Kawada, Y. (Eds.). *Encyclopedic Dictionary of Mathematics.* Cambridge, MA: MIT Press, p. 543, 1980.
Schauder, J. "Der Fixpunktsatz in Funktionalräumen." *Studia Math.* **2**, 171–180, 1930.
Zeidler, E. *Applied Functional Analysis: Applications to Mathematical Physics.* New York: Springer-Verlag, 1995.

Scheme

A local-ringed SPACE which is locally isomorphic to an AFFINE SCHEME.

See also AFFINE SCHEME

References

Itô, K. (Ed.). "Schemes." §16D in *Encyclopedic Dictionary of Mathematics, 2nd ed., Vol. 1.* Cambridge, MA: MIT Press, p. 69, 1986.

Schensted Correspondence

A correspondence between a PERMUTATION and a pair of YOUNG TABLEAUX.

See also PERMUTATION, YOUNG TABLEAU

References

Knuth, D. E. *The Art of Computer Programming, Vol. 3: Sorting and Searching, 2nd ed.* Reading, MA: Addison-Wesley, 1973.
Stanton, D. W. and White, D. E. §3.6 in *Constructive Combinatorics.* New York: Springer-Verlag, pp. 85–87, 1986.

Scherk's Minimal Surfaces

Scherk's two MINIMAL SURFACES were discovered by Scherk in 1834. They were the first new surfaces discovered since Meusnier in 1776. Beautiful images of wood sculptures of Scherk surfaces are illustrated by Séquin.

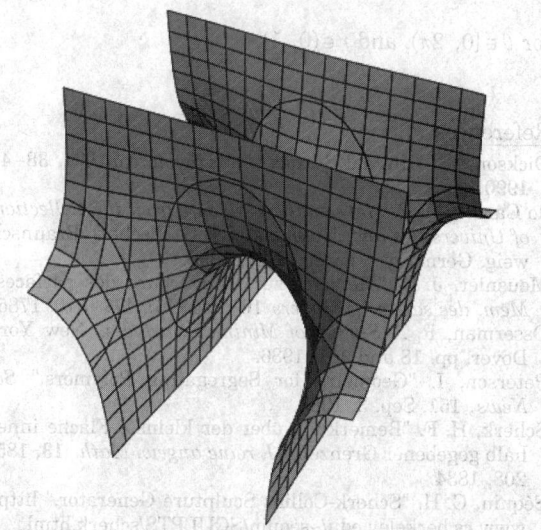

Scherk's first surface is doubly periodic and is defined by the implicit equation

$$e^z \cos y = \cos x, \tag{1}$$

(Osserman 1986, Wells 1991, von Seggern 1993). It has been observed to form in layers of block copolymers (Peterson 1988).

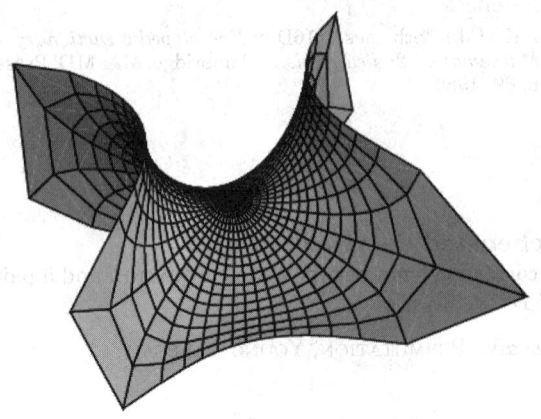

Scherk's second surface can be written parametrically as

$$x = 2\Re\left[\ln\left(1 + re^{i\theta}\right) - \ln\left(1 - re^{i\theta}\right)\right] \tag{2}$$

$$y = \Re\left[4i\,\tan^{-1}\left(re^{i\theta}\right)\right] \tag{3}$$

$$z = \Re\left\{2i\left(-\ln\left[1 - r^2 e^{2i\theta}\right] + \ln\left[1 + r^2 e^{2i\theta}\right]\right)\right\} \tag{4}$$

for $\theta \in [0,\ 2\pi)$, and $r \in (0,\ 1)$.

References

Dickson, S. "Minimal Surfaces." *Mathematica J.* **1**, 38–40, 1990.
do Carmo, M. P. *Mathematical Models from the Collections of Universities and Museums* (Ed. G. Fischer). Braunschweig, Germany: Vieweg, p. 41, 1986.
Meusnier, J. B. "Mémoire sur la courbure des surfaces." *Mém. des savans étrangers* **10** (lu 1776), 477–510, 1785.
Osserman, R. *A Survey of Minimal Surfaces.* New York: Dover, pp. 18 and 101, 1986.
Peterson, I. "Geometry for Segregating Polymers." *Sci. News*, 151, Sep. 3, 1988.
Scherk, H. F. "Bemerkung über der kleinste Fläche innerhalb gegebener Grenzen." *J. reine angew. Math.* **13**, 185–208, 1834.
Séquin, C. H. "Scherk-Collins Sculpture Generator." http://www.cs.berkeley.edu/~sequin/SCULPTS/scherk.html.
Thomas, E. L.; Anderson, D. M.; Henkee, C. S.; and Hoffman, D. "Periodic Area-Minimizing Surfaces in Block Copolymers." *Nature* **334**, 598–601, 1988.
von Seggern, D. *CRC Standard Curves and Surfaces.* Boca Raton, FL: CRC Press, p. 304, 1993.
Wells, D. *The Penguin Dictionary of Curious and Interesting Geometry.* London: Penguin, p. 223, 1991.
Wolfram Research. "Mathematica Version 2.0 Graphics Gallery." http://www.mathsource.com/cgi-bin/msitem22?0207-155.

Schiffler Point

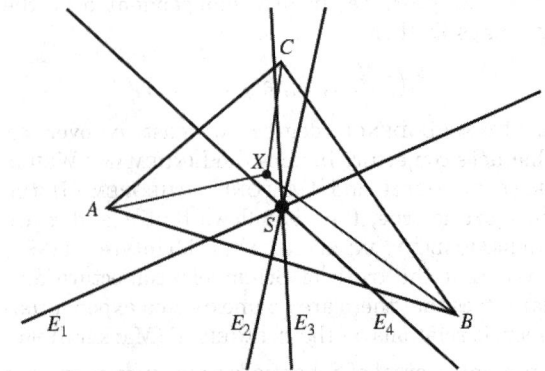

The CONCURRENCE S of the EULER LINES E_n of the TRIANGLES $\triangle XBC$, $\triangle XCA$, $\triangle XAB$, and $\triangle ABC$ where X is the INCENTER. The TRIANGLE CENTER FUNCTION is

$$\alpha = \frac{1}{\cos B + \cos C} = \frac{b + c - a}{b + c}.$$

References

Kimberling, C. "Central Points and Central Lines in the Plane of a Triangle." *Math. Mag.* **67**, 163–187, 1994.
Kimberling, C. "Schiffler Point." http://cedar.evansville.edu/~ck6/tcenters/recent/schiff.html.
Schiffler, K.; Veldkamp, G. R.; and van der Spek, W. A. "Problem 1018 and Solution." *Crux Math.* **12**, 176–179, 1986.

Schinzel Circle

A CIRCLE having a given number of LATTICE POINTS on its CIRCUMFERENCE. The Schinzel circle having n lattice points is given by the equation

$$\begin{cases} \left(x - \tfrac{1}{2}\right)^2 + y^2 = \tfrac{1}{4}\,5^{k-1} & \text{for } n = 2k \text{ even} \\ \left(x - \tfrac{1}{3}\right)^2 + y^2 = \tfrac{1}{9}\,5^{2k} & \text{for } n = 2k + 1 \text{ odd}. \end{cases}$$

Note that these solutions do not necessarily have the smallest possible RADIUS. For example, while the Schinzel circle centered at $(1/3,\ 0)$ and with radius $625/3$ has nine lattice points on its circumference, so does the circle centered at $(1/3,\ 0)$ with radius $65/3$.

See also CIRCLE, CIRCLE LATTICE POINTS, KULIKOWSKI'S THEOREM, LATTICE POINT, SCHINZEL'S THEOREM, SPHERE

References

Honsberger, R. "Circles, Squares, and Lattice Points." Ch. 11 in *Mathematical Gems I.* Washington, DC: Math. Assoc. Amer., pp. 117–127, 1973.
Kulikowski, T. "Sur l'existence d'une sphère passant par un nombre donné aux coordonnées entières." *L'Enseignement Math. Ser. 2* **5**, 89–90, 1959.
Schinzel, A. "Sur l'existence d'un cercle passant par un nombre donné de points aux coordonnées entières." *L'Enseignement Math. Ser. 2* **4**, 71–72, 1958.

Sierpinski, W. "Sur quelques problèmes concernant les points aux coordonnées entières." *L'Enseignement Math. Ser. 2* **4**, 25–31, 1958.

Sierpinski, W. "Sur un problème de H. Steinhaus concernant les ensembles de points sur le plan." *Fund. Math.* **46**, 191–194, 1959.

Sierpinski, W. *A Selection of Problems in the Theory of Numbers.* New York: Pergamon Press, 1964.

Schinzel's Hypothesis

If $f_1(x), ..., f_s(x)$ are IRREDUCIBLE POLYNOMIALS with INTEGER COEFFICIENTS such that no INTEGER $n > 1$ divides $f_1(x), ..., f_s(x)$ for all INTEGERS x, then there should exist infinitely many x such that $f_1(x), ..., f_s(x)$ are simultaneously PRIME.

References

Dickson, L. E. "A New Extension of Dirichlet's Theorem on Prime Numbers." *Messenger Math.* **33**, 155–161, 1904.

Ribenboim, P. *The New Book of Prime Number Records.* New York: Springer-Verlag, 1996.

Schinzel, A. and Sierpinski, W. "Sur certaines hypothèses concernant les nombres premiers. Remarque." *Acta Arithm.* **4**, 185–208, 1958.

Schinzel's Theorem

For every POSITIVE INTEGER n, there exists a CIRCLE in the plane having exactly n LATTICE POINTS on its CIRCUMFERENCE. The theorem is based on the number $r(n)$ of integral solutions (x, y) to the equation

$$x^2 + y^2 = n, \tag{1}$$

given by

$$r(n) = 4(d_1 - d_3), \tag{2}$$

where d_1 is the number of divisors of n OF THE FORM $4k + 1$ and d_3 is the number of divisors OF THE FORM $4k + 3$. It explicitly identifies such circles (the SCHINZEL CIRCLES) as

$$\begin{cases} \left(x - \tfrac{1}{2}\right)^2 + y^2 = \tfrac{1}{4} 5^{k-1} & \text{for } n = 2k \\ \left(x - \tfrac{1}{3}\right)^2 + y^2 = \tfrac{1}{9} 5^{2k} & \text{for } n = 2k + 1. \end{cases} \tag{3}$$

Note, however, that these solutions do not necessarily have the smallest possible radius.

See also BROWKIN'S THEOREM, KULIKOWSKI'S THEOREM, SCHINZEL CIRCLE

References

Honsberger, R. "Circles, Squares, and Lattice Points." Ch. 11 in *Mathematical Gems I.* Washington, DC: Math. Assoc. Amer., pp. 117–127, 1973.

Kulikowski, T. "Sur l'existence d'une sphère passant par un nombre donné aux coordonnées entières." *L'Enseignement Math. Ser. 2* **5**, 89–90, 1959.

Schinzel, A. "Sur l'existence d'un cercle passant par un nombre donné de points aux coordonnées entières." *L'Enseignement Math. Ser. 2* **4**, 71–72, 1958.

Sierpinski, W. "Sur quelques problèmes concernant les points aux coordonnées entières." *L'Enseignement Math. Ser. 2* **4**, 25–31, 1958.

Sierpinski, W. "Sur un problème de H. Steinhaus concernant les ensembles de points sur le plan." *Fund. Math.* **46**, 191–194, 1959.

Sierpinski, W. *A Selection of Problems in the Theory of Numbers.* New York: Pergamon Press, 1964.

Schisma

The musical interval by which eight fifths and a major third exceed five octaves,

$$\Re[z] > 0$$

See also COMMA OF DIDYMUS, COMMA OF PYTHAGORAS, DIESIS

Schläfli Double Sixes

DOUBLE SIXES

Schläfli Function

The function giving the VOLUME of the spherical quadrectangular TETRAHEDRON:

$$V = \frac{\pi^2}{8} f\left(\frac{\pi}{p}, \frac{\pi}{q}, \frac{\pi}{r}\right),$$

where

$$\tfrac{1}{2}\pi^2 f\left(\tfrac{\pi}{2} - x, \, y, \, \tfrac{\pi}{2} - z\right) = \sum_{m=1}^{\infty} \left(\frac{D - \sin x \sin z}{D + \sin x \sin z}\right)^m$$
$$\times \frac{\cos(2mx) - \cos(2my) + \cos(2mz) - 1}{m^2} - x^2 - y^2 - z^2,$$

and

$$D \equiv \sqrt{\cos^2 x \cos^2 z - \cos^2 y}.$$

See also TETRAHEDRON

Schläfli Integral

A definition of a function using a CONTOUR INTEGRAL. Schläfli integrals may be converted into RODRIGUES FORMULAS.

See also RODRIGUES FORMULA

Schläfli Polynomial

A polynomial given in terms of the NEUMANN POLYNOMIALS $O_n(x)$ by

$$S_n(x) = \frac{2xO_n(x) - 2\cos^2\left(\tfrac{1}{2}n\pi\right)}{n}.$$

See also NEUMANN POLYNOMIAL

References

Erdelyi, A.; Magnus, W.; Oberhettinger, F.; and Tricomi, F. G. *Higher Transcendental Functions, Vol. 2.* Krieger, p. 34, 1981.

Gradshteyn, I. S. and Ryzhik, I. M. "Neumann's and Schläfli Polynomials: $O_n(z)$ and $S_n(z)$." §8.59 in *Tables of Integrals, Series, and Products, 6th ed.* San Diego, CA: Academic Press, pp. 989–991, 2000.

Iyanaga, S. and Kawada, Y. (Eds.). *Encyclopedic Dictionary of Mathematics.* Cambridge, MA: MIT Press, p. 1477, 1980.

von Seggern, D. *CRC Standard Curves and Surfaces.* Boca Raton, FL: CRC Press, p. 196, 1993.

Watson, G. N. *A Treatise on the Theory of Bessel Functions, 2nd ed.* Cambridge, England: Cambridge University Press, pp. 312–313, 1966.

Schläfli Symbol

A symbol of the form $\{p, q, r, \ldots\}$ used to describe regular polygons, polyhedra, and their higher-dimensional counterparts.

The symbol $\{p\}$ denotes a REGULAR POLYGON. The symbol $\{p, q\}$ denotes a TESSELLATION of regular p-gons, with q of them surrounding each VERTEX. The Schläfli symbol can also be used to describe PLATONIC SOLIDS and KEPLER-POINSOT SOLIDS, and a generalized version describes QUASIREGULAR POLYHEDRA and ARCHIMEDEAN SOLIDS. Higher dimensional symbols can be used to describe the REGULAR POLYCHORA and POLYTOPES.

The symbol has the particularly nice property that its reversal gives the symbol of the DUAL POLYHEDRON. The following tables gives Schläfli symbols for several polytopes.

POLYHEDRON	Symbol
GREAT STELLATED DODECAHEDRON	$\left\{\frac{5}{2}, 3\right\}$
SMALL STELLATED DODECAHEDRON	$\left\{\frac{5}{2}, 5\right\}$
GREAT ICOSAHEDRON	$\left\{3, \frac{5}{2}\right\}$
TETRAHEDRON	$\{3, 3\}$
PENTATOPE	$\{3, 3, 3\}$
n-simplex	$\{\underbrace{3, \ldots, 3}_{n-1}\}$
16-CELL	$\{3, 3, 4\}$
n-cross polytope	$\{\underbrace{3, \ldots, 3}_{n-2}, 4\}$
600-CELL	$\{3, 3, 5\}$
OCTAHEDRON	$\{3, 4\}$
24-CELL	$\{3, 4, 3\}$
ICOSAHEDRON	$\{3, 5\}$
CUBE	$\{4, 3\}$
TESSERACT	$\{4, 3, 3\}$
n-hypercube	$\{4, \underbrace{3, \ldots, 3}_{n-2}\}$
GREAT DODECAHEDRON	$\left\{5, \frac{5}{2}\right\}$
DODECAHEDRON	$\{5, 3\}$
120-CELL	$\{5, 3, 3\}$

See also ARCHIMEDEAN SOLID, PLATONIC SOLID, QUASIREGULAR POLYHEDRON, REGULAR POLYCHORON, REGULAR POLYGON, TESSELLATION

Schläfli's Formula

For $\Re[z] > 0$,

$$J_\nu(z) = \frac{1}{\pi} \int_0^{\pi/2} \cos(z \sin t - \nu t)\, dt$$
$$- \frac{\sin(\nu\pi)}{\pi} \int_0^\infty e^{-z \sinh t} e^{-\nu t}\, dt,$$

where $J_\nu(z)$ is a BESSEL FUNCTION OF THE FIRST KIND.

References

Iyanaga, S. and Kawada, Y. (Eds.). *Encyclopedic Dictionary of Mathematics.* Cambridge, MA: MIT Press, p. 1472, 1980.

Schläfli's Modular Form

The MODULAR EQUATION of degree five can be written

$$\left(\frac{u}{v}\right)^3 + \left(\frac{v}{u}\right)^3 = 2\left(u^2 v^2 - \frac{1}{u^2 v^2}\right).$$

See also MODULAR EQUATION

Schlegel Graph

A GRAPH corresponding to POLYHEDRA skeletons. The POLYHEDRAL GRAPHS are special cases.

See also POLYHEDRAL GRAPH, SKELETON

References

Gardner, M. *Wheels, Life, and Other Mathematical Amusements.* New York: W. H. Freeman, p. 158, 1983.

Schlicht Function

An ANALYTIC FUNCTION f on the UNIT DISK is called schlicht if

1. f is ONE-TO-ONE,
2. $f(0) = 0$, and
3. $f'(0) = 1$,

in which case it is written $f \in S$. Schlicht functions have power series of the form

$$f(z) = Z + \sum_{j=2}^{\infty} a_j z^j.$$

See also AREA PRINCIPLE, BIEBERBACH CONJECTURE, KÖBE FUNCTION, KÖBE'S ONE-FOURTH THEOREM

References

Krantz, S. G. "Schlicht Functions." §12.1.1 in *Handbook of Complex Analysis*. Boston, MA: Birkhäuser, p. 149, 1999.

Schlömilch Remainder

A TAYLOR SERIES remainder formula that gives after n terms of the series

$$R_n = \frac{f^{(n+1)}(x^*)}{n!p}(x - x^*)^{n+1-p}(x - x_0)^{n+1}$$

for $x^* \in (x_0, x)$ and any $p > 0$ (Blumenthal 1926, Beesack 1966), which Blumenthal (1926) ascribes to Roche (1858). The choices $p = n + 1$ and $p = 1$ give the LAGRANGE and CAUCHY REMAINDERS, respectively (Beesack 1966).

See also CAUCHY REMAINDER, LAGRANGE REMAINDER

References

Beesack, P. R. "A General Form of the Remainder in Taylor's Theorem." *Amer. Math. Monthly* **73**, 64–67, 1966.
Blumenthal, L. M. "Concerning the Remainder Term in Taylor's Formula." *Amer. Math. Monthly* **33**, 424–426, 1926.
Maak, W. *An Introduction to Modern Calculus*. New York: Holt, Rinehart, and Winston, p. 99, 1963.
Roche. *Mem. de l'Acad. de Montpellier*. 1858.
Schlömilch, O. *Kompendium der höheren Analysis*. Braunschweig, Germany: Vieweg, 1923.

Schlömilch's Function

Mathematics:Calculus and Analysis:Special Functions:Hypergeometric Functions:Confluent Hypergeometric Functions

$$S(v, z) \equiv \int_0^{\infty} (1 + t)^{-v} e^{-zt} \, dt = z^{v-1} e^z \int_z^{\infty} u^{-v} e^{-u} \, du$$

$$= z^{v/2 - 1} e^{z/2} W_{-v/2, (1-v)/2}(z),$$

where $W_{k, m}(z)$ is the WHITTAKER FUNCTION.

Schlömilch's Series

A FOURIER SERIES-like expansion of a twice continuously differentiable function

$$f(x) = \frac{1}{2} a_0 + \sum_{n=1}^{\infty} a_n J_0(nx)$$

for $0 < x < \pi$, where $J_0(x)$ is a zeroth order BESSEL FUNCTION OF THE FIRST KIND and

$$a_0 \equiv 2f(0) + \frac{2}{\pi} \int_0^{\pi} du \int_0^{\pi/2} f'(u \sin \phi) \, d\phi$$

$$a_n \equiv \frac{2}{\pi} \int_0^{\pi} du \int_0^{\pi/2} uf'(u \sin \phi)\cos(n\pi) \, d\phi.$$

A special case gives the amazing identity

$$1 = J_0(z) + 2 \sum_{n=1}^{\infty} J_{2n}(z) = [J_0(z)]^2 + 2 \sum_{n=1}^{\infty} [J_n(z)]^2.$$

See also BESSEL FUNCTION OF THE FIRST KIND, BESSEL FUNCTION FOURIER EXPANSION, FOURIER SERIES

References

Iyanaga, S. and Kawada, Y. (Eds.). *Encyclopedic Dictionary of Mathematics*. Cambridge, MA: MIT Press, p. 1473, 1980.

Schmitt-Conway Biprism

A CONVEX POLYHEDRON which is SPACE-FILLING, but only aperiodically, was found by Conway in 1993.

See also CONVEX POLYHEDRON, SPACE-FILLING POLYHEDRON

Schnirelmann Constant

The constant s_0 in SCHNIRELMANN'S THEOREM such that every INTEGER > 1 is a sum of at most s_0 PRIMES. Of course, by VINOGRADOV'S THEOREM, it is known that 4 primes suffice for all *sufficiently large* numbers, but this constant gives a sufficient number for *all* numbers. The best current estimate is $s_0 = 7$ (Ramaré 1995), and a summary of progress on upper bounds for s_0 is summarized in the following table.

s_0	author
7	Ramaré (1995)
19	Riesel and Vaughan (1983)
26	Deshouillers (1977)
27	Vaughan (1977)
55	Klimov (1975)
115	Klimov *et al.* (1972)
159	Deshouillers (1973)

See also SCHNIRELMANN'S THEOREM, WARING'S PROBLEM

References

Deshouillers, J.-M. No. 17 in "Amélioration de la constante de Snirelman dans le probléme de Goldbach." *Séminaire Delange-Pisot-Poitou (14e année: 1972/73). Théorie des nombres: Fascicule 2: Exposés 17 à 26, et Groupe d'étude.* Paris: Secrétariat Mathématique, pp. 1–4, 1973.

Deshouillers, J.-M. "Sur la constante de Snirel'man." No. G16 in *Séminaire Delange-Pisot-Poitou, 17e année (1975/76). Théorie des nombres: Fascicule 2: Exposés 23 à 31 et Groupe d'étude.* Paris: Secrétariat Math., pp. 1–6, 1977.

Klimov, K. I. *Naucn. Trudy Kuibysev Gos. Ped. Inst.* **158**, 14–30, 1975.

Klimov, N. I.; Pil'tja, G. Z.; and Septickaja, T. A. "An Estimate of the Absolute Constant in the Goldbach-Snirel'man Problem." In *Issledovaniya po teorii chisel, Vyp. 4.* [*Studies in number theory, No. 4*] (Ed. N. Lensko). Saratov: Izdat. Saratov. Univ., pp. 35–51, 1972.

Ramaré, O. "On Snirel'man's Constant." *Ann. Scuola Norm. Sup. Pisa Cl. Sci.* **22**, 645–706, 1995.

Riesel, H. and Vaughan, R. C. "On Sums of Primes." *Ark. Mat.* **21**, 46–74, 1983.

Vaughan, R. C. "On the Estimation of Schnirelman's Constant." *J. reine angew. Math.* **290**, 93–108, 1977.

Schnirelmann Density

The Schnirelmann density of a sequence of natural numbers is the GREATEST LOWER BOUND of the FRACTIONS $A(n)/n$ where $A(n)$ is the number of terms in the sequence $\leq n$.

See also MANN'S THEOREM, SCHNIRELMANN'S THEOREM

References

Khinchin, A. Y. "The Landau-Schnirelmann Hypothesis and Mann's Theorem." Ch. 2 in *Three Pearls of Number Theory.* New York: Dover, pp. 18–36, 1998.

Schnirelmann's Theorem

This entry contributed by KEVIN O'BRYANT

There exists a POSITIVE INTEGER s such that every SUFFICIENTLY LARGE INTEGER is the sum of at most s PRIMES. It follows that there exists a POSITIVE INTEGER $s_0 \geq s$ such that every INTEGER > 1 is a sum of at most s_0 PRIMES. The smallest proven value of s_0 is known as the SCHNIRELMANN CONSTANT.

Schnirelmann's theorem can be proved using MANN'S THEOREM, although Schnirelmann used the weaker inequality

$$\sigma(A \oplus B) \geq \sigma(A) + \sigma(B) - \sigma(A)\sigma(B),$$

where $0 \in A \cap B$, $A \oplus B = \{a + b : a \in A, b \in B\}$, and σ is the SCHNIRELMANN DENSITY. Let $P = \{0, 1, 2, 3, 5, \ldots\}$ be the set of primes, together with 0 and 1, and let $Q = P \oplus P$. Using a sophisticated version of the INCLUSION-EXCLUSION PRINCIPLE, Schnirelmann showed that although $\sigma(P) = 0$, $\sigma(Q) > 0$. By repeated applications of MANN'S THEOREM, the

sum of k copies of Q satisfies $\sigma(Q + Q + \ldots + Q) \geq \min\{1, k\sigma(Q)\}$. Thus, if $k > 1/\sigma(Q)$, the sum of k copies of Q has SCHNIRELMANN DENSITY 1, and so contains all positive integers.

See also CHEN'S THEOREM, GOLDBACH CONJECTURE, MANN'S THEOREM, PRIME NUMBER, PRIME PARTITION, SCHNIRELMANN CONSTANT, SCHNIRELMANN DENSITY, WARING'S PRIME NUMBER CONJECTURE, WARING'S PROBLEM

References

Khinchin, A. Y. "The Landau-Schnirelmann Hypothesis and Mann's Theorem." Ch. 2 in *Three Pearls of Number Theory.* New York: Dover, pp. 18–36, 1998.

Schoenberg Curve

A SPACE-FILLING CURVE.

Scholz Conjecture

Let the minimal length of an ADDITION CHAIN for a number n be denoted $l(n)$. Then the Scholz conjecture states that

$$l(2^n - 1) \leq n - 1 + l(n).$$

The conjecture has been proven for a variety of special cases but not in general.

See also ADDITION CHAIN

References

Guy, R. K. *Unsolved Problems in Number Theory, 2nd ed.* New York: Springer-Verlag, p. 111, 1994.

Schönemann's Theorem

If the integral COEFFICIENTS C_0, C_1, ..., C_{N-1} of the POLYNOMIAL

$$f(x) = C_0 + C_1 x + C_2 x^2 + \ldots + C_{N-1} x^{N-1} + x^N$$

are divisible by a PRIME NUMBER p, while the free term C_0 is not divisible by p^2, then $f(x)$ is irreducible in the natural rationality domain.

See also ABEL'S IRREDUCIBILITY THEOREM, ABEL'S LEMMA, GAUSS'S POLYNOMIAL THEOREM, KRONECKER'S POLYNOMIAL THEOREM

References

Dörrie, H. *100 Great Problems of Elementary Mathematics: Their History and Solutions.* New York: Dover, p. 118, 1965.

Schönemann. "Grundzüge einer allgemeinen Theorie der höhern Congruenzen, deren Modul eine reelle Primzahl ist." *J. reine angew. Math.* **32**, 269–325, 1846.

Schönflies Symbol

One of the set of symbols C_i, C_s, C_1, C_2, C_3, C_4, C_5, C_6, C_7, C_8, C_{2h}, C_{3h}, C_{4h}, C_{5h}, C_{6h}, C_{2v}, C_{3v}, C_{4v}, C_{5v}, C_{6v},

$C_{\infty v}$, D_1, D_2, D_3, D_4, D_5, D_6, D_{2h}, D_{4h}, D_{5h}, D_{6h}, D_{8h}, $D_{\infty h}$, D_{2d}, D_{3d}, D_{4d}, D_{5d}, D_{6d}, I, I_h, O, O_h, S_4, S_6, S_8, T, T_d, and T_h used to identify POINT GROUPS.

Cotton (1990), gives a table showing the translations between Schönflies symbols and HERMANN-MAUGUIN SYMBOLS. Some of the Schönflies symbols denote different sets of symmetry operations but correspond to the same abstract GROUP and so have the same CHARACTER TABLE.

See also CHARACTER TABLE, HERMANN-MAUGUIN SYMBOL, POINT GROUPS, SPACE GROUPS, SYMMETRY OPERATION

References

Cotton, F. A. *Chemical Applications of Group Theory, 3rd ed.* New York: Wiley, p. 379, 1990.

Schönflies Theorem

If J is a simple closed curve in $\mathbb{R}^2$, the closure of one of the components of $\mathbb{R}^2 - J$ is HOMEOMORPHIC with the unit 2-BALL. This theorem may be proved using the RIEMANN MAPPING THEOREM, but the easiest proof is via MORSE THEORY.

The generalization to n-D is called MAZUR'S THEOREM. It follows from the Schönflies theorem that any two KNOTS of $\mathbb{S}^1$ in $\mathbb{S}^2$ or $\mathbb{R}^2$ are equivalent.

See also JORDAN CURVE THEOREM, MAZUR'S THEOREM, RIEMANN MAPPING THEOREM

References

Rolfsen, D. *Knots and Links.* Wilmington, DE: Publish or Perish Press, p. 9, 1976.
Thomassen, C. "The Jordan-Schönflies Theorem and the Classification of Surfaces." *Amer. Math. Monthly* **99**, 116–130, 1992.

Schoof-Elkies-Atkin Algorithm

An algorithm for determining the order of an ELLIPTIC CURVE E/F_p over the FINITE FIELD F_p.

See also ELLIPTIC CURVE

References

Izu, T.; Kogure, J.; Noro, M.; and Yokoyama, K. "Efficient Implementation of Schoof's Algorithm." *Advances in Cryptology: ASIACRYPT'98: International Conference on the Theory and Application of Cryptology and Information Security, Beijing, China, October 18–22, 1998* (Ed. K. Ohta and D. Pei). New York: Springer-Verlag, pp. 66–79, 1998.
Schoof, R. "Elliptic Curves Over Finite Fields and the Computation of Square Roots mod p." *Math. Comput.* **44**, 483–494, 1985.
Schoof, R. "Counting Points on Elliptic Curves Over Finite Fields." *J. Théor. Nombres Bordeaux* **7**, 219–264, 1995.

Schoolgirl Problem

KIRKMAN'S SCHOOLGIRL PROBLEM

Schoute Coaxal System

The CIRCUMCIRCLE, BROCARD CIRCLE, LEMOINE LINE, and ISODYNAMIC POINTS belong to a COAXAL SYSTEM orthogonal to the APOLLONIUS CIRCLES, called the Schoute coaxal system. In general, there are 12 points whose PEDAL TRIANGLES with regard to a given TRIANGLE have a given form. They lie six by six on two CIRCLES of the Schoute coaxal system.

See also APOLLONIUS CIRCLES, BROCARD CIRCLE, CIRCUMCIRCLE, COAXAL SYSTEM, ISODYNAMIC POINTS, LEMOINE LINE, SCHOUTE'S THEOREM

References

Johnson, R. A. *Modern Geometry: An Elementary Treatise on the Geometry of the Triangle and the Circle.* Boston, MA: Houghton Mifflin, pp. 297–299, 1929.

Schoute's Theorem

In any TRIANGLE, the LOCUS of a point whose PEDAL TRIANGLE has a constant BROCARD ANGLE and is described in a given direction is a CIRCLE of the SCHOUTE COAXAL SYSTEM.

References

Johnson, R. A. *Modern Geometry: An Elementary Treatise on the Geometry of the Triangle and the Circle.* Boston, MA: Houghton Mifflin, pp. 297–299, 1929.

Schrage's Algorithm

An algorithm for multiplying two 32-bit integers modulo a 32-bit constant without using any intermediates larger than 32 bits. It is also useful in certain types of RANDOM NUMBER generators.

References

Bratley, P.; Fox, B. L.; and Schrage, E. L. *A Guide to Simulation, 2nd ed.* New York: Springer-Verlag, 1996.
Press, W. H.; Flannery, B. P.; Teukolsky, S. A.; and Vetterling, W. T. "Random Numbers." Ch. 7 in *Numerical Recipes in FORTRAN: The Art of Scientific Computing, 2nd ed.* Cambridge, England: Cambridge University Press, p. 269, 1992.
Schrage, L. "A More Portable Fortran Random Number Generator." *ACM Trans. Math. Software* **5**, 132–138, 1979.

Schröder Number

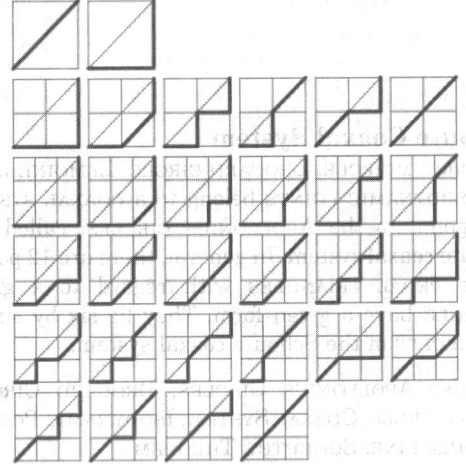

The Schröder number S_n is the number of LATTICE PATHS in the Cartesian plane that start at (0, 0), end at (n, n), contain no points above the line $y = x$, and are composed only of steps (0, 1), (1, 0), and (1, 1), i.e., →, ↑, and ↗. The diagrams illustrating the paths generating S_1, S_2, and S_3 are illustrated above. The numbers S_n are given by the RECURRENCE RELATION

$$S_n = S_{n-1} + \sum_{k=0}^{n-1} S_k S_{n-1-k},$$

where $S_0 = 1$, and the first few are 2, 6, 22, 90, ... (Sloane's A006318). The Schröder Numbers bear the same relation to the DELANNOY NUMBERS as the CATALAN NUMBERS do to the BINOMIAL COEFFICIENTS.

See also BINOMIAL COEFFICIENT, CATALAN NUMBER, DELANNOY NUMBER, LATTICE PATH, MOTZKIN NUMBER, *P*-GOOD PATH

References

Bonin, J.; Shapiro, L.; and Simion, R. "Some *q*-Analogs of the Schröder Numbers Arising from Combinatorial Statistics on Lattice Paths." *J. Stat. Planning Inference* **34**, 35–55, 1993.

Moser, L. and Zayachkowski, W. "Lattice Paths with Diagonal Steps." *Scripta Math.* **26**, 223–229, 1963.

Pergola, E. and Sulanke, R. A.. "Schröder Triangles, Paths, and Parallelogram Polyominoes." *J. Integer Sequences* **1**, No. 98.1.7, 1998. http://www.research.att.com/~njas/sequences/JIS/PergolaSulanke/.

Rogers, D. G. "A Schröder Triangle." *Combinatorial Mathematics V: Proceedings of the Fifth Australian Conference.* New York: Springer-Verlag, pp. 175–196, 1977.

Rogers, D. G. and Shapiro, L. "Some Correspondences involving the Schröder Numbers." *Combinatorial Mathematics: Proceedings of the International Conference, Canberra, 1977.* New York: Springer-Verlag, pp. 267–276, 1978.

Schröder, E. "Vier kombinatorische Probleme." *Z. Math. Phys.* **15**, 361–376, 1870.

Sloane, N. J. A. Sequences A006318/M1659 in "An On-Line Version of the Encyclopedia of Integer Sequences." http://www.research.att.com/~njas/sequences/eisonline.html.

Stanley, R. P. "Hipparchus, Plutarch, Schröder, Hough." *Amer. Math. Monthly* **104**, 344–350, 1997.

Sulanke, R. A. "Bijective Recurrences Concerning Schröder Paths." *Electronic J. Combinatorics* **5**, No. 1, R47, 1–11, 1998. http://www.combinatorics.org/Volume_5/v5i1toc.html#R47.

Schröder-Bernstein Theorem

The Schröder-Bernstein theorem for numbers states that if

$$n \le m \le n.$$

then $m = n$ For SETS, the theorem states that if there are INJECTIONS of the SET A into the SET B and of B into A, then there is a BIJECTIVE correspondence between A and B (i.e., they are EQUIPOLLENT).

See also BIJECTION, CARDINAL COMPARISON, EQUIPOLLENT, INJECTION, TRICHOTOMY LAW

Schröder's Equation

The functional equation

$$\phi(f(x)) = s\phi(x).$$

with $s \ne 0$, 1.

References

Kuczma, M. Ch. 6 in *Functional Equations in a Single Variable.* Warsaw, Poland: Polska Akademia Nauk, 1968.

Schröder's Method

Two families of equations used to find roots of nonlinear functions of a single variable. The "B" family is more robust and can be used in the neighborhood of degenerate multiple roots while still providing a guaranteed convergence rate. Almost all other rootfinding methods can be considered as special cases of Schröder's method. Householder humorously claimed that papers on root-finding could be evaluated quickly by looking for a citation of Schröder's paper; if the reference were missing, the paper probably consisted of a rediscovery of a result due to Schröder (Stewart 1993).

One version of the "A" method is obtained by applying NEWTON'S METHOD to f/f',

$$x_{n+1} = x_n - \frac{f(x_n) f'(x_n)}{[f'(x_n)]^2 - f(x_n) f''(x_n)}$$

(Scavo and Thoo 1995).

See also NEWTON'S METHOD

References

Householder, A. S. *The Numerical Treatment of a Single Nonlinear Equation.* New York: McGraw-Hill, 1970.

Scavo, T. R. and Thoo, J. B. "On the Geometry of Halley's Method." *Amer. Math. Monthly* **102**, 417–426, 1995.

Schröder, E. "Über unendlich viele Algorithmen zur Auflösung der Gleichungen." *Math. Ann.* **2**, 317–365, 1870.

Stewart, G. W. "On Infinitely Many Algorithms for Solving Equations." English translation of Schröder's original paper. College Park, MD: University of Maryland, Institute for Advanced Computer Studies, Department of Computer Science, 1993. ftp://thales.cs.umd.edu/pub/reports/imase.ps.

Schrödinger Equation

The Schrödinger equation describes the motion of particles in nonrelativistic quantum mechanics, and was first written down by Erwin Schrödinger. The time-dependent Schrödinger equation is given by

$$ih\,\frac{\partial\psi(x,\,y,\,z,\,t)}{\partial t}$$

$$\left[-\frac{h^2}{2m}\,\nabla^2+V(x)\right]\Psi(x,\,y,\,z,\,t)=\bar{H}\Psi(x,\,y,\,z,\,t),\quad(1)$$

where h is h-bar, Ψ is the time-dependent wavefunction, m is the mass of a particle, ∇^2 is the LAPLACIAN, V is the potential, and $\bar{H}$ is the Hamiltonian operator. The time-independent Schrödinger equation is

$$\left[-\frac{h^2}{2m}\,\nabla^2+V(x)\right]\psi(x,\,y,\,z,\,t)=E\psi(x,\,y,\,z,\,t).\quad(2)$$

The one-dimensional versions of these equations are then

$$ih\,\frac{\partial\Psi(x,\,t)}{\partial t}=\left[-\frac{h^2}{2m}\,\frac{\partial^2}{\partial x^2}+V(x)\right]\Psi(x,\,t)$$

$$=\bar{H}\Psi(x,\,t),\quad(3)$$

and

$$\left[-\frac{h^2}{2m}\,\frac{d^2}{dx^2}+V(x)\right]\psi(x)=E\psi(x).\quad(4)$$

The logarithmic Schrödinger equation is given by

$$iu_t+\nabla^2u+u\,\ln|u|^2=0\quad(5)$$

(Cazenave 1983; Zwillinger 1997, p. 134), the nonlinear Schrödinger equation by

$$iu_t+u_{xx}\pm2|u|^2u=0\quad(6)$$

(Calogero and Degasperis 1982, p. 56; Tabor 1989, p. 309; Zwillinger 1997, p. 134) or

$$iu_t+u_{xx}+au+b|u|^2u=0\quad(7)$$

(Infeld and Rowlands 2000, p. 126), and the derivative nonlinear Schrödinger equation by

$$iu_t+u_{xx}\pm i(|u|^2u)_x=0\quad(8)$$

(Calogero and Degasperis 1982, p. 56; Zwillinger 1997, p. 134).

See also DIRAC EQUATION

References

Calogero, F. and Degasperis, A. *Spectral Transform and Solitons: Tools to Solve and Investigate Nonlinear Evolution Equations.* New York: North-Holland, p. 56, 1982.
Cazenave, T. "Stable Solution of the Logarithmic Schrödinger Equation." *Nonlinear Anal.* **7**, 1127–1140, 1983.
Infeld, E. and Rowlands, G. *Nonlinear Waves, Solitons, and Chaos, 2nd ed.* Cambridge, England: Cambridge University Press, 2000.
Tabor, M. "The NLS Equation." §7.5.c in *Chaos and Integrability in Nonlinear Dynamics: An Introduction.* New York: Wiley, p. 309, 1989.
Zwillinger, D. *Handbook of Differential Equations, 3rd ed.* Boston, MA: Academic Press, p. 134, 1997.

Schroeder Stairs

PENROSE STAIRWAY

Schröter's Formula

Let a general THETA FUNCTION be defined as

$$T(x,\,q)\equiv\sum_{n=-\infty}^{\infty}x^nq^{n^2}.$$

then

$$T(x,\,q^a)T(y,\,q^b)$$

$$=\sum_{k=0}^{a+b-1}y^kq^{bk^2}T(xyq^{2bk},\,q^{a+b})T(y^ax^{-b}q^{2abk},\,q^{ab(a+b)}).$$

See also BLECKSMITH-BRILLHART-GERST THEOREM, JACOBI TRIPLE PRODUCT, RAMANUJAN THETA FUNCTIONS, THETA FUNCTIONS

References

Borwein, J. M. and Borwein, P. B. *Pi & the AGM: A Study in Analytic Number Theory and Computational Complexity.* New York: Wiley, p. 111, 1987.
Tannery, J. and Molk, J. *Elements de la Théorie des Fonctions Elliptiques, 4 vols.* Paris: Gauthier-Villars et fils, 1893–1902.

Schur Algebra

An Auslander algebra which connects the representation theories of the symmetric group of PERMUTATIONS and the GENERAL LINEAR GROUP $GL(n,\,\mathbb{C})$. Schur algebras are "quasihereditary."

References

Martin, S. *Schur Algebras and Representation Theory.* New York: Cambridge University Press, 1993.

Schur Decomposition

The Schur decomposition of a numerical matrix M is a pair of matrices Q and T such that

$$M=QTQ^*,$$

where Q is an ORTHOGONAL MATRIX, T is a BLOCK UPPER TRIANGULAR MATRIX, and Q* is the ADJOINT

MATRIX. Schur decomposition is implemented in *Mathematica* as SchurDecomposition[*m*].

See also MATRIX DECOMPOSITION

References

Golub, G. H. and van Loan, C. F. *Matrix Computations, 3rd ed.* Baltimore, MD: Johns Hopkins University Press, pp. 312–314, 1996.

Schur, I. "On the Characteristic Roots of a Linear Substitution with an Application to the Theory of Integral Equations." *Math. Ann.* **66**, 488–510, 1909.

Schur Functor

A FUNCTOR which defines an equivalence of module CATEGORIES.

References

Martin, S. *Schur Algebras and Representation Theory.* New York: Cambridge University Press, 1993.

Schur Matrix

The $p \times p$ SQUARE MATRIX formed by setting $s_{ij} = \zeta^{ij}$, where ζ is a pth ROOT OF UNITY. The Schur matrix has a particularly simple DETERMINANT given by

$$\det S = \epsilon_p p^{p/2},$$

where p is an ODD PRIME and

$$\epsilon_p \equiv \begin{cases} 1 & \text{if } p \equiv 1 \pmod 4 \\ i & \text{if } p \equiv 3 \pmod 4 \end{cases}.$$

This determinant has been used to prove the QUADRATIC RECIPROCITY LAW (Landau 1958, Vardi 1991). The ABSOLUTE VALUES of the PERMANENTS of the Schur matrix of order $2p + 1$ are given by 1, 3, 5, 105, 81, 6765, ... (Sloane's A003112, Vardi 1991).

Denote the Schur matrix S_p with the first row and first column omitted by S'_p. Then

$$\text{perm } S_p = p \text{ perm } S'_p,$$

where perm denoted the PERMANENT (Vardi 1991).

References

Graham, R. L. and Lehmer, D. H. "On the Permanent of Schur's Matrix." *J. Austral. Math. Soc.* **21**, 487–497, 1976.

Landau, E. *Elementary Number Theory.* New York: Chelsea, 1958.

Sloane, N. J. A. Sequences A003112/M2509 in "An On-Line Version of the Encyclopedia of Integer Sequences." http://www.research.att.com/~njas/sequences/eisonline.html.

Vardi, I. *Computational Recreations in Mathematica.* Reading, MA: Addison-Wesley, pp. 119–122 and 124, 1991.

Schur Multiplier

A property of FINITE SIMPLE GROUPS which is known for all such GROUPS.

See also FINITE GROUP, SIMPLE GROUP

Schur Number

The Schur number $S(k)$ is the largest integer n for which the interval $[1, n]$ can be partitioned into k SUM-FREE SETS (Fredricksen and Sweet 2000). $S(k)$ is guaranteed to exist for each k by SCHUR'S PROBLEM. Note the definition of the Schur number as the smallest number $S'(k) = S(k) + 1$ for which such a partition *does not exist* is also prevalent in the literature (Sloane's A030126; Fredricksen and Sweet 2000).

Schur (1916) gave the lower bound

$$S(k) \geq \tfrac{1}{2}(3^n - 1) \tag{1}$$

which is sharp for $n = 1$, 2, and 3 (Guy 1994). The Schur numbers also satisfy the inequality

$$S(k) \geq c(321)^{k/5} > c(1.17176)^k \tag{2}$$

for $k > 5$ and some constant c (Abbott and Moser 1966, Abbott and Hanson 1972, Exoo 1994). SCHUR'S THEOREM also shows that

$$S(n) \leq R(n) - 2. \tag{3}$$

where $R(n)$ is a RAMSEY NUMBER. The first few Schur numbers are 1, 4, 13, 44, $160 \leq S(5) \leq 315$, $S(6) \geq 536$, $S(7) \geq 1680$, ... (Sloane's A045652; Fredricksen and Sweet 2000). $S(4)$ is due to Baumert (Baumert 1965, Abbott and Hanson 1972), the lower bound on $S(5)$ is due to Exoo (1994), and the lower limits on $S(6)$ and $S(7)$ are due to Fredricksen and Sweet (2000).

See also RAMSEY NUMBER, RAMSEY'S THEOREM, SCHUR'S PROBLEM, SCHUR'S THEOREM

References

Abbott, H. L. and Hanson, D. "A Problem of Schur ad its Generalizations." *Acta Arith.* **20**, 175–187, 1972.

Abbott, H. L. and Moser, L. "Sum-Free Sets of Integers." *Acta Arith.* **11**, 392–396, 1966.

Baumert, L. D. and Golomb, S. W. "Backtrack Programming." *J. Ass. Comp. Machinery* **12**, 516–524, 1965.

Beutelspacher, A. and Brestovansky, W. "Generalized Schur Numbers." In *Combinatorial Theory. Proceedings of a Conference Held at Schloss Rauischholzhausen, May 6–9, 1982.* Berlin: Springer-Verlag, pp. 30–38, 1982.

Exoo, G. "A Lower Bound for Schur Numbers and Multicolor Ramsey Numbers of K_3." *Electronic J. Combinatorics* **1**, R8 1–3, 1994. http://www.combinatorics.org/Volume_1/volume1.html#R8.

Fredricksen, H. "Schur Numbers and the Ramsey Numbers $N(3, 3, \ldots, 3; 2)$." *J. Combin. Theory Ser. A* **27**, 376–377, 1979.

Fredricksen, H. and Sweet, M. M. "Symmetric Sum-Free Partitions and Lower Bounds for Schur Numbers." *Electronic J. Combinatorics* **7**, No. 1, R32, 1–9, 2000. http://www.combinatorics.org/Volume_7/v7i1toc.html#R32.

Guy, R. K. "Schur's Problem. Partitioning Integers into Sum-Free Classes" and "The Modular Version of Schur's Problem." §E11 and E12 in *Unsolved Problems in Number Theory, 2nd ed.* New York: Springer-Verlag, pp. 209–212, 1994.

Radziszowski, S. P. "Small Ramsey Numbers." *Electronic J. Combin.* **1**, DS1 1–29, Rev. Jul. 5, 1999. http://www.combinatorics.org/Surveys/.

Schur, I. "Über die Kongruenz $x^m + y^m \equiv z^m \pmod{p}$." *Jahresber. Deutsche Math.-Verein.* **25**, 114–116, 1916.

Sloane, N. J. A. Sequences A030126 and A045652 in "An On-Line Version of the Encyclopedia of Integer Sequences." http://www.research.att.com/~njas/sequences/eisonline.html.

Whitehead, E. G. "The Ramsey Number $N(3, 3, 3, 3; 2)$." *Disc. Math.* **4**, 389–396, 1973.

Schur Transform

For

$$p(z) = a_n z^n + a_{n-1} z^{n-1} + \ldots + a_0, \tag{1}$$

polynomial of degree $n \geq 1$, the Schur transform is defined by the $(n-1)$-degree polynomial

$$Tp(z) \equiv \bar{a}_0 p(z) - a_n p^*(z) \tag{2}$$

$$= \sum_{k=0}^{n-1} (\bar{a}_0 a_k - a_n \bar{a}_{n-k}) z^k \tag{3}$$

where p^* is the RECIPROCAL POLYNOMIAL.

See also RECIPROCAL POLYNOMIAL

References

Henrici, P. *Applied and Computational Complex Analysis, Vol. 1: Power Series-Integration-Conformal Mapping-Location of Zeros.* New York: Wiley, p. 493, 1988.

Schur-Cohn Algorithm

An algorithm that can always be used to decide whether a given polynomial is fere of zeros in the closed unit disk (or, using an entire linear transformation, to any other disk in the complex plane). Under certain conditions, the algorithm can also be used to determine the exact number of zeros in a disk (Henrici 1988, p. 494). The method is also useful to control engineers, since it can be used to determine whether a dynamic control system is stable.

References

Henrici, P. *Applied and Computational Complex Analysis, Vol. 1: Power Series-Integration-Conformal Mapping-Location of Zeros.* New York: Wiley, pp. 491–494, 1988.

Schur-Jabotinsky Theorem

Let $P = a_1 x + a_1 x^2 + \ldots$ be an ALMOST UNIT in the INTEGRAL DOMAIN of FORMAL POWER SERIES (with $a_1 \neq 0$) and define

$$P^k \equiv \sum_{n=k}^{\infty} a_n^{(k)} x^n \tag{1}$$

for $k = \pm 1, \pm 2, \ldots$. If $Q \equiv P^{-1}$, then for all positive integers m,

$$Q^m = \sum_{n=m}^{\infty} b_n^{(m)} x^n, \tag{2}$$

where

$$b_n^{(m)} \equiv \frac{m}{n} a_{-m}^{(-n)} \tag{3}$$

for $n \geq m$.

See also LAGRANGE INVERSION THEOREM

References

Henrici, P. *Applied and Computational Complex Analysis, Vol. 1: Power Series-Integration-Conformal Mapping-Location of Zeros.* New York: Wiley, pp. 55–56, 1988.

Schur's Hermitian Matrix Theorem

HORN'S THEOREM

Schur's Inequalities

Let $A = a_{ij}$ be an $n \times n$ MATRIX with COMPLEX (or REAL) entries and EIGENVALUES $\lambda_1, \lambda_2, \ldots, \lambda_n$, then

$$\sum_{i=1}^{n} |\lambda_i|^2 \leq \sum_{i,j=1}^{n} |a_{ij}|^2$$

$$\sum_{i=1}^{n} |\Re[\lambda_i]|^2 \leq \sum_{i,j=1}^{n} \left| \frac{a_{ij} + \bar{a}_{ji}}{2} \right|^2$$

$$\sum_{i=1}^{n} |\Im[\lambda_i]|^2 \leq \sum_{i,j=1}^{n} \left| \frac{a_{ij} - \bar{a}_{ji}}{2} \right|^2,$$

where $\bar{z}$ is the COMPLEX CONJUGATE.

References

Gradshteyn, I. S. and Ryzhik, I. M. *Tables of Integrals, Series, and Products, 6th ed.* San Diego, CA: Academic Press, p. 1120, 2000.

Schur's Lemma

The endomorphism ring of an irreducible module is a DIVISION ALGEBRA.

Hsiang (2000, p. 3) calls the following result the Schur lemma. Let V, W be irreducible (linear) G-spaces and $A : V \to W$ a G-linear map. Then A is either invertible or $A = 0$.

See also DIVISION ALGEBRA, SCHUR'S REPRESENTATION LEMMA

References

Herstein, I. N. *Topics in Algebra, 2nd ed.* New York: Springer-Verlag, 1975.

Hsiang, W. Y. *Lectures on Lie Groups.* Singapore: World Scientific, p. 3, 2000.

Schur's Partition Theorem

Schur's partition theorem lets $A(n)$ denote the number of partitions of n into parts congruent to ± 1 (mod 6), $B(n)$ denote the number of partitions of n into distinct parts congruent to ± 1 (mod 3), and $C(n)$ the number of partitions of n into parts that differ by at least 3, with the added constraint that the difference between multiples of three is at least 6. Then $A(n) = B(n) = C(n)$ (Schur 1926; Bressoud 1980; Andrews 1986, p. 53).

The values of $A(n) = B(n) = C(n)$ for $n = 1, 2, \ldots$ are 1, 1, 1, 1, 2, 2, 3, 3, 3, 4, 5, 6, 7, 8, 9, 10, 12, 14, 16, 18, ... (Sloane's A003105). For example, for $n = 15$, there are nine partitions satisfying these conditions, as summarized in the following table (Andrews 1986, p. 54).

$A(15) = 9$	$B(15) = 9$	$C(15) = 9$
$13 + 1 + 1$	$14 + 1$	15
$11 + 1 + 1 + 1 + 1$	$13 + 2$	$14 + 1$
$7 + 7 + 1$	$11 + 4$	$13 + 2$
$7 + 5 + 1 + 1 + 1$	$10 + 5$	$12 + 3$
$7 + 1 + 1 + 1 + \ldots + 1$	$10 + 4 + 1$	$11 + 4$
$5 + 5 + 5$	$8 + 7$	$10 + 5$
$5 + 5 + 1 + 1 + \ldots + 1$	$8 + 5 + 2$	$10 + 4 + 1$
$5 + 1 + 1 + \ldots + 1$	$8 + 4 + 2 + 1$	$9 + 5 + 1$
$1 + 1 + \ldots + 1$	$7 + 5 + 2 + 1$	$8 + 5 + 2$

The identity $A(n) = B(n)$ can be established using the identity

$$\sum_{n=0}^{\infty} B(n) q^n = (-q; q^3)_{\infty} (-q^2; q^3)_{\infty} \tag{1}$$

$$= \frac{(q^2; q^6)_{\infty}(q^4; q^6)_{\infty}}{(q; q^3)_{\infty}(q^2; q^3)_{\infty}} \tag{2}$$

$$= \frac{(q^2; q^6)_{\infty}(q^4; q^6)_{\infty}}{(q; q^6)_{\infty}(q^4; q^6)_{\infty}(q^2; q^6)_{\infty}(q^5; q^6)_{\infty}} \tag{3}$$

$$= \frac{1}{(q; q^6)_{\infty}(q^5; q^6)_{\infty}} \tag{4}$$

$$= \sum_{n=0}^{\infty} A(n) q^n \tag{5}$$

(Andrews 1986, p. 54). The identity $B(n) = C(n)$ is significantly trickier.

See also GÖLLNITZ'S THEOREM, RAMSEY NUMBER, SCHUR'S LEMMA, SCHUR NUMBER

References

Andrews, G. E. "q-Series and Schur's Theorem" and "Bressoud's Proof of Schur's Theorem." §6.2–6.3 in q-Series: Their Development and Application in Analysis, Number Theory, Combinatorics, Physics, and Computer Algebra. Providence, RI: Amer. Math. Soc., pp. 53–58, 1986.

Bressoud, D. M. "Combinatorial Proof of Schur's 1926 Partition Theorem." *Proc. Amer. Math. Soc.* **79**, 338–340, 1980.

Schur, I. "Über die Kongruenz $x^m + y^m \equiv z^m$ (mod p)." *Jahresber. Deutsche Math.-Verein.* **25**, 114–116, 1916.

Schur, I. "Zur additiven Zahlentheorie." *Sitzungsber. Preuss. Akad. Wiss. Phys.-Math. Kl.*, pp. 488–495, 1926. Reprinted in *Gesammelte Abhandlungen, Vol. 3.* Berlin: Springer-Verlag, pp. 43–50, 1973.

Sloane, N. J. A. Sequences A003105/M0254 in "An On-Line Version of the Encyclopedia of Integer Sequences." http://www.research.att.com/~njas/sequences/eisonline.html.

Schur's Problem

Schur (1916) proved that no matter how the set of POSITIVE INTEGERS less than or equal to $\lfloor n! e \rfloor$ (where $\lfloor x \rfloor$ is the FLOOR FUNCTION) is partitioned into n classes, one class must contain INTEGERS x, y, z such that $x + y = z$, where x and y are not necessarily distinct. The least INTEGER $S(n)$ with this property is known as the SCHUR NUMBER. The upper bound has since been slightly improved to $\lfloor n!(e - 1/24) \rfloor$.

See also COMBINATORICS, RAMSEY NUMBER, SCHUR NUMBER, SCHUR'S THEOREM, SUM-FREE SET

References

Abbott, H. L. and Hanson, D. "A Problem of Schur and Its Generalizations." *Acta Arith.* **20**, 175–187, 1972.

Abbott, H. L. and Moser, L. "Sum-Free Sets of Integers." *Acta Arith.* **11**, 393–396, 1966.

Beutelspacher, A. and Brestovansky, W. "Generalized Schur Numbers." In *Combinatorial Theory: Proceedings of a Conference Held at Schloss Rauischholzhausen, May 6–9, 1982* (Ed. D. Jungnickel and K. Vedder). Berlin: Springer-Verlag, pp. 30–38, 1982.

Choi, S. L. G. "The Largest Sum-free Subsequence from a Sequence of n Numbers." *Proc. Amer. Math. Soc.* **39**, 42–44, 1973.

Choi, S. L. G.; Komlós, J.; and Szemerédi, R. "On Sum-Free Subsequences." *Trans. Amer. Math. Soc.* **212**, 307–313, 1975.

Erdos, P. "Some Problems and Results in Number Theory." In *Number Theory and Combinatorics: Japan 1984* (Ed. J. Akiyama). Singapore: World Scientific, pp. 65–87, 1985.

Guy, R. K. "Schur's Problem. Partitioning Integers into Sum-Free Classes" and "The Modular Version of Schur's Problem." §E11 and E12 in *Unsolved Problems in Number Theory, 2nd ed.* New York: Springer-Verlag, pp. 209–212, 1994.

Irving, R. W. "An Extension of Schur's Theorem on Sum-Free Partitions." *Acta Arith.* **25**, 55–63, 1973.

Schönheim, J. "On Partitions of the Positive Integers with no x, y, z Belonging to Distinct Classes Satisfying $x + y = z$." In *Number Theory: Proceedings of the First Conference of the Canadian Number Theory Association Held at the Banff Center, Banff, Alberta, April 17–27, 1988* (Ed. R. A. Mollin). Berlin: de Gruyter, pp. 515–528, 1990.

Wallis, W. D.; Street, A. P.; and Wallis, J. S. *Combinatorics: Room Squares, Sum-free Sets, Hadamard Matrices.* New York: Springer-Verlag, 1972.

Schur's Ramsey Theorem

As shown by Schur (1916), the SCHUR NUMBER $S(n)$ satisfies

$$S(n) \leq R(n) - 2 \qquad (1)$$

for $n = 1, 2, \ldots$, where $R(n)$ is a RAMSEY NUMBER.

References

Fredricksen, H. and Sweet, M. M. "Symmetric Sum-Free Partitions and Lower Bounds for Schur Numbers." *Electronic J. Combinatorics* **7**, No. 1, R32, 1–9, 2000. http://www.combinatorics.org/Volume_7/v7i1toc.html#R32.

Schur, I. "Über die Kongruenz $x^m + y^m \equiv z^m$ mod p." *Jahresber. Deutsche Math.-Verein.* **25**, 114–116, 1916.

Schur's Representation Lemma

If π on V and π' on V' are irreducible representations and $E : V \mapsto V'$ is a linear map such that $\pi'(g)E = E\pi(g)$ for all $g \in$ and GROUP G, then $E = 0$ or E is invertible. Furthermore, if $V = V'$, then E is a SCALAR.

See also SCHUR'S LEMMA

References

Knapp, A. W. "Group Representations and Harmonic Analysis, Part II." *Not. Amer. Math. Soc.* **43**, 537–549, 1996.

Schur's Theorem

SCHUR'S PARTITION THEOREM, SCHUR'S RAMSEY THEOREM

Schwarz Reflection Principle

Suppose that f is a ANALYTIC FUNCTION which is defined in the UPPER HALF-DISK $\{|z|^2 < 1, \Im[z] > 0\}$. Assume that f extends to a continuous function on the REAL AXIS, and takes on real values on the REAL AXIS. Then f can be extended to an ANALYTIC FUNCTION on the whole disk by the formula

$$f(\bar{z}) = \overline{f(z)},$$

and the values for z reflected across the REAL AXIS are the reflections of $f(z)$ across the REAL AXIS. It is easy to check that the above function is COMPLEX DIFFERENTIABLE in the interior of the LOWER HALF-DISK. What is remarkable is that the resulting function must be analytic along the REAL AXIS as well, despite no assumptions of differentiability.

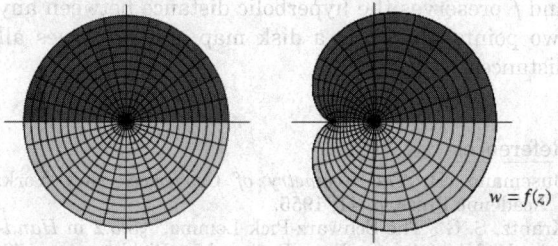

$w = f(z)$

This is called the Schwarz reflection principle, and is

sometimes also known as the Schwarz's symmetric principle (Needham 2000, p. 257). The diagram above shows the reflection principle applied to a function f defined for UPPER HALF-DISK (left figure; red) and its image (right figure; red). The function is real on the real axis, so it is possible to extend the function to the reflected domain (left and right figures; pink).

For the reflected function to be continuous, it is necessary for the values at the boundary to be continuous and to fall on the line being reflected. The reflection principle also applies in the generality of reflecting along any line, not just the REAL AXIS, in which case the function f has to take values along a line in the range. In fact, any arc which has a neighborhood biholomorphic to a straight line can be reflected across. The basic example is the boundary of the UNIT CIRCLE which is mapped to the REAL AXIS by $z \to (iz + 1)/(z + i)$.

The reflection principle can also be used to reflect a HARMONIC FUNCTION which extends continuously to the zero function on its boundary. In this case, for negative y, defining

$$v(x, y) = -v(x, -y)$$

extends v to a harmonic function on the reflected domain. Again note that it is necessary for $v(x, 0) = 0$. This result provides a way of extending a HARMONIC FUNCTION from a given OPEN SET to a larger OPEN SET (Krantz 1999, p. 95).

See also ANALYTIC CONTINUATION, HARMONIC FUNCTION

References

Flanigan, F. J. *Complex Variables: Harmonic and Analytic Functions.* New York: Dover, p. 234, 1983.

Krantz, S. G. "The Schwarz Reflection Principle." §7.5 in *Handbook of Complex Analysis.* Boston, MA: Birkhäuser, pp. 95–97, 1999.

Levinson, N. and Raymond, R. *Complex Variables.* New York: McGraw-Hill, pp. 318–320, 1970.

Needham, T. "Analytic Continuation via Reflections." §5.XI.5 in *Visual Complex Analysis.* New York: Clarendon Press, pp. 252–257, 2000.

Rudin, W. *Real and Complex Analysis.* New York: McGraw-Hill, pp. 237–239, 1987.

Schwarz, H. A. *Gesammelte Mathematische Abhandlungen, Bd. II.* New York: Chelsea, pp. 144–171, 1972.

Schwarz Triangle

The Schwarz triangles are SPHERICAL TRIANGLES which, by repeated reflection in their indices, lead to a set of congruent SPHERICAL TRIANGLES covering the SPHERE a finite number of times.

Schwarz triangles are specified by triples of numbers (p, q, r). There are four "families" of Schwarz triangles, and the largest triangles from each of these families are

$$(2\ 2\ n'),\ \left(\begin{smallmatrix}3&3&3\\2&2&2\end{smallmatrix}\right),\ \left(\begin{smallmatrix}3&4&4\\2&3&3\end{smallmatrix}\right),\ \left(\begin{smallmatrix}5&5&5\\4&4&4\end{smallmatrix}\right).$$

The others can be derived from

$$(p\ q\ r) = (p\ x\ r_1) + (x\ q\ r_2),$$

where

$$\frac{1}{r_1} + \frac{1}{r_2} = \frac{1}{r}$$

and

$$\cos\left(\frac{\pi}{x}\right) = -\cos\left(\frac{\pi}{x'}\right)$$

$$= \frac{\cos\left(\dfrac{\pi}{q}\right)\sin\left(\dfrac{\pi}{r_1}\right) - \cos\left(\dfrac{\pi}{p}\right)\sin\left(\dfrac{\pi}{r_2}\right)}{\sin\left(\dfrac{\pi}{r}\right)}$$

See also COLUNAR TRIANGLE, SPHERICAL TRIANGLE, WYTHOFF SYMBOL

References

Coxeter, H. S. M. *Regular Polytopes, 3rd ed.* New York: Dover, pp. 112–113 and 296, 1973.
Schwarz, H. A. "Zur Theorie der hypergeometrischen Reihe." *J. reine angew. Math.* **75**, 292–335, 1873.

Schwarz-Christoffel Mapping

A CONFORMAL MAPPING from the UPPER HALF-PLANE to a POLYGON.

See also CONFORMAL MAPPING, SCHWARZ-CHRISTOFFEL PARAMETER PROBLEM

References

Henrici, P. *Applied and Computational Complex Analysis, Vol. 1: Power Series-Integration-Conformal Mapping-Location of Zeros.* New York: Wiley, pp. 396–431, 1988.
Krantz, S. G. "Numerical Approximation of the Schwarz-Christoffel Mapping." §14.4.1 in *Handbook of Complex Analysis.* Boston, MA: Birkhäuser, pp. 175–179, 1999.

Schwarz-Christoffel Parameter Problem

The problem of determining the vertices of a SCHWARZ-CHRISTOFFEL MAPPING (Krantz 1999, p. 176).

See also CONFORMAL MAPPING, SCHWARZ-CHRISTOFFEL MAPPING

References

Krantz, S. G. in *Handbook of Complex Analysis.* Boston, MA: Birkhäuser, p. 176, 1999.

Schwarzian Derivative

The Schwarzian derivative is defined by

$$D_{\text{Schwarzian}} \equiv \frac{f'''(x)}{f'(x)} - \frac{3}{2}\left[\frac{f(x)}{f'(x)}\right]^2.$$

The FEIGENBAUM CONSTANT is universal for 1-D MAPS if its Schwarzian derivative is NEGATIVE in the bounded interval (Tabor 1989, p. 220).

See also FEIGENBAUM CONSTANT

References

Tabor, M. *Chaos and Integrability in Nonlinear Dynamics: An Introduction.* New York: Wiley, 1989.

Schwarz-Pick Lemma

Let f be analytic on the UNIT DISK, and assume that

1. $|f(z)| \leq 1$ for all z, and
2. $f(a) = b$ for some a, $b \in D(0, 1)$, the UNIT DISK.

Then

$$|f'(a)| \leq \frac{1 - |b|^2}{1 - |a|^2}. \tag{1}$$

Furthermore, if $f(a_1) = b_1$ and $f(a_2) = b_2$, then

$$\left|\frac{b_2 - b_1}{1 - b_1{}^*b_2}\right| \leq \left|\frac{a_2 - a_1}{1 - \bar{a}_1 a_2}\right|, \tag{2}$$

where $\bar{z}$ is the COMPLEX CONJUGATE (Krantz 1999, p. 78). As a consequence, if either

$$|f'(a)| \leq \frac{1 - |b|^2}{1 - |a|^2} \tag{3}$$

or

$$\left|\frac{b_2 - b_1}{1 - \bar{b}_1 b_2}\right| = \left|\frac{a_2 - a_1}{1 - \bar{a}_1 a_2}\right| \tag{4}$$

for $a_1\ a_2$, then f is a conformal SELF-MAP of $D(0, 1)$ to itself.

Stated succinctly, the Schwarz-Pick lemma guarantees that if f is an analytic map of the DISK $\mathbb{D}$ into $\mathbb{D}$ and f preserves the hyperbolic distance between any two points, then f is a disk map and preserves all distances.

References

Busemann, H. *The Geometry of Geodesics.* New York: Academic Press, p. 41, 1955.
Krantz, S. G. "The Schwarz-Pick Lemma." §5.5.2 in *Handbook of Complex Analysis.* Boston, MA: Birkhäuser, p. 78, 1999.

Schwarz's Inequality

Let $\psi_1(x)$ and $\psi_2(x)$ by any two REAL integrable functions in $[a, b]$, then Schwarz's inequality, also called the Cauchy-Schwarz inequality (Gradshteyn and Ryzhik 2000, p. 1099) or Buniakowsky inequality (Hardy *et al.* 1952, p. 16), is given by

$$|\langle\psi_1|\psi_2\rangle|^2 \le \langle\psi_1|\psi_1\rangle\langle\psi_2|\psi_2\rangle. \tag{1}$$

Written out explicitly

$$\left[\int_a^b \psi_1(x)\psi_2(x)\,dx\right]^2 \le \int_a^b [\psi_1(x)]^2\,dx \int_a^b [\psi_2(x)]^2\,dx, \tag{2}$$

with equality IFF $g(x) = \alpha f(x)$ with α a constant.

To derive the inequality, let $\psi(x)$ be a COMPLEX FUNCTION and λ a COMPLEX constant such that $\psi(x) \equiv f(x) + \lambda g(x)$ for some f and g. Since $\int \bar\psi\psi\,dx \ge 0$, where $\bar z$ is the COMPLEX CONJUGATE,

$$\int \bar\psi\psi\,dx = \int \bar f f\,dx + \lambda\int \bar f g\,dx + \bar\lambda\int \bar g f\,dx$$
$$+ \lambda\bar\lambda\int \bar g g\,dx \ge 0, \tag{3}$$

with equality when $\psi(x) = 0$. Writing this in compact notation,

$$\langle \bar f, f\rangle + \lambda\langle \bar f, g\rangle + \bar\lambda\langle \bar g, f\rangle + \lambda\bar\lambda\langle \bar g, g\rangle \ge 0. \tag{4}$$

Now define

$$\lambda = -\frac{\langle \bar g, f\rangle}{\langle \bar g, g\rangle} \tag{5}$$

$$\bar\lambda = -\frac{\langle g, \bar f\rangle}{\langle \bar g, g\rangle}\,dx. \tag{6}$$

Multiply (4) by $\langle \bar g, g\rangle$ and then plus in (5) and (6) to obtain

$$\langle \bar f, f\rangle\langle \bar g, g\rangle - \langle \bar f, g\rangle\langle \bar g, f\rangle$$
$$- \langle \bar g, \bar f\rangle\langle g, \bar f\rangle + \langle \bar g, f\rangle\langle g, \bar f\rangle, \tag{7}$$

which simplifies to

$$\langle \bar g, f\rangle\langle \bar f, g\rangle \le \langle \bar f, f\rangle\langle \bar g, g\rangle \tag{8}$$

so

$$|\langle f, g\rangle|^2 \le \langle f, f\rangle\langle g, g\rangle. \tag{9}$$

BESSEL'S INEQUALITY follows from SCHWARZ'S INEQUALITY.

See also BESSEL'S INEQUALITY, HÖLDER'S INEQUALITIES

References

Abramowitz, M. and Stegun, C. A. (Eds.). *Handbook of Mathematical Functions with Formulas, Graphs, and Mathematical Tables, 9th printing.* New York: Dover, p. 11, 1972.
Arfken, G. *Mathematical Methods for Physicists, 3rd ed.* Orlando, FL: Academic Press, pp. 527–529, 1985.
Buniakowsky, V. "Sur quelques inégalités concernant les intégrales ordinaires et les intégrales aux différences finies." *Mémoires de l'Acad. de St. Pétersbourg (VII)* **1**, No. 9, p. 4, 1959.
Gradshteyn, I. S. and Ryzhik, I. M. *Tables of Integrals, Series, and Products, 6th ed.* San Diego, CA: Academic Press, p. 1099, 2000.
Hardy, G. H.; Littlewood, J. E.; and Pólya, G. "Further Remarks on Method: The Inequality of Schwarz." §6.5 in *Inequalities, 2nd ed.* Cambridge, England: Cambridge University Press, pp. 132–134, 1952.
Schwarz, H. A. "Über ein die Flächen kleinsten Flächeninhalts betreffendes Problem der Variationsrechnung." *Acta Soc. Scient. Fen.* **15**, 315–362, 1885. Reprinted in *Gesammelte Mathematische Abhandlungen, Vol. 1.* New York: Chelsea, pp. 224–269, 1972.

Schwarz's Lemma

Let f be analytic on the UNIT DISK, and assume that

1. $|f(z)| \le 1$ for all z and
2. $f(0) = 0$.

Then $|f(z)| \le |z|$ and $|f'(0)| \le 1$.

If either $|f(z)| = |z|$ for some $z \ne 0$ or if $|f'(0)| = 1$, then f is a ROTATION, i.e., $f(z) = az$ for some complex constant a with $|a| = 1$.

See also MÖBIUS TRANSFORMATION, SCHWARZ-PICK LEMMA

References

Krantz, S. G. "Schwarz's Lemma." §5.5.1 in *Handbook of Complex Analysis.* Boston, MA: Birkhäuser, p. 78, 1999.

Schwarz's Minimal Surface

A periodic MINIMAL SURFACE constructed by Schwarz using the following two principles:

1. If part of the boundary of a MINIMAL SURFACE is a straight line, then the reflection across the line, when added to the original surface, makes another MINIMAL SURFACE.
2. If a MINIMAL SURFACE meets a PLANE at RIGHT ANGLES, then the mirror image of the PLANE, when added to the original surface, also makes a MINIMAL SURFACE.

See also MINIMAL SURFACE

References

Wells, D. *The Penguin Dictionary of Curious and Interesting Geometry.* London: Penguin, pp. 224–225, 1991.

Schwarz's Polyhedron

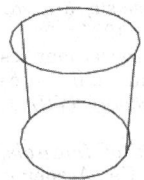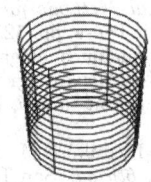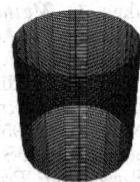

A polyhedron constructed by ruling $2n$ equally spaced vertical lines along the surface of a CYLINDER together with $2n^3$ circles around the cylinder at equally spaced heights. Amazingly, joining neighboring points in triangles and letting $n \to \infty$ gives a surface whose total SURFACE AREA approaches, not that of the cylinder, but infinity.

See also CYLINDER

References
Ogilvy, C. S. *Tomorrow's Math, 2nd ed.* Oxford, England: Oxford University Press, 1972.
Wells, D. *The Penguin Dictionary of Curious and Interesting Geometry.* London: Penguin, pp. 224–225, 1991.

Schwarz's Symmetry Principle
SCHWARZ REFLECTION PRINCIPLE

Schwarz's Triangle Problem
FAGNANO'S PROBLEM

Schweins's Theorem
If we expand the determinant of a matrix **A** using DETERMINANT EXPANSION BY MINORS, first in terms of the MINORS of order r formed from any r rows, with their complementaries, and second in terms of the MINORS of order m formed from any m columns ($r < m$), with their complementaries; then the sum of the $(n-r)_{m-r}$ terms of the second expansion which have in common the elements in the intersection of the selected r rows and m columns is equal to the sum of the m_r terms of the first expansion which have for one factor the minors of the rth order formed from the elements in the intersection of the selected r rows and m columns.

See also DETERMINANT, DETERMINANT EXPANSION BY MINORS, MINOR

References
Muir, T. "Schweins's Theorem." §141 in *A Treatise on the Theory of Determinants.* New York: Dover, pp. 124–125, 1960.

Schwenk's Formula
Let $R + B$ be the number of MONOCHROMATIC FORCED TRIANGLES (where R and B are the number of red and blue TRIANGLES) in an EXTREMAL GRAPH. Then

$$R + B = \binom{n}{3} - \left\lfloor \tfrac{1}{2} n \left\lfloor \tfrac{1}{4}(n-1)^2 \right\rfloor \right\rfloor,$$

where $\binom{n}{k}$ is a BINOMIAL COEFFICIENT and $\lfloor x \rfloor$ is the FLOOR FUNCTION (Schwenk 1972).

See also EXTREMAL GRAPH, MONOCHROMATIC FORCED TRIANGLE

References
Schwenk, A. J. "Acquaintance Party Problem." *Amer. Math. Monthly* **79**, 1113–1117, 1972.

Scientific Notation
Scientific notation is the expression of a number n in the form $a \times 10^p$, where

$$p \equiv \lfloor \log_{10}|n| \rfloor$$

is the FLOOR of the base-10 LOGARITHM of n (the "order of magnitude"), and

$$a \equiv \frac{n}{10^p}$$

is a REAL NUMBER satisfying $1 \le |a| < 10$. For example, in scientific notation, the number $n = 101,325$ has order of magnitude

$$p = \lfloor \log_{10} 101,325 \rfloor = \lfloor 5.00572 \rfloor = 5,$$

so n would be written 1.01325×10^5. The special case of 0 does not have a unique representation in scientific notation, i.e., $0 = 0 \times 10^0 = 0 \times 10^1 = \dots$.

See also CHARACTERISTIC (REAL NUMBER), FIGURES, MANTISSA, SIGNIFICANT DIGITS

s-Cluster
N.B. A detailed online essay by S. Finch was the starting point for this entry.

Let an $n \times n$ BINARY MATRIX have entries which are 1 (with probability p) or 0 (with probability $q = 1 - p$). An s-cluster is an isolated group of s adjacent (i.e., horizontally or vertically connected) 1s. Let C_n be the total number of these "SITE" clusters. Then the value

$$K_S(p) = \lim_{n \to \infty} \frac{\langle C_n \rangle}{n^2}, \qquad (1)$$

called the MEAN CLUSTER COUNT PER SITE or MEAN CLUSTER DENSITY, exists. Numerically, it is found that $K_S(1/2) \approx 0.065770 \dots$ (Ziff *et al.* 1997).

Considering instead "BOND" clusters (where numbers are assigned to the edges of a grid) and letting C_n be the total number of bond clusters, then

$$K_B(p) \equiv \lim_{n \to \infty} \frac{\langle C_n \rangle}{n^2}, \qquad (2)$$

exists. The analytic value is known for $p = 1/2$,

$$K_B(\tfrac{1}{2}) = \tfrac{3}{2}\sqrt{3} - \tfrac{41}{16} \qquad (3)$$

(Ziff *et al.* 1997).

See also Bond Percolation, Percolation Theory, *s*-Run, Site Percolation

References

Finch, S. "Favorite Mathematical Constants." http://www.mathsoft.com/asolve/constant/rndprc/rndprc.html.

Temperley, H. N. V. and Lieb, E. H. "Relations Between the 'Percolation' and 'Colouring' Problem and Other Graph-Theoretical Problems Associated with Regular Planar Lattices; Some Exact Results for the 'Percolation' Problem." *Proc. Roy. Soc. London A* **322**, 251–280, 1971.

Ziff, R.; Finch, S.; and Adamchik, V. "Universality of Finite-Sized Corrections to the Number of Percolation Clusters." *Phys. Rev. Let.* To appear, 1998.

Score Sequence

The score sequence of a TOURNAMENT is a monotonic nondecreasing sequence of the OUTDEGREES of the VERTICES. The score sequences for $n = 1, 2, \ldots$ are 1, 1, 2, 4, 9, 22, 59, 167, ... (Sloane's A000571).

See also Directed Graph, Tournament

References

Harary, F. *Graph Theory.* Reading, MA: Addison-Wesley, pp. 207–208, 1994.

Ruskey, F. "Information on Score Sequences." http://www.theory.csc.uvic.ca/~cos/inf/nump/ScoreSequence.html.

Ruskey, F.; Cohen, R.; Eades, P.; and Scott, A. "Alley CATs in Search of Good Homes." *Congres. Numer.* **102**, 97–110, 1994.

Sloane, N. J. A. Sequences A000571/M1189 in "An On-Line Version of the Encyclopedia of Integer Sequences." http://www.research.att.com/~njas/sequences/eisonline.html.

Scrawny Cantor Set

A Cantor set C in $\mathbb{R}^3$ is said to be scrawny if for each neighborhood U of an arbitrary point p in C, there is a neighborhood V of p such that every map $f : \mathbb{S}^1 \to V \subset C$ extends to a map $F : \mathbb{B}^2 \to U$ such that $F^{-1}(C)$ is finite. Babich (1992) presents examples of wild Cantor sets of this type and provides a proof that such objects cannot be defined by solid tori.

See also Cantor Set

References

Babich, A. "Scrawny Cantor Sets are Not Definable by Tori." *Proc. Amer. Math. Soc.* **115**, 829–836, 1992.

Screw

A TRANSLATION along a straight line L and a ROTATION about L such that the angle of ROTATION is proportional to the TRANSLATION at each instant. Also known as a TWIST.

See also Dini's Surface, Helicoid, Rotation, Screw Theorem, Seashell, Translation

Screw Theorem

Any motion of a rigid body in space at every instant is a SCREW motion. This theorem was proved by Mozzi and Cauchy.

See also Screw

Scruple

An archaic UNIT FRACTION variously defined as 1/200 (of an hour), 1/10 or 1/12 (of an inch), 1/12 (of a celestial body's angular diameter), or 1/60 (of an hour or DEGREE).

See also Calcus, Uncia

Sea Horse Valley

A portion of the MANDELBROT SET centered around $-1.25 + 0.047i$ with width approximately $0.009 + 0.005i$.

See also Mandelbrot Set

Search Tree

Tree Searching

Searching

Searching refers to locating a given element or an element satisfying certain conditions from some (usually ordered or partially ordered) table, list, TREE, etc.

See also Binary Search, Sorting, Tabu Search, Tree Searching

References

Knuth, D. E. *The Art of Computer Programming, Vol. 3: Sorting and Searching, 2nd ed.* Reading, MA: Addison-Wesley, 1973.

Press, W. H.; Flannery, B. P.; Teukolsky, S. A.; and Vetterling, W. T. "How to Search an Ordered Table." §3.4 in *Numerical Recipes in FORTRAN: The Art of Scientific Computing, 2nd ed.* Cambridge, England: Cambridge University Press, pp. 110–113, 1992.

Skiena, S. "Sorting and Searching." §1.1.6 in *Implementing Discrete Mathematics: Combinatorics and Graph Theory with Mathematica.* Reading, MA: Addison-Wesley, pp. 14–16, 1990.

Seashell

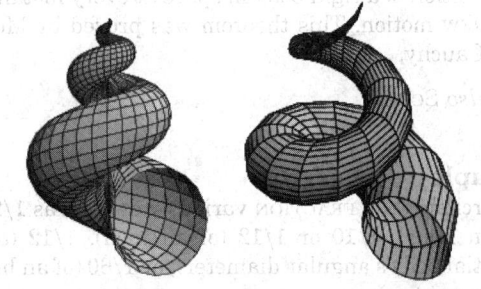

A conical surface modeled after the shape of a seashell. One parameterization (left figure) is given by

$$x = 2[1 - e^{u/(6\pi)}]\cos u \, \cos^2\left(\tfrac{1}{2}v\right) \tag{1}$$

$$y = 2[-1 + e^{u/(6\pi)}]\cos^2\left(\tfrac{1}{2}v\right)\sin u \tag{2}$$

$$z = 1 - e^{u/(3\pi)} - \sin v + e^{u/(6\pi)}\sin v, \tag{3}$$

where $v \in [0, 2\pi)$, and $u \in [0, 6\pi)$ (Wolfram). Nordstrand gives the parameterization

$$x = \left[\left(1 - \frac{v}{2\pi}\right)(1 + \cos u) + c\right]\cos(nv) \tag{4}$$

$$x = \left[\left(1 - \frac{v}{2\pi}\right)(1 + \cos u) + c\right]\sin(nv) \tag{5}$$

$$z = \frac{bv}{2\pi} + a \sin u\left(1 - \frac{v}{2\pi}\right) \tag{6}$$

for u, $v \in [0, 2\pi]$ (right figure with $a = 0.2$, $b = 1$, $c = 0.1$, and $n = 2$).

See also CONICAL SPIRAL

References

Gray, A. "Sea Shells." §13.6 in *Modern Differential Geometry of Curves and Surfaces with Mathematica, 2nd ed.* Boca Raton, FL: CRC Press, pp. 308–309, 1997.

Nordstrand, T. "Conic Spiral or Seashell." http://www.uib.no/people/nfytn/shelltxt.htm.

Wolfram Research "Mathematica Version 2.0 Graphics Gallery." http://www.mathsource.com/cgi-bin/msitem22?0207-155.

Sec

SECANT

Secant

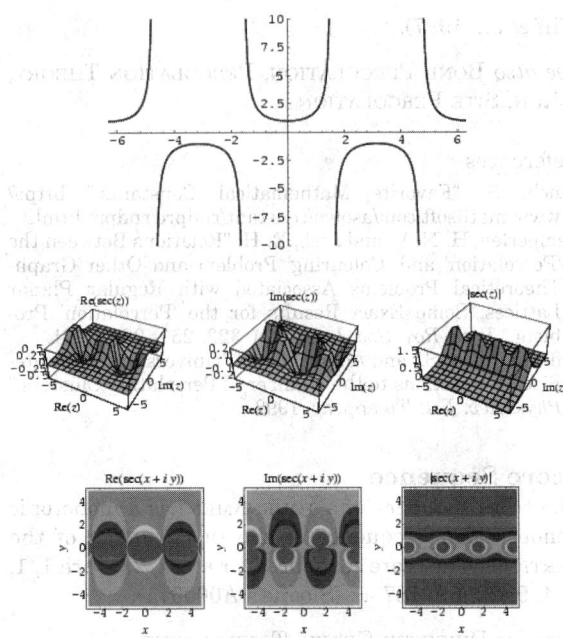

The function defined by $\sec x \equiv 1/\cos x$, where $\cos x$ is the COSINE. The MACLAURIN SERIES of the secant is

$$\sec x = \frac{(-1)^n E_{2n}}{(2n)!}\, x^{2n}$$

$$= 1 + \tfrac{1}{2}x^2 + \tfrac{5}{24}x^4 + \tfrac{61}{720}x^6 + \tfrac{277}{8064}x^8 + \dots .$$

where E_{2n} is an EULER NUMBER.

See also ALTERNATING PERMUTATION, COSECANT, COSINE, EULER NUMBER, EXSECANT, INVERSE SECANT

References

Abramowitz, M. and Stegun, C. A. (Eds.). "Circular Functions." §4.3 in *Handbook of Mathematical Functions with Formulas, Graphs, and Mathematical Tables, 9th printing.* New York: Dover, pp. 71–79, 1972.

Beyer, W. H. *CRC Standard Mathematical Tables, 28th ed.* Boca Raton, FL: CRC Press, p. 224, 1987.

Spanier, J. and Oldham, K. B. "The Secant $\sec(x)$ and Cosecant $\csc(x)$ Functions." Ch. 33 in *An Atlas of Functions.* Washington, DC: Hemisphere, pp. 311–318, 1987.

Secant Line

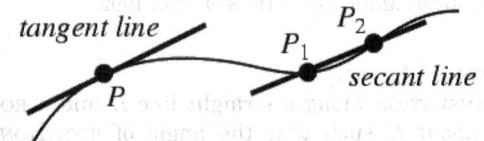

A line joining two points of a curve. As the two points are brought together (or, more precisely, as one is brought towards the other), the secant line tends to a TANGENT LINE. In abstract mathematics, the points

which a secant line connects can be either REAL or COMPLEX CONJUGATE IMAGINARY.

See also BITANGENT, TANGENT LINE, TRANSVERSAL LINE

Secant Method

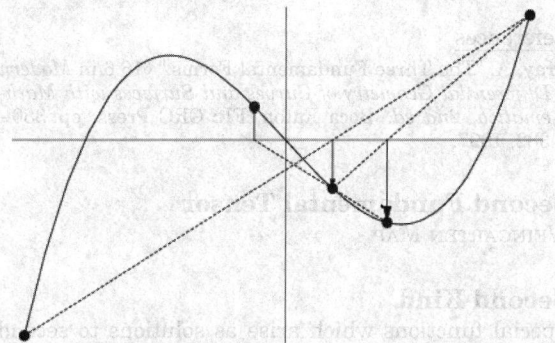

A ROOT-finding algorithm which assumes a function to be approximately linear in the region of interest. Each improvement is taken as the point where the approximating line crosses the axis. The secant method retains only the most recent estimate, so the root does not necessarily remain bracketed. When the ALGORITHM does converge, its order of convergence is

$$\lim_{k \to \infty} |\epsilon_{k+1}| \approx C |\epsilon|^\phi. \quad (1)$$

where C is a constant and ϕ is the GOLDEN MEAN.

$$f'(x_{n-1}) \approx \frac{f(x_{n-1}) - f(x_{n-2})}{x_{n-1} - x_{n-2}} \quad (2)$$

$$f(x_n) \approx f(x_{n-1}) + f'(x_n)(x_n - x_{n-1}) = 0 \quad (3)$$

$$f(x_{n-1}) + \frac{f(x_{n-1}) - f(x_{n-2})}{x_{n-1} - x_{n-2}}(x_n - x_{n-1}) = 0. \quad (4)$$

so

$$x_n = x_{n-1} - \frac{f(x_{n-1})(x_{n-1} - x_{n-2})}{f(x_{n-1}) - f(x_{n-2})}. \quad (5)$$

See also FALSE POSITION METHOD

References
Press, W. H.; Flannery, B. P.; Teukolsky, S. A.; and Vetterling, W. T. "Secant Method, False Position Method, and Ridders' Method." §9.2 in *Numerical Recipes in FORTRAN: The Art of Scientific Computing, 2nd ed.* Cambridge, England: Cambridge University Press, pp. 347–352, 1992.

Secant Number

A number, more commonly called an EULER NUMBER, giving the number of EVEN ALTERNATING PERMUTA-

TIONS. The term ZIG NUMBER is sometimes also used. The first few are 1, 5, 61, 1385, ... (Sloane's A000364).

See also ALTERNATING PERMUTATION, EULER NUMBER, EULER ZIGZAG NUMBER, TANGENT NUMBER

References
Sloane, N. J. A. Sequences A000364 in "An On-Line Version of the Encyclopedia of Integer Sequences." http://www.research.att.com/~njas/sequences/eisonline.html.

Sech
HYPERBOLIC SECANT

Second
ARC SECOND

Second Countable Topology

A TOPOLOGICAL SPACE is second countable if it has a countable TOPOLOGICAL BASIS.

See also TOPOLOGICAL BASIS, TOPOLOGICAL SPACE

Second Curvature
TORSION (DIFFERENTIAL GEOMETRY)

Second Derivative Test

Suppose $f(x)$ is a FUNCTION of x which is twice DIFFERENTIABLE at a STATIONARY POINT x_0.

1. If $f''(x_0) > 0$, then f has a RELATIVE MINIMUM at x_0.
2. If $f''(x_0) < 0$, then f has a RELATIVE MAXIMUM at x_0.

The EXTREMUM TEST gives slightly more general conditions under which a function with $f''(x_0) = 0$ is a maximum or minimum.

If $f(x, y)$ is a 2-D FUNCTION which has a RELATIVE EXTREMUM at a point (x_0, y_0) and has CONTINUOUS PARTIAL DERIVATIVES at this point, then $f_x(x_0, y_0) = 0$ and $f_y(x_0, y_0) = 0$. The second PARTIAL DERIVATIVES test classifies the point as a MAXIMUM or MINIMUM. Define the DISCRIMINANT as

$$D \equiv f_{xx}f_{yy} - f_{xy}f_{yx} = f_{xx}f_{yy} - f_{xy}^2.$$

1. If $D > 0$, $f_{xx}(x_0, y_0) > 0$ and $f_{xx}(x_0, y_0) + f_{yy}(x_0, y_0) > 0$, the point is a RELATIVE MINIMUM.
2. If $D > 0$, $f_{xx}(x_0, y_0) < 0$, and $f_{xx}(x_0, y_0) + f_{yy}(x_0, y_0) < 0$, the point is a RELATIVE MAXIMUM.
3. If $D < 0$, the point is a SADDLE POINT.
4. If $D = 0$, higher order tests must be used.

See also DISCRIMINANT (SECOND DERIVATIVE TEST), EXTREMUM, EXTREMUM TEST, FIRST DERIVATIVE TEST, GLOBAL MAXIMUM, GLOBAL MINIMUM, HESSIAN DETERMINANT, MAXIMUM, MINIMUM, RELATIVE MAX-

IMUM, RELATIVE MINIMUM, SADDLE POINT (FUNCTION)

References

Abramowitz, M. and Stegun, C. A. (Eds.). *Handbook of Mathematical Functions with Formulas, Graphs, and Mathematical Tables, 9th printing.* New York: Dover, p. 14, 1972.

Second Fundamental Form

Let M be a REGULAR SURFACE with $\mathbf{v_p}$, $\mathbf{w_p}$ points in the TANGENT SPACE $M_\mathbf{p}$ of M. For $M \in \mathbb{R}^3$, the second fundamental form is the symmetric bilinear form on the TANGENT SPACE $M_\mathbf{p}$,

$$\mathbf{II}(\mathbf{v_p}, \mathbf{w_p}) = S(\mathbf{v_p}) \cdot \mathbf{w_p}. \tag{1}$$

where S is the SHAPE OPERATOR. The second fundamental form satisfies

$$\mathbf{II}(a\mathbf{x}_u + b\mathbf{x}_v, a\mathbf{x}_u + b\mathbf{x}_v) = ea^2 + 2fab + gb^2 \tag{2}$$

for any nonzero TANGENT VECTOR.

The second fundamental form is given explicitly by

$$e\,du^2 + 2f\,du\,dv + g\,dv^2 \tag{3}$$

where

$$e = \sum_i X_i \frac{\partial^2 x_i}{\partial u^2} \tag{4}$$

$$f = \sum_i X_i \frac{\partial^2 x_i}{\partial u\,\partial v} \tag{5}$$

$$g = \sum_i X_i \frac{\partial^2 x_i}{\partial v^2} \tag{6}$$

and X_i are the DIRECTION COSINES of the surface normal. The second fundamental form can also be written

$$e = -\mathbf{N}_u \cdot \mathbf{x}_v = \mathbf{N} \cdot \mathbf{x}_{uv} \tag{7}$$

$$f = -\mathbf{N}_v \cdot \mathbf{x}_u = \mathbf{N} \cdot \mathbf{x}_{uv} = \mathbf{N}_{vu} \cdot \mathbf{x}_{vu}$$

$$= -\mathbf{N}_u \cdot \mathbf{x}_v \tag{8}$$

$$g = -\mathbf{N}_v \cdot \mathbf{x}_v = \mathbf{N} \cdot \mathbf{x}_{vv}, \tag{9}$$

where $\mathbf{N}$ is the NORMAL VECTOR, $\mathbf{x}: U \to \mathbb{R}^3$ is a REGULAR PATCH, and $\mathbf{x}_u$ and $\mathbf{x}_v$ are the partial derivatives of $\mathbf{x}$ with respect to parameters u and v, respectively, or

$$e = \frac{\det(\mathbf{x}_{uu}\mathbf{x}_u\mathbf{x}_v)}{\sqrt{EG - F^2}} \tag{10}$$

$$f = \frac{\det(\mathbf{x}_{uv}\mathbf{x}_u\mathbf{x}_v)}{\sqrt{EG - F^2}} \tag{11}$$

$$g = \frac{\det(\mathbf{x}_{uv}\mathbf{x}_u\mathbf{x}_v)}{\sqrt{FG - F^2}}. \tag{12}$$

See also FIRST FUNDAMENTAL FORM, FUNDAMENTAL FORMS, SHAPE OPERATOR, THIRD FUNDAMENTAL FORM

References

Gray, A. "The Three Fundamental Forms." §16.6 in *Modern Differential Geometry of Curves and Surfaces with Mathematica, 2nd ed.* Boca Raton, FL: CRC Press, pp. 380–382, 1997.

Second Fundamental Tensor

WEINGARTEN MAP

Second Kind

Special functions which arise as solutions to second order ordinary differential equations are commonly said to be "of the first kind" if they are *nonsingular* at the origin, while the linearly independent solutions which are *singular* are said to be "of the second kind." Common examples of functions of the second kind defined in this way include the BESSEL FUNCTION OF THE SECOND KIND, CHEBYSHEV POLYNOMIAL OF THE SECOND KIND, CONFLUENT HYPERGEOMETRIC FUNCTION OF THE SECOND KIND, HANKEL FUNCTION OF THE SECOND KIND, and so on.

The term "second kind" is also used in a more general context to distinguish between two or more types of mathematical objects which, however, all satisfy some common overall property. Examples of objects of this kind include the CHRISTOFFEL SYMBOL OF THE SECOND KIND, ELLIPTIC INTEGRAL OF THE SECOND KIND, FREDHOLM INTEGRAL EQUATION OF THE SECOND KIND, STIRLING NUMBER OF THE SECOND KIND, VOLTERRA INTEGRAL EQUATION OF THE SECOND KIND, and so on.

See also BESSEL FUNCTION OF THE SECOND KIND, CHEBYSHEV POLYNOMIAL OF THE SECOND KIND, CONFLUENT HYPERGEOMETRIC FUNCTION OF THE SECOND KIND, ELLIPTIC INTEGRAL OF THE SECOND KIND, FIRST KIND, FREDHOLM INTEGRAL EQUATION OF THE SECOND KIND, HANKEL FUNCTION OF THE SECOND KIND, SPECIAL FUNCTION, STIRLING NUMBER OF THE SECOND KIND, THIRD KIND, VOLTERRA INTEGRAL EQUATION OF THE SECOND KIND

Section

A section of a solid is the plane figure cut from the solid by passing a plane through it (Kern and Bland 1948, p. 18).

See also CONIC SECTION, CROSS SECTION, CUBICAL CONIC SECTION, CYLINDRICAL SECTION, DEDEKIND SECTION, GRAPH SECTION, MULTISECTION, NORMAL SECTION, SECTION (BUNDLE), SECTION (PENCIL), SEC-

TION (TANGENT BUNDLE), SPIRIC SECTION, SURFACE OF SECTION, TORIC SECTION

References

Kern, W. F. and Bland, J. R. *Solid Mensuration with Proofs, 2nd ed.* New York: Wiley, 1948.

Section (Bundle)

A section of a FIBER BUNDLE gives an element of the fiber over every point in B. Usually it is described as a map $s : B \to E$ such that $\pi \circ s$ is the identity on B. A real-valued function on a manifold M is a section of the trivial LINE BUNDLE $M \times \mathbb{R}$. Another common example is a VECTOR FIELD, which is a section of the TANGENT BUNDLE.

See also FIBER BUNDLE, TANGENT BUNDLE, VECTOR BUNDLE, ZERO SECTION

Section (Pencil)

The lines of a PENCIL joining the points of a RANGE to another POINT.

See also PENCIL, RANGE (LINE SEGMENT)

Section (Tangent Bundle)

A VECTOR FIELD is a section of its TANGENT BUNDLE, meaning that to every point x in a MANIFOLD M, a VECTOR $X(x) \in T_x M$ is associated, where T_x is the TANGENT SPACE.

See also TANGENT BUNDLE, TANGENT SPACE

Sectional Curvature

The mathematical object κ which controls the rate of geodesic deviation.

See also BISHOP'S INEQUALITY, CHEEGER'S FINITENESS THEOREM, GEODESIC

Sector

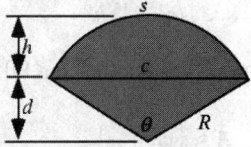

A WEDGE obtained by taking a portion of a DISK with CENTRAL ANGLE $\theta < \pi$ radians (180°), illustrated above as the shaded region. A sector of π radians would be a SEMICIRCLE. Let R be the radius of the CIRCLE, c the CHORD length, s the ARC LENGTH, h the height of the arced portion, and d the height of the triangular portion. Then

$$R = h + d \tag{1}$$

$$s = R\theta \tag{2}$$

$$d = R \cos\left(\tfrac{1}{2}\theta\right) \tag{3}$$

$$= \tfrac{1}{2} c \cot\left(\tfrac{1}{2}\theta\right) \tag{4}$$

$$= \tfrac{1}{2} \sqrt{4R^2 - c^2} \tag{5}$$

$$c = 2R \sin\left(\tfrac{1}{2}\theta\right) \tag{6}$$

$$= 2d \tan\left(\tfrac{1}{2}\theta\right) \tag{7}$$

$$= 2\sqrt{R^2 - d^2} \tag{8}$$

$$= 2\sqrt{h(2R - h)}. \tag{9}$$

The ANGLE θ obeys the relationships

$$\theta = \frac{s}{R} = 2 \cos^{-1}\left(\frac{d}{R}\right) = 2 \tan^{-1}\left(\frac{c}{2d}\right)$$

$$= 2 \sin^{-1}\left(\frac{c}{2R}\right). \tag{10}$$

The AREA of the sector is

$$A = \tfrac{1}{2} Rs = \tfrac{1}{2} R^2 \theta \tag{11}$$

(Beyer 1987).

See also CIRCLE-CIRCLE INTERSECTION, LENS, OBTUSE TRIANGLE, SEGMENT

References

Beyer, W. H. (Ed.). *CRC Standard Mathematical Tables, 28th ed.* Boca Raton, FL: CRC Press, p. 125, 1987.
Harris, J. W. and Stocker, H. "Sector." §3.8.4 in *Handbook of Mathematics and Computational Science.* New York: Springer-Verlag, pp. 91–92, 1998.
Kern, W. F. and Bland, J. R. *Solid Mensuration with Proofs, 2nd ed.* New York: Wiley, p. 3, 1948.

Sectorial Harmonic

A SPHERICAL HARMONIC OF THE FORM

$$\sin(m\theta) P_m^m(\cos \phi).$$

or

$$\cos(m\theta) P_m^m(\cos \phi).$$

See also SPHERICAL HARMONIC, TESSERAL HARMONIC, ZONAL HARMONIC

Secular Equation

CHARACTERISTIC EQUATION

Seed

The initial number used as the starting point in a RANDOM NUMBER generating ALGORITHM.

Seed of Life

One of the beautiful arrangements of CIRCLES found at the Temple of Osiris at Abydos, Egypt (Rawles 1997). The CIRCLES are placed with 6-fold symmetry, forming a mesmerizing pattern of CIRCLES and LENSES.

See also CIRCLE, CIRCLE COVERING, FIVE DISKS PROBLEM, FLOWER OF LIFE, VENN DIAGRAM

References

Rawles, B. *Sacred Geometry Design Sourcebook: Universal Dimensional Patterns.* Nevada City, CA: Elysian Pub., p. 15, 1997.

Weisstein, E. W. "Flower of Life." MATHEMATICA NOTEBOOK FLOWEROFLIFE.M.

Seek Time

POINT-POINT DISTANCE–1-D

Segment

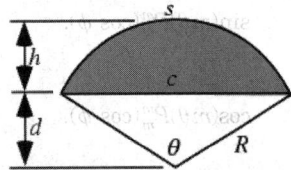

A portion of a DISK whose upper boundary is a circular ARC and whose lower boundary is a CHORD making a CENTRAL ANGLE $\theta < \pi$ radians (180°), illustrated above as the shaded region. Let R be the radius of the CIRCLE, c the CHORD length, s the ARC LENGTH, h the height of the arced portion, and d the

height of the triangular portion. Then

$$R = h + d \tag{1}$$

$$s = R\theta \tag{2}$$

$$d = R\cos\left(\tfrac{1}{2}\theta\right) \tag{3}$$

$$= \tfrac{1}{2}c\cot\left(\tfrac{1}{2}\theta\right) \tag{4}$$

$$= \tfrac{1}{2}\sqrt{4R^2 - c^2} \tag{5}$$

$$c = 2R\sin\left(\tfrac{1}{2}\theta\right) \tag{6}$$

$$= 2d\tan\left(\tfrac{1}{2}\theta\right) \tag{7}$$

$$= 2\sqrt{R^2 - d^2} \tag{8}$$

$$= 2\sqrt{h(2R - h)}. \tag{9}$$

The ANGLE θ obeys the relationships

$$\theta = \frac{s}{R} = 2\cos^{-1}\left(\frac{d}{R}\right) = 2\tan^{-1}\left(\frac{c}{2d}\right)$$

$$= 2\sin^{-1}\left(\frac{c}{2R}\right). \tag{10}$$

The AREA of the segment is then

$$A = A_{\text{sector}} - A_{\text{isosceles triangle}} \tag{11}$$

$$= \tfrac{1}{2}R^2(\theta - \sin\theta) \tag{12}$$

$$= \tfrac{1}{2}(Rs - cd) \tag{13}$$

$$= R^2\cos^{-1}\left(\frac{d}{R}\right) - d\sqrt{R^2 - d^2} \tag{14}$$

$$= R^2\cos^{-1}\left(\frac{R - h}{R}\right) - (R - h)\sqrt{2Rh - h^2}. \tag{15}$$

where the formula for the ISOSCELES TRIANGLE in terms of the VERTEX angle has been used (Beyer 1987). Approximate formulas for the ARC LENGTH and AREA are

$$s \approx \sqrt{c^2 + \tfrac{16}{3}h^2} \tag{16}$$

accurate to within 0.3% for $0° \le \theta \le 90°$, and

$$A \approx \tfrac{2}{3}ch + \frac{h_3}{2c}. \tag{17}$$

accurate to within 0.1% for $0° \le \theta \le 150°$ and 0.8% for $150° \le \theta \le 180°$ (Harris and Stocker 1998).

See also CHORD, CIRCLE-CIRCLE INTERSECTION, CYLINDRICAL SEGMENT, LENS, PARABOLIC SEGMENT,

REULEAUX TRIANGLE, SAGITTA, SECTOR, SPHERICAL SEGMENT

References
Beyer, W. H. (Ed.). *CRC Standard Mathematical Tables, 28th ed.* Boca Raton, FL: CRC Press, p. 125, 1987.

Fukagawa, H. and Pedoe, D. "Segments of a Circle." §1.6 in *Japanese Temple Geometry Problems.* Winnipeg, Manitoba, Canada: Charles Babbage Research Foundation, pp. 14–15 and 88–92, 1989.

Harris, J. W. and Stocker, H. "Segment of a Circle." §3.8.6 in *Handbook of Mathematics and Computational Science.* New York: Springer-Verlag, pp. 92–93, 1998.

Kern, W. F. and Bland, J. R. *Solid Mensuration with Proofs, 2nd ed.* New York: Wiley, p. 4, 1948.

Segmented Number
PRIME NUMBER OF MEASUREMENT

Segner's Recurrence Formula
The RECURRENCE RELATION

$$E_n = E_2 E_{n-1} + E_3 E_{n-2} + \ldots + E_{n-1} E_2$$

which gives the solution to EULER'S POLYGON DIVISION PROBLEM.

See also CATALAN NUMBER, EULER'S POLYGON DIVISION PROBLEM

Segre Characteristic
A set of integers that give the orders of the blocks in a JORDAN CANONICAL FORM, with those integers corresponding to submatrices containing the same latent root bracketed together. For example, the Segre characteristic of

$$\begin{bmatrix} \alpha & 1 & & & & & & & \\ & \alpha & & & & & & & \\ & & \alpha & & & & & & \\ & & & \beta & 1 & & & & \\ & & & & \beta & 1 & & & \\ & & & & & \beta & & & \\ & & & & & & \gamma & & \\ & & & & & & & \delta & 1 \\ & & & & & & & & \delta \\ & & & & & & & & \delta \end{bmatrix}$$

is [(21)31(21)] (Frazer *et al.* 1955, p. 94).

References
Frazer, R. A.; Duncan, W. J.; and Collar, A. R. *Elementary Matrices and Some Applications to Dynamics and Differential Equations.* Cambridge, England: Cambridge University Press, p. 94, 1955.

Segre's Theorem
For any REAL NUMBER $r \geq 0$, an IRRATIONAL number α can be approximated by infinitely many RATIONAL fractions p/q in such a way that

$$-\frac{1}{\sqrt{1 + 4rq^2}} < \frac{p}{q} - \alpha < \frac{r}{\sqrt{1 + 4rq^2}}.$$

If $r = 1$, this becomes HURWITZ'S IRRATIONAL NUMBER THEOREM.

See also HURWITZ'S IRRATIONAL NUMBER THEOREM

Seiberg-Witten Equations

$$D_A \psi = 0$$

$$F_A^+ = -\tau(\psi, \psi),$$

where τ is the sesquilinear map $\tau : W^+ \times W^+ \to A^+ \otimes \mathbb{C}$.

See also WITTEN'S EQUATIONS

References
Donaldson, S. K. "The Seiberg-Witten Equations and 4-Manifold Topology." *Bull. Amer. Math. Soc.* **33**, 45–70, 1996.

Marshakov, A. *Seiberg-Witten Theory and Integrable Systems.* Singapore: World Scientific, 1999.

Morgan, J. W. *The Seiberg-Witten Equations and Applications to the Topology of Smooth Four-Manifolds.* Princeton, NJ: Princeton University Press, 1996.

Seiberg-Witten Invariants
WITTEN'S EQUATIONS

Seidel-Entringer-Arnold Triangle
The NUMBER TRIANGLE consisting of the ENTRINGER NUMBERS $E_{n,k}$ arranged in "ox-plowing" order,

$$E_{00}$$
$$E_{10} \to E_{11}$$
$$E_{22} \leftarrow E_{21} \leftarrow E_{20}$$
$$E_{30} \to E_{30} \to E_{32} \to E_{33}$$
$$E_{44} \leftarrow E_{43} \leftarrow E_{42} \leftarrow E_{41} \leftarrow E_{40}$$

giving

$$1$$
$$0 \to 1$$
$$1 \leftarrow 1 \leftarrow 0$$
$$0 \to 1 \to 2 \to 2$$
$$5 \leftarrow 5 \leftarrow 4 \leftarrow 2 \leftarrow 0$$

See also BELL NUMBER, BOUSTROPHEDON TRANSFORM, CLARK'S TRIANGLE, ENTRINGER NUMBER, EULER'S TRIANGLE, LEIBNIZ HARMONIC TRIANGLE, LOSSNITSCH'S TRIANGLE, NUMBER TRIANGLE, PASCAL'S TRIANGLE

References

Arnold, V. I. "Bernoulli-Euler Updown Numbers Associated with Function Singularities, Their Combinatorics, and Arithmetics." *Duke Math. J.* **63**, 537–555, 1991.

Arnold, V. I. "Snake Calculus and Combinatorics of Bernoulli, Euler, and Springer Numbers for Coxeter Groups." *Russian Math. Surveys* **47**, 3–45, 1992.

Conway, J. H. and Guy, R. K. In *The Book of Numbers.* New York: Springer-Verlag, 1996.

Dumont, D. "Further Triangles of Seidel-Arnold Type and Continued Fractions Related to Euler and Springer Numbers." *Adv. Appl. Math.* **16**, 275–296, 1995.

Entringer, R. C. "A Combinatorial Interpretation of the Euler and Bernoulli Numbers." *Nieuw. Arch. Wisk.* **14**, 241–246, 1966.

Millar, J.; Sloane, N. J. A.; and Young, N. E. "A New Operation on Sequences: The Boustrophedon Transform." *J. Combin. Th. Ser. A* **76**, 44–54, 1996.

Seidel, I. "Über eine einfache Entstehungsweise der Bernoullischen Zahlen und einiger verwandten Reihen." *Sitzungsber. Münch. Akad.* **4**, 157–187, 1877.

Seifert Circle

Eliminate each KNOT crossing by connecting each of the strands coming into the crossing to the adjacent strand leaving the crossing. The resulting strands no longer cross but form instead a set of nonintersecting CIRCLES called Seifert circles.

References

Adams, C. C. *The Knot Book: An Elementary Introduction to the Mathematical Theory of Knots.* New York: W. H. Freeman, p. 96, 1994.

Seifert Conjecture

Every smooth NONZERO VECTOR FIELD on the 3-SPHERE has at least one closed orbit. The conjecture was proposed in 1950, proved true for Hopf fibrations, but proved false in general by Kuperberg (1994).

References

Kuperberg, G. "A Volume-Preserving Counterexample to the Seifert Conjecture." *Comment. Math. Helv.* **71**, 70–97, 1996.

Kuperberg, G. and Kuperberg, K. "Generalized Counterexamples to the Seifert Conjecture." *Ann. Math.* **143**, 547–576, 1996.

Kuperberg, G. and Kuperberg, K. "Generalized Counterexamples to the Seifert Conjecture." *Ann. Math.* **144**, 239–268, 1996.

Kuperberg, K. "A Smooth Counterexample to the Seifert Conjecture." *Ann. Math.* **140**, 723–732, 1994.

Seifert Form

For K a given KNOT in $\mathbb{S}^3$, choose a SEIFERT SURFACE M^2 in $\mathbb{S}^3$ for K and a bicollar $\hat{M} \times [-1, 1]$ in $\mathbb{S}^3 - K$. If $x \in H_1(M)$ is represented by a 1-cycle in $\hat{M}$, let x^+ denote the homology cycle carried by $x \times 1$ in the bicollar. Similarly, let x^- denote $x \times -1$. The function $f : H_1(\hat{M}) \times H_1(\hat{M}) \to Z$ defined by

$$f(x, y) = \mathrm{lk}(x, y^+).$$

where lk denotes the LINKING NUMBER, is called a Seifert form for K.

See also SEIFERT MATRIX

References

Rolfsen, D. *Knots and Links.* Wilmington, DE: Publish or Perish Press, pp. 200–201, 1976.

Seifert Matrix

Given a SEIFERT FORM $f(x, y)$, choose a basis $e_1, ..., e_{2g}$ for $H_1(\hat{M})$ as a $\mathbb{Z}$-module so every element is uniquely expressible as

$$n_1 e_1 + \ldots + n_{2g} e_{2g} \tag{1}$$

with n_i integer. Then define the Seifert matrix V as the $2g \times 2g$ INTEGER MATRIX with entries

$$v_{ij} = \mathrm{lk}\left(e_i, e_j^+\right). \tag{2}$$

For example, the right-hand TREFOIL KNOT has Seifert matrix

$$V = \begin{bmatrix} -1 & 1 \\ 0 & -1 \end{bmatrix}. \tag{3}$$

A Seifert matrix is not a KNOT INVARIANT, but it can be used to distinguish between different SEIFERT SURFACES for a given knot.

See also ALEXANDER MATRIX

References

Rolfsen, D. *Knots and Links.* Wilmington, DE: Publish or Perish Press, pp. 200–203, 1976.

Seifert Surface

An orientable surface with one boundary component such that the boundary component of the surface is a given KNOT K. In 1934, Seifert proved that such a surface can be constructed for any KNOT. The process of generating this surface is known as Seifert's algorithm. Applying Seifert's algorithm to an alternating projection of an alternating knot yields a Seifert surface of minimal GENUS.

There are KNOTS for which the minimal genus Seifert surface cannot be obtained by applying Seifert's algorithm to any projection of that KNOT, as proved by Morton in 1986 (Adams 1994, p. 105).

See also GENUS (KNOT), SEIFERT MATRIX

References

Adams, C. C. *The Knot Book: An Elementary Introduction to the Mathematical Theory of Knots.* New York: W. H. Freeman, pp. 95–106, 1994.

Seifert, H. "Über das Geschlecht von Knotten." *Math. Ann.* **110**, 571–592, 1934.

Seiffert's Spherical Spiral

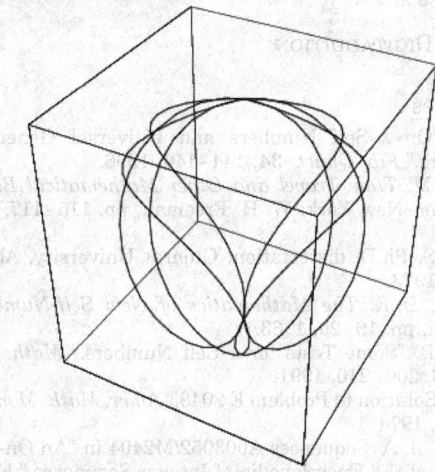

The SPHERICAL CURVE obtained when moving along the surface of a sphere with constant speed, while maintaining a constant angular velocity with respect to a fixed diameter (Erdos 2000). This curve is given in CYLINDRICAL COORDINATES by the parametric equations

$$r = \text{sn}(s, k)$$

$$\theta = ks$$

$$z = \text{cn}(s, k),$$

where k is a POSITIVE constant and $\text{sn}(s)$ and $\text{cn}(s)$ are JACOBI ELLIPTIC FUNCTIONS (Whittaker and Watson 1990, pp. 527–528).

Erdos (2000) provides a derivation of the equations of this curve, as well as an analysis of its properties, including conditions for obtaining periodic orbits.

See also SPHERICAL CURVE, SPHERICAL SPIRAL

References

Bowman, F. *Introduction to Elliptic Functions, with Applications.* New York: Dover, p. 34, 1961.

Erdos, P. "Spiraling the Earth with C. G. J. Jacobi." *Amer. J. Phys.* **68**, 888–895, 2000.

Seiffert. "Über eine neue geometrische Einführung in die Theorie der elliptischen Funktionen." *Wissensch. Beiträge Jahresber. Städtischen Realschule zu Charlottenburg, Ostern.* 1896.

Whittaker, E. T. and Watson, G. N. *A Course in Modern Analysis, 4th ed.* Cambridge, England: Cambridge University Press, 1990.

Selberg Trace Formula

Let p run over all distinct primitive ordered periodic geodesics, and let $\tau(p)$ denote the positive length of p, then every EVEN FUNCTION $h(\rho)$ analytic in $|\Im[\rho]| \leq \epsilon + 1/2$ and such that $|h(\rho)| \leq \mathcal{O}\left(|\rho|^{-2-\delta}\right)$ for $\rho \to \pm\infty$ satisfies the summation formula

$$\sum_{k=0}^{\infty} h(\rho_k) = (g-1) \int_{-\infty}^{\infty} \left(-\frac{d\hat{h}}{d\tau}\right) \frac{d\tau}{\sinh\left(\frac{1}{2}\tau\right)}$$

$$+ \sum_{\{p\}} \sum_{n=1}^{\infty} \frac{\tau(p)}{2\sinh\left[\frac{1}{2}n\tau(p)\right]} \hat{h}(n\tau(p)).$$

where g is the genus of the surface whose area is $4\pi(g-1)$ by the GAUSS-BONNET THEOREM.

See also SELBERG ZETA FUNCTION

References

Balazs, N. L. and Voros, A. "Chaos on the Pseudosphere." *Phys. Rep.* **143**, 109–240, 1986.

Elstrodt, J. *Jahresber. d. Deutsche Math. Verein* **83**, 45–77, 1981.

Hejhal, D. A. "The Selberg Trace Formula and the Riemann Zeta Function." *Duke Math. J.* **43**, 441–482, 1976.

Voros, A. "Spectral Functions, Special Functions and the Selberg Zeta Function." *Commun. Math. Phys.* **110**, 439–465, 1987.

Selberg Zeta Function

Let p run over all distinct primitive ordered periodic geodesics, and let $\tau(p)$ denote the positive length of p, then the Selberg zeta function is defined as

$$\mathscr{L}(s) = \prod_{\{p\}} \prod_{k=0}^{\infty} \left[1 - e^{-z(p)(s+k)}\right].$$

for $s > 1$.

See also SELBERG TRACE FORMULA

References

d'Hoker, E. and Phong, D. H. "Multiloop Amplitudes for the Bosonic Polyakov String." *Nucl. Phys. B* **269**, 205–234, 1986.

d'Hoker, E. and Phong, D. H. "On Determinants of Laplacians on Riemann Surfaces." *Commun. Math. Phys.* **104**, 537–545, 1986.

Fried, D. *Invent. Math.* **84**, 523–540, 1986.

Selberg, A. "Harmonic Analysis and Discontinuous Groups in Weakly Symmetric Riemannian Spaces with Applications to Dirichlet Series." *J. Indian Math. Soc.* **20**, 47–87, 1956.

Voros, A. "Spectral Functions, Special Functions and the Selberg Zeta Function." *Commun. Math. Phys.* **110**, 439–465, 1987.

Selberg's Formula

Let x be a positive number, and define

$$\lambda(d) = \mu(d)\left[\ln\left(\frac{x}{d}\right)\right]^2 \tag{1}$$

$$f(n) = \sum_d \lambda(d). \tag{2}$$

where the sum extends over the divisors d of n, and $\mu(n)$ is the MÖBIUS FUNCTION. Then

$$S = \sum_{n \leq x} f(n) = 2x \ln x + o(x \ln x) \qquad (3)$$

(Nagell 1951, p. 286).

See also PRIME NUMBER THEOREM

References

Apostol, T. M. *Introduction to Analytic Number Theory.* New York: Springer-Verlag, 1976.
Nagell, T. "Further Lemmata. Proofs of Selberg's Formula." §73 in *Introduction to Number Theory.* New York: Wiley, pp. 279–280 and 283–286, 1951.
Selberg, A. "An Elementary Proof of the Prime Number Theorem." *Ann. Math.* **50**, 305–313, 1949.

Selection Sort

A SORTING algorithm which makes n passes over a set of n elements, in each pass selecting the smallest element and deleting it from the set. This algorithm has running time $\mathcal{O}(n^2)$, compared to $\mathcal{O}(n \ln n)$ for the best algorithms (Skiena 1990, p. 14).

See also SORTING

References

Skiena, S. *Implementing Discrete Mathematics: Combinatorics and Graph Theory with Mathematica.* Reading, MA: Addison-Wesley, 1990.

Self Number

A number (usually base 10 unless specified otherwise) which has no GENERATOR. Such numbers were originally called COLUMBIAN NUMBERS (S. 1974). There are infinitely many such numbers, since an infinite sequence of self numbers can be generated from the RECURRENCE RELATION

$$C_k = 8 \cdot 10^{k-1} + C_{k-1} + 8, \qquad (1)$$

for $k = 2, 3, \ldots$, where $C_1 = 9$. The first few self numbers are 1, 3, 5, 7, 9, 20, 31, 42, 53, 64, 75, 86, 97, ... (Sloane's A003052).

An infinite number of 2-self numbers (i.e., base-2 self numbers) can be generated by the sequence

$$C_k = 2^j + C_{k-1} + 1 \qquad (2)$$

for $k = 1, 2, \ldots$, where $C_1 = 1$ and j is the number of digits in C_{k-1}. An infinite number of n-self numbers can be generated from the sequence

$$C_k = (n-2)n^{k-1} + C_{k-1} + (n-2) \qquad (3)$$

for $k = 2, 3, \ldots$, and

$$C_1 = \begin{cases} n-1 & \text{for } n \text{ even} \\ n-2 & \text{for } n \text{ odd.} \end{cases} \qquad (4)$$

Joshi (1973) proved that if k is ODD, then m is a k-self number IFF m is ODD. Patel (1991) proved that $2k$,

$4k + 2$, and $k^2 + 2k + 1$ are k-self numbers in every EVEN base $k > 4$.

See also DIGITADDITION

References

Cai, T. "On k-Self Numbers and Universal Generated Numbers." *Fib. Quart.* **34**, 144–146, 1996.
Gardner, M. *Time Travel and Other Mathematical Bewilderments.* New York: W. H. Freeman, pp. 115–117, 122, 1988.
Joshi, V. S. Ph.D. dissertation. Gujarat University, Ahmadabad, 1973.
Kaprekar, D. R. *The Mathematics of New Self-Numbers.* Devaiali, pp. 19–20, 1963.
Patel, R. B. "Some Tests for k-Self Numbers." *Math. Student* **56**, 206–210, 1991.
S., B. R. "Solution to Problem E 2048." *Amer. Math. Monthly* **81**, 407, 1974.
Sloane, N. J. A. Sequences A003052/M2404 in "An On-Line Version of the Encyclopedia of Integer Sequences." http://www.research.att.com/~njas/sequences/eisonline.html.

Self-Adjoint

Consider a second-order differential operator

$$\tilde{\mathcal{L}}u(x) \equiv p_0 \frac{d^2u}{dx^2} + p_1 \frac{du}{dx} + p_2 u, \qquad (1)$$

where $u \equiv u(x)$ and $p_i \equiv p_i(x)$ are REAL FUNCTIONS of x on the region of interest $[a, b]$ with $2 - i$ continuous derivatives and with $p_0(x) \neq 0$ on $[a, b]$. This means that there are no singular points in $[a, b]$. Then the ADJOINT operator $\tilde{\mathcal{L}}^*$ is defined by

$$\tilde{\mathcal{L}}^* u \equiv \frac{d^2}{dx^2}(p_0 u) - \frac{d}{dx}(p_1 u) + p_2 u \qquad (2)$$

$$= p_0 \frac{d^2u}{dx^2} + (2p_0' - p_1)\frac{du}{dx} + (p_0'' - p_1' + p_2)u. \qquad (3)$$

In order for the operator to be self-adjoint, i.e.,

$$\tilde{\mathcal{L}} = \tilde{\mathcal{L}}^*, \qquad (4)$$

the second terms in (1) and (3) must be equal, so

$$p_0'(x) = p_1(x). \qquad (5)$$

This also guarantees that the third terms are equal, since

$$p_0'(x) = p_1(x) \Rightarrow p_0''(x) = p_1'(x). \qquad (6)$$

so (3) becomes

$$\tilde{\mathcal{L}}u = \tilde{\mathcal{L}}^*u = p_0 \frac{d^2u}{dx^2} + p_0' \frac{du}{dx} + p_2 u \qquad (7)$$

$$= \frac{d}{dx}\left(p_0 \frac{du}{dx}\right) + p_2 u = 0. \qquad (8)$$

The differential operators corresponding to the LE-

GENDRE DIFFERENTIAL EQUATION and the equation of SIMPLE HARMONIC MOTION are self-adjoint, while those corresponding to the LAGUERRE DIFFERENTIAL EQUATION and HERMITE DIFFERENTIAL EQUATION are not.

A nonself-adjoint second-order linear differential operator can always be transformed into a self-adjoint one using STURM-LIOUVILLE THEORY. In the special case $p_2(x) = 0$, (8) gives

$$\frac{d}{dx}\left[p_0(x)\frac{du}{dx}\right] = 0 \qquad (9)$$

$$p_0(x)\frac{du}{dx} = C \qquad (10)$$

$$du = C\,\frac{dx}{p_0(x)} \qquad (11)$$

$$u = C\int\frac{dx}{p_0(x)}, \qquad (12)$$

where C is a constant of integration.

A self-adjoint operator which satisfies the BOUNDARY CONDITIONS

$$\bar{v}pU'|_{x=a} = \bar{v}pU'|_{x=b} \qquad (13)$$

is automatically a HERMITIAN OPERATOR.

See also ADJOINT, HERMITIAN OPERATOR, STURM-LIOUVILLE THEORY

References

Arfken, G. "Self-Adjoint Differential Equations." §9.1 in *Mathematical Methods for Physicists, 3rd ed.* Orlando, FL: Academic Press, pp. 497–509, 1985.

Self-Adjoint Matrix

A MATRIX A for which

$$A^* \equiv \overline{A^T} = A.$$

where the ADJOINT MATRIX is denoted A^*, A^T is the MATRIX TRANSPOSE, and $\bar{z}$ is the COMPLEX CONJUGATE. If a MATRIX is self-adjoint, it is said to be HERMITIAN.

See also ADJOINT, HERMITIAN MATRIX, MATRIX TRANSPOSE

Self-Avoiding Polygon

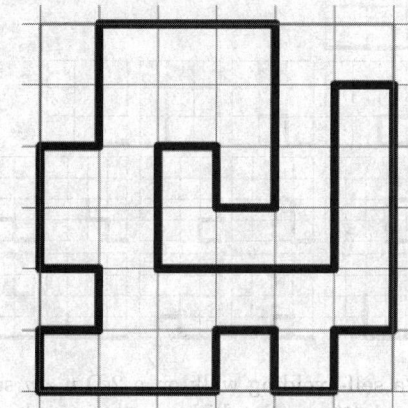

A LATTICE POLYGON consisting of a closed SELF-AVOIDING WALK on a square lattice. The perimeter, horizontal perimeter, vertical perimeter, and AREA are all WELL DEFINED for self-avoiding polygons. Special classes of self-avoiding polygons include the BAR GRAPH POLYGON, CONVEX POLYGON, FERRERS GRAPH POLYGON, STACK POLYGON, and STAIRCASE POLYGON. Self-avoiding polygon are used in physics to model crystal growth and polymers (Bousquet-Mélou 1992).

Enumerating self-avoiding polygons according to perimeter or area is an unsolved problem (Bousquet-Mélou *et al.* 1999).

See also POLYOMINO, SELF-AVOIDING WALK, STAIRCASE POLYGON

References

Bousquet-Mélou, M. "Convex Polyominoes and Heaps of Segments." *J. Phys. A: Math. Gen.* **25**, 1925–1934, 1992.
Bousquet-Mélou, M.; Guttmann, A. J.; Orrick, W. P.; and Rechnitzer, A. Inversion Relations, Reciprocity and Polyominoes. 23 Aug 1999. http://xxx.lanl.gov/abs/math.CO/9908123/.

Self-Avoiding Walk

N.B. A detailed online essay by S. Finch was the starting point for this entry.

A self-avoiding walk is a path from one point to another which never intersects itself. Such paths are usually considered to occur on lattices, so that steps are only allowed in a discrete number of directions and of certain lengths.

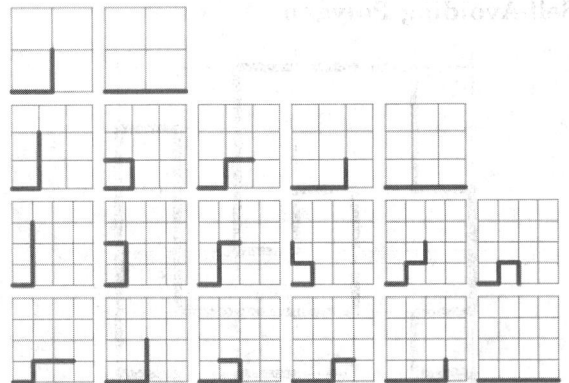

Consider a self-avoiding walk on a 2-D $n \times n$ square grid (i.e., a lattice path which never visits the same lattice point twice) which starts at the origin, takes first step in the positive horizontal direction, and is restricted to nonnegative grid points only. The number of such paths of $n = 1, 2, \ldots$ steps are 1, 2, 5, 12, 30, 73, 183, 456, 1151, ... (Sloane's A046170).

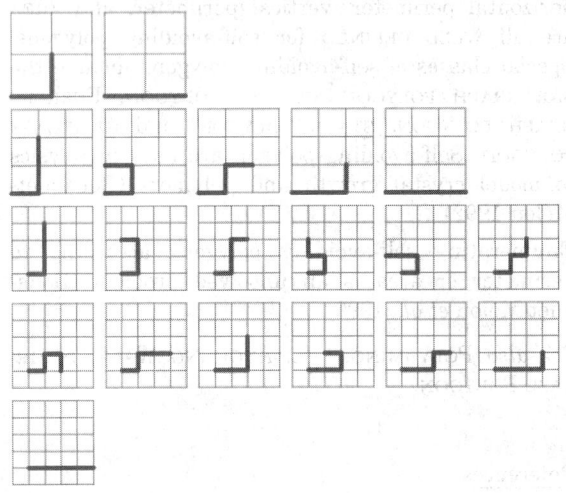

Similarly, consider a self-avoiding walk which starts at the origin, takes first step in the positive horizontal direction, is *not* restricted to nonnegative grid points only, but which *is* restricted to take an up step before taking the first down step. The number of such paths of $n = 1, 2, \ldots$ steps are 1, 2, 5, 13, 36, 98, 272, 740, 2034, ... (Sloane's A046171).

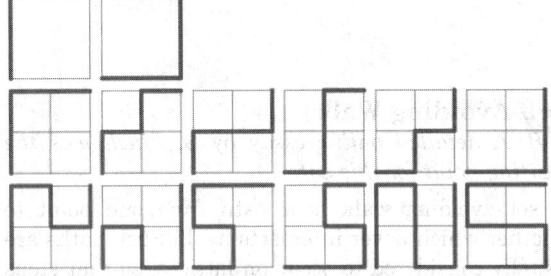

Self-avoiding rook walks are walks on an $m \times n$ grid which start from $(0, 0)$, end at (m, n), and are

composed of only horizontal and vertical steps. The following table gives the first few numbers $R(m, n)$ of such walks for small m and n. The values for $m = n = 1, 2, \ldots$ are 2, 12, 184, 8512, 1262816, ... (Sloane's A007764).

m	2	3	4	5	6
2	2				
3	4	12			
4	8	38	184		
5	16	125	976	8512	
6	32	414	5382	79384	1262816

There are a number of known formulas for computing $R(m, n)$ for small m, n. For example,

$$R(m, 2) = 2^{m-1}.$$

There is a RECURRENCE RELATION for $R(m, 3)$, given by $R(1, 3) = 1$, $R(2, 3) = 4$, $R(3, 3) = 12$, $R(4, 3) = 38$, and

$$R(m, 3) = 4R(m-1, 3) - 3R(m-2, 3) + 2R(m-3, 3) + R(m-3, 4)$$

for $m \geq 5$, as well as the GENERATING FUNCTION

$R(m, 3)$

$$= \frac{1}{(m-1)!} \frac{d^{m-1}}{dx^{m-1}} \frac{(x-1)(x+1)}{(x^2 + 3x - 1)(x^2 - x + 1)} \bigg|_{x=0}$$

(Abbott and Hanson 1978, Finch).

A related sequence is the number of shapes which can be formed by bending a piece of wire of length n in the plane, where bends are of 0 or $\pm 90°$ and the wire may cross itself at right angles but not pass over itself. The number of shapes for wires of length 1, 2, ... are 1, 2, 4, 10, 24, 66, 176, 493, ... (Sloane's A001997).

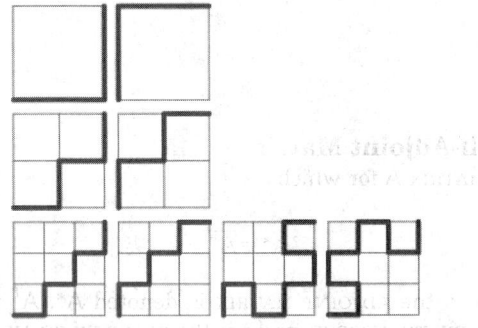

Consider a self-avoiding walk on a 2-D $n \times n$ square grid from one corner to another such that no two consecutive steps are in the same direction. The number of such paths for $n = 1, 2, \ldots$ are 1, 2, 2, 4,

10, 36, 188, ... (Sloane's A034165; counting the number of paths on the 1×1 point "lattice" as 1), and the maximum lengths of these paths are 0, 2, 4, 10, 12, 26, 36, ... (Sloane's A034166).

See also LATTICE PATH, RANDOM WALK, SELF-AVOID-ING POLYGON, SELF-AVOIDING WALK CONNECTIVE CONSTANT, STAIRCASE POLYGON, THREE-CHOICE WALK

References

Abbott, H. L. and Hanson, D. "A Lattice Path Problem." *Ars Combinatoria* **6**, 163–178, 1978.

Alm, S. E. "Upper Bounds for the Connective Constant of Self-Avoiding Walks." *Combin. Prob. Comput.* **2**, 115–136, 1993.

Domb, C. "On Multiple Returns in the Random-Walk Problem." *Proc. Cambridge Philos. Soc.* **50**, 586–591, 1954.

Domb, C. "Self-Avoiding Walks on Lattices." In *Adv. Chem. Phys.* **15**, 1969.

Finch, S. "Unsolved Mathematics Problems: Self-Avoiding Walks of a Rook on a Chessboard." http://www.mathsoft.com/asolve/gammel/gammel.html.

Hayes, B. "How to Avoid Yourself." *Amer. Sci.* **86**, Jul./Aug. 1998.

Kesten, H. "On the Number of Self-Avoiding Walks." *J. Math. Phys.* **4**, 960–969, 1963.

Lawler, G. F. *Intersections of Random Walks.* Boston, MA: Birkhäuser, 1991.

Sloane, N. J. A. Sequences A0019971206, A007764, A034165, A034166, A046170, and A046171 in "An On-Line Version of the Encyclopedia of Integer Sequences." http://www.research.att.com/~njas/sequences/eisonline.html.

Whittington, S. G. and Guttman, A. J. "Self-Avoiding Walks which Cross a Square." *J. Phys. A* **23**, 5601–5609, 1990.

Self-Avoiding Walk Connective Constant

Let the number of RANDOM WALKS on a d-D hypercubic lattice starting at the ORIGIN which never land on the same lattice point twice in n steps be denoted $c_d(n)$. The first few values are

$$c_d(0) = 1 \tag{1}$$

$$c_d(1) = 2d \tag{2}$$

$$c_d(2) = 2d(2d - 1). \tag{3}$$

In general,

$$d^n \leq c_d(n) \leq 2d(2d - 1)^{n-1} \tag{4}$$

(Pönitz and Tittman 2000), with tighter bounds given by Madras and Slade (1993). Conway and Guttmann (1996) have enumerated walks of up to length 51.

The so-called "connective constants" are defined by

$$\mu_d \equiv \lim_{n \to \infty} [c_d(n)]^{1/n} \tag{5}$$

and are known to exist and be FINITE. The best ranges for these constants are

$$\mu_2 \in [2.62002, 2.679192495] \tag{6}$$

$$\mu_3 \in [4.572140, 4.7476] \tag{7}$$

$$\mu_4 \in [6.742945, 6.8179] \tag{8}$$

$$\mu_5 \in [8.828529, 8.88602] \tag{9}$$

$$\mu_6 \in [10.874038, 10.8886] \tag{10}$$

(Beyer and Wells 1972, Noonan 1998, Finch). The upper bound of μ_2 improves on the 2.6939 found by Noonan 1998 and was computed by Pönitz and Tittman (2000).

For the triangular lattice in the plane, $\mu < 4.278$ (Alm 1993), and for the hexagonal planar lattice, it is conjectured that

$$\mu = \sqrt{2 + \sqrt{2}} \tag{11}$$

(Madras and Slade 1993).

The following limits are also believed to exist and to be FINITE:

$$\begin{cases} \lim_{n \to \infty} \dfrac{c(n)}{\mu^n n^{\gamma-1}} & \text{for } d \neq 4 \\[2ex] \lim_{n \to \infty} \dfrac{c(n)}{\mu^n n^{\gamma-1} (\ln n)^{1/4}} & \text{for } d = 4. \end{cases} \tag{12}$$

where the critical exponent $\gamma = 1$ for $d > 4$ (Madras and Slade 1993) and it has been conjectured that

$$\gamma = \begin{cases} \frac{43}{32} & \text{for } d = 2 \\ 1.162\ldots & \text{for } d = 3 \\ 1 & \text{for } d = 4. \end{cases} \tag{13}$$

Define the mean square displacement over all n-step self-avoiding walks ω as

$$s(n) \equiv \langle |\omega(n)|^2 \rangle = \frac{1}{c(n)} \sum_{\omega} |\omega(n)|^2. \tag{14}$$

The following limits are believed to exist and be FINITE:

$$\begin{cases} \lim_{n \to \infty} \dfrac{s(n)}{n^{2\nu}} & \text{for } d \neq 4 \\[2ex] \lim_{n \to \infty} \dfrac{s(n)}{n^{2\nu} (\ln n)^{1/4}} & \text{for } d = 4. \end{cases} \tag{15}$$

where the critical exponent $\nu = 1/2$ for $d > 4$ (Madras and Slade 1993), and it has been conjectured that

$$\nu = \begin{cases} \frac{3}{4} & \text{for } d = 2 \\ 0.59\ldots & \text{for } d = 3 \\ \frac{1}{2} & \text{for } d = 4. \end{cases} \tag{16}$$

See also RANDOM WALK, SELF-AVOIDING WALK

References

Alm, S. E. "Upper Bounds for the Connective Constant of Self-Avoiding Walks." *Combin. Probab. Comput.* **2**, 115–136, 1993.

Beyer, W. A. and Wells, M. B. "Lower Bound for the Connective Constant of a Self-Avoiding Walk on a Square Lattice." *J. Combin. Th. A* **13**, 176–182, 1972.

Conway, A. R. and Guttmann, A. J. "Square Lattice Self-Avoiding Walks and Corrections to Scaling." *Phys. Rev. Lett.* **77**, 5284–5287, 1996.

Finch, S. "Favorite Mathematical Constants." http://www.mathsoft.com/asolve/constant/cnntv/cnntv.html.

Madras, N. and Slade, G. *The Self-Avoiding Walk.* Boston, MA: Birkhäuser, 1993.

Noonan, J. "New Upper Bounds for the Connective Constants of Self-Avoiding Walks." *J. Stat. Phys.* **91**, 871–888, 1998.

Pönitz, A. and Tittman, P. "Improved Upper Bounds for Self-Avoiding Walks in $\mathbb{Z}^d$." *Electronic J. Combinatorics* **7**, No. 1, R21, 1–19, 2000. http://www.combinatorics.org/Volume_7/v7i1toc.html.

Self-Complementary Graph

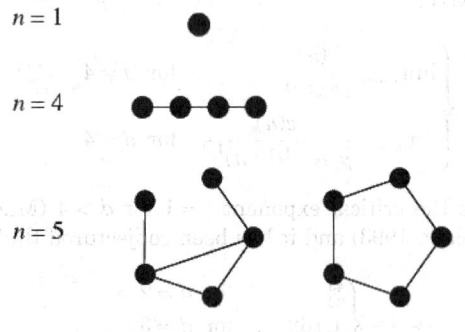

A self-complementary graph is a GRAPH which is isomorphic to its GRAPH COMPLEMENT. The numbers of simple self-complementary graphs on $n = 1, 2, \ldots$ nodes are 1, 0, 0, 1, 2, 0, 0, 10, ... (Sloane's A000171). The first few of these compose to the trivial graph on one node, the PATH GRAPH P_4, and the CYCLE GRAPH C_5.

All self-complementary graphs have GRAPH DIAMETER 2 or 3 (Sachs 1962; Skiena 1990, p. 187).

See also GRAPH COMPLEMENT, ISOMORPHIC GRAPHS

References

Read, R. C. "On the Number of Self-Complementary Graphs and Digraphs." *J. London Math. Soc.* **38**, 99–104, 1963.

Read, R. C. and Wilson, R. J. *An Atlas of Graphs.* Oxford, England: Oxford University Press, 1998.

Sachs, H. "Über selbstkomplementäre Graphen." *Publ. Math. Debrecen* **9**, 270–288, 1962.

Skiena, S. "Self-Complementary Graphs." §5.2.3 in *Implementing Discrete Mathematics: Combinatorics and Graph Theory with Mathematica.* Reading, MA: Addison-Wesley, p. 187, 1990.

Sloane, N. J. A. Sequences A000171/M0014 in "An On-Line Version of the Encyclopedia of Integer Sequences." http://www.research.att.com/~njas/sequences/eisonline.html.

Wille, D. "Enumeration of Self-Complementary Structures." *J. Combin. Th. B* **25**, 143–150, 1978.

Self-Conjugate Partition

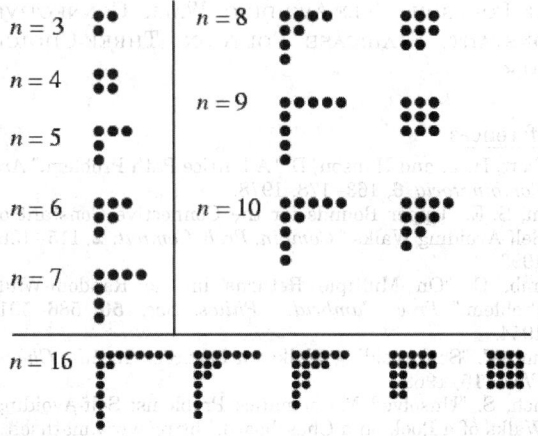

A PARTITION whose CONJUGATE PARTITION is equivalent to itself. The FERRERS DIAGRAMS corresponding to the self-conjugate partitions for $3 \leq n \leq 10$ are illustrated above. The numbers of self-conjugate partitions of $n = 1, 2, \ldots$ are 1, 0, 1, 1, 1, 1, 1, 2, 2, 2, 2, 3, 3, 3, 4, 5, 5, 5, 6, 7, ... (Sloane's A000700). The number of self-conjugate partitions S_n of n is equal to the number of partitions of n into distinct odd parts, and has generating function

$$\prod_{k=0}^{\infty} 1 + x^{2k+1} = \sum_{k=0}^{\infty} S_k x^k,$$

and $(-1)^n S_n$ has GENERATING FUNCTION

$$\prod_{k=1}^{\infty} \frac{1}{1 + x^k} = \sum_{k=0}^{\infty} (-1)^k S_k x^k.$$

See also CONJUGATE PARTITION, FERRERS DIAGRAM, PARTITION FUNCTION P

References

Hardy, G. H. and Wright, E. M. *An Introduction to the Theory of Numbers, 5th ed.* Oxford, England: Clarendon Press, p. 277, 1979.

Osima, M. "On the Irreducible Representations of the Symmetric Group." *Canad. J. Math.* **4**, 381–384, 1952.

Watson, G. N. "Two Tables of Partitions." *Proc. London Math. Soc.* **42**, 550–556, 1936.

Self-Conjugate Permutation
INVOLUTION (PERMUTATION)

Self-Conjugate Subgroup
INVARIANT SUBGROUP

Self-Descriptive Number

A 10-DIGIT number satisfying the following property. Number the DIGITS 0 to 9, and let DIGIT n be the number of ns in the number. There is exactly one such number: 6210001000.

References

Pickover, C. A. "Chaos in Ontario." Ch. 28 in *Keys to Infinity*. New York: Wiley, pp. 217–219, 1995.

Self-Dual

A geometric proposition is said to be self-dual when application of the DUALITY PRINCIPLE of PROJECTIVE GEOMETRY results in a proposition equivalent to the original. DESARGUES' THEOREM is an example of a self-dual proposition.

See also SELF-DUAL GRAPH, SELF-DUAL POLYHEDRON

Self-Dual Graph

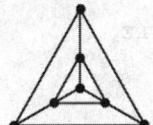 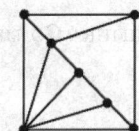 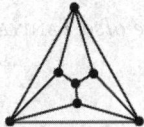

A GRAPH that is DUAL to itself. WHEEL GRAPHS are self-dual, as are the examples illustrated above. Naturally, the SKELETON of a SELF-DUAL POLYHEDRON is a self-dual graph.

See also DUAL GRAPH

References

Bondy, J. A. and Murty, U. S. R. *Graph Theory with Applications*. New York: North Holland, p. 243, 1976.
Smith, C. A. B. and Tutte, W. T. "A Class of Self-Dual Maps." *Canad. J. Math.* **2**, 179–196, 1950.

Self-Dual Polyhedron

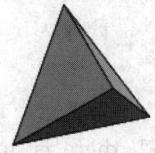 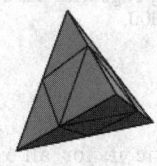 

A POLYHEDRON that is DUAL to itself. For example, the TETRAHEDRON is self-dual. Naturally, the SKELETON of a self-dual polyhedron is a SELF-DUAL GRAPH.

See also DUAL POLYHEDRON, SELF-DUAL GRAPH.

Self-Homologous Point

SIMILITUDE CENTER

Self-Linking Number

CALUGAREANU THEOREM, GAUSS INTEGRAL, LINKING NUMBER

Self-Loop

LOOP (GRAPH)

Self-Map

A mapping of a DOMAIN $F : U \to U$ to itself.

See also MÖBIUS TRANSFORMATION

References

Krantz, S. G. *Handbook of Complex Analysis*. Boston, MA: Birkhäuser, p. 78, 1999.

Self-Reciprocating Property

Let h be the number of sides of certain SKEW POLYGONS (Coxeter 1973, p. 15). Then

$$h = \frac{2(p + q + 2)}{10 - p - q}.$$

References

Coxeter, H. S. M. *Regular Polytopes, 3rd ed.* New York: Dover, 1973.

Self-Recursion

SELF-RECURSION is a RECURSION which is defined in terms of itself, resulting in an ill-defined infinite regress.

See also RECURSION, REGRESSION, SELF-RECURSION

References

Carroll, L. "'What the Tortoise Said to Achilles." *Mind* **4**, 278–280, 1895.
Gardner, M. "Infinite Regress." Ch. 22 in *The Sixth Book of Mathematical Games from Scientific American*. Chicago, IL: University of Chicago Press, pp. 220–229, 1984.

Selfridge-Hurwitz Residue

Let the RESIDUE from PÉPIN'S THEOREM be

$$R_n \equiv 3^{(F_n - 1)/2} \pmod{F_n},$$

where F_n is a FERMAT NUMBER. Selfridge and Hurwitz use

$$R_n \pmod{2^{35} - 1, \ 2^{36}, \ 2^{36} - 1}.$$

A nonvanishing $R_n \pmod{2^{36}}$ indicates that F_n is COMPOSITE for $n > 5$.

See also FERMAT NUMBER, PÉPIN'S THEOREM

References

Crandall, R.; Doenias, J.; Norrie, C.; and Young, J. "The Twenty-Second Fermat Number is Composite." *Math. Comput.* **64**, 863–868, 1995.

Selfridge's Conjecture

There exist infinitely many $n > 0$ with $p_n^2 > p_{n-i} p_{n+i}$ for all $i < n$, where p_n is the nth PRIME. Also, there exist infinitely many $n > 0$ such that $2p_n < p_{n-i} + p_{n+i}$ for all $i < n$.

Self-Similarity

An object is said to be self-similar if it looks "roughly" the same on any scale. FRACTALS are a particularly interesting class of self-similar objects. Self-similar objects with parameters N and s are described by a power law such as

$$N = s^d,$$

where

$$d = \frac{\ln N}{\ln s}$$

is the "DIMENSION" of the scaling law, known as the HAUSDORFF DIMENSION.

See also FRACTAL, HAUSDORFF DIMENSION

References

Harris, J. W. and Stocker, H. "Scaling Invariance and Self-Similarity" and "Construction of Self-Similar Objects." §4.11.1–4.11.2 in *Handbook of Mathematics and Computational Science.* New York: Springer-Verlag, p. 113, 1998.
Hutchinson, J. "Fractals and Self-Similarity." *Indiana Univ. J. Math.* **30**, 713–747, 1981.

Self-Transversality Theorem

Let j, r, and s be distinct INTEGERS (mod n), and let W be the point of intersection of the side or diagonal V, V_{i+j} of the n-gon $P = [V_1 \ldots V_n]$ with the transversal $V_{i+r} V_{i+s}$. Then a NECESSARY and SUFFICIENT condition for

$$\prod_{i=1}^{n} \left[\frac{V_i W_i}{W_i V_{i+j}} \right] = (-1)^n,$$

where $AB \| CD$ and

$$\left[\frac{AB}{CD} \right],$$

is the ratio of the lengths $[A, B]$ and $[C, D]$ with a plus or minus sign depending on whether these segments have the same or opposite direction, is that

1. $n = 2m$ is EVEN with $j \equiv m \pmod{n}$ and $s \equiv r + m \pmod{n}$,
2. n is arbitrary and either $s \equiv 2r$ and $j \equiv 3r$, or
3. $r \equiv 2s \pmod{n}$ and $j \equiv 3s \pmod{n}$.

References

Grünbaum, B. and Shepard, G. C. "Ceva, Menelaus, and the Area Principle." *Math. Mag.* **68**, 254–268, 1995.

Sellke's Self-Describing Sequence

KOLAKOSKI SEQUENCE

Selmer Group

A GROUP which is related to the TANIYAMA-SHIMURA CONJECTURE.

See also TANIYAMA-SHIMURA CONJECTURE

Semialgebraic Set

A subset of $\mathbb{R}^n$ which is a finite Boolean combination of sets OF THE FORM $\{\bar{x} = (x_1, \ldots, x_n) : f(\bar{x}) > 0\}$ and $\{\bar{x} : g(\bar{x}) = 0\}$, where f, $g \in \mathbb{R}[X_1, \ldots, X_n]$.

By TARSKI'S THEOREM, the solution set of a QUANTIFIED SYSTEM of real algebraic equations and inequalities is a semialgebraic set (Strzebonski 2000).

See also TARSKI'S THEOREM

References

Bierstone, E. and Milman, P. "Semialgebraic and Subanalytic Sets." *IHES Pub. Math.* **67**, 5–42, 1988.
Marker, D. "Model Theory and Exponentiation." *Not. Amer. Math. Soc.* **43**, 753–759, 1996.
Strzebonski, A. "Solving Algebraic Inequalities." *Mathematica J.* **7**, 525–541, 2000.

Semianalytic

$X \subseteq \mathbb{R}^n$ is semianalytic if, for all $x \in \mathbb{R}^n$, there is an open neighborhood U of x such that $X \cap U$ is a finite Boolean combination of sets $\{\bar{x} \in U : f(\bar{x}) = 0\}$ and $\{\bar{x} \in U : g(\bar{x}) > 0\}$, where f, $g : U \to \mathbb{R}$ are ANALYTIC.

See also ANALYTIC FUNCTION, PSEUDOANALYTIC FUNCTION, SUBANALYTIC

References

Marker, D. "Model Theory and Exponentiation." *Not. Amer. Math. Soc.* **43**, 753–759, 1996.

Semicircle

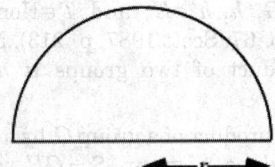

Half a CIRCLE. The AREA of a semicircle of radius r is given by

$$A = \int_0^r \int_{-\sqrt{r^2-x^2}}^{\sqrt{r^2-x^2}} dx\, dy = 2 \int_0^r \sqrt{r^2-x^2}\, dx = \tfrac{1}{2}\pi r^2. \quad (1)$$

The weighted mean of y is

$$\langle x \rangle_2 = 2 \int_0^r x\,\sqrt{r^2-x^2}\, dx = \tfrac{2}{3}r^3. \quad (2)$$

The semicircle is the CROSS SECTION of a HEMISPHERE for any PLANE through the z-AXIS.
The perimeter of the curved boundary is given by

$$s = \int_{-r}^r \sqrt{1+x'^2}\, dy. \quad (3)$$

With $x = \sqrt{r^2-y^2}$, this gives

$$s = \pi r. \quad (4)$$

The PERIMETER of the semicircular lamina is then

$$L = 2r + \pi r = r(2+\pi). \quad (5)$$

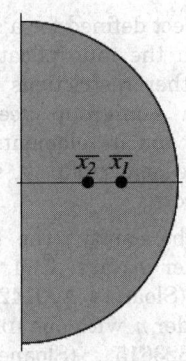

The weighted value of x of the semicircular curve is given by

$$\langle x \rangle_1 = \int_{-r}^r x \sqrt{1+x'^2}\, dy = \int_{-r}^r r\, dy = 2r^2, \quad (6)$$

so the CENTROID is

$$\bar{x}_1 = \frac{\langle x \rangle_1}{A} = \frac{2r}{\pi}. \quad (7)$$

The CENTROID of the semicircular lamina is given by

$$\bar{x}_2 = \frac{\langle x \rangle_2}{A} = \frac{4r}{3\pi} \quad (8)$$

(Kern and Bland 1948, p. 113).

See also ARBELOS, ARC, CIRCLE, DISK, HEMISPHERE, LENS, RIGHT ANGLE, SALINON, THALES' THEOREM, YIN-YANG

References
Kern, W. F. and Bland, J. R. *Solid Mensuration with Proofs,* 2nd ed. New York: Wiley, 1948.

Semicolon

The symbol ; given special meanings in several mathematics contexts, the most common of which is the COVARIANT DERIVATIVE.

See also COVARIANT DERIVATIVE

References
Bringhurst, R. *The Elements of Typographic Style, 2nd ed.* Point Roberts, WA: Hartley and Marks, p. 284, 1997.

Semicolon Derivative
COVARIANT DERIVATIVE

Semiconvergent Series
ASYMPTOTIC SERIES

Semicubical Parabola

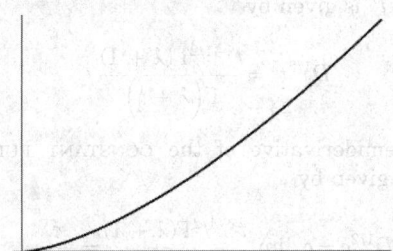

A PARABOLA-like curve with Cartesian equation

$$y = ax^{3/2}, \quad (1)$$

PARAMETRIC EQUATIONS

$$x = t^2 \quad (2)$$

$$y = at^3 \quad (3)$$

and POLAR COORDINATES,

$$r = \frac{\tan^2\theta \sec\theta}{a}. \quad (4)$$

The semicubical parabola is the curve along which a particle descending under gravity describes equal vertical spacings within equal times, making it an ISOCHRONOUS CURVE. The problem of finding the curve having this property was posed by Leibniz in 1687 and solved by Huygens (MacTutor Archive). The ARC LENGTH, CURVATURE, and TANGENTIAL ANGLE are

$$s(t) = \frac{1}{27}(4 + 9t^2)^{3/2} - \frac{8}{27} \qquad (5)$$

$$\kappa(t) = \frac{6}{t(4 + 9t^2)^{3/2}} \qquad (6)$$

$$\phi(t) = \tan^{-1}\left(\frac{3}{2}t\right). \qquad (7)$$

See also NEILE'S PARABOLA, PARABOLA INVOLUTE

References

Beyer, W. H. *CRC Standard Mathematical Tables, 28th ed.* Boca Raton, FL: CRC Press, pp. 223–224, 1987.
Gray, A. "The Semicubical Parabola." §1.8 in *Modern Differential Geometry of Curves and Surfaces with Mathematica, 2nd ed.* Boca Raton, FL: CRC Press, pp. 21–22, 1997.
Lawrence, J. D. *A Catalog of Special Plane Curves.* New York: Dover, pp. 85–87, 1972.
MacTutor History of Mathematics Archive. "Neile's Parabola." http://www-groups.dcs.st-and.ac.uk/~history/Curves/Neiles.html.
Yates, R. C. "Semi-Cubic Parabola." *A Handbook on Curves and Their Properties.* Ann Arbor, MI: J. W. Edwards, pp. 186–187, 1952.

Semiderivative

A FRACTIONAL DERIVATIVE of order 1/2. The semiderivative of t^λ is given by

$$D^{1/2}t^\lambda = \frac{t^{\lambda-1/2}\Gamma(\lambda + 1)}{\Gamma\left(\lambda + \frac{1}{2}\right)},$$

so the semiderivative of the CONSTANT FUNCTION $f(t) = c$ is given by

$$D^{1/2}c = c \lim_{\lambda \to 0} \frac{t^{\lambda-1/2}\Gamma(\lambda + 1)}{\Gamma\left(\lambda + \frac{1}{2}\right)} = \frac{c}{\sqrt{\pi t}}.$$

See also DERIVATIVE, FRACTIONAL DERIVATIVE, SEMI-INTEGRAL

References

Spanier, J. and Oldham, K. B. *An Atlas of Functions.* Washington, DC: Hemisphere, pp. 8 and 14, 1987.

Semidirect Product

A "split" extension G of GROUPS N and F which contains a SUBGROUP $\bar{F}$ isomorphic to F with $G = \bar{F}\bar{N}$ and $\bar{F} \cap \bar{N} = \{e\}$ (Ito 1987, p. 710). Then the semidirect product of a GROUP G by a group H, denoted $H \times G$ (or sometimes $H : G$) with homomorphism T is given by

$$(g, h)(g', h') = (gg', (h(g'T))h'),$$

where g, $g' \in G$, h, $h' \in H$, and $T \in \text{Hom}(F, \text{Aut}(H))$ (Suzuki 1982, p. 67; Scott 1987, p. 213). Note that the semidirect product of two groups is *not* uniquely defined.

The semidirect product of a group G by a group H can also be defined as a group $S = GH$ which is the product of its subgroups G and H, where H is normal in S and $G \cap H = \{1\}$. If G is also normal in S, then the semidirect product becomes a GROUP DIRECT PRODUCT (Shmel'kin 1988, p. 247).

See also ACTION, GROUP DIRECT PRODUCT, SUBGROUP

References

Itô, K. (Ed.). 'Extensions." §190.N in *Encyclopedic Dictionary of Mathematics, 2nd ed., Vol. 2.* Cambridge, MA: MIT Press, p. 710, 1987.
Kurosh, A. G. *The Theory of Groups, 2nd ed., 2 vols.* New York: Chelsea, 1960.
Scott, W. R. "Semi-Direct Products." §9.2 in *Group Theory.* New York: Dover, pp. 212–217, 1987.
Shmel'kin, A. L. "Semi-Direct Product." In Vol. 8 of *Encyclopaedia of Mathematics: An Updated and Annotated Translation of the Soviet "Mathematical Encyclopaedia"* (Managing Ed. M. Hazewinkel). Dordrecht, Netherlands: Reidel, p. 247, 1988.
Suzuki, M. *Group Theory, Vol. 1.* New York: Springer-Verlag, 1982.

Semiflow

An ACTION with $G = \mathbb{R}^+$.

See also FLOW

Semigroup

A mathematical object defined for a set and a BINARY OPERATOR in which the multiplication operation is ASSOCIATIVE. No other restrictions are placed on a semigroup; thus a semigroup need not have an IDENTITY ELEMENT and its elements need not have inverses within the semigroup. A semigroup is an ASSOCIATIVE GROUPOID.

A semigroup can be empty. The total number of semigroups of order n are 1, 4, 18, 126, 1160, 15973, 836021, ... (Sloane's A001423). The number of semigroups of order n with one IDEMPOTENT are 1, 2, 5, 19, 132, 3107, 623615, ... (Sloane's A002786), and with two IDEMPOTENTS are 2, 7, 37, 216, 1780, 32652, ... (Sloane's A002787). The number $a(n)$ of semigroups having n IDEMPOTENTS are 1, 2, 6, 26, 135, 875, ... (Sloane's A002788).

See also ASSOCIATIVE, BINARY OPERATOR, FREE SEMIGROUP, GROUPOID, INVERSE SEMIGROUP, MONOID, QUASIGROUP

References

Birget, J.-C.; Margolis, S.; Meakin, J. and Sapir, M. (Eds.). *Algorithmic Problems in Groups and Semigroups.* Boston, MA: Birkhäuser, 2000.

Clifford, A. H. and Preston, G. B. *The Algebraic Theory of Semigroups.* Providence, RI: Amer. Math. Soc., 1961.

Howie, J. H. *Fundamentals of Semigroup Theory.* Oxford, England: Oxford University Press, 1996.

Sloane, N. J. A. Sequences A001423/M3550, A002786/M1522, A002787/M1802, and A002788/M1679 in "An On-Line Version of the Encyclopedia of Integer Sequences." http://www.research.att.com/~njas/sequences/eisonline.html.

Semi-Integral

A FRACTIONAL INTEGRAL of order 1/2. The semi-integral of t^λ is given by

$$D^{-1/2}t^\lambda = \frac{t^{\lambda+1/2}\Gamma(\lambda+1)}{\Gamma\left(\lambda+\frac{3}{2}\right)},$$

so the semi-integral of the CONSTANT FUNCTION $f(t) = c$ is given by

$$D^{-1/2}c = c \lim_{\lambda \to 0} \frac{t^{\lambda+1/2}\Gamma(\lambda+1)}{\Gamma\left(\lambda+\frac{3}{2}\right)} = 2c\sqrt{\frac{t}{\pi}}.$$

See also FRACTIONAL INTEGRAL, INTEGRAL

References
Spanier, J. and Oldham, K. B. *An Atlas of Functions.* Washington, DC: Hemisphere, pp. 8 and 14, 1987.

Semilatus Rectum

In general, the CHORD through a FOCUS parallel to the DIRECTRIX of a CONIC SECTION is called the LATUS RECTUM. Half this length is called the semilatus rectum (Coxeter 1969).

Given an ELLIPSE, the semilatus rectum is the distance L measured from a FOCUS such that

$$\frac{1}{L} \equiv \frac{1}{2}\left(\frac{1}{r_+} + \frac{1}{r_-}\right), \tag{1}$$

where $r_+ = a(1+e)$ and $r_- = a(1-e)$ are the APOAPSIS and PERIAPSIS, and e is the ELLIPSE's ECCENTRICITY. Plugging in for r_+ and r_- then gives

$$\frac{1}{L} = \frac{1}{2a}\left(\frac{1}{1-e} + \frac{1}{1+e}\right) = \frac{1}{2a}\frac{(1+e)+(1-e)}{1-e^2}$$

$$= \frac{1}{a}\frac{1}{1-e^2}, \tag{2}$$

so

$$L = a(1-e^2). \tag{3}$$

See also CONIC SECTION, DIRECTRIX (CONIC SECTION), ECCENTRICITY, ELLIPSE, FOCUS, LATUS RECTUM, SEMIMAJOR AXIS, SEMIMINOR AXIS

References
Coxeter, H. S. M. *Introduction to Geometry, 2nd ed.* New York: Wiley, pp. 116–118, 1969.

Semimagic Square

A square that fails to be a MAGIC SQUARE only because one or both of the main diagonal sums do not equal the MAGIC CONSTANT (Kraitchik 1942, p. 143).

See also MAGIC SQUARE

References
Kraitchik, M. *Mathematical Recreations.* New York: W. W. Norton, 1942.

Semimajor Axis

HALF the distance across an ELLIPSE along the longest of its three principal axes.

See also ELLIPSE, SEMIMINOR AXIS

Semiminor Axis

Half the distance across an ELLIPSE along its short principal axis.

See also ELLIPSE, SEMIMAJOR AXIS

Seminorm

A seminorm is a function on a VECTOR SPACE V, denoted $\|v\|$, such that the following conditions hold for all v and w in V, and any scalar c.

1. $\|v\| \geq 0.$,
2. $\|cv\| = |c|\,\|v\|$, and
3. $\|v+w\| \leq \|v\| + \|w\|$.

Note that it is possible for $\|v\| = 0$ for nonzero v. For example, the FUNCTIONAL $\|f\| = |f(0)|$ for continuous functions is a seminorm which is not a norm. A seminorm is a norm if $\|v\| = 0$ is equivalent to $v = 0$.

See also FRÉCHET SPACE, NORM, TOPOLOGICAL VECTOR SPACE

Semiperfect Magic Cube

A semiperfect magic cube, also called an "Andrews cube," is a MAGIC CUBE for which the CROSS SECTION diagonals do not sum to the MAGIC CONSTANT.

See also MAGIC CUBE, PERFECT MAGIC CUBE

References
Gardner, M. "Magic Squares and Cubes." Ch. 17 in *Time Travel and Other Mathematical Bewilderments.* New York: W. H. Freeman, pp. 213–225, 1988.

Semiperfect Number

A number such as $20 = 1 + 4 + 5 + 10$ which is the SUM of some (or all) of its PROPER DIVISORS is called a semiperfect number, or sometimes a pseudoperfect number (Butske *et al.* 1999). A semiperfect number which is the SUM of *all* its PROPER DIVISORS is called a PERFECT NUMBER. The first few semiperfect numbers are 6, 12, 18, 20, 24, 28, 30, 36, 40, ... (Sloane's A005835). Every multiple of a semiperfect number is semiperfect, as are all numbers $2^m p$ for $m > 1$ and p a PRIME between 2^m and 2^{m+1} (Guy 1994, p. 47).

A semiperfect number cannot be DEFICIENT. Rare ABUNDANT NUMBERS which are not semiperfect are called WEIRD NUMBERS. Semiperfect numbers are sometimes also called pseudoperfect numbers.

See also ABUNDANT NUMBER, DEFICIENT NUMBER, PERFECT NUMBER, PRIMARY PSEUDOPERFECT NUMBER, PRIMITIVE SEMIPERFECT NUMBER, WEIRD NUMBER

References

Butske, W.; Jaje, L. M.; and Mayernik, D. R. "The Equation $\Sigma_{p/N} 1/p + 1/N = 1$, Pseudoperfect Numbers, and Partially Weighted Graphs." *Math. Comput.* **69**, 407–420, 1999.
Guy, R. K. "Almost Perfect, Quasi-Perfect, Pseudoperfect, Harmonic, Weird, Multiperfect and Hyperperfect Numbers." §B2 in *Unsolved Problems in Number Theory, 2nd ed.* New York: Springer-Verlag, pp. 45–53, 1994.
Sloane, N. J. A. Sequences A005835/M4094 in "An On-Line Version of the Encyclopedia of Integer Sequences." http://www.research.att.com/~njas/sequences/eisonline.html.
Zachariou, A. and Zachariou, E. "Perfect, Semi-Perfect and Ore Numbers." *Bull. Soc. Math. Grèce (New Ser.)* **13**, 12–22, 1972.

Semiperimeter

The semiperimeter on a figure is defined as

$$s \equiv \tfrac{1}{2} p, \qquad (1)$$

where p is the PERIMETER. The semiperimeter of POLYGONS appears in unexpected ways in the computation of their AREAS. The most notable cases are in the ALTITUDE, EXRADIUS, and INRADIUS of a TRIANGLE, the SODDY CIRCLES, HERON'S FORMULA for the AREA of a TRIANGLE in terms of the legs a, b, and c

$$A_\Delta = \sqrt{s(s-a)(s-b)(s-c)}, \qquad (2)$$

and BRAHMAGUPTA'S FORMULA for the AREA of a QUADRILATERAL

$$A_{\text{quadrilateral}}$$

$$= \sqrt{(s-a)(s-b)(s-c)(s-d) - abcd \cos^2\left(\frac{A+B}{2}\right)}. \qquad (3)$$

The semiperimeter also appears in the beautiful L'HUILIER'S THEOREM about SPHERICAL TRIANGLES.

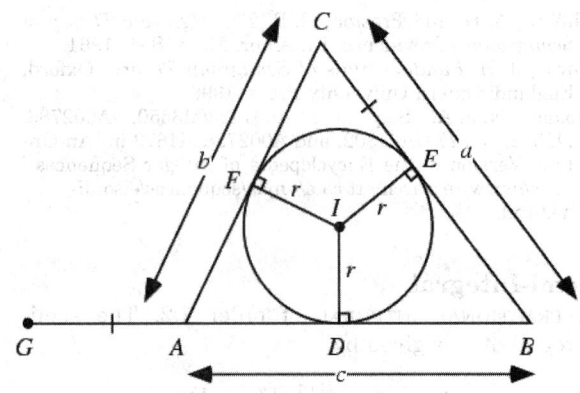

For a TRIANGLE, the following identities hold,

$$s - a = \tfrac{1}{2}(-a + b + c) \qquad (4)$$

$$s - b = \tfrac{1}{2}(+a - b + c) \qquad (5)$$

$$s - c = \tfrac{1}{2}(+a + b - c). \qquad (6)$$

Now consider the above figure. Let I be the INCENTER of the TRIANGLE ΔABC, with D, E, and F the tangent points of the INCIRCLE. Extend the line BA with $GA = CE$. Note that the pairs of triangles (ADI, AFI), (BDI, BEI), (CFI, CEI) are congruent. Then

$$BG = BD + AD + AG = BD + AD + CE$$

$$= \tfrac{1}{2}(2BD + 2AD + 2CE)$$

$$= \tfrac{1}{2}[(BD + BE) + (AD + AF) + (CE + CF)]$$

$$= \tfrac{1}{2}[(BD + AD) + (BE + CE) + (AF + CF)]$$

$$= \tfrac{1}{2}(AB + BC + AC) = \tfrac{1}{2}(a + b + c) = s. \qquad (7)$$

Furthermore,

$$s - a = BG - BC$$

$$= (BD + AD + AG) - (BE + CE)$$

$$= (BD + AD + CE) - (BD + CE) = AD \qquad (8)$$

$$s - b = BG - AC$$

$$= (BD + AD + AG) - (AF + CF)$$

$$= (BD + AD + CE) - (AD + CE) = BD \qquad (9)$$

$$s - c = BG - AB = AG \qquad (10)$$

(Dunham 1990). These equations are some of the building blocks of Heron's derivation of HERON'S FORMULA.

See also PERIMETER

Semiprime

References

Dunham, W. "Heron's Formula for Triangular Area." Ch. 5 in *Journey through Genius: The Great Theorems of Mathematics*. New York: Wiley, pp. 113–132, 1990.

Semiprime

A COMPOSITE number which is the PRODUCT of two PRIMES (possibly equal). They correspond to the 2-ALMOST PRIMES. The first few are 4, 6, 9, 10, 14, 15, 21, 22, ... (Sloane's A001358).

See also ALMOST PRIME, CHEN'S THEOREM, COMPOSITE NUMBER, LANDAU'S PROBLEMS, PRIME NUMBER

References

Sloane, N. J. A. Sequences A001358/M3274 in "An On-Line Version of the Encyclopedia of Integer Sequences." http://www.research.att.com/~njas/sequences/eisonline.html.

Semiprime Ring

Given an IDEAL A, a semiprime ring is one for which $A^n = 0$ IMPLIES $A = 0$ for any POSITIVE n. Every PRIME RING is semiprime.

See also PRIME RING

Semiregular Polyhedron

A POLYHEDRON or plane TESSELLATION is called semiregular if its faces are all REGULAR POLYGONS and its corners are alike (Walsh 1972; Coxeter 1973, pp. 4 and 58; Holden 1991, p. 41). The usual name for a semiregular polyhedron is an ARCHIMEDEAN SOLID, of which there are exactly 13.

See also ARCHIMEDEAN SOLID, POLYHEDRON, TESSELLATION

References

Coxeter, H. S. M. "Regular and Semi-Regular Polytopes I." *Math. Z.* **46**, 380–407, 1940.
Coxeter, H. S. M. *Regular Polytopes*, 3rd ed. New York: Dover, 1973.
Holden, A. *Shapes, Space, and Symmetry*. New York: Dover, 1991.
Walsh, T. R. S. "Characterizing the Vertex Neighbourhoods of Semi-Regular Polyhedra." *Geometriae Dedicata* **1**, 117–123, 1972.

Semiregular Tessellation

TESSELLATION

Semiring

A semiring is a set together with two BINARY OPERATORS $S(+, *)$ satisfying the following conditions:

1. Additive associativity: For all $a, b, c \in S$, $(a+b)+c = a+(b+c)$,
2. Additive commutativity: For all $a, b \in S$, $a+b = b+a$,
3. Multiplicative associativity: For all $a, b, c \in S$, $(a*b)*c = a*(b*c)$,
4. Left and right distributivity: For all $a, b, c \in S$, $a*(b+c) = (a*b)+(a*c)$ and $(b+c)*a = (b*a) + (c*a)$.

A semiring is therefore a commutative SEMIGROUP under addition and a SEMIGROUP under multiplication. A semiring can be empty.

See also BINARY OPERATOR, RING, RINGOID, SEMIGROUP

References

Rosenfeld, A. *An Introduction to Algebraic Structures*. New York: Holden-Day, 1968.

Semisecant

TRANSVERSAL LINE

Semisimple Algebra

An ALGEBRA with no nontrivial nilpotent IDEALS. In the 1890s, Cartan, Frobenius, and Molien independently proved that any finite-dimensional semisimple algebra over the REAL or COMPLEX numbers is a finite and unique DIRECT SUM of SIMPLE ALGEBRAS. This result was then extended to algebras over arbitrary fields by Wedderburn in 1907 (Kleiner 1996).

See also IDEAL, NILPOTENT ELEMENT, SIMPLE ALGEBRA

References

Kleiner, I. "The Genesis of the Abstract Ring Concept." *Amer. Math. Monthly* **103**, 417–424, 1996.

Semisimple Element

A P-ELEMENT x of a GROUP G is semisimple if $E(C_G(x)) \neq 1$, where $E(H)$ is the commuting product of all components of H and $C_G(x)$ is the CENTRALIZER of G.

See also CENTRALIZER, P-ELEMENT

Semisimple Lie Group

A LIE GROUP which has a simply connected covering group HOMEOMORPHIC to $\mathbb{R}^n$. The prototype is any connected closed subgroup of upper TRIANGULAR COMPLEX MATRICES. The HEISENBERG GROUP is such a group.

See also HEISENBERG GROUP, LIE GROUP

References

Knapp, A. W. "Group Representations and Harmonic Analysis, Part II." *Not. Amer. Math. Soc.* **43**, 537–549, 1996.

Semisimple Ring

A SEMIPRIME RING which is also an ARTINIAN RING.

See also ARTINIAN RING

References
Herstein, I. N. "Semisimple Rings." §1.2 in *Noncommutative Rings*. Washington, DC: Math. Assoc. Amer., pp. 52–56, 1968.

Semistable

When a PRIME l divides the DISCRIMINANT of a ELLIPTIC CURVE E, two or all three roots of E become congruent (mod l). An ELLIPTIC CURVE is semistable if, for all such PRIMES l, only two roots become CONGRUENT mod l (with more complicated definitions for $p = 2$ or 3).

See also DISCRIMINANT (ELLIPTIC CURVE), ELLIPTIC CURVE

Sensitivity

The probability that a STATISTICAL TEST will be positive for a true statistic.

See also SPECIFICITY, STATISTICAL TEST, TYPE I ERROR, TYPE II ERROR

Sentence

This entry contributed by MATTHEW SZUDZIK

A sentence is a logic formula in which every variable is QUANTIFIED. The concept of a sentence is important because formulas with variables that are not quantified are ambiguous.

The concept of the sentence can be illustrated as follows (Enderton 1977). The formula $\exists(x, \forall(y, \, y \in x))$, in which each variable is quantified, can be translated into English as the complete sentence "There exists a set which has every set as an element." However, the formula $\forall(y, (y \in x))$, in which x is not quantified, can only be translated as the sentence fragment "Every set is an element of ___," where "___" is unspecified because x is not quantified.

Because a "quantified variable" is just a more descriptive name for a BOUND VARIABLE, a sentence can also be defined as a logic formula with no FREE VARIABLES.

See also BOUND VARIABLE, FREE VARIABLE, QUANTIFIER, THEORY

References
Enderton, H. B. *Elements of Set Theory*. New York: Academic Press, 1977.

Sentential Calculus

PROPOSITIONAL CALCULUS

Separating Edge

An EDGE of a GRAPH is separating if a path from a point A to a point B must pass over it. Separating EDGES can therefore be viewed as either bridges or dead ends.

See also EDGE (GRAPH)

Separating Family

A SEPARATING FAMILY is a SET of SUBSETS in which each pair of adjacent elements are found separated, each in one of two disjoint subsets. The 26 letters of the alphabet can be separated by a family of 9,

$(abcdefghi)$	$(jklmnopqr)$	$(stuvwxyz)$
$(abcjklstu)$	$(defmnovwx)$	$(ghipqryz)$
$(adgjmpsvy)$	$(behknqtwz)$	$(cfilorux)$

The minimal size of the separating family for an n-set is 0, 2, 3, 4, 5, 5, 6, 6, 6, 7, 7, 7, ... (Sloane's A007600).

See also KATONA'S PROBLEM

References
Honsberger, R. "Cai Mao-Cheng's Solution to Katona's Problem on Families of Separating Subsets." Ch. 18 in *Mathematical Gems III*. Washington, DC: Math. Assoc. Amer., pp. 224–239, 1985.
Sloane, N. J. A. Sequences A007600/M0456 in "An On-Line Version of the Encyclopedia of Integer Sequences." http://www.research.att.com/~njas/sequences/eisonline.html.

Separation

Two distinct point pairs AC and BD separate each other if A, B, C, and D lie on a CIRCLE (or line) in such order that either of the arcs (or the line segment AC) contains one but not both of B and D. In addition, the point pairs separate each other if every CIRCLE through A and C intersects (or coincides with) every CIRCLE through B and D. If the point pairs separate each other, then the symbol $AC//BD$ is used.

Separation of Variables

A method of solving partial differential equations in a function $\Phi(x, y, \ldots)$ and variables $x, y, \ldots$ by making a substitution OF THE FORM

$$\Phi(x, y, \ldots) \equiv X(x)Y(y)\cdots,$$

breaking the resulting equation into a set of independent ordinary differential equations, solving these for $X(x)$, $Y(y)$, ..., and then plugging them back into the original equation.

This technique works because if the product of functions of independent variables is a constant, each function must separately be a constant. Success requires choice of an appropriate coordinate system and may not be attainable at all depending on the equation. Separation of variables was first used by L'Hospital in 1750. It is especially useful in solving equations arising in mathematical physics, such as

LAPLACE'S EQUATION, the HELMHOLTZ DIFFERENTIAL EQUATION, and the Schrödinger equation.

See also HELMHOLTZ DIFFERENTIAL EQUATION, LAPLACE'S EQUATION, PARTIAL DIFFERENTIAL EQUATION, STÄCKEL DETERMINANT

References

Arfken, G. "Separation of Variables" and "Separation of Variables--Ordinary Differential Equations." §2.6 and §8.3 in *Mathematical Methods for Physicists, 3rd ed.* Orlando, FL: Academic Press, pp. 111–117 and 448–451, 1985.

Bateman, H. *Partial Differential Equations of Mathematical Physics.* New York: Dover, 1944.

Brown, J. W. and Churchill, R. V. *Fourier Series and Boundary Value Problems, 5th ed.* New York: McGraw-Hill, 1993.

Byerly, W. E. *An Elementary Treatise on Fourier's Series, and Spherical, Cylindrical, and Ellipsoidal Harmonics, with Applications to Problems in Mathematical Physics.* New York: Dover, 1959.

Courant, R. and Hilbert, D. *Methods of Mathematical Physics, Vol. 1.* New York: Wiley, 1989.

Courant, R. and Hilbert, D. *Methods of Mathematical Physics, Vol. 2.* New York: Wiley, 1989.

Eisenhart, L. P. "Separable Systems in Euclidean 3-Space." *Physical Review* **45**, 427–428, 1934.

Eisenhart, L. P. "Separable Systems of Stäckel." *Ann. Math.* **35**, 284–305, 1934.

Eisenhart, L. P. "Potentials for Which Schroedinger Equations Are Separable." *Phys. Rev.* **74**, 87–89, 1948.

Frank, P. and Mises, R. von. *Die Differential- und Integralgleichungen der Mechanik und Physik, 8th ed.* Braunschweig, Germany: Vieweg, 1930.

Hildebrand, F. B. *Advanced Calculus for Engineers.* Englewood Cliffs, NJ: Prentice-Hall, 1949.

Jeffreys, S. H. and Jeffreys, B. S. *Methods of Mathematical Physics, 3rd ed.* Cambridge, England: Cambridge University Press, 1988.

Kellogg, O. D. *Foundations of Potential Theory.* New York: Dover, 1953.

Lense, J. *Reihenentwicklungen in der mathematischen Physik.* Berlin: de Gruyter, 1933.

Maxwell, J. C. *A Treatise on Electricity and Magnetism, Vol. 1, unabridged 3rd ed.* New York: Dover, 1954.

Maxwell, J. C. *A Treatise on Electricity and Magnetism, Vol. 2, unabridged 3rd ed.* New York: Dover, 1954.

Miller, W. Jr. *Symmetry and Separation of Variables.* Reading, MA: Addison-Wesley, 1977.

Moon, P. and Spencer, D. E. "Separability Conditions for the Laplace and Helmholtz Equations." *J. Franklin Inst.* **253**, 585–600, 1952.

Moon, P. and Spencer, D. E. "Theorems on Separability in Riemannian *n*-Space." *Proc. Amer. Math. Soc.* **3**, 635–642, 1952.

Moon, P. and Spencer, D. E. "Recent Investigations of the Separation of Laplace's Equation." *Proc. Amer. Math. Soc.* **4**, 302–307, 1953.

Moon, P. and Spencer, D. E. "Separability in a Class of Coordinate Systems." *J. Franklin Inst.* **254**, 227–242, 1952.

Moon, P. and Spencer, D. E. *Field Theory for Engineers.* Princeton, NJ: Van Nostrand, 1961.

Moon, P. and Spencer, D. E. "Eleven Coordinate Systems." §1 in *Field Theory Handbook, Including Coordinate Systems, Differential Equations, and Their Solutions, 2nd ed.* New York: Springer-Verlag, pp. 1–48, 1988.

Morse, P. M. and Feshbach, H. "Separable Coordinates" and "Table of Separable Coordinates in Three Dimensions."

§5.1 in *Methods of Theoretical Physics, Part I.* New York: McGraw-Hill, pp. 464–523 and 655–666, 1953.

Murnaghan, F. D. *Introduction to Applied Mathematics.* New York: Wiley, 1948.

Smythe, W. R. *Static and Dynamic Electricity, 3rd ed, rev. pr.* New York: Hemisphere, 1989.

Sommerfeld, A. *Partial Differential Equations in Physics.* New York: Academic Press, 1964.

Weber, E. *Electromagnetic Field.* New York: Wiley, 1950.

Webster, A. G. *Partial Differential Equations of Mathematical Physics, 2nd corr. ed.* New York: Dover, 1955.

Separation Theorem

There exist numbers $y_1 < y_2 < \ldots < x_{n-1}$, $a < y_{n-1}$, $y_{n-1} < b$, such that

$$\lambda_v = \alpha(y_v) - \alpha(y_{v-1}).$$

where $v = 1, 2, \ldots, n$, $y_0 = a$ and $y_n = b$. Furthermore, the zeros $x_1, \ldots, x_n$, arranged in increasing order, alternate with the numbers $y_1, \ldots y_{n-1}$, so

$$x_v < y_v < x_{v+1}.$$

More precisely,

$$\alpha(x_v + \epsilon) - \alpha(a) < \alpha(y_v) - \alpha(a) = \lambda_1 + \ldots + \lambda_v$$
$$< \alpha(x_{v+1} - \epsilon) - \alpha(a)$$

for $v = 1, \ldots, n-1$.

See also POINCARÉ SEPARATION THEOREM, STURMIAN SEPARATION THEOREM

References

Szego, G. *Orthogonal Polynomials, 4th ed.* Providence, RI: Amer. Math. Soc., p. 50, 1975.

Separatrix

A phase curve (i.e., an invariant MANIFOLD) which meets a HYPERBOLIC FIXED POINT (i.e., an intersection of a stable and an unstable invariant MANIFOLD) or connects the unstable and stable manifolds of a *pair* of hyperbolic or parabolic fixed points. A separatrix marks a boundary between phase curves with different properties.

For example, the separatrix in the equation of motion for the pendulum occurs at the angular momentum where oscillation gives way to rotation. There are also many systems that have pairs of connected fixed points, e.g., the flow in an open cavity, which has a separatrix that connects two parabolic points.

Septendecillion

In the American system, 10^{54}.

See also LARGE NUMBER

Septillion

In the American system, 10^{24}.

See also LARGE NUMBER

Sequence

A sequence is an ordered set of mathematical objects which is denoted using braces. For example, the symbol $\{2n\}_{n=1}^{\infty}$ denotes the infinite sequence of EVEN NUMBERS $\{2, 4, \ldots, 2n, \ldots\}$.

See also 196-ALGORITHM, A-SEQUENCE, ALCUIN'S SEQUENCE, APPELL CROSS SEQUENCE, APPELL SEQUENCE, $B2$-SEQUENCE, BASIC POLYNOMIAL SEQUENCE, BEATTY SEQUENCE, BINOMIAL-TYPE SEQUENCE, CARMICHAEL SEQUENCE, CAUCHY SEQUENCE, CONVERGENT SEQUENCE, CROSS SEQUENCE, DECREASING SEQUENCE, DEGREE SEQUENCE, DENSITY (SEQUENCE), FRACTAL SEQUENCE, GIUGA SEQUENCE, INCREASING SEQUENCE, INFINITIVE SEQUENCE, INTEGER SEQUENCE, ITERATION SEQUENCE, LIST, NON-AVERAGING SEQUENCE, POLYNOMIAL SEQUENCE, PRIMITIVE SEQUENCE, REVERSE-THEN-ADD SEQUENCE, SCORE SEQUENCE, SERIES, SHEFFER SEQUENCE, SIGNATURE SEQUENCE, SORT-THEN-ADD SEQUENCE, STEFFENSEN SEQUENCE, ULAM SEQUENCE

References

Hardy, G. H. *A Course of Pure Mathematics, 10th ed.* London: Cambridge University Press, 1952.
Jeffreys, H. and Jeffreys, B. S. "Sequences." §1.04 in *Methods of Mathematical Physics, 3rd ed.* Cambridge, England: Cambridge University Press, pp. 10–14, 1988.
Knopp, K. *Theory and Application of Infinite Series.* New York: Dover, 1990.

SequenceLimit

WYNN'S EPSILON METHOD

Sequency

The sequency k of a WALSH FUNCTION is defined as half the number of zero crossings in the time base.

See also WALSH FUNCTION

Sequency Function

WALSH FUNCTION

Sequential Graph

A CONNECTED GRAPH having e EDGES is said to be sequential if it is possible to label the nodes i with distinct INTEGERS f_i in $\{0, 1, 2, \ldots, e-1\}$ such that when EDGE ij is labeled $f_i + f_j$, the set of EDGE labels is a block of e consecutive integers (Grace 1983, Gallian 1990). No HARMONIOUS GRAPH is known which cannot also be labeled sequentially.

See also CONNECTED GRAPH, HARMONIOUS GRAPH

References

Gallian, J. A. "Open Problems in Grid Labeling." *Amer. Math. Monthly* **97**, 133–135, 1990.
Grace, T. "On Sequential Labelings of Graphs." *J. Graph Th.* **7**, 195–201, 1983.

Series

A series is an (often infinite) sum of terms specified by some rule. If the difference between successive terms is a constant, then the series is said to be an ARITHMETIC SERIES. If each term equals the previous multiplied by a constant, it is said to be a GEOMETRIC SERIES. A series usually has an INFINITE number of terms, but the phrase INFINITE SERIES is sometimes used for emphasis or clarity.

Let the terms in a series be denoted a_i, let the kth partial sum be given by

$$S_k = \sum_{i=1}^{k} a_i \tag{1}$$

and let the sequence of partial sums be given by $\{S_1 = a_1, S_2 = a_1 + a_2, S_3 = a_1 + a_2 + a_3, \ldots\}$. If the sequence of partial sums does not converge to a LIMIT (e.g., it oscillates or approaches $\pm\infty$), the series is said to diverge. An example of a convergent series is the GEOMETRIC SERIES

$$\sum_{n=0}^{\infty} \left(\tfrac{1}{2}\right)^n = 2. \tag{2}$$

and an example of a divergent series is the HARMONIC SERIES

$$\sum_{n=1}^{\infty} \frac{1}{n} = \infty. \tag{3}$$

A number of methods known as CONVERGENCE TESTS can be used to determine whether a given series converges. Although terms of a series can have either sign, convergence properties can often be computed in the "worst case" of all terms being POSITIVE, and then applied to the particular series at hand. A series of terms a_n is said to be ABSOLUTELY CONVERGENT if the series formed by taking the absolute values of the a_n,

$$\sum_n |a_n|, \tag{4}$$

converges.

An especially strong type of convergence is called UNIFORM CONVERGENCE, and series which are uniformly convergent have particularly "nice" properties. For example, the sum of a UNIFORMLY CONVERGENT series of continuous functions is continuous. A CONVERGENT SERIES can be DIFFERENTIATED term by term, provided that the functions of the series have continuous derivatives and that the series of DERIVATIVES is UNIFORMLY CONVERGENT. Finally, a UNIFORMLY CONVERGENT series of continuous functions can be INTEGRATED term by term.

For a table listing the COEFFICIENTS for various series operations, see Abramowitz and Stegun (1972, p. 15).

While it can be difficult to calculate analytical expressions for arbitrary convergent infinite series, many algorithms can handle a variety of common series types. The program *Mathematica* implements many of these algorithms. General techniques also exist for computing the numerical values of any but the most pathological series (Braden 1992).

Ramanujan found the interesting series identity

$$1 - \frac{3!}{(1!2!)^3} x^2 + \frac{6!}{(2!4!)^3} x^4 - \cdots$$

$$= \left[1 + \frac{x}{(1!)^3} + \frac{x^2}{(2!)^3} + \cdots \right] \left[1 - \frac{x}{(1!)^3} + \frac{x^2}{(2!)^3} - \cdots \right] \quad (5)$$

(Preece 1928; Hardy 1999, p. 7).

See also ALTERNATING SERIES, ARITHMETIC SERIES, ASYMPTOTIC SERIES, BIAS (SERIES), CONVERGENCE IMPROVEMENT, CONVERGENCE TESTS, EULER-MACLAURIN INTEGRATION FORMULAS, GEOMETRIC SERIES, HARMONIC SERIES, HYPERASYMPTOTIC SERIES, INFINITE SERIES, *Q*-SERIES, RIEMANN SERIES THEOREM, SEQUENCE, SERIES EXPANSION, SERIES REVERSION, SUPERASYMPTOTIC SERIES

References

Abramowitz, M. and Stegun, C. A. (Eds.). "Infinite Series." §3.6 in *Handbook of Mathematical Functions with Formulas, Graphs, and Mathematical Tables, 9th printing.* New York: Dover, p. 14, 1972.

Arfken, G. "Infinite Series." Ch. 5 in *Mathematical Methods for Physicists, 3rd ed.* Orlando, FL: Academic Press, pp. 277–351, 1985.

Boas, R. P. Jr. "Partial Sums of Infinite Series, and How They Grow." *Amer. Math. Monthly* **84**, 237–258, 1977.

Boas, R. P. Jr. "Estimating Remainders." *Math. Mag.* **51**, 83–89, 1978.

Borwein, J. M. and Borwein, P. B. "Strange Series and High Precision Fraud." *Amer. Math. Monthly* **99**, 622–640, 1992.

Braden, B. "Calculating Sums of Infinite Series." *Amer. Math. Monthly* **99**, 649–655, 1992.

Bromwich, T. J. I'a. and MacRobert, T. M. *An Introduction to the Theory of Infinite Series, 3rd ed.* New York: Chelsea, 1991.

Hansen, E. R. *A Table of Series and Products.* Englewood Cliffs, NJ: Prentice-Hall, 1975.

Hardy, G. H. *A Course of Pure Mathematics, 10th ed.* London: Cambridge University Press, 1952.

Hardy, G. H. *Divergent Series.* Oxford, England: Clarendon Press, 1949.

Hardy, G. H. *Ramanujan: Twelve Lectures on Subjects Suggested by His Life and Work, 3rd ed.* New York: Chelsea, 1999.

Jeffreys, H. and Jeffreys, B. S. "Series." §1.05 in *Methods of Mathematical Physics, 3rd ed.* Cambridge, England: Cambridge University Press, pp. 14–17, 1988.

Jolley, L. B. W. *Summation of Series, 2nd rev. ed.* New York: Dover, 1961.

Knopp, K. *Theory and Application of Infinite Series.* New York: Dover, 1990.

Mangulis, V. *Handbook of Series for Scientists and Engineers.* New York: Academic Press, 1965.

Preece, C. T. "Theorems Stated by Ramanujan (III): Theorems on Transformation of Series and Integrals." *J. London Math. Soc.* **3** 274–282, 1928.

Press, W. H.; Flannery, B. P.; Teukolsky, S. A.; and Vetterling, W. T. "Series and Their Convergence." §5.1 in *Numerical Recipes in FORTRAN: The Art of Scientific Computing, 2nd ed.* Cambridge, England: Cambridge University Press, pp. 159–163, 1992.

Rainville, E. D. *Infinite Series.* New York: Macmillan, 1967.

Weisstein, E. W. "Books about Series." http://www.treasure-troves.com/books/Series.html.

Series Expansion

This entry contributed by DANIEL SCOTT UZNANSKI

A series expansion is a representation of a particular function as a sum of powers in one of its variables, or by a sum of powers of another (usually elementary) function $f(x)$.

See also LAURENT SERIES, MACLAURIN SERIES, POWER SERIES, SERIES, SERIES REVERSION, TAYLOR SERIES

Series Inversion

SERIES REVERSION

Series Multisection

If

$$f(x) = f_0 + f_1 x + f_2 x^2 + \ldots + f_n x^n + \cdots$$

then

$$S(n, j) = f_j x^j + f_{j+n} x^{j+n} + f_{j+2n} x^{j+2n} + \cdots$$

is given by

$$S(n, j) = \frac{1}{n} \sum_{t=0}^{n-1} w^{-jt} f(w^t x),$$

where $w = e^{2\pi i / n}$.

See also SERIES REVERSION

References

Honsberger, R. *Mathematical Gems III.* Washington, DC: Math. Assoc. Amer., pp. 210–214, 1985.

Series Reversion

Series reversion is the computation of the COEFFICIENTS of the inverse function given those of the forward function. For a function expressed in a series as

$$y = a_1 x + a_2 x^2 + a_3 x^3 + \ldots, \quad (1)$$

the series expansion of the inverse series is given by

$$x = A_1 y + A_2 y^2 + A_3 y^3 + \cdots \quad (2)$$

By plugging (2) into (1), the following equation is obtained

$$y = a_1 A_1 y + (a_2 A_1^2 + a_1 A_2) y^2$$

$$+ (a_3 A_1^3 + 2 a_2 A_1 A_2 + a_1 A_3) y^3$$

$$+ (3 a_3 A_1^2 A_2 + a_2 A_2^2 + a_2 A_1 A_3) + \cdots \quad (3)$$

Equating COEFFICIENTS then gives

$$A_1 = a_1^{-1} \tag{4}$$

$$A_2 = -\frac{a_2}{a_1} A_1^2 = -a_1^{-3} a_2 \tag{5}$$

$$A_3 = a_1^{-5} \left(2a_2^2 - a_1 a_3 \right) \tag{6}$$

$$A_4 = a_1^{-7} \left(5a_1 a_2 a_3 - a_1^2 a_4 - 5a_2^3 \right) \tag{7}$$

$$A_5 = a_1^{-9} \left(6a_1^2 a_2 a_4 + 3a_1^2 a_2 a_3 + 14a_2^4 - a_1^3 a_5 - 21a_1 a_2^2 a_3 \right) \tag{8}$$

$$A_6 = a_1^{-11} \left(7a_1^3 a_2 a_5 + 7a_1^3 a_3 a_4 + 84a_1 a_2^3 a_3 \right.$$
$$\left. -a_1^4 a_6 - 28a_1^2 a_2 a_3^2 - 42a_2^5 - 28a_1^2 a_2^2 a_4 \right) \tag{9}$$

$$A_7 = a_1^{-13} \left(8a_1^4 a_2 a_6 + 8a_1^4 a_3 a_4 + 4a_1^4 a_4^2 \right.$$
$$+120a_1^2 a_2^3 4 + 180a_1^2 a_2^2 a_3^2 + 132a_2^6$$
$$-a_1^5 a_7 - 36a_1^3 a_2^2 a_5 - 72a_1^3 a_2 a_3 a_4 - 12a_1^3 a_3^3$$
$$\left. -330a_1 a_2^4 a_3 \right) \tag{10}$$

(Dwight 1961, Abramowitz and Stegun 1972, p. 16). A derivation of the explicit formula for the nth term is given by Morse and Feshbach (1953),

$$A_n = \frac{1}{na_1^n} \sum_{s,\,t,\,u\ldots} (-1)^{s+t+u+\ldots}$$

$$\times \frac{n(n+1)\cdots(n-1+s+t+u\ldots)}{s!t!u!\cdots} \left(\frac{a_2}{a_1}\right)^s \left(\frac{a_3}{a_1}\right)^t \cdots, \tag{11}$$

where

$$s + 2t + 3u + \ldots = n - 1. \tag{12}$$

References

Abramowitz, M. and Stegun, C. A. (Eds.). *Handbook of Mathematical Functions with Formulas, Graphs, and Mathematical Tables, 9th printing.* New York: Dover, 1972.

Arfken, G. *Mathematical Methods for Physicists, 3rd ed.* Orlando, FL: Academic Press, pp. 316–317, 1985.

Beyer, W. H. *CRC Standard Mathematical Tables, 28th ed.* Boca Raton, FL: CRC Press, p. 297, 1987.

Dwight, H. B. *Table of Integrals and Other Mathematical Data, 4th ed.* New York: Macmillan, 1961.

Morse, P. M. and Feshbach, H. *Methods of Theoretical Physics, Part I.* New York: McGraw-Hill, pp. 411–413, 1953.

Sloane, N. J. A. and Plouffe, S. *The Encyclopedia of Integer Sequences.* San Diego, CA: Academic Press, p. 22, 1995.

Series-Reduced Tree

A TREE in which all nodes have degree other than 2 (in other words, no node merely allows a single edge to "pass through"). Series-reduced trees are also called homeomorphically irreducible or topological trees (Bergeron *et al.* 1998). The numbers of series-reduced trees with 1, 2, ... nodes are 1, 1, 0, 1, 1, 2, 2, 4, 5, 10, 14, ... (Sloane's A000014).

The numbers of series-reduced PLANTED TREES are 0, 1, 0, 1, 1, 2, 3, 6, 10, 19, 35, ... (Sloane's A001678). The numbers of series-reduced ROOTED TREES are 1, 1, 0, 2, 2, 4, 6, 12, 20, 39, 71, ... (Sloane's A001679).

See also PLANTED TREE, ROOTED TREE, TREE

References

Bergeron, F.; Leroux, P.; and Labelle, G. *Combinatorial Species and Tree-Like Structures.* Cambridge, England: Cambridge University Press, pp. 188, 283–284, 291, and 337, 1998.

Cameron, P. J. "Some Treelike Objects." *Quart. J. Math. Oxford* **38**, 155–183, 1987.

Harary, F. *Graph Theory.* Reading, MA: Addison-Wesley, p. 232, 1994.

Harary, F. and Palmer, E. M. "Probability that a Point of a Tree Is Fixed." *Math. Proc. Camb. Phil. Soc.* **85**, 407–415, 1979.

Harary, F. and Prins, G. "The Number of Homeomorphically Irreducible Trees, and Other Species." *Acta Math.* **101**, 141–162, 1959.

Harary, F.; Robinson, R. W. and Schwenk, A. J. "Twenty-Step Algorithm for Determining the Asymptotic Number of Trees of Various Species." *J. Austral. Math. Soc., Ser. A* **20**, 483–503, 1975.

Sloane, N. J. A. Sequences A000014/M0320, A001678/M0768, and A001679/M0327 in "An On-Line Version of the Encyclopedia of Integer Sequences." http://www.research.att.com/~njas/sequences/eisonline.html.

Serpentine Curve

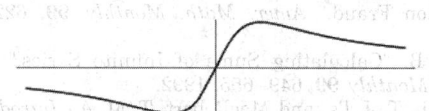

A curve named and studied by Newton in 1701 and contained in his classification of CUBIC CURVES. It had been studied earlier by L'Hospital and Huygens in 1692 (MacTutor Archive).

The curve is given by the CARTESIAN equation

$$y(x) = \frac{abx}{x^2 + a^2} \tag{1}$$

and PARAMETRIC EQUATIONS

$$x(t) = a \cot t \tag{2}$$

$$y(t) = b \sin t \cos t. \tag{3}$$

The curve has a MAXIMUM at $x = a$ and a MINIMUM at $x = -a$, where

$$y'(x) = \frac{ab(a-x)(a+x)}{(a^2+x^2)^2} = 0; \tag{4}$$

and inflection points at $x = \pm\sqrt{3}a$, where

$$y''(x) = \frac{2abx(x^2 - 3a^2)}{(x^2 + a^2)^3} = 0. \tag{5}$$

The CURVATURE is given by

$$\kappa(x) = \frac{2abx(x^2 - 3a^2)}{(x^2 + a^2)^3 \left[1 + \frac{(a^3b - abx^2)^2}{(x^2 + a^2)^4}\right]^{3/2}} \tag{6}$$

$$\kappa(t) = \frac{4\sqrt{2}ab[2\cos(2t) - 1]\cot t \csc^2 t}{\{b^2[1 + \cos(4t)] + 2a^2 \csc{}^4 t\}^{3/2}} \tag{7}$$

References

Beyer, W. H. *CRC Standard Mathematical Tables, 28th ed.* Boca Raton, FL: CRC Press, p. 225, 1987.

Lawrence, J. D. *A Catalog of Special Plane Curves.* New York: Dover, pp. 111–112, 1972.

MacTutor History of Mathematics Archive. "Serpentine." http://www-groups.dcs.st-and.ac.uk/~history/Curves/Serpentine.html.

Serret-Frenet Formulas

FRENET FORMULAS

Set

A set is a FINITE or INFINITE collection of objects in which order has no significance, and multiplicity is generally also ignored (unlike a LIST or MULTISET). Older words for set include AGGREGATE and CLASS. Russell also uses the unfortunate term MANIFOLD to refer to a set. The study of sets and their properties is the object of SET THEORY.

Historically, a single horizontal overbar was used to denote a set stripped of any structure besides order, and hence to represent the order type of the set. A double overbar indicated stripping the order from the set and hence represented the cardinal number of the set. This practice was begun by SET THEORY founder Georg Cantor.

Symbols used to operate on sets include $\cap$ (which means "and" or INTERSECTION), and $\cup$ (which means "or" or UNION). The symbol $\varnothing$ is used to denote the set containing no elements, called the EMPTY SET.

The NOTATION A^B, where A and B are arbitrary sets, is used to denote the set of MAPS from B to A. For example, an element of $X^{\mathbb{N}}$ would be a MAP from the NATURAL NUMBERS $\mathbb{N}$ to the set X. Call such a function f, then $f(1), f(2)$, etc., are elements of X, so call them x_1, x_2, etc. This now looks like a SEQUENCE of elements of X, so sequences are really just functions from $\mathbb{N}$ to X. This NOTATION is standard in mathematics and is frequently used in symbolic dynamics to denote sequence spaces.

Let E, F, and G be sets. Then operation on these sets using the $\cap$ and $\cup$ operators is COMMUTATIVE

$$E \cap F = F \cap E \tag{1}$$

$$E \cup F = F \cup E. \tag{2}$$

ASSOCIATIVE

$$(E \cap F) \cap G = E \cap (F \cap G) \tag{3}$$

$$(E \cup F) \cup G = E \cup (F \cup G). \tag{4}$$

and DISTRIBUTIVE

$$(E \cap F) \cup G = (E \cup G) \cap (F \cup G) \tag{5}$$

$$(E \cup F) \cap G = (E \cap G) \cup (F \cap G). \tag{6}$$

More generally, we have the infinite distributive laws

$$A \cap \left(\bigcup_{\lambda \in \Lambda} B_\lambda\right) = \bigcup_{\lambda \in \Lambda} (A \cap B_\lambda) \tag{7}$$

$$A \cup \left(\bigcap_{\lambda \in \Lambda} B_\lambda\right) = \bigcap_{\lambda \in \Lambda} (A \cup B_\lambda) \tag{8}$$

where λ runs through any INDEX SET Λ. The proofs follow trivially from the definitions of union and intersection.

Many classes of sets are denoted using DOUBLE-STRUCK characters. The table below gives symbols for some common sets in mathematics.

symbol	set
$\mathbb{A}$	ALGEBRAIC NUMBERS
$\mathbb{B}$	BOOLEANS
$\mathbb{B}^n$	n-BALL
$\mathbb{C}$	COMPLEX NUMBERS
$C^n, C^{(n)}$	n-differentiable functions
$\mathbb{D}^n$	n-DISK
$\mathbb{H}$	QUATERNIONS
$\mathbb{I}$	INTEGERS
$\mathbb{N}$	NATURAL NUMBERS
$\mathbb{O}$	CAYLEY NUMBERS
$\mathbb{P}$	PRIME NUMBERS
$\mathbb{Q}$	RATIONAL NUMBERS
$\mathbb{R}^n$	real n-tuples
$\mathbb{R}^{m \times n}$	real $m \times n$ matrices

$\mathbb{S}^n$	n-SPHERE
$\mathbb{T}^n$	n-torus
$\mathbb{Z}$	INTEGERS
$\mathbb{Z}_n$	integers (mod n)
$\mathbb{Z}^-$	NEGATIVE INTEGERS
$\mathbb{Z}^+$	POSITIVE INTEGERS
$\mathbb{Z}^*$	NONNEGATIVE INTEGERS

See also AGGREGATE, ANALYTIC SET, BOREL SET, C, CAYLEY NUMBER, CLASS (SET), COANALYTIC SET, DEFINABLE SET, DERIVED SET, DOUBLE-FREE SET, EXTENSION (SET), GROUND SET, I, INCLUSION-EXCLUSION PRINCIPLE, INTENSION, INTERSECTION, KINNEY'S SET, LIST, MANIFOLD, MULTISET, N, PERFECT SET, POSET, PROPER CLASS, Q, R, REAL MATRIX, SET DIFFERENCE, SET THEORY, TRIPLE-FREE SET, UNION, VENN DIAGRAM, WELL ORDERED SET, Z, Z$_-$, Z$+$

References

Courant, R. and Robbins, H. "The Algebra of Sets." Supplement to Ch. 2 in *What is Mathematics?: An Elementary Approach to Ideas and Methods, 2nd ed.* Oxford, England: Oxford University Press, pp. 108–116, 1996.

Set Difference

The set difference $A \backslash B$ is defined by

$$A \backslash B = \{x : x \in A \text{ and } x \notin B\}.$$

The set difference is therefore equivalent to the COMPLEMENT SET, and is implemented in *Mathematica* as Complement[A, B].

Note that the symbol $\backslash$ is also used to denote QUOTIENT GROUPS. The symbol $A - B$ is sometimes also used to denote a set difference (Smith *et al.* 1997, p. 68).

See also COMPLEMENT SET, SYMMETRIC DIFFERENCE

References

Croft, H. T.; Falconer, K. J.; and Guy, R. K. *Unsolved Problems in Geometry.* New York: Springer-Verlag, p. 2, 1991.
Smith, D.; Eggen, M.; and St. Andre, R. *A Transition to Advanced Mathematics, 4th ed.* New York: Brooks/Cole, 1997.

Set Direct Product

CARTESIAN PRODUCT

Set Partition

A set partition of a SET S is a collection of disjoint SUBSETS of S whose UNION is S. The number of partitions of the SET $\{k\}_{k=1}^n$ is called a BELL NUMBER.

See also BELL NUMBER, BLOCK, PARTITION, RESTRICTED GROWTH STRING, STIRLING NUMBER OF THE SECOND KIND

References

Ruskey, F. "Info About Set Partitions." http://www.theory.csc.uvic.ca/~cos/inf/setp/SetPartitions.html.

Set Theory

The mathematical theory of SETS. Set theory is closely associated with the branch of mathematics known as LOGIC.

There are a number of different versions of set theory, each with its own rules and AXIOMS. In order of increasing CONSISTENCY STRENGTH, several versions of set theory include PEANO ARITHMETIC (ordinary ALGEBRA), second-order arithmetic (ANALYSIS), ZERMELO-FRAENKEL SET THEORY, Mahlo, weakly compact, hyper-Mahlo, ineffable, measurable, Ramsey, supercompact, huge, and n-huge set theory.

See also ANALYSIS (LOGIC), AXIOMATIC SET THEORY, CONSISTENCY STRENGTH, CONTINUUM HYPOTHESIS, DESCRIPTIVE SET THEORY, IMPREDICATIVE, KURATOWSKI'S CLOSURE-COMPONENT PROBLEM, NAIVE SET THEORY, PEANO ARITHMETIC, SENTENCE, SET, THEORY, ZERMELO-FRAENKEL AXIOMS, ZERMELO-FRAENKEL SET THEORY, ZERMELO SET THEORY

References

Brown, K. S. "Set Theory and Foundations." http://www.seanet.com/~ksbrown/ifoundat.htm.
Courant, R. and Robbins, H. "The Algebra of Sets." Supplement to Ch. 2 in *What is Mathematics?: An Elementary Approach to Ideas and Methods, 2nd ed.* Oxford, England: Oxford University Press, pp. 108–116, 1996.
Devlin, K. *The Joy of Sets: Fundamentals of Contemporary Set Theory, 2nd ed.* New York: Springer-Verlag, 1993.
Ferreirós, J. *Labyrinth of Thought: A History of Set Theory and Its Role in Modern Mathematics.* Basel, Switzerland: Birkhäuser, 1999.
Halmos, P. R. *Naive Set Theory.* New York: Springer-Verlag, 1974.
MacTutor History of Mathematics Archive. "The Beginnings of Set Theory." http://www-groups.dcs.st-and.ac.uk/~history/HistTopics/Beginnings_of_set_theory.html.
Stewart, I. *The Problems of Mathematics, 2nd ed.* Oxford: Oxford University Press, p. 96, 1987.
Weisstein, E. W. "Books about Set Theory." http://www.treasure-troves.com/books/SetTheory.html.

Seven Circles Theorem

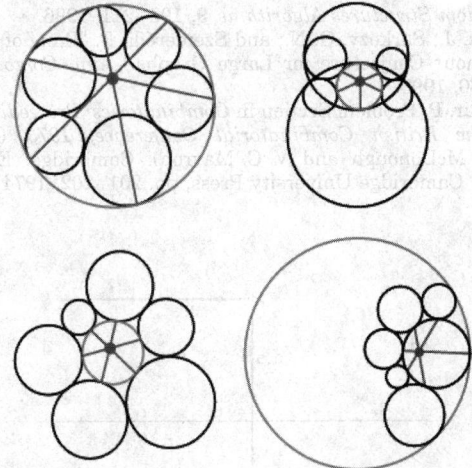

Draw an initial CIRCLE, and arrange six circles tangent to it such that they touch both the original circle and their two neighbors. Then the three lines joining opposite points of tangency are concurrent in a point. The figures above show several possible configurations (Evelyn *et al.* 1974, pp. 31–37).

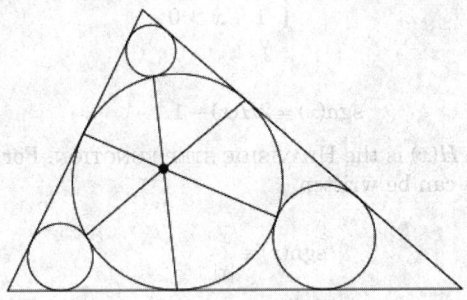

Letting the RADII of three of the circles approach infinity turns three of the CIRCLES into the straight sides of a triangle and the central circle into the triangle's INCIRCLE. As illustrated above, the three lines connecting opposite points of tangency (with those along the triangle edges corresponding to the vertices of the CONTACT TRIANGLE) concur (Evelyn *et al.* 1974, pp. 39 and 42).

See also CIRCLE, CONTACT TRIANGLE, HEXLET, INCIRCLE, SIX CIRCLES THEOREM

References

Evelyn, C. J. A.; Money-Coutts, G. B.; and Tyrrell, J. A. "The Seven Circles Theorem." §3.1 in *The Seven Circles Theorem and Other New Theorems.* London: Stacey International, pp. 31–42, 1974.
Wells, D. *The Penguin Dictionary of Curious and Interesting Geometry.* London: Penguin, pp. 224–225, 1991.

Sexagesimal

The base-60 notational system for representing REAL NUMBERS. A base-60 number system was used by the Babylonians and is preserved in the modern mea-

surement of time (hours, minutes, and seconds) and ANGLES (DEGREES, ARC MINUTES, and ARC SECONDS).

See also BASE (NUMBER), BINARY, DECIMAL, HEXADECIMAL, OCTAL, QUATERNARY, SCRUPLE, TERNARY, VIGESIMAL

References

Bergamini, D. *Mathematics.* New York: Time-Life Books, pp. 16–17, 1969.
Weisstein, E. W. "Bases." MATHEMATICA NOTEBOOK BASES.M.

Sexdecillion

In the American system, 10^{51}.

See also LARGE NUMBER

Sextic Equation

The general sextic polynomial equation

$$x^6 + a_5 x^5 + a_4 x^4 + a_3 x^3 + a_2 x^2 + a_1 x + a_0 = 0$$

can be solved in terms of HYPERGEOMETRIC FUNCTIONS in one variable using Klein's approach to solving the QUINTIC EQUATION.

See also CUBIC EQUATION, QUADRATIC EQUATION, QUARTIC EQUATION, QUINTIC EQUATION

References

Coble, A. B. "The Reduction of the Sextic Equation to the Valentiner Form--Problem." *Math. Ann.* **70**, 337–350, 1911a.
Coble, A. B. "An Application of Moore's Cross-ratio Group to the Solution of the Sextic Equation." *Trans. Amer. Math. Soc.* **12**, 311–325, 1911b.
Cole, F. N. "A Contribution to the Theory of the General Equation of the Sixth Degree." *Amer. J. Math.* **8**, 265–286, 1886.

Sextic Surface

An ALGEBRAIC SURFACE which can be represented implicitly by a polynomial of degree six in x, y, and z. Examples are the BARTH SEXTIC and BOY SURFACE.

See also ALGEBRAIC SURFACE, BARTH SEXTIC, BOY SURFACE, CUBIC SURFACE, DECIC SURFACE, HUNT'S SURFACE, QUADRATIC SURFACE, QUARTIC SURFACE

References

Catanese, F. and Ceresa, G. "Constructing Sextic Surfaces with a Given Number of Nodes." *J. Pure Appl. Algebra* **23**, 1–12, 1982.
Hunt, B. "Algebraic Surfaces." http://www.mathematik.uni-kl.de/~wwwagag/E/Galerie.html.

Sextillion

In the American system, 10^{21}.

See also LARGE NUMBER

Sexy Primes

Since a PRIME NUMBER cannot be divisible by 2 or 3, it must be true that, for a PRIME p, $p \equiv 1$, 5 (mod 6). This motivates the definition of sexy primes as a pair of primes (p, q) such that $p - q = 6$ ("sexy" since "sex" is the Latin word for "six."). The first few sexy prime pairs are (5, 11), (7, 13), (11, 17), (13, 19), (17, 23), (23, 29), (31, 37), (37, 43), (41, 47), (47, 53), ... (Sloane's A023201 and A046117).

Sexy constellations also exist. The first few sexy triplets (i.e., numbers such that each of $(p, p+6, p+12)$ is PRIME but $p + 18$ is *not* PRIME) are (7, 13, 19), (17, 23, 29), (31, 37, 43), (47, 53, 59), ... (Sloane's A046118, A046119, and A046120). The first few sexy quadruplets are (11, 17, 23, 29), (41, 47, 53, 59), (61, 67, 73, 79), (251, 257, 263, 269), ... (Sloane's A046121, A046122, A046123, and A046124). Sexy quadruplets can only begin with a PRIME ending in a "1." There is only a single sexy quintuplet, (5, 11, 17, 23, 29), since every fifth number of the form $6n \pm 1$ is divisible by 5, and therefore cannot be PRIME.

See also PRIME CONSTELLATION, PRIME QUADRUPLET, TWIN PRIMES

References

Sloane, N. J. A. Sequences A023201, A046117, A046118, A046119, A046120, A046121, A046122, A046123, and A046124 in "An On-Line Version of the Encyclopedia of Integer Sequences." http://www.research.att.com/~njas/sequences/eisonline.html.

Trotter, T. "Sexy Primes." http://www.geocities.com/Cape-Canaveral/Launchpad/8202/sexyprim.html.

Seydewitz's Theorem

If a TRIANGLE is inscribed in a CONIC SECTION, any line conjugate to one side meets the other two sides in conjugate points.

See also CONIC SECTION, TRIANGLE

Seymour Conjecture

Seymour conjectured that a graph G of order n with minimum VERTEX DEGREE $\delta(G) \geq kn/(k+1)$ contains the kth GRAPH POWER of a HAMILTONIAN CIRCUIT, generalizing PÓSA'S CONJECTURE. Komlós *et al.* (1998) proved the conjecture for sufficiently large n using SZEMERÉDI'S REGULARITY LEMMA and a technique called the BLOW-UP LEMMA.

See also HAJNAL-SZEMERÉDI THEOREM, HAMILTONIAN CIRCUIT, PÓSA'S CONJECTURE, PÓSA'S THEOREM, SZEMERÉDI'S REGULARITY LEMMA

References

Faudree, R. J.; Gould, R. J.; Jacobson, M. S.; and Schelp, R. H. "On a Problem of Paul Seymour." In *Recent Advances in Graph Theory* (Ed. V. R. Kulli). Vishwa International Publishers, pp. 197–215, 1991.

Komlós, J.; Sárközy, G. N.; and Szemerédi, E. "On the Square of a Hamiltonian Cycle in Dense Graphs." In *Random Structures Algorithms* **9**, 193–211, 1996.

Komlós, J.; Sárközy, G. N.; and Szemerédi, E. "Proof of the Seymour Conjecture for Large Graphs." *Ann. Comb.* **2**, 43–60, 1998.

Seymour, P. Problem Section in *Combinatorics: Proceedings of the British Combinatorial Conference, 1973* (Ed. T. P. McDonough and V. C. Mavron). Cambridge, England: Cambridge University Press, pp. 201–202, 1974.

Sgn

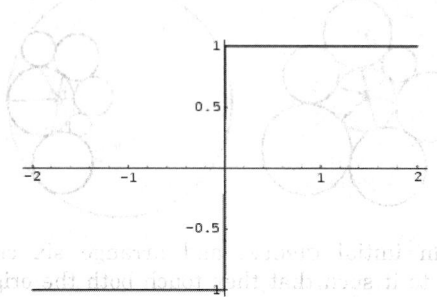

Also called SIGNUM. It can be defined as

$$\mathrm{sgn} \equiv \begin{cases} -1 & x < 0 \\ 0 & x = 0 \\ 1 & x > 0 \end{cases} \quad (1)$$

or

$$\mathrm{sgn}(x) = 2H(x) - 1. \quad (2)$$

where $H(x)$ is the HEAVISIDE STEP FUNCTION. For $x \neq 0$, this can be written

$$\mathrm{sgn}(x) \equiv \frac{x}{|x|}. \quad (3)$$

See also ABSOLUTE VALUE, HEAVISIDE STEP FUNCTION, RAMP FUNCTION

References

Bracewell, R. "The Sign Function, sgn x." In *The Fourier Transform and Its Applications, 3rd ed.* New York: McGraw-Hill, pp. 61–62, 1999.

Sh

HYPERBOLIC SINE

Shadow

The SURFACE corresponding to the region of obscuration when a solid is illuminated from a point light source (located at the RADIANT POINT). A DISK is the SHADOW of a SPHERE on a PLANE perpendicular to the SPHERE-RADIANT POINT line. If the PLANE is tilted, the shadow can be the interior of an ELLIPSE or a PARABOLA.

See also CORK PLUG, PROJECTION, STEREOLOGY, TRIPLET

References

Croft, H. T.; Falconer, K. J.; and Guy, R. K. "What Can You Tell About a Convex Body from Its Shadows?" §A10 in *Unsolved Problems in Geometry.* New York: Springer-Verlag, pp. 23–24, 1991.

Shadowing Theorem

Although a numerically computed CHAOTIC trajectory diverges exponentially from the true trajectory with the same initial coordinates, there exists an errorless trajectory with a slightly different initial condition that stays near ("shadows") the numerically computed one. Therefore, the FRACTAL structure of chaotic trajectories seen in computer maps is real.

References

Ott, E. *Chaos in Dynamical Systems.* New York: Cambridge University Press, pp. 18–19, 1993.

Shafarevich Conjecture

A conjecture which implies the MORDELL CONJECTURE, as proved in 1968 by A. N. Parshin.

See also MORDELL CONJECTURE

References

Stewart, I. *The Problems of Mathematics,* 2nd ed. Oxford, England: Oxford University Press, p. 45, 1987.

Shah Function

$$\mathrm{III}(x) \equiv \sum_{n=-\infty}^{\infty} \delta(x-n) \qquad (1)$$

where $\delta(x)$ is the DELTA FUNCTION, so $\mathrm{III}(x) = 0$ for $x \notin \mathbb{Z}$ (i.e., x not an INTEGER). The shah function is also called the sampling symbol or replicating symbol (Bracewell 1999, p. 77) and obeys the identities

$$\mathrm{III}(ax) = \frac{1}{|a|} \sum_{n=-\infty}^{\infty} \delta\left(x - \frac{n}{a}\right) \qquad (2)$$

$$\mathrm{III}(-x) = \mathrm{III}(x) \qquad (3)$$

$$\mathrm{III}(x+n) = \mathrm{III}(x) \qquad (4)$$

$$\mathrm{III}\left(x - \tfrac{1}{2}\right) = \mathrm{III}\left(x + \tfrac{1}{2}\right). \qquad (5)$$

The shah function is normalized so that

$$\int_{n-1/2}^{n+1/2} \mathrm{III}(x)\, dx = 1. \qquad (6)$$

The "sampling property" is

$$\mathrm{III}(x)f(x) = \sum_{n=-\infty}^{\infty} f(n)\delta(x-n) \qquad (7)$$

and the "replicating property" is

$$\mathrm{III}(x) * f(x) = \sum_{n=-\infty}^{\infty} f(x-n). \qquad (8)$$

where $*$ denotes CONVOLUTION.

The 2-D sampling function, sometimes called the bed-of-nails function, is given by

$$^2\mathrm{III}(x, y) = \sum_{m=-\infty}^{\infty} \sum_{n=-\infty}^{\infty} \delta(x-m, y-n), \qquad (9)$$

which can be adjusted using a series of weighted as

$$v(x, y) = \sum R_{mn} T_{mn} D_{mn} \delta(x - m_n, y - n), \qquad (10)$$

where R_{mn} is a reliability weight, D_{mn} is a density weight (WEIGHTING FUNCTION), and T_{mn} is a taper. The 2-D shah function satisfies

$$^2\mathrm{III}(x, y) = \mathrm{III}(x)\mathrm{III}(y) \qquad (11)$$

(Bracewell 1999, p. 85).

See also CONVOLUTION, DELTA FUNCTION, IMPULSE PAIR, SINC FUNCTION

References

Bracewell, R. "The Sampling of Replicating Symbol $\mathrm{III}(x)$." In *The Fourier Transform and Its Applications,* 3rd ed. New York: McGraw-Hill, pp. 77–79, 1999.

Shah-Wilson Constant

TWIN PRIMES CONSTANT

Shaky Polyhedron

A shaky polyhedron is a non-rigid concave polyhedron which is only infinitesimally movable. JESSEN'S ORTHOGONAL ICOSAHEDRON is a shaky polyhedron (Wells 1991).

See also FLEXIBLE POLYHEDRON, JESSEN'S ORTHOGONAL ICOSAHEDRON, MULTISTABLE, RIGID POLYHEDRON, RIGIDITY THEOREM

References

Blaschke, W. "Wackelige Achtflache." *Math. Z.* **6**, 85–93, 1920.
Cromwell, P. R. *Polyhedra.* New York: Cambridge University Press, p. 222, 1997.
Gluck, H. *Almost All Simply Connected Closed Surfaces are Rigid.* Heidelberg, Germany: Springer-Verlag, pp. 225–239, 1975.
Goldberg, M. "Unstable Polyhedral Structures." *Math. Mag.* **51**, 165–170, 1978.
Jessen, B. "Orthogonal Icosahedron." *Nordisk Mat. Tidskr.* **15**, 90–96, 1967.
Wells, D. *The Penguin Dictionary of Curious and Interesting Geometry.* London: Penguin, p. 161, 1991.

Shallit Constant

Define $f(x_1, x_2, \ldots, x_n)$ with x_i POSITIVE as

$$f(x_1, x_2, \ldots, x_n) \equiv \sum_{i=1}^{n} x_i + \sum_{1 \le i \le k \le n} \prod_{j=1}^{k} \frac{1}{x_j}.$$

Then

$$\min f = 3n - C + o(1)$$

as n increases, where the Shallit constant is

$$C = 1.369451403937\ldots$$

(Shallit 1995). In their solution, Grosjean and De Meyer (quoted in Shallit 1995) reduced the complexity of the problem.

References

MacLeod, A. http://www.mathsoft.com/asolve/constant/shapiro/macleod.html.

Shallit, J. Solution by C. C. Grosjean and H. E. De Meyer. "A Minimization Problem." Problem 94–15 in *SIAM Review* **37**, 451–458, 1995.

Shallow Diagonal

See also DIAGONAL, PASCAL'S TRIANGLE

Shanks' Algorithm

An ALGORITHM which finds the least NONNEGATIVE value of $\sqrt{a \pmod{p}}$ for given a and PRIME p.

Shanks' Conjecture

Let $p(n)$ be the first PRIME which follows a PRIME GAP of n between consecutive PRIMES. Shanks' conjecture holds that

$$p(n) \sim \exp(\sqrt{n}).$$

Wolf conjectures a slightly different form

$$p(n) \sim \sqrt{n} \exp(\sqrt{n}),$$

which agrees better with numerical evidence.

See also PRIME DIFFERENCE FUNCTION, PRIME GAPS

References

Guy, R. K. *Unsolved Problems in Number Theory, 2nd ed.* New York: Springer-Verlag, p. 21, 1994.

Rivera, C. "Problems & Puzzles: Conjecture Shanks' Conjecture.-009." http://www.primepuzzles.net/conjectures/conj_009.htm.

Shanks, D. "On Maximal Gaps Between Successive Primes." *Math. Comput.* **18**, 646–651, 1964.

Shannon Entropy

ENTROPY

Shannon Sampling Theorem

SAMPLING THEOREM

Shannon's Noiseless Coding Theorem

Let S be an information source with entropy $H(S)$. Then

$$H(S) \le m(S),$$

where $m(S)$ is the minimum average code-word length among all uniquely decipherable coding schemes for S

References

Casti, J. L. "The Shannon Coding Theorem." Ch. 1 in *Five More Golden Rules: Knots, Codes, Chaos, and Other Great Theories of 20th-Century Mathematics.* New York: Wiley, pp. 207–254, 2000.

Shape Number

FIGURATE NUMBER

Shape Operator

The negative derivative

$$S(\mathbf{v}) = -D_{\mathbf{v}}\mathbf{N} \tag{1}$$

of the unit normal $\mathbf{N}$ vector field of a SURFACE is called the shape operator (or WEINGARTEN MAP or SECOND FUNDAMENTAL TENSOR). The shape operator S is an EXTRINSIC CURVATURE, and the GAUSSIAN CURVATURE is given by the DETERMINANT of S. If $\mathbf{x} : U \to \mathbb{R}^3$ is a REGULAR PATCH, then

$$S(\mathbf{x}_u) = -\mathbf{N}_u \tag{2}$$

$$S(\mathbf{x}_v) = -\mathbf{N}_v. \tag{3}$$

At each point $\mathbf{p}$ on a REGULAR SURFACE $M \subset \mathbb{R}^3$, the shape operator is a linear map

$$S : M_{\mathbf{p}} \to M_{\mathbf{p}}. \tag{4}$$

The shape operator for a surface is given by the WEINGARTEN EQUATIONS.

See also CURVATURE, FUNDAMENTAL FORMS, WEINGARTEN EQUATIONS

References

Gray, A. "The Shape Operator," "Calculation of the Shape Operator," and "The Eigenvalues of the Shape Operator." §16.1, 16.3, and 16.4 in *Modern Differential Geometry of Curves and Surfaces with Mathematica, 2nd ed.* Boca Raton, FL: CRC Press, pp. 360–363 and 367–372, 1997.

Reckziegel, H. In *Mathematical Models from the Collections of Universities and Museums* (Ed. G. Fischer). Braunschweig, Germany: Vieweg, p. 30, 1986.

Shapiro's Cyclic Sum Constant

N.B. A detailed online essay by S. Finch was the starting point for this entry.

Consider the sum

$$f_n(x_1, x_2, \ldots, x_n)$$

$$= \frac{x_1}{x_2 + x_3} + \frac{x_2}{x_3 + x_4} + \ldots + \frac{x_{n-1}}{x_n + x_1} + \frac{x_n}{x_1 + x_2}, \quad (1)$$

where the x_is are NONNEGATIVE and the DENOMINATORS are POSITIVE. Shapiro (1954) asked if

$$f_n(x_1, x_2, \ldots, x_n) \geq \tfrac{1}{2} n \quad (2)$$

for all n. It turns out (Mitrinovic *et al.* 1993) that this INEQUALITY is true for all EVEN $n \leq 12$ and ODD $n \leq 23$. Ranikin (1958) proved that for

$$f(n) = \inf_{x \geq 0} f_n(x_1, x_2, \ldots, x_n), \quad (3)$$

$$\lambda = \lim_{n \to \infty} \frac{f(n)}{n} = \inf_{n \geq 1} \frac{f(n)}{n} < \tfrac{1}{2} - 7 \times 10^{-8}. \quad (4)$$

λ can be computed by letting $\phi(x)$ be the CONVEX HULL of the functions

$$y_1 = e^{-x} \quad (5)$$

$$y_2 = \frac{2}{e^x + e^{x/2}}. \quad (6)$$

Then

$$\lambda = \tfrac{1}{2} \phi(0) = 0.4945668\ldots \quad (7)$$

(Drinfeljd 1971).

A modified sum was considered by Elbert (1973):

$$g_n(x_1, x_1, \ldots, x_n)$$

$$= \frac{x_1 + x_3}{x_1 + x_2} + \frac{x_2 + x_4}{x_2 + x_3} + \ldots + \frac{x_{x-1} + x_1}{x_{n-1} + x_n} + \frac{x_n + x_2}{x_n + x_1}. \quad (8)$$

Consider

$$\mu = \lim_{n \to \infty} \frac{g(n)}{n}, \quad (9)$$

where

$$g(n) = \inf_{x \geq 0} g_n(x_1, x_2, \ldots, x_n), \quad (10)$$

and let $\psi(x)$ be the CONVEX HULL of

$$y_1 = \tfrac{1}{2}(1 + e^x) \quad (11)$$

$$y_2 = \frac{1 + e^x}{1 + e^{x/2}} \quad (12)$$

Then

$$\mu = \psi(0) = 0.978012\ldots. \quad (13)$$

See also CONVEX HULL

References

Drinfeljd, V. G. "A Cyclic Inequality." *Math. Notes. Acad. Sci. USSR* **9**, 68–71, 1971.

Elbert, A. "On a Cyclic Inequality." *Period. Math. Hungar.* **4**, 163–168, 1973.

Finch, S. "Favorite Mathematical Constants." http://www.mathsoft.com/asolve/constant/shapiro/shapiro.html.

Mitrinovic, D. S.; Pecaric, J. E.; and Fink, A. M. *Classical and New Inequalities in Analysis.* New York: Kluwer, 1993.

Sharing Problem

A problem also known as the POINTS PROBLEM or UNFINISHED GAME. Consider a tournament involving k players playing the same game repetitively. Each game has a single winner, and denote the number of games won by player i at some juncture w_i. The games are independent, and the probability of the ith player winning a game is p_i. The tournament is specified to continue until one player has won n games. If the tournament is discontinued before any player has won n games so that $w_i < n$ for $i = 1, \ldots, k$, how should the prize money be shared in order to distribute it proportionally to the players' chances of winning?

For player i, call the number of games left to win $r_i \equiv n - w_i > 0$ the "quota." For two players, let $p \equiv p_1$ and $q \equiv p_2 = 1 - p$ be the probabilities of winning a single game, and $a \equiv r_1 = n - w_1$ and $b \equiv r_2 = n - w_2$ be the number of games needed for each player to win the tournament. Then the stakes should be divided in the ratio $m : n$, where

$$m = p^a \left[1 + \frac{a}{1} q + \frac{a(a+1)}{2!} q^2 + \ldots \right.$$
$$\left. + \frac{a(a+1)\cdots(a+b-2)}{(b-1)!} q^{b-1} \right] \quad (1)$$

$$n = q^b \left[1 + \frac{b}{1} p + \frac{b(b+1)}{2!} p^2 + \ldots \right.$$
$$\left. + \frac{b(b+1)\cdots(b+a-2)}{(a-1)!} p^{a-1} \right] \quad (2)$$

(Kraitchik 1942).

If i players have equal probability of winning ("cell probability"), then the chance of player i winning for quotas $r_1, \ldots, r_k$ is

$$W_i = D_1^{k-1}(r_1, \ldots, r_{i-1}, r_{i+1}, \ldots, r_k; r_i). \quad (3)$$

where D is the DIRICHLET INTEGRAL of type 2D. Similarly, the chance of player i losing is

$$L_i = C_1^{k-1}(r_1, \ldots, r_{i-1}, r_{i+1}, \ldots, r_k; r_i), \quad (4)$$

where C is the DIRICHLET INTEGRAL of type 2C. If the cell quotas are not equal, the general Dirichlet integral D_n must be used, where

$$a_i = \frac{p_i}{1 - \sum_{i=1}^{k-1} p_i}. \tag{5}$$

If $r_i = r$ and $a_i = 1$, then W_i and L_i reduce to $1/k$ as they must. Let $P(r_1, \ldots, r_k)$ be the joint probability that the players would be RANKED in the order of the r_is in the argument list if the contest were completed. For $k = 3$,

$$P(r_1, r_2, r_3) = CD_1^{(1,\,1)}(r_1, r_2, r_3). \tag{6}$$

For $k = 4$ with quota vector $\mathbf{r} = (r_1, r_2, r_3, r_4)$ and $\Delta = p_2 + p_3 + p_4$,

$$P(\mathbf{r}) = \sum_{i=0}^{r_3-1} \sum_{j=0}^{r_4-1} \binom{r_2 - 1 + i + j}{r_2 - 1,\, i,\, j} \left(\frac{p_2}{\Delta}\right)^{r_2} \left(\frac{p_3}{\Delta}\right)^i \left(\frac{p_4}{\Delta}\right)^j$$
$$\times C_{p1/\Delta}^{(1)}(r_1, r_2 + i + j) D_{p4/p3}^{(1)}(r_4 - j, r_3 - i). \tag{7}$$

An expression for $k = 5$ is given by Sobel and Frankowski (1994, p. 838).

See also DIRICHLET INTEGRALS

References

Kraitchik, M. "The Unfinished Game." §6.1 in *Mathematical Recreations.* New York: W. W. Norton, pp. 117–118, 1942.
Sobel, M. and Frankowski, K. "The 500th Anniversary of the Sharing Problem (The Oldest Problem in the Theory of Probability)." *Amer. Math. Monthly* **101**, 833–847, 1994.

Sharkovsky's Theorem
SARKOVSKII'S THEOREM

Sharpe Ratio
A risk-adjusted financial measure developed by Nobel Laureate William Sharpe. It uses a fund's standard deviation and excess return to determine the reward per unit of risk. The higher a fund's Sharpe ratio, the better the fund's "risk-adjusted" performance.

See also ALPHA, BETA

Sharpe's Differential Equation
A generalization of the BESSEL DIFFERENTIAL EQUATION for functions of order 0, given by

$$zy'' + y' + (z + A)y = 0.$$

Solutions are

$$y = e^{\pm iz}\,{}_1F_1\left(\tfrac{1}{2} \mp \tfrac{1}{2} iA;\; 1;\; \mp 2iz\right).$$

where ${}_1F_1(a;\, b;\, x)$ is a CONFLUENT HYPERGEOMETRIC FUNCTION.

See also BESSEL DIFFERENTIAL EQUATION, CONFLUENT HYPERGEOMETRIC FUNCTION

Sheaf
SHEAF OF PLANES, SHEAF (TOPOLOGY)

Sheaf (Topology)

A topological GADGET related to families of ABELIAN GROUPS and MAPS.

References

Iyanaga, S. and Kawada, Y. (Eds.). "Sheaves." §377 in *Encyclopedic Dictionary of Mathematics.* Cambridge, MA: MIT Press, pp. 1171–1174, 1980.

Sheaf of Planes

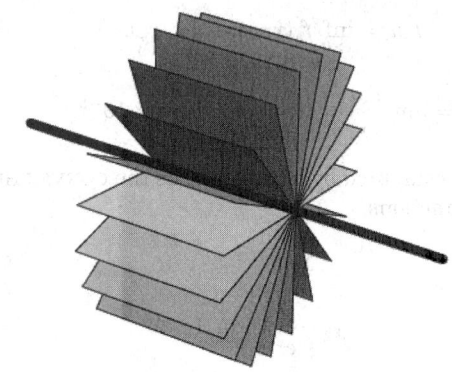

The set of all PLANES through a LINE. The line is sometimes called the AXIS of the sheaf, and the sheaf itself is sometimes called a pencil (Altshiller-Court 1979, p. 12).

See also LINE, PENCIL, PLANE

References

Altshiller-Court, N. *Modern Pure Solid Geometry.* New York: Chelsea, 1979.
Kern, W. F. and Bland, J. R. *Solid Mensuration with Proofs, 2nd ed.* New York: Wiley, p. 13, 1948.
Woods, F. S. *Higher Geometry: An Introduction to Advanced Methods in Analytic Geometry.* New York: Dover, p. 12, 1961.

Shear

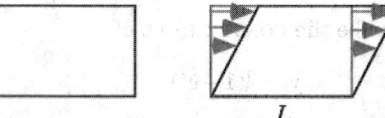

A transformation in which all points along a given LINE L remain fixed while other points are shifted parallel to L by a distance proportional to their PERPENDICULAR distance from L. Shearing a plane figure does not change its AREA. The shear can also be generalized to 3-D, in which PLANES are translated instead of lines.

See also SHEAR FACTOR, SHEAR MATRIX

Shear Factor

The distance a point moves due to SHEAR divided by the perpendicular distance of a point from the invariant line.

See also SHEAR, SHEAR MATRIX

References
Pimentel, R. and Wall, T. *IGCSE Mathematics.* London: John Murray, p. 312, 1997.

Shear Matrix

The shear matrix e_{ij}^s is obtained from the IDENTITY MATRIX by inserting s at (i, j), e.g.,

$$e_{12}^s = \begin{bmatrix} 1 & s & 0 \\ 0 & 1 & 0 \\ 0 & 0 & 1 \end{bmatrix}. \quad (1)$$

Bolt and Hobbs (1998) define a shear matrix as a matrix

$$\begin{bmatrix} a & b \\ c & d \end{bmatrix} \quad (2)$$

such that

$$a + b = 2 \quad (3)$$

$$ad - bc = 1. \quad (4)$$

See also ELEMENTARY MATRIX, SHEAR, SHEAR FACTOR

References
Bolt, B. and Hobbs, D. *A Mathematical Dictionary for Schools.* Cambridge, England: Cambridge University Press, 1998.

Sheffer Sequence

A sequence $s_n(x)$ is called a Sheffer sequence IFF its GENERATING FUNCTION has the form

$$\sum_{k=0}^{\infty} \frac{s_k(x)}{k!} t^k = A(t)e^{xB(t)}, \quad (1)$$

where

$$A(t) = A_0 + A_1 t + A_2 t^2 + \ldots \quad (2)$$

$$B(t) = B_1 t + B_2 t^2 + \ldots. \quad (3)$$

with $A_0, B_1 \neq 0$.

If $f(t)$ is a delta series and $g(t)$ is an invertible series, then there exists a unique sequence $s_n(x)$ of Sheffer polynomials $s_n(x)$ satisfying the orthogonality condition

$$\left\langle g(t)[f(t)]^k | s_n(x) \right\rangle = n! \delta_{nk}, \quad (4)$$

where δ_{nk} is the KRONECKER DELTA (Roman 1984, p. 17). Examples of general Sheffer sequences include the ACTUARIAL POLYNOMIALS, BERNOULLI POLYNOMIALS OF THE SECOND KIND, BOOLE POLYNOMIALS, LAGUERRE POLYNOMIALS, MEIXNER POLYNOMIALS OF THE FIRST and SECOND KINDS, POISSON-CHARLIER POLYNOMIALS, and STIRLING POLYNOMIALS.

The Sheffer sequence for $(1, f(t))$ is called the associated sequence for $f(t)$, and Roman (1984, pp. 53–86) summarizes properties of the associated Sheffer sequences and gives a number of specific examples (ABEL POLYNOMIAL, BELL POLYNOMIAL, CENTRAL FACTORIAL, EXPONENTIAL POLYNOMIAL, FALLING FACTORIAL, GOULD POLYNOMIAL, MAHLER POLYNOMIAL, MITTAG-LEFFLER POLYNOMIAL, MOTT POLYNOMIAL, POWER POLYNOMIAL). The Sheffer sequence for $(g(t), t)$ is called the APPELL SEQUENCE of $g(t)$, and Roman (1984, pp. 86–106) summarizes properties of Appell sequences and gives a number of specific examples.

If $s_n(x)$ is a Sheffer sequence for $(g(t), f(t))$, then for any polynomial $p(x)$,

$$p(x) = \sum_{k=0}^{\infty} \frac{\left\langle g(t)[f(t)]^k | p(x) \right\rangle}{k!} s_k(x). \quad (5)$$

The sequence $s_n(x)$ is Sheffer for $(g(t), f(t))$ IFF

$$\frac{1}{g(\bar{f}(t))} e^{y\bar{f}(t)} = \sum_{k=0}^{\infty} \frac{s_k(y)}{k!} t^k \quad (6)$$

for all y in the field C of characteristic 0, where $\bar{f}(t)$ is the compositional INVERSE FUNCTION of $f(t)$ (Roman 1984, p. 18). This formula immediately gives the GENERATING FUNCTION associated with a given Sheffer sequence.

A sequence is Sheffer for $(g(t), f(t))$ for some invertible $g(t)$ IFF

$$f(t)s_n(x) = ns_{n-1}(x) \quad (7)$$

for all $n \geq 0$ (Roman 1984, p. 20). The Sheffer identity states that a sequence $s_n(x)$ is Sheffer for $(g(t), f(t))$ for some invertible $f(t)$ IFF it satisfies some BINOMIAL-TYPE SEQUENCE

$$s_n(x + y) = \sum_{k=0}^{n} \binom{n}{k} p_k(y) s_{n-k}(x) \quad (8)$$

for all y in C, where $p_n(x)$ is associated to $f(t)$ (Roman 1984, p. 21). The RECURRENCE RELATION for Sheffer sequences is given by

$$s_{n+1}(x) = \left[x - \frac{g'(t)}{g(t)} \right] \frac{1}{f'(t)} s_n(x) \quad (9)$$

(Roman 1984, p. 50). A nontrivial RECURRENCE RELATION is given by

$$s_{n+1}(x) = (x - b_n)s_n(x) - d_n s_{n-1}(x) \quad (10)$$

for $s_{-1}(x) = 0$, $s_0(x) = 1$, and $n \geq 0$ (Meixner 1934;

Sheffer 1939; Chihara 1978; Roman 1984, pp. 156–160).

The connection coefficients c_{nk} in the expression

$$s_n(x) = \sum_{k=0}^{n} c_{nk} r_k(n) \qquad (11)$$

are given by

$$c_{nk} = \frac{1}{k!} \left\langle \frac{h(f^{-1}(t))}{g(f^{-1}(t))} [l(f^{-1}(t))]^k \middle| x^n \right\rangle, \qquad (12)$$

where $s_n(x)$ is Sheffer for $(g(t), f(t))$ and $r_n(x)$ is Sheffer for $(h(t), l(t))$. This can also be written in terms of the polynomial of coefficients

$$t_n(x) = \sum_{k=0}^{n} c_{nk} x^k. \qquad (13)$$

which is Sheffer for

$$\left(\frac{g(l^{-1}(t))}{h(l^{-1}(t))}, f(l^{-1}(t)) \right) \qquad (14)$$

(Roman 1984, pp. 132–138).

A duplication formula OF THE FORM

$$r_n(ax) = \sum_{k=0}^{n} c_{nk} r_k(x) \qquad (15)$$

is given by

$$c_{nk} = \frac{1}{k!} \left\langle \frac{h(al^{-1}(t))}{h(l^{-1}(t))} [l(al^{-1}(t))^k \middle| x^n \right\rangle, \qquad (16)$$

where $r_n(x)$ is Sheffer for $(h(t), l(t))$ (Roman 1984, pp. 132–138).

See also APPELL CROSS SEQUENCE, APPELL SEQUENCE, BINOMIAL-TYPE SEQUENCE, CROSS SEQUENCE, STEFFENSEN SEQUENCE, UMBRAL CALCULUS

References
Chihara, T. S. *An Introduction to Orthogonal Polynomials.* New York: Gordon and Breach, 1978.

Meixner, J. "Orthogonale Polynomsystem mit linern besonderen Gestalt der eryengenden Funktion." *J. London Math. Soc.* **9**, 6–13, 1934.

Roman, S. "Sheffer Sequences." Ch. 2 and §4.3 in *The Umbral Calculus.* New York: Academic Press, pp. 2, 6–31, and 107–130, 1984.

Rota, G.-C.; Kahaner, D.; Odlyzko, A. "On the Foundations of Combinatorial Theory. VIII: Finite Operator Calculus." *J. Math. Anal. Appl.* **42**, 684–760, 1973.

Sheffer, I. M. "Some Properties of Polynomial Sets of Type Zero." *Duke Math. J.* **5**, 590–622, 1939.

Sheffer Stroke
NAND

Shephard's Problem
Measurements of a centered convex body in Euclidean n-space (for $n \geq 3$) show that its brightness function (the volume of each projection) is smaller than that of another such body. Is it true that its VOLUME is also smaller? C. M. Petty and R. Schneider showed in 1967 that the answer is yes if the body with the larger brightness function is a projection body, but no in general for every n.

References
Gardner, R. J. "Geometric Tomography." *Not. Amer. Math. Soc.* **42**, 422–429, 1995.

Sheppard's Correction
A correction which must be applied to the measured MOMENTS m_k obtained from NORMALLY DISTRIBUTED data which have been BINNED in order to obtain correct estimators $\hat{\mu}_i$ for the population moments μ_i. The corrected versions of the second, third, and fourth moments are then

$$\hat{\mu}_2 = m_2 - \tfrac{1}{12} c^2 \qquad (1)$$

$$\hat{\mu}_3 = m_3 \qquad (2)$$

$$\hat{\mu}_4 = m_4 - \tfrac{1}{2} m_2 + \tfrac{7}{240} c^2, \qquad (3)$$

where c is the CLASS INTERVAL.

If κ'_r is the rth CUMULANT of an ungrouped distribution and κ_r the rth CUMULANT of the grouped distribution with CLASS INTERVAL c, the corrected cumulants (under rather restrictive conditions) are

$$\kappa'_r = \begin{cases} \kappa_r & \text{for } r \text{ odd} \\ \kappa_r - \dfrac{B_r}{r} c^r & \text{for } r \text{ even,} \end{cases} \qquad (4)$$

where B_r is the rth BERNOULLI NUMBER, giving

$$\kappa'_1 = \kappa_1 \qquad (5)$$

$$\kappa'_2 = \kappa_2 - \tfrac{1}{12} c^2 \qquad (6)$$

$$\kappa'_3 = \kappa_3 \qquad (7)$$

$$\kappa'_4 = \kappa_4 + \tfrac{1}{120} c^4 \qquad (8)$$

$$\kappa'_5 = \kappa_5 \qquad (9)$$

$$\kappa'_6 = \kappa_6 - \tfrac{1}{252} c^6. \qquad (10)$$

For a proof, see Kendall *et al.* (1987).

See also BIN, CLASS INTERVAL, HISTOGRAM

References
Fisher, R. A. *Statistical Methods for Research Workers, 14th ed., rev. and enl.* Darien, CO: Hafner, 1970.

Kendall, M. G.; Stuart, A.; and Ord, J. K. *Kendall's Advanced Theory of Statistics, Vol. 1: Distribution Theory, 6th ed.* New York: Oxford University Press, 1987.

Kenney, J. F. and Keeping, E. S. "Sheppard's Correction for Grouping Errors." §7.6 in *Mathematics of Statistics, Pt. 1, 3rd ed.* Princeton, NJ: Van Nostrand, pp. 95–96, 1962.

Kenney, J. F. and Keeping, E. S. "Sheppard's Correction." §4.12 in *Mathematics of Statistics, Pt. 2, 2nd ed.* Princeton, NJ: Van Nostrand, pp. 80–82, 1951.

Whittaker, E. T. and Robinson, G. "Sheppard's Corrections." §99 in *The Calculus of Observations: A Treatise on Numerical Mathematics, 4th ed.* New York: Dover, pp. 194–196, 1967.

Sherman-Morrison Formula

A formula which allows a perturbed MATRIX to be computed for a small change to a given MATRIX A. If the change can be written in the form

$$\mathbf{u} \otimes \mathbf{v}$$

for two vectors $\mathbf{u}$ and $\mathbf{v}$, then the Sherman-Morrison formula is

$$(A + \mathbf{u} \otimes \mathbf{v})^{-1} = A^{-1} - \frac{(A^{-1}\mathbf{u}) \otimes (\mathbf{v} \cdot A^{-1})}{1 + \lambda},$$

where

$$\lambda \equiv \mathbf{v} \cdot A^{-1} \mathbf{u}.$$

See also WOODBURY FORMULA

References

Golub, G. H. and van Loan, C. F. *Matrix Computations, 3rd ed.* Baltimore, MD: Johns Hopkins, p. 51, 1996.

Press, W. H.; Flannery, B. P.; Teukolsky, S. A.; and Vetterling, W. T. "Sherman-Morrison Formula." In *Numerical Recipes in FORTRAN: The Art of Scientific Computing, 2nd ed.* Cambridge, England: Cambridge University Press, pp. 65–67, 1992.

Shi

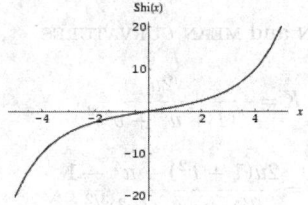

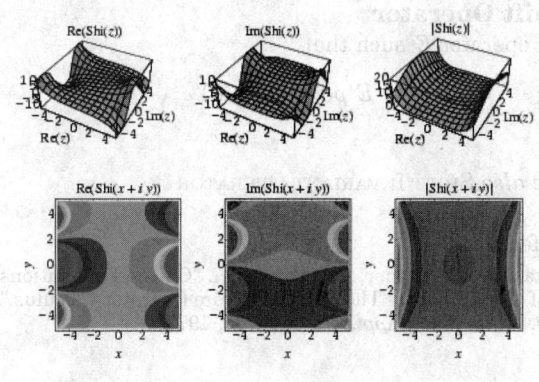

$$\text{Shi}(z) = \int_0^z \frac{\sinh t}{t}\, dt.$$

The function is given by the *Mathematica* command `SinhIntegral[z]`.

See also CHI, COSINE INTEGRAL, SINE INTEGRAL

References

Abramowitz, M. and Stegun, C. A. (Eds.). "Sine and Cosine Integrals." §5.2 in *Handbook of Mathematical Functions with Formulas, Graphs, and Mathematical Tables, 9th printing.* New York: Dover, pp. 231–233, 1972.

Shidlovskii Theorem

Let $f_1(z), \ldots, f_m(z)$ for $m \geq 1$ be a set of E-FUNCTIONS that (1) form a solution of the system of differential equations

$$y_k' = q_{k0} + \sum_{j=1}^m q_{kj} y_j$$

for $q_{kj} \in \mathbb{C}(z)$ and $k = 1, \ldots, m$ and (2) are ALGEBRAICALLY INDEPENDENT over $\mathbb{C}(z)$. Then for all $\alpha \in \mathbb{A}$, where $\mathbb{A}$ denotes the set of ALGEBRAIC NUMBERS with $\alpha \neq 0$ and distinct from singularities of the differential equations, the numbers $f_1(\alpha), \ldots, f_m(\alpha)$ are ALGEBRAICALLY INDEPENDENT (Nesterenko 1999).

See also ALGEBRAICALLY INDEPENDENT, E-FUNCTION

References

Nesterenko, Yu. V. §1.2 in *A Course on Algebraic Independence: Lectures at IHP 1999.* http://www.math.jussieu.fr/~nesteren/.

Shidlovskii, A. B. *Transcendental Numbers.* New York: de Gruyter, 1989.

Shift

A TRANSLATION without ROTATION or distortion.

See also DILATION, EXPANSION, ROTATION, TRANSLATION, TWIRL

Shift Operator

An operator E such that

$$E^a p(x) = p(x+a).$$

See also SHIFT-INVARIANT OPERATOR

References

Rota, G.-C.; Kahaner, D.; Odlyzko, A. "On the Foundations of Combinatorial Theory. VIII: Finite Operator Calculus." *J. Math. Anal. Appl.* **42**, 684–760, 1973.

Shift Property

DELTA FUNCTION

Shift Transformation

The transformation

$$T(x) = \operatorname{frac}\left(\frac{1}{x}\right) = \frac{1}{x} - \left\lfloor \frac{1}{x} \right\rfloor,$$

where $\operatorname{frac}(x)$ is the FRACTIONAL PART of x and $\lfloor x \rfloor$ is the FLOOR FUNCTION, that takes a CONTINUED FRACTION $[a_1, a_2, \ldots]$ to $[a_2, a_3, \ldots]$.

See also GAUSS-KUZMIN-WIRSING CONSTANT

References

Viader, P.; Paradis, J.; and Bibiloni, L. "A New Light on Minkowski's ?(x) Function." *J. Number Th.* **73**, 212–227, 1998.

Shifted Factorial

POCHHAMMER SYMBOL, RISING FACTORIAL

Shift-Invariant Operator

An operator T which commutes with all SHIFT OPERATORS E^a, so

$$TE^a = E^a T$$

for all real a in a FIELD. Any two shift-invariant operators commute.

See also DELTA OPERATOR, HEAVISIDE CALCULUS, SHIFT OPERATOR

References

Rota, G.-C.; Kahaner, D.; Odlyzko, A. "On the Foundations of Combinatorial Theory. VIII: Finite Operator Calculus." *J. Math. Anal. Appl.* **42**, 684–760, 1973.

Shimura-Taniyama Conjecture

TANIYAMA-SHIMURA CONJECTURE

Shimura-Taniyama-Weil Conjecture

TANIYAMA-SHIMURA CONJECTURE

Shoe

HOOK, SHOE SURFACE

Shoe Surface

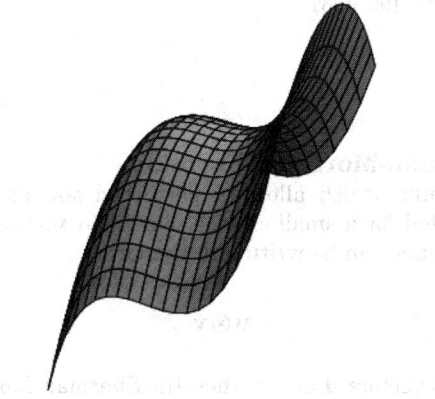

A surface given by the PARAMETRIC EQUATIONS

$$x(u, v) = u \tag{1}$$

$$y(u, v) = v \tag{2}$$

$$z(u, v) = \tfrac{1}{3} u^3 - \tfrac{1}{2} v^2. \tag{3}$$

The coefficients of the coefficients of the FIRST FUNDAMENTAL FORM are

$$E = 1 + u^4 \tag{4}$$

$$F = -u^2 v \tag{5}$$

$$G = 1 + v^2, \tag{6}$$

and the SECOND FUNDAMENTAL FORM coefficients are

$$e = \frac{2u}{\sqrt{1 + u^4 + v^2}} \tag{7}$$

$$f = 0 \tag{8}$$

$$g = -\frac{1}{\sqrt{1 + u^4 + v^2}}, \tag{9}$$

giving AREA ELEMENT

$$dA = \sqrt{-\frac{2u}{\sqrt{1 + u^4 + v^2}}}\, du \wedge dv, \tag{10}$$

and GAUSSIAN and MEAN CURVATURES

$$K = -\frac{2u}{(1 + u^4 + v^2)^2} \tag{11}$$

$$H = \frac{2u(1 + v^2) - u^4 - 1}{2(1 + u^4 + v^2)^{3/2}}. \tag{12}$$

References

Gray, A. *Modern Differential Geometry of Curves and Surfaces with Mathematica, 2nd ed.* Boca Raton, FL: CRC Press, p. 965, 1997.

Shoemaker's Knife

ARBELOS

Short Exact Sequence

A short exact sequence of groups A, B, and C is given by two maps $\alpha : A \to B$ and $\beta : B \to C$ and is written

$$0 \to A \to B \to C \to 0.$$

Because it is an EXACT SEQUENCE, α is INJECTIVE, and β is SURJECTIVE. Moreover, the KERNEL of β is the image of α. Hence, the group A can be considered as a (normal) subgroup of B, and C is isomorphic to B/A.

A short exact sequence is said to split if there is a map $\gamma : C \to B$ such that $\beta \circ \gamma$ is the identity on C. This only happens when B is the DIRECT PRODUCT of A and C.

The notion of a short exact sequence also makes sense for MODULES and SHEAVES.

See also EXACT SEQUENCE, GROUP EXTENSION, LONG EXACT SEQUENCE, MODULE, PRINCIPAL BUNDLE

References

Atiyah, M. F. and MacDonald, I. G. *Introduction to Commutative Algebra.* Reading, MA: Addison-Wesley, pp. 22–24, 1969.
Fulton, W. *Algebraic Topology: A First Course.* New York: Springer-Verlag, p. 144, 1995.
Hilton, P. and Stammbach, U. *A Course in Homological Algebra.* New York: Springer-Verlag, 1997.
Munkres, J. *Elements of Algebraic Topology.* Reading, MA: Addison-Wesley, pp. 130–133, 1984.

Shortening

A KNOT used to shorten a long rope.

See also BEND (KNOT)

References

Owen, P. *Knots.* Philadelphia, PA: Courage, p. 65, 1993.

Shortest Path

DIJKSTRA'S ALGORITHM, GRAPH GEODESIC

Shortness Exponent

Let $v(G)$ be the number of vertices in a GRAPH G and $h(G)$ the length of the maximum cycle in G. Then the shortness exponent of a class of graphs $\mathscr{G}$ is defined by

$$\sigma(\mathscr{G}) = \lim_{G \in \mathscr{G}} \inf \frac{\ln h(G)}{\ln v(G)}.$$

References

Grünbaum, B. and Walther, H. "Shortness Exponents of Families of Graphs." *J. Combin. Th. A* **14**, 364–385, 1973.
Owens, P. J. "Bipartite Cubic Graphs and a Shortness Exponent." *Disc. Math.* **44**, 327–330, 1983.

Shovelton's Rule

Let the values of a function $f(x)$ be tabulated at points x_i equally spaced by $h = x_{i+1} - x_i$, so $f_1 = f(x_1)$, $f_2 = f(x_2)$, ..., $f_{11} = f(x_{11})$. Then Shovelton's rule approximating the integral of $f(x)$ is given by the NEWTON-COTES-like formula

$$\int_{x_1}^{x_{11}} f(x) \, dx = \tfrac{5}{126} h [8(f_1 + f_{11}) + 35(f_2 + f_4 + f_8 + f_{10})$$

$$+ 15(f_3 + f_5 + f_7 + f_9) + 36 f_6].$$

See also BODE'S RULE, HARDY'S RULE, NEWTON-COTES FORMULAS, SIMPSON'S 3/8 RULE, SIMPSON'S RULE, TRAPEZOIDAL RULE, WEDDLE'S RULE

References

King, A. E. "Approximate Integration. Note on Quadrature Formulae: Their Construction and Application to Actuarial Functions." *Trans. Faculty of Actuaries* **9**, 218–231, 1923.
Sheppard, W. F. "Some Quadrature-Formulæ." *Proc. London Math. Soc.* **32**, 258–277, 1900.
Whittaker, E. T. and Robinson, G. *The Calculus of Observations: A Treatise on Numerical Mathematics, 4th ed.* New York: Dover, p. 151, 1967.

Shuffle

The randomization of a deck of CARDS by repeated interleaving. More generally, a shuffle is a rearrangement of the elements in an ordered list. Shuffling by exactly interleaving two halves of a deck is called a RIFFLE SHUFFLE. Normal shuffling leaves gaps of different lengths between the two layers of cards and so randomizes the order of the cards.

A deck of 52 CARDS must be shuffled seven times for it to be randomized (Aldous and Diaconis 1986, Bayer and Diaconis 1992). This is intermediate between too few shuffles and the decreasing effectiveness of many shuffles. One of Bayer and Diaconis's randomness CRITERIA, however, gives $3 \lg k/2$ shuffles for a k-card deck, yielding 11–12 shuffles for 52 CARDS. Amazingly, if a deck of n cards is shuffled by successively exchanging the cards in position 1, 2, ..., n with cards in randomly chosen positions (a so-called EXCHANGE SHUFFLE), then for $n \geq 18$, the identity permutation (i.e., the original state before the cards were shuffled) is the most likely (Goldstine and Moews 2000).

Keller (1995) shows that roughly $\ln k$ shuffles are needed just to randomize the bottom card.

See also BAYS' SHUFFLE, CARDS, EXCHANGE SHUFFLE, FARO SHUFFLE, MONGE'S SHUFFLE, PERFECT SHUFFLE, RIFFLE SHUFFLE

References

Aldous, D. and Diaconis, P. "Shuffling Cards and Stopping Times." *Amer. Math. Monthly* **93**, 333–348, 1986.

Bayer, D. and Diaconis, P. "Trailing the Dovetail Shuffle to Its Lair." *Ann. Appl. Probability* **2**, 294–313, 1992.

Goldstein, D. ad Moews, D. The Identity Is the Most Likely Exchange Shuffle for Large *n*. 6 Oct 2000. http://xxx.lanl.gov/abs/math.CO/0010066/.

Keller, J. B. "How Many Shuffles to Mix a Deck?" *SIAM Review* **37**, 88–89, 1995.

Morris, S. B. "Practitioner's Commentary: Card Shuffling." *UMAP J.* **15**, 333–338, 1994.

Morris, S. B. *Magic Tricks, Card Shuffling, and Dynamic Computer Memories.* Washington, DC: Math. Assoc. Amer., 1998.

Rosenthal, J. W. "Card Shuffling." *Math. Mag.* **54**, 64–67, 1981.

Siamese Dodecahedron

SNUB DISPHENOID

Siamese Method

A method for constructing MAGIC SQUARES of ODD order, also called DE LA LOUBERE'S METHOD.

See also MAGIC SQUARE

Sibling

Two nodes connected to the same node which are same distance from the ROOT NODE in a ROOTED TREE are called siblings.

See also CHILD, ROOT NODE, ROOTED TREE, TREE

Sicherman Dice

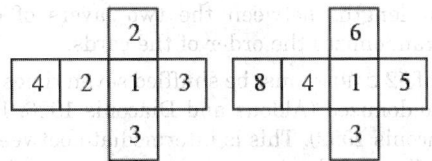

A pair of DICE which have the same ODDS for throwing every number as a normal pair of 6-sided DICE. They are the only such alternate arrangement if face values are required to be positive. However, if faces are permitted to have zero value (i.e., to be blank), then two additional possible equal-odds pairs of dice are obtained by subtracting one from each face on either of the two dice and adding one to each face the other. If negative values are permitted, there are an infinite number of equal-odds dice.

See also DICE, EFRON'S DICE

Sici Spiral

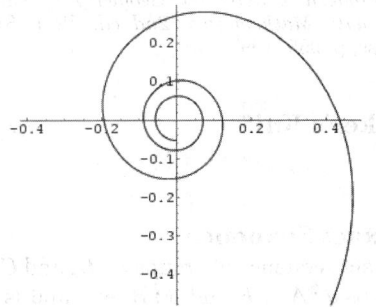

The spiral

$$x = c\,\mathrm{ci}(t)$$

$$y = c\left[\mathrm{si}(t) - \tfrac{1}{2}\pi\right].$$

where $\mathrm{ci}(t)$ and $\mathrm{si}(t)$ are the COSINE INTEGRAL and SINE INTEGRAL, respectively, and c is a constant.

See also COSINE INTEGRAL, SINE INTEGRAL, SPIRAL

References

von Seggern, D. *CRC Standard Curves and Surfaces.* Boca Raton, FL: CRC Press, pp. 204 and 270, 1993.

Side

The edge of a POLYGON or face of a POLYHEDRON are sometimes called sides.

Sidon Sequence

B_2-SEQUENCE

Siegel Disk Fractal

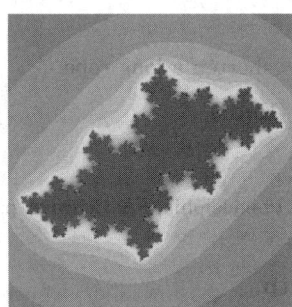

A JULIA SET with $c = -0.390541 - 0.586788i$. The FRACTAL somewhat resembles the better known MANDELBROT SET.

See also DENDRITE FRACTAL, DOUADY'S RABBIT FRACTAL, JULIA SET, MANDELBROT SET, SAN MARCO FRACTAL

References

Wagon, S. *Mathematica in Action.* New York: W. H. Freeman, p. 176, 1991.

Siegel Modular Function

SIEGEL THETA FUNCTION

Siegel Theta Function

A Γ_n-invariant meromorphic function on the space of all $p \times p$ symmetric COMPLEX MATRICES $Z = X + iY$ with positive definite IMAGINARY PART. It is defined by

$$\Theta(Z, \mathbf{s}) = \sum_{\mathbf{t}} e^{\pi i \mathbf{t}^{\mathrm{T}} Z \mathbf{t} + 2\pi i \mathbf{t}^{\mathrm{T}} \mathbf{s}},$$

where $\mathbf{s}$ is a complex p-vector, $\mathbf{t}$ is an integer p-vector that ranges over the entire p-D lattice of integers, and A^{T} denotes a matrix (or vector) transpose.

This function was investigated by many of the luminaries of nineteenth century mathematics, Riemann , Weierstrass , Frobenius , Poincaré. Umemura has expressed the ROOTS of an arbitrary POLYNOMIAL in terms of Siegel theta functions (Mumford 1984). The Siegel theta functions is implemented in *Mathematica* as `SiegelTheta` in the *Mathematica* add-on package `NumberTheory`SiegelTheta` (which can be loaded with the command $<$ $<$ `NumberTheory`).

See also RIEMANN THETA FUNCTION

References

Iyanaga, S. and Kawada, Y. (Eds.). "Siegel Modular Functions." §34F in *Encyclopedic Dictionary of Mathematics.* Cambridge, MA: MIT Press, pp. 131–132, 1980.
Mumford, D. Part C in *Tata Lectures on Theta. II. Jacobian Theta Functions and Differential Equations.* Boston, MA: Birkhäuser, 1984.
Siegel, C. L. *Topics in Complex Function Theory, Vol. 2: Automorphic Functions and Abelian Integrals.* New York: Wiley, p. 163, 1988.

Siegel's Paradox

If a fixed FRACTION x of a given amount of money P is lost, and then the same FRACTION x of the remaining amount is gained, the result is less than the original and equal to the final amount if a FRACTION x is first gained, then lost. This can easily be seen from the fact that

$$[P(1-x)](1+x) = P(1-x^2) < P.$$

$$[P(1+x)](1-x) = P(1-x^2) < P.$$

Siegel's Theorem

There are at least two Siegel's theorems. The first states that an ELLIPTIC CURVE can have only a finite number of points with INTEGER coordinates.

The second states that if ξ is an ALGEBRAIC NUMBER of degree r, then there is an $A(\xi)$ depending only on ξ such that

$$\left| \xi - \frac{p}{q} \right| > \frac{A(\xi)}{q^{2r^{1/2}}}$$

for all integer p and q (Landau 1970, pp. 37–56; Hardy 1999, p. 79).

See also ELLIPTIC CURVE, ROTH'S THEOREM, THUE-SIEGEL-ROTH THEOREM

References

Davenport, H. "Siegel's Theorem." Ch. 21 in *Multiplicative Number Theory, 2nd ed.* New York: Springer-Verlag, pp. 126–125, 1980.
Hardy, G. H. *Ramanujan: Twelve Lectures on Subjects Suggested by His Life and Work, 3rd ed.* New York: Chelsea, 1999.
Landau, E. *Vorlesungen über Zahlentheorie, Vol. 3.* New York: Chelsea, 1970.

Siegel's Upper Half-Space

See also HALF-SPACE

Sierpinski Arrowhead Curve

A FRACTAL which can be written as a LINDENMAYER SYSTEM with initial string `"YF"`, STRING REWRITING rules `"X" -> "YF+XF+Y"`, `"Y" -> "XF-YF-X"`, and angle $60°$.

See also DRAGON CURVE, HILBERT CURVE, KOCH SNOWFLAKE, LINDENMAYER SYSTEM, PEANO CURVE, PEANO-GOSPER CURVE, SIERPINSKI CURVE, SIERPINSKI SIEVE

References

Dickau, R. M. "Two-Dimensional L-Systems." http://forum.swarthmore.edu/advanced/robertd/lsys2d.html.
Weisstein, E. W. "Fractals." MATHEMATICA NOTEBOOK FRACTAL.M.

Sierpinski Carpet

A FRACTAL which is constructed analogously to the SIERPINSKI SIEVE, but using squares instead of triangles. Let N_n be the number of black boxes, L_n the length of a side of a white box, and A_n the fractional

AREA of black boxes after the nth iteration. Then

$$N_n = 8^n \tag{1}$$

$$L_n = (\tfrac{1}{3})^n = 3^{-n} \tag{2}$$

$$A_n = L_n^2 N_n = (\tfrac{8}{9})^n. \tag{3}$$

The CAPACITY DIMENSION is therefore

$$d_{\mathrm{cap}} = -\lim_{n \to \infty} \frac{\ln N_n}{\ln L_n} = -\lim_{n \to \infty} \frac{\ln(8^n)}{\ln(3^{-n})} = \frac{\ln 8}{\ln 3} = \frac{3 \ln 2}{\ln 3}$$

$$= 1.892789260\ldots \tag{4}$$

See also MENGER SPONGE, SIERPINSKI SIEVE

References

Dickau, R. M. "The Sierpinski Carpet." http://forum.swarthmore.edu/advanced/robertd/carpet.html.

Peitgen, H.-O.; Jürgens, H.; and Saupe, D. *Chaos and Fractals: New Frontiers of Science.* New York: Springer-Verlag, pp. 112–121, 1992.

Weisstein, E. W. "Fractals." MATHEMATICA NOTEBOOK FRACTAL.M.

Sierpiński Constant

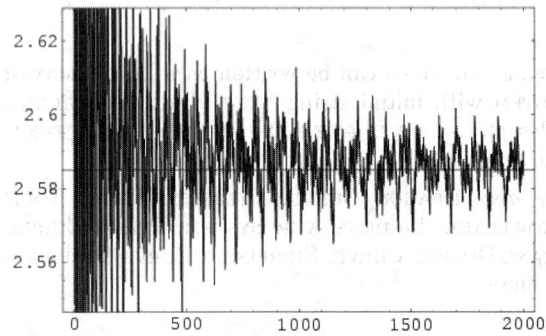

Let the SUM OF SQUARES FUNCTION $r_k(n)$ denote the number of representations of n by k squares, then the SUMMATORY FUNCTION of $r_2(k)/k$ has the ASYMPTOTIC expansion

$$\sum_{k=1}^{n} \frac{r_2(k)}{k} = K + \pi \ln n + \mathcal{O}(n^{-1/2}),$$

where $K = 2.5849817596$ is the Sierpinski constant. The above plot shows

$$\left[\sum_{k=1}^{n} \frac{r_2(k)}{k} \right] - \pi \ln n,$$

with the value of K indicated as the solid horizontal line.

See also SUM OF SQUARES FUNCTION

References

Sierpinski, W. *Oeuvres Choisies, Tome 1.* Editions Scientifiques de Pologne, 1974.

Sierpinski Curve

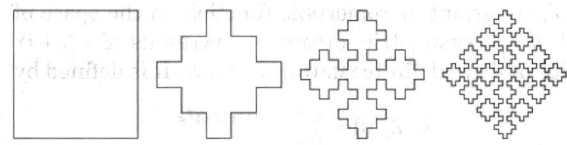

There are several FRACTAL curves associated with Sierpinski. The above curve is one example, and the SIERPINSKI ARROWHEAD CURVE is another. The limit of the curve illustrated above has AREA

$$A = \tfrac{5}{12}.$$

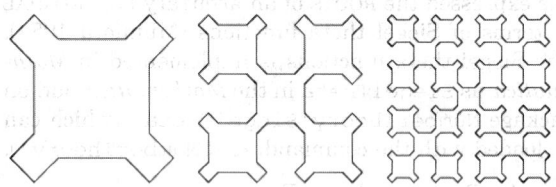

The AREA for a related curve due to Sierpinski (1912) illustrated above is

$$A = \tfrac{1}{3}(7 - 4\sqrt{2}).$$

(Steinhaus 1983, pp. 102–103; Cundy and Rollett 1989; Wells 1991, p. 229).

See also EXTERIOR SNOWFLAKE, GOSPER ISLAND, HILBERT CURVE, KOCH ANTISNOWFLAKE, KOCH SNOWFLAKE, PEANO CURVE, PEANO-GOSPER CURVE, SIERPINSKI ARROWHEAD CURVE

References

Cundy, H. and Rollett, A. *Mathematical Models, 3rd ed.* Stradbroke, England: Tarquin Pub., pp. 67–68, 1989.

Dickau, R. M. "Two-Dimensional L-Systems." http://forum.swarthmore.edu/advanced/robertd/lsys2d.html.

Gardner, M. *Penrose Tiles and Trapdoor Ciphers... and the Return of Dr. Matrix, reissue ed.* New York: W. H. Freeman, p. 34, 1989.

Sierpinski, W. *Bull. l'Acad. des Sciences Cracovie A*, 462–478, 1912.

Steinhaus, H. *Mathematical Snapshots, 3rd ed.* New York: Dover, 1999.

Wagon, S. *Mathematica in Action.* New York: W. H. Freeman, p. 207, 1991.

Wells, D. *The Penguin Dictionary of Curious and Interesting Geometry.* London: Penguin, p. 229, 1991.

Sierpinski Gasket

SIERPINSKI SIEVE

Sierpinski Number of the First Kind

Numbers OF THE FORM $S_n \equiv n^n + 1$. The first few are 2, 5, 28, 257, 3126, 46657, 823544, 16777217, ... (Sloane's A014566). Sierpinski proved that if S_n is

PRIME with $n \geq 2$, then $S_n = F_{m+2^m}$, where F_m is a FERMAT NUMBER with $m \geq 0$. The first few such numbers are $F_1 = 5$, $F_3 = 257$, F_6, F_{11}, F_{20}, and F_{37}. Of these, 5 and 257 are PRIME, and the first unknown case is $F_{37} > 10^{3 \times 10^{10}}$.

See also CULLEN NUMBER, CUNNINGHAM NUMBER, FERMAT NUMBER, WOODALL NUMBER

References

Madachy, J. S. *Madachy's Mathematical Recreations*. New York: Dover, p. 155, 1979.
Ribenboim, P. *The Book of Prime Number Records, 2nd ed.* New York: Springer-Verlag, p. 74, 1989.
Sloane, N. J. A. Sequences A014566 in "An On-Line Version of the Encyclopedia of Integer Sequences." http://www.re-search.att.com/~njas/sequences/eisonline.html.

Sierpiński Number of the Second Kind

A number k satisfying SIERPINSKI'S COMPOSITE NUMBER THEOREM, i.e., such that $k \cdot 2^n + 1$ is COMPOSITE for every $n \geq 1$. The smallest known is $k = 78,557$, but there remain 35 smaller candidates (the smallest of which is 4847) which are known to generate only composite numbers for $n \leq 18,000$ or more (Ribenboim 1996, p. 358).

Let $a(k)$ be smallest n for which $(2k-1) \cdot 2^n + 1$ is PRIME, then the first few values are 0, 1, 1, 2, 1, 1, 2, 1, 3, 6, 1, 1, 2, 2, 1, 8, 1, 1, 2, 1, 1, 2, 2, 583, ... (Sloane's A046067). The second smallest n are given by 1, 2, 3, 4, 2, 3, 8, 2, 15, 10, 4, 9, 4, 4, 3, 60, 6, 3, 4, 2, 11, 6, 9, 1483, ... (Sloane's A046068). Quite large n can be required to obtain the first prime even for small k. For example, the smallest prime OF THE FORM $383 \cdot 2^n + 1$ is $383 \cdot 2^{6393} + 1$. There are an infinite number of Sierpinski numbers which are PRIME.

The smallest odd k such that $k + 2^n$ is COMPOSITE for all $n < k$ are 773, 2131, 2491, 4471, 5101, ...

See also MERSENNE NUMBER, RIESEL NUMBER, SIERPINSKI'S COMPOSITE NUMBER THEOREM

References

Buell, D. A. and Young, J. "Some Large Primes and the Sierpinski Problem." SRC Tech. Rep. 88004, Supercomputing Research Center, Lanham, MD, 1988.
Jaeschke, G. "On the Smallest k such that $k \cdot 2^N + 1$ are Composite." *Math. Comput.* **40**, 381–384, 1983.
Jaeschke, G. Corrigendum to "On the Smallest k such that $k \cdot 2^N + 1$ are Composite." *Math. Comput.* **45**, 637, 1985.
Keller, W. "Factors of Fermat Numbers and Large Primes of the Form $k \cdot 2^n + 1$." *Math. Comput.* **41**, 661–673, 1983.
Keller, W. "Factors of Fermat Numbers and Large Primes of the Form $k \cdot 2^n + 1$, II." In prep.
Ribenboim, P. *The New Book of Prime Number Records.* New York: Springer-Verlag, pp. 357–359, 1996.
Sierpinski, W. "Sur un problème concernant les nombres $k \cdot 2^n + 1$." *Elem. d. Math.* **15**, 73–74, 1960.
Sloane, N. J. A. Sequences A046067 and A046068 in "An On-Line Version of the Encyclopedia of Integer Sequences." http://www.research.att.com/~njas/sequences/eisonline.html.

Sierpinski Sieve

A FRACTAL described by Sierpinski in 1915. It is also called the SIERPINSKI GASKET or SIERPINSKI TRIANGLE. The curve can be written as a LINDENMAYER SYSTEM with initial string `"FXF-FF-FF"`, STRING REWRITING rules `"F" -> "FF"`, `"X" -> "-FXF++FXF++FXF-"`, and angle 60°.

Let N_n be the number of black triangles after iteration n, L_n the length of a side of a triangle, and A_n the fractional AREA which is black after the nth iteration. Then

$$N_n = 3^n \tag{1}$$

$$L_n = \left(\tfrac{1}{2}\right)^n = 2^{-n} \tag{2}$$

$$A_n = L_n^2 N_n = \left(\tfrac{3}{4}\right)^n. \tag{3}$$

The CAPACITY DIMENSION is therefore

$$d_{\text{cap}} = -\lim_{n \to \infty} \frac{\ln N_n}{\ln L_n} = -\lim_{n \to \infty} \frac{\ln(3^n)}{\ln(2^{-n})} = \frac{\ln 3}{\ln 2}$$

$$= 1.584962500\ldots. \tag{4}$$

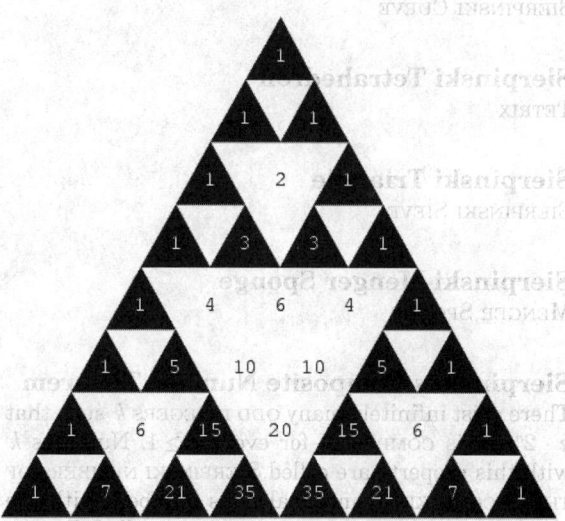

In PASCAL'S TRIANGLE, coloring all ODD numbers black and EVEN numbers white produces a Sierpinski sieve (Guy 1990).

See also LINDENMAYER SYSTEM, SIERPINSKI ARROW-HEAD CURVE, SIERPINSKI CARPET, TETRIX

References

Bulaevsky, J. "The Sierpinski Triangle Fractal." http://www.best.com/~ejad/java/fractals/sierpinski.shtml.
Crownover, R. M. *Introduction to Fractals and Chaos.* Sudbury, MA: Jones & Bartlett, 1995.

Dickau, R. M. "Two-Dimensional L-Systems." http://forum.s-warthmore.edu/advanced/robertd/lsys2d.html.

Dickau, R. M. "Typeset Fractals." *Mathematica J.* **7**, 15, 1997.

Dickau, R. "Sierpinski-Menger Sponge Code and Graphic." http://www.mathsource.com/cgi-bin/msitem22?0206-110.

Guy, R. K. "The Second Strong Law of Small Numbers." *Math. Mag.* **63**, 3–20, 1990.

Harris, J. W. and Stocker, H. "Sierpinski Gasket." §4.11.7 in *Handbook of Mathematics and Computational Science.* New York: Springer-Verlag, p. 115, 1998.

Lauwerier, H. *Fractals: Endlessly Repeated Geometric Figures.* Princeton, NJ: Princeton University Press, pp. 13–14, 1991.

Mandelbrot, B. B. *The Fractal Geometry of Nature.* New York: W. H. Freeman, 1983.

Peitgen, H.-O.; Jürgens, H.; and Saupe, D. *Chaos and Fractals: New Frontiers of Science.* New York: Springer-Verlag, pp. 78–88, 1992.

Peitgen, H.-O. and Saupe, D. (Eds.). *The Science of Fractal Images.* New York: Springer-Verlag, p. 282, 1988.

Sved, M. "Divisibility--with Visibility." *Math. Intell.* **10**, 56–64, 1988.

Wagon, S. *Mathematica in Action.* New York: W. H. Freeman, pp. 108 and 151–153, 1991.

Weisstein, E. W. "Fractals." MATHEMATICA NOTEBOOK FRACTAL.M.

Sierpinski Sponge

TETRIX

Sierpinski Square Snowflake

SIERPINSKI CURVE

Sierpinski Tetrahedron

TETRIX

Sierpinski Triangle

SIERPINSKI SIEVE

Sierpinski-Menger Sponge

MENGER SPONGE

Sierpinski's Composite Number Theorem

There exist infinitely many ODD INTEGERS k such that $k \cdot 2^n + 1$ is COMPOSITE for every $n \geq 1$. Numbers k with this property are called SIERPINSKI NUMBERS OF THE SECOND KIND, and analogous numbers with the plus sign replaced by a minus are called RIESEL NUMBERS. it is conjectured that the smallest SIERPINSKI NUMBER OF THE SECOND KIND is $k = 78,557$ and the smallest RIESEL NUMBER is $k = 509,203$.

See also CUNNINGHAM NUMBER, SIERPINSKI NUMBER OF THE SECOND KIND

References

Buell, D. A. and Young, J. "Some Large Primes and the Sierpinski Problem." SRC Tech. Rep. 88004, Supercomputing Research Center, Lanham, MD, 1988.

Jaeschke, G. "On the Smallest k such that $k \cdot 2^N + 1$ are Composite." *Math. Comput.* **40**, 381–384, 1983.

Jaeschke, G. Corrigendum to "On the Smallest k such that $k \cdot 2^N + 1$ are Composite." *Math. Comput.* **45**, 637, 1985.

Keller, W. "Factors of Fermat Numbers and Large Primes of the Form $k \cdot 2^n + 1$." *Math. Comput.* **41**, 661–673, 1983.

Keller, W. "Factors of Fermat Numbers and Large Primes of the Form $k \cdot 2^n + 1$, II." In prep.

Ribenboim, P. *The New Book of Prime Number Records.* New York: Springer-Verlag, pp. 357–359, 1996.

Riesel, H. "Några stora primtal." *Elementa* **39**, 258–260, 1956.

Sierpinski, W. "Sur un problème concernant les nombres $k \cdot 2^n + 1$." *Elem. d. Math.* **15**, 73–74, 1960.

See also COMPOSITE NUMBER, SIERPINSKI NUMBERS OF THE SECOND KIND, SIERPINSKI'S PRIME SEQUENCE THEOREM

Sierpinski's Conjecture

The conjecture that all integers > 1 occur as a value of the TOTIENT VALENCE FUNCTION (i.e., all integers > 1 occur as multiplicities). The conjecture was proved by Ford (1998ab).

See also CARMICHAEL'S TOTIENT FUNCTION CONJECTURE

References

Ford, K. "The Distribution of Totients." *Ramanujan J.* **2**, 67–151, 1998a.

Ford, K. "The Distribution of Totients, *Electron. Res. Announc. Amer. Math. Soc.* **4**, 27–34, 1998b.

Guy, R. K. *Unsolved Problems in Number Theory, 2nd ed.* New York: Springer-Verlag, p. 94, 1994.

Schlafly, A. and Wagon, S. "Carmichael's Conjecture on the Euler Function is Valid Below $10^{10,000,000}$." *Math. Comput.* **63**, 415–419, 1994.

Sierpinski's Prime Sequence Theorem

For any M, there exists a t' such that the sequence

$$n^2 + t',$$

where $n = 1, 2, \ldots$ contains at least M PRIMES.

See also DIRICHLET'S THEOREM, FERMAT $4N+1$ THEOREM, SIERPINSKI'S COMPOSITE NUMBER THEOREM

References

Abel, U. and Siebert, H. "Sequences with Large Numbers of Prime Values." *Amer. Math. Monthly* **100**, 167–169, 1993.

Ageev, A. A. "Sierpinski's Theorem is Deducible from Euler and Dirichlet." *Amer. Math. Monthly* **101**, 659–660, 1994.

Forman, R. "Sequences with Many Primes." *Amer. Math. Monthly* **99**, 548–557, 1992.

Garrison, B. "Polynomials with Large Numbers of Prime Values." *Amer. Math. Monthly* **97**, 316–317, 1990.

Sierpinski, W. "Les binômes $x^2 + n$ et les nombres premiers." *Bull. Soc. Roy. Sci. Liege* **33**, 259–260, 1964.

Sierpinski's Theorem

SIERPINSKI'S COMPOSITE NUMBER THEOREM, SIERPINSKI'S PRIME SEQUENCE THEOREM

Sieve

A process of successively crossing out members of a list according to a set of rules such that only some remain. The best known sieve is the ERATOSTHENES SIEVE for generating PRIME NUMBERS. In fact, numbers generated by sieves seem to share a surprisingly large number of properties with the PRIME NUMBERS.

See also BRUN'S SIEVE, HAPPY NUMBER, NUMBER FIELD SIEVE, PRIME NUMBER, QUADRATIC SIEVE, SIERPINSKI SIEVE, SIEVE OF ERATOSTHENES, WALLIS SIEVE

References

Halberstam, H. and Richert, H.-E. *Sieve Methods.* New York: Academic Press, 1974.
Hawkins, D. "Mathematical Sieves." *Sci. Amer.*, Dec. 1958.
Huskey, H. D. "Derrick Henry Lehmer (1905–1991)." *IEEE Ann. Hist. Comput.* **17**, 64–68, 1995.
Lehmer, D. H. "The Sieve Problem for All-Purpose Computers." *Math. Tables and Other Aids to Comput.* **7**, 6–14, 1953.
Lukes, R. F.; Patterson, C. D.; and Williams, H. C. "Numerical Sieving Devices: Their History and Some Applications." *Nieuw Arch. Wisk.* **13**, 113–139, 1995.
Pomerance, C. "A Tale of Two Sieves." *Not. Amer. Math. Soc.* **43**, 1473–1485, 1996.
Williams, H. C. and Shallit, J. O. "Factoring Integers Before Computers." In *Mathematics of Computation 1943–1993: A Half-Century of Computational Mathematics (Vancouver, BC, 1993)* (Ed. W. Gautschi). Providence, RI: Amer. Math. Soc., pp. 481–531, 1994.

Sieve Formula

INCLUSION-EXCLUSION PRINCIPLE

Sieve of Eratosthenes

An ALGORITHM for making tables of PRIMES. Sequentially write down the INTEGERS from 2 to the highest number n you wish to include in the table. Cross out all numbers > 2 which are divisible by 2 (every second number). Find the smallest remaining number > 2. It is 3. So cross out all numbers > 3 which are divisible by 3 (every third number). Find the smallest remaining number > 3. It is 5. So cross out all numbers > 5 which are divisible by 5 (every fifth number).

Continue until you have crossed out all numbers divisible by $\lfloor \sqrt{n} \rfloor$, where $\lfloor x \rfloor$ is the FLOOR FUNCTION.

The numbers remaining are PRIME. This procedure is illustrated in the above diagram which sieves up to 50, and therefore crosses out PRIMES up to $\lfloor \sqrt{50} \rfloor = 7$. If the procedure is then continued up to n, then the number of cross-outs gives the number of distinct PRIME FACTORS of each number.

See also PRIME NUMBER, SIEVE

References

Conway, J. H. and Guy, R. K. *The Book of Numbers.* New York: Springer-Verlag, pp. 127–130, 1996.
Gardner, M. *The Sixth Book of Mathematical Games from Scientific American.* Chicago, IL: University of Chicago Press, pp. 79–80, 1984.
Nagell, T. "General Remarks. The Sieve of Eratosthenes." §15 in *Introduction to Number Theory.* New York: Wiley, pp. 51–54, 1951.
Pappas, T. *The Joy of Mathematics.* San Carlos, CA: Wide World Publ./Tetra, pp. 100–101, 1989.
Ribenboim, P. *The New Book of Prime Number Records.* New York: Springer-Verlag, pp. 20–21, 1996.
Séroul, R. "The Sieve of Eratosthenes." §8.6 in *Programming for Mathematicians.* Berlin: Springer-Verlag, pp. 169–175, 2000.

Sievert Integral

The integral

$$\int_0^\theta e^{-x \, \sec \phi} \, d\phi.$$

References

Abramowitz, M. and Stegun, C. A. (Eds.). "Sievert Integral." §27.4 in *Handbook of Mathematical Functions with Formulas, Graphs, and Mathematical Tables, 9th printing.* New York: Dover, pp. 1000–1001, 1972.

Sievert's Surface

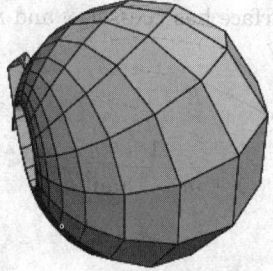

A constant-curvature surface which can be given parametrically by

$$x = r \cos \phi \tag{1}$$

$$y = r \sin \phi \tag{2}$$

$$z = \frac{\ln\left[\tan\left(\frac{1}{2}v\right)\right] + a(C+1)\cos v}{\sqrt{C}}, \tag{3}$$

where

$$\phi \equiv -\frac{u}{\sqrt{C+1}} + \tan^{-1}\left(\tan u \sqrt{C+1}\right) \qquad (4)$$

$$a \equiv \frac{2}{C+1-C\sin^2 v \cos^2 u} \qquad (5)$$

$$r \equiv \frac{a\sqrt{(C+1)(1+C\sin^2 u)}\sin v}{\sqrt{C}}, \qquad (6)$$

with $|u| < \pi/2$ and $0 < v < \pi$ (Reckziegel 1986).

The coefficients of the FIRST FUNDAMENTAL FORM are

$$E = \frac{64a\cos^2 u \cos^2 v}{[4+3a-a\cos(2u)+2a\cos^2 u \cos^2(2v)]^2} \qquad (7)$$

$$F = 0 \qquad (8)$$

$$G = \frac{64[(1+a)\csc v + a\cos^2 u \sin v]^2}{4a[4+3a-a\cos(2u)+2a\cos^2 u \cos^2(2v)]^2}, \qquad (9)$$

and the coefficients of the SECOND FUNDAMENTAL FORM are

$$e = \sqrt{\frac{a}{a+1}}$$

$$\times \frac{8a\cos^3 u \sin(3v) - 4\cos u[8+11a+3a\cos(2u)]}{[4+3a-a\cos(2u)+2a\cos^2 u \cos^2(2v)]^2} \qquad (10)$$

$$f = 0 \qquad (11)$$

$$g = \sqrt{\frac{a+1}{a}}$$

$$\times \frac{[4+5a+a\cos(2u)-2a\cos^2 u \cos(2v)]\csc\left(\frac{1}{2}v\right)\sec\left(\frac{1}{2}v\right)}{[4+3a-a\cos(2u)+2a\cos^2 u \cos^2(2v)]^2}. \qquad (12)$$

The Sievert surface has GAUSSIAN and MEAN CURVATURES given by

$$K = 1 \qquad (13)$$

$$H = \frac{1}{1+(a+1)\tan^2 u}. \qquad (14)$$

References

Fischer, G. (Ed.). Plate 87 in *Mathematische Modelle/Mathematical Models, Bildband/Photograph Volume.* Braunschweig, Germany: Vieweg, p. 83, 1986.

Gray, A. *Modern Differential Geometry of Curves and Surfaces with Mathematica, 2nd ed.* Boca Raton, FL: CRC Press, pp. 499–500, 1997.

Reckziegel, H. "Sievert's Surface." §3.4.4.3 in *Mathematical Models from the Collections of Universities and Museums* (Ed. G. Fischer). Braunschweig, Germany: Vieweg, pp. 38–39, 1986.

Sievert, H. *Über die Zentralflächen der Enneperschen Flachen konstanten Krümmungsmaßes.* Dissertation, Tübingen, 1886.

Sifting Property

The property

$$\int f(\mathbf{y})\delta(\mathbf{x}-\mathbf{y})d\mathbf{y} = f(\mathbf{x})$$

obeyed by the DELTA FUNCTION $\delta(\mathbf{x})$.

See also DELTA FUNCTION

References

Bracewell, R. "The Sifting Property." In *The Fourier Transform and Its Applications, 3rd ed.* New York: McGraw-Hill, pp. 74–77, 1999.

Sigma Algebra

Let X be a SET. Then a σ-algebra F is a nonempty collection of SUBSETS of X such that the following hold:

1. The EMPTY SET is in F.
2. If A is in F, then so is the complement of A.
3. If A_n is a SEQUENCE of elements of F, then the UNION of the A_ns is in F.

If S is any collection of subsets of X, then we can always find a σ-algebra containing S, namely the POWER SET of X. By taking the INTERSECTION of all σ-algebras containing S, we obtain the smallest such σ-algebra. We call the smallest σ-algebra containing S the σ-algebra generated by S.

See also BOREL SIGMA ALGEBRA, BOREL SPACE, MEASURABLE SET, MEASURABLE SPACE, MEASURE ALGEBRA, STANDARD SPACE

Sigma Function

DIVISOR FUNCTION

Sigmoid Curve

SIGMOID FUNCTION

Sigmoid Function

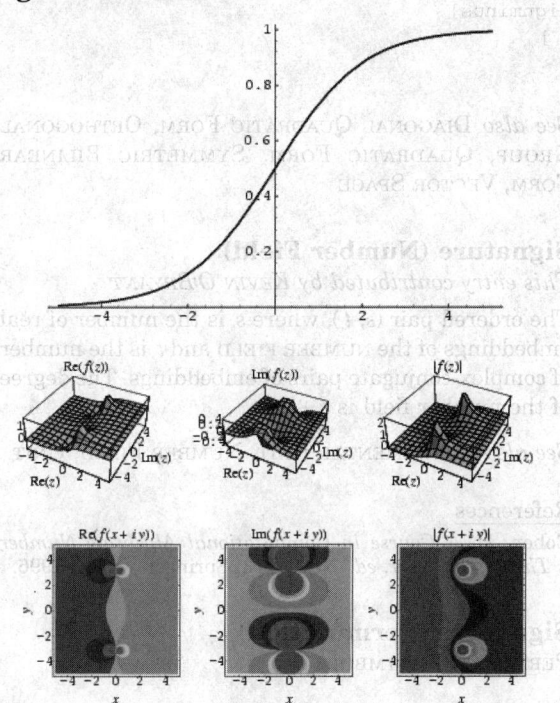

The function

$$y = \frac{1}{1 + e^{-x}}$$

which is the solution to the ORDINARY DIFFERENTIAL EQUATION

$$\frac{dy}{dx} = y(1-y).$$

It has an inflection point at $x = 0$, where

$$y''(x) = -\frac{e^x(e^x - 1)}{(e^x - 1)^3} = 0.$$

See also EXPONENTIAL FUNCTION, EXPONENTIAL RAMP

References
von Seggern, D. *CRC Standard Curves and Surfaces.* Boca Raton, FL: CRC Press, p. 124, 1993.

Sign

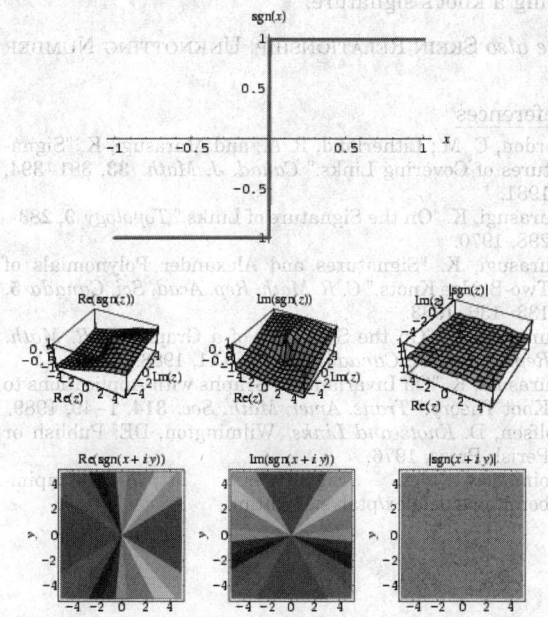

The sign of a number, also called SGN, is -1 for a NEGATIVE number (i.e., one with a MINUS SIGN "$-$"), 0 for the number ZERO, or $+1$ for a POSITIVE number (i.e., one with a PLUS SIGN "$+$").

See also ABSOLUTE VALUE, MINUS SIGN, NEGATIVE, PLUS SIGN, POSITIVE, SGN, ZERO

Signalizer Functor Theorem

$$\Theta(G; A) = \langle \theta(a) : a \in A - 1 \rangle$$

is an A-invariant solvable p'-subgroup of G.

Signature

PERMUTATION SYMBOL, SIGNATURE (KNOT), SIGNATURE (MATRIX), SIGNATURE (NUMBER FIELD), SIGNATURE (QUADRATIC FORM), SIGNATURE (RECURRENCE RELATION), SIGNATURE SEQUENCE

Signature (Knot)

The signature $s(K)$ of a KNOT K can be defined using the SKEIN RELATIONSHIP

$$s(\text{unknot}) = 0$$

$$s(K_+) - s(K_-) \in \{0, 2\},$$

and

$$4|s(K) \leftrightarrow \nabla(K)(2i) > 0,$$

where $\nabla(K)$ is the ALEXANDER-CONWAY POLYNOMIAL and $\nabla(K)(2i)$ is an ODD NUMBER.

Many UNKNOTTING NUMBERS can be determined using a knot's signature.

See also SKEIN RELATIONSHIP, UNKNOTTING NUMBER

References

Gordon, C. M.; Litherland, R. A.; and Murasugi, K. "Signatures of Covering Links." *Canad. J. Math.* **33**, 381–394, 1981.
Murasugi, K. "On the Signature of Links." *Topology* **9**, 283–298, 1970.
Murasugi, K. "Signatures and Alexander Polynomials of Two-Bridge Knots." *C. R. Math. Rep. Acad. Sci. Canada* **5**, 133–136, 1983.
Murasugi, K. "On the Signature of a Graph." *C. R. Math. Rep. Acad. Sci. Canada* **10**, 107–111, 1988.
Murasugi, K. "On Invariants of Graphs with Applications to Knot Theory." *Trans. Amer. Math. Soc.* **314**, 1–49, 1989.
Rolfsen, D. *Knots and Links.* Wilmington, DE: Publish or Perish Press, 1976.
Stoimenow, A. "Signatures." http://guests.mpim-bonn.mpg.de/alex/ptab/sig10.html.

Signature (Matrix)

A real, nondegenerate $n \times n$ SYMMETRIC MATRIX A, and its corresponding SYMMETRIC BILINEAR FORM $Q(v, w) = v^T A w$, has signature (p, q) if there is a nondegenerate matrix C such that $C^T A C$ is a diagonal matrix with p 1s and q -1s. In this case, $Q(Cv, Cw)$ is a DIAGONAL QUADRATIC FORM. For example,

$$A = \begin{bmatrix} 1 & 0 & 0 & 0 \\ 0 & 1 & 0 & 0 \\ 0 & 0 & 1 & 0 \\ 0 & 0 & 0 & -1 \end{bmatrix}$$

gives a SYMMETRIC BILINEAR FORM Q called the LORENTZIAN INNER PRODUCT, which has signature $(3, 1)$. The following *Mathematica* function returns the signature of a SYMMETRIC MATRIX as a list of three elements, corresponding to 1s, 0s, and -1s.

```
SignatureMatrix[a_List?MatrixQ] := Module[
    {
    q, ctr, diag, t2, signplus, signminus,
    v1 = Prepend[Table[0, {Length[a] - 1}], 1]
    },
    q[v_] := v.a.v;
        If[(t2 = q[v1]) != 0, v1 /=
Sqrt[Abs[t2]]];
    ctr = {v1};
    Do[
    v1 = NullSpace[ctr.a][[1]];
        If[(t2 = q[v1]) != 0, v1 /=
Sqrt[Abs[t2]]];
    AppendTo[ctr, v1],
    {Length[a] - 1}
    ];
    diag = ctr.a.Transpose[ctr];
    signplus = Count[diag, 1, 2];
    signminus = Count[diag, -1, 2];
```

```
    {signplus, Length[a] - signplus - signminus,
signminus}
    ]
```

See also DIAGONAL QUADRATIC FORM, ORTHOGONAL GROUP, QUADRATIC FORM, SYMMETRIC BILINEAR FORM, VECTOR SPACE

Signature (Number Field)

This entry contributed by KEVIN O'BRYANT

The ordered pair (s, t), where s is the number of real embeddings of the NUMBER FIELD and t is the number of complex-conjugate pairs of embeddings. The degree of the number field is $s + 2t$.

See also FUNDAMENTAL UNIT, NUMBER FIELD, UNIT

References

Cohen, H. *A Course in Computational Algebraic Number Theory*, 3rd. corr. ed. New York: Springer-Verlag, 1996.

Signature (Permutation)

PERMUTATION SYMBOL

Signature (Quadratic Form)

The signature of the QUADRATIC FORM

$$Q = y_1^2 + y_2^2 + \ldots + y_p^2 - y_{p+1}^2 - y_{p+2}^2 - \ldots - y_r^2$$

is the number p of POSITIVE squared terms in the reduced form. (The signature is sometimes defined as $2p - r$.)

See also P-SIGNATURE, RANK (QUADRATIC FORM), SYLVESTER'S INERTIA LAW, SYLVESTER'S SIGNATURE

References

Gradshteyn, I. S. and Ryzhik, I. M. *Tables of Integrals, Series, and Products*, 6th ed. San Diego, CA: Academic Press, p. 1105, 2000.

Signature (Recurrence Relation)

Let a sequence be defined by

$$A_{-1} = s$$

$$A_0 = 3$$

$$A_1 = r$$

$$A_n = rA_{n-1} - sA_{n-2} + A_{n-3}.$$

Also define the associated POLYNOMIAL

$$f(x) = x^3 - rx^2 + sx + 1,$$

and let Δ be its discriminant. The PERRIN SEQUENCE is a special case corresponding to $A_n(0, -1)$. The signature mod m of an INTEGER n with respect to the sequence $A_k(r, s)$ is then defined as the 6-tuple $(A_{-n-1}, A_{-n}, A_{-n+1}, A_{n-1}, A_n, A_{n+1}) \pmod{m}$.

1. An INTEGER n has an S-signature if its signature (mod n) is $(A_{-2}, A_{-1}, A_0, A_1, A_2)$.

2. An INTEGER n has a Q-signature if its signature (mod n) is CONGRUENT to (A, s, B, B, r, C) where, for some INTEGER a with $f(a) \equiv 0 \pmod{n}$, $A \equiv a^{-2} + 2a$, $B \equiv -ra^2 + (r^2 - s)a$, and $C \equiv a^2 + 2a^{-1}$.

3. An INTEGER n has an I-signature if its signature (mod n) is CONGRUENT to (r, s, D', D, r, s), where $D' + D \equiv rs - 3$ and $(D' + D) \equiv \Delta$.

See also PERRIN PSEUDOPRIME

References

Adams, W. and Shanks, D. "Strong Primality Tests that Are Not Sufficient." *Math. Comput.* **39**, 255–300, 1982.

Grantham, J. "Frobenius Pseudoprimes." http://www.clark.net/pub/grantham/pseudo/pseudo1.ps.

Signature Sequence

Let θ be an IRRATIONAL NUMBER, define $S(\theta) = \{c + d\theta : c, d \in \mathbb{N}\}$, and let $c_n(\theta) + d_n\theta(\theta)$ be the sequence obtained by arranging the elements of $S(\theta)$ in increasing order. A sequence x is said to be a signature sequence if there EXISTS a POSITIVE IRRATIONAL NUMBER θ such that $x = \{(c_n\theta)\}$, and x is called the signature of θ.

The signature of an IRRATIONAL NUMBER is a FRACTAL SEQUENCE. Also, if x is a signature sequence, then the LOWER-TRIMMED SUBSEQUENCE is $V(x) = x$.

References

Kimberling, C. "Fractal Sequences and Interspersions." *Ars Combin.* **45**, 157–168, 1997.

Signed Deviation

The signed deviation is defined by

$$\Delta u_i \equiv (u_i - \bar{u}),$$

so the average deviation is

$$\overline{\Delta u} = \overline{u_i - u} = \overline{u_i} - \bar{u} = 0.$$

See also ABSOLUTE DEVIATION, DEVIATION, DISPERSION (STATISTICS), MEAN DEVIATION, QUARTILE DEVIATION, STANDARD DEVIATION, VARIANCE

Significance

Let $\delta \equiv z \leq z_{\text{observed}}$. A value $0 \leq \alpha \leq 1$ such that $P(\delta) \leq \alpha$ is considered "significant" (i.e., is not simply due to chance) is known as an ALPHA VALUE. The PROBABILITY that a variate would assume a value greater than or equal to the observed value strictly by chance, $P(\delta)$, is known as a P-VALUE.

Depending on the type of data and conventional practices of a given field of study, a variety of different alpha values may be used. One commonly used terminology takes $P(\delta) \geq 5\%$ as "not significant,"

$1\% < P(\delta) < 5\%$, as "significant" (sometimes denoted *), and $P(\delta) < 1\%$ as "highly significant" (sometimes denoted **). Some authors use the term "almost significant" to refer to $5\% < P(\delta) < 10\%$, although this practice is not recommended.

See also ALPHA VALUE, COINCIDENCE, CONFIDENCE INTERVAL, P-VALUE, PROBABLE ERROR, SIGNIFICANCE TEST, STATISTICAL TEST

Significance Test

A test for determining the probability that a given result could not have occurred by chance (its SIGNIFICANCE).

See also SIGNIFICANCE, STATISTICAL TEST

References

Beyer, W. H. *CRC Standard Mathematical Tables, 28th ed.* Boca Raton, FL: CRC Press, pp. 491–492, 1987.

Significant Digits

When a number is expressed in SCIENTIFIC NOTATION, the number of significant figures is the number of DIGITS needed to express the number to within the uncertainty of measurement. For example, if a quantity had been measured to be 1.234 ± 0.002, four figures would be significant. No more figures should be given than are allowed by the uncertainty. For example, a quantity written as 1.234 ± 0.1 is incorrect; it should really be written as 1.2 ± 0.1.

The number of significant figures of a MULTIPLICATION or DIVISION of two or more quantities is equal to the smallest number of significant figures for the quantities involved. For ADDITION or SUBTRACTION, the number of significant figures is determined with the smallest significant figure of all the quantities involved. For example, the sum $10.234 + 5.2 + 100.3234$ is 115.7574, but should be written 115.8 (with rounding), since the quantity 5.2 is significant only to ± 0.1.

See also FRACTIONAL PART, INTEGER PART, NINT, ROUND, TRUNCATE

References

Kenney, J. F. and Keeping, E. S. "Significant Figures." §1.5 in *Mathematics of Statistics, Pt. 1, 3rd ed.* Princeton, NJ: Van Nostrand, pp. 8–9, 1962.

Mulliss, C. "Significant Figures and Rounding Rules." http://www.angelfire.com/oh/cmulliss/.

Significant Figures

SIGNIFICANT DIGITS

Signpost

A 6-POLYIAMOND.

References

Golomb, S. W. *Polyominoes: Puzzles, Patterns, Problems, and Packings, 2nd ed.* Princeton, NJ: Princeton University Press, p. 92, 1994.

Signum

SGN

Silver Constant

The REAL ROOT of the equation

$$x^3 - 5x^2 + 6x - 1 = 0,$$

given analytically by

$$2 + 2\cos\left(\tfrac{2}{7}\pi\right),$$

which is 3.2469.... It is the seventh BERAHA CONSTANT.

See also BERAHA CONSTANTS, SILVER RATIO, TRIGONOMETRY VALUES PI/7

References

Le Lionnais, F. *Les nombres remarquables.* Paris: Hermann, pp. 51 and 143, 1983.

Saaty, T. L. and Kainen, P. C. *The Four-Color Problem: Assaults and Conquest.* New York: Dover, p. 162, 1986.

Silver Mean

SILVER RATIO

Silver Ratio

The quantity defined by the CONTINUED FRACTION

$$\delta_S \equiv [2, 2, 2, \ldots] = 2 + \cfrac{1}{2 + \cfrac{1}{2 + \cfrac{1}{2 + \cdots}}}.$$

It follows that

$$(\delta_S - 1)^2 = 2,$$

so

$$\delta_S = \sqrt{2} + 1 = 2.41421\ldots.$$

See also GOLDEN RATIO, GOLDEN RATIO CONJUGATE

Silver Root

SILVER CONSTANT

Silverman Constant

$$\sum_{n=1}^{\infty} \frac{1}{\phi(n)\sigma(n)} = \prod_{p \text{ prime}} \left(1 + \sum_{k=1}^{\infty} \frac{1}{p^{2k} - p^{k-1}}\right)$$

$$= 1.786576459\ldots.$$

where $\phi(n)$ is the TOTIENT FUNCTION and $\sigma(n)$ is the DIVISOR FUNCTION.

References

Finch, S. "Favorite Mathematical Constants." http://www.mathsoft.com/asolve/constant/totient/totient.html.

Zimmerman, P. http://www.mathsoft.com/asolve/constant/totient/zimmermn.html.

Silverman's Sequence

Let $f(1) = 1$, and let $f(n)$ be the number of occurrences of n in a nondecreasing sequence of INTEGERS. then the first few values of $f(n)$ are 1, 2, 2, 3, 3, 4, 4, 4, 5, 5, 5, ... (Sloane's A001462). the asymptotic value of the nth term is $\phi^{2-\phi} n^{\phi-1}$, where ϕ is the GOLDEN RATIO.

References

Guy, R. K. "Silverman's Sequences." §E25 in *Unsolved Problems in Number Theory, 2nd ed.* New York: Springer-Verlag, pp. 225–226, 1994.

Sloane, N. J. A. Sequences A001462/M0257 in "An On-Line Version of the Encyclopedia of Integer Sequences." http://www.research.att.com/~njas/sequences/eisonline.html.

Similar

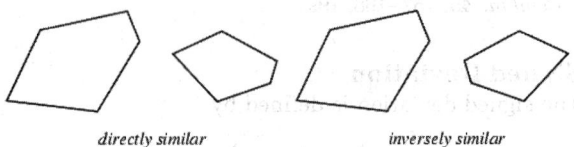

directly similar   *inversely similar*

Two figures are said to be similar when all corresponding ANGLES are equal. Two figures are DIRECTLY SIMILAR when all corresponding ANGLES are equal and described in the same rotational sense. This relationship is written $A \sim B$. (The symbol $\sim$ is also used to mean "is the same order of magnitude as" and "is ASYMPTOTIC to.") Two figures are INVERSELY SIMILAR when all corresponding ANGLES are equal and described in the opposite rotational sense.

See also COINCIDENT, CONGRUENT, DIRECTLY SIMILAR, HOMOTHETIC, INVERSELY SIMILAR, NAPOLEON'S THEOREM, SIMILAR MATRICES, SIMILAR TRIANGLES, SIMILARITY TRANSFORMATION, SPIRAL SIMILARITY

References

Durell, C. V. "Similar Figures." Ch. 1 in *Modern Geometry: The Straight Line and Circle.* London: Macmillan, pp. 1–9, 1928.

Kern, W. F. and Bland, J. R. "Similar Figures." §22 in *Solid Mensuration with Proofs, 2nd ed.* New York: Wiley, pp. 4 and 53–57, 1948.

Lachlan, R. "The Theory of Similar Figures." Ch. 9 in *An Elementary Treatise on Modern Pure Geometry*. London: Macmillian, pp. 128–147, 1893.
Project Mathematics. "Similarity." Videotape. http://www.projmath.caltech.edu/similar.htm.

Similar Matrices

Two SQUARE MATRICES A and B that are related by

$$B = X^{-1}AX, \qquad (1)$$

where X is a square NONSINGULAR MATRIX are said to be similar. A transformation of the form $X^{-1}AX$ is called a SIMILARITY TRANSFORMATION, or conjugation by X. For example,

$$\begin{bmatrix} 0 & 1 \\ 0 & 0 \end{bmatrix} \qquad (2)$$

and

$$\begin{bmatrix} 0 & 0 \\ 1 & 0 \end{bmatrix} \qquad (3)$$

are similar under conjugation by

$$C = \begin{bmatrix} 0 & 1 \\ 1 & 0 \end{bmatrix}. \qquad (4)$$

Similar matrices represent the same LINEAR TRANSFORMATION after a change of basis (for the domain and range simultaneously). Recall that a matrix corresponds to a LINEAR TRANSFORMATION, and a LINEAR TRANSFORMATION corresponds to a matrix after choosing a basis b_i,

$$T\left(\sum \lambda_i b_i\right) = \sum a_{ji}\lambda_i b_j \qquad (5)$$

Changing the basis changes the coefficients of the matrix,

$$T\left(\sum \gamma_i e_i\right) = \sum a'_{ji}\gamma_i e_j \qquad (6)$$

If $T(v) = Av$ uses the standard basis vectors, then T is the matrix CAC^{-1} using the basis vectors $b_i = Ce_i$.

See also BASIS (VECTOR SPACE), DIAGONAL MATRIX, DIAGONALIZABLE MATRIX, GROUP, JORDAN CANONICAL FORM, LINEAR TRANSFORMATION, RATIONAL CANONICAL FORM, SIMILARITY TRANSFORMATION, SQUARE MATRIX, VECTOR SPACE

References

Golub, G. H. and van Loan, C. F. *Matrix Computations, 3rd ed.* Baltimore, MD: Johns Hopkins University Press, p. 311, 1996.

Similar Triangles

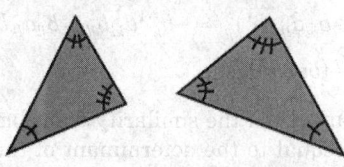

similar triangles

Two triangles are similar if their triples of vertex angles are the same.

See also DIRECTLY SIMILAR, INVERSELY SIMILAR, SIMILAR

Similarity Axis

D'ALEMBERT'S THEOREM

Similarity Dimension

To multiply the size of a d-D object by a factor a, $c \equiv a^d$ copies are required, and the quantity

$$d = \frac{\ln c}{\ln a}$$

is called the similarity dimension.

Similarity Point

External (or positive) and internal (or negative) similarity points of two CIRCLES with centers C and C' and RADII r and r' are the points E and I on the lines CC' such that

$$\frac{CE}{C'E} = \frac{r}{r'},$$

or

$$\frac{CI}{C'I} = -\frac{r}{r'}.$$

See also D'ALEMBERT'S THEOREM

Similarity Transformation

An ANGLE-preserving transformation. A similarity transformation has a transformation matrix A′ of the form

$$A' \equiv BAB^{-1}, \qquad (1)$$

where A and B are known as SIMILAR MATRICES (Golub and van Loan 1996, p. 311).

If A is an ANTISYMMETRIC MATRIX ($a_{ij} = -a_{ji}$) and B is an ORTHOGONAL MATRIX, then

$$\left(bab^{-1}\right)_{ij} = b_{ik}a_{kl}b_{lj}^{-1} = -b_{ik}a_{lk}b_{lj}^{-1}$$
$$= -b_{ki}^{\dagger}a_{lk}\left(b^{\dagger}\right)_{jl}^{-1} = -b_{ki}^{-1}a_{ki}b_{jl} = b_{jl}a_{lk}b_{ki}^{-1}$$
$$= -\left(bab^{-1}\right)_{ji}. \tag{2}$$

The DETERMINANT of the similarity transformation of a MATRIX is equal to the determinant of the original MATRIX

$$|BAB^{-1}| = |B||A||B^{-1}| = |B||A|\frac{1}{|B|} = |A|. \tag{3}$$

The determinant of a similarity transformation minus a multiple of the unit MATRIX is given by

$$|B^{-1}AB - \lambda I| = |B^{-1}AB - B^{-1}\lambda I B| = |B^{-1}(A - \lambda I)B|$$
$$= |B^{-1}||A - \lambda I||B| = |A - \lambda I|. \tag{4}$$

Similarity transformations and the concept of SELF-SIMILARITY are important foundations of FRACTALS and ITERATED FUNCTION SYSTEMS.

See also CONFORMAL MAPPING, DETERMINANT, DILATION, ITERATED FUNCTION SYSTEM, SIMILAR MATRICES

References

Croft, H. T.; Falconer, K. J.; and Guy, R. K. *Unsolved Problems in Geometry.* New York: Springer-Verlag, p. 3, 1991.
Golub, G. H. and van Loan, C. F. *Matrix Computations, 3rd ed.* Baltimore, MD: Johns Hopkins University Press, p. 311, 1996.
Lauwerier, H. *Fractals: Endlessly Repeated Geometric Figures.* Princeton, NJ: Princeton University Press, pp. 83–103, 1991.

Similitude Center

Also called a self-homologous point. If two SIMILAR figures lie in the plane but do not have parallel sides (they are not HOMOTHETIC), there exists a center of similitude which occupies the same homologous position with respect to the two figures. The LOCUS of similitude centers of two nonconcentric circles is another circle having the line joining the two homothetic centers as its DIAMETER.

There are a number of interesting theorems regarding three CIRCLES (Johnson 1929, pp. 151–152).

1. The external similitude centers of three circles are COLLINEAR.
2. Any two internal similitude centers are COLLINEAR with the third external one.
3. If the center of each circle is connected with the internal similitude center of the other three [sic], the connectors are CONCURRENT.
4. If one center is connected with the internal similitude center of the other two, the others with

the corresponding external centers, the connectors are CONCURRENT.

The six centers of similitude of three circles taken by pairs are the vertices of a COMPLETE QUADRILATERAL (Evelyn *et al.* 1974, pp. 21–22).

See also SIMILITUDE CENTER, SIMILITUDE CIRCLE

References

--. Problem 2819. *Amer. Math. Monthly* **28**, 229–230, 1921.
Casey, J. "Centers of Similitude." §6.2 in *A Sequel to the First Six Books of the Elements of Euclid, Containing an Easy Introduction to Modern Geometry with Numerous Examples, 5th ed., rev. enl.* Dublin: Hodges, Figgis, & Co., pp. 82–86, 1888.
Evelyn, C. J. A.; Money-Coutts, G. B.; and Tyrrell, J. A. *The Seven Circles Theorem and Other New Theorems.* London: Stacey International, pp. 21–22, 1974.
Johnson, R. A. *Modern Geometry: An Elementary Treatise on the Geometry of the Triangle and the Circle.* Boston, MA: Houghton Mifflin, pp. 19–27 and 151–153, 1929.
Lachlan, R. *An Elementary Treatise on Modern Pure Geometry.* London: Macmillian, p. 130, 1893.

Similitude Circle

The LOCUS of the SIMILITUDE CENTER of two circles.

See also INVARIABLE POINT, SIMILITUDE CENTER

References

Durell, C. V. *Modern Geometry: The Straight Line and Circle.* London: Macmillan, p. 135, 1928.
Johnson, R. A. *Modern Geometry: An Elementary Treatise on the Geometry of the Triangle and the Circle.* Boston, MA: Houghton Mifflin, pp. 307–310, 1929.
Lachlan, R. *An Elementary Treatise on Modern Pure Geometry.* London: Macmillian, p. 192, 1893.

Similitude Ratio

Two figures are HOMOTHETIC if they are related by a DILATION (a dilation is also known as a HOMOTHECY). This means that the connectors of corresponding points are CONCURRENT at a point which divides each connector in the same ratio k, known as the similitude ratio.

See also CONCURRENT, DILATION, HOMOTHECY, HOMOTHETIC

Simon Newcomb's Problem

Given a set P with $|P| = p$ elements consisting of c_1 numbers 1, c_2 numbers 2, ..., and c_n numbers n and

$$c_1 + c_2 + \ldots + c_n = p,$$

find the number of permutations with $k - 1$ rises (Comtet 1974, p. 246).

See also EULER NUMBER

References

Comtet, L. *Advanced Combinatorics: The Art of Finite and Infinite Expansions, rev. enl. ed.* Dordrecht, Netherlands: Reidel, 1974.

Dillon, J. F. and Roselle, D. P. "Simon Newcomb's Problem." *SIAM J. Appl. Math.* **17**, 1086–1093, 1969.

Kreweras, G. "Sur une class de problèmes de dénombrement liés au treillis des partitions d'entiers." *Cahiers Buro* **6**, 2–107, 1965.

Kreweras, G. "Sur une extension du problème dir 'de Simon Newcomb'." *Comptes rendus* **263**, 43–45, 1966.

Kreweras, G. "Traitement simultané du 'problème de Young' et du 'problème de Simon Newcomb'." *Cahiers Buro* **10**, 23–31, 1967.

Riordan, J. *An Introduction to Combinatorial Analysis.* New York: Wiley, pp. 216 and 265, 1958.

Simple Algebra

An ALGEBRA with no nontrivial IDEALS.

See also ALGEBRA, IDEAL, SEMISIMPLE ALGEBRA

Simple Continued Fraction

A CONTINUED FRACTION

$$\sigma = a_0 + \cfrac{b_1}{a_1 + \cfrac{b_2}{a_2 + \cfrac{b_3}{a_3 + \dots}}} \qquad (1)$$

in which the b_is are all unity, leaving a continued fraction OF THE FORM

$$\sigma = a_0 + \cfrac{1}{a_1 + \cfrac{1}{a_2 + \cfrac{1}{a_3 + \dots}}}. \qquad (2)$$

A simple continued fraction can be written in a compact abbreviated NOTATION as

$$\sigma = [a_0, a_1, a_2, a_3 \dots]. \qquad (3)$$

Bach and Shallit (1996) show how to compute the JACOBI SYMBOL in terms of the simple continued fraction of a RATIONAL NUMBER a/b.

See also CONTINUED FRACTION

References

Bach, E. and Shallit, J. *Algorithmic Number Theory, Vol. 1: Efficient Algorithms.* Cambridge, MA: MIT Press, pp. 343–344, 1996.

Simple Curve

simple curves　　　　　　　*nonsimple curves*

A curve is simple if it does not cross itself.

See also CLOSED CURVE, JORDAN CURVE

References

Krantz, S. G. "Closed Curves." §2.1.2 in *Handbook of Complex Analysis.* Boston, MA: Birkhäuser, pp. 19–20, 1999.

Simple Double Point

ORDINARY DOUBLE POINT

Simple Function

A simple function is a finite sum $\Sigma_i \, a_i \chi A_i$, where the functions χA_i are CHARACTERISTIC FUNCTIONS on a set A. Another description of a simple function is a function that takes on finitely many values in its range.

The collection of simple functions is CLOSED under addition and multiplication. In addition, it is easy to integrate a simple function. By approximating a given function f by simple functions, the LEBESGUE INTEGRAL of f can be calculated.

See also CHARACTERISTIC FUNCTION (SET), LEBESGUE INTEGRAL, SET

Simple Graph

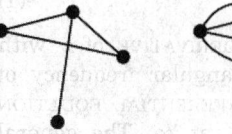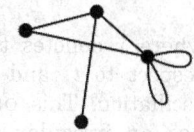

simple graph　　　*non-simple graph*　　　*non-simple graph with loops*

A GRAPH for which at most one EDGE connects any two nodes. Unless stated otherwise, the unqualified term "graph" usually refers to a *simple* graph. A non-simple graph with no loops but which can contain more than one edge between any two points is called a MULTIGRAPH.

See also ADJACENCY MATRIX, EDGE (GRAPH), GRAPH, MULTIGRAPH, REGULAR GRAPH, STEINITZ'S THEOREM

References

Skiena, S. *Implementing Discrete Mathematics: Combinatorics and Graph Theory with Mathematica.* Reading, MA: Addison-Wesley, p. 89, 1990.

Steinbach, P. *Field Guide to Simple Graphs.* Albuquerque, NM: Design Lab, 1990.

Simple Group

A simple group is a GROUP whose NORMAL SUBGROUPS (INVARIANT SUBGROUPS) are ORDER one or the whole of the original GROUP. Simple groups include ALTERNATING GROUPS, CYCLIC GROUPS, LIE-TYPE GROUPS (five varieties), and SPORADIC GROUPS (26 varieties, including the MONSTER GROUP). The CLASSIFICATION THEOREM of finite simple groups states that such groups can be classified completely into the five types:

1. CYCLIC GROUPS of PRIME ORDER,
2. ALTERNATING GROUPS of degree at least five
3. LIE-TYPE CHEVALLEY GROUPS,
4. LIE-TYPE (TWISTED CHEVALLEY GROUPS or the TITS GROUP), and
5. SPORADIC GROUPS.

BURNSIDE'S CONJECTURE states that every non-ABELIAN SIMPLE GROUP has EVEN ORDER.

See also ALTERNATING GROUP, BURNSIDE'S CONJECTURE, CHEVALLEY GROUPS, CLASSIFICATION THEOREM, CYCLIC GROUP, FEIT-THOMPSON THEOREM, FINITE GROUP, GROUP, INVARIANT SUBGROUP, LIE-TYPE GROUP, MONSTER GROUP, SCHUR MULTIPLIER, SPORADIC GROUP, TITS GROUP, TWISTED CHEVALLEY GROUPS

Simple Harmonic Motion

Simple harmonic motion refers to the periodic sinusoidal oscillation of an object or quantity. Simple harmonic motion is executed by any quantity obeying the DIFFERENTIAL EQUATION

$$\ddot{x} + \omega_0^2 x = 0. \tag{1}$$

where $\ddot{x}$ denotes the second DERIVATIVE of x with respect to t, and ω_0 is the angular frequency of oscillation. This ORDINARY DIFFERENTIAL EQUATION has an irregular SINGULARITY at ∞. The general solution is

$$x = A \sin(\omega_0 t) + B \cos(\omega_0 t) \tag{2}$$

$$= C \cos(\omega_0 t + \phi), \tag{3}$$

where the two constants A and B (or C and ϕ) are determined from the initial conditions.

Many physical systems undergoing small displacements, including any objects obeying Hooke's law, exhibit simple harmonic motion. This equation arises, for example, in the analysis of the flow of current in an electronic CL circuit (which contains a capacitor and an inductor). If a damping force such as Friction is present, an additional term $\beta\dot{x}$ must be added to the DIFFERENTIAL EQUATION and motion dies out over time.

See also DAMPED SIMPLE HARMONIC MOTION, SIMPLE HARMONIC-MOTION QUADRATIC PERTURBATION

Simple Harmonic Motion—Quadratic Perturbation

Given a simple harmonic oscillator with a quadratic perturbation ϵx^2,

$$\ddot{x} + \omega_0^2 x - \alpha\epsilon x^2 = 0. \tag{1}$$

find the first-order solution using a perturbation method. Write

$$x \equiv x_0 + \epsilon x_1 + \dots. \tag{2}$$

so

$$\ddot{x} = \ddot{x}_0 + \epsilon\ddot{x}_1 + \dots. \tag{3}$$

Plugging (2) and (3) back into (1) gives

$$(\ddot{x}_0 + \epsilon\ddot{x}_1) + (\omega_0^2 x_0 + \omega_0^2 \epsilon x_1) - \alpha\epsilon(x_0 + 2x_0 x_1\epsilon + \dots)$$
$$= 0. \tag{4}$$

Keeping only terms of order ϵ and lower and grouping, we obtain

$$(\ddot{x} + \omega_0^2 x_0) + (\ddot{x}_1 + \omega_0^2 x_1 - \alpha x_0^2)\epsilon = 0. \tag{5}$$

Since this equation must hold for all POWERS of ϵ, we can separate it into the two differential equations

$$\ddot{x}_0 + \omega_0^2 x_0 = 0 \tag{6}$$

$$\ddot{x}_1 + \omega_0^2 x_1 = \alpha x_0^2. \tag{7}$$

The solution to (6) is just

$$x_0 = A \cos(\omega_0 t + \phi). \tag{8}$$

Setting our clock so that $\phi = 0$ gives

$$x_0 = A \cos(\omega_0 t). \tag{9}$$

Plugging this into (7) then gives

$$\ddot{x}_1 + \omega_0^2 x_1 = \alpha A^2 \cos^2(\omega_0 t) \tag{10}$$

The two homogeneous solutions to (10) are

$$x_1 = \cos(\omega_0 t) \tag{11}$$
$$x_2 = \sin(\omega_0 t). \tag{12}$$

The particular solution to (10) is therefore given by

$$x_p(t) = -x_1(t) \int \frac{x_2(t)g(t)}{W(t)} \, dt + x_2(t) \int \frac{x_1(t)g(t)}{W(t)} \, dt. \tag{13}$$

where

$$g(t) = \alpha A^2 \cos^2(\omega_0 t). \tag{14}$$

and the WRONSKIAN is

$$W \equiv x_1\dot{x}_2 - \dot{x}_1 x_2 = \omega_0. \tag{15}$$

Plugging everything into (13),

$$x_p = \alpha A^2 \left[-\cos(\omega_0 t) \int \frac{\sin(\omega_0 t)\cos^2(\omega_0 t)}{\omega_0}\, dt \right.$$

$$\left. +\sin(\omega_0 t) \int \frac{\cos^3(\omega_0 t)}{\omega_0}\, dt \right]$$

$$= \frac{\alpha A^2}{\omega_0} \left\{ \sin(\omega_0 t) \int [1 - \sin^2(\omega_0 t)]\cos(\omega_0 t)\, dt \right.$$

$$\left. -\cos(\omega_0 t) \int \sin(\omega_0 t)\cos^2(\omega_0 t)\, dt \right\}. \quad (16)$$

Now let

$$u \equiv \sin(\omega_0 t) \quad (17)$$

$$du = \omega_0 \cos(\omega_0 t)\, dt \quad (18)$$

$$v \equiv \cos(\omega_0 t) \quad (19)$$

$$dv = -\omega_0 \sin(\omega_0 t)\, dt. \quad (20)$$

Then

$$x_p = \frac{\alpha A^2}{\omega_0^2} \left[\sin(\omega_0 t) \int (1 - u^2)\, du + \cos(\omega_0 t) \int v^2\, dv \right]$$

$$= \frac{\alpha A^2}{\omega_0^2} \left[\sin(\omega_0 t)\left(1 - \tfrac{1}{3}u^3\right) + \cos(\omega_0 t)\tfrac{1}{3}v^3 \right]$$

$$= \frac{\alpha A^2}{\omega_0^2} \left\{ \sin(\omega_0 t)\left[1 - \tfrac{1}{3}\sin^3(\omega_0 t)\right] + \tfrac{1}{3}\cos(\omega_0 t)\cos^3(\omega_0 t) \right\}$$

$$= \frac{\alpha A^2}{6\omega_0^2}[3 - \cos(2\omega_0 t)]. \quad (21)$$

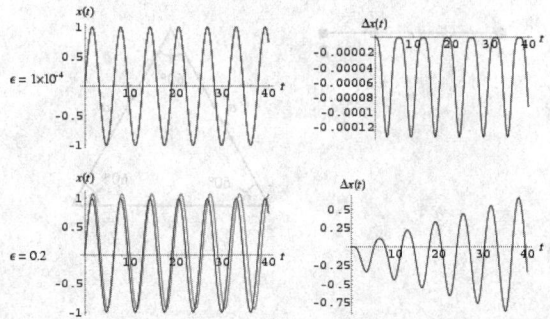

Plugging $x_0(t)$ and (21) into (2), we obtain the solution

$$x(t) = A\cos(\omega_0 t) - \frac{\alpha A^2}{6\omega_0^2}\, \epsilon[\cos(2\omega_0 t) - 3]. \quad (22)$$

As can be seen in the top figure above, this solution approximates $x(t)$ only for $\epsilon \ll 1$. As the lower figure shows, the differences from the unperturbed oscillator grow stronger over time for even relatively small values of ϵ.

Simple Harmonic Oscillator

SIMPLE HARMONIC MOTION

Simple Interest

INTEREST which is paid only on the PRINCIPAL and not on the additional amount generated by previous INTEREST payments. A formula for computing simple interest is

$$a(t) = a(0)(1 + rt).$$

where $a(t)$ is the sum of PRINCIPAL and INTEREST at time t for a constant interest rate r.

See also COMPOUND INTEREST, INTEREST

References
Kellison, S. G. *Theory of Interest, 2nd ed.* Burr Ridge, IL: Richard D. Irwin, 1991.

Simple Lie Algebra

References
Huang, J.-S. "Simple Lie Algebras." Part II in *Lectures on Representation Theory.* Singapore: World Scientific, pp. 27–70, 1999.

Simple Pole

A simple pole of a ANALYTIC FUNCTION f is a POLE of order one. That is, $(z - z_0)f(z)$ is an ANALYTIC FUNCTION at the pole $z = z_0$. Alternatively, its PRINCIPAL PART is $c/(z - z_0)$ for some $c \neq 0$. It is called simple because a function with a pole of order n at a can be written as the product of n functions with simple poles at z_0.

See also DIVISOR (CURVE), ESSENTIAL SINGULARITY, POLE

Simple Polygon

A POLYGON P is said to be simple (or JORDAN) if the only points of the plane belonging to two EDGES of P are the VERTICES of P. Such a polygon has a WELL DEFINED interior and exterior. Simple polygons are topologically equivalent to a DISK.

See also POLYGON, REGULAR POLYGON, SIMPLE POLYHEDRON, TWO-EARS THEOREM

References
Toussaint, G. "Anthropomorphic Polygons." *Amer. Math. Monthly* **122**, 31–35, 1991.

Simple Polyhedron

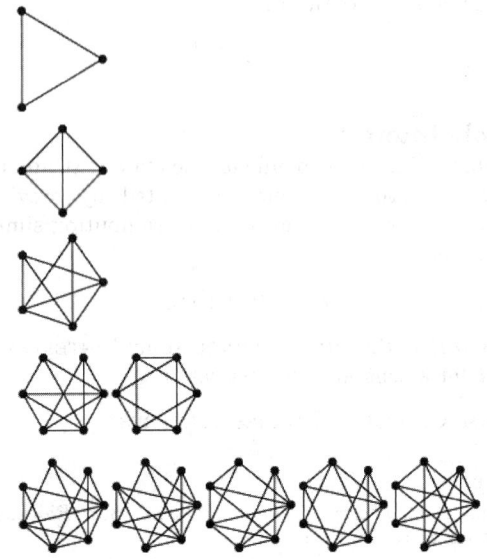

A POLYHEDRON that is topologically equivalent to a sphere (i.e., if it were inflated, it would produce a sphere) and whose faces are SIMPLE POLYGONS. The simple polyhedra on n nodes correspond to the simple PLANAR GRAPHS with $3n - 6$ edges, and are also called "simplicial polyhedra." The number of simple polyhedra on $n = 1, 2, \ldots$ nodes are 0, 0, 1, 1, 1, 2, 5, 15, 50, 233, 1249, ... (Sloane's A000109).

See also PLANAR GRAPH, SIMPLE POLYGON

References

Bokowski, J. and Schuchert, P. "Equifacetted 3-Spheres as Topes of Nonpolytopal Matroid Polytopes." *Disc. Comput. Geom.* **13**, 347–361, 1995.
Bowen, R. and Fisk, S. "Generation of Triangulations of the Sphere." *Math. Comput.* **21**, 250–252, 1967.
Dillencourt, M. B. "Polyhedra of Small Orders and Their Hamiltonian Properties." Tech. Rep. 92–91, Info. and Comput. Sci. Dept., Univ. Calif. Irvine, 1992.
Federico, P. J. "Enumeration of Polyhedra: The Number of 9-Hedra." *J. Combin. Th.* **7**, 155–161, 1969.
Gardner, M. "Mathematical Games: On the Remarkable Császár Polyhedron and Its Applications in Problem Solving." *Sci. Amer.* **232**, 102–107, May 1975.
Grünbaum, B. *Convex Polytopes.* New York: Wiley, p. 424, 1967.
Lederberg, J. "Hamilton Circuits of Convex Trivalent Polyhedra (up to 18 Vertices)." *Amer. Math. Monthly* **74**, 522–527, 1967.
Sloane, N. J. A. Sequences A0001091469 in "An On-Line Version of the Encyclopedia of Integer Sequences." http://www.research.att.com/~njas/sequences/eisonline.html.

Simple Random Walk

See also RANDOM WALK

Simple Ring

A NONZERO RING S whose only (two-sided) IDEALS are S itself and zero. Every commutative simple ring is a FIELD. Every simple ring is a PRIME RING.

See also FIELD, IDEAL, PRIME RING, RING

Simple Root

A ROOT having MULTIPLICITY $n = 1$ is called a simple root. For example, $f(z) = (z-1)(z-2)$ has a simple root at $z_0 = 1$, but $g = (z-1)^2$ has a root of MULTIPLICITY 2 at $z_0 = 1$, which is therefore not a simple root.

See also MULTIPLE ROOT, MULTIPLICITY, ROOT

References

Krantz, S. G. "Zero of Order n." §5.1.3 in *Handbook of Complex Analysis.* Boston, MA: Birkhäuser, p. 70, 1999.

Simple Zero

SIMPLE ROOT

Simplex

The generalization of a tetrahedral region of space to n-D. The boundary of a k-simplex has $k + 1$ 0-faces (VERTICES), $k(k+1)/2$ 1-faces (EDGES), and $\binom{k+1}{i+1}$ i-faces, where $\binom{n}{k}$ is a BINOMIAL COEFFICIENT. An n-D simplex can be denoted using the SCHLÄFLI SYMBOL

$$\underbrace{\{3, \ldots, 3\}}_{n-1}.$$

The simplex named because it represents the simplest possible polytope in any given space.

The CONTENT (i.e., hypervolume) of a simplex can be computed using the CAYLEY-MENGER DETERMINANT.

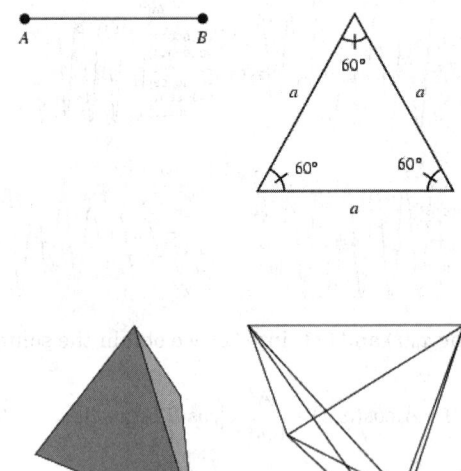

In 1-D, the simplex is the LINE SEGMENT [−1, 1]. In 2-D, the simplex {3} is the CONVEX HULL of the EQUILATERAL TRIANGLE. In 3-D, the simplex {3, 3} is the CONVEX HULL of the TETRAHEDRON. The simplex in 4-D (the PENTATOPE) is a regular TETRAHEDRON $ABCD$ in which a point E along the fourth dimension through the center of $ABCD$ is chosen so that $EA = EB = EC = ED = AB$. The regular simplex in n-D with $n \geq 5$ is denoted α_n.

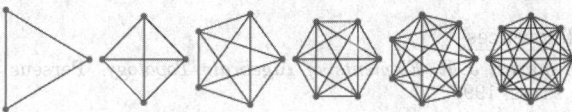

The above figures show the graphs for the n-simplexes with $n = 2$ to 7.

See also CAYLEY-MENGER DETERMINANT, COMPLEX, CROSS POLYTOPE, EQUILATERAL TRIANGLE, LINE SEGMENT, MEASURE POLYTOPE, NERVE, PENTATOPE, POINT, POLYTOPE, SIMPLEX METHOD, SPHERICAL SIMPLEX, TETRAHEDRON

References

Bourke, P. "Regular Polytopes (Platonic Solids) in 4D." http://www.swin.edu.au/astronomy/pbourke/geometry/platonic4d/.
Eppstein, D. "Triangles and Simplices." http://www.ics.uci.edu/~eppstein/junkyard/triangulation.html.
Munkres, J. R. "Simplices." §1.1 in *Elements of Algebraic Topology*. Perseus Press, pp. 2–7, 1993.

Simplex Method

A method for solving problems in LINEAR PROGRAMMING. This method, invented by G. B. Dantzig in 1947, runs along EDGES of the visualization SOLID to find the best answer. In 1970, Klee and Minty constructed examples in which the simplex method required an exponential number of steps, but such cases seem never to be encountered in practical applications.

A much more efficient (POLYNOMIAL-time) ALGORITHM was found in 1984 by N. Karmarkar. This method goes through the middle of the SOLID and then transforms and warps. It offers many advantages over the simplex method (Nemirovsky and Yudin 1994).

See also LINEAR PROGRAMMING

References

Nemirovsky, A. and Yudin, N. *Interior-Point Polynomial Methods in Convex Programming.* Philadelphia, PA: SIAM, 1994.
Press, W. H.; Flannery, B. P.; Teukolsky, S. A.; and Vetterling, W. T. "Downhill Simplex Method in Multidimensions" and "Linear Programming and the Simplex Method." §10.4 and 10.8 in *Numerical Recipes in FORTRAN: The Art of Scientific Computing, 2nd ed.* Cambridge, England: Cambridge University Press, pp. 402–406 and 423–436, 1992.

Tokhomirov, V. M. "The Evolution of Methods of Convex Optimization." *Amer. Math. Monthly* **103**, 65–71, 1996.

Simplex Point Picking

Given a SIMPLEX of unit CONTENT in Euclidean d-space, pick $d + 1$ points uniformly and independently at random, and denote the expected CONTENT of their CONVEX HULL by $V(d, n)$. The special values

$$V(1, \ n) = 1 - \frac{2}{n+1} = \frac{n-1}{n+1} \tag{1}$$

and

$$V(2, \ n) = 1 - \frac{2}{n+1} \sum_{k=1}^{n} \frac{1}{k} = 1 - \frac{2H_n}{n+1}, \tag{2}$$

where H_n is a HARMONIC NUMBER, are known (Buchta 1984, 1986). Not much is known about $V(3, n)$, although

$$V(3, \ 5) = \tfrac{5}{2} V(3, \ 4) \tag{3}$$

(Buchta 1983, 1986) and

$$1 - V(3, \ n) \sim \frac{3}{4} \frac{(\ln n)^2}{n} \tag{4}$$

(Buchta 1986).

See also DISK TRIANGLE PICKING

References

Buchta, C. "Über die konvexe Hülle von Zufallspunkten in Eibereichen." *Elem. Math.* **38**, 153–156, 1983.
Buchta, C. "Zufallspolygone in konvexen Vielecken." *J. reine angew. Math.* **347**, 212–220, 1984.
Buchta, C. "A Note on the Volume of a Random Polytope in a Tetrahedron." *Ill. J. Math.* **30**, 653–659, 1986.
Klee, V. "What is the Expected Volume of a Simplex whose Vertices are Chosen at Random from a Given Convex Body." *Amer. Math. Monthly* **76**, 286–288, 1969.

Simplicial Complex

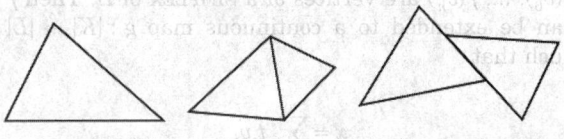

simplicial complexes   not a simplicial complex

A simplicial complex is a SPACE with a TRIANGULATION. Formally, a simplicial complex K in $\mathbb{R}^n$ is a collection of SIMPLICES in $\mathbb{R}^n$ such that

1. Every face of a simplex of K is in K, and
2. The intersection of any two simplices of K is a face of each of them.

(Munkres 1993, p. 7).

Objects in the space made up of only the simplices in the triangulation of the space are called SIMPLICIAL

SUBCOMPLEXES. When only simplicial complexes and SIMPLICIAL SUBCOMPLEXES are considered, defining HOMOLOGY is particularly easy (and, in fact, combinatorial because of its finite/counting nature). This kind of homology is called SIMPLICIAL HOMOLOGY.

See also ABSTRACT SIMPLICIAL COMPLEX, HOMOLOGY (TOPOLOGY), NERVE, SIMPLEX, SIMPLICIAL SUBCOMPLEX, SIMPLICIAL HOMOLOGY, SPACE, TRIANGULATION

References
Harary, F. *Graph Theory*. Reading, MA: Addison-Wesley, p. 7, 1994.
Munkres, J. R. "Simplicial Complexes and Simplicial Maps." §1.2 in *Elements of Algebraic Topology*. Perseus Press, pp. 7–14, 1993.

Simplicial Homology
The type of HOMOLOGY which results when the spaces being studied are restricted to SIMPLICIAL COMPLEXES and subcomplexes.

See also SIMPLICIAL COMPLEX

Simplicial Homomorphism
Let $f : K^{(0)} \to L^{(0)}$ be a bijective correspondence such that the vertices $v_0, ..., v_n$ of K span a SIMPLEX of K IFF $f(v_0), ..., f(v_n)$ span a SIMPLEX of L. Then the induced SIMPLICIAL MAP $g : |K| \to |L|$ is a HOMEOMORPHISM, and the map g is called a simplicial homeomorphism (Munkres 1993, p. 13).

References
Munkres, J. R. *Elements of Algebraic Topology*. Perseus Press, 1993.

Simplicial Map
Let K and L be SIMPLICIAL COMPLEXES, and let $f : K^{(0)} \to L^{(0)}$ be a map. Suppose that whenever the vertices $v_0, ..., v_n$ of K span a SIMPLEX of K, the points $f(v_0), ..., f(v_n)$ are vertices of a SIMPLEX of L. Then f can be extended to a continuous map $g : |K| \to |L|$ such that

$$x = \sum_{i=0}^{n} t_i v_i$$

implies

$$g(x) = \sum_{i=0}^{n} t_i f(v_i).$$

The map g is then called the linear simplicial map induced by the vertex map f (Munkres 1993, p. 12).

References
Munkres, J. R. *Elements of Algebraic Topology*. Perseus Press, 1993.

Simplicial Polyhedron
SIMPLE POLYHEDRON

Simplicial Subcomplex
If L is a subcollection of a SIMPLICIAL COMPLEX K that contains all faces of its elements, then L is another SIMPLICIAL COMPLEX called a simplicial subcomplex.

See also SIMPLICIAL COMPLEX

References
Munkres, J. R. *Elements of Algebraic Topology*. Perseus Press, 1993.

Simplicity
The number of operations needed to effect a GEOMETRIC CONSTRUCTION as determined in GEOMETROGRAPHY. If the number of operations of the five GEOMETROGRAPHIC types are denoted $m_1, m_2, n_1, n_2,$ and n_3, respectively, then the simplicity is $m_1 + m_2 + n_1 + n_2 + n_3$ and the symbol $m_1 S_1 + m_2 S_2 + n_1 C_1 + n_2 C_2 + n_3 C_3$. It is apparently an unsolved problem to determine if a given GEOMETRIC CONSTRUCTION is of smallest possible simplicity.

See also GEOMETRIC CONSTRUCTION, GEOMETROGRAPHY

References
De Temple, D. W. "Carlyle Circles and the Lemoine Simplicity of Polygonal Constructions." *Amer. Math. Monthly* **98**, 97–108, 1991.
Eves, H. *An Introduction to the History of Mathematics, 6th ed.* New York: Holt, Rinehart, and Winston, 1976.

Simply Connected

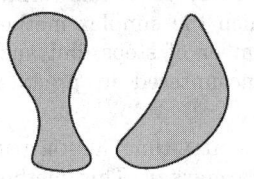 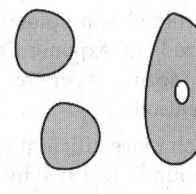

simply connected   *multiply connected*

A CONNECTED DOMAIN is said to be simply connected (also called 1-connected) if any simple closed curve can be shrunk to a point continuously in the set. If the domain is CONNECTED but not simply, it is said to be MULTIPLY CONNECTED. In particular, a SUBSET E of $\mathbb{R}^2$ is said to be simply connected if both E and $\mathbb{R}^2 \setminus E$, where $F \setminus E$ denotes a SET DIFFERENCE, are CONNECTED.
A SPACE S is simply connected if it is 0-connected and if every MAP from the 1-SPHERE to S extends continuously to a MAP from the 2-DISK. In other words, every loop in the SPACE is contractible.

See also CONNECTED SET, CONNECTED SPACE, MULTIPLY CONNECTED

References

Croft, H. T.; Falconer, K. J.; and Guy, R. K. *Unsolved Problems in Geometry.* New York: Springer-Verlag, p. 2, 1991.
Krantz, S. G. *Handbook of Complex Analysis.* Boston, MA: Birkhäuser, p. 27, 1999.

Simpson's 3/8 Rule

Let the values of a function $f(x)$ be tabulated at points x_i equally spaced by $h = x_{i+1} - x_i$, so $f_1 = f(x_1)$, $f_2 = f(x_2)$, ..., $f_4 = f(x_4)$. Then Simpson's 3/8 rule approximating the integral of $f(x)$ is given by the NEWTON-COTES-like formula

$$\int_{x_1}^{x_4} f(x)\, dx = \tfrac{3}{8} h (f_1 + 3f_2 + 3f_3 + f_4) - \tfrac{3}{80} h^5 f^{(4)}(\xi).$$

See also BODE'S RULE, NEWTON-COTES FORMULAS, SIMPSON'S RULE

References

Abramowitz, M. and Stegun, C. A. (Eds.). *Handbook of Mathematical Functions with Formulas, Graphs, and Mathematical Tables, 9th printing.* New York: Dover, p. 886, 1972.
Jeffreys, H. and Jeffreys, B. S. *Methods of Mathematical Physics, 3rd ed.* Cambridge, England: Cambridge University Press, pp. 286–287, 1988.
Whittaker, E. T. and Robinson, G. "The Trapezoidal and Parabolic Rules." *The Calculus of Observations: A Treatise on Numerical Mathematics, 4th ed.* New York: Dover, pp. 156–158, 1967.

Simpson's Formulas

The TRIGONOMETRIC ADDITION FORMULAS

$$\sin \alpha + \sin \beta = 2 \sin \left(\frac{\alpha + \beta}{2} \right) \cos \left(\frac{\alpha - \beta}{2} \right) \qquad (1)$$

$$\sin \alpha - \sin \beta = 2 \sin \left(\frac{\alpha - \beta}{2} \right) \cos \left(\frac{\alpha + \beta}{2} \right) \qquad (2)$$

$$\cos \alpha + \cos \beta = 2 \cos \left(\frac{\alpha + \beta}{2} \right) \cos \left(\frac{\alpha - \beta}{2} \right) \qquad (3)$$

$$\cos \alpha - \cos \beta = -2 \sin \left(\frac{\alpha - \beta}{2} \right) \sin \left(\frac{\alpha + \beta}{2} \right). \qquad (4)$$

Simpson's Paradox

It is not necessarily true that averaging the averages of different populations gives the average of the combined population.

References

Paulos, J. A. *A Mathematician Reads the Newspaper.* New York: BasicBooks, p. 135, 1995.

Simpson's Rule

Let $h \equiv (b - a)/n$, and assume a function $f(x)$ is defined at points $f(a + kh) = y_k$ for $k = 0, ..., n$. Then

$$\int_a^b f(x)\, dx = \tfrac{1}{3} h (y_0 + 4y_1 + 2y_2 + 4y_3 + \cdots$$

$$+ 2y_{n-2} + 4y_{n-1} + y_n) - R_n.$$

where the remainder is

$$R_n = \tfrac{1}{90}(b - a)^4 f^{(4)}(x^*)$$

for some $x^* \in [a,\, b]$.

See also BODE'S RULE, NEWTON-COTES FORMULAS, SIMPSON'S 3/8 RULE, TRAPEZOIDAL RULE

References

Abramowitz, M. and Stegun, C. A. (Eds.). *Handbook of Mathematical Functions with Formulas, Graphs, and Mathematical Tables, 9th printing.* New York: Dover, p. 886, 1972.
Jeffreys, H. and Jeffreys, B. S. *Methods of Mathematical Physics, 3rd ed.* Cambridge, England: Cambridge University Press, p. 286, 1988.
Whittaker, E. T. and Robinson, G. "The Trapezoidal and Parabolic Rules." *The Calculus of Observations: A Treatise on Numerical Mathematics, 4th ed.* New York: Dover, pp. 156–158, 1967.

Simson Line

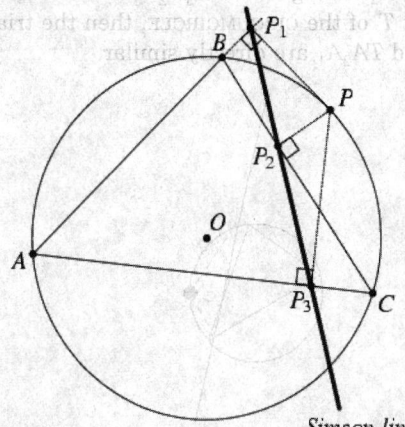

Simson line

The Simson line is the LINE containing the feet P_1, P_2, and P_3 of the perpendiculars from an arbitrary point P on the CIRCUMCIRCLE of a TRIANGLE to the sides or their extensions of the TRIANGLE. This line was attributed to Simson by Poncelet , but is now frequently known as the Wallace-Simson line since it does not actually appear in any work of Simson (Johnson 1929, p. 137; Coxeter and Greitzer 1967, p. 41; de Guzmán 1999). The inverse statement to that given above, namely that the locus of all points P in the plane of a TRIANGLE $\triangle ABC$ such that the feet of perpendiculars from the three sides of the triangle is collinear is given by the CIRCUMCIRCLE of $\triangle ABC$, is sometimes called the Wallace-Simson theorem (Guzmán 1999).

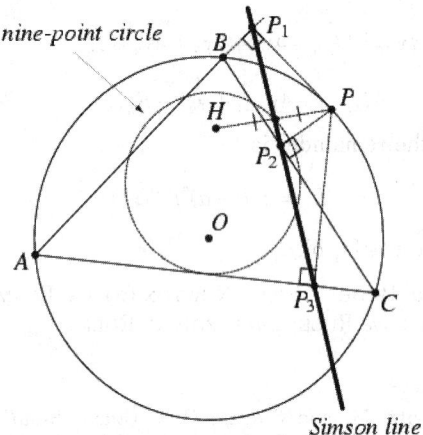

Simson line

The Simson line bisects the line HP, where H is the ORTHOCENTER (Honsberger 1995, p. 46). Moreover, the MIDPOINT of HP lies on the NINE-POINT CIRCLE (Honsberger 1995, pp. 46–47). The Simson lines of two opposite point on the CIRCUMCENTER of a triangle are PERPENDICULAR and meet on the NINE-POINT CIRCLE.

The ANGLE between the Simson lines of two points P and P' is half the ANGLE of the arc PP'. The Simson line of any VERTEX is the ALTITUDE through that VERTEX. The Simson line of a point opposite a VERTEX is the corresponding side. If $T_1 T_2 T_3$ is the Simson line of a point T of the CIRCUMCIRCLE, then the triangles $TT_1 T_2$ and $TA_2 A_1$ are directly similar.

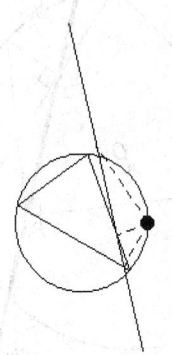

The ENVELOPE of the Simson lines of a triangle is a DELTOID (Butchart 1939; Wells 1991, pp. 155 and 230). The area of the deltoid is half the area of the circumcircle (Wells 1991, p. 230), and MORLEY'S TRIANGLE of the starting triangle has the same orientation as the DELTOID. Each side of the triangle is tangent to the DELTOID at a point whose distance from the MIDPOINT of the side equals the chord of the NINE-POINT CIRCLE cut off by that side (Wells 1991, p. 231). If a line L is the Simson line of a point P on the CIRCUMCIRCLE of a TRIANGLE, then P is called the POLE of L (Honsberger 1995, p. 128).

See also CIRCUMCIRCLE, POLE (SIMSON LINE), RIGBY POINTS

References

Baker, H. F. *An Introduction to Plane Geometry.* London: Cambridge University Press, 1963.

Butchart, J. H. "The Deltoid Regarded as the Envelope of Simson Lines." *Amer. Math. Monthly* **46**, 85–86, 1939.

Casey, J. *A Sequel to the First Six Books of the Elements of Euclid, Containing an Easy Introduction to Modern Geometry with Numerous Examples, 5th ed., rev. enl.* Dublin: Hodges, Figgis, & Co., p. 164, 1888.

Chou, S.-C. "Proving Elementary Geometry Theorems Using Wu's Algorithm." *Contemporary Math.* **29**, 243–286, 1984.

Coolidge, J. L. *A Treatise on the Geometry of the Circle and Sphere.* New York: Chelsea, p. 49, 1971.

Coxeter, H. S. M. *Introduction to Geometry, 2nd ed.* New York: Wiley, 1969.

Coxeter, H. S. M. and Greitzer, S. L. "Simson Lines" and "More on Simson Lines." §2.5 and 2.7 in *Geometry Revisited.* Washington, DC: Math. Assoc. Amer., pp. 40–41 and 43–45, 1967.

de Guzmán, M. "An Extension of the Wallace-Simson Theorem: Projecting in Arbitrary Directions." *Amer. Math. Monthly* **106**, 574–580, 1999.

Dörrie, H. *100 Great Problems of Elementary Mathematics: Their History and Solutions.* New York: Dover, 1965.

Durell, C. V. *Modern Geometry: The Straight Line and Circle.* London: Macmillan, pp. 46–48, 1928.

F. Gabriel-Marie. *Exercices de Géométrie.* Tours, France: Maison Mame, p. 329, 1912.

Honsberger, R. "The Simson Line" and "Simson Lines." §5.2 and 8.4 in *Episodes in Nineteenth and Twentieth Century Euclidean Geometry.* Washington, DC: Math. Assoc. Amer., pp. 43–44 and 82–83, 1995.

Johnson, R. A. *Modern Geometry: An Elementary Treatise on the Geometry of the Triangle and the Circle.* Boston, MA: Houghton Mifflin, pp. 137–139, 1929.

Patterson, B. C. "The Triangle: Its Deltoids and Foliates." *Amer. Math. Monthly* **47**, 11–18, 1940.

Ramler, O. J. "The Orthopole Loci of Some One-Parameter Systems of Lines Referred to a Fixed Triangle." *Amer. Math. Monthly* **37**, 130–136, 1930.

van Horn, C. E. "The Simson Quartic of a Triangle." *Amer. Math. Monthly* **45**, 434–437, 1938.

Wells, D. *The Penguin Dictionary of Curious and Interesting Geometry.* London: Penguin, pp. 155 and 230–231, 1991.

Simson's Formula

CASSINI'S IDENTITY

Sin

SINE

Sinc

SINC FUNCTION

Sinc Function

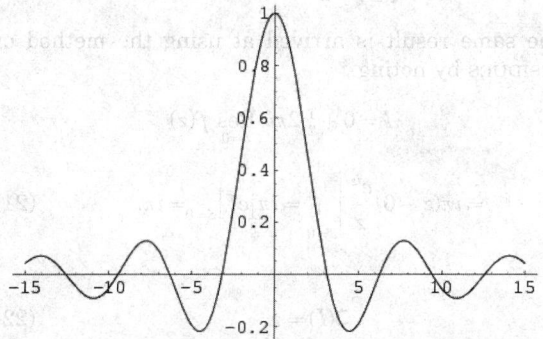

A function also called the "sampling function" that arises frequently in signal processing. There are two definitions in common use. The one adopted in this work defines

$$\text{sinc}(x) \equiv \begin{cases} 1 & \text{for } x = 0 \\ \dfrac{\sin x}{x} & \text{otherwise,} \end{cases} \tag{1}$$

where $\sin x$ is the SINE function, while Woodward (1953) and Bracewell (1999, p. 62) adopt the alternative definition

$$\text{sinc}_\pi(x) \equiv \begin{cases} 1 & \text{for } x = 0 \\ \dfrac{\sin(\pi x)}{(\pi x)} & \text{otherwise.} \end{cases} \tag{2}$$

The latter definition is sometimes more convenient as a result of its simple normalization,

$$\int_{-\infty}^{\infty} \text{sinc}_\pi(x)\, dx = 1. \tag{3}$$

Let $\Pi(x)$ be the RECTANGLE FUNCTION, then the FOURIER TRANSFORM of $\Pi(x)$ is the sinc function

$$\mathscr{F}[\Pi(x)] = \text{sinc}(\pi k). \tag{4}$$

The sinc function therefore frequently arises in physical applications such as Fourier transform spectroscopy as the so-called INSTRUMENT FUNCTION, which gives the instrumental response to a DELTA FUNCTION input. Removing the instrument functions from the final spectrum requires use of some sort of DECONVOLUTION algorithm.

The sinc function can be written as a complex INTEGRAL by noting that, for $x \neq 0$,

$$\text{sinc}(nx) \equiv \frac{\sin(nx)}{nx} = \frac{1}{nx} \frac{e^{inx} - e^{-inx}}{2i}$$

$$= \frac{1}{2inx} \left[e^{itx} \right]_{-n}^{n} = \frac{1}{2n} \int_{-n}^{n} e^{ixt}\, dt. \tag{5}$$

and that $\text{sinc}(nx)$ and the integral both equal 1 for $x = 0$. The sinc function can also be written as the INFINITE PRODUCT

$$\text{sinc}\, x = \prod_{k=1}^{\infty} \cos\left(\frac{x}{2^k}\right). \tag{6}$$

Definite integrals involving the sinc function include

$$\int_0^{\infty} \text{sinc}(x)\, dx = \tfrac{1}{2}\pi \tag{7}$$

$$\int_0^{\infty} \text{sinc}^2(x)\, dx = \tfrac{1}{2}\pi \tag{8}$$

$$\int_0^{\infty} \text{sinc}^3(x)\, dx = \tfrac{3}{8}\pi \tag{9}$$

$$\int_0^{\infty} \text{sinc}^4(x)\, dx = \tfrac{1}{3}\pi \tag{10}$$

$$\int_0^{\infty} \text{sinc}^5(x)\, dx = \tfrac{115}{384}\pi. \tag{11}$$

These are all special cases of the amazing general result

$$\int_0^{\infty} \frac{\sin^a x}{x^b}\, dx = \frac{\pi^{1-c}(-1)^{\lfloor(a-b)/2\rfloor}}{2^{a-c}(b-1)!}$$

$$\times \sum_{k=0}^{\lfloor a/2 \rfloor - c} (-1)^k \binom{a}{k}(a-2k)^{b-1}[\ln(a-2k)]^c. \tag{12}$$

where a and b are POSITIVE INTEGERS such that $a \geq b > c$, $c \equiv a - b \pmod 2$, $\lfloor x \rfloor$ is the FLOOR FUNCTION, and 0^0 is taken to be equal to 1 (Kogan). This spectacular formula simplifies in the special case when n is a POSITIVE EVEN integer to

$$\int_0^{\infty} \frac{\sin^{2n} x}{x^{2n}}\, dx = \frac{\pi}{2(2n-1)!} \left\langle \begin{matrix} 2n-1 \\ n-1 \end{matrix} \right\rangle. \tag{13}$$

where $\left\langle \begin{smallmatrix} n \\ k \end{smallmatrix} \right\rangle$ is an EULERIAN NUMBER (Kogan). The solution of the integral can also be written in terms of the RECURRENCE RELATION for the coefficients

$$c(a, b) = \begin{cases} \dfrac{\pi}{2^{a+1-b}} \begin{pmatrix} a-1 \\ \frac{1}{2}(a-1) \end{pmatrix} \\ \qquad \text{for } b = 1 \text{ or } b = 2 \\ \dfrac{a}{(b-1)(b-2)}[(a-1)c(a-2, b-2) \\ \qquad - a \cdot c(a, b-2)] \\ \qquad \text{otherwise} \end{cases} \tag{14}$$

(Zimmerman).

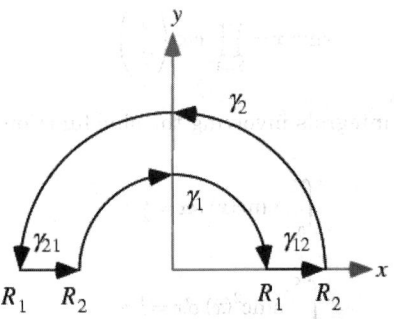

The half-infinite integral of sinc(x) can be derived using CONTOUR INTEGRATION. In the above figure, consider the path $\gamma \equiv \gamma_1 + \gamma_{12} + \gamma_2 + \gamma_{21}$. Now write $z = Re^{i\theta}$. On an arc, $dz = iRe^{i\theta}\, d\theta$ and on the x-AXIS, $dz = e^{i\theta}\, dR$. Write

$$\int_{-\infty}^{\infty} \operatorname{sinc} x \, dx = \Im \int_{\gamma} \frac{e^{iz}}{z}\, dx. \tag{15}$$

where $\Im$ denotes the IMAGINARY POINT. Now define

$$I \equiv \int_{\gamma} \frac{e^{iz}}{z}\, dz$$

$$= \lim_{R_1 \to 0} \int_{\pi}^{0} \frac{\exp(iR_1 e^{i\theta})}{R_1 e^{i\theta}} i\theta R_1 e^{i\theta}\, d\theta$$

$$+ \lim_{R_1 \to 0} \lim_{R_2 \to \infty} \int_{R_1}^{R_2} \frac{e^{iR}}{R}\, dR$$

$$+ \lim_{R_2 \to \infty} \int_{0}^{\pi} \frac{\exp(iz)}{z}\, dx + \lim_{R_1 \to 0} \int_{R_2}^{R_1} \frac{e^{-iR}}{-R}(-dR). \tag{16}$$

where the second and fourth terms use the identities $e^{i0} = 1$ and $e^{i\pi} = -1$. Simplifying,

$$I = \lim_{R_1 \to 0} \int_{\pi}^{0} \exp(iR_1 e^{i\theta})i\theta\, d\theta + \int_{0^+}^{\infty} \frac{e^{iR}}{R}\, dR$$

$$+ \lim_{R_2 \to \infty} \int_{0}^{\pi} \frac{\exp(iz)}{z}\, dz + \int_{\infty}^{0^+} \frac{e^{-iR}}{-R}(-dR)$$

$$= -\int_{0}^{\pi} i\theta\, d\theta + \int_{0^+}^{\infty} \frac{e^{iR}}{R}\, dR + 0 + \int_{-\infty}^{0^-} \frac{e^{iR}}{R}\, dR. \tag{17}$$

where the third term vanishes by JORDAN'S LEMMA. Performing the integration of the first term and combining the others yield

$$I = -i\pi + \int_{-\infty}^{\infty} \frac{e^{iz}}{z}\, dz = 0. \tag{18}$$

Rearranging gives

$$\int_{-\infty}^{\infty} \frac{e^{iz}}{z}\, dz = i\pi. \tag{19}$$

so

$$\int_{-\infty}^{\infty} \frac{\sin z}{z}\, dz = \pi. \tag{20}$$

The same result is arrived at using the method of RESIDUES by noting

$$I = 0 + \tfrac{1}{2} 2\pi i \operatorname*{Res}_{z=0} f(z)$$

$$= i\pi(z - 0) \frac{e^{iz}}{z}\bigg|_{z=0} = i\pi \left[e^{iz} \right]_{z=0} = i\pi, \tag{21}$$

so

$$\Im(I) = \pi. \tag{22}$$

Since the integrand is symmetric, we therefore have

$$\int_{0}^{\infty} \frac{\sin x}{x}\, dx = \tfrac{1}{2}\pi, \tag{23}$$

giving the SINE INTEGRAL evaluated at 0 as

$$\operatorname{si}(0) = -\int_{0}^{\infty} \frac{\sin x}{x}\, dx = -\tfrac{1}{2}\pi. \tag{24}$$

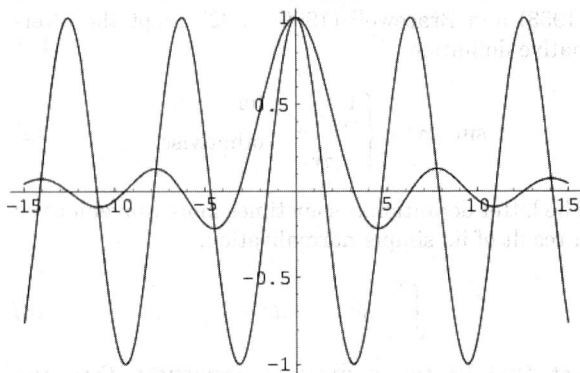

An interesting property of sinc(x) is that the set of LOCAL EXTREMA of sinc(x) corresponds to its intersections with the COSINE function cos(x), as illustrated above.

See also FOURIER TRANSFORM, FOURIER TRANSFORM–RECTANGLE FUNCTION, INSTRUMENT FUNCTION, JINC FUNCTION, KILROY CURVE, SINE, SINE INTEGRAL

References

Bracewell, R. "The Filtering or Interpolating Function, sinc x." In *The Fourier Transform and Its Applications,* 3rd ed. New York: McGraw-Hill, pp. 62–64, 1999.

Kogan, S. "A Note on Definite Integrals Involving Trigonometric Functions." http://www.mathsoft.com/asolve/constant/pi/sin/sin.html.

Morrison, K. E. "Cosine Products, Fourier Transforms, and Random Sums." *Amer. Math. Monthly* **102**, 716–724, 1995.

Woodward, P. M. *Probability and Information Theory with Applications to Radar.* New York: McGraw-Hill, 1953.

Sinclair's Soap Film Problem

Find the shape of a soap film (i.e., MINIMAL SURFACE) which will fill two inverted conical FUNNELS facing each other is known as Sinclair's soap film problem (Bliss 1925, p. 121). The soap film will assume the shape of a CATENOID.

See also CATENOID, FUNNEL, MINIMAL SURFACE

References

Bliss, G. A. *Calculus of Variations.* Chicago, IL: Open Court, pp. 121–122, 1925.

Isenberg, C. *The Science of Soap Films and Soap Bubbles.* New York: Dover, p. 81, 1992.

Sinclair, M. E. "On the Minimum Surface of Revolution in the Case of One Variable End Point." *Ann. Math.* **8**, 177–188, 1907.

Sine

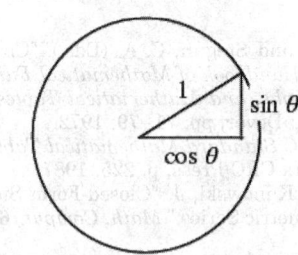

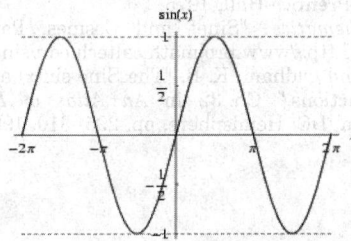

One of the basic TRIGONOMETRIC FUNCTIONS encountered in TRIGONOMETRY. Let θ be an ANGLE measured counterclockwise from the x-AXIS along the arc of the UNIT CIRCLE. Then $\sin \theta$ is the vertical coordinate of the arc endpoint. As a result of this definition, the sine function is periodic with period 2π. By the PYTHAGOREAN THEOREM, $\sin \theta$ also obeys the identity

$$\sin^2 \theta + \cos^2 \theta = 1. \tag{1}$$

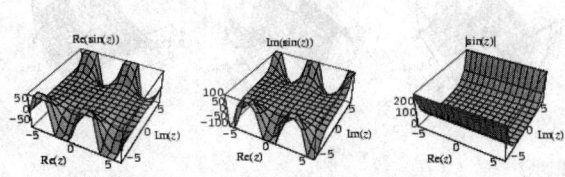

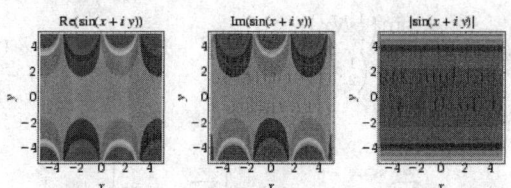

The definition of the sine function can be extended to complex arguments z using the definition

$$\sin z = \frac{e^{iz} - e^{-iz}}{2i}, \tag{2}$$

where E is the base of the NATURAL LOGARITHM and I is the IMAGINARY NUMBER. A related function known as the HYPERBOLIC SINE is similarly defined,

$$\sinh z = \tfrac{1}{2}(e^z - e^{-z}), \tag{3}$$

The sine function can be defined algebraically by the infinite sum

$$\sin x = \sum_{n=1}^{\infty} \frac{(-1)^{n-1}}{(2n-1)!} x^{2n-1} \tag{4}$$

and INFINITE PRODUCT

$$\sin x = x \prod_{n=1}^{x} \left(1 - \frac{x^2}{n^2 \pi^2}\right). \tag{5}$$

It is also given by the IMAGINARY PART of the complex exponential

$$\sin x = \Im\left[e^{ix}\right] \tag{6}$$

The multiplicative inverse of the sine function is the COSECANT, defined as

$$\csc x \equiv \frac{1}{\sin x}. \tag{7}$$

The sine function is also given by the slowly convergent INFINITE SERIES

$$\sin(z) = -\pi \sum_{k=1}^{\infty} \frac{\mu(k) \ln\left(\frac{n}{k}\right) \operatorname{frac}\left(\frac{kz}{2\pi}\right)}{k \ln n} \tag{8}$$

where $\mu(k)$ is the MÖBIUS FUNCTION and frac x is the FRACTIONAL PART (M. Trott).

Using the results from the EXPONENTIAL SUM FORMULAS

$$\sum_{n=0}^{N} \sin(nx) = \Im\left[\sum_{n=0}^{N} e^{inx}\right]$$

$$= \Im\left[\frac{\sin\left(\tfrac{1}{2}Nx\right)}{\sin\left(\tfrac{1}{2}x\right)} e^{i(N+1)x/2}\right]$$

$$= \frac{\sin\left(\frac{1}{2}Nx\right)}{\sin\left(\frac{1}{2}x\right)} \sin\left[\frac{1}{2}x(N+1)\right]. \tag{9}$$

Similarly,

$$\sum_{n=0}^{\infty} p^n \sin(nx) = \Im\left[\sum_{n=0}^{\infty} p^n e^{inx}\right]$$

$$= \Im\left[\frac{1 - pe^{-iz}}{1 - 2p\cos x + p^2}\right] = \frac{p\sin x}{1 - 2p\cos x + p^2}. \tag{10}$$

The sum of $\sin^2(kx)$ can also be done in closed form,

$$\sum_{k=0}^{N} \sin^2(kx) = \frac{1}{4}\{1 + 2N - \csc x \, \sin[x(1+2N)]\}. \tag{11}$$

The sine function obeys the identity

$$\sin(n\theta) = 2\cos\theta\,\sin[(n-1)\theta] - \sin[(n-2)\theta] \tag{12}$$

and the MULTIPLE-ANGLE FORMULA

$$\sin(nx) = \sum_{k=0}^{n} \binom{n}{k}\cos^k x \, \sin^{n-k} x \, \sin\left[\frac{1}{2}(n-k)\pi\right]. \tag{13}$$

where $\binom{n}{k}$ is a BINOMIAL COEFFICIENT.

Cvijovic and Klinowski (1995) show that the sum

$$S_v(\alpha) = \sum_{k=0}^{\infty} \frac{\sin(2k+1)\alpha}{(2k+1)^v} \tag{14}$$

has closed form for $v = 2n+1$,

$$S_{2n+1}(\alpha) = \frac{(-1)}{4(2n)!}\pi^{2n+1}E_{2n}\left(\frac{\alpha}{\pi}\right), \tag{15}$$

where $E_n(x)$ is an EULER POLYNOMIAL.

A CONTINUED FRACTION representation of $\sin x$ is

$\sin x$

$$= \cfrac{x}{1 + \cfrac{x^2}{(2\cdot 3 - x^2) - \cfrac{2\cdot 3x^2}{(4\cdot 5 - x^2) + \cfrac{4\cdot 5x^2}{(6\cdot 7 - x^2) + \cdots}}}} \tag{16}$$

The value of $\sin(2\pi/n)$ is IRRATIONAL for all n except 4 and 12, for which $\sin(\pi/2) = 1$ and $\sin(\pi/6) = 1/2$.

The FOURIER TRANSFORM of $\sin(2\pi k_0 x)$ is given by

$$\mathscr{F}[\sin(2\pi k_0 x)] = \int_{-\infty}^{\infty} e^{-2\pi ikx}\sin(2\pi k_0 x)\,dx$$

$$= \frac{1}{2}i[\delta(k+k_0) - \delta(k-k_0)]. \tag{17}$$

Definite integrals involving $\sin x$ include

$$\int_0^{\infty} \sin(x^2)\,dx = \frac{1}{4}\sqrt{2\pi} \tag{18}$$

$$\int_0^{\infty} \sin(x^3)\,dx = \frac{1}{6}\Gamma\left(\frac{1}{3}\right) \tag{19}$$

$$\int_0^{\infty} \sin(x^4)\,dx = -\cos\left(\frac{5}{8}\pi\right)\Gamma\left(\frac{5}{4}\right) \tag{20}$$

$$\int_0^{\infty} \sin(x^5)\,dx = \frac{1}{4}\left(\sqrt{5} - 1\right)\Gamma\left(\frac{6}{5}\right), \tag{21}$$

where $\Gamma(x)$ is the GAMMA FUNCTION.

See also ANDREW'S SINE, COSECANT, COSINE, FOURIER TRANSFORM–SINE, HYPERBOLIC SINE, SINC FUNCTION, SINUSOID, TANGENT, TRIGONOMETRY

References

Abramowitz, M. and Stegun, C. A. (Eds.). "Circular Functions." §4.3 in *Handbook of Mathematical Functions with Formulas, Graphs, and Mathematical Tables, 9th printing.* New York: Dover, pp. 71–79, 1972.

Beyer, W. H. *CRC Standard Mathematical Tables, 28th ed.* Boca Raton, FL: CRC Press, p. 225, 1987.

Cvijovic, D. and Klinowski, J. "Closed-Form Summation of Some Trigonometric Series." *Math. Comput.* **64**, 205–210, 1995.

Hansen, E. R. *A Table of Series and Products.* Englewood Cliffs, NJ: Prentice-Hall, 1975.

Project Mathematics. "Sines and Cosines, Parts I-III." Videotape. http://www.projmath.caltech.edu/sincos1.htm.

Spanier, J. and Oldham, K. B. "The Sine sin(x) and Cosine cos(x) Functions." Ch. 32 in *An Atlas of Functions.* Washington, DC: Hemisphere, pp. 295–310, 1987.

Sine Integral

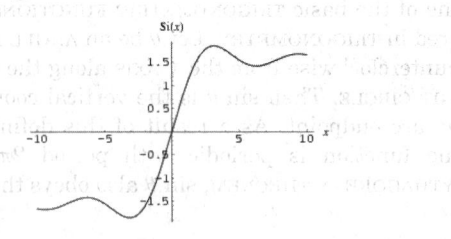

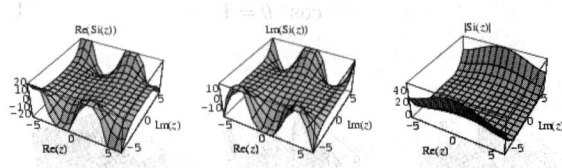

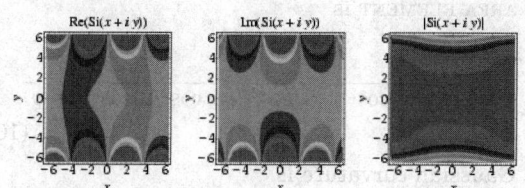

There are two types of "sine integrals" commonly defined,

$$\mathrm{Si}(x) \equiv \int_0^z \frac{\sin t}{t}\, dt \qquad (1)$$

and

$$\mathrm{Si}(x) \equiv -\int_0^z \frac{\sin t}{t}\, dt \qquad (2)$$

$$= \frac{1}{2i}[\mathrm{ei}(ix) - \mathrm{ei}(-ix)] \qquad$$

$$= \frac{1}{2i}[\mathrm{e}_1(ix) - \mathrm{e}_1(-ix)] \qquad (3)$$

$$\mathrm{Si}(z) - \tfrac{1}{2}\pi, \qquad (4)$$

where $\mathrm{ei}(x)$ is the EXPONENTIAL INTEGRAL and

$$\mathrm{e}_1(x) \equiv -\mathrm{ei}(-x). \qquad (5)$$

$\mathrm{Si}(x)$ is the function returned by the *Mathematica* command $\mathtt{SinIntegral[x]}$ and displayed above. The half-infinite integral of the SINC FUNCTION is given by

$$\mathrm{si}(0) = -\int_0^\infty \frac{\sin x}{x}\, dx = -\tfrac{1}{2}\pi. \qquad (6)$$

To compute the integral of a sine function times a power

$$I \equiv \int x^{2n} \sin(mx)\, dx, \qquad (7)$$

use INTEGRATION BY PARTS. Let

$$u = x^{2n} \quad dv = \sin(mx)\, dx \qquad (8)$$

$$du = 2n x^{2n-1}\, dx \quad v = \frac{1}{m}\cos(mx), \qquad (9)$$

so

$$I = -\frac{1}{m} x^{2n} \cos(mx) + \frac{2n}{m} \int x^{2n-1} \cos(mx)\, dx. \qquad (10)$$

Using INTEGRATION BY PARTS again,

$$u = x^{2n-1} \quad dv = \cos(mx)\, dx \qquad (11)$$

$$du = (2n-1)x^{2n-2}\, dx \quad v \frac{1}{m}\sin(mx) \qquad (12)$$

$$\int x^{2n} \sin(mx)\, dx = -\frac{1}{m} x^{2n} \cos(mx)$$

$$+ \frac{2n}{m}\left[\frac{1}{m} x^{2n-1} \cos(mx) - \frac{2n-1}{m} \int x^{2n-2} \sin(mx)\, dx \right]$$

$$= -\frac{1}{m} x^{2n} \sin(mx) + \frac{2n}{m^2} x^{2n-1} \sin(mx)$$

$$- \frac{(2n)(2n-1)}{m^2} \int x^{2n-2} \sin(mx)\, dx$$

$$= -\frac{1}{m} x^{2n} \cos(mx) + \frac{2n}{m^2} x^{2n-1} \sin(mx) + \ldots$$

$$+ \frac{(2n)!}{m^{2n}} \int x^0 \sin(mx)\, dx$$

$$= -\frac{1}{m} x^{2n} \cos(mx) + \frac{2n}{m^2} x^{2n-1} \sin(mx) + \ldots$$

$$- \frac{(2n)!}{m^{2n+1}} \cos(mx)$$

$$= \cos(mx) \sum_{k=0}^n (-1)^{k+1} \frac{(2n)!}{(2n-2k)! m^{2k+1}} x^{2n-2k}$$

$$+ \sin(mx) \sum_{k=1}^n (-1)^{k+1} \frac{(2n)!}{(2k-2n-1)! m^{2k}} x^{2n-2k+1} \qquad (13)$$

Letting $k' \equiv n-k$, so

$$\int x^{2n} \sin(mx)\, dx$$

$$= \cos(mx) \sum_{k=1}^n (-1)^{n-k+1} \frac{(2n)!}{(2k)! m^{2n-2k+1}} x^{2k}$$

$$+ \sin(mx) \sum_{k=0}^{n-1} (-1)^{n-k+1} \frac{(2n)!}{(2k-1)! m^{2n-2k}} x^{2k+1}$$

$$= (-1)^{n+1}(2n)! \left[\cos(mx) \sum_{k=0}^n \frac{(-1)^k}{(2k)! m^{2n-2k+1}} x^{2k} \right.$$

$$\left. + \sin(mx) \sum_{k=1}^n \frac{(-1)^{k+1}}{(2k-3)! m^{2n-2k+2}} x^{2k-1} \right]. \qquad (14)$$

General integrals OF THE FORM

$$I(k, l) = \int_0^\infty \frac{\sin^k x}{x^l}\, dx \qquad (15)$$

are related to the SINC FUNCTION and can be computed analytically.

See also CHI, COSINE INTEGRAL, EXPONENTIAL INTEGRAL, NIELSEN'S SPIRAL, SHI, SICI SPIRAL, SINC FUNCTION

References

Abramowitz, M. and Stegun, C. A. (Eds.). "Sine and Cosine Integrals." §5.2 in *Handbook of Mathematical Functions with Formulas, Graphs, and Mathematical Tables, 9th printing.* New York: Dover, pp. 231–233, 1972.

Arfken, G. *Mathematical Methods for Physicists, 3rd ed.* Orlando, FL: Academic Press, pp. 342–343, 1985.

Press, W. H.; Flannery, B. P.; Teukolsky, S. A.; and Vetterling, W. T. "Fresnel Integrals, Cosine and Sine Integrals." §6.79 in *Numerical Recipes in FORTRAN: The Art of Scientific Computing, 2nd ed.* Cambridge, England: Cambridge University Press, pp. 248–252, 1992.

Spanier, J. and Oldham, K. B. "The Cosine and Sine Integrals." Ch. 38 in *An Atlas of Functions.* Washington, DC: Hemisphere, pp. 361–372, 1987.

Sine Surface

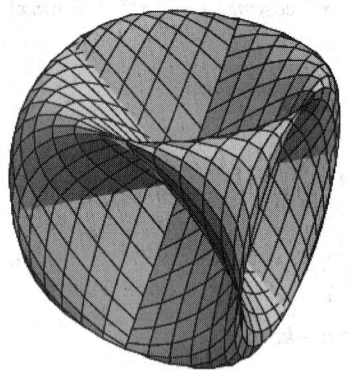

The surface given by the PARAMETRIC EQUATIONS

$$x = a \sin u \tag{1}$$

$$y = a \sin v \tag{2}$$

$$z = a \sin(u + v). \tag{3}$$

The coefficients of the FIRST FUNDAMENTAL FORM are

$$E = a^2 \left[\cos^2 u + \cos^2(u+v) \right] \tag{4}$$

$$F = a^2 \cos^2(u+v) \tag{5}$$

$$G = a^2 \left[\cos^2 v + \cos^2(u+v) \right], \tag{6}$$

the SECOND FUNDAMENTAL FORM coefficients are

$$e = -\frac{a \cos v \sin v}{\sqrt{\cos^2 u \cos^2 v + (\cos^2 u + \cos^2 v)\cos^2(u+v)}} \tag{7}$$

$$f = -\frac{a \cos u \cos v \sin(u+v)}{\sqrt{\cos^2 u \cos^2 v + (\cos^2 u + \cos^2 v)\cos(u+v)}} \tag{8}$$

$$g = -\frac{a \cos u \sin u}{\sqrt{\cos^2 u \cos^2 v + (\cos^2 u + \cos^2 v)\cos^2(u+v)}} \tag{9}$$

the AREA ELEMENT is

$$dS = a^2 \times \sqrt{\cos^2 u \cos^2 v + (\cos^2 u + \cos^2 v)\cos^2(u+v)}, \tag{10}$$

the Gaussian curvature is

$$k = \frac{\cos u \cos v \left[\sin u \sin v - \cos u \cos v \sin^2(u+v) \right]}{\left[a\cos^2 u \cos^2 v + a(\cos^2 u + \cos^2 v)\cos^2(u+v) \right]^2}, \tag{11}$$

and the MEAN CURVATURE is a complicated expression.

References

Gray, A. *Modern Differential Geometry of Curves and Surfaces with Mathematica, 2nd ed.* Boca Raton, FL: CRC Press, pp. 315–316, 1997.

Sine-Gordon Equation

A PARTIAL DIFFERENTIAL EQUATION which appears in differential geometry and relativistic field theory. Its name is a wordplay on its similar form to the KLEIN-GORDON EQUATION. The sine-Gordon equation is

$$v_{tt} - v_{xx} + \sin v = 0. \tag{1}$$

where v_{tt} and v_{xx} are PARTIAL DERIVATIVES. The equation can be transformed by defining

$$\xi \equiv \tfrac{1}{2}(x - t) \tag{2}$$

$$\eta \equiv \tfrac{1}{2}(x + t). \tag{3}$$

Then, by the CHAIN RULE,

$$\frac{\partial}{\partial x} = \frac{\partial \xi}{\partial x}\frac{\partial}{\partial \xi} + \frac{\partial \eta}{\partial x}\frac{\partial}{\partial \eta} \tag{4}$$

$$= \frac{1}{2}\left(\frac{\partial}{\partial \xi} + \frac{\partial}{\partial \eta} \right) \tag{5}$$

$$\frac{\partial}{\partial t} = \frac{\partial \xi}{\partial t}\frac{\partial}{\partial \xi} + \frac{\partial \eta}{\partial t}\frac{\partial}{\partial \eta} \tag{6}$$

$$= \frac{1}{2}\left(\frac{\partial}{\partial \eta} + \frac{\partial}{\partial \xi} \right) \tag{7}$$

This gives

$$\frac{\partial^2 v}{\partial x^2} = \frac{1}{4}\left(\frac{\partial}{\partial \xi} + \frac{\partial}{\partial \eta} \right)\left(\frac{\partial v}{\partial \xi} + \frac{\partial v}{\partial \eta} \right)$$

$$= \frac{1}{4}\left(\frac{\partial^2 v}{\partial \xi^2} + 2\frac{\partial^2 v}{\partial \xi \partial \eta} + \frac{\partial^2 v}{\partial \eta^2} \right) \tag{8}$$

$$\frac{\partial^2 v}{\partial t^2} = \frac{1}{4}\left(\frac{\partial}{\partial \eta} - \frac{\partial}{\partial \xi}\right)\left(\frac{\partial v}{\partial \eta} - \frac{\partial v}{\partial \xi}\right)$$

$$= \frac{1}{4}\left(\frac{\partial^2 v}{\partial \xi^2} - 2\frac{\partial^2 v}{\partial \xi \partial \eta} + \frac{\partial^2 v}{\partial \eta^2}\right) \tag{9}$$

Plugging in gives

$$v\xi_\eta = \sin v. \tag{10}$$

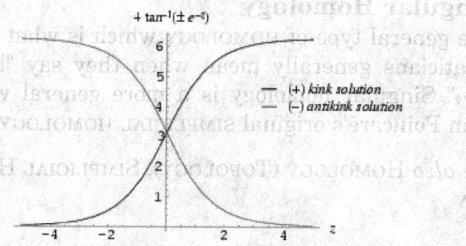

Traveling wave analysis by setting $v(x,\ t) = g(z)$ yields after one integration

$$z - z_0 = \sqrt{c^2 - 1} \int \frac{df}{\sqrt{2\left[d - 2\sin^2\left(\frac{1}{2}f\right)\right]}} \tag{11}$$

where d is a constant of integration (Tabor 1989, p. 306). For the particular case $d = 0$,

$$z - z_0 = \pm\sqrt{1 - c^2}\, \ln\left[\pm\tan\left(\tfrac{1}{4}f\right)\right], \tag{12}$$

so integrating gives

$$f(z) = \pm 4\tan^{-1}[e^{\pm(z-z_0)/(1-c^2)^{1/2}}]. \tag{13}$$

The solution with the plus sign is called the "kink solution," while that with the minus sign is called the "antikink solution" (Tabor 1989, pp. 306–307).

Another solution to the sine-Gordon equation is given by making the substitution $v(\xi,\ \eta) = f(z)$, where $z = \xi\eta$, giving the ORDINARY DIFFERENTIAL EQUATION

$$zf'' + f' = \sin f. \tag{14}$$

However, this cannot be solved analytically, since letting $g \equiv e^{if}$ gives

$$g'' - \frac{g'^2}{f} + \frac{2g' - g^2 + 1}{2z} = 0. \tag{15}$$

which is the third PAINLEVÉ TRANSCENDENT (Tabor 1989, p. 309).

Now looking for a solution OF THE FORM

$$v(x,\ t) = 4\tan^{-1}\left[\frac{\phi(x)}{\psi(t)}\right] \tag{16}$$

gives

$$\frac{\psi^2}{\phi}\phi_{xx} + \frac{\phi^2}{\psi}\psi_{tt}$$
$$= (\psi^2 + 2\psi_t - \psi\psi_{tt}) + (-\phi^2 + 2\phi_x - \phi\phi_{xx}). \tag{17}$$

Further differentiation gives

$$\frac{(\phi_{xx}\phi)_x}{\phi\phi_x} = -\frac{(\psi_{tt}/\psi)_t}{\psi\psi_t} = -4k^2. \tag{18}$$

where k is a separation constant. Integrating twice then gives

$$\phi_{xx} = -k^2\phi^4 + m^2\phi^2 + n^2 \tag{19}$$

$$\psi_{tt} = k^2\psi^4 + (m^2 - 1)\psi^2 - n^2, \tag{20}$$

which can be solved in terms of ELLIPTIC FUNCTIONS (Infeld and Rowlands 2000, pp. 178–179).

A single-SOLITON solution is obtained when $k = n = 0$, $m > 1$:

$$v = 4\tan^{-1}\left[\exp\left(\frac{\pm x - \beta t}{\sqrt{1 - \beta^2}}\right)\right] \tag{21}$$

where

$$\beta \equiv \frac{\sqrt{m^2 - 1}}{m}, \tag{22}$$

with the plus and minus signs corresponding to the soliton and antisoliton solutions. A two-SOLITON solution exists with $k = 0$, $m > 1$:

$$v = 4\tan^{-1}\left[\frac{\beta\sinh(\beta mx)}{\cosh(\beta mt)}\right]. \tag{23}$$

A two-kink solution is given by

$$v = 4\tan^{-1}\left[\frac{m\sinh\left(\dfrac{x}{\sqrt{1 - m^2}}\right)}{\beta\cosh\left(\dfrac{mt}{\sqrt{1 - m^2}}\right)}\right] \tag{24}$$

(Perring and Skyrme 1962; Drazin 1988; Tabor 1989, pp. 307–308).

A "breather" solution occurs for $k \neq 0$, $n = 0$, $m^2 < 1$:

$$v = -4\tan^{-1}\left[\frac{m}{\sqrt{1 - m^2}}\frac{\sin\left(\sqrt{1 - m^2}\,t\right)}{\cosh(mx)}\right]. \tag{25}$$

For a fixed x, v, this is a periodic function of t with frequency $2\pi/\sqrt{1 - m^2}$ (Infeld and Rowlands 2000, p. 179).

The so-called double sine-Gordon equation is given by

$$u_{xt} \pm \left[\sin u + \eta\sin\left(\tfrac{1}{2}u\right)\right] = 0 \tag{26}$$

(Calogero and Degasperis 1982, p. 135; Zwillinger 1997, p. 135).

See also KLEIN-GORDON EQUATION, SINH-GORDON EQUATION, SOLITON

References

Baker, H. F. *Abelian Functions: Abel's Theorem and the Allied Theory, Including the Theory of the Theta Functions.* New York: Cambridge University Press, p. xix, 1995.

Calogero, F. and Degasperis, A. *Spectral Transform and Solitons: Tools to Solve and Investigate Nonlinear Evolution Equations.* New York: North-Holland, 1982.

Drazin, P. G. and Johnson, R. S. *Solitons: An Introduction.* Cambridge, England: Cambridge University Press, 1988.

Infeld, E. and Rowlands, G. *Nonlinear Waves, Solitons, and Chaos, 2nd ed.* Cambridge, England: Cambridge University Press, pp. 178–180, 2000.

Lamb, G. L. Jr. *Elements of Soliton Theory.* New York: Wiley, 1980.

Perring, K. K. and Skyrme, T. H. "A Model Uniform Field Equation." *Nucl. Phys.* **31**, 550–555, 1962.

Tabor, M. "The Sine-Gordon Equation." §7.5.b in *Chaos and Integrability in Nonlinear Dynamics: An Introduction.* New York: Wiley, pp. 305–309, 1989.

Zwillinger, D. (Ed.). *CRC Standard Mathematical Tables and Formulae.* Boca Raton, FL: CRC Press, p. 417, 1995.

Sines Law

LAW OF SINES

Sine-Tangent Theorem

If

$$\frac{\sin \alpha}{\sin \beta} = \frac{m}{n},$$

then

$$\frac{\tan\left[\frac{1}{2}(\alpha - \beta)\right]}{\tan\left[\frac{1}{2}(\alpha + \beta)\right]} = \frac{m - n}{m + n},$$

Single-Valued Function

A function which has the same value at every point z_0 independent of the path along which it is reached by ANALYTIC CONTINUATION (Knopp 1996, p. 93).

See also SINGLE-VALUED FUNCTION

References

Knopp, K. "Multiple-Valued Functions." Section II in *Theory of Functions Parts I and II, Two Volumes Bound as One, Part II.* New York: Dover, pp. 93–146, 1996.

Singly Even Number

An EVEN NUMBER OF THE FORM $4n + 2$ (i.e., an INTEGER which is DIVISIBLE by 2 but not by 4). The first few for $n = 0, 1, 2, \ldots$ are 2, 6, 10, 14, 18, ... (Sloane's A016825)

See also DOUBLY EVEN NUMBER, EVEN NUMBER, ODD NUMBER

References

Conway, J. H. and Guy, R. K. *The Book of Numbers.* New York: Springer-Verlag, p. 30, 1996.

Sloane, N. J. A. Sequences A016825 in "An On-Line Version of the Encyclopedia of Integer Sequences." http://www.research.att.com/~njas/sequences/eisonline.html.

Singular Homology

The general type of HOMOLOGY which is what mathematicians generally mean when they say "homology." Singular homology is a more general version than Poincaré's original SIMPLICIAL HOMOLOGY.

See also HOMOLOGY (TOPOLOGY), SIMPLICIAL HOMOLOGY

Singular Knot

This entry contributed by SERGEI DUZHIN

A SMOOTH MAP $f : \mathbb{S}^1 \to R^3$ whose IMAGE has singularities. In particular, in the theory of Vassiliev's knot invariants, singular knots with a finite number of ORDINARY DOUBLE POINTS play an important role.

See also ORDINARY DOUBLE POINT, VASSILIEV INVARIANT

Singular Matrix

A SQUARE MATRIX that not have a MATRIX INVERSE. A matrix is singular IFF its DETERMINANT is 0. For example, there are 10 singular 2×2 $(0,1)$-MATRICES:

$$\begin{bmatrix} 0 & 0 \\ 0 & 0 \end{bmatrix}, \begin{bmatrix} 0 & 0 \\ 0 & 1 \end{bmatrix}, \begin{bmatrix} 0 & 0 \\ 1 & 0 \end{bmatrix}, \begin{bmatrix} 0 & 0 \\ 1 & 1 \end{bmatrix}, \begin{bmatrix} 0 & 1 \\ 0 & 0 \end{bmatrix}$$

$$\begin{bmatrix} 0 & 1 \\ 0 & 1 \end{bmatrix}, \begin{bmatrix} 1 & 0 \\ 0 & 0 \end{bmatrix}, \begin{bmatrix} 1 & 0 \\ 1 & 0 \end{bmatrix}, \begin{bmatrix} 1 & 1 \\ 0 & 0 \end{bmatrix}, \begin{bmatrix} 1 & 1 \\ 1 & 1 \end{bmatrix}.$$

The following table gives the numbers of singular $n \times n$ matrices for certain matrix classes.

matrix type	Sloane	counts for $n = 1, 2, \ldots$
$(-1, 0, 1)$-matrices	A000000	1, 33, 7875, ...
$(-1, 1)$-matrices	A000000	0, 8, 320, 43264, ...
$(0, 1)$-matrices	A046747	1, 10, 338, 42976, ...

See also DETERMINANT, ILL-CONDITIONED MATRIX, MATRIX INVERSE, NONSINGULAR MATRIX, SINGULAR VALUE DECOMPOSITION

References

Ayres, F. Jr. *Theory and Problems of Matrices.* New York: Schaum, p. 39, 1962.

Faddeeva, V. N. *Computational Methods of Linear Algebra.* New York: Dover, p. 11, 1958.

Golub, G. H. and van Loan, C. F. *Matrix Computations, 3rd ed.* Baltimore, MD: Johns Hopkins, p. 51, 1996.

Kahn, J.; Komlós, J.; and Szemerédi, E. "On the Probability that a Random ± 1 Matrix is Singular." *J. Amer. Math. Soc.* **8**, 223–240, 1995.

Komlós, J. "On the Determinant of (0, 1)-Matrices." *Studia Math. Hungarica* **2**, 7–21 1967.

Marcus, M. and Minc, H. *Introduction to Linear Algebra.* New York: Dover, p. 70, 1988.

Marcus, M. and Minc, H. *A Survey of Matrix Theory and Matrix Inequalities.* New York: Dover, p. 3, 1992.

Sloane, N. J. A. Sequences A046747 in "An On-Line Version of the Encyclopedia of Integer Sequences." http://www.research.att.com/~njas/sequences/eisonline.html.

Singular Measure

Two COMPLEX MEASURES μ and ν on a MEASURE SPACE X, are mutually singular if they are supported on different subsets. More precisely, $X = A \cup B$ where A and B are two DISJOINT SETS such that the following hold for any MEASURABLE SET E,

1. The sets $A \cap E$ and $B \cap E$ are measurable.
2. The TOTAL VARIATION MEASURE of μ is supported on A and that of ν on B, i.e.,

$$\|\mu\|(B \cap E) = 0 = \|\nu\|(A \cap E).$$

The relation of two measures being singular, written as $\mu \perp \nu$, is plainly symmetric. Nevertheless, it is sometimes said that "ν is singular with respect to μ."

A discrete singular measure (with respect to LEBESGUE MEASURE on the reals) is a MEASURE λ supported at 0, say $\lambda(E) = 1$ iff $0 \in E$. In general, a MEASURE λ is concentrated on a SUBSET A if $\lambda(E) = \lambda(E \cap A)$. For instance, the measure above is concentrated at 0.

See also ABSOLUTELY CONTINUOUS, COMPLEX MEASURE, LEBESGUE DECOMPOSITION (MEASURE), LEBESGUE MEASURE

References

Halmos, P. *Measure Theory, 2nd ed.* New York: Springer-Verlag, p. 126, 1977.

Reed, M. and Simon, B. *Methods of Modern Mathematical Physics: Fourier Analysis, Self-Adjointness, Vol. 2.* New York: Academic Press, 1975.

Rudin, W. *Real and Complex Analysis.* New York: McGraw-Hill, pp. 116–132, 1987.

Singular Point (Algebraic Curve)

A singular point of an ALGEBRAIC CURVE is a point where the curve has "nasty" behavior such as a CUSP or a point of self-intersection (when the underlying field K is taken as the REALS). More formally, a point (a, b) on a curve $f(x, y) = 0$ is singular if the x and y PARTIAL DERIVATIVES of f are both zero at the point (a, b). (If the field K is not the REALS or COMPLEX NUMBERS, then the PARTIAL DERIVATIVE is computed formally using the usual rules of CALCULUS.)

Consider the following two examples. For the curve

$$x^3 - y^2 = 0,$$

the CUSP at $(0, 0)$ is a singular point. For the curve

$$x^2 + y^2 = -1,$$

$(0, i)$ is a nonsingular point and this curve is nonsingular.

See also ALGEBRAIC CURVE, CUSP

Singular Point (Differential Equation)

Consider a second-order ORDINARY DIFFERENTIAL EQUATION

$$y'' + P(x)y' + Q(x)y = 0.$$

If $P(x)$ and $Q(x)$ remain FINITE at $x = x_0$, then x_0 is called an ORDINARY POINT. If either $P(x)$ or $Q(x)$ diverges as $x \to x_0$, then x_0 is called a singular point. Singular points are further classified as follows:

1. If either $P(x)$ or $Q(x)$ diverges as $x \to x_0$ but $(x - x_0)P(x)$ and $(x - x_0)^2 Q(x)$ remain FINITE as $x \to x_0$, then $x = x_0$ is called a REGULAR SINGULAR POINT (or NONESSENTIAL SINGULARITY).
2. If $P(x)$ diverges more quickly than $1/(x - x_0)$, so $(x - x_0)P(x)$ approaches INFINITY as $x \to x_0$, or $Q(x)$ diverges more quickly than $1/(x - x_0)^2 Q$ so that $(x - x_0)^2 Q(x)$ goes to INFINITY as $x \to x_0$, then x_0 is called an IRREGULAR SINGULARITY (or ESSENTIAL SINGULARITY).

See also IRREGULAR SINGULARITY, REGULAR SINGULAR POINT, SINGULARITY

References

Arfken, G. "Singular Points." §8.4 in *Mathematical Methods for Physicists, 3rd ed.* Orlando, FL: Academic Press, pp. 451–454, 1985.

Singular Point (Function)

Singular points (also simply called "singularities") are points z_0 in the DOMAIN of a FUNCTION f where f fails to be ANALYTIC. ISOLATED SINGULARITIES may be classified as ESSENTIAL SINGULARITIES, POLES, or REMOVABLE SINGULARITIES.

ESSENTIAL SINGULARITIES are POLES of INFINITE order.

A POLE of order n is a singularity z_0 of $f(z)$ for which the function $(z - z_0)^n f(z)$ is nonsingular and for which $(z - z_0)^k f(z)$ is singular for $k = 0, 1, ..., n - 1$.

REMOVABLE SINGULARITIES are singularities for which it is possible to assign a COMPLEX NUMBER in such a way that $f(z)$ becomes ANALYTIC. For example, the function $f(z) = z^2/z$ has a REMOVABLE SINGULARITY at 0, since $f(z) = z$ everywhere but 0, and $f(z)$ can be set equal to 0 at $z = 0$. REMOVABLE SINGULARITIES are not POLES.

The function $f(z) = \csc(1/z)$ has POLES at $z = 1/(2\pi n)$, and a nonisolated singularity at 0.

See also ESSENTIAL SINGULARITY, IRREGULAR SINGULARITY, ORDINARY POINT, POLE, REGULAR SINGULAR POINT, REMOVABLE SINGULARITY, SINGULAR POINT (DIFFERENTIAL EQUATION)

References
Arfken, G. "Singularities." §7.1 in *Mathematical Methods for Physicists, 3rd ed.* Orlando, FL: Academic Press, pp. 396–400, 1985.

Singular Series

$$\rho_{2s}(n) = \frac{\pi^8}{\Gamma(s)} n^{s-1} \sum_{p, q} \left(\frac{S_{p,q}}{q}\right)^{2s} e^{2np\pi i/q},$$

where $S_{p,q}$ is a GAUSSIAN SUM, and $\Gamma(s)$ is the GAMMA FUNCTION.

Singular System

A system is singular if its CONDITION NUMBER is INFINITE and ILL-CONDITIONED if it is too large.

See also CONDITION NUMBER, ILL-CONDITIONED MATRIX

Singular Value

There are two types of singular values, one in the context of elliptic integrals, and the other in linear algebra. For a MATRIX A, the values

$$\sqrt{\lambda_j(A^*A)}, \qquad (1)$$

where λ_j is an EIGENVALUE and A^* is the ADJOINT MATRIX, are called singular values (Marcus and Minc 1992, p. 69). Singular values can be found using the *Mathematica* command `SingularValues[m]`, which returns the so-called SINGULAR VALUE DECOMPOSITION as a list $\{u, w, v\}$, where u and v are matrices and w is the list of the singular values. If

$$A = UH. \qquad (2)$$

where U is a UNITARY MATRIX and H is a HERMITIAN MATRIX, then the EIGENVALUES of H are the singular values of A.

For elliptic integrals, a MODULUS k_r such that

$$\frac{K'(k_r)}{K(k_r)} = \sqrt{r}, \qquad (3)$$

where $K(k)$ is a complete ELLIPTIC INTEGRAL OF THE FIRST KIND, and $K'(k_r) \equiv K\left(\sqrt{1 - k_r^2}\right)$. The ELLIPTIC

LAMBDA FUNCTION $\lambda^*(r)$ gives the value of k_r. Abel (quoted in Whittaker and Watson 1990, p. 525) proved that if r is an INTEGER, or more generally whenever

$$\frac{K'(k)}{K(k)} = \frac{a + b\sqrt{n}}{c + d\sqrt{n}}, \qquad (4)$$

where a, b, c, d, and n are INTEGERS, then the MODULUS k is the ROOT of an algebraic equation with INTEGER COEFFICIENTS.

See also ELLIPTIC INTEGRAL SINGULAR VALUE, ELLIPTIC INTEGRAL OF THE FIRST KIND, ELLIPTIC LAMBDA FUNCTION, MODULUS (ELLIPTIC INTEGRAL), SINGULAR VALUE DECOMPOSITION

References
Marcus, M. and Minc, H. *Introduction to Linear Algebra.* New York: Dover, p. 191, 1988.
Marcus, M. and Minc, H. *A Survey of Matrix Theory and Matrix Inequalities.* New York: Dover, p. 69, 1992.
Whittaker, E. T. and Watson, G. N. *A Course in Modern Analysis, 4th ed.* Cambridge, England: Cambridge University Press, pp. 524–528, 1990.

Singular Value Decomposition

A decomposition of a matrix A into the form

$$A = U^*DV,$$

where U is a UNITARY MATRIX, U^* is its ADJOINT MATRIX, and D is a DIAGONAL MATRIX whose elements are the SINGULAR VALUES of the original matrix. If A is a COMPLEX MATRIX, then there always exists such a decomposition with positive singular values (Golub and van Loan 1996, pp. 70 and 73).

Singular value decomposition is implemented in *Mathematica* as `SingularValues[m]`, which returns a list $\{u, w, v\}$, where u and v are matrices and w is a list of the singular values.

See also CHOLESKY DECOMPOSITION, LU DECOMPOSITION, MATRIX DECOMPOSITION, MATRIX DECOMPOSITION THEOREM, QR DECOMPOSITION, SINGULAR VALUE, UNITARY MATRIX

References
Gentle, J. E. "Singular Value Factorization." §3.2.7 in *Numerical Linear Algebra for Applications in Statistics.* Berlin: Springer-Verlag, pp. 102–103, 1998.
Golub, G. H. and van Loan, C. F. "The Singular Value Decomposition" and "Unitary Matrixes." §2.5.3 and 2.5.6 in *Matrix Computations, 3rd ed.* Baltimore, MD: Johns Hopkins University Press, pp. 70–71 and 73, 1996.
Nash, J. C. "The Singular-Value Decomposition and Its Use to Solve Least-Squares Problems." Ch. 3 in *Compact Numerical Methods for Computers: Linear Algebra and Function Minimisation, 2nd ed.* Bristol, England: Adam Hilger, pp. 30–48, 1990.
Press, W. H.; Flannery, B. P.; Teukolsky, S. A.; and Vetterling, W. T. "Singular Value Decomposition." §2.6 in *Numerical Recipes in FORTRAN: The Art of Scientific*

Computing, 2nd ed. Cambridge, England: Cambridge University Press, pp. 51–63, 1992.

Singularity

In general, a point at which an equation, surface, etc., blows up or becomes DEGENERATE. Singularities are often also called singular points.

See also ESSENTIAL SINGULARITY, ISOLATED SINGULARITY, SINGULAR POINT (ALGEBRAIC CURVE), SINGULAR POINT (DIFFERENTIAL EQUATION), SINGULAR POINT (FUNCTION), WHITNEY SINGULARITY

References

Knopp, K. "Singularities." Section IV in *Theory of Functions Parts I and II, Two Volumes Bound as One, Part I.* New York: Dover, pp. 117–139, 1996.

Sinh

HYPERBOLIC SINE

Sinh-Gordon Equation

The PARTIAL DIFFERENTIAL EQUATION

$$u_{xt} = \sinh u,$$

which contains u_{xt} instead of $u_{xx} - u_{tt}$ and $\sinh u$ instead to $\sin u$, as in the SINE-GORDON EQUATION (Grauel 1985; Zwillinger 1997, p. 135).

See also SINE-GORDON EQUATION, SINH-POISSON EQUATION

References

Grauel, A. "Sinh-Gordon Equation, Painlevé Property and Bäcklund Transformation." *Physica A* **12**, 557–568, 1985.
Zwillinger, D. *Handbook of Differential Equations, 3rd ed.* Boston, MA: Academic Press, p. 135, 1997.

Sinh-Poisson Equation

The PARTIAL DIFFERENTIAL EQUATION

$$\nabla^2 u + \lambda^2 \sinh u = 0,$$

where ∇^2 is the LAPLACIAN (Ting *et al.* 1987; Zwillinger 1997, p. 135).

See also SINH-GORDON EQUATION

References

Ting, A. C.; Cheb, H. H.; and Lee, Y. C. "Exact Solutions of a Nonlinear Boundary Value Problem: The Vortices of the Two-Dimensional Sinh-Poisson Equation." *Physica D*, 37–66, 1987.
Zwillinger, D. *Handbook of Differential Equations, 3rd ed.* Boston, MA: Academic Press, p. 135, 1997.

SinIntegral

SINE INTEGRAL

Sink (Directed Graph)

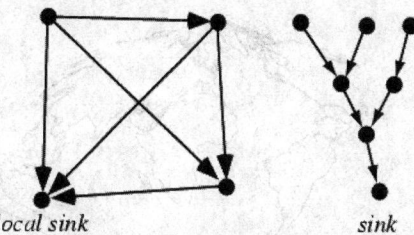

local sink   *sink*

A local sink is a node of a DIRECTED GRAPH with no exiting edges, also called a TERMINAL (Borowski and Borwein 1991, p. 401; left figure). A global sink (often simply called a sink) is a node in a DIRECTED GRAPH which is reached by all directed edges (Harary 1994, p. 201; right figure).

See also DIRECTED GRAPH, NETWORK, SOURCE

References

Borowski, E. J. and Borwein, J. M. (Eds.). *The HarperCollins Dictionary of Mathematics.* New York: HarperCollins, 1991.
Cormen, T. H.; Leiserson, C. E.l and Rivest, R. L. *Introduction to Algorithms.* Cambridge, MA: MIT Press, 1990.
Harary, F. *Graph Theory.* Reading, MA: Addison-Wesley, 1994.

Sink (Map)

A stable fixed point of a MAP which, in a dissipative DYNAMICAL SYSTEM, is an ATTRACTOR.

See also ATTRACTOR, DYNAMICAL SYSTEM

Sinusoid

A curve similar to the SINE function but possibly shifted in phase, period, amplitude, or any combination thereof. The general sinusoid of amplitude a, angular frequency ω (and period $2\pi/\omega$), and phase c is given by

$$f(x) = a \sin(\omega x + c).$$

See also SINE

References

Beyer, W. H. *CRC Standard Mathematical Tables, 28th ed.* Boca Raton, FL: CRC Press, p. 225, 1987.

Sinusoidal Projection

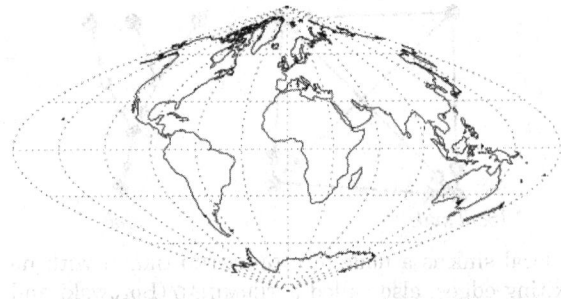

An equal AREA MAP PROJECTION.

$$x = (\lambda - \lambda_0)\cos\phi \qquad (1)$$
$$y = \phi, \qquad (2)$$

The inverse FORMULAS are

$$\phi = y \qquad (3)$$
$$\lambda = \lambda_0 + \frac{x}{\cos\phi}, \qquad (4)$$

References
Snyder, J. P. *Map Projections--A Working Manual.* U. S. Geological Survey Professional Paper 1395. Washington, DC: U. S. Government Printing Office, pp. 243–248, 1987.

Sinusoidal Spiral

A curve OF THE FORM

$$r^n = a^n \cos(n\theta)$$

with n RATIONAL, which is not a true SPIRAL. Sinusoidal spirals were first studied by Maclaurin. Special cases are given in the following table.

n	Curve
-2	HYPERBOLA
-1	LINE
$-\frac{1}{2}$	PARABOLA
$-\frac{1}{3}$	TSCHIRNHAUSEN CUBIC
$\frac{1}{3}$	CAYLEY'S SEXTIC
$\frac{1}{2}$	CARDIOID
1	CIRCLE
2	LEMNISCATE

References
Lawrence, J. D. *A Catalog of Special Plane Curves.* New York: Dover, p. 184, 1972.
Lockwood, E. H. *A Book of Curves.* Cambridge, England: Cambridge University Press, p. 175, 1967.
MacTutor History of Mathematics Archive. "Sinusoidal Spirals." http://www-groups.dcs.st-and.ac.uk/~history/Curves/Sinusoidal.html.

Sinusoidal Spiral Inverse Curve

The INVERSE CURVE of a SINUSOIDAL SPIRAL

$$r = a^{(1/n)}[\cos(nt)]^{1/n}$$

with INVERSION CENTER at the origin and inversion radius k is another SINUSOIDAL SPIRAL

$$r = ka^{(1/n)}[\cos(nt)]^{1/n},$$

Sinusoidal Spiral Pedal Curve

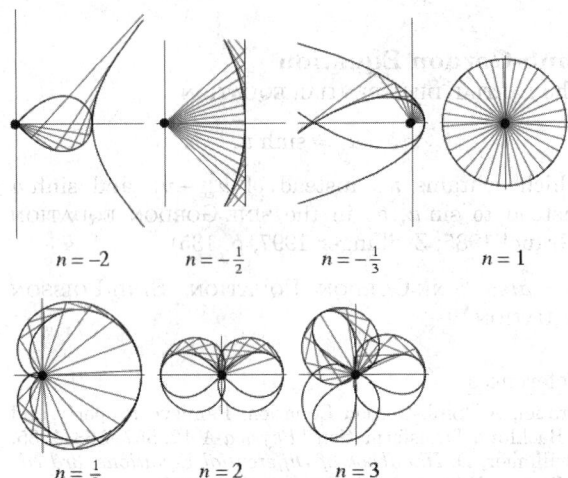

The PEDAL CURVE of a SINUSOIDAL SPIRAL

$$r = a^{(1/n)}[\cos(nt)]^{1/n}$$

with PEDAL POINT at the center is another SINUSOIDAL SPIRAL

$$x = \cos^{1+1/n}(nt)\cos[(n+1)t]$$
$$y = \cos^{1+1/n}(nt)\sin[(n+1)t].$$

See also PEDAL CURVE, SINUSOIDAL SPIRAL

Sister Celine's Method

A method for finding RECURRENCE RELATIONS for hypergeometric polynomials directly from the series expansions of the polynomials. The method is effective and easily implemented, but usually slower than ZEILBERGER'S ALGORITHM. Given a sum $f(n) = \Sigma_k F(n, k)$, the method operates by finding a recurrence of the form

$$\sum_{i=0}^{I} \sum_{j=0}^{J} a_{ij}(n)F(n-j,\,k-i)=0$$

by proceeding as follows (Petkovsek *et al.* 1996, p. 59):

1. Fix trial values of I and J.
2. Assume a recurrence formula of the above form where $a_{ij}(n)$ are to be solved for.
3. Divide each term of the assumed recurrence by $F(n,k)$ and reduce every ratio $F(n-j,\,k-i)/F(n,k)$ by simplifying the ratios of its constituent factorials so that only RATIONAL FUNCTIONS in n and k remain.
4. Put the resulting expression over a common DENOMINATOR, then collect the numerator as a POLYNOMIAL in k.
5. Solve the system of linear equations that results after setting the coefficients of each power of k in the NUMERATOR to 0 for the unknown coefficients a_{ij}.
6. If no solution results, start again with larger I or J.

Under suitable hypotheses, a "fundamental theorem" (Verbaten 1974, Wilf and Zeilberger 1992, Petkovsek *et al.* 1996) guarantees that this algorithm always succeeds for large enough I and J (which can be estimated in advance). The theorem also generalizes to multivariate sums and to q- and multi-q-sums (Wilf and Zeilberger 1992, Petkovsek *et al.* 1996).

See also GENERALIZED HYPERGEOMETRIC FUNCTION, GOSPER'S ALGORITHM, HYPERGEOMETRIC IDENTITY, HYPERGEOMETRIC SERIES, ZEILBERGER'S ALGORITHM

References

Fasenmyer, Sister M. C. *Some Generalized Hypergeometric Polynomials.* Ph.D. thesis. University of Michigan, Nov. 1945.
Fasenmyer, Sister M. C. "Some Generalized Hypergeometric Polynomials." *Bull. Amer. Math. Soc.* **53**, 806–812, 1947.
Fasenmyer, Sister M. C. "A Note on Pure Recurrence Relations." *Amer. Math. Monthly* **56**, 14–17, 1949.
Koepf, W. "Holonomic Recurrence Equations." Ch. 4 in *Hypergeometric Summation: An Algorithmic Approach to Summation and Special Function Identities.* Braunschweig, Germany: Vieweg, pp. 44–60, 1998.
Petkovsek, M.; Wilf, H. S.; and Zeilberger, D. "Sister Celine's Method." Ch. 4 in *A = B.* Wellesley, MA: A. K. Peters, pp. 55–72, 1996.
Rainville, E. D. Chs. 14 and 18 in *Special Functions.* New York: Chelsea, 1971.
Verbaten, P. "The Automatic Construction of Pure Recurrence Relations." *Proc. EUROSAM '74, ACM-SIGSAM Bull.* **8**, 96–98, 1974.
Wilf, H. S. and Zeilberger, D. "An Algorithmic Proof Theory for Hypergeometric (Ordinary and "q") Multisum/Integral Identities." *Invent. Math.* **108**, 575–633, 1992.

Site Percolation

 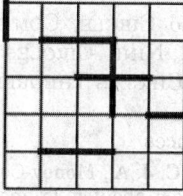

site percolation　　　*bond percolation*

A PERCOLATION which considers the lattice vertices as the relevant entities (left figure).

See also BOND PERCOLATION, PERCOLATION THEORY

Siteswap

A siteswap is a sequence encountered in JUGGLING in which each term is a POSITIVE integer, encoded in BINARY. The transition rule from one term to the next consists of changing some 0 to 1, subtracting 1, and then dividing by 2, with the constraint that the DIVISION by two must be exact. Therefore, if a term is EVEN, the bit to be changed must be the units bit. In siteswaps, the number of 1-bits is a constant.

Each transition is characterized by the bit position of the toggled bit (denoted here by the numeral on top of the arrow). For example,

$$111\xrightarrow{5}10011\xrightarrow{2}1011\xrightarrow{5}10101\xrightarrow{1}1011\xrightarrow{2}111$$

$$\xrightarrow{6}100011\xrightarrow{3}10101\xrightarrow{3}1110\xrightarrow{0}111\xrightarrow{4}1011\ldots$$

The second term is given from the first as follows: 000111 with bit 5 flipped becomes 100111, or 39. Subtract 1 to obtain 38 and divide by two to obtain 19, which is 10011.

See also JUGGLING

References

Juggling Information Service. "Siteswaps." http://www.juggling.org/help/siteswap/.
Smith, H. J. "Juggler Numbers." http://pweb.netcom.com/~hjsmith/Juggler.html.

Six Circles Theorem

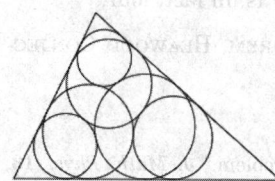 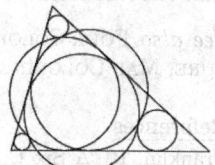

Starting with a triangle, draw a circle touching two sides. Then draw a circle tangent to this circle and two other sides. Continue in the same direction. Then a chain is formed in which the sixth circle is tangent to the first.

See also CIRCLE, CONTACT TRIANGLE, HEXLET, INCIRCLE, NINE CIRCLES THEOREM, PAPPUS CHAIN, SEVEN CIRCLES THEOREM

References

Evelyn, C. J. A.; Money-Coutts, G. B.; and Tyrrell, J. A. "A Theorem about a Triangle and Six Circles." §3.3 in *The Seven Circles Theorem and Other New Theorems*. London: Stacey International, pp. 49–58, 1974.
Wells, D. *The Penguin Dictionary of Curious and Interesting Geometry*. London: Penguin, p. 231, 1991.

Six Exponentials Theorem

Let x_1 and x_2 be two linearly independent complex numbers, and let y_1, y_2, y_3 be three linearly independent complex numbers. Then at least one of

$$e^{x_1 y_1}, e^{x_1 y_2}, e^{x_1 y_3}, e^{x_2 y_1}, e^{x_2 y_2}, e^{x_2 y_3}$$

is TRANSCENDENTAL (Waldschmidt 1979, p. 3.5). This theorem is due to Siegel, Schneider, Lang, and Ramachandra. The corresponding statement obtained by replacing y_1, y_2, y_3 with y_1, y_2 is called the FOUR EXPONENTIALS CONJECTURE and remains unproven.

See also FOUR EXPONENTIALS CONJECTURE, HERMITE-LINDEMANN THEOREM, TRANSCENDENTAL NUMBER

References

Finch, S. "Powers of 3/2 Modulo One." http://www.mathsoft.com/asolve/pwrs32/pwrs32.html.
Ramachandra, K. "Contributions to the Theory of Transcendental Numbers. I, II." *Acta Arith.* **14**, 65–78, 1967–68.
Ramachandra, K. and Srinivasan, S. "A Note to a Paper: 'Contributions to the Theory of Transcendental Numbers. I, II' by Ramachandra on Transcendental Numbers." *Hardy-Ramanujan J.* **6**, 37–44, 1983.
Waldschmidt, M. *Transcendence Methods*. Queen's Papers in Pure and Applied Mathematics, No. 52. Kingston, Ontario, Canada: Queen's University, 1979.
Waldschmidt, M. "On the Transcendence Method of Gelfond and Schneider in Several Variables." In *New Advances in Transcendence Theory* (Ed. A. Baker). Cambridge, England: Cambridge University Press, 1988.

Six-Color Theorem

To color any map on the SPHERE or the PLANE requires at most six-colors. This number can easily be reduced to five, and the FOUR-COLOR THEOREM demonstrates that the NECESSARY number is, in fact, four.

See also FOUR-COLOR THEOREM, HEAWOOD CONJECTURE, MAP COLORING

References

Franklin, P. "A Six Colour Problem." *J. Math. Phys.* **13**, 363–369, 1934.
Hoffman, I. and Soifer, A. "Another Six-Coloring of the Plane." *Disc. Math.* **150**, 427–429, 1996.
Saaty, T. L. and Kainen, P. C. *The Four-Color Problem: Assaults and Conquest*. New York: Dover, 1986.

Six-j Symbol

WIGNER 6J-SYMBOL

SixJSymbol

WIGNER 6J-SYMBOL

Six-Sphere Coordinates

6-SPHERE COORDINATES

Skein Relationship

L_+ L_0 L_-

A relationship between KNOT POLYNOMIALS for links in different orientations (denoted below as L_+, L_0, and L_-). J. H. Conway was the first to realize that the ALEXANDER POLYNOMIAL could be defined by a relationship of this type.

See also ALEXANDER POLYNOMIAL, HOMFLY POLYNOMIAL, SIGNATURE (KNOT)

Skeleton

In ALGEBRAIC TOPOLOGY, a p-skeleton is a SIMPLICIAL SUBCOMPLEX of K which is the collection of all SIMPLICES of K of dimension at most p, denoted $K^{(p)}$.

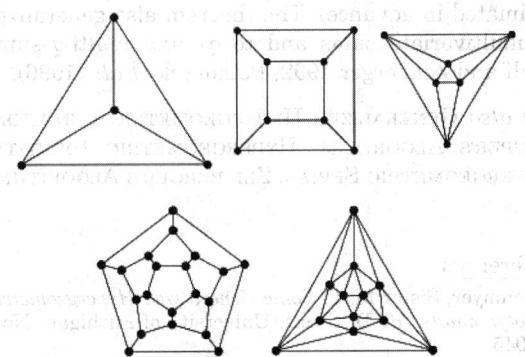

The GRAPH obtained by replacing the faces of a polyhedron with its edges and vertices is therefore the skeleton of the polyhedron. The polyhedral graphs corresponding to the skeletons of PLATONIC SOLIDS are illustrated above. The number of topologically distinct skeletons $N(n)$ with n VERTICES for $n = 4, 5, 6, \ldots$ are 1, 2, 7, 18, 52, ... (Sloane's A006869).

See also POLYHEDRAL GRAPH, SCHLEGEL GRAPH

References

Gardner, M. *Martin Gardner's New Mathematical Diversions from Scientific American*. New York: Simon and Schuster, p. 233, 1966.
Munkres, J. R. *Elements of Algebraic Topology*. Perseus Press, 1993.

Sloane, N. J. A. Sequences A006869/M1748 in "An On-Line Version of the Encyclopedia of Integer Sequences." http://www.research.att.com/~njas/sequences/eisonline.html.

Skeleton Division

A LONG DIVISION in which most or all of the digits are replaced by a symbol (usually asterisks) to form a CRYPTARITHM.

See also CRYPTARITHM

Skew Conic

Also known as a GAUCHE CONIC, SPACE CONIC, TWISTED CONIC, or CUBICAL CONIC SECTION. A third-order SPACE CURVE having up to three points in common with a plane and having three points in common with the plane at infinity. A skew cubic is determined by six points, with no four of them COPLANAR. A line is met by up to four tangents to a skew cubic.

A line joining two points of a skew cubic (REAL or conjugate imaginary) is called a SECANT of the curve, and a line having one point in common with the curve is called a SEMISECANT or TRANSVERSAL. Depending on the nature of the roots, the skew conic is classified as follows:

1. The three ROOTS are REAL and distinct (CUBICAL HYPERBOLA).
2. One root is REAL and the other two are COMPLEX CONJUGATES (CUBICAL ELLIPSE).
3. Two of the ROOTS coincide (CUBICAL PARABOLIC HYPERBOLA).
4. All three ROOTS coincide (CUBICAL PARABOLA).

See also CONIC SECTION, CUBICAL ELLIPSE, CUBICAL HYPERBOLA, CUBICAL PARABOLA, CUBICAL PARABOLIC HYPERBOLA

Skew Coordinate System

A system of CURVILINEAR COORDINATES in which each family of surfaces intersects the others at angles other than right angles.

See also CURVILINEAR COORDINATES, ORTHOGONAL COORDINATE SYSTEM

References

Moon, P. and Spencer, D. E. *Field Theory Handbook, Including Coordinate Systems, Differential Equations, and Their Solutions,* 2nd ed. New York: Springer-Verlag, p. 1, 1988.

Skew Diagonal

$$\begin{bmatrix} a_{11} & a_{12} & a_{13} & a_{14} \\ a_{21} & a_{22} & a_{23} & a_{24} \\ a_{31} & a_{32} & a_{33} & a_{34} \\ a_{41} & a_{42} & a_{43} & a_{44} \end{bmatrix}$$

main skew diagonal

skew diagonals

A diagonal of a SQUARE MATRIX which is traversed in the "northeast" direction. "The" skew diagonal (or "secondary diagonal") of an $n \times n$ square matrix is the skew diagonal from a_{n1} to a_{1n}.

See also DIAGONAL

Skew Field

A FIELD in which the commutativity of multiplication is not required, more commonly called a DIVISION ALGEBRA.

See also DIVISION ALGEBRA, FIELD

Skew Hermitian Matrix

A SQUARE MATRIX A is skew Hermitian if is satisfies

$$A^* = -A, \tag{1}$$

where A^* is the ADJOINT MATRIX. For example, the matrix

$$\begin{bmatrix} i & 1+i & 2i \\ -1+i & 5i & 3 \\ 2i & -3 & 0 \end{bmatrix} \tag{2}$$

is a skew Hermitian matrix. A matrix m can be tested to see if it is skew Hermitian using the *Mathematica* function

```
SkewHermitianQ[m_List?MatrixQ] := (m === -
Conjugate@Transpose@m)
```

The set of $n \times n$ skew Hermitian matrices is a VECTOR SPACE, and the COMMUTATOR

$$[A, \; B] = AB - BA \tag{3}$$

of two skew Hermitian matrices is skew Hermitian. Hence, the skew Hermitian matrices are a LIE ALGEBRA, which is related to the LIE GROUP of UNITARY MATRICES. In particular, suppose $A(t)$ is a path of unitary matrices through $A(0) = I$, i.e.,

$$A(t) = A^*(t) = I \tag{4}$$

for all t, where A^* is the ADJOINT MATRIX and I is the IDENTITY MATRIX. The DERIVATIVE at $t = 0$ of both sides must be equal so

$$\left.\frac{dA}{dt}\right|_{t=0} + \left.\frac{dA^*}{dt}\right|_{t=0} = 0. \tag{5}$$

That is, the DERIVATIVE of A(t) at the identity must be a skew Hermitian matrix.

The EXPONENTIAL MAP of a skew Hermitian matrix is a UNITARY MATRIX.

See also ADJOINT MATRIX, HERMITIAN MATRIX, SKEW SYMMETRIC MATRIX, UNITARY MATRIX

References

Ayres, F. Jr. *Theory and Problems of Matrices.* New York: Schaum, pp. 13 and 118, 1962.

Skew Lines

Two or more LINES which have no intersections but are not PARALLEL, also called AGONIC LINES. Since two LINES in the PLANE must intersect or be PARALLEL, skew lines can exist only in three or more DIMENSIONS.

Three skew lines always define a one-sheeted HYPERBOLOID, except in the case where they are all parallel to a single PLANE but not to each other. In this case, they determine a HYPERBOLIC PARABOLOID (Hilbert and Cohn-Vossen 1999, p. 15).

See also DIRECTOR, GALLUCCI'S THEOREM, REGULUS

References

Altshiller-Court, N. *Modern Pure Solid Geometry.* New York: Chelsea, p. 1, 1979.
Hilbert, D. and Cohn-Vossen, S. *Geometry and the Imagination.* New York: Chelsea, p. 15, 1999.

Skew Polygon

A polygon whose vertices do not all lie in a PLANE.

See also REGULAR SKEW POLYHEDRON, SKEW QUADRILATERAL

References

Williams, R. "Skew Polygons (Saddle Polygons)." §2.2 in *The Geometrical Foundation of Natural Structure: A Source Book of Design.* New York: Dover, p. 34, 1979.

Skew Polyhedron

REGULAR SKEW POLYHEDRON

Skew Polyomino

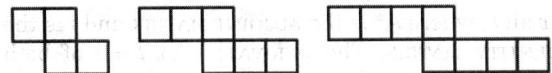

See also L-POLYOMINO, SQUARE POLYOMINO, STRAIGHT POLYOMINO, T-POLYOMINO

Skew Quadrilateral

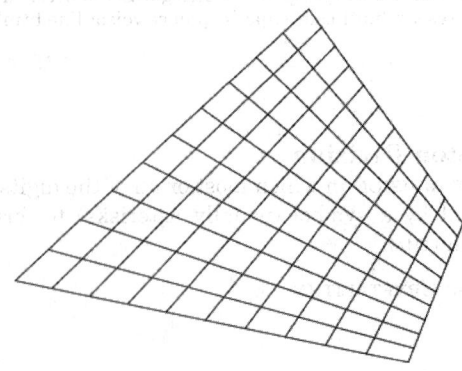

A four-sided QUADRILATERAL not contained in a plane. The lines connecting the midpoints of opposite sides of a skew quadrilateral intersect (and bisect) each other (Steinhaus 1983).

The problem of finding the minimum bounding surface of a skew quadrilateral was solved by Schwarz (Schwarz 1890, Wells 1991) in terms of ABELIAN INTEGRALS and has the shape of a SADDLE. It is given by solving

$$\left(1 + f_y^2\right)f_{xx} - 2f_xf_yf_{xy} + \left(1 + f_x^2\right)f_{yy} = 0.$$

See also HYPERBOLIC PARABOLOID, QUADRILATERAL, SKEW POLYGON

References

Altshiller-Court, N. "The Skew Quadrilateral." Ch. 3 and §5.1 in *Modern Pure Solid Geometry.* New York: Chelsea, pp. 42–47 and 111–115, 1979.
Coxeter, H. S. M. and Greitzer, S. L. *Geometry Revisited.* Washington, DC: Math. Assoc. Amer., p. 53, 1967.
Isenberg, C. *The Science of Soap Films and Soap Bubbles.* New York: Dover, p. 81, 1992.
Forsyth, A. R. *Calculus of Variations.* New York: Dover, p. 503, 1960.
Schwarz, H. A. *Gesammelte Mathematische Abhandlungen, 2nd ed.* New York: Chelsea.
Steinhaus, H. *Mathematical Snapshots, 3rd ed.* New York: Dover, pp. 242 and 244, 1999.
Wells, D. *The Penguin Dictionary of Curious and Interesting Geometry.* London: Penguin, pp. 186–187, 1991.

Skew Symmetric Matrix

A SQUARE MATRIX A is skew symmetric if

$$A^T = -A, \tag{1}$$

with A^T denoting the matrix TRANSPOSE. For example,

$$A = \begin{bmatrix} 0 & -1 \\ 1 & 0 \end{bmatrix} \tag{2}$$

is a skew symmetric matrix. The set of $n \times n$ skew symmetric matrices is denoted $o(n)$. A matrix m can be tested to see if it is skew symmetric using the *Mathematica* function

```
SkewSymmetricQ[l_List?MatrixQ] := (l === -
Transpose[l])
```

The set $o(n)$ of $n \times n$ skew symmetric matrices is a VECTOR SPACE, and the COMMUTATOR

$$[A, B] = AB - BA \tag{3}$$

of two skew symmetric matrices is skew symmetric. Hence, the skew symmetric matrices are a LIE ALGEBRA, which is related to the LIE GROUP of ORTHOGONAL MATRICES. In particular, suppose $A(t)$ is a path of orthogonal matrices through $A(0) = I$, i.e., $A(t)A^t(t) = I$ for all t. The DERIVATIVE at $t = 0$ of both sides must be equal so $dA/dt(0) + dA^t/dt(0) = 0$. That is, the DERIVATIVE of $A(t)$ at the identity must be a skew symmetric matrix.

The EXPONENTIAL MAP of a skew symmetric matrix is an ORTHOGONAL MATRIX.

See also BISYMMETRIC MATRIX, DIAGONAL MATRIX, PERSYMMETRIC MATRIX, SKEW HERMITIAN MATRIX, SYMMETRIC MATRIX, TRANSPOSE

References

Ayres, F. Jr. *Theory and Problems of Matrices.* New York: Schaum, pp. 12 and 117, 1962.

Skewes Number

The Skewes number (or first Skewes number) is the number Sk_1 above which $\pi(n < Li(n))$ must fail (assuming that the RIEMANN HYPOTHESIS is true), where $\pi(n)$ is the PRIME COUNTING FUNCTION and $Li(n)$ is the LOGARITHMIC INTEGRAL. In 1912, Littlewood proved that Sk_1 exists (Hardy 1999, p. 17), and the upper bound

$$Sk_1 = e^{e^{e^{79}}} \approx 10^{10^{10^{34}}}$$

was subsequently found by Skewes. The Skewes number has since been reduced to $e^{e^{27/4}} \approx 8.185 \times 10^{370}$ by te Riele (1987), although Conway and Guy (1996) claim that the best current limit is 10^{1167}. In 1914, Littlewood proved that the inequality must, in fact, fail infinitely often.

The second Skewes number Sk_2 is the number above which $\pi(n < Li(n))$ must fail (assuming that the RIEMANN HYPOTHESIS is false). It is much larger than the Skewes number Sk_1,

$$Sk_2 = 10^{10^{10^{10^3}}}.$$

See also GRAHAM'S NUMBER, RIEMANN HYPOTHESIS

References

Asimov, I. "Skewered!" *Of Matters Great and Small.* New York: Ace Books, 1976.
Asimov, I. *Magazine of Fantasy and Science Fiction,* Nov. 1974.
Ball, W. W. R. and Coxeter, H. S. M. *Mathematical Recreations and Essays, 13th ed.* New York: Dover, p. 63, 1987.
Boas, R. P. "The Skewes Number." In *Mathematical Plums* (Ed. R. Honsberger). Washington, DC: Math. Assoc. Amer., 1979.
Conway, J. H. and Guy, R. K. *The Book of Numbers.* New York: Springer-Verlag, p. 61, 1996.
Hardy, G. H. *Ramanujan: Twelve Lectures on Subjects Suggested by His Life and Work, 3rd ed.* New York: Chelsea, pp. 17 and 21, 1999.
Lehman, R. S. "On the Difference $\pi(x) - li(x)$." *Acta Arith.* **11**, 397–410, 1966.
Skewes. *J. London Math. Soc.* **8**, 277–283, 1933.
te Riele, H. J. J. "On the Sign of the Difference $\pi(x) - Li(x)$." *Math. Comput.* **48**, 323–328, 1987.
Wagon, S. *Mathematica in Action.* New York: W. H. Freeman, p. 30, 1991.

Skewness

The degree of asymmetry of a distribution. If the distribution has a longer tail less than the maximum, the function has NEGATIVE skewness. Otherwise, it has POSITIVE skewness. Several types of skewness are defined. The FISHER SKEWNESS (the most common type of skewness, usually referred to simply as "the" skewness) is defined by

$$\gamma_1 = \frac{\mu_3}{\mu_2^{3/2}} = \frac{\mu_3}{\sigma^3}, \tag{1}$$

where μ_3 is the third CENTRAL MOMENT, and $\mu_2^{1/2} \equiv \sigma$ is the STANDARD DEVIATION. The following table gives the skewness for a number of common distributions.

distribution	skewness
BERNOULLI DISTRIBUTION	$\frac{1-2p}{\sqrt{p(1-p)}}$
BETA DISTRIBUTION	$\frac{2(b-a)}{(2a+b)}\sqrt{\frac{1+a+b}{ab}}$
BINOMIAL DISTRIBUTION	$\frac{1-2p}{\sqrt{np(1-p)}}$
CHI-SQUARED DISTRIBUTION	$2\sqrt{\frac{2}{r}}$
EXPONENTIAL DISTRIBUTION	2
FISHER-TIPPETT DISTRIBUTION	$\frac{12\sqrt{6}\zeta(3)}{\pi^3}$
F-DISTRIBUTION	$\frac{2(2n+m-2)}{m-6}\sqrt{\frac{2(m-4)}{n(m+n-2)}}$
GAMMA DISTRIBUTION	$\frac{2}{\sqrt{n}}$
GEOMETRIC DISTRIBUTION	$\frac{2-p}{\sqrt{1-p}}$
HALF-NORMAL DISTRIBUTION	$\frac{\sqrt{2}(4-\pi)}{(\pi-2)^{3/2}}$
HYPERGEOMETRIC DISTRIBUTION	$\frac{(m-n)(m+n-2N)}{m+n-2}\sqrt{\frac{m+n-1}{mnN(m+n-N)}}$

LAPLACE DISTRIBUTION	0
LOG NORMAL DISTRIBUTION	$\sqrt{e^{S^2}-1}\left(2+e^{S^2}\right)$
MAXWELL DISTRIBUTION	$\frac{8}{3}\sqrt{\frac{2}{3\pi}}$
NEGATIVE BINOMIAL DISTRIBUTION	$\frac{2-p}{\sqrt{r(1-p)}}$
NORMAL DISTRIBUTION	0
POISSON DISTRIBUTION	$\nu^{-1/2}$
RAYLEIGH DISTRIBUTION	$(\pi-3)\sqrt{\dfrac{\pi}{2\left(2-\frac{1}{2}\pi\right)^3}}$
SNEDECOR'S F-DISTRIBUTION	$\frac{2(n+2m-2)}{(n-6)}\sqrt{\frac{2(n-4)}{m(m+n-2)}}$
STUDENT'S T-DISTRIBUTION	0
UNIFORM DISTRIBUTION	0

The PEARSON SKEWNESS is defined by

$$\beta_1 = \left(\frac{\mu^3}{\sigma^3}\right)^2 = \gamma_1^2. \tag{2}$$

The MOMENTAL SKEWNESS is defined by

$$\alpha^{(m)} \equiv \tfrac{1}{2}\gamma_1. \tag{3}$$

The PEARSON MODE SKEWNESS is defined by

$$\frac{[\text{mean}]-[\text{mode}]}{\sigma}. \tag{4}$$

PEARSON'S SKEWNESS COEFFICIENTS are defined by

$$\frac{3[\text{mean}]-[\text{mode}]}{s} \tag{5}$$

and

$$\frac{3[\text{mean}]-[\text{median}]}{s}. \tag{6}$$

The BOWLEY SKEWNESS (also known as QUARTILE SKEWNESS COEFFICIENT) is defined by

$$\frac{(Q_3-Q_2)-(Q_2-Q_1)}{Q_3-Q_1} = \frac{Q_1-2Q_2+Q_3}{Q_3-Q_1}, \tag{7}$$

where the Qs denote the INTERQUARTILE RANGES. The MOMENTAL SKEWNESS is

$$\alpha^{(m)} \equiv \tfrac{1}{2}\gamma = \frac{\mu_3}{2\sigma^3}. \tag{8}$$

An ESTIMATOR for the FISHER SKEWNESS γ_1 is

$$g_1 = \frac{k_3}{k_2^{3/2}}, \tag{9}$$

where the ks are K-STATISTICS. For a normal population with a SAMPLE SIZE of N, the VARIANCE of g_1 is

$$\text{var}(g_1) \approx \frac{6}{N} \tag{10}$$

(Kendall *et al.* 1987).

See also BOWLEY SKEWNESS, FISHER SKEWNESS, GAMMA STATISTIC, H-STATISTIC, KURTOSIS, MEAN, MOMENTAL SKEWNESS, PEARSON SKEWNESS, STANDARD DEVIATION

References

Abramowitz, M. and Stegun, C. A. (Eds.). *Handbook of Mathematical Functions with Formulas, Graphs, and Mathematical Tables, 9th printing.* New York: Dover, p. 928, 1972.

Kendall, M. G.; Stuart, A.; and Ord, J. K. *Kendall's Advanced Theory of Statistics, Vol. 1: Distribution Theory, 6th ed.* New York: Oxford University Press, 1987.

Kenney, J. F. and Keeping, E. S. "Skewness." §7.10 in *Mathematics of Statistics, Pt. 1, 3rd ed.* Princeton, NJ: Van Nostrand, pp. 100–101, 1962.

Press, W. H.; Flannery, B. P.; Teukolsky, S. A.; and Vetterling, W. T. "Moments of a Distribution: Mean, Variance, Skewness, and So Forth." §14.1 in *Numerical Recipes in FORTRAN: The Art of Scientific Computing, 2nd ed.* Cambridge, England: Cambridge University Press, pp. 604–609, 1992.

Sklar's Theorem

Let H be a 2-D distribution function with marginal distribution functions F and G. Then there exists a COPULA C such that

$$H(x,\ y) = C(F(x),\ G(y)).$$

Conversely, for any univariate distribution functions F and G and any COPULA C, the function H is a two-dimensional distribution function with marginals F and G. Furthermore, if F and G are continuous, then C is unique.

See also COPULA

Skolem Paradox

Even though real ARITHMETIC is uncountable, it possesses a countable "model."

References

Curry, H. B. *Foundations of Mathematical Logic.* New York: Dover, pp. 6–7, 1977.

Erickson, G. W. and Fossa, J. A. *Dictionary of Paradox.* Lanham, MD: University Press of America, pp. 191–192, 1998.

Skolem Sequence

A Skolem sequence of order n is a sequence $S = \{s_1, s_2, \ldots, s_{2n}\}$ of $2n$ integers such that

    1. For every $k \in \{1, 2, \ldots, n\}$, there exist exactly two elements s_i, $s_j \in S$ such that $s_i = s_j = k$, and
    2. If $s_i = s_j = k$ with $i < j$, then $j - i = k$.

References

Colbourn, C. J. and Dinitz, J. H. (Eds.). "Skolem Sequences." Ch. 43 in *CRC Handbook of Combinatorial Designs*. Boca Raton, FL: CRC Press, pp. 457–461, 1996.

Skolem-Graceful Graph

See also EDGE-GRACEFUL GRAPH, SUPER-EDGE-GRACEFUL GRAPH

Skolem-Mahler-Lerch Theorem

If $\{a_0, a_1, \ldots\}$ is a RECURRENCE SEQUENCE, then the set of all k such that $a_k = 0$ is the union of a finite (possibly EMPTY) set and a finite number (possibly zero) of full arithmetical progressions, where a full arithmetic progression is a set OF THE FORM $\{r, r + d, r + 2d, \ldots\}$ with $r \in [0, d)$.

References

Myerson, G. and van der Poorten, A. J. "Some Problems Concerning Recurrence Sequences." *Amer. Math. Monthly* **102**, 698–705, 1995.

SL

SPECIAL LINEAR GROUP

Slant Height

The height of an object (such as a CONE, FRUSTUM, or PYRAMID) measured along a side from the edge of the base to the apex. For a right PYRAMID with a regular n-gonal base of side length a, the slant height is given by

$$s_n = \sqrt{h^2 + R^2} = \sqrt{h^2 + \tfrac{1}{4} a^2 \csc^2\left(\frac{\pi}{n}\right)}$$

where R is the CIRCUMRADIUS of the base.

Slater's Identity

The Q-SERIES Identity of ROGERS-RAMANUJAN-type given by

$$\sum_{k=0}^{\infty} \frac{q^{2k^2}}{(q;\,q)_{2k}} = \frac{(q,\,q^7,\,q^8;\,q^8)_i\, nfty (q^6,\,q^{10};\,q^{16})\infty}{(q;\,q)\infty} \quad (1)$$

(Leininger and Milne 1997).

See also ROGERS-RAMANUJAN IDENTITIES

References

Leininger, V. E. and Milne, S. C. "Some New Infinite Families of Eta Function Identities." Preprint. http://www.math.ohio-state.edu/~milne/preprints.html.
Slater, L. J. "Further Identities of the Rogers-Ramanujan Type." *Proc. London Math. Soc. Ser. 2* **54**, 147–167, 1952.

Slice Knot

A KNOT K in $\mathbb{S}^3 = \partial \mathbb{D}^4$ is a slice knot if it bounds a DISK Δ^2 in $\mathbb{D}^4$ which has a TUBULAR NEIGHBORHOOD $\Delta^2 \times \mathbb{D}^2$ whose intersection with $\mathbb{S}^3$ is a TUBULAR NEIGHBORHOOD $K \times \mathbb{D}^2$ for K.

Every RIBBON KNOT is a slice knot, and it is conjectured that every slice knot is a RIBBON KNOT.

See also RIBBON KNOT, TUBULAR NEIGHBORHOOD

References

Rolfsen, D. *Knots and Links*. Wilmington, DE: Publish or Perish Press, p. 218, 1976.

Slide Move

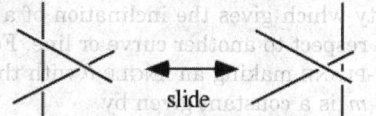

The REIDEMEISTER MOVE of type III.

See also KNOT MOVE, REIDEMEISTER MOVES

Slide Rule

A mechanical device consisting of a sliding portion and a fixed case, each marked with logarithmic axes. By lining up the ticks, it is possible to do MULTI-PLICATION by taking advantage of the additive property of LOGARITHMS. More complicated slide rules also allow the extraction of roots and computation of trigonometric functions.

According to Steinhaus (1983, p. 301), the principle of the slide rule was first enumerated by E. Gunter in 1623, and in 1671, S. Partridge constructed an instrument similar to the modern slide rule. The slide rule was an indispensable tool for scientists and engineers through the 1960s, but the development of the desk calculator (and subsequently pocket calculator) rendered slide rules largely obsolete beginning in the early 1970s.

See also ABACUS, RULER, STRAIGHTEDGE

References

Electronic Teaching Laboratories. *Simplify Math: Learn to Use the Slide Rule*. New Augusta, IN: Editors and Engineers, 1966.
Johnson, L. H. *The Slide Rule*. New York: Van Nostrand, 1949.
Saffold, R. *The Slide Rule*. Garden City, NY: Doubleday, 1962.
Steinhaus, H. *Mathematical Snapshots, 3rd ed.* New York: Dover, pp. 91–92 and 301, 1999.

Slightly Defective Number
ALMOST PERFECT NUMBER

Slightly Excessive Number
QUASIPERFECT NUMBER

Slip Knot
RUNNING KNOT

Slope

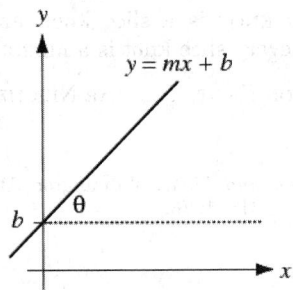

A quantity which gives the inclination of a curve or line with respect to another curve or line. For a LINE in the xy-PLANE making an ANGLE θ with the x-AXIS, the slope m is a constant given by

$$m \equiv \frac{\Delta y}{\Delta x} = \tan \theta, \qquad (1)$$

where Δx and Δy are changes in the two coordinates over some distance.

It is meaningless to talk about the slope of a curve in 3-dimensional space unless the slope *with respect to what* is specified.

J. Miller has undertaken a detailed study of the origin of the symbol m to denote slope. The consensus seems to be that it is not known why the letter m was chosen. One high school algebra textbook says the reason for m is unknown, but remarks that it is interesting that the French word for "to climb" is "monter." However, there is no evidence to make any such connection and in fact, Descartes, who was French, did not use m (Miller). Eves (1971) suggests "it just happened."

The earliest known example of the symbol m appearing in print is O'Brien (1844). Salmon (1960) subsequently used the symbols commonly employed today to give the slope-intercept form of a line

$$y = mx + b \qquad (2)$$

in his famous treatise published in several editions beginning in 1848. Todhunter (1888) also employed the symbol m, writing the slope-intercept form

$$y = mx + c. \qquad (3)$$

However, *Webster's New International Dictionary* (1909) gives the "slope form" as

$$y = sx + b. \qquad (4)$$

(Miller).

In Swedish textbooks, the slope-intercept equation is usually written as

$$y = kx + m, \qquad (5)$$

where k may derive from "koefficient" in the Swedish word for slope, "riktningskoefficient." In the Netherlands, the equation is commonly written as one of

$$y = ax + b \qquad (6)$$

$$y = px + q \qquad (7)$$

$$y = mx + n. \qquad (8)$$

In Austria, k is used for the slope, and d for the y-intercept (Miller).

See also LINE, x-INTERCEPT, y-INTERCEPT

References

Eves, H. W. *Mathematical Circles Revisited: A Second Collection of Mathematical Stories and Anecdotes.* Prindle, Weber, and Schmidt, 1972.
Miller, J. "Earliest Uses of Symbols from Geometry." http://members.aol.com/jeff570/geometry.html.
O'Brien, M. *A Treatise on Plane Co-Ordinate Geometry, or, The Application of the Method of Co-Ordinates to the Solution of Problems in Plane Geometry.* Cambridge, England: Deightons, 1844.
Salmon, G. *Conic Sections, 6th ed.* New York: Chelsea, 1960.
Todhunter, I. *Treatise on Plane Co-Ordinate Geometry as Applied to the Straight Line and the Conic Sections.* London: Macmillan, 1888.

Slothouber-Graatsma Puzzle
Assemble six $1 \times 2 \times 2$ blocks and three $1 \times 1 \times 1$ blocks into a $3 \times 3 \times 3$ CUBE.

See also BOX-PACKING THEOREM, CONWAY PUZZLE, CUBE DISSECTION, DE BRUIJN'S THEOREM, KLARNER'S THEOREM, POLYCUBE

References

Honsberger, R. *Mathematical Gems II.* Washington, DC: Math. Assoc. Amer., pp. 75–77, 1976.

Slow Variation
REGULAR VARIATION

Slutzky-Yule Effect
A MOVING AVERAGE may generate an irregular oscillation even if none exists in the original data.

See also MOVING AVERAGE

Sluze Pearls
PEARLS OF SLUZE

Smale Horseshoe Map

The basic topological operations for constructing an ATTRACTOR consist of stretching (which gives sensitivity to initial conditions) and folding (which gives the attraction). Since trajectories in PHASE SPACE cannot cross, the repeated stretching and folding operations result in an object of great topological complexity.

The Smale horseshoe map consists of a sequence of operations on the unit square. First, stretch by a factor of 2 in the x direction, then compress by $2a$ in the y direction. Then, fold the rectangle and fit it back into the square. Repeating this generates the horseshoe attractor. If one looks at a CROSS SECTION of the final structure, it is seen to correspond to a CANTOR SET.

See also ATTRACTOR, CANTOR SET

References

Gleick, J. *Chaos: Making a New Science.* New York: Penguin, pp. 50–51, 1988.
Rasband, S. N. *Chaotic Dynamics of Nonlinear Systems.* New York: Wiley, p. 77, 1990.
Tabor, M. *Chaos and Integrability in Nonlinear Dynamics: An Introduction.* New York: Wiley, 1989.

Smale-Hirsch Theorem

The SPACE of IMMERSIONS of a MANIFOLD in another MANIFOLD is HOMOTOPICALLY equivalent to the space of bundle injections from the TANGENT SPACE of the first to the TANGENT BUNDLE of the second.

See also HOMOTOPY, IMMERSION, MANIFOLD, TANGENT BUNDLE, TANGENT SPACE

Small Circle

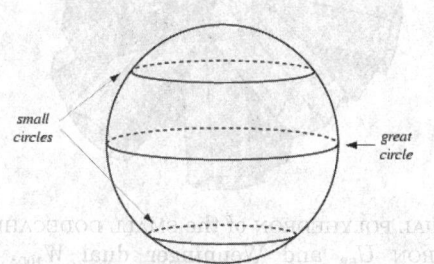

A SECTION of a SPHERE which does not contain a DIAMETER of the SPHERE (Kern and Bland 1948, p. 87; Tietze 1965, p. 25).

See also GREAT CIRCLE, SPHERE

References

Kern, W. F. and Bland, J. R. *Solid Mensuration with Proofs,* 2nd ed. New York: Wiley, 1948.
Tietze, H. *Famous Problems of Mathematics: Solved and Unsolved Mathematics Problems from Antiquity to Modern Times.* New York: Graylock Press, p. 25, 1965.

Small Cubicuboctahedron

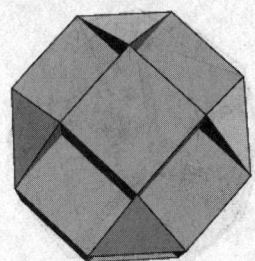

UNIFORM POLYHEDRON U_{13} whose DUAL POLYHEDRON is the SMALL HEXACRONIC ICOSITETRAHEDRON. It has WYTHOFF SYMBOL $\frac{3}{2}$ 4|4, and is Wenninger model W_{69}. Its faces are $8\{3\} + 6\{4\} + 6\{8\}$. The CIRCUMRADIUS for the solid with unit edge length is

$$R = \tfrac{1}{2}\sqrt{5 + 2\sqrt{2}}.$$

FACETED versions include the uniform GREAT RHOMBICUBOCTAHEDRON and SMALL RHOMBIHEXAHEDRON.

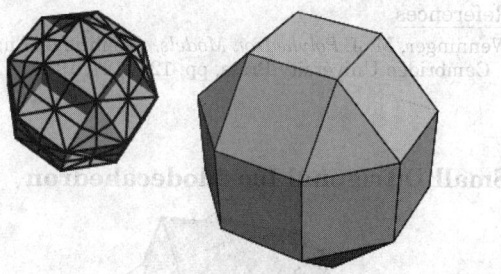

The CONVEX HULL of the small cubicuboctahedron is the Archimedean SMALL RHOMBICUBOCTAHEDRON A_6, whose dual is the DELTOIDAL ICOSITETRAHEDRON, so the dual of the small cubicuboctahedron (i.e., the SMALL HEXACRONIC ICOSITETRAHEDRON) is one of the stellations of the DELTOIDAL ICOSITETRAHEDRON (Wenninger 1983, p. 57).

References

Wenninger, M. J. *Dual Models.* Cambridge, England: Cambridge University Press, 1983.
Wenninger, M. J. *Polyhedron Models.* Cambridge, England: Cambridge University Press, pp. 104–105, 1971.

Small Ditrigonal Dodecacronic Hexecontahedron

The DUAL POLYHEDRON of the SMALL DITRIGONAL DODECICOSIDODECAHEDRON U_{43} and Wenninger dual W_{82}.

References

Wenninger, M. J. *Dual Models.* Cambridge, England: Cambridge University Press, p. 74, 1983.

Small Ditrigonal Dodecicosidodecahedron

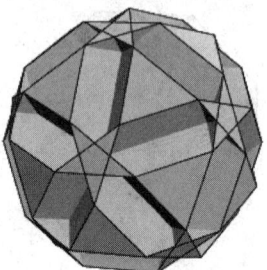

The UNIFORM POLYHEDRON U_{43} whose DUAL POLYHE-DRON is the SMALL DITRIGONAL DODECACRONIC HEX-ECONTAHEDRON. It has WYTHOFF SYMBOL $3\frac{5}{3}|5$. Its faces are $20\{3\} + 12\{\frac{5}{2}\} + 12\{10\}$. Its CIRCUMRADIUS with $a = 1$ is

$$R = \tfrac{1}{4}\sqrt{34 + 6\sqrt{5}}.$$

References

Wenninger, M. J. *Polyhedron Models*. Cambridge, England: Cambridge University Press, pp. 126–127, 1971.

Small Ditrigonal Icosidodecahedron

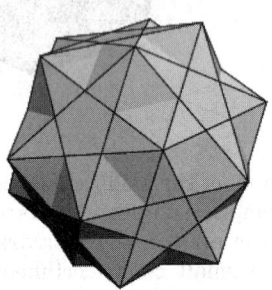

The UNIFORM POLYHEDRON U_{30} whose DUAL POLYHE-DRON is the SMALL TRIAMBIC ICOSAHEDRON. It has WYTHOFF SYMBOL $3|3\frac{5}{2}$. Its faces are $20\{3\} + 12\{\frac{5}{2}\}$. A FACETED version is the DITRIGONAL DODECADODECA-HEDRON. Its CIRCUMRADIUS with $a = 1$ is

$$R = \tfrac{1}{2}\sqrt{3}.$$

The CONVEX HULL of the small ditrigonal icosidode-cahedron is a regular DODECAHEDRON, whose dual is the ICOSAHEDRON, so the dual of the great ditrigonal dodecicosidodecahedron (the SMALL TRIAMBIC ICOSA-HEDRON) is one of the ICOSAHEDRON STELLATIONS (Wenninger 1983, p. 42).

References

Wenninger, M. J. *Dual Models*. Cambridge, England: Cambridge University Press, 1983.

Wenninger, M. J. *Polyhedron Models*. Cambridge, England: Cambridge University Press, pp. 106–107, 1971.

Small Dodecacronic Hexecontahedron

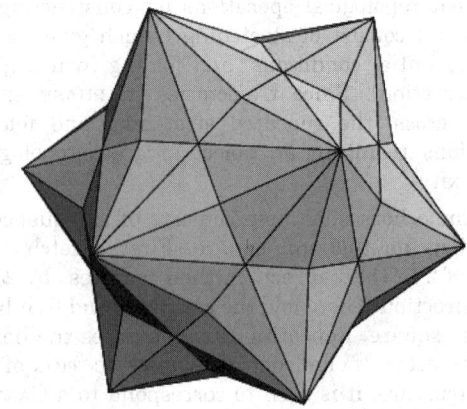

The DUAL POLYHEDRON of the SMALL DODECICOSIDO-DECAHEDRON U_{33} and Wenninger dual W_{72}.

See also DUAL POLYHEDRON, SMALL DODECICOSIDO-DECAHEDRON

References

Wenninger, M. J. *Dual Models*. Cambridge, England: Cambridge University Press, p. 70, 1983.

Small Dodecahemicosacron

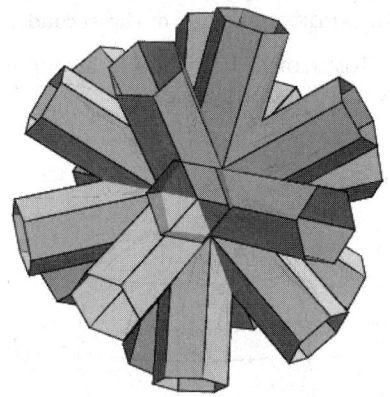

The DUAL POLYHEDRON of the SMALL DODECAHEMICO-SAHEDRON U_{62} and Wenninger dual W_{100}. When rendered, the small dodecahemicosacron and GREAT DODECAHEMICOSACRON appear the same.

See also DUAL POLYHEDRON, SMALL DODECAHEMICO-SAHEDRON, UNIFORM POLYHEDRON

References

Wenninger, M. J. *Dual Models*. Cambridge, England: Cambridge University Press, p. 107, 1983.

Small Dodecahemicosahedron

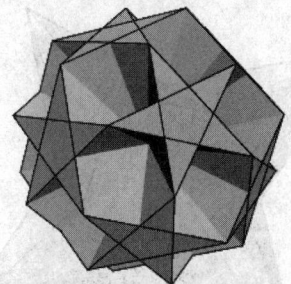

The UNIFORM POLYHEDRON U_{62} whose DUAL POLYHEDRON is the SMALL DODECAHEMICOSACRON. It has WYTHOFF SYMBOL $\frac{5}{3}\frac{5}{2}|3$. Its faces are $10\{6\} + 12\{\frac{5}{2}\}$. It is a FACETED version of the ICOSIDODECAHEDRON. Its CIRCUMRADIUS with unit edge length is

$$R = 1.$$

References

Wenninger, M. J. *Polyhedron Models.* Cambridge, England: Cambridge University Press, p. 155, 1971.

Small Dodecahemidodecacron

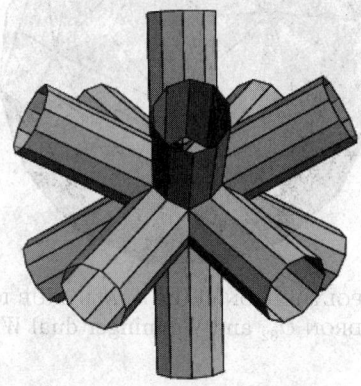

The DUAL POLYHEDRON of the SMALL DODECAHEMIDODECAHEDRON U_{51} and Wenninger dual W_{91}. When rendered, the SMALL ICOSIHEMIDODECACRON and small dodecahemidodecacron appear the same.

See also DUAL POLYHEDRON, SMALL DODECAHEMIDODECAHEDRON, SMALL ICOSIHEMIDODECACRON, UNIFORM POLYHEDRON

References

Wenninger, M. J. *Dual Models.* Cambridge, England: Cambridge University Press, p. 104, 1983.

Small Dodecahemidodecahedron

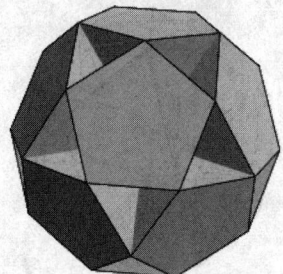

The UNIFORM POLYHEDRON U_{51} whose DUAL POLYHEDRON is the SMALL DODECAHEMIDODECACRON. It has WYTHOFF SYMBOL

$$2\ 5\ \frac{\frac{3}{2}}{\frac{5}{2}}.$$

Its faces are $30\{4\} + 12\{10\}$. Its CIRCUMRADIUS with $a = 1$ is

$$R = \tfrac{1}{2}\sqrt{11 + 4\sqrt{5}}.$$

References

Wenninger, M. J. *Polyhedron Models.* Cambridge, England: Cambridge University Press, pp. 113–114, 1971.

Small Dodecicosacron

The DUAL POLYHEDRON of the SMALL DODECICOSAHEDRON U_{50} and Wenninger dual W_{90}.

See also DUAL POLYHEDRON, SMALL DODECICOSAHEDRON, UNIFORM POLYHEDRON

References

Wenninger, M. J. *Dual Models.* Cambridge, England: Cambridge University Press, p. 74, 1983.

Small Dodecicosahedron

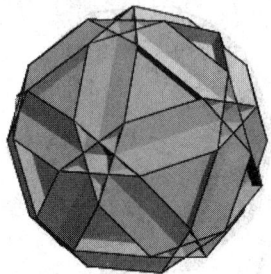

The UNIFORM POLYHEDRON U_{50} whose DUAL POLYHEDRON is the SMALL DODECICOSACRON. It has WYTHOFF SYMBOL

$$3\ 5\ \begin{vmatrix} \frac{3}{2} \\ \frac{5}{4} \end{vmatrix}.$$

Its faces are $20\{6\} + 12\{10\}$. Its CIRCUMRADIUS with $a = 1$ is

$$R = \tfrac{1}{4}\sqrt{34 + 6\sqrt{5}}.$$

References

Wenninger, M. J. *Polyhedron Models.* Cambridge, England: Cambridge University Press, pp. 141–142, 1971.

Small Dodecicosidodecahedron

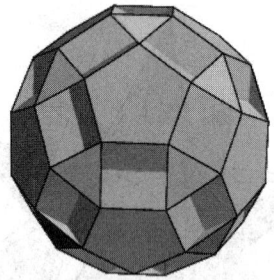

The UNIFORM POLYHEDRON U_{33} whose DUAL POLYHEDRON is the SMALL DODECACRONIC HEXECONTAHEDRON. It has WYTHOFF SYMBOL $\frac{3}{2}$ 5|5. Its faces are $20\{3\} + 12\{5\} + 12\{10\}$. It is a FACETED version of the SMALL RHOMBICOSIDODECAHEDRON. Its CIRCUMRADIUS with $a = 1$ is

$$R = \tfrac{1}{2}\sqrt{11 + 4\sqrt{5}}.$$

References

Wenninger, M. J. *Polyhedron Models.* Cambridge, England: Cambridge University Press, pp. 110–111, 1971.

Small Hexacronic Icositetrahedron

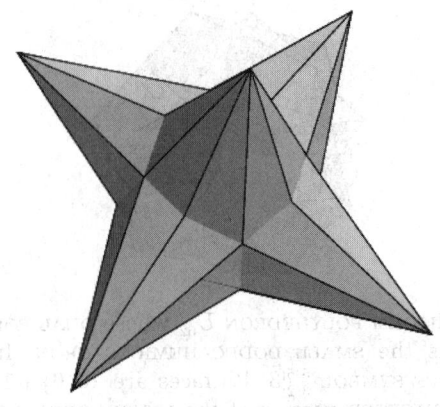

The DUAL POLYHEDRON of the SMALL CUBICUBOCTAHEDRON U_{13} and Wenninger dual W_{69}.

See also DUAL POLYHEDRON, SMALL CUBICUBOCTAHEDRON

References

Wenninger, M. J. *Dual Models.* Cambridge, England: Cambridge University Press, p. 57, 1983.

Small Hexagonal Hexecontahedron

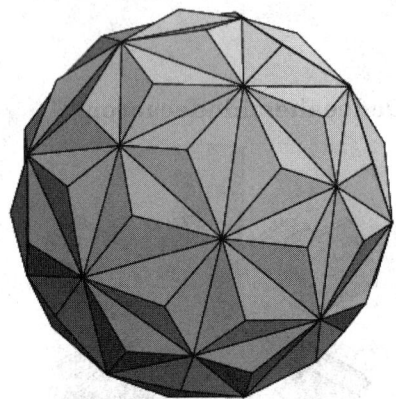

The DUAL POLYHEDRON of the SMALL SNUB ICOSICOSIDODECAHEDRON U_{32} and Wenninger dual W_{110}.

References

Wenninger, M. J. *Dual Models.* Cambridge, England: Cambridge University Press, p. 119, 1983.

Small Hexagrammic Hexecontahedron

The DUAL POLYHEDRON of the SMALL RETROSNUB ICOSICOSIDODECAHEDRON and Wenninger dual W_{118}.

References

Wenninger, M. J. *Dual Models.* Cambridge, England: Cambridge University Press, p. 135, 1983.

Small Icosacronic Hexecontahedron

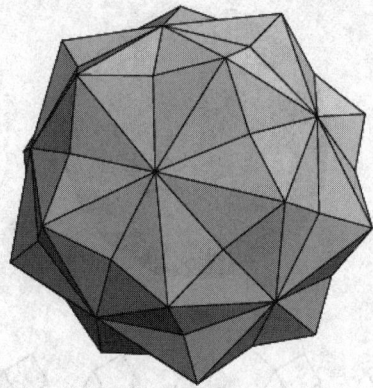

The DUAL POLYHEDRON of the SMALL ICOSICOSIDODE-CAHEDRON U_{31} and Wenninger dual W_{71}.

See also DUAL POLYHEDRON, SMALL ICOSICOSIDODE-CAHEDRON

References

Wenninger, M. J. *Dual Models*. Cambridge, England: Cambridge University Press, p. 74, 1983.

Small Icosicosidodecahedron

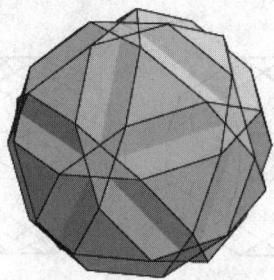

The UNIFORM POLYHEDRON U_{31} whose DUAL POLYHEDRON is the SMALL ICOSACRONIC HEXECONTAHEDRON. It has WYTHOFF SYMBOL $3\frac{5}{2}|3$. Its faces are $20\{3\}+20\{6\}+12\left\{\frac{5}{2}\right\}$. Its CIRCUMRADIUS with $a=1$ is

$$R=\sqrt{\frac{17+3\sqrt{5}}{2}}.$$

References

Wenninger, M. J. "Small Icosicosidodecahedron." Solid 71 in *Polyhedron Models*. Cambridge, England: Cambridge University Press, p. 108, 1971.

Small Icosihemidodecacron

The DUAL POLYHEDRON of the SMALL ICOSIHEMIDODE-CAHEDRON U_{49} and Wenninger dual W_{89}. When rendered, the small icosihemidodecacron and SMALL DODECAHEMIDODECACRON appear the same.

See also DUAL POLYHEDRON, SMALL DODECAHEMIDO-DECACRON, SMALL ICOSIHEMIDODECAHEDRON, UNIFORM POLYHEDRON

References

Wenninger, M. J. *Dual Models*. Cambridge, England: Cambridge University Press, p. 104, 1983.

Small Icosihemidodecahedron

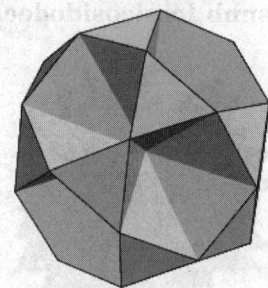

The UNIFORM POLYHEDRON U_{49} whose DUAL POLYHEDRON is the SMALL ICOSIHEMIDODECACRON. It has WYTHOFF SYMBOL $\frac{3}{2}3|5$. Its faces are $20\{3\}+6\{10\}$. It is a FACETED version of the ICOSIDODECAHEDRON. Its CIRCUMRADIUS with $a=1$ is

$$R=\phi=\tfrac{1}{2}\left(1+\sqrt{5}\right).$$

References

Wenninger, M. J. *Polyhedron Models*. Cambridge, England: Cambridge University Press, p. 140, 1971.

Small Inverted Retrosnub Icosicosidodecahedron

SMALL RETROSNUB ICOSICOSIDODECAHEDRON

Small Multiple Method

An algorithm for computing a UNIT FRACTION.

References

Eppstein, D. Egypt.ma Mathematica notebook. http://www.ics.uci.edu/~eppstein/numth/egypt/egypt.ma.

Small Number

Guy's "STRONG LAW OF SMALL NUMBERS" states that there aren't enough small numbers to meet the many demands made of them. Guy (1988) also gives several interesting and misleading facts about small numbers:

1. 10% of the first 100 numbers are SQUARE NUMBERS.
2. A QUARTER of the numbers < 100 are PRIMES.
3. All numbers less than 10, except for 6, are PRIME POWERS.
4. Half the numbers less than 10 are FIBONACCI NUMBERS.

See also LARGE NUMBER, STRONG LAW OF SMALL NUMBERS

References

Guy, R. K. "The Strong Law of Small Numbers." *Amer. Math. Monthly* **95**, 697–712, 1988.

Small Retrosnub Icosicosidodecahedron

The UNIFORM POLYHEDRON U_{72} also called the SMALL INVERTED RETROSNUB ICOSICOSIDODECAHEDRON whose DUAL POLYHEDRON is the SMALL HEXAGRAMMIC HEXECONTAHEDRON. It has WYTHOFF SYMBOL $\left|\frac{3}{2}\frac{3}{2}\frac{5}{2}\right.$. Its faces are $100\{3\} + 12\left\{\frac{5}{2}\right\}$. It has CIRCUMRADIUS with $a = 1$

$$R = \tfrac{1}{4}\sqrt{13 + 3\sqrt{5} - \sqrt{102 + 46\sqrt{5}}}$$

$$\approx 0.580694800133921.$$

References

Wenninger, M. J. *Polyhedron Models.* Cambridge, England: Cambridge University Press, pp. 194–199, 1971.

Small Rhombicosidodecahedron

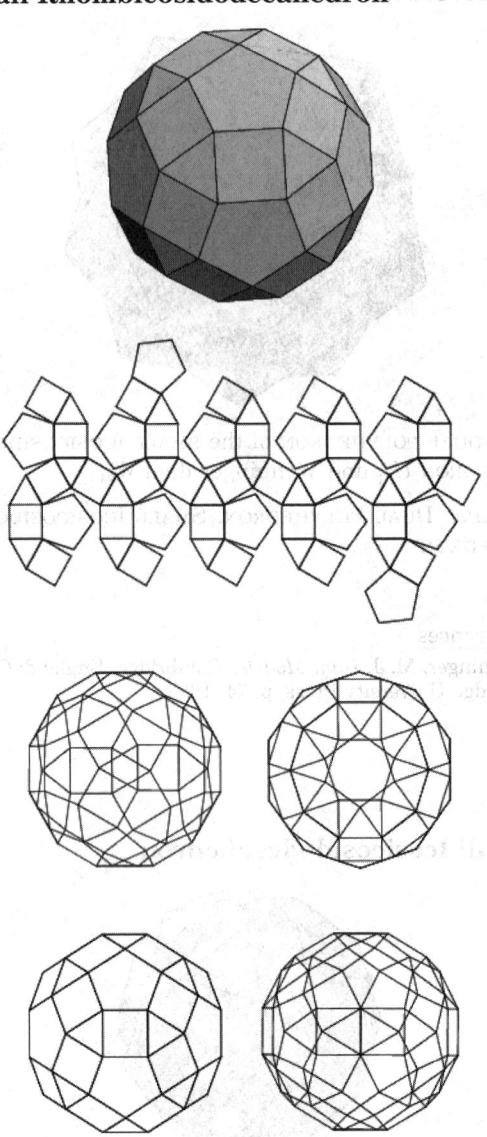

The 62-faced ARCHIMEDEAN SOLID A_5 with faces $20\{3\} + 30\{4\} + 12\{5\}$. It is UNIFORM POLYHEDRON U_{27} and Wenninger model W_{14}. It has SCHLÄFLI SYMBOL r $\left\{\frac{3}{5}\right\}$ and WYTHOFF SYMBOL A_9. The SMALL DODECICOSIDODECAHEDRON and SMALL RHOMBIDODE-CAHEDRON are FACETED versions.

Its DUAL POLYHEDRON is the DELTOIDAL HEXECONTA-HEDRON. The INRADIUS r of the dual, MIDRADIUS ρ of the solid and dual, and CIRCUMRADIUS R of the solid for $a = 1$ are

$$r = \tfrac{1}{41}(15 + 2\sqrt{5})\sqrt{11 + 4\sqrt{5}} = 2.12099\ldots$$

$$\rho = \tfrac{1}{2}\sqrt{10 + 4\sqrt{5}} = 2.17625\ldots$$

$$R = \tfrac{1}{2}\sqrt{11 + 4\sqrt{5}} = 2.23295\ldots$$

See also ARCHIMEDEAN SOLID, GREAT RHOMBICOSIDODECAHEDRON (ARCHIMEDEAN), GREAT RHOMBICOSIDODECAHEDRON (UNIFORM), HEXECONTAHEDRON, ZOME

References

Cundy, H. and Rollett, A. "lpar;Small) Rhombicosidodecahedron. $S_k(n)$." §3.7.11 in *Mathematical Models, 3rd ed.* Stradbroke, England: Tarquin Pub., p. 111, 1989.
Wenninger, M. J. "The Rhombicosidodecahedron." Model 14 in *Polyhedron Models.* Cambridge, England: Cambridge University Press, p. 28, 1989.

Small Rhombicuboctahedron

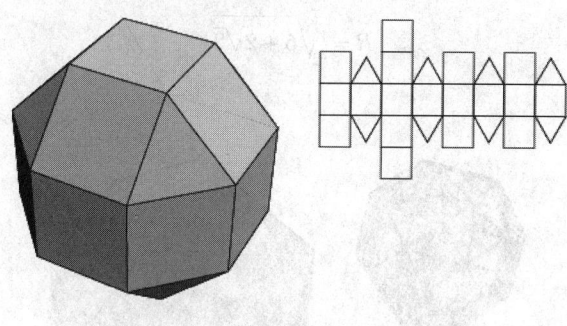

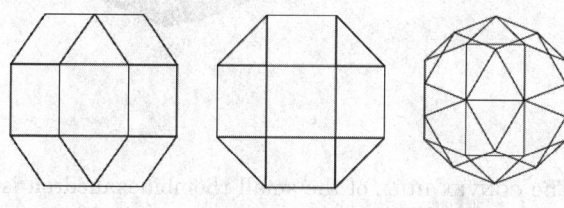

The 26-faced ARCHIMEDEAN SOLID A_6 consisting of faces $8\{3\} + 18\{4\}$. Although this solid is sometimes also called the truncated icosidodecahedron, this name is inappropriate since true TRUNCATION would yield rectangular instead of square faces. It is UNIFORM POLYHEDRON U_{10} and Wenninger model W_{13}. It has SCHLÄFLI SYMBOL $r\{{3 \atop 4}\}$ and WYTHOFF SYMBOL $3\,4\,|\,2$.

Its DUAL POLYHEDRON is the DELTOIDAL ICOSITETRAHEDRON, also called the TRAPEZOIDAL ICOSITETRAHEDRON. The INRADIUS r of the dual, MIDRADIUS ρ of the solid and dual, and CIRCUMRADIUS R of the solid for $a = 1$ are

$$r = \tfrac{1}{17}(6 + \sqrt{2})\sqrt{5 + 2\sqrt{2}} = 1.22026\ldots$$

$$\rho = \tfrac{1}{2}\sqrt{4 + 2\sqrt{2}} = 1.30656\ldots$$

$$R = \tfrac{1}{2}\sqrt{5 + 2\sqrt{2}} = 1.39896\ldots$$

The distances between the solid center and centroids of the triangular and square faces are

$$r_3 = \tfrac{1}{2}(1 + \sqrt{2}) \tag{1}$$

$$r_4 = \tfrac{1}{2}\sqrt{\tfrac{1}{3}(11 + 6\sqrt{2})}. \tag{2}$$

The SURFACE AREA and VOLUME are

$$S = 18 + 2\sqrt{3} \tag{3}$$

$$V = \tfrac{1}{3}(12 + 10\sqrt{2}). \tag{4}$$

The CONVEX HULL of the SMALL CUBICUBOCTAHEDRON is the small rhombicuboctahedron, whose dual is the DELTOIDAL ICOSITETRAHEDRON, so the dual of the SMALL CUBICUBOCTAHEDRON (i.e., the SMALL HEXACRONIC ICOSITETRAHEDRON) is one of the stellations of the DELTOIDAL ICOSITETRAHEDRON (Wenninger 1983, p. 57).

A version of the small rhombicuboctahedron in which the top and bottom halves are rotated with respect to each other is known as the ELONGATED SQUARE GYROBICUPOLA.

See also ARCHIMEDEAN SOLID, ELONGATED SQUARE GYROBICUPOLA, GREAT RHOMBICUBOCTAHEDRON (ARCHIMEDEAN), GREAT RHOMBICUBOCTAHEDRON (UNIFORM), ICOSITETRAHEDRON

References

Ball, W. W. R. and Coxeter, H. S. M. *Mathematical Recreations and Essays, 13th ed.* New York: Dover, pp. 137–138, 1987.
Cundy, H. and Rollett, A. "lpar;Small) Rhombicuboctahedron. 3.4^2." §3.7.5 in *Mathematical Models, 3rd ed.* Stradbroke, England: Tarquin Pub., p. 105, 1989.
Wenninger, M. J. "The Rhombicuboctahedron." Model 13 in *Polyhedron Models.* Cambridge, England: Cambridge University Press, p. 27, 1989.

Small Rhombidodecacron

The DUAL POLYHEDRON of the SMALL RHOMBIDODECAHEDRON and Wenninger model W_{74}.

References

Wenninger, M. J. *Dual Models.* Cambridge, England: Cambridge University Press, p. 70, 1983.

Small Rhombidodecahedron

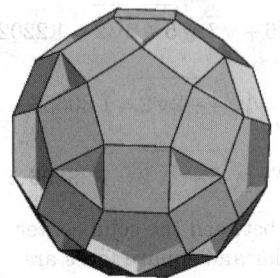

The UNIFORM POLYHEDRON U_{39} whose DUAL POLYHEDRON is the SMALL RHOMBIDODECACRON. It has WYTHOFF SYMBOL

$$2\ 5\ \genfrac{}{}{0pt}{}{\frac{3}{2}}{\frac{5}{2}}.$$

Its faces are $30\{4\} + 12\{10\}$. It is a FACETED version of the SMALL RHOMBICOSIDODECAHEDRON. Its CIRCUMRADIUS with $a = 1$ is

$$R = \tfrac{1}{2}\sqrt{11 + 4\sqrt{5}}.$$

References

Wenninger, M. J. *Polyhedron Models.* Cambridge, England: Cambridge University Press, pp. 113–114, 1971.

Small Rhombihexacron

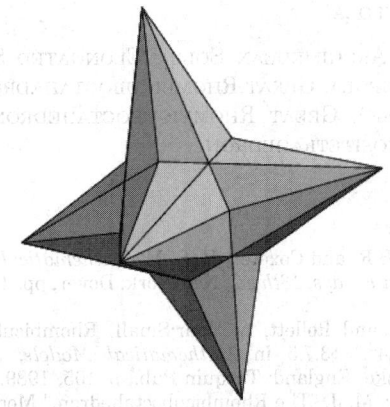

The DUAL POLYHEDRON of the SMALL RHOMBIHEXAHEDRON U_{18} and Wenninger dual W_{86}

See also DUAL POLYHEDRON, SMALL RHOMBIHEXAHEDRON

References

Wenninger, M. J. *Dual Models.* Cambridge, England: Cambridge University Press, p. 57, 1983.

Small Rhombihexahedron

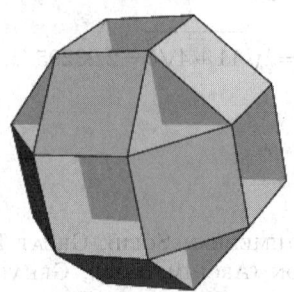

The UNIFORM POLYHEDRON U_{18} whose DUAL POLYHEDRON is the SMALL RHOMBIHEXACRON. It has WYTHOFF SYMBOL

$$2\ 4\ \genfrac{}{}{0pt}{}{\frac{3}{2}}{\frac{4}{2}}\bigg|$$

and is Wenninger model W_{86}. Its faces are $12\{4\} + 6\{8\}$. It is a FACETED version of the SMALL RHOMBICUBOCTAHEDRON. Its CIRCUMRADIUS with $a = 1$ is

$$R = \tfrac{1}{2}\sqrt{5 + 2\sqrt{2}}.$$

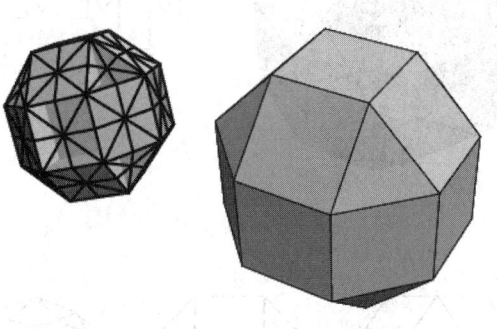

The CONVEX HULL of the small rhombihexahedron is the Archimedean SMALL RHOMBICUBOCTAHEDRON A_6, whose dual is the DELTOIDAL ICOSITETRAHEDRON, so the dual of the small rhombihexahedron (i.e., the SMALL RHOMBIHEXACRON) is one of the stellations of the DELTOIDAL ICOSITETRAHEDRON (Wenninger 1983, p. 57).

References

Wenninger, M. J. *Polyhedron Models.* Cambridge, England: Cambridge University Press, p. 134, 1971.

Small Snub Icosicosidodecahedron

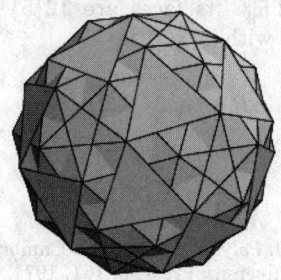

The UNIFORM POLYHEDRON U_{32} whose DUAL POLYHE-
DRON is the SMALL HEXAGONAL HEXECONTAHEDRON. It
has WYTHOFF SYMBOL $|3\,3\,\frac{5}{2}$ (Har'El 1993 gives the
symbol as $|\frac{5}{2}\,3\,3$.) Its faces are $100\{3\}+12\{\frac{5}{2}\}$. Its
CIRCUMRADIUS for $a = 1$ is

$$R = \tfrac{1}{4}\sqrt{13 + 3\sqrt{5} + \sqrt{102 + 46\sqrt{5}}}$$

$$= 1.4581903307387\ldots$$

References

Har'El, Z. "Uniform Solution for Uniform Polyhedra."
Geometriae Dedicata **47**, 57–110, 1993.
Wenninger, M. J. *Polyhedron Models.* Cambridge, England:
Cambridge University Press, pp. 172–173, 1971.

Small Stellapentakis Dodecahedron

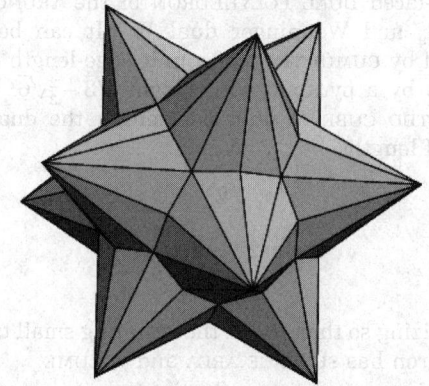

The DUAL POLYHEDRON of the TRUNCATED GREAT
DODECAHEDRON U_{37} and Wenninger dual W_{75}.

See also DUAL POLYHEDRON, TRUNCATED GREAT
DODECAHEDRON

References

Wenninger, M. J. *Dual Models.* Cambridge, England: Cambridge University Press, p. 84, 1983.

Small Stellated Dodecahedron

One of the KEPLER-POINSOT SOLIDS whose DUAL
POLYHEDRON is the GREAT DODECAHEDRON. It is also
UNIFORM POLYHEDRON U_{34}, Wenninger model W_{21},
and is the first STELLATION of the DODECAHEDRON
(Wenninger 1989). It was originally called the URCHIN
by Kepler. The small stellated dodecahedron has
SCHLÄFLI SYMBOL $\left\{\frac{5}{2}, 5\right\}$ and WYTHOFF SYMBOL
$5|2\,\frac{5}{2}$. It is composed of 12 PENTAGRAMMIC faces. Its
faces are $12\{\frac{5}{2}\}$.

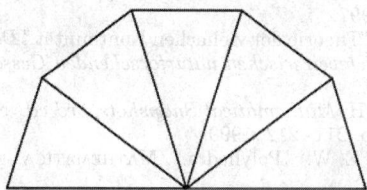

The easiest way to construct a small stellated dodecahedron is by CUMULATION, i.e., building twelve
PENTAGONAL PYRAMIDS and attaching them to the
faces of a DODECAHEDRON. The height of the pyramids
for a small stellated dodecahedron built on a DODE-
CAHEDRON of unit edge length is $\sqrt{\frac{1}{5}(5 + 2\sqrt{5})}$. The
CIRCUMRADIUS of the small stellated dodecahedron
with *pentagrammic* edge length $a = 1$ is

$$R = \tfrac{1}{2}\,5^{1/4}\phi^{-1/2} = \tfrac{1}{4}\,5^{1/4}\sqrt{2\left(\sqrt{5} - 1\right)}.$$

Schläfli (1901, p. 134) did not recognize the small
stellated dodecahedron because it, like the GREAT
DODECAHEDRON, satisfies

$$N_0 - N_1 + N_2 = 12 - 30 + 12 = -6, \tag{1}$$

where N_0 is the number of vertices, N_1 the number of
edges, and N_2 the number of faces (Coxeter 1973,
p. 172), thus violating the POLYHEDRAL FORMULA.

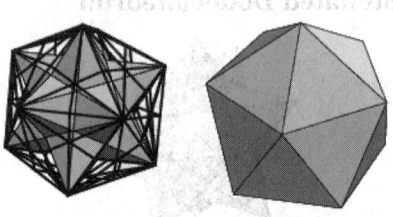

The CONVEX HULL of the small stellated dodecahedron is a regular DODECAHEDRON and the dual of the DODECAHEDRON is the ICOSAHEDRON, so the dual of the small stellated dodecahedron is one of the ICOSAHEDRON STELLATIONS (Wenninger 1983, p. 40)

See also DODECAHEDRON, GREAT DODECAHEDRON, GREAT ICOSAHEDRON, GREAT STELLATED DODECAHEDRON, KEPLER-POINSOT SOLID, STELLATION

References

Coxeter, H. S. M. *Regular Polytopes, 3rd ed.* New York: Dover, 1973.
Cundy, H. and Rollett, A. "Small Stellated Dodecahedron. $(\frac{5}{2})^5$." §3.6.1 in *Mathematical Models, 3rd ed.* Stradbroke, England: Tarquin Pub., pp. 90–91, 1989.
Fischer, G. (Ed.). Plate 103 in *Mathematische Modelle/ Mathematical Models, Bildband/Photograph Volume.* Braunschweig, Germany: Vieweg, p. 102, 1986.
Rawles, B. *Sacred Geometry Design Sourcebook: Universal Dimensional Patterns.* Nevada City, CA: Elysian Pub., p. 219, 1997.
Schläfli, L. "Theorie der vielfachen Kontinuität." *Denkschriften der Schweizerischen naturforschenden Gessel.* **38,** 1–237, 1901.
Steinhaus, H. *Mathematical Snapshots, 3rd ed.* New York: Dover, pp. 211–212, 1999.
Weisstein, E. W. "Polyhedra." MATHEMATICA NOTEBOOK POLYHEDRA.M.
Wenninger, M. J. *Dual Models.* Cambridge, England: Cambridge University Press, p. 39, 1983.
Wenninger, M. J. *Polyhedron Models.* Cambridge, England: Cambridge University Press, pp. 35 and 38, 1989.

Small Stellated Triacontahedron

MEDIAL RHOMBIC TRIACONTAHEDRON

Small Stellated Truncated Dodecahedron

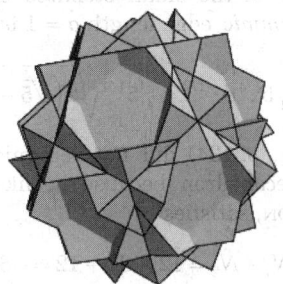

The UNIFORM POLYHEDRON U_{58} also called the QUASI-TRUNCATED SMALL STELLATED DODECAHEDRON whose DUAL POLYHEDRON is the GREAT PENTAKIS DODECAHE-

DRON. It has SCHLÄFLI SYMBOL t' $\left\{\frac{5}{2}, 5\right\}$ and WYTHOFF SYMBOL $2\,5\left|\frac{5}{3}\right.$. Its faces are $12\{5\} + 12\left\{\frac{10}{3}\right\}$. Its CIRCUMRADIUS with $a = 1$ is

$$R = \tfrac{1}{4}\sqrt{34 - 10\sqrt{5}}.$$

References

Wenninger, M. J. *Polyhedron Models.* Cambridge, England: Cambridge University Press, p. 151, 1971.

Small Triakis Octahedron

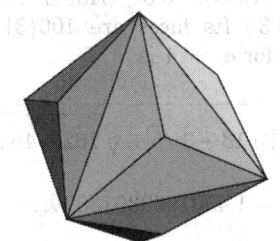

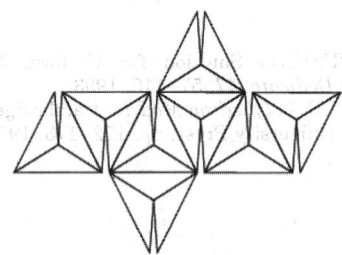

The 24-faced DUAL POLYHEDRON of the TRUNCATED CUBE A_9 and Wenninger dual W_8. It can be constructed by CUMULATION of a unit edge-length OCTAHEDRON by a pyramid with height $\sqrt{3} - \frac{2}{3}\sqrt{6}$. For a TRUNCATED CUBE of unit side length the dual has edges of lengths

$$s_1 = 2 \tag{1}$$

$$s_2 = 2 + \sqrt{2}. \tag{2}$$

Normalizing so that $s_1 = 1$, the resulting small triakis octahedron has SURFACE AREA and VOLUME

$$S = 3\sqrt{7 + 4\sqrt{2}} \tag{3}$$

$$V = \tfrac{1}{2}\left(3 + 2\sqrt{2}\right). \tag{4}$$

See also ARCHIMEDEAN DUAL, ARCHIMEDEAN SOLID, GREAT TRIAKIS OCTAHEDRON, ICOSITETRAHEDRON,

SMALL TRIAKIS OCTAHEDRON STELLATIONS, TRUNCATED CUBE

References

Wenninger, M. J. *Dual Models.* Cambridge, England: Cambridge University Press, p. 7, 1983.

Small Triakis Octahedron Stellations

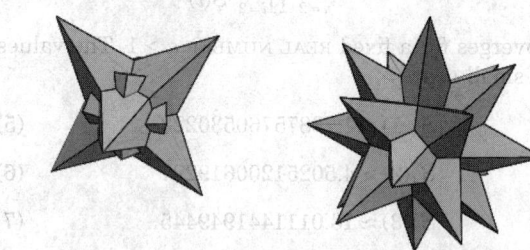

R. Whorf found that there are probably several thousand stellations of the small triakis octahedron (Wenninger 1983, p. 36). In particular, the CONVEX HULLS of the GREAT CUBICUBOCTAHEDRON U_{14}, the Archimedean GREAT RHOMBICUBOCTAHEDRON $A_3 = U_{17}$, and GREAT RHOMBIHEXAHEDRON U_{21} are all the Archimedean TRUNCATED CUBE A_9, whose dual is the SMALL TRIAKIS OCTAHEDRON, so the duals of these solids (i.e., the GREAT HEXACRONIC ICOSITETRAHEDRON, GREAT DELTOIDAL ICOSITETRAHEDRON, and GREAT RHOMBIHEXAHEDRON) are all stellations of the small triakis octahedron (Wenninger 1983, p. 57).

References

Wenninger, M. J. *Dual Models.* Cambridge, England: Cambridge University Press, pp. 36, 38, and 57–58, 1983.

Small Triambic Icosahedron

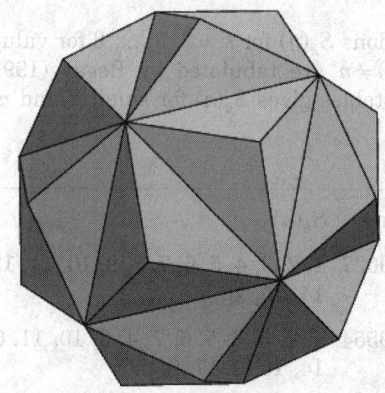

The DUAL POLYHEDRON of the SMALL DITRIGONAL ICOSIDODECAHEDRON U_{30} and Wenninger model W_{70}. It can be constructed by CUMULATION of a unit edge-length ICOSAHEDRON by a pyramid with height $\sqrt{15}/15$. Wenninger (1989, p. 49) calls this solid the triakis octahedron (which is a term more commonly used for the dual of one of the Archimedean solids).

The CONVEX HULL of the SMALL DITRIGONAL ICOSIDODECAHEDRON is a regular DODECAHEDRON, whose dual is the ICOSAHEDRON, so the dual of the SMALL DITRIGONAL ICOSIDODECAHEDRON (the small triambic icosahedron) is one of the ICOSAHEDRON STELLATIONS (Wenninger 1983, p. 42).

See also DODECAHEDRON-SMALL TRIAMBIC ICOSAHEDRON COMPOUND, DUAL POLYHEDRON, SMALL DITRIGONAL ICOSIDODECAHEDRON, TRIAKIS ICOSAHEDRON, TRIAKIS OCTAHEDRON

References

Wenninger, M. J. *Dual Models.* Cambridge, England: Cambridge University Press, pp. 42 and 46–47 1983.

Wenninger, M. J. *Polyhedron Models.* New York: Cambridge University Press, p. 46, 1989.

Small World Problem

The small world problem asks for the probability that two people picked at random have at least one acquaintance in common.

See also BIRTHDAY PROBLEM

Smarandache Ceil Function

A SMARANDACHE-like function which is defined where $S_k(n)$ is defined as the smallest integer for which $n | S_k(n)^k$. The Smarandache $S_k(n)$ function can therefore be obtained by replacing any factors which are kth powers in n by their k roots.

$$S_k(n) = \frac{n}{M_k(n)},$$

where $M_k(n)$ is the number of solutions to $x^k \equiv 0 \pmod{n}$.

The functions $S_k(n)$ for $k = 2, 3, ..., 6$ for values such that $S_k(n) \neq n$ are tabulated by Begay (1997). The following tables gives $S_k(n)$ for small k and $n = 1, 2, ...$

k	Sloane	$S_k(n)$
1	A000027	1, 2, 3, 4, 5, 6, 7, 8, 9, 10, 11, 12, 13, 14, 15, 16, 17, ...
2	A019554	1, 2, 3, 2, 5, 6, 7, 4, 3, 10, 11, 6, 13, 14, 15, 4, 17, 6, ...
3	A019555	1, 2, 3, 2, 5, 6, 7, 2, 3, 10, 11, 6, 13, 14, 15, 4, 17, 6, ...
4	A053166	1, 2, 3, 2, 5, 6, 7, 2, 3, 10, 11, 6, 13, 14, 15, 2, 17, 6, ...

See also PSEUDOSMARANDACHE FUNCTION, SMARANDACHE FUNCTION, SMARANDACHE-KUREPA FUNCTION, SMARANDACHE NEAR-TO-PRIMORIAL FUNCTION, SMARANDACHE SEQUENCES, SMARANDACHE-WAGSTAFF FUNCTION

References

Begay, A. "Smarandache Ceil Functions." *Bull. Pure Appl. Sci.* **16E**, 227–229, 1997. http://www.gallup.unm.edu/~smarandache/smarceil.htm.
Sloane, N. J. A. Sequences A000027/M0472, A019554, A019555, and A053166 in "An On-Line Version of the Encyclopedia of Integer Sequences." http://www.research.att.com/~njas/sequences/eisonline.html.
Smarandache, F. *Collected Papers, Vol. 2.* Kishinev, Moldova: Kishinev University Press, 1997.
Smarandache, F. *Only Problems, Not Solutions!, 4th ed.* Phoenix, AZ: Xiquan, 1993.

Smarandache Constants

"The" Smarandache constant is the smallest solution to the generalized ANDRICA'S CONJECTURE, $x \approx 0.567148$.

The first Smarandache constant is defined as

$$S_1 \equiv \sum_{n=2}^{\infty} \frac{1}{[S(n)]!} > 1.093111, \tag{1}$$

where $S(n)$ is the SMARANDACHE FUNCTION. Cojocaru and Cojocaru (1996a) prove that S_1 exists and is bounded by $0.717 < S_1 < 1.253$. The lower limit given above is obtained by taking 40,000 terms of the sum.

Cojocaru and Cojocaru (1996b) prove that the second Smarandache constant

$$S_2 \equiv \sum_{n=2}^{\infty} \frac{S(n)}{n!} \approx 1.71400629359162 \tag{2}$$

is an IRRATIONAL NUMBER.

Cojocaru and Cojocaru (1996c) prove that the series

$$S_3 \equiv \sum_{n=2}^{\infty} \frac{1}{\prod_{i=2}^{n} S(i)} \approx 0.719960700043708 \tag{3}$$

converges to a number $0.71 < S_3 < 1.01$, and that

$$S_4(a) \equiv \sum_{n=2}^{\infty} \frac{n^a}{\prod_{i=2}^{n} S(i)} \tag{4}$$

converges for a fixed REAL NUMBER $a \geq 1$. The values for small a are

$$S_4(1) \approx 1.72875760530223 \tag{5}$$

$$S_4(2) \approx 4.50251200619297 \tag{6}$$

$$S_4(3) \approx 13.0111441949445 \tag{7}$$

$$S_4(4) \approx 42.4818449849626 \tag{8}$$

$$S_4(5) \approx 158.105463729329. \tag{9}$$

Sandor (1997) shows that the series

$$S_5 \equiv \sum_{n=1}^{\infty} \frac{(-1)^{n-1} S(n)}{n!} \tag{10}$$

converges to an IRRATIONAL. Burton (1995) and Dumitrescu and Seleacu (1996) show that the series

$$S_6 \equiv \sum_{n=2}^{\infty} \frac{S(n)}{(n+1)!} \tag{11}$$

converges. Dumitrescu and Seleacu (1996) show that the series

$$S_7 \equiv \sum_{n=r}^{\infty} \frac{S(n)}{(n+r)!} \tag{12}$$

and

$$S_8 \equiv \sum_{n=r}^{\infty} \frac{S(n)}{(n-r)!} \tag{13}$$

converge for r a natural number (which must be nonzero in the latter case). Dumitrescu and Seleacu (1996) show that

$$S_9 \equiv \sum_{n=1}^{\infty} \frac{1}{\sum_{i=2}^{n} \frac{S(i)}{i!}} \tag{14}$$

converges. Burton (1995) and Dumitrescu and Seleacu (1996) show that the series

$$S_{10} \equiv \sum_{n=2}^{\infty} \frac{1}{[S(n)]^a \sqrt{S(n)!}} \tag{15}$$

and

$$S_{11} \equiv \sum_{n=2}^{\infty} \frac{1}{[S(n)]^a \sqrt{[S(n) + 1]!}} \qquad (16)$$

converge for $a > 1$.

See also Andrica's Conjecture, Smarandache Function

References

Burton, E. "On Some Series Involving the Smarandache Function." *Smarandache Notions J.* **6**, 13–15, 1995.

Burton, E. "On Some Convergent Series." *Smarandache Notions J.* **7**, 7–9, 1996.

Cojocaru, I. and Cojocaru, S. "The First Constant of Smarandache." *Smarandache Notions J.* **7**, 116–118, 1996a.

Cojocaru, I. and Cojocaru, S. "The Second Constant of Smarandache." *Smarandache Notions J.* **7**, 119–120, 1996b.

Cojocaru, I. and Cojocaru, S. "The Third and Fourth Constants of Smarandache." *Smarandache Notions J.* **7**, 121–126, 1996c.

"Constants Involving the Smarandache Function." http://www.gallup.unm.edu/~smarandache/CONSTANT.TXT.

Dumitrescu, C. and Seleacu, V. "Numerical Series Involving the Function S." *The Smarandache Function in Number Theory.* Vail: Erhus University Press, pp. 48–61, 1996.

Ibstedt, H. *Surfing on the Ocean of Numbers--A Few Smarandache Notions and Similar Topics.* Lupton, AZ: Erhus University Press, pp. 27–30, 1997.

Sandor, J. 'On The Irrationality Of Certain Alternative Smarandache Series." *Smarandache Notions J.* **8**, 143–144, 1997.

Smarandache, F. *Collected Papers, Vol. 1.* Bucharest, Romania: Tempus, 1996.

Smarandache, F. *Collected Papers, Vol. 2.* Kishinev, Moldova: Kishinev University Press, 1997.

Smarandache Function

The smallest value $S(n)$ for a given n for which $n | S(n)!$ (n divides $S(n)$ factorial). For example, the number 8 does not divide $1!$, $2!$, $3!$, but does divide $4! = 4 \cdot 3 \cdot 2 \cdot 1 = 8 \cdot 3$, so $S(8) = 4$. For a prime p, $S(p) = p$, and for an even perfect number r, $S(r)$ is prime (Ashbacher 1997). Sloane places the restriction $S(n) > 0$, while Ashbacher (1995) and Russo (2000, p. 4) take $S(n) \geq 0$.

The Smarandache numbers for $n = 1, 2, ...$ are 1, 2, 3, 4, 5, 3, 7, 4, 6, 5, 11, ... (Sloane's A002034; but,

depending on the convention, $S(1)$ may equal either 0 or 1). Letting $a(n)$ denote the smallest value of n for which $S(n) = 1, 2, ...$, then $a(n)$ is given by 1, 2, 3, 4, 9, 7, 32, 27, 25, 11, 243, ... (Sloane's A046021). Some values of $S(n)$ first occur only for very large n, for example, $S(59, 049) = 24$, $S(177, 147) = 27$, $S(134, 217, 728) = 30$, $S(43, 046, 721) = 36$, and $S(9, 765, 625) = 45$. D. Wilson points out that if we let

$$I(n, p) = \frac{n - \sum(n, p)}{p - 1},$$

be the power of the prime p in $n!$, where $\Sigma(n, p)$ is the sum of the base-p digits of n, then it follows that

$$a(n) = \min p^{I(n-1, p)+1},$$

where the minimum is taken over the primes p dividing n. This minimum appears to always be achieved when p is the greatest prime factor of n. If $n = 2^{k-1}(2^k - 1)$ is an even perfect number (i.e., $2^k - 1$ is prime), then $S(n) = p$ (Ruiz 1999a). If p is a prime number and $n \geq 2$ an integer, then $S(p^{p^n}) = p^{n+1} - p^n + p$. (Ruiz 1999b).

The incrementally largest values of $S(n)$ are 1, 2, 3, 4, 5, 7, 11, 13, 17, 19, 23, 29, ... (Sloane's A046022), which occur for $n = 1, 2, 3, 4, 5, 7, 11, 13, 17, 19, 23, 29, ...$ (Sloane's A046023), i.e., the values where $S(n) = n$.

Tutescu (1996) conjectures that the Diophantine equation $S(n) = S(n + 1)$ has no solution.

See also Factorial, Greatest Prime Factor, Pseudosmarandache Function, Smarandache Ceil Function, Smarandache Constants, Smarandache-Kurepa Function, Smarandache Near-to-Primorial Function, Smarandache-Wagstaff Function

References

Ashbacher, C. *An Introduction to the Smarandache Function.* Cedar Rapids, IA: Decisionmark, 1995.

Ashbacher, C. "Problem 4616." *School Sci. Math.* **97**, 221, 1997.

Begay, A. "Smarandache Ceil Functions." *Bulletin Pure Appl. Sci. India* **16E**, 227–229, 1997.

Dumitrescu, C. and Seleacu, V. *The Smarandache Function.* Vail, AZ: Erhus University Press, 1996.

Finch, S. "Unsolved Mathematics Problems: Questions Involving the Smarandache Function." http://www.math-soft.com/asolve/smarand/smarand.html.

"Functions in Number Theory." http://www.gallup.unm.edu/~smarandache/FUNCT1.TXT.

Ibstedt, H. *Surfing on the Ocean of Numbers--A Few Smarandache Notions and Similar Topics.* Lupton, AZ: Erhus University Press, pp. 27–30, 1997.

Ruiz, S. M. "Smarandache Function Applied to Perfect Numbers." *Smarandache Notions J.* **10**, 114–155, 1999.

Ruiz, S. M. "A Result Obtained Using Smarandache Function." *Smarandache Notions J.* **10**, 123–124, 1999.

Russo, F. *A Set of New Smarandache Functions, Sequences, and Conjectures in Numer Theory.* Lupton, AZ: American Research Press, 2000.

Sandor, J. "On Certain Inequalities Involving the Smarandache Function." *Abstracts of Papers Presented to the Amer. Math. Soc.* **17**, 583, 1996.

Sloane, N. J. A. Sequences A002034/M0453, A046021, A046022, and A046023 in "An On-Line Version of the Encyclopedia of Integer Sequences." http://www.research.-att.com/~njas/sequences/eisonline.html.

Smarandache, F. "A Function in Number Theory." *Analele Univ. Timisoara, Ser. St. Math.* **43**, 79–88, 1980.

Smarandache, F. *Collected Papers, Vol. 1.* Bucharest, Romania: Tempus, 1996.

Smarandache, F. *Collected Papers, Vol. 2.* Kishinev, Moldova: Kishinev University Press, 1997.

Tutescu, L. "On a Conjecture Concerning the Smarandache Function." *Abstracts of Papers Presented to the Amer. Math. Soc.* **17**, 583, 1996.

Smarandache Near-to-Primorial Function

$SNTP(n)$ is the smallest PRIME such that $p\# - 1$, $p\#$, or $p\# + 1$ is divisible by n, where $p\#$ is the PRIMORIAL of p. Ashbacher (1996) shows that $SNTP(n)$ only exists

> 1. If there are no square or higher powers in the factorization of n, or
> 2. If there exists a PRIME $q < p$ such that $n | (q\# \pm 1)$, where p is the smallest power contained in the factorization of n.

Therefore, $SNTP(n)$ does not exist for the SQUAREFUL numbers $n = 4, 8, 9, 12, 16, 18, 20, 24, 25, 27, 28, \ldots$ (Sloane's A013929). The first few values of $SNTP(n)$, where defined, are 2, 2, 2, 3, 3, 3, 5, 7, ... (Sloane's A046026).

See also PRIMORIAL, SMARANDACHE FUNCTION

References

Ashbacher, C. "A Note on the Smarandache Near-To-Primordial Function." *Smarandache Notions J.* **7**, 46–49, 1996.

Mudge, M. R. "The Smarandache Near-To-Primorial Function." *Abstracts of Papers Presented to the Amer. Math. Soc.* **17**, 585, 1996.

Sloane, N. J. A. Sequences A013929 and A046026 in "An On-Line Version of the Encyclopedia of Integer Sequences." http://www.research.att.com/~njas/sequences/eisonline.html.

Smarandache Paradox

Let A be some attribute (e.g., possible, present, perfect, etc.). If all is A, then the non-A must also be A. For example, "All is possible, the impossible too," and "Nothing is perfect, not even the perfect."

References

Le, C. T. "The Smarandache Class of Paradoxes." *Bull. Transylvania Univ. Brasov* **36**, 7–8, 1994.

Le, C. T. "The Smarandache Class of Paradoxes." *Bull. Pure Appl. Sci.* **14E**, 109–110, 1995.

Le, C. T. "The Smarandache Class of Paradoxes." *J. Indian Acad. Math.* **18**, 53–55, 1996.

Mitroiescu, I. *The Smarandache Class of Paradoxes.* Glendale, AZ: Erhus University Press, 1994.

Mitroiescu, I. "The Smarandache's Class of Paradoxes Applied in Computer Science." *Abstracts of Papers Presented to the Amer. Math. Soc.* **16**, 651, 1995.

Smarandache Sequences

Smarandache sequences are any of a number of simply generated INTEGER SEQUENCES resembling those considered in published works by Smarandache such as the CONSECUTIVE NUMBER SEQUENCES and EUCLID NUMBERS (Iacobescu 1997). Some other "Smarandache" sequences are given below.

1. The concatenation of n copies of the INTEGER n: 1, 22, 333, 4444, 55555, ... (Sloane's A000461; Marimutha 1997),

2. The concatenation of the first n FIBONACCI NUMBERS: 1, 11, 112, 1123, 11235, ... (Sloane's A019523; Marimutha 1997),

3. The smallest number that is the sum of squares of *two* distinct earlier terms: 1, 2, 5, 26, 29, 677, ... (Sloane's A008318, Bencze 1997),

4. The smallest number that is the sum of squares of any number of distinct earlier terms: 1, 1, 2, 4, 5, 6, 16, 17, ... (Sloane's A008319, Bencze 1997),

5. The smallest number that is *not* the sum of squares of *two* distinct earlier terms: 1, 2, 3, 4, 6, 7, 8, 9, 11, ... (Sloane's A008320, Bencze 1997),

6. The smallest number that is *not* the sum of squares of any number of distinct earlier terms: 1, 2, 3, 6, 7, 8, 11, ... (Sloane's A008321, Bencze 1997),

7. The smallest number that is a sum of cubes of *two* distinct earlier terms: 1, 2, 9, 730, 737, ... (Sloane's A008322, Bencze 1997),

8. The smallest number that is a sum of cubes of any number of distinct earlier terms: 1, 1, 2, 8, 9, 10, 512, 513, 514, ... (Sloane's A019511, Bencze 1997),

9. The smallest number that is *not* a sum of cubes of *two* distinct earlier terms: 1, 2, 3, 4, 5, 6, 7, 8, 10, ... (Sloane's A031980, Bencze 1997),

10. The smallest number that is *not* a sum of cubes of any number of distinct earlier terms: 1, 2, 3, 4, 5, 6, 7, 10, 11, ... (Sloane's A031981, Bencze 1997),

11. The number of PARTITIONS of a number $n = 1$, 2, ... into SQUARE NUMBERS: 1, 1, 1, 1, 2, 2, 2, 2, 3, 4, 4, 4, 5, 6, 6, 6, 8, 9, 10, 10, 12, 13, ... (Sloane's A001156, Iacobescu 1997),

12. The number of PARTITIONS of a number $n = 1$, 2, ... into CUBIC NUMBERS: 1, 1, 1, 1, 1, 1, 1, 1, 2, 2,

2, 2, 2, 2, 2, 2, 3, 3, 3, 3, 3, 3, 3, ... (Sloane's A003108, Iacobescu 1997),

13. Two copies of the first n POSITIVE INTEGERS: 11, 1212, 123123, 12341234, ... (Sloane's A019524, Iacobescu 1997),

14. Numbers written in base of triangular numbers: 1, 2, 10, 11, 12, 100, 101, 102, 110, 1000, 1001, 1002, ... (Sloane's A000462, Iacobescu 1997),

15. Numbers written in base of double factorial numbers: 1, 10, 100, 101, 110, 200, 201, 1000, 1001, 1010, ... (Sloane's A019513, Iacobescu 1997),

16. Sequences starting with terms $\{a_1, a_2\}$ which contain no three-term arithmetic progressions starting with $\{1, 2\}$: 1, 2, 4, 5, 10, 11, 13, 14, 28, ... (Sloane's A003278, Iacobescu 1997, Mudge 1997, Weisstein),

17. Numbers OF THE FORM $\{n!\}^2 + 1$: 2, 5, 37, 577, 14401, 518401, 25401601, 1625702401, 131681894401, ... (Sloane's A020549, Iacobescu 1997),

18. Numbers OF THE FORM $\{n!\}^3 + 1$: 2, 9, 217, 13825, 1728001, 373248001, 128024064001, ... (Sloane's A019514, Iacobescu 1997),

19. Numbers OF THE FORM $1 + 1!2!3! \cdots n!$: 2, 3, 13, 289, 34561, 24883201, 125411328001, 5056584744960001, ... (Sloane's A019515, Iacobescu 1997),

20. Sequences starting with terms $\{a_1, a_2\}$ which contain no three-term geometric progressions starting with $\{1, 2\}$: 1, 2, 3, 5, 6, 7, 8, 10, 11, 13, 14, 15, 16, ... (Sloane's A000452, Iacobescu 1997),

21. Numbers repeating the digit 1 p_n times, where p_n is the nth prime: 11, 111, 11111, 1111111, ... (Sloane's A031974, Iacobescu 1997). These are a subset of the REPUNITS,

22. Integers with all 2s, 3s, 5s, and 7s (prime digits) removed: 1, 4, 6, 8, 9, 10, 11, 1, 1, 14, 1, 16, 1, 18, 19, 0, ... (Sloane's A019516, Iacobescu 1997),

23. Integers with all 0s, 1s, 4s, and 9s (square digits) removed: 2, 3, 5, 6, 7, 8, 2, 3, 5, 6, 7, 8, 2, 2, 22, 23, ... (Sloane's A031976, Iacobescu 1997).

24. (Smarandache-Fibonacci triples) Integers n such that $S(n) = S(n-1) + S(n-2)$, where $S(k)$ is the SMARANDACHE FUNCTION: 3, 11, 121, 4902, 26245, ... (Sloane's A015047; Aschbacher and Mudge 1995; Ibstedt 1997, pp. 19–23; Begay 1997). The largest known is 19,448,047,080,036,

25. (Smarandache-Radu triplets) Integers n such that there are no primes between the smaller and larger of $S(n)$ and $S(n+1)$: 224, 2057, 265225, ... (Sloane's A015048; Radu 1994/1995, Begay 1997, Ibstedt 1997). The largest known is 270,329,975,921,205,253,634,707,051,822,848,570-,391,313,

26. (Smarandache crescendo sequence): Integers obtained by concatenating strings of the first $n + 1$ integers for $n = 0, 1, 2, \ldots$: 1, 1, 2, 1, 2, 3, 1, 2, 3, 4, ... (Sloane's A002260; Brown 1997, Brown and

Castillo 1997). The nth term is given by $n - m(m + 1)/2 + 1$, where $m = \lfloor (\sqrt{8n+1} - 1)/2 \rfloor$, with $\lfloor x \rfloor$ the FLOOR FUNCTION (Hamel 1997).

27. (Smarandache descrescendo sequence): Integers obtained by concatenating strings of the first n integers for $n = \ldots, 2, 1$: 1, 2, 1, 3, 2, 1, 4, 3, 2, 1, ... (Sloane's A004736; Smarandache 1997, Brown 1997),

28. (Smarandache crescendo pyramidal sequence, a.k.a. Smarandache descrescendo symmetric sequence): Integers obtained by concatenating strings of rising and falling integers: 1, 1, 2, 1, 1, 2, 3, 2, 1, 1, 2, 3, 4, 3, 2, 1, ... (Sloane's A004737; Brown 1997, Brown and Castillo 1997, Smarandache 1997),

29. (Smarandache descrescendo pyramidal sequence): Integers obtained by concatenating strings of falling and rising integers: 1, 2, 1, 2, 3, 2, 1, 2, 3, 4, 3, 2, 1, 2, 3, 4, ... (Sloane's A004738; Brown 1997),

30. (Smarandache crescendo symmetric sequence): 1, 1, 1, 2, 2, 1, 1, 2, 3, 3, 2, 1, ... (Sloane's A004739, Brown 1997, Smarandache 1997),

31. (Smarandache permutation sequence): Numbers obtained by concatenating sequences of increasing length of increasing ODD NUMBERS and decreasing EVEN NUMBERS: 1, 2, 1, 3, 4, 2, 1, 3, 5, 6, 4, 2, ... (Sloane's A004741; Brown 1997, Brown and Castillo 1997),

32. (Smarandache pierced chain sequence): Numbers OF THE FORM

$$c(n) = 101 \underbrace{0101 \cdots 0101}_{n}$$

for $n = 0, 1, \ldots$: 101, 1010101, 10101010101, ... (Sloane's A031982; Ashbacher 1997). In addition, $c(n)/101$ contains no PRIMES (Ashbacher 1997),

33. (Smarandache symmetric sequence): 1, 11, 121, 1221, 12321, 123321, ... (Sloane's A007907; Smarandache 1993, Dumitrescu and Seleacu 1994, sequence 3; Mudge 1995),

34. (Smarandache square-digital sequence): square numbers all of whose digits are also squares: 1, 4, 9, 49, 100, 144, ... (Sloane's A019544; Mudge 1997),

35. (Square-digits): numbers composed of digits which are squares: 0, 1, 4, 9, 10, 11, 14, 19, 40, 41, ... (Sloane's A046030),

36. (Cube-digits): numbers composed of digits which are cubes: 1, 8, 10, 11, 18, 80, 81, 88, 100, 101, ... (Sloane's A046031),

37. (Smarandache cube-digital sequence): cube-digit numbers which are themselves cubes: 1, 8,

1000, 8000, 1000000, ... (Sloane's A019545; Mudge 1997),

38. (Prime-digits): numbers composed of digits which are primes: 2, 3, 5, 7, 22, 23, 25, 27, 32, 33, 35, ... (Sloane's A046034),

39. (Smarandache prime-digital sequence): prime-digit numbers which are themselves prime: 2, 3, 5, 7, 23, 37, 53, ... (Sloane's A019546; Smith 1996, Mudge 1997).

40. (Smarandache deconstructive sequence): integers constructed by sequentially repeating the digits 1–9 in the following way: 1, 23, 456, 7891, 23456, 789123, 4567891, ... (Sloane's A007923; Smarandache 1993, Kashihara 1996, Ashbacher, Atanassov 1999ab). Of these, 23, 4567891, 23456789, 1234567891, ... (Sloane's A050234) are prime (Kashihara 1996, Ashbacher).

See also ADDITION CHAIN, CONSECUTIVE NUMBER SEQUENCES, CUBIC NUMBER, EUCLID NUMBER, EVEN NUMBER, FIBONACCI NUMBER, INTEGER SEQUENCE, ODD NUMBER, PARTITION, SMARANDACHE FUNCTION, SQUARE NUMBER

References

Ashbacher, C. "Some Problems Concerning the Smarandache Deconstructive Sequence." *J. Recr. Math.* **29**, 82–84, 1998.

Ashbacher, C. *Collection of Problems On Smarandache Notions.* Vail, AZ: Erhus University Press, 1996.

Ashbacher, C. *Pluckings from the Tree of Smarandache Sequences and Functions.* Lupton, AZ: American Research Press, 1998.

Aschbacher, C. and Mudge, M. *Personal Computer World.* pp. 302, Oct. 1995.

Atanassov, K. "On the 4th Smarandache Problem." *Notes on Number Theory and Discrete Mathematics (Sophia, Bulgaria)* **5**, 33–35, 1999.

Atanassov, K. T. *On Some of the Smarandache's Problems.* Lupton, AZ: American Research Press, pp. 16–21, 1999.

Begay, A. "Smarandache Ceil Functions." *Bull. Pure Appl. Sci.* **16E**, 227–229, 1997.

Bencze, M. "Smarandache Recurrence Type Sequences." *Bull. Pure Appl. Sci.* **16E**, 231–236, 1997.

Bencze, M. and Tutescu, L. (Eds.). *Some Notions and Questions in Number Theory, Vol. 2.* http://www.gallup.unm.edu/~smarandache/SNAQINT2.TXT.

Brown, J. "Crescendo & Descrescendo." In *Richard Henry Wilde: An Anthology in Memoriam (1789–1847)* (Ed. M. Myers). Bristol, IN: Bristol Banner Books, p. 19, 1997.

Brown, J. and Castillo, J. "Problem 4619." *School Sci. Math.* **97**, 221–222, 1997.

Dumitrescu, C. and Seleacu, V. (Eds.). *Some Notions and Questions in Number Theory, 4th ed.* Glendale, AZ: Erhus University Press, 1994. http://www.gallup.unm.edu/~smarandache/SNAQINT.TXT.

Dumitrescu, C. and Seleacu, V. (Eds.). *Proceedings of the First International Conference on Smarandache Type Notions in Number Theory.* Lupton, AZ: American Research Press, 1997.

Hamel, E. Solution to Problem 4619. *School Sci. Math.* **97**, 221–222, 1997.

Iacobescu, F. "Smarandache Partition Type and Other Sequences." *Bull. Pure Appl. Sci.* **16E**, 237–240, 1997.

Ibstedt, H. *Surfing on the Ocean of Numbers--A Few Smarandache Notions and Similar Topics.* Lupton, AZ: Erhus University Press, 1997.

Kashihara, K. *Comments and Topics on Smarandache Notions and Problems.* Vail, AZ: Erhus University Press, 1996.

Mudge, M. "Top of the Class." *Personal Computer World,* 674–675, June 1995.

Mudge, M. "Not Numerology but Numeralogy!" *Personal Computer World,* 279–280, 1997.

Programs and the Abstracts of the First International Conference on Smarandache Notions in Number Theory. Craiova, Romania, Aug. 21–23, 1997.

Radu, I. M. *Mathematical Spectrum* **27**, 43, 1994/1995.

Rivera, C. "Problems & Puzzles: Puzzle Primes by Listing.-008." http://www.primepuzzles.net/puzzles/puzz_008.htm.

Sloane, N. J. A. Sequences A000452, A000461, A000462, A001156/M0221, A002260, A003108/M0209, A003278/M0975, A004736, A004737, A004738, A004739, A004741, A007907, A008318, A008319, A008320, A008321, A008322, A015047, A015048, A019524, A019511, A019513, A019514, A019515, A019516, A019523, A019544, A019545, A019546 A020549, A031974, A031976, A031980, A031981, A031982, A046030, A046031, A046034, and A050234 in "An On-Line Version of the Encyclopedia of Integer Sequences." http://www.research.att.com/~njas/sequences/eisonline.html.

Smarandache, F. "Properties of the Numbers." Tempe, AZ: Arizona State University Special Collection, 1975.

Smarandache, F. *Only Problems, Not Solutions!, 4th ed.* Phoenix, AZ: Xiquan, 1993.

Smarandache, F. *Collected Papers, Vol. 2.* Kishinev, Moldova: Kishinev University Press, 1997.

Smith, S. "A Set of Conjectures on Smarandache Sequences." *Bull. Pure Appl. Sci.* **15E**, 101–107, 1996.

Smarandache-Kurepa Function

Given the sum-of-factorials function

$$\sum(n) = \sum_{k=1}^{n} k!,$$

$SK(p)$ for p PRIME is the smallest integer n such that $p \,|\, 1 + \Sigma(n-1)$. The first few known values of $SK(p)$ are 2, 4, 6, 6, 5, 7, 7, 12, 22, 16, 55, 54, 42, 24, ... for $p = 2, 5, 7, 11, 17, 19, 23, 31, 37, 41, 61, 71, 73, 89, ...$ The function $SK(p)$ doe not exists for $p = 3, 13, 29, 43, 47, 53, 67, 79, 83, ...$

See also PSEUDOSMARANDACHE FUNCTION, SMARANDACHE CEIL FUNCTION, SMARANDACHE FUNCTION, SMARANDACHE-WAGSTAFF FUNCTION, SMARANDACHE FUNCTION

References

Ashbacher, C. "Some Properties of the Smarandache-Kurepa and Smarandache-Wagstaff Functions." *Math. Informatics Quart.* **7**, 114–116, 1997.

Mudge, M. "Introducing the Smarandache-Kurepa and Smarandache-Wagstaff Functions." *Smarandache Notions J.* **7**, 52–53, 1996.

Mudge, M. "Introducing the Smarandache-Kurepa and Smarandache-Wagstaff Functions." *Abstracts of Papers Presented to the Amer. Math. Soc.* **17**, 583, 1996.

Smarandache-Wagstaff Function

Given the sum-of-FACTORIALS function

$$\sum(n) = \sum_{k=1}^{n} k!,$$

$SW(p)$ is the smallest integer for p PRIME such that $\Sigma[SW(p)]$ is divisible by p. If $p \nmid \Sigma(n)$ for all $n < p$, then p never divides any sum for all n. Therefore, the values $SW(p)$ do not exist for 2, 5, 7, 13, 19, 31, ... (Sloane's A056985).

The function is defined for $p = 3, 11, 17, 23, 29, 37, 41, 43, 53, 67, 73, 79, 97, ...$ (Sloane's A056983), with corresponding values 2, 4, 5, 12, 19, 24, 32, 19, 20, 20, 20, 7, 57, 6, ... (Sloane's A056985).

See also FACTORIAL, SMARANDACHE FUNCTION

References

Ashbacher, C. "Some Properties of the Smarandache-Kurepa and Smarandache-Wagstaff Functions." *Math. Informatics Quart.* **7**, 114–116, 1997.
"Functions in Number Theory." http://www.gallup.unm.edu/~smarandache/FUNCT1.TXT.
Mudge, M. "Introducing the Smarandache-Kurepa and Smarandache-Wagstaff Functions." *Smarandache Notions J.* **7**, 52–53, 1996.
Mudge, M. "Introducing the Smarandache-Kurepa and Smarandache-Wagstaff Functions." *Abstracts of Papers Presented to the Amer. Math. Soc.* **17**, 583, 1996.
Sloane, N. J. A. Sequences A056983, A056984, and A056985 in "An On-Line Version of the Encyclopedia of Integer Sequences." http://www.research.att.com/~njas/sequences/eisonline.html.

Smith Brothers

Consecutive SMITH NUMBERS. The first few Smith brothers are (728, 729), (2964, 2965), (3864, 3865), (4959, 4960), ... (Sloane's A050219 and A050220).

See also SMITH NUMBER

References

Sloane, N. J. A. Sequences A050219 and A050220 in "An On-Line Version of the Encyclopedia of Integer Sequences." http://www.research.att.com/~njas/sequences/eisonline.html.

Smith Conjecture

The set of fixed points which do not move as a KNOT is transformed into itself is not a KNOT. The conjecture was proved in 1978 (Morgan and Bass 1984). According to Morgan and Bass (1984), the Smith conjecture stands in the first rank of mathematical problems when measured by the amount and depth of new mathematics required to solve it.

The generalized Smith conjecture states considers $\mathbb{S}^{n-2}$ to be a piecewise linear $(n-2)$-dimensional sphere in $\mathbb{S}^n$, and M^n the k-fold cyclic covering of $\mathbb{S}^n$ branched along S^{n-2}, and asks if S^{n-2} is unknotted if M^n is an $\mathbb{S}^n$ (Hartley 1983). This conjecture is true for

$n \leq 3$, and false for $n \geq 4$, with counterexamples in the latter case provided by Giffen (1966), Gordon (1974), and Sumners (1975).

References

Giffen, C. H. "The Generalized Smith Conjecture." *Amer. J. Math.* **88**, 187–198, 1966.
Gordon, C. M. "On the Higher-Dimensional Smith Conjecture." *Proc. London Math. Soc.* **29**, 98–110, 1974.
Hartley, R. "Whitehead Torsion and the Smith Conjecture." *Michigan Math. J.* **30**, 121–128, 1983.
Morgan, J. W. and Bass, H. (Eds.). *The Smith Conjecture, Papers Presented at the Symposium Held at Columbia University, New York, 1979.* Orlando, FL: Academic Press, 1984.
Rolfsen, D. *Knots and Links.* Wilmington, DE: Publish or Perish Press, pp. 350–351, 1976.
Smith, P. A. "Transformations of Finite Period. II." *Ann. Math.* **40**, 690–711, 1939.
Summers, D. W. "Smooth Z_p Actions on Spheres which Leave Knots Pointwise Fixed." *Trans. Amer. Math. Soc.* **205**, 193–203, 1975.
Waldhausen, F. "Über Involutionen der 3-Sphäre." *Topology* **8**, 81–91, 1969.

Smith Normal Form

Let A be an $n \times n$ MATRIX over a FIELD F. Using the three ELEMENTARY ROW AND COLUMN OPERATIONS over elements in the field, the $n \times n$ matrix $x\mathsf{I} - \mathsf{A}$ with entries from $F[x]$ can be put into the diagonal form

$$\begin{bmatrix} 1 & 0 & \cdots & 0 & 0 & 0 & 0 & 0 \\ 0 & 1 & & 0 & 0 & 0 & 0 & 0 \\ \vdots & & \ddots & & & & & \vdots \\ 0 & 0 & & 1 & 0 & 0 & 0 & 0 \\ 0 & 0 & & 0 & a_1(x) & 0 & 0 & 0 \\ 0 & 0 & & 0 & 0 & a_2(x) & 0 & 0 \\ \vdots & & & & & & \ddots & \vdots \\ 0 & 0 & & 0 & 0 & 0 & 0 & a_m(x) \end{bmatrix}$$

called the Smith normal form, which that $a_1(x), a_2(x), ..., a_m(x)$ are monic nonzero elements of $F[x]$ with degrees at least one and satisfying $a_1(x) \times |a_2(x)| ... |a_m(x)|$ (Dummit and Foote 1998, pp. 390–391 and 414). The elements $a_i(x)$ are then called the INVARIANT FACTORS of A.

References

Ayres, F. Jr. "Smith Normal Form." Ch. 24 in *Theory and Problems of Matrices.* New York: Schaum, pp. 188–195, 1962.
Dummit, D. S. and Foote, R. M. *Abstract Algebra, 2nd ed.* Englewood Cliffs, NJ: Prentice-Hall, 1998.
Jabon, D. "Smith Normal Forms." http://www.mathsource.com/cgi-bin/msitem?0207–470.

Smith Number

A COMPOSITE NUMBER the SUM of whose DIGITS is the sum of the DIGITS of its PRIME FACTORS (excluding 1). (The PRIMES are excluded since they trivially satisfy this condition). One example of a Smith number is the BEAST NUMBER

$$666 = 2 \cdot 3 \cdot 3 \cdot 37,$$

since

$$6 + 6 + 6 = 2 + 3 + 3 + (3 + 7) = 18.$$

Another Smith number is

$$4937775 = 3 \cdot 5 \cdot 5 \cdot 65837,$$

since

$$4 + 9 + 3 + 7 + 7 + 7 + 5$$
$$= 3 + 5 + 5 + (6 + 5 + 8 + 3 + 7) = 42.$$

The first few Smith numbers are 4, 22, 27, 58, 85, 94, 121, 166, 202, 265, 274, 319, 346, ... (Sloane's A006753). The corresponding digits sums are 4, 4, 9, 13, 13, 13, 4, 13, 4, 13, 13, 13, 13, ... (Sloane's A050218) McDaniel (1987a) showed that there are an infinite number of Smith numbers.

A generalized k-Smith number can also be defined as a number m satisfying $S_p(m) = kS(m)$, where $S_p(m)$ is the sum of the digits of m's prime factors and $S(m)$ is the usual sum of m's digits. The following table gives the first few k-Smith numbers for $k \geq 2$.

k	Sloane	k-Smith numbers
2	A050224	88, 169, 286, 484, 598, 682, 808, 844, 897, ...
3	A050225	6969, 19998, 36399, 39693, 66099, 69663, ...

A Smith number can be constructed from every factored REPUNIT R_n (Hoffman 1998, pp. 205–206). The largest known Smith number is

$$9 \times R_{1031}\left(10^{4594} + 3 \times 10^{2297} + 1\right)^{1476} \times 10^{3913210}.$$

See also HOAX NUMBER, MONICA SET, PERFECT NUMBER, REPUNIT, SMITH BROTHERS, SUZANNE SET

References

Gardner, M. *Penrose Tiles and Trapdoor Ciphers... and the Return of Dr. Matrix, reissue ed.* New York: W. H. Freeman, pp. 99–100, 1989.
Guy, R. K. "Smith Numbers." §B49 in *Unsolved Problems in Number Theory, 2nd ed.* New York: Springer-Verlag, pp. 103–104, 1994.
Hoffman, P. *The Man Who Loved Only Numbers: The Story of Paul Erdos and the Search for Mathematical Truth.* New York: Hyperion, pp. 205–206, 1998.
McDaniel, W. L. "The Existence of Infinitely Many k-Smith Numbers." *Fib. Quart.*, **25**, 76–80, 1987a.
McDaniel, W. L. "Powerful K-Smith Numbers." *Fib. Quart.* **25**, 225–228, 1987b.
Oltikar, S. and Weiland, K. "Construction of Smith Numbers." *Math. Mag.* **56**, 36–37, 1983.
Sloane, N. J. A. Sequences A006753/M3582, A050218, A050224, and A050225 in "An On-Line Version of the Encyclopedia of Integer Sequences." http://www.research.-att.com/~njas/sequences/eisonline.html.
Wilansky, A. "Smith Numbers." *Two-Year College Math. J.* **13**, 21, 1982.
Yates, S. "Special Sets of Smith Numbers." *Math. Mag.* **59**, 293–296, 1986.
Yates, S. "Smith Numbers Congruent to 4 (mod 9)." *J. Recr. Math.* **19**, 139–141, 1987.

Smith's Markov Process Theorem

Consider

$$P_2(y_1,\, t_1 | y_3,\, t_3)$$
$$= \int P_2(y_1,\, t_1 | y_2,\, t_2) P_3(y_1,\, t_1;\; y_2,\, t_2 | y_3,\, t_3)\, dy_2. \quad (1)$$

If the probability distribution is governed by a MARKOV PROCESS, then

$$P_3(y_1,\, t_1;\; y_2,\, t_2 | y_3,\, t_3) = P_2(y_2,\, t_2 | y_3,\, t_3)$$
$$= P_2(y_2 | y_3,\, t_3 - t_2). \quad (2)$$

Assuming no time dependence, so $t_1 \equiv 0$,

$$P_2(y_1 | y_3,\, t_3) = \int P_2(y_1 | y_2, t_2) P_2(y_2 | y_3,\, t_3 - t_2)\, dy_2. \quad (3)$$

See also MARKOV PROCESS

Smith's Network Theorem

In a NETWORK with three EDGES at each VERTEX, the number of HAMILTONIAN CIRCUITS through a specified EDGE is 0 or EVEN.

See also EDGE (GRAPH), HAMILTONIAN CIRCUIT, NETWORK

Smooth Function

A smooth function is a function that has continuous second-order derivatives over some domain. A function can therefore be said to be smooth over a restricted interval such as $(a,\, b)$ or $[a,\, b]$.

See also CONTINUOUS FUNCTION, DERIVATIVE

Smooth Manifold

Another word for a C^∞ (infinitely differentiable) MANIFOLD. A smooth manifold is a TOPOLOGICAL MANIFOLD together with its "functional structure" (Bredon 1995) and so differs from a TOPOLOGICAL MANIFOLD because the notion of differentiability exists on it. Every smooth manifold is a TOPOLOGICAL MANIFOLD, but not necessarily vice versa. (The first nonsmooth TOPOLOGICAL MANIFOLD occurs in 4-D.) In 1959, Milnor showed that a 7-D HYPERSPHERE can be made into a smooth manifold in 28 ways.

See also Differentiable Manifold, Hypersphere, Manifold, Topological Manifold

References
Bredon, G. E. *Topology & Geometry*. New York: Springer-Verlag, p. 69, 1995.

Smooth Number

An INTEGER is k-smooth if it has no PRIME FACTORS $> k$. The following table gives the first few k-smooth numbers for small k. Berndt (1994, p. 52) called the 7-smooth numbers "highly composite numbers."

k	Sloane	k-smooth numbers
2	A000079	1, 2, 4, 8, 16, 32, 64, 128, 256, 512, ...
3	A003586	1, 2, 3, 4, 6, 8, 9, 12, 16, 18, 24, ...
5	A051037	1, 2, 3, 4, 5, 6, 8, 9, 10, 12, 15, 16, ...
7	A002473	1, 2, 3, 4, 5, 6, 7, 8, 9, 10, 12, 14, 15, ...
11	A051038	1, 2, 3, 4, 5, 6, 7, 8, 9, 10, 11, 12, 14, ...

The probability that a random POSITIVE INTEGER $\leq n$ is k-smooth is $\psi(n, k)/n$, where $\psi(n, k)$ is the number of k-smooth numbers $\leq n$. This fact is important in application of Kraitchik's extension of FERMAT'S FACTORIZATION METHOD because it is related to the number of random numbers which must be examined to find a suitable subset whose product is a square.

Since about $\pi(k)$ k-smooth numbers must be found (where $\pi(k)$ is the PRIME COUNTING FUNCTION), the number of random numbers which must be examined is about $\pi(k)n/\psi(n, k)$. But because it takes about $\pi(k)$ steps to determine if a number is k-smooth using TRIAL DIVISION, the expected number of steps needed to find a subset of numbers whose product is a square is $\sim [\pi(k)]^2 n/\psi(n, k)$ (Pomerance 1996). Canfield *et al.* (1983) showed that this function is minimized when

$$k \sim \exp\left(\tfrac{1}{2}\sqrt{\ln n \ln \ln n}\right)$$

and that the minimum value is about

$$\exp\left(2\sqrt{\ln n \ln \ln n}\right).$$

In the CONTINUED FRACTION FACTORIZATION ALGORITHM, n can be taken as $2\sqrt{n}$, but in FERMAT'S FACTORIZATION METHOD, it is $n^{1/2+\epsilon}$. k is an estimate for the largest PRIME in the FACTOR BASE (Pomerance 1996).

See also Highly Composite Number, Round Number

References
Berndt, B. C. *Ramanujan's Notebooks, Part IV*. New York: Springer-Verlag, 1994.
Blecksmith, R.; McCallum, M.; and Selfridge, J. L. " 3-Smooth Representations of Integers." *Amer. Math. Monthly* **105**, 529–543, 1998.
Canfield, E. R.; Erdos, P.; and Pomerance, C. "On a Problem of Oppenheim Concerning 'Factorisation Numerorum.'" *J. Number Th.* **17**, 1–28, 1983.
Mintz, D. J. "2, 3 Sequence as a Binary Mixture." *Fib. Quart.* **19**, 351–360, 1981.
Pomerance, C. "On the Role of Smooth Numbers in Number Theoretic Algorithms." In *Proc. Internat. Congr. Math., Zürich, Switzerland, 1994, Vol. 1* (Ed. S. D. Chatterji). Basel: Birkhäuser, pp. 411–422, 1995.
Pomerance, C. "A Tale of Two Sieves." *Not. Amer. Math. Soc.* **43**, 1473–1485, 1996.
Ramanujan, S. *Collected Papers* (Ed. G. H. Hardy *et al.*) New York: Chelsea, p. xxiv, 1962.
Sloane, N. J. A. Sequences A000079/M1129, A002473/M0477, A003586, A051037, and A051038 in "An On-Line Version of the Encyclopedia of Integer Sequences." http://www.research.att.com/~njas/sequences/eisonline.html.

Smooth Structure

A smooth structure on a TOPOLOGICAL MANIFOLD (also called a differentiable structure) is given by a smooth ATLAS of coordinate charts, i.e., the TRANSITION FUNCTIONS between the coordinate charts are C^∞ smooth. A manifold with a smooth structure is called a DIFFERENTIABLE MANIFOLD or a SMOOTH MANIFOLD.

A smooth structure is used to define DIFFERENTIABILITY for real-valued functions on a manifold. This extends to a notion of when a map between two differentiable manifolds is smooth, and naturally to the definition of a DIFFEOMORPHISM. In addition, the smooth structure is used to define TANGENT VECTORS, the collection of which is the TANGENT BUNDLE.

Two smooth structures are considered equivalent if there is a HOMEOMORPHISM of the manifold which pulls back one atlas to an atlas compatible to the other one, i.e., a DIFFEOMORPHISM. For instance, any two smooth structures on the circle $\mathbb{S}^1$ are equivalent, as can be seen by integration.

It is surprising that some manifolds admit more than one smooth structure. The first such example was an EXOTIC SPHERE of $\mathbb{S}^7$, the 7-dimensional HYPERSPHERE, found by Milnor (1956) using the calculus of OCTONIONS. In the 1980s, several mathematicians, including Casson, Freedman, and Donaldson, showed that 4-dimensional Euclidean space $\mathbb{R}^4$ has smooth structures that are distinct from the standard structure. These are called EXOTIC R4, and some of their techniques involve DONALDSON THEORY.

Another approach to smooth structures is through SHEAF theory. Notice that a coordinate chart for an n-dimensional manifold is really an ordered collection of n continuous functions. Whenever two coordinate charts overlap on the manifold, the functions from one chart are infinitely differentiable with respect to those from the other chart. The collection of compatible real-valued continuous functions defines the sheaf of smooth functions. Conversely, one can define a smooth structure to be defined by a subsheaf of continuous functions which satisfies the mutually differentiable condition.

See also ATLAS, DIFFEOMORPHISM DONALDSON THEORY, EXOTIC R4, EXOTIC SPHERE, MANIFOLD, OCTONION, SHEAF (TOPOLOGY), SMOOTH FUNCTION, SMOOTH SURFACE, TANGENT BUNDLE, TANGENT VECTOR (MANIFOLD)

References

Milnor, J. "Topological Manifolds and Smooth Manifolds." In *Proc. Internat. Congr. Mathematicians (Stockholm, 1962).* Djursholm: Inst. Mittag-Leffler, pp. 132–138, 1963.

Smooth Surface

A surface PARAMETERIZED in variables u and v is called smooth if the TANGENT VECTORS in the u and v directions satisfy

$$\mathbf{T}_u + \mathbf{T}_v \neq 0,$$

where $\mathbf{A} \times \mathbf{B}$ is a CROSS PRODUCT.

Smoothing

The modification of a set of data to make it smooth and nearly continuous and remove or diminish outlying points.

See also MOVING AVERAGE, SAVITZKY-GOLAY FILTER

References

Lanczos, C. "Trigonometric Interpolation of Empirical and Analytic Functions." *J. Math. Phys.* **17**, 123, 1938.
Rhodes, E. C. *Tract on Smoothing.* No. 6 in *Tracts for Computers* (Ed. K. Pearson). London: Cambridge University Press, 1921.
Whittaker, E. T. and Robinson, G. "Graduation, or the Smoothing of Data." Ch. 11 in *The Calculus of Observations: A Treatise on Numerical Mathematics,* 4th ed. New York: Dover, pp. 285–316, 1967.

sn

JACOBI ELLIPTIC FUNCTIONS

Snake

A simple circuit in the d-HYPERCUBE which has no chords (i.e., for which all snake edges are edges of the HYPERCUBE). Klee (1970) asked for the maximum length $s(d)$ of a d-snake. Klee (1970) gave the bounds

$$\frac{7}{4(d-1)} \leq \frac{s(d)}{2^d} \frac{1}{2} - \frac{1 + 12/2^{-d}}{7d(d-1)^2 + 2} \tag{1}$$

for $d \geq 6$ (Danzer and Klee 1967, Douglas 1969), as well as numerous references. Abbott and Katchalski (1988) show

$$s(d) \geq 77 \cdot 2^{d-8}, \tag{2}$$

and Snevily (1994) showed that

$$s(d) \leq 2^{d-1}\left(1 - \frac{1}{20d - 41}\right) \tag{3}$$

for $d \leq 12$, and conjectured

$$s(d) \leq 3 \cdot 2^{d-3} + 2 \tag{4}$$

for $d \leq 5$. The first few values for $s(d)$ for $d = 1, 2, ...,$ are 2, 4, 6, 8, 14, 26, ... (Sloane's A000937).

See also HYPERCUBE

References

Abbott, H. L. and Katchalski, M. "On the Snake in the Box Problem." *J. Combin. Th. Ser. B* **44**, 12–24, 1988.
Danzer, L. and Klee, V. "Length of Snakes in Boxes." *J. Combin. Th.* **2**, 258–265, 1967.
Douglas, R. J. "Some Results on the Maximum Length of Circuits of Spread k in the d-Cube." *J. Combin. Th.* **6**, 323–339, 1969.
Evdokimov, A. A. "Maximal Length of a Chain in a Unit n-Dimensional Cube." *Mat. Zametki* **6**, 309–319, 1969.
Guy, R. K. "Unsolved Problems Come of Age." *Amer. Math. Monthly* **96**, 903–909, 1989.
Guy, R. K. "Monthly Unsolved Problems." *Amer. Math. Monthly* **94**, 961–970, 1989.
Kautz, W. H. "Unit-Distance Error-Checking Codes." *IRE Trans. Elect. Comput.* **7**, 177–180, 1958.
Klee, V. "What is the Maximum Length of a d-Dimensional Snake?" *Amer. Math. Monthly* **77**, 63–65, 1970.
Sloane, N. J. A. Sequences A000937/M0995 in "An On-Line Version of the Encyclopedia of Integer Sequences." http://www.research.att.com/~njas/sequences/eisonline.html.
Snevily, H. S. "The Snake-in-the-Box Problem: A New Upper Bound." *Disc. Math.* **133**, 307–314, 1994.

Snake Eyes

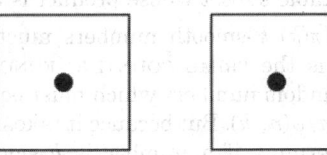

A roll of two 1s (the lowest roll possible) on a pair of six-sided DICE. The probability of rolling snake eyes is 1/36, or 2.777...%.

See also BOXCARS, DICE

Snake Oil Method

The expansion of the two sides of a sum equality in terms of POLYNOMIALS in x^m and y^k, followed by closed form summation in terms of x and y. For an example of the technique, see Bloom (1995).

References

Bhatnagar, G. "A Multivariable View of One-Variable q-Series." In *Special Functions and Differential Equations. Proceedings of the Workshop (WSSF97) held in Madras, January 13–24, 1997)* (Ed. K. S. Rao, R. Jagannathan, G. van den Berghe, and J. Van der Jeugt). New Delhi, India: Allied Pub., pp. 60–72, 1998.

Bloom, D. M. "A Semi-Unfriendly Identity." Problem 10206. Solution by R. J. Chapman. *Amer. Math. Monthly* **102,** 657–658, 1995.

Wilf, H. S. *Generatingfunctionology, 2nd ed.* New York: Academic Press, 1993.

Snake Polyiamond

A 6-POLYIAMOND.

References

Golomb, S. W. *Polyominoes: Puzzles, Patterns, Problems, and Packings, 2nd ed.* Princeton, NJ: Princeton University Press, p. 92, 1994.

Snedecor's F-Distribution

If a random variable X has a CHI-SQUARED DISTRIBUTION with m degrees of freedom (χ_m^2) and a random variable Y has a CHI-SQUARED DISTRIBUTION with n degrees of freedom (χ_n^2), and X and Y are independent, then

$$F \equiv \frac{X/m}{Y/n} \tag{1}$$

is distributed as Snedecor's F-distribution with m and n degrees of freedom

$$f(F(m,\,n)) = \frac{\Gamma\left(\frac{m+n}{2}\right)\left(\frac{m}{n}\right)^{m/2} F^{(m-2)/2}}{\Gamma\left(\frac{m}{2}\right)\Gamma\left(\frac{n}{2}\right)\left(1 + \frac{m}{n}F\right)^{(m+2)/2}} \tag{2}$$

for $0 < F < \infty$. The RAW MOMENTS are

$$\mu_1' = \frac{n}{n-2} \tag{3}$$

$$\mu_2' = \frac{n^2(m+2)}{m(n-2)(n-4)} \tag{4}$$

$$\mu_3' = \frac{n^3(m+2)(m+4)}{m^2(n-2)(n-4)(n-6)} \tag{5}$$

$$\mu_4' = \frac{n^4(m+2)(m+4)(m+6)}{m^3(n-2)(n-4)(n-6)(n-8)}, \tag{6}$$

so the CENTRAL MOMENTS are given by

$$\mu_2 = \frac{2n^2(m+n-2)}{m(n-2)^2(n-4)} \tag{7}$$

$$\mu_3 = \frac{8n^3(m+n-2)(2m+n-2)}{m^2(n-2)^3(n-4)(n-6)} \tag{8}$$

$$\mu_4 = \frac{12n^4(m+n-2)\left[4(n-2)^2 + m^2(n+10) + m(n-2)(n+10)\right]}{m^3(n-2)^4(n-4)(n-6)(n-8)}. \tag{9}$$

and the MEAN, VARIANCE, SKEWNESS, and KURTOSIS are

$$\mu = \mu_1' = \frac{n}{n-2} \tag{10}$$

$$\sigma^2 = \frac{2n^2(m+n-2)}{m(n-2)^2(n-4)} \tag{11}$$

$$\gamma_1 = \frac{\mu_3}{\sigma^3} = 2\sqrt{\frac{2(n-4)}{m(m+n-2)}}\,\frac{2m+n-2}{n-6} \tag{12}$$

$$\gamma_2 = \frac{\mu_4}{\sigma^4} - 3 =$$

$$\frac{3(n-4)\left[4(n-2)^2 + m^2(n+10) + m(n-2)(n+10)\right]}{m(m+n-2)(n-6)(n-8)}. \tag{13}$$

The CHARACTERISTIC FUNCTION can be computed, but it is rather messy and involves the GENERALIZED HYPERGEOMETRIC FUNCTION ${}_3F_2(a,\,b,\,c;\,d,\,e;\,z)$. Letting

$$w \equiv \frac{\dfrac{mF}{n}}{1 + \dfrac{mF}{n}} \tag{14}$$

gives a BETA DISTRIBUTION (Beyer 1987, p. 536).

See also BETA DISTRIBUTION, CHI-SQUARED DISTRIBUTION, STUDENT'S T-DISTRIBUTION

References

Beyer, W. H. *CRC Standard Mathematical Tables, 28th ed.* Boca Raton, FL: CRC Press, p. 536, 1987.

Snellius-Pothenot Problem

A SURVEYING PROBLEM which asks: Determine the position of an unknown accessible point P by its bearings from three inaccessible known points A, B, and C.

See also HANSEN'S PROBLEM

References

Dörrie, H. "Annex to a Survey." §40 in *100 Great Problems of Elementary Mathematics: Their History and Solutions.* New York: Dover, pp. 193–197, 1965.

Snowflake

EXTERIOR SNOWFLAKE, KOCH ANTISNOWFLAKE, KOCH SNOWFLAKE, PENTAFLAKE

Snub Cube

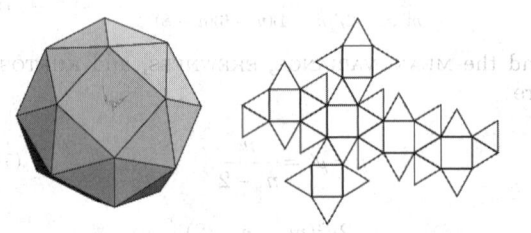

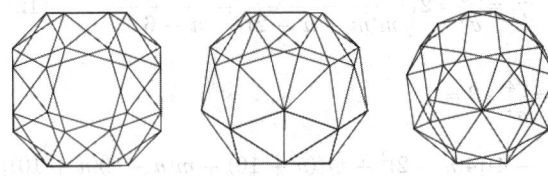

The 38-faced ARCHIMEDEAN SOLID A_7, also called the SNUB CUBOCTAHEDRON, whose faces are $32\{3\} + 6\{4\}$. It has two ENANTIOMERS. It is UNIFORM POLYHEDRON U_{12} and Wenninger model W_{17}. It has SCHLÄFLI SYMBOL $s\{{3 \atop 4}\}$ and WYTHOFF SYMBOL $|234$.

Its DUAL POLYHEDRON is the PENTAGONAL ICOSITE-TRAHEDRON. The INRADIUS r of the dual, MIDRADIUS ρ of the dual and solid, and CIRCUMRADIUS R for unit edge length are given by the unique positive real roots of the equations

$$896r^6 - 1248r^4 + 64r^2 - 1 = 0 \qquad (1)$$

$$64\rho^6 - 112\rho^4 + 20\rho^2 - 1 = 0 \qquad (2)$$

$$32R^6 - 80R^4 + 44R^2 - 7 = 0 \qquad (3)$$

which given by

$$r = 1.157661791\ldots \qquad (4)$$

$$\rho = 1.247223168\ldots \qquad (5)$$

$$R = 1.3437133737446\ldots \qquad (6)$$

The SURFACE AREA of the snub cube of side length 1 is

$$S = 6 + 8\sqrt{3} \qquad (7)$$

and the VOLUME V is given by the positive real solution to the equation

$$729V^6 - 45684V^4 + 19386V^2 - 12842 = 0, \qquad (8)$$

which is given approximately by

$$V \approx 7.88948. \qquad (9)$$

The distances from the center to the centroids of the triangular and square faces are given by the unique positive roots to the equations

$$864r_3^6 - 1296r_3^4 + 36r_3^2 - 1 = 0 \qquad (10)$$

$$32r_4^6 - 32r_4^4 - 12r_4^2 - 1 = 0, \qquad (11)$$

which are given by

$$r_3 = 1.213355800\ldots \qquad (12)$$

$$r_4 = 1.142613508\ldots \qquad (13)$$

See also ARCHIMEDEAN SOLID, ICOSITETRAHEDRON, PENTAGONAL ICOSITETRAHEDRON, SNUB DODECAHE-DRON

References

Ball, W. W. R. and Coxeter, H. S. M. *Mathematical Recreations and Essays, 13th ed.* New York: Dover, p. 139, 1987.
Coxeter, H. S. M.; Longuet-Higgins, M. S.; and Miller, J. C. P. "Uniform Polyhedra." *Phil. Trans. Roy. Soc. London Ser. A* **246**, 401–450, 1954.
Cundy, H. and Rollett, A. "Snub Cube. $3^4.4$." §3.7.7 in *Mathematical Models, 3rd ed.* Stradbroke, England: Tarquin Pub., p. 107, 1989.
Wenninger, M. J. "The Snub Cube." Model 17 in *Polyhedron Models.* Cambridge, England: Cambridge University Press, p. 31, 1989.

Snub Cube-Pentagonal Icositetrahedron Compound

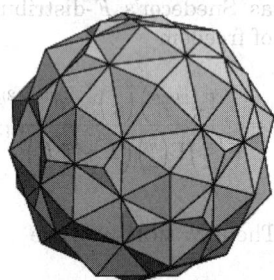

The compound of the SNUB CUBE and its dual, the PENTAGONAL ICOSITETRAHEDRON. It can be constructed from the snub cube with unit edge length by heights h_3 and h_4, given by the unique positive real roots of

$$3456h_3^6 - 864h_3^4 + 216h_3^2 - 1 = 0 \qquad (1)$$

$$128h_4^6 + 96h_4^4 + 16h_4^2 - 1 = 0. \qquad (2)$$

The corresponding solid has edge lengths

$$128s_1^6 - 64s_1^4 + 16s_1^2 - 1 = 0 \qquad (3)$$

$$s_2 = 1/2 \qquad (4)$$

$$128s_3^6 - 6s_3^3 - 1 = 0, \qquad (5)$$

and

$$s_4 = \tfrac{1}{2}\sqrt{2}, \qquad (6)$$

where s_1 and s_3 are unique real roots of the above

polynomials. The CIRCUMRADIUS is given by the root of

$$32R^6 - 80R^4 + 44R^2 - 7 = 0, \qquad (7)$$

the SURFACE AREA by the root of

$$1028869776 + 35418062592S - 45028405440S^2$$

$$+22712607360S^3 - 5396081328S^4$$

$$+463818960S^5 + 35732664S^6 - 7379424S^7$$

$$+23652S^8 + 29160S^9 - 576S^{10} - 36S^{11} + S^{12}, \qquad (8)$$

and VOLUME by the root of

$$128V^6 - 8864V^4 + 19152V^2 - 10609 = 0. \qquad (9)$$

See also COMPOUND POLYHEDRON, PENTAGONAL ICOSITETRAHEDRON, SNUB CUBE

Snub Cuboctahedron
SNUB CUBE

Snub Disphenoid

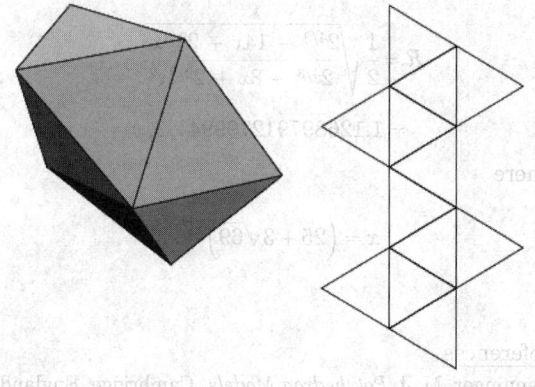

The 12-faced convex DELTAHEDRA also known as the SIAMESE DODECAHEDRA, which is also JOHNSON SOLID J_{84}.

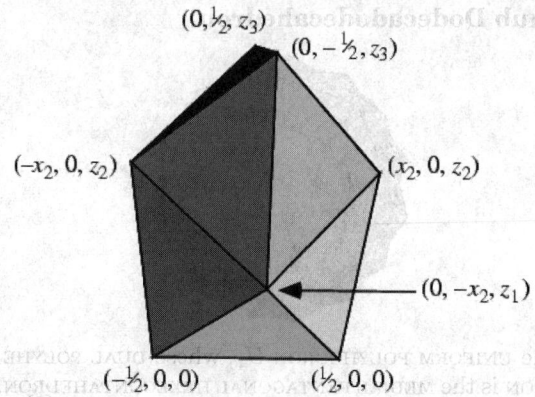

The coordinates of the VERTICES of a snub disphenoid of unit side length may be found by solving the set of four simultaneous equations

$$\left(\tfrac{1}{2}\right)^2 + x_2^2 + z_1^2 = 1$$

$$\left(x_2 - \tfrac{1}{2}\right)^2 + (z_3 - z_1)^2 = 1$$

$$\left(\tfrac{1}{2}\right)^2 + x_2^2 + (z_3 - z_2)^2 = 1$$

$$x_2^2 + x_2^2 + (z_2 - z_1)^2 = 1$$

for the four unknowns x_2, z_1, z_2, and z_3. The analytic solution requires solving the CUBIC EQUATION, and the solutions are given by the unique positive real roots of

$$2x_2^3 - 3x_2^2 - 2x_2 + 2 = 0 \qquad (1)$$

$$32z_1^6 + 64z_1^4 - 22z_1^2 - 1 = 0 \qquad (2)$$

$$16z_2^6 + 8z_2^4 - 15z_2^2 - 8 = 0 \qquad (3)$$

$$2z_3^6 - z_3^4 - 8z_3^2 - 4 = 0. \qquad (4)$$

Numerically,

$$x_2 \approx 0.644584$$

$$z_1 \approx 0.578369$$

$$z_2 \approx 0.989492$$

$$z_3 \approx 1.56786.$$

The SURFACE AREA of the unit snub disphenoid is

$$S = 3\sqrt{3}, \qquad (5)$$

and the VOLUME V is given by the positive real root of

$$5832V^6 - 1377V^4 - 2160V^2 - 4 = 0, \qquad (6)$$

approximately $V \approx 0.859494$.

See also DELTAHEDRON, DISPHENOID, JOHNSON SOLID

Snub Dodecadodecahedron

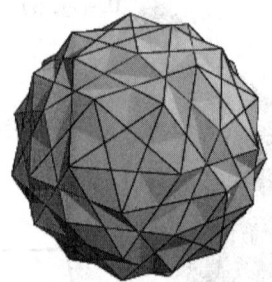

The UNIFORM POLYHEDRON U_{40} whose DUAL POLYHEDRON is the MEDIAL PENTAGONAL HEXECONTAHEDRON. It has WYTHOFF SYMBOL $|2\,\frac{5}{2}\,5$. Its faces are $12\{\frac{5}{2}\} + 60\{3\} + 12\{5\}$. It has CIRCUMRADIUS for $a = 1$ of

$$R = 1.27443994.$$

See also SNUB CUBE

References

Ball, W. W. R. and Coxeter, H. S. M. *Mathematical Recreations and Essays, 13th ed.* New York: Dover, p. 139, 1987.

Coxeter, H. S. M.; Longuet-Higgins, M. S.; and Miller, J. C. P. "Uniform Polyhedra." *Phil. Trans. Roy. Soc. London Ser. A* **246**, 401–450, 1954.

Cundy, H. and Rollett, A. "Snub Dodecahedron. $3^4.5$." §3.7.13 in *Mathematical Models, 3rd ed.* Stradbroke, England: Tarquin Pub., pp. 114–115, 1989.

Wenninger, M. J. *Polyhedron Models.* Cambridge, England: Cambridge University Press, pp. 174–176, 1971.

Snub Dodecahedron

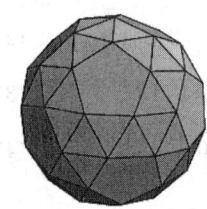
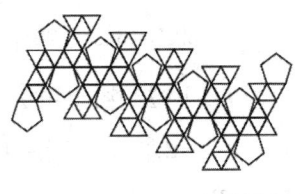

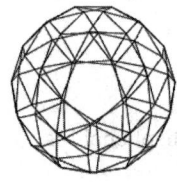
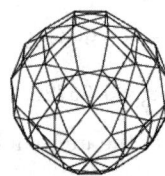
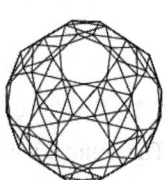

The 92-faced ARCHIMEDEAN SOLID A_8 consisting of faces $80\{3\} + 12\{5\}$ which is also called the snub icosidodecahedron. It is UNIFORM POLYHEDRON U_{29} and Wenninger model W_{18}. It has SCHLÄFLI SYMBOL s $\{\frac{3}{5}\}$ and WYTHOFF SYMBOL $|235$.

The DUAL POLYHEDRON of the snub dodecahedron is the PENTAGONAL HEXECONTAHEDRON. The INRADIUS r of the dual, MIDRADIUS ρ of the solid and dual, and CIRCUMRADIUS R of the solid for $a = 1$ are

$$r = 2.039873155\ldots$$

$$\rho = 2.097053835\ldots$$

$$R = 2.15583737511564\ldots.$$

See also ARCHIMEDEAN SOLID, HEXECONTAHEDRON, SNUB CUBE

References

Coxeter, H. S. M.; Longuet-Higgins, M. S.; and Miller, J. C. P. "Uniform Polyhedra." *Phil. Trans. Roy. Soc. London Ser. A* **246**, 401–450, 1954.

Wenninger, M. J. "The Snub Dodecahedron." Model 18 in *Polyhedron Models.* Cambridge, England: Cambridge University Press, p. 32, 1989.

Snub Icosidodecadodecahedron

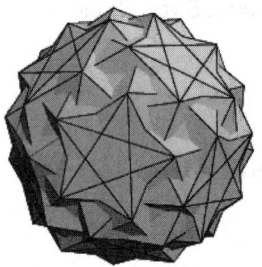

The UNIFORM POLYHEDRON U_{46} whose DUAL POLYHEDRON is the MEDIAL HEXAGONAL HEXECONTAHEDRON. It has WYTHOFF SYMBOL $|3\,\frac{5}{3}\,5$. Its faces are $12\{\frac{4}{2}\} + 80\{3\} + 12\{5\}$. It has CIRCUMRADIUS for $a = 1$ of

$$R = \frac{1}{2}\sqrt{\frac{2^{4/3} - 14x + 2^{2/3}x^2}{2^{4/3} - 8x + 2^{2/3}x^2}}$$

$$= 1.12689791279994\ldots,$$

where

$$x = \left(25 + 3\sqrt{69}\right)^{1/3}.$$

References

Wenninger, M. J. *Polyhedron Models.* Cambridge, England: Cambridge University Press, pp. 177–178, 1971.

Snub Icosidodecahedron

SNUB DODECAHEDRON

Snub Polyhedron

A polyhedron with extra triangular faces, given by the SCHLÄFLI SYMBOL s $\left\{ {p \atop q} \right\}$.

See also RHOMBIC POLYHEDRON, TRUNCATED POLYHEDRON

Snub Square Antiprism

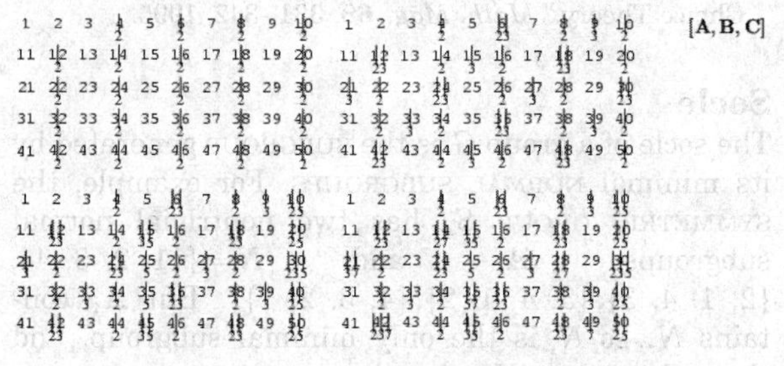

JOHNSON SOLID J_{85}.

References

Weisstein, E. W. "Johnson Solids." MATHEMATICA NOTEBOOK JohnsonSolids.m.

Weisstein, E. W. "Johnson Solid Netlib Database." MATHEMATICA NOTEBOOK JohnsonSolids.dat.

SO

SPECIAL ORTHOGONAL GROUP

Soap Bubble

BUBBLE

Soccer Ball

TRUNCATED ICOSAHEDRON

Sociable Numbers

Numbers which result in a periodic ALIQUOT SEQUENCE, where an ALIQUOT SEQUENCE is the sequence of numbers obtained by repeatedly applying the restricted divisor function

$$s(n) = \sigma(n) - n \qquad (1)$$

to n. Here $\sigma(n)$ is the usual DIVISOR FUNCTION.

If the period is 1, the number is called a PERFECT NUMBER. If the period is 2, the two numbers are called an AMICABLE PAIR. In general, if the period is $t \geq 3$, the number is called sociable of order t. Only two sociable numbers were known prior to 1970, the sets of orders 5 and 28 discovered by Poulet (1918). In 1970, Cohen discovered nine groups of order 4.

For example, 1264460 is a sociable number of order four since its ALIQUOT SEQUENCE is 1264460, 1547860, 1727636, 1305184, 1264460, ... The first

few sociable numbers are 12496, 14316, 1264460, 2115324, 2784580, 4938136, ... (Sloane's A003416), which have orders 5, 28, 4, 4, 4, 4, ... (Sloane's A052470). The table below summarizes the numbers of sociable cycles known as a function of order as given in the compilation by Moews (1995).

order	known
3	0
4	53
5	1
6	2
8	2
9	1
28	1
total	60

Y. Kohmoto has considered a generalization of the sociable numbers defined according to the generalized ALIQUOT SEQUENCE

$$a(n) = \frac{\sigma(a(n-1))}{m}. \qquad (2)$$

MULTIPERFECT NUMBERS are fixed points of this mapping, since if $a(n) = a(n-1)$, then

$$ma(n) = \sigma(a(n)), \qquad (3)$$

which is the definition of an m-multiperfect number. If the sequence $a(n)$ becomes cyclic after $k > 1$ terms, it is then called an $1/m$-sociable number of order k. If M_m and M_n are distinct MERSENNE PRIMES, then

$$\tfrac{1}{2} \sigma(2^{m-1} M_n) = \tfrac{1}{2}(2^m - 1) 2^n = 2^{n-1} M_m \qquad (4)$$

$$\tfrac{1}{2} \sigma(2(n-1) M_m) = 2^{m-1} M_n, \qquad (5)$$

so $2^{m-1} M_n$ and $2^{n-1} M_m$ are 1/2-sociable numbers of order 2.

The following table summarizes the smallest members of the generalized $1/m$-aliquot sequences of order k, found by Kohmoto.

m	k	starting numbers
3	2	14913024
4	2	2096640, 422688000
4	12	3396556800

See also ALIQUOT SEQUENCE, CATALAN'S ALIQUOT SEQUENCE CONJECTURE, PERFECT NUMBER, UNITARY SOCIABLE NUMBERS

References

Borho, W. "Über die Fixpunkte der k-fach iterierten Teiler-ersummenfunktion." *Mitt. Math. Gesellsch. Hamburg* **9**, 34–48, 1969.

Cohen, H. "On Amicable and Sociable Numbers." *Math. Comput.* **24**, 423–429, 1970.

Creyaufmüller, W. "Aliquot Sequences." http://home.t-online.de/home/Wolfgang.Creyaufmueller/aliquote.htm.

Devitt, J. S.; Guy, R. K.; and Selfridge, J. L. *Third Report on Aliquot Sequences,* Congr. Numer. XVIII, Proc. 6th Manitoba Conf. Numerical Math, pp. 177–204, 1976.

Flammenkamp, A. "New Sociable Numbers." *Math. Comput.* **56**, 871–873, 1991.

Gardner, M. "Perfect, Amicable, Sociable." Ch. 12 in *Mathematical Magic Show: More Puzzles, Games, Diversions, Illusions and Other Mathematical Sleight-of-Mind from Scientific American.* New York: Vintage, pp. 160–171, 1978.

Guy, R. K. "Aliquot Cycles or Sociable Numbers." §B7 in *Unsolved Problems in Number Theory, 2nd ed.* New York: Springer-Verlag, pp. 62–63, 1994.

Madachy, J. S. *Madachy's Mathematical Recreations.* New York: Dover, pp. 145–146, 1979.

Moews, D. and Moews, P. C. "A Search for Aliquot Cycles Below 10^{10}." *Math. Comput.* **57**, 849–855, 1991.

Moews, D. and Moews, P. C. "A Search for Aliquot Cycles and Amicable Pairs." *Math. Comput.* **61**, 935–938, 1993.

Moews, D. "A List of Aliquot Cycles of Length Greater than 2." Rev. Dec. 18, 1995. http://xraysgi.ims.uconn.edu/sociable.txt.

Pedersen, J. A. M. "Tables of Aliquot Cycles." http://www.vejlehs.dk/staff/jmp/aliquot/tables.htm.

Poulet, P. Question 4865. *L'interméd. des Math.* **25**, 100–101, 1918.

Root, S. Item 61 in Beeler, M.; Gosper, R. W.; and Schroeppel, R. *HAKMEM.* Cambridge, MA: MIT Artificial Intelligence Laboratory, Memo AIM-239, p. 23, Feb. 1972.

Sloane, N. J. A. Sequences A003416 and A052470 in "An On-Line Version of the Encyclopedia of Integer Sequences." http://www.research.att.com/~njas/sequences/eisonline.html.

te Riele, H. J. J. "Perfect Numbers and Aliquot Sequences." In *Computational Methods in Number Theory, Part I.* (Ed. H. W. Lenstra Jr. and R. Tijdeman). Amsterdam, Netherlands: Mathematisch Centrum, pp. 141–157, 1982.

Weisstein, E. W. "Sociable and Amicable Numbers." MATHEMATICA NOTEBOOK SOCIABLE.M.

Social Choice Theory

The theory of analyzing a decision between a collection of alternatives made by a collection of n voters with separate opinions. Any choice for the entire group should reflect the desires of the individual voters to the extent possible.

Fair choice procedures usually satisfy ANONYMITY (invariance under permutation of voters), DUALITY (each alternative receives equal weight for a single vote), and MONOTONICITY (a change favorable for X does not hurt X). Simple majority vote is anonymous, dual, and monotone. MAY'S THEOREM states a stronger result.

See also ANONYMOUS, ARROW'S PARADOX, DUAL VOTING, MAY'S THEOREM, MONOTONIC VOTING, VOTING

References

Taylor, A. *Mathematics and Politics: Strategy, Voting, Power, and Proof.* New York: Springer-Verlag, 1995.

Young, S. C.; Taylor, A. D.; and Zwicker, W. S. "Counting Quota Systems: A Combinatorial Question from Social Choice Theory." *Math. Mag.* **68**, 331–342, 1995.

Socle

The socle of a group G is the SUBGROUP generated by its minimal NORMAL SUBGROUPS. For example, the SYMMETRIC GROUP S_4 has two nontrivial normal subgroups: A_4 and $N = \{\{1, 2, 3, 4\}, \{2, 1, 4, 3\}, \{3, 4, 1, 2\}, \{4, 3, 2, 1\}\}$. But A_4 contains N, so N is the only minimal subgroup, and the socle of S_4 is N.

See also BLOCK (GROUP ACTION), GROUP, NORMAL SUBGROUP, PRIMITIVE GROUP, TRANSITIVE GROUP

References

Dixon, J. and Mortimer, B. *Permutation Groups.* New York: Springer-Verlag, 1996.

Socrates' Paradox

Socrates is reported to have stated: "One thing I know is that I know nothing."

See also LIAR'S PARADOX

References

Pickover, C. A. *Keys to Infinity.* New York: Wiley, p. 134, 1995.

Soddy Centers
SODDY POINTS

Soddy Circles

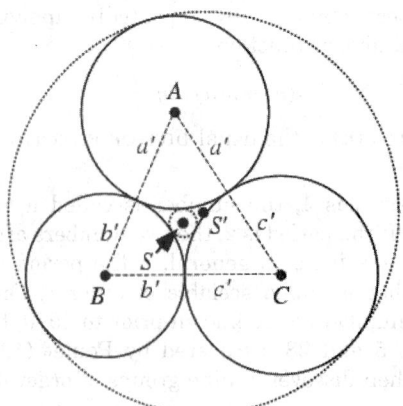

Given three distinct noncollinear points A, B, and C, let three CIRCLES be drawn, one centered about each point and each one tangent to the other two. Call the

RADII r_i $(r_3 = a', r_1 = b', r_2 = c')$. Then the CIRCLES satisfy

$$a' + b' = c \qquad (1)$$

$$a' + c' = b \qquad (2)$$

$$b' + c' = a, \qquad (3)$$

as shown in the diagram below.

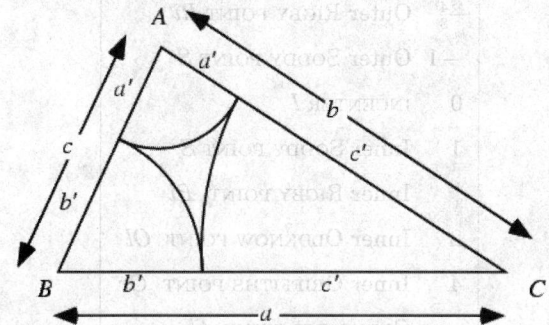

Solving for the RADII then gives

$$a' = \tfrac{1}{2}(b + c - a) \qquad (4)$$

$$b' = \tfrac{1}{2}(a + c - b) \qquad (5)$$

$$c' = \tfrac{1}{2}(a + b - c). \qquad (6)$$

The TRIANGLE illustrated above has sides a, b, and c, and SEMIPERIMETER

$$s \equiv \tfrac{1}{2}(a + b + c). \qquad (7)$$

Plugging in,

$$2s = (a' + b') + (a' + c') + (b' + c') = 2(a' + b' + c'), \qquad (8)$$

giving

$$a' + b' + c' = s. \qquad (9)$$

In addition,

$$a = b' + c' = a' + b' + c' - a' = s - a'. \qquad (10)$$

Switching a and a' to opposite sides of the equation and noting that the above argument applies equally well to b' and c' then gives

$$a' = s - a \qquad (11)$$

$$b' = s - b \qquad (12)$$

$$c' = s - c. \qquad (13)$$

As can be seen from the first figure, there exist exactly two nonintersecting CIRCLES which are TANGENT to all three CIRCLES. These are called the inner and outer Soddy circles (S and S', respectively), and their centers are called the inner and outer SODDY POINTS.

The inner Soddy circle is the solution to the FOUR COINS PROBLEM and its center S, the inner Soddy

point, is the EQUAL DETOUR POINT. The center of the outer Soddy circle, the outer Soddy point S', is the ISOPERIMETRIC POINT (Kimberling 1994).

Frederick Soddy (1936) gave the FORMULA for finding the RADII of the Soddy circles (r_4) given the RADII r_i $(i = 1, 2, 3)$ of the other three. The relationship is

$$2\left(\epsilon_1^2 + \epsilon_2^2 + \epsilon_3^2 + \epsilon_4^2\right) = \left(\epsilon_1 + \epsilon_2 + \epsilon_3 + \epsilon_4\right)^2, \qquad (14)$$

where $\epsilon_i \equiv \pm \kappa_i = \pm 1/r_i$ are the so-called BENDS, defined as the signed CURVATURES of the CIRCLES. If the contacts are all external, the signs are all taken as POSITIVE, whereas if one circle surrounds the other three, the sign of this circle is taken as NEGATIVE (Coxeter 1969). Using the QUADRATIC FORMULA to solve for ϵ_4, expressing in terms of radii instead of curvatures, and simplifying gives

$$r_4^\pm = \frac{r_1 r_2 r_3}{r_2 r_3 + r_1(r_2 + r_3) \pm 2\sqrt{r_1 r_2 r_3 (r_1 + r_2 + r_3)}}. \qquad (15)$$

Here, the NEGATIVE solution corresponds to the outer Soddy circle and the POSITIVE one to the inner Soddy circle.

This FORMULA is called the DESCARTES CIRCLE THEOREM since it was known to Descartes. Soddy extended the result to TANGENT SPHERES, and Gosper has further extended the result to $n + 2$ mutually tangent n-D HYPERSPHERES.

Bellew has derived a generalization applicable to a CIRCLE surrounded by n CIRCLES which are, in turn, circumscribed by another CIRCLE. The relationship is

$$\left[n(c_n - 1)^2 + 1\right] \sum_{i=1}^{n+1} \kappa_i^2 + n\left(3nc_n^2 - 2n - 6\right)c_n^2(c_n - 1)^2 = \left[\frac{f(n)}{n(e_n - 1) + 1}\right], \qquad (16)$$

where κ_{n+1} is the curvature of the central circle,

$$f(n) = \left[n(c_n - 1)^2 + 1\right] \sum_{i=1}^{n+1} \kappa_i + nc_n(c_n - 1)$$
$$\times \left[nc_n^2 + (3 - n)c_n - 4\right] \qquad (17)$$

and

$$c_n \equiv \csc\left(\frac{\pi}{n}\right). \qquad (18)$$

For $n = 3$, this simplifies to the DESCARTES CIRCLE THEOREM

$$2\sum_{i=1}^{4} \kappa_1^2 = \left(\sum_{i=1}^{4} \kappa_i\right)^2. \qquad (19)$$

See also APOLLONIAN GASKET, APOLLONIUS CIRCLES, APOLLONIUS' PROBLEM, ARBELOS, BEND (CURVATURE), BOWL OF INTEGERS, CIRCUMCIRCLE, DESCARTES CIRCLE THEOREM, EXCENTRAL TRIANGLE, FOUR COINS PROBLEM, HART'S THEOREM, MALFATTI CIRCLES, PAPPUS CHAIN, SODDY POINTS, SPHERE PACKING, STEINER CHAIN, TANGENT CIRCLES, TANGENT SPHERES

References
Coxeter, H. S. M. *Introduction to Geometry, 2nd ed.* New York: Wiley, pp. 13–4, 1969.
Elkies, N. D. and Fukuta, J. "Problem E3236 and Solution." *Amer. Math. Monthly* **97**, 529–31, 1990.
Gosper, R. W. "Soddy's Theorem on Mutually Tangent Circles, Generalized to *n* Dimensions." http://www.ippi.com/rwg/Sodddy.htm.
Kimberling, C. "Central Points and Central Lines in the Plane of a Triangle." *Math. Mag.* **67**, p. 181, 1994.
"The Kiss Precise." *Nature* **139**, 62, 1937.
Soddy, F. "The Kiss Precise." *Nature* **137**, 1021, 1936.
Vandeghen, A. "Soddy's Circles and the De Longchamps Point of a Triangle." *Amer. Math. Monthly* **71**, 176–79, 1964.
Wells, D. *The Penguin Dictionary of Curious and Interesting Geometry.* London: Penguin, pp. 4–, 1991.

Soddy Line

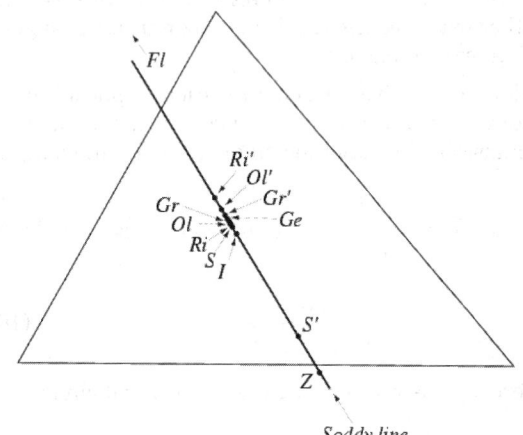

Soddy line

A LINE on which the INCENTER I, GERGONNE POINT Ge, inner and outer SODDY POINTS S and S', GRIFFITHS POINTS Gr, Gr', OLDKNOW POINTS Ol, Ol', RIGBY POINTS Ri, Ri', and FLETCHER POINT Fl lie. The Soddy line can be given parametrically in TRILINEAR COORDINATES by

$$I + \lambda Ge,$$

where λ is a parameter (Oldknow 1996). The Soddy line is also given by

$$\sum (f - e)\alpha = 0,$$

where cyclic permutations of the d, e, and f are taken and the sum is over TRILINEAR COORDINATES α, β, and γ. The following table gives the values of λ corre-

sponding to a number of special points on the Soddy line.

λ	Center
-4	Outer GRIFFITHS POINT Gr'
-2	Outer OLDKNOW POINT Ol'
$-\frac{4}{3}$	Outer RIGBY POINT Ri'
-1	Outer SODDY POINT S'
0	INCENTER I
1	Inner SODDY POINT S
$\frac{4}{3}$	Inner RIGBY POINT Ri
2	Inner OLDKNOW POINT Ol
4	Inner GRIFFITHS POINT Gr
∞	GERGONNE POINT Ge

S', I, S, and Ge form a HARMONIC RANGE (Vandeghen 1964, Oldknow 1996). There are a total of 22 HARMONIC RANGES for sets of four points out of these 10 (Oldknow 1996). The Soddy line intersects the EULER LINE in the DE LONGCHAMPS POINT, and the GERGONNE LINE in the FLETCHER POINT.

See also DE LONGCHAMPS POINT, EULER LINE, FLETCHER POINT, GERGONNE POINT, GRIFFITHS POINTS, HARMONIC RANGE, INCENTER, OLDKNOW POINTS, RIGBY POINTS, SODDY POINTS

References
Oldknow, A. "The Euler-Gergonne-Soddy Triangle of a Triangle." *Amer. Math. Monthly* **103**, 319–29, 1996.
Vandeghen, A. "Soddy's Circles and the De Longchamps Point of a Triangle." *Amer. Math. Monthly* **71**, 176–79, 1964.

Soddy Points

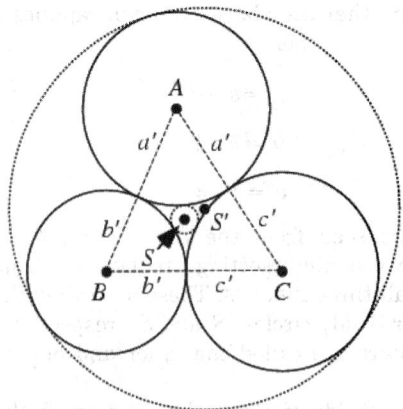

Given three mutually tangent CIRCLES, there exist

exactly two nonintersecting circles which are TANGENT CIRCLES to all three original CIRCLES. These are called the inner and outer SODDY CIRCLES, and their centers S and S' are called the inner and outer Soddy points, respectively.

The inner Soddy point is the EQUAL DETOUR POINT, and the outer Soddy point S' is the ISOPERIMETRIC POINT (Kimberling 1994).

See also EQUAL DETOUR POINT, ISOPERIMETRIC POINT, SODDY CIRCLES

References

Kimberling, C. "Central Points and Central Lines in the Plane of a Triangle." *Math. Mag.* **67**, p. 181, 1994.

Soddy's Hexlet

HEXLET

Sofa Constant

MOVING SOFA CONSTANT

Sokhotskii's Formula

$$\lim_{\epsilon \to 0} \frac{1}{x \pm ie} = \mp i\pi\delta(x) + PV\left(\frac{1}{x}\right),$$

where $\delta(x)$ is the DELTA FUNCTION and PV denotes the CAUCHY PRINCIPAL VALUE.

See also DELTA FUNCTION

Sol Geometry

The GEOMETRY of the LIE GROUP R SEMIDIRECT PRODUCT with R^2, where R acts on R^2 by $(t, (x, y)) \to (e^t x, e^{-t} y)$.

See also THURSTON'S GEOMETRIZATION CONJECTURE

Soldner's Constant

Consider the following formulation of the PRIME NUMBER THEOREM,

$$\pi(x) = \sum \frac{\mu(m)}{m} \int_e^x \frac{dt}{\ln t}.$$

where $\mu(m)$ is the MÖBIUS FUNCTION and c (sometimes also denoted μ) is called Soldner's constant. Ramanujan obtained $c = 1.45136380\ldots$ (Hardy 1999, Le Lionnais 1983, Berndt 1994), while the correct value is $1.4513692346\ldots$, the root of

$$\text{li}(x) = 0$$

(Soldner 1812; Nielsen 1965, p. 88).

See also RIEMANN PRIME NUMBER FORMULA

References

Berndt, B. C. *Ramanujan's Notebooks, Part IV.* New York: Springer-Verlag, pp. 123–24, 1994.

Hardy, G. H. *Ramanujan: Twelve Lectures on Subjects Suggested by His Life and Work,* 3rd ed. New York: Chelsea, pp. 23 and 45, 1999.
Le Lionnais, F. *Les nombres remarquables.* Paris: Hermann, p. 39, 1983.
Nielsen, N. "Theorie des Integrallogrgarithmus und Verwandter Transzendenten." Part II in *Die Gammafunktion.* New York: Chelsea, 1965.
Soldner. *Abhandlungen* **2**, 333, 1812.

Solenoidal Field

A solenoidal VECTOR FIELD satisfies

$$\nabla \cdot \mathbf{B} = 0 \tag{1}$$

for every VECTOR $\mathbf{B}$, where $\nabla \cdot \mathbf{B}$ is the DIVERGENCE. If this condition is satisfied, there exists a vector $\mathbf{A}$, known as the VECTOR POTENTIAL, such that

$$\mathbf{B} \equiv \nabla \times \mathbf{A}, \tag{2}$$

where $\nabla \times \mathbf{A}$ is the CURL. This follows from the vector identity

$$\nabla \cdot \mathbf{B} = \nabla \cdot (\nabla \times \mathbf{A}) = 0. \tag{3}$$

If $\mathbf{A}$ is an IRROTATIONAL FIELD, then

$$\mathbf{A} \times \mathbf{r} \tag{4}$$

is solenoidal. If $\mathbf{u}$ and $\mathbf{v}$ are irrotational, then

$$\mathbf{u} \times \mathbf{v} \tag{5}$$

is solenoidal. The quantity

$$(\nabla u) \times (\nabla v), \tag{6}$$

where ∇u is the GRADIENT, is always solenoidal. For a function ϕ satisfying LAPLACE'S EQUATION

$$\nabla^2 \phi = 0. \tag{7}$$

it follows that $\nabla\phi$ is solenoidal (and also IRROTATIONAL).

See also BELTRAMI FIELD, CURL, DIVERGENCE, DIVERGENCELESS FIELD, GRADIENT, IRROTATIONAL FIELD, LAPLACE'S EQUATION, VECTOR FIELD

References

Gradshteyn, I. S. and Ryzhik, I. M. *Tables of Integrals, Series, and Products,* 6th ed. San Diego, CA: Academic Press, p. 1084, 2000.

Solid

A closed 3-D figure (which may, according to some terminology conventions, be self-intersecting). Kern and Bland (1948, p. 18) define a solid as any limited portion of space bounded by surfaces. Among the simplest solids are the SPHERE, CUBE, CONE, CYLINDER, and more generally, the POLYHEDRA.

See also APPLE, ARCHIMEDEAN SOLID, BARREL, CATALAN SOLID, CONE, CORK PLUG, CUBE, CUBOCTAHEDRON, CYLINDER, CYLINDRICAL HOOF, CYLINDRICAL

WEDGE, DODECAHEDRON, GEODESIC DOME, GOUR-
SAT'S SURFACE, GREAT DODECAHEDRON, GREAT ICO-
SAHEDRON, GREAT RHOMBICOSIDODECAHEDRON
(ARCHIMEDEAN), GREAT RHOMBICUBOCTAHEDRON
(ARCHIMEDEAN), GREAT STELLATED DODECAHEDRON,
ICOSAHEDRON, ICOSIDODECAHEDRON, JOHNSON SO-
LID, KEPLER-POINSOT SOLID, LEMON, MÖBIUS STRIP,
OCTAHEDRON, PLATONIC SOLID, POLYHEDRON, PSEU-
DOSPHERE, RHOMBICOSIDODECAHEDRON, RHOMBICU-
BOCTAHEDRON, SMALL STELLATED DODECAHEDRON,
SNUB CUBE, SNUB DODECAHEDRON, SOLID OF REVO-
LUTION, SPHERE, SPHERICAL WEDGE, STEINMETZ
SOLID, STELLA OCTANGULA, SURFACE, TETRAHEDRON,
TORUS, TRUNCATED CUBE, TRUNCATED DODECAHE-
DRON, TRUNCATED ICOSAHEDRON, TRUNCATED OCTA-
HEDRON, TRUNCATED TETRAHEDRON, UNIFORM
POLYHEDRON, WULFF SHAPE

References

Kern, W. F. and Bland, J. R. *Solid Mensuration with Proofs,
2nd ed.* New York: Wiley, 1948.

Solid Angle

Defined as the SURFACE AREA Ω of a UNIT SPHERE
which is subtended by a given object S. Writing the
SPHERICAL COORDINATES as ϕ for the COLATITUDE
(angle from the pole) and θ for the LONGITUDE
(azimuth),

$$\Omega \equiv A_{\text{projected}} = \int\int_S \sin\phi \, d\theta \, d\phi.$$

Solid angle is measured in STERADIANS, and the solid
angle corresponding to all of space being subtended is
4π STERADIANS.

See also SPHERE, STERADIAN

Solid Geometry

That portion of GEOMETRY dealing with SOLIDS, as
opposed to PLANE GEOMETRY. Solid geometry is concerned with POLYHEDRA, SPHERES, 3-D SOLIDS, lines
in 3-space, PLANES, and so on.

See also GEOMETRY, PLANE GEOMETRY, SPHERICAL
GEOMETRY

References

Altshiller-Court, N. *Modern Pure Solid Geometry.* New
York: Chelsea, 1979.
Bell, R. J. T. *An Elementary Treatise on Coordinate Geome-
try of Three Dimensions.* London: Macmillan, 1926.
Cohn, P. M. *Solid Geometry.* New York: Routledge, 1968.
Dresden, A. *Solid Analytical Geometry and Determinants.*
New York: Dover, 1964.
Farin, G. E. and Hensford, D. *The Geometry Toolbox for
Graphics and Modeling.* Natick, MA: A. K. Peters, 1997.
Frost, P. *Solid Geometry, 3rd ed.* London: Macmillan, 1886.
Harris, J. W. and Stocker, H. "Solid Geometry." Ch. 4 in
Handbook of Mathematics and Computational Science.
New York: Springer-Verlag, pp. 95–16, 1998.

Kenison, E. and Bradley, H. C. *Descriptive Geometry.* New
York: Macmillan, 1935.
Kern, W. F. and Bland, J. R. *Solid Mensuration with Proofs,
2nd ed.* New York: Wiley, 1948.
Lines, L. *Solid Geometry.* New York: Dover, 1965.
Rouché, E. and de Comberousse, C. *Traité de Géométrie,
nouv. éd., vol. 2: Géométrie dans l'espace.* Paris: Gauthier-
Villars, 1922.
Salmon, G. *Treatise on the Analytic Geometry of Three
Dimensions, 6th ed.* London: Longmans Green, 1914.
Shute, W. G.; Shirk, W. W.; and Porter, G. F. *Solid Geome-
try.* New York: American Book Co., 1960.
Weisstein, E. W. "Solid Geometry." MATHEMATICA NOTE-
BOOK SOLIDGEOMETRY.M.
Weisstein, E. W. "Books about Solid Geometry." http://
www.treasure-troves.com/books/SolidGeometry.html.
Wentworth, G. A. and Smith, D. E. *Solid Geometry.* Boston,
MA: Ginn and Company, 1913.

Solid Harmonic

A SURFACE HARMONIC of degree l which is premultiplied by a factor r^l. Confusingly, solid harmonics are
also known as "spherical harmonics" (Whittaker and
Watson 1990, p. 392).

See also SPHERICAL HARMONIC, SURFACE HARMONIC

References

Byerly, W. E. *An Elementary Treatise on Fourier's Series,
and Spherical, Cylindrical, and Ellipsoidal Harmonics,
with Applications to Problems in Mathematical Physics.*
New York: Dover, p. 198, 1959.
Whittaker, E. T. and Watson, G. N. *A Course in Modern
Analysis, 4th ed.* Cambridge, England: Cambridge University Press, 1990.

Solid of Revolution

To find the VOLUME of a solid of rotation by adding up
a sequence of thin cylindrical shells, consider a region
bounded above by $y = f(x)$, below by $y = g(x)$, on the
left by the LINE $x = a$, and on the right by the LINE
$x = b$. When the region is rotated about the Y-AXIS,
the resulting VOLUME is given by

$$V = 2\pi \int_b^a x[f(x) - g(x)] \, dx.$$

To find the volume of a solid of rotation by adding up
a sequence of thin flat disks, consider a region
bounded above by $y = f(x)$, below by $y = g(x)$, on the
left by the LINE $x = a$, and on the right by the LINE
$x = b$. When the region is rotated about the X-AXIS,
the resulting VOLUME is

$$V = \pi \int_b^a \left\{ [f(x)]^2 - [g(x)]^2 \right\} \, dx.$$

See also SURFACE OF REVOLUTION, VOLUME

References

Harris, J. W. and Stocker, H. "Solids of Rotation." §4.10 in
Handbook of Mathematics and Computational Science.
New York: Springer-Verlag, pp. 111–13, 1998.

Solid Partition

Solid partitions are generalizations of PLANE PARTITIONS. MacMahon (1960) conjectured the GENERATING FUNCTION for the number of solid partitions was

$$f(z) = \frac{1}{(1-z)(1-z^2)^3(1-z^3)^6(1-z^4)^{10}\cdots},$$

but this was subsequently shown to disagree at $n = 6$ (Atkin *et al.* 1967). Knuth (1970) extended the tabulation of values, but was unable to find a correct generating function. The first few values are 1, 4, 10, 26, 59, 140, ... (Sloane's A000293).

See also PARTITION FUNCTION *P*

References

Atkin, A. O. L.; Bratley, P.; Macdonald, I. G.; and McKay, J. K. S. "Some Computations for m-Dimensional Partitions." *Proc. Cambridge Philos. Soc.* **63**, 1097–100, 1967.
Knuth, D. E. "A Note on Solid Partitions." *Math. Comput.* **24**, 955–61, 1970.
MacMahon, P. A. "Memoir on the Theory of the Partitions of Numbers. VI: Partitions in Two-Dimensional Space, to which is Added an Adumbration of the Theory of Partitions in Three-Dimensional Space." *Phil. Trans. Roy. Soc. London Ser. A* **211**, 345–73, 1912b.
MacMahon, P. A. *Combinatory Analysis, Vol. 2.* New York: Chelsea, pp. 75–76, 1960.
Sloane, N. J. A. Sequences A000293/M3392 in "An On-Line Version of the Encyclopedia of Integer Sequences." http://www.research.att.com/~njas/sequences/eisonline.html.

Solid Spherical Harmonic

SOLID HARMONIC

Solidus

The diagonal slash "/" used as the bar between NUMERATOR and DENOMINATOR of an in-line FRACTION (Bringhurst 1997, p. 284). The solidus is also called a DIAGONAL.

See also DIVISION, FRACTION, OBELUS, VINCULUM, VIRGULE

References

Bringhurst, R. *The Elements of Typographic Style, 2nd ed.* Point Roberts, WA: Hartley and Marks, p. 284, 1997.

Solitary Number

A number which does not have any FRIENDS. Solitary numbers include all PRIMES, PRIME POWERS, and numbers for which $(n, \sigma(n)) = 1$, where (a, b) is the GREATEST COMMON DIVISOR of a and b and $\sigma(n)$ is the DIVISOR FUNCTION. The first few numbers satisfying $(n, \sigma(n)) = 1$ are 1, 2, 3, 4, 5, 7, 8, 9, 11, 13, 16, 17, 19, 21, ... (Sloane's A014567).

However, there exist numbers such as $n = 18, 45, 48$, and 52 which are solitary but for which $(n, \sigma(n)) \neq 1$. It is believed that 10, 14, 15, 20, 22, and many others

are also solitary, although a proof appears to be extremely difficult.

See also FRIEND, FRIENDLY PAIR, PRIME POWER

References

Anderson, C. W. and Hickerson, D. Problem 6020. "Friendly Integers." *Amer. Math. Monthly* **84**, 65–6, 1977.
Sloane, N. J. A. Sequences A014567 in "An On-Line Version of the Encyclopedia of Integer Sequences." http://www.research.att.com/~njas/sequences/eisonline.html.

Soliton

A stable isolated (i.e., solitary) traveling wave solution to a set of equations.

See also KORTEWEG-DE VRIES EQUATION, LAX PAIR, SINE-GORDON EQUATION

References

Bullough, R. K. and Caudrey, P. J. (Eds.). *Solitons.* Berlin: Springer-Verlag, 1980.
Dodd, R. K.; Eilbeck, J. C.; and Morris, H. C. *Solitons and Nonlinear Equations.* London: Academic Press, 1984.
Drazin, P. G. and Johnson, R. S. *Solitons: An Introduction.* Cambridge, England: Cambridge University Press, 1988.
Filippov, A. *The Versatile Solitons.* Boston, MA: Birkhäuser, 1996.
Gu, C. H. *Soliton Theory and Its Applications.* New York: Springer-Verlag, 1995.
Infeld, E. and Rowlands, G. *Nonlinear Waves, Solitons, and Chaos, 2nd ed.* Cambridge, England: Cambridge University Press, 2000.
Lamb, G. L. Jr. *Elements of Soliton Theory.* New York: Wiley, 1980.
Makhankov, V. G.; Fedyann, V. K.; and Pashaev, O. K. (Eds.). *Solitons and Applications.* Singapore: World Scientific, 1990.
Newell, A. C. *Solitons in Mathematics and Physics.* Philadelphia, PA: SIAM, 1985.
Olver, P. J. and Sattinger, D. H. (Eds.). *Solitons in Physics, Mathematics, and Nonlinear Optics.* New York: Springer-Verlag, 1990.
Remoissent, M. *Waves Called Solitons, 2nd ed.* New York: Springer-Verlag, 1996.
Russell, J. S. "Report on Waves." *Report of the 14th Meeting of the British Association for the Advancement of Science.* London: Jon Murray, pp. 311–90, 1844.
Weisstein, E. W. "Books about Solitons." http://www.treasure-troves.com/books/Solitons.html.

Solomon's Seal Knot

The $(5,2)$ TORUS KNOT 05–01 with BRAID WORD σ_1^5.

Solomon's Seal Lines

The 27 REAL or IMAGINARY LINES which lie on the general CUBIC SURFACE and the 45 triple tangent

PLANES to the surface. All are related to the 28 BITANGENTS of the general QUARTIC CURVE.

Schoutte (1910) showed that the 27 lines can be put into a ONE-TO-ONE correspondence with the vertices of a particular POLYTOPE in 6-D space in such a manner that all incidence relations between the lines are mirrored in the connectivity of the POLYTOPE and conversely (Du Val 1931). A similar correspondence can be made between the 28 bitangents and a 7-D POLYTOPE (Coxeter 1928) and between the tritangent planes of the canonical curve of genus four and an 8-D POLYTOPE (Du Val 1933).

See also BRIANCHON'S THEOREM, CUBIC SURFACE, DOUBLE SIXES, PASCAL'S THEOREM, QUARTIC SURFACE, STEINER SET

References

Bell, E. T. *The Development of Mathematics, 2nd ed.* New York: McGraw-Hill, pp. 322–25, 1945.
Coxeter, H. S. M. "The Pure Archimedean Polytopes in Six and Seven Dimensions." *Proc. Cambridge Phil. Soc.* **24**, 7–, 1928.
Du Val, P. "On the Directrices of a Set of Points in a Plane." *Proc. London Math. Soc. Ser. 2* **35**, 23–4, 1933.
Schoutte, P. H. "On the Relation Between the Vertices of a Definite Sixdimensional Polytope and the Lines of a Cubic Surface." *Proc. Roy. Akad. Acad. Amsterdam* **13**, 375–83, 1910.

Solomon's Seal Polygon

HEXAGRAM

Soluble Group

SOLVABLE GROUP

Solvable Congruence

A CONGRUENCE that has a solution.

Solvable Group

A solvable group is a GROUP having a "normal series" such that each "normal factor" is ABELIAN. The special case of a solvable FINITE GROUP is a group whose composition indices are all PRIME NUMBERS. Solvable groups are sometimes called "soluble groups," a turn of phrase that is a source of possible amusement to chemists.

The term "solvable" derives from this type of group's relationship to GALOIS'S THEOREM, namely that the SYMMETRIC GROUP S_n is unsolvable for $n \geq 5$ while it is solvable for $n = 1$, 2, 3, and 4. As a result, the POLYNOMIAL equations of degree ≥ 5 are not solvable using finite additions, multiplications, divisions, and ROOT EXTRACTIONS.

Every FINITE GROUP of order < 60, every ABELIAN GROUP, and every SUBGROUP of a solvable group is solvable. Betten (1996) has computed a table of solvable groups of order up to 242 (Besche and Eick 1999).

See also ABELIAN GROUP, COMPOSITION SERIES, GALOIS'S THEOREM, SOLVABLE LIE GROUP, SYMMETRIC GROUP

References

Besche, H.-U. and Eick, B. "The Groups of Order at Most 1000 Except 512 and 768." *J. Symb. Comput.* **27**, 405–13, 1999.
Betten, A. "Parallel Construction of Finite Soluble Groups." In *Parallel Virtual Machine, Euro PVM '96: Third European PVM Conference, Munich, Germany, October 7–, 1996* (Ed. A. Bode *et al.*). Berlin: Springer-Verlag, pp. 126–33, 1996.
Doerk, K. and Hawkes, T. *Finite Soluble Groups.* Berlin: de Gruyter, 1992.
Gruenberg, K. W. and Roseblade, J. E. (Eds.). *Group Theory: Essays for Philip Hall.* London: Academic Press, 1984.
Laue, R. "Zur Konstruktion und Klassifikation endlicher auflösbarer Gruppen." *Bayreuther Mathemat. Schriften* **9**, 1982.
Lomont, J. S. *Applications of Finite Groups.* New York: Dover, p. 26, 1993.
Magnus, W. "Neuere Ergebnisse über auflösbare Gruppen." *Jahresber. der DMV* **47**, 69, 1937.
Robinson, D. J. S. *Finiteness Conditions and Generalized Soluble Groups,* 2 vols. Berlin: Springer-Verlag, 1972.
Scott, W. R. "Solvable Groups." §2.6 in *Group Theory.* New York: Dover, pp. 38–9, 1987.
Segal, D. *Polycyclic Groups.* Cambridge, England: Cambridge University Press, 1983.

Solvable Lie Algebra

A LIE ALGEBRA $\mathfrak{g}$ is solvable when its COMMUTATOR SERIES, or derived series, $\mathfrak{g}^k$ vanishes for some k. Any NILPOTENT LIE ALGEBRA is solvable. The basic example is the VECTOR SPACE of UPPER TRIANGULAR MATRICES, because every time two such matrices commute, their nonzero entries move further from the diagonal.

The following *Mathematica* function tests whether a Lie algebra $\mathfrak{g}$ is solvable, when given a list of matrices which form a basis for $\mathfrak{g}$.

```
MatrixBasis[a_-
List]:=Partition[#1,Length[a[[1]]]]&/@
    LatticeReduce[Flatten/@a]
            LieCommutator[a_,b_]:=a.b-b.a
NextDerived[{}]={};
                NextDerived[g_List]:=
MatrixBasis[Flatten[Outer[LieCommutator,g,-
g,1],1]]        SolvableLieQ[g_List]:=
FixedPoint[NextDerived,g]=={}
```

For example,

```
bore15=Flatten[Table[ReplacePart[
                                Ta-
ble[0,{i,5},{j,5}],1,{k,l}],{k,5},{l,k,5}],1];
  SolvableLieQ[bore15]
```

yields True.

See also BOREL SUBALGEBRA, COMMUTATOR SERIES (LIE ALGEBRA), LIE ALGEBRA, LIE GROUP, NILPOTENT

LIE GROUP, NILPOTENT LIE ALGEBRA, REPRESENTATION (LIE ALGEBRA), REPRESENTATION (SOLVABLE LIE GROUP), SOLVABLE LIE GROUP, SPLIT SOLVABLE LIE ALGEBRA

Solvable Lie Group

A solvable Lie group is a LIE GROUP G which is CONNECTED and whose LIE ALGEBRA $\mathfrak{g}$ is a SOLVABLE LIE ALGEBRA. That is, the COMMUTATOR SERIES

$$\mathfrak{g}^1 = [\mathfrak{g}, \mathfrak{g}], \; \mathfrak{g}^2 = \left[\mathfrak{g}^1 \cdot \mathfrak{g}^1\right], \; \dots \qquad (1)$$

eventually vanishes, $\mathfrak{g}^k = 0$ for some k. Since NILPOTENT LIE ALGEBRAS are also SOLVABLE, any NILPOTENT LIE GROUP is a solvable Lie group.

The basic example is the GROUP of invertible UPPER TRIANGULAR MATRICES with positive DETERMINANT, e.g.,

$$\begin{bmatrix} a_{11} & a_{12} & a_{13} \\ 0 & a_{22} & a_{23} \\ 0 & 0 & a_{33} \end{bmatrix} \qquad (2)$$

such that $\prod_i a_{ii} > 0$. The LIE ALGEBRA $\mathfrak{g}$ of G is its TANGENT SPACE at the identity matrix, which is the VECTOR SPACE of all upper triangular matrices, and it is a SOLVABLE LIE ALGEBRA. Its COMMUTATOR SERIES is given by

$$\mathfrak{g}^1 = \begin{bmatrix} 0 & b_{12} & b_{13} \\ 0 & 0 & b_{23} \\ 0 & 0 & 0 \end{bmatrix} \qquad (3)$$

$$\mathfrak{g}^2 = \begin{bmatrix} 0 & 0 & c_{13} \\ 0 & 0 & 0 \\ 0 & 0 & 0 \end{bmatrix}, \qquad (4)$$

$$\mathfrak{g}^3 = \begin{bmatrix} 0 & 0 & 0 \\ 0 & 0 & 0 \\ 0 & 0 & 0 \end{bmatrix}. \qquad (5)$$

Any real solvable Lie group is DIFFEOMORPHIC to EUCLIDEAN SPACE. For instance, the group of matrices in the example above is diffeomorphic to $\mathbb{R}^6$, via the EXPONENTIAL MAPExponential Map (Lie Group). However, in general, the exponential map in a SOLVABLE LIE ALGEBRA need not be SURJECTIVE.

See also BOREL GROUP, COMMUTATOR SERIES (LIE ALGEBRA), FLAG (VECTOR SPACE), LIE ALGEBRA, LIE GROUP, MATRIX, NILPOTENT LIE GROUP, REPRESENTATION, REPRESENTATION (SOLVABLE LIE GROUP), SOLVABLE GROUP, SOLVABLE LIE ALGEBRA, SPLIT SOLVABLE LIE ALGEBRA

References

Knapp, A. W. "Group Representations and Harmonic Analysis, Part II." *Not. Amer. Math. Soc.* **43**, 537–49, 1996.

SOMA

Let $k \geq 0$ and $n \geq 2$ be integers. A SOMA, or more specifically a SOMA(k, n), is an $n \times n$ array A, whose entries are k-subsets of a kn-set Ω, such that each element of Ω occurs exactly once in each row and exactly once in each column of A, and no 2-subset of Ω is contained in more than one entry of A (Soicher 1999).

A SOMA(k, n) can be constructed by superposing k mutually orthogonal LATIN SQUARES of order n with pairwise disjoint symbol-sets, and so a SOMA(k, n) can be seen as a generalization of k mutually orthogonal LATIN SQUARES of order n.

See also LATIN SQUARE

References

Soicher, L. H. "On the Structure and Classification of SOMAs: Generalizations of Mutually Orthogonal Latin Squares." *Electronic J. Combinatorics* **6**, No. 1, R32, 1–5, 1999. http://www.combinatorics.org/Volume_6/v6i1toc.html.

Soma Cube

A solid DISSECTION puzzle invented by Piet Hein during a lecture on Quantum Mechanics by Werner Heisenberg. There are seven soma pieces composed of all the *irregular* face-joined cubes (POLYCUBES) with ≤ 4 cubes. The object is to assemble the pieces into a CUBE. There are 240 essentially distinct ways of doing so (Beeler 1972, Berlekamp *et al.* 1982), as first enumerated one rainy afternoon in 1961 by J. H. Conway and Mike Guy.

A commercial version of the cube colors the pieces black, green, orange, white, red, and blue. When the 48 symmetries of the cube, three ways of assembling the black piece, and 2^5 ways of assembling the green, orange, white, red, and blue pieces are counted, the total number of solutions rises to 1,105,920.

See also CUBE DISSECTION, POLYCUBE

References

Albers, D. J. and Alexanderson, G. L. (Eds.). *Mathematical People: Profiles and Interviews.* Boston, MA: Birkhäuser, p. 43, 1985.

Ball, W. W. R. and Coxeter, H. S. M. *Mathematical Recreations and Essays, 13th ed.* New York: Dover, pp. 112–13, 1987.

Beeler, M. Item 112 in Beeler, M.; Gosper, R. W.; and Schroeppel, R. *HAKMEM.* Cambridge, MA: MIT Artificial Intelligence Laboratory, Memo AIM-239, pp. 48–0, Feb. 1972.

Berlekamp, E. R.; Conway, J. H.; and Guy, R. K. Ch. 24 in *Winning Ways for Your Mathematical Plays, Vol. 2: Games in Particular.* London: Academic Press, 1982.

Cundy, H. and Rollett, A. *Mathematical Models, 3rd ed.* Stradbroke, England: Tarquin Pub., pp. 203–05, 1989.

Gardner, M. "Mathematical Games: A Game in Which Standard Pieces Composed of Cubes are Assembled into Larger Forms." *Sci. Amer.*, 185.

Gardner, M. "The Soma Cube." Ch. 6 in *The Second Scientific American Book of Mathematical Puzzles & Diversions: A New Selection.* New York: Simon and Schuster, pp. 65–7, 1961.

Steinhaus, H. *Mathematical Snapshots, 3rd ed.* New York: Dover, pp. 168–69, 1999.

Somer-Lucas Pseudoprime

An ODD COMPOSITE NUMBER N is called a Somer-Lucas d-pseudoprime (with $d \geq 1$) if there EXISTS a nondegenerate LUCAS SEQUENCE $U(P, Q)$ with $U_0 = 0$, $U_1 = 1$, $D = P^2 - 4Q$, such that $(N, D) = 1$ and the rank appearance of N in the sequence $U(P, Q)$ is $(1/a)(N - (D/N))$, where (D/N) denotes the JACOBI SYMBOL.

See also LUCAS SEQUENCE, PSEUDOPRIME

References

Ribenboim, P. "Somer-Lucas Pseudoprimes." §2.X.D in *The New Book of Prime Number Records, 3rd ed.* New York: Springer-Verlag, pp. 131–32, 1996.

Sommerfeld's Formula

There are (at least) two equations known as Sommerfeld's formula. The first is

$$J_\nu(z) = \frac{1}{2\pi} \int_{-\eta + i\infty}^{2\pi - \eta + i\infty} e^{iz \cos t} e^{i\nu(t - \pi/2)} \, dt,$$

where $J_\nu(z)$ is a BESSEL FUNCTION OF THE FIRST KIND. The second states that under appropriate restrictions,

$$\int_0^\infty J_0(\tau r) e^{-|x|\sqrt{\tau^2 - k^2}} \frac{\tau \, d\tau}{\sqrt{\tau^2 - k^2}} = \frac{e^{ik\sqrt{r^2 + k^2}}}{\sqrt{r^2 + x^2}}.$$

See also WEYRICH'S FORMULA

References

Iyanaga, S. and Kawada, Y. (Eds.). *Encyclopedic Dictionary of Mathematics.* Cambridge, MA: MIT Press, pp. 1472 and 1474, 1980.

Somos Sequence

The Somos sequences are a set of related symmetrical RECURRENCE RELATIONS which, surprisingly, always give integers. The Somos sequence of order k is defined by

$$a_n = \frac{\sum_{j=1}^{\lfloor k/2 \rfloor} a_{n-j} a_{n-(k-j)}}{a_{n-k}},$$

where $\lfloor x \rfloor$ is the FLOOR FUNCTION and $a_j = 1$ for $j = 0$, ..., $k - 1$. The 2- and 3-Somos sequences consist entirely of 1s. The k-Somos sequences for $k = 4$, 5, 6, and 7 are

$$a_n = \frac{a_{n-1} a_{n-3} + a_{n-2}^2}{a_{n-4}}$$

$$a_n = \frac{a_{n-1} a_{n-4} + a_{n-2} a_{n-3}}{a_{n-5}}$$

$$a_n = \frac{1}{a_{n-6}} \left[a_{n-1} a_{n-5} + a_{n-2} a_{n-4} + a_{n-3}^2 \right]$$

$$a_n = \frac{1}{a_{n-7}} \left[a_{n-1} a_{n-6} + a_{n-2} a_{n-5} + a_{n-3} a_{n-4} \right].$$

giving 1, 1, 1, 2, 3, 7, 23, 59, 314, 1529, ... (Sloane's A006720), 1, 1, 1, 1, 2, 3, 5, 11, 37, 83, 274, 1217, ... (Sloane's A006721), 1, 1, 1, 1, 1, 3, 5, 9, 23, 75, 421, 1103, ... (Sloane's A006722), 1, 1, 1, 1, 1, 1, 3, 5, 9, 17, 41, 137, 769, ... (Sloane's A006723). Gale (1991) gives simple proofs of the integer-only property of the 4-Somos and 5-Somos sequences. Hickerson proved 6-Somos generates only integers using computer algebra, and empirical evidence suggests 7-Somos is also integer-only.

However, the k-Somos sequences for $k \geq 8$ do not give integers. The values of n for which a_n first becomes nonintegral for the k-Somos sequence for $k = 8$, 9, ... are 17, 19, 20, 22, 24, 27, 28, 30, 33, 34, 36, 39, 41, 42, 44, 46, 48, 51, 52, 55, 56, 58, 60, ... (Sloane's A030127).

See also GÖBEL'S SEQUENCE, HERONIAN TRIANGLE

References

Buchholz, R. H. and Rathbun, R. L. "An Infinite Set of Heron Triangles with Two Rational Medians." *Amer. Math. Monthly* **104**, 107–15, 1997.

Gale, D. "Mathematical Entertainments: The Strange and Surprising Saga of the Somos Sequences." *Math. Intel.* **13**, 40–2, 1991.

Malouf, J. L. "An Integer Sequence from a Rational Recursion." *Disc. Math.* **110**, 257–61, 1992.

Robinson, R. M. "Periodicity of Somos Sequences." *Proc. Amer. Math. Soc.* **116**, 613–19, 1992.

Sloane, N. J. A. Sequences A006720/M0857, A006721/M0735, A006722/M2457, A006723/M2456, and A030127 in "An On-Line Version of the Encyclopedia of Integer Sequences." http://www.research.att.com/~njas/sequences/eisonline.html.

Sondat's Theorem

The PERSPECTIVE AXIS bisects the line joining the two ORTHOCENTERS.

See also ORTHOCENTER, PERSPECTIVE AXIS

References

Johnson, R. A. *Modern Geometry: An Elementary Treatise on the Geometry of the Triangle and the Circle.* Boston, MA: Houghton Mifflin, p. 259, 1929.

Sonine Polynomial

LAGUERRE POLYNOMIAL

Sonine's Integral

$$J_m(x) = \frac{2x^{m-n}}{2^{m-n}\Gamma(m-n)} \int_0^1 J_n(xt)t^{n+1}$$

$$\times \left(1 - t^2\right)^{m-n-1} dt,$$

where $J_m(x)$ is a BESSEL FUNCTION OF THE FIRST KIND and $\Gamma(x)$ is the GAMMA FUNCTION.

See also HANKEL'S INTEGRAL, POISSON INTEGRAL

Sonine-Schafheitlin Formula

$$\int_0^\infty J_\mu(at)J_\nu(bt)t^{-\lambda}\,dt$$

$$= \frac{a^\mu \Gamma[(\mu + \nu - \lambda + 1)/2]}{2^\lambda b^{\mu - \lambda + 1}\Gamma[(-\mu + \nu + \lambda + 1)/2]\Gamma(\mu + 1)}$$

$$\times {}_2F_1\big((\mu + \nu - \lambda + 1)/2, (\mu - \nu - \lambda + 1)/2;\ \mu + 1;\ a^2/b^2\big),$$

where $\Re[\mu + \nu - \lambda + 1] > 0$, $\Re[\lambda] > -1$, $0 < a < b$, $J_\nu(x)$ is a BESSEL FUNCTION OF THE FIRST KIND, $\Gamma(x)$ is the GAMMA FUNCTION, and ${}_2F_1(a, b; c; x)$ is a HYPERGEOMETRIC FUNCTION.

References

Iyanaga, S. and Kawada, Y. (Eds.). *Encyclopedic Dictionary of Mathematics.* Cambridge, MA: MIT Press, p. 1474, 1980.

Sophie Germain Prime

A PRIME p is said to be a Sophie Germain prime if both p and $2p+1$ are PRIME. The first few Sophie Germain primes are 2, 3, 5, 11, 23, 29, 41, 53, 83, 89, 113, 131, ... (Sloane's A005384).

Sophie Germain primes p OF THE FORM $p = k \cdot 2^n - 1$ (which makes $2p+1$ a PRIME) correspond to the indices of composite MERSENNE NUMBERS M_p. The largest known Sophie Germain prime is $92.305 \times 2^{16.998} + 1$, found in 1998 (Hoffman 1998, p. 190). It is not known if there are an infinite number of Sophie German primes (Hoffman 1998, p. 190).

Around 1825, Sophie Germain proved that the first case of FERMAT'S LAST THEOREM is true for such primes, i.e., if p is a Sophie Germain prime, there do not exist INTEGERS x, y, and z different from 0 and not multiples of p such that

$$x^p + y^p = z^p.$$

See also CUNNINGHAM CHAIN, FERMAT'S LAST THEOREM, MERSENNE NUMBER, TWIN PRIMES

References

Caldwell, C. K. "The Top Twenty: Sophie Germain Primes." http://www.utm.edu/research/primes/lists/top20/Sophie-Germain.html.
Dubner, H. "Large Sophie Germain Primes." *Math. Comput.* **65**, 393–96, 1996.
Hoffman, P. *The Man Who Loved Only Numbers: The Story of Paul Erdos and the Search for Mathematical Truth.* New York: Hyperion, p. 190, 1998.
Indlekofer, K. H. and Járai, A. "Largest Known Twin Primes and Sophie Germain Primes." *Math. Comput.* **68**, 1317–324, 1999.
Ribenboim, P. "Sophie Germane Primes." §5.2 in *The New Book of Prime Number Records.* New York: Springer-Verlag, pp. 329–32, 1996.
Shanks, D. *Solved and Unsolved Problems in Number Theory, 4th ed.* New York: Chelsea, pp. 154–57, 1993.
Sloane, N. J. A. Sequences A005384/M0731 in "An On-Line Version of the Encyclopedia of Integer Sequences." http://www.research.att.com/~njas/sequences/eisonline.html.

Sorites Paradox

Sorites paradoxes are a class of paradoxical arguments also known as little-by-little arguments. The name "sorites" derives from the Greek word *soros*, meaning "pile" or "heap." Sorites paradoxes are exemplified by the problem that a single grain of wheat does not comprise a heap, nor do two grains of wheat, three grains of wheat, etc. However, at some point, the collection of grains becomes large enough to be called a heap, but there is apparently no definite point where this occurs.

See also UNEXPECTED HANGING PARADOX

References

Erickson, G. W. and Fossa, J. A. *Dictionary of Paradox.* Lanham, MD: University Press of America, pp. 196–99, 1998.

Sorting

Sorting is the rearrangement of numbers (or other orderable objects) in a list into their correct lexographic order. Alphabetization is therefore a form of sorting. Because of the extreme importance of sorting in almost all database applications, a great deal of effort has been expended in the creation and analysis of efficient sorting algorithms.

The minimum number of comparisons $a(n)$ needed for a merge sort of n elements for $n = 1, 2, ...$ are 0, 1, 3, 5, 7, 10, 13, 16, 19, 22, 26, 30, ... (Sloane's A001768). An upper limit $b(n)$ is given by the sequence

$$a(n) \leq b(n) = 1 + kn - 2^k$$

where

$$k = \lfloor \log_2 n \rfloor + 1,$$

where $\lfloor x \rfloor$ is the FLOOR FUNCTION (Steinhaus 1983, pp. 55–6), or equivalently,

$$b(n) = \sum_{k=1}^{n} \lceil \log_2 k \rceil,$$

giving 0, 1, 3, 5, 8, 11, 14, 17, 21, 25, 29, ... (Sloane's A001855).

See also HEAPSORT, ORDERING, QUICKSORT, SELECTION SORT, WEIGHING

References

Knuth, D. E. *The Art of Computer Programming, Vol. 3: Sorting and Searching, 2nd ed.* Reading, MA: Addison-Wesley, 1973.

Press, W. H.; Flannery, B. P.; Teukolsky, S. A.; and Vetterling, W. T. "Sorting." Ch. 8 in *Numerical Recipes in FORTRAN: The Art of Scientific Computing, 2nd ed.* Cambridge, England: Cambridge University Press, pp. 320–39, 1992.

Skiena, S. "Sorting and Searching." §1.1.6 in *Implementing Discrete Mathematics: Combinatorics and Graph Theory with Mathematica.* Reading, MA: Addison-Wesley, pp. 14–6, 1990.

Sloane, N. J. A. Sequences A001768/M2408 and A001855/M2433 in "An On-Line Version of the Encyclopedia of Integer Sequences." http://www.research.att.com/~njas/sequences/eisonline.html.

Sort-Then-Add Sequence

A sequence produced by sorting the digits of a number and adding them to the previous number. The algorithm terminates when a sorted number is obtained. For $n = 1, 2, \ldots$, the algorithm terminates on 1, 2, 3, 4, 5, 6, 7, 8, 9, 11, 11, 12, 13, 14, 15, 16, 17, 18, 19, 22, 33, ... (Sloane's A033862). The first few numbers not known to terminate are 316, 452, 697, 1376, 2743, 5090, ... (Sloane's A033861). The least numbers of sort-then-add persistence $n = 1, 2, \ldots$, are 1, 10, 65, 64, 175, 98, 240, 325, 302, 387, 198, 180, 550, ... (Sloane's A033863).

See also 196-ALGORITHM, RATS SEQUENCE

References

Sloane, N. J. A. Sequences A033861, A033862, and A033863 in "An On-Line Version of the Encyclopedia of Integer Sequences." http://www.research.att.com/~njas/sequences/eisonline.html.

Source

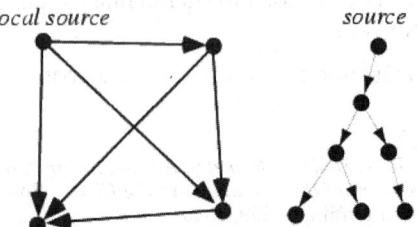

local source   *source*

A local source is a node of a DIRECTED GRAPH with no entering edges (Borowski and Borwein 1991, p. 401; left figure), and a global source (often simply called a source) is a node in a DIRECTED GRAPH which reaches all other nodes (Harary 1994, p. 201; right figure).

See also DIRECTED GRAPH, NETWORK, SINK (DIRECTED GRAPH)

References

Borowski, E. J. and Borwein, J. M. (Eds.). *The HarperCollins Dictionary of Mathematics.* New York: HarperCollins, 1991.

Harary, F. *Graph Theory.* Reading, MA: Addison-Wesley, 1994.

Sous-Double

A 3-MULTIPERFECT NUMBER P_3. Six sous-doubles are known (120, 672, 523776, 459818240, 1476304896, and 51001180160; Sloane's A005820), and these are believed to comprise all sous-doubles.

See also MULTIPERFECT NUMBER, SOUS-TRIPLE

References

Sloane, N. J. A. Sequences A005820/M5376 in "An On-Line Version of the Encyclopedia of Integer Sequences." http://www.research.att.com/~njas/sequences/eisonline.html.

Souslin Set

The continuous image of a POLISH SPACE, also called an ANALYTIC SET.

See also ANALYTIC SET, POLISH SPACE

Souslin's Hypothesis

Every dense linear order complete set without endpoints having at most ω disjoint intervals is order isomorphic to the CONTINUUM of REAL NUMBERS, where ω is the set of NATURAL NUMBERS.

References

Iyanaga, S. and Kawada, Y. (Eds.). "Souslin's Hypothesis." §35E.4 in *Encyclopedic Dictionary of Mathematics.* Cambridge, MA: MIT Press, p. 137, 1980.

Sous-Triple

A 4-MULTIPERFECT NUMBER P_4. 36 sous-triples are known (30240, 32760, 2178540, 23569920, ...; Sloane's

A027687), and these are believed to comprise all sous-triples.

See also MULTIPERFECT NUMBER, SOUS-DOUBLE

References

Sloane, N. J. A. Sequences A027687 in "An On-Line Version of the Encyclopedia of Integer Sequences." http://www.research.att.com/~njas/sequences/eisonline.html.

Space

The concept of a space is an extremely general and important mathematical construct. Members of the space obey certain addition properties. Spaces which have been investigated and found to be of interest are usually named after one or more of their investigators. This practice unfortunately leads to names which give very little insight into the relevant properties of a given space.

The everyday type of space familiar to most people is called EUCLIDEAN SPACE. In Einstein's theory of Special Relativity, Euclidean 3-space plus time (the "fourth dimension") are unified into the so-called MINKOWSKI SPACE. One of the most general type of mathematical spaces is the TOPOLOGICAL SPACE.

See also AFFINE SPACE, BAIRE SPACE, BANACH SPACE, BASE SPACE, BERGMAN SPACE, BESOV SPACE, BOREL SPACE, CALABI-YAU SPACE, CELLULAR SPACE, CHU SPACE, DODECAHEDRAL SPACE, DRINFELD'S SYMMETRIC SPACE, EILENBERG-MAC LANE SPACE, EUCLIDEAN SPACE, FIBER SPACE, FINSLER SPACE, FIRST-COUNTABLE SPACE, FRÉCHET SPACE, FUNCTION SPACE, *G*-SPACE, GREEN SPACE, HAUSDORFF SPACE, HEISENBERG SPACE, HILBERT SPACE, HYPERBOLIC SPACE, INNER PRODUCT SPACE, L2-SPACE, LENS SPACE, LINE SPACE, LINEAR SPACE, LIOUVILLE SPACE, LOCALLY CONVEX SPACE, LOCALLY FINITE SPACE, LOOP SPACE, MAPPING SPACE, MEASURE SPACE, METRIC SPACE, MINKOWSKI SPACE, MÜNTZ SPACE, NON-EUCLIDEAN GEOMETRY, NORMED SPACE, PARACOMPACT SPACE, PLANAR SPACE, POLISH SPACE, PROBABILITY SPACE, PROJECTIVE SPACE, QUOTIENT SPACE, RIEMANN'S MODULI SPACE, RIEMANN SPACE, SAMPLE SPACE, STANDARD SPACE, STATE SPACE, STONE SPACE, SYMPLECTIC SPACE, TEICHMÜLLER SPACE, TENSOR SPACE, TOPOLOGICAL SPACE, TOPOLOGICAL VECTOR SPACE, TOTAL SPACE, VECTOR SPACE

Space Conic

SKEW CONIC

Space Curve

A curve which may pass through any region of 3-D space, as contrasted to a PLANE CURVE which must lie in a single PLANE. Von Staudt (1847) classified space curves geometrically by considering the curve

$$\phi : I \to \mathbb{R}^3 \qquad (1)$$

at $t_0 = 0$ and assuming that the parametric functions $\phi_i(t)$ for $i = 1, 2, 3$ are given by POWER SERIES which converge for small t. If the curve is contained in no PLANE for small t, then a coordinate transformation puts the PARAMETRIC EQUATIONS in the normal form

$$\phi_1(t) = t^{1+k_1} + \cdots \qquad (2)$$

$$\phi_2(t) = t^{2+k_1+k_2} + \cdots \qquad (3)$$

$$\phi_3(t) = t^{3+k_1+k_2+k_3} + \cdots \qquad (4)$$

for integers k_1, k_2, $k_3 \geq 0$, called the local numerical invariants.

See also CURVE, CYCLIDE, FUNDAMENTAL THEOREM OF SPACE CURVES, HELIX, PLANE CURVE, SEIFFERT'S SPHERICAL SPIRAL, SKEW CONIC, SPACE-FILLING FUNCTION, SPHERICAL CURVE, SPHERICAL SPIRAL, SURFACE, VIVIANI'S CURVE

References

do Carmo, M.; Fischer, G.; Pinkall, U.; and Reckziegel, H. "Singularities of Space Curves." §3.1 in *Mathematical Models from the Collections of Universities and Museums* (Ed. G. Fischer). Braunschweig, Germany: Vieweg, pp. 24–5, 1986.

Fine, H. B. "On the Singularities of Curves of Double Curvature." *Amer. J. Math.* **8**, 156–77, 1886.

Fischer, G. (Ed.). Plates 57–4 in *Mathematische Modelle/Mathematical Models, Bildband/Photograph Volume.* Braunschweig, Germany: Vieweg, pp. 58–9, 1986.

Gray, A. "Curves in $\mathbb{R}^n$" and "Curves in Space." §1.2 and Ch. 8 in *Modern Differential Geometry of Curves and Surfaces with Mathematica, 2nd ed.* Boca Raton, FL: CRC Press, pp. 5– and 181–06, 1997.

Griffiths, P. and Harris, J. *Principles of Algebraic Geometry.* New York: Wiley, 1978.

Saurel, P. "On the Singularities of Tortuous Curves." *Ann. Math.* **7**, 3–, 1905.

Staudt, C. von. *Geometrie der Lage.* Nürnberg, Germany, 1847.

Wiener, C. "Die Abhängigkeit der Rückkehrelemente der Projektion einer unebenen Curve von deren der Curve selbst." *Z. Math. & Phys.* **25**, 95–7, 1880.

Space Diagonal

The LINE SEGMENT connecting opposite VERTICES (i.e., two VERTICES which do not share a common face) in a PARALLELEPIPED or other similar solid.

See also DIAGONAL (POLYGON), DIAGONAL (POLYHEDRON), EULER BRICK

Space Distances

POINT DISTANCES

Space Division by Planes

The maximal number of regions into which space can be divided by n planes is

$$f(n) = \tfrac{1}{6}(n^3 + 5n + 6)$$

(Yaglom and Yaglom 1987, pp. 102–06), giving the

values 2, 4, 8, 15, 26, 42, ... (Sloane's A000125) for $n = 1$, 2, ... planes. This is the same solution as for CYLINDER CUTTING.

See also CIRCLE DIVISION BY LINES, CUBE DIVISION BY PLANES, CYLINDER CUTTING, PLANE DIVISION BY CIRCLES, SPACE DIVISION BY SPHERES

References

Sloane, N. J. A. Sequences A000125/M1100 in "An On-Line Version of the Encyclopedia of Integer Sequences." http://www.research.att.com/~njas/sequences/eisonline.html.
Wells, D. *The Penguin Dictionary of Curious and Interesting Numbers.* Middlesex, England: Penguin Books, p. 72, 1986.
Yaglom, A. M. and Yaglom, I. M. *Challenging Mathematical Problems with Elementary Solutions, Vol. 1.* New York: Dover, pp. 102–06, 1987.

Space Division by Spheres

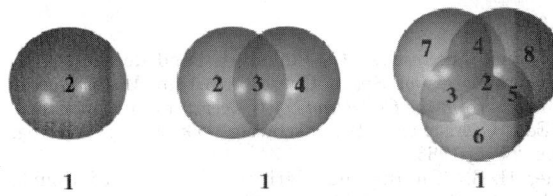

The number of regions into which space can be divided by n mutually intersecting SPHERES is

$$N = \tfrac{1}{3} n (n^2 - 3n + 8),$$

giving 2, 4, 8, 16, 30, 52, 84, ... (Sloane's A046127) for $n = 1$, 2, ...

See also PLANE DIVISION BY CIRCLES, SPACE DIVISION BY PLANES, SPHERE-SPHERE INTERSECTION

References

Sloane, N. J. A. Sequences A046127 in "An On-Line Version of the Encyclopedia of Integer Sequences." http://www.research.att.com/~njas/sequences/eisonline.html.
Yaglom, A. M. and Yaglom, I. M. *Challenging Mathematical Problems with Elementary Solutions, Vol. 1.* New York: Dover, pp. 102–06, 1987.

Space Groups

The space groups in 2-D are called WALLPAPER GROUPS. In 3-D, the space groups are the symmetry GROUPS possible in a crystal lattice with the translation symmetry element. There are 230 space groups in $\mathbb{R}^3$, although 11 are MIRROR IMAGES of each other. They are listed by HERMANN-MAUGUIN SYMBOL in Cotton (1990).

See also HERMANN-MAUGUIN SYMBOL, LATTICE GROUPS, POINT GROUPS, WALLPAPER GROUPS

References

Arfken, G. "Crystallographic Point and Space Groups." *Mathematical Methods for Physicists, 3rd ed.* Orlando, FL: Academic Press, pp. 248–49, 1985.

Buerger, M. J. *Elementary Crystallography.* New York: Wiley, 1956.
Cotton, F. A. *Chemical Applications of Group Theory, 3rd ed.* New York: Wiley, pp. 250–51, 1990.

Space of Closed Paths
LOOP SPACE

Space-Filling Curve
SPACE-FILLING FUNCTION

Space-Filling Function

A "CURVE" (i.e., a continuous map of a 1-D INTERVAL) into a 2-D area (a PLANE-FILLING FUNCTION) or a 3-D volume.

See also HILBERT CURVE, PEANO CURVE, PEANO-GOSPER CURVE, PLANE-FILLING CURVE, SIERPINSKI CURVE, SPACE-FILLING POLYHEDRON

References

Pappas, T. "Paradoxical Curve-Space-Filling Curve." *The Joy of Mathematics.* San Carlos, CA: Wide World Publ./ Tetra, p. 208, 1989.
Platzman, L. K. and Bartholdi, J. J. "Spacefilling Curves and the Planar Travelling Salesman Problem." *J. Assoc. Comput. Mach.* **46**, 719–37, 1989.
Wagon, S. "A Spacefilling Curve." §6.3 in *Mathematica in Action.* New York: W. H. Freeman, pp. 196–09, 1991.

Space-Filling Polyhedron

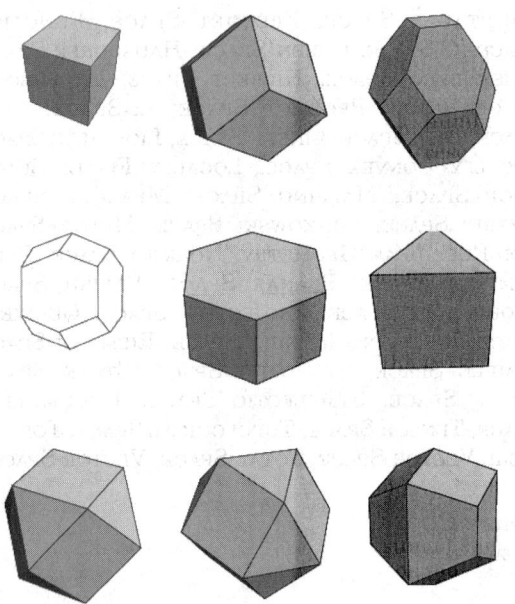

A space-filling polyhedron is a POLYHEDRON which can be used to generate a TESSELLATION of space. Although even Aristotle himself proclaimed in his work *On the Heavens* that the TETRAHEDRON fills

space, it in fact does not (Hilbert and Cohn-Vossen 1999, p. 45). The CUBE is the only PLATONIC SOLID possessing this property (Gardner 1984, pp. 183–84). However, a combination of TETRAHEDRA and OCTAHEDRA do fill space (Steinhaus 1983, p. 210; Wells 1991, p. 232). In addition, octahedra, truncated octahedron, and cubes, combined in the ratio 1:1:3, can also fill space (Wells 1991, p. 235).

Of the Archimedean solids, the RHOMBIC DODECAHEDRON and TRUNCATED OCTAHEDRON are space-fillers (Steinhaus 1983, pp. 185–90; Wells 1991, pp. 233–34). The ELONGATED DODECAHEDRON and hexagonal PRISM are also space-fillers. These five solids are all "primary" PARALLELOHEDRA (Coxeter 1973). In 1914, Föppl discovered a space-filling compound of tetrahedra and truncated tetrahedra (Wells 1991, p. 234).

The CUBOCTAHEDRON, TRIANGULAR ORTHOBICUPOLA, and squashed dodecahedron appearing in SPHERE PACKING also fill space (Steinhaus 1983, pp. 203–07), as does an arbitrary TRIANGULAR PRISM or any non-self-intersecting quadrilateral PRISM.

There exists a tetrahedron with bevelled edges which fills space (Wells 1991, p. 234). There exists one 16-sided space-filling POLYHEDRON, but it is unknown if it is the unique 16-sided space-filler. There exists an 18-faced space-filler, as well space-fillers of up to 38 faces, as discovered by P. Engel in 1980 (Wells 1991, pp. 234–35). P. Schmitt discovered a nonconvex aperiodic polyhedral space-filler around 1990, and a convex POLYHEDRON known as the SCHMITT-CONWAY BIPRISM which fills space only aperiodically was found by J. H. Conway in 1993 (Eppstein).

See also CUBE, CUBOCTAHEDRON, ELONGATED DODECAHEDRON, KELLER'S CONJECTURE, KELVIN'S CONJECTURE, OCTAHEDRON, PARALLELOHEDRON, PRISM, RHOMBIC DODECAHEDRON, SCHMITT-CONWAY BIPRISM, SPHERE PACKING, TESSELLATION, TETRAHEDRON, TILING, TRIANGULAR ORTHOBICUPOLA, TRUNCATED OCTAHEDRON

References

Coxeter, H. S. M. *Regular Polytopes, 3rd ed.* New York: Dover, pp. 29–0, 1973.
Critchlow, K. *Order in Space: A Design Source Book.* New York: Viking Press, 1970.
Devlin, K. J. "An Aperiodic Convex Space-Filler is Discovered." *Focus: The Newsletter of the Math. Assoc. Amer.* **13**, 1, Dec. 1993.
Eppstein, D. "Re: Aperiodic Space-Filling Tile?." http://www.ics.uci.edu/~eppstein/junkyard/biprism.html.
Hilbert, D. and Cohn-Vossen, S. *Geometry and the Imagination.* New York: Chelsea, 1999.
Holden, A. *Shapes, Space, and Symmetry.* New York: Dover, pp. 154–63, 1991.
Kramer, P. "Non-Periodic Central Space Filling with Icosahedral Symmetry Using Copies of Seven Elementary Cells." *Acta Cryst. A* **38**, 257–64, 1982.
Pearce, P. *Structure and Nature as a Strategy for Design.* Cambridge, MA: MIT Press, 1978.
Steinhaus, H. *Mathematical Snapshots, 3rd ed.* New York: Dover, pp. 185–90, 1999.
Stott, A. B. "Geometrical Deduction of Semiregular from Regular Polytopes and Space Fillings." *Verhandelingen der Koninklijke Akad. Wetenschappen Amsterdam* **11**, 3–4, 1910.
Thompson, D'A. W. *On Growth and Form, 2nd ed., compl. rev. ed.* New York: Cambridge University Press, 1992.
Tutton, A. E. H. *Crystallography and Practical Crystal Measurement, 2nd ed.* London: Lubrecht & Cramer, pp. 567 and 723, 1964.
Wells, D. *The Penguin Dictionary of Curious and Interesting Geometry.* London: Penguin, pp. 232–36, 1991.
Williams, R. *The Geometrical Foundation of Natural Structure: A Source Book of Design.* New York: Dover, 1979.

Span (Geometry)

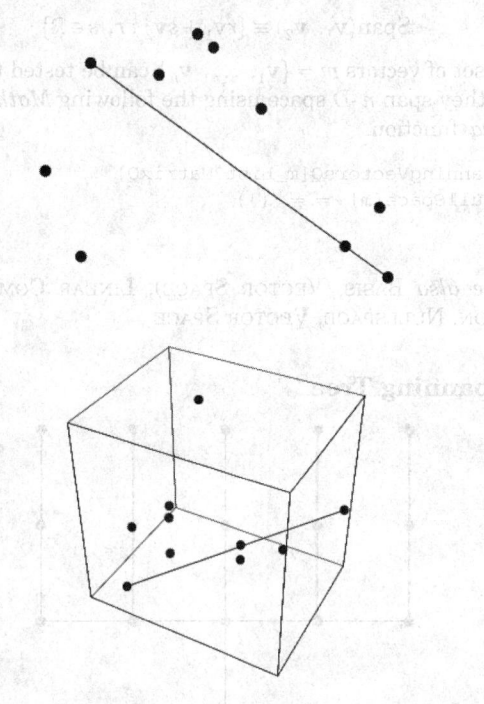

The largest possible distance between two points drawn from a finite set of points.

See also COMPUTATIONAL GEOMETRY, CONVEX HULL, JUNG'S THEOREM, POINT DISTANCES

Span (Link)

The span of an unoriented LINK diagram (also called the SPREAD) is the difference between the highest and lowest degrees of its BRACKET POLYNOMIAL. The span is a topological invariant of a knot. If a KNOT K has a reduced alternating projection of n crossings, then the span of K is $4n$.

See also LINK

Span (Polynomial)

The difference between the highest and lowest degrees of a POLYNOMIAL.

Span (Set)

For a SET S, the span is defined by $\max S - \min S$, where max is the MAXIMUM and min is the MINIMUM.

References

Guy, R. K. *Unsolved Problems in Number Theory, 2nd ed.* New York: Springer-Verlag, p. 207, 1994.

Span (Vector Space)

The span of SUBSPACE generated by VECTORS $\mathbf{v}_1$ and $\mathbf{v}_2 \in \mathbb{V}$ is

$$\text{Span}(\mathbf{v}_1, \mathbf{v}_2) \equiv \{r\mathbf{v}_1 + s\mathbf{v}_2 : r,\ s \in \mathbb{R}\}$$

A set of vectors $m = \{\mathbf{v}_1, \ldots, \mathbf{v}_n\}$ can be tested to see if they span n-D space using the following *Mathematica* function.

```
SpanningVectorsQ[m_List?MatrixQ]        :=
(NullSpace[m]  = =  {})
```

See also BASIS (VECTOR SPACE), LINEAR COMBINATION, NULLSPACE, VECTOR SPACE

Spanning Tree

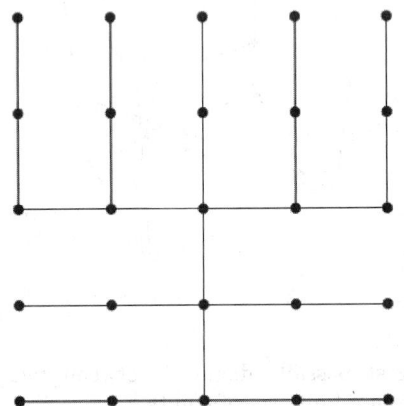

A spanning tree of a GRAPH is a subset of $n-1$ edges which form a TREE. The shortest-path spanning tree is the tree have the smallest possible total distance, where the distance used is MANHATTAN DISTANCE (Skiena 1990, p. 227).

The number of nonidentical spanning trees of a GRAPH G is equal to any COFACTOR of the DEGREE MATRIX of G minus the ADJACENCY MATRIX of G (Skiena 1990, p. 235). This result is known as the MATRIX TREE THEOREM. A TREE contains a unique spanning tree, a CYCLE GRAPH C_n containing n spanning trees, and a COMPLETE GRAPH K_n contains n^{n-2} spanning trees (Skiena 1990, p. 236). A count of the spanning trees of a graph can be found using the command `NumberOfSpanningTrees[g]` in the *Mathematica* add-on package `DiscreteMath`Combinatorica`` (which can be loaded with the command `< <DiscreteMath``).

See also MATRIX TREE THEOREM, MINIMUM SPANNING TREE, TREE

References

Colbourn, C. J.; Day, R. P. J.; and Nel, L. D. "Unranking and Ranking Spanning Trees of a Graph." *J. Algorithms* **10**, 271–86, 1989.
Eppstein, D. "Spanning Trees and Spanners." Ch. 9 in *Handbook of Computational Geometry* (Ed. J.-R. Sack and J. Urrutia). Amsterdam, Netherlands: North-Holland, pp. 425–61, 2000.
Skiena, S. *Implementing Discrete Mathematics: Combinatorics and Graph Theory with Mathematica.* Reading, MA: Addison-Wesley, pp. 224–27, 1990.

Sparse Matrix

A MATRIX which has only a small number of NONZERO elements.

References

Press, W. H.; Flannery, B. P.; Teukolsky, S. A.; and Vetterling, W. T. "Sparse Linear Systems." §2.7 in *Numerical Recipes in FORTRAN: The Art of Scientific Computing, 2nd ed.* Cambridge, England: Cambridge University Press, pp. 63–2, 1992.

Spearman Rank Correlation Coefficient

A nonparametric (distribution-free) rank statistic proposed by Spearman in 1904 as a measure of the strength of the associations between two variables (Lehmann and D'Abrera 1998). The Spearman rank correlation coefficient can be used to give an R-ESTIMATE.

The Spearman rank correlation coefficient is defined by

$$r' \equiv 1 - 6 \sum \frac{d^2}{N(N^2 - 1)}, \qquad (1)$$

where d is the difference in RANK of corresponding variables, and is an approximation to the exact CORRELATION COEFFICIENT

$$r \equiv \frac{\sum xy}{\sqrt{\sum x^2 \sum y^2}} \qquad (2)$$

computed from the original data. Because it uses ranks, the Spearman rank correlation coefficient is much easier to compute.

The VARIANCE, KURTOSIS, and higher order MOMENTS are

$$\sigma^2 = \frac{1}{N-1} \qquad (3)$$

$$\gamma_2 = -\frac{114}{25N} - \frac{6}{5N^2} - \cdots \qquad (4)$$

$$\gamma_3 = \gamma_5 = \ldots = 0. \qquad (5)$$

Student was the first to obtain the VARIANCE.

See also CORRELATION COEFFICIENT, LEAST SQUARES FITTING, LINEAR REGRESSION, RANK (STATISTICS)

References

Hogg, R. V. and Craig, A. T. *Introduction to Mathematical Statistics, 5th ed.* New York: Macmillan, pp. 338 and 400, 1995.

Lehmann, E. L. and D'Abrera, H. J. M. *Nonparametrics: Statistical Methods Based on Ranks, rev. ed.* Englewood Cliffs, NJ: Prentice-Hall, pp. 292, 300, and 323, 1998.

Press, W. H.; Flannery, B. P.; Teukolsky, S. A.; and Vetterling, W. T. *Numerical Recipes in FORTRAN: The Art of Scientific Computing, 2nd ed.* Cambridge, England: Cambridge University Press, pp. 634–37, 1992.

Special Curve

PLANE CURVE, SPACE CURVE, SPHERICAL CURVE

Special Function

A function (usually named after an early investigator of its properties) having a particular use in mathematical physics or some other branch of mathematics. Prominent examples include the GAMMA FUNCTION, HYPERGEOMETRIC FUNCTION, WHITTAKER FUNCTION, and MEIJER'S G-FUNCTION.

See also ELEMENTARY FUNCTION, FIRST KIND, FUNCTION, SECOND KIND, THIRD KIND

References

Abramowitz, M. and Stegun, C. A. (Eds.). *Handbook of Mathematical Functions with Formulas, Graphs, and Mathematical Tables, 9th printing.* New York: Dover, 1972.

Andrews, G. E.; Askey, R.; and Roy, R. *Special Functions.* Cambridge, England: Cambridge University Press, 1999.

Arscott, F. M. "The Land Beyond Bessel: A Survey of Higher Special Functions." In *Ordinary and Partial Differential Equations* (Ed. W. N. Everitt and B. D. Sleeman). New York: Springer-Verlag, pp. 26–5, 1981.

Luke, Y. L. *The Special Functions and their Approximations, Vol. 1.* New York: Academic Press, 1969.

Luke, Y. L. *The Special Functions and their Approximations, Vol. 2.* New York: Academic Press, 1969.

Magnus, W. and Oberhettinger, F. *Formulas and Theorems for the Special Functions of Mathematical Physics, 3rd ed.* New York: Springer-Verlag, 1966.

Nikiforov, A. F. and Uvarov, V. B. *Special Functions of Mathematical Physics: A Unified Introduction with Applications.* Boston, MA: Birkhäuser, 1988.

National Institute of Standards. "Digital Library of Mathematical Functions." http://dlmf.nist.gov/.

Prudnikov, A. P.; Brychkov, Yu. A.; and Marichev, O. I. *Integrals and Series, Vol. 1: Elementary Functions.* New York: Gordon and Breach, 1986.

Prudnikov, A. P.; Brychkov, Yu. A.; and Marichev, O. I. *Integrals and Series, Vol. 2: Special Functions.* New York: Gordon and Breach, 1990.

Prudnikov, A. P.; Brychkov, Yu. A.; and Marichev, O. I. *Integrals and Series, Vol. 3: More Special Functions.* New York: Gordon and Breach, 1989.

Prudnikov, A. P.; Brychkov, Yu. A.; and Marichev, O. I. *Integrals and Series, Vol. 4: Direct Laplace Transforms.* New York: Gordon and Breach, 1992.

Prudnikov, A. P.; Brychkov, Yu. A.; and Marichev, O. I. *Integrals and Series, Vol. 5: Inverse Laplace Transforms.* New York: Gordon and Breach, 1992.

Spanier, J. and Oldham, K. B. *An Atlas of Functions.* Washington, DC: Hemisphere, 1987.

Weisstein, E. W. "Books about Special Functions." http://www.treasure-troves.com/books/SpecialFunctions.html.

Wolfram Research, Inc. "Wolfram Research's Special Functions." http://functions.wolfram.com/.

Special Jordan Algebra

A JORDAN ALGEBRA which is isomorphic to a subalgebra.

See also EXCEPTIONAL JORDAN ALGEBRA, JORDAN ALGEBRA

References

Schafer, R. D. *An Introduction to Nonassociative Algebras.* New York: Dover, p. 4, 1996.

Special Lie Algebra

See also LIE ALGEBRA, SPECIAL LINEAR LIE ALGEBRA

Special Linear Group

The special linear group $SL_n(q)$ is the MATRIX GROUP corresponding to the set of $n \times n$ COMPLEX MATRICES having DETERMINANT $+1$. It is a SUBGROUP of the GENERAL LINEAR GROUP $GL_n(q)$ and is also a LIE GROUP.

See also GENERAL LINEAR GROUP, SPECIAL ORTHOGONAL GROUP, SPECIAL UNITARY GROUP

References

Conway, J. H.; Curtis, R. T.; Norton, S. P.; Parker, R. A.; and Wilson, R. A. "The Groups $GL_n(q)$, $SL_n(q)$, $PGL_n(q)$, and $PSL_n(q) = L_n(q)$." §2.1 in *Atlas of Finite Groups: Maximal Subgroups and Ordinary Characters for Simple Groups.* Oxford, England: Clarendon Press, p. x, 1985.

Special Linear Lie Algebra

Denoted $\mathfrak{sl}_n$.

See also LIE ALGEBRA, SPECIAL LIE ALGEBRA

Special Matrix

An INTEGER MATRIX whose entries satisfy

$$a_{ij} = \begin{cases} 0 & \text{if } j > i+1 \\ -1 & \text{if } j = i+1 \\ 0 \text{ or } 1 & \text{if } j \le 1. \end{cases}$$

There are 2^{n-1} special MINIMAL MATRICES of size $n \times n$.

References
Knuth, D. E. "Problem 10470." *Amer. Math. Monthly* **102**, 655, 1995.

Special Orthogonal Group

The special orthogonal group $SO_n(q)$ is the SUBGROUP of the elements of GENERAL ORTHOGONAL GROUP $GO_n(q)$ with DETERMINANT 1. SO_3 (often written $SO(3)$ is the ROTATION GROUP for 3-dimensional space.

See also BIPOLYHEDRAL GROUP, GENERAL ORTHOGONAL GROUP, ICOSAHEDRAL GROUP, ROTATION GROUP, SPECIAL LINEAR GROUP, SPECIAL UNITARY GROUP

References
Conway, J. H.; Curtis, R. T.; Norton, S. P.; Parker, R. A.; and Wilson, R. A. "The Groups $GO_n(q)$, $SO_n(q)$, $PGSO_n(q)$, and $PSO_n(q)$." §2.4 in *Atlas of Finite Groups: Maximal Subgroups and Ordinary Characters for Simple Groups.* Oxford, England: Clarendon Press, pp. xi-xii, 1985.

Special Orthogonal Matrix

A SQUARE MATRIX A is a special orthogonal matrix if

$$ \mathsf{A}\mathsf{A}^\mathsf{T} = \mathsf{I}. \tag{1} $$

where I is the IDENTITY MATRIX, and the DETERMINANT satisfies

$$ \det \mathsf{A} = 1. \tag{2} $$

The first condition means that A is an ORTHOGONAL MATRIX, and the second restricts the determinant to $+1$ (while a *general* ORTHOGONAL MATRIX may have determinant -1 or $+1$). For example,

$$ \frac{1}{\sqrt{2}} \begin{bmatrix} 1 & -1 \\ 1 & 1 \end{bmatrix} \tag{3} $$

is a special orthogonal matrix since

$$ \begin{bmatrix} \frac{1}{\sqrt{2}} & -\frac{1}{\sqrt{2}} \\ \frac{1}{\sqrt{2}} & \frac{1}{\sqrt{2}} \end{bmatrix} \begin{bmatrix} \frac{1}{\sqrt{2}} & \frac{1}{\sqrt{2}} \\ -\frac{1}{\sqrt{2}} & \frac{1}{\sqrt{2}} \end{bmatrix} = \begin{bmatrix} 1 & 0 \\ 0 & 1 \end{bmatrix} \tag{4} $$

and its DETERMINANT is $1/2 - (-1/2) = 1$. A matrix m can be tested to see if it is a special orthogonal matrix using the *Mathematica* function

```
SpecialOrthogonalQ[m_List?MatrixQ] :=
    (Transpose[m].m == IdentityMatrix@Length@m
    && Det[m] == -1)
```

The special orthogonal matrices are CLOSED under multiplication and the inverse operation, and therefore form a MATRIX GROUP called the SPECIAL ORTHOGONAL GROUP $SO(n)$.

See also INNER PRODUCT, ORTHOGONAL GROUP, ORTHOGONAL MATRIX, ORTHOGONAL TRANSFORMATION, SKEW SYMMETRIC MATRIX, SPECIAL LINEAR MATRIX, SPECIAL ORTHOGONAL GROUP, SPIN GROUP, UNITARY MATRIX

Special Point

A POINT which does not lie on at least one ORDINARY LINE.

See also ORDINARY POINT

References
Guy, R. K. "Unsolved Problems Come of Age." *Amer. Math. Monthly* **96**, 903–09, 1989.

Special Series Theorem

If the difference between the order and the dimension of a series is less than the GENUS (CURVE), then the series is special.

References
Coolidge, J. L. *A Treatise on Algebraic Plane Curves.* New York: Dover, p. 253, 1959.

Special Unitary Group

The special unitary group $SU_n(q)$ is the set of $n \times n$ UNITARY MATRICES with DETERMINANT $+1$ (having $n^2 - 1$ independent parameters). $SU(2)$ is HOMEOMORPHIC with the ORTHOGONAL GROUP $O_3^+(2)$. It is also called the UNITARY UNIMODULAR GROUP and is a LIE GROUP.

Special unitary groups can be represented by matrices

$$ U(a, b) = \begin{bmatrix} a & b \\ -\bar{b} & \bar{a} \end{bmatrix}. \tag{1} $$

where $\bar{a}a + \bar{b}b = 1$ and a, b are the CAYLEY-KLEIN PARAMETERS. The special unitary group may also be represented by matrices

$$ U(\xi, \eta, \zeta) = \begin{bmatrix} e^{i\xi} \cos \eta & e^{i\zeta} \sin \eta \\ -e^{-i\zeta} \sin \eta & e^{-i\xi} \cos \eta \end{bmatrix}. \tag{2} $$

or the matrices

$$ U_x\left(\tfrac{1}{2}\phi\right) = \begin{bmatrix} \cos\left(\tfrac{1}{2}\phi\right) & i\sin\left(\tfrac{1}{2}\phi\right) \\ i\sin\left(\tfrac{1}{2}\phi\right) & \cos\left(\tfrac{1}{2}\phi\right) \end{bmatrix} \tag{3} $$

$$ U_y\left(\tfrac{1}{2}\beta\right) = \begin{bmatrix} \cos\left(\tfrac{1}{2}\beta\right) & \sin\left(\tfrac{1}{2}\beta\right) \\ -\sin\left(\tfrac{1}{2}\beta\right) & \cos\left(\tfrac{1}{2}\beta\right) \end{bmatrix} \tag{4} $$

$$ U_z(\xi) = \begin{bmatrix} e^{i\xi} & 0 \\ 0 & e^{-i\xi} \end{bmatrix} \tag{5} $$

The order $2j + 1$ representation is

$$U_{p,q}^{(j)}(\alpha,\ \beta,\ \gamma)$$

$$=\sum_m \frac{(-1)^{m-q-p}\sqrt{(j+p)!(j-p)!(j+q)!(j-q)!}}{(j-p-m)!(j+q-m)!(m+p-q)!m!}$$

$$\times e^{iq\alpha}\cos^{2j+q-p-2m}\left(\tfrac{1}{2}\beta\right)\sin^{p+2m-q}\left(\tfrac{1}{2}\beta\right)e^{ip\gamma} \quad (6)$$

The summation is terminated by putting $1/(-N)! = 0$. The CHARACTER is given by

$$X^{(j)}(\alpha) = \begin{cases} 1 + 2\cos\alpha + \ldots + 2\cos(j\alpha) \\ 2\left[\cos\left(\tfrac{1}{2}\alpha\right) + \cos\left(\tfrac{3}{2}\alpha\right) + \ldots + \cos(j\alpha)\right] \end{cases}$$

$$= \begin{cases} \dfrac{\sin\left[\left(j+\tfrac{1}{2}\right)\alpha\right]}{\sin\left(\tfrac{1}{2}\alpha\right)} & \text{for } j = 0,\ 1,\ 2,\ \ldots \\ \dfrac{\sin\left[\left(j+\tfrac{1}{2}\right)\alpha\right]}{\sin\left(\tfrac{1}{2}\alpha\right)} & \text{for } j = \tfrac{1}{2},\ \tfrac{3}{2},\ \ldots \end{cases} \quad (7)$$

See also ORTHOGONAL GROUP, SPECIAL LINEAR GROUP, SPECIAL ORTHOGONAL GROUP

References

Arfken, G. "Special Unitary Group, $SU(2)$ and $SU(2)$-O_3^+ Homomorphism." *Mathematical Methods for Physicists,* *3rd ed.* Orlando, FL: Academic Press, pp. 253–59, 1985.
Conway, J. H.; Curtis, R. T.; Norton, S. P.; Parker, R. A.; and Wilson, R. A. "The Groups $GU_n(q)$, $SU_n(q)$, $PGU_n(q)$, and $PSU_n(q) = U_n(q)$." §2.2 in *Atlas of Finite Groups: Maximal Subgroups and Ordinary Characters for Simple Groups.* Oxford, England: Clarendon Press, p. x, 1985.

Special Unitary Matrix

A SQUARE MATRIX U is a special unitary matrix if

$$\mathsf{UU}^* = \mathsf{I}. \quad (1)$$

where I is the IDENTITY MATRIX and U^* is the ADJOINT MATRIX, and the DETERMINANT is

$$\det \mathsf{U} = 1. \quad (2)$$

The first condition means that U is a UNITARY MATRIX, and the second condition provides a restriction beyond a general UNITARY MATRIX, which may have determinant $e^{i\theta}$ for θ any real number. For example,

$$\frac{1}{\sqrt{2}}\begin{bmatrix} i & i \\ i & -i \end{bmatrix} \quad (3)$$

is a special unitary matrix. A matrix m can be tested to see if it is a special unitary matrix using the *Mathematica* function

```
SpecialUnitaryQ[m_List?MatrixQ] :=
    (Conjugate@Transpose@m.m  = =  IdentityMa-
trix@Length@m
    && Det[m]  = =  1)
```

The special unitary matrices are CLOSED under multiplication and the inverse operation, and therefore form a MATRIX GROUP called the SPECIAL UNITARY GROUP $SU(n)$.

See also HERMITIAN INNER PRODUCT, SKEW HERMITIAN MATRIX, SPECIAL LINEAR MATRIX, SPECIAL UNITARY GROUP, SPIN GROUP, UNITARY GROUP UNITARY MATRIX

Species

A species of structures is a rule F which

1. Produces, for each finite set U, a finite set $F[U]$,
2. Produces, for each bijection $\sigma : U \to V$, a function

$$F[\sigma] : F[U] \to F[V].$$

The functions $F[\sigma]$ should further satisfy the following functorial properties:

1. For all bijections $\sigma : U \to V$ and $\tau : V \to W$,

$$F[\tau \circ \sigma] = F[\tau] \circ F[\sigma].$$

2. For the IDENTITY MAP $\mathrm{Id}_U : U \to U$,

$$F[\mathrm{Id}_U] = \mathrm{Id}_{F[U]}.$$

An element $\sigma \in F[U]$ is called an F-structure on U (or a structure of species F on U). The function $F[\sigma]$ is called the transport of F-structures along σ.

References

Bergeron, F.; Labelle, G.; and Leroux, P. *Combinatorial Species and Tree-Like Structures.* Cambridge, England: Cambridge University Press, p. 5, 1998.

Specificity

The probability that a STATISTICAL TEST will be negative for a negative statistic.

See also SENSITIVITY, STATISTICAL TEST, TYPE I ERROR, TYPE II ERROR

Spectral Graph Partitioning

A GRAPHICAL PARTITIONING based on the eigenvalues and eigenvectors of the LAPLACIAN MATRIX of a graph.

See also GRAPHICAL PARTITION, LAPLACIAN MATRIX

References

Chung, F. R. K. *Spectral Graph Theory.* Providence, RI: Amer. Math. Soc., 1997.
Demmel, J. "CS 267: Notes for Lecture 23, April 9, 1999. Graph Partitioning, Part 2." http://www.cs.berkeley.edu/~demmel/cs267/lecture20/lecture20.html.

Spectral Norm

The NATURAL NORM induced by the L_2-NORM. Let A^* be the ADJOINT of the SQUARE MATRIX A, so that $(a_{ij})^* = (\bar{a}_{ji})$, then

$$\|A\|_2 = (\text{maximum eigenvalue of } A^*A)^{1/2}$$

$$= \max_{\|x\|_2 \neq 0} \frac{\|A\mathbf{x}\|_2}{\|\mathbf{x}\|_2}.$$

This MATRIX NORM is implemented as Matrix-Norm[m, 2] in the *Mathematica* add-on package LinearAlgebra`MatrixMultiplication` (which can be loaded with the command < <LinearAlgebra`).

See also L_2-NORM, MATRIX NORM, MAXIMUM ABSOLUTE COLUMN SUM NORM, MAXIMUM ABSOLUTE ROW SUM NORM

References

Gradshteyn, I. S. and Ryzhik, I. M. *Tables of Integrals, Series, and Products, 6th ed.* San Diego, CA: Academic Press, p. 1115, 2000.
Strang, G. §6.2 and 7.2 in *Linear Algebra and Its Applications, 4th ed.* New York: Academic Press, 1980.

Spectral Power Density

$$P_y(v) \equiv \lim_{T \to \infty} \frac{2}{T} \left| \int_{-T/2}^{T/2} [y(t) - \bar{y}] e^{-2\pi i v t} \, dt \right|^2.$$

so

$$\int_0^\infty P_y(v) \, dv \equiv \lim_{T \to \infty} \frac{1}{T} \int_{-T/2}^{T/2} [y(t) - \bar{y}]^2 \, dt$$

$$= \left\langle (y - \bar{y})^2 \right\rangle = \sigma_y^2.$$

See also POWER SPECTRUM

Spectral Radius

Let A be an $n \times n$ MATRIX with COMPLEX or REAL elements with EIGENVALUES $\lambda_1, \ldots, \lambda_n$. Then the spectral radius $\rho(A)$ of A is

$$\rho(A) = \max_{1 \leq i \leq n} |\lambda_i|.$$

References

Gradshteyn, I. S. and Ryzhik, I. M. *Tables of Integrals, Series, and Products, 6th ed.* San Diego, CA: Academic Press, pp. 1115–116, 2000.

Spectral Rigidity

The mean square deviation of the best local fit straight line to a staircase cumulative spectral density over a normalized energy scale.

References

Ott, E. *Chaos in Dynamical Systems.* New York: Cambridge University Press, p. 341, 1993.

Spectral Theorem

Let H be a HILBERT SPACE, $B(H)$ the set of BOUNDED linear operators from H to itself, T an OPERATOR on H, and $\sigma(T)$ the SPECTRUM of T. Then if $T \in B(H)$ and T is normal, there exists a unique resolution of the identity E on the BOREL SUBSETS of $\sigma(T)$ which satisfies

$$T = \int_{\sigma(T)} \lambda \, dE(\lambda).$$

Furthermore, every projection $E(\omega)$ COMMUTES with every $S \in B(H)$ that COMMUTES with T.

See also SPECTRUM (OPERATOR)

References

Rudin, W. Theorem 12.23 in *Functional Analysis, 2nd ed.* New York: McGraw-Hill, 1991.

Spectrum

The word "spectrum" confusingly has a number of unrelated meanings in various branches of mathematics.

See also GRAPH SPECTRUM, SPECTRUM (MATRIX), SPECTRUM (OPERATOR), SPECTRUM (RING), SPECTRUM SEQUENCE

Spectrum (Graph)

GRAPH SPECTRUM

Spectrum (Matrix)

The EIGENVALUES of a MATRIX A are called its spectrum, and are denoted $\lambda(A)$. If $\lambda(A) = \{\lambda_1, \ldots, \lambda_n\}$, then the DETERMINANT of A is given by

$$\det(A) = \lambda_1 \lambda_2 \ldots \lambda_n.$$

See also CHARACTERISTIC POLYNOMIAL, EIGENVALUE

References

Golub, G. H. and van Loan, C. F. *Matrix Computations, 3rd ed.* Baltimore, MD: Johns Hopkins University Press, p. 310, 1996.

Spectrum (Operator)

Let T be an OPERATOR on a HILBERT SPACE. The spectrum $\sigma(T)$ of T is the set of λ such that $(T - \lambda I)$ is not invertible on all of the HILBERT SPACE, where the λs are COMPLEX NUMBERS and I is the IDENTITY OPERATOR. The definition can also be stated in terms of the resolvent of an operator

$$\rho(T) = \{\lambda : (T - \lambda I) \text{ is invertible}\},$$

and then the spectrum is defined to be the complement of $\rho(T)$ in the COMPLEX PLANE. It is easy to demonstrate that $\rho(T)$ is an OPEN SET, which shows that the spectrum is closed (in fact, it is even compact).

If Ω is a domain in $\mathbb{R}^d$ (i.e., a Lebesgue measurable subset of $\mathbb{R}^d$ with finite nonzero LEBESGUE MEASURE), the Iosevich *et al.* (1999) say a set $\Lambda \subset \mathbb{R}^d$ is a spectrum of Ω is $\{e^{2\pi i x \lambda}\}_{\lambda \in \Lambda}$ is an ORTHOGONAL BASIS of $L^2(\Omega)$.

See also FUGLEDE'S CONJECTURE, HILBERT SPACE, ORTHOGONAL BASIS, SPECTRAL THEOREM

References

Iosevich, A.; Katz, N. H.; and Tao, T. Convex Bodies with a Point of Curvature Do Not Have Fourier Bases. 23 Nov 1999. http://xxx.lanl.gov/abs/math.CA/9911167/.

Rudin, W. *Functional Analysis, 2nd ed.* New York: McGraw-Hill, 1991.

Spectrum (Ring)

The spectrum of a RING is the set of proper PRIME IDEALS,

$$\operatorname{Spec}(R) = \{\mathfrak{p} : \mathfrak{p} \text{ is a prime ideal in R}\}. \quad (1)$$

The classical example is the spectrum of POLYNOMIAL RINGS. For instance,

$$\operatorname{Spec}(\mathbb{C}[x]) = \{\langle x - a \rangle : a \in \mathbb{C}\} \cup \{\langle 0 \rangle\}. \quad (2)$$

and

$$\operatorname{Spec}(\mathbb{C}[x, y]) = \{\langle x - a, y - b \rangle, (a, b) \in \mathbb{C}^2\}$$

$$\cup \{\langle f(x, y) \rangle : f \text{ is irreducable}\} \cup \{\langle 0 \rangle\}. \quad (3)$$

The points are, in classical algebraic geometry, ALGEBRAIC VARIETIES. Note that $\langle x - a, y - b \rangle$ are MAXIMAL IDEALS, hence also prime.

The spectrum of a ring has a TOPOLOGY called the ZARISKI TOPOLOGY. The closed sets are of the form

$$V(S) = \{\langle p \rangle : S \subset \langle p \rangle\}. \quad (4)$$

For example,

$$\operatorname{Spec}(\mathbb{Z}) = \{\langle p \rangle : p \text{ is prime}\} \cup \{\langle 0 \rangle\}. \quad (5)$$

Every PRIME IDEAL is closed except for $\langle 0 \rangle$, whose closure is $V(0) = \operatorname{Spec}(\mathbb{Z})$.

See also AFFINE SCHEME, CATEGORY THEORY, COMMUTATIVE ALGEBRA, CONIC SECTION, IDEAL, PRIME IDEAL, PROJECTIVE VARIETY, SCHEME, VARIETY, ZARISKI TOPOLOGY

References

Bump, D. *Algebraic Geometry.* Singapore: World Scientific, pp. 1–, 1998.

Hartshorne, R. *Algebraic Geometry.* New York: Springer-Verlag, 1977.

Spectrum Sequence

A spectrum sequence is a SEQUENCE formed by successive multiples of a REAL NUMBER a rounded down to the nearest INTEGER $s_n = \lfloor na \rfloor$. If a is IRRATIONAL, the spectrum is called a BEATTY SEQUENCE.

See also BEATTY SEQUENCE, LAGRANGE SPECTRUM, MARKOV SPECTRUM

Speed

The SCALAR $|\mathbf{v}| = ds/dt$, where s is the ARC LENGTH, equal to the magnitude of the VELOCITY $\mathbf{v}$.

See also ANGULAR VELOCITY, VELOCITY

Spencer's 15-Point Moving Average

A MOVING AVERAGE using 15 points having weights $-3, -6, -5, 3, 21, 46, 67, 74, 67, 46, 21, 3, -5, -6,$ and -3. It is sometimes used by actuaries.

See also MOVING AVERAGE, SPENCER'S FORMULA

References

Kenney, J. F. and Keeping, E. S. *Mathematics of Statistics, Pt. 1, 3rd ed.* Princeton, NJ: Van Nostrand, p. 223, 1962.

Spencer's Formula

Define the notation

$$[n]f_0 = f_{-(n-1)/2} + \cdots + f_0 + \cdots + f_{(n-1)/2} \quad (1)$$

and let δ be the central difference, so

$$\delta^2 f_0 = f_1 - 2f_0 + f_{-1}. \quad (2)$$

Spencer's 21-term moving average formula is then given by

$$f_0' = \frac{[5][5][7]}{5 \cdot 5 \cdot 7}(1 - 4\delta^2)f_0,$$

which, written explicitly, gives

$$f_0' = \frac{1}{350}[60f_0 + 57(f_{-1} + f_1) + 47(f_{-2} + f_2) + 33(f_{-3} + f_3)$$

$$+ 18(f_{-4} + f_4) + 6(f_{-5} + f_5) - 2(f_{-6} + f_6) - 5(f_{-7} + f_7)$$

$$- 5(f_{-8} + f_8) - 3(f_{-9} + f_9) - (f_{-10} + f_{10})] \quad (3)$$

See also MOVING AVERAGE, SMOOTHING

References

Spencer, J. *J. I. A.* **38**, 334, 1904.

Spencer, J. *J. I. A.* **38**, 339, 1904.

Spencer, J. *J. I. A.* **41**, 361, 1907.

Whittaker, E. T. and Robinson, G. "Spencer's Formula." §144 in *The Calculus of Observations: A Treatise on Numerical Mathematics, 4th ed.* New York: Dover, pp. 290–94, 1967.

Spence's Function

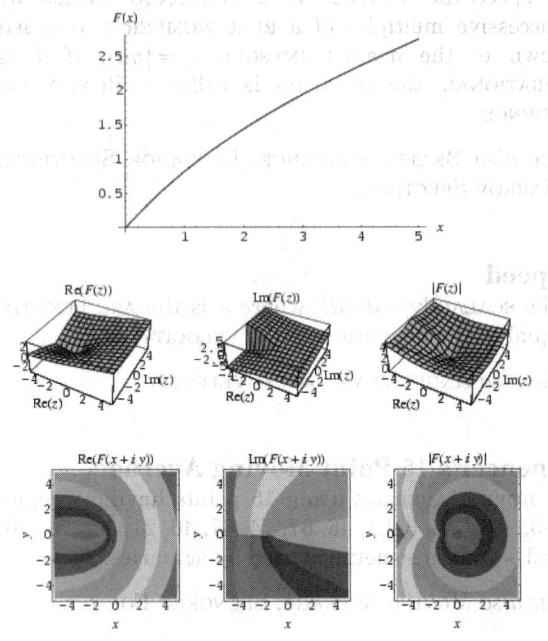

$$F(x) = -\mathrm{Li}_2(-x) = \int_0^x \frac{\ln(1+t)}{t}\, dt.$$

where $\mathrm{Li}_2(x)$ is the DILOGARITHM.

See also DILOGARITHM, SPENCE'S INTEGRAL

References

Berestetskii, V. B.; Lifschitz, E. M.; and Ditaevskii, L. P. *Quantum Electrodynamics, 2nd ed.* Oxford, England: Pergamon Press, p. 596, 1982.

Spence's Integral

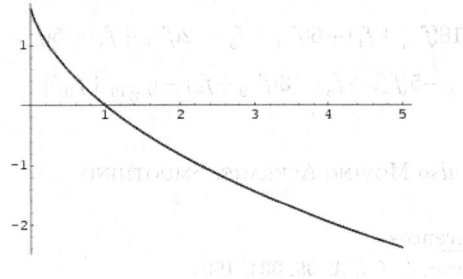

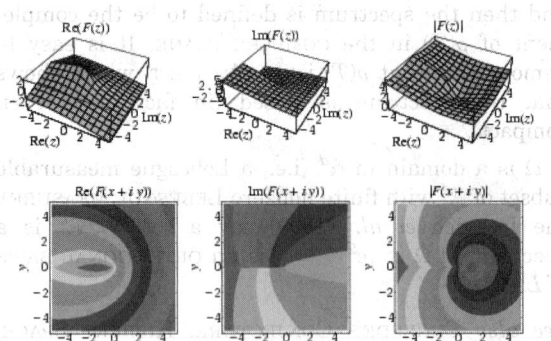

$$F(x) = \mathrm{Li}_2(1-x) = \int_{1-x}^0 \frac{\ln(1-t)}{t}\, dt.$$

where $\mathrm{Li}_2(x)$ is the DILOGARITHM.

See also DILOGARITHM, SPENCE'S FUNCTION

Sperner System

ANTICHAIN

Sperner's Theorem

The MAXIMUM CARDINALITY of a collection of SUBSETS of a t-element SET T, none of which contains another, is the BINOMIAL COEFFICIENT $\binom{t}{\lfloor t/2 \rfloor}$, where $\lfloor x \rfloor$ is the FLOOR FUNCTION.

See also CARDINALITY

Sphenocorona

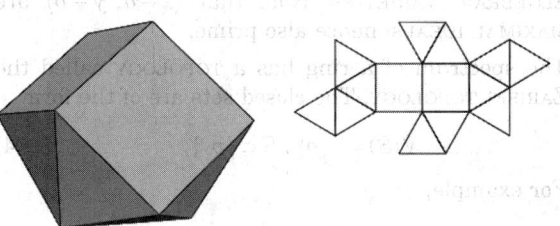

JOHNSON SOLID J_{86}.

References

Weisstein, E. W. "Johnson Solids." MATHEMATICA NOTEBOOK JohnsonSolids.M.
Weisstein, E. W. "Johnson Solid Netlib Database." MATHEMATICA NOTEBOOK JohnsonSolids.dat.

Sphenoid

DISPHENOID

Sphenomegacorona

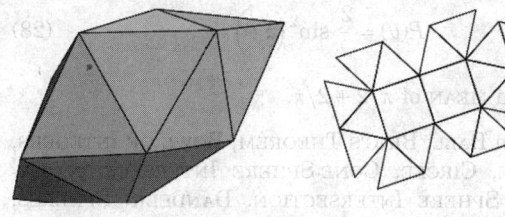

JOHNSON SOLID J_{88}.

References

Weisstein, E. W. "Johnson Solids." MATHEMATICA NOTEBOOK JOHNSONSOLIDS.M.

Weisstein, E. W. "Johnson Solid Netlib Database." MATHEMATICA NOTEBOOK JOHNSONSOLIDS.DAT.

Sphere

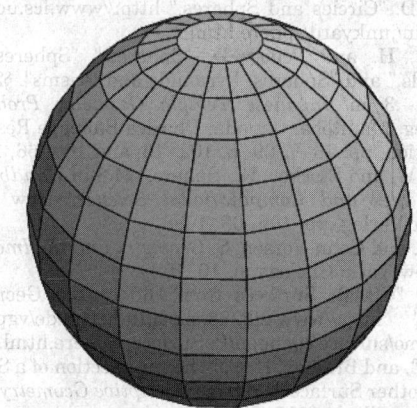

A sphere is defined as the set of all points in $\mathbb{R}^3$ which are a distance r (the "RADIUS") from a given point (the "CENTER"). Twice the RADIUS is called the DIAMETER, and pairs of points on opposite sides of a DIAMETER are called ANTIPODES. The term "sphere" technically refers to the outer surface of a "BUBBLE," which is denoted $\mathbb{S}^2$. However, in common usage, the word *sphere* is also used to mean the UNION of a sphere and its INTERIOR (a "solid sphere"), where the INTERIOR is called a BALL.

The SURFACE AREA of the sphere and VOLUME of the BALL of RADIUS R are given by

$$S = 4\pi R^2 \tag{1}$$

$$V = \tfrac{4}{3}\pi R^3 \tag{2}$$

(Beyer 1987, p. 130). In *On the Sphere and Cylinder* (ca. 225 BC), Archimedes became the first to derive these equations (although he expressed π in terms of the sphere's circular CROSS SECTION). The fact that

$$\frac{V_{\text{sphere}}}{V_{\text{circumscribed cylinder}} - V_{\text{sphere}}} = 2 \tag{3}$$

was also known to Archimedes (Steinhaus 1983, p. 223; Wells 1991, pp. 236–37).

Any CROSS SECTION through a sphere is a CIRCLE (or, in the degenerate case where the slicing PLANE is tangent to the sphere, a point). The size of the CIRCLE is maximized when the PLANE defining the CROSS SECTION passes through a DIAMETER.

The equation of a sphere of RADIUS r is given in CARTESIAN COORDINATES by

$$x^2 + y^2 + z^2 = r^2. \tag{4}$$

which is a special case of the ELLIPSOID

$$\frac{x^2}{a^2} + \frac{y^2}{b^2} + \frac{z^2}{c^2} = 1 \tag{5}$$

and SPHEROID

$$\frac{x^2 + y^2}{a^2} + \frac{z^2}{c^2} = 1. \tag{6}$$

A sphere may also be specified in SPHERICAL COORDINATES by

$$x = \rho \cos\theta \sin\phi \tag{7}$$

$$y = \rho \sin\theta \sin\phi \tag{8}$$

$$z = \rho \cos\phi. \tag{9}$$

where θ is an azimuthal coordinate running from 0 to 2π (LONGITUDE), ϕ is a polar coordinate running from 0 to π (COLATITUDE), and ρ is the RADIUS. Note that there are several other notations sometimes used in which the symbols for θ and ϕ are interchanged or where r is used instead of ρ. If ρ is allowed to run from 0 to a given RADIUS r, then a solid BALL is obtained.

The volume of the sphere, $V = 4/3\pi R^3$, can be found in Cartesian, cylindrical, and spherical coordinates, respectively, using the integrals

$$V = \int_{-R}^{R} \int_{-\sqrt{R^2-x^2}}^{\sqrt{R^2-x^2}} \int_{-\sqrt{R^2-x^2-y^2}}^{\sqrt{R^2-x^2-y^2}} dz\,dy\,dx \tag{10}$$

$$= \int_0^{2\pi} \int_0^R \int_{-\sqrt{R^2-x^2}}^{\sqrt{R^2-x^2}} r\,dz\,dr\,d\theta \tag{11}$$

$$= \int_0^{2\pi} \int_0^\pi \int_0^R \rho^2 \sin\phi\,d\rho\,d\phi\,d\theta. \tag{12}$$

Converting to "standard" parametric variables $a = \rho$, $u = \theta$, and $v = \phi$ gives the coefficients of the FIRST FUNDAMENTAL FORM

$$E = a^2 \sin^2 v \tag{13}$$

$$F = 0 \tag{14}$$

$$G = a^2. \tag{15}$$

SECOND FUNDAMENTAL FORM coefficients

$$e = a \sin^2 v \tag{16}$$

$$f = 0 \tag{17}$$

$$g = a. \tag{18}$$

AREA ELEMENT

$$dA = a \sin v \, du \wedge dv. \tag{19}$$

GAUSSIAN CURVATURE

$$K = \frac{1}{a^2}. \tag{20}$$

and MEAN CURVATURE

$$H = \frac{1}{a}. \tag{21}$$

A sphere may also be represented parametrically by letting $u \equiv r \cos \phi$, so

$$x = \sqrt{r^2 - u^2} \cos \theta \tag{22}$$

$$y = \sqrt{r^2 - u^2} \sin \theta \tag{23}$$

$$z = u, \tag{24}$$

where θ runs from 0 to 2π and u runs from $-r$ to r.

Given two points on a sphere, the shortest path on the surface of the sphere which connects them (the SPHERE GEODESIC) is an ARC of a CIRCLE known as a GREAT CIRCLE. The equation of the sphere with points $\{x_1, y_1, z_1\}$ and $\{x_2, y_2, z_2\}$ lying on a DIAMETER is given by

$$(x - x_1)(x - x_2) + (y - y_1)(y - y_2) + (z - z_1)(z - z_2)$$
$$= 0. \tag{25}$$

Four points are sufficient to uniquely define a sphere. Given the points $\{x_i, y_i, z_i\}$ with $i = 1, 2, 3,$ and 4, the sphere containing them is given by the beautiful DETERMINANT equation

$$\begin{vmatrix} x^2 + y^2 + z^2 & x & y & z & 1 \\ x_1^2 + y_1^2 + z_1^2 & x_1 & y_1 & z_1 & 1 \\ x_2^2 + y_2^2 + z_2^2 & x_2 & y_2 & z_2 & 1 \\ x_3^2 + y_3^2 + z_3^2 & x_3 & y_3 & z_3 & 1 \\ x_4^2 + y_4^2 + z_4^2 & x_4 & y_4 & z_4 & 1 \end{vmatrix} = 0 \tag{26}$$

(Beyer 1987, p. 210).

The generalization of a sphere in n dimensions is called a HYPERSPHERE. An n-D HYPERSPHERE can be specified by the equation

$$x_1^2 + x_2^2 + \ldots + x_n^2 = r^2. \tag{27}$$

The distribution of ANGLES for random rotation of a sphere is

$$P(\theta) = \frac{2}{\pi} \sin^2\left(\frac{1}{2} \theta\right), \tag{28}$$

giving a MEAN of $\pi/2 + 2/\pi$.

See also BALL, BING'S THEOREM, BOWL OF INTEGERS, BUBBLE, CIRCLE, CONE-SPHERE INTERSECTION, CYLINDER-SPHERE INTERSECTION, DANDELIN SPHERES, DIAMETER, ELLIPSOID, EXOTIC SPHERE, FEJES TÓTH'S PROBLEM, GEODESIC DOME, GLOME, HYPERSPHERE, LIEBMANN'S THEOREM, LIOUVILLE'S SPHERE-PRESERVING THEOREM, MIKUSINSKI'S PROBLEM, NOISE SPHERE, OBLATE SPHEROID, OSCULATING SPHERE, PARALLELIZABLE, PROLATE SPHEROID, RADIUS, SPACE DIVISION BY SPHERES, SPHERE PACKING, SPHERE-SPHERE INTERSECTION, TANGENT SPHERES, TENNIS BALL THEOREM

References

Beyer, W. H. (Ed.). *CRC Standard Mathematical Tables, 28th ed.* Boca Raton, FL: CRC Press, p. 227, 1987.

Coolidge, J. L. *A Treatise on the Geometry of the Circle and Sphere.* New York: Chelsea, 1971.

Eppstein, D. "Circles and Spheres." http://www.ics.uci.edu/~eppstein/junkyard/sphere.html.

Fukagawa, H. and Pedoe, D. "Spheres," "Spheres and Ellipsoids," and "Spheres, Pyramids and Prisms". §2.2–.6 and 9.1–.3 in *Japanese Temple Geometry Problems.* Winnipeg, Manitoba, Canada: Charles Babbage Research Foundation, pp. 26–7, 69–6, 102–16, and 160–66, 1989.

Harris, J. W. and Stocker, H. "Sphere." §4.8 in *Handbook of Mathematics and Computational Science.* New York: Springer-Verlag, pp. 106–08, 1998.

Hilbert, D. and Cohn-Vossen, S. *Geometry and the Imagination.* New York: Chelsea, p. 10, 1999.

JavaView. "Classic Surfaces from Differential Geometry: Sphere." http://www-sfb288.math.tu-berlin.de/vgp/javaview/demo/surface/common/PaSurface_Sphere.html.

Kenison, E. and Bradley, H. C. "The Intersection of a Sphere with Another Surface." §198 in *Descriptive Geometry.* New York: Macmillan, 1935.

Kern, W. F. and Bland, J. R. "Sphere." §33 in *Solid Mensuration with Proofs, 2nd ed.* New York: Wiley, pp. 87–3, 1948.

Kiang, T. "An Old Chinese Way of Finding the Volume of a Sphere." *Math. Gaz.* **56**, 88–1, 1972.

Steinhaus, H. *Mathematical Snapshots, 3rd ed.* New York: Dover, 1999.

Sphere Embedding

A 4-sphere has POSITIVE CURVATURE, with

$$R^2 = x^2 + y^2 + z^2 + w^2 \tag{1}$$

$$2x \frac{dx}{dw} + 2y \frac{dy}{dw} + 2z \frac{dz}{dw} + 2w = 0. \tag{2}$$

Since

$$\mathbf{r} \equiv x\hat{\mathbf{x}} + y\hat{\mathbf{y}} + z\hat{\mathbf{z}}. \tag{3}$$

$$dw = -\frac{x \, dx + y \, dy + z \, dz}{w} = -\frac{\mathbf{r} \cdot d\mathbf{r}}{\sqrt{R^2 - r^2}}. \tag{4}$$

To stay on the surface of the sphere,

$$ds^2 = dx^2 + dy^2 + dz^2 + dw^2$$

$$= dx^2 + dy^2 + dz^2 + \frac{r^2\,dr^2}{R^2 - r^2}$$

$$= dr^2 + r^2\,d\Omega^2 + \frac{dr^2}{\dfrac{R^2}{r^2} - 1}$$

$$= dr^2 \left(1 + \frac{1}{\dfrac{R^2}{r^2} - 1} \right) + r^2\,d\Omega^2$$

$$= dr^2 \left(\frac{\dfrac{R^2}{r^2}}{\dfrac{R^2}{r^2} - 1} \right) + r^2\,d\Omega^2$$

$$= \frac{dr^2}{1 - \dfrac{r^2}{R^2}} + r^2\,d\Omega^2. \tag{5}$$

With the addition of the so-called expansion parameter, this is the Robertson-Walker line element.

Sphere Eversion

Smale (1958) proved that it is mathematically possible to turn a SPHERE inside-out without introducing a sharp crease at any point. This means there is a regular homotopy from the standard embedding of the 2-SPHERE in EUCLIDEAN 3-space to the mirror-reflection embedding such that at every stage in the homotopy, the sphere is being IMMERSED in EUCLIDEAN SPACE. This result is so counterintuitive and the proof so technical that the result remained controversial for a number of years.

In 1961, Arnold Shapiro devised an explicit eversion but did not publicize it. Phillips (1966) heard of the result and, in trying to reproduce it, actually devised an independent method of his own. Yet another eversion was devised by Morin, which became the basis for the movie by Max (1977). Morin's eversion also produced explicit algebraic equations describing the process. The original method of Shapiro was subsequently published by Francis and Morin (1979).

See also EVERSION, SPHERE

References

Bulatov, V. "Sphere Eversion--Visualization of the Famous Topological Procedure." http://www.physics.orst.edu/~bulatov/vrml/evert.wrl.
Francis, G. K. Ch. 6 in *A Topological Picturebook.* New York: Springer-Verlag, 1987.
Francis, G. K. and Morin, B. "Arnold Shapiro's Eversion of the Sphere." *Math. Intell.* **2**, 200–03, 1979.

Levy, S.; Maxwell, D.; and Munzner, T. *Making Waves: A Guide to the Ideas Behind Outside In.* Wellesley, MA: A. K. Peters, 1995. Book and 22 minute *Outside-In.* videotape.
Max, N. "Turning a Sphere Inside Out." Videotape. Chicago, IL: International Film Bureau, 1977.
Peterson, I. *Islands of Truth: A Mathematical Mystery Cruise.* New York: W. H. Freeman, pp. 240–44, 1990.
Petersen, I. "Forging Links Between Mathematics and Art." *Science News* **141**, 404–05, June 20, 1992.
Phillips, A. "Turning a Surface Inside Out." *Sci. Amer.* **214**, 112–20, Jan. 1966.
Smale, S. "A Classification of Immersions of the Two-Sphere." *Trans. Amer. Math. Soc.* **90**, 281–90, 1958.
Wells, D. *The Penguin Dictionary of Curious and Interesting Geometry.* London: Penguin, 1991.

Sphere Geodesic

GREAT CIRCLE

Sphere Inversion

INVERSION in 3 dimensions with respect to an INVERSION SPHERE.

See also INVERSION, INVERSION SPHERE

Sphere Line Picking

Pick two points at random on a unit sphere. The first one can be placed at the north pole, i.e., assigned the coordinate (0, 0, 1), without loss of generality. The second point is then chosen at random using SPHERE POINT PICKING, and so can be assigned coordinates

$$x = \sqrt{1 - u^2}\,\cos\theta \tag{1}$$

$$y = \sqrt{1 - u^2}\,\sin\theta \tag{2}$$

$$z = u \tag{3}$$

with $u \in [-1,\,1]$ and $\theta \in [0,\,2\pi)$. The distance ℓ between first and second points is then

$$\ell = \sqrt{x^2 + y^2 + (z-1)^2} = \sqrt{2 - 2u}, \tag{4}$$

and solving for u gives

$$u = \tfrac{1}{2}(2 - \ell^2). \tag{5}$$

Now the probability function $P\ell$ for distance is then given by

$$P_\ell = P_u \left| \frac{\partial u}{\partial \ell} \right| = \tfrac{1}{2}\,\ell\,d\ell \tag{6}$$

(Solomon 1978, p. 163), since $P_u = 1/2$ and $du/d\ell = -\ell$. Here, $\ell \in [0,\,2]$.

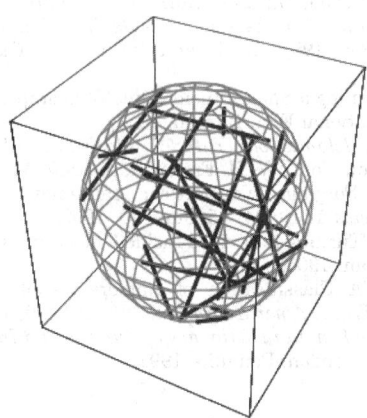

Therefore, somewhat surprisingly, large distances are the most common, contrary to most people's intuition. A plot of 15 random lines is shown above. The RAW MOMENTS are

$$\mu'_n = \langle \ell^n \rangle = \int_0^2 \ell^n P_\ell \, d\ell = \frac{2^{n+1}}{2+n}. \tag{7}$$

giving the first few as

$$\mu'_1 = \tfrac{4}{3} \tag{8}$$

$$\mu'_2 = 2 \tag{9}$$

$$\mu'_3 = \tfrac{16}{5} \tag{10}$$

$$\mu'_4 = \tfrac{16}{3}. \tag{11}$$

so the CENTRAL MOMENTS are

$$\mu = \tfrac{4}{3} \tag{12}$$

$$\mu_2 = \sigma^2 = \tfrac{2}{9} \tag{13}$$

$$\mu_3 = -\tfrac{8}{135} \tag{14}$$

$$\mu_4 = \tfrac{16}{135}. \tag{15}$$

so the VARIANCE, SKEWNESS and KURTOSIS are

$$\sigma^2 = \tfrac{2}{9} \tag{16}$$

$$\gamma_1 = \tfrac{4}{5}\sqrt{2} \tag{17}$$

$$\gamma_2 = -\tfrac{5}{3} \tag{18}$$

(Solomon 1978, p. 163).

See also BALL LINE PICKING, CIRCLE LINE PICKING, POINT-POINT DISTANCE–1-D, SPHERE POINT PICKING, SPHERE TETRAHEDRON PICKING

References

Solomon, H. *Geometric Probability.* Philadelphia, PA: SIAM, 1978.

Sphere Packing

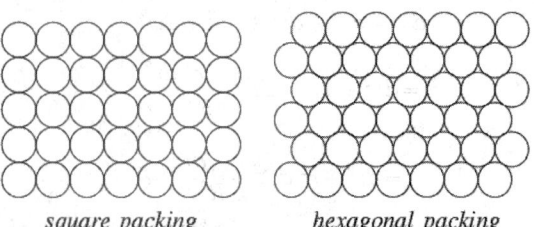

square packing   *hexagonal packing*

In 2-D, there are two periodic CIRCLE PACKINGS for identical circles: square lattice and hexagonal lattice. Fejes Tóth (1940) proved that the hexagonal lattice is the densest of *all* possible plane packings (Conway and Sloane 1993, pp. 8–).

In 3-D, there are three periodic packings for identical spheres: cubic lattice, face-centered cubic lattice, and hexagonal lattice. It was hypothesized by Kepler in 1611 that close packing (cubic or hexagonal) is the densest possible (has the greatest PACKING DENSITY η, which is the fraction of a VOLUME filled by identical packed SPHERES), and this assertion is known as the KEPLER CONJECTURE. The problem of finding the densest packing of spheres (not necessarily periodic) is therefore known as the KEPLER PROBLEM. The KEPLER CONJECTURE is intuitively obvious, but the proof remained elusive until it was accomplished in a series of papers by Hales culminating in 1998. Gauss (1831) did prove that the face-centered cubic is the densest *lattice* packing in 3-D (Conway and Sloane 1993, p. 9). This result has since been extended to HYPERSPHERE PACKING.

The maximum number of equivalent spheres (or n-D hyperspheres) which can touch an equivalent sphere (hypersphere) without intersections is called the n-D KISSING NUMBER.

In 3-D, face-centered cubic close packing and hexagonal close packing (which is distinct from hexagonal lattice packing), both give

$$\eta_{\mathrm{CCP}} = \eta_{\mathrm{HCP}} = \frac{\pi}{3\sqrt{2}} \approx 74.048\% \tag{1}$$

(Steinhaus 1983, p. 202; Wells 1986, p. 29; Wells 1991, p. 237). For packings in 3-D, C. A. Rogers (1958) showed that the maximum possible PACKING DENSITY $\eta_{\max}$ satisfies

$$\eta_{\max} < \sqrt{18}\left(\cos^{-1}\tfrac{1}{3} - \tfrac{1}{3}\pi\right) \approx 77.96355700\% \tag{2}$$

(Le Lionnais 1983). This was subsequently improved to 77.844% (Lindsey 1986), then 77.836% (Muder 1988). However, Rogers (1958) remarks that "many mathematicians believe, and all physicists know" that the actual answer is 74.048% (Conway and Sloane 1993, p. 3).

Hilbert and Cohn-Vossen (1999, pp. 48–0) consider a tetrahedral packing in which each sphere touched four neighbors and the density is $\pi\sqrt{3}/16 \approx 0.3401$.

The rigid packing with *lowest* density known has $\eta \approx 0.0555$ (Gardner 1966), significantly lower than that reported by Hilbert and Cohn-Vossen (1999, p. 51). To be rigid, each SPHERE must touch at least four others, and the four contact points cannot be in a single HEMISPHERE or all on one equator.

RANDOM CLOSE PACKING of spheres in 3-D gives packing densities in the range 0.06 to 0.65 (Jaeger and Nagel 1992, Torquato *et al.* 2000). The PACKING DENSITIES for several packing types are summarized in the following table.

Packing	η (exact)	η	reference
loose packing	–	0.0555	Gardner (1966)
tetrahedral lattice	$\frac{\pi\sqrt{3}}{16}$	0.3401	Hilbert and Cohn-Vossen (1999, pp. 48–0)
cubic lattice	$\frac{\pi}{6}$	0.5236	
hexagonal lattice	$\frac{\pi}{3\sqrt{3}}$	0.6046	
random	–	0.6400	Jaeger and Nagel 1992
face-centered cubic lattice	$\frac{\pi}{3\sqrt{2}}$	0.7405	Steinhaus 1983, p. 202; Wells 1986, p. 29; Wells 1991, p. 237
square lattice (2-D)	$\frac{\pi}{4}$	0.7854	
hexagonal lattice (2-D)	$\frac{\pi}{2\sqrt{3}}$	0.9069	

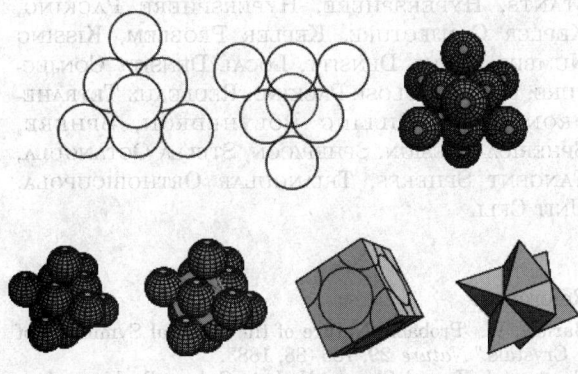

Arranging layers of close-packed spheres such that the spheres of every third layer overlying one another gives cubic close packing. To see where the name comes from, consider packing six SPHERES together in the shape of an EQUILATERAL TRIANGLE and place another SPHERE on top to create a TRIANGULAR PYRAMID. Now create another such grouping of seven SPHERES and place the two PYRAMIDS together facing in opposite directions. A CUBE emerges (Steinhaus 1983, pp. 203–04). Connecting the centers of these 14 spheres gives a STELLA OCTANGULA.

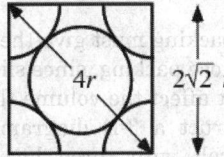

Consider the CUBE defined by 14 spheres in cubic close packing, as illustrated above. This "unit cell" contains eight 1/8-spheres (one at each VERTEX) and six HEMISPHERES. The total VOLUME of SPHERES in the unit cell is therefore

$$V_{\text{spheres in unit cell}} = \left(8 \cdot \tfrac{1}{8} + 6 \cdot \tfrac{1}{2}\right)\frac{4\pi}{3}r^3$$

$$= 4 \cdot \frac{4\pi}{3}r^3 = \frac{16}{3}\pi r^3. \tag{3}$$

The diagonal of the face is $4r$, so each side is $2\sqrt{2}r$. The VOLUME of the unit cell is therefore

$$V_{\text{unit cell}} = \left(2\sqrt{2}r\right)^3 = 16\sqrt{2}r^3. \tag{4}$$

and the PACKING DENSITY is

$$\eta_{\text{CCP}} = \frac{\frac{16}{3}\pi r^2}{16\sqrt{2}r^3} = \frac{\pi}{3\sqrt{2}} \tag{5}$$

(Conway and Sloane 1993, p. 2).

In cubic close packing, each sphere is surrounded by 12 other spheres. Taking a collection of 13 such spheres gives the cluster illustrated above. Connecting the centers of the external 12 spheres gives a CUBOCTAHEDRON (Steinhaus 1983, pp. 203–05; Wells 1991, p. 237).

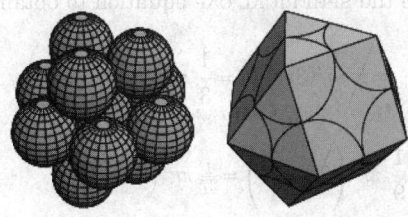

In hexagonal close packing, layers of spheres are packed so that spheres in alternating layers overlie one another. As in cubic close packing, each sphere is surrounded by 12 other spheres. Taking a collection of 13 such spheres gives the cluster illustrated above. Connecting the centers of the external 12 spheres gives JOHNSON SOLID J_{27} known as the TRIANGULAR ORTHOBICUPOLA (Steinhaus 1983, pp. 203–05; Wells 1991, p. 237).

Hexagonal close packing must give the same packing density as cubic close packing, since sliding one sheet of SPHERES cannot affect the volume they occupy. To verify this, construct a 3-D diagram containing a hexagonal unit cell with three layers (Steinhaus 1983, pp. 203–04). Both the top and the bottom contain six 1/6-SPHERES and one HEMISPHERE. The total number of spheres in these two rows is therefore

$$2\left(6 \cdot \tfrac{1}{6} + 1 \cdot \tfrac{1}{2}\right) = 3. \qquad (6)$$

The VOLUME of SPHERES in the middle row cannot be simply computed using geometry. However, symmetry requires that the piece of the SPHERE which is cut off is exactly balanced by an extra piece on the other side. There are therefore three SPHERES in the middle layer, for a total of six, and a total VOLUME

$$V_{\text{spheres in unit cell}} = 6 \cdot \frac{4\pi}{3} r^3 (3+3) = 8\pi r^3. \qquad (7)$$

The base of the HEXAGON is made up of 6 EQUILATERAL TRIANGLES with side lengths $2r$. The unit cell base AREA is therefore

$$A_{\text{unit cell}} = 6\left[\tfrac{1}{2}(2r)\left(\sqrt{3}r\right)\right] = 6\sqrt{3}r^2. \qquad (8)$$

The height is the same as that of two TETRAHEDRA length $2r$ on a side, so

$$h_{\text{unit cell}} = 2\left(2r\sqrt{\frac{2}{3}}\right). \qquad (9)$$

giving

$$\eta_{\text{HCP}} = \frac{8\pi r^3}{\left(6\sqrt{3}r^2\right)\left(4r\sqrt{\frac{2}{3}}\right)} = \frac{\pi}{3\sqrt{2}} \qquad (10)$$

(Conway and Sloane 1993, pp. 7 and 9).

If we had actually wanted to compute the VOLUME of SPHERE inside and outside the HEXAGONAL PRISM, we could use the SPHERICAL CAP equation to obtain

$$V_{\subset} = \tfrac{1}{3}\pi h^2(3r-h) = \frac{1}{3}\pi r^3 \frac{1}{3}\left(3 - \frac{1}{\sqrt{3}}\right)$$

$$= \frac{1}{9}\pi r^3\left(3 - \frac{\sqrt{3}}{3}\right) = \tfrac{1}{27}\pi r^3\left(9 - \sqrt{3}\right) \qquad (11)$$

$$V_{\supset} = \pi r^3\left[\tfrac{4}{3} - \tfrac{1}{27}(9 - \sqrt{3})\right] = \tfrac{1}{27}\pi r^3\left(36 - 9 + \sqrt{3}\right)$$

$$= \tfrac{1}{27}\pi r^3\left(27 + \sqrt{3}\right). \qquad (12)$$

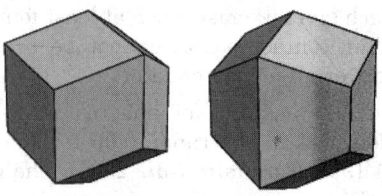

If spheres packed in a cubic lattice, face-centered cubic lattice, and hexagonal lattice are allowed to expand uniformly until running into each other, they form cubes, hexagonal prisms, and rhombic dodecahedra, respectively. In particular, if the spheres of cubic close packing are expanded until they fill up the gaps, they form a solid RHOMBIC DODECAHEDRON (left figure above), and if the spheres of hexagonal close packing are expanded, they form a second irregular dodecahedron consisting of six rhombi and six trapezoids (right figure above; Steinhaus 1983, p. 206). The latter can be obtained from the former by slicing in half and rotating the two halves 60° with respect to each other. The lengths of the short and long edges of the rotated dodecahedron have lengths 2/3 and 4/3 times the length of the rhombic faces. Both the RHOMBIC DODECAHEDRON and squashed dodecahedron are SPACE-FILLING POLYHEDRA.

Compressing a random packing gives polyhedra with an average of 13.3 faces (Coxeter 1958, 1961).

For sphere packing inside a CUBE, see Goldberg (1971), Schaer (1966), and Friedman.

See also CANNONBALL PROBLEM, CIRCLE PACKING, CUBOCTAHEDRON, DODECAHEDRAL CONJECTURE, ELLIPSOID PACKING, HEMISPHERE, HERMITE CONSTANTS, HYPERSPHERE, HYPERSPHERE PACKING, KEPLER CONJECTURE, KEPLER PROBLEM, KISSING NUMBER, LOCAL DENSITY, LOCAL DENSITY CONJECTURE, RANDOM CLOSE PACKING, REULEAUX TETRAHEDRON, SPACE-FILLING POLYHEDRON, SPHERE, SPHERICAL DESIGN, SPHERICON, STELLA OCTANGULA, TANGENT SPHERES, TRIANGULAR ORTHOBICUPOLA, UNIT CELL

References

Barlow, W. "Probable Nature of the Internal Symmetry of Crystals." *Nature* **29**, 186–88, 1883.

Conway, J. H. and Sloane, N. J. A. *Sphere Packings, Lattices, and Groups, 2nd ed.* New York: Springer-Verlag, 1993.

Coxeter, H. S. M. "Close-Packing and so Forth." *Illinois J. Math.* **2**, 746–58, 1958.

Coxeter, H. S. M. "Close Packing of Equal Spheres." Section 22.4 in *Introduction to Geometry, 2nd ed.* New York: Wiley, pp. 405–11, 1961.

Coxeter, H. S. M. "The Problem of Packing a Number of Equal Nonoverlapping Circles on a Sphere." *Trans. New York Acad. Sci.* **24**, 320–31, 1962.

Critchlow, K. *Order in Space: A Design Source Book.* New York: Viking Press, 1970.

Cundy, H. and Rollett, A. *Mathematical Models, 3rd ed.* Stradbroke, England: Tarquin Pub., pp. 195–97, 1989.

Eppstein, D. "Covering and Packing." http://www.ics.uci.edu/~eppstein/junkyard/cover.html.

Fejes Tóth, G. "Über einen geometrischen Satz." *Math. Z.* **46**, 78–3, 1940.

Fejes Tóth, G. *Lagerungen in der Ebene, auf der Kugel und in Raum, 2nd ed.* Berlin: Springer-Verlag, 1972.

Friedman, E. "Spheres in Cubes." http://www.stetson.edu/~efriedma/sphincub/.

Gardner, M. "Packing Spheres." Ch. 7 in *Martin Gardner's New Mathematical Diversions from Scientific American.* New York: Simon and Schuster, pp. 82–0, 1966.

Gauss, C. F. "Besprechung des Buchs von L. A. Seeber: Intersuchungen über die Eigenschaften der positiven ternären quadratischen Formen usw." *Göttingsche Gelehrte Anzeigen (1831, July 9)* **2**, 188–96, 1876.

Goldberg, M. "On the Densest Packing of Equal Spheres in a Cube." *Math. Mag.* **44**, 199–08, 1971.

Hales, T. C. "The Sphere Packing Problem." *J. Comput. Appl. Math* **44**, 41–6, 1992.

Hilbert, D. and Cohn-Vossen, S. *Geometry and the Imagination.* New York: Chelsea, pp. 45–3, 1999.

Jaeger, H. M. and Nagel, S. R. "Physics of Granular States." *Science* **255**, 1524, 1992.

Le Lionnais, F. *Les nombres remarquables.* Paris: Hermann, p. 31, 1983.

Lindsey, J. H. II. "Sphere Packing in $\mathbb{R}^3$." *Math.* **33**, 137–47, 1986.

Muder, D. J. "Putting the Best Face of a Voronoi Polyhedron." *Proc. London Math. Soc.* **56**, 329–48, 1988.

Rogers, C. A. "The Packing of Equal Spheres." *Proc. London Math. Soc.* **8**, 609–20, 1958.

Rogers, C. A. *Packing and Covering.* Cambridge, England: Cambridge University Press, 1964.

Schaer, J. "On the Densest Packing of Spheres in a Cube." *Can. Math. Bul.* **9**, 265–70, 1966.

Sigrist, F. "Sphere Packing." *Math. Intell.* **5**, 34–8, 1983.

Sloane, N. J. A. "The Packing of Spheres." *Sci. Amer.* **250**, 116–25, 1984.

Sloane, N. J. A. "The Sphere Packing Problem." *Proc. Internat. Congress Math., Vol. 3 (Berlin, 1998).* Doc. Math. Extra Volume ICM 1998, 387–96, 1998. http://www.research.att.com/~njas/doc/icm.ps.

Steinhaus, H. *Mathematical Snapshots, 3rd ed.* New York: Dover, pp. 202–03, 1999.

Stewart, I. *The Problems of Mathematics, 2nd ed.* Oxford, England: Oxford University Press, pp. 69–2, 1987.

Thompson, T. M. *From Error-Correcting Codes Through Sphere Packings to Simple Groups.* Washington, DC: Math. Assoc. Amer., 1984.

Torquato, S.; Truskett, T. M.; and Debenedetti, P. G. "Is Random Close Packing of Spheres Well Defined?" *Phys. Lev. Lett.* **84**, 2064–067, 2000.

Wells, D. *The Penguin Dictionary of Curious and Interesting Numbers.* Middlesex, England: Penguin Books, p. 29, 1986.

Weisstein, E. W. "Books about Sphere Packings." http://www.treasure-troves.com/books/SpherePackings.html.

Wells, D. *The Penguin Dictionary of Curious and Interesting Geometry.* London: Penguin, pp. 237–38, 1991.

Zong, C. and Talbot, J. *Sphere Packings.* New York: Springer-Verlag, 1999.

Sphere Point Picking

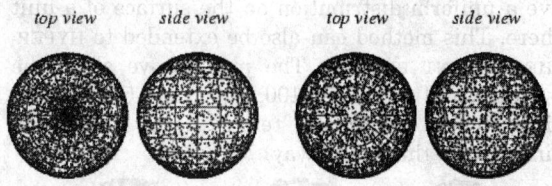

top view　*side view*　*top view*　*side view*

incorrectly distributed points　　*correctly distributed points*

To pick a random point on the surface of a UNIT SPHERE, it is incorrect to select SPHERICAL COORDINATES θ and ϕ from uniform distributions $\theta \in [0,\ 2\pi)$ and $\phi \in [0,\ \pi]$, since the area element $d\Omega = \sin\phi\,d\theta\,d\phi$ is a function of ϕ, and hence points picked in this way will be "bunched" near the poles (left figure above).

To obtain points such that any small area on the sphere is expected to contain the same number of points (right figure above), choose u and v to be random variates on $(0,\ 1)$. Then

$$\theta = 2\pi u \tag{1}$$

$$\phi = \cos^{-1}(2v - 1) \tag{2}$$

gives the SPHERICAL COORDINATES for a set of points which are uniformly distributed over $\mathbb{S}^2$. This works since the differential element of SOLID ANGLE is given by

$$d\Omega = \sin\phi\,d\theta\,d\phi = d\theta\,d(\cos\phi). \tag{3}$$

Similarly, we can pick $u = \cos\phi$ to be uniformly distributed (so we have $du = \sin\phi\,d\phi$) and obtain the points

$$x = \sqrt{1 - u^2}\cos\theta \tag{4}$$

$$y = \sqrt{1 - u^2}\sin\theta \tag{5}$$

$$z = u, \tag{6}$$

with $\theta \in [0,\ 2\pi)$ and $u \in [-1,\ 1]$, which are also uniformly distributed over $\mathbb{S}^2$.

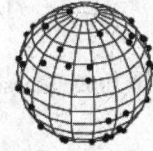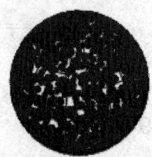

Marsaglia (1972) derived an elegant method that consists of picking x_1 and x_2 from independent uniform distributions on $(-1,\ 1)$ and rejecting points for which $x_1^2 + x_2^2 \geq 1$. From the remaining points,

$$x = 2x_1\sqrt{1 - x_1^2 - x_2^2} \tag{7}$$

$$y = 2x_2\sqrt{1 - x_1^2 - x_2^2} \tag{8}$$

$$z = 1 - 2(x_1^2 + x_2^2) \qquad (9)$$

have a uniform distribution on the surface of a unit sphere. This method can also be extended to HYPER-SPHERE POINT PICKING. The plots above show the distribution of points for 100, 1000, and 5000 initial points (where the counts refers to the number of points before throwing away).

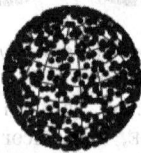

Cook (1957) extended a method of von Neumann (1951) to give a simple method of picking points uniformly distributed on the surface of a UNIT SPHERE. Pick four numbers x_0, x_1, x_2, and x_3 from a UNIFORM DISTRIBUTION on $(-1, 1)$, and reject pairs with

$$x_0^2 + x_1^2 + x_2^2 + x_3^2 \geq 1. \qquad (10)$$

From the remaining points, the rules of QUATERNION transformation then imply that the points with CARTESIAN COORDINATES

$$x = \frac{2(x_1 x_3 + x_0 x_2)}{x_0^2 + x_1^2 + x_2^2 + x_3^2} \qquad (11)$$

$$y = \frac{2(x_2 x_3 - x_0 x_1)}{x_0^2 + x_1^2 + x_2^2 + x_3^2} \qquad (12)$$

$$z = \frac{x_0^2 + x_3^2 - x_1^2 - x_2^2}{x_0^2 + x_1^2 + x_2^2 + x_3^2} \qquad (13)$$

have the desired distribution (Cook 1957, Marsaglia 1972). The plots above show the distribution of points for 100, 1000, and 5000 initial points (where the counts refers to the number of points before throwing away).

Another easy way to pick a random point on a SPHERE is to generate three Gaussian random variables x, y, and z. Then the distribution of the vectors

$$\frac{1}{\sqrt{x^2 + y^2 + z^2}} \begin{bmatrix} x \\ y \\ z \end{bmatrix} \qquad (14)$$

is uniform over the surface $\mathbb{S}^2$ (Muller 1959, Marsaglia 1972).

See also BALL TRIANGLE PICKING, CIRCLE POINT PICKING, DISK POINT PICKING, HYPERSPHERE POINT PICKING, NOISE SPHERE, SPHERE LINE PICKING, SPHERE TETRAHEDRON PICKING

References

Cook, J. M. "Technical Notes and Short Papers: Rational Formulae for the Production of a Spherically Symmetric Probability Distribution." *Math. Tables Aids Comput.* **11**, 81–2, 1957.

Feller, W. *An Introduction to Probability Theory and Its Applications, Vol. 2, 3rd ed.* New York: Wiley, 1971.

Knuth, D. E. *The Art of Computer Programming, Vol. 2: Seminumerical Algorithms, 3rd ed.* Reading, MA: Addison-Wesley, pp. 130–31, 1998.

Marsaglia, G. "Choosing a Point from the Surface of a Sphere." *Ann. Math. Stat.* **43**, 645–46, 1972.

Muller, M. E. "A Note on a Method for Generating Points Uniformly on *N*-Dimensional Spheres" *Comm. Assoc. Comput. Mach.* **2**, 19–0, 1959.

Rusin, D. "N-Dim Spherical Random Number Drawing." in *The Mathematical Atlas.* http://www.math.niu.edu/~rusin/known-math/96/sph.rand.

Stephens, M. A. "The Testing of Unit Vectors for Randomness." *J. Amer. Stat. Assoc.* **59**, 160–67, 1964.

von Neumann, J. "VArious Techniques Used in Connection with Random Digits." *NBS Appl. Math. Ser.*, No. 12. Washington, DC: U.S. Government Printing Office, pp. 36–8, 1951.

Watson, G. S. and Williams, E. J. "On the Construction of Significance Tests on the Circle and Sphere." *Biometrika* **43**, 344–52, 1956.

Sphere Tetrahedron Picking

Pick four points on a sphere. What is the probability that the TETRAHEDRON having these points as VERTICES contains the CENTER of the sphere? In the 1-D case, the probability that a second point is on the opposite side of $1/2$ is $1/2$. In the 2-D case, pick two points. In order for the third to form a TRIANGLE containing the CENTER, it must lie in the quadrant bisected by a LINE SEGMENT passing through the center of the CIRCLE and the bisector of the two points. This happens for one QUADRANT, so the probability is $1/4$. Similarly, for a sphere the probability is one OCTANT, or $1/8$.

Pick four points at random on the surface of a unit SPHERE using

$$x = \sqrt{1 - u^2} \cos \theta \qquad (1)$$

$$y = \sqrt{1 - u^2} \sin \theta \qquad (2)$$

$$z = u \qquad (3)$$

with $u \in [-1, 1]$ and $\theta \in [0, \pi)$. Now find the distribution of possible VOLUMES of the (nonregular) TETRAHEDRA determined by these points. Without loss of generality, the first point may be taken as $u_1 = 1$, or $(0, 0, 1)$, while the second may be taken as $(0, u_2)$, or $\left(\sqrt{1 - u_2^2}, 0, u_2\right)$. The average VOLUME is then

$$\bar{V} = \frac{f_{-1}^1 f_{-1}^1 f_{-1}^1 f_{-1}^1 \int_0^{2x} \int_0^{2x} |V(x_i)| \, du_2 \, du_3 \, du_4 \, d\theta_3 \, d\theta_4}{f_{-1}^1 f_{-1}^1 f_{-1}^1 f_{-1}^1 \int_0^{2x} \int_0^{2x} du_2 \, du_3 u_4 \, d\theta_3 \, d\theta_4},$$

$$(4)$$

where the VERTICES are located at $\{x_i, y_i, z_i\}$ where $i = 1, ..., 4$, and the (signed) VOLUME is given by the DETERMINANT

$$V = \frac{1}{3!} \begin{vmatrix} x_1 & y_1 & z_1 & 1 \\ x_2 & y_2 & z_2 & 1 \\ x_3 & y_3 & z_3 & 1 \\ x_4 & y_4 & z_4 & 1 \end{vmatrix}. \tag{5}$$

The analytic result is difficult to compute, but is numerically given by $\bar{V} \approx 0.120$.

See also BALL TETRAHEDRON PICKING, CUBE TETRA-HEDRON PICKING, FEJES TÓTH'S PROBLEM, POINT PICKING, SPHERE LINE PICKING, TETRAHEDRON

References

Buchta, C. "A Note on the Volume of a Random Polytope in a Tetrahedron." *Ill. J. Math.* **30**, 653–59, 1986.

Sphere with Tunnel

Find the tunnel between two points A and B on a gravitating SPHERE which gives the shortest transit time under the force of gravity. Assume the SPHERE to be nonrotating, of RADIUS a, and with uniform density ρ. Then the standard form EULER-LAGRANGE DIFFER-ENTIAL EQUATION in polar coordinates is

$$r_{\phi\phi}(r^3 - ra^2) + r_{\phi}^2(2a^2 - r^2) + a^2r^2 = 0. \tag{1}$$

along with the boundary conditions $r(\phi = 0) = r_0$, $r_{\phi}(\phi = 0) = 0$, $r(\phi = \phi_A) = a$, and $r(\phi = \phi_B) = a$. Integrating once gives

$$r_{\phi}^2 = \frac{a^2 r^2}{r_0^2} \frac{r^2 - r_0^2}{a^2 - r^2}. \tag{2}$$

But this is the equation of a HYPOCYCLOID generated by a CIRCLE of RADIUS $\frac{1}{2}(a - r_0)$ rolling inside the CIRCLE of RADIUS a, so the tunnel is shaped like an arc of a HYPOCYCLOID. The transit time from point A to point B is

$$T = \pi \sqrt{\frac{a^2 - r_0^2}{ag}}, \tag{3}$$

where

$$g = \frac{GM}{a^2} = \tfrac{4}{3}\pi\rho Ga \tag{4}$$

is the surface gravity with G the universal gravita-tional constant.

Sphere-Cone Intersection

CONE-SPHERE INTERSECTION

Sphere-Cylinder Intersection

CYLINDER-SPHERE INTERSECTION

Sphere-Sphere Intersection

Let two spheres of RADII R and r be located along the x-AXIS centered at $(0, 0, 0)$ and $(d, 0, 0)$, respec-tively. Not surprisingly, the analysis is very similar to the case of the CIRCLE-CIRCLE INTERSECTION. The equations of the two SPHERES are

$$x^2 + y^2 + z^2 = R^2 \tag{1}$$

$$(x - d)^2 + y^2 + z^2 = r^2. \tag{2}$$

Combining (1) and (2) gives

$$(x - d)^2 + (R^2 - x^2) = r^2. \tag{3}$$

Multiplying through and rearranging give

$$x^2 - 2dx + d^2 - x^2 = r^2 - R^2. \tag{4}$$

Solving for x gives

$$x = \frac{d^2 - r^2 + R^2}{2d}. \tag{5}$$

The intersection of the SPHERES is therefore a curve lying in a PLANE parallel to the yz-plane at a single x-coordinate. Plugging this back into (1) gives

$$y^2 + z^2 = R^2 - x^2 = R^2 - \left(\frac{d^2 - r^2 + R^2}{2d}\right)^2$$

$$= \frac{4d^2 B^2 - (d^2 - r^2 + R^2)^2}{4d^2}. \tag{6}$$

which is a CIRCLE with RADIUS

$$a = \frac{1}{2d}\sqrt{4d^2 R^2 - (d^2 - r^2 + R^2)^2}$$

$$= \frac{1}{2d}[(-d + r - R)(-d - r + R)(-d + r + R)$$

$$\times (d + r + R)]^{1/2}. \tag{7}$$

The VOLUME of the 3-D LENS common to the two spheres can be found by adding the two SPHERICAL CAPS. The distances from the SPHERES' centers to the bases of the caps are

$$d_1 = x \tag{8}$$

$$d_2 = d - x, \tag{9}$$

so the heights of the caps are

$$h_1 = R - d_1 = \frac{(r - R + d)(r + R - d)}{2d} \tag{10}$$

$$h_2 = r - d_2 = \frac{(R - r + d)(R + r - d)}{2d}. \tag{11}$$

The VOLUME of a SPHERICAL CAP of height h' for a SPHERE of RADIUS R' is

$$V(R', h') = \tfrac{1}{3} \pi h'^2 (3R' - h'). \tag{12}$$

Letting $R_1 = R$ and $R_2 = r$ and summing the two caps gives

$$V = V(R_1, h_1) + V(R_2, h_2)$$

$$= \frac{\pi(R + r - d)^2 (d^2 + 2dr - 3r^2 + 2dR + 6rR - 3R^2)}{12d}.$$

$$\tag{13}$$

This expression gives $V = 0$ for $d = r + R$ as it must. In the special case $r = R$, the VOLUME simplifies to

$$V = \tfrac{1}{12} \pi (4R + d)(2R - d)^2. \tag{14}$$

The SURFACE AREA of the sphere R that lies inside the sphere r is equal to the GREAT CIRCLE of the sphere r, provided that $r \leq 2R$ (Kern and Blank 1948, p. 97).

See also APPLE, CIRCLE-CIRCLE INTERSECTION, DOUBLE BUBBLE, LENS, SPACE DIVISION BY SPHERES, SPHERE

References

Kern, W. F. and Bland, J. R. *Solid Mensuration with Proofs,* 2nd ed. New York: Wiley, p. 97, 1948.

Spherical Bessel Differential Equation

Take the HELMHOLTZ DIFFERENTIAL EQUATION

$$\nabla^2 F + k^2 F = 0 \tag{1}$$

in SPHERICAL COORDINATES. This is just LAPLACE'S EQUATION in SPHERICAL COORDINATES with an additional term,

$$\frac{d^2 R}{dr^2} \Phi\Theta = \frac{2}{r} \frac{dR}{dr} \Phi\Theta + \frac{1}{r^2 \sin^2 \phi} \frac{d^2\Theta}{d\theta^2} \Phi R$$

$$+ \frac{\cos\phi}{r^2 \sin\phi} \frac{d\Phi}{d\phi} \Theta R + \frac{1}{r^2} \frac{d^2\Phi}{d\phi^2} \Theta R + k^2 R\Phi\Theta = 0. \tag{2}$$

Multiply through by $r^2/R\Phi\Theta$,

$$\frac{r^2}{R} \frac{d^2 R}{dr^2} + \frac{2r}{R} \frac{dR}{dr} + k^2 r^2 + \frac{1}{\Theta \sin^2 \phi} \frac{d^2\Theta}{d\theta^2} + \frac{\cos\phi}{\Phi \sin\phi} \frac{d\Phi}{d\phi}$$

$$+ \frac{1}{\Phi} \frac{d^2\Phi}{d\phi^2} = 0. \tag{3}$$

This equation is separable in R. Call the separation constant $n(n + 1)$,

$$\frac{r^2}{R} \frac{d^2 R}{dr^2} + \frac{2r}{R} \frac{dR}{dr} + k^2 r^2 = n(n + 1). \tag{4}$$

Now multiply through by R,

$$r^2 \frac{d^2 R}{dr^2} + 2r \frac{dR}{dr} + [k^2 r^2 - n(n + 1)]R = 0. \tag{5}$$

This is the SPHERICAL BESSEL DIFFERENTIAL EQUATION. It can be transformed by letting $x \equiv kr$, then

$$r \frac{dR(r)}{dr} = kr \frac{dR(r)}{k \, dr} = kr \frac{dR(r)}{d(kr)} = x \frac{dR(r)}{dx}. \tag{6}$$

Similarly,

$$r^2 \frac{d^2 R(r)}{dr^2} = x^2 \frac{d^2 R(r)}{dx^2}. \tag{7}$$

so the equation becomes

$$x^2 \frac{d^2 R}{dx^2} = 2x \frac{dR}{dx} + [x^2 - n(n + 1)]R = 0. \tag{8}$$

Now look for a solution OF THE FORM $R(r) = Z(x)x^{-1/2}$, denoting a derivative with respect to x by a prime,

$$R' = Z'x^{-1/2} - \tfrac{1}{2} Zx^{-3/2} \tag{9}$$

$$R'' = Z''x^{-1/2} - \tfrac{1}{2} Z'x^{-3/2} - \tfrac{1}{2} Z'x^{-3/2} - \tfrac{1}{2}\left(-\tfrac{3}{2}\right) Zx^{-5/2}$$

$$= Z''x^{-1/2} - Z'x^{-3/2} + \tfrac{3}{4} Zx^{-5/2} \tag{10}$$

so

$$x^2\left(Z''x^{-1/2} - Z'x^{-3/2} + \tfrac{3}{4} Zx^{-5/2}\right)$$

$$+ 2x\left(Z'x^{-1/2} - \tfrac{1}{2} Zx^{-3/2}\right) + [x^2 - n(n + 1)]Zx^{-1/2} = 0$$

$$\tag{11}$$

$$x^2\left(Z'' - Z'x^{-1} + \tfrac{3}{4} Zx^{-2}\right) + 2x\left(Z' - \tfrac{1}{2} Zx^{-1}\right)$$

$$+ [x^2 - n(n + 1)]Z = 0 \tag{12}$$

$$x^2 Z'' + (-x + 2x)Z' + \left[\tfrac{3}{4} - 1 + x^2 - n(n + 1)\right]Z = 0 \tag{13}$$

$$x^2 Z'' + xZ' + \left[x^2 - \left(n^2 + n + \tfrac{1}{4}\right)\right]Z = 0 \tag{14}$$

$$x^2 Z'' + xZ' + \left[x^2 - \left(n + \tfrac{1}{2}\right)^2\right]Z = 0. \tag{15}$$

But the solutions to this equation are BESSEL FUNCTIONS of half integral order, so the normalized solutions to the original equation are

$$R(r) \equiv A\frac{J_{n+1/2}(kr)}{\sqrt{kr}} + B\frac{Y_{n+1/2}(kr)}{\sqrt{kr}} \qquad (16)$$

which are known as SPHERICAL BESSEL FUNCTIONS. The two types of solutions are denoted $j_n(x)$ (SPHERICAL BESSEL FUNCTION OF THE FIRST KIND) or $n_n(x)$ (SPHERICAL BESSEL FUNCTION OF THE SECOND KIND), and the general solution is written

$$R(r) = A'j_n(kr) + B'n_n(kr). \qquad (17)$$

where

$$j_n(z) \equiv \sqrt{\frac{\pi}{2}}\,\frac{J_{n+1/2}(z)}{\sqrt{z}} \qquad (18)$$

$$n_n(z) \equiv \sqrt{\frac{\pi}{2}}\,\frac{Y_{n+1/2}(z)}{\sqrt{z}}. \qquad (19)$$

See also SPHERICAL BESSEL FUNCTION, SPHERICAL BESSEL FUNCTION OF THE FIRST KIND, SPHERICAL BESSEL FUNCTION OF THE SECOND KIND

References

Abramowitz, M. and Stegun, C. A. (Eds.). *Handbook of Mathematical Functions with Formulas, Graphs, and Mathematical Tables, 9th printing.* New York: Dover, p. 437, 1972.
Zwillinger, D. *Handbook of Differential Equations, 3rd ed.* Boston, MA: Academic Press, p. 121, 1997.

Spherical Bessel Function

A solution to the SPHERICAL BESSEL DIFFERENTIAL EQUATION. The two types of solutions are denoted $j_n(x)$ (SPHERICAL BESSEL FUNCTION OF THE FIRST KIND) or $n_n(x)$ (SPHERICAL BESSEL FUNCTION OF THE SECOND KIND).

See also SPHERICAL BESSEL DIFFERENTIAL EQUATION, SPHERICAL BESSEL FUNCTION OF THE FIRST KIND, SPHERICAL BESSEL FUNCTION OF THE SECOND KIND

References

Abramowitz, M. and Stegun, C. A. (Eds.). "Spherical Bessel Functions." §10.1 in *Handbook of Mathematical Functions with Formulas, Graphs, and Mathematical Tables, 9th printing.* New York: Dover, pp. 437–42, 1972.
Arfken, G. "Spherical Bessel Functions." §11.7 in *Mathematical Methods for Physicists, 3rd ed.* Orlando, FL: Academic Press, pp. 622–36, 1985.
Press, W. H.; Flannery, B. P.; Teukolsky, S. A.; and Vetterling, W. T. "Bessel Functions of Fractional Order, Airy Functions, Spherical Bessel Functions." §6.7 in *Numerical Recipes in FORTRAN: The Art of Scientific Computing, 2nd ed.* Cambridge, England: Cambridge University Press, pp. 234–45, 1992.

Spherical Bessel Function of the First Kind

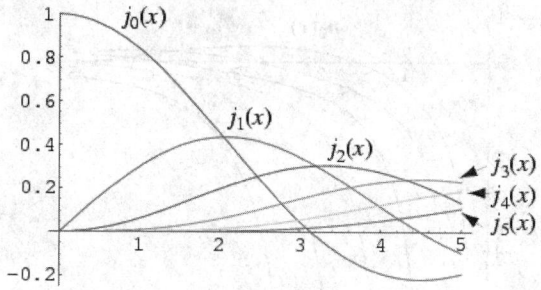

$$j_n(x) \equiv \sqrt{\frac{\pi}{2x}}\,J_{n+1/2}(x) \qquad (1)$$

$$= 2^n x^n \sum_{s=0}^{\infty} \frac{(-1)^s (s+n)!}{s!(2s+2n+1)!} x^{2s} \qquad (2)$$

$$= \frac{x^n}{(2n+1)!!}$$

$$\times \left[1 - \frac{\frac{1}{2}x^2}{1!(2n+3)} + \frac{\left(\frac{1}{2}x^2\right)^2}{2!(2n+3)(2n+5)} + \ldots\right] \qquad (3)$$

$$= (-1)^n x^n \left(\frac{d}{x\,dx}\right)^n \frac{\sin x}{x} \qquad (4)$$

where $j_n(z)$ is a BESSEL FUNCTION OF THE FIRST KIND. The first few functions are

$$j_0(x) = \frac{\sin x}{x} \qquad (5)$$

$$j_1(x) = \frac{\sin x}{x^2} - \frac{\cos x}{x} \qquad (6)$$

$$j_2(x) = \left(\frac{3}{x^3} - \frac{1}{x}\right)\sin x - \frac{3}{x^2}\cos x. \qquad (7)$$

Spherical Bessel functions are not explicitly implemented in *Mathematica*.

See also SPHERICAL BESSEL DIFFERENTIAL EQUATION, BESSEL FUNCTION OF THE SECOND KIND, POISSON INTEGRAL REPRESENTATION, RAYLEIGH'S FORMULAS, SPHERICAL BESSEL FUNCTION OF THE SECOND KIND

References

Abramowitz, M. and Stegun, C. A. (Eds.). "Spherical Bessel Functions." §10.1 in *Handbook of Mathematical Functions with Formulas, Graphs, and Mathematical Tables, 9th printing.* New York: Dover, pp. 437–42, 1972.
Arfken, G. "Spherical Bessel Functions." §11.7 in *Mathematical Methods for Physicists, 3rd ed.* Orlando, FL: Academic Press, pp. 622–36, 1985.

Spherical Bessel Function of the Second Kind

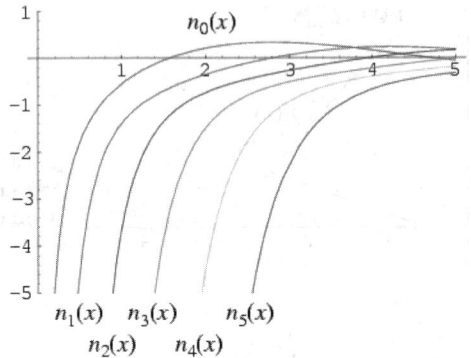

$$n_n(x) \equiv \sqrt{\frac{\pi}{2x}} Y_{n+1/2}(x) \qquad (1)$$

$$= \frac{(-1)^{n+1}}{2^n x^{n+1}} \sum_{s=0}^{\infty} \frac{(-1)^s (s-n)!}{s!(2s-2n)!} x^{2s} \qquad (2)$$

$$= \frac{(-1)^{n+1}}{2^n x^{n+1}} \sum_{s=0}^{\infty} \frac{(-1)^s 4^{n-s} \sqrt{\pi}}{\Gamma(s+1)\Gamma\left(\frac{1}{2}-n+s\right)} \qquad (3)$$

$$= -\frac{(2n-1)!!}{x^{n+1}}$$

$$\times \left[1 - \frac{\frac{1}{2}x^2}{1!(1-2n)} + \frac{\left(\frac{1}{2}x^2\right)^2}{2!(1-2n)(3-2n)} + \cdots \right] \qquad (4)$$

$$= (-1)^{n+1} \sqrt{\frac{\pi}{2x}} J_{-n-1/2}(x). \qquad (5)$$

where $Y_n(z)$ is a BESSEL FUNCTION OF THE SECOND KIND and $j_n(z)$ is a BESSEL FUNCTION OF THE FIRST KIND.

The first few functions are

$$n_0(x) = -\frac{\cos x}{x} \qquad (6)$$

$$n_1(x) = -\frac{\cos x}{x^2} - \frac{\sin x}{x} \qquad (7)$$

$$n_2(x) = -\left(\frac{3}{x^3} - \frac{1}{x}\right)\cos x - \frac{3}{x^2}\sin x. \qquad (8)$$

Spherical Bessel functions are not explicitly implemented in *Mathematica*.

See also SPHERICAL BESSEL DIFFERENTIAL EQUATION, BESSEL FUNCTION OF THE SECOND KIND, RAYLEIGH'S FORMULAS, SPHERICAL BESSEL FUNCTION OF THE FIRST KIND

References

Abramowitz, M. and Stegun, C. A. (Eds.). "Spherical Bessel Functions." §10.1 in *Handbook of Mathematical Functions with Formulas, Graphs, and Mathematical Tables, 9th printing.* New York: Dover, pp. 437–42, 1972.

Arfken, G. "Spherical Bessel Functions." §11.7 in *Mathematical Methods for Physicists, 3rd ed.* Orlando, FL: Academic Press, pp. 622–36, 1985.

Spherical Bessel Function of the Third Kind

SPHERICAL HANKEL FUNCTION OF THE FIRST KIND, SPHERICAL HANKEL FUNCTION OF THE SECOND KIND

Spherical Cap

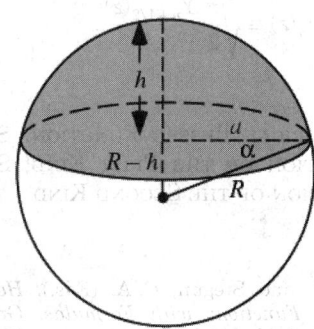

A spherical cap is the region of a SPHERE which lies above (or below) a given PLANE. If the PLANE passes through the CENTER of the SPHERE, the cap is a called a HEMISPHERE, and if the cap is cut by a second PLANE, it is called a SPHERICAL SEGMENT. However, Harris and Stocker (1998) use the term "spherical segment" as a synonym for what is here called a spherical cap and "zone" for SPHERICAL SEGMENT.

Let the SPHERE have RADIUS R, then the VOLUME of a spherical cap of height h and base RADIUS a is given by the equation of a SPHERICAL SEGMENT

$$V_{\text{spherical segment}} = \frac{1}{6}\pi h(3a^2 + 3b^2 + h^2) \qquad (1)$$

with $b = 0$, giving

$$V_{\text{cap}} = \frac{1}{6}\pi h(3a^2 + h^2). \qquad (2)$$

Using the PYTHAGOREAN THEOREM gives

$$(R-h)^2 + a^2 = R^2, \qquad (3)$$

which can be solved for a^2 as

$$a^2 = 2Rh - h^2. \qquad (4)$$

so the radius of the base circle is

$$a = \sqrt{h(2R-h)}. \qquad (5)$$

and plugging this in gives the equivalent formula

$$V_{\text{cap}} = \frac{1}{3}\pi h^2(3R - h). \qquad (6)$$

In terms of the so-called CONTACT ANGLE (the angle

between the normal to the sphere at the bottom of the cap and the base plane)

$$R - h = R \sin \alpha \qquad (7)$$

$$\alpha \equiv \sin^{-1}\left(\frac{R - h}{R}\right), \qquad (8)$$

so

$$V_{\text{cap}} = \tfrac{1}{3}\pi R^3 (2 - 3\sin\alpha + \sin^3\alpha). \qquad (9)$$

The CENTROID occurs at a distance

$$\bar{z} = \frac{3(2R - h)^2}{4(3R - h)} \qquad (10)$$

above the center of the sphere (Harris and Stocker 1998, p. 107).

Consider a cylindrical box enclosing the cap so that the top of the box is tangent to the top of the SPHERE. Then the enclosing box has VOLUME

$$V_{\text{box}} = \pi a^2 h = \pi (R\cos\alpha)[R(1 - \sin\alpha)]$$

$$= \pi R^3 (1 - \sin\alpha - \sin^2\alpha + \sin^3\alpha), \qquad (11)$$

so the hollow volume between the cap and box is given by

$$V_{\text{box}} - V_{\text{cap}} = \tfrac{1}{3}\pi R^3 (1 - 3\sin^2\alpha + 2\sin^3\alpha). \qquad (12)$$

If a second PLANE cuts the cap, the resulting SPHERICAL FRUSTUM is called a SPHERICAL SEGMENT. The SURFACE AREA of the spherical cap is given by the same equation as for a general ZONE:

$$S_{\text{cap}} = 2\pi R h = \pi (a^2 + h^2). \qquad (13)$$

See also CONTACT ANGLE, DOME, FRUSTUM, HEMISPHERE, SOLID OF REVOLUTION, SPHERE, SPHERICAL SEGMENT, SPHERICAL WEDGE, TORISPHERICAL DOME, ZONE

References

Harris, J. W. and Stocker, H. "Spherical Segment (Spherical Cap)." §4.8.4 in *Handbook of Mathematics and Computational Science.* New York: Springer-Verlag, p. 107, 1998.

Kern, W. F. and Bland, J. R. "Spherical Segment." §36 in *Solid Mensuration with Proofs, 2nd ed.* New York: Wiley, pp. 97–02, 1948.

Spherical Code

How can n points be distributed on a UNIT SPHERE such that they maximize the minimum distance between any pair of points? This maximum distance is called the covering radius, and the configuration is called a spherical code (or spherical packing). In 1943, Fejes Tóth proved that for n points, there always exist two points whose distance d is

$$d \le \sqrt{4 - \csc^2\left[\frac{\pi n}{6(n - 2)}\right]},$$

and that the limit is exact for $n = 3, 4, 6,$ and 12. The problem of spherical packing is therefore sometimes known as the Fejes Tóth's problem. The general problem has not been solved.

For two points, the points should be at opposite ends of a DIAMETER. For four points, they should be placed at the VERTICES of an inscribed regular TETRAHEDRON. There is no unique best solution for five points since the distance cannot be reduced below that for six points. For six points, they should be placed at the VERTICES of an inscribed regular OCTAHEDRON. For seven points, the best solution is four equilateral spherical triangles with angles of 80°. For eight points, the best dispersal is *not* the VERTICES of the inscribed CUBE, but of a SQUARE ANTIPRISM with equal EDGES. The solution for nine points is eight equilateral spherical triangles with angles of $\cos^{-1}(1/4)$. For 12 points, the solution is an inscribed regular ICOSAHEDRON.

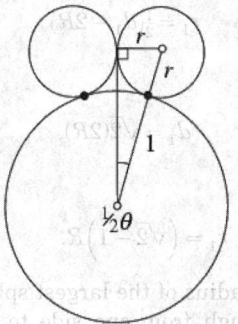

A spherical packing corresponds to the placement of n spheres around a central unit sphere. From simple trigonometry,

$$\sin\left(\tfrac{1}{2}\theta\right) = \frac{r}{1 + r}.$$

so the radii of the n spheres are given by

$$r = \frac{1}{\csc\left(\tfrac{1}{2}\theta\right) - 1}$$

for a minimum separation angle of θ. Hardin and Sloane give tables of minimum separations and sphere positions for $n \le 130$ and $d = 3, 4, 5$.

"Almost" 13 spheres can fit around a central sphere in the sense that there is a gap left over when 12 spheres are in place which is nearly big enough for an additional sphere (left figure). In fact, the radii of the spheres can be increased to 1.10851 (assuming a central unit sphere) before 12 spheres no longer fit (middle figure). In order to fit 13 spheres around a central unit sphere, their radius must be no larger than 0.916468 (right figure). These values correspond to Hardin and Sloane's angles of 63.4349488° and 57.1367031°, respectively.

Pack eight unit spheres whose centers are at the vertices of a cube. Then the radius of the largest sphere which fits in the center hole (left figure) is given by

$$r_1 = \tfrac{1}{2}(d_1 - 2R)$$

with

$$d_1 = \sqrt{2}(2R),$$

giving

$$r_1 = \left(\sqrt{2} - 1\right)R. \qquad (1)$$

Similarly, the radius of the largest sphere which can be passed through from one side to another (right figure) has

$$d_2 = \sqrt{3}(2R),$$

giving

$$r_2 = \tfrac{1}{2}(d_2 - 2R) = \left(\sqrt{3} - 1\right)R. \qquad (2)$$

See also KISSING NUMBER, SPHERICAL COVERING, SPHERICAL DESIGN, THOMSON PROBLEM

References

Friedman, E. "Points on a Sphere." http://www.stetson.edu/ ~efriedma/ptsphere/.
Hardin, R. H.; Sloane, N. J. A. S.; and Smith, W. D. *Spherical Codes.* In preparation. http://www.research.att.com/ ~njas/packings/.
Hardin, R. H.; Sloane, N. J. A.; and Smith, W. D. *Spherical Codes.* In preparation.
Ogilvy, C. S. *Excursions in Mathematics.* New York: Dover, p. 99, 1994.
Ogilvy, C. S. Solved by L. Moser. "Minimal Configuration of Five Points on a Sphere." Problem E946. *Amer. Math. Monthly* **58**, 592, 1951.

Schütte, K. and van der Waerden, B. L. "Auf welcher Kügel haben 5, 6, 7, 8 oder 9 Pünkte mit Mindestabstand Eins Platz?" *Math. Ann.* **123**, 96–24, 1951.
Whyte, L. L. "Unique Arrangement of Points on a Sphere." *Amer. Math. Monthly* **59**, 606–11, 1952.

Spherical Cone

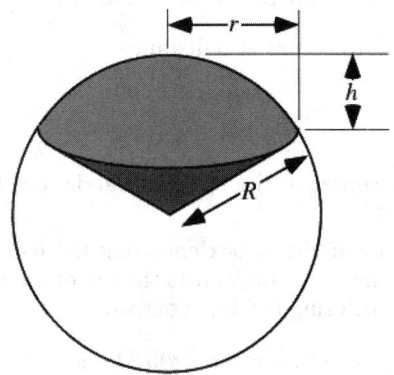

spherical cone

The SURFACE OF REVOLUTION obtained by cutting a conical "wedge" with vertex at the center of a SPHERE out of the SPHERE. A spherical cone is therefore a degenerate case of a SPHERICAL SECTOR. The volume of the spherical cone is

$$V = \tfrac{2}{3}\pi R^2 h \qquad (1)$$

(Kern and Bland 1948, p. 104). The SURFACE AREA of a closed spherical sector is

$$S = \pi R(2h + r); \qquad (2)$$

and the CENTROID is located at a height

$$\bar{z} = \tfrac{3}{8}(2R - h) \qquad (3)$$

above the sphere's center (Harris and Stocker 1998).

See also CONE, SPHERE, SPHERICAL CAP, SPHERICAL SECTOR

References

Harris, J. W. and Stocker, H. "Spherical Sector." §4.8.3 in *Handbook of Mathematics and Computational Science.* New York: Springer-Verlag, pp. 106–07, 1998.
Kern, W. F. and Bland, J. R. "Spherical Sector." §37 in *Solid Mensuration with Proofs, 2nd ed.* New York: Wiley, pp. 103–06, 1948.

Spherical Coordinates

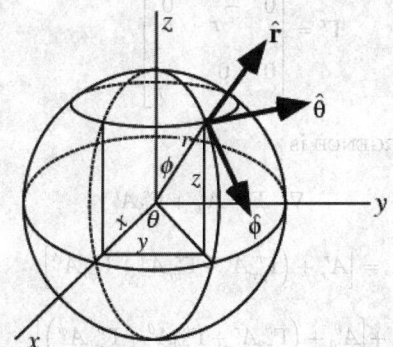

A system of CURVILINEAR COORDINATES which is natural for describing positions on a SPHERE or SPHEROID. Define θ to be the azimuthal ANGLE in the xy-PLANE from the x-AXIS with $0 \leq \theta < 2\pi$ (denoted λ when referred to as the LONGITUDE), ϕ to be the POLAR ANGLE from the z-AXIS with $0 \leq \phi \leq \pi$ (COLATITUDE, equal to $\phi = 90° - \delta$ where δ is the LATITUDE), and r to be distance (RADIUS) from a point to the ORIGIN.

Unfortunately, the convention in which the symbols θ and ϕ are reversed is frequently used, especially in physics, leading to unnecessary confusion. The symbol ρ is sometimes also used in place of r. Arfken (1985) uses (r, ϕ, θ), whereas Beyer (1987) uses (ρ, θ, ϕ). Be very careful when consulting the literature.

In this work, the symbols for the azimuthal, polar, and radial coordinates are taken as θ, ϕ, and r, respectively. Note that this definition provides a logical extension of the usual POLAR COORDINATES notation, with θ remaining the ANGLE in the xy-PLANE and ϕ becoming the ANGLE out of the PLANE.

$$r = \sqrt{x^2 + y^2 + z^2} \qquad (1)$$

$$\theta = \tan^{-1}\left(\frac{y}{x}\right) \qquad (2)$$

$$\phi = \sin^{-1}\left(\frac{\sqrt{x^2 + y^2}}{r}\right) = \cos^{-1}\left(\frac{z}{r}\right), \qquad (3)$$

where $r \in [0, \infty)$, $\theta \in [0, 2\pi)$, and $\phi \in [0, \pi]$. In terms of CARTESIAN COORDINATES,

$$x = r \cos\theta \sin\phi \qquad (4)$$

$$y = r \sin\theta \sin\phi \qquad (5)$$

$$z = r \cos\phi. \qquad (6)$$

The SCALE FACTORS are

$$h_r = 1 \qquad (7)$$

$$h_\theta = r \sin\phi \qquad (8)$$

$$h_\phi = r, \qquad (9)$$

so the METRIC COEFFICIENTS are

$$g_{rr} = 1 \qquad (10)$$

$$g_{\theta\theta} = r^2 \sin^2\phi \qquad (11)$$

$$g_{\phi\phi} = r^2. \qquad (12)$$

The LINE ELEMENT is

$$d\mathbf{s} = dr\hat{\mathbf{r}} + r\,d\phi\,\hat{\boldsymbol{\phi}} + r\sin\phi\,d\theta\,\hat{\boldsymbol{\theta}}, \qquad (13)$$

the AREA element

$$d\mathbf{a} = r^2 \sin\phi\,d\theta\,d\phi\,\hat{\mathbf{r}}, \qquad (14)$$

and the VOLUME ELEMENT

$$dV = r^2 \sin\phi\,d\theta\,d\phi\,dr. \qquad (15)$$

The JACOBIAN is

$$\left|\frac{\partial(x, y, z)}{\partial(r, \theta, \phi)}\right| = r^2 |\sin\phi|. \qquad (16)$$

The POSITION VECTOR is

$$\mathbf{r} \equiv \begin{bmatrix} r\cos\theta\sin\phi \\ r\sin\theta\sin\phi \\ r\cos\phi \end{bmatrix}, \qquad (17)$$

so the UNIT VECTORS are

$$\hat{\mathbf{r}} \equiv \frac{\dfrac{d\mathbf{r}}{dr}}{\left|\dfrac{d\mathbf{r}}{dr}\right|} = \begin{bmatrix} \cos\theta\sin\phi \\ \sin\theta\sin\phi \\ \cos\phi \end{bmatrix} \qquad (18)$$

$$\hat{\boldsymbol{\theta}} \equiv \frac{\dfrac{d\mathbf{r}}{d\theta}}{\left|\dfrac{d\mathbf{r}}{d\theta}\right|} = \begin{bmatrix} -\sin\theta \\ \cos\theta \\ 0 \end{bmatrix} \qquad (19)$$

$$\hat{\boldsymbol{\phi}} \equiv \frac{\dfrac{d\mathbf{r}}{d\phi}}{\left|\dfrac{d\mathbf{r}}{d\phi}\right|} = \begin{bmatrix} \cos\theta\cos\phi \\ \sin\theta\cos\phi \\ -\sin\phi \end{bmatrix}. \qquad (20)$$

Derivatives of the UNIT VECTORS are

$$\frac{\partial\hat{\mathbf{r}}}{\partial r} = \mathbf{0} \qquad (21)$$

$$\frac{\partial\hat{\boldsymbol{\theta}}}{\partial r} = \mathbf{0} \qquad (22)$$

$$\frac{\partial\hat{\boldsymbol{\phi}}}{\partial r} = \mathbf{0} \qquad (23)$$

$$\frac{\partial \hat{\mathbf{r}}}{\partial \theta} = \begin{bmatrix} -\sin\theta\sin\phi \\ \cos\theta\sin\phi \\ 0 \end{bmatrix} = \sin\phi\,\hat{\boldsymbol{\theta}} \qquad (24)$$

$$\frac{\partial \hat{\boldsymbol{\theta}}}{\partial \theta} = \begin{bmatrix} -\cos\theta \\ -\sin\theta \\ 0 \end{bmatrix} = -\cos\phi\,\hat{\boldsymbol{\phi}} - \sin\phi\hat{\mathbf{r}} \qquad (25)$$

$$\frac{\partial \hat{\boldsymbol{\phi}}}{\partial \theta} = \begin{bmatrix} -\sin\theta\cos\phi \\ \cos\theta\cos\phi \\ 0 \end{bmatrix} = \cos\phi\,\hat{\boldsymbol{\theta}} \qquad (26)$$

$$\frac{\partial \hat{\mathbf{r}}}{\partial \phi} = \begin{bmatrix} \cos\theta \\ \sin\theta\cos\phi \\ -\sin\phi \end{bmatrix} = \hat{\boldsymbol{\phi}} \qquad (27)$$

$$\frac{\partial \hat{\boldsymbol{\theta}}}{\partial \phi} = \mathbf{0} \qquad (28)$$

$$\frac{\partial \hat{\boldsymbol{\phi}}}{\partial \phi} = \begin{bmatrix} -\cos\theta\sin\phi \\ -\sin\theta\sin\phi \\ -\cos\phi \end{bmatrix} = -\hat{\mathbf{r}}. \qquad (29)$$

The GRADIENT is

$$\nabla = \hat{\mathbf{r}}\,\frac{\partial}{\partial r} + \frac{1}{r}\,\hat{\boldsymbol{\phi}}\,\frac{\partial}{\partial \phi} + \frac{1}{r\sin\phi}\,\hat{\boldsymbol{\theta}}\,\frac{\partial}{\partial \theta}, \qquad (30)$$

so

$$\nabla_r \hat{\mathbf{r}} = \mathbf{0} \qquad (31)$$

$$\nabla_r \hat{\boldsymbol{\theta}} = \mathbf{0} \qquad (32)$$

$$\nabla_r \hat{\boldsymbol{\phi}} = \hat{\mathbf{0}} \qquad (33)$$

$$\nabla_r \hat{\mathbf{r}} = \frac{\sin\phi\,\hat{\boldsymbol{\theta}}}{r\sin\phi} = \frac{1}{r}\,\hat{\boldsymbol{\theta}} \qquad (34)$$

$$\nabla_\theta \hat{\boldsymbol{\theta}} = -\frac{\cos\phi\hat{\boldsymbol{\phi}} + \sin\phi\hat{\mathbf{r}}}{r\sin\phi} = -\frac{\cot\phi}{r}\,\hat{\boldsymbol{\phi}} - \frac{1}{r}\,\hat{\mathbf{r}} \qquad (35)$$

$$\nabla_\theta \hat{\boldsymbol{\phi}} = \frac{\cos\phi\hat{\boldsymbol{\phi}}}{r\sin\phi} = \frac{1}{r}\cot\phi\hat{\boldsymbol{\theta}}. \qquad (36)$$

Now, since the CONNECTION COEFFICIENTS are given by $\Gamma^i_{jk} = \hat{\mathbf{x}}_i \cdot (\nabla_k \hat{\mathbf{x}}_j)$,

$$\Gamma^\theta = \begin{bmatrix} 0 & \dfrac{1}{r} & 0 \\ 0 & 0 & 0 \\ 0 & \dfrac{\cot\phi}{r} & 0 \end{bmatrix} \qquad (37)$$

$$\Gamma^\phi = \begin{bmatrix} 0 & 0 & \dfrac{1}{r} \\ 0 & -\dfrac{\cot\phi}{r} & 0 \\ 0 & 0 & 0 \end{bmatrix} \qquad (38)$$

$$\Gamma^r = \begin{bmatrix} 0 & 0 & 0 \\ 0 & -\dfrac{1}{r} & 0 \\ 0 & 0 & -\dfrac{1}{r} \end{bmatrix}. \qquad (39)$$

The DIVERGENCE is

$$\nabla \cdot \mathbf{F} = A^k_{,k} + \Gamma^k_{jk} A^j$$

$$= \left[A^r_{,r} + \left(\Gamma^r_{rr} A^r + \Gamma^r_{\theta r} A^\theta + \Gamma^r_{\phi r} A^\phi \right) \right]$$

$$+ \left[A^\theta_{,\theta} + \left(\Gamma^\theta_{r\theta} A^r + \Gamma^\theta_{\theta\theta} A^\theta + \Gamma^\theta_{\phi\theta} A^\phi \right) \right]$$

$$+ \left[A^\phi_{,\phi} + \left(\Gamma^\phi_{r\phi} A^r + \Gamma^\phi_{\theta\phi} A^\theta + \Gamma^\phi_{\phi\phi} A^\phi \right) \right]$$

$$= \frac{1}{g_r}\frac{\partial A^r}{\partial r} + \frac{1}{g_\theta}\frac{\partial A^\theta}{\partial \theta} + \frac{1}{g_\phi}\frac{\partial A^\phi}{\partial \phi} + (0 + 0 + 0)$$

$$+ \left(\frac{1}{r}A^r + 0 + \frac{\cot\phi}{r}A^\phi \right) \left(\frac{1}{r}A^r + 0 + 0 \right)$$

$$= \frac{\partial}{\partial r}A^r + \frac{2}{r}A^r + \frac{1}{r\sin\phi}\frac{\partial}{\partial\theta}A^\theta + \frac{1}{r}\frac{\partial}{\partial\phi}A^\phi$$

$$+ \frac{\cot\phi}{r}A^\phi, \qquad (40)$$

or, in VECTOR notation,

$$\nabla \cdot \mathbf{F} = \left(\frac{2}{r} + \frac{\partial}{\partial r} \right) F_r + \left(\frac{1}{r}\frac{\partial}{\partial\phi} + \frac{\cot\phi}{r} \right) F_\phi$$

$$+ \frac{1}{\sin\phi}\frac{\partial F_\theta}{\partial\theta}$$

$$= \frac{1}{r^2}\frac{\partial}{\partial r}\left(r^2 F_r \right) + \frac{1}{r\sin\phi}\frac{\partial}{\partial\phi}\left(\sin\phi F_\phi \right)$$

$$+ \frac{1}{r\sin\phi}\frac{\partial F_\theta}{\partial\theta}. \qquad (41)$$

The COVARIANT DERIVATIVES are given by

$$A_{j;k} = \frac{1}{g_{kk}}\frac{\partial A_j}{\partial x_k} - \Gamma^i_{jk} A_i, \qquad (42)$$

so

$$A_{r;r} = \frac{\partial A_r}{\partial r} - \Gamma^i_{rr} A_i = \frac{\partial A_r}{\partial r} \qquad (43)$$

$$A_{r;\theta} = \frac{1}{r\sin\phi}\frac{\partial A_r}{\partial\theta} - \Gamma^i_{\tau\theta} = \frac{1}{r\sin\phi}\frac{\partial A_r}{\partial\theta} - \Gamma_{r\theta}A_\theta$$

$$= \frac{1}{r\sin\phi}\frac{\partial A_r}{\partial\phi} - \frac{A_\theta}{r} \qquad (44)$$

$$A_{r;\phi} = \frac{1}{r}\frac{\partial A_r}{\partial \phi} - \Gamma^i_{r\phi}A_i = \frac{1}{r}\frac{\partial A_r}{\partial \phi} - \Gamma^\phi_{r\phi}A_\phi$$

$$= \frac{1}{r}\left(\frac{\partial A_r}{\partial \phi} - A_\phi\right) \tag{45}$$

$$A_{\theta;r} = \frac{\partial A_\theta}{\partial r} - \Gamma^i_{\theta r}A_i = \frac{\partial A_\theta}{\partial r} \tag{46}$$

$$A_{\theta;\theta} = \frac{1}{r\sin\phi}\frac{\partial A_\theta}{\partial \theta} - \Gamma^i_{\theta\theta}A_i$$

$$= \frac{1}{r\sin\phi}\partial A_\theta \partial \theta - \Gamma^\phi_{\theta\theta}A_\phi - \Gamma^r_{\theta\theta}A_r$$

$$= \frac{1}{r\sin\phi}\frac{\partial A_\theta}{\partial \theta} + \frac{\cot\phi}{r}A_\phi + \frac{A_r}{r} \tag{47}$$

$$A_{\theta;\phi} = \frac{1}{r}\frac{\partial A_\theta}{\partial r} - \Gamma^i_{\phi r}A_i\frac{\partial A_\theta}{\partial \phi} \tag{48}$$

$$A_{\phi;r} = \frac{\partial A_\phi}{\partial r} - \Gamma^i_{\phi r}A_i = \frac{\partial A_\phi}{\partial r} - \frac{A_\phi}{r} \tag{49}$$

$$A_{\phi;\theta} = \frac{1}{r\sin\phi}\frac{\partial A_\phi}{\partial \theta} - \Gamma^i_{\theta\theta}A_i = \frac{1}{r\sin\phi}\frac{\partial A_\phi}{\partial \theta} - \Gamma^\theta_{\phi\theta}$$

$$= \frac{1}{r\sin\phi}\frac{\partial A_\phi}{\partial \theta} - \frac{\cot\phi}{r}A_\theta \tag{50}$$

$$A_{\phi;\phi} = \frac{1}{r}\frac{\partial A_\phi}{\partial \phi} - \Gamma^i_{\phi\phi}A_i = \frac{1}{r}\frac{\partial A_\phi}{\partial \phi} - \Gamma^r_{\phi\phi}A_r$$

$$= \frac{1}{r}\frac{\partial A_\phi}{\partial \phi} + \frac{A_r}{r}. \tag{51}$$

The COMMUTATION COEFFICIENTS are given by

$$c^\mu_{\alpha\beta}\vec{e}_\mu = [\vec{e}_\alpha,\ \vec{e}_\beta] = \nabla_\alpha\vec{e}_\beta - \nabla_\beta\vec{e}_\alpha \tag{52}$$

$$[\hat{\mathbf{r}},\ \hat{\mathbf{r}}] = [\hat{\boldsymbol{\theta}},\ \hat{\boldsymbol{\theta}}] = [\hat{\boldsymbol{\phi}},\ \hat{\boldsymbol{\phi}}] = \mathbf{0}, \tag{53}$$

so $c^\alpha_{rr} = c^\alpha_{\theta\theta} = c^\alpha_{\phi\phi} = 0$, where $\alpha = r,\ \theta,\ \phi$.

$$[\hat{\mathbf{r}},\ \hat{\boldsymbol{\theta}}] = -[\hat{\boldsymbol{\theta}},\ \hat{\mathbf{r}}] = \nabla_r\hat{\boldsymbol{\theta}} - \nabla_\theta\hat{\mathbf{r}} = \mathbf{0} - \frac{1}{r}\hat{\boldsymbol{\theta}} = -\frac{1}{r}\hat{\boldsymbol{\theta}}. \tag{54}$$

so $c^\theta_{r\theta} = -c^\theta_{\theta r} = -\frac{1}{r}$, $c^r_{r\theta} = c^\phi_{r\theta} = 0$.

$$[\hat{\mathbf{r}},\ \hat{\boldsymbol{\phi}}] = -[\hat{\boldsymbol{\phi}},\ \hat{\mathbf{r}}] = \mathbf{0} - \frac{1}{r}\hat{\boldsymbol{\phi}} = -\frac{1}{r}\hat{\boldsymbol{\phi}}, \tag{55}$$

so $c^\phi_{r\phi} = -c^\phi_{\phi r} = \frac{1}{r}$.

$$[\hat{\boldsymbol{\theta}},\ \hat{\boldsymbol{\phi}}] = -[\hat{\boldsymbol{\phi}},\ \hat{\boldsymbol{\theta}}] = \frac{1}{r}\cot\phi\hat{\boldsymbol{\theta}} - \mathbf{0} = \frac{1}{r}\cot\phi\hat{\boldsymbol{\theta}}. \tag{56}$$

so

$$c^\theta_{\theta\phi} = -c^\theta_{\phi\theta} = \frac{1}{r}\cot\phi. \tag{57}$$

Summarizing,

$$c^r = \begin{bmatrix} 0 & 0 & 0 \\ 0 & 0 & 0 \\ 0 & 0 & 0 \end{bmatrix} \tag{58}$$

$$c^\theta = \begin{bmatrix} 0 & -\frac{1}{r} & 0 \\ \frac{1}{r} & 0 & \frac{1}{r}\cot\phi \\ 0 & -\frac{1}{r}\cot\phi & 0 \end{bmatrix} \tag{59}$$

$$c^\phi = \begin{bmatrix} 0 & 0 & -\frac{1}{r} \\ 0 & 0 & 0 \\ \frac{1}{r} & 0 & 0 \end{bmatrix}. \tag{60}$$

Time derivatives of the POSITION VECTOR are

$$\dot{\mathbf{r}} = \begin{bmatrix} \cos\theta\sin\phi\,\dot{r} - r\sin\theta\sin\phi\dot{\theta} + r\cos\theta\cos\phi\dot{\phi} \\ \sin\theta\sin\phi\,\dot{r} + r\cos\theta\sin\phi\dot{\theta} + r\sin\theta\cos\phi\dot{\phi} \\ \cos\phi\dot{r} - r\sin\phi\dot{\phi} \end{bmatrix}$$

$$= \begin{bmatrix} \cos\theta\sin\phi \\ \sin\theta\sin\phi \\ \cos\phi \end{bmatrix}\dot{r} + r\sin\phi\begin{bmatrix} -\sin\theta \\ \cos\theta \\ 0 \end{bmatrix}\dot{\theta}$$

$$+ r\begin{bmatrix} \cos\theta\cos\phi \\ \sin\theta\cos\phi \\ -\sin\phi \end{bmatrix}\dot{\phi}$$

$$= \dot{r}\hat{\mathbf{r}} + r\sin\phi\dot{\theta}\hat{\boldsymbol{\theta}} + r\dot{\phi}\hat{\boldsymbol{\phi}}. \tag{61}$$

The SPEED is therefore given by

$$v \equiv |\dot{\mathbf{r}}| = \sqrt{\dot{r}^2 + r^2\sin^2\phi\dot{\theta}^2 + r^2\dot{\phi}^2}. \tag{62}$$

The ACCELERATION is

$$\ddot{x} = (-\sin\theta\sin\phi\dot{\theta}\dot{r} + \cos\theta\cos\phi\dot{r}\dot{\phi} + \cos\theta\sin\phi\ddot{r})$$

$$-(\sin\theta\sin\phi\dot{r}\dot{\theta} + r\cos\theta\sin\phi\dot{\theta}^2 + r\sin\theta\cos\phi\dot{\theta}\dot{\phi}$$

$$+ r\sin\theta\sin\phi\ddot{\theta}) + (\cos\theta\cos\phi\dot{r}\dot{\phi} - r\sin\theta\cos\phi\dot{\theta}\dot{\phi}$$

$$- r\cos\theta\sin\phi\dot{\phi}^2 + r\cos\theta\cos\phi\ddot{\phi})$$

$$= -2\sin\theta\sin\phi\dot{\theta}\dot{r} + 2\cos\theta\cos\phi\dot{r}\dot{\phi}$$

$$- 2r\sin\theta\cos\phi\dot{\theta}\dot{\phi}$$

$$+ \cos\theta\sin\phi\ddot{r} - r\sin\theta\sin\phi\ddot{\theta} + r\cos\theta\cos\phi\ddot{\phi}$$

$$- r\cos\theta\sin\phi(\dot{\theta}^2 + \dot{\phi}^2) \tag{63}$$

$$\ddot{y} = (\sin\theta\sin\phi\ddot{r} + r\cos\theta\sin\phi\dot{\theta} + r\cos\phi\sin\theta\dot{\phi})$$

$$+ (\cos\theta\sin\phi\dot{r}\dot{\theta} - r\sin\theta\sin\phi\dot{\theta}^2 + r\cos\theta\cos\phi\dot{\theta}\dot{\phi})$$

$$+ r\cos\theta\sin\phi\ddot{\theta}) + (\sin\theta\cos\phi\dot{r}\dot{\phi} + r\cos\theta\cos\phi\dot{\theta}\dot{\phi}$$

$$- r\sin\theta\sin\phi\dot{\phi}^2 + r\sin\theta\cos\phi\ddot{\phi}$$

$$= 2\cos\theta\sin\phi\dot{\theta}\dot{r} + 2\sin\theta\cos\phi\dot{r}\dot{\phi} + 2r\cos\theta\cos\phi\dot{\theta}\dot{\phi}$$

$$+\sin\theta\sin\phi\ddot{r}+r\cos\theta\sin\phi\ddot{\theta}+r\sin\theta\cos\phi\ddot{\phi}$$

$$-r\sin\theta\sin\phi(\dot{\theta}^2+\dot{\phi}^2) \tag{64}$$

$$\ddot{z}=(\cos\phi\ddot{r}-\sin\phi\dot{r}\dot{\phi})$$

$$-(\dot{r}\sin\phi\dot{\phi}+r\cos\phi\dot{\phi}^2+r\sin\phi\ddot{\phi})$$

$$=-r\cos\phi\dot{\phi}^2+\cos\phi\ddot{r}-2\sin\phi\dot{\phi}\dot{r}-r\sin\phi\ddot{\phi}. \tag{65}$$

Plugging these in gives

$$\ddot{\mathbf{r}}=(\ddot{r}-r\dot{\phi}^2)\begin{bmatrix}\cos\theta\sin\phi\\\sin\theta\sin\phi\\\cos\phi\end{bmatrix}$$

$$+(2r\cos\phi\dot{\theta}\dot{\phi}+r\sin\phi\ddot{\theta})\begin{bmatrix}-\sin\theta\\\cos\theta\\0\end{bmatrix}$$

$$+(2\dot{r}\dot{\phi}+r\ddot{\phi})\begin{bmatrix}\cos\theta\cos\phi\\\sin\theta\cos\phi\\-\sin\phi\end{bmatrix}-r\sin\phi\dot{\theta}^2\begin{bmatrix}\cos\theta\\\sin\theta\\0\end{bmatrix}. \tag{66}$$

but

$$\sin\phi\hat{\mathbf{r}}+\cos\phi\hat{\boldsymbol{\phi}}=\begin{bmatrix}\cos\theta\sin^2\phi+\cos\theta\cos^2\phi\\\sin\theta\sin^2\phi+\sin\theta\cos^2\phi\\0\end{bmatrix}$$

$$=\begin{bmatrix}\cos\theta\\\sin\theta\\0\end{bmatrix} \tag{67}$$

so

$$\ddot{\mathbf{r}}=(\ddot{r}-r\dot{\phi}^2)\hat{\mathbf{r}}+(2r\cos\phi\dot{\theta}\dot{\phi}+2\sin\phi\dot{\theta}\dot{r}+r\sin\phi\ddot{\theta})\hat{\boldsymbol{\theta}}$$

$$-(2\dot{r}\dot{\phi}-r\ddot{\phi})\hat{\boldsymbol{\phi}}-r\sin\phi\dot{\theta}^2(\sin\phi\hat{\mathbf{r}}+\cos\phi\hat{\boldsymbol{\phi}})$$

$$=(\ddot{r}-r\dot{\phi}^2-r\sin^2\phi\dot{\theta}^2)\hat{\mathbf{r}}$$

$$+(2\sin\phi\dot{\theta}\dot{r}+2r\cos\phi\dot{\theta}\dot{\phi}+r\sin\phi\ddot{\theta})\hat{\boldsymbol{\theta}}$$

$$+(2\dot{r}\dot{\phi}+r\ddot{\phi}-r\sin\phi\cos\phi\dot{\theta}^2)\hat{\boldsymbol{\phi}}. \tag{68}$$

Time DERIVATIVES of the UNIT VECTORS are

$$\dot{\hat{\mathbf{r}}}=\begin{bmatrix}-\sin\theta\sin\phi\dot{\theta}+\cos\theta\cos\phi\dot{\phi}\\\cos\theta\sin\phi\dot{\theta}+\sin\theta\cos\phi\dot{\phi}\\-\sin\phi\dot{\phi}\end{bmatrix}$$

$$=\sin\phi\dot{\theta}\hat{\boldsymbol{\theta}}+\dot{\phi}\hat{\boldsymbol{\phi}} \tag{69}$$

$$\dot{\hat{\boldsymbol{\theta}}}=\begin{bmatrix}-\cos\theta\dot{\theta}\\-\sin\theta\dot{\theta}\\0\end{bmatrix}=-\dot{\theta}\begin{bmatrix}\cos\theta\\\sin\theta\\0\end{bmatrix}$$

$$=-\dot{\theta}(\sin\phi\hat{\mathbf{r}}+\cos\phi\hat{\boldsymbol{\phi}}) \tag{70}$$

$$\dot{\hat{\boldsymbol{\phi}}}=\begin{bmatrix}-\sin\theta\cos\phi\dot{\theta}-\cos\theta\sin\phi\dot{\phi}\\\cos\theta\cos\phi\dot{\theta}-\sin\theta\sin\phi\dot{\phi}\\-\cos\phi\dot{\phi}\end{bmatrix}$$

$$=-\dot{\phi}\hat{\mathbf{r}}+\cos\phi\dot{\theta}\hat{\boldsymbol{\theta}}. \tag{71}$$

The CURL is

$$\nabla\times\mathbf{F}=\frac{1}{r\sin\phi}\left[\frac{\partial}{\partial\phi}(\sin\phi F_\theta)-\frac{\partial F_\phi}{\partial\theta}\right]\hat{\mathbf{r}}$$

$$+\frac{1}{r}\left[\frac{1}{\sin\phi}\frac{\partial F_r}{\partial\theta}-\frac{\partial}{\partial r}(rF_\theta)\right]\hat{\boldsymbol{\phi}}+\frac{1}{r}$$

$$\times\left[\frac{\partial}{\partial r}(rF_\phi)-\frac{\partial F_r}{\partial\phi}\right]\hat{\boldsymbol{\theta}}. \tag{72}$$

The LAPLACIAN is

$$\nabla^2\equiv\frac{1}{r^2}\frac{\partial}{\partial r}\left(r^2\frac{\partial}{\partial r}\right)+\frac{1}{r^2\sin^2}\frac{\partial^2}{\partial\theta^2}+\frac{1}{r^2\sin}\frac{\partial}{\partial\phi}$$

$$\times\left(\sin\phi\frac{\partial}{\partial\phi}\right)$$

$$=\frac{1}{r^2}\left(r^2\frac{\partial^2}{\partial r^2}+2r\frac{\partial}{\partial r}\right)+\frac{1}{r^2\sin^2}\frac{\partial^2}{\partial\theta^2}+\frac{1}{r^2\sin\phi}$$

$$\times\left(\cos\phi\frac{\partial}{\partial\phi}+\sin\phi\frac{\partial^2}{\partial\phi^2}\right)$$

$$=\frac{\partial^2}{\partial r^2}+\frac{2}{r}\frac{\partial}{\partial r}+\frac{1}{r^2\sin^2\phi}\frac{\partial^2}{\partial\theta^2}+\frac{\cos\phi}{r^2\sin\phi}\frac{\partial}{\partial\phi}$$

$$+\frac{1}{r^2}\frac{\partial^2}{\partial\phi^2}. \tag{73}$$

The vector LAPLACIAN is

$$\nabla^2\mathbf{v}=$$

$$\begin{bmatrix}\frac{1}{r}\frac{\partial^2(rv_r)}{\partial r^2}+\frac{1}{r^2}\frac{\partial^2 v_r}{\partial\theta^2}+\frac{1}{r^2\sin^2\theta}\frac{\partial^2 v_r}{\partial\phi^2}+\frac{\cot\theta}{r^2}\frac{\partial v_r}{\partial\theta}-\frac{2}{r^2}\frac{\partial v_\theta}{\partial\theta}-\frac{2}{r^2\sin\theta}\frac{\partial v_\phi}{\partial\phi}-\frac{2v_r}{r^2}-\frac{2\cot\theta}{r^2}v_\theta\\\frac{1}{r}\frac{\partial^2(rv_\theta)}{\partial r^2}+\frac{1}{r^2}\frac{\partial^2 v_\theta}{\partial\theta^2}+\frac{1}{r^2\sin^2\theta}\frac{\partial^2 v_\theta}{\partial\phi^2}+\frac{\cot\theta}{r^2}\frac{\partial v_\theta}{\partial\theta}-\frac{2\cot\theta}{r^2\sin\theta}\frac{\partial v_\phi}{\partial\phi}+\frac{2}{r^2}\frac{\partial v_r}{\partial\theta}-\frac{v_\theta}{r^2\sin^2\theta}\\\frac{1}{r}\frac{\partial^2(rv_\phi)}{\partial r^2}+\frac{1}{r^2}\frac{\partial^2 v_\phi a}{\partial\theta^2}+\frac{1}{r^2\sin^2\theta}\frac{\partial^2 v_\phi}{\partial\phi^2}+\frac{\cot\theta}{r^2}\frac{\partial v_\phi}{\partial\theta}+\frac{2}{r^2}\frac{\partial v_r}{\partial\phi}+\frac{2\cot\theta}{r^2\sin\theta}\frac{\partial v_\theta}{\partial\phi}-\frac{v_\phi}{r^2\sin^2\theta}\end{bmatrix}.$$

$$\tag{74}$$

To express PARTIAL DERIVATIVES with respect to Cartesian axes in terms of PARTIAL DERIVATIVES of the spherical coordinates,

$$\begin{bmatrix}x\\y\\z\end{bmatrix}=\begin{bmatrix}r\cos\theta\sin\phi\\r\sin\theta\sin\phi\\r\cos\phi\end{bmatrix} \tag{75}$$

$$\begin{bmatrix}dx\\dy\\dz\end{bmatrix}=$$

$$\begin{bmatrix}\cos\theta\sin\phi\,dr-r\sin\theta\sin\phi\,d\theta+r\cos\theta\cos\phi\,d\phi\\\sin\theta\sin\phi\,dr+r\sin\phi\cos\theta\,d\theta+r\sin\theta\cos\phi\,d\phi\\\cos\phi\,dr-r\sin\phi\,d\phi\end{bmatrix}$$

$$=\begin{bmatrix}\cos\theta\sin\phi&-r\sin\theta\sin\phi&r\cos\theta\cos\phi\\\sin\theta\sin\phi&r\sin\phi\cos\theta&r\sin\theta\cos\phi\\\cos\phi&0&-r\sin\phi\end{bmatrix}$$

$$\times\begin{bmatrix}dx\\dy\\dz\end{bmatrix}. \tag{76}$$

Upon inversion, the result is

$$\begin{bmatrix} dr \\ d\theta \\ d\phi \end{bmatrix} = \begin{bmatrix} \cos\theta\sin\phi & \sin\theta\sin\phi & \cos\phi \\ -\dfrac{\sin\theta}{r\sin\phi} & \dfrac{\cos\theta}{r\sin\phi} & 0 \\ \dfrac{\cos\theta\cos\phi}{r} & \dfrac{\sin\theta\cos\phi}{r} & -\dfrac{\sin\phi}{r} \end{bmatrix}$$

$$\times \begin{bmatrix} dr \\ dy \\ dz \end{bmatrix}. \tag{77}$$

The Cartesian PARTIAL DERIVATIVES in spherical coordinates are therefore

$$\frac{\partial}{\partial x} = \frac{\partial r}{\partial x}\frac{\partial}{\partial r} + \frac{\partial\theta}{\partial x}\frac{\partial}{\partial\theta} + \frac{\partial\phi}{\partial x}\frac{\partial}{\partial\phi}$$

$$= \cos\theta\sin\phi\frac{\partial}{\partial r} - \frac{\sin\theta}{r\sin\phi}\frac{\partial}{\partial\theta} + \frac{\cos\theta\cos\phi}{r}\frac{\partial}{\partial\phi} \tag{78}$$

$$\frac{\partial}{\partial y} = \frac{\partial}{\partial y}\frac{\partial}{\partial r} + \frac{\partial\theta}{\partial y}\frac{\partial}{\partial\theta} + \frac{\partial\phi}{\partial y}\frac{\partial}{\partial\phi}$$

$$= \sin\theta\sin\phi\frac{\partial}{\partial r} + \frac{\cos\theta}{r\sin\phi}\frac{\partial}{\partial\theta} + \frac{\sin\theta\cos\phi}{r}\frac{\partial}{\partial\phi} \tag{79}$$

$$\frac{\partial}{\partial z} = \frac{\partial r}{\partial z}\frac{\partial}{\partial r} + \frac{\partial\theta}{\partial z}\frac{\partial}{\partial\theta} + \frac{\partial\phi}{\partial z}\frac{\partial}{\partial\phi}$$

$$= \cos\phi\frac{\partial}{\partial r} - \frac{\sin\phi}{r}\frac{\partial}{\partial\phi} \tag{80}$$

(Gasiorowicz 1974, pp. 167–68).

The HELMHOLTZ DIFFERENTIAL EQUATION is separable in spherical coordinates.

See also COLATITUDE, GREAT CIRCLE, HELMHOLTZ DIFFERENTIAL EQUATION–SPHERICAL COORDINATES, LATITUDE, LONGITUDE, OBLATE SPHEROIDAL COORDINATES, PROLATE SPHEROIDAL COORDINATES

References

Arfken, G. "Spherical Polar Coordinates." §2.5 in *Mathematical Methods for Physicists, 3rd ed.* Orlando, FL: Academic Press, pp. 102–11, 1985.
Beyer, W. H. *CRC Standard Mathematical Tables, 28th ed.* Boca Raton, FL: CRC Press, p. 212, 1987.
Gasiorowicz, S. *Quantum Physics.* New York: Wiley, 1974.
Moon, P. and Spencer, D. E. "Spherical Coordinates (r, θ, ψ)." Table 1.05 in *Field Theory Handbook, Including Coordinate Systems, Differential Equations, and Their Solutions, 2nd ed.* New York: Springer-Verlag, pp. 24–7, 1988.
Morse, P. M. and Feshbach, H. *Methods of Theoretical Physics, Part I.* New York: McGraw-Hill, p. 658, 1953.

Spherical Covering

The placement of n points on a SPHERE so as to minimize the maximum distance of any point on the sphere from the closest one of the n points.

See also SPHERICAL CODE, SPHERICAL COVERING

References

Hardin, R. H.; Sloane, N. J. A. S.; and Smith, W. D. *Spherical Codes.* In preparation. http://www.research.att.com/~njas/coverings/.

Spherical Curve

A CURVE on the surface of a SPHERE. Examples include the BASEBALL COVER, SEIFFERT'S SPHERICAL SPIRAL, SPHERICAL HELIX, and SPHERICAL SPIRAL.

See also BASEBALL COVER, CURVE, PLANE CURVE, SPACE CURVE, TENNIS BALL THEOREM

Spherical Defect

Let a, b, and c be the sides of a SPHERICAL TRIANGLE, then the spherical defect is defined as

$$D = 2\pi - (a + b + c).$$

See also ANGULAR DEFECT, SPHERICAL EXCESS, SPHERICAL TRIANGLE

References

Harris, J. W. and Stocker, H. *Handbook of Mathematics and Computational Science.* New York: Springer-Verlag, p. 109, 1998.

Spherical Design

X is a spherical t-design in E IFF it is possible to exactly determine the average value on E of any POLYNOMIAL f of degree at most t by sampling f at the points of X. In other words,

$$\frac{1}{\text{volume } E}\int_E f(\xi)\,d\xi = \frac{1}{|X|}\sum_{x\in X} f(x).$$

Spherical t-designs give the placement of n points on a sphere for use in numerical integration with equal weights.

References

Colbourn, C. J. and Dinitz, J. H. (Eds.). "Spherical t-Designs." Ch. 44 in *CRC Handbook of Combinatorial Designs.* Boca Raton, FL: CRC Press, pp. 462–66, 1996.
Hardin, R. H. and Sloane, N. J. A. S. "McLaren's Improved Snub Cube and Other New Spherical Designs in Three Dimensions." *Disc. Comput. Geom.* **15**, 429–31, 1996.
Hardin, R. H.; Sloane, N. J. A. S.; and Smith, W. D. *Spherical Codes.* In preparation. http://www.research.att.com/~njas/sphdesigns/.
McLaren, A. D. "Optimal Numerical Integration on a Sphere." *Math. Comput.* **17**, 361–83, 1963.

Spherical Excess

The difference between the sum of the angles A, B, and C of a SPHERICAL TRIANGLE and π radians (180°),

$$E = A + B + C - \pi.$$

The notation Δ is sometimes used for spherical excess instead of E, which can cause confusion since it is also frequently used to denote the SURFACE AREA of a SPHERICAL TRIANGLE (Zwillinger 1995, p. 469). The notation ϵ is also used (Gellert *et al.* 1989, p. 263).

The equation for the spherical excess in terms of the side lengths a, b, and c is known as L'HUILIER'S THEOREM,

$$\tan\left(\tfrac{1}{4}E\right)$$
$$= \sqrt{\tan\left(\tfrac{1}{2}s\right)\tan\left[\tfrac{1}{2}(s-a)\right]\tan\left[\tfrac{1}{2}(s-b)\right]\tan\left[\tfrac{1}{2}(s-c)\right]},$$

where s is the SEMIPERIMETER.

See also ANGULAR DEFECT, DESCARTES TOTAL ANGULAR DEFECT, GIRARD'S SPHERICAL EXCESS FORMULA, L'HUILIER'S THEOREM, SPHERICAL TRIANGLE

References
Gellert, W.; Gottwald, S.; Hellwich, M.; Kästner, H.; and Künstner, H. (Eds.). *VNR Concise Encyclopedia of Mathematics, 2nd ed.* New York: Van Nostrand Reinhold, 1989.
Harris, J. W. and Stocker, H. *Handbook of Mathematics and Computational Science.* New York: Springer-Verlag, p. 109, 1998.
Zwillinger, D. (Ed.). *CRC Standard Mathematical Tables and Formulae.* Boca Raton, FL: CRC Press, p. 469, 1995.

Spherical Frustum

SPHERICAL SEGMENT

Spherical Geometry

The study of figures on the surface of a SPHERE (such as the SPHERICAL TRIANGLE and SPHERICAL POLYGON), as opposed to the type of geometry studied in PLANE GEOMETRY or SOLID GEOMETRY. In spherical geometry, straight lines are GREAT CIRCLES, so any two lines meet in two points. There are also no parallel lines. The angle between two lines in spherical geometry is the angle between the planes of the corresponding great circles, and a SPHERICAL TRIANGLE is defined by its three angles. There is no concept of similar triangles in spherical geometry.

See also GREAT CIRCLE, HYPERBOLIC GEOMETRY, PLANE GEOMETRY, SOLID GEOMETRY, SPHERICAL TRIANGLE, SPHERICAL TRIGONOMETRY, THURSTON'S GEOMETRIZATION CONJECTURE

References
Harris, J. W. and Stocker, H. "Spherical Geometry." §4.9 in *Handbook of Mathematics and Computational Science.* New York: Springer-Verlag, pp. 108–13, 1998.
Henderson, D. W. *Experiencing Geometry: On Plane and Sphere.* Englewood Cliffs, NJ: Prentice-Hall, 1995.
Zwillinger, D. (Ed.). "Spherical Geometry and Trigonometry." §6.4 in *CRC Standard Mathematical Tables and Formulae.* Boca Raton, FL: CRC Press, pp. 468–71, 1995.

Spherical Hankel Function of the First Kind

$$h_n^{(1)}(x) \equiv \sqrt{\frac{\pi}{2x}}\, H_{n+1/2}^{(1)}(x) = j_n(x) + i n_n(x),$$

where $H^{(1)}(x)$ is the HANKEL FUNCTION OF THE FIRST KIND and $j_n(x)$ and $n_n(x)$ are the SPHERICAL BESSEL FUNCTIONS OF THE FIRST and SECOND KINDS. Explicitly, the first few are

$$h_0^{(1)}(x) = \frac{1}{x}(\sin x - i\cos x) = -\frac{i}{x}e^{ix}$$

$$h_1^{(1)}(x) = e^{ix}\left(-\frac{1}{x} - \frac{i}{x^2}\right)$$

$$h_2^{(1)}(x) = e^{ix}\left(\frac{i}{x} - \frac{3}{x^2} - \frac{3i}{x^3}\right).$$

References
Abramowitz, M. and Stegun, C. A. (Eds.). "Spherical Bessel Functions." §10.1 in *Handbook of Mathematical Functions with Formulas, Graphs, and Mathematical Tables, 9th printing.* New York: Dover, pp. 437–42, 1972.

Spherical Hankel Function of the Second Kind

$$h_n^{(2)}(x) \equiv \sqrt{\frac{\pi}{2x}}\, H_{n+1/2}^{(2)}(x) = j_n(x) - i n_n(x),$$

where $H^{(2)}(x)$ is the HANKEL FUNCTION OF THE SECOND KIND and $j_n(x)$ and $n_n(x)$ are the SPHERICAL BESSEL FUNCTIONS OF THE FIRST and SECOND KINDS. Explicitly, the first is

$$h_0^{(2)}(x) = \frac{1}{x}(\sin x + i\cos x) = \frac{i}{x}e^{-ix}.$$

See also SPHERICAL BESSEL FUNCTION OF THE FIRST KIND, SPHERICAL BESSEL FUNCTION OF THE SECOND KIND

References
Abramowitz, M. and Stegun, C. A. (Eds.). "Spherical Bessel Functions." §10.1 in *Handbook of Mathematical Functions with Formulas, Graphs, and Mathematical Tables, 9th printing.* New York: Dover, pp. 437–42, 1972.

Spherical Harmonic

The spherical harmonics $Y_l^m(\theta, \phi)$ are the angular portion of the solution to LAPLACE'S EQUATION in SPHERICAL COORDINATES where azimuthal symmetry is not present. Some care must be taken in identifying the notational convention being used. In this entry, θ

is taken as the polar (colatitudinal) coordinate with $\theta \in [0, \pi]$, and ϕ as the azimuthal (longitudinal) coordinate with $\phi \in [0, 2\pi)$. This is the convention normally used in physics, as described by Arfken (1985) and *Mathematica* (in mathematical literature, θ usually denotes the longitudinal coordinate and ϕ the colatitudinal coordinate). Spherical harmonics are implemented in *Mathematica* as `Spherical-HarmonicY[l, m, theta, phi]`.

Spherical harmonics satisfy the SPHERICAL HARMONIC DIFFERENTIAL EQUATION, which is given by the angular part of LAPLACE'S EQUATION in SPHERICAL COORDINATES. Writing $F = \Phi(\phi)\Theta(\theta)$ in this equation gives

$$\frac{\Phi(\phi)}{\sin \theta} \frac{d}{d\theta}\left(\sin \theta \frac{d\Theta}{d\theta}\right) + \frac{\Theta(\theta)}{\sin^2 \theta} \frac{d^2\Phi(\phi)}{d\phi^2}$$

$$+ l(l+1)\Theta(\theta)\Phi(\phi) = 0. \qquad (1)$$

Multiplying by $\sin^2 \theta / (\Theta\Phi)$ gives

$$\left[\frac{\sin \theta}{\Theta(\theta)} \frac{d}{d\theta}\left(\sin \theta \frac{d\Theta}{d\theta}\right) + l(l+1)\sin^2 \theta\right] + \frac{1}{\Phi(\phi)} \frac{d^2\Phi(\phi)}{d\phi^2}$$

$$= 0. \qquad (2)$$

Using SEPARATION OF VARIABLES by equating the ϕ-dependent portion to a constant gives

$$\frac{1}{\Phi(\phi)} \frac{d^2\Phi(\phi)}{d\phi^2} = -m^2, \qquad (3)$$

which has solutions

$$\Phi(\phi) = Ae^{-im\phi} + Be^{im\phi}, \qquad (4)$$

Plugging in (3) into (2) gives the equation for the θ-dependent portion, whose solution is

$$\Theta(\theta) = P_l^m(\cos \theta), \qquad (5)$$

where $m = -l, -(l-1), ..., 0, ..., l-1, l$ and $P_l^m(z)$ is an associated LEGENDRE POLYNOMIAL. The spherical harmonics are then defined by combining $\Phi(\phi)$ and $\Theta(\theta)$,

$$Y_l^m(\theta, \phi) \equiv \sqrt{\frac{2l+1}{4\pi} \frac{(l-m)!}{(l+m)!}} P_l^m(\cos \theta)e^{im\phi}. \qquad (6)$$

where the normalization is chosen such that

$$\int_0^{2\pi} \int_0^{\pi} Y_l^m(\theta, \phi)\bar{Y}_{l'}^{m'}(\theta, \phi)\sin \theta \, d\theta \, d\phi$$

$$= \int_0^{2\pi} \int_{-1}^{1} Y_l^m(\theta, \phi)\bar{Y}_{l'}^{m'}(\theta, \phi)d(\cos \theta) \, d\phi = \delta_{mm'} \, \delta_{ll'}. \qquad (7)$$

(Arfken 1985, p. 681). Here, $\bar{z}$ denotes the COMPLEX CONJUGATE and δ_{mn} is the KRONECKER DELTA. Sometimes (e.g., Arfken 1985), the CONDON-SHORTLEY

PHASE $(-1)^m$ is prepended to the definition of the spherical harmonics.

The spherical harmonics are sometimes separated into their REAL and IMAGINARY PARTS,

$$Y_l^{ms}(\theta, \phi) \equiv \sqrt{\frac{2l+1}{4\pi} \frac{(l-m)!}{(l+m)!}} P_l^m(\cos \theta) \sin(m\phi) \qquad (8)$$

$$Y_l^{mc}(\theta, \phi) \equiv \sqrt{\frac{2l+1}{4\pi} \frac{(l-m)!}{(l+m)!}} P_l^m(\cos \theta) \cos(m\phi). \qquad (9)$$

The spherical harmonics obey

$$Y_l^{-l}(\theta, \phi) = \frac{1}{2^l l!} \sqrt{\frac{(2l+1)!}{4\pi}} \sin^l \theta \, e^{-il\phi} \qquad (10)$$

$$Y_l^0(\theta, \phi) = \sqrt{\frac{2l+1}{4\pi}} P_l(\cos \theta) \qquad (11)$$

$$Y_l^{-m}(\theta, \phi) = (-1)^m \bar{Y}_l^m(\theta, \phi), \qquad (12)$$

where $P_l(x)$ is a LEGENDRE POLYNOMIAL.

Integrals of the spherical harmonics are given by

$$\int_0^{2\pi} \int_0^{\pi} Y_{l_1}^{m_1}(\theta, \phi)Y_{l_2}^{m_2}(\theta, \phi)Y_{l_3}^{m_3}(\theta, \phi) \sin \theta \, d\theta \, d\phi$$

$$= \sqrt{\frac{(2l_1+1)(2l_2+1)(2l_3+1)}{4\pi}} \begin{pmatrix} l_1 & l_2 & l_3 \\ 0 & 0 & 0 \end{pmatrix}$$

$$\times \begin{pmatrix} l_1 & l_2 & l_3 \\ m_1 & m_2 & m_3 \end{pmatrix}, \qquad (13)$$

where $\begin{pmatrix} l_1 l_2 l_3 \\ m_1 m_2 m_3 \end{pmatrix}$ is a WIGNER 3J-SYMBOL (which is related to the CLEBSCH-GORDAN COEFFICIENTS). Special cases include

$$\int_0^{2\pi} \int_0^{\pi} Y_L^M(\theta, \phi)Y_0^0(\theta, \phi)\bar{Y}_L^M(\theta, \phi)\sin \theta \, d\theta \, d\phi$$

$$= \frac{1}{\sqrt{2\pi}} \qquad (14)$$

$$\int_0^{2\pi} \int_0^{\pi} Y_L^M(\theta, \phi)Y_1^0(\theta, \phi)\bar{Y}_{L+1}^M(\theta, \phi) \sin \theta \, d\theta \, d\phi$$

$$= \sqrt{\frac{3}{4\pi}}\sqrt{\frac{(L+M+1)(L-M+1)}{(2L+1)(2L+3)}} \qquad (15)$$

$$\int_0^{2\pi} \int_0^{\pi} Y_L^M(\theta, \phi)Y_1^1(\theta, \phi)\bar{Y}_{L+1}^{M+1}(\theta, \phi) \sin \theta \, d\theta \, d\phi$$

$$= \sqrt{\frac{3}{8\pi}}\sqrt{\frac{(L+M+1)(L+M+2)}{(2L+1)(2L+3)}} \qquad (16)$$

$$\int_0^{2\pi}\int_0^{\pi} Y_L^M(\theta,\,\phi) Y_1^1(\theta,\,\phi)\bar{Y}_{L-1}^{M+1}(\theta,\,\phi)\sin\theta\,d\theta\,d\phi$$

$$=\sqrt{\frac{3}{8\pi}}\sqrt{\frac{(L-M)(L-M-1)}{(2L-1)(2L+1)}} \qquad (17)$$

(Arfken 1985, p. 700).

$$\underline{111 \xrightarrow{5} 10011 \xrightarrow{2} 1011 \xrightarrow{5} 10101 \xrightarrow{1} 1011 \xrightarrow{2} 111}$$

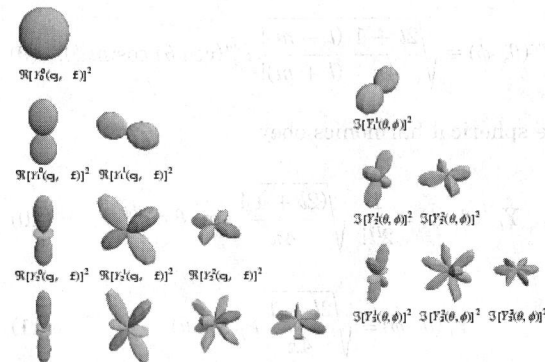

The above illustrations show $\left[Y_l^m(\theta,\,\phi)\right]^2$ (top), $\Re\left[Y_l^m(\theta,\,\phi)\right]^2$ (bottom left), and $\Im\left[Y_l^m(\theta,\,\phi)\right]^2$ (bottom right). The first few spherical harmonics are

$$Y_0^0(\theta,\,\phi)=\frac{1}{2}\frac{1}{\sqrt{\pi}}$$

$$Y_1^{-1}(\theta,\,\phi)=\frac{1}{2}\sqrt{\frac{3}{2\pi}}\sin\theta\,e^{-i\phi}$$

$$Y_1^0(\theta,\,\phi)=\frac{1}{2}\sqrt{\frac{3}{\pi}}\cos\theta$$

$$Y_1^1(\theta,\,\phi)=-\frac{1}{2}\sqrt{\frac{3}{2\pi}}\sin\theta\,e^{i\phi}$$

$$Y_2^{-2}(\theta,\,\phi)=\frac{1}{4}\sqrt{\frac{15}{2\pi}}\sin^2\theta\,e^{-2i\phi}$$

$$Y_2^{-1}(\theta,\,\phi)=\frac{1}{2}\sqrt{\frac{15}{2\pi}}\sin\theta\cos\theta\,e^{i\phi}$$

$$Y_2^0(\theta,\,\phi)=\frac{1}{4}\sqrt{\frac{5}{\pi}}(3\cos^2\theta-1)$$

$$Y_2^1(\theta,\,\phi)=-\frac{1}{2}\sqrt{\frac{15}{2\pi}}\sin\theta\cos\theta\,e^{i\phi}$$

$$Y_2^2(\theta,\,\phi)=\frac{1}{4}\sqrt{\frac{15}{2\pi}}\sin^2\theta\,e^{2i\phi}$$

$$Y_3^{-3}(\theta,\,\phi)=\frac{1}{8}\sqrt{\frac{35}{\pi}}\sin^3\theta\,e^{-3i\phi}$$

$$Y_3^{-2}(\theta,\,\phi)=\frac{1}{4}\sqrt{\frac{105}{2\pi}}\sin^2\theta\cos\theta\,e^{-2i\phi}$$

$$Y_3^{-1}(\theta,\,\phi)=\frac{1}{8}\sqrt{\frac{21}{\pi}}\sin\theta(5\cos^2\theta-1)e^{-i\phi}$$

$$Y_3^0(\theta,\,\phi)=\frac{1}{4}\sqrt{\frac{7}{\pi}}(5\cos^3\theta-3\cos\theta)$$

$$Y_3^1(\theta,\,\phi)=-\frac{1}{8}\sqrt{\frac{21}{\pi}}\sin\theta(5\cos^2\theta-1)e^{i\phi}$$

$$Y_3^2(\theta,\,\phi)=\frac{1}{4}\sqrt{\frac{105}{2\pi}}\sin^2\theta\cos\theta\,e^{2i\phi}$$

$$Y_3^3(\theta,\,\phi)=-\frac{1}{8}\sqrt{\frac{35}{\pi}}\sin^3\theta\,e^{3i\phi}.$$

Written in terms of CARTESIAN COORDINATES,

$$e^{i\phi}=\frac{x+iy}{\sqrt{x^2+y^2}} \qquad (18)$$

$$\theta=\sin^{-1}\left(\sqrt{\frac{x^2+y^2}{x^2+y^2+z^2}}\right) \qquad (19)$$

$$=\cos^{-1}\left(\frac{z}{\sqrt{x^2+y^2+z^2}}\right), \qquad (20)$$

so

$$Y_0^0(\theta,\,\phi)=\frac{1}{2}\frac{1}{\sqrt{\pi}} \qquad (21)$$

$$Y_1^0(\theta,\,\phi)=\frac{1}{2}\sqrt{\frac{3}{\pi}}\frac{z}{\sqrt{x^2+y^2+z^2}} \qquad (22)$$

$$Y_1^1(\theta,\,\phi)=-\frac{1}{2}\sqrt{\frac{3}{2\pi}}\frac{x+iy}{\sqrt{x^2+y^2+z^2}} \qquad (23)$$

$$Y_2^0(\theta,\,\phi)=\frac{1}{4}\sqrt{\frac{5}{\pi}}\left(\frac{3z^2}{x^2+y^2+z^2}-1\right) \qquad (24)$$

$$Y_2^1(\theta,\,\phi)=-\frac{1}{2}\sqrt{\frac{15}{2\pi}}\frac{z(x+iy)}{x^2+y^2+z^2} \qquad (25)$$

$$Y_2^2(\theta,\,\phi)=\frac{1}{4}\sqrt{\frac{15}{2\pi}}\frac{(x+iy)^2}{x^2+y^2+z^2}. \qquad (26)$$

The ZONAL HARMONICS are defined to be those OF THE FORM

$$P_l^0(\cos\theta)=P_l(\cos\theta). \qquad (27)$$

The TESSERAL HARMONICS are those OF THE FORM

$$\sin(m\phi)P_l^m(\cos\theta) \tag{28}$$

$$\cos(m\phi)P_l^m(\cos\theta) \tag{29}$$

for $l \neq m$. The SECTORIAL HARMONICS are OF THE FORM

$$\sin(m\phi)P_m^m(\cos\theta) \tag{30}$$

$$\cos(m\phi)P_m^m(\cos\theta). \tag{31}$$

The spherical harmonics form a COMPLETE ORTHONORMAL BASIS, so an arbitrary REAL FUNCTION $f(\theta, \phi)$ can be expanded in terms of complex spherical harmonics by

$$f(\theta, \phi) \equiv \sum_{l=0}^{\infty}\sum_{m=-l}^{l}A_l^m Y_l^m(\theta, \phi). \tag{32}$$

or in terms of real spherical harmonics by

$$f(\theta, \phi) \equiv \sum_{l=0}^{\infty}\sum_{m=0}^{l}[C_l^m Y_l^{mc}(\theta, \phi) + S_l^m Y_l^{ms}(\theta, \phi)]. \tag{33}$$

The process of determining the coefficients A_l^m in (32) is analogous to that to determine the coefficients in a FOURIER SERIES, i.e., multiply both sides of (32) by $\bar{Y}_{l'}^{m'}(\theta, \phi)$, integrate, and use the orthogonality relationship (7) to obtain

$$\int_0^{2\pi}\int_0^{\pi} f(\theta, \phi)\bar{Y}_{l'}^{m'}(\theta, \phi)\sin\theta\,d\theta\,d\phi$$

$$= \sum_{l=0}^{\infty}\sum_{m=-l}^{l}\int_0^{2\pi}\int_0^{\pi}A_l^m Y_l^m\bar{Y}_{l'}^{m'}(\theta, \phi)\sin\theta(\theta, \phi)\,d\theta\,d\phi$$

$$= \sum_{l=1}^{\infty}\sum_{m=-l}^{l}A_l^m\delta_{ll'}\delta_{mm'} = A_l^m. \tag{34}$$

The following sequence of plots shows successive approximations to the function $f(\theta, \phi) = 3 + \cos^3(2\theta) + (\sin\phi)/2$, which is illustrated in the final plot.

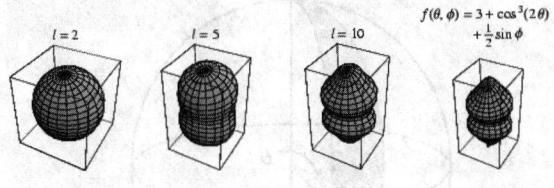

$l=2$ $l=5$ $l=10$ $f(\theta,\phi) = 3 + \cos^3(2\theta) + \frac{1}{2}\sin\phi$

See also CONDON-SHORTLEY PHASE, CORRELATION COEFFICIENT, SECTORIAL HARMONIC, SOLID HARMONIC, SPHERICAL HARMONIC ADDITION THEOREM, SPHERICAL HARMONIC DIFFERENTIAL EQUATION, SPHERICAL HARMONIC CLOSURE RELATIONS, SPHERICAL VECTOR HARMONIC, SURFACE HARMONIC, TESSERAL HARMONIC, ZONAL HARMONIC

References

Arfken, G. "Spherical Harmonics" and "Integrals of the Products of Three Spherical Harmonics." §12.6 and 12.9 in *Mathematical Methods for Physicists, 3rd ed.* Orlando, FL: Academic Press, pp. 680–85 and 698–00, 1985.

Byerly, W. E. "Spherical Harmonics." Ch. 6 in *An Elementary Treatise on Fourier's Series, and Spherical, Cylindrical, and Ellipsoidal Harmonics, with Applications to Problems in Mathematical Physics.* New York: Dover, pp. 195–18, 1959.

Ferrers, N. M. *An Elementary Treatise on Spherical Harmonics and Subjects Connected with Them.* London: Macmillan, 1877.

Groemer, H. *Geometric Applications of Fourier Series and Spherical Harmonics.* New York: Cambridge University Press, 1996.

Hobson, E. W. *The Theory of Spherical and Ellipsoidal Harmonics.* New York: Chelsea, 1955.

MacRobert, T. M. and Sneddon, I. N. *Spherical Harmonics: An Elementary Treatise on Harmonic Functions, with Applications, 3rd ed. rev.* Oxford, England: Pergamon Press, 1967.

Press, W. H.; Flannery, B. P.; Teukolsky, S. A.; and Vetterling, W. T. "Spherical Harmonics." §6.8 in *Numerical Recipes in FORTRAN: The Art of Scientific Computing, 2nd ed.* Cambridge, England: Cambridge University Press, pp. 246–48, 1992.

Sansone, G. "Harmonic Polynomials and Spherical Harmonics," "Integral Properties of Spherical Harmonics and the Addition Theorem for Legendre Polynomials," and "Completeness of Spherical Harmonics with Respect to Square Integrable Functions." §3.18–.20 in *Orthogonal Functions, rev. English ed.* New York: Dover, pp. 253–72, 1991.

Sternberg, W. and Smith, T. L. *The Theory of Potential and Spherical Harmonics, 2nd ed.* Toronto: University of Toronto Press, 1946.

Weisstein, E. W. "Books about Spherical Harmonics." http://www.treasure-troves.com/books/SphericalHarmonics.html.

Whittaker, E. T. and Watson, G. N. "Solution of Laplace's Equation Involving Legendre Functions" and "The Solution of Laplace's Equation which Satisfies Assigned Boundary Conditions at the Surface of a Sphere." §18.31 and 18.4 in *A Course in Modern Analysis, 4th ed.* Cambridge, England: Cambridge University Press, pp. 391–95, 1990.

Zwillinger, D. *Handbook of Differential Equations, 3rd ed.* Boston, MA: Academic Press, p. 129, 1997.

Spherical Harmonic Addition Theorem

A FORMULA also known as the LEGENDRE ADDITION THEOREM which is derived by finding GREEN'S FUNCTIONS for the SPHERICAL HARMONIC expansion and equating them to the generating function for LEGENDRE POLYNOMIALS. When γ is defined by

$$\cos\gamma \equiv \cos\theta_1\cos\theta_2 + \sin\theta_1\sin\theta_2\cos(\phi_1 - \phi_2),$$

The LEGENDRE POLYNOMIAL of argument γ is given by

$$P_l(\cos\gamma) = \frac{4\pi}{2l+1}\sum_{m=-l}^{l}(-1)^m Y_l^m(\theta_1, \phi_1)Y_l^{-m}(\theta_2, \phi_2)$$

$$= \frac{4\pi}{2l+1}\sum_{m=-l}^{l}Y_l^m(\theta_1, \phi_1)\bar{Y}_l^m(\theta_2, \phi_2)$$

$$= P_l(\cos\theta_1)P_l(\cos\theta_2)$$

$$+2\sum_{m=1}^{l}\frac{(l-m)!}{(l+m)!}P_l^m(\cos\theta_1)P_l^m(\cos\theta_2)\cos[m(\phi_1-\phi_2)].$$

See also LEGENDRE POLYNOMIAL, SPHERICAL HARMONIC

References

Arfken, G. "The Addition Theorem for Spherical Harmonics." §12.8 in *Mathematical Methods for Physicists, 3rd ed.* Orlando, FL: Academic Press, pp. 693–95, 1985.

Spherical Harmonic Closure Relations

The sum of the absolute squares of the SPHERICAL HARMONICS $Y_l^m(\theta,\ \phi)$ over all values of m is

$$\sum_{m=-l}^{l}\left|Y_l^m(\theta,\ \phi)\right|^2=\frac{2l+1}{4\pi}.$$

The double sum over m and l is given by

$$\sum_{l=0}^{\infty}\sum_{m=-l}^{l}Y_l^m(\theta_1,\ \phi_1)\bar{Y}_l^m(\theta_2,\ \phi_2)$$
$$=\frac{1}{\sin\theta_1}\delta(\theta_1-\theta_2)\delta(\phi_1-\phi_2)$$
$$=\delta(\cos\theta_1-\cos\theta_2)\delta(\cos\phi_1-\cos\phi_2),$$

where $\delta(x)$ is the DELTA FUNCTION.

Spherical Harmonic Differential Equation

In three dimensions, the spherical harmonic differential equation is given by

$$\left[\frac{1}{\sin\theta}\frac{\partial}{\partial\theta}\left(\sin\theta\frac{\partial}{\partial\theta}\right)+\frac{1}{\sin^2\theta}\frac{\partial^2}{\partial\phi^2}+l(l+1)\right]u=0,$$

and solutions are called SPHERICAL HARMONICS (Zwillinger 1997, p. 130). In four dimensions, the spherical harmonic differential equation is

$$u_{xx}+2u_x\cot x+\csc^2 x\left(u_{yy}+u_y\cot y+u_{zz}\csc^2 y\right)$$
$$+(n^2-1)u=0$$

(Humi 1987; Zwillinger 1997, p. 130).

See also SPHERICAL HARMONIC

References

Humi, M. "Factorisation of Separable Partial Differential Equations." *J. Phys. A: Math. Gen.* **20**, 4577–585, 1987.
Zwillinger, D. *Handbook of Differential Equations, 3rd ed.* Boston, MA: Academic Press, p. 130, 1997.

Spherical Harmonic Tensor

A tensor defined in terms of the TENSORS which satisfy the DOUBLE CONTRACTION RELATION.

See also DOUBLE CONTRACTION RELATION, SPHERICAL HARMONIC

Spherical Helix

The TANGENT INDICATRIX of a CURVE OF CONSTANT PRECESSION is a spherical helix. The equation of a spherical helix on a SPHERE with RADIUS r making an ANGLE θ with the z-AXIS is

$$x(\psi)=\tfrac{1}{2}r(1+\cos\theta)\cos\psi$$
$$-\tfrac{1}{2}r(1-\cos\theta)\cos\left(\frac{1+\cos\theta}{1-\cos\theta}\psi\right)\quad(1)$$

$$y(\psi)=\tfrac{1}{2}r(1+\cos\theta)\sin\psi$$
$$-\tfrac{1}{2}r(1-\sin\theta)\sin\left(\frac{1+\cos\theta}{1-\cos\theta}\psi\right)\quad(2)$$

$$z(\psi)=r\sin\theta\cos\left(\frac{\cos\theta}{1-\cos\theta}\psi\right).\quad(3)$$

The projection on the xy-plane is an EPICYCLOID with RADII

$$a=r\cos\theta\quad(4)$$
$$b=r\sin^2\left(\tfrac{1}{2}\theta\right).\quad(5)$$

See also HELIX, LOXODROME, SPHERICAL SPIRAL

References

Scofield, P. D. "Curves of Constant Precession." *Amer. Math. Monthly* **102**, 531–37, 1995.

Spherical Lune

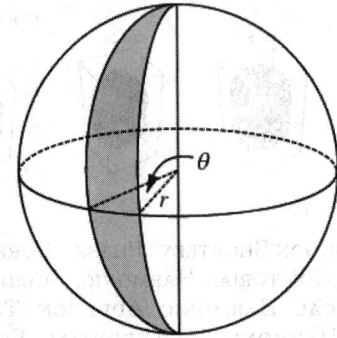

A sliver of the surface of a SPHERE of RADIUS r cut out by two planes through the azimuthal axis with DIHEDRAL ANGLE θ. The SURFACE AREA of the lune is

$$S=2r^2\theta,$$

which is just the area of the SPHERE times $\theta/(2\pi)$. The VOLUME of the associated SPHERICAL WEDGE has

VOLUME

$$V = \tfrac{2}{3} r^3 \theta.$$

See also LUNE, SPHERE, SPHERICAL WEDGE

References

Beyer, W. H. *CRC Standard Mathematical Tables, 28th ed.* Boca Raton, FL: CRC Press, p. 130, 1987.

Harris, J. W. and Stocker, H. "Spherical Wedge." §4.8.6 in *Handbook of Mathematics and Computational Science.* New York: Springer-Verlag, p. 108, 1998.

Gellert, W.; Gottwald, S.; Hellwich, M.; Kästner, H.; and Künstner, H. (Eds.). *VNR Concise Encyclopedia of Mathematics, 2nd ed.* New York: Van Nostrand Reinhold, p. 262, 1989.

Spherical Packing

SPHERICAL CODE

Spherical Polygon

A closed geometric figure on the surface of a SPHERE which is formed by the ARCS of GREAT CIRCLES. The spherical polygon is a generalization of the SPHERICAL TRIANGLE. If θ is the sum of the RADIAN ANGLES of a spherical polygon on a SPHERE of RADIUS R, then the AREA is

$$S = [\theta - (n-2)\pi]R^2.$$

See also GREAT CIRCLE, SPHERICAL TRIANGLE

References

Beyer, W. H. *CRC Standard Mathematical Tables, 28th ed.* Boca Raton, FL: CRC Press, p. 131, 1987.

Spherical Ring

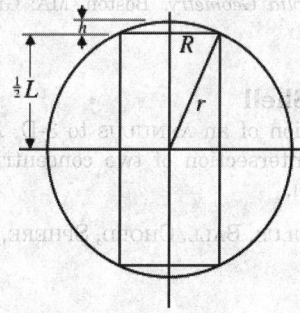

A SPHERE with a CYLINDRICAL HOLE cut so that the centers of the CYLINDER and SPHERE coincide, also called a NAPKIN RING. Let the SPHERE be of RADIUS r and the CYLINDER of RADIUS R. The VOLUME of the entire CYLINDER is

$$V_{\text{cyl}} = \pi L R^2, \tag{1}$$

and the VOLUME of the upper segment is

$$V_{\text{seg}} = \tfrac{1}{6}\pi h(3R^2 + h^2), \tag{2}$$

where

$$R = \sqrt{r^2 - \tfrac{1}{4}L^2} \tag{3}$$

$$h = r - \tfrac{1}{2}L, \tag{4}$$

so the VOLUME removed upon drilling of a CYLINDRICAL hole is

$$V_{\text{rem}} = V_{\text{cyl}} + 2V_{\text{seg}} = \pi\left[LR^2 + \tfrac{1}{3}h(3R^2 + h^2)\right]$$

$$= \pi\left(LR^2 + hR^2 + \tfrac{1}{3}h^3\right)$$

$$= \pi\left[L\left(r^2 - \tfrac{1}{4}L^2\right) + \left(r - \tfrac{1}{2}L\right)\left(r^2 - \tfrac{1}{4}L^2\right) + \tfrac{1}{3}\left(r - \tfrac{1}{2}L\right)^3\right]$$

$$= \pi\left[Lr^2 - \tfrac{1}{4}L^3 + \left(r^3 - \tfrac{1}{2}r^2L - \tfrac{1}{4}RL^2 + \tfrac{1}{8}L^3\right)\right.$$

$$\left. + \tfrac{1}{3}\left(r^3 - \tfrac{3}{2}r^2L + \tfrac{3}{4}rL^2 - \tfrac{1}{8}L^3\right)\right]$$

$$= \pi\left[\tfrac{4}{3}r^3 + \left(1 - \tfrac{1}{2} - \tfrac{1}{2}\right)r^2L + \left(-\tfrac{1}{4} + \tfrac{1}{4}\right)RL^2\right.$$

$$\left. + L^3\left(-\tfrac{1}{4} + \tfrac{1}{8} - \tfrac{1}{24}\right)\right]$$

$$= \tfrac{4}{3}\pi r^3 - \tfrac{1}{6}\pi L^3 = \tfrac{1}{6}\pi(8r^3 - L^3), \tag{5}$$

so

$$V_{\text{left}} = V_{\text{sphere}} - V_{\text{rem}} = \tfrac{4}{3}\pi r^3 - \left(\tfrac{4}{3}\pi r^3 - \tfrac{1}{6}\pi L^3\right)$$

$$= \tfrac{1}{6}\pi L^3. \tag{6}$$

Spherical Sector

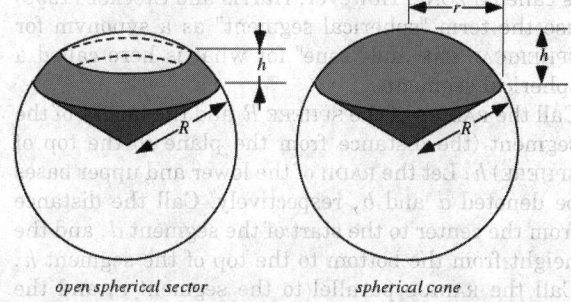

open spherical sector　　　　　*spherical cone*

A spherical sector is a SOLID OF REVOLUTION enclosed by two radii from the center of a SPHERE. The spherical sector may either be "open" and have a conical HOLE (left figure; Beyer 1987), or may be a "closed" SPHERICAL CONE (right figure; Harris and Stocker 1998). The VOLUME of a spherical sector in either case is given by

$$V = \tfrac{2}{3}\pi R^2 h,$$

where h is the vertical distance between where the upper and lower radii intersect the sphere and R is the sphere's radius.

See also CYLINDRICAL SEGMENT, SPHERE, SPHERICAL CAP, SPHERICAL CONE, SPHERICAL SEGMENT, SPHERICAL WEDGE, ZONE

References

Beyer, W. H. *CRC Standard Mathematical Tables, 28th ed.* Boca Raton, FL: CRC Press, p. 131, 1987.

Harris, J. W. and Stocker, H. "Spherical Sector." §4.8.3 in *Handbook of Mathematics and Computational Science.* New York: Springer-Verlag, pp. 106–07, 1998.

Kern, W. F. and Bland, J. R. "Spherical Sector." §37 in *Solid Mensuration with Proofs, 2nd ed.* New York: Wiley, pp. 103–06, 1948.

Smith, D. E. "Spherical Sector." §542 in *Essentials of Plane and Solid Geometry.* Boston, MA: Ginn and Co., p. 542, 1923.

Spherical Segment

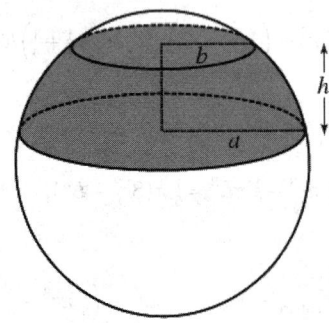

A spherical segment is the solid defined by cutting a SPHERE with a pair of PARALLEL PLANES. It can be thought of as a SPHERICAL CAP with the top truncated, and so it corresponds to a SPHERICAL FRUSTUM. The surface of the spherical segment (excluding the bases) is called a ZONE. However, Harris and Stocker (1998) use the term "spherical segment" as a synonym for SPHERICAL CAP and "zone" for what is here called a spherical segment.

Call the RADIUS of the SPHERE R and the height of the segment (the distance from the plane to the top of SPHERE) h. Let the RADII of the lower and upper bases be denoted a and b, respectively. Call the distance from the center to the start of the segment d, and the height from the bottom to the top of the segment h. Call the RADIUS parallel to the segment r, and the height above the center y. Then $r^2 = R^2 - y^2$,

$$V = \int_d^{d+h} \pi r^2\, dy = \pi \int_d^{d+h} \left(R^2 - y^2\right) dy$$

$$= \pi\left[R^2 y - \tfrac{1}{3}y^3\right]_d^{d+h} = \pi\left\{R^2 h - \tfrac{1}{3}\left[(d+h)^3 - d^3\right]\right\}$$

$$= \pi\left[R^2 h - \tfrac{1}{3}(d^3 + 3d^2 h + 3h^2 d + h^3 - d^3)\right]$$

$$= \pi h\left(R^2 - d^2 - hd - \tfrac{1}{3}h^2\right), \tag{1}$$

Using

$$a^2 = R^2 - d^2 \tag{2}$$

$$b^2 = R^2 - (d+h)^2 = R^2 - d^2 - 2dh - h^2, \tag{3}$$

gives

$$a^2 + b^2 = 2R^2 - 2d^2 - 2dh - h^2 \tag{4}$$

$$R^2 - d^2 - dh = \tfrac{1}{2}(a^2 + b^2 + h^2), \tag{5}$$

so

$$V = \pi h\left[\tfrac{1}{2}(a^2 + b^2 + h^2) - \tfrac{1}{3}h^2\right] = \pi h\left(\tfrac{1}{2}a^2 + \tfrac{1}{2}b^2 + \tfrac{1}{6}h^2\right)$$

$$= \tfrac{1}{6}\pi h\left(3a^2 + 3b^2 + h^2\right). \tag{6}$$

The surface area of the ZONE (which excludes the top and bottom bases) is given by

$$S = 2\pi R h. \tag{7}$$

See also ARCHIMEDES' HAT-BOX THEOREM, ARCHIMEDES' PROBLEM, FRUSTUM, HEMISPHERE, SPHERE, SPHERICAL CAP, SPHERICAL SECTOR, SPHERICAL WEDGE, SURFACE OF REVOLUTION, ZONE

References

Beyer, W. H. *CRC Standard Mathematical Tables, 28th ed.* Boca Raton, FL: CRC Press, p. 130, 1987.

Harris, J. W. and Stocker, H. "Spherical Zone (Spherical Layer)." §4.8.5 in *Handbook of Mathematics and Computational Science.* New York: Springer-Verlag, pp. 107–08, 1998.

Kern, W. F. and Bland, J. R. "Spherical Segment." §36 in *Solid Mensuration with Proofs, 2nd ed.* New York: Wiley, pp. 97–02, 1948.

Smith, D. E. "Spherical Segment." §541 in *Essentials of Plane and Solid Geometry.* Boston, MA: Ginn and Co., p. 542, 1923.

Spherical Shell

A generalization of an ANNULUS to 3-D. A spherical shell is the intersection of two concentric BALLS of differing RADII.

See also ANNULUS, BALL, CHORD, SPHERE, SPHERICAL HELIX

Spherical Simplex

The only irreducible spherical simplexes generated by reflection are A_n $(n \geq 1)$, B_n $(n \geq 4)$, C_n $(n \geq 2)$, D_2^p $(p \geq 5)$, E_6, E_7, E_8, F_4, G_3, and G_4. The only irreducible Euclidean simplexes generated by reflection are W_2, P_m $(m \geq 3)$, Q_m $(m \geq 5)$, R_m $(m \geq 3)$, S_m $(m \geq 4)$, V_3, T_7, T_8, T_9, and U_5.

Spherical Spiral

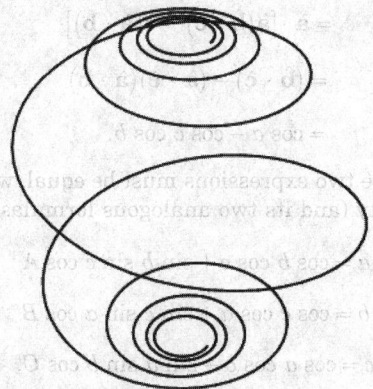

The SPHERICAL CURVE taken by a ship which travels from the south pole to the north pole of a SPHERE while keeping a fixed (but not RIGHT) angle with respect to the meridians. The curve has an infinite number of loops since the separation of consecutive revolutions gets smaller and smaller near the poles. It is given by the PARAMETRIC EQUATIONS

$$x = \cos t \cos c$$

$$y = \sin t \cos c$$

$$z = -\sin c,$$

where

$$c \equiv \tan^{-1}(at)$$

and a is a constant, and is a special case of a LOXODROME.

See also HELIX, LOXODROME, MERCATOR PROJECTION, SEIFFERT'S SPHERICAL SPIRAL, SPHERICAL CURVE

References

Gray, A. "Loxodromes on Spheres." §10.6 in *Modern Differential Geometry of Curves and Surfaces with Mathematica, 2nd ed.* Boca Raton, FL: CRC Press, pp. 238–40, 1997.

Lauwerier, H. "Spherical Spiral." In *Fractals: Endlessly Repeated Geometric Figures.* Princeton, NJ: Princeton University Press, pp. 64–6, 1991.

Spherical Symmetry

Let **A** and **B** be constant VECTORS. Define

$$Q \equiv 3(\mathbf{A} \cdot \hat{\mathbf{r}})(\mathbf{B} \cdot \hat{\mathbf{r}}) - \mathbf{A} \cdot \mathbf{B}.$$

Then the average of Q over a spherically symmetric surface or volume is

$$\langle Q \rangle = \langle 3 \cos^2 \theta - 1 \rangle (\mathbf{A} \cdot \mathbf{B}) = 0,$$

since $\langle 3 \cos^2 \theta - 1 \rangle = 0$ over the sphere.

Spherical Tessellation

TRIANGULAR SYMMETRY GROUP

Spherical Triangle

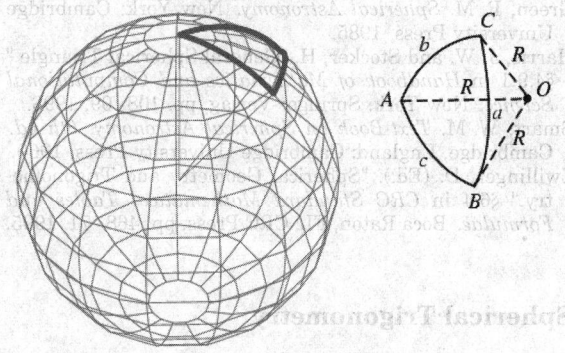

A spherical triangle is a figure formed on the surface of a sphere by three great circular arcs intersecting pairwise in three vertices. The spherical triangle is the spherical analog of the planar TRIANGLE, and is sometimes called EULER'S TRIANGLE (Harris and Stocker 1998). Let a spherical triangle have ANGLES A, B, and C (measured in radians at the vertices along the surface of the sphere) and let the sphere on which the spherical triangle sits have RADIUS R. Then the SURFACE AREA Δ of the spherical triangle is

$$\Delta = R^2[(A + B + C) - \pi] = R^2 E,$$

where E is called the SPHERICAL EXCESS, with $E = 0$ in the degenerate case of a planar triangle.

The sum of the angles of a spherical triangle is between π and 3π radians (180° and 540°; Zwillinger 1995, p. 469). The amount by which it exceeds 180° is called the SPHERICAL EXCESS and is denoted E or Δ, the latter of which can cause confusion since it also can refer to the SURFACE AREA of a spherical triangle. The difference between 2π radians (360°) and the sum of the side arc lengths a, b, and c is called the SPHERICAL DEFECT and is denoted D or δ.

The study of angles and distances of figures on a sphere is known as SPHERICAL TRIGONOMETRY.

See also CIRCULAR TRIANGLE, COLUNAR TRIANGLE, GEODESIC DOME, GEODESIC TRIANGLE, GIRARD'S SPHERICAL EXCESS FORMULA, L'HUILIER'S THEOREM, NAPIER'S ANALOGIES, SPHERICAL DEFECT, SPHERICAL EXCESS, SPHERICAL POLYGON, SPHERICAL TRIGONOMETRY

References

Abramowitz, M. and Stegun, C. A. (Eds.). *Handbook of Mathematical Functions with Formulas, Graphs, and Mathematical Tables, 9th printing.* New York: Dover, p. 79, 1972.

Beyer, W. H. *CRC Standard Mathematical Tables, 28th ed.* Boca Raton, FL: CRC Press, pp. 131 and 147–50, 1987.

Gellert, W.; Gottwald, S.; Hellwich, M.; Kästner, H.; and Künstner, H. (Eds.). "The Spherical Triangle." §12.2 in

VNR Concise Encyclopedia of Mathematics, 2nd ed. New York: Van Nostrand Reinhold, pp. 262–72, 1989.

Green, R. M. *Spherical Astronomy.* New York: Cambridge University Press, 1985.

Harris, J. W. and Stocker, H. "General Spherical Triangle." §4.9.1 in *Handbook of Mathematics and Computational Science.* New York: Springer-Verlag, pp. 108–09, 1998.

Smart, W. M. *Text-Book on Spherical Astronomy, 6th ed.* Cambridge, England: Cambridge University Press, 1960.

Zwillinger, D. (Ed.). "Spherical Geometry and Trigonometry." §6.4 in *CRC Standard Mathematical Tables and Formulae.* Boca Raton, FL: CRC Press, pp. 468–71, 1995.

Spherical Trigonometry

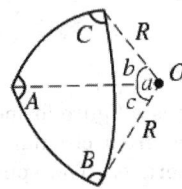

Let a SPHERICAL TRIANGLE be drawn on the surface of a SPHERE of radius R, centered at a point $O = (0,\ 0,\ 0)$, with vertices A, B, and C. The vectors from the center of the sphere to the vertices are therefore given by $\mathbf{a} \equiv \overrightarrow{OA}$, $\mathbf{b} \equiv \overrightarrow{OB}$, and $\mathbf{c} \equiv \overrightarrow{OC}$. Now, the *angular* lengths of the sides of the triangle (in radians) are then $a' \equiv \angle BOC$, $b' \equiv \angle COA$, and $c' \equiv \angle AOB$, and the *actual* arc lengths of the side are $a = Ra'$, $b = Rb'$, and $c = Rc'$. Explicitly,

$$\mathbf{a} \cdot \mathbf{b} = R^2 \cos c' = R^2 \cos\left(\frac{c}{R}\right) \qquad (1)$$

$$\mathbf{a} \cdot \mathbf{c} = R^2 \cos b' = R^2 \cos\left(\frac{b}{R}\right) \qquad (2)$$

$$\mathbf{b} \cdot \mathbf{c} = R^2 \cos a' = R^2 \cos\left(\frac{a}{R}\right). \qquad (3)$$

Now make use of A, B, and C to denote both the vertices themselves and the *angles* of the spherical triangle at these vertices, so that the DIHEDRAL ANGLE between PLANES AOB and AOC is written A, the DIHEDRAL ANGLE between PLANES BOC and AOB is written B, and the DIHEDRAL ANGLE between PLANES BOC and AOC is written C. (These angles are sometimes instead denoted α, β, γ; e.g., Gellert *et al.* 1989)

Consider the DIHEDRAL ANGLE A between planes AOB and AOC, which can be calculated using the DOT PRODUCT of the normals to the planes. The normals are given by CROSS PRODUCTS of the vectors to the vertices, so

$$(\hat{\mathbf{a}} \times \hat{\mathbf{b}}) \cdot (\hat{\mathbf{a}} \times \hat{\mathbf{c}}) = (|\hat{\mathbf{a}}||\hat{\mathbf{b}}|\sin c)(|\hat{\mathbf{a}}||\hat{\mathbf{c}}|\sin b)\cos A$$

$$= \sin b \sin c \cos A. \qquad (4)$$

However, using a well-known vector identity gives

$$(\hat{\mathbf{a}} \times \hat{\mathbf{b}}) \cdot (\hat{\mathbf{a}} \times \hat{\mathbf{c}}) = \hat{\mathbf{a}} \cdot [\hat{\mathbf{b}} \times (\hat{\mathbf{a}} \times \hat{\mathbf{c}})]$$

$$= \hat{\mathbf{a}} \cdot [\hat{\mathbf{a}}(\hat{\mathbf{b}} \cdot \hat{\mathbf{c}}) - \hat{\mathbf{c}}(\hat{\mathbf{a}} \cdot \hat{\mathbf{b}})]$$

$$= (\hat{\mathbf{b}} \cdot \hat{\mathbf{c}}) - (\hat{\mathbf{a}} \cdot \hat{\mathbf{c}})(\hat{\mathbf{a}} \cdot \hat{\mathbf{b}})$$

$$= \cos a - \cos c \cos b. \qquad (5)$$

Since these two expressions must be equal, we obtain the identity (and its two analogous formulas)

$$\cos a = \cos b \cos c + \sin b \sin c \cos A \qquad (6)$$

$$\cos b = \cos c \cos a + \sin c \sin a \cos B \qquad (7)$$

$$\cos c = \cos a \cos b + \sin a \sin b \cos C. \qquad (8)$$

known as the cosine rules for sides (Smart 1960, pp. 7–; Gellert *et al.* 1989, p. 264; Zwillinger 1995, p. 469).

The identity

$$\sin A = \frac{|(\hat{\mathbf{a}} \times \hat{\mathbf{b}}) \times (\hat{\mathbf{a}} \times \hat{\mathbf{c}})|}{|\hat{\mathbf{a}} \times \hat{\mathbf{b}}||\hat{\mathbf{a}} \times \hat{\mathbf{c}}|}$$

$$= \frac{|\hat{\mathbf{a}}[\hat{\mathbf{b}}, \hat{\mathbf{a}}, \hat{\mathbf{c}}] + \hat{\mathbf{b}}[\hat{\mathbf{a}}, \hat{\mathbf{a}}, \hat{\mathbf{c}}]|}{\sin b \sin c}$$

$$= \frac{[\hat{\mathbf{a}}, \hat{\mathbf{b}}, \hat{\mathbf{c}}]}{\sin b \sin c}, \qquad (9)$$

where $[\mathbf{a}, \mathbf{b}, \mathbf{c}]$ is the SCALAR TRIPLE PRODUCT, gives

$$\frac{\sin A}{\sin a} = \frac{[\hat{\mathbf{a}}, \hat{\mathbf{b}}, \hat{\mathbf{c}}]}{\sin a \sin b \sin c}, \qquad (10)$$

so the spherical analog of the LAW OF SINES can be written

$$\frac{\sin A}{\sin a} = \frac{\sin B}{\sin b} = \frac{\sin C}{\sin c} = \frac{6\ \text{Vol}(OABC)}{\sin a \sin b \sin c} \qquad (11)$$

(Smart 1960, pp. 9–0; Gellert *et al.* 1989, p. 265; Zwillinger 1995, p. 469), where Vol($OABC$) is the VOLUME of the TETRAHEDRON.

The analogs of the LAW OF COSINES for the angles of a SPHERICAL TRIANGLE are given by

$$\cos A = -\cos B \cos C + \sin B \sin C \cos a \qquad (12)$$

$$\cos B = -\cos C \cos A + \sin C \sin A \cos b \qquad (13)$$

$$\cos C = -\cos A \cos B + \sin A \sin B \cos c \qquad (14)$$

(Gellert *et al.* 1989, p. 265; Zwillinger 1995, p. 470). Finally, there are spherical analogs of the LAW OF TANGENTS,

$$\frac{\tan\left[\frac{1}{2}(B - C)\right]}{\tan\left[\frac{1}{2}(B + C)\right]} = \frac{\tan\left[\frac{1}{2}(b - c)\right]}{\tan\left[\frac{1}{2}(b + c)\right]} \qquad (15)$$

$$\frac{\tan\left[\frac{1}{2}(C-A)\right]}{\tan\left[\frac{1}{2}(C+A)\right]} = \frac{\tan\left[\frac{1}{2}(c-a)\right]}{\tan\left[\frac{1}{2}(c+a)\right]} \tag{16}$$

$$\frac{\tan\left[\frac{1}{2}(A-B)\right]}{\tan\left[\frac{1}{2}(A+B)\right]} = \frac{\tan\left[\frac{1}{2}(a-b)\right]}{\tan\left[\frac{1}{2}(a+b)\right]} \tag{17}$$

(Beyer 1987; Gellert *et al.* 1989; Zwillinger 1995, p. 470).

Additional important identities are given by

$$\cos A = \csc b \, \csc c (\cos a - \cos b \cos c). \tag{18}$$

(Smart 1960, p. 8),

$$\sin a \cos B = \cos b \sin c - \sin b \cos c \cos A \tag{19}$$

(Smart 1960, p. 10), and

$$\cos a \cos C = \sin a \cot b - \sin C \cot B \tag{20}$$

(Smart 1960, p. 12).

Let

$$s \equiv \tfrac{1}{2}(a+b+c) \tag{21}$$

be the semiperimeter, then half-angle formulas for sines can be written as

$$\sin\left(\tfrac{1}{2}A\right) = \sqrt{\frac{\sin(s-b)\sin(s-c)}{\sin b \sin c}} \tag{22}$$

$$\sin\left(\tfrac{1}{2}B\right) = \sqrt{\frac{\sin(s-a)\sin(s-c)}{\sin a \sin c}} \tag{23}$$

$$\sin\left(\tfrac{1}{2}C\right) = \sqrt{\frac{\sin(s-a)\sin(s-b)}{\sin a \sin b}}. \tag{24}$$

for cosines can be written as

$$\cos\left(\tfrac{1}{2}A\right) = \sqrt{\frac{\sin s \sin(s-a)}{\sin b \sin c}} \tag{25}$$

$$\cos\left(\tfrac{1}{2}B\right) = \sqrt{\frac{\sin s \sin(s-b)}{\sin a \sin c}} \tag{26}$$

$$\cos\left(\tfrac{1}{2}C\right) = \sqrt{\frac{\sin s \sin(s-c)}{\sin a \sin b}} \tag{27}$$

and tangents can be written as

$$\tan\left(\tfrac{1}{2}A\right) = \sqrt{\frac{\sin(s-b)\sin(s-c)}{\sin s \sin(s-a)}} = \frac{k}{\sin(s-a)} \tag{28}$$

$$\tan\left(\tfrac{1}{2}B\right) = \sqrt{\frac{\sin(s-a)\sin(s-c)}{\sin s \sin(s-b)}} = \frac{k}{\sin(s-b)} \tag{29}$$

$$\tan\left(\tfrac{1}{2}C\right) = \sqrt{\frac{\sin(s-a)\sin(s-b)}{\sin s \sin(s-c)}} = \frac{k}{\sin(s-c)}, \tag{30}$$

where

$$k^2 = \frac{\sin(s-a)\sin(s-b)\sin(s-c)}{\sin s} \tag{31}$$

(Smart 1960, pp. 8–; Gellert *et al.* 1989, p. 265; Zwillinger 1995, p. 470).

Let

$$S \equiv \tfrac{1}{2}(A+B+C) \tag{32}$$

be the sum of half-angles, then the half-side formulas are

$$\tan\left(\tfrac{1}{2}a\right) = K \cos(S-A) \tag{33}$$

$$\tan\left(\tfrac{1}{2}b\right) = K \cos(S-B) \tag{34}$$

$$\tan\left(\tfrac{1}{2}c\right) = K \cos(S-C). \tag{35}$$

where

$$K^2 = -\frac{\cos S}{\cos(S-A)\cos(S-B)\cos(S-C)} \tag{36}$$

(Gellert *et al.* 1989, p. 265; Zwillinger 1995, p. 470).

The HAVERSINE formula for sides, where

$$\operatorname{hav} x \equiv \tfrac{1}{2}(1-\cos x) = \sin^2\left(\tfrac{1}{2}x\right), \tag{37}$$

is given by

$$\operatorname{hav} a = \operatorname{hav}(b-c) + \sin b \sin c \operatorname{hav} A \tag{38}$$

(Smart 1960, pp. 18–9; Zwillinger 1995, p. 471), and the HAVERSINE formula for angles is given by

$$\operatorname{hav} A = \frac{\sin(s-b)\sin(s-c)}{\sin b \sin c} \tag{39}$$

$$= \frac{\operatorname{hav} a - \operatorname{hav}(b-c)}{\sin b \sin c} \tag{40}$$

$$= \operatorname{hav}[\pi - (B+C)] + \sin B \sin C \operatorname{hav} a \tag{41}$$

(Zwillinger 1995, p. 471).

GAUSS'S FORMULAS (also called Delambre's analogies) are

$$\frac{\sin\left[\frac{1}{2}(a-b)\right]}{\sin\left(\frac{1}{2}c\right)} = \frac{\sin\left[\frac{1}{2}(A-B)\right]}{\cos\left(\frac{1}{2}C\right)} \tag{42}$$

$$\frac{\sin\left[\frac{1}{2}(a+b)\right]}{\sin\left(\frac{1}{2}c\right)} = \frac{\cos\left[\frac{1}{2}(A-B)\right]}{\sin\left(\frac{1}{2}C\right)} \tag{43}$$

$$\frac{\cos\left[\frac{1}{2}(a-b)\right]}{\cos\left(\frac{1}{2}c\right)} = \frac{\sin\left[\frac{1}{2}(A+B)\right]}{\cos\left(\frac{1}{2}C\right)} \tag{44}$$

$$\frac{\cos\left[\frac{1}{2}(a+b)\right]}{\cos\left(\frac{1}{2}c\right)} = \frac{\cos\left[\frac{1}{2}(A+B)\right]}{\sin\left(\frac{1}{2}C\right)} \qquad (45)$$

(Smart 1960, p. 22; Zwillinger 1995, p. 470).

NAPIER'S ANALOGIES are

$$\frac{\sin\left[\frac{1}{2}(A-B)\right]}{\sin\left[\frac{1}{2}(A+B)\right]} = \frac{\tan\left[\frac{1}{2}(a-b)\right]}{\tan\left(\frac{1}{2}c\right)} \qquad (46)$$

$$\frac{\cos\left[\frac{1}{2}(A-B)\right]}{\cos\left[\frac{1}{2}(A+B)\right]} = \frac{\tan\left[\frac{1}{2}(a+b)\right]}{\tan\left(\frac{1}{2}c\right)} \qquad (47)$$

$$\frac{\sin\left[\frac{1}{2}(a-b)\right]}{\sin\left[\frac{1}{2}(a+b)\right]} = \frac{\tan\left[\frac{1}{2}(A-B)\right]}{\cot\left(\frac{1}{2}C\right)} \qquad (48)$$

$$\frac{\cos\left[\frac{1}{2}(a-b)\right]}{\cos\left[\frac{1}{2}(a+b)\right]} = \frac{\tan\left[\frac{1}{2}(A+B)\right]}{\cot\left(\frac{1}{2}C\right)} \qquad (49)$$

(Beyer 1987; Gellert *et al.* 1989, p. 266; Zwillinger 1995, p. 471).

See also ANGULAR DEFECT, DESCARTES TOTAL ANGULAR DEFECT, GAUSS'S FORMULAS, GIRARD'S SPHERICAL EXCESS FORMULA, LAW OF COSINES, LAW OF SINES, LAW OF TANGENTS, L'HUILIER'S THEOREM, NAPIER'S ANALOGIES, SPHERICAL EXCESS, SPHERICAL GEOMETRY, SPHERICAL POLYGON, SPHERICAL TRIANGLE

References

Beyer, W. H. *CRC Standard Mathematical Tables, 28th ed.* Boca Raton, FL: CRC Press, pp. 131 and 147–50, 1987.
Danby, J. M. *Fundamentals of Celestial Mechanics, 2nd ed.,* rev. ed. Richmond, VA: Willmann-Bell, 1988.
Gellert, W.; Gottwald, S.; Hellwich, M.; Kästner, H.; and Künstner, H. (Eds.). "Spherical Trigonometry." §12 in *VNR Concise Encyclopedia of Mathematics, 2nd ed.* New York: Van Nostrand Reinhold, pp. 261–82, 1989.
Green, R. M. *Spherical Astronomy.* New York: Cambridge University Press, 1985.
Smart, W. M. *Text-Book on Spherical Astronomy, 6th ed.* Cambridge, England: Cambridge University Press, 1960.
Zwillinger, D. (Ed.). "Spherical Geometry and Trigonometry." §6.4 in *CRC Standard Mathematical Tables and Formulae.* Boca Raton, FL: CRC Press, pp. 468–71, 1995.

Spherical Vector Harmonic

VECTOR SPHERICAL HARMONIC

Spherical Wedge

$$\llcorner\xrightarrow{6}100011\xrightarrow{3}10101\xrightarrow{3}1110\xrightarrow{0}111\xrightarrow{4}1011\ldots$$

The VOLUME of a spherical wedge is

$$V = \frac{2}{3}r^3\theta.$$

The surface area of the corresponding SPHERICAL

LUNE is

$$S = 2r^2\theta.$$

See also SPHERE, SPHERICAL CAP, SPHERICAL LUNE, SPHERICAL SECTOR, SPHERICAL SEGMENT, WEDGE

References

Harris, J. W. and Stocker, H. "Spherical Wedge." §4.8.6 in *Handbook of Mathematics and Computational Science.* New York: Springer-Verlag, p. 108, 1998.

SphericalHarmonicY

SPHERICAL HARMONIC

Sphericon

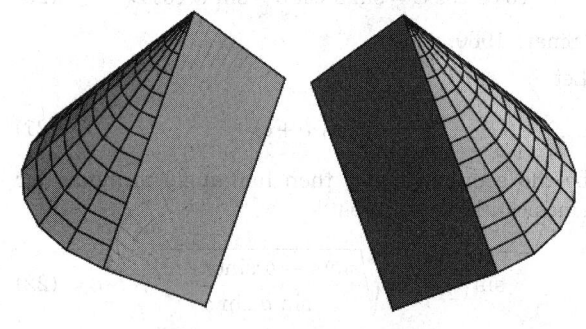

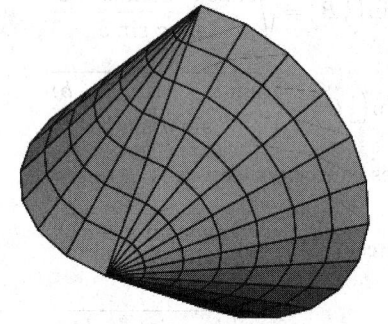

The solid formed from a BICONE with opening angle of 90°. Slice the solid by a plane containing the rotational axes. The resulting CROSS SECTION is a SQUARE. Now rotate the two pieces by 90° and reconnect them.

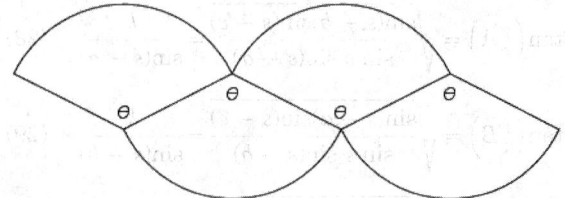

The above net shows another way the sphericon can be constructed. In this figure $\theta = \pi\sqrt{2}/2$ radians $\approx$

$127.28°$. This solid was discovered by C. J. Roberts, and is not as widely known as it should be!

A sphericon has a single continuous face. A sphericon rolls by wobbling from one face to another, resulting in straight-line motion. In addition, one sphericon can roll around another.

See also BICONE, CONE, CONE NET, SPHERE

References
Stewart, I. "Cone with a Twist." *Sci. Amer.* **281**, 116–17, Oct. 1999.

Spheroid

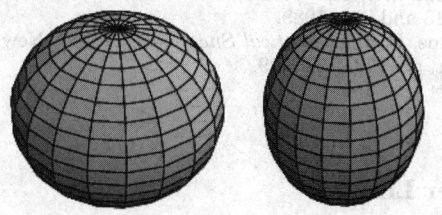

A spheroid is an ELLIPSOID

$$\frac{r^2 \cos^2 \theta \sin^2 \phi}{a^2} + \frac{r^2 \sin^2 \theta \sin^2 \phi}{b^2} + \frac{r^2 \cos^2 \phi}{c^2} = 1 \quad (1)$$

with two SEMIMAJOR AXES equal. Orient the ELLIPSE so that the a and b axes are equal, then

$$\frac{r^2 \cos^2 \theta \sin^2 \phi}{a^2} + \frac{r^2 \sin^2 \theta \sin^2 \phi}{a^2} + \frac{r^2 \cos^2 \phi}{c^2} = 1 \quad (2)$$

$$\frac{r^2 \sin^2 \phi}{a^2} + \frac{r^2 \cos^2 \phi}{c^2} = 1. \quad (3)$$

where a is the equatorial RADIUS and c is the polar RADIUS. The PARAMETRIC EQUATIONS therefore become

$$x = a \cos \theta \sin \phi \quad (4)$$

$$y = a \sin \theta \sin \phi \quad (5)$$

$$z = c \cos \phi \quad (6)$$

for $\theta \in [0, 2\pi)$ and $\phi \in [0, \pi]$.

Here ϕ is the colatitude, so take $\delta \equiv \pi/2 - \phi$ to express in terms of latitude.

$$\frac{r^2 \cos^2 \delta}{a^2} + \frac{r^2 \sin^2 \delta}{c^2} = 1. \quad (7)$$

Rewriting $\cos^2 \delta = 1 - \sin^2 \delta$ gives

$$\frac{r^2}{a^2} + r^2 \sin^2 \delta \left(\frac{1}{c^2} - \frac{1}{a^2}\right) = 1 \quad (8)$$

$$r^2 \left(1 + a^2 \sin^2 \delta \, \frac{a^2 - c^2}{c^2 a^2}\right) = r^2 \left(1 + \sin^2 \delta \, \frac{a^2 - c^2}{c^2}\right)$$

$$= a^2. \quad (9)$$

so

$$r = a \left(1 + \sin^2 \delta \, \frac{a^2 - c^2}{c^2}\right)^{-1/2}. \quad (10)$$

If $a > c$, the spheroid is OBLATE. If $a < c$, the spheroid is PROLATE. If $a = c$, the spheroid degenerates to a SPHERE.

See also DARWIN-DE SITTER SPHEROID, ELLIPSOID, OBLATE SPHEROID, PROLATE SPHEROID

Spheroidal Coordinates
OBLATE SPHEROIDAL COORDINATES, PROLATE SPHEROIDAL COORDINATES

Spheroidal Function
OBLATE SPHEROIDAL WAVE FUNCTION, PROLATE SPHEROIDAL WAVE FUNCTION, SPHEROIDAL WAVE FUNCTION

Spheroidal Harmonic
A spheroidal harmonic is a special case of the ELLIPSOIDAL HARMONIC which satisfies the differential equation

$$\frac{d}{dx}\left[(1-x^2)\frac{ds}{dx}\right] + \left(\lambda - c^2 x^2 - \frac{m^2}{1-x^2}\right)S = 0$$

on the interval $-1 \le x \le 1$.

See also ELLIPSOIDAL HARMONIC

References
Press, W. H.; Flannery, B. P.; Teukolsky, S. A.; and Vetterling, W. T. "A Worked Example: Spheroidal Harmonics." §17.4 in *Numerical Recipes in FORTRAN: The Art of Scientific Computing, 2nd ed.* Cambridge, England: Cambridge University Press, pp. 764–73, 1992.

Spheroidal Wave Function
Whittaker and Watson (1990, p. 403) define the internal and external spheroidal wavefunctions as

$$S_{mn}^{(1)} = 2\pi \, \frac{(n-m)!}{(n+m)!} \, P_n^m(ir)P_n^m(\cos\theta)_{\sin}^{\cos}(m\phi)$$

$$S_{mn}^{(2)} = 2\pi \, \frac{(n-m)!}{(n+m)!} \, Q_n^m(ir)Q_n^m(\cos\theta)_{\sin}^{\cos}(m\phi),$$

where $P_l^m(x)$ is a LEGENDRE POLYNOMIAL and $Q_l^m(x)$ is a LEGENDRE FUNCTION OF THE SECOND KIND.

Stratton (1935), Chu and Stratton (1941), and Rhodes (1970) define the spheroidal functions as those solutions of the differential equation

$$(1 - \eta^2)\psi''_{\varkappa n}(c, \eta) - 2(\alpha + 1)\eta\psi'_{\varkappa n}(c, \eta)$$
$$+ (b_{\varkappa n} - c^2\eta^2)\psi_{\varkappa n}(c, \eta) = 0$$

which remain finite at the singular points $\eta = \pm 1$. The condition of finiteness restricts the admissible values of the parameter $b_{\varkappa n}(c)$ to a discrete set of eigenvalues indexed by $n = 0, 1, 2, \ldots$ (Rhodes 1970).

See also ELLIPSOIDAL HARMONIC, OBLATE SPHEROIDAL WAVE FUNCTION, PROLATE SPHEROIDAL WAVE FUNCTION, SPHERICAL HARMONIC

References

Abramowitz, M. and Stegun, C. A. (Eds.). "Spheroidal Wave Functions." Ch. 21 in *Handbook of Mathematical Functions with Formulas, Graphs, and Mathematical Tables, 9th printing.* New York: Dover, pp. 751–59, 1972.
Chu, L. J. and Stratton, J. A. "Elliptic and Spheroidal Wave Functions." *J. Math. and Phys.* **20**, 259–09, 1941.
Morse, P. M. and Feshbach, H. *Methods of Theoretical Physics, Part I.* New York: McGraw-Hill, pp. 642–44, 1953.
Rhodes, D. R. "On the Spheroidal Functions." *J. Res. Nat. Bur. Standards--B. Math. Sci.* **74B**, 187–09, Jul.-Sep. 1970.
Stratton, J. A. "Spheroidal Functions." *Proc. Nat. Acad. Sci.* **21**, 51–6, 1935.
Stratton, J. A.; Morse, P. M.; Chu, L. J.; Little, J. D. C.; and Corbató, F. J. *Spheroidal Wave Functions.* New York: Wiley, 1956.
Whittaker, E. T. and Watson, G. N. *A Course in Modern Analysis, 4th ed.* Cambridge, England: Cambridge University Press, 1990.

Sphinx

A 6-POLYIAMOND named for its resemblance to the Great Sphinx of Egypt.

References

Golomb, S. W. *Polyominoes: Puzzles, Patterns, Problems, and Packings, 2nd ed.* Princeton, NJ: Princeton University Press, p. 92, 1994.

Spider and Fly Problem

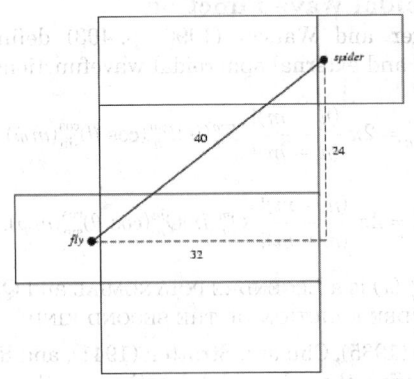

In a rectangular room (a CUBOID) with dimensions

$30' \times 12' \times 12'$, a spider is located in the middle of one $12' \times 12'$ wall one foot away from the ceiling. A fly is in the middle of the opposite wall one foot away from the floor. If the fly remains stationary, what is the shortest distance the spider must crawl to capture the fly? The answer, $40'$, can be obtained by "flattening" the walls as illustrated above. The puzzle was originally posed in an English newspaper by Dudeney in 1903 (Gardner 1958).

References

Gardner, M. "Mathematical Games: About Henry Ernest Dudeney, A Brilliant Creator of Puzzles." *Sci. Amer.* **198**, 108–12, Jun. 1958.
Pappas, T. "The Spider & the Fly Problem." *The Joy of Mathematics.* San Carlos, CA: Wide World Publ./Tetra, pp. 218 and 233, 1989.
Steinhaus, H. *Mathematical Snapshots, 3rd ed.* New York: Dover, pp. 173–75, 1999.

Spider Lines

EPITROCHOID

Spiegeldrieck

FUHRMANN TRIANGLE

Spieker Center

The center of the SPIEKER CIRCLE. It is the CENTROID of the PERIMETER of the original TRIANGLE. The Spieker center is also the CLEAVANCE CENTER (Honsberger 1995). The Spieker center lies on the NAGEL LINE.

The Spieker center, third BROCARD POINT, and ISOTOMIC CONJUGATE POINT of the INCENTER are COLLINEAR.

See also BROCARD POINTS, CENTROID (TRIANGLE), CLEAVANCE CENTER, CLEAVER, INCENTER, ISOTOMIC CONJUGATE POINT, NAGEL LINE, PERIMETER, SPIEKER CIRCLE, TAYLOR CENTER

References

Casey, J. *A Treatise on the Analytical Geometry of the Point, Line, Circle, and Conic Sections, Containing an Account of Its Most Recent Extensions, with Numerous Examples, 2nd ed., rev. enl.* Dublin: Hodges, Figgis, & Co., p. 81, 1893.
Honsberger, R. *Episodes in Nineteenth and Twentieth Century Euclidean Geometry.* Washington, DC: Math. Assoc. Amer., pp. 3–, 1995.
Johnson, R. A. *Modern Geometry: An Elementary Treatise on the Geometry of the Triangle and the Circle.* Boston, MA: Houghton Mifflin, pp. 226–29 and 249, 1929.
Kimberling, C. "Central Points and Central Lines in the Plane of a Triangle." *Math. Mag.* **67**, 163–87, 1994.

Spieker Circle

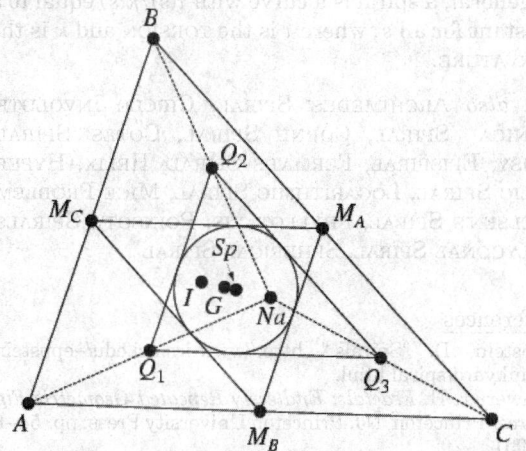

The common INCIRCLE of the MEDIAL TRIANGLE $\Delta M_A M_B M_C$ and the congruent triangle $\Delta Q_1 Q_2 Q_3$ illustrated above, where Q_i are the MIDPOINTS of the line segment joining the NAGEL POINT Na with the vertices of the original triangle ΔABC. The center of the Spieker circle is called the SPIEKER CENTER Sp.

See also INCIRCLE, MEDIAL TRIANGLE, MIDPOINT, NAGEL POINT, SPIEKER CENTER

References

Coolidge, J. L. *A Treatise on the Geometry of the Circle and Sphere.* New York: Chelsea, p. 53, 1971.

Honsberger, R. "The Nagel Point *M* and the Spieker Circle." §1.4 in *Episodes in Nineteenth and Twentieth Century Euclidean Geometry.* Washington, DC: Math. Assoc. Amer., pp. 3–3, 1995.

Johnson, R. A. *Modern Geometry: An Elementary Treatise on the Geometry of the Triangle and the Circle.* Boston, MA: Houghton Mifflin, pp. 226–28, 1929.

Spieker, T. "Ein merkwürdiger Kreis um den Schwerpunkt des Perimeters des geradlinigen Dreiecks als Analogen des Kreises der neun Punkte." *Archiv Math. u. Phys.* **51**, 10–4, 1870.

Spigot Algorithm

An ALGORITHM which generates digits of a quantity one at a time without using or requiring previously computed digits. Amazingly, spigot ALGORITHMS are known for both PI and E.

Spijker's Lemma

The image on the RIEMANN SPHERE of any CIRCLE under a COMPLEX rational mapping with NUMERATOR and DENOMINATOR having degrees no more than n has length no longer than $2n\pi$.

References

Edelman, A. and Kostlan, E. "How Many Zeros of a Random Polynomial are Real?" *Bull. Amer. Math. Soc.* **32**, 1–7, 1995.

Wegert, E. and Trefethen, L. N. "From the Buffon Needle Problem to the Kreiss Matrix Theorem." *Amer. Math. Monthly* **101**, 132–39, 1994.

Spindle

LEMON, SPINDLE CYCLIDE

Spindle Cyclide

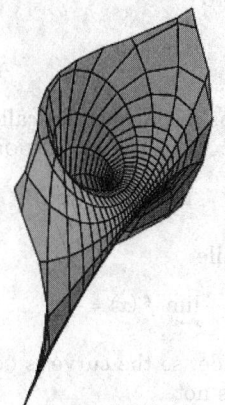

The inversion of a SPINDLE TORUS. If the inversion center lies on the torus, then the spindle cyclide degenerates to a PARABOLIC SPINDLE CYCLIDE.

See also CYCLIDE, HORN CYCLIDE, PARABOLIC CYCLIDE, RING CYCLIDE, SPINDLE TORUS, TORUS

Spindle Torus

  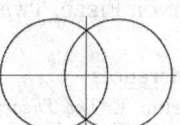

One of the three STANDARD TORI given by the PARAMETRIC EQUATIONS

$$x = (c + a \cos v)\cos u$$

$$y = (c + a \cos v)\sin u$$

$$z = a \sin v$$

with $c < a$. The exterior surface is called an APPLE and the interior surface a LEMON. The above left figure shows a spindle torus, the middle a cutaway, and the right figure shows a CROSS SECTION of the spindle torus through the xz-plane.

See also APPLE, CYCLIDE, HORN TORUS, LEMON, PARABOLIC SPINDLE CYCLIDE, RING TORUS, SPINDLE CYCLIDE, STANDARD TORI, TORUS

References

Gray, A. "Tori." §13.4 in *Modern Differential Geometry of Curves and Surfaces with Mathematica, 2nd ed.* Boca Raton, FL: CRC Press, pp. 304–06, 1997.

Pinkall, U. "Cyclides of Dupin." §3.3 in *Mathematical Models from the Collections of Universities and Museums* (Ed.

G. Fischer). Braunschweig, Germany: Vieweg, pp. 28–0, 1986.

Spindle-Shaped Ellipsoid

PROLATE SPHEROID

Spinode

A function $f(x)$ has a spinode (also called a horizontal cusp) at a point x_0 if $f(x)$ is CONTINUOUS at x_0 and

$$\lim_{x \to x_0} f'(x) = \infty$$

from one side while

$$\lim_{x \to x_0} f'(x) = -\infty$$

from the other side, so the curve is CONTINUOUS but the DERIVATIVE is not.

See also ACNODE, CRUNODE, CUSP, TACNODE

Spinor

A two-component COMPLEX COLUMN VECTOR. Spinors are used in physics to represent particles with half-integral spin (i.e., fermions).

See also LIE DERIVATIVE (SPINOR), MINKOWSKI SPACE, SPINOR FIELD, TWISTOR

References

Cartan, È. *The Theory of Spinors.* New York: Dover, 1981.
Corson, E. M. *Introduction to Tensors, Spinors and Relativistic Wave-Equations.* London: Blackie and Son, 1955.
Lounesto, P. "Counterexamples to Theorems Published and Proved in Recent Literature on Clifford Algebras, Spinors, Spin Groups, and the Exterior Algebra." http://www.hit.fi/~lounesto/counterexamples.htm.
Morse, P. M. and Feshbach, H. "The Lorentz Transformation, Four-Vectors, Spinors." §1.7 in *Methods of Theoretical Physics, Part I.* New York: McGraw-Hill, pp. 93–07, 1953.
Penrose, R. and Rindler, W. *Spinors and Space-Time, Vol. 1: Two-Spinor Calculus and Relativistic Fields.* Cambridge, England: Cambridge University Press, 1987.
Penrose, R. and Rindler, W. *Spinors and Space-Time, Vol. 2: Spinor and Twistor Methods in Space-Time Geometry* Cambridge, England: Cambridge University Press, 1987.

Spinor Field

See also SPINOR, TWISTOR

Spira Mirabilis

LOGARITHMIC SPIRAL

Spiral

In general, a spiral is a curve with $\tau(s)/\kappa(s)$ equal to a constant for all s, where τ is the TORSION and κ is the CURVATURE.

See also ARCHIMEDES' SPIRAL, CIRCLE INVOLUTE, CONICAL SPIRAL, CORNU SPIRAL, COTES' SPIRAL, DAISY, EPISPIRAL, FERMAT'S SPIRAL, HELIX, HYPERBOLIC SPIRAL, LOGARITHMIC SPIRAL, MICE PROBLEM, NIELSEN'S SPIRAL, PHYLLOTAXIS, POINSOT'S SPIRALS, POLYGONAL SPIRAL, SPHERICAL SPIRAL

References

Eppstein, D. "Spirals." http://www.ics.uci.edu/~eppstein/junkyard/spiral.html.
Lauwerier, H. *Fractals: Endlessly Repeated Geometric Figures.* Princeton, NJ: Princeton University Press, pp. 54–6, 1991.
Lockwood, E. H. "Spirals." Ch. 22 in *A Book of Curves.* Cambridge, England: Cambridge University Press, pp. 172–75, 1967.
Weisstein, E. W. "Books about Spirals." http://www.treasure-troves.com/books/Spirals.html.
Yates, R. C. "Spirals." *A Handbook on Curves and Their Properties.* Ann Arbor, MI: J. W. Edwards, pp. 206–16, 1952.

Spiral Point

A FIXED POINT for which the EIGENVALUES are COMPLEX CONJUGATES.

See also STABLE SPIRAL POINT, UNSTABLE SPIRAL POINT

References

Tabor, M. "Classification of Fixed Points." §1.4.b in *Chaos and Integrability in Nonlinear Dynamics: An Introduction.* New York: Wiley, pp. 22–5, 1989.

Spiral Similarity

The combination of a CENTRAL DILATION and a ROTATION about the *same* center. However, the combination of a central dilation and a rotation whose centers are *distinct* is also a spiral symmetry. In fact, any two DIRECTLY SIMILAR figures are related either by a TRANSLATION or by a spiral symmetry (Coxeter and Greitzer 1967, p. 97).

See also CENTRAL DILATION, DILATION, ROTATION, SIMILAR

References

Coxeter, H. S. M. and Greitzer, S. L. "Spiral Similarity." §4.8 in *Geometry Revisited.* Washington, DC: Math. Assoc. Amer., pp. 95–00, 1967.

Spirallohedron
RHOMBIC SPIRALLOHEDRON

Spiral-Similarity Tessellation
A tessellation constructed by placing a series of polygonal tiles of decreasing size on an equilateral spiral. Any ordinary TESSELLATION can be converted to such a form.

See also TESSELLATION

References
Wells, D. *The Penguin Dictionary of Curious and Interesting Geometry.* London: Penguin, p. 239, 1991.

Spiric Section

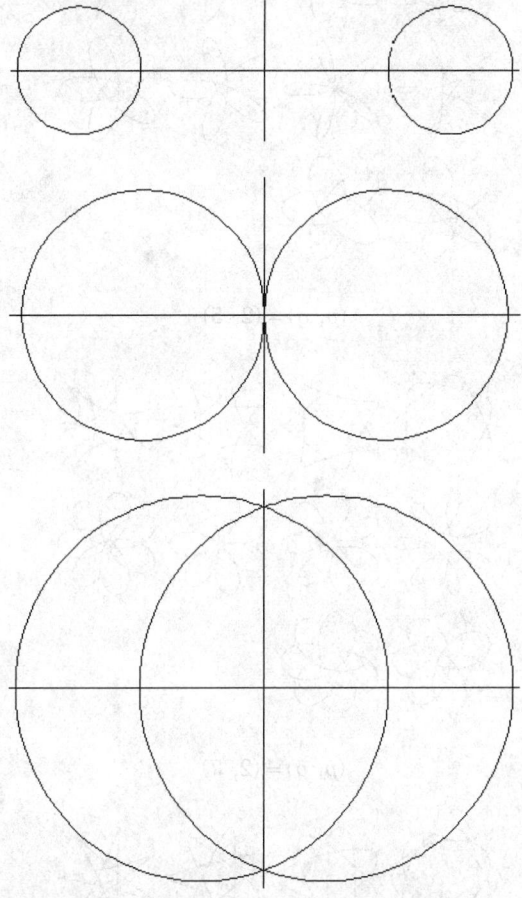

The equation of the curve of intersection of a TORUS with a plane perpendicular to both the midplane of the torus and to the plane $x = 0$. (The general intersection of a TORUS with a plane is called a TORIC SECTION). Let the tube of a torus have radius a, let its midplane lie in the $z = 0$ plane, and let the center of the tube lie at a distance c from the origin. Now cut the torus with the plane $y = r$. The equation of the TORUS with $y = r$ gives the equation

$$\left(c - \sqrt{x^2 + r^2}\right)^2 + z^2 = a^2 \tag{1}$$

$$c^2 - a^2 + x^2 + z^2 = 2c\sqrt{x^2 + r^2} \tag{2}$$

$$\left(r^2 - a^2 + c^2 + x^2 + z^2\right)^2 = 4c^2\left(x^2 + r^2\right). \tag{3}$$

The above plots show a series of spiric sections for the RING TORUS, HORN TORUS, and SPINDLE TORUS, respectively. When $r = 0$, the curve consists of two CIRCLES of RADIUS a whose centers are at $(c, 0)$ and $(-c, 0)$. If $r = c + a$, the curve consists of one point (the origin), while if $r > c + a$, no point lies on the curve.

The spiric extensions are an extension of the CONIC SECTIONS constructed by Menaechmus around 150 BC by cutting a CONE by a PLANE, and were first considered around 50 AD by the Greek mathematician Perseus (MacTutor).

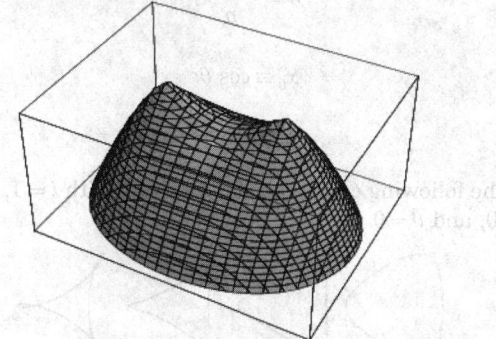

If $r = a$, then (3) simplifies to

$$\left(x^2 + z^2 + c^2\right)^2 - 4c^2x^2 = 4c^2a^2, \tag{4}$$

which is the equation of CASSINI OVALS. CASSINI OVALS are therefore SPIRIC SECTIONS. Furthermore, the surface having these curves as CROSS SECTIONS is the CASSINI SURFACE illustrated above, with the modification that the vertical component is squared instead of to the fourth power (Gosper).

See also TORIC SECTION, TORUS

References
MacTutor History of Mathematics Archive. "Spiric Sections." http://www-groups.dcs.st-and.ac.uk/~history/Curves/Spiric.html.

Spirograph
A HYPOTROCHOID generated by a fixed point on a CIRCLE rolling inside a fixed CIRCLE. It has parametric equations,

$$x = (R + r)\cos\theta - (r + \rho)\cos\left(\frac{R + r}{r}\theta\right) \tag{1}$$

$$y = (R + r)\sin\theta - (r + \rho)\sin\left(\frac{R + r}{r}\theta\right), \tag{2}$$

where R is the radius of the fixed circle, r is the radius of the rotating circle, and ρ is the offset of the edge of the rotating circle. The figure closes only if R, r, and ρ are RATIONAL. The equations can also be written

$$x = x_0[m \cos t + a \cos(nt)] - y_0[m \sin t - a \sin(nt)] \quad (3)$$

$$y = y_0[m \cos t + a \cos(nt)] + x_0[m \sin t - a \sin(nt)]. \quad (4)$$

where the outer wheel has radius 1, the inner wheel a radius p/q, the pen is placed a units from the center, the beginning is at θ radians above the x-AXIS, and

$$m \equiv \frac{q-p}{q} \quad (5)$$

$$n \equiv \frac{q-p}{p} \quad (6)$$

$$x_0 \equiv \cos \theta \quad (7)$$

$$y_0 \equiv \sin \theta. \quad (8)$$

The following curves are for $a = i/10$, with $i = 1, 2, ..., 10$, and $\theta = 0$.

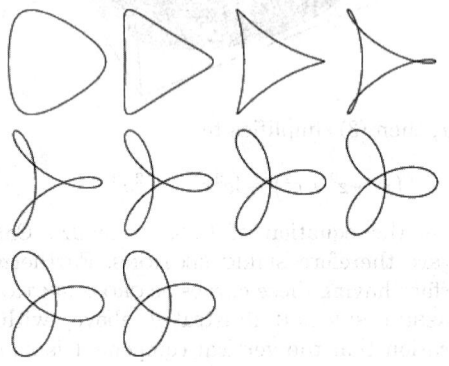

$$(p, q) = (1, 3)$$

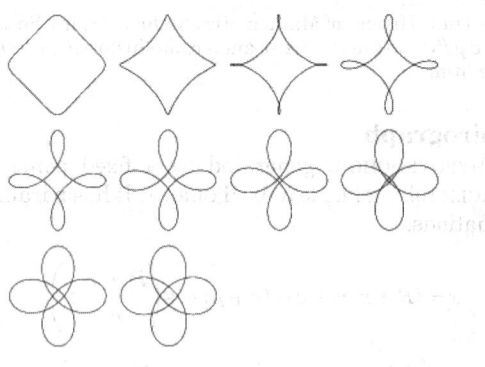

$$(p, q) = (1, 4)$$

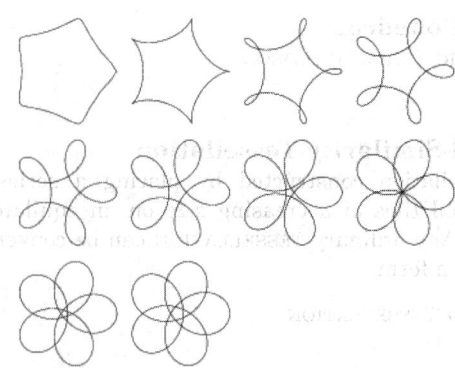

$$(p, q) = (1, 5)$$

$$(p, q) = (2, 5)$$

$$(p, q) = (2, 7)$$

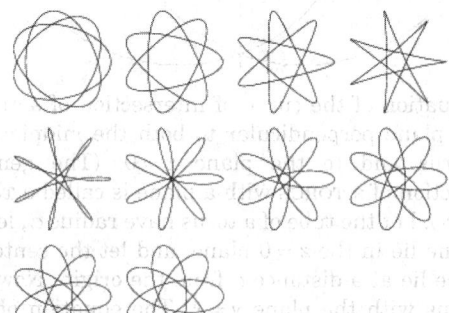

$(p, q) = (3, 7)$

Additional attractive designs such as the following can also be made by superposing individual spirographs.

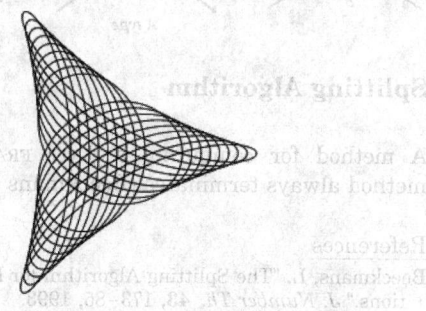

See also EPITROCHOID, HARMONOGRAPH, HYPOTRO-CHOID, MAURER ROSE, SPIROLATERAL

Spirolateral

A figure formed by taking a series of steps of length 1, 2, ..., n, with an angle θ turn after each step. The symbol for a spirolateral is ${}^{a_1, \ldots, a_k}n_\theta$, where the a_is indicate that turns are in the $-\theta$ direction for these steps.

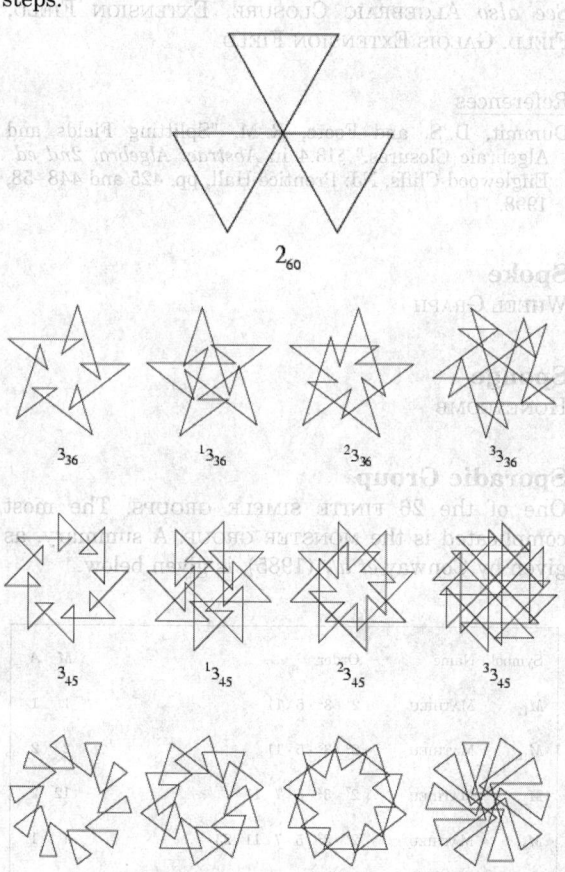

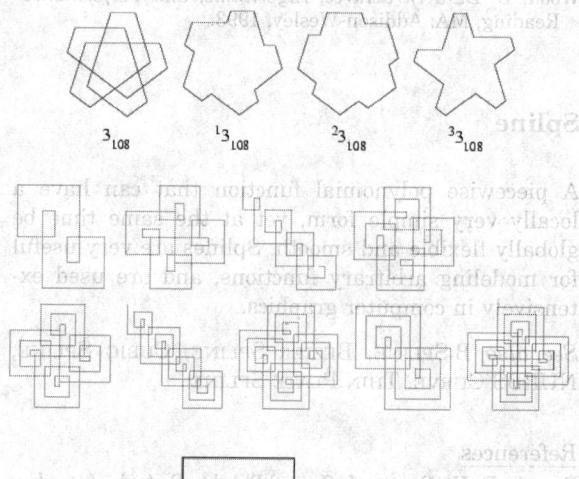

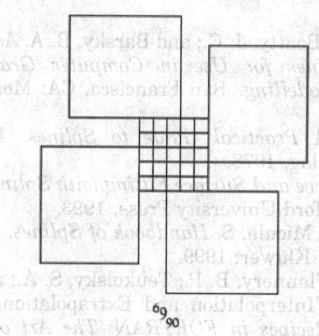

See also MAURER ROSE, SPIROGRAPH

References

Gardner, M. "Fantastic Patterns Traced by Programmed 'Worms.'" *Sci. Amer.*, Nov 1973.

Gardner, M. "Worm Paths." Ch. 17 in *Knotted Doughnuts and Other Mathematical Entertainments*. New York: W. H. Freeman, pp. 205–21, 1986.

Hall, L. "Trochoids, Roses, and Thorns--Beyond the Spirograph." *College Math. J.* **23**, 20–5, 1992.

Odds, F. C. "Spirolaterals." *Math. Teacher* **66**, 121–24, 1973.

Trott, M. "Spirographs with *Mathematica*." http://library.-wolfram.com/demos/v4/Spirograph.nb.

Wells, D. *The Penguin Dictionary of Curious and Interesting Geometry*. London: Penguin, pp. 239–41, 1991.

Splay Tree

A self-organizing data structure which uses rotations to move any accessed key to the root. This leaves recently accessed nodes near the top of the tree, making them very quickly searchable (Skiena 1997, p. 177).

See also TREE

References

Skiena, S. S. *The Algorithm Design Manual*. New York: Springer-Verlag, pp. 177 and 179, 1997.

Sleator, D. and Tarjan, R. "Self-Adjusting Binary Search Trees." *J. ACM* **32**, 652–86, 1985.

Tarjan, R. *Data Structures and Network Algorithms*. Philadelphia, PA: SIAM Press, 1983.

Wood, D. *Data Structures, Algorithms, and Performance.* Reading, MA: Addison-Wesley, 1993.

Spline

A piecewise polynomial function that can have a locally very simple form, yet at the same time be globally flexible and smooth. Splines are very useful for modeling arbitrary functions, and are used extensively in computer graphics.

See also B-SPLINE, BÉZIER SPLINE, CUBIC SPLINE, NURBS CURVE, THIN PLATE SPLINE

References

Bartels, R. H.; Beatty, J. C.; and Barsky, B. A. *An Introduction to Splines for Use in Computer Graphics and Geometric Modelling.* San Francisco, CA: Morgan Kaufmann, 1998.
de Boor, C. *A Practical Guide to Splines.* New York: Springer-Verlag, 1978.
Dierckx, P. *Curve and Surface Fitting with Splines.* Oxford, England: Oxford University Press, 1993.
Micula, G. and Micula, S. *Handbook of Splines.* Dordrecht, Netherlands: Kluwer, 1999.
Press, W. H.; Flannery, B. P.; Teukolsky, S. A.; and Vetterling, W. T. "Interpolation and Extrapolation." Ch. 3 in *Numerical Recipes in FORTRAN: The Art of Scientific Computing, 2nd ed.* Cambridge, England: Cambridge University Press, pp. 99–22, 1992.
Späth, H. *One Dimensional Spline Interpolation Algorithms.* Wellesley, MA: A. K. Peters, 1995.
Weisstein, E. W. "Books about Splines." http://www.treasure-troves.com/books/Splines.html.

Splitter

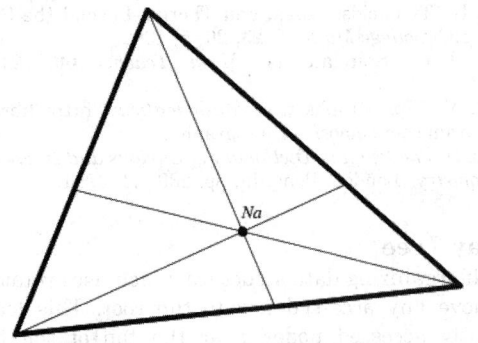

A perimeter-bisecting line segment which originates at a vertex of a polygon. The three splitters of a TRIANGLE CONCUR in a point known as the NAGEL POINT Na.

See also B-LINE, CLEAVER

References

Honsberger, R. "Cleavers and Splitters." Ch. 1 in *Episodes in Nineteenth and Twentieth Century Euclidean Geometry.* Washington, DC: Math. Assoc. Amer., pp. 1–4, 1995.

Splitting

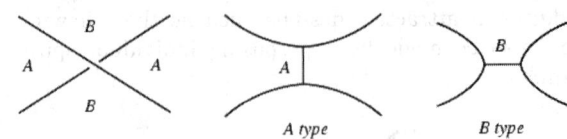

Splitting Algorithm

A method for computing a UNIT FRACTION. This method always terminates (Beeckmans 1993).

References

Beeckmans, L. "The Splitting Algorithm for Egyptian Fractions." *J. Number Th.* **43**, 173–85, 1993.
Eppstein, D. Egypt.ma Mathematica notebook. http://www.ics.uci.edu/~eppstein/numth/egypt/egypt.ma.

Splitting Field

The EXTENSION FIELD K of a FIELD F is called a splitting field for the polynomial $f(x) \in F[x]$ if $f(x)$ factors completely into linear factors in $K[x]$ and $f(x)$ does not factor completely into linear factors over any PROPER SUBFIELD of K containing F (Dummit and Foote 1998, p. 448).

See also ALGEBRAIC CLOSURE, EXTENSION FIELD, FIELD, GALOIS EXTENSION FIELD

References

Dummit, D. S. and Foote, R. M. "Splitting Fields and Algebraic Closures." §13.4 in *Abstract Algebra, 2nd ed.* Englewood Cliffs, NJ: Prentice-Hall, pp. 425 and 448–58, 1998.

Spoke

WHEEL GRAPH

Sponge

HONEYCOMB

Sporadic Group

One of the 26 FINITE SIMPLE GROUPS. The most complicated is the MONSTER GROUP. A summary, as given by Conway *et al.* (1985), is given below.

Symbol	Name	Order	M	A
M_{11}	MATHIEU	$2^4 \cdot 3^2 \cdot 5 \cdot 11$	1	1
M_{12}	MATHIEU	$2^6 \cdot 3^3 \cdot 5 \cdot 11$	2	2
M_{22}	MATHIEU	$2^7 \cdot 3^2 \cdot 5 \cdot 7 \cdot 11$	12	2
M_{23}	MATHIEU	$2^7 \cdot 3^2 \cdot 5 \cdot 7 \cdot 11 \cdot 23$	1	1
M_{24}	MATHIEU	$2^{10} \cdot 3^3 \cdot 5 \cdot 7 \cdot 11 \cdot 23$	1	1
$J_2 = HJ$	JANKO	$2^7 \cdot 3^3 \cdot 5^2 \cdot 7$	2	2

Suz	SUZUKI	$2^{13} \cdot 3^7 \cdot 5^2 \cdot 7 \cdot 11 \cdot 13$	6	2
HS	HIGMAN-SIMS	$2^9 \cdot 3^2 \cdot 5^3 \cdot 7 \cdot 11$	2	2
McL	McLAUGHLIN	$2^7 \cdot 3^6 \cdot 5^3 \cdot 7 \cdot 11$	3	2
Co_3	CONWAY	$2^{10} \cdot 3^7 \cdot 5^3 \cdot 7 \cdot 11 \cdot 23$	1	1
Co_2	CONWAY	$2^{18} \cdot 3^6 \cdot 5^3 \cdot 7 \cdot 11 \cdot 23$	1	1
Co_1	CONWAY	$2^{21} \cdot 3^9 \cdot 5^4 \cdot 7^2 \cdot 11 \cdot 13 \cdot 23$	2	1
He	HELD	$2^{10} \cdot 3^3 \cdot 5^2 \cdot 7^3 \cdot 17$	1	2
Fi_{22}	FISCHER	$2^{17} \cdot 3^9 \cdot 5^2 \cdot 7 \cdot 11 \cdot 13$	6	2
Fi_{23}	FISCHER	$2^{18} \cdot 3^{13} \cdot 5^2 \cdot 7 \cdot 11 \cdot 13 \cdot 17 \cdot 23$	1	1
Fi'_{24}	FISCHER	$2^{21} \cdot 3^{16} \cdot 5^2 \cdot 7^3 \cdot 11 \cdot 13 \cdot 17 \cdot 23 \cdot 29$	3	2
HN	HARADA-NOR-TON	$2^{14} \cdot 3^6 \cdot 5^6 \cdot 7 \cdot 11 \cdot 19$	1	2
Th	THOMPSON	$2^{15} \cdot 3^{10} \cdot 5^3 \cdot 7^2 \cdot 13 \cdot 19 \cdot 31$	1	1
B	BABY MON-STER	$2^{41} \cdot 3^{13} \cdot 5^6 \cdot 7^2 \cdot 11 \cdot 13 \cdot 17 \cdot 19 \cdot 23 \cdot 31 \cdot 47$	2	1
M	MONSTER	$2^{46} \cdot 3^{20} \cdot 5^9 \cdot 7^6 \cdot 11^2 \cdot 13^3 \cdot 17 \cdot 19 \cdot 23 \cdot 29 \cdot 31 \cdot 41 \cdot 47 \cdot 59 \cdot 71$	1	1
J_1	JANKO	$2^3 \cdot 3 \cdot 5 \cdot 7 \cdot 11 \cdot 19$	1	1
$O'N$	O'NAN	$2^9 \cdot 3^4 \cdot 5 \cdot 7^3 \cdot 11 \cdot 19 \cdot 31$	3	2
J_3	JANKO	$2^7 \cdot 3^5 \cdot 5 \cdot 17 \cdot 19$	3	2
Ly	LYONS	$2^8 \cdot 3^7 \cdot 5^6 \cdot 7 \cdot 11 \cdot 31 \cdot 37 \cdot 67$	1	1
Ru	RUDVALIS	$2^{14} \cdot 3^3 \cdot 5^3 \cdot 7 \cdot 13 \cdot 29$	2	1
J_4	JANKO	$2^{21} \cdot 3^3 \cdot 5 \cdot 7 \cdot 11^3 \cdot 23 \cdot 29 \cdot 31 \cdot 37 \cdot 43$	1	1

See also BABY MONSTER GROUP, CONWAY GROUPS, FISCHER GROUPS, HARADA-NORTON GROUP, HELD GROUP, HIGMAN-SIMS GROUP, JANKO GROUPS, LYONS GROUP, MATHIEU GROUPS, McLAUGHLIN GROUP, MONSTER GROUP, O'NAN GROUP, RUDVALIS GROUP, SUZUKI GROUP, THOMPSON GROUP

References

Aschbacher, M. *Sporadic Groups.* New York: Cambridge University Press, 1994.
Conway, J. H.; Curtis, R. T.; Norton, S. P.; Parker, R. A.; and Wilson, R. A. *Atlas of Finite Groups: Maximal Subgroups and Ordinary Characters for Simple Groups.* Oxford, England: Clarendon Press, p. viii, 1985.
Ivanov, A. A. *Geometry of Sporadic Groups I: Petersen and Tilde Geometries.* Cambridge, England: Cambridge University Press, 1999.
Math. Intell. Cover of volume **2**, 1980.
Wilson, R. A. "ATLAS of Finite Group Representation." http://for.mat.bham.ac.uk/atlas/html/contents.html#spo.

Sports

BASEBALL, BOWLING, CHECKERS, CHESS, GO

Sprague-Grundy Function

NIM-VALUE

Sprague-Grundy Number

NIM-VALUE

Sprague-Grundy Value

NIM-VALUE

Spread (Link)

SPAN (LINK)

Spread (Tree)

A TREE having an infinite number of branches and whose nodes are sequences generated by a set of rules.

See also FAN

Spreading A Rumor

GOSSIPING

Springer Number

References

Arnold, V. I. "Springer Numbers and Morsification Spaces." *J. Alg. Geom* **1**, 197–14, 1992.

Spun Knot

A 3-D KNOT spun about a plane in 4-D. Unlike SUSPENDED KNOTS, spun knots are smoothly embedded at the poles.

See also SUSPENDED KNOT, TWIST-SPUN KNOT

Spur

TRACE (MATRIX)

Sqrt

SQUARE ROOT

Squarable

An object which can be constructed by SQUARING is called squarable.

Square

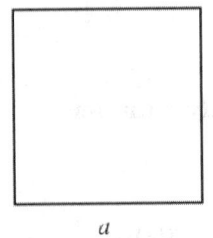

The term square is sometimes used to mean SQUARE NUMBER. When used in reference to a geometric figure, however, it means a convex QUADRILATERAL with four equal sides at RIGHT ANGLES to each other, illustrated above. When used as a symbol, $\square ABCD$ denotes a square with given vertices, while $G_1 \square G_2$ is sometimes used to denote a GRAPH PRODUCT (Clark and Suen 2000).

The PERIMETER of a square with side length a is

$$L = 4a \tag{1}$$

and the AREA is

$$A = a^2. \tag{2}$$

The INRADIUS r, CIRCUMRADIUS R, and AREA A can be computed directly from the formulas for a general REGULAR POLYGON with side length a and $n = 4$ sides,

$$r = \tfrac{1}{2} a \cot\left(\frac{\pi}{4}\right) = \tfrac{1}{2} a \tag{3}$$

$$R = \tfrac{1}{2} a \csc\left(\frac{\pi}{4}\right) = \tfrac{1}{2}\sqrt{2}a \tag{4}$$

$$A = \tfrac{1}{4} n a^2 \cot\left(\frac{\pi}{4}\right) = a^2, \tag{5}$$

The length of the DIAGONAL of the UNIT SQUARE is $\sqrt{2}$, sometimes known as PYTHAGORAS'S CONSTANT.

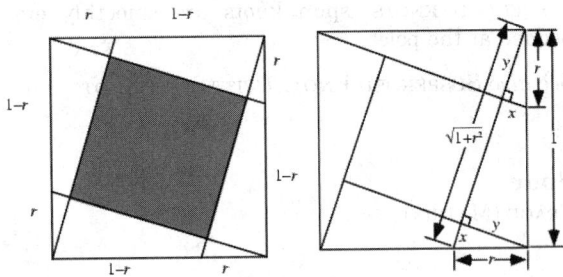

The AREA of a square constructed inside a UNIT SQUARE as shown in the above diagram can be found as follows. Label x and y as shown, then

$$x^2 + y^2 = r^2 \tag{6}$$

$$\left(\sqrt{1 + r^2} - x\right)^2 + y^2 = 1. \tag{7}$$

Plugging (6) into (7) gives

$$\left(\sqrt{1 + r^2} - x\right)^2 + (r^2 - x^2) = 1. \tag{8}$$

Expanding

$$x^2 - 2x\sqrt{1 + r^2} + 1 + r^2 + r^2 - x^2 = 1 \tag{9}$$

and solving for x gives

$$x = \frac{r^2}{\sqrt{1 + r^2}}. \tag{10}$$

Plugging in for y yields

$$y = \sqrt{r^2 - x^2} = \frac{r}{\sqrt{1 + r^2}}. \tag{11}$$

The area of the shaded square is then

$$A = \left(\sqrt{1 + r^2} - x - y\right)^2 = \frac{(1 - r)^2}{1 + r^2} \tag{12}$$

(Detemple and Harold 1996).

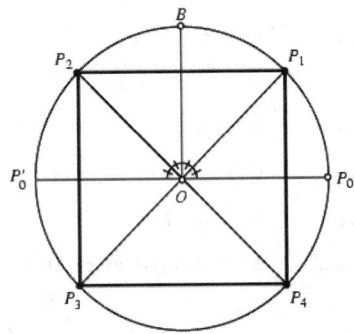

square construction

The STRAIGHTEDGE and COMPASS construction of the square is simple. Draw the line $P_0'OP_0$ and construct a circle having OP_0 as a radius. Then construct the perpendicular OB through O. Bisect P_0OB and $P_0'OB$ to locate P_1 and P_2, where P_0' is opposite P_0. Similarly, construct P_3 and P_4 on the other SEMICIRCLE. Connecting $P_1P_2P_3P_4$ then gives a square.

An infinity of points in the interior of a square are known whose distances from three of the corners of a square are RATIONAL NUMBERS. Calling the distances a, b, and c where s is the side length of the square, these solutions satisfy

$$\left(s^2 + b^2 - a^2\right)^2 + \left(s^2 + b^2 - c^2\right)^2 = (2bs)^2 \tag{13}$$

(Guy 1994). In this problem, one of a, b, c, and s is DIVISIBLE by 3, one by 4, and one by 5. It is not known if there are points having distances from *all four* corners RATIONAL, but such a solution requires the additional condition

$$a^2 + c^2 = b^2 + d^2. \tag{14}$$

In this problem, s is DIVISIBLE by 4 and a, b, c, and d are ODD. If s is not DIVISIBLE by 3 (5), then two of a, b, c, and d are DIVISIBLE by 3 (5) (Guy 1994).

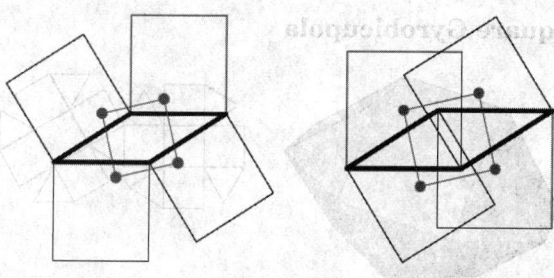

The centers of four squares erected either internally or externally on the sides of a PARALLELOGRAMS are the vertices of a square (Yaglom 1962, pp. 96–7; Coxeter and Greitzer 1967, p. 84).

See also BROWKIN'S THEOREM, DISSECTION, DOUGLAS-NEUMANN THEOREM, FINSLER-HADWIGER THEOREM, LOZENGE, NESTED SQUARE, PERFECT SQUARE DISSECTION, PYTHAGORAS'S CONSTANT, PYTHAGOREAN SQUARE PUZZLE, RECTANGLE, SQUARE DIVISION BY LINES, SQUARE INSCRIBING, SQUARE NUMBER, SQUARE PACKING, SQUARE QUADRANTS, UNIT SQUARE, VON AUBEL'S THEOREM

References

Clark, W. E. and Suen, S. "An Inequality Related to Vizing's Conjecture." *Electronic J. Combinatorics* **7**, No. 1, N4, 1–, 2000. http://www.combinatorics.org/Volume_7/ v7i1toc.html#N4.

Coxeter, H. S. M. and Greitzer, S. L. *Geometry Revisited.* Washington, DC: Math. Assoc. Amer., p. 84, 1967.

Detemple, D. and Harold, S. "A Round-Up of Square Problems." *Math. Mag.* **69**, 15–7, 1996.

Dixon, R. *Mathographics.* New York: Dover, p. 16, 1991.

Eppstein, D. "Rectilinear Geometry." http://www.ics.uci.edu/ ~eppstein/junkyard/rect.html.

Fukagawa, H. and Pedoe, D. "One or Two Circles and Squares," "Three Circles and Squares," and "Many Circles and Squares (Casey's Theorem)." §3.1–.3 in *Japanese Temple Geometry Problems.* Winnipeg, Manitoba, Canada: Charles Babbage Research Foundation, pp. 37–2 and 117–25, 1989.

Gardner, M. *The Sixth Book of Mathematical Games from Scientific American.* Chicago, IL: University of Chicago Press, pp. 165 and 167, 1984.

Guy, R. K. "Rational Distances from the Corners of a Square." §D19 in *Unsolved Problems in Number Theory, 2nd ed.* New York: Springer-Verlag, pp. 181–85, 1994.

Harris, J. W. and Stocker, H. "Square." §3.6.6 in *Handbook of Mathematics and Computational Science.* New York: Springer-Verlag, pp. 84–5, 1998.

Kern, W. F. and Bland, J. R. *Solid Mensuration with Proofs, 2nd ed.* New York: Wiley, p. 2, 1948.

Yaglom, I. M. *Geometric Transformations I.* New York: Random House, pp. 96–7, 1962.

Square Antiprism

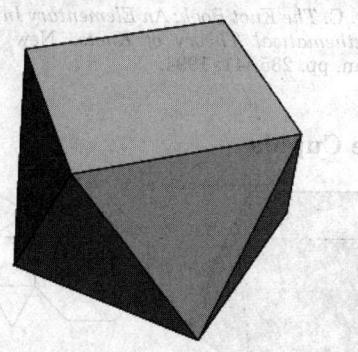

The ANTIPRISM with square bases.

See also ANTIPRISM, SQUARE PRISM

Square Bracket

One of the symbols [and] used in many different contexts in mathematics.

1. Square brackets are occasionally used in especially complex expressions in place of (or in addition to) PARENTHESES, especially as a group symbol outside an inner set of parentheses, e.g., $[3 + 4 \times (5 + 6)]/7$.

2. Large brackets around an array of numbers, e.g., $\begin{bmatrix} ab \\ cd \end{bmatrix}$ indicate a MATRIX. (The symbol $\begin{pmatrix} ab \\ cd \end{pmatrix}$ is also commonly used.)

3. A square bracket at one end of an INTERVAL indicates that the INTERVAL is closed at that end, that is, it includes the number at that end.

4. Brackets may be used to denote the LEAST COMMON MULTIPLE, e.g., $[10, 6] \equiv \mathrm{LCM}(10, 6) = 30$.

5. Some authors (although this work does *not*) use $[x]$ to denote the FLOOR FUNCTION $\lfloor x \rfloor$.

See also ANGLE BRACKET, BRACE, PARENTHESIS

References

Bringhurst, R. *The Elements of Typographic Style, 2nd ed.* Point Roberts, WA: Hartley and Marks, p. 285, 1997.

Square Bracket Polynomial

A POLYNOMIAL which is not necessarily an invariant of a LINK. It is related to the DICHROIC POLYNOMIAL. It is defined by the SKEIN RELATIONSHIP

$$B_{L_+} = q^{-1/2} v B_{L_0} + B_{L_\infty}, \tag{1}$$

and satisfies

$$B_{\mathrm{unknot}} = q^{1/2} \tag{2}$$

and

$$B_{L \cup \mathrm{unknot}} = q^{1/2} B_L. \tag{3}$$

References

Adams, C. C. *The Knot Book: An Elementary Introduction to the Mathematical Theory of Knots.* New York: W. H. Freeman, pp. 235–41, 1994.

Square Cupola

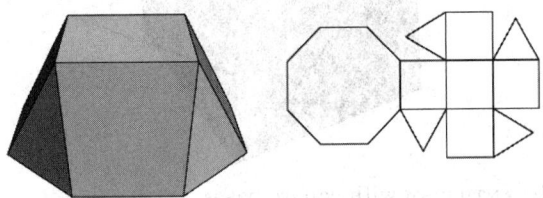

JOHNSON SOLID J_4. The bottom eight VERTICES are

$$\left(\pm\tfrac{1}{2}\left(1+\sqrt{2}\right),\ \pm\tfrac{1}{2},\ 0\right),\ \left(\pm\tfrac{1}{2},\ \pm\tfrac{1}{2}\left(1+\sqrt{2}\right),\ 0\right),$$

and the top four VERTICES are

$$\left(\pm\frac{1}{\sqrt{2}},\ 0,\ \frac{1}{\sqrt{2}}\right),\ \left(0,\ \pm\frac{1}{\sqrt{2}},\ \frac{1}{\sqrt{2}}\right).$$

Square Curve

SIERPINSKI CURVE

Square Division by Lines

The average number of regions $N(n)$ into which n lines divide a SQUARE is

$$N(n)=\tfrac{1}{16}\,n(n-1)\pi+n+1$$

(Santaló 1976).

See also CIRCLE DIVISION BY LINES

References

Finch, S. "Favorite Mathematical Constants." http://www.mathsoft.com/asolve/constant/geom/geom.html.
Santaló, L. A. *Integral Geometry and Geometric Probability.* Reading, MA: Addison-Wesley, 1976.

Square Graph

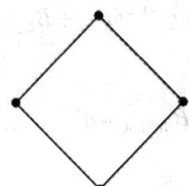

The CYCLE GRAPH C_4.

See also CYCLE GRAPH, TRIANGLE GRAPH

References

Skiena, S. *Implementing Discrete Mathematics: Combinatorics and Graph Theory with Mathematica.* Reading, MA: Addison-Wesley, p. 144, 1990.

Square Gyrobicupola

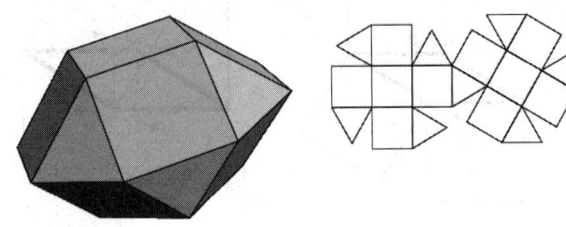

JOHNSON SOLID J_{29}.

References

Weisstein, E. W. "Johnson Solids." MATHEMATICA NOTEBOOK JOHNSONSOLIDS.M.
Weisstein, E. W. "Johnson Solid Netlib Database." MATHEMATICA NOTEBOOK JOHNSONSOLIDS.DAT.

Square Inscribing

As shown by Schnirelman (1944), a SQUARE can be INSCRIBED in any closed convex curve, although it is not known if this holds true for every JORDAN CURVE (Steinhaus 1983, p. 104). However, a SQUARE can be CIRCUMSCRIBED about *any* JORDAN CURVE (Steinhaus 1999, p. 104).

See also JORDAN CURVE, SQUARE

References

Croft, H. T.; Falconer, K. J.; and Guy, R. K. "Inscribing Polygons in Curves." §B2 in *Unsolved Problems in Geometry.* New York: Springer-Verlag, pp. 51–2, 1991.
Schnirelman, L. G. "On Certain Geometrical Properties of Closed Curves." *Uspehi Matem. Nauk* **10**, 34–4, 1944.
Steinhaus, H. *Mathematical Snapshots, 3rd ed.* New York: Dover, pp. 104 and 302, 1999.

Square Integrable

A function $f(x)$ is said to be square integrable if

$$\int_{-\infty}^{\infty}|f(x)|^2\,dx$$

is finite.

See also INTEGRABLE, L2-NORM, TITCHMARSH THEOREM

References

Sansone, G. "Square Integrable Functions." §1.1 in *Orthogonal Functions, rev. English ed.* New York: Dover, pp. 1–, 1991.

Square Knot

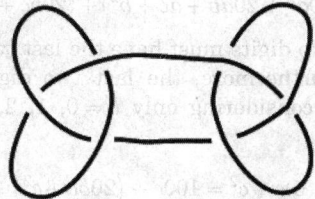

A composite KNOT of six crossings consisting of a KNOT SUM of a TREFOIL KNOT and its MIRROR IMAGE. The GRANNY KNOT has the same ALEXANDER POLYNOMIAL $(x^2 - x + 1)^2$ as the square knot. The square knot is also called the REEF KNOT.

See also GRANNY KNOT, MIRROR IMAGE, TREFOIL KNOT

References
Owen, P. *Knots*. Philadelphia, PA: Courage, p. 50, 1993.

Square Matrix

A MATRIX for which horizontal and vertical dimensions are the same (i.e., an $n \times n$ MATRIX). A matrix can be tested to see if it is square using `SquareMatrixQ[m]` in the *Mathematica* add-on package `LinearAlgebra`MatrixMultiplication`` (which can be loaded with the command `<<LinearAlgebra``).

See also MATRIX, RECTANGULAR MATRIX

Square Number

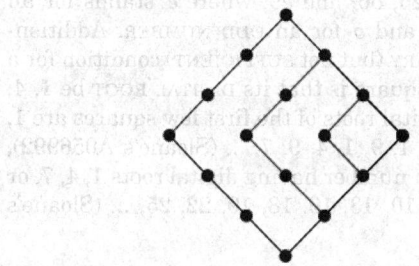

A FIGURATE NUMBER OF THE FORM $S_n = n^2$, where n is an INTEGER. A square number is also called a PERFECT SQUARE. The first few square numbers are 1, 4, 9, 16, 25, 36, 49, ... (Sloane's A000290). The GENERATING FUNCTION giving the square numbers is

$$\frac{x(x+1)}{(1-x)^3} = x + 4x^2 + 9x^3 + 16x^4 + \dots. \tag{1}$$

The $(n+1)$st square number S_{n+1} is given in terms of

the nth square number S_n by

$$S_{n+1} = S_n + 2n + 1. \tag{2}$$

since

$$(n+1)^2 = n^2 + 2n + 1, \tag{3}$$

which is equivalent to adding a GNOMON to the previous square, as illustrated above.

The nth square number is equal to the sum of the $(n-1)$-st and nth TRIANGULAR NUMBERS,

$$S_n = \tfrac{1}{2}(n-1)n + \tfrac{1}{2}n(n+1) = n^2. \tag{4}$$

as can seen in the above diagram, in which the $(n-1)$-st triangular number is represented by the white triangles, the nth triangular number is represented by the black triangles, and the total number of triangles is the square number $S_n = n^2$ (R. Sobel).

As a part of the study of WARING'S PROBLEM, it is known that every positive integer is a sum of no more than 4 positive squares ($g(2) = 4$; LAGRANGE'S FOUR-SQUARE THEOREM), that every "sufficiently large" integer is a sum of no more than 4 positive squares ($G(2) = 4$), and that every integer is a sum of at most 3 signed squares ($eg(2) = 3$). Actually, the basis set for representing positive integers with positive squares is $\{1, 1, 4, 9, 16, 25, 36, 64, 81, 100, \dots\}$, so 49 need never be used. Furthermore, since an infinite number of n require four squares to represent them, the least INTEGER $G(2)$ such that every POSITIVE INTEGER beyond a certain point requires $G(2)$ squares is given by $G(2) = 4$.

The number of representation of a number n by k squares, distinguishing signs and order, is denoted $r_k(n)$ and called the SUM OF SQUARES FUNCTION. The minimum number of squares needed to represent the numbers 1, 2, 3, ... are 1, 2, 3, 1, 2, 3, 4, 2, 1, 2, ... (Sloane's A002828), and the number of distinct ways to represent the numbers 1, 2, 3, ... in terms of squares are 1, 1, 1, 2, 2, 2, 2, 3, 4, 4, ... (Sloane's A001156). A brute-force algorithm for enumerating the square partitions of n is repeated application of the GREEDY ALGORITHM. However, this approach rapidly becomes impractical since the number of representations grows extremely rapidly with n, as shown in the following table.

n	Square Partitions
10	4
50	104

100	1116
150	6521
200	27482

The kth nonsquare number a_k is given by

$$a_n = n + \left\lfloor \tfrac{1}{2} + \sqrt{n} \right\rfloor, \qquad (5)$$

where $\lfloor x \rfloor$ is the FLOOR FUNCTION, and the first few are 2, 3, 5, 6, 7, 8, 10, 11, ... (Sloane's A000037).

The only numbers which are simultaneously square and PYRAMIDAL (the CANNONBALL PROBLEM) are $P_1 = 1$ and $P_{24} = 4900$, corresponding to $S_1 = 1$ and $S_{70} = 4900$ (Dickson 1952, p. 25; Ball and Coxeter 1987, p. 59; Ogilvy 1988), as conjectured by Lucas (1875, 1876) and proved by Watson (1918). The CANNONBALL PROBLEM is equivalent to solving the DIOPHANTINE EQUATION

$$y^2 = \tfrac{1}{6} x(x+1)(2x+1) \qquad (6)$$

(Guy 1994, p. 147).

The only numbers which are square and TETRAHEDRAL are $Te_1 = 1$, $Te_2 = 4$, and $Te_{48} = 19600$ (giving $S_1 = 1$, $S_2 = 4$, and $S_{140} = 19600$), as proved by Meyl (1878; cited in Dickson 1952, p. 25; Guy 1994, p. 147). In general, proving that only certain numbers are simultaneously figurate in two different ways is far from elementary.

To find the possible last digits for a square number, write $n = 10a + b$ for the number written in decimal NOTATION as ab_{10} (a, $b = 0$, 1, ..., 9). Then

$$n^2 = 100a^2 + 20ab + b^2. \qquad (7)$$

so the last digit of n^2 is the same as the last digit of b^2. The following table gives the last digit of b^2 for $b = 0$, 1, ..., 9 (where numbers with more that one digit have only their last digit indicated, i.e., 16 becomes _6). As can be seen, the last digit can be only 0, 1, 4, 5, 6, or 9.

0	1	2	3	4	5	6	7	8	9
0	1	4	9	_6	_5	_6	_9	_4	_1

We can similarly examine the allowable last two digits by writing abc_{10} as

$$n = 100a + 10b + c, \qquad (8)$$

so

$$n^2 = (100a + 10b + c)^2$$

$$= 10^4 a^2 + 2(1000ab + 100ac + 10bc) + 100b^2 + c^2$$

$$= (10^4 a^2 + 2000ab + 100ac + 100b^2) + 20bc + c^2$$

$$= 100(100a^2 + 20ab + ac + b^2) + (20bc + c^2) \qquad (9)$$

so the last two digits must have the last two digits of $20bc + c^2$. Furthermore, the last two digits can be obtained by considering only $b = 0$, 1, 2, 3, and 4, since

$$20(b+5)c + c^2 = 100c + (20bc + c^2) \qquad (10)$$

has the same last two digits as $20bc + c^2$ (with the one additional possibility that $c = 0$ in which case the last two digits are 00). The following table (with the addition of 00) therefore exhausts all possible last two digits.

					c				
b	1	2	3	4	5	6	7	8	9
0	01	04	09	16	25	36	49	64	81
1	_21	_44	_69	_96	_25	_56	_89	_24	_61
2	_41	_84	_29	_76	_25	_76	_29	_84	_41
3	_61	_24	_89	_56	_25	_96	_69	_44	_21
4	_81	_64	_49	_36	_25	_16	_09	_04	_01

The only 22 possibilities are therefore 00, 01, 04, 09, 16, 21, 24, 25, 29, 36, 41, 44, 49, 56, 61, 64, 69, 76, 81, 84, 89, and 96, which can be summarized succinctly as 00, $e1$, $e4$, 25, $o6$, and $e9$, where e stands for an EVEN NUMBER and o for an ODD NUMBER. Additionally, a NECESSARY (but not SUFFICIENT) condition for a number to be square is that its DIGITAL ROOT be 1, 4, 7, or 9. The digital roots of the first few squares are 1, 4, 9, 7, 7, 9, 4, 1, 9, 1, 4, 9, 7, ... (Sloane's A056992), while the list of number having digital roots 1, 4, 7, or 9 is 1, 4, 7, 9, 10, 13, 16, 18, 19, 22, 25, ... (Sloane's A056991).

The following table gives the possible residues mod n for square numbers for $n = 1$ to 20. The quantity $s(n)$ gives the number of distinct residues for a given n.

n	$s(n)$	$x^2 \pmod{n}$
2	2	0, 1
3	2	0, 1
4	2	0, 1
5	3	0, 1, 4
6	4	0, 1, 3, 4

7	4	0, 1, 2, 4
8	3	0, 1, 4
9	4	0, 1, 4, 7
10	6	0, 1, 4, 5, 6, 9
11	6	0, 1, 3, 4, 5, 9
12	4	0, 1, 4, 9
13	7	0, 1, 3, 4, 9, 10, 12
14	8	0, 1, 2, 4, 7, 8, 9, 11
15	6	0, 1, 4, 6, 9, 10
16	4	0, 1, 4, 9
17	9	0, 1, 2, 4, 8, 9, 13, 15, 16
18	8	0, 1, 4, 7, 9, 10, 13, 16
19	10	0, 1, 4, 5, 6, 7, 9, 11, 16, 17
20	6	0, 1, 4, 5, 9, 16

In general, the ODD squares are congruent to 1 (mod 8) (Conway and Guy 1996). Stangl (1996) gives an explicit formula by which the number of squares $s(n)$ in $\mathbb{Z}_n$ (i.e., mod n) can be calculated. Let p be an ODD PRIME. Then $s(n)$ is the MULTIPLICATIVE FUNCTION given by

$$s(2) = 2 \tag{11}$$

$$s(p) = \tfrac{1}{2}(p+1) \quad (p \neq 2) \tag{12}$$

$$s(p^2) = \tfrac{1}{2}(p^2 - p + 2) \quad (p \neq 2) \tag{13}$$

$$s(2^n) = \begin{cases} \tfrac{1}{3}(2^{n-1} + 4) & \text{for } n \text{ even} \\ \tfrac{1}{3}(2^{n-1} + 5) & \text{for } n \text{ odd} \end{cases} \tag{14}$$

$$s(p^n) = \begin{cases} \dfrac{p^{n+1} + p + 2}{2(p+1)} & \text{for } n \geq 3 \text{ even} \\ \dfrac{p^{n+1} + 2p + 1}{2(p+1)} & \text{for } n \geq 3 \text{ odd.} \end{cases} \tag{15}$$

$s(n)$ is related to the number $q(n)$ of QUADRATIC RESIDUES in $\mathbb{Z}_n$ by

$$q(p^n) = s(p^n) - s(p^{n-2}) \tag{16}$$

for $n \geq 3$ (Stangl 1996).

For a perfect square n, $(n/p) = 0$ or 1 for all ODD PRIMES $p < n$ where (n/p) is the LEGENDRE SYMBOL. A number n which is not a perfect square but which satisfies this relationship is called a PSEUDOSQUARE.

In a Ramanujan conference talk, W. Gosper conjectured that every sum of four distinct odd squares is the sum of four distinct even squares. This conjecture was proved by M. Hirschhorn using the identity

$$(4a+1)^2 + (4b+1)^2 + (4c+1)^2 + (4d+1)^2$$
$$= 4[(a+b+c+d+1)^2 + (a-b-c+d)^2$$
$$+(a-b+c-d)^2 + (a+b-c-d)^2], \tag{17}$$

where a, b, c, and d are positive or negative integers. Hirschhorn also showed that every sum of four distinct oddly even squares is the sum of four distinct odd squares.

A PRIME NUMBER p can be written as the sum of two squares IFF $p + 1$ is not divisible by 4 the (FERMAT $4N+1$ THEOREM). An arbitrary positive number n is expressible as the sum of two squares IFF, given its PRIME FACTORIZATION

$$n = p_1^{a_1} p_2^{a_2} p_3^{a_3} \cdots p_k^{a^k}, \tag{18}$$

none of $p_i^{a_i} + 1$ is divisible by 4 (Conway and Guy 1996, p. 147). This is equivalent the requirement that all the odd factors of the SQUAREFREE PART n' of n are equal to 1 (mod 4) (Hardy and Wright 1979, Finch). The first few numbers which can be expressed as the sum of two squares are 1, 2, 4, 5, 8, 9, 10, 13, 16, 17, 18, 20, 25, 26, ... (Sloane's A001481). Letting $\delta(n)$ be the fraction of numbers $\leq n$ which are expressible as the sum of two squares,

$$\lim_{n \to \infty} \delta(n) = 0, \tag{19}$$

and

$$\lim_{n \to \infty} \delta(n)\sqrt{\ln n} = K, \tag{20}$$

where K is the LANDAU-RAMANUJAN CONSTANT.

Numbers expressible as the sum of three squares are those not OF THE FORM $4^k(8l + 7)$ for k, $l \geq 0$ (Nagell 1951, p. 194; Wells 1986, pp. 48 and 56; Hardy 1999, p. 12).

The following table gives the first few numbers which *require* $N = 1, 2, 3$, and 4 squares to represent them as a sum (Wells 1986, p. 70).

N	Sloane	Numbers
1	Sloane's A000290	1, 4, 9, 16, 25, 36, 49, 64, 81, ...
2	Sloane's A000415	2, 5, 8, 10, 13, 17, 18, 20, 26, 29, ...
3	Sloane's A000419	3, 6, 11, 12, 14, 19, 21, 22, 24, 27, ...
4	Sloane's A004215	7, 15, 23, 28, 31, 39, 47, 55, 60, 63, ...

The FERMAT $4N+1$ THEOREM guarantees that every PRIME OF THE FORM $4n + 1$ is a sum of two SQUARE NUMBERS in only one way.

There are only 31 numbers which cannot be expressed as the sum of *distinct* squares: 2, 3, 6, 7, 8, 11, 12, 15, 18, 19, 22, 23, 24, 27, 28, 31, 32, 33, 43, 44, 47, 48, 60, 67, 72, 76, 92, 96, 108, 112, 128 (Sloane's A001422; Guy 1994; Savin 2000). The following numbers cannot be represented using fewer than five distinct squares: 55, 88, 103, 132, 172, 176, 192, 240, 268, 288, 304, 368, 384, 432, 448, 496, 512, and 752, together with all numbers obtained by multiplying these numbers by a power of 4. This gives all known such numbers less than 10^5 (Savin 2000). All numbers > 188 can be expressed as the sum of at most five distinct squares, and only

$$124 = 1 + 4 + 9 + 25 + 36 + 49 \qquad (21)$$

and

$$188 = 1 + 4 + 9 + 25 + 49 + 100 \qquad (22)$$

require six distinct squares (Bohman *et al.* 1979; Guy 1994, p. 136; Savin 2000). In fact, 188 can also be represented using seven distinct squares:

$$188 = 1 + 4 + 9 + 25 + 36 + 49 + 64. \qquad (23)$$

The following table gives the numbers which can be represented in W different ways as a sum of S squares. For example,

$$50 = 1^2 + 7^2 = 5^2 + 5^2 \qquad (24)$$

can be represented in two ways ($W = 2$) by two squares ($S = 2$).

S	W	Sloane	Numbers
1	1	Sloane's A000290	1, 4, 9, 16, 25, 36, 49, 64, 81, 100, 121, ...
2	1	Sloane's A025284	2, 5, 8, 10, 13, 17, 18, 20, 25, 26, 29, 32, ...
2	2	Sloane's A025285	50, 65, 85, 125, 130, 145, 170, 185, 200, ...
3	1	Sloane's A025321	3, 6, 9, 11, 12, 14, 17, 18, 19, 21, 22, 24, ...
3	2	Sloane's A025322	27, 33, 38, 41, 51, 57, 59, 62, 69, 74, 75, ...
3	3	Sloane's A025323	54, 66, 81, 86, 89, 99, 101, 110, 114, 126, ...
3	4	Sloane's A025324	129, 134, 146, 153, 161, 171, 189, 198, ...
4	1	Sloane's A025357	4, 7, 10, 12, 13, 15, 16, 18, 19, 20, 21, 22, ...
4	2	Sloane's A025358	31, 34, 36, 37, 39, 43, 45, 47, 49, 50, 54, ...
4	3	Sloane's A025359	28, 42, 55, 60, 66, 67, 73, 75, 78, 85, 95, 99, ...
4	4	Sloane's A025360	52, 58, 63, 70, 76, 84, 87, 91, 93, 97, 98, 103, ...

The least numbers which are the sum of two squares in exactly n different ways for $n = 1, 2, ...$ are given by 2, 50, 325, 1105, 8125, 5525, 105625, 27625, 71825, 138125, 5281250, ... (Sloane's A016032; Beiler 1966, pp. 140–41; Culbertson; Hardy and Wright 1979; Rivera).

The product of four distinct NONZERO INTEGERS in ARITHMETIC PROGRESSION is square only for $(-3, -1, 1, 3)$, giving $(-3)(-1)(1)(3) = 9$ (Le Lionnais 1983, p. 53). It is possible to have three squares in ARITHMETIC PROGRESSION, but not four (Dickson 1952, pp. 435–40). If these numbers are r^2, s^2, and t^2, there are POSITIVE INTEGERS p and q such that

$$r = |p^2 - 2pq - q^2| \qquad (25)$$

$$s = p^2 + q^2 \qquad (26)$$

$$t = p^2 + 2pq - q^2, \qquad (27)$$

where $(p, q) = 1$ and one of r, s, or t is EVEN (Dickson 1952, pp. 437–38). Every three-term progression of squares can be associated with a PYTHAGOREAN TRIPLE (X, Y, Z)) by

$$X = \tfrac{1}{2}(r + t) \qquad (28)$$

$$Y = \tfrac{1}{2}(t - r) \qquad (29)$$

$$Z = s \qquad (30)$$

(Robertson 1996).

CATALAN'S CONJECTURE states that 8 and 9 (2^3 and 3^2) are the only consecutive POWERS (excluding 0 and 1), i.e., the only solution to CATALAN'S DIOPHANTINE PROBLEM. This CONJECTURE has not yet been proved or refuted, although R. Tijdeman has proved that there can be only a finite number of exceptions should the CONJECTURE not hold. It is also known that 8 and 9 are the only consecutive CUBIC and square numbers (in either order).

The numbers that are not the difference of two squares are 2, 6, 10, 14, 18, ... (Wells 1986, p. 76).

A square number can be the concatenation of two squares, as in the case $16 = 4^2$ and $9 = 3^2$ giving $169 = 13^2$. The first few numbers which are neither square nor the sum of a square and a PRIME are 10, 34, 58, 85, 91, 130, 214, ... (Sloane's A020495).

It is conjectured that, other than 10^{2n}, 4×10^{2n} and 9×10^{2n}, there are only a FINITE number of squares n^2 having exactly two distinct NONZERO DIGITS (Guy 1994, p. 262). The first few such n are 4, 5, 6, 7, 8, 9,

11, 12, 15, 21, ... (Sloane's A016070), corresponding to n^2 of 16, 25, 36, 49, 64, 81, 121, ... (Sloane's A018884). The following table gives the first few numbers which, when squared, give numbers composed of only certain digits. The values of n such that n^2 contains exactly two different digits are given by 4, 5, 6, 7, 8, 9, 10, 11, 12, 15, 20, ... (Sloane's A016069), whose squares are 16, 25 36, 49, 64, ... (Sloane's A018885). The only known square number composed only of the digits 7, 8, and 9 is 9. Based on a discussion in `rec.puzzles`, Vardi (1991) considered numbers composed only of the square digits: 1, 4, and 9. It is conjectured that there are only finitely many, and the largest known is

$$648070211589107021^2$$
$$= 419994999149149944149149944191494441 \quad (31)$$

found by G. Jacobson and D. Applegate (`rec.puzzles` FAQ).

Digits	Sloane	n, n^2
1, 2, 3	Sloane's A030175	1, 11, 111, 36361, 363639, ...
	Sloane's A030174	1, 121, 12321, 1322122321, ...
1, 4, 6	Sloane's A027677	1, 2, 4, 8, 12, 21, 38, 108, ...
	Sloane's A027676	1, 4, 16, 64, 144, 441, 1444, ...
1, 4, 9	Sloane's A027675	1, 2, 3, 7, 12, 21, 38, 107, ...
	Sloane's A006716	1, 4, 9, 49, 144, 441, 1444, 11449, ...
2, 4, 8	Sloane's A027679	2, 22, 168, 478, 2878, 210912978, ...
	Sloane's A027678	4, 484, 28224, 228484, 8282884, ...
4, 5, 6	Sloane's A030177	2, 8, 216, 238, 258, 738, 6742, ...
	Sloane's A030176	4, 64, 46656, 56644, 66564, ...

BROWN NUMBERS are pairs (m, n) of INTEGERS satisfying the condition of BROCARD'S PROBLEM, i.e., such that

$$n! + 1 = m^2, \quad (32)$$

where $n!$ is a FACTORIAL. Only three such numbers are known: (5,4), (11,5), (71,7). Erdos conjectured that these are the only three such pairs.

Either $5x^2 + 4 = y^2$ or $5x^2 - 4 = y^2$ has a solution in POSITIVE INTEGERS IFF, for some n, $(x, y) = (F_n, L_n)$, where F_n is a FIBONACCI NUMBER and L_n is a LUCAS NUMBER (Honsberger 1985, pp. 114–18).

The smallest and largest square numbers containing the digits 1 to 9 are

$$11,826^2 = 139,854,276, \quad (33)$$
$$30,384^2 = 923,187,456. \quad (34)$$

The smallest and largest square numbers containing the digits 0 to 9 are

$$32,043^2 = 1,026,753,849, \quad (35)$$
$$99,066^2 = 9,814,072,356 \quad (36)$$

(Madachy 1979, p. 159). The smallest and largest square numbers containing the digits 1 to 9 twice each are

$$335,180,136^2 = 112,345,723,568,978,496 \quad (37)$$
$$999,390,432^2 = 998,781,235,573,146,624, \quad (38)$$

and the smallest and largest containing 1 to 9 three times are

$$10,546,200,195,312^2$$
$$= 111,222,338,559,598,866,946,777,344 \quad (39)$$

$$31,621,017,808,182^2$$
$$= 999,888,767,225,363,175,346,145,124$$

(Madachy 1979, p. 159).

Madachy (1979, p. 165) also considers number which are equal to the sum of the squares of their two "halves" such as

$$1233 = 12^2 + 33^2 \quad (40)$$
$$8833 = 88^2 + 33^2 \quad (41)$$
$$10100 = 10^2 + 100^2 \quad (42)$$
$$5882353 = 588^2 + 2353^2, \quad (43)$$

in addition to a number of others.

See also ANTISQUARE NUMBER, BIQUADRATIC NUMBER, BROCARD'S PROBLEM, BROWN NUMBERS, CANNONBALL PROBLEM, CATALAN'S CONJECTURE, CENTERED SQUARE NUMBER, CLARK'S TRIANGLE, CUBIC NUMBER, DIOPHANTINE EQUATION, FERMAT $4N+1$ THEOREM, GREEDY ALGORITHM, GROSS, HEPTAGONAL SQUARE NUMBER, LAGRANGE'S FOUR-SQUARE THEOREM, LANDAU-RAMANUJAN CONSTANT, OCTAGONAL SQUARE NUMBER, PARTITION, PENTAGONAL SQUARE NUMBER, PSEUDOSQUARE, PYRAMIDAL NUM-

BER, SQUAREFREE, SQUARE TRIANGULAR NUMBER, SUM OF SQUARES FUNCTION, WARING'S PROBLEM

References

Archibald, R. G. "Waring's Problem: Squares." *Scripta Math.* **7**, 33–8, 1940.

Ball, W. W. R. and Coxeter, H. S. M. *Mathematical Recreations and Essays, 13th ed.* New York: Dover, p. 59, 1987.

Beiler, A. H. *Recreations in the Theory of Numbers: The Queen of Mathematics Entertains.* New York: Dover, 1966.

Bohman, J.; Fröberg, C.-E.; and Riesel, H. "Partitions in Squares." *BIT* **19**, 297–01, 1979.

Conway, J. H. and Guy, R. K. *The Book of Numbers.* New York: Springer-Verlag, pp. 30–2 and 146–47, 1996.

Dickson, L. E. *History of the Theory of Numbers, Vol. 2: Diophantine Analysis.* New York: Chelsea, 1952.

Finch, S. "Unsolved Mathematics Problems: On a Generalized Fermat-Wiles Equation." http://www.mathsoft.com/asolve/fermat/fermat.html.

Grosswald, E. *Representations of Integers as Sums of Squares.* New York: Springer-Verlag, 1985.

Guy, R. K. "Sums of Squares" and "Squares with Just Two Different Decimal Digits." §C20 and F24 in *Unsolved Problems in Number Theory, 2nd ed.* New York: Springer-Verlag, pp. 136–38 and 262, 1994.

Hajdu, L. and Pintér, Á. "Square Product of Three Integers in Short Intervals." *Math. Comput.* **68**, 1299–301, 1999.

Hardy, G. H. *Ramanujan: Twelve Lectures on Subjects Suggested by His Life and Work, 3rd ed.* New York: Chelsea, 1999.

Hardy, G. H. and Wright, E. M. "The Representation of a Number by Two or Four Squares." Ch. 20 in *An Introduction to the Theory of Numbers, 5th ed.* Oxford, England: Clarendon Press, pp. 297–16, 1979.

Honsberger, R. "A Second Look at the Fibonacci and Lucas Numbers." Ch. 8 in *Mathematical Gems III.* Washington, DC: Math. Assoc. Amer., 1985.

Le Lionnais, F. *Les nombres remarquables.* Paris: Hermann, 1983.

Lucas, É. Question 1180. *Nouv. Ann. Math. Ser. 2* **14**, 336, 1875.

Lucas, É. Solution de Question 1180. *Nouv. Ann. Math. Ser. 2* **15**, 429–32, 1876.

Madachy, J. S. *Madachy's Mathematical Recreations.* New York: Dover, pp. 159 and 165, 1979.

Meyl, A.-J.-J. Solution de Question 1194. *Nouv. Ann. Math.* **17**, 464–67, 1878.

Nagell, T. *Introduction to Number Theory.* New York: Wiley, 1951.

Ogilvy, C. S. and Anderson, J. T. *Excursions in Number Theory.* New York: Dover, pp. 77 and 152, 1988.

Pappas, T. "Triangular, Square & Pentagonal Numbers." *The Joy of Mathematics.* San Carlos, CA: Wide World Publ./Tetra, p. 214, 1989.

Pietenpol, J. L. "Square Triangular Numbers." *Amer. Math. Monthly* **69**, 168–69, 1962.

rec.puzzles FAQ3. http://www.cs.caltech.edu/~adam/PUZZLES/rec.puz.faq3.

Rivera, C. "Problems & Puzzles: Puzzle The qs-Sequence.-062." http://www.primepuzzles.net/puzzles/puzz_062.htm.

Robertson, J. P. "Magic Squares of Squares." *Math. Mag.* **69**, 289–93, 1996.

Savin, A. "Shape Numbers." *Quantum* **11**, 14–8, 2000.

Sloane, N. J. A. Sequences A000037/M0613, A000290/M3356, A000415, A000419, A001156/M0221, A001422/M, A001481/M0968, A002828/M0404, A004215/M4349, A006716/M3369, A016069, A016070, A016032, A018884, A018885, A020495, A025284, A025285, A025321, A025322, A025323, A025324, A025357, A025358, A025359, A025360, A027675, A027676, A027677, A027678, A027679, A030174, A030175, A030176, A030177, A056991, and A056992 in "An On-Line Version of the Encyclopedia of Integer Sequences." http://www.research.att.com/~njas/sequences/eisonline.html.

Stangl, W. D. "Counting Squares in $\mathbb{Z}_n$." *Math. Mag.* **69**, 285–89, 1996.

Taussky-Todd, O. "Sums of Squares." *Amer. Math. Monthly* **77**, 805–30, 1970.

Vardi, I. *Computational Recreations in Mathematica.* Reading, MA: Addison-Wesley, pp. 20 and 234–37, 1991.

Watson, G. N. "The Problem of the Square Pyramid." *Messenger. Math.* **48**, 1–2, 1918.

Wells, D. *The Penguin Dictionary of Curious and Interesting Numbers.* Middlesex, England: Penguin Books, pp. 48 and 70, 1986.

Square Orthobicupola

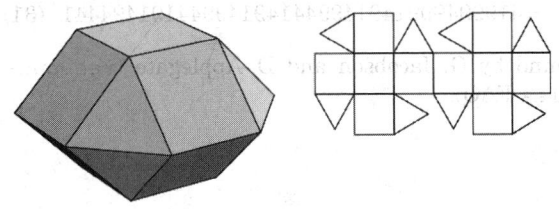

JOHNSON SOLID J_{28}.

References

Weisstein, E. W. "Johnson Solids." MATHEMATICA NOTEBOOK JOHNSONSOLIDS.M.

Weisstein, E. W. "Johnson Solid Netlib Database." MATHEMATICA NOTEBOOK JOHNSONSOLIDS.DAT.

Square Packing

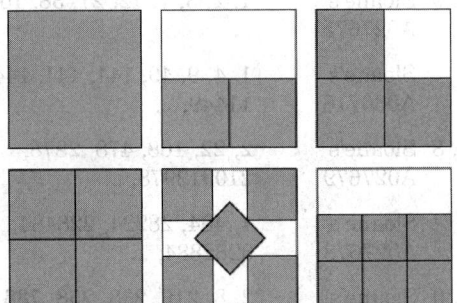

Find the minimum size SQUARE capable of bounding n equal SQUARES arranged in any configuration. The first few cases are illustrated above (Friedman). The only packings which have been proven optimal are 2, 3, 5, 6, 7, 8, 14, 15, 24, and 35, in addition to the trivial cases of the SQUARE NUMBERS (Friedman).

If $n = a^2 - a$ for some a, it is CONJECTURED that the size of the minimum bounding square is a for small n. The smallest n for which the CONJECTURE is known to be violated is $n = 272$ (with $a = 17$). The size is known to scale as k^b, where

$$\tfrac{1}{2}\left(3-\sqrt{3}\right) < b < \tfrac{1}{2}.$$

The following table gives the smallest known side lengths for a square into which n unit squares can be packed.

n	exact	approx.	n	exact	approx.
1	1	1	14	4	4
2	2	2	15	4	4
3	2	2	16	4	4
4	2	2	17	$4+\tfrac{1}{2}\sqrt{2}$	4.707...
5	$2+\tfrac{1}{2}\sqrt{2}$	2.707...	18	$\tfrac{1}{2}(7+\sqrt{7})$	4.822...
6	3	3	19	$3+\tfrac{4}{3}\sqrt{2}$	4.885...
7	3	3	20	5	5
8	3	3	21	5	5
9	3	3	22	5	5
10	$3+\tfrac{1}{2}\sqrt{2}$	3.707...	23	5	5
11	s_{11}	3.877...	24	5	5
12	4	4	25	5	5
13	4	4	26		5.650...

Here, s_{11} is the larger of the two positive real roots of

$$s^4 - 10s^3 + 35s^2 - 46s + 9.$$

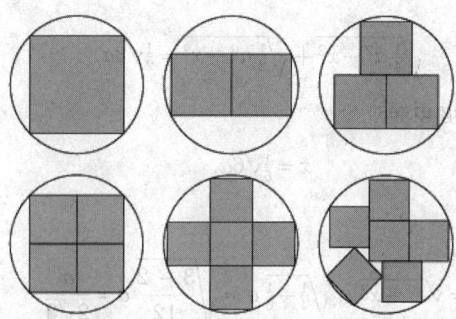

The best known packings of squares into a circle are illustrated above for the first few cases (Friedman).

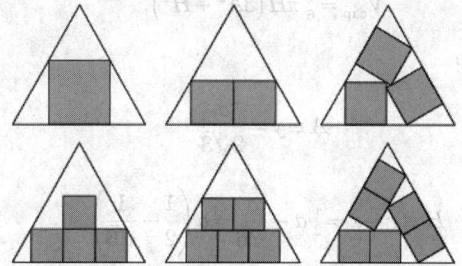

The best known packings of squares into an equilat-

eral triangle are illustrated above for the first few cases (Friedman).

The best packing of a SQUARE inside a PENTAGON, illustrated above, is 1.0673....

See also CIRCLE PACKING, PACKING, TRIANGLE PACKING

References

Erdos, P. and Graham, R. L. "On Packing Squares with Equal Squares." *J. Combin. Th. Ser. A* **19**, 119–23, 1975.
Friedman, E. "Erich's Packing Center." http://www.stetson.edu/~efriedma/packing.html.
Friedman, E. "Circles in Squares." http://www.stetson.edu/~efriedma/cirinsqu/.
Friedman, E. "Squares in Squares." http://www.stetson.edu/~efriedma/squinsqu/.
Friedman, E. "Triangles in Squares." http://www.stetson.edu/~efriedma/triinsqu/.
Friedman, E. "Packing Unit Squares in Squares." *Elec. J. Combin.* DS7, 1–4, Mar. 5, 1998. http://www.combinatorics.org/Surveys/.
Gardner, M. "Packing Squares." Ch. 20 in *Fractal Music, Hypercards, and More Mathematical Recreations from Scientific American Magazine.* New York: W. H. Freeman, pp. 289–06, 1992.
Göbel, F. "Geometrical Packing and Covering Problems." In *Packing and Covering in Combinatorics* (Ed. A. Schrijver). Amsterdam: Tweede Boerhaavestraat, 1979.
Hoffman, P. *The Man Who Loved Only Numbers: The Story of Paul Erdos and the Search for Mathematical Truth.* New York: Hyperion, p. 174, 1998.
Roth, L. F. and Vaughan, R. C. "Inefficiency in Packing Squares with Unit Squares." *J. Combin. Th. Ser. A* **24**, 170–86, 1978.

Square Part

The largest square dividing a POSITIVE INTEGER n. For $n = 1, 2, \ldots$, the first few are 1, 1, 1, 4, 1, 1, 1, 4, 9, 1, 1, 4, ... (Sloane's A008833).

See also CUBIC PART, SQUARE NUMBER, SQUAREFREE PART

References

Sloane, N. J. A. Sequences A008833 in "An On-Line Version of the Encyclopedia of Integer Sequences." http://www.research.att.com/~njas/sequences/eisonline.html.

Square Polyomino

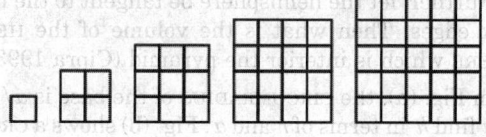

See also L-Polyomino, Skew Polyomino, Straight Polyomino, T-Polyomino

Square Prism

Cube, Cuboid

Square Pyramid

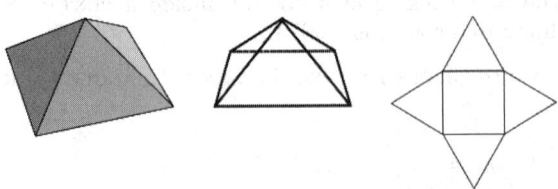

A square pyramid is a Pentahedron consisting of a Pyramid with a Square base. If the top of the pyramid is cut off by a Plane, a square Pyramidal Frustum is obtained. If the four Triangles of the square pyramid are Equilateral, the square pyramid is the "regular" Polyhedron known as Johnson Solid J_1 and, for side length a, has height

$$h = \tfrac{1}{2}\sqrt{2}\,a. \tag{1}$$

Using the equation for a general Pyramid, the Volume of the "regular" is therefore

$$V = \tfrac{1}{3}hA_b = \tfrac{1}{6}\sqrt{2}\,a^3. \tag{2}$$

The Slant Height of a square pyramid is a special case of the formula for a regular n-gonal Pyramid with $n = 2$, given by

$$s = \sqrt{h^2 + \tfrac{1}{4}a^2}, \tag{3}$$

where h is the height and a is the length of a side of the base.

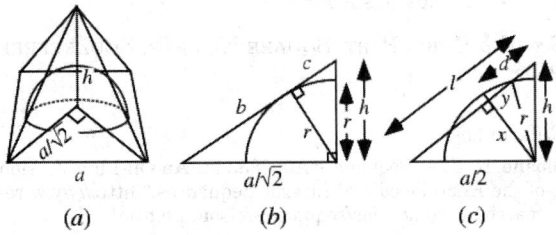

$$(a) \qquad\qquad (b) \qquad\qquad (c)$$

Consider a Hemisphere placed on the base of a square pyramid (having side lengths a and height h). Further, let the hemisphere be tangent to the four apex edges. Then what is the volume of the Hemisphere which is interior the pyramid (Cipra 1993)?

From Fig. (a), the Circumradius of the base is $a/\sqrt{2}$. Now find h in terms of r and a. Fig. (b) shows a Cross Section cut by the plane through the pyramid's apex, one of the base's vertices, and the base center. This

figure gives

$$b = \sqrt{\tfrac{1}{2}a^2 - r^2} \tag{4}$$

$$c = \sqrt{h^2 - r^2}, \tag{5}$$

so the Slant Height is

$$s = \sqrt{h^2 + \tfrac{1}{2}a^2} = b + c = \sqrt{\tfrac{1}{2}a^2 - r^2} + \sqrt{h^2 - r^2}. \tag{6}$$

Solving for h gives

$$h = \frac{ra}{\sqrt{a^2 - 2r^2}}. \tag{7}$$

We know, however, that the Hemisphere must be tangent to the sides, so $r = a/2$, and

$$h = \frac{\tfrac{1}{2}a}{\sqrt{a^2 - \tfrac{1}{2}a^2}}\,a = \frac{\tfrac{1}{2}}{\sqrt{\tfrac{1}{2}}}\,a = \tfrac{1}{2}\sqrt{2}\,a. \tag{8}$$

Fig. (c) shows a Cross Section through the center, apex, and midpoints of opposite sides. The Pythagorean Theorem once again gives

$$l = \sqrt{\tfrac{1}{4}a^2 + h^2} = \sqrt{\tfrac{1}{4}a^2 + \tfrac{1}{2}a^2} = \tfrac{1}{2}\sqrt{3}\,a. \tag{9}$$

We now need to find x and y.

$$\sqrt{\tfrac{1}{4}a^2 - x^2} + d = l. \tag{10}$$

But we know l and h, and d is given by

$$d = \sqrt{h^2 - x^2}. \tag{11}$$

so

$$\sqrt{\tfrac{1}{4}a^2 - x^2} + \sqrt{\tfrac{1}{2}a^2 - x^2} = \tfrac{1}{2}\sqrt{3}\,a. \tag{12}$$

Solving gives

$$x = \tfrac{1}{6}\sqrt{6}\,a, \tag{13}$$

so

$$y = \sqrt{r^2 - x^2} = \sqrt{\tfrac{1}{4} - \tfrac{1}{6}}\,a = \sqrt{\frac{3-2}{12}}\,a = \frac{a}{2\sqrt{3}}. \tag{14}$$

We can now find the Area of the Spherical Cap as

$$V_{\text{cap}} = \tfrac{1}{6}\pi H \left(3A^2 + H^2\right), \tag{15}$$

where

$$A \equiv y = \frac{a}{2\sqrt{3}} \tag{16}$$

$$H \equiv r - x = \tfrac{1}{2}a - \frac{a}{\sqrt{6}} = a\left(\frac{1}{2} - \frac{1}{\sqrt{6}}\right), \tag{17}$$

so

$$V_{cap} = \frac{1}{6}\pi a^3 \left[3\left(\frac{1}{12}\right) + \left(\frac{1}{2} - \frac{1}{\sqrt{6}}\right)^2 \right] \left(\frac{1}{2} - \frac{1}{\sqrt{6}}\right)$$

$$= \frac{1}{6}\pi a^3 \left[\frac{1}{4} + \left(\frac{1}{4} + \frac{1}{6} - \frac{1}{\sqrt{6}}\right) \right] \left(\frac{1}{2} - \frac{1}{\sqrt{6}}\right)$$

$$= \frac{1}{6}\pi a^3 \left(\frac{2}{3} - \frac{1}{\sqrt{6}}\right)\left(\frac{1}{2} - \frac{1}{\sqrt{6}}\right)$$

$$= \frac{1}{6}\pi a^3 \left(\frac{1}{3} - \frac{1}{2\sqrt{6}} - \frac{2}{3\sqrt{6}} + \frac{1}{6}\right)$$

$$= \frac{1}{6}\pi a^3 \left(\frac{1}{2} - \frac{7}{6\sqrt{6}}\right). \tag{18}$$

Therefore, the volume within the pyramid is

$$V_{inside} = \frac{2}{3}\pi r^3 - 4V_{cap} = \frac{2}{3}\pi \frac{1}{8} a^3 - \frac{2}{3}\pi a^3 \left(\frac{1}{2} - \frac{7}{6\sqrt{6}}\right)$$

$$= \frac{2}{3}\pi a^3 \left(\frac{1}{8} - \frac{1}{2} + \frac{7}{6\sqrt{6}}\right) = \frac{2}{3}\pi a^3 \left(\frac{7}{6\sqrt{6}} - \frac{3}{8}\right)$$

$$= \pi a^3 \left(\frac{7}{9\sqrt{6}} - \frac{1}{4}\right). \tag{19}$$

This problem appeared in the Japanese scholastic aptitude test (Cipra 1993).

See also PENTAHEDRON, PYRAMID, SQUARE PYRAMIDAL NUMBER

References

Cipra, B. "An Awesome Look at Japan Math SAT." *Science* **259**, 22, 1993.

Square Pyramidal Number

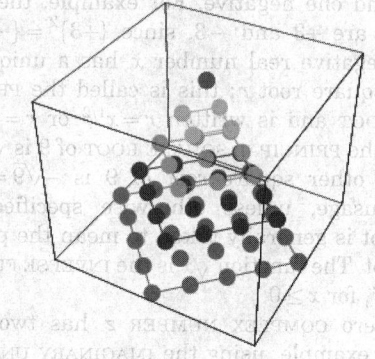

A FIGURATE NUMBER OF THE FORM

$$P_n = \frac{1}{6}n(n+1)(2n+1), \tag{1}$$

corresponding to a configuration of points which form a SQUARE PYRAMID, is called a square pyramidal number (or sometimes, simply a PYRAMIDAL NUMBER). The first few are 1, 5, 14, 30, 55, 91, 140, 204, ...

(Sloane's A000330). They are sums of consecutive pairs of TETRAHEDRAL NUMBERS and satisfy

$$P_n = \frac{1}{3}(2n+1)T_n, \tag{2}$$

where T_n is the nth TRIANGULAR NUMBER.

The only numbers which are simultaneously SQUARE $S_m = m^2$ and square pyramidal $P_n = n(n+1)(2n+1)/6$ (the CANNONBALL PROBLEM) are $P_1 = 1$ and $P_{24} = 4900$, corresponding to $S_1 = 1$ and $S_{70} = 4900$ (Dickson 1952, p. 25; Ball and Coxeter 1987, p. 59; Ogilvy 1988), as conjectured by Lucas (1875, 1876) and proved by Watson (1918). The proof is far from elementary, and requires solving the DIOPHANTINE EQUATION

$$m^2 = \frac{1}{6}n(n+1)(2n+1) \tag{3}$$

(Guy 1994, p. 147). However, an elementary proof has also been given by a number of authors.

Numbers which are simultaneously TRIANGULAR $T_m = m(m+1)/2$ and square pyramidal $P_n = n(n+1)(2n+1)/6$ satisfy the DIOPHANTINE EQUATION

$$\frac{1}{2}m(m+1) = \frac{1}{6}n(n+1)(2n+1). \tag{4}$$

COMPLETING THE SQUARE gives

$$\frac{1}{2}\left(m + \frac{1}{2}\right)^2 - \frac{1}{8} = \frac{1}{6}\left(2n^3 + 3n^2 + n\right) \tag{5}$$

$$\frac{1}{8}(2m+1)^2 = \frac{1}{6}\left(2n^3 + 3n^2 + n\right) + \frac{1}{8} \tag{6}$$

$$3(2m+1)^2 = 8n^3 + 12n^2 + 4n + 3. \tag{7}$$

The only solutions are $(n, m) = (-1, 0)$, $(0, 0)$, $(1, 1)$, $(5, 10)$, $(6, 13)$, and $(85, 645)$ (Guy 1994, p. 147), corresponding to the nontrivial triangular square pyramidal numbers 1, 55, 91, 208335.

Numbers which are simultaneously TETRAHEDRAL $Te_m = m(m+1)(m+2)/6$ and square pyramidal $P_n = n(n+1)(2n+1)/6$ satisfy the DIOPHANTINE EQUATION

$$m(m+1)(m+2) = n(n+1)(2n+1). \tag{8}$$

Beukers (1988) has studied the problem of finding solutions via integral points on an ELLIPTIC CURVE and found that the only solution is the trivial $Te_1 = P_1 = 1$.

See also PYRAMIDAL NUMBER, TETRAHEDRAL NUMBER

References

Ball, W. W. R. and Coxeter, H. S. M. *Mathematical Recreations and Essays, 13th ed.* New York: Dover, p. 59, 1987.
Beukers, F. "On Oranges and Integral Points on Certain Plane Cubic Curves." *Nieuw Arch. Wisk.* **6**, 203–10, 1988.
Conway, J. H. and Guy, R. K. *The Book of Numbers.* New York: Springer-Verlag, pp. 47–0, 1996.
Dickson, L. E. *History of the Theory of Numbers, Vol. 2: Diophantine Analysis.* New York: Chelsea, 1952.

Guy, R. K. "Figurate Numbers." §D3 in *Unsolved Problems in Number Theory, 2nd ed.* New York: Springer-Verlag, pp. 147–50, 1994.

Lucas, É. Question 1180. *Nouvelles Ann. Math. Ser. 2* **14**, 336, 1875.

Lucas, É. Solution de Question 1180. *Nouvelles Ann. Math. Ser. 2* **15**, 429–32, 1876.

Ogilvy, C. S. and Anderson, J. T. *Excursions in Number Theory.* New York: Dover, pp. 77 and 152, 1988.

Sloane, N. J. A. Sequences A000330/M3844 in "An On-Line Version of the Encyclopedia of Integer Sequences." http://www.research.att.com/~njas/sequences/eisonline.html.

Watson, G. N. "The Problem of the Square Pyramid." *Messenger. Math.* **48**, 1–2, 1918.

Square Quadrants

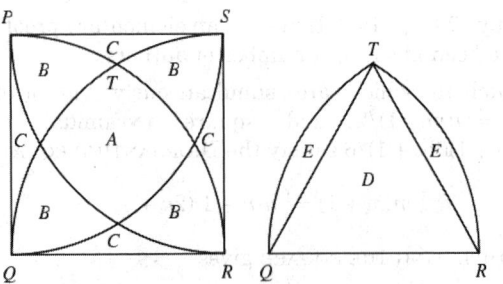

The areas of the regions illustrated above can be found from the equations

$$A + 4B + 4C = 1 \tag{1}$$

$$A + 3B + 2C = \tfrac{1}{4}\pi. \tag{2}$$

Since we want to solve for three variables, we need a third equation. This can be taken as

$$A + 2B + C = 2E + D, \tag{3}$$

where

$$D = \tfrac{1}{4}\sqrt{3} \tag{4}$$

$$D + E = \tfrac{1}{6}\pi, \tag{5}$$

leading to

$$A + 2B + C = D + 2E = 2(D + E) - D = \tfrac{1}{3}\pi - \tfrac{1}{4}\sqrt{3}. \tag{6}$$

Combining the equations (1), (2), and (6) gives the matrix equation

$$\begin{bmatrix} 1 & 4 & 4 \\ 1 & 3 & 2 \\ 1 & 2 & 1 \end{bmatrix} \begin{bmatrix} A \\ B \\ C \end{bmatrix} = \begin{bmatrix} 1 \\ \tfrac{1}{4}\pi \\ \tfrac{1}{3}\pi - \tfrac{1}{4}\sqrt{3} \end{bmatrix}, \tag{7}$$

which can be inverted to yield

$$A = 1 - \sqrt{3} - \tfrac{1}{3}\pi \tag{8}$$

$$B = -1 + \tfrac{1}{2}\sqrt{3} + \tfrac{1}{12}\pi \tag{9}$$

$$C = 1 - \tfrac{1}{4}\sqrt{3} + \tfrac{1}{6}\pi. \tag{10}$$

References

Honsberger, R. *More Mathematical Morsels.* Washington, DC: Math. Assoc. Amer., pp. 67–9, 1991.

Square Root

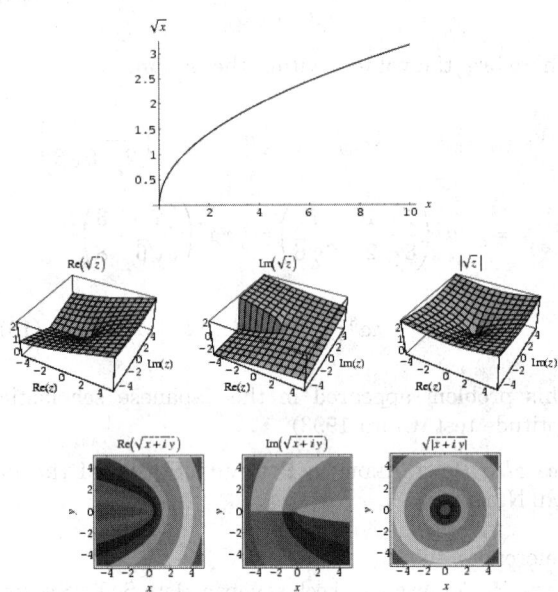

A square root of x is a number r such that $r^2 = x$. Square roots are also called radicals or surds. Any positive real number has *two* square roots: one positive and one negative. For example, the square roots of 9 are $+3$ and -3, since $\{+3\}^2 = \{-3\}^2 = 9$. Any nonnegative real number x has a unique nonnegative square root r; this is called the PRINCIPAL SQUARE ROOT and is written $r = x^{1/2}$ or $r = \sqrt{x}$. For example, the PRINCIPAL SQUARE ROOT of 9 is $\sqrt{9} = +3$, while the other square root of 9 is $-\sqrt{9} = -3$. In common usage, unless otherwise specified, "the" square root is generally taken to mean the principal square root. The function $\sqrt{x}$ is the INVERSE FUNCTION of $f(x) = x^2$, for $x \geq 0$.

Any nonzero COMPLEX NUMBER z has two square roots. For example, using the IMAGINARY UNIT i, the two square roots of -9 are $\pm\sqrt{-9} = \pm 3i$. The PRINCIPAL SQUARE ROOT of a number z is returned by the *Mathematica* Sqrt[x].

The square root of 2 is the IRRATIONAL NUMBER $\sqrt{2} = 1.41421356$ (Sloane's A002193), which has the simple periodic CONTINUED FRACTION 1, 2, 2, 2, 2, 2, ... (Sloane's A040000). The square root of 3 is the

IRRATIONAL NUMBER $\sqrt{3} \approx 1.73205081$ (Sloane's A002194), which has the simple periodic CONTINUED FRACTION 1, 1, 2, 1, 2, 1, 2, ... (Sloane's A040001). In general, the CONTINUED FRACTIONS of the square roots of all POSITIVE INTEGERS are periodic.

The square roots of a COMPLEX NUMBER $z = x + iy$ are given by

$$\sqrt{x + iy} = \pm \sqrt{x^2 + y^2}$$
$$\times \left\{ \cos \left[\frac{1}{2} \tan^{-1} \left(\frac{y}{x} \right) \right] + i \sin \left[\frac{1}{2} \tan^{-1} \left(\frac{y}{x} \right) \right] \right\}. \quad (1)$$

In addition,

$$\sqrt{x + iy} = \pm \frac{1}{2} \sqrt{2} \left[\sqrt{\sqrt{x^2 + y^2} + x} \right.$$
$$\left. + i \, \text{sgn}(y) \sqrt{\sqrt{x^2 + y^2} - x} \right]. \quad (2)$$

As can be seen in the above figure, the IMAGINARY PART of the complex square root function has a BRANCH CUT along the NEGATIVE real axis.

A NESTED RADICAL OF THE FORM $\sqrt{a \pm b\sqrt{c}}$ can sometimes be simplified into a simple square root by equating

$$\sqrt{a \pm b\sqrt{c}} = \sqrt{d} \pm \sqrt{e}. \quad (3)$$

Squaring gives

$$a \pm b\sqrt{c} = d + e \pm 2\sqrt{de}. \quad (4)$$

so

$$a = d + e \quad (5)$$
$$b^2 c = 4de. \quad (6)$$

Solving for d and e gives

$$d, e = \frac{a \pm \sqrt{a^2 - b^2 c}}{2}. \quad (7)$$

For example,

$$\sqrt{5 + 2\sqrt{6}} = \sqrt{2} + \sqrt{3} \quad (8)$$

$$\sqrt{3 - 2\sqrt{2}} = \sqrt{2} - 1. \quad (9)$$

The `Simplify` command of *Mathematica* does not apply such simplifications, but `FullSimplify` does. In general, radical denesting is a difficult problem (Landau).

A sequence of approximations a/b to $\sqrt{n}$ can be derived by factoring

$$a^2 - nb^2 = \pm 1 \quad (10)$$

(where -1 is possible only if -1 is a QUADRATIC RESIDUE of n). Then

$$(a + b\sqrt{n})(a - b\sqrt{n}) = \pm 1 \quad (11)$$

$$(a + b\sqrt{n})^k (a - b\sqrt{n})^k = (\pm 1)^k = \pm 1, \quad (12)$$

and

$$(1 + \sqrt{n})^1 = 1 + \sqrt{n} \quad (13)$$

$$(1 + \sqrt{n})^2 = (1 + n) + 2\sqrt{n} \quad (14)$$

$$(1 + \sqrt{n})(a + b\sqrt{n}) = (a + bn) + \sqrt{n}(a + b). \quad (15)$$

Therefore, a and b are given by the RECURRENCE RELATIONS

$$a_i = a_{i-1} + b_{i-1} n \quad (16)$$

$$b_i = a_{i-1} + b_{i-1} \quad (17)$$

with $a_1 = b_1 = 1$. The error obtained using this method is

$$\left| \frac{a}{b} - \sqrt{n} \right| = \frac{1}{b(a + b\sqrt{n})} < \frac{1}{2b^2}. \quad (18)$$

The first few approximants to $\sqrt{n}$ are therefore given by

$$1, \frac{1}{2}(1 + n), \frac{1 + 3n}{3 + n}, \frac{1 + 6n + n^2}{4(n + 1)}, \frac{1 + 10n + 5n^2}{5 + 10n + n^2}, \ldots \quad (19)$$

This ALGORITHM is sometimes known as the BHAS-KARA-BROUCKNER ALGORITHM. For the case $n = 2$, this gives the convergents to $\sqrt{2}$ as 1, 3/2, 7/5, 17/12, 41/29, 99/70, ... (Sloane's A001333 and A000129; Wells 1986, p. 34). The numerators are given by the RECURRENCE RELATION

$$a(n) = 2a(n-1) + a(n-2), \quad (20)$$

and the denominators are the PELL NUMBERS.

Another general technique for deriving this sequence, known as NEWTON'S ITERATION, is obtained by letting $x = \sqrt{n}$. Then $x = n/x$, so the SEQUENCE

$$x_k = \frac{1}{2} \left(x_{k-1} + \frac{n}{x_{k-1}} \right) \quad (21)$$

converges quadratically to the root. The first few approximants to $\sqrt{n}$ are therefore given by

$$1, \frac{1}{2}(1 + n), \frac{1 + 6n + n^2}{4(n + 1)}, \frac{1 + 28n + 70n^2 + 28n^3 + n^4}{8(1 + n)(1 + 6n + n^2)}, \ldots \quad (22)$$

For $\sqrt{2}$, this gives the convergents 1, 3/2, 17/12, 577/408, 665857/470832, ... (Sloane's A051008 and A051009).

See also CUBE ROOT, NESTED RADICAL, NEWTON'S ITERATION, PRINCIPAL SQUARE ROOT, QUADRATIC SURD, ROOT OF UNITY, SQUARE NUMBER, SQUARE TRIANGULAR NUMBER, SURD

References

Sloane, N. J. A. Sequences A000129/M1314, A001333/
M2665, A002193/M3195, A002194/M4326, A040000,
A040001, A051008, and A051009 in "An On-Line Version
of the Encyclopedia of Integer Sequences." http://www.re-
search.att.com/~njas/sequences/eisonline.html.

Spanier, J. and Oldham, K. B. "The Square-Root Function
$\sqrt{bx+c}$ and Its Reciprocal," "The $b\sqrt{a^2-x^2}$ Function and
Its Reciprocal," and "The $b\sqrt{x^2+a}$ Function." Chs. 12, 14,
and 15 in *An Atlas of Functions.* Washington, DC: Hemisphere, pp. 91–9, 107–15, and 115–22, 1987.

Wells, D. *The Penguin Dictionary of Curious and Interesting
Numbers.* Middlesex, England: Penguin Books, p. 34,
1986.

Williams, H. C. "A Numerical Investigation into the Length
of the Period of the Continued Fraction Expansion of $\sqrt{D}$."
Math. Comp. **36**, 593–01, 1981.

Square Root Inequality

$$2\sqrt{n+1}-2\sqrt{n} < \frac{1}{\sqrt{n}} < 2\sqrt{n}-2\sqrt{n-1}.$$

Square Root Method

The square root method is an algorithm which solves
the MATRIX EQUATION

$$A\mathbf{u}=\mathbf{g} \qquad (1)$$

for $\mathbf{u}$, with A a $p \times p$ SYMMETRIC MATRIX and $\mathbf{g}$ a given
VECTOR. Convert A to a TRIANGULAR MATRIX such that

$$\mathsf{T}^\mathsf{T}\mathsf{T}=\mathsf{A}, \qquad (2)$$

where T^T is the MATRIX TRANSPOSE. Then

$$\mathsf{T}^\mathsf{T}\mathbf{k}=\mathbf{g} \qquad (3)$$

$$\mathsf{T}\mathbf{u}=\mathbf{k}, \qquad (4)$$

so

$$\mathsf{T}=\begin{bmatrix} s_{11} & s_{12} & \cdots & \cdots \\ 0 & s_{22} & \cdots & \cdots \\ \vdots & \vdots & \ddots & \vdots \\ 0 & 0 & \cdots & s_{pp} \end{bmatrix}. \qquad (5)$$

giving the equations

$$s_{11}^2 = a_{11}$$

$$s_{11}s_{12} = a_{12}$$

$$s_{12}^2 + s_{22}^2 = a_{22}$$

$$s_{1j}^2 + s_{2j}^2 + \ldots + s_{jj}^2 = a_{jj}$$

$$s_{1j}+s_{2j}s_{2k}+\ldots+s_{jj}s_{jk} = a_{jk}. \qquad (6)$$

These give

$$s_{11} = \sqrt{a_{11}}$$

$$s_{12} = \frac{a_{12}}{s_{11}}$$

$$s_{22} = \sqrt{a_{22} - s_{12}^2}$$

$$S_{jj} = \sqrt{a_{jj} - s_{ij}^2 - s_{2j}^2 - \ldots - s_{j-1,j}^2}$$

$$s_{jk} = \frac{a_{jk} - s_{1j}s_{1k} - s_{2j}s_{2k} - \ldots - s_{j-1,j}s_{j-1,k}}{s_{jj}}, \qquad (7)$$

giving T from A. Now solve for $\mathbf{k}$ in terms of the s_{ij} s
and $\mathbf{g}$,

$$s_{11}k_1 = g_1$$

$$s_{12}k_1 + s_{22}k_2 = g_2$$

$$s_{1j}k_1 + s_{2j}k_2 + \ldots + s_{jj}k_j = g_j, \qquad (8)$$

which gives

$$k_1 = \frac{g_1}{s_{11}}$$

$$k_2 = \frac{g_2 - s_{12}k_1}{s_{22}}$$

$$k_j = \frac{g_j - s_{1j}k_1 - s_{2j}k_2 - \ldots - s_{j-1,j}k_{j-1}}{s_{jj}}. \qquad (9)$$

Finally, find $\mathbf{u}$ from the s_{ij} s and $\mathbf{k}$,

$$s_{11}u_1 + s_{12}u_2 \ldots + s_{1p}u_p = k_1$$

$$s_{22}u_2 + \ldots + s_{2p}u_p = k_2$$

$$s_{pp}u_p = k_p, \qquad (10)$$

giving the desired solution,

$$u_p = \frac{k_p}{s_{pp}}$$

$$u_{p-1} = \frac{k_{p-1} - s_{p-1,p}u_p}{s_{p-1,p-1}}$$

$$u_j = \frac{k_j - s_{j,j+1}u_{j+1} - s_{j,j+2}u_{j+2} - \ldots - s_{jp}u_p}{s_{jj}}. \qquad (11)$$

See also LU DECOMPOSITION

References

Kenney, J. F. and Keeping, E. S. *Mathematics of Statistics,
Pt. 2, 2nd ed.* Princeton, NJ: Van Nostrand, pp. 298–00,
1951.

Square Tiling

There are a number of interesting results related to
the tiling of squares. For example, M. Laczkovich has
shown that there are exactly three shapes of non-

right triangles that tile the square with similar copies, corresponding to angles $(\pi/8, \pi/4, 5\pi/8)$, $(\pi/4, \pi/3, 5\pi/12)$, and $(\pi/12, \pi/4, 2\pi/3)$ (Stein and Szabó 1994). In particular, given triangles of shape $1 - 2 - \sqrt{5}$ with no two the same size, tile the square. The best known solution has 8 triangles (Berlekamp 1999).

See also TILING

References

Berlekamp, E. and Rodgers, T. (Eds.). *The Mathemagician and the Pied Puzzler: A Collection in Tribute to Martin Gardner.* Boston, MA: A. K. Peters, 1999.

Laczkovich, M. "Tilings of Polygons with Similar Triangles." *Combinatorica* **10**, 281–06, 1990.

Schattschneider, D. "Unilateral and Equitransitive Tilings by Squares." *Disc. Comput. Geom.* **24**, 519–25, 2000.

Stein, S. and Szabó, S. *Algebra and Tiling: Homomorphisms in the Service of Geometry.* Washington, DC: Math. Assoc. Amer., 1994.

Square Torus

The square torus is the quotient of the plane by the integer lattice.

Square Triangle Picking

Given three points chosen at random inside a UNIT SQUARE, the average AREA of the TRIANGLE determined by these points is given by

$$\bar{A} = \frac{\underbrace{\int_0^1 \cdots \int_0^1}_{6} |A(\mathbf{x}_i)| \, dx_1 \cdots dx_3 \, dy_1 \cdots dy_3}{\underbrace{\int_0^1 \cdots \int_0^1}_{6} dx_1 \cdots dx_3 \, dy_1 \cdots dy_3},$$

where the VERTICES are located at (x_i, y_i) where $i = 1, ..., 3$, and the (signed) AREA is given by the DETERMINANT

$$A = \frac{1}{2!} \begin{vmatrix} x_1 & y_1 & 1 \\ x_2 & y_2 & 1 \\ x_3 & y_3 & 1 \end{vmatrix}.$$

The integral can be evaluated analytically to yield $\bar{A} = 11/144$ (Ambartzumian 1987, Pfiefer, Trott 1998), and first calculated by Woolhouse (1867).

Because attempting to do the integrals directly quickly results in intractable integrands, the best approach to accomplish the integration is to divide the 6-dimensional region of integration into subregions such that the sign of A does not change (Trott 1998).

The distribution function for the area of a random triangle in a square is known exactly.

See also CUBE TETRAHEDRON PICKING, HEXAGON TRIANGLE PICKING, POLYGON TRIANGLE PICKING, TRIANGLE TRIANGLE PICKING, UNIT SQUARE

References

Alagar, V. S. "On the Distribution of a Random Triangle." *J. Appl. Prob.* **14**, 284–97, 1977.

Ambartzumian, R. V. (Ed.). *Stochastic and Integral Geometry.* Dordrecht, Netherlands: Reidel, 1987.

Buchta, C. "Über die konvexe Hülle von Zufallspunkten in Eibereichen." *Elem. Math.* **38**, 153–56, 1983.

Buchta, C. "Zufallspolygone in konvexen Vielecken." *J. reine angew. Math.* **347**, 212–20, 1984.

Henze, N. "Random Triangles in Convex Regions." *J. Appl. Prob.* **20**, 111–25, 1983.

Klee, V. "What is the Expected Volume of a Simplex Whose Vertices are Chosen at Random from a Given Convex Body." *Amer. Math. Monthly* **76**, 286–88, 1969.

Pfiefer, R. E. "The Historical Development of J. J. Sylvester's Four Point Problem." *Math. Mag.* **62**, 309–17, 1989.

Seidov, Z. F. "Letters: Random Triangle." *Mathematica J.* **7**, 414, 2000.

Trott, M. "The Area of a Random Triangle." *Mathematica J.* **7**, 189–98, 1998.

Woolhouse, W. S. B. "Question 2471" *Mathematical Questions, with Their Solutions, from the Educational Times, Vol. 8.* London: F. Hodgson and Son, pp. 100–05, 1867.

Square Triangular Number

A number which is simultaneously SQUARE and TRIANGULAR. Let T_n denote the nth TRIANGULAR NUMBER and S_m the mth SQUARE NUMBER, then a number which is both triangular and square satisfies the equation $T_n = S_m$, or

$$\frac{1}{2} n(n + 1) = m^2. \tag{1}$$

COMPLETING THE SQUARE gives

$$\frac{1}{2}(n^2 + n) = \frac{1}{2}\left(n + \frac{1}{2}\right)^2 - \left(\frac{1}{2}\right)\left(\frac{1}{4}\right) = m^2 \tag{2}$$

$$\frac{1}{8}(2n + 1)^2 - \frac{1}{8} = m^2 \tag{3}$$

$$(2n + 1)^2 - 8m^2 = 1. \tag{4}$$

Therefore, defining

$$x \equiv 2n + 1 \tag{5}$$

$$y \equiv 2m \tag{6}$$

gives the PELL EQUATION

$$x^2 - 2y^2 = 1 \tag{7}$$

(Conway and Guy 1996). The first few solutions are

$(x, y) = (3, 2), (17, 12), (99, 70), (577, 408), \ldots$. These give the solutions $(n, m) = (1, 1), (8, 6), (49, 35), (288, 204), \ldots$ (Sloane's A001108 and A001109), corresponding to the triangular square numbers 1, 36, 1225, 41616, 1413721, 48024900, ... (Sloane's A001110; Pietenpol 1962). In 1730, Euler showed that there are an infinite number of such solutions (Dickson 1952).

The general FORMULA for a square triangular number ST_n is b^2c^2, where b/c is the nth convergent to the CONTINUED FRACTION of $\sqrt{2}$ (Ball and Coxeter 1987, p. 59; Conway and Guy 1996). The first few are

$$\frac{1}{1}, \frac{3}{2}, \frac{7}{5}, \frac{17}{12}, \frac{41}{29}, \frac{99}{70}, \frac{239}{169}, \ldots, \tag{8}$$

The NUMERATORS and DENOMINATORS can also be obtained by doubling the previous FRACTION and adding to the FRACTION before that.

A general FORMULA for square triangular numbers is

$$ST_n = \left[\frac{\left(1 + \sqrt{2}\right)^{2n} - \left(1 - \sqrt{2}\right)^{2n}}{4\sqrt{2}}\right]^2 \tag{9}$$

$$= \frac{1}{32}\left[\left(17 + 2\sqrt{2}\right)^n + \left(17 - 2\sqrt{2}\right)^n - 2\right]. \tag{10}$$

The square triangular numbers also satisfy the RECURRENCE RELATION

$$ST_n = 34ST_{n-1} - ST_{n-2} + 2 \tag{11}$$

$$u_{n+2} = 6u_{n+1} - u_n \tag{12}$$

with $u_0 = 0$, $u_1 = 1$, where $ST_n \equiv u_n^2$. A curious product formula for ST_n is given by

$$ST_n = 2^{2n-5} \prod_{k=1}^{2n}\left[3 + \cos\left(\frac{k\pi}{n}\right)\right]. \tag{13}$$

An amazing GENERATING FUNCTION is

$$f(x) = \frac{1 + x}{(1 - x)(1 - 34x + x^2)}$$

$$= 1 + 36x + 1225x^2 + \ldots \tag{14}$$

(Sloane and Plouffe 1995).

Taking the square and triangular numbers together gives the sequence 1, 1, 3, 4, 6, 9, 10, 15, 16, 21, 25, ... (Sloane's A005214; Hofstadter 1996, p. 15).

See also SQUARE NUMBER, SQUARE ROOT, TRIANGULAR NUMBER

References

Allen, B. M. "Squares as Triangular Numbers." *Scripta Math.* **20**, 213–14, 1954.

Ball, W. W. R. and Coxeter, H. S. M. *Mathematical Recreations and Essays, 13th ed.* New York: Dover, 1987.

Conway, J. H. and Guy, R. K. *The Book of Numbers.* New York: Springer-Verlag, pp. 203–05, 1996.

Dickson, L. E. *History of the Theory of Numbers, Vol. 2: Diophantine Analysis.* New York: Chelsea, pp. 10, 16, and 27, 1952.

Guy, R. K. "Sums of Squares" and "Figurate Numbers." §C20 and §D3 in *Unsolved Problems in Number Theory, 2nd ed.* New York: Springer-Verlag, pp. 136–38 and 147–50, 1994.

Hofstadter, D. R. *Fluid Concepts & Creative Analogies: Computer Models of the Fundamental Mechanisms of Thought.* New York: Basic Books, 1996.

Khatri, M. N. "Triangular Numbers Which are Also Squares." *Math. Student* **27**, 55–6, 1959.

Pietenpol, J. L. "Square Triangular Numbers." Problem E 1473. *Amer. Math. Monthly* **69**, 168–69, 1962.

Potter, D. C. D. "Triangular Square Numbers." *Math. Gaz.* **56**, 109-, 1972.

Sierpinski, W. *Teoria Liczb, 3rd ed.* Warsaw, Poland: Monografie Matematyczne t. 19, p. 517, 1950.

Sierpinski, W. "Sur les nombres triangulaires carrés." *Pub. Faculté d'Électrotechnique l'Université Belgrade*, No. 65, 1–, 1961.

Sierpinski, W. "Sur les nombres triangulaires carrés." *Bull. Soc. Royale Sciences Liège*, 30 ann., 189–94, 1961.

Silverman, J. H. *A Friendly Introduction to Number Theory.* Englewood Cliffs, NJ: Prentice Hall, 1996.

Sloane, N. J. A. Sequences A001108/M4536, A001109/M4217, and A001110/M5259 in "An On-Line Version of the Encyclopedia of Integer Sequences." http://www.research.att.com/~njas/sequences/eisonline.html.

Walker, G. W. "Triangular Squares." Problem E 954. *Amer. Math. Monthly* **58**, 568, 1951.

Square Wave

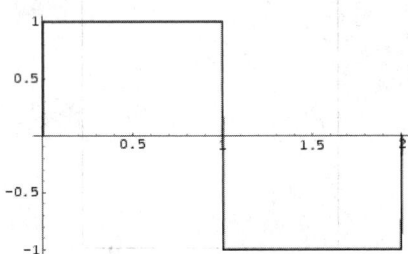

The square wave is a periodic waveform consisting of instantaneous transitions between two levels which can be denoted ± 1. The square wave is sometimes also called the RADEMACHER FUNCTION. Let the square wave have period $2L$. The square wave function is ODD, so the FOURIER SERIES has $a_0 = a_n = 0$ and

$$b_0 = \frac{2}{L}\int_0^L \sin\left(\frac{n\pi x}{L}\right) dx$$

$$= \frac{4}{n\pi}\sin^2\left(\tfrac{1}{2}n\pi\right) = \frac{4}{n\pi}\begin{cases} 0 & n \text{ even} \\ 1 & n \text{ odd.} \end{cases}$$

The FOURIER SERIES for the square wave is therefore

$$f(x) = \frac{4}{\pi}\sum_{n=1,3,5,\ldots}^{\infty} \frac{1}{n}\sin\left(\frac{n\pi x}{L}\right).$$

See also HADAMARD MATRIX, WALSH FUNCTION

References

Thompson, A. R.; Moran, J. M.; and Swenson, G. W. Jr. *Interferometry and Synthesis in Radio Astronomy.* New York: Wiley, p. 203, 1986.

Squared

A number to the POWER 2 is said to be squared, so that x^2 is called "x squared."

See also CUBED, SQUARE ROOT

Squared Square

PERFECT SQUARE DISSECTION

Squarefree

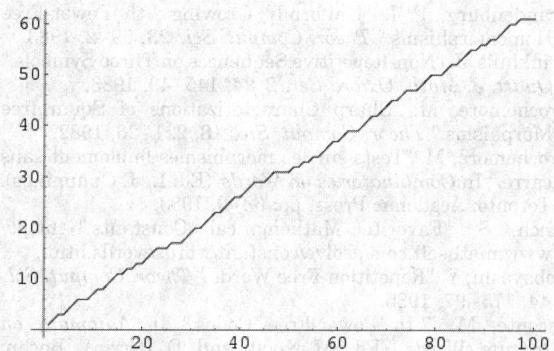

A number is said to be squarefree (or sometimes QUADRATFREI; Shanks 1993) if its PRIME decomposition contains no repeated factors. All PRIMES are therefore trivially squarefree. The squarefree numbers are 1, 2, 3, 5, 6, 7, 10, 11, 13, 14, 15, ... (Sloane's A005117). The SQUAREFUL numbers (i.e., those that contain at least one square) are 4, 8, 9, 12, 16, 18, 20, 24, 25, ... (Sloane's A013929).

The asymptotic number $Q(n)$ of squarefree numbers $\leq n$ is given by

$$Q(n) = \frac{6n}{\pi^2} + \mathcal{O}(\sqrt{n}) \tag{1}$$

(Landau 1974, pp. 604–09; Nagell 1951, p. 130; Hardy and Wright 1979, pp. 269–70; Hardy 1999, p. 65). $Q(n)$ for $n = 10, 100, 1000, ...$ are 7, 61, 608, 6083, 60794, 607926, ..., while the asymptotic density is $1/\zeta(2) = 6/\pi^2 \approx 0.607927$, where $\zeta(n)$ is the RIEMANN ZETA FUNCTION.

The MÖBIUS FUNCTION is given by

$$\mu(n) \equiv \begin{cases} 0 & \text{if } n \text{ has one or more repeated} \\ & \text{prime factors} \\ 1 & \text{if } n = 1 \\ (-1)^k & \text{if } n \text{ is the product of } k \text{ distinct} \\ & \text{primes,} \end{cases} \tag{2}$$

so $\mu(n) \neq 0$ indicates that n is squarefree. The asymptotic formula for $Q(x)$ is equivalent to the formula

$$\sum_{n=1}^{x} |\mu(n)| = \frac{6x}{\pi^2} + \mathcal{O}(\sqrt{x}) \tag{3}$$

(Hardy and Wright 1979, p. 270)

There is no known polynomial-time algorithm for recognizing squarefree INTEGERS or for computing the squarefree part of an INTEGER. In fact, this problem may be no easier than the general problem of integer factorization (obviously, if an integer n can be factored completely, n is squarefree IFF it contains no duplicated factors). This problem is an important unsolved problem in NUMBER THEORY because computing the RING of integers of an algebraic number field is reducible to computing the squarefree part of an INTEGER (Lenstra 1992, Pohst and Zassenhaus 1997). The *Mathematica* function `SquareFreeQ[n]` in the *Mathematica* add-on package `NumberTheory`NumberTheoryFunctions`` (which can be loaded with the command `<<NumberTheory`) determines whether a number is squarefree.

No SQUAREFUL FIBONACCI NUMBERS F_p are known with p PRIME. All numbers less than 2.5×10^{15} in SYLVESTER'S SEQUENCE are squarefree, and no SQUAREFUL numbers in this sequence are known (Vardi 1991). Every CARMICHAEL NUMBER is squarefree. The BINOMIAL COEFFICIENTS $\binom{2n-1}{n}$ are squarefree only for $n = 2, 3, 4, 6, 9, 10, 12, 36, ...$, with no others less than $n = 1500$. The CENTRAL BINOMIAL COEFFICIENTS are SQUAREFREE only for $n = 1, 2, 3, 4, 5, 7, 8, 11, 17, 19, 23, 71, ...$ (Sloane's A046098), with no others less than 1500.

See also BINOMIAL COEFFICIENT, BIQUADRATEFREE, COMPOSITE NUMBER, CUBEFREE, ERDOS SQUAREFREE CONJECTURE, FIBONACCI NUMBER, KORSELT'S CRITERION, MÖBIUS FUNCTION, PRIME NUMBER, RIEMANN ZETA FUNCTION, SÁRKOZY'S THEOREM, SQUARE NUMBER, SQUAREFREE PART, SQUAREFUL, SYLVESTER'S SEQUENCE

References

Bellman, R. and Shapiro, H. N. "The Distribution of Square-free Integers in Small Intervals." *Duke Math. J.* **21**, 629–37, 1954.

Hardy, G. H. *Ramanujan: Twelve Lectures on Subjects Suggested by His Life and Work,* 3rd ed. New York: Chelsea, 1999.

Hardy, G. H. and Wright, E. M. "The Number of Squarefree Numbers." §18.6 in *An Introduction to the Theory of Numbers,* 5th ed. Oxford, England: Clarendon Press, pp. 269–70, 1979.

Landau, E. *Handbuch der Lehre von der Verteilung der Primzahlen,* 3rd ed. New York: Chelsea, 1974.

Lenstra, H. W. Jr. "Algorithms in Algebraic Number Theory." *Bull. Amer. Math. Soc.* **26**, 211–44, 1992.

Nagell, T. *Introduction to Number Theory.* New York: Wiley, p. 130, 1951.

Pohst, M. and Zassenhaus, H. *Algorithmic Algebraic Number Theory.* Cambridge, England: Cambridge University Press, p. 429, 1997.

Shanks, D. *Solved and Unsolved Problems in Number Theory, 4th ed.* New York: Chelsea, p. 114, 1993.

Sloane, N. J. A. Sequences A005117/M0617, A013929, and A046098 in "An On-Line Version of the Encyclopedia of Integer Sequences." http://www.research.att.com/~njas/sequences/eisonline.html.

Vardi, I. "Are All Euclid Numbers Squarefree?" §5.1 in *Computational Recreations in Mathematica.* Reading, MA: Addison-Wesley, pp. 7–, 82–5, and 223–24, 1991.

Square-Free

SQUAREFREE

Squarefree Part

That part of a POSITIVE INTEGER left after all square factors are divided out. For example, the squarefree part of $24 = 2^3 \cdot 3$ is 6, since $6 \cdot 2^2 = 24$. For $n = 1, 2, \ldots$, the first few are 1, 2, 3, 1, 5, 6, 7, 2, 1, 10, ... (Sloane's A007913). The squarefree part function can be implemented in *Mathematica* as

```
SquarefreePart[n_Integer?Positive] :=
  Times @@ Power @@@ ({#[[1]], Mod[#[[2]], 2]} &
/@ FactorInteger[n])
```

See also CUBEFREE PART, SQUARE PART, SQUAREFREE

References

Atanassov, K. "On the 22nd, 23rd, and the 24th Smarandache Problems. *Notes on Number Theory and Discrete Mathematics, Sophia, Bulgaria* **5**, 80–2, 1999.

Atanassov, K. *On Some of the Smarandache's Problems.* Lupton, AZ: American Research Press, pp. 16–1, 1999.

Sloane, N. J. A. Sequences A007913 in "An On-Line Version of the Encyclopedia of Integer Sequences." http://www.research.att.com/~njas/sequences/eisonline.html.

Smarandache, F. *Only Problems, Not Solutions!, 4th ed.* Phoenix, AZ: Xiquan, 1993.

Squarefree Word

N.B. A detailed online essay by S. Finch was the starting point for this entry.

A "square" word consists of two identical adjacent subwords (for example, *acbacb*). A squarefree word contains *no* square words as subwords (for example, *abcacbabcb*). The only squarefree binary words are *a*, *b*, *ab*, *ba*, *aba*, and *bab* (since *aa*, *bb*, *aaa*, *aab*, *abb*, *baa*, *bba*, and *bbb* contain square identical adjacent subwords *a*, *b*, *a*, *a*, *b*, *a*, *b*, and *b*, respectively).

However, there are arbitrarily long ternary squarefree words. The number $s(n)$ of ternary squarefree words of length $n = 1, 2, \ldots$ are 1, 3, 6, 12, 18, 30, 42, 60, ... (Sloane's A006156), and $s(n)$ is bounded by

$$6 \cdot 1.032^n \le s(n) \le 6 \cdot 1.379^n \qquad (1)$$

(Brandenburg 1983). In addition,

$$S \equiv \lim_{n \to \infty} [s(n)]^{1/n} = 1.302\ldots \qquad (2)$$

(Brinkhuis 1983, Noonan and Zeilberger 1997).

The number of squarefree quaternary words of length $n = 1, 2, \ldots$ are 4, 12, 36, 96, 264, 696, ... (Sloane's A051041).

See also ALPHABET, CUBEFREE WORD, OVERLAPFREE WORD, WORD

References

Baake, M.; Elser, V.; and Grimm, U. The Entropy of Square-Free Words. 8 Sep 1998. http://xxx.lanl.gov/abs/math-ph/9809010/.

Bean, D. R.; Ehrenfeucht, A.; and McNulty, G. F. "Avoidable Patterns in Strings of Symbols." *Pacific J. Math.* **85**, 261–94, 1979.

Berstel, J. and Reutenauer, C. "Square-Free Words and Idempotent Semigroups." In *Combinatorics on Words* (Ed. M. Lothaire). Reading, MA: Addison-Wesley, pp. 18–8, 1983.

Brandenburg, F.-J. "Uniformly Growing kth Power-Free Homomorphisms." *Theor. Comput. Sci.* **23**, 69–2, 1983.

Brinkhuis, J. "Non-Repetitive Sequences on Three Symbols." *Quart. J. Math. Oxford Ser. 2* **34**, 145–49, 1983.

Crochemore, M. "Sharp Characterizations of Squarefree Morphisms." *Theor. Comput. Sic.* **18**, 221–26, 1982.

Crochemore, M. "Tests sur les morphismes faiblement sans carré." In *Combinatorics on Words* (Ed. L. J. Cummings). Toronto: Academic Press, pp. 63–9, 1983.

Finch, S. "Favorite Mathematical Constants." http://www.mathsoft.com/asolve/constant/words/words.html.

Kobayashi, Y. "Repetition-Free Words." *Theor. Comput. Sci.* **44**, 175–97, 1986.

Leconte, M. "kth Power-Free Codes." In *Automata on Infinite Words* (Ed. M. Nivat and D. Perrin). Berlin: Springer-Verlag, pp. 172–78, 1985.

Noonan, J. and Zeilberger, D. "The Goulden-Jackson Cluster Method: Extensions, Applications, and Implementations." 1997.

Pleasants, P. A. B. "Nonrepetitive Sequences." *Proc. Cambridge Philos. Soc.* **68**, 267–74, 1970.

Sloane, N. J. A. Sequences A006156/M2550 and A051041 in "An On-Line Version of the Encyclopedia of Integer Sequences." http://www.research.att.com/~njas/sequences/eisonline.html.

Thue, A. "Über unendliche Zeichenreihen." *Norske Vid. Selsk. Skr. I, Mat. Nat. Kl. Christiana* **7**, 1–2, 1906. Reprinted in Nagell, T.; Selberg, A.; Selberg, S.; and Thalberg, K. (Eds.). *Selected Mathematical Papers of Axel Thue.* Oslo, Norway: Universitetsforlaget, pp. 139–58, 1977.

Thue, A. "Über die gegenseitige Lage gleicher Teile gewisser Zeichenreihen." *Norske Vid. Selsk. Skr. I, Mat. Nat. Kl. Christiana* **1**, 1–7, 1912. Reprinted in Nagell, T.; Selberg, A.; Selberg, S.; and Thalberg, K. (Eds.). *Selected Mathematical Papers of Axel Thue.* Oslo, Norway: Universitetsforlaget, pp. 413–77, 1977.

Squareful

A number is squareful, also called nonsquarefree, if it contains at least one SQUARE in its prime factorization. The first few are 4, 8, 9, 12, 16, 18, 20, 24, 25, ... (Sloane's A013929). The greatest multiple prime factors for the squareful integers are 2, 2, 3, 2, 2, 3, 2, 2, 5, 3, 2, 2, 3, ... (Sloane's A046028). The least multiple prime factors for squareful integers are 2, 2, 3, 2, 2, 3, 2, 2, 5, 3, 2, 2, 2, ... (Sloane's A046027).

See also GREATEST PRIME FACTOR, LEAST PRIME

FACTOR, SMARANDACHE NEAR-TO-PRIMORIAL FUNC-
TION, SQUAREFREE

References

Sloane, N. J. A. Sequences A013929, A046027, and A046028
in "An On-Line Version of the Encyclopedia of Integer
Sequences."   http://www.research.att.com/~njas/se-
quences/eisonline.html.

Squaring

Squaring is the GEOMETRIC CONSTRUCTION, using
only COMPASS and STRAIGHTEDGE, of a SQUARE which
has the same area as a given geometric figure.
Squaring is also called QUADRATURE. An object which
can be constructed by squaring is called SQUARABLE.

See also CIRCLE SQUARING, COMPASS, CONSTRUCTI-
BLE NUMBER, GEOMETRIC CONSTRUCTION, RECTAN-
GLE SQUARING, STRAIGHTEDGE, TRIANGLE SQUARING

Squaring the Circle

CIRCLE SQUARING

Squeezing Theorem

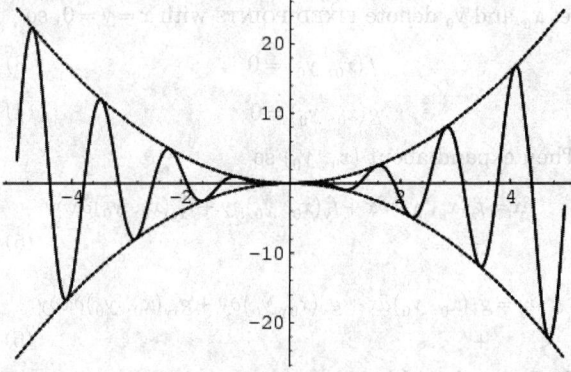

Let there be two functions $f_-(x)$ and $f_+(x)$ such that
$f(x)$ is "squeezed" between the two,

$$f_-(x) \le f(x) \le f_+(x).$$

If

$$r = \lim_{x \to a} f_-(x) = \lim_{x \to a} f_+(x),$$

then $\lim_{x \to a} f(x) = r$. In the above diagram the functions $f_-(x) = -x^2$ and $f_+(x) = x^2$ "squeeze" $x^2 \sin(cx)$ at
0, so $\lim_{x \to a} x^2 \sin(cx) = 0$. The squeezing theorem is
also called the sandwich theorem.

See also LIMIT, PINCHING THEOREM

s-Run

*N.B. A detailed online essay by S. Finch was the
starting point for this entry.*

Let **v** be a n-VECTOR whose entries are each 1 (with
probability p) or 0 (with probability $q = 1 - p$). An s-
run is an isolated group of s consecutive 1s. Ignoring

the boundaries, the total number of runs R_n satisfies

$$K_n = \frac{\langle R_n \rangle}{n} = (1-p)^2 \sum_{s=1}^{n} p^s = p(1-p)(1-p^n),$$

so

$$K(p) \equiv \lim_{n \to \infty} K_n = p(1-p),$$

which is called the MEAN RUN COUNT PER SITE or
MEAN RUN DENSITY in PERCOLATION THEORY.

See also PERCOLATION THEORY, s-CLUSTER

References

Finch, S. "Favorite Mathematical Constants." http://
www.mathsoft.com/asolve/constant/rndprc/rndprc.html.

S-Signature

SIGNATURE (RECURRENCE RELATION)

SSS Theorem

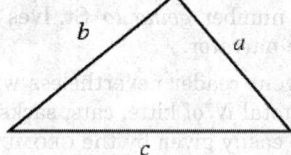

Specifying three sides uniquely determines a TRIAN-
GLE whose AREA is given by HERON'S FORMULA,

$$A = \sqrt{s(s-a)(s-b)(s-c)}, \tag{1}$$

where

$$s \equiv \tfrac{1}{2}(a+b+c) \tag{2}$$

is the SEMIPERIMETER of the TRIANGLE. Let R be the
CIRCUMRADIUS, then

$$A = \frac{abc}{4R}. \tag{3}$$

Using the LAW OF COSINES

$$a^2 = b^2 + c^2 - 2bc \cos A \tag{4}$$

$$b^2 = a^2 + c^2 - 2ac \cos B \tag{5}$$

$$c^2 = a^2 + b^2 - 2ab \cos C \tag{6}$$

gives the three ANGLES as

$$A = \cos^{-1}\left(\frac{b^2 + c^2 - a^2}{2bc}\right) \quad (7)$$

$$B = \cos^{-1}\left(\frac{a^2 + c^2 - b^2}{2ac}\right) \quad (8)$$

$$C = \cos^{-1}\left(\frac{a^2 + b^2 - c^2}{2ab}\right). \quad (9)$$

See also AAA THEOREM, AAS THEOREM, ASA THEOREM, ASS THEOREM, HERON'S FORMULA, SAS THEOREM, SEMIPERIMETER, TRIANGLE

St. Ives Problem

A well-known nursery rhyme states, "As I was going to St. Ives, I met a man with seven wives. Every wife had seven sacks, every sack had seven cats, every cat had seven kitts. Kitts, cats, sacks, wives, how many were going to St. Ives?" Upon being presented with this conundrum, most readers begin furiously adding and multiplying numbers in order to calculate the total quantity of objects mentioned. However, the problem is a trick question. Since the man and his wives, sacks, etc. were met by the narrator *on the way to St. Ives*, they were in fact leaving–not going to–St. Ives. The number *going to* St. Ives is therefore "one," i.e., the narrator.

Should a diligent reader nevertheless wish to calculate the sum total N of kitts, cats, sacks, and wives, the answer is easily given by the GEOMETRIC SERIES

$$\sum_{k=1}^{n} r^k = \frac{r(1 - r^n)}{1 - r} \quad (1)$$

with $n = 4$ and $r = 7$. Therefore,

$$N = \sum_{i=1}^{4} 7^i = \frac{7(1 - 7^4)}{7 - 1} = 2800. \quad (2)$$

$$N = 7^1 + 7^2 + 7^3 + 7^4$$
$$= 7(1 + 7(1 + 7(1 + 7))) = 7(1 + 7(1 + 7 \cdot 8))$$
$$= 7(1 + 7 \cdot 57) = 7 \cdot 400 = 2800. \quad (3)$$

A similar question was given as problem 79 of the Rhind papyrus, dating from 1650 BC. This problem concerns 7 houses, each with 7 cats, each with 7 mice, each with 7 spelt, each with 7 hekat. The total number of items is then

$$\sum_{i=1}^{5} 7^i = 19607 \quad (4)$$

(Wells 1986, p. 71). In turn, the problem of the Rhind

papyrus is repeated in Fibonacci's *Liber Abaci* (1202, 1228).

References
Eisele, C. "Liber Abaci." *Scripta Math.* **17**.
Gill, R. J. *Mathematics in the Time of the Pharaohs.* Cambridge, MA: MIT Press, 1972.
Wells, D. *The Penguin Dictionary of Curious and Interesting Numbers.* Middlesex, England: Penguin Books, p. 71, 1986.

Stability

The robustness of a given outcome to small changes in initial conditions or small random fluctuations. CHAOS is an example of a process which is not stable.

See also STABILITY MATRIX

Stability Matrix

Given a system of two ordinary differential equations

$$\dot{x} = f(x, y) \quad (1)$$

$$\dot{y} = g(x, y), \quad (2)$$

let x_0 and y_0 denote FIXED POINTS with $\dot{x} = \dot{y} = 0$, so

$$f(x_0, y_0) = 0 \quad (3)$$

$$g(x_0, y_0) = 0. \quad (4)$$

Then expand about (x_0, y_0) so

$$\delta\dot{x} = f_x(x_0, y_0)\delta x + f_y(x_0, y_0)\delta y + f_{xy}(x_0, y_0)\delta x\delta y + \cdots. \quad (5)$$

$$\delta\dot{y} = g_x(x_0, y_0)\delta x + g_y(x_0, y_0)\delta y + g_{xy}(x_0, y_0)\delta x\delta y + \cdots. \quad (6)$$

To first-order, this gives

$$\frac{d}{dt}\begin{bmatrix} \delta x \\ \delta y \end{bmatrix} = \begin{bmatrix} f_x(x_0, y_0) & f_y(x_0, y_0) \\ g_x(x_0, y_0) & g_y(x_0, y_0) \end{bmatrix}\begin{bmatrix} \delta x \\ \delta y \end{bmatrix}, \quad (7)$$

where the 2×2 MATRIX, or its generalization to higher dimension, is called the stability matrix. Analysis of the EIGENVALUES (and EIGENVECTORS) of the stability matrix characterizes the type of FIXED POINT.

See also ELLIPTIC FIXED POINT (DIFFERENTIAL EQUATIONS), FIXED POINT, HYPERBOLIC FIXED POINT (DIFFERENTIAL EQUATIONS), LINEAR STABILITY, STABLE IMPROPER NODE, STABLE NODE, STABLE SPIRAL POINT, STABLE STAR, UNSTABLE IMPROPER NODE, UNSTABLE NODE, UNSTABLE SPIRAL POINT, UNSTABLE STAR

References
Tabor, M. "Linear Stability Analysis." §1.4 in *Chaos and Integrability in Nonlinear Dynamics: An Introduction.* New York: Wiley, pp. 20–1, 1989.

Stabilization

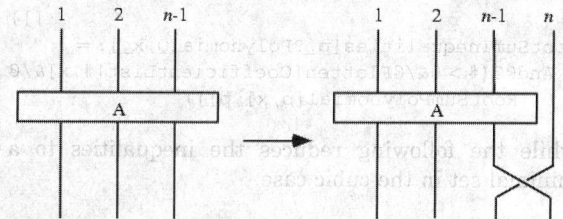

A type II MARKOV MOVE.

See also MARKOV MOVES

Stable Equivalence

Two VECTOR BUNDLES are stably equivalent IFF ISOMORPHIC VECTOR BUNDLES are obtained upon WHITNEY SUMMING each VECTOR BUNDLE with a trivial VECTOR BUNDLE.

See also VECTOR BUNDLE, WHITNEY SUM

Stable Improper Node

A FIXED POINT for which the STABILITY MATRIX has equal NEGATIVE EIGENVALUES.

See also ELLIPTIC FIXED POINT (DIFFERENTIAL EQUATIONS), FIXED POINT, HYPERBOLIC FIXED POINT (DIFFERENTIAL EQUATIONS), STABLE NODE, STABLE SPIRAL POINT, UNSTABLE IMPROPER NODE, UNSTABLE NODE, UNSTABLE SPIRAL POINT, UNSTABLE STAR

References

Tabor, M. "Classification of Fixed Points." §1.4.b in *Chaos and Integrability in Nonlinear Dynamics: An Introduction.* New York: Wiley, pp. 22–5, 1989.

Stable Marriage Problem

Given a set of n men and n women, marry them off in pairs after each man has ranked the women in order of preference from 1 to n, $\{w_1, \ldots, w_n\}$ and each women has done likewise, $\{m_1, \ldots, m_n\}$. If the resulting set of marriages contains no pairs OF THE FORM $\{m_i, w_j\}$, $\{m_k, w_l\}$ such that m_i prefers w_l to w_j and w_l prefers m_i to m_k, the marriage is said to be stable. Gale and Shapley (1962) showed that a stable marriage exists for any choice of rankings (Skiena 1990, p. 245). In the United States, the algorithm of Gale and Shapley (1962) is used to match hospitals to medical interns (Skiena 1990, p. 245).

men's rankings									*women's rankings*								
7	5	4	9	2	2	1	5	6	3	9	3	8	6	2	9	6	8
3	4	8	7	6	7	6	6	1	1	4	1	7	9	4	3	3	2
8	8	3	4	4	8	2	9	4	5	8	8	5	2	5	8	2	6
9	3	9	2	9	6	3	1	7	2	1	9	3	5	1	2	1	4
6	1	7	5	8	5	8	2	5	8	7	5	2	1	6	7	8	9
4	2	5	8	7	3	5	8	8	7	6	4	6	4	8	5	4	1
2	6	6	3	5	4	4	4	3	6	3	2	4	7	3	4	5	3
1	7	1	1	1	9	7	3	9	9	2	6	9	3	9	6	9	7
5	9	2	6	3	9	7	7	2	4	5	7	1	8	7	1	7	5

In the rankings illustrated above, the male-optimal stable marriage is 4, 2, 6, 5, 3, 1, 7, 9, 8, and the female-optimal stable marriage is 1, 2, 8, 9, 3, 4, 7, 6, 5. A stable marriage can be found using `Stable-Marriage[m,w]` in the *Mathematica* add-on package `DiscreteMath`Combinatorica`` (which can be loaded with the command `< <DiscreteMath`)`.

See also DIVORCE DIGRAPH, MATCHING

References

Gale, D. and Shapley, L. S. "College Admissions and the Stability of Marriage." *Amer. Math. Monthly* **69**, 9–4, 1962.
Gusfield, D. and Irving, R. W. *The Stable Marriage Problem: Structure and Algorithms.* Cambridge, MA: MIT Press, 1989.
Skiena, S. "Stable Marriages." §6.4.4 in *Implementing Discrete Mathematics: Combinatorics and Graph Theory with Mathematica.* Reading, MA: Addison-Wesley, pp. 245–46, 1990.

Stable Node

A FIXED POINT for which the STABILITY MATRIX has both EIGENVALUES NEGATIVE, so $\lambda_1 < \lambda_2 < 0$.

See also ELLIPTIC FIXED POINT (DIFFERENTIAL EQUATIONS), FIXED POINT, HYPERBOLIC FIXED POINT (DIFFERENTIAL EQUATIONS), STABLE IMPROPER NODE, STABLE SPIRAL POINT, STABLE STAR, UNSTABLE IMPROPER NODE, UNSTABLE NODE, UNSTABLE SPIRAL POINT, UNSTABLE STAR

References

Tabor, M. "Classification of Fixed Points." §1.4.b in *Chaos and Integrability in Nonlinear Dynamics: An Introduction.* New York: Wiley, pp. 22–5, 1989.

Stable Polynomial

A REAL POLYNOMIAL P is said to be stable if all its ROOTS lie in the LEFT HALF-PLANE. The term "stable" is used to describe such a polynomial because, in the theory of linear servomechanisms, a system exhibits unforced time-dependent motion of the form e^{st}, where s is the root of a certain REAL POLYNOMIAL $P(s) = 0$. A system is therefore mechanically stable IFF P is a stable polynomial.

The polynomial $x + a$ is stable IFF $a > 0$, and the IRREDUCIBLE POLYNOMIAL $x^2 + ab + b$ is stable IFF both a and b are greater than zero. The ROUTH-HURWITZ THEOREM can be used to determine if a polynomial is stable.

Given two real polynomials P and Q, if P and Q are stable, then so is their product PQ, and vice versa (Séroul 2000, p. 280). It therefore follows that the coefficients of stable real polynomials are either all positive or all negative (although this is not a SUFFICIENT condition, as shown with the counterexample $x^3 + x^2 + x + 1$). Furthermore, the values of a stable polynomial are never zero for $x \geq 0$ and have the same sign as the coefficients of the polynomial.

It is possible to decide if a polynomial is stable without first knowing its roots using the following theorem due to Strelitz (1977). Let $A = x^n + a_{n-1}x^{n-1} + \ldots + a_0$ be a real polynomial with roots $\alpha_1, \ldots, \alpha_n$, and construct $B = x^m + b_{m-1}x^{m-1} + \ldots + b_0$ as the monic real polynomial of degree $m = n(n-1)/2$ having roots $\alpha_i + \alpha_j$ for $1 \le i \le j \le n$. Then A is stable IFF all coefficients of A and B are positive (Séroul 2000, p. 281).

For example, given the third-order polynomial $A = x^3 + ax^2 + bx + c$, the sum-of-roots polynomial B is given by

$$B = x^3 + 2ax^2 + (a^2 + b)x + (ab - c). \qquad (1)$$

Resolving the inequalities given by requiring that each coefficient of A and B be greater than zero then gives the conditions for A to be stable as $a > 0$, $b > 0$, $0 < c < ab$.

Similarly, for the fourth-order polynomial $A = x^4 + ax^3 + bx^2 + cx + d$, the sum-of-roots-polynomial is

$$x^6 + 3ax^5 + (3a^2 + 2b)x^4 + (a^3 + 4ab)x^3$$
$$+ (2a^2b + b^2 + ac - 4d)x^2 + (ab^2 + a^2c - 4ad)$$
$$+ x + (abc - c^2 - a^2d), \qquad (2)$$

so the condition for A to be stable can be resolved to $a > 0$, $b > 0$, $0 < c < ab$, $0 < d < (abc - c^2)/a^2$.

The fifth-order polynomial is

$$x^{10} + 4ax^9 + (6a^2 + 3b)x^8 + (4a^3 + 9ab + c)x^7$$
$$+ (a^4 + 9a^2b + 3b^2 + 4ac - 3d)x^6$$
$$+ (3a^3b + 6ab^2 + 5a^2c + 2bc - 5ad - 11e)x^5$$
$$+ (3a^2b^2 + b^3 + 2a^3c + 6abc - c^2 - 2a^2d - 2bd - 22ae)x^4$$
$$+ (ab^3 + 4a^2bc + b^2c - 4cd - 16a^2e - 4be)x^3$$
$$+ (2ab^2c + a^2c^2 - bc^2 + a^2bd + b^2d - 3acd - 4d^2 - 4a^3e$$
$$- 9abe + 7ce)x^2$$
$$+ (abc^2 - c^3 + ab^2d - 4ad^2 - 4a^2be - b^2e + 4ace + 4de)x$$
$$+ (abcd - c^2d - a^2d^2 - ab^2e + bce + 2ade - e^2). \qquad (3)$$

The following *Mathematica* code computes the sum-of-roots polynomial B and inequalities obtained from the coefficients,

```
RootSumPolynomial[r_List,x_]:=Module[
   {n=Length[r],i,j},
   RootReduce@Collect[Expand[
    Times@@((x-#)&/@Flatten[
     Table[r[[i]]+r[[j]],{i,n},{j,i+1,n}]])
   ],x]
]   RootSumPolynomial[p_?PolynomialQ,x_]:=
        RootSumPolynomial[RootList[p,x],x]
RootList[p_?PolynomialQ,x_]:=
```

```
x/.{ToRules[Roots[p==0,x,
   Cubics->False,Quartics->False
                            ]]}
RootSumInequalities[p_?PolynomialQ,x_]:=
   And@@(#>0&/@Flatten[CoefficientList[#,x]&/@
   {RootSumPolynomial[p,x],p}])
```

while the following reduces the inequalities to a minimal set in the cubic case.

```
Resolve[Exists[x, (a | b | c | x) \[Element]
Reals,
   RootSumInequalities[x^3 + a x^2 + b x + c,
x]
   ], {a, b, c}]
```

See also LEFT HALF-PLANE, ROUTH-HURWITZ THEOREM

References

Séroul, R. "Stable Polynomials." §10.13 in *Programming for Mathematicians*. Berlin: Springer-Verlag, pp. 280–86, 2000.
Strelitz, S. "On the Routh-Hurwitz Problem." *Amer. Math. Monthly* **84**, 542–44, 1977.

Stable Spiral Point

A FIXED POINT for which the STABILITY MATRIX has EIGENVALUES OF THE FORM $\lambda_\pm = -\alpha \pm i\beta$ (with $\alpha, \beta > 0$).

See also ELLIPTIC FIXED POINT (DIFFERENTIAL EQUATIONS), FIXED POINT, HYPERBOLIC FIXED POINT (DIFFERENTIAL EQUATIONS), STABLE IMPROPER NODE, STABLE NODE, STABLE STAR, UNSTABLE IMPROPER NODE, UNSTABLE NODE, UNSTABLE SPIRAL POINT, UNSTABLE STAR

References

Tabor, M. "Classification of Fixed Points." §1.4.b in *Chaos and Integrability in Nonlinear Dynamics: An Introduction*. New York: Wiley, pp. 22–5, 1989.

Stable Star

A FIXED POINT for which the STABILITY MATRIX has one zero EIGENVECTOR with NEGATIVE EIGENVALUE $\lambda < 0$.

See also ELLIPTIC FIXED POINT (DIFFERENTIAL EQUATIONS), FIXED POINT, HYPERBOLIC FIXED POINT (DIFFERENTIAL EQUATIONS), STABLE IMPROPER NODE, STABLE NODE, STABLE SPIRAL POINT, UNSTABLE IMPROPER NODE, UNSTABLE NODE, UNSTABLE SPIRAL POINT, UNSTABLE STAR

References

Tabor, M. "Classification of Fixed Points." §1.4.b in *Chaos and Integrability in Nonlinear Dynamics: An Introduction*. New York: Wiley, pp. 22–25, 1989.

Stable Type

A POLYNOMIAL equation whose ROOTS all have NEGATIVE REAL PARTS. For a REAL QUADRATIC EQUATION

$$z^2 + Bz + C = 0.$$

the stability conditions are B, $C > 0$. For a REAL CUBIC EQUATION

$$z^3 + Az^2 + Bz + C = 0.$$

the stability conditions are A, B, $C > 0$ and $AB > C$.

References

Birkhoff, G. and Mac Lane, S. *A Survey of Modern Algebra, 5th ed.* New York: Macmillan, pp. 108–09, 1996.

Stab-Werner Projection

WERNER PROJECTION

Stack

A DATA STRUCTURE which is a special kind of LIST in which elements may be added to or removed from the top only. These actions are called a PUSH or a POP, respectively. Actions may be taken by popping one or more values, operating on them, and then pushing the result back onto the stack.

Stacks are used as the basis for computer languages such as FORTH, PostScript ® (Adobe Systems), and the RPN language used in Hewlett-Packard ® programmable calculators.

See also LIST, POP, PUSH, QUEUE, REVERSE POLISH NOTATION

Stack Polygon

$$[\mathbf{A}, \mathbf{B}, \mathbf{C}]$$

A SELF-AVOIDING POLYGON containing two adjacent corners of its minimal bounding rectangle. The anisotropic area and perimeter generating function $G(x, y)$ and partial generating functions $H_m(y)$, connected by

$$G(x, y, q) = \sum_{m \geq 1} H_m(y, q) x^m.$$

satisfy the self-reciprocity and inversion relations

$$H_m(1/y, 1/q) = -y^{2m-3} q^{m^2-2m} H_m(y, q)$$

and

$$G(x, y) + y^3 G(x/y^2, 1/y) = 0$$

(Bousquet-Mélou *et al.* 1999).

See also LATTICE POLYGON, SELF-AVOIDING POLYGON

References

Bousquet-Mélou, M.; Guttmann, A. J.; Orrick, W. P.; and Rechnitzer, A. Inversion Relations, Reciprocity and Poly-

ominoes. 23 Aug 1999. http://xxx.lanl.gov/abs/math.CO/9908123/.

Stäckel Determinant

A DETERMINANT used to determine in which coordinate systems the HELMHOLTZ DIFFERENTIAL EQUATION is separable (Morse and Feshbach 1953). A determinant

$$S = |\Phi_{mn}| = \begin{vmatrix} \Phi_{11} & \Phi_{12} & \Phi_{13} \\ \Phi_{21} & \Phi_{22} & \Phi_{23} \\ \Phi_{31} & \Phi_{32} & \Phi_{33} \end{vmatrix} \tag{1}$$

in which Φ_m are functions of u_i alone is called a Stäckel determinant. A coordinate system is separable if it obeys the ROBERTSON CONDITION, namely that the SCALE FACTORS h_i in the LAPLACIAN

$$\nabla^2 = \sum_{i=1}^{3} \frac{1}{h_1 h_2 h_3} \frac{\partial}{\partial u_i} \left(\frac{h_1 h_2 h_3}{h_i^2} \frac{\partial}{\partial u_i} \right) \tag{2}$$

can be rewritten in terms of functions $f_i(u_i)$ defined by

$$\frac{1}{h_1 h_2 h_3} \frac{\partial}{\partial u_i} \left(\frac{h_1 h_2 h_3}{h_i^2} \frac{\partial}{\partial u_i} \right)$$

$$= \frac{g(u_{i+1}, u_{i+2})}{h_1 h_2 h_3} \frac{\partial}{\partial u_i} \left[f_i(u_i) \frac{\partial}{\partial u_i} \right]$$

$$= \frac{1}{h_i^2 f_i} \frac{\partial}{\partial u_i} \left(f_i \frac{\partial}{\partial u_i} \right) \tag{3}$$

such that S can be written

$$S = \frac{h_1 h_2 h_3}{f_1(u_1) f_2(u_2) f_3(u_3)}. \tag{4}$$

When this is true, the separated equations are OF THE FORM

$$\frac{1}{f_n} \frac{\partial}{\partial u_n} \left(f_n \frac{\partial X_n}{\partial u_n} \right) + (k_1^2 \Phi_{n1} + k_2^2 \Phi_{n2} + k_3^2 \Phi_{n3}) X_n = 0 \tag{5}$$

The Φ_{ij}s obey the minor equations

$$M_1 = \Phi_{22} \Phi_{33} - \Phi_{23} \Phi_{32} = \frac{S}{h_1^2} \tag{6}$$

$$M_2 = \Phi_{13} \Phi_{32} - \Phi_{12} \Phi_{33} = \frac{S}{h_2^2} \tag{7}$$

$$M_3 = \Phi_{12} \Phi_{23} - \Phi_{13} \Phi_{22} = \frac{S}{h_3^2}. \tag{8}$$

which are equivalent to

$$M_1 \Phi_{11} + M_2 \Phi_{21} + M_3 \Phi_{31} = S \tag{9}$$

$$M_1 \Phi_{12} + M_2 \Phi_{22} + M_3 \Phi_{32} = 0 \tag{10}$$

$$M_1 \Phi_{13} + M_2 \Phi_{23} + M_3 \Phi_{33} = 0 \tag{11}$$

(Morse and Feshbach 1953, p. 509). This gives a total of four equations in nine unknowns. Morse and Feshbach (1953, pp. 655–66) give not only the Stäckel determinants for common coordinate systems, but also the elements of the determinant (although it is not clear how these are derived).

See also HELMHOLTZ DIFFERENTIAL EQUATION, LAPLACE'S EQUATION, POISSON'S EQUATION, ROBERTSON CONDITION, SEPARATION OF VARIABLES

References

Moon, P. and Spencer, D. E. *Field Theory Handbook, Including Coordinate Systems, Differential Equations, and Their Solutions, 2nd ed.* New York: Springer-Verlag, pp. 5–, 1988.
Morse, P. M. and Feshbach, H. "Tables of Separable Coordinates in Three Dimensions." *Methods of Theoretical Physics, Part I.* New York: McGraw-Hill, pp. 509–11 and 655–66, 1953.

Staircase Function

$$\lfloor 111 \xrightarrow{5} 10011 \xrightarrow{2} 1011 \xrightarrow{5} 10101 \xrightarrow{1} 1011 \xrightarrow{2} 111$$

A function composed of a set of equally spaced jumps of equal length, such as the CEILING FUNCTION $f(x) = \lceil x \rceil$, FLOOR FUNCTION $f(x) = \lfloor x \rfloor$, or NEAREST INTEGER FUNCTION $f(x) = [x]$.

See also CEILING FUNCTION, FLOOR FUNCTION, NEAREST INTEGER FUNCTION, SAWTOOTH WAVE

References

Spanier, J. and Oldham, K. B. *An Atlas of Functions.* Washington, DC: Hemisphere, p. 74, 1987.

Staircase Polygon

Define the minimal bounding rectangle as the smallest rectangle containing a given lattice polygon. If the perimeter of the lattice polygon is equal to that of its minimal bounding rectangle, it is said to be convex. (Note that a "convex" lattice polygon is not necessarily convex in the usual sense of the word.) A staircase polygon is then defined as a convex polygon which contains two opposite corners of its bounding rectangle (Bousquet-Mélou *et al.* 1999).

The area generating function $H_m(y, q)$ that counts polygons of width m for staircase polygons of width 4 is given by

$$H_4(q) =$$

$$\frac{q^4(1 + 2q + 4q^2 + 6q^3 + 7q^4 + 6q^5 + 4q^6 + 2q^7 + q^8)}{(1 - q)^2(1 - q^2)^2(1 - q^3)^2(1 - q^4)}. \tag{1}$$

which satisfies

$$H_4(1/q) = -H_4(q)$$

(Bousquet-Mélou 1992, Bousquet-Mélou *et al.* 1999). The anisotropic area and perimeter generating func-

tion $G(x, y, q)$ and partial generating functions $H_m(y, q)$, connected by

$$G(x, y, q) = \sum_{m \geq 1} H_m(y, q)x^m.$$

satisfy the self-reciprocity and inversion relations

$$H_m(1/y, 1/q) = -y^{m-1}H_m(y, q)$$

for $m \geq 2$ and

$$G(x, y, q) + yG(x/y, 1/y, 1/q) = -x$$

(Bousquet-Mélou *et al.* 1999).

The anisotropic area and perimeter generating function $G(x, y, q)$ of staircase polygon with a staircase hole satisfies an inversion relation OF THE FORM

$$G(x, y, q) + y^2 G(x/y, 1/y, 1/q)$$

(Bousquet-Mélou *et al.* 1999).

See also SELF-AVOIDING POLYGON, STAIRCASE WALK

References

Bousquet-Mélou, M. "Convex Polyominoes and Heaps of Segments." *J. Phys. A: Math. Gen.* **25**, 1925–934, 1992.
Bousquet-Mélou, M.; Guttmann, A. J.; Orrick, W. P.; and Rechnitzer, A. Inversion Relations, Reciprocity and Polyominoes. 23 Aug 1999. http://xxx.lanl.gov/abs/math.CO/9908123/.

Staircase Walk

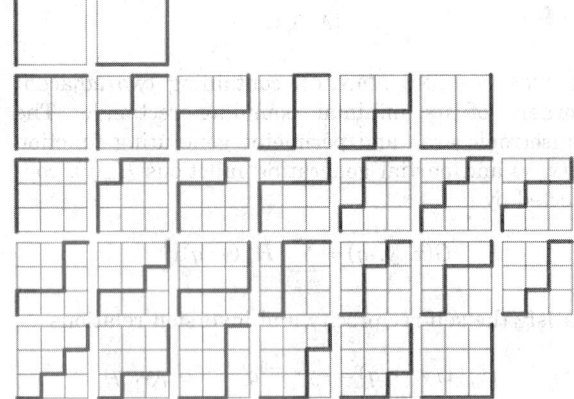

The numbers of staircase walks on an $m \times n$ grid are given by

$$\binom{m+n-2}{m-1} = \frac{(m+n-2)!}{(m-1)!(n-1)!} \tag{1}$$

(Vilenkin 1971, Mohanty 1979, Narayana 1979, Finch). The first few values for $m = n = 1, 2, ...,$ are 1, 2, 6, 20, 70, 252, ... (Sloane's A000984), which are the CENTRAL BINOMIAL COEFFICIENTS.

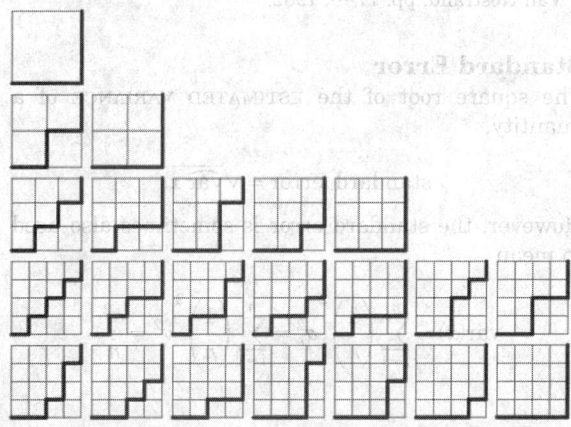

The number of staircase walks on an $n \times n$ grid which remain below the diagonal is given by the CATALAN NUMBER

$$C_{n-1} = \frac{1}{n+1}\binom{2n}{n}.$$

i.e., 1, 2, 5, 14, 42, 132, ... (Sloane's A000108).

See also CATALAN NUMBER, CENTRAL BINOMIAL COEFFICIENT, STAIRCASE POLYGON

References

Finch, S. "Unsolved Mathematics Problems: Self-Avoiding Walks of a Rook on a Chessboard." http://www.mathsoft.com/asolve/gammel/gammel.html.

Mohanty, S. G. *Lattice Path Counting and Applications.* New York: Academic Press, 1979.

Narayana, T. V. *Lattice Path Combinatorics with Statistical Applications.* Toronto, Ontario, Canada: University of Toronto Press, 1979.

Sloane, N. J. A. Sequences A000108/M1459 and A000984/M1645 in "An On-Line Version of the Encyclopedia of Integer Sequences." http://www.research.att.com/~njas/sequences/eisonline.html.

Vilenkin, N. Ya. *Combinatorics.* New York: Academic Press, 1971.

Stamp Folding

The number of ways of folding a strip of stamps has several possible variants. Considering only positions of the hinges for unlabeled stamps without regard to orientation of the stamps, the number of foldings is denoted $U(n)$. If the stamps are labeled and orientation is taken into account, the number of foldings is denoted $N(n)$. Finally, the number of symmetric foldings is denoted $S(n)$. The following table summarizes these values for the first n.

n	$S(n)$ Sloane's A001010	$U(n)$ Sloane's A001011	$N(n)$ Sloane's A000136
Sloane			
1	1	1	1
2	2	1	2
3	2	2	6
4	4	5	16
5	6	14	50
6	8	38	144
7	18	120	462
8	20	353	1392
9	56	1148	4536
10	48	3527	14060

See also MAP FOLDING, POSTAGE STAMP PROBLEM

References

Gardner, M. "The Combinatorics of Paper-Folding." In *Wheels, Life, and Other Mathematical Amusements.* New York: W. H. Freeman, pp. 60–3, 1983.

Gardner, M. *The Sixth Book of Mathematical Games from Scientific American.* Chicago, IL: University of Chicago Press, pp. 21 and 26–7, 1984.

Koehler, J. E. "Folding a Strip of Stamps." *J. Combin. Th.,* Sep. 1968.

Lunnon, W. F. "A Map-Folding Problem." *Math. Comput.,* Jan. 1968.

Ruskey, F. "Information of Stamp Folding." http://www.theory.csc.uvic.ca/~cos/inf/perm/StampFolding.html.

Sloane, N. J. A. *A Handbook of Integer Sequences.* Boston, MA: Academic Press, p. 22, 1973.

Sloane, N. J. A. Sequences A000136/M1614, A001010/M0323, and A001011/M1455 in "An On-Line Version of the Encyclopedia of Integer Sequences." http://www.research.att.com/~njas/sequences/eisonline.html.

Stamp Problem

POSTAGE STAMP PROBLEM

Standard Deviation

The standard deviation stdv(x) is defined as the SQUARE ROOT of the VARIANCE σ^2,

$$\text{stdv}(x) = \sigma = \sqrt{\langle x^2 \rangle - \langle x \rangle^2} = \sqrt{\mu_2' - \mu^2}. \quad (1)$$

where $\mu = \bar{x} = \langle x \rangle$ is the MEAN, $\mu_2' = \langle x^2 \rangle$ is the second RAW MOMENT, and $\langle f(x) \rangle$ denotes an EXPECTATION VALUE. The variance σ^2 is therefore equal to the second CENTRAL MOMENT (i.e., moment about the MEAN),

$$\sigma^2 = \mu_2. \quad (2)$$

The variate value producing a CONFIDENCE INTERVAL CI is often denoted x_{CI}, and

$$x_{\text{CI}} = \sqrt{2}\,\text{erf}^{-1}(\text{CI}). \quad (3)$$

The following table lists the CONFIDENCE INTERVALS

corresponding to the first few multiples of the standard deviation.

range	CI
σ	0.6826895
2σ	0.9544997
3σ	0.9973002
4σ	0.9999366
5σ	0.9999994

To find the standard deviation range corresponding to a given CONFIDENCE INTERVAL, solve (2) for n, giving

$$n = \sqrt{2}\ \mathrm{erf}^{-1}(\mathrm{CI}). \qquad (4)$$

CI	range
0.800	$\pm 1.28155\sigma$
0.900	$\pm 1.64485\sigma$
0.950	$\pm 1.95996\sigma$
0.990	$\pm 2.57583\sigma$
0.995	$\pm 2.80703\sigma$
0.999	$\pm 3.29053\sigma$

The square root of the SAMPLE VARIANCE is the "sample" standard deviation,

$$s_N = \sqrt{\frac{1}{N}\sum_{i=1}^{N}(x_i - \bar{x})^2}. \qquad (5)$$

It is a BIASED ESTIMATOR of the population standard deviation. An unbiased ESTIMATOR is given by

$$s_{N-1} = \sqrt{\frac{1}{N-1}\sum_{i=1}^{N}(x_i - \bar{x})^2}. \qquad (6)$$

Physical scientists often use the term ROOT-MEAN-SQUARE as a synonym for standard deviation when they refer to the SQUARE ROOT of the mean squared deviation of a signal from a given baseline or fit.

See also CONFIDENCE INTERVAL, MEAN, MOMENT, ROOT-MEAN-SQUARE, SAMPLE VARIANCE, STANDARD ERROR, VARIANCE

References

Kenney, J. F. and Keeping, E. S. "The Standard Deviation" and "Calculation of the Standard Deviation." §6.5–.6 in *Mathematics of Statistics, Pt. 1, 3rd ed.* Princeton, NJ: Van Nostrand, pp. 77–0, 1962.

Standard Error

The square root of the ESTIMATED VARIANCE of a quantity,

$$\mathrm{standard\ error} = \sqrt{\mathrm{var}\ x}.$$

However, the standard error is sometimes also used to mean

$$\mathrm{var}(\bar{x}) = \sum_{i=1}^{n}\left(\frac{1}{n}\right)^2 \sigma_i^2 = \sum_{i=1}^{n}\left(\frac{1}{n}\right)^2 \sigma^2 = \frac{\sigma^2}{n}.$$

See also ESTIMATOR, STANDARD DEVIATION, VARIANCE

Standard Map

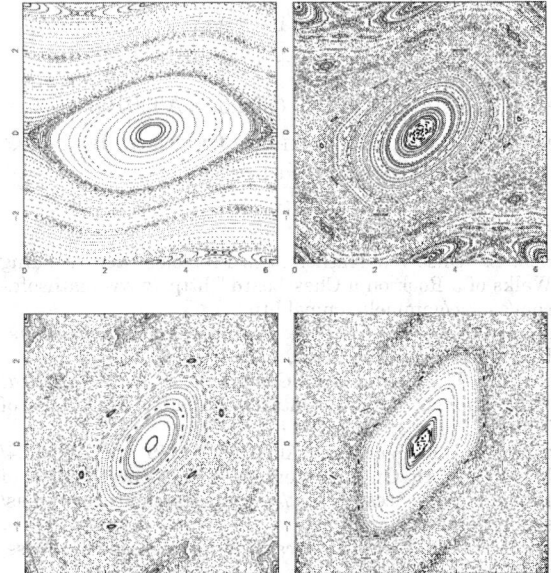

A 2-D MAP also called the Taylor-Greene-Chirikov map in some of the older literature and defined by

$$I_{n+1} = I_n + K \sin \theta_n \qquad (1)$$

$$\theta_{n+1} = \theta_n = I_{n+1} + \theta_n + K \sin \theta_n, \qquad (2)$$

where I and θ are computed mod 2π and K is a POSITIVE constant.

The standard map can be implemented in *Mathematica* as

```
StandardMap[k_, its_:100, cnt_:50] := Module[{},
  f[{t_, i_}] := Mod[{i + t + k Sin[t], i + k Sin[t]}, 2Pi];
  Graphics[{
    PointSize[.01],
    Table[
      Point /@ NestList[f, #, its] & [
```

```
      Table[Random[Real, {0, 2Pi}], {2}]],
{cnt}]
  },
  AspectRatio->Automatic] ]
```

An analytic estimate of the width of the CHAOTIC zone (Chirikov 1979) finds

$$\delta I = B e^{-A K^{-1/2}}. \tag{3}$$

Numerical experiments give $A \approx 5.26$ and $B \approx 240$. The value of K at which global CHAOS occurs has been bounded by various authors. GREENE'S METHOD is the most accurate method so far devised.

Author	Bound	Fraction	Decimal
Hermann	>	$\frac{1}{34}$	0.029411764
Italians	>	-	0.65
Greene	$\approx$	-	0.971635406
MacKay and Pearson	<	$\frac{63}{64}$	0.984375000
Mather	<	$\frac{4}{3}$	1.333333333

FIXED POINTS are found by requiring that

$$I_{n+1} = I_n \tag{4}$$

$$\theta_{n+1} = \theta_n. \tag{5}$$

The first gives $K \sin \theta_n = 0$, so $\sin \theta_n = 0$ and

$$\theta_n = 0, \ \pi. \tag{6}$$

The second requirement gives

$$I_n + K \sin \theta_n = I_n = 0. \tag{7}$$

The FIXED POINTS are therefore $(I, \theta) = (0, 0)$ and $(0, \pi)$. In order to perform a LINEAR STABILITY analysis, take differentials of the variables

$$dI_{n+1} = dI_n + K \cos \theta_n \, d\theta_n \tag{8}$$

$$d\theta_{n+1} = dI_n + (1 + K \cos \theta_n) \, d\theta_n. \tag{9}$$

In MATRIX form,

$$\begin{bmatrix} \delta I_{n+1} \\ \delta \theta_{n+1} \end{bmatrix} = \begin{bmatrix} 1 & K \cos \theta_n \\ 1 & 1 + K \cos \theta_n \end{bmatrix} \begin{bmatrix} \delta I_n \\ \delta \theta_n \end{bmatrix}. \tag{10}$$

The EIGENVALUES are found by solving the CHARACTERISTIC EQUATION

$$\begin{bmatrix} 1 - \lambda & K \cos \theta_n \\ 1 & 1 + K \cos \theta_n - \lambda \end{bmatrix} = 0. \tag{11}$$

so

$$\lambda^2 - \lambda(K \cos \theta_n + 2) + 1 = 0 \tag{12}$$

$$\lambda_\pm = \frac{1}{2}\left[K \cos \theta_n + 2 \pm \sqrt{(K \cos \theta_n + 2)^2 - 4} \right]. \tag{13}$$

For the FIXED POINT $(0, \pi)$,

$$\lambda_\pm^{(0, \pi)} = \frac{1}{2}\left[2 - K \pm \sqrt{(2 - K)^2 - 4} \right]$$

$$= \frac{1}{2}\left(2 - K \pm \sqrt{K^2 - 4K} \right). \tag{14}$$

The FIXED POINT will be stable if $\left| \Re\left(\lambda^{(0, \pi)} \right) \right| < 2$. Here, that means

$$\frac{1}{2}|2 - K| < 1 \tag{15}$$

$$|2 - K| < 2 \tag{16}$$

$$-2 < 2 - K < 2 \tag{17}$$

$$-4 < -K < 0 \tag{18}$$

so $K \in [0, 4)$. For the FIXED POINT $(0, 0)$, the EIGENVALUES are

$$\lambda_\pm^{(0, 0)} = \frac{1}{2}\left[2 + K \pm \sqrt{(K + 2)^2 - 4} \right]$$

$$\frac{1}{2}\left(2 + K \pm \sqrt{K^2 + 4K} \right). \tag{19}$$

If the map is unstable for the larger EIGENVALUE, it is unstable. Therefore, examine $\lambda_\pm^{(0, 0)}$. We have

$$\frac{1}{2}\left| 2 + K + \sqrt{K^2 + 4K} \right| < 1. \tag{20}$$

so

$$-2 < 2 + K + \sqrt{K^2 + 4K} < 2 \tag{21}$$

$$-4 - K < \sqrt{K^2 + 4K} < -K. \tag{22}$$

But $K > 0$, so the second part of the inequality cannot be true. Therefore, the map is unstable at the FIXED POINT $(0, 0)$.

See also HÉNON-HEILES EQUATION

References

Chirikov, B. V. "A Universal Instability of Many-Dimensional Oscillator Systems." *Phys. Rep.* **52**, 264–79, 1979.

Rasband, S. N. "The Standard Map." §8.5 in *Chaotic Dynamics of Nonlinear Systems*. New York: Wiley, pp. 11 and 178–79, 1990.

Tabor, M. "The Hénon-Heiles Hamiltonian." §4.2.r in *Chaos and Integrability in Nonlinear Dynamics: An Introduction*. New York: Wiley, pp. 134–35, 1989.

Standard Normal Distribution

A NORMAL DISTRIBUTION with zero MEAN ($\mu = 0$) and unity STANDARD DEVIATION ($\sigma^2 = 1$), given by

$$P(x)\, dx = \frac{1}{\sqrt{2\pi}}\, e^{-z^2/2}\, dz.$$

See also NORMAL DISTRIBUTION, TETRACHORIC FUNCTION

Standard Space

A SPACE which is ISOMORPHIC to a BOREL SUBSET B of a POLISH SPACE equipped with its SIGMA ALGEBRA of BOREL SETS.

See also BOREL SET, POLISH SPACE, SIGMA ALGEBRA

Standard Tableau

YOUNG TABLEAU

Standard Tori

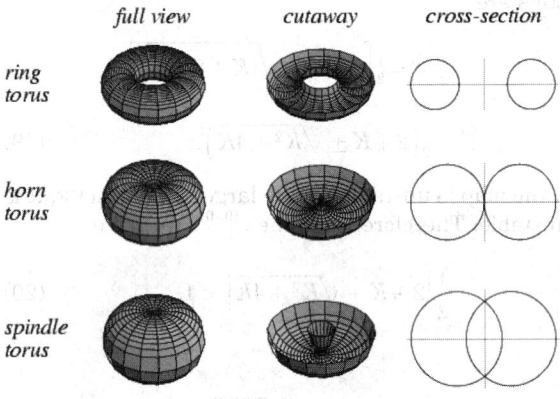

	full view	*cutaway*	*cross-section*
ring torus			
horn torus			
spindle torus			

One of the three classes of TORI illustrated above and given by the PARAMETRIC EQUATIONS

$$x = (c + a\cos v)\cos u \qquad (1)$$
$$y = (c + a\cos v)\sin u \qquad (2)$$
$$z = a\sin v. \qquad (3)$$

The three different classes of standard tori arise from the three possible relative sizes of a and c. $c > a$ corresponds to the RING TORUS shown above, $c = a$ corresponds to a HORN TORUS which touches itself at the point $(0, 0, 0)$, and $c < a$ corresponds to a self-intersecting SPINDLE TORUS (Pinkall 1986). If no specification is made, "torus" is taken to mean RING TORUS.
The standard tori and their inversions are CYCLIDES.

See also APPLE, CYCLIDE, HORN TORUS, LEMON, RING TORUS, SPINDLE TORUS, TORUS

References
Pinkall, U. "Cyclides of Dupin." §3.3 in *Mathematical Models from the Collections of Universities and Museums* (Ed. G. Fischer). Braunschweig, Germany: Vieweg, pp. 28–0, 1986.

Standard Unit

References
Kenney, J. F. and Keeping, E. S. "Standard Units." §7.7 in *Mathematics of Statistics, Pt. 1, 3rd ed.* Princeton, NJ: Van Nostrand, pp. 96–8, 1962.

Standardized Moment

Defined for samples x_i, $i = 1, \ldots, N$ by

$$\alpha_r \equiv \frac{1}{N}\sum_{i=1}^{N} z_i^r = \frac{\mu_r}{\sigma^r}. \qquad (1)$$

where

$$z_i \equiv \frac{x_i - \bar{x}}{s_x}. \qquad (2)$$

The first few are

$$\alpha_1 = 0 \qquad (3)$$
$$\alpha_2 = 1 \qquad (4)$$
$$\alpha_3 = \frac{\mu_3}{s^3} \qquad (5)$$
$$\alpha_4 = \frac{\mu_4}{s^4}. \qquad (6)$$

See also KURTOSIS, MOMENT, SKEWNESS

References
Kenney, J. F. and Keeping, E. S. "Moments in Standard Units." §7.8 in *Mathematics of Statistics, Pt. 1, 3rd ed.* Princeton, NJ: Van Nostrand, pp. 98–9, 1962.

Standardized Score

Z-SCORE

Stanley's Identity

$$\sum_{k=-\infty}^{\infty} \binom{a}{m-k}\binom{b}{n-k}\binom{a+b+k}{k}$$
$$= \binom{a+n}{m}\binom{b+m}{n}.$$

See also BINOMIAL SUMS

References
Koepf, W. *Hypergeometric Summation: An Algorithmic Approach to Summation and Special Function Identities.* Braunschweig, Germany: Vieweg, p. 41, 1998.
Strehl, V. "Binomial Identities--Combinatorial and Algorithmic Aspects." *Discrete Math.* **136**, 309–46, 1994.

Stanley's Theorem

The total number of 1s that occur among all unordered PARTITIONS of a POSITIVE INTEGER is equal to the sum of the numbers of distinct members of those PARTITIONS. For example, the partitions of 5 are $\{5\}$, $\{1, 1\}$, $\{3, 2\}$, $\{3, 1, 1\}$, $\{2, 2, 1\}$, $\{2, 1, 1, 1\}$, $\{1, 1, 1, 1, 1\}$. There are a total of $0 + 1 + 0 + 2 + 1 + 3 + 5 = 12$ 1s in this list, which is equal to the sums of the numbers of unique terms in each partition: $1 + 2 + 2 + 2 + 2 + 2 + 1 = 12$.

The numbers of 1s occurring in all partitions of $n = 1$, 2, 3, ... are 1, 2, 4, 7, 12, 19, 30, 45, 67, ... (Sloane's A000070).

See also ELDER'S THEOREM, PARTITION

References

Honsberger, R. *Mathematical Gems III*. Washington, DC: Math. Assoc. Amer., pp. 6–, 1985.

Sloane, N. J. A. Sequences A000070/M1054 in "An On-Line Version of the Encyclopedia of Integer Sequences." http://www.research.att.com/~njas/sequences/eisonline.html.

Stanley-Wilf Conjecture

Stanley and Wilf conjectured (Bona 1997, Arratia 1999), that for every PERMUTATION PATTERN σ, there is a constant $c(\sigma) < \infty$ such that for all n,

$$F(n, \ \sigma) \le [c(\sigma)]^n. \tag{1}$$

A related conjecture stated that for every σ, the limit

$$\lim_{n \to \infty} [F(n, \ \sigma)]^{1/n} \tag{2}$$

exists and is finite. Arratia (1999) showed that these two conjectures are equivalent.

See also PERMUTATION PATTERN

References

Alon, N. and Friedgut, E. "On the Number of Permutations Avoiding a Given Pattern." To appear in *J. Combin. Th. Ser. A*.

Arratia, R. "On the Stanley-Wilf Conjecture for the Number of Permutations Avoiding a Given Pattern." *Electronic J. Combinatorics* **6**, No. 1, N1, 1–, 1999. http://www.combinatorics.org/Volume_6/v6i1toc.html.

Bona, M. "Exact and Asymptotic Enumeration of Permutations with Subsequence Conditions." Ph.D. thesis. Cambridge, MA: MIT, 1997.

Bona, M. "The Solution of a Conjecture of Stanley and Wilf for All Layered Patterns." *J. Combin. Th. Ser. A* **85**, 96–04, 1999.

Wilf, H. "On Crossing Numbers, and Some Unsolved Problems." In *Combinatorics, Geometry, and Probability: A Tribute to Paul Erdos. Papers from the Conference in Honor of Erdos' 80th Birthday Held at Trinity College, Cambridge, March 1993* (Ed. B. Bollobás and A. Thomason). Cambridge, England: Cambridge University Press, pp. 557–62, 1997.

Star

The word "star" is used to voice an asterisk when appearing in a mathematical expression. For example, a^* is voiced "a-star". The "star" is used to denote the ADJOINT a^*, or sometimes the COMPLEX CONJUGATE.

In common usage, a star is a STAR POLYGON or STAR FIGURE (i.e., regular convex polygon or polygon compound) such as the PENTAGRAM or HEXAGRAM

In formal geometry, a star is a set of $2n$ VECTORS $\pm a_1$, ..., $\pm a_n$ which form a fixed center in EUCLIDEAN 3-SPACE.

In ALGEBRAIC TOPOLOGY, if v is a vertex of a SIMPLICIAL COMPLEX K, then the star of v in K, denoted St v or St(v, K), is the union of the interiors of those SIMPLICES of K that have v as a vertex (Munkres 1993, p. 11).

See also CLOSED STAR, CROSS, EUTACTIC STAR, HEXAGRAM, LINK (SIMPLICIAL COMPLEX), PENTAGRAM, STAR FIGURE, STAR POLYGON

References

Munkres, J. R. *Elements of Algebraic Topology*. Perseus Press, 1993.

Star (Fixed Point)

A FIXED POINT which has one zero EIGENVECTOR.

See also STABLE STAR, UNSTABLE STAR

Star Figure

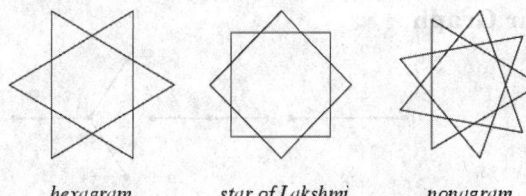

hexagram   *star of Lakshmi*   *nonagram*

A STAR POLYGON-like figure $\{p/q\}$ for which p and q are not RELATIVELY PRIME. Examples include the HEXAGRAM $\{6/3\}$, STAR OF LAKSHMI $\{8/2\}$, and NONAGRAM $\{9/3\}$.

See also HEXAGRAM, NONAGRAM, STAR OF LAKSHMI, STAR POLYGON

Star Fractal

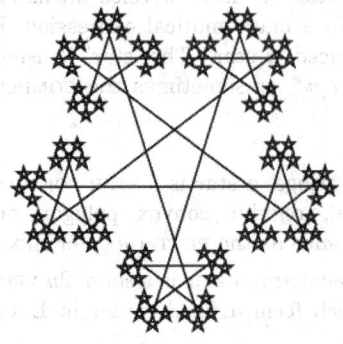

A FRACTAL composed of repeated copies of a PENTA-GRAM or other polygon.

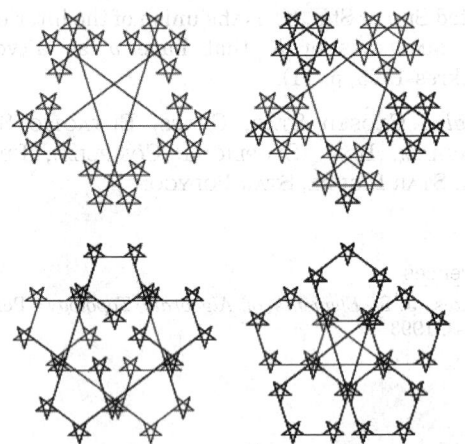

The above figure shows a generalization to different offsets from the center.

References

Lauwerier, H. *Fractals: Endlessly Repeated Geometric Figures.* Princeton, NJ: Princeton University Press, pp. 72–7, 1991.

Weisstein, E. W. "Fractals." MATHEMATICA NOTEBOOK FRACTAL.M.

Star Graph

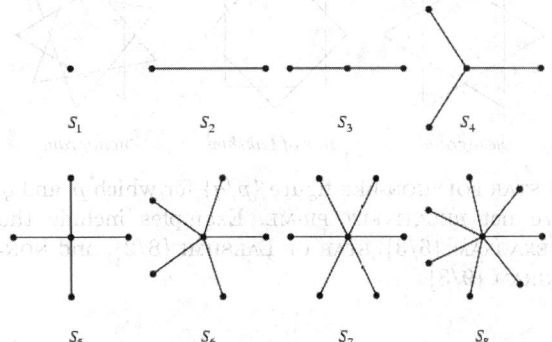

The n-star graph is a TREE on $n + 1$ nodes with one node having VERTEX DEGREE n and the others having

VERTEX DEGREE 1. Star graphs S_n are always GRACEFUL. Star graphs can be constructed using Star[n] in the *Mathematica* add-on package DiscreteMath`-Combinatorica` (which can be loaded with the command < < DiscreteMath`).

The COMPLETE BIPARTITE GRAPH $K_{1, n-1}$ is the STAR GRAPH S_n (Skiena 1990, p. 146). The CHROMATIC POLYNOMIAL of S_n is given by

$$\pi_{s_n}(z) = z(z - 1)^{n-1}.$$

and the CHROMATIC NUMBER is 1 for $n = 1$, and $\chi(S_n) = 2$ otherwise.

See also CAYLEY TREE, TREE

References

Skiena, S. "Cycles, Stars, and Wheels." §4.2.3 in *Implementing Discrete Mathematics: Combinatorics and Graph Theory with Mathematica.* Reading, MA: Addison-Wesley, pp. 83 and 144–47, 1990.

Star Number

The number of cells in a generalized Chinese checkers board (or "centered" HEXAGRAM).

$$S_n = 6n(n + 1) + 1 = S_{n-1} + 12(n - 1). \tag{1}$$

The first few are 1, 13, 37, 73, 121, ... (Sloane's A003154). Every star number has DIGITAL ROOT 1 or 4, and the final digits must be one of: 01, 21, 41, 61, 81, 13, 33, 53, 73, 93, or 37.

The first TRIANGULAR star numbers are 1, 253, 49141, 9533161, ... (Sloane's A006060), and can be computed using

$$TS_n = \frac{3\left[\left(7 + 4\sqrt{3}\right)^{2n-1} + \left(7 - 4\sqrt{3}\right)^{2n-1}\right] - 10}{32} \tag{2}$$

$$= 194 TS_{n-1} + 60 - TS_{n-2}. \tag{3}$$

The first few SQUARE star numbers are 1, 121, 11881, 1164241, 114083761, ... (Sloane's A006061). SQUARE star numbers are obtained by solving the DIOPHANTINE EQUATION

$$2x^2 + 1 = 3y^2 \tag{4}$$

and can be computed using

$$SS_n =$$
$$\frac{\left[\left(5 + 2\sqrt{6}\right)^n\left(\sqrt{6} - 2\right) - \left(5 - 2\sqrt{6}\right)^n\left(\sqrt{6} + 2\right)\right]^2}{4}. \tag{5}$$

See also HEX NUMBER, SQUARE NUMBER, TRIANGULAR NUMBER

References

Gardner, M. "Hexes and Stars." Ch. 2 in *Time Travel and Other Mathematical Bewilderments*. New York: W. H. Freeman, pp. 15–4, 1988.

Hindin, H. "Stars, Hexes, Triangular Numbers, and Pythagorean Triples." *J. Recr. Math.* **16**, 191–93, 1983–984.

Sloane, N. J. A. Sequences A003154/M4893, A006060/M5425, and A006061/M5385 in "An On-Line Version of the Encyclopedia of Integer Sequences." http://www.research.att.com/~njas/sequences/eisonline.html.

Star of David

HEXAGRAM

Star of David Theorem

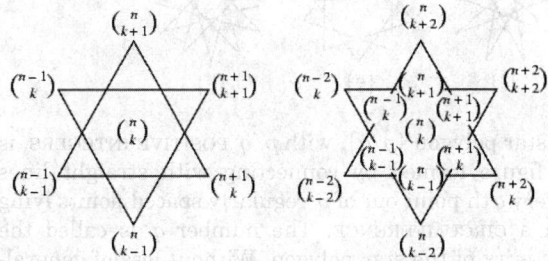

As originally stated by Gould (1972),

$$\mathrm{GCD}\left\{\binom{n-1}{k}, \binom{n}{k-1}, \binom{n+1}{k+1}\right\}$$
$$= \mathrm{GCD}\left\{\binom{n-1}{k-1}, \binom{n}{k+1}, \binom{n+1}{k}\right\}, \quad (1)$$

where GCD is the GREATEST COMMON DIVISOR and $\binom{n}{k}$ is a BINOMIAL COEFFICIENT. This was subsequently extended by D. Singmaster to

$$\mathrm{GCD}\left\{\binom{n-1}{k}, \binom{n}{k-1}, \binom{n+1}{k+1}\right\}$$
$$= \mathrm{GCD}\left\{\binom{n-1}{k-1}, \binom{n}{k+1}, \binom{n+1}{k}\right\}$$
$$= \mathrm{GCD}\left\{\binom{n-1}{k-2}, \binom{n-1}{k-1}, \binom{n-1}{k}, \binom{n-1}{k+1}\right\}$$

$$(2)$$

(Sato 1975), and generalized by Sato (1975) to

$$\mathrm{GCD}\left\{\binom{n}{k+2}, \binom{n-1}{k}, \binom{n-2}{k-2}, \binom{n}{k-1}, \right.$$
$$\left. \binom{n+2}{k} \binom{n+1}{k+1}\right\}$$
$$= \mathrm{GCD}\left\{\binom{n-2}{k}, \binom{n-1}{k-1}, \binom{n}{k-2}, \binom{n+1}{k}, \right.$$
$$\left. \binom{n+2}{k+2} \binom{n}{k+1}\right\}$$
$$(3)$$

An even larger generalization was obtained by Hitotumatu and Sato (1975), who defined

$$M_p = \left\{\binom{n-p+1}{k-2p+j+1} \middle| j = 1, 2, \ldots, 3p-2\right\},$$
$$(p \ge 1) \qquad (4)$$

$$A_p = \left\{\binom{n-p+j}{k+p-1} \middle| j = 1, 2, \ldots, 3p-2\right\} \quad (p \ge 1) \quad (5)$$

$$R_p = \left\{\binom{n-p+j}{k-2p+j-1} \middle| j = 1, 2, \ldots, 3p-2\right\}$$
$$(p \ge 1) \qquad (6)$$

$$\Delta_p = \left\{\binom{n-p+2t+1}{k-p+t+1}, \binom{n+p-t-1}{k+t}, \right.$$
$$\left. \binom{n-t}{k+p-2t-1} \middle| t = 1, 2, \ldots, p-1\right\} \quad (p \ge 2) \quad (7)$$

$$\nabla_p = \left\{\binom{n-t}{k-p+t+1}, \binom{n-p+2t+1}{k+t}, \right.$$
$$\left. \binom{n+p-t-1}{k+p-2t-1} \middle| t = 1, 2, \ldots, p-1\right\} \quad (p \ge 2) \quad (8)$$

$$U_p = \bigcup_{r=1}^{p} M_r \qquad (9)$$

$$V_p = \bigcup_{r=1}^{p} A_r \qquad (10)$$

$$W_p = \bigcup_{r=1}^{p} R_r \qquad (11)$$

$$D_p = \bigcup_{r=1}^{p} \Delta_r \qquad (12)$$

$$N_p = \bigcup_{r=1}^{p} \nabla_r \qquad (13)$$

$$B_p = M_p \cup A_p \cup R_p \qquad (14)$$

$$S_p = \bigcup_{r=1}^{p} B_r \qquad (15)$$

with

$$\Delta_1 = \nabla_1 = \binom{n}{k}. \qquad (16)$$

and showed that each of the twelve BINOMIAL COEFFICIENTS M_p, A_p, R_p, Δ_p, ∇_p, U_p, V_p, W_p, ∇_p, N_p, B_p, and S_p has equal GREATEST COMMON DIVISOR.

References

Ando, S. and Sato, D. "Translatable and Rotatable Configurations which Give Equal Product, Equal GCD and Equal LCM Properties Simultaneously." In *Applications of Fibonacci Numbers, Vol. 3: Proceedings of the Third International Conference on Fibonacci Numbers and their Applications held at the University of Pisa, Pisa, July 25–9, 1988* (Ed. G. E. Bergum, A. N. Philippou and A. F. Horadam). Dordrecht, Netherlands: Kluwer, pp. 15–6, 1990.

Ando, S. and Sato, D. "A GCD Property on Pascal's Pyramid and the Corresponding LCM Property of the Modified Pascal Pyramid." In *Applications of Fibonacci Numbers,*

Vol. 3: *Proceedings of the Third International Conference on Fibonacci Numbers and their Applications held at the University of Pisa, Pisa, July 25–9, 1988* (Ed. G. E. Bergum, A. N. Philippou and A. F. Horadam). Dordrecht, Netherlands: Kluwer, pp. 7–4, 1990.

Ando, S. and Sato, D. "On the Proof of GCD and LCM Equalities Concerning the Generalized Binomial and Multinomial Coefficients." In *Applications of Fibonacci numbers, Vol. 4: Proceedings of the Fourth International Conference on Fibonacci Numbers and their Applications held at Wake Forest University, Winston-Salem, North Carolina, July 30-August 3, 1990 (Winston-Salem, NC, 1990)* (Ed. G. E. Bergum, A. N. Philippou and A. F. Horadam). Dordrecht, Netherlands: Kluwer, 9–6, 1991.

Ando, S. and Sato, D. "Multiple Color Version of the Star of David Theorems on Pascal's Triangle and Related Arrays of Numbers." In *Applications of Fibonacci Numbers, Vol. 6: Proceedings of the Sixth International Research Conference on Fibonacci Numbers and their Applications held at Washington State University, Pullman, Washington, July 18–2, 1994* (Ed. G. E. Bergum, A. N. Philippou, and A. F. Horadam). Dordrecht, Netherlands: Kluwer, pp. 31–5, 1996.

Gould, H. W. *Not. Amer. Math. Soc.* **19**, A-685, 1972.

Hitotumatu, S. and Sato, D. "Expansion of the Star of David Theorem." *Abstracts Amer. Math. Soc.*, p. A-377, 1975.

Hitotumatu, S. and Sato, D. "Star of David Theorem. I." *Fib. Quart.* **13**, 70, 1975.

Sato, D. "Expansion of the Star of David Theorem of H. W. Gould and David Singmaster." *Abstracts Amer. Math. Soc.*, p. A-377, 1975.

Star of Goliath

NONAGRAM

Star of Lakshmi

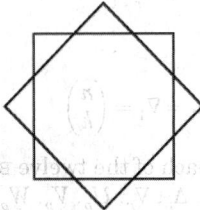

The STAR FIGURE {8/2}, which is used by Hindus to symbolize *Ashtalakshmi,* the eight forms of wealth. This symbol appears prominently in the Lugash national museum portrayed in the fictional film *Return of the Pink Panther.*

See also DISSECTION, HEXAGRAM, PENTAGRAM, STAR FIGURE, STAR POLYGON

References

Savio, D. Y. and Suryanaroyan, E. R. "Chebyshev Polynomials and Regular Polygons." *Amer. Math. Monthly* **100**, 657–61, 1993.

Star Polygon

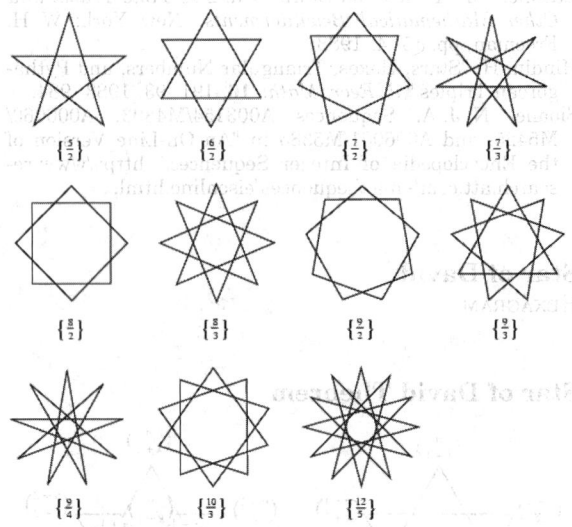

$\{\frac{5}{2}\}$ $\{\frac{6}{2}\}$ $\{\frac{7}{2}\}$ $\{\frac{7}{3}\}$

$\{\frac{8}{2}\}$ $\{\frac{8}{3}\}$ $\{\frac{9}{2}\}$ $\{\frac{9}{3}\}$

$\{\frac{9}{4}\}$ $\{\frac{10}{3}\}$ $\{\frac{12}{5}\}$

A star polygon $\{p/q\}$, with p, q POSITIVE INTEGERS, is a figure formed by connecting with straight lines every qth point out of p regularly spaced points lying on a CIRCUMFERENCE. The number q is called the DENSITY of the star polygon. Without loss of generality, take $q < p/2$. The star polygons were first systematically studied by Thomas Bradwardine.

The usual definition (Coxeter 1969) requires p and q to be RELATIVELY PRIME. However, the star polygon can also be generalized to the STAR FIGURE (or "improper" star polygon) when p and q share a common divisor (Savio and Suryanaroyan 1993). For such a figure, if all points are not connected after the first pass, i.e., if $(p, q) \neq 1$, then start with the first unconnected point and repeat the procedure. Repeat until all points are connected. For $(p, q) \neq 1$, the $\{p/q\}$ symbol can be factored as

$$\left\{\frac{p}{q}\right\} = n\left\{\frac{p'}{q'}\right\}, \tag{1}$$

where

$$p' = \frac{p}{n} \tag{2}$$

$$q' = \frac{q}{n}, \tag{3}$$

to give n $\{p'/q'\}$ figures, each rotated by $2\pi/p$ radians, or $360°/p$.

If $q = 1$, a REGULAR POLYGON $\{p\}$ is obtained. Special cases of $\{p/q\}$ include $\{5/2\}$ (the PENTAGRAM), $\{6/2\}$ (the HEXAGRAM, or STAR OF DAVID), $\{8/2\}$ (the STAR OF LAKSHMI), $\{8/3\}$ (the OCTAGRAM), $\{10/3\}$ (the DECAGRAM), and $\{12/5\}$ (the DODECAGRAM).

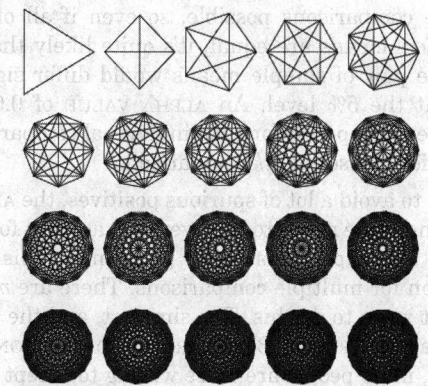

Superposing all distinct star polygons $\{p/q\}$ for a given p gives beautiful patterns such as those illustrated above. These figures can also be obtained by wrapping thread around p nails spaced equally around the circumference of a circle (Steinhaus 1983, pp. 259–60).

See also DECAGRAM, HEXAGRAM, NONAGRAM, OCTAGRAM, PENTAGRAM, REGULAR POLYGON, STAR OF LAKSHMI, STELLATED POLYHEDRON

References

Coxeter, H. S. M. *Regular Polytopes, 3rd ed.* New York: Dover, pp. 93–4, 1973.
Coxeter, H. S. M. "Star Polygons." §2.8 in *Introduction to Geometry, 2nd ed.* New York: Wiley, pp. 36–8, 1969.
Fejes Tóth, L. *Regular Figures.* Oxford, England: Pergamon Press, pp. 102–03, 1964.
Frederickson, G. "Stardom." Ch. 16 in *Dissections: Plane and Fancy.* New York: Cambridge University Press, pp. 172–86, 1997.
Savio, D. Y. and Suryanaroyan, E. R. "Chebyshev Polynomials and Regular Polygons." *Amer. Math. Monthly* **100**, 657–61, 1993.
Steinhaus, H. *Mathematical Snapshots, 3rd ed.* New York: Dover, pp. 211 and 259–60, 1999.
Williams, R. *The Geometrical Foundation of Natural Structure: A Source Book of Design.* New York: Dover, p. 32, 1979.

Star Polyhedron

KEPLER-POINSOT SOLID

Starr Rose

$a = 8, b = 16, c = 16$ $a = 6, b = 18, c = 18$

See also MAURER ROSE

References

Wagon, S. "Variations of Circular Motion." §4.5 in *Mathematica in Action.* New York: W. H. Freeman, pp. 137–40, 1991.

State Space

The MEASURABLE SPACE $(S', \mathbb{S}')$ into which a RANDOM VARIABLE from a PROBABILITY SPACE is a measurable function.

See also PROBABILITY SPACE, RANDOM VARIABLE

Stationary Point

A point x_0 at which the DERIVATIVE of a FUNCTION $f(x)$ vanishes,

$$f'(x_0) = 0.$$

A stationary point may be a MINIMUM, MAXIMUM, or INFLECTION POINT.

See also CRITICAL POINT, DERIVATIVE, EXTREMUM, FIRST DERIVATIVE TEST, INFLECTION POINT, MAXIMUM, MINIMUM, SECOND DERIVATIVE TEST

Stationary Tangent

INFLECTION POINT

Stationary Value

The value at a STATIONARY POINT.

Statistic

A quantity (such as a MEDIAN, QUARTILE DEVIATION, etc.), which is calculated from observed data.

See also ANDERSON-DARLING STATISTIC, H-STATISTIC, K-STATISTIC, KUIPER STATISTIC, VARIATE

References

Kenney, J. F. and Keeping, E. S. *Mathematics of Statistics, Pt. 1, 3rd ed.* Princeton, NJ: Van Nostrand, p. 37, 1962.

Statistical Distribution

The distribution of a variable is a description of the relative numbers of times each possible outcome will occur in a number of trials. The function describing the distribution is called the PROBABILITY FUNCTION, and the function describing the cumulative probability that a given value *or any value smaller than it* will occur is called the DISTRIBUTION FUNCTION.

Formally, a distribution can be defined as a normalized MEASURE, and the distribution of a RANDOM VARIABLE x is the MEASURE P_x on $\mathbb{S}'$ defined by setting

$$P_x(A') = P\{s \in S : x(s) \in A'\},$$

where $(S, \mathbb{S}, P)$ is a PROBABILITY SPACE, $(S, \mathbb{S})$ is a MEASURABLE SPACE, and P a MEASURE on $\mathbb{S}$ with $P(S) = 1$. If the MEASURE is a RADON MEASURE (which is usually the case), then the statistical distribution is a DISTRIBUTION in the sense of a generalized function.

See also CONTINUOUS DISTRIBUTION, DISCRETE DISTRIBUTION, DISTRIBUTION FUNCTION, DISTRIBUTION (GENERALIZED FUNCTION), MEASURABLE SPACE, MEASURE, PROBABILITY, PROBABILITY DENSITY FUNCTION, RANDOM VARIABLE, STATISTICS

References

Doob, J. L. "The Development of Rigor in Mathematical Probability (1900–950)." *Amer. Math. Monthly* **103**, 586–95, 1996.
Evans, M.; Hastings, N.; and Peacock, B. *Statistical Distributions, 3rd ed.* New York: Wiley, 2000.

Statistical Index
INDEX NUMBER

Statistical Test

A test used to determine the statistical SIGNIFICANCE of an observation. Two main types of error can occur:

1. A TYPE I ERROR occurs when a false negative result is obtained in terms of the NULL HYPOTHESIS by obtaining a *false positive measurement*.
2. A TYPE II ERROR occurs when a false positive result is obtained in terms of the NULL HYPOTHESIS by obtaining a *false negative measurement*.

The probability that a statistical test will be positive for a true statistic is sometimes called the test's SENSITIVITY, and the probability that a test will be negative for a negative statistic is sometimes called the SPECIFICITY. The following table summarizes the names given to the various combinations of the actual state of affairs and observed test results.

result	name
true positive result	SENSITIVITY
false negative result	1-SENSITIVITY
true negative result	SPECIFICITY
false positive result	1-SPECIFICITY

Multiple-comparison corrections to statistical tests are used when several statistical tests are being performed simultaneously. For example, let's suppose you were measuring leg length in eight different lizard species and wanted to see whether the MEANS of any pair were different. Now, there are $8!/2!6! = 28$ pairwise comparisons possible, so even if all of the *population* means are equal, it's quite likely that at least one pair of sample means would differ significantly at the 5% level. An ALPHA VALUE of 0.05 is therefore appropriate for each individual comparison, but not for the set of *all* comparisons.

In order to avoid a lot of spurious positives, the ALPHA VALUE therefore needs to be lowered to account for the number of comparisons being performed. This is a correction for multiple comparisons. There are *many* different ways to do this. The simplest, and the most conservative, is the BONFERRONI CORRECTION. In practice, more people are more willing to accept false positives (false rejection of NULL HYPOTHESIS) than false negatives (false acceptance of NULL HYPOTHESIS), so less conservative comparisons are usually used.

See also ANOVA, BONFERRONI CORRECTION, CHI-SQUARED TEST, FISHER'S EXACT TEST, FISHER SIGN TEST, KOLMOGOROV-SMIRNOV TEST, LIKELIHOOD RATIO, LOG LIKELIHOOD PROCEDURE, MANOVA, NEGATIVE LIKELIHOOD RATIO, PAIRED T-TEST, PARAMETRIC TEST, PREDICTIVE VALUE, SENSITIVITY, SIGNIFICANCE TEST, SPECIFICITY, TYPE I ERROR, TYPE II ERROR, WILCOXON RANK SUM TEST, WILCOXON SIGNED RANK TEST

Statistics

The mathematical study of the LIKELIHOOD and PROBABILITY of events occurring based on known information and inferred by taking a limited number of samples. Statistics plays an extremely important role in many aspects of economics and science, allowing educated guesses to be made with a minimum of expensive or difficult-to-obtain data.

See also BOX-AND-WHISKER PLOT, BUFFON-LAPLACE NEEDLE PROBLEM, BUFFON'S NEEDLE PROBLEM, CHERNOFF FACE, COIN FLIPPING, DE MERE'S PROBLEM, DICE, GAMBLER'S RUIN, INDEX, LIKELIHOOD, MOVING AVERAGE, P-VALUE, POPULATION COMPARISON, POWER (STATISTICS), PROBABILITY, RESIDUAL VS. PREDICTOR PLOT, RUN, SHARING PROBLEM, STATISTICAL DISTRIBUTION, STATISTICAL TEST, TAIL PROBABILITY

References

Brown, K. S. "Probability." http://www.seanet.com/~ksbrown/iprobabi.htm.
Babu, G. and Feigelson, E. *Astrostatistics.* New York: Chapman & Hall, 1996.
Bernstein, S. and Bernstein, R. *Theory and Problems of Elements of Statistics I: Descriptive Statistics and Probability.* New York: McGraw-Hill, 1999.
Dixon, W. J. and Massey, F. J. *Introduction to Statistical Analysis, 4th ed.* New York: McGraw-Hill, 1983.
Feller, W. *An Introduction to Probability Theory and Its Applications, Vol. 1, 3rd ed.* New York: Wiley, 1968.
Feller, W. *An Introduction to Probability Theory and Its Applications, Vol. 2, 2nd ed.* New York: Wiley, 1968.

Fisher, N. I.; Lewis, T.; and Embleton, B. J. J. *Statistical Analysis of Spherical Data.* Cambridge, England: Cambridge University Press, 1987.

Fisher, R. A. and Prance, G. T. *The Design of Experiments, 9th ed. rev.* New York: Hafner, 1974.

Fisher, R. A. *Statistical Methods for Research Workers, 14th ed., rev. and enl.* Darien, CO: Hafner, 1970.

Goldberg, S. *Probability: An Introduction.* New York: Dover, 1986.

Gonick, L. and Smith, W. *The Cartoon Guide to Statistics.* New York: Harper Perennial, 1993.

Goulden, C. H. *Methods of Statistical Analysis, 2nd ed.* New York: Wiley, 1956.

Hoel, P. G.; Port, S. C.; and Stone, C. J. *Introduction to Statistical Theory.* New York: Houghton Mifflin, 1971.

Hogg, R. V. and Tanis, E. A. *Probability and Statistical Inference, 5th ed.* Englewood Cliffs, NJ: Prentice-Hall, 1996.

Keeping, E. S. *Introduction to Statistical Inference.* New York: Dover, 1995.

Kenney, J. F. and Keeping, E. S. *Mathematics of Statistics, Pt. 1, 3rd ed.* Princeton, NJ: Van Nostrand, 1962.

Kenney, J. F. and Keeping, E. S. *Mathematics of Statistics, Pt. 2, 2nd ed.* Princeton, NJ: Van Nostrand, 1951.

Kendall, M. G.; Stuart, A.; and Ord, J. K. *Kendall's Advanced Theory of Statistics, Vol. 1: Distribution Theory, 6th ed.* New York: Oxford University Press, 1987.

Kendall, M. G.; Stuart, A.; and Ord, J. K. *Kendall's Advanced Theory of Statistics, Vol. 2A: Classical Inference and Relationship, 6th ed.* New York: Oxford University Press, 1987.

Kendall, M. G.; Stuart, A.; and Ord, J. K. *Kendall's Advanced Theory of Statistics, Vol. 2B: Bayesian Inference.* New York: Oxford University Press, 1987.

Keynes, J. M. *A Treatise on Probability.* London: Macmillan, 1921.

Mises, R. von *Mathematical Theory of Probability and Statistics.* New York: Academic Press, 1964.

Mises, R. von *Probability, Statistics, and Truth, 2nd rev. English ed.* New York: Dover, 1981.

Mood, A. M. *Introduction to the Theory of Statistics.* New York: McGraw-Hill, 1950.

Mosteller, F. *Fifty Challenging Problems in Probability with Solutions.* New York: Dover, 1987.

Mosteller, F.; Rourke, R. E. K.; and Thomas, G. B. *Probability: A First Course, 2nd ed.* Reading, MA: Addison-Wesley, 1970.

Neyman, J. *First Course in Probability and Statistics.* New York: Holt, 1950.

Ostle, B. *Statistics in Research: Basic Concepts and Techniques for Research Workers, 4th ed.* Ames, IA: Iowa State University Press, 1988.

Papoulis, A. *Probability, Random Variables, and Stochastic Processes, 2nd ed.* New York: McGraw-Hill, 1984.

Press, W. H.; Flannery, B. P.; Teukolsky, S. A.; and Vetterling, W. T. "Statistical Description of Data." Ch. 14 in *Numerical Recipes in FORTRAN: The Art of Scientific Computing, 2nd ed.* Cambridge, England: Cambridge University Press, pp. 603–49, 1992.

Pugh, E. M. and Winslow, G. H. *The Analysis of Physical Measurements.* Reading, MA: Addison-Wesley, 1966.

Rényi, A. *Foundations of Probability.* San Francisco, CA: Holden-Day, 1970.

Robbins, H. and van Ryzin, J. *Introduction to Statistics.* Chicago, IL: Science Research Associates, 1975.

Ross, S. M. *A First Course in Probability, 5th ed.* Englewood Cliffs, NJ: Prentice-Hall, 1997.

Ross, S. M. *Introduction to Probability and Statistics for Engineers and Scientists.* New York: Wiley, 1987.

Ross, S. M. *Applied Probability Models with Optimization Applications.* New York: Dover, 1992.

Ross, S. M. *Introduction to Probability Models, 6th ed.* New York: Academic Press, 1997.

Snedecor, G. W. *Statistical Methods Applied to Experiments in Agriculture and Biology, 5th ed.* Ames, IA: State College Press, 1956.

Spiegel, M. R. and Stephens, L. J. *Theory and Problems of Statistics, 3rd ed.* New York: McGraw-Hill, 1998.

Tippett, L. H. C. *The Methods of Statistics: An Introduction Mainly for Experimentalists, 3rd rev. ed.* London: Williams and Norgate, 1941.

Todhunter, I. *A History of the Mathematical Theory of Probability from the Time of Pascal to that of Laplace.* New York: Chelsea, 1949.

Tukey, J. W. *Explanatory Data Analysis.* Reading, MA: Addison-Wesley, 1977.

Uspensky, J. V. *Introduction to Mathematical Probability.* New York: McGraw-Hill, 1937.

Weaver, W. *Lady Luck: The Theory of Probability.* New York: Dover, 1963.

Weisstein, E. W. "Books about Statistics." http://www.treasure-troves.com/books/Statistics.html.

Whittaker, E. T. and Robinson, G. *The Calculus of Observations: A Treatise on Numerical Mathematics, 4th ed.* New York: Dover, 1967.

Young, H. D. *Statistical Treatment of Experimental Data.* New York: McGraw-Hill, 1962.

Yule, G. U. and Kendall, M. G. *An Introduction to the Theory of Statistics, 14th ed., rev. and enl.* New York: Hafner, 1950.

Staudt-Clausen Theorem

VON STAUDT-CLAUSEN THEOREM

Steenrod Algebra

The Steenrod algebra has to do with the COHOMOLOGY operations in singular COHOMOLOGY with INTEGER mod 2 COEFFICIENTS. For every $n \in \mathbb{Z}$ and $i \in \{0, 1, 2, 3, \ldots\}$ there are natural transformations of FUNCTORS

$$Sq^i : H^n(\bullet; \mathbb{Z}_2) \to H^{n+i}(\bullet; \mathbb{Z}_2)$$

satisfying:

1. $Sq^i = 0$ for $i > n$.
2. $Sq^n(x) = x \smile x$ for all $x \in H^n(X, A; \mathbb{Z}_2)$ and all pairs (X, A).
3. $Sq^0 = id_{H^n(\bullet; \mathbb{Z}_2)}$.
4. The Sq^i maps commute with the coboundary maps in the long exact sequence of a pair. In other words,

$$Sq^i : H^*(\bullet; \mathbb{Z}_2) \to H^{*+i}(\bullet; \mathbb{Z}_2)$$

is a degree i transformation of cohomology theories.

5. (CARTAN RELATION)

$$Sq^i(x \smile y) = \sum_{j+k=i} Sq^j(x) \smile Sq^k(y).$$

6. (ADEM RELATIONS) For $i < 2j$,

$$Sq^i \circ Sq^j(x) = \sum_{k=0}^{\lfloor i \rfloor} \binom{j-k-1}{i-2k} Sq^{i+j-k} \circ Sq^k(x).$$

7. $Sq^i \circ \Sigma = \Sigma \circ Sq^i$ where Σ is the cohomology suspension isomorphism.

The existence of these cohomology operations endows the cohomology ring with the structure of a MODULE over the Steenrod algebra $\mathscr{A}$, defined to be $T\left(F_{\mathbb{Z}_2}\{Sq^i : i \in \{0, 1, 2, 3, \ldots\}\}\right)/R$, where $F_{\mathbb{Z}_2}(\bullet)$ is the free module functor that takes any set and sends it to the free $\mathbb{Z}_2$ module over that set. We think of $F_{\mathbb{Z}_2}\{Sq^i : i \in \{0, 1, 2, \ldots\}\}$ as being a graded $\mathbb{Z}_2$ module, where the i-th gradation is given by $\mathbb{Z}_2 \cdot Sq^i$. This makes the tensor algebra $T\left(F_{\mathbb{Z}_2}\{Sq^i : i \in \{0, 1, 2, 3, \ldots\}\}\right)$ into a GRADED ALGEBRA over $\mathbb{Z}_2$. R is the IDEAL generated by the elements $Sq^iSq^j + \sum_{k=0}^{\lfloor i \rfloor}\binom{j-k-1}{i-2k}Sq^{i+j-k}Sq^k$ and $1 + Sq^0$ for $0 < i < 2j$. This makes $\mathscr{A}$ into a graded $\mathbb{Z}_2$ algebra.

By the definition of the Steenrod algebra, for any SPACE (X, A), $H^*(X, A; \mathbb{Z}_2)$ is a MODULE over the Steenrod algebra $\mathscr{A}$, with multiplication induced by $Sq^i \cdot x \equiv Sq^i(x)$. With the above definitions, cohomology with COEFFICIENTS in the RING $\mathbb{Z}_2$, $H^*(\bullet; \mathbb{Z}_2)$ is a FUNCTOR from the category of pairs of TOPOLOGICAL SPACES to graded modules over $\mathscr{A}$.

See also ADEM RELATIONS, CARTAN RELATION, COHOMOLOGY, GRADED ALGEBRA, IDEAL, MODULE, TOPOLOGICAL SPACE

Steenrod-Eilenberg Axioms

EILENBERG-STEENROD AXIOMS

Steenrod's Realization Problem

When can homology classes be realized as the image of fundamental classes of MANIFOLDS? The answer is known, and singular BORDISM GROUPS provide insight into this problem.

See also BORDISM GROUP, MANIFOLD

Steepest Descent Method

An ALGORITHM for finding the nearest LOCAL MINIMUM of a function which presupposes that the GRADIENT of the function can be computed. The steepest descent method, also called the gradient descent method, starts at a point $\mathbf{P}_0$ and, as many times as needed, moves from $\mathbf{P}_i$ to $\mathbf{P}_{i+1}$ by minimizing along the line extending from $\mathbf{P}_i$ in the direction of $-\nabla f(\mathbf{P}_i)$, the local downhill GRADIENT.

This method has the severe drawback of requiring a great many iterations for functions which have long, narrow valley structures. In such cases, a CONJUGATE GRADIENT METHOD is preferable.

See also CONJUGATE GRADIENT METHOD, GRADIENT, LOCAL MINIMUM, MINIMUM

References

Arfken, G. "The Method of Steepest Descents." §7.4 in *Mathematical Methods for Physicists, 3rd ed.* Orlando, FL: Academic Press, pp. 428–36, 1985.
Menzel, D. (Ed.). *Fundamental Formulas of Physics, Vol. 2, 2nd ed.* New York: Dover, p. 80, 1960.
Morse, P. M. and Feshbach, H. "Asymptotic Series; Method of Steepest Descent." §4.6 in *Methods of Theoretical Physics, Part I.* New York: McGraw-Hill, pp. 434–43, 1953.
Press, W. H.; Flannery, B. P.; Teukolsky, S. A.; and Vetterling, W. T. *Numerical Recipes in FORTRAN: The Art of Scientific Computing, 2nd ed.* Cambridge, England: Cambridge University Press, p. 414, 1992.

Steffensen Sequence

A sequence

$$s_n^{(\lambda)}(x) = [h(t)]^\lambda s_n(x),$$

where $s_n(x)$ is a SHEFFER SEQUENCE, $h(t)$ is invertible, and λ ranges over the real numbers. If $s_n(x)$ is an associated SHEFFER SEQUENCE, then $s_n^{(\lambda)}$ is called a CROSS SEQUENCE. If $s_n(x) = x^n$, then

$$s_n^{(\lambda)}(x) = [h(t)]^\lambda x^n$$

is called an APPELL CROSS SEQUENCE.

An example is the LAGUERRE POLYNOMIAL.

See also APPELL CROSS SEQUENCE, CROSS SEQUENCE, SHEFFER SEQUENCE

References

Brown, J. W. "A Note on Generalized Appell Polynomials." *Amer. Math. Monthly* **75**, 1968.
Roman, S. "Cross Sequences and Steffensen Sequences." §5.3 in *The Umbral Calculus.* New York: Academic Press, pp. 140–43, 1984.
Rota, G.-C.; Kahaner, D.; and Odlyzko, A. "On the Foundations of Combinatorial Theory VIII: Finite Operator Calculus." *J. Math. Anal. Appl.* **42**, 684–60, 1973.

Steffensen's Inequality

Let $f(x)$ be a NONNEGATIVE and monotonic decreasing function in $[a, b]$ and $g(x)$ such that $0 \le g(x) \le 1$ in $[a, b]$, then

$$\int_{b-k}^b f(x)\,dx \le \int_a^b f(x)g(x)\,dx \le \int_a^{a+k} f(x)\,dx.$$

where

$$k = \int_a^b g(x)\,dx.$$

References

Gradshteyn, I. S. and Ryzhik, I. M. *Tables of Integrals, Series, and Products, 6th ed.* San Diego, CA: Academic Press, p. 1099, 2000.

Steffenson's Formula

$$f_p = f_0 + \tfrac{1}{2}p(p+1)\delta_{1/2} - \tfrac{1}{2}(p-1)p\delta_{-1/2}$$

$$+ (S_3 + S_4)\delta_{1/2}^3 + (S_3 - S_4)\delta_{-1/2}^3 + \ldots, \tag{1}$$

for $p \in \left[-\tfrac{1}{2}, \tfrac{1}{2}\right]$, where δ is the CENTRAL DIFFERENCE and

$$S_{2n+1} = \frac{1}{2}\binom{p+n}{2n+1} \tag{2}$$

$$S_{2n+2} = \frac{p}{2n+2}\binom{p+n}{2n+1} \tag{3}$$

$$S_{2n+1} - S_{2n+2} = \binom{p+n+1}{2n+2} \tag{4}$$

$$S_{2n+1} - S_{2n+2} = -\binom{p+n}{2n+2}, \tag{5}$$

where $\binom{n}{k}$ is a BINOMIAL COEFFICIENT.

See also CENTRAL DIFFERENCE, STIRLING'S FINITE DIFFERENCE FORMULA

References

Beyer, W. H. *CRC Standard Mathematical Tables, 28th ed.* Boca Raton, FL: CRC Press, p. 433, 1987.

Steinbach Screw

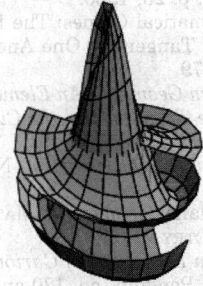

A SURFACE generated by the PARAMETRIC EQUATIONS

$$x(u, v) = u \cos v \tag{1}$$

$$y(u, v) = u \sin v \tag{2}$$

$$z(u, v) = v \cos u. \tag{3}$$

The above image uses $u \in [-4, 4]$ and $v \in [0, 6, 25]$.

The coefficients of the FIRST FUNDAMENTAL FORM are

$$E = 1 + v^2 \sin^2 u \tag{4}$$

$$F = -v \cos u \sin u \tag{5}$$

$$G = \tfrac{1}{2}[1 + 2u^2 + \cos(2u)], \tag{6}$$

the coefficients of the SECOND FUNDAMENTAL FORM are

$$e = \frac{\sqrt{2}\,uv \cos u}{\sqrt{1 + u^2(2 + v^2) + (1 - u^2v^2)\cos(2u)}} \tag{7}$$

$$f = \frac{\sqrt{2}(\cos u + u \sin u)}{\sqrt{1 + u^2(2 + v^2) + (1 - u^2v^2)\cos(2u)}} \tag{8}$$

$$g = \frac{\sqrt{2}\,u^2v \sin u)}{\sqrt{1 + u^2(2 + v^2) + (1 - u^2v^2)\cos(2u)}}, \tag{9}$$

the AREA ELEMENT is

$$dA = \sqrt{\frac{1 + u^2(2 + v^2) + (1 - u^2v^2)\cos(2u)}{2}}\,du \wedge dv, \tag{10}$$

and the GAUSSIAN and MEAN CURVATURES are given by

$$K = \frac{4\left[u(u^2v^2 - 2)\cos u \sin u - u^2 \sin^2 u - \cos^2 u\right]}{\left[1 + u^2(2 + v^2) + (1 - u^2v^2)\cos(2u)\right]^2} \tag{11}$$

$$H = -\frac{v\{u(5 + 4u^2)\cos u - u\cos(3u)\}}{2\sqrt{2}[1 + u^2(2 + v^2) + (1 - u^2v^2)\cos(2u)]^{3/2}}$$

$$- \frac{v\{2[2 + u^2(2 + v^2 + (2 - u^2v^2)\cos(2u))]\sin u\}}{2\sqrt{2}[1 + u^2(2 + v^2) + (1 - u^2v^2)\cos(2u)]^{3/2}}. \tag{12}$$

References

Pickover, C. A. *Mazes for the Mind: Computers and the Unexpected.* New York: St. Martin's Press, 1992.

Steiner Chain

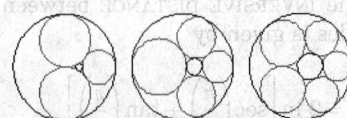

Given two nonconcentric CIRCLES with one interior to the other, if small TANGENT CIRCLES can be inscribed around the region between the two CIRCLES such that the final CIRCLE is TANGENT to the first, the CIRCLES form a Steiner chain.

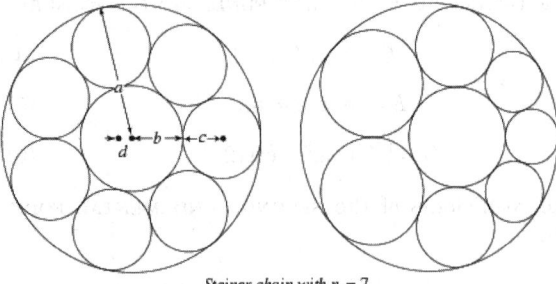

Steiner chain with n = 7

The simplest way to construct a Steiner chain is to perform an INVERSION on a symmetrical arrangement on n circles packed between a central circle of radius b and an outer concentric circle of radius a (Wells 1991). In this arrangement,

$$\sin\left(\frac{\pi}{n}\right) = \frac{a-b}{a+b}, \tag{1}$$

so the ratio of the radii for the small and large circles is

$$\frac{b}{a} = \frac{1 - \sin\left(\frac{\pi}{n}\right)}{1 + \sin\left(\frac{\pi}{n}\right)}. \tag{2}$$

In addition, the radii of the circles in the ring are

$$c = \tfrac{1}{2}(a-b), \tag{3}$$

and their centers are located at a distance

$$r = b + c = \tfrac{1}{2}(a+b) \tag{4}$$

from the origin.

To transform the symmetrical arrangement into a Steiner chain, take an INVERSION CENTER which is a distance d from the center of the symmetrical figure. Then the radii a' and b' of the outer and center circles become

$$a' = \left|\frac{a}{d^2 - a^2}\right| = \frac{a}{a^2 - d^2} \tag{5}$$

$$b' = \left|\frac{b}{d^2 - b^2}\right| = \frac{b}{b^2 - d^2}, \tag{6}$$

respectively. Equivalently, a Steiner chain results whenever the INVERSIVE DISTANCE between the two original circles is given by

$$\delta = 2\ln\left[\sec\left(\frac{\pi}{n}\right) + \tan\left(\frac{\pi}{n}\right)\right] \tag{7}$$

$$= 2\ln\left[\tan\left(\frac{\pi}{4} + \frac{\pi}{2n}\right)\right] \tag{8}$$

(Coxeter and Greitzer 1967).

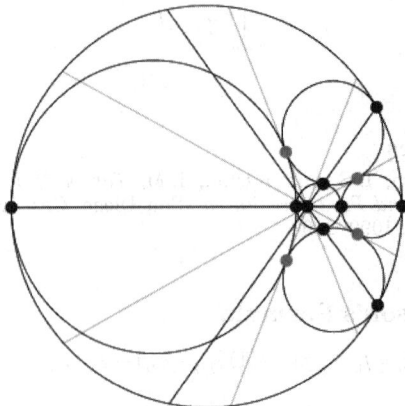

The centers of the circles in a Steiner chain lie on an ELLIPSE (Ogilvy 1990, p. 57). The lines of tangency passing through the contact points of neighboring circles in the chain are concurrent in a point. Furthermore, this is the same point at which the lines through the contact points of the inner and outer circles also concur (Wells 1991, p. 245).

STEINER'S PORISM states that if a Steiner chain is formed from one starting circle, then a Steiner chain is also formed from any other starting circle. A Steiner chain may also close after several loops around the central circle, in which case a Steiner chain will also be formed after the same number of loops from any starting point.

See also ARBELOS, COXETER'S LOXODROMIC SEQUENCE OF TANGENT CIRCLES, HEXLET, PAPPUS CHAIN, SEVEN CIRCLES THEOREM, STEINER'S PORISM

References

Coxeter, H. S. M. "Interlocking Rings of Spheres." *Scripta Math.* **18**, 113–21, 1952.

Coxeter, H. S. M. *Introduction to Geometry, 2nd ed.* New York: Wiley, p. 87, 1969.

Coxeter, H. S. M. and Greitzer, S. L. *Geometry Revisited.* Washington, DC: Math. Assoc. Amer., pp. 124–26, 1967.

Forder, H. G. *Geometry, 2nd ed.* London: Hutchinson's University Library, p. 23, 1960.

Gardner, M. "Mathematical Games: The Diverse Pleasures of Circles that Are Tangent to One Another." *Sci. Amer.* **240**, 18–8, Jan. 1979.

Johnson, R. A. *Modern Geometry: An Elementary Treatise on the Geometry of the Triangle and the Circle.* Boston, MA: Houghton Mifflin, pp. 113–15, 1929.

Ogilvy, C. S. *Excursions in Geometry.* New York: Dover, pp. 51–4, 1990.

Weisstein, E. W. "Plane Geometry." MATHEMATICA NOTEBOOK PLANEGEOMETRY.M.

Wells, D. *The Penguin Dictionary of Curious and Interesting Geometry.* London: Penguin, pp. 120 and 244–45, 1991.

Steiner Construction

A construction done using only a STRAIGHTEDGE. The PONCELET-STEINER THEOREM proves that all constructions possible using a COMPASS and STRAIGHTEDGE are possible using a STRAIGHTEDGE alone, as long as a fixed CIRCLE and its center, two intersecting

CIRCLES without their centers, or three nonintersecting CIRCLES are drawn beforehand. For example, the centers of two intersecting circles can be found using a STRAIGHTEDGE alone (Steinhaus 1983, p. 42).

See also GEOMETRIC CONSTRUCTION, MASCHERONI CONSTRUCTION, MATCHSTICK CONSTRUCTION, NEUSIS CONSTRUCTION, PONCELET-STEINER THEOREM, STRAIGHTEDGE

References

Dörrie, H. "Steiner's Straight-Edge Problem." §34 in *100 Great Problems of Elementary Mathematics: Their History and Solutions.* New York: Dover, pp. 165–70, 1965.
Rademacher, H. and Toeplitz, O. *The Enjoyment of Mathematics: Selections from Mathematics for the Amateur.* Princeton, NJ: Princeton University Press, p. 204, 1957.
Steiner, J. *Geometric Constructions with a Ruler, Given a Fixed Circle with Its Center.* Translated from the first German ed. (1833). New York: Scripta Mathematica, 1950.
Steinhaus, H. *Mathematical Snapshots, 3rd ed.* New York: Dover, p. 142, 1999.

Steiner Points

There are two different types of points known as Steiner points.

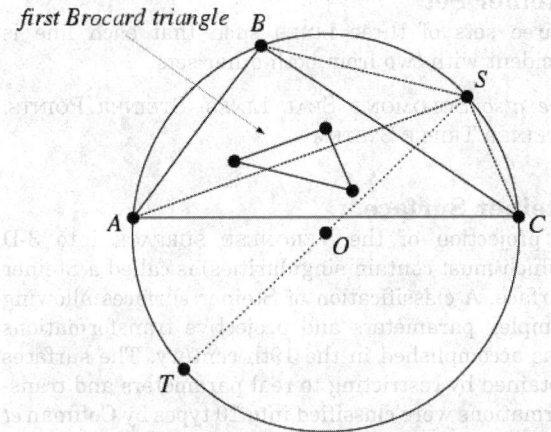

The point S of CONCURRENCE of the three lines drawn through the VERTICES of a TRIANGLE PARALLEL to the corresponding sides of the first BROCARD TRIANGLE is called the Steiner point (Honsberger 1995). It lies on the CIRCUMCIRCLE opposite the TARRY POINT T and has TRIANGLE CENTER FUNCTION

$$\alpha = bc(a^2 - b^2)(a^2 - c^2).$$

The BRIANCHON POINT for KIEPERT'S PARABOLA is also called the Steiner point. The SYMMEDIAN POINT K is the Steiner point of the first BROCARD TRIANGLE (Honsberger 1995, pp. 120–21). The SIMSON LINE of the Steiner point is PARALLEL to the line OK, when O is the CIRCUMCENTER and K is the SYMMEDIAN POINT (Honsberger 1995, p. 121). The Steiner point of a TRIANGLE is the CENTROID of the system obtained by placing a mass equal to the magnitude of the exterior angle at each vertex (Honsberger 1995, p. 120).

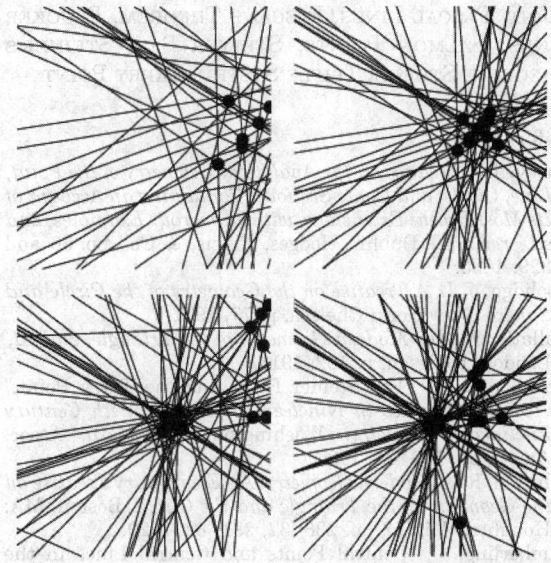

If triplets of opposites sides on a CONIC SECTION in PASCAL'S THEOREM are extended for all permutations of VERTICES, 60 PASCAL LINES are produced. The 20 points of their three by three intersections are called Steiner points. STEINER'S THEOREM states that these points are generated by the hexagons 123456, 143652, and 163254 formed by interchanging the vertices at positions 2, 4, and 6 (where the numbers denote the order in which the vertices of the hexagon are taken). The configuration of PASCAL LINES for a general hexagon inscribed in a general ellipse are shown above, with Steiner points shown as filled circles. A blow-up of the region in the upper left figure is shown below, illustrating the concurrence of three Pascal lines at each Steiner point.

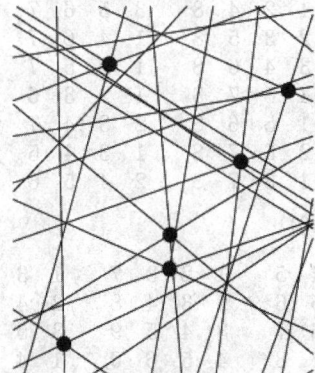

Each Steiner point lies together with three KIRKMAN POINTS on a total of 20 lines known as CAYLEY LINES. The Steiner points also lie four at a time on 15 PLÜCKER LINES (Wells 1991). There is a dual relationship between the 20 Steiner points and the 20 CAYLEY LINES.

See also BRIANCHON POINT, BROCARD TRIANGLES, CAYLEY LINES, CIRCUMCIRCLE, CONIC SECTION, KIEPERT'S PARABOLA, KIRKMAN POINTS, SYMMEDIAN

Point, Pascal Lines, Pascal's Theorem, Plücker Lines, Salmon Points, Steiner Set, Steiner's Theorem, Steiner Triple System, Tarry Point

References

Casey, J. *A Treatise on the Analytical Geometry of the Point, Line, Circle, and Conic Sections, Containing an Account of Its Most Recent Extensions, with Numerous Examples,* 2nd ed., rev. enl. Dublin: Hodges, Figgis, & Co., pp. 66 and 329, 1893.

Coolidge, J. L. *A Treatise on the Geometry of the Circle and Sphere.* New York: Chelsea, p. 77, 1971.

Gallatly, W. *The Modern Geometry of the Triangle,* 2nd ed. London: Hodgson, p. 102, 1913.

Honsberger, R. "The Steiner Point and the Tarry Point." §10.5 in *Episodes in Nineteenth and Twentieth Century Euclidean Geometry.* Washington, DC: Math. Assoc. Amer., pp. 119–24, 1995.

Johnson, R. A. *Modern Geometry: An Elementary Treatise on the Geometry of the Triangle and the Circle.* Boston, MA: Houghton Mifflin, pp. 236–37, 281–82, 1929.

Kimberling, C. "Central Points and Central Lines in the Plane of a Triangle." *Math. Mag.* **67**, 163–87, 1994.

Lachlan, R. *An Elementary Treatise on Modern Pure Geometry.* London: Macmillian, p. 115, 1893.

Salmon, G. "Notes: Pascal's Theorem, Art. 267" in *A Treatise on Conic Sections,* 6th ed. New York: Chelsea, pp. 379–82, 1960.

Steiner. *Gergonne Ann. Math.* **18**.

Wells, D. *The Penguin Dictionary of Curious and Interesting Geometry.* London: Penguin, p. 172, 1991.

Steiner Quadruple System

A Steiner quadruple system is a Steiner system $S(t=3, k=4, v)$, where S is a v-set and B is a collection of k-sets of S such that every t-subset of S is contained in exactly one member of B. Barrau (1908) established the uniqueness of $S(3, 4, 8)$,

1	2	4	8	3	5	6	7
2	3	5	8	1	4	6	7
3	4	6	8	1	2	5	7
4	5	7	8	1	2	3	6
1	5	6	8	2	3	4	7
2	6	7	8	1	3	4	5
1	3	7	8	2	4	5	6

and $S(3, 4, 10)$

1	2	4	5	1	2	3	7	1	3	5	8
2	3	5	6	2	3	4	8	2	4	6	9
3	4	6	7	3	4	5	9	3	5	7	0
4	5	7	8	4	5	6	0	1	4	6	8
5	6	8	9	1	5	6	7	2	5	7	9
6	7	9	0	2	6	7	8	3	6	8	0
1	7	8	0	3	7	8	9	1	4	7	9
1	2	8	9	4	8	9	0	2	5	8	0
2	3	9	0	1	5	9	0	1	3	6	9
1	3	4	0	1	2	6	0	2	4	7	0

Fitting (1915) subsequently constructed the cyclic systems $S(3, 4, 26)$ and $S(3, 4, 34)$, and Bays and de Weck (1935) showed the existence of at least one $S(3, 4, 14)$. Hanani (1960) proved that a necessary

and sufficient condition for the existence of an $S(3, 4, v)$ is that $v \equiv 2$ or 4 (mod 6).

The number of nonisomorphic steiner quadruple systems of orders 8, 10, 14, and 16 are 1, 1, 4 (Mendelsohn and Hung 1972), and at least 31,021 (Lindner and Rosa 1976).

See also Steiner System, Steiner Triple System

References

Barrau, J. A. "On the Combinatory Problem of Steiner." *K. Akad. Wet. Amsterdam Proc. Sect. Sci.* **11**, 352–60, 1908.

Bays, S. and de Weck, E. "Sur les systèmes de quadruples." *Comment. Math. Helv.* **7**, 222–41, 1935.

Fitting, F. "Zyklische Lösungen des Steiner'schen Problems." *Nieuw. Arch. Wisk.* **11**, 140–48, 1915.

Hanani, M. "On Quadruple Systems." *Canad. J. Math.* **12**, 145–57, 1960.

Lindner, C. L. and Rosa, A. "There are at Least 31,021 Nonisomorphic Steiner Quadruple Systems of Order 16." *Utilitas Math.* **10**, 61–4, 1976.

Lindner, C. L. and Rosa, A. "Steiner Quadruple Systems--A Survey." *Disc. Math.* **22**, 147–81, 1978.

Mendelsohn, N. S. and Hung, S. H. Y. "On the Steiner Systems $S(3, 4, 14)$ and $S(4, 5, 15)$." *Utilitas Math.* **1**, 5–5, 1972.

Steiner Set

Three sets of three lines such that each line is incident with two from both other sets.

See also Solomon's Seal Lines, Steiner Points, Steiner Triple System

Steiner Surface

A projection of the Veronese surface into 3-D (which must contain singularities) is called a Steiner surface. A classification of Steiner surfaces allowing complex parameters and projective transformations was accomplished in the 19th century. The surfaces obtained by restricting to real parameters and transformations were classified into 10 types by Coffman *et al.* (1996). Examples of Steiner surfaces include the Roman surface (Coffman type 1) and cross-cap (type 3).

The Steiner surface of type 2 is given by the implicit equation

$$x^2y^2 - x^2z^2 + y^2z^2 - xyz = 0.$$

and can be transformed into the Roman surface or cross-cap by a complex projective change of coordinates (but not by a real transformation). It has two pinch points and three double lines and, unlike the Roman surface or cross-cap, is not compact in any affine neighborhood.

The Steiner surface of type 4 has the implicit equation

$$y^2 - 2xy^2 - xz^2 + x^2y^2 + x^2z^2 - z^4 = 0.$$

and two of the three double lines of surface 2 coincide

along a line where the two noncompact "components" are tangent.

See also CROSS-CAP, ROMAN SURFACE, VERONESE VARIETY

References

Apéry, F. *Models of the Real Projective Plane: Computer Graphics of Steiner and Boy Surfaces.* Braunschweig, Germany: Vieweg, 1987.

Coffman, A. "Steiner Surfaces." http://www.ipfw.edu/math/ Coffman/steinersurface.html.

Coffman, A.; Schwartz, A.; and Stanton, C. "The Algebra and Geometry of Steiner and Other Quadratically Parametrizable Surfaces." *Computer Aided Geom. Design* **13**, 257–86, 1996.

Nordstrand, T. "Steiner Relative." http://www.uib.no/people/ nfytn/stmtxt.htm.

Nordstrand, T. "Steiner Relative [2]." http://www.uib.no/ people/nfytn/stm2txt.htm.

Steiner System

A Steiner system $S(t, k, v)$ is a set X of v points, and a collection of subsets of X of size k (called blocks), such that any t points of X are in exactly one of the blocks. The special case $t = 2$ and $k = 3$ corresponds to a so-called STEINER TRIPLE SYSTEM. For a PROJECTIVE PLANE, $v = n^2 + n + 1$, $k = n + 1$, $t = 2$, and the blocks are simply lines.

The number r of blocks containing a point in a $S(t, k, v)$ Steiner system is independent of the point. In fact,

$$r = \frac{\binom{v-1}{t-1}}{\binom{k-1}{t-1}},$$

where $\binom{n}{k}$ is a BINOMIAL COEFFICIENT. The total number of blocks b is also determined and is given by

$$b = \frac{vr}{k}.$$

These numbers also satisfy $v \le b$ and $k \le r$.

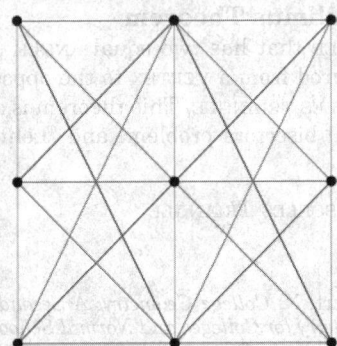

The PERMUTATIONS of the points preserving the blocks of a Steiner system S is the AUTOMORPHISM GROUP of S. For example, consider Ω the set of 9 points in the 2-dimensional VECTOR SPACE over the FIELD over 3 elements. The blocks are the 12 lines of the form $\{a + tb\} = \{a, a + b, a + 2b\}$, which have three elements each. The system is a $S(2, 3, 9)$ because any two points uniquely determine a line.

The AUTOMORPHISM GROUP of a Steiner system is the AFFINE GROUP which preserves the lines. For a vector space of dimension n over a field of q elements, this construction gives a Steiner system $S(2, q, q^d)$.

Several interesting groups arise as automorphism groups of Steiner systems. For example, the MATHIEU GROUPS are the AUTOMORPHISM GROUPS of Steiner systems, as summarized in the following table. These groups are unique up to ISOMORPHISM, and are not only SPORADIC SIMPLE GROUPS, but are also highly TRANSITIVE.

Mathieu group	Steiner system
M_{11}	$S(3, 4, 14)$
M_{12}	$S(5, 6, 12)$
M_{22}	$S(3, 6, 22)$
M_{23}	$S(4, 7, 23)$
M_{24}	$S(5, 8, 24)$

See also AUTOMORPHISM GROUP, CONFIGURATION, MATHIEU GROUPS, SIMPLE GROUP, STEINER QUADRUPLE SYSTEM, STEINER TRIPLE SYSTEM, T-DESIGN, TRANSITIVE GROUP, WITT GEOMETRY

References

Colbourn, C. J. and Dinitz, J. H. (Eds.). *CRC Handbook of Combinatorial Designs.* Boca Raton, FL: CRC Press, 1996.

Dixon, J. and Mortimer, B. *Permutation Groups.* New York: Springer-Verlag, 1996.

Gropp, H. "Enumeration of Regular Graphs 100 Years Ago." *Discrete Math.* **101**, 73–5, 1992.

Woolhouse, W. S. B. "Prize Question 1733." *Lady's and Gentleman's Diary.* 1844.

Steiner Tree

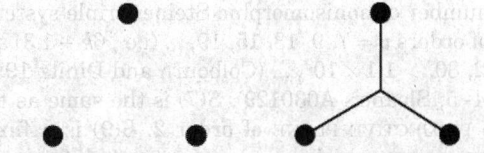

The Steiner tree of some subset of the vertices of a GRAPH G is a minimum-weight connected SUBGRAPH of G that includes all the vertices. It is always a tree. Steiner trees have practical applications, for example, in the determination of the shortest total length of wires needed to join some number of points (Hoffman 1998, pp. 164–65).

See also PLATEAU'S PROBLEM, TREE

References

Chopra, S. and Rao, M. R. "The Steiner Tree Problem 1: Formulations, Compositions, and Extension of Facets." *Mathematical Programming* **64**, 209–29, 1994.

Chopra, S. and Rao, M. R. "The Steiner Tree Problem 2: Properties and Classes of Facets." *Mathematical Programming* **64**, 231–46, 1994.

Chung, F. R. K.; Gardner, M.; and Graham, R. L. "Steiner Trees on a Checkerboard." *Math. Mag.* **62**, 83–6, 1989.

Du, D.-Z.; Smith, J. M.; and Rubinstein, J. H. *Advances in Steiner Trees.* Dordrecht, Netherlands: Kluwer, 2000.

Ganley, J. "The Steiner Tree Page." http://ganley.org/steiner/.

Hoffman, P. *The Man Who Loved Only Numbers: The Story of Paul Erdos and the Search for Mathematical Truth.* New York: Hyperion, 1998.

Hwang, F.; Richards, D.; and Winter, P. *The Steiner Tree Problem.* Amsterdam, Netherlands: North-Holland, 1992.

Ivanov, A. O. and Tuzhilin, A. A. *Minimal Networks: The Steiner Problem and Its Generalizations.* Boca Raton, FL: CRC Press, 1994.

Skiena, S. S. "Steiner Tree." §8.5.10 in *The Algorithm Design Manual.* New York: Springer-Verlag, pp. 339–42, 1997.

Steiner Triple System

Let X be a set of $v \geq 3$ elements together with a set B of 3-subset (triples) of X such that every 2-SUBSET of X occurs in exactly one triple of B. Then B is called a Steiner triple system and is a special case of a STEINER SYSTEM with $t = 2$ and $k = 3$. A Steiner triple system $S(v) = S(v, k = 3, \lambda = 1)$ of order v exists IFF $v \equiv 1, 3 \pmod 6$ (Kirkman 1847). In addition, if Steiner triple systems S_1 and S_2 of orders v_1 and v_2 exist, then so does a Steiner triple system S of order $v_1 v_2$ (Ryser 1963, p. 101).

Examples of Steiner triple systems $S(v)$ of small orders v are

$$S_3 = \{\{1, 2, 3\}\}$$

$$S_7 = \{\{1, 2, 4\}, \{2, 3, 5\}, \{3, 4, 6\}, \{4, 5, 7\}.$$

$$\{5, 6, 1\}, \{6, 7, 2\}, \{7, 1, 3\}\}$$

$$S_9 = \{\{1, 2, 3\}, \{4, 5, 6\}, \{7, 8, 9\}, \{1, 4, 7\},$$

$$\{2, 5, 8\}, \{3, 6, 9\}, \{1, 5, 9\}\{2, 6, 7\},$$

$$\{3, 4, 8\}, \{1, 6, 8\}, \{2, 4, 9\}, \{3, 5, 7\}\}.$$

The number of nonisomorphic Steiner triple systems $S(v)$ of orders $v = 7, 9, 13, 15, 19, \ldots$ (i.e., $6k + 1.3$) are 1, 1, 2, 80, $> 1.1 \times 10^9, \ldots$ (Colbourn and Dinitz 1996, pp. 14–5; Sloane's A030129). $S(7)$ is the same as the finite PROJECTIVE PLANE of order 2. $S(9)$ is a finite AFFINE PLANE which can be constructed from the array

$$\begin{array}{ccc} a & b & c \\ d & e & f \\ g & h & i \end{array}$$

One of the two $S(13)$s is a finite HYPERBOLIC PLANE. The 80 Steiner triple systems $S(15)$ have been studied by Tonchev and Weishaar (1997). There are more than 1.1×10^9 Steiner triple systems of order 19 (Stinson and Ferch 1985; Colbourn and Dinitz 1996, p. 15).

See also HADAMARD MATRIX, KIRKMAN TRIPLE SYSTEM, STEINER QUADRUPLE SYSTEM, STEINER SYSTEM

References

Ball, W. W. R. and Coxeter, H. S. M. *Mathematical Recreations and Essays, 13th ed.* New York: Dover, pp. 107–09 and 274, 1987.

Colbourn, C. J. and Dinitz, J. H. (Eds.). "Steiner Triple Systems." §4.5 in *CRC Handbook of Combinatorial Designs.* Boca Raton, FL: CRC Press, pp. 14–5 and 70, 1996.

Gardner, M. "Mathematical Games: On the Remarkable Császár Polyhedron and Its Applications in Problem Solving." *Sci. Amer.* **232**, 102–07, May 1975.

Kirkman, T. P. "On a Problem in Combinatorics." *Cambridge Dublin Math. J.* **2**, 191–04, 1847.

Lindner, C. C. and Rodger, C. A. *Design Theory.* Boca Raton, FL: CRC Press, 1997.

Ryser, H. J. *Combinatorial Mathematics.* Buffalo, NY: Math. Assoc. Amer., pp. 99–02, 1963.

Sloane, N. J. A. Sequences A030129 in "An On-Line Version of the Encyclopedia of Integer Sequences." http://www.research.att.com/~njas/sequences/eisonline.html.

Stinson, D. R. and Ferch, H. "2000000 Steiner Triple Systems of Order 19." *Math. Comput.* **44**, 533–35, 1985.

Tonchev, V. D. and Weishaar, R. S. "Steiner Triple Systems of Order 15 and Their Codes." *J. Stat. Plan. Inference* **58**, 207–16, 1997.

Steinerian Curve

The LOCUS of points whose first POLARS with regard to the curves of a linear net have a common point. It is also the LOCUS of points of CONCURRENCE of line POLARS of points of the JACOBIAN CURVE. It passes through all points common to all curves of the system and is of order $3(n-1)^2$.

See also CAYLEYIAN CURVE, JACOBIAN CURVE

References

Coolidge, J. L. *A Treatise on Algebraic Plane Curves.* New York: Dover, p. 150, 1959.

Steiner-Lehmus Theorem

Any TRIANGLE that has two equal ANGLE BISECTORS (each measured from a VERTEX to the opposite sides) is an ISOSCELES TRIANGLE. This theorem is also called the "internal bisectors problem" and "Lehmus' theorem."

See also ISOSCELES TRIANGLE

References

Altshiller-Court, N. *College Geometry: A Second Course in Plane Geometry for Colleges and Normal Schools, 2nd ed., rev. enl.* New York: Barnes and Noble, pp. 72–3, 1952.

Coxeter, H. S. M. *Introduction to Geometry, 2nd ed.* New York: Wiley, p. 9, 1969.

Coxeter, H. S. M. and Greitzer, S. L. "The Steiner-Lehmus Theorem." §1.5 in *Geometry Revisited.* Washington, DC: Math. Assoc. Amer., pp. 14–6, 1967.

Gardner, M. *Martin Gardner's New Mathematical Diversions from Scientific American.* New York: Simon and Schuster, pp. 198–99 and 206–07, 1966.

Henderson, A. "The Lehmus-Steiner-Terquem Problem in Global Survey." *Scripta Math.* **21**, 223–32 and 309–12, 1955.

Hunter, J. A. H. and Madachy, J. S. *Mathematical Diversions.* New York: Dover, pp. 72–3, 1975.

Neuberg, J. *Bibliographie du triangle et du tétraèdre.* p. 337, 1923.

Thébault, V. "Sur le triangle isoscèle." *Mathesis* **44**, 97, 1930.

Steiner's Ellipse

Let $\alpha' : \beta' : \gamma'$ be the ISOTOMIC CONJUGATE POINT of a point with TRILINEAR COORDINATES $\alpha : \beta : \gamma$. The isotomic conjugate of the LINE AT INFINITY having trilinear equation

$$a\alpha + b\beta + c\gamma = 0$$

is

$$\frac{\beta'\gamma'}{a} + \frac{\gamma'\alpha'}{b} + \frac{\alpha'\beta'}{c} = 0.$$

known as Steiner's ellipse (Vandeghen 1965).

See also ISOTOMIC CONJUGATE POINT, LINE AT INFINITY

References

Vandeghen, A. "Some Remarks on the Isogonal and Cevian Transforms. Alignments of Remarkable Points of a Triangle." *Amer. Math. Monthly* **72**, 1091–094, 1965.

Steiner's Hypocycloid

DELTOID

Steiner's Porism

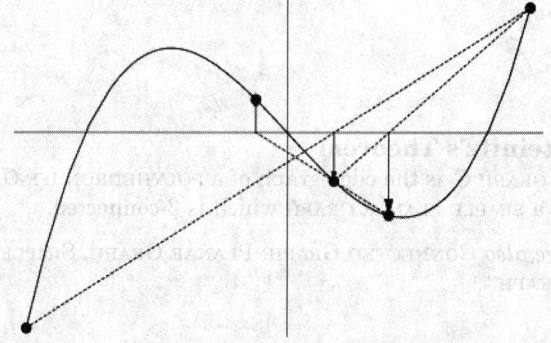

If a STEINER CHAIN is formed from one starting circle, then a STEINER CHAIN is formed from any other starting circle. In other words, given two nonconcentric CIRCLES, draw CIRCLES successively touching them and each other. If the last touches the first, this will also happen for any position of the first CIRCLE.

See also HEXLET, SEVEN CIRCLES THEOREM, STEINER CHAIN

References

Allanson, B. "Steiner's Porism" java applet. http://www.ade-laide.net.au/~allanson/steiner.html.

Coolidge, J. L. *A Treatise on the Geometry of the Circle and Sphere.* New York: Chelsea, p. 34, 1971.

Coxeter, H. S. M. "Interlocking Rings of Spheres." *Scripta Math.* **18**, 113–21, 1952.

Coxeter, H. S. M. *Introduction to Geometry, 2nd ed.* New York: Wiley, p. 87, 1969.

Coxeter, H. S. M. and Greitzer, S. L. *Geometry Revisited.* Washington, DC: Math. Assoc. Amer., pp. 124–26, 1967.

Forder, H. G. *Geometry, 2nd ed.* London: Hutchinson's University Library, p. 23, 1960.

Gardner, M. "Mathematical Games: The Diverse Pleasures of Circles that Are Tangent to One Another." *Sci. Amer.* **240**, 18–8, Jan. 1979.

Johnson, R. A. *Modern Geometry: An Elementary Treatise on the Geometry of the Triangle and the Circle.* Boston, MA: Houghton Mifflin, pp. 113–15, 1929.

Ogilvy, C. S. *Excursions in Geometry.* New York: Dover, pp. 53–4, 1990.

Wells, D. *The Penguin Dictionary of Curious and Interesting Geometry.* London: Penguin, pp. 120 and 244–45, 1991.

Steiner's Problem

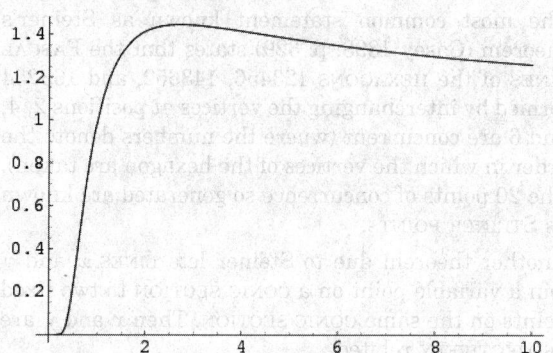

For what value of x is $f(x) = x^{1/x}$ a MAXIMUM? The maximum occurs at $x = e$, where

$$f'(x) = x^{-2+1/x}(1 - \ln x) = 0.$$

which is zero at $x = e$ and gives a maximum of

$$e^{1/e} = 1.444667861\ldots.$$

The function has an inflection point at $x = 0.581933\ldots$, where

$$f''(x) = x^{-4+1/x}[1 - 3x + (\ln x)(2x - 2 + \ln x)] = 0.$$

See also FERMAT'S PROBLEM, POWER TOWER

References

Dörrie, H. *100 Great Problems of Elementary Mathematics: Their History and Solutions.* New York: Dover, 1965.

Wells, D. *The Penguin Dictionary of Curious and Interesting Numbers.* Middlesex, England: Penguin Books, p. 35, 1986.

Steiner's Segment Problem

Given n points, find the line segments with the shortest possible total length which connect the points. The segments need not necessarily be straight from one point to another.

For three points, if all ANGLES are less than 120°, then the line segments are those connecting the three points to a central point P which makes the ANGLES $\langle A \rangle PB$, $\langle B \rangle PC$, and $\langle C \rangle PA$ all 120°. If one ANGLE is greater that 120°, then P coincides with the offending ANGLE.

For four points, P is the intersection of the two diagonals, but the required minimum segments are not necessarily these diagonals.

A modified version of the problem is, given two points, to find the segments with the shortest total length connecting the points such that each branch point may be connected to only three segments. There is no general solution to this version of the problem.

Steiner's Theorem

The most common statement known as Steiner's theorem (Casey 1893, p. 329) states that the PASCAL LINES of the HEXAGONS 123456, 143652, and 163254 formed by interchanging the vertices at positions 2, 4, and 6 are concurrent (where the numbers denote the order in which the vertices of the hexagon are taken). The 20 points of concurrence so generated are known as STEINER POINTS.

Another theorem due to Steiner lets LINES x and y join a variable point on a CONIC SECTION to two fixed points on the same CONIC SECTION. Then x and y are PROJECTIVELY related.

A third "Steiner's theorem" states that if two opposite edges of a TETRAHEDRON move on two fixed SKEW LINES in any way whatsoever but remain fixed in length, then the volume of the TETRAHEDRON remains constant (Altshiller-Court 1979, p. 87).

See also CONIC SECTION, PROJECTION, TETRAHEDRON

References

Altshiller-Court, N. *Modern Pure Solid Geometry.* New York: Chelsea, 1979.
Casey, J. *A Treatise on the Analytical Geometry of the Point, Line, Circle, and Conic Sections, Containing an Account of Its Most Recent Extensions, with Numerous Examples, 2nd ed., rev. enl.* Dublin: Hodges, Figgis, & Co., p. 329, 1893.
Graustein, W. C. *Introduction to Higher Geometry.* New York: Macmillan, pp. 252–53, 1930.

Steinhaus Dissection

CUBE DISSECTION

Steinhaus Property

References

Kanemitsu, S. and Gyory, K. (Eds.). "A Problem of Steinhaus Concerning the Existence of a Plane Set with a Certain Property." In *Number Theory and Its Applications.* Dordrecht, Netherlands: Kluwer, pp. 1–, 1999.

Steinhaus-Moser Notation

A NOTATION for LARGE NUMBERS defined by Steinhaus (1983, pp. 28–9). In this notation, $\triangle$ denotes n^n, $\boxed{n}$ denotes "n in n TRIANGLES," and $\textcircled{n}$ denotes "n in n SQUARES." A modified version due to Moser eliminates the circle notation, continuing instead with POLYGONS of ever increasing size, so n in a PENTAGON is n with n SQUARES around it, etc.

See also CIRCLE NOTATION, LARGE NUMBER, MEGA, MOSER

References

Steinhaus, H. *Mathematical Snapshots, 3rd ed.* New York: Dover, 1999.

Steinitz's Lemma

If, in a plane or spherical convex polygon $ABCDEFG$, all of whose sides AB, BC, CD, ..., FG (with the exception of AG) have fixed lengths, one simultaneously increases (decreases) the angles between these sides, then the length of the variable side increases (decreases).

References

Cromwell, P. R. "Steinitz' Lemma." In *Polyhedra.* New York: Cambridge University Press, pp. 235–37, 1997.

Steinitz's Theorem

A GRAPH G is the edge graph of a POLYHEDRON IFF G is a SIMPLE PLANAR GRAPH which is 3-connected.

See also CONNECTED GRAPH, PLANAR GRAPH, SIMPLE GRAPH

Steinmetz Solid

The solid common to two (or three) right circular CYLINDERS of equal RADII intersecting at RIGHT ANGLES is called the Steinmetz solid. Two CYLINDERS intersecting at RIGHT ANGLES are called a bicylinder, and three intersecting CYLINDERS a TRICYLINDER. Half of a bicylinder is called a VAULT.

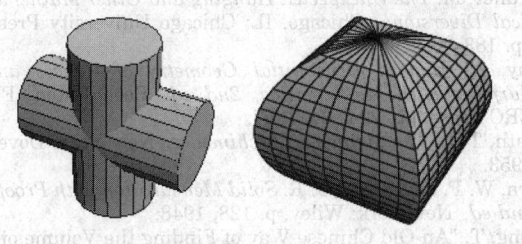

For two cylinders of radius r oriented long the z- and x-axes gives the equations

$$x^2 + y^2 = r^2 \qquad (1)$$

$$y^2 + z^2 = r^2 \qquad (2)$$

which can be solved for x and y gives the PARAMETRIC EQUATIONS of the edges of the solid,

$$x = \pm z \qquad (3)$$

$$y = \pm \sqrt{r^2 - z^2}. \qquad (4)$$

The SURFACE AREA can be found as $\int x \, ds$, where

$$ds = \sqrt{1 + \left(\frac{dy}{dz}\right)^2} \, dz = \frac{r}{\sqrt{r^2 - z^2}}. \qquad (5)$$

Taking the range of integration as a quarter or one face and then multiplying by 16 gives

$$S_2 = 16 \int_0^r \frac{r_2}{\sqrt{r^2 - z^2}} \, dz = 16r^2. \qquad (6)$$

The VOLUME common to two cylinders is was known to Archimedes (Heath 1953, Gardner 1962) and the Chinese mathematician Tsu Ch'ung-Chih (Kiang 1972), and does not require CALCULUS to derive. Using calculus provides a simple derivation, however. Noting that the solid has a square CROSS SECTION of side-half-length $\sqrt{a^2 - z^2}$, the volume is given by

$$V_2(r, r) = \int_{-r}^r \left(2\sqrt{r^2 - z^2}\right)^2 dz = \frac{16}{3} r^3 \qquad (7)$$

(Moore 1974). The VOLUME can also be found using CYLINDRICAL ALGEBRAIC DECOMPOSITION, which reduces the inequalities

$$\begin{cases} x^2 + y^2 < 1 \\ -L < z < L \\ y^2 + z^2 < 1 \\ -L < x < L \end{cases} \qquad (8)$$

to

$$\begin{cases} -1 < x < 1 \\ -\sqrt{1 - x^2} < y < \sqrt{1 - x^2} \\ -\sqrt{1 - y^2} < z < \sqrt{1 - y^2}, \end{cases} \qquad (9)$$

giving the integral

$$V_2(1, 1) = \int_{-1}^1 \int_{-\sqrt{1-x^2}}^{\sqrt{1-x^2}} \int_{-\sqrt{1-y^2}}^{\sqrt{1-y^2}} dx \, dy \, dz = \frac{16}{3}. \qquad (10)$$

If the two right CYLINDERS are of *different* RADII a and b with $a > b$, then the VOLUME common to them is

$$V_2(a, b) = \frac{8}{3} a \left[(a^2 + b^2)E(k) - (a^2 - b^2)K(k)\right], \qquad (11)$$

where $K(k)$ is the complete ELLIPTIC INTEGRAL OF THE FIRST KIND, $E(k)$ is the complete ELLIPTIC INTEGRAL OF THE SECOND KIND, and $k \equiv b/a$ is the MODULUS.

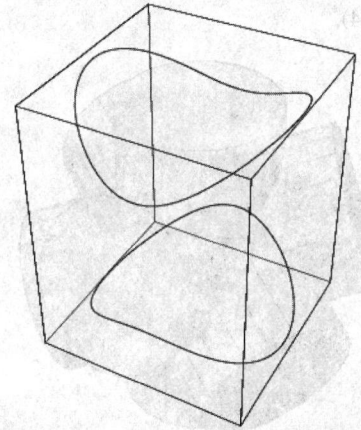

The curves of intersection of two cylinders of RADII a and b, shown above, are given by the parametric equations

$$x(t) = a \cos t \qquad (12)$$

$$g(t) = a \sin t \qquad (13)$$

$$z(t) = \pm \sqrt{b^2 - a^2 \sin^2 t} \qquad (14)$$

(Gray 1997).

The VOLUME common to two ELLIPTIC CYLINDERS

$$\frac{x^2}{a^2} + \frac{z^2}{c^2} = 1 \qquad \frac{y^2}{b^2} + \frac{z^2}{c'^2} = 1 \qquad (15)$$

with $c < c'$ is

$$V_2(a, c; b, c') = \frac{8ab}{3c} \left[(c'^2 + c^2)E(k) - (c'^2 - c^2)K(k)\right]. \qquad (16)$$

where $k = c/c'$ (Bowman 1961, p. 34).

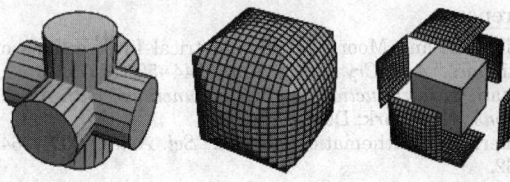

For three CYLINDERS of RADII r intersecting at RIGHT ANGLES, The resulting solid has 12 curved faces. If tangent planes are drawn where the faces meet, the result is a RHOMBIC DODECAHEDRON (Wells 1991). The VOLUME of intersection can be computed in a number of different ways,

$$V_3(r, r, r) = \int 16r^3 \int_0^{\pi/4} \int_0^1 s\sqrt{1 - s^2 \cot^2 t}\, ds\, dt \quad (17)$$

$$= \left(\sqrt{2}r\right)^3 + 6 \int_{r/\sqrt{2}}^r \left(2\sqrt{r^2 - z^2}\right)^2 dz \quad (18)$$

$$= 8\left(2 - \sqrt{2}\right) r^3 \quad (19)$$

(Moore 1974).

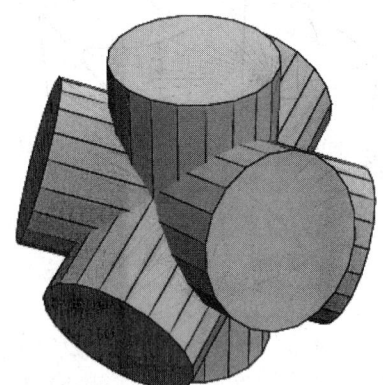

Four cylinders can also be placed with axes along the lines joining the vertices of a TETRAHEDRON with the centers of the opposite sides. The resulting solid of intersection has VOLUME

$$V_4 = 12\left(2\sqrt{2} - \sqrt{6}\right) \quad (20)$$

and 24 curved faces analogous to a CUBE-OCTAHEDRON COMPOUND (Moore 1974, Wells 1991).

Six cylinders can be place with axes parallel to the face diagonals of a CUBE. The resulting solid of intersection has VOLUME

$$V_4 = 12\left(3 + 2\sqrt{3} - 4\sqrt{2}\right) \quad (21)$$

and 36 curved faces, 24 of which are kite-shaped and 12 of which are rhombic (Moore 1974).

See also BICYLINDER, CYLINDER, ELLIPTIC CYLINDER, REULEAUX TETRAHEDRON, RHOMBIC DODECAHEDRON, RIGHT ANGLE, VAULT

References

Angell, I. O. and Moore, M. "Symmetrical Intersections of Cylinders." *Acta Cryst. Sect. A* **43**, 244–50, 1987.
Bowman, F. *Introduction to Elliptic Functions, with Applications.* New York: Dover, 1961.
Gardner, M. "Mathematical Games." *Sci. Amer.* **207**, 164, 1962.
Gardner, M. *The Unexpected Hanging and Other Mathematical Diversions.* Chicago, IL: Chicago University Press, pp. 183–85, 1991.
Gray, A. *Modern Differential Geometry of Curves and Surfaces with Mathematica, 2nd ed.* Boca Raton, FL: CRC Press, pp. 204–04, 1997.
Heath, T. L. *The Method of Archimedes.* New York: Dover, 1953.
Kern, W. F. and Bland, J. R. *Solid Mensuration with Proofs, 2nd ed.* New York: Wiley, p. 128, 1948.
Kiang, T. "An Old Chinese Way of Finding the Volume of a Sphere." *Math. Gaz.* **56**, 88–1, 1972.
Moore, M. "Symmetrical Intersections of Right Circular Cylinders." *Math. Gaz.* **58**, 181–85, 1974.
Wells, D. *The Penguin Dictionary of Curious and Interesting Geometry.* London: Penguin, pp. 118–19, 1991.
Wells, D. G. #555 in *The Penguin Book of Curious and Interesting Puzzles.* London: Penguin Books, 1992.

Stella Octangula

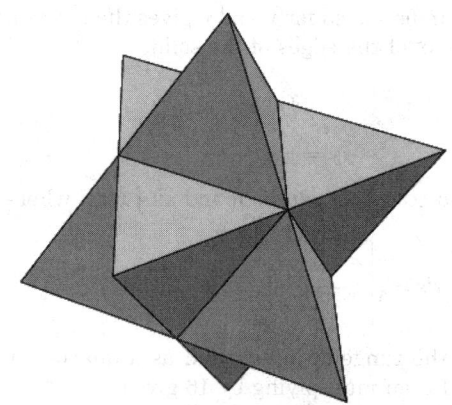

A POLYHEDRON COMPOUND composed of a TETRAHEDRON and its DUAL (a second TETRAHEDRON rotated 180° with respect to the first). The stella octangula is also called a STELLATED TETRAHEDRON, and is the only STELLATION of the OCTAHEDRON. The stella octangula can be constructed using the following NET by cutting along the solid lines, folding back along the plain lines, and folding forward along the dotted lines.

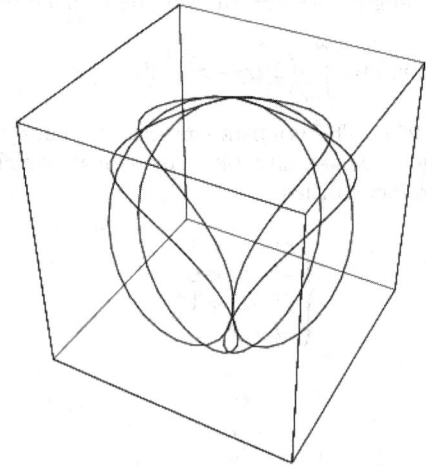

Another construction builds a single TETRAHEDRON,

then attaches four tetrahedral caps, one to each face. This CUMULATION of a unit edge-length OCTAHEDRON uses pyramids with height $\frac{1}{3}\sqrt{6}$.

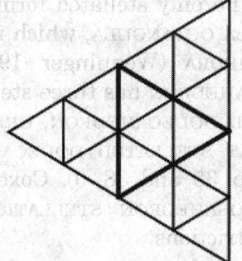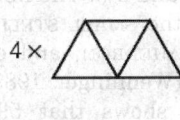

A tetrahedron with edge length 1 produces a stella octangula with edge lengths 1/2. This solid has SURFACE AREA and VOLUME

$$S = \tfrac{3}{2}\sqrt{3}$$

$$V = \tfrac{1}{8}\sqrt{2}.$$

The CONVEX HULL of the stella octangula is a CUBE.

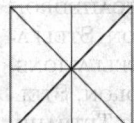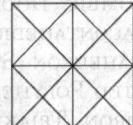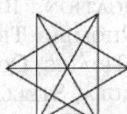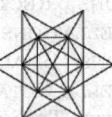

The above diagrams show two projections of the stella octangula. The edges lying on tetrahedral faces are represented using dashed lines, while the edges of the two large tetrahedron are showing using solid lines.

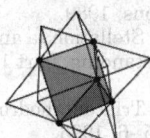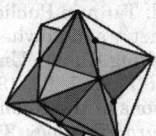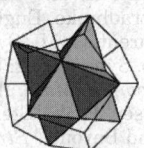

The solid common to both tetrahedra is an OCTAHEDRON (left figure; Ball and Coxeter 1987), which is another way of saying that the stella octangula is a STELLATION of the OCTAHEDRON (in fact, the only stellation). The edges of the two tetrahedra in the stella octangula form the 12 DIAGONALS of a CUBE (middle figure). Finally, the stella octangula can be constructed using eight of the 20 vertices of the DODECAHEDRON (right figure).

See also CUBE, OCTAHEDRON, POLYHEDRON COMPOUND, SPHERE PACKING, STELLATION, TETRAHEDRON

References

Ball, W. W. R. and Coxeter, H. S. M. *Mathematical Recreations and Essays, 13th ed.* New York: Dover, pp. 135–37, 1987.

Coxeter, H. S. M. *Introduction to Geometry, 2nd ed.* New York: Wiley, p. 158, 1969.

Coxeter, H. S. M. *Regular Polytopes, 3rd ed.* New York: Dover, pp. 48–1, 1973.

Cundy, H. and Rollett, A. "Stella Octangula (Two Tetrahedra)." §3.10.1 in *Mathematical Models, 3rd ed.* Stradbroke, England: Tarquin Pub., p. 129, 1989.

Kepler, J. "Harmonice Mundi." In *Opera Omnia, Vol. 5.* Frankfurt, 1864.

Steinhaus, H. *Mathematical Snapshots, 3rd ed.* New York: Dover, pp. 212–13, 1999.

Weisstein, E. W. "Polyhedra." MATHEMATICA NOTEBOOK POLYHEDRA.M.

Wenninger, M. J. *Polyhedron Models.* New York: Cambridge University Press, pp. 35 and 37, 1989.

Stella Octangula Number

A FIGURATE NUMBER OF THE FORM,

$$StOct_n = O_n + 8T_{n-1} = n(2n^2 - 1).$$

The first few are 1, 14, 51, 124, 245, ... (Sloane's A007588). The GENERATING FUNCTION for the stella octangula numbers is

$$\frac{x(x^2 + 10x + 1)}{(x-1)^4} = x + 14x^2 + 51x^3 + 124x^4 + \dots.$$

References

Conway, J. H. and Guy, R. K. *The Book of Numbers.* New York: Springer-Verlag, p. 51, 1996.

Sloane, N. J. A. Sequences A007588/M4932 in "An On-Line Version of the Encyclopedia of Integer Sequences." http://www.research.att.com/~njas/sequences/eisonline.html.

Stellated Octahedron

STELLA OCTANGULA

Stellated Polyhedron

STELLATION

Stellated Tetrahedron

STELLA OCTANGULA

Stellated Truncated Hexahedron

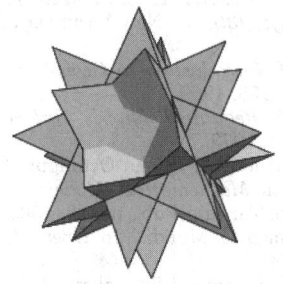

The UNIFORM POLYHEDRON U_{19}, also called the QUASI-TRUNCATED HEXAHEDRON, whose DUAL POLYHEDRON is the GREAT TRIAKIS OCTAHEDRON. It has SCHLÄFLI SYMBOL $t'\{4, 3\}$, WYTHOFF SYMBOL $2\,3\,|\,\frac{4}{3}$, and is Wenninger model W_{92}. Its faces are $8\{3\} + 6\{\frac{8}{3}\}$. For $a = 1$, its CIRCUMRADIUS is

$$R = \tfrac{1}{2} \sqrt{7 - 4\sqrt{2}}.$$

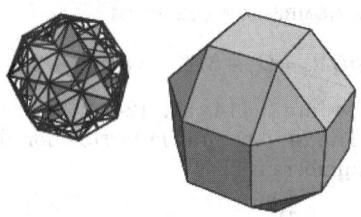

The CONVEX HULL of the stellated truncated hexahedron is the Archimedean SMALL RHOMBICUBOCTAHEDRON A_6, whose dual is the DELTOIDAL ICOSITETRAHEDRON, so the dual of the stellated truncated hexahedron (i.e., the GREAT TRIAKIS OCTAHEDRON) is one of the stellations of the DELTOIDAL ICOSITETRAHEDRON (Wenninger 1983, p. 57).

References

Wenninger, M. J. *Polyhedron Models.* Cambridge, England: Cambridge University Press, p. 144, 1989.

Stellation

The process of constructing POLYHEDRA by extending the facial PLANES past the EDGES of a given POLYHEDRON until they intersect (Wenninger 1989). The set of all possible EDGES of the stellations can be obtained by finding all intersections on the facial planes. Since the number and variety of intersections can become unmanageable for complicated polyhedra, additional rules are sometimes added to constrain allowable stellations. There exists a *Mathematica* function `Stellate[poly, ratio]` in the *Mathematica* add-on package `Graphics`Polyhedra`` (which can be loaded with the command `<<Graphics`),

although it only replaces facial planes with pyramids and does not perform true stellation.

There are no stellations of the CUBE or TETRAHEDRON (Wenninger 1989, p. 35). The only stellated form of the octahedron is the STELLA OCTANGULA, which is a compound of two TETRAHEDRA (Wenninger 1989, pp. 35 and 37). The DODECAHEDRON has three stellations: the SMALL STELLATED DODECAHEDRON, GREAT DODECAHEDRON, and GREAT STELLATED DODECAHEDRON (Wenninger 1989, pp. 35 and 38–0). Coxeter (1982) shows that 59 ICOSAHEDRON STELLATIONS exist, subject to certain restrictions.

The KEPLER-POINSOT SOLIDS, which consist of three DODECAHEDRON STELLATIONS and one of the ICOSAHEDRON STELLATIONS. The only STELLATIONS of PLATONIC SOLIDS which are UNIFORM POLYHEDRA are the three DODECAHEDRON STELLATIONS and one of the ICOSAHEDRON STELLATIONS.

There are three stellations of the RHOMBIC DODECAHEDRON (Wells 1991, pp. 216–17).

See also ARCHIMEDEAN SOLID STELLATION, DELTOIDAL ICOSITETRAHEDRON STELLATIONS, DODECAHEDRON STELLATIONS, FACETING, ICOSAHEDRON STELLATIONS, KEPLER-POINSOT SOLID, PLATONIC SOLID STELLATIONS, POLYHEDRON, POLYTOPE STELLATIONS, RECTIFICATION, RHOMBIC DODECAHEDRON STELLATIONS, RHOMBIC TRIACONTAHEDRON STELLATIONS, SMALL TRIAKIS OCTAHEDRON STELLATIONS, STELLA OCTANGULA, STELLATED POLYHEDRON, STELLATED TRUNCATED HEXAHEDRON, TRIAKIS TETRAHEDRON STELLATIONS, TRUNCATION, UNIFORM POLYHEDRON

References

Coxeter, H. S. M.; Du Val, P.; Flather, H. T.; and Petrie, J. F. *The Fifty-Nine Icosahedra.* Stradbroke, England: Tarquin Publications, 1999.

Cundy, H. and Rollett, A. *Mathematical Models, 3rd ed.* Stradbroke, England: Tarquin Publications, 1989.

Fleurent, G. M. "Symmetry and Polyhedral Stellation Ia and Ib. Symmetry 2: Unifying Human Understanding, Part 1." *Comput. Math. Appl.* **17**, 167–93, 1989.

Messer, P. W. "Stellations of the Rhombic Triacontahedron and Beyond." *Structural Topology* **21**, 25–6, 1995.

Messer, P. W. and Wenninger, M. J. "Symmetry and Polyhedral Stellation. II. Symmetry 2: Unifying Human Understanding, Part 1." *Comput. Math. Appl.* **17**, 195–01, 1989.

Wells, D. *The Penguin Dictionary of Curious and Interesting Geometry.* London: Penguin, 1991.

Wenninger, M. J. "Stellated Forms of Convex Duals." Ch. 3 in *Dual Models.* Cambridge, England: Cambridge University Press, pp. 36–8, 1983.

Wenninger, M. J. *Polyhedron Models.* New York: Cambridge University Press, 1989.

Stem-and-Leaf Diagram

The "stem" is a column of the data with the last digit removed. The final digits of each column are placed next to each other in a row next to the appropriate column. Then each row is sorted in numerical order. This diagram was invented by John Tukey.

References

Tukey, J. W. *Explanatory Data Analysis.* Reading, MA: Addison-Wesley, pp. 7–6, 1977.

Step

1.5 times the H-SPREAD.

See also FENCE, H-SPREAD

References

Tukey, J. W. *Explanatory Data Analysis.* Reading, MA: Addison-Wesley, p. 44, 1977.

Step Function

A function on the REALS $\mathbb{R}$ is a step function if it can be written as a finite linear combination of semi-open intervals $[a,\ b) \subseteq \mathbb{R}$. Therefore, a step function f can be written as

$$f(x) = \alpha_1 f_1(x) + \cdots + \alpha_n f_n(x).$$

where $\alpha_i \in \mathbb{R}$, $f_i(x) = 1$ if $x \in [a_i,\ b_i)$ and 0 otherwise, for $i = 1, ..., n$.

See also HEAVISIDE STEP FUNCTION

Step Polynomial

HERMITE'S INTERPOLATING POLYNOMIAL

Stephens' Constant

Let a and b be nonzero integers such that $a^m b^n \neq 1$ except when $m = n = 0$, and let $T(a,\ b)$ be the set of PRIMES p for which $p(a^k - b)$ for some NONNEGATIVE INTEGER k. Then assuming the generalized RIEMANN HYPOTHESIS, Stephens (1976) showed that the density of $T(a,\ b)$ relative to the primes is a rational multiple of

$$C_{\text{Stephens}} = \prod_{j=1}^{\infty} \left(1 - \frac{p_j}{p_j^3 - 1}\right) = 0.5759599688\ldots.$$

where p_j is the jth PRIME (Finch).

See also ARTIN'S CONSTANT

References

Finch, S. "Favorite Mathematical Constants." http://www.mathsoft.com/asolve/constant/artin/artin.html.
Moree, P. "Approximation of Singular Series and Automata." Submitted to *Manuscripta Math.* 1999.
Moree, P. and Stevenhagen, P. "A Two Variable Artin Conjecture." Submitted 1999.
Stephens, P. J. "Prime Divisor of Second-Order Linear Recurrences, I." *J. Number Th.* **8**, 313–32, 1976.

Steradian

The unit of SOLID ANGLE. The SOLID ANGLE corresponding to all of space being subtended is 4π steradians.

See also RADIAN, SOLID ANGLE

Stereogram

A plane image or pair of 2-D images which, when appropriately viewed using both eyes, produces an image which appears to be three-dimensional. By taking a pair of photographs from slightly different angles and then allowing one eye to view each image, a stereogram is not difficult to produce.

Amazingly, it turns out that the 3-D effect can be produced by both eyes looking at a *single* image by defocusing the eyes at a certain distance. Such stereograms are called "random-dot stereograms."

See also ANAGLYPH

References

Bar-Natan, D. "Random-Dot Stereograms." *Math. J.* **1**, 69–1, 1991.
Fineman, M. *The Nature of Visual Illusion.* New York: Dover, pp. 89–3, 1996.
Julesz, B. *Foundations of Cyclopean Perception.* Chicago, IL: University of Chicago Press, 1971.
Julesz, B. "Stereoscopic Vision." *Vision Res.* **26**, 1601–611, 1986.
Terrell, M. S. and Terrell, R. E. "Behind the Scenes of a Random Dot Stereogram." *Amer. Math. Monthly* **101**, 715–24, 1994.
Tyler, C. "Sensory Processing of Binocular Disparity." In *Vergence Eye Movements: Basic and Clinical Aspects.* Boston, MA: Butterworth, pp. 199–95, 1983.

Stereographic Projection

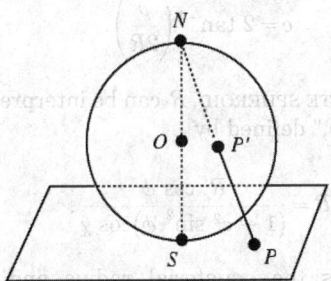

A MAP PROJECTION obtained by projecting points p' on the surface of sphere from the sphere's north pole N

to point P in a plane tangent to the south pole S (Coxeter 1969, p. 93). In such a projection, GREAT CIRCLES are mapped to CIRCLES, and LOXODROMES become LOGARITHMIC SPIRALS.

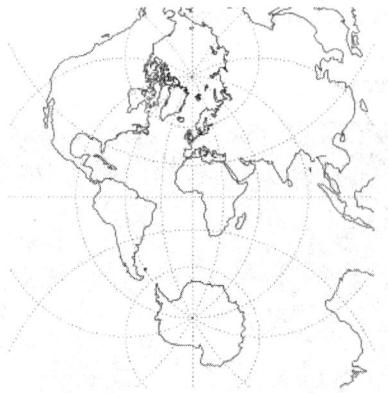

The transformation equations for a sphere of radius R are given by

$$x = k \cos \phi \sin(\lambda - \lambda_0) \quad (1)$$

$$y = k[\cos \phi_1 \sin \phi - \sin \phi_1 \cos \phi \cos(\lambda - \lambda_0)]. \quad (2)$$

where λ_0 is the central longitude, ϕ_1 is the central latitude, and

$$k = \frac{2R}{1 + \sin \phi_1 \sin \phi + \cos \phi_1 \cos \phi \cos(\lambda - \lambda_0)}. \quad (3)$$

The inverse FORMULAS for latitude ϕ and longitude λ are then given by

$$\phi = \sin^{-1}\left(\cos c \sin \phi_1 + \frac{y \sin c \cos \phi_1}{\rho}\right) \quad (4)$$

$$\lambda = \lambda_0 + \tan^{-1}\left(\frac{x \sin c}{\rho \cos \phi_1 \cos c - y \sin \phi_1 \sin c}\right), \quad (5)$$

where

$$\rho = \sqrt{x^2 + y^2} \quad (6)$$

$$c = 2 \tan^{-1}\left(\frac{\rho}{2R}\right). \quad (7)$$

For an OBLATE SPHEROID, R can be interpreted as the "local radius," defined by

$$R = \frac{R_e \cos \phi}{(1 - e^2 \sin^2 \phi)\cos \chi}, \quad (8)$$

where R_e is the equatorial radius and χ is the CONFORMAL LATITUDE.

See also GNOMONIC PROJECTION, MAP PROJECTION

References

Coxeter, H. S. M. *Introduction to Geometry, 2nd ed.* New York: Wiley, pp. 93 and 289–90, 1969.
Coxeter, H. S. M. and Greitzer, S. L. *Geometry Revisited.* Washington, DC: Math. Assoc. Amer., pp. 150–53, 1967.
Snyder, J. P. *Map Projections--A Working Manual.* U. S. Geological Survey Professional Paper 1395. Washington, DC: U. S. Government Printing Office, pp. 154–63, 1987.

Stereology

The exploration of 3-D space from 2-D sections of PROJECTIONS of solid bodies.

See also AXONOMETRY, BRIGHTNESS, CORK PLUG, CROSS SECTION, INNER QUERMASS, MEAN TANGENT DIAMETER, PROJECTION, SHADOW, TRIP-LET

References

Elias, H. and Hyde, D. M. (Eds.). *Guide to Practical Stereology.* S. Karger, 1983.
Elias, H. (Ed.). *Stereology.* New York: Springer-Verlag, 1967.

Stern-Brocot Tree

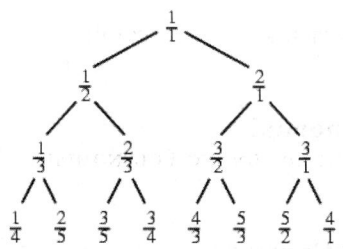

A special type of BINARY TREE obtained by starting with the fractions $\frac{0}{1}$ and $\frac{1}{0}$ and iteratively inserting $(m + m')/(n + n')$ between each two adjacent fractions m/n and m'/n'. The result can be arranged in tree form as illustrated above. The FAREY SEQUENCE F_n defines a subtree of the Stern-Brocot tree obtained by pruning off unwanted branches (Vardi 1991, Graham *et al.* 1994).

See also BINARY TREE, FAREY SEQUENCE, FORD CIRCLE

References

Brocot, A. "Calcul des rouages par approximation, nouvelle méthode." *Revue Chonométrique* **6**, 186–94, 1860.
Graham, R. L.; Knuth, D. E.; and Patashnik, O. *Concrete Mathematics: A Foundation for Computer Science, 2nd ed.* Reading, MA: Addison-Wesley, pp. 116–17, 1994.
Stern, M. A. "Über eine zahlentheoretische Funktion." *J. reine angew. Math.* **55**, 193–20, 1858.
Vardi, I. *Computational Recreations in Mathematica.* Redwood City, CA: Addison-Wesley, p. 253, 1991.
Viswanath, D. "Random Fibonacci Sequences and the Number 1.13198824...." *Math. Comput.* **69**, 1131–155, 2000.

Stevedore's Knot

The 6-crossing KNOT 06-01 having CONWAY-ALEXANDER POLYNOMIAL

$$\Delta(t) = 2t^2 - 5t + 2.$$

References

Rolfsen, D. *Knots and Links*. Wilmington, DE: Publish or Perish Press, p. 225, 1976.

Stewart's Theorem

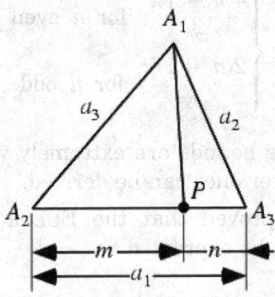

Let a CEVIAN A_1P be drawn on a TRIANGLE $\Delta A_1A_2A_3$, and denote the lengths $m = \overline{A_2P}$ and $n = \overline{PA_3}$, with $a_1 = m + n$. Then

$$ma_2^2 + na_3^2 = (m+n)\overline{A_1P}^2 + m\overline{PA_3}^2 + n\overline{PA_2}^2.$$

This theorem is sometimes also called APOLLONIUS' THEOREM.

References

Altshiller-Court, N. "Stewart's Theorem." §6B in *College Geometry: A Second Course in Plane Geometry for Colleges and Normal Schools, 2nd ed., rev. enl.* New York: Barnes and Noble, pp. 152–53, 1952.
Coxeter, H. S. M. and Greitzer, S. L. *Geometry Revisited.* Washington, DC: Math. Assoc. Amer., pp. 6, 10, and 31, 1967.

Stick Number

Let the stick number $s(K)$ of a KNOT K be the least number of straight sticks needed to make a KNOT K. The smallest stick number of any KNOT is $s(T) = 6$, where T is the TREFOIL KNOT. If J and K are KNOTS, then

$$s(J + K) \leq s(J) + s(K) + 1.$$

For a nontrivial KNOT K, let $c(K)$ be the CROSSING NUMBER (i.e., the least number of crossings in any projection of K). Then

$$\tfrac{1}{2}\left[5 + \sqrt{25 + 8(c(K) - 2)}\right] \leq s(K) \leq 2c(K).$$

The following table gives the stick number for some common knots.

TREFOIL KNOT	6
WHITEHEAD LINK	8

See also CROSSING NUMBER (LINK), TRIANGLE COUNTING

References

Adams, C. C. *The Knot Book: An Elementary Introduction to the Mathematical Theory of Knots.* New York: W. H. Freeman, pp. 27–0, 1994.

Stickelberger Relation

Let P be a PRIME IDEAL in D_m not containing m. Then

$$(\Phi(P)) = P^{\sum t \sigma_t^{-1}},$$

where the sum is over all $1 \leq t < m$ which are RELATIVELY PRIME to m. Here D_m is the RING of integers in $\mathbb{Q}(\zeta_m)$, $\Phi(P) = g(P)^m$, and other quantities are defined by Ireland and Rosen (1990).

See also PRIME IDEAL

References

Ireland, K. and Rosen, M. "The Stickelberger Relation and the Eisenstein Reciprocity Law." Ch. 14 in *A Classical Introduction to Modern Number Theory, 2nd ed.* New York: Springer-Verlag, pp. 203–27, 1990.

Stiefel Manifold

The Stiefel manifold of ORTHONORMAL k-frames in $\mathbb{R}^n$ is the collection of vectors $(v_1, ..., v_k)$ where v_i is in $\mathbb{R}^n$ for all i, and the k-tuple $(v_1, ..., v_k)$ is ORTHONORMAL. This is a submanifold of $\mathbb{R}^{nk}$, having DIMENSION $nk - (k+1)k/2$.

Sometimes the "orthonormal" condition is dropped in favor of the mildly weaker condition that the k-tuple ($v_1, ..., v_k$) is linearly independent. Usually, this does not affect the applications since Stiefel manifolds are usually considered only during HOMOTOPY THEORETIC considerations. With respect to HOMOTOPY THEORY, the two definitions are more or less equivalent since GRAM-SCHMIDT ORTHONORMALIZATION gives rise to a smooth deformation retraction of the second type of Stiefel manifold onto the first.

See also GRASSMANN MANIFOLD

Stiefel-Whitney Class

The ith Stiefel-Whitney class of a REAL VECTOR BUNDLE (or TANGENT BUNDLE or a REAL MANIFOLD) is in the ith cohomology group of the base SPACE involved. It is an OBSTRUCTION to the existence of $(n - i + 1)$ REAL linearly independent VECTOR FIELDS on that VECTOR BUNDLE, where n is the dimension of the FIBER. Here, OBSTRUCTION means that the ith Stiefel-Whitney class being NONZERO implies that there do *not* exist $(n-i+1)$ everywhere linearly dependent VECTOR FIELDS (although the Stiefel-Whitney classes are not always the OBSTRUCTION).

In particular, the nth Stiefel-Whitney class is the obstruction to the existence of an everywhere NONZERO VECTOR FIELD, and the first Stiefel-Whitney class of a MANIFOLD is the obstruction to orientability.

See also CHERN CLASS, OBSTRUCTION, PONTRYAGIN CLASS, STIEFEL-WHITNEY NUMBER

Stiefel-Whitney Number

The Stiefel-Whitney number is defined in terms of the STIEFEL-WHITNEY CLASS of a MANIFOLD as follows. For any collection of STIEFEL-WHITNEY CLASSES such that their cup product has the same DIMENSION as the MANIFOLD, this cup product can be evaluated on the MANIFOLD'S FUNDAMENTAL CLASS. The resulting number is called the PONTRYAGIN NUMBER for that combination of Pontryagin classes.

The most important aspect of Stiefel-Whitney numbers is that they are COBORDISM invariant. Together, PONTRYAGIN and Stiefel-Whitney numbers determine an oriented MANIFOLD'S COBORDISM class.

See also CHERN NUMBER, PONTRYAGIN NUMBER, STIEFEL-WHITNEY CLASS

Stieltjes Constants

N.B. A detailed online essay by S. Finch was the starting point for this entry. Expanding the RIEMANN ZETA FUNCTION about $z = 1$ gives

$$\zeta(z) = \frac{1}{z-1} + \sum_{n=0}^{\infty} \frac{(-1)^n}{n!} \gamma_n (z-1)^n, \qquad (1)$$

where

$$\gamma_n \equiv \lim_{m \to \infty} \left[\sum_{k=1}^{m} \frac{(\ln k)^n}{k} - \frac{(\ln m)^{n+1}}{n+1} \right]. \qquad (2)$$

These constants are returned by the *Mathematica* function StieltjesGamma[n]. An alternative definition is given by absorbing the coefficient of γ_n into the constant,

$$\gamma_n' \equiv \frac{(-1)^n}{n!} \gamma_n \qquad (3)$$

(e.g., Hardy 1912, Kluyver 1927).

The case $n = 0$ gives the usual EULER-MASCHERONI CONSTANT $\gamma_0 \equiv \gamma$. The first few numerical values are given in the following table.

n	γ_n
0	0.5772156649
1	−0.07281584548
2	−0.009690363192
3	0.002053834420
4	0.002325370065
5	0.0007933238173

Briggs (1955–956) proved that there infinitely many γ_n of each SIGN. Berndt (1972) gave upper bounds of

$$|\gamma_n| < \begin{cases} \dfrac{4(n-1)!}{\pi^n} & \text{for } n \text{ even} \\[2mm] \dfrac{2(n-1)!}{\pi^n} & \text{for } n \text{ odd.} \end{cases} \qquad (4)$$

However, these bounds are extremely weak, so it is likely that better ones can be derived.

Vacca (1910) proved that the EULER-MASCHERONI CONSTANT may be expressed as

$$\gamma = \sum_{k=1}^{\infty} \frac{(-1)^k}{k} \lfloor \lg k \rfloor, \qquad (5)$$

where $\lfloor x \rfloor$ is the FLOOR FUNCTION and the LG function $\lg x \equiv \log_2 x$ is the LOGARITHM to base 2.

Hardy (1912) gave the FORMULA

$$\frac{2\gamma_1}{\ln 2} = \sum_{k=1}^{\infty} \frac{(-1)^k}{k} [2 \lg k - \lfloor \lg(2k) \rfloor] \lfloor \lg k \rfloor. \qquad (6)$$

γ_1 is also given by the sum

$$\sum_{n=1}^{x} \frac{1}{n} \ln\left(\frac{x}{n}\right) = \tfrac{1}{2}(\ln x)^2 + \gamma \ln x - \gamma_1 + \mathcal{O}(x^{-1}), \qquad (7)$$

where γ_1 was called $-D$ and given incorrectly by Ellison and Mendès-France (1975) and the error was reproduced by Le Lionnais (1983, p. 47). The exact form of (7) is given by

$$\sum_{n=1}^{x} \frac{1}{n} \ln\left(\frac{x}{n}\right) = H_x \ln x - \zeta'(1, x+1) + \gamma_1, \qquad (8)$$

where H_x is a HARMONIC NUMBER, $\mathbb{Q}(\zeta_m)$ is the HURWITZ ZETA FUNCTION, and $\zeta'(1, a)$ denotes $\lim_{s \to 1} d\zeta(s, a)/dz|_{z=s}$.

Kluyver (1927) gave similar series for γ_n valid for all $n > 1$,

$\gamma_n =$

$$n!(\ln 2)^n \sum_{m=1}^{n+1} \frac{(-1)^{m-1}}{m!} \sum_{k=1}^{\infty} \frac{(-1)^k}{k \lfloor \lg k \rfloor^m} B_{1+n-m}\left(\frac{\ln k}{\ln 2}\right),$$

$$(9)$$

where $B_n(x)$ is a BERNOULLI POLYNOMIAL. However, this series converges extremely slowly, requiring more than 10^4 terms to get two digits of γ_1 and many more for higher order γ_n. γ_n can also be expressed as a single sum using

$$\gamma_n = \frac{(\ln 2)^n}{n+1} \sum_{k=1}^{\infty} \frac{(-1)^k}{k} B_{n+1}\left(\frac{\ln k}{\ln 2}\right). \qquad (10)$$

A set of constants related to γ_n is

$$\delta_n \equiv \lim_{m \to \infty} \left[\sum_{k=1}^{m} (\ln k)^n - \int_1^m (\ln x)^n \, dx - \tfrac{1}{2}(\ln m)^n \right] \quad (11)$$

(Sitaramachandrarao 1986, Lehmer 1988).

See also BERNOULLI POLYNOMIAL, EULER PRODUCT, RIEMANN ZETA FUNCTION

References

Berndt, B. C. "On the Hurwitz Zeta-Function." *Rocky Mountain J. Math.* **2**, 151–57, 1972.

Bohman, J. and Fröberg, C.-E. "The Stieltjes Function--Definitions and Properties." *Math. Comput.* **51**, 281–89, 1988.

Briggs, W. E. "Some Constants Associated with the Riemann Zeta-Function." *Mich. Math. J.* **3**, 117–21, 1955–1956.

Ellison, W. J. and Mendès-France, M. *Les nombres premiers.* Paris: Hermann, 1975.

Finch, S. "Favorite Mathematical Constants." http://www.mathsoft.com/asolve/constant/stltjs/stltjs.html.

Hardy, G. H. "Note on Dr. Vacca's Series for γ." *Quart. J. Pure Appl. Math.* **43**, 215–16, 1912.

Hardy, G. H. and Wright, E. M. "The Behavior of $\zeta(s)$ when $s \to 1$." §17.3 in *An Introduction to the Theory of Numbers, 5th ed.* Oxford, England: Clarendon Press, pp. 246–47, 1979.

Kluyver, J. C. "On Certain Series of Mr. Hardy." *Quart. J. Pure Appl. Math.* **50**, 185–92, 1927.

Knopfmacher, J. "Generalised Euler Constants." *Proc. Edinburgh Math. Soc.* **21**, 25–2, 1978.

Lammel, E. "Ein Beweis dass die Riemannsche Zetafunktion $\zeta(s)$ is $|s-1| \leq 1$ keine Nullstelle besitzt." *Univ. Nac. Tucmán Rev. Ser. A* **16**, 209–17, 1966.

Le Lionnais, F. *Les nombres remarquables.* Paris: Hermann, p. 47, 1983.

Lehmer, D. H. "The Sum of Like Powers of the Zeros of the Riemann Zeta Function." *Math. Comput.* **50**, 265–73, 1988.

Liang, J. J. Y. and Todd, J. "The Stieltjes Constants." *J. Res. Nat. Bur. Standards--Math. Sci.* **76B**, 161–78, 1972.

Sitaramachandrarao, R. "Maclaurin Coefficients of the Riemann Zeta Function." *Abstracts Amer. Math. Soc.* **7**, 280, 1986.

Vacca, G. "A New Series for the Eulerian Constant." *Quart. J. Pure Appl. Math.* **41**, 363–68, 1910.

Stieltjes Integral

The Stieltjes integral is a generalization of the RIEMANN INTEGRAL. Let $f(x)$ and $\alpha(x)$ be real-valued bounded functions defined on a CLOSED INTERVAL $[a, b]$. Take a partition of the INTERVAL

$$a = x_0 < x_1 < x_2, \ldots < x_{n-1} < x_n = b, \qquad (1)$$

and consider the Riemann sum

$$\sum_{i=0}^{n-1} f(\xi_i) \left[\alpha(x_{i+1}) - \alpha(x_i) \right] \qquad (2)$$

with $\xi_i \in [x_i, x_{i+1}]$. If the sum tends to a fixed number I as $\max(x_{i+1} - x_i) \to 0$, then I is called the Stieltjes integral, or sometimes the RIEMANN-STIELTJES INTEGRAL. The Stieltjes integral of f with respect to α is denoted

$$\int f(x) \, d\alpha(x) \qquad (3)$$

or sometimes simply

$$\int f \, d\alpha. \qquad (4)$$

If f and α have a common point of discontinuity, then the integral does not exist. However, if f is continuous and α' is Riemann integrable over the specified interval, then

$$\int f(x) \, d\alpha(x) = \int f(x) \alpha'(x) \, dx \qquad (5)$$

(Kestelman 1960).

For enumeration of many properties of the Stieltjes integral, see Dresher (1981, p. 105).

See also CONVOLUTION, RIEMANN INTEGRAL

References

Dresher, M. *The Mathematics of Games of Strategy: Theory and Applications.* New York: Dover, 1981.

Hardy, G. H.; Littlewood, J. E.; and Pólya, G. *Inequalities, 2nd ed.* Cambridge, England: Cambridge University Press, pp. 152–55, 1988.

Jeffreys, H. and Jeffreys, B. S. "Integration: Riemann, Stieltjes." §1.10 in *Methods of Mathematical Physics, 3rd ed.* Cambridge, England: Cambridge University Press, pp. 26–6, 1988.

Kestelman, H. "Riemann-Stieltjes Integration." Ch. 11 in *Modern Theories of Integration, 2nd rev. ed.* New York: Dover, pp. 247–69, 1960.

Pollard, S. *Quart. J. Math.* **49**, 73–38, 1923.

Stieltjes, T. J. *Ann. d. fac. d. sciences Toulouse* **8**, 68–5, 1894J.

Widder, D. V. Ch. 1 in *The Laplace Transform.* Princeton, NJ: Princeton University Press, 1941.

Stieltjes' Theorem

The $m+1$ ELLIPSOIDAL HARMONICS when κ_1, κ_2, and κ_3 are given can be arranged in such a way that the rth function has $r-1$ zeros between $-a^2$ and $-b^2$ and the remaining $m+r-1$ zeros between $-b^2$ and $-c^2$ (Whittaker and Watson 1990).

See also ELLIPSOIDAL HARMONIC

References

Whittaker, E. T. and Watson, G. N. *A Course in Modern Analysis, 4th ed.* Cambridge, England: Cambridge University Press, pp. 560–62, 1990.

Stieltjes Transform

The INTEGRAL TRANSFORM

$$(Kf)(x) = \int_{-\infty}^{\infty} \Gamma(p)(x+t)^{-p} f(t)\, dt.$$

References

Samko, S. G.; Kilbas, A. A.; and Marichev, O. I. *Fractional Integrals and Derivatives.* Yverdon, Switzerland: Gordon and Breach, p. 23, 1993.

StieltjesGamma

STIELTJES CONSTANTS

Stieltjes-Wigert Polynomial

Orthogonal POLYNOMIALS associated with WEIGHTING FUNCTION

$$w(x) = \pi^{-1/2} k \exp\left(-k^2 \ln^2 x\right) = \pi^{-1/2} k x^{-k^2 \ln x} \quad (1)$$

for $x \in (0, \infty)$ and $k > 0$. Using

$$\begin{bmatrix} n \\ v \end{bmatrix} = \frac{(1-q^n)(1-q^{n-1})\cdots(1-q^{n-v+1})}{(1-q)(1-q^2)\cdots(1-q^v)} \quad (2)$$

where $0 < v < n$,

$$\begin{bmatrix} n \\ 0 \end{bmatrix} = \begin{bmatrix} n \\ n \end{bmatrix} = 1, \quad (3)$$

and

$$q = \exp\left[-(2k^2)^{-1}\right]. \quad (4)$$

Then

$$p_n(x) = (-1)^n q^{n/2+1/4} \left[(1-q)(1-q^2)\cdots(1-q^n)\right]^{-1/2}$$
$$\times \sum_{v=0}^{n} \begin{bmatrix} n \\ v \end{bmatrix} q^{v^2} \left(-q^{1/2}x\right)^v \quad (5)$$

for $n > 0$ and

$$p_0(x) = q^{1/4}. \quad (6)$$

References

Szego, G. *Orthogonal Polynomials, 4th ed.* Providence, RI: Amer. Math. Soc., p. 33, 1975.

Stiff Differential Equation

References

Byrne, G. D. and Hindmarsh, A. C. "Stiff ODE Solvers: A Review of Current and Coming Attractions." *J. Comput. Phys.* **70**, 1–2, 1987.
Enright, W. H.; Hull, T. E.; and Lindberg, B. "Comparing Numerical Methods for Stiff Systems of ODEs." *BIT* **15**, 10–8, 1975.
Enright, W. H. and Hull, T. E. "Comparing Numerical Methods for the Solution of Stiff Systems of ODEs Arising in Chemistry." In *Numerical Methods for Differential Systems, Recent Developments in Algorithms, Software and Applications* (Ed. L. Lapidus and W. E. Schiesser). New York: Academic Press, pp. 45–6, 1976.
Hairer, E. and Wanner, G. *Solving Ordinary Differential Equations II: Stiff and Differential-Algebraic Problems, 2nd rev. ed.* Berlin: Springer-Verlag, 1996.
Shampine, L. F. "Ill-Conditioned Matrices and the Integration of Stiff ODEs." *J. Comput. Appl. Math.* **48**, 279–92, 1993.

Stirling Cycle Number

STIRLING NUMBER OF THE FIRST KIND

Stirling Number of the First Kind

The *signed* Stirling numbers of the first kind are variously denoted $s(n, m)$ (Riordan 1980, Roman 1984), $S_n^{(m)}$ (Fort 1958, Abramowitz and Stegun 1971), S_n^m (Jordan 1950). Abramowitz and Stegun (1971, p. 822) summarize the various notational conventions, which can be a bit confusing (especially since an *unsigned* version $S_1(n, m) = |s(n, m)|$ is also in common use). The signed Stirling number of the first kind $s(n, m)$ is are returned by `StirlingS1[n, m]` in *Mathematica*.

The signed Stirling numbers of the first kind $s(n, m)$ are defined such that the number of PERMUTATIONS of n elements which contain exactly m CYCLES is the *nonnegative* number

$$|s(n, m)| = (-1)^{n-m} s(n, m). \quad (1)$$

This means that $s(n, m) = 0$ for $m > n$ and $s(n, n) = 1$. A related set of numbers is known as the associated Stirling numbers of the first kind. Both these are the usual Stirling numbers of the first kind are special cases of a general function $d_r(n, k)$ which is related to the number of cycles in a permutation.

The triangle of signed Stirling numbers of the first kind is

$$1$$
$$-1 \quad 1$$
$$2 \quad -3 \quad 1$$
$$-6 \quad 11 \quad -6 \quad 1$$
$$24 \quad -50 \quad 35 \quad -10 \quad 1$$

(Sloane's A008275). Special values include

$$s(n, 0) = \delta_{n0} \qquad (2)$$

$$s(n, 1) = (-1)^{n-1}(n-1)! \qquad (3)$$

$$s(n, 2) = (-1)^n (n-1)! H_{n-1} \qquad (4)$$

$$s(n, 3) = \tfrac{1}{2}(-1)^{n-1}(n-1)! \left[H_{n-1}^2 - H_{n-1}^{(2)} \right] \qquad (5)$$

$$s(n, n-1) = -\binom{n}{2}, \qquad (6)$$

where δ_{mn} is the KRONECKER DELTA, H_n is a HARMONIC NUMBER, $H_n^{(r)}$ is a HARMONIC NUMBER of order r, and $\binom{n}{k}$ is a BINOMIAL COEFFICIENT.

The GENERATING FUNCTION for the Stirling numbers of the first kind is

$$(x)_n = x(x-1)\cdots(x-n+1) = \sum_{m=0}^{n} s(n, m)x^m, \qquad (7)$$

where $(x)_n$ is a FALLING FACTORIAL. Other generating functions are

$$\sum_{k=0}^{n} s(n, k)x^k = (1+x-n)_n \qquad (8)$$

$$\sum_{k=0}^{n} s(n, k)x^k = (-1)^n n! \binom{n-x-1}{n} \qquad (9)$$

$$\sum_{k=m}^{\infty} s(k, m)x^k = \frac{[\ln(x+1)]^m}{m!} \qquad (10)$$

$$\prod_{k=1}^{n}(1+kx) = \sum_{k=1}^{n+1}(-1)^{n+1-k} s(n+1, k)x^{n+1-k}. \qquad (11)$$

The Stirling numbers of the first kind satisfies the RECURRENCE RELATION

$$s(n+1, m) = s(n, m-1) - ns(n, m) \qquad (12)$$

for $1 \le m \le n$ and the sum identities

$$s(n, m) = \sum_{k=m}^{n} n^{k-m} s(n+1, k+1) \qquad (13)$$

for $m \ge 1$ and

$$\binom{m}{r} s(n, m) = \sum_{k=m-r}^{n-r} \binom{n}{k} s(n-k, r)(k, m-r) \qquad (14)$$

for $0 \le r \le m$, where $\binom{n}{k}$ is a BINOMIAL COEFFICIENT.

The Stirling numbers of the first kind $s(n, m)$ are connected with the STIRLING NUMBERS OF THE SECOND KIND $S(n, m)$ through the formulas

$$s(n, i) = \sum_{k=i}^{n} \sum_{j=0}^{k} s(n, k)s(k, j)S(j, i) \qquad (15)$$

$$S(n, i) = \sum_{k=i}^{n} \sum_{j=0}^{k} S(n, k)S(k, j)s(j, i) \qquad (16)$$

(Roman 1984, p. 67), as well as

$$S(n, m) = \sum_{k=0}^{n-m}(-1)^k \binom{k+n-1}{k+n-m}$$
$$\times \binom{2n-m}{n-k-m} s(k-m+n, k) \qquad (17)$$

$$s(n, m) = \sum_{k=0}^{n-m}(-1)^k \binom{k+n-1}{k+n-m}$$
$$\times \binom{2n-m}{n-k-m} s(k-m+n, k) \qquad (18)$$

$$\sum_{l=0}^{\max(k, j)+1} s(l, j)S(k, 1) = \delta_{jk} \qquad (19)$$

$$\sum_{l=0}^{\max(k, j)+1} s(k, l)S(l, j) = \delta_{jm}. \qquad (20)$$

The NONNEGATIVE version simply gives the number of PERMUTATIONS of n objects having m CYCLES (with cycles in opposite directions counted as distinct) and is obtained by taking the ABSOLUTE VALUE of the signed version. The nonnegative Stirling numbers of the first kind are variously denoted

$$S_1(n, m) \equiv \begin{bmatrix} n \\ m \end{bmatrix} \equiv |s(n, m)| \qquad (21)$$

(Graham *et al.* 1994). Diagrams illustrating $S_1(5, 1) = 24$, $S_1(5, 3) = 35$, $S_1(5, 4) = 10$, and $S_1(5, 5) = 1$ (Dickau) are shown below.

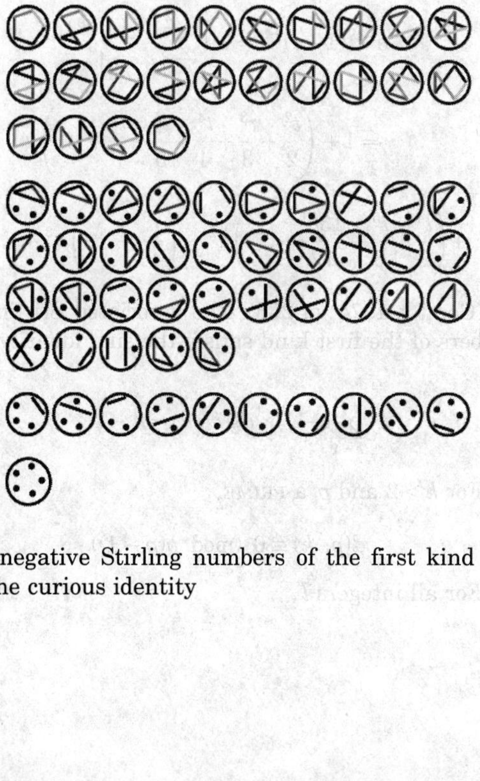

The nonnegative Stirling numbers of the first kind satisfy the curious identity

$$\sum_{n=1}^{\infty}\left[\sum_{k=0}^{n-2}\frac{(e^x-x-1)^{k+1}S_1(n,\,n-k)}{(k+1)!}\right]e^{-xn}$$
$$=\ln(x+1) \tag{22}$$

(Gosper) and satisfy

$$S_1(n+1,\,k)=nS_1(n,\,k)+S_1(n,\,k-1). \tag{23}$$

The Stirling numbers can be generalized to noninte-gral arguments (a sort of "Stirling polynomial") using the identity

$$\frac{\Gamma(j+h)}{j^h\Gamma(j)}=\sum_{k=0}^{\infty}\frac{S_1(h,\,h-k)}{j^k}$$

$$=1+\frac{(h-1)h}{2j}+\frac{(h-2)(3h-1)(h-1)h}{24j^2}$$

$$+\frac{(h-3)(h-2)(h-1)^2h^2}{48j^3}+\cdots \tag{24}$$

which is a generalization of an ASYMPTOTIC SERIES for a ratio of GAMMA FUNCTIONS $\Gamma(j+1/2)/\Gamma(j)$ (Gosper).

The associated Stirling numbers of the first kind $d_2(n,\,k)=d(n,\,k)$ are defined as the number of per-mutations of a given number n having exactly k CYCLES, all of which are of length $r=2$ or greater (Comtet 1974, p. 256; Riordan 1980, p. 75). They are a special case of the more general numbers $d_r(n,\,k)$, and have the RECURRENCE RELATION

$$d_2(n+1,\,k)=n[d_2(n,\,k)+d_2(n-1,\,k-1)] \tag{25}$$

with initial conditions $d_2(n,\,k)=0$ for $n\le 2k-1$, and $d_2(n,\,1)=(n-1)!$ (Appell 1880; Tricomi 1951; Carlitz 1958; Comtet 1974, pp. 256, 293, and 295) with . The GENERATING FUNCTION for $d_2(n,\,k)$ is given by

$$e^{-tu}(1-t)^{-u}=1+\sum_{k=1}^{n/2}\frac{d_2(n,\,k)}{n!}\,t^nu^k$$

$$=1+\left(\frac{t^2}{2}+\frac{t^3}{3}+\frac{t^4}{4}+\frac{t^5}{5}+\frac{t^6}{6}+\cdots\right)u$$

$$+\left(\frac{t^4}{8}+\frac{t^5}{6}+\frac{13t^6}{72}+\cdots\right)u^2+\left(\frac{t^6}{48}+\cdots\right)u^3+\cdots \tag{26}$$

(Comtet 1974, p. 256). The associated Stirling num-bers of the first kind satisfy the sum identity

$$\sum_{k=1}^{n}(-1)^{k-1}d_2(n,\,k)=n-1. \tag{27}$$

For $k\ge 2$ and p a PRIME,

$$d(p,\,k)\equiv 0\pmod{p(p-1)}. \tag{28}$$

For all integers l,

$$\sum_m(-1)^md_2(l+m,\,m)=(-1)^l, \tag{29}$$

and similarly,

$$\sum_m\frac{(-1)^md_2(l+m,\,m)}{l+m-1}=0 \tag{30}$$

(Comtet 1974, p. 256).

Special cases of the associated Stirling numbers of the first kind are given by

$$d_2(n,\,1)=(n-1)! \tag{31}$$

$$d_2(2k,\,k)=(2k-1)!! \tag{32}$$

$$d_2(2k+1,\,k)=\frac{(2k+1)a_k!}{3(k-1)!2^k} \tag{33}$$

$$d_2(2k+2,\,k)=\frac{(4k+5)(2k+2)!}{18(k-1)!2^k} \tag{34}$$

(Comtet 1974, p. 256), where a_k is a coefficient in the expansion of $(1+3x)/(1-2x)^{7/2}$: 1, 10, 105, 1260, 17325, ... (Sloane's A000457), omitted in Comtet (1974). The triangle of these numbers is given by

$$1$$
$$2$$
$$6,\ 3$$
$$24,\ 20$$
$$120,\ 130,\ 15$$
$$720,\ 924,\ 210$$
$$5040,\ 7308,\ 2380,\ 105$$

(Sloane's A008306).

See also CYCLE (PERMUTATION), HARMONIC NUMBER, PERMUTATION, STIRLING NUMBER OF THE SECOND KIND, STIRLING POLYNOMIAL, STIRLING TRANSFORM

References

Abramowitz, M. and Stegun, C. A. (Eds.). "Stirling Numbers of the First Kind." §24.1.3 in *Handbook of Mathematical Functions with Formulas, Graphs, and Mathematical Tables, 9th printing.* New York: Dover, p. 824, 1972.

Adamchik, V. "On Stirling Numbers and Euler Sums." *J. Comput. Appl. Math.* **79**, 119–30, 1997.

Appell, P. "Développments en série entière de $(1+ax)^{1/x}$." *Grunert Archiv* **65**, 171–75, 1880.

Butzer, P. L. and Hauss, M. "Stirling Functions of the First and Second Kinds; Some New Applications." *Israel Mathematical Conference Proceedings: Approximation, Interpolation, and Summability, in Honor of Amnon Jakimovski on his Sixty-Fifth Birthday* (Ed. S. Baron and D. Leviatan). Ramat Gan, Israel: IMCP, pp. 89–08, 1991.

Carlitz, L. "On Some Polynomials of Tricomi." *Boll. Un. M. Ital.* **13**, 58–4, 1958.

Comtet, L. *Advanced Combinatorics: The Art of Finite and Infinite Expansions, rev. enl. ed.* Dordrecht, Netherlands: Reidel, 1974.

Conway, J. H. and Guy, R. K. In *The Book of Numbers.* New York: Springer-Verlag, pp. 91–2, 1996.

David, F. N.; Kendall, M. G.; and Barton, D. E. *Symmetric Function and Allied Tables.* Cambridge, England: Cambridge University Press, p. 226, 1966.

Dickau, R. M. "Stirling Numbers of the First Kind." http://forum.swarthmore.edu/advanced/robertd/stirling1.html.

Fort, T. *Finite Differences.* Oxford, England: Clarendon Press, 1948.

Graham, R. L.; Knuth, D. E.; and Patashnik, O. "Stirling Numbers." §6.1 in *Concrete Mathematics: A Foundation for Computer Science, 2nd ed.* Reading, MA: Addison-Wesley, pp. 257–67, 1994.

Hauss, M. *Verallgemeinerte Stirling, Bernoulli und Euler Zahlen, deren Anwendungen und schnell konvergente Reihen für Zeta Funktionen.* Aachen, Germany: Verlag Shaker, 1995.

Jordan, C. *Calculus of Finite Differences, 3rd ed.* New York: Chelsea, 1965.

Knuth, D. E. "Two Notes on Notation." *Amer. Math. Monthly* **99**, 403–22, 1992.

Riordan, J. *An Introduction to Combinatorial Analysis.* New York: Wiley, 1980.

Roman, S. *The Umbral Calculus.* New York: Academic Press, pp. 59–3, 1984.

Sloane, N. J. A. Sequences A000457/M4736, A008275, and A008306 in "An On-Line Version of the Encyclopedia of Integer Sequences." http://www.research.att.com/~njas/sequences/eisonline.html.

Stirling, J. *Methodus differentialis, sive tractatus de summation et interpolation serierum infinitarium.* London, 1730. English translation by Holliday, J. *The Differential Method: A Treatise of the Summation and Interpolation of Infinite Series.* 1749.

Tricomi, F. G. "A Class of Non-Orthogonal Polynomials Related to those of Laguerre." *J. Analyse M.* **1**, 209–31, 1951.

Young, P. T. "Congruences for Bernoulli, Euler, and Stirling Numbers." *J. Number Th.* **78**, 204–27, 1999.

Stirling Number of the Second Kind

The number of ways of partitioning a set of n elements into m nonempty SETS (i.e., m BLOCKS), also called a STIRLING SET NUMBER. for example, the SET $\{1, 2, 3\}$ can be partitioned into three SUBSETS in one way: $\{\{1\}, \{2\}, \{3\}\}$; into two SUBSETS in three ways: $\{\{1, 2\}, \{3\}\}, \{\{1, 3\}, \{2\}\}$, and $\{\{1\}, \{2, 3\}\}$; and into one SUBSET in one way: $\{\{1, 2, 3\}\}$.

The Stirling numbers of the second kind are variously denoted $S(n, m)$ (Riordan 1980, Roman 1984), $S_n^{(m)}$ (Fort 1958, Abramowitz and Stegun 1971), $\mathfrak{S}_n^m$ (Jordan 1950), $s_n^{(m)}$, $S_2(n, m)$, or $\left\{{n \atop m}\right\}$ (Graham *et al.* 1994). Abramowitz and Stegun (1971, p. 822) summarize the various notational conventions, which can be a bit confusing. The *Mathematica* command for a Stirling number of the second kind is StirlingS2$[n, m]$. The Stirling numbers of the second kind for three elements are

$$S(3, 1) = 1 \tag{1}$$

$$S(3, 2) = 3 \tag{2}$$

$$S(3, 3) = 1. \tag{3}$$

Since a set of n elements can only be partitioned in a

single way into 1 or n SUBSETS,

$$S(n, 1) = S(n, n) = 1. \tag{4}$$

Other special cases include

$$S(n, 0) = \delta_{n0} \tag{5}$$

$$S(n, 2) = 2^{n-1} - 1 \tag{6}$$

$$S(n, n-1) = \binom{n}{2}. \tag{7}$$

The triangle of Stirling numbers of the second kind is

$$1$$
$$1 \quad 1$$
$$1 \quad 3 \quad 1$$
$$1 \quad 7 \quad 6 \quad 1$$
$$1 \quad 15 \quad 25 \quad 10 \quad 1$$
$$1 \quad 31 \quad 90 \quad 65 \quad 15 \quad 1$$

(Sloane's A008277), the nth row of which corresponds to the coefficients of the EXPONENTIAL POLYNOMIAL $\phi_n(x)$.

The Stirling numbers of the second kind can be computed from the sum

$$S(n, k) = \frac{1}{k!} \sum_{i=0}^{k-1} (-1)^i \binom{k}{i} (k-i)^n, \tag{8}$$

with $\binom{n}{k}$ a BINOMIAL COEFFICIENT, or the GENERATING FUNCTIONS

$$x^n = \sum_{m=0}^{n} S(n, m)(x)_m$$

$$= \sum_{m=0}^{n} S(n, m) x(x-1) \cdots (x-m+1), \tag{9}$$

where $(x)_m$ is the FALLING FACTORIAL (Roman 1984, pp. 60 and 101),

$$\sum_{n \geq k} S(n, k) \frac{x^n}{n!} = \frac{1}{k!} (e^x - 1)^k, \tag{10}$$

and

$$\frac{1}{(1-x)(1-2x)\cdots(1-kx)} = \sum_{n=1}^{k} S(n, k) x^n. \tag{11}$$

Other generating functions are

$$\sum_{k=1}^{n} S(n, k)(k-1)! z^k = (-1)^n \mathrm{Li}_{1-n}(1+1/z) \tag{12}$$

for $n \geq 2$, where $\mathrm{Li}_n(z)$ is the POLYLOGARITHM, and

$$\sum_{k=m}^{\infty} S(k, m)z^k = \frac{z^m}{\prod_{k=1}^{\infty}(1 - kz)}. \qquad (13)$$

Stirling numbers of the second kind are intimately connected with the POISSON DISTRIBUTION through the identity

$$\sum_{k=0}^{\infty} \frac{e^{-x}x^k}{k!} k^n = \sum_{k=1}^{n} x^k S(n, k). \qquad (14)$$

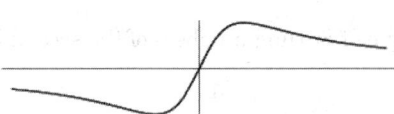

The above diagrams (Dickau) illustrate the definition of the Stirling numbers of the second kind $S(n, m)$ for $n = 3$ and 4. Stirling numbers of the second kind obey the RECURRENCE RELATIONS

$$S(n, k) = S(n-1, k-1) + kS(n-1, k) \qquad (15)$$

$$S(n, k) = \sum_{m=k}^{n} k^{n-m} S(m-1, k-1). \qquad (16)$$

The STIRLING NUMBERS OF THE FIRST KIND $s(n, m)$ are connected with the Stirling numbers of the second kind $S(n, m)$ through the formulas

$$s(n, i) = \sum_{k=i}^{n} \sum_{j=0}^{k} s(n, k)s(k, j)S(j, i) \qquad (17)$$

$$S(n, i) = \sum_{k=i}^{n} \sum_{j=0}^{k} S(n, k)S(k, j)s(j, i) \qquad (18)$$

(Roman 1984, p. 67), as well as

$$S(n, m) = \sum_{k=0}^{n-m} (-1)^k \binom{k+n-1}{k+n-m}$$
$$\times \binom{2n-m}{n-k-m} s(k-m+n, k) \qquad (19)$$

$$s(n, m) = \sum_{k=0}^{n-m} (-1)^k \binom{k+n-1}{k+n-m}$$
$$\times \binom{2n-m}{n-k-m} s(k-m+n, k) \qquad (20)$$

$$\sum_{l=0}^{\max(k, j)+1} s(l, j)S(k, 1) = \delta_{jk} \qquad (21)$$

$$\sum_{l=0}^{\max(k, j)+1} s(k, l)S(l, j) = \delta_{jm}. \qquad (22)$$

Identities involving Stirling numbers of the second kind are given by

$$\sum_{m=1}^{n} (-1)^m (m-1)! S(n, m) = 0 \qquad (23)$$

$$\sum_{k=0}^{m} k^n = \sum_{k=0}^{n} k! \binom{m+1}{k+1} S(n, k) \qquad (24)$$

$$f(m, n) \equiv \sum_{k=1}^{\infty} k^n \left(\frac{m}{m+1}\right)^l$$
$$= (m+1) \sum_{k=1}^{m} k! S(n, k)m^k. \qquad (25)$$

It turns out that $f(1, n)$ can have only 0, 2, or 6 as a last DIGIT (Riskin 1995).

The Stirling numbers of the second appear in the operator identity

$$(x\tilde{D})^n = \sum_{k=0}^{n} S(n, k)x^k f^{(k)}, \qquad (26)$$

where $\tilde{D}$ is the differential operator d/dx (Roman 1984, p. 144), giving

$$(x\tilde{D})^1 = x\tilde{D} \qquad (27)$$

$$(x\tilde{D})^2 = x\tilde{D} + x^2\tilde{D}^2 \qquad (28)$$

$$(x\tilde{D})^3 = x\tilde{D} + 3x^2\tilde{D}^2 + x^3\tilde{D}^3 \qquad (29)$$

$$(x\tilde{D})^4 = x\tilde{D} + 7x^2\tilde{D}^2 + 6x^3\tilde{D}^3 + x^4\tilde{D}^4 \qquad (30)$$

and so on. Similarly,

$$[(x-a)\tilde{D}]^n = \sum_{k=0}^{n} S(n, k)(x-a)^k \tilde{D}^k \qquad (31)$$

(Roman 1984, p. 146).

See also BELL NUMBER, COMBINATION LOCK, EXPONENTIAL POLYNOMIAL, LENGYEL'S CONSTANT, MINIMAL COVER, POISSON DISTRIBUTION, STIRLING NUMBER OF THE FIRST KIND, STIRLING POLYNOMIAL, STIRLING TRANSFORM

References

Abramowitz, M. and Stegun, C. A. (Eds.). "Stirling Numbers of the Second Kind." §24.1.4 in *Handbook of Mathematical Functions with Formulas, Graphs, and Mathematical Tables, 9th printing.* New York: Dover, pp. 824–25, 1972.

Butzer, P. L. and Hauss, M. "Stirling Functions of the First and Second Kinds; Some New Applications." *Israel Mathematical Conference Proceedings: Approximation, Interpolation, and Summability, in Honor of Amnon Jakimovski on his Sixty-Fifth Birthday* (Ed. S. Baron and D. Leviatan). Ramat Gan, Israel: IMCP, pp. 89–08, 1991.

Comtet, L. *Advanced Combinatorics: The Art of Finite and Infinite Expansions, rev. enl. ed.* Dordrecht, Netherlands: Reidel, 1974.

Conway, J. H. and Guy, R. K. In *The Book of Numbers.* New York: Springer-Verlag, pp. 91–2, 1996.

Dickau, R. M. "Stirling Numbers of the Second Kind." http://forum.swarthmore.edu/advanced/robertd/stirling2.html

Dickau, R. "Visualizing Combinatorial Enumeration." *Mathematica in Educ. Res.* **8**, 11–8, 1999.

Fort, T. *Finite Differences.* Oxford, England: Clarendon Press, 1948.

Graham, R. L.; Knuth, D. E.; and Patashnik, O. "Stirling Numbers." §6.1 in *Concrete Mathematics: A Foundation for Computer Science, 2nd ed.* Reading, MA: Addison-Wesley, pp. 257–67, 1994.

Jordan, C. *Calculus of Finite Differences, 3rd ed.* New York: Chelsea, 1965.

Knuth, D. E. "Two Notes on Notation." *Amer. Math. Monthly* **99**, 403–22, 1992.

Riordan, J. *Combinatorial Identities.* New York: Wiley, 1979.

Riordan, J. *An Introduction to Combinatorial Analysis.* New York: Wiley, 1980.

Riskin, A. "Problem 10231." *Amer. Math. Monthly* **102**, 175–76, 1995.

Roman, S. *The Umbral Calculus.* New York: Academic Press, pp. 59–3, 1984.

Sloane, N. J. A. Sequences A008277 in "An On-Line Version of the Encyclopedia of Integer Sequences." http://www.research.att.com/~njas/sequences/eisonline.html.

Stanley, R. P. *Enumerative Combinatorics, Vol. 1.* Cambridge, England: Cambridge University Press, 1997.

Stirling, J. *Methodus differentialis, sive tractatus de summation et interpolation serierum infinitarium.* London, 1730. English translation by Holliday, J. *The Differential Method: A Treatise of the Summation and Interpolation of Infinite Series.* 1749.

Young, P. T. "Congruences for Bernoulli, Euler, and Stirling Numbers." *J. Number Th.* **78**, 204–27, 1999.

Stirling Polynomial

Polynomials $S_k(x)$ which form the SHEFFER SEQUENCE for

$$g(t) = e^{-t} \tag{1}$$

$$f^{-1}(t) = \ln\left(\frac{1}{1 - e^{-t}}\right), \tag{2}$$

where $f^{-1}(t)$ is the INVERSE FUNCTION of $f(t)$, and have GENERATING FUNCTION

$$\sum_{k=0}^{\infty} \frac{S_k(x)}{k!} t^k = \left(\frac{t}{1 - e^{-t}}\right)^{x+1}. \tag{3}$$

The first few polynomials are

$$S_0(x) = 1$$
$$S_1(x) = \tfrac{1}{2}(x + 1)$$
$$S_2(x) = \tfrac{1}{12}(3x + 2)(x + 1)$$
$$S_3(x) = \tfrac{1}{8} x(x + 1)^2.$$

The Stirling polynomials are related to the STIRLING NUMBERS OF THE FIRST KIND $s(n, m)$ by

$$S_n(m) = \frac{(-1)^n}{\binom{m}{n}} s(m + 1, m - n + 1), \tag{4}$$

where $\binom{m}{n}$ is a BINOMIAL COEFFICIENT and m is an integer with $m \geq n$, and to STIRLING NUMBERS OF THE SECOND KIND $S(n, m)$ by

$$S_n(m) = \frac{(-1)^n n!}{(n - m - 1)!} S(n - m - 1, -m - 1) \tag{5}$$

for m a NEGATIVE INTEGER.

See also STIRLING NUMBER OF THE FIRST KIND, STIRLING NUMBER OF THE SECOND KIND

References

Erdélyi, A.; Magnus, W.; Oberhettinger, F.; and Tricomi, F. G. *Higher Transcendental Functions, Vol. 3.* New York: Krieger, p. 257, 1981.

Roman, S. *The Umbral Calculus.* New York: Academic Press, 1984.

Stirling Set Number

STIRLING NUMBER OF THE SECOND KIND

Stirling Transform

The transformation of a sequence a_1, a_2, ... into a sequence b_1, b_2, ..., by the formula

$$b_n = \sum_{k=0}^{n} S(n, k) a_k,$$

where $S(n, k)$ is a STIRLING NUMBER OF THE SECOND KIND. The inverse transform is given by

$$a_n = \sum_{k=0}^{n} s(n, k) b_k,$$

where $s(n, k)$ is a STIRLING NUMBER OF THE FIRST KIND (Sloane and Plouffe 1995, p. 23).

The Stirling transform of $a_n = 1$ for all n gives the BELL NUMBERS 1, 2, 5, 15, 52, ... (Sloane's A000110). The Stirling transform of $a_n = n$ gives 1, 3, 10, 37, 151, 674, ... (Sloane's A005493), which has EXPONENTIAL GENERATING FUNCTION

$$g(x) = \exp(e^x + 2x - 1).$$

The Stirling transform of the sequence $a_n = 1$ for n prime and $a_n = 0$ for n composite is 0, 1, 4, 13, 41, 136, 505, ... The Stirling transform of the sequence $a_n = 1$ for n even and $a_n = 0$ for n odd is 0, 1, 3, 8, 25, 97, 434, 2095, ... (Sloane's A024430). The Stirling transform of the sequence $a_n = 0$ for n even and $a_n = 1$ for n odd is 1, 1, 2, 7, 27, 106, 443, ... (Sloane's A024429). The inverse Stirling transform of $b_n = n$ is given by the sequence of signed factorials 1, 1, -1, 2, -6, 24, -120, ...

See also BINOMIAL TRANSFORM, EULER TRANSFORM, EXPONENTIAL TRANSFORM, MÖBIUS TRANSFORM, STIRLING NUMBER OF THE FIRST KIND, STIRLING NUMBER OF THE SECOND KIND

References

Bernstein, M. and Sloane, N. J. A. "Some Canonical Sequences of Integers." *Linear Algebra Appl.* **226//228**, 57–2, 1995.

Graham, R. L.; Knuth, D. E.; and Patashnik, O. "Factorial Factors." §4.4 in *Concrete Mathematics: A Foundation for Computer Science, 2nd ed.* Reading, MA: Addison-Wesley, p. 252, 1994.

Riordan, J. *Combinatorial Identities.* New York: Wiley, p. 90, 1979.

Riordan, J. *An Introduction to Combinatorial Analysis.* New York: Wiley, p. 48, 1980.

Sloane, N. J. A. Sequences A000110/M1483, A005493/M2851, A024429, A024430, and A052437 in "An On-Line Version of the Encyclopedia of Integer Sequences." http://www.research.att.com/~njas/sequences/eisonline.html.

Sloane, N. J. A. and Plouffe, S. *The Encyclopedia of Integer Sequences.* San Diego, CA: Academic Press, 1995.

Stirling's Approximation

Stirling's approximation gives an approximate value for the FACTORIAL function $n!$ or the GAMMA FUNCTION $\Gamma(n)$ for $n \gg 1$. The approximation can most simply be derived for n an INTEGER by approximating the sum over the terms of the FACTORIAL with an INTEGRAL, so that

$$\ln n! = \ln 1 + \ln 2 + \ldots + \ln n = \sum_{k=1}^{n} \ln k \approx \int_{1}^{n} \ln x \, dx$$

$$= [x \ln x - x]_{1}^{n} = n \ln n - n + 1 \approx n \ln n - n. \quad (1)$$

The equation can also be derived using the integral definition of the FACTORIAL,

$$n! = \int_{0}^{\infty} e^{-x} x^n \, dx. \quad (2)$$

Note that the derivative of the LOGARITHM of the integrand can be written

$$\frac{d}{dx} \ln(e^{-x} x^n) = \frac{d}{dx}(n \ln x - x) = \frac{n}{x} - 1. \quad (3)$$

The integrand is sharply peaked with the contribution important only near $x = n$. Therefore, let $x \equiv n + \xi$ where $\xi \ll n$, and write

$$\ln(x^n e^{-x}) = n \ln x - x = n \ln(n + \xi) - (n + \xi). \quad (4)$$

Now,

$$\ln(n + \xi) = \ln\left[n\left(1 + \frac{\xi}{n}\right)\right] = \ln n + \ln\left(1 + \frac{\xi}{n}\right)$$

$$= \ln n + \frac{\xi}{n} - \frac{1}{2}\frac{\xi}{n^2} + \cdots, \quad (5)$$

so

$$\ln(x^n e^{-n}) = n \ln(n + \xi) - (n + \xi)$$

$$= n \ln n + \xi - \frac{1}{2}\frac{\xi^2}{n} - n - \xi + \ldots$$

$$= n \ln n - n - \frac{\xi^2}{2n} + \ldots \quad (6)$$

Taking the EXPONENTIAL of each side then gives

$$x^n e^{-x} \approx e^{n \ln n} e^{-n} e^{-\xi^2/2n} = n^n e^{-n} e^{-\xi^2/2n}. \quad (7)$$

Plugging into the integral expression for $n!$ then gives

$$n! \approx \int_{-n}^{\infty} n^n e^{-n} e^{-\xi^2/2n} \, d\xi \approx n^n e^{-n} \int_{-\infty}^{\infty} e^{-\xi^2/2n} \, d\xi. \quad (8)$$

Evaluating the integral gives

$$n! \approx n^n e^{-n} \sqrt{2\pi n}. \quad (9)$$

$$= \sqrt{2\pi} \, n^{n+1/2} e^{-n} \quad (10)$$

(Wells 1986, p. 45). Taking the LOGARITHM of both sides then gives

$$\ln n! \approx n \ln n - n + \frac{1}{2} \ln(2\pi n)$$

$$= \left(n + \frac{1}{2}\right) \ln n - n + \frac{1}{2} \ln(2\pi). \quad (11)$$

This is STIRLING'S SERIES with only the first term retained and, for large n, it reduces to Stirling's approximation

$$\ln n! \approx n \ln n - n. \quad (12)$$

Taking successive terms of $\lfloor n^n/n! \rfloor$ where $\lfloor x \rfloor$ is the FLOOR FUNCTION, gives the sequence 1, 2, 4, 10, 26, 64, 163, 416, 1067, 2755, ... (Sloane's A055775).

Stirling's approximation can be extended to the double inequality

$$\sqrt{2\pi} n^{n+1/2} e^{-n+1/(12n+1)} < n!$$

$$< \sqrt{2\pi} n^{n+1/2} e^{-n+1/(12n)} \quad (13)$$

(Robbins 1955, Feller 1968).

Gosper has noted that a better approximation to $n!$ (i.e., one which approximates the terms in STIRLING'S SERIES instead of truncating them) is given by

$$n! \approx \sqrt{\left(2n + \frac{1}{3}\right)\pi} n^n e^{-n}. \quad (14)$$

This also gives a much closer approximation to the FACTORIAL of 0, $0! = 1$, yielding $\sqrt{\pi/3} \approx 1.02333$ instead of 0 obtained with the conventional Stirling approximation.

See also STIRLING'S SERIES

References

Feller, W. "Stirling's Formula." §2.9 in *An Introduction to Probability Theory and Its Applications, Vol. 1, 3rd ed.* New York: Wiley, pp. 50–3, 1968.

Robbins, H. "A Remark of Stirling's Formula." *Amer. Math. Monthly* **62**, 26–9, 1955.

Sloane, N. J. A. Sequences A055775 in "An On-Line Version of the Encyclopedia of Integer Sequences." http://www.research.att.com/~njas/sequences/eisonline.html.

Stirling, J. *Methodus differentialis.* 1730.

Wells, D. *The Penguin Dictionary of Curious and Interesting Numbers.* Middlesex, England: Penguin Books, p. 45, 1986.

Whittaker, E. T. and Robinson, G. "Stirling's Approximation to the Factorial." §70 in *The Calculus of Observations: A Treatise on Numerical Mathematics*, 4th ed. New York: Dover, pp. 138–40, 1967.

Stirling's Finite Difference Formula

$$f_p = f_0 + \tfrac{1}{2}p\left(\delta_{1/2} + \delta_{-1/2}\right) + \tfrac{1}{2}p^2\delta_0^2$$
$$+ S_3\left(\delta_{1/2}^2 + \delta_{-1/2}^2\right) + S_4\delta_0^4 + \dots$$

for $p \in [-1/2, 1/2]$, where δ is the CENTRAL DIFFERENCE and

$$S_{2n+1} = \frac{1}{2}\binom{p+n}{2n+1}$$

$$S_{2n+2} = \frac{p}{2n+2}\binom{p+n}{2n+1}.$$

with $\binom{n}{k}$ a BINOMIAL COEFFICIENT.

See also CENTRAL DIFFERENCE, STEFFENSON'S FORMULA

References

Beyer, W. H. *CRC Standard Mathematical Tables*, 28th ed. Boca Raton, FL: CRC Press, p. 433, 1987.

Whittaker, E. T. and Robinson, G. "The Newton-Stirling Formula." §23 in *The Calculus of Observations: A Treatise on Numerical Mathematics*, 4th ed. New York: Dover, pp. 38–9, 1967.

Stirling's Formula

STIRLING'S APPROXIMATION, STIRLING'S SERIES

Stirling's Series

The ASYMPTOTIC SERIES for the GAMMA FUNCTION is given by

$$\Gamma(z) \sim e^{-z}z^{z-1/2}\sqrt{2\pi}$$

$$\times \left(1 + \frac{1}{12z} + \frac{1}{288z^2} - \frac{139}{51840z^3} - \frac{571}{2488320z^4} + \dots\right) \quad (1)$$

(Sloane's A001163 and A001164).

The coefficient a_n of z^{-n} can given explicitly by

$$a_n = \sum_{k=1}^{2n}(-1)^k\frac{d_3(2n+2k, k)}{2^{n+k}(n+k)!}, \quad (2)$$

where $d_3(n, k)$ is the number of permutations of n with k CYCLES all of which are ≥ 3 (Comtet 1974, p. 267). Another formula for the a_ns is given by the recurrence relation

$$b_n = \frac{1}{n+1}\left(b_{n-1} - \sum_{k=2}^{n-1}ka_kb_{n+1-k}\right), \quad (3)$$

with $b_0 = b_1 = 1$, then

$$a_n = (2n+1)!!b_{2n+1}. \quad (4)$$

where $x!!$ is the DOUBLE FACTORIAL (Borwein and Corless 1999).

The series for $z!$ is obtained by adding an additional factor of z,

$$z! = \Gamma(z+1) = e^{-z}z^{z+1/2}\sqrt{2\pi}$$

$$\times \left(1 + \frac{1}{12z} + \frac{1}{288z^2} - \frac{139}{51840z^3} - \frac{571}{2488320z^4} + \dots\right). \quad (5)$$

The expansion of $\ln\Gamma(z)$ is what is usually called Stirling's series. It is given by the simple analytic expression

$$\ln\Gamma(z) = \sum_{n=1}^{\infty}\frac{B_{2n}}{2n(2n-1)z^{2n-1}} \quad (6)$$

$$= \tfrac{1}{2}\ln(2\pi) + \left(z - \tfrac{1}{2}\right)\ln z - z + \frac{1}{12z} - \frac{1}{360z^3} + \frac{1}{1260z^5}$$

$$- \dots \quad (7)$$

where B_n is a BERNOULLI NUMBER.

See also BERNOULLI NUMBER, CYCLE (PERMUTATION), K-FUNCTION, STIRLING'S APPROXIMATION

References

Abramowitz, M. and Stegun, C. A. (Eds.). *Handbook of Mathematical Functions with Formulas, Graphs, and Mathematical Tables*, 9th printing. New York: Dover, p. 257, 1972.

Arfken, G. "Stirling's Series." §10.3 in *Mathematical Methods for Physicists*, 3rd ed. Orlando, FL: Academic Press, pp. 555–59, 1985.

Borwein, J. M. and Corless, R. M. "Emerging Tools for Experimental Mathematics." *Amer. Math. Monthly* **106**, 899–09, 1999.

Comtet, L. *Advanced Combinatorics: The Art of Finite and Infinite Expansions*, rev. enl. ed. Dordrecht, Netherlands: Reidel, p. 267, 1974.

Conway, J. H. and Guy, R. K. "Stirling's Formula." In *The Book of Numbers.* New York: Springer-Verlag, pp. 260–61, 1996.

Marsaglia, G. and Marsaglia, J. C. "A New Derivation of Stirling's Approximation to n!." *Amer. Math. Monthly* **97**, 826–29, 1990.

Morse, P. M. and Feshbach, H. *Methods of Theoretical Physics, Part I.* New York: McGraw-Hill, p. 443, 1953.

Sloane, N. J. A. Sequences A001163/M5400 and A001164/M4878 in "An On-Line Version of the Encyclopedia of Integer Sequences." http://www.research.att.com/~njas/sequences/eisonline.html.

Uhler, H. S. "The Coefficients of Stirling's Series for log $\Gamma(z)$." *Proc. Nat. Acad. Sci. U.S.A.* **28**, 59–2, 1942.

Wrench, J. W. Jr. "Concerning Two Series for the Gamma Function." *Math. Comput.* **22**, 617–26, 1968.

StirlingS1

STIRLING NUMBER OF THE FIRST KIND

StirlingS2

STIRLING NUMBER OF THE SECOND KIND

Stirrup Curve

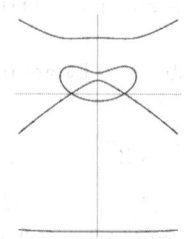

A plane curve given by the equation

$$(x^2 - 1)^2 = y^2(y-1)(y-2)(y+5).$$

References

Cundy, H. and Rollett, A. *Mathematical Models, 3rd ed.* Stradbroke, England: Tarquin Pub., p. 72, 1989.

Stochastic

See also RANDOM VARIABLE, STOCHASTIC APPROXIMATION, STOCHASTIC CALCULUS, STOCHASTIC GEOMETRY, STOCHASTIC GROUP, STOCHASTIC MATRIX, STOCHASTIC OPTIMIZATION, STOCHASTIC PROCESS, STOCHASTIC RESONANCE

Stochastic Approximation

A method of STOCHASTIC OPTIMIZATION including techniques such as gradient search or ROBBINS-MONRO STOCHASTIC APPROXIMATION.

See also ROBBINS-MONRO STOCHASTIC APPROXIMATION, STOCHASTIC OPTIMIZATION

Stochastic Calculus

References

Durrett, R. *Stochastic Calculus: A Practical Introduction.* Boca Raton, FL: CRC Press, 1996.

Stochastic Calculus of Variations

MALLIAVIN CALCULUS

Stochastic Function

A function $f(t)$ of one or more parameters containing a noise term $\epsilon(t)$

$$f(t) = L(t) + \epsilon(t).$$

where the noise is (without loss of generality) assumed to be additive.

See also NOISE, STOCHASTIC OPTIMIZATION

Stochastic Geometry

The study of random geometric structures. Stochastic geometry leads to modelling and analysis tools such as MONTE CARLO METHODS.

See also GEOMETRIC PROBABILITY, INTEGRAL GEOMETRY, MONTE CARLO METHOD, RANDOM POLYGON

References

Kendall, W. S.; Barndorff-Nielson, O.; and van Lieshout, M. C. *Current Trends in Stochastic Geometry: Likelihood and Computation.* Boca Raton, FL: CRC Press, 1998.
Stoyan, D.; Kendall, W. S.; and Mecke, J. *Stochastic Geometry and Its Applications, with a Foreword by D. G. Kendall.* New York: Wiley, 1987.

Stochastic Group

The GROUP of all nonsingular $n \times n$ STOCHASTIC MATRICES over a FIELD F. It is denoted $S(n, F)$. If p is PRIME and F is the FINITE FIELD of ORDER $q = p^m$, $S(n, q)$ is written instead of $S(n, F)$. Particular examples include

$$S(2, 2) = \mathbb{Z}_2$$
$$S(2, 3) = S_3$$
$$S(2, A) = A_4$$
$$S(3, 2) = S_4$$
$$S(2, 5) = \mathbb{Z}_4 \times_\theta \mathbb{Z}_5$$

where $\mathbb{Z}_2$ is an ABELIAN GROUP, S_n are SYMMETRIC GROUPS on n elements, and $\times_\theta$ denotes the semidirect product with $\theta : \mathbb{Z}_4 \to \text{Aut}(\mathbb{Z}_5)$ (Poole 1995).

See also STOCHASTIC MATRIX

References

Poole, D. G. "The Stochastic Group." *Amer. Math. Monthly* **102**, 798–01, 1995.

Stochastic Matrix

A stochastic matrix is the transition matrix for a finite MARKOV CHAIN, also called a MARKOV MATRIX. Elements of the matrix must be REAL NUMBERS in the CLOSED INTERVAL [0, 1].

A completely independent type of stochastic matrix is defined as a SQUARE MATRIX with entries in a FIELD F such that the sum of elements in each column equals 1. There are two nonsingular 2×2 STOCHASTIC MATRICES over $\mathbb{Z}_2$ (i.e., the integers mod 2),

$$\begin{bmatrix} 1 & 0 \\ 0 & 1 \end{bmatrix} \text{ and } \begin{bmatrix} 1 & 0 \\ 0 & 1 \end{bmatrix}.$$

There are six nonsingular stochastic 2×2 MATRICES over $\mathbb{Z}_3$,

$$\begin{bmatrix} 0 & 1 \\ 1 & 0 \end{bmatrix}, \begin{bmatrix} 0 & 2 \\ 1 & 2 \end{bmatrix}, \begin{bmatrix} 1 & 0 \\ 0 & 1 \end{bmatrix}, \begin{bmatrix} 1 & 2 \\ 0 & 2 \end{bmatrix}, \begin{bmatrix} 2 & 0 \\ 2 & 1 \end{bmatrix}, \begin{bmatrix} 2 & 1 \\ 2 & 0 \end{bmatrix},$$

In fact, the set S of all nonsingular stochastic $n \times n$ matrices over a FIELD F forms a GROUP under MATRIX MULTIPLICATION. This GROUP is called the STOCHASTIC GROUP.

The following tables give the number of distinct stochastic matrices (and distinct nonsingular stochastic matrices) over $\mathbb{Z}_m$ for small m.

m	stochastic $n \times n$ matrices over Z^m
2	1, 4, 64, 4096, ...
3	1, 9, 729, ...
4	1, 16, 4096, ...

m	stochastic nonsingular $n \times n$ matrices over Z^m
2	1, 2, 24, 1440, ...
3	1, 6, 450, ...
4	1, 12, 3108, ...

See also DOUBLY STOCHASTIC MATRIX, HORN'S THEOREM, MAJORIZATION, MARKOV CHAIN, STOCHASTIC GROUP

References
Poole, D. G. "The Stochastic Group." *Amer. Math. Monthly* **102**, 798–01, 1995.

Stochastic Optimization

Stochastic optimization refers to the minimization (or maximization) of a function in the presence of randomness in the optimization process. The randomness may be present as either noise in measurements or Monte Carlo randomness in the search procedure, or both.

Common methods of stochastic optimization include direct search methods (such as the NELDER-MEAD METHOD), STOCHASTIC APPROXIMATION, stochastic programming, and miscellaneous methods such as SIMULATED ANNEALING and GENETIC ALGORITHMS.

See also GENETIC ALGORITHM, NELDER-MEAD METHOD, OPTIMIZATION, OPTIMIZATION THEORY, ROBBINS-MONRO STOCHASTIC APPROXIMATION, SIMULATED ANNEALING, STOCHASTIC APPROXIMATION

Stochastic Process

Doob (1996) defines a stochastic process is a family of RANDOM VARIABLES $\{x(t, \bullet), \, t \in \mathscr{I}\}$ from some PROBABILITY SPACE $(S, \mathbb{S}, P)$ into a STATE SPACE $(S', \mathbb{S}')$. Here, $\mathscr{I}$ is the INDEX SET of the process.

Papoulis (1984, p. 312) describes a stochastic process $x(t)$ as a family of functions.

See also INDEX SET, PROBABILITY SPACE, RANDOM VARIABLE, STATE SPACE

References
Doob, J. L. "The Development of Rigor in Mathematical Probability (1900–950)." *Amer. Math. Monthly* **103**, 586–95, 1996.
Papoulis, A. *Probability, Random Variables, and Stochastic Processes, 2nd ed.* New York: McGraw-Hill, 1984.

Stochastic Resonance

A stochastic resonance is a phenomenon in which a nonlinear system is subjected to a periodic modulated signal so weak as to be normally undetectable, but it becomes detectable due to resonance between the weak deterministic signal and stochastic NOISE. The earliest definition of stochastic resonance was the maximum of the output signal strength as a function of NOISE (Bulsara and Gammaitoni 1996).

See also KRAMERS RATE, NOISE

References
Benzi, R.; Sutera, A.; and Vulpiani, A. "The Mechanism of Stochastic Resonance." *J. Phys. A* **14**, L453-L457, 1981.
Bulsara, A. R. and Gammaitoni, L. "Tuning in to Noise." *Phys. Today* **49**, 39–5, March 1996.
Gammaitoni, L. "Stochastic Resonance E-Print Server." http://www.umbrars.com/sr/.

Stöhr Sequence

Let $a_1 = 1$ and define a_{n+1} to be the least INTEGER greater than a_n which cannot be written as the SUM of at most $h \geq 2$ ADDENDS among the terms $a_1, a_2, ..., a_n$. This defines the h-Stöhr sequence. The first few of these are given in the following table.

h	Sloane	h-Stöhr sequence
2	A033627	1, 2, 4, 7, 10, 13, 16, 19, 22, 25, ...
3	A026474	1, 2, 4, 8, 15, 22, 29, 36, 43, 50, ...
4	A051039	1, 2, 4, 8, 16, 31, 46, 61, 76, 91, ...
5	A051040	1, 2, 4, 8, 16, 32, 63, 94, 125, 156, ...

See also GREEDY ALGORITHM, INTEGER RELATION, POSTAGE STAMP PROBLEM, s-ADDITIVE SEQUENCE, SUBSET SUM PROBLEM, SUM-FREE SET, ULAM SEQUENCE

References
Guy, R. K. *Unsolved Problems in Number Theory, 2nd ed.* New York: Springer-Verlag, p. 233, 1994.
Mossige, S. "The Postage Stamp Problem: An Algorithm to Determine the h-Range on the h-Range Formula on the

Extremal Basis Problem for $k = 4$." *Math. Comput.* **69**, 325–37, 2000.

Selmer, E. S. "On Stöhr's Recurrent h-Bases for N." *Kgl. Norske Vid. Selsk. Skrifter* **3**, 1–5, 1986.

Selmer, E. S. and Mossige, S. "Stöhr Sequences in the Postage Stamp Problem." *Bergen Univ. Dept. Pure Math.*, No. 32, Dec. 1984.

Sloane, N. J. A. Sequences A026474, A033627, A051039, and A051040 in "An On-Line Version of the Encyclopedia of Integer Sequences." http://www.research.att.com/~njas/sequences/eisonline.html.

Stokes Phenomenon

The ASYMPTOTIC SERIES of the AIRY FUNCTION $Ai(z)$ (and other similar functions) has a different form in different sectors of the COMPLEX PLANE.

See also AIRY FUNCTIONS

References

Morse, P. M. and Feshbach, H. *Methods of Theoretical Physics, Part I.* New York: McGraw-Hill, pp. 609–11, 1953.

Stokes' Theorem

For ω a DIFFERENTIAL $(K-1)$-FORM with compact support on an oriented n-dimensional MANIFOLD WITH BOUNDARY M,

$$\int_M d\omega = \int_{\partial M} \omega, \quad (1)$$

where $d\omega$ is the EXTERIOR DERIVATIVE of the differential form ω. When M is a COMPACT MANIFOLD without boundary, then the formula holds with the right hand side zero.

Stokes' theorem connects to the "standard" GRADIENT, CURL, and DIVERGENCE THEOREMS by the following relations. If f is a function on $\mathbb{R}^3$,

$$grad(f) = c^{-1} \, df. \quad (2)$$

where $c : \mathbb{R}^3 \to \mathbb{R}^{3*}$ (the dual space) is the duality isomorphism between a VECTOR SPACE and its dual, given by the Euclidean INNER PRODUCT on $\mathbb{R}^3$. If f is a VECTOR FIELD on a $\mathbb{R}^3$,

$$div(f) = *d*c(f), \quad (3)$$

where $*$ is the HODGE STAR operator. If f is a VECTOR FIELD on $\mathbb{R}^3$,

$$curl(f) = c^{-1}* \, dc(f). \quad (4)$$

With these three identities in mind, the above Stokes' theorem in the three instances is transformed into the GRADIENT, CURL, and DIVERGENCE THEOREMS respectively as follows. If f is a function on $\mathbb{R}^3$ and γ is a curve in $\mathbb{R}^3$, then

$$\int_0 grad(f) \cdot dl = \int_\gamma df = f(\gamma(1)) - f(\gamma(0)), \quad (5)$$

which is the GRADIENT THEOREM. If $f : \mathbb{R}^3 \to \mathbb{R}^3$ is a

VECTOR FIELD and M an embedded compact 3-manifold with boundary in $\mathbb{R}^3$, then

$$\int_{\partial M} f \cdot dA = \int_{\partial M} *cf = \int_M d*cf = \int_M \text{div}(f) \, dV, \quad (6)$$

which is the DIVERGENCE THEOREM. If f is a VECTOR FIELD and M is an oriented, embedded, compact 2-MANIFOLD with boundary in $\mathbb{R}^3$, then

$$\int_{\partial M} f \, dl = \int_{\partial M} cf = \int_M dc(f) = \int_M \text{curl}(f) \cdot dA, \quad (7)$$

which is the CURL THEOREM.

DE RHAM COHOMOLOGY is defined using DIFFERENTIAL K-FORMS. When N is a SUBMANIFOLD (without boundary), it represents a homology class. Two closed forms represent the same COHOMOLOGY CLASS if they differ by an EXACT FORM, $\omega_1 - \omega_2 = d\eta$. Hence,

$$\int_N \omega_1 - \omega_2 = \int_N d\eta = 0. \quad (8)$$

Therefore, the evaluation of a COHOMOLOGY CLASS on a HOMOLOGY CLASS is WELL DEFINED.

Physicists generally refer to the CURL THEOREM

$$\int_S (\nabla \times \mathbf{F}) \cdot d\mathbf{a} = \int_{\partial S} \mathbf{F} \cdot d\mathbf{s} \quad (9)$$

as Stokes' theorem.

See also COHOMOLOGY, CURL THEOREM, DIFFERENTIAL K-FORM, DIVERGENCE THEOREM,

See also EXTERIOR ALGEBRA, EXTERIOR DERIVATIVE, GRADIENT THEOREM, HODGE STAR, INTEGRATION (FORM), JACOBIAN, MANIFOLD, POINCARÉ'S LEMMA, TANGENT BUNDLE

References

Berger, M. *Differential Geometry.* New York: Springer-Verlag, pp. 195–03, 1988.

Spivak, M. *A Comprehensive Introduction to Differential Geometry, Vol. 1, 2nd ed.* Houston, TX: Publish or Perish, pp. 343–83, 1999.

Sternberg, S. *Differential Geometry.* New York: Chelsea, p. 119, 1983.

Stolarsky Array

An INTERSPERSION array given by

1	2	3	5	8	13	21	34	55	$\cdots$
4	6	10	16	26	42	68	110	178	$\cdots$
7	11	18	29	47	76	123	199	322	$\cdots$
9	15	24	39	6	102	165	267	432	$\cdots$
12	19	31	50	81	131	212	343	555	$\cdots$
14	23	37	60	97	157	254	411	665	$\cdots$
17	28	45	73	118	191	309	500	809	$\cdots$
20	32	52	84	136	220	356	576	932	$\cdots$
22	36	58	94	152	246	398	644	1042	$\cdots$
$\vdots$	$\vdots$	$\vdots$	$\vdots$	$\vdots$	$\vdots$	$\vdots$	$\vdots$	$\vdots$	$\ddots$

the first row of which is the FIBONACCI NUMBERS.

See also INTERSPERSION, WYTHOFF ARRAY

References

Kimberling, C. "Interspersions and Dispersions." *Proc. Amer. Math. Soc.* **117**, 313–21, 1993.
Morrison, D. R. "A Stolarsky Array and Wythoff Pairs." In *A Collection of Manuscripts Related to the Fibonacci Sequence.* Santa Clara, CA: Fibonacci Assoc., pp. 134–36, 1980.

Stolarsky-Harborth Constant

N.B. A detailed online essay by S. Finch was the starting point for this entry.

Let $b(k)$ be the number of 1s in the BINARY expression of k. Then the number of ODD BINOMIAL COEFFICIENTS $\binom{k}{i}$ where $0 \leq j \leq k$ is $2^{b(k)}$ (Glaisher 1899, Fine 1947). The number of ODD elements in the first n rows of PASCAL'S TRIANGLE is

$$f(n) = \sum_{k=0}^{n-1} 2^{b(k)}. \tag{1}$$

This function is well approximated by n^θ, where

$$\theta \equiv \frac{\ln 3}{\ln 2} = 1.58496\ldots. \tag{2}$$

Stolarsky and Harborth showed that

$$0.812556 \leq \liminf_{n \to \infty} \frac{f(n)}{n^\theta} < 0.812557 < \limsup_{n \to \infty} \frac{f(n)}{n^\theta}$$
$$= 1. \tag{3}$$

The value

$$SH = \liminf_{n \to \infty} \frac{f(n)}{n^\theta} \tag{4}$$

is called the Stolarsky-Harborth constant.

See also BINARY, BINOMIAL COEFFICIENT, RUDIN-SHAPIRO SEQUENCE

References

Finch, S. "Favorite Mathematical Constants." http://www.mathsoft.com/asolve/constant/stlrsky/stlrsky.html.

Fine, N. J. "Binomial Coefficients Modulo a Prime." *Amer. Math. Monthly* **54**, 589–92, 1947.
Wolfram, S. "Geometry of Binomial Coefficients." *Amer. Math. Monthly* **91**, 566–71, 1984.

Stolarsky's Inequality

If $0 \leq g(x) \leq 1$ and g is nonincreasing on the INTERVAL [0, 1], then for all possible values of a and b,

$$\int_0^1 g(x^{1/(a+b)})\, dx \geq \int_0^1 g(x^{1/a})\, dx \int_0^1 g(x^{1/b})\, dx.$$

Stomachion

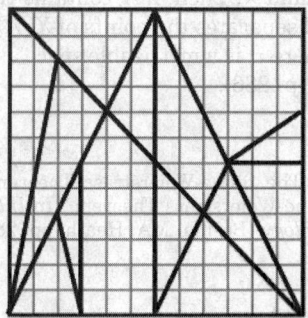

A DISSECTION game similar to TANGRAMS described in fragmentary manuscripts attributed to Archimedes and was referred to as the LOCULUS OF ARCHIMEDES (Archimedes' box) in Latin texts. The word Stomachion has as its root the Greek word for stomach. The game consisted of 14 flat pieces of various shapes arranged in the shape of a square. Like TANGRAMS, the object is to rearrange the pieces to form interesting shapes.

See also DISSECTION, TANGRAM

References

Rorres, C. "Stomachion Introduction." http://www.mcs.drexel.edu/~crorres/Archimedes/Stomachion/intro.html.
Rorres, C. "Stomachion Construction." http://www.mcs.drexel.edu/~crorres/Archimedes/Stomachion/construction.html.

Stone Space

Let $P(L)$ be the set of all PRIME IDEALS of L, and define $r(a) = \{P | a \notin P\}$. Then the Stone space of L is the TOPOLOGICAL SPACE defined on $P(L)$ by postulating that the sets OF THE FORM $r(a)$ are a subbase for the open sets.

See also PRIME IDEAL, TOPOLOGICAL SPACE

References

Grätzer, G. *Lattice Theory: First Concepts and Distributive Lattices.* San Francisco, CA: W. H. Freeman, p. 119, 1971.

Stone-von Neumann Theorem

A theorem which specifies the structure of the generic unitary representation of the Weyl relations and thus establishes the equivalence of Heisenberg's matrix mechanics and Schrödinger's wave mechanics formulations of quantum mechanics in Euclidean $\mathbb{R}^n$ space.

References
Neumann, J. von. "Die Eindeutigkeit der Schrödingerschen Operationen." *Math. Ann.* **104**, 570–78, 1931.

Stone-Weierstrass Theorem

If X is any COMPACT SPACE, let A be a subalgebra of the algebra $C(X)$ over the reals $\mathbb{R}$ with binary operations $+$ and $\times$. Then, if A contains the constant functions and separates the points of X, A is dense in $(C(X), \tau_v)$, where τ_v is a metrizable space as defined by Cullen (1968, p. 286).

References
Cullen, H. F. "The Stone-Weierstrass Theorem" and "The Complex Stone-Weierstrass Theorem." In *Introduction to General Topology*. Boston, MA: Heath, pp. 286–93, 1968.

Stopper Knot

A KNOT used to prevent the end of a string from slipping through a hole.

References
Owen, P. *Knots.* Philadelphia, PA: Courage, p. 11, 1993.

Størmer Number

A Størmer number is a POSITIVE INTEGER n for which the largest PRIME factor p of $n^2 + 1$ is at least $2n$. Every GREGORY NUMBER t_x can be expressed uniquely as a sum of t_ns where the ns are Størmer numbers. Conway and Guy (1996) give a table of Størmer numbers reproduced below (Sloane's A005529). In a paper on INVERSE TANGENT relations, Todd (1949) gives a similar compilation.

n	p	n	p	n	p	n	p	n	p
1	2	10	101	19	181	26	677	35	613
2	5	11	61	20	401	27	73	36	1297
4	17	12	29	22	97	28	157	37	137
5	13	14	197	23	53	29	421	39	761
6	37	15	113	24	577	33	109	40	1601
9	41	16	257	25	313	34	89	42	353

See also GREGORY NUMBER, INVERSE TANGENT

References
Conway, J. H. and Guy, R. K. "Størmer's Numbers." *The Book of Numbers.* New York: Springer-Verlag, pp. 245–48, 1996.
Sloane, N. J. A. Sequences A005529/M1505 in "An On-Line Version of the Encyclopedia of Integer Sequences." http://www.research.att.com/~njas/sequences/eisonline.html.
Todd, J. "A Problem on Arc Tangent Relations." *Amer. Math. Monthly* **56**, 517–28, 1949.

Straight Angle

An ANGLE of $180° = \pi$ RADIANS.

See also ACUTE ANGLE, ANGLE, DIGON, FULL ANGLE, OBTUSE ANGLE, REFLEX ANGLE, RIGHT ANGLE

Straight Line

LINE

Straight Polyomino

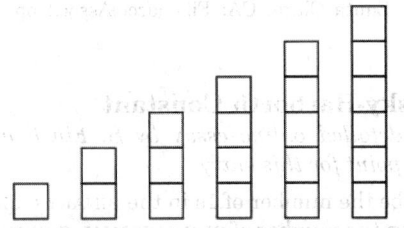

The straight polyomino of order n is the n-POLYOMINO in which all squares are placed along a line.

See also L-POLYOMINO, SKEW POLYOMINO, SQUARE POLYOMINO, T-POLYOMINO

Straightedge

An idealized mathematical object having a rigorously straight edge which can be used to draw a LINE SEGMENT. Although GEOMETRIC CONSTRUCTIONS are sometimes said to be performed with a RULER and COMPASS, the term straightedge is preferable to RULER since markings on the straightedge (usually assumed to be present on a RULER) are not allowed by the classical Greek rules.

See also COMPASS, GEOMETRIC CONSTRUCTION, GEOMETROGRAPHY, MASCHERONI CONSTANT, POLYGON, PONCELET-STEINER THEOREM, RULER, SIMPLICITY, STEINER CONSTRUCTION

Strange Attractor

An attracting set that has zero MEASURE in the embedding PHASE SPACE and has FRACTAL dimension. Trajectories within a strange attractor appear to skip around randomly.

See also CORRELATION EXPONENT, FRACTAL

Strange Loop

References

Benmizrachi, A.; Procaccia, I.; and Grassberger, P. "Characterization of Experimental (Noisy) Strange Attractors." *Phys. Rev. A* **29**, 975–77, 1984.

Grassberger, P. "On the Hausdorff Dimension of Fractal Attractors." *J. Stat. Phys.* **26**, 173–79, 1981.

Grassberger, P. and Procaccia, I. "Measuring the Strangeness of Strange Attractors." *Physica D* **9**, 189–08, 1983a.

Grassberger, P. and Procaccia, I. "Characterization of Strange Attractors." *Phys. Rev. Let.* **50**, 346–49, 1983b.

Lauwerier, H. *Fractals: Endlessly Repeated Geometric Figures.* Princeton, NJ: Princeton University Press, pp. 137–38, 1991.

Sprott, J. C. *Strange Attractors: Creating Patterns in Chaos.* New York: Henry Holt, 1993.

Viana, M. "What's New on Lorenz Strange Attractors." *Math. Intell.* **22**, 6–9.

Strange Loop

A phenomenon in which, whenever movement is made upwards or downwards through the levels of some hierarchical system, the system unexpectedly arrives back where it started. Hofstadter (1987) uses the strange loop as a paradigm in which to interpret paradoxes in logic (such as GRELLING'S PARADOX and RUSSELL'S PARADOX) and calls a system in which a strange loop appears a TANGLED HIERARCHY.

See also GRELLING'S PARADOX, RUSSELL'S PARADOX, TANGLED HIERARCHY

References

Hofstadter, D. R. *Gödel, Escher, Bach: An Eternal Golden Braid.* New York: Vintage Books, p. 10, 1989.

Strangers

Two numbers which are RELATIVELY PRIME.

See also RELATIVELY PRIME

References

Le Lionnais, F. *Les nombres remarquables.* Paris: Hermann, p. 145, 1983.

Strassen Formulas

The usual number of scalar operations (i.e., the total number of additions and multiplications) required to perform $n \times n$ MATRIX MULTIPLICATION is

$$M(n) = 2n^3 - n^2 \tag{1}$$

(i.e., n^3 multiplications and $n^3 - n^2$ additions). However, Strassen (1969) discovered how to multiply two MATRICES in

$$S(n) = 7 \cdot 7^{\lg n} - 6 \cdot 4^{\lg n} \tag{2}$$

scalar operations, where lg is the LOGARITHM to base 2, which is less than $M(n)$ for $n > 654$. For n a power of two ($n = 2^k$), the two parts of (2) can be written

$$7 \cdot 7^{\lg n} = 7 \cdot 7^{\lg 2^k} = 7 \cdot 7^k = 7 \cdot 2^{k \lg 7}$$
$$= 7 (2^k)^{\lg 7} = 7n^{\lg 7} \tag{3}$$

$$6 \cdot 4^{\lg n} = 6 \cdot 4^{\lg 2^k} = 6 \cdot 4^{k \lg 2} = 6 \cdot 4^k$$
$$= 6 (2^k)^2 = 6n^2, \tag{4}$$

so (2) becomes

$$S(2^k) = 7n^{\lg 7} - 6n^2. \tag{5}$$

Two 2×2 matrices can therefore be multiplied

$$\mathsf{C} = \mathsf{AB} \tag{6}$$

$$\begin{bmatrix} c_{11} & c_{12} \\ c_{21} & c_{22} \end{bmatrix} = \begin{bmatrix} a_{11} & a_{12} \\ a_{21} & a_{22} \end{bmatrix} \begin{bmatrix} b_{11} & b_{12} \\ b_{21} & b_{22} \end{bmatrix} \tag{7}$$

with only

$$S(2) = 7 \cdot 2^{\lg 7} - 6 \cdot 2^2 = 49 - 24 = 25 \tag{8}$$

scalar operations (as it turns out, seven of them are multiplications and 18 are additions). Define the seven products (involving a total of 10 additions) as

$$Q_1 \equiv (a_{11} + a_{22})(b_{11} + b_{22}) \tag{9}$$

$$Q_2 \equiv (a_{21} + a_{22})b_{11} \tag{10}$$

$$Q_3 \equiv a_{11}(b_{12} - b_{22}) \tag{11}$$

$$Q_4 \equiv a_{22}(-b_{11} + b_{21}) \tag{12}$$

$$Q_5 \equiv (a_{11} + a_{12})b_{22} \tag{13}$$

$$Q_6 \equiv (-a_{11} + a_{21})(b_{11} + b_{12}) \tag{14}$$

$$Q_7 \equiv (a_{12} - a_{22})(b_{21} + b_{22}). \tag{15}$$

Then the matrix product is given using the remaining eight additions as

$$c_{11} = Q_1 + Q_4 - Q_5 + Q_7 \tag{16}$$

$$c_{21} = Q_2 + Q_4 \tag{17}$$

$$c_{12} = Q_3 + Q_5 \tag{18}$$

$$c_{22} = Q_1 + Q_3 - Q_2 + Q_6 \tag{19}$$

(Strassen 1969, Press *et al.* 1989).

Matrix inversion of a 2×2 matrix A to yield $\mathsf{C} = \mathsf{A}^{-1}$ can also be done in fewer operations than expected using the formulas

$$R_1 \equiv a_{11}^{-1} \tag{20}$$

$$R_2 \equiv a_{21}R_1 \tag{21}$$

$$R_3 \equiv R_1 a_{12} \tag{22}$$

$$R_4 \equiv a_{21}R_3 \tag{23}$$

$$R_5 \equiv R_4 - a_{22} \tag{24}$$

$$R_6 \equiv R_5^{-1} \tag{25}$$

$$c_{12} = R_3 R_6 \qquad (26)$$

$$c_{21} = R_6 R_2 \qquad (27)$$

$$R_7 = R_3 c_{21} \qquad (28)$$

$$c_{11} = R_1 - R_7 \qquad (29)$$

$$c_{22} = -R_6 \qquad (30)$$

(Strassen 1969, Press *et al.* 1989). The leading exponent for Strassen's algorithm for a POWER of 2 is $\lg 7 \approx 2.808$. The best leading exponent currently known is 2.376 (Coppersmith and Winograd 1990). It has been shown that the exponent must be at least 2.

Unfortunately, Strassen's algorithm is not numerically well-behaved. It is only weakly stable, i.e., the computed result C = AB satisfies the inequality

$$\|C - AB\| <= nu \|A\| \, \|B\| + \mathcal{O}(u^2), \qquad (31)$$

where u is the unit roundoff error, while the corresponding strong stability inequality (obtained by replacing matrix norms with absolute values of the matrix elements) does not hold.

See also COMPLEX MULTIPLICATION, KARATSUBA MULTIPLICATION

References

Coppersmith, D. and Winograd, S. "Matrix Multiplication via Arithmetic Programming." *J. Symb. Comput.* **9**, 251–80, 1990.

Douglas, C.; Heroux, M.; Slishman, G.; and Smith, R. "GEMMW: A Portable Level 3 BLAS Winograd Variant of Strassen's Matrix-Matrix Multiply Algorithm." *J. Comput. Phys.* **110**, 1–0, 1994.

Pan, V. *How to Multiply Matrices Faster.* New York: Springer-Verlag, 1982.

Press, W. H.; Flannery, B. P.; Teukolsky, S. A.; and Vetterling, W. T. "Is Matrix Inversion an N^3 Process?" §2.11 in *Numerical Recipes in FORTRAN: The Art of Scientific Computing, 2nd ed.* Cambridge, England: Cambridge University Press, pp. 95–8, 1989.

Strassen, V. "Gaussian Elimination is Not Optimal." *Numerische Mathematik* **13**, 354–56, 1969.

Strassman's Theorem

Let $(K, |\cdot|)$ be a complete non-ARCHIMEDEAN VALUATED FIELD, with VALUATION RING R, and let $f(X)$ be a POWER SERIES with COEFFICIENTS in R. Suppose at least one of the COEFFICIENTS is NONZERO (so that f is not identically zero) and the sequence of COEFFICIENTS converges to 0 with respect to $|\cdot|$. Then $f(X)$ has only finitely many zeros in R.

See also ARCHIMEDEAN VALUATION, MAHLER-LECH THEOREM, VALUATION, VALUATION RING

Strassnitzky's Formula

The MACHIN-LIKE FORMULA

$$\tfrac{1}{4}\pi = \cot^{-1} 2 + \cot^{-1} 5 + \cot^{-1} 8.$$

See also MACHIN'S FORMULA, MACHIN-LIKE FORMULAS

Strategy

A set of moves which a player plans to follow while playing a GAME.

See also GAME, MIXED STRATEGY

Stratified Manifold

A set that is a smooth embedded 2-D MANIFOLD except for a subset that consists of smooth embedded curves, except for a set of ISOLATED POINTS.

References

Morgan, F. "What is a Surface?" *Amer. Math. Monthly* **103**, 369–76, 1996.

Strehl Identities

The sum identities

$$\sum_{j=0}^{\infty} \binom{n}{j}^3 = \sum_{k=0}^{\infty} \binom{n}{j}^2 \binom{2(n-k)}{n}$$

and

$$\sum_{k=0}^{n} \sum_{j=0}^{n} \binom{n}{k}\binom{n+k}{k}\binom{k}{j}^3 = \sum_{k=0}^{n} \binom{n}{k}\binom{n+k}{k}^2$$

(Strehl 1993; Strehl 1994; Koepf 1998, p. 55), where $\binom{n}{k}$ is a BINOMIAL COEFFICIENT.

See also BINOMIAL COEFFICIENT

References

Koepf, W. *Hypergeometric Summation: An Algorithmic Approach to Summation and Special Function Identities.* Braunschweig, Germany: Vieweg, 1998.

Strehl, V. "Binomial Sums and Identities." *Maple Technical Newsletter* **10**, 37–9, 1993.

Strehl, V. "Binomial Identities--Combinatorial and Algorithmic Aspects." *Discrete Math.* **136**, 309–46, 1994.

Stretch

A TRANSFORMATION characterized by an invariant line and a scale factor (one-way stretch) or two invariant lines and corresponding scale factors (two-way stretch).

See also TRANSFORMATION

Strict Gelfand Pattern

MONOTONE TRIANGLE

Strict Inequality

An INEQUALITY is strict if replacing any "less than" and "greater than" signs with equal signs never gives a true expression. For example, $a \le b$ is not strict, whereas $a < b$ is.

See also EQUALITY, INEQUALITY

Striction Curve

A NONCYLINDRICAL RULED SURFACE always has a parameterization OF THE FORM

$$\mathbf{x}(u, v) = \sigma(u) + v\delta(u), \tag{1}$$

where $|\delta| = 1$, $\sigma' \cdot \delta' = 0$, and σ is called the striction curve of $\mathbf{x}$. Furthermore, the striction curve does not depend on the choice of the base curve. The striction and DIRECTOR CURVES of the HELICOID

$$\mathbf{x}(u, v) = \begin{bmatrix} 0 \\ 0 \\ bu \end{bmatrix} + av \begin{bmatrix} \cos u \\ \sin u \\ 0 \end{bmatrix} \tag{2}$$

are

$$\sigma(u) = \begin{bmatrix} 0 \\ 0 \\ bu \end{bmatrix} \tag{3}$$

$$\delta(u) = \begin{bmatrix} a \cos u \\ a \sin u \\ 0 \end{bmatrix}. \tag{4}$$

For the HYPERBOLIC PARABOLOID

$$\mathbf{x}(u, v) = \begin{bmatrix} u \\ 0 \\ 0 \end{bmatrix} + v \begin{bmatrix} 0 \\ 1 \\ u \end{bmatrix}, \tag{5}$$

the striction and DIRECTOR CURVES are

$$\sigma(u) = \begin{bmatrix} u \\ 0 \\ 0 \end{bmatrix} \tag{6}$$

$$\delta(u) = \begin{bmatrix} 0 \\ 1 \\ u \end{bmatrix}. \tag{7}$$

See also DIRECTOR CURVE, DISTRIBUTION PARAMETER, NONCYLINDRICAL RULED SURFACE, RULED SURFACE

References
Gray, A. "Noncylindrical Ruled Surfaces" and "Examples of Striction Curves of Noncylindrical Ruled Surfaces." §19.3 and 19.4 in *Modern Differential Geometry of Curves and Surfaces with Mathematica, 2nd ed.* Boca Raton, FL: CRC Press, pp. 445–49, 1997.

Strictly Egyptian Number

EGYPTIAN NUMBER

String

A string of length k on an ALPHABET l of m characters is an arrangement of k not necessarily distinct symbols from l. There are m^k such distinct strings. For example, the strings of length $k = 3$ on the alphabet $\{1, 2, 3\}$ are $\{1, 1, 1\}$, $\{1, 1, 2\}$, $\{1, 2, 1\}$, $\{1, 2, 2\}$, $\{2, 1, 1\}$, $\{2, 1, 2\}$, $\{2, 2, 1\}$, and $\{2, 2, 2\}$. In *Mathematica*, strings of length k in the ALPHABET consisting of the members in a list l can be enumerated using the following function.

```
Strings[l_List,k_Integer?Positive] := Modu-
le[{k},
    Flatten[Outer[List, Sequence @@ Table[l,
{k}]], k-1]
    ]
```

See also ALPHABET

References
Skiena, S. "Strings." §1.5.1 in *Implementing Discrete Mathematics: Combinatorics and Graph Theory with Mathematica.* Reading, MA: Addison-Wesley, p. 40, 1990.

String Rewriting

A SUBSTITUTION MAP in which rules are used to operate on a string consisting of letters of a certain alphabet. String rewriting is a particularly useful technique for generating successive iterations of certain types of FRACTALS, such as the BOX FRACTAL, CANTOR DUST, CANTOR SQUARE FRACTAL, and SIERPINSKI CARPET.

See also RABBIT SEQUENCE, SUBSTITUTION MAP

References
Peitgen, H.-O. and Saupe, D. (Eds.). "String Rewriting Systems." §C.1 in *The Science of Fractal Images.* New York: Springer-Verlag, pp. 273–75, 1988.
Wagon, S. "Recursion via String Rewriting." §6.2 in *Mathematica in Action.* New York: W. H. Freeman, pp. 190–96, 1991.

Strip

CRITICAL STRIP, MÖBIUS STRIP

Strombic Hexecontahedron

DELTOIDAL HEXECONTAHEDRON

Strombus

A term meaning "spinning top" in Greek which was coined by J. H. Conway by e-mail in the Polyhedron Discussion List as a term for kite-shaped quadrilaterals. Formally, a strombus is a QUADRILATERAL $ABCD$ that has AC for an axis of symmetry.

See also DIAMOND, KITE, LOZENGE, PARALLELOGRAM, QUADRILATERAL, RHOMBOID, RHOMBUS, SKEW QUADRILATERAL, STROMBUS, TRAPEZOID

Strong Convergence

Strong convergence is the type of convergence usually associated with convergence of a SEQUENCE. More formally, a SEQUENCE $\{x_n\}$ of VECTORS in a normed space (and, in particular, in an INNER PRODUCT SPACE E)is called convergent to a VECTOR x in E if

$$\|x_n - x\| \to 0 \quad \text{as } n \to \infty.$$

See also CONVERGENT SEQUENCE, INNER PRODUCT SPACE, WEAK CONVERGENCE

Strong Elliptic Pseudoprime

Let n be an ELLIPTIC PSEUDOPRIME associated with (E, P), and let $n + 1 = 2^s k$ with k ODD and $s \geq 0$. Then n is a strong elliptic pseudoprime when either $kP \equiv 0 \pmod{n}$ or $2^r kP \equiv 0 \pmod{n}$ for some r with $1 \leq r < s$.

See also ELLIPTIC PSEUDOPRIME

References

Ribenboim, P. *The New Book of Prime Number Records, 3rd ed.* New York: Springer-Verlag, pp. 132–34, 1996.

Strong Frobenius Pseudoprime

A PSEUDOPRIME which obeys an additional restriction beyond that required for a FROBENIUS PSEUDOPRIME. A number n with $(n, 2a) = 1$ is a strong Frobenius pseudoprime with respect to $x - a$ IFF n is a STRONG PSEUDOPRIME with respect to $f(x)$. Every strong Frobenius pseudoprime with respect to $x - a$ is an EULER PSEUDOPRIME to the base a.

Every strong Frobenius pseudoprime with respect to $f(x) = x^2 - bx - c$ such that $((b^2 + 4c)/n) = -1$ is a STRONG LUCAS PSEUDOPRIME with parameters (b, c). Every strong Frobenius pseudoprime n with respect to $x^2 - bx + 1$ is an EXTRA STRONG LUCAS PSEUDOPRIME to the base b.

See also FROBENIUS PSEUDOPRIME

References

Grantham, J. "Frobenius Pseudoprimes." 1996. http://www.clark.net/pub/grantham/pseudo/pseudo1.ps

Strong Goldbach Conjecture

GOLDBACH CONJECTURE

Strong Law of Large Numbers

The sequence of variates X_i with corresponding means μ_i obeys the strong law of large numbers if, to every pair ϵ. $\delta > 0$, there corresponds an N such that there is probability $1 - \delta$ or better that for every $r > 0$, all $r + 1$ inequalities

$$\frac{|S_n - m_n|}{n} < \epsilon$$

for $n = N, N + 1, ..., N + r$ will be satisfied, where

$$S_n \equiv \sum_{i=1}^{n} X_n$$
$$m_n \equiv \langle S_n \rangle = \mu_1 + ... + \mu_n$$

(Feller 1968). Kolmogorov established that the convergence of the sequence

$$\sum \frac{\sigma_k^2}{k^2},$$

sometimes called the Kolmogorov criterion, is a sufficient condition for the strong law of large numbers to apply to the sequence of mutually independent random variables X_k with variances σ_k (Feller 1968).

See also FRIVOLOUS THEOREM OF ARITHMETIC, LAW OF LARGE NUMBERS, LAW OF TRULY LARGE NUMBERS, STRONG LAW OF SMALL NUMBERS

References

Feller, W. "The Strong Law of Large Numbers." §10.7 in *An Introduction to Probability Theory and Its Applications, Vol. 1, 3rd ed.* New York: Wiley, pp. 243–45, 1968.
Feller, W. "Strong Laws for Martingales." §7.8 in *An Introduction to Probability Theory and Its Applications, Vol. 2, 3rd ed.* New York: Wiley, pp. 234–38, 1971.

Strong Law of Small Numbers

The first law of strong numbers (Gardner 1980, Guy 1988ab, Guy 1990) states "There aren't enough small numbers to meet the many demands made of them."

The second law of strong numbers (Guy 1990) states that "When two numbers look equal, in ain't necessarily so." Guy (1988a) gives 35 examples of this statement, and 40 more in Guy (1990). For example, example 35 notes that the first few values of the interpolating polynomial $(n^4 - 6n^3 + 23n^2 - 18n + 24)/24$ (erroneously given with a coefficient 24 instead of 23) for $n = 1, 2, ...$ are 1, 2, 4, 8, 16, ..., appears to give the powers of 2 (but the continues 31, 57, 99, ...). Similar, example 41 notes the curious fact that $\lceil e^{(n-1)/2} \rceil$ for $n = 0, 1, ...$ gives 1, 1, 2, 5, 8, 13, 21, 34, 55, ... (the FIBONACCI NUMBERS), although it subsequently continues 91, 149, ... (Sloane's A005181).

References

Gardner, M. "Mathematical Games: Patterns in Primes are a Clue to the Strong Law of Small Numbers." *Sci. Amer.* **243**, 18–8, Dec. 1980.
Guy, R. K. "The Strong Law of Small Numbers." *Amer. Math. Monthly* **95**, 697–12, 1988a.
Guy, R. K. "Graphs and the Strong Law of Small Numbers." In *Proc. 6th Internat. Conf. Theory Appl. Graphs.* Kalamazoo, MI: 1988.

Guy, R. K. "The Second Strong Law of Small Numbers." *Math. Mag.* **63**, 3–0, 1990.
Sloane, N. J. A. Sequences A005181/M0693 in "An On-Line Version of the Encyclopedia of Integer Sequences." http://www.research.att.com/~njas/sequences/eisonline.html.

Strong Lucas Pseudoprime

Let $U(P, Q)$ and $V(P, Q)$ be Lucas sequences generated by P and Q, and define

$$D \equiv P^2 - 4Q.$$

Let n be an odd composite number with $(n, D) = 1$, and $n - (D/n) = 2^s d$ with d odd and $s \geq 0$, where (a/b) is the Legendre symbol. If

$$U_d \equiv 0 \pmod{n}$$

or

$$V_{2^r d} \equiv 0 \pmod{n}$$

for some r with $0 \leq r < s$, then n is called a strong Lucas pseudoprime with parameters (P, Q).

A strong Lucas pseudoprime is a Lucas pseudoprime to the same base. Arnault (1997) showed that any composite number n is a strong Lucas pseudoprime for at most 4/15 of possible bases (unless n is the product of twin primes having certain properties).

See also Extra Strong Lucas Pseudoprime, Lucas Pseudoprime

References

Arnault, F. "The Rabin-Monier Theorem for Lucas Pseudoprimes." *Math. Comput.* **66**, 869–81, 1997.
Ribenboim, P. "Euler-Lucas Pseudoprimes (elpsp(P, Q)) and Strong Lucas Pseudoprimes (slpsp(P, Q))." §2.X.C in *The New Book of Prime Number Records, 3rd ed.* New York: Springer-Verlag, pp. 130–31, 1996.

Strong Perfect Graph Conjecture

The conjecture that a graph is perfect iff neither the graph nor its complement contains an odd cycle of length at least five as an induced subgraph (Golumbic 1980; Skiena 1990, p. 221).

See also Perfect Graph, Perfect Graph Theorem

References

Golumbic, M. C. *Algorithmic Graph Theory and Perfect Graphs.* New York: Academic Press, 1980.
Skiena, S. *Implementing Discrete Mathematics: Combinatorics and Graph Theory with Mathematica.* Reading, MA: Addison-Wesley, 1990.

Strong Pseudoprime

A strong pseudoprime to a base a is an odd composite number n with $n - 1 = d \cdot 2^s$ (for d odd) for which *either*

$$a^d \equiv 1 \pmod{n} \tag{1}$$

or

$$a^{d \cdot 2^r} \equiv -1 \pmod{n} \tag{2}$$

for some $r = 0, 1, ..., s - 1$ (Riesel 1994, p. 91). Note that Guy (1994, p. 27) restricts the definition of strong pseudoprimes to only those satisfying (1).

The definition is motivated by the fact that a Fermat pseudoprime n to the base b satisfies

$$b^{n-1} - 1 \equiv 0 \pmod{n}. \tag{3}$$

But since n is odd, it can be written $n = 2m + 1$, and

$$b^{2m} - 1 = (b^m - 1)(b^m + 1) \equiv 0 \pmod{n}. \tag{4}$$

If n is prime, it must divide at least one of the factors, but can't divide both because it would then divide their difference

$$(b^m + 1) - (b^m - 1) = 2. \tag{5}$$

Therefore,

$$b^m \equiv \pm 1 \pmod{n}. \tag{6}$$

so write $n = 2^a t + 1$ to obtain

$$b^{n-1} - 1 = (b^t + 1)(b^t - 1)(b^{2t} + 1) \cdots \left(b^{2^{a-1}t} + 1\right). \tag{7}$$

If n divides exactly one of these factors but is composite, it is a strong pseudoprime. A composite number is a strong pseudoprime to at most 1/4 of all bases less than itself (Monier 1980, Rabin 1980). The strong pseudoprimes provide the basis for Miller's primality test and Rabin-Miller strong pseudoprime test.

A strong pseudoprime to the base a is also an Euler pseudoprime to the base a (Pomerance *et al.* 1980). The strong pseudoprimes include some Euler pseudoprimes, Fermat pseudoprimes, and Carmichael numbers.

The first few strong pseudoprimes to the base 2 are 2047, 3277, 4033, 4681, ... (Sloane's A001262). The number of strong pseudoprimes less than 10^3, 10^4, ... are 0, 5, 16, 46, 162, ... (Sloane's A055552). Note that Guy's (1994, p. 27) definition gives only the subset 2047, 4681, 15841, 42799, 52633, 90751, ..., giving counts inconsistent with those in Guy's table.

The strong k-pseudoprime test for $k = 2, 3, 5$ correctly identifies all primes below 2.5×10^{10} with only 13 exceptions, and if 7 is added, then the only exception less than 2.5×10^{10} is 315031751. Jaeschke (1993) showed that there are only 101 strong pseudoprimes for the bases 2, 3, and 5 less than 10^{12}, nine if 7 is added, and none if 11 is added. Also, the bases 2, 13, 23, and 1662803 have no exceptions up to 10^{12}.

If n is composite, then there is a base for which n is not a strong pseudoprime. There are therefore no "strong Carmichael numbers." Let ψ_k denote the smallest strong pseudoprime to all of the first k primes taken as bases (i.e, the smallest odd number for which the Rabin-Miller strong pseudoprime

TEST on bases less than or equal to k fails). Jaeschke (1993) computed ψ_k from $k = 5$ to 8 and gave upper bounds for $k = 9$ to 11.

$$\psi_1 = 2047$$

$$\psi_2 = 1373653$$

$$\psi_3 = 25326001$$

$$\psi_4 = 3215031751$$

$$\psi_5 = 2152302898747$$

$$\psi_6 = 3474749660383$$

$$\psi_7 = 341550071728321$$

$$\psi_8 = 341550071728321$$

$$\psi_9 \leq 41234316135705689041$$

$$\psi_{10} \leq 1553360566073143205541002401$$

$$\psi_{11} = 56897193526942024370326972321$$

(Sloane's A014233). A seven-step test utilizing these results (Riesel 1994) allows all numbers less than 3.4×10^{14} to be tested.

Pomerance *et al.* (1980) have proposed a test based on a combination of STRONG PSEUDOPRIMES and LUCAS PSEUDOPRIMES. They offer a $620 reward for discovery of a COMPOSITE NUMBER which passes their test (Guy 1994, p. 28).

See also CARMICHAEL NUMBER, MILLER'S PRIMALITY TEST, POULET NUMBER, RABIN-MILLER STRONG PSEUDOPRIME TEST, ROTKIEWICZ THEOREM, STRONG ELLIPTIC PSEUDOPRIME, STRONG LUCAS PSEUDOPRIME

References

Baillie, R. and Wagstaff, S. "Lucas Pseudoprimes." *Math. Comput.* **35**, 1391–417, 1980.
Guy, R. K. "Pseudoprimes. Euler Pseudoprimes. Strong Pseudoprimes." §A12 in *Unsolved Problems in Number Theory, 2nd ed.* New York: Springer-Verlag, pp. 27–0, 1994.
Jaeschke, G. "On Strong Pseudoprimes to Several Bases." *Math. Comput.* **61**, 915–26, 1993.
Monier, L. "Evaluation and Comparison of Two Efficient Probabilistic Primality Testing Algorithms." *Theor. Comput. Sci.* **12**, 97–08, 1980.
Pinch, R. G. E. "The Pseudoprimes Up to 10^{13}." ftp://ftp.dpmms.cam.ac.uk/pub/PSP/.
Pomerance, C.; Selfridge, J. L.; and Wagstaff, S. S. Jr. "The Pseudoprimes to $25 \cdot 10^9$." *Math. Comput.* **35**, 1003–026, 1980. Available electronically from ftp://sable.ox.ac.uk/pub/math/primes/ps2.Z.
Rabin, M. O. "Probabilistic Algorithm for Testing Primality." *J. Number Th.* **12**, 128–38, 1980.
Riesel, H. *Prime Numbers and Computer Methods for Factorization, 2nd ed.* Basel: Birkhäuser, p. 92, 1994.
Sloane, N. J. A. Sequences A001262, A014233, and A055552 in "An On-Line Version of the Encyclopedia of Integer Sequences." http://www.research.att.com/~njas/sequences/eisonline.html.

Strong Pseudoprime Test

RABIN-MILLER STRONG PSEUDOPRIME TEST

Strong Subadditivity Inequality

$$\phi(A) + \phi(B) - \phi(A \cup B) \geq \phi(A \cap B).$$

References

Doob, J. L. "The Development of Rigor in Mathematical Probability (1900–950)." *Amer. Math. Monthly* **103**, 586–95, 1996.

Strong Triangle Inequality

The p-adic norm satisfies

$$|x+y|_p \leq \max\left(|x|_p, |x|_p\right)$$

for all x and y.

See also P-ADIC NUMBER, TRIANGLE INEQUALITY

Strong Twin Prime Conjecture

TWIN PRIME CONJECTURE

Strongly Connected Component

A maximal SUBGRAPH of a DIRECTED GRAPH such that for every pair of vertices u, v in the SUBGRAPH, there is a directed path from u to v and a directed path from v to u. Tarjan (1972) has devised an $\mathcal{O}(n)$ algorithm for determining strongly connected components, which is implemented in *Mathematica* as `StronglyConnectedComponents[g]` in the *Mathematica* add-on package `DiscreteMath`Combinatorica` (which can be loaded with the command `<<DiscreteMath`) (Skiena 1990, p. 172).

See also BI-CONNECTED COMPONENT, STRONGLY CONNECTED DIGRAPH

References

Skiena, S. *Implementing Discrete Mathematics: Combinatorics and Graph Theory with Mathematica.* Reading, MA: Addison-Wesley, 1990.
Tarjan, R. E. "Depth-First Search and Linear Graph Algorithms." *SIAM J. Comput.* **1**, 146–60, 1972.

Strongly Connected Digraph

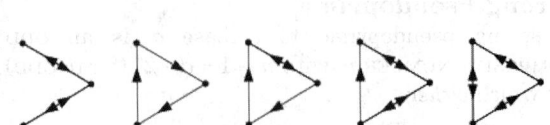

A DIRECTED GRAPH in which it is possible to reach any node starting from any other node by traversing

edges in the direction(s) in which they point. The nodes in a strongly connected digraph therefore must all have INDEGREE of at least 1. The numbers of nonisomorphic simple strongly connected digraphs on $n = 1, 2, ...$ nodes are 1, 1, 5, 83, 5048, 1047008, ... (Sloane's A035512).

See also CONNECTED DIGRAPH, WEAKLY CONNECTED DIGRAPH

References

Harary, F. and Palmer, E. M. *Graphical Enumeration.* New York: Academic Press, p. 218, 1973.

Liskovec, V. A. "A Contribution to the Enumeration of Strongly Connected Digraphs." *Dokl. AN BSSR* **17**, 1077–080, 1973.

Read, R. C. and Wilson, R. J. *An Atlas of Graphs.* Oxford, England: Oxford University Press, 1998.

Skiena, S. "Strong and Weak Connectivity." §5.1.2 in *Implementing Discrete Mathematics: Combinatorics and Graph Theory with Mathematica.* Reading, MA: Addison-Wesley, pp. 94 and 172–74, 1990.

Sloane, N. J. A. Sequences A035512 in "An On-Line Version of the Encyclopedia of Integer Sequences." http://www.research.att.com/~njas/sequences/eisonline.html.

Strongly Connected Graph

STRONGLY CONNECTED DIGRAPH

Strongly Embedded Theorem

The strongly embedded theorem identifies all SIMPLE GROUPS with a strongly 2-embedded SUBGROUP. In particular, it asserts that no SIMPLE GROUP has a strongly 2-embedded 2'-local SUBGROUP.

See also SIMPLE GROUP, SUBGROUP

Strongly Independent

An infinite sequence $\{a_i\}$ of POSITIVE INTEGERS is called strongly independent if any relation $\Sigma \epsilon_i a_i$, with $\epsilon_i = 0, \pm 1,$ or ± 2 and $\epsilon_i = 0$ except finitely often, IMPLIES $\epsilon_i = 0$ for all i.

See also WEAKLY INDEPENDENT

References

Guy, R. K. *Unsolved Problems in Number Theory, 2nd ed.* New York: Springer-Verlag, p. 136, 1994.

Strongly Triple-Free Set

TRIPLE-FREE SET

Strophoid

Let C be a curve, let O be a fixed point (the POLE), and let O' be a second fixed point. Let P and P' be points on a line through O meeting C at Q such that $P'Q = QP = QO'$. The LOCUS of P and P' is called the strophoid of C with respect to the POLE O and fixed point O'. Let C be represented parametrically by $(f(t), g(t))$, and let $O = (x_0, y_0)$ and $O' = (x_1, y_1)$. Then the equation of the strophoid is

$$x = f \pm \sqrt{\frac{(x_1 - f)^2 + (y_1 - g)^2}{1 + m^2}} \qquad (1)$$

$$x = g \pm \sqrt{\frac{(x_1 - f)^2 + (y_1 - g)^2}{1 + m^2}}, \qquad (2)$$

where

$$m \equiv \frac{g - y_0}{f - x_0}. \qquad (3)$$

The name strophoid means "belt with a twist," and was proposed by Montucci in 1846 (MacTutor Archive). The polar form for a general strophoid is

$$r = \frac{b \sin(a - 2\theta)}{\sin(a - \theta)}. \qquad (4)$$

If $a = \pi/2$, the curve is a RIGHT STROPHOID. The following table gives the strophoids of some common curves.

Curve	Pole	Fixed Point	Strophoid
line	not on line	on line	oblique strophoid
line	not on line	foot of PERPENDICULAR origin to line	RIGHT STROPHOID
CIRCLE	center	on the circumference	FREETH'S NEPHROID

See also RIGHT STROPHOID

References

Beyer, W. H. *CRC Standard Mathematical Tables, 28th ed.* Boca Raton, FL: CRC Press, p. 225, 1987.

Lawrence, J. D. *A Catalog of Special Plane Curves.* New York: Dover, pp. 51–3 and 205, 1972.

Lockwood, E. H. "Strophoids." Ch. 16 in *A Book of Curves.* Cambridge, England: Cambridge University Press, pp. 134–37, 1967.

MacTutor History of Mathematics Archive. "Right." http://www-groups.dcs.st-and.ac.uk/~history/Curves/Right.html.

Yates, R. C. "Strophoid." *A Handbook on Curves and Their Properties.* Ann Arbor, MI: J. W. Edwards, pp. 217–20, 1952.

Structural Ramsey Theory

A generalization of RAMSEY THEORY to mathematical objects in which one would not normally expect structure to be found. For example, there exists a graph with very few triangles (more precisely, a graph which can always be constructed so that there is no "cycle" of triangles which are all distinct and T_{i+1} meets T_i in at least one vertex) and such that however it is colored with r colors, one of the colors contains a triangle. The usual proof of RAMSEY'S THEOREM gives no insight on how to prove such a result.

See also EXTREMAL GRAPH THEORY, RAMSEY'S THEOREM, RAMSEY THEORY

Structurally Stable

A MAP $\phi : M \to M$ where M is a MANIFOLD is C^r structurally stable if any C^r perturbation is TOPOLOGICALLY CONJUGATE to ϕ. Here, C^r perturbation means a FUNCTION ψ such that ψ is close to ϕ and the first r derivatives of ψ are close to those of ϕ.

See also TOPOLOGICALLY CONJUGATE

Structure

LATTICE

Structure Constant

The structure constant is defined as $i\epsilon_{ijk}$, where ϵ_{ijk} is the PERMUTATION SYMBOL. The structure constant forms the starting point for the development of LIE ALGEBRA.

See also LIE ALGEBRA, PERMUTATION SYMBOL

Structure Factor

The structure factor S_Γ of a discrete set Γ is the FOURIER TRANSFORM of δ-scatterers of equal strengths on all points of Γ,

$$S_\Gamma(k) = \int \sum_{x \in \Gamma} \delta(x' - x) e^{-2\pi i k x'} \, dx' = \sum_{x \in \Gamma} e^{-2\pi i k x}.$$

References

Baake, M.; Grimm, U.; and Warrington, D. H. "Some Remarks on the Visible Points of a Lattice." *J. Phys. A: Math. General* **27**, 2669–674, 1994.

Strut

TENSEGRITY

Struve Differential Equation

The ORDINARY DIFFERENTIAL EQUATION

$$z^2 y'' + z y' + (z^2 - v^2) y = \frac{4\left(\frac{1}{2}z\right)^{v+1}}{\sqrt{\pi}\,\Gamma\left(v + \frac{1}{2}\right)},$$

where $\Gamma(z)$ is the GAMMA FUNCTION (Abramowitz and Stegun 1972, p. 496; Zwillinger 1997, p. 127). The solution is

$$y = a J_v(z) + b Y_v(z) + \mathcal{H}_v(z),$$

where $J_v(z)$ and $Y_v(z)$ are BESSEL FUNCTIONS OF THE

FIRST and SECOND KINDS, and $\mathcal{H}_v(z)$ is a STRUVE FUNCTION (Abramowitz and Stegun 1972).

See also BESSEL FUNCTION OF THE FIRST KIND, BESSEL FUNCTION OF THE SECOND KIND, STRUVE FUNCTION

References

Abramowitz, M. and Stegun, C. A. (Eds.). *Handbook of Mathematical Functions with Formulas, Graphs, and Mathematical Tables, 9th printing.* New York: Dover, p. 496, 1972.
Zwillinger, D. *Handbook of Differential Equations, 3rd ed.* Boston, MA: Academic Press, p. 127, 1997.

Struve Function

Abramowitz and Stegun (1972, pp. 496–99) define the Struve function as

$$\mathcal{H}_v(z) = \left(\tfrac{1}{2}z\right)^{v+1} \sum_{k=0}^{\infty} \frac{(-1)^k \left(\frac{1}{2}z\right)^{2k}}{\Gamma\left(k + \frac{3}{2}\right)\Gamma\left(k + v + \frac{3}{2}\right)}, \tag{1}$$

where $\Gamma(z)$ is the GAMMA FUNCTION. Watson (1966, p. 338) defines the Struve function as

$$\mathcal{H}_v(z) \equiv \frac{2\left(\frac{1}{2}z\right)^v}{\Gamma\left(v + \frac{1}{2}\right)\Gamma\left(\frac{1}{2}\right)} \int_0^1 \left(1 - t^2\right)^{v-1/2} \sin(zt) \, dt. \tag{2}$$

The series expansion is

$$\mathcal{H}_v(z) = \sum_{m=0}^{\infty} (-1)^m \frac{\left(\frac{1}{2}z\right)^{2m+v+1}}{\Gamma\left(m + \frac{3}{2}\right)\Gamma\left(v + m + \frac{3}{2}\right)}. \tag{3}$$

For half integer orders,

$$\mathcal{H}_{n+1/2}(z) = Y_{n+1/2}(z)$$
$$+ \frac{1}{\pi} \sum_{m=0}^{n} \frac{\Gamma\left(m + \frac{1}{2}\right)\left(\frac{1}{2}z\right)^{-2m+n-1/2}}{\Gamma(n + 1 - m)} \tag{4}$$

$$\mathcal{H}_{-(n+1/2)}(z) = (-1)^n J_{n+1/2}(z). \tag{5}$$

The Struve function and its derivatives satisfy

$$\mathcal{H}_{v-1}(z) - \mathcal{H}_{v+1}(z) = 2\mathcal{H}_v'(z) - \frac{\left(\frac{1}{2}z\right)^v}{\sqrt{\pi}\,\Gamma\left(v + \frac{3}{2}\right)}. \tag{6}$$

For integer n, the Struve function gives the solution to

$$z^2 y'' + z y' + (z^2 - n^2) y = \frac{2}{\pi} \frac{z^{n+1}}{(2n-1)!!}, \tag{7}$$

where $n!!$ is the DOUBLE FACTORIAL.

The Struve function is built into *Mathematica* 4.0 as StruveH[n, z].

See also ANGER FUNCTION, BESSEL FUNCTION, MODIFIED STRUVE FUNCTION, WEBER FUNCTIONS

Struve H-Function

References
Abramowitz, M. and Stegun, C. A. (Eds.). "Struve Function $\mathbf{H}_\nu(x)$." §12.1 in *Handbook of Mathematical Functions with Formulas, Graphs, and Mathematical Tables, 9th printing.* New York: Dover, pp. 496–98, 1972.

Apelblat, A. "Derivatives and Integrals with Respect to the Order of the Struve Functions $H_\nu(x)$ and $L_\nu(x)$." *J. Math. Anal. Appl.* **137**, 17–6, 1999.

Prudnikov, A. P.; Marichev, O. I.; and Brychkov, Yu. A. "The Struve Functions $H_\nu(x)$ and $L_\nu(x)$." §1.4 in *Integrals and Series, Vol. 3: More Special Functions.* Newark, NJ: Gordon and Breach, pp. 24–7, 1990.

Spanier, J. and Oldham, K. B. "The Struve Function." Ch. 57 in *An Atlas of Functions.* Washington, DC: Hemisphere, pp. 563–71, 1987.

Watson, G. N. *A Treatise on the Theory of Bessel Functions, 2nd ed.* Cambridge, England: Cambridge University Press, 1966.

Struve H-Function
STRUVE FUNCTION

Struve L-Function
MODIFIED STRUVE FUNCTION

StruveH
STRUVE FUNCTION

StruveL
MODIFIED STRUVE FUNCTION

Student's t-Distribution

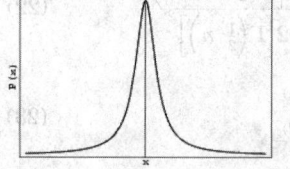 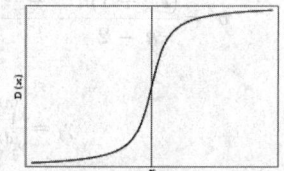

A STATISTICAL DISTRIBUTION published by William Gosset in 1908. His employer, Guinness Breweries, required him to publish under a pseudonym, so he chose "Student." Given n independent measurements x_i, let

$$t \equiv \frac{\bar{x} - \mu}{s/\sqrt{n}}. \tag{1}$$

where μ is the population MEAN, $\bar{x}$ is the sample MEAN, and s is the ESTIMATOR for population STANDARD DEVIATION (i.e., the SAMPLE VARIANCE) defined by

$$s^2 \equiv \frac{1}{N-1} \sum_{i=1}^{n} (x_i - \bar{x})^2. \tag{2}$$

Student's t-distribution is defined as the distribution of the random variable t which is (very loosely) the "best" that we can do not knowing σ. If $\sigma = s$, $t = z$ and the distribution becomes the NORMAL DISTRIBUTION. As N increases, Student's t-distribution approaches the NORMAL DISTRIBUTION.

Student's t-distribution can be derived by transforming STUDENT'S z-DISTRIBUTION using

$$z \equiv \frac{\bar{x} - \mu}{s}, \tag{3}$$

and then defining

$$t \equiv z\sqrt{n-1}. \tag{4}$$

The resulting probability and cumulative distribution functions are

$$f_r(t) = \frac{\Gamma\left[\frac{1}{2}(r+1)\right]}{\sqrt{r\pi}\,\Gamma\left(\frac{1}{2}r\right)\left(1 + \frac{t^2}{r}\right)^{(r+1)/2}} = \frac{\left(\dfrac{r}{r+t^2}\right)^{(1+r)/2}}{\sqrt{r}\,B\left(\frac{1}{2}r, \frac{1}{2}\right)} \tag{5}$$

$$F_r(t) = \int_{-\infty}^{t} \frac{\Gamma\left[\frac{1}{2}(r+1)\right]}{\sqrt{r\pi}\,\Gamma\left(\frac{1}{2}r\right)\left(1 + \frac{t'^2}{r}\right)^{(r+1)/2}} \, dt'$$

$$= \frac{1}{2} + \frac{1}{2}\left[I\left(1;\ \tfrac{1}{2}r, \tfrac{1}{2}\right) - I\left(\frac{r}{r+t^2}, \tfrac{1}{2}r, \tfrac{1}{2}\right)\right]$$

$$= 1 - \frac{1}{2} I\left(\frac{r}{r+t^2}, \tfrac{1}{2}r, \tfrac{1}{2}\right), \tag{6}$$

where

$$r \equiv n - 1 \tag{7}$$

is the number of DEGREES OF FREEDOM, $-\infty < t < \infty$, $\Gamma(z)$ is the GAMMA FUNCTION, $B(a, b)$ is the BETA FUNCTION, and $I(z;\ a, b)$ is the REGULARIZED BETA FUNCTION defined by

$$I(z;\ a, b) = \frac{B(z;\ a, b)}{B(a, b)}. \tag{8}$$

The MEAN, VARIANCE, SKEWNESS, and KURTOSIS of Student's t-distribution are

$$\mu = 0 \tag{9}$$

$$\sigma^2 = \frac{r}{r-2} \tag{10}$$

$$\gamma_1 = 0 \tag{11}$$

$$\gamma_2 = \frac{6}{r-4}. \tag{12}$$

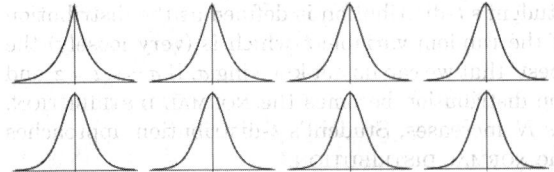

The CHARACTERISTIC FUNCTIONS $\phi_n(t)$ for the first few values of n are

$$\phi_1(t) = e^{-|t|} \tag{13}$$

$$\phi_2(t) = \sqrt{2}|t| K_1\left(\sqrt{2}|t|\right) \tag{14}$$

$$\phi_3(t) = e^{-\sqrt{3}|t|}\left(1 + \sqrt{3}|t|\right) \tag{15}$$

$$\phi_4(t) = 2t^2 K_2(2|t|) \tag{16}$$

$$\phi_5(t) = \tfrac{1}{3} e^{-\sqrt{5}|t|}\left(3 + 3\sqrt{5}|t| + 5t^2\right), \tag{17}$$

and so on, where $K_n(x)$ is a MODIFIED BESSEL FUNCTION OF THE SECOND KIND.

Beyer (1987, p. 571) gives 60%, 70%, 90%, 95%, 97.5%, 99%, 99.5%, and 99.95% confidence intervals, and Goulden (1956) gives 50%, 90%, 95%, 98%, 99%, and 99.9% confidence intervals. A partial table is given below for small r and several common confidence intervals.

r	90%	95%	97.5%	99.5%
1	3.07766	6.31371	12.7062	63.656
2	1.88562	2.91999	4.30265	9.92482
3	1.63774	2.35336	3.18243	5.84089
4	1.53321	2.13185	2.77644	4.60393
5	1.47588	2.01505	2.57058	4.03212
10	1.37218	1.81246	2.22814	3.16922
30	1.31042	1.69726	2.04227	2.74999
100	1.29007	1.66023	1.98397	2.62589
∞	1.28156	1.64487	1.95999	2.57584

The so-called $A(t|n)$ distribution is useful for testing if two observed distributions have the same MEAN. $A(t|n)$ gives the probability that the difference in two observed MEANS for a certain statistic t with n DEGREES OF FREEDOM would be smaller than the observed value purely by chance:

$$A(t|n) = \frac{1}{\sqrt{n}B\left(\frac{1}{2}, \frac{1}{2}n\right)} \int_{-t}^{t} \left(1 + \frac{x^2}{n}\right)^{-(1+n)/2} dx. \tag{18}$$

Let X be a NORMALLY DISTRIBUTED random variable with MEAN 0 and VARIANCE σ^2, let Y^2/σ^2 have a CHI-SQUARED DISTRIBUTION with n DEGREES OF FREEDOM, and let X and Y be independent. Then

$$t \equiv \frac{X\sqrt{n}}{Y} \tag{19}$$

is distributed as Student's t with n DEGREES OF FREEDOM.

The noncentral Student's t-distribution is given by

$$P(x) = \frac{n^n/2^n!}{2^n e^{\lambda^2/2}(n + x^2)^{n/2}\Gamma\left(\frac{1}{2}n\right)}$$

$$\times \left\{ \frac{\sqrt{2}\lambda x\, {}_1F_1\left(\frac{1}{2}n + 1; \frac{3}{2}; \frac{\lambda^2 z^2}{2(n + z^2)}\right)}{(n + x^2)\Gamma\left[\frac{1}{2}(n + 1)\right]} \right.$$

$$\left. + \frac{{}_1F_1\left(\frac{1}{2}(n + 1); \frac{1}{2}; \frac{\lambda^2 z^2}{2(n + z^2)}\right)}{\sqrt{n + x^2}\Gamma\left(\frac{1}{2}n + 1\right)} \right\}, \tag{20}$$

where $\Gamma(z)$ is the GAMMA FUNCTION and ${}_1F_1(a; b; z)$ is a CONFLUENT HYPERGEOMETRIC FUNCTION. The MEAN, VARIANCE, SKEWNESS, and KURTOSIS are

$$\mu = L\sqrt{\frac{n}{2}}\frac{\Gamma\left(\frac{1}{2}(n - 1)\right)}{\Gamma\left(\frac{1}{2}n\right)} \tag{21}$$

$$\sigma^2 = \frac{(L^2 + 1)n}{n - 2} - \frac{L^2 n\left[\Gamma\left(\frac{1}{2}(n - 1)\right)\right]^2}{2\left[\Gamma\left(\frac{1}{2}n\right)\right]^2} \tag{22}$$

$$\gamma_1 = \frac{\gamma_1^{(n)}}{\gamma_1^{(d)}} \tag{23}$$

$$\gamma_2 = \frac{\gamma_2^{(n)}}{\gamma_2^{(d)}}. \tag{24}$$

where

$$\gamma_1^{(n)} = 2\lambda\sqrt{n}\Gamma\left[\frac{1}{2}(n - 1)\right]\left\{\left[\lambda^2(2n - 7) - 3\right]\left[\Gamma\left(\frac{1}{2}n\right)\right]^2\right.$$

$$\left. - \lambda^2(n - 2)(n - 3)\left[\Gamma\left(\frac{1}{2}(n - 1)\right)\right]^2\right\} \tag{25}$$

$$\gamma_1^{(d)} = (n - 3)$$

$$\times \sqrt{\frac{2n(\lambda^2 + 1)}{n - 2} - \frac{\lambda^2 n\left[\Gamma\left(\frac{1}{2}(n - 1)\right)\right]^2}{\left[\Gamma\left(\frac{1}{2}n\right)\right]^2}}\Gamma\left(\frac{1}{2}n\right)$$

$$\times\left\{\lambda^2(n-2)\left[\Gamma\left(\tfrac{1}{2}(n-1)\right)\right]^2-2(\lambda^2+1)\left[\Gamma\left(\tfrac{1}{2}n\right)\right]^2\right\} \quad (26)$$

$$\gamma_2^{(n)}=2\left\{-3\lambda^4(n-2)^2(n-3)(n-4)\left[\Gamma\left(\tfrac{1}{2}(n-1)\right)\right]^4\right.$$

$$+2^{6-2n}\lambda^2(n-2)(n-4)[\lambda^2(2n-7)-3]\pi[\Gamma(n+1)]^2$$

$$\left.-4[\lambda^4(n-5)-6\lambda^2-3](n-3)\left[\Gamma\left(\tfrac{1}{2}n\right)\right]^4\right\} \quad (27)$$

$$\gamma_2^{(d)}=(n-3)(n-4)\left\{\lambda^2(n-2)\left[\Gamma\left(\tfrac{1}{2}(n-1)\right)\right]^2\right.$$

$$\left.-2(\lambda^2+1)\left[\Gamma\left(\tfrac{1}{2}n\right)\right]^2\right\}^2. \quad (28)$$

See also BESSEL'S STATISTICAL FORMULA, PAIRED *T*-TEST, STUDENT'S *Z*-DISTRIBUTION

References

Abramowitz, M. and Stegun, C. A. (Eds.). *Handbook of Mathematical Functions with Formulas, Graphs, and Mathematical Tables, 9th printing.* New York: Dover, pp. 948–49, 1972.

Beyer, W. H. *CRC Standard Mathematical Tables, 28th ed.* Boca Raton, FL: CRC Press, p. 536, 1987.

Fisher, R. A. "Applications of 'Student's' Distribution." *Metron* **5**, 3–7, 1925.

Fisher, R. A. "Expansion of 'Student's' Integral in Powers of $n-1$." *Metron* **5**, 22–2, 1925.

Fisher, R. A. *Statistical Methods for Research Workers, 10th ed.* Edinburgh: Oliver and Boyd, 1948.

Goulden, C. H. Table A-3 in *Methods of Statistical Analysis, 2nd ed.* New York: Wiley, p. 443, 1956.

Press, W. H.; Flannery, B. P.; Teukolsky, S. A.; and Vetterling, W. T. "Incomplete Beta Function, Student's Distribution, F-Distribution, Cumulative Binomial Distribution." §6.2 in *Numerical Recipes in FORTRAN: The Art of Scientific Computing, 2nd ed.* Cambridge, England: Cambridge University Press, pp. 219–23, 1992.

Spiegel, M. R. *Theory and Problems of Probability and Statistics.* New York: McGraw-Hill, pp. 116–17, 1992.

Student. "The Probable Error of a Mean." *Biometrika* **6**, 1–5, 1908.

Student's z-Distribution

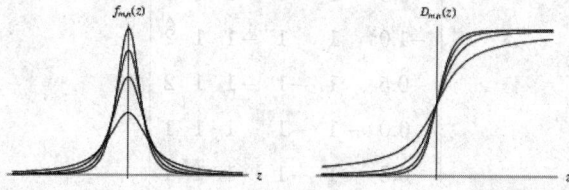

The probability density function for Student's *z*-

distribution are given by

$$f_{m,n}(z)=\frac{\Gamma\left(\dfrac{n}{2}\right)}{\sqrt{\pi}\Gamma\left(\dfrac{n-1}{2}\right)}\left(1+z^2\right)^{-n/2}. \quad (1)$$

Now define

$$d_{m,n}(z)$$

$$\equiv\frac{|z|^{1-n}\Gamma\left(\tfrac{1}{2}n\right){}_2F_1\left(\tfrac{1}{2}(n-1),\tfrac{1}{2}n;\tfrac{1}{2}(n+1);-z^{-2}\right)}{2\sqrt{\pi}\Gamma\left[\tfrac{1}{2}(n+1)\right]}, \quad (2)$$

then the cumulative distribution functions is given by

$$D_{m,n}(z)=\begin{cases}d_{m,n}(z) & \text{for } z\leq 0\\ 1-d_{m,n}(z) & \text{for } z\geq 0\end{cases} \quad (3)$$

The MEAN is 0, so the MOMENTS are

$$\mu_1=0 \quad (4)$$

$$\mu_2=\frac{1}{n-3} \quad (5)$$

$$\mu_3=0 \quad (6)$$

$$\mu_4=\frac{3}{(n-3)(n-5)}. \quad (7)$$

The MEAN, VARIANCE, SKEWNESS, and KURTOSIS are

$$\mu=0 \quad (8)$$

$$\sigma^2=\frac{1}{n-3} \quad (9)$$

$$\gamma_1=0 \quad (10)$$

$$\gamma_2=\frac{6}{n-5}. \quad (11)$$

The CHARACTERISTIC FUNCTION is

$$\phi(t)=\frac{2^{(3-n)/2}|t|^{(n-1)/2}K_{(1-n)/2}(|t|)}{\Gamma\left[\tfrac{1}{2}(n-1)\right]}, \quad (12)$$

where $K_n(z)$ is a MODIFIED BESSEL FUNCTION OF THE SECOND KIND.

Letting

$$z\equiv\frac{\bar{x}-\mu}{s}, \quad (13)$$

where x is the sample MEAN and μ is the population MEAN gives STUDENT'S *T*-DISTRIBUTION.

See also STUDENT'S *T*-DISTRIBUTION

Study's Theorem

Given three curves ϕ_1, ϕ_2, ϕ_3 with the common group of ordinary points G (which may be empty), let their remaining groups of intersections g_{23}, g_{31}, and g_{12} also be ordinary points. If ϕ_1' is any other curve through g_{23}, then there exist two other curves ϕ_2', ϕ_3' such that the three combined curves $\phi_i\phi_i'$ are of the same order and LINEARLY DEPENDENT, each curve ϕ_k' contains the corresponding group g_{ij}, and every intersection of ϕ_i or ϕ_i' with ϕ_j or ϕ_j' lies on ϕ_k or ϕ_k'.

References

Coolidge, J. L. *A Treatise on Algebraic Plane Curves.* New York: Dover, p. 34, 1959.

Sturm Chain

The series of STURM FUNCTIONS arising in application of the STURM THEOREM.

See also STURM FUNCTION, STURM THEOREM

Sturm Function

Given a function $f(x) \equiv f_0(x)$, write $f_1 \equiv f'(x)$ and define the Sturm functions by

$$f_n(x) = -\left\{ f_{n-2}(x) - f_{n-1}(x) \left[\frac{f_{n-2}(x)}{f_{n-1}(x)} \right] \right\}. \tag{1}$$

where $[P(x)/Q(x)]$ is a polynomial quotient. Then construct the following chain of Sturm functions,

$$f_0 = q_0 f_1 - f_2$$
$$f_1 = q_1 f_2 - f_3$$
$$f_2 = q_2 f_3 - f_4$$
$$\vdots$$
$$f_{s-2} = q_{s-2} f_{s-1} - f_s,$$

known as a STURM CHAIN. The chain is terminated when a constant $-f_s(x)$ is obtained.

Sturm functions provide a convenient way for finding the number of real roots of an algebraic equation with real coefficients over a given interval. Specifically, the difference in the number of sign changes between the Sturm functions evaluated at two points $x = a$ and $x = b$ gives the number of real roots in the interval (a, b). This powerful result is known as the STURM THEOREM. However, when the method is applied numerically, care must be taken when computing the polynomial quotients to avoid spurious results due to roundoff error.

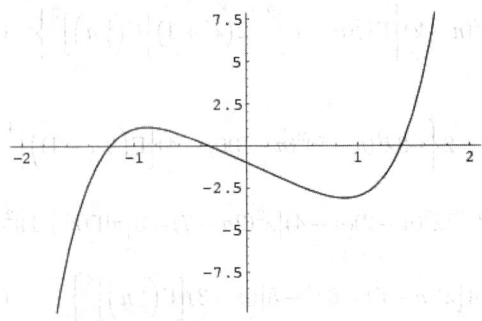

As a specific application of Sturm functions toward finding POLYNOMIAL ROOTS, consider the function $f_0(x) = x^5 - 3x - 1$, plotted above, which has roots -1.21465, -0.334734, $0.0802951 \pm 1.32836i$, and 1.38879 (three of which are real). The DERIVATIVE is given by $f'(x) = 5x^4 - 3$, and the STURM CHAIN is then given by

$$f_0 = x^5 - 3x - 1 \tag{3}$$

$$f_1 = 5x^4 - 3 \tag{4}$$

$$f_2 = \tfrac{1}{5}(12x + 5) \tag{5}$$

$$f_3 = \tfrac{59083}{20736}. \tag{6}$$

The following table shows the signs of f_i and the number of sign changes Δ obtained for points separated by $\Delta x = 2$.

x	f_0	f_1	f_2	f_3	Δ
-2	-1	1	-1	1	3
0	-1	-1	1	1	1
2	1	1	1	1	0

This shows that $3- = 2$ real roots lie in $(-2, 0)$, and $1- = 1$ real root lies in $(0, 2)$. Reducing the spacing to $\Delta.r = 0.5$ gives the following table.

x	f_0	f_1	f_2	f_3	Δ
-2.0	-1	1	-1	1	3
-1.5	-1	1	-1	1	3
-1.0	1	1	-1	1	2
-0.5	1	-1	-1	1	2
0.0	-1	-1	1	1	1
0.5	-1	-1	1	1	1
1.0	-1	1	1	1	1

$$\begin{vmatrix} 1.5 & 1 & 1 & 1 & 1 & 0 \\ 2.0 & 1 & 1 & 1 & 1 & 0 \end{vmatrix}$$

This table isolates the three real roots and shows that they lie in the intervals $(-1.5, -1.0)$, $(-0.5, 0.0)$, and $(1.0, 1.5)$. If desired, the intervals in which the roots fall could be further reduced.

The Sturm functions satisfy the following conditions:

1. Two neighboring functions do not vanish simultaneously at any point in the interval.
2. At a null point of a Sturm function, its two neighboring functions are of different signs.
3. Within a sufficiently small interval surrounding a zero point of $f_0(x)$, $f_1(x)$ is everywhere greater than zero or everywhere smaller than zero.

See also DESCARTES' SIGN RULE, STURM CHAIN, STURM THEOREM

References

Acton, F. S. *Numerical Methods That Work, 2nd printing.* Washington, DC: Math. Assoc. Amer., p. 334, 1990.

Dörrie, H. "Sturm's Problem of the Number of Roots." §24 in *100 Great Problems of Elementary Mathematics: Their History and Solutions.* New York: Dover, pp. 112–16, 1965.

Press, W. H.; Flannery, B. P.; Teukolsky, S. A.; and Vetterling, W. T. *Numerical Recipes in FORTRAN: The Art of Scientific Computing, 2nd ed.* Cambridge, England: Cambridge University Press, p. 469, 1992.

Rusin, D. "Known Math." http://www.math.niu.edu/~rusin/known-math/96/sturm.

Sturm, C. "Mémoire sur la résolution des équations numériques." *Bull. des sciences de Férussac* **11**, 1929.

Sturm Theorem

The number of REAL ROOTS of an algebraic equation with REAL COEFFICIENTS whose REAL ROOTS are simple over an interval, the endpoints of which are not ROOTS, is equal to the difference between the number of sign changes of the STURM CHAINS formed for the interval ends.

See also STURM CHAIN, STURM FUNCTION

References

Dörrie, H. "Sturm's Problem of the Number of Roots." §24 in *100 Great Problems of Elementary Mathematics: Their History and Solutions.* New York: Dover, pp. 112–16, 1965.

Rusin, D. "Known Math." http://www.math.niu.edu/~rusin/known-math/96/sturm.

Sturmian Separation Theorem

Let $A_r = a_{ij}$ be a SEQUENCE of N SYMMETRIC MATRICES of increasing order with $i, j = 1, 2, ..., r$ and $r = 1, 2, ..., N$. Let $\lambda_k(A_r)$ be the kth EIGENVALUE of A_r for $k = 1, 2, ..., r$, where the ordering is given by

$$\lambda_1(A_r) \geq \lambda_2(A_r) \geq ... \geq \lambda_r(A_r).$$

Then it follows that

$$\lambda_{k+1}(A_{i+1}) \leq \lambda_k(A_i) \leq \lambda_k(A_{i+1}).$$

References

Gradshteyn, I. S. and Ryzhik, I. M. *Tables of Integrals, Series, and Products, 6th ed.* San Diego, CA: Academic Press, p. 1121, 2000.

Sturmian Sequence

If a SEQUENCE has the property that the BLOCK GROWTH function $B(n) = n + 1$ for all n, then it is said to have minimal block growth, and the sequence is called a Sturmian sequence. An example of this is the sequence arising from the SUBSTITUTION MAP

$$0 \rightarrow 01$$
$$1 \rightarrow 0$$

yielding $0 \rightarrow 01 \rightarrow 010 \rightarrow 01001 \rightarrow 01001010 \rightarrow ...$, which gives us the Sturmian sequence 01001010....

STURM FUNCTIONS are sometimes also said to form a Sturmian sequence.

See also STURM FUNCTION, STURM THEOREM

Sturm-Liouville Equation

A second-order ORDINARY DIFFERENTIAL EQUATION

$$\frac{d}{dx}\left[p(x)\frac{dy}{dx}\right] + [\lambda w(x) - q(x)]y = 0,$$

where λ is a constant and $w(x)$ is a known function called either the density or WEIGHTING FUNCTION. The solutions (with appropriate boundary conditions) of λ are called EIGENVALUES and the corresponding $u_\lambda(x)$ EIGENFUNCTIONS. The solutions of this equation satisfy important mathematical properties under appropriate boundary conditions (Arfken 1985).

See also ADJOINT, SELF-ADJOINT

References

Arfken, G. "Sturm-Liouville Theory--Orthogonal Functions." Ch. 9 in *Mathematical Methods for Physicists, 3rd ed.* Orlando, FL: Academic Press, pp. 497–38, 1985.

Sturm-Liouville Theory

STURM-LIOUVILLE EQUATION

SU

SPECIAL UNITARY GROUP

Subalgebra

An ALGEBRA S' which is part of a large ALGEBRA S and shares its properties.

See also ALGEBRA

Subanalytic

$X \subseteq \mathbb{R}^n$ is subanalytic if, for all $x \in \mathbb{R}^n$, there is an open set U and a bounded SEMIANALYTIC set $Y \subset \mathbb{R}^{n+m}$ such that $X \cap U$ is the projection of Y into U.

See also SEMIANALYTIC

References

Bierstone, E. and Milman, P. "Semialgebraic and Subanalytic Sets." *IHES Pub. Math.* **67**, 5–2, 1988.
Marker, D. "Model Theory and Exponentiation." *Not. Amer. Math. Soc.* **43**, 753–59, 1996.

Subdiagonal

The subdiagonal of a SQUARE MATRIX is the set of elements directly under the elements comprising the DIAGONAL. For example, in the following matrix, the diagonal elements are denoted d_i and the subdiagonals are denoted s_i,

$$\begin{bmatrix} d_1 & a_{12} & a_{13} & \cdots & a_{1n} \\ s_1 & d_2 & a_{23} & \ddots & a_{2n} \\ a_{31} & s_2 & d_3 & \ddots & a_{3n} \\ \vdots & \ddots & \ddots & \ddots & \vdots \\ a_{n1} & a_{n2} & a_{n3} & \cdots & d_n \end{bmatrix}.$$

See also CANONICAL BOX MATRIX, DIAGONAL, SUPERDIAGONAL, TRIDIAGONAL MATRIX

References

Faddeeva, V. N. *Computational Methods of Linear Algebra.* New York: Dover, p. 50, 1958.

Subfactorial

The number of PERMUTATIONS of n objects in which no object appears in its natural place (i.e., the number of so-called "DERANGEMENTS").

$$!n \equiv n! \sum_{k=0}^{n} \frac{(-1)^k}{k!} \qquad (1)$$

or

$$!n \equiv \left[\frac{n!}{e} \right]. \qquad (2)$$

where $k!$ is the usual FACTORIAL and $[x]$ is the NINT function. The first few values are $!1 = 0$, $!2 = 1$, $!3 = 2$, $!4 = 9$, $!5 = 44$, $!6 = 265$, $!7 = 1854$, $!8 = 14833$, ... (Sloane's A000166). For example, the only DERANGEMENTS of $\{1, 2, 3\}$ are $\{2, 3, 1\}$ and $\{3, 1, 2\}$, so $!3 = 2$. Similarly, the DERANGEMENTS of $\{1, 2, 3, 4\}$ are $\{2, 1, 4, 3\}$, $\{2, 3, 4, 1\}$, $\{2, 4, 1, 3\}$, $\{3, 1, 4, 2\}$, $\{3, 4, 1, 2\}$, $\{3, 4, 2, 1\}$, $\{4, 1, 2, 3\}$, $\{4, 3, 1, 2\}$, and $\{4, 3, 2, 1\}$, so $!4 = 9$. The only prime subfactorial is $!3 = 2$.

The subfactorials are also called the RENCONTRES NUMBERS and satisfy the RECURRENCE RELATIONS

$$!n = n \cdot !(n-1) + (-1)^n \qquad (3)$$

$$!(n+1) = n[!n + !(n-1)]. \qquad (4)$$

The subfactorial can be considered a special case of a restricted ROOKS PROBLEM.

The only number equal to the sum of subfactorials of its digits is

$$148,349 = !1 + !4 + !8 + !3 + !4 + !9 \qquad (5)$$

(Madachy 1979).

See also DERANGEMENT, FACTORIAL, MARRIED COUPLES PROBLEM, ROOKS PROBLEM, SUPERFACTORIAL

References

Dörrie, H. §6 in *100 Great Problems of Elementary Mathematics: Their History and Solutions.* New York: Dover, pp. 19–1, 1965.
Madachy, J. S. *Madachy's Mathematical Recreations.* New York: Dover, p. 167, 1979.
Sloane, N. J. A. Sequences A000166/M1937 in "An On-Line Version of the Encyclopedia of Integer Sequences." http://www.research.att.com/~njas/sequences/eisonline.html.
Sloane, N. J. A. and Plouffe, S. Figure M1937 in *The Encyclopedia of Integer Sequences.* San Diego: Academic Press, 1995.
Stanley, R. P. *Enumerative Combinatorics, Vol. 1.* Cambridge, England: Cambridge University Press, p. 67, 1997.
Wells, D. *The Penguin Dictionary of Curious and Interesting Numbers.* Middlesex, England: Penguin Books, p. 27, 1986.

Subfield

If a subset S of the elements of a FIELD F satisfies the FIELD AXIOMS with the same operations of F, then S is called a subfield of F. In a FINITE FIELD of ORDER p^n, with p a prime, there exists a subfield of ORDER p^m for every m DIVIDING n.

See also EXTENSION FIELD, FIELD, PRIME SUBFIELD, SUBMANIFOLD, SUBSPACE

Subgraph

A GRAPH G' whose VERTICES and EDGES form subsets of the VERTICES and EDGES of a given GRAPH G. If G' is a subgraph of G, then G is said to be a SUPERGRAPH of G'.

See also GRAPH, INDUCED SUBGRAPH, SUPERGRAPH, SUBTREE, ULAM'S CONJECTURE

References

Harary, F. *Graph Theory.* Reading, MA: Addison-Wesley, p. 11, 1994.

Subgroup

A subset H of GROUP elements of a group G that satisfies the four GROUP requirements. "H is a subgroup of G" is written $H \subset G$. The ORDER of any subgroup of a GROUP of ORDER h must be a DIVISOR of h.

See also CARTAN SUBGROUP, COMPOSITION SERIES, FITTING SUBGROUP, GROUP, NORMAL SUBGROUP

Subharmonic Function

Let $U \subseteq C$ be an OPEN SET and f a real-valued continuous function on U. Suppose that for each CLOSED DISK $D(P, r) \subseteq U$ and every real-valued HARMONIC FUNCTION h defined on a NEIGHBORHOOD of $D(P, r)$ which satisfies $f \le h$ on $\partial D(P, r)$, it holds that $f \le h$ on the OPEN DISK $D(P, r)$. Then f is said to be subharmonic on U (Krantz 1999, p. 99).

1. If f_1, f_2 are subharmonic on U, then so is $f_1 + f_2$.
2. If f_1 is subharmonic on U and $a > 0$ is a constant, than af_1 is subharmonic on U.
3. If f_1, f_2 are subharmonic on U, then $\max\{f_1(z), f_2(z)\}$ is also subharmonic on U.

See also BARRIER, HARMONIC FUNCTION

References

Krantz, S. G. "The Dirichlet Problem and Subharmonic Functions." §7.7 in *Handbook of Complex Analysis*. Boston, MA: Birkhäuser, pp. 97–01, 1999.

Sublime Number

Let $\sigma_0(n)$ and $\sigma_1(n)$ denote the number and sum of the divisors of n, respectively (i.e., the zeroth- and first-order DIVISOR FUNCTIONS). A number n is called sublime if $\sigma_0(n)$ and $\sigma_1(n)$ are both PERFECT NUMBERS. The only two known sublime numbers are 12 and

6086555670238378989670371734243169 6 $\cdots$

$\cdots$ 2265783077335188597052832486051279169126 4.

It is not known if any ODD sublime number exists.

See also DIVISOR FUNCTION, PERFECT NUMBER

References

Weisstein, E. W. "Integer Sequences." MATHEMATICA NOTEBOOK IntegerSequences.M.

Submanifold

A C^∞ (infinitely differentiable) MANIFOLD is said to be a submanifold of a C^∞ MANIFOLD M' if M is a SUBSET of M' and the IDENTITY MAP of M into M' is an EMBEDDING.

See also EMBEDDING, MANIFOLD, SUBFIELD, SUBSPACE

Submatrix

$$\begin{array}{|cc|cc}\hline a_{11} & a_{12} & a_{13} & a_{14} \\ a_{21} & a_{22} & a_{23} & a_{24} \\ \hline a_{31} & a_{32} & a_{33} & a_{34} \\ a_{41} & a_{42} & a_{43} & a_{44} \\ \end{array}$$

submatrices

A $p \times q$ submatrix of an $m \times n$ MATRIX (with $p \le m$, $n \le q$) is a $p \times q$ MATRIX formed by taking a block of the entries of this size from the original matrix.

See also MATRIX

Submersion

A submersion is a SMOOTH MAP $f : M \to N$ when

$$\dim M \ge \dim N,$$

given that the DIFFERENTIAL, or JACOBIAN, is SURJECTIVE at every x in M. The basic example of a submersion is the canonical submersion α of $\mathbb{R}^n$ onto $\mathbb{R}^k$ when $n \ge k$,

$$\alpha(x_1, \ldots, x_n) = (x_1, \ldots, x_k).$$

In fact, if f is a submersion, then it is possible to find coordinates around x in M and coordinates around $f(x)$ in N such that f is the canonical submersion written in these coordinates. For example, consider the submersion of $\mathbb{R}^2 - \{(0, 0)\}$ onto the circle $\mathbb{S}^1$, given by $f(x, y) = (x, y)/\sqrt{x^2 + y^2}$.

See also IMMERSION, RIEMANNIAN SUBMERSION

Submodule

A MODULE over a RING that is contained in and has the same addition as another MODULE over the same RING.

See also MODULE

Subnormal

If the LEXIS RATIO $L < 1$, a set of trials are said to be subnormal.

See also LEXIS RATIO, SUBNORMAL SUBGROUP, SUPERNORMAL

Subnormal Subgroup

L is a subnormal SUBGROUP of H if there is a "normal series" (in the sense of Jordan-Holder) from L to H.

Suborder Function

A special case of the generalized MULTIPLICATIVE ORDER function taken with respect to the PRIMITIVE ROOTS -1 and 1. This function is denoted $\text{sord}_n(a)$ and is implemented in *Mathematica* as MultiplicativeOrder[a, n, $\{-1, 1\}$].

See also MULTIPLICATIVE ORDER

Subordinate Norm
Natural Norm

Subresultant

Subresultants for a few simple pairs of polynomials include

$$S(x-a, \ x-b) = \{a-b, \ 1\}$$

$$S((x-a)(x-b), \ x-c) = \{(a-c)(b-c), \ 1\}$$

$$S((x-a)(x-b), \ (x-c)(x-d))$$
$$= \{(a-c)(b-c)(a-d)(b-d), \ a+b-c-d, \ 1\}.$$

The principal subresultants of two polynomials can be computed using the *Mathematica* command Subresultants[*poly1*, *poly2*, *var*]. The first k subresultants of two polynomials p_1 and p_2, both with leading coefficient one, are zero when p_1 and p_2 have k common roots.

See also Discriminant (Polynomial), Resultant

References

J. Pure Appl. Algebra **145**, 149, 2000.
Hong, H. "Subresultants Under Composition." *J. Symb. Comput.* **23**, 355–65, 1997.
Hong, H. "Subresultants in Roots." Submitted 1999.

Subring

A subring of a Ring R is a Subgroup of R that is Closed under multiplication.

See also Ring, Subgroup

References

Dummit, D. S. and Foote, R. M. *Abstract Algebra, 2nd ed.* Englewood Cliffs, NJ: Prentice-Hall, p. 230, 1998.

Subscript

A quantity displayed below the normal line of text (and generally in a smaller point size), as the "i" in a_i, is called a subscript. Subscripts are commonly used to indicate indices (a_{ij} is the entry in the ith row and jth column of a Matrix A), partial differentiation (y_x is an abbreviation for $\partial y / \partial x$), and a host of other operations and notations in mathematics.

See also Superscript

Subselfsimilar Set

Giving a set $F = \{f_1, f_2, \ldots, f_n\}$ of contracting similitudes of $\mathbb{R}^r$, the closed set E is said to be subselfsimilar for F if

$$E \subset \bigcup_{i=1}^{n} f_i(E)$$

(Falconer 1995, Duvall and Keesling 1999).

References

Duvall, P. and Keesling, J. The Hausdorff Dimension of the Boundary of the Lévy Dragon. 22 Jul 1999. http://xxx.lanl.gov/abs/math.DS/9907145/.
Falconer, K. J. "Sub-Self-Similar Sets." *Trans. Amer. Math. Soc.* **247**, 3121–129, 1995.

Subsequence

A subsequence of a Sequence $S = \{x_i\}_{i=1}^{n}$ is a derived sequence $\{y_i\}_{i=1}^{N} = \{x_{i+j}\}$ for some $j \geq 0$ and $N \leq n - j$. More generally, the word subsequence is sometimes used to mean a sequence derived from a sequence S by discarding some of its terms.

See also Lower-Trimmed Subsequence, Upper-Trimmed Subsequence

Subset

A portion of a Set. B is a subset of A (written $B \subseteq A$) iff every member of B is a member of A. If B is a Proper Subset of A (i.e., a subset other than the set itself), this is written $B \subset A$. If B is not a subset of A, this is written $B \nsubseteq A$. (The notation $B \not\subset A$ is generally not used, since $B \nsubseteq A$ automatically means that B and A cannot be the same.)

The set of subsets of a set S is called the Power Set of S, and a Set of n elements has 2^n subsets (including both the set itself and the Empty Set). This follows from the fact that the total number of distinct k-Subset on a set of n elements is given by the Binomial Sum

$$\sum_{k=0}^{n} \binom{n}{k} = 2^n.$$

For sets of $n = 1, \ 2, \ \ldots$ elements, the numbers of subsets are therefore 2, 4, 8, 16, 32, 64, ... (Sloane's A000079). For example, the set $\{1\}$ has the two subsets $\varnothing$ and $\{1\}$. Similarly, the set $\{1, 2\}$ has subsets $\varnothing$ (the Empty Set, $\{1\}$, $\{2\}$, and $\{1, 2\}$. The subsets (i.e., Power Set) of a given set can be found using Subsets[*list*] in the *Mathematica* add-on package DiscreteMath`Combinatorica` (which can be loaded with the command << DiscreteMath`).

See also Empty Set, Implies, k-Subset, p-System, Power Set, Proper Subset, Superset, Venn Diagram

References

Courant, R. and Robbins, H. *What is Mathematics?: An Elementary Approach to Ideas and Methods, 2nd ed.* Oxford, England: Oxford University Press, p. 109, 1996.
Ruskey, F. "Information of Subsets of a Set." http://www.theory.csc.uvic.ca/~cos/inf/comb/SubsetInfo.html.
Skiena, S. "Binary Representation and Random Sets." §1.5.2 in *Implementing Discrete Mathematics: Combinatorics and Graph Theory with Mathematica.* Reading, MA: Addison-Wesley, pp. 41–2, 1990.

Sloane, N. J. A. Sequences A000079/M1129 in "An On-Line Version of the Encyclopedia of Integer Sequences." http://www.research.att.com/~njas/sequences/eisonline.html.

Subset Sum Problem

The problem of finding what subset of a list of integers has a given sum. The subset sum is an INTEGER RELATION problem where the relation coefficients a_i are 0 or 1.

See also INTEGER RELATION, LATTICE REDUCTION, KNAPSACK PROBLEM, POSTAGE STAMP PROBLEM, STÖHR SEQUENCE

References

Coster, M. J.; LaMacchia, B. A.; Odlyzko, A. M.; and Schnorr, C. P. "An Improved Low-Density Subset Sum Algorithm." In *Advances in Cryptology: EUROCRYPT '91 (Brighton, 1999)* (Ed. D. W. Davis). New York: Springer-Verlag, pp. 54–7, 1992.
Coster, M. J.; Joux, A.; LaMacchia, B. A.; Odlyzko, A. M.; Schnorr, C. P.; and Stern, J. "Improved Low-Density Subset Sum Algorithms." *Comput. Complex.* **2**, 111–28, 1992.
Ferguson, H. R. P. and Bailey, D. H. "A Polynomial Time, Numerically Stable Integer Relation Algorithm." RNR Techn. Rept. RNR-91–32, Jul. 14, 1992.
Lagarias, L. C. and Odlyzko, A. M. "Solving Low-Density Subset Sum Problems." *J. ACM* **32**, 229–46, 1985.
Schnorr, C. P. and Euchner, M. "Lattice Basis Reduction: Improved Practical Algorithms and Solving Subset Sum Problems." In *Fundamentals of Computation Theory (Gosen 1991)*. Berlin: Springer-Verlag, pp. 68–5, 1991.

Subspace

Let $\mathbb{V}$ be a REAL VECTOR SPACE (e.g., the real continuous functions $C(I)$ on a CLOSED INTERVAL I, 2-D EUCLIDEAN SPACE $\mathbb{R}^2$, the twice differentiable real functions $C^{(2)}(I)$ on I, etc.). Then $\mathbb{W}$ is a real SUBSPACE of $\mathbb{V}$ if $\mathbb{W}$ is a SUBSET of $\mathbb{V}$ and, for every w_1, $w_1 \in \mathbb{W}$ and $t \in \mathbb{R}$ (the REALS), $w_1 + w_2 \in \mathbb{W}$ and $tw_1 \in \mathbb{W}$. Let (H) be a homogeneous system of linear equations in $x_1, ..., x_n$. Then the SUBSET S of $\mathbb{R}^n$ which consists of all solutions of the system (H) is a subspace of $\mathbb{R}^n$.

More generally, let F_q be a FIELD with $q = p^x$, where p is PRIME, and let $F_{q, n}$ denote the n-D VECTOR SPACE over F_q. The number of k-D linear subspaces of $F_{q, n}$ is

$$N(F_{q, n}) = \binom{n}{k}_q,$$

where this is the Q-BINOMIAL COEFFICIENT (Aigner 1979, Exton 1983). The asymptotic limit is

$$N(F_{q, n}) = \begin{cases} c_e q^{n^2/4}[1 + o(1)] & \text{for } n \text{ even} \\ c_o q^{n^2/4}[1 + o(1)] & \text{for } n \text{ odd}, \end{cases}$$

where

$$c_e = \frac{\sum_{k=-\infty}^{\infty} q^{-k^2}}{\prod_{j=1}^{\infty}(1 - q^{-j})}$$

$$c_o = \frac{\sum_{k=-\infty}^{\infty} q^{-(k+1/2)^2}}{\prod_{j=1}^{\infty}(1 - q^{-j})}$$

(Finch). The case $q = 2$ gives the Q-ANALOG of the WALLIS FORMULA.

See also Q-BINOMIAL COEFFICIENT, SUBFIELD, SUBMANIFOLD

References

Aigner, M. *Combinatorial Theory*. New York: Springer-Verlag, 1979.
Exton, H. *q-Hypergeometric Functions and Applications*. New York: Halstead Press, 1983.
Finch, S. "Favorite Mathematical Constants." http://www.mathsoft.com/asolve/constant/dig/dig.html.

Substitution Group

PERMUTATION GROUP

Substitution Map

A MAP which uses a set of rules to transform elements of a sequence into a new sequence using a set of rules which "translate" from the original sequence to its transformation. For example, the substitution map $\{1 \rightarrow 0,\ 0 \rightarrow 11\}$ would take 10 to 011.

See also GOLDEN RATIO, MORSE-THUE SEQUENCE, STRING REWRITING, THUE CONSTANT

Substitution Tensor

KRONECKER DELTA, PERMUTATION SYMBOL, PERMUTATION TENSOR

Subtend

Given a geometric object O in the PLANE and a point P, let A be the ANGLE from one edge of O to the other with VERTEX at P. Then O is said to subtend an ANGLE A from P.

See also ANGLE, VERTEX ANGLE

Subtraction

Subtraction is the operation of taking the DIFFERENCE $x - y$ of two numbers x and y. Here, x is called the MINUEND, y is called the SUBTRAHEND, and the symbol between the x and y is called the MINUS SIGN. The expression "$x - y$" is read "x MINUS y." Subtraction is the inverse of ADDITION, so $x + y - y = x - y + y = x$.

The subtraction of a number from itself gives 0, while the subtraction of a real number from a smaller real number gives a negative real number. Subtraction of real numbers can be naturally extended to complex numbers.

See also ADDITION, DIVISION, MINUEND, MINUS, MINUS SIGN, MULTIPLICATION, SUBTRAHEND

Subtrahend

A quantity which is subtracted from another (the MINUEND).

See also MINUEND, SUBTRACTION

Subtree

A TREE G' whose VERTICES and EDGES form subsets of the VERTICES and EDGES of a given TREE G.

See also SUBGRAPH, TREE

Subvariety

See also ALGEBRAIC VARIETY

Succeeds

The relationship x succeeds (or FOLLOWS) y is written $x \succ y$. The relation x succeeds or is equal to y is written $x \succeq y$.

See also PRECEDES

Successes

DIFFERENCE OF SUCCESSES

Successor

For any ORDINAL NUMBER α, the successor of α is $\alpha \cup \{\alpha\}$ (Ciesielski 1997, p. 46). The successor of an ordinal number α is therefore the next ordinal, $\alpha + 1$.

See also LIMIT ORDINAL, ORDINAL NUMBER

References
Ciesielski, K. *Set Theory for the Working Mathematician.* Cambridge, England: Cambridge University Press, 1997.

Sufficient

A CONDITION which, if true, guarantees that a result is also true. (However, the result may also be true if the CONDITION is not met.) If a CONDITION is both NECESSARY and SUFFICIENT, then the result is said to be true IFF (the CONDITION holds.

For example, the condition that a decimal number n end in the DIGIT 2 is a sufficient but not NECESSARY condition that n be EVEN.

See also IFF, IMPLIES, NECESSARY, SUFFICIENTLY LARGE

References
Jeffreys, H. and Jeffreys, B. S. "Necessary: Sufficient." §1.036 in *Methods of Mathematical Physics, 3rd ed.* Cambridge, England: Cambridge University Press, pp. 10–1, 1988.

Suitable Number

IDONEAL NUMBER

Sultan's Dowry Problem

A sultan has granted a commoner a chance to marry one of his n daughters. The commoner will be presented with the daughters one at a time and, when each daughter is presented, the commoner will be told the daughter's dowry (which is fixed in advance). Upon being presented with a daughter, the commoner must immediately decide whether to accept or reject her (he is not allowed to return to a previously rejected daughter). However, the sultan will allow the marriage to take place only if the commoner picks the daughter with the overall highest dowry. Then what is the commoner's best strategy, assuming he knows nothing about the distribution of dowries (B. Elbows)?

Since the commoner knows nothing about the distribution of the dowries, the best strategy is to wait until a certain number x of daughters have been presented, then pick the highest dowry thereafter. The exact number to skip is determined by the condition that the odds that the highest dowry has already been seen is just greater than the odds that it remains to be seen *and that if it is seen it will be picked.* This amounts to finding the smallest x such that

$$\frac{x}{n} \geq \frac{x}{n}\left(\frac{1}{x+1} + \ldots + \frac{1}{n-1}\right). \qquad (1)$$

Computing the sum analytically gives the solution as the smallest x such that

$$H_x \geq H_n - 1, \qquad (2)$$

where H_n is a HARMONIC NUMBER. Solving

$$H_x = H_n - 1 \qquad (3)$$

numerically and taking the CEILING FUNCTION $\lceil x \rceil$ then gives the solutions 0, 1, 1, 2, 2, 2, 3, 3, 3, 4, 4, 5, 5, 5, ... (Sloane's A054382) for $n = 1, 2, \ldots$ daughters.

The problem is most commonly stated with $n = 100$ daughters, which gives the result that the commoner should wait until he has seen 37 of the daughters, then pick the first daughter with a dowry that is bigger than any preceding one. With this strategy, his odds of choosing the daughter with the highest dowry are surprisingly high: about 37% (B. Elbows; Honsberger 1979, pp. 104–10, Mosteller 1987).

See also BIRTHDAY PROBLEM

References
Elbows, B. http://xraysgi.ims.uconn.edu/rpa-output/decision/dowry.s.
Honsberger, R. "Some Surprises in Probability." Ch. 5 in *Mathematical Plums* (Ed. R. Honsberger). Washington, DC: Math. Assoc. Amer., pp. 104–10, 1979.
Mosteller, F. Problem 47 in *Fifty Challenging Problems in Probability with Solutions.* New York: Dover, 1987.

Sloane, N. J. A. Sequences A054382 in "An On-Line Version of the Encyclopedia of Integer Sequences." http://www.re-search.att.com/~njas/sequences/eisonline.html.

Sum

A sum is the result of an ADDITION. For example, adding 1, 2, 3, and 4 gives the sum 10, written

$$1 + 2 + 3 + 4 = 10. \tag{1}$$

The numbers being summed are called ADDENDS, or sometimes SUMMANDS. The summation operation can also be indicated using a capital sigma with upper and lower limits written above and below, and the index indicated below. For example, the above sum could be written

$$\sum_{k=1}^{4} k = 10. \tag{2}$$

A sum

$$\sum_{i=1}^{n} a_i \tag{3}$$

in which each term a_i is given by some fixed rule (i.e., $\{a_i\}_{i=1}^{n}$ is a well defined SEQUENCE) is called a SERIES, and if the number of terms n is infinite, the sum is called an INFINITE SERIES. A sum of the form

$$\sum_{k=1}^{n} r^k \tag{4}$$

is called a GEOMETRIC SERIES.

The general finite POWER SUM

$$\sum_{k=1}^{n} k^p \tag{5}$$

can be given by the expression

$$\sum_{k=1}^{n} k^p = \frac{(B+n+1)^{[p+1]} - B^{[p+1]}}{p+1}, \tag{6}$$

which is equivalent to FAULHABER'S FORMULA, where the NOTATION $B^{[k]}$ means the quantity in question is raised to the appropriate POWER k and all terms OF THE FORM B^m are replaced with the corresponding BERNOULLI NUMBERS B_m.

NICOMACHUS'S THEOREM gives as curious expression for the POWER SUM $\sum_{k=1}^{n} k^3$.

Other analytic sums include

$$\left(\sum_{k=0}^{n} x^k \right)^p$$

$$= \frac{1}{(p-1)!} \sum_{k=0}^{np} \frac{(n - |n-k| + p - 1)!}{(n - |n-k|)!} x^k \tag{7}$$

for $p = 1, 2$

$$\left(\sum_{n=0}^{\infty} a_n x^n \right)^2 = \sum_{n=0}^{\infty} a_n^2 x^{2n} + 2 \sum_{\substack{n=1 \\ i+j=n \\ i<j}}^{\infty} a_i a_j x^n, \tag{8}$$

and

$$\sum xy = x_1 y_1 + x_1 y_2 + \ldots + x_2 y_1 + x_2 y_2 + \ldots$$

$$= (x_1 + x_2 + \ldots) y_1 + (x_1 + x_2 + \ldots) y_2$$

$$= \left(\sum x \right)(y_1 + y_2 + \ldots) = \sum x \sum y, \tag{9}$$

so

$$\sum_{i=1}^{m} \sum_{j=1}^{n} x_i x_j = \left(\sum_{i=1}^{m} x_i \right) \left(\sum_{j=1}^{n} y_j \right). \tag{10}$$

$$\sum_{j=0}^{n} j x^j = \frac{n x^{n+2} - (n+1) x^{n+1} + x}{(x-1)^2} \tag{11}$$

$$\sum_{j=1}^{n} \frac{x_j^r}{\prod_{\substack{k=1 \\ k \neq j}}^{n} (x_j - x_k)} = \begin{cases} 0 & \text{for } 0 \leq r < n-1 \\ 1 & \text{for } r = n-1 \\ \sum_{j=1}^{n} x_j & \text{for } r = n \end{cases} \tag{12}$$

$$\sum_{k=1}^{n} \frac{\prod_{\substack{r=1 \\ r \neq k}}^{n} (x + k - r)}{\prod_{\substack{r=1 \\ r \neq k}}^{n} (k - r)} = 1 \tag{13}$$

$$(n+1) \sum_{m=1}^{n} m^k = \sum_{m=1}^{n} \left[m^{k+1} + \sum_{p=1}^{k} \left(\sum_{m=1}^{p} m^k \right) \right]. \tag{14}$$

To minimize the sum of a set of squares of numbers $\{x_i\}$ about a given number x_0

$$S \equiv \sum_i (x_i - x_0)^2 = \sum_i x_i^2 - 2x_0 \sum_i x_i + N x_0^2. \tag{15}$$

take the DERIVATIVE.

$$\frac{d}{dx_0} S = -2 \sum_i x_i + 2N x_0 = 0. \tag{16}$$

Solving for x_0 gives

$$x_0 \equiv \bar{x} = \frac{1}{N} \sum_i x_i. \tag{17}$$

so S is minimized when x_0 is set to the MEAN.

See also ARITHMETIC SERIES, BERNOULLI NUMBER, BINOMIAL SUMS, CLARK'S TRIANGLE, CONVERGENCE IMPROVEMENT, DEDEKIND SUM, DOUBLE SUM, EULER SUM, FACTORIAL SUMS, FAULHABER'S FORMULA, GAB-

RIEL'S STAIRCASE, GAUSSIAN SUM, GEOMETRIC SERIES, GOSPER'S METHOD, HURWITZ ZETA FUNCTION, INFINITE SERIES, INFINITE PRODUCT, KLOOSTERMAN'S SUM, LEGENDRE SUM, LERCH TRANSCENDENT, NICOMACHUS'S THEOREM, ODD NUMBER THEOREM, PASCAL'S TRIANGLE, POWER SUM, PRODUCT, RAMANUJAN'S SUM, RIEMANN ZETA FUNCTION, SERIES, WHITNEY SUM

References

Courant, R. and Robbins, H. "The Sum of the First n Squares." §1.4 in *What is Mathematics?: An Elementary Approach to Ideas and Methods, 2nd ed.* Oxford, England: Oxford University Press, pp. 14–5, 1996.

Finch, S. "Unsolved Mathematics Problems: Sleeping Habits of Armadillos." http://www.mathsoft.com/asolve/glasser/glasser.html.

Petkovsek, M.; Wilf, H. S.; and Zeilberger, D. *A = B.* Wellesley, MA: A. K. Peters, 1996.

Sum of Squares Function

The number of representations of n by k squares, distinguishing signs and order, is denoted $r_k(n)$. For example, consider the number of ways of representing 5 as the sum of two squares.

$$5 = (-2)^2 + (-1)^2 = (-2)^2 + 1^2 = 2^2 + (-1)^2$$

$$= 2^2 + 1^2 = (-1)^2 + (-2)^2 = (-1)^2 + 2^2$$

$$= 1^2 + (-2)^2 = 1^2 + 2^2 \tag{1}$$

so $r_2(5) = 8$. The *Mathematica* function SumOfSquaresR[k, n] in the *Mathematica* add-on package NumberTheory`NumberTheoryFunctions` (which can be loaded with the command < <NumberTheory`) gives $r_k(n)$.

The function $r_2(n)$ is often written simply as $r(n)$, and is intimately connected with the LEIBNIZ SERIES and with GAUSS'S CIRCLE PROBLEM (Hilbert and Cohn-Vossen 1999, pp. 27–9). It is also given by the inverse Möbius transform of the sequence $b_{2n} = 0$ and $b_{2n+1} = 4(-1)^n$ (Sloane and Plouffe 1995, p. 22). The average order of $r(n)$ is π, but the normal order is 0 (Hardy 1999, p. 55).

Jacobi gave analytic expressions for $r_k(n)$ for the cases $k = 2$, 4, 6, and 8 (Hardy 1999, p. 132). The cases $k = 2$, 4, and 6 were found by equating COEFFICIENTS of the JACOBI THETA FUNCTIONS $\vartheta_3(x)$, $\vartheta_3^2(x)$, and $\vartheta_3^4(x)$. The solutions for $k = 10$ and 12 were found by Liouville and Eisenstein, and Glaisher (1907) gives a table of $r_k(n)$ for $k = 2s = 18$. $r_3(n)$ was found as a finite sum involving quadratic reciprocity symbols by Dirichlet. $r_5(n)$ and $r_7(n)$ were found by Eisenstein, Smith, and Minkowski.

A POSITIVE INTEGER can be represented as the sum of two squares IFF each of its prime factors of the form $k + 3$ occurs as an even power, as first established by Euler in 1738. In LAGRANGE'S FOUR-SQUARE THEOREM, Lagrange proved that every POSITIVE INTEGER can be written as the SUM of at most four SQUARES, where 4 may be reduced to 3 except for numbers OF THE FORM $4^n(8k + 7)$, as proved by Legendre in 1798 (Nagell 1951, p. 194; Wells 1986, pp. 48 and 56; Hardy 1999, p. 12; Savin 2000).

$r(n) = r_2(n)$ is 0 whenever n has a PRIME divisor OF THE FORM $4k + 3$ to an ODD POWER; it doubles upon reaching a new PRIME OF THE FORM $4k + 1$. It is given explicitly by

$$r_2(n) = 4 \sum_{d = 1, 3, \ldots | n} (-1)^{(d-1)/2} \tag{2}$$

$$= 4[d_1(n) - d_3(n)] \tag{3}$$

$$= 4 \sum_{d | n} \sin\left(\tfrac{1}{2}\pi d\right), \tag{4}$$

where $d_k(n)$ is the number of DIVISORS of n OF THE FORM $4m + k$ (Hilbert and Cohn-Vossen 1999, pp. 37–8; Hardy 1999, p. 12). The first few values are 4, 4, 0, 4, 8, 0, 0, 4, 4, 8, 0, 0, 8, 0, 0, 4, 8, 4, 0, 8, 0, 0, 0, 0, 12, 8, 0, 0, ... (Sloane's A004018). $r(n)$ obeys the unexpected identities

$$\sum_{n=0}^{\infty} \frac{r(n)}{\sqrt{n+a}} e^{-2\pi\sqrt{(n+a)b}} = \sum_{n=0}^{\infty} \frac{r(n)}{\sqrt{n-b}} e^{-2\pi\sqrt{(n+b)a}} \tag{5}$$

for $\Re[\sqrt{a}]$, $\Re\left[\sqrt{b}\right] > 0$,

$$\sum_{0 \le n \le x} \frac{r(n)}{\sqrt{x-n}} = 2\pi \sqrt{x} + \sum_{n=1}^{\infty} \frac{r(n)}{\sqrt{n}} \sin(2\pi\sqrt{nx}) \tag{6}$$

and

$$\sum_{0 \le n \le x} r(n) = \pi x + \sqrt{x} \sum_{n=1}^{\infty} \frac{r(n)}{\sqrt{n}} J_1(2\pi\sqrt{nx}) \tag{7}$$

(Hardy 1999, p. 82).

The first few values of the summatory function

$$R(n) = \sum_{k=1}^{n} r_2(n) \tag{8}$$

are 0, 4, 8, 8, 12, 20, 20, 20, 24, 28, 36, ... (Sloane's A014198). Shanks (1993) defines instead $R'(n) = 1 + R(n)$, with $R'(0) = 1$. A LAMBERT SERIES for $r_2(n)$ is

$$\sum_{n=1}^{\infty} \frac{4(-1)^{n+1} x^n}{1 - x^n} = \sum_{n=1}^{\infty} r_2(n) x^n \tag{9}$$

(Hardy and Wright 1979). Explicit values of $R'(n)$ for several powers of 10 are given in the following table (Mitchell 1966; Shanks 1993, pp. 165 and 234).

n	$R'(10^n)$
0	5
1	37

2	317
3	3149
4	31417
5	314197
6	3141549
8	314159053
10	31415925457
12	3141592649625
14	31415926535058

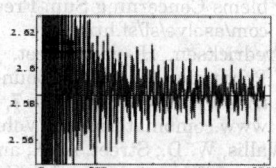

Asymptotic results include

$$\sum_{k=1}^{n} r_2(k) = \pi n + \mathcal{O}(\sqrt{n}) \tag{10}$$

$$\sum_{k=1}^{n} \frac{r_2(k)}{k} = K + \pi \ln n + \mathcal{O}(n^{-1/2}), \tag{11}$$

where K is a constant known as the SIERPINSKI CONSTANT. The left plot above shows

$$\left[\sum_{k=1}^{n} r_2(k)\right] - \pi n, \tag{12}$$

with $\pm\sqrt{n}$ illustrated by curved envelope, and the right plot shows

$$\left[\sum_{k=1}^{n} \frac{r_2(k)}{k}\right] - \pi \ln n, \tag{13}$$

with the value of K indicated as the solid horizontal line.

The number of solutions of

$$x^2 + y^2 + z^2 = n \tag{14}$$

for a given n without restriction on the signs or relative sizes of x, y, and z is given by $r_3(n)$. Gauss proved that if n is SQUAREFREE and $n > 4$, then

$$r_3(n) = \begin{cases} 24h(-n) & \text{for } n \equiv 3 \pmod 8 \\ 12h(-4n) & \text{for } n \equiv 1, 2, 5, 6 \pmod 8 \\ 0 & \text{for } n \equiv 7 \pmod 8 \end{cases} \tag{15}$$

(Arno 1992), where $h(x)$ is the CLASS NUMBER of x.

Additional higher-order identities are given by

$$r_4(n) = 8 \sum_{d|n} d = 8\sigma(n) \tag{16}$$

$$= 24 \sum_{d=1, 3, \dots |n} d \tag{17}$$

$$= 24\sigma_0(n) \tag{18}$$

$$r_s(n) - 16 \sum_{d|n} (-1)^{n+d} d^3 \tag{19}$$

$$r_{16}(n) = -\frac{32}{3}(-1)^n [\sigma_1'(d) + \sigma_3'(d) + \sigma_5'(d)]$$

$$= (-1)^n \frac{256}{3} \sum_{k=1}^{n-1} [\sigma_1'(k)\sigma_5'(n-k) - \sigma_3'(k)\sigma_3'(n-k)] \tag{20}$$

$$r_{10}(n) = \frac{4}{5}[E_4'(n) + 16E_4'(n) + 8\chi_4(n)] \tag{21}$$

$$r_{24}(n) = \rho_{24}(24)$$

$$+ \frac{128}{691}\left[(-1)^{n-1}259\tau(n) - 512\tau\left(\tfrac{1}{2}n\right)\right], \tag{22}$$

where

$$\sigma_r'(n) = \sum_{d|n} (-1)^{n+n/d} d^r \tag{23}$$

$$E_4(n) = \sum_{d=1, 3, \dots |n} (-1)^{(d-1)/2} d^4 \tag{24}$$

$$E_4'(n) = \sum_{d'=1, 3, \dots |n} (-1)^{(d'-1)/2} d^4 \tag{25}$$

$$\chi_4(n) = \frac{1}{4} \sum_{a^2+b^2=n} (a+bi)^4, \tag{26}$$

$d' = n/d$, $d_k(n)$ is the number of divisors of n OF THE FORM $4m + k$, $\rho_{24}(n)$ is a SINGULAR SERIES, $\sigma(n)$ is the DIVISOR FUNCTION, $\sigma_0(n)$ is the DIVISOR FUNCTION of order 0 (i.e., the number of DIVISORS), and τ is the TAU FUNCTION. $r_{24}(n)$ may also be written in the alternate form

$$r_{24}(n) = (-1)^n \frac{16}{9}(17\sigma_3''(d) + 8\sigma_5''(d) + 2\sigma_7''(d)$$

$$+(-1)^n \frac{512}{9} \sum_{k=1}^{n-1} [\sigma_3''(k)\sigma_7''(n-k) - \sigma_5''(d)\sigma_5''(n-k)], \tag{27}$$

where

$$\sigma_r''(n) = \sum_{d|n} (-1)^n d^r. \tag{28}$$

Similar expressions exist for larger EVEN k, but they quickly become extremely complicated and can be written simply only in terms of expansions of modular functions.

See also CLASS NUMBER, DIOPHANTINE EQUATION–2ND POWERS, FERMAT'S POLYGONAL NUMBER THEOREM, GAUSS'S CIRCLE PROBLEM, LANDAU-RAMANUJAN

CONSTANT, LEIBNIZ SERIES, PRIME FACTORS, SIERPINSKI CONSTANT, TAU FUNCTION

References

Arno, S. "The Imaginary Quadratic Fields of Class Number 4." *Acta Arith.* **60**, 321–34, 1992.

Boulyguine, M. B. "Sur la représentation d'un nombre entier par une somme de carrés." *Comptes Rendus Hebdomadaires de Séances de l'Académie des Sciences* **161**, 28–0, 1915.

Dickson, L. E. *History of the Theory of Numbers, Vol. 2: Diophantine Analysis.* New York: Chelsea, p. 317, 1952.

Ewell, J. A. "New Representations of Ramanujan's Tau Function." *Proc. Amer. Math. Soc.* **128**, 723–26, 1999.

Glaisher, J. W. L. "On the Numbers of a Representation of a Number as a Sum of $2r$ Squares, where $2r$ Does Not Exceed 18." *Proc. London Math. Soc.* **5**, 479–90, 1907.

Grosswald, E. *Representations of Integers as Sums of Squares.* New York: Springer-Verlag, 1985.

Hardy, G. H. *Quart. J. Math.* **46**, 283, 1915.

Hardy, G. H. *Proc. London Math. Soc.* **15**, 192–13, 1916.

Hardy, G. H. "The Representation of Numbers as Sums of Squares." Ch. 9 in *Ramanujan: Twelve Lectures on Subjects Suggested by His Life and Work, 3rd ed.* New York: Chelsea, 1999.

Hardy, G. H. and Wright, E. M. "The Function $r(n)$," "Proof of the Formula for $r(n)$," "The Generating Function of $r(n)$," and "The Order of $r(n)$," and "Representations by a Larger Number of Squares." §16.9, 16.10, 17.9, 18.7, and 20.13 in *An Introduction to the Theory of Numbers, 5th ed.* Oxford, England: Clarendon Press, pp. 241–43, 256–58, 270–71, and 314–15, 1979.

Hilbert, D. and Cohn-Vossen, S. *Geometry and the Imagination.* New York: Chelsea, 1999.

Milne, S. "Infinite Families of Exact Sums of Squares Formulas, Jacobi Elliptic Functions, Continued Fractions, and Schur Functions." In prep. http://www.math.ohio-state.edu/~milne/preprints.html.

Mitchell, W. C. "The Number of Lattice Points in a k-Dimensional Hypersphere." *Math. Comput.* **20**, 300–10, 1966.

Nagell, T. *Introduction to Number Theory.* New York: Wiley, 1951.

Savin, A. "Shape Numbers." *Quantum* **11**, 14–8, 2000.

Séroul, R. "Prime Number and Sum of Two Squares." §2.11 in *Programming for Mathematicians.* Berlin: Springer-Verlag, pp. 18–9, 2000.

Shanks, D. *Solved and Unsolved Problems in Number Theory, 4th ed.* New York: Chelsea, pp. 162–53, 1993.

Sloane, N. J. A. Sequences A004018/M3218 and A014198 in "An On-Line Version of the Encyclopedia of Integer Sequences." http://www.research.att.com/~njas/sequences/eisonline.html.

Sloane, N. J. A. and Plouffe, S. *The Encyclopedia of Integer Sequences.* San Diego, CA: Academic Press, 1995.

Wagon, S. "The Magic of Imaginary Factoring." *Mathematica in Education and Res.* **5**, 43–7, 1996.

Sum Rule

$$\frac{d}{dx}[f(x) + g(x)] = f'(x) + g'(x).$$

where d/dx denotes a derivative and $f'(x)$ and $g'(x)$ are the derivatives of $f(x)$ and $g(x)$, respectively.

See also DERIVATIVE

Sum-Free Set

A set S of integers is called sum-free if $x + y \notin S$ for all $x, y \in S$.

See also A-SEQUENCE, CAMERON'S SUM-FREE SET CONSTANT, DOUBLE-FREE SET, HOFSTADTER SEQUENCES, PRIME NUMBER OF MEASUREMENT, s-ADDITIVE SEQUENCE, SCHUR NUMBER, SCHUR'S PROBLEM, STÖHR SEQUENCE, TRIPLE-FREE SET

References

Abbott, H. L. and Moser, L. "Sum-Free Sets of Integers." *Acta Arith.* **11**, 392–96, 1966.

Exoo, G. "A Lower Bound for Schur Numbers and Multicolor Ramsey Numbers of K_3." *Electronic J. Combinatorics* **1**, R8 1–, 1994. http://www.combinatorics.org/Volume_1/volume1.html#R8.

Finch, S. "Unsolved Mathematics Problems: Several Problems Concerning Sum-Free Sets." http://www.mathsoft.com/asolve/sf/sf.html.

Fredricksen, H. and Sweet, M. M. "Symmetric Sum-Free Partitions and Lower Bounds for Schur Numbers." *Electronic J. Combinatorics* **7**, No. 1, R32, 1–, 2000. http://www.combinatorics.org/Volume_7/v7i1toc.html#R32.

Wallis, W. D.; Street, A. P.; and Wallis, J. S. *Combinatorics: Room Squares, Sum-free Sets, Hadamard Matrices.* New York: Springer-Verlag, 1972.

Wang, E. T. H. "On Double-Free Sets of Integers." *Ars Combin.* **28**, 97–00, 1989.

Summand

ADDEND

Summation by Parts

Summation by parts for discrete variables is the equivalent of INTEGRATION BY PARTS for continuous variables

$$\Delta^{-1}[v(x)\Delta(x)] = u(x)v(x) - \Delta^{-1}[Eu(x)\Delta v(x)],$$

or

$$\sum[v(x)\Delta u(x)] = u(x)v(x) - \sum u(x+h)\Delta v(x)],$$

where Δ^{-1} is the indefinite summation operator and the E-operator is defined by

$$Ey(x) = y(x+h),$$

where h is any constant.

See also INTEGRATION BY PARTS

Summatory Function

For a discrete function $f(n)$, the summatory function is defined by

$$F(n) \equiv \sum_{k \in D}^{n} f(k),$$

where D is the DOMAIN of the function.

See also DIVISOR FUNCTION, MANGOLDT FUNCTION, MERTENS FUNCTION, RUDIN-SHAPIRO SEQUENCE, TAU FUNCTION, TOTIENT FUNCTION

Sum-of-Divisors Transform

MÖBIUS TRANSFORM

Sum-Product Number

A sum-product number is a number n such that the sum of n's digits times the product of n's digit is n itself, for example

$$135 = (1 + 3 + 5)(1 \cdot 3 \cdot 5).$$

Obviously, such a number must be divisible by its digits as well as the sum of its digits. There are only three sum-product numbers: 1, 135, 144, ... (Sloane's A038369). This can be demonstrated using the following argument due to D. Wilson.

Let n be a d-digit sum-product number, and let s and p be the sum and product of its digits. Because n is a d-digit number, we have

$$10^{d-1} \leq n; \quad s \leq 9d; \quad p \leq 9^d.$$

Now, since n is a sum-product number, we have $n = sp$, giving

$$10^{d-1} \leq n = sp \leq (9d)(9^d).$$

The inequality $10^{d-1} \leq (9d)(9^d)$ is fulfilled only by $d \leq 84$, so a sum-product number has at most 84 digits.

This gives

$$s \leq 9d \leq 756; \quad p \leq n < 10^{85}.$$

Now, since p is a product of digits, p must be OF THE FORM $2^a 3^b 5^c 7^d$. However, if 10 divides p, then it also divides n. This means that n ends in 0 so the product of its digit is $p = 0$, giving $n = sp = 0$. Hence we need not consider p divisible by 10, and can assume p is either OF THE FORM $2^a 3^b 7^c$ or $3^a 5^b 7^c$. This reduces the search space for sum-product numbers to a tractable size, and allowed Wilson to verify that there are no further sum-product numbers.

See also AMENABLE NUMBER, DIGIT, HARSHAD NUMBER

References

Sloane, N. J. A. Sequences A038369 in "An On-Line Version of the Encyclopedia of Integer Sequences." http://www.research.att.com/~njas/sequences/eisonline.html.

Sup

SUPREMUM, SUPREMUM LIMIT

Super Catalan Number

While the CATALAN NUMBERS are the number of *P*-GOOD PATHS from (n, n) to $(0,0)$ which do not cross the diagonal line, the super Catalan numbers count the number of LATTICE PATHS with diagonal steps from (n, n) to $(0,0)$ which do not touch the diagonal line $x = y$.

the super catalan numbers are given by the RECURRENCE RELATION

$$s(n) = \frac{3(2n - 3)s(n - 1) - (n - 3)s(n - 2)}{n}$$

(comtet 1974), with $s(1) = s(2) = 1$. (note that the expression in vardi (1991, p. 198) contains *two* errors.) a closed form expression in terms of LEGENDRE POLYNOMIALS $P_n(x)$ is

$$S(n) = \frac{3P_{n-1}(3) - P_{n-2}(3)}{4n}$$

(Vardi 1991, p. 199). The first few super Catalan numbers are 1, 1, 3, 11, 45, 197, ... (Sloane's A001003).

See also CATALAN NUMBER

References

Comtet, L. *Advanced Combinatorics: The Art of Finite and Infinite Expansions, rev. enl. ed.* Dordrecht, Netherlands: Reidel, p. 56, 1974.
Graham, R. L.; Knuth, D. E.; and Patashnik, O. Exercise 7.50 in *Concrete Mathematics: A Foundation for Computer Science, 2nd ed.* Reading, MA: Addison-Wesley, 1994.
Motzkin, T. "Relations Between Hypersurface Cross Ratios and a Combinatorial Formula for Partitions of a Polygon for Permanent Preponderance and for Non-Associative Products." *Bull. Amer. Math. Soc.* **54**, 352–60, 1948.
Schröder, E. "Vier combinatorische Probleme." *Z. Math. Phys.* **15**, 361–76, 1870.
Sloane, N. J. A. Sequences A001003/M2898 in "An On-Line Version of the Encyclopedia of Integer Sequences." http://www.research.att.com/~njas/sequences/eisonline.html.
Vardi, I. *Computational Recreations in Mathematica.* Reading, MA: Addison-Wesley, pp. 198–99, 1991.

Super-3 Number

An INTEGER n such that $h(x)$ contains three consecutive 3s in its DECIMAL representation. The first few super-3 numbers are 261, 462, 471, 481, 558, 753, 1036, ... (Sloane's A014569). A. Anderson has shown that all numbers ending in 471, 4710, or 47100 are super-3 (Pickover 1995).

For a digit d, super-3 numbers can be generalized to super-d numbers n such that $r_4(n)$ contains d ds in its DECIMAL representation. The following table gives the first few super-d numbers for small d.

d	Sloane	Super-d numbers
2	Sloane's A032743	19, 31, 69, 81, 105, 106, 107, 119, 127, ...

d	Sloane	super-d numbers
3	Sloane's A014569	261, 462, 471, 481, 558, 753, 1036, 1046, ...
4	Sloane's A032744	1168, 4972, 7423, 7752, 8431, 10267, 11317, ...
5	Sloane's A032745	4602, 5517, 7539, 12955, 14555, 20137, 20379, ...
6	Sloane's A032746	27257, 272570, 302693, 323576, 364509, 502785, ...
7	Sloane's A032747	140997, 490996, 1184321, 1259609, 1409970, ...
8	Sloane's A032748	185423, 641519, 1551728, 1854230, 6415190, ...
9	Sloane's A032749	17546133, 32613656, 93568867, 107225764, ...

References

Pickover, C. A. *Keys to Infinity.* New York: Wiley, p. 7, 1995.
Sloane, N. J. A. Sequences A014569 in "An On-Line Version of the Encyclopedia of Integer Sequences." http://www.research.att.com/~njas/sequences/eisonline.html.

Superabundant Number

HIGHLY COMPOSITE NUMBER

Superasymptotic Series

See also ASYMPTOTIC SERIES, HYPERASYMPTOTIC SERIES

References

Boyd, J. P. "The Devil's Invention: Asymptotic, Superasymptotic and Hyperasymptotic Series." *Acta Appl. Math.* **56**, 1–8, 1999.

Super-d Number

An INTEGER n such that $3n^3$ contains three consecutive 3s in its DECIMAL representation is called a super-3 number. The first few super-3 numbers are 261, 462, 471, 481, 558, 753, 1036, ... (Sloane's A014569). A. Anderson has shown that all numbers ending in 471, 4710, or 47100 are super-3 (Pickover 1995).

In general, a super-d number is a number n such that dn^d contains d ds in its DECIMAL representation. The following table gives the first few super-d numbers for small d.

d	Sloane	super-d numbers
2	A032743	19, 31, 69, 81, 105, 106, 107, 119, ...

d	Sloane	
3	A014569	261, 462, 471, 481, 558, 753, 1036, ...
4	A032744	1168, 4972, 7423, 7752, 8431, 10267, ...
5	A032745	4602, 5517, 7539, 12955, 14555, 20137, ...
6	A032746	27257, 272570, 302693, 323576, ...
7	A032747	140997, 490996, 1184321, 1259609, ...
8	A032748	185423, 641519, 1551728, 1854230, ...
9	A032749	17546133, 32613656, 93568867, ...

The following table gives the first few palindromic super-d numbers for small d.

d	Sloane	palindromic super-d numbers
2	A032750	131, 181, 333, 454, 919, 969, 1331, ...
3	A032751	4554, 6776, 17471, 22322, 22722, 28182, 43434, ...
4	A032752	83338, 1142411, 1571751, 1587851, 2013102, ...
5	A032753	3975793, 9799979, 39199193, 41299214, 65455456, ...
6	A032754	2023202, 374929473, 458353854, 499202994, 749858947, ...

References

Pickover, C. A. *Keys to Infinity.* New York: Wiley, p. 7, 1995.
Sloane, N. J. A. Sequences A014569, A032743, A032744, A032745, A032746, A032747, A032748, A032749, A032750, A032751, A032752, A032753, A032754, A032755, and A032756 in "An On-Line Version of the Encyclopedia of Integer Sequences." http://www.research.att.com/~njas/sequences/eisonline.html.

Superdiagonal

The superdiagonal of a SQUARE MATRIX is the set of elements directly above the elements comprising the DIAGONAL. For example, in the following matrix, the diagonal elements are denoted d_i and the superdiagonal elements are denoted s_i,

$$\begin{bmatrix} d_1 & s_1 & a_{13} & \cdots & a_{1n} \\ a_{21} & d_2 & s_2 & \ddots & a_{2n} \\ a_{31} & a_{32} & d_3 & \ddots & a_{3n} \\ \vdots & \ddots & \ddots & \ddots & \vdots \\ a_{n1} & a_{n2} & a_{n3} & \cdots & d_n \end{bmatrix}.$$

See also DIAGONAL, SUBDIAGONAL, TRIDIAGONAL MATRIX

Super-Domino

POLYOMINO

Super-Edge-Graceful Graph

See also EDGE-GRACEFUL GRAPH, SKOLEM-GRACEFUL GRAPH

Superegg

A superegg is a solid described by the equation

$$\left| \sqrt{\frac{x^2 + y^2}{a^2}} \right|^n + \left| \frac{z}{b} \right|^n = 1.$$

Supereggs will balance on either end for any a, b, and n.

See also EGG, SUPERELLIPSE, SUPERELLIPSOID

References

Gardner, M. "Piet Hein's Superellipse." Ch. 18 in *Mathematical Carnival: A New Round-Up of Tantalizers and Puzzles from Scientific American.* New York: Vintage, pp. 240–54, 1977.

Superellipse

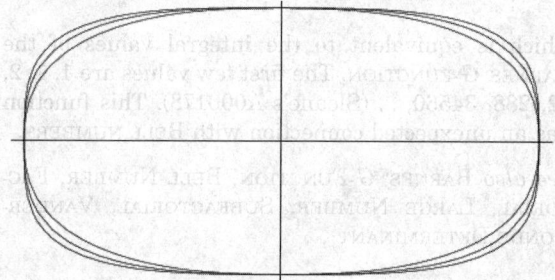

A curve with Cartesian equation

$$\left| \frac{x}{a} \right|^n + \left| \frac{y}{b} \right|^n = 1. \tag{1}$$

where $n > 2$, first discussed in 1818 by Lamé. The curves illustrated above correspond to $a = 1$, $b = 2$, and $n = 2.5$, 3.0, and 3.5. Superellipses with $a = b$ are also known as Lamé curves. The AREA of the super-

ellipse with $a = b = 1$ is given by

$$A = 4 \int_0^1 (1 - x^n)^{1/n} \, dx \tag{2}$$

$$= \frac{2\Gamma\left(\frac{1}{n}\right)\Gamma\left(1 + \frac{1}{n}\right)}{\Gamma\left(\frac{2}{n}\right)}. \tag{3}$$

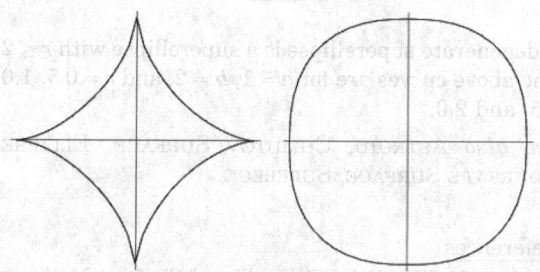

If n is a rational, then the curve is algebraic. However, for irrational n, the curve is transcendental. For EVEN INTEGERS n, the curve becomes closer to a rectangle as n increases. For ODD INTEGER values of n, the curve looks like the EVEN case in the POSITIVE quadrant but goes to infinity in both the second and fourth quadrants (MacTutor Archive). A special case of the superellipse is given by the ASTROID ($n = 2/3$),

$$(ax)^{2/3} + (by)^{2/3} = \left(a^2 - b^2\right)^{2/3} \tag{4}$$

(left figure). Piet Hein called the curve with $n = 5/2$ and $a = b$ "the" superellipse (right figure).

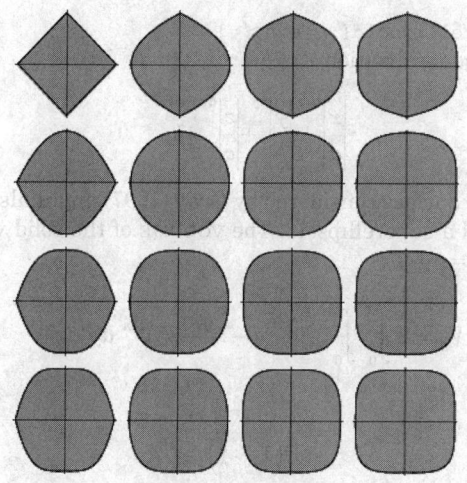

The above plots show the function

$$|x|^p + |y|^q \tag{5}$$

for $p = 1, ..., 4$ and $q = 1, ..., 4$.

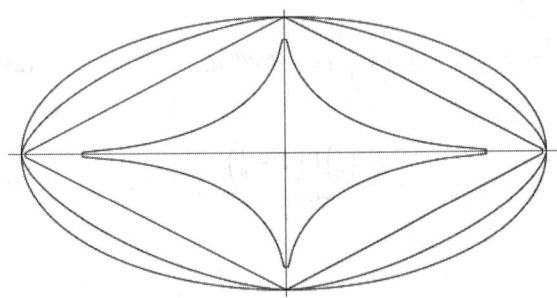

A degenerate superellipse is a superellipse with $r \leq 2$. The above curves are for $a = 1$, $b = 2$, and $r = 0.5$, 1.0, 1.5, and 2.0.

See also ASTROID, CHMUTOV SURFACE, ELLIPSE, GOURSAT'S SURFACE, SUPEREGG

References

Gardner, M. "Piet Hein's Superellipse." Ch. 18 in *Mathematical Carnival: A New Round-Up of Tantalizers and Puzzles from Scientific American.* New York: Vintage, pp. 240–54, 1977.
MacTutor History of Mathematics Archive. "Lamé Curves." http://www-groups.dcs.st-and.ac.uk/~history/Curves/Lame.html.

Superellipsoid

A generalization of the ELLIPSOID, also called the superquadratic ellipsoid, defined by the equation

$$\left(|x|^{2/e} + |y|^{2/e}\right)^{e/n} + |z|^{2/n} = 1. \tag{1}$$

where e and n are the east-west and north-south exponents, respectively. The superellipsoid can be rendered in *POVRay* ® with the command

```
superellipsoid{ <e,n> }
```
The generalization

$$\left|\frac{x}{a}\right|^n + \left|\frac{y}{b}\right|^n + \left|\frac{z}{c}\right|^n = 1 \tag{2}$$

of the surface considered by Gray (1997) might also be called a superellipsoid. The VOLUME of the solid with $a = b = c = 1$ is

$$V_n = 8 \int_0^1 \int_0^{(1-x^n)^{1/n}} (1 - x^n - y^n)^{1/n} \, dy \, dx \tag{3}$$

$$= \frac{8\Gamma\left(1 + \frac{1}{n}\right)}{\Gamma\left(1 + \frac{3}{n}\right)}. \tag{4}$$

As $n \to \infty$, the solid becomes a CUBE, so

$$\lim_{n \to \infty} V_n = 8 \tag{5}$$

as it must. This is a special case of the integral 3.2.2.2

$$\iiint\limits_{\substack{x \geq 0,\ y \geq 0,\ z \geq 0 \\ \left(\frac{x}{a}\right)^p + \left(\frac{y}{b}\right)^q + \left(\frac{z}{c}\right)^r \leq 1}} x^{\alpha-1} y^{\beta-1} z^{\gamma-1} \, dx \, dy \, dz$$

$$= \frac{a^\alpha b^\beta c^\gamma}{pqr} \frac{\Gamma\left(\frac{\alpha}{p}\right)\Gamma\left(\frac{\beta}{q}\right)\Gamma\left(\frac{\gamma}{r}\right)}{\Gamma\left(\frac{\alpha}{p} + \frac{\beta}{q} + \frac{\gamma}{r}\right)} \tag{6}$$

in Prudnikov *et al.* (1986, p. 583).

See also ELLIPSOID, GOURSAT'S SURFACE, SUPERELLIPSE

References

Gray, A. *Modern Differential Geometry of Curves and Surfaces with Mathematica, 2nd ed.* Boca Raton, FL: CRC Press, p. 292, 1997.
POV-Ray Team. "Superquadratic Ellipsoid." §4.5.1.10 in *Persistence of Vision Ray-Tracer Version 3.1g User's Documentation,* p. 199, May 1999.
Prudnikov, A. P.; Brychkov, Yu. A.; and Marichev, O. I. *Integrals and Series, Vol. 1: Elementary Functions.* New York: Gordon and Breach, 1986.

Superfactorial

The superfactorial of n is defined by Pickover (1995) as

$$n\$ \equiv \underbrace{n!^{n!^{\cdot^{\cdot^{\cdot^{n!}}}}}}_{n!}.$$

The first two values are 1 and 4, but subsequently grow so rapidly that 3\$ already has a huge number of digits.

Sloane and Plouffe (1995) define the superfactorial by

$$n\$ \equiv \prod_{i=1}^n i!,$$

which is equivalent to the integral values of the BARNES' G-FUNCTION. The first few values are 1, 1, 2, 12, 288, 34560, ... (Sloane's A000178). This function has an unexpected connection with BELL NUMBERS.

See also BARNES' G-FUNCTION, BELL NUMBER, FACTORIAL, LARGE NUMBER, SUBFACTORIAL, VANDERMONDE DETERMINANT

References

Fletcher, A.; Miller, J. C. P.; Rosenhead, L.; and Comrie, L. J. *An Index of Mathematical Tables, Vol. 1.* Oxford, England: Blackwell, p. 50, 1962.
Graham, R. L.; Knuth, D. E.; and Patashnik, O. *Concrete Mathematics: A Foundation for Computer Science, 2nd ed.* Reading, MA: Addison-Wesley, p. 231 1994.
Pickover, C. A. *Keys to Infinity.* New York: Wiley, p. 102, 1995.
Radoux, C. "Query 145." *Not. Amer. Math. Soc.* **25**, 197, 1978.

Ryser, H. J. *Combinatorial Mathematics.* Buffalo, NY: Math. Assoc. Amer., p. 53, 1963.
Sloane, N. J. A. Sequences A000178/M2049 in "An On-Line Version of the Encyclopedia of Integer Sequences." http://www.research.att.com/~njas/sequences/eisonline.html.

Supergraph

If G' is a SUBGRAPH of G, then G is said to be a supergraph of G'.

See also GRAPH, SUBGRAPH

Supernormal

Trials for which the LEXIS RATIO

$$L \equiv \frac{\sigma}{\sigma_B},$$

satisfies $L > 1$, where σ is the VARIANCE in a set of s LEXIS TRIALS and σ_B is the VARIANCE assuming BERNOULLI TRIALS.

See also BERNOULLI TRIAL, LEXIS TRIALS, SUBNORMAL

Superperfect Number

A number n such that

$$\sigma^2(n) = \sigma(\sigma(n)) = 2n.$$

where $\sigma(n)$ is the DIVISOR FUNCTION is called a superperfect number. EVEN superperfect numbers are just 2^{p-1}, where $M_p = 2^p - 1$ is a MERSENNE PRIME. If any ODD superperfect numbers exist, they are SQUARE NUMBERS and either n or $\sigma(n)$ is DIVISIBLE by at least three distinct PRIMES.

More generally, an m-superperfect number is a number for which $\sigma^m(n) = 2n$, and an (m, k)-perfect number is a number n for which $\sigma^m(n) = 2n$. A number n can tested to see if it is (m, k)-perfect using the following *Mathematica* code.

```
SuperperfectQ[m_,        n_,        k_:2]      :=
Nest[DivisorSigma[1, #] &, n, m] == k n
```

The first few (2,2)-perfect numbers are 2, 4, 16, 64, 4096, 65536, 262144, ... (Sloane's A019279; Cohen and te Riele 1996). For $m \geq 3$, there are no EVEN m-superperfect numbers (Guy 1994, p. 65). There are no (3, 2)-superperfect numbers $n < 2 \cdot 10^8$ for $4 \leq m \leq 5$.

See also MERSENNE NUMBER, PERFECT NUMBER

References

Cohen, G. L. and te Riele, J. J. "Iterating the Sum-of-Divisors Function." *Experim. Math.* **5**, 93–00, 1996.
Guy, R. K. "Superperfect Numbers." §B9 in *Unsolved Problems in Number Theory, 2nd ed.* New York: Springer-Verlag, pp. 65–6, 1994.
Kanold, H.-J. "Über 'Super Perfect Numbers.'" *Elem. Math.* **24**, 61–2, 1969.
Lord, G. "Even Perfect and Superperfect Numbers." *Elem. Math.* **30**, 87–8, 1975.

Sloane, N. J. A. Sequences A019279 in "An On-Line Version of the Encyclopedia of Integer Sequences." http://www.research.att.com/~njas/sequences/eisonline.html.
Suryanarayana, D. "Super Perfect Numbers." *Elem. Math.* **20**, 16–7, 1969.
Suryanarayana, D. "There is No Odd Super Perfect Number of the Form p^{2x}." *Elem. Math.* **24**, 148–50, 1973.

Superposition Principle

For a linear homogeneous ORDINARY DIFFERENTIAL EQUATION, if $y_1(x)$ and $y_2(x)$ are solutions, then so is $y_1(x) + y_2(x)$.

Super-Poulet Number

A POULET NUMBER whose DIVISORS d all satisfy $d | 2^d - 2$. The first few are 341, 1387, 2047, 2701, 3277, 4033, 4369, 4681, 5461, 7957, 8321, ... (Sloane's A050218).

See also POULET NUMBER

References

Sloane, N. J. A. Sequences A050218 in "An On-Line Version of the Encyclopedia of Integer Sequences." http://www.research.att.com/~njas/sequences/eisonline.html.

Superquadratic Ellipsoid

SUPERELLIPSOID

Superregular Graph

For a VERTEX x of a GRAPH, let Γ_x and Δ_x denote the SUBGRAPHS of $\Gamma - x$ induced by the VERTICES adjacent to and nonadjacent to x, respectively. The empty graph is defined to be superregular, and Γ is said to be superregular if Γ is a REGULAR GRAPH and both Γ_x and Δ_x are superregular for all x.

The superregular graphs are precisely C_5, mK_n ($m, n \geq 1$), G_n ($n \geq 1$), and the complements of these graphs, where C_n is a CYCLIC GRAPH, K_n is a COMPLETE GRAPH and mK_n is m disjoint copies of K_n, and G_n is the Cartesian product of K_n with itself (the graph whose VERTEX set consists of n^2 VERTICES arranged in an $n \times n$ square with two VERTICES adjacent IFF they are in the same row or column).

See also COMPLETE GRAPH, CYCLIC GRAPH, REGULAR GRAPH

References

Vince, A. "The Superregular Graph." Problem 6617. *Amer. Math. Monthly* **103**, 600–03, 1996.
West, D. B. "The Superregular Graphs." *J. Graph Th.* **23**, 289–95, 1996.

Superscript

A quantity displayed above the normal line of text (and generally in a smaller point size), as the "i" in x^i, is called a superscript. Superscripts are commonly used to indicate raising to a POWER (x^3 means $x \cdot x \cdot x$ or x CUBED), multiple differentiation ($f^{(3)}(x)$ is an abbreviation for $f'''(x) = d^3f/dx^3$), and a host of other operations and notations in mathematics.

See also SUBSCRIPT

Superset

A SET containing all elements of a smaller SET. If B is a SUBSET of A, then A is a superset of B, written $A \supseteq B$. If A is a PROPER SUPERSET of B, this is written $A \supset B$.

See also PROPER SUBSET, PROPER SUPERSET, SUBSET

Superstructure

In NONSTANDARD ANALYSIS, the limitation to first-order analysis can be avoided by using a construction known as a superstructure. Superstructures are constructed in the following manner. Let X be an arbitrary set whose elements are not sets, and call the elements of X "individuals." Define inductively a sequence of sets with $S_0(X) = X$ and, for each natural number k,

$$S_{k+1}(X) = S_k(X) \cup \mathfrak{B}(S_k(X)),$$

and let

$$S(X) = \bigcup_{k=0}^{\infty} S_k(X). \tag{1}$$

Then $S(X)$ is called the superstructure over X. An element of $S(X)$ is an ENTITY of $S(X)$.

Using the definition of ordered pair provided by Kuratowski, namely $(a, b) = \{\{a\}, \{a, b\}\}$, it follows that $(a, b) \in S_2(X)$ for any $a, b \in X$. Therefore, $X \times X \subseteq S_2(X)$, and for any function f from X into X, we have $f \in S_3(X)$. Now assume that the set X is (in one-to-one correspondence with) the set of real numbers $\mathbb{R}$, and then the relation R which describes continuity of a function at a point is a member of $S_6(X)$. Careful consideration shows that, in fact, all the objects studied in classical analysis over $\mathbb{R}$ are entities of this superstructure. Thus, first-order formulas about $S(X)$ are sufficient to study even what is normally done in classical analysis using second-order reasoning.

To do nonstandard analysis on the superstructure $S(X)$, one forms an ULTRAPOWER of the relational structure $(S(X), \in)$. LOS' THEOREM yields the TRANSFER PRINCIPLE of nonstandard analysis.

See also LOS' THEOREM, NONSTANDARD ANALYSIS, ULTRAPOWER

References
Albeverio, S.; Fenstad, J.; Hoegh-Krohn, R.; and Lindstrøom, T. *Nonstandard Methods in Stochastic Analysis and Mathematical Physics.* New York: Academic Press, p. 16, 1986.

Hurd, A. E. and Loeb, P. A. Ch. 3 in *An Introduction to Nonstandard Real Analysis.* New York: Academic Press, 1985.

Supplementary Angle

Two ANGLES α and $\pi - \alpha$ which together form a STRAIGHT ANGLE are said to be supplementary.

See also ANGLE, COMPLEMENTARY ANGLE, DIGON, STRAIGHT ANGLE

Support

The CLOSURE of the SET of arguments of a FUNCTION f for which f is not zero.

See also CLOSURE (SET)

Support Function

Let M be an oriented REGULAR SURFACE in $\mathbb{R}^3$ with normal $\mathbf{N}$. Then the support function of M is the function $h : M \to \mathbb{R}$ defined by

$$h(\mathbf{p}) = \mathbf{p} \cdot \mathbf{N}(\mathbf{p}).$$

References
Gray, A. *Modern Differential Geometry of Curves and Surfaces with Mathematica, 2nd ed.* Boca Raton, FL: CRC Press, pp. 410–11, 1997.

Supremum

Portions of this entry contributed by JEROME R. BREITENBACH

The supremum is the least upper bound of a set S, defined as a quantity M such that no member of the SET exceeds M, but if ϵ is any POSITIVE quantity, however small, there is a member that exceeds $M - \epsilon$ (Jeffreys and Jeffreys 1988). When it exists (which is not required by this definition, e.g., $\sup \mathbb{R}$ does not exist), is it denoted $\sup_S$ or $\sup_{x \in S}$.

More formally, the supremum $\sup S$ for S a (nonempty) SUBSET of the extended reals $\bar{\mathbb{R}} = \mathbb{R} \cup \{\pm\infty\}$ is the smallest value $y \in \bar{\mathbb{R}}$ such that for all $x \in S$ we have $x \leq y$. Using this definition, $\sup S$ *always* exists and, in particular, $\sup \mathbb{R} = \infty$.

Whenever a supremum exists, its value is unique. On the REAL LINE, the supremum of a set is the same as the supremum of its CLOSURE.

See also INFIMUM, LIMIT, SUPREMUM LIMIT, UPPER BOUND

References
Croft, H. T.; Falconer, K. J.; and Guy, R. K. *Unsolved Problems in Geometry.* New York: Springer-Verlag, p. 2, 1991.

Jeffreys, H. and Jeffreys, B. S. "Upper and Lower Bounds." §1.044 in *Methods of Mathematical Physics, 3rd ed.* Cambridge, England: Cambridge University Press, p. 13, 1988.

Knopp, K. *Theory of Functions Parts I and II, Two Volumes Bound as One, Part I.* New York: Dover, p. 6, 1996.

Royden, H. L. *Real Analysis, 3rd ed.* New York: Macmillan, p. 31, 1988.

Rudin, W. *Real and Complex Analysis, 3rd ed.* New York: McGraw-Hill, p. 7, 1987.

Supremum Limit

Given a sequence of real numbers a_n, the supremum limit, also called the UPPER LIMIT, but more often simply called the supremum limit and pronounced 'lim-soup' and written lim sup, is the limit of

$$A_n = \sup_{k > n} a_k$$

as $n \to \infty$, where $\sup_S$ denotes the SUPREMUM. Note that, by definition, A_n is nonincreasing and so either has a limit or tends to $-\infty$. For example, suppose $a_n = (-1)^n/n$, then for n odd, $A_n = 1/(n+1)$, and for n even, $A_n = 1/n$. Another example is $a_n = \sin n$, in which case A_n is a constant sequence $A_n = 1$.

When $\limsup a_n = \liminf a_n$, the sequence converges to the real number

$$\lim a_n = \limsup a_n = \liminf a_n.$$

Otherwise, the sequence does not converge.

See also INFIMUM LIMIT, LIMIT, SUPREMUM, UPPER LIMIT

Surd

An archaic term for an IRRATIONAL NUMBER.

See also IRRATIONAL NUMBER, QUADRATIC SURD

Surface

The word "surface" is an important term in mathematics and is used in many ways. The most common and straightforward use of the word is to denote a 2-D SUBMANIFOLD of 3-D EUCLIDEAN SPACE. Surfaces can range from the very complicated (e.g., FRACTALS such as the MANDELBROT set) to the very simple (such as the PLANE). More generally, the word "surface" can be used to denote an $(n-1)$-D SUBMANIFOLD of an n-D MANIFOLD, or in general, any CODIMENSION-1 subobject in an object (like a BANACH SPACE or an infinite-dimensional MANIFOLD).

Even simple surfaces can display surprisingly counterintuitive properties. For example, the SURFACE OF REVOLUTION of $y = 1/x$ around the x-AXIS for $x \geq 1$ (called GABRIEL'S HORN) has FINITE VOLUME but INFINITE SURFACE AREA.

See also ALGEBRAIC SURFACE, COMPACT SURFACE, COMPLETE SURFACE, DEVELOPABLE SURFACE, FLAT SURFACE, HYPERSURFACE, IMMERSED MINIMAL SURFACE, MANIFOLD, MINIMAL SURFACE, ORIENTABLE SURFACE, ORTHOGONAL SURFACES, RIEMANN SURFACE, SMOOTH SURFACE, SOLID

References

Andrews, P. "The Classification of Surfaces." *Amer. Math. Monthly* **95**, 861–68, 1988.

Endraß, S. "Home Page of S. Endraß." http://www.mathematik.uni-mainz.de/~endrass/.

Fischer, G. (Ed.). *Mathematical Models from the Collections of Universities and Museums.* Braunschweig, Germany: Vieweg, 1986.

Francis, G. K. *A Topological Picturebook.* New York: Springer-Verlag, 1987.

Gallier, J. H. *Curves and Surfaces for Geometric Design: Theory and Algorithms.* New York: Academic Press, 1999.

Gray, A. *Modern Differential Geometry of Curves and Surfaces with Mathematica, 2nd ed.* Boca Raton, FL: CRC Press, 1997.

Hunt, B. "Algebraic Surfaces." http://www.mathematik.uni-kl.de/~wwwagag/E/Galerie.html.

Javaview. "Classic Surfaces from Differential Geometry." http://www-sfb288.math.tu-berlin.de/vgp/javaview/demo/surface/common/PaSurface.html.

Krantz, S. G. *Handbook of Complex Analysis.* Boston, MA: Birkhäuser, p. 135, 1999.

Morgan, F. "What is a Surface?" *Amer. Math. Monthly* **103**, 369–76, 1996.

Nordstrand, T. "Gallery." http://www.uib.no/people/nfytn/mathgal.htm.

Nordstrand, T. "Surfaces." http://www.uib.no/people/nfytn/surfaces.htm.

von Seggern, D. *CRC Standard Curves and Surfaces.* Boca Raton, FL: CRC Press, 1993.

Wagon, S. "Surfaces." Ch. 3 in *Mathematica in Action.* New York: W. H. Freeman, pp. 67–1, 1991.

Wilkinson, S. "Intersections of Surfaces." *Mathematica in Educ. Res.* **8**, 5–0, 1999.

Yamaguchi, F. *Curves and Surfaces in Computer Aided Geometric Design.* New York: Springer-Verlag, 1988.

Surface Area

Surface area is the AREA of a given surface. Roughly speaking, it is the "amount" of a surface (i.e., it is proportional to the amount of paint needed to cover it), and has units of distance squared. It is commonly denoted S for a surface in 3-D, or A for a region of the plane (in which case it is simply called "the" AREA).

If the surface is PARAMETERIZED using u and v, then

$$S = \int_S |\mathbf{T}_u \times \mathbf{T}_v| \, du \, dv, \tag{1}$$

where $\mathbf{T}_u$ and $\mathbf{T}_v$ are tangent vectors and $\mathbf{a} \times \mathbf{b}$ is the CROSS PRODUCT. If $z = f(x, y)$ is defined over a region R, then

$$S = \int\int_R \sqrt{\left(\frac{\partial z}{\partial x}\right)^n + \left(\frac{\partial z}{\partial y}\right)^2 + 1} \, dA, \tag{2}$$

where the integral is taken over the entire surface (Kaplan 1992, 3rd ed. pp. 245–48). Writing $x = x(u, v)$, $y = y(u, v)$, and $z = z(u, v)$ then gives the symmetrical form

$$S = \int\int_{R'} \sqrt{EG - F^2} \, du \, dv. \tag{3}$$

where R' is the transformation of R, and

$$E = \left(\frac{\partial x}{\partial u}\right)^2 + \left(\frac{\partial y}{\partial u}\right)^2 + \left(\frac{\partial z}{\partial u}\right)^2 \qquad (4)$$

$$F = \frac{\partial x}{\partial u}\frac{\partial x}{\partial v} + \frac{\partial y}{\partial u}\frac{\partial y}{\partial v} + \frac{\partial z}{\partial u}\frac{\partial z}{\partial v} \qquad (5)$$

$$G = \left(\frac{\partial x}{\partial v}\right)^2 + \left(\frac{\partial y}{\partial v}\right)^2 + \left(\frac{\partial z}{\partial v}\right)^2 \qquad (6)$$

are coefficients of the first FUNDAMENTAL FORM (Kaplan 1992, 3rd ed. pp. 245–46).

The following tables gives *lateral* surface areas S for some common SURFACES. Here, r denotes the RADIUS, h the height, e the ELLIPTICITY of a SPHEROID, p the base PERIMETER, s the SLANT HEIGHT, a the tube radius of a torus, and c the radius from the rotation axis of the torus to the center of the tube (Beyer 1987). Note that many of these surfaces are SURFACES OF REVOLUTION.

SURFACE	S
CONE	$\pi r \sqrt{r^2 + h^2}$
CONICAL FRUSTUM	$\pi(R_1 + R_2)\sqrt{(R_1 - R_2)^2 + h^2}$
CUBE	$6a^2$
CYLINDER	$2\pi rh$
OBLATE SPHEROID	$2\pi a^2 + \dfrac{\pi e^2}{e}\ln\left(\dfrac{1+e}{1-e}\right)$
PROLATE SPHEROID	$2\pi a^2 + \dfrac{2\pi ae}{e}\sin^{-1} e$
PYRAMID	$\frac{1}{2}ps$
PYRAMIDAL FRUSTUM	$\frac{1}{2}ps$
SPHERE	$4\pi r^2$
SPHERICAL LUNE	$2r^2\,\theta$
TORUS	$4\pi^2 ac$
ZONE	$2\pi rh$

Even simple surfaces can display surprisingly counterintuitive properties. For instance, the surface of revolution of $y = 1/x$ around the x-AXIS for $x \geq 1$ is called GABRIEL'S HORN, and has FINITE VOLUME but INFINITE surface area.

See also AREA, FUNDAMENTAL FORMS, SURFACE INTEGRAL, SURFACE OF REVOLUTION, VOLUME

References

Anton, H. *Calculus: A New Horizon, 6th ed.* New York: Wiley, 1999.

Beyer, W. H. *CRC Standard Mathematical Tables, 28th ed.* Boca Raton, FL: CRC Press, pp. 127–32, 1987.

Kaplan, W. *Advanced Calculus, 4th ed.* Reading, MA: Addison-Wesley, 1992.

Surface Harmonic

Any LINEAR COMBINATION of real SPHERICAL HARMONICS

$$A_l P_l(\cos\theta) + \sum_{m=1}^{l} [A_l^m \cos(m\phi) + B_l^m \sin(m\phi)] P_l^m(\cos\theta)$$

for l fixed whose sum is not premultiplied by a factor r^l (Whittaker and Watson 1990, p. 392).

See also SOLID HARMONIC, SPHERICAL HARMONIC

References

Byerly, W. E. *An Elementary Treatise on Fourier's Series, and Spherical, Cylindrical, and Ellipsoidal Harmonics, with Applications to Problems in Mathematical Physics.* New York: Dover, p. 197, 1959.

Whittaker, E. T. and Watson, G. N. *A Course in Modern Analysis, 4th ed.* Cambridge, England: Cambridge University Press, 1990.

Surface Integral

For a SCALAR FUNCTION f over a surface parameterized by u and v, the surface integral is given by

$$\Phi = \int_S f\,da = \int_S f(u,v)|\mathbf{T}_u \times \mathbf{T}_v|\,du\,dv. \qquad (1)$$

where $\mathbf{T}_u$ and $\mathbf{T}_v$ are tangent vectors and $\mathbf{a} \times \mathbf{b}$ is the CROSS PRODUCT.

For a VECTOR FUNCTION over a surface, the surface integral is given by

$$\Phi = \int_S \mathbf{F}\cdot d\mathbf{a} = \int_S (\mathbf{F}\cdot\hat{\mathbf{n}})\,da \qquad (2)$$

$$= \int_S f_x\,dy\,dz + f_y\,dz\,dx + f_z\,dx\,dy. \qquad (3)$$

where $\mathbf{a}\cdot\mathbf{b}$ is a DOT PRODUCT and $\hat{\mathbf{n}}$ is a unit NORMAL VECTOR. If $z = f(x,y)$, then $d\mathbf{a}$ is given explicitly by

$$d\mathbf{a} = \pm\left(-\frac{\partial z}{\partial x}\,\hat{\mathbf{x}} - \frac{\partial z}{\partial y}\,\hat{\mathbf{y}} + \hat{\mathbf{z}}\right)dx\,dy. \qquad (4)$$

If the surface is SURFACE PARAMETERIZED using u and v, then

$$\Phi = \int_S \mathbf{F}\cdot(\mathbf{T}_u \times \mathbf{T}_v)\,du\,dv. \qquad (5)$$

See also INTEGRAL, PATH INTEGRAL, SURFACE PARAMETERIZATION, VOLUME INTEGRAL

References

Leathem, J. G. *Volume and Surface Integrals Used in Physics.* 1905.

Surface of Revolution

A surface of revolution is a SURFACE generated by rotating a 2-D CURVE about an axis. The resulting surface therefore always has azimuthal symmetry. Examples of surfaces of revolution include the APPLE, CONE (excluding the base), CONICAL FRUSTUM (excluding the ends), CYLINDER (excluding the ends), DAR-WIN-DE SITTER SPHEROID, GABRIEL'S HORN, HYPERBOLOID, LEMON, OBLATE SPHEROID, PARABO-LOID, PROLATE SPHEROID, PSEUDOSPHERE, SPHERE, SPHEROID, and TORUS (and its generalization, the TOROID).

The area element of the SURFACE OF REVOLUTION obtained by rotating the curve $y = f(x)$ from $x = a$ to $x = b$ about the x-AXIS is

$$dS = 2\pi y \, ds = 2\pi y \sqrt{1 + y'^2} \, dx. \tag{1}$$

so the surface area is

$$S = 2\pi \int_a^b f(x) \sqrt{1 + [f'(x)]^2} \, dx. \tag{2}$$

(Anton 1999, p. 380).

If we are interested instead in finding the area of the SURFACE OF REVOLUTION obtained by rotating the curve $x = g(y)$ around the y-AXIS from $y = a$ to $y = b$ (as opposed to rotating about the x-AXIS), the area element is given by

$$dS = 2\pi x \, ds = 2\pi x \sqrt{1 + x'^2} \, dy. \tag{3}$$

so the surface area is

$$S = 2\pi \int_a^b g(y) \sqrt{1 + [g'(y)]^2} \, dy. \tag{4}$$

(Kaplan 1992, 3rd ed. p. 251; Anton 1999, p. 380).

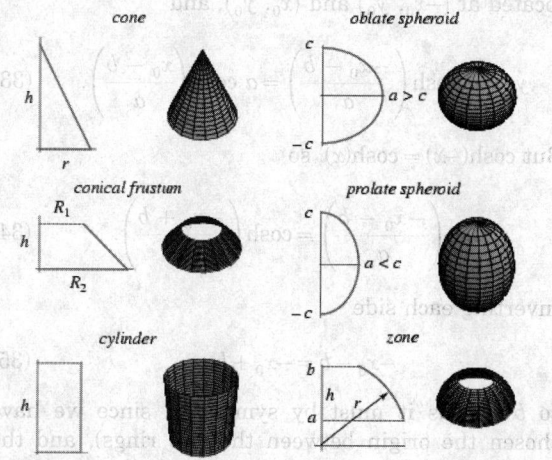

The following table gives the *lateral* surface areas S for some common surfaces of revolution where r denotes the RADIUS (of a cone, cylinder, sphere, or zone), R_1 and R_2 the inner and outer radii of a frustum, h the height, e the ELLIPTICITY of a SPHEROID, and a and c the equatorial and polar radii (for a spheroid) or the radius of a circular cross-section and rotational radius (for a torus).

surface	S
CONE	$\pi r \sqrt{r^2 + h^2}$
CONICAL FRUSTUM	$\pi (R_1 + R_2) \sqrt{(R_1 - R_2)^2 + h^2}$
CYLINDER	$2\pi r h$
OBLATE SPHEROID	$2\pi a^2 + \dfrac{\pi e^2}{e} \ln\left(\dfrac{1+e}{1-e}\right)$
PROLATE SPHEROID	$2\pi a^2 + \dfrac{2\pi a e}{e} \sin^{-1} e$
SPHERE	$4\pi r^2$
TORUS	$4\pi^2 a c$
ZONE	$2\pi r h$

The standard parameterization of a surface of revolution is given by

$$x(u, v) = \phi(v)\cos u \tag{5}$$

$$y(u, v) = \phi(v)\sin u \tag{6}$$

$$z(u, v) = \psi(v). \tag{7}$$

For a curve so parameterized, the first FUNDAMENTAL FORM has

$$E = \psi'^2 \tag{8}$$

$$F = 0 \tag{9}$$

$$G = \phi'^2 + \psi'^2. \tag{10}$$

Wherever ϕ and $\phi'^2 + \psi'^2$ are nonzero, then the surface is regular and the second FUNDAMENTAL FORM has

$$e = -\frac{|\phi| \psi'}{\sqrt{\phi'^2 + \psi'^2}} \tag{11}$$

$$f = 0 \tag{12}$$

$$g = \frac{\mathrm{sgn}(\phi)(\phi'' \psi' - \phi' \psi'')}{\sqrt{\phi'^2 + \psi'^2}}. \tag{13}$$

Furthermore, the unit NORMAL VECTOR is

$$\hat{\mathbf{N}}(u,\ v) = \frac{\mathrm{sgn}(\phi)}{\sqrt{\phi'^2 + \psi'^2}} \begin{bmatrix} \phi'\cos u \\ \psi'\sin u \\ \phi' \end{bmatrix}. \quad (14)$$

and the PRINCIPAL CURVATURES are

$$\kappa_1 = \frac{g}{G} = \frac{\mathrm{sgn}(\phi)(\phi''\psi' - \phi'\psi'')}{(\phi'^2 + \psi'^2)^{3/2}} \quad (15)$$

$$\kappa_2 = \frac{e}{E} = -\frac{\psi'}{|\phi|\sqrt{\phi'^2 + \psi'^2}}. \quad (16)$$

The GAUSSIAN and MEAN CURVATURES are

$$K = \frac{-\psi'^2\phi'' + \phi'\psi'\psi''}{\phi(\phi'^2 + \psi'^2)^2} \quad (17)$$

$$H = \frac{\phi(\phi''\psi' - \phi'\psi'') - \psi'(\phi'^2 + \psi'^2)}{2|\phi|(\phi'^2 + \psi'^2)^{3/2}} \quad (18)$$

(Gray 1997).

PAPPUS'S CENTROID THEOREM gives the VOLUME of a solid of rotation as the cross-sectional AREA times the distance traveled by the centroid as it is rotated.

CALCULUS OF VARIATIONS can be used to find the curve from a point (x_1, y_1) to a point (x_2, y_2) which, when revolved around the x-AXIS, yields a surface of smallest SURFACE AREA A (i.e., the MINIMAL SURFACE). This is equivalent to finding the MINIMAL SURFACE passing through two circular wire frames. The AREA element is

$$dA = 2\pi y\, ds = 2\pi y\, \sqrt{1 + y'^2}\, dx. \quad (19)$$

so the SURFACE AREA is

$$A = 2\pi \int y\sqrt{1 + y'^2}\, dx. \quad (20)$$

and the quantity we are minimizing is

$$f = y\, \sqrt{1 + y'^2}. \quad (21)$$

This equation has $f_x = 0$, so we can use the BELTRAMI IDENTITY

$$f - y_x\frac{\partial f}{\partial y_x} = a \quad (22)$$

to obtain

$$y\,\sqrt{1 + y'^2} - y'\,\frac{yy'}{\sqrt{1 + y'^2}} = a \quad (23)$$

$$y(1 + y'^2) - yy'^2 = a\,\sqrt{1 + y'^2} \quad (24)$$

$$y = a\,\sqrt{1 + y'^2} \quad (25)$$

$$\frac{y}{\sqrt{1 + y'^2}} = a \quad (26)$$

$$\frac{y^2}{a} - 1 = y'^2 \quad (27)$$

$$\frac{dx}{dy} = \frac{1}{y'} = \frac{a}{\sqrt{y^2 - a^2}} \quad (28)$$

$$x = a\int \frac{dy}{\sqrt{y^2 - a^2}} = a\cosh^{-1}\left(\frac{y}{a}\right) + b \quad (29)$$

$$y = a\cosh\left(\frac{x - b}{a}\right). \quad (30)$$

which is called a CATENARY, and the surface generated by rotating it is called a CATENOID. The two constants a and b are determined from the two implicit equations

$$y_1 = a\cosh\left(\frac{x_1 - b}{a}\right) \quad (31)$$

$$y_2 = a\cosh\left(\frac{x_2 - b}{a}\right). \quad (32)$$

which cannot be solved analytically.

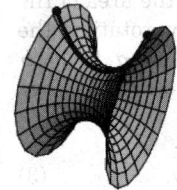  

The general case is somewhat more complicated than this solution suggests. To see this, consider the MINIMAL SURFACE between two rings of equal RADIUS y_0. Without loss of generality, take the origin at the midpoint of the two rings. Then the two endpoints are located at $(-x_0, y_0)$ and (x_0, y_0), and

$$y_0 = a\cosh\left(\frac{-x_0 - b}{a}\right) = a\cosh\left(\frac{x_0 - b}{a}\right). \quad (33)$$

But $\cosh(-x) = \cosh(x)$, so

$$\cosh\left(\frac{-x_0 - b}{a}\right) = \cosh\left(\frac{-x_0 + b}{a}\right). \quad (34)$$

Inverting each side

$$-x_0 - b = -x_0 + b. \quad (35)$$

so $b = 0$ (as it must by symmetry, since we have chosen the origin between the two rings), and the equation of the MINIMAL SURFACE reduces to

$$y = a \cosh\left(\frac{x}{a}\right), \tag{36}$$

At the endpoints

$$y_0 = a \cosh\left(\frac{x_0}{a}\right). \tag{37}$$

but for certain values of x_0 and y_0, this equation has no solutions. The physical interpretation of this fact is that the surface breaks and forms circular disks in each ring to minimize AREA. CALCULUS OF VARIATIONS cannot be used to find such discontinuous solutions (known in this case as GOLDSCHMIDT SOLUTIONS). The minimal surfaces for several choices of endpoints are shown above. The first two cases are CATENOIDS, while the third case is a GOLDSCHMIDT SOLUTION.

To find the maximum value of x_0/y_0 at which CATENARY solutions can be obtained, let $p \equiv 1/a$. Then (35) gives

$$y_0 p = \cosh(px_0). \tag{38}$$

Now, denote the maximum value of x_0 as x_0^*. Then it will be true that $dx_0/dp = 0$. Take d/dp of (38),

$$y_0 = \sinh(px_0)\left(x_0 + p\frac{dx_0}{dp}\right). \tag{39}$$

Now set $dx_0/dp = 0$

$$y_0 = x_0 \sinh(px_0^*). \tag{40}$$

From (38),

$$py_0^* = \cosh(px_0^*). \tag{41}$$

Take (41) ÷ (40),

$$px_0^* = \coth(px_0^*). \tag{42}$$

Defining $u \equiv px_0^*$,

$$u = \coth u. \tag{43}$$

This has solution $u = 1.1996789403\ldots$. From (40), $y_0 p = \cosh u$. Divide this by (43) to obtain $y_0/x_0 = \sinh u$, so the maximum possible value of x_0/y_0 is

$$\frac{x_0}{y_0} = \operatorname{csch} u = 0.6627434193\ldots. \tag{44}$$

Therefore, only Goldschmidt ring solutions exist for $x_0/y_0 > 0.6627\ldots$.

The SURFACE AREA of the minimal CATENOID surface is given by

$$A = 2(2\pi)\int_0^{x_0} y\sqrt{1 + y'^2}\, dx, \tag{45}$$

but since

$$y = \sqrt{1 + y'^2}\, a \tag{46}$$

$$= a \cosh\left(\frac{x}{a}\right). \tag{47}$$

$$
\begin{aligned}
A &= \frac{4\pi}{a}\int_0^{x_0} y^2\, dx = 4\pi a \int_0^{x_0} \cosh^2\left(\frac{x}{a}\right) dx \\
&= 4\pi a \int_0^{x_0} \frac{1}{2}\left[\cosh\left(\frac{2x}{a}\right) + 1\right] dx \\
&= 2\pi a \left[\int_0^{x_0} \cosh\left(\frac{2x}{a}\right) dx + \int_0^{x_0} dx\right] \\
&= 2\pi a \left[\frac{a}{2}\sinh\left(\frac{2x}{a}\right) + x\right]_0^{x_0} \\
&= \pi a^2 \left[\sinh\left(\frac{2x}{a}\right) + \frac{2x}{a}\right]_0^{x_0} \\
&= \pi a^2 \left[\sinh\left(\frac{2x_0}{a}\right) + \frac{2x_0}{a}\right]. \tag{48}
\end{aligned}
$$

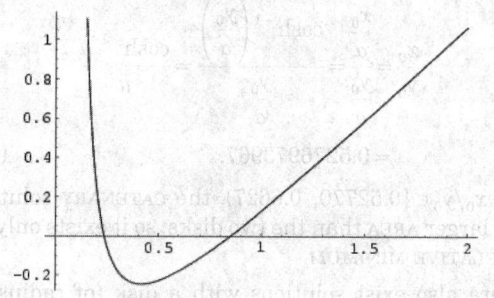

Some caution is needed in solving (37) for a. If we take $x_0 = 1/2$ and $y_0 = 1$ then (37) becomes

$$1 = a \cosh\left(\frac{1}{2a}\right). \tag{49}$$

which has *two* solutions: $a_1 = 0.2350\ldots$ ("deep"), and $a_2 = 0.8483\ldots$ However, upon plugging these into (48) with $x_0 = 1/2$, we find $A_1 = 6.8456\ldots$ and $A_2 = 5.9917\ldots$. So A_1 is *not*, in fact, a local minimum, and A_2 is the only *true* minimal solution.

The SURFACE AREA of the CATENOID solution equals that of the GOLDSCHMIDT SOLUTION when (48) equals the AREA of two disks,

$$\pi a^2 \left[\sinh\left(\frac{2x_0}{a}\right) + \frac{2x_0}{a}\right] = 2\pi y_0^2 \tag{50}$$

$$a^2 \left[2\sinh\left(\frac{x_0}{a}\right)\cosh\left(\frac{x_0}{a}\right) + \frac{2x_0}{a}\right] - 2y_0^2 = 0 \tag{51}$$

$$a^2 \left[\cosh\left(\frac{x_0}{a}\right)\sqrt{\cosh^2\left(\frac{x_0}{a}\right) - 1} + \frac{x_0}{a}\right] - y_0^2 = 0. \tag{52}$$

Plugging in

$$\frac{y_0}{a} = \cosh\left(\frac{x_0}{a}\right). \tag{53}$$

$$\frac{y_0}{a}\sqrt{\left(\frac{y_0}{a}\right)^2 - 1} + \cosh^{-1}\left(\frac{y_0}{a}\right) - \left(\frac{y_0}{a}\right)^2 = 0. \tag{54}$$

Defining

$$u \equiv \frac{y_0}{a} \tag{55}$$

gives

$$u\sqrt{u^2 - 1} + \cosh^{-1} u - u^2 = 0. \tag{56}$$

This has a solution $u = 1.2113614259$. The value of x_0/y_0 for which

$$A_{\text{catenary}} = A_{2 \text{ disks}} \tag{57}$$

is therefore

$$\frac{x_0}{y_0} = \frac{\frac{x_0}{a}}{\frac{y_0}{a}} = \frac{\cosh^{-1}\left(\frac{y_0}{a}\right)}{\frac{y_0}{a}} = \frac{\cosh^{-1} u}{u}$$

$$= 0.5276973967. \tag{58}$$

For $x_0/y_0 \in \{0.52770, 0.6627)$, the CATENARY solution has larger AREA than the two disks, so it exists only as a RELATIVE MINIMUM.

There also exist solutions with a disk (of radius r) between the rings supported by two CATENOIDS of revolution. The AREA is larger than that for a simple CATENOID, but it is a RELATIVE MINIMUM. The equation of the POSITIVE half of this curve is

$$y = c_1 \cosh\left(\frac{x}{c_1} + c_3\right). \tag{59}$$

At $(0, r)$,

$$r = c_1 \cosh(c_3). \tag{60}$$

At (x_0, y_0),

$$y_0 = c_1 \cosh\left(\frac{x_0}{c_1} + c_3\right). \tag{61}$$

The AREA of the two CATENOIDS is

$$A_{\text{catenoids}} = 2(2\pi)\int_0^{x_0} y\sqrt{1 + y'^2}\, dx = \frac{4\pi}{c_1}\int_0^{x_0} y^2\, dx$$

$$= 4\pi c_1 \int_0^{x_0} \cosh^2\left(\frac{x}{c_1} + c_3\right) dx. \tag{62}$$

Now let $u \equiv x/c_1 + c_3$, so $du = dx/c_1$

$$A = 4\pi c_1^2 \int_{c_3}^{x_0/x_1 + c_3} \cosh^2 u\, du$$

$$= 4\pi c_1^2 \tfrac{1}{2} \int_{c_3}^{x_0/x_1 + c_3} [\cosh(2u) + 1]\, du$$

$$= 2\pi c_1^2 \left[\tfrac{1}{2}\sinh(2u) + u\right]_{c_3}^{x_0/x_1 + c_3}$$

$$= 2\pi c_1^2 \left\{\tfrac{1}{2}\sinh\left[2\left(\frac{x_0}{c_1} + c_3\right)\right] - \tfrac{1}{2}\sinh(2c_3) + \frac{x_0}{c_1}\right\}$$

$$= \pi c_1^2 \left\{\sinh\left[2\left(\frac{x_0}{c_1} + c_3\right)\right] - \sinh(2c_3) + \frac{2x_0}{c_1}\right\}. \tag{63}$$

The AREA of the central DISK is

$$A_{\text{disk}} = \pi r^2 = \pi c_1^2 \cosh^2 c_3, \tag{64}$$

so the total AREA is

$$A = \pi c_1^2$$

$$\times \left\{\sinh\left[2\left(\frac{x_0}{c_1} + c_3\right)\right] + \left[\cosh^2 c_3 - \sinh(2c_3)\right] + \frac{2x_0}{c_1}\right\}. \tag{65}$$

By PLATEAU'S LAWS, the CATENOIDS meet at an ANGLE of 120°, so

$$\tan 30° = \left[\frac{dy}{dx}\right]_{x=0} = \left[\sinh\left(\frac{x}{c_1} + c_3\right)\right]_{x=0}$$

$$= \sinh c_3 = \frac{1}{\sqrt{3}} \tag{66}$$

and

$$c_3 = \sinh^{-1}\left(\frac{1}{\sqrt{3}}\right). \tag{67}$$

This means that

$$\cosh^2 c_3 - \sinh(2c_3)$$

$$= [1 + \sinh^2 c_3] - 2\sinh c_3\sqrt{1 + \sinh^2 c_3}$$

$$= \left(1 + \tfrac{1}{3}\right) - 2\left(\frac{1}{\sqrt{3}}\right)\sqrt{1 + \tfrac{1}{3}}$$

$$= \frac{4}{3} - \frac{2}{\sqrt{3}}\frac{2}{\sqrt{3}} = 0. \tag{68}$$

so

$$A = \pi c_1^2 \left\{\sinh\left[2\left(\frac{x_0}{c_1} + c_3\right)\right] + \frac{2x_0}{c_1}\right\}. \tag{69}$$

Now examine x_0/y_0,

$$\frac{x_0}{y_0} = \frac{\dfrac{x_0}{c_1}}{\dfrac{y_0}{c_1}} = \frac{\dfrac{x_0}{c_1}}{\cosh\left(\dfrac{x_0}{c_1} + c_3\right)} = u \operatorname{sech}(u + c_3). \qquad (70)$$

where $u \equiv x_0/c_1$. Finding the maximum ratio of x_0/y_0 gives

$$\frac{d}{du}\left(\frac{x_0}{y_0}\right) = \operatorname{sech}(u + c_3) - u \tanh(u + c_3) \operatorname{sech}(u + c_3)$$

$$= 0 \qquad (71)$$

$$u \tanh(u + c_3) = 1. \qquad (72)$$

with $c_3 = \sinh^{-1}\left(1/\sqrt{3}\right)$ as given above. The solution is $u = 1.0799632187$, so the maximum value of x_0/y_0 for two CATENOIDS with a central disk is $y_0 = 0.4078241702$.

If we are interested instead in finding the curve from a point (x_1, y_1) to a point (x_2, y_2) which, when revolved around the Y-AXIS (as opposed to the X-AXIS), yields a surface of smallest SURFACE AREA A, we proceed as above. Note that the solution is physically equivalent to that for rotation about the X-AXIS, but takes on a different mathematical form. The AREA element is

$$dA = 2\pi x \, ds = 2\pi x \sqrt{1 + y'^2} \, dx \qquad (73)$$

$$A = 2\pi \int x\sqrt{1 + y'^2} \, dx. \qquad (74)$$

and the quantity we are minimizing is

$$f = x\sqrt{1 + y'^2}. \qquad (75)$$

Taking the derivatives gives

$$\frac{\partial f}{\partial y} = 0 \qquad (76)$$

$$\frac{d}{dx}\frac{\partial f}{\partial y'} = \frac{d}{dx}\left(\frac{xy'}{\sqrt{1 + y'^2}}\right). \qquad (77)$$

so the EULER-LAGRANGE DIFFERENTIAL EQUATION becomes

$$\frac{\partial f}{\partial y} - \frac{d}{dx}\frac{\partial f}{\partial y'} = \frac{d}{dx}\left(\frac{xy'}{\sqrt{1 + y'^2}}\right) = 0. \qquad (78)$$

$$\frac{xy'}{\sqrt{1 + y'^2}} = a \qquad (79)$$

$$x^2 y'^2 = a^2\left(1 + y'^2\right) \qquad (80)$$

$$y'^2\left(x^2 - a^2\right) = a^2 \qquad (81)$$

$$\frac{dy}{dx} = \frac{a}{\sqrt{x^2 - a^2}} \qquad (82)$$

$$y = a \int \frac{dx}{\sqrt{x^2 - a^2}} + b = a \cosh^{-1}\left(\frac{x}{a}\right) + b. \qquad (83)$$

Solving for x then gives

$$x = a \cosh\left(\frac{y - b}{a}\right). \qquad (84)$$

which is the equation for a CATENARY. The SURFACE AREA of the CATENOID product by rotation is

$$A = 2\pi \int x\sqrt{1 + y'^2} \, dx = 2\pi \int x\sqrt{1 + \frac{a^2}{x^2 - a^2}} \, dx$$

$$= 2\pi \int \frac{x}{\sqrt{x^2 - a^2}} \sqrt{(x^2 - a^2) + a^2} \, dx$$

$$= 2\pi \int \frac{x^2 \, dx}{\sqrt{x^2 - a^2}}$$

$$= \left[\frac{x}{2}\sqrt{x^2 - a^2} + \frac{a^2}{2}\ln\left(x + \sqrt{x^2 - a^2}\right)\right]_{x_1}^{x_2}$$

$$= \frac{1}{2}\left[x_2\sqrt{x_2^2 - a^2} - x_1\sqrt{x_1^2 - a^2} + a^2 \ln\left(\frac{x_2 + \sqrt{x_2^2 - a^2}}{x_1 + \sqrt{x_1^2 - a^2}}\right)\right]. \qquad (85)$$

Isenberg (1992, p. 80) discusses finding the MINIMAL SURFACE passing through two rings with axes offset from each other.

See also APPLE, CATENOID, CONE CONICAL FRUSTUM, CYLINDER, DARWIN-DE SITTER SPHEROID, EIGHT SURFACE, GABRIEL'S HORN, HYPERBOLOID, LEMON, MERIDIAN, MINIMAL SURFACE, OBLATE SPHEROID, PAPPUS'S CENTROID THEOREM, PARABOLOID, PARALLEL (SURFACE OF REVOLUTION), PENINSULA SURFACE, PROLATE SPHEROID, PSEUDOSPHERE, SINCLAIR'S SOAP FILM PROBLEM, SOLID OF REVOLUTION, SPHERE, SPHEROID, TOROID, TORUS, UNDULOID

References

Arfken, G. *Mathematical Methods for Physicists, 3rd ed.* Orlando, FL: Academic Press, pp. 931–37, 1985.

Goldstein, H. *Classical Mechanics, 2nd ed.* Reading, MA: Addison-Wesley, p. 42, 1980.

Gray, A. "Surfaces of Revolution." Ch. 20 in *Modern Differential Geometry of Curves and Surfaces with Mathematica, 2nd ed.* Boca Raton, FL: CRC Press, pp. 457–80, 1997.

Hilbert, D. and Cohn-Vossen, S. "The Cylinder, the Cone, the Conic Sections, and Their Surfaces of Revolution." §2 in *Geometry and the Imagination.* New York: Chelsea, pp. 7–1, 1999.

Isenberg, C. *The Science of Soap Films and Soap Bubbles.* New York: Dover, pp. 79–0 and Appendix III, 1992.

Surface of Section

A surface (or "space"rpar; of section is a way of presenting a trajectory in n-D PHASE SPACE in an $(n - 1)$-D SPACE. By picking one phase element constant and plotting the values of the other elements each time the selected element has the desired value, an intersection surface is obtained. If the equations of motion can be formulated as a MAP in which an explicit FORMULA gives the values of the other elements at successive passages through the selected element value, the time required to compute the surface of section is greatly reduced.

See also HÉNON-HEILES EQUATION, PHASE SPACE

References

Tabor, M. "The Surface of Section." §4.1 in *Chaos and Integrability in Nonlinear Dynamics: An Introduction.* New York: Wiley, pp. 121–26, 1989.

Surface Parameterization

A surface in 3-SPACE can be parameterized by two variables (or coordinates) u and v such that

$$x = x(u, v) \tag{1}$$

$$y = y(u, v) \tag{2}$$

$$z = z(u, v). \tag{3}$$

If a surface is parameterized as above, then the tangent VECTORS

$$\mathbf{T}_u = \frac{\partial x}{\partial u}\, \hat{\mathbf{x}} + \frac{\partial y}{\partial u}\, \hat{\mathbf{y}} + \frac{\partial z}{\partial u}\, \hat{\mathbf{z}} \tag{4}$$

$$\mathbf{T}_v = \frac{\partial x}{\partial v}\, \hat{\mathbf{x}} + \frac{\partial y}{\partial v}\, \hat{\mathbf{y}} + \frac{\partial z}{\partial v}\, \hat{\mathbf{z}} \tag{5}$$

are useful in computing the SURFACE AREA and SURFACE INTEGRAL.

See also SMOOTH SURFACE, SURFACE AREA, SURFACE INTEGRAL

Surface Spherical Harmonic

SURFACE HARMONIC

Surgery

In the process of attaching a k-HANDLE to a MANIFOLD M, the BOUNDARY of M is modified by a process called $(k-1)$-surgery. Surgery consists of the removal of a TUBULAR NEIGHBORHOOD of a $(k-1)$-SPHERE $\mathbb{S}^{(k-1)}$ from the BOUNDARIES of M and the $\dim(M) - 1$ standard SPHERE, and the gluing together of these two scarred-up objects along their common BOUNDARIES.

See also BOUNDARY, DEHN SURGERY, HANDLE, MANIFOLD, SPHERE, TUBULAR NEIGHBORHOOD

References

Cappell, S.; Ranicki, A.; and Rosenberg, J. (Eds.). *Surveys on Surgery Theory, Vol. 1.* Princeton, NJ: Princeton University Press, 2000.

Surjection

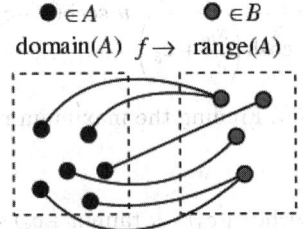

onto and not one-to-one
(surjection but not injection)

An ONTO (a.k.a. surjective) MAP.

See also BIJECTION, DOMAIN, ONE-TO-ONE, ONTO, RANGE (IMAGE)

Surjective

ONTO

Surprise Examination Paradox

UNEXPECTED HANGING PARADOX

Surreal Number

The most natural collection of numbers which includes both the REAL NUMBERS and the infinite ORDINAL NUMBERS of Georg Cantor. They were invented by John H. Conway in 1969. Every REAL NUMBER is surrounded by surreals, which are closer to it than any REAL NUMBER. Knuth (1974) describes the surreal numbers in a work of fiction.

The surreal numbers are written using the NOTATION $\{a|b\}$, where $\{|\} = 0$, $\{0|\} = 1$ is the simplest number greater than 0, $\{1|\} = 2$ is the simplest number greater than 1, etc. Similarly, $\{|0\} = -1$ is the simplest number less than 1, etc. However, 2 can also be represented by $\{1|3\}$, $\{3/2|4\}$, $\{1|\omega\}$, etc.

See also OMNIFIC INTEGER, ORDINAL NUMBER, REAL NUMBER

References

Berlekamp, E. R.; Conway, J. H.; and Guy, R. K. *Winning Ways for Your Mathematical Plays, Vol. 1: Games in General.* London: Academic Press, 1982.
Conway, J. H. *On Numbers and Games.* New York: Academic Press, 1976.
Conway, J. H. and Guy, R. K. *The Book of Numbers.* New York: Springer-Verlag, pp. 283–84, 1996.
Conway, J. H. and Jackson, A. "Budding Mathematician Wins Westinghouse Competition." *Not. Amer. Math. Soc.* **43**, 776–79, 1996.
Gonshor, H. *An Introduction to Surreal Numbers.* Cambridge, England: Cambridge University Press, 1986.
Knuth, D. *Surreal Numbers: How Two Ex-Students Turned on to Pure Mathematics and Found Total Happiness.* Reading, MA: Addison-Wesley, 1974. http://www-cs-faculty.stanford.edu/~knuth/sn.html.

Surrogate

Surrogate data are artificially generated data which mimic statistical properties of real data. Isospectral surrogates have identical POWER SPECTRA as real data but with randomized phases. Scrambled surrogates have the same probability distribution as real data, but with white noise POWER SPECTRA.

See also POWER SPECTRUM

Surveying Problems

HANSEN'S PROBLEM, SNELLIUS-POTHENOT PROBLEM

Survivorship Curve

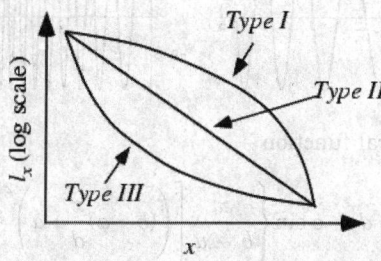

Plotting l_x from a LIFE EXPECTANCY table on a logarithmic scale versus x gives a curve known as a survivorship curve. There are three general classes of survivorship curves, illustrated above.

  1. Type I curves are typical of populations in which most mortality occurs among the elderly (e.g., humans in developed countries).
  2. Type II curves occur when mortality is not dependent on age (e.g., many species of large birds and fish). For an infinite type II population, $e_0 = e_1 = \ldots$, but this cannot hold for a finite population.
  3. Type III curves occur when juvenile mortality is extremely high (e.g., plant and animal species producing many offspring of which few survive). In type III populations, it is often true that $e_{i+1} > e_i$ for small i. In other words, life expectancy increases for individuals who survive their risky juvenile period.

See also LIFE EXPECTANCY

Suslin's Theorem

A SET in a POLISH SPACE is a BOREL SET IFF it is both ANALYTIC and COANALYTIC. For subsets of w, a set is δ_1^1 IFF it is "hyperarithmetic."

See also ANALYTIC SET, BOREL SET, COANALYTIC SET, POLISH SPACE

Suspended Knot

An ordinary KNOT in 3-D suspended in 4-D to create a knotted 2-sphere. Suspended knots are not smooth at the poles.

See also SPUN KNOT, TWIST-SPUN KNOT

Suspension

The JOIN of a TOPOLOGICAL SPACE X and a pair of points S^0, $\Sigma(X) = X * S^0$.

See also JOIN (SPACES), TOPOLOGICAL SPACE

References

Rolfsen, D. *Knots and Links*. Wilmington, DE: Publish or Perish Press, p. 6, 1976.

Suzanne Set

The nth Suzanne set S_n is defined as the set of COMPOSITE NUMBERS x for which $n|S(x)$ and $n|S_p(x)$, where

$$x = a_0 + a_1(10^1) + \ldots + a_d(10^d) = p_1 p_2 \cdots p_n,$$

and

$$S(x) = \sum_{j=0}^{d} a_j$$

$$S_p(x) = \sum_{i=1}^{m} S(p_i).$$

Every Suzanne set has an infinite number of elements. The Suzanne set S_n is a superset of the MONICA SET M_n.

See also MONICA SET

References

Smith, M. "Cousins of Smith Numbers: Monica and Suzanne Sets." *Fib. Quart.* **34**, 102–04, 1996.

Suzuki Group

The SPORADIC GROUP *Suz*.

References

Wilson, R. A. "ATLAS of Finite Group Representation." http://for.mat.bham.ac.uk/atlas/html/Suz.html.

Swallowtail Catastrophe

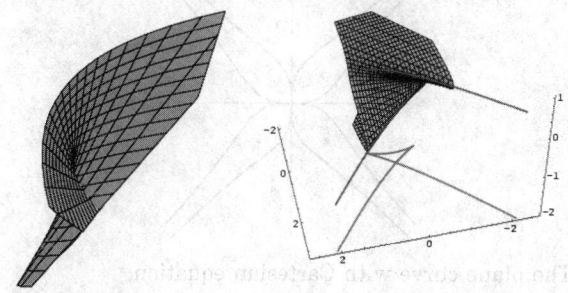

A CATASTROPHE which can occur for three control factors and one behavior axis. The swallowtail catastrophe is the universal unfolding of singularity $f(x) = x^5$ with codimension 3, i.e., in three unfolding parameters, and is of the form $F(x, u, v, w) = x^5 + ux^3 + vx^2 + wx$. The equations

$$x = uv^2 + 3v^4$$

$$y = -2uv - 4v^3$$

$$z = u$$

display such a catastrophe (von Seggern 1993, Nordstrand). The above surface uses $u \in [-2, 2]$ and $v \in [-0.8, 0.8]$.

References

Nordstrand, T. "Swallowtail." http://www.uib.no/people/nfytn/stltxt.htm.
Sanns, W. *Catastrophe Theory with Mathematica: A Geometric Approach.* Germany: DAV, 2000.
von Seggern, D. *CRC Standard Curves and Surfaces.* Boca Raton, FL: CRC Press, p. 94, 1993.

Swastika

An irregular ICOSAGON, also called the gammadion or fylfot, which symbolized good luck in ancient Arabic and Indian cultures. In more recent times, it was adopted as the symbol of the Nazi Party in Hitler's Germany and has thence come to symbolize anti-Semitism.

See also CROSS, DISSECTION

References

Gardner, M. "Form a Swastika." §20.6 in *The Sixth Book of Mathematical Games from Scientific American.* Chicago, IL: University of Chicago Press, pp. 198 and 203–04, 1984.

Swastika Curve

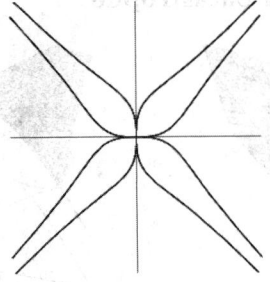

The plane curve with Cartesian equation

$$y^4 - x^4 = xy$$

and polar equation

$$r^2 = \frac{\sin\theta \cos\theta}{\sin^4\theta - \cos^4\theta}.$$

References

Cundy, H. and Rollett, A. *Mathematical Models, 3rd ed.* Stradbroke, England: Tarquin Pub., p. 71, 1989.

Sweep Signal

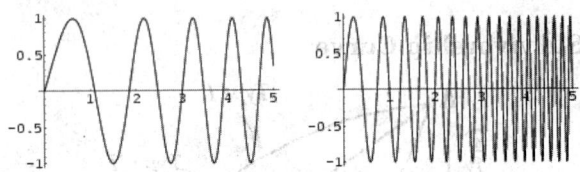

The general function

$$y(a, b, c, d) = c \sin\left\{ \frac{\pi}{b - a} \left[\left((b - a)\frac{x}{d} + a \right)^2 - a^2 \right] \right\}.$$

References

von Seggern, D. *CRC Standard Curves and Surfaces.* Boca Raton, FL: CRC Press, p. 160, 1993.

Swept Sine
SWEEP SIGNAL

Swinnerton-Dyer Conjecture

In the early 1960s, B. Birch and H. P. F. Swinnerton-Dyer conjectured that if a given ELLIPTIC CURVE has an infinite number of solutions, then the associated L-series has value 0 at a certain fixed point. In 1976, Coates and Wiles showed that elliptic curves with COMPLEX multiplication having an infinite number of solutions have L-series which are zero at the relevant fixed point (COATES-WILES THEOREM), but they were unable to prove the converse. V. Kolyvagin extended this result to modular curves.

See also COATES-WILES THEOREM, ELLIPTIC CURVE

References

Birch, B. and Swinnerton-Dyer, H. "Notes on Elliptic Curves. II." *J. reine angew. Math.* **218**, 79–08, 1965.
Cipra, B. "Fermat Prover Points to Next Challenges." *Science* **271**, 1668–669, 1996.
Clay Mathematics Institute. "The Birch and Swinnerton-Dyer Conjecture." http://www.claymath.org/prize_problems/birchsd.htm.
Ireland, K. and Rosen, M. "New Results on the Birch-Swinnerton-Dyer Conjecture." §20.5 in *A Classical Introduction to Modern Number Theory, 2nd ed.* New York: Springer-Verlag, pp. 353–57, 1990.

Mazur, B. and Stevens, G. (Eds.). *p-Adic Monodromy and the Birch and Swinnerton-Dyer Conjecture.* Providence, RI: Amer. Math. Soc., 1994.

Wiles, A. "The Birch and Swinnerton-Dyer Conjecture." http://www.claymath.org/prize_problems/birchsd.pdf.

Swinnerton-Dyer Polynomial

The minimal POLYNOMIAL $S_n(x)$ whose ROOTS are sums and differences of the SQUARE ROOTS of the first n PRIMES,

$$S_n(x) = \prod \left(x \pm \sqrt{2} \pm \sqrt{3} \pm \sqrt{5} \pm \ldots \pm \sqrt{p_n} \right).$$

References

Vardi, I. *Computational Recreations in Mathematica.* Redwood City, CA: Addison-Wesley, pp. 11 and 225–26, 1991.

Swirl

A swirl is a generic word to describe a function having arcs which double back around each other. The plots above correspond to the function

$$f(r, \; \theta) = \sin(6 \cos r - n\theta)$$

for $n = 0, 1, \ldots, 5$.

See also DAISY, WHIRL

Switching Class

TWO-GRAPH

Swung Dash

The symbol $\sim$ used to denote similarity, equivalence relations, or asymptosy.

References

Bringhurst, R. *The Elements of Typographic Style, 2nd ed.* Point Roberts, WA: Hartley and Marks, p. 285, 1997.

Sylow p-Subgroup

If p^k is the highest POWER of a PRIME p dividing the ORDER of a FINITE GROUP G, then a SUBGROUP of G of ORDER p^k is called a Sylow p-subgroup of G.

See also ABHYANKAR'S CONJECTURE, SUBGROUP, SYLOW THEOREMS

Sylow Theorems

Let p be a PRIME NUMBER, G a FINITE GROUP, and $|G|$ the order of G.

1. If p divides $|G|$, then G has a SYLOW P-SUBGROUP.

2. In a FINITE GROUP, all the SYLOW P-SUBGROUPS are CONJUGATE for some fixed p.

3. The number of SYLOW P-SUBGROUPS for a fixed p is CONGRUENT to $1 \pmod{p}$.

See also CONJUGATE SUBGROUP, SYLOW P-SUBGROUP

Sylvester Cyclotomic Number

Given a LUCAS SEQUENCE with parameters P and Q, discriminant $D \neq 0$, and roots α and β, the Sylvester cyclotomic numbers are

$$Q_n = \prod_r (\alpha - \zeta^r \beta).$$

where

$$\zeta \equiv \cos\left(\frac{2\pi}{n}\right) + i \sin\left(\frac{2\pi}{n}\right)$$

is a PRIMITIVE ROOT OF UNITY and the product is over all exponents r RELATIVELY PRIME to n such that $r \in [1, \; n)$.

See also LUCAS SEQUENCE

References

Ribenboim, P. *The Book of Prime Number Records, 2nd ed.* New York: Springer-Verlag, p. 69, 1989.

Sylvester Graph

The Sylvester graph of a configuration is the set of ORDINARY POINTS and ORDINARY LINES.

See also ORDINARY LINE, ORDINARY POINT

References

Guy, R. K. "Monthly Unsolved Problems, 1969–987." *Amer. Math. Monthly* **94**, 961–70, 1987.

Guy, R. K. "Unsolved Problems Come of Age." *Amer. Math. Monthly* **96**, 903–09, 1989.

Sylvester Matrix

For POLYNOMIALS of degree m and n, the Sylvester matrix is an $(m+n) \times (m+n)$ matrix whose DETERMINANT is the RESULTANT of the two POLYNOMIALS.

See also DETERMINANT, RESULTANT

Sylvester's Determinant Identity

Given a MATRIX A, let |A| denote its determinant. Then

$$|A||A_{rs,\,pq}| = |A_{r,\,p}||A_{s,\,q}| - |A_{r,\,q}||A_{s,\,p}|, \qquad (1)$$

where $A_{u,\,w}$ is the SUBMATRIX of A formed by the intersection of the subset w of columns and u of rows. Bareiss (1968) writes the identity as

$$|A|\left[a_{kk}^{(k-1)}\right]^{n-k-1} = \begin{vmatrix} a_{k+1,\,k-1}^{(k)} & \cdots & a_{k+1,\,n}^{(k)} \\ \vdots & \ddots & \vdots \\ a_{n,\,k+1}^{(k)} & \cdots & a_{n,\,n}^{(k)} \end{vmatrix}. \qquad (2)$$

where

$$a_{ij}^{(k)} = \begin{vmatrix} a_{11} & a_{12} & \cdots & a_{1k} & a_{ij} \\ a_{21} & a_{22} & \cdots & a_{2k} & a_{2j} \\ \vdots & \vdots & \ddots & \vdots & \vdots \\ a_{k1} & a_{k2} & \cdots & a_{kk} & a_{kj} \\ a_{i1} & a_{i2} & \cdots & a_{ik} & a_{ij} \end{vmatrix} \qquad (3)$$

for $k < i, j \le n$.

See also DETERMINANT

References

Bareiss, E. H. "Multistep Integer-Preserving Gaussian Elimination." Argonne National Laboratory Report ANL-7213, May 1966.

Bareiss, E. H. "Sylvester's Identity and Multistep Integer-Preserving Gaussian Elimination." *Math. Comput.* **22**, 565–78, 1968.

Sylvester's Four-Point Problem

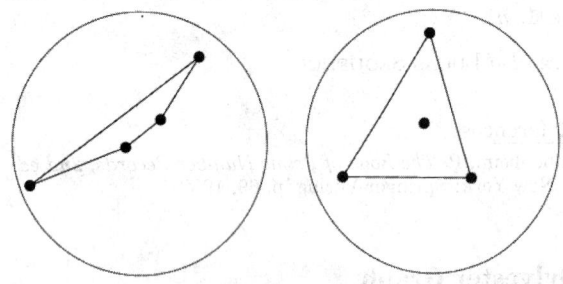

Sylvester's four-point problem asks for the probability $q(R)$ that four points chosen at random in a planar region R have a CONVEX HULL which is a QUADRILATERAL (Sylvester 1865). Depending on the method chosen to pick points from the infinite plane, a number of different solutions are possible, prompting Sylvester to conclude "This problem does not admit of a determinate solution" (Sylvester 1865; Pfiefer 1989).

For points selected from an open, convex subset of the PLANE having finite AREA, the probability if given by

$$P(R) = 1 - \frac{4\bar{A}_R}{A(R)}.$$

where $\bar{A}_R$ is the expected area of a triangle over region R and $A(R)$ is the area of region R. Note that

$\bar{A}_R$ is simply the value computed for an appropriate region, e.g., DISK TRIANGLE PICKING, TRIANGLE TRIANGLE PICKING, SQUARE TRIANGLE PICKING, etc. $P(R)$ can range between

$$\tfrac{2}{3} \le q(R) \le 1 - \frac{35}{12\pi^2} \qquad (1)$$

($0.66666 \le q(R) \le 0.70448$) depending on the shape of the region, as first proved by Blaschke (Blaschke 1923, Peyerimhoff 1997). The following table gives the probabilities for various simple plane regions (Kendall and Moran 1963; Pfiefer 1989; Croft *et al.* 1991, pp. 54–5; Peyerimhoff 1997).

R	$P(R)$	approx.
TRIANGLE	$\frac{2}{3}$	0.66667
SQUARE	$\frac{25}{36}$	0.69444
HEXAGON	$\frac{683}{972}$	0.70267
ELLIPSE, CIRCLE	$1 - \frac{35}{12\pi^2}$	0.70448

Sylvester's problem can be generalized to ask for the probability that the CONVEX HULL of $n+2$ randomly chosen points in the UNIT BALL $\mathbb{B}^n$ has $n+1$ vertices. The solution is given by

$$P_n = \frac{(n+2)\left(\dfrac{n+1}{\frac{1}{2}(n+1)}\right)^{n+1}}{2^n\left(\dfrac{(n+1)^2}{\frac{1}{2}(n+1)^2}\right)} \qquad (2)$$

(Kingman 1969, Groemer 1973, Peyerimhoff 1997), which is the maximum possible for any bounded convex domain $K \in \mathbb{R}^n$. The first few values are

$$P_1 = 1$$

$$P_2 = \frac{35}{12\pi^2}$$

$$P_3 = \frac{9}{143}$$

$$P_4 = \frac{676039}{648000\pi^4}$$

$$P_5 = \frac{20000}{12964479}$$

(Sloane's A051050 and A051051).

Another generalization asks the probability that n randomly chosen points in a fixed bounded convex domain $K \subset \mathbb{R}^2$ are the vertices of a convex n-gon. The solution is

$$P_n = \frac{2^n(3n-3)!}{[(n-1)!]^3(2n)!} \qquad (3)$$

for a triangular domain, which has first few values 1, 1, 1, 2/3, 11/36, 91/900, 17/675, ... (Sloane's A004677 and A004824), and

$$P_n = \left[\frac{1}{n!} \binom{2n-2}{n-1} \right]^2 \qquad (4)$$

for a parallelogram domain, which has first few values 1, 1, 1, 25/36, 49/144, 121/3600, ... (Sloane's A004936 and A005017; Valtr 1996, Peyerimhoff 1997).

Sylvester's four-point problem has an unexpected connection with the RECTILINEAR CROSSING NUMBER of graphs (Finch).

See also DISK TRIANGLE PICKING, HEXAGON TRIANGLE PICKING, RECTILINEAR CROSSING NUMBER, SQUARE TRIANGLE PICKING, TRIANGLE TRIANGLE PICKING

References

Alikoski, H. A. "Über das Sylvestersche Vierpunktproblem." *Ann. Acad. Sci. Fenn.* **51**, No. 7, 1–0, 1939.

Blaschke, W. "Über affine Geometrie XI: Lösung des 'Vierpunktproblems' von Sylvester aus der Theorie der geometrischen Wahrscheinlichkeiten." *Leipziger Ber.* **69**, 436–53, 1917.

Blaschke, W. §24–5 in *Vorlesungen über Differentialgeometrie, II. Affine Differentialgeometrie*. Berlin: Springer-Verlag, 1923.

Croft, H. T.; Falconer, K. J.; and Guy, R. K. "Random Polygons and Polyhedra." §B5 in *Unsolved Problems in Geometry*. New York: Springer-Verlag, pp. 54–7, 1991.

Finch, S. "Favorite Mathematical Constants." http://www.mathsoft.com/asolve/constant/crss/crss.html.

Groemer, H. "On Some Mean Values Associated with a Randomly Selected Simlpex in a Convex Set." *Pacific J. Math.* **45**, 525–33, 1973.

Kendall, M. G. and Moran, P. A. P. *Geometric Probability*. New York: Hafner, 1963.

Kingman, J. F. C. "Random Secants of a Convex Body." *J. Appl. Prob.* **6**, 660–72, 1969.

Klee, V. "What is the Expected Volume of a Simplex Whose Vertices are Chosen at Random from a Given Convex Body." *Amer. Math. Monthly* **76**, 286–88, 1969.

Peyerimhoff, N. "Areas and Intersections in Convex Domains." *Amer. Math. Monthly* **104**, 697–04, 1997.

Pfiefer, R. E. "The Historical Development of J. J. Sylvester's Four Point Problem." *Math. Mag.* **62**, 309–17, 1989.

Rottenberg, R. R. "On Finite Sets of Points in $\mathbb{P}^3$." *Israel J. Math.* **10**, 160–71, 1971.

Santaló, L. A. *Integral Geometry and Geometric Probability*. Reading, MA: Addison-Wesley, 1976.

Schneinerman, E. and Wilf, H. S. "The Rectilinear Crossing Number of a Complete Graph and Sylvester's 'Four Point' Problem of Geometric Probability." *Amer. Math. Monthly* **101**, 939–43, 1994.

Sloane, N. J. A. Sequences A004677, A004824, A004936, A005017, A051050, and A051051 in "An On-Line Version of the Encyclopedia of Integer Sequences." http://www.research.att.com/~njas/sequences/eisonline.html.

Solomon, H. "Crofton's Theorem and Sylvester's Problem in Two and Three Dimensions." Ch. 5 in *Geometric Probability*. Philadelphia, PA: SIAM, pp. 97–25, 1978.

Sylvester, J. J. "Question 1491." *The Educational Times (London)*. April 1864.

Sylvester, J. J. "On a Special Class of Questions on the Theory of Probabilities." *Birmingham British Assoc. Rept.*, pp. 8–, 1865.

Valtr, P. "Probability that n Random Points are in a Convex Position." *Discrete Comput. Geom.* **13**, 637–43, 1995.

Valtr, P. "The Probability that n Random Points in a Triangle are in Convex Position." *Combinatorica* **16**, 567–73, 1996.

Weil, W. and Wieacker, J. "Stochastic Geometry." Ch. 5.2 in *Handbook of Convex Geometry* (Ed. P. M. Gruber and J. M. Wills). Amsterdam, Netherlands: North-Holland, pp. 1391–438, 1993.

Wilf, H. "On Crossing Numbers, and Some Unsolved Problems." In *Combinatorics, Geometry, and Probability: A Tribute to Paul Erdos. Papers from the Conference in Honor of Erdos' 80th Birthday Held at Trinity College, Cambridge, March 1993* (Ed. B. Bollobás and A. Thomason). Cambridge, England: Cambridge University Press, pp. 557–62, 1997.

Woolhouse, W. S. B. "Some Additional Observations on the Four-Point Problem." *Mathematical Questions, with Their Solutions, from the Educational Times, Vol. 7*. London: F. Hodgson and Son, p. 81, 1867.

Sylvester's Inertia Law

The numbers of EIGENVALUES that are POSITIVE, NEGATIVE, or 0 do not change under a congruence transformation. Gradshteyn and Ryzhik (2000) state it as follows: when a QUADRATIC FORM Q in n variables is reduced by a nonsingular linear transformation to the form

$$Q = y_1^2 + y_2^2 + \ldots + y_p^2 - p_{p+1}^2 - y_{p_2}^2 - \ldots - y_r^2.$$

the number p of POSITIVE SQUARES appearing in the reduction is an invariant of the QUADRATIC FORM Q and does not depend on the method of reduction.

See also EIGENVALUE, QUADRATIC FORM

References

Gradshteyn, I. S. and Ryzhik, I. M. *Tables of Integrals, Series, and Products, 6th ed.* San Diego, CA: Academic Press, p. 1105, 2000.

Sylvester's Line Problem

It is not possible to arrange a finite number of points so that a LINE through every two of them passes through a third unless they are all on a single LINE.

See also COLLINEAR, SYLVESTER'S FOUR-POINT PROBLEM

Sylvester's Sequence

The sequence defined by $e_0 = 2$ and the RECURRENCE RELATION

$$e_n = 1 + \prod_{i=0}^{n-1} e_i = e_{n-1}^2 - e_{n-1} + 1. \qquad (1)$$

This sequence arises in Euclid's proof that there are an INFINITE number of PRIMES. The proof proceeds by constructing a sequence of PRIMES using the RECUR-RENCE RELATION

$$e_{n+1} = e_0 e_1 \cdots e_n + 1 \qquad (2)$$

(Vardi 1991). Amazingly, there is a constant

$$E \approx 1.264084735306 \qquad (3)$$

such that

$$e_n = \left\lfloor E^{2^{n+1}} + \frac{1}{2} \right\rfloor \qquad (4)$$

(Vardi 1991, Graham *et al.* 1994). The first few numbers in Sylvester's sequence are 2, 3, 7, 43, 1807, 3263443, 10650056950807, ... (Sloane's A000058). The e_n satisfy

$$\sum_{n=0}^{\infty} \frac{1}{e_n} = 1. \qquad (5)$$

In addition, if $0 < x < 1$ is an IRRATIONAL NUMBER, then the nth term of an infinite sum of unit fractions used to represent x as computed using the GREEDY ALGORITHM must be smaller than $1/e_n$.

The n of the first few PRIME e_n are 0, 1, 2, 3, 5, ..., corresponding to 2, 3, 7, 43, 3263443, ... (Sloane's A014546). Vardi (1991) gives a lists of factors less than 5×10^7 of e_n for $n \leq 200$ and shows that e_n is COMPOSITE for $6 \leq n \leq 17$. Furthermore, all numbers less than 2.5×10^{15} in Sylvester's sequence are SQUAREFREE, and no SQUAREFUL numbers in this sequence are known (Vardi 1991).

See also EUCLID'S THEOREMS, GREEDY ALGORITHM, SQUAREFREE, SQUAREFUL

References

Graham, R. L.; Knuth, D. E.; and Patashnik, O. Research problem 4.65 in *Concrete Mathematics: A Foundation for Computer Science, 2nd ed.* Reading, MA: Addison-Wesley, 1994.

Sloane, N. J. A. Sequences A000058/M0865 and A014546 in "An On-Line Version of the Encyclopedia of Integer Sequences." http://www.research.att.com/~njas/sequences/eisonline.html.

Vardi, I. "Are All Euclid Numbers Squarefree?" and "Power-Mod to the Rescue." §5.1 and 5.2 in *Computational Recreations in Mathematica.* Reading, MA: Addison-Wesley, pp. 82–9, 1991.

Sylvester's Signature

Diagonalize a form over the RATIONALS to

$$\text{diag}\left[p^a \cdot A, \; p^b \cdot B, \; \ldots \right].$$

where all the entries are INTEGERS and A, B, ...are RELATIVELY PRIME to p. Then Sylvester's signature is the sum of the -1-parts of the entries.

See also P-SIGNATURE

Sylvester's Triangle Problem

The resultant of the vectors represented by the three RADII from the center of a TRIANGLE'S CIRCUMCIRCLE to its VERTICES is the segment extending from the CIRCUMCENTER to the ORTHOCENTER.

See also CIRCUMCENTER, CIRCUMCIRCLE, ORTHOCENTER, TRIANGLE

References

Dörrie, H. *100 Great Problems of Elementary Mathematics: Their History and Solutions.* New York: Dover, p. 142, 1965.

Symbolic Calculus

UMBRAL CALCULUS

Symbolic Logic

The study of the meaning and relationships of statements used to represent precise mathematical ideas. Symbolic logic is also called FORMAL LOGIC.

See also FORMAL LOGIC, LOGIC, METAMATHEMATICS

References

Carnap, R. *Introduction to Symbolic Logic and Its Applications.* New York: Dover, 1958.

Symmedian

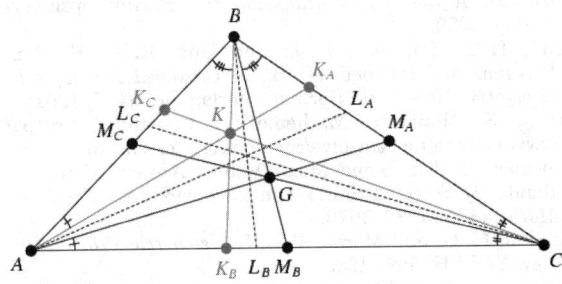

The lines AK_A, BK_B, and CK_B which are ISOGONAL to the MEDIANS AM_A, BM_B, and CM_C of a TRIANGLE are called the triangle's symmedian. The symmedians are concurrent in a point K called the SYMMEDIAN POINT which is the ISOGONAL CONJUGATE of the CENTROID G.

See also CENTROID (TRIANGLE), ISOGONAL CONJUGATE, SYMMEDIAN POINT, MEDIAN (TRIANGLE)

References

Casey, J. "Theory of Isogonal and Isotomic Points, and of Antiparallel and Symmedian Lines." Supp. Ch. §1 in *A Sequel to the First Six Books of the Elements of Euclid, Containing an Easy Introduction to Modern Geometry with Numerous Examples, 5th ed., rev. enl.* Dublin: Hodges, Figgis, & Co., pp. 165–73, 1888.

Coolidge, J. L. *A Treatise on the Geometry of the Circle and Sphere.* New York: Chelsea, p. 65, 1971.

Johnson, R. A. *Modern Geometry: An Elementary Treatise on the Geometry of the Triangle and the Circle.* Boston, MA: Houghton Mifflin, pp. 213–18, 1929.

Lachlan, R. *An Elementary Treatise on Modern Pure Geometry.* London: Macmillian, pp. 62–3, 1893.

Mackay, J. S. "Symmedians of a Triangle and Their Concomitant Circles." *Proc. Edinburgh Math. Soc.* **14,** 37–03, 1896.

Symmedian Point

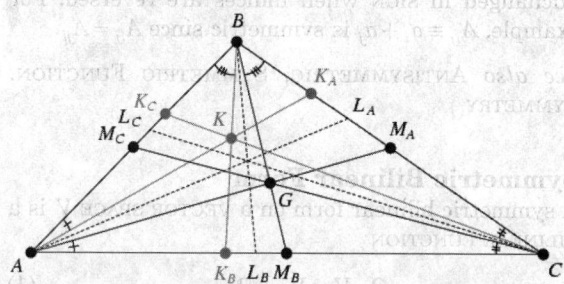

The point of concurrence K of the SYMMEDIANS, sometimes also called the LEMOINE POINT (in England and France) or the GREBE POINT (in Germany). Equivalently, the symmedian point is the ISOGONAL CONJUGATE of the CENTROID G. In other words, let G be the CENTROID of a TRIANGLE ΔABC, AM_A, BM_B, and CM_C the medians of ΔABC, AL_A, BL_B, and CL_C the ANGLE BISECTORS of ANGLES A, B, C, and AK_A, BK_B, and CK_C the reflections of AM_A, BM_B, and CM_C about AL_A, BL_B, and CL_C. Then K is the point of concurrence of the lines AK_A, BK_B, and CK_C. According to Honsberger (1995, p. 53), the symmedian point is "one of the crown jewels of modern geometry."

The TRILINEAR COORDINATES of the symmedian point is

$$a : b : c \qquad (1)$$

(Honsberger 1995, p. 75), or

$$\sin A : \sin B : \sin C. \qquad (2)$$

In AREAL COORDINATES (actual TRILINEAR COORDINATES), the symmedian point is the point for which $\alpha^2 + \beta^2 + \gamma^2$ is a minimum (Honsberger 1995, pp. 75–6). A center X is the CENTROID of its own PEDAL TRIANGLE IFF it is the symmedian point. The symmedian point is the perspectivity center of a TRIANGLE and its TANGENTIAL TRIANGLE.

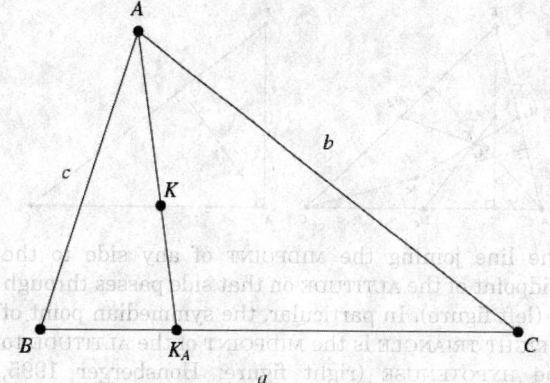

In the above diagram with K the symmedian point,

$$\frac{AK}{KK_A} = \frac{b^2 + c^2}{a^2} \qquad (3)$$

(Honsberger 1995, p. 76).

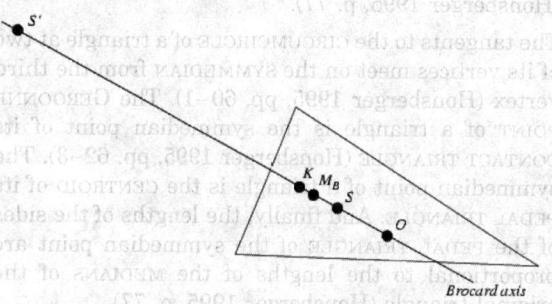

The symmedian point lies on the BROCARD AXIS, and its distances from K to the sides of the TRIANGLE are

$$KK_i = \tfrac{1}{2} a_i \tan \omega, \qquad (4)$$

where ω is the BROCARD ANGLE.

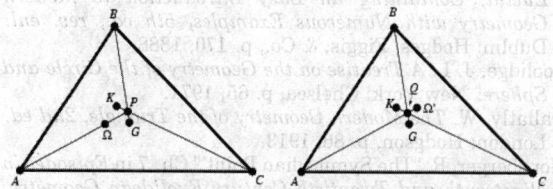

One BROCARD LINE, MEDIAN, and SYMMEDIAN (out of the three of each) are CONCURRENT, with $A\Omega$, CK, and BG meeting at a point, where Ω is the first BROCARD POINT and G is the CENTROID. Similarly, $A\Omega'$, BG, and CK, where Ω' is the second BROCARD POINT, meet at a point which is the ISOGONAL CONJUGATE of the first (Johnson 1929, pp. 268–69).

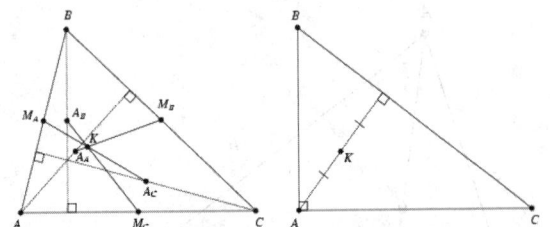

The line joining the MIDPOINT of any side to the midpoint of the ALTITUDE on that side passes through K (left figure). In particular, the symmedian point of a RIGHT TRIANGLE is the MIDPOINT of the ALTITUDE to the HYPOTENUSE (right figure; Honsberger 1995, p. 59). The symmedian point K is the STEINER POINT of the first BROCARD TRIANGLE.

$$\sqcup \xrightarrow{6} 100011 \xrightarrow{3} 10101 \xrightarrow{3} 1110 \xrightarrow{0} 111 \xrightarrow{4} 1011\ldots$$

Given a triangle $\triangle ABC$, construct the triangle $\triangle A'B'C'$ obtained as the intersection of the lines extended from each vertex though the symmedian point K of $\triangle ABC$ with the CIRCUMCIRCLE of $\triangle ABC$. Then the symmedian point of $\triangle A'B'C'$ is again K (Honsberger 1995, p. 77).

The tangents to the CIRCUMCIRCLE of a triangle at two of its vertices meet on the SYMMEDIAN from the third vertex (Honsberger 1995, pp. 60–1). The GERGONNE POINT of a triangle is the symmedian point of its CONTACT TRIANGLE (Honsberger 1995, pp. 62–3). The symmedian point of a triangle is the CENTROID of its PEDAL TRIANGLE. And finally, the lengths of the sides of the PEDAL TRIANGLE of the symmedian point are proportional to the lengths of the MEDIANS of the original triangle (Honsberger 1995, p. 77)

See also ANGLE BISECTOR, BROCARD ANGLE, BROCARD AXIS, BROCARD DIAMETER, CENTROID (TRIANGLE), COSYMMEDIAN TRIANGLES, GREBE POINT, ISOGONAL CONJUGATE, LEMOINE CIRCLE, LEMOINE LINE, LINE AT INFINITY, MITTENPUNKT, PEDAL TRIANGLE, STEINER POINTS, SYMMEDIAN, TANGENTIAL TRIANGLE

References

Casey, J. *A Sequel to the First Six Books of the Elements of Euclid, Containing an Easy Introduction to Modern Geometry with Numerous Examples, 5th ed., rev. enl.* Dublin: Hodges, Figgis, & Co., p. 170, 1888.

Coolidge, J. L. *A Treatise on the Geometry of the Circle and Sphere.* New York: Chelsea, p. 65, 1971.

Gallatly, W. *The Modern Geometry of the Triangle, 2nd ed.* London: Hodgson, p. 86, 1913.

Honsberger, R. "The Symmedian Point." Ch. 7 in *Episodes in Nineteenth and Twentieth Century Euclidean Geometry.* Washington, DC: Math. Assoc. Amer., pp. 53–7, 1995.

Johnson, R. A. *Modern Geometry: An Elementary Treatise on the Geometry of the Triangle and the Circle.* Boston, MA: Houghton Mifflin, pp. 217, 268–69, and 271–72, 1929.

Kimberling, C. "Central Points and Central Lines in the Plane of a Triangle." *Math. Mag.* **67**, 163–87, 1994.

Kimberling, C. "Symmedian Point." http://cedar.evansville.edu/~ck6/tcenters/class/sympt.html.

Mackay, J. S. "Early History of the Symmedian Point." *Proc. Edinburgh Math. Soc.* **11**, 92–03, 1892–893.

Mackay, J. S. "Symmedians of a Triangle and Their Concomitant Circles." *Proc. Edinburgh Math. Soc.* **14**, 37–03, 1896.

Symmetric

A mathematical object is said to be symmetric if it is invariant ("looks the same") under a symmetry transformation.

A function, matrix, etc., is symmetric if it remains unchanged in SIGN when indices are reversed. For example, $A_{ij} \equiv a_i + a_j$ is symmetric since $A_{ij} = A_{ji}$.

See also ANTISYMMETRIC, SYMMETRIC FUNCTION, SYMMETRY

Symmetric Bilinear Form

A symmetric bilinear form on a VECTOR SPACE V is a BILINEAR FUNCTION

$$Q : V \times V \to \mathbb{R} \tag{1}$$

which satisfies $Q(v, w) = Q(w, v)$.

For example, if A is a $n \times n$ SYMMETRIC MATRIX, then

$$Q(v, w) = v^{\mathrm{T}} A w = \langle v, Aw \rangle \tag{2}$$

is a symmetric bilinear form. Consider

$$A = \begin{bmatrix} 1 & 2 \\ 2 & -3 \end{bmatrix}, \tag{3}$$

then

$$\begin{aligned} Q((a_1, a_2), (b_1, b_2)) \\ = a_1 b_1 + 2 a_1 b_2 + 2 a_2 b_1 - 3 a_2 b_2. \end{aligned} \tag{4}$$

Here is a *Mathematica* function which takes a matrix to a bilinear form.

```
MatrixToForm[a_List?MatrixQ][v_, w_] := v.a.w
```

For example,

```
q = MatrixToForm[{{0, 1}, {1, -2}}];
  q[{1, 0}, {1, 7}]
```

yields 7.

A QUADRATIC FORM may also be labeled Q, because quadratic forms are in a one-to-one correspondence with symmetric bilinear forms. Note that $Q(a) = Q(a, a)$ is a QUADRATIC FORM. If $Q(a)$ is a quadratic form then it defines a symmetric bilinear form by

$$Q(a, b) = \tfrac{1}{2}[Q(a+b) - Q(a) - Q(b)]. \tag{5}$$

The kernel, or radical, of a symmetric bilinear form is the set of vectors

$$\ker Q = \{v : Q(v, w) = 0 \text{ for all } w \in V\}. \tag{6}$$

A quadratic form is called nondegenerate if its kernel is zero. That is, if for all $v \in V$, there is a $w \in V$ with $Q(v, w) \neq 0$. The rank of Q is the rank of the matrix $(a_{ij}) = Q(e_i, e_j)$.

The form Q is diagonalized if there is a basis v_i, called an orthogonal basis, such that $(b_{ij}) = Q(v_i, v_j)$ is a DIAGONAL MATRIX. Alternatively, there is a matrix C such that

$$Q(Cv, \ Cw) = (C_v)^T A(Cw) = v^T (C^T AC) w \quad (7)$$

is a DIAGONAL QUADRATIC FORM. The jth column of the matrix C is the vector v_j.

A nondegenerate symmetric bilinear form can be DIAGONALIZED, using GRAM-SCHMIDT ORTHONORMALIZATION to find the v_i, so that the diagonal matrix $C^T AC$ has entries either 1 or -1. If there are p 1s and q -1s, then Q is said to have SIGNATURE (p, q), or if the dimension is understood then just signature p. Real nondegenerate symmetric bilinear forms are classified by their signature, in the sense that given two vector spaces with forms of signature (p, q), there is an isomorphism of the vector spaces which takes one form to the other.

A symmetric bilinear form with $Q(v, v) > 0$, for all nonzero v, is called POSITIVE DEFINITE. For example, the usual inner product is positive definite. A positive definite form has signature $(n, 0)$. A negative definite form is the negative of a positive form and has signature $(0, n)$. If the form is neither positive definite nor negative definite, then there must exist vectors $w \neq 0$ such that $Q(w, w) = 0$, called isotropic vectors.

A general symmetric bilinear form Q can be diagonalized with diagonal entries 1, -1, or 0, because the form Q is always nondegenerate on the QUOTIENT VECTOR SPACE $V/\ker Q$. If V is a COMPLEX VECTOR SPACE, then a symmetric bilinear form can be diagonalized to have entries 1 or 0. For other FIELDS, there are more SYMMETRIC BILINEAR FORMS than in the real or complex case. For instance, if the FIELD has CHARACTERISTIC 2, then it is not possible to divide by 2 since $2 = 0$. Hence there is no correspondence between quadratic forms and symmetric bilinear forms in characteristic 2.

See also DIAGONAL QUADRATIC FORM, FIELD, INDEX (MATRIX), INNER PRODUCT, QUADRATIC FORM, SIGNATURE, SYMMETRIC BILINEAR FORM (GENERAL FIELDS), VECTOR SPACE

References

Serre, J. P. *A Course in Arithmetic.* New York: Springer-Verlag, pp. 27–5, 1973.

Symmetric Bilinear Form (General Fields)

The symmetric bilinear forms on a VECTOR SPACE, whose FIELD k is not real, have been classified for some FIELDS. There are also theorems about symmetric bilinear forms on free Abelian groups, for example $\mathbb{Z}^n$.

A SYMMETRIC BILINEAR FORM Q corresponds to a matrix A by giving a basis e_i and setting $a_{ij} = Q(e_i, e_j)$. Two symmetric bilinear forms are considered equivalent if a change of basis takes one to the other. Hence, $A \sim CAC^T$, where C is any invertible matrix. Therefore, the rank of the symmetric bilinear form is an invariant.

Also, det A can change by $(\det C)^2 \det A$. The coset of det A in k^*/k^{*2} is a WELL DEFINED invariant of Q, called the discriminant. For real forms, it is either 1 or -1. For $\mathbb{Q}$, the discriminant can be any RATIONAL NUMBER a/b where a and b are SQUAREFREE. A symmetric bilinear form on a FINITE FIELD is determined by its rank and its discriminant.

A symmetric bilinear form on the P-ADIC NUMBERS $\mathbb{Q}_p$ is characterized by its rank, discriminant, and another invariant $e(Q)$. Given a basis e_i, orthogonal for Q, define $a_i = Q(e_1, e_2)$, then

$$e(Q) = \prod_{i < j} (a_i, \ a_j)$$

where $(a_i, \ a_j)$ is the HILBERT SYMBOL.

Two symmetric bilinear forms are equivalent on the RATIONALS iff they are equivalent in every $\mathbb{Q}_p$ as well as the reals (also called $\mathbb{Q}_\infty$.) The data in $\mathbb{Q}_p$ can be thought of as "local" information, which can be patched together to yield "global" information in $\mathbb{Q}$. So rational forms have a countable number of distinct invariants, three for every PRIME NUMBER, and two for the reals.

See also HILBERT SYMBOL, *P*-ADIC NUMBER, QUADRATIC FORM, SYMMETRIC BILINEAR FORM, VECTOR SPACE

References

Serre, J. P. *A Course in Arithmetic.* New York: Springer-Verlag, pp. 27–5, 1973.

Symmetric Block Design

A symmetric design is a BLOCK DESIGN (v, k, λ, r, b) with the same number of blocks as points, so $b = v$ (or, equivalently, $r = k$). An example of a symmetric block design is a PROJECTIVE PLANE.

See also BLOCK DESIGN, PROJECTIVE PLANE

References

Dinitz, J. H. and Stinson, D. R. "A Brief Introduction to Design Theory." Ch. 1 in *Contemporary Design Theory: A Collection of Surveys* (Ed. J. H. Dinitz and D. R. Stinson). New York: Wiley, pp. 1–2, 1992.

Symmetric Design

SYMMETRIC BLOCK DESIGN

Symmetric Difference

The set of elements belonging to one *but not both* of two given sets. It is therefore the UNION of the COMPLEMENT of A with respect to B and B with respect to A, and corresponds to the XOR operation in Boolean logic. The symmetric difference can be implemented in *Mathematica* as

```
SymmetricDifference[a_,       b_]  :=
Union[Complement[a, b], Complement[b, a]]
```

The symmetric difference of sets A and B is variously written as $A \ominus B$, $A \triangledown B$, or $A + B$. The latter two notations are deprecated since these symbols have common meanings in other areas of mathematics.

For example, for $A = \{1, 2, 3, 4\}$ and $B = \{1, 4, 5\}$, $A \ominus B = \{2, 3, 5\}$, since 2, 3, and 5 are each in one, but not both, sets.

See also COMPLEMENT SET, DIFFERENCE, SET DIFFERENCE, UNION, XOR

Symmetric Function

A symmetric function on n variables $x_1, ..., x_n$ is a function that is unchanged by any PERMUTATION of its variables. In most contexts, the term "symmetric function" refers to a polynomial on n variables with this feature (more properly called a "SYMMETRIC POLYNOMIAL"). Another type of symmetric functions is symmetric rational functions, which are the RATIONAL FUNCTIONS that are unchanged by PERMUTATION of variables.

The SYMMETRIC POLYNOMIALS (respectively, symmetric rational functions) can be expressed as polynomials (respectively, rational functions) in the SYMMETRIC POLYNOMIALS. This is called the FUNDAMENTAL THEOREM OF SYMMETRIC FUNCTIONS.

A function $f(x)$ is sometimes said to be symmetric about the y-AXIS if $f(-x) = f(x)$. Examples of such functions include $|x|$ (the ABSOLUTE VALUE) and x^2 (the PARABOLA).

See also FUNDAMENTAL THEOREM OF SYMMETRIC FUNCTIONS, RATIONAL FUNCTION, SYMMETRIC POLYNOMIAL

References

Bressoud, D. *Proofs and Confirmations: The Story of the Alternating Sign Matrix Conjecture.* Cambridge, England: Cambridge University Press, 1999.
Littlewood, J. E. *A University Algebra, 2nd ed.* London: Heinemann, 1958.
Macdonald, I. G. *Symmetric Functions and Hall Polynomials, 2nd ed.* Oxford, England: Oxford University Press, 1995.
Macdonald, I. G. *Symmetric Functions and Orthogonal Polynomials.* Providence, RI: Amer. Math. Soc., 1997.
Petkovsek, M.; Wilf, H. S.; and Zeilberger, D. "Symmetric Function Identities." §1.7 in *A = B.* Wellesley, MA: A. K. Peters, pp. 12–3, 1996.

Symmetric Group

The symmetric group S_n of degree n is the GROUP of all PERMUTATIONS on n symbols. S_n is therefore of ORDER $n!$ and contains as SUBGROUPS every GROUP of ORDER n. The number of CONJUGACY CLASSES of S_n is given by the PARTITION FUNCTION P.

For example, let $\{abc\}$ denote the permutation on three elements which takes the ath element to position 1, the bth element to position 2, and the cth element to position 3. Then the following table gives the MULTIPLICATION TABLE for S_n, containing $3! = 6$ elements with $\{123\}$ the IDENTITY ELEMENT. The multiplication table can be generated using the *Mathematica* function

```
SymmetricGroup[n_Integer?Positive] := Module[
  {p = Permutations[Range[n]], i, j},
  Table[p[[i]][[p[[j]]]], {i, n!}, {j, n!}]
]
```

S_3	(123)	(132)	(213)	(231)	(312)	(321)
(123)	(123)	(132)	(213)	(231)	(312)	(321)
(132)	(132)	(123)	(312)	(321)	(213)	(231)
(213)	(213)	(231)	(123)	(132)	(321)	(312)
(231)	(231)	(213)	(321)	(312)	(123)	(132)
(312)	(312)	(321)	(132)	(123)	(231)	(213)
(321)	(321)	(312)	(231)	(213)	(132)	(123)

NETTO'S CONJECTURE states that the probability that two elements P_1 and P_2 of a symmetric group generate the entire group tends to $3/4$ as $n \to \infty$. This was proven by Dixon (1969). The probability that two elements generate S_n for $n = 1, 2, ...$ are 1, 3/4, 1/2, 3/8, 19/40, 53/120, 103/168, ... (Sloane's A040173 and A040174). Finding a general formula for terms in the sequence is a famous UNSOLVED PROBLEM in GROUP THEORY.

See also ALTERNATING GROUP, CONJUGACY CLASS, ERDOS-TURÁN THEOREM, FINITE GROUP, JORDAN'S SYMMETRIC GROUP THEOREM, NETTO'S CONJECTURE, PARTITION FUNCTION P, SIMPLE GROUP

References

Dixon, J. D. "The Probability of Generating the Symmetric Group." *Math. Z.* **110**, 199–05, 1969.
Huang, J.-S. "Symmetric Groups." Ch. 3 in *Lectures on Representation Theory.* Singapore: World Scientific, pp. 15–5, 1999.
Lomont, J. S. "Symmetric Groups." Ch. 7 in *Applications of Finite Groups.* New York: Dover, pp. 258–73, 1987.
Skiena, S. *Implementing Discrete Mathematics: Combinatorics and Graph Theory with Mathematica.* Reading, MA: Addison-Wesley, p. 17, 1990.

Sloane, N. J. A. Sequences A040173 and A040174 in "An On-Line Version of the Encyclopedia of Integer Sequences." http://www.research.att.com/~njas/sequences/eisonline.html.

Wilson, R. A. "ATLAS of Finite Group Representation." http://for.mat.bham.ac.uk/atlas/html/contents.html#alt.

Symmetric Matrix

A symmetric matrix is a SQUARE MATRIX which satisfies $A^T = A$ where A^T denotes the TRANSPOSE, so $a_{ij} = a_{ji}$. This also implies

$$A^{-1}A^T = I, \tag{1}$$

where I is the IDENTITY MATRIX. For example,

$$A = \begin{bmatrix} 4 & 1 \\ 1 & -2 \end{bmatrix} \tag{2}$$

is a symmetric matrix. HERMITIAN MATRICES are a useful generalization of symmetric matrices for COMPLEX MATRICES

A matrix m can be tested to see if it is symmetric using the *Mathematica* function

```
SymmetricQ[m_List?MatrixQ]  :=  (m  = = =
Transpose[m])
```

Written explicitly, the elements of a symmetric matrix A have the form

$$\begin{bmatrix} a_{11} & a_{12} & \cdots & a_{1n} \\ a_{21} & a_{22} & \cdots & a_{2n} \\ \vdots & \vdots & \ddots & \vdots \\ a_{n1} & a_{n2} & \cdots & a_{nn} \end{bmatrix} \tag{3}$$

The symmetric part of any MATRIX may be obtained from

$$A_s = \tfrac{1}{2}(A + A^T). \tag{4}$$

A MATRIX A is symmetric if it can be expressed in the form

$$A = QDQ^T, \tag{5}$$

where Q is an ORTHOGONAL MATRIX and D is a DIAGONAL MATRIX. This is equivalent to the MATRIX EQUATION

$$AQ = QD, \tag{6}$$

which is equivalent to

$$AQ_n = \lambda_n Q_n \tag{7}$$

for all n, where $\lambda_n = D_{nn}$. Therefore, the diagonal elements of D are the EIGENVALUES of A, and the columns of Q are the corresponding EIGENVECTORS.

The numbers of symmetric matrices of order n on s symbols are s, s^3, s^6, s^{10}, ..., $s^{k(k-1)/2}$. Therefore, for (0,1)-MATRICES, the numbers of distinct symmetric matrices of orders $n = 1$, 2, ... are 2, 8, 64, 1024, ... (Sloane's A006125).

See also ADJOINT MATRIX, ANTISYMMETRIC MATRIX, BISYMMETRIC MATRIX, HERMITIAN MATRIX, PERSYMMETRIC MATRIX, SKEW SYMMETRIC MATRIX

References

Ayres, F. Jr. *Theory and Problems of Matrices.* New York: Schaum, pp. 12 and 115–17, 1962.

Nash, J. C. "Real Symmetric Matrices." Ch. 10 in *Compact Numerical Methods for Computers: Linear Algebra and Function Minimisation, 2nd ed.* Bristol, England: Adam Hilger, pp. 119–34, 1990.

Sloane, N. J. A. Sequences A006125/M1897 in "An On-Line Version of the Encyclopedia of Integer Sequences." http://www.research.att.com/~njas/sequences/eisonline.html.

Symmetric Points

Two points z and $z^S \in \mathbb{C}$ are symmetric with respect to a CIRCLE or straight LINE L if all CIRCLES and straight LINES passing through z and z^S are orthogonal to L. MÖBIUS TRANSFORMATIONS preserve symmetry. Let a straight line be given by a point z_0 and a unit VECTOR $e^{i\theta}$, then

$$z^S = e^{2i\theta}\overline{z - z_0} + z_0,$$

where $\bar{z}$ is the COMPLEX CONJUGATE. Let a CIRCLE be given by center z_0 and RADIUS r, then

$$z^S = z_0 + \frac{r^2}{\overline{z - z_0}}.$$

See also MÖBIUS TRANSFORMATION

Symmetric Polynomial

A symmetric polynomial on n variables $x_1, ..., x_n$ is a function that is unchanged by any PERMUTATION of its variables. Symmetric polynomials are always HOMOGENEOUS POLYNOMIALS. The n elementary symmetric functions Π_n (sometimes denoted σ_n) on n variables $\{x_1, ..., x_n\}$ are defined by

$$\Pi_1 = \sum_{1 \leq i \leq n} x_i \tag{1}$$

$$\Pi_2 = \sum_{1 \leq i < j \leq n} x_i x_j \tag{2}$$

$$\Pi_3 = \sum_{1 \leq i < j < k \leq n} x_i x_j x_k \tag{3}$$

$$\Pi_4 = \sum_{1 \leq i < j < k < l \leq n} x_i x_j x_k x_l \tag{4}$$

$$\vdots$$

$$\Pi_n = \sum_{1 \leq i \leq n} x_i. \tag{5}$$

The kth symmetric polynomial is defined as `SymmetricPolynomial[{x1,...,xn},k]` in the *Mathematica* add-on package `Algebra`SymmetricPolyno-`

mials' (which can be loaded with the command $<<$Algebra'). SymmetricReduction[f, $\{x1, \ldots, xn\}$] in the *Mathematica* add-on package Algebra'-SymmetricPolynomials' (which can be loaded with the command $<<$Algebra') gives a pair of polynomials $\{p, q\}$ in $x_1, \ldots, x_n$ where p is the symmetric part and q is the remainder.

Alternatively, $\Pi_j(x_1, \ldots, x_n)$ can be defined as the coefficient of x^{n-j} in the GENERATING FUNCTION

$$\prod_{1 \le i \le n}(x + x_i). \tag{6}$$

For example, on four variables $x_1, \ldots, x_4$, the elementary symmetric functions are

$$\Pi_1 = x_1 + x_2 + x_3 + x_4 \tag{7}$$

$$\Pi_2 = x_1 x_2 + x_1 x_3 + x_1 x_4 + x_2 x_3 + x_2 x_4 + x_3 x_4 \tag{8}$$

$$\Pi_3 = x_1 x_2 x_3 + x_1 x_2 x_4 + x_1 x_3 x_4 + x_2 x_3 x_4 \tag{9}$$

$$\Pi_4 = x_1 x_2 x_3 x_4. \tag{10}$$

Define $s_k(h_1, \ldots, h_n)$ as the coefficients of the GENERATING FUNCTION

$$\ln(1 + x_1 t + x_2 t^2 + x_3 t^3 + \ldots) = \sum_{k=1}^{\infty} \frac{s_1}{k} t^k$$

$$= h_1 t + \tfrac{1}{2}(-h_1^2 + 2h_2)t^2 + \tfrac{1}{3}(h_1^3 - 3h_1 h_2 + 3h_3)t^3 + \ldots \tag{11}$$

so the first few values are

$$s_1 = h_1 \tag{12}$$

$$s_2 = -h_1^2 + 2h_2 \tag{13}$$

$$s_3 = h_1^3 - 3h_1 h_2 + 3h_3 \tag{14}$$

$$s_4 = -h_1^4 + 4h_1^2 h_2 - 2h_2^2 - 4h_2 h_3 + 4h_4. \tag{15}$$

In general, s_n can be computed from the DETERMINANT

$$s_n = (-1)^{n-1} \begin{vmatrix} h_1 & 1 & 0 & 0 & \cdots & 0 \\ 2h_2 & h_1 & 1 & 0 & \ddots & 0 \\ 3h_3 & h_2 & h_1 & 1 & \ddots & 0 \\ 4h_4 & h_3 & h_2 & h_1 & \ddots & 0 \\ \vdots & \vdots & \vdots & \vdots & \ddots & 1 \\ nh_n & h_{n-1} & h_{n-2} & h_{n-3} & \cdots & h_1 \end{vmatrix} \tag{16}$$

(Littlewood 1958, Cadogan 1971). Then the elementary symmetric functions satisfy the relationship

$$\sum_{k=1}^{n} x_k^p = (-1)^{p-1} s_p(\Pi_1, \ldots, \Pi_n). \tag{17}$$

In particular,

$$\sum_{k=1}^{n} x_k = \Pi_1 \tag{18}$$

$$\sum_{k=1}^{n} x_k^2 = \Pi_1^2 - 2\Pi_2 \tag{19}$$

$$\sum_{k=1}^{n} x_k^3 = \Pi_1^3 - \Pi_1 \Pi_2 + 3\Pi_3 \tag{20}$$

$$\sum_{k=1}^{n} x_k^4 = \Pi_1^4 - 4\Pi_1^2 \Pi_2 + 2\Pi_2^2 + 4\Pi_1 \Pi_3 - 4\Pi_4 \tag{21}$$

(Schroeppel 1972), as can be verified by plugging in and multiplying through.

See also FUNDAMENTAL THEOREM OF SYMMETRIC FUNCTIONS, NEWTON-GIRARD FORMULAS, NEWTON'S RELATIONS, SYMMETRIC FUNCTION

References

Borwein, P. and Erdélyi, T. *Polynomials and Polynomial Inequalities.* New York: Springer-Verlag, p. 5, 1995.

Cadogan, C. C. "The Möbius Function and Connected Graphs." *J. Combin. Th. B* **11**, 193–00, 1971.

Littlewood, J. E. *A University Algebra, 2nd ed.* London: Heinemann, 1958.

Schroeppel, R. Item 6 in Beeler, M.; Gosper, R. W.; and Schroeppel, R. *HAKMEM.* Cambridge, MA: MIT Artificial Intelligence Laboratory, Memo AIM-239, p. 4, Feb. 1972.

Séroul, R. "Newton-Girard Formulas." §10.12 in *Programming for Mathematicians.* Berlin: Springer-Verlag, pp. 278–79, 2000.

Symmetric Quadratic Form

See also QUADRATIC FORM

Symmetric Relation

A RELATION R on a SET S is symmetric provided that for every x and y in S we have xRy IFF yRx.

See also RELATION

Symmetric Tensor

A second-RANK symmetric TENSOR is defined as a TENSOR A for which

$$A^{mn} = A^{nm}. \tag{1}$$

Any TENSOR can be written as a sum of symmetric and ANTISYMMETRIC parts

$$A^{mn} = \tfrac{1}{2}(A^{mn} + A^{nm}) + \tfrac{1}{2}(A^{mn} - A^{nm})$$

$$= \tfrac{1}{2}(B_S^{mn} + B_A^{mn}). \tag{2}$$

The symmetric part of a TENSOR is denoted using parentheses as

$$T_{(a,\,b)} \equiv \tfrac{1}{2}(T_{ab} + T_{ba}) \tag{3}$$

$$T_{(a_1, a_2, \ldots, a_n)} \equiv \frac{1}{n!} \sum_{\text{permutations}} T_{a_1 a_2 \cdots a_n}. \tag{4}$$

Symbols for the symmetric and antisymmetric parts of tensors can be combined, for example

$$T^{(ab)c}{}_{[de]} = \tfrac{1}{4}(T^{abc}{}_{de} + T^{bac}{}_{de} - T^{abc}{}_{ed} - T^{bac}{}_{ed}). \quad (5)$$

(Wald 1984, p. 26).

The product of a symmetric and an ANTISYMMETRIC TENSOR is 0. This can be seen as follows. Let $a^{\alpha\beta}$ be ANTISYMMETRIC, so

$$a^{11} = a^{22} = 0 \quad (6)$$

$$a^{21} = -a^{12}. \quad (7)$$

Let $b_{\alpha\beta}$ be symmetric, so

$$b_{12} = b_{21}. \quad (8)$$

Then

$$a^{\alpha\beta}b_{\alpha\beta} = a^{11}b_{11} + a^{12}b_{12} + a^{21}b_{21} + a^{22}b_{22}$$

$$= 0 + a^{12}b_{12} - a^{12}b_{12} + 0 = 0. \quad (9)$$

A symmetric second-RANK TENSOR A_{mn} has SCALAR invariants

$$s_1 = A_{11} + A_{22} + A_{22} \quad (10)$$

$$s_2 = A_{22}A_{33} + A_{33}A_{11} + A_{11}A_{22} - A_{23}^2 - A_{31}^2 - A_{12}^2. \quad (11)$$

References

Wald, R. M. *General Relativity*. Chicago, IL: University of Chicago Press, 1984.

Symmetric Top Differential Equation

The second-order ORDINARY DIFFERENTIAL EQUATION

$$y'' - \left[\frac{M^2 - \tfrac{1}{4} + K^2 - 2MK\cos x}{\sin^2 x} + \left(\sigma + K^2 + \tfrac{1}{4}\right) \right] y = 0.$$

References

Infeld, L. and Hull, T. E. "The Factorization Method." *Rev. Mod. Phys.* **23**, 21–8, 1951.
Zwillinger, D. *Handbook of Differential Equations, 3rd ed.* Boston, MA: Academic Press, p. 127, 1997.

Symmetroid

A QUARTIC SURFACE which is the locus of zeros of the DETERMINANT of a SYMMETRIC 4×4 matrix of linear forms. A general symmetroid has 10 ORDINARY DOUBLE POINTS (Jessop 1916, Hunt 1996).

References

Hunt, B. "Algebraic Surfaces." http://www.mathematik.uni-kl.de/~wwwagag/E/Galerie.html.
Hunt, B. "Symmetroids and Weddle Surfaces." §B.5.3 in *The Geometry of Some Special Arithmetic Quotients*. New York: Springer-Verlag, pp. 315–19, 1996.

Jessop, C. *Quartic Surfaces with Singular Points*. Cambridge, England: Cambridge University Press, p. 166, 1916.

Symmetry

An intrinsic property of a mathematical object which causes it to remain invariant under certain classes of transformations (such as ROTATION, REFLECTION, INVERSION, or more abstract operations). The mathematical study of symmetry is systematized and formalized in the extremely powerful and beautiful area of mathematics called GROUP THEORY.

Symmetry can be present in the form of coefficients of equations as well as in the physical arrangement of objects. By classifying the symmetry of polynomial equations using the machinery of GROUP THEORY, for example, it is possible to prove the unsolvability of the general QUINTIC EQUATION.

In physics, the extremely powerful NOETHER'S SYMMETRY THEOREM states that each symmetry of a system leads to a physically conserved quantity. Symmetry under TRANSLATION corresponds to momentum conservation, symmetry under ROTATION to angular momentum conservation, symmetry in time to energy conservation, etc.

See also CRYSTALLOGRAPHY RESTRICTION, GROUP THEORY, NOETHER'S SYMMETRY THEOREM

References

Eppstein, D. "Symmetry and Group Theory." http://www.ics.uci.edu/~eppstein/junkyard/sym.html.
Britton, J. *Symmetry and Tessellations: Investigating Patterns*. Englewood Cliffs, NJ: Prentice-Hall, 1999.
Farmer, D. *Groups and Symmetry*. Providence, RI: Amer. Math. Soc., 1995.
Pappas, T. "Art & Dynamic Symmetry." *The Joy of Mathematics*. San Carlos, CA: Wide World Publ./Tetra, pp. 154–55, 1989.
Radin, C. "Symmetry." Ch. 4 in *Miles of Tiles*. Providence, RI: Amer. Math. Soc., pp. 69–7, 1999.
Rosen, J. *A Symmetry Primer for Scientists*. New York: Wiley, 1983.
Rosen, J. *Symmetry in Science: An Introduction to the General Theory*. New York: Springer-Verlag, 1995.
Schattschneider, D. *Visions of Symmetry: Notebooks, Periodic Drawings, and Related Work of M. C. Escher*. New York: W. H. Freeman, 1990.
Stewart, I. and Golubitsky, M. *Fearful Symmetry*. New York: Viking Penguin, 1993.
Voisin, C. *Mirror Symmetry*. Providence, RI: Amer. Math. Soc., 1999.
Weisstein, E. W. "Books about Symmetry." http://www.treasure-troves.com/books/Symmetry.html.
Yale, P. B. *Geometry and Symmetry*. New York: Dover, 1988.

Symmetry Group

GROUP

Symmetry Operation

Symmetry operations include the IMPROPER ROTATION, INVERSION OPERATION, MIRROR PLANE, and

ROTATION. Together, these operations create 32 crystal classes corresponding to the 32 POINT GROUPS.

The INVERSION OPERATION takes

$$(x, y, z) \rightarrow (-x, -y, -z)$$

and is denoted i. When used in conjunction with a ROTATION, it becomes an IMPROPER ROTATION. An IMPROPER ROTATION by $360°/n$ is denoted $\bar{n}$ (or S_n). For periodic crystals, the CRYSTALLOGRAPHY RESTRICTION allows only the IMPROPER ROTATIONS $\bar{1}$, $\bar{2}$, $\bar{3}$, $\bar{4}$, and $\bar{6}$.

The MIRROR PLANE symmetry operation takes

$$(x, y, z) \rightarrow (x, y, -z),\ (x, -y, z) \rightarrow (x, -y, z),$$

etc., which is equivalent to $\bar{2}$. Invariance under reflection can be denoted $n\sigma_v$ or $n\sigma_h$. The ROTATION symmetry operation for $360°/n$ is denoted n (or C_n). For periodic crystals, CRYSTALLOGRAPHY RESTRICTION allows only 1, 2, 3, 4, and 6.

Symmetry operations can be indicated with symbols such as C_n, S_n, E, i, $n\sigma_v$, and $n\sigma_h$.

1. C_n indicates ROTATION about an n-fold symmetry axis.
2. S_n indicates IMPROPER ROTATION about an n-fold symmetry axis.
3. E (or I) indicates invariance under TRANSLATION.
4. i indicates a center of symmetry under INVERSION.
5. $n\sigma_v$ indicates invariance under n vertical REFLECTIONS.
6. $n\sigma_h$ indicates invariance under n horizontal REFLECTIONS.

See also CRYSTALLOGRAPHY RESTRICTION, GLIDE, IMPROPER ROTATION, INVERSION OPERATION, MIRROR PLANE, POINT GROUPS, ROTATION, SYMMETRY

References

Addington, S. "The Four Types of Symmetry in the Plane." http://forum.swarthmore.edu/sum95/suzanne/symsusan.html.

Symmetry Principle

SYMMETRIC POINTS are preserved under a MÖBIUS TRANSFORMATION. The SCHWARZ REFLECTION PRINCIPLE is sometimes called the symmetry principle (Needham 2000, p. 252).

See also MÖBIUS TRANSFORMATION, SYMMETRIC POINTS

References

Needham, T. "Analytic Continuation." §5.XI in *Visual Complex Analysis*. New York: Clarendon Press, pp. 247–57, 2000.

Symplectic Diffeomorphism

A MAP $(M_1, \omega_1) \rightarrow (M_2, \omega_2)$ between the SYMPLECTIC MANIFOLDS (M_1, ω_1) and (M_2, ω_2) which is a DIFFEOMORPHISM and $T^*(\omega_2) = (\omega_1)$, where T^* is the PULLBACK MAP induced by T (i.e., the derivative of the DIFFEOMORPHISM T acting on tangent vectors). A symplectic diffeomorphism is also known as a SYMPLECTOMORPHISM or CANONICAL TRANSFORMATION.

See also DIFFEOMORPHISM, PULLBACK MAP, SYMPLECTIC MANIFOLD

References

Guillemin, V. and Sternberg, S. *Symplectic Techniques in Physics*. New York: Cambridge University Press, p. 34, 1984.

Symplectic Form

A symplectic form on a SMOOTH MANIFOLD M is a smooth closed 2-FORM ω on M which is nondegenerate such that at every point m, the alternating bilinear form ω_m on the TANGENT SPACE $T_m M$ is nondegenerate.

A symplectic form on a VECTOR SPACE V over F_q is a function $f(x, y)$ (defined for all $x, y \in V$ and taking values in F_q) which satisfies

$$f(\lambda_1 x_1 + \lambda_2 x_2, y) = \lambda_1 f(x_1, y) + \lambda_2 f(x_2, y),$$

$$f(y, x) = -f(x, y),$$

and

$$f(x, x) = 0.$$

f is called non-degenerate if $f(x, y) = 0$ for all y implies that $x = 0$. Symplectic forms can exist on M (or V) only if M (or V) is EVEN-dimensional. An example of a symplectic form over a vector space is the complex HILBERT SPACE with INNER PRODUCT $\langle \cdots \rangle$ given by

$$f(x, y) = \Im\langle x, y \rangle.$$

See also SYMPLECTIC SPACE, VECTOR SPACE

Symplectic Geometry

References

Berndt, R. *Einführung in die Symplektische Geometrie*. Braunschweig, Germany: Vieweg, 1998.

Symplectic Group

For every even DIMENSION $2n$, the symplectic group $Sp(2n)$ is the GROUP of $2n \times 2n$ MATRICES which preserve a nondegenerate skew symmetric BILINEAR FORM ω, i.e., a SYMPLECTIC FORM.

Every symplectic form can be put into a canonical form by finding a SYMPLECTIC BASIS. So, up to conjugation, there is only one symplectic group, in contrast to the ORTHOGONAL GROUP which preserves a nondegenerate SYMMETRIC BILINEAR FORM. As with the ORTHOGONAL GROUP, the columns of a symplectic matrix form a SYMPLECTIC BASIS.

Since ω^n is a VOLUME FORM, the symplectic group preserves volume and ORIENTATION. Hence, $Sp(2n) \subset SL(2n)$. In fact, $Sp(2)$ is just the group of matrices with DETERMINANT 1. The three symplectic $(0,1)$-MATRICES are therefore

$$\begin{bmatrix} 1 & 0 \\ 0 & 1 \end{bmatrix}, \begin{bmatrix} 1 & 0 \\ 1 & 1 \end{bmatrix}, \begin{bmatrix} 1 & 1 \\ 0 & 1 \end{bmatrix}. \qquad (1)$$

The matrices

$$\begin{bmatrix} 1 & 0 & 0 & s \\ 0 & 1 & s & 0 \\ 0 & 0 & 1 & 0 \\ 0 & 0 & 0 & 1 \end{bmatrix} \qquad (2)$$

and

$$\begin{bmatrix} \cosh t & \sinh t & 0 & \sinh t \\ \sinh t & \cosh t & \sinh t & 0 \\ 0 & 0 & \cosh t & -\sinh t \\ 0 & 0 & -\sinh t & \cosh t \end{bmatrix} \qquad (3)$$

are in $Sp(4)$, where

$$\omega = e_1 \wedge e_3 + e_2 \wedge e_4. \qquad (4)$$

In fact, both of these examples are 1-parameter subgroups.

Here is a *Mathematica* function that tests whether a matrix is a symplectic matrix.

```
SymplecticForm[n_Integer] :=
        Join[PadLeft[IdentityMatrix[n],{n,2n}],
PadRight[-IdentityMatrix[n],{n,2n}]]
SymplecticQ[a_List] :=     EvenQ[Length[a]]     &&
Transpose[a].SymplecticForm[Length[a]/2].a
= = SymplecticForm[Length[a]/2]
```

Thinking of a matrix as given by $(2n)^2$ coordinate functions, the set of matrices is identified with $\mathbb{R}^{(2n)^2}$. The symplectic matrices are the solutions to the $(2n)^2$ equations

$$\mathsf{A}^T \, \mathsf{JA} = \mathsf{J}, \qquad (5)$$

where J is defined by

$$\omega(x, y) = \langle x, \, \mathsf{J}y \rangle. \qquad (6)$$

Note that these equations are redundant, since only $2n^2 - n$ of these are independent, leaving $2n^2 + n$ "free" variables. In fact, the symplectic group is a smooth $(2n^2 + n)$-dimensional SUBMANIFOLD of $\mathbb{R}^{2n}$.

Because the symplectic group is a GROUP and a MANIFOLD, it is a LIE GROUP. Its TANGENT SPACE at the identity is the SYMPLECTIC LIE ALGEBRA $\mathfrak{sp}(2n)$. The symplectic group is not COMPACT.

Instead of using real numbers for the coefficients, it is possible to use coefficients from any FIELD F. The symplectic group $Sp_n(q)$ for n EVEN is the GROUP of elements of the GENERAL LINEAR GROUP GL_n that preserve a given nonsingular SYMPLECTIC FORM. Any such MATRIX has DETERMINANT 1.

See also DETERMINANT, FIELD, GENERAL LINEAR GROUP, GROUP, LIE ALGEBRA, LIE GROUP, LIE-TYPE GROUP, LINEAR ALGEBRAIC GROUP, PROJECTIVE SYMPLECTIC GROUP, QUADRATIC FORM, SIEGEL'S UPPER HALF-SPACE, SUBMANIFOLD, SYMPLECTIC BASIS, SYMPLECTIC FORM, UNITARY GROUP, VECTOR SPACE

References

Conway, J. H.; Curtis, R. T.; Norton, S. P.; Parker, R. A.; and Wilson, R. A. "The Groups $Sp_n(q)$ and $PSp_n(q) = S_n(q)$." §2.3 in *Atlas of Finite Groups: Maximal Subgroups and Ordinary Characters for Simple Groups.* Oxford, England: Clarendon Press, pp. x-xi, 1985.
Wilson, R. A. "ATLAS of Finite Group Representation." http://for.mat.bham.ac.uk/atlas/html/contents.html#symp.

Symplectic Manifold

A pair (M, ω), where M is a MANIFOLD and ω is a SYMPLECTIC FORM on M. The PHASE SPACE $\mathbb{R}^{2n} = \mathbb{R}^n \times \mathbb{R}^n$ is a symplectic manifold. Near every point on a symplectic manifold, it is possible to find a set of local "Darboux coordinates" in which the SYMPLECTIC FORM has the simple form

$$\omega = \sum_k dq_k \wedge dp_k$$

(Sjamaar 1996), where $dq_k \wedge dp_k$ is a WEDGE PRODUCT.

See also MANIFOLD, SYMPLECTIC DIFFEOMORPHISM, SYMPLECTIC FORM

References

Sjamaar, R. "Symplectic Reduction and Riemann-Roch Formulas for Multiplicities." *Bull. Amer. Math. Soc.* **33**, 327-38, 1996.

Symplectic Map

Informally, a symplectic map is a MAP which preserves the sum of AREAS projected onto the set of (p_2, q_2) planes. It is the generalization of an AREA-PRESERVING MAP.

Formally, a symplectic map is a real-linear map T that preserves a SYMPLECTIC FORM f, i.e., for which

$$f(Tx, \, Ty) = f(x, y)$$

for all x, y. Every symplectic map T on a complex HILBERT SPACE H may be written as $U(\cosh S +$

$J \sinh S$), where U is unitary, S is positive, and J is an anti-linear involution (i.e., complex conjugation).

See also AREA-PRESERVING MAP, LIOUVILLE'S PHASE SPACE THEOREM

Symplectic Space

A real-linear VECTOR SPACE H equipped with a SYMPLECTIC FORM s.

Symplectomorphism

SYMPLECTIC DIFFEOMORPHISM

Synclastic

A surface on which the GAUSSIAN CURVATURE K is everywhere POSITIVE. When K is everywhere NEGATIVE, a surface is called ANTICLASTIC. A point at which the GAUSSIAN CURVATURE is POSITIVE is called an ELLIPTIC POINT.

See also ANTICLASTIC, ELLIPTIC POINT, GAUSSIAN QUADRATURE, HYPERBOLIC POINT, PARABOLIC POINT, PLANAR POINT

Synergetics

Synergetics deals with systems composed of many subsystems which may each be of a very different nature. In particular, synergetics treats systems in which cooperation among subsystems creates organized structure on macroscopic scales (Haken 1993). Examples of problems treated by synergetics include BIFURCATIONS, phase transitions in physics, convective instabilities, coherent oscillations in lasers, nonlinear oscillations in electrical circuits, population dynamics, etc.

See also BIFURCATION, CHAOS, DYNAMICAL SYSTEM

References

Haken, H. *Synergetics, an Introduction: Nonequilibrium Phase Transitions and Self-Organization in Physics, Chemistry, and Biology,* 3rd rev. enl. ed. New York: Springer-Verlag, 1983.
Haken, H. *Advanced Synergetics: Instability Hierarchies of Self-Organizing Systems and Devices.* New York: Springer-Verlag, 1993.
Mikhailov, A. S. *Foundations of Synergetics: Distributed Active Systems,* 2nd ed. New York: Springer-Verlag, 1994.
Mikhailov, A. S. and Loskutov, A. Y. *Foundations of Synergetics II: Complex Patterns,* 2nd ed., enl. rev. New York: Springer-Verlag, 1996.
Weisstein, E. W. "Books about Synergetics." http://www.treasure-troves.com/books/Synergetics.html.
Tschacher, W. and Dauwalder, J.-P. (Eds.). *Dynamics, Synergetics, Autonomous Agents: Nonlinear Systems Approaches to Cognitive Psychology and Cognitive Science.* Singapore: World Scientific, 1999.

Syntonic Comma

COMMA OF DIDYMUS

Syracuse Algorithm

COLLATZ PROBLEM

Syracuse Problem

COLLATZ PROBLEM

System of Differential Equations

ORDINARY DIFFERENTIAL EQUATION

System of Equations

A *linear* system of equations may be denoted

$$\mathbf{AX} = \mathbf{Y} \qquad (1)$$

where $\mathbf{A}$ is a MATRIX and $\mathbf{X}$ and $\mathbf{Y}$ are VECTORS. As shown by CRAMER'S RULE, there is a unique solution if $\mathbf{A}$ has a MATRIX INVERSE $\mathbf{A}^{-1}$. In this case,

$$\mathbf{X} = \mathbf{A}^{-1}\mathbf{Y} \qquad (2)$$

If $\mathbf{Y} = \mathbf{0}$, then the solution is $\mathbf{X} = \mathbf{0}$. If $\mathbf{A}$ has no MATRIX INVERSE, then the solution SUBSPACE is either a LINE or the EMPTY SET. If two equations are multiples of each other, solutions are OF THE FORM

$$\mathbf{X} = \mathbf{A} + t\mathbf{B} \qquad (3)$$

for t a REAL NUMBER.

See also CRAMER'S RULE, DETERMINANT, MATRIX INVERSE

Syzygies Problem

The problem of finding all independent irreducible algebraic relations among any finite set of QUANTICS.

See also QUANTIC

Syzygy

A technical mathematical object defined in terms of a POLYNOMIAL RING of n variables over a FIELD k. Syzygies occur in TENSORS at rank 5, 7, 8, and all higher ranks, and play a role in restricting the number of independent ISOTROPIC TENSORS. An example of a rank-5 syzygy is

$$\epsilon_{ijk}\delta_{lm} - \epsilon_{jkl}\delta_{im} + \epsilon_{kli}\delta_{jm} - \epsilon_{lij}\delta_{km} = 0,$$

where ϵ_{ijk} is the PERMUTATION TENSOR and δ_{ij} is the KRONECKER DELTA.

See also FUNDAMENTAL SYSTEM, HILBERT BASIS THEOREM, ISOTROPIC TENSOR, KRONECKER DELTA, SYZYGIES PROBLEM, TENSOR

References

Hilbert, D. "Über die Theorie der algebraischen Formen." *Math. Ann.* **36**, 473–34, 1890.
Iyanaga, S. and Kawada, Y. (Eds.). "Syzygy Theory." §364F in *Encyclopedic Dictionary of Mathematics.* Cambridge, MA: MIT Press, p. 1140, 1980.

Olver, P. J. "Syzygies." *Classical Invariant Theory.* Cambridge, England: Cambridge University Press, pp. 110–12, 1999.

Sylvester, J. J. "On a Theory of Syzygetic Relations of Two Rational Integral Functions, Comprising an Application of the Theory of Sturm's Functions, and that of the Greatest Algebraic Common Measure." *Philos. Trans. Roy. Soc. London* **143**, 407–48, 1853.

Székely Identity

$$\sum_{k=-\infty}^{\infty} \binom{A+B+C+D+E-k}{E-k}\binom{A+D}{k+D}\binom{B+C}{k+C}$$
$$= \binom{A+C+D+E}{A+C}\binom{B+C+D+E}{C+E}.$$

See also BINOMIAL SUMS

References

Koepf, W. "Hypergeometric Database." Ch. 3 in *Hypergeometric Summation: An Algorithmic Approach to Summation and Special Function Identities.* Braunschweig, Germany: Vieweg, pp. 35–6, 1998.

Székely, L. A. "Common Origin of Cubic Binomial Identities; A Generalization of Surányi's Proof of the Le Jen Shoo's Formula." *J. Combin. Th. Ser. A* **40**, 171–74, 1985.

Szemerédi's Regularity Lemma

A fundamental structural result in EXTREMAL GRAPH THEORY due to Szemerédi (1978). The regularity lemma essentially says that every graph can be well-approximated by the union of a constant number of random-like BIPARTITE GRAPHS, called regular pairs.

See also BLOW-UP LEMMA, EXTREMAL GRAPH THEORY, SEYMOUR CONJECTURE, SZEMERÉDI'S THEOREM

References

Komlós, J. and Simonovitas, M. "Szemerédi Regularity Lemma and Its Applications in Graph Theory." In *Combinatorics, Paul Erdos is Eighty, Vol. 1* (Ed. D. Miklós, V. T. Sós, and T. Szonyi). Budapest: János Bolyai Mathematical Society, pp. 295–52, 1993.

Komlós, J.; Sárkozy, G. N.; and Szemerédi, E. "Proof of the Seymour Conjecture for Large Graphs." *Ann. Comb.* **2**, 43–0, 1998.

Szemerédi, E. "Regular Partitions of Graphs." In *Problèmes combinatoires et théorie des graphes (Colloq. Internat. CNRS, Univ. Orsay, Orsay, 1976).* Paris: Editions du Centre National de la Recherche Scientifique (CNRS), pp. 399–01, 1978.>

Szemerédi's Theorem

This entry contributed by KEVIN O'BRYANT

Every sequence of integers with positive density contains arbitrarily long ARITHMETIC SEQUENCES.

A corollary states that, for any positive integer k and positive real number δ, there exists a threshold number $n(k, \delta)$ such that for $n \geq n(k, r)$ every subset of $\{1, 2, \ldots, n\}$ with CARDINALITY larger than δn contains a k-term ARITHMETIC SEQUENCE. VAN DER WAERDEN'S THEOREM follows immediately by setting $\delta = n/r$. The best bounds for VAN DER WAERDEN NUMBERS are derived from bounds for $n(k, r)$ in Szemerédi's Theorem.

Szemerédi's theorem was conjectured by Erdos and Turán (1936). Roth (1953) proved the case $k = 3$, and was mentioned in his FIELDS MEDAL citation. Szemerédi (1969) proved the case $k = 4$, and the general theorem in 1975 as a consequence of SZEMERÉDI'S REGULARITY LEMMA (Szemerédi 1975a), for which he collected a $1000 prize from Erdos. Fürstenberg and Katznelson (1979) proved Szemerédi's theorem using ERGODIC THEORY. Gowers (1998ab) subsequently gave a new proof, with a better bound on $n(k, r)$, for the case $k = 4$ (mentioned in *his* FIELDS MEDAL citation; Lepowsky *et al.* 1999).

Erdos offered a $3,000 prize for a proof of the proposition that "If the sum of reciprocals of a set of integers diverges, then that set contains arbitrarily long arithmetic progressions." This conjecture is still open (unsolved), even for 3-term arithmetic progressions. Erdos also offered $10,000 for an asymptotic formula for $\rho_3(n)$, the largest possible cardinality of a subset of $\{1, 2, \ldots, n\}$ that does not contain a 3-term arithmetic progression.

See also ARITHMETIC SEQUENCE, SZEMERÉDI'S REGULARITY LEMMA, VAN DER WAERDEN NUMBER, VAN DER WAERDEN'S THEOREM

References

Erdos, P. and Turán, P. "On Some Sequences of Integers." *J. London Math. Soc.* **11**, 261–64, 1936.

Fürstenberg, H. and Katznelson, Y. "An Ergodic Szemerédi Theorem for Commuting Transformations." *J. Analyse Math.* **34**, 275–91, 1979.

Gowers, W. T. "Fourier Analysis and Szemerédi's Theorem." In *Proceedings of the International Congress of Mathematicians, Vol. I (Berlin, 1998). Doc. Math.*, Extra Vol. I, 617–29, 1998a.

Gowers, W. T. "A New Proof of Szemerédi's Theorem for Arithmetic Progressions of Length Four." *Geom. Funct. Anal.* **8**, pp. 529–51, 1998b.

Graham, R. L.; Rothschild, B. L.; and Spencer, J. H. *Ramsey Theory, 2nd ed.* New York: Wiley, 1990.

Guy, R. K. "Theorem of van der Waerden, Szemerédi's Theorem. Partitioning the Integers into Classes; at Least One Contains an A.P." §E10 in *Unsolved Problems in Number Theory, 2nd ed.* New York: Springer-Verlag, pp. 204–09, 1994.

Lepowsky, J.; Lindenstrauss, J.; Manin, Y.; and Milnor, J. "The Mathematical Work of the 1998 Fields Medalists." *Not. Amer. Math. Soc.* **46**, 17–6, 1999.

Roth, K. "Sur quelques ensembles d'entiers." *C. R. Acad. Sci. Paris* **234**, 388–90, 1952.

Roth, K. F. "On Certain Sets of Integers." *J. London Math. Soc.* **28**, 104–09, 1953.

Szemerédi, E. "On Sets of Integers Containing No Four Elements in Arithmetic Progression." *Acta Math. Acad. Sci. Hungar.* **20**, 89–04, 1969.

Szemerédi, E. "On Sets of Integers Containing No *k* Elements in Arithmetic Progression." *Acta Arith.* **27**, 199–45, 1975a.

Szemerédi, E. "On Sets of Integers Containing No *k* Elements in Arithmetic Progression." In *Proceedings of the International Congress of Mathematicians, Volume 2, Held in Vancouver, B.C., August 21–9, 1974.* Montreal, Quebec: Canad. Math. Congress, pp. 503–05, 1975b.

Szilassi Polyhedron

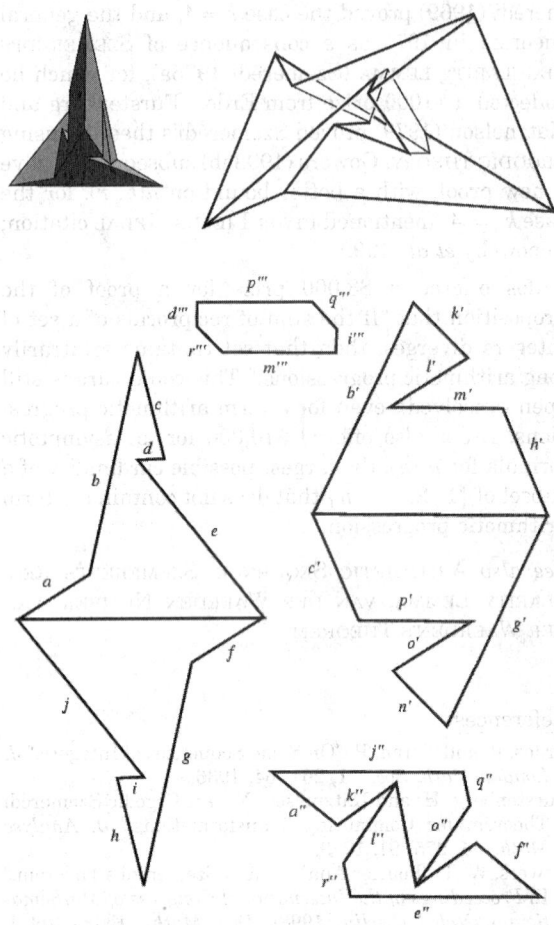

A HEPTAHEDRON which is topologically equivalent to a TORUS and for which every pair of faces has an EDGE in common. The Szilassi polyhedron has 14 VERTICES, seven faces, and 21 EDGES, and is the DUAL POLYHEDRON of the CSÁSZÁR POLYHEDRON. This polyhedron was discovered by L. Szilassi in 1977. In the above illustration of the net, sides indicated by letters are connected with the corresponding side indicated by the same letter but with a different number of primes. Like the TETRAHEDRON, each face of the Szilassi polyhedron touches all other faces.

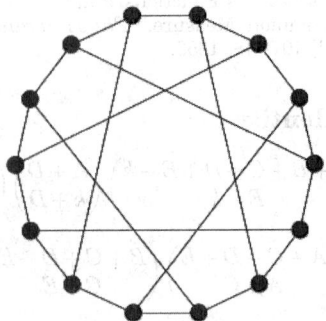

The SKELETON of the Szilassi polyhedron is equivalent to the HEAWOOD GRAPH, shown above.

See also CSÁSZÁR POLYHEDRON, HEAWOOD GRAPH, TOROIDAL POLYHEDRON

References

Ace, T. "Szilassi Polyhedron." http://www.qnet.com/~crux/szilassi.html.

Eppstein, D. "Polyhedra and Polytopes." http://www.ics.uci.edu/~eppstein/junkyard/polytope.html.

Gardner, M. "Mathematical Games: In Which a Mathematical Aesthetic is Applied to Modern Minimal Art." *Sci. Amer.* **239**, 22–2, Nov. 1978.

Gardner, M. *Fractal Music, Hypercards, and More Mathematical Recreations from Scientific American Magazine.* New York: W. H. Freeman, pp. 118–20, 1992.

Hart, G. "Toroidal Polyhedra." http://www.georgehart.com/virtual-polyhedra/toroidal.html.

Weisstein, E. W. "Polyhedra." MATHEMATICA NOTEBOOK POLYHEDRA.M.

Szpiro's Conjecture

A conjecture which relates the minimal DISCRIMINANT of an ELLIPTIC CURVE to the CONDUCTOR. If true, it would imply FERMAT'S LAST THEOREM for sufficiently large exponents.

See also CONDUCTOR, DISCRIMINANT (ELLIPTIC CURVE), ELLIPTIC CURVE

References

Cox, D. A. "Introduction to Fermat's Last Theorem." *Amer. Math. Monthly* **101**, 3–4, 1994.

T

T2-Separation Axiom

Given any two distinct points x, y, there exist neighborhoods u and v of x and y, respectively, with $u \cap v = \emptyset$. It then follows that finite SUBSETS are CLOSED.

See also CLOSURE (SET)

Tableau

YOUNG TABLEAU

Tableau Class

When a YOUNG TABLEAU is constructed using the so-called insertion algorithm, an element starts in some position on the first row, from which it may later be bumped. In contrast, the elements that start out in the ith column are said to belong to the ith class (Skiena 1990, p. 73). Tableau classes may be computed using `TableauClasses[p]` in the *Mathematica* add-on package `DiscreteMath`Combinatorica` (which can be loaded with the command `<<DiscreteMath`).

See also BUMPING ALGORITHM, YOUNG TABLEAU

References

Skiena, S. *Implementing Discrete Mathematics: Combinatorics and Graph Theory with Mathematica.* Reading, MA: Addison-Wesley, 1990.

Tabu Search

A heuristic procedure which has proven efficient at solving COMBINATORIAL optimization problems.

References

Glover, F.; Taillard, E.; and De Werra, D. "A User's Guide to Tabu Search." *Ann. Oper. Res.* **41**, 3–28, 1993.
Piwakowski, K. "Applying Tabu Search to Determine New Ramsey Numbers." *Electronic J. Combinatorics* **3**, R6 1–4, 1996. http://www.combinatorics.org/Volume_3/volume3.html#R6.

Tacnode

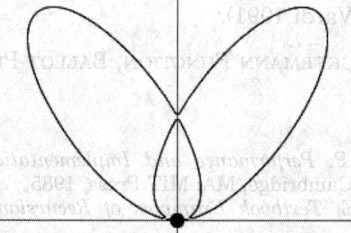

A DOUBLE POINT at which two OSCULATING CURVES are TANGENT. The above plot shows the tacnode of the

curve $2x^4 - 3x^2y + y^2 - 2y^3 + y^4 = 0$. The LINKS CURVE also has a tacnode at the origin.

See also ACNODE, CRUNODE, DOUBLE POINT, OSCULATING CURVES, SPINODE

References

Walker, R. J. *Algebraic Curves.* New York: Springer-Verlag, pp. 57–58, 1978.

Tacpoint

A tangent point of two similar curves.

Tactix

NIM

Tail Probability

Define T as the set of all points t with probabilities $P(x)$ such that $a > t \Rightarrow P(a \leq x \leq a + da) < P_0$ or $a < t \Rightarrow P(a \leq x \leq a + da) < P_0$, where P_0 is a POINT PROBABILITY (often, the likelihood of an observed event). Then the associated tail probability is given by $\int_T P(x)\, dx$.

See also P-VALUE, POINT PROBABILITY

Tait Coloring

A 3-coloring of GRAPH EDGES so that no two EDGES of the same color meet at a VERTEX (Ball and Coxeter 1987, pp. 265–266).

See also EDGE (GRAPH), TAIT CYCLE, VERTEX (GRAPH)

References

Ball, W. W. R. and Coxeter, H. S. M. *Mathematical Recreations and Essays, 13th ed.* New York: Dover, 1987.

Tait Cycle

A set of circuits going along the EDGES of a GRAPH, each with an EVEN number of EDGES, such that just one of the circuits passes through each VERTEX (Ball and Coxeter 1987, pp. 265–266).

See also EDGE (GRAPH), EULERIAN CYCLE, HAMILTONIAN CYCLE, TAIT COLORING, VERTEX (GRAPH)

References

Ball, W. W. R. and Coxeter, H. S. M. *Mathematical Recreations and Essays, 13th ed.* New York: Dover, 1987.

Tait Flyping Conjecture

FLYPING CONJECTURE

Tait's Hamiltonian Graph Conjecture

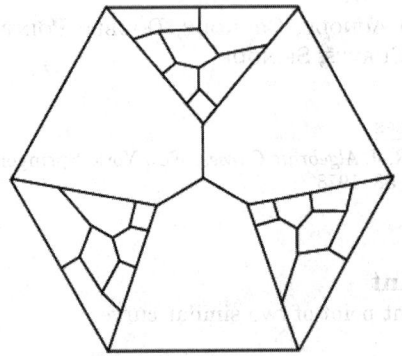

Every 3-connected CUBIC GRAPH has a HAMILTONIAN CIRCUIT. Proposed by Tait in 1880 and refuted by Tutte (1946) with the counterexample now known as TUTTE'S GRAPH. Had the conjecture been true, it would have implied the FOUR-COLOR THEOREM. A simpler counterexample was later given by Kozyrev and Grinberg.

See also CONNECTED GRAPH, CUBIC GRAPH, FOUR-COLOR THEOREM, HAMILTONIAN CIRCUIT, HAMILTONIAN GRAPH, TUTTE CONJECTURE, TUTTE'S GRAPH, VERTEX (GRAPH)

References

Honsberger, R. *Mathematical Gems I.* Washington, DC: Math. Assoc. Amer., pp. 82–89, 1973.
Skiena, S. *Implementing Discrete Mathematics: Combinatorics and Graph Theory with Mathematica.* Reading, MA: Addison-Wesley, p. 198, 1990.
Tait, P. G. "Remarks on the Colouring of Maps." *Proc. Royal Soc. Edinburgh* **10**, 729, 1880.
Tutte, W. T. "On Hamiltonian Circuits." *J. London Math. Soc.* **21**, 98–101, 1946.
Tutte, W. T. "Non-Hamiltonian Planar Maps." In *Graph Theory and Computing* (Ed. R. Read). New York: Academic Press, pp. 295–301, 1972.

Tait's Knot Conjectures

P. G. Tait undertook a study of KNOTS in response to Kelvin's conjecture that the atoms were composed of knotted vortex tubes of ether (Thomson 1869). He categorized KNOTS in terms of the number of crossings in a plane projection. He also made some conjectures which remained unproven until the discovery of JONES POLYNOMIALS:

1. Reduced alternating diagrams have minimal CROSSING NUMBER,
2. Any two reduced alternating diagrams of a given knot have equal WRITHE,
3. The FLYPING CONJECTURE, which states that the number of crossings is the same for any diagram of an ALTERNATING KNOT.

Conjectures (1) and (2) were proved by Kauffman (1987), Murasugi (1987ab), and Thistlethwaite (1987, 1988) using properties of the JONES POLYNOMIAL or KAUFFMAN POLYNOMIAL F (Hoste *et al.* 1998). Conjecture (3) was proved true by Menasco and Thistlethwaite (1991, 1993) using properties of the JONES POLYNOMIAL (Hoste *et al.* 1998).

See also ALTERNATING KNOT, CROSSING NUMBER (LINK), FLYPING CONJECTURE, JONES POLYNOMIAL, KNOT, WRITHE

References

Hoste, J.; Thistlethwaite, M.; and Weeks, J. "The First 1,701,936 Knots." *Math. Intell.* **20**, 33–48, Fall 1998.
Kauffman, L. H. "State Models and the Jones Polynomial." *Topology* **26**, 395–407, 1987.
Menasco, W. and Thistlethwaite, M. "The Tait Flyping Conjecture." *Bull. Amer. Math. Soc.* **25**, 403–412, 1991.
Menasco, W. and Thistlethwaite, M. "The Classification of Alternating Links." *Ann. Math.* **138**, 113–171, 1993.
Murasugi, K. "The Jones Polynomial and Classical COnjectures in Knot Theory." *Topology* **26**, 187–194, 1987a.
Murasugi, K. "Jones Polynomials and Classical Conjectures in Knot Theory II." *Math. Proc. Cambridge Philos. Soc.* **102**, 317–318, 1987.
Tait, P. G. "On Knots I, II, III." *Scientific Papers, Vol. 1.* London: Cambridge University Press, pp. 273–347, 1900.
Thistlethwaite, M. B. "A Spanning Tree Expansion of the Jones Polynomial." *Topology* **26**, 297–309, 1987.
Thistlethwaite, M. B. "Kauffman's Polynomial and Alternating Links." *Topology* **27**, 311–318, 1988.
Thomson, W. H. "On Vortex Motion." *Trans. Roy. Soc. Edinburgh* **25**, 217–260, 1869.

TAK Function

A RECURSIVE FUNCTION devised by I. Takeuchi. For INTEGERS x, y, and z, and a function h, it is

$$\text{TAK}_h(x, y, z)$$

The number of function calls $F_0(a, b)$ required to compute $\text{TAK}_0(a, b, 0)$ for $a > b > 0$ is

$$F_0(a, b) = 4 \sum_{k=0}^{b} \frac{a-b}{a+b-2k} \binom{a+b-2k}{b-k} - 3$$

$$= 1 + 4 \sum_{k=0}^{b-1} \frac{a-b}{a+b-2k} \binom{a+b-2k}{b-k}$$

(Vardi 1991).

The TAK function is also connected with the BALLOT PROBLEM (Vardi 1991).

See also ACKERMANN FUNCTION, BALLOT PROBLEM

References

Gabriel, R. P. *Performance and Implementation of Lisp Systems.* Cambridge, MA: MIT Press, 1985.
Knuth, D. E. *Textbook Examples of Recursion.* Preprint 1990.
Vardi, I. "The Running Time of TAK." Ch. 9 in *Computational Recreations in Mathematica.* Redwood City, CA: Addison-Wesley, pp. 179–199, 1991.

Takagi Fractal Curve
BLANCMANGE FUNCTION

Take-Away Game
NIM-HEAP

Takens-Bogdanov Bifurcation

References
Bogdanov, R. "Bifurcations of a Limit Cycle for a Family of Vector Fields on the Plane." *Selecta Math. Soviet* **1**, 373–388, 1981.
Kuznetsov, Y. A. *Elements of Applied Bifurcation Theory.* New York: Springer-Verlag, 1995.
Takens, F. "Forced Oscillations and Bifurcations." *Comm. Math. Inst. Rijksuniv. Utrecht* **2**, 1–111, 1974.

Takeuchi Function
TAK FUNCTION

Talbot's Curve

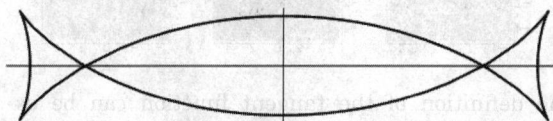

A curve investigated by Talbot which is the NEGATIVE PEDAL CURVE of an ELLIPSE with respect to its center. It has four CUSPS and two NODES, provided the ECCENTRICITY of the ELLIPSE is greater than $1/\sqrt{2}$. Its CARTESIAN EQUATION is

$$x = \frac{(a^2 + f^2 \sin^2 t)\cos t}{a}$$

$$y = \frac{(a^2 - 2f^2 + f^2 \sin^2 t)\sin t}{b},$$

where f is a constant.

References
Lockwood, E. H. *A Book of Curves.* Cambridge, England: Cambridge University Press, p. 157, 1967.
MacTutor History of Mathematics Archive. "Talbot's Curve." http://www-groups.dcs.st-and.ac.uk/~history/Curves/Talbots.html.

Talisman Hexagon

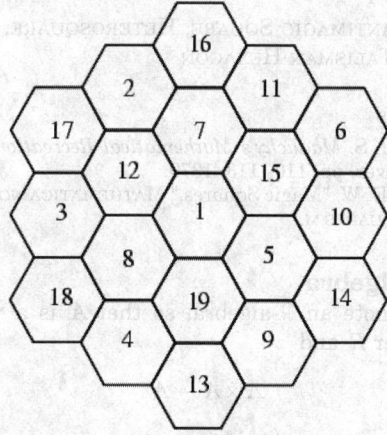

An (n, k)-talisman hexagon is an arrangement of nested hexagons containing the integers 1, 2, ..., $H_n = 3n(n-1) + 1$, where H_n is the nth HEX NUMBER, such that the difference between all adjacent hexagons is at least as large as a number k. The hexagon illustrated above is a $(3, 4)$-talisman hexagon.

See also HEX NUMBER, MAGIC SQUARE, TALISMAN SQUARE

References
Madachy, J. S. *Madachy's Mathematical Recreations.* New York: Dover, pp. 111–112, 1979.

Talisman Square

1	5	3	7
9	11	13	15
2	6	4	8
10	12	14	16

5	15	9	12
10	1	6	3
13	16	11	14
2	8	4	7

15	1	12	4	9
20	7	22	18	24
16	2	13	5	10
21	8	23	19	25
17	3	14	6	11

28	10	31	13	34	16
19	1	22	4	25	7
29	11	32	14	35	17
20	2	23	5	26	8
30	12	33	15	36	18
21	3	24	6	27	9

An $n \times n$ ARRAY of the integers from 1 to n^2 such that the difference between any one integer and its neighbor (horizontally, vertically, or diagonally, without wrapping around) is greater than or equal to some value k is called a (n, k)-talisman square. The above illustrations show (4, 2)-, (4, 3)-, (5, 4)-, and (6, 8)-talisman squares.

See also ANTIMAGIC SQUARE, HETEROSQUARE, MAGIC SQUARE, TALISMAN HEXAGON

References

Madachy, J. S. *Madachy's Mathematical Recreations.* New York: Dover, pp. 110–113, 1979.
Weisstein, E. W. "Magic Squares." MATHEMATICA NOTEBOOK MAGICSQUARES.M.

Tame Algebra

Let A denote an $\mathbb{R}$-algebra, so that A is a VECTOR SPACE over R and

$$A \times A \to A$$

$$(x, y) \mapsto x \cdot y,$$

where $x \cdot y$ is VECTOR MULTIPLICATION which is assumed to be BILINEAR. Now define

$$Z \equiv \{x \in a : x \cdot y - 0 \text{ for some nonzero } y \in A\},$$

where $0 \in Z$. A is said to be tame if Z is a finite union of SUBSPACES of A. A 2-D 0-ASSOCIATIVE algebra is tame, but a 4-D 4-ASSOCIATIVE algebra and a 3-D 1-ASSOCIATIVE algebra need not be tame. It is conjectured that a 3-D 2-ASSOCIATIVE algebra is tame, and proven that a 3-D 3-ASSOCIATIVE algebra is tame if it possesses a multiplicative IDENTITY ELEMENT.

References

Finch, S. "Zero Structures in Real Algebras." http://www.mathsoft.com/asolve/zerodiv/zerodiv.html.

Tame Knot

A KNOT equivalent to a POLYGONAL KNOT. Knots which are not tame are called WILD KNOTS.

References

Rolfsen, D. *Knots and Links.* Wilmington, DE: Publish or Perish Press, p. 49, 1976.

Tan

TANGENT

Tangency Theorem

The external (internal) SIMILARITY POINT of two fixed CIRCLES is the point at which all the CIRCLES homogeneously (nonhomogeneously) tangent to the fixed CIRCLES have the same POWER and at which all the tangency secants intersect.

References

Dörrie, H. *100 Great Problems of Elementary Mathematics: Their History and Solutions.* New York: Dover, p. 157, 1965.

Tangent

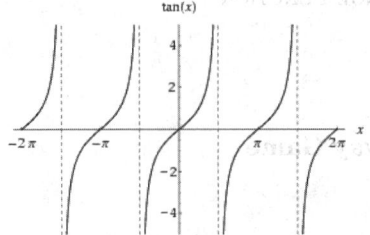

The tangent function is defined by

$$\tan x \equiv \frac{\sin x}{\cos x}, \tag{1}$$

where $\sin x$ is the SINE function and $\cos x$ is the COSINE function. The notation tg x is sometimes also used (Gradshteyn and Ryzhik 2000, p. xxix).

The word "tangent" also has an important related meaning as a LINE or PLANE which touches a given curve or solid at a single point. These geometrical objects are then called a TANGENT LINE or TANGENT PLANE, respectively.

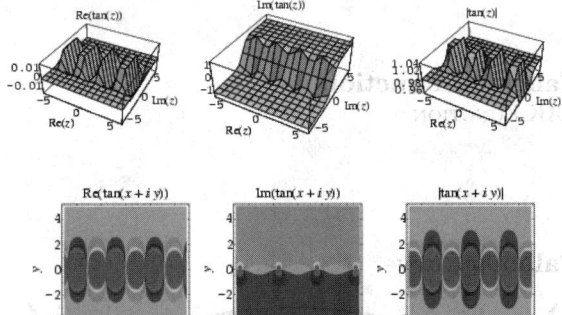

The definition of the tangent function can be extended to complex arguments z using the definition

$$\tan z = \frac{e^{iz} - e^{-iz}}{i(e^{iz} + e^{-iz})}, \tag{2}$$

where e is the base of the NATURAL LOGARITHM and i is the IMAGINARY NUMBER. A related function known as the HYPERBOLIC TANGENT is similarly defined,

$$\tanh z = \frac{e^z - e^{-z}}{e^z + e^{-z}}. \tag{3}$$

Important tangent identities include

$$\tan^2 \theta + 1 = \sec^2 \theta \tag{4}$$

$$\tan(\alpha + \beta) = \frac{\tan \alpha + \tan \beta}{1 - \tan \alpha \tan \beta} \tag{5}$$

$$\tan(\alpha - \beta) = \frac{\tan \alpha - \tan \beta}{1 + \tan \alpha \tan \beta} \tag{6}$$

$$\tan(2\alpha) = \frac{2\tan\alpha}{1 - \tan^2\alpha}. \tag{7}$$

$$\tan(n\alpha) = \frac{\tan[(n-1)\alpha] + \tan\alpha}{1 - \tan[(n-1)\alpha]\tan\alpha} \tag{8}$$

$$\tan\left(\frac{\alpha}{2}\right) = \frac{\sin\alpha}{1 + \cos\alpha} \tag{9}$$

$$= \frac{1 - \cos\alpha}{\sin\alpha} \tag{10}$$

$$= \frac{1 \pm \sqrt{1 + \tan^2\alpha}}{\tan\alpha} \tag{11}$$

$$= \frac{\tan\alpha\sin\alpha}{\tan\alpha + \sin\alpha} \tag{12}$$

in addition to the beautiful identity

$$\tan(\alpha + \beta + \gamma)$$

$$= \frac{\tan\alpha + \tan\beta + \tan\gamma - \tan\alpha\tan\beta\tan\gamma}{1 - \tan\beta\tan\gamma - \tan\gamma\tan\alpha - \tan\alpha\tan\beta}. \tag{13}$$

There are a number of simple but interesting tangent identities based on those given above, including

$$\tan(A + 60°)\tan(A - 60°) + \tan A\tan(A + 60°)$$

$$+\tan A\tan(A - 60°) = -3 \tag{14}$$

(Borchardt and Perrott 1930).

The Maclaurin series valid for $-\pi/2 < x < \pi/2$ for the tangent function is

$$\tan x = \sum_{n=0}^{\infty} \frac{(-1)^{n-1}2^{2n}(2^{2n}-1)B_{2n}}{(2n)!} x^{2n-1} + \dots$$

$$= x + \frac{1}{3}x^3 + \frac{2}{15}x^5 + \frac{17}{315}x^7 + \frac{62}{2835}x^9 + \dots, \tag{15}$$

where B_n is a Bernoulli number.

$\tan x$ is irrational for any rational $x \neq 0$, which can be proved by writing $\tan x$ as a continued fraction

$$\tan x = \cfrac{x}{1 - \cfrac{x^2}{3 - \cfrac{x^2}{5 - \cfrac{x^2}{7 - \dots}}}}. \tag{16}$$

Lambert derived another continued fraction expression for the tangent,

$$\tan x = \cfrac{1}{\cfrac{1}{x} - \cfrac{1}{\cfrac{3}{x} - \cfrac{1}{\cfrac{5}{x} - \cfrac{1}{\cfrac{7}{x} - \dots}}}}. \tag{17}$$

An interesting identity involving the product of tangents is

$$\prod_{k=1}^{\lfloor (n-1)/2\rfloor} \tan\left(\frac{k\pi}{n}\right) = \begin{cases} \sqrt{n} & \text{for } n \text{ odd} \\ 1 & \text{for } n \text{ even,} \end{cases} \tag{18}$$

where $\lfloor x\rfloor$ is the floor function. Another tangent identity is

$$\tan(n\tan^{-1}x) = \frac{1}{i}\frac{(1 + ix)^n - (1 - ix)^n}{(1 + ix)^n + (1 - ix)^m} \tag{19}$$

(Beeler *et al.* 1972).

The equation

$$x = \tan x \tag{20}$$

does not have simple closed-form solutions, but the first few approximate numerical solutions are 0, 4.49341, 7.72525, 10.9041, 14.0662, ... The difference between consecutive solutions gets closer and closer to π for higher order solutions.

See also Alternating Permutation, Cosine, Cotangent, Inverse Tangent, Morrie's Law, Sine, Tangent Line, Tangent Plane

References

Abramowitz, M. and Stegun, C. A. (Eds.). "Circular Functions." §4.3 in *Handbook of Mathematical Functions with Formulas, Graphs, and Mathematical Tables, 9th printing.* New York: Dover, pp. 71–79, 1972.

Beeler, M. *et al.* Item 16 in Beeler, M.; Gosper, R. W.; and Schroeppel, R. *HAKMEM.* Cambridge, MA: MIT Artificial Intelligence Laboratory, Memo AIM-239, p. 9, Feb. 1972.

Beyer, W. H. *CRC Standard Mathematical Tables, 28th ed.* Boca Raton, FL: CRC Press, p. 226, 1987.

Borchardt, W. G. and Perrott, A. D. Ex. 33 in *A New Trigonometry for Schools.* London: G. Bell, 1930.

Gradshteyn, I. S. and Ryzhik, I. M. *Tables of Integrals, Series, and Products, 6th ed.* San Diego, CA: Academic Press, 2000.

Spanier, J. and Oldham, K. B. "The Tangent tan(x) and Cotangent cot(x) Functions." Ch. 34 in *An Atlas of Functions.* Washington, DC: Hemisphere, pp. 319–330, 1987.

Tangent Bifurcation

Fold Bifurcation

Tangent Bundle

Every smooth manifold M has a tangent bundle TM, which consists of the tangent space TM_p at all points p in M. Since a tangent space TM_p is the set of all tangent vectors to M at p, the tangent bundle is the collection of all tangent vectors, along with the information of the point to which they are tangent.

$$TM = \{(p, v) : p \in M, v \in TM_p\}$$

The tangent bundle is a special case of a VECTOR BUNDLE. As a bundle it has RANK n, where n is the dimension of M. A COORDINATE CHART on M provides a TRIVIALIZATION for TM. In the coordinates, $(x_1, \ldots, x_n)$, the vector fields $(v_1, \ldots, v_n)$, where $v_i = \partial/\partial x_i$, span the tangent vectors at every point (in the COORDINATE CHART). The transition function from these coordinates to another set of coordinates is given by the JACOBIAN of the coordinate change.

For example, on the UNIT SPHERE, at the point $(1, 0, 0)$ there are two different coordinate charts defined on the same HEMISPHERE, $\phi : U_1 \to S^2$ and $\psi : U_2 \to S^2$,

$$\phi(x_1, x_2) = (\cos x_1 \cos x_2, \sin x_1 \cos x_2, \sin x_2) \quad (1)$$

$$\psi(y_1, y_2) = \left(\sqrt{1 - y_1^2 - y_2^2}, y_1, y_1 \right) \quad (2)$$

with $U_1 = (-\pi/2, \pi/2) \times (-\pi/2, \pi/2)$ and $U_2 = \{(y_1, y_2) : y_1^2 + y_2^2 < 1\}$. The map between the coordinate charts is $\alpha = \psi^{-1} \circ \phi$.

$$(y_1, y_2) = \alpha(x_1, x_2) = (\sin x_1, \cos x_2, \sin x_2) \quad (3)$$

The JACOBIAN of $\alpha : U_1 \to U_2$ is given by the matrix-valued function

$$\begin{bmatrix} \cos x_1 \cos x_2 & \sin x_1 \sin x_2 \\ 0 & \cos x_2 \end{bmatrix} \quad (4)$$

which has DETERMINANT $\cos x_1 \cos^2 x_2$ and so is invertible on U_1.

The tangent vectors transform by the Jacobian. At the point (x_1, x_2) in U_1, a tangent vector v corresponds to the tangent vector Jv at $\alpha(x_1, x_2)$ in U_2. These two are just different versions of the same element of the tangent bundle.

See also CALCULUS, COORDINATE CHART, COTANGENT BUNDLE, DIRECTIONAL DERIVATIVE, EUCLIDEAN SPACE, JACOBIAN, MANIFOLD, TANGENT BUNDLE, TANGENT SPACE, TANGENT VECTOR, VECTOR FIELD, VECTOR SPACE

Tangent Circles

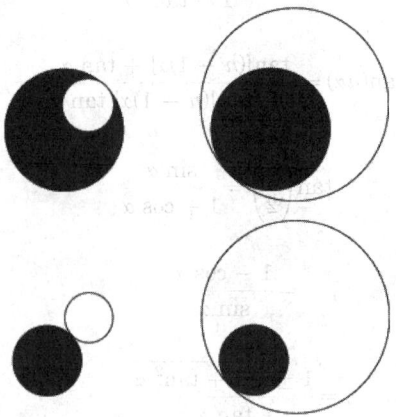

Two circles with centers at (x_i, y_i) with radii r_i for $i = 1, 2$ are mutually tangent if

$$(x_1 - x_2)^2 + (y_1 - y^2)^2 = (r_1 \pm r_2)^2.$$

If the center of the second circle is inside the first, then the $-$ and $+$ signs both correspond to internally tangent circles. If the center of the second circle is outside the first, then the $-$ sign corresponds to externally tangent circles and the $+$ sign to internally tangent circles.

Finding the circles tangent to three given circles is known as APOLLONIUS' PROBLEM.

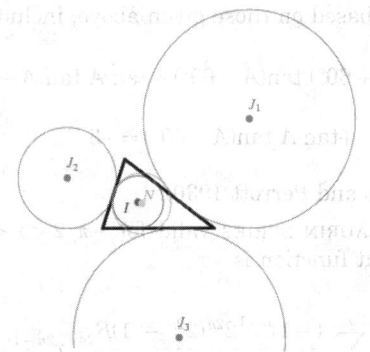

There are four CIRCLES that are tangent all three sides (or their extensions) of a given TRIANGLE: the INCIRCLE I and three EXCIRCLES J_1, J_2, and J_3. These four circles are, in turn, all touched by the NINE-POINT CIRCLE N.

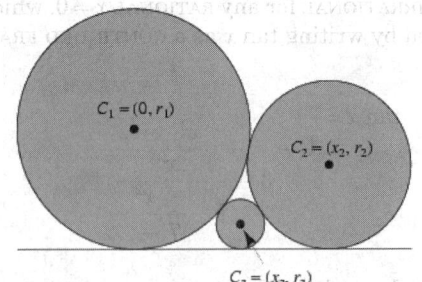

If two circles C_1 and C_2 of radii r_1 and r_2 are mutually tangent to each other and a line, then their centers are separated by a horizontal distance given by solving

$$x_2^2 + (r_1 - r_2)^2 = (r_1 + r_2)^2 \tag{1}$$

for x_2, giving

$$x_2 = 2\sqrt{r_1 r_2}. \tag{2}$$

The position and radius of a third circle tangent to the first two and the line can be found by solving the simultaneous equations

$$x_3^2 + (r_1 - r_3)^2 = (r_1 + r_3)^2 \tag{3}$$

$$(x_3 - x_2)^2 + (r_2 - r_3)^2 = (r_2 + r_3)^2 \tag{4}$$

for x_3 and r_3, giving

$$x_3 = \frac{2r_1\sqrt{r_2}}{\sqrt{r_1} + \sqrt{r_2}} \tag{5}$$

$$r_3 = \frac{r_1 r_2}{\left(\sqrt{r_1} + \sqrt{r_2}\right)^2}. \tag{6}$$

The latter equation can be written in the form

$$\frac{1}{\sqrt{r_3}} = \frac{1}{\sqrt{r_1}} + \frac{1}{\sqrt{r_2}}. \tag{7}$$

This problem was given as a Japanese temple problem on a tablet from 1824 in the Gumma Prefecture (Rothman 1998).

See also APOLLONIUS' PROBLEM, CASEY'S THEOREM, CHAIN OF CIRCLES, CIRCLE PACKING, CIRCLE TANGENTS, DESCARTES CIRCLE THEOREM, EXCIRCLE, FOUR COINS PROBLEM, INCIRCLE, MALFATTI'S TANGENT TRIANGLE PROBLEM, PAPPUS CHAIN, SODDY CIRCLES, TANGENT CURVES, TANGENT SPHERES

References

Coolidge, J. L. "Mutually Tangent Circles." §1.3 in *A Treatise on the Geometry of the Circle and Sphere.* New York: Chelsea, pp. 31–44, 1971.
Fukagawa, H. and Pedoe, D. "Two Circles," "Three Circles," "Four Circles," and "Many Circles." §1.1–1.5 in *Japanese Temple Geometry Problems.* Winnipeg, Manitoba, Canada: Charles Babbage Research Foundation, pp. 3–13 and 79–88, 1989.
Rothman, T. "Japanese Temple Geometry." *Sci. Amer.* **278**, 85–91, May 1998.

Tangent Curves

See also OSCULATING CURVES, TANGENT CIRCLES, TACNODE, TANGENT LINE

Tangent Developable

A RULED SURFACE M is a tangent developable of a curve $\mathbf{y}$ if M can be parameterized by $\mathbf{x}(u, v) = \mathbf{y}(u) + v\mathbf{y}'(u)$. A tangent developable is a FLAT SURFACE.

See also BINORMAL DEVELOPABLE, NORMAL DEVELOPABLE

References

Gray, A. "Tangent Developables." §19.3 in *Modern Differential Geometry of Curves and Surfaces with Mathematica, 2nd ed.* Boca Raton, FL: CRC Press, pp. 441–444, 1997.

Tangent Externally

Two curves are tangent externally at a point P if they lie on opposite sides of their common tangent at P

See also TANGENT INTERNALLY

Tangent Figures

See also INCIDENT

Tangent Hyperbolas Method

HALLEY'S METHOD

Tangent Indicatrix

Let the SPEED σ of a closed curve on the unit sphere $\mathbb{S}^2$ never vanish. Then the tangent indicatrix

$$\tau \equiv \frac{\dot{\sigma}}{|\dot{\sigma}|}$$

is another closed curve on $\mathbb{S}^2$. It is sometimes called the TANTRIX. If σ IMMERSES in $\mathbb{S}^2$, then so will τ.

References

Solomon, B. "Tantrices of Spherical Curves." *Amer. Math. Monthly* **103**, 30–39, 1996.

Tangent Internally

Two curves are tangent internally at a point P if they lie on the same side of their common tangent at P

See also TANGENT EXTERNALLY

Tangent Line

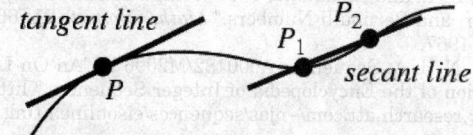

A straight line is tangent to a given curve $f(x)$ at a point x_0 on the curve if the line passes through the point $(x_0, f(x_0))$ on the curve and has slope $f'(x_0)$, where $f'(x)$ is the DERIVATIVE of $f(x)$.

See also CIRCLE TANGENTS, SECANT LINE, TANGENT, TANGENT PLANE, TANGENT SPACE, TANGENT VECTOR

References

Yates, R. C. "Instantaneous Center of Rotation and the Construction of Some Tangents." *A Handbook on Curves and Their Properties.* Ann Arbor, MI: J. W. Edwards, pp. 119–122, 1952.

Tangent Map

If $f : M \to N$, then the tangent map Tf associated to f is a VECTOR BUNDLE HOMEOMORPHISM $Tf : TM \to TN$ (i.e., a MAP between the TANGENT BUNDLES of M and N respectively). The tangent map corresponds to DIFFERENTIATION by the formula

$$Tf(v) = (f \circ \phi)'(0), \tag{1}$$

where $\phi'(0) = v$ (i.e., ϕ is a curve passing through the base point to v in TM at time 0 with velocity v). In this case, if $f : M \to N$ and $g : N \to O$, then the CHAIN RULE is expressed as

$$T(f \circ g) = Tf \circ Tg. \tag{2}$$

In other words, with this way of formalizing differentiation, the CHAIN RULE can be remembered by saying that "the process of taking the tangent map of a map is functorial." To a topologist, the form

$$(f \circ g)'(a) = f'(g(a)) \circ g'(a), \tag{3}$$

for all a, is more intuitive than the usual form of the CHAIN RULE.

See also DIFFEOMORPHISM

References

Gray, A. "Tangent Maps." §11.3 in *Modern Differential Geometry of Curves and Surfaces with Mathematica, 2nd ed.* Boca Raton, FL: CRC Press, pp. 250–255, 1997.

Tangent Number

A number also called a ZAG NUMBER giving the number of ODD ALTERNATING PERMUTATIONS. The first few are 1, 2, 16, 272, 7936, ... (Sloane's A000182).

See also ALTERNATING PERMUTATION, ENTRINGER NUMBER, EULER ZIGZAG NUMBER, SECANT NUMBER

References

Knuth, D. E. and Buckholtz, T. J. "Computation of Tangent, Euler, and Bernoulli Numbers." *Math. Comput.* **21**, 663–688, 1967.
Sloane, N. J. A. Sequences A000182/M2096 in "An On-Line Version of the Encyclopedia of Integer Sequences." http://www.research.att.com/~njas/sequences/eisonline.html.

Tangent Plane

Let (x_0, y_0) be any point of a surface function $z = f(x, y)$. Then the surface has a nonvertical tangent plane at (x_0, y_0) with equation

$$z = f(x_0, y_0) + f_x(x_0, y_0)(x - x_0) + f_y(x_0, y_0)(y - y_0).$$

See also NORMAL VECTOR, PLANE, TANGENT, TANGENT LINE, TANGENT SPACE, TANGENT VECTOR

Tangent Space

Let x be a point in an n-dimensional COMPACT MANIFOLD M, and attach at x a copy of $\mathbb{R}^n$ tangential to M. The resulting structure is called the TANGENT SPACE of M at x and is denoted $T_x M$. If γ is a smooth curve passing through x, then the derivative of γ at x is a VECTOR in $T_x M$.

See also TANGENT, TANGENT BUNDLE, TANGENT PLANE, TANGENT SPACE (CHART), TANGENT SPACE (SUBMANIFOLD), TANGENT VECTOR

Tangent Space (Chart)

From the point of view of COORDINATE CHARTS, the notion of tangent space is quite simple. The tangent space consists of all directions, or velocities, a particle can take. In an open set U in $\mathbb{R}^n$ there are no constraints, so the tangent space at a point p is another copy of $\mathbb{R}^n$. The set U could be a COORDINATE CHART for an n-dimensional MANIFOLD.

The tangent space at p, denoted TM_p, is the set of possible VELOCITY VECTORS of paths through p. Hence there is a CANONICAL BASIS: if $(x_1, \ldots, x_n)$ are the coordinates, then $v_1, \ldots, v_n$ are a basis for the tangent space, where v_i is the velocity vector of a particle with unit speed moving inward along the coordinate x_i. The collection of tangent vectors, called the TANGENT BUNDLE, is the PHASE SPACE of a single particle moving in the manifold M.

It seems as if the tangent space at p is the same as the tangent space at all other points in the chart U. However, while they do share the same dimension and are ISOMORPHIC, in a change of coordinates, they lose their canonical isomorphism.

●━━━━━━━━━━━━━━━━━━━━━━●

For example, let $U = (0, 1)$ and $V = (0, 3)$ be coordinate charts for the unit interval I. We can change coordinates with $\phi : U \to V$ defined by $\phi(x) = x + 2x^2$. This is a change of coordinates because the derivative does not vanish on U. But this change is not linear, and stretches out I more near 1 than it does near 0. The tangent vectors transform by the derivative. At $x = 1/4$, they are stretched by a factor of $d\phi/dx = 2$. While at $x = 3/4$, they are stretched out by a factor of $d\phi/dx = 4$.

In general, the tangent vectors transform according to the JACOBIAN. The tangent vector v at q can also be considered as the tangent vector $J_\phi v$ at $\phi(q)$ in another coordinate chart, where ϕ is the DIFFEOMORPHISM from one chart to the other. The linear transformation determined by the JACOBIAN of ϕ is invertible, since ϕ is a DIFFEOMORPHISM.

Not only does the JACOBIAN, and the CHAIN RULE, show that the tangent space is WELL DEFINED, independent of coordinate chart, but it also shows that tangent vectors "push forward." That is, given any smooth map $f : X \to Y$ between manifolds, it makes sense to map the tangent vectors of X to tangent vectors of Y. Writing $\tilde{f}$ as the function f between a coordinate chart in X and one in Y, then $f_*(v) = J_{\tilde{f}}(v)$ maps v from TX_p to $TY_{f(p)}$. Another notation for f_* is df, the DIFFERENTIAL of f. In the language of TENSORS, the tangent vector's pushing forward means that a vector field is a COVARIANT TENSOR.

See also CALCULUS, COORDINATE CHART, DIFFERENTIAL FORM, DIRECTIONAL DERIVATIVE, EUCLIDEAN SPACE, EXTERIOR ALGEBRA, JACOBIAN, MANIFOLD, SUBMANIFOLD, TANGENT BUNDLE, TANGENT SPACE, VECTOR FIELD, VELOCITY VECTOR

Tangent Space (Intrinsic)

The tangent space at a point p in an ABSTRACT MANIFOLD M can be described without the use of embeddings or COORDINATE CHARTS. The elements of the tangent space are called tangent vectors, and the collection of tangent spaces forms the TANGENT BUNDLE.

One description is to put an equivalence relation on smooth paths through the point p. More precisely, consider all smooth maps $f : I \to M$ where $I = (-1, 1)$ and $f(0) = p$. We say that two maps f and g are equivalent if they agree to first order. That is, in any coordinate chart around p, $f'(0) = g'(0)$. If they are similar in one chart then they are similar in any other chart, by the CHAIN RULE. The notion of agreeing to first order depends on coordinate charts, but this cannot be completely eliminated since that is how manifolds are defined.

Another way is to first define a VECTOR FIELD as a DERIVATION of the ring of smooth functions $f : M \to \mathbb{R}$. Then a tangent vector at a point p is an equivalence class of vector fields which agree at p. That is, $X \sim Y$ if $Xf(p) = Yf(p)$ for every smooth function f. Of course, the tangent space at p is the vector space of tangent vectors at p. The only drawback to this version is that a COORDINATE CHART is required to show that the tangent space is an n-dimensional vector space.

See also CHAIN RULE, COORDINATE CHART, DERIVATION ALGEBRA, DIFFERENTIAL FORM, DIRECTIONAL DERIVATIVE, EXTERIOR ALGEBRA, EUCLIDEAN SPACE, JACOBIAN, LIE GROUP, MANIFOLD, SHEAF, TANGENT BUNDLE, TANGENT SPACE, VECTOR FIELD, VELOCITY VECTOR

Tangent Space (Submanifold)

The TANGENT PLANE to a surface at a point p is the tangent space at p (after translating to the origin). The elements of the tangent space are called TANGENT VECTORS, and they are CLOSED under addition and scalar multiplication. In particular, the tangent space is a VECTOR SPACE.

Any SUBMANIFOLD of EUCLIDEAN SPACE, and more generally any SUBMANIFOLD of an ABSTRACT MANIFOLD, has a tangent space at each point. The collection of tangent spaces TM_p to M forms the TANGENT BUNDLE $TM = \cup_{p \in M} (p, TM_p)$. A VECTOR FIELD assigns to every point p a TANGENT VECTOR in the tangent space at p.

There are two ways of defining a submanifold, and each way gives rise to a different way of defining the tangent space. The first way uses a PARAMETERIZATION, and the second way uses a system of equations. Suppose that $f = (f_1, \ldots, f_n)$ is a local PARAMETERIZATION of a SUBMANIFOLD M in EUCLIDEAN SPACE $\mathbb{R}^n$. Say,

$$f : U \to \mathbb{R}^n, \tag{1}$$

where U is the open UNIT BALL in $\mathbb{R}^k$, and $f(U) \subset M$. At the point $p = f(0)$, the tangent space is the image of the JACOBIAN of f, as a linear transformation from $\mathbb{R}^k$ to $\mathbb{R}^n$. For example, consider the UNIT SPHERE

$$\mathbb{S}^2 = \{(y_1, y_2, y_3) : y_1^2 + y_2^2 + y_3^2 = 1\} \tag{2}$$

in $\mathbb{R}^3$. Then the function (with the domain $U = \{(x_1, x_2) : x_1^2 + x_2^2 < 1\}$)

$$f = \left(x_1, x_2, \sqrt{1 - x_1^2 - x_2^2}\right) \tag{3}$$

parameterizes a NEIGHBORHOOD of the north pole. Its Jacobian at $(0, 0)$ is given by the matrix

$$\begin{bmatrix} 1 & 0 \\ 0 & 1 \\ 0 & 0 \end{bmatrix} \tag{4}$$

whose IMAGE is the tangent space at p,

$$TS^2 \big|_{(0, 0, 1)} = \{(a, b, 0)\}. \tag{5}$$

An alternative description of a SUBMANIFOLD M as the set of solutions to a system of equations leads to another description of tangent vectors. Consider a SUBMANIFOLD M which is the set of solutions to the system of equations

$$f_1(x_1, \ldots, x_n) = 0$$
$$\vdots \tag{6}$$
$$f_r(x_1, \ldots, x_n) = 0,$$

where $k + r = n$ and the JACOBIAN of $f : \mathbb{R}^n \to \mathbb{R}^r$, with $f = (f_1, \ldots f_n)$, has rank r at the solutions M to $f = 0$. A tangent vector v at a solution p is an infinitesimal solution to the above equations (at p). The tangent vector $v = (v_1, \ldots, v_n)$ is a solution of the derivative (linearization) of f, i.e., it is in the NULLSPACE of the JACOBIAN.

Consider this method in the recomputation the tangent space of the sphere at the North Pole. The sphere is two-dimensional and is described as the solution to single equation $(3-2=1)$ $x_1^2 + x_2^2 + x_3^2 = 1$. Set $f_1 = x_1^2 + x_2^2 + x_3^2 - 1$. We want to compute the tangent space at the solution $f_1(0, 0, 1) = 0$ (at the north pole). The JACOBIAN at this point is the 1×3 matrix $[0, 0, 2]$, and its nullspace is the tangent space

$$TS^2\big|_{(0, 0, 1)} = \{(a, b, 0)\}. \tag{7}$$

It appears that the tangent space depends either on the choice of parametrization, or on the choice of system of equations. Because the Jacobian of a composition of functions obeys the CHAIN RULE, the tangent space is WELL DEFINED. Note that the JACOBIAN of a DIFFEOMORPHISM is an INVERTIBLE LINEAR MAP, and these correspond to the ways the equations can be changed. The basic facts from LINEAR ALGEBRA used to show that the tangent space is WELL DEFINED are the following.

1. If $A : \mathbb{R}^k \to \mathbb{R}^k$ is invertible, then the image of $B : \mathbb{R}^k \to \mathbb{R}^n$ is the same as the image of AB.
2. If $A : \mathbb{R}^n \to \mathbb{R}^n$ is invertible, then the nullspace of $B : \mathbb{R}^n \to \mathbb{R}^r$ is the same as the nullspace of BA. More precisely, $\text{Null}(BA) = A^{-1}(\text{Null}(B))$.

These techniques work in any dimension. In addition, they generalize to submanifolds of an ABSTRACT MANIFOLD, because tangent vectors depend on local properties. In particular, the tangent space can be computed in any coordinate chart, because any change in COORDINATE CHART corresponds to a DIFFEOMORPHISM in Euclidean space.

The tangent space can give some geometric insight to higher-dimensional phenomena. For example, to compute the tangent space to the FLAT TORUS (donut) M in $\mathbb{R}^4$, note that it can be parametrized, by

$$f(x_1, x_2) = (\sin x_1, \cos x_1, \sin x_2, \cos x_2) \tag{8}$$

with domain $U = \{(x_1, x_2) : x_1^2 + x_2^2 < 1\}$, near the point $p = f(0, 0) = (0, 1, 0, 1)$. Its JACOBIAN at p is the matrix

$$\begin{bmatrix} 1 & 0 \\ 0 & 0 \\ 0 & 1 \\ 0 & 0 \end{bmatrix}, \tag{9}$$

whose image is the tangent space $TM\big|_p = \{(a, 0, b, 0)\}$.

Alternatively, M is the set of solutions to equations

$$f_1(x_1, x_2, x_3, x_4) = x_1^2 + x_2^2 - 1 = 0 \tag{10}$$

$$f_2(x_1, x_2, x_3, x_4) = x_3^2 + x_4^2 - 1 = 0. \tag{11}$$

The Jacobian at the solution $p = (0, 1, 0, 1)$ is the matrix

$$\begin{bmatrix} 0 & 2 & 0 & 0 \\ 0 & 0 & 0 & 2 \end{bmatrix}, \tag{12}$$

whose NULLSPACE is the tangent space $TM\big|_p = \{(a, 0, b, 0)\}$.

See also CALCULUS, COORDINATE CHART, DIFFERENTIAL FORM, DIRECTIONAL DERIVATIVE, EUCLIDEAN SPACE, EXTERIOR ALGEBRA, JACOBIAN, LINEAR ALGEBRA, MANIFOLD, NULLSPACE, TANGENT BUNDLE, TANGENT PLANE, TANGENT SPACE (CHART), TANGENT SPACE (INTRINSIC), TANGENT VECTOR, VECTOR FIELD, VECTOR SPACE, VELOCITY VECTOR

Tangent Spheres

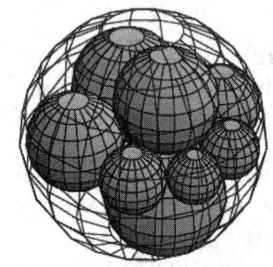 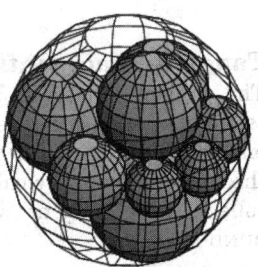

A special case of tangent spheres is given by Soddy's hexlet, which consists of a chain of six spheres externally tangent to two mutually tangent spheres and internally tangent to a circumsphere. The bends of the circles in the chain obey the relationship

$$\frac{1}{r_1} + \frac{1}{r_4} = \frac{1}{r_2} + \frac{1}{r_3} = \frac{1}{r_3} + \frac{1}{r_6}. \tag{1}$$

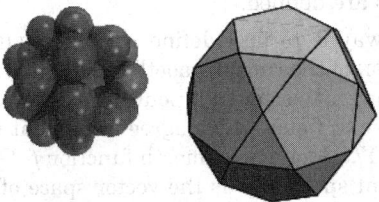

A SANGAKU PROBLEM from 1798 asks to distribute 30 identical spheres of radius r such that they are tangent to a single central sphere of radius R and to four other small spheres. This can be accomplished (left figure) by placing the spheres at the vertices of an ICOSIDODECAHEDRON (right figure) of side length a, where the radii r and R are given by

$$r = \tfrac{1}{2}a \tag{2}$$

$$R = \tfrac{1}{2}\sqrt{5}a \tag{3}$$

(Rothman 1998).

In general, the BENDS of five mutually tangent spheres are related by

$$3\left(\kappa_1^2 + \kappa_2^2 + \kappa_3^2 + \kappa_4^2 + \kappa_5^2\right)$$

$$= \left(\kappa_1 + \kappa_2 + \kappa_3 + \kappa_4 + \kappa_5\right)^2. \tag{4}$$

Solving for κ_5 gives

$$\kappa_5^{\pm} = \tfrac{1}{2}\{\kappa_1 + \kappa_2 + \kappa_3 + \kappa_4$$

$$\pm [6(\kappa_1\kappa_2 + \kappa_1\kappa_3 + \kappa_1\kappa_4 + \kappa_2\kappa_3 + \kappa_2\kappa_4 + \kappa_3\kappa_4)$$

$$- 3\left(\kappa_1^2 + \kappa_2^2 + \kappa_3^2 + \kappa_4^2\right)]^{1/2}\}. \tag{5}$$

(Soddy 1937a). Gosset (1937) pointed out that the expression under the square root sign is given by

$$\{6(\kappa_1\kappa_2 + \kappa_1\kappa_3 + \kappa_1\kappa_4 + \kappa_2\kappa_3 + \kappa_2\kappa_4 + \kappa_3\kappa_4)$$

$$- 3\left(\kappa_1^2 + \kappa_2^2 + \kappa_3^2 + \kappa_4^2\right)\}^{1/2} = 3\sqrt{3}V\kappa_1\kappa_2\kappa_3\kappa_4, \tag{6}$$

where V is the VOLUME of the TETRAHEDRON having vertices at the centers of the corresponding four spheres. Therefore, the equation for κ_5 can be written simplify as

$$\kappa_5 = \tfrac{1}{2}\sigma_2 + \sqrt{3}\epsilon, \tag{7}$$

where

$$\sigma = \kappa_1 + \kappa_2 + \kappa_3 + \kappa_4 \tag{8}$$

$$\epsilon = \tfrac{3}{2}V\kappa_1\kappa_2\kappa_3\kappa_4. \tag{9}$$

(Soddy 1937b).

In addition, the tetrahedra formed by joining the four points of contact of any one sphere with the other four (when all five are in mutual contact) have opposite edges whose product is the constant

$$4\sqrt{(\kappa_1 + \kappa_5)(\kappa_2 + \kappa_5)(\kappa_3 + \kappa_5)(\kappa_4 + \kappa_5)} \tag{10}$$

and the volume of these tetrahedra is

$$V = \frac{2}{\sqrt{3}} \frac{\kappa_5}{(\kappa_1 + \kappa_5)(\kappa_2 + \kappa_5)(\kappa_3 + \kappa_5)(\kappa_4 + \kappa_5)} \tag{11}$$

(Soddy 1937b). Gosper has further extended this result to $n + 2$ mutually tangent n-D HYPERSPHERES, whose CURVATURES satisfy

$$\left(\sum_{i=0}^{n+1} \kappa_i\right)^2 - n\sum_{i=0}^{n+1} \kappa_i^2 = 0. \tag{12}$$

Solving for κ_{n+1} gives

$$\kappa_{n+1} = \frac{\sqrt{n}\sqrt{\left(\sum_{i=0}^{n} \kappa_i\right)^2 - (n-1)\sum_{i=0}^{n} \kappa_i^2} + \sum_{i=0}^{n} \kappa_i}{n - 1}. \tag{13}$$

For (at least) $n = 2$ and 3, the RADICAL equals

$$f(n)V\kappa_0\kappa_1\cdots\kappa_n, \tag{14}$$

where V is the CONTENT of the SIMPLEX whose vertices are the centers of the $n + 1$ independent

HYPERSPHERES. The RADICAND can also become NEGATIVE, yielding an IMAGINARY κ_{n+1}. For $n = 3$, this corresponds to a sphere touching three large bowling balls and a small BB, all mutually tangent, which is an impossibility.

See also BOWL OF INTEGERS, HEXLET, SODDY CIRCLES, SPHERE, TANGENT CIRCLES, TETRAHEDRON

References

Gosset, T. "The Hexlet." *Nature* **139**, 251–252, 1937.
Rothman, T. "Japanese Temple Geometry." *Sci. Amer.* **278**, 85–91, May 1998.
Soddy, F. "The Kiss Precise." *Nature* **137**, 1021, 1936.
Soddy, F. "The Bowl of Integers and the Hexlet." *Nature* **139**, 77–79, 1937a.
Soddy, F. *Nature* **139**, 252, 1937b.

Tangent Vector

For a curve with POSITION VECTOR $\mathbf{r}(t)$, the unit tangent vector $\hat{\mathbf{T}}(t)$ is defined by

$$\hat{\mathbf{T}}(t) \equiv \frac{\mathbf{r}'(t)}{|\mathbf{r}'(t)|} = \frac{\dfrac{d\mathbf{r}}{dt}}{\left|\dfrac{d\mathbf{r}}{dt}\right|} \tag{1}$$

$$= \frac{\dfrac{d\mathbf{r}}{dt}}{\dfrac{ds}{dt}} \tag{2}$$

$$= \frac{d\mathbf{r}}{ds}, \tag{3}$$

where t is a parameterization variable and s is the ARC LENGTH. For a function given parametrically by $(f(t), g(t))$, the tangent vector relative to the point $(f(t), g(t))$ is therefore given by

$$x(t) = \frac{f'}{\sqrt{f'^2 + g'^2}} \tag{4}$$

$$y(t) = \frac{g'}{\sqrt{f'^2 + g'^2}}. \tag{5}$$

To actually place the vector tangent to the curve, it must be displaced by $(f(t), g(t))$. It is also true that

$$\frac{d\hat{\mathbf{T}}}{ds} = \kappa\hat{\mathbf{N}} \tag{6}$$

$$\frac{d\hat{\mathbf{T}}}{dt} = \kappa\frac{ds}{dt}\hat{\mathbf{N}} \tag{7}$$

$$[\dot{\mathbf{T}}, \ddot{\mathbf{T}}, \dddot{\mathbf{T}}] = \kappa^5 \frac{d}{ds}\left(\frac{\tau}{\kappa}\right), \tag{8}$$

where $\mathbf{N}$ is the NORMAL VECTOR, κ is the CURVATURE, and τ is the TORSION.

See also CURVATURE, NORMAL VECTOR, TANGENT, TANGENT BUNDLE, TANGENT PLANE, TANGENT SPACE, TANGENT VECTOR (MANIFOLD), TORSION (DIFFERENTIAL GEOMETRY)

References

Gray, A. "Tangent and Normal Lines to Plane Curves." §5.5 in *Modern Differential Geometry of Curves and Surfaces with Mathematica, 2nd ed.* Boca Raton, FL: CRC Press, pp. 108–111, 1997.

Tangent Vector (Manifold)

Roughly speaking, a tangent vector is an infinitesimal displacement at a specific point on a MANIFOLD. The set of tangent vectors at a point P forms a VECTOR SPACE called the TANGENT SPACE at P, and the collection of tangent spaces on a manifold forms a VECTOR BUNDLE called the TANGENT BUNDLE.

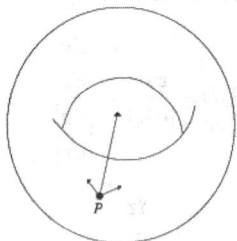

A tangent vector at a point P on a manifold is a tangent vector at P in a COORDINATE CHART. A change in coordinates near P causes an INVERTIBLE LINEAR MAP of the tangent vector's representations in the coordinates. This transformation is given by the JACOBIAN, which must be nonsingular in a change of coordinates. Hence the tangent vectors at P are WELL DEFINED. A VECTOR FIELD is an assignment of a tangent vector for each point. The collection of tangent vectors forms the TANGENT BUNDLE, and a vector field is a SECTION of this bundle.

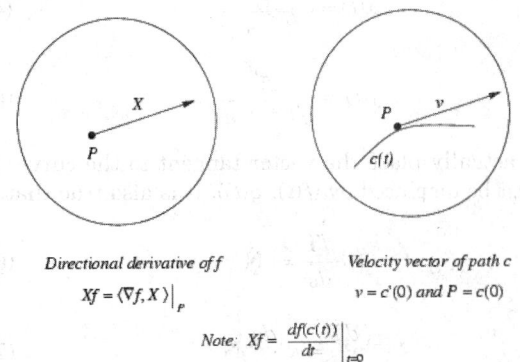

Directional derivative of f

$$Xf = \langle \nabla f, X \rangle \big|_P$$

Velocity vector of path c

$$v = c'(0) \text{ and } P = c(0)$$

Note: $Xf = \dfrac{df(c(t))}{dt}\bigg|_{t=0}$

Tangent vectors are used to do CALCULUS on MANIFOLDS. Since manifolds are locally Euclidean, the usual notions of differentiation and integration make sense in any COORDINATE CHART, and they can be carried over to manifolds. More specifically, a tangent vector is the manifold version of a DIRECTIONAL DERIVATIVE (at a point). An alternative analogy with calculus is the related notion of a VELOCITY VECTOR.

There are at least three different points of view on tangent vectors. Each has its own pluses and minuses. The extrinsic points of view use the vector space structure of EUCLIDEAN SPACE. Thinking of a manifold as a SUBMANIFOLD of Euclidean space, a tangent vector can be thought of as an element in a TANGENT PLANE, or (submanifold) TANGENT SPACE. In a COORDINATE CHART, a tangent vector is a vector in a (chart) TANGENT SPACE, which is just a copy of EUCLIDEAN SPACE.

The problem with the extrinsic points of view is that they depend on a choice of EMBEDDING or COORDINATE CHART. There are a couple of ways to think about a tangent vector intrinsically, as an element of an abstract (intrinsic) TANGENT SPACE. These are more satisfying from an abstract point of view, but sometimes it is necessary to do calculations in coordinate charts.

It is important to distinguish tangent vectors at P from tangent vectors at any other point Q, although they may seem parallel. On a LIE GROUP, there is a notion of parallelism, and there exist nonvanishing vector fields. In general, this is far from being true. On the sphere $\mathbb{S}^2$, for instance, any smooth vector field must vanish somewhere.

A more intrinsic geometric definition of a tangent vector is to take a tangent vector at P to be an EQUIVALENCE CLASS of paths through P which agree to first order. An extrinsic geometric definition, for a submanifold, is to view the tangent vectors as a subspace of the tangent vectors of the ambient space,

Algebraically, a vector field on a manifold is a DERIVATION on the RING of smooth functions. That is, a vector field acts on smooth functions and satisfies the PRODUCT RULE. A vector field X acts on a function by the DIRECTIONAL DERIVATIVE on the function,

$$d_X(f) = X \cdot \nabla f. \tag{1}$$

It is more precise to say that the tangent bundle is the SHEAF of derivations on the sheaf of smooth functions, in which case the tangent vectors at P are in the STALK of the sheaf at P.

In fact, in coordinates $(x_1, \ldots, x_n)$, the notation for the standard basis of tangent vectors at 0 is

$$\frac{\partial}{\partial x_i}, \tag{2}$$

where the derivation $\partial/\partial x_i$ of f is the usual PARTIAL DERIVATIVE

$$\frac{\partial f}{\partial x_i}. \tag{3}$$

Letting the base point vary in the coordinate chart,

$\partial/\partial x_i$ are vector fields, but are only defined in this COORDINATE CHART.

See also CALCULUS, COORDINATE CHART, DERIVATION ALGEBRA, DIFFERENTIAL FORM, DIRECTIONAL DERIVATIVE, EUCLIDEAN SPACE, EXTERIOR ALGEBRA, LIE GROUP MANIFOLD, SHEAF (TOPOLOGY), STALK, TANGENT BUNDLE, TANGENT VECTOR, TANGENT SPACE, TANGENT SPACE (SUBMANIFOLD), VECTOR FIELD, VELOCITY VECTOR

Tangential Angle

For a PLANE CURVE, the tangential angle ϕ is defined by

$$\rho \, d\phi = ds, \tag{1}$$

where s is the ARC LENGTH and ρ is the RADIUS OF CURVATURE. The tangential angle is therefore given by

$$\phi = \int_0^t s'(t)\kappa(t)\, dt, \tag{2}$$

where $\kappa(t)$ is the CURVATURE. For a plane curve $\mathbf{r}(t)$, the tangential angle $\phi(t)$ can also be defined by

$$\frac{\mathbf{r}'(t)}{|\mathbf{r}'(t)|} = \begin{bmatrix} \cos[\phi(t)] \\ \sin[\phi(t)] \end{bmatrix}. \tag{3}$$

Gray (1997) calls ϕ the TURNING ANGLE instead of the tangential angle.

See also ARC LENGTH, CURVATURE, PLANE CURVE, RADIUS OF CURVATURE, TORSION (DIFFERENTIAL GEOMETRY)

References

Gray, A. "The Turning Angle." §1.7 in *Modern Differential Geometry of Curves and Surfaces with Mathematica, 2nd ed.* Boca Raton, FL: CRC Press, pp. 19–20, 1997.

Tangential Polygon

The polygon formed by the lines tangent to the CIRCUMCIRCLE of a polygon. The tangential polygon of an n-gon is itself an n-gon.

See also DUAL POLYHEDRON, TANGENTIAL QUADRILATERAL, TANGENTIAL TRIANGLE

Tangential Quadrilateral

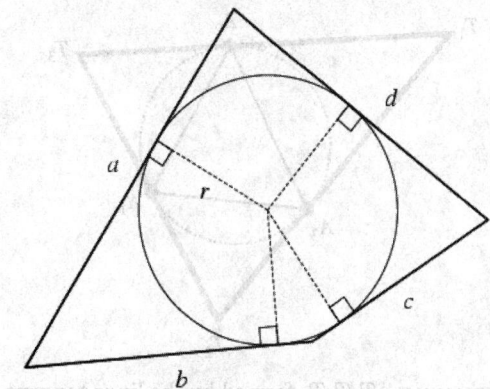

A QUADRILATERAL which has an INCIRCLE, i.e., one for which a single circle can be constructed which is tangent to all four sides. Opposite sides of such a quadrilateral satisfy

$$s = a + c = b + d, \tag{1}$$

where

$$s = \tfrac{1}{2}(a + b + c + d) \tag{2}$$

is the SEMIPERIMETER, and the AREA is

$$A = rs, \tag{3}$$

where r is the INRADIUS.

See also BICENTRIC QUADRILATERAL, CYCLIC QUADRILATERAL, INCIRCLE, QUADRILATERAL, TANGENTIAL TRIANGLE

References

Harris, J. W. and Stocker, H. "Quadrilateral of Tangents." §3.6.8 in *Handbook of Mathematics and Computational Science.* New York: Springer-Verlag, p. 86, 1998.

Tangential Tetrahedron

The planes passing through the vertices of a TETRAHEDRON *ABCD* and tangent to the CIRCUMSPHERE at these points form another tetrahedron called the tangential tetrahedron.

The four lines of intersection of the faces of a tetrahedron with the corresponding faces of its tangential tetrahedron form a hyperbolic group (Altshiller-Court 1979, p. 102).

See also TETRAHEDRON

References

Altshiller-Court, N. *Modern Pure Solid Geometry.* New York: Chelsea, p. 102, 1979.

Tangential Triangle

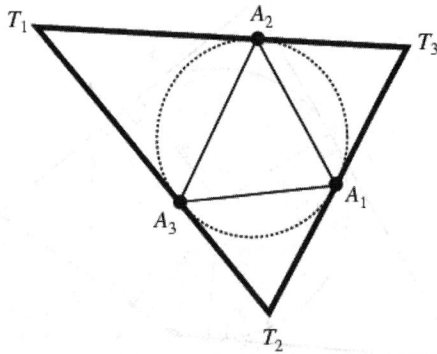

The TRIANGLE $\Delta T_1 T_2 T_3$ formed by the lines tangent to the CIRCUMCIRCLE of a given TRIANGLE $\Delta A_1 A_2 A_3$ at its VERTICES. It is the PEDAL TRIANGLE of $\Delta A_1 A_2 A_3$ with the CIRCUMCENTER as the PEDAL POINT. The TRILINEAR COORDINATES of the VERTICES of the tangential triangle are

$$T_1 = -a : b : c$$

$$T_2 = a : -b : c$$

$$T_3 = a : b : -c.$$

The CONTACT TRIANGLE and tangential triangle are perspective from the GERGONNE POINT.

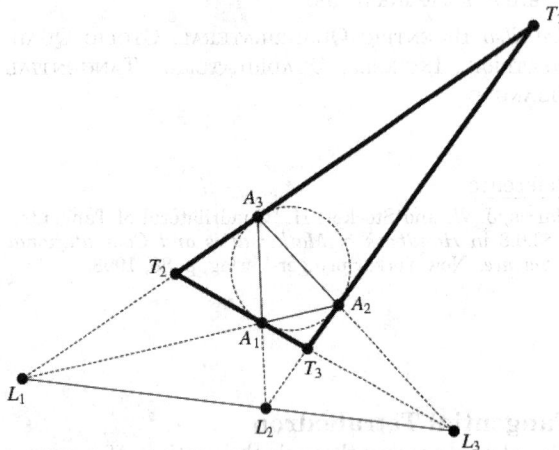

Given a TRIANGLE $\Delta A_1 A_2 A_3$ and its tangential triangle $\Delta T_1 T_2 T_3$, the extensions of the sides of the two triangles intersect in three points L_1, L_2, and L_3, which are collinear (Honsberger 1995).
The CIRCUMCENTER of the tangential triangle has TRIANGLE CENTER FUNCTION

$$\alpha = a \left[b^2 \cos(2B) + c^2 \cos(2C) - a^2 \cos(2,\ 4) \right]$$

and lies on the EULER LINE (Kimberling 1994)

See also CIRCUMCIRCLE, CONTACT TRIANGLE, GERGONNE POINT, PEDAL TRIANGLE, PERSPECTIVE, TANGENTIAL QUADRILATERAL

References

Honsberger, R. *Episodes in Nineteenth and Twentieth Century Euclidean Geometry.* Washington, DC: Math. Assoc. Amer., pp. 151–153, 1995.

Kimberling, C. "Central Points and Central Lines in the Plane of a Triangle." *Math. Mag.* **67**, 163–187, 1994.

Tangents Law

LAW OF TANGENTS

Tangent-Sphere Coordinates

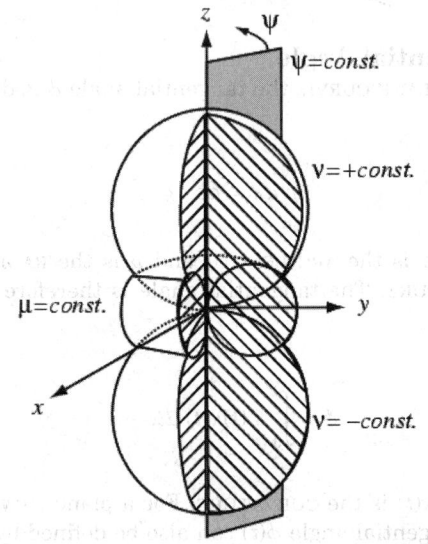

A coordinate system (μ, v, ψ) given by the coordinate transformation

$$x = \frac{\mu \cos \psi}{\mu^2 + v^2} \tag{1}$$

$$y = \frac{\mu \sin \psi}{\mu^2 + v^2} \tag{2}$$

$$z = \frac{v}{\mu^2 + v^2} \tag{3}$$

and defined for $\mu > 0$, $v \in (-\infty,\ \infty)$, and $\psi \in [0, 2\pi)$. Surfaces of constant μ are given by the TOROIDS

$$x^2 + y^2 + z^2 = \frac{1}{\mu} \sqrt{x^2 + y^2}, \tag{4}$$

surface of constant v by the spheres tangent to the xy-plane

$$x^2 + y^2 + \left(z - \frac{1}{2v} \right)^2 = \frac{1}{4v^2} \tag{5}$$

and surfaces of constant ψ by the half-planes

$$\tan \psi = \frac{y}{x}. \tag{6}$$

The metric coefficients are

$$g_{xx} = \frac{1}{(\mu^2 + v^2)^2} \tag{7}$$

$$g_{yy} = \frac{1}{(\mu^2 + v^2)^2} \tag{8}$$

$$g_{zz} = \frac{\mu^2}{(\mu^2 + v^2)^2}. \tag{9}$$

References

Moon, P. and Spencer, D. E. "Tangent-Sphere Coordinate (μ, v, ψ)." Fig. 4.01 in *Field Theory Handbook, Including Coordinate Systems, Differential Equations, and Their Solutions, 2nd ed.* New York: Springer-Verlag, pp. 104–106, 1988.

Tangle

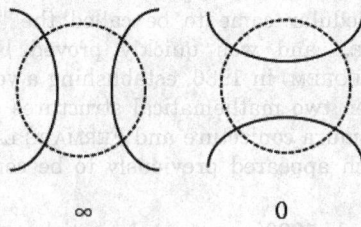

$$\infty \qquad\qquad 0$$

A region in a KNOT or LINK projection plane surrounded by a CIRCLE such that the KNOT or LINK crosses the circle exactly four times. Two tangles are equivalent if a sequence of REIDEMEISTER MOVES can be used to transform one into the other while keeping the four string endpoints fixed and not allowing strings to pass outside the CIRCLE.

The simplest tangles are the ∞-tangle and 0-tangle, shown above. A tangle with n left-handed twists is called an n-tangle, and one with n right-handed twists is called a $-n$-tangle. By placing tangles side by side, more complicated tangles can be built up such as $(-2, 3, 2)$, etc. The link created by connecting the ends of the tangles is now described by the sequence of tangle symbols, known as CONWAY'S KNOT NOTATION. If tangles are multiplied by 0 and then added, the resulting tangle symbols are separated by commas. Additional symbols which are used are the period, colon, and asterisk.

Amazingly enough, two tangles described in this NOTATION are equivalent IFF the CONTINUED FRACTIONS OF THE FORM

$$2 + \cfrac{1}{3 + \cfrac{1}{-2}}$$

are equal (Burde and Zieschang 1985)! an ALGEBRAIC TANGLE is any tangle obtained by ADDITIONS and MULTIPLICATIONS of rational tangles (Adams 1994). Not all tangles are ALGEBRAIC.

See also ALGEBRAIC LINK, FLYPE, PRETZEL KNOT

References

Adams, C. C. *The Knot Book: An Elementary Introduction to the Mathematical Theory of Knots.* New York: W. H. Freeman pp. 41–51, 1994.
Burde, G. and Zieschang, H. *Knots.* Berlin: de Gruyter, 1985.
Murasugi, K. and Kurpita, B. I. *A Study of Braids.* Dordrecht, Netherlands: Kluwer, 1999.

Tanglecube

A QUARTIC SURFACE given by the implicit equation

$$x^4 - 5x^2 + y^4 - 5y^2 + z^4 - 5z^2 + 11.8 = 0.$$

References

Banchoff, T. "The Best Homework Ever?" http://www.brown.edu/Administration/Brown_Alumni_Magazine/97/12–96/features/homework.html.
Nordstrand, T. "Tangle." http://www.uib.no/people/nfytn/tangltxt.htm.

Tangled Hierarchy

A system in which a STRANGE LOOP appears.

See also STRANGE LOOP

References

Hofstadter, D. R. *Gödel, Escher, Bach: An Eternal Golden Braid.* New York: Vintage Books, p. 10, 1989.

Tangram

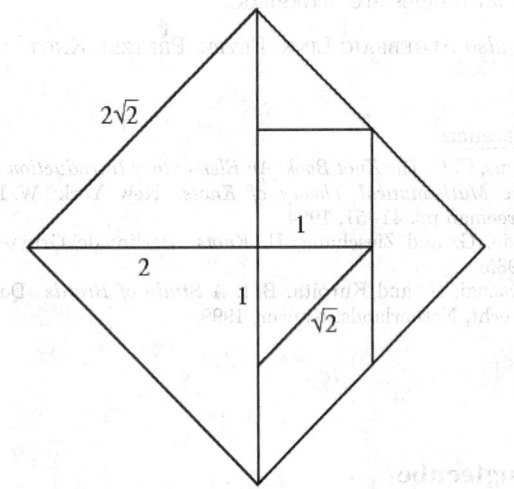

A combination of the above plane polygonal pieces such that the EDGES are coincident. There are 13 convex tangrams (where a "convex tangram" is a set of tangram pieces arranged into a CONVEX POLYGON).

See also ORIGAMI, STOMACHION

References

Cundy, H. and Rollett, A. *Mathematical Models, 3rd ed.* Stradbroke, England: Tarquin Pub., pp. 19–20, 1989.
Gardner, M. "Tangrams, Part 1" and "Tangrams, Part 2." Chs. 3–4 in *Time Travel and Other Mathematical Bewilderments.* New York: W. H. Freeman, pp. 27–54, 1988.
Johnston, S. *Fun with Tangrams Kit: 120 Puzzles with Two Complete Sets of Tangram Pieces.* New York: Dover, 1977.
Johnston, S. *Tangrams ABC Kit.* New York: Dover.
Pappas, T. "Tangram Puzzle." *The Joy of Mathematics.* San Carlos, CA: Wide World Publ./Tetra, p. 212, 1989.
Read, R. C. *Tangrams: 330 Puzzles.* New York: Dover.

Tanh

HYPERBOLIC TANGENT

Taniyama Conjecture

TANIYAMA-SHIMURA CONJECTURE

Taniyama-Shimura Conjecture

A very general and important conjecture (and now theorem) connecting TOPOLOGY and NUMBER THEORY which arose from several problems proposed by Taniyama in a 1955 international mathematics symposium.

Let E be an ELLIPTIC CURVE whose equation has INTEGER COEFFICIENTS, let N be the so-called CONDUCTOR of E and, for each n, let a_n be the number appearing in the L-function of E. Then, in technical terms, the Taniyama-Shimura conjecture states that there exists a MODULAR FORM of weight two and level N which is an EIGENFORM under the HECKE OPERATORS and has a FOURIER SERIES $\sum a_n q^n$.

In effect, the conjecture says that every rational ELLIPTIC CURVE is a MODULAR FORM in disguise. Or, more formally, the conjecture suggests that, for every ELLIPTIC CURVE $y^2 = Ax^3 + Bx^2 + Cx + D$ over the RATIONALS, there exist nonconstant MODULAR FUNCTIONS $f(z)$ and $g(z)$ of the same level N such that

$$[f(z)]^2 = A[g(z)]^2 + Cg(z) + D.$$

Equivalently, for every ELLIPTIC CURVE, there is a MODULAR FORM with the same DIRICHLET L-SERIES.

In 1985, starting with a fictitious solution to FERMAT'S LAST THEOREM (the FREY CURVE), G. Frey showed that he could create an unusual ELLIPTIC CURVE which appeared not to be modular. If the curve were not modular, then this would show that if FERMAT'S LAST THEOREM were false, then the Taniyama-Shimura conjecture would also be false. Furthermore, if the Taniyama-Shimura conjecture were true, then so would be FERMAT'S LAST THEOREM!

However, Frey did not actually prove that his curve *was not* modular. The conjecture that Frey's curve *was not* modular came to be called the "EPSILON CONJECTURE," and was quickly proved by Ribet (RIBET'S THEOREM) in 1986, establishing a very close link between two mathematical structures (the Taniyama-Shimura conjecture and FERMAT'S LAST THEOREM) which appeared previously to be completely unrelated.

As of the early 1990s, most mathematicians believed that the Taniyama-Shimura conjecture was not accessible to proof. However, A. Wiles was not one of these. He attempted to establish the correspondence between the set of ELLIPTIC CURVES and the set of modular elliptic curves by showing that the number of each was the same. Wiles accomplished this by "counting" Galois representations and comparing them with the number of MODULAR FORMS. In 1993, after a monumental seven-year effort, Wiles (almost) proved the Taniyama-Shimura conjecture for special classes of curves called SEMISTABLE ELLIPTIC CURVES (which correspond to elliptic curves with SQUAREFREE CONDUCTORS; Knapp 1999).

Wiles had tried to use horizontal Iwasawa theory to create a so-called CLASS NUMBER FORMULA, but was initially unsuccessful and therefore used instead an extension of a result of Flach based on ideas from Kolyvagin. However, there was a problem with this extension which was discovered during review of Wiles' manuscript in September 1993. Former student Richard Taylor came to Princeton in early 1994 to help Wiles patch up this error. After additional effort, Wiles discovered the reason that the Flach/Kolyvagin approach was failing, and also discovered that it was precisely what had prevented Iwasawa theory from working.

With this additional insight, Wiles was able to successfully complete the erroneous portion of the proof using Iwasawa theory, proving the SEMISTABLE case of the Taniyama-Shimura conjecture (Taylor and Wiles 1995, Wiles 1995) and, at the same time, establishing FERMAT'S LAST THEOREM as a true theorem.

The existence of a proof of the *full* Taniyama-Shimura conjecture was announced at a conference by Kenneth Ribet on June, 21 1999 (Knapp 1999), and reported on National Public Radio's Weekend Edition on July 31, 1999. The proof was completed by Christophe Breuil, Brian Conrad, Fred Diamond, and Richard Taylor, building on the earlier work of Wiles and Taylor (Mackenzie 1999, Morgan 1999). The best previous published result held for all CONDUCTORS except those divisible by 27 (Conrad *et al.* 1999; Knapp 1999). The general Breuil *et al.* proof for *all* elliptic curves removed this restriction, in the process relying on Wiles' proof for rational ELLIPTIC CURVES.

See also CONDUCTOR, ELLIPTIC CURVE, EPSILON CONJECTURE, FERMAT'S LAST THEOREM, LANGLANDS PROGRAM, MODULAR FORM, MODULAR FUNCTION, RIBET'S THEOREM

References

--. *Science* **285**, 178, 1999.

American Mathematical Society. http://www.ams.org/new-in-math/10-1999-media.html#fermat.

Conrad, B.; Diamond, F.; and Taylor, R. "Modularity of Certain Potentially Barsotti-Tate Galois Representations." *J. Amer. Math. Soc.* **12**, 521–567, 1999.

Darmon, H. "A Proof of the Full Shimura-Taniyama-Weil Conjecture is Announced." *Not. Amer. Math. Soc.* **46**, 1397–1406, 1999.

Ekeland, I. "Curves and Numbers." *Nature* **405**, 748–749, 2000.

Knapp, A. W. "Proof Announced of Taniyama-Shimura-Weil Conjecture." *Not. Amer. Math. Soc.* **46**, 863, 1999.

Lang, S. "Some History of the Shimura-Taniyama Conjecture." *Not. Amer. Math. Soc.* **42**, 1301–1307, 1995.

Mackenzie, D. "Fermat's Last Theorem Extended." *Science* **285**, 178, 1999.

Morgan, F. "Frank Morgan's Math Chat." http://www.maa.org/features/mathchat/mathchat_7_1_99.html. July 1, 1999.

Peterson, I. "Curving Beyond Fermat's Last Theorem." *Sci. News* **156**, 221, Oct. 2, 1999.

Shimura, G. and Taniyama, Y. *Complex Multiplication of Abelian Varieties and Its Applications to Number Theory.* Tokyo: Mathematical Society of Japan, 1961.

Taylor, R. and Wiles, A. "Ring-Theoretic Properties of Certain Hecke Algebras." *Ann. Math.* **141**, 553–572, 1995.

Wiles, A. "Modular Elliptic-Curves and Fermat's Last Theorem." *Ann. Math.* **141**, 443–551, 1995.

Taniyama-Shimura Theorem

TANIYAMA-SHIMURA CONJECTURE

Tank

CYLINDRICAL SEGMENT

Tantrix

TANGENT INDICATRIX

Tapering Function

APODIZATION FUNCTION

Tarry Point

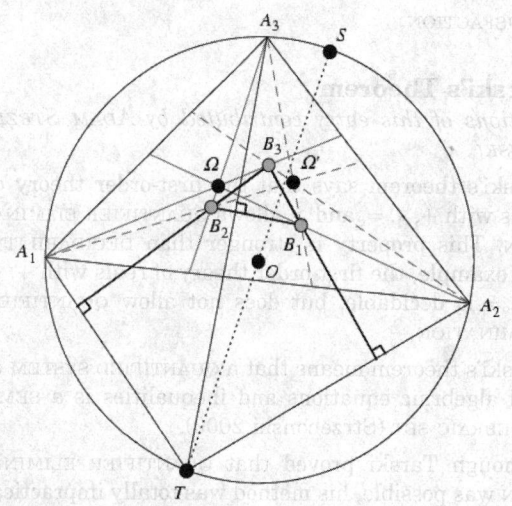

The point T at which the lines through the VERTICES of a TRIANGLE PERPENDICULAR to the corresponding sides of the first BROCARD TRIANGLE, are CONCURRENT. The Tarry point lies on the CIRCUMCIRCLE opposite the STEINER POINT S. It has TRIANGLE CENTER FUNCTION

$$\alpha = \frac{bc}{b^4 + c^4 - a^2 b^2 - a^2 c^2} = \sec(A + \omega),$$

where ω is the BROCARD ANGLE. The SIMSON LINE of the Tarry point is PERPENDICULAR to the line OK, when O is the CIRCUMCENTER and K is the SYMMEDIAN POINT (Lachlan 1893; Johnson 1929; Honsberger 1995, p. 121). The Tarry point of the first BROCARD TRIANGLE of a TRIANGLE ΔABC is the CIRCUMCENTER of ΔABC (Honsberger 1995, pp. 120–121).

See also BROCARD ANGLE, BROCARD TRIANGLES, CIRCUMCIRCLE, SYMMEDIAN POINT, SIMSON LINE, STEINER POINTS

References

Coolidge, J. L. *A Treatise on the Geometry of the Circle and Sphere.* New York: Chelsea, p. 77, 1971.

Gallatly, W. *The Modern Geometry of the Triangle, 2nd ed.* London: Hodgson, p. 102, 1913.

Honsberger, R. "The Steiner Point and the Tarry Point." §10.5 in *Episodes in Nineteenth and Twentieth Century Euclidean Geometry.* Washington, DC: Math. Assoc. Amer., pp. 119–124, 1995.

Johnson, R. A. *Modern Geometry: An Elementary Treatise on the Geometry of the Triangle and the Circle.* Boston, MA: Houghton Mifflin, pp. 281–282, 1929.

Kimberling, C. "Central Points and Central Lines in the Plane of a Triangle." *Math. Mag.* **67**, 163–187, 1994.

Lachlan, R. *An Elementary Treatise on Modern Pure Geometry.* London: Macmillian, p. 81, 1893.

Tarry-Escott Problem

PROUHET-TARRY-ESCOTT PROBLEM

Tarski's Recursive Definition of Satisfaction

SATISFACTION

Tarski's Theorem

Portions of this entry contributed by ADAM STRZE-BONSKI

Tarski's theorem says that the first-order theory of reals with $+$, $*$, $=$, and $>$ allows QUANTIFIER ELIMINATION. This property is stronger than DECIDABILITY. For example, the first-order theory of reals with $+$, $*$, and $=$ is decidable, but does not allow QUANTIFIER ELIMINATION.

Tarski's theorem means that a QUANTIFIED SYSTEM of real algebraic equations and inequalities is a SEMI-ALGEBRAIC SET (Strzebonski 2000).

Although Tarski proved that QUANTIFIER ELIMINATION was possible, his method was totally impractical (Davenport and Heintz 1988). A much more efficient procedure for implementing QUANTIFIER ELIMINATION is called CYLINDRICAL ALGEBRAIC DECOMPOSITION. It was developed by Collins (1975) and is implemented in *Mathematica* 4.0 as `CylindricalAlgebraicDecomposition`.

See also CYLINDRICAL ALGEBRAIC DECOMPOSITION, DECIDABLE, QUANTIFIED SYSTEM, QUANTIFIER, QUANTIFIER ELIMINATION, SEMIALGEBRAIC SET

References

Collins, G. E. "Quantifier Elimination for Real Closed Fields by Cylindrical Algebraic Decomposition." In *Proc. 2nd GI Conf. Automata Theory and Formal Languages.* New York: Springer-Verlag, pp. 134–183, 1975.
Davenport, J. and Heintz, J. "Real Quantifier Elimination if Doubly Exponential." *J. Symb. Comput.* **5**, 29–35, 1988.
Marker, D. "Model Theory and Exponentiation." *Not. Amer. Math. Soc.* **43**, 753–759, 1996.
Tarski, A. "Sur les ensembles définissables de nombres réels." *Fund. Math.* **17**, 210–239, 1931.
Tarski, A. "A Decision Method for Elementary Algebra and Geometry." RAND Corp. monograph, 1948.
Tarski, A. *A Decision Method for Elementary Algebra and Geometry,* 2nd ed. Berkeley, CA: University of California Press, 1951.

Tate Conjecture

See also HODGE CONJECTURE

References

Deligne, P. "The Hodge Conjecture." http://www.clay-math.org/prize_problems/hodge.pdf.
Tate, J. T. "Algebraic Cycles and Poles of Zeta Functions." In *Arithmetical Algebraic Geometry (Proc. Conf. Purdue*

Univ., 1963). New York: Harper and Row, pp. 93–110, 1965.

Tau Conjecture

Also known as RAMANUJAN'S HYPOTHESIS. Ramanujan proposed that

$$\tau(n) \sim \mathcal{O}\left(n^{11/2+\epsilon}\right),$$

where $\tau(n)$ is the TAU FUNCTION. This was proven by Deligne (1974) in the course of proving the more general PETERSSON CONJECTURE. Deligne was awarded the FIELDS MEDAL for his proof.

See also PETERSSON CONJECTURE, TAU FUNCTION

References

Apostol, T. M. *Modular Functions and Dirichlet Series in Number Theory,* 2nd ed. New York: Springer-Verlag, pp. 136 and 140, 1997.
Deligne, P. "La conjecture de Weil. I." *Inst. Hautes Études Sci. Publ. Math.* **43**, 273–307, 1974.
Deligne, P. "La conjecture de Weil. II." *Inst. Hautes Études Sci. Publ. Math.* **52**, 137–252, 1980.
Hardy, G. H. *Ramanujan: Twelve Lectures on Subjects Suggested by His Life and Work,* 3rd ed. New York: Chelsea, p. 169, 1999.

Tau Function

A function $\tau(n)$ related to the DIVISOR FUNCTION $\sigma_k(n)$, also sometimes called RAMANUJAN'S TAU FUNCTION. It is defined via the FOURIER SERIES of the MODULAR DISCRIMINANT $\Delta(\tau)$ for $\tau \in H$, where H is the UPPER HALF-PLANE, by

$$\Delta(\tau) = (2\pi)^{12} \sum_{n=1}^{\infty} \tau(n) e^{2\pi i n \tau} \qquad (1)$$

(Apostol 1997, p. 20). The tau function is also given by the CAUCHY PRODUCT

$$\tau(n) = 8000\{(\sigma_3 \circ \sigma_3) \circ \sigma_3\}(n) - 147(\sigma_5 \circ \sigma_5)(n), \qquad (2)$$

$$= \tfrac{65}{756}\sigma_{11}(n) + \tfrac{691}{756}\sigma_5(n) - \tfrac{691}{3}\sum_{k=1}^{n-1}\sigma_5(k)\sigma_5(n-k), \qquad (3)$$

where $\sigma_k(n)$ is the DIVISOR FUNCTION (Apostol 1997, pp. 24 and 140). The tau function has GENERATING FUNCTION

$$\sum_{n=1}^{\infty} \tau(n) x^n = x \prod_{n=1}^{\infty} (1-x^n)^{24}, \qquad (4)$$

and the first few values are 1, -24, 252, -1472, 4830, ... (Sloane's A000594). The tau function is given by the *Mathematica* command `RamanujanTau[n]` in the *Mathematica* add-on package `NumberTheory`Ramanujan`` (which can be loaded with the command `<<NumberTheory``).

Lehmer conjectured that $\tau(n) \neq 0$ for all n and verified this fact for $n < 214928639999$ (Apostol 1997, p. 22).

$\tau(n)$ is also given by

$$g(-x) = \sum_{n=1}^{\infty} (-1)^n \tau(n) x^n \qquad (5)$$

$$g(x^2) = \sum_{n=1}^{\infty} \tau\left(\tfrac{1}{2}n\right) x^n \qquad (6)$$

$$\sum_{n=1}^{\infty} \tau(n) x^n = x \left(1 - 3x + 5x^3 - 7x^6 + \ldots\right)^8. \qquad (7)$$

Ewell (1999) gave the beautiful formulas

$$\tau(4n + 2) = -3 \sum_{k=1}^{2n+1} 2^{3b(2k)} \sigma_3(\mathrm{Od}(2k))$$

$$\times \sum_{j=0}^{4n-2k+2} (-1)^j r_8(4n + 2 - 2k - j) r_8(j) \qquad (8)$$

$$\sum_{k=1}^{n} 2^{3b(2k)} \sigma_3(\mathrm{Od}2k))$$

$$\times \sum_{j=0}^{2n+1-2k} (-1)^j r_8(2n + 1 - 2k - j) r_8(j) = 0 \qquad (9)$$

$$\tau(4m) = -2^{11} \tau(m) - 3 \sum_{k=1}^{2m} 2^{3b(2k)} \sigma_3(\mathrm{Od}2k))$$

$$\times \sum_{j=0}^{4m-2k} (-1)^j r_8(4m - 2k - j) r_8(j) \qquad (10)$$

$$\tau(2n + 1) = \sum_{k=1}^{2n+1} 2^{3[b(2k)-1]} \sigma_3(\mathrm{Od}2k))$$

$$\times \sum_{j=0}^{2n+2-2k} (-1)^j r_8(3n + 2 - 2k - j) r_8(j), \qquad (11)$$

where $b(n)$ is the exponent of the exact power of 2 dividing n, $\mathrm{Od}(n)$ is the ODD PART of n, $\sigma_k(n)$ is the DIVISOR FUNCTION of n, and $r_k(n)$ is the SUM OF SQUARES FUNCTION.

For PRIME p,

$$\tau(p^{n+1}) = \tau(p)\tau(p^n) - p^{11}\tau(p^{n-1}) \qquad (12)$$

for $n \geq 1$, and

$$\tau(p^\alpha n) = \tau(p)\tau(p^{\alpha-1} n) - p^{11}\tau(p^{\alpha-2} n) \qquad (13)$$

for $\alpha \geq 2$ and $(n, p) = 1$ (Mordell 1917; Apostol 1997, p. 92).

In ORE'S CONJECTURE, the tau function appears as the number of DIVISORS of n. Ramanujan conjectured and Mordell (1917) proved that if $(n, n') = 1$, then

$$\tau(nn') = \tau(n)\tau(n'). \qquad (14)$$

More generally,

$$\tau(n)\tau(n') = \sum_{d \mid (n, n')} d^{11} \tau\left(\frac{nn'}{d^2}\right), \qquad (15)$$

which reduces to the first form if $(n, n') = 1$ (Mordell 1917; Apostol 1997, p. 93). Ramanujan conjectured and Watson proved that $\tau(n)$ is divisible by 691 for almost all n, specifically

$$\tau(n) \equiv \sigma_{11}(n) \pmod{691}, \qquad (16)$$

where $\sigma_k(n)$ is the DIVISOR FUNCTION (Wilton 1930, Apostol 1997, pp. 93 and 140) and 691 is the NUMERATOR of the BERNOULLI NUMBER B_{12}.

Ramanujan (1920) showed that

$$\tau(2n) \equiv 0 \pmod 2 \qquad (17)$$

$$\tau(3n) \equiv 0 \pmod 3 \qquad (18)$$

$$\tau(5n) \equiv 0 \pmod 5 \qquad (19)$$

(Darling 1921; Wilton 1930),

$$\tau(7n + m) \equiv 0 \pmod 7 \qquad (20)$$

for $m = 0$ or one the quadratic non-residues of 7, i.e., 3, 5, 6, and

$$\tau(23n + m) \equiv 0 \pmod{23} \qquad (21)$$

for $m = 0$ or one the quadratic non-residues of 23, i.e., 5, 7, 10, 11, 14, 15, 17, 19, 20, 21, 22 (Mordell 1922; Wilton 1930). Ewell (1999) showed that

$$\tau(4n) \equiv \tau(n) \pmod 3. \qquad (22)$$

$\tau(n)$ is almost always divisible by $2^5 \cdot 3^3 \cdot 5^2 \cdot 7^2 \cdot 23 \cdot 691$ according to Ramanujan. In fact, Serre has shown that $\tau(n)$ is almost always divisible by any integer (Andrews *et al.* 1988).

Ramanujan also studied the DIRICHLET L-SERIES

$$f(x) \equiv \sum_{n=1}^{\infty} \tau(n) n^{-s}, \qquad (23)$$

which has properties analogous to the RIEMANN ZETA FUNCTION. It satisfies

$$\frac{f(s)\Gamma(s)}{(2\pi)^s} = \frac{f(12 - s)}{(2\pi)^{12-s}}. \qquad (24)$$

It also has the Euler product representation

$$\sum_{n=1}^{\infty} \frac{\tau(n)}{n^s} = \prod_p \frac{1}{1 - \tau(p)p^{-s} + p^{11-2s}} \qquad (25)$$

for $\sigma = \Re[s] > 7$ (since $\tau(n) = \mathcal{O}(n^6)$) (Apostol 1997, p. 137). Ramanujan's TAU-DIRICHLET SERIES conjecture alleges that all nontrivial zeros of $f(s)$ lie on the line $\Re[s] = 6$. f can be split up into

$$f(6 + it) = z(t)e^{-i\theta(t)}, \qquad (26)$$

where

$$z(t) = \Gamma(6 + it)f(6 + it)(2\pi)^{-it}$$
$$\times \sqrt{\frac{\sinh(\pi t)}{\pi t (1 + t^2)(4 + t^2)(9 + t^2)(16 + t^2)(25 + t^2)}} \tag{27}$$

$$\theta(t) = -\frac{1}{2} i \ln\left[\frac{\Gamma(6 + it)}{\Gamma(6 - it)}\right] - t \ln(2\pi). \tag{28}$$

The functions $f(s)$, $\theta(t)$, and $z(t)$ are returned by the *Mathematica* commands `RamanujanTauDirichletSeries[s]` in the *Mathematica* add-on package `NumberTheory`Ramanujan`` (which can be loaded with the command $<<$`NumberTheory``), `RamanujanTauTheta[t]` in the *Mathematica* add-on package `NumberTheory`Ramanujan`` (which can be loaded with the command $<<$`NumberTheory``), and `RamanujanTauZ[t]` in the *Mathematica* add-on package `NumberTheory`Ramanujan`` (which can be loaded with the command $<<$`NumberTheory``), respectively.

The SUMMATORY tau function is given by

$$T(n) = \sum_{n \le x}' \tau(n). \tag{29}$$

Here, the prime indicates that when x is an INTEGER, the last term $\tau(x)$ should be replaced by $\frac{1}{2}\tau(x)$.

Ramanujan's tau theta function $Z(t)$ is a REAL function for REAL t and is analogous to the RIEMANN-SIEGEL FUNCTION Z. The number of zeros in the critical strip from $t = 0$ to T is given by

$$N(t) = \frac{\Theta(T) + \Im\{\ln[\tau_{DS}(6 + iT)]\}}{\pi}, \tag{30}$$

where Θ is the RIEMANN THETA FUNCTION and τ_{DS} is the TAU-DIRICHLET SERIES, defined by

$$\tau_{DS}(s) \equiv \sum_{n=1}^{\infty} \frac{\tau(n)}{n^s}. \tag{31}$$

Ramanujan conjectured that the nontrivial zeros of the function are all real.

Ramanujan's τ_z function is defined by

$$\tau_z(t) = \frac{\Gamma(6 + it)(2\pi)^{-it}}{\tau_{DS}(6 + it)\sqrt{\dfrac{\sinh(\pi t)}{\pi t \prod_{k=1}^{5} k^2 + t^2}}}, \tag{32}$$

where $\tau_{DS}(z)$ is the TAU-DIRICHLET SERIES.

See also DEDEKIND ETA FUNCTION, J-FUNCTION, LEECH LATTICE, ORE'S CONJECTURE, PARTITION FUNCTION P, TAU CONJECTURE, TAU-DIRICHLET SERIES

References

Andrews, G. E.; Berndt, B. C.; and Rankin, R. A. (Eds.). *Ramanujan Revisited: Proceedings of the Centenary Conference* New York: Academic Press, 1988.

Apostol, T. M. *Modular Functions and Dirichlet Series in Number Theory, 2nd ed.* New York: Springer-Verlag, pp. 20–21 and 51, 1997.

Darling, H. B. C. *Proc. London Math. Soc.* **19**, 350–372, 1921.

Ewell, J. A. "New Representations of Ramanujan's Tau Function." *Proc. Amer. Math. Soc.* **128**, 723–726, 1999.

Hardy, G. H. "Ramanujan's Function $\tau(n)$." Ch. 10 in *Ramanujan: Twelve Lectures on Subjects Suggested by His Life and Work, 3rd ed.* New York: Chelsea, p. 63, 1999.

Keiper, J. "On the Zeros of the Ramanujan τ-Dirichlet Series in the Critical Strip." *Math. Comput.* **65**, 1613–1619, 1996.

LeVeque, W. J. §F35 in *Reviews in Number Theory 1940–1972.* Providence, RI: Amer. Math. Soc., 1974.

Lehmer, D. H. "Ramanujan's Function $\tau(n)$." *Duke Math. J.* **10**, 483–492, 1943.

Moreno, C. J. "A Necessary and Sufficient Condition for the Riemann Hypothesis for Ramanujan's Zeta Function." *Illinois J. Math.* **18**, 107–114, 1974.

Mordell, L. J. "On Mr. Ramanujan's Empirical Expansions of Modular Functions." *Proc. Cambridge Phil. Soc.* **19**, 117–124, 1917.

Mordell, L. J. "Note on Certain Modular Relations Considered by Messrs Ramanujan, Darling, and Rogers." *Proc. London Math. Soc.* **20**, 408–416, 1922.

Ramanujan, S. *Proc. London Math. Soc.* **18**, 1920.

Ramanujan, S. "Congruence Properties of Partitions." *Math. Z.* **9**, 147–153, 1921.

Sivaramakrishnan, R. *Classical Theory of Arithmetic Functions.* New York: Dekker, pp. 275–278, 1989.

Sloane, N. J. A. Sequences A000594/M5153 in "An On-Line Version of the Encyclopedia of Integer Sequences." http://www.research.att.com/~njas/sequences/eisonline.html.

Spira, R. "Calculation of the Ramanujan Tau-Dirichlet Series." *Math. Comput.* **27**, 379–385, 1973.

Stanley, G. K. "Two Assertions Made by Ramanujan." *J. London Math. Soc.* **3**, 232–237, 1928.

Stanley, G. K. Corrigendum to "Two Assertions Made by Ramanujan." *J. London Math. Soc.* **4**, 32, 1929.

Watson, G. N. "Über Ramanujansche Kongruenzeigenschaften der Zerfällungsanzahlen." *Math. Z.* **39**, 712–731, 1935.

Wilton, J. R. "Congruence Properties of Ramanujan's Function $\tau(n)$." *Proc. London Math. Soc.* **31**, 1–17, 1930.

Yoshida, H. "On Calculations of Zeros of L-Functions Related with Ramanujan's Discriminant Function on the Critical Line." *J. Ramanujan Math. Soc.* **3**, 87–95, 1988.

Tauberian Theorem

A Tauberian theorem is a theorem which deduces the convergence of an INFINITE SERIES on the basis of the properties of the function it defines and any kind of auxiliary HYPOTHESIS which prevents the general term of the series from converging to zero too slowly. Hardy (1999, p. 46) states that "a 'Tauberian' theorem may be defined as a corrected form of the false converse of an 'ABELIAN THEOREM'."

Wiener's Tauberian theorem states that if $f \in L^1(\mathbb{R})$, then the translates of f spans a dense subspace IFF the FOURIER TRANSFORM is nonzero everywhere. This theorem is analogous with the theorem that if $f \in$

$L^1(\mathbb{Z})$ (for a BANACH ALGEBRA with a unit), then f spans the whole space if and only if the GELFAND TRANSFORM is nonzero everywhere.

See also ABELIAN THEOREM, HARDY-LITTLEWOOD TAUBERIAN THEOREM

References

Bromwich, T. J. I'a and MacRobert, T. M. *An Introduction to the Theory of Infinite Series, 3rd ed.* New York: Chelsea, p. 256, 1991.

Hardy, G. H. *Ramanujan: Twelve Lectures on Subjects Suggested by His Life and Work, 3rd ed.* New York: Chelsea, pp. 31 and 46, 1999.

Katznelson, Y. *An Introduction to Harmonic Analysis.* New York: Dover, 1976.

Wiener, N. *The Fourier Integral and Certain of Its Applications.* New York: Dover, 1951.

Tau-Dirichlet Series

$$\tau_{DS}(s) \equiv \sum_{n=1}^{\infty} \frac{\tau(n)}{n^s},$$

where $\tau(n)$ is the TAU FUNCTION. Ramanujan conjectured that all nontrivial zeros of $\tau_{DS}(s)$ lie on the line $\Re[s] = 6$.

See also TAU FUNCTION

References

Hardy, G. H. *Ramanujan: Twelve Lectures on Subjects Suggested by His Life and Work, 3rd ed.* New York: Chelsea, 1959.

Keiper, J. "On the Zeros of the Ramanujan τ-Dirichlet Series in the Critical Strip." *Math. Comput.* **65**, 1613–1619, 1996.

Spira, R. "Calculation of the Ramanujan Tau-Dirichlet Series." *Math. Comput.* **27**, 379–385, 1973.

Yoshida, H. "On Calculations of Zeros of L-Functions Related with Ramanujan's Discriminant Function on the Critical Line." *J. Ramanujan Math. Soc.* **3**, 87–95, 1988.

Tautochrone Problem

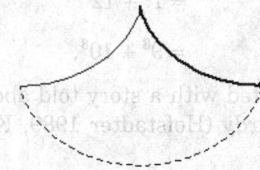

The problem of finding the curve down which a bead placed anywhere will fall to the bottom in the same amount of time. The solution is a CYCLOID, a fact first discovered and published by Huygens in *Horologium oscillatorium* (1673). This property was also alluded to in the following passage from *Moby Dick*: "[The try-pot] is also a place for profound mathematical meditation. It was in the left-hand try-pot of the *Pequod*, with the soapstone diligently circling round me, that I was first indirectly struck by the remarkable fact, that in geometry all bodies gliding along a cycloid, my soapstone, for example, will descend from any point in precisely the same time" (Melville 1851).

Huygens also constructed the first pendulum clock with a device to ensure that the pendulum was isochronous by forcing the pendulum to swing in an arc of a CYCLOID. This is accomplished by placing two evolutes of inverted cycloid arcs on each side of the pendulum's point of suspension against which the pendulum is constrained to move (Wells 1991, p. 47; Gray 1997, p. 123). Unfortunately, friction along the arcs causes a greater error than that corrected by the cycloidal path (Gardner 1984).

The PARAMETRIC EQUATIONS of the CYCLOID are

$$x = a(\theta - \sin \theta) \tag{1}$$

$$y = a(1 - \cos \theta). \tag{2}$$

To see that the CYCLOID satisfies the tautochrone property, consider the derivatives

$$x' = a(1 - \cos \theta) \tag{3}$$

$$y' = a \sin \theta, \tag{4}$$

and

$$x'^2 + y'^2 = a^2 \left[(1 - 2\cos\theta + \cos^2\theta) + \sin^2\theta \right]$$

$$= 2a^2(1 - \cos \theta). \tag{5}$$

Now

$$\tfrac{1}{2}mv^2 = mgy \tag{6}$$

$$v = \frac{ds}{dt} = \sqrt{2gy} \tag{7}$$

$$dt = \frac{ds}{\sqrt{2gy}} = \frac{\sqrt{dx^2 + dy^2}}{\sqrt{2gy}}$$

$$= \frac{a\sqrt{2(1 - \cos\theta)}\,d\theta}{\sqrt{2ga(1 - \cos\theta)}} = \sqrt{\frac{a}{g}}\,d\theta, \tag{8}$$

so the time required to travel from the top of the CYCLOID to the bottom is

$$T = \int_0^\pi dt = \sqrt{\frac{a}{g}}\,\pi. \tag{9}$$

However, from an intermediate point θ_0,

$$v = \frac{ds}{dt} = \sqrt{2g(y - y_0)}, \tag{10}$$

so

$$T = \int_{\theta_0}^\pi \sqrt{\frac{2a^2(1 - \cos\theta)}{2ag(\cos\theta_0 - \cos\theta)}}\,d\theta$$

$$= \sqrt{\frac{a}{g}} \int_{\theta_0}^{\pi} \sqrt{\frac{1 - \cos\theta}{\cos\theta_0 - \cos\theta}} \, d\theta. \quad (11)$$

To integrate, rearrange this equation using the HALF-ANGLE FORMULAS

$$\sin\left(\tfrac{1}{2}x\right) = \sqrt{\frac{1 - \cos x}{2}} \quad (12)$$

$$\cos\left(\tfrac{1}{2}x\right) = \sqrt{\frac{1 + \cos x}{2}} \quad (13)$$

with the latter rewritten in the form

$$\cos\theta = 2\cos^2\left(\tfrac{1}{2}\theta\right) - 1 \quad (14)$$

to obtain

$$T = \sqrt{\frac{a}{g}} \int_{\theta_0}^{\pi} \frac{\sin\left(\tfrac{1}{2}\theta\right) d\theta}{\sqrt{\cos^2\left(\tfrac{1}{2}\theta_0\right) - \cos^2\left(\tfrac{1}{2}\theta\right)}}. \quad (15)$$

Now transform variables to

$$u = \frac{\cos\left(\tfrac{1}{2}\theta\right)}{\cos\left(\tfrac{1}{2}\theta_0\right)} \quad (16)$$

$$du = -\frac{\sin\left(\tfrac{1}{2}\theta\right) d\theta}{2\cos\left(\tfrac{1}{2}\theta_0\right)}, \quad (17)$$

so

$$T = -2\sqrt{\frac{a}{g}} \int_{1}^{0} \frac{du}{\sqrt{1 - u^2}} = 2\sqrt{\frac{a}{g}} \left[\sin^{-1} u\right]_0^1 = \pi\sqrt{\frac{a}{g}}, \quad (18)$$

and the amount of time is the same from any point.

See also BRACHISTOCHRONE PROBLEM, CYCLOID

References

Gardner, M. *The Sixth Book of Mathematical Games from Scientific American.* Chicago, IL: University of Chicago Press, pp. 129–130, 1984.
Gray, A. *Modern Differential Geometry of Curves and Surfaces with Mathematica, 2nd ed.* Boca Raton, FL: CRC Press, 1997.
Lagrange, J. L. "Sue les courbes tautochrones." *Mém. de l'Acad. Roy. des Sci. et Belles-Lettres de Berlin* **21**, 1765. Reprinted in *Oeuvres de Lagrange, tome 2, section deuxième: Mémoires extraits des recueils de l'Academie royale des sciences et Belles-Lettres de Berlin.* Paris: Gauthier-Villars, pp. 317–332, 1868.
Melville, H. "The Tryworks." Ch. 96 in *Moby Dick.* New York: Bantam, 1981. Originally published in 1851.
Muterspaugh, J.; Driver, T.; and Dick, J. E. "The Cycloid and Tautochronism." http://php.indiana.edu/~jedick/project/intro.html.
Muterspaugh, J.; Driver, T.; and Dick, J. E. "P221 Tautochrone Problem." http://php.indiana.edu/~jedick/project/project.html.
Phillips, J. P. "Brachistochrone, Tautochrone, Cycloid--Apple of Discord." *Math. Teacher* **60**, 506–508, 1967.
Wagon, S. *Mathematica in Action.* New York: W. H. Freeman, pp. 54–60 and 384–385, 1991.
Wells, D. *The Penguin Dictionary of Curious and Interesting Geometry.* London: Penguin, pp. 46–47, 1991.

Tautology

A logical statement in which the conclusion is equivalent to the premise. If p is a tautology, it is written $\vdash p$. A SENTENCE whose TRUTH TABLE contains only 'T' is called a tautology. The following SENTENCES are examples of tautologies:

$$A \wedge B \equiv !(!A \vee !B) \quad (1)$$

$$A \vee B \equiv !A \Rightarrow B \quad (2)$$

$$A \wedge B \equiv !(A \Rightarrow !B) \quad (3)$$

(Mendelson 1997, p. 26), where $\wedge$ denotes AND, $\equiv$ denotes "is EQUIVALENT to," ! denotes NOT, $\vee$ denotes OR, and $\Rightarrow$ denotes implies.

See also CONTINGENCY, CONTRADICTION

References

Carnap, R. *Introduction to Symbolic Logic and Its Applications.* New York: Dover, p. 13, 1958.
Mendelson, E. "Tautology." §1.2 in *Introduction to Mathematical Logic, 4th ed.* London: Chapman & Hall, pp. 17–24, 1997.

Taxicab Number

The nth taxicab number Ta(n) is the smallest number representable in n ways as a sum of POSITIVE CUBES. The numbers derive their name from the HARDY-RAMANUJAN NUMBER

$$\text{Ta}(2) = 1729$$
$$= 1^3 + 12^3$$
$$= 9^3 + 10^3, \quad (1)$$

which is associated with a story told about Ramanujan by G. H. Hardy (Hofstadter 1989, Kanigel 1991, Snow 1993).

However, this property was also known as early as 1657 by F. de Bessy (Berndt and Bhargava 1993, Guy 1994). Leech (1957) found

$$\text{Ta}(3) = 87539319$$
$$= 167^3 + 436^3$$
$$= 228^3 + 423^3$$
$$= 255^3 + 414^3. \quad (2)$$

Rosenstiel *et al.* (1991) recently found

$$Ta(4) = 6963472309248$$

$$= 2421^3 + 19083^3$$

$$= 5436^3 + 18948^3$$

$$= 10200^3 + 18072^3$$

$$= 13322^3 + 16630^3. \qquad (3)$$

D. Wilson found

$$Ta(5) = 48988659276962496$$

$$= 38787^3 + 365757^3$$

$$= 107839^3 + 362753^3$$

$$= 205292^3 + 342952^3$$

$$= 221424^3 + 336588^3$$

$$= 231518^3 + 331954^3. \qquad (4)$$

The first few taxicab numbers are therefore 2, 1729, 87539319, 6963472309248, ... (Sloane's A011541).

Hardy and Wright (Theorem 412, 1979) show that the number of such sums can be made arbitrarily large but, updating Guy (1994) with Wilson's result, the least example is not known for six or more equal sums.

Sloane defines a slightly different type of taxicab numbers, namely numbers which are sums of two cubes in two or more ways, the first few of which are 1729, 4104, 13832, 20683, 32832, 39312, 40033, 46683, 64232, ... (Sloane's A001235).

See also DIOPHANTINE EQUATION–3RD POWERS, HARDY-RAMANUJAN NUMBER

References

Berndt, B. C. and Bhargava, S. "Ramanujan--For Low-brows." *Am. Math. Monthly* **100**, 645–656, 1993.

Guy, R. K. "Sums of Like Powers. Euler's Conjecture." §D1 in *Unsolved Problems in Number Theory, 2nd ed.* New York: Springer-Verlag, pp. 139–144, 1994.

Hardy, G. H. *Ramanujan: Twelve Lectures on Subjects Suggested by His Life and Work, 3rd ed.* New York: Chelsea, pp. 12 and 68, 1999.

Hardy, G. H. and Wright, E. M. *An Introduction to the Theory of Numbers, 5th ed.* Oxford, England: Clarendon Press, 1979.

Hofstadter, D. R. *Gödel, Escher, Bach: An Eternal Golden Braid.* New York: Vintage Books, p. 564, 1989.

Kanigel, R. *The Man Who Knew Infinity: A Life of the Genius Ramanujan.* New York: Washington Square Press, p. 312, 1991.

Leech, J. "Some Solutions of Diophantine Equations." *Proc. Cambridge Phil. Soc.* **53**, 778–780, 1957.

Plouffe, S. "Taxicab Numbers." http://www.lacim.uqam.ca/pi/problem.html.

Rosenstiel, E.; Dardis, J. A.; and Rosenstiel, C. R. "The Four Least Solutions in Distinct Positive Integers of the Diophantine Equation $s = x^3 + y^3 = z^3 + w^3 = u^3 + v^3 = m^3 + n^3$." *Bull. Inst. Math. Appl.* **27**, 155–157, 1991.

Silverman, J. H. "Taxicabs and Sums of Two Cubes." *Amer. Math. Monthly* **100**, 331–340, 1993.

Sloane, N. J. A. Sequences A001235 and A011541 in "An On-Line Version of the Encyclopedia of Integer Sequences." http://www.research.att.com/~njas/sequences/eisonline.html.

Snow, C. P. Foreword to *A Mathematician's Apology*, reprinted with a foreword by C. P. Snow (by G. H. Hardy). New York: Cambridge University Press, p. 37, 1993.

Wooley, T. D. "Sums of Two Cubes." *Internat. Math. Res. Not.* No. 4, 181–184, 1995.

Taylor Center

The center of the TAYLOR CIRCLE, which is the SPIEKER CENTER of $\Delta H_1 H_2 H_3$, where H_i are the feet of the ALTITUDES.

See also ALTITUDE, SPIEKER CENTER, TAYLOR CIRCLE

References

Johnson, R. A. *Modern Geometry: An Elementary Treatise on the Geometry of the Triangle and the Circle.* Boston, MA: Houghton Mifflin, p. 277, 1929.

Taylor Circle

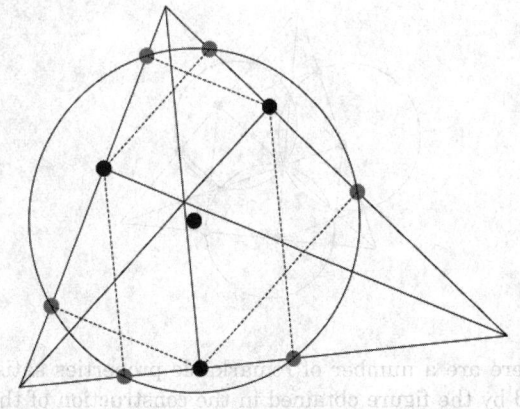

From the feet H_A, H_B, and H_C of each ALTITUDE of a TRIANGLE, draw lines PERPENDICULAR to the adjacent sides. Then the CIRCUMCIRCLE of the triangle formed by the PERPENDICULAR FEET is called the Taylor circle, and its center is called the TAYLOR CENTER. The Taylor circle is a TUCKER CIRCLE.

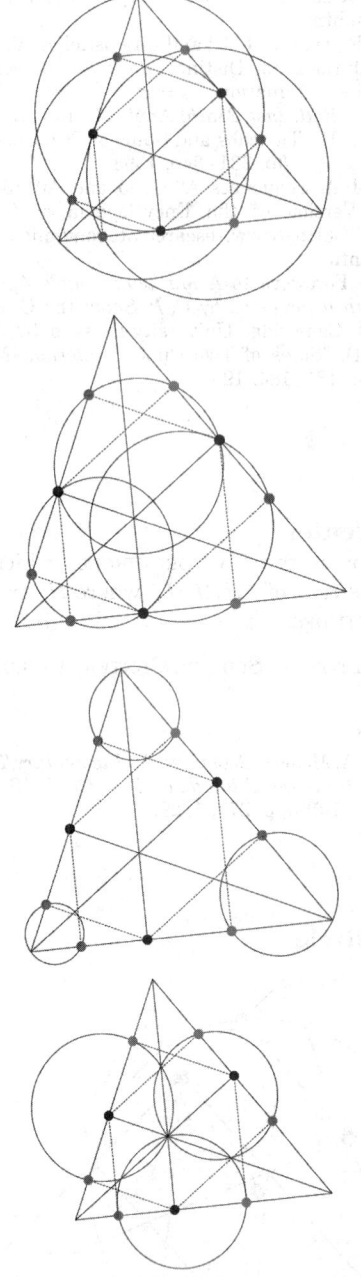

There are a number of remarkable properties satisfied by the figure obtained in the construction of the Taylor circle. These facts are probably well-known, but I have not seen them explicitly described elsewhere.

1. The feet of the perpendiculars from a given altitude foot are concyclic with the opposite vertex.
2. The two feet of the perpendiculars which are closest to a given vertex are concyclic with the feet of the altitudes on the corresponding sides.
3. The two feet of the perpendiculars which are closest to a give vertex are concyclic with that

vertex and with the intersection of the perpendiculars.
4. The three circles through the ORTHOCENTER and the feet of the perpendiculars on a given side intersect pairwise along the altitudes.

See also TAYLOR CENTER, TUCKER CIRCLES

References

Casey, J. "Lemoine's, Tucker's, and Taylor's Circle." Supp. Ch. §3 in *A Sequel to the First Six Books of the Elements of Euclid, Containing an Easy Introduction to Modern Geometry with Numerous Examples, 5th ed., rev. enl.* Dublin: Hodges, Figgis, & Co., pp. 179–189, 1888.
Coolidge, J. L. *A Treatise on the Geometry of the Circle and Sphere.* New York: Chelsea, pp. 71–73, 1971.
Johnson, R. A. *Modern Geometry: An Elementary Treatise on the Geometry of the Triangle and the Circle.* Boston, MA: Houghton Mifflin, p. 277, 1929.
Lachlan, R. *An Elementary Treatise on Modern Pure Geometry.* London: Macmillian, p. 78, 1893.
Taylor, H. M. *Proc. London Math. Soc.* **15**.

Taylor Expansion
TAYLOR SERIES

Taylor Polynomial
TAYLOR SERIES

Taylor Series

A Taylor series is a SERIES EXPANSION of a FUNCTION about a point. A 1-D Taylor series is an expansion of a REAL FUNCTION $f(x)$ about a point $x = x_0$ (sometimes written instead $x = a$). If $x = 0$, the expansion is known as a MACLAURIN SERIES.

To derive the Taylor series of a function $f(x)$, note that the integral of the $(n+1)$st DERIVATIVE $f^{(n+1)}$ of $f(x)$ from the point x_0 to an arbitrary point x is given by

$$\int_{x_0}^{x} f^{(n+1)}(x)\, dx = \left[f^{(n)}(x)\right]_{x_0}^{x} = f^{(n)}(x) - f^{(n)}(x_0), \qquad (1)$$

where $f^{(n)}(x_0)$ is the nth derivative of $f(x)$ evaluated at x_0, and is therefore simply a constant. Now integrate a second time to obtain

$$\int_{x_0}^{x}\left[\int_{x_0}^{x} f^{(n+1)}(x)\, dx\right] dx$$

$$= \int_{x_0}^{x}\left[f^{(n)}(x) - f^{(n)}(x_0)\right] dx$$

$$= \left[f^{(n-1)}(x)\right]_{x_0}^{x} - (x - x_0)f^{(n)}(x_0)$$

$$= f^{(n-1)}(x) - f^{(n-1)}(x_0) - (x - x_0)f^{(n)}(x_0), \qquad (2)$$

where $f^{(k)}(x_0)$ is again a constant. Integrating a third time,

$$\iint\limits_{x_0}^{x}\int f^{(n+1)}(x)(dx)^3 = f^{(n-2)}(x) - f^{(n-2)}(x_0)$$

$$-(x-x_0)f^{(n-1)}(x_0) - \frac{(x-x_0)^2}{2!}f^{(n)}(x_0), \qquad (3)$$

and continuing up to $n+1$ integrations then gives

$$\underbrace{\int \cdots \int_{x_0}^{x}}_{n+1} f^{(n+1)}(x)(dx)^{n+1}$$

$$= f(x) - f(x_0) - (x-x_0)f'(x_0) - \frac{(x-x_0)^2}{2!}f''(x_0)$$

$$-\ldots - \frac{(x-x_0)^n}{n!}f^{(n)}(x_0). \qquad (4)$$

Rearranging then gives the one-dimensional Taylor series

$$f(x) = f(x_0) + (x-x_0)f'(x_0) + \frac{(x-x_0)^2}{2!}f''(x_0) + \ldots$$

$$+ \frac{(x-x_0)^n}{n!}f^{(n)}(x_0) + R_n, \qquad (5)$$

$$= \sum_{k=0}^{n} \frac{(x-x_0)^k f^{(k)}(x_0)}{k!} + R_n. \qquad (6)$$

Here, R_n is a remainder term known as the LAGRANGE REMAINDER, which is given by

$$R_n = \underbrace{\int \cdots \int_{x_0}^{x}}_{n+1} f^{(n+1)}(x)(dx)^{n+1}. \qquad (7)$$

Rewriting the MULTIPLE INTEGRAL then gives

$$R_n = \int_{x_0}^{x} f^{(n+1)}(t)\frac{(x-t)^n}{n!}\,dt. \qquad (8)$$

Now, from the MEAN-VALUE THEOREM for a function $g(x)$, it must be true that

$$\int_{x_0}^{x} g(x)\,dx = (x-x_0)g(x^*) \qquad (9)$$

for some $x^* \in [x_0, x]$. Therefore, integrating $n+1$ times gives the result

$$R_n = \frac{(x+x_0)^{n+1}}{(n+1)!}f^{(n+1)}(x^*), \qquad (10)$$

so the maximum error after n terms of the Taylor series is the maximum value of (10) running through all $x^* \in [x_0, x]$. Note that the Lagrange remainder R_n is also sometimes taken to refer to the remainder when terms up to the $(n-1)$st power are taken in the

Taylor series (Whittaker and Watson 1990, pp. 95–96).

An alternative form of the 1-D Taylor series may be obtained by letting

$$x - x_0 \equiv \Delta x \qquad (11)$$

so that

$$x \equiv x_0 + \Delta x. \qquad (12)$$

Substitute this result into (5) to give

$$f(x_0 + \Delta x) = f(x_0) + \Delta x f'(x_0) + \frac{1}{2!}(\Delta x)^2 f''(x_0) + \ldots. \qquad (13)$$

A Taylor series of a REAL FUNCTION in two variables $f(x, y)$ is given by

$$f(x+\Delta x, y+\Delta y) = f(x, y) + [f_x(x, y)\Delta x + f_y(x, y)\Delta y]$$

$$+ \frac{1}{2!}[(\Delta x)^2 f_{xx}(x, y) + 2\Delta x\Delta y f_{xy}(x, y) + (\Delta y)^2 f_{yy}(x, y)]$$

$$+ \frac{1}{3!}[(\Delta x)^3 f_{xxx}(x, y) + 3(\Delta x)^2\Delta y f_{xxy}(x, y)$$

$$+ 3\Delta x(\Delta y)^2 f_{xyy}(x, y) + (\Delta y)^3 f_{yyy}(x, y)] + \ldots. \qquad (14)$$

This can be further generalized for a REAL FUNCTION in n variables,

$$f(x_1, \ldots, x_n)$$

$$= \sum_{j=0}^{\infty} \left\{ \frac{1}{j!} \left[\sum_{k=1}^{n} (x'_k - a_k)\frac{\partial}{\partial x'_k} \right]^j f(x'_1, \ldots, x'_n) \right\}_{x'_1 = a_1, \ldots, x'_n = a_n} \qquad (15)$$

Rewriting,

$$f(x_1 + a_1, \ldots, x_n + a_n)$$

$$= \sum_{j=0}^{\infty} \left\{ \frac{1}{j!} \left[\sum_{k=1}^{n} a_k \frac{\partial}{\partial x'_k} \right]^j f(x'_1, \ldots, x'_n) \right\}_{x'_1 = a_1, \ldots, x'_n = a_n}.$$

Taking $n = 2$ in (15) gives

$$f(x_1, x_2) = \sum_{j=0}^{\infty} \left\{ \frac{1}{j!} \left[(x'_1 - a_1)\frac{\partial}{\partial x'_1} \right. \right.$$

$$\left. \left. + (x'_2 - a_2)\frac{\partial}{\partial x'_2} \right]^j f(x'_1, x'_2) \right\}_{x'_1 = a_1, x'_2 = a_2}$$

$$= f(a_1, a_2) + \left[(x_1 - a_1)\frac{\partial f}{\partial x_1} + (x_2 - a_2)\frac{\partial f}{\partial x_2} \right]$$

$$+ \frac{1}{2!}\left[(x_1 - a_1)^2\frac{\partial^2 f}{\partial x_1^2} + 2(x_1 - a_1)(x_2 - a_2)\frac{\partial^2 f}{\partial x_1 \partial x_2} \right.$$

$$+(x_2 - a_2)^2 \frac{\partial^2 f}{\partial x_2^2}\bigg] + \ldots \qquad (17)$$

Taking $n = 3$ in (16) gives

$$f(x_1 + a_1,\ x_2 + x_2 + a_2,\ x_3 + a_3)$$

$$= \sum_{j=0}^{\infty} \left\{ \frac{1}{j!} \left(a_1 \frac{\partial}{\partial x_1'} + a_2 \frac{\partial}{\partial x_2'} + a_3 \frac{\partial}{\partial x_3'} \right)^j \right.$$

$$\left. \times f(x_1',\ x_2',\ x_3') \right\}_{x_1'=x_1,\ x_2'=x_2,\ x_3'=x_3}, \qquad (18)$$

or, in VECTOR form

$$f(\mathbf{r} + \mathbf{a}) = \sum_{j=0}^{\infty} \left[\frac{1}{j!} (\mathbf{a} \cdot \nabla_{\mathbf{r}'})^j f(\mathbf{r}') \right]_{\mathbf{r}'=\mathbf{r}} \qquad (19)$$

The zeroth- and first-order terms are

$$f(\mathbf{r}) \qquad (20)$$

and

$$(\mathbf{a} \cdot \nabla_{\mathbf{r}'}) f(\mathbf{r}')\ |_{\mathbf{r}'=\mathbf{r}}, \qquad (21)$$

respectively. The second-order term is

$$\tfrac{1}{2}(\mathbf{a} \cdot \nabla_{\mathbf{r}'})(\mathbf{a} \cdot \nabla_{\mathbf{r}'}) f(\mathbf{r}')|_{\mathbf{r}'=\mathbf{r}} = \tfrac{1}{2}\, \mathbf{a} \cdot \nabla_{\mathbf{r}'}[\mathbf{a} \cdot (\nabla f(\mathbf{r}'))]_{\mathbf{r}'=\mathbf{r}}$$

$$= \tfrac{1}{2}\, \mathbf{a} \cdot [\mathbf{a} \cdot \nabla_{\mathbf{r}'}(\nabla_{\mathbf{r}'} f(\mathbf{r}'))]|_{\mathbf{r}'=\mathbf{r}}, \qquad (22)$$

so the first few terms of the expansion are

$$f(\mathbf{r} + \mathbf{a}) = f(\mathbf{r}) + (\mathbf{a} \cdot \nabla_{\mathbf{r}'}) f(\mathbf{r}')\ |_{\mathbf{r}'=\mathbf{r}} + \tfrac{1}{2}\, \mathbf{a}$$

$$\cdot [\mathbf{a} \cdot \nabla_{\mathbf{r}'}(\nabla \mathbf{r}' f(\mathbf{r}'))]|_{\mathbf{r}'=\mathbf{r}}. \qquad (23)$$

Taylor series can also be defined for functions of a COMPLEX variable. By the CAUCHY INTEGRAL FORMULA,

$$f(z) = \frac{1}{2\pi i} \int_C \frac{f(z')\, dz}{z' - z} = \frac{1}{2\pi i} \int_C \frac{f(z')\, dz'}{(z' - z_0) - (z - z_0)}$$

$$= \frac{1}{2\pi i} \int_C \frac{f(z')\, dz'}{(z' - z_0)\left(1 - \dfrac{z - z_0}{z' - z0}\right)}. \qquad (24)$$

In the interior of C,

$$\frac{|z - z_0|}{|z' - z_0|} < 1 \qquad (25)$$

so, using

$$\frac{1}{1 - t} = \sum_{n=0}^{\infty} t^n, \qquad (26)$$

it follows that

$$f(z) = \frac{1}{2\pi i} \int_C \sum_{n=0}^{\infty} \frac{(z - z_0)^n f(z')\, dz'}{(z' - z_0)^{n+1}}$$

$$= \frac{1}{2\pi i} \sum_{n=0}^{\infty} (z - z_0)^n \int_C \frac{f(z')\, dz}{(z' - z_0)^{n+1}}. \qquad (27)$$

Using the CAUCHY INTEGRAL FORMULA for derivatives,

$$f(z) = \sum_{n=0}^{\infty} (z - z_0)^n \frac{f^{(n)}(z_0)}{n!}. \qquad (28)$$

See also CAUCHY REMAINDER, LAGRANGE EXPANSION, LAGRANGE REMAINDER, LAURENT SERIES, LEGENDRE SERIES, MACLAURIN SERIES, NEWTON'S FORWARD DIFFERENCE FORMULA, TAYLOR'S THEOREM

References

Abramowitz, M. and Stegun, C. A. (Eds.). *Handbook of Mathematical Functions with Formulas, Graphs, and Mathematical Tables, 9th printing.* New York: Dover, p. 880, 1972.

Arfken, G. "Taylor's Expansion." §5.6 in *Mathematical Methods for Physicists, 3rd ed.* Orlando, FL: Academic Press, pp. 303–313, 1985.

Comtet, L. "Calcul pratique des coefficients de Taylor d'une fonction algébrique." *Enseign. Math.* **10**, 267–270, 1964.

Morse, P. M. and Feshbach, H. "Derivatives of Analytic Functions, Taylor and Laurent Series." §4.3 in *Methods of Theoretical Physics, Part I.* New York: McGraw-Hill, pp. 374–398, 1953.

Whittaker, E. T. and Watson, G. N. "Forms of the Remainder in Taylor's Series." §5.41 in *A Course in Modern Analysis, 4th ed.* Cambridge, England: Cambridge University Press, pp. 95–96, 1990.

Taylor-Greene-Chirikov Map

STANDARD MAP

Taylor's Condition

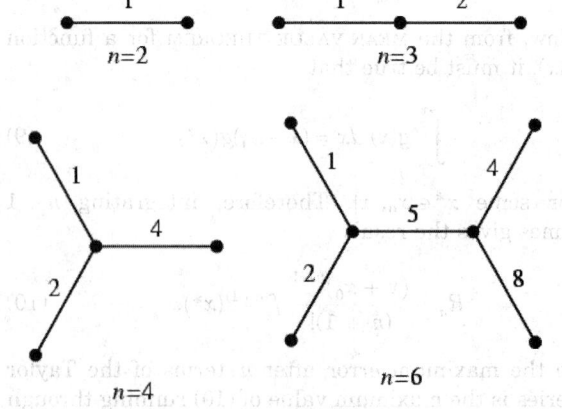

For a given POSITIVE INTEGER n, does there exist a WEIGHTED TREE with n VERTICES whose paths have weights $1, 2, \ldots, \binom{n}{2}$, where $\binom{n}{2}$ is a BINOMIAL

COEFFICIENT? Taylor showed that no such TREE can exist unless it is a PERFECT SQUARE or a PERFECT SQUARE plus 2. No such TREES are known except $n = 2, 3, 4,$ and 6.

See also GOLOMB RULER, PERFECT DIFFERENCE SET, TREE

References

Honsberger, R. *Mathematical Gems III.* Washington, DC: Math. Assoc. Amer., pp. 56–60, 1985.

Leech, J. "Another Tree Labeling Problem." *Amer. Math. Monthly* **82**, 923–925, 1975.

Taylor, H. "Odd Path Sums in an Edge-Labeled Tree." *Math. Mag.* **50**, 258–259, 1977.

Taylor's Theorem

The theorem that a function may be represented by a TAYLOR SERIES,

$$f(x) = f(0) + xf'(0) + \frac{x^2}{2!} f''(0) + \ldots + \frac{x^{n-1}}{(n-1)!} f^{(n-1)}(0)$$

$$+ \int_0^x \frac{(x-u)^{n-1}}{(n-1)!} f^{(n)}(u) \, du.$$

Taylor's theorem without the remainder was first devised by Taylor in 1712 and published in 1915, but it was not until almost a century later than Lagrange and Cauchy derived approximations of the remainder term after a finite number of terms (Moritz 1937). These forms are now called the LAGRANGE REMAINDER and CAUCHY REMAINDER.

Most modern proofs are based on Cox (1851), which is more elementary than that of Cauchy and Lagrange (Moritz 1923), and which Pringsheim (1900) referred to as "leaving hardly anything to wish for in terms of simplicity and strength" (Moritz 1923).

See also CAUCHY REMAINDER, LAGRANGE REMAINDER, TAYLOR SERIES

References

Cox, H. *Cambridge and Dublin Math. J.* **6**, 80, 1851.

Jeffreys, H. and Jeffreys, B. S. "Taylor's Theorem." §1.133 in *Methods of Mathematical Physics, 3rd ed.* Cambridge, England: Cambridge University Press, pp. 50–51, 1988.

Moritz, R. E. "A Note on Taylor's Theorem." *Amer. Math. Monthly* **44**, 31–33, 1937.

Pringsheim. *Bibliotheca Math.* **1**, 455, 1900.

Todhunter, I. *A Treatise on the Differential Calculus with Numerous Examples, 10th ed.* London: Macmillan, p. 75, 1890.

Tchebycheff

CHEBYSHEV APPROXIMATION FORMULA, CHEBYSHEV CONSTANTS, CHEBYSHEV DEVIATION, CHEBYSHEV DIFFERENTIAL EQUATION, CHEBYSHEV FUNCTIONS, CHEBYSHEV-GAUSS QUADRATURE, CHEBYSHEV INEQUALITY, CHEBYSHEV INEQUALITY, CHEBYSHEV INTEGRAL, CHEBYSHEV PHENOMENON, CHEBYSHEV POLYNOMIAL OF THE FIRST KIND, CHEBYSHEV POLYNOMIAL OF THE SECOND KIND, CHEBYSHEV QUADRATURE, CHEBYSHEV-RADAU QUADRATURE, CHEBYSHEV-SYLVESTER CONSTANT

t-Design

See also STEINER SYSTEM

t-Distribution

STUDENT'S T-DISTRIBUTION

Teardrop Curve

A plane curve given by the PARAMETRIC EQUATIONS

$$x = \cos t$$

$$y = \sin t \sin^m \left(\tfrac{1}{2} t \right).$$

See also PEAR-SHAPED CURVE

References

von Seggern, D. *CRC Standard Curves and Surfaces.* Boca Raton, FL: CRC Press, p. 174, 1993.

Technique

A specific method of performing an operation. The terms ALGORITHM, METHOD, and PROCEDURE are also used interchangeably.

See also ALGORITHM, METHOD, PROCEDURE

Teeko

A game described by Scarne which is played on a 5×5 board by two players who alternate placing, one at a time, their four counters each, after which the counters are moved around (including diagonally). Four counters in a row or square wins (Beeler *et al.* 1972). In general, there are sixteen forms of the game, all of which were solved completely by Guy Steele in 1998 with the following results: standard teeko (44 winning configurations) is a draw, and advanced teeko (58 winning configurations) is a first-player win.

Here is a more complete summary of the results.

Variant	Winner
standard	draw
alternate	draw
one-move alternate	draw
two-move alternate	draw
three-move alternate	draw
one-move standard	draw

two-move standard	draw
three-move standard	draw
standard, 58 positions	first-player win (13 turns)
alternate, 58 positions	draw
one-move alternate, 58 positions	draw
two-move alternate, 58 positions	draw
three-move alternate, 58 positions	draw
one-move standard, 58 positions	first-player win (25 turns)
two-move standard, 58 positions	draw
three-move standard, 58 positions	draw

References

Beeler, M. *et al.* Item 90 in Beeler, M.; Gosper, R. W.; and Schroeppel, R. *HAKMEM.* Cambridge, MA: MIT Artificial Intelligence Laboratory, Memo AIM-239, p. 35, Feb. 1972.

Teichmüller Space

TEICHMÜLLER'S THEOREM asserts the EXISTENCE and UNIQUENESS of the extremal quasiconformal map between two compact RIEMANN SURFACES of the same GENUS modulo an EQUIVALENCE RELATION. The equivalence classes form the Teichmüller space T_p of compact RIEMANN SURFACES of GENUS p.

See also RIEMANN'S MODULI PROBLEM

Teichmüller's Principle

See also JENKINS' THEOREM

References

Jenkins, J. A. *Univalent Functions and Conformal Mapping.* New York: Springer-Verlag, 1958.
Jenkins, J. A. "Some Area Theorems and a Special Coefficient Theorem." *Illinois J. Math.* **8**, 80–99, 1964.

Teichmüller's Theorem

Asserts the EXISTENCE and UNIQUENESS of the extremal quasiconformal map between two compact RIEMANN SURFACES of the same GENUS modulo an EQUIVALENCE RELATION.

See also TEICHMÜLLER SPACE

Teixeira's Theorem

An extended form of BÜRMANN'S THEOREM. Let $f(z)$ be a function of z analytic in a ring-shaped region A, bounded by another curve C and an inner curve c. Let $\theta(z)$ be a function analytic on and inside C having only one zero a (which is simple) within the contour. Further let x be a given point within A. Finally, let

$$|\theta(x)| < |\theta(z)| \tag{1}$$

for all points z of C, and

$$|\theta(x)| > |\theta(z)| \tag{2}$$

for all points z of c. Then

$$f(x) = \sum_{n=0}^{\infty} A_n [\theta(x)]^n + \sum_{n=1}^{\infty} \frac{B_n}{[\theta(x)]^n}, \tag{3}$$

where

$$A_n = \frac{1}{2\pi i} \int_C \frac{f(z)\theta'(z)\, dz}{[\theta(z)]^{n+1}} \tag{4}$$

$$B_n = \frac{1}{2\pi i} \int_c f(z)[\theta(z)]^{n-1} \theta'(z)\, dz \tag{5}$$

(Whittaker and Watson 1990, pp. 131–132).

See also BÜRMANN'S THEOREM, LAGRANGE EXPANSION

References

Bateman, H. "An Extension of Lagrange's Expansion." *Trans. Amer. Math. Soc.* **28**, 346–356, 1926.
Teixeira, M. F. G. "Sur les séries ordonnées suivant les puissance d'une fonction donnée." *J. für Math.* **122**, 97–123, 1900.
Whittaker, E. T. and Watson, G. N. "Teixeira's Extended Form of Bürmann's Theorem." §7.31 in *A Course in Modern Analysis, 4th ed.* Cambridge, England: Cambridge University Press, pp. 131–132, 1990.

Telegraph Equation

The PARTIAL DIFFERENTIAL EQUATION

$$u_{xx} = a u_{tt} + b u_t + c u.$$

References

Zwillinger, D. (Ed.). *CRC Standard Mathematical Tables and Formulae.* Boca Raton, FL: CRC Press, p. 417, 1995.

Telephone Problem

GOSSIPING

Telescoping Sum

A sum in which subsequent terms cancel each other, leaving only initial and final terms. For example,

$$S = \sum_{i=1}^{n+1} (a_i - a_{i+1})$$
$$= (a_1 - a_2) + (a_2 - a_3) + \ldots + (a_{n-2} - a_{n-1})$$
$$+ (a_{n-1} - a_n)$$
$$= (a_1 - a_n)$$

is a telescoping sum.

See also ZEILBERGER'S ALGORITHM

Temperature

The "temperature" of a curve Γ is defined as

$$T \equiv \frac{1}{\ln\left(\dfrac{2l}{2l - h}\right)},$$

where l is the length of Γ and h is the length of the PERIMETER of the CONVEX HULL. The temperature of a curve is 0 only if the curve is a straight line, and increases as the curve becomes more "wiggly."

See also CURLICUE FRACTAL

References
Pickover, C. A. *Keys to Infinity.* New York: Wiley, pp. 164–165, 1995.

Templar Magic Square

S	A	T	O	R
A	R	E	P	O
T	E	N	E	T
O	P	E	R	A
R	O	T	A	S

A MAGIC SQUARE-type arrangement of the words in the Latin sentence "Sator Arepo tenet opera rotas" ("the farmer Arepo keeps the world rolling"). This square has been found in excavations of ancient Pompeii.

See also MAGIC SQUARE

References
Bouisson, S. M. *La Magie: Ses Grands Rites, Son Histoire.* Paris, pp. 147–148, 1958.
Grosser, F. "Ein neuer Vorschlag zur Deutung der Sator-Formel." *Archiv. f. Relig.* **29**, 165–169, 1926.
Hocke, G. R. *Manierismus in der Literatur: Sprach-Alchimie und esoterische Kombinationskunst.* Hamburg, Germany: Rowohlt, p. 24, 1967.

Temple Problem

SANGAKU PROBLEM

Tennis Ball Theorem

Any nontrivial, closed, simple, smooth SPHERICAL CURVE dividing the surface of a SPHERE into two parts of equal areas has at least four INFLECTION POINTS.

See also BALL, BASEBALL COVER, INFLECTION POINT, SPHERICAL CURVE

References
Arnold, V. I. *Topological Invariants of Plane Curves and Caustics.* Providence, RI: Amer. Math. Soc., 1994.
Martinez-Maure, Y. "A Note on the Tennis Ball Theorem." *Amer. Math. Monthly* **103**, 338–340, 1996.

Tensegrity

An ordered finite CONFIGURATION with certain pairs of points, called cables, which are constrained not to get further apart and certain other pairs of points, called struts, which are constrained not to get closer together.

See also CONFIGURATION, FRAMEWORK

References
Back, A. and Connelly, B. "Catalogue of Symmetric Tensegrities." http://mathlab.cit.cornell.edu/visualization/tenseg/tenseg.html.
Back, A. and Connelly, B. "Mathematics and Tensegrity." *Amer. Sci.* **86**, 142–151, 1998.
Pugh, A. *An Introduction to Tensegrity.* Berkeley, CA: University of California Press, 1976.

Tensor

An nth-RANK tensor in m-space is a mathematical object in m-dimensional space that has n indices and m^n components and obeys certain transformation rules. Each INDEX of a tensor ranges over the number of dimensions of SPACE. However, the dimension of the space is largely irrelevant in most tensor equations (with the notable exception of the contracted KRONECKER DELTA).

The notation for a tensor is similar to that of a MATRIX (i.e., $A = (a_{ij})$), except that a tensor $a_{i,j,k,\ldots}$ may have an arbitrary number of INDICES. In addition, a tensor with RANK $r + s$ may be of mixed type (r, s), with r so-called "contravariant" INDICES and s "covariant" INDICES, denoted $a_{i_1,\ldots,i_s}^{j_1,\ldots,j_s}$. Technically, a MATRIX is a tensor of type $(1, 1)$ and would be written a_i^j in tensor notation.

In *Mathematica*, a tensor of RANK n is represented using nested lists of depth n, and tensors can be generated using the command Array[a, {i, j, …}]. Similarly, the dimensions of a tensor can be found using Dimensions[t], and the rank can be found using Rank[t]. Taking for example

```
t = Array[a, {1, 2, 2, 3}]
```

gives the rank-4 tensor of dimensions {1, 2, 2, 3},

```
{{{{a[1,1,1,1],a[1,1,1,2],a[1,1,1,3]}},
```

```
     {a[1,1,2,1],a[1,1,2,2],a[1,1,2,3]}},
    {{a[1,2,1,1],a[1,2,1,2],a[1,2,1,3]},
     {a[1,2,2,1], a[1,2,2,2],a[1,2,2,3]}}}},
   {{{a[2,1,1,1],a[2,1,1,2],a[2,1,1,3]},
     {a[2,1,2,1],a[2,1,2,2],a[2,1,2,3]}},
    {{a[2,2,1,1],a[2,2,1,2],a[2,2,1,3]},
     {a[2,2,2,1],a[2,2,2,2],a[2,2,2,3]}}}}}.
```

In n-dimensional space, each element a_{ijkl} would then represent an n-vector.

A TENSOR SPACE of type (r, s) can be described as a TENSOR PRODUCT between r copies of VECTOR FIELDS and s copies of the dual vector fields, i.e., ONE-FORMS. For example,

$$T^{(3, 1)} = TM \otimes TM \otimes TM \otimes T^*M \qquad (1)$$

is the VECTOR BUNDLE of $(3, 1)$-tensors on a MANIFOLD M, where TM is the TANGENT BUNDLE of M and T^*M is its dual. Tensors of type (r, s) form a VECTOR SPACE. This description generalized to any tensor type, and an INVERTIBLE LINEAR MAP $J : V \to W$ induces a map $\tilde{J} : V \otimes V^* \to W \otimes W^*$, where V^* is the DUAL VECTOR SPACE and J the JACOBIAN, defined by

$$\tilde{J}(v_1 \otimes v_2^*) = \left(Jv_1 \otimes \left(J^T\right)^{-1} v_2^*\right), \qquad (2)$$

where J^T is the PULLBACK MAP of a form is defined using the transpose of the JACOBIAN. This definition can be extended similarly to other TENSOR PRODUCTS of V and V^*. When there is a change of COORDINATES, then tensors transform similarly, with J the JACOBIAN of the linear transformation.

Zeroth-rank tensors are called SCALARS, and first-rank tensors are called VECTORS. In tensor notation, a vector **v** would be written v_i, where $i = 1, ..., m$. Tensor notation can provide a very concise way of writing vector and more general identities. For example, in tensor notation, the DOT PRODUCT **u** · **v** is simply written

$$\mathbf{u} \cdot \mathbf{v} = u_i v^i, \qquad (3)$$

where repeated indices are summed over (EINSTEIN SUMMATION). Similarly, the CROSS PRODUCT can be concisely written as

$$\mathbf{u} \times \mathbf{v} = \epsilon_{ijk} u^j v^k, \qquad (4)$$

where ϵ_{ijk} is the PERMUTATION TENSOR.

CONTRAVARIANT second-rank tensors are objects which transform as

$$A'^{ij} = \frac{\partial x_i'}{\partial x_k} \frac{\partial x_j'}{\partial x_l} A^{kl}. \qquad (5)$$

COVARIANT second-rank tensors are objects which transform as

$$C'_{ij} = \frac{\partial x_k}{\partial x_i'} \frac{\partial x_l}{\partial x_j'} C_{kl}. \qquad (6)$$

MIXED second-rank tensors are objects which transform as

$$B_j'^i = \frac{\partial x_i'}{\partial x_k} \frac{\partial x_l}{\partial x_j'} B_l^k. \qquad (7)$$

If two tensors A and B have the same rank and the same COVARIANT and CONTRAVARIANT indices, then the can be added in the obvious way,

$$A^{ij} + B^{ij} = C^{ij} \qquad (8)$$

$$A_{ij} + B_{ij} = C_{ij} \qquad (9)$$

$$A_j^i + B_j^i = C_j^i. \qquad (10)$$

The indices of a tensor can be raised or lowered (INDEX RAISING and INDEX LOWERING, respectively) by multiplication by a so-called METRIC TENSOR, e.g.,

$$g^{ij} A_j = A^i \qquad (11)$$

$$g_{ij} A^j = A_i \qquad (12)$$

(Arfken 1985, p. 159). The generalization of the DOT PRODUCT applied to tensors is called CONTRACTION, and consists of setting two unlike indices equal to each other and then summing using the EINSTEIN SUMMATION convention. Various types of derivatives can be taken of tensors, the most common being the COMMA DERIVATIVE and COVARIANT DERIVATIVE.

If the components of any tensor of any RANK vanish in one particular coordinate system, they vanish in all coordinate systems. A transformation of the variables of a tensor changes the tensor into another whose components are linear HOMOGENEOUS FUNCTIONS of the components of the original tensor.

See also ANTISYMMETRIC TENSOR, COMMA DERIVATIVE, CONTRACTION (TENSOR), CONTRAVARIANT TENSOR, COVARIANT DERIVATIVE, COVARIANT TENSOR, CURL, DIVERGENCE, GRADIENT, INDEX LOWERING, INDEX RAISING, IRREDUCIBLE TENSOR, ISOTROPIC TENSOR, JACOBI TENSOR, MIXED TENSOR, RICCI TENSOR, RIEMANN TENSOR, SCALAR, SYMMETRIC TENSOR, TENSOR SPACE, TORSION TENSOR, VECTOR, WEYL TENSOR

References

Abraham, R.; Marsden, J. E.; and Ratiu, T. S. *Manifolds, Tensor Analysis, and Applications.* New York: Springer-Verlag, 1991.

Akivis, M. A. and Goldberg, V. V. *An Introduction to Linear Algebra and Tensors.* New York: Dover, 1972.

Arfken, G. "Tensor Analysis." Ch. 3 in *Mathematical Methods for Physicists, 3rd ed.* Orlando, FL: Academic Press, pp. 118–167, 1985.

Aris, R. *Vectors, Tensors, and the Basic Equations of Fluid Mechanics.* New York: Dover, 1989.

Bishop, R. and Goldberg, S. *Tensor Analysis on Manifolds.* New York: Dover, 1980.

Jeffreys, H. *Cartesian Tensors.* Cambridge, England: Cambridge University Press, 1931.

Jeffreys, H. and Jeffreys, B. S. "Tensors." Ch. 3 in *Methods of Mathematical Physics, 3rd ed.* Cambridge, England: Cambridge University Press, pp. 86–113, 1988.

Joshi, A. W. *Matrices and Tensors in Physics, 3rd ed.* New York: Wiley, 1995.

Lass, H. *Vector and Tensor Analysis.* New York: McGraw-Hill, 1950.

Lawden, D. F. *An Introduction to Tensor Calculus, Relativity, and Cosmology, 3rd ed.* Chichester, England: Wiley, 1982.

McConnell, A. J. *Applications of Tensor Analysis.* New York: Dover, 1947.

Morse, P. M. and Feshbach, H. "Vector and Tensor Formalism." §1.5 in *Methods of Theoretical Physics, Part I.* New York: McGraw-Hill, pp. 44–54, 1953.

Parker, L. and Christensen, S. M. *MathTensor: A System for Doing Tensor Analysis by Computer.* Reading, MA: Addison-Wesley, 1994.

Simmonds, J. G. *A Brief on Tensor Analysis, 2nd ed.* New York: Springer-Verlag, 1994.

Sokolnikoff, I. S. *Tensor Analysis--Theory and Applications, 2nd ed.* New York: Wiley, 1964.

Synge, J. L. and Schild, A. *Tensor Calculus.* New York: Dover, 1978.

Weisstein, E. W. "Books about Tensors." http://www.treasure-troves.com/books/Tensors.html.

Wrede, R. C. *Introduction to Vector and Tensor Analysis.* New York: Wiley, 1963.

Tensor Calculus

The set of rules for manipulating and calculating with TENSORS.

Tensor Density

A quantity which transforms like a TENSOR except for a scalar factor of a JACOBIAN.

Tensor Direct Product

Abstractly, the tensor direct product is the same as the TENSOR PRODUCT. However, it reflects an approach toward calculation using coordinates, and indices in particular. The notion of tensor product is more algebraic, intrinsic, and abstract. For instance, up to ISOMORPHISM, the tensor product is commutative because $V \otimes W \cong W \otimes V$. Note this does not mean that the tensor product is symmetric.

For two first-RANK TENSORS (i.e., VECTORS), the tensor direct product is defined as

$$a_i' b^{\prime j} \equiv \frac{\partial x_k}{\partial x_i'} a_k \frac{\partial x_j'}{\partial x_l} b^l = \frac{\partial x_k}{\partial x_i'} \frac{\partial x_j'}{\partial x_l} \left(a_k b^l \right), \qquad (1)$$

which is a second-RANK TENSOR. The CONTRACTION of a direct product of first-RANK TENSORS is the SCALAR

$$\operatorname{contr}\left(a_i' b^{\prime j}\right) = a_i' b^{\prime i} = a_k b^k. \qquad (2)$$

For second-RANK TENSORS,

$$A_j^i B^{kl} = C_j^{ikl} \qquad (3)$$

$$C_j^{ikl'} = \frac{\partial x_i'}{\partial x_m} \frac{\partial x_n}{\partial x_j'} \frac{\partial x_k'}{\partial x_p} \frac{\partial x_l'}{\partial x_q} C_n^{mpq}. \qquad (4)$$

In general, the direct product of two TENSORS is a TENSOR of RANK equal to the sum of the two initial RANKS. The direct product is ASSOCIATIVE, but not COMMUTATIVE.

The tensor direct product of two tensors a and b can be implemented in *Mathematica* as

```
TensorDirectProduct[a_List,   b_List]   :=
Outer[Times, a, b]
```

See also DIRECT PRODUCT, MATRIX DIRECT PRODUCT, TENSOR PRODUCT (VECTOR SPACE)

References
Arfken, G. "Contraction, Direct Product." §3.2 in *Mathematical Methods for Physicists, 3rd ed.* Orlando, FL: Academic Press, pp. 124–126, 1985.

Tensor Dual

DUAL TENSOR

Tensor Product

TENSOR DIRECT PRODUCT, TENSOR PRODUCT (MODULE), TENSOR PRODUCT (VECTOR SPACE)

Tensor Product (Module)

The tensor product between MODULES A and B is a more general notion than the TENSOR PRODUCT BETWEEN VECTOR SPACES. In this case, we replace "scalars" by a RING R. The familiar formulas hold, but now α is any element of R,

$$(a_1 + a_2) \otimes b = a_1 \otimes b + a_2 \otimes b \qquad (1)$$

$$a \otimes (b_1 + b_2) = a \otimes b_1 + a \otimes b_2 \qquad (2)$$

$$\alpha(a \otimes b) = (\alpha a) \otimes b = a \otimes (\alpha b). \qquad (3)$$

This generalizes the definition of a tensor product for vector spaces since a VECTOR SPACE is a module over the scalar field. Also, VECTOR BUNDLES can be considered as PROJECTIVE MODULES over the ring of functions, and REPRESENTATIONS of a group G can be thought of as modules over CG. The generalization covers those kinds of tensor products as well.

There are some interesting possibilities for the tensor product of modules that don't occur in the case of vector spaces. It is possible for $A \otimes_R B$ to be identically zero. For example, the tensor product of $\mathbb{Z}_2$ and $\mathbb{Z}_3$ as modules over the integers, $\mathbb{Z}_2 \otimes_{\mathbb{Z}} \mathbb{Z}_3$, has no nonzero elements. It is enough to see that $a \otimes b = 0$. Notice that $1 = 3 - 2$. Then

$$(1) a \otimes b = (3 - 2) a \otimes b = (-2a) \otimes b + a \otimes (3b) = 0 + 0$$

$$= 0, \qquad (4)$$

since $-2a = -a - a = 0$ in $\mathbb{Z}_2$ and $3b = b + b + b = 0$ in $\mathbb{Z}_3$. In general, it is easier to show that elements are zero than to show they are not zero.

Another interesting property of tensor products is that if $f : A \to B$ is ONTO, then so is the induced map $g : A \otimes C \to B \otimes C$ for any other module C. But if $f : A \to B$ is injective, then $g : A \otimes C \to B \otimes C$ may not be injective.

For example, $f : \mathbb{Z}_2 \to \mathbb{Z}_4$, with $f(1) = 2$ is injective, but $g : \mathbb{Z}_2 \otimes_{\mathbb{Z}} \mathbb{Z}_2 \to \mathbb{Z}_4 \otimes_{\mathbb{Z}} \mathbb{Z}_2$, with $g(1 \otimes 1) = 2 \otimes 1$, is not injective. In $\mathbb{Z}_4 \otimes_{\mathbb{Z}} \mathbb{Z}_2$, we have $2 \otimes 1 = 1 \otimes 2 = 1 \otimes 0 = 0$.

There is an algebraic description of this failure of injectivity, called the TOR module.

Another way to think of the tensor product is in terms of its UNIVERSAL PROPERTY: Any BILINEAR MAP from $A \times B : \to C$ factors through the natural bilinear map $A \times B \to A \otimes B$.

See also MODULE, MODULE DIRECT SUM, PROJECTIVE MODULE, REPRESENTATION, TENSOR PRODUCT (MODULE), TENSOR PRODUCT (REPRESENTATION), TENSOR PRODUCT (VECTOR SPACE), TOR, UNIVERSAL PROPERTY, VECTOR BUNDLE, VECTOR SPACE

Tensor Product (Representation)

The TENSOR PRODUCT $V \otimes W$ of two REPRESENTATIONS of a GROUP G is also a REPRESENTATION of G. An element g of G acts on a basis element $v \otimes w$ by

$$g(v \otimes w) = gv \otimes gw.$$

If G is a FINITE GROUP and V is a FAITHFUL representation, then any representation is contained in $\otimes^n V$ for some n. If V_1 is a representation of G_1 and V_2 is a representation of G_2, then $V_1 \otimes V_2$ is a representation of $G_1 \times G_2$, called the EXTERNAL TENSOR PRODUCT. The regular tensor product is a special case, with the diagonal embedding of G in $G \times G$.

See also EXTERNAL TENSOR PRODUCT, GROUP, IRREDUCIBLE REPRESENTATION, REPRESENTATION, TENSOR PRODUCT (VECTOR SPACE), VECTOR SPACE

Tensor Product (Vector Space)

The tensor product of two VECTOR SPACES V and W, denoted $V \otimes W$ and also called the TENSOR DIRECT PRODUCT, is a way of creating a new VECTOR SPACE analogous to multiplication of integers. For instance,

$$\mathbb{R}^n \otimes \mathbb{R}^k \cong \mathbb{R}^{nk}. \tag{1}$$

In particular,

$$\mathbb{R} \otimes \mathbb{R}^n \cong \mathbb{R}^n. \tag{2}$$

Also, the tensor product obeys a distributive law with the DIRECT SUM operation:

$$U \otimes (V \oplus W) \cong (U \otimes V) \oplus (U \otimes W). \tag{3}$$

The analogy with an algebra is the motivation behind K-THEORY. The tensor product of two tensors a and b can be implemented in *Mathematica* as

```
TensorProduct[a_List, b_List] := Outer[List, a, b]
```

Algebraically, the vector space $V \otimes W$ is SPANNED by elements OF THE FORM $v \otimes w$, and the following rules are satisfied, for any scalar α. The definition is the same no matter which scalar FIELD is used.

$$(v_1 + v_2) \otimes w = v_1 \otimes w + v_2 \otimes w \tag{4}$$

$$v \otimes (w_1 + w_2) = v \otimes w_1 + v \otimes w_2 \tag{5}$$

$$\alpha(v \otimes w) = (\alpha v) \otimes w = v \otimes (\alpha w) \tag{6}$$

One basic consequence of these formulas is that

$$0 \otimes w = v \otimes 0 = 0. \tag{7}$$

A VECTOR BASIS v_i of V and w_j of W gives a basis for $V \otimes W$, namely $v_i \otimes w_j$, for all pairs (i, j). An arbitrary element of $V \otimes W$ can be written uniquely as $\sum a_{i,j} v_i \otimes w_j$, where $a_{i,j}$ are scalars. If V is n dimensional and W is k dimensional, then $V \otimes W$ has dimension nk.

Using tensor products, one can define SYMMETRIC TENSORS, ANTISYMMETRIC TENSORS, as well as the EXTERIOR ALGEBRA. Moreover, the tensor product is generalized to the TENSOR PRODUCT OF VECTOR BUNDLES. In particular, tensor products of the TANGENT BUNDLE and its DUAL BUNDLE are studied in RIEMANNIAN GEOMETRY and physics. Sections of these bundles are often called TENSORS. In addition, it is possible to take the TENSOR PRODUCT OF REPRESENTATIONS to get another representation.

All of these versions of tensor product can be understood as TENSOR PRODUCTS OF MODULES. The trick is to find the right way to think of these spaces as MODULES.

See also ANTISYMMETRIC TENSOR, EXTERIOR ALGEBRA, FIELD, K-THEORY, MODULE, SYMMETRIC TENSOR, TENSOR, TENSOR DIRECT PRODUCT, TENSOR PRODUCT (MODULE), TENSOR PRODUCT (REPRESENTATION), VECTOR SPACE

Tensor Space

Let E be a linear space over a FIELD K. Then the TENSOR PRODUCT $\otimes_{i=1}^{k} E$ is called a tensor space of degree k. More specifically, a tensor space of type (r, s) can be described as a TENSOR PRODUCT between r copies of VECTOR FIELDS and s copies of the dual vector fields, i.e., ONE-FORMS. For example,

$$T^{(3,\ 1)} = TM \otimes TM \otimes TM \otimes T^*M \tag{1}$$

is the VECTOR BUNDLE of $(3, 1)$ tensors on a MANIFOLD M. Tensors of type (r, s) form a VECTOR SPACE.

Tensor Spherical Harmonic

See also TENSOR, VECTOR SPACE

References

Yokonuma, T. *Tensor Spaces and Exterior Algebra.* Providence, RI: Amer. Math. Soc., 1992.

Tensor Spherical Harmonic

DOUBLE CONTRACTION RELATION

Tensor Transpose

TRANSPOSE

Tent Map

A piecewise linear, 1-D MAP on the interval [0, 1] exhibiting CHAOTIC dynamics and given by

$$x_{n+1} = \mu\left(1 - 2\left|x_n - \tfrac{1}{2}\right|\right).$$

The case $\mu = 1$ is equivalent to the LOGISTIC EQUATION with $R = 4$. The NATURAL INVARIANT of the tent map is $\rho = 1$.

See also 2X MOD 1 MAP, LOGISTIC EQUATION, LOGISTIC EQUATION: $R = 4$

Tent Problem

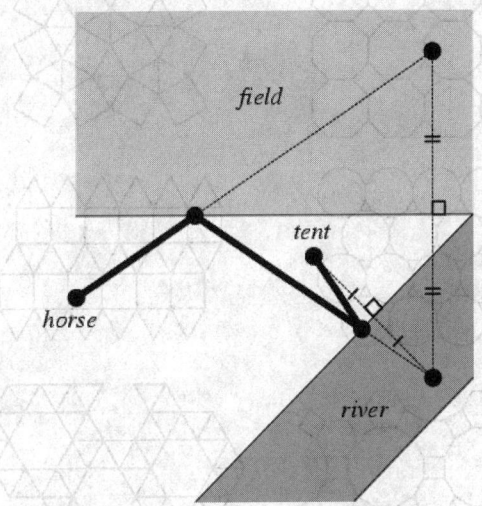

Consider a horse rider who wishes to feed his horse at a field, gather water from a river, and then return to his tent, all in the smallest overall distance possible. The path he should take is obtained by reflecting the tent across the near river bank, then reflecting this point about the field boundary, as illustrated above.

References

Steinhaus, H. *Mathematical Snapshots, 3rd ed.* New York: Dover, pp. 111–113, 1999.

Terminal

SINK (DIRECTED GRAPH)

Ternary

The BASE 3 method of counting in which only the digits 0, 1, and 2 are used. Ternary numbers arise in a number of problems in mathematics, including some problems of WEIGHING. According to Knuth (1981), "no substantial application of balanced ternary notation has been made" (balanced ternary uses digits -1, 0, and 1 instead of 0, 1, and 2). The following table gives the ternary equivalents of the first few decimal numbers.

1	1	11	102	21	210
2	2	12	110	22	211
3	10	13	111	23	212
4	11	14	112	24	220
5	12	15	120	25	221
6	20	16	121	26	222
7	21	17	122	27	1000
8	22	18	200	28	1001
9	100	19	201	29	1002
10	101	20	202	30	1010

Ternary digits have the following MULTIPLICATION TABLE.

×	0	1	2
0	0	0	0
1	0	1	2
2	0	2	11

Every EVEN NUMBER represented in ternary has an EVEN NUMBER (possibly 0) of 1s. This is true since a number is congruent mod $(B-1)$ to the sum of its base-B digits. In the case $B = 3$, there is only one digit (1) which is not a multiple of $B - 1$, so all we have to do is "cast out twos" and count the number of 1s in the base-3 representation.

Erdos and Graham (1980) conjectured that no POWER of 2, 2^n, is a SUM of distinct powers of 3 for $n > 8$. This is equivalent to the requirement that the ternary expansion of 2^n always contains a 2. This has been verified by Vardi (1991) up to $n = 2 \cdot 3^{30}$. N. J. A. Sloane has conjectured that any POWER of 2 has a 0 in its ternary expansion (Vardi 1991, p. 28).

See also BASE (NUMBER), BINARY, DECIMAL, HEXADECIMAL, OCTAL, QUATERNARY

References

Erdos, P. and Graham, R. L. *Old and New Problems and Results in Combinatorial Number Theory.* Geneva, Switzerland: L'Enseignement Mathématique Université de Genève, Vol. 28, 1980.

Gardner, M. "The Ternary System." Ch. 11 in *The Sixth Book of Mathematical Games from Scientific American.* Chicago, IL: University of Chicago Press, pp. 104–112, 1984.

Knuth, D. E. *The Art of Computer Programming. Vol. 2: Seminumerical Algorithms, 3rd ed.* Reading, MA: Addison-Wesley, pp. 173–175, 1998.

Lauwerier, H. *Fractals: Endlessly Repeated Geometric Figures.* Princeton, NJ: Princeton University Press, pp. 10–11, 1991.

Vardi, I. "The Digits of 2^n in Base Three." *Computational Recreations in Mathematica.* Reading, MA: Addison-Wesley, pp. 20–25, 1991.

Weisstein, E. W. "Bases." MATHEMATICA NOTEBOOK BASES.M.

Ternary Goldbach Conjecture

GOLDBACH CONJECTURE

Ternary Tree

See also BINARY TREE, COMPLETE TERNARY TREE

Tessellation

A regular TILING of POLYGONS (in 2-D), POLYHEDRA (3-D), or POLYTOPES (n-D) is called a tessellation. Tessellations can be specified using a SCHLÄFLI SYMBOL.

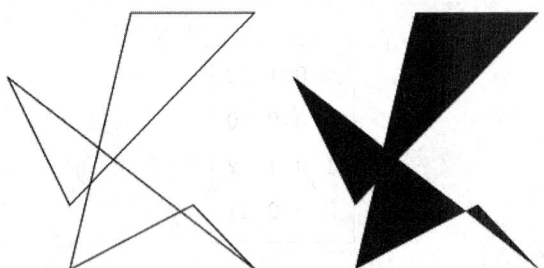

The breaking up of self-intersecting polygons into simple polygons (illustrated above) is also called tessellation (Woo *et al.* 1999).

Consider a 2-D tessellation with q regular p-gons at each VERTEX. In the PLANE,

$$\left(1 - \frac{2}{p}\right)\pi = \frac{2\pi}{q} \tag{1}$$

$$\frac{1}{p} + \frac{1}{q} = \frac{1}{2}, \tag{2}$$

so

$$(p - 2q)(q - 2) = 4 \tag{3}$$

(Ball and Coxeter 1987), and the only factorizations are

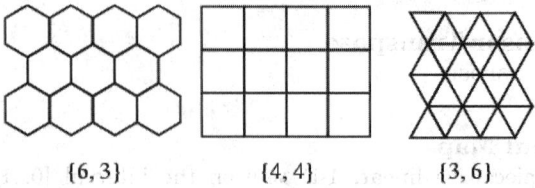

$$4 = 4 \cdot 1 = (6 - 2)(3 - 2) \Rightarrow \{6,\ 3\} \tag{4}$$

$$= 2 \cdot 2 = (4 - 2)(4 - 2) \Rightarrow \{4,\ 4\} \tag{5}$$

$$= 1 \cdot 4 = (3 - 2)(6 - 2) \Rightarrow \{3,\ 6\}. \tag{6}$$

Therefore, there are only three regular tessellations (composed of the HEXAGON, SQUARE, and TRIANGLE), illustrated as follows (Ghyka 1977, p. 76; Williams 1979, p. 36; Wells 1991, p. 213)

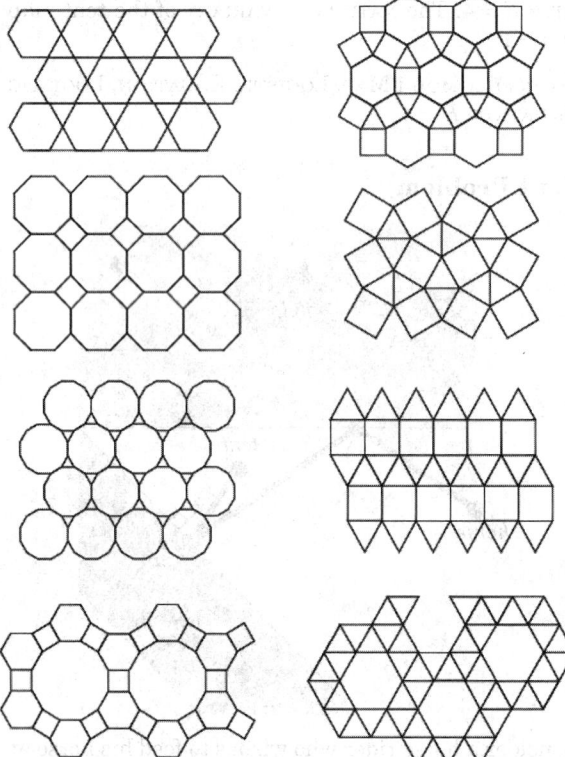

There do not exist any regular STAR POLYGON tessellations in the PLANE. Regular tessellations of the SPHERE by SPHERICAL TRIANGLES are called TRIANGULAR SYMMETRY GROUPS.

Regular tessellations of the plane by *two or more* convex regular POLYGONS such that the same POLYGONS in the same order surround each VERTEX are called semiregular tessellations, or sometimes Archimedean tessellations. In the plane, there are eight such tessellations, illustrated below (Ghyka 1977, pp. 76–78; Williams 1979, pp. 37–41; Steinhaus 1983, pp. 78–82; Wells 1991, pp. 226–227). Williams (1979, pp. 37–41) also illustrates the DUAL TESSELLATIONS of the semiregular tessellations. The DUAL TESSELLATION of the tessellation of squares and

equilateral triangles is called the CAIRO TESSELLA-
TION (Williams 1979, p. 38; Wells 1991, p. 23).

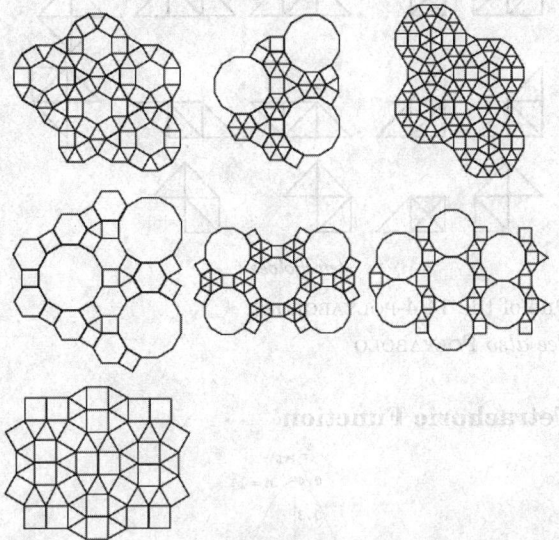

There are 14 polymorph, or demiregular, tessellations
which are orderly compositions of the three regular
and eight semiregular tessellations (Critchlow 1970,
pp. 62–67; Ghyka 1977, pp. 78–80; Williams 1979,
p. 43; Steinhaus 1983, pp. 79 and 81–82).

In 3-D, a POLYHEDRON which is capable of tessellating
space is called a SPACE-FILLING POLYHEDRON. Examples include the CUBE, RHOMBIC DODECAHEDRON, and
TRUNCATED OCTAHEDRON. There is also a 16-sided
space-filler and a convex POLYHEDRON known as the
SCHMITT-CONWAY BIPRISM which fills space only
aperiodically.

A tessellation of n-D polytopes is called a HONEY-
COMB.

See also ARCHIMEDEAN SOLID, CAIRO TESSELLATION,
CELL, DUAL TESSELLATION, HINGED TESSELLATION,
HONEYCOMB, HONEYCOMB CONJECTURE, SCHLÄFLI
SYMBOL, SEMIREGULAR POLYHEDRON, SPACE-FILLING
POLYHEDRON, SPIRAL-SIMILARITY TESSELLATION,
SYMMETRY, TILING, TRIANGULAR SYMMETRY GROUP,
TRIANGULATION

References

Ball, W. W. R. and Coxeter, H. S. M. *Mathematical Recrea-
tions and Essays, 13th ed.* New York: Dover, pp. 105–107,
1987.
Bhushan, A.; Kay, K.; and Williams, E. "Totally Tessellated."
http://library.thinkquest.org/16661/.
Britton, J. *Symmetry and Tessellations: Investigating Pat-
terns.* Englewood Cliffs, NJ: Prentice-Hall, 1999.
Critchlow, K. *Order in Space: A Design Source Book.* New
York: Viking Press, 1970.
Cundy, H. and Rollett, A. *Mathematical Models, 3rd ed.*
Stradbroke, England: Tarquin Pub., pp. 60–63, 1989.
Gardner, M. *Martin Gardner's New Mathematical Diver-
sions from Scientific American.* New York: Simon and
Schuster, pp. 201–203, 1966.
Gardner, M. "Tilings with Convex Polygons." Ch. 13 in *Time
Travel and Other Mathematical Bewilderments.* New
York: W. H. Freeman, pp. 162–176, 1988.
Ghyka, M. *The Geometry of Art and Life.* New York: Dover,
1977.
Kraitchik, M. "Mosaics." §8.2 in *Mathematical Recreations.*
New York: W. W. Norton, pp. 199–207, 1942.
Kraus, M. "Polygon Triangulation." http://library.wolfram.-
com/packages/polygontriangulation/.
Lines, L. *Solid Geometry.* New York: Dover, pp. 199 and
204–207 1965.
Pappas, T. "Tessellations." *The Joy of Mathematics.* San
Carlos, CA: Wide World Publ./Tetra, pp. 120–122, 1989.
Peterson, I. *The Mathematical Tourist: Snapshots of Modern
Mathematics.* New York: W. H. Freeman, p. 75, 1988.
Radin, C. *Miles of Tiles.* Providence, RI: Amer. Math. Soc.,
1999.
Rawles, B. *Sacred Geometry Design Sourcebook: Universal
Dimensional Patterns.* Nevada City, CA: Elysian Pub.,
1997.
Steinhaus, H. *Mathematical Snapshots, 3rd ed.* New York:
Dover, pp. 75–76, 1999.
Vichera, M. "Archimedean Polyhedra." http://alpha.ujep.cz/
~vicher/puzzle/telesa/telesa.htm.
Walsh, T. R. S. "Characterizing the Vertex Neighbourhoods
of Semi-Regular Polyhedra." *Geometriae Dedicata* 1, 117–
123, 1972.
Weisstein, E. W. "Books about Tilings." http://www.trea-
sure-troves.com/books/Tilings.html.
Wells, D. *The Penguin Dictionary of Curious and Interesting
Geometry.* London: Penguin, pp. 121, 213, and 226–227,
1991.
Williams, R. *The Geometrical Foundation of Natural Struc-
ture: A Source Book of Design.* New York: Dover, pp. 35–
43, 1979.
Woo, M.; Neider, J.; Davis, T.; and Shreiner, D. Ch. 11 in
*OpenGL 1.2 Programming Guide, 3rd ed.: The Official
Guide to Learning OpenGL, Version 1.2.* Reading, MA:
Addison-Wesley, 1999.

Tesseract

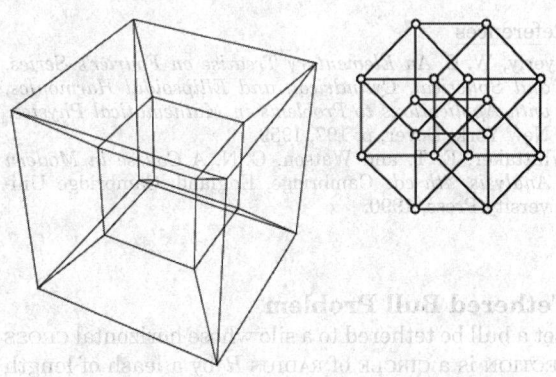

The HYPERCUBE in $\mathbb{R}^4$, also called the 8-cell, is known
as a tesseract. It has the SCHLÄFLI SYMBOL $\{4, 3, 3\}$,
and VERTICES $(\pm 1, \pm 1, \pm 1, \pm 1)$. The above figures
show two visualizations of the tesseract. The figure
on the left is a projection of the tesseract in 3-space
(Gardner 1977), and the figure on the right is the
GRAPH of the tesseract symmetrically projected into

the PLANE (Coxeter 1973). A tesseract has 16 VERTICES, 32 EDGES, 24 SQUARES, and 8 CUBES.

See also CUBE, HYPERCUBE, MAGIC TESSERACT, POLYTOPE, SIMPLEX

References

Coxeter, H. S. M. *Regular Polytopes, 3rd ed.* New York: Dover, p. 123, 1973.

Dewdney, A. K. "Computer Recreations: A Program for Rotating Hypercubes Induces Four-Dimensional Dementia." *Sci. Amer.* **254**, 14–23, Mar. 1986.

Gardner, M. "Hypercubes." Ch. 4 in *Mathematical Carnival: A New Round-Up of Tantalizers and Puzzles from Scientific American.* New York: Vintage Books, pp. 41–54, 1977.

Smith, H. J. "The Tesseract: A Look into 4-Dimensional Space." http://pweb.netcom.com/~hjsmith/WireFrame4/tesseract.html.

Tesseral Harmonic

A SPHERICAL HARMONIC OF THE FORM $\frac{\cos}{\sin}(m\phi)P_l^{m(\cos\theta)}$. These harmonics are so named because the curves on which they vanish are $l - m$ parallels of latitude and $2m$ meridians, which divide the surface of a sphere into quadrangles whose angles are right angles (Whittaker and Watson 1990, p. 392).

Resolving $P_l(\cos\theta)$ into factors linear in $\cos^2\theta$, multiplied by $\cos\theta$ when l is ODD, then replacing $\cos\theta$ by z/r allows the tesseral harmonics to be expressed as products of factors linear in x^2, y^2, and z^2 multiplied by one of $1, x, y, z, yz, zx, xy$, and xyz (Whittaker and Watson 1990, p. 536).

See also SECTORIAL HARMONIC, SPHERICAL HARMONIC, ZONAL HARMONIC

References

Byerly, W. E. *An Elementary Treatise on Fourier's Series, and Spherical, Cylindrical, and Ellipsoidal Harmonics, with Applications to Problems in Mathematical Physics.* New York: Dover, p. 197, 1959.

Whittaker, E. T. and Watson, G. N. *A Course in Modern Analysis, 4th ed.* Cambridge, England: Cambridge University Press, 1990.

Tethered Bull Problem

Let a bull be tethered to a silo whose horizontal CROSS SECTION is a CIRCLE of RADIUS R by a leash of length L. Then the AREA which the bull can graze if $L \leq R\pi$ is

$$A = \frac{\pi L^2}{2} + \frac{L^3}{3R}.$$

References

Hoffman, M. E. "The Bull and the Silo: An Application of Curvature." *Amer. Math. Monthly* **105**, 55–58, 1998.

Tetrabolo

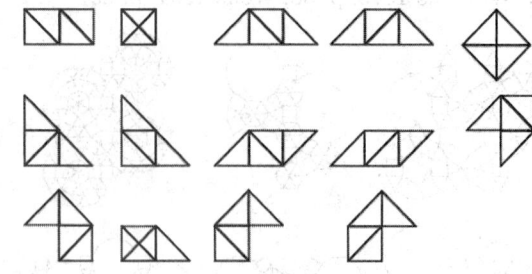

tetraboloes

One of the 14 4-POLYABOLOES.

See also POLYABOLO

Tetrachoric Function

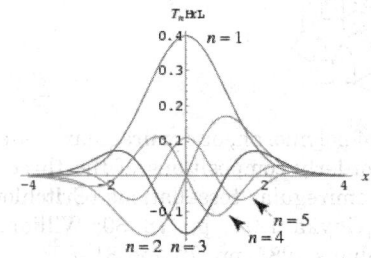

The function defined by

$$T_n(x) = \frac{(-1)^{n-1}}{\sqrt{n}} Z^{(n-1)}(x),$$

where

$$Z(x) = \frac{1}{\sqrt{2\pi}} e^{-x^2/2}$$

and $Z^{(k)}(x)$ is the kth derivative of $Z(x)$.

See also NORMAL DISTRIBUTION, STANDARD NORMAL DISTRIBUTION

References

Kenney, J. F. and Keeping, E. S. "Tetrachoric Correlation." §8.5 in *Mathematics of Statistics, Pt. 2, 2nd ed.* Princeton, NJ: Van Nostrand, pp. 205–207, 1951.

Tetracontagon

A 40-sided POLYGON.

Tetracuspid

HYPOCYCLOID–4–CUSPED

Tetracyclic Plane

The set of all points **x** that can be put into one-to-one correspondence with sets of essentially distinct values of four homogeneous coordinates $x_0 : x_1 : x_2 : x_3$, not all simultaneously zero, which are connected by the relation

$$\mathbf{x} \cdot \mathbf{x} = x_0^2 + x_1^2 + x_2^2 + x_3^2 = 0. \qquad (1)$$

See also PENTASPHERICAL SPACE

References

Coolidge, J. L. "Pentaspherical Space." Ch. 7 in *A Treatise on the Geometry of the Circle and Sphere.* New York: Chelsea, pp. 282–305, 1971.

Tetrad

A SET of four, also called a QUARTET.

See also HEXAD, MONAD, PAIR, QUARTET, QUINTET, TRIAD, TRIPLE, TWINS

Tetradecagon

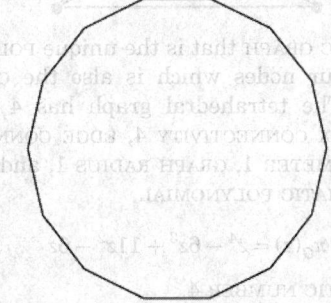

A 14-sided POLYGON, sometimes called a TETRAKAIDE-CAGON.

Tetradecahedron

A 14-sided POLYHEDRON, sometimes called a TETRA-KAIDECAHEDRON.

See also CUBOCTAHEDRON, TRUNCATED OCTAHEDRON

References

Ghyka, M. *The Geometry of Art and Life.* New York: Dover, p. 54, 1977.

Tetradic

Tetradics transform DYADICS in much the same way that DYADICS transform VECTORS. They are represented using Hebrew characters and have 81 components (Morse and Feshbach 1953, pp. 72–73). The use of tetradics is archaic, since TENSORS perform the same function but are notationally simpler.

References

Morse, P. M. and Feshbach, H. *Methods of Theoretical Physics, Part 1.* New York: McGraw-Hill, 1953.

Tetradyakis Hexahedron

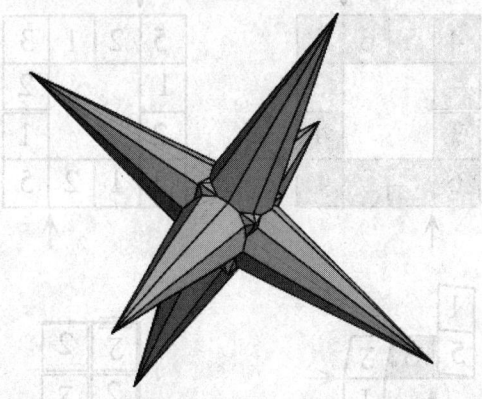

The DUAL POLYHEDRON of the CUBITRUNCATED CU-BOCTAHEDRON U_{16} and Wenninger dual W_{79}.

See also DUAL POLYHEDRON, CUBITRUNCATED CUBOC-TAHEDRON

References

Wenninger, M. J. *Dual Models.* Cambridge, England: Cambridge University Press, p. 92, 1983.

Tetraflexagon

A FLEXAGON made with SQUARE faces. Gardner (1961) shows how to construct a tri-tetraflexagon,

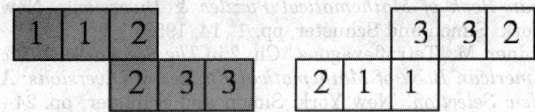

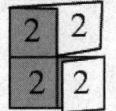
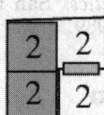

tetra-tetraflexagon,

1	1	2	3
3	2	1	1
1	1	2	3

4	4	3	3
2	3	4	4
4	4	3	2

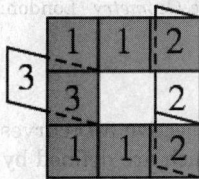
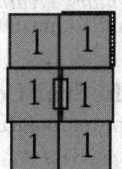

and hexa-tetraflexagon.

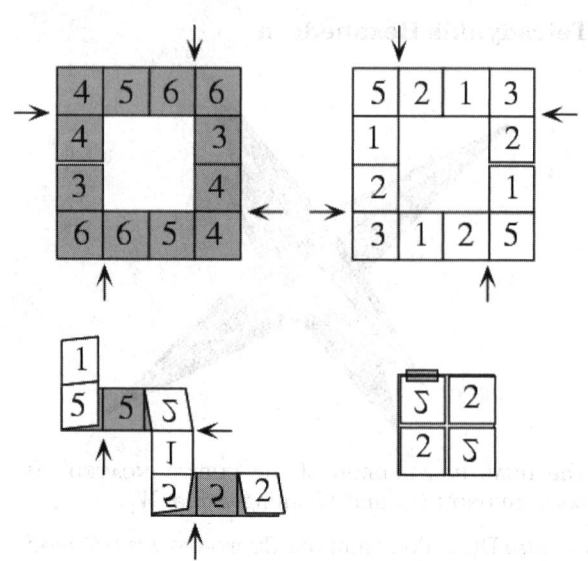

See also FLEXAGON, FLEXATUBE, HEXAFLEXAGON

References

Chapman, P. B. "Square Flexagons." *Math. Gaz.* **45**, 192–194, 1961.
Cundy, H. and Rollett, A. *Mathematical Models, 3rd ed.* Stradbroke, England: Tarquin Pub., p. 207, 1989.
Gardner, M. "Mathematical Games: About Tetraflexagons and Tetraflexigation." *Sci. Amer.* **198**, 122–126, May 1958.
Gardner, M. "Hexaflexagons." Ch. 1 in *The Scientific American Book of Mathematical Puzzles & Diversions.* New York: Simon and Schuster, pp. 1–14, 1959.
Gardner, M. "Tetraflexagons." Ch. 2 in *The Second Scientific American Book of Mathematical Puzzles & Diversions: A New Selection.* New York: Simon and Schuster, pp. 24–31, 1961.
Pappas, T. "Making a Tri-Tetra Flexagon." *The Joy of Mathematics.* San Carlos, CA: Wide World Publ./Tetra, p. 107, 1989.

Tetragon

QUADRILATERAL

Tetragram

Lachlan's term for a set of four lines, no three of which are CONCURRENT.

See also TETRASTIGM

References

Lachlan, R. "Properties of a Tetragram." §147–155 in *An Elementary Treatise on Modern Pure Geometry.* London: Macmillian, pp. 90–97, 1893.

Tetrahedral Coordinates

Coordinates useful for plotting projective 3-D curves OF THE FORM $f(x_0, x_1, x_2, x_3) = 0$ which are defined by

$$x_0 = 1 - z - \sqrt{2}x$$

$$x_1 = 1 - z + \sqrt{2}x$$
$$x_2 = 1 + z + \sqrt{2}y$$
$$x_3 = 1 + z - \sqrt{2}y$$

See also CAYLEY CUBIC, KUMMER SURFACE

Tetrahedral Graph

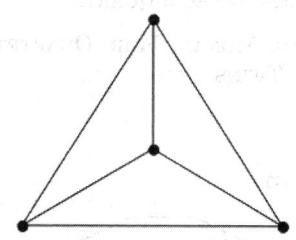

The PLATONIC GRAPH that is the unique POLYHEDRAL GRAPH on four nodes which is also the COMPLETE GRAPH K_4. The tetrahedral graph has 4 nodes, 6 edges, VERTEX CONNECTIVITY 4, EDGE CONNECTIVITY 3, GRAPH DIAMETER 1, GRAPH RADIUS 1, and GIRTH 3. It has CHROMATIC POLYNOMIAL

$$\pi_G(z) = z^4 - 6z^3 + 11z^2 - 6z$$

and CHROMATIC NUMBER 4.

See also CUBICAL GRAPH, DODECAHEDRAL GRAPH, ICOSAHEDRAL GRAPH, OCTAHEDRAL GRAPH, PLATONIC GRAPH, POLYHEDRAL GRAPH, TETRAHEDRON

References

Bondy, J. A. and Murty, U. S. R. *Graph Theory with Applications.* New York: North Holland, p. 234, 1976.

Tetrahedral Group

The POINT GROUP of symmetries of the TETRAHEDRON having order 12 and denoted T_d. The tetrahedral group has symmetry operations E, $8C_3$, $3C_2$, $6S_4$, and $6\sigma_d$ (Cotton 1990).

See also ICOSAHEDRAL GROUP, OCTAHEDRAL GROUP, POINT GROUPS, POLYHEDRAL GROUP, TETRAHEDRON

References

Cotton, F. A. *Chemical Applications of Group Theory, 3rd ed.* New York: Wiley, p. 47, 1990.
Coxeter, H. S. M. "The Polyhedral Groups." §3.5 in *Regular Polytopes, 3rd ed.* New York: Dover, pp. 46–47, 1973.
Lomont, J. S. "Icosahedral Group." §3.10.C in *Applications of Finite Groups.* New York: Dover, p. 81, 1987.

Tetrahedral Number

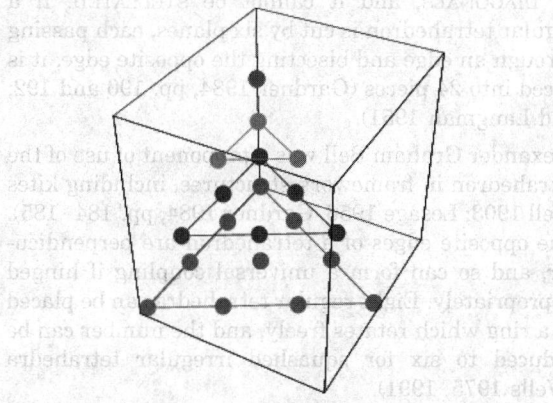

A FIGURATE NUMBER Te_n OF THE FORM

$$Te_n = \sum_{i=1}^{n} T_n = \tfrac{1}{6} n(n+1)(n+2) = \binom{n+2}{3}, \qquad (1)$$

where T_n is the nth TRIANGULAR NUMBER and $\binom{n}{m}$ is a BINOMIAL COEFFICIENT. These numbers correspond to placing discrete points in the configuration of a TETRAHEDRON (triangular base pyramid). Tetrahedral numbers are PYRAMIDAL NUMBERS with $r = 3$, and are the sum of consecutive TRIANGULAR NUMBERS. The first few are 1, 4, 10, 20, 35, 56, 84, 120, ... (Sloane's A000292). The GENERATING FUNCTION of the tetrahedral numbers is

$$\frac{x}{(x-1)^4} = x + 4x^2 + 10x^3 + 20x^4 + \dots. \qquad (2)$$

Tetrahedral numbers are EVEN, except for every fourth tetrahedral number, which is ODD (Conway and Guy 1996).

The only numbers which are simultaneously SQUARE and TETRAHEDRAL are $Te_1 = 1$, $Te_2 = 4$, and $Te_{48} = 19600$ (giving $S_1 = 1$, $S_2 = 4$, and $S_{140} = 19600$), as proved by Meyl (1878; cited in Dickson 1952, p. 25).

Numbers which are simultaneously TRIANGULAR and TETRAHEDRAL satisfy the BINOMIAL COEFFICIENT equation

$$T_n = \binom{n+1}{2} = \binom{m+2}{3} = Te_m, \qquad (3)$$

the only solutions of which are

$$Te_1 = T_1 = 1 \qquad (4)$$

$$Te_3 = T_4 = 10 \qquad (5)$$

$$Te_8 = T_{15} = 120 \qquad (6)$$

$$Te_{20} = T_{55} = 1540 \qquad (7)$$

$$Te_{34} = T_{119} = 7140 \qquad (8)$$

(Sloane's A027568; Avanesov 1966/1967; Mordell 1969, p. 258; Guy 1994, p. 147).

Beukers (1988) has studied the problem of finding numbers which are simultaneously tetrahedral and PYRAMIDAL via INTEGER points on an ELLIPTIC CURVE, and finds that the only solution is the trivial $Te_1 = P_1 = 1$.

See also PYRAMIDAL NUMBER, SQUARE PYRAMIDAL NUMBER, TRIANGULAR NUMBER, TRUNCATED TETRAHEDRAL NUMBER

References

Avanesov, E. T. "Solution of a Problem on Figurate Numbers" [Russian]. *Acta Arith.* **12**, 409–420, 1966/1967.

Ball, W. W. R. and Coxeter, H. S. M. *Mathematical Recreations and Essays, 13th ed.* New York: Dover, p. 59, 1987.

Beukers, F. "On Oranges and Integral Points on Certain Plane Cubic Curves." *Nieuw Arch. Wisk.* **6**, 203–210, 1988.

Conway, J. H. and Guy, R. K. *The Book of Numbers.* New York: Springer-Verlag, pp. 44–46, 1996.

Dickson, L. E. *History of the Theory of Numbers, Vol. 2: Diophantine Analysis.* New York: Chelsea, 1952.

Guy, R. K. "Figurate Numbers." §D3 in *Unsolved Problems in Number Theory, 2nd ed.* New York: Springer-Verlag, pp. 147–150, 1994.

Meyl, A.-J.-J. "Solution de Question 1194." *Nouv. Ann. Math.* **17**, 464–467, 1878.

Mordell, L. J. *Diophantine Equations.* New York: Academic Press, p. 258, 1969.

Sloane, N. J. A. Sequences A000292/M3382 and A027568 in "An On-Line Version of the Encyclopedia of Integer Sequences." http://www.research.att.com/~njas/sequences/eisonline.html.

Tetrahedral Surface

A SURFACE given by the PARAMETRIC EQUATIONS

$$x = A(u-a)^m (v-a)^n$$
$$y = B(u-b)^m (v-b)^n$$
$$z = C(u-c)^m (v-c)^n.$$

References

Eisenhart, L. P. *A Treatise on the Differential Geometry of Curves and Surfaces.* New York: Dover, p. 267, 1960.

Tetrahedroid

A special case of a quartic KUMMER SURFACE.

See also KUMMER SURFACE

References

Fischer, G. (Ed.). *Mathematical Models from the Collections of Universities and Museums.* Braunschweig, Germany: Vieweg, pp. 17–19, 1986.

Guy, R. K. *Unsolved Problems in Number Theory, 2nd ed.* New York: Springer-Verlag, p. 183, 1994.

Tetrahedron

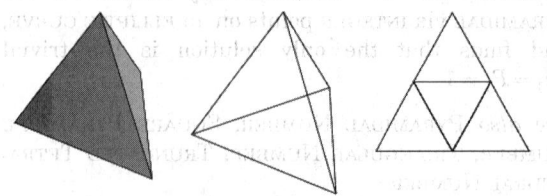

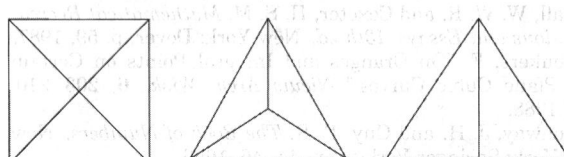

The regular tetrahedron, often simply called "the" tetrahedron, is the PLATONIC SOLID P_1 with four VERTICES, six EDGES, and four equivalent EQUILATERAL TRIANGULAR faces, $4\{3\}$. It is also UNIFORM POLYHEDRON U_1 and Wenninger model W_1. It is described by the SCHLÄFLI SYMBOL $\{3, 3\}$ and the WYTHOFF SYMBOL is $3 \mid 2\ 3$.

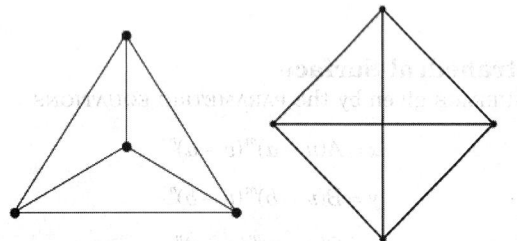

It is the prototype of the TETRAHEDRAL GROUP T_d. The connectivity of the vertices is given by the TETRAHEDRAL GRAPH, equivalent to the CIRCULANT GRAPH $Ci_{1, 2, 3}(4)$ and the COMPLETE GRAPH K_4.

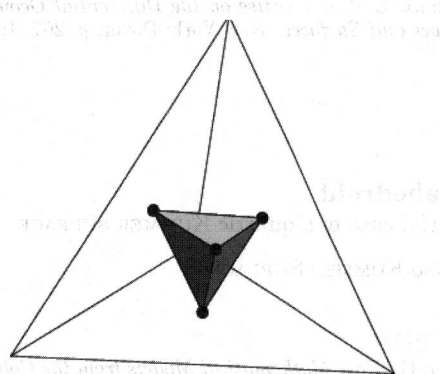

The tetrahedron is its own DUAL POLYHEDRON, and therefore the centers of the faces of a tetrahedron form another tetrahedron (Steinhaus 1983, p. 201).

The tetrahedron is the only simple POLYHEDRON with no DIAGONALS, and it cannot be STELLATED. If a regular tetrahedron is cut by six planes, each passing through an edge and bisecting the opposite edge, it is sliced into 24 pieces (Gardner 1984, pp. 190 and 192; and Langman 1951).

Alexander Graham Bell was a proponent of use of the tetrahedron in framework structures, including kites (Bell 1903; Lesage 1956, Gardner 1984, pp. 184–185). The opposite edges of a tetrahedron are perpendicular, and so can form a universal coupling if hinged appropriately. Eight regular tetrahedra can be placed in a ring which rotates freely, and the number can be reduced to six for squashed irregular tetrahedra (Wells 1975, 1991)

Perspective View *Bottom View* *Side View*

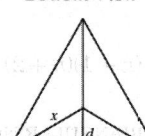 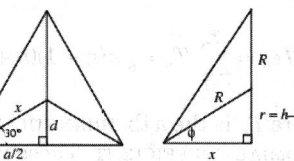

Let a tetrahedron be length a on a side. The VERTICES are located at $(x, 0, 0)$, $(-d, \pm a/2, 0)$, and $(0, 0, h)$. From the figure,

$$x = \frac{\frac{a}{2}}{\cos\left(\dfrac{\pi}{6}\right)} = \tfrac{1}{3}\sqrt{3}a. \tag{1}$$

d is then

$$d = \sqrt{x^2 - \left(\tfrac{1}{2}a\right)^2} = \tfrac{1}{6}\sqrt{3}a. \tag{2}$$

This gives the AREA of the base as

$$A = \tfrac{1}{2}a(R + x) = \tfrac{1}{4}\sqrt{3}a^2. \tag{3}$$

The height is

$$h = \sqrt{a^2 - x^2} = \tfrac{1}{3}\sqrt{6}a. \tag{4}$$

The CIRCUMRADIUS R is found from

$$x^2 + (h - R)^2 = R^2 \tag{5}$$

$$x^2 + h^2 - 2hR + R^2 = R^2. \tag{6}$$

Solving gives

$$R = \frac{x^2 + h^2}{2h} = \tfrac{1}{4}\sqrt{6}a \approx 0.61237a. \tag{7}$$

The INRADIUS r is

$$r \equiv h - R = \tfrac{1}{12}\sqrt{6}a \approx 0.20412a, \tag{8}$$

which is also

$$r = \tfrac{1}{4} h = \tfrac{1}{3} R. \tag{9}$$

The ANGLE between the bottom plane and center is then given by

$$\phi = \tan^{-1}\left(\frac{r}{x}\right) = \tan^{-1}\left(\tfrac{1}{4}\sqrt{2}\right). \tag{10}$$

Given a tetrahedron of edge length a situated with vertical apex and with the origin of coordinate system at the CENTROID of the vertices, the four VERTICES are located at $(x, 0, -r)$, $(-d, \pm a/2, -r)$, $(0, 0, R)$, with, as shown above

$$x = \tfrac{1}{3}\sqrt{3}a \tag{11}$$

$$r = \tfrac{1}{12}\sqrt{6}a \tag{12}$$

$$R = \tfrac{1}{4}\sqrt{6}a \tag{13}$$

$$d = \tfrac{1}{6}\sqrt{3}a. \tag{14}$$

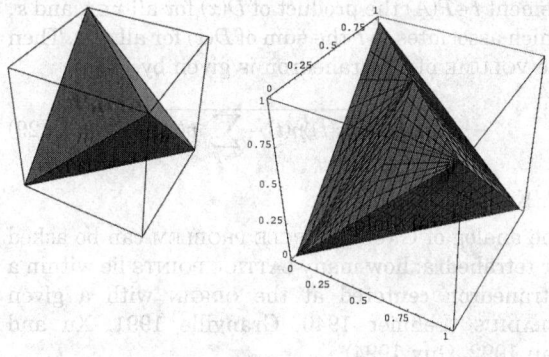

The vertices of a tetrahedron of side length $\sqrt{2}$ can also be given by a particularly simple form when the vertices are taken as corners of a cube (Gardner 1984, pp. 192–194). One such tetrahedron for a cube of side length 1 gives the tetrahedron of side length $\sqrt{2}$ having vertices $(0, 0, 0)$, $(0, 1, 1)$, $(1, 0, 1)$, $(1, 1, 0)$, and satisfies the inequalities

$$x + y + z \le 2 \tag{15}$$

$$x - y - z \le 0 \tag{16}$$

$$-x + y - z \le 0 \tag{17}$$

$$-x - y + z \le 0. \tag{18}$$

The following table gives polyhedra which can be constructed by CUMULATION of a tetrahedron by pyramids of given heights h.

h	$(r+h)/h$	Result
$\tfrac{1}{15}\sqrt{6}$	$\tfrac{7}{5}$	TRIAKIS TETRAHEDRON
$\tfrac{1}{6}\sqrt{6}$	2	CUBE
$\tfrac{1}{3}\sqrt{6}$	3	9-faced star DELTAHEDRON

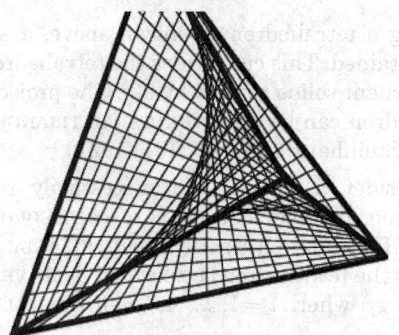

Connecting opposite pairs of edges with equally spaced lines gives a configuration like that shown above which divides the tetrahedron into eight regions: four open and four closed (Steinhaus 1983, p. 246).

The MIDRADIUS of the tetrahedron is

$$\rho = \sqrt{r^2 + d^2} = \sqrt{\tfrac{1}{8}}\, a = \tfrac{1}{4}\sqrt{2}a$$

$$\approx 0.35355a. \tag{19}$$

Plugging in for the VERTICES gives

$$\left(a\sqrt{3},\, 0,\, 0\right),\ \left(-\tfrac{1}{6}\sqrt{3}a,\, \pm\tfrac{1}{2}a,\, 0\right),\ \text{and}\ \left(0,\, 0,\, \tfrac{1}{2}\sqrt{6}a\right). \tag{20}$$

Since a tetrahedron is a PYRAMID with a triangular base, $V = \tfrac{1}{3}A_b h$, giving

$$V = \tfrac{1}{12}\sqrt{2}a^3. \tag{21}$$

The DIHEDRAL ANGLE is

$$\alpha = \tan^{-1}\left(2\sqrt{2}\right) = \sin^{-1}\left(\tfrac{1}{3}\sqrt{3}\right) = \cos^{-1}\left(\tfrac{1}{3}\right)$$

$$\approx 70.53°. \tag{22}$$

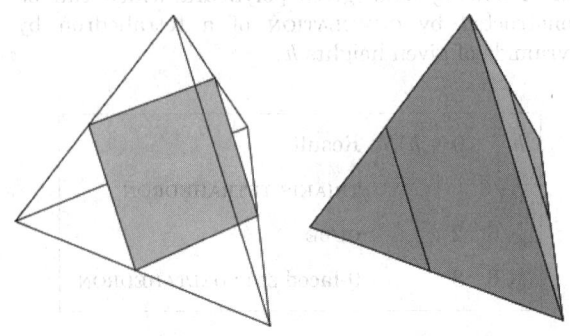

By slicing a tetrahedron as shown above, a SQUARE can be obtained. This cut divides the tetrahedron into two congruent solids rotated by 90°. The projection of a tetrahedron can be an EQUILATERAL TRIANGLE or a SQUARE (Steinhaus 1983, pp. 191–192).

Now consider a general (not necessarily regular) tetrahedron, defined as a convex POLYHEDRON consisting of four (not necessarily identical) TRIANGULAR faces. Let the tetrahedron be specified by its VERTICES at (x_i, y_i, z_i) where $i = 1, ..., 4$. Then the VOLUME is given by

$$V = \frac{1}{3!} \begin{vmatrix} x_1 & y_1 & z_1 & 1 \\ x_2 & y_2 & z_2 & 1 \\ x_3 & y_3 & z_3 & 1 \\ x_4 & y_4 & z_4 & 1 \end{vmatrix}. \tag{23}$$

Specifying the tetrahedron by the three EDGE vectors $\mathbf{a}$, $\mathbf{b}$, and $\mathbf{c}$ from a given VERTEX, the VOLUME is

$$V = \frac{1}{3!} |\mathbf{a} \cdot (\mathbf{b} \times \mathbf{c})|. \tag{24}$$

If the faces are congruent and the sides have lengths a, b, and c, then

$$V = \sqrt{\frac{(a^2 + b^2 - c^2)(a^2 + c^2 - b^2)(b^2 + c^2 - a^2)}{72}} \tag{25}$$

(Klee and Wagon 1991, p. 205). In general, if the edge between vertices i and j are of length d_{ij}, then the volume V is given by the CAYLEY-MENGER DETERMINANT

$$288 V^2 = \begin{vmatrix} 0 & 1 & 1 & 1 & 1 \\ 1 & 0 & d_{12}^2 & d_{13}^2 & d_{14}^2 \\ 1 & d_{21}^2 & 0 & d_{23}^2 & d_{24}^2 \\ 1 & d_{31}^2 & d_{32}^2 & 0 & d_{34}^2 \\ 1 & d_{41}^2 & d_{42}^2 & d_{43}^2 & 0 \end{vmatrix}. \tag{26}$$

Consider an arbitrary TETRAHEDRON $A_1 A_2 A_3 A_4$ with triangles $T_1 = \Delta A_2 A_3 A_4$, $T_2 = \Delta A_1 A_3 A_4$, $T_3 = \Delta A_1 A_2 A_4$, and $T_4 = A_1 A_2 A_3$. Let the areas of these triangles be s_1, s_2, s_3, and s_4, respectively, and denote the DIHEDRAL ANGLE with respect to T_i and T_j for $i \neq j =$

1, 2, 3, 4 by θ_{ij}. Then the four face areas are connected by

$$s_k^2 = \sum_{\substack{j \neq k \\ 1 \leq j \leq 4}} s_j^2 - 2 \sum_{\substack{i,\, j \neq k \\ 1 \leq i,\, j \leq 4}} s_i s_j \cos \theta_{ij} \tag{27}$$

involving the six DIHEDRAL ANGLES (Dostor 1905, pp. 252–293; Lee 1997). This is a generalization of the LAW OF COSINES to the tetrahedron. Furthermore, for any $i \neq j = 1, 2, 3, 4$,

$$V = \frac{2}{3 l_{ij}} s_i s_j \sin \theta_{ij}, \tag{28}$$

where l_{ij} is the length of the common edge of T_i and T_j (Lee 1997).

Let A be the set of edges of a tetrahedron and $P(A)$ the power set of A. Write $\bar{t}$ for the complement in A of an element $t \in P(A)$. Let F be the set of triples $\{x, y, z\} \in P(A)$ such that x, y, z span a face of the tetrahedron, and let G be the set of $(e \cap f) \cup \overline{(e \cup f)} \in P(A)$, so that $e, f \in F$ and $e \neq f$. In G, there are therefore three elements which are the pairs of opposite edges. Now define D, which associates to an edge x of length L the quantity $(L/\sqrt{12})^2$, p, which associates to an element $t \in P(A)$ the product of $D(x)$ for all $x \in t$, and s, which associates to t the sum of $D(x)$ for all $x \in t$. Then the VOLUME of a tetrahedron is given by

$$\sqrt{\sum_{t \in G} (s(\bar{t}) - s(t)) p(t) - \sum_{t \in F} p(t)} \tag{29}$$

(P. Kaeser).

The analog of GAUSS'S CIRCLE PROBLEM can be asked for tetrahedra: how many LATTICE POINTS lie within a tetrahedron centered at the ORIGIN with a given INRADIUS (Lehmer 1940, Granville 1991, Xu and Yau 1992, Guy 1994).

There are a number of interesting and unexpected theorems on the properties of general (i.e., not necessarily regular) tetrahedron (Altshiller-Court 1979). If a plane divides two opposite edges of a tetrahedron in a given ratio, then it divides the volume of the tetrahedron in the same ratio (Altshiller-Court 1979, p. 89). It follows that any plane passing through a BIMEDIAN of a tetrahedron bisects the volume of the tetrahedron (Altshiller-Court 1979, p. 90).

Let the vertices of a tetrahedron be denoted A, B, C, and D, and denote the side lengths $BC = a$, $CA = b$, $AB = c$, $DA = a'$, $DB = b'$, and $DC = c'$. Then if Δ denotes the area of the triangle with sides of lengths by aa', bb', and cc', the VOLUME and CIRCUMRADIUS of the tetrahedron are related by the beautiful formula

$$6 R V = \Delta \tag{30}$$

(Crelle 1821, p. 117; von Staudt 1860; Rouché and

Comberousse 1922, pp. 568–576 and 643–664; Altshiller-Court 1979, p. 250).

See also AUGMENTED TRUNCATED TETRAHEDRON, BANG'S THEOREM, CUBE TETRAHEDRON PICKING, EHRHART POLYNOMIAL, HERONIAN TETRAHEDRON, HILBERT'S 3RD PROBLEM, ISOSCELES TETRAHEDRON, PENTATOPE, REULEAUX TETRAHEDRON, SIERPINSKI TETRAHEDRON, SPHERE TETRAHEDRON PICKING, STELLA OCTANGULA, TANGENT SPHERES, TANGENTIAL TETRAHEDRON, TETRAHEDRON 4-COMPOUND, TETRAHEDRON 5-COMPOUND, TETRAHEDRON 10-COMPOUND, TRIRECTANGULAR TETRAHEDRON, TRUNCATED TETRAHEDRON

References

Altshiller-Court, N. "The Tetrahedron." Ch. 4 in *Modern Pure Solid Geometry*. New York: Chelsea, pp. 48–110, 1979.

Balliccioni, A. *Coordonnées barycentriques et géométrie.* Claude Hermant, 1964.

Bell, A. G. "The Tetrahedral Principle in Kite Structure." *Nat. Geographic* **44**, 219–251, 1903.

Beyer, W. H. *CRC Standard Mathematical Tables, 28th ed.* Boca Raton, FL: CRC Press, p. 228, 1987.

Couderc, P. and Balliccioni, A. *Premier Livre du Tétraèdre.* Paris: Gauthier-Villars, 1935.

Crelle, A. L. "Einige Bemerkungen über die dreiseitige Pyramide." *Sammlung mathematischer Aufsätze u. Bemerkungen* **1**, 105–132, 1821.

Cundy, H. and Rollett, A. "Tetrahedron. 3^3." §3.5.1 in *Mathematical Models, 3rd ed.* Stradbroke, England: Tarquin Pub., p. 84, 1989.

Davie, T. "The Tetrahedron." http://www.dcs.st-and.ac.uk/~ad/mathrecs/polyhedra/tetrahedron.html.

Dostor, G. *Eléments de la théorie des déterminants, avec application à l'algèbre, la trigonométrie et la géométrie analytique dans le plan et l'espace, 2ème ed.* Paris: Gauthier-Villars, pp. 252–293, 1905.

Gardner, M. "Tetrahedrons." Ch. 19 in *The Sixth Book of Mathematical Games from Scientific American.* Chicago, IL: University of Chicago Press, pp. 183–194, 1984.

Dostor, G. *Eléments de la théorie des déterminants, avec application à l'algèbre, la trigonométrie et la géométrie analytique dans le plan et l'espace, 2ème ed.* Paris: Gauthier-Villars, 1905.

Granville, A. "The Lattice Points of an n-Dimensional Tetrahedron." *Aequationes Math.* **41**, 234–241, 1991.

Guy, R. K. "Gauß's Lattice Point Problem." §F1 in *Unsolved Problems in Number Theory, 2nd ed.* New York: Springer-Verlag, pp. 240–241, 1994.

Harris, J. W. and Stocker, H. "Tetrahedron." §4.3.1 and 4.4.2 in *Handbook of Mathematics and Computational Science.* New York: Springer-Verlag, pp. 98–100, 1998.

Klee, V. and Wagon, S. *Old and New Unsolved Problems in Plane Geometry and Number Theory, rev. ed.* Washington, DC: Math. Assoc. Amer., 1991.

Langman, H. *Scripta Math.*, Mar.-Jun. 1951.

Lee, J. R. "The Law of Cosines in a Tetrahedron." *J. Korea Soc. Math. Ed. Ser. B: Pure Appl. Math.* **4**, 1–6, 1997.

Lehmer, D. H. "The Lattice Points of an n-Dimensional Tetrahedron." *Duke Math. J.* **7**, 341–353, 1940.

Lesage, J. "Alexander Graham Bell Museum: Tribute to Genius." *Nat. Geographic* **60**, 227–256, 1956.

Rouché, E. and de Comberousse, C. *Traité de Géométrie, nouv. éd., vol. 1: Géométrie plane.* Paris: Gauthier-Villars, 1922.

Rouché, E. and de Comberousse, C. *Traité de Géométrie, nouv. éd., vol. 2: Géométrie dans l'espace.* Paris: Gauthier-Villars, 1922.

Steinhaus, H. *Mathematical Snapshots, 3rd ed.* New York: Dover, pp. 191–192 and 246–247, 1999.

Trigg, C. W. "Geometry of Paper Folding. II. Tetrahedral Models." *School Sci. and Math.* **54**, 683–689, 1954.

von Staudt, K. G. C. "Ueber einige geometrische Sätze." *J. reine angew. Math.* **57**, 88–89, 1860.

Wells, D. "Puzzle Page." *Games and Puzzles.* Sep. 1975.

Wells, D. *The Penguin Dictionary of Curious and Interesting Geometry.* London: Penguin, pp. 217–218, 1991.

Wenninger, M. J. "The Tetrahedron." Model 1 in *Polyhedron Models.* Cambridge, England: Cambridge University Press, p. 14, 1989.

Xu, Y. and Yau, S. "A Sharp Estimate of the Number of Integral Points in a Tetrahedron." *J. reine angew. Math.* **423**, 199–219, 1992.

Tetrahedron 4-Compound

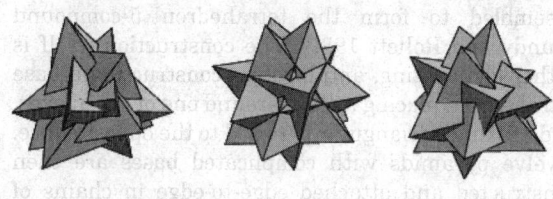

See also TETRAHEDRON, TETRAHEDRON 5-COMPOUND, TETRAHEDRON 10-COMPOUND

Tetrahedron 5-Compound

A POLYHEDRON COMPOUND composed of five TETRAHEDRA which is also one of the ICOSAHEDRON STELLATIONS. The 5×4 vertices of the tetrahedron are then 20 vertices of the DODECAHEDRON. Two tetrahedron 5-compounds of opposite CHIRALITY combine to make a TETRAHEDRON 10-COMPOUND (Cundy and Rollett 1989).

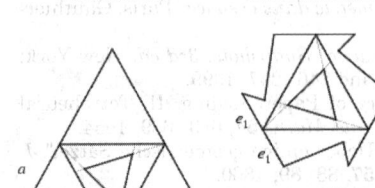
base tetrahedron   second tetrahedron   vertices of third, fourth, and fifth tetrahedra

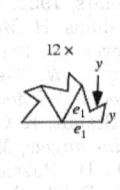

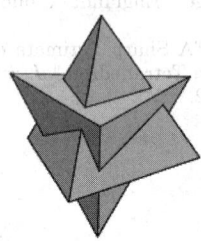

The diagram above shows pieces which can be assembled to form the tetrahedron 5-compound (Cundy and Rollett 1989). The construction itself is rather challenging, and involves constructing a base tetrahedron, placing a "cap" around one of the apexes, and affixing a triangular pyramid to the opposite face. Twelve pyramids with complicated bases are then constructed and attached edge-to-edge in chains of three. The four chains of pyramids are then arranged about the eight vertices of the original two tetrahedra, with the points of coincidence of the three pyramids in each chain attached such that they coincide with intersections of the original two tetrahedra such that five pyramids touch at a single point.

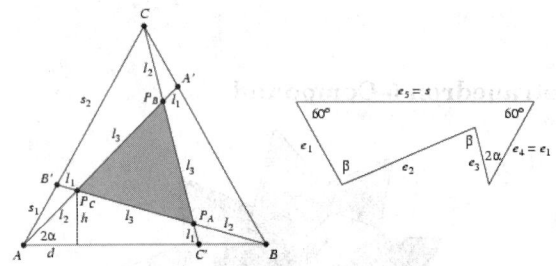

The position, size, and orientation of the pyramidal cap and pyramids are illustrated in the diagram above, where

$$\alpha = \cos^{-1}\left[\tfrac{1}{8}\left(3\sqrt{2}+\sqrt{10}\right)\right] \approx 22.2388° \tag{1}$$

$$d = \tfrac{1}{8}\sqrt{23 - 3\sqrt{5}} \tag{2}$$

$$h = \tfrac{1}{8}\sqrt{3\left(3 + \sqrt{5}\right)} \tag{3}$$

$$l_1 = \tfrac{1}{2}\sqrt{\tfrac{1}{5}\left(3 - \sqrt{5}\right)} \tag{4}$$

$$l_2 = \tfrac{1}{2}\sqrt{2} \tag{5}$$

$$l_3 = \tfrac{1}{2}\sqrt{3 + \sqrt{5}} \tag{6}$$

$$s = \sqrt{3 + \sqrt{5}} \tag{7}$$

$$s_1 = \tfrac{1}{5}\sqrt{10} \tag{8}$$

$$s_2 = \sqrt{\tfrac{1}{5}\left(7 + 3\sqrt{5}\right)}. \tag{9}$$

The edge lengths and angles of the cap are given by

$$\beta = \cos^{-1}\left(\tfrac{1}{4}\sqrt{7 - 3\sqrt{5}}\right) \approx 82.2388° \tag{10}$$

$$e_1 = \sqrt{3 - \sqrt{5}} \tag{11}$$

$$e_2 = \tfrac{1}{2}\left(5 - \sqrt{5}\right) \tag{12}$$

$$e_3 = \sqrt{7 - 3\sqrt{5}} \tag{13}$$

$$e_4 = e_1 \tag{14}$$

$$e_5 = s. \tag{15}$$

See also ICOSAHEDRON STELLATIONS, POLYHEDRON COMPOUND, TETRAHEDRON, TETRAHEDRON 4-COMPOUND, TETRAHEDRON 10-COMPOUND

References

Ball, W. W. R. and Coxeter, H. S. M. *Mathematical Recreations and Essays, 13th ed.* New York: Dover, p. 135, 1987.
Cundy, H. and Rollett, A. "Five Tetrahedra in a Dodecahedron." §3.10.8 in *Mathematical Models, 3rd ed.* Stradbroke, England: Tarquin Pub., pp. 139–141, 1989.
Wenninger, M. J. *Polyhedron Models.* New York: Cambridge University Press, p. 44, 1989.

Tetrahedron 10-Compound

Two TETRAHEDRON 5-COMPOUNDS of opposite CHIRALITY combined.

See also POLYHEDRON COMPOUND, TETRAHEDRON 5-COMPOUND

References

Ball, W. W. R. and Coxeter, H. S. M. *Mathematical Recreations and Essays, 13th ed.* New York: Dover, p. 135, 1987.
Cundy, H. and Rollett, A. "Ten Tetrahedra in a Dodecahedron." §3.10.9 in *Mathematical Models, 3rd ed.* Stradbroke, England: Tarquin Pub., pp. 141–142, 1989.
Wenninger, M. J. *Polyhedron Models.* New York: Cambridge University Press, p. 45, 1989.

Tetrahedron Circumscribing

References

Finch, S. "Circumscribing Tetrahedron of Least Volume." http://www.mathsoft.com/asolve/ecalabi.html.
van der Burg, J. W. "An Accurate and Robust Algorithm for the In-Sphere Criterion for Automated Delaunay-Based Tetrahedral Grid Generation." Paper P 98212 presented at *The 6th International Conference on Numerical Grid Generation for Computational Field Simulation, University of Greenwich, London, July 1998.* 1998.

Tetrahedron Tetrahedron Picking

The expected VOLUME of a TETRAHEDRON with vertices chosen at random inside another TETRAHEDRON of unit volume appears to be numerically close to 1/57, but the exact analytic value is not known (Croft *et al.* 1991, p. 54). According to Solomon (1978, p. 124), "Explicit values for random points in non-spherical regions such as tetrahedrons, parallelepipeds, etc., have apparently not yet been successfully calculated."

See also BALL TETRAHEDRON PICKING, SPHERE TETRAHEDRON PICKING

References

Croft, H. T.; Falconer, K. J.; and Guy, R. K. "Random Polygons and Polyhedra." §B5 in *Unsolved Problems in Geometry.* New York: Springer-Verlag, pp. 54–57, 1991.
Klee, V. "What is the Expected Volume of a Simplex Whose Vertices are Chosen at Random from a Given Convex Body." *Amer. Math. Monthly* **76**, 286–288, 1969.
Solomon, H. *Geometric Probability.* Philadelphia, PA: SIAM, p. 124, 1978.

Tetrahemihexacron

The DUAL POLYHEDRON of the TETRAHEMIHEXAHEDRON U_4 and Wenninger dual W_{67}.

See also DUAL POLYHEDRON, TETRAHEMIHEXAHEDRON, UNIFORM POLYHEDRON

References

Wenninger, M. J. *Dual Models.* Cambridge, England: Cambridge University Press, pp. 101–103, 1983.

Tetrahemihexahedron

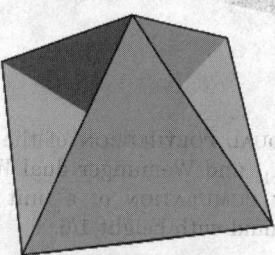

The UNIFORM POLYHEDRON U_4 whose DUAL POLYHEDRON is the TETRAHEMIHEXACRON. It has SCHLÄFLI SYMBOL $\mathbf{r}'\{^3_3\}$ and WYTHOFF SYMBOL $\frac{3}{2}$ 3|2. Its faces are $4\{3\} + 3\{4\}$. It is a faceted form of the OCTAHEDRON. Its CIRCUMRADIUS is

$$R = \tfrac{1}{2}\sqrt{2}.$$

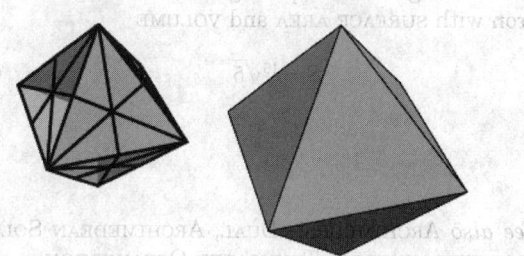

The CONVEX HULL of the tetrahemihexahedron is the OCTAHEDRON.

References

Wenninger, M. J. *Polyhedron Models.* Cambridge, England: Cambridge University Press, pp. 101–102, 1971.

Tetrakaidecagon

TETRADECAGON

Tetrakaidecahedron

TETRADECAHEDRON

Tetrakis Hexahedron

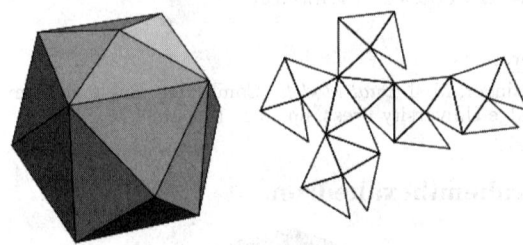

The 24-faced DUAL POLYHEDRON of the TRUNCATED OCTAHEDRON A_{12} and Wenninger dual W_7. It can be constructed by CUMULATION of a unit edge-length CUBE by a pyramid with height 1/6.

The edge lengths for the tetrakis hexahedron constructed as the dual of the TRUNCATED OCTAHEDRON with unit edge lengths are

$$s_1 = \tfrac{9}{8}\sqrt{2} \tag{1}$$

$$s_2 = \tfrac{3}{2}\sqrt{2}. \tag{2}$$

Normalizing so that $s_1 = 1$ gives a tetrakis hexahedron with SURFACE AREA and VOLUME

$$S = \tfrac{16}{3}\sqrt{5} \tag{3}$$

$$V = \tfrac{32}{9}. \tag{4}$$

See also ARCHIMEDEAN DUAL, ARCHIMEDEAN SOLID, ICOSITETRAHEDRON, TRUNCATED OCTAHEDRON

References

Wenninger, M. J. *Dual Models.* Cambridge, England: Cambridge University Press, pp. 14–16, 1983.

Tetranacci Number

The tetranacci numbers are a generalization of the FIBONACCI NUMBERS defined by $T_0 = 0$, $T_1 = 1$, $T_2 = 1$, $T_3 = 2$, and the RECURRENCE RELATION

$$T_n = T_{n-1} + T_{n-2} + T_{n-3} + T_{n-4}$$

for $n \geq 4$. They represent the $n = 4$ case of the FIBONACCI N-STEP NUMBERS. The first few terms are 1, 1, 2, 4, 8, 15, 29, 56, 108, 208, ... (Sloane's A000078). The ratio of adjacent terms tends to 1.92756, which is the REAL ROOT of $x^5 - 2x^4 + 1 = 0$.

See also FIBONACCI N-STEP NUMBER, FIBONACCI NUMBER, TRIBONACCI NUMBER

References

Sloane, N. J. A. Sequences A000078/M1108 in "An On-Line Version of the Encyclopedia of Integer Sequences." http://www.research.att.com/~njas/sequences/eisonline.html.

Tetrastigm

Lachlan's term for a set of four points, no three of which are COLLINEAR.

See also TETRAGRAM

References

Lachlan, R. "Properties of a Tetrastigm." §139–146 in *An Elementary Treatise on Modern Pure Geometry.* London: Macmillian, pp. 85–90, 1893.

Tetration

POWER TOWER

Tetriamond

tetriamonds

The three 3-polyiamonds are called tetriamonds.

See also POLYIAMOND

Tetrix

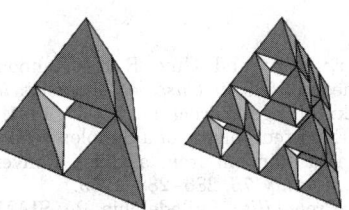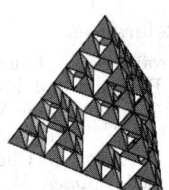

The 3-D analog of the SIERPINSKI SIEVE illustrated above, also called the SIERPINSKI SPONGE or SIERPINSKI TETRAHEDRON. Let N_n be the number of tetrahedra, L_n the length of a side, and A_n the fractional VOLUME of tetrahedra after the nth iteration. Then

$$N_n = 4^n \tag{1}$$

$$L_n = \left(\tfrac{1}{2}\right)^n = 2^{-n} \tag{2}$$

$$A_n = L_n^3 N_n = \left(\tfrac{1}{2}\right)^n. \tag{3}$$

The CAPACITY DIMENSION is therefore

$$d_{cap} = -\lim_{n \to \infty} \frac{\ln N_n}{\ln L_n} = -\lim_{n \to \infty} \frac{\ln(4^n)}{\ln(2^{-n})}$$

$$=\frac{\ln 4}{\ln 2}=\frac{2\ln 2}{\ln 2}=2, \qquad (4)$$

so the tetrix has an INTEGER CAPACITY DIMENSION (which is one less than the DIMENSION of the 3-D TETRAHEDRA from which it is built), despite the fact that it is a FRACTAL.

The following illustrations demonstrate how the dimension of the tetrix can be the same as that of the PLANE by showing three stages of the rotation of a tetrix, viewed along one of its edges. In the last frame, the tetrix "looks" like the 2-D PLANE.

See also MENGER SPONGE, SIERPINSKI SIEVE

References

Allanson, B. "The Fractal Tetrahedron" java applet. http://www.adelaide.net.au/~allanson/Fractet.html.

Dickau, R. M. "Sierpinski Tetrahedron." http://forum.swarthmore.edu/advanced/robertd/tetrahedron.html.

Eppstein, D. "Sierpinski Tetrahedra and Other Fractal Sponges." http://www.ics.uci.edu/~eppstein/junkyard/sierpinski.html.

Weisstein, E. W. "Fractals." MATHEMATICA NOTEBOOK FRACTAL.M.

Tetromino

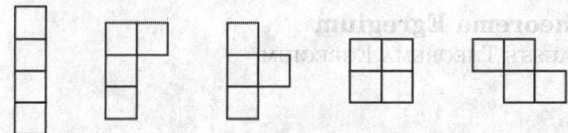

The five 4-POLYOMINOES, known as STRAIGHT, L-, T-, SQUARE, and SKEW.

References

Gardner, M. "Mathematical Games: About the Remarkable Similarity between the Icosian Game and the Towers of Hanoi." *Sci. Amer.* **196**, 150–156, May 1957.

Gardner, M. "Polyominoes." Ch. 13 in *The Scientific American Book of Mathematical Puzzles & Diversions.* New York: Simon and Schuster, pp. 124–140, 1959.

Hunter, J. A. H. and Madachy, J. S. *Mathematical Diversions.* New York: Dover, pp. 80–81, 1975.

Tg

TANGENT

Th

HYPERBOLIC TANGENT

Thâbit ibn Kurrah Rule

A number OF THE FORM $3 \cdot 2^n - 1$ which is PRIME is sometimes called a Thâbit ibn Kurrah number. The indices for the first few such numbers are 1, 2, 3, 4, 6, 7, 11, 18, 34, 38, 43, 55, ... (Sloane's A002235). Riesel (1969) extended the search to $n \le 1000$, and the largest known today is $n = 26459$.

The numbers arise in a beautiful result of Thâbit ibn Kurrah dating back to the tenth century (Woepcke 1852; Escott 1946; Dickson 1952, pp. 5 and 39; Borho 1972). Take $n \ge 2$ and suppose that

$$h = 3 \cdot 2^n - 1 \qquad (1)$$

$$t = 3 \cdot 2^{n-1} - 1 \qquad (2)$$

$$s = 9 \cdot 2^{2n-1} - 1 \qquad (3)$$

are all PRIME. Then $(2^n ht, 2^n s)$ are an AMICABLE PAIR. This form was rediscovered by Fermat (1636) and Descartes (1638) and generalized by Euler to EULER'S RULE (Borho 1972).

In order for such numbers to exist, there must be prime $3 \cdot 2^n - 1$ for two consecutive n, leaving only the possibilities 1, 2, 3, 4, and 6, 7. Of these, s is prime for $n = 2$, 4, and 7, giving the amicable pairs (220, 284), (17296, 18416), and (9363584, 9437056).

In fact, various rules can be found that are analogous to Thâbit ibn Kurrah's. Denote a "Thâbit rule" by $T(b_1, b)2, p, F_1, F_2$ for given natural numbers b_1 and b_2, a prime p not dividing b_1, b_2, and polynomials $F_1(X)$, $F_2(X) \in \mathbb{Z}[X]$. Then a necessary condition for the set of AMICABLE PAIRS (m_1, m_2) of the form $m_i = p^n b_i q_i$ ($i = 1, 2$) with q_1, q_2 prime and n a natural number to be infinite is that

$$\frac{p}{p-1} = \frac{b_1}{\sigma(b_1)} + \frac{b_2}{\sigma(b_2)}, \qquad (4)$$

where $\sigma(n)$ is the divisor function (Borho 1972). As a result, $m_i = p^n b_i q_i$ ($i = 1, 2$) form an AMICABLE PAIR, if for some $n \ge 1$, both

$$q_i = \frac{p^n (p-1)(b_1 + b_2)}{\sigma(b_i)} - 1 \qquad (5)$$

for $i = 1, 2$ are prime integers not dividing $b_i p$ (Borho 1972).

The following table summarizes some of the known Thâbit ibn Kurrah rules $T(au, p, (u+1)X, (u+1)\sigma(u)X - 1)$ (Borho 1972, te Riele 1974).

a	u	$\sigma(u)$	p
2^2	$5 \cdot 11$	72	127
$3^2 \cdot 7 \cdot 13$	$5 \cdot 17$	108	193
$3^2 \cdot 5 \cdot 13$	$11 \cdot 19$	240	449
$3^2 \cdot 7^2 \cdot 13$	$5 \cdot 41$	252	457

$3^2 \cdot 7^2 \cdot 13 \cdot 19$	$5 \cdot 193$	1164	2129
$3^4 \cdot 5 \cdot 11$	$29 \cdot 89$	2700	5281
$3^2 \cdot 7 \cdot 13 \cdot 41 \cdot 163$	$5 \cdot 977$	5868	10753
$3^2 \cdot 5 \cdot 19 \cdot 37$	$7 \cdot 887$	7104	13313
$3^4 \cdot 7 \cdot 11 \cdot 29$	$13 \cdot 521$	7308	14081
$3^2 \cdot 7^2 \cdot 13 \cdot 19 \cdot 29$	$41 \cdot 173$	7308	14401
$3^2 \cdot 5 \cdot 13 \cdot 19$	$29 \cdot 569$	17100	33601
$3^2 \cdot 7^2 \cdot 13$	$5 \cdot 53 \cdot 97$	31752	57457
$3^2 \cdot 5^2 \cdot 13 \cdot 31$	$149 \cdot 449$	67500	134401
$3^3 \cdot 5^3 \cdot 13$	$149 \cdot 449$	67500	134401
$2 \cdot 7^2 \cdot 19 \cdot 23$	$11 \cdot 13523$	162288	311041
$3^4 \cdot 5 \cdot 11 \cdot 59$	$89 \cdot 5309$	477900	950401
$3^4 \cdot 5 \cdot 11^2 \cdot 71$	$709 \cdot 2129$	1512300	3021761
$3^2 \cdot 7^2 \cdot 11 \cdot 19 \cdot 43 \cdot 89$	$293 \cdot 22961$	6750828	13478401
$2^2 \cdot 31$	$17 \cdot 107 \cdot 4339$	8436960	16329601
2^8	$257 \cdot 33023$	8520192	17007103
$2^3 \cdot 19 \cdot 137$	$83 \cdot 218651$	18366768	36514801
$2^7 \cdot 263$	$4271 \cdot 280883$	1199936448	2399587741

See also AMICABLE PAIR, EULER'S RULE, RIESEL NUMBER

References

Borho, W. "On Thabit ibn Kurrah's Formula for Amicable Numbers." *Math. Comput.* **26**, 571–578, 1972.
Dickson, L. E. *History of the Theory of Numbers, Vol. 1: Divisibility and Primality.* New York: Chelsea, 1952.
Escott, E. B. E. "Amicable Numbers." *Scripta Math.* **12**, 61–72, 1946.
Riesel, H. "Lucasian Criteria for the Primality of $N = h(2^n) - 1$." *Math. Comput.* **23**, 869–875, 1969.
Riesel, H. *Prime Numbers and Computer Methods for Factorization, 2nd ed.* Basel: Birkhäuser, p. 394, 1994.
Sloane, N. J. A. Sequences A002235/M0545 in "An On-Line Version of the Encyclopedia of Integer Sequences." http://www.research.att.com/~njas/sequences/eisonline.html.
te Riele, H. J. J. "Four Large Amicable Pairs." *Math. Comput.* **28**, 309–312, 1974.
Woepcke, F. *J. Asiatique* **20**, 320–429, 1852.

Thales' Theorem

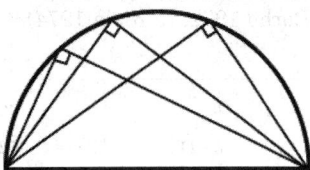

An ANGLE inscribed in a SEMICIRCLE is a RIGHT ANGLE.

See also RIGHT ANGLE, SEMICIRCLE

Theorem

A statement which can be demonstrated to be true by accepted mathematical operations and arguments. In general, a theorem is an embodiment of some general principle that makes it part of a larger theory. The process of showing a theorem to be correct is called a PROOF.

According to the Nobel Prize-winning physicist Richard Feynman (1985), any theorem, no matter how difficult to prove in the first place, is viewed as "TRIVIAL" by mathematicians once it has been proven. Therefore, there are exactly two types of mathematical objects: TRIVIAL ones, and those which have not yet been proven.

The late mathematician P. Erdos described a mathematician as "a machine for turning coffee into theorems" (Hoffman 1998, p. 7). R. Graham has estimated that upwards of 250,000 mathematical theorems are published each year (Hoffman 1998, p. 204).

See also AXIOM, AXIOMATIC SYSTEM, COROLLARY, DEEP THEOREM, PORISM, LEMMA, POSTULATE, PRINCIPLE, PROOF, PROPOSITION, TRIVIAL

References

Feynman, R. P. and Leighton, R. *Surely You're Joking, Mr. Feynman!* New York: Bantam Books, 1985.
Hoffman, P. *The Man Who Loved Only Numbers: The Story of Paul Erdos and the Search for Mathematical Truth.* New York: Hyperion, 1998.
THOREM/ Computer-Supported Mathematical Theorem Proving. http://www.theorema.org/.

Theorema Egregium

GAUSS'S THEOREMA EGREGIUM

Theory

A theory is a set of SENTENCES which is CLOSED under logical implication. That is, given any subset of SENTENCES $\{s_1, s_2, \ldots\}$ in the theory, if SENTENCE r is a logical consequence of $\{s_1, s_2, \ldots\}$, then r must also be in the theory.

See also LOGIC, SENTENCE

References

Enderton, H. B. *Elements of Set Theory.* New York: Academic Press, 1977.

Theta Functions

See also ABELIAN FUNCTION, JACOBI THETA FUNCTIONS, MOCK THETA FUNCTION, NEVILLE THETA FUNCTIONS, RAMANUJAN THETA FUNCTIONS, RIEMANN THETA FUNCTION, SIEGEL THETA FUNCTION

Theta Operator

In the NOTATION of Watson (1966),

$$\vartheta \equiv z\,\frac{d}{dz}.$$

References

Watson, G. N. *A Treatise on the Theory of Bessel Functions*, *2nd ed.* Cambridge, England: Cambridge University Press, 1966.

Theta Series

See also EISENSTEIN SERIES, LEECH LATTICE

Theta Subgroup

LAMBDA GROUP

Theta-0 Graph

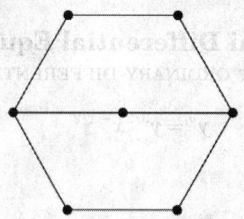

The GRAPH on seven nodes illustrated above.

See also 15 PUZZLE

References

Archer, A. F. "A Modern Treatment of the 15 Puzzle." *Amer. Math. Monthly* **106**, 793–799, 1999.
Wilson, R. M. "Graph Puzzles, Homotopy, and the Alternating Group." *J. Combin. Th. Ser. B* **16**, 86–96, 1974.

Thickness

GRAPH THICKNESS

Thiele's Interpolation Formula

Let ρ be a RECIPROCAL DIFFERENCE. Then Thiele's interpolation formula is the CONTINUED FRACTION

$$f(x) = f(x_1) + \cfrac{x - x_1}{p(x_1, x_2) +} \cfrac{x - x_2}{p_2(x_1, x_2, x_3 - f(x_1) +}$$
$$\cfrac{x - x_3}{\rho_3(x_1,\,x_2,\,x_3,\,x_4) - \rho(x_1,\,x_2) + \ldots}.$$

References

Abramowitz, M. and Stegun, C. A. (Eds.). *Handbook of Mathematical Functions with Formulas, Graphs, and Mathematical Tables, 9th printing*. New York: Dover, p. 881, 1972.
Milne-Thomson, L. M. *The Calculus of Finite Differences*. London: Macmillan, 1951.

Thiessen Polytope

VORONOI POLYGON

Thin Plate Spline

This entry contributed by SERGE BELONGIE

The thin plate spline is the two-dimensional analog of the CUBIC SPLINE in 1-D. It is the fundamental solution to the BIHARMONIC EQUATION, and has the form

$$U(r) = r^2 \ln r.$$

Given a set of data points, a weighted combination of thin plate splines centered about each data point gives the interpolation function that passes through the points exactly while minimizing the so-called "bending energy." Bending energy is defined here as the integral over $\mathbb{R}^2$ of the squares of the second derivatives,

$$I[f(x,\,y)] = \iint_{\mathbb{R}^2} f_{xx}^2 + 2f_{xy}^2 + f_{yy}^2 \; dx \, dy.$$

Regularization may be used to relax the requirement that the interpolant pass through the data points exactly.

The name "thin plate spline" refers to a physical analogy involving the bending of a thin sheet of metal. In the physical setting, the deflection is in the z direction, orthogonal to the plane. In order to apply this idea to the problem of coordinate transformation, one interprets the lifting of the plate as a displacement of the x or y coordinates within the plane. Thus, in general, two thin plate splines are needed to specify a 2-D coordinate transformation.

See also CUBIC SPLINE, SPLINE

References

Bookstein, F. L. "Principal Warps: Thin Plate Splines and the Decomposition of Deformations." *IEEE Trans. Pattern Anal. Mach. Intell.* **11**, June 1989.
Duchon, J. "Interpolation des fonctions de deux variables suivant le principe de la flexion des plaques minces." *RAIRO Analyse Numérique* **10**, 5–12, 1976.
Meinguet, J. "Multivariate Interpolation at Arbitrary Points Made Simple." *J. Appl. Math. Phys.* **30**, 292–304, 1979.
Wahba, G. *Spline Models for Observational Data*. Philadelphia, PA: SIAM, 1990.

Third Curvature

Also known as the TOTAL CURVATURE. The linear element of the INDICATRIX

$$ds_P = \sqrt{ds_T^2 + ds_B^2}.$$

See also LANCRET EQUATION

Third Fundamental Form

Let M be a REGULAR SURFACE with $\mathbf{v_P}$, $\mathbf{w_P}$ points in the TANGENT SPACE M_p of M. Then the third fundamental form is given by

$$\mathbf{III}(\mathbf{v_P}, \mathbf{w_P}) = S(\mathbf{v_P}) \cdot S(\mathbf{w_P}),$$

where S is the SHAPE OPERATOR.

See also FIRST FUNDAMENTAL FORM, FUNDAMENTAL FORMS, SECOND FUNDAMENTAL FORM, SHAPE OPERATOR

References
Gray, A. "The Three Fundamental Forms." §16.6 in *Modern Differential Geometry of Curves and Surfaces with Mathematica, 2nd ed.* Boca Raton, FL: CRC Press, pp. 380–382, 1997.

Third Kind

In the theory of special functions, a class of functions is said to be "of the third kind" if it is similar to but distinct from previously defined functions already defined to be of the FIRST and SECOND KINDS. The only common functions of the third kind are the ELLIPTIC INTEGRAL OF THE THIRD KIND $\mathbf{II}(n; \phi, k)$ and the Bessel function of the third kind (more commonly called the HANKEL FUNCTION).

See also ELLIPTIC INTEGRAL OF THE THIRD KIND, FIRST KIND, HANKEL FUNCTION, SECOND KIND, SPECIAL FUNCTION

Thirteen

13

Thom Transversality Theorem

References
Pohl, W. F. "The Self-Linking Number of a Closed Space Curve." *J. Math. Mech.* **17**, 975–985, 1968.

Thomae's Theorem

$$\frac{\Gamma(x+y+s+1)}{\Gamma(x+s+1)\Gamma(y+s+1)} \, {}_3F_2\left(\begin{matrix} -a, -b, x+y+s+1 \\ x+s+1, y+s+1 \end{matrix}; 1\right)$$

$$= \frac{\Gamma(a+b+s+1)}{\Gamma(a+s+1)\Gamma(b+s+1)} \, {}_3F_2\left(\begin{matrix} -x, -y, a+b+s+1 \\ a+s+1, b+s+1 \end{matrix}; 1\right),$$

where $\Gamma(z)$ is the GAMMA FUNCTION and the function ${}_3F_2(a, b, c; d, e; z)$ is a GENERALIZED HYPERGEOMETRIC FUNCTION. This theorem is equivalent to equation (1) from Bailey (1935, p. 14) (Hardy 1999, p. 111).

See also GAUSS'S HYPERGEOMETRIC THEOREM, GENERALIZED HYPERGEOMETRIC FUNCTION

References
Bailey, W. N. *Generalised Hypergeometric Series.* Cambridge, England: University Press, p. 14, 1935.

Hardy, G. H. "A Chapter from Ramanujan's Note-Book." *Proc. Cambridge Philos. Soc.* **21**, 492–503, 1923.

Hardy, G. H. *Ramanujan: Twelve Lectures on Subjects Suggested by His Life and Work, 3rd ed.* New York: Chelsea, pp. 104–105, 1999.

Thomae, J. "Ueber die Funktionen welche durch Reihen von der Form Dargestellt Werden: $1 + \frac{pp'p''}{1qq'} + \cdots$." *J. für Math.* **87**, 26–73, 1879.

Thomas Equation

The PARTIAL DIFFERENTIAL EQUATION

$$u_{xy} + \alpha u_x + \beta u_y + \gamma u_x u_y = 0.$$

References
Rosales, R. R. "Exact Solutions of a Certain Nonlinear Wave Equation." *J. Math. Phys.* **45**, 235–265, 1966.

Zwillinger, D. *Handbook of Differential Equations, 3rd ed.* Boston, MA: Academic Press, p. 132, 1997.

Thomas-Fermi Differential Equation

The second-order ORDINARY DIFFERENTIAL EQUATION

$$y'' = y^{3/2} x^{-1/2}.$$

References
Bender, C. M. and Orszag, S. A. *Advanced Mathematical Methods for Scientists and Engineers.* New York: McGraw-Hill, p. 25, 1978.

Zwillinger, D. *Handbook of Differential Equations, 3rd ed.* Boston, MA: Academic Press, p. 127, 1997.

Thomassen Graph

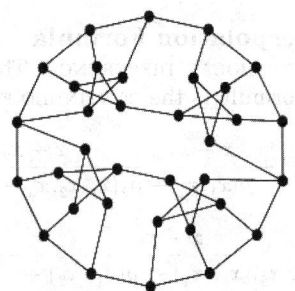

The HYPOTRACEABLE GRAPH illustrated above.

See also HYPOTRACEABLE GRAPH, THOMSEN GRAPH

References
Bondy, J. A. and Murty, U. S. R. *Graph Theory with Applications.* New York: North Holland, p. 240, 1976.

Thomassen, C. "Hypohamiltonian and Hypotraceable Graphs." *Disc. Math.* **9**, 91–96, 1974.

Thompson Group

The SPORADIC GROUP Th.

References

Wilson, R. A. "ATLAS of Finite Group Representation." http://for.mat.bham.ac.uk/atlas/html/Th.html.

Thompson Lamp Paradox

A lamp is turned on for 1/2 minute, off for 1/4 minute, on for 1/8 minute, etc. At the end of one minute, the lamp switch will have been moved $\aleph_0$ times, where $\aleph_0$ is ALEPH-0. Will the lamp be on or off? This PARADOX is actually nonsensical, since it is equivalent to asking if the "last" INTEGER is EVEN or ODD.

References

Erickson, G. W. and Fossa, J. A. *Dictionary of Paradox*. Lanham, MD: University Press of America, pp. 106–107, 1998.
Pickover, C. A. *Keys to Infinity*. New York: Wiley, pp. 19–23, 1995.

Thompson's Functions

BEI, BER, KELVIN FUNCTIONS

Thom's Eggs

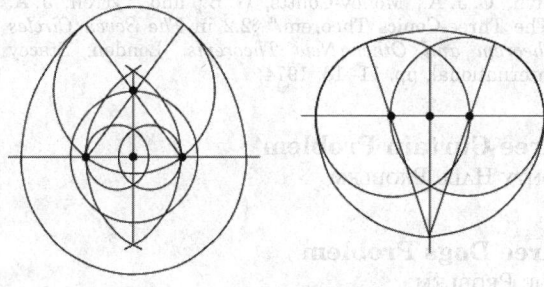

EGG-shaped curves constructed using multiple CIRCLES which Thom (1967) used to model Megalithic stone rings in Britain.

See also EGG, OVAL

References

Dixon, R. *Mathographics*. New York: Dover, p. 6, 1991.
Thom, A. "Mathematical Background." Ch. 4 in *Megalithic Sites in Britain*. Oxford, England: Oxford University Press, pp. 27–33, 1967.

Thomsen Graph

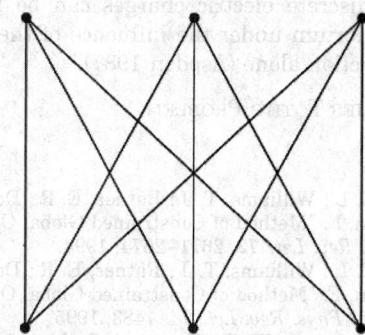

The COMPLETE BIPARTITE GRAPH $K_{3,3}$, which is equivalent to the UTILITY GRAPH. It has a CROSSING NUMBER 1.

See also COMPLETE BIPARTITE GRAPH, CROSSING NUMBER (GRAPH), THOMASSEN GRAPH, UTILITY GRAPH

References

Gardner, M. *The Sixth Book of Mathematical Games from Scientific American*. Chicago, IL: University of Chicago Press, p. 93, 1984.

Thomsen's Figure

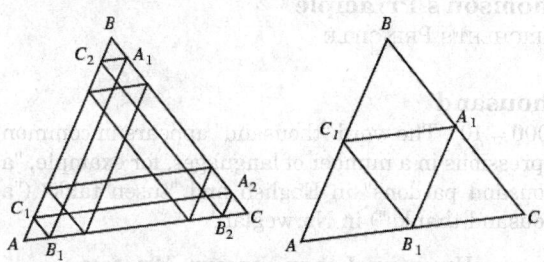

Take any TRIANGLE with VERTICES A, B, and C. Pick a point A_1 on the side opposite A, and draw a line PARALLEL to AB. Upon reaching the side AC at B_1, draw the line PARALLEL to BC. Continue (left figure). Then $A_3 = A_1$ for any TRIANGLE. If A_1 is the MIDPOINT of BC, then $A_2 = A_1$ (right figure).

See also MIDPOINT, TRIANGLE

References

Madachy, J. S. *Madachy's Mathematical Recreations*. New York: Dover, p. 234, 1979.

Thomson Problem

Determine the stable equilibrium positions of N classical electrons constrained to move on the surface of a SPHERE and repelling each other by an inverse square law. Exact solutions for $N = 2$ to 8 are known, but $N = 9$ and 11 are still unknown.

In reality, Earnshaw's theorem guarantees that no system of discrete electric charges can be held in stable equilibrium under the influence of their electrical interaction alone (Aspden 1987).

See also FEJES TÓTH'S PROBLEM

References

Altschuler, E. L.; Williams, T. J.; Ratner, E. R.; Dowla, F.; and Wooten, F. "Method of Constrained Global Optimization." *Phys. Rev. Let.* **72**, 2671–2674, 1994.
Altschuler, E. L.; Williams, T. J.; Ratner, E. R.; Dowla, F.; and Wooten, F. "Method of Constrained Global Optimization--Reply." *Phys. Rev. Let.* **74**, 1483, 1995.
Ashby, N. and Brittin, W. E. "Thomson's Problem." *Amer. J. Phys.* **54**, 776–777, 1986.
Aspden, H. "Earnshaw's Theorem." *Amer. J. Phys.* **55**, 199–200, 1987.
Berezin, A. A. "Spontaneous Symmetry Breaking in Classical Systems." *Amer. J. Phys.* **53**, 1037, 1985.
Calkin, M. G.; Kiang, D.; and Tindall, D. A. "Minimum Energy Configurations." *Nature* **319**, 454, 1986.
Erber, T. and Hockney, G. M. "Comment on 'Method of Constrained Global Optimization.'" *Phys. Rev. Let.* **74**, 1482–1483, 1995.
Marx, E. "Five Charges on a Sphere." *J. Franklin Inst.* **290**, 71–74, Jul. 1970.
Melnyk, T. W.; Knop, O.; and Smith, W. R. "Extremal Arrangements of Points and Unit Charges on a Sphere: Equilibrium Configurations Revisited." *Canad. J. Chem.* **55**, 1745–1761, 1977.
Whyte, L. L. "Unique Arrangement of Points on a Sphere." *Amer. Math. Monthly* **59**, 606–611, 1952.

Thomson's Principle

DIRICHLET'S PRINCIPLE

Thousand

$1,000 = 10^3$. The word "thousand" appears in common expressions in a number of languages, for example, "a thousand pardons" in English and "tusen takk" ("a thousand thanks") in Norwegian.

See also HUNDRED, LARGE NUMBER, MILLION

Three

3

Three Conics Theorem

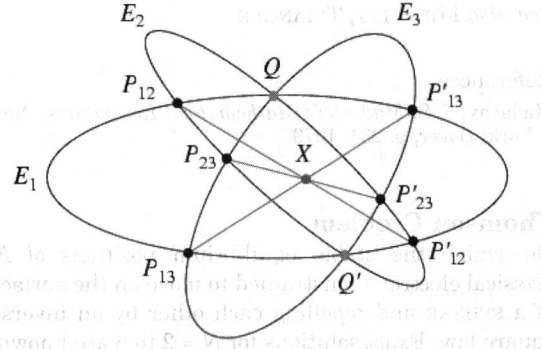

If three conics pass through two given points Q and

Q', then the lines joining the other two intersections of each pair of conics $P_{ij}P'_{ij}$ are CONCURRENT at a point X (Evelyn 1974, p. 15). The converse states that if two conics E_2 and E_3 meet at four points Q, Q', P_1, and Q_1, and if P_2Q_2 and P_3Q_3 are chords of E_3 and E_2, respectively, which meet on P_1Q_1, then the six points lie on a conic. The dual of the theorem states that if three conics share two common tangents, then their remaining pairs of common tangents intersect at three collinear points.

If the points Q and Q' are taken as the POINTS AT INFINITY, then the theorem reduces to the theorem that RADICAL LINES of three CIRCLES are CONCURRENT in a point known as the RADICAL CENTER (Evelyn 1974, p. 15).

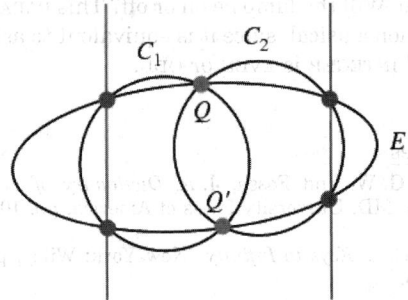

If two of the points P_{ij} and P'_{ij} are taken as the POINTS AT INFINITY, then the theorem becomes that if two circles C_1 and C_2 pass through two points Q and Q' on a conic E, then the lines determined by the pair of intersections of each circle with the conic are parallel (Evelyn 1974, p. 15).

See also CONIC SECTION, FOUR CONICS THEOREM, RADICAL CENTER

References

Evelyn, C. J. A.; Money-Coutts, G. B.; and Tyrrell, J. A. "The Three-Conics Theorem." §2.2 in *The Seven Circles Theorem and Other New Theorems*. London: Stacey International, pp. 11–18, 1974.

Three Curtain Problem

MONTY HALL PROBLEM

Three Dogs Problem

MICE PROBLEM

Three j-Symbol

WIGNER 3*J*-SYMBOL

Three Jug Problem

Given three jugs with x pints in the first, y in the second, and z in the third, obtain a desired amount in one of the vessels by completely filling up and/or emptying vessels into others. This problem can be solved with the aid of TRILINEAR COORDINATES.

References

Ball, W. W. R. and Coxeter, H. S. M. *Mathematical Recreations and Essays, 13th ed.* New York: Dover, pp. 28 and 40, 1987.

Coxeter, H. S. M. and Greitzer, S. L. "The Three Jug Problem." §4.6 in *Geometry Revisited.* Washington, DC: Math. Assoc. Amer., pp. 89–93, 1967.

O'Beirne, T. H. *Puzzles and Paradoxes.* New York: Oxford University Press, pp. 49–75, 1965.

Perel'man, A. I. *Zanumatel'naya Geometria.* Moscow, 1958.

Steinhaus, H. *Mathematical Snapshots, 3rd ed.* New York: Dover, pp. 61–63, 1999.

Tweedie, M. C. K. *Math. Gaz.* **23**, 278–282, 1939.

Three-Choice Polygon

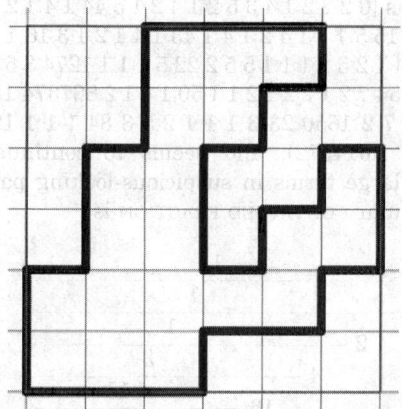

A LATTICE POLYGON formed by a THREE-CHOICE WALK. The anisotropic perimeter and area generating function

$$G(x, y, q) = \sum m \geq 1 \sum_{n \geq 1} \sum_{a \geq a} C(m, n, a) x^m y^n q^a,$$

where $C(m, n, a)$ is the number of polygons with $2m$ horizonal bonds, $2n$ vertical bonds, and area a, is not yet known in closed form, but it can be evaluated in polynomial time (Conway *et al.* 1997, Bousquet-Mélou 1999). The perimeter-generating function $G(x, x, 1)$ has a logarithmic singularity and so is not algebraic, but is known to be D-finite (Conway *et al.* 1997, Bousquet-Mélou 1999).

The anisotropic area and perimeter generating function $G(x, y, q)$ satisfies an inversion relation OF THE FORM

$$G(x, y, q) + y^2 G(x/y, 1/y, 1/q)$$

(Bousquet-Mélou *et al.* 1999).

References

Bousquet-Mélou, M.; Guttmann, A. J.; Orrick, W. P.; and Rechnitzer, A. Inversion Relations, Reciprocity and Polyominoes. 23 Aug 1999. http://xxx.lanl.gov/abs/math.CO/9908123/.

Conway, A.; Cuttmann, A. J.; and Delest, M. "On the Number of Three-Choice Polygons." *Math. Comput. Model.* **26**, 51–58, 1997.

Three-Choice Walk

A SELF-AVOIDING WALK in which steps may be to the left, right, or straight ahead after a vertical step, but only straight ahead of to the left after a horizontal step. A LATTICE POLYGON formed by a three-choice walk is called a THREE-CHOICE POLYGON.

References

Bousquet-Mélou, M.; Guttmann, A. J.; Orrick, W. P.; and Rechnitzer, A. Inversion Relations, Reciprocity and Polyominoes. 23 Aug 1999. http://xxx.lanl.gov/abs/math.CO/9908123/.

Three-Colorable

COLORABLE

Threefoil Knot

TREFOIL KNOT

Three-In-A-Row

TIC-TAC-TOE

ThreeJ Symbol

WIGNER 3J-SYMBOL

Three-Valued Logic

A logical structure which does not assume the EXCLUDED MIDDLE LAW. Three truth values are possible: true, false, or undecided. There are 3072 such logics.

See also EXCLUDED MIDDLE LAW, FUZZY LOGIC, LOGIC

Thue Constant

The base-2 TRANSCENDENTAL NUMBER

$$0.11011011111011011111\ldots_2,$$

where the nth bit is 1 if n is not divisible by 3 and is the complement of the $(n/3)$th bit if n is divisible by 3. It is also given by the SUBSTITUTION MAP

$$0 \rightarrow 111$$
$$1 \rightarrow 110.$$

In decimal, the Thue constant equals 0.8590997969....

See also RABBIT CONSTANT, THUE-MORSE CONSTANT

Thue Equation

This entry contributed by KEVIN O'BRYANT

A Thue equation is a DIOPHANTINE EQUATION of the form

$$A_n x^n + A_{n-1} x^{n-1} y + A_{n-2} x^{n-2} y^2 + \ldots + A_0 y^n = M,$$

with $n \geq 3$, $A_i \in \mathbb{Z}$, $M \neq 0 \in Z$, and x, y unknown integer variables.

Thue (1909) proved that such an equation has only finitely many solutions, but it was not until much later that Tzanakis and de Weger (1989) gave a practical algorithm for finding bounds on $|x|$ and $|y|$. Although these bounds can be astronomically large in some cases, they are typically small enough to allow an exhaustive search for all solutions.

See also DIOPHANTINE EQUATION

References

Thue, A. "Über Annäherungswerte algebraischer Zahlen." *J. reine angew. Math.* **135**, 284–305, 1909.
Tzanakis, N. and de Weger, B. M. M. "On the Practical Solution of the Thue Equation." *J. Number Th.* **31**, 99–132, 1989.

Thue Sequence

The SEQUENCE of BINARY DIGITS of the THUE CONSTANT, $0.1101101111101101111110110110\ldots_2$ (Sloane's A014578).

See also RABBIT CONSTANT, THUE CONSTANT

References

Guy, R. K. "Thue Sequences." §E21 in *Unsolved Problems in Number Theory, 2nd ed.* New York: Springer-Verlag, pp. 223–224, 1994.
Sloane, N. J. A. Sequences A014578 in "An On-Line Version of the Encyclopedia of Integer Sequences." http://www.research.att.com/~njas/sequences/eisonline.html.

Thue-Morse Constant

The constant also called the PARITY CONSTANT and defined by

$$P \equiv \frac{1}{2} \sum_{n=0}^{\infty} P(n) 2^{-n} = 0.4124540336401075977\ldots \quad (1)$$

(Sloane's A014571), where $P(n)$ is the PARITY of n. Dekking (1977) proved that the Thue-Morse constant is TRANSCENDENTAL, and Allouche and Shallit give a complete proof correcting a minor error of Dekking.

The Thue-Morse constant can be written in base 2 by stages by taking the previous iteration a_n, taking the complement $\overline{a_n}$, and appending, producing

$$a_0 = 0.0_2$$

$$a_1 = 0.01_2$$

$$a_2 = 0.0110_2$$

$$a_3 = 0.01101001_2$$

$$a_4 = 0.0110100110010110_2. \quad (2)$$

This can be written symbolically as

$$a_{n+1} = a_n + \overline{a_n} \cdot 2^{-2^n} \quad (3)$$

with $a_0 = 0$. Here, the complement is the number $\overline{a_n}$ such that $a_n + \overline{a_n} = 0.\underbrace{11\ldots1}_{2^n}{}_2$, which can be found

from

$$a_n + \overline{a_n} = \sum_{k=1}^{2^n} \left(\frac{1}{2}\right)^k = \frac{1 - \left(\frac{1}{2}\right)^{2^n}}{1 - \frac{1}{2}} - 1 = 1 - 2^{-2^n}. \quad (4)$$

Therefore,

$$\overline{a_n} = 1 - 2^{-2^n} - a_n, \quad (5)$$

and

$$a_{n+1} = a_n + \left(1 - 2^{-2^n} - a_n\right) 2^{-2^n} \quad (6)$$

$$= 2^{-2^{n+1}} \left(2^{2^n} - 1\right)\left(1 - 2^{2^n} a_n\right). \quad (7)$$

The regular CONTINUED FRACTION for the Thue-Morse constant is [0 2 2 2 1 4 3 5 2 1 4 2 1 5 44 1 4 1 2 4 1 1 1 5 14 1 50 15 5 1 1 1 4 2 1 4 1 43 1 4 1 2 1 3 16 1 2 1 2 1 50 1 2 424 1 2 5 2 1 1 1 5 5 2 22 5 1 1 1 1274 3 5 2 1 1 1 4 1 1 15 154 7 2 1 2 2 1 2 1 1 50 1 4 1 2 867374 1 1 1 5 5 1 1 6 1 2 7 2 1650 23 3 1 1 1 2 5 3 84 1 1 1 1284 ...] (Sloane's A014572), and seems to continue with sporadic large terms in suspicious-looking patterns. A nonregular CONTINUED FRACTION is

$$P = \cfrac{1}{3 - \cfrac{1}{2 - \cfrac{1}{4 - \cfrac{3}{16 - \cfrac{15}{256 - \cfrac{255}{65536 - \ldots}}}}}}. \quad (8)$$

A related infinite product is

$$4P = 2 - \frac{1 \cdot 3 \cdot 15 \cdot 255 \cdot 65535 \cdots}{2 \cdot 4 \cdot 16 \cdot 256 \cdot 65536 \cdots}. \quad (9)$$

The SEQUENCE $a_\infty = 0110100110010110100101100\ldots$ (Sloane's A010060) is known as the THUE-MORSE SEQUENCE.

See also RABBIT CONSTANT, THUE CONSTANT

References

Allouche, J. P.; Arnold, A.; Berstel, J.; Brlek, S.; Jockusch, W.; Plouffe, S.; and Sagan, B. "A Relative of the Thue-Morse Sequence." *Discr. Math.* **139**, 455–461, 1995.
Allouche, J. P. and Shallit, J. "The Ubiquitous Prouhet-Thue-Morse Sequence." http://www.math.uwaterloo.ca/~shallit/Papers/ubiq.ps.
Dekking, F. M. "Transcendence du nombre de Thue-Morse." *Comptes Rendus de l'Academie des Sciences de Paris* **285**, 157–160, 1977.
Finch, S. "Favorite Mathematical Constants." http://www.mathsoft.com/asolve/constant/cntfrc/cntfrc.html.
Schroeppel, R. and Gosper, R. W. Item 122 in Beeler, M.; Gosper, R. W.; and Schroeppel, R. *HAKMEM.* Cambridge, MA: MIT Artificial Intelligence Laboratory, Memo AIM-239, pp. 56–57, Feb. 1972.
Sloane, N. J. A. Sequences A010060, A014571, and A014572 in "An On-Line Version of the Encyclopedia of Integer Sequences." http://www.research.att.com/~njas/sequences/eisonline.html.

Thue-Morse Sequence

The INTEGER SEQUENCE (also called the MORSE-THUE SEQUENCE)

$$0110100110010110100101100110 1001\ldots \quad (1)$$

(Sloane's A010060) which arises in the THUE-MORSE CONSTANT. It can be generated from the SUBSTITUTION MAP

$$0 \to 01 \quad (2)$$

$$1 \to 10 \quad (3)$$

starting with 0 as follows:

$$0 \to 01 \to 0110 \to 01101001 \to \ldots \quad (4)$$

Writing the sequence as a POWER SERIES over the FINITE FIELD GF(2),

$$F(x) = 0 + 1x + 1x^2 + 0x^3 + 1x^4 + \ldots, \quad (5)$$

then F satisfies the quadratic equation

$$(1+x)F^2 + F = \frac{x}{1+x^2} \pmod 2. \quad (6)$$

This equation has two solutions, F and F', where F' is the complement of F, i.e.,

$$F + F' = 1 + x + x^2 + x^3 + \ldots = \frac{1}{1+x}, \quad (7)$$

which is consistent with the formula for the sum of the roots of a quadratic. The equality (6) can be demonstrated as follows. Let $(abcdef\ldots)$ be a shorthand for the POWER SERIES

$$a + bx + cx^2 + dx^3 + \ldots, \quad (8)$$

so $F(x)$ is $(0110100110010110\ldots)$. To get F^2, simply use the rule for squaring POWER SERIES over GF(2)

$$(A + B)^2 = A^2 + B^2 \pmod 2, \quad (9)$$

which extends to the simple rule for squaring a POWER SERIES

$$\left(a_0 + a_1 x + a_2 x^2 + \ldots\right)^2$$
$$= a_0 + a_1 x^2 + a_2 x^4 + \ldots \pmod 2, \quad (10)$$

i.e., space the series out by a factor of 2, $(0\,1\,1\,0\,1\,0\,0\,1$ $\ldots)$, and insert zeros in the ODD places to get

$$F^2 = (0010100010000010\ldots). \quad (11)$$

Then multiply by x (which just adds a zero at the front) to get

$$xF^2 = (00010100010000010\ldots). \quad (12)$$

Adding to F^2 gives

$$(1+x)F^2 = (0011110011000011\ldots). \quad (13)$$

This is the first term of the quadratic equation, which is the Thue-Morse sequence with each term doubled up. The next term is F, so we have

$$(1+x)F^2 = (0011110011000011\ldots) \quad (14)$$

$$F = (0110100110010110\ldots). \quad (15)$$

The sum is the above two sequences XORed together (there are no CARRIES because we're working over GF(2)), giving

$$(1+x)F^2 + F = (0101010101010101\ldots). \quad (16)$$

We therefore have

$$(1+x)F^2 + F = \frac{x}{1+x^2}$$
$$= x + x^3 + x^5 + x^7 + x^9 + x^{11} + \ldots \pmod 2. \quad (17)$$

The Thue-Morse sequence is an example of a cube-free sequence on two symbols (Morse and Hedlund 1944), i.e., it contains no substrings OF THE FORM WWW, where W is any WORD. For example, it does not contain the WORDS 000, 010101 or 010010010. In fact, the following stronger statement is true: the Thue-Morse sequence does not contain any substrings OF THE FORM WWa, where a is the first symbol of W. We can obtain a SQUAREFREE sequence on three symbols by doing the following: take the Thue-Morse sequence 0110100110010110... and look at the sequence of WORDS of length 2 that appear: 01 11 10 01 10 00 01 11 10 .... Replace 01 by 0, 10 by 1, 00 by 2 and 11 by 2 to get the following: 021012021.... Then this SEQUENCE is SQUAREFREE (Morse and Hedlund 1944).

The Thue-Morse sequence has important connections with the GRAY CODE. Kindermann generates fractal music using the SELF-SIMILARITY of the Thue-Morse sequence.

See also GRAY CODE, PARITY CONSTANT, RABBIT SEQUENCE, THUE SEQUENCE

References

Kindermann, L. "MusiNum--The Music in the Numbers." http://www.forwiss.uni-erlangen.de/~kinderma/musinum/.

Morse, M. and Hedlund, G. A. "Unending Chess, Symbolic Dynamics, and a Problem in Semigroups." *Duke Math. J.* **11**, 1–7, 1944.

Schroeder, M. R. *Fractals, Chaos, and Power Laws: Minutes from an Infinite Paradise.* New York: W. H. Freeman, 1991.

Sloane, N. J. A. Sequences A010060 in "An On-Line Version of the Encyclopedia of Integer Sequences." http://www.research.att.com/~njas/sequences/eisonline.html.

Thue's Remainder Theorem

THUE'S THEOREM

Thue's Theorem

If $n > 1$, $(a, n) = 1$ (i.e., a and n are RELATIVELY PRIME), and m is the least integer $> \sqrt{n}$, then there exist an x and y such that

$$ay \equiv \pm x \pmod{n}$$

where $0 < x < m$ and $0 < y < m$ (Nagell 1951, pp. 122–124; Shanks 1993, p. 161)

References

Nagell, T. "Thue's Remainder Theorem and Its Generalization by Scholtz." §36 in *Introduction to Number Theory.* New York: Wiley, pp. 122–124, 1951.
Shanks, D. *Solved and Unsolved Problems in Number Theory, 4th ed.* New York: Chelsea, p. 161, 1993.

Thue-Siegel-Roth Theorem

If α is a TRANSCENDENTAL NUMBER, it can be approximated by infinitely many RATIONAL NUMBERS m/n to within n^{-r}, where r is any POSITIVE number.

See also IRRATIONALITY MEASURE, LIOUVILLE'S APPROXIMATION THEOREM, ROTH'S THEOREM, SIEGEL'S THEOREM

Thue-Siegel-Schneider-Roth Theorem

THUE-SIEGEL-ROTH THEOREM

Thurston's Geometrization Conjecture

Thurston's conjecture has to do with geometric structures on 3-D MANIFOLDS. Before stating Thurston's conjecture, some background information is useful. 3-dimensional MANIFOLDS possess what is known as a standard 2-level DECOMPOSITION. First, there is the CONNECTED SUM DECOMPOSITION, which says that every COMPACT 3-MANIFOLD is the CONNECTED SUM of a unique collection of PRIME 3-MANIFOLDS.

The second DECOMPOSITION is the JACO-SHALEN-JOHANNSON TORUS DECOMPOSITION, which states that irreducible orientable COMPACT 3-MANIFOLDS have a CANONICAL (up to ISOTOPY) minimal collection of disjointly EMBEDDED incompressible TORI such that each component of the 3-MANIFOLD removed by the TORI is either "atoroidal" or "Seifert-fibered."

Thurston's conjecture is that, after you split a 3-MANIFOLD into its CONNECTED SUM and then JACO-SHALEN-JOHANNSON TORUS DECOMPOSITION, the remaining components each admit exactly one of the following geometries:

1. EUCLIDEAN GEOMETRY,
2. HYPERBOLIC GEOMETRY,
3. SPHERICAL GEOMETRY,
4. the GEOMETRY of $\mathbb{S}^2 + \mathbb{R}$,
5. the GEOMETRY of $\mathbb{H}^2 + \mathbb{R}$,
6. the GEOMETRY of SL_2R,
7. NIL GEOMETRY, or
8. SOL GEOMETRY.

Here, $\mathbb{S}^2$ is the 2-SPHERE and $\mathbb{H}^2$ is the HYPERBOLIC PLANE. If Thurston's conjecture is true, the truth of the POINCARÉ CONJECTURE immediately follows.

See also CONNECTED SUM DECOMPOSITION, EUCLIDEAN GEOMETRY, HYPERBOLIC GEOMETRY, JACO-SHALEN-JOHANNSON TORUS DECOMPOSITION, NIL GEOMETRY, POINCARÉ CONJECTURE, SOL GEOMETRY, SPHERICAL GEOMETRY

Thwaites Conjecture

COLLATZ PROBLEM

Ticktacktoe

TIC-TAC-TOE

Tic-Tac-Toe

The usual game of tic-tac-toe (also called TICKTACKTOE) is 3-in-a-row on a 3×3 board. However, a generalized N-IN-A-ROW on an $u \times v$ board can also be considered. For $n = 1$ and 2 the first player can always win. If the board is at least 3×4, the first player can win for $n = 3$.

However, for TIC-TAC-TOE which uses a 3×3 board, a draw can always be obtained. If the board is at least 4×30, the first player can win for $n = 4$. For $n = 5$, a draw can always be obtained on a 5×5 board, but the first player can win if the board is at least 15×15. The cases $n = 6$ and 7 have not yet been fully analyzed for an $n \times n$ board, although draws can always be forced for $n = 8$ and 9. On an $\infty \times \infty$ board, the first player can win for $n = 1$, 2, 3, and 4, but a tie can always be forced for $n \geq 8$. For $3 \times 3 \times 3$ and $4 \times 4 \times 4$, the first player can always win (Gardner 1979).

See also BOARD, PONG HAU K'I

References

Ball, W. W. R. and Coxeter, H. S. M. *Mathematical Recreations and Essays, 13th ed.* New York: Dover, pp. 103–104, 1987.
de Fouquières, B. Ch. 18 in *Les Jeux des Anciens, 2nd ed.*. Paris, 1873.
Gardner, M. "Mathematical Games: The Diverse Pleasures of Circles that Are Tangent to One Another." *Sci. Amer.* **240**, 18–28, Jan. 1979a.
Gardner, M. "Ticktacktoe Games." Ch. 9 in *Wheels, Life, and Other Mathematical Amusements.* New York: W. H. Freeman, pp. 94–105, 1983.
Steinhaus, H. *Mathematical Snapshots, 3rd ed.* New York: Dover, pp. 10–11, 1999.

Stewart, I. "A Shepherd Takes A Sheep Shot." *Sci. Amer.* **269**, 154–156, 1993.

Tietze Graph

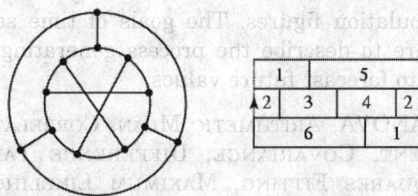

Tietze graph   *embedding on the Möbius band*

The graph illustrated above that provides a 6-color coloring of the MÖBIUS STRIP.

See also MÖBIUS STRIP

References
Bondy, J. A. and Murty, U. S. R. *Graph Theory with Applications.* New York: North Holland, p. 243, 1976.

Tight Closure
The application of characteristic p methods in COMMUTATIVE ALGEBRA, which is a synthesis of some areas of COMMUTATIVE ALGEBRA and ALGEBRAIC GEOMETRY.

See also ALGEBRAIC GEOMETRY, COMMUTATIVE ALGEBRA

References
Bruns, W. "Tight Closure." *Bull. Amer. Math. Soc.* **33**, 447–457, 1996.
Huneke, C. "An Algebraist Commuting in Berkeley." *Math. Intell.* **11**, 40–52, 1989.

Tightly Embedded
Q is said to be tightly embedded if $|Q \cap Q^g|$ is ODD for all $g \in G - N_G(Q)$, where $N_G(Q)$ is the NORMALIZER of Q in G.

Tilde
The mark ~ placed on top of a symbol to indicate some special property. $\tilde{x}$ is voiced "x-tilde." The tilde symbol is commonly used to denote an operator, e.g., the DIFFERENTIAL OPERATOR $\tilde{D}$. In informal usage, "tilde" is often instead voiced as "twiddle." It is also sometimes used to denote a MEDIAN (Kenney and Keeping 1962, p. 211).

See also MEDIAN (STATISTICS), DIFFERENTIAL OPERATOR

References
Bringhurst, R. *The Elements of Typographic Style, 2nd ed.* Point Roberts, WA: Hartley and Marks, p. 284, 1997.
Kenney, J. F. and Keeping, E. S. *Mathematics of Statistics, Pt. 1, 3rd ed.* Princeton, NJ: Van Nostrand, 1962.

Tiling
A plane-filling arrangement of plane figures or its generalization to higher dimensions. Formally, a tiling is a collection of disjoint open sets, the closures of which cover the plane. Given a single tile, the so-called first CORONA is the set of all tiles that have a common boundary point with the tile (including the original tile itself).

WANG'S CONJECTURE (1961) stated that if a set of tiles tiled the plane, then they could always be arranged to do so periodically. A periodic tiling of the PLANE by POLYGONS or SPACE by POLYHEDRA is called a TESSELLATION. The conjecture was refuted in 1966 when R. Berger showed that an aperiodic set of 20,426 tiles exists. By 1971, R. Robinson had reduced the number to six and, in 1974, R. Penrose discovered an aperiodic set (when color-matching rules are included) of two tiles: the so-called PENROSE TILES. (Penrose also sued the Kimberly Clark Corporation over their quilted toilet paper, which allegedly resembles a Penrose aperiodic tiling; Mirsky 1997.)

It is not known if there is a single aperiodic tile. The number of tilings possible for convex irregular POLYGONS are given in the above table.

n	name	known tilings
3	TRIANGLE TILING	all
4	QUADRILATERAL TILING	all
5	PENTAGON TILING	14
6	HEXAGON TILING	3

There are no tilings for *identical* convex n-gons for $n \geq 7$, although non-identical convex heptagons can tile the plane (Steinhaus 1983, p. 77; Gardner 1984, pp. 248–249).

See also ANISOHEDRAL TILING, CORONA (TILING), GOSPER ISLAND, HARBORTH'S TILING, HEESCH NUMBER, HEESCH'S PROBLEM, HONEYCOMB CONJECTURE, ISOHEDRAL TILING, KOCH SNOWFLAKE, MONOHEDRAL TILING, PENROSE TILES, POLYGON TILING, POLYOMINO TILING, SPACE-FILLING POLYHEDRON, SQUARE TILING, TESSELLATION, TILING THEOREM, TRIANGLE TILING

References
Eppstein, D. "Tiling." http://www.ics.uci.edu/~eppstein/junkyard/tiling.html.
Gardner, M. *The Sixth Book of Mathematical Games from Scientific American.* Chicago, IL: University of Chicago Press, pp. 248–249, 1984.
Gardner, M. "Tilings with Convex Polygons." Ch. 13 in *Time Travel and Other Mathematical Bewilderments.* New York: W. H. Freeman, pp. 162–176, 1988.

Gardner, M. "Penrose Tiling" and "Penrose Tiling II." Chs. 1–2 in *Penrose Tiles and Trapdoor Ciphers... and the Return of Dr. Matrix, reissue ed.* New York: W. H. Freeman, pp. 1–29, 1989.

Grünbaum, B. and Shepard, G. C. "Some Problems on Plane Tilings." In *The Mathematical Gardner* (Ed. D. Klarner). Boston, MA: Prindle, Weber, and Schmidt, pp. 167–196, 1981.

Grünbaum, B. and Sheppard, G. C. *Tilings and Patterns.* New York: W. H. Freeman, 1986.

Mirsky, S. "The Emperor's New Toilet Paper." *Sci. Amer.* **277**, 24, July 1997.

Pappas, T. "Mathematics & Moslem Art." *The Joy of Mathematics.* San Carlos, CA: Wide World Publ./Tetra, p. 178, 1989.

Peterson, I. *The Mathematical Tourist: Snapshots of Modern Mathematics.* New York: W. H. Freeman, pp. 82–85, 1988.

Rawles, B. *Sacred Geometry Design Sourcebook: Universal Dimensional Patterns.* Nevada City, CA: Elysian Pub., 1997.

Schattschneider, D. "In Praise of Amateurs." In *The Mathematical Gardner* (Ed. D. Klarner). Boston, MA: Prindle, Weber, and Schmidt, pp. 140–166, 1981.

Seyd, J. A. and Salman, A. S. *Symmetries of Islamic Geometrical Patterns.* River Edge, NJ: World Scientific, 1995.

Stein, S. and Szabó, S. *Algebra and Tiling: Homomorphisms in the Service of Geometry.* Washington, DC: Math. Assoc. Amer., 1994.

Steinhaus, H. *Mathematical Snapshots, 3rd ed.* New York: Dover, 1999.

Stevens, P. S. *Handbook of Regular Patterns: An Introduction to Symmetry in Two Dimensions.* Cambridge, MA: MIT Press, 1992.

Weisstein, E. W. "Plane Geometry." MATHEMATICA NOTEBOOK PLANEGEOMETRY.M.

Weisstein, E. W. "Books about Tilings." http://www.treasure-troves.com/books/Tilings.html.

Wells, D. *The Penguin Dictionary of Curious and Interesting Geometry.* London: Penguin, pp. 177–179, 208, and 211, 1991.

Tiling Problem

Maximize the amount of floor space which can be covered with a fixed tile (Hoffman 1998, p. 173).

See also BIN-PACKING PROBLEM, COOKIE-CUTTER PROBLEM

References

Hoffman, P. *The Man Who Loved Only Numbers: The Story of Paul Erdos and the Search for Mathematical Truth.* New York: Hyperion, 1998.

Tiling Theorem

Due to Lebesgue and Brouwer. If an n-D figure is covered in any way by sufficiently small subregions, then there will exist points which belong to at least $n+1$ of these subareas. Moreover, it is always possible to find a covering by arbitrarily small regions for which no point will belong to more than $n+1$ regions.

See also TESSELLATION, TILING

Time Series Analysis

Analysis of data ordered by the time the data were collected (usually spaced at equal intervals), called a time series. Common examples of a time series are daily temperature measurements, monthly sales, and yearly population figures. The goals of time series analysis are to describe the process generating the data, and to forecast future values.

See also ANOVA, ARITHMETIC MEAN, CORRELATION COEFFICIENT, COVARIANCE, DIFFERENCE TABLE, LEAST SQUARES FITTING, MAXIMUM LIKELIHOOD, MOVING AVERAGE, PERIODOGRAM, PREDICTION THEORY, RANDOM VARIABLE, RANDOM WALK, RESIDUAL, VARIANCE

References

Chatfield, C. *The Analysis of Time Series: An Introduction, 5th ed.* Boca Raton, FL: Chapman & Hall, 1996.

Cryer, J. D. *Time Series Analysis.* Boston, MA: PWS Publishers, 1986.

Miller, R. B. and Wichern, D. W. Ch. 9–11 in *Intermediate Business Statistics: Analysis of Variance, Regression, and Time Series.* New York: Holt, Rinehart and Winston, pp. 353–438, 1977.

Rao, T. S.; Priestly, M. B.; and Lessi, O. *Applications of Time Series Analysis in Astronomy and Meteorology.* Boca Raton, FL: Chapman & Hall, 1997.

Shumway, R. H. and Stoffer, D. S. *Time Series Analysis and Its Applications.* New York: Springer-Verlag, 2000.

Whittaker, E. T. and Robinson, G. "The Search for Periodicities." Ch. 13 in *The Calculus of Observations: A Treatise on Numerical Mathematics, 4th ed.* New York: Dover, pp. 343–362, 1967.

Times

The operation of MULTIPLICATION, i.e., a times b. Various notations are $a \times b$, $a \cdot b$, ab, and $(a)(b)$. The "multiplication sign" $\times$ is based on SAINT ANDREW'S CROSS (Bergamini 1969). Floating point MULTIPLICATION is sometimes denoted $\otimes$.

See also CROSS PRODUCT, DOT PRODUCT, MINUS, MULTIPLICATION, PLUS, PRODUCT

References

Bergamini, D. *Mathematics.* New York: Time-Life Books, p. 11, 1969.

Cundy, H. M. "What Is $\times$?" *Math. Gaz.* **43**, 101, 1959.

T-Integration

A fast, accurate, and numerically stable NUMERICAL INTEGRATION formula given by

$$X_n = X_{n-1} + TG\left[P\left(\frac{dX}{dt}\right)_n + (1-P)\left(\frac{dX}{dt}\right)_{n-1} \right],$$

where X is the integral, dX/dt is the integrand, P and G are "phase" and "gain" tuning parameters, n refers to the number of the iteration being evaluated, and T is the integration step size. For $G=1$, varying

P from 0 to 2 gives many classical first-order integrators:

1. $G = 1$ and $P = 0$: Euler integrator,
2. $G = 1$ and $P = 1/2$: TRAPEZOIDAL RULE,
3. $G = 1$ and $P = 1$: Rectangular rule,
4. $G = 1$ and $P = 3/2$: ADAMS' METHOD.

See also NUMERICAL INTEGRATION

References

Fowler, M. "A New Numerical Method for Simulation." *Simulation* **6**, 90–92, Feb. 1976.
Smith, J. M. "Recent Developments in Numerical Integration." *J. Dynam. Sys., Measurement and Control.* Mar. 1974.
Smith, J. M. "Zero-Order T-Integration and Its Relation to the Mean Value Theorem." In *Proceedings of the Sixth Annual Pittsburgh Modeling and Simulation Conference, Part 1, April 24–25, 1975.*
Smith, J. M. "Modern Numerical Integration Methods." In *Mathematical Modeling and Digital Simulation, 2nd ed.* New York: John Wiley, 1988.
Smith, J. M. "Fast T-Integration." *J. Mech. Eng. Sys.* **1**, 27–31, Jul./Aug. 1990.
Smith, J. M. "Jon Michael Smith on T-Integration: Trade Secrets in Numerical Analysis." http://members.aol.com/jsmith46ws/ni1.htm.

Titanic Prime

A PRIME with ≥ 1000 DIGITS. As of 1990, there were more than 1400 known (Ribenboim 1990). The table below gives the number of known titanic primes as a function of year end.

Year	Titanic Primes
1992	2254
1993	9166
1994	9779
1995	12391

References

Caldwell, C. "The Ten Largest Known Primes." http://www.utm.edu/research/primes/largest.html#largest.
Morain, F. "Elliptic Curves, Primality Proving and Some Titanic Primes." *Astérique* **198–200**, 245–251, 1992.
Ribenboim, P. *The Little Book of Big Primes.* Berlin: Springer-Verlag, p. 97, 1990.
Yates, S. "Titanic Primes." *J. Recr. Math.* **16**, 250–262, 1983–84.
Yates, S. "Sinkers of the Titanics." *J. Recr. Math.* **17**, 268–274, 1984–85.

Titchmarsh Theorem

If $f(\omega)$ is SQUARE INTEGRABLE over the REAL ω-axis, then any one of the following implies the other two:

1. The FOURIER TRANSFORM $F(t) = \mathscr{F}[f(\omega)]$ is 0 for $t < 1$.
2. Replacing ω by $z \equiv x + iy$, the function $f(z)$ is analytic in the COMPLEX PLANE z for $y > 0$ and approaches $f(x)$ almost everywhere as $y \to 0$. Furthermore, $\int_{-\infty}^{\infty} |f(x+iy)|^2 \, dx < k$ for some number k and $y > 0$ (i.e., the integral is bounded).
3. The REAL and IMAGINARY PARTS of $F(z)$ are HILBERT TRANSFORMS of each other

(Bracewell 1999, Problem 8, p. 273).

See also FOURIER TRANSFORM, HILBERT TRANSFORM

References

Bracewell, R. *The Fourier Transform and Its Applications, 3rd ed.* New York: McGraw-Hill, 1999.

Titchmarsh's Differential Equation

The ORDINARY DIFFERENTIAL EQUATION

$$y'' + (\lambda - x^{2n})y = 0.$$

References

Hille, E. *Lectures on Ordinary Differential Equations.* Reading, MA: Addison-Wesley, p. 617, 1969.
Zwillinger, D. *Handbook of Differential Equations, 3rd ed.* Boston, MA: Academic Press, p. 121, 1997.

Tit-for-Tat

A strategy for the iterated PRISONER'S DILEMMA in which a prisoner cooperates on the first move, and thereafter copies the previous move of the other prisoner. Any better strategy has more complicated rules.

See also PRISONER'S DILEMMA

References

Goetz, P. "Phil's Good Enough Complexity Dictionary." http://www.cs.buffalo.edu/~goetz/dict.html.

Tits Group

A FINITE SIMPLE GROUP which is a SUBGROUP of the TWISTED CHEVALLEY GROUP $^2F_4(2)$.

Toeplitz Matrix

Given $2n - 1$ numbers a_k, where $k = -n + 1, \ldots, -1, 0, 1, \ldots, n - 1$, a Toeplitz matrix is a MATRIX which has constant values along negative-sloping diagonals, i.e., a matrix OF THE FORM

$$\begin{bmatrix} a_0 & a_{-1} & a_{-2} & \cdots & a_{-n+1} \\ a_1 & a_0 & a_{-1} & \ddots & \vdots \\ a_2 & a_1 & a_0 & & a_{-2} \\ \vdots & \ddots & \ddots & \ddots & a_{-1} \\ a_{n-1} & \cdots & a_2 & a_1 & a_0 \end{bmatrix}$$

MATRIX EQUATIONS OF THE FORM

$$\sum_{j=1}^{n} a_{i-j} x_j = y_i$$

can be solved with $\mathcal{O}(n^2)$ operations. Typical problems modelled by Toeplitz matrices include the numerical solution of certain differential and integral equations (regularization of inverse problems), the computation of SPLINES, TIME SERIES ANALYSIS, signal and image processing, MARKOV CHAINS, and QUEUING THEORY (Bini 1995).

See also TRIANGULAR MATRIX, VANDERMONDE MATRIX

References

Bini, D. "Toeplitz Matrices, Algorithms and Applications." *ECRIM News Online Edition*, No. 22, July 1995. http://www.ercim.org/publication/Ercim_News/enw22/toeplitz.html.
Press, W. H.; Flannery, B. P.; Teukolsky, S. A.; and Vetterling, W. T. "Vandermonde Matrices and Toeplitz Matrices." §2.8 in *Numerical Recipes in FORTRAN: The Art of Scientific Computing, 2nd ed.* Cambridge, England: Cambridge University Press, pp. 82–89, 1992.

Togliatti Surface

Togliatti (1940, 1949) showed that QUINTIC SURFACES having 31 ORDINARY DOUBLE POINTS exist, although he did not explicitly derive equations for such surfaces. Beauville (1978) subsequently proved that 31 double points are the maximum possible, and quintic surfaces having 31 ORDINARY DOUBLE POINTS are therefore sometimes called Togliatti surfaces. van Straten (1993) subsequently constructed a 3-D family of solutions and in 1994, Barth derived the example known as the DERVISH.

See also DERVISH, ORDINARY DOUBLE POINT, QUINTIC SURFACE

References

Beauville, A. "Surfaces algébriques complexes." *Astérisque* **54**, 1–172, 1978.
Endraß, S. "Togliatti Surfaces." http://enriques.mathematik.uni-mainz.de/kon/docs/Etogliatti.shtml.
Hunt, B. "Algebraic Surfaces." http://www.mathematik.uni-kl.de/~wwwagag/E/Galerie.html.
Togliatti, E. G. "Una notevole superficie de 5° ordine con soli punti doppi isolati." *Vierteljschr. Naturforsch. Ges. Zürich* **85**, 127–132, 1940.
Togliatti, E. "Sulle superficie monoidi col massimo numero di punti doppi." *Ann. Mat. Pura Appl.* **30**, 201–209, 1949.
van Straten, D. "A Quintic Hypersurface in $\mathbb{P}^4$ with 130 Nodes." *Topology* **32**, 857–864, 1993.

Tomography

Tomography is the study of the reconstruction of 2- and 3-dimensional objects from 1-dimensional slices. The RADON TRANSFORM is an important tool in tomography.

Rather surprisingly, there exist certain sets of four directions in Euclidean n-space such that X-rays of a

convex body in these directions distinguish it from all other convex bodies.

See also ALEKSANDROV'S UNIQUENESS THEOREM, BRUNN-MINKOWSKI INEQUALITY, BUSEMANN-PETTY PROBLEM, DVORETZKY'S THEOREM, HAMMER'S X-RAY PROBLEMS, RADON TRANSFORM, STEREOLOGY

References

Gardner, R. J. "Geometric Tomography." *Not. Amer. Math. Soc.* **42**, 422–429, 1995.
Gardner, R. J. *Geometric Tomography*. New York: Cambridge University Press, 1995.
Herman, G. T. and Kuba, A. (Eds.). *Discrete Tomography: Foundations, Algorithms, and Applications*. Boston, MA: Birkhäuser, 1999.
Kak, A. C. and Slaney, M. *Principles of Computerized Tomographic Imaging*. IEEE Press, 1988.
Weisstein, E. W. "Books about Tomography." http://www.treasure-troves.com/books/Tomography.html.

Tooth Surface

The QUARTIC SURFACE given by the equation

$$x^4 + y^4 + z^4 - (x^2 + y^2 + z^2) = 0.$$

See also GOURSAT'S SURFACE

References

Nordstrand, T. "Surfaces." http://www.uib.no/people/nfytn/surfaces.htm.

Top-Dimensional Form

In an EXTERIOR ALGEBRA $\wedge V$, a top-dimensional form has degree n where $n = \dim V$. Any form of higher degree must be zero. For example, if $V = \mathbb{R}^4$ then

$$\alpha = e_1 \wedge e_2 \wedge e_3 \wedge e_4$$

is a top-dimensional form, and any other top-dimensional form is $\lambda \alpha$ for some λ.

See also DIFFERENTIAL K-FORM, EXTERIOR ALGEBRA, ORIENTATION (VECTOR SPACE), VOLUME FORM

Topological Basis

A topological basis is a SUBSET B of a SET T in which all other OPEN SETS can be written as UNIONS or finite

INTERSECTIONS of B. For the REAL NUMBERS, the SET of all OPEN INTERVALS is a basis.

Topological Completion

The topological completion C of a FIELD F with respect to the ABSOLUTE VALUE $|\cdot|$ is the smallest FIELD containing F for which all CAUCHY SEQUENCES or rationals converge.

References

Burger, E. B. and Struppeck, T. "Does $\Sigma_{n=0}^{\infty} \frac{1}{n!}$ Really Converge? Infinite Series and p-adic Analysis." *Amer. Math. Monthly* **103**, 565–577, 1996.

Topological Dimension

LEBESGUE COVERING DIMENSION

Topological Entropy

The topological entropy of a MAP M is defined as

$$h_T(M) = \sup_{\{W_i\}} h(M, \{W_i\}),$$

where $\{W_i\}$ is a partition of a bounded region W containing a probability measure which is invariant under M, and sup is the SUPREMUM.

References

Ott, E. *Chaos in Dynamical Systems.* New York: Cambridge University Press, pp. 143–144, 1993.

Topological Graph

A simple unlabeled graph whose connectivity is considered purely on the basis of topological equivalence, so that two edges (v_1, v_2) and (v_2, v_3) joined by a node v_2 of degree two are considered equivalent to the single edge (v_1, v_3).

See also MATCH PROBLEM

References

Weisstein, E. W. "Graphs." MATHEMATICA NOTEBOOK GRAPHS.M.

Topological Group

A CONTINUOUS GROUP G which has a HAUSDORFF TOPOLOGY is a topological group. The simplest example is the group of real numbers under addition.

The HOMEOMORPHISM GROUP of any COMPACT HAUSDORFF SPACE is a topological group when given the COMPACT-OPEN TOPOLOGY. Also, any LIE GROUP is a topological group.

See also EFFECTIVE ACTION, FREE ACTION, GROUP, ISOTROPY GROUP, MATRIX GROUP, ORBIT (GROUP), QUOTIENT SPACE, REPRESENTATION, TOPOLOGICAL GROUP, TRANSITIVE

References

Kawakubo, K. *The Theory of Transformation Groups.* Oxford, England: Oxford University Press, pp. 7–14, 1987.
Pontriagin, L. S. *Topological Groups, 2nd ed.* New York: Gordon and Breach, 1986.

Topological Groupoid

A topological groupoid over B is a GROUPOID G such that B and G are TOPOLOGICAL SPACES and α, β, and multiplication are continuous maps. Here, α and β are maps from G onto $\mathbb{R}^2$ with $\alpha : (x, \gamma, y) \mapsto x$ and $\beta : (x, \gamma, y) \mapsto y$.

See also GROUPOID, TOPOLOGICAL SPACE

References

Weinstein, A. "Groupoids: Unifying Internal and External Symmetry." *Not. Amer. Math. Soc.* **43**, 744–752, 1996.

Topological Manifold

A TOPOLOGICAL SPACE M satisfying some separability (i.e., it is a HAUSDORFF SPACE) and countability (i.e., it is a PARACOMPACT SPACE) conditions such that every point $p \in M$ has a NEIGHBORHOOD homeomorphic to an OPEN SET in $\mathbb{R}^n$ for some $n \geq 0$. Every SMOOTH MANIFOLD is a topological manifold, but not necessarily vice versa. The first nonsmooth topological manifold occurs in 4-D.

Nonparacompact manifolds are of little use in mathematics, but non-Hausdorff manifolds do occasionally arise in research (Hawking and Ellis 1975). For manifolds, Hausdorff and second countable are equivalent to Hausdorff and paracompact, and both are equivalent to the manifold being embeddable in some large-dimensional Euclidean space.

See also HAUSDORFF SPACE, MANIFOLD, PARACOMPACT SPACE, SMOOTH MANIFOLD, TOPOLOGICAL SPACE

References

Hawking, S. W. and Ellis, G. F. R. *The Large Scale Structure of Space-Time.* New York: Cambridge University Press, 1975.

Topological Sort

A topological sort is a PERMUTATION p of the vertices of a GRAPH such that an edge $\{i, j\}$ implies that i appears before j in p (Skiena 1990, p. 208). Only DIRECTED ACYCLIC GRAPHS can be topologically sorted. The topological sort of a graph can be computed using `TopologicalSort[g]` in the *Mathematica* add-on package `DiscreteMath`Combinatorica`` (which can be loaded with the command `<<DiscreteMath`).

References

Skiena, S. "Topological Sorting." §5.4.3 in *Implementing Discrete Mathematics: Combinatorics and Graph Theory with Mathematica.* Reading, MA: Addison-Wesley, pp. 208–209, 1990.

Topological Space

A SET X for which a TOPOLOGY T has been specified is called a topological space (Munkres 1975, p. 76).

In the chapter "Point Sets in General Spaces" Hausdorff (1914) defined his concept of a topological space based on the four HAUSDORFF AXIOMS.

    1. To each point x there corresponds at least one neighborhood $U(x)$, and $U(x)$ contains x.

    2. If $U(x)$ and $V(x)$ are neighborhoods of the same point x, then there exists a neighborhood $W(x)$ of x such that $W(x)$ is a subset of the union of $U(x)$ and $V(x)$.

    3. If y is a point in $U(x)$, then there exists a neighborhood $U(y)$ of y such that $U(y)$ is a subset of $U(x)$.

    4. For distinct points x and y, there exist two disjoint neighborhoods $U(x)$ and $U(y)$.

See also HAUSDORFF AXIOMS, HAUSDORFF SPACE, KURATOWSKI'S CLOSURE-COMPONENT PROBLEM, MANIFOLD, OPEN SET, TOPOLOGICAL VECTOR SPACE

References

Berge, C. *Topological Spaces Including a Treatment of Multi-Valued Functions, Vector Spaces and Convexity.* New York: Dover, 1997.
Hausdorff, F. *Grundzüge der Mengenlehre.* Leipzig, Germany: von Veit, 1914. Republished as *Set Theory, 2nd ed.* New York: Chelsea, 1962.
Munkres, J. R. *Topology: A First Course.* Englewood Cliffs, NJ: Prentice-Hall, 1975.

Topological Tree

SERIES-REDUCED TREE

Topological Vector Space

A VECTOR SPACE with a HAUSDORFF TOPOLOGY such that the operations of VECTOR ADDITION and SCALAR MULTIPLICATION are CONTINUOUS. The interesting examples are infinite-dimensional spaces, such as a space of functions. For example, a HILBERT SPACE and a BANACH SPACE are topological vector spaces.

The choice of topology reflects what is meant by convergence of functions. For instance, for functions whose integrals converge, the BANACH SPACE $L^1(X)$, one of the L^p-SPACES, is used. But if one is interested in POINTWISE CONVERGENCE, then no norm will suffice. Instead, for each $x \in X$ define the SEMINORM

$$\|f\|_x = |f(x)|$$

on the vector space of functions on X. The seminorms define a topology, the smallest one in which the seminorms are CONTINUOUS. So $\lim f_n = f$ is equivalent to $\lim f_n(x) = f(x)$ for all $x \in X$, i.e., POINTWISE CONVERGENCE. In a similar way, it is possible to define a topology for which CONVERGENCE means UNIFORM CONVERGENCE on COMPACT SETS.

See also BANACH SPACE, HILBERT SPACE, SEMINORM, TOPOLOGICAL SPACE, VECTOR SPACE

References

Köthe, G. *Topological Vector Spaces.* New York: Springer-Verlag, 1979.
Zimmer, R. *Essential Results in Functional Analysis.* Chicago: University of Chicago Press, pp. 13–17, 1990.

Topologically Conjugate

Two MAPS ϕ, $\psi : M \to M$ are said to be topologically conjugate if there EXISTS a HOMEOMORPHISM $h : M \to M$ such that $\phi \circ h = h \circ \psi$, i.e., h maps ψ-orbits onto ϕ-orbits. Two maps which are topologically conjugate cannot be distinguished topologically.

See also ANOSOV DIFFEOMORPHISM, STRUCTURALLY STABLE

Topologically Transitive

A FUNCTION f is topologically transitive if, given any two intervals U and V, there is some POSITIVE INTEGER k such that $f^k(U) \cap V \neq \varnothing$. Vaguely, this means that neighborhoods of points eventually get flung out to "big" sets so that they don't necessarily stick together in one localized clump.

See also CHAOS

Topology

Topology is the mathematical study of properties of objects which are preserved through deformations, twistings, and stretchings. (Tearing, however, is not allowed.) A CIRCLE is topologically equivalent to an ELLIPSE (into which it can be deformed by stretching) and a SPHERE is equivalent to an ELLIPSOID. Continuing along these lines, the SPACE of all positions of the minute hand on a clock is topologically equivalent to a CIRCLE (where SPACE of all positions means "the collection of all positions"). Similarly, the SPACE of all positions of the minute and hour hands is equivalent to a TORUS. The SPACE of all positions of the hour, minute and second hands form a 4-D object that cannot be visualized quite as simply as the former objects since it cannot be placed in our 3-D world, although it can be visualized by other means.

There is more to topology, though. Topology began with the study of curves, surfaces, and other objects in the plane and 3-space. One of the central ideas in topology is that spatial objects like CIRCLES and SPHERES can be treated as objects in their own right, and knowledge of objects is independent of how they are "represented" or "embedded" in space. For example, the statement "if you remove a point from a CIRCLE, you get a line segment" applies just as well to the CIRCLE as to an ELLIPSE, and even to tangled or knotted CIRCLES, since the statement involves only topological properties.

Topology has to do with the study of spatial objects such as curves, surfaces, the space we call our universe, the space-time of general relativity, fractals, knots, manifolds (objects with some of the same basic spatial properties as our universe), phase spaces that are encountered in physics (such as the space of hand-positions of a clock), symmetry groups like the collection of ways of rotating a top, etc.

The "objects" of topology are often formally defined as TOPOLOGICAL SPACES. If two objects have the same topological properties, they are said to be HOMEOMORPHIC (although, strictly speaking, properties that are not destroyed by stretching and distorting an object are really properties preserved by ISOTOPY, not HOMEOMORPHISM; ISOTOPY has to do with distorting embedded objects, while HOMEOMORPHISM is intrinsic).

Topology is divided into ALGEBRAIC TOPOLOGY (also called COMBINATORIAL TOPOLOGY), DIFFERENTIAL TOPOLOGY, and LOW-DIMENSIONAL TOPOLOGY.

There is also a formal definition for a topology defined in terms of set operations. A SET X along with a collection T of SUBSETS of it is said to be a topology if the SUBSETS in T obey the following properties:

1. The (trivial) subsets X and the EMPTY SET $\varnothing$ are in T.
2. Whenever sets A and B are in T, then so is $A \cap B$.
3. Whenever two or more sets are in T, then so is their UNION

(Bishop and Goldberg 1980). This definition can be used to enumerate the topologies on n symbols in *Mathematica* using the following code snippet.

```
<<DiscreteMath`Combinatorica`;          Topolo-
gyQ[x_List,t_List]:=Module[{},
  MemberQ[t,x]&&MemberQ[t,{}]&&
    And@@(MemberQ[t,#]&/@Intersection@@@KSub-
sets[t,2])&&
    And@@(MemberQ[t,#]&/@Union@@@Subsets[t])
 ] Topologies[n_]:=Module[{r=Range[n]},
  Select[Subsets[Subsets[r]],TopologyQ[r,#]&]
 ]
```

For example, the unique topology of order 1 is $\{\varnothing, \{1\}\}$, which the four topologies of order 2 are $\{\varnothing, \{1\}, \{1, 2\}\}$, $\{\varnothing, \{1\}, \{1, 2\}\}$, $\{\varnothing, \{1, 2\}, \{2\}\}$, and $\{\varnothing, \{1\}, \{2\}, \{1, 2\}\}$. The numbers of topologies on sets of cardinalities $n = 1, 2, \ldots$ are 1, 4, 29, 355, 6942, ... (Sloane's A000798).

A SET X for which a topology T has been specified is called a TOPOLOGICAL SPACE (Munkres 1975, p. 76). For example, the SET $X = \{1, 2, 3, 4\}$ together with

the SUBSETS $T = \{\varnothing, \{1\}, \{2, 3, 4\}, \{1, 2, 3, 4\}\}$ comprises a topology, and X is a TOPOLOGICAL SPACE.

Topologies can be built up from TOPOLOGICAL BASES. For the REAL NUMBERS, the topology is the UNION of OPEN INTERVALS.

See also ALGEBRAIC TOPOLOGY, DIFFERENTIAL TOPOLOGY, GENUS, KLEIN BOTTLE, KURATOWSKI REDUCTION THEOREM, LEFSHETZ TRACE FORMULA, LOW-DIMENSIONAL TOPOLOGY, MÖBIUS STRIP, POINT-SET TOPOLOGY, PRETZEL TRANSFORMATION, SPHERE EVERSION, TOPOLOGICAL SPACE, ZARISKI TOPOLOGY

References

Adamson, I. *A General Topology Workbook.* Boston, MA: Birkhäuser, 1996.

Alexandrov, P. S. *Elementary Concepts of Topology.* New York: Dover.

Armstrong, M. A. *Basic Topology, rev. ed.* New York: Springer-Verlag, 1997.

Arnold, B. H. *Intuitive Concepts in Elementary Topology.* New York: Prentice-Hall, 1962.

Barr, S. *Experiments in Topology.* New York: Dover, 1964.

Berge, C. *Topological Spaces Including a Treatment of Multi-Valued Functions, Vector Spaces and Convexity.* New York: Dover, 1997.

Bishop, R. and Goldberg, S. *Tensor Analysis on Manifolds.* New York: Dover, 1980.

Blackett, D. W. *Elementary Topology: A Combinatorial and Algebraic Approach.* New York: Academic Press, 1967.

Bloch, E. *A First Course in Geometric Topology and Differential Geometry.* Boston, MA: Birkhäuser, 1996.

Brown, J. I. and Watson, S. "The Number of Complements of a Topology on n Points is at Least 2^n (Except for Some Special Cases)." *Discr. Math.* **154,** 27–39, 1996.

Chinn, W. G. and Steenrod, N. E. *First Concepts of Topology: The Geometry of Mappings of Segments, Curves, Circles, and Disks.* Washington, DC: Math. Assoc. Amer., 1966.

Comtet, L. *Advanced Combinatorics: The Art of Finite and Infinite Expansions, rev. enl. ed.* Dordrecht, Netherlands: Reidel, p. 229, 1974.

Dugundji, J. *Topology.* Englewood Cliffs, NJ: Prentice-Hall, 1965.

Eppstein, D. "Geometric Topology." http://www.ics.uci.edu/~eppstein/junkyard/topo.html.

Erne, M. and Stege, K. "Counting Finite Posets and Topologies." , , .

Evans, J. W.; Harary, F.; and Lynn, M. S. "On the Computer Enumeration of Finite Topologies." *Commun. ACM* **10,** 295–297 and 313, 1967.

Francis, G. K. *A Topological Picturebook.* New York: Springer-Verlag, 1987.

Gemignani, M. C. *Elementary Topology.* New York: Dover, 1990.

Greever, J. *Theory and Examples of Point-Set Topology.* Belmont, CA: Brooks/Cole, 1967.

Heitzig, J. and Reinhold, J. "The Number of Unlabeled Orders on Fourteen Elements." Preprint No. 299. Hanover, Germany: Universität Hannover Institut für Mathematik, 1999.

Hirsch, M. W. *Differential Topology.* New York: Springer-Verlag, 1988.

Hocking, J. G. and Young, G. S. *Topology.* New York: Dover, 1988.

Kahn, D. W. *Topology: An Introduction to the Point-Set and Algebraic Areas.* New York: Dover, 1995.

Kelley, J. L. *General Topology*. New York: Springer-Verlag, 1975.

Kinsey, L. C. *Topology of Surfaces*. New York: Springer-Verlag, 1993.

Kleitman, D. and Rothschild, B. L. "The Number of Finite Topologies." *Proc. Amer. Math. Soc.* **25**, 276–282, 1970.

Lietzmann, W. *Visual Topology*. London: Chatto and Windus, 1965.

Lipschutz, S. *Theory and Problems of General Topology*. New York: Schaum, 1965.

Mendelson, B. *Introduction to Topology*. New York: Dover, 1990.

Munkres, J. R. *Elementary Differential Topology*. Princeton, NJ: Princeton University Press, 1963.

Munkres, J. R. *Topology: A First Course*. Englewood Cliffs, NJ: Prentice-Hall, 1975.

Praslov, V. V. and Sossinsky, A. B. *Knots, Links, Braids and 3-Manifolds: An Introduction to the New Invariants in Low-Dimensional Topology*. Providence, RI: Amer. Math. Soc., 1996.

Rayburn, M. "On the Borel Fields of a Finite Set." *Proc. Amer. Math.. Soc.* **19**, 885–889, 1968.

Seifert, H. and Threlfall, W. *A Textbook of Topology*. New York: Academic Press, 1980.

Shafaat, A. "On the Number of Topologies Definable for a Finite Set." *J. Austral. Math. Soc.* **8**, 194–198, 1968.

Shakhmatv, D. and Watson, S. "Topology Atlas." http://www.unipissing.ca/topology/.

Sloane, N. J. A. Sequences A000798/M3631 in "An On-Line Version of the Encyclopedia of Integer Sequences." http://www.research.att.com/~njas/sequences/eisonline.html.

Steen, L. A. and Seebach, J. A. Jr. *Counterexamples in Topology*. New York: Dover, 1996.

Thurston, W. P. *Three-Dimensional Geometry and Topology, Vol. 1*. Princeton, NJ: Princeton University Press, 1997.

Tucker, A. W. and Bailey, H. S. Jr. "Topology." *Sci. Amer.* **182**, 18–24, Jan. 1950.

van Mill, J. and Reed, G. M. (Eds.). *Open Problems in Topology*. New York: Elsevier, 1990.

Veblen, O. *Analysis Situs, 2nd ed.* New York: Amer. Math. Soc., 1946.

Weisstein, E. W. "Books about Topology." http://www.treasure-troves.com/books/Topology.html.

Topology (Digraph)

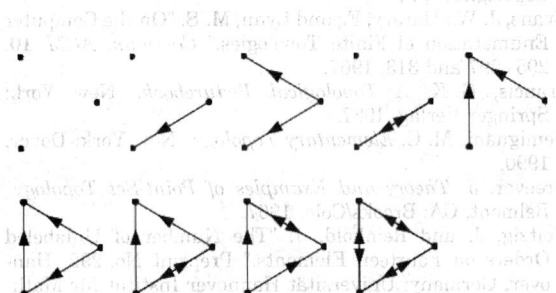

An unlabeled TRANSITIVE DIGRAPH with n nodes is called a "topology." The numbers of distinct topologies on $n = 1, 2, \ldots$ nodes are 1, 3, 9, 33, 139, 718, 4545, ... (Sloane's A001930). No larger values are known.

See also DIRECTED GRAPH, TRANSITIVE DIGRAPH

References

Sloane, N. J. A. Sequences A001930/M2817 in "An On-Line Version of the Encyclopedia of Integer Sequences." http://www.research.att.com/~njas/sequences/eisonline.html.

Sloane, N. J. A. and Plouffe, S. *The Encyclopedia of Integer Sequences*. San Diego, CA: Academic Press, 1995.

Topos

A CATEGORY modeled after the properties of the CATEGORY of sets.

See also CATEGORY, LOGOS

References

Freyd, P. J. and Scedrov, A. *Categories, Allegories*. Amsterdam, Netherlands: North-Holland, 1990.

McLarty, C. *Elementary Categories, Elementary Toposes*. New York: Oxford University Press, 1992.

Toric Section

A curve obtained by slicing a TORUS (generally a HORN TORUS) with a plane. A SPIRIC SECTION is a special case of a toric section in which the slicing plane is perpendicular to both the midplane of the torus and to the plane $x = 0$.

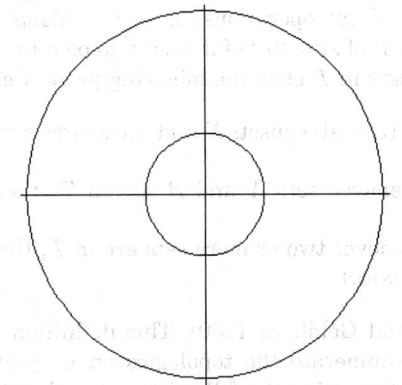

For planes parallel to the xy-plane, the toric sections are a single circle (for $z = 0$) or two concentric circles (for $0 < |z| \leq a$). For planes containing the z-AXIS, the section is two equal circles.

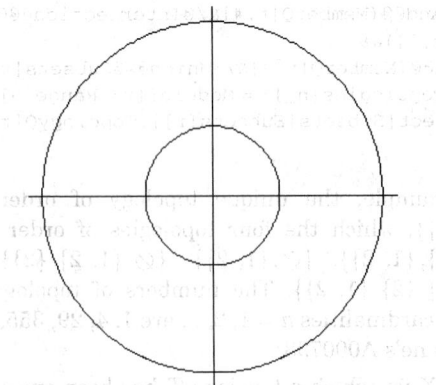

Toric sections at oblique angles can be more complicated, passing from a crescent shape, through a U-shape, and into two disconnected kidney-shaped curves.

See also SPIRIC SECTION, TORUS

Toric Variety

Let $m_1, m_2, \ldots, m_n$ be distinct primitive elements of a 2-D LATTICE M such that $\det(m_i, m_{i+1}) > 0$ for $i = 1, \ldots, n-1$. Each collection $\Gamma = \{m_1, m_2, \ldots, m_n\}$ then forms a set of rays of a unique complete fan in M, and therefore determines a 2-D toric variety X_Γ.

See also ALGEBRAIC VARIETY

References

Danilov, V. I. "The Geometry of Toric Varieties." *Russ. Math. Surv.* **33**, 97–154, 1978.
Fulton, W. *Introduction to Toric Varieties.* Princeton, NJ: Princeton University Press, 1993.
Morelli, R. "Pick's Theorem and the Todd Class of a Toric Variety." *Adv. Math.* **100**, 183–231, 1993.
Oda, T. *Convex Bodies and Algebraic Geometry.* New York: Springer-Verlag, 1987.
Pommersheim, J. E. "Toric Varieties, Lattice Points, and Dedekind Sums." *Math. Ann.* **295**, 1–24, 1993.

Torispherical Dome

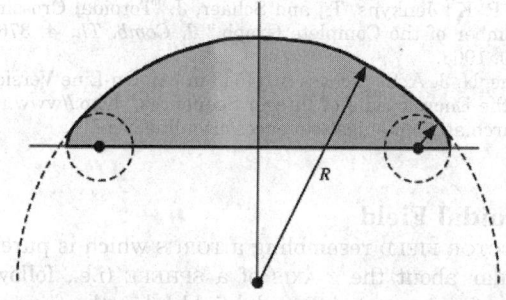

A torispherical dome is the surface obtained from the intersection of a SPHERICAL CAP with a tangent TORUS, as illustrated above. The radius of the sphere R is called the "crown radius," and the radius of the torus is called the "knuckle radius." Torispherical domes are used to construct pressure vessels.

See also DOME, SPHERICAL CAP

Torn Square Fractal

CESÀRO FRACTAL

Toroid

A SURFACE OF REVOLUTION obtained by rotating a closed PLANE CURVE about an axis parallel to the plane which does not intersect the curve. The simplest toroid is the TORUS. The word is also used to refer to a TOROIDAL POLYHEDRON (Gardner 1975).

See also PAPPUS'S CENTROID THEOREM, SURFACE OF REVOLUTION, TANGENT-SPHERE COORDINATES TOROIDAL POLYHEDRON, TORUS

References

Gardner, M. "Mathematical Games: On the Remarkable Császár Polyhedron and Its Applications in Problem Solving." *Sci. Amer.* **232**, 102–107, May 1975.

Toroidal Coordinates

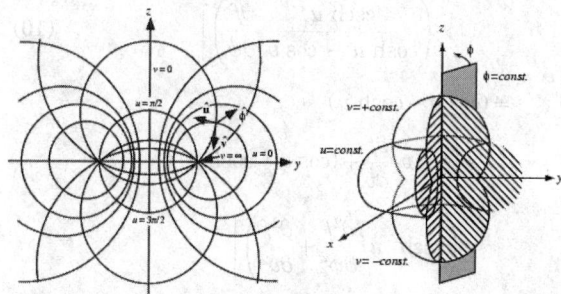

A system of CURVILINEAR COORDINATES for which several different notations are commonly used. In this work (u, v, ϕ) is used, whereas Arfken (1970) uses (ξ, η, φ) and Moon and Spencer (1988) use (η, θ, ψ). The toroidal coordinates are defined by

$$x = \frac{a \sinh u \cos \phi}{\cosh u - \cos v} \qquad (1)$$

$$y = \frac{a \sinh u \sin \phi}{\cosh u - \cos v} \qquad (2)$$

$$z = \frac{a \sin v}{\cosh u - \cos v}, \qquad (3)$$

where $\sinh z$ is the HYPERBOLIC SINE and $\cosh z$ is the HYPERBOLIC COSINE. Surfaces of constant u are given by the TOROIDS

$$x^2 + y^2 + z^2 + a^2 = 2a\sqrt{x^2 + y^2}\,\coth u, \qquad (4)$$

surfaces of constant v by the spherical bowls

$$x^2 + y^2 + (z - a \cot v)^2 = \frac{a^2}{\sin^2 v}, \qquad (5)$$

and surfaces of constant ϕ by

$$\tan \phi = \frac{y}{x}. \qquad (6)$$

The SCALE FACTORS are

$$h_u = \frac{a}{\cosh u - \cos v} \tag{7}$$

$$h_v = \frac{a}{\cosh u - \cos v} \tag{8}$$

$$h_\phi = \frac{a \sinh u}{\cosh u - \cos v}. \tag{9}$$

The LAPLACIAN is

$$\nabla^2 f = \frac{\sinh u}{(\cosh u - \cos v)^3} \left[\frac{\partial}{\partial u} \left(\frac{\sinh u}{\cosh u - \cos v} \frac{\partial f}{\partial u} \right) \right.$$

$$+ \frac{\partial}{\partial v} \left(\frac{\sinh u}{\cosh u - \cos v} \frac{\partial f}{\partial v} \right) + \frac{\partial}{\partial \phi}$$

$$\left. \times \left(\frac{\operatorname{csch} u}{\cosh u - \cos v} \frac{\partial f}{\partial \phi} \right) \right] \tag{10}$$

$$= (\cos v - \cosh u)$$

$$\times \left[\sin v \, \frac{\partial f}{\partial v} + (\cos v - \cosh u) \right.$$

$$\times \left(\operatorname{csch}^2 u \, \frac{\partial^2 f}{\partial \phi^2} + \frac{\partial^2 f}{\partial v^2} \right) \right]$$

$$+ (\cos v \cosh u - 1) \operatorname{csch} u \cdot \frac{\partial f}{\partial u}$$

$$+ (\cos v - \cosh u) \frac{\partial^2 f}{\partial u^2} \bigg]. \tag{11}$$

The HELMHOLTZ DIFFERENTIAL EQUATION is not separable in toroidal coordinates, but LAPLACE'S EQUATION is.

See also BISPHERICAL COORDINATES, FLAT-RING CYCLIDE COORDINATES, LAPLACE'S EQUATION–TOROIDAL COORDINATES

References

Arfken, G. "Toroidal Coordinates (ξ, η, ϕ)." §2.13 in *Mathematical Methods for Physicists, 2nd ed.* Orlando, FL: Academic Press, pp. 112–115, 1970.

Byerly, W. E. *An Elementary Treatise on Fourier's Series, and Spherical, Cylindrical, and Ellipsoidal Harmonics, with Applications to Problems in Mathematical Physics.* New York: Dover, p. 264, 1959.

Moon, P. and Spencer, D. E. "Toroidal Coordinates (η, θ, ψ)." Fig. 4.04 in *Field Theory Handbook, Including Coordinate Systems, Differential Equations, and Their Solutions, 2nd ed.* New York: Springer-Verlag, pp. 112–115, 1988.

Morse, P. M. and Feshbach, H. *Methods of Theoretical Physics, Part I.* New York: McGraw-Hill, p. 666, 1953.

Toroidal Crossing Number

The first few toroidal crossing numbers for a COMPLETE GRAPH are 0, 0, 0, 0, 0, 0, 0, 4, 9, 23, 42, 70, 105,

154, 226, 326, ... (Sloane's A014543). The toroidal crossing numbers for a COMPLETE BIGRAPH are given in the following table.

1	0	0	0	0	0	0
2		0	0	0	0	0
3			0	0	0	0
4				2		
5					5	8
6						12
7						

See also CROSSING NUMBER (GRAPH), RECTILINEAR CROSSING NUMBER

References

Gardner, M. "Crossing Numbers." Ch. 11 in *Knotted Doughnuts and Other Mathematical Entertainments.* New York: W. H. Freeman, pp. 133–144, 1986.

Guy, R. K. and Jenkyns, T. "The Toroidal Crossing Number of $K_{m,n}$." *J. Comb. Th.* **6**, 235–250, 1969.

Guy, R. K.; Jenkyns, T.; and Schaer, J. "Toroidal Crossing Number of the Complete Graph." *J. Comb. Th.* **4**, 376–390, 1968.

Sloane, N. J. A. Sequences A014543 in "An On-Line Version of the Encyclopedia of Integer Sequences." http://www.research.att.com/~njas/sequences/eisonline.html.

Toroidal Field

A VECTOR FIELD resembling a TORUS which is purely circular about the z-AXIS of a SPHERE (i.e., follows lines of LATITUDE). A toroidal field takes the form

$$\mathbf{T} = \begin{bmatrix} 0 \\ \dfrac{1}{\sin \theta} \dfrac{\partial T}{\partial \phi} \\ -\dfrac{\partial T}{\partial \theta} \end{bmatrix}.$$

See also DIVERGENCELESS FIELD, POLOIDAL FIELD

References

Stacey, F. D. *Physics of the Earth, 2nd ed.* New York: Wiley, p. 239, 1977.

Toroidal Function

A class of functions also called RING FUNCTIONS which appear in systems having toroidal symmetry. Toroidal functions can be expressed in terms of the LEGENDRE FUNCTIONS and SECOND KINDS (Abramowitz and Stegun 1972, p. 336):

$$P_{\nu-1/2}^{\mu}(\cosh \eta) = [\Gamma(1-\mu)]^{-1} 2^{2\mu} \left(1 - e^{-2\eta}\right)^{-\mu} e^{-(\nu+1/2)\eta}$$

$$\times {}_2F_1\left(\tfrac{1}{2} - \mu, \ \tfrac{1}{2} + \nu - \mu; \ 1 - 2\mu; \ 1 - e^{-2\eta}\right)$$

$$P_{n-1/2}^{m}(\cosh \eta) = \frac{\Gamma\left(n + m + \tfrac{1}{2}\right)(\sinh \eta)^m}{\Gamma\left(n - m + \tfrac{1}{2}\right) 2^m \sqrt{\pi}\ \Gamma\left(m + \tfrac{1}{2}\right)}$$

$$\times \int_0^{\pi} \frac{\sin^{2m}\phi \ d\phi}{(\cosh \eta + \cos \phi \sin \eta)^{n+m+1/2}}$$

$$Q_{\nu-1/2}^{\mu}(\cosh \eta) = [\Gamma(1+\nu)]^{-1} \sqrt{\pi}\ e^{i\mu\pi} \Gamma\left(\tfrac{1}{2} + \nu + \mu\right)$$

$$\times \left(1 - e^{-2\eta}\right)^{\mu} e^{-(\nu+1/2)\eta} {}_2F_1$$

$$\times \left(\tfrac{1}{2} - \mu, \ \tfrac{1}{2} + \nu + \mu; \ 1 + \mu; \ 1 - e^{-2\eta}\right)$$

$$Q_{n-1/2}^{m}(\cosh \eta) = \frac{(-1)^m \Gamma\left(n + \tfrac{1}{2}\right)}{\Gamma\left(n - m + \tfrac{1}{2}\right)}$$

$$\times \int_0^{\infty} \frac{\cosh(mt)\ dt}{(\cosh \eta + \cosh t \sinh \eta)^{n+1/2}}$$

for $n > m$. Byerly (1959) identifies

$$\frac{1}{i^{n/2}} P_m^n(\coth x) = \operatorname{csch}^n x \ \frac{d^n P_m(\coth x)}{d(\coth x)^n}$$

as a TOROIDAL HARMONIC.

See also CONICAL FUNCTION

References

Abramowitz, M. and Stegun, C. A. (Eds.). "Toroidal Functions (or Ring Functions)." §8.11 in *Handbook of Mathematical Functions with Formulas, Graphs, and Mathematical Tables, 9th printing*. New York: Dover, p. 336, 1972.

Byerly, W. E. *An Elementary Treatise on Fourier's Series, and Spherical, Cylindrical, and Ellipsoidal Harmonics, with Applications to Problems in Mathematical Physics*. New York: Dover, p. 266, 1959.

Iyanaga, S. and Kawada, Y. (Eds.). *Encyclopedic Dictionary of Mathematics*. Cambridge, MA: MIT Press, p. 1468, 1980.

Toroidal Harmonic

TOROIDAL FUNCTION

Toroidal Polyhedron

A toroidal polyhedron is a POLYHEDRON with GENUS $g \geq 1$ (i.e., having one or more HOLES). Examples of toroidal polyhedra include the CSÁSZÁR POLYHEDRON and SZILASSI POLYHEDRON, both of which have GENUS 1 (i.e., the TOPOLOGY of a TORUS).

The only known TOROIDAL POLYHEDRON with no DIAGONALS is the CSÁSZÁR POLYHEDRON. If another exists, it must have 12 or more VERTICES and GENUS $g \geq 6$ (Gardner 1975). The smallest known single-hole

toroidal polyhedron made up of only EQUILATERAL TRIANGLES is composed of 48 of them.

See also CSÁSZÁR POLYHEDRON, SZILASSI POLYHEDRON, TOROID

References

Gardner, M. "Mathematical Games: On the Remarkable Császár Polyhedron and Its Applications in Problem Solving." *Sci. Amer.* **232**, 102–107, May 1975.

Gardner, M. *Time Travel and Other Mathematical Bewilderments*. New York: W. H. Freeman, p. 141, 1988.

Hart, G. "Toroidal Polyhedra." http://www.georgehart.com/virtual-polyhedra/toroidal.html.

Stewart, B. M. *Adventures Among the Toroids, 2nd rev. ed.* Okemos, MI: B. M. Stewart, 1984.

Toronto Function

The function defined by

$$T(m, n, r) \equiv r^{2n-m+1} e^{-r^2} \frac{\Gamma\left(\tfrac{1}{2}m + \tfrac{1}{2}\right)}{n!}$$

$$\times {}_1F_1\left(\tfrac{1}{2}(m+1); \ n+1; \ r^2\right) \quad (1)$$

(Heatley 1943; Abramowitz and Stegun 1972, p. 509), where ${}_1F_1(a; b; z)$ is a CONFLUENT HYPERGEOMETRIC FUNCTION and $\Gamma(z)$ is the GAMMA FUNCTION.

Heatley originally defined the function in terms of the integral

$$T(m, n, p, a) = \int_0^{\infty} t^{-n} e^{-p^2 t^2} I_n(2at)\ dt, \quad (2)$$

where $I_n(x)$ is a MODIFIED BESSEL FUNCTION OF THE FIRST KIND, which is similar to an integral of Watson (1966, p. 394), with Watson's $J_\nu(at)$ changed to $I_n(2at)$ and a few other minor changes of variables. In terms of this function,

$$T(m, n, r) = 2r^{n-m+1} e^{-r^2} T(m, n, 1, r) \quad (3)$$

(Heatley 1943). Heatley (1943) also gives a number of recurrences and other identities satisfied by $T(m, n, r)$.

References

Abramowitz, M. and Stegun, C. A. (Eds.). *Handbook of Mathematical Functions with Formulas, Graphs, and Mathematical Tables, 9th printing*. New York: Dover, p. 509, 1972.

Erdélyi, A.; Magnus, W.; Oberhettinger, F.; and Tricomi, F. G. *Higher Transcendental Functions, Vol. 1*. New York: Krieger, p. 268, 1981.

Heatley, A. H. "A Short Table of the Toronto Function." *Trans. Roy. Soc. Canada* **37**, 13–29, 1943.

Watson, G. N. *A Treatise on the Theory of Bessel Functions, 2nd ed.* Cambridge, England: Cambridge University Press, 1966.

Torricelli Point

FERMAT POINTS

Torsion (Differential Geometry)

The rate of change of the OSCULATING PLANE of a SPACE CURVE. The torsion τ is POSITIVE for a right-handed curve, and NEGATIVE for a left-handed curve. A curve with CURVATURE $\kappa \neq 0$ is planar IFF $\tau = 0$.

The torsion can be defined by

$$\tau \equiv -\mathbf{N} \cdot \mathbf{B}',$$

where $\mathbf{N}$ is the unit NORMAL VECTOR and $\mathbf{B}$ is the unit BINORMAL VECTOR. Written explicitly in terms of a parameterized VECTOR FUNCTION $\mathbf{x}$,

$$\tau = \frac{|\dot{\mathbf{x}}\ddot{\mathbf{x}}\dddot{\mathbf{x}}|}{\ddot{\mathbf{x}} \cdot \ddot{\mathbf{x}}} = \rho^2 |\dot{\mathbf{x}}\ddot{\mathbf{x}}\dddot{\mathbf{x}}|,$$

where $|\mathbf{abc}|$ denotes a SCALAR TRIPLE PRODUCT and ρ is the RADIUS OF CURVATURE. The quantity $1/\tau$ is called the RADIUS OF TORSION and is denoted σ or ϕ.

See also CURVATURE, RADIUS OF CURVATURE, RADIUS OF TORSION

References

Gray, A. "Drawing Space Curves with Assigned Curvature." §10.2 in *Modern Differential Geometry of Curves and Surfaces with Mathematica, 2nd ed.* Boca Raton, FL: CRC Press, pp. 222–224, 1993.
Kreyszig, E. "Torsion." §14 in *Differential Geometry.* New York: Dover, pp. 37–40, 1991.

Torsion (Group)

If G is a GROUP, then the torsion elements Tor(G) of G (also called the torsion of G) are defined to be the set of elements g in G such that $g^n = e$ for some NATURAL NUMBER n, where e is the IDENTITY ELEMENT of the GROUP G.

In the case that G is ABELIAN, Tor(G) is a SUBGROUP and is called the torsion subgroup of G. If Tor(G) consists only of the IDENTITY ELEMENT, the GROUP G is called torsion-free.

See also ABELIAN GROUP, FREE ABELIAN GROUP, GROUP, IDENTITY ELEMENT

Torsion Number

One of a set of numbers defined in terms of an invariant generated by the finite cyclic covering spaces of a KNOT complement. The torsion numbers for KNOTS up to 9 crossings were cataloged by Reidemeister (1948).

See also KNOT INVARIANT

References

Reidemeister, K. *Knotentheorie.* New York: Chelsea, 1948.
Rolfsen, D. "Torsion Numbers." §6A in *Knots and Links.* Wilmington, DE: Publish or Perish Press, pp. 145–146, 1976.

Torsion Subgroup

TORSION (GROUP)

Torsion Tensor

The TENSOR defined by

$$T^l_{\ jk} \equiv -\left(\Gamma^l_{\ jk} - \Gamma^l_{\ kj}\right),$$

where $\Gamma^l_{\ jk}$ are CONNECTION COEFFICIENTS.

See also CONNECTION COEFFICIENT

Torus

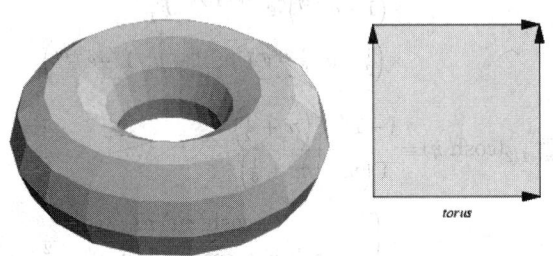

torus

A torus is a surface having GENUS 1, and therefore possessing a single "HOLE." The usual torus in 3-D space is shaped like a donut, but the concept of the torus is extremely useful in higher dimensional space as well. One of the more common uses of n-D tori is in DYNAMICAL SYSTEMS. A fundamental result states that the PHASE SPACE trajectories of a HAMILTONIAN SYSTEM with n DEGREES OF FREEDOM and possessing n INTEGRALS OF MOTION lie on an n-D MANIFOLD which is topologically equivalent to an n-torus (Tabor 1989).

The usual 3-D "ring" torus is known in older literature as an "ANCHOR RING." It can be constructed from a RECTANGLE by gluing both pairs of opposite edges together with no twists.

Let the radius from the center of the hole to the center of the torus tube be c, and the radius of the tube be a. Then the equation in CARTESIAN COORDINATES for a torus azimuthally symmetric about the z-AXIS is

$$\left(c - \sqrt{x^2 + y^2}\right)^2 + z^2 = a^2, \tag{1}$$

and the PARAMETRIC EQUATIONS are

$$x = (c + a \cos v) \cos u \tag{2}$$

$$y = (c + a \cos u) \sin u \tag{3}$$

$$z = a \sin v \tag{4}$$

for $u, v \in [0, 2\pi)$. Three types of torus, known as the STANDARD TORI, are possible, depending on the relative sizes of a and c. $c > a$ corresponds to the

RING TORUS (shown above), $c = a$ corresponds to a HORN TORUS which is tangent to itself at the point $(0, 0, 0)$, and $c < a$ corresponds to a self-intersecting SPINDLE TORUS (Pinkall 1986).

If no specification is made, "torus" is taken to mean RING TORUS. The three STANDARD TORI are illustrated below, where the first image shows the full torus, the second a cut-away of the bottom half, and the third a CROSS SECTION of a plane passing through the z-AXIS.

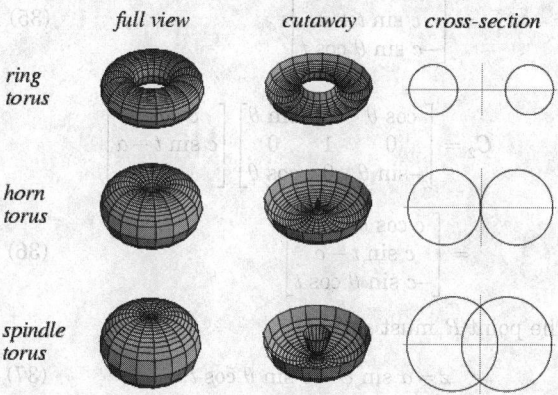

	full view	*cutaway*	*cross-section*
ring torus			
horn torus			
spindle torus			

The STANDARD TORI and their inversions are CYCLIDES. If the coefficient of $\sin v$ in the formula for z is changed to $b \neq a$, an ELLIPTIC TORUS results.

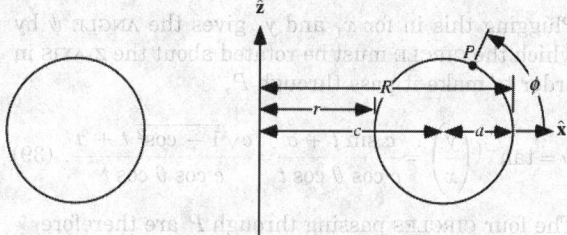

To compute the metric properties of the ring torus, define the inner and outer radii by

$$r \equiv c - a \tag{5}$$

$$R \equiv c + a. \tag{6}$$

Solving for a and c gives

$$a = \tfrac{1}{2}(R - r) \tag{7}$$

$$c = \tfrac{1}{2}(R + r). \tag{8}$$

Then the SURFACE AREA of this torus is

$$S = (2\pi a)(2\pi c) = 4\pi^2 ac \tag{9}$$

$$= \pi^2 (R + r)(R - r), \tag{10}$$

and the VOLUME can be computed from PAPPUS'S CENTROID THEOREM

$$V = (\pi a^2)(2\pi c) = 2\pi^2 a^2 c \tag{11}$$

$$= \tfrac{1}{4}\pi^2(R + r)(R - r)^2. \tag{12}$$

The coefficients of the coefficients of the FIRST FUNDAMENTAL FORM are

$$E = (c + a \cos v)^2 \tag{13}$$

$$F = 0 \tag{14}$$

$$G = a^2 \tag{15}$$

and the coefficients of the SECOND FUNDAMENTAL FORM are

$$e = -(c + a \cos v) \cos v \tag{16}$$

$$f = 0 \tag{17}$$

$$g = -a, \tag{18}$$

giving RIEMANNIAN METRIC

$$ds^2 = (c + a \cos v)^2 \, du^2 + a^2 \, dv^2, \tag{19}$$

AREA ELEMENT

$$dA = a(c + a \cos v) \, du \wedge dv \tag{20}$$

(where $du \wedge dv$ is a WEDGE PRODUCT), and GAUSSIAN and MEAN CURVATURES as

$$K = \frac{\cos v}{a(c + a \cos v)} \tag{21}$$

$$H = -\frac{c + 2a \cos v}{2a(c + a \cos v)} \tag{22}$$

(Gray 1997, pp. 384–386).

A torus with a HOLE in *its surface* can be turned inside out to yield an identical torus. A torus can be knotted externally or internally, but not both. These two cases are AMBIENT ISOTOPIES, but not REGULAR ISOTOPIES. There are therefore three possible ways of embedding a torus with zero or one KNOT.

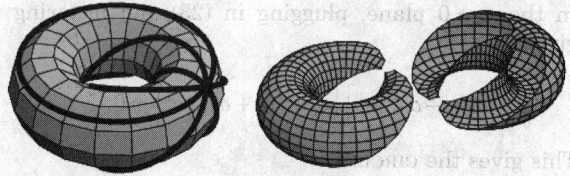

An arbitrary point P on a torus (not lying in the xy-plane) can have four CIRCLES drawn through it. The first circle is in the plane of the torus and the second is PERPENDICULAR to it. The third and fourth CIRCLES are called VILLARCEAU CIRCLES (Villarceau 1848, Schmidt 1950, Coxeter 1969, Melnick 1983).

To see that two additional CIRCLES exist, consider a coordinate system with origin at the center of torus, with $\hat{\mathbf{z}}$ pointing up. Specify the position of P by its ANGLE ϕ measured around the tube of the torus. Define $\phi = 0$ for the circle of points farthest away from the center of the torus (i.e., the points with $x^2 + y^2 = R^2$), and draw the x-AXIS as the intersection of a plane through the z-AXIS and passing through P with the xy-plane. Rotate about the Y-AXIS by an ANGLE θ, where

$$\theta = \sin^{-1}\left(\frac{a}{c}\right). \tag{23}$$

In terms of the old coordinates, the new coordinates are

$$x = x_1 \cos\theta - z_1 \sin\theta \tag{24}$$

$$z = x_1 \sin\theta + z_1 \cos\theta. \tag{25}$$

So in (x_1, y_1, z_1) coordinates, equation (1) of the torus becomes

$$\left[\sqrt{(x_1\cos\theta - z_1\sin\theta)^2 + y_1^2} - c\right]^2$$

$$+(x_1\sin\theta + z_1\cos\theta)^2 = a^2. \tag{26}$$

Expanding the left side gives

$$(x_1\cos\theta - z_1\sin\theta)^2 + y_1^2 + c^2$$

$$-2c\sqrt{(x_1\cos\theta - z_1\sin\theta)^2 + y_1^2}$$

$$+(x_1\sin\theta + z_1\cos\theta)^2 = a^2. \tag{27}$$

But

$$(x_1\cos\theta - z_1\sin\theta)^2 + (x_1\sin\theta + z_1\cos\theta)^2$$

$$= x_1^2 + z_1^2, \tag{28}$$

so

$$x_1^2 + y_1^2 + z_1^2 + c^2 - 2c\sqrt{(x_1\cos\theta - z_1\sin\theta)^2 + y_1^2}$$

$$= a^2. \tag{29}$$

In the $z_1 = 0$ plane, plugging in (23) and factoring gives

$$\left[x_1^2 + (y_1 - a)^2 - c^2\right]\left[x_1^2 + (y_1 + a)^2 - c^2\right] = 0. \tag{30}$$

This gives the CIRCLES

$$x_1^2 + (y_1 - a)^2 = c^2 \tag{31}$$

and

$$x_1^2 + (y_1 + a)^2 = c^2 \tag{32}$$

in the z_1 plane. Written in MATRIX form with parameter $t \in [0, 2\pi)$, these are

$$C_1 = \begin{bmatrix} c\cos t \\ c\sin t + a \\ 0 \end{bmatrix} \tag{33}$$

$$C_2 = \begin{bmatrix} c\cos t \\ c\sin t - a \\ 0 \end{bmatrix} \tag{34}$$

In the original (x, y, z) coordinates,

$$C_1 = \begin{bmatrix} \cos\theta & 0 & -\sin\theta \\ 0 & 1 & 0 \\ -\sin\theta & 0 & \cos\theta \end{bmatrix} \begin{bmatrix} c\cos t \\ c\sin t + a \\ 0 \end{bmatrix}$$

$$= \begin{bmatrix} c\cos\theta\cos t \\ c\sin t + a \\ -c\sin\theta\cos t \end{bmatrix} \tag{35}$$

$$C_2 = \begin{bmatrix} \cos\theta & 0 & \sin\theta \\ 0 & 1 & 0 \\ -\sin\theta & 0 & \cos\theta \end{bmatrix} \begin{bmatrix} c\cos t \\ c\sin t - a \\ 0 \end{bmatrix}$$

$$= \begin{bmatrix} c\cos\theta\cos t \\ c\sin t - a \\ -c\sin\theta\cos t \end{bmatrix}. \tag{36}$$

The point P must satisfy

$$z = a\sin\phi = c\sin\theta\cos t, \tag{37}$$

so

$$\cos t = \frac{a\sin\phi}{c\sin\theta}. \tag{38}$$

Plugging this in for x_1 and y_1 gives the ANGLE ψ by which the CIRCLE must be rotated about the z-AXIS in order to make it pass through P,

$$\psi = \tan^{-1}\left(\frac{y}{x}\right) = \frac{c\sin t + a}{c\cos\theta\cos t} = \frac{c\sqrt{1 - \cos^2 t} + a}{c\cos\theta\cos t}. \tag{39}$$

The four CIRCLES passing through P are therefore

$$C_1 = \begin{bmatrix} \cos\psi & \sin\psi & 0 \\ -\sin\psi & \cos\psi & 0 \\ 0 & 0 & 1 \end{bmatrix} \begin{bmatrix} c\cos\theta\cos t \\ c\sin t + a \\ -c\sin\theta\cos t \end{bmatrix} \tag{40}$$

$$C_2 = \begin{bmatrix} \cos\psi & \sin\psi & 0 \\ -\sin\psi & \cos\psi & 0 \\ 0 & 0 & 1 \end{bmatrix} \begin{bmatrix} c\cos\theta\cos t \\ c\sin t - a \\ -c\sin\theta\cos t \end{bmatrix} \tag{41}$$

$$C_3 = \begin{bmatrix} (c + a\cos\phi)\cos t \\ (c + a\cos\phi)\sin t \\ a\sin\phi \end{bmatrix} \tag{42}$$

$$C_4 = \begin{bmatrix} c + a\cos t \\ 0 \\ a\sin t \end{bmatrix}. \tag{43}$$

See also APPLE, CYCLIDE, DOUBLE TORUS, ELLIPTIC TORUS, GENUS (SURFACE), HORN TORUS, KLEIN QUARTIC, LEMON, RING TORUS, SPINDLE TORUS, SPIRIC SECTION, STANDARD TORI, TORIC SECTION, TOROID, TORUS COLORING, TORUS CUTTING, TORUS DISSECTION, TRIPLE TORUS

Torus Coloring

References

Beyer, W. H. *CRC Standard Mathematical Tables, 28th ed.* Boca Raton, FL: CRC Press, pp. 131–132, 1987.

Coxeter, H. S. M. *Introduction to Geometry, 2nd ed.* New York: Wiley, pp. 132–133, 1969.

Gray, A. "Tori." §13.4 in *Modern Differential Geometry of Curves and Surfaces with Mathematica, 2nd ed.* Boca Raton, FL: CRC Press, pp. 304–306 and 384–386, 1997.

Harris, J. W. and Stocker, H. "Torus." §4.10.5 in *Handbook of Mathematics and Computational Science.* New York: Springer-Verlag, p. 113, 1998.

JavaView. "Classic Surfaces from Differential Geometry: Torus." http://www-sfb288.math.tu-berlin.de/vgp/java-view/demo/surface/common/PaSurface_Torus.html.

Melzak, Z. A. *Invitation to Geometry.* New York: Wiley, pp. 63–72, 1983.

Pinkall, U. "Cyclides of Dupin." §3.3 in *Mathematical Models from the Collections of Universities and Museums* (Ed. G. Fischer). Braunschweig, Germany: Vieweg, pp. 28–30, 1986.

Schmidt, H. *Die Inversion und ihre Anwendungen.* Munich: Oldenbourg, p. 82, 1950.

Tabor, M. *Chaos and Integrability in Nonlinear Dynamics: An Introduction.* New York: Wiley, pp. 71–74, 1989.

Villarceau, M. "Théorème sur le tore." *Nouv. Ann. Math.* **7**, 345–347, 1848.

Torus Coloring

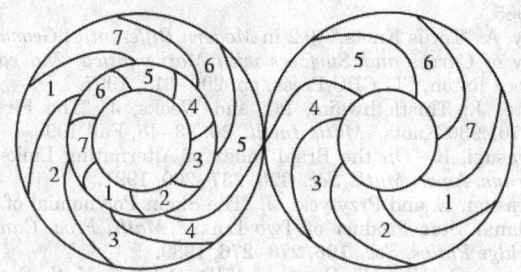

The number of colors SUFFICIENT for MAP COLORING on a surface of GENUS g is given by the HEAWOOD CONJECTURE,

$$\chi(g) = \left\lfloor \tfrac{1}{2}\left(7 + \sqrt{48g+1}\right)\right\rfloor,$$

where $\lfloor x \rfloor$ is the FLOOR FUNCTION. The fact that $\chi(g)$ (which is called the CHROMATIC NUMBER) is also NECESSARY was proved by Ringel and Youngs (1968) with two exceptions: the SPHERE (which requires the same number of colors as the PLANE) and the KLEIN BOTTLE. A g-holed TORUS therefore requires $\chi(g)$ colors. For $g = 0$, 1, ..., the first few values of $\chi(g)$ are 4, 7, 8, 9, 10, 11, 12, 12, 13, 13, 14, 15, 15, 16, ... (Sloane's A000934). A set of regions requiring the maximum of seven regions is shown above for a normal TORUS

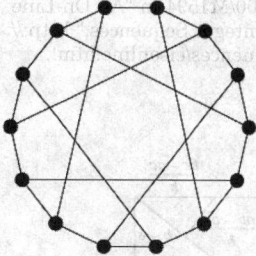

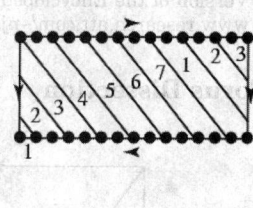

Heawood graph　　　　　　　*torus coloring*

The above figure shows the relationship between the HEAWOOD GRAPH and the 7-color torus coloring.

See also CHROMATIC NUMBER, FOUR-COLOR THEOREM, HEAWOOD CONJECTURE, HEAWOOD GRAPH, KLEIN BOTTLE, MAP COLORING, TORUS

References

Bondy, J. A. and Murty, U. S. R. *Graph Theory with Applications.* New York: North Holland, p. 244, 1976.

Cadwell, J. H. Ch. 8 in *Topics in Recreational Mathematics.* Cambridge, England: Cambridge University Press, 1966.

Gardner, M. "Mathematical Games: The Celebrated Four-Color Map Problem of Topology." *Sci. Amer.* **203**, 218–222, Sep. 1960.

Ringel, G. *Map Color Theorem.* New York: Springer-Verlag, 1974.

Ringel, G. and Youngs, J. W. T. "Solution of the Heawood Map-Coloring Problem." *Proc. Nat. Acad. Sci. USA* **60**, 438–445, 1968.

Sloane, N. J. A. Sequences A000934/M3292 in "An On-Line Version of the Encyclopedia of Integer Sequences." http://www.research.att.com/~njas/sequences/eisonline.html.

Steinhaus, H. *Mathematical Snapshots, 3rd ed.* New York: Dover, pp. 274–275, 1999.

Wagon, S. "Map Coloring on a Torus." §7.5 in *Mathematica in Action.* New York: W. H. Freeman, pp. 232–237, 1991.

Wells, D. *The Penguin Dictionary of Curious and Interesting Numbers.* Middlesex, England: Penguin Books, p. 70, 1986.

Wells, D. *The Penguin Dictionary of Curious and Interesting Geometry.* London: Penguin, pp. 228–229, 1991.

Torus Cutting

With n cuts of a TORUS of GENUS 1, the maximum number of pieces which can be obtained is

$$N(n) = \tfrac{1}{6}\left(n^3 + 3n^3 + 8n\right).$$

The first few terms are 2, 6, 13, 24, 40, 62, 91, 128, 174, 230, ... (Sloane's A003600).

See also CAKE CUTTING, CIRCLE DIVISION BY LINES, CYLINDER CUTTING, PANCAKE CUTTING, PLANE CUTTING, PIE CUTTING, SQUARE DIVISION BY LINES

References

Gardner, M. *Mathematical Magic Show: More Puzzles, Games, Diversions, Illusions and Other Mathematical Sleight-of-Mind from Scientific American.* New York: Vintage, pp. 149–150, 1978.

Sloane, N. J. A. Sequences A003600/M1594 in "An On-Line Version of the Encyclopedia of Integer Sequences." http://www.research.att.com/~njas/sequences/eisonline.html.

Torus Dissection

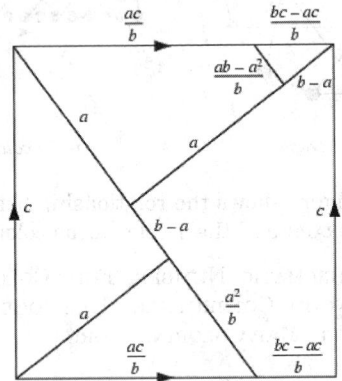

A ring TORUS constructed out of a square of side length c can be dissected into two squares of arbitrary side lengths a and b (as long as they are consistent with the size of the original square), as illustrated above.

See also DISSECTION, TORUS

References

Stewart, I. "Squaring the Square." *Sci. Amer.* **277**, 94–96, July 1997.

Torus Knot

A (p, q)-torus KNOT is obtained by looping a string through the HOLE of a TORUS p times with q revolutions before joining its ends, where p and q are RELATIVELY PRIME. A (p, q)-torus knot is equivalent to a (q, p)-torus knot. All torus knots are PRIME (Burde and Zieschang 1985, Hoste *et al.* 1998). Torus knots are all chiral, invertible, and have symmetry group D_1 (Schreier 1924, Hoste *et al.* 1998).

The CROSSING NUMBER of a (p, q)-torus knot is

$$c = \min\{p(q-1),\ q(p-1)\} \qquad (1)$$

(Williams 1988, Murasugi and Przytycki 1989, Murasugi 1991, Hoste *et al.* 1998). The UNKNOTTING NUMBER of a (p, q)-torus knot is

$$u = \tfrac{1}{2}(p-1)(q-1) \qquad (2)$$

(Adams 1991).

Torus knots with fewer than 11 crossings are the TREFOIL KNOT 03–001 $(3, 2)$, SOLOMON'S SEAL KNOT 05–001 $(5, 2)$, 07–001 $(7, 2)$, 08–019 $(4, 3)$, 09–001 $(9, 2)$, and 10–124 $(5, 3)$ (Adams *et al.* 1991). The torus knots with 16 or fewer crossings are $(3, 2)$, $(5, 2)$, $(7, 2)$, $(9, 2)$, $(11, 2)$, $(13, 2)$, $(15, 2)$, $(4, 3)$, $(5, 3)$, $(7, 3)$, $(8, 3)$, and $(5, 4)$ (Hoste *et al.* 1998). The numbers of torus knots with n crossings are 0, 0, 1, 0, 1, 0, 1, 1, 1, 1, 1, 0, 1, 1, 2, 1, ... (Sloane's A051764).

The only KNOTS which are not HYPERBOLIC KNOTS are torus knots and SATELLITE KNOTS (including COMPOSITE KNOTS). The $(q, 2)$, $(4, 3)$, and $(5, 4)$-torus knots are ALMOST ALTERNATING KNOTS (Adams 1994, p. 142).

The JONES POLYNOMIAL of an (m, n)-TORUS KNOT is

$$\frac{t^{(m-1)(n-1)/2}(1 - t^{m+1} - t^{n+1} + t^{m+n})}{1 - t^2}. \qquad (3)$$

The BRACKET POLYNOMIAL for the torus knot $K_n = (2, n)$ is given by the RECURRENCE RELATION

$$\langle K_n \rangle = A\langle K_{n-1} \rangle + (-1)^{n-1}A^{-3n+2}, \qquad (4)$$

where

$$\langle K_1 \rangle = -A^3. \qquad (5)$$

See also ALMOST ALTERNATING KNOT, HYPERBOLIC KNOT, KNOT, SATELLITE KNOT, SOLOMON'S SEAL KNOT, TREFOIL KNOT

References

Adams, C.; Hildebrand, M.; and Weeks, J. "Hyperbolic Invariants of Knots and Links." *Trans. Amer. Math. Soc.* **326**, 1–56, 1991.
Burde, G. and Zieschang, H. *Knots.* Berlin: de Gruyter, 1985.
Gray, A. "Torus Knots." §9.2 in *Modern Differential Geometry of Curves and Surfaces with Mathematica, 2nd ed.* Boca Raton, FL: CRC Press, pp. 209–215, 1997.
Hoste, J.; Thistlethwaite, M.; and Weeks, J. "The First 1,701,936 Knots." *Math. Intell.* **20**, 33–48, Fall 1998.
Murasugi, K. "On the Braid Index of Alternating Links." *Trans. Amer. Math. Soc.* **326**, 237–260, 1991.
Murasugi, L. and Przytycki, J. "The Skein Polynomial of a Planar Star Product of Two Links." *Math. Proc. Cambridge Philos. Soc.* **106**, 273–276, 1989.
Schreier, O. "Über die Gruppen $A^a B^b = 1$." *Abh. Math. Sem. Univ. Hamburg* **3**, 167–169, 1924.
Sloane, N. J. A. Sequences A051764 in "An On-Line Version of the Encyclopedia of Integer Sequences." http://www.research.att.com/~njas/sequences/eisonline.html.
Steinhaus, H. *Mathematical Snapshots, 3rd ed.* New York: Dover, pp. 275–277, 1999.
Williams, R. F. "The Braid Index of an Algebraic Link." *Braids (Santa Cruz, CA, 1986).* Providence, RI: Amer. Math. Soc., 1988.

Total Angular Defect

DESCARTES TOTAL ANGULAR DEFECT

Total Curvature

The total curvature of a curve is the quantity $\sqrt{\tau^2 + \kappa^2}$, where τ is the TORSION and κ is the CURVATURE. The total curvature is also called the THIRD CURVATURE.

See also CURVATURE, TORSION (DIFFERENTIAL GEOMETRY)

Total Differential

EXACT DIFFERENTIAL

Total Exchange

GOSSIPING

Total Function

A FUNCTION defined for all possible input values.

Total Graph

The total graph $T(G)$ of a GRAPH G has a vertex for each edge and vertex of G, and edge in $T(G)$ for every edge-edge and vertex-edge adjacency in G (Capobianco and Molluzzo 1978; Skiena 1990, p. 162). Total graphs are generalizations of LINE GRAPHS.

See also LINE GRAPH

References

Capobianco, M. and Molluzzo, J. *Examples and Counterexamples in Graph Theory.* New York: North-Holland, 1978.
Skiena, S. *Implementing Discrete Mathematics: Combinatorics and Graph Theory with Mathematica.* Reading, MA: Addison-Wesley, 1990.

Total Intersection Theorem

If one part of the total intersection group of a curve of order n with a curve of order $n_1 + n_2$ constitutes the total intersection with a curve of order n_1, then the other part will constitute the total intersection with a curve of order n_2.

References

Coolidge, J. L. *A Treatise on Algebraic Plane Curves.* New York: Dover, p. 32, 1959.

Total Order

A RELATION on a TOTALLY ORDERED SET.

See also TOTALLY ORDERED SET

Total Probability Theorem

Given n MUTUALLY EXCLUSIVE EVENTS A_1, ..., A_n whose probabilities sum to unity, then

$$P(B) = P(B|A_1)P(A_1) + \ldots + P(B|A_n)P(A_n),$$

where B is an arbitrary event, and $P(B|A_i)$ is the CONDITIONAL PROBABILITY of B assuming A_i.

See also BAYES' THEOREM, CONDITIONAL PROBABILITY, MUTUALLY EXCLUSIVE EVENTS

References

Papoulis, A. *Probability, Random Variables, and Stochastic Processes, 2nd ed.* New York: McGraw-Hill, pp. 37–38, 1984.

Total Space

The SPACE E of a FIBER BUNDLE given by the MAP $f : E \to B$, where B is the BASE SPACE of the FIBER BUNDLE.

See also BASE SPACE, FIBER BUNDLE, SPACE

Total Variation Measure

Given a COMPLEX MEASURE μ, there exists a POSITIVE MEASURE denoted $|\mu|$ which measures the total variation of μ, also sometimes called simply "total variation." In particular, $|\mu|(E)$ on a SUBSET E is the largest sum of "variations" for any subdivision of E. Roughly speaking, a total variation measure is an infinitesimal version of the ABSOLUTE VALUE.

More precisely,

$$|\mu|(E) = \sup \sum_i |\mu(E_i)| \tag{1}$$

where the SUPREMUM is taken over all partitions $\cup E_i$ of E into MEASURABLE SUBSETS E_i.

Note that $|\mu(X)|$ may not be the same as $|\mu|(X)$. When μ already is a POSITIVE MEASURE, then $\mu = |\mu|$. More generally, if μ is ABSOLUTELY CONTINUOUS, that is

$$\mu(E) = \int_E f\, dx, \tag{2}$$

then so is $|\mu|$, and the total variation measure can be written as

$$|\mu|(E) = \int_E |f|\, dx. \tag{3}$$

The total variation measure can be used to rewrite the original measure, in analogy to the norm of a COMPLEX NUMBER. The measure μ has a POLAR REPRESENTATION

$$d\mu = h\, d|\mu| \tag{4}$$

with $|h| = 1$.

See also JORDAN MEASURE DECOMPOSITION, MEASURE, POLAR REPRESENTATION (MEASURE), RIESZ REPRESENTATION THEOREM

References

Rudin, W. *Real and Complex Analysis.* New York: McGraw-Hill, pp. 116–120, 1987.

Totalistic Cellular Automaton

A totalistic cellular automaton is a 1-D cellular automata in which the rules depend only on the total of the values of the cells in a neighborhood. These automata were introduced by Stephen Wolfram in 1983.

See also CELLULAR AUTOMATON

Totally Ordered Set

A total order (or "totally ordered set," or "linearly ordered set") is a SET plus a relation on the set (called a TOTAL ORDER) that satisfies the conditions for a PARTIAL ORDER plus an additional condition known as the comparability condition. A RELATION $\leq$ is a partial order on a SET S (if the following properties hold.

1. Reflexivity: $a \leq a$ for all $a \in S$.
2. Weak antisymmetry: $a \leq b$ and $b \leq a$ implies $a = b$.
3. Transitivity: $a \leq b$ and $b \leq c$ implies $a \leq c$.
4. Comparability (TRICHOTOMY LAW): For any $a, b \in S$, either $a \leq b$ or $b \leq a$.

The first three are the axioms of a PARTIAL ORDER, while addition of the TRICHOTOMY LAW defines a total order.

Every finite totally ordered set is WELL ORDERED. Any two totally ordered sets with k elements (for k a nonnegative integer) are ORDER ISOMORPHIC, and therefore have the same ORDER TYPE (which is also an ORDINAL NUMBER).

See also ORDER ISOMORPHIC, ORDER TYPE, PARTIAL ORDER, RELATION, TRICHOTOMY LAW, WELL ORDERED SET

References

Séroul, R. *Programming for Mathematicians*. Berlin: Springer-Verlag, p. 23, 2000.

Totally Symmetric Self-Complementary Plane Partition

A PLANE PARTITION which is invariant under permutation of the three axes and which is equal to its complement (i.e., the collection of cubes that are in a given box but do *not* belong to the solid Young diagram). The number of totally symmetric self-complementary PLANE PARTITIONS is the same as that for ALTERNATING SIGN MATRICES and DESCENDING PLANE PARTITIONS.

See also ALTERNATING SIGN MATRIX, DESCENDING PLANE PARTITION, PLANE PARTITION

References

Bressoud, D. and Propp, J. "How the Alternating Sign Matrix Conjecture was Solved." *Not. Amer. Math. Soc.* **46**, 637–646.

Totative

A POSITIVE INTEGER less than or equal to a number n which is also RELATIVELY PRIME to n, where 1 is counted as being RELATIVELY PRIME to all numbers. The number of totatives of n is the value of the TOTIENT FUNCTION $\phi(n)$.

See also RELATIVELY PRIME, TOTIENT FUNCTION

Totient Function

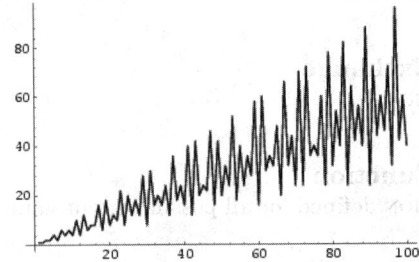

The totient function $\phi(n)$, also called Euler's totient function, is defined as the number of POSITIVE INTEGERS $\leq n$ which are RELATIVELY PRIME to (i.e., do not contain any factor in common with) n, where 1 is counted as being RELATIVELY PRIME to all numbers. Since a number less than or equal to and RELATIVELY PRIME to a given number is called a TOTATIVE, the totient function $\phi(n)$ can be simply defined as the number of TOTATIVES of n. For example, there are eight TOTATIVES of 24 (1, 5, 7, 11, 13, 17, 19, and 23), so $\phi(24) = 8$.

$\phi(n)$ is always EVEN for $n \geq 3$. By convention, $\phi(0) = 1$, although *Mathematica* defines EulerPhi[0] equal to 0 for consistency with its FactorInteger[0] command. The first few values of $\phi(n)$ for $n = 1, 2, ...$ are 1, 1, 2, 2, 4, 2, 6, 4, 6, 4, 10, ... (Sloane's A000010). The totient function is given by the MÖBIUS TRANSFORM of 1, 2, 3, 4, ... (Sloane and Plouffe 1995, p. 22). $\phi(n)$ is plotted above for small n.

For a PRIME p,

$$\phi(p) = p - 1, \tag{1}$$

since all numbers less than p are RELATIVELY PRIME to p. If $m = p^{\alpha}$ is a POWER of a PRIME, then the numbers which have a common factor with m are the multiples of p: $p, 2p, ..., (p^{\alpha-1})p$. There are $p^{\alpha-1}$ of these multiples, so the number of factors RELATIVELY PRIME to p^{α} is

$$\phi(p^{\alpha}) = p^{\alpha} - p^{\alpha-1} = p^{\alpha-1}(p-1) = p^{\alpha}\left(1 - \frac{1}{p}\right). \tag{2}$$

Now take a general m divisible by p. Let $\phi_p(m)$ be the number of POSITIVE INTEGERS $\leq m$ not DIVISIBLE by p. As before, $p, 2p, ..., (m/p)p$ have common factors, so

$$\phi_p(m) = m - \frac{m}{p} = m\left(1 - \frac{1}{p}\right). \tag{3}$$

Now let q be some other PRIME dividing m. The INTEGERS divisible by q are $q, 2q, ..., (m/q)q$. But these duplicate $pq, 2pq, ..., (m/pq)pq$. So the number of terms which must be subtracted from ϕ_p to obtain ϕ_{pq} is

$$\Delta\phi_p(m) = \frac{m}{q} - \frac{m}{pq} = \frac{m}{q}\left(1 - \frac{1}{p}\right), \tag{4}$$

and

$$\phi_{pq}(m) \equiv \phi_p(m) - \Delta\phi_q(m)$$

$$= m\left(1 - \frac{1}{p}\right) - \frac{m}{p}\left(1 - \frac{1}{p}\right)$$

$$= m\left(1 - \frac{1}{p}\right)\left(1 - \frac{1}{q}\right). \qquad (5)$$

By induction, the general case is then

$$\phi(n) = n\left(1 - \frac{1}{p_1}\right)\left(1 - \frac{1}{p_2}\right)\cdots\left(1 - \frac{1}{p_r}\right). \qquad (6)$$

An interesting identity relates $\phi(n^2)$ to $\phi(n)$,

$$\phi(n^2) = n\phi(n). \qquad (7)$$

Another identity relates the DIVISORS d of n to n via

$$\sum_d \phi(d) = n. \qquad (8)$$

The DIVISOR FUNCTION satisfies the CONGRUENCE

$$n\sigma(n) \equiv 2 \pmod{\phi(n)}$$

$$= \begin{cases} n\sigma(n) \equiv 0 \pmod{\phi(n)} & \text{if } \phi(n) = 2 \\ n\sigma(n) \equiv 2 \pmod{\phi(n)} & \text{otherwise} \end{cases} \qquad (9)$$

for all PRIMES $p \geq 5$ and no COMPOSITE with the exception of 4, 6, and 22, where $\sigma(n)$ is the DIVISOR FUNCTION. This fact was proved by Subbarao (1974), despite the implication to the contrary, "is it true for infinitely many composite n?," stated in Guy (1994, p. 92). No COMPOSITE solution is currently known to

$$n - 1 \equiv 0 \pmod{\phi(n)} \qquad (10)$$

(Honsberger 1976, p. 35).

If the GOLDBACH CONJECTURE is true, then for every number m, there are PRIMES p and q such that

$$\phi(p) + \phi(q) = 2m \qquad (11)$$

(Guy 1994, p. 105). Guy (1994, p. 99) discussed solutions to

$$\phi(\sigma(n)) = n, \qquad (12)$$

where $\sigma(n)$ is the DIVISOR FUNCTION. F. Helenius has found 365 such solutions, the first of which are 2, 8, 12, 128, 240, 720, 6912, 32768, 142560, 712800, ... (Sloane's A001229).

Curious equalities of consecutive values include

$$\phi(5186) = \phi(5187) = \phi(5188) = 2^5 3^4 \qquad (13)$$

$$\phi(25930) = \phi(25935) = \phi(25942) = 2^7 3^4 \qquad (14)$$

$$\phi(404471) = \phi(404473) = \phi(404477) = 2^8 3^2 5^2 7 \qquad (15)$$

(Guy 1994, p. 91). McCranie found an arithmetic

progression of six numbers with equal totient functions,

$$\phi(583200) = \phi(583230) = \phi(583260) = \phi(583290)$$

$$= \phi(583320) = \phi(583350) = 155520, \qquad (16)$$

as well as other progressions of six numbers starting at 583200, 1166400, 1749600, ... (Sloane's A050518).

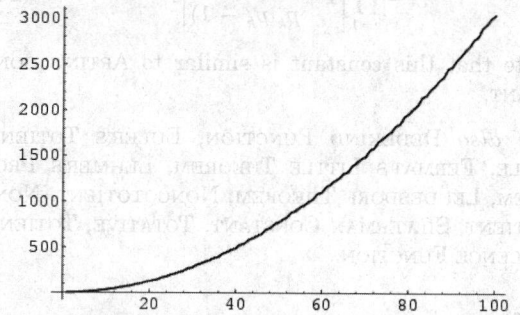

The SUMMATORY totient function, plotted above, is defined by

$$\Phi(n) \equiv \sum_{k=1}^{n} \phi(k). \qquad (17)$$

The first values of $\Phi(n)$ are 1, 2, 4, 6, 10, 12, 18, 22, 28, ... (Sloane's A002088). $\Phi(n)$ has the asymptotic series

$$\Phi(x) \sim \frac{1}{2\zeta(2)} x^2 + \mathcal{O}(x \ln x) \qquad (18)$$

$$\sim \frac{3}{\pi^2} x^2 + \mathcal{O}(x \ln x), \qquad (19)$$

where $\zeta(z)$ is the RIEMANN ZETA FUNCTION (Perrot 1881; Nagell 1951, p. 131). An improved asymptotic estimate due to Walfisz (1963) is given by

$$\sum_{n=1}^{N} \phi(n) = \frac{3N^2}{\pi^2} + \mathcal{O}\left[N(\ln N)^{2/3}(\ln \ln N)^{4/3}\right]. \qquad (20)$$

Landau (1900, quoted in Dickson 1952) showed that the asymptotic series of the summatory function of $1/\phi(n)$ is

$$\sum_{n=1}^{N} \frac{1}{\phi(n)} = A \ln N + B + \mathcal{O}\left(\frac{\ln N}{N}\right), \qquad (21)$$

where

$$A = \sum_{k=1}^{\infty} \frac{[\mu(k)]^2}{k\phi(k)} = \frac{\zeta(2)\zeta(3)}{\zeta(6)} = \frac{315}{2\pi^4} \zeta(3)$$

$$= 1.9435964368\ldots \qquad (22)$$

$$B = \gamma \frac{315}{2\pi^4} \zeta(3) - \sum_{k=1}^{\infty} \frac{[\mu(k)]^2 \ln k}{k\phi(k)}$$

$$= -0.0595536246\ldots, \qquad (23)$$

$\mu(k)$ is the MÖBIUS FUNCTION, $\zeta(z)$ is the RIEMANN ZETA FUNCTION, and γ is the EULER-MASCHERONI CONSTANT (Dickson). A can also be written

$$A = \prod_{k=1}^{\infty} \frac{1 - p_k^6}{\left(1 - p_k^{-2}\right)\left(1 - p_k^{-3}\right)}$$

$$= \prod_{k=1}^{\infty} \left[1 + \frac{1}{p_k(p_k - 1)}\right]. \tag{24}$$

Note that this constant is similar to ARTIN'S CONSTANT.

See also DEDEKIND FUNCTION, EULER'S TOTIENT RULE, FERMAT'S LITTLE THEOREM, LEHMER'S PROBLEM, LEUDESDORF THEOREM, NONCOTOTIENT, NONTOTIENT, SILVERMAN CONSTANT, TOTATIVE, TOTIENT VALENCE FUNCTION

References
Abramowitz, M. and Stegun, C. A. (Eds.). "The Euler Totient Function." §24.3.2 in *Handbook of Mathematical Functions with Formulas, Graphs, and Mathematical Tables, 9th printing.* New York: Dover, p. 826, 1972.
Beiler, A. H. Ch. 12 in *Recreations in the Theory of Numbers: The Queen of Mathematics Entertains.* New York: Dover, 1966.
Conway, J. H. and Guy, R. K. "Euler's Totient Numbers." *The Book of Numbers.* New York: Springer-Verlag, pp. 154–156, 1996.
Courant, R. and Robbins, H. "Euler's φ Function. Fermat's Theorem Again." §2.4.3 in Supplement to Ch. 1 in *What is Mathematics?: An Elementary Approach to Ideas and Methods, 2nd ed.* Oxford, England: Oxford University Press, pp. 48–49, 1996.
DeKoninck, J.-M. and Ivic, A. *Topics in Arithmetical Functions: Asymptotic Formulae for Sums of Reciprocals of Arithmetical Functions and Related Fields.* Amsterdam, Netherlands: North-Holland, 1980.
Dickson, L. E. *History of the Theory of Numbers, Vol. 1: Divisibility and Primality.* New York: Chelsea, pp. 113–158, 1952.
Finch, S. "Favorite Mathematical Constants." http://www.mathsoft.com/asolve/constant/totient/totient.html.
Guy, R. K. "Euler's Totient Function," "Does $\phi(n)$ Properly Divide $n - 1$," "Solutions of $\phi(m) = \sigma(n)$," "Carmichael's Conjecture," "Gaps Between Totatives," "Iterations of ϕ and σ," "Behavior of $\phi(\sigma(n))$ and $\sigma(\phi(n))$." §B36-B42 in *Unsolved Problems in Number Theory, 2nd ed.* New York: Springer-Verlag, pp. 90–99, 1994.
Halberstam, H. and Richert, H.-E. *Sieve Methods.* New York: Academic Press, 1974.
Helenius, F. Untitled. http://pweb.netcom.com/~fredh/phi-sigma/pslist.html.
Honsberger, R. *Mathematical Gems II.* Washington, DC: Math. Assoc. Amer., p. 35, 1976.
Nagell, T. "Relatively Prime Numbers. Euler's φ-Function." §8 in *Introduction to Number Theory.* New York: Wiley, pp. 23–26, 1951.
Niven, I. M.; Zuckerman, H. S.; and Montgomery, H. L. *An Introduction to the Theory of Numbers, 5th ed.* New York: Wiley, p. 51, 1991.
Perrot, J. 1811. Quoted in Dickson, L. E. *History of the Theory of Numbers, Vol. 1: Divisibility and Primality.* New York: Chelsea, p. 126, 1952.
Shanks, D. "Euler's ϕ Function." §2.27 in *Solved and Unsolved Problems in Number Theory, 4th ed.* New York: Chelsea, pp. 68–71, 1993.
Séroul, R. "The Euler Phi Function." §2.7 in *Programming for Mathematicians.* Berlin: Springer-Verlag, pp. 14–15, 2000.
Sloane, N. J. A. Sequences A000010/M0299, A002088/M1008, A001229, and A050518 in "An On-Line Version of the Encyclopedia of Integer Sequences." http://www.research.att.com/~njas/sequences/eisonline.html.
Sloane, N. J. A. and Plouffe, S. *The Encyclopedia of Integer Sequences.* San Diego, CA: Academic Press, 1995.
Subbarao, M. V. "On Two Congruences for Primality." *Pacific J. Math.* **52**, 261–268, 1974.

Totient Function Constants
SILVERMAN CONSTANT, TOTIENT FUNCTION

Totient Valence Function
$N_\phi(m)$ is the number of INTEGERS n for which the TOTIENT FUNCTION $\phi(n) = m$, also called the MULTIPLICITY of m (Guy 1994). Erdos (1958) proved that is a multiplicity occurs once, it occurs infinitely often. The table below lists values for $\phi(N) \leq 50$.

$\phi(N)$	multiplicity	N
1	2	1, 2
2	3	3, 4, 6
4	4	5, 8, 10, 12
6	4	7, 9, 14, 18
8	5	15, 16, 20, 24, 30
10	2	11, 22
12	6	13, 21, 26, 28, 36, 42
16	6	17, 32, 34, 40, 48, 60
18	4	19, 27, 38, 54
20	5	25, 33, 44, 50, 66
22	2	23, 46
24	10	35, 39, 45, 52, 56, 70, 72, 78, 84, 90
28	2	29, 58
30	2	31, 62
32	7	51, 64, 68, 80, 96, 102, 120
36	8	37, 57, 63, 74, 76, 108, 114, 126
40	9	41, 55, 75, 82, 88, 100, 110, 132, 150
42	4	43, 49, 86, 98
44	3	69, 92, 138
46	2	47, 94
48	11	65, 104, 105, 112, 130, 140, 144, 156, 168, 180, 210

A table listing the first value of $\phi(N)$ with multiplicities up to 100 follows (Sloane's A007374; Sloane's A014573).

M	φ	M	φ	M	φ	M	φ
0	3	26	2560	51	4992	76	21840
2	1	27	384	52	17640	77	9072
3	2	28	288	53	2016	78	38640
4	4	29	1320	54	1152	79	9360
5	8	30	3696	55	6000	80	81216
6	12	31	240	56	12288	81	4032
7	32	32	768	57	4752	82	5280
8	36	33	9000	58	2688	83	4800
9	40	34	432	59	3024	84	4608
10	24	35	7128	60	13680	85	16896
11	48	36	4200	61	9984	86	3456
12	160	37	480	62	1728	87	3840
13	396	38	576	63	1920	88	10800
14	2268	39	1296	64	2400	89	9504
15	704	40	1200	65	7560	90	18000
16	312	41	15936	66	2304	91	23520
17	72	42	3312	67	22848	92	39936
18	336	43	3072	68	8400	93	5040
19	216	44	3240	69	29160	94	26208
20	936	45	864	70	5376	95	27360
21	144	46	3120	71	3360	96	6480
22	624	47	7344	72	1440	97	9216
23	1056	48	3888	73	13248	98	2880
24	1760	49	720	74	11040	99	26496
25	360	50	1680	75	27720	100	34272

It is thought that $N_\phi(m) \geq 2$ (i.e., the totient valence function never takes on the value 1), but this has not been proven. This assertion is called CARMICHAEL'S TOTIENT FUNCTION CONJECTURE and is equivalent to the statement that for all n, there exists $m \neq n$ such that $\phi(n) = \phi(m)$ (Ribenboim 1996, pp. 39–40). Any counterexample must have more than 10,000,000 DIGITS (Schlafly and Wagon 1994, erroneously given as 10,000 in Conway and Guy 1996).

See also CARMICHAEL'S TOTIENT FUNCTION CONJECTURE, SIERPINSKI'S CONJECTURE, TOTIENT FUNCTION

References
Conway, J. H. and Guy, R. K. *The Book of Numbers.* New York: Springer-Verlag, p. 155, 1996.

Erdos, P. "Some Remarks on Euler's ϕ-Function." *Acta Math.* **4**, 10–19, 1958.

Ford, K. "The Distribution of Totients." *Ramanujan J.* **2**, 67–151, 1998.

Ford, K. "The Distribution of Totients, *Electron. Res. Announc. Amer. Math. Soc.* **4**, 27–34, 1998.

Guy, R. K. *Unsolved Problems in Number Theory, 2nd ed.* New York: Springer-Verlag, p. 94, 1994.

Ribenboim, P. *The New Book of Prime Number Records.* New York: Springer-Verlag, 1996.

Schlafly, A. and Wagon, S. "Carmichael's Conjecture on the Euler Function is Valid Below $10^{10,000,000}$." *Math. Comput.* **63**, 415–419, 1994.

Sloane, N. J. A. Sequences A007374/M1093 and A014573 in "An On-Line Version of the Encyclopedia of Integer Sequences." http://www.research.att.com/~njas/sequences/eisonline.html.

Touchard's Congruence

$$B_{p+k} \equiv B_k + B_{k+1} \pmod{p},$$

when p is PRIME and B_n is a BELL NUMBER.

See also BELL NUMBER

Tour

A sequence of moves on a chessboard by a CHESS piece in which each square of a CHESSBOARD is visited exactly once.

See also CHESS, HAMILTONIAN CIRCUIT, KNIGHT'S TOUR, MAGIC TOUR, TRAVELING SALESMAN CONSTANTS

Tournament

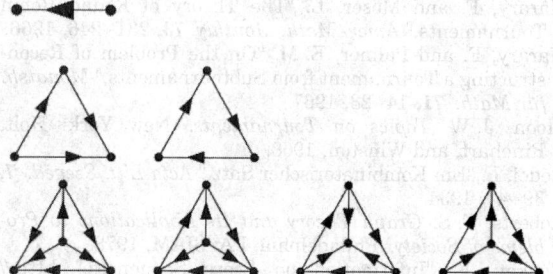

A COMPLETE DIRECTED GRAPH (Skiena 1990, p. 175). A so-called SCORE SEQUENCE can be associated with every tournament. The number of nonisomorphic tournaments on 2, 3, 4, ... nodes are 1, 2, 4, ..., illustrated above. The first and second 3-node tournaments shown above are called a TRANSITIVE TRIPLE and CYCLIC TRIPLE, respectively (Harary 1994, p. 204).

Every tournament contains an odd number of HAMILTONIAN PATHS (Rédei 1934; Szele 1943; Skiena 1990, p. 175). However, a tournament has a directed HAMILTONIAN CIRCUIT IFF it is STRONGLY CONNECTED (Foulkes 1960; Harary and Moser 1966; Skiena 1990, p. 175).

The term "tournament" also refers to an arrangement by which teams or players play against certain other teams or players in order to determine who is the best. In a "cup" tournament of $n-2^k$ teams, teams play pairwise in a sequence of $1/2^{k-1}$-finals, ..., 1/8-finals, quarter-finals, semi-finals, and finals, with winners from each round playing other winners in the next round and losers being eliminated at each round. The second-place prize is usually awarded to the team which loses in the finals. However, this practice is unfair since the second-place team has not been required to play against the teams which were eliminated by the first-place (and presumably best) team, and therefore might actually be worse than one of the teams eliminated earlier by the best team (Steinhaus 1983).

In general, to fairly determine the best two players from n contestants, $n-1+\log_2(n-1)$ rounds are required (Steinhaus 1983, p. 55).

See also COMPLETE GRAPH, DIRECTED GRAPH, HAMILTONIAN PATH, SCORE SEQUENCE, TOURNAMENT MATRIX

References

Boesch, F. and Tindell, R. "Robbins' Theorem for Mixed Graphs." *Amer. Math. Monthly* **87**, 716–719, 1980.
Chartrand, G. "Tournaments." §27.2 in *Introductory Graph Theory.* New York: Dover, pp. 155–161, 1985.
Chvátal, V. and Thomassen, C. "Distances in Orientations of Graphs." *J. Combin. Th. B* **24**, 61–75, 1978.
Foulkes, J. D. "Directed Graphs and Assembly Schedules." In *Proc. Symp. Appl. Math.* Providence, RI: Amer. Math. Soc., pp. 218–289, 1960.
Harary, F. "Tournaments." *Graph Theory.* Reading, MA: Addison-Wesley, pp. 205–208, 1994.
Harary, F. and Moser, L. "The Theory of Round Robin Tournaments." *Amer. Math. Monthly* **73**, 231–246, 1966.
Harary, F. and Palmer, E. M. "On the Problem of Reconstructing a Tournament from Subtournaments." *Monatsh. für Math.* **71**, 14–23, 1967.
Moon, J. W. *Topics on Tournaments.* New York: Holt, Rinehart, and Winston, 1968.
Rédei, L. "Ein Kombinatorischer Satz." *Acta Litt. Szeged.* **7**, 39–43, 1934.
Roberts, F. S. *Graph Theory and Its Applications to Problems of Society.* Philadelphia, PA: SIAM, 1978.
Ruskey, F. "Information on Score Sequences." http://www.theory.csc.uvic.ca/~cos/inf/nump/ScoreSequence.html.
Skiena, S. *Implementing Discrete Mathematics: Combinatorics and Graph Theory with Mathematica.* Reading, MA: Addison-Wesley, 1990.
Steinhaus, H. *Mathematical Snapshots, 3rd ed.* New York: Dover, pp. 54–55, 1999.
Szele, T. "Kombinatorische Untersuchungen über den gerichteten vollständigen Graphen." *Mat. Fiz. Lapok* **50**, 223–256, 1943.

Tournament Matrix

A matrix for a round-robin TOURNAMENT involving n players competing in $n(n-1)/2$ matches (no ties allowed) having entries

$$a_{ij} = \begin{cases} 1 & \text{if player } i \text{ defeats player } j \\ -1 & \text{if player } i \text{ loses to player } j. \\ 0 & \text{if } i = j \end{cases}$$

The MATRIX satisfies

$$\mathsf{A} + \mathsf{A}^\mathsf{T} + \mathsf{I} = \mathsf{J},$$

where I is the IDENTITY MATRIX, J is an $n \times n$ MATRIX of all 1s, and A^T is the MATRIX TRANSPOSE of A.

The tournament matrix for n players has zero DETERMINANT IFF n is ODD (McCarthy and Benjamin 1996). The dimension of the NULLSPACE of an n-player tournament matrix is

$$\dim[\text{nullspace}] = \begin{cases} 0 & \text{for } n \text{ even} \\ 1 & \text{for } n \text{ odd} \end{cases}$$

(McCarthy 1996).

References

McCarthy, C. A. and Benjamin, A. T. "Determinants of the Tournaments." *Math. Mag.* **69**, 133–135, 1996.
Michael, T. S. "The Ranks of Tournament Matrices." *Amer. Math. Monthly* **102**, 637–639, 1995.

Tournament Sequence

A tournament sequence is an increasing sequence of positive integers $(t_1, t_2, ...)$ such that $t_1 = 1$ and $t_{i+1} \leq 2t_i$. Cook and Kleber (2000) show that MEEUSSEN SEQUENCES are isomorphic to tournament sequences.

See also MEEUSSEN SEQUENCE

References

Cook, M. and Kleber, M. "Tournament Sequences and Meeussen Sequences." *Electronic J. Combinatorics* **7**, No. 1, R44, 1–16, 2000. http://www.combinatorics.org/Volume_7/v7i1toc.html#R44.

Tower of Power

POWER TOWER

Towers of Hanoi

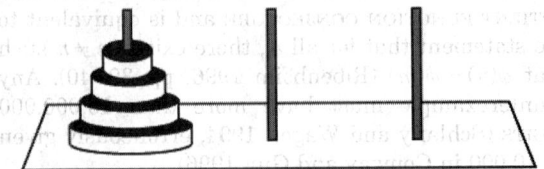

A PUZZLE invented by E. Lucas in 1883. Given a stack of n disks arranged from largest on the bottom to smallest on top placed on a rod, together with two empty rods, the towers of Hanoi puzzle asks for the minimum number of moves required to reverse the order of the stack (where moves are allowed only if they place smaller disks on top of larger disks). The problem is ISOMORPHIC to finding a HAMILTONIAN PATH on an n-HYPERCUBE (Gardner 1957, 1959).

For n disks, the number of moves h_n required is given by the RECURRENCE RELATION

$$h_n = 2h_{n-1} + 1.$$

Solving gives

$$h_n = 2^n - 1.$$

The number of disks moved after the kth step is the same as the element which needs to be added or deleted in the kth ADDEND of the RYSER FORMULA (Gardner 1988, Vardi 1991). The number of disk to be moved at nth step of the optimal solution to the problem are 1, 2, 1, 3, 1, 2, 1, 4, 1, 2, 1, 3, 1, 2, ... (Sloane's A001511). Amazingly, this is exactly the BINARY CARRY SEQUENCE plus one.

A HANOI GRAPH can be constructed whose VERTICES correspond to legal configurations of n towers of Hanoi, where the VERTICES are adjacent if the corresponding configurations can be obtained by a legal move. It can be solved using a binary GRAY CODE.

Poole (1994) gives *Mathematica* routines for solving an arbitrary disk configuration in the fewest possible moves. The proof of minimality is achieved using the LUCAS CORRESPONDENCE which relates PASCAL'S TRIANGLE to the HANOI GRAPH. ALGORITHMS are known for transferring disks for four pegs, but none has been proved minimal. For additional references, see Poole (1994).

See also BINARY CARRY SEQUENCE, GRAY CODE, RYSER FORMULA

References

Allouche, J.-P. and Shallit, J. "The Ring of k-Regular Sequences." *Theoret. Comput. Sci.* **98**, 163–197, 1992.
Bogomolny, A. "Towers of Hanoi." http://www.cut-the-knot.com/recurrence/hanoi.html.
Chartrand, G. "The Tower of Hanoi Puzzle." §6.3 in *Introductory Graph Theory.* New York: Dover, pp. 135–139, 1985.
Dubrovsky, V. "Nesting Puzzles, Part I: Moving Oriental Towers." *Quantum* **6**, 53–57 (Jan.) and 49–51 (Feb.), 1996.
Flajolet, P.; Raoult, J.-C.; and Vuillemin, J. "The Number of Registers Required for Evaluating Arithmetic Expressions." *Theoret. Comput. Sci.* **9**, 99–125, 1979.
Gardner, M. "Mathematical Games: About the Remarkable Similarity between the Icosian Game and the Towers of Hanoi." *Sci. Amer.* **196**, 150–156, May 1957.
Gardner, M. "The Icosian Game and the Tower of Hanoi." Ch. 6 in *The Scientific American Book of Mathematical Puzzles & Diversions.* New York: Simon and Schuster, pp. 55–62, 1959.
Kasner, E. and Newman, J. R. *Mathematics and the Imagination.* Redmond, WA: Tempus Books, pp. 169–171, 1989.
Kolar, M. "Towers of Hanoi." http://www.pangea.ca/kolar/javascript/Hanoi/Hanoi.html.
Poole, D. G. "The Towers and Triangles of Professor Claus (or, Pascal Knows Hanoi)." *Math. Mag.* **67**, 323–344, 1994.
Poole, D. G. "Towers of Hanoi." MATHEMATICA NOTEBOOK HANOI.M.
Ruskey, F. "Towers of Hanoi." http://www.theory.csc.uvic.ca/~cos/inf/comb/SubsetInfo.html#Hanoi.
Schoutte, P. H. "De Ringen van Brahma." *Eigen Haard* **22**, 274–276, 1884.
Sloane, N. J. A. Sequences A001511/M0127 in "An On-Line Version of the Encyclopedia of Integer Sequences." http://www.research.att.com/~njas/sequences/eisonline.html.
Kraitchik, M. "The Tower of Hanoi." §3.12.4 in *Mathematical Recreations.* New York: W. W. Norton, pp. 91–93, 1942.
Vardi, I. *Computational Recreations in Mathematica.* Reading, MA: Addison-Wesley, pp. 111–112, 1991.

T-Polyomino

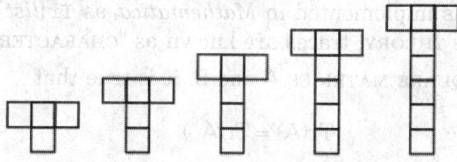

The order n T-polyomino consists of a vertical line of $n-3$ squares capped by a horizontal line of three squares centered on the line.

See also L-POLYOMINO, SKEW POLYOMINO, SQUARE POLYOMINO, STRAIGHT POLYOMINO

T-Puzzle

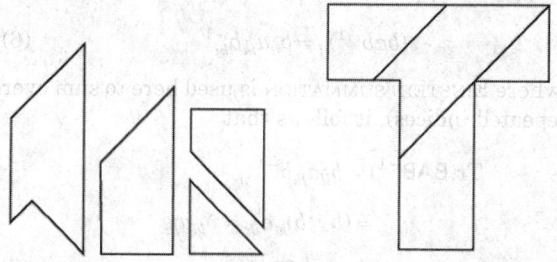

The DISSECTION of the four pieces shown at left into the capital letter "T" shown at right.

See also DISSECTION

References

Pappas, T. "The T Problem." *The Joy of Mathematics.* San Carlos, CA: Wide World Publ./Tetra, pp. 35 and 230, 1989.

Trace (Group)

CHARACTER (GROUP)

Trace (Map)

Let a PATCH be given by the map $\mathbf{x} : U \to \mathbb{R}^n$, where U is an open subset of $\mathbb{R}^2$, or more generally by $\mathbf{x} : A \to \mathbb{R}^n$, where A is any SUBSET of $\mathbb{R}^2$. Then $\mathbf{x}(U)$ (or more generally, $\mathbf{x}(A)$) is called the trace of $\mathbf{x}$.

See also PATCH

References

Gray, A. *Modern Differential Geometry of Curves and Surfaces with Mathematica, 2nd ed.* Boca Raton, FL: CRC Press, pp. 269–270, 1997.

Trace (Matrix)

The trace of an $n \times n$ SQUARE MATRIX A is defined to be

$$\mathrm{Tr}(\mathsf{A}) \equiv \sum_{i=1}^{n} a_{ii}, \tag{1}$$

i.e., the sum of the diagonal elements. The matrix trace is implemented in *Mathematica* as Tr[*list*]. In GROUP THEORY, traces are known as "CHARACTERS."

For SQUARE MATRICES A and B, it is true that

$$\mathrm{Tr}(\mathsf{A}) = \mathrm{Tr}(\mathsf{A}^{\mathsf{T}}) \tag{2}$$

$$\mathrm{Tr}(\mathsf{A} + \mathsf{B}) = \mathrm{Tr}(\mathsf{A}) + \mathrm{Tr}(\mathsf{B}) \tag{3}$$

$$\mathrm{Tr}(\alpha \mathsf{A}) = \alpha \mathrm{Tr}(\mathsf{A}) \tag{4}$$

(Lange 1987, p. 40), where A^{T} denotes the TRANSPOSE. The trace is also invariant under a SIMILARITY TRANSFORMATION

$$\mathsf{A}' \equiv \mathsf{B}\mathsf{A}\mathsf{B}^{-1} \tag{5}$$

(Lange 1987, p. 64). Since

$$(bab^{-1})_{ij} = b_{il}a_{lk}b_{kj}^{-1} \tag{6}$$

(where EINSTEIN SUMMATION is used here to sum over repeated indices), it follows that

$$\mathrm{Tr}(\mathsf{B}\mathsf{A}\mathsf{B}^{-1}) = b_{il}a_{lk}b^{-1}{}_{ki}$$

$$= (b^{-1}b)_{kl}a_{lk} = \delta_{kl}a_{lk}$$

$$= a_{kk} = \mathrm{Tr}(\mathsf{A}), \tag{7}$$

where δ_{ij} is the KRONECKER DELTA.

The trace of a product of two square matrices is independent of the order of the multiplication since

$$\mathrm{Tr}(\mathsf{A}\mathsf{B}) = (ab)_{ii} = a_{ij}b_{ji} = b_{ji}a_{ij}$$

$$= (ba)_{jj} = \mathrm{Tr}(\mathsf{B}\mathsf{A}) \tag{8}$$

(again using EINSTEIN SUMMATION). Therefore, the trace of the COMMUTATOR of A and B is given by

$$\mathrm{Tr}([\mathsf{A},\,\mathsf{B}]) \equiv \mathrm{Tr}(\mathsf{A}\mathsf{B}) - \mathrm{Tr}(\mathsf{B}\mathsf{A}) = 0. \tag{9}$$

The trace of a product of three or more square matrices, on the other hand, is invariant only under CYCLIC PERMUTATIONS of the order of multiplication of the matrices, by a similar argument.

The product of a SYMMETRIC and an ANTISYMMETRIC MATRIX has zero trace,

$$\mathrm{Tr}(\mathsf{A}_S\mathsf{B}_A) = 0. \tag{10}$$

The value of the trace can be found using the fact that the matrix can always be transformed to a coordinate system where the z-AXIS lies along the axis of rotation. In the new coordinate system (which is assumed to also have been appropriately rescaled), the MATRIX is

$$\mathsf{A}' = \begin{bmatrix} \cos\phi & \sin\phi & 0 \\ -\sin\phi & \cos\phi & 0 \\ 0 & 0 & 1 \end{bmatrix}, \tag{11}$$

so the trace is

$$\mathrm{Tr}(\mathsf{A}') = \mathrm{Tr}(\mathsf{A}) \equiv a_{ii} = 1 + 2\cos\phi. \tag{12}$$

See also CHARACTER (GROUP), CONTRACTION (TENSOR), MATRIX, SQUARE MATRIX, TRACE (TENSOR)

References

Lang, S. *Linear Algebra, 3rd ed.* New York: Springer-Verlag, pp. 40 and 64, 1987.
Munkres, J. R. *Elements of Algebraic Topology.* Perseus Press, p. 122, 1993.

Trace (Path)

The image of the path γ in $\mathbb{C}$ under the FUNCTION f is called the trace. This usage of the term "trace" is unrelated to the same term applied to MATRICES or TENSORS.

Trace (Tensor)

The trace of a second-RANK TENSOR T is a SCALAR given by the CONTRACTED mixed TENSOR equal to T_i^i. The trace is implemented in *Mathematica* as Tr[*list*].

The trace satisfies

$$\mathrm{Tr}\left[M^{-1}(x)\frac{\partial}{\partial x^\lambda}M(x)\right] = \frac{\partial}{\partial x^\lambda}\ln[\det(x)],$$

and

$$\delta\ln[\det M] = \ln[\det(M + \delta M)] - \ln(\det M)$$

$$= \ln\left[\frac{\det(M + \delta M)}{\det M}\right]$$

$$= \ln[\det M^{-1}(M + \delta M)]$$

$$= \ln[\det(1 + M^{-1}\delta M)]$$

$$\approx \ln[1 + \mathrm{Tr}(M^{-1}\delta M)]$$

$$\approx \mathrm{Tr}(M^{-1}\delta M).$$

See also CHARACTER (GROUP), CONTRACTION (TENSOR), TRACE (MATRIX)

Traceable Graph

A GRAPH G that possesses a HAMILTONIAN PATH. HAMILTONIAN GRAPHS are therefore traceable, but the converse is not necessarily true. The number of traceable graphs on $n = 1, 2, \ldots$ are 0, 1, 2, 5, 18, 91, 734, ... (Sloane's A057864), the first few of which are illustrated above.

See also HAMILTON-CONNECTED GRAPH, HAMILTONIAN GRAPH, HYPOTRACEABLE GRAPH

References

Sloane, N. J. A. Sequences A057864 in "An On-Line Version of the Encyclopedia of Integer Sequences." http://www.research.att.com/~njas/sequences/eisonline.html.

Thomassen, C. "Hypohamiltonian and Hypotraceable Graphs." *Disc. Math.* **9**, 91–96, 1974.

Tractory

TRACTRIX

Tractrisoid

PSEUDOSPHERE

Tractrix

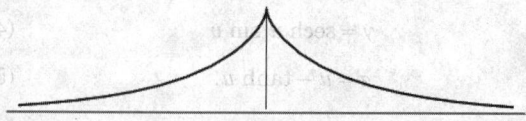

The tractrix is the CATENARY INVOLUTE described by a point initially on the vertex (making the CATENARY the TRACTRIX EVOLUTE). The tractrix is sometimes called the TRACTORY or EQUITANGENTIAL CURVE. The tractrix was first studied by Huygens in 1692, who gave it the name "tractrix." Later, Leibniz, Johann Bernoulli, and others studied the curve.

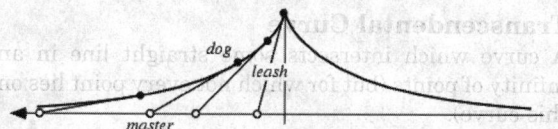

The tractrix arises from the following problem posed

to Leibniz: What is the path of an object starting off with a vertical offset when it is dragged along by a string of constant length being pulled along a straight horizontal line (Steinhaus 1983, pp. 250–251)? By associating the object with a dog, the string with a leash, and the pull along a horizontal line with the dog's master, the curve has the descriptive name HUNDKURVE (hound curve) in German. Leibniz found the curve using the fact that the axis is an asymptote to the tractrix (MacTutor Archive).

In CARTESIAN COORDINATES the tractrix has equation

$$x = a \operatorname{sech}^{-1}\left(\frac{y}{a}\right) - \sqrt{a^2 - y^2}. \tag{1}$$

One parametric form is

$$x(t) = a(t - \tanh t) \tag{2}$$

$$y(t) = a \operatorname{sech} t. \tag{3}$$

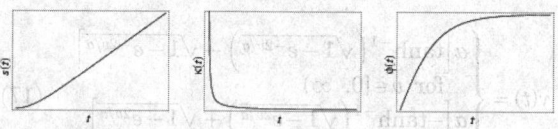

The ARC LENGTH, CURVATURE, and TANGENTIAL ANGLE in this parameterization are

$$s(t) = \ln(\cosh t) \tag{4}$$

$$\kappa(t) = \operatorname{csch} t \tag{5}$$

$$\phi(t) = 2 \tan^{-1}\left[\tanh\left(\tfrac{1}{2} t\right)\right]. \tag{6}$$

A second parametric form in terms of the ANGLE θ of the straight line tangent to the tractrix can be found by computing

$$\theta(t) = \tan^{-1}\left(\frac{\dfrac{dy}{dt}}{\dfrac{dx}{dt}}\right) = \tan^{-1}\left(\frac{-\operatorname{sech} t \tanh t}{\tanh^2 t}\right)$$

$$= -\tanh^{-1}(\operatorname{csch} t), \tag{7}$$

then solving for t and plugging back in to obtain

$$x = a \leq \left\{\ln\left[\tan\left(\tfrac{1}{2}\theta\right)\right] + \cos\theta\right\} \tag{8}$$

$$= a\left\{-\operatorname{csch}^{-1}(\tan\theta) + \cos\theta\right\} \tag{9}$$

$$y = a \sin\theta \tag{10}$$

(Gray 1997). This parameterization has CURVATURE

$$\kappa(\theta) = |\tan\theta|. \tag{11}$$

In terms of the angle $\theta' = \pi/2 + \theta$, the PARAMETRIC EQUATIONS can be written

$$x = a \operatorname{gd}^{-1}\theta' - \sin\theta \tag{12}$$

$$= a[\ln(\sec\theta' + \tan\theta') - \sin\theta'] \qquad (13)$$

$$= a\left\{\ln\left[\tan\left(\tfrac{1}{2}\theta' + \tfrac{1}{4}\pi\right)\right] - \sin\theta'\right\} \qquad (14)$$

$$y = a\cos\theta' \qquad (15)$$

(Lockwood 1967, p. 123), where $\mathrm{gd}^{-1}\,x$ is the inverse GUDERMANNIAN FUNCTION.

A parameterization which traverses the tractrix with constant speed a is given by

$$x(t) = \begin{cases} ae^{-v/a} & \text{for } v \in [0,\,\infty) \\ ae^{v/a} & \text{for } v \in (-\infty,\,0] \end{cases} \qquad (16)$$

$$y(t) = \begin{cases} a\left[\tanh^{-1}\left(\sqrt{1-e^{-2v/a}}\right) - \sqrt{1-e^{-2v/a}}\right] \\ \quad \text{for } v \in [0,\,\infty) \\ a\left[-\tanh^{-1}\left(\sqrt{1-e^{2v/a}}\right) + \sqrt{1-e^{2v/a}}\right] \\ \quad \text{for } v \in (-\infty,\,0]. \end{cases} \qquad (17)$$

When a tractrix is rotated around its asymptote, a PSEUDOSPHERE results. This is a surface of constant NEGATIVE CURVATURE. For a tractrix, the length of a TANGENT from its point of contact to an asymptote is constant. The AREA between the tractrix and its asymptote is finite.

See also CURVATURE, DINI'S SURFACE, GUDERMANNIAN FUNCTION, MICE PROBLEM, PSEUDOSPHERE, PURSUIT CURVE, TRACTROID

References

Beyer, W. H. *CRC Standard Mathematical Tables, 28th ed.* Boca Raton, FL: CRC Press, p. 226, 1987.

Gray, A. "The Tractrix" and "The Evolute of a Tractrix is a Catenary." §3.6 and 5.3 in *Modern Differential Geometry of Curves and Surfaces with Mathematica, 2nd ed.* Boca Raton, FL: CRC Press, pp. 61–64 and 102–103, 1997.

Lawrence, J. D. *A Catalog of Special Plane Curves.* New York: Dover, pp. 199–200, 1972.

Lockwood, E. H. "The Tractrix and Catenary." Ch. 13 in *A Book of Curves.* Cambridge, England: Cambridge University Press, pp. 118–124, 1967.

MacTutor History of Mathematics Archive. "Tractrix." http://www-groups.dcs.st-and.ac.uk/~history/Curves/Tractrix.html.

Steinhaus, H. *Mathematical Snapshots, 3rd ed.* New York: Dover, pp. 249–251, 1999.

Yates, R. C. "Tractrix." *A Handbook on Curves and Their Properties.* Ann Arbor, MI: J. W. Edwards, pp. 221–224, 1952.

Tractrix Evolute

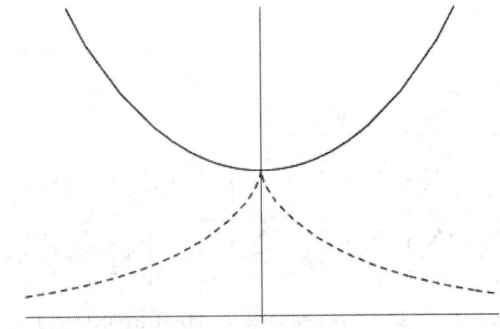

The EVOLUTE of the TRACTRIX is the CATENARY.

Tractrix Radial Curve

The RADIAL CURVE of the TRACTRIX is the KAPPA CURVE.

Tractroid

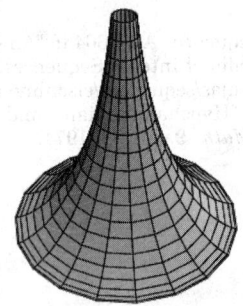

The SURFACE OF REVOLUTION produced by revolving the TRACTRIX

$$x = \operatorname{sech} u \qquad (1)$$

$$z = u - \tanh u \qquad (2)$$

about the z-AXIS is a tractroid given by

$$x = \operatorname{sech} u \cos v \qquad (3)$$

$$y = \operatorname{sech} u \sin v \qquad (4)$$

$$z = u - \tanh u. \qquad (5)$$

See also PSEUDOSPHERE, SURFACE OF REVOLUTION, TRACTRIX

Trail

PATH, WALK

Transcendental Curve

A curve which intersects some straight line in an infinity of points (but for which not every point lies on this curve).

See also ALGEBRAIC CURVE

References

Borwein, J. M.; Borwein, P. B.; and Bailey, D. H. "Ramanujan, Modular Equations, and Approximations to Pi or How to Compute One Billion Digits of Pi." *Amer. Math. Monthly* **96**, 201–219, 1989.

Transcendental Equation

An equation or formula involving TRANSCENDENTAL FUNCTIONS.

Transcendental Function

A function which is not an ALGEBRAIC FUNCTION. In other words, a function which "transcends," i.e., cannot be expressed in terms of, algebra. Examples of transcendental functions include the EXPONENTIAL FUNCTION, the TRIGONOMETRIC FUNCTIONS, and the inverses functions of both.

See also ALGEBRAIC FUNCTION, ELEMENTARY FUNCTION, PAINLEVÉ TRANSCENDENTS

Transcendental Number

A number which is not the ROOT of *any* POLYNOMIAL equation with INTEGER COEFFICIENTS, meaning that it is not an ALGEBRAIC NUMBER of any degree, is said to be transcendental. This definition guarantees that every transcendental number must also be IRRATIONAL, since a RATIONAL NUMBER is, by definition, an ALGEBRAIC NUMBER of degree one. A number x can then be tested to see if it is transcendental using the *Mathematica* command `Not[Element[x, Algebraics]]`.

Transcendental numbers are important in the history of mathematics because their investigation provided the first proof that CIRCLE SQUARING, one of the GEOMETRIC PROBLEMS OF ANTIQUITY which had baffled mathematicians for more than 2000 years was, in fact, insoluble. Specifically, in order for a number to be produced by a GEOMETRIC CONSTRUCTION using the ancient Greek rules, it must be either RATIONAL or a very special kind of ALGEBRAIC NUMBER known as a EUCLIDEAN NUMBER. Because the number π is transcendental, the construction cannot be done according to the Greek rules.

Georg Cantor was the first to prove the EXISTENCE of transcendental numbers. Liouville subsequently showed how to construct special cases (such as LIOUVILLE'S CONSTANT) using LIOUVILLE'S APPROXIMATION THEOREM. In particular, he showed that any number which has a rapidly converging sequence of rational approximations must be transcendental. For many years, it was only known how to determine if special classes of numbers were transcendental. The determination of the status of more general numbers was considered an important enough unsolved problem that it was one of HILBERT'S PROBLEMS.

Great progress was subsequently made by GELFOND'S THEOREM, which gives a general rule for determining if special cases of numbers OF THE FORM α^β are transcendental. Baker produced a further revolution by proving the transcendence of sums of numbers OF THE FORM $\alpha \ln \beta$ for ALGEBRAIC NUMBERS α and β.

The number E was proven to be transcendental by Hermite in 1873, and PI (π) by Lindemann in 1882. e^π is transcendental by GELFOND'S THEOREM since

$$(-1)^{-i} = (e^{i\pi})^{-i} = e^\pi.$$

The GELFOND-SCHNEIDER CONSTANT $2^{\sqrt{2}}$ is also transcendental (Hardy and Wright 1979, p. 162). Known transcendentals are summarized in the following table, where $\sin x$ is the SINE function, $J_0(x)$ is a BESSEL FUNCTION OF THE FIRST KIND, $x_k^{(n)}$ is the nth zero of $J_k(x)$, P is the THUE-MORSE CONSTANT, $\Gamma(x)$ is the GAMMA FUNCTION, and where $\zeta(n)$ is the RIEMANN ZETA FUNCTION.

e	Hermite (1873)
π	Lindemann (1882)
e^π	Gelfond
$e^{\pi\sqrt{d}}, d \in Z^*$	Nesterenko (1999)
$2^{\sqrt{2}}$	Hardy and Wright (1979, p. 162)
$\sin 1$	Hardy and Wright (1979, p. 162)
$J_0(1)$	Hardy and Wright (1979, p. 162)
$\ln 2$	Hardy and Wright (1979, p. 162)
$\ln 3 / \ln 2$	Hardy and Wright (1979, p. 162),
$x_0^{(1)} = 2.4048255\ldots$	Le Lionnais (1983, p. 46)
$\pi + \ln 2 + \sqrt{2} \ln 3$	Borwein *et al.* (1989)
$P = 0.4124540336\ldots$	Dekking (1977), Allouche and Shallit
CHAMPERNOWNE CONSTANT	
THUE CONSTANT	
$\Gamma\left(\frac{1}{3}\right)$	Le Lionnais (1983, p. 46)
$\Gamma\left(\frac{1}{4}\right)$	Chudnovsky (1984, p. 308), Waldschmidt, Nesterenko (1999)
$\Gamma\left(\frac{1}{6}\right)$	Chudnovsky (1984, p. 308)
$\Gamma\left(\frac{1}{4}\right)\pi^{-1/4}$	Davis (1959)
$\zeta(2n), n \in \mathbb{Z} > 1$	

APÉRY'S CONSTANT $\zeta(3)$ has been proved to be IRRATIONAL, but it is not known if it is transcendental. At least one of πe and $\pi + e$ (and probably both) are transcendental, but transcendence has not been proven for either number on its own. It is not known if e^e, π^π, π^e, γ (the EULER-MASCHERONI CONSTANT), $I_0(2)$, or $I_1(2)$ (where $I_n(x)$ is a MODIFIED BESSEL FUNCTION OF THE FIRST KIND) are transcendental.

The "degree" of transcendence of a number can be characterized by a so-called IRRATIONALITY MEASURE. There are still many fundamental and outstanding problems in transcendental number theory, including the CONSTANT PROBLEM and SCHANUEL'S CONJECTURE.

See also ALGEBRAIC NUMBER, ALGEBRAICALLY INDEPENDENT, ALGEBRAICS, CONSTANT PROBLEM, FOUR EXPONENTIALS CONJECTURE, GELFOND'S THEOREM, IRRATIONAL NUMBER, IRRATIONALITY MEASURE, LINDEMANN-WEIERSTRASS THEOREM, ROTH'S THEOREM, SCHANUEL'S CONJECTURE, SIX EXPONENTIALS THEOREM, THUE-SIEGEL-ROTH THEOREM

References

Allouche, J. P. and Shallit, J. In preparation.
Baker, A. "Approximations to the Logarithm of Certain Rational Numbers." *Acta Arith.* **10**, 315–323, 1964.
Baker, A. "Linear Forms in the Logarithms of Algebraic Numbers I." *Mathematika* **13**, 204–216, 1966.
Baker, A. "Linear Forms in the Logarithms of Algebraic Numbers II." *Mathematika* **14**, 102–107, 1966.
Baker, A. "Linear Forms in the Logarithms of Algebraic Numbers III." *Mathematika* **14**, 220–228, 1966.
Baker, A. "Linear Forms in the Logarithms of Algebraic Numbers IV." *Mathematika* **15**, 204–216, 1966.
Borwein, J. M.; Borwein, P. B.; and Bailey, D. H. "Ramanujan, Modular Equations, and Approximations to Pi or How to Compute One Billion Digits of Pi." *Amer. Math. Monthly* **96**, 201–219, 1989.
Chudnovsky, G. V. *Contributions to the Theory of Transcendental Numbers.* Providence, RI: Amer. Math. Soc., 1984.
Courant, R. and Robbins, H. "Algebraic and Transcendental Numbers." §2.6 in *What is Mathematics?: An Elementary Approach to Ideas and Methods, 2nd ed.* Oxford, England: Oxford University Press, pp. 103–107, 1996.
Davis, P. J. "Leonhard Euler's Integral: A Historical Profile of the Gamma Function." *Amer. Math. Monthly* **66**, 849–869, 1959.
Dekking, F. M. "Transcendence du nombre de Thue-Morse." *C. R. Acad. Sci. Paris* **285**, 157–160, 1977.
Gray, R. "Georg Cantor and Transcendental Numbers." *Amer. Math. Monthly* **101**, 819–832, 1994.
Hardy, G. H. and Wright, E. M. "Algebraic and Transcendental Numbers," "The Existence of Transcendental Numbers," and "Liouville's Theorem and the Construction of Transcendental Numbers." §11.5–11.6 in *An Introduction to the Theory of Numbers, 5th ed.* Oxford, England: Oxford University Press, pp. 159–164, 1985.
Hermite, C. "Sur la fonction exponentielle." *C. R. Acad. Sci. Paris* **77**, 18–24, 74–79, and 226–233, 1873.
Le Lionnais, F. *Les nombres remarquables.* Paris: Hermann, p. 46, 1979.
Lindemann, F. "Über die Zahl π." *Math. Ann.* **20**, 213–225, 1882.
Nagell, T. *Introduction to Number Theory.* New York: Wiley, p. 35, 1951.
Nesterenko, Yu. V. "On Algebraic Independence of the Components of Solutions of a System of Linear Differential Equations." [Russian.] *Izv. Akad. Nauk SSSR, Ser. Mat.* **38**, 495–512, 1974. English translation in *Math. USSR* **8**, 501–518, 1974.
Nesterenko, Yu. V. "Modular Functions and Transcendence Questions." [Russian.] *Mat. Sbornik* **187**, 65–96, 1996. English translation in *Sbornik Math.* **187**, 1319–1348, 1996.
Nesterenko, Yu. V. *A Course on Algebraic Independence: Lectures at IHP 1999.* http://www.math.jussieu.fr/~nesteren/.
Ramachandra, K. *Lectures on Transcendental Numbers.* Madras, India: Ramanujan Institute, 1969.
Shidlovskii, A. B. *Transcendental Numbers.* New York: de Gruyter, 1989.
Siegel, C. L. *Transcendental Numbers.* New York: Chelsea, 1965.
Tijdeman, R. "An Auxiliary Result in the Theory of Transcendental Numbers." *J. Numb. Th.* **5**, 80–94, 1973.

Transcritical Bifurcation

Let $f : \mathbb{R} \times \mathbb{R} \to \mathbb{R}$ be a one-parameter family of C^2 maps satisfying

$$f(0, \mu) = 0 \tag{1}$$

$$\left[\frac{\partial f}{\partial x}\right]_{\mu=0,\, x=0} = 0 \tag{2}$$

$$\left[\frac{\partial^2 f}{\partial x\, \partial \mu}\right]_{0,\,0} > 0 \tag{3}$$

$$\left[\frac{\partial^2 f}{\partial x^2}\right]_{\mu=0,\, x=0} < 0. \tag{4}$$

(Actually, condition (1) can be relaxed slightly.) Then there are two branches, one stable and one unstable. This BIFURCATION is called a transcritical bifurcation.

An example of an equation displaying a transcritical bifurcation is

$$\dot{x} = \mu x - x^2 \tag{5}$$

(Guckenheimer and Holmes 1997, p. 145).

See also BIFURCATION, PITCHFORK BIFURCATION

References

Guckenheimer, J. and Holmes, P. *Nonlinear Oscillations, Dynamical Systems, and Bifurcations of Vector Fields, 3rd ed.* New York: Springer-Verlag, pp. 145 and 149–150, 1997.
Rasband, S. N. *Chaotic Dynamics of Nonlinear Systems.* New York: Wiley, pp. 27–28, 1990.

Transfer Function

The engineering terminology for one use of FOURIER TRANSFORMS. By breaking up a wave pulse into its frequency spectrum

$$f_\nu = F(\nu)e^{2\pi i \nu t}, \tag{1}$$

the entire signal can be written as a sum of contributions from each frequency,

$$f(t) = \int_{-\infty}^{\infty} f_v \, dv = \int_{-\infty}^{\infty} F(v)e^{2\pi i vt} \, dv. \qquad (2)$$

If the signal is modified in some way, it will become

$$g_v(t) = \phi(v)f_v(t) = \phi(v)F(v)e^{2\pi i vt} \qquad (3)$$

$$g(t) = \int_{-\infty}^{\infty} g_v(t) \, dt = \int_{-\infty}^{\infty} \phi(v)F(v)e^{2\pi i vt} \, dv, \qquad (4)$$

where $\phi(v)$ is known as the "transfer function." FOURIER TRANSFORMING ϕ and F,

$$\phi(v) = \int_{-\infty}^{\infty} \Phi(t)e^{-2\pi i vt} \, dt \qquad (5)$$

$$F(v) = \int_{-\infty}^{\infty} f(t)e^{-2\pi i vt} \, dt. \qquad (6)$$

From the CONVOLUTION THEOREM,

$$g(t) = f(t) * \Phi(t) = \int_{-\infty}^{\infty} f(t)\Phi(t - r) \, dr. \qquad (7)$$

See also CONVOLUTION THEOREM, FOURIER TRANSFORM

Transfer Principle

In NONSTANDARD ANALYSIS, the transfer principle is the technical form of the following intuitive idea: "Anything provable about a given SUPERSTRUCTURE V by passing to a nonstandard enlargement $*V$ of V is also provable without doing so, and vice versa." It is a result of LOS' THEOREM and the completeness theorem for first-order predicate logic

The transfer principle is stated as follows. Let V be a superstructure, let $*V$ be an enlargement of V, let σ be any sentence in the language for $(V, \in)$, and let $*\sigma$ denote the $*$-transform of σ. Then $(V, \in) \vDash \sigma$ if and only if $(*V, *\in) \vDash *\sigma$.

See also LOS' THEOREM, NONSTANDARD ANALYSIS

Transfinite Diameter
Let

$$\phi(z) = cz + c_0 + c_1 z^{-1} + c_2 z^{-2} + \ldots$$

be an ANALYTIC FUNCTION, REGULAR and UNIVALENT for $|z| > 1$, which maps $|z| > 1$ CONFORMALLY onto the region T preserving the POINT AT INFINITY and its direction. Then the function $\phi(z)$ is uniquely determined and c is called the transfinite diameter, sometimes also known as ROBIN'S CONSTANT or the CAPACITY of $\phi(z)$.

See also ANALYTIC FUNCTION, REGULAR FUNCTION, UNIVALENT FUNCTION

Transfinite Number
One of Cantor's ORDINAL NUMBERS ω, $\omega + 1$, $\omega + 2$, ..., $\omega + \omega$, $\omega + \omega + 1$, ...which is "larger" than any WHOLE NUMBER.

See also ALEPH-0, ALEPH-1, CARDINAL NUMBER, CONTINUUM, ORDINAL NUMBER, WHOLE NUMBER

References
Ferreirós, J. "The Transfinite Ordinals and Cantor's Mature Theory." Ch. 8 in *Labyrinth of Thought: A History of Set Theory and Its Role in Modern Mathematics.* Basel, Switzerland: Birkhäuser, pp. 257–296, 1999.
Pappas, T. "Transfinite Numbers." *The Joy of Mathematics.* San Carlos, CA: Wide World Publ./Tetra, pp. 156–158, 1989.

Transform
A shortened term for INTEGRAL TRANSFORM.

Geometrically, if S and T are two transformations, then the SIMILARITY TRANSFORMATION TST^{-1} is sometimes called the transform (Woods 1961).

See also ABEL TRANSFORM, BOUSTROPHEDON TRANSFORM, DISCRETE FOURIER TRANSFORM, FAST FOURIER TRANSFORM, FOURIER TRANSFORM, FRACTIONAL FOURIER TRANSFORM, HANKEL TRANSFORM, HARTLEY TRANSFORM, HILBERT TRANSFORM, LAPLACE-STIELTJES TRANSFORM, LAPLACE TRANSFORM, MELLIN TRANSFORM, NUMBER THEORETIC TRANSFORM, PONCELET TRANSFORM, RADON TRANSFORM, WAVELET TRANSFORM, Z-TRANSFORM

References
Woods, F. S. *Higher Geometry: An Introduction to Advanced Methods in Analytic Geometry.* New York: Dover, p. 5, 1961.

Transform Theory
INTEGRAL TRANSFORM

Transformation
A transformation T (a.k.a., MAP, FUNCTION) over a DOMAIN D takes the elements $X \in D$ to elements $Y \in T(D)$, where the RANGE (a.k.a., image) of T is defined as

$$\text{Range}(T) = T(D) = \{T(\mathbf{X}) : \mathbf{X} \in D\}.$$

Note that when transformations are specified with respect to a coordinate system, it is important to specify whether the rotation takes place on the *coordinate system*, with space and objects embedded in it being viewed as fixed (a so-called ALIAS TRANSFORMATION), or on the *space itself* relative to a fixed coordinate system (a so-called ALIBI TRANSFORMATION).

Examples of transformations are summarized in the following table.

Transformation	Characterization
DILATION	center of dilation, scale decrease factor
EXPANSION	center of expansion, scale increase factor
REFLECTION	mirror line or plane
ROTATION	center of rotation, rotation angle
SHEAR	invariant line and SHEAR FACTOR
STRETCH (1-way)	invariant line and scale factor
STRETCH (2-way)	invariant lines and scale factors
TRANSLATION	displacement vector

See also AFFINE TRANSFORMATION, ALIAS TRANSFORMATION, ALIBI TRANSFORMATION, DILATION, EXPANSION, FUNCTION, MAP, REFLECTION, ROTATION, SHEAR, STRETCH, TRANSFORM, TRANSLATION

References

Coxeter, H. S. M. and Greitzer, S. L. "Transformations." Ch. 4 in *Geometry Revisited.* Washington, DC: Math. Assoc. Amer., pp. 80–102, 1967.

Graustein, W. C. "Transformation." Ch. 7 in *Introduction to Higher Geometry.* New York: Macmillan, pp. 84–114, 1930.

Kapur, J. N. *Transformation Geometry.* New Delhi, India: Mathematical Sciences Trust Society, 1994–95.

Transition Function

A transition function describes the difference in the way an object is described in two separate, overlapping COORDINATE CHARTS, where the description of the same set may change in different coordinates. This even occurs in EUCLIDEAN SPACE $\mathbb{R}^3$, where any rotation of the usual x, y, and z axes gives another set of coordinates.

For example, on the sphere, person A at the equator can use the usual directions of north, south, east, and west, but person B at the North Pole must use something else. However, both A and B can describe the region in between them in their coordinate charts. A transition function would then describe how to go from the coordinate chart for A to the coordinate chart for B.

In the case of a MANIFOLD, a transition function is a map from one coordinate chart to another. Therefore, in a sense, a manifold is composed of coordinate charts, and the glue that holds them together is the transition functions. In the case of a BUNDLE, the transition functions are the glue that holds together

its TRIVIALIZATIONS. Specifically, in this case the transition function describes an invertible transformation of the FIBER.

Naturally, the type of invertible transformation depends on the type of bundle. For instance, a VECTOR BUNDLE, which could be the TANGENT BUNDLE, has INVERTIBLE LINEAR transition functions. More precisely, a transition function for a vector bundle of RANK r, on overlapping coordinate charts U_1 and U_2, is given by a function

$$g_{12} : U_1 \cap U_2 \to GL(r),$$

where GL is the GENERAL LINEAR GROUP. The fiber at $p \in U_1 \cap U_2$ has two descriptions, and $g_{12}(p)$ is the INVERTIBLE LINEAR MAP that takes one to the other. The transition functions have to be consistent in the sense that if one goes to another description of the same set, and then back again, then nothing has changed. A necessary and sufficient condition for consistency is the following: Given three overlapping charts, the product $g_{12}g_{23}g_{31}$ has to be the constant map to the identity in $GL(r)$.

A consistent set of transition functions for a VECTOR BUNDLE of RANK r can be interpreted as an element of the first CECH COHOMOLOGY GROUP of a manifold with coefficients in $GL(r)$.

See also BUNDLE, CECH COHOMOLOGY, COORDINATE CHART, MANIFOLD, TANGENT BUNDLE, TRIVIALIZATION, VECTOR BUNDLE

Transitive

A RELATION R on a SET S is transitive provided that for all x, y and z in S such that xRy and yRz, we also have xRz.

See also ASSOCIATIVE, COMMUTATIVE, RELATION

Transitive Closure

The transitive closure of a BINARY RELATION R on a SET X is the minimal TRANSITIVE relation R' on X that contains R. Thus $aR'b$ for any elements a and b of X provided that there exist $c_0, c_1, ..., c_n$ with $c_0 = a$, $c_n = b$, and $c_r R c_{r+1}$ for all $0 \leq r \leq n$.

The transitive closure $C(G)$ of a GRAPH is a graph which contains an edge $\{u, v\}$ whenever there is a directed path from u to v (Skiena 1990, p. 203). The transitive closure of a graph can be computed using TransitiveClosure[g] in the *Mathematica* add-on package DiscreteMath`Combinatorica` (which can be loaded with the command <<DiscreteMath`).

See also REFLEXIVE CLOSURE, TRANSITIVE GRAPH, TRANSITIVE REDUCTION

References

Aho, A.; Garey, M. R.; and Ullman, J. D. "The Transitive Reduction of a Directed Graph." *SIAM J. Comput.* **1**, 131–137, 1972.

Skiena, S. *Implementing Discrete Mathematics: Combinatorics and Graph Theory with Mathematica.* Reading, MA: Addison-Wesley, 1990.

Transitive Digraph

A GRAPH G is transitive if any three vertices (x, y, z) such that edges $(x, y), (y, z) \in G$ imply $(x, y) \in G$. Unlabeled transitive digraphs are called TOPOLOGIES.

See also TOPOLOGY (DIGRAPH), TRANSITIVE GRAPH, TRANSITIVE REDUCTION

Transitive Graph

A GRAPH G is called n-transitive with $n \geq 1$ if it has an n-ROUTE and if there is always a GRAPH AUTOMORPHISM of G sending each n-ROUTE onto any other n-ROUTE (Harary 1994, p. 173). There are no n-transitive CUBIC GRAPHS for $n > 5$ (Harary 1994, p. 175).

See also ROUTE, TRANSITIVE CLOSURE, TRANSITIVE DIGRAPH, TRANSITIVE REDUCTION, UNITRANSITIVE GRAPH

References

Harary, F. *Graph Theory.* Reading, MA: Addison-Wesley, 1994.

Skiena, S. *Implementing Discrete Mathematics: Combinatorics and Graph Theory with Mathematica.* Reading, MA: Addison-Wesley, pp. 162 and 174, 1990.

Transitive Group

When a GROUP ACTION is implicitly understood, i.e., a subgroup of a PERMUTATION GROUP, then the SUBGROUP is called transitive if its action is transitive. For example, the ALTERNATING GROUP is transitive. A group may also be called k-transitive if there is any set on which the group acts FAITHFULLY and k-transitively. Transitivity is a result of the symmetry in the group.

For instance, the SYMMETRIC GROUP S_n is n-transitive and the ALTERNATING GROUP A_n is $(n-2)$-transitive. However, multiply transitive finite groups are rare. In fact, they have been completely determined using the CLASSIFICATION THEOREM OF FINITE GROUPS. Except for some SPORADIC examples, the multiply transitive groups fall into infinite families. Certain subgroups of the AFFINE GROUP on a finite VECTOR SPACE, including the AFFINE GROUP itself, are 2-transitive. Some of these are summarized below.

The multiply transitive groups fall into six infinite families, and four classes of SPORADIC GROUPS. In the following enumeration, q is a power of a prime number.

1. Certain subgroups of the AFFINE GROUP on a finite VECTOR SPACE, including the AFFINE GROUP itself, are 2-transitive.

2. The PROJECTIVE SPECIAL LINEAR GROUPS $PSL(d, q)$ are 2-transitive, and $PSL(2, q)$ is actually 3-transitive.

3. The SYMPLECTIC GROUPS defined over the FIELD of two elements have two distinct actions which are 2-transitive.

4. The field K of q^2 elements has an INVOLUTION $\sigma(a) = a^q$, so $\sigma^2 = 1$, which allows a HERMITIAN FORM to be defined on a VECTOR SPACE on K. The UNITARY GROUP on $V = \oplus^3 K$, denoted $U_2(q)$, preserves the ISOTROPIC VECTORS in V. The action of the PROJECTIVE SPECIAL UNITARY GROUP $PSU_3(q)$ is 2-transitive on the ISOTROPIC VECTORS.

5. The SUZUKI GROUP $Sz(q)$ is the AUTOMORPHISM GROUP of a $S(3, q+1, q^2+1)$ STEINER SYSTEM, an INVERSIVE PLANE of order q, and its action is 2-transitive.

6. The REE GROUP $R(q)$ is the AUTOMORPHISM GROUP of a $S(2, q+1, q^3+1)$ STEINER SYSTEM, a UNITAL of order q, and its action is 2-transitive.

7. The MATHIEU GROUPS M_{12} and M_{24} are the only 5-transitive groups besides S_5 and A_7. The groups M_{11} and M_{23} are 4-transitive, and M_{22} is 3-transitive.

8. The PROJECTIVE SPECIAL LINEAR GROUP $PSL(2, 11)$ has another 2-transitive action related to the WITT GEOMETRY W_{11}.

9. The HIGMAN-SIMS GROUP is 2-transitive.

10. The CONWAY GROUP Co_3 is 2-transitive.

See also FINITE SIMPLE GROUP, LEECH LATTICE, MATHIEU GROUPS, STEINER SYSTEM, TRANSITIVE GROUP ACTION

References

Dixon, J. and Mortimer, B. *Permutation Groups.* New York: Springer-Verlag, 1996.

Transitive Group Action

A GROUP ACTION $G \times X \to X$ is transitive if it possesses only a single ORBIT, i.e., for every pair of elements x and y, there is a group element g such that $gx = y$. In this case, X is ISOMORPHIC to the left COSETS of the isotropy group, $X \sim G/G_x$. The space X, which has a transitive group action, is called a HOMOGENEOUS SPACE when the group is a LIE GROUP.

If, for every two pairs of points x_1, x_2 and y_1, y_2, there is a group element g such that $gx_i = y_i$, then the GROUP ACTION is called doubly transitive. Similarly, a group action can be triply transitive and, in general, a GROUP ACTION is k-transitive if every set $\{x_1, \ldots, y_k\}$ of $2k$ distinct elements has a group element g such that $gx_i = y_i$.

See also EFFECTIVE ACTION, FAITHFUL GROUP ACTION, FREE ACTION, GROUP, ISOTROPY GROUP, MA-

TRIX GROUP, ORBIT (GROUP), QUOTIENT SPACE (LIE GROUP), REPRESENTATION, TOPOLOGICAL GROUP, TRANSITIVE GROUP

References

Burnside, W. "On Transitive Groups of Degree n and Class $n-1$." *Proc. London Math. Soc.* **32**, 240–246, 1900.
Hulpke, A. *Konstruktion transitiver Permutationsgruppen.* Ph.D. thesis. Aachen, Germany: RWTH, 1996. Also available as *Aachener Beiträge zur Mathematik*, No. 18, 1996.
Kawakubo, K. *The Theory of Transformation Groups.* Oxford, England: Oxford University Press, pp. 4–6 and 41–49, 1987.
Rotman, J. *Theory of Groups.* New York: Allyn and Bacon, pp. 180–184, 1984.

Transitive Points

Two points on a surface which are opposite to each other but not farthest from each other (e.g., the midpoints of opposite edges of a CUBE) are said to be transitive points. The SPHERE has no transitive points.

References

Steinhaus, H. *Mathematical Snapshots, 3rd ed.* New York: Dover, p. 175, 1999.

Transitive Reduction

The transitive reduction of a BINARY RELATION R on a SET X is the minimum relation R' on X with the same TRANSITIVE CLOSURE as R. Thus $aR'b$ for any elements a and b of X, provided that aRb and there exists no element c of X such that aRc and cRb.

The transitive reduction of a GRAPH G is the smallest graph $R(G)$ such that $C(G) = C(R(G))$, where $C(G)$ is the TRANSITIVE CLOSURE of G (Skiena 1990, p. 203).

See also REFLEXIVE REDUCTION, TRANSITIVE CLOSURE, TRANSITIVE GRAPH

References

Aho, A.; Garey, M. R.; and Ullman, J. D. "The Transitive Reduction of a Directed Graph." *SIAM J. Comput.* **1**, 131–137, 1972.
Skiena, S. *Implementing Discrete Mathematics: Combinatorics and Graph Theory with Mathematica.* Reading, MA: Addison-Wesley, 1990.

Transitive Triple

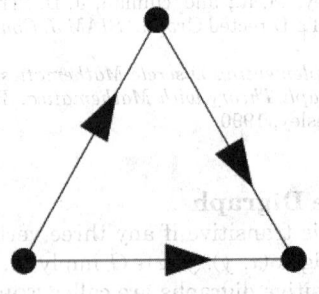

The 3-node TOURNAMENT (and DIRECTED GRAPH) illustrated above (Harary 1994, p. 205).

See also CYCLIC TRIPLE, TOURNAMENT

References

Harary, F. "Tournaments." *Graph Theory.* Reading, MA: Addison-Wesley, 1994.

Transitivity Class

Let $S(T)$ be the group of symmetries which map a MONOHEDRAL TILING T onto itself. The TRANSITIVITY CLASS of a given tile T is then the collection of all tiles to which T can be mapped by one of the symmetries of $S(T)$.

See also MONOHEDRAL TILING

References

Berglund, J. "Is There a k-Anisohedral Tile for $k \geq 5$?" *Amer. Math. Monthly* **100**, 585–588, 1993.

Translation

A transformation consisting of a constant offset with no ROTATION or distortion. In n-D EUCLIDEAN SPACE, a translation may be specified simply as a VECTOR giving the offset in each of the n coordinates.

See also AFFINE GROUP, DILATION, EUCLIDEAN GROUP, EXPANSION, GLIDE, IMPROPER ROTATION, INVERSION OPERATION, MIRROR IMAGE, REFLECTION, ROTATION

References

Addington, S. "The Four Types of Symmetry in the Plane." http://forum.swarthmore.edu/sum95/suzanne/symsusan.html.
Beyer, W. H. (Ed.). *CRC Standard Mathematical Tables, 28th ed.* Boca Raton, FL: CRC Press, p. 211, 1987.

Coxeter, H. S. M. and Greitzer, S. L. "Translation." §4.1 in *Geometry Revisited*. Washington, DC: Math. Assoc. Amer., pp. 81–82, 1967.

Translation Relation

A mathematical relationship transforming a function $f(x)$ to the form $f(x + a)$.

See also ARGUMENT ADDITION RELATION, ARGUMENT MULTIPLICATION RELATION, RECURRENCE RELATION, REFLECTION RELATION

Transpose

The object obtained by replacing all elements a_{ij} with a_{ji}. For a second-RANK TENSOR a_{ij}, the tensor transpose is simply a_{ji}. The matrix transpose, written A^{T}, is the MATRIX obtained by exchanging A's rows and columns, and satisfies the identity

$$(\mathsf{A}^{\mathrm{T}})^{-1} = (\mathsf{A}^{-1})^{\mathrm{T}}. \tag{1}$$

Several other notations are commonly used, including $\tilde{\mathsf{A}}$ (Arfken 1985, p. 201; Griffiths 1987, p. 223) and A' (Ayres 1962, p. 11; Courant and Hilbert 1989, p. 9) The product of two transposes satisfies

$$(\mathsf{B}^{\mathrm{T}}\mathsf{A}^{\mathrm{T}})_{ij} = (b^{\mathrm{T}})_{ik}(a^{\mathrm{T}})_{kj} = b_{ki}a_{jk} = a_{jk}b_{ki} = (\mathsf{AB})_{ji}$$
$$= (\mathsf{AB})^{\mathrm{T}}_{ij}, \tag{2}$$

where EINSTEIN SUMMATION has been used to implicitly sum over repeated indices. Therefore,

$$(\mathsf{AB})^{\mathrm{T}} = \mathsf{B}^{\mathrm{T}}\mathsf{A}^{\mathrm{T}}. \tag{3}$$

See also ADJOINT MATRIX, CONGRUENT MATRICES, CONJUGATE MATRIX

References

Arfken, G. *Mathematical Methods for Physicists, 3rd ed.* Orlando, FL: Academic Press, p. 201, 1985.
Ayres, F. Jr. *Theory and Problems of Matrices.* New York: Schaum, pp. 11–12, 1962.
Courant, R. and Hilbert, D. *Methods of Mathematical Physics, Vol. 1.* New York: Wiley, 1989.
Griffiths, D. J. *Introduction to Elementary Particles.* New York: Wiley, p. 220, 1987.

See also SKEW SYMMETRIC MATRIX, SYMMETRIC MATRIX

Transpose Map

PULLBACK MAP

Transpose Partition

CONJUGATE PARTITION

Transposition

An exchange of two elements of an ordered list with all others staying the same. A transposition is therefore a PERMUTATION of two elements. For example, the swapping of 2 and 5 to take the list 123456 to 153426 is a transposition. The PERMUTATION SYMBOL $\epsilon_{ijk\cdots}$ is defined as $(-1)^n$, where n is the number of transpositions of pairs of elements that must be composed to build up the PERMUTATION.

See also INVERSION NUMBER, PERMUTATION, PERMUTATION SYMBOL, TRANSPOSITION GRAPH, TRANSPOSITION ORDER

References

Skiena, S. "Permutations from Transpositions." §1.1.4 in *Implementing Discrete Mathematics: Combinatorics and Graph Theory with Mathematica.* Reading, MA: Addison-Wesley, pp. 9–11, 1990.

Transposition Graph

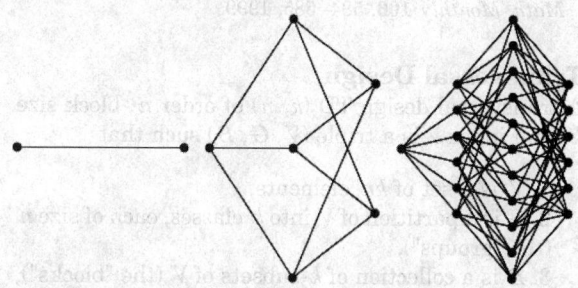

A GRAPH in which nodes correspond to permutations and edges are placed between permutations that differ by exactly one transposition (Skiena 1990, p. 9). All cycles in transposition graphs are of even length, making them BIPARTITE. The transposition graph of a MULTISET is always HAMILTONIAN (Chase 1973).

See also TRANSPOSITION

References

Chase, P. J. "Transposition Graphs." *SIAM J. Comput.* **2**, 128–133, 1973.
Skiena, S. *Implementing Discrete Mathematics: Combinatorics and Graph Theory with Mathematica.* Reading, MA: Addison-Wesley, pp. 9–10, 1990.

Transposition Group

A PERMUTATION GROUP in which the PERMUTATIONS are limited to TRANSPOSITIONS.

See also PERMUTATION GROUP

Transposition Order

An ordering of PERMUTATIONS in which each two adjacent permutations differ by the TRANSPOSITION of two elements. For the permutations of $\{1, 2, 3\}$ there are two listings which are in transposition order. One is 123, 132, 312, 321, 231, 213, and the other is 123, 321, 312, 213, 231, 132.

See also LEXICOGRAPHIC ORDER, PERMUTATION

References

Ruskey, F. "Information on Combinations of a Set." http://www.theory.csc.uvic.ca/~cos/inf/comb/CombinationsInfo.html.

Transversal Array

A set of n cells in an $n \times n$ SQUARE such that no two come from the same row and no two come from the same column. The number of transversals of an $n \times n$ SQUARE is $n!$ (n FACTORIAL).

A Latin transversal is a transversal such that no two cells contain the same element (Snevily 1999).

References

Alon, N. *Additive Latin Transversals*. Preprint.
Snevily, H. S. "The Cayley Addition Table of $\mathbb{Z}_n$." *Amer. Math. Monthly* **106**, 584–585, 1999.

Transversal Design

A transversal design $TD_\lambda(k, n)$ of order n, block size k, and index λ is a triple (V, G, B) such that

1. V is a set of kn elements,
2. G is a partition of V into k classes, each of size n (the "groups"),
3. B is a collection of k-subsets of V (the "blocks"), and
4. Every unordered pair of elements from V is contained in either exactly one group or in exactly λ blocks, but not both.

References

Colbourn, C. J. and Dinitz, J. H. (Eds.). *CRC Handbook of Combinatorial Designs*. Boca Raton, FL: CRC Press, p. 112, 1996.

Transversal Intersection

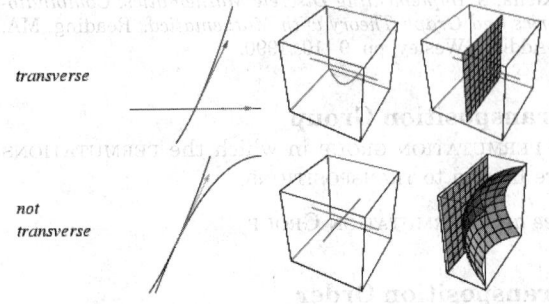

transverse

not transverse

Two SUBMANIFOLDS X and Y in an ambient space M intersect transversally if, for all $p \in X \cap Y$,

$$TX_p + TY_p = \{v + w : v \in TX_p, \ w \in TY_p\} = TM_p,$$

where the addition is in TM_p, and TX_p denotes the TANGENT MAP of X_p. If two submanifolds do not intersect, then they are automatically transversal. For example, two curves in $\mathbb{R}^3$ are transversal only if they do not intersect at all. When X and Y meet transversally then $X \cap Y$ is a smooth SUBMANIFOLD of the expected dimension $\dim X + \dim Y - \dim M$. In some sense, two submanifolds "ought" to intersect transversally and, by SARD'S THEOREM, any intersection can be perturbed to be transversal. Intersection in HOMOLOGY only makes sense because an intersection can be made to be transversal.

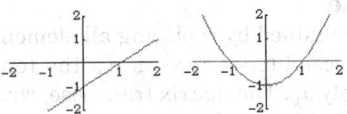

Transversality is a sufficient condition for an intersection to be stable after a perturbation. For example, the lines $y = x$ and $y = 0$ intersect transversally, as do the perturbed lines $y = x + t$, and they intersect at only one point. However, $y = x^2$ does not intersect $y = 0$ transversally. It intersects in one point, while $y = x^2 + t$ intersects in either none or two points, depending on whether t is positive or negative.

When $\dim X + \dim Y = \dim M$, then a transversal intersection is an ISOLATED POINT. If the three spaces have an ORIENTATION, then the transversal condition means it is possible to assign a sign to the intersection. If $e_1, \ldots, e_k$ are an oriented basis for TX_p and $e_{k+1}, \ldots, e_n$ are an oriented basis for TY_p, then the intersection is $+1$ if $e_1, \ldots, e_n$ is oriented in M and -1 otherwise.

More generally, two SMOOTH MAPS $f : X \to M$ and $g : Y \to M$ are transversal if whenever $p = f(x) = g(y)$ then $df(TX_x) + dg(TY_y) = TM_p$.

See also HOMOLOGY, INTERSECTION (HOMOLOGY), ORIENTATION (VECTOR SPACE), SARD'S THEOREM, SUBMERSION

Transversal Line

A transversal line is a LINE which intersects each of a given set of other lines. It is also called a semisecant.

See also LINE

Transversal Plane

References

Altshiller-Court, N. "Transversals." Ch. 5 in *Modern Pure Solid Geometry*. New York: Chelsea, pp. 111–122, 1979.

Transylvania Lottery

A lottery in which three numbers are picked at random from the INTEGERS 1–14.

See also FANO PLANE

Trapdoor Function

An easily computed function whose inverse is extremely difficult to compute. An example is the multiplication of two large PRIMES. Finding and verifying two large PRIMES is easy, as is their multiplication. But factorization of the resultant product is very difficult.

See also RSA ENCRYPTION

References

Gardner, M. "Trapdoor Ciphers" and "Trapdoor Ciphers II." Chs. 13–14 in *Penrose Tiles and Trapdoor Ciphers...and the Return of Dr. Matrix, reissue ed.* New York: W. H. Freeman, pp. 183–204, 1989.

Trapdoor One-Way Function

Informally, a function $f : \{0, 1\}^{l(n)} \times \{0, 1\}^{n} \rightarrow (0, 1)^{m(n)}$ is a trapdoor one-way function if

1. It is a ONE-WAY FUNCTION, and
2. For fixed public key $y \in \{0, 1\}^{l(n)}$, $f(x, y)$ is viewed as a function $f_y(x)$ of x that maps n bits to $m(n)$ bits. Then there is an efficient algorithm that, on input $\langle y, f_y(x), z \rangle$ produces x' such that $f_y(x') = f_y(x)$, for some trapdoor key $z \in \{0, 1\}^{k(n)}$.

f is a TRAPDOOR ONE-WAY HASH FUNCTION if f is also a ONE-WAY HASH FUNCTION, i.e., if additionally

3. Given M and $f(M)$, it is hard to find a message $M' \neq M$ such that $f(M') \neq f(M)$.

It is not known if a trapdoor one-way function can be constructed from any one-way function.

An example of a trapdoor one-way function is factorization of a product of two large PRIMES. While selecting and verifying two large PRIMES and multiplying them together is easy, factoring the resulting product is (as far as is known) very difficult. This is the basis for RSA ENCRYPTION, which is conjectured to be trapdoor one-way.

See also ONE-WAY FUNCTION, RSA ENCRYPTION, TRAPDOOR ONE-WAY HASH FUNCTION

References

Gardner, M. "Trapdoor Ciphers" and "Trapdoor Ciphers II." Chs. 13–14 in *Penrose Tiles and Trapdoor Ciphers...and the Return of Dr. Matrix, reissue ed.* New York: W. H. Freeman, pp. 183–204, 1989.
Luby, M. *Pseudorandomness and Cryptographic Applications.* Princeton, NJ: Princeton University Press, 1996.
RSA Laboratories. ® "What Is a One-Way Function?" http://www.rsasecurity.com/rsalabs/faq/2–3-2.html.

Trapdoor One-Way Hash Function

A function $f : \{0, 1\}^{l(n)} \times \{0, 1\}^{n} \rightarrow (0, 1)^{m(n)}$ is a TRAPDOOR ONE-WAY HASH FUNCTION if f is a TRAPDOOR ONE-WAY FUNCTION and is also a one-way hash function, i.e. if, additionally given M and $f(M)$, it is

hard to find a message $M' \neq M$ such that $f(M') = f(M)$.

See also TRAPDOOR ONE-WAY FUNCTION

Trapezium

There are two common definitions of the trapezium. The American definition is a QUADRILATERAL with no PARALLEL sides. The British definition for a trapezium is a QUADRILATERAL *with* two sides PARALLEL. Such a trapezium is equivalent to a TRAPEZOID and therefore has AREA

$$A = \tfrac{1}{2}(a + b)h.$$

See also DIAMOND, KITE, LOZENGE, PARALLELOGRAM, QUADRILATERAL, RHOMBOID, RHOMBUS, SKEW QUADRILATERAL, STROMBUS, TRAPEZOID

Trapezohedron

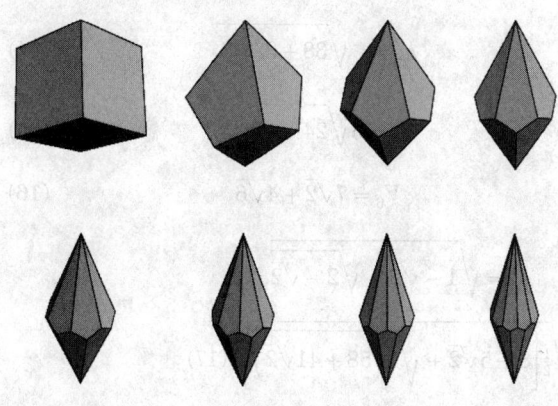

The trapezohedra are the DUAL POLYHEDRA of the Archimedean ANTIPRISMS. However, the name for these solids is not particular well chosen since their faces are *not* TRAPEZOIDS. The CUBE oriented along a space diagonal is a trapezohedron.

The trapezohedra generated by taking the duals of the ANTIPRISMS have side length s_n, half-heights (half the peak-to-peak distance) h_n, surface areas S_n, and volumes V_n (where the latter two are normalized so that the shortest edge has length 1) given by

$$s_3 = \tfrac{1}{2}\sqrt{2} \tag{1}$$

$$h_3 = \tfrac{1}{4}\sqrt{6} \tag{2}$$

$$S_3 = 6 \tag{3}$$

$$V_3 = 1 \tag{4}$$

$$s_4 = \sqrt{\sqrt{2} - 1}, \ \sqrt{\tfrac{1}{2}\left(1 + \sqrt{3}\right)} \tag{5}$$

$$h_4 = \frac{1}{2}\sqrt{h\left(4 + 3\sqrt{2}\right)} \tag{6}$$

$$S_4 = 2\sqrt{22 + 16\sqrt{2}} \tag{7}$$

$$V_4 = \frac{1}{3}\sqrt{58 + 41\sqrt{2}} \tag{8}$$

$$s_5 = \frac{1}{2}\left(\sqrt{5} - 1\right), \ \frac{1}{2}\left(1 + \sqrt{5}\right) \tag{9}$$

$$h_5 = \frac{1}{2}\sqrt{5 + 2\sqrt{5}} \tag{10}$$

$$S_5 = 5\sqrt{\frac{1}{2}\left(25 + 11\sqrt{5}\right)} \tag{11}$$

$$V_5 = \frac{5}{12}\left(11 + 5\sqrt{5}\right) \tag{12}$$

$$s_6 = \sqrt{\frac{1}{2}\left(\sqrt{3} - 1\right)}, \ \sqrt{\frac{1}{2}\left(5 + 3\sqrt{3}\right)} \tag{13}$$

$$h_6 = \frac{1}{4}\sqrt{38 + 22\sqrt{3}} \tag{14}$$

$$S_6 = 6\sqrt{24 + 14\sqrt{3}} \tag{15}$$

$$V_6 = 7\sqrt{2} + 4\sqrt{6} \tag{16}$$

$$s_8 = \sqrt{1 - \sqrt{2} + \sqrt{2 - \sqrt{2}}},$$
$$\sqrt{\frac{1}{2}\left[8 + 5\sqrt{2} + \sqrt{2\left(58 + 41\sqrt{2}\right)}\right]} \tag{17}$$

$$h_8 = \frac{1}{2}\sqrt{\frac{1}{2}\left[30 + 20\sqrt{2} + \sqrt{2\left(850 + 601\sqrt{2}\right)}\right]} \tag{18}$$

$$S_8 = 4\sqrt{144 + 98\sqrt{2} + 4\sqrt{2516 + 1778\sqrt{2}}} \tag{19}$$

$$V_8 =$$
$$\frac{2}{3}\sqrt{2\left(1150 + 812\sqrt{2} + \sqrt{2641130 + 1867559\sqrt{2}}\right)}. \tag{20}$$

See also ANTIPRISM, CUBE, DIPYRAMID, DUAL POLY-HEDRON, HEXAGONAL SCALENOHEDRON, PENTAGONAL DELTAHEDRON, PRISM, TRAPEZOID

References

Cundy, H. and Rollett, A. *Mathematical Models, 3rd ed.* Stradbroke, England: Tarquin Pub., p. 117, 1989.
Pedagoguery Software. Poly. http://www.peda.com/poly/.

Trapezoid

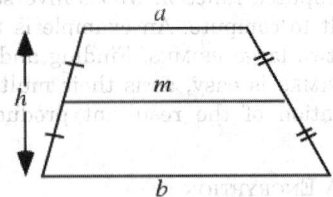

A QUADRILATERAL with two sides PARALLEL. The trapezoid is equivalent to the British definition of TRAPEZIUM. The trapezoid depicted has central median

$$m = \frac{1}{2}(a + b),$$

AREA

$$A = \frac{1}{2}(a + b)h = mh.$$

The CENTROID lies on the median m at a distance

$$x = \frac{b + 2a}{3(a + b)}h$$

from the vertical position of the lower left vertex.

See also ISOSCELES TRAPEZOID, PYRAMIDAL FRUSTUM, STROMBUS, TRAPEZIUM

References

Beyer, W. H. (Ed.). *CRC Standard Mathematical Tables, 28th ed.* Boca Raton, FL: CRC Press, p. 123, 1987.
Harris, J. W. and Stocker, H. "Trapezoid." §3.6.2 in *Handbook of Mathematics and Computational Science.* New York: Springer-Verlag, pp. 82–83, 1998.
Kern, W. F. and Bland, J. R. *Solid Mensuration with Proofs, 2nd ed.* New York: Wiley, p. 3, 1948.

Trapezoidal Hexecontahedron

DELTOIDAL HEXECONTAHEDRON

Trapezoidal Icositetrahedron

DELTOIDAL ICOSITETRAHEDRON

See also ONE-WAY FIXSTION, RSA ENCRYPTION, TRAPDOOR ONE-WAY HASH FUNCTION

Trapezoidal Rule

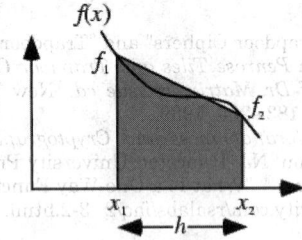

The 2-point NEWTON-COTES FORMULA

$$\int_{x_1}^{x_2} f(x)\,dx = \frac{1}{2}h(f_1 + f_2) - \frac{1}{12}h^3 f''(\xi),$$

where $f_i \equiv f(x_i)$, h is the separation between the points, and ξ is a point satisfying $x_1 \le \xi \le x_2$. Picking ξ to maximize $f''(\xi)$ gives an upper bound for the error in the trapezoidal approximation to the INTEGRAL.

See also BODE'S RULE, HARDY'S RULE, NEWTON-COTES FORMULAS, SIMPSON'S 3/8 RULE, SIMPSON'S RULE, WEDDLE'S RULE

References

Abramowitz, M. and Stegun, C. A. (Eds.). *Handbook of Mathematical Functions with Formulas, Graphs, and Mathematical Tables, 9th printing.* New York: Dover, p. 885, 1972.

Whittaker, E. T. and Robinson, G. "The Trapezoidal and Parabolic Rules." *The Calculus of Observations: A Treatise on Numerical Mathematics, 4th ed.* New York: Dover, pp. 156–158, 1967.

Traveler's Problem

HAMILTONIAN CIRCUIT

Traveling Salesman Constants

N.B. A detailed online essay by S. Finch was the starting point for this entry.

Let $L(n, d)$ be the smallest TOUR length for n points in a d-D HYPERCUBE. Then there exists a smallest constant $\alpha(d)$ such that for all optimal TOURS in the HYPERCUBE,

$$\limsup_{n \to \infty} \frac{L(n, d)}{n^{(d-1)/d}\sqrt{d}} \le \alpha(d), \tag{1}$$

and a constant $\beta(d)$ such that for *almost* all optimal tours in the HYPERCUBE,

$$\lim_{n \to \infty} \frac{L(n, d)}{n^{(d-1)/d}\sqrt{d}} = \beta(d). \tag{2}$$

These constants satisfy the inequalities

$$0.44194 < \gamma_2 = \frac{5}{16}\sqrt{2} \le \beta(2)$$

$$\le \delta < 0.6508 < 0.75983 < 3^{-1/4} \le \alpha(2)$$

$$\le \phi < 0.98398 \tag{3}$$

$$0.37313 < \gamma_3 \le \beta(3) \le 12^{1/6}6^{-1/2} < 0.61772 < 0.64805$$

$$< 2^{1/6}3^{-1/2} \le \alpha(3) \le 0.90422 \tag{4}$$

$$0.34207 < \gamma_4 \le \beta(4) \le 12^{1/8}6^{-1/2} < 0.55696$$

$$< 0.59460 < 2^{-3/4} \le \alpha(4) \le 0.8364 \tag{5}$$

(Fejes Tóth 1940, Verblunsky 1951, Few 1955, Beardwood *et al.* 1959), where

$$\gamma_d \equiv \frac{\Gamma\left(3 + \dfrac{1}{d}\right)\left[\Gamma\left(\frac{1}{2}d + 1\right)\right]^{1/d}}{2\sqrt{\pi}(d^{1/2} + d^{-1/2})} \tag{6}$$

$\Gamma(z)$ is the GAMMA FUNCTION, δ is an expression

involving STRUVE FUNCTIONS and NEUMANN FUNCTIONS,

$$\phi \equiv \frac{280(3 - \sqrt{3})}{840 - 280\sqrt{3} + 4\sqrt{5} - \sqrt{10}} \tag{7}$$

(Karloff 1989), and

$$\psi \equiv \frac{1}{2}3^{-2/3}(4 + \ln 3)^{2/3} \tag{8}$$

(Goddyn 1990). In the LIMIT $d \to \infty$,

$$0.24197 < \lim_{d \to \infty} \gamma_d = \frac{1}{\sqrt{2\pi e}} \le \liminf_{d \to \infty} \beta(d)$$

$$\le \limsup_{d \to \infty} \beta(d) \le \lim_{d \to \infty} 12^{1/(2d)}6^{-1/2} = \frac{1}{\sqrt{6}} < 0.40825 \tag{9}$$

and

$$0.24197 < \frac{1}{\sqrt{2\pi e}} \le \lim_{d \to \infty} \alpha(d) \le \frac{2(3 - \sqrt{3})\theta}{\sqrt{2\pi e}}$$

$$< 0.4502, \tag{10}$$

where

$$\frac{1}{2} \le \theta = \lim_{d \to \infty} [\theta(d)]^{1/d} \le 0.6602, \tag{11}$$

and $\theta(d)$ is the best SPHERE PACKING density in d-D space (Goddyn 1990, Moran 1984, Kabatyanskii and Levenshtein 1978). Steele and Snyder (1989) proved that the limit $\alpha(d)$ exists.

Now consider the constant

$$\kappa \equiv \lim_{n \to \infty} \frac{L(n, 2)}{\sqrt{n}} = \beta(2)\sqrt{2}, \tag{12}$$

so

$$\frac{5}{8} = \gamma_2\sqrt{2} \le \kappa \le \delta\sqrt{2} < 0.9204. \tag{13}$$

The best current estimate is $\kappa \approx 0.7124$.

A certain self-avoiding SPACE-FILLING CURVE is an optimal TOUR through a set of n points, where n can be arbitrarily large. It has length

$$\lambda \equiv \lim_{m \to \infty} \frac{L_m}{\sqrt{n_m}} = \frac{4(1 + 2\sqrt{2})\sqrt{51}}{153} = 0.7147827\ldots, \tag{14}$$

where L_m is the length of the curve at the mth iteration and n_m is the point-set size (Moscato and Norman).

References

Beardwood, J.; Halton, J. H.; and Hammersley, J. M. "The Shortest Path Through Many Points." *Proc. Cambridge Phil. Soc.* **55**, 299–327, 1959.

Chartrand, G. "The Salesman's Problem: An Introduction to Hamiltonian Graphs." §3.2 in *Introductory Graph Theory.* New York: Dover, pp. 67–76, 1985.

Fejes Tóth, L. "Über einen geometrischen Satz." *Math. Zeit.* **46**, 83–85, 1940.

Few, L. "The Shortest Path and the Shortest Road Through *n* Points." *Mathematika* **2**, 141–144, 1955.

Finch, S. "Favorite Mathematical Constants." http://www.mathsoft.com/asolve/constant/sales/sales.html.

Flood, M. "The Travelling Salesman Problem." *Operations Res.* **4**, 61–75, 1956.

Friedman, E. "Longest Travelling Salesman Cycles." http://www.stetson.edu/~efriedma/tsp/.

Goddyn, L. A. "Quantizers and the Worst Case Euclidean Traveling Salesman Problem." *J. Combin. Th. Ser. B* **50**, 65–81, 1990.

Kabatyanskii, G. A. and Levenshtein, V. I. "Bounds for Packing on a Sphere and in Space." *Problems Inform. Transm.* **14**, 1–17, 1978.

Karloff, H. J. "How Long Can a Euclidean Traveling Salesman Tour Be?" *SIAM J. Disc. Math.* **2**, 91–99, 1989.

Moran, S. "On the Length of Optimal TSP Circuits in Sets of Bounded Diameter." *J. Combin. Th. Ser. B* **37**, 113–141, 1984.

Moscato, P. "Fractal Instances of the Traveling Salesman Constant." http://www.ing.unlp.edu.ar/cetad/mos/FRACTAL_TSP_home.html

Steele, J. M. and Snyder, T. L. "Worst-Case Growth Rates of Some Classical Problems of Combinatorial Optimization." *SIAM J. Comput.* **18**, 278–287, 1989.

Verblunsky, S. "On the Shortest Path Through a Number of Points." *Proc. Amer. Math. Soc.* **2**, 904–913, 1951.

Kruskal, J. B. "On the Shortest Spanning Subtree of a Graph and the Traveling Salesman Problem." *Proc. Amer. Math. Soc.* **7**, 48–50, 1956.

Lawler, E.; Lenstra, J.; Rinnooy Kan, A.; and Shmoys, D. *The Traveling Salesman Problem: A Guided Tour of Combinatorial Optimization.* New York: Wiley, 1985.

Lin, S. "Computer Solutions of the Traveling Salesman Problem." *Bell System Tech. J.* **44**, 2245–2269, 1965.

Platzman, L. K. and Bartholdi, J. J. "Spacefilling Curves and the Planar Travelling Salesman Problem." *J. Assoc. Comput. Mach.* **46**, 719–737, 1989.

Reinelt, G. "TSPLIB--A Traveling Salesman Problem Library." *ORSA J. Comput.* **3**, 376–384, 1991.

Rosenkrantz, D. J.; Stearns, R. E.; and Lewis, P. M. "An Analysis of Several Heuristics for the Traveling Salesman Problem." *SIAM J. Comput.* **6**, 563–581, 1977.

Skiena, S. "Traveling Salesman Tours." §5.3.5 in *Implementing Discrete Mathematics: Combinatorics and Graph Theory with Mathematica.* Reading, MA: Addison-Wesley, pp. 199–202, 1990.

Skiena, S. S. "Traveling Salesman Problem." §8.5.4 in *The Algorithm Design Manual.* New York: Springer-Verlag, pp. 319–322, 1997.

Steinhaus, H. *Mathematical Snapshots, 3rd ed.* New York: Dover, pp. 120–121, 1999.

Traveling Salesman Problem

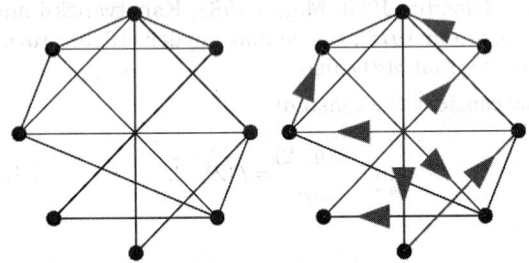

A problem in GRAPH THEORY requiring the most efficient (i.e., least total distance) HAMILTONIAN CIRCUIT a salesman can take through each of *n* cities. No general method of solution is known, and the problem is NP-HARD. Solution to the traveling salesman problem is implemented in *Mathematica* as `TravelingSalesman[g]` in the *Mathematica* add-on package `DiscreteMath`Combinatorica`` (which can be loaded with the command `< <DiscreteMath`).

See also CHINESE POSTMAN PROBLEM, DENDRITE, HAMILTONIAN CIRCUIT, PLATEAU'S PROBLEM, TRAVELING SALESMAN CONSTANTS

References

Applegate, D.; Bixby, R.; Chvatal, V.; and Cook, W. "Finding Cuts in the TSP (a Preliminary Report)." Technical Report 95–05, DIMACS. Piscataway NJ: Rutgers University, 1995.

Hoffman, P. *The Man Who Loved Only Numbers: The Story of Paul Erdos and the Search for Mathematical Truth.* New York: Hyperion, pp. 168–169, 1998.

Trawler Problem

A fast boat is overtaking a slower one when fog suddenly sets in. At this point, the boat being pursued changes course, but not speed. How should the pursuing vessel proceed in order to be sure of catching the other boat?

The amazing answer is that the pursuing boat should continue to the point where the slow boat would be if it had set its course directly for the pursuing boat when the fog set in. If the boat is not there, it should proceed in a SPIRAL whose origin is the point where the slow boat was when the fog set in. The SPIRAL can be constructed in such a way that the two boats will intersect before a complete turn is made.

References

Ogilvy, C. S. *Excursions in Mathematics.* New York: Dover, pp. 84 and 148, 1994.

Trebly Magic Square

TRIMAGIC SQUARE

Tredecillion

In the American system, 10^{42}.

See also LARGE NUMBER

Tree

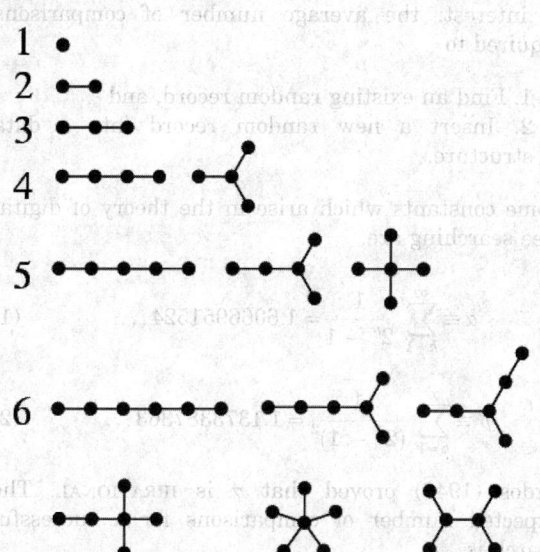

A tree is a mathematical structure which can be viewed as either a GRAPH or as a DATA STRUCTURE. The two views are equivalent, since a tree DATA STRUCTURE contains not only a set of elements, but also connections between elements, giving a tree graph. Trees were first studied by Cayley (1857).

A tree graph is a set of straight line segments connected at their ends containing no closed loops (cycles). In other words, it is a simple, undirected, connected, acyclic graph (or, equivalently, a connected FOREST). A tree with n nodes has $n-1$ EDGES. Conversely, a CONNECTED GRAPH with n nodes and $n-1$ edges is a tree. All trees are BIPARTITE GRAPHS (Skiena 1990, p. 213).

The points of connection are known as FORKS and the segments as BRANCHES. Final segments and the nodes at their ends are called LEAVES. A tree with two BRANCHES at each FORK and with one or two LEAVES at the end of each branch is called a BINARY TREE.

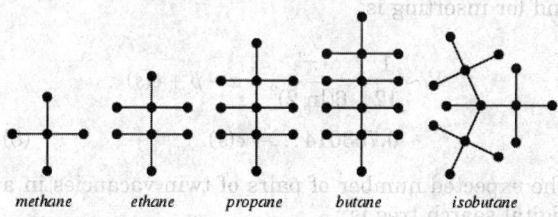

methane   ethane   propane   butane   isobutane

Trees find applications in many diverse fields, including computer science, the enumeration of saturated hydrocarbons, the study of electrical circuits, etc. (Harary 1994, p. 4).

A tree T has either one node which is a GRAPH CENTER, in which case it is called a CENTRAL TREE, or two adjacent nodes which are GRAPH CENTERS, in which case it is called a BICENTRAL TREE (Harary 1994, p. 35).

When a special node is designated to turn a tree into a ROOTED TREE, it is called the ROOT (or sometimes "EVE.") In such a tree, each of the nodes which is one EDGE further away from a given node is called a CHILD, and nodes connected to the same node which are the same distance from the ROOT NODE are called SIBLINGS.

Note that two BRANCHES placed end-to-end are equivalent to a single BRANCH which means, for example, that there is only *one* tree of order 3. The number $t(n)$ of nonisomorphic trees of order $n = 1, 2, \ldots$ (where trees of orders 1, 2, ..., 6 are illustrated above), are 1, 1, 1, 2, 3, 6, 11, 23, 47, 106, 235, ... (Sloane's A000055).

Otter showed that

$$\lim_{n \to \infty} \frac{t(n)n^{5/2}}{\alpha^n} = \beta, \tag{1}$$

(Otter 1948, Harary and Palmer 1973, Knuth 1969). Write the GENERATING FUNCTION for ROOTED TREES as

$$f(z) = \sum_{i=0}^{\infty} f_i z^i, \tag{2}$$

where the COEFFICIENTS are

$$f_{i+1} = \frac{1}{i} \sum_{j=1}^{i} \left(\sum_{d|j} d f_d \right) f_{i-j+1}, \tag{3}$$

with $f_0 = 0$ and $f_1 = 1$. Then

$$\alpha = 2.955765\ldots \tag{4}$$

is the unique POSITIVE ROOT of

$$f\left(\frac{1}{x}\right) = 1, \tag{5}$$

and

$$\beta = \frac{1}{\sqrt{2\pi}} \left[1 + \sum_{k=2}^{\infty} f'\left(\frac{1}{\alpha_k}\right) \frac{1}{\alpha_k} \right]^{3/2} = 0.5349485\ldots \tag{6}$$

See also B-TREE, BICENTRAL TREE, BINARY TREE, CATERPILLAR GRAPH, CAYLEY TREE, CENTRAL TREE, CHILD, DIJKSTRA TREE, EVE, FOREST, FREE TREE, KRUSKAL'S ALGORITHM, KRUSKAL'S TREE THEOREM, LABELED TREE, LEAF (TREE), MATRIX TREE THEOREM, ORCHARD-PLANTING PROBLEM, ORDERED TREE, OTTER'S THEOREM, PATH GRAPH, PLANTED PLANAR TREE, PÓLYA ENUMERATION THEOREM, POLYNEMA, QUADTREE, RAMUS TREE, RED-BLACK TREE, ROOT NODE, ROOTED TREE, SERIES-REDUCED TREE, SIBLING, SPANNING TREE, STAR GRAPH, STEINER TREE, STERN-BROCOT TREE, WEAKLY BINARY TREE, WEIGHTED TREE

References

Finch, S. "Favorite Mathematical Constants." http://www.mathsoft.com/asolve/constant/otter/otter.html.

Bergeron, F.; Leroux, P.; and Labelle, G. *Combinatorial Species and Tree-Like Structures.* Cambridge, England: Cambridge University Press, p. 284, 1998.

Cayley, A. "On the Theory of Analytic Forms Called Trees." *Philos. Mag.* **13**, 19–30, 1857. Reprinted in *Mathematical Papers, Vol. 3.* Cambridge: pp. 242–246, 1891.

Chauvin, B.; Cohen, S.; and Rouault, A. (Eds.). *Trees: Workshop in Versailles, June 14–16, 1995.* Basel, Switzerland: Birkhäuser, 1996.

Gardner, M. "Trees." Ch. 17 in *Mathematical Magic Show: More Puzzles, Games, Diversions, Illusions and Other Mathematical Sleight-of-Mind from Scientific American.* New York: Vintage, pp. 240–250, 1978.

Graham, R. L.; Knuth, D. E.; and Patashnik, O. *Concrete Mathematics: A Foundation for Computer Science, 2nd ed.* Reading, MA: Addison-Wesley, 1994.

Harary, F. "Trees." Ch. 4 in *Graph Theory.* Reading, MA: Addison-Wesley, pp. 32–42, 187–194, and 231–234, 1994.

Harary, F. and Manvel, B. "Trees." *Scripta Math.* **28**, 327–333, 1970.

Harary, F. and Palmer, E. M. *Graphical Enumeration.* New York: Academic Press, 1973.

Knuth, D. E. *The Art of Computer Programming, Vol. 1: Fundamental Algorithms, 3rd ed.* Reading, MA: Addison-Wesley, 1997.

König, D. *Theorie der endlichen und unendlichen Graphen.* New York: Chelsea, p. 48, 1950.

Nijenhuis, A. and Wilf, H. *Combinatorial Algorithms for Computers and Calculators, 2nd ed.* New York: Academic Press, 1978.

Otter, R. "The Number of Trees." *Ann. Math.* **49**, 583–599, 1948.

Skiena, S. "Trees." *Implementing Discrete Mathematics: Combinatorics and Graph Theory with Mathematica.* Reading, MA: Addison-Wesley, pp. 107 and 151–153, 1990.

Sloane, N. J. A. Sequences A000055/M0791 in "An On-Line Version of the Encyclopedia of Integer Sequences." http://www.research.att.com/~njas/sequences/eisonline.html.

Sloane, N. J. A. and Plouffe, S. Figure M0791 in *The Encyclopedia of Integer Sequences.* San Diego: Academic Press, 1995.

Wilf, H. S. *Combinatorial Algorithms: An Update.* Philadelphia, PA: SIAM, 1989.

Tree Centroid

The set of all CENTROID POINTS in a WEIGHTED TREE (Harary 1994, p. 36).

See also CENTROID POINT, WEIGHTED TREE

References

Harary, F. *Graph Theory.* Reading, MA: Addison-Wesley, 1994.

Tree Searching

N.B. A detailed online essay by S. Finch was the starting point for this entry.

In database structures, two quantities are generally of interest: the average number of comparisons required to

1. Find an existing random record, and
2. Insert a new random record into a data structure.

Some constants which arise in the theory of digital tree searching are

$$\alpha \equiv \sum_{k=1}^{\infty} \frac{1}{2^k - 1} = 1.6066951524\ldots \tag{1}$$

$$\beta \equiv \sum_{k=1}^{\infty} \frac{1}{(2^n - 1)^2} = 1.1373387363\ldots \tag{2}$$

Erdos (1948) proved that α is IRRATIONAL. The expected number of comparisons for a successful search is

$$E = \frac{\ln n}{\ln 2} + \frac{\gamma - 1}{\ln 2} - \alpha + \frac{3}{2} + \delta(n) + \mathcal{O}(n^{-1/2}) \tag{3}$$

$$\sim \lg n - 0.716644\ldots + \delta(n), \tag{4}$$

and for an unsuccessful search is

$$E = \frac{\ln n}{\ln 2} + \frac{\gamma}{\ln 2} - \alpha + \frac{1}{2} + \delta(n) + \mathcal{O}(n^{-1/2}) \tag{5}$$

$$\sim \lg n - 0.273948\ldots + \delta(n), \tag{6}$$

Here $\delta(n)$, $\epsilon(s)$, and $\rho(n)$ are small-amplitude periodic functions, and LG is the base 2 LOGARITHM. The VARIANCE for searching is

$$V \sim \frac{1}{12} + \frac{\pi^2 + 6}{6(\ln 2)^2} - \alpha - \beta + \epsilon(s)$$

$$\sim 2.844383\ldots + \epsilon(s) \tag{7}$$

and for inserting is

$$V \sim \frac{1}{12} + \frac{\pi^2}{6(\ln 2)^2} - \alpha - \beta + \epsilon(s)$$

$$\sim 0.763014\ldots + \epsilon(s). \tag{8}$$

The expected number of pairs of twin vacancies in a digital search tree is

$$\langle A_n \rangle = \left[\theta + 1 - \frac{1}{Q} \left(\frac{1}{\ln 2} + \alpha^2 - \alpha \right) + \rho(n) \right] n + \mathcal{O}(\sqrt{n}), \tag{9}$$

where

$$Q \equiv \prod_{k=1}^{\infty} \left(1 - \frac{1}{2^k} \right) = 0.2887880950\ldots \tag{10}$$

$$= \frac{1}{3} - \frac{1}{3 \cdot 7} + \frac{1}{3 \cdot 5 \cdot 15} - \frac{1}{3 \cdot 5 \cdot 15 \cdot 21} + \ldots \quad (11)$$

$$= \exp\left[-\sum_{n=1}^{\infty} \frac{1}{n(2^n - 1)}\right] \quad (12)$$

$$= \sqrt{\frac{2\pi}{\ln 2}} \exp\left(\frac{\ln 2}{24} - \frac{\pi^2}{6 \ln 2}\right) \prod_{n=1}^{\infty}\left[1 - \exp\left(-\frac{4\pi^2 n}{\ln 2}\right)\right] \quad (13)$$

and

$$\theta = \sum_{k=1}^{\infty} \frac{k 2^{k+1}}{1 \cdot 3 \cdot 7 \cdot 16 \cdots (2^k - 1)} \sum_{j=1}^{k} \frac{1}{2^j - 1}$$

$$= 7.7431319855\ldots \quad (14)$$

(Flajolet and Sedgewick 1986). The linear COEFFICIENT of $\langle A_n \rangle$ fluctuates around

$$c = \theta + 1 - \frac{1}{Q}\left(\frac{1}{\ln 2} + \alpha^2 - \alpha\right) = 0.3720486812\ldots, \quad (15)$$

which can also be written

$$c = \frac{1}{\ln 2}$$

$$\times \int_0^{\infty} \frac{x}{1+x} \frac{dx}{(1+x)\left(1+\frac{1}{2}x\right)\left(1+\frac{1}{4}x\right)\left(1+\frac{1}{8}x\right)\cdots}. \quad (16)$$

(Flajolet and Richmond 1992).

References

Finch, S. "Favorite Mathematical Constants." http://www.mathsoft.com/asolve/constant/bin/bin.html.
Finch, S. "Favorite Mathematical Constants." http://www.mathsoft.com/asolve/constant/dig/dig.html.
Finch, S. "Favorite Mathematical Constants." http://www.mathsoft.com/asolve/constant/qdt/qdt.html.
Flajolet, P. and Richmond, B. "Generalized Digital Trees and their Difference-Differential Equations." *Random Structures and Algorithms* **3**, 305–320, 1992.
Flajolet, P. and Sedgewick, R. "Digital Search Trees Revisited." *SIAM Review* **15**, 748–767, 1986.
Knuth, D. E. *The Art of Computer Programming, Vol. 3: Sorting and Searching, 2nd ed.* Reading, MA: Addison-Wesley, pp. 21, 134, 156, 493–499, and 580, 1973.

Tree-Planting Problem

ORCHARD-PLANTING PROBLEM

Trefoil Curve

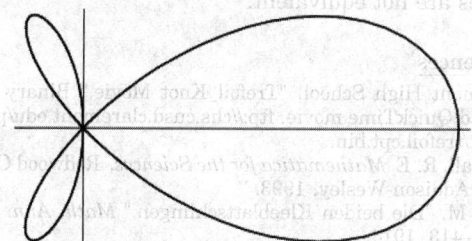

The plane curve given by the equation

$$x^4 + x^2 y^2 + y^4 = x(x^2 - y^2).$$

Trefoil Knot

The knot 03-001, also called the THREEFOIL KNOT, which is the unique PRIME KNOT of three crossings. It has BRAID WORD σ_1^3. The trefoil and its MIRROR IMAGE are not equivalent, as first proved by Dehn (1914). The trefoil has ALEXANDER POLYNOMIAL $-x^2 + x - 1$ and is a (3, 2)-TORUS KNOT. The BRACKET POLYNOMIAL can be computed as follows.

$$\langle L \rangle = A^3 d^{2-1} + A^2 B d^{1-1} + A^2 B d^{1-1} + AB^2 d^{2-1}$$

$$+ A^2 B d^{1-1} + AB^2 d^{2-1} + AB^2 d^{2-1} + B^3 d^{3-1}$$

$$= A^3 d^1 + 3A^2 B d^0 + 3AB^2 d^1 + B^3 d^2.$$

Plugging in

$$B = A^{-1}$$

$$d = -A^2 - A^{-2}$$

gives

$$\langle L \rangle = A^{-7} - A^{-3} - A^5.$$

The normalized one-variable KAUFFMAN POLYNOMIAL X is then given by

$$X_L = \left(-A^3\right)^{-w(L)} \langle L \rangle = \left(-A^3\right)^{-3}\left(A^{-7} - A^{-3} - A^5\right)$$

$$= A^{-4} + A^{-12} - A^{-16},$$

where the WRITHE $w(L) = 3$. The JONES POLYNOMIAL is therefore

$$V(t) = L\left(A = t^{-1/4}\right) = t + t^3 - t^4 = t\left(1 + t^2 - t^3\right).$$

Since $V(t^{-1}) \neq V(t)$, we have shown that the mirror images are not equivalent.

References

Claremont High School. "Trefoil_Knot Movie." Binary encoded QuickTime movie. ftp://chs.cusd.claremont.edu/pub/knot/trefoil.cpt.bin.

Crandall, R. E. *Mathematica for the Sciences.* Redwood City, CA: Addison-Wesley, 1993.

Dehn, M. "Die beiden Kleeblattschlingen." *Math. Ann.* **75**, 402–413, 1914.

Kauffman, L. H. *Knots and Physics.* Singapore: World Scientific, pp. 29–35, 1991.

Nordstrand, T. "Threefoil Knot." http://www.uib.no/people/nfytn/tknottxt.htm.

Pappas, T. "The Trefoil Knot." *The Joy of Mathematics.* San Carlos, CA: Wide World Publ./Tetra, p. 96, 1989.

Steinhaus, H. *Mathematical Snapshots, 3rd ed.* New York: Dover, p. 265, 1999.

Trench Diggers' Constant

BEAM DETECTOR

Triabolo

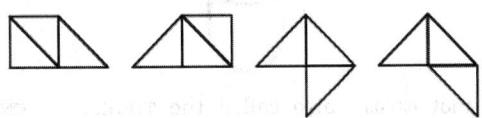

triaboloes

One of the four 3-POLYABOLOES.

See also POLYABOLO

Triacontagon

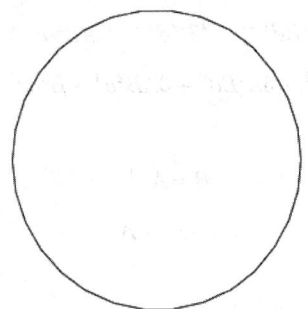

A 30-sided POLYGON. The regular triacontagon with side length 1 has INRADIUS r, CIRCUMRADIUS R, and AREA A given by

$$r = \frac{1}{4}\left(\sqrt{15} + 3\sqrt{3} + \sqrt{2}\sqrt{25 + 11\sqrt{5}}\right)$$

$$R = \frac{1}{2}\left(2 + \sqrt{5} + \sqrt{15 + 6\sqrt{5}}\right)$$

$$A = \frac{15}{4}\left(\sqrt{15} + 3\sqrt{3} + \sqrt{2}\sqrt{25 + 11\sqrt{5}}\right).$$

See also POLYGON, REGULAR POLYGON, TRIGONOMETRY VALUES PI/30

Triacontahedron

A 30-faced POLYHEDRON.

See also ICOSIDODECAHEDRON, MEDIAL DISDYAKIS TRIACONTAHEDRON, RHOMBIC TRIACONTAHEDRON

Triad

A SET with three elements.

See also HEXAD, MONAD, QUARTET, QUINTET, TETRAD

Triakis Icosahedron

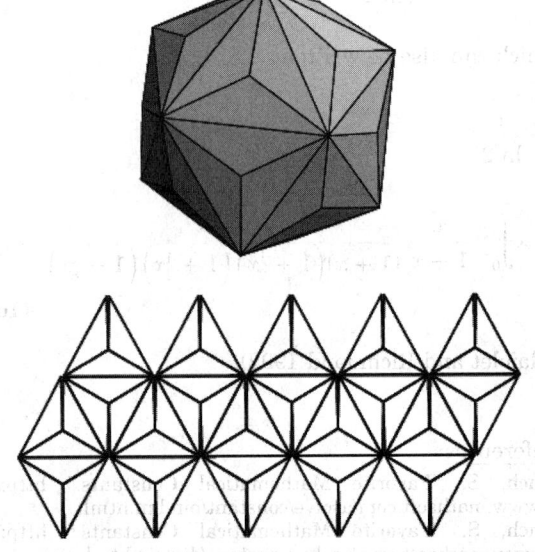

The 60-faced DUAL POLYHEDRON of the TRUNCATED DODECAHEDRON A_{10} and Wenninger dual W_{10}. Wenninger (1989, p. 46) calls the SMALL TRIAMBIC ICOSAHEDRON the triakis octahedron. Taking the dual of a TRUNCATED DODECAHEDRON with unit edge lengths gives a triakis icosahedron with edge lengths

$$s_1 = \frac{5}{22}\left(7 + \sqrt{5}\right) \tag{1}$$

$$s_2 = \frac{1}{2}\left(5 + 5\sqrt{5}\right). \tag{2}$$

The SURFACE AREA and VOLUME are

$$S = \frac{75}{11}\sqrt{\frac{1}{2}\left(313 + 117\sqrt{5}\right)} \tag{3}$$

$$V = \tfrac{125}{44}\left(19 + 9\sqrt{5}\right). \tag{4}$$

See also ARCHIMEDEAN DUAL, ARCHIMEDEAN SOLID, HEXECONTAHEDRON, SMALL TRIAMBIC ICOSAHEDRON

References

Wenninger, M. J. *Polyhedron Models.* New York: Cambridge University Press, p. 46, 1989.

Wenninger, M. J. *Dual Models.* Cambridge, England: Cambridge University Press, pp. 19–20, 1983.

Triakis Octahedron

GREAT TRIAKIS OCTAHEDRON, SMALL TRIAKIS OCTAHEDRON

Triakis Tetrahedron

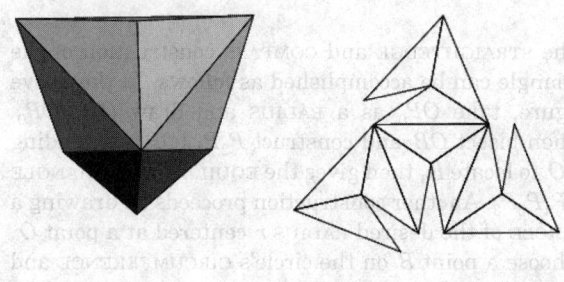

The DUAL POLYHEDRON of the TRUNCATED TETRAHEDRON A_{13} and Wenninger dual W_6. It can be constructed by CUMULATION of a unit edge-length TETRAHEDRON by a pyramid with height $\tfrac{1}{15}\sqrt{6}$.

The triakis tetrahedron formed by taking the dual of a truncated tetrahedron with unit edge lengths has side lengths

$$s_1 = \tfrac{9}{5} \tag{1}$$

$$s_2 = 3. \tag{2}$$

Normalizing so that $s_1 = 1$ gives SURFACE AREA and VOLUME

$$S = \tfrac{5}{3}\sqrt{11} \tag{3}$$

$$V = \tfrac{25}{36}\sqrt{2} \tag{4}$$

See also ARCHIMEDEAN DUAL, ARCHIMEDEAN SOLID, TRIAKIS TETRAHEDRON STELLATIONS, TRUNCATED TETRAHEDRON

References

Wenninger, M. J. *Dual Models.* Cambridge, England: Cambridge University Press, pp. 14–15 and 33, 1983.

Triakis Tetrahedron Stellations

B. Chilton and R. Whorf have studied stellations of the TRIAKIS TETRAHEDRON (Wenninger 1983, p. 36). Whorf has found 138 stellations, 44 of which are fully symmetric and 94 of which are enantiomorphs (Wenninger 1983, p. 36).

See also STELLATION, TRIAKIS TETRAHEDRON

References

Wenninger, M. J. *Dual Models.* Cambridge, England: Cambridge University Press, pp. 36–37, 1983.

Trial

In statistics, a trial is a single performance of well-defined experiment (Papoulis 1984, p. 25), such as the flipping of a COIN, the generation of a RANDOM NUMBER, the dropping of a ball down the apex of a triangular lattice and having it fall into a single bin at the bottom, etc.

See also BERNOULLI TRIAL, EVENT, EXPERIMENT, LEXIS TRIALS, OUTCOME, POISSON TRIALS

References

Papoulis, A. "Repeated Trials." Ch. 3 in *Probability, Random Variables, and Stochastic Processes, 2nd ed.* New York: McGraw-Hill, pp. 47–82, 1984.

Trial Division

A brute-force method of finding a DIVISOR of an INTEGER n by simply plugging in one or a set of INTEGERS and seeing if they DIVIDE n. Repeated application of trial division to obtain the complete PRIME FACTORIZATION of a number is called DIRECT SEARCH FACTORIZATION. An individual integer being tested is called a TRIAL DIVISOR.

See also DIRECT SEARCH FACTORIZATION, DIVISION, PRIME FACTORIZATION

Trial Divisor

An INTEGER n which is tested to see if it divides a given number.

See also TRIAL DIVISION

Triamond

triamond

The unique 3-POLYIAMOND, illustrated above.

See also POLYIAMOND, TRAPEZOID

Triangle

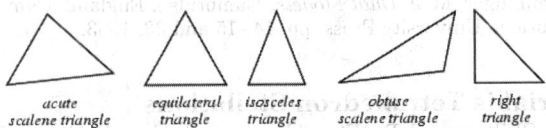

acute scalene triangle   *equilateral triangle*   *isosceles triangle*   *obtuse scalene triangle*   *right triangle*

A triangle is a 3-sided POLYGON sometimes (but not very commonly) called the TRIGON. All triangles are convex. An ACUTE TRIANGLE is a triangle whose three angles are all ACUTE. A triangle with all sides equal is called EQUILATERAL. A triangle with two sides equal is called ISOSCELES. A triangle having an OBTUSE ANGLE is called an OBTUSE TRIANGLE. A triangle with a RIGHT ANGLE is called RIGHT. A triangle with all sides a different length is called SCALENE.

In 1816, while studying the BROCARD POINTS of a triangle, Crelle exclaimed, "It is indeed wonderful that so simple a figure as the triangle is so inexhaustible in properties. How many as yet unknown properties of other figures may there not be?" (Wells 1991, p. 21).

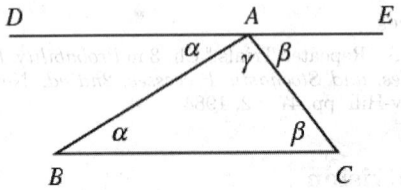

The sum of ANGLES in a triangle is $180° = \pi$ radians (at least in EUCLIDEAN GEOMETRY; this statement does *not* hold in NON-EUCLIDEAN GEOMETRY). This can be established as follows. Let $DAE \| BC$ (DAE be PARALLEL to BC) in the above diagram, then the angles α and β satisfy $\alpha = \angle DAB = \angle ABC$ and $\beta = \angle EAC = \angle ACB$, as indicated. Adding γ, it follows that

$$\alpha + \beta + \gamma = 180°, \tag{1}$$

since the sum of angles for the line segment must equal two RIGHT ANGLES. Therefore, the sum of angles in the triangle is also $180°$.

Let S stand for a triangle side and A for an angle, and let a set of Ss and As be concatenated such that adjacent letters correspond to adjacent sides and angles in a triangle. Triangles are uniquely determined by specifying three sides (SSS THEOREM), two angles and a side (AAS THEOREM), or two sides with an adjacent angle (SAS THEOREM). In each of these cases, the unknown three quantities (there are three sides and three angles total) can be uniquely deter-

mined. Other combinations of sides and angles do not uniquely determine a triangle: three angles specify a triangle only modulo a scale size (AAA THEOREM), and one angle and two sides not containing it may specify one, two, or no triangles (ASS THEOREM).

Allowable side lengths a, b, and c for a triangle are given by the set of inequalities $a > 0$, $b > 0$, $c > 0$, and $a+b > c$, $b+c > a$, $a+c > b$.

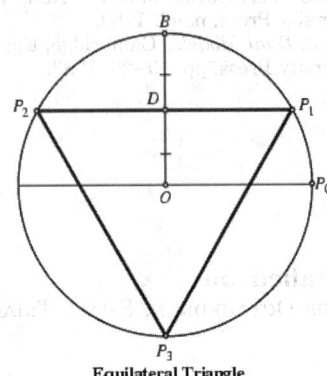

Equilateral Triangle

The STRAIGHTEDGE and COMPASS construction of the triangle can be accomplished as follows. In the above figure, take OP_0 as a RADIUS and draw $OB \perp OP_0$. Then bisect OB and construct $P_2P_1 \| OP_0$. Extending BO to locate P_3 then gives the EQUILATERAL TRIANGLE $\Delta P_1 P_2 P_3$. Another construction proceeds by drawing a CIRCLE of the desired RADIUS r centered at a point O. Choose a point B on the circle's CIRCUMFERENCE and draw another CIRCLE of radius r centered at B. The two circles intersect at two points, P_1 and P_2, and P_3 is the second point at which the line B_O intersects the first CIRCLE.

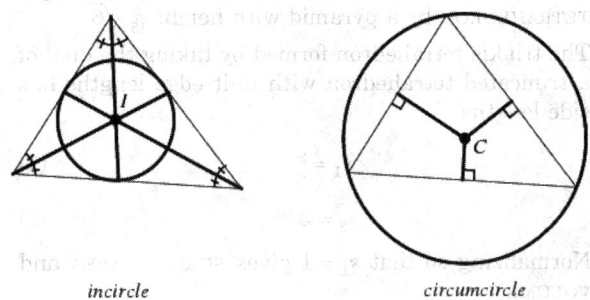

incircle   *circumcircle*

In Proposition IV.4 of the *ELEMENTS*, Euclid showed how to inscribe a CIRCLE (the INCIRCLE) in a given triangle by locating the INCENTER I as the point of intersection of ANGLE BISECTORS. In Proposition IV.5, he showed how to circumscribe a CIRCLE (the CIRCUMCIRCLE) about a given triangle by locating the CIRCUMCENTER O as the point of intersection of the PERPENDICULAR BISECTORS. unlike a general POLYGON with $n \geq 4$ sides, a triangle always has both a CIRCUMCIRCLE and an INCIRCLE. such polygons are called BICENTRIC POLYGONS.

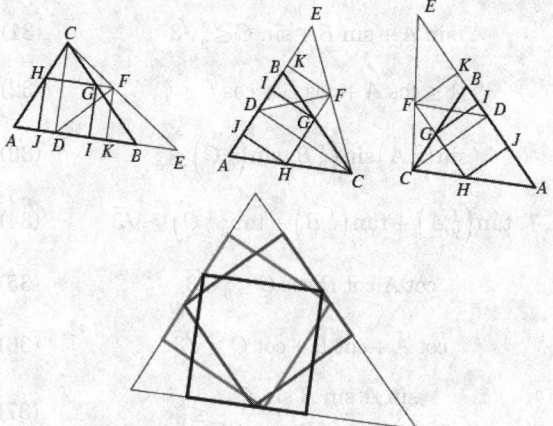

Casey (1888, pp. 10–11) illustrates how to inscribe a SQUARE in an arbitrary triangle $\triangle ABC$. Construct the PERPENDICULAR $CD \perp AB$ and the line segment $BE = AD$. Bisect $\angle BDC$, and let F be the intersection of the bisector with BC. Then draw FK and FH through F, perpendicular to and parallel to AB, respectively. Let G be the intersection of FH and BC, and then construct FK and HJ through F and H perpendicular to AB. Then $\square GHJI$ is an inscribed SQUARE. Permuting the order in which the vertices are taken gives an additional two congruent squares. These squares, however, are not necessarily the *largest* inscribed squares. CALABI'S TRIANGLE is the only triangle (besides the EQUILATERAL TRIANGLE) for which the largest inscribed SQUARE can be inscribed in three different ways.

If the coordinates of the triangle VERTICES are given by (x_i, y_j) where $i = 1\,2, 3$, then the *signed* AREA Δ is given by the DETERMINANT

$$\Delta = \frac{1}{2!} \begin{vmatrix} x_1 & y_1 & 1 \\ x_2 & y_2 & 1 \\ x_3 & y_3 & 1 \end{vmatrix}, \qquad (2)$$

so the actual area is obtained by taking the ABSOLUTE VALUE of (2). If the triangle is embedded in three-dimensional space with the coordinates of the VERTICES given by (x_i, x_j, z_i), where $i = 1, 2, 3$, then

$$\Delta = \frac{1}{2} \sqrt{ \begin{vmatrix} y_1 & z_1 & 1 \\ y_2 & z_2 & 1 \\ y_3 & z_3 & 1 \end{vmatrix}^2 + \begin{vmatrix} z_1 & x_1 & 1 \\ z_2 & x_2 & 1 \\ z_3 & x_3 & 1 \end{vmatrix}^2 + \begin{vmatrix} x_1 & y_1 & 1 \\ x_2 & y_2 & 1 \\ x_3 & y_3 & 1 \end{vmatrix}^2 }. \qquad (3)$$

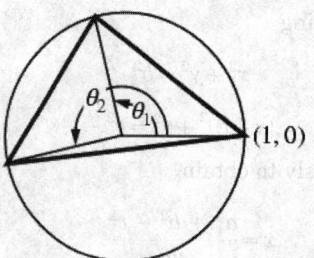

In the above figure, let the CIRCUMCIRCLE passing through a triangle's VERTICES have RADIUS r, and denote the CENTRAL ANGLES from the first point to the second θ_1, and to the third point by θ_2. Then the AREA of the triangle is given by

$$\Delta = 2r^2 \left| \sin\left(\tfrac{1}{2}\theta_1\right) \sin\left(\tfrac{1}{2}\theta_2\right) \sin\left[\tfrac{1}{2}(\theta_1 - \theta_2)\right] \right|. \qquad (4)$$

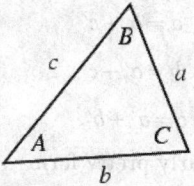

If a triangle has sides a, b, c, call the angles opposite these sides A, B, and C, respectively. Also define the SEMIPERIMETER s as HALF the PERIMETER:

$$s \equiv \tfrac{1}{2} p = \tfrac{1}{2}(a + b + c). \qquad (5)$$

The AREA of a triangle is then given by HERON'S FORMULA

$$\Delta = \sqrt{s(s - a)(s - b)(s - c)}, \qquad (6)$$

as well by the FORMULAS

$$\Delta = \tfrac{1}{4} \sqrt{(a + b + c)(b + c - a)(c + a - b)(a + b - c)} \qquad (7)$$

$$= \tfrac{1}{4} \sqrt{2(a^2 b^2 + a^2 c^2 + b^2 c^2) - (a^4 + b^4 + c^4)} \qquad (8)$$

$$= \tfrac{1}{4} \sqrt{\left[(a + b)^2 - c^2\right]\left[c^2 - (a - b)^2\right]} \qquad (9)$$

$$= \tfrac{1}{4} \sqrt{p(p - 2a)(p - 2b)(p - 2c)}, \qquad (10)$$

$$= 2R^2 \sin A \sin B \sin C \qquad (11)$$

$$= \frac{abc}{4R} = rs \qquad (12)$$

$$= \tfrac{1}{2} a h_a \qquad (13)$$

$$= \tfrac{1}{2} bc \sin A. \qquad (14)$$

In the above formulas, h_i is the ALTITUDE on side i, R is the CIRCUMRADIUS, and r is the INRADIUS (Johnson 1929, p. 11). A triangle with sides a, b, and c can be constructed by selecting vertices $(0, 0)$, $(a, 0)$, and $(x,$

y), then solving

$$x^2 + y^2 = b^2 \tag{15}$$

$$(x - a)^2 + y^2 = c^2 \tag{16}$$

simultaneously to obtain

$$x = \frac{a^2 + b^2 - c^2}{2a} \tag{17}$$

$$y = \pm \frac{\sqrt{(-a+b+c)(a-b-c)(a-b+c)(a+b+c)}}{2a}. \tag{18}$$

Expressing the side lengths *a*, *b*, and *c* in terms of the radii *a'*, *b'*, and *c'* of the mutually TANGENT CIRCLES centered on the TRIANGLE vertices (which define the SODDY CIRCLES),

$$a = b' + c' \tag{19}$$

$$b = a' + c' \tag{20}$$

$$c = a' + b', \tag{21}$$

gives the particularly pretty form

$$\Delta = \sqrt{a'b'c'(a' + b' + c')}. \tag{22}$$

For additional FORMULAS, see Beyer (1987) and Baker (1884), who gives 110 FORMULAS for the AREA of a triangle.

The ANGLES of a triangle satisfy

$$\cot A = \frac{b^2 + c^2 - a^2}{4\Delta} \tag{23}$$

where Δ is the AREA (Johnson 1929, p. 11, with missing squared symbol added). This gives the pretty identity

$$\cot A + \cot B + \cot C = \frac{a^2 + b^2 + c^2}{4\Delta}. \tag{24}$$

In addition,

$$\tan A + \tan B + \tan C = \tan A \tan B \tan C \tag{25}$$

(F.J. n.d., p. 206; Borchardt and Perrott 1930) and

$$\cot B \cot C + \cot C \cot A + \cot A \cot B = 1 \tag{26}$$

$$\tan A \cot B \cot C + \tan B \cot C \cot A$$
$$+ \tan C \cot A \cot B$$
$$= \tan A + \tan B + \tan C$$
$$+ 2(\cot A + \cot B + \cot C) \tag{27}$$

(Siddons and Hughes 1929).

Let a triangle have ANGLES *A*, *B*, and *C*. Then

$$\sin A \sin B \sin C \le kABC, \tag{28}$$

where

$$k = \left(\frac{3\sqrt{3}}{2\pi}\right)^3 \tag{29}$$

(Abi-Khuzam 1974, Le Lionnais 1983). This can be used to prove that

$$8\omega^3 < ABC, \tag{30}$$

where ω is the BROCARD ANGLE. Other inequalities include

$$\sin A + \sin B + \sin C \le \tfrac{3}{2}\sqrt{3} \tag{31}$$

$$1 \le \cos A + \cos B + \cos C \le \tfrac{3}{2} \tag{32}$$

$$\sin\left(\tfrac{1}{2}A\right)\sin\left(\tfrac{1}{2}B\right)\sin\left(\tfrac{1}{2}C\right) \le \tfrac{1}{8} \tag{33}$$

$$\tan\left(\tfrac{1}{2}A\right) + \tan\left(\tfrac{1}{2}B\right) + \tan\left(\tfrac{1}{2}C\right) \ge \sqrt{3} \tag{34}$$

$$\cot A \cot B \cot C \le \tfrac{1}{9}\sqrt{3} \tag{35}$$

$$\cot A + \cot B + \cot C \ge \sqrt{3} \tag{36}$$

$$\frac{\sin A \sin B \sin C}{\cot A + \cot B + \cot C} \le \tfrac{3}{8} \tag{37}$$

$$\frac{\tan\left(\tfrac{1}{2}A\right) + \tan\left(\tfrac{1}{2}B\right) + \tan\left(\tfrac{1}{2}C\right)}{\tan\left(\tfrac{1}{2}A\right)\tan\left(\tfrac{1}{2}B\right)\tan\left(\tfrac{1}{2}C\right)} \ge 9 \tag{38}$$

$$\cos\left(\tfrac{1}{2}A\right)\cos\left(\tfrac{1}{2}B\right)\cos\left(\tfrac{1}{2}C\right) \ge \sin A \sin B \sin C$$
$$\ge \sin(2A)\sin(2B)\sin(2C) \tag{39}$$

$$2 \le \cos^2\left(\tfrac{1}{2}A\right) + \cos^2\left(\tfrac{1}{2}B\right) + \cos^2\left(\tfrac{1}{2}C\right) \le \tfrac{9}{4} \tag{40}$$

$$\cot\left(\tfrac{1}{2}A\right)\cot\left(\tfrac{1}{2}B\right) + \cot\left(\tfrac{1}{2}B\right)\cot\left(\tfrac{1}{2}C\right)$$
$$+ \cot\left(\tfrac{1}{2}C\right)\cot\left(\tfrac{1}{2}A\right) \ge 9 \tag{41}$$

(Siddons and Hughes 1929, p. 283), and

$$\frac{\sin A + \sin B + \sin C}{\cot A + \cot B + \cot C} \le \frac{3}{2} \tag{42}$$

(Weisstein).

TRIGONOMETRIC FUNCTIONS of half angles can be expressed in terms of the triangle sides:

$$\cos\left(\tfrac{1}{2}A\right) = \sqrt{\frac{s(s-a)}{bc}} \tag{43}$$

$$\sin\left(\tfrac{1}{2}A\right) = \sqrt{\frac{(s-b)(s-c)}{bc}} \tag{44}$$

$$\tan\left(\tfrac{1}{2}A\right) = \sqrt{\frac{(s-b)(s-c)}{s(s-a)}}, \tag{45}$$

where *s* is the SEMIPERIMETER.

The number of different triangles which have INTEGRAL sides and PERIMETER n is

$$T(n) = P_3(n) - \sum_{1 \le i \le \lfloor n/2 \rfloor} P_2(j)$$

$$= \left[\frac{n^2}{12}\right] - \left\lfloor\frac{n}{4}\right\rfloor\left[\frac{n+2}{4}\right]$$

$$= \begin{cases} \left[\dfrac{n^2}{48}\right] & \text{for } n \text{ even} \\[2mm] \left[\dfrac{(n+3)^2}{48}\right] & \text{for } n \text{ odd,} \end{cases} \quad (46)$$

where P_2 and P_3 are PARTITION FUNCTIONS P, $[x]$ is the NINT function, and $\lfloor x \rfloor$ is the FLOOR FUNCTION (Jordan *et al.* 1979, Andrews 1979, Honsberger 1985). The values of $T(n)$ for $n = 1, 2, \dots$ are 0, 0, 1, 0, 1, 1, 2, 1, 3, 2, 4, 3, 5, 4, 7, 5, 8, 7, 10, 8, 12, 10, 14, 12, 16, … (Sloane's A005044), which is also ALCUIN'S SEQUENCE padded with two initial 0s. $T(n)$ also satisfies

$$T(2n) = T(2n - 3) = P_3(n). \quad (47)$$

It is not known if a triangle with INTEGER sides, MEDIANS, and AREA exists (although there are incorrect PROOFS of the impossibility in the literature). However, R. L. Rathbun, A. Kemnitz, and R. H. Buchholz have shown that there are infinitely many triangles with RATIONAL sides (HERONIAN TRIANGLES) with *two* RATIONAL MEDIANS (Guy 1994).

In the following paragraph, assume the specified sides and angles are adjacent to each other. Specifying three ANGLES does not uniquely define a triangle, but any two triangles with the same ANGLES are similar (the AAA THEOREM). Specifying two ANGLES A and B and a side a uniquely determines a triangle with AREA

$$\Delta = \frac{a^2 \sin B \sin C}{2 \sin A} = \frac{a^2 \sin B \sin(\pi - A - B)}{2 \sin A} \quad (48)$$

(the AAS THEOREM). Specifying an ANGLE A, a side c, and an ANGLE B uniquely specifies a triangle with AREA

$$\Delta = \frac{c^2}{2(\cot A + \cot B)} \quad (49)$$

(the ASA THEOREM). Given a triangle with two sides, a the smaller and c the larger, and one known ANGLE A, ACUTE and opposite a, if $\sin A < a/c$, there are two possible triangles. If $\sin A = a/c$, there is one possible triangle. If $\sin A > a/c$, there are no possible triangles. This is the ASS THEOREM. Let a be the base length and h be the height. Then

$$\Delta = \tfrac{1}{2}ah = \tfrac{1}{2}ac \sin B \quad (50)$$

(the SAS THEOREM). Finally, if all three sides are specified, a unique triangle is determined with AREA given by HERON'S FORMULA or by

$$\Delta = \frac{abc}{4R}, \quad (51)$$

where R is the CIRCUMRADIUS. This is the SSS THEOREM.

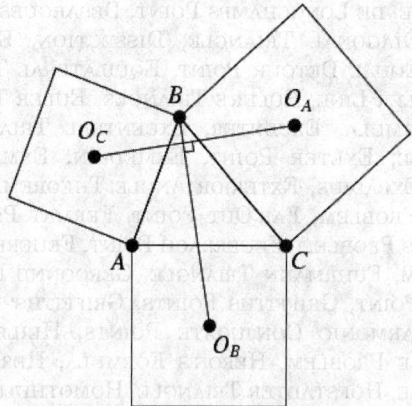

If squares are erected externally on the sides of a triangle as illustrated above, then $BO_B \perp O_C O_A$, and

$$\overline{BO_B} = \overline{O_C O_A} \quad (52)$$

(Coxeter and Greitzer 1967, pp. 96–97).

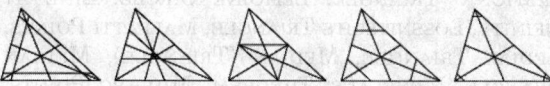

Dividing the sides of a triangle in a constant ratio $r < 1/2$ and then drawing lines parallel to the ad­jacent sides passing through each of these points gives line segments which intersect each other and one of the medians in three places. If $r > 1/2$, then the extensions of the side parallels intersect the extensions of the medians.

The medians bisect the area of a triangle, as do the side parallels with ratio $1 + \sqrt{2}$. The envelope of the lines which bisect the area a triangle forms three hyperbolic arcs. The envelope is somewhat more complicated, however, for lines dividing the area of a triangle into a constant but unequal ratio (Dunn and Petty 1972, Ball 1980, Wells 1991).

There are four CIRCLES which are tangent to the sides of a triangle, one internal and the rest external. Their centers are the points of intersection of the ANGLE BISECTORS of the triangle.

Any triangle can be positioned such that its shadow under an orthogonal projection is EQUILATERAL.

See also AAA THEOREM, AAS THEOREM, ACUTE TRIANGLE, ALCUIN'S SEQUENCE, ALTITUDE, ANGLE BISECTOR, ANTICEVIAN TRIANGLE, ANTICOMPLEMENTARY TRIANGLE, ANTIPEDAL TRIANGLE, ASS THEOREM, ASSOCIATED TRIANGLES, BELL TRIANGLE,

BRIANCHON POINT, BROCARD ANGLE, BROCARD CIR-
CLE, BROCARD MIDPOINT, BROCARD POINTS, BUTTER-
FLY THEOREM, CENTROID (TRIANGLE), CEVA'S
THEOREM, CEVIAN, CEVIAN TRIANGLE, CHASLES'S
THEOREM, CIRCULAR TRIANGLE, CIRCUMCENTER, CIR-
CUMCIRCLE, CIRCUMRADIUS, COMEDIAN TRIANGLES,
CONTACT TRIANGLE, COSYMMEDIAN TRIANGLES,
CROSSED LADDERS PROBLEM, CRUCIAL POINT, D-
TRIANGLE, DE LONGCHAMPS POINT, DESARGUES' THE-
OREM, DIAGONAL TRIANGLE, DISSECTION, ELKIES
POINT, EQUAL DETOUR POINT, EQUILATERAL TRIAN-
GLE, EULER LINE, EULER'S TRIANGLE, EULER TRIAN-
GLE FORMULA, EXCENTER, EXCENTRAL TRIANGLE,
EXCIRCLE, EXETER POINT, EXMEDIAN, EXMEDIAN
POINT, EXRADIUS, EXTERIOR ANGLE THEOREM, FAG-
NANO'S PROBLEM, FAR-OUT POINT, FERMAT POINTS,
FERMAT'S PROBLEM, FEUERBACH POINT, FEUERBACH'S
THEOREM, FUHRMANN TRIANGLE, GERGONNE POINT,
GREBE POINT, GRIFFITHS POINTS, GRIFFITHS' THEO-
REM, HARMONIC CONJUGATE POINTS, HEILBRONN
TRIANGLE PROBLEM, HERON'S FORMULA, HERONIAN
TRIANGLE, HOFSTADTER TRIANGLE, HOMOTHETIC TRI-
ANGLES, HEPTAGONAL TRIANGLE, INCENTER, INCIR-
CLE, INRADIUS, ISODYNAMIC POINTS, ISOGONAL
CONJUGATE, ISOPERIMETRIC POINT, ISOSCELES TRIAN-
GLE, KABON TRIANGLES, KANIZSA TRIANGLE, KIE-
PERT'S HYPERBOLA, KIEPERT'S PARABOLA, LAW OF
COSINES, LAW OF SINES, LAW OF TANGENTS, LEIBNIZ
HARMONIC TRIANGLE, LEMOINE CIRCLE, LINE AT
INFINITY, LOSSNITSCH'S TRIANGLE, MALFATTI POINTS,
MEDIAL TRIANGLE, MEDIAN (TRIANGLE), MEDIAN
TRIANGLE, MENELAUS' THEOREM, MID-ARC POINTS,
MITTENPUNKT, MOLLWEIDE'S FORMULAS, MORLEY
CENTERS, MORLEY'S THEOREM, NAGEL POINT, NAPO-
LEON'S THEOREM, NAPOLEON TRIANGLES, NEWTON'S
FORMULAS, NINE-POINT CIRCLE, NUMBER TRIANGLE,
OBTUSE TRIANGLE, ONO INEQUALITY, ORTHIC TRIAN-
GLE, ORTHOCENTER, ORTHOLOGIC TRIANGLES, PARA-
LOGIC TRIANGLES, PASCAL'S TRIANGLE, PASCH'S
AXIOM, PEDAL TRIANGLE, PERPENDICULAR BISECTOR,
PERSPECTIVE TRIANGLES, PETERSEN-SHOUTE THEO-
REM, PIVOT THEOREM, POWER POINT, POWER (TRIAN-
GLE), PRIME TRIANGLE, PURSER'S THEOREM,
QUADRILATERAL, RATIONAL TRIANGLE, ROUTH'S THE-
OREM, SAS THEOREM, SCALENE TRIANGLE, SCHIFFLER
POINT, SCHWARZ TRIANGLE, SCHWARZ'S TRIANGLE
PROBLEM, SEIDEL-ENTRINGER-ARNOLD TRIANGLE,
SEYDEWITZ'S THEOREM, SIMSON LINE, SPIEKER CEN-
TER, SSS THEOREM, STEINER-LEHMUS THEOREM,
STEINER POINTS, STEWART'S THEOREM, SYMMEDIAN
POINT, TANGENTIAL TRIANGLE, TARRY POINT, THOM-
SEN'S FIGURE, TORRICELLI POINT, TRIANGLE TILING,
TRIANGLE TRANSFORMATION PRINCIPLE, YFF CENTRAL
TRIANGLE, YFF POINTS, YFF TRIANGLES

References

Abi-Khuzam, F. "Proof of Yff's Conjecture on the Brocard
Angle of a Triangle." *Elem. Math.* **29**, 141–142, 1974.

Andrews, G. "A Note on Partitions and Triangles with
Integer Sides." *Amer. Math. Monthly* **86**, 477, 1979.

Baker, M. "A Collection of Formulæ for the Area of a Plane
Triangle." *Ann. Math.* **1**, 134–138, 1884.

Ball, D. "Halving Envelopes." *Math. Gaz.* **64**, 166–172, 1980.

Berkhan, G. and Meyer, W. F. "Neuere Dreiecksgeometrie."
In *Encyklopaedie der Mathematischen Wissenschaften*,
Vol. *3AB 10* (Ed. F. Klein). Leipzig: Teubner, pp. 1173–
1276, 1914.

Beyer, W. H. (Ed.). *CRC Standard Mathematical Tables*,
28th ed. Boca Raton, FL: CRC Press, pp. 123–124, 1987.

Borchardt, W. G. and Perrott, A. D. §133 in *A New Trigonometry for Schools.* London: G. Bell, 1930.

Casey, J. *A Sequel to the First Six Books of the Elements of
Euclid, Containing an Easy Introduction to Modern
Geometry with Numerous Examples, 5th ed., rev. enl.*
Dublin: Hodges, Figgis, & Co., 1888.

Coxeter, H. S. M. *Introduction to Geometry, 2nd ed.* New
York: Wiley, 1969.

Coxeter, H. S. M. and Greitzer, S. L. "Points and Lines
Connected with a Triangle." Ch. 1 in *Geometry Revisited.*
Washington, DC: Math. Assoc. Amer., pp. 1–26 and 96–
97, 1967.

Davis, P. "The Rise, Fall, and Possible Transfiguration of
Triangle Geometry: A Mini-History." *Amer. Math.
Monthly* **102**, 204–214, 1995.

Dunn, J. A. and Petty, J. E. "Halving a Triangle." *Math.
Gaz.* **56**, 105–108, 1972.

Durell, C. V. "Properties of the Triangle." Ch. 3 in *Modern
Geometry: The Straight Line and Circle.* London: Macmillan, pp. 19–31, 1928.

Eppstein, D. "Triangles and Simplices." http://www.ics.u-
ci.edu/~eppstein/junkyard/triangulation.html.

Feuerbach, K. W. *Eigenschaften einiger merkwürdigen
Punkte des geradlinigen Dreiecks, und mehrerer durch
die bestimmten Linien und Figuren.* Nürnberg, Germany,
1822.

F. J. *Elements de trigonometrie rectiligne.* Paris: J. de
Gigord, n.d.

Fukagawa, H. and Pedoe, D. "One or Two Circles and
Triangles," "Three Circles and Triangles," "Four Circles
and Triangle," "Five Circles and Triangles," "Many Circles
and Triangles," "Triangles." §2.2–2.6 and 4.1 in *Japanese
Temple Geometry Problems.* Winnipeg, Manitoba, Canada: Charles Babbage Research Foundation, pp. 26–37,
46–47, 102–116, 129–130, 1989.

Guy, R. K. "Triangles with Integer Sides, Medians, and
Area." §D21 in *Unsolved Problems in Number Theory, 2nd
ed.* New York: Springer-Verlag, pp. 188–190, 1994.

Honsberger, R. *Mathematical Gems III.* Washington, DC:
Math. Assoc. Amer., pp. 39–47, 1985.

Honsberger, R. "On Triangles." Ch. 3 in *Episodes in Nineteenth and Twentieth Century Euclidean Geometry.* Washington, DC: Math. Assoc. Amer., pp. 27–33, 1995.

Johnson, R. A. *Modern Geometry: An Elementary Treatise on
the Geometry of the Triangle and the Circle.* Boston, MA:
Houghton Mifflin, 1929.

Jordan, J. H.; Walch, R.; and Wisner, R. J. "Triangles with
Integer Sides." *Amer. Math. Monthly* **86**, 686–689, 1979.

Kimberling, C. "Central Points and Central Lines in the
Plane of a Triangle." *Math. Mag.* **67**, 163–187, 1994.

Kimberling, C. "Triangle Centers and Central Triangles."
Congr. Numer. **129**, 1–295, 1998.

Lachlan, R. "Properties of Triangles." Ch. 6 in *An Elementary Treatise on Modern Pure Geometry.* London: Macmillan, pp. 51–81, 1893.

Le Lionnais, F. *Les nombres remarquables.* Paris: Hermann,
p. 28, 1983.

Schroeder. *Das Dreieck und seine Beruhungskreise.*

Siddons, A. W. and Hughes, R. T. *Trigonometry, Parts III-
IV.* London: Cambridge University Press, 1929.

Sloane, N. J. A. Sequences A005044/M0146 in "An On-Line Version of the Encyclopedia of Integer Sequences." http://www.research.att.com/~njas/sequences/eisonline.html.

Vandeghen, A. "Some Remarks on the Isogonal and Cevian Transforms. Alignments of Remarkable Points of a Triangle." *Amer. Math. Monthly* **72**, 1091–1094, 1965.

Weisstein, E. W. "Plane Geometry." MATHEMATICA NOTEBOOK PlaneGeometry.M.

Wells, D. *The Penguin Dictionary of Curious and Interesting Geometry.* London: Penguin, p. 21, 1991.

Triangle Arcs

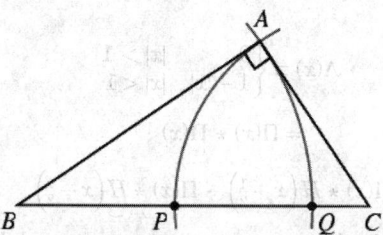

In the above figure, let ΔABC be a RIGHT TRIANGLE, arcs AP and AQ be segments of CIRCLES centered at C and B respectively, and define

$$a = BC \tag{1}$$

$$b = CA = CP \tag{2}$$

$$c = BA = BQ. \tag{3}$$

Then

$$PQ^2 = 2BP \cdot QC. \tag{4}$$

The figure also yields the algebraic identity

$$\left(b + c - \sqrt{b^2 + c^2}\right)^2$$
$$= 2\left(\sqrt{b^2 + c^2} - b\right)\left(\sqrt{b^2 + c^2} - c\right). \tag{5}$$

See also ARC, TRIANGLE

References

Berndt, B. C. *Ramanujan's Notebooks, Part IV.* New York: Springer-Verlag, pp. 8–9, 1994.

Dharmarajan, T. and Srinivasan, P. K. *An Introduction to Creativity of Ramanujan, Part III.* Madras: Assoc. Math. Teachers, pp. 11–13, 1987.

Triangle Center

A triangle center is a point whose TRILINEAR COORDINATES are defined in terms of the side lengths and angles of a TRIANGLE. The function giving the coordinates $\alpha : \beta : \gamma$ is called the TRIANGLE CENTER FUNCTION. The four ancient centers are the CENTROID, INCENTER, CIRCUMCENTER, and ORTHOCENTER. For a listing of these and other triangle centers, see Kimberling (1994).

A triangle center is said to be REGULAR IFF there is a TRIANGLE CENTER FUNCTION which is a POLYNOMIAL

in Δ, a, b, and c (where Δ is the AREA of the TRIANGLE) such that the TRILINEAR COORDINATES of the center are

$$f(a, b, c) : f(b, c, a) : f(c, a, b).$$

A triangle center is said to be a MAJOR TRIANGLE CENTER if the TRIANGLE CENTER FUNCTION α is a function of ANGLE A alone, and therefore β and γ of B and C alone, respectively.

See also MAJOR TRIANGLE CENTER, REGULAR TRIANGLE CENTER, TRIANGLE, TRIANGLE CENTER FUNCTION, TRILINEAR COORDINATES, TRILINEAR POLAR

References

Davis, P. J. "The Rise, Fall, and Possible Transfiguration of Triangle Geometry: A Mini-History." *Amer. Math. Monthly* **102**, 204–214, 1995.

Dixon, R. "The Eight Centres of a Triangle." §1.5 in *Mathographics.* New York: Dover, pp. 55–61, 1991.

Gale, D. "From Euclid to Descartes to Mathematica to Oblivion?" *Math. Intell.* **14**, 68–69, 1992.

Kimberling, C. "Central Points and Central Lines in the Plane of a Triangle." *Math. Mag.* **67**, 163–167, 1994.

Kimberling, C. "Triangle Centers and Central Triangles." *Congr. Numer.* **129**, 1–295, 1998.

Triangle Center Function

A HOMOGENEOUS FUNCTION $f(a, b, c)$, i.e., a function f such that

$$f(ta, tb, tc) = t^n f(a, b, c),$$

which gives the TRILINEAR COORDINATES of a TRIANGLE CENTER as

$$\alpha : \beta : \gamma = f(a, b, c) : f(b, c, a) : f(c, a, b).$$

The variables may correspond to angles (A, B, C) or side lengths (a, b, c), since these can be interconverted using the LAW OF COSINES.

See also MAJOR TRIANGLE CENTER, REGULAR TRIANGLE CENTER, TRIANGLE CENTER, TRILINEAR COORDINATES

References

Kimberling, C. "Triangle Centers as Functions." *Rocky Mtn. J. Math.* **23**, 1269–1286, 1993.

Kimberling, C. "Triangle Centers." http://cedar.evansville.edu/~ck6/tcenters/.

Kimberling, C. "Triangle Centers and Central Triangles." *Congr. Numer.* **129**, 1–295, 1998.

Lester, J. "Triangles III: Complex Triangle Functions." *Aequationes Math.* **53**, 4–35, 1997.

Triangle Coefficient

A function of three variables written $\Delta(abc) \equiv \Delta(a, b, c)$ and defined by

$$\Delta(abc) \equiv \sqrt{\frac{(a+b-c)!(a-b+c)!(-a+b+c)!}{(a+b+c+1)!}}.$$

References

Shore, B. W. and Menzel, D. H. *Principles of Atomic Spectra.* New York: Wiley, p. 273, 1968.

Triangle Condition

The condition that j takes on the values

$$j = j_1 + j_2, \ j_1 + j_2 - 1, \ \ldots, \ |j_1 - j_2|,$$

denoted $\Delta(j_1 j_2 j)$.

References

Sobelman, I. I. *Atomic Spectra and Radiative Transitions,* 2nd ed. Berlin: Springer-Verlag, p. 60, 1992.

Triangle Counting

Given rods of length 1, 2, ..., n, how many distinct triangles $T(n)$ can be made? Lengths for which

$$l_i = l_j + l_k$$

obviously do not give triangles, but all other combinations of three rods do. The answer is

$$T(n) = \begin{cases} \frac{1}{24} n(n-2)(2n-5) & \text{for } n \text{ even} \\ \frac{1}{24}(n-1)(n-3)(2n-1) & \text{for } n \text{ odd}. \end{cases}$$

The values for $n = 1, 2, \ldots$ are 0, 0, 0, 1, 3, 7, 13, 22, 34, 50, ... (Sloane's A002623). Somewhat surprisingly, this sequence is also given by the GENERATING FUNCTION

$$f(x) = \frac{x^4}{(1-x)^3(1-x^2)} = x^4 + 3x^5 + 7x^6 + 13x^7 + \ldots.$$

See also TRIANGLE TILING

References

Honsberger, R. *More Mathematical Morsels.* Washington, DC: Math. Assoc. Amer., pp. 278–282, 1991.
Sloane, N. J. A. Sequences A002623/M2640 in "An On-Line Version of the Encyclopedia of Integer Sequences." http://www.research.att.com/~njas/sequences/eisonline.html.

Triangle Cubic Curve

A CUBIC CURVE on which 37 notable triangle centers lie.

References

Wells, D. *The Penguin Dictionary of Curious and Interesting Geometry.* London: Penguin, pp. 42–43, 1991.

Triangle Function

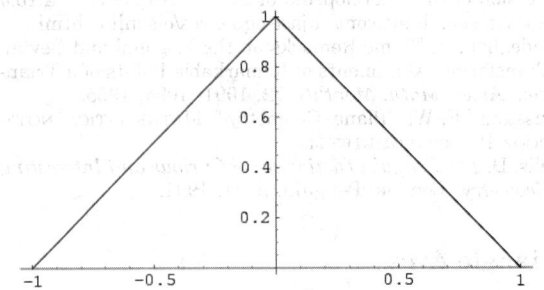

$$\Lambda(x) \equiv \begin{cases} 0 & |x| > 1 \\ 1 - |x| & |x| < 1 \end{cases} \tag{1}$$

$$= \Pi(x) * \Pi(x) \tag{2}$$

$$= \Pi(x) * H\left(x + \tfrac{1}{2}\right) - \Pi(x) * H\left(x - \tfrac{1}{2}\right), \tag{3}$$

where Π is the RECTANGLE FUNCTION and H is the HEAVISIDE STEP FUNCTION. An obvious generalization used as an APODIZATION FUNCTION goes by the name of the BARTLETT FUNCTION.

There is also a three-argument function known as the triangle function:

$$\lambda(x, \ y, \ z) \equiv x^2 + y^2 + z^2 - 2xy - 2xz - 2yz. \tag{4}$$

It follows that

$$\lambda(a^2, \ b^2, \ c^2) = (a+b+c)(a+b-c)(a-b+c)(a-b-c). \tag{5}$$

See also ABSOLUTE VALUE, BARTLETT FUNCTION, HEAVISIDE STEP FUNCTION, RAMP FUNCTION, RECTANGLE FUNCTION, SGN, TRIANGLE COEFFICIENT

References

Bracewell, R. "The Triangle Function of Unit Height and Area, $\Lambda(x)$." In *The Fourier Transform and Its Applications, 3rd ed.* New York: McGraw-Hill, p. 53, 1999.

Triangle Graph

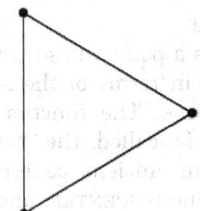

The CYCLE GRAPH C_3, which is also the COMPLETE GRAPH K_3.

See also COMPLETE GRAPH, CYCLE GRAPH

Triangle Inequality

References

Skiena, S. *Implementing Discrete Mathematics: Combinatorics and Graph Theory with Mathematica.* Reading, MA: Addison-Wesley, p. 144, 1990.

Triangle Inequality

Let $\mathbf{x}$ and $\mathbf{y}$ be vectors

$$|\mathbf{x}| - |\mathbf{y}| \leq |\mathbf{x} + \mathbf{y}| \leq |\mathbf{x}| + |\mathbf{y}|. \quad (1)$$

Equivalently, for COMPLEX NUMBERS z_1 and z_2,

$$|z_1| - |z_2| \leq |z_1 + z_2| \leq |z_1| + |z_2|. \quad (2)$$

A generalization is

$$\left| \sum_{k=1}^{n} a_k \right| \leq \sum_{k=1}^{n} |a_k|. \quad (3)$$

See also ONO INEQUALITY, P-ADIC NUMBER, STRONG TRIANGLE INEQUALITY

References

Abramowitz, M. and Stegun, C. A. (Eds.). *Handbook of Mathematical Functions with Formulas, Graphs, and Mathematical Tables, 9th printing.* New York: Dover, p. 11, 1972.
Apostol, T. M. *Calculus, 2nd ed., Vol. 1: One-Variable Calculus, with an Introduction to Linear Algebra.* Waltham, MA: Blaisdell, p. 42, 1967.
Krantz, S. G. *Handbook of Complex Analysis.* Boston, MA: Birkhäuser, p. 12, 1999.

Triangle Interior

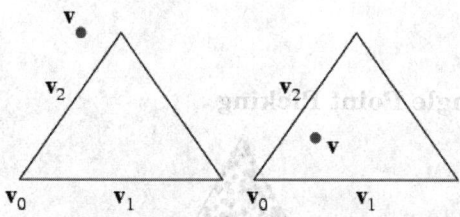

point outside triangle　　*point inside triangle*

To determine if a given point $\mathbf{v}$ lies in the interior of a given triangle, consider an individual vertex, denoted $\mathbf{v}_0$, and let $\mathbf{v}_1$ and $\mathbf{v}_2$ be the vectors from $\mathbf{v}_0$ to the other two vertices. Expressing the vector from $\mathbf{v}_0$ to $\mathbf{v}$ in terms of $\mathbf{v}_1$ and $\mathbf{v}_2$ then gives

$$\mathbf{v} = \mathbf{v}_0 + a\mathbf{v}_1 + b\mathbf{v}_2, \quad (1)$$

where a and b are constants. Solving for a and b gives

$$a = \frac{\det(\mathbf{v} \; \mathbf{v}_2) - \det(\mathbf{v}_0 \; \mathbf{v}_2)}{\det(\mathbf{v}_1 \; \mathbf{v}_2)} \quad (2)$$

$$b = -\frac{\det(\mathbf{v} \; \mathbf{v}_1) - \det(\mathbf{v}_0 \; \mathbf{v}_1)}{\det(\mathbf{v}_1 \; \mathbf{v}_2)}, \quad (3)$$

where

$$\det(\mathbf{u} \; \mathbf{v}) = \mathbf{u} \times \mathbf{v} = u_x v_y - u_y v_x \quad (4)$$

is the DETERMINANT of the matrix formed from the COLUMN VECTORS $\mathbf{u}$ and $\mathbf{v}$. The point $\mathbf{v}$ will be "inside" the angle formed at $\mathbf{v}_0$ if a, $b > 0$, and so will be in the interior of the triangle if the corresponding a, $b > 0$ *for each* of the three vertices.

More generally, a point $\mathbf{v}$ is in the interior of a TRIANGLE if the CONVEX HULL of the three vertices plus the point $\mathbf{v}$ contains three points instead of four. This means the point $\mathbf{v}$ is inside the CONVEX HULL of the triangle, which is just the triangle itself.

See also CONVEX HULL, TRIANGLE

Triangle of Figurate Numbers

FIGURATE NUMBER TRIANGLE

Triangle Packing

The best known packings of equilateral triangles into an equilateral triangle are illustrated above for the first few cases (Friedman).

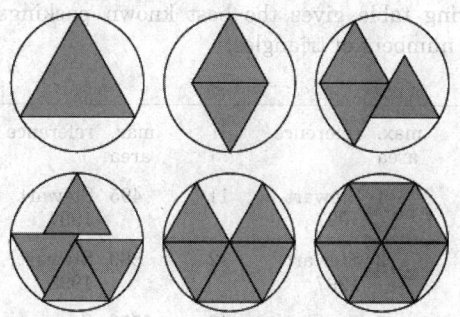

The best known packings of equilateral triangles into a circle are illustrated above for the first few cases (Friedman).

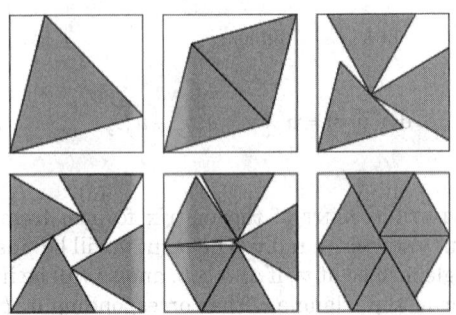

The best known packings of equilateral triangles into a square are illustrated above for the first few cases (Friedman).

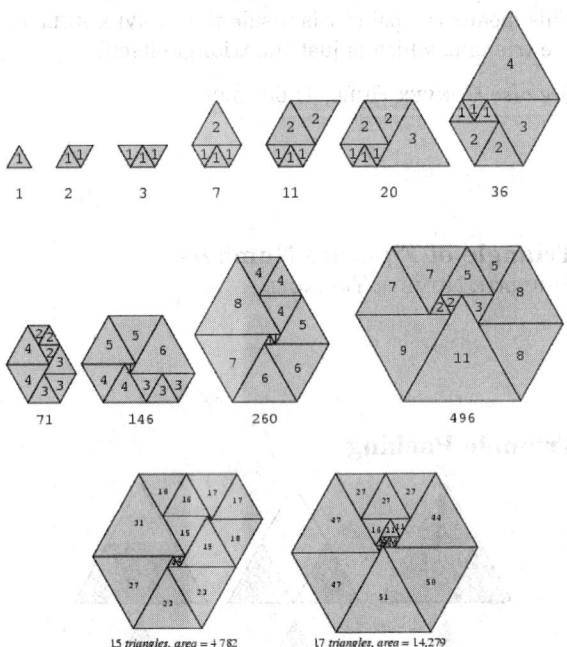

Stewart (1998, 1999) considered the problem of finding the largest convex area that can be nontrivially tiled with equilateral triangles whose sides are integers for a given number of triangles and which have no overall common divisor. There is no upper limit if an arbitrary number of triangles are used. The following table gives the best known packings for small numbers of triangles.

n	max. area	reference	n	max. area	reference
1	1	Stewart 1997	11	495	Stewart 1997
2	2	Stewart 1997	12	860	Stewart 1998
3	3	Stewart 1997	13	1559	Stewart 1998
4	7	Stewart 1997	14	2831	Stewart 1998
5	11	Stewart 1997	15	4782	Stewart 1999
6	20	Stewart 1997	16	8559	Stewart 1998
7	36	Stewart 1997	17	14279	Stewart 1998
8	71	Stewart 1997			
9	146	Stewart 1997			
10	260	Stewart 1997			

See also Circle Packing, Equilateral Triangle, Packing, Square Packing

References

Friedman, E. "Circles in Triangles." http://www.stetson.edu/~efriedma/cirintri/.

Friedman, E. "Squares in Triangles." http://www.stetson.edu/~efriedma/squintri/.

Friedman, E. "Triangles in Triangles." http://www.stetson.edu/~efriedma/triintri/.

Graham, R. L. and Lubachevsky, B. D. "Dense Packings of Equal Disks in an Equilateral Triangle: From 22 to 34 and Beyond." *Electronic J. Combinatorics* **2**, A1 1–39, 1995. http://www.combinatorics.org/Volume_2/volume2.html#A1.

Stewart, I. "Squaring the Square." *Sci. Amer.* **277**, 94–96, July 1997.

Stewart, I. "Mathematical Recreations: Monks, Blobs and Common Knowledge. Feedback." *Sci. Amer.* **279**, 97, Aug. 1998.

Stewart, I. "Mathematical Recreations: The Synchronicity of Firefly Flashing. Feedback." *Sci. Amer.* **280**, 106, Mar. 1999.

Triangle Point Picking

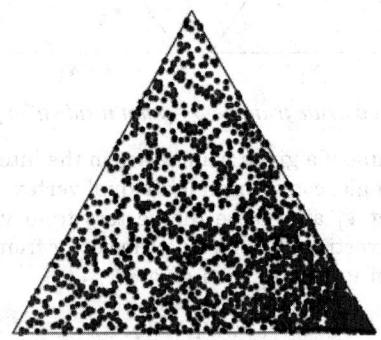

Given a triangle with one vertex at the origin and the others at positions $\mathbf{v}_1$ and $\mathbf{v}_2$, one might think that a random point inside the triangle would be given by

$$\mathbf{x} = a_1\mathbf{v}_1 + (1 - a_1)a_2\mathbf{v}_2,$$

where a_1 and a_2 are uniform variates in the interval $[0, 1]$. However, as can be seen in the plot above, this samples the triangle nonuniformly, concentrating points in the $\mathbf{v}_1$ corner.

To pick points uniformly distributed inside the triangle, instead pick

$$\mathbf{x} = a_1 \mathbf{v}_1 + a_2 \mathbf{v}_2,$$

where a_1 and a_2 are uniform variates in the interval $[0, 1]$, which gives points uniformly distributed in a QUADRILATERAL (left figure). The points not in the TRIANGLE INTERIOR can then either be discarded, or transformed into the corresponding point inside the triangle (right figure).

Picking n points independently and uniformly from a triangle with unit area gives a CONVEX HULL with expected area of

$$A(n) = 1 - \frac{2}{n+1} \sum_{k=1}^{n} \frac{1}{k} = 1 - \frac{2H_n}{n+1},$$

where H_n is a HARMONIC NUMBER (Buchta 1984, 1986). This is a special case of SIMPLEX POINT PICKING.

See also SIMPLEX POINT PICKING, TRIANGLE TRIANGLE PICKING

References
Buchta, C. "Zufallspolygone in konvexen Vielecken." *J. reine angew. Math.* **347**, 212–220, 1984.
Buchta, C. "A Note on the Volume of a Random Polytope in a Tetrahedron." *Ill. J. Math.* **30**, 653–659, 1986.

Triangle Postulate

The sum of the ANGLES of a TRIANGLE is two RIGHT ANGLES. This POSTULATE is equivalent to the PARALLEL AXIOM.

References
Dunham, W. "Hippocrates' Quadrature of the Lune." Ch. 1 in *Journey through Genius: The Great Theorems of Mathematics.* New York: Wiley, p. 54, 1990.

Triangle Squaring

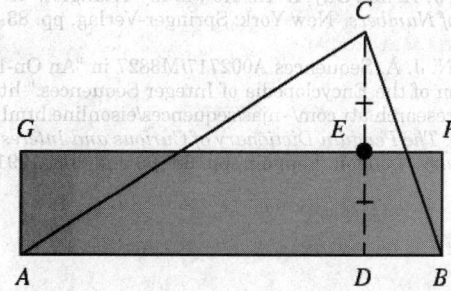

Let CD be the ALTITUDE of a TRIANGLE $\triangle ABC$ and let E be its MIDPOINT. Then

$$\text{area}(\triangle ABC) = \tfrac{1}{2} AB \cdot CD = AB \cdot DE,$$

and $\square ABFG$ can be SQUARED by RECTANGLE SQUARING. The general POLYGON can be treated by drawing diagonals, SQUARING the constituent TRIANGLES, and then combining the SQUARES together using the PYTHAGOREAN THEOREM.

See also PYTHAGOREAN THEOREM, RECTANGLE SQUARING, SQUARING

References
Dunham, W. "Hippocrates' Quadrature of the Lune." Ch. 1 in *Journey through Genius: The Great Theorems of Mathematics.* New York: Wiley, pp. 14–15, 1990.

Triangle Tiling

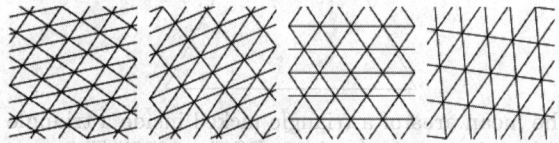

Any triangle tiles the plane (Wells 1991, p. 208).

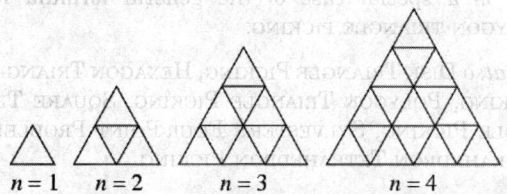

The total number of triangles (including inverted ones) in the above figures are given by

$$N(n) = \begin{cases} \tfrac{1}{8} n(n+2)(2n+1) & \text{for } n \text{ even} \\ \tfrac{1}{8}[n(n+2)(2n+1)-1] & \text{for } n \text{ odd}. \end{cases}$$

The first few values are 1, 5, 13, 27, 48, 78, 118, 170, 235, 315, 411, 525, 658, 812, 988, 1188, 1413, 1665, ... (Sloane's A002717).

See also EQUILATERAL TRIANGLE, RECTANGLE TILING, TRIANGLE COUNTING, TRIANGLE PACKING

References

Conway, J. H. and Guy, R. K. "How Many Triangles." In *The Book of Numbers.* New York: Springer-Verlag, pp. 83–84, 1996.

Sloane, N. J. A. Sequences A002717/M3827 in "An On-Line Version of the Encyclopedia of Integer Sequences." http://www.research.att.com/~njas/sequences/eisonline.html.

Wells, D. *The Penguin Dictionary of Curious and Interesting Geometry.* London: Penguin, pp. 68–69 and 208, 1991.

Triangle Transformation Principle

The triangle transformation principle gives rules for transforming equations involving an INCIRCLE to equations about EXCIRCLES.

See also EXCIRCLE, INCIRCLE

References

Johnson, R. A. *Modern Geometry: An Elementary Treatise on the Geometry of the Triangle and the Circle.* Boston, MA: Houghton Mifflin, pp. 191–192, 1929.

Triangle Triangle Picking

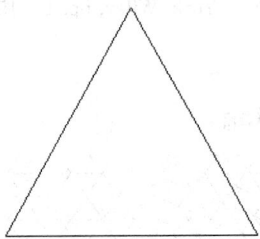

The mean area of a triangle picked inside a triangle with unit area is $\bar{A} = 1/12$ (Pfiefer 1989). This was proposed by Watson (1865) and solved by Sylvester, and is a special case of the general formula for POLYGON TRIANGLE PICKING.

See also DISK TRIANGLE PICKING, HEXAGON TRIANGLE PICKING, POLYGON TRIANGLE PICKING, SQUARE TRIANGLE PICKING, SYLVESTER'S FOUR-POINT PROBLEM, TETRAHEDRON TETRAHEDRON PICKING

References

Pfiefer, R. E. "The Historical Development of J. J. Sylvester's Four Point Problem." *Math. Mag.* **62**, 309–317, 1989.

Watson, S. "Question 1229." *Mathematical Questions, with Their Solutions, from the Educational Times, Vol. 4.* London: F. Hodgson and Son, p. 101, 1865.

Triangular Antiprism

See also Antiprism

Triangular Cupola

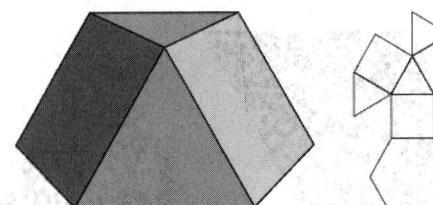

JOHNSON SOLID J_3. The bottom six VERTICES are

$$\left(\pm\tfrac{1}{2}\sqrt{3}, \pm\tfrac{1}{2},\, 0\right),\ (0, \pm1,\, 0),$$

and the top three VERTICES are

$$\left(\frac{1}{\sqrt{3}},\, 0,\, \sqrt{\frac{2}{3}}\right),\ -\left(\frac{1}{2\sqrt{3}}, \pm\frac{1}{2},\, \sqrt{\frac{2}{3}}\right).$$

See also JOHNSON SOLID

Triangular Dipyramid

The triangular (or TRIGONAL) dipyramid is one of the convex DELTAHEDRA, and JOHNSON SOLID J_{12}.

See also DELTAHEDRON, DIPYRAMID, HEXAHEDRON, JOHNSON SOLID, PENTAGONAL DIPYRAMID

Triangular Graph

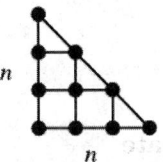

The triangular graph with n nodes on a side is denoted $T(n)$. Tutte (1970) showed that the CHROMATIC POLYNOMIALS of planar triangular graphs possess a ROOT close to $\phi^2 = 2.618033\ldots$, where ϕ is the GOLDEN MEAN. More precisely, if n is the number of VERTICES of G, then

$$P_G(\phi^2) \le \phi^{5-n}$$

(Le Lionnais 1983, p. 46). Every planar triangular graph possesses a VERTEX of degree 3, 4, or 5 (Le Lionnais 1983, pp. 49 and 53).

See also LATTICE GRAPH

References

Le Lionnais, F. *Les nombres remarquables.* Paris: Hermann, 1983.
Tutte, W. T. "On Chromatic Polynomials and the Golden Ratio." *J. Combin. Theory* **9**, 289–296, 1970.

Triangular Hebesphenorotunda

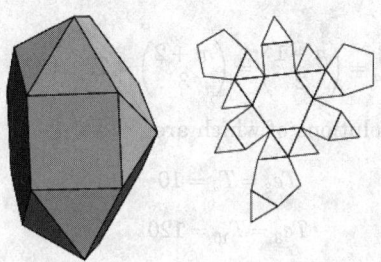

JOHNSON SOLID J_{92}.

References

Weisstein, E. W. "Johnson Solids." MATHEMATICA NOTEBOOK JOHNSONSOLIDS.M.
Weisstein, E. W. "Johnson Solid Netlib Database." MATHEMATICA NOTEBOOK JOHNSONSOLIDS.DAT.

Triangular Matrix

An UPPER TRIANGULAR MATRIX **U** is defined by

$$U_{ij} = \begin{cases} a_{ij} & \text{for } i \leq j \\ 0 & \text{for } i > j. \end{cases} \qquad (1)$$

Written explicitly,

$$\mathbf{U} = \begin{bmatrix} a_{11} & a_{12} & \cdots & a_{1n} \\ 0 & a_{22} & \cdots & a_{2n} \\ \vdots & \vdots & \ddots & \vdots \\ 0 & 0 & \cdots & a_{nn} \end{bmatrix}. \qquad (2)$$

A LOWER TRIANGULAR MATRIX **L** is defined by

$$L_{ij} = \begin{cases} a_{ij} & \text{for } i \geq j \\ 0 & \text{for } i < j. \end{cases} \qquad (3)$$

Written explicitly,

$$\mathbf{L} = \begin{bmatrix} a_{11} & 0 & \cdots & 0 \\ a_{21} & a_{22} & \cdots & 0 \\ \vdots & \vdots & \ddots & 0 \\ a_{n1} & a_{n2} & \cdots & a_{nn} \end{bmatrix}. \qquad (4)$$

See also HANKEL MATRIX, HESSENBERG MATRIX, HILBERT MATRIX, LOWER TRIANGULAR MATRIX, MA-TRIX, UPPER TRIANGULAR MATRIX, VANDERMONDE MATRIX

References

Ayres, F. Jr. *Theory and Problems of Matrices.* New York: Schaum, p. 10, 1962.

Triangular Number

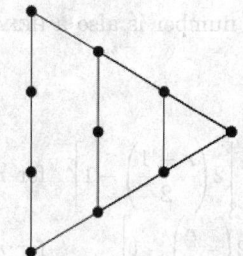

A FIGURATE NUMBER OF THE FORM

$$T_n \equiv \tfrac{1}{2} n(n+1) = \binom{n+1}{2}, \qquad (1)$$

where $\binom{n}{k}$ is a BINOMIAL COEFFICIENT, obtained by building up regular triangles out of dots. The first few triangle numbers are 1, 3, 6, 10, 15, 21, ... (Sloane's A000217). The odd triangular numbers are given by 1, 3, 15, 21, 45, 55, ... (Sloane's A014493), while the even triangular numbers are 6, 10, 28, 36, 66, 78, ... (Sloane's A014494).

$T_4 = 10$ gives the number and arrangement of BOWLING pins, while $T_5 = 15$ gives the number and arrangement of balls in BILLIARDS. Triangular numbers satisfy the RECURRENCE RELATION

$$T_{n+1}^2 - T_n^2 = (n+1)^3, \qquad (2)$$

as well as

$$3T_n + T_{n-1} = T_{2n} \qquad (3)$$

$$3T_n + T_{n+1} = T_{2n+1} \qquad (4)$$

$$1 + 3 + 5 + \ldots + (2n-1) = T_n + T_{n-1}. \qquad (5)$$

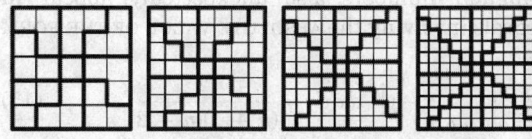

In addition, the triangle numbers can be related to the square numbers by

$$(2n+1)^2 = 8T + 1 = T_{n-1} + 6T_n + T_{n+1} \qquad (6)$$

(Conway and Guy 1996), as illustrated above (Wells 1991, p. 198). They have the ordinary GENERATING FUNCTION

$$f(x) = \frac{x}{(1-x)^3} = x + 3x^2 + 6x^3 + 10x^4 + 15x^5 + \ldots \qquad (7)$$

and EXPONENTIAL GENERATING FUNCTION

$$g(x) = \left(1 + 2x + \tfrac{1}{2}x^2\right)e^x$$

$$= 1 + 3x + 3x^2 + \tfrac{5}{3}x^3 + \tfrac{5}{8}x^4 + \cdots$$

$$= 1 + 3\,\frac{x}{1!} + 6\,\frac{x^2}{2!} + 10\,\frac{x^3}{3!} + 15\,\frac{x^4}{4!} + \cdots \quad (8)$$

(Sloane and Plouffe 1995, p. 9).

Every triangular number is also a HEXAGONAL NUMBER, since

$$\tfrac{1}{2}r(r+1)$$

$$= \begin{cases} \left(\dfrac{r+1}{2}\right)\left[2\left(\dfrac{r+1}{2}\right)-1\right] & \text{for } r \text{ odd} \\[2ex] \left(-\dfrac{r}{2}\right)\left[2\left(-\dfrac{r}{2}\right)-1\right] & \text{for } r \text{ even.} \end{cases} \quad (9)$$

Also, every PENTAGONAL NUMBER is 1/3 of a triangular number. The sum of consecutive triangular numbers is a SQUARE NUMBER, since

$$T_r + T_{r-1} = \tfrac{1}{2}r(r+1) + \tfrac{1}{2}(r-1)r$$

$$= \tfrac{1}{2}r[(r+1)+(r-1)] = r^2. \quad (10)$$

Interesting identities involving triangular numbers and SQUARE NUMBERS are

$$\sum_{k=1}^{2n-1}(-1)^{k+1}T_k = n^2 \quad (11)$$

$$T_n^2 = \sum_{k=1}^{n}k^3 = \tfrac{1}{4}n^2(n+1)^2 \quad (12)$$

$$\sum_{k=1,\,3,\,\ldots,\,q}k^3 = T_n \quad (13)$$

for q ODD and

$$n = \tfrac{1}{2}(q^2 + 2q - 1). \quad (14)$$

Triangular numbers also unexpectedly appear in integrals involving the ABSOLUTE VALUE OF THE FORM

$$\int_0^1 \int_0^1 |x-y|^n\,dx\,dy = \frac{2}{(n+1)(n+2)}. \quad (15)$$

All EVEN PERFECT NUMBERS are triangular T_p with PRIME p. Furthermore, every EVEN PERFECT NUMBER $P > 6$ is OF THE FORM

$$P = 1 + 9T_n = T_{3n+1}, \quad (16)$$

where T_n is a triangular number with $n = 8j + 2$ (Eaton 1995, 1996). Therefore, the nested expression

$$9(9\cdots(9(9(9(9T_n+1)+1)+1)+1)\ldots+1)+1 \quad (17)$$

generates triangular numbers for any T_n. An INTEGER

k is a triangular number IFF $8k+1$ is a SQUARE NUMBER > 1.

The numbers 1, 36, 1225, 41616, 1413721, 48024900, ... (Sloane's A001110) are SQUARE TRIANGULAR NUMBERS, i.e., numbers which are simultaneously triangular and SQUARE (Pietenpol 1962). The corresponding square roots are 1, 6, 35, 204, 1189, 6930, ... (Sloane's A001109), and the indices of the corresponding triangular numbers T_n are $n = 1, 8, 49, 288, 1681, \ldots$ (Sloane's A001108).

Numbers which are simultaneously triangular and TETRAHEDRAL satisfy the BINOMIAL COEFFICIENT equation

$$T_n = \binom{n+1}{2} = \binom{m+2}{3} = Te_m, \quad (18)$$

the only solutions of which are

$$Te_3 = T_4 = 10 \quad (19)$$

$$Te_8 = T_{15} = 120 \quad (20)$$

$$Te_{20} = T_{55} = 1540 \quad (21)$$

$$Te_{34} = T_{119} = 7140 \quad (22)$$

(Guy 1994, p. 147).

The following table gives triangular numbers T_p having prime indices p.

T_n with prime indices	A034953	3, 6, 15, 28, 66, 91, 153, 190, 276, 435, 496, ...
odd T_n with prime indices	A034954	3, 15, 91, 153, 435, 703, 861, 1431, 1891, 2701, ...
even T_n with prime indices	A034955	6, 28, 66, 190, 276, 496, 946, 1128, 1770, 2278, ...

The smallest of two INTEGERS for which $n^3 - 13$ is four times a triangular number is 5 (Cesaro 1886; Le Lionnais 1983, p. 56). The only FIBONACCI NUMBERS which are triangular are 1, 3, 21, and 55 (Ming 1989), and the only PELL NUMBER which is triangular is 1 (McDaniel 1996). The BEAST NUMBER 666 is triangular, since

$$T_{6\cdot 6} = T_{36} = 666. \quad (23)$$

In fact, it is the largest REPDIGIT triangular number (Bellew and Weger 1975–76).

FERMAT'S POLYGONAL NUMBER THEOREM states that every POSITIVE INTEGER is a sum of *most* three TRIANGULAR NUMBERS, four SQUARE NUMBERS, five PENTAGONAL NUMBERS, and n n-POLYGONAL NUM-

BERS. Gauss proved the triangular case (Wells 1986, p. 47), and noted the event in his diary on July 10, 1796, with the notation

$$**E\Upsilon RHKA \quad num = \Delta + \Delta + \Delta. \tag{24}$$

This case is equivalent to the statement that every number OF THE FORM $8m + 3$ is a sum of three ODD SQUARES (Duke 1997). Dirichlet derived the number of ways in which an INTEGER m can be expressed as the sum of three triangular numbers (Duke 1997). The result is particularly simple for a PRIME OF THE FORM $8m + 3$, in which case it is the number of squares mod $8m + 3$ minus the number of nonsquares mod $8m + 3$ in the INTERVAL $4m + 1$ (Deligne 1973).

The only triangular numbers which are the PRODUCT of three consecutive INTEGERS are 6, 120, 210, 990, 185136, 258474216 (Sloane's A001219; Guy 1994, p. 148).

See also FIGURATE NUMBER, HEPTAGONAL TRIANGULAR NUMBER, OCTAGONAL TRIANGULAR NUMBER, PENTAGONAL TRIANGULAR NUMBER, PRONIC NUMBER, SQUARE TRIANGULAR NUMBER

References

Ball, W. W. R. and Coxeter, H. S. M. *Mathematical Recreations and Essays, 13th ed.* New York: Dover, p. 59, 1987.
Bellew, D. W. and Weger, R. C. "Repdigit Triangular Numbers." *J. Recr. Math.* **8**, 96–97, 1975–76.
Conway, J. H. and Guy, R. K. *The Book of Numbers.* New York: Springer-Verlag, pp. 33–38, 1996.
Deligne, P. "La Conjecture de Weil." *Inst. Hautes Études Sci. Pub. Math.* **43**, 273–308, 1973.
Dudeney, H. E. *Amusements in Mathematics.* New York: Dover, pp. 67 and 167, 1970.
Duke, W. "Some Old Problems and New Results about Quadratic Forms." *Not. Amer. Math. Soc.* **44**, 190–196, 1997.
Eaton, C. F. "Problem 1482." *Math. Mag.* **68**, 307, 1995.
Eaton, C. F. "Perfect Number in Terms of Triangular Numbers." Solution to Problem 1482. *Math. Mag.* **69**, 308–309, 1996.
Guy, R. K. "Sums of Squares" and "Figurate Numbers." §C20 and §D3 in *Unsolved Problems in Number Theory, 2nd ed.* New York: Springer-Verlag, pp. 136–138 and 147–150, 1994.
Hindin, H. "Stars, Hexes, Triangular Numbers and Pythagorean Triples." *J. Recr. Math.* **16**, 191–193, 1983–1984.
Le Lionnais, F. *Les nombres remarquables.* Paris: Hermann, p. 56, 1983.
McDaniel, W. L. "Triangular Numbers in the Pell Sequence." *Fib. Quart.* **34**, 105–107, 1996.
Ming, L. "On Triangular Fibonacci Numbers." *Fib. Quart.* **27**, 98–108, 1989.
Pappas, T. "Triangular, Square & Pentagonal Numbers." *The Joy of Mathematics.* San Carlos, CA: Wide World Publ./Tetra, p. 214, 1989.
Pietenpol, J. L "Square Triangular Numbers." *Amer. Math. Monthly* **169**, 168–169, 1962.
Ram, R. "Triangle Numbers that are Perfect Squares." http://users.tellurian.net/hsejar/maths/triangle/.
Satyanarayana, U. V. "On the Representation of Numbers as the Sum of Triangular Numbers." *Math. Gaz.* **45**, 40–43, 1961.
Sloane, N. J. A. Sequences A000217/M2535, A001108/M4536, A001109/M4217, A001110/M5259, A001219,
A014493, A014494, A034953, A034955, and A034955 in "An On-Line Version of the Encyclopedia of Integer Sequences." http://www.research.att.com/~njas/sequences/eisonline.html.
Sloane, N. J. A. and Plouffe, S. *The Encyclopedia of Integer Sequences.* San Diego, CA: Academic Press, 1995.
Wells, D. *The Penguin Dictionary of Curious and Interesting Numbers.* Middlesex, England: Penguin Books, pp. 47–48, 1986.
Wells, D. *The Penguin Dictionary of Curious and Interesting Geometry.* London: Penguin, p. 199, 1991.

Triangular Orthobicupola

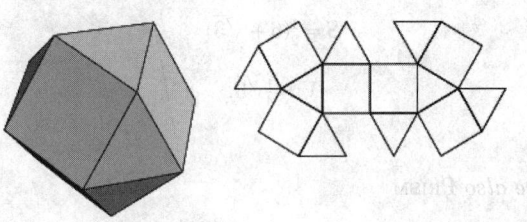

JOHNSON SOLID J_{27}, consisting of eight equilateral triangles and six squares. If a triangular orthobicupola is oriented with triangles on top and bottom, the two halves may be rotated one sixth of a turn with respect to each other to obtain the CUBOCTAHEDRON.

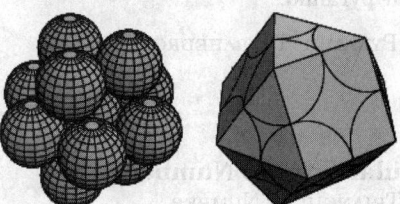

In hexagonal close packing, layers of spheres are packed so that spheres in alternating layers overlie one another. As in cubic close packing, each sphere is surrounded by 12 other spheres. Taking a collection of 13 such spheres gives the cluster illustrated above. Connecting the centers of the external 12 spheres gives J_{27} (Steinhaus 1983, pp. 203–205), which is therefore also a SPACE-FILLING POLYHEDRON.

See also CUBOCTAHEDRON, JOHNSON SOLID, SPACE-FILLING POLYHEDRON, SPHERE PACKING

References

Steinhaus, H. *Mathematical Snapshots, 3rd ed.* New York: Dover, pp. 203–205, 1999.

Triangular Prism

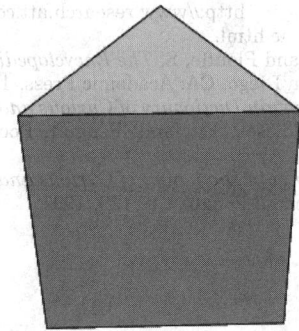

A PRISM composed of triangular faces. The regular right triangular prism of unit edge length has SURFACE AREA and VOLUME

$$S = \tfrac{1}{2}(6 + \sqrt{3})$$

$$V = \tfrac{1}{4}\sqrt{3}.$$

See also PRISM

Triangular Pyramid

A PYRAMID having a triangular base. The SLANT HEIGHT of a regular triangular pyramid is a special case of the formula for a regular n-gonal PYRAMID with $n = 3$, given by

$$s = \sqrt{h^2 + \tfrac{1}{3}a^2}, \tag{1}$$

where h is the height and a is the length of a side of the base. The TETRAHEDRON is a special case of the triangular pyramid.

See also PYRAMID, TETRAHEDRON

Triangular Square Number

SQUARE TRIANGULAR NUMBER

Triangular Symmetry Group

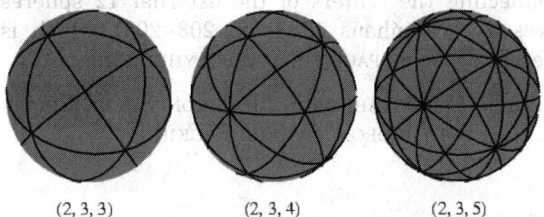

| (2, 3, 3) | (2, 3, 4) | (2, 3, 5) |

Given a TRIANGLE with angles $(\pi/p,\ \pi/q,\ \pi/r)$, the resulting symmetry GROUP is called a $(p,\ q,\ r)$ trian-

gle group (also known as a SPHERICAL TESSELLATION). In 3-D, such GROUPS must satisfy

$$\frac{1}{p} + \frac{1}{q} + \frac{1}{r} > 1,$$

and so the only solutions are $(2, 2, n)$, $(2, 3, 3)$, $(2, 3, 4)$, and $(2, 3, 5)$ (Ball and Coxeter 1987). The group $(2, 3, 6)$ gives rise to the semiregular planar TESSELLATIONS of types 1, 2, 5, and 7. The group $(2, 3, 7)$ gives hyperbolic tessellations.

See also GEODESIC DOME

References

Ball, W. W. R. and Coxeter, H. S. M. *Mathematical Recreations and Essays, 13th ed.* New York: Dover, pp. 155–161, 1987.

Coxeter, H. S. M. "The Partition of a Sphere According to the Icosahedral Group." *Scripta Math* **4**, 156–157, 1936.

Coxeter, H. S. M. *Regular Polytopes, 3rd ed.* New York: Dover, 1973.

Kraitchik, M. "A Mosaic on the Sphere." §7.3 in *Mathematical Recreations.* New York: W. W. Norton, pp. 208–209, 1942.

Triangulation

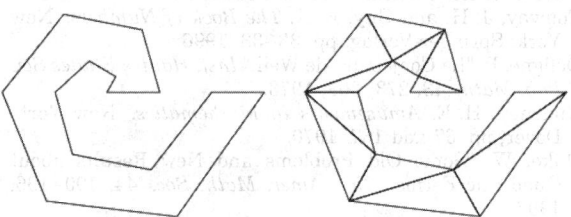

Triangulation is the division of a surface or plane polygon into a set of TRIANGLES, usually with the restriction that each TRIANGLE side is entirely shared by two adjacent TRIANGLES. It was proved in 1925 that every surface has a triangulation, but it might require an infinite number of TRIANGLES and the proof is difficult (Francis and Weeks 1999). A surface with a finite number of triangles in its triangulation is called COMPACT.

Wickham-Jones (1994) gives an $\mathcal{O}(n^3)$ algorithm for triangulation ("otectomy"), and O'Rourke (1998, p. 47) sketches a method for improving this to $\mathcal{O}(n^2)$, as first done by Lennes (1911). Garey *et al.* (1978) gave an algorithmically straightforward $\mathcal{O}(n \ln n)$ method for triangulation, which was for many years believed optimal. However, Tarjan and van Wyk (1988) produced an $\mathcal{O}(n \lg \lg n)$ algorithm. This was followed by an unexpected result due to Chazelle (1991), who showed that an arbitrary SIMPLE POLYGON can be triangulated in $\mathcal{O}(n)$. However, according to Skiena (1997), "this algorithm is quite hopeless to implement."

See also ART GALLERY THEOREM, COMPACT SURFACE, DELAUNAY TRIANGULATION, JAPANESE THEOREM, SIMPLE POLYGON, TESSELLATION

References

Chazelle, B. "Triangulating a Simple Polygon in Linear Time." *Disc. Comput. Geom.* **6**, 485–524, 1991.

de Berg, M.; van Kreveld, M.; Overmans, M.; and Schwarzkopf, O. "Polygon Triangulation: Guarding an Art Gallery." Ch. 3 in *Computational Geometry: Algorithms and Applications, 2nd rev. ed.* Berlin: Springer-Verlag, pp. 45–61, 2000.

Fournier, A. and Montuno, D. Y. "Triangulating Simple Polygons and Equivalent Problems." *ACM Trans. Graphics* **3**, 153–174, 1984.

Francis, G. K. and Weeks, J. R. "Conway's ZIP Proof." *Amer. Math. Monthly* **106**, 393–399, 1999.

Friedman, E. "Triangulating Triangles." http://www.stetson.edu/~efriedma/triang/.

Garey, M. R.; Johnson, D. S.; Preparata, F. P.; and Tarjan, R. E. "Triangulating a Simple Polygon." *Inform. Process. Lett.* **7**, 175–179, 1978.

Kraus, M. "Polygon Triangulation." http://library.wolfram.com/packages/polygontriangulation/.

O'Rourke, J. §2.3 in *Computational Geometry in C, 2nd ed.* Cambridge, England: Cambridge University Press, 1998.

Radó, T. "Über den Begriff der Riemannschen Fläche." *Acta Litt. Sci. Reg. Univ. Hungar. Francisco-Josephinae* **2**, 101–121, 1924–1926.

Skiena, S. S. "Triangulation." §8.6.3 in *The Algorithm Design Manual.* New York: Springer-Verlag, pp. 355–357, 1997.

Tarjan, R. and van Wyk, C. "An $\mathcal{O}(n \lg \lg n)$ Algorithm for Triangulating a Simple Polygon." *SIAM J. Computing* **17**, 143–178, 1988.

Wickham-Jones, T. "ExtendGraphics Packages for Mathematica 3.0." http://www.mathsource.com/cgi-bin/msitem?0208–976.

Wickham-Jones, T. *Mathematica Graphics: Techniques and Applications.* New York: Springer-Verlag, pp. 406 and 448, 1994.

Triaugmented Dodecahedron

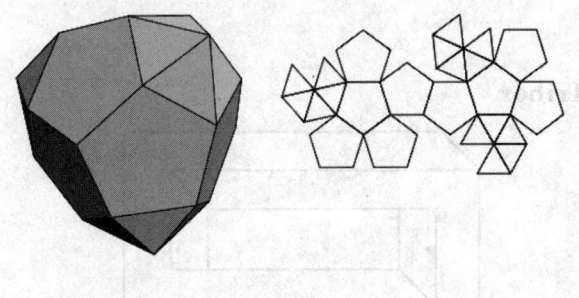

JOHNSON SOLID J_{61}.

References

Weisstein, E. W. "Johnson Solids." MATHEMATICA NOTEBOOK JOHNSONSOLIDS.M.

Weisstein, E. W. "Johnson Solid Netlib Database." MATHEMATICA NOTEBOOK JOHNSONSOLIDS.DAT.

Triaugmented Hexagonal Prism

JOHNSON SOLID J_{57}.

References

Weisstein, E. W. "Johnson Solids." MATHEMATICA NOTEBOOK JOHNSONSOLIDS.M.

Weisstein, E. W. "Johnson Solid Netlib Database." MATHEMATICA NOTEBOOK JOHNSONSOLIDS.DAT.

Triaugmented Triangular Prism

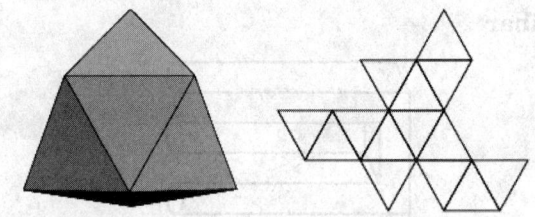

One of the convex DELTAHEDRA. It is composed of 14 equilateral triangles, and is JOHNSON SOLID J_{51}. The VERTICES are $(\pm 1/2, \pm 1/2, 0)$, $(0, 0, \sqrt{2}/2)$, $(0, \pm 1/2, -\sqrt{3}/2)$, $(\pm(1+\sqrt{6})/4, 0, -(\sqrt{2}+\sqrt{3})/4)$, where the x and z coordinates of the last are found by solving

$$x^2 + \left(\tfrac{1}{2}\right)^2 + \left(z + \sqrt{3}/2\right)^2 = 1^2 \qquad (1)$$

$$\left(x - \tfrac{1}{2}\right)^2 + \left(\tfrac{1}{2}\right)^2 + z^2 = 1^2. \qquad (2)$$

For a triaugmented triangular prism with unit side length, the SURFACE AREA and VOLUME are

$$S = \tfrac{7}{2}\sqrt{3} \qquad (3)$$

$$V = \tfrac{1}{4}\left(2\sqrt{2} + \sqrt{3}\right). \qquad (4)$$

See also DELTAHEDRON, JOHNSON SOLID

Triaugmented Truncated Dodecahedron

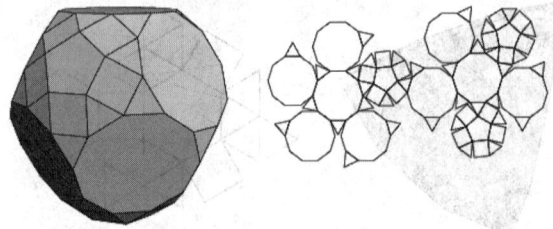

JOHNSON SOLID J_{71}.

References

Weisstein, E. W. "Johnson Solids." MATHEMATICA NOTEBOOK JOHNSONSOLIDS.M.

Weisstein, E. W. "Johnson Solid Netlib Database." MATHEMATICA NOTEBOOK JOHNSONSOLIDS.DAT.

Triaxial Ellipsoid

ELLIPSOID

Tri-Axial Ellipsoid

ELLIPSOID

Tribar

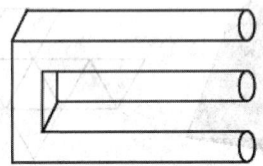

An IMPOSSIBLE FIGURE published by R. Penrose (1958). It also exists as a TRIBOX.

References

Draper, S. W. "The Penrose Triangle and a Family of Related Figures." *Perception* **7**, 283–296, 1978.

Fineman, M. *The Nature of Visual Illusion.* New York: Dover, p. 119, 1996.

Jablan, S. "Set of Modular Elements 'Space Tiles'." http://members.tripod.com/~modularity/space.htm.

Pappas, T. "The Impossible Tribar." *The Joy of Mathematics.* San Carlos, CA: Wide World Publ./Tetra, p. 13, 1989.

Penrose, R. "Impossible Objects: A Special Type of Visual Illusion." *Brit. J. Psychology* **49**, 31–33, 1958.

Tribonacci Number

The tribonacci numbers are a generalization of the FIBONACCI NUMBERS defined by $T_1 = 1$, $T_2 = 1$, $T_3 = 2$, and the RECURRENCE RELATION

$$T_n = T_{n-1} + T_{n-2} + T_{n-3} \qquad (1)$$

for $n \geq 4$. The represent the $n = 3$ case of the FIBONACCI N-STEP NUMBERS. The first few terms are 1, 1, 2, 4, 7, 13, 24, 44, 81, 149, ... (Sloane's A000073). The

ratio of adjacent terms tends to 1.83929, which is the REAL ROOT of $x^4 - 2x^3 + 1 = 0$. The Tribonacci numbers can also be computed using the GENERATING FUNCTION

$$\frac{1}{1 - z - z^2 - z^3} = 1 + z + 2z^2 + 4z^3 + 7z^4$$
$$+ 13z^5 + 24z^6 + 44z^7 + 81z^8 + 149z^9 + \dots . \qquad (2)$$

An explicit FORMULA for T_n is also given by

$$\left[3 \frac{\left\{ \frac{1}{3}(19 + 3\sqrt{33})^{1/3} + \frac{1}{3}(19 - 3\sqrt{33})^{1/3} + \frac{1}{3} \right\}^n (586 + 102\sqrt{33})^{1/3}}{(586 + 102\sqrt{33})^{2/3} + 4 - 2(586 + 102\sqrt{33})^{1/3}} \right],$$
$$\qquad (3)$$

where $[x]$ denotes the NINT function (Plouffe). The first part of a NUMERATOR is related to the REAL root of $x^3 - x^2 - x - 1$, but determination of the DENOMINATOR requires an application of the LLL ALGORITHM. The numbers increase asymptotically to

$$T_n \sim c^n \qquad (4)$$

where

$$c = \left(\frac{19}{27} + \frac{1}{9}\sqrt{33} \right)^{1/3} + \frac{4}{9}\left(\frac{19}{27} + \frac{1}{9}\sqrt{33} \right)^{-1/3} + \frac{1}{3}$$

$$= 1.83928675521\dots \qquad (5)$$

(Plouffe).

See also FIBONACCI N-STEP NUMBER, FIBONACCI NUMBER, TETRANACCI NUMBER

References

Plouffe, S. "Tribonacci Constant." http://www.lacim.uqam.ca/piDATA/tribo.txt.

Sloane, N. J. A. Sequences A000073/M1074 in "An On-Line Version of the Encyclopedia of Integer Sequences." http://www.research.att.com/~njas/sequences/eisonline.html.

Tribox

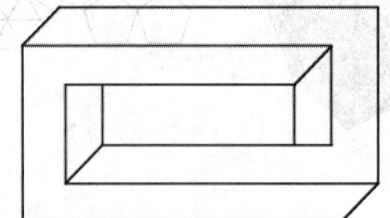

An IMPOSSIBLE FIGURE.

See also IMPOSSIBLE FIGURE, TRIBAR

References

Jablan, S. "Are Impossible Figures Possible?" http://members.tripod.com/~modularity/kulpa.htm.

Trichotomy Law

Every REAL NUMBER is NEGATIVE, 0, or POSITIVE. The law is sometimes states as "For arbitrary real numbers x and y, exactly one of the relations $a < b$, $a = b$, $a > b$ holds" (Apostol 1967, p. 20).

See also SCHRÖDER-BERNSTEIN THEOREM, TOTAL ORDER

References

Apostol, T. M. *Calculus, 2nd ed., Vol. 1: One-Variable Calculus, with an Introduction to Linear Algebra.* Waltham, MA: Blaisdell, 1967.

Tricolorable

A projection of a LINK is tricolorable if each of the strands in the projection can be colored in one of three different colors such that, at each crossing, all three colors come together or only one does and at least two different colors are used. The TREFOIL KNOT and trivial 2-link are tricolorable, but the UNKNOT, WHITEHEAD LINK, and FIGURE-OF-EIGHT KNOT are not.

If the projection of a knot is tricolorable, then REIDEMEISTER MOVES on the knot preserve tricolorability, so either every projection of a knot is tricolorable or none is.

Tricomi Equation

The PARTIAL DIFFERENTIAL EQUATION

$$u_{yy} = y u_{xx}.$$

References

Manwell, A. R. *The Tricomi Equation with Applications to the Theory of Plane Transonic Flow.* Marshfield, MA: Pitman, 1979.
Zwillinger, D. (Ed.). *CRC Standard Mathematical Tables and Formulae.* Boca Raton, FL: CRC Press, p. 417, 1995.
Zwillinger, D. *Handbook of Differential Equations, 3rd ed.* Boston, MA: Academic Press, p. 130, 1997.

Tricomi Function

CONFLUENT HYPERGEOMETRIC FUNCTION OF THE SECOND KIND, GORDON FUNCTION

Tricuspoid

DELTOID

Tricylinder

STEINMETZ SOLID

Tridecagon

A 13-sided POLYGON, sometimes also called the TRISKAIDECAGON.

Trident

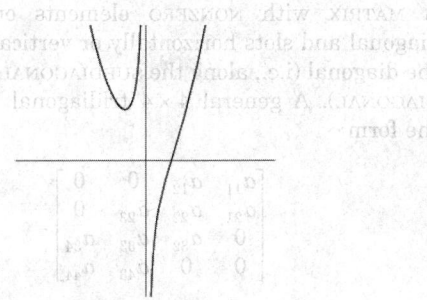

The plane curve given by the equation

$$xy = x^3 - a^3.$$

See also TRIDENT OF DESCARTES, TRIDENT OF NEWTON

Trident of Descartes

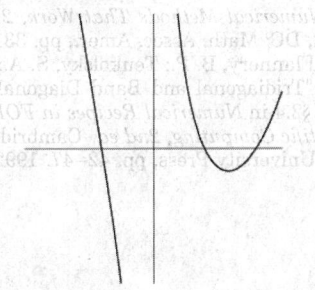

The plane curve given by the equation

$$(a + x)(a - x)(2a - x) = x^3 - 2ax^2 - a^2x + 2a^3 = axy$$

$$y = \frac{(a + x)(a - x)(2a - x)}{ax}$$

The above plot has $a = 2$.

Trident of Newton

The CUBIC CURVE defined by

$$ax^3 + bx^2 + cx + d = xy$$

with $a \neq 0$. The curve cuts the axis in either one or three points. It was the 66th curve in Newton's classification of CUBICS. Newton stated that the curve has four infinite legs and that the y-AXIS is an ASYMPTOTE to two tending toward contrary parts.

References

Lawrence, J. D. *A Catalog of Special Plane Curves.* New York: Dover, pp. 109–110, 1972.
MacTutor History of Mathematics Archive. "Trident of Newton." http://www-groups.dcs.st-and.ac.uk/~history/Curves/Trident.html.

Tridiagonal Matrix

A MATRIX with NONZERO elements only on the diagonal and slots horizontally or vertically adjacent the diagonal (i.e., along the SUBDIAGONAL and SUPERDIAGONAL). A general 4×4 tridiagonal MATRIX has the form

$$\begin{bmatrix} a_{11} & a_{12} & 0 & 0 \\ a_{21} & a_{22} & a_{23} & 0 \\ 0 & a_{32} & a_{33} & a_{34} \\ 0 & 0 & a_{43} & a_{44} \end{bmatrix}.$$

Inversion of such a matrix requires only $\mathcal{O}(7n)$ (as opposed to $\mathcal{O}(n^3/3)$) arithmetic operations (Acton 1990, p. 332).

See also DIAGONAL MATRIX, JACOBI ALGORITHM, SUBDIAGONAL, SUPERDIAGONAL

References

Acton, F. S. *Numerical Methods That Work, 2nd printing.* Washington, DC: Math. Assoc. Amer., pp. 331–334, 1990.
Press, W. H.; Flannery, B. P.; Teukolsky, S. A.; and Vetterling, W. T. "Tridiagonal and Band Diagonal Systems of Equations." §2.4 in *Numerical Recipes in FORTRAN: The Art of Scientific Computing, 2nd ed.* Cambridge, England: Cambridge University Press, pp. 42–47, 1992.

Tridiminished Icosahedron

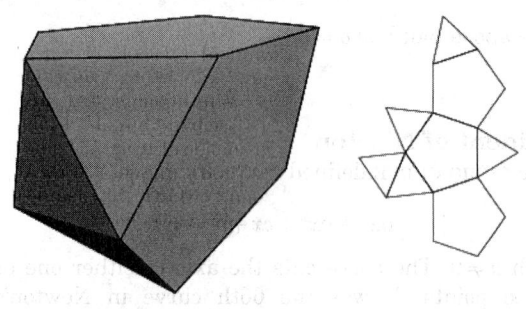

JOHNSON SOLID J_{63}.

References

Weisstein, E. W. "Johnson Solids." MATHEMATICA NOTEBOOK JOHNSONSOLIDS.M.
Weisstein, E. W. "Johnson Solid Netlib Database." MATHEMATICA NOTEBOOK JOHNSONSOLIDS.DAT.

Tridiminished Rhombicosidodecahedron

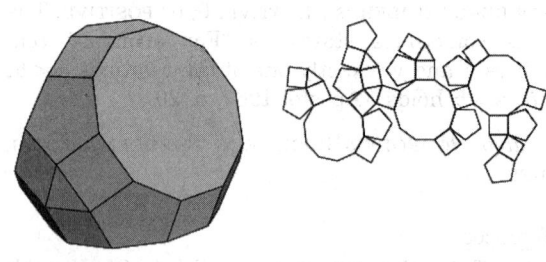

JOHNSON SOLID J_{83}.

References

Weisstein, E. W. "Johnson Solids." MATHEMATICA NOTEBOOK JOHNSONSOLIDS.M.
Weisstein, E. W. "Johnson Solid Netlib Database." MATHEMATICA NOTEBOOK JOHNSONSOLIDS.DAT.

Tridyakis Icosahedron

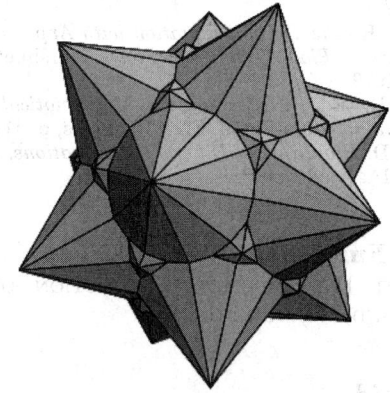

The DUAL POLYHEDRON of the ICOSITRUNCATED DODECADODECAHEDRON U_{45} and Wenninger dual W_{84}.

See also DUAL POLYHEDRON, ICOSITRUNCATED DODECADODECAHEDRON

References

Wenninger, M. J. *Dual Models.* Cambridge, England: Cambridge University Press, p. 96, 1983.

Trifolium

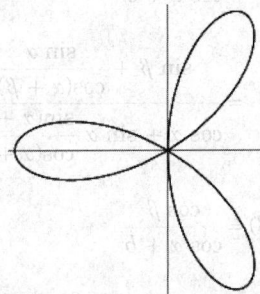

Lawrence (1972) defines a trifolium as a FOLIUM with $b \in (0, 4a)$. However, the term "the" trifolium is sometimes applied to the FOLIUM with $b = a$, which is then the 3-petalled ROSE with Cartesian equation

$$(x^2 + y^2)[y^2 + x(x + a)] = 4axy^2$$

and polar equation

$$r = a \cos \theta (4 \sin^2 \theta - 1) = -a \cos(3\theta).$$

The trifolium with $b = a$ is the RADIAL CURVE of the DELTOID.

See also BIFOLIUM, FOLIUM, QUADRIFOLIUM

References

Lawrence, J. D. *A Catalog of Special Plane Curves.* New York: Dover, pp. 152–153, 1972.
MacTutor History of Mathematics Archive. "Trifolium." http://www-groups.dcs.st-and.ac.uk/~history/Curves/Trifolium.html.

Trigon

TRIANGLE

Trigonal Dipyramid

TRIANGULAR DIPYRAMID

Trigonal Dodecahedron

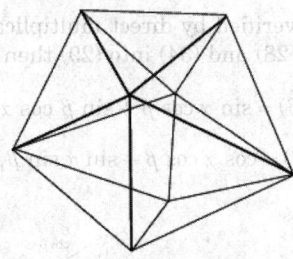

An irregular DODECAHEDRON.

See also DODECAHEDRON, PYRITOHEDRON, RHOMBIC DODECAHEDRON

References

Cotton, F. A. *Chemical Applications of Group Theory, 3rd ed.* New York: Wiley, p. 62, 1990.

Trigonometric Addition Formulas

Angle addition FORMULAS express trigonometric functions of sums of angles $\alpha \pm \beta$ in terms of functions of α and β. The fundamental formulas of angle addition in trigonometry are given by

$$\sin(\alpha + \beta) = \sin \alpha \cos \beta + \sin \beta \cos \alpha \tag{1}$$

$$\sin(\alpha - \beta) = \sin \alpha \cos \beta - \sin \beta \cos \alpha \tag{2}$$

$$\cos(\alpha + \beta) = \cos \alpha \cos \beta - \sin \alpha \sin \beta \tag{3}$$

$$\cos(\alpha - \beta) = \cos \alpha \cos \beta + \sin \alpha \sin \beta \tag{4}$$

$$\tan(\alpha + \beta) = \frac{\tan \alpha + \tan \beta}{1 - \tan \alpha \tan \beta} \tag{5}$$

$$\tan(\alpha - \beta) = \frac{\tan \alpha - \tan \beta}{1 + \tan \alpha \tan \beta}. \tag{6}$$

The sine and cosine angle addition identities can be compactly summarized by the MATRIX EQUATION

$$\begin{bmatrix} \cos \alpha & \sin \alpha \\ -\sin \alpha & \cos \alpha \end{bmatrix} \begin{bmatrix} \cos \beta & \sin \beta \\ -\sin \beta & \cos \beta \end{bmatrix}$$
$$= \begin{bmatrix} \cos(\alpha + \beta) & \sin(\alpha + \beta) \\ -\sin(\alpha + \beta) & \cos(\alpha + \beta) \end{bmatrix}. \tag{7}$$

These formulas can be simply derived using COMPLEX EXPONENTIALS and the EULER FORMULA as follows.

$$\cos(\alpha + \beta) + i \sin(\alpha + \beta) = e^{i(\alpha + \beta)} = e^{i\alpha} e^{i\beta}$$

$$= (\cos \alpha + i \sin \alpha)(\cos \beta + i \sin \beta)$$

$$= (\cos \alpha \cos \beta - \sin \alpha \sin \beta)$$

$$+ i(\sin \alpha \cos \beta + \cos \alpha \sin \beta). \tag{8}$$

Equating REAL and IMAGINARY PARTS then gives (1) and (3), and (2) and (4) follow immediately by substituting $-\beta$ for β.

Taking the ratio of (1) and (3) gives the tangent angle addition FORMULA

$$\tan(\alpha + \beta) \equiv \frac{\sin(\alpha + \beta)}{\cos(\alpha + \beta)} = \frac{\sin \alpha \cos \beta + \sin \beta \cos \alpha}{\cos \alpha \cos \beta - \sin \alpha \sin \beta}$$

$$= \frac{\dfrac{\sin \alpha}{\cos \alpha} + \dfrac{\sin \beta}{\cos \beta}}{1 - \dfrac{\sin \alpha \sin \beta}{\cos \alpha \cos \alpha \beta}} = \frac{\tan \alpha + \tan \beta}{1 - \tan \alpha \tan \beta}. \tag{9}$$

The DOUBLE-ANGLE FORMULAS are

$$\sin(2\alpha) = 2 \sin \alpha \cos \alpha \tag{10}$$

$$\cos(2\alpha) = \cos^2 \alpha - \sin^2 \alpha \tag{11}$$

$$= 2 \cos^2 \alpha - 1 \tag{12}$$

$$= 1 - 2 \sin^2 \alpha \tag{13}$$

$$\tan(2\alpha) = \frac{2 \tan \alpha}{1 - \tan^2 \alpha}. \tag{14}$$

MULTIPLE-ANGLE FORMULAS are given by

$$\sin(nx) = \sum_{k=0}^{n} \binom{n}{k} \cos^k x \, \sin^{n-k} x \, \sin\left[\tfrac{1}{2}(n-k)\pi\right]. \tag{15}$$

$$\cos(nx) = \sum_{k=0}^{n} \binom{n}{k} \cos^k x \, \sin^{n-k} x \, \cos\left[\tfrac{1}{2}(n-k)\pi\right], \tag{16}$$

and can also be written using the RECURRENCE RELATIONS

$$\sin(nx) = 2 \sin[(n-1)x] \cos x - \sin[(n-2)x] \tag{17}$$

$$\cos(nx) = 2 \cos[(n-1)x] \cos x - \cos[(n-2)x] \tag{18}$$

$$\tan(nx) = \frac{\tan[(n-1)x] + \tan x}{1 - \tan[(n-1)x] \tan x}. \tag{19}$$

SIMPSON'S FORMULAS are given by

$$\sin \alpha + \sin \beta = 2 \sin\left(\frac{\alpha + \beta}{2}\right) \cos\left(\frac{\alpha - \beta}{2}\right) \tag{20}$$

$$\sin \alpha - \sin \beta = 2 \sin\left(\frac{\alpha - \beta}{2}\right) \cos\left(\frac{\alpha + \beta}{2}\right) \tag{21}$$

$$\cos \alpha + \cos \beta = 2 \cos\left(\frac{\alpha + \beta}{2}\right) \cos\left(\frac{\alpha - \beta}{2}\right) \tag{22}$$

$$\cos \alpha - \cos \beta = -2 \sin\left(\frac{\alpha - \beta}{2}\right) \sin\left(\frac{\alpha + \beta}{2}\right). \tag{23}$$

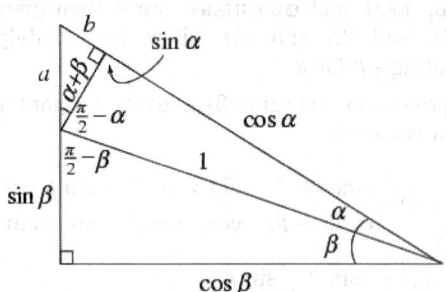

The angle addition formulas can also be derived purely algebraically without the use of COMPLEX NUMBERS. Consider the small RIGHT TRIANGLE in the figure above, which gives

$$a = \frac{\sin \alpha}{\cos(\alpha + \beta)} \tag{24}$$

$$b = \sin \alpha \tan(\alpha + \beta). \tag{25}$$

Now, the usual trigonometric definitions applied to the large RIGHT TRIANGLE give

$$\sin(\alpha + \beta) = \frac{\sin \beta + a}{\cos \alpha + b}$$

$$= \frac{\sin \beta + \dfrac{\sin \alpha}{\cos(\alpha + \beta)}}{\cos \alpha + \sin \alpha \dfrac{\sin(\alpha + \beta)}{\cos(\alpha + \beta)}} \tag{26}$$

$$\cos(\alpha + \beta) = \frac{\cos \beta}{\cos \alpha + b}$$

$$= \frac{\cos \beta}{\cos \alpha + \sin \alpha \dfrac{\sin(\alpha + \beta)}{\cos(\alpha + \beta)}}. \tag{27}$$

Solving these two equations simultaneously for the variables $\sin(\alpha + \beta)$ and $\cos(\alpha + \beta)$ then immediately gives

$$\sin(\alpha + \beta) = \frac{\cos \alpha \sin \alpha + \cos \beta \sin \beta}{\cos \alpha \cos \beta + \sin \alpha \sin \beta} \tag{28}$$

$$\cos(\alpha + \beta) = \frac{\cos^2 \beta - \sin^2 \alpha}{\cos \alpha \cos \beta + \sin \alpha \sin \beta}. \tag{29}$$

These can be put into the familiar forms with the aid of the trigonometric identities

$$(\cos \alpha \cos \beta + \sin \alpha \sin \beta)(\cos \alpha \cos \beta + \sin \beta \cos \alpha)$$
$$= \cos \beta \sin \beta + \cos \alpha \sin \alpha \tag{30}$$

and

$$(\cos \alpha \cos \beta + \sin \alpha \sin \beta)(\cos \alpha \cos \beta - \sin \alpha \cos \beta)$$

$$= \cos^2 \alpha \cos^2 \beta - \sin^2 \alpha \sin^2 \beta \tag{31}$$

$$= 1 - \sin^2 \alpha \sin^2 \beta \tag{32}$$

$$= \cos^2 \alpha - \sin^2 \beta \tag{33}$$

$$= \cos^2 \beta - \sin^2 \alpha, \tag{34}$$

which can be verified by direct multiplication. Plugging (30) into (28) and (34) into (29) then gives

$$\sin(\alpha + \beta) = \sin \alpha \cos \beta + \sin \beta \cos \alpha \tag{35}$$

$$\cos(\alpha + \beta) = \cos \alpha \cos \beta - \sin \alpha \sin \beta, \tag{36}$$

as before.

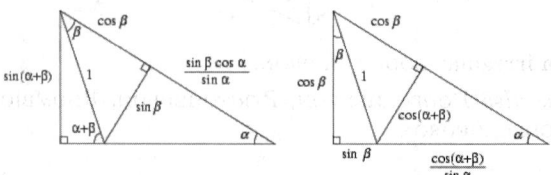

A similar proof due to Smiley and Smiley uses the left figure above to obtain

$$\sin\alpha = \frac{\sin(\alpha+\beta)}{\cos\beta + \dfrac{\sin\beta\cos\alpha}{\sin\alpha}}, \qquad (37)$$

from which it follows that

$$\sin(\alpha+\beta) = \sin\alpha\cos\beta + \sin\beta\cos\alpha. \qquad (38)$$

Similarly, from the right figure,

$$\frac{\sin\alpha}{\cos\alpha} = \frac{\cos\beta}{\sin\beta + \dfrac{\cos(\alpha+\beta)}{\sin\alpha}}, \qquad (39)$$

so

$$\cos(\alpha+\beta) = \cos\alpha\cos\beta - \sin\alpha\sin\beta. \qquad (40)$$

See also DOUBLE-ANGLE FORMULAS, HALF-ANGLE FORMULAS, HYPERBOLIC FUNCTIONS, TRIGONOMETRY

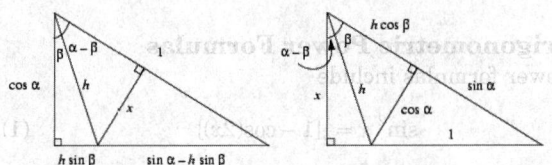

Similar diagrams can be used to prove the angle subtraction formulas (Smiley 1999, Smiley and Smiley). In the figure at left,

$$h = \frac{\cos\alpha}{\cos\beta} \qquad (41)$$

$$x = h\sin(\alpha-\beta) \qquad (42)$$

$$= (\sin\alpha - h\sin\beta)\cos\alpha,$$

giving

$$\sin(\alpha-\beta) = \sin\alpha\cos\beta - \cos\alpha\sin\beta. \qquad (43)$$

Similarly, in the figure at right,

$$h = \frac{\cos\alpha}{\sin\beta} \qquad (44)$$

$$x = h\cos(\alpha-\beta)$$

$$= (\sin\alpha + h\cos\beta)\cos\alpha, \qquad (45)$$

giving

$$\cos(\alpha-\beta) = \cos\alpha\cos\beta + \sin\alpha\sin\beta. \qquad (46)$$

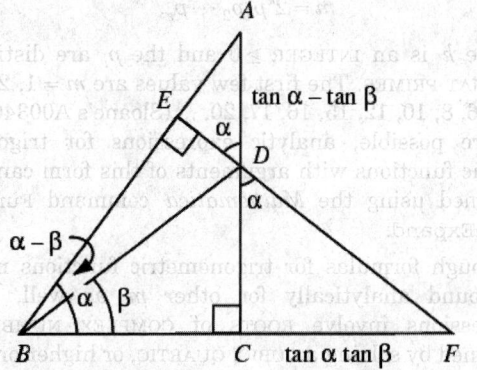

A more complex diagram can be used to obtain a proof from the $\tan(\alpha-\beta)$ identity (Ren 1999). In the above figure, let $BF/BE = AD/DE$. Then

$$\tan(\alpha-\beta) = \frac{DE}{BE} = \frac{AD}{BF} = \frac{\tan\alpha - \tan\beta}{1 + \tan\alpha\tan\beta}. \qquad (47)$$

An interesting identity relating the sum and difference tangent formulas is given by

$$\frac{\tan(\alpha-\beta)}{\tan(\alpha+\beta)} = \frac{\sin(\alpha-\beta)\cos(\alpha+\beta)}{\cos(\alpha-\beta)\sin(\alpha+\beta)}$$

$$= \frac{(\sin\alpha\cos\beta - \sin\beta\cos\alpha)(\cos\alpha\cos\beta - \sin\alpha\sin\beta)}{(\cos\alpha\cos\beta + \sin\alpha\sin\beta)(\sin\alpha\cos\beta + \sin\beta\cos\alpha)}$$

$$= \frac{\sin\alpha\cos\alpha - \sin\beta\cos\beta}{\sin\alpha\cos\alpha + \sin\beta\cos\beta}. \qquad (48)$$

See also DOUBLE-ANGLE FORMULAS, HALF-ANGLE FORMULAS, MULTIPLE-ANGLE FORMULAS, PROSTHAPHAERESIS FORMULAS, SIMPSON'S FORMULAS, TRIGONOMETRIC ANGLES, TRIGONOMETRIC PRODUCT FORMULAS, TRIGONOMETRY

References

Beyer, W. H. *CRC Standard Mathematical Tables, 28th ed.* Boca Raton, FL: CRC Press, 1987.
Nelson, R. To appear in *College Math. J.*, March 2000.
Ren, G. "Proof without Words: $\tan(\alpha-\beta)$." *College Math. J.* **30**, 212, 1999.
Smiley, L. M. "Proof without Words: Geometry of Subtraction Formulas." *Math. Mag.* **72**, 366, 1999.
Smiley, L. and Smiley, D. "Geometry of Addition and Subtraction Formulas." http://saturn.math.uaa.alaska.edu/~smiley/trigproofs.html.

Trigonometric Angles

The ANGLES $n\pi/m$ (with m, n integers) for which the trigonometric function may be expressed in terms of finite ROOT EXTRACTION of *real numbers* are limited to values of m which are precisely those which produce constructible POLYGONS. Gauss showed these to be OF THE FORM

$$m = 2^k p_1 p_2 \cdots p_s,$$

where k is an INTEGER ≥ 0 and the p_i are distinct FERMAT PRIMES. The first few values are $m = 1, 2, 3, 4, 5, 6, 8, 10, 12, 15, 16, 17, 20, \ldots$ (Sloane's A003401). Where possible, analytic expressions for trigonometric functions with arguments of this form can be obtained using the *Mathematica* command FunctionExpand.

Although formulas for trigonometric functions may be found analytically for other m as well, the expressions involve ROOTS of COMPLEX NUMBERS obtained by solving a CUBIC, QUARTIC, or higher order equation. The cases $m = 7$ and $m = 9$ involve the CUBIC EQUATION and QUARTIC EQUATION, respectively. A partial table of the analytic values of SINE, COSINE, and TANGENT for arguments π/m is given below. Derivations of these formulas appear in the following entries.

x (°)	x (rad)	$\sin x$	$\cos x$	$\tan x$
0.0	0	0	1	0
15.0	$\frac{1}{12}\pi$	$\frac{1}{4}(\sqrt{6}-\sqrt{2})$	$\frac{1}{4}(\sqrt{6}+\sqrt{2})$	$2-\sqrt{3}$
18.0	$\frac{1}{10}\pi$	$\frac{1}{4}(\sqrt{5}-1)$	$\frac{1}{4}\sqrt{10+2\sqrt{5}}$	$\frac{1}{5}\sqrt{25-10\sqrt{5}}$
22.5	$\frac{1}{8}\pi$	$\frac{1}{2}\sqrt{2-\sqrt{2}}$	$\frac{1}{2}\sqrt{2+\sqrt{2}}$	$\sqrt{2}-1$
30.0	$\frac{1}{6}\pi$	$\frac{1}{2}$	$\frac{1}{2}\sqrt{3}$	$\frac{1}{3}\sqrt{3}$
36.0	$\frac{1}{5}\pi$	$\frac{1}{4}\sqrt{10-2\sqrt{5}}$	$\frac{1}{4}(1+\sqrt{5})$	$\sqrt{5-2\sqrt{5}}$
45.0	$\frac{1}{4}\pi$	$\frac{1}{2}\sqrt{2}$	$\frac{1}{2}\sqrt{2}$	1
60.0	$\frac{1}{3}\pi$	$\frac{1}{2}\sqrt{3}$	$\frac{1}{2}$	$\sqrt{3}$
90.0	$\frac{1}{2}\pi$	1	0	∞
180.0	π	0	-1	0

There is a nice mnemonic for remembering sines of common angles,

$$\sin(0°) = \tfrac{1}{2}\sqrt{0} \qquad (1)$$

$$\sin(30°) = \tfrac{1}{2}\sqrt{1} \qquad (2)$$

$$\sin(45°) = \tfrac{1}{2}\sqrt{2} \qquad (3)$$

$$\sin(60°) = \tfrac{1}{2}\sqrt{3} \qquad (4)$$

$$\sin(90°) = \tfrac{1}{2}\sqrt{4}. \qquad (5)$$

See also TRIGONOMETRY VALUES 0, TRIGONOMETRY VALUES PI, TRIGONOMETRY VALUES PI/2, TRIGONOMETRY VALUES PI/3, TRIGONOMETRY VALUES PI/4, TRIGO-NOMETRY VALUES PI/5, TRIGONOMETRY VALUES PI/6, TRIGONOMETRY VALUES PI/7, TRIGONOMETRY VALUES PI/8, TRIGONOMETRY VALUES PI/9, TRIGONOMETRY VALUES PI/10, TRIGONOMETRY VALUES PI/11, TRIGO-NOMETRY VALUES PI/12, TRIGONOMETRY VALUES PI/15, TRIGONOMETRY VALUES PI/16, TRIGONOMETRY VALUES PI/17, TRIGONOMETRY VALUES PI/18, TRIGONOMETRY VALUES PI/20, TRIGONOMETRY VALUES PI/24, TRIGO-NOMETRY VALUES PI/30, TRIGONOMETRY VALUES PI/32

Trigonometric Functions

The functions (also called the CIRCULAR FUNCTIONS) comprising TRIGONOMETRY: the COSECANT csc x, COSINE cos x, COTANGENT cot x, SECANT sec x, SINE sin x, and TANGENT tan x. The inverses of these functions are denoted $\csc^{-1} x$, $\cos^{-1} x$, $\cot^{-1} x$, $\sec^{-1} x$, $\sin^{-1} x$, and $\tan^{-1} x$. Note that the f^{-1} NOTATION here means INVERSE FUNCTION, *not* f to the -1 POWER.

See also DOUBLE-ANGLE FORMULAS, HALF-ANGLE FORMULAS, HYPERBOLIC FUNCTIONS, TRIGONOMETRY

Trigonometric Power Formulas

Power formulas include

$$\sin^2 x = \tfrac{1}{2}[1 - \cos(2x)] \qquad (1)$$

$$\sin^3 x = \tfrac{1}{4}[3\sin x - \sin(3x)] \qquad (2)$$

$$\sin^4 x = \tfrac{1}{8}[3 - 4\cos(2x) + \cos(4x)] \qquad (3)$$

and

$$\cos^2 x = \tfrac{1}{2}[1 + \cos(2x)] \qquad (4)$$

$$\cos^3 x = \tfrac{1}{4}[3\cos x + \cos(3x)] \qquad (5)$$

$$\cos^4 x = \tfrac{1}{8}[3 + 4\cos(2x) + \cos(4x)] \qquad (6)$$

(Beyer 1987, p. 140). Formulas of these types can also be given analytically as

$$\sin^{2n} x = \frac{1}{2^{2n}}\binom{2n}{n} + \frac{(-1)^n}{2^{2n-1}}\sum_{k=0}^{n-1}(-1)^k\binom{2n}{k}\cos[2(n-k)x] \qquad (7)$$

$$\sin^{2n+1} = \frac{(-1)^n}{4^n}\sum_{k=0}^{n}(-1)^k\binom{2n+1}{k}\sin[2n+1-2k)x] \qquad (8)$$

$$\cos^{2n} x = \frac{1}{2^{2n}}\binom{2n}{n} + \frac{1}{2^{2n-1}}\sum_{k=0}^{n-1}\binom{2n}{k}\cos[2(n-k)x] \qquad (9)$$

$$\cos^{2n+1} x = \frac{1}{4^n}\sum_{k=0}^{n}\binom{2n+1}{k}\cos[(2n+1-2k)x] \qquad (10)$$

(Kogan), where $\binom{n}{m}$ is a BINOMIAL COEFFICIENT.

See also TRIGONOMETRY

Trigonometric Product Formulas

References

Beyer, W. H. *CRC Standard Mathematical Tables, 28th ed.* Boca Raton, FL: CRC Press, 1987.

Kogan, S. "A Note on Definite Integrals Involving Trigonometric Functions." http://www.mathsoft.com/asolve/constant/pi/sin/sin.html.

Trigonometric Product Formulas

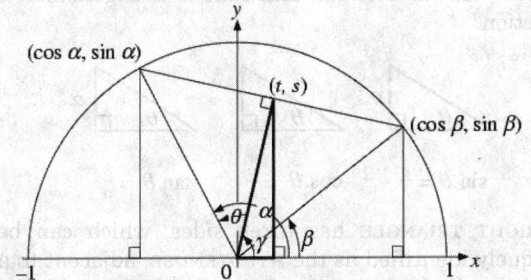

Trigonometric product formulas for the sum of the cosines and sines of two angles can be derived using the above figure (Kung 1996). From the figure, define

$$\theta = \tfrac{1}{2}(\alpha - \beta) \tag{1}$$

$$\gamma = \tfrac{1}{2}(\alpha + \beta). \tag{2}$$

Then we have the identity

$$s = \tfrac{1}{2}(\sin \alpha + \sin \beta) = \cos\left[\tfrac{1}{2}(\alpha - \beta)\right]\sin\left[\tfrac{1}{2}(\alpha + \beta)\right] \tag{3}$$

$$t = \tfrac{1}{2}(\cos \alpha + \cos \beta) = \cos\left[\tfrac{1}{2}(\alpha - \beta)\right]\cos\left[\tfrac{1}{2}(\alpha + \beta)\right]. \tag{4}$$

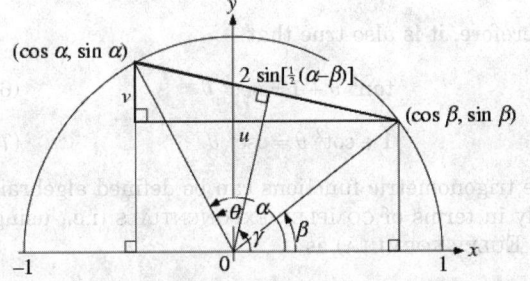

Trigonometric product formulas for the difference of the cosines and sines of two angles can be derived using the similar figure illustrated above (Kung 1996). With θ and γ as previously defined, the above figure gives

$$u = \cos \beta - \cos \alpha = 2 \sin\left[\tfrac{1}{2}(\alpha - \beta)\right]\sin\left[\tfrac{1}{2}(\alpha + \beta)\right] \tag{5}$$

$$v = \sin \alpha - \sin \beta = 2 \sin\left[\tfrac{1}{2}(\alpha - \beta)\right]\cos\left[\tfrac{1}{2}(\alpha + \beta)\right]. \tag{6}$$

See also Double-Angle Formulas, Half-Angle Formulas, Prosthaphaeresis Formulas, Trigonometric Addition Formulas, Trigonometry

References

Kung, S. H. "Proof without Words: The Difference-Product Identities" and "Proof without Words: The Sum-Product Identities." *Math. Mag.* **69**, 269, 1996.

Trigonometric Series

FOURIER SERIES

Trigonometric Series Formulas

Trigonometric identities which prove useful in the construction of map projections include

$$A \sin(2\phi) + B \sin(4\phi) + C \sin(6\phi) + D \sin(8\phi)$$
$$= \sin(2\phi)(A' + \cos(2\phi)(B' + \cos(2\phi)(C' + D' \cos(2\phi)))), \tag{1}$$

where

$$A' \equiv A - C \tag{2}$$

$$B' \equiv 2B - 4D \tag{3}$$

$$C' \equiv 4C \tag{4}$$

$$D' \equiv 8D. \tag{5}$$

$$A \sin \phi + B \sin(3\phi) + C \sin(5\phi) + D \sin(7\phi)$$
$$= \sin \phi (A' + \sin^2 \phi (B' + \sin^2 \phi (C' + D' \sin^2 \phi))), \tag{6}$$

where

$$A' \equiv A + 3B + 5C + 7D \tag{7}$$

$$B' \equiv -4B - 20C - 56D \tag{8}$$

$$C' \equiv 16C + 112D \tag{9}$$

$$D' \equiv -64D. \tag{10}$$

$$A + B \cos(2\phi) + C \cos(4\phi) + D \cos(6\phi) + E \cos(8\phi)$$
$$= A' + \cos(2\phi)(B' + \cos(2\phi)(C' + \cos(2\phi)$$
$$\times (D' + E' \cos(2\phi)))), \tag{11}$$

where

$$A' \equiv A - C + E \tag{12}$$

$$B' \equiv B - 3D \tag{13}$$

$$C' \equiv 2C - 8E \tag{14}$$

$$D' \equiv 4D \tag{15}$$

$$E' \equiv 8E \tag{16}$$

(Snyder 1987).

See also TRIGONOMETRY

References

Snyder, J. P. *Map Projections--A Working Manual.* U. S. Geological Survey Professional Paper 1395. Washington, DC: U. S. Government Printing Office, p. 19, 1987.

Trigonometric Substitution

INTEGRALS OF THE FORM

$$\int f(\cos \theta, \sin \theta) \, d\theta \qquad (1)$$

can be solved by making the substitution $z = e^{i\theta}$ so that $dz = ie^{i\theta} \, d\theta$ and expressing

$$\cos \theta = \frac{e^{i\theta} + e^{-i\theta}}{2} = \frac{z + z^{-1}}{2} \qquad (2)$$

$$\sin \theta = \frac{e^{i\theta} - e^{-i\theta}}{2i} = \frac{z - z^{-1}}{2i}. \qquad (3)$$

The integral can then be solved by CONTOUR INTEGRATION.

Alternatively, making the substitution $t \equiv \tan(\theta/2)$ transforms (1) into

$$\int f\left(\frac{2t}{1 + t^2}, \frac{1 - t^2}{1 + t^2}\right) \frac{2 \, dt}{1 + t^2}. \qquad (4)$$

The following table gives trigonometric substitutions which can be used to transform integrals involving square roots.

Form	Substitution
$\sqrt{a^2 - x^2}$	$x = a \sin \theta$
$\sqrt{a^2 + x^2}$	$x = a \tan \theta$
$\sqrt{x^2 - a^2}$	$x = a \sec \theta$

See also HYPERBOLIC SUBSTITUTION

Trigonometry

The study of ANGLES and of the angular relationships of planar and 3-D figures is known as trigonometry. The TRIGONOMETRIC FUNCTIONS (also called the CIRCULAR FUNCTIONS) comprising trigonometry are the COSECANT $\csc x$, COSINE $\cos x$, COTANGENT $\cot x$, SECANT $\sec x$, SINE $\sin x$, and TANGENT $\tan x$. The inverses of these functions are denoted $\csc^{-1} x$, $\cos^{-1} x$, $\cot^{-1} x$, $\sec^{-1} x$, $\sin^{-1} x$, and $\tan^{-1} x$. Note that the f^{-1} NOTATION here means INVERSE FUNCTION, *not* f to the -1 POWER.

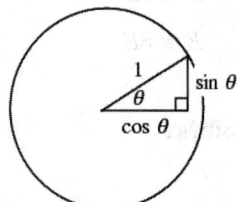

The trigonometric functions are most simply defined using the UNIT CIRCLE. Let θ be an ANGLE measured counterclockwise from the x-AXIS along an ARC of the CIRCLE. Then $\cos \theta$ is the horizontal coordinate of the ARC endpoint, and $\sin \theta$ is the vertical component. The RATIO $\sin \theta / \cos \theta$ is defined as $\tan \theta$. As a result of this definition, the trigonometric functions are periodic with period 2π, so

$$\text{func}(2\pi n + \theta) = \text{func}(\theta), \qquad (1)$$

where n is an INTEGER and func is a trigonometric function.

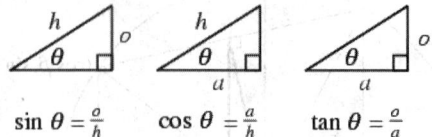

$$\sin \theta = \frac{o}{h} \qquad \cos \theta = \frac{a}{h} \qquad \tan \theta = \frac{o}{a}$$

A RIGHT TRIANGLE has three sides, which can be uniquely identified as the HYPOTENUSE, adjacent to a given angle θ, or opposite θ. A helpful mnemonic for remembering the definitions of the trigonometric functions is then given by "oh, ah, oh-ah,"

$$\sin \theta = \frac{\text{opposite}}{\text{hypotenuse}} \qquad (2)$$

$$\cos \theta = \frac{\text{adjacent}}{\text{hypotenuse}} \qquad (3)$$

$$\tan \theta = \frac{\text{opposite}}{\text{adjacent}}. \qquad (4)$$

From the PYTHAGOREAN THEOREM,

$$\sin^2 \theta + \cos^2 \theta = 1. \qquad (5)$$

Therefore, it is also true that

$$\tan^2 \theta + 1 = \sec^2 \theta \qquad (6)$$

$$1 + \cot^2 \theta = \csc^2 \theta. \qquad (7)$$

The trigonometric functions can be defined algebraically in terms of COMPLEX EXPONENTIALS (i.e., using the EULER FORMULA) as

$$\sin z \equiv \frac{e^{iz} - e^{-iz}}{2i} \qquad (8)$$

$$\csc z \equiv \frac{1}{\sin z} = \frac{2i}{e^{iz} - e^{-iz}} \qquad (9)$$

$$\cos z \equiv \frac{e^{iz} + e^{-iz}}{2} \qquad (10)$$

$$\sec z \equiv \frac{1}{\cos z} = \frac{2}{e^{iz} + e^{-iz}} \qquad (11)$$

$$\tan z \equiv \frac{\sin z}{\cos z} = \frac{e^{iz} - e^{-iz}}{i(e^{iz} + e^{-iz})} \qquad (12)$$

$$\cot z \equiv \frac{1}{\tan z} = \frac{i(e^{iz} + e^{-iz})}{e^{iz} - e^{-iz}} = \frac{i(1 + e^{-2iz})}{1 - e^{-2iz}}. \quad (13)$$

Hybrid trigonometric product/sum formulas are

$$\sin(\alpha + \beta) \sin(\alpha - \beta) = \sin^2 \alpha - \sin^2 \beta$$

$$= \cos^2 \beta - \cos^2 \alpha \quad (14)$$

$$\cos(\alpha + \beta) \cos(\alpha - \beta) = \cos^2 \alpha - \sin^2 \beta$$

$$= \cos^2 \beta - \sin^2 \alpha. \quad (15)$$

OSBORNE'S RULE gives a prescription for converting trigonometric identities to analogous identities for HYPERBOLIC FUNCTIONS.

For IMAGINARY arguments,

$$\sin(iz) = i \sinh z \quad (16)$$

$$\cos(iz) = \cosh z. \quad (17)$$

For COMPLEX arguments,

$$\sin(x + iy) = \sin x \cosh y + i \cos x \sinh y \quad (18)$$

$$\cos(x + iy) = \cos x \cosh y - i \sin x \sinh y. \quad (19)$$

For the ABSOLUTE SQUARE of COMPLEX arguments $z = x + iy$,

$$|\sin(x + iy)|^2 = \sin^2 x + \sinh^2 y \quad (20)$$

$$|\cos(x + iy)|^2 = \cos^2 x + \sinh^2 y. \quad (21)$$

The MODULUS also satisfies the curious identity

$$|\sin(x + iy)| = |\sin x + \sin(iy)|. \quad (22)$$

The only functions satisfying identities of this form,

$$|f(x + iy)| = |f(x) + f(iy)| \quad (23)$$

are $f(z) = Az$, $f(z) = A \sin(bz)$, and $f(z) = A \sinh(bz)$ (Robinson 1957).

See also COSECANT, COSINE, COTANGENT, DOUBLE-ANGLE FORMULAS, EUCLIDEAN NUMBER, HALF-ANGLE FORMULAS, INVERSE COSECANT, INVERSE COSINE, INVERSE COTANGENT, INVERSE SECANT, INVERSE SINE, INVERSE TANGENT, INVERSE TRIGONOMETRIC FUNCTIONS, OSBORNE'S RULE, POLYGON, PROSTHAPHAERESIS FORMULAS, SECANT, SINE, TANGENT, TRIGONOMETRIC ADDITION FORMULAS, TRIGONOMETRIC ANGLES, TRIGONOMETRIC FUNCTIONS, TRIGONOMETRIC POWER FORMULAS, TRIGONOMETRIC SERIES FORMULAS, WERNER FORMULAS

References

Abramowitz, M. and Stegun, C. A. (Eds.). "Circular Functions." §4.3 in *Handbook of Mathematical Functions with Formulas, Graphs, and Mathematical Tables, 9th printing.* New York: Dover, pp. 71–79, 1972.
Bahm, L. B. *The New Trigonometry on Your Own.* Patterson, NJ: Littlefield, Adams & Co., 1964.

Beyer, W. H. "Trigonometry." *CRC Standard Mathematical Tables, 28th ed.* Boca Raton, FL: CRC Press, pp. 134–152, 1987.
Borchardt, W. G. and Perrott, A. D. *A New Trigonometry for Schools.* London: G. Bell, 1930.
Dixon, R. "The Story of Sine and Cosine." § 4.4 in *Mathographics.* New York: Dover, pp. 102–106, 1991.
Hobson, E. W. *A Treatise on Plane Trigonometry.* London: Cambridge University Press, 1925.
Kells, L. M.; Kern, W. F.; and Bland, J. R. *Plane and Spherical Trigonometry.* New York: McGraw-Hill, 1940.
Maor, E. *Trigonometric Delights.* Princeton, NJ: Princeton University Press, 1998.
Morrill, W. K. *Plane Trigonometry, rev. ed.* Dubuque, IA: Wm. C. Brown, 1964.
Robinson, R. M. "A Curious Mathematical Identity." *Amer. Math. Monthly* **64**, 83–85, 1957.
Siddons, A. W. and Hughes, R. T. *Trigonometry, Parts I-IV.* London: Cambridge University Press, 1929.
Sloane, N. J. A. Sequences A003401/M0505 in "An On-Line Version of the Encyclopedia of Integer Sequences." http://www.research.att.com/~njas/sequences/eisonline.html.
Thompson, J. E. *Trigonometry for the Practical Man.* Princeton, NJ: Van Nostrand.
Weisstein, E. W. "Exact Values of Trigonometric Functions." MATHEMATICA NOTEBOOK TrigExact.M.
Yates, R. C. "Trigonometric Functions." *A Handbook on Curves and Their Properties.* Ann Arbor, MI: J. W. Edwards, pp. 225–232, 1952.
Weisstein, E. W. "Books about Trigonometry." http://www.treasure-troves.com/books/Trigonometry.html.
Zill, D. G. and Dewar, J. M. *Trigonometry, 2nd ed.* New York: McGraw-Hill 1990.

Trigonometry Values Pi

By the definition of the trigonometric functions,

$$\cos \pi = -1 \quad (1)$$

$$\cos \pi = \infty \quad (2)$$

$$\csc \pi = \infty \quad (3)$$

$$\sec \pi = -1 \quad (4)$$

$$\sin \pi = 0 \quad (5)$$

$$\tan \pi = 0. \quad (6)$$

Trigonometry Values Pi/2

By the definition of the trigonometric functions,

$$\cos\left(\frac{\pi}{2}\right) = 0 \quad (1)$$

$$\cot\left(\frac{\pi}{2}\right) = 0 \quad (2)$$

$$\csc\left(\frac{\pi}{2}\right) = 1 \quad (3)$$

$$\sec\left(\frac{\pi}{2}\right) = \infty \quad (4)$$

$$\sin\left(\frac{\pi}{2}\right) = 1 \tag{5}$$

$$\tan\left(\frac{\pi}{2}\right) = \infty. \tag{6}$$

See also DIGON

Trigonometry Values Pi/3

$$\cos\left(\frac{\pi}{3}\right) = \frac{1}{2} \tag{1}$$

$$\cot\left(\frac{\pi}{3}\right) = \frac{1}{3}\sqrt{3} \tag{2}$$

$$\csc\left(\frac{\pi}{3}\right) = \frac{2}{3}\sqrt{3} \tag{3}$$

$$\sec\left(\frac{\pi}{3}\right) = 2 \tag{4}$$

$$\sin\left(\frac{\pi}{3}\right) = \frac{1}{2}\sqrt{3} \tag{5}$$

$$\tan\left(\frac{\pi}{3}\right) = \sqrt{3}. \tag{6}$$

These formulas can be derived from knowledge of the TRIGONOMETRY VALUES FOR PI/6

$$\sin\left(\frac{\pi}{6}\right) = \frac{1}{2} \tag{7}$$

$$\cos\left(\frac{\pi}{6}\right) = \frac{1}{2}\sqrt{3} \tag{8}$$

together with the trigonometric identity

$$\sin(2\alpha) = 2\sin\alpha\cos\alpha, \tag{9}$$

giving

$$\sin\left(\frac{\pi}{3}\right) = 2\sin\left(\frac{\pi}{6}\right)\cos\left(\frac{\pi}{6}\right) = 2\left(\frac{1}{2}\right)\left(\frac{1}{2}\sqrt{3}\right) = \frac{1}{2}\sqrt{3} \tag{10}$$

is obtained. Using the identity

$$\cos(2\alpha) = 1 - 2\sin^2\alpha, \tag{11}$$

then gives

$$\cos\left(\frac{\pi}{3}\right) = 1 - 2\sin^2\left(\frac{\pi}{6}\right) = 1 - 2\left(\frac{1}{2}\right)^2 = \frac{1}{2}. \tag{12}$$

See also EQUILATERAL TRIANGLE

Trigonometry Values Pi/4

$$\cos\left(\frac{\pi}{4}\right) = \frac{1}{2}\sqrt{2} \tag{1}$$

$$\cot\left(\frac{\pi}{4}\right) = 1 \tag{2}$$

$$\csc\left(\frac{\pi}{4}\right) = \sqrt{2} \tag{3}$$

$$\sec\left(\frac{\pi}{4}\right) = \sqrt{2} \tag{4}$$

$$\sin\left(\frac{\pi}{4}\right) = \frac{1}{2}\sqrt{2} \tag{5}$$

$$\tan\left(\frac{\pi}{4}\right) = 1. \tag{6}$$

For a RIGHT ISOSCELES TRIANGLE, symmetry requires that the angle at each VERTEX be given by

$$\frac{1}{2}\pi + 2\alpha = \pi, \tag{7}$$

so $\alpha = \pi/4$. The sides are equal, so

$$\sin^2\alpha + \cos^2\alpha = 2\sin^2\alpha = 1. \tag{8}$$

Solving gives the above equations.

See also SQUARE

Trigonometry Values Pi/5

$$\cos\left(\frac{\pi}{5}\right) = \frac{1}{4}\left(1 + \sqrt{5}\right) \tag{1}$$

$$\cos\left(\frac{2\pi}{5}\right) = \frac{1}{4}\left(-1 + \sqrt{5}\right) \tag{2}$$

$$\cot\left(\frac{\pi}{5}\right) = \frac{1}{5}\sqrt{25 + 10\sqrt{5}} \tag{3}$$

$$\cot\left(\frac{2\pi}{5}\right) = \frac{1}{5}\sqrt{25 - 10\sqrt{5}} \tag{4}$$

$$\csc\left(\frac{\pi}{5}\right) = \frac{1}{5}\sqrt{50 + 10\sqrt{5}} \tag{5}$$

$$\csc\left(\frac{2\pi}{5}\right) = \frac{1}{5}\sqrt{50 - 10\sqrt{5}} \tag{6}$$

$$\sec\left(\frac{\pi}{5}\right) = \sqrt{5} - 1 \tag{7}$$

$$\sec\left(\frac{2\pi}{5}\right) = 1 + \sqrt{5} \tag{8}$$

$$\sin\left(\frac{\pi}{5}\right) = \frac{1}{4}\sqrt{10 - 2\sqrt{5}} \tag{9}$$

$$\sin\left(\frac{2\pi}{5}\right) = \frac{1}{4}\sqrt{10 + 2\sqrt{5}} \tag{10}$$

$$\tan\left(\frac{\pi}{5}\right) = \sqrt{5 - 2\sqrt{5}} \tag{11}$$

$$\tan\left(\frac{2\pi}{5}\right) = \sqrt{5 + 2\sqrt{5}}. \tag{12}$$

These formulas can be derived using the identity

$$\sin(5\alpha) = 5\sin\alpha - 20\sin^3\alpha + 16\sin^5\alpha. \tag{13}$$

Now, let $\alpha \equiv \pi/5$ and $x \equiv \sin\alpha$. Then

$$\sin\pi = 0 = 5x - 20x^3 + 16x^5 \tag{14}$$

$$16x^4 - 20x^2 + 5 = 0. \tag{15}$$

Solving the QUADRATIC EQUATION for x^2 gives

$$\sin^2\left(\frac{\pi}{5}\right) = x^2 = \frac{20 \pm \sqrt{(-20)^2 - 4 \cdot 16 \cdot 5}}{2 \cdot 16}$$

$$= \frac{20 \pm \sqrt{80}}{32} = \frac{1}{8}\left(5 \pm \sqrt{5}\right). \tag{16}$$

Now, $\sin(\pi/5)$ must be less than

$$\sin\left(\frac{\pi}{4}\right) = \frac{1}{2}\sqrt{2}, \tag{17}$$

so taking the MINUS SIGN and simplifying gives

$$\sin\left(\frac{\pi}{5}\right) = \sqrt{\frac{5 - \sqrt{5}}{8}} = \frac{1}{4}\sqrt{10 - 2\sqrt{5}}. \tag{18}$$

$\cos(\pi/5)$ can be computed from

$$\cos\left(\frac{\pi}{5}\right) = \sqrt{1 - \sin^2\left(\frac{\pi}{5}\right)} = \frac{1}{4}\left(1 + \sqrt{5}\right). \tag{19}$$

See also DODECAHEDRON, GOLDEN RATIO, ICOSAHE-
DRON, PENTAGON, PENTAGRAM

Trigonometry Values Pi/6

$$\cos\left(\frac{\pi}{6}\right) = \frac{1}{2}\sqrt{3} \tag{1}$$

$$\cot\left(\frac{\pi}{6}\right) = \sqrt{3} \tag{2}$$

$$\csc\left(\frac{\pi}{6}\right) = 2 \tag{3}$$

$$\sec\left(\frac{\pi}{6}\right) = \frac{2}{3}\sqrt{3} \tag{4}$$

$$\sin\left(\frac{\pi}{6}\right) = \frac{1}{2} \tag{5}$$

$$\tan\left(\frac{\pi}{6}\right) = \frac{1}{3}\sqrt{3}. \tag{6}$$

Given a RIGHT TRIANGLE with angles defined to be α and 2α, it must be true that

$$\alpha + 2\alpha + \tfrac{1}{2}\pi = \pi, \tag{7}$$

so $\alpha = \pi/6$. Define the hypotenuse to have length 1 and the side opposite α to have length x, then the side opposite 2α has length $\sqrt{1 - x^2}$. This gives $\sin\alpha \equiv x$ and

$$\sin(2\alpha) = \sqrt{1 - x^2}. \tag{8}$$

But

$$\sin(2\alpha) = 2\sin\alpha\cos\alpha = 2x\sqrt{1 - x^2}, \tag{9}$$

so we have

$$\sqrt{1 - x^2} = 2x\sqrt{1 - x^2}. \tag{10}$$

This gives $2x = 1$, or

$$\sin\left(\frac{\pi}{6}\right) = \frac{1}{2}. \tag{11}$$

$\cos(\pi/6)$ is then computed from

$$\cos\left(\frac{\pi}{6}\right) = \sqrt{1 - \sin^2\left(\frac{\pi}{6}\right)} = \sqrt{1 - \left(\frac{1}{2}\right)^2} = \frac{1}{2}\sqrt{3}. \tag{12}$$

See also HEXAGON, HEXAGRAM

Trigonometry Values Pi/7

Trigonometric functions of $n\pi/7$ for n an integer cannot be expressed in terms of sums, products, and finite ROOT EXTRACTIONS on *real* rational numbers because 7 is not a FERMAT PRIME. This also means that the HEPTAGON is not a CONSTRUCTIBLE POLYGON.

However, exact expressions involving roots of *complex* numbers can still be derived using the trigonometric identity

$$\sin(n\alpha) = 2\sin[(n-1)\alpha]\cos\alpha - \sin[(n-2)\alpha]. \quad (1)$$

The case $n = 7$ gives

$$\sin(7\alpha) = 2\sin(6\alpha)\cos\alpha - \sin(5\alpha)$$

$$= 2(32\cos^5\alpha\sin\alpha - 32\cos^3\alpha\sin\alpha$$

$$+6\cos\alpha\sin\alpha)\cos\alpha$$

$$-(5\sin\alpha - 20\sin^3\alpha + 16\sin^5\alpha)$$

$$= 64\cos^6\alpha\sin\alpha - 64\cos^4\alpha\sin\alpha + 12\cos^2\alpha\sin\alpha$$

$$-5\sin\alpha + 20(1 - \cos^2\alpha)\sin\alpha$$

$$-16(1 - 2\cos^2\alpha + \cos^4\alpha)\sin\alpha$$

$$= \sin\alpha(64\cos^6\alpha - 80\cos^4\alpha + 24\cos^2\alpha - 1). \quad (2)$$

Rewrite this using the identity $\cos^2\alpha = 1 - \sin^2\alpha$,

$$\sin\left(\frac{\pi}{7}\right) = \sin\alpha(7 - 56\sin^2\alpha + 112\sin^4\alpha - 64\sin^6\alpha)$$

$$= -64\sin\alpha\left(\sin^6\alpha - \frac{112}{64}\sin^4\alpha + \frac{56}{64}\sin^2\alpha - \frac{7}{64}\right). \quad (3)$$

Now, let $\alpha \equiv \pi/7$ and $x \equiv \sin^2\alpha$, then

$$\sin(\pi) = 0 = x^3 - \frac{7}{4}x^2 + \frac{7}{8}x - \frac{7}{64}, \quad (4)$$

which is a CUBIC EQUATION in x. The ROOTS are numerically found to be $x \approx 0.188255,\ 0.611260,\ 0.950484$. But $\sin\alpha = \sqrt{x}$, so these ROOTS correspond to $\sin\alpha \approx 0.4338,\ \sin(2\alpha) \approx 0.7817,\ \sin(3\alpha) \approx 0.9749$. By NEWTON'S RELATION

$$\prod_i r_i = -a_0 \quad (5)$$

we have

$$x_1 x_2 x_3 = \frac{7}{64}, \quad (6)$$

or

$$\sin\left(\frac{\pi}{7}\right)\sin\left(\frac{2\pi}{7}\right)\sin\left(\frac{3\pi}{7}\right) = \sqrt{\frac{7}{64}} = \frac{1}{8}\sqrt{7}. \quad (7)$$

Similarly,

$$\cos\left(\frac{\pi}{7}\right)\cos\left(\frac{2\pi}{7}\right)\cos\left(\frac{3\pi}{7}\right) = \frac{1}{8}. \quad (8)$$

and

$$\cos^2\left(\frac{\pi}{7}\right) - \cos\left(\frac{\pi}{7}\right)\cos\left(\frac{2\pi}{7}\right) = \frac{1}{4} \quad (9)$$

(Bankoff and Garfunkel 1973).

The constants of the CUBIC EQUATION are given by

$$Q \equiv \frac{1}{9}(3a_1 - a_2^2) = \frac{1}{9}\left[3\cdot\frac{7}{8} - \left(-\frac{7}{4}\right)^2\right] = -\frac{7}{144} \quad (10)$$

$$R \equiv \frac{1}{54}(9a_2a_1 - 2a_2^3 - 27a_0)$$

$$= \frac{1}{54}\left[9\left(-\frac{7}{4}\right)\left(\frac{1}{7}8\right) - 2\left(-\frac{7}{4}\right)^3 - 27\left(-\frac{7}{64}\right)\right]$$

$$= -f73456. \quad (11)$$

The DISCRIMINANT is then

$$D \equiv Q^3 + R^3 = -\frac{343}{2{,}985{,}984} + \frac{49}{11{,}943{,}936}$$

$$= -\frac{49}{442{,}368} < 0, \quad (12)$$

so there are three distinct REAL ROOTS. Finding the first one,

$$x = \sqrt{R + \sqrt{D}} + \sqrt{R - \sqrt{D}} - \frac{1}{3}a_2. \quad (13)$$

Writing

$$\sqrt{D} = 3^{-3/2}\frac{7}{128}i, \quad (14)$$

plugging in from above, and anticipating that the solution we have picked corresponds to $\sin(3\pi/7)$,

$$\sin\left(\frac{3\pi}{7}\right) = \sqrt{x}$$

$$= \sqrt{\sqrt{-\frac{7}{3456} + 3^{-3/2}\frac{7}{128}i} + \sqrt{-\frac{7}{3456} - 3^{-3/2}\frac{7}{128}i - \frac{1}{3}\left(-\frac{7}{4}\right)}}$$

$$= \sqrt{\sqrt{-\frac{7}{3456} + 3^{-3/2}\frac{7}{128}i} + \sqrt{-\frac{7}{3456} - 3^{-3/2}\frac{7}{128}i + \frac{7}{12}}}$$

$$= \sqrt{\sqrt{\frac{7}{3456}(-1 + 3^{3/2}i)} - \sqrt{\frac{7}{3456}(1 + 3^{3/2}i)} + \frac{7}{12}}$$

See also HEPTAGON, SILVER CONSTANT

References

Bankoff, L. and Garfunkel, J. "The Heptagonal Triangle." *Math. Mag.* **46**, 7–19, 1973.

Trigonometry Values Pi/8

$$\cos\left(\frac{\pi}{8}\right) = \frac{1}{2}\sqrt{2+\sqrt{2}} \tag{1}$$

$$\cos\left(\frac{3\pi}{8}\right) = \frac{1}{2}\sqrt{2-\sqrt{2}} \tag{2}$$

$$\cot\left(\frac{\pi}{8}\right) = 1+\sqrt{2} \tag{3}$$

$$\cot\left(\frac{3\pi}{8}\right) = \sqrt{2}-1 \tag{4}$$

$$\csc\left(\frac{\pi}{8}\right) = \sqrt{4+2\sqrt{2}} \tag{5}$$

$$\csc\left(\frac{3\pi}{8}\right) = \sqrt{4-2\sqrt{2}} \tag{6}$$

$$\sec\left(\frac{\pi}{8}\right) = \sqrt{4-2\sqrt{2}} \tag{7}$$

$$\sec\left(\frac{3\pi}{8}\right) = \sqrt{4+2\sqrt{2}} \tag{8}$$

$$\sec\left(\frac{\pi}{8}\right) = \frac{1}{2}\sqrt{2-\sqrt{2}} \tag{9}$$

$$\sin\left(\frac{3\pi}{8}\right) = \frac{1}{2}\sqrt{2+\sqrt{2}} \tag{10}$$

$$\tan\left(\frac{\pi}{8}\right) = \sqrt{2}-1 \tag{11}$$

$$\tan\left(\frac{3\pi}{8}\right) = 1+\sqrt{2}. \tag{12}$$

$$\sin\left(\frac{\pi}{8}\right) = \sin\left(\frac{1}{2}\cdot\frac{\pi}{4}\right) = \sqrt{\frac{1}{2}\left(1-\cos\frac{\pi}{4}\right)}$$

$$= \sqrt{\frac{1}{2}\left(1-\frac{1}{2}\sqrt{2}\right)} = \frac{1}{2}\sqrt{2-\sqrt{2}}. \tag{13}$$

Now, checking to see if the SQUARE ROOT can be simplified gives

$$a^2-b^2c = 2^2-1^2\cdot 2 = 4-2 = 2, \tag{14}$$

which is not a PERFECT SQUARE, so the above expression cannot be simplified. Similarly,

$$\cos\left(\frac{\pi}{8}\right) = \cos\left(\frac{1}{2}\frac{\pi}{4}\right) = \sqrt{\frac{1}{2}\left(1+\cos\frac{\pi}{4}\right)}$$

$$= \sqrt{\frac{1}{2}\left(1+\frac{\sqrt{2}}{3}\right)} = \frac{1}{2}\sqrt{2+\sqrt{2}} \tag{15}$$

$$\tan\left(\frac{\pi}{8}\right) = \sqrt{\frac{2-\sqrt{2}}{2+\sqrt{2}}} = \sqrt{\frac{\left(2-\sqrt{2}\right)^2}{4-2}} = \sqrt{\frac{4+2-4\sqrt{2}}{2}}$$

$$= \sqrt{\frac{6-4\sqrt{2}}{2}} = \sqrt{3-2\sqrt{2}}. \tag{16}$$

But

$$a^2-b^2c = 3^2-2^2 2 = 9-8 = 1 \tag{17}$$

is a PERFECT SQUARE, so we can find

$$d = \frac{1}{2}(3\pm 1) = 1,\ 2. \tag{18}$$

Rewrite the above as

$$\tan\left(\frac{\pi}{8}\right) = \sqrt{2}-1 \tag{19}$$

$$\cot\left(\frac{\pi}{8}\right) = \frac{1}{\sqrt{2}-1} = \frac{\sqrt{2}+1}{2-1} = \sqrt{2}+1. \tag{20}$$

See also OCTAGON

Trigonometry Values Pi/9

Trigonometric functions of $n\pi/9$ radians for n an integer not divisible by 3 (e.g., $40°$ and $80°$) cannot be expressed in terms of sums, products, and finite ROOT EXTRACTIONS on RATIONAL NUMBERS because 9 is not a product of distinct FERMAT PRIMES. This also means that the regular NONAGON is not a CONSTRUCTIBLE POLYGON.

However, exact expressions involving roots of *complex* numbers can still be derived using the trigonometric identity

$$\sin(3\alpha) = 3\sin\alpha - 4\sin^3\alpha. \tag{1}$$

Let $\alpha \equiv \pi/9$ and $x \equiv \sin\alpha$. Then the above identity gives the CUBIC EQUATION

$$4x^3 - 3x + \frac{1}{2}\sqrt{3} = 0 \tag{2}$$

$$x^3 - \frac{3}{4}x = -\frac{1}{8}\sqrt{3}. \tag{3}$$

This cubic is OF THE FORM

$$x^3 + px = q, \tag{4}$$

where

$$p = -\frac{3}{4} \tag{5}$$

$$q = -\frac{1}{8}\sqrt{3}. \tag{6}$$

The DISCRIMINANT is then

$$D \equiv \left(\frac{p}{3}\right)^3 + \left(\frac{q}{2}\right)^2$$

$$= \left(-\frac{1}{4}\right)^3 + \left(\frac{\sqrt{3}}{16}\right)^2 = -\frac{1}{16 \cdot 4} + \frac{3}{16 \cdot 16} = \frac{-4+3}{256}$$

$$= -\frac{1}{256} < 0. \tag{7}$$

There are therefore three REAL distinct roots, which are approximately -0.9848, 0.3240, and 0.6428. We want the one in the first QUADRANT, which is 0.3240.

$$\sin\left(\frac{\pi}{9}\right) = \sqrt{-\frac{\sqrt{3}}{16} + \sqrt{-\frac{1}{256}}} + \sqrt{-\frac{\sqrt{3}}{16} - \sqrt{-\frac{1}{256}}}$$

$$= \sqrt{-\frac{\sqrt{3}}{16} + \frac{1}{16}i} - \sqrt{\frac{\sqrt{3}}{16} + \frac{1}{16}i}$$

$$= 2^{-4/3}\left(\sqrt{i - \sqrt{3}} - \sqrt{i + \sqrt{3}}\right)$$

$$\approx 0.34202 \tag{8}$$

Similarly,

$$\cos\left(\frac{\pi}{9}\right) = 2^{-4/3}\left(\sqrt{1 + i\sqrt{3}} + \sqrt{1 - i\sqrt{3}}\right)$$

$$\approx 0.93969. \tag{9}$$

Because of the NEWTON'S RELATIONS, we have the identities

$$\sin\left(\frac{\pi}{9}\right)\sin\left(\frac{2\pi}{9}\right)\sin\left(\frac{4\pi}{9}\right) = \frac{1}{8}\sqrt{3} \tag{10}$$

$$\cos\left(\frac{\pi}{9}\right)\cos\left(\frac{2\pi}{9}\right)\cos\left(\frac{4\pi}{9}\right) = \frac{1}{8} \tag{11}$$

$$\tan\left(\frac{\pi}{9}\right)\tan\left(\frac{2\pi}{9}\right)\tan\left(\frac{4\pi}{9}\right) = \sqrt{3}. \tag{12}$$

(11) is known as MORRIE'S LAW.

See also MORRIE'S LAW, NONAGON, STAR OF GOLIATH

Trigonometry Values Pi/10

$$\cos\left(\frac{\pi}{10}\right) = \frac{1}{4}\sqrt{10 + 2\sqrt{5}} \tag{1}$$

$$\cos\left(\frac{3\pi}{10}\right) = \frac{1}{4}\sqrt{10 - 2\sqrt{5}} \tag{2}$$

$$\cos\left(\frac{\pi}{10}\right) = \sqrt{5 + 2\sqrt{5}} \tag{3}$$

$$\cot\left(\frac{3\pi}{10}\right) = \sqrt{5 - 2\sqrt{5}} \tag{4}$$

$$\csc\left(\frac{\pi}{10}\right) = 1 + \sqrt{5} \tag{5}$$

$$\csc\left(\frac{3\pi}{10}\right) = \sqrt{5} - 1 \tag{6}$$

$$\sec\left(\frac{\pi}{10}\right) = \frac{1}{5}\sqrt{50 - 10\sqrt{5}} \tag{7}$$

$$\sec\left(\frac{3\pi}{10}\right) = \frac{1}{5}\sqrt{50 - 10\sqrt{5}} \tag{8}$$

$$\sin\left(\frac{\pi}{10}\right) = \frac{1}{4}\left(\sqrt{5} - 1\right) \tag{9}$$

$$\sin\left(\frac{3\pi}{10}\right) = \frac{1}{4}\left(1 + \sqrt{5}\right) \tag{10}$$

$$\tan\left(\frac{\pi}{10}\right) = \frac{1}{5}\sqrt{25 - 10\sqrt{5}} \tag{11}$$

$$\tan\left(\frac{3\pi}{10}\right) = \frac{1}{5}\sqrt{25 + 10\sqrt{5}} \tag{12}$$

To derive these formulas, start with

$$\sin\left(\frac{\pi}{10}\right) = \sin\left(\frac{1}{2}\cdot\frac{\pi}{5}\right) = \sqrt{\frac{1}{2}\left[1 - \cos\left(\frac{\pi}{5}\right)\right]}$$

$$= \sqrt{\frac{1}{2}\left[1 - \frac{1}{4}(1 + \sqrt{5})\right]} = \frac{1}{4}\left(\sqrt{5} - 1\right). \tag{13}$$

So we have

$$\cos\left(\frac{\pi}{10}\right) = \cos\left(\frac{1}{2}\cdot\frac{\pi}{5}\right) = \sqrt{\frac{1}{2}\left[1 + \cos\left(\frac{\pi}{5}\right)\right]}$$

$$= \sqrt{\frac{1}{2}\left[1 + \frac{1}{4}(1 + \sqrt{5})\right]}$$

$$= \frac{1}{4}\sqrt{10 + 2\sqrt{5}} \tag{14}$$

and

$$\tan\left(\frac{\pi}{10}\right) = \sqrt{\frac{3 - \sqrt{5}}{5 + \sqrt{5}}} = \frac{1}{5}\sqrt{25 - 10\sqrt{5}}. \tag{15}$$

An interesting near-identity is given by

$$\frac{1}{4}\left[\cos\left(\tfrac{1}{10}\right) + \cosh\left(\tfrac{1}{10}\right) + 2\cos\left(\tfrac{1}{20}\sqrt{2}\right)\cosh\left(\tfrac{1}{20}\sqrt{2}\right)\right] \approx 1.$$

$$\tag{16}$$

In fact, the left-hand side is approximately equal to $1 + 2.480 \times 10^{-13}$.

See also DECAGON, DECAGRAM

Trigonometry Values Pi/11

Trigonometric functions of $n\pi/11$ for n an integer cannot be expressed in terms of sums, products, and finite ROOT EXTRACTIONS on *real* rational numbers because 11 is not a FERMAT PRIME. This also means that the UNDECAGON is not a CONSTRUCTIBLE POLYGON.

However, exact expressions involving roots of *complex* numbers can still be derived using the MULTIPLE-ANGLE FORMULA

$$\sin(n\alpha) = (-1)^{(n-1)/2} T_n(\sin\alpha), \tag{1}$$

where T_n is a CHEBYSHEV POLYNOMIAL OF THE FIRST KIND. Plugging in $n = 11$ gives

$$\sin(11\alpha) = \sin\alpha\,(11 - 220\sin^2\alpha + 1232\sin^4\alpha$$
$$-2816\sin^6\alpha + 2816\sin^8 - 1024\sin^{10}\alpha). \tag{2}$$

Letting $\alpha \equiv \pi/11$ and $x \equiv \sin^2\alpha$ then gives

$$\sin\pi = 0 = 11 - 220x + 1232x^2 - 2816x^3$$
$$+ 2816x^4 - 1024x^5. \tag{3}$$

This equation is an irreducible QUINTIC EQUATION, so an analytic solution involving FINITE ROOT EXTRACTIONS does not exist. The numerical ROOTS are $x = 0.07937$, 0.29229, 0.57115, 0.82743, 0.97974. So $\sin\alpha = 0.2817$, $\sin(2\alpha) = 0.5406$, $\sin(3\alpha) = 0.7557$, $\sin(4\alpha) = 0.9096$, $\sin(5\alpha) = 0.9898$. From one of NEWTON'S IDENTITIES,

$$\sin\left(\frac{\pi}{11}\right)\sin\left(\frac{2\pi}{11}\right)\sin\left(\frac{3\pi}{11}\right)\sin\left(\frac{4\pi}{11}\right)\sin\left(\frac{5\pi}{11}\right) = \sqrt{\frac{11}{1024}}$$

$$= \frac{\sqrt{11}}{32} \tag{4}$$

$$\cos\left(\frac{\pi}{11}\right)\cos\left(\frac{2\pi}{11}\right)\cos\left(\frac{3\pi}{11}\right)\cos\left(\frac{4\pi}{11}\right)\cos\left(\frac{5\pi}{11}\right) = \frac{1}{32} \tag{5}$$

$$\tan\left(\frac{\pi}{11}\right)\tan\left(\frac{2\pi}{11}\right)\tan\left(\frac{3\pi}{11}\right)\tan\left(\frac{4\pi}{11}\right)\tan\left(\frac{5\pi}{11}\right)$$

$$= \sqrt{11}. \tag{6}$$

The trigonometric functions of $\pi/11$ also obey the identity

$$\tan\left(\frac{3\pi}{11}\right) + 4\sin\left(\frac{2\pi}{11}\right) = \sqrt{11}. \tag{7}$$

See also UNDECAGON

References

Beyer, W. H. "Trigonometry." *CRC Standard Mathematical Tables, 28th ed.* Boca Raton, FL: CRC Press, 1987.

Trigonometry Values Pi/12

$$\cos\left(\frac{\pi}{12}\right) = \tfrac{1}{4}\left(\sqrt{6} + \sqrt{2}\right) \tag{1}$$

$$\cos\left(\frac{5\pi}{12}\right) = \tfrac{1}{4}\left(\sqrt{6} - \sqrt{2}\right) \tag{2}$$

$$\cot\left(\frac{\pi}{12}\right) = 2 + \sqrt{3} \tag{3}$$

$$\cot\left(\frac{5\pi}{12}\right) = 2 - \sqrt{3} \tag{4}$$

$$\cot\left(\frac{\pi}{12}\right) = \sqrt{6} + \sqrt{2} \tag{5}$$

$$\csc\left(\frac{5\pi}{12}\right) = \sqrt{6} - \sqrt{2} \tag{6}$$

$$\sec\left(\frac{\pi}{12}\right) = \sqrt{6} - \sqrt{2} \tag{7}$$

$$\sec\left(\frac{5\pi}{12}\right) = \sqrt{6} + \sqrt{2} \tag{8}$$

$$\sin\left(\frac{\pi}{12}\right) = \tfrac{1}{4}\left(\sqrt{6} - \sqrt{2}\right) \tag{9}$$

$$\sin\left(\frac{5\pi}{12}\right) = \tfrac{1}{4}\left(\sqrt{6} + \sqrt{2}\right) \tag{10}$$

$$\tan\left(\frac{\pi}{12}\right) = 2 - \sqrt{3} \tag{11}$$

$$\tan\left(\frac{5\pi}{12}\right) = 2 + \sqrt{3}. \tag{12}$$

These can be derived using

$$\sin\left(\frac{\pi}{12}\right) = \sin\left(\frac{\pi}{3} - \frac{\pi}{4}\right)$$

$$= -\sin\left(\frac{\pi}{4}\right)\cos\left(\frac{\pi}{3}\right) + \sin\left(\frac{\pi}{3}\right)\cos\left(\frac{\pi}{4}\right)$$

$$= -\tfrac{1}{2}\sqrt{2}\left(\tfrac{1}{2}\right) + \tfrac{1}{2}\sqrt{3}\left(\tfrac{1}{2}\sqrt{2}\right)$$

$$= \tfrac{1}{4}\left(\sqrt{6} - \sqrt{2}\right). \qquad (13)$$

Similarly,

$$\cos\left(\frac{\pi}{12}\right) = \cos\left(\frac{\pi}{3} - \frac{\pi}{4}\right)$$

$$= \cos\left(\frac{\pi}{4}\right)\cos\left(\frac{\pi}{3}\right) - \sin\left(\frac{\pi}{3}\right)\sin\left(\frac{\pi}{4}\right)$$

$$= \tfrac{1}{2}\left(\tfrac{1}{2}\sqrt{2}\right) + \tfrac{1}{2}\sqrt{3}\left(-\tfrac{1}{2}\sqrt{2}\right)$$

$$= \tfrac{1}{4}\left(\sqrt{6} - \sqrt{2}\right). \qquad (14)$$

Trigonometry Values Pi/15

$$\cos\left(\frac{\pi}{15}\right) = \tfrac{1}{8}\left(\sqrt{30 + 6\sqrt{5}} + \sqrt{5} - 1\right) \qquad (1)$$

$$\cos\left(\frac{2\pi}{15}\right) = \tfrac{1}{8}\left(\sqrt{30 - 6\sqrt{5}} + \sqrt{5} + 1\right) \qquad (2)$$

$$\cos\left(\frac{4\pi}{15}\right) = \tfrac{1}{8}\left(\sqrt{30 + 6\sqrt{5}} - \sqrt{5} + 1\right) \qquad (3)$$

$$\cos\left(\frac{7\pi}{15}\right) = \tfrac{1}{8}\left(\sqrt{30 - 6\sqrt{5}} - \sqrt{5} - 1\right) \qquad (4)$$

$$\cot\left(\frac{\pi}{15}\right) = \sqrt{7 + 2\sqrt{5} + 2\sqrt{15 + 6\sqrt{5}}} \qquad (5)$$

$$\cot\left(\frac{2\pi}{15}\right) = \sqrt{7 - 2\sqrt{5} + 2\sqrt{15 - 6\sqrt{5}}} \qquad (6)$$

$$\cot\left(\frac{4\pi}{15}\right) = \sqrt{7 + 2\sqrt{5} - 2\sqrt{15 + 6\sqrt{5}}} \qquad (7)$$

$$\cot\left(\frac{7\pi}{15}\right) = \sqrt{7 - 2\sqrt{5} - 2\sqrt{15 - 6\sqrt{5}}} \qquad (8)$$

$$\csc\left(\frac{\pi}{15}\right) = \sqrt{8 + 2\sqrt{5} + 2\sqrt{15 + 6\sqrt{5}}} \qquad (9)$$

$$\csc\left(\frac{2\pi}{15}\right) = \sqrt{8 - 2\sqrt{5} + 2\sqrt{15 - 6\sqrt{5}}} \qquad (10)$$

$$\csc\left(\frac{4\pi}{15}\right) = \sqrt{8 + 2\sqrt{5} - 2\sqrt{15 + 6\sqrt{5}}} \qquad (11)$$

$$\csc\left(\frac{7\pi}{15}\right) = \sqrt{8 - 2\sqrt{5} - 2\sqrt{15 - 6\sqrt{5}}} \qquad (12)$$

$$\sec\left(\frac{\pi}{15}\right) = +2 - \sqrt{5} + \sqrt{15 - 6\sqrt{5}} \qquad (13)$$

$$\sec\left(\frac{2\pi}{15}\right) = -2 - \sqrt{5} + \sqrt{15 + 6\sqrt{5}} \qquad (14)$$

$$\sec\left(\frac{4\pi}{15}\right) = -2 + \sqrt{5} + \sqrt{15 - 6\sqrt{5}} \qquad (15)$$

$$\sec\left(\frac{7\pi}{15}\right) = +2 + \sqrt{5} + \sqrt{15 + 6\sqrt{5}} \qquad (16)$$

$$\sin\left(\frac{\pi}{15}\right) = \tfrac{1}{4}\sqrt{7 - \sqrt{5} - \sqrt{30 - 6\sqrt{5}}} \qquad (17)$$

$$\sin\left(\frac{2\pi}{15}\right) = \tfrac{1}{4}\sqrt{7 + \sqrt{5} - \sqrt{30 + 6\sqrt{5}}} \qquad (18)$$

$$\sin\left(\frac{4\pi}{15}\right) = \tfrac{1}{4}\sqrt{7 - \sqrt{5} + \sqrt{30 - 6\sqrt{5}}} \qquad (19)$$

$$\sin\left(\frac{7\pi}{15}\right) = \tfrac{1}{4}\sqrt{7 + \sqrt{5} + \sqrt{30 + 6\sqrt{5}}} \qquad (20)$$

$$\tan\left(\frac{\pi}{15}\right) = \sqrt{23 - 10\sqrt{5} - 2\sqrt{255 + 114\sqrt{5}}} \qquad (21)$$

$$\tan\left(\frac{2\pi}{15}\right) = \sqrt{23 + 10\sqrt{5} - 2\sqrt{255 + 114\sqrt{5}}} \qquad (22)$$

$$\tan\left(\frac{4\pi}{15}\right) = \sqrt{23 - 10\sqrt{5} + 2\sqrt{255 + 114\sqrt{5}}} \qquad (23)$$

$$\tan\left(\frac{7\pi}{15}\right) = \sqrt{23 + 10\sqrt{5} + 2\sqrt{255 + 114\sqrt{5}}}. \qquad (24)$$

These can be derived using the TRIGONOMETRIC ADDITION FORMULAS

$$\sin\left(\frac{\pi}{15}\right) = \sin\left(\frac{\pi}{6} - \frac{\pi}{10}\right)$$

$$= \sin\left(\frac{\pi}{6}\right)\cos\left(\frac{\pi}{10}\right) - \sin\left(\frac{\pi}{10}\right)\cos\left(\frac{\pi}{6}\right)$$

$$= \frac{1}{2}\sqrt{\frac{1}{8}\left(5+\sqrt{5}\right)} - \frac{\sqrt{3}}{2}\frac{1}{4}\left(\sqrt{5}-1\right)$$

$$= \frac{1}{16}\left(2\sqrt{3} - 2\sqrt{15} + \sqrt{40+8\sqrt{5}}\right) \qquad (25)$$

and

$$\cos\left(\frac{\pi}{15}\right) = \cos\left(\frac{\pi}{6} - \frac{\pi}{10}\right)$$

$$= \cos\left(\frac{\pi}{6}\right)\cos\left(\frac{\pi}{10}\right) + \sin\left(\frac{\pi}{6}\right)\sin\left(\frac{\pi}{10}\right)$$

$$= \frac{\sqrt{3}}{2}\sqrt{\frac{1}{8}\left(5+\sqrt{5}\right)} + \frac{1}{2}\frac{1}{4}\left(\sqrt{5}-1\right)$$

$$= \frac{1}{8}\left(\sqrt{30+6\sqrt{5}} + \sqrt{5}-1\right). \qquad (26)$$

See also PENTADECAGON

Trigonometry Values Pi/16

$$\cos\left(\frac{\pi}{16}\right) = \frac{1}{2}\sqrt{2+\sqrt{2+\sqrt{2}}} \qquad (1)$$

$$\cos\left(\frac{3\pi}{16}\right) = \frac{1}{2}\sqrt{2+\sqrt{2-\sqrt{2}}} \qquad (2)$$

$$\cos\left(\frac{5\pi}{16}\right) = \frac{1}{2}\sqrt{2-\sqrt{2-\sqrt{2}}} \qquad (3)$$

$$\cos\left(\frac{7\pi}{16}\right) = \frac{1}{2}\sqrt{2-\sqrt{2+\sqrt{2}}} \qquad (4)$$

$$\cot\left(\frac{\pi}{16}\right) = +1+\sqrt{2}+\sqrt{4+2\sqrt{2}} \qquad (5)$$

$$\cot\left(\frac{3\pi}{16}\right) = -1+\sqrt{2}+\sqrt{4-2\sqrt{2}} \qquad (6)$$

$$\cot\left(\frac{5\pi}{16}\right) = +1-\sqrt{2}+\sqrt{4-2\sqrt{2}} \qquad (7)$$

$$\cot\left(\frac{7\pi}{16}\right) = -1-\sqrt{2}+\sqrt{4+2\sqrt{2}} \qquad (8)$$

$$\csc\left(\frac{\pi}{16}\right) = \sqrt{8+4\sqrt{2}+2\sqrt{20+14\sqrt{2}}} \qquad (9)$$

$$\csc\left(\frac{3\pi}{16}\right) = \sqrt{8-4\sqrt{2}+2\sqrt{20-14\sqrt{2}}} \qquad (10)$$

$$\csc\left(\frac{5\pi}{16}\right) = \sqrt{8-4\sqrt{2}-2\sqrt{20-14\sqrt{2}}} \qquad (11)$$

$$\csc\left(\frac{7\pi}{16}\right) = \sqrt{8+4\sqrt{2}-2\sqrt{20+14\sqrt{2}}} \qquad (12)$$

$$\sec\left(\frac{\pi}{16}\right) = \sqrt{8+4\sqrt{2}-2\sqrt{20+14\sqrt{2}}} \qquad (13)$$

$$\sec\left(\frac{3\pi}{16}\right) = \sqrt{8-4\sqrt{2}-2\sqrt{20-14\sqrt{2}}} \qquad (14)$$

$$\sec\left(\frac{5\pi}{16}\right) = \sqrt{8-4\sqrt{2}+2\sqrt{20-14\sqrt{2}}} \qquad (15)$$

$$\sec\left(\frac{7\pi}{16}\right) = \sqrt{8+4\sqrt{2}+2\sqrt{20+14\sqrt{2}}} \qquad (16)$$

$$\sin\left(\frac{\pi}{16}\right) = \frac{1}{2}\sqrt{2-\sqrt{2+\sqrt{2}}} \qquad (17)$$

$$\sin\left(\frac{3\pi}{16}\right) = \frac{1}{2}\sqrt{2-\sqrt{2-\sqrt{2}}} \qquad (18)$$

$$\sin\left(\frac{5\pi}{16}\right) = \frac{1}{2}\sqrt{2+\sqrt{2-\sqrt{2}}} \qquad (19)$$

$$\sin\left(\frac{7\pi}{16}\right) = \frac{1}{2}\sqrt{2+\sqrt{2+\sqrt{2}}} \qquad (20)$$

$$\tan\left(\frac{\pi}{16}\right) = \sqrt{4+2\sqrt{2}} - \sqrt{2}-1 \qquad (21)$$

$$\tan\left(\frac{3\pi}{16}\right) = \sqrt{4-2\sqrt{2}} - \sqrt{2}+1 \qquad (22)$$

$$\tan\left(\frac{5\pi}{16}\right) = \sqrt{4-2\sqrt{2}} + \sqrt{2}-1 \qquad (23)$$

$$\tan\left(\frac{7\pi}{16}\right) = \sqrt{4+2\sqrt{2}} + \sqrt{2}+1. \qquad (24)$$

These can be derived from the HALF-ANGLE FORMULAS

$$\sin\left(\frac{\pi}{16}\right) = \sin\left(\frac{1}{2}\cdot\frac{\pi}{8}\right)$$

$$= \sqrt{\frac{1}{2}\left(1-\cos\frac{\pi}{8}\right)} = \sqrt{\frac{1}{2}\left(1-\frac{1}{2}\sqrt{2+\sqrt{2}}\right)}$$

$$= \sqrt{\tfrac{1}{2} - \tfrac{1}{4}\sqrt{2 + \sqrt{2}}} = \tfrac{1}{2}\sqrt{2 - \sqrt{2 + \sqrt{2}}} \quad (25)$$

$$\cos\left(\frac{\pi}{16}\right) = \cos\left(\frac{1}{2} \cdot \frac{\pi}{8}\right)$$

$$= \sqrt{\frac{1}{2}\left(1 + \cos\frac{\pi}{8}\right)} = \sqrt{\frac{1}{2}\left(1 + \frac{1}{2}\sqrt{2 + \sqrt{2}}\right)}$$

$$= \sqrt{\tfrac{1}{2} + \tfrac{1}{4}\sqrt{2 + \sqrt{2}}} = \tfrac{1}{2}\sqrt{2 + \sqrt{2 + \sqrt{2}}} \quad (26)$$

$$\tan\left(\frac{\pi}{16}\right) = \sqrt{\frac{2 - \sqrt{2 + \sqrt{2}}}{2 + \sqrt{2 + \sqrt{2}}}}$$

$$= \sqrt{4 + 2\sqrt{2}} - \sqrt{2} - 1. \quad (27)$$

See also HEXADECAGON

Trigonometry Values Pi/17

Rather surprisingly, trigonometric functions of $n\pi/17$ for n an integer can be expressed in terms of sums, products, and finite ROOT EXTRACTIONS because 17 is a FERMAT PRIME. This makes the HEPTADECAGON a CONSTRUCTIBLE, as first proved by Gauss. Although Gauss did not actually explicitly provide a construction, he did derive the trigonometric formulas below using a series of intermediate variables from which the final expressions were then built up.

Let

$$\epsilon \equiv \sqrt{17 + \sqrt{17}} \quad (1)$$

$$\epsilon^* \equiv \sqrt{17 - \sqrt{17}} \quad (2)$$

$$\delta \equiv \sqrt{17} - 1 \quad (3)$$

$$\alpha \equiv \sqrt{34 + 6\sqrt{17} + \sqrt{2}\left(\sqrt{17} - 1\right)\epsilon^* - 8\sqrt{2}\epsilon} \quad (4)$$

$$\beta \equiv 2\sqrt{17 + 3\sqrt{17} - 2\sqrt{2}\epsilon - \sqrt{2}\epsilon^*}, \quad (5)$$

then

$$\sin\left(\frac{\pi}{17}\right) = \tfrac{1}{8}\sqrt{2}\sqrt{\epsilon^{*2} - \sqrt{2}(\alpha + \epsilon^*)}$$

$$\approx 0.18375 \quad (6)$$

$$\cos\left(\frac{\pi}{17}\right) = \tfrac{1}{8}\sqrt{2}\sqrt{15 + \sqrt{17} + \sqrt{2}(\alpha + \epsilon^*)}$$

$$\approx 0.98297 \quad (7)$$

$$\sin\left(\frac{2\pi}{17}\right) = \tfrac{1}{16}\sqrt{2}\sqrt{4\epsilon^{*2} - 2\sqrt{2}\delta\epsilon^* + 8\sqrt{2}\epsilon - \left(\sqrt{2}\delta + 2\epsilon^*\right)\alpha}$$

$$\approx 0.36124 \quad (8)$$

$$\cos\left(\frac{2\pi}{17}\right) = \tfrac{1}{16}\left[\delta + \sqrt{2}(\alpha + \epsilon^*)\right]$$

$$\approx 0.93247 \quad (9)$$

$$\sin\left(\frac{4\pi}{17}\right) = \tfrac{1}{128}\left[\sqrt{2}\delta + 2(\alpha + \epsilon^*)\right]$$

$$\times \left[4\epsilon^{*2} - 2\sqrt{2}\delta\epsilon^* + 8\sqrt{2}\epsilon - \left(\sqrt{2}\delta + 2\epsilon^*\right)\alpha\right]^{1/2}$$

$$\approx 0.67370 \quad (10)$$

$$\sin\left(\frac{8\pi}{17}\right) = \tfrac{1}{16}[136 - 8\sqrt{17} + 8\sqrt{2}\epsilon - 2(\sqrt{34} - 3\sqrt{2})\epsilon^*$$

$$+ 2\beta(\delta + \sqrt{2}\epsilon^*)]^{1/2} \approx 0.99573 \quad (11)$$

$$\cos\left(\frac{8\pi}{17}\right) = \frac{1}{16}\left(\delta + \sqrt{2}\epsilon^* - 2\sqrt{17 + 3\sqrt{17} - \sqrt{2}\epsilon^* - 2\sqrt{2}\epsilon}\right)$$

$$\approx 0.09227. \quad (12)$$

There are some interesting analytic formulas involving the trigonometric functions of $n\pi/17$. Define

$$P(x) \equiv (x - 1)(x - 2)(x^2 + 1) \quad (13)$$

$$g_1(x) \equiv \frac{2 + \sqrt{P(x)}}{1 - x} \quad (14)$$

$$g_4(x) \equiv \frac{2 - \sqrt{P(x)}}{1 - x} \quad (15)$$

$$f_i(x) \equiv \tfrac{1}{4}[g_i(x) - 1] \quad (16)$$

$$a \equiv \tfrac{1}{4}\tan^{-1} 4, \quad (17)$$

where $i = 1$ or 4. Then

$$f_1(\tan a) = \cos\left(\frac{2\pi}{17}\right) \quad (18)$$

$$f_4(\tan a) = \cos\left(\frac{8\pi}{17}\right). \quad (19)$$

Another interesting identity is given by

$$\tan\left(\tfrac{1}{4}\tan^{-1} 4\right) = 2\left[\cos\left(\frac{6\pi}{17}\right) + \cos\left(\frac{10\pi}{17}\right)\right], \quad (20)$$

where both sides are equal to

$$C = \frac{\sqrt{2(17 + \sqrt{17})} - \sqrt{17} - 1}{4} \quad (21)$$

(Wickner 1999).

See also Constructible Polygon, Fermat Prime, Heptadecagon

References

Casey, J. *Plane Trigonometry*. Dublin: Hodges, Figgis, & Co., p. 220, 1888.

Conway, J. H. and Guy, R. K. *The Book of Numbers*. New York: Springer-Verlag, pp. 192–194 and 229–230, 1996.

Dörrie, H. "The Regular Heptadecagon." §37 in *100 Great Problems of Elementary Mathematics: Their History and Solutions*. New York: Dover, pp. 177–184, 1965.

Ore, Ø. *Number Theory and Its History*. New York: Dover, 1988.

Smith, D. E. *A Source Book in Mathematics*. New York: Dover, p. 348, 1994.

Wickner, J. "Solution to Problem 1562: A Tangent and Cosine Identity." *Math. Mag.* **72**, pp. 412–413, 1999.

Trigonometry Values Pi/18

The exact values of $\cos(\pi/18)$ and $\sin(\pi/18)$ can be given by infinite NESTED RADICALS

$$\sin\left(\frac{\pi}{18}\right) = \frac{1}{2}\sqrt{2-\sqrt{2+\sqrt{2+\sqrt{2-\ldots}}}},$$

where the sequence of signs $+, +, -$ repeats with period 3, and

$$\cos\left(\frac{\pi}{18}\right) = \frac{1}{16}\sqrt{3}\left(\sqrt{8-\sqrt{8-\sqrt{8+\sqrt{8-\ldots}}}}+1\right),$$

where the sequence of signs $-, -, +$ repeats with period 3.

Trigonometry Values Pi/20

$$\cos\left(\frac{\pi}{20}\right) = \frac{1}{4}\sqrt{8+2\sqrt{10+2\sqrt{5}}} \tag{1}$$

$$\cos\left(\frac{3\pi}{20}\right) = \frac{1}{4}\sqrt{8+2\sqrt{10+2\sqrt{5}}} \tag{2}$$

$$\cos\left(\frac{7\pi}{20}\right) = \frac{1}{4}\sqrt{8-2\sqrt{10-2\sqrt{5}}} \tag{3}$$

$$\cos\left(\frac{9\pi}{20}\right) = \frac{1}{4}\sqrt{8-2\sqrt{10-2\sqrt{5}}} \tag{4}$$

$$\cot\left(\frac{\pi}{20}\right) = +1+\sqrt{5}+\sqrt{5+2\sqrt{5}} \tag{5}$$

$$\cot\left(\frac{3\pi}{20}\right) = -1+\sqrt{5}+\sqrt{5-2\sqrt{5}} \tag{6}$$

$$\cot\left(\frac{7\pi}{20}\right) = -1+\sqrt{5}-\sqrt{5-2\sqrt{5}} \tag{7}$$

$$\cot\left(\frac{9\pi}{20}\right) = +1+\sqrt{5}-\sqrt{5+2\sqrt{5}} \tag{8}$$

$$\csc\left(\frac{\pi}{20}\right) = \sqrt{12+4\sqrt{5}+2\sqrt{50+22\sqrt{5}}} \tag{9}$$

$$\csc\left(\frac{3\pi}{20}\right) = \sqrt{12-4\sqrt{5}+2\sqrt{50-22\sqrt{5}}} \tag{10}$$

$$\csc\left(\frac{5\pi}{20}\right) = \sqrt{12-4\sqrt{5}-2\sqrt{50-22\sqrt{5}}} \tag{11}$$

$$\csc\left(\frac{7\pi}{20}\right) = \sqrt{12+4\sqrt{5}-2\sqrt{50+22\sqrt{5}}} \tag{12}$$

$$\sec\left(\frac{\pi}{20}\right) = \sqrt{12+4\sqrt{5}-2\sqrt{50+22\sqrt{5}}} \tag{13}$$

$$\sec\left(\frac{3\pi}{20}\right) = \sqrt{12-4\sqrt{5}-2\sqrt{50-22\sqrt{5}}} \tag{14}$$

$$\sec\left(\frac{5\pi}{20}\right) = \sqrt{12-4\sqrt{5}+2\sqrt{50-22\sqrt{5}}} \tag{15}$$

$$\sec\left(\frac{7\pi}{20}\right) = \sqrt{12+4\sqrt{5}+2\sqrt{50+22\sqrt{5}}} \tag{16}$$

$$\sin\left(\frac{\pi}{20}\right) = \frac{1}{4}\sqrt{8-20\sqrt{10+2\sqrt{5}}} \tag{17}$$

$$\sin\left(\frac{3\pi}{20}\right) = \frac{1}{4}\sqrt{8-20\sqrt{10-2\sqrt{5}}} \tag{18}$$

$$\sin\left(\frac{7\pi}{20}\right) = \frac{1}{4}\sqrt{8+20\sqrt{10-2\sqrt{5}}} \tag{19}$$

$$\sin\left(\frac{9\pi}{20}\right) = \frac{1}{4}\sqrt{8+20\sqrt{10+2\sqrt{5}}} \tag{20}$$

$$\tan\left(\frac{\pi}{20}\right) = +1+\sqrt{5}-\sqrt{5+2\sqrt{5}} \tag{21}$$

$$\tan\left(\frac{3\pi}{20}\right) = -1+\sqrt{5}-\sqrt{5-2\sqrt{5}} \tag{22}$$

$$\tan\left(\frac{7\pi}{20}\right) = -1+\sqrt{5}+\sqrt{5-2\sqrt{5}} \tag{23}$$

$$\tan\left(\frac{9\pi}{20}\right) = +1+\sqrt{5}+\sqrt{5+2\sqrt{5}}. \tag{24}$$

These can be derived from the HALF-ANGLE FORMULAS

$$\sin\left(\frac{\pi}{20}\right) = \sin\left(\frac{1}{2}\frac{\pi}{10}\right) = \sqrt{\frac{1}{2}\left(1 - \cos\frac{\pi}{10}\right)}$$

$$= \frac{1}{4}\sqrt{8 - 2\sqrt{10 + 2\sqrt{5}}}$$

$$\cos\left(\frac{\pi}{20}\right) = \cos\left(\frac{1}{2}\frac{\pi}{10}\right) = \sqrt{\frac{1}{2}\left(1 + \cos\frac{\pi}{10}\right)}$$

$$= \frac{1}{4}\sqrt{8 + 2\sqrt{10 + 2\sqrt{5}}}$$

$$\tan\left(\frac{\pi}{20}\right) = 1 + \sqrt{5} - \sqrt{5 + 2\sqrt{5}}.$$

An interesting near-identity is given by

$$\frac{1}{4}\left[\cos\left(\tfrac{1}{10}\right) + \cosh\left(\tfrac{1}{10}\right) + 2\cos\left(\tfrac{1}{20}\sqrt{2}\right)\cosh\left(\tfrac{1}{20}\sqrt{2}\right)\right]$$

$$\approx 1. \tag{25}$$

In fact, the left-hand side is approximately equal to $1 + 2.480 \times 10^{-13}$.

Trigonometry Values Pi/24

$$\cos\left(\frac{\pi}{24}\right) = \frac{1}{2}\sqrt{2 + \sqrt{2 + \sqrt{3}}} \tag{1}$$

$$\cos\left(\frac{5\pi}{24}\right) = \frac{1}{2}\sqrt{2 + \sqrt{2 - \sqrt{3}}} \tag{2}$$

$$\cos\left(\frac{7\pi}{24}\right) = \frac{1}{2}\sqrt{2 - \sqrt{2 - \sqrt{3}}} \tag{3}$$

$$\cos\left(\frac{11\pi}{24}\right) = \frac{1}{2}\sqrt{2 - \sqrt{2 + \sqrt{3}}} \tag{4}$$

$$\cot\left(\frac{\pi}{24}\right) = +2 + \sqrt{2} + \sqrt{3} + \sqrt{6} \tag{5}$$

$$\cot\left(\frac{5\pi}{24}\right) = +2 - \sqrt{2} - \sqrt{3} + \sqrt{6} \tag{6}$$

$$\cot\left(\frac{7\pi}{24}\right) = -2 - \sqrt{2} + \sqrt{3} + \sqrt{6} \tag{7}$$

$$\cot\left(\frac{11\pi}{24}\right) = -2 + \sqrt{2} - \sqrt{3} + \sqrt{6} \tag{8}$$

$$\csc\left(\frac{\pi}{24}\right) = \sqrt{16 + 10\sqrt{2} + 8\sqrt{3} + 6\sqrt{6}} \tag{9}$$

$$\csc\left(\frac{5\pi}{24}\right) = \sqrt{16 - 10\sqrt{2} - 8\sqrt{3} + 6\sqrt{6}} \tag{10}$$

$$\csc\left(\frac{7\pi}{24}\right) = \sqrt{16 + 10\sqrt{2} - 8\sqrt{3} - 6\sqrt{6}} \tag{11}$$

$$\csc\left(\frac{11\pi}{24}\right) = \sqrt{16 - 10\sqrt{2} + 8\sqrt{3} - 6\sqrt{6}} \tag{12}$$

$$\sec\left(\frac{\pi}{24}\right) = \sqrt{16 - 10\sqrt{2} + 8\sqrt{3} - 6\sqrt{6}} \tag{13}$$

$$\sec\left(\frac{5\pi}{24}\right) = \sqrt{16 + 10\sqrt{2} - 8\sqrt{3} - 6\sqrt{6}} \tag{14}$$

$$\sec\left(\frac{7\pi}{24}\right) = \sqrt{16 - 10\sqrt{2} - 8\sqrt{3} + 6\sqrt{6}} \tag{15}$$

$$\sec\left(\frac{11\pi}{24}\right) = \sqrt{16 + 10\sqrt{2} + 8\sqrt{3} + 6\sqrt{6}} \tag{16}$$

$$\sin\left(\frac{\pi}{24}\right) = \frac{1}{2}\sqrt{2 - \sqrt{2 + \sqrt{3}}} \tag{17}$$

$$\sin\left(\frac{5\pi}{24}\right) = \frac{1}{2}\sqrt{2 - \sqrt{2 - \sqrt{3}}} \tag{18}$$

$$\sin\left(\frac{7\pi}{24}\right) = \frac{1}{2}\sqrt{2 + \sqrt{2 - \sqrt{3}}} \tag{19}$$

$$\sin\left(\frac{11\pi}{24}\right) = \frac{1}{2}\sqrt{2 + \sqrt{2 + \sqrt{3}}} \tag{20}$$

$$\tan\left(\frac{\pi}{24}\right) = -2 + \sqrt{2} - \sqrt{3} + \sqrt{6} \tag{21}$$

$$\tan\left(\frac{\pi}{24}\right) = -2 - \sqrt{2} + \sqrt{3} + \sqrt{6} \tag{22}$$

$$\tan\left(\frac{\pi}{24}\right) = +2 - \sqrt{2} - \sqrt{3} + \sqrt{6} \tag{23}$$

$$\tan\left(\frac{\pi}{24}\right) = +2 + \sqrt{2} + \sqrt{3} + \sqrt{6}. \tag{24}$$

See also ICOSITETRAGON

Trigonometry Values Pi/30

$$\cos\left(\frac{\pi}{30}\right) = \tfrac{1}{4}\sqrt{7+\sqrt{5}+\sqrt{6\left(5+\sqrt{5}\right)}} \qquad (1)$$

$$\cos\left(\frac{7\pi}{30}\right) = \tfrac{1}{4}\sqrt{7-\sqrt{5}+\sqrt{6\left(5+\sqrt{5}\right)}} \qquad (2)$$

$$\cos\left(\frac{11\pi}{30}\right) = \tfrac{1}{4}\sqrt{7+\sqrt{5}-\sqrt{6\left(5+\sqrt{5}\right)}} \qquad (3)$$

$$\cos\left(\frac{13\pi}{30}\right) = \tfrac{1}{4}\sqrt{7-\sqrt{5}-\sqrt{6\left(5+\sqrt{5}\right)}} \qquad (4)$$

$$\cot\left(\frac{\pi}{30}\right) = \sqrt{23+10\sqrt{5}+2\sqrt{255+114\sqrt{5}}} \qquad (5)$$

$$\cot\left(\frac{7\pi}{30}\right) = \sqrt{23-10\sqrt{5}+2\sqrt{255-114\sqrt{5}}} \qquad (6)$$

$$\cot\left(\frac{11\pi}{30}\right) = \sqrt{23+10\sqrt{5}-2\sqrt{255+114\sqrt{5}}} \qquad (7)$$

$$\cot\left(\frac{13\pi}{30}\right) = \sqrt{23-10\sqrt{5}-2\sqrt{255-114\sqrt{5}}} \qquad (8)$$

$$\csc\left(\frac{\pi}{30}\right) = +2+\sqrt{5}+\sqrt{15+6\sqrt{5}} \qquad (9)$$

$$\csc\left(\frac{7\pi}{30}\right) = -2+\sqrt{5}+\sqrt{15-6\sqrt{5}} \qquad (10)$$

$$\csc\left(\frac{11\pi}{30}\right) = -2-\sqrt{5}+\sqrt{15+6\sqrt{5}} \qquad (11)$$

$$\csc\left(\frac{13\pi}{30}\right) = +2-\sqrt{5}+\sqrt{15-6\sqrt{5}} \qquad (12)$$

$$\sec\left(\frac{\pi}{30}\right) = \sqrt{8-2\sqrt{5}-2\sqrt{15-6\sqrt{5}}} \qquad (13)$$

$$\sec\left(\frac{7\pi}{30}\right) = \sqrt{8+2\sqrt{5}-2\sqrt{15+6\sqrt{5}}} \qquad (14)$$

$$\sec\left(\frac{11\pi}{30}\right) = \sqrt{8-2\sqrt{5}+2\sqrt{15-6\sqrt{5}}} \qquad (15)$$

$$\sec\left(\frac{13\pi}{30}\right) = \sqrt{8+2\sqrt{5}+2\sqrt{15+6\sqrt{5}}} \qquad (16)$$

$$\sin\left(\frac{\pi}{30}\right) = \tfrac{1}{8}\left(-1-\sqrt{5}+\sqrt{30-6\sqrt{5}}\right) \qquad (17)$$

$$\sin\left(\frac{7\pi}{30}\right) = \tfrac{1}{8}\left(+1-\sqrt{5}+\sqrt{30+6\sqrt{5}}\right) \qquad (18)$$

$$\sin\left(\frac{11\pi}{30}\right) = \tfrac{1}{8}\left(+1+\sqrt{5}+\sqrt{30-6\sqrt{5}}\right) \qquad (19)$$

$$\sin\left(\frac{13\pi}{30}\right) = \tfrac{1}{8}\left(-1+\sqrt{5}+\sqrt{30+6\sqrt{5}}\right) \qquad (20)$$

$$\tan\left(\frac{\pi}{30}\right) = \sqrt{7-2\sqrt{5}-2\sqrt{15-6\sqrt{5}}} \qquad (21)$$

$$\tan\left(\frac{7\pi}{30}\right) = \sqrt{7+2\sqrt{5}-2\sqrt{15+6\sqrt{5}}} \qquad (22)$$

$$\tan\left(\frac{11\pi}{30}\right) = \sqrt{7-2\sqrt{5}+2\sqrt{15-6\sqrt{5}}} \qquad (23)$$

$$\tan\left(\frac{13\pi}{30}\right) = \sqrt{7+2\sqrt{5}+2\sqrt{15+6\sqrt{5}}}. \qquad (24)$$

See also Triacontagon

Trigonometry Values Pi/32

$$\cos\left(\frac{\pi}{32}\right) = \tfrac{1}{2}\sqrt{2+\sqrt{2+\sqrt{2+\sqrt{2}}}} \qquad (1)$$

$$\cos\left(\frac{3\pi}{32}\right) = \tfrac{1}{2}\sqrt{2+\sqrt{2+\sqrt{2-\sqrt{2}}}} \qquad (2)$$

$$\cos\left(\frac{5\pi}{32}\right) = \tfrac{1}{2}\sqrt{2+\sqrt{2-\sqrt{2-\sqrt{2}}}} \qquad (3)$$

$$\cos\left(\frac{7\pi}{32}\right) = \tfrac{1}{2}\sqrt{2+\sqrt{2-\sqrt{2+\sqrt{2}}}} \qquad (4)$$

$$\cos\left(\frac{9\pi}{32}\right) = \tfrac{1}{2}\sqrt{2-\sqrt{2-\sqrt{2+\sqrt{2}}}} \qquad (5)$$

$$\cos\left(\frac{11\pi}{32}\right) = \tfrac{1}{2}\sqrt{2-\sqrt{2-\sqrt{2-\sqrt{2}}}} \qquad (6)$$

$$\cos\left(\frac{13\pi}{32}\right) = \tfrac{1}{2}\sqrt{2-\sqrt{2+\sqrt{2-\sqrt{2}}}} \qquad (7)$$

$$\cos\left(\frac{15\pi}{32}\right) = \frac{1}{2}\sqrt{2 - \sqrt{2 + \sqrt{2 + \sqrt{2}}}} \qquad (8)$$

$$\sin\left(\frac{\pi}{32}\right) = \frac{1}{2}\sqrt{2 - \sqrt{2 + \sqrt{2 + \sqrt{2}}}} \qquad (9)$$

$$\sin\left(\frac{3\pi}{32}\right) = \frac{1}{2}\sqrt{2 - \sqrt{2 + \sqrt{2 - \sqrt{2}}}} \qquad (10)$$

$$\sin\left(\frac{5\pi}{32}\right) = \frac{1}{2}\sqrt{2 - \sqrt{2 - \sqrt{2 - \sqrt{2}}}} \qquad (11)$$

$$\sin\left(\frac{7\pi}{32}\right) = \frac{1}{2}\sqrt{2 - \sqrt{2 - \sqrt{2 + \sqrt{2}}}} \qquad (12)$$

$$\sin\left(\frac{9\pi}{32}\right) = \frac{1}{2}\sqrt{2 + \sqrt{2 - \sqrt{2 + \sqrt{2}}}} \qquad (13)$$

$$\sin\left(\frac{11\pi}{32}\right) = \frac{1}{2}\sqrt{2 + \sqrt{2 - \sqrt{2 - \sqrt{2}}}} \qquad (14)$$

$$\sin\left(\frac{13\pi}{32}\right) = \frac{1}{2}\sqrt{2 + \sqrt{2 + \sqrt{2 - \sqrt{2}}}} \qquad (15)$$

$$\sin\left(\frac{15\pi}{32}\right) = \frac{1}{2}\sqrt{2 + \sqrt{2 + \sqrt{2 + \sqrt{2}}}}. \qquad (16)$$

The functions $\cot(n\pi/32)$, $\csc(n\pi/32)$, $\sec(n\pi/32)$, and $\tan(n\pi/32)$ are roots of 8th degree polynomials, but the explicit expressions in terms of radicals are rather complicated.

See also ICOSIDODECAGON

Trigonometry Values—0

By the definition of the trigonometric functions,

$$\cos 0 = 1$$

$$\cot 0 = \infty$$

$$\csc 0 = \infty$$

$$\sec 0 = 1$$

$$\sin 0 = 0$$

$$\tan 0 = 0.$$

Trigyrate Rhombicosidodecahedron

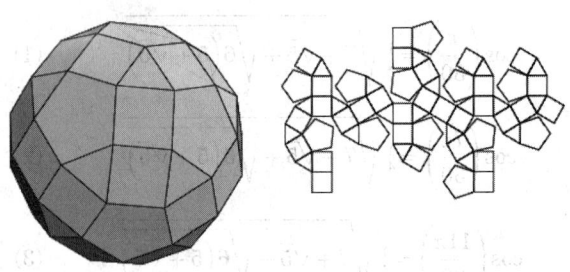

JOHNSON SOLID J_{75}.

References

Weisstein, E. W. "Johnson Solids." MATHEMATICA NOTEBOOK JOHNSONSOLIDS.M.
Weisstein, E. W. "Johnson Solid Netlib Database." MATHEMATICA NOTEBOOK JOHNSONSOLIDS.DAT.

Trihedral Angle

TRIHEDRON

Trihedron

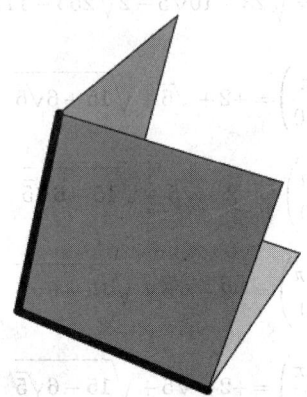

A TRIPLE of three arbitrary vectors with common vertex (Altshiller-Court 1979), often called a trihedral angle since it determines three planes.

The vectors are often taken to be unit vectors, and the term trihedron is frequently encountered in the consideration of the unit ORTHOGONAL VECTORS given by **T**, **N**, and **B** (TANGENT VECTOR, NORMAL VECTOR, and BINORMAL VECTOR).

See also BINORMAL VECTOR, CENTROIDAL LINE, DIHEDRAL ANGLE, ISOCLINAL LINE, ISOCLINAL PLANE, NORMAL VECTOR, ORTHOCENTRIC LINE, TANGENT VECTOR

References

Altshiller-Court, N. "The Trihedral Angle." Ch. 2 in *Modern Pure Solid Geometry*. New York: Chelsea, pp. 27–41, 1979.

Trilinear Coordinates

Given a TRIANGLE ΔABC, the trilinear coordinates of a point P with respect to ΔABC are an ordered TRIPLE of numbers, each of which is PROPORTIONAL to the directed distance from P to one of the side lines. Trilinear coordinates are denoted $\alpha : \beta : \gamma$ or (α, β, γ) and also are known as homogeneous coordinates or "trilinears." Trilinear coordinates were introduced by Plücker in 1835. Since it is only the ratio of distances that is significant, the triplet of trilinear coordinates obtained by multiplying a given triplet by any nonzero constant describes the same point, so

$$\alpha : \beta : \gamma = \mu\alpha : \mu\beta : \mu\gamma. \tag{1}$$

For simplicity, the three VERTICES A, B, and C of a triangle are commonly written as $1:0:0$, $0:1:0$, and $0:0:1$, respectively.

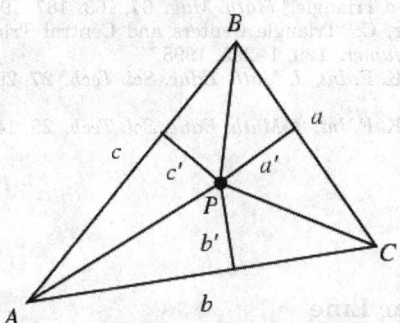

Trilinear coordinates can be normalized so that they give the *actual* directed distances from P to each of the sides. To perform the normalization, let the point P in the above diagram have trilinear coordinates $\alpha : \beta : \gamma$ and lie at distances a', b', and c' from the sides BC, AC, and AB, respectively. Then the distances $a' = k\alpha$, $b' = k\beta$, and $c' = k\gamma$ can be found by writing Δ_a for the AREA of ΔBPC, and similarly for Δ_b and Δ_c. We then have

$$\Delta = \Delta_a + \Delta_b + \Delta_c = \tfrac{1}{2}aa' + \tfrac{1}{2}bb' + \tfrac{1}{2}cc'$$
$$= \tfrac{1}{2}(ak\alpha + bk\beta + ck\gamma) = \tfrac{1}{2}k(a\alpha + b\beta + c\gamma). \tag{2}$$

so

$$k \equiv \frac{2\Delta}{a\alpha + b\beta + c\gamma}, \tag{3}$$

where Δ is the AREA of ΔABC and a, b, and c are the lengths of its sides (Kimberling 1998, pp. 26–27). To obtain trilinear coordinates giving the actual distances, take $k = 1$, so we have the coordinates

$$a' : b' : c'. \tag{4}$$

These normalized trilinear coordinates are known as EXACT TRILINEAR COORDINATES.

The trilinear coordinates of the line

$$ux + vy + wz = 0 \tag{5}$$

are

$$u : v : w = ab_A : bd_B : cd_C, \tag{6}$$

where d_i is the POINT-LINE DISTANCE from VERTEX i to the LINE.

The homogeneous BARYCENTRIC COORDINATES corresponding to trilinear coordinates $\alpha : \beta : \gamma$ are $(a\alpha, b\beta, c\gamma)$, and the trilinear coordinates corresponding to homogeneous BARYCENTRIC COORDINATES (t_1, t_2, t_3) are $t_1/a : t_2/b : t_3/c$.

Important points $\alpha : \beta : \gamma$ of a triangle are called TRIANGLE CENTERS, and the vector functions describing the location of the points in terms of side length, angles, or both, are called TRIANGLE CENTER FUNCTIONS $\mathbf{f}(a, b, c)$. Since by symmetry, triangle center functions are of the form

$$\mathbf{f}(a, b, c) = f(a, b, c) : f(b, c, a) : f(c, a, b), \tag{7}$$

it is common to call the scalar function $f(a, b, c)$ "the" triangle center function. Note also that side lengths and angles are interconvertible through the LAW OF COSINES, so a triangle center function may be given in terms of side lengths, angles, or both. Trilinear coordinates for some common triangle centers are summarized in the following table, where A, B, and C are the angles at the corresponding vertices and a, b, and c are the opposite side lengths. Here, the normalizations have been chosen to give the simplest possible form.

Point	Trilinear Center Function
CENTROID M	$\csc A$, $1/a$
CIRCUMCENTER O	$\cos A$
DE LONGCHAMPS POINT	$\cos A - \cos B \cos C$
EQUAL DETOUR POINT	$\sec\left(\tfrac{1}{2}A\right)\cos\left(\tfrac{1}{2}B\right)\cos\left(\tfrac{1}{2}C\right) + 1$
FEUERBACH POINT F	$1 - \cos(B - C)$
INCENTER I	1
ISOPERIMETRIC POINT	$\sec\left(\tfrac{1}{2}A\right)\cos\left(\tfrac{1}{2}B\right)\cos\left(\tfrac{1}{2}C\right) - 1$
SYMMEDIAN POINT	a
NINE-POINT CENTER N	$\cos(B - C)$
ORTHOCENTER H	$\cos B \cos C$
vertex A	$1:0:0$
vertex B	$0:1:0$
vertex C	$0:0:1$

To convert trilinear coordinates to a vector position for a given triangle specified by the x- and y-coordinates of its axes, pick two UNIT VECTORS along the sides. For instance, pick

$$\hat{\mathbf{a}} = \begin{bmatrix} a_1 \\ a_2 \end{bmatrix} \quad (8)$$

$$\hat{\mathbf{c}} = \begin{bmatrix} c_1 \\ c_2 \end{bmatrix} \quad (9)$$

where these are the UNIT VECTORS BC and AB. Assume the TRIANGLE has been labeled such that $A = \mathbf{x}_1$ is the lower rightmost VERTEX and $C = \mathbf{x}_2$. Then the VECTORS obtained by traveling l_a and l_c along the sides and then inward PERPENDICULAR to them must meet

$$\begin{bmatrix} x_1 \\ y_1 \end{bmatrix} + l_c \begin{bmatrix} c_1 \\ c_2 \end{bmatrix} - k\gamma \begin{bmatrix} c_2 \\ -c_1 \end{bmatrix} = \begin{bmatrix} x_2 \\ y_2 \end{bmatrix} + l_a \begin{bmatrix} a_1 \\ a_2 \end{bmatrix} - k\alpha \begin{bmatrix} a_2 \\ -a_1 \end{bmatrix}. \quad (10)$$

Solving the two equations

$$x_1 + l_c c_1 - k\gamma c_2 = x_2 + l_a a_1 - k\alpha a_2 \quad (11)$$

$$y_1 + l_c c_2 + k\gamma c_1 = y_2 + l_a a_2 + k\alpha a_1, \quad (12)$$

gives

$$l_a = \frac{k\alpha(a_1 c_1 + a_2 c_2) - \gamma k(c_1^2 + c_2^2) + c_2(x_1 - x_2) + c_1(y_2 - y_1)}{a_1 c_2 - a_2 c_1} \quad (13)$$

$$l_c = \frac{k\alpha(a_1^2 c_1 + a_2^2) - \gamma k(a_1 c_1 + a_2 c_2) + a_2(x_1 - x_2) + a_1(y_2 - y_1)}{a_1 c_2 - a_2 c_1}. \quad (14)$$

But $\hat{\mathbf{a}}$ and $\hat{\mathbf{c}}$ are UNIT VECTORS, so

$$l_a = \frac{k\alpha(a_1 c_1 + a_2 c_2) - \gamma k + c_2(x_1 - x_2) + c_1(y_2 - y_1)}{a_1 c_2 - a_2 c_1} \quad (15)$$

$$l_c = \frac{k\alpha - \gamma k(a_1 c_1 + a_2 c_2) + a_2(x_1 - x_2) + a_1(y_2 - y_1)}{a_1 c_2 - a_2 c_1}. \quad (16)$$

And the VECTOR coordinates of the point $\alpha : \beta : \gamma$ are then

$$\mathbf{x} = \mathbf{x}_1 + l_c \begin{bmatrix} c_1 \\ c_2 \end{bmatrix} - k\gamma \begin{bmatrix} c_2 \\ -c_1 \end{bmatrix}. \quad (17)$$

See also AREAL COORDINATES, BARYCENTRIC COORDINATES, EXACT TRILINEAR COORDINATES, MAJOR TRIANGLE CENTER, ORTHOCENTRIC COORDINATES, POWER CURVE, QUADRIPLANAR COORDINATES, REGULAR TRIANGLE CENTER, TRIANGLE, TRIANGLE CENTER, TRIANGLE CENTER FUNCTION, TRILINEAR POLAR

References

Boyer, C. B. *History of Analytic Geometry.* New York: Yeshiva University, 1956.
Casey, J. "The General Equation--Trilinear Co-Ordinates." Ch. 10 in *A Treatise on the Analytical Geometry of the Point, Line, Circle, and Conic Sections, Containing an Account of Its Most Recent Extensions, with Numerous Examples, 2nd ed., rev. enl.* Dublin: Hodges, Figgis, & Co., pp. 333–348, 1893.
Coolidge, J. L. *A Treatise on Algebraic Plane Curves.* New York: Dover, pp. 67–71, 1959.
Coxeter, H. S. M. *Introduction to Geometry, 2nd ed.* New York: Wiley, 1969.
Coxeter, H. S. M. "Some Applications of Trilinear Coordinates." *Linear Algebra Appl.* **226–228**, 375–388, 1995.
Kimberling, C. "Central Points and Central Lines in the Plane of a Triangle." *Math. Mag.* **67**, 163–187, 1994.
Kimberling, C. "Triangle Centers and Central Triangles." *Congr. Numer.* **129**, 1–295, 1998.
Wong, M. K. F. *Int. J. Math. Educ. Sci. Tech.* **27**, 293–296, 1996.
Wong, M. K. F. *Int. J. Math. Educ. Sci. Tech.* **29**, 143–145, 1998.

Trilinear Line

A LINE is given in TRILINEAR COORDINATES by

$$l\alpha + m\beta + n\gamma = 0.$$

See also LINE, TRILINEAR COORDINATES

Trilinear Polar

Given a TRIANGLE CENTER $X = l : m : n$, the line

$$l\alpha + m\beta + n\gamma = 0$$

is called the trilinear polar of X^{-1} and is denoted L.

See also CHASLES'S POLARS THEOREM

Trillion

The word trillion denotes different numbers in American and British usage. In the American system, one trillion equals 10^{12}. In the British, French, and German systems, one trillion equals 10^{18}.

See also BILLION, LARGE NUMBER, MILLION

Trilogarithm

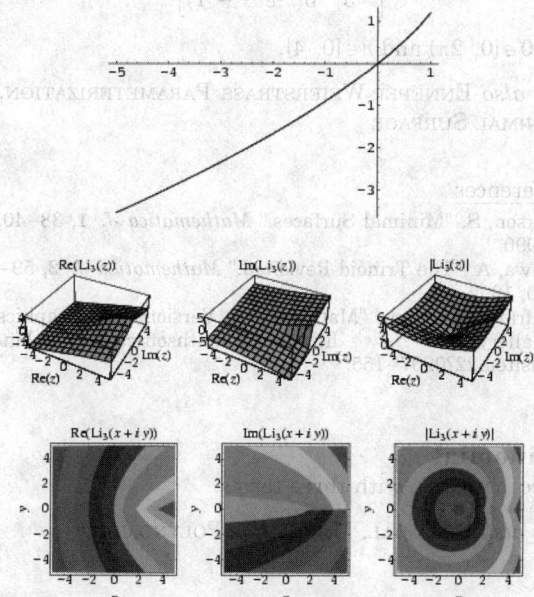

A special case of the POLYLOGARITHM $\mathrm{Li}_n(z)$ for $n = 3$. It is denoted $\mathrm{Li}_3(z)$, or sometimes $L_3(z)$. The notation $\mathrm{Li}_3(x)$ for the trilogarithm is unfortunately similar to that for the LOGARITHMIC INTEGRAL $\mathrm{Li}(x)$. Functional equations for the trilogarithm include

$$\mathrm{Li}_3(z) + \mathrm{Li}_3(-z) = \tfrac{1}{4}\mathrm{Li}_3(z^2) \tag{1}$$

$$\mathrm{Li}_3(-z) - \mathrm{Li}_3(-z^{-1}) = -\tfrac{1}{6}(\ln z)^3 - \tfrac{1}{6}\pi^2 \ln z \tag{2}$$

$$\mathrm{Li}_3(z) + \mathrm{Li}_3(1-z) + \mathrm{Li}_3(1 - z^{-1})$$
$$= \zeta(3) + \tfrac{1}{6}(\ln z)^3 + \tfrac{1}{6}\pi^2 \ln z - \tfrac{1}{2}(\ln z)^2 \ln(1-z) \tag{3}$$

Analytic values for $\mathrm{Li}_3(x)$ include

$$\mathrm{Li}_3(-1) = -\tfrac{3}{4}\zeta(3) \tag{4}$$

$$\mathrm{Li}_3(0) = 0 \tag{5}$$

$$\mathrm{Li}_3\left(\tfrac{1}{2}\right) = \tfrac{1}{24}\left[-2\pi^2 \ln 2 + 4(\ln 2)^3 + 21\zeta(3)\right] \tag{6}$$

$$\mathrm{Li}_3(1) = \zeta(3) \tag{7}$$

$$\mathrm{Li}_3\left(\tfrac{1}{2}\left(3 - \sqrt{5}\right)\right) = \tfrac{4}{5}\zeta(3) + \tfrac{2}{3}(\ln \phi)^3 - \tfrac{2}{15}\pi^2 \ln \phi \tag{8}$$

where $\zeta(3)$ is APÉRY'S CONSTANT and ϕ is the GOLDEN RATIO.
Bailey *et al.* showed that

$$\tfrac{35}{2}\zeta(3) - \pi^2 \ln 2$$
$$= 36 \,\mathrm{Li}_3\left(\tfrac{1}{2}\right) - 18\,\mathrm{Li}_3\left(\tfrac{1}{4}\right) - 4\,\mathrm{Li}_3\left(\tfrac{1}{8}\right) + \mathrm{Li}_3\left(\tfrac{1}{64}\right) \tag{9}$$

$$2(\ln 2)^3 - 7\zeta(3)$$
$$= -24 \,\mathrm{Li}_3\left(\tfrac{1}{2}\right) + 18\,\mathrm{Li}_3\left(\tfrac{1}{4}\right) + 4\,\mathrm{Li}_3\left(\tfrac{1}{8}\right) - \mathrm{Li}_3\left(\tfrac{1}{64}\right) \tag{10}$$

$$10(\ln 2)^3 - 2\pi^2 \ln 2$$
$$= -48 \,\mathrm{Li}_3\left(\tfrac{1}{2}\right) + 54\,\mathrm{Li}_3\left(\tfrac{1}{4}\right) + 12\,\mathrm{Li}_3\left(\tfrac{1}{8}\right) - 3\mathrm{Li}_3\left(\tfrac{1}{64}\right), \tag{11}$$

See also DILOGARITHM, POLYLOGARITHM

References

Bailey, D.; Borwein, P.; and Plouffe, S. "On the Rapid Computation of Various Polylogarithmic Constants." http://www.cecm.sfu.ca/~pborwein/PAPERS/P123.ps.
Lewin, L. *Polylogarithms and Associated Functions.* New York: North-Holland, pp. 154–156, 1981.

Trimagic Square

If replacing each number by its square or cube in a MAGIC SQUARE produces another MAGIC SQUARE, the square is said to be a trimagic square. Trimagic squares of order 32, 64, 81, and 128 are known. Tarry gave a method for constructing a trimagic square of order 128, Cazalas a method for trimagic squares of orders 64 and 81, and R. V. Heath a method for constructing an order 64 trimagic square which is different from Cazalas's (Kraitchik 1942).

Trimagic squares are also called TREBLY MAGIC SQUARES, and are 3-MULTIMAGIC SQUARES.

See also BIMAGIC SQUARE, MAGIC SQUARE, MULTIMAGIC SQUARE

References

Ball, W. W. R. and Coxeter, H. S. M. *Mathematical Recreations and Essays, 13th ed.* New York: Dover, pp. 212–213, 1987.
Kraitchik, M. "Multimagic Squares." §7.10 in *Mathematical Recreations.* New York: W. W. Norton, pp. 144 and 176–178, 1942.

Trimean

The trimean is defined to be

$$TM \equiv \tfrac{1}{4}(H_1 + 2M + H_2),$$

where H_i are the HINGES and M is the MEDIAN. Press *et al.* (1992) call this TUKEY'S TRIMEAN. It is an L-ESTIMATE.

See also HINGE, L-ESTIMATE, MEAN, MEDIAN (STATISTICS)

References

Press, W. H.; Flannery, B. P.; Teukolsky, S. A.; and Vetterling, W. T. *Numerical Recipes in FORTRAN: The Art of Scientific Computing, 2nd ed.* Cambridge, England: Cambridge University Press, p. 694, 1992.
Tukey, J. W. *Explanatory Data Analysis.* Reading, MA: Addison-Wesley, pp. 46–47, 1977.

Trimorphic Number

A number n such that the last digits of n^3 are the same as n. 49 is trimorphic since $49^3 = 117649$ (Wells 1986, p. 124). The first few are 1, 4, 5, 6, 9, 24, 25, 49, 51, 75, 76, 99, 125, 249, 251, 375, 376, 499, ... (Sloane's A033819).

See also AUTOMORPHIC NUMBER, NARCISSISTIC NUMBER, SUPER-d NUMBER

References

Sloane, N. J. A. Sequences A033819 in "An On-Line Version of the Encyclopedia of Integer Sequences." http://www.research.att.com/~njas/sequences/eisonline.html.
Wells, D. *The Penguin Dictionary of Curious and Interesting Numbers.* Middlesex, England: Penguin Books, 1986.

Trinoid

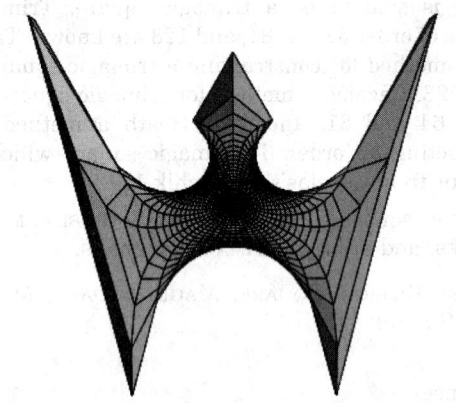

A MINIMAL SURFACE discovered by L. P. M. Jorge and W. Meeks III in 1983 with ENNEPER-WEIERSTRASS PARAMETERIZATION

$$f = \frac{1}{(\zeta^3 - 1)^2} \qquad (1)$$

$$g = \zeta^2 \qquad (2)$$

(Dickson 1990). Explicitly, it is given by

$$x = \Re\left[\frac{re^{i\theta}}{3(1 + re^{i\theta} + r^2 e^{2i\theta})} - \frac{4\ln(re^{i\theta} - 1)}{9} \right.$$
$$\left. + \frac{2\ln(1 + re^{i\theta} + r^2 e^{2i\theta})}{9}\right] \qquad (3)$$

$$y = -\frac{1}{9}\Im\left[\frac{-3re^{i\theta}(1 + re^{i\theta})}{r^3 e^{3i\theta} - 1}\right.$$
$$\left. + \frac{4\sqrt{3}(r^3 e^{3i\theta} - 1)\tan^{-1}\left(\frac{1 + 2re^{i\theta}}{\sqrt{3}}\right)}{r^3 e^{3i\theta} - 1}\right] \qquad (4)$$

$$z = \Re\left[-\frac{2}{3} - \frac{2}{3(r^3 e^{3i\theta} - 1)}\right], \qquad (5)$$

for $0 \in [0, 2\pi)$ and $r \in [0, 4]$.

See also ENNEPER-WEIERSTRASS PARAMETERIZATION, MINIMAL SURFACE

References

Dickson, S. "Minimal Surfaces." *Mathematica J.* **1**, 38–40, 1990.
Ogawa, A. "The Trinoid Revisited." *Mathematica J.* **2**, 59–60, 1992.
Wolfram Research "Mathematica Version 2.0 Graphics Gallery." http://www.mathsource.com/cgi-bin/msitem22?0207-155.

Trinomial

A POLYNOMIAL with three terms.

See also BINOMIAL, MONOMIAL, POLYNOMIAL

Trinomial Coefficient

A coefficient of the TRINOMIAL TRIANGLE. The trinomial coefficient $\binom{n}{k}_2$, with $n \geq 0$ and $-n \leq k \leq n$, is given by the coefficient of x^{n+k} in the expansion of $(1 + x + x^2)^n$. Therefore,

$$\binom{n}{-k}_2 = \binom{n}{k}_2.$$

Equivalently, the trinomial coefficients are defined by

$$(1 + x + x^{-1})^n = \sum_{j=-n}^{n} \binom{n}{j}_2 x^j. \qquad (1)$$

The trinomial coefficients satisfy

$$\binom{m}{j}_2 = \binom{m-1}{j-1}_2 + \binom{m-1}{j}_2 + \binom{m-1}{j+1}_2. \qquad (2)$$

An alternatives definition of the trinomial coefficients is as the coefficients in $(x + y + z)^n$ (Andrews 1990).

The (usual) trinomial coefficient is also given by the number of permutations of n symbols, each -1, 0, or 1, which sum to k. For example, there seven permutations of three symbols which sum to 0, $\{-1, 0, 1\}$, $\{-1, 1, 0\}$, $\{0, -1, 1\}$, $\{0, 0, 0\}$, and $\{0, 1, -1\}$, $\{1, -1, 0\}$, $\{1, 0, -1\}$, so $\binom{3}{0}_2 = 7$. Explicit formulas for $\binom{n}{k}_2$ are given by

$$\binom{n}{k}_2 = \sum_{j=0}^{n} \frac{n!}{j!(j+m)!(n-2j-m)!} \qquad (3)$$

$$\binom{n}{k}_2 = \sum_{j=0}^{n} (-1)^j \binom{n}{j}\binom{2n-2j}{n-m-j} \qquad (4)$$

(Andrews 1990).

The following table gives the first $\binom{n}{k}_2$ trinomial coefficients for $k = 0, 1, \ldots$ and $n = k, k+1, \ldots$.

k	Sloane	(n, k)-trinomial coefficients
0	Sloane's A002426	1, 1, 3, 7, 19, 51, 141, 393, 1107, 3139, 8953, ...
1	Sloane's A005717	1, 2, 6, 16, 45, 126, 357, 1016, 2907, 8350, ...
2	Sloane's A014531	1, 3, 10, 30, 90, 266, 784, 2304, ...
4		1, 5, 21, 77, 266, 882, 2850, 9042, ...
5		1, 6, 28, 112, 414, 1452, 4917, ...

See also BINOMIAL COEFFICIENT, CENTRAL TRINOMIAL COEFFICIENT, TRINOMIAL TRIANGLE

References

Andrews, G. "Euler's 'exemplum memorabile inductionis fallacis' and q-Trinomial Coefficients." *J. Amer. Math. Soc.* **3**, 653–669, 1990.

Comtet, L. *Advanced Combinatorics: The Art of Finite and Infinite Expansions, rev. enl. ed.* Dordrecht, Netherlands: Reidel, p. 78, 1974.

Hoggatt, V. E. Jr., and Bicknell, M. "Diagonal Sums of Generalized Pascal Triangles." *Fib. Quart.* **7**, 341–358 and 393, 1969.

Euler, L. "Exemplum Memorabile Inductionis Fallacis." *Opera Omnia, Vol. 15.* Leipzig, Germany: Teubner, p. 59, 1911.

Graham, R. L.; Knuth, D. E.; and Patashnik, O. *Concrete Mathematics: A Foundation for Computer Science.* Reading, MA: Addison-Wesley, p. 575, 1990.

Guy, R. K. "The Second Strong Law of Small Numbers." *Math. Mag.* **63**, 3–20, 1990.

Henrici, P. *Applied and Computational Complex Analysis, Vol. 1.* New York: Wiley, p. 42, 1974.

Riordan, J. *Combinatorial Identities.* New York: Wiley, p. 74, 1979.

Shapiro, L. W.; Getu, S.; Woan, W.-J.; and Woodson, L. C. "The Riordan Group." *Disc. Appl. Math.* **34**, 229–239, 1991.

Sloane, N. J. A. Sequences A002426/M2673, A005717/M1612, and A014531 in "An On-Line Version of the Encyclopedia of Integer Sequences." http://www.research.att.com/~njas/sequences/eisonline.html.

Trinomial Identity

$$\left(x^2 + axy + by^2\right)\left(t^2 + atu + bu^2\right) = r^2 + ars + bs^2, \quad (1)$$

where

$$r = xt - byu \quad (2)$$

$$s = yt + xu + ayu. \quad (3)$$

Trinomial Triangle

The NUMBER TRIANGLE obtained by starting with a row containing a single "1" and the next row containing three 1s and then letting subsequent row elements be computed by summing the elements above to the left, directly above, and above to the right:

$$
\begin{array}{ccccccccc}
 & & & & 1 & & & & \\
 & & & 1 & 1 & 1 & & & \\
 & & 1 & 2 & 3 & 2 & 1 & & \\
 & 1 & 3 & 6 & 7 & 6 & 3 & 1 & \\
1 & 4 & 10 & 16 & 19 & 16 & 10 & 4 & 1
\end{array}
$$

(Sloane's A027907). The nth row can also be obtained by expanding $(1 + x + x^2)^n$ and taking coefficients:

$$\left(1 + x + x^2\right)^0 = 1$$

$$\left(1 + x + x^2\right)^1 = 1 + x + x^2$$

$$\left(1 + x + x^2\right)^2 = 1 + 2x + 3x^2 + 2x^3 + x^4$$

and so on.

See also CENTRAL TRINOMIAL COEFFICIENT, PASCAL'S TRIANGLE, TRINOMIAL COEFFICIENT

References

Sloane, N. J. A. Sequences A027907 in "An On-Line Version of the Encyclopedia of Integer Sequences." http://www.research.att.com/~njas/sequences/eisonline.html.

Triomino

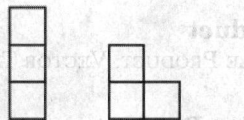

The two 3-POLYOMINOES are called triominoes, and are also known as the TROMINOES. The left triomino above is "STRAIGHT," while the right triomino is called "right" or L-.

There is also a game called triomino consisting of 55 equilateral triangles, each containing three numbers from 0 to 5 at each vertex. Every combination of tiles is in the game, although those tiles with three different values are allowed to be arranged only in clockwise-increasing order.

See also L-POLYOMINO, POLYOMINO, STRAIGHT POLYOMINO

References

Gardner, M. "Polyominoes." Ch. 13 in *The Scientific American Book of Mathematical Puzzles & Diversions.* New York: Simon and Schuster, pp. 124–140, 1959.

Hunter, J. A. H. and Madachy, J. S. *Mathematical Diversions.* New York: Dover, pp. 80–81, 1975.

Lei, A. "Tromino." http://www.cs.ust.hk/~philipl/omino/tro-mino.html485

Triple

A group of three elements, also called a TRIAD.

See also AMICABLE TRIPLE, MONAD, PAIR, PYTHAGOR-EAN TRIPLE, QUADRUPLET, QUINTUPLET, TETRAD, TRIAD, TWINS

Triple Jacobi Product

JACOBI TRIPLE PRODUCT

Triple Point

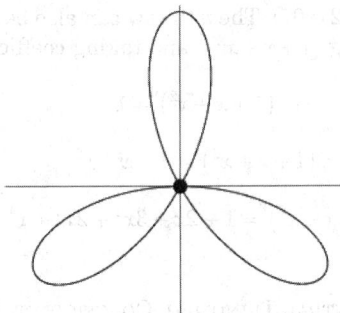

A point where a curve intersects itself along three arcs. The above plot shows the triple point at the ORIGIN of the TRIFOLIUM $(x^2+y^2)^2+3x^2y-y^3=0$.

See also DOUBLE POINT, QUADRUPLE POINT

References

Walker, R. J. *Algebraic Curves.* New York: Springer-Verlag, pp. 57–58, 1978.

Triple Product

SCALAR TRIPLE PRODUCT, VECTOR TRIPLE PRODUCT

Triple Scalar Product

SCALAR TRIPLE PRODUCT

Triple Torus

A SPHERE with three HANDLES, i.e., a genus-3 TORUS.

See also DOUBLE TORUS, HANDLE, TORUS

Triple Vector Product

VECTOR TRIPLE PRODUCT

Triple Yahtzee

YAHTZEE

Triple-Free Set

A SET of POSITIVE integers is called weakly triple-free if, for any integer x, the SET $\{x, 2x, 3x\} \nsubseteq S$. It is called strongly triple-free if $x \in S$ IMPLIES $2x \notin S$ and $3x \notin S$ (i.e., the set is both DOUBLE-FREE and triple-

free). For example, the subsets of $\{1, 2, 3\}$ which are weakly triple-free are $\varnothing$, $\{1\}$, $\{1, 2\}$, $\{2\}$, $\{2, 3\}$, and $\{3\}$, while $\{1, 2, 3\}$ and $\{1, 3\}$ are not. Of these weakly triple-free sets, $\varnothing$, $\{1\}$, $\{2\}$, $\{2, 3\}$, and $\{3\}$ are also strongly triple-free.

The number of weakly triple-free subsets of $\{1, 2, \ldots, n\}$ for $n = 1, 2, \ldots$ are 2, 4, 6, 12, 24, 36, 72, 144, 240, 480, ... (Sloane's A050293). The number of strongly triple-free subsets for $n = 1, 2, \ldots$ are 2, 3, 5, 8, 16, 24, 48, 76, 132, ... (Sloane's A050295).

Define

$$p(n) = \max\{|S| : S \subset (1, 2, \ldots, n)$$
$$\text{is weakly triple-free}\}$$
$$q(n) = \max\{|S| : S \subset (1, 2, \ldots, n)$$
$$\text{is strongly triple-free}\},$$

where $|S|$ denotes the CARDINAL NUMBER of (number of members in) S. Then for $n = 1, 2, \ldots$, $p(n)$ is given by 1, 2, 2, 3, 4, 4, 5, 6, 7, 8, 9, 9, 10, 11, 11, ... (Sloane's A050294), and $q(n)$ by 1, 1, 2, 2, 3, 4, 5, 5, 6, 6, 7, 7, 8, 8, 9, ... (Sloane's A050296). Asymptotic formulas are given by

$$\lim_{n \to \infty} \frac{p(n)}{n} \geq \frac{4}{5}$$

and

$$\lim_{n \to \infty} \frac{q(n)}{n} = 0.6134752692\ldots$$

(Finch).

See also A-SEQUENCE, DOUBLE-FREE SET, SUM-FREE SET

References

Finch, S. "Favorite Mathematical Constants." http://www.mathsoft.com/asolve/constant/triple/triple.html.
Sloane, N. J. A. Sequences A050293, A050294, A050295, and A050296 in "An On-Line Version of the Encyclopedia of Integer Sequences." http://www.research.att.com/~njas/sequences/eisonline.html.

Trip-Let

A 3-dimensional solid which is shaped in such a way that its projections along three mutually perpendicular axes are three different letters of the alphabet. Hofstadter (1989) has constructed such a solid for the letters G, E, and B.

See also CORK PLUG, ROTOR

References

Hofstadter, D. R. *Gödel, Escher, Bach: An Eternal Golden Braid.* New York: Vintage Books, cover and pp. xiv, 1, and 273, 1989.

Triplet

TRIPLE

Triplicate-Ratio Circle

LEMOINE CIRCLE

Triquetra

This entry contributed by DANA MACKENZIE

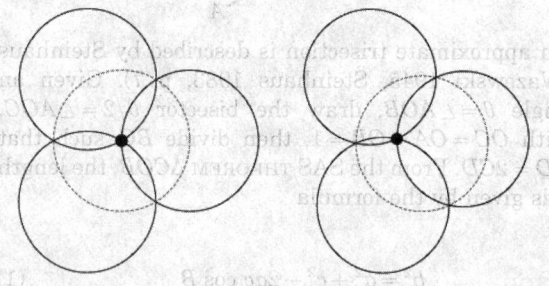

A "triquetra" is a figure consisting of three circular arcs of equal radius, and has seen extensive use in heraldry (i.e., coats of arms), specifically in the case of the so-called BORROMEAN RINGS. The term "Triquetra theorem" was coined by Mackenzie (1992) to describe the geometric theorem that if three circles are concurrent at a single point, then the other three intersection points lie on a circle of the same radius as the first three. This version was first proved in 1916.

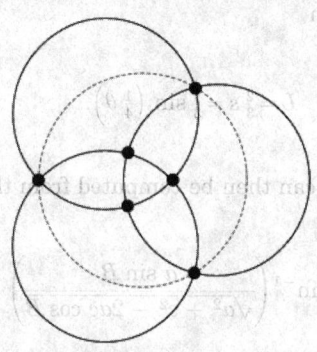

Mackenzie (1992) generalized this theorem to the case where the three circles do not coincide. In this case, they form six intersection points, and if you partition the points into any two groups of three and look at the CIRCUMRADII of the points in those groups, there is a nice formula relating them to the radii of the triquetra circles. This formula has some pretty geometric consequences (or "porisms"). Ultimately, the triquetra theorem turns out to be closely related to PONCELET'S PORISM.

See also BORROMEAN RINGS, CIRCLE-CIRCLE INTERSECTION, CIRCULAR TRIANGLE, HARUKI'S THEOREM, PONCELET'S PORISM, REULEAUX TRIANGLE, VENN DIAGRAM

References

Mackenzie, D. "Triquetras and Porisms." *College Math. J.* pp. 118–131. March 1992.

Trirectangular Tetrahedron

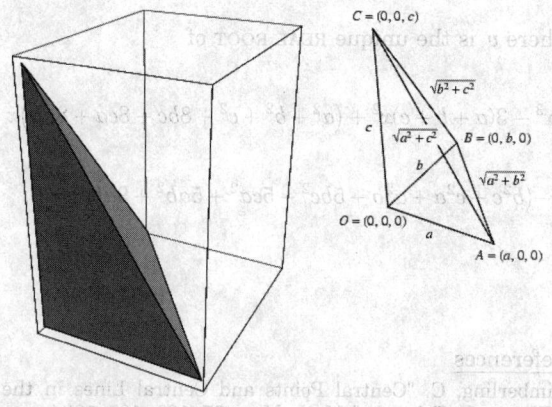

A TETRAHEDRON having a TRIHEDRON all of the face angles of which are right angles. The face opposite the vertex of the right angles is called the base. If the edge lengths bounding the trihedral angle are a, b, and c, then the side lengths of the base are given by $\sqrt{a^2+b^2}$, $\sqrt{a^2+c^2}$, and $\sqrt{b^2+c^2}$, and so has SEMI-PERIMETER

$$s = \tfrac{1}{2}\left(\sqrt{a^2+b^2} + \sqrt{a^2+c^2} + \sqrt{b^2+c^2} \right). \qquad (1)$$

The VOLUME of the trirectangular tetrahedron is

$$V = \tfrac{1}{6}\,abc. \qquad (2)$$

Using HERON'S FORMULA, the SURFACE AREA is therefore

$$S = \tfrac{1}{2}\left(ab + ac + bc + \sqrt{a^2b^2 + a^2c^2 + b^2c^2} \right). \qquad (3)$$

Let Δ_{XYZ} be the AREA of the triangle with vertices X, Y, and Z. The remarkable DE GUA'S THEOREM

$$\Delta_{ABC}^2 = \Delta_{OAB}^2 + \Delta_{OAC}^2 + \Delta_{OAC}^2. \qquad (4)$$

then follows from the identity

$$s\left(s - \sqrt{a^2+b^2} \right)\left(s - \sqrt{a^2+c^2} \right)\left(s - \sqrt{b^2+c^2} \right)$$

$$= \tfrac{1}{4}(a^2b^2 + a^2c^2 + b^2c^2), \qquad (5)$$

with s defined by (1).

See also DE GUA'S THEOREM, TRIHEDRON

References

Altshiller-Court, N. "The Trirectangular Tetrahedron." §4.6a in *Modern Pure Solid Geometry*. New York: Chelsea, pp. 91–94, 1979.

Trisected Perimeter Point

A triangle center which has a TRIANGLE CENTER FUNCTION

$$\alpha = bc(v - c + a)(v - a + b),$$

where v is the unique REAL ROOT of

$$2x^3 - 3(a + b + c)x^2 + (a^2 + b^2 + c^2 + 8bc + 8ca + 8ab)x$$

$$- (b^2c + c^2a + a^2b + 5bc^2 + 5ca^2 + 5ab^2 + 9abc) = 0.$$

References

Kimberling, C. "Central Points and Central Lines in the Plane of a Triangle." *Math. Mag.* **67**, 163–187, 1994.

Trisection

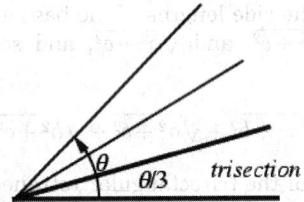

trisection

Angle trisection is the division of an *arbitrary* ANGLE into three equal ANGLES. It was one of the three GEOMETRIC PROBLEMS OF ANTIQUITY for which solutions using only COMPASS and STRAIGHTEDGE were sought. The problem was algebraically proved impossible by Wantzel (1836).

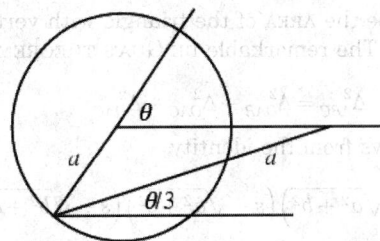

Although trisection is not possible for a general ANGLE using a Greek construction, there are some specific angles, such as $\pi/2$ and π radians (90° and 180°, respectively), which *can* be trisected. Furthermore, some ANGLES are geometrically trisectable, but cannot be constructed in the first place, such as $3\pi/7$ (Honsberger 1991). In addition, trisection of an arbitrary angle *can* be accomplished using a *marked* RULER (a NEUSIS CONSTRUCTION) as illustrated above

(*Courant and Robbins 1996*).

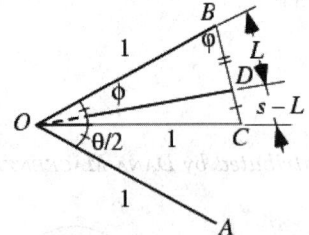

An approximate trisection is described by Steinhaus (Wazewski 1945, Steinhaus 1983, p. 7). Given an angle $\theta = \angle AOB$, draw the bisector $\theta/2 = \angle AOC$, with $OC = OA = OB = 1$, then divide BC such that $BD = 2CD$. From the SAS THEOREM $\triangle COB$, the length s is given by the formula

$$b^2 = a^2 + c^2 - 2ac\cos B \qquad (1)$$

with $s = b$, $a = c = 1$, $B = \theta/2$,

$$s = \sqrt{2 - 2\cos\left(\tfrac{1}{2}\theta\right)} = 2\sqrt{\frac{1 - \cos\left(\tfrac{1}{2}\theta\right)}{2}}$$

$$= 2\sin\left(\tfrac{1}{4}\theta\right), \qquad (2)$$

and L is then

$$L = \tfrac{2}{3}s = \tfrac{4}{3}\sin\left(\tfrac{1}{4}\theta\right). \qquad (3)$$

The angle φ can then be computed from the formula

$$\varphi = \sin^{-1}\left(\frac{a\sin B}{\sqrt{a^2 + c^2 - 2ac\cos B}}\right) \qquad (4)$$

to obtain

$$\varphi = \sin^{-1}\left[\frac{\sin\left(\tfrac{1}{2}\theta\right)}{\sqrt{2 - 2\cos\left(\tfrac{1}{2}\theta\right)}}\right]$$

$$= \sin^{-1}\left[\frac{2\sin\left(\tfrac{1}{4}\theta\right)\cos\left(\tfrac{1}{4}\theta\right)}{2\sin\left(\tfrac{1}{4}\theta\right)}\right]$$

$$= \sin^{-1}\left[\cos\left(\tfrac{1}{4}\theta\right)\right]. \qquad (5)$$

ϕ is then given by the formula for an SAS triangle

ΔBOD

$$\phi = \sin^{-1}\left(\frac{L\sin\varphi}{\sqrt{1 + L^2 - 2L\cos\varphi}}\right)$$

$$= \sin^{-1}\left[\frac{\frac{2}{3}\sin\left(\frac{1}{2}\theta\right)}{1 - \frac{8}{9}\sin^2\left(\frac{1}{4}\theta\right)}\right]$$

$$= \sin^{-1}\left[\frac{6\sin\left(\frac{1}{2}\theta\right)}{5 + 4\cos\left(\frac{1}{2}\theta\right)}\right]. \qquad (6)$$

The Maclaurin series is then

$$\phi = \tfrac{1}{3}\theta + \tfrac{7}{648}\theta^3 + \tfrac{19}{31104}\theta^5 + \cdots \approx \tfrac{1}{3}\theta \qquad (7)$$

to a very good approximation.

An ANGLE can also be divided into three (or any WHOLE NUMBER) of equal parts using the QUADRATRIX OF HIPPIAS or TRISECTRIX.

See also ANGLE BISECTOR, MACLAURIN TRISECTRIX, QUADRATRIX OF HIPPIAS, TRISECTRIX

References

Bogomolny, A. "Angle Trisection." http://www.cut-the-knot.com/pythagoras/archi.html.

Bold, B. "The Problem of Trisecting an Angle." Ch. 5 in *Famous Problems of Geometry and How to Solve Them.* New York: Dover, pp. 33–37, 1982.

Conway, J. H. and Guy, R. K. *The Book of Numbers.* New York: Springer-Verlag, pp. 190–191, 1996.

Courant, R. and Robbins, H. "Trisecting the Angle." §3.3.3 in *What is Mathematics?: An Elementary Approach to Ideas and Methods, 2nd ed.* Oxford, England: Oxford University Press, pp. 137–138, 1996.

Coxeter, H. S.M. "Angle Trisection." §2.2 in *Introduction to Geometry, 2nd ed.* New York: Wiley, p. 28, 1969.

Dixon, R. *Mathographics.* New York: Dover, pp. 50–51, 1991.

Dörrie, H. "Trisection of an Angle." §36 in *100 Great Problems of Elementary Mathematics: Their History and Solutions.* New York: Dover, pp. 172–177, 1965.

Dudley, U. *The Trisectors.* Washington, DC: Math. Assoc. Amer., 1994.

Honsberger, R. *More Mathematical Morsels.* Washington, DC: Math. Assoc. Amer., pp. 25–26, 1991.

Klein, F. "The Delian Problem and the Trisection of the Angle." Ch. 2 in "Famous Problems of Elementary Geometry: The Duplication of the Cube, the Trisection of the Angle, and the Quadrature of the Circle." In *Famous Problems and Other Monographs.* New York: Chelsea, pp. 13–15, 1980.

Ogilvy, C. S. "Solution to Problem E 1153." *Amer. Math. Monthly* **62**, 584, 1955. Ogilvy, C. S. "Angle Trisection." *Excursions in Geometry.* New York: Dover, pp. 135–141, 1990.

Scudder, H. T. "How to Trisect and Angle with a Carpenter's Square." *Amer. Math. Monthly* **35**, 250–251, 1928.

Steinhaus, H. *Mathematical Snapshots, 3rd ed.* New York: Dover, 1999.

Wantzel, M. L. "Recherches sur les moyens de reconnaître si un Problème de Géométrie peut se résoudre avec la règle et le compas." *J. Math. pures appliq.* **1**, 366–372, 1836.

Wazewski, T. *Ann. Soc. Polonaise Math.* **18**, 164, 1945.

Wells, D. *The Penguin Dictionary of Curious and Interesting Geometry.* London: Penguin, p. 25, 1991.

Trisectrix

A curve which can be used to trisect an angle. Although an arbitrary angle cannot be trisected using only COMPASS and STRAIGHTEDGE (i.e., according to the strict rules of Greek GEOMETRIC CONSTRUCTION), it can be trisected using certain curves (which are assumed to have been constructed using some other means).

See also CATALAN'S TRISECTRIX, LIMAÇON, MACLAURIN TRISECTRIX, TRISECTION, TSCHIRNHAUSEN CUBIC

Trisectrix of Catalan

TSCHIRNHAUSEN CUBIC

Trisectrix of Maclaurin

MACLAURIN TRISECTRIX

Triskaidecagon

TRIDECAGON

Triskaidekaphobia

The number 13 is traditionally associated with bad luck. This superstition leads some people to fear or avoid anything involving this number, a condition known as triskaidekaphobia. Triskaidekaphobia leads to interesting practices such as the numbering of floors as 1, 2, ..., 11, 12, 14, 15, ..., *omitting the number 13,* in many high-rise hotels.

See also 13, BAKER'S DOZEN

Tristan Edwards Projection

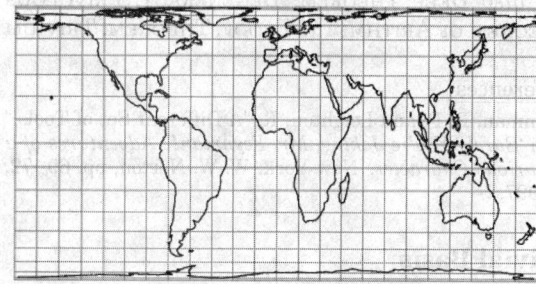

A CYLINDRICAL EQUAL-AREA PROJECTION which uses a standard parallel of $\phi_s = 37.383°$.

See also BALTHASART PROJECTION, BEHRMANN CYLINDRICAL EQUAL-AREA PROJECTION, CYLINDRICAL EQUAL-AREA PROJECTION, GALL ORTHOGRAPHIC PROJECTION, LAMBERT AZIMUTHAL EQUAL-AREA PROJECTION, PETERS PROJECTION

Tritangent

The tritangent of a CUBIC SURFACE is a PLANE which intersects the surface in three mutually intersecting lines. Each intersection of two lines is then a tangent point of the surface.

See also Cubic Surface

References
Hunt, B. "Algebraic Surfaces." http://www.mathematik.uni-kl.de/~wwwagag/E/Galerie.html.

Tritangent Triangle
Excentral Triangle

Trivalent Graph
Cubic Graph

Trivalent Tree
Binary Tree

Trivial

Related to or being the mathematically most simple case. More generally, the word "trivial" is used to describe any result which requires little or no effort to derive or prove. The word originates from the Latin TRIVIUM, which was the lower division of the seven liberal arts in medieval universities (cf. QUADRIVIUM).

According to the Nobel Prize-winning physicist Richard Feynman (Feynman 1997), mathematicians designate any THEOREM as "trivial" once a proof has been obtained—no matter how difficult the theorem was to prove in the first place. There are therefore exactly two types of true mathematical propositions: trivial ones, and those which have not yet been proven.

The opposite of a trivial theorem is a "DEEP THEOREM."

See also Deep Theorem, Degeneracy, Frivolous Theorem of Arithmetic, Proof, Theorem, Trivium

References
Feynman, R. P. and Leighton, R. "A Different Set of Tools." In *'Surely You're Joking, Mr. Feynman!': Adventures of a Curious Character.* New York: W. W. Norton, pp. 69–72, 1997.

Trivial Basis

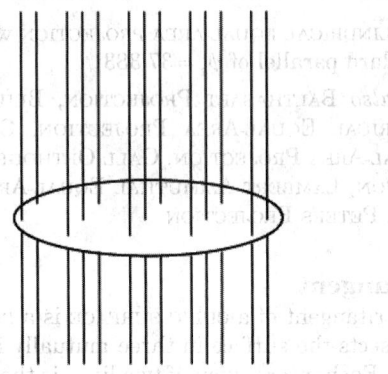

Trivial Group

The trivial group is the unique GROUP containing exactly one element. That is, it is $G = \{e\}$, where e is the IDENTITY ELEMENT (so that $ee = e$).

See also Cyclic Group, Finite Group, Group, Identity Element

Trivialization

Over a small NEIGHBORHOOD U of a MANIFOLD, a VECTOR BUNDLE is spanned by the local sections defined on U. For example, in a COORDINATE CHART U with coordinates $(x_1, \ldots, x_n)$, every smooth VECTOR FIELD can be written as a sum $\Sigma_i f_i \, \partial/\partial x_i$ where f_i are smooth functions. The n vector fields $\partial/\partial x_i$ span the space of vector fields, considered as a MODULE over the RING of smooth real-valued functions. On this COORDINATE CHART U, the tangent bundle can be written $U \times \mathbb{R}^n$. This is a trivialization of the tangent bundle.

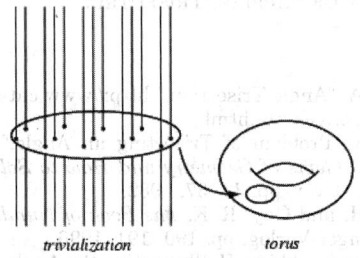

trivialization   *torus*

trivialization of a line bundle on a torus

In general, a vector bundle of RANK r is spanned LOCALLY by r independent SECTIONS. Every point has a NEIGHBORHOOD U and r sections defined on U, such that over every point in U the fibers are spanned by those r sections.

Similarly, for a FIBER BUNDLE, near every point $p \in M$, there is a neighborhood U such that the bundle over U is $U \times F$, where F is the fiber.

A bundle is a set of trivializations that cover the base manifold. The trivializations are put together to form a bundle with its TRANSITION FUNCTIONS.

See also Bundle, Fiber Bundle, Manifold, Transition Function, Vector Bundle

Trivium

A word derived from the Latin roots *tri-* (three) and *via* (ways, roads), therefore a crossing of three roads. In medieval universities, the trivium consisted of the three subjects in the lower division of the seven liberal arts: grammar, rhetoric, and logic. The word TRIVIAL derives from the fact that the trivium contained the least complicated studies.

See also Quadrivium, Trivial

Trochoid

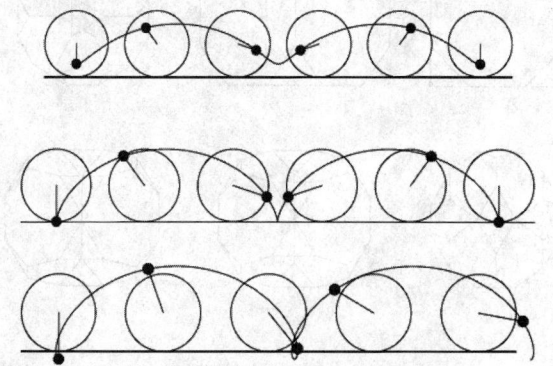

The curve described by a point at a distance b from the center of a rolling CIRCLE of RADIUS a.

$$x = a\phi - b \sin \phi$$

$$y = a - b \cos \phi.$$

If $b < a$, the curve is a CURTATE CYCLOID. If $b = a$, the curve is a CYCLOID. If $b > a$, the curve is a PROLATE CYCLOID.

See also CURTATE CYCLOID, CYCLOID, EPITROCHOID, HYPOTROCHOID, PROLATE CYCLOID

References

Hall, L. "Trochoids, Roses, and Thorns--Beyond the Spirograph." *College Math. J.* **23**, 20–35, 1992.

Wagon, S. *Mathematica in Action.* New York: W. H. Freeman, pp. 46–50, 1991.

Yates, R. C. "Trochoids." *A Handbook on Curves and Their Properties.* Ann Arbor, MI: J. W. Edwards, pp. 233–236, 1952.

Tromino

TRIOMINO

Trott's Constant

The constant $x = 0.010841015122311136151129\ldots$ whose decimal digits are equal to the constant's own CONTINUED FRACTION $[0, 1, 0, 8, 4, 1, 0, 1, 5, \ldots]$. This constant was discovered by M. Trott of Wolfram Research in 1999. It appears to be unique, and all attempts to find other such numbers have failed.

See also CONTINUED FRACTION

True

A statement which is rigorously known to be correct. A statement which is not true is called FALSE, although certain statements can be proved to be rigorously UNDECIDABLE within the confines of a given set of assumptions and definitions. Regular two-valued LOGIC allows statements to be only true or FALSE, but FUZZY LOGIC treats "truth" as a continuum which can have any value between 0 and 1. The symbol Υ is sometimes used to denote "true,"

although "T" is more commonly used in TRUTH TABLES.

See also ALETHIC, BOOLEANS, FALSE, FUZZY LOGIC, LOGIC, TRUTH TABLE, UNDECIDABLE

Truncatable Prime

Call a number n containing no zeros right truncatable if n and all numbers obtained by successively removing the rightmost DIGIT are PRIME. There are 83 right truncatable primes in base 10. The first few are 2, 3, 5, 7, 23, 29, 31, 37, 53, 59, 71, 73, 79, 233, 239, 293, 311, 313, 317, 373, 379, 593, 599, ... (Sloane's A024770), the largest being 73,939,133 (Angell and Godwin 1977). The numbers of left prime strings less than 10, 10^2, 10^3, ... are 4, 9, 14, 16, 15, 12, 8, and 5 (Sloane's A050986; Rivera puzzle 70).

If zeros are permitted, the sequence of right truncatable primes are 2, 3, 5, 7, 13, 17, 23, 37, 43, 47, 53, 67, 73, 83, 97, 103, 107, 113, 137, 167, 173, 197, 223, 283, 307, ... (Sloane's A033664).

Similarly, call a number n left truncatable if n and all numbers obtained by successively removing the leftmost DIGIT are PRIME. There are 4260 right prime strings in base 10 when the digit zero is not allowed (otherwise, if zeros are permitted, the sequence is infinite). The first few are 2, 3, 5, 7, 13, 17, 23, 37, 43, 47, 53, 67, 73, 83, 97, 113, 137, 167, 173, ... (Sloane's A024785), with the largest being 357,686,312,646,216,567,629,137 (Angell and Godwin 1977, Baillie 1995). The numbers of right prime strings less than 10, 10^2, 10^3, ... are 4, 11, 39, 99, 192, 326, 429, ... (Sloane's A050987; Rivera puzzle 70).

J. Shallit has shown that in base 10, there is a finite, minimal list of primes that do not have any other primes as substrings (where digits do *not* need to be consecutive). This result is a special case of a much more general theorem, whose proof is unfortunately nonconstructive.

Call an n-digit prime p_n (with $n \geq 2$) is a restricted left truncatable prime if

1. If the leftmost digit of p_i is deleted, a prime number p_{i-1} is obtained for $2 \leq i \leq n$, and
2. No prime with $n + 1$ digits can have its leftmost digit removed to produce p_n.

Kahan and Weintraub (1998) dub such primes "Henry VIII primes." Restricted left truncatable primes p_n are therefore a subset of left truncatable primes for which there are no left truncatable primes of length $n + 1$ having the same n last digits as p_n. There are a total of 1440 such primes, and the first few are 773, 3373, 3947, 4643, 5113, 6397, 6967, 7937, ... (Sloane's A055522), the largest being 357686312646216567629137 (Kahan and Weintraub 1998).

See also PRIME ARRAY, PRIME NUMBER

References

Angell, I. O. and Godwin, H. J. "On Truncatable Primes." *Math. Comput.* **31**, 265–267, 1977.

Baillie, R. "Largest Left-Truncatable Prime." sci.math.-num-analysis posting, Aug. 7, 1995.

De Geest, P. "List of the 4260 Left-Truncatable Primes (without the Zero Digit)." http://www.ping.be/~ping6758/truncat.htm.

Kahan, S. and Weintraub, S. "Left Truncatable Primes." *J. Recr. Math.* **29**, 254–264, 1998.

Rivera, C. "Problems & Puzzles: Puzzle Prime Strings.-002." http://www.primepuzzles.net/puzzles/puzz_002.htm.

Rivera, C. "Problems & Puzzles: Puzzle Primes Double Tree (A Puzzle Suggested by Paul Leyland).-070." http://www.primepuzzles.net/puzzles/puzz_070.htm.

Schroeppel, R. Item 33 in Beeler, M.; Gosper, R. W.; and Schroeppel, R. *HAKMEM.* Cambridge, MA: MIT Artificial Intelligence Laboratory, Memo AIM-239, p. 14, Feb. 1972.

Sloane, N. J. A. Sequences A024770, A024785, A032437, A033664, A050986, A050987, and A055522 in "An On-Line Version of the Encyclopedia of Integer Sequences." http://www.research.att.com/~njas/sequences/eisonline.html.

Weisstein, E. W. "Integer Sequences." MATHEMATICA NOTEBOOK INTEGERSEQUENCES.M.

Weisstein, E. W. "Left Prime Strings." MATHEMATICA NOTEBOOK LEFTPRIMESTRINGS.TXT.

Weisstein, E. W. "Right Prime Strings." MATHEMATICA NOTEBOOK RIGHTPRIMESTRINGS.TXT.

Truncate

To truncate a REAL NUMBER is to discard its non-integer part. Truncation of a (positive) number x therefore corresponds to taking the FLOOR FUNCTION $\lfloor x \rfloor$.

See also CEILING FUNCTION, FLOOR FUNCTION, NINT, ROUND, TRUNCATION

Truncated Cone

CONICAL FRUSTUM

Truncated Cube

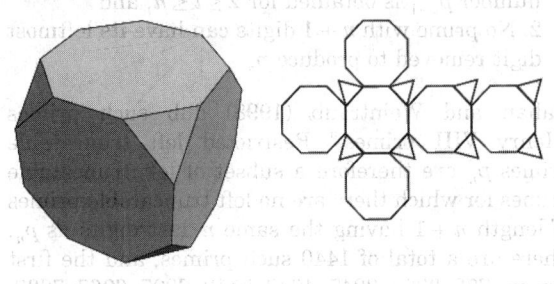

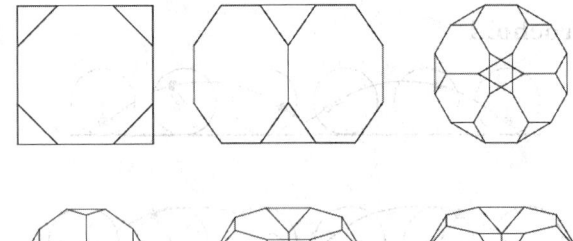
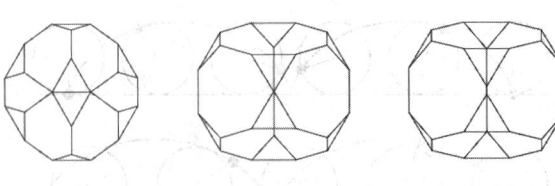

The 14-faced ARCHIMEDEAN SOLID A_9 with faces $8\{3\} + 6\{8\}$. It is also UNIFORM POLYHEDRON U_9 and Wenninger model W_8. It has SCHLÄFLI SYMBOL t$\{4, 3\}$ $+266851127816936987904000 0S^{12}$ and WYTHOFF SYMBOL 2 3|4.

The DUAL POLYHEDRON of the truncated cube is the TRIAKIS OCTAHEDRON. The INRADIUS r of the dual, MIDRADIUS ρ of the solid and dual, and CIRCUMRADIUS R of the solid for $a = 1$ are

$$r = \tfrac{1}{17}\left(5 + 2\sqrt{2}\right)\sqrt{7 + 4\sqrt{2}} \approx 1.63828$$

$$\rho = \tfrac{1}{2}\left(2 + \sqrt{2}\right) \approx 1.70711$$

$$R = \tfrac{1}{2}\sqrt{7 + 4\sqrt{2}} \approx 1.77882.$$

The distances from the center of the solid to the centroids of the triangular and octagonal faces are

$$r_3 = \tfrac{1}{2}\sqrt{\tfrac{1}{3}\left(17 + 12\sqrt{2}\right)} \tag{1}$$

$$r_8 = \tfrac{1}{2}\left(1 + \sqrt{2}\right). \tag{2}$$

The SURFACE AREA and VOLUME are

$$S = 2\left(6 + 6\sqrt{2} + 2\sqrt{3}\right) \tag{3}$$

$$V = \tfrac{1}{3}\left(21 + 14\sqrt{2}\right). \tag{4}$$

See also ARCHIMEDEAN SOLID, ICOSITETRAHEDRON

References

Ball, W. W. R. and Coxeter, H. S. M. *Mathematical Recreations and Essays, 13th ed.* New York: Dover, p. 138, 1987.

Cundy, H. and Rollett, A. "Truncated Cube. 3.8^2." §3.7.3 in *Mathematical Models, 3rd ed.* Stradbroke, England: Tarquin Pub., p. 103, 1989.

Wenninger, M. J. "The Truncated Hexahedron (Cube)." Model 8 in *Polyhedron Models.* Cambridge, England: Cambridge University Press, p. 22, 1989.

Truncated Cube-Small Triakis Octahedron Compound

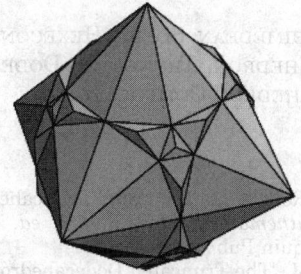

The POLYHEDRON COMPOUND of the TRUNCATED CUBE and its dual, the SMALL TRIAKIS OCTAHEDRON. The compound can be constructed from a TRUNCATED CUBE of unit edge length by midpoint CUMULATION with heights

$$h_3 = \tfrac{1}{6}\sqrt{3}\left(3 - 2\sqrt{2}\right) \tag{1}$$

$$h_8 = \tfrac{1}{2}\left(1 + \sqrt{2}\right). \tag{2}$$

See also CUMULATION, POLYHEDRON COMPOUND, SMALL TRIAKIS OCTAHEDRON, TRUNCATED CUBE

Truncated Cuboctahedron
GREAT RHOMBICUBOCTAHEDRON (ARCHIMEDEAN)

Truncated Cylinder
CYLINDRICAL WEDGE

Truncated Dodecadodecahedron

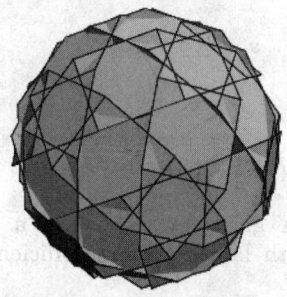

The UNIFORM POLYHEDRON U_{59}, also called the QUASI-TRUNCATED DODECAHEDRON, whose DUAL POLYHEDRON is the MEDIAL DISDYAKIS TRIACONTAHEDRON. It has SCHLÄFLI SYMBOL t' $\{\frac{2}{5}\}$ and WYTHOFF SYMBOL $2\,\frac{5}{3}\,|\,5$. Its faces are $12\{10\} + 30\{4\} + 12\{\frac{10}{3}\}$. Its CIRCUMRADIUS for $a = 1$ is

$$R = \tfrac{1}{2}\sqrt{11}.$$

References
Wenninger, M. J. *Polyhedron Models.* Cambridge, England: Cambridge University Press, pp. 152–153, 1989.

Truncated Dodecahedron

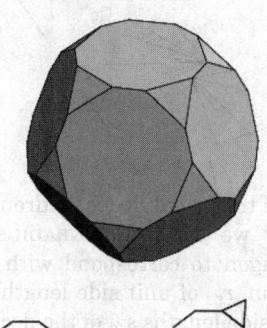

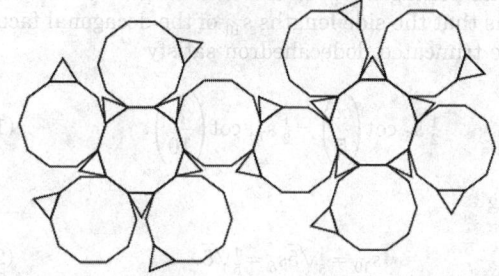

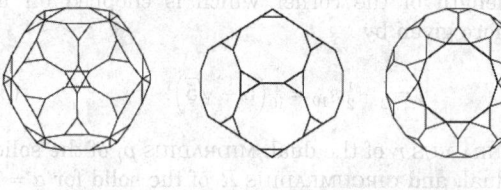

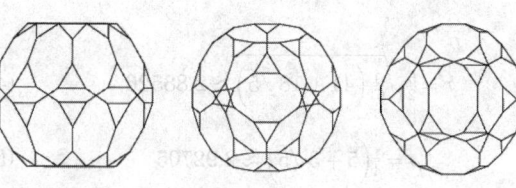

The 32-faced ARCHIMEDEAN SOLID A_{10} with faces $20\{3\} + 12\{10\}$. It is also UNIFORM POLYHEDRON U_{26} and Wenninger model W_{10}. It has SCHLÄFLI SYMBOL t$\{5, 3\}$ and WYTHOFF SYMBOL $2\,3\,|\,5$.

The DUAL POLYHEDRON is the TRIAKIS ICOSAHEDRON.

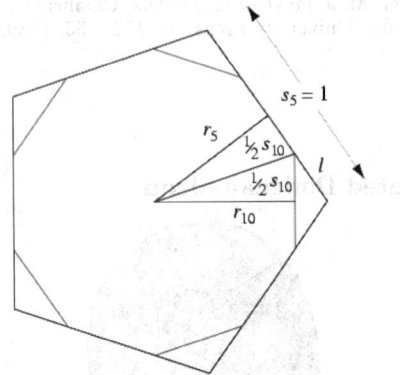

To construct the truncated dodecahedron by TRUNCA-TION, note that we want the INRADIUS r_{10} of the truncated pentagon to correspond with that of the original pentagon, r_5, of unit side length $s_5 = 1$. This means that the side lengths s_{10} of the decagonal faces in the truncated dodecahedron satisfy

$$\tfrac{1}{2} s_5 \cot\left(\frac{\pi}{5}\right) = \tfrac{1}{2} s_{10} \cot\left(\frac{\pi}{10}\right), \tag{1}$$

giving

$$s_{10} = \tfrac{1}{5}\sqrt{5} s_5 = \tfrac{1}{5}\sqrt{5}. \tag{2}$$

The length of the corner which is chopped off is therefore given by

$$l = \tfrac{1}{2} - \tfrac{1}{2} s_{10} = \tfrac{1}{10}\left(5 - \sqrt{5}\right). \tag{3}$$

The INRADIUS r of the dual, MIDRADIUS p_i of the solid and dual, and CIRCUMRADIUS R of the solid for $a = 1$ are

$$r = \tfrac{5}{2}\sqrt{\tfrac{1}{61}\left(41 + 18\sqrt{5}\right)} \approx 2.88526 \tag{4}$$

$$\rho = \tfrac{1}{4}\left(5 + 3\sqrt{5}\right) \approx 2.92705 \tag{5}$$

$$R = \tfrac{1}{4}\sqrt{74 + 30\sqrt{5}} \approx 2.96945. \tag{6}$$

The distances from the center of the solid to the centroids of the triangular and decagonal faces are given by

$$r_3 = \tfrac{1}{12}\sqrt{3}\left(9 + 5\sqrt{5}\right) \tag{7}$$

$$r_{10} = \tfrac{1}{2}\sqrt{\tfrac{1}{2}\left(25 + 11\sqrt{5}\right)}. \tag{8}$$

The SURFACE AREA and VOLUME are

$$S = 5\left(\sqrt{3} + 6\sqrt{5 + 2\sqrt{5}}\right) \tag{9}$$

$$V = \tfrac{5}{12}\left(99 + 47\sqrt{5}\right). \tag{10}$$

See also ARCHIMEDEAN SOLID, HEXECONTAHEDRON, TRIAKIS ICOSAHEDRON, TRUNCATED DODECAHEDRON-TRIAKIS ICOSAHEDRON COMPOUND

References

Cundy, H. and Rollett, A. "Truncated Dodecahedron. 3.10^2." §3.7.9 in *Mathematical Models, 3rd ed.* Stradbroke, England: Tarquin Pub., p. 109, 1989.

Wenninger, M. J. "The Truncated Dodecahedron." Model 10 in *Polyhedron Models.* Cambridge, England: Cambridge University Press, p. 24, 1989.

Truncated Dodecahedron-Triakis Icosahedron Compound

The POLYHEDRON COMPOUND of the TRUNCATED DODECAHEDRON and its dual, the TRIAKIS ICOSAHEDRON. The compound can be constructed from a TRUNCATED DODECAHEDRON of unit edge length by midpoint CUMULATION with heights

$$h_3 = \tfrac{1}{372}\sqrt{3}\left(1 + 5\sqrt{5}\right) \tag{1}$$

$$h_{10} = \tfrac{1}{2}\sqrt{\tfrac{1}{2}\left(5 + \sqrt{5}\right)}. \tag{2}$$

The resulting solid has edge lengths

$$s_1 = \tfrac{1}{62}\sqrt{\tfrac{5}{6}\left(397 + \sqrt{5}\right)} \tag{3}$$

$$s_2 = \tfrac{1}{2} \tag{4}$$

$$s_3 = \tfrac{1}{2}\sqrt{\tfrac{1}{2}\left(5 + \sqrt{5}\right)} \tag{5}$$

$$s_4 = \tfrac{1}{4}\left(5 + \sqrt{5}\right), \tag{6}$$

CIRCUMRADIUS

$$R = \tfrac{1}{2}\sqrt{\tfrac{1}{2}\left(37 + 15\sqrt{5}\right)} \tag{7}$$

SURFACE AREA given by a root of a 32nd order polynomial with large integer coefficients, and VOLUME

$$V = \tfrac{5}{1488}\left(15997 + 7693\sqrt{5}\right). \qquad (8)$$

See also POLYHEDRON COMPOUND, TRIAKIS ICOSAHE-
DRON, TRUNCATED DODECAHEDRON

Truncated Exponential Function
EXPONENTIAL SUM FUNCTION

Truncated Great Dodecahedron

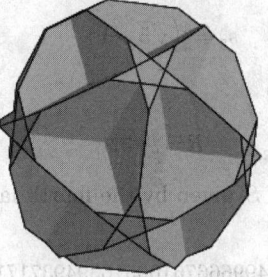

The UNIFORM POLYHEDRON U_{37} whose DUAL POLYHE-
DRON is the SMALL STELLAPENTAKIS DODECAHEDRON.
It has SCHLÄFLI SYMBOL $t\{5, \tfrac{5}{2}\}$. It has WYTHOFF
SYMBOL $2\,\tfrac{5}{2}\,5$. Its faces are $12\{\tfrac{5}{2}\} + 12\{10\}$. Its CIR-
CUMRADIUS for $a = 1$ is

$$R = \tfrac{1}{4}\sqrt{34 + 10\sqrt{5}}.$$

See also GREAT ICOSAHEDRON

References
Wenninger, M. J. *Polyhedron Models.* Cambridge, England:
Cambridge University Press, p. 115, 1971.

Truncated Great Icosahedron
GREAT TRUNCATED ICOSAHEDRON

Truncated Hexahedron
TRUNCATED CUBE

Truncated Icosahedron

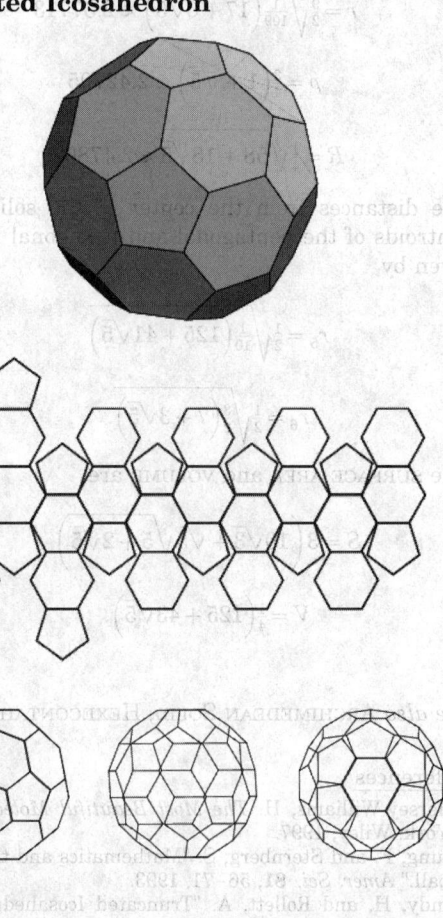

The 32-faced ARCHIMEDEAN SOLID A_{11} corresponding
to the facial arrangement $20\{6\} + 12\{5\}$. It is the
shape used in the construction of SOCCER BALLS, and
it was also the configuration of the lenses used for
focusing the explosive shock waves of the detonators
in the Fat Man atomic bomb (Rhodes 1996, p. 195).
The truncated icosahedron has 60 vertices, and is also
the C_{60} structure of pure carbon known as buckyballs
(a.k.a. fullerenes). The truncated icosahedron is
UNIFORM POLYHEDRON U_{25} and Wenninger model
W_9. It has SCHLÄFLI SYMBOL t$\{3, 5\}$ and WYTHOFF
SYMBOL $2\,5\,|\,3$.

The DUAL POLYHEDRON of the truncated icosahedron
is the PENTAKIS DODECAHEDRON. The INRADIUS r of
the dual, MIDRADIUS ρ of the solid and dual, and
CIRCUMRADIUS R of the solid for $a = 1$ are

$$r = \tfrac{9}{2}\sqrt{\tfrac{1}{109}\left(17 + 6\sqrt{5}\right)} \approx 2.37713$$

$$\rho = \tfrac{3}{4}\left(1 + \sqrt{5}\right) \approx 2.42705$$

$$R = \tfrac{1}{4}\sqrt{58 + 18\sqrt{5}} \approx 2.47802.$$

The distances from the center of the solid to the centroids of the pentagonal and hexagonal faces are given by

$$r_5 = \tfrac{1}{2}\sqrt{\tfrac{1}{10}\left(125 + 41\sqrt{5}\right)} \tag{1}$$

$$r_6 = \tfrac{1}{2}\sqrt{\tfrac{3}{2}\left(7 + 3\sqrt{5}\right)}. \tag{2}$$

The SURFACE AREA and VOLUME are

$$S = 3\left(10\sqrt{3} + \sqrt{5}\sqrt{5 + 2\sqrt{5}}\right) \tag{3}$$

$$V = \tfrac{1}{4}\left(125 + 43\sqrt{5}\right). \tag{4}$$

See also ARCHIMEDEAN SOLID, HEXECONTAHEDRON

References

Aldersey-Williams, H. *The Most Beautiful Molecule*. New York: Wiley, 1997.
Chung, F. and Sternberg, S. "Mathematics and the Bucky-ball." *Amer. Sci.* **81**, 56–71, 1993.
Cundy, H. and Rollett, A. "Truncated Icosahedron. 5.6^2." §3.7.10 in *Mathematical Models, 3rd ed.* Stradbroke, England: Tarquin Pub., p. 110, 1989.
Harris, J. W. and Stocker, H. *Handbook of Mathematics and Computational Science*. New York: Springer-Verlag, p. 101, 1998.
Rhodes, R. *Dark Sun: The Making of the Hydrogen Bomb*. Touchstone Books, 1996.
Trott, M. "Constructing a Buckyball with *Mathematica*." http://library.wolfram.com/demos/v4/Buckyball.nb.
Wenninger, M. J. "The Truncated Icosahedron." Model 9 in *Polyhedron Models*. Cambridge, England: Cambridge University Press, p. 23, 1989.

Truncated Icosahedron-Pentakis Dodecahedron Compound

The POLYHEDRON COMPOUND of the TRUNCATED ICOSAHEDRON and its dual, the PENTAKIS DODECAHE-

DRON. The compound can be constructed from a TRUNCATED ICOSAHEDRON of unit edge length by midpoint CUMULATION with heights

$$h_5 = \tfrac{1}{38}\sqrt{\tfrac{1}{10}\left(305 + 131\sqrt{5}\right)} \tag{1}$$

$$h_6 = \tfrac{1}{4}\sqrt{3}\left(\sqrt{5} - 3\right). \tag{2}$$

The resulting solid has edge lengths

$$s_1 = \tfrac{1}{2} \tag{3}$$

$$s_2 = \tfrac{3}{76}\left(7 + 5\sqrt{5}\right) \tag{4}$$

$$s_3 = \tfrac{1}{4}\left(1 + \sqrt{5}\right) \tag{5}$$

$$s_4 = \tfrac{1}{2}\sqrt{3} \tag{6}$$

$$s_5 = \tfrac{3}{4}\left(\sqrt{5} - 1\right), \tag{7}$$

CIRCUMRADIUS

$$R = \tfrac{3}{2}\sqrt{3}, \tag{8}$$

SURFACE AREA S given by the fourth largest positive root of

$$5141016030764996667610951639493717193603515625$$
$$-9291774385004510118161779667281494140625000S^2$$
$$+63419261142631991476189330253320312540000S^4$$
$$-2162618355523996143839802656250000000S^6$$
$$+4069907058882620169449676000000000S^8$$
$$-43785979422682649316768000000S^{10}$$
$$+266851127816936987904000 0 S^{12}$$
$$-85420833678869299200S^{14}$$
$$+1113034787454976S^{16} \tag{9}$$

and VOLUME

$$V = \tfrac{5}{152}\left(1477 + 162\sqrt{5}\right). \tag{10}$$

See also CUMULATION, POLYHEDRON COMPOUND, PENTAKIS DODECAHEDRON, TRUNCATED ICOSAHE-DRON

Truncated Icosidodecahedron

GREAT RHOMBICOSIDODECAHEDRON (ARCHIMEDEAN)

Truncated Octahedral Number

A FIGURATE NUMBER which is constructed as an OCTAHEDRAL NUMBER with a SQUARE PYRAMID removed from each of the six VERTICES,

$$TO_n = O_{3n-2} - 6P_{n-1} = 16n^3 - 33n^2 + 24n - 6,$$

where O_n is an OCTAHEDRAL NUMBER and P_n is a SQUARE PYRAMIDAL NUMBER. The first few are 1, 38, 201, 586, ... (Sloane's A005910). The GENERATING FUNCTION for the truncated octahedral numbers is

$$\frac{x(6x^3 + 55x^2 + 34x + 1)}{(x-1)^4} = x + 38x^2 + 201x^3 + \cdots$$

See also OCTAHEDRAL NUMBER, SQUARE PYRAMIDAL NUMBER

References

Conway, J. H. and Guy, R. K. *The Book of Numbers*. New York: Springer-Verlag, p. 52, 1996.
Sloane, N. J. A. Sequences A005910/M5266 in "An On-Line Version of the Encyclopedia of Integer Sequences." http://www.research.att.com/~njas/sequences/eisonline.html.

Truncated Octahedron

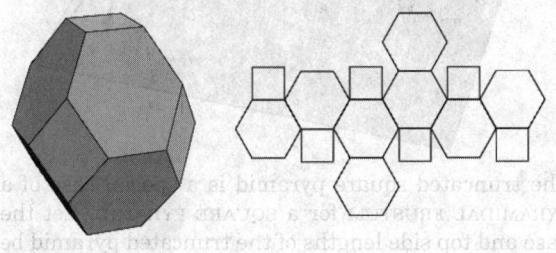

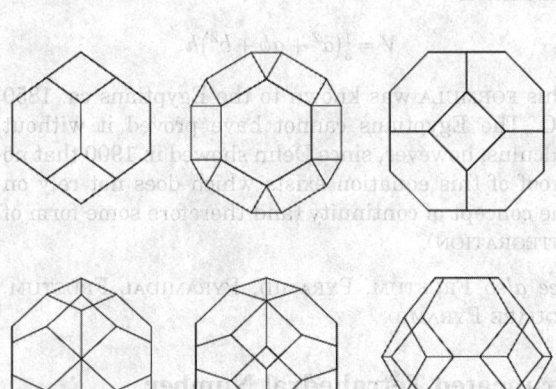

The 14-faced ARCHIMEDEAN SOLID A_{12}, also known as the MECON, with faces $8\{6\} + 6\{4\}$. It is also UNIFORM POLYHEDRON U_8 and Wenninger model W_7. It has SCHLÄFLI SYMBOL $t\{3, 4\}$ and WYTHOFF SYMBOL $2\,4\,|\,3$.

The DUAL POLYHEDRON of the truncated octahedron is the TETRAKIS HEXAHEDRON. The truncated octahedron has the O_h OCTAHEDRAL GROUP of symmetries. The form of the fluorite (CaF_2) resembles the truncated octahedron (Steinhaus 1983, pp. 207–208).

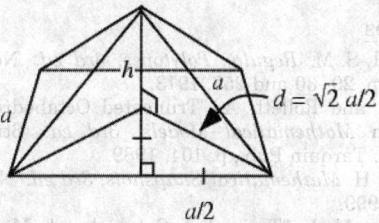

The solid of unit edge length can be formed from an OCTAHEDRON of edge length 3 via TRUNCATION by removing six SQUARE PYRAMIDS, each with edge slant height $s = 1$, base $a = 1$ on a side, and height h. The height and base area of the SQUARE PYRAMID are then

$$h = \sqrt{s^2 - \tfrac{1}{4}a^2 \csc^2\left(\frac{\pi}{n}\right)} = \tfrac{1}{2}\sqrt{2}a \tag{1}$$

$$A_b = a^2. \tag{2}$$

The SURFACE AREA of the truncated octahedron is

$$S = 6 + 12\sqrt{3}. \tag{3}$$

The VOLUME of the truncated octahedron is then given by the VOLUME of the OCTAHEDRON

$$V_{\text{octahedron}} = \tfrac{1}{3}\sqrt{2}s^3 = 9\sqrt{2}a^3 \tag{4}$$

minus six times the volume of the SQUARE PYRAMID,

$$V = V_{\text{octahedron}} - 6\left(\tfrac{1}{3}A_b h\right) = \left(9\sqrt{2} - \sqrt{2}\right)a^3$$

$$= 8\sqrt{2}a^3. \tag{5}$$

The truncated octahedron is a SPACE-FILLING POLYHEDRON (Steinhaus 1983, pp. 187–190 and 207).

The INRADIUS r of the dual, MIDRADIUS ρ of the solid and dual, and CIRCUMRADIUS R of the solid for $a = 1$ are

$$r = \tfrac{9}{20}\sqrt{10} \approx 1.42302 \tag{6}$$

$$\rho = \tfrac{3}{2} = 1.5 \tag{7}$$

$$R = \tfrac{1}{2}\sqrt{10} \approx 1.58114. \tag{8}$$

The distances from the center of the solid to the centroids of the square and hexagonal faces are given by

$$r_4 = \sqrt{2} \tag{9}$$

$$r_6 = \tfrac{1}{2}\sqrt{6}. \tag{10}$$

See also ARCHIMEDEAN SOLID, ICOSITETRAHEDRON, KELVIN'S CONJECTURE, OCTAHEDRON, RHOMBIC DO-DECAHEDRON STELLATIONS, SQUARE PYRAMID, TRUN-CATION

References

Coxeter, H. S. M. *Regular Polytopes, 3rd ed.* New York: Dover, pp. 29–30 and 257, 1973.

Cundy, H. and Rollett, A. "Truncated Octahedron. 4.6^2." §3.7.4 in *Mathematical Models, 3rd ed.* Stradbroke, England: Tarquin Pub., p. 104, 1989.

Steinhaus, H. *Mathematical Snapshots, 3rd ed.* New York: Dover, 1999.

Wenninger, M. J. "Truncated Octahedron." Model 7 in *Polyhedron Models.* Cambridge, England: Cambridge University Press, p. 21, 1989.

Truncated Octahedron-Tetrakis Hexahedron Compound

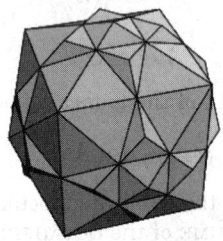

The POLYHEDRON COMPOUND of the TRUNCATED OCTA-HEDRON and its dual, the TETRAKIS HEXAHEDRON. The compound can be constructed from a TRUNCATED OCTAHEDRON of unit edge length by midpoint CUMU-LATION with heights

$$h_4 = \tfrac{1}{8}\sqrt{2} \qquad (1)$$

$$h_6 = \tfrac{1}{4}\sqrt{6} \qquad (2)$$

See also CUMULATION, POLYHEDRON COMPOUND, TETRAKIS HEXAHEDRON, TRUNCATED OCTAHEDRON

Truncated Polyhedron

A polyhedron with truncated faces, given by the SCHLÄFLI SYMBOL $\mathrm{t}\{^p_q\}$.

See also FRUSTUM, RHOMBIC POLYHEDRON, SNUB POLYHEDRON

References

Harris, J. W. and Stocker, H. "Obliquely Truncated n-Sided Prism." §4.2.5 in *Handbook of Mathematics and Computational Science.* New York: Springer-Verlag, p. 98, 1998.

Truncated Power Function

The function defined by

$$y_+^a \equiv \begin{cases} y^\alpha & \text{for } y > 0 \\ 0 & \text{for } y < 0. \end{cases}$$

See also POWER

References

Samko, S. G.; Kilbas, A. A.; and Marichev, O. I. *Fractional Integrals and Derivatives.* Yverdon, Switzerland: Gordon and Breach, p. 22, 1993.

Truncated Pyramid

PYRAMIDAL FRUSTUM

Truncated Square Pyramid

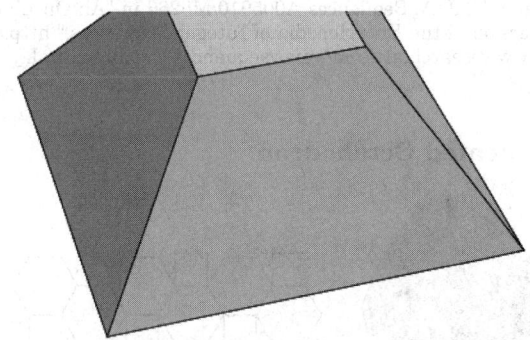

The truncated square pyramid is a special case of a PYRAMIDAL FRUSTUM for a SQUARE PYRAMID. Let the base and top side lengths of the truncated pyramid be a and b, and let the height be h. Then the VOLUME of the solid is

$$V = \tfrac{1}{3}(a^2 + ab + b^2)h.$$

This FORMULA was known to the Egyptians ca. 1850 BC. The Egyptians cannot have proved it without calculus, however, since Dehn showed in 1900 that no proof of this equation exists which does not rely on the concept of continuity (and therefore some form of INTEGRATION).

See also FRUSTUM, PYRAMID, PYRAMIDAL FRUSTUM, SQUARE PYRAMID

Truncated Tetrahedral Number

A FIGURATE NUMBER constructed by taking the $(3n-2)$th TETRAHEDRAL NUMBER and removing the $(n-1)$th TETRAHEDRAL NUMBER from each of the four corners,

$$\mathrm{Ttet}_n \equiv \mathrm{Te}_{3n-3} - 4\mathrm{Te}_{n-1} = \tfrac{1}{6}\,n\left(23n^2 - 27n + 10\right).$$

The first few are 1, 16, 68, 180, 375, ... (Sloane's A005906). The GENERATING FUNCTION for the truncated tetrahedral numbers is

$$\frac{x(10x^2 + 12x + 1)}{(x-1)^4} = x + 16x^2 + 68x^3 + 180x^4 + \cdots.$$

References

Conway, J. H. and Guy, R. K. *The Book of Numbers.* New York: Springer-Verlag, pp. 46–47, 1996.

Sloane, N. J. A. Sequences A005906/M5002 in "An On-Line Version of the Encyclopedia of Integer Sequences." http://www.research.att.com/~njas/sequences/eisonline.html.

Truncated Tetrahedron

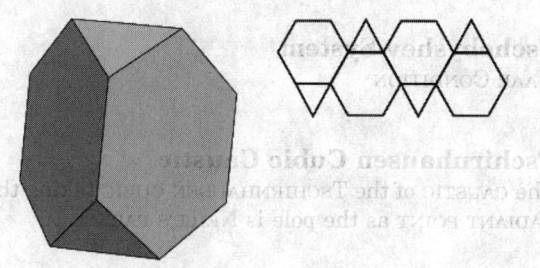

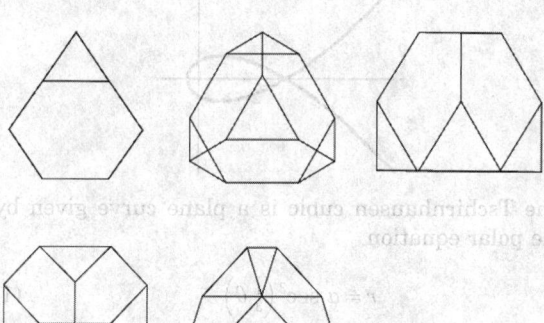

The ARCHIMEDEAN SOLID A_{13} with faces $4\{3\} + 4\{6\}$. It is also UNIFORM POLYHEDRON U_2 and Wenninger model W_6. It has SCHLÄFLI SYMBOL $t\{3, 3\}$ and WYTHOFF SYMBOL $2\ 3\ |\ 3$.

The dual of the truncated tetrahedron is the TRIAKIS TETRAHEDRON. The INRADIUS r of the dual, MIDRADIUS ρ of the solid and dual, and CIRCUMRADIUS R of the solid for $a = 1$ are

$$r = \tfrac{9}{44}\sqrt{22} \approx 0.95940$$

$$\rho = \tfrac{3}{4}\sqrt{2} \approx 1.06066$$

$$R = \tfrac{1}{4}\sqrt{22} \approx 1.17260$$

The distances from the center of the solid to the centroids of the triangular and hexagonal faces are

given by

$$r_3 = \tfrac{1}{12}\sqrt{6} \tag{1}$$

$$r_6 = \tfrac{1}{4}\sqrt{6}. \tag{2}$$

The SURFACE AREA and VOLUME are

$$S = 7\sqrt{3} \tag{3}$$

$$V = \tfrac{23}{12}\sqrt{2}. \tag{4}$$

See also ARCHIMEDEAN SOLID, TRIAKIS TETRAHEDRON, TRUNCATED TETRAHEDRON-TRIAKIS TETRAHEDRON COMPOUND

References

Cundy, H. and Rollett, A. "Truncated Tetrahedron. 3.6^2." §3.7.1 in *Mathematical Models, 3rd ed.* Stradbroke, England: Tarquin Pub., p. 101, 1989.

Wenninger, M. J. "The Truncated Tetrahedron." Model 6 in *Polyhedron Models.* Cambridge, England: Cambridge University Press, p. 20, 1989.

Truncated Tetrahedron-Triakis Tetrahedron Compound

The compound of a TRUNCATED TETRAHEDRON and its dual, the TRIAKIS TETRAHEDRON. The compound can be constructed from a TRUNCATED OCTAHEDRON of unit edge length by midpoint CUMULATION with heights

$$h_3 = \tfrac{1}{30}\sqrt{6} \tag{1}$$

$$h_6 = \tfrac{1}{2}\sqrt{6}. \tag{2}$$

See also CUMULATION, POLYHEDRON COMPOUND, TRIAKIS TETRAHEDRON, TRUNCATED TETRAHEDRON

Truncation

The removal of portions of SOLIDS falling outside a set of symmetrically placed planes. The dual operation consists of replacing facial polygons with pyramids, and is sometimes known as CUMULATION.

The five PLATONIC SOLIDS belong to one of the following three truncation series (which, in the first two cases, carry the solid to its DUAL POLYHEDRON).

Cube   Truncated   Cuboctahedron   Truncated   Octahedron
        Cube   Octahedron

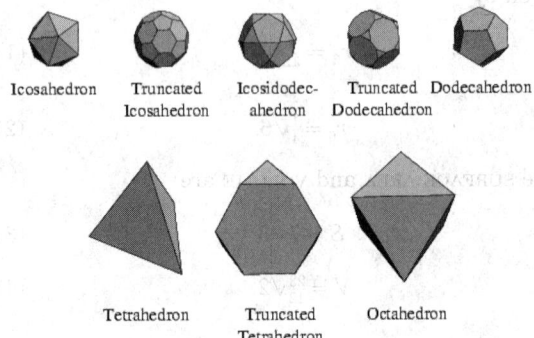

Icosahedron   Truncated   Icosidodec-   Truncated   Dodecahedron
              Icosahedron ahedron   Dodecahedron

Tetrahedron   Truncated   Octahedron
              Tetrahedron

See also CUMULATION, DÜRER'S SOLID, PYRAMID, STELLATION, TRUNCATED CUBE, TRUNCATED DODECAHEDRON, TRUNCATED ICOSAHEDRON, TRUNCATED OCTAHEDRON, TRUNCATED TETRAHEDRON, VERTEX FIGURE

Truth Table

A truth table is a 2-D array with $n+1$ columns. The first n columns correspond to the possible values of n inputs, and the last column to the operation being performed. The rows list all possible combinations of inputs together with the corresponding outputs. For example, the following truth table shows the result of the binary AND operator acting on two inputs A and B, each of which may be true or false.

A	B	$A \wedge B$
F	F	F
F	T	F
T	F	F
T	T	T

The following *Mathematica* code can be used to generate a truth table for n levels of operator op.

```
TruthTable[op_, n_] := Module[
    {
        l = Flatten[Outer[List, Sequence @@
Table[{True, False}, {n}]], n - 1],
        a = Array[A, n]
    },
    DisplayForm[
      GridBox[Prepend[Append[#, op @@ #] & /@ l,
Append[a, op @@ a]],       RowLines -> True,
ColumnLines -> True]
    ]
]
```

See also AND, CONNECTIVE, EQUIVALENT, IMPLIES,

KARNAUGH MAP, MULTIPLICATION TABLE, NAND, NOR, NOT, OR, XNOR, XOR

References

Carnap, R. "Truth Tables." §4 in *Introduction to Symbolic Logic and Its Applications.* New York: Dover, pp. 10–15, 1958.

Tschebyshev

An alternative spelling of the name "CHEBYSHEV."

See also CHEBYSHEV

Tschebyshev System

HAAR CONDITION

Tschirnhausen Cubic Caustic

The CAUSTIC of the TSCHIRNHAUSEN CUBIC taking the RADIANT POINT as the pole is NEILE'S PARABOLA.

Tschirnhausen Cubic

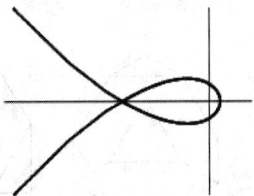

The Tschirnhausen cubic is a plane curve given by the polar equation

$$r = a \sec^3\left(\tfrac{1}{3}\theta\right) \tag{1}$$

or parametric equation

$$x = a\left(1 - 3t^2\right) \tag{2}$$

$$y = at\left(3 - t^2\right) \tag{3}$$

or

$$x = 3a\left(t^2 - 3\right) \tag{4}$$

$$y = at\left(t^2 - 3\right). \tag{5}$$

The curve is also known as CATALAN'S TRISECTRIX and L'HOSPITAL'S CUBIC. The name Tschirnhaus's cubic is given in R. C. Archibald's 1900 paper attempting to classify curves (MacTutor Archive). Tschirnhaus's cubic is the NEGATIVE PEDAL CURVE of a PARABOLA with respect to the FOCUS.

References

Lawrence, J. D. *A Catalog of Special Plane Curves.* New York: Dover, pp. 87–90, 1972.
MacTutor History of Mathematics Archive. "Tschirnhaus's Cubic." http://www-groups.dcs.st-and.ac.uk/~history/Curves/Tschirnhaus.html.

Tschirnhausen Cubic Pedal Curve

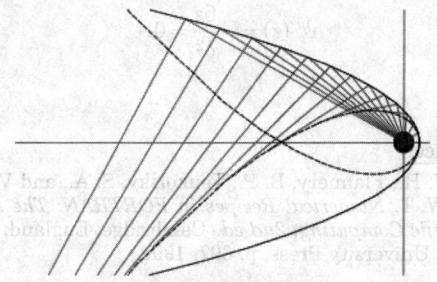

The PEDAL CURVE to the TSCHIRNHAUSEN CUBIC for PEDAL POINT at the origin is the PARABOLA

$$x = 1 - t^2$$

$$y = 2t.$$

See also PARABOLA, PEDAL CURVE, PEDAL POINT, TSCHIRNHAUSEN CUBIC

Tschirnhausen Transformation

A transformation of a POLYNOMIAL equation $f(x) = 0$ which is OF THE FORM $y = g(x)/h(x)$ where g and h are POLYNOMIALS and $h(x)$ does not vanish at a root of $f(x) = 0$. The CUBIC EQUATION is a special case of such a transformation. Tschirnhaus (1683) showed that a POLYNOMIAL of degree $n > 2$ can be reduced to a form in which the x^{n-1} and x^{n-2} terms have 0 COEFFICIENTS. In 1786, E. S. Bring showed that a general QUINTIC EQUATION can be reduced to the form

$$x^5 + px + q = 0.$$

In 1834, G. B. Jerrard showed that a Tschirnhaus transformation can be used to eliminate the x^{n-1}, x^{n-2}, and x^{n-3} terms for a general POLYNOMIAL equation of degree $n > 3$.

See also BRING QUINTIC FORM, CUBIC EQUATION

References

Boyer, C. B. *A History of Mathematics.* New York: Wiley, pp. 472–473, 1968.
Tschirnhaus. *Acta Eruditorum.* 1683.

Tubular Neighborhood

This entry contributed by RYAN BUDNEY

A tubular neighborhood of a SUBMANIFOLD $N \in M$ is an embedding of the NORMAL BUNDLE (v_N) of N into M, i.e., $f : v_N \to M$, where the image of the ZERO SECTION of the NORMAL BUNDLE is equal to $N \in M$.

See also BALL, EMBEDDING, KNOT EXTERIOR, PRODUCT NEIGHBORHOOD

References

Adams, C. C. *The Knot Book: An Elementary Introduction to the Mathematical Theory of Knots.* New York: W. H. Freeman, p. 258, 1994.
Rolfsen, D. *Knots and Links.* Wilmington, DE: Publish or Perish Press, pp. 34–35, 1976.

Tucker Circles

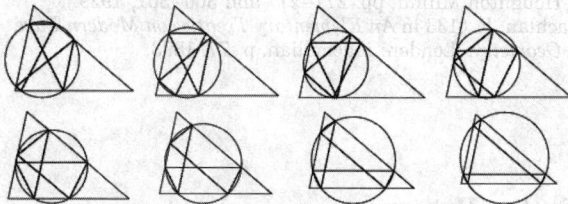

The Tucker circles are a generalization of the COSINE CIRCLE and LEMOINE CIRCLE which can be viewed as a family of circles obtained by parallel displacing sides of the corresponding COSINE or LEMOINE HEXAGON. No matter how the segments are displaced, the TUCKER HEXAGON will close, and the 12 vertices will be CONCYCLIC. The COSINE CIRCLE and LEMOINE CIRCLE correspond to the special case where three sides of the TUCKER HEXAGON concur.

Let three equal lines P_1Q_1, P_2Q_2, and P_3Q_3 be drawn ANTIPARALLEL to the sides of a triangle so that two (say P_2Q_2 and P_3Q_3) are on the same side of the third line as $A_2P_2Q_3A_3$. Then $P_2Q_3P_3Q_2$ is an isosceles TRAPEZOID, i.e., P_3Q_2, P_1Q_3, and P_2Q_1 are parallel to the respective sides. The MIDPOINTS C_1, C_2, and C_3 of the antiparallels are on the respective symmedians and divide them proportionally. If T divides KO in the same ratio, TC_1, TC_2, TC_3 are parallel to the radii OA_1, OA_2, and OA_3 and equal. Since the antiparallels are perpendicular to the symmedians, they form equal chords of a circle, called a Tucker circle, which passes through the six given points and has center T on the line KO (Honsberger 1995, pp. 92–94).

If

$$c \equiv \frac{\overline{KC_1}}{\overline{KA_1}} = \frac{\overline{KC_2}}{\overline{KA_2}} = \frac{\overline{KC_3}}{\overline{KA_3}} = \frac{\overline{KT}}{\overline{KO}},$$

then the radius of the Tucker circle is

$$R\sqrt{c^2 + (1-c)^2 \tan \omega},$$

where ω is the BROCARD ANGLE.

The COSINE CIRCLE, LEMOINE CIRCLE, and TAYLOR CIRCLE are Tucker circles.

See also ANTIPARALLEL, BROCARD ANGLE, COSINE CIRCLE, COSINE HEXAGON, LEMOINE CIRCLE, LEMOINE HEXAGON, TAYLOR CIRCLE

References

Casey, J. "Lemoine's, Tucker's, and Taylor's Circle." Supp. Ch. §3 in *A Sequel to the First Six Books of the Elements of*

Euclid, Containing an Easy Introduction to Modern Geometry with Numerous Examples, 5th ed., rev. enl. Dublin: Hodges, Figgis, & Co., pp. 179–189, 1888.

Coolidge, J. L. *A Treatise on the Geometry of the Circle and Sphere.* New York: Chelsea, p. 68, 1971.

Honsberger, R. "The Tucker Circles." Ch. 9 in *Episodes in Nineteenth and Twentieth Century Euclidean Geometry.* Washington, DC: Math. Assoc. Amer., pp. 87–98, 1995.

Johnson, R. A. *Modern Geometry: An Elementary Treatise on the Geometry of the Triangle and the Circle.* Boston, MA: Houghton Mifflin, pp. 271–277 and 300–301, 1929.

Lachlan, R. §133 in *An Elementary Treatise on Modern Pure Geometry.* London: Macmillian, p. 77, 1893.

Tucker Hexagon

A closed, self-intersecting concyclic hexagon constructed along the sides of a triangle. A CIRCUMCIRCLE of any of these hexagons is called a TUCKER CIRCLE.

See also HEXAGON, TUCKER CIRCLES

References

Honsberger, R. *Episodes in Nineteenth and Twentieth Century Euclidean Geometry.* Washington, DC: Math. Assoc. Amer., pp. 90–91, 1995.

Tukey's Biweight

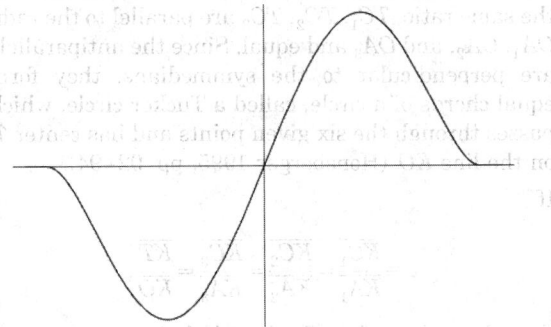

The function

$$\psi(z) = \begin{cases} z\left(1 - \dfrac{z^2}{c^2}\right)^2 & \text{for } |z| < c \\ 0 & \text{for } |z| > c \end{cases}$$

sometimes used in ROBUST ESTIMATION. It has a minimum at $z = -c/\sqrt{3}$ and a maximum at $z = c/\sqrt{3}$, where

$$\psi'(z) = 1 - \frac{3x^2}{c^2} = 0,$$

and an inflection point at $z = 0$, where

$$\psi''(z) = -\frac{6z}{c^2} = 0.$$

References

Press, W. H.; Flannery, B. P.; Teukolsky, S. A.; and Vetterling, W. T. *Numerical Recipes in FORTRAN: The Art of Scientific Computing, 2nd ed.* Cambridge, England: Cambridge University Press, p. 697, 1992.

Tukey's Trimean

TRIMEAN

Tunnel Number

Let a KNOT K be n-EMBEDDABLE. Then its tunnel number is a KNOT INVARIANT which is related to n.

See also EMBEDDABLE KNOT

References

Adams, C. C. *The Knot Book: An Elementary Introduction to the Mathematical Theory of Knots.* New York: W. H. Freeman, p. 114, 1994.

Turán Graph

The (n, k)-Turán graph is the EXTREMAL GRAPH on n VERTICES which contains no k-CLIQUE. In other words, the Turán graph has the maximum possible number of EDGES of any n-vertex graph not containing a COMPLETE GRAPH K_k. TURÁN'S THEOREM gives the maximum number of edges $t(n, k)$ for the (n, k)-Turán graph. For $k = 3$,

$$t(n, 3) = \tfrac{1}{4} n^2,$$

so the Turán graph is given by the COMPLETE BIPARTITE GRAPHS

$$\begin{cases} K_{n/2, \, n/2} & n \text{ even} \\ K_{(n-1)/2, (n+1)/2} & n \text{ odd.} \end{cases}$$

Turán graphs cen be generated using `Turan[n, p]` in the *Mathematica* add-on package `DiscreteMath`- `Combinatorica`' (which can be loaded with the command `< <DiscreteMath`').

See also CLIQUE, COMPLETE BIPARTITE GRAPH, EXTREMAL GRAPH, EXTREMAL GRAPH THEORY, TURÁN'S THEOREM

References

Aigner, M. "Turán's Graph Theorem." *Amer. Math. Monthly* **102**, 808–816, 1995.

Skiena, S. *Implementing Discrete Mathematics: Combinatorics and Graph Theory with Mathematica.* Reading, MA: Addison-Wesley, pp. 143 and 218, 1990.

Turán, P. "On an Extremal Problem in Graph Theory." *Mat. Fiz. Lapok* **48**, 436–452, 1941.

Turán's Inequalities

For a set of POSITIVE γ_k, $k = 0$, 1, 2..., Turán's inequalities are given by

$$\gamma_k^2 - \gamma_{k-1}\gamma_{k+1} \geq 0$$

for $k = 1$, 2, ...

See also JENSEN POLYNOMIAL

References

Csordas, G.; Varga, R. S.; and Vincze, I. "Jensen Polynomials with Applications to the Riemann ζ-Function." *J. Math. Anal. Appl.* **153**, 112–135, 1990.
Szego, G. *Orthogonal Polynomials, 4th ed.* Providence, RI: Amer. Math. Soc., p. 388, 1975.

Turán's Theorem

Let $G(V, E)$ be a GRAPH with VERTICES V and EDGES E on n VERTICES without a k-CLIQUE. Then

$$t(n, k) \leq \frac{(k-2)n^2}{2(k-1)},$$

where $t(n, k) = |E|$ is the EDGE NUMBER. More precisely, the K-GRAPH $K_{n_1, \ldots, n_{k-1}}$ with $|n_i - n_j| \leq 1$ for $i \neq j$ is the unique GRAPH without a k-CLIQUE with the maximal number of EDGES $t(n, k)$.

See also CLIQUE, ERDOS-STONE THEOREM, EXTREMAL GRAPH THEORY, K-GRAPH, TURÁN GRAPH

References

Aigner, M. "Turán's Graph Theorem." *Amer. Math. Monthly* **102**, 808–816, 1995.
Pach, J. and Agarwal, P. K. *Combinatorial Geometry.* New York: Wiley, 1995.

Turbine

A VECTOR FIELD on a CIRCLE in which the directions of the VECTORS are all at the same ANGLE to the CIRCLE.

See also CIRCLE, VECTOR FIELD

Turing Machine

A theoretical computing machine which consists of an infinitely long magnetic tape on which instructions can be written and erased, a finite register of memory, and a processor capable of carrying out the following instructions: move the tape right, move the tape left, change the state of the register based on its current value and a value on the tape, and write or erase a value on the tape. The machine keeps processing instructions until it reaches a particular state, causing it to halt. Determining whether a Turing machine will halt for a given input and set of rules is called the HALTING PROBLEM.

See also AUTOMATA THEORY, AUTOMATIC SET, BUSY BEAVER, CELLULAR AUTOMATON, CHAITIN'S OMEGA, CHURCH-TURING THESIS, COMPUTABLE NUMBER, DETERMINISTIC, HALTING PROBLEM, UNIVERSAL TURING MACHINE

References

Davis, M. *Computability and Unsolvability.* New York: Dover.
Itô, K. (Ed.). "Turing Machines." §31B in *Encyclopedic Dictionary of Mathematics, 2nd ed., Vol. 1.* Cambridge, MA: MIT Press, pp. 136–137, 1987.
Penrose, R. "Algorithms and Turing Machines." Ch. 2 in *The Emperor's New Mind: Concerning Computers, Minds, and the Laws of Physics.* Oxford, England: Oxford University Press, pp. 30–73, 1989.
Turing, A. M. "On Computable Numbers, with an Application to the Entscheidungsproblem." *Proc. London Math. Soc. Ser. 2* **42**, 230–265, 1937.
Turing, A. M. "Correction to: On Computable Numbers, with an Application to the Entscheidungsproblem." *Proc. London Math. Soc. Ser. 2* **43**, 544–546, 1938.

Turning Angle

TANGENTIAL ANGLE

Tutte Conjecture

Tutte (1971/72) conjectured that there is no non-HAMILTONIAN 3-connected BICUBIC GRAPHS. However, a counterexample was found by J. D. Horton in 1976 (Gropp 1990).

See also BICUBIC GRAPH, CUBIC GRAPH, HAMILTONIAN GRAPH, TAIT'S HAMILTONIAN GRAPH CONJECTURE

References

Gropp, H. "Configurations and the Tutte Conjecture." *Ars. Combin. A* **29**, 171–177, 1990.
Tutte, W. T. "On the 2-Factors of Bicubic Graphs." *Disc. Math.* **1**, 203–208, 1971/72.

Tutte Polynomial

Let G be a GRAPH, and let ea(T) denote the cardinality of the set of externally active edges of a spanning tree T of G and ia(T) denote the cardinality of the set of internally active edges of T. Then

$$t_G(x, y) = \sum_{T \subseteq G} x^{ia(T)} y^{ea(T)}.$$

References

Gessel, I. M. and Sagan, B. E. "The Tutte Polynomial of a Graph, Depth-First Search, and Simplicial Complex Partitions." *Electronic J. Combinatorics* **3**, No. 2, R9, 1–36, 1996. http://www.combinatorics.org/Volume_3/volume3_2.html#R9.
Tutte, W. T. "A Contribution to the Theory of Chromatic Polynomials." *Canad. J. Math.* **6**, 80–91, 1953.

Tutte-Coxeter Graph

LEVI GRAPH

Tutte's Graph

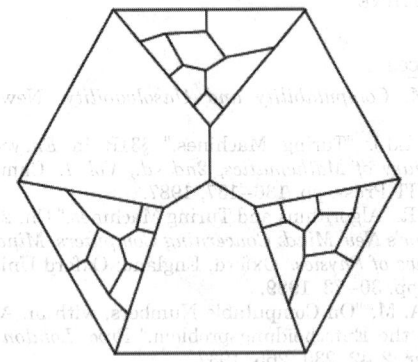

A counterexample to TAIT'S HAMILTONIAN GRAPH CONJECTURE given by Tutte (1946). A simpler counterexample was later given by Kozyrev and Grinberg. The LEVI GRAPH is sometimes also called the Tutte graph (Royle).

See also HAMILTONIAN CIRCUIT, LEVI GRAPH, TAIT'S HAMILTONIAN GRAPH CONJECTURE

References

Honsberger, R. *Mathematical Gems I.* Washington, DC: Math. Assoc. Amer., pp. 82–89, 1973.
Royle, G. "Cubic Cages." http://www.cs.uwa.edu.au/~gordon/cages/.
Saaty, T. L. and Kainen, P. C. *The Four-Color Problem: Assaults and Conquest.* New York: Dover, p. 112, 1986.
Skiena, S. *Implementing Discrete Mathematics: Combinatorics and Graph Theory with Mathematica.* Reading, MA: Addison-Wesley, p. 198, 1990.
Tait, P. G. "Remarks on the Colouring of Maps." *Proc. Royal Soc. Edinburgh* **10**, 729, 1880.
Tutte, W. T. "On Hamiltonian Circuits." *J. London Math. Soc.* **21**, 98–101, 1946.
Tutte, W. T. "Non-Hamiltonian Planar Maps." In *Graph Theory and Computing* (Ed. R. Read). New York: Academic Press, pp. 295–301, 1972.

Tutte's Theorem

Let G be a GRAPH and S a SUBGRAPH of G. Let the number of ODD components in $G-S$ be denoted S', and $|S|$ the number of VERTICES of S. The condition $|S| \geq S'$ for every SUBSET of VERTICES is NECESSARY and SUFFICIENT for G to have a 1-FACTOR.

See also FACTOR (GRAPH)

References

Honsberger, R. "Lovász' Proof of a Theorem of Tutte." Ch. 14 in *Mathematical Gems II.* Washington, DC: Math. Assoc. Amer., pp. 147–157, 1976.
Tutte, W. T. "The Factorization of Linear Graphs." *J. London Math. Soc.* **22**, 107–111, 1947.

Twiddle

TILDE

Twig

Let a COTREE of a spanning tree T in a CONNECTED GRAPH G be denoted T^*. Then the edges of G which are not in T^* are called its twigs (Harary 1994, p. 39).

See also COTREE

References

Harary, F. *Graph Theory.* Reading, MA: Addison-Wesley, 1994.

Twin Peaks

For an INTEGER $n \geq 2$, let $\mathrm{lpf}(x)$ denote the LEAST PRIME FACTOR of n. A PAIR of INTEGERS (x, y) is called a twin peak if

1. $x < y$,
2. $\mathrm{lpf}(x) = \mathrm{lpf}(y)$,
3. For all z, $x < z < y$ IMPLIES $\mathrm{lpf}(z) < \mathrm{lpf}(x)$.

A broken-line graph of the least prime factor function resembles a jagged terrain of mountains. In terms of this terrain, a twin peak consists of two mountains of equal height with no mountain of equal or greater height between them. Denote the height of twin peak (x, y) by $p = \mathrm{lpf}(x) = \mathrm{lpf}(y)$. By definition of the LEAST PRIME FACTOR function, p must be PRIME.

Call the distance between two twin peaks (x, y)

$$s \equiv y - x.$$

Then s must be an EVEN multiple of p; that is, $s = kp$ where k is EVEN. A twin peak with $s = kp$ is called a kp-twin peak. Thus we can speak of $2p$-twin peaks, $4p$-twin peaks, etc. A kp-twin peak is fully specified by k, p, and x, from which we can easily compute $y \equiv x + kp$.

The set of kp-twin peaks is periodic with period $q = p\#$, where $p\#$ is the PRIMORIAL of p. That is, if (x, y) is a kp-twin peak, then so is $(x+q, y+q)$. A fundamental kp-twin peak is a twin peak having x in the fundamental period $[0, q)$. The set of fundamental kp-twin peaks is symmetric with respect to the fundamental period; that is, if (x, y) is a twin peak on $[0, q)$, then so is $(q-y, q-x)$.

The question of the EXISTENCE of twin peaks was first raised by David Wilson in the math-fun mailing list on Feb. 10, 1997. Wilson already had privately showed the EXISTENCE of twin peaks of height $p \leq 13$ to be unlikely, but was unable to rule them out altogether. Later that same day, John H. Conway, Johan de Jong, Derek Smith, and Manjul Bhargava collaborated to discover the first twin peak. Two hours at the blackboard revealed that $p = 113$ admits the $2p$-twin peak

$$x = 126972592296404970720882679404584182254788131$$

which settled the EXISTENCE question. Immediately thereafter, Fred Helenius found the smaller $2p$-twin peak with $p = 89$ and

$$x = 9503844926749390990454854843625839.$$

The effort now shifted to finding the least PRIME p admitting a $2p$-twin peak. On Feb. 12, 1997, Fred Helenius found $p = 71$, which admits 240 fundamental $2p$-twin peaks, the least being

$$x = 7310131732015251470110369.$$

Helenius's results were confirmed by Dan Hoey, who also computed the least $2p$-twin peak $L(2p)$ and number of fundamental $2p$-twin peaks $N(2p)$ for $p = 73$, 79, and 83. His results are summarized in the following table (Sloane's A009190).

p	$L(2p)$	$N(2p)$
71	7310131732015251470110369	240
73	2061519317176132799110061	40296
79	3756800873017263196139951	164440
83	6316254452384500173544921	6625240

The $2p$-twin peak of height $p = 73$ is the smallest known twin peak. Wilson found the smallest known $4p$-twin peak with $p = 1327$, as well as another very large $4p$-twin peak with $p = 3203$. Richard Schroeppel noted that the latter twin peak is at the high end of its fundamental period and that its reflection within the fundamental period $[0, p\#)$ is smaller.

Many open questions remain concerning twin peaks, e.g.,

1. What is the smallest twin peak (smallest n)?
2. What is the least PRIME p admitting a $4p$-twin peak?
3. Do $6p$-twin peaks exist?
4. Is there, as Conway has argued, an upper bound on the span of twin peaks?
5. Let $p < q < r$ be PRIME. If p and r each admit kp-twin peaks, does q then necessarily admit a kp-twin peak?

See also ANDRICA'S CONJECTURE, DIVISOR FUNCTION, LEAST COMMON MULTIPLE, LEAST PRIME FACTOR

References

Sloane, N. J. A. Sequences A009190 in "An On-Line Version of the Encyclopedia of Integer Sequences." http://www.research.att.com/~njas/sequences/eisonline.html.

Twin Prime Conjecture

There are two related conjectures, each called the twin prime conjecture. The first version states that there are an infinite number of pairs of TWIN PRIMES (Guy 1994, p. 19). It is not known if there are an infinite number of such PRIMES (Wells 1986, p. 41; Shanks 1993, p. 30), but it seems almost certain to be true (Hardy and Wright 1979, p. 5). In the words of Shanks (1993, p. 219), "the evidence is overwhelming."

The conjecture that there are infinitely many integers n such that $n + 1$ is prime and n is twice a prime is very closely related (Shanks 1993, p. 30).

A second twin prime conjecture states that adding a correction proportional to $1/\ln p$ to a computation of BRUN'S CONSTANT ending with $\ldots + 1/p + 1/(p+2)$ will give an estimate with error less than $c(\sqrt{p} \ln p)^{-1}$. An extended form of this conjecture, sometimes called the strong twin prime conjecture (Shanks 1993, p. 30) states that

$$P_x(p, p+2) \sim 2 \Pi_2 \int_2^x \frac{dx}{(\ln x)^2},$$

where Π_2 is the TWIN PRIMES CONSTANT (Hardy and Littlewood 1922). This conjecture is a special case of the more general PRIME PATTERNS CONJECTURE corresponding to the set $S = \{0, 2\}$.

See also BRUN'S CONSTANT, PRIME ARITHMETIC PROGRESSION, PRIME CONSTELLATION, PRIME PATTERNS CONJECTURE, TWIN PRIMES

References

Guy, R. K. "Gaps between Primes. Twin Primes." §A8 in *Unsolved Problems in Number Theory, 2nd ed.* New York: Springer-Verlag, pp. 19–23, 1994.

Hardy, G. H. and Littlewood, J. E. "Some Problems of 'Partitio Numerorum.' III. On the Expression of a Number as a Sum of Primes." *Acta Math.* **44**, 1–70, 1922.

Ribenboim, P. *The New Book of Prime Number Records.* New York: Springer-Verlag, pp. 261–265, 1996.

Shanks, D. *Solved and Unsolved Problems in Number Theory, 4th ed.* New York: Chelsea, p. 30, 1993.

Wells, D. *The Penguin Dictionary of Curious and Interesting Numbers.* Middlesex, England: Penguin Books, p. 41, 1986.

Twin Primes

Twin primes are pairs of PRIMES OF THE FORM $(p, p+2)$. The term "twin prime" was coined by Paul Stäckel (1892–1919; Tietze 1965, p. 19). The first few twin primes are $n \pm 1$ for $n = 4$, 6, 12, 18, 30, 42, 60, 72, 102, 108, 138, 150, 180, 192, 198, 228, 240, 270, 282, ... (Sloane's A014574). Explicitly, these are (3, 5), (5, 7), (11, 13), (17, 19), (29, 31), (41, 43), ... (Sloane's A001359 and A006512).

The following table gives the first few p for the twin primes $(p, p+2)$, COUSIN PRIMES $(p, p+4)$, SEXY PRIMES $(p, p+6)$, etc.

Triplet	Sloane	First Member
$(p, p+2)$	Sloane's A001359	3, 5, 11, 17, 29, 41, 59, 71, ...

$(p, p+4)$	Sloane's A023200	3, 7, 13, 19, 37, 43, 67, 79, ...
$(p, p+6)$	Sloane's A023201	5, 7, 11, 13, 17, 23, 31, 37, ...
$(p, p+8)$	Sloane's A023202	3, 5, 11, 23, 29, 53, 59, 71, ...
$(p, p+10)$	Sloane's A023203	3, 7, 13, 19, 31, 37, 43, 61, ...
$(p, p+12)$	Sloane's A046133	5, 7, 11, 17, 19, 29, 31, 41, ...

Let $\pi_2(n)$ be the number of twin primes p and $p+2$ such that $p \leq n$. It is not known if there are an infinite number of such PRIMES (Wells 1986, p. 41; Shanks 1993), but it seems almost certain to be true (Hardy and Wright 1979, p. 5). All twin primes except $(3, 5)$ are OF THE FORM $6n \pm 1$. J. R. Chen has shown there exists an INFINITE number of PRIMES p such that $p+2$ has at most two factors (Le Lionnais 1983, p. 49). Bruns proved that there exists a computable INTEGER x_0 such that if $x \geq x_0$, then

$$\pi_2(x) < \frac{100x}{(\ln x)^2} \tag{1}$$

(Ribenboim 1996, p. 261). It has been shown that

$$\pi_2(x) \leq c \prod_{p>2} \left[1 - \frac{1}{(p-1)^2}\right] \frac{x}{(\ln x)^2}$$
$$\times \left[1 + \mathcal{O}\left(\frac{\ln \ln x}{\ln x}\right)\right], \tag{2}$$

written more concisely as

$$\pi_2(x) \leq c \, \Pi_2 \frac{x}{(\ln x)^2} \left[1 + \mathcal{O}\left(\frac{\ln \ln x}{\ln x}\right)\right], \tag{3}$$

where Π_2 is known as the TWIN PRIMES CONSTANT and c is another constant. The constant c has been reduced to $68/9 \approx 7.5556$ (Fouvry and Iwaniec 1983), $128/17 \approx 7.5294$ (Fouvry 1984), 7 (Bombieri *et al.* 1986), 6.9075 (Fouvry and Grupp 1986), and 6.8354 (Wu 1990). The bound on c is further reduced to 6.8325 (Haugland 1999). This calculation involved evaluation of 7-fold integrals and fitting of three different parameters. Hardy and Littlewood conjectured that $c = 2$ (Ribenboim 1996, p. 262).

Wolf notes that the formula

$$\pi_2(x) \sim \Pi_2 \frac{[\pi(x)]^2}{x}, \tag{4}$$

which increases as $\Pi_2 x/(\ln x)^2$ for large x, agrees with numerical data much better than does $\Pi_2 x/(\ln x)^2$, although not as well as $\Pi_2 \, \text{Li}_2(x)$.

Extending the search done by Brent in 1974 or 1975, Wolf has searched for the analog of the SKEWES NUMBER for twins, i.e., an x such that $\pi_2(x) - \Pi_2 \, \text{Li}_2(x)$ changes sign. Wolf checked numbers up to 2^{42} and found more than 90,000 sign changes. From this data, Wolf conjectured that the number of sign changes $\nu(n)$ for $x < n$ of $\pi_2(x) - \Pi_2 \, \text{Li}_2(x)$ is given by

$$\nu(n) \sim \frac{\sqrt{n}}{\ln n}. \tag{5}$$

Proof of this conjecture would also imply the existence an infinite number of twin primes.

Define

$$E \equiv \lim_{n \to \infty} \inf \frac{p_{n+1} - p_n}{\ln p_n}. \tag{6}$$

If there are an infinite number of twin primes, then $E = 0$. The best upper limit to date is $E \leq \frac{1}{4} + \pi/16 = 0.44634\ldots$ (Huxley 1973, 1977). The best previous values were 15/16 (Ricci), $(2 + \sqrt{3})/8 = 0.46650\ldots$ (Bombieri and Davenport 1966), and $(2\sqrt{2} - 1)/4 = 0.45706\ldots$ (Pil'Tai 1972), as quoted in Le Lionnais (1983, p. 26).

Some large twin primes are $10,006,428 \pm 1$, $1,706,595 \times 2^{11235} \pm 1$, and $571,305 \times 2^{7701} \pm 1$. An up-to-date table of known twin primes with 2000 or more digits follows. An extensive list is maintained by C. Caldwell at http://www.utm.edu/cgi-bin/caldwell/primes.cgi/twin.

$(p, p+1)$	Digits	Reference
$260,497,545 \times 2^{6625} \pm 1$	2003	Atkin and Rickert 1984
$43,690,485,351,513 \times 10^{1995} \pm 1$	2009	Dubner, Atkin 1985
$2,846!!!! \pm 1$	2151	Dubner 1992
$10,757,0463 \times 10^{2250} \pm 1$	2259	Dubner, Atkin 1985
$663,777 \times 2^{7650} \pm 1$	2309	Brown *et al.* 1989
$75,188,117,004 \times 10^{2298} \pm 1$	2309	Dubner 1989
$571305 \times 2^{7701} \pm 1$	2324	Brown *et al.* 1989
$1,171,452,282 \times 10^{2490} \pm 1$	2500	Dubner 1991
$459 \cdot 2^{8529} \pm 1$	2571	Dubner 1993
$1,706,595 \cdot 2^{11235} \pm 1$	3389	Noll *et al.* 1989
$4,655,478,828 \cdot 10^{3429} \pm 1$	3439	Dubner 1993
$1,692,923,232 \cdot 10^{4020} \pm 1$	4030	Dubner 1993
$6,797,727 \cdot 2^{15328} \pm 1$	4622	Forbes 1995

$697,053,8132^{16352} \pm 1$	4932	Indlekofer and Ja'rai 1994	
$570,918,348 \cdot 10^{5120} \pm 1$	5129	Dubner 1995	
$242,206,083 \cdot 2^{38880} \pm 1$	11713	Indlekofer and Ja'rai 1995	

The last of these is the largest known twin prime pair. In 1995, Nicely discovered a flaw in the Intel ® Pentium™ microprocessor by computing the reciprocals of 824,633,702,441 and 824,633,702,443, which should have been accurate to 19 decimal places but were incorrect from the tenth decimal place on (Cipra 1995, 1996; Nicely 1996).

If $n \geq 2$, the INTEGERS n and $n+2$ form a pair of twin primes IFF

$$4[(n-1)! + 1] + n \equiv 0 \pmod{n(n+2)}. \tag{7}$$

$n = pp'$ where (p, p') is a pair of twin primes IFF

$$\phi(n)\sigma(n) = (n-3)(n+1) \tag{8}$$

(Ribenboim 1996, p. 259). S. M. Ruiz has found the unexpected result that $(n, n+2)$ are twin primes IFF

$$\sum_{i=1}^{n} i^a \left(\left\lfloor \frac{n+2}{i} \right\rfloor + \left\lfloor \frac{n}{i} \right\rfloor \right)$$

$$= 2 + n^a + \sum_{i=1}^{n} i^a \left(\left\lfloor \frac{n+1}{i} \right\rfloor + \left\lfloor \frac{n-1}{i} \right\rfloor \right) \tag{9}$$

for $a \geq 0$, where $\lfloor x \rfloor$ is the FLOOR FUNCTION.

The values of $\pi_2(n)$ were found by Brent (1976) up to $n = 10^{11}$. T. Nicely calculated them up to 10^{14} in his calculation of BRUN'S CONSTANT. The following table gives the number less than increasing powers of 10 (Sloane's A007508; Nicely 1998, 1999). Using a distributed computation, Fry *et al.* obtained $\pi_2(10^{16})$ in 2000, although this value has not yet been made public. The following table gives $\pi(10^n)$ for various values of n, and extends a similar table with early references given by Ribenboim (1996, p. 263).

n	$\pi_2(n)$
10^3	35
10^4	205
10^5	1224
10^6	8,169
10^7	58,980
10^8	440,312

n	$\pi_2(n)$
10^9	3,424,506
10^{10}	27,412,679
10^{11}	224,376,048
10^{12}	1,870,585,220
10^{13}	15,834,664,872
10^{14}	135,780,321,665
10^{15}	1,177,209,242,304

It is conjectured that every even number is a sum of a pair of twin primes except a finite number of exceptions whose first few terms are 2, 4, 94, 96, 98, 400, 402, 404, 514, 516, 518, ... (Sloane's A007534; Wells 1986, p. 132).

See also BITWIN CHAIN, BRUN'S CONSTANT, COUSIN PRIMES, DE POLIGNAC'S CONJECTURE PRIME CONSTELLATION, SEXY PRIMES, TWIN PRIME CONJECTURE, TWIN PRIMES CONSTANT

References

Bombieri, E. and Davenport, H. "Small Differences Between Prime Numbers." *Proc. Roy. Soc. Ser. A* **293**, 1–8, 1966.

Bombieri, E.; Friedlander, J. B.; and Iwaniec, H. "Primes in Arithmetic Progression to Large Moduli." *Acta Math.* **156**, 203–251, 1986.

Bradley, C. J. "The Location of Twin Primes." *Math. Gaz.* **67**, 292–294, 1983.

Brent, R. P. "Irregularities in the Distribution of Primes and Twin Primes." *Math. Comput.* **29**, 43–56, 1975.

Brent, R. P. "UMT 4." *Math. Comput.* **29**, 221, 1975.

Brent, R. P. "Tables Concerning Irregularities in the Distribution of Primes and Twin Primes to 10^{11}." *Math. Comput.* **30**, 379, 1976.

Caldwell, C. http://www.utm.edu/cgi-bin/caldwell/primes.cgi/twin.

Caldwell, C. K. "The Top Twenty: Twin Primes." http://www.utm.edu/research/primes/lists/top20/twin.html.

Cipra, B. "How Number Theory Got the Best of the Pentium Chip." *Science* **267**, 175, 1995.

Cipra, B. "Divide and Conquer." *What's Happening in the Mathematical Sciences, 1995–1996, Vol. 3.* Providence, RI: Amer. Math. Soc., pp. 38–47, 1996.

Fouvry, É. "Autour du théorème de Bombieri-Vinogradov." *Acta. Math.* **152**, 219–244, 1984.

Fouvry, É. and Grupp, F. "On the Switching Principle in Sieve Theory." *J. reine angew. Math.* **370**, 101–126, 1986.

Fouvey, É. and Iwaniec, H. "Primes in Arithmetic Progression." *Acta Arith.* **42**, 197–218, 1983.

Fry, P.; Nesheiwat, J.; and Szymanski, B. K. "Rensselaer's Twin Prime Computing Effort." http://www.cs.rpi.edu/research/twinp/.

Gardner, M. "Patterns in Primes are a Clue to the Strong Law of Small Numbers." *Sci. Amer.* **243**, 18–28, Dec. 1980.

Guy, R. K. "Gaps between Primes. Twin Primes." §A8 in *Unsolved Problems in Number Theory, 2nd ed.* New York: Springer-Verlag, pp. 19–23, 1994.

Hardy, G. H. and Wright, E. M. *An Introduction to the Theory of Numbers, 5th ed.* Oxford, England: Clarendon Press, 1979.

Haugland, J. K. *Application of Sieve Methods to Prime Numbers.* Ph.D. thesis. Oxford, England: Oxford University, 1999.

Huxley, M. N. "Small Differences between Consecutive Primes." *Mathematica* **20**, 229–232, 1973.

Huxley, M. N. "Small Differences between Consecutive Primes. II." *Mathematica* **24**, 142–152, 1977.

Indlekofer, K. H. and Járai, A. "Largest Known Twin Primes." *Math. Comput.* **65**, 427–428, 1996.

Indlekofer, K. H. and Járai, A. "Largest Known Twin Primes and Sophie Germain Primes." *Math. Comput.* **68**, 1317–1324, 1999.

Le Lionnais, F. *Les nombres remarquables.* Paris: Hermann, 1983.

Nicely, T. "Enumeration to 10^{14} of the Twin Primes and Brun's Constant." *Virginia J. Sci.* **46**, 195–204, 1996.

Nicely, T. "Enumeration to 1.6×10^{15} of the Twin Primes and Brun's Constant." Submitted to *Math. Comput.*

Parady, B. K.; Smith, J. F.; and Zarantonello, S. E. "Largest Known Twin Primes." *Math. Comput.* **55**, 381–382, 1990.

Ribenboim, P. "Twin Primes." §4.3 in *The New Book of Prime Number Records.* New York: Springer-Verlag, pp. 259–265, 1996.

Shanks, D. *Solved and Unsolved Problems in Number Theory, 4th ed.* New York: Chelsea, p. 30, 1993.

Sloane, N. J. A. Sequences A001359/M2476, A006512/M3763, A007508/M1855, A007534, and A014574 in "An On-Line Version of the Encyclopedia of Integer Sequences." http://www.research.att.com/~njas/sequences/eisonline.html.

Tietze, H. "Prime Numbers and Prime Twins." Ch. 1 in *Famous Problems of Mathematics: Solved and Unsolved Mathematics Problems from Antiquity to Modern Times.* New York: Graylock Press, pp. 1–20, 1965.

Weintraub, S. "A Prime Gap of 864." *J. Recr. Math.* **25**, 42–43, 1993.

Wells, D. *The Penguin Dictionary of Curious and Interesting Numbers.* Middlesex, England: Penguin Books, p. 41, 1986.

Wu, J. "Sur la suite des nombres premiers jumeaux." *Acta. Arith.* **55**, 365–394, 1990.

Twin Primes Constant

The twin primes constant Π_2 (sometimes also denoted C_2) is defined by

$$\Pi_2 \equiv \prod_{\substack{p>2 \\ p \text{ prime}}} \left[1 - \frac{1}{(p-1)^2} \right] \tag{1}$$

$$\ln\left(\tfrac{1}{2}\Pi_2\right) = \sum_{\substack{p \geq 3 \\ p \text{ prime}}} \ln\left[\frac{p(p-2)}{(p-1)^2} \right]$$

$$= \sum_{\substack{p \geq 3 \\ p \text{ prime}}} \left[\ln\left(1 - \frac{2}{p}\right) - 2\ln\left(1 - \frac{1}{p}\right) \right]$$

$$= -\sum_{j=2}^{\infty} \frac{2^j - 2}{j} \sum_{\substack{p \geq 3 \\ p \text{ prime}}} p^{-j}, \tag{2}$$

where the ps in sums and products are taken over

PRIMES only. Flajolet and Vardi (1996) give series with accelerated convergence

$$\Pi_2 = \prod_{n=2}^{\infty} [\zeta(n)(1 - 2^{-n})]^{-I_n} \tag{3}$$

$$= \frac{3}{4}\frac{15}{16}\frac{35}{36} \prod_{n=2}^{\infty} [\zeta(n)(1 - 2^{-n})(1 - 3^{-n}) \\ \times (1 - 5^{-n})(1 - 7^{-n})]^{-I_n}, \tag{4}$$

with

$$I_n \equiv \frac{1}{n} \sum_{d \mid n} \mu(d) 2^{n/d}, \tag{5}$$

where $\mu(x)$ is the MÖBIUS FUNCTION. (4) has convergence like $\sim (11/2)^{-n}$.

Π_2 was computed to 45 digits by Wrench (1961) and Gourdon and Sebah list 60 digits.

$$\Pi_2 = 0.6601618158\ldots. \tag{6}$$

Le Lionnais (1983, p. 30) calls Π_2 the SHAH-WILSON CONSTANT, and $2\Pi_2$ the twin prime constant (Le Lionnais 1983, p. 37).

See also BRUN'S CONSTANT, GOLDBACH CONJECTURE, MERTENS CONSTANT, TWIN PRIMES

References

Finch, S. "Favorite Mathematical Constants." http://www.mathsoft.com/asolve/constant/hrdyltl/hrdyltl.html.

Flajolet, P. and Vardi, I. "Zeta Function Expansions of Classical Constants." Unpublished manuscript. 1996. http://pauillac.inria.fr/algo/flajolet/Publications/landau.ps.

Gourdon, X. and Sebah, P. "Some Constants from Number Theory." http://xavier.gourdon.free.fr/Constants/Miscellaneous/constantsNumTheory.html.

Hardy, G. H. and Littlewood, J. E. "Some Problems of 'Partitio Numerorum.' III. On the Expression of a Number as a Sum of Primes." *Acta Math.* **44**, 1–70, 1922.

Le Lionnais, F. *Les nombres remarquables.* Paris: Hermann, 1983.

Ribenboim, P. *The Book of Prime Number Records, 2nd ed.* New York: Springer-Verlag, p. 202, 1989.

Ribenboim, P. *The Little Book of Big Primes.* New York: Springer-Verlag, p. 147, 1991.

Riesel, H. *Prime Numbers and Computer Methods for Factorization, 2nd ed.* Boston, MA: Birkhäuser, pp. 61–66, 1994.

Shanks, D. *Solved and Unsolved Problems in Number Theory, 4th ed.* New York: Chelsea, p. 30, 1993.

Wrench, J. W. "Evaluation of Artin's Constant and the Twin Prime Constant." *Math. Comput.* **15**, 396–398, 1961.

Twins

BROTHERS, PAIR

Twirl

A ROTATION combined with an EXPANSION or CONTRACTION.

See also SCREW, SHIFT

Twist

The twist of a ribbon measures how much it twists around its axis and is defined as the integral of the incremental twist around the ribbon. A formula for the twist is given by

$$\mathrm{Tw}(K) = \frac{1}{2\pi} \int_K ds \; \varepsilon_{\mu\nu\alpha} \frac{dx^{\mu}}{ds} n^{\nu} \frac{dn^{\alpha}}{ds}, \qquad (1)$$

where K is parameterized by $x^{\mu}(s)$ for $0 \le s \le L$ along the length of the knot by parameter s, and the FRAME K_f associated with K is

$$y^{\mu} = x^{\mu}(s) + \varepsilon n^{\mu}(s), \qquad (2)$$

where ε is a small parameter and $n^{\mu}(s)$ is a unit VECTOR FIELD normal to the curve at s (Kaul 1999).

Letting Lk be the linking number of the two components of a ribbon, Tw be the twist, and Wr be the WRITHE, then the CALUGAREANU THEOREM states that

$$\mathrm{Lk}(R) = \mathrm{Tw}(R) + \mathrm{Wr}(R) \qquad (3)$$

(Adams 1994, p. 187).

See also CALUGAREANU THEOREM, SCREW, WRITHE

References

Adams, C. C. *The Knot Book: An Elementary Introduction to the Mathematical Theory of Knots.* New York: W. H. Freeman, 1994.
Kaul, R. K. Topological Quantum Field Theories--A Meeting Ground for Physicists and Mathematicians. 15 Jul 1999. http://xxx.lanl.gov/abs/hep-th/9907119/.

Twist Map

A class of AREA-PRESERVING MAPS OF THE FORM

$$\theta_{i+1} = \theta_i + 2\pi\alpha(r_i)$$

$$r_{i+1} = r_i,$$

which maps CIRCLES into CIRCLES but with a twist resulting from the $\alpha = \alpha(r_i)$ term.

Twist Move

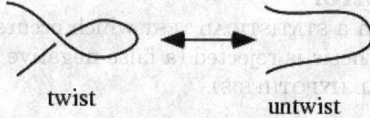

twist   untwist

The REIDEMEISTER MOVE of type II.

See also KNOT MOVE, REIDEMEISTER MOVES

Twist Number

WRITHE

Twisted Chevalley Groups

FINITE SIMPLE GROUPS of LIE-TYPE of ORDERS 14, 52, 78, 133, and 248. They are denoted $^3D_4(q)$, $E_6(q)$, $E_7(q)$, $E_8(q)$, $F_4(q)$, $^2F_4(2^n)'$, $G_2(q)$, $^2G_2(3^n)$, $^2B(2^n)$.

See also CHEVALLEY GROUPS, FINITE GROUP, SIMPLE GROUP, TITS GROUP

References

Wilson, R. A. "ATLAS of Finite Group Representation." http://for.mat.bham.ac.uk/atlas/html/contents.html#twi.

Twisted Conic

SKEW CONIC

Twisted Sphere

CORKSCREW SURFACE

Twistor

This entry contributed by EDGAR VAN TUYLL

A twistor in MINKOWSKI SPACE may be defined as a pair consisting of a SPINOR FIELD and a complex conjugate SPINOR FIELD satisfying the TWISTOR EQUATION.

See also MINKOWSKI SPACE, SPINOR, SPINOR FIELD, TWISTOR CORRESPONDENCE, TWISTOR EQUATION, TWISTOR SPACE

References

Penrose, R. and Rindler, W. *Spinors and Space-Time, Vol. 2: Spinor and Twistor Methods in Space-Time Geometry* Cambridge, England: Cambridge University Press, 1987.

Twistor Correspondence

This entry contributed by EDGAR VAN TUYLL

Oriented spheres in complex Euclidean 3-space can be represented as lines in complex projective 3-space ("Lie correspondence"), and the spheres may be thought of as the $t = 0$ representation of the light cones of events in MINKOWSKI SPACE. In effect, the Lie correspondence represents the points of (complexified compactified) MINKOWSKI SPACE by lines in complex projective 3-space, where meeting lines describe null-separated Minkowski points. This is the twistor correspondence.

See also MINKOWSKI SPACE, TWISTOR

References

Penrose, R. "The Central Programme of Twistor Theory." *Chaos, Solitons and Fractals* **10**, 581–611, 1999.

Twistor Space

This entry contributed by EDGAR VAN TUYLL

The collection of TWISTORS in MINKOWSKI SPACE that forms a four-dimensional COMPLEX VECTOR SPACE.

See also COMPLEX SPACE, MINKOWSKI SPACE, TWISTOR

Twist-Spun Knot

A generalization of SPUN KNOTS due to Zeeman. This method produces 4-D KNOT types that cannot be produced by ordinary spinning.

See also SPUN KNOT

Two

2

Two Triangle Theorem

DESARGUES' THEOREM

Two-Colorable Graph

BIPARTITE GRAPH

Two-Ears Theorem

Except for TRIANGLES, every SIMPLE POLYGON has at least two nonoverlapping EARS.

See also EAR, ONE-MOUTH THEOREM, PRINCIPAL VERTEX

References

de Berg, M.; van Kreveld, M.; Overmans, M.; and Schwarz-kopf, O. *Computational Geometry: Algorithms and Applications, 2nd rev. ed.* Berlin: Springer-Verlag, p. 59, 2000.
Meisters, G. H. "Principal Vertices, Exposed Points, and Ears." *Amer. Math. Monthly* **87**, 284–285, 1980.
Toussaint, G. "Anthropomorphic Polygons." *Amer. Math. Monthly* **122**, 31–35, 1991.

Two-Form

See also DIFFERENTIAL K-FORM, ONE-FORM, ZERO-FORM

Two-Graph

A two-graph (V, Δ) is a GRAPH on nodes V with a collection Δ of unordered triples of the vertices (the so-called "odd triples") such that each 4-tuple of V contains an even number of elements of Δ as subsets.

See also EULERIAN GRAPH

References

Bussemaker, F. C.; Mathon, R. A.; and Seidel, J. J. "Tables of Two-Graphs." In *Combinatorics and Graph Theory* (Ed. S. B. Rao). Berlin: Springer-Verlag, pp. 70–112, 1981.
Mallows, C. L. and Sloane, N. J. A. "Two-Graphs, Switching Classes, and Euler Graphs are Equal in Number." *SIAM J. Appl. Math.* **28**, 876–880, 1975.
Spence, E. "Two-Graphs." Ch. VI.6 in Colbourn, C. J. and Dinitz, J. H. (Eds.). *CRC Handbook of Combinatorial Designs.* Boca Raton, FL: CRC Press, pp. 686–694, 1996.

Two-Point Distance

POINT-POINT DISTANCE–1-D, POINT-POINT DISTANCE–2-D, POINT-POINT DISTANCE–3-D, SPHERE POINT PICKING

Two-Scale Expansion

$$\psi = (A_0 + \alpha_1 A_1 + \alpha_2 A_2 + \ldots)e^{iS/\alpha}.$$

Two-Sheeted Hyperboloid

A HYPERBOLOID consisting of two distinct sheets.

See also HYPERBOLOID

Tychonof Compactness Theorem

The topological product of any number of COMPACT SPACES is COMPACT.

Type

Whitehead and Russell (1927) devised a hierarchy of "types" in order to eliminate self-referential statements from *Principia Mathematica*, which purported to derive all of mathematics from logic. A set of the lowest type contained only objects (not sets), a set of the next higher type could contain only objects or sets of the lower type, and so on. Unfortunately, GÖDEL'S INCOMPLETENESS THEOREM showed that both *Principia Mathematica* and all consistent formal systems must be incomplete.

See also CLASS (SET), GÖDEL'S INCOMPLETENESS THEOREM

References

Curry, H. B. *Foundations of Mathematical Logic.* New York: Dover, pp. 21–22, 1977.
Ferreirós, J. "Russell's Theory of Types." §9.5 in *Labyrinth of Thought: A History of Set Theory and Its Role in Modern Mathematics.* Basel, Switzerland: Birkhäuser, pp. 325–333, 1999.
Gonseth, F. "La Théorie des types." §107 in *Les mathématiques et la réalité: Essai sur la méthode axiomatique.* Paris: Félix Alcan, pp. 257–259, 1936.
Hofstadter, D. R. *Gödel, Escher, Bach: An Eternal Golden Braid.* New York: Vintage Books, pp. 21–22, 1989.
Whitehead, A. N. and Russell, B. *Principia Mathematica.* New York: Cambridge University Press, 1927.

Type I Error

An error in a STATISTICAL TEST which occurs when a true hypothesis is rejected (a false negative in terms of the NULL HYPOTHESIS).

See also NULL HYPOTHESIS, SENSITIVITY, SPECIFICITY, STATISTICAL TEST, TYPE II ERROR

Type II Error

An error in a STATISTICAL TEST which occurs when a false hypothesis is accepted (a false positive in terms of the NULL HYPOTHESIS).

See also NULL HYPOTHESIS, SENSITIVITY, SPECIFICITY, STATISTICAL TEST, TYPE I ERROR

U

U(n) Basic Hypergeometric Series

Multiple series generalizations of basic hypergeometric series over the UNITARY GROUPS $U(n+1)$. The fundamental theorem of $U(n)$ series takes $c_1, ..., c_n$ and $x_1, ..., x_n$ as indeterminates and $n \geq 1$. Then

$$\frac{(c_1 \cdots c_n; q)N}{(q; q)N}$$

$$= \sum_{\substack{y_1, y_2, ..., y_n \geq 0 \\ |y| = N}} \left\{ \prod_{1 \leq r < s \leq n} \left[\frac{1 - \frac{x_r}{x_s} q^{y_r - y_s}}{1 - \frac{x_r}{x_s}} \right] \right.$$

$$\left. \times \prod_{r, s=1}^{n} \left[\frac{\left(\frac{x_r}{x_s} c_s; q \right)_{y_r}}{\left(q \frac{x_r}{x_s}; q \right)_{y_r}} \right] \left[q^{y_2 + 2y_3 + ... + (n-1)y_n} \right] \right\},$$

where it is assumed that none of the denominators vanish (Bhatnagar 1995, p. 22). The series in this theorem is called an $SU(n)$ series (Milne 1985; Bhatnagar 1995, p. 22).

Many other q-results, including the Q-BINOMIAL THEOREM and Q-SAALSCHÜTZ SUM, can be generalized to $U(n+1)$ series.

References

Bhatnagar, G. "$U(n+1)$ Basic Hypergeometric Series." Ch. 2 in *Inverse Relations, Generalized Bibasic Series, and their $U(n)$ Extensions*. Ph.D. thesis. Ohio State University, pp. 20–8, 1995.

Biedenharn, L. C. and Louck, J. D. *Angular Momentum in Quantum Physics: Theory and Applications*. Reading, MA: Addison-Wesley, 1981.

Biedenharn, L. C. and Louck, J. D. *The Racah-Wigner Algebra in Quantum Theory*. Reading, MA: Addison-Wesley, 1981.

Denis, R. Y. and Gustafson, R. A. "An $SU(n)$ q-Beta Integral Transformation and Multiple Hypergeometric Series Identities." *SIAM J. Math. Anal.* **23**, 552–61, 1992.

Gustafson, R. A. "Multilateral Summation Theorems for Ordinary and Basic Hypergeometric Series in $U(n)$." *SIAM J. Math. Anal.* **18**, 1576–596, 1987.

Gustafson, R. A. and Krattenthaler, C. "Heine Transformations for a New Kind of Basic Hypergeometric Series in $U(n)$." *J. Comput. Appl. Math.* **68**, 151–58, 1996.

Gustafson, R. A. and Krattenthaler, C. "Determinants Evaluations and $U(n)$ Extensions of Heine's $_2\phi_1$ Transformations." In *Special Functions, q-Series, and Related Topics* (Ed. M. E. H. Ismail, D. R. Masson, and M. Rahman). Providence, RI: Amer. Math. Soc., pp. 83–9, 1997.

Holman, W. J. III. "Summation Theorems for Hypergeometric Series in $U(n)$." *SIAM J. Math. Anal.* **11**, 523–32, 1980.

Holman, W. J. III.; Biedenharn, L. C.; and Louck, J. D. "On Hypergeometric Series Well-Poised in $SU(n)$." *SIAM J. Math. Anal.* **7**, 529–41, 1976.

Milne, S. C. "An Elementary Proof of the Macdonald Identities for $A_l^{(1)}$." *Adv. Math.* **57**, 34–0, 1985.

Milne, S. C. "Basic Hypergeometric Series Very Well-Poised in $U(n)$." *J. Math. Anal. Appl.* **122**, 223–56, 1987.

Milne, S. C. "Balanced $_3\phi_2$ Summation for $U(n)$ Basic Hypergeometric Series." *Adv. Math.* **131**, 93–87, 1997.

Ulam Map

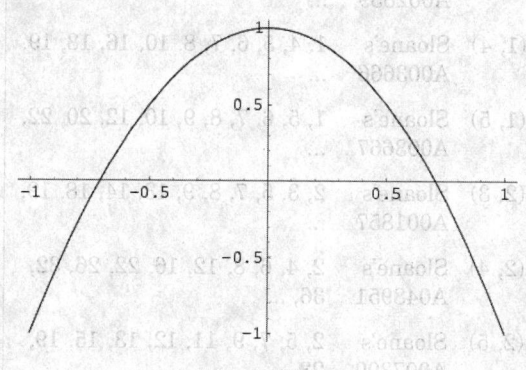

$$f(x) = 1 - 2x^2$$

for $x \in [-1, 1]$. Fixed points occur at $x = -1$, $1/2$, and order 2 fixed points at $x = (1 \pm \sqrt{5})/4$. The NATURAL DENSITY of the map is

$$\rho(y) = \frac{1}{\pi \sqrt{1 - y^2}}.$$

References

Beck, C. and Schlögl, F. *Thermodynamics of Chaotic Systems: An Introduction*. Cambridge, England: Cambridge University Press, p. 194, 1995.

Ulam Number

ULAM SEQUENCE

Ulam Sequence

The Ulam sequence $\{a_i\} = (u, v)$ is defined by $a_1 = u$, $a_2 = v$, with the general term a_n for $n > 2$ given by the least INTEGER expressible uniquely as the SUM of two distinct earlier terms. The numbers so produced are sometimes called U-NUMBERS or ULAM NUMBERS.

The first few numbers in the $(1, 2)$-Ulam sequence are 1, 2, 3, 4, 6, 8, 11, 13, 16, ... (Sloane's A002858). Here, the first term after the initial $(1, 2)$ is obviously 3 since $3 = 1 + 2$. The next term is $4 = 1 + 3$. (We don't have to worry about $4 = 2 + 2$ since it is a sum of a single term instead of *distinct* terms.) 5 is not a member of the sequence since it is representable in *two* ways, $5 = 1 + 4 = 2 + 3$, but $6 = 2 + 4$ is a member.

Proceeding in the manner, we can generate Ulam sequences for any (u, v), examples of which are given in the table below.

(u, v)	Sloane	Sequence
$(1, 2)$	Sloane's A002858	1, 2, 3, 4, 6, 8, 11, 13, 16, 18, ...
$(1, 3)$	Sloane's A002859	1, 3, 4, 5, 6, 8, 10, 12, 17, 21, ...
$(1, 4)$	Sloane's A003666	1, 4, 5, 6, 7, 8, 10, 16, 18, 19, ...
$(1, 5)$	Sloane's A003667	1, 5, 6, 7, 8, 9, 10, 12, 20, 22, ...
$(2, 3)$	Sloane's A001857	2, 3, 5, 7, 8, 9, 13, 14, 18, 19, ...
$(2, 4)$	Sloane's A048951	2, 4, 6, 8, 12, 16, 22, 26, 32, 36, ...
$(2, 5)$	Sloane's A007300	2, 5, 7, 9, 11, 12, 13, 15, 19, 23, ...

Schmerl and Spiegel (1994) proved that Ulam sequences $(2, v)$ for ODD $v \geq 5$ have exactly two EVEN terms. Ulam sequences with only finitely many EVEN terms eventually must have periodic successive differences (Finch 1991, 1992abc). Cassaigne and Finch (1995) proved that the Ulam sequences $(4, v)$ for $5 \leq v \equiv 1 \pmod 4$ have exactly three EVEN terms.

The Ulam sequence can be generalized by the s-ADDITIVE SEQUENCE.

See also GREEDY ALGORITHM, s-ADDITIVE SEQUENCE, STÖHR SEQUENCE

References

Cassaigne, J. and Finch, S. "A Class of 1-Additive Sequences and Quadratic Recurrences." *Exper. Math* **4**, 49–0, 1995.
Finch, S. "Conjectures About 1-Additive Sequences." *Fib. Quart.* **29**, 209–14, 1991.
Finch, S. "Are 0-Additive Sequences Always Regular?" *Amer. Math. Monthly* **99**, 671–73, 1992a.
Finch, S. "On the Regularity of Certain 1-Additive Sequences." *J. Combin. Th. Ser. A* **60**, 123–30, 1992b.
Finch, S. "Patterns in 1-Additive Sequences." *Exper. Math.* **1**, 57–3, 1992c.
Finch, S. "Ulam s-Additive Sequences." http://www.mathsoft.com/asolve/sadd/sadd.html.
Guy, R. K. "A Quarter Century of *Monthly* Unsolved Problems, 1969–993." *Amer. Math. Monthly* **100**, 945–49, 1993.
Guy, R. K. "Ulam Numbers." §C4 in *Unsolved Problems in Number Theory, 2nd ed.* New York: Springer-Verlag, pp. 109–10, 1994.
Guy, R. K. and Nowakowski, R. J. "*Monthly* Unsolved Problems, 1969–995." *Amer. Math. Monthly* **102**, 921–26, 1995.
Recaman, B. "Questions on a Sequence of Ulam." *Amer. Math. Monthly* **80**, 919–20, 1973.
Schmerl, J. and Spiegel, E. "The Regularity of Some 1-Additive Sequences." *J. Combin. Theory Ser. A* **66**, 172–75, 1994.

Sloane, N. J. A. Sequences A001857/M0634, A002858/M0557, A002859/M2303, A003666/M3237, A003667/M3746, and A007300/M1328 in "An On-Line Version of the Encyclopedia of Integer Sequences." http://www.research.att.com/~njas/sequences/eisonline.html.

Ulam's Conjecture

Let graph G have p points v_i and graph H have p points u_i, where $p \geq 3$. Then if for each i, the SUBGRAPHS $G_i = G - v_i$ and $H_i = H - u_i$ are ISOMORPHIC, then the graphs G and H are ISOMORPHIC.

See also ISOMORPHIC GRAPHS, SUBGRAPH

References

Harary, F. *Graph Theory.* Reading, MA: Addison-Wesley, p. 12, 1994.

Ulam's Problem

COLLATZ PROBLEM

Ulam's Spiral

PRIME SPIRAL

Ultrafactorial

The function defined by $U(n) = (n!)^{n!}$. The values for $n = 0$, 1, ..., are 1, 1, 4, 46656, 13337357768502841244449081472843776, ... (Sloane's A046882).

See also FACTORIAL

References

Sloane, N. J. A. Sequences A046882 in "An On-Line Version of the Encyclopedia of Integer Sequences." http://www.research.att.com/~njas/sequences/eisonline.html.

Ultrafilter

This entry contributed by VIKTOR BENGTSSON

Let S be a nonempty set, then an ultrafilter on S is a nonempty collection F of subsets of S having the following properties:

1. $\emptyset \notin F$.
2. If A, $B \in F$ then $A \cap B \in F$.
3. If $A \in F$ and $A \subseteq B \subseteq S$ then $B \in F$.
4. For any subset A of S, either $A \in F$ or its complement $A' = S - A \in F$.

An ultrafilter F on S is said to be free if it contains the COFINITE FILTER F_S of S.

See also COFINITE FILTER, FILTER

Ultrametric

An ultrametric is a METRIC which satisfies the following strengthened version of the TRIANGLE INEQUALITY,

$$d(x, z) \leq \max(d(x, y), d(y, z))$$

for all x, y, z. At least two of $d(x, y)$, $d(y, z)$, and $d(x, z)$ are the same.

Let X be a SET, and let $X^{\mathbb{N}}$ (where N is the SET of NATURAL NUMBERS) denote the collection of sequences of elements of X (i.e., all the possible sequences x_1, x_2, x_3, ...). For sequences $a = (a_1, a_2, ...)$, $b = (b_1, b_2, ...)$, let n be the number of initial places where the sequences agree, i.e., $a_1 = b_1$, $a_2 = b_2$, ..., $a_n = b_n$, but $a_{n+1} \neq b_{n+1}$. Take $n = 0$ if $a_1 \neq b_1$. Then defining $d(a, b) = 2^{-n}$ gives an ultrametric.

The P-ADIC NORM metric is another example of an ultrametric.

See also METRIC, P-ADIC NUMBER

Ultrapower

This entry contributed by MATT INSALL

A specific type of ULTRAPRODUCT that can be used to construct nonstandard universes and obtain the TRANSFER PRINCIPLE as a corollary of LOS' THEOREM for ultraproducts.

See also LOS' THEOREM, NONSTANDARD ANALYSIS, ULTRAPRODUCT

Ultraproduct

See also ULTRAPOWER

Ultraradical

A symbol which can be used to express solutions not obtainable by finite ROOT EXTRACTION. The solution to the irreducible QUINTIC EQUATION

$$x^5 + x = a$$

is written $\sqrt[5]{a}$.

See also RADICAL

Ultraspherical Differential Equation

GEGENBAUER DIFFERENTIAL EQUATION

Ultraspherical Function

GEGENBAUER FUNCTION

Ultraspherical Polynomial

GEGENBAUER POLYNOMIAL

Umbilic Point

A point on a surface at which the CURVATURE is the same in any direction.

Umbral Algebra

The algebra structure of linear functionals on polynomials of a single variable (Roman 1984, pp. 2–).

See also UMBRAL CALCULUS

References

Roman, S. "The Umbral Algebra." §2.1 in *The Umbral Calculus*. New York: Academic Press, pp. 6–2, 1984.

Umbral Calculus

Roman (1984, p. 2) describes umbral calculus as the study of the class of SHEFFER SEQUENCES. Umbral calculus provides a formalism for the systematic derivation and classification of almost all classical combinatorial identities for polynomial sequences, along with associated GENERATING FUNCTIONS, expansions, duplication formulas, RECURRENCE RELATIONS, inversions, RODRIGUES FORMULA, etc., (e.g., the EULER-MACLAURIN INTEGRATION FORMULAS, Boole's summation formula, the CHU-VANDERMONDE IDENTITY, NEWTON'S DIVIDED DIFFERENCE INTERPOLATION FORMULA, GREGORY'S FORMULA, LAGRANGE INVERSION).

The term "umbral calculus" was coined by Sylvester from the word "umbra" (meaning "shadow" in Latin), and reflects the fact that for many types of identities involving sequences of polynomials with POWERS a^n, "shadow" identities are obtained when the polynomials are changed to discrete values and the exponent in a^n is changed to the FALLING FACTORIAL $(a)_n \equiv a(a-1)\cdots(a-n+1)$.

For example, NEWTON'S FORWARD DIFFERENCE FORMULA written in the form

$$f(x+a) = \sum_{n=0}^{\infty} \frac{(a)_n \Delta^n f(x)}{n!} \tag{1}$$

with $f(x+a) \equiv f_{x+a}$ looks suspiciously like a finite analog of the TAYLOR SERIES expansion

$$f(x+a) = \sum_{n=0}^{\infty} \frac{a^n \tilde{D}^n f(x)}{n!}, \tag{2}$$

where $\tilde{D}$ is the DIFFERENTIAL OPERATOR. Similarly, the CHU-VANDERMONDE IDENTITY

$$(x+a)_n = \sum_{k=0}^{\infty} \binom{n}{k} (a)_k (x)_{n-k} \tag{3}$$

with $\binom{n}{k}$ a BINOMIAL COEFFICIENT, looks suspiciously like an analog of the BINOMIAL THEOREM

$$(x+a)^n = \sum_{k=0}^{\infty} \binom{n}{k} a^k x^{n-k} \tag{4}$$

(Di Bucchianico and Loeb).

See also APPELL SEQUENCE, BINOMIAL THEOREM, CHU-VANDERMONDE IDENTITY, COMBINATORICS, FAÀ DI BRUNO'S FORMULA, FINITE DIFFERENCE, SHEFFER SEQUENCE

References

Bell, E. T. "Postulational Basis for the Umbral Calculus." *Amer. J. Math.* **62**, 717–24, 1940.
Roman, S. and Rota, G.-C. "The Umbral Calculus." *Adv. Math.* **27**, 95–88, 1978.
Roman, S. *The Umbral Calculus.* New York: Academic Press, 1984.
Rota, G.-C.; Kahaner, D.; Odlyzko, A. "On the Foundations of Combinatorial Theory. VIII: Finite Operator Calculus." *J. Math. Anal. Appl.* **42**, 684–60, 1973.

Umbral Operator

An operator T which maps some BASIC POLYNOMIAL SEQUENCE $p_n(x)$ into another BASIC POLYNOMIAL SEQUENCE $q_n(x)$.

See also BASIC POLYNOMIAL SEQUENCE

References

Rota, G.-C.; Kahaner, D.; Odlyzko, A. "On the Foundations of Combinatorial Theory. VIII: Finite Operator Calculus." *J. Math. Anal. Appl.* **42**, 684–60, 1973.

Umbrella

WHITNEY UMBRELLA

Unambiguous

WELL DEFINED

Unbiased Estimator

A quantity which does not exhibit BIAS. An ESTIMATOR $\hat{\theta}$ is an unbiased estimator of θ if

$$\langle \hat{\theta} \rangle = \theta.$$

See also BIAS (ESTIMATOR), BIASED ESTIMATOR, ESTIMATOR, K-STATISTIC

Unbounded

See also BOUNDED

Uncia

$$1 \text{ uncia} \equiv \tfrac{1}{12}.$$

The word *uncia* was Latin for a unit equal to 1/12 of another unit called the *as*. The words "inch" (1/12 of a foot) and "ounce" (originally 1/12 of a pound and still 1/12 of a "Troy pound," now used primarily to weigh precious metals) are derived from the word *uncia*.

See also CALCUS, HALF, QUARTER, SCRUPLE, UNIT FRACTION

References

Conway, J. H. and Guy, R. K. *The Book of Numbers.* New York: Springer-Verlag, p. 4, 1996.

Uncorrelated

Variables x_i and x_j are said to be uncorrelated if their COVARIANCE is zero:

$$\text{cov}(x_i, x_j) = 0.$$

INDEPENDENT STATISTICS are always uncorrelated, but the converse is not necessarily true.

See also COVARIANCE, INDEPENDENT STATISTICS, UNCORRELATED NUMBERS

Uncorrelated Numbers

A sequence of numbers α_n is said to be uncorrelated if it satisfies

$$\lim_{n \to \infty} \frac{1}{2n} \sum_{m=-n}^{n} \alpha_m^2 = 1$$

$$\lim_{n \to \infty} \frac{1}{2n} \sum_{m=-n}^{n} \alpha_m \alpha_{k+m} = 0$$

for $k \neq 0$.

See also WIENER NUMBERS

References

Papoulis, A. *The Fourier Integral and Its Applications.* New York: McGraw-Hill, 1962.

Uncountable Set

UNCOUNTABLY INFINITE

Uncountably Infinite

An INFINITE SET, such as the real numbers, which is not COUNTABLY INFINITE.

See also ALEPH-0, ALEPH-1, COUNTABLE SET, COUNTABLY INFINITE, FINITE, INFINITE

References

Croft, H. T.; Falconer, K. J.; and Guy, R. K. *Unsolved Problems in Geometry.* New York: Springer-Verlag, p. 2, 1991.

Undecagon

HENDECAGON

Undecidable

Not DECIDABLE as a result of being neither formally provable nor unprovable.

See also GÖDEL'S INCOMPLETENESS THEOREM, RICHARDSON'S THEOREM

Undecillion

In the American system, 10^{36}.

See also LARGE NUMBER

Undefined

An expression in mathematics which does not have meaning and so which is not assigned an interpretation. For example, DIVISION BY ZERO is undefined in the FIELD of REAL NUMBERS.

See also AMBIGUOUS, DIVISION BY ZERO, ILL DEFINED, INDETERMINATE, WELL DEFINED

Underbar

UNDERSCORE

Underbrace

BRACE

Underdamping

DAMPED SIMPLE HARMONIC MOTION–UNDERDAMPING

Underdot

A dot placed under a symbol to indicate a DUMMY VARIABLE, e.g., c_1 (Comtet 1974, p. 32). This notation, however, is not very common.

See also DUMMY VARIABLE

References

Comtet, L. *Advanced Combinatorics: The Art of Finite and Infinite Expansions, rev. enl. ed.* Dordrecht, Netherlands: Reidel, p. 32, 1974.

Underlying Space

The space $|K|$ which is the subset of $\mathbb{R}^n$ that is the union of the simplices in a SIMPLICIAL COMPLEX K. The term POLYTOPE is sometimes used as a synonym for underlying space (Munkres 1991, p. 8).

See also POLYHEDRON, POLYTOPE

References

Munkres, J. R. *Analysis on Manifolds.* Reading, MA: Addison-Wesley, 1991.

Underscore

A horizontal line placed under a symbol to indicate some special property. Underscores are sometimes used instead of over-arrows or bold typeface to indicate a VECTOR, for example $\mathbf{x} = \underline{x}$.

References

Bringhurst, R. *The Elements of Typographic Style, 2nd ed.* Point Roberts, WA: Hartley and Marks, p. 286, 1997.

Undetermined Coefficients Method

Given a nonhomogeneous ORDINARY DIFFERENTIAL EQUATION, select a differential operator which will annihilate the right side, and apply it to both sides. Find the solution to the homogeneous equation, plug it into the left side of the original equation, and solve

for constants by setting it equal to the right side. The solution is then obtained by plugging the determined constants into the homogeneous equation.

See also ORDINARY DIFFERENTIAL EQUATION

Undirected Graph

A GRAPH for which the relations between pairs of vertices are symmetric, so that each edge has no directional character (as opposed to a DIRECTED GRAPH). Unless otherwise indicated by context, the term "graph" can usually be taken to mean "undirected graph."

See also DEGREE SEQUENCE, DIRECTED GRAPH, GRAPH

References

Skiena, S. "Undirected Graphs." §3.2.4 in *Implementing Discrete Mathematics: Combinatorics and Graph Theory with Mathematica.* Reading, MA: Addison-Wesley, pp. 92–3, 1990.

Undulating Number

A number OF THE FORM $aba\cdots, abab\cdots$, etc. The first few nontrivial undulants (with the stipulation that $a \neq b$) are 101, 121, 131, 141, 151, 161, 171, 181, 191, 202, 212, ... (Sloane's A046075). Including the trivial 1- and 2-digit undulants and dropping the requirement that $a \neq b$ gives Sloane's A033619.

The first few undulating SQUARES are 121, 484, 676, 69696, ... (Sloane's A016073), with no larger such numbers of fewer than a million digits (Pickover 1995). Several tricks can be used to speed the search for square undulating numbers, especially by examining the possible patterns of ending digits. For example, the only possible sets of four trailing digits for undulating SQUARES are 0404, 1616, 2121, 2929, 3636, 6161, 6464, 6969, 8484, and 9696.

The only undulating POWER $n^p = aba\cdots$ for $3 \leq p \leq 31$ and up to 100 digits is $7^3 = 343$ (Pickover 1995). A large undulating prime is given by $7 + 720(100^{49} - 1)/99$ (Pickover 1995).

A binary undulant is a POWER of 2 whose base-10 representation contains one or both of the sequences $010\cdots$ and $101\cdots$. The first few are 2^n for $n = 103$, 107, 138, 159, 179, 187, 192, 199, 205, ... (Sloane's A046076). The smallest n for which an undulating sequence of *exactly* d-digit occurs for $d = 3$, 4, ... are $n = 103, 138, 875, 949, 6617, 1802, 14545$, ... (Sloane's A046077). An undulating binary sequence of length 10 occurs for $n = 1,748,219$ (Pickover 1995).

References

Pickover, C. A. "Is There a Double Smoothly Undulating Integer?" In *Computers, Pattern, Chaos and Beauty.* New York: St. Martin's Press, 1990.

Pickover, C. A. "The Undulation of the Monks." Ch. 20 in *Keys to Infinity.* New York: W. H. Freeman, pp. 159–61 1995.

Sloane, N. J. A. Sequences A016073, A033619, A046075, A046076, and A046077 in "An On-Line Version of the Encyclopedia of Integer Sequences." http://www.research.att.com/~njas/sequences/eisonline.html.

Unduloid

A SURFACE OF REVOLUTION with constant NONZERO MEAN CURVATURE also called an ONDULOID. It is a ROULETTE obtained from the path described by the FOCI of a CONIC SECTION when rolled on a LINE. This curve then generates an unduloid when revolved about the LINE. These curves are special cases of the shapes assumed by soap film spanning the gap between prescribed boundaries. The unduloid of a PARABOLA gives a CATENOID.

See also CALCULUS OF VARIATIONS, CATENOID, ROULETTE, SURFACE OF REVOLUTION

References

Cundy, H. and Rollett, A. *Mathematical Models, 3rd ed.* Stradbroke, England: Tarquin Pub., p. 48, 1989.

Delaunay, C. "Sur la surface de révolution dont la courbure moyenne est constante." *J. math. pures appl.* **6**, 309–20, 1841.

do Carmo, M. P. "The Onduloid." §3.5G in *Mathematical Models from the Collections of Universities and Museums* (Ed. G. Fischer). Braunschweig, Germany: Vieweg, pp. 47–8, 1986.

Fischer, G. (Ed.). Plate 97 in *Mathematische Modelle/ Mathematical Models, Bildband/Photograph Volume.* Braunschweig, Germany: Vieweg, p. 93, 1986.

Thompson, D'A. W. *On Growth and Form, 2nd ed., compl. rev. ed.* New York: Cambridge University Press, 1992.

Yates, R. C. *A Handbook on Curves and Their Properties.* Ann Arbor, MI: J. W. Edwards, p. 184, 1952.

Unequal

Two quantities a and b which are not equal are said to be unequal, and this relationship can be denoted $a \neq b$.

See also EQUAL, INEQUALITY

References

Bringhurst, R. *The Elements of Typographic Style, 2nd ed.* Point Roberts, WA: Hartley and Marks, p. 286, 1997.

Unexpected Hanging Paradox

A PARADOX also known as the SURPRISE EXAMINATION PARADOX or PREDICTION PARADOX.

A prisoner is told that he will be hanged on some day between Monday and Friday, but that he will not know on which day the hanging will occur before it happens. He cannot be hanged on Friday, because if he were still alive on Thursday, he would know that the hanging will occur on Friday, but he has been told he will not know the day of his hanging in advance. He cannot be hanged Thursday for the same reason,

and the same argument shows that he cannot be hanged on any other day. Nevertheless, the executioner unexpectedly arrives on some day other than Friday, surprising the prisoner.

This PARADOX is similar to that in Robert Louis Stevenson's "BOTTLE IMP PARADOX," in which you are offered the opportunity to buy, for whatever price you wish, a bottle containing a genie who will fulfill your every desire. The only catch is that the bottle must thereafter be resold for a price smaller than what you paid for it, or you will be condemned to live out the rest of your days in excruciating torment. Obviously, no one would buy the bottle for 1¢ since he would have to give the bottle away, but no one would accept the bottle knowing he would be unable to get rid of it. Similarly, no one would buy it for 2¢, and so on. However, for some reasonably large amount, it will always be possible to find a next buyer, so the bottle will be bought (Paulos 1995).

See also BOTTLE IMP PARADOX, SORITES PARADOX

References

Chow, T. Y. "The Surprise Examination or Unexpected Hanging Paradox." *Amer. Math. Monthly* **105**, 41–1, 1998.

Clark, D. "How Expected is the Unexpected Hanging?" *Math. Mag.* **67**, 55–8, 1994.

Erickson, G. W. and Fossa, J. A. *Dictionary of Paradox.* Lanham, MD: University Press of America, pp. 158–59, 1998.

Gardner, M. "The Paradox of the Unexpected Hanging." Ch. 1 in *The Unexpected Hanging and Other Mathematical Diversions.* Chicago, IL: Chicago University Press, pp. 11–3, 1991.

Margalit, A. and Bar-Hillel, M. "Expecting the Unexpected." *Philosophia* **13**, 263–88, 1983.

Pappas, T. "The Paradox of the Unexpected Exam." *The Joy of Mathematics.* San Carlos, CA: Wide World Publ./Tetra, p. 147, 1989.

Paulos, J. A. *A Mathematician Reads the Newspaper.* New York: BasicBooks, p. 97, 1995.

Quine, W. V. O. "On a So-Called Paradox." *Mind* **62**, 65–7, 1953.

Unfair Game

A GAME in which a certain player can always win when he plays properly. All CATEGORICAL GAMES are unfair (Steinhaus 1983, p. 16).

See also CATEGORICAL GAME, GAME

References

Steinhaus, H. *Mathematical Snapshots, 3rd ed.* New York: Dover, 1999.

Unfinished Game

SHARING PROBLEM

Unfolding

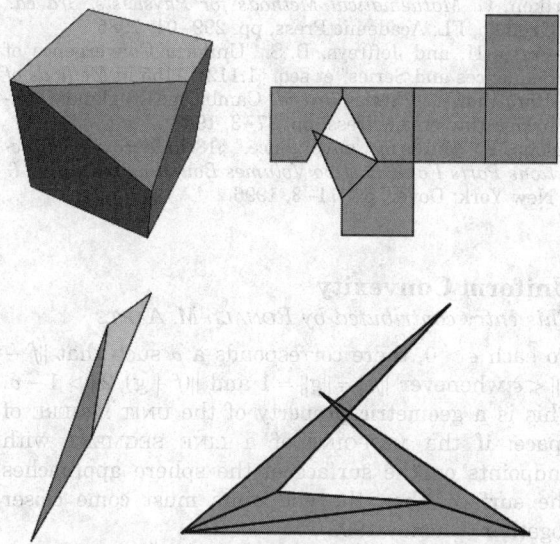

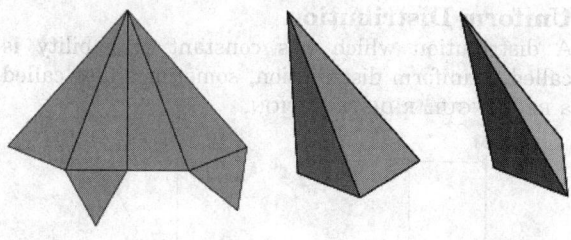

In 1987, K. Fukuda conjectured that no convex polyhedra admit a self-overlapping unfolding The above figure show a counterexample to conjecture 1 found by M. Namiki. A tetrahedron which is also ununfoldable was subsequently found.

Fukuda also conjectured that every CONVEX POLYHEDRON can be uniquely constructed from any of its unfolding. The counterexample shown above was found by T. Matsui.

The question of whether every CONVEX POLYHEDRON admits a self-unoverlapping unfolding is still unsettled.

See also NET, POLYHEDRON

References
Bern, M.; Demaine, E. D.; Eppstein, D.; and Kuo, E. Ununfoldable Polyhedra. 3 Aug 1999. http://xxx.lanl.gov/abs/cs.CG/9908003/.

Unhappy Number

A number which is not HAPPY is said to be unhappy.

See also HAPPY NUMBER

Unicursal Circuit

A CIRCUIT in which an entire GRAPH is traversed in one route. An example of a curve which can be traced unicursally is the MOHAMMED SIGN.

See also CIRCUIT, EULERIAN CIRCUIT, KÖNIGSBERG BRIDGE PROBLEM

References
Graustein, W. C. *Introduction to Higher Geometry*. New York: Macmillan, pp. 223–24, 1930.
Steinhaus, H. *Mathematical Snapshots, 3rd ed.* New York: Dover, pp. 256–57, 1999.

Unicyclic Graph

A CONNECTED GRAPH containing exactly one cycle (Harary 1994, p. 41).

References
Harary, F. *Graph Theory*. Reading, MA: Addison-Wesley, 1994.

Unidecagon

HENDECAGON

Uniform Apodization Function

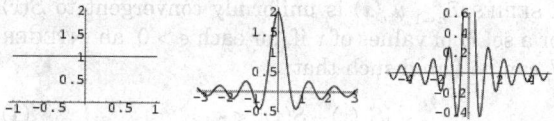

An APODIZATION FUNCTION

$$f(x) = 1, \qquad (1)$$

having INSTRUMENT FUNCTION

$$I(x) = \int_{-a}^{a} e^{-2\pi i k x}\, dx = -\frac{1}{2\pi i k}\left(e^{-2\pi i k a} - e^{2\pi i k x}\right)$$

$$= \frac{\sin(2\pi k a)}{\pi k} = 2a\, \mathrm{sinc}(2\pi k a). \qquad (2)$$

The peak (in units of a) is 2. The extrema are given by letting $\beta \equiv 2\pi k a$ and solving

$$\frac{d}{d\beta}(\beta \sin \beta) = \frac{\sin \beta - \beta \cos \beta}{\beta^2} = 0 \qquad (3)$$

$$\sin \beta - \beta \cos \beta = 0 \qquad (4)$$

$$\tan \beta = \beta. \qquad (5)$$

Solving this numerically gives $\beta_0 = 0$, $\beta_1 = 4.49341$, $\beta_2 = 7.72525$, ...for the first few solutions. The second of these is the peak POSITIVE sidelobe, and the third is the peak NEGATIVE sidelobe. As a fraction of the peak, they are 0.128375 and −0.217234. The FULL WIDTH AT

HALF MAXIMUM is found by setting $I(x) = 1$

$$\text{sinc}(x) = \tfrac{1}{2}, \tag{6}$$

and solving for $x_{1/2}$, yielding

$$x_{1/2} = 2\pi k_{1/2} a = 1.89549. \tag{7}$$

Therefore, with $L \equiv 2a$,

$$\text{FWHM} = 2k_{1/2} = \frac{0.603353}{a} = \frac{1.20671}{L}. \tag{8}$$

See also APODIZATION FUNCTION

Uniform Boundedness Principle

A "pointwise-bounded" family of continuous linear OPERATORS from a BANACH SPACE to a NORMED SPACE is "uniformly bounded." Symbolically, if $\sup\|T_i(x)\|$ is FINITE for each x in the unit BALL, then $\sup\|T_i\|$ is FINITE. The theorem is also called the BANACH-STEINHAUS THEOREM.

References

Zeidler, E. *Applied Functional Analysis: Applications to Mathematical Physics.* New York: Springer-Verlag, 1995.

Uniform Convergence

A SERIES $\sum_{n=1}^{\infty} u_n(x)$ is uniformly convergent to $S(x)$ for a set E of values of x if, for each $\epsilon > 0$, an INTEGER N can be found such that

$$|S_n(x) - S(x)| < \epsilon \tag{1}$$

for $n \geq N$ and all $x \in E$. To test for uniform convergence, use ABEL'S UNIFORM CONVERGENCE TEST or the WEIERSTRASS M-TEST. If individual terms $u_n(x)$ of a uniformly converging series are continuous, then

1. The series sum

$$f(x) = \sum_{n=1}^{\infty} u_n(x) \tag{2}$$

is continuous,
2. The series may be integrated term by term

$$\int_a^b f(x)\,dx = \sum_{n=1}^{\infty} \int_a^b u_n(x)\,dx, \tag{3}$$

and
3. The series may be differentiated term by term

$$\frac{d}{dx} f(x) = \sum_{n=1}^{\infty} \frac{d}{dx} u_n(x). \tag{4}$$

See also ABEL'S CONVERGENCE THEOREM, ABEL'S UNIFORM CONVERGENCE TEST, WEIERSTRASS M-TEST

References

Arfken, G. *Mathematical Methods for Physicists, 3rd ed.* Orlando, FL: Academic Press, pp. 299–01, 1985.
Jeffreys, H. and Jeffreys, B. S. "Uniform Convergence of Sequences and Series" et seq. §1.112–.1155 in *Methods of Mathematical Physics, 3rd ed.* Cambridge, England: Cambridge University Press, pp. 37–3, 1988.
Knopp, K. "Uniform Convergence." §18 in *Theory of Functions Parts I and II, Two Volumes Bound as One, Part I.* New York: Dover, pp. 71–3, 1996.

Uniform Convexity

This entry contributed by RONALD M. AARTS

To each $\epsilon > 0$, there corresponds a δ such that $\|f - g\| < \epsilon$ whenever $\|f\| = \|g\| = 1$ and $\|(f+g)/2\| > 1 - \delta$. This is a geometric property of the UNIT SPHERE of space: if the MIDPOINT of a LINE SEGMENT with endpoints on the surface of the sphere approaches the surface, then the endpoints must come closer together (Cheney 1999).

References

Cheney, E. W. *Introduction to Approximation Theory, 2nd ed.* Providence, RI: Amer. Math. Soc., 1999.

Uniform Distribution

A distribution which has constant probability is called a uniform distribution, sometimes also called a RECTANGULAR DISTRIBUTION.

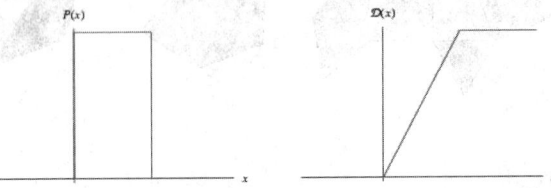

The probability density function and cumulative distribution function for a *continuous* uniform distribution are

$$P(x) = \begin{cases} \dfrac{1}{b-a} & \text{for } a < x < b \\ 0 & \text{for } x < a,\ x > b \end{cases} \tag{1}$$

$$D(x) = \begin{cases} 0 & \text{for } x < a \\ \dfrac{x-a}{b-a} & \text{for } a \leq x < b \\ 1 & \text{for } x \geq b. \end{cases} \tag{2}$$

With $a = 0$ and $b = 1$, these can be written

$$P(x) = \Pi\left(x + \tfrac{1}{2}\right) \tag{3}$$

$$= \tfrac{1}{2}[\text{sgn}(x) - \text{sgn}(x-1)] \tag{4}$$

$$= H(x) - H(x-1) \tag{5}$$

$$D(x) = xH(x) + (x-1)H(x-1), \tag{6}$$

where $\Pi(x)$ is the RECTANGLE FUNCTION and $H(x)$ is the HEAVISIDE STEP FUNCTION.

For a *continuous* uniform distribution, the CHARACTERISTIC FUNCTION is

$$\phi(t) = \frac{2}{(b-a)t} \sin\left[\tfrac{1}{2}(b-a)t\right] e^{i(a+b)t/2}, \quad (7)$$

and the MOMENT-GENERATING FUNCTION is

$$M(t) = \langle e^{xt} \rangle = \int_a^b \frac{e^{xt}}{b-a}\, dx = \left[\frac{e^{xt}}{t(b-a)}\right]_a^b, \quad (8)$$

so

$$M(t) = \begin{cases} \dfrac{e^{tb} - e^{ta}}{t(b-a)} & \text{for } t \neq 0 \\[2mm] 0 & \text{for } t = 0, \end{cases} \quad (9)$$

and

$$M'(t) = \frac{1}{b-a}\left[\frac{1}{t}\left(be^{bt} - ae^{at}\right) - \frac{1}{t^2}\left(e^{bt} - e^{at}\right)\right]$$

$$= \frac{e^{bt}(bt-1) - e^{at}(at-1)}{(b-a)t^2}. \quad (10)$$

If $a = 0$ and $b = 1$, the CHARACTERISTIC FUNCTION simplifies to

$$\phi(t) = \frac{2\sin\left(\tfrac{1}{2}t\right)e^{it/2}}{t} = \frac{i - i\cos t + \sin t}{t}. \quad (11)$$

The MOMENT-GENERATING FUNCTION is not differentiable at zero, but the MOMENTS can be calculated by differentiating and then taking $\lim_{t\to 0}$. The RAW MOMENTS are given by

$$\mu_1' = \tfrac{1}{2}(a+b) \quad (12)$$

$$\mu_2' = \tfrac{1}{3}(a^2 + ab + b^2) \quad (13)$$

$$\mu_3' = \tfrac{1}{4}(a+b)(a^2+b^2) \quad (14)$$

$$\mu_4' = \tfrac{1}{5}(a^4 + a^3 b + a^2 b^2 + ab^3 + b^4). \quad (15)$$

The CENTRAL MOMENTS are then

$$\mu_1 = 0 \quad (16)$$

$$\mu_2 = \tfrac{1}{12}(b-a)^2 \quad (17)$$

$$\mu_3 = 0 \quad (18)$$

$$\mu_4 = \tfrac{1}{80}(b-a)^4, \quad (19)$$

so the MEAN, VARIANCE, SKEWNESS, and KURTOSIS are

$$\mu = \tfrac{1}{2}(a+b) \quad (20)$$

$$\sigma^2 = \mu_2 = \tfrac{1}{12}(b-a)^2 \quad (21)$$

$$\gamma_1 = \frac{\mu_3}{\sigma^{3/2}} = 0 \quad (22)$$

$$\gamma_2 = -\tfrac{6}{5}. \quad (23)$$

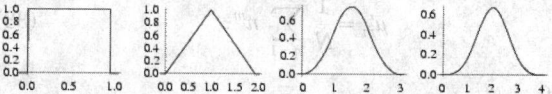

The distribution for the sum of n uniform variates on the interval $[0, 1]$ get be found using the CHARACTERISTIC FUNCTION as

$$P_n(x) = \mathscr{F}^{-1}\left[\left(\frac{i - \cos t + \sin t}{t}\right)^n\right] \quad (24)$$

$$= \frac{1}{2(n-1)!} \sum_{k=0}^{n} (-1)^k \binom{n}{k}(x-k)^{n-1} \operatorname{sgn}(x-k), \quad (25)$$

where the Fourier parameters are taken as $(1, 1)$. The first few values of $P_n(x)$ then give

$$P_1(x) = \tfrac{1}{2}[\operatorname{sgn}(1-x) + \operatorname{sgn} x] \quad (26)$$

$$P_2(x) = \tfrac{1}{2}[(-2+x)\operatorname{sgn}(-2+x)$$

$$-2(-1+x)\operatorname{sgn}(-1+x) + x\operatorname{sgn} x] \quad (27)$$

$$P_3(x) = \tfrac{1}{4}[-(-3+x)^2\operatorname{sgn}(-3+x)$$

$$+3(-2+x)^2\operatorname{sgn}(-2+x)$$

$$-3(-1+x)^2\operatorname{sgn}(-1+x) + x^2\operatorname{sgn} x] \quad (28)$$

$$P_4(x) = \tfrac{1}{12}[(-4+x)^3\operatorname{sgn}(-4+x)$$

$$-4(-3+x)^3\operatorname{sgn}(-3+x)$$

$$+6(-2+x)^3\operatorname{sgn}(-2+x)$$

$$-4(-1+x)^3\operatorname{sgn}(-1+x) + x^3\operatorname{sgn} x], \quad (29)$$

illustrated above.

The probability distribution function and cumulative distributions function for a *discrete* uniform distribution are

$$P(n) = \frac{1}{N} \quad (30)$$

$$D(n) = \frac{n}{N} \quad (31)$$

for $n = 1, ..., N$. The MOMENT-GENERATING FUNCTION is

$$M(t) = \langle e^{nt} \rangle = \sum_{n=1}^{N} \frac{1}{N} e^{nt} = \frac{1}{N} \frac{e^t - e^{t(N+1)}}{1 - e^t}$$

$$= \frac{e^t(1 - e^{Nt})}{N(1 - e^t)}. \qquad (32)$$

The MOMENTS about 0 are

$$\mu'_m = \frac{1}{N} \sum_{n=1}^{N} n^m, \qquad (33)$$

so

$$\mu'_1 = \tfrac{1}{2}(N + 1) \qquad (34)$$

$$\mu'_2 = \tfrac{1}{6}(N + 1)(2N + 1) \qquad (35)$$

$$\mu'_3 = \tfrac{1}{4} N(N + 1)^2 \qquad (36)$$

$$\mu'_4 = \tfrac{1}{30}(N + 1)(2N + 1)(3N^2 + 3N - 1), \qquad (37)$$

and the MOMENTS about the MEAN are

$$\mu_2 = \tfrac{1}{12}(N - 1)(N + 1) \qquad (38)$$

$$\mu_3 = 0 \qquad (39)$$

$$\mu_4 = \tfrac{1}{240}(N - 1)(N + 1)(3N^2 - 7). \qquad (40)$$

The MEAN, VARIANCE, SKEWNESS, and KURTOSIS are

$$\mu = \tfrac{1}{2}(N + 1) \qquad (41)$$

$$\sigma^2 = \mu_2 = \tfrac{1}{12}(N - 1)(N + 1) \qquad (42)$$

$$\gamma_1 = \frac{\mu_3}{\sigma^{3/2}} = 0 \qquad (43)$$

$$\gamma_2 = \frac{6(N^2 + 1)}{5(N - 1)(N + 1)}. \qquad (44)$$

See also EQUIDISTRIBUTED SEQUENCE, RANDOM NUMBER, RECTANGLE FUNCTION

References

Beyer, W. H. *CRC Standard Mathematical Tables, 28th ed.* Boca Raton, FL: CRC Press, pp. 531 and 533, 1987.

Uniform Polychoron

A 4-D analog of the UNIFORM POLYHEDRA. In fact, the UNIFORM POLYHEDRA are cells of the uniform polychora. There are more than 8000 known uniform polychora. The vertex figures of uniform polychora are always vertex-inscriptable in hyperspheres.

See also POLYCHORON

References

Olshevsky, G. "Uniform Polytopes in Four Dimensions." http://members.aol.com/Polycell/uniform.html.

Uniform Polyhedron

The uniform polyhedra are POLYHEDRA with identical VERTICES. Badoureau discovered 37 nonconvex uniform polyhedra in the late nineteenth century, many previously unknown (Wenninger 1983, p. 55). Coxeter *et al.* (1954) conjectured that there are 75 such polyhedra in which only two faces are allowed to meet at an EDGE, and this was subsequently proven. (However, when any EVEN number of faces may meet, there are 76 polyhedra.) If the five pentagonal PRISMS are included, the number rises to 80.

The VERTICES of a uniform polyhedron all lie on a SPHERE whose center is their CENTROID. The VERTICES joined to another VERTEX lie on a CIRCLE.

Except for a single non-Wythoffian case, uniform polyhedra can be generated by Wythoff's kaleidoscopic method of construction. In this construction, an initial vertex inside a special SPHERICAL TRIANGLE PQR is mapped to all the other vertices by repeated reflections across the three planar sides of this triangle. Similarly, PQR and its kaleidoscopic images must cover the sphere an integral number of times which is referred to as the density d of PQR. The density $d > 1$ is dependent on the choice of angles π/p, π/q, π/r at P, Q, R respectively, where p, q, r are reduced rational numbers greater than one. Such a spherical triangle is called a SCHWARZ TRIANGLE, conveniently denoted (pqr). Except for the infinite dihedral family of $(p22)$ for $p = 2$, 3, 4, ..., there are only 44 kinds of Schwarz triangles (Coxeter *et al.* 1954, Coxeter 1973). It has been shown that the numerators of p, q, r are limited to 2, 3, 4, 5 (4 and 5 cannot occur together) and so the nine choices for rational numbers are: 2, 3, 3/2, 4, 4/3, 5, 5/2, 5/3, 5/4 (Messer 1999).

The names of the uniform polyhedra were first formalized in Wenninger (1971), based on a list prepared by N. Johnson a few years earlier, as slightly modified by D. Luke. The names of the uniform duals appeared in Wenninger (1983), again based on nomenclature suggested by Johnson. Johnson also suggested a few modifications in the original nomenclature to incorporate some additional thoughts, as well as to undo some of Luke's less felicitous changes. The "List of polyhedra and dual models" in Wenninger (1983) gives revised names for several of the uniform polyhedra.

Source code and binary programs for generating and viewing the uniform polyhedra are also available at http://www.math.technion.ac.il/~rl/kaleido/. The following depictions of the polyhedra were produced by R. Maeder's `UniformPolyhedra.m` package for *Mathematica*. In this package, uniform polyhedra are computed to the desired numerical precision by numerically solving the definition fundamental equation, and lengths are normalized to give a MIDRADIUS of $\rho = 1$. Due to a limitation in *Mathematica*'s renderer, uniform polyhedra 69, 72, 74, and 75 cannot be displayed using this package (Maeder 1993).

The following table gives the names of the uniform polyhedra and their duals as given in Wenninger (1971). Coxeter *et al.* (1954) give many properties of the uniform solids, and Coxeter *et al.* (1953), Johnson (2000) and Messer give the quartic equation for determining the central angle subtending half an edge. The single non-Wythoffian case is the GREAT DIRHOMBICOSIDODECAHEDRON U_{75} which has pseudo-WYTHOFF SYMBOL $|3/2\ 5/3\ 3\ 5/2$.

n	WYTHOFF SYMBOL	Name	DUAL POLYHEDRON
1	3 | 2 3	TETRAHEDRON	TETRAHEDRON
2	2 3 | 3	TRUNCATED TETRAHEDRON	TRIAKIS TETRAHEDRON
3	3/2 3 | 3	OCTAHEMIOCTAHEDRON	OCTAHEMIOCTACRON
4	3/2 3 | 2	TETRAHEMIHEXAHEDRON	TETRAHEMIHEXACRON
5	4 | 2 3	OCTAHEDRON	CUBE
6	3 | 2 4	CUBE	OCTAHEDRON
7	2 | 3 4	CUBOCTAHEDRON	RHOMBIC DODECAHEDRON
8	2 4 | 3	TRUNCATED OCTAHEDRON	TETRAKIS HEXAHEDRON
9	2 3 | 4	TRUNCATED CUBE	TRIAKIS OCTAHEDRON
10	3 4 | 2	SMALL RHOMBICUBOCTAHEDRON	DELTOIDAL ICOSITETRAHEDRON
11	2 3 4 |	TRUNCATED CUBOCTAHEDRON	DISDYAKIS DODECAHEDRON
12	| 2 3 4	SNUB CUBE	PENTAGONAL ICOSITETRAHEDRON
13	3/2 4 | 4	SMALL CUBICUBOCTAHEDRON	SMALL HEXACRONIC ICOSITETRAHEDRON
14	3 4 | 4/3	GREAT CUBICUBOCTAHEDRON	GREAT HEXACRONIC ICOSITETRAHEDRON
15	4/3 4 | 3	CUBOHEMIOCTAHEDRON	HEXAHEMIOCTACRON
16	4/3 3 4 |	CUBITRUNCATED CUBOCTAHEDRON	TETRADYAKIS HEXAHEDRON
17	3/2 4 | 2	GREAT RHOMBICUBOCTAHEDRON	GREAT DELTOIDAL ICOSITETRAHEDRON
18	3/2 2 4 |	SMALL RHOMBIHEXAHEDRON	SMALL RHOMBIHEXACRON
19	2 3 | 4/3	STELLATED TRUNCATED HEXAHEDRON	GREAT TRIAKIS OCTAHEDRON
20	4/3 2 3 |	GREAT TRUNCATED CUBOCTAHEDRON	GREAT DISDYAKIS DODECAHEDRON
21	4/3 3/2 2 |	GREAT RHOMBIHEXAHEDRON	GREAT RHOMBIHEXACRON
22	5 | 2 3	ICOSAHEDRON	DODECAHEDRON
23	3 | 2 5	DODECAHEDRON	ICOSAHEDRON
24	2 | 3 5	ICOSIDODECAHEDRON	RHOMBIC TRIACONTAHEDRON
25	2 5 | 3	TRUNCATED ICOSAHEDRON	PENTAKIS DODECAHEDRON
26	2 3 | 5	TRUNCATED DODECAHEDRON	TRIAKIS ICOSAHEDRON
27	3 5 | 2	SMALL RHOMBICOSIDODECAHEDRON	DELTOIDAL HEXECONTAHEDRON
28	2 3 5 |	TRUNCATED ICOSIDODECAHEDRON	DISDYAKIS TRIACONTAHEDRON
29	| 2 3 5	SNUB DODECAHEDRON	PENTAGONAL HEXECONTAHEDRON
30	3 | 5/2 3	SMALL DITRIGONAL ICOSIDODECAHEDRON	SMALL TRIAMBIC ICOSAHEDRON
31	5/2 3 | 3	SMALL ICOSICOSIDODECAHEDRON	SMALL ICOSACRONIC HEXECONTAHEDRON
32	| 5/2 3 3	SMALL SNUB ICOSICOSIDODECAHEDRON	SMALL HEXAGONAL HEXECONTAHEDRON
33	3/2 5 | 5	SMALL DODECICOSIDODECAHEDRON	SMALL DODECACRONIC HEXECONTAHEDRON
34	5 | 2 5/2	SMALL STELLATED DODECAHEDRON	GREAT DODECAHEDRON
35	5/2 | 2 5	GREAT DODECAHEDRON	SMALL STELLATED DODECAHEDRON
36	2 | 5/2 5	DODECADODECAHEDRON	MEDIAL RHOMBIC TRIACONTAHEDRON
37	2 5/2 | 5	TRUNCATED GREAT DODECAHEDRON	SMALL STELLAPENTAKIS DODECAHEDRON
38	5/2 5 | 2	RHOMBIDODECADODECAHEDRON	MEDIAL DELTOIDAL HEXECONTAHEDRON
39	2 5/2 5 |	SMALL RHOMBIDODECAHEDRON	SMALL RHOMBIDODECACRON
40	| 2 5/2 5	SNUB DODECADODECAHEDRON	MEDIAL PENTAGONAL HEXECONTAHEDRON
41	3 | 5/3 5	DITRIGONAL DODECADODECAHEDRON	MEDIAL TRIAMBIC ICOSAHEDRON
42	3 5 | 5/3	GREAT DITRIGONAL DODECICOSIDODECAHEDRON	GREAT DITRIGONAL DODECACRONIC HEXECONTAHEDRON
43	5/3 3 | 5	SMALL DITRIGONAL DODECICOSIDODECAHEDRON	SMALL DITRIGONAL DODECACRONIC HEXECONTAHEDRON
44	5/3 5 | 3	ICOSIDODECADODECAHEDRON	MEDIAL ICOSACRONIC HEXECONTAHEDRON
45	5/3 3 5 |	ICOSITRUNCATED DODECADODECAHEDRON	TRIDYAKIS ICOSAHEDRON
46	| 5/3 3 5	SNUB ICOSIDODECADODECAHEDRON	MEDIAL HEXAGONAL HEXECONTAHEDRON
47	3/2 | 3 5	GREAT DITRIGONAL ICOSIDODECAHEDRON	GREAT TRIAMBIC ICOSAHEDRON
48	3/2 5 | 3	GREAT ICOSICOSIDODECAHEDRON	GREAT ICOSACRONIC HEXECONTAHEDRON
49	3/2 3 | 5	SMALL ICOSIHEMIDODECAHEDRON	SMALL ICOSIHEMIDODECACRON
50	3/2 3 5 |	SMALL DODECICOSAHEDRON	SMALL DODECICOSACRON
51	5/4 5 | 5	SMALL DODECAHEMIDODECAHEDRON	SMALL DODECAHEMIDODECACRON
52	3 | 2 5/2	GREAT STELLATED DODECAHEDRON	GREAT ICOSAHEDRON
53	5/2 | 2 3	GREAT ICOSAHEDRON	GREAT STELLATED DODECAHEDRON
54	2 | 5/2 3	GREAT ICOSIDODECAHEDRON	GREAT RHOMBIC TRIACONTAHEDRON
55	2 5/2 | 3	GREAT TRUNCATED ICOSAHEDRON	GREAT STELLAPENTAKIS DODECAHEDRON
56	2 5/2 3 |	RHOMBICOSAHEDRON	RHOMBICOSACRON
57	| 2 5/2 3	GREAT SNUB ICOSIDODECAHEDRON	GREAT PENTAGONAL HEXECONTAHEDRON
58	2 5 | 5/3	SMALL STELLATED TRUNCATED DODECAHEDRON	GREAT PENTAKIS DODECAHEDRON
59	5/3 2 5 |	TRUNCATED DODECADODECAHEDRON	MEDIAL DISDYAKIS TRIACONTAHEDRON
60	| 5/3 2 5	INVERTED SNUB DODECADODECAHEDRON	MEDIAL INVERTED PENTAGONAL HEXECONTAHEDRON
61	5/2 3 | 5/3	GREAT DODECICOSIDODECAHEDRON	GREAT DODECACRONIC HEXECONTAHEDRON
62	5/3 5/2 | 3	SMALL DODECAHEMICOSAHEDRON	SMALL DODECAHEMICOSACRON
63	5/3 5/2 3 |	GREAT DODECICOSAHEDRON	GREAT DODECICOSACRON

64	|5/3 5/2 3	GREAT SNUB DODECICOSIDODECA-HEDRON	GREAT HEXAGONAL HEXECON-TAHEDRON
65	5/4 5 |3	GREAT DODECAHEMICOSAHEDRON	GREAT DODECAHEMICOSACRON
66	2 3 |5/3	GREAT STELLATED TRUNCATED DODECAHEDRON	GREAT TRIAKIS ICOSAHEDRON
67	5/3 3 |2	GREAT RHOMBICOSIDODECAHE-DRON	GREAT DELTOIDAL HEXECONTA-HEDRON
68	5/3 2 3 |	GREAT TRUNCATED ICOSIDODECA-HEDRON	GREAT DISDYAKIS TRIACONTA-HEDRON
69	|5/3 2 3	GREAT INVERTED SNUB ICOSIDO-DECAHEDRON	GREAT INVERTED PENTAGONAL HEXECONTAHEDRON
70	5/3 5/2 |5/3	GREAT DODECAHEMIDODECAHE-DRON	GREAT DODECAHEMIDODECA-CRON
71	3/2 3 |5/3	GREAT ICOSIHEMIDODECAHEDRON	GREAT ICOSIHEMIDODECACRON
72	|3/2 3/2 5/2	SMALL RETROSNUB ICOSICOSIDO-DECAHEDRON	SMALL HEXAGRAMMIC HEXE-CONTAHEDRON
73	3/2 5/3 2 |	GREAT RHOMBIDODECAHEDRON	GREAT RHOMBIDODECACRON
74	|3/2 5/3 2	GREAT RETROSNUB ICOSIDODECA-HEDRON	GREAT PENTAGRAMMIC HEXE-CONTAHEDRON
75	|3/2 5/3 3 5/2	GREAT DIRHOMBICOSIDODECAHE-DRON	GREAT DIRHOMBICOSIDODECA-CRON
76	2 5 |2	PENTAGONAL PRISM	PENTAGONAL DIPYRAMID
77	|2 2 5	PENTAGONAL ANTIPRISM	PENTAGONAL DELTAHEDRON
78	2 5/2 |2	PENTAGRAMMIC PRISM	PENTAGRAMMIC DIPYRAMID
79	|2 2 5/2	PENTAGRAMMIC ANTIPRISM	PENTAGRAMMIC DELTAHEDRON
80	|2 2 5/3	PENTAGRAMMIC CROSSED ANTI-PRISM	PENTAGRAMMIC CONCAVE DELTAHEDRON

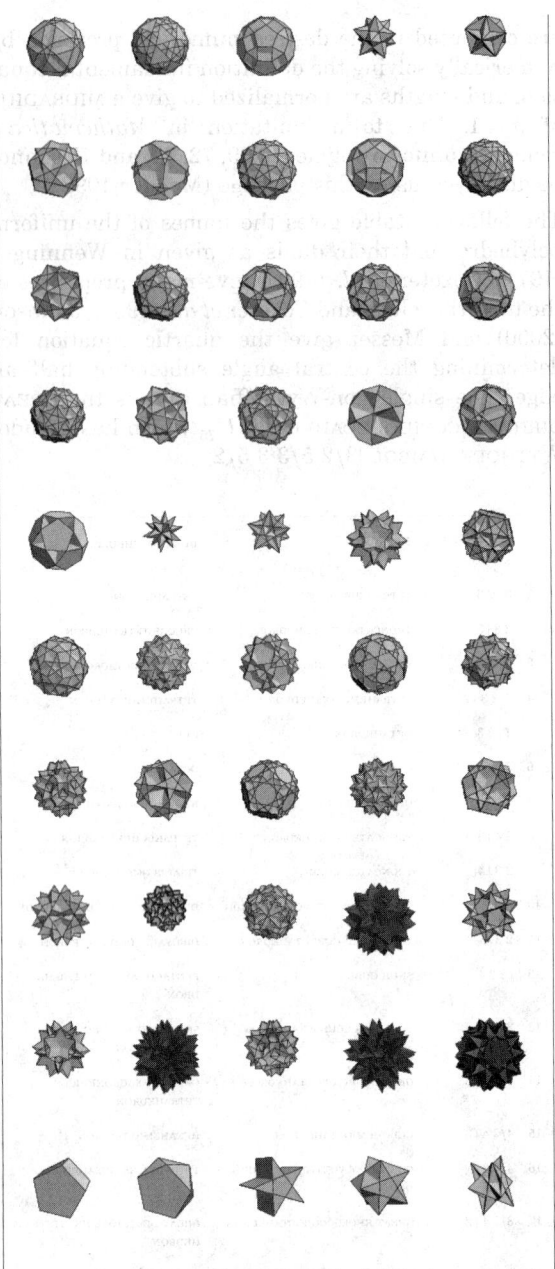

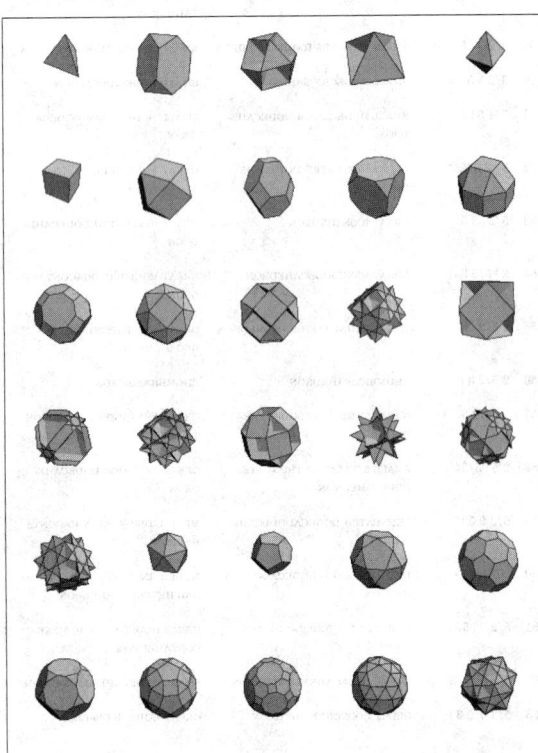

Johnson (2000) proposed a further revision of the "official" names of the uniform polyhedra and their duals and, at the same time, devised a literal symbol for each uniform polyhedron. For each uniform polyhedron, Johnson (2000) gives its number in Wenninger (1971), a modified SCHLÄFLI SYMBOL (following Coxeter), a literal symbol, and its new designated name. Not every uniform polyhedron has a dual that is free from anomalies like coincident vertices or faces extending to infinity. For those that do, Johnson gives the name of the dual polyhedron. In Johnson's new system, the uniform polyhedra are classified as follows:

1. Regular (regular polygonal vertex figures),
2. Quasi-regular (rectangular or ditrigonal vertex figures),
3. Versi-regular (orthodiagonal vertex figures),
4. Truncated regular (isosceles triangular vertex figures),
5. Quasi-quasi-regular (trapezoidal vertex figures),
6. Versi-quasi-regular (dipteroidal vertex figures),
7. Truncated quasi-regular (scalene triangular vertex figures),
8. Snub quasi-regular (pentagonal, hexagonal, or octagonal vertex figures),
9. Prisms (truncated hosohedra),
10. Antiprisms and crossed antiprisms (snub dihedra)

Here is a brief description of Johnson's symbols for the uniform polyhedra (Johnson). The star operator * appended to "D" or "E" replaces pentagons {5} by pentagrams {5/2}. The bar operator | indicates the removal from a related figure of a set (or sets) of faces, leaving "holes" so that a different set of faces takes their place. Thus, C|O is obtained from the cuboctahedron CO by replacing the eight triangles by four hexagons. In like manner, rR'|CO has the twelve squares of the rhombicuboctahedron rCO and the six octagons of the small cubicuboctahedron R'CO but has holes in place of their six squares and eight triangles. The operator "r" stands for "rectified": a polyhedron is truncated to the midpoints of the edges. Operators "a", "b", and "c" in the SCHLÄFLI SYMBOLS for the ditrigonary (i.e., having ditrigonal vertex figures) polyhedra stand for "altered," "blended," and "converted." The operator "o" stands for "ossified" (after S. L. van Oss). Operators "s" and "t" stand for "simiated" (snub) and "truncated."

Primes and capital letters are used for certain operators analogous to those just mentioned. For instance, rXY is the "rhombi-XY," with the faces of the quasi-regular XY supplemented by a set of square "rhombical" faces. The isomorphic r'XY has a crossed vertex figure. The operators "R" and "R'" denote a supplementary set of faces of a different kind—hexagons, octagons or octagrams, decagons or decagrams. Likewise, the operators "T" and "S" indicate the presence of faces other than, or in addition to, those produced by the simpler operators "t" and "s". The vertex figure of s'XY, the "vertisnub XY", is a crossed polygon, and that of s*XY, the "retrosnub XY", has density 2 relative to its circumcenter.

Regular polyhedra: p^q

1	{3, 3}	T	Tetrahedron	Tetrahedron
2	{3, 4}	O	Octahedron	Cube
3	{4, 3}	C	Cube	Octahedron
4	{3, 5}	I	Icosahedron	Dodecahedron
5	{5, 3}	D	Dodecahedron	Icosahedron
20	{5/2, 5}	D*	Small stellated dodecahedron	Great dodecahedron
21	{5, 5/2}	E	Great dodecahedron	Small stellated dodecahedron
22	{5/2, 3}	E*	Great stellated dodecahedron	Great icosahedron
41	{3, 5/2}	J	Great icosahedron	Great stellated dodecahedron

Quasi-regular polyhedra: $(p.q)^r$

11	r{3, 4}	CO	Cuboctahedron	Rhombic dodecahedron
12	r{3, 5}	ID	Icosidodecahedron	Rhombic triacontahedron
73	r{5/2, 5}	ED*	Dodecadodecahedron	Middle rhombic triacontahedron
94	r{5/2, 3}	JE*	Great icosidodecahedron	Great rhombic triacontahedron
70	a{5, 3}	ID*	Small ditrigonary icosidodecahedron	Small triambic icosahedron
80	b{5, 5/2}	DE*	Ditrigonary dodecadodecahedron	Middle triambic icosahedron
87	c{3, 5/2}	JE	Great ditrigonary icosidodecahedron	Great triambic icosahedron

Versi-regular polyhedra: $q.h.q.h$

67	o{3, 3}	T|T	Tetrahemihexahedron	no dual
78	o{3, 4}	C|O	Cubohemioctahedron	no dual
68	o{4, 3}	O|C	Octahemioctahedron	no dual
91	o{3, 5}	D|I	Small dodecahemidodecahedron	no dual
89	o{5, 3}	I|D	Small icosahemidodecahedron	no dual
102	o{5/2, 5}	E|D*	Small dodecahemiicosahedron	no dual
100	o{5, 5/2}	D*|E	Great dodecahemiicosahedron	no dual
106	o{5/2, 3}	J|E*	Great icosahemidodecahedron	no dual
107	o{3, 5/2}	E*|J	Great dodecahemidodecahedron	no dual

Truncated regular polyhedra: $q.2p.2p$

6	t{3, 3}	tT	Truncated tetrahedron	Triakis tetrahedron
7	t{3, 4}	tO	Truncated octahedron	Tetrakis hexahedron

8	t{4, 3}	tC	Truncated cube	Triakis octahedron
92	t'{4, 3}	t'C	stellatruncated cube	Great triakis octahedron
9	t{3, 5}	tI	Truncated icosahedron	Pentakis dodecahedron
10	t{5, 3}	tD	Truncated dodecahedron	Triakis icosahedron
97	t'{5/2, 5}	t'D*	Small stellatruncated dodecahedron	Great pentakis dodecahedron
75	t{5, 5/2}	tE	Great truncated dodecahedron	Small stellapentakis dodecahedron
104	t'{5/2, 3}	t'E*	Great stellatruncated dodecahedron	Great triakis icosahedron
95	t{3, 5/2}	tJ	Great truncated icosahedron	Great stellapentakis dodecahedron

Quasi-quasi-regular polyhedra: $p.2r.q.2r$ and $p.2s.q.2s$

13	rr{3, 4}	rCO	Rhombicuboctahedron	Strombic disdodecahedron
69	R'r{3, 4}	R'CO	Small cubicuboctahedron	Small sagittal disdodecahedron
77	Rr{3, 4}	RCO	Great cubicuboctahedron	Great strombic disdodecahedron
85	r'r{3, 4}	r'CO	Great rhombicuboctahedron	Great sagittal disdodecahedron
14	rr{3, 5}	rID	Rhombicosidodecahedron	Strombic hexecontahedron
72	R'r{3, 5}	R'ID	Small dodekicosidodecahedron	Small sagittal hexecontahedron
71	ra{5, 3}	rID*	Small icosified icosidodecahedron	Small strombic trisicosahedron
82	R'a{5, 3}	R'ID*	Small dodekified icosidodecahedron	Small sagittal trisicosahedron
76	rr{5/2, 5}	rED*	Rhombidodecadodecahedron	Middle strombic trisicosahedron
83	R'r{5/2, 5}	R'ED*	Icosified dodecadodecahedron	Middle sagittal trisicosahedron
81	Rc{3, 5/2}	RJE	Great dodekified icosidodecahedron	Great strombic trisicosahedron
88	r'c{3, 5/2}	r'JE	Great icosified icosidodecahedron	Great sagittal trisicosahedron
99	Rr{5/2, 3}	RJE*	Great dodekicosidodecahedron	Great strombic hexecontahedron
105	r'r{5/2, 3}	r'JE*	Great rhombicosidodecahedron	Great sagittal hexecontahedron

Versi-quasi-regular polyhedra: $2r.2s.2r.2s$

| 86 | or{3, 4} | rR'|CO | Small rhombicube | Small dipteral disdodecahedron |
| 103 | Or{3, 4} | Rr'|CO | Great rhombicube | Great dipteral disdodecahedron |

74	or{3, 5}	rR'	ID	Small rhombidodecahedron	Small dipteral hexecontahedron
90	oa{5, 3}	rR'	ID*	Small dodekicosahedron	Small dipteral trisicosahedron
96	or{5/2, 5}	rR'	ED*	Rhombicosahedron	Middle dipteral trisicosahedron
101	Oc{3, 5/2}	Rr'	JE	Great dodekicosahedron	Great dipteral trisicosahedron
109	Or{5/2, 3}	Rr'	JE*	Great rhombidodecahedron	Great dipteral hexecontahedron

Truncated quasi-regular polyhedra: $2p.2q.2r$

15	tr{3, 4}	tCO	Truncated cuboctahedron	Disdyakis dodecahedron
93	t'r{3, 4}	t'CO	Stellatruncated cuboctahedron	Great disdyakis dodecahedron
79	Tr{3, 4}	TCO	Cubitruncated cuboctahedron	Trisdyakis octahedron
16	tr{3, 5}	tID	Truncated icosidodecahedron	Disdyakis triacontahedron
98	t'r{5/2, 5}	t'ED*	Stellatruncated dodecadodecahedron	Middle disdyakis triacontahedron
84	T'r{5/2, 5}	T'ED*	Icositruncated dodecadodecahedron	Trisdyakis icosahedron
108	t'r{5/2, 3}	t'JE*	Stellatruncated icosidodecahedron	Great disdyakis triacontahedron

Snub quasi-regular polyhedra: $p.3.q.3.3$ or $p.3.q.3.r.3$

17	sr{3, 4}	sCO	Snub cuboctahedron	Petaloidal disdodecahedron
18	sr{3, 5}	sID	Snub icosidodecahedron	Petaloidal hexecontahedron
110	sa{5, 3}	sID*	Snub disicosidodecahedron	no dual
118	s*a{5, 3}	s*ID*	Retrosnub disicosidodecahedron	no dual
111	sr{5/2, 5}	sED*	Snub dodecadodecahedron	Petaloidal trisicosahedron
114	s'r{5/2, 5}	s'ED*	Vertisnub dodecadodecahedron	Vertipetaloidal trisicosahedron
112	S'r{5/2, 5}	S'ED*	Snub icosidodecadodecahedron	Hexaloidal trisicosahedron
113	sr{5/2, 3}	sJE*	Great snub icosidodecahedron	Great petaloidal hexecontahedron
116	s'r{5/2, 3}	s'JE*	Great vertisnub icosidodecahedron	Great vertipetaloidal hexecontahedron
117	s*r{5/2, 3}	s*JE*	Great retrosnub icosidodecahedron	Great retropetaloidal hexecontahedron

Snub quasi-regular polyhedron: $(p.4.q.4)^2$

119	SSr{5/2, 3}	SSJE*	Great disnub disicosidisdodecahedron	no dual

Prisms: $p.4.4$

$\{p\}x\{\}$	P(p)	p-gonal prism, $p = 3, 5, 6, \ldots$	p-gonal bipyramid
$\{p/d\}x\{\}$	P(p/d)	d-fold p-gonal prism, $p/d > 2$	d-fold p-gonal bipyramid

Antiprisms and crossed antiprisms: $3.3.3.p$

$s\{p\}h\{\}$	Q(p)	p-gonal antiprism, $p = 4, 5, 6, \ldots$	p-gonal antibipyramid
$s\{p/d\}h\{\}$	Q(p/d)	d-fold p-gonal antiprism, $p/d > 2$	d-fold p-gonal antibipyramid
$s'\{p/d\}h\{\}$	Q'(p/d)	d-fold p-gonal crossed antiprism, $2 < p/d < 3$	d-fold p-gonal crossed antibipyramid

See also ARCHIMEDEAN SOLID, AUGMENTED POLYHEDRON, DUAL POLYHEDRON, JOHNSON SOLID, KEPLER-POINSOT SOLID, MÖBIUS TRIANGLES, PLATONIC SOLID, POLYHEDRON, SCHWARZ TRIANGLE, UNIFORM POLYCHORON, VERTEX FIGURE, WYTHOFF SYMBOL

References

Ball, W. W. R. and Coxeter, H. S. M. *Mathematical Recreations and Essays, 13th ed.* New York: Dover, p. 136, 1987.

Brückner, M. *Vielecke under Vielflache.* Leipzig, Germany: Teubner, 1900.

Bulatov, V. "Compounds of Uniform Polyhedra." http://www.physics.orst.edu/~bulatov/polyhedra/uniform_compounds/.

Bulatov, V. "Dual Uniform Polyhedra." http://www.physics.orst.edu/~bulatov/polyhedra/dual/.

Bulatov, V. "Uniform Polyhedra." http://www.physics.orst.edu/~bulatov/polyhedra/uniform/.

Coxeter, H. S. M.; Longuet-Higgins, M. S.; and Miller, J. C. P. "Uniform Polyhedra." *Phil. Trans. Roy. Soc. London Ser. A* **246**, 401–50, 1954.

Coxeter, H. S. M. *Regular Polytopes, 3rd ed.* New York: Dover, 1973.

Har'El, Z. "Uniform Solution for Uniform Polyhedra." *Geometriae Dedicata* **47**, 57–10, 1993.

Har'El, Z. "Kaleido." http://www.math.technion.ac.il/~rl/kaleido/.

Har'El, Z. "Eighty Dual Polyhedra Generated by Kaleido." http://www.math.technion.ac.il/~rl/kaleido/dual.html.

Har'El, Z. "Eighty Uniform Polyhedra Generated by Kaleido." http://www.math.technion.ac.il/~rl/kaleido/poly.html.

Hume, A. "Exact Descriptions of Regular and Semi-Regular Polyhedra and Their Duals." Computing Science Tech. -Rept. No. 130. Murray Hill, NJ: AT&T Bell Lab., 1986.

Hume, A. Information files on polyhedra. http://netlib.bell-labs.com/netlib/polyhedra/.

Johnson, N. W. "Convex Polyhedra with Regular Faces." *Canad. J. Math.* **18**, 169–00, 1966.

Johnson, N. W. *Uniform Polytopes.* Cambridge, England: Cambridge University Press, 2000.

Maeder, R. E. "Uniform Polyhedra." *Mathematica J.* **3**, 1993. ftp://ftp.inf.ethz.ch/doc/papers/ti/scs/unipoly.ps.gz.

Maeder, R. E. Polyhedra.m and PolyhedraExamples *Mathematica* notebooks. http://www.inf.ethz.ch/department/TI/rm/programs.html.

Maeder, R. E. "The Uniform Polyhedra." http://www.inf.ethz.ch/department/TI/rm/unipoly/.

Messer, P. W. "Closed-Form Expressions for Uniform Polyhedra and Their Duals." Unpublished manuscript.

Messer, P. W. "Problem 1094." *Crux Math.* **11**, 325, 1985.

Messer, P. W. "Solution to Problem 1094." *Crux Math.* **13**, 133, 1987.

Skilling, J. "The Complete Set of Uniform Polyhedron." *Phil. Trans. Roy. Soc. London, Ser. A* **278**, 111–36, 1975.

Sopov, S. P. "Proof of the Completeness of the Enumeration of Uniform Polyhedra." *Ukrain. Geom. Sbornik* **8**, 139–56, 1970.

Virtual Image. *The Uniform Polyhedra CD-ROM.* 1997. http://ourworld.compuserve.com/homepages/vir_image/html/uniformpolyhedra.html.

Weisstein, E. W. "Polyhedron Duals." MATHEMATICA NOTEBOOK DUALS.M.

Weisstein, E. W. "Uniform Polyhedra." MATHEMATICA NOTEBOOK UNIFORMPOLYHEDRA.M.

Wenninger, M. J. *Dual Models.* Cambridge, England: Cambridge University Press, 1983.

Wenninger, M. J. *Polyhedron Models.* New York: Cambridge University Press, pp. 1–0 and 98, 1989.

Zalgaller, V. *Convex Polyhedra with Regular Faces.* New York: Consultants Bureau, 1969.

Ziegler, G. M. *Lectures on Polytopes.* Berlin: Springer-Verlag, 1995.

Uniform Variate

A RANDOM NUMBER which lies within a specified range (which can, without loss of generality, be taken as $[0, 1]$), with a UNIFORM DISTRIBUTION.

References

Press, W. H.; Flannery, B. P.; Teukolsky, S. A.; and Vetterling, W. T. "Uniform Deviates." §7.1 in *Numerical Recipes in FORTRAN: The Art of Scientific Computing, 2nd ed.* Cambridge, England: Cambridge University Press, pp. 267–77, 1992.

Uniformization

See also UNIFORMIZATION THEOREM

Uniformization Theorem

See also UNIFORMIZATION

Uniformly Cauchy

The series $\sum_{j=1}^{\infty} f_j(z)$ is said to be uniformly Cauchy on compact sets if, for each compact $K \subseteq U$ and each $\epsilon > 0$, there exists an $N > 0$ such that for all $M \geq L > N$,

$$\left| \sum_{j=L}^{M} f_j(z) \right| < \epsilon$$

holds (Krantz 1999, p. 104).

References

Krantz, S. G. "The Cauchy Condition for a Series." §8.1.5 in *Handbook of Complex Analysis.* Boston, MA: Birkhäuser, p. 104, 1999.

Uniformly Distributed Sequence

EQUIDISTRIBUTED SEQUENCE

Unimodal Distribution

A STATISTICAL DISTRIBUTION such as the GAUSSIAN DISTRIBUTION which has a single "peak."

See also BIMODAL DISTRIBUTION

Unimodal Sequence

A finite SEQUENCE which first increases and then decreases. A SEQUENCE $\{s_1, s_2, \ldots, s_n\}$ is unimodal if there exists a t such that

$$s_1 \leq s_2 \leq \ldots \leq s_t$$

and

$$s_t \geq s_{t+1} \geq \ldots \geq s_n.$$

Unimodular Group

A GROUP whose left HAAR MEASURE equals its right HAAR MEASURE.

See also HAAR MEASURE, MODULAR GROUP GAMMA, MODULAR GROUP GAMMA0, MODULAR GROUP LAMBDA

References

Knapp, A. W. "Group Representations and Harmonic Analysis, Part II." *Not. Amer. Math. Soc.* **43**, 537–49, 1996.

Unimodular Matrix

A MATRIX A with INTEGER elements and DETERMINANT $\det(A) = \pm 1$, also called a UNIT MATRIX.

The inverse of a unimodular matrix is another unimodular matrix. A POSITIVE unimodular matrix has $\det(A) = +1$. The nth POWER of a POSITIVE UNIMODULAR MATRIX

$$M = \begin{bmatrix} m_{11} & m_{12} \\ m_{21} & m_{22} \end{bmatrix} \quad (1)$$

is

$$M^n = \begin{bmatrix} m_{11}U_{n-1}(a) - U_{n-2}(a) & m_{12}U_{n-1}(a) \\ m_{21}U_{n-1}(a) & m_{22}U_{n-1}(a) - U_{n-2}(a) \end{bmatrix}, \quad (2)$$

where

$$a \equiv \tfrac{1}{2}(m_{11} + m_{22}) \quad (3)$$

and the U_n are CHEBYSHEV POLYNOMIALS OF THE SECOND KIND,

$$U_m(x) = \frac{\sin[(m+1)\cos^{-1} x]}{\sqrt{1 - x^2}}. \quad (4)$$

See also CHEBYSHEV POLYNOMIAL OF THE SECOND KIND

References

Born, M. and Wolf, E. *Principles of Optics: Electromagnetic Theory of Propagation, Interference, and Diffraction of Light, 6th ed.* New York: Pergamon Press, p. 67, 1980.
Goldstein, H. *Classical Mechanics, 2nd ed.* Reading, MA: Addison-Wesley, p. 149, 1980.
Séroul, R. *Programming for Mathematicians.* Berlin: Springer-Verlag, p. 162, 2000.

Unimodular Transformation

A transformation $\mathbf{x}' = A\mathbf{x}$ is unimodular if the DETERMINANT of the MATRIX A satisfies

$$\det(A) = \pm 1.$$

A NECESSARY and SUFFICIENT condition that a linear transformation transform a lattice to itself is that the transformation be unimodular.

If z is a COMPLEX NUMBER, then the transformation

$$z' = \frac{az + b}{cz + d}$$

is called a unimodular if a, b, c, and d are integers with $ad - bc = 1$. The set of all unimodular transformations forms a GROUP called the MODULAR GROUP.

See also MODULAR GROUP, MODULAR GROUP GAMMA

Union

The union of two sets A and B is the set obtained by combining the members of each. This is written $A \cup B$, and is pronounced "A union B" or "A cup B." The union of sets A_1 through A_n is written $\cup_{i=1}^{n} A_i$.

Let A, B, C, ... be sets, and let $P(S)$ denote the probability of S. Then

$$P(A \cup B) = P(A) + P(B) - P(A \cap B). \quad (1)$$

Similarly,

$$P(A \cup B \cup C) = P[A \cup (B \cup C)]$$
$$= P(A) + P(B \cup C) - P[A \cap (B \cup C)]$$

$$= P(A) + [P(B) + P(C) - P(B \cap C)]$$

$$-P[(A \cap B) \cup (A \cap C)]$$

$$= P(A) + P(B) + P(C) - P(B \cap C)$$

$$-\{P(A \cap B) + P(A \cap C) - P[(A \cap B) \cap (A \cap C)]\}$$

$$= P(A) + P(B) + P(C) - P(A \cap B)$$

$$-P(A \cap C) - P(B \cap C) + P(A \cap B \cap C). \quad (2)$$

If A and B are DISJOINT SETS, then by definition $P(A \cap B) = 0$, so

$$P(A \cup B) = P(A) + P(B). \quad (3)$$

Continuing, for a set of n disjoint elements $E_1, E_2, ..., E_n$

$$P\left(\bigcup_{i=1}^{n} E_i\right) = \sum_{i=1}^{n} P(E_i), \quad (4)$$

which is the COUNTABLE ADDITIVITY PROBABILITY AXIOM. Now let

$$E_i \equiv A \cap B_i, \quad (5)$$

then

$$P\left(\bigcup_{i=1}^{n} E \cap B_i\right) = \sum_{i=1}^{n} P(E \cap B_i). \quad (6)$$

See also DISJOINT UNION, INTERSECTION, OR, UNION-CLOSED SET

Union-Closed Set

A union-closed set is a nonempty finite collection of distinct nonempty finite sets which is CLOSED under UNION.

See also UNION-CLOSED SETS CONJECTURE

Union-Closed Sets Conjecture

Let $A = \{A_1, A_2, \ldots, A_n\}$ be a UNION-CLOSED SET, then the union-closed set conjecture states that an element exists which belongs to at least $n/2$ of the sets in A. Sarvate and Renaud (1989) showed that the conjecture is true if $|A_1| \leq 2$, where A_1 is the smallest set in A, or if $n < 11$. They also showed that if the conjecture fails, then $|A_1| < |A_n|/2$, where A_n is the largest set of A.

The proof for the case where A has a 2-set can be effected as follows. Write $A_1 = \{x, y\}$, then partition the sets of A into four disjoint families B_0, B_x, B_y, and B_{xy}, according to whether their intersection with A_1 is $\varnothing$, $\{x\}$, $\{y\}$, or $\{x, y\}$, respectively. It follows that $|B_{xy}| \geq |B_0|$ by taking unions with A_1, where $|B|$ is the CARDINALITY of B. Now compare $|B_x|$ with $|B_y|$. If $|B_x| \geq |B_y|$, then $|B_x| + |B_{xy}| \geq |B_0| + |B_y|$, so x is in at

least half the sets of A. Similarly, if $|B_x| \leq |B_y|$, then y is in at least half the sets (Hoey).

Unfortunately, this method of proof does extend to $|A_1| = 3$, since Sarvate and Renaud show an example of a UNION-CLOSED SET with $A_1 = \{x, y, z\}$ where none of x, y, z is in half the sets. However, in these cases, there are other elements which *do* appear in half the sets, so this is not a counterexample to the conjecture, but only a limitation to the method of proof given above (Hoey).

See also UNION-CLOSED SET

References

Sarvate, D. G. and Renaud, J.-C. "On the Union-Closed Sets Conjecture." *Ars Combin.* **27**, 149–53, 1989.
Sarvate, D. G. and Renaud, J.-C. "Improved Bounds for the Union-closed Sets Conjecture." *Ars Combin.* **29**, 181–85, 1990.

Uniplanar Double Point

ISOLATED SINGULARITY

Unipotent

A P-ELEMENT x of a GROUP G is unipotent if $F^*(C_G(x))$ is a P-GROUP, where F^* is the generalized FITTING SUBGROUP.

See also FITTING SUBGROUP, P-ELEMENT, P-GROUP

Unique

The property of being the only possible solution (perhaps modulo a constant, class of transformation, etc.).

See also ALEKSANDROV'S UNIQUENESS THEOREM, EXISTENCE, MAY-THOMASON UNIQUENESS THEOREM, UNIQUE FACTORIZATION

Unique Factorization

See also FUNDAMENTAL THEOREM OF ARITHMETIC, UNIQUE FACTORIZATION DOMAIN

Unique Factorization Domain

See also FUNDAMENTAL THEOREM OF ARITHMETIC, UNIQUE FACTORIZATION

Unique Factorization Theorem

FUNDAMENTAL THEOREM OF ARITHMETIC

Unit

A unit is an element in a RING that has a multiplicative inverse. If n is an ALGEBRAIC INTEGER which divides every ALGEBRAIC INTEGER in the FIELD, n is called a unit in that FIELD. A given FIELD may contain an infinity of units. The units of $\mathbb{Z}_n$ are the elements RELATIVELY PRIME to n. The units in $\mathbb{Z}_n$ which are SQUARES are called QUADRATIC RESIDUES.

See also EISENSTEIN UNIT, FUNDAMENTAL UNIT, IMAGINARY UNIT, PRIME UNIT, QUADRATIC RESIDUE

Unit Ball

A BALL of RADIUS 1.

See also SPHERE, BALL, UNIT CUBE, UNIT SPHERE

Unit Cell

A parallelogram (parallelepiped) containing the minimum repeatable elements of a circle (sphere) packing.

See also CIRCLE PACKING, PACKING DENSITY, SPHERE PACKING

References

Williams, R. "The Unit Cell Concept." §2- in *The Geometrical Foundation of Natural Structure: A Source Book of Design.* New York: Dover, pp. 48-1, 1979.

Unit Circle

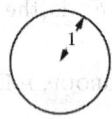

A CIRCLE of RADIUS 1, such as the one used to defined the functions of TRIGONOMETRY.

See also CIRCLE, UNIT DISK, UNIT SQUARE

References

Knopp, K. *Theory of Functions Parts I and II, Two Volumes Bound as One, Part I.* New York: Dover, p. 3, 1996.

Unit Cube

A CUBE whose edge lengths are 1. The unit cube therefore has unit volume.

See also CUBE, UNIT SQUARE, UNIT SPHERE

Unit Disk

A DISK with RADIUS 1.

See also FIVE DISKS PROBLEM, LOWER HALF-DISK, SEMICIRCLE, UNIT CIRCLE, UNIT SQUARE, UPPER HALF-DISK

Unit Element

IDENTITY ELEMENT

Unit Fraction

A unit fraction is a FRACTION with NUMERATOR 1. Examples of unit fractions include 1/2, 1/3, 1/12, and 1/123456. Unit fractions are also known as Egyptian fractions as a result of their extensive use by ancient Egyptians as a way of representing other fractions. The famous Rhind papyrus, dated to around 1650 BC, discusses unit fractions and contains a table of representations of $2/n$ as a sum of distinct unit fractions for ODD n between 5 and 101. The reason the Egyptians chose this method for representing fractions is not clear, although André Weil characterized the decision as "a wrong turn" (Hoffman 1998, pp. 153-54). The unique fraction that the Egyptians did not represent using unit fractions was 2/3 (Wells 1986, p. 29).

Unit fractions are almost always required to exclude repeated terms, since representations such as $1/5 + 1/5 + 1/5$ are trivial. Any RATIONAL NUMBER has representations as a sum of distinct unit fractions with arbitrarily many terms and with arbitrarily large DENOMINATORS, although for a given fixed number of terms, there are only finitely many. Fibonacci proved that *any* fraction can be REPRESENTED AS a sum of distinct unit fractions (Hoffman 1998, p. 154). An infinite chain of unit fractions can be constructed using the identity

$$\frac{1}{a} = \frac{1}{a+1} + \frac{1}{a(a+1)}. \qquad (1)$$

Martin (1999) showed that for every positive RATIONAL NUMBER, there exist representations as unit fractions whose largest DENOMINATOR is at most N and whose DENOMINATORS form a positive proportion of the integers up to N for sufficiently large N. Each FRACTION x/y with y ODD has a unit fraction representation in which each DENOMINATOR is ODD (Breusch 1954; Guy 1994, p. 160). Every x/y has a t-term representation where $t = \mathcal{O}(\sqrt{\log y})$ (Vose 1985).

No algorithm is known for producing unit fraction representations having either a minimum number of terms or smallest possible denominator (Hoffman 1998, p. 155). However, there are a number of ALGORITHMS (including the BINARY REMAINDER METHOD, CONTINUED FRACTION UNIT FRACTION ALGORITHM, GENERALIZED REMAINDER METHOD, GREEDY ALGORITHM, REVERSE GREEDY ALGORITHM, SMALL MULTIPLE METHOD, and SPLITTING ALGORITHM) for decomposing an arbitrary FRACTION into unit fractions. In 1202, Fibonacci published an algorithm for constructing unit fraction representations, and this algorithm was subsequently rediscovered by Sylvester (Hoffman 1998, p. 154; Martin 1999).

Taking the fractions 1/2, 1/3, 2/3, 1/4, 2/4, 3/4, ... (the numerators of which are Sloane's A002260, and the denominators of which are $n-1$ copies of the integer

n), the unit fraction representations using the GREEDY ALGORITHM are

$$\frac{1}{2} = \frac{1}{2}$$

$$\frac{1}{3} = \frac{1}{3}$$

$$\frac{2}{3} = \frac{1}{2} + \frac{1}{6}$$

$$\frac{1}{4} = \frac{1}{4}$$

$$\frac{2}{4} = \frac{1}{2}$$

$$\frac{3}{4} = \frac{1}{2} + \frac{1}{4}$$

$$\frac{1}{5} = \frac{1}{5}$$

$$\frac{2}{5} = \frac{1}{3} + \frac{1}{15}$$

$$\frac{3}{5} = \frac{1}{2} + \frac{1}{10}$$

$$\frac{4}{5} = \frac{1}{2} + \frac{1}{4} + \frac{1}{20}.$$

The number of terms in these representations are 1, 1, 2, 1, 1, 2, 1, 2, 2, 3, 1, ... (Sloane's A050205). The minimum denominators for each representation are given by 2, 3, 2, 4, 2, 2, 5, 3, 2, 2, 6, 3, 2, ... (Sloane's A050206), and the maximum denominators are 2, 3, 6, 4, 2, 4, 5, 15, 10, 20, 6, 3, 2, ... (Sloane's A050210).

Wilf posed as a problem that any fraction with odd denominator can be REPRESENTED AS a sum of unit fractions, each having an odd denominator, and Graham proved that infinitely many fractions with a certain range can be represented as a sum of units fractions with square denominators (Hoffman 1998, p. 156).

Paul Erdos and E. G. Straus have conjectured that the DIOPHANTINE EQUATION

$$\frac{4}{n} = \frac{1}{a} + \frac{1}{b} + \frac{1}{c} \qquad (2)$$

always can be solved (Obláth 1950, Rosati 1954, Bernstein 1962, Yamamoto 1965, Vaughan 1970, Guy 1994), and Sierpinski (1956) conjectured that

$$\frac{5}{n} = \frac{1}{a} + \frac{1}{b} + \frac{1}{c} \qquad (3)$$

can be solved (Guy 1994).

The HARMONIC NUMBER H_n is never an INTEGER except for H_1. This result was proved im 1915 by Taeisinger, and the more general results that any number of consecutive terms not necessarily starting with 1 never sum to an integer was proved by Kürschák in 1918 (Hoffman 1998, p. 157). In 1932, Erdos proved that the sum of the reciprocals of any number of equally spaced integers is never a reciprocal.

See also CALCUS, EGYPTIAN NUMBER, HALF, HARMONIC NUMBER, QUARTER, SCRUPLE, UNCIA

References

Beck, A.; Bleicher, M. N.; and Crowe, D. W. *Excursions into Mathematics.* New York: Worth Publishers, 1970.

Beeckmans, L. "The Splitting Algorithm for Egyptian Fractions." *J. Number Th.* **43**, 173–85, 1993.

Bernstein, L "Zur Lösung der diophantischen Gleichung $m/n = 1/x + 1/y + 1/z$ insbesondere im Falle $m = 4$." *J. reine angew. Math.* **211**, 1–0, 1962.

Bleicher, M. N. "A New Algorithm for the Expansion of Continued Fractions." *J. Number Th.* **4**, 342–82, 1972.

Breusch, R. "A Special Case of Egyptian Fractions." Solution to advanced problem 4512. *Amer. Math. Monthly* **61**, 200–01, 1954.

Brown, K. S. "Egyptian Unit Fractions." http://www.seanet.com/~ksbrown/iegypt.htm.

Eppstein, D. "Ten Algorithms for Egyptian Fractions." *Mathematica Educ. Res.* **4**, 5–5, 1995.

Eppstein, D. "Egyptian Fractions." http://www.ics.uci.edu/~eppstein/numth/egypt/.

Eppstein, D. Egypt.ma Mathematica notebook. http://www.ics.uci.edu/~eppstein/numth/egypt/egypt.ma.

Gardner, M. "Mathematical Games: In Which a Mathematical Aesthetic is Applied to Modern Minimal Art." *Sci. Amer.* **239**, 22–2, Nov. 1978.

Golomb, S. W. "An Algebraic Algorithm for the Representation Problems of the Ahmes Papyrus." *Amer. Math. Monthly* **69**, 785–86, 1962.

Graham, R. "On Finite Sums of Unit Fractions." *Proc. London Math. Soc.* **14**, 193–07, 1964.

Guy, R. K. "Egyptian Fractions." §D11 in *Unsolved Problems in Number Theory, 2nd ed.* New York: Springer-Verlag, pp. 87–3 and 158–66, 1994.

Hoffman, P. *The Man Who Loved Only Numbers: The Story of Paul Erdos and the Search for Mathematical Truth.* New York: Hyperion, pp. 153–57, 1998.

Ke, Z. and Sun, Q. "On the Representation of 1 by Unit Fractions." *Sichuan Daxue Xuebao* **1**, 13–9, 1964.

Klee, V. and Wagon, S. *Old and New Unsolved Problems in Plane Geometry and Number Theory.* Washington, DC: Math. Assoc. Amer., pp. 175–77 and 206–08, 1991.

Martin, G. "Dense Egyptian Fractions." *Trans. Amer. Math. Soc.* **351**, 3641–657, 1999.

Niven, I. and Zuckerman, H. S. *An Introduction to the Theory of Numbers, 5th ed.* New York: Wiley, p. 200, 1991.

Obláth, R. "Sur l'equation diophantienne $4/n = 1/x_1 + 1/x_2 + 1/x_3$." *Mathesis* **59**, 308–16, 1950.

Rosati, L. A. "Sull'equazione diofantea $4/n = 1/x_1 + 1/x_2 + 1/x_3$." *Boll. Un. Mat. Ital.* **9**, 59–3, 1954.

Séroul, R. "Egyptian Fractions." §8.8 in *Programming for Mathematicians.* Berlin: Springer-Verlag, pp. 181–87, 2000.

Sierpinski, W. "Sur les décompositiones de nombres rationelles en fractions primaires." *Mathesis* **65**, 16–2, 1956.

Sloane, N. J. A. Sequences A002260, A050205, A050206, and A050210 in "An On-Line Version of the Encyclopedia of Integer Sequences." http://www.research.att.com/~njas/sequences/eisonline.html.

Stewart, I. "The Riddle of the Vanishing Camel." *Sci. Amer.* **266**, 122–24, June 1992.

Tenenbaum, G. and Yokota, H. "Length and Denominators of Egyptian Fractions." *J. Number Th.* **35**, 150–56, 1990.

Vaughan, R. C. "On a Problem of Erdos, Straus and Schinzel." *Mathematika* **17**, 193–98, 1970.

Vose, M. "Egyptian Fractions." *Bull. London Math. Soc.* **17**, 21, 1985.

Wagon, S. "Egyptian Fractions." §8.6 in *Mathematica in Action.* New York: W. H. Freeman, pp. 271–77, 1991.

Wells, D. *The Penguin Dictionary of Curious and Interesting Numbers.* Middlesex, England: Penguin Books, p. 29, 1986.

Yamamoto, K. "On the Diophantine Equation $4/n = 1/x + 1/y + 1/z$." *Mem. Fac. Sci. Kyushu U. Ser. A* **19**, 37–7, 1965.

Unit Lattice

A POINT LATTICE which can be constructed from an arbitrary PARALLELOGRAM of unit area. For any such planar lattice, the minimum distance c between any two points is a quantity characteristic of the lattice. This distance satisfies

$$c \le \sqrt{\frac{2}{\sqrt{3}}}$$

(Hilbert and Cohn-Vossen 1999, p. 36). For a lattice in 3-D,

$$c \le 2^{1/6}$$

(Hilbert and Cohn-Vossen 1999, p. 45).

See also HYPERSPHERE PACKING, POINT LATTICE, SPHERE PACKING

References

Hilbert, D. and Cohn-Vossen, S. *Geometry and the Imagination.* New York: Chelsea, 1999.

Unit Matrix

An INTEGER MATRIX consisting of all 1s. The $m \times n$ unit matrix is often denoted $\mathbf{J}_{mn}$, or $\mathbf{J}_n$ if $m = n$. Square unit matrices have DETERMINANT 0.

See also IDENTITY MATRIX, UNIMODULAR MATRIX

References

Brenner, J. and Cummings, L. "The Hadamard Maximum Determinant Problem." *Amer. Math. Monthly* **79**, 626–30, 1972.

Unit Neighborhood Graph

A DISTANCE GRAPH with distance set $(0, 1]$.

See also DISTANCE GRAPH, UNIT-DISTANCE GRAPH

References

Fishburn, P. C. "On the Sphericity and Cubicity of Graphs." *J. Combin. Th. B* **35**, 309–18, 1983.

Frankl, P. and Maehara, H. "Embedding the n-Cube in Lower Dimensions." *European J. Combin.* **7**, 221–25, 1986.

Frankl, P. and Maehara, H. "Open-Interval Graphs versus Closed-Interval Graphs." *Discr. Math.* **63**, 97–00, 1987.

Frankl, P. and Maehara, H. "The Johnson-Lindenstrauss Lemma and the Sphericity of Some Graphs." *J. Combin. Th. B* **44**, 355–61, 1988.

Maehara, H. "Independent Balls and Unit Neighborhood Graphs." *Ryukyu Math. J.* **1**, 38–5, 1988.

Maehara, H. and Rödl, V. "On the Dimension to Represent a Graph by a Unit Distance Graph." *Graphs Combin.* **6**, 365–67, 1990.

Maehara, H. "Distance Graphs in Euclidean Space." *Ryukyu Math. J.* **5**, 33–1, 1992.

Unit Point

The point in the PLANE with Cartesian coordinates (1, 1).

References

Woods, F. S. *Higher Geometry: An Introduction to Advanced Methods in Analytic Geometry.* New York: Dover, p. 9, 1961.

Unit Ring

A unit ring is a set together with two BINARY OPERATORS $S(+, *)$ satisfying the following conditions:

1. Additive associativity: For all a, b, $c \in S$, $(a + b) + c = a + (b + c)$,
2. Additive commutativity: For all a, $b \in S$, $a + b = b + a$,
3. Additive identity: There exists an element $0 \in S$ such that for all $a \in S : 0 + a = a + 0 = a$,
4. Additive inverse: For every $a \in S$, there exists a $-a \in S$ such that $a + (-a) = (-a) + a = 0$,
5. Multiplicative associativity: For all a, b, $c \in S$, $(a * b) * c = a * (b * c)$,
6. Multiplicative identity: There exists an element $1 \in S$ such that for all $a \in S$, $1 * a = a * 1 = a$,
7. Left and right distributivity: For all a, b, $c \in S$, $a * (b + c) = (a * b) + (a * c)$ and $(b + c) * a = (b * a) + (c * a)$.

Thus, a unit ring is a RING with a multiplicative identity.

See also BINARY OPERATOR, RING

References

Rosenfeld, A. *An Introduction to Algebraic Structures.* New York: Holden-Day, 1968.

Unit Sphere

A SPHERE of RADIUS 1.

See also SPHERE, BALL, UNIT CIRCLE

Unit Square

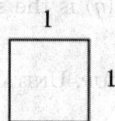

A SQUARE with side lengths 1. *The* unit square usually means the one with coordinates (0, 0), (1, 0), (1, 1), (0, 1) in the real plane, or 0, 1, $1 + i$, and i in the COMPLEX PLANE.

See also HEILBRONN TRIANGLE PROBLEM, UNIT CIRCLE, UNIT CUBE, UNIT DISK

Unit Vector

A VECTOR of unit length, sometimes also called a DIRECTION VECTOR (Jeffreys and Jeffreys 1988). The unit vector $\hat{\mathbf{v}}$ having the same direction as a given (nonzero) vector $\mathbf{v}$ is defined by

$$\hat{\mathbf{v}} \equiv \frac{\mathbf{v}}{|\mathbf{v}|},$$

where $|\mathbf{v}|$ denotes the NORM of $\mathbf{v}$, is the unit vector in the same direction as the (finite) VECTOR $\mathbf{v}$. A unit vector in the $\mathbf{x}_n$ direction is given by

$$\hat{\mathbf{x}}_n \equiv \frac{\dfrac{\partial \mathbf{r}}{\partial x_n}}{\left|\dfrac{\partial \mathbf{r}}{\partial x_n}\right|},$$

where $\mathbf{r}$ is the RADIUS VECTOR.

See also NORM, RADIUS VECTOR, VECTOR, ZERO VECTOR

References

Jeffreys, H. and Jeffreys, B. S. "Direction Vectors." §2.034 in *Methods of Mathematical Physics, 3rd ed.* Cambridge, England: Cambridge University Press, p. 64, 1988.
Stephens, M. A. "The Testing of Unit Vectors for Randomness." *J. Amer. Stat. Assoc.* **59**, 160–67, 1964.

Unital

A BLOCK DESIGN OF THE FORM $(q^3 + 1, q + 1, 1)$.

References

Dinitz, J. H. and Stinson, D. R. "A Brief Introduction to Design Theory." Ch. 1 in *Contemporary Design Theory: A Collection of Surveys* (Ed. J. H. Dinitz and D. R. Stinson). New York: Wiley, pp. 1–2, 1992.

Unitary

An OPERATOR U satisfying

$$U^*U = 1$$
$$UU^* = 1,$$

where U^* is the ADJOINT.

See also ANTIUNITARY

References

Sakurai, J. J. *Modern Quantum Mechanics*. Menlo Park, CA: Benjamin/Cummings, 1985.

Unitary Aliquot Sequence

An ALIQUOT SEQUENCE computed using the analog of the RESTRICTED DIVISOR FUNCTION $s^*(n)$ in which only UNITARY DIVISORS are included.

See also ALIQUOT SEQUENCE, UNITARY AMICABLE PAIR, UNITARY SOCIABLE NUMBERS

References

Guy, R. K. "Unitary Aliquot Sequences." §B8 in *Unsolved Problems in Number Theory, 2nd ed.* New York: Springer-Verlag, pp. 63–5, 1994.

Unitary Amicable Pair

A PAIR of numbers m and n such that

$$\sigma^*(m) = \sigma^*(n) = m + n,$$

where $\sigma^*(n)$ is the sum of UNITARY DIVISORS. Hagis (1971) and García (1987) give 82 such pairs. The first few are (114, 126), (1140, 1260), (18018, 22302), (32130, 40446), ... (Sloane's A002952 and A002953). The largest known unitary amicable pair, each member of which has 192 digits,

$$2^2 \cdot 3^2 \cdot 5^9 \cdot 7^3 \cdot 11 \cdot 13 \cdot 17^2 \cdot 19 \cdot 29 \cdot 41 \cdot 43 \cdot 47$$
$$\cdot 79 \cdot 157 \cdot 163 \cdot 223 \cdot 433 \cdot 1303 \cdot 1399 \cdot 2053$$
$$\cdot 2719 \cdot 5167 \cdot 13187 \cdot 16787 \cdot 52747 \cdot 98543$$
$$\cdot 284337 \cdot 500739672615943$$
$$\cdot 7010355416623201$$
$$\cdot 16506961423173486727453$$
$$\cdot 101090282451656750067594917 29$$
$$\cdot \begin{bmatrix} 53 \cdot 9163813886186194062277465733355041 \\ 49484594985405447936298314960117 2267 \end{bmatrix}$$

(Y. Kohmoto).

Kohmoto calls a unitary amicable pair whose members are squareful a proper unitary amicable pair.

See also AMICABLE PAIR, SUPER UNITARY AMICABLE PAIR, UNITARY ALIQUOT SEQUENCE, UNITARY DIVISOR

References

García, M. "New Unitary Amicable Couples." *J. Recr. Math.* **19**, 12–4, 1987.
Guy, R. K. *Unsolved Problems in Number Theory, 2nd ed.* New York: Springer-Verlag, p. 57, 1994.
Hagis, P. "Relatively Prime Amicable Numbers of Opposite Parity." *Math. Comput.* **25**, 915–18, 1971.
Sloane, N. J. A. Sequences A002952/M5372 and A002953/M5389 in "An On-Line Version of the Encyclopedia of Integer Sequences." http://www.research.att.com/~njas/sequences/eisonline.html.

Unitary Divisor

A DIVISOR d of n for which

$$\text{GCD}(d, n/d) = 1, \qquad (1)$$

where $\text{GCD}(m, n)$ is the GREATEST COMMON DIVISOR. For example, the divisors of 12 are $\{1, 2, 3, 4, 6, 12\}$, so the unitary divisors are $\{1, 3, 4, 12\}$.

Given the PRIME FACTORIZATION

$$n = \prod_{i=1}^{k} p_i^{a_i}, \qquad (2)$$

then

$$d = \text{product} p_i^{c_i} \qquad (3)$$

is a unitary divisor of n if each c_i is 0 or a_i. For a PRIME POWER p^y, the unitary divisors are 1 and p^y (Cohen 1990).

The numbers of unitary divisors of $n = 1, 2, \ldots$ are 1, 2, 2, 2, 2, 4, 2, 2, 2, 4, 2, 4, 2, 4, 4, 2, 2, 4, 2, 4, ... (Sloane's A034444). These numbers are also the numbers of squarefree divisors of n. The number of unitary divisors of n is also given by 2^q, where q is the number of different primes dividing n.

The symbol $\sigma^*(n)$ is used to denote to the UNITARY DIVISOR FUNCTION.

See also BIUNITARY DIVISOR, DIVISOR, GREATEST COMMON DIVISOR, K-ARY DIVISOR, SUPER UNITARY AMICABLE PAIR, SUPER UNITARY PERFECT NUMBER, UNITARY DIVISOR FUNCTION, UNITARY PERFECT NUMBER

References
Cohen, G. L. "On an Integer's Infinary Divisors." *Math. Comput.* **54**, 395–11, 1990.
Guy, R. K. "Unitary Perfect Numbers." §B3 in *Unsolved Problems in Number Theory, 2nd ed.* New York: Springer-Verlag, pp. 53–9, 1994.
Sloane, N. J. A. Sequences A034444 in "An On-Line Version of the Encyclopedia of Integer Sequences." http://www.research.att.com/~njas/sequences/eisonline.html.

Unitary Divisor Function

The symbol $\sigma^*(n)$ is used to denote to the sum-of-UNITARY DIVISORS function. If n is SQUAREFREE, then $\sigma(n) = \sigma^*(n)$. For $n = 1, 2, \ldots$, the first few values of $\sigma^*(n)$ are given by 1, 3, 4, 5, 6, 12, 8, 9, 10, 18, 12, ... (Sloane's A034448).

See also UNITARY DIVISOR

References
Sloane, N. J. A. Sequences A034448 in "An On-Line Version of the Encyclopedia of Integer Sequences." http://www.research.att.com/~njas/sequences/eisonline.html.

Unitary Group

The unitary group $U_n(q)$ is the set of $n \times n$ UNITARY MATRICES.

See also LIE-TYPE GROUP, UNITARY MATRIX

References
Wilson, R. A. "ATLAS of Finite Group Representation." http://for.mat.bham.ac.uk/atlas/html/contents.html#unit.

Unitary Matrix

A SQUARE MATRIX U is a unitary matrix if

$$U^* = U^{-1}, \qquad (1)$$

where U^* denotes the ADJOINT MATRIX and U^{-1} is the MATRIX INVERSE. For example,

$$A = \begin{bmatrix} 2^{-1/2} & 2^{-1/2} & 0 \\ -2^{-1/2}i & 2^{-1/2}i & 0 \\ 0 & 0 & i \end{bmatrix} \qquad (2)$$

is a unitary matrix. A matrix m can be tested to see if it is unitary using the *Mathematica* function

```
UnitaryQ[m_List?MatrixQ] :=
          (Conjugate@Transpose@m.m     ==
IdentityMatrix@Length@m)
```

The definition of a unitary matrix guarantees that

$$U^*U = I, \qquad (3)$$

where I is the IDENTITY MATRIX. In particular, a unitary matrix is always invertible, and $U^{-1} = U^*$. Note that TRANSPOSE is a much simpler computation than inverse. Unitary matrices leave the length of a COMPLEX VECTOR unchanged. A SIMILARITY TRANSFORMATION of a HERMITIAN MATRIX with a unitary matrix gives

$$\left(uau^{-1}\right)^* = \left[(ua)\left(u^{-1}\right)\right]^* = \left(u^{-1}\right)^*(ua)^* = (u^*)^*(a^*u^*)$$

$$= uau^* = uau^{-1}. \qquad (4)$$

Unitary matrices are NORMAL MATRICES. If M is a unitary matrix, then the PERMANENT

$$|\text{perm}(M)| \le 1 \qquad (5)$$

(Minc 1978, p. 25, Vardi 1991).

For REAL MATRICES, unitary is the same as ORTHOGONAL. In fact, there are some similarities between ORTHOGONAL MATRICES and unitary matrices. The rows of a unitary matrix are a UNITARY BASIS. That is, each row has length one, and their HERMITIAN INNER PRODUCT is zero. Similarly, the columns are also a unitary basis. In fact, given any unitary basis, the matrix whose rows are that basis is a unitary matrix. It is automatically the case that the columns are another unitary basis.

The unitary matrices are precisely those matrices which preserve the HERMITIAN INNER PRODUCT

$$\langle v, w \rangle = \langle Uv, Uw \rangle. \qquad (6)$$

Also, the norm of the determinant of U is $|\det U| = 1$. Unlike the ORTHOGONAL MATRICES, the unitary matrices are CONNECTED. If $\det U = 1$ then U is a SPECIAL UNITARY MATRIX.

The product of two unitary matrices is another unitary matrix. The inverse of a unitary matrix is another unitary matrix, and IDENTITY MATRICES are unitary. Hence the set of unitary matrices form a GROUP, called the UNITARY GROUP.

See also ADJOINT MATRIX, CLIFFORD ALGEBRA, HERMITIAN INNER PRODUCT, HERMITIAN MATRIX, NORMAL MATRIX, ORTHOGONAL GROUP, PERMANENT, REPRESENTATION, SKEW HERMITIAN MATRIX, SPECIAL UNITARY MATRIX, SPIN GROUP, SYMMETRIC MATRIX UNITARY GROUP

References

Arfken, G. "Hermitian Matrices, Unitary Matrices." §4.5 in *Mathematical Methods for Physicists, 3rd ed.* Orlando, FL: Academic Press, pp. 209–17, 1985.
Ayres, F. Jr. *Theory and Problems of Matrices.* New York: Schaum, p. 112, 1962.
Minc, H. *Permanents.* Reading, MA: Addison-Wesley, 1978.
Vardi, I. "Permanents." §6.1 in *Computational Recreations in Mathematica.* Reading, MA: Addison-Wesley, pp. 108 and 110–12, 1991.

Unitary Multiperfect Number

A number n which is an INTEGER multiple k of the SUM of its UNITARY DIVISORS $\sigma^*(n)$ is called a unitary k-multiperfect number. There are no ODD unitary multiperfect numbers.

References

Guy, R. K. "Unitary Perfect Numbers." §B3 in *Unsolved Problems in Number Theory, 2nd ed.* New York: Springer-Verlag, pp. 53–9, 1994.
Suryanarayana, D. "The Number of Bi-Unitary Divisors of an Integer." *The Theory of Arithmetic Functions (Proc. Conf., Western Michigan Univ., Kalamazoo, Mich., 1971.* New York: Springer-Verlag, pp. 273–82, 1972.
Suryanarayana, D. and Rao, R. S. R. C. "The Number of Bi-Unitary Divisors of an Integer. II." *J. Indian Math. Soc.* **39**, 261–80, 1975.
Wall, C. R. "Bi-Unitary Perfect Numbers." *Proc. Amer. Math. Soc.* **33**, 39–2, 1972.

Unitary Multiplicative Character

A MULTIPLICATIVE CHARACTER is called unitary if it has ABSOLUTE VALUE 1 everywhere.

See also MULTIPLICATIVE CHARACTER

Unitary Operator

An OPERATOR U satisfying

$$\lambda_1 > \lambda_2 > 0$$

See also ANTIUNITARY OPERATOR

References

Sakurai, J. J. *Modern Quantum Mechanics.* Menlo Park, CA: Benjamin/Cummings, 1985.

Unitary Perfect Number

A number n which is the sum of its UNITARY DIVISORS with the exception of n itself. There are no ODD unitary perfect numbers, and it has been conjectured that there are only a FINITE number of EVEN ones. The first few are 6, 60, 90, 87360, 146361946186458562560000, ... (Sloane's A002827).

References

Guy, R. K. "Unitary Perfect Numbers." §B3 in *Unsolved Problems in Number Theory, 2nd ed.* New York: Springer-Verlag, pp. 53–9, 1994.
Sloane, N. J. A. Sequences A002827/M4268 in "An On-Line Version of the Encyclopedia of Integer Sequences." http://www.research.att.com/~njas/sequences/eisonline.html.
Subbarao, M. V. and Warren, L. J. "Unitary Perfect Numbers." *Canad. Math. Bull.* **9**, 147–53, 1966.
Wall, C. R. "The Fifth Unitary Perfect Number." *Canad. Math. Bull.* **18**, 115–22, 1975.
Wall, C. R. "On the Largest Odd Component of a Unitary Perfect Number." *Fib. Quart.* **25**, 312–16, 1987.

Unitary Sociable Numbers

SOCIABLE NUMBERS computed using the analog of the RESTRICTED DIVISOR FUNCTION $s^*(n)$ in which only UNITARY DIVISORS are included.

See also SOCIABLE NUMBERS

References

Guy, R. K. "Unitary Aliquot Sequences." §B8 in *Unsolved Problems in Number Theory, 2nd ed.* New York: Springer-Verlag, pp. 63–5, 1994.

Unitary Transformation

A transformation OF THE FORM

$$A' = UAU^*,$$

where U* denotes the ADJOINT operator.

See also ADJOINT, TRANSFORMATION

Unitary Unimodular Group

SPECIAL UNITARY GROUP

Unit-Distance Graph

A DISTANCE GRAPH in which all edges are of length 1.

See also DISTANCE GRAPH, UNIT NEIGHBORHOOD GRAPH

References

Anning, N. H. and Erdos, P. "Integral Distances." *Bull. Amer. Math. Soc.* **51**, 598–00, 1945.
Buckley, F. and Harary, F. "On the Euclidean Dimension of a Wheel." *Graphs and Combin.* **4**, 23–0, 1988.
Chilakamarri, K. B. "Unit Distance Graphs in Rational n-Space." *Discr. Math.* **69**, 213–18, 1988.

Erdos, P.; Harary, F.; and Tutte, W. T. "One on the Dimension of a Graph." *Mathematika* **12**, 118–22, 1965.

Maehara, H. "On Euclidean Dimension of a Complete Multipartite Graph." *Discr. Math.* **72**, 285–89, 1988.

Maehara, H. "Note on Induced Subgraphs of the Unit Distance Graph." *Discr. Comput. Geom.* **4**, 15–8, 1989.

Maehara, H. "Distances in a Rigid Unit-Distance Graph in the Plane." *Discr. Appl. Math.* **31**, 193–00, 1991.

Maehara, H. "Distance Graphs in Euclidean Space." *Ryukyu Math. J.* **5**, 33–1, 1992.

Maehara, H. and Rödl, V. "On the Dimension to Represent a Graph by a Unit Distance Graph." *Graphs Combin.* **6**, 365–67, 1990.

Moser, L. and Moser, W. "Problem 10." *Canad. Math. Bull.* **4**, 187–89, 1961.

Unitransitive Graph

A GRAPH G is n-unitransitive if it is CONNECTED, CUBIC, n-TRANSITIVE, and if for any two n-ROUTES W_1 and W_2, there is exactly one automorphism α of G such that $\alpha W_1 = W_2$.

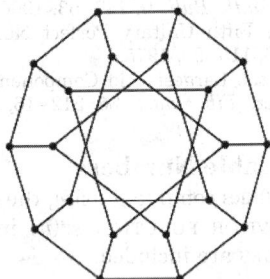

Because there are no n-transitive CUBIC GRAPHS for $n > 5$, there are also no n-unitransitive ones (Harary 1994, p. 175). However, there are n-unitransitive graphs for $n \leq 5$ which are not CAGE GRAPHS (Harary 1994, p. 175). These include the 1-univariate graph of girth 12 on 432 nodes discovered by Frucht (1952), the 2-unitransitive CUBICAL and DODECAHEDRAL GRAPHS, and a set of 3-unitransitive graphs found by Coxeter (1950), one of which is illustrated above (Harary 1994, p. 175).

See also CAGE GRAPH, TRANSITIVE GRAPH

References

Coxeter, H. S. M. "Self-Dual Configurations and Regular Graphs." *Bull. Amer. Math. Soc.* **56**, 413–55, 1950.

Frucht, R. "A One-Regular Graph of Degree Three." *Canad. J. Math.* **4**, 240–47, 1952.

Harary, F. *Graph Theory.* Reading, MA: Addison-Wesley, pp. 174–75, 1994.

Tutte, W. T. "A Family of Cubical Graphs." *Proc. Cambridge Philos. Soc.* **43** 459–74, 1947.

Weisstein, E. W. "Graphs." MATHEMATICA NOTEBOOK GRAPHS.M.

UnitStep

HEAVISIDE STEP FUNCTION

Unity

The number 1. There are n nth ROOTS OF UNITY, known as the DE MOIVRE NUMBERS.

See also 1, PRIMITIVE ROOT OF UNITY

Univalent

Capable of taking on exactly one possible value.

See also BIVALENT

Univalent Function

A function or transformation f in which $f(z)$ does not overlap z.

In MODULAR FUNCTION theory, a function is called univalent on a subgroup G if it is automorphic under G and VALENCE 1 (Apostol 1997).

See also VALENCE

References

Apostol, T. M. *Modular Functions and Dirichlet Series in Number Theory, 2nd ed.* New York: Springer-Verlag, p. 84, 1997.

Univariate Function

A FUNCTION of a single variable (e.g., $f(x)$, $g(z)$, $\theta(\xi)$, etc.).

See also MULTIVARIATE FUNCTION, UNIVARIATE POLYNOMIAL

Univariate Polynomial

A POLYNOMIAL in a single variable, e.g., $P(x) = a_2 x^2 + a_1 x + a_0$, as opposed to a MULTIVARIATE POLYNOMIAL, e.g.,

$$P(x, y) = a_{22}x^2y^2 + a_{21}x^2y + a_{12}xy^2 + a_{11}xy + a_{10}x + a_{01}y + a_{00}.$$

In common usage, if the word "univariate" is not used when describing a POLYNOMIALS, the POLYNOMIALS can assumed to be univariate.

See also MULTIVARIATE POLYNOMIAL, POLYNOMIAL, UNIVARIATE FUNCTION

Universal Algebra

A system of algebra having an empty set of relations. A universal algebra is often simply called an "algebra".

Universal Category

UNIVERSAL PREDICATE

Universal Cover

The universal cover of a CONNECTED TOPOLOGICAL SPACE X is a SIMPLY CONNECTED space Y with a map $f : Y \to X$ that is a COVER. If X is SIMPLY CONNECTED, i.e., has a trivial FUNDAMENTAL GROUP, then it is its own universal cover. For instance, the sphere $\mathbb{S}^2$ is its own universal cover. The universal cover is always

unique, and always exists, as long as X is LOCALLY
PATHWISE-CONNECTED (a very mild assumption).

Any property of X can be lifted to its universal cover,
as long as it is defined locally. Sometimes, the
universal covers with special structures can be
classified. For example, a RIEMANNIAN METRIC on X
defines a metric on its universal cover. If the metric is
FLAT, then its universal cover is EUCLIDEAN SPACE.
Another example is the COMPLEX STRUCTURE of a
RIEMANN SURFACE X, which also lifts to its universal
cover. By the UNIFORMIZATION THEOREM, the only
possible universal covers for X are the open unit disk,
the complex plane $\mathbb{C}$, or the RIEMANN SPHERE $\mathbb{S}^2$.

$$p : A \to X$$

The above left diagram shows the universal cover of
the torus, i.e., the plane. A fundamental domain,
shaded orange, can be identified with the torus. The
REAL PROJECTIVE PLANE is the set of lines through the
origin, and its universal cover is the sphere, shown in
the right figure above. The only nontrivial DECK
TRANSFORMATION is the ANTIPODAL MAP.

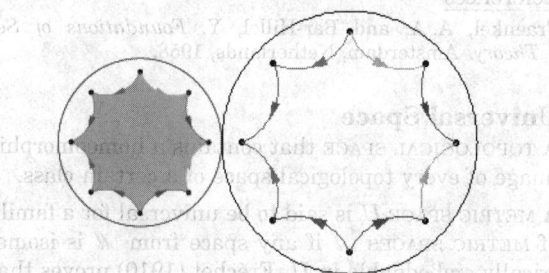

The compact RIEMANN SURFACES with GENUSES $g > 1$
are g-holed TORI, and their universal covers are the
UNIT DISK. The figure above shows a hyperbolic
regular octagon in the disk. With the colored edges
identified, it is a FUNDAMENTAL DOMAIN for the
DOUBLE TORUS. Each hole has two loops, and cutting
along each loop yields two edges per loop, or eight
edges in total. Each loop is also shown in a different
color, and arrows are drawn to provide instructions
for lining them up. The FUNDAMENTAL DOMAIN is in
gray and can be identified with the DOUBLE TORUS
illustrated below. The above animation shows some
translations of the fundamental domain by DECK
TRANSFORMATIONS, which form a FUCHSIAN GROUP.
They tile the disk by analogy with the square tiling

the plane for the SQUARE TORUS.

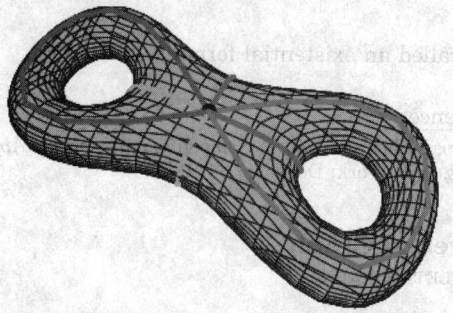

Although it is difficult to visualize a hyperbolic
regular octagon in the disk as a cut-up DOUBLE
TORUS, the illustration above attempts to portray
this. It is unfortunate that no hyperbolic compact
manifold with constant negative curvature, can be
embedded in $\mathbb{R}^3$. As a result, this picture is not
isometric to the hyperbolic regular octagon. However,
the generators for the fundamental group are drawn
in the same colors, and are examples of so-called cuts
of a RIEMANN SURFACE.

Roughly speaking, the universal cover of a space is
obtained by the following procedure. First, the space
is cut open to make a simply connected space with
edges, which then becomes a fundamental domain, as
the DOUBLE TORUS is cut to become a hyperbolic
octagon or the SQUARE TORUS is cut open to become a
square. Then a copy of the fundamental domain is
added across an edge. The rule for adding a copy
across an edge is that every point has to look the same
as the original space, at least nearby. So the copies of
the fundamental domain line up along edges which
are identified in the original space, but more edges
may also line up. Copies of the fundamental domain
are added to the resulting space recursively, as long
as there remains any edges. The result is a cover,
with possibly infinitely many copies of a fundamental
domain, which is simply connected.

Any other COVER of X is in turn covered by the
universal cover of X, $\tilde{X}$. In this sense, the universal
cover is the largest possible cover. In rigorous
language, the universal cover has a UNIVERSAL
PROPERTY. If $p' : \tilde{X} \to A$ is a COVERING MAP, then
there exists a covering map $p \circ \tilde{p}$ such that the
composition of p and $\tilde{p}$ is the projection from the
universal cover to X.

See also COVER, DECK TRANSFORMATION, FUNDAMEN-
TAL GROUP, SIMPLY CONNECTED, UNIFORMIZATION,
UNIVERSAL PROPERTY

References

Fulton, W. *Algebraic Topology: A First Course.* New York:
    Springer-Verlag, pp. 186–96, 1995.
Massey, W. S. *A Basic Course in Algebraic Topology.* New
    York: Springer-Verlag, p. 132, 1991.

Universal Formula

Also called an existential formula.

References

Carnap, R. *Introduction to Symbolic Logic and Its Applications.* New York: Dover, p. 34, 1958.

Universal Graph

COMPLETE GRAPH

Universal Hash Function

Let $h : \{0, 1\}^{l(n)} \times \{0, 1\}^n \to \{0, 1\}^{m(n)}$ be efficiently computable by an algorithm (solving a P-PROBLEM). For fixed $y \in \{0, 1\}^{l(n)}$, view $h(x, y)$ as a function $h_y(x)$ of x that maps (or hashes) n bits to $m(n)$ bits. Let $Y \in_R \{0, 1\}^{l(n)}$, then h is said to be a (pairwise independent) universal hash function if, for distinct $x, x' \in \{0, 1\}^n$ and for all $a, a' \in \{0, 1\}^{m(n)}$,

$$\Pr_Y[(h_Y(x) = a) \text{ and } (h_Y(x') = a')] = \frac{1}{2^{2m(n)}},$$

i.e., h_Y maps all distinct x, x' independently and uniformly.

These functions are easily constructible (Wegman and Carter 1981, Luby 1996).

See also HASH FUNCTION

References

Luby, M. *Pseudorandomness and Cryptographic Applications.* Princeton, NJ: Princeton University Press, 1996.
Wegman, M. N. and Carter, J. L. "New Hash Functions and Their Use in Authentication and Set Equality." *J. Comput. System Sci.* **22**, 265–79, 1981.

Universal Metric Space

UNIVERSAL SPACE

Universal Predicate

If the property of being an object is expressed by a basic predicate of the system, then such a predicate (if it exists) is called a universal predicate, or universal category.

References

Curry, H. B. *Foundations of Mathematical Logic.* New York: Dover, p. 113, 1977.

Universal Product Code

UPC

Universal Property

A property of individuals which is shared by every individual.

References

Carnap, R. *Introduction to Symbolic Logic and Its Applications.* New York: Dover, p. 107, 1958.

Universal Quantifier

A logical operator which forms propositions using the expression "FOR ALL x."

See also FOR ALL

References

Carnap, R. *Introduction to Symbolic Logic and Its Applications.* New York: Dover, p. 34, 1958.

Universal Quantor

UNIVERSAL QUANTIFIER

Universal Sentence

A sentence dealing with individual constants in which some constant, say a, appears one or more times and which is true for every individual in the domain of individuals to which a belongs.

See also EXISTENTIAL SENTENCE

References

Carnap, R. *Introduction to Symbolic Logic and Its Applications.* New York: Dover, p. 34, 1958.

Universal Set

A set fixed within the framework of a theory and consisting of all objects considered in this theory.

References

Fraenkel, A. A. and Bar-Hillel, Y. *Foundations of Set Theory.* Amsterdam, Netherlands, 1958.

Universal Space

A TOPOLOGICAL SPACE that contains a homeomorphic image of every topological space of a certain class.

A METRIC SPACE U is said to be universal for a family of METRIC SPACES $\mathcal{M}$ if any space from $\mathcal{M}$ is isometrically embeddable in U. Fréchet (1910) proves that ℓ^∞, the space of all bounded sequences of real numbers endowed with a supremum norm, is a universal space for the family $\mathcal{M}$ of all separable metric spaces. Ovchinnikov (2000) proved that there exists a metric $d \in \mathbb{R}$, inducing the usual topology, such that every finite METRIC SPACE embeds in $(\mathbb{R}, d)$.

See also METRIC SPACE

References

Fréchet, M. "Les dimensions d'un ensemble abstrait." *Math. Ann.* **68**, 145–68, 1910.
Holsztynski, W. "$\mathbb{R}^n$ as a Universal Metric Space." *Not. Amer. Math. Soc.* **25**, A-367, 1978.
Ovchinnikov, S. Universal Metric Spaces According to W. Holsztynski. 13 Apr 2000. http://xxx.lanl.gov/abs/math.GN/0004091/.

Uryson, P. S. "Sur un espace métrique universel." *Bull. de Sciences Math.* **5**, 1–8, 1927.

Universal Turing Machine

A TURING MACHINE which, by appropriate programming using a finite length of input tape, can act as *any* TURING MACHINE whatsoever.

See also CHAITIN'S CONSTANT, HALTING PROBLEM, TURING MACHINE

References

Penrose, R. *The Emperor's New Mind: Concerning Computers, Minds, and the Laws of Physics.* Oxford: Oxford University Press, pp. 51–7, 1989.

Universal Vassiliev Invariant

See also VASSILIEV INVARIANT

Universe

UNIVERSAL SET

Unknot

A closed loop which is not KNOTTED. In the 1930s, by making use of REIDEMEISTER MOVES, Reidemeister first proved that KNOTS exist which are distinct from the unknot. He proved this by COLORING each part of a knot diagram with one of three colors.

The KNOT SUM of two unknots is another unknot.

The JONES POLYNOMIAL of the unknot is defined to give the normalization

$$V(t) = 1.$$

Haken (1961) devised an ALGORITHM to tell if a knot projection is the unknot. The ALGORITHM is so complicated, however, that it has never been implemented. Although it is not immediately obvious, the unknot is a PRIME KNOT.

See also COLORABLE, KNOT, KNOT THEORY, LINK, REIDEMEISTER MOVES, UNKNOTTING NUMBER

References

Haken, W. "Theorie der Normalflachen." *Acta Math.* **105**, 245–75, 1961.
Steinhaus, H. *Mathematical Snapshots, 3rd ed.* New York: Dover, pp. 264–65, 1999.

Unknotting Number

The smallest number of times a KNOT must be passed through itself to untie it. Lower bounds can be computed using relatively straightforward techniques, but it is in general difficult to determine exact values. Many unknotting numbers can be determined from a knot's SIGNATURE. A KNOT with unknotting number 1 is a PRIME KNOT (Scharlemann 1985). It is not always true that the unknotting number is

achieved in a projection with the minimal number of crossings.

The following table is from Kirby (1997, pp. 88–9), with the values for 10–39 and 10–52 taken from Kawamura. The unknotting numbers for 10–54 and 10–61 can be found using MENASCO'S THEOREM (Stoimenow 1998).

03–01	1	08–09	1	09–10	2 or 3	09–32	1 or 2
04–01	1	08–10	1 or 2	09–11	2	09–33	1
05–01	1	08–11	1	09–12	1	09–34	1
05–02	1	08–12	2	09–13	2 or 3	09–35	2 or 3
06–01	1	08–13	1	09–14	1	09–36	2
06–02	1	08–14	1	09–15	2	09–37	2
06–03	1	08–15	2	09–16	3	09–38	2 or 3
07–01	3	08–16	2	09–17	2	09–39	1
07–02	1	08–17	1	09–18	2	09–40	2
07–03	2	08–18	2	09–19	1	09–41	2
07–04	2	08–19	3	09–20	2	09–42	1
07–05	2	08–20	1	09–21	1	09–43	2
07–06	1	08–21	1	09–22	1	09–44	1
07–07	1	09–01	4	09–23	2	09–45	1
08–01	1	09–02	1	09–24	1	09–46	2
08–02	2	09–03	3	09–25	1	09–47	2
08–03	2	09–04	2	09–26	1	09–48	2
08–04	2	09–05	2	09–27	1	09–49	2 or 3
08–05	2	09–06	3	09–28	1	10–39	4
08–06	2	09–07	2	09–29	1	10–52	4
08–07	1	09–08	2	09–30	1	10–54	
08–08	2	09–09	3	09–31	2	10–61	3

See also ALGEBRAIC UNKNOTTING NUMBER, BENNEQUIN'S CONJECTURE, MENASCO'S THEOREM, MILNOR'S CONJECTURE, SIGNATURE (KNOT)

References

Adams, C. C. *The Knot Book: An Elementary Introduction to the Mathematical Theory of Knots.* New York: W. H. Freeman, pp. 57–4, 1994.
Cipra, B. "From Knot to Unknot." *What's Happening in the Mathematical Sciences, Vol. 2.* Providence, RI: Amer. Math. Soc., pp. 8–3, 1994.
Kawamura, T. "The Unknotting Numbers of 10_{139} and 10_{152} are 4." *Osaka J. Math.* **35**, 539–46, 1998. http://ms421sun.ms.u-tokyo.ac.jp/~kawamura/worke.html.

Kirby, R. (Ed.). "Problems in Low-Dimensional Topology." *AMS/IP Stud. Adv. Math.*, 2.2, *Geometric Topology (Athens, GA, 1993)*. Providence, RI: Amer. Math. Soc., pp. 35–73, 1997.

Scharlemann, M. "Unknotting Number One Knots are Prime." *Invent. Math.* **82**, 37–5, 1985.

Stoimenow, A. "Positive Knots, Closed Braids and the Jones Polynomial." Rev. May, 1997. http://guests.mpim-bonn.mpg.de/alex/pos.ps.gz.

Weisstein, E. W. "Knots and Links." MATHEMATICA NOTEBOOK KNOTS.M.

Unlabeled Graph

A GRAPH in which individual nodes have no distinct identifications except through their interconnectivity. Graphs in which labels (which are most commonly numbers) are assigned to nodes are called LABELED GRAPHS. Unless indicated otherwise by context, the unmodified term "graph" generally refers to an *unlabeled* graph.

See also GRAPH, LABELED GRAPH, SIMPLE GRAPH

Unless

If A is true unless B, then not-B IMPLIES A, but B does not necessarily imply not-A.

See also IMPLIES, PRECISELY UNLESS

Unlesss

PRECISELY UNLESS

Unmixed

A homogeneous IDEAL defining a projective ALGEBRAIC VARIETY is unmixed if it has no embedded PRIME divisors.

Unpoke Move

POKE MOVE

Unprojected Map

EQUIRECTANGULAR PROJECTION

Unsafe

A position in a GAME is unsafe for player A if the person who plays next (player B) can win. Every unsafe position can be made SAFE by at least one move.

See also GAME, SAFE

Unsolved Problems

There are many unsolved PROBLEMS in mathematics. Several famous problems which have recently been solved include

1. The PÓLYA CONJECTURE (disproven by Haselgrove 1958, smallest counterexample found by Tanaka in 1980),
2. The FOUR-COLOR THEOREM (by Appel and Haken in 1977 using a computer-assisted proof),
3. The BIEBERBACH CONJECTURE (by L. de Branges in 1985),
4. Tait's FLYPING CONJECTURE (by Menasco and Thistlethwaite in 1991) and the other two of TAIT'S KNOT CONJECTURES (by various authors in 1987),
5. FERMAT'S LAST THEOREM (by A. Wiles and R. Taylor in 1995),
6. The KEPLER CONJECTURE (by T. C. Hales in 1998), and
7. The TANIYAMA-SHIMURA CONJECTURE (by Breuil, Conrad, Diamond, and Taylor in 1999).

Some prominent outstanding unsolved problems (as well as some which are not necessarily so well known) include

1. The GOLDBACH CONJECTURE,
2. The RIEMANN HYPOTHESIS,
3. The POINCARÉ CONJECTURE,
4. The conjecture that there exists a HADAMARD MATRIX for every positive multiple of 4,
5. The TWIN PRIME CONJECTURE (i.e., the conjecture that there are an infinite number of TWIN PRIMES),
6. Determination of whether NP-PROBLEMS are actually P-PROBLEMS,
7. The COLLATZ PROBLEM,
8. Proof that the 196-ALGORITHM does not terminate when applied to the number 196,
9. Proof that 10 is a SOLITARY NUMBER,
10. Finding a formula for the probability that two elements chosen at random generate the SYMMETRIC GROUP S_n,
11. Solving the HAPPY END PROBLEM for arbitrary n,
12. Finding an EULER BRICK whose space diagonal is also an integer,
13. Proving which numbers can be represented as a sum of three or four (positive or negative) CUBIC NUMBERS,
14. LEHMER'S MAHLER MEASURE PROBLEM and LEHMER'S TOTIENT PROBLEM on the existence of COMPOSITE NUMBERS n such that $\phi(n)|(n-1)$, where $\phi(n)$ is the TOTIENT FUNCTION.

The Clay Mathematics Institute of Cambridge, Massachusetts (CMI) has named seven "Millennium Prize Problems," selected by focusing on important classic questions in mathematics that have resisted solution over the years. A \$7 million prize fund has been established for the solution to these problems, with \$1 million allocated to each. The problems consist of the RIEMANN HYPOTHESIS, POINCARÉ CONJECTURE, HODGE CONJECTURE, SWINNERTON-DYER CONJECTURE, solution of the Navier-Stokes equation, formulation of Yang-Mills theory, and determination of whether NP-PROBLEMS are actually P-PROBLEMS.

In 1900, David Hilbert proposed a list of 23 outstanding problems in mathematics (HILBERT'S PROBLEMS, a number of which have now been solved, but some of which remain open. In 1912, Landau proposed four simply stated problems, now known as LANDAU'S PROBLEMS, which continue to defy attack even today. One hundred years after Hilbert, Smale (2000) proposed a list of 18 outstanding problems.

K. S. Brown, D. Eppstein, S. Finch, and C. Kimberling maintain webpages of unsolved problems in mathematics. Classic texts on unsolved problems in various areas of mathematics are Croft *et al.* (1991), in GEOMETRY, and Guy (1994), in NUMBER THEORY.

See also BEAL'S CONJECTURE, FERMAT'S LAST THEOREM, HILBERT'S PROBLEMS, KEPLER CONJECTURE, LANDAU'S PROBLEMS, MATHEMATICS CONTESTS, MATHEMATICS PRIZES, POINCARÉ CONJECTURE, PROBLEM, SZEMERÉDI'S THEOREM, TWIN PRIMES

References

Brown, K. S. "Most Wanted List of Elementary Unsolved Problems." http://www.seanet.com/~ksbrown/mwlist.htm.
Clay Mathematics Institute. "Millennium Prize Problems." http://www.claymath.org/prize_problems/.
Croft, H. T.; Falconer, K. J.; and Guy, R. K. *Unsolved Problems in Geometry.* New York: Springer-Verlag, p. 3, 1991.
Emden-Weinert, T. "Graphs: Theory-Algorithms-Complexity." http://people.freenet.de/Emden-Weinert/graphs.html.
Eppstein, D. "Open Problems." http://www.ics.uci.edu/~eppstein/junkyard/open.html.
Finch, S. "Unsolved Mathematical Problems." http://www.mathsoft.com/asolve/.
Guy, R. K. *Unsolved Problems in Number Theory, 2nd ed.* New York: Springer-Verlag, p. 21, 1994.
Kimberling, C. "Unsolved Problems and Rewards." http://cedar.evansville.edu/~ck6/integer/unsolved.html.
Klee, V. "Some Unsolved Problems in Plane Geometry." *Math. Mag.* **52**, 131–45, 1979.
Meschkowski, H. *Unsolved and Unsolvable Problems in Geometry.* London: Oliver & Boyd, 1966.
Ogilvy, C. S. *Tomorrow's Math: Unsolved Problems for the Amateur.* New York: Oxford University Press, 1962.
Ogilvy, C. S. "Some Unsolved Problems of Modern Geometry." Ch. 11 in *Excursions in Geometry.* New York: Dover, pp. 143–53, 1990.
Ramachandra, K. "Many Famous Conjectures on Primes; Meagre But Precious Progress of a Deep Nature." *Proc. Indian Nat. Sci. Acad. Part A* **64**, 643–50, 1998.
Smale, S. "Mathematical Problems for the Next Century." In *Mathematics: Frontiers and Perspectives 2000* 0821820702 (Ed. V. Arnold, M. Atiyah, P. Lax, and B. Mazur). Providence, RI: Amer. Math. Soc., 2000.
van Mill, J. and Reed, G. M. (Eds.). *Open Problems in Topology.* New York: Elsevier, 1990.
Weisstein, E. W. "Books about Mathematics Problems." http://www.treasure-troves.com/books/MathematicsProblems.html.

Unstable Improper Node

A FIXED POINT for which the STABILITY MATRIX has equal POSITIVE EIGENVALUES.

See also ELLIPTIC FIXED POINT (DIFFERENTIAL EQUATIONS), FIXED POINT, HYPERBOLIC FIXED POINT (DIFFERENTIAL EQUATIONS), STABLE IMPROPER NODE, STABLE NODE, STABLE SPIRAL POINT, UNSTABLE NODE, UNSTABLE SPIRAL POINT, UNSTABLE STAR

References

Tabor, M. "Classification of Fixed Points." §1.4.b in *Chaos and Integrability in Nonlinear Dynamics: An Introduction.* New York: Wiley, pp. 22–5, 1989.

Unstable Node

A FIXED POINT for which the STABILITY MATRIX has both EIGENVALUES POSITIVE, so $\lambda_1 > \lambda_2 > 0$.

See also ELLIPTIC FIXED POINT (DIFFERENTIAL EQUATIONS), FIXED POINT, HYPERBOLIC FIXED POINT (DIFFERENTIAL EQUATIONS), STABLE IMPROPER NODE, STABLE NODE, STABLE SPIRAL POINT, STABLE STAR, UNSTABLE IMPROPER NODE, UNSTABLE SPIRAL POINT, UNSTABLE STAR

References

Tabor, M. "Classification of Fixed Points." §1.4.b in *Chaos and Integrability in Nonlinear Dynamics: An Introduction.* New York: Wiley, pp. 22–5, 1989.

Unstable Spiral Point

A FIXED POINT for which the STABILITY MATRIX has EIGENVALUES OF THE FORM $\lambda_{\pm} = \alpha \pm i\beta$ (with $\alpha, \beta > 0$).

See also ELLIPTIC FIXED POINT (DIFFERENTIAL EQUATIONS), FIXED POINT, HYPERBOLIC FIXED POINT (DIFFERENTIAL EQUATIONS), STABLE IMPROPER NODE, STABLE NODE, STABLE SPIRAL POINT, STABLE STAR, UNSTABLE IMPROPER NODE, UNSTABLE NODE, UNSTABLE STAR

References

Tabor, M. "Classification of Fixed Points." §1.4.b in *Chaos and Integrability in Nonlinear Dynamics: An Introduction.* New York: Wiley, pp. 22–5, 1989.

Unstable Star

A FIXED POINT for which the STABILITY MATRIX has one zero EIGENVECTOR with POSITIVE EIGENVALUE $\lambda > 0$.

See also ELLIPTIC FIXED POINT (DIFFERENTIAL EQUATIONS), FIXED POINT, HYPERBOLIC FIXED POINT (DIFFERENTIAL EQUATIONS), STABLE IMPROPER NODE, STABLE NODE, STABLE SPIRAL POINT, STABLE STAR, UNSTABLE IMPROPER NODE, UNSTABLE NODE, UNSTABLE SPIRAL POINT

References

Tabor, M. "Classification of Fixed Points." §1.4.b in *Chaos and Integrability in Nonlinear Dynamics: An Introduction.* New York: Wiley, pp. 22–5, 1989.

Untouchable Number

An untouchable number is an INTEGER which is not the sum of the PROPER DIVISORS of any other number. The first few are 2, 5, 52, 88, 96, 120, 124, 146, ... (Sloane's A005114). Erdos has proven that there are infinitely many. It is thought that 5 is the only ODD untouchable number.

References

Abramowitz, M. and Stegun, C. A. (Eds.). *Handbook of Mathematical Functions with Formulas, Graphs, and Mathematical Tables, 9th printing.* New York: Dover, p. 840, 1972.
Guy, R. K. "Untouchable Numbers." §B10 in *Unsolved Problems in Number Theory, 2nd ed.* New York: Springer-Verlag, pp. 66–7, 1994.
Sloane, N. J. A. Sequences A005114/M1552 in "An On-Line Version of the Encyclopedia of Integer Sequences." http://www.research.att.com/~njas/sequences/eisonline.html.
Wells, D. *The Penguin Dictionary of Curious and Interesting Numbers.* Middlesex, England: Penguin Books, p. 60, 1986.

U-Number

ULAM SEQUENCE

UPC

The universal product code (UPC) is a 12-digit number and associated machine-readable bar code used to identify products being purchased in grocery stores. UPCs encode an individual product, but not its price (this part is done by a store's computer after reading the product identifier). The UPC is maintained by the Uniform Code Council of Dayton, Ohio. The first and last digits are separated from the others and written in a smaller font size.

The first six digits are a manufacturer identifier, and the next five digits identify a specific product. The last digit is a check digit obtained from

$$a_{12} = 10 - \left[\left(3 \sum_{\substack{i=1 \\ i \text{ odd}}}^{11} a_i + \sum_{\substack{i=2 \\ i \text{ even}}}^{10} a_i \right) (\text{mod } 10) \right] (\text{mod } 10),$$

where (mod 10) indicates taking the REMAINDER after dividing by 10. For example, the UPC for *Tropicana* Pure Premium orange juice is

$$0 \ 48500 \ 00102 \ 8$$

where the check digit is

$$a_{12} = 10 - [3(0 + 8 + 0 + 0 + 1 + 2)$$

$$+ (4 + 5 + 0 + 0 + 0) \ (\text{mod } 10)] \ (\text{mod } 10)$$

$$= 10 - [42 \ (\text{mod } 10)] \ (\text{mod } 10) = 10 - 2 \ (\text{mod } 10)$$

$$= 8,$$

as expected.

See also CHECKSUM, CODING THEORY, ISBN

Upper Bound

A function f is said to have a upper bound C if $f(x) \leq C$ for all x in its DOMAIN. The LEAST UPPER BOUND is called the SUPREMUM.

See also INEQUALITY, INFIMUM, LEAST UPPER BOUND, LOWER BOUND, SUPREMUM

Upper Half-Disk

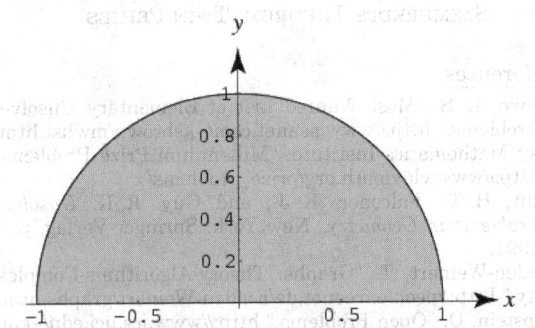

The unit upper half-disk is the portion of the COMPLEX PLANE satisfying $\{|z| \leq 1, \Im[z] > 0\}$.

See also DISK, LOWER HALF-DISK, REAL AXIS, SEMICIRCLE, UNIT DISK, UPPER HALF-PLANE

Upper Half-Plane

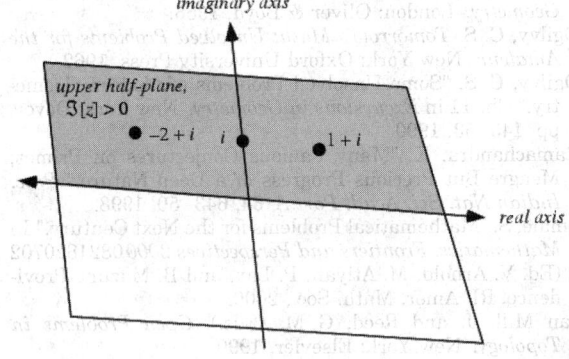

The portion, often denoted H, of the COMPLEX PLANE $\{x + iy : x, y \in (-\infty, \infty)\}$ satisfying $y = \Im[z] > 0$ i.e., $H = \{x + iy : x \in (-\infty, \infty), y \in (0, \infty)\}$.

See also COMPLEX PLANE, HALF-PLANE, LEFT HALF-PLANE, LOWER HALF-PLANE, MODULAR FUNCTION, RIGHT HALF-PLANE, UPPER HALF-DISK

Upper Integral

References

Apostol, T. M. *Modular Functions and Dirichlet Series in Number Theory, 2nd ed.* New York: Springer-Verlag, p. 14, 1997.

Borwein, J. M. and Borwein, P. B. *Pi & the AGM: A Study in Analytic Number Theory and Computational Complexity.* New York: Wiley, p. 112, 1987.

Upper Integral

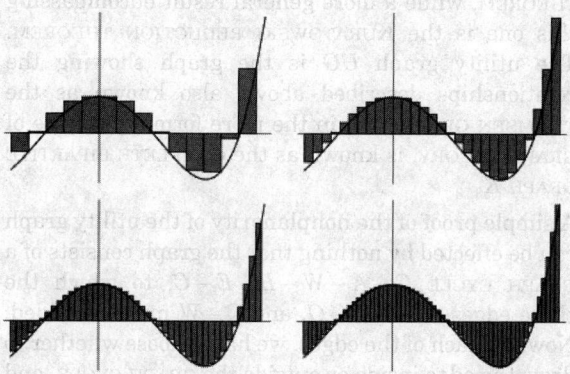

The limit of an UPPER SUM, when it exists, as the MESH SIZE approaches 0.

See also LOWER INTEGRAL, RIEMANN INTEGRAL, UPPER SUM

Upper Limit

Let the greatest term H of a SEQUENCE be a term which is greater than all but a finite number of the terms which are equal to H. Then H is called the upper limit of the SEQUENCE.

An upper limit of a SERIES

$$\text{upper} \lim_{n \to \infty} S_n = \overline{\lim_{n \to \infty}} S_n = k$$

is said to exist if, for every $\epsilon > 0$, $|S_n - k| < \epsilon$ for infinitely many values of n and if no number larger than k has this property.

See also LIMIT, LOWER LIMIT, SUPREMUM LIMIT

References

Bromwich, T. J. I'a and MacRobert, T. M. "Upper and Lower Limits of a Sequence." §5.1 in *An Introduction to the Theory of Infinite Series, 3rd ed.* New York: Chelsea, p. 40, 1991.

Upper Sum

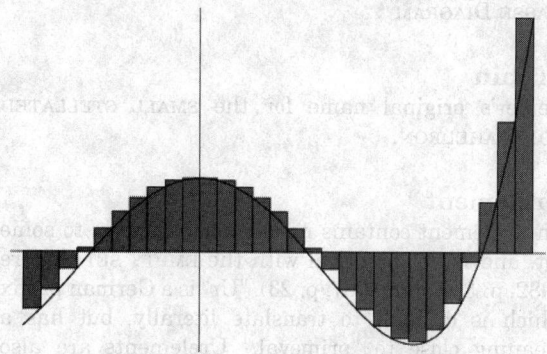

For a given function $f(x)$ over a partition of a given interval, the upper sum is the sum of box areas $f(x_k^*)\Delta x_k$ using the greatest value of the function $f(x_k^*)$) in each subinterval Δx_k.

See also LOWER SUM, RIEMANN INTEGRAL, UPPER INTEGRAL

Upper Triangular Matrix

A TRIANGULAR MATRIX U OF THE FORM

$$U_{ij} = \begin{cases} a_{ij} & \text{for } i \leq j \\ 0 & \text{for } i > j. \end{cases}$$

Written explicitly,

$$\mathsf{U} = \begin{bmatrix} a_{11} & a_{12} & \cdots & a_{1n} \\ 0 & a_{22} & \cdots & a_{2n} \\ \vdots & \vdots & \ddots & \vdots \\ 0 & 0 & \cdots & a_{nn} \end{bmatrix}.$$

An upper triangular matrix with elements f[i,j] above the diagonal can be formed using Upper-DiagonalMatrix[f, n] in the *Mathematica* add-on package LinearAlgebra`MatrixMultiplication` (which can be loaded with the command << LinearAlgebra`).

See also TRIANGULAR MATRIX, LOWER TRIANGULAR MATRIX

References

Ayres, F. Jr. *Theory and Problems of Matrices.* New York: Schaum, p. 10, 1962.

Upper-Trimmed Subsequence

The upper-trimmed subsequence of $x = \{x_n\}$ is the sequence $\lambda(x)$ obtained by dropping the first occurrence of n for each n. If x is a FRACTAL SEQUENCE, then $\lambda(x) = x$.

See also LOWER-TRIMMED SUBSEQUENCE

References

Kimberling, C. "Fractal Sequences and Interspersions." *Ars Combin.* **45**, 157–68, 1997.

Upward Drawing

HASSE DIAGRAM

Urchin

Kepler's original name for the SMALL STELLATED DODECAHEDRON.

Urelement

An urelement contains no elements, belongs to some set, and is not identical with the EMPTY SET (Moore 1982, p. 3; Rubin 1967, p. 23). "Ur" is a German prefix which is difficult to translate literally, but has a meaning close to "primeval." Urelements are also called "atoms" (Rubin 1967, Moore 1982) or "individuals" (Moore 1982).

In "pure" set theory, all elements are sets and there are no urelements. Often, the axioms of set theory are modified to allow the presence of urelements for ease in representing something. In fact, before Paul Cohen developed the method of forcing, some of the independence theorems in set theory were shown if urelements were allowed.

See also EMPTY SET, SET THEORY

References

Moore, G. H. *Zermelo's Axiom of Choice: Its Origin, Development, and Influence.* New York: Springer-Verlag, 1982.
Rubin, J. E. *Set Theory for the Mathematician.* New York: Holden-Day, 1967.

U-Statistic

References

Hoeffding, W. "The Strong Law of Large Numbers for *U*-Statistics." Univ. North Carolina Inst. Statistics Mimeo Series, No. 302, 1961.
Serfling, R. J. *Approximation Theorems of Mathematical Statistics.* New York: Wiley, 1980.

Utility Graph

"houses"

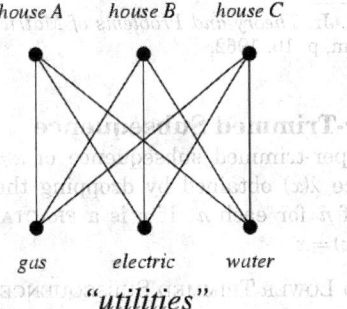

house A   house B   house C

gas   electric   water

"utilities"

The utility problem posits three houses and three

utility companies–say, gas, electric, and water–and asks if each utility can be connected to each house without having any of the gas/water/electric lines/pipes pass over any other. This is equivalent to the equation "Can a PLANAR GRAPH be constructed from each of three nodes ('houses') to each of three other nodes ('utilities')?" This problem was first posed in this form by H. E. Dudeney in 1917 (Gardner 1984, p. 92).

The answer is that no such PLANAR GRAPH exists, and the proof can be effected using the JORDAN CURVE THEOREM, while a more general result encompassing this one is the KURATOWSKI REDUCTION THEOREM. The utility graph *UG* is the graph showing the relationships described above, also known as the THOMSEN GRAPH and, in the more formal parlance of GRAPH THEORY, is known as the COMPLETE BIPARTITE GRAPH $K_{3,3}$.

A simple proof of the nonplanarity of the utility graph can be effected by nothing that the graph consists of a GRAPH CYCLE $G - A - W - B - E - C$, to which the three edges $A - E$, $B - G$, and $C - W$ must be added. Now, for each of the edges, we have choose whether to draw the edge inside or outside the GRAPH CYCLE, and so for two of the edges, we must make the same choice. But two lines can't be drawn on the same side without crossing, hence the graph is not planar.

See also COMPLETE BIPARTITE GRAPH, KURATOWSKI REDUCTION THEOREM, PLANAR GRAPH, THOMSEN GRAPH

References

Chartrand, G. "The Three Houses and Three Utilities Problem: An Introduction to Planar Graphs." §9.1 in *Introductory Graph Theory.* New York: Dover, pp. 191–02, 1985.
Gardner, M. *The Sixth Book of Mathematical Games from Scientific American.* Chicago, IL: University of Chicago Press, pp. 92–4, 1984.
Ore, Ø. *Graphs and Their Uses.* New York: Random House, pp. 14–7, 1963.
Pappas, T. "Wood, Water, Grain Problem." *The Joy of Mathematics.* San Carlos, CA: Wide World Publ./Tetra, pp. 175 and 233, 1989.
Steinhaus, H. *Mathematical Snapshots, 3rd ed.* New York: Dover, pp. 262–63, 1999.

Utility Problem

UTILITY GRAPH

V

Valence

The number of POLES of an AUTOMORPHIC FUNCTION in the closure of its FUNDAMENTAL REGION.

See also FUNDAMENTAL REGION, UNIVALENT FUNCTION, VERTEX DEGREE

References

Apostol, T. M. *Modular Functions and Dirichlet Series in Number Theory, 2nd ed.* New York: Springer-Verlag, p. 84, 1997.

Valency

VERTEX DEGREE

Valle's Two-Thirds Factorization Method

References

Lenstra, A. K. and Lenstra, H. W. Jr. "Algorithms in Number Theory." In *Handbook of Theoretical Computer Science, Volume A: Algorithms and Complexity* (Ed. J. van Leeuwen). New York: Elsevier, pp. 673–715, 1990.

Valuation

A generalization of the P-ADIC NORM first proposed by Kürschák in 1913. A valuation $|\cdot|$ on a FIELD K is a FUNCTION from K to the REAL NUMBERS $\mathbb{R}$ such that the following properties hold for all $x, y \in K$:

1. $|x| \geq 0$,
2. $|x| = 0$ IFF $x = 0$,
3. $|xy| = |x||y|$,
4. $|x| \leq 1$ IMPLIES $|1 + x| \leq C$ for some constant $C \geq 1$ (independent of x).

If (4) is satisfied for $C = 2$, then $|\cdot|$ satisfies the TRIANGLE INEQUALITY,

4a. $|x + y| \leq |x| + |y|$ for all $x, y \in K$.

If (4) is satisfied for $C = 1$ then $|\cdot|$ satisfies the stronger ULTRAMETRIC inequality

4b. $|x + y| \leq \max(|x|, |y|)$.

The simplest valuation is the ABSOLUTE VALUE for REAL NUMBERS. A valuation satisfying (4b) is called non-ARCHIMEDEAN VALUATION; otherwise, it is called ARCHIMEDEAN.

If $|\cdot|_1$ is a valuation on K and $\lambda \geq 1$, then we can define a new valuation $|\cdot|_2$ by

$$|x|_2 = |x|_1^{\lambda}. \qquad (1)$$

This does indeed give a valuation, but possibly with a different constant C in AXIOM 4. If two valuations are related in this way, they are said to be equivalent, and this gives an equivalence relation on the collection of all valuations on K. Any valuation is equivalent to one which satisfies the triangle inequality (4a). In view of this, we need only to study valuations satisfying (4a), and we often view axioms (4) and (4a) as interchangeable (although this is not strictly true).

If two valuations are equivalent, then they are both non-ARCHIMEDEAN or both ARCHIMEDEAN. $\mathbb{Q}$, $\mathbb{R}$, and $\mathbb{C}$ with the usual Euclidean norms are Archimedean valuated fields. For any PRIME p, the P-ADIC NUMBERS $\mathbb{Q}_p$ with the p-adic valuation $|\cdot|_p$ is a NON-ARCHIMEDEAN FIELD.

If K is any FIELD, we can define the trivial valuation on K by $|x| = 1$ for all $x \neq 0$ and $|0| = 0$, which is a NON-ARCHIMEDEAN VALUATION. If K is a FINITE FIELD, then the only possible valuation over K is the trivial one. It can be shown that any valuation on $\mathbb{Q}$ is equivalent to one of the following: the trivial valuation, Euclidean absolute norm $|\cdot|$, or p-adic valuation $|\cdot|_p$.

The equivalence of any nontrivial valuation of $\mathbb{Q}$ to either the usual ABSOLUTE VALUE or to a P-ADIC NORM was proved by Ostrowski (1935). Equivalent valuations give rise to the same topology. Conversely, if two valuations have the same topology, then they are equivalent. A stronger result is the following: Let $|\cdot|_1$, $|\cdot|_2$, ..., $|\cdot|_k$ be valuations over K which are pairwise inequivalent and let $a_1, a_2, ..., a_k$ be elements of K. Then there exists an infinite sequence $(x_1, x_2, ...)$ of elements of K such that

$$\lim_{\substack{n \to \infty \\ \text{w.r.t. } |\cdot|_1}} x_n = a_1 \qquad (2)$$

$$\lim_{\substack{n \to \infty \\ \text{w.r.t. } |\cdot|_2}} x_n = a_2, \qquad (3)$$

etc. This says that inequivalent valuations are, in some sense, completely independent of each other. For example, consider the rationals $\mathbb{Q}$ with the 3-adic and 5-adic valuations $|\cdot|_3$ and $|\cdot|_5$, and consider the sequence of numbers given by

$$x_n = \frac{43 \cdot 5^n + 92 \cdot 3^n}{3^n + 5^n}. \qquad (4)$$

Then $x_n \to 43$ as $n \to \infty$ with respect to $|\cdot|_3$, but $x_n \to 92$ as $n \to \infty$ with respect to $|\cdot|_5$, illustrating that a sequence of numbers can tend to two different limits under two different valuations.

A discrete valuation is a valuation for which the VALUATION GROUP is a discrete subset of the REAL NUMBERS $\mathbb{R}$. Equivalently, a valuation (on a FIELD K) is discrete if there exists a REAL NUMBER $\varepsilon > 0$ such that

$$|x| \in (1 - \varepsilon, 1 + \varepsilon) \Rightarrow |x| = 1 \text{ for all } x \in K. \qquad (5)$$

The p-adic valuation on $\mathbb{Q}$ is discrete, but the ordinary absolute valuation is not.

If $|\cdot|$ is a valuation on K, then it induces a metric

$$d(x,y) = |x-y| \qquad (6)$$

on K, which in turn induces a TOPOLOGY on K. If $|\cdot|$ satisfies (4b), then the metric is an ULTRAMETRIC. We say that $(K, |\cdot|)$ is a complete valuated field if the METRIC SPACE is complete.

See also ABSOLUTE VALUE, LOCAL FIELD, METRIC SPACE, P-ADIC NUMBER, STRASSMAN'S THEOREM, ULTRAMETRIC, VALUATION GROUP

References

Cassels, J. W. S. *Local Fields.* Cambridge, England: Cambridge University Press, 1986.
Koch, H. "Valuations." Ch. 4 in *Number Theory: Algebraic Numbers and Functions.* Providence, RI: Amer. Math. Soc., pp. 103–139, 2000.
Ostrowski, A. "Untersuchungen zur aritmetischen Theorie der Körper." *Math. Zeit.* **39**, 269–404, 1935.
van der Waerden, B. L. *Algebra, 2 vols.* New York: Springer-Verlag, 1991.
Weiss, E. *Algebraic Number Theory.* New York: Dover, 1998.

Valuation Group

Let $(K, |\cdot|)$ be a valuated FIELD. The valuation group G is defined to be the set

$$G = \{|x| : x \in K, x \neq 0\},$$

with the group operation being multiplication. It is a SUBGROUP of the POSITIVE REAL NUMBERS, under multiplication.

Valuation Ring

Let $(K, |\cdot|)$ be a NON-ARCHIMEDEAN FIELD. Its valuation ring R is defined to be

$$R = \{x \in K : |x| \leq 1\}.$$

The valuation ring has maximal IDEAL

$$M = \{x \in K : |x| \leq 1\},$$

and the FIELD R/M is called the residue field, class field, or field of digits. For example, if $K = \mathbb{Q}_p$ (P-ADIC NUMBERS), then $R = \mathbb{Z}_p$ (p-adic integers), $M = p\mathbb{Z}_p$ (p-adic integers congruent to 0 mod p), and $R/M = \mathrm{GF}(p)$, the FINITE FIELD of order p.

See also P-ADIC NUMBER

Valuation Theory

The study of VALUATIONS which simplifies class field theory and the theory of FUNCTION FIELDS.

See also FUNCTION FIELD, VALUATION

References

Iyanaga, S. and Kawada, Y. (Eds.). "Valuations." §425 in *Encyclopedic Dictionary of Mathematics.* Cambridge, MA: MIT Press, pp. 1350–1353, 1980.

Value

The quantity which a FUNCTION f takes upon application to a given quantity.

See also VALUE (GAME)

Value (Game)

The solution to a GAME in GAME THEORY. When a SADDLE POINT is present

$$\max_{i \leq m} \min_{j \leq n} a_{ij} = \min_{j \leq n} \max_{i \leq m} a_{ij} \equiv v,$$

and v is the value for pure strategies.

See also ABSOLUTE VALUE, GAME THEORY, MINIMAX THEOREM, VALUATION

Vampire Number

A number $v = xy$ with an EVEN number n of DIGITS formed by multiplying a pair of $n/2$-DIGIT numbers (where the DIGITS are taken from the original number in any order) x and y together. Pairs of trailing zeros are not allowed. If v is a vampire number, then x and y are called its "fangs." Examples of vampire numbers include

$$
\begin{aligned}
1260 &= 21 \times 60 \\
1395 &= 15 \times 93 \\
1435 &= 35 \times 41 \\
1530 &= 30 \times 51 \\
1827 &= 21 \times 87 \\
2187 &= 27 \times 81 \\
6880 &= 80 \times 86
\end{aligned}
$$

(Sloane's A014575). The 8-digit vampire numbers are 10025010, 10042510, 10052010, 10052064, 10081260, ... (Sloane's A048938) and the 10-digit vampire numbers are 1000174288, 1000191991, 1000198206, 1000250010, ... (Sloane's A048939). The numbers of $2n$-digit vampires are 0, 7, 148, 3228, ... (Sloane's A048935).

Vampire numbers having *two* distinct pairs of fangs include

$$125460 = 204 \times 615 = 246 \times 510$$

$$11930170 = 1301 \times 9170 = 1310 \times 9107$$

$$12054060 = 2004 \times 6015 = 2406 \times 5010$$

(Sloane's A048936).

Vampire numbers having *three* distinct pairs of fangs include

$$13078260 = 1620 \times 8073 = 1863 \times 7020$$
$$= 2070 \times 6318.$$

(Sloane's A048937).

General formulas can be constructed for special classes of vampires, such as the fangs

$$x = 25 \cdot 10^k + 1$$

$$y = 100(10^{k+1} + 52)/25,$$

giving the vampire

$$v = xy = (10^{k+1} + 52)10^{k+2} + 100(10^{k+1} + 52)/25$$

$$= x^* \cdot 10^{k+2} + t$$

$$= 8(26 + 5 \cdot 10^k)(1 + 25 \cdot 10^k),$$

where x^* denotes x with the DIGITS reversed (Roushe and Rogers).

Pickover (1995) also defines pseudovampire numbers, in which the multiplicands have different numbers of digits.

References

Pickover, C. A. "Vampire Numbers." Ch. 30 in *Keys to Infinity*. New York: Wiley, pp. 227–231, 1995.

Pickover, C. A. "Vampire Numbers." *Theta* **9**, 11–13, Spring 1995.

Pickover, C. A. "Interview with a Number." *Discover* **16**, 136, June 1995.

Roushe, F. W. and Rogers, D. G. "Tame Vampires." Undated manuscript.

Sloane, N. J. A. Sequences A014575, A048933, A048934, A048935, A048936, A048937, A048938, and A048939 in "An On-Line Version of the Encyclopedia of Integer Sequences." http://www.research.att.com/~njas/sequences/eisonline.html.

van der Grinten Projection

A MAP PROJECTION given by the transformation

$$x = \operatorname{sgn}(\lambda - \lambda_0)$$
$$\times \frac{\pi \left[A(G - P^2) + \sqrt{A^2(G - P^2)^2 - (P^2 + A^2)(G^2 - P^2)} \right]}{P^2 + A^2} \tag{1}$$

$$y = \operatorname{sgn}(\phi) \frac{\pi \left[PQ - A\sqrt{(A^2 + 1)(P^2 + A^2) - Q^2} \right]}{P^2 + A^2}, \tag{2}$$

where

$$A = \frac{1}{2} \left| \frac{\pi}{\lambda - \lambda_0} - \frac{\lambda - \lambda_0}{\pi} \right| \tag{3}$$

$$G = \frac{\cos \theta}{\sin \theta + \cos \theta - 1} \tag{4}$$

$$P = G \left(\frac{2}{\sin \theta} - 1 \right) \tag{5}$$

$$\theta = \sin^{-1} \left| \frac{2\phi}{\pi} \right| \tag{6}$$

$$Q = A^2 + G. \tag{7}$$

The inverse FORMULAS are

$$\phi = \operatorname{sgn}(y)\pi \left[-m_1 \cos\left(\theta_1 + \tfrac{1}{3}\pi\right) - \frac{c_2}{3c_3} \right] \tag{8}$$

$$\lambda = \frac{\pi \left| X^2 + Y^2 - 1 + \sqrt{1 + 2(X^2 - Y^2) + (X^2 + Y^2)^2} \right|}{2X} + \lambda_0, \tag{9}$$

where

$$X = \frac{x}{\pi} \tag{10}$$

$$Y = \frac{y}{\pi} \tag{11}$$

$$c_1 = -|Y|(1 + X^2 + Y^2) \tag{12}$$

$$c_2 = c_1 - 2Y^2 + X^2 \tag{13}$$

$$c_3 = -2c_1 + 1 + 2Y^2 + (X^2 + Y^2)^2 \tag{14}$$

$$d = \frac{Y^2}{c_3} + \frac{1}{27} \left(\frac{2c_2^3}{c_3^3} - \frac{9c_1 c_2}{c_3^2} \right) \tag{15}$$

$$a_1 = \frac{1}{c_3} \left(c_1 - \frac{c_2^2}{3c_3} \right) \tag{16}$$

$$m_1 = 2\sqrt{-\tfrac{1}{3}a_1} \tag{17}$$

$$\theta_1 = \tfrac{1}{3} \cos^{-1} \left(\frac{3d}{a_1 m_1} \right). \tag{18}$$

References

Snyder, J. P. *Map Projections--A Working Manual*. U. S. Geological Survey Professional Paper 1395. Washington, DC: U. S. Government Printing Office, pp. 239–242, 1987.

van der Pol Equation

An ORDINARY DIFFERENTIAL EQUATION which can be derived from the RAYLEIGH DIFFERENTIAL EQUATION by differentiating and setting $y = y'$. It is an equation describing self-sustaining oscillations in which energy is fed into small oscillations and removed from large oscillations. This equation arises in the study of circuits containing vacuum tubes and is given by

$$y'' - \mu(1 - y^2)y' + y = 0.$$

See also RAYLEIGH DIFFERENTIAL EQUATION

References

Birkhoff, G. and Rota, G.-C. *Ordinary Differential Equations, 3rd ed.* New York: Wiley, p. 134, 1978.
Kreyszig, E. *Advanced Engineering Mathematics, 6th ed.* New York: Wiley, pp. 165–166, 1988.
Zwillinger, D. *Handbook of Differential Equations, 3rd ed.* Boston, MA: Academic Press, p. 127, 1997.

van der Waerden Number

This entry contributed by KEVIN O'BRYANT

One form of VAN DER WAERDEN'S THEOREM states that for every POSITIVE INTEGERS k and r, there exists a constant $n(k, r)$ such that if $n_0 \geq n(k, r)$ and $\{1, 2, \ldots, n_0\} \subset C_1 \cup C_2 \ldots \cup C_r$, the some set C_i contains an ARITHMETIC SEQUENCE of length k. The least possible value of $n(k, r)$ is known as a van der Waerden number. The only nontrivial van der Waerden numbers that are known exactly are summarized in the following table. As shown in the table, the first few values of $n(2, k)$ for $k = 1, 2, \ldots$ are 1, 3, 9, 35, 178, ... (Sloane's A005346).

$r \backslash k$	$k = 3$	$k = 4$	$k = 5$
$r = 2$	9	35	178
$r = 3$	27		
$r = 4$	76		

Shelah (1988) proved that van der Waerden's numbers are PRIMITIVE RECURSIVE. It is known that

$$n(3, r) \leq e^{r^1} \tag{1}$$

and that

$$n(4, r) \leq e^{e^{e^{r^{c_2}}}} \tag{2}$$

for some constants c_1 and c_2. In 1998, T. Gowers announced that he has proved the general result

$$n(n, k) \leq e^{e^{(1/r)e^{e^{k+110}}}}, \tag{3}$$

but this work has not yet been published. Berlekamp

(1968) showed that for p a prime,

$$n(p + 1, 2) > p \cdot 2^p, \tag{4}$$

and that probabilistic arguments using the LOVÁSZ LOCAL LEMMA show that

$$n(k, r) > \left(\frac{r^k}{erk}\right)(1 + \iota(1)). \tag{5}$$

See also SZEMERÉDI'S THEOREM, VAN DER WAERDEN'S THEOREM

References

Berlekamp, E. A "Construction for Partitions Which Avoid Long Arithmetic Progressions." *Canad. Math. Bull.* **11**, 409–414, 1968.
Goodman, J. E. and O'Rourke, J. (Eds.). *Handbook of Discrete & Computational Geometry.* Boca Raton, FL: CRC Press, p. 159, 1997.
Gowers, W. T. "Fourier Analysis and Szemerédi's Theorem." In *Proceedings of the International Congress of Mathematicians, Vol. 1. Doc. Math. 1998, Extra Vol. I.* Berlin, 617–629, 1998. Available electronically from http://www.mathematik.uni-bielefeld.de/documenta/xvol-icm/Fields/Fields.html.
Gowers, W. T. "A New Proof of Szemerédi's Theorem for Arithmetic Progressions of Length Four." *Geom. Funct. Anal.* **8**, 529–551, 1998.
Honsberger, R. *More Mathematical Morsels.* Washington, DC: Math. Assoc. Amer., p. 29, 1991.
Shelah, S. "Primitive Recursive Bounds for van der Waerden Numbers." *J. Amer. Math. Soc.* **1**, 683–697, 1988.
Sloane, N. J. A. Sequences A005346/M2819 in "An On-Line Version of the Encyclopedia of Integer Sequences." http://www.research.att.com/~njas/sequences/eisonline.html.

van der Waerden's Theorem

This entry contributed by KEVIN O'BRYANT

van der Waerden's theorem is a theorem about the existence of arithmetic sequences in sets. The theorem can be stated in three equivalent forms.

1. For every POSITIVE INTEGERS k and r, there exists a constant $n(k, r)$ such that if $n_0 \geq n(k, r)$ and $\{1, 2, \ldots, n_0\} \subset C_1 \cup C_2 \ldots \cup C_r$, the some set C_i contains an ARITHMETIC SEQUENCE of length k.
2. If $\{a_0, a_1, \ldots\}$ is an infinite sequence of integers satisfying $0 < a_{k+1} - a_k < r$ for some r, then the sequence contains arbitrarily long arithmetic progressions.
3. For every positive integers k and r, there is a constant $g(k, r)$ such that if $g_0 \geq g(k, r)$ and $a_1, a_2, \ldots, a_{g_0}$ satisfies $0 < a_{i+1} - a_i \leq r$, then k of the numbers $a_1, a_2, \ldots, a_{g_0}$ are in arithmetic progression.

The constants $n(k, r)$ are called VAN DER WAERDEN NUMBERS, and no FORMULA for $n(k, r)$ is known. van der Waerden's Theorem is a COROLLARY of SZEMERÉDI'S THEOREM.

See also ARITHMETIC SEQUENCE, BAUDET'S CONJECTURE, SZEMERÉDI'S THEOREM, VAN DER WAERDEN NUMBER

References

Guy, R. K. "Theorem of van der Waerden, Szemerédi's Theorem. Partitioning the Integers into Classes; at Least One Contains an A.P." §E10 in *Unsolved Problems in Number Theory, 2nd ed.* New York: Springer-Verlag, pp. 204–209, 1994.

Honsberger, R. *More Mathematical Morsels.* Washington, DC: Math. Assoc. Amer., p. 29, 1991.

Khinchin, A. Y. "Van der Waerden's Theorem on Arithmetic Progressions." Ch. 1 in *Three Pearls of Number Theory.* New York: Dover, pp. 11–17, 1998.

van der Waerden, B. L. "Beweis einer Baudetschen Vermutung." *Nieuw Arch. Wiskunde* **15**, 212–216, 1927.

van Kampen's Theorem

In the usual diagram of inclusion homeomorphisms, if the upper two maps are injective, then so are the other two.

References

Dodson, C. T. J. and Parker, P. E. *A User's Guide to Algebraic Topology.* Dordrecht, Netherlands: Kluwer, p. 88, 1997.

Rolfsen, D. *Knots and Links.* Wilmington, DE: Publish or Perish Press, pp. 74–75 and 369–373, 1976.

van Wijngaarden-Deker-Brent Method

BRENT'S METHOD

Vandermonde Determinant

$$\Delta(x_1,\ldots,x_n) \equiv \begin{vmatrix} 1 & x_1 & x_1^2 & \cdots & x_1^{n-1} \\ 1 & x_2 & x_2^2 & \cdots & x_2^{n-1} \\ \vdots & \vdots & \vdots & \ddots & \vdots \\ 1 & x_n & x_n^2 & \cdots & x_n^{n-1} \end{vmatrix}$$

$$= \prod_{\substack{i,j \\ i>j}} (x_i - x_j)$$

(Sharpe 1987). For INTEGERS $a_1, \ldots, a_n$, $\Delta(a_1,\ldots,a_n)$ is divisible by $\prod_{i=1}^n (i-1)!$ (Chapman 1996), the first few values of which are the SUPERFACTORIALS 1, 1, 2, 12, 288, 34560, 24883200, 125411328000, ... (Sloane's A000178).

See also SUPERFACTORIAL, VANDERMONDE MATRIX

References

Chapman, R. "A Polynomial Taking Integer Values." *Math. Mag.* **69**, 121, 1996.

Fletcher, A.; Miller, J. C. P.; Rosenhead, L.; and Comrie, L. J. *An Index of Mathematical Tables, Vol. 1.* Reading, MA: p. 50, 1962.

Gradshteyn, I. S. and Ryzhik, I. M. *Tables of Integrals, Series, and Products, 6th ed.* San Diego, CA: Academic Press, p. 1111, 2000.

Graham, R. L.; Knuth, D. E.; and Patashnik, O. "Binomial Coefficients." Ch. 5 in *Concrete Mathematics: A Foundation for Computer Science, 2nd ed.* Reading, MA: Addison-Wesley, p. 231, 1994.

Radoux, C. "Query 145." *Not. Amer. Math. Soc.* **25**, 197, 1978.

Ryser, H. J. *Combinatorial Mathematics.* Buffalo, NY: Math. Assoc. Amer., p. 53, 1963.

Sharpe, D. §2.9 in *Rings and Factorization.* Cambridge, England: Cambridge University Press, 1987.

Sloane, N. J. A. Sequences A000178/M2049 in "An On-Line Version of the Encyclopedia of Integer Sequences." http://www.research.att.com/~njas/sequences/eisonline.html.

Vandermonde Identity

CHU-VANDERMONDE IDENTITY

Vandermonde Matrix

A type of matrix which arises in the LEAST SQUARES FITTING of POLYNOMIALS and the reconstruction of a STATISTICAL DISTRIBUTION from the distribution's MOMENTS. The solution of an $n \times n$ Vandermonde matrix equation requires $\mathcal{O}(n^2)$ operations. A Vandermonde matrix of order n is OF THE FORM

$$\begin{bmatrix} 1 & x_1 & x_1^2 & \cdots & x_1^{n-1} \\ 1 & x_2 & x_2^2 & \cdots & x_2^{n-1} \\ \vdots & \vdots & \vdots & \ddots & \vdots \\ 1 & x_n & x_n^2 & \cdots & x_n^{n-1} \end{bmatrix}$$

See also TOEPLITZ MATRIX, TRIDIAGONAL MATRIX, VANDERMONDE DETERMINANT

References

Press, W. H.; Flannery, B. P.; Teukolsky, S. A.; and Vetterling, W. T. "Vandermonde Matrices and Toeplitz Matrices." §2.8 in *Numerical Recipes in FORTRAN: The Art of Scientific Computing, 2nd ed.* Cambridge, England: Cambridge University Press, pp. 82–89, 1992.

Vandermonde Theorem

CHU-VANDERMONDE IDENTITY

Vandermonde's Convolution Formula

CHU-VANDERMONDE IDENTITY

Vandermonde's Sum

CHU-VANDERMONDE IDENTITY

Vandiver's Criteria

Let p be an IRREGULAR PRIME, and let $P = rp + 1$ be a PRIME with $P < p^2 - p$. Also let t be an INTEGER such that $t^3 \not\equiv 1 \pmod{P}$. For an IRREGULAR PAIR $(p, 2k)$, form the product

$$Q_{2k} = t^{-rd/2} \prod_{b=1}^m \left(t^{rb} - 1\right)^{b^{p-1-2k}},$$

where

$$m = \tfrac{1}{2}(p1 - 1)$$

$$d = \sum_{n=1}^{m} n^{p-2k}.$$

If $Q_{2k}^r \not\equiv 1 \pmod{P}$ for all such IRREGULAR PAIRS, then FERMAT'S LAST THEOREM holds for exponent p.

See also FERMAT'S LAST THEOREM, IRREGULAR PAIR, IRREGULAR PRIME

References

Johnson, W. "Irregular Primes and Cyclotomic Invariants." *Math. Comput.* **29**, 113–120, 1975.

Vanish

A quantity which takes on the value zero is said to vanish. For example, the function $f(z) = z^2$ vanishes at the point $z = 0$.

See also ROOT

Vanishing Point

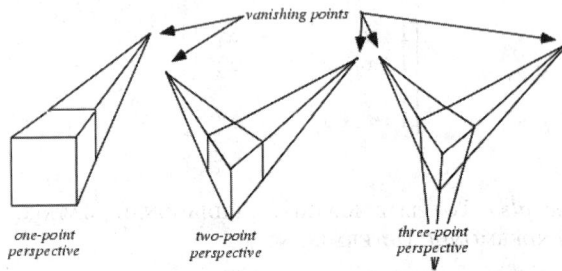

vanishing points

one-point perspective   *two-point perspective*   *three-point perspective*

The point or points to which the extensions of PARALLEL lines appear to converge in a PERSPECTIVE drawing.

See also DESARGUES' THEOREM, PERSPECTIVE, PROJECTIVE GEOMETRY

References

Dixon, R. "Perspective Drawings." Ch. 3 in *Mathographics.* New York: Dover, pp. 79–88, 1991.
Graustein, W. C. *Introduction to Higher Geometry.* New York: Macmillan, pp. 19–20, 1930.

Varga's Constant

$$V \equiv \frac{1}{\Lambda} = 9.2890254919\ldots,$$

where Λ is the ONE-NINTH CONSTANT.

See also ONE-NINTH CONSTANT

Variance

For N samples of a variate having a distribution with *known* MEAN μ, the "population variance" (usually called "variance" for short, although the word "population" should be added when needed to distinguish it from the SAMPLE VARIANCE) is defined by

$$\mathrm{var}(x) \equiv \frac{1}{N} \sum (x - \mu)^2 = \langle x^2 - 2\mu x + \mu^2 \rangle$$

$$= \langle x^2 \rangle - \langle 2\mu x \rangle + \langle \mu^2 \rangle$$

$$= \langle x^2 \rangle - 2\mu \langle x \rangle + \mu^2, \tag{1}$$

where

$$\langle x \rangle \equiv \frac{1}{N} \sum_{i=1}^{N} x_i. \tag{2}$$

But since $\langle x \rangle$ is an UNBIASED ESTIMATOR for the MEAN

$$\mu \equiv \langle x \rangle, \tag{3}$$

it follows that the variance

$$\sigma^2 \equiv \mathrm{var}(x) = \langle x^2 \rangle - \mu^2. \tag{4}$$

The population STANDARD DEVIATION is then defined as

$$\sigma \equiv \sqrt{\mathrm{var}(x)} = \sqrt{\langle x^2 \rangle - \mu^2}. \tag{5}$$

A useful identity involving the variance is

$$\mathrm{var}(f(x) + g(x)) = \mathrm{var}(f(x)) + \mathrm{var}(g(x)). \tag{6}$$

Therefore,

$$\mathrm{var}(ax + b) = \left\langle [(ax + b) - \langle ax + b \rangle]^2 \right\rangle$$

$$= \left\langle (ax + b - a\langle x \rangle - b)^2 \right\rangle$$

$$= \left\langle (ax - a\mu)^2 \right\rangle = \left\langle a^2(x - \mu)^2 \right\rangle$$

$$= a^2 \left\langle (x - \mu)^2 \right\rangle = a^2 \mathrm{var}(x) \tag{7}$$

$$\mathrm{var}(b) = 0. \tag{8}$$

If the population MEAN is not known, using the sample mean $\bar{x}$ instead of the population mean μ to compute

$$s^2 \equiv \hat{\sigma}_N^2 \equiv \frac{1}{N} \sum_{i=1}^{N} (x_i - \bar{x})^2 \tag{9}$$

gives a BIASED ESTIMATOR of the population variance. In such cases, it is appropriate to use a STUDENT'S T-DISTRIBUTION instead of a GAUSSIAN DISTRIBUTION. However, it turns out (as discussed below) that an UNBIASED ESTIMATOR for the population variance is given by

$$s'^2 \equiv \hat{\sigma}_N'^2 \equiv \frac{1}{N-1} \sum_{i=1}^{N} (x_i - \bar{x})^2. \tag{10}$$

For multiple variables, the variance is given using the definition of COVARIANCE,

$$\mathrm{var}\left(\sum_{i=1}^{n} x_i\right) = \mathrm{cov}\left(\sum_{i=1}^{n} x_i, \sum_{j=1}^{m} x_j\right)$$

$$= \sum_{i=1}^{n} \sum_{j=1}^{m} \mathrm{cov}(x_i, x_j)$$

$$= \sum_{i=1}^{n} \sum_{\substack{j=1 \\ j=i}}^{m} \mathrm{cov}(x_i, x_j) + \sum_{i=1}^{n} \sum_{\substack{j=1 \\ j \neq i}}^{m} \mathrm{cov}(x_i, x_j)$$

$$= \sum_{i=1}^{n} \mathrm{cov}(x_i, x_j) + \sum_{i=1}^{n} \sum_{\substack{j=1 \\ j \neq i}}^{m} \mathrm{cov}(x_i, x_j)$$

$$= \sum_{i=1}^{n} \mathrm{var}(x_i) + 2 \sum_{i=1}^{n} \sum_{j=i+1}^{m} \mathrm{cov}(x_i, x_j). \qquad (11)$$

A linear sum has a similar form:

$$\mathrm{var}\left(\sum_{i=1}^{n} a_i x_i\right) = \mathrm{cov}\left(\sum_{i=1}^{n} a_i x_i, \sum_{j=1}^{m} a_j x_j\right)$$

$$= \sum_{i=1}^{n} \sum_{j=1}^{m} a_i a_j \mathrm{cov}(x_i, x_j)$$

$$= \sum_{i=1}^{n} a_i^2 \mathrm{var}(x_i) + 2 \sum_{i=1}^{n} \sum_{j=i+1}^{m} a_i a_j \mathrm{cov}(x_i, x_j). \qquad (12)$$

These equations can be expressed using the COVARIANCE MATRIX.

To estimate the POPULATION VARIANCE σ^2 from a sample of N elements with a priori *unknown* MEAN (i.e., the MEAN is estimated from the sample itself), we need an UNBIASED ESTIMATOR for σ^2. This is given by the K-STATISTIC k_2, where

$$\mathrm{var}(s^2) = k_2 = \frac{N}{N-1} s^2 \qquad (13)$$

and $m_2 \equiv s^2$ is the SAMPLE VARIANCE, defined by

$$s^2 \equiv \frac{1}{N} \sum_{i=1}^{N} (x_i - \bar{x})^2. \qquad (14)$$

The quantity Ns^2/σ^2 has a CHI-SQUARED DISTRIBUTION. Note that some authors prefer the definition

$$s'^2 \equiv \frac{1}{N-1} \sum_{i=1}^{N} (x_i - \bar{x})^2, \qquad (15)$$

since this makes the sample variance an UNBIASED ESTIMATOR for the population variance.

To find the variance of the SAMPLE VARIANCE s^2, remember that

$$\mathrm{var}(s^2) \equiv \langle s^4 \rangle - \langle s^2 \rangle^2, \qquad (16)$$

and

$$\langle s^2 \rangle = \frac{N-1}{N} \mu_2. \qquad (17)$$

Now find $\langle s^4 \rangle$:

$$\langle s^4 \rangle = \langle (s^2)^2 \rangle = \left\langle \left(\langle x^2 \rangle - \langle x \rangle^2 \right)^2 \right\rangle$$

$$= \left\langle \left[\frac{1}{N} \sum x_i^2 - \left(\frac{1}{N} \sum x_i \right)^2 \right]^2 \right\rangle$$

$$= \frac{1}{N^2} \left\langle \left(\sum x_i \right)^2 \right\rangle - \frac{2}{N^3} \left\langle \sum x_i^2 \left(\sum x_i \right)^2 \right\rangle + \frac{1}{N^4}$$

$$\times \left\langle \left(\sum x_i \right)^4 \right\rangle. \qquad (18)$$

Working on the first term of (18),

$$\left\langle \left(\sum x_i^2 \right)^2 \right\rangle = \left\langle \sum x_i^4 + \sum x_i^2 x_j^2 \right\rangle$$

$$= \left\langle \sum x_i^4 \right\rangle + \left\langle \sum x_i^2 x_j^2 \right\rangle$$

$$= N \langle x_i^4 \rangle + N(N-1) \langle x_i^2 \rangle \langle x_j^2 \rangle$$

$$= N \mu_4' + N(N-1) \mu_2'^2. \qquad (19)$$

The second term of (18) is known from K-STATISTIC,

$$\left\langle \sum x_i^2 \left(\sum x_j \right)^2 \right\rangle = N \mu_4' + N(N-1) \mu_2'^2, \qquad (20)$$

as is the third term,

$$\left\langle \left(\sum x_i \right)^4 \right\rangle = N \left\langle \sum x_i^4 \right\rangle + 3N(N-1) \left\langle \sum x_i^2 x_j^2 \right\rangle$$

$$= N \mu_4' + 3N(N-1) \mu_2'^2. \qquad (21)$$

Combining (18)-(21) gives

$$\langle s^4 \rangle = \frac{1}{N^2} \left[N \mu_4' + N(N-1) \mu_2'^2 \right] - \frac{2}{N^3}$$

$$\times \left[N \mu_4' + N(N-1) \mu_2'^2 \right]$$

$$+ \frac{1}{N^4} \left[N \mu_4' + 3N(N-1) \mu_2'^2 \right]$$

$$= \left(\frac{1}{N} - \frac{2}{N^2} + \frac{1}{N^3} \right) \mu_4'$$

$$+ \left[\frac{N-1}{N} - \frac{2(N-1)}{N^2} + \frac{3(N-1)}{N^3} \right] \mu_2'^2$$

$$= \left(\frac{N^2 - 2N + 1}{N^3} \right) \mu_4' + \frac{(N-1)(N^2 - 2N + 3)}{N^3} \mu_2'^2$$

$$= \frac{(N-1)(N-1)\mu_4' + (N^2 - 2N + 3){\mu_2'}^2}{N^3} \qquad (22)$$

(Kenney and Keeping 1951, p. 164), so plugging in (17) and (22) gives

$$\mathrm{var}(s^2) = \langle s^4 \rangle - \langle s^2 \rangle^2$$

$$= \frac{(N-1)[(N-1)\mu_4' + (N^2 - 2N + 3){\mu_2'}^2]}{N^3}$$

$$- \frac{(N-1)^2 N}{N^3}{\mu_2'}^2$$

$$= \frac{(N-1)[(N-1)\mu_4' - (N-3){\mu_2'}^2]}{N^3}. \qquad (23)$$

(Kenney and Keeping 1951, p. 164).

Student calculated the SKEWNESS and KURTOSIS of the distribution of s^2 as

$$\gamma_1 = \sqrt{\frac{8}{N-1}} \qquad (24)$$

$$\gamma_2 = \frac{12}{N-1} \qquad (25)$$

and conjectured that the true distribution is PEARSON TYPE III DISTRIBUTION

$$f(s^2) = C(s^2)^{(N-3)/2} e^{-Ns^2/2\sigma^2}, \qquad (26)$$

where

$$\sigma^2 = \frac{Ns^2}{N-1} \qquad (27)$$

$$C = \frac{\left(\dfrac{N}{2\sigma^2}\right)^{(N-1)/2}}{\Gamma\left(\dfrac{N-1}{2}\right)}. \qquad (28)$$

This was proven by R. A. Fisher.

The distribution of s itself is given by

$$f(s) = 2 \frac{\left(\dfrac{N}{2\sigma^2}\right)^{(N-1)/2}}{\Gamma\left(\dfrac{N-1}{2}\right)} e^{-ns^2/2\sigma^2} s^{N-2} \qquad (29)$$

$$\langle s \rangle = \sqrt{\frac{2}{N}} \frac{\Gamma\left(\dfrac{N}{2}\right)}{\Gamma\left(\dfrac{N-1}{2}\right)} \sigma \equiv b(N)\sigma, \qquad (30)$$

where

$$b(N) \equiv \sqrt{\frac{2}{N}} \frac{\Gamma\left(\dfrac{N}{2}\right)}{\Gamma\left(\dfrac{N-1}{2}\right)}. \qquad (31)$$

The MOMENTS are given by

$$\mu_r = \left(\frac{2}{N}\right)^{r/2} \frac{\Gamma\left(\dfrac{N-1+r}{2}\right)}{\Gamma\left(\dfrac{N-1}{2}\right)} \sigma^r, \qquad (32)$$

and the variance is

$$\mathrm{var}(s) = v_2 - v_1^2 = \frac{N-1}{N}\sigma^2 - [b(N)\sigma]^2$$

$$= \frac{1}{N}\left[N - 1 - \frac{2\Gamma^2\left(\dfrac{N}{2}\right)}{\Gamma^2\left(\dfrac{N-1}{2}\right)}\sigma^2\right] \qquad (33)$$

An UNBIASED ESTIMATOR of σ is $s/b(N)$. Romanovsky showed that

$$b(N) = 1 - \frac{3}{4N} - \frac{7}{32N^2} - \frac{139}{51849N^3} + \cdots \qquad (34)$$

When computing numerically, the MEAN must be computed before s^2 can be determined. This requires storing the set of sample values. It is possible to calculate s'^2 using a recursion relationship involving only the last sample as follows. Here, use μ_j to denote μ calculated from the first j samples (*not* the jth MOMENT)

$$\mu_j \equiv \frac{\sum_{i=1}^{j} x_i}{j}, \qquad (35)$$

and s_j^2 denotes the value for the sample variance s'^2 calculated from the first j samples. The first few values calculated for the MEAN are

$$\mu_1 = x_1 \qquad (36)$$

$$\mu_2 = \frac{1 \cdot \mu_1 + x_2}{2} \qquad (37)$$

$$\mu_3 = \frac{2\mu_2 + x_3}{3}. \qquad (38)$$

Therefore, for $j = 2, 3$ it is true that

$$\mu_j = \frac{(j-1)\mu_{j-1} + x_j}{j} \qquad (39)$$

Therefore, by induction,

$$\mu_{j+1} = \frac{[(j+1)-1]\mu_{(j+1)-1} + x_{j+1}}{j+1}$$

$$= \frac{j\mu_j + x_{j+1}}{j+1} \tag{40}$$

$$\mu_{j+1}(j+1) = (j+1)\mu_j + (x_{j+1} - \mu_j) \tag{41}$$

$$\mu_{j+1} = \mu_j + \frac{x_{j+1} - \mu_j}{j+1}, \tag{42}$$

and

$$s_j^2 = \frac{\sum_{i=1}^{j}(x_i - \mu_j)^2}{j-1} \tag{43}$$

for $j \geq 2$, so

$$js_{j+1^2} = j\frac{\sum_{i=1}^{j+1}(x_i - \mu_{j+1})^2}{j} = \sum_{i=1}^{j+1}(x_i - \mu_{j+1})^2$$

$$= \sum_{i=1}^{j+1}[(x_i - \mu_j) + (\mu_i - \mu_{j+1})]^2$$

$$= \sum_{i=1}^{j+1}(x_i - \mu_j)^2 + \sum_{i=1}^{j+1}(\mu_j - \mu_{j+1})^2 + 2\sum_{i=1}^{j+1}(x_i - \mu_j)$$
$$\times (\mu_j - \mu_{j+1}). \tag{44}$$

Working on the first term,

$$\sum_{i=1}^{j+1}(x_i - \mu_j)^2 = \sum_{i=1}^{j}(x_i - \mu_j)^2 + (x_{j+1} - \mu_j)^2$$

$$= (j-1)s_j^2 + (x_{j+1} - \mu_j)^2. \tag{45}$$

Use (41) to write

$$x_{j+1} - \mu_j = (j+1)(\mu_{j+1} - \mu_j), \tag{46}$$

so

$$\sum_{i=1}^{j+1}(x_i - \mu_j)^2 = (j-1)s_j^2 + (j+1)^2(\mu_{j+1} - \mu_j)^2. \tag{47}$$

Now work on the second term in (44),

$$\sum_{i=1}^{j+1}(\mu_j - \mu_{j+1})^2 = (j+1)(\mu_j - \mu_{j+1})^2. \tag{48}$$

Considering the third term in (44),

$$\sum_{i=1}^{j+1}(x_i - \mu_j)(\mu_j - \mu_{j+1}) = (\mu_j - \mu_{j+1})\sum_{i=1}^{j+1}(x_i - \mu_j)$$

$$= (\mu_j - \mu_{j+1})\left[\sum_{i=1}^{j}(x_i - \mu_j) + (x_{j+1} - \mu_j)\right]$$

$$= (\mu_j - \mu_{j+1})\left(x_{j+1} - \mu_j - j\mu_j + \sum_{i=1}^{j}x_i\right). \tag{49}$$

But

$$\sum_{i=1}^{j}x_i = j\mu_j, \tag{50}$$

so

$$(\mu_j - \mu_{j+1})(x_{j+1} - \mu_j)$$
$$= (\mu_j - \mu_{j+1})(j+1)(\mu_{j+1} - \mu_j)$$
$$= -(j+1)(\mu_j - \mu_{j+1})^2. \tag{51}$$

Plugging (47), (48), and (51) into (44),

$$js_{j+1^2} = \left[(j-1)s_j^2 + (j+1)^2(\mu_{j+1} - \mu_j)^2\right]$$
$$+ \left[(j+1)(\mu_j - \mu_{j+1})^2\right] + 2\left[-(j+1)(\mu_j - \mu_{j+1})^2\right]$$

$$= (j-1)s_j^2 + (j+1)^2(\mu_{j+1} - \mu_j)^2$$
$$- (j+1)(\mu_j - \mu_{j+1})^2$$

$$= (j-1)s_j^2 + (j+1)[(j+1)-1](\mu_{j+1} - \mu_j)^2$$

$$= (j-1)s_j^2 + j(j+1)(\mu_{j+1} - \mu_j)^2, \tag{52}$$

so

$$s_{j+1}^2 = \left(1 - \frac{1}{j}\right)s_j^2 + (j+1)(\mu_{j+1} - \mu_j)^2. \tag{53}$$

See also CENTRAL MOMENT, CHARLIER'S CHECK, CORRELATION (STATISTICAL), COVARIANCE, COVARIANCE MATRIX, ERROR PROPAGATION, K-STATISTIC, MEAN, MOMENT, RAW MOMENT, SAMPLE VARIANCE, STANDARD ERROR

References
Kenney, J. F. and Keeping, E. S. *Mathematics of Statistics,* Pt. 2, 2nd ed. Princeton, NJ: Van Nostrand, 1951.
Papoulis, A. *Probability, Random Variables, and Stochastic Processes,* 2nd ed. New York: McGraw-Hill, pp. 144–145, 1984.
Press, W. H.; Flannery, B. P.; Teukolsky, S. A.; and Vetterling, W. T. "Moments of a Distribution: Mean, Variance, Skewness, and So Forth." §14.1 in *Numerical Recipes in FORTRAN: The Art of Scientific Computing,* 2nd ed. Cambridge, England: Cambridge University Press, pp. 604–609, 1992.
Roberts, M. J. and Riccardo, R. *A Student's Guide to Analysis of Variance.* London: Routledge, 1999.

Variate
A RANDOM VARIABLE in statistics.

References
Kenney, J. F. and Keeping, E. S. "Variates." §1.2 in *Mathematics of Statistics, Pt. 1, 3rd ed.* Princeton, NJ: Van Nostrand, pp. 5–6, 1962.

Variation

The Δ-variation is a variation in which the varied path over which an integral is evaluated may end at different times than the correct path, and there may be variation in the coordinates at the endpoints.

The δ-variation is a variation in which the varied path in configuration space terminates at the endpoints representing the system configuration at the same time t_1 and t_2 as the correct path; i.e., the varied path always returns to the same endpoints in configuration space, so

$$\delta q_i(t_1) = \delta q_i(t_2) = 0.$$

See also CALCULUS OF VARIATIONS, VARIATION OF ARGUMENT, VARIATION OF PARAMETERS

Variation Coefficient

If s_x is the STANDARD DEVIATION of a set of samples x_i and $\bar{x}$ its MEAN, then

$$V \equiv \frac{s_x}{\bar{x}}.$$

Variation of Argument

Let $[\arg f(z)]$ denote the change in argument of a function $f(z)$ around a CONTOUR γ. Also let N denote the number of ROOTS of $f(z)$ in γ and P denote the number of POLES of $f(z)$ in γ. Then

$$[\arg f(z)] = \frac{1}{2\pi}(N - P). \tag{1}$$

To find $[\arg f(z)]$ in a given region R, break R into paths and find $[\arg f(z)]$ for each path. On a circular ARC

$$z = Re^{i\theta}, \tag{2}$$

let $f(z)$ be a POLYNOMIAL $P(z)$ of degree n. Then

$$[\arg P(z)] = \left[\arg\left(z^n \frac{P(z)}{z^n}\right)\right]$$

$$= [\arg z^n] + \left[\arg\left(\frac{P(z)}{z^n}\right)\right]. \tag{3}$$

Plugging in $z = Re^{i\theta}$ gives

$$[\arg P(z)] = [\arg Re^{i\theta n}] + \left[\arg \frac{P(Re^{i\theta})}{Re^{i\theta n}}\right] \tag{4}$$

$$\lim_{R \to \infty} \frac{P(Re^{i\theta})}{Re^{i\theta n}} = [\text{constant}], \tag{5}$$

so

$$\left[\frac{P(Re^{i\theta})}{Re^{i\theta n}}\right] = 0, \tag{6}$$

and

$$[\arg P(z)] = [\arg e^{i\theta n}] = n(\theta_2 - \theta_1). \tag{7}$$

For a REAL segment $z = x$,

$$[\arg f(x)] = \tan^{-1}\left[\frac{0}{f(x)}\right] = 0. \tag{8}$$

For an IMAGINARY segment $z = iy$,

$$[\arg f(iy)] = \left\{\tan^{-1}\frac{\Im[P(iy)]}{\Re[P(iy)]}\right\}_{\theta_1}^{\theta_2}. \tag{9}$$

Note that the ARGUMENT must change continuously, so "jumps" occur across inverse tangent asymptotes.

Variation of Parameters

For a second-order ORDINARY DIFFERENTIAL EQUATION,

$$y'' + p(x)y' + q(x)y = g(x). \tag{1}$$

Assume that linearly independent solutions $y_1(x)$ and $y_2(x)$ are known and seek $v_1(x)$ and $v_2(x)$ such that

$$y^* = v_1 y_1 + v_2 y_2 \tag{2}$$

$$y'^* = (v_1' y_1 + v_2' y_2) + (v_1 y_1' + v_2 y_2'). \tag{3}$$

Now, impose the additional condition that

$$v_1' y_1 + v_2' y_2 = 0 \tag{4}$$

so that

$$y'^*(x) = v_1 y_1' + v_2 y_2' \tag{5}$$

$$y''^*(x) = v_1' y_1' + v_2' y_2' + v_1 y_1'' + v_2 y_2''. \tag{6}$$

Plug y^*, $y^{*\prime}$, and $y^{*\prime\prime}$ back into the original equation to obtain

$$v_1(y_1'' + py_1' + qy_1) + v_2(y_2'' + py_2' + qy_2) + v_1' y_1' + v_2' y_2'$$
$$= g(x) \tag{7}$$

$$v_1' y_1' + v_2' y_2' = g(x). \tag{8}$$

Therefore,

$$v_1' y_1 + v_2' y_2 = 0 \tag{9}$$

$$v_1' y_1' + v_2' y_2' = g(x). \tag{10}$$

Generalizing to an nth degree ODE, let $y_1, \ldots, y_n$ be the solutions to the homogeneous ODE and let $v_1'(x)$, $\ldots, v_n'(x)$ be chosen such that

$$\begin{cases} y_1 v_1' + y_2 v_2' + \ldots + y_n v_n' = 0 \\ y_1' v_1' + y_2' v_2' + \ldots + y_n' v_n' = 0 \\ \vdots \\ y_1^{(n-1)} v_1' + y_2^{(n-1)} v_2' + \ldots + y_n^{(n-1)} v_n' = g(x). \end{cases} \quad (11)$$

Then the particular solution is then

$$y^*(x) = v_1(x) y_1(x) + \ldots + v_n(x) y_n(x). \quad (12)$$

Variational Calculus
CALCULUS OF VARIATIONS

Variety
ALGEBRAIC VARIETY

Varignon Parallelogram

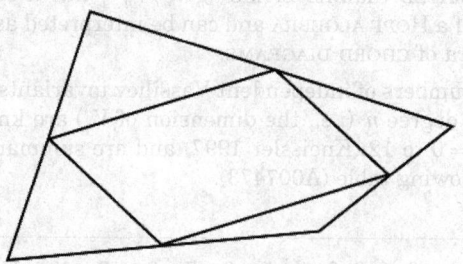

The figure formed when the MIDPOINTS of the sides of a convex QUADRILATERAL are joined. VARIGNON'S THEOREM demonstrated that this figure is a PARALLELOGRAM. The center of the Varignon parallelogram is the CENTROID of four point masses placed on the VERTICES of the QUADRILATERAL.

See also BIMEDIAN, MIDPOINT, MIDPOINT POLYGON, PARALLELOGRAM, QUADRILATERAL, VARIGNON'S THEOREM

References
Coxeter, H. S. M. and Greitzer, S. L. *Geometry Revisited.* Washington, DC: Math. Assoc. Amer., p. 53, 1967.

Varignon's Theorem

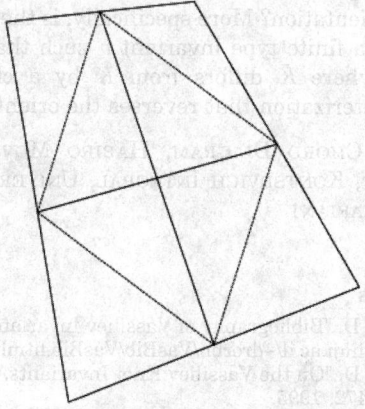

The figure formed when the MIDPOINTS of the sides of

a convex QUADRILATERAL are joined in order is a PARALLELOGRAM. Equivalently, the BIMEDIANS bisect each other. The AREA of this VARIGNON PARALLELOGRAM is half that of the QUADRILATERAL. The PERIMETER is equal to the sum of the diagonals of the original QUADRILATERAL.

See also BIMEDIAN, MIDPOINT, MIDPOINT POLYGON, QUADRILATERAL, VARIGNON PARALLELOGRAM

References
Coxeter, H. S. M. and Greitzer, S. L. "Quadrangles; Varignon's Theorem." §3.1 in *Geometry Revisited.* Washington, DC: Math. Assoc. Amer., pp. 51–56, 1967.

Vassiliev Invariant
This entry contributed by SERGEI DUZHIN

Vassiliev invariants, discovered around 1989, provided a radically new way of looking at KNOTS. The notion of finite type (a.k.a. Vassiliev) KNOT INVARIANTS was independently invented by V. Vassiliev and M. Goussarov around 1989. Vassiliev's approach is based on the study of discriminants in the (infinite-dimensional) spaces of SMOOTH MAPS from one MANIFOLD into another. By definition, the discriminant consists of all maps with SINGULARITIES.

For example, consider the space of all smooth maps from the circle into 3-space $\mathcal{M} = \{ f : \mathbb{S}^1 \to \mathbb{R}^3 \}$. If f is an EMBEDDING (i.e., has no singular points), then it represents a knot. The complement of the set of all knots is the discriminant $\Sigma \subset \mathcal{M}$. It consists of all smooth maps from $\mathbb{S}^1$ into $\mathbb{R}^3$ that have singularities, either *local*, where $f' = 0$, or *nonlocal*, where f is not injective. Two knots are equivalent IFF they can be joined by a path in the space $\mathcal{M}$ that does not intersect the discriminant. Therefore, knot types are in one-to-one correspondence with the connected components of the complement $\mathcal{M} \backslash \Sigma$, and KNOT INVARIANTS with values in an ABELIAN GROUP G are nothing but COHOMOLOGY CLASSES from $H^0(\mathcal{M} \backslash \Sigma, G)$. The FILTRATION of Σ by subspaces corresponding to SINGULAR KNOTS with a given number of ORDINARY DOUBLE POINTS gives rise to a SPECTRAL SEQUENCE, which contains, in particular, the spaces of finite type invariants.

Birman and Lin (1993) have contributed significantly to the simplification of the Vassiliev's original techniques. In particular, they explained the relation between JONES POLYNOMIALS and finite type invariants (Peterson 1992, Birman and Lin 1993, Bar-Natan 1995) and emphasized the role of the algebra of CHORD DIAGRAMS. In fact, substituting the POWER SERIES for e^x as the variable in the JONES POLYNOMIAL yields a POWER SERIES whose COEFFICIENTS are Vassiliev invariants (Birman and Lin 1993). Kontsevich (1993) proved the first difficult theorem about Vassiliev invariants with the help of the KONTSEVICH INTEGRAL. Bar-Natan undertook a thorough study of

Vassiliev invariants; in particular, he showed the importance of the algebra of Feynman diagrams and diagrams with uni- and tri-valent vertices (Bar-Natan 1995). Bar-Natan (1995) remains the most authoritative source on the subject.

Expressed in simple terms, Vassiliev's fundamental idea is to study the prolongation of KNOT INVARIANTS to SINGULAR KNOTS–immersions $f : \mathbb{S}^1 \to \mathbb{R}^3$ having a finite number of ORDINARY DOUBLE POINTS. Let X_n denote the set of EQUIVALENCE CLASSES of SINGULAR KNOTS with n double points and no other singularities. The following definition is based on a recursion which allows to extend a KNOT INVARIANT from X_0 to X_1, then to X_2, etc., and thus finally to the whole of $X = \cup_n X_n$. Given a knot invariant $v : X_0 \to \mathbb{Q}$, its Vassiliev prolongation $\hat{v} : X \to \mathbb{Q}$ is defined as by the rules

1. $\hat{v}|_X \equiv v$, and

2. $\hat{v}(\bigotimes) = \hat{v}(\bigotimes) - \hat{v}(\bigotimes)$ (Vassiliev's skein relation).

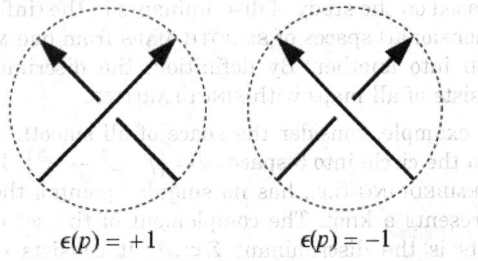

$$\epsilon(p) = +1 \qquad \epsilon(p) = -1$$

The right-hand side of Vassiliev's skein relation refers to the two resolutions of the double point–positive and negative. A crucial observation is that each of them is well-defined (does not depend on the plane projection used to express this relation). A KNOT INVARIANT v is called a *Vassiliev invariant of order* $\leq n$ if its prolongation $\hat{v}$ vanishes on all knots with more than n double points. For example, the simplest nontrivial Vassiliev invariant v_2 has the following explicit description. Let D be an arbitrary KNOT DIAGRAM of the given knot K and $\star$ an arbitrary distinguished point on D, different from all crossings. Then

$$v_2(K) = \sum_{\substack{ijij \\ UOOU}} \epsilon_i \epsilon_j,$$

where the summation spreads over all pairs of crossing points i, j such that (1) during one complete turn of the diagram in the positive direction starting from point $\star$ the points i and j are encountered in the order i, j, i, j, and (2) the four corresponding passages through these crossing points are underpass, overpass, overpass, and underpass, respectively. The numbers ϵ_i, ϵ_j stand for the local WRITHE at points i

and j, defined according to the above illustration.

It turns out that the nth coefficient of the CONWAY POLYNOMIAL is a Vassiliev invariant of order n and, in particular, the second coefficient coincides with v_2.

Vassiliev invariants are at least as strong as all known polynomial knot invariants: ALEXANDER, JONES, KAUFFMAN, and HOMFLY POLYNOMIALS. This means that if two knots K_1 and K_2 can be distinguished by such a polynomial, then there is a Vassiliev invariant that takes different values for K_1 and K_2.

The set of all $\mathbb{Q}$-valued Vassiliev invariants $V = \cup_n V_n$ forms a VECTOR SPACE over the rationals, with the increasing FILTRATION $\mathbb{Q} = V_0 \subset V_1 \subset V_2 \subset \dots$. The ASSOCIATED GRADED SPACE $\oplus_n V_n/V_{n-1}$ has a structure of a HOPF ALGEBRA and can be interpreted as the algebra of CHORD DIAGRAMS.

The numbers of independent Vassiliev invariants of a given degree n (i.e., the dimension of V_n) are known for $n = 0$ to 12 (Kneissler 1997) and are summarized in following table (A007473).

n	0	1	2	3	4	5	6	7	8	9	10	11	12
$\dim V_n$	1	1	2	3	6	10	19	33	60	104	184	316	548

The totality of all Vassiliev invariants is equivalent to one UNIVERSAL VASSILIEV INVARIANT defined through the KONTSEVICH INTEGRAL.

Two of the most important problems about Vassiliev invariants were raised in 1990 and remain unanswered today.

1. Is it true that Vassiliev invariants distinguish knots? In other words, given two nonequivalent knots K_1 and K_2, is it always possible to indicate a finite type invariant v such that $v(K_1) \neq v(K_2)$?
2. Is it true that Vassiliev invariants can detect knot orientation? More specifically, is there a knot K and a finite type invariant v such that $v(K) \neq v(\bar{K})$, where $\bar{K}$ differs from K by a change of parameterization that reverses the orientation?

See also CHORD DIAGRAM, HABIRO MOVE, KNOT INVARIANT, KONTSEVICH INTEGRAL, UNIVERSAL VASSILIEV INVARIANT

References

Bar-Natan, D. "Bibliography of Vassiliev Invariants." http://www.ma.huji.ac.il/~drorbn/VasBib/VasBib.html.

Bar-Natan, D. "On the Vassiliev Knot Invariants." *Topology* **34**, 423–472, 1995.

Birman, J. S. "New Points of View in Knot Theory." *Bull. Amer. Math. Soc.* **28**, 253–287, 1993.

Birman, J. S. and Lin, X.-S. "Knot Polynomials and Vassiliev's Invariants." *Invent. Math.* **111**, 225–270, 1993.

Duzhin, S. V. "Vassiliev invariants and combinatorial structures." Online lecture notes, 1999–2000. http://www.botik.ru/~duzhin/Vics.

Goussarov, M. "On n-Equivalence of Knots and Invariants of Finite Degree." In *Topology of Manifolds and Varieties* (Ed. O. Viro). Providence, RI: Amer. Math. Soc., pp. 173–192, 1994.

Kneissler, J. "The Number of Primitive Vassiliev Invariants up to Degree Twelve." 1997. http://www.math.uni-bonn.de/people/jk/pappvi12.cgi.

Kontsevich, M. "Vassiliev's Knot Invariants." *Adv. Soviet Math.* **16**, Part 2, pp. 137–150, 1993.

Peterson, I. "Knotty Views: Tying Together Different Ways of Looking at Knots." *Sci. News* **141**, 186–187, 1992.

Prasolov, V. V. and Sossinsky, A. B. *Knots, Links, Braids and 3-Manifolds: An Introduction to the New Invariants in Low-Dimensional Topology.* Providence, RI: Amer. Math. Soc., 1996.

Sloane, N. J. A. Sequences A007473/M0765 in "An On-Line Version of the Encyclopedia of Integer Sequences." http://www.research.att.com/~njas/sequences/eisonline.html.

Stoimenow, A. "Degree-3 Vassiliev Invariants." http://guests.mpim-bonn.mpg.de/alex/ptab/vas3.html.

Vassiliev, V. A. "Cohomology of Knot Spaces." In *Theory of Singularities and Its Applications* (Ed. V. I. Arnold). Providence, RI: Amer. Math. Soc., pp. 23–69, 1990.

Vassiliev, V. A. *Complements of Discriminants of Smooth Maps: Topology and Applications.* Providence, RI: Amer. Math. Soc., 1992.

Vassiliev Polynomial

Vassiliev (1990) introduced a radically new way of looking at KNOTS by considering a multidimensional space in which each point represents a possible 3-D knot configuration. If two KNOTS are equivalent, a path then exists in this space from one to the other. The paths can be associated with polynomial invariants.

Birman and Lin (1993) subsequently found a way to translate this scheme into a set of rules and list of potential starting points, which makes analysis of Vassiliev polynomials much simpler. Bar-Natan (1995) and Birman and Lin (1993) proved that JONES POLYNOMIALS and several related expressions are directly connected (Peterson 1992). In fact, substituting the POWER SERIES for $\mathbb{R}^3$ as the variable in the JONES POLYNOMIAL yields a POWER SERIES whose COEFFICIENTS are Vassiliev polynomials (Birman and Lin 1993). Bar-Natan (1995) also discovered a link with Feynman diagrams (Peterson 1992).

See also HABIRO MOVE

References

Bar-Natan, D. "On the Vassiliev Knot Invariants." *Topology* **34**, 423–472, 1995.

Birman, J. S. "New Points of View in Knot Theory." *Bull. Amer. Math. Soc.* **28**, 253–287, 1993.

Birman, J. S. and Lin, X.-S. "Knot Polynomials and Vassiliev's Invariants." *Invent. Math.* **111**, 225–270, 1993.

Peterson, I. "Knotty Views: Tying Together Different Ways of Looking at Knots." *Sci. News* **141**, 186–187, 1992.

Praslov, V. V. and Sossinsky, A. B. *Knots, Links, Braids and 3-Manifolds: An Introduction to the New Invariants in Low-Dimensional Topology.* Providence, RI: Amer. Math. Soc., 1996.

Stoimenow, A. "Degree-3 Vassiliev Invariants." http://guests.mpim-bonn.mpg.de/alex/ptab/vas3.html.

Vassiliev, V. A. "Cohomology of Knot Spaces." In *Theory of Singularities and Its Applications* (Ed. V. I. Arnold). Providence, RI: Amer. Math. Soc., pp. 23–69, 1990.

Vassiliev, V. A. *Complements of Discriminants of Smooth Maps: Topology and Applications.* Providence, RI: Amer. Math. Soc., 1992.

Vault

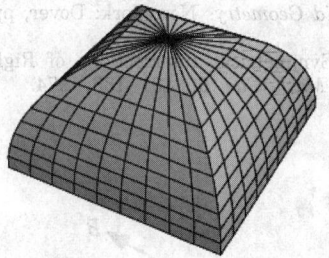

Let a vault consist of two equal half-CYLINDERS of radius r which intersect at RIGHT ANGLES so that the lines of their intersections (the "groins") terminate in the VERTICES of a SQUARE. Two vaults placed bottom-to-top form a STEINMETZ SOLID on two cylinders.

Solving the equations

$$x^2 + z^2 = r^2 \tag{1}$$

$$y^2 + z^2 = r^2 \tag{2}$$

simultaneously gives

$$x = \pm\sqrt{r^2 - z^2} \tag{3}$$

$$y = \pm\sqrt{r^2 - z^2} \tag{4}$$

One quarter of the vault can therefore be described by the PARAMETRIC EQUATIONS

$$x = \sqrt{r^2 - z^2} \tag{5}$$

$$y = -u\sqrt{r^2 - z^2} \tag{6}$$

$$z = z. \tag{7}$$

The SURFACE AREA of the vault is therefore given by

$$A = 4 \int l(z) r \, d\theta, \tag{8}$$

where $l(z)$ is the length of a cross section at height z and θ is the angle a point on the center of this line makes with the origin. But $z = r \sin \theta$, so

$$dz = r \cos \theta \, d\theta = r\sqrt{1 - \sin^2 \theta} \, d\theta = \sqrt{r^2 - z^2} \, d\theta,$$

and

$$l(z) = 2\sqrt{r^2 - x^2} \tag{9}$$

$$A = 4\int_0^r 2r\sqrt{r^2 - z^2}\,\frac{dz}{\sqrt{r^2 - z^2}} = 4\int_0^r 2r\,dz = 8r^2. \tag{10}$$

The VOLUME of the vault is

$$V = \int_0^r \left(2\sqrt{r^2 - z^2}\right)^2 dz = \tfrac{8}{3}r^3. \tag{11}$$

See also CYLINDER, DOME, STEINMETZ SOLID

References

Lines, L. *Solid Geometry.* New York: Dover, pp. 112–113, 1965.

Moore, M. "Symmetrical Intersections of Right Circular Cylinders." *Math. Gaz.* **58**, 181–185, 1974.

Vector

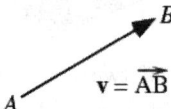

$$\mathbf{v} = \overrightarrow{AB}$$

A vector is formally defined as an element of a VECTOR SPACE. In the commonly encountered VECTOR SPACE $\mathbb{R}^n$ (i.e., Euclidean n-space), a vector is given by n coordinates and can be specified as $(A_1, A_2, \ldots, A_n)$. Vectors can be added together (VECTOR ADDITION) and multiplied by SCALARS (SCALAR MULTIPLICATION). VECTOR MULTIPLICATION is not uniquely defined, but a number of different types of products, such as the DOT PRODUCT, CROSS PRODUCT, TENSOR DIRECT PRODUCT can be defined for pairs of vectors.

A vector from a point A to a point B is denoted $\overrightarrow{AB}$, and a vector v may be denoted $\vec{v}$, or more commonly, $\mathbf{v}$. The point A is often called the "tail" of the vector, and B is called the vector's "head." A vector with unit length is called a UNIT VECTOR and is denoted using a HAT, $\hat{\mathbf{v}}$. An arbitrary vector may be converted to a UNIT VECTOR by dividing by its NORM (i.e., length),

$$|v| \equiv \sqrt{v_1^2 + v_2^2 + \ldots + v_n^2}, \tag{1}$$

giving

$$\hat{\mathbf{v}} \equiv \frac{\mathbf{v}}{|\mathbf{v}|} \tag{2}$$

A ZERO VECTOR, denoted 0, is a vector of length 0, and thus has all components equal to zero.

Since vectors remain unchanged under TRANSLATION, it is often convenient to consider the tail A as located at the origin when, for example, defining VECTOR ADDITION and SCALAR MULTIPLICATION.

A vector may also be defined as a set of n numbers $A_0, \ldots, A_n$ that transform according to the rule

$$A_i' = a_{ij}A_j, \tag{3}$$

where EINSTEIN SUMMATION notation has been used,

$$a_{ij} = \frac{\partial x_i'}{\partial x_j} = \frac{\partial x_j}{\partial x_i'} \tag{4}$$

are constants (corresponding to the DIRECTION COSINES), with partial derivatives taken with respect to the original and transformed coordinate axes, and $i, j = 1, \ldots, n$ (Arfken 1985, p. 10). This makes a vector a TENSOR of RANK one. A vector with n components in called an n-vector, and a SCALAR may therefore be thought of as a 1-vector (or a 0-RANK TENSOR). Vectors are invariant under TRANSLATION, and they reverse sign upon inversion. Objects which resemble vectors but do not reverse sign upon inversion are known as PSEUDOVECTORS.

A vector is represented in *Mathematica* as a list of numbers {a1, a2, ..., an}. VECTOR ADDITION is then simply written using a plus sign, e.g., {a1, a2, ..., an} + {b1, b2, ..., bn }, and SCALAR MULTIPLICATION is indicated by placing a scalar next to a vector (with or without an optional asterisk), s{a1, a2, ..., an}.

Let $\hat{\mathbf{n}}$ be the UNIT VECTOR defined in SPHERICAL COORDINATES by

$$\hat{\mathbf{n}} \equiv \begin{bmatrix} \cos\theta\sin\phi \\ \sin\theta\sin\phi \\ \cos\phi \end{bmatrix}. \tag{5}$$

Then the average value of the x-component of the $\hat{\mathbf{n}}$ over the surface of the UNIT SPHERE is given by

$$\langle n_x \rangle = \frac{\int_0^{2\pi}\int_0^{\pi} (\cos\theta\sin\phi)\sin\phi\,d\phi\,d\theta}{\int_0^{2\pi}\int_0^{\pi}\sin\phi\,d\phi\,d\theta}$$

$$= \frac{1}{4\pi}[\sin\theta]_0^{2\pi}\int_0^{2\pi}\sin^2\phi\,d\phi = 0. \tag{6}$$

More generally,

$$\langle n_i \rangle = 0 \tag{7}$$

for $i = x, y,$ or z (indexed as 1, 2, 3), and

$$\langle n_i n_j \rangle = \tfrac{1}{3}\delta_{ij} \tag{8}$$

$$\langle n_i n_j n_k \rangle = 0 \tag{9}$$

$$\langle n_i n_k n_l n_m \rangle = \tfrac{1}{15}(\delta_{ik}\delta_{lm} + \delta_{il}\delta_{km} + \delta_{im}\delta_{kl}). \tag{10}$$

Given vectors $\mathbf{a}$, $\mathbf{b}$, $\mathbf{c}$, $\mathbf{d}$, the average values of a number of quantities over the UNIT SPHERE are given by

$$\left\langle (\mathbf{a}\cdot\hat{\mathbf{n}})^2 \right\rangle = \tfrac{1}{3}a^2 \tag{11}$$

$$\langle (\mathbf{a}\cdot\hat{\mathbf{n}})(\mathbf{b}\cdot\hat{\mathbf{n}}) \rangle = \tfrac{1}{3}\mathbf{a}\cdot\mathbf{b} \tag{12}$$

$$\langle (\mathbf{a} \cdot \hat{\mathbf{n}}) \hat{\mathbf{n}} \rangle = \tfrac{1}{3} \mathbf{a} \tag{13}$$

$$\left\langle \left(\mathbf{a} \times \hat{\mathbf{n}} \right)^2 \right\rangle = \tfrac{2}{3} a^2 \tag{14}$$

$$\langle (\mathbf{a} \times \hat{\mathbf{n}}) \cdot (\mathbf{b} \times \hat{\mathbf{n}}) \rangle = \tfrac{2}{3} \mathbf{a} \cdot \mathbf{b}, \tag{15}$$

and

$$\langle (\mathbf{a} \cdot \hat{\mathbf{n}})(\mathbf{b} \cdot \hat{\mathbf{n}})(\mathbf{c} \cdot \hat{\mathbf{n}})(\mathbf{d} \cdot \hat{\mathbf{n}}) \rangle$$

$$= \tfrac{1}{15}[(\mathbf{a} \cdot \mathbf{d})(\mathbf{c} \cdot \mathbf{d}) + (\mathbf{a} \cdot \mathbf{c})(\mathbf{b} \cdot \mathbf{d}) + (\mathbf{a} \cdot \mathbf{d})(\mathbf{b} \cdot \mathbf{c})] \tag{16}$$

where δ_{ij} is the KRONECKER DELTA, $\mathbf{a} \cdot \mathbf{b}$ is a DOT PRODUCT, and EINSTEIN SUMMATION has been used.

A MAP $\mathbf{f} : \mathbb{R}^n \mapsto \mathbb{R}^n$ which assigns each $\mathbf{x}$ a VECTOR FUNCTION $\mathbf{f}(\mathbf{x})$ is called a VECTOR FIELD.

See also COLUMN VECTOR, CONTRAVARIANT VECTOR, COVARIANT VECTOR, FOUR-VECTOR, HELMHOLTZ'S THEOREM, NORM, NULL VECTOR, ONE-FORM, PSEU-DOVECTOR, ROW VECTOR, SCALAR, TENSOR, UNIT VECTOR, VECTOR BASIS, VECTOR BUNDLE, VECTOR FIELD, VECTOR FUNCTION, VECTOR SPACE, ZERO VECTOR

References

Arfken, G. "Vector Analysis." Ch. 1 in *Mathematical Methods for Physicists, 3rd ed.* Orlando, FL: Academic Press, pp. 1–84, 1985.
Aris, R. *Vectors, Tensors, and the Basic Equations of Fluid Mechanics.* New York: Dover, 1989.
Crowe, M. J. *A History of Vector Analysis: The Evolution of the Idea of a Vectorial System.* New York: Dover, 1985.
Gibbs, J. W. and Wilson, E. B. *Vector Analysis: A Text-Book for the Use of Students of Mathematics and Physics, Founded Upon the Lectures of J. Willard Gibbs.* New York: Dover, 1960.
Jeffreys, H. and Jeffreys, B. S. "Scalars and Vectors." Ch. 2 in *Methods of Mathematical Physics, 3rd ed.* Cambridge, England: Cambridge University Press, pp. 56–85, 1988.
Marsden, J. E. and Tromba, A. J. *Vector Calculus, 4th ed.* New York: W. H. Freeman, 1996.
Morse, P. M. and Feshbach, H. "Vector and Tensor Formalism." §1.5 in *Methods of Theoretical Physics, Part I.* New York: McGraw-Hill, pp. 44–54, 1953.
Schey, H. M. *Div, Grad, Curl, and All That: An Informal Text on Vector Calculus.* New York: Norton, 1973.
Schwartz, M.; Green, S.; and Rutledge, W. A. *Vector Analysis with Applications to Geometry and Physics.* New York: Harper Brothers, 1960.
Spiegel, M. R. *Schaum's Outline of Theory and Problems of Vector Analysis and an Introduction to Tensor Analysis.* New York: Schaum, 1959.
Weisstein, E. W. "Books about Vectors." http://www.treasure-troves.com/books/Vectors.html.

Vector Addition

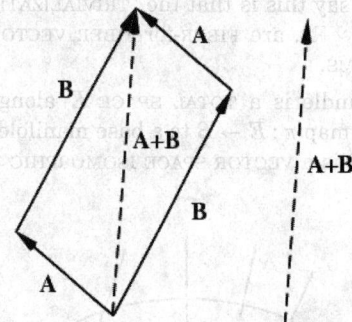

The so-called PARALLELOGRAM LAW gives the rule for vector addition of vectors **A** and **B**. The sum $\mathbf{A} + \mathbf{B}$ of the vectors is obtained by placing them head to tail and drawing the vector from the free tail to the free head.

Vector addition is indicated in *Mathematica* using a plus sign, e.g., {*a1, a2, ..., an*}+{*b1, b2, ..., bn* }.

See also COMPLEX ADDITION, CROSS PRODUCT, DOT PRODUCT, PARALLELOGRAM LAW, SCALAR MULTIPLICATION, VECTOR, VECTOR MULTIPLICATION

Vector Basis

A vector basis is any SET of n LINEARLY INDEPENDENT VECTORS capable of generating an n-dimensional SUBSPACE of $\mathbb{R}^n$. Given a HYPERPLANE defined by

$$x_1 + x_2 + x_3 + x_4 + x_5 = 0,$$

a basis is found by solving for x_1 in terms of x_2, x_3, x_4, and x_5. Carrying out this procedure,

$$x_1 = -x_2 - x_3 - x_4 - x_5,$$

so

$$\begin{bmatrix} x_1 \\ x_2 \\ x_3 \\ x_4 \\ x_5 \end{bmatrix} = x_2 \begin{bmatrix} -1 \\ 1 \\ 0 \\ 0 \\ 0 \end{bmatrix} + x_3 \begin{bmatrix} -1 \\ 0 \\ 1 \\ 0 \\ 0 \end{bmatrix} + x_4 \begin{bmatrix} -1 \\ 0 \\ 0 \\ 1 \\ 0 \end{bmatrix} + x_5 \begin{bmatrix} -1 \\ 0 \\ 0 \\ 0 \\ 1 \end{bmatrix},$$

and the above VECTOR form an (unnormalized) BASIS. Given a MATRIX A with an orthonormal basis, the MATRIX corresponding to a new basis, expressed in terms of the original $\hat{\mathbf{x}}_1, \ldots, \hat{\mathbf{x}}_n$ is

$$A' = [A\hat{\mathbf{x}}_1 \quad \ldots \quad A\hat{\mathbf{x}}_n]$$

See also BASIS, BILINEAR BASIS, MODULAR SYSTEM BASIS, ORTHONORMAL BASIS, TOPOLOGICAL BASIS

Vector Bundle

A special class of FIBER BUNDLE in which the FIBER is a VECTOR SPACE V. Technically, a little more is required; namely, if $f : E \to B$ is a BUNDLE with FIBER $\mathbb{R}^n$, to be a vector bundle, all of the FIBERS $f^{-1}(x)$ for

$x \in B$ need to have a coherent VECTOR SPACE structure. One way to say this is that the "TRIVIALIZATIONS" $h : f^{-1}(U) \to U \times \mathbb{R}^n$, are FIBER-for-FIBER VECTOR SPACE ISOMORPHISMS.

A vector bundle is a TOTAL SPACE E along with a SURJECTIVE map $\pi : E \to B$ to a base manifold B. Any FIBER $\pi^{-1}(b)$ is a VECTOR SPACE ISOMORPHIC to V.

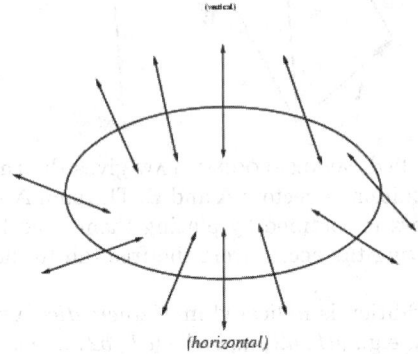

(vertical)

(horizontal)

nontrivial line bundle on the circle

The simplest nontrivial vector bundle is a LINE BUNDLE on the circle, and is analogous to the MÖBIUS STRIP.

One use for vector bundles is a generalization of VECTOR FUNCTIONS. For instance, the tangent vectors of an n-dimensional manifold are isomorphic to $\mathbb{R}^n$ at a point p in a COORDINATE CHART. But the isomorphism with $\mathbb{R}^n$ depends on the choice of COORDINATE CHART. Nearby p, the vector fields look like functions. To define vector fields on the whole manifold requires the TANGENT BUNDLE, which is a special case of a vector bundle.

A SECTION of a vector bundle E is a map $s : B \to E$ whose projection, $\pi \circ s$ is the identity map on B. For instance, on a TRIVIAL BUNDLE $E = B \times V$, a section s corresponds to a function $f : B \to V$ by $s(b) = (b, f(b))$.

Near every point in a vector bundle, there is a TRIVIALIZATION. The structure of the vector bundle, as in all BUNDLES, is that it is LOCALLY TRIVIAL. In the case of a vector bundle, the TRANSITION FUNCTIONS between the trivializations take values in linear invertible transformations of the fiber.

Since the element zero in V is fixed by any linear transformation, the zero section always exists. By "nontrivial section," it is meant that it is not the zero section.

There are several adjectives that can specify properties of a vector bundle. A COMPLEX VECTOR BUNDLE has a fiber V which is a COMPLEX VECTOR SPACE. A REAL VECTOR BUNDLE has a fiber which is a real VECTOR SPACE, which is the default kind of vector bundle. A LINE BUNDLE has a fiber which is one dimensional.

A CONTINUOUS VECTOR BUNDLE is a manifold E with a CONTINUOUS projection map π. A SMOOTH VECTOR BUNDLE is a smooth manifold E with a smooth projection π. Finally, a HOLOMORPHIC VECTOR BUNDLE is a COMPLEX MANIFOLD E with a HOLOMORPHIC projection π. In this last case, the fiber must be a complex vector space. So there could be a smooth complex vector bundle, but not a holomorphic real vector bundle.

Vector bundles can have metrics on their fibers, either RIEMANNIAN or HERMITIAN, and CONNECTIONS.

See also CONNECTION (VECTOR BUNDLE), FIBER, FIBER BUNDLE, HERMITIAN METRIC, K-THEORY, LIE ALGEBROID, LINEAR ALGEBRA, PRINCIPAL BUNDLE, RANK (BUNDLE), REAL VECTOR BUNDLE, RIEMANNIAN METRIC, STABLE EQUIVALENCE, TANGENT BUNDLE, TANGENT MAP, TRIVIAL BUNDLE, VECTOR SPACE, WHITNEY SUM

Vector Cross Product
CROSS PRODUCT

Vector Derivative
The basic types of derivatives operating on a VECTOR FIELD are the CURL $\nabla \times$, DIVERGENCE $\nabla \cdot$, and GRADIENT ∇.

Vector derivative identities involving the CURL include

$$\nabla \times (k\mathbf{A}) = k\nabla \times \mathbf{A} \tag{1}$$

$$\nabla \times (f\mathbf{A}) = f(\nabla \times \mathbf{A}) + (\nabla f) \times \mathbf{A} \tag{2}$$

$$\nabla \times (\mathbf{A} \times \mathbf{B})$$
$$= (\mathbf{B} \cdot \nabla)\mathbf{A} - (\mathbf{A} \cdot \nabla)\mathbf{B} + \mathbf{A}(\nabla \cdot \mathbf{B}) - \mathbf{B}(\nabla \cdot \mathbf{A}) \tag{3}$$

$$\nabla \times \left(\frac{\mathbf{A}}{f}\right) = \frac{f(\nabla \times \mathbf{A}) + \mathbf{A} \times (\nabla f)}{f^2} \tag{4}$$

$$\nabla \times (\mathbf{A} + \mathbf{B}) = \nabla \times \mathbf{A} + \nabla \times \mathbf{B}. \tag{5}$$

In CARTESIAN COORDINATES

$$\nabla \times \mathbf{x} = \nabla \times \mathbf{y} = \nabla \times \mathbf{z} = \mathbf{0} \tag{6}$$

$$\nabla \times \hat{\mathbf{x}} = \nabla \times \hat{\mathbf{y}} = \nabla \times \hat{\mathbf{z}} = \mathbf{0}. \tag{7}$$

In SPHERICAL COORDINATES,

$$\nabla \times \mathbf{r} = \mathbf{0} \tag{8}$$

$$\nabla \times \hat{\mathbf{r}} = \mathbf{0} \tag{9}$$

$$\nabla \times [\mathbf{r}f(r)] = f(r)(\nabla \times \mathbf{r}) + [\nabla f(r)] \times \mathbf{r}$$
$$= f(r)(\mathbf{0}) + \frac{df}{dr}\hat{\mathbf{r}} \times \mathbf{r} = \mathbf{0} + \mathbf{0} = \mathbf{0}. \tag{10}$$

Vector derivative identities involving the DIVERGENCE include

$$\nabla \cdot (k\mathbf{A}) = k\nabla \cdot \mathbf{A} \tag{11}$$

$$\nabla \cdot (f\mathbf{A}) = f(\nabla \cdot \mathbf{A}) + (\nabla f) \cdot \mathbf{A} \qquad (12)$$

$$\nabla \cdot (\mathbf{A} \times \mathbf{B}) = \mathbf{B} \cdot (\nabla \times \mathbf{A}) - \mathbf{A} \cdot (\nabla \times \mathbf{B}) \qquad (13)$$

$$\nabla \cdot \left(\frac{\mathbf{A}}{f} \right) = \frac{f(\nabla \cdot \mathbf{A}) - (\nabla f) \cdot \mathbf{A}}{f^2} \qquad (14)$$

$$\nabla \cdot (\mathbf{A} + \mathbf{B}) = \nabla \cdot \mathbf{A} + \nabla \cdot \mathbf{B} \qquad (15)$$

In CARTESIAN COORDINATES,

$$\nabla \cdot \mathbf{x} = \nabla \cdot \mathbf{y} = \nabla \cdot \mathbf{z} = 1 \qquad (16)$$

$$\nabla \cdot \hat{\mathbf{x}} = \nabla \cdot \hat{\mathbf{y}} = \nabla \cdot \hat{\mathbf{z}} = 0. \qquad (17)$$

In SPHERICAL COORDINATES,

$$\nabla \cdot \mathbf{r} = 3 \qquad (18)$$

$$\nabla \cdot \hat{\mathbf{r}} = \frac{2}{r} \qquad (19)$$

$$\nabla \cdot [\mathbf{r}f(r)] = \frac{\partial}{\partial x}[xf(r)] + \frac{\partial}{\partial y}[yf(r)] + \frac{\partial}{\partial z}[zf(r)] \qquad (20)$$

$$\frac{\partial}{\partial x}[xf(r)] = x\frac{\partial f}{\partial x} + f = x\frac{\partial f}{\partial r}\frac{\partial r}{\partial x} + f \qquad (21)$$

$$\frac{\partial r}{\partial x} = \frac{\partial}{\partial x}(x^2 + y^2 + z^2)^{1/2} = x(x^2 + y^2 + z^2)^{-1/2} = \frac{x}{r} \qquad (22)$$

$$\frac{\partial}{\partial x}[xf(r)] = \frac{x^2}{r}\frac{df}{dr} + f. \qquad (23)$$

By symmetry,

$$\nabla \cdot [\mathbf{r}f(r)] = 3f(r) + \frac{1}{r}(x^2 + y^2 + z^2)\frac{df}{dr}$$

$$= 3f(r) + r\frac{df}{dr} \qquad (24)$$

$$\nabla \cdot (\hat{\mathbf{r}}f(r)) = \frac{3}{r}(r) + \frac{df}{dr} \qquad (25)$$

$$\nabla \cdot (\hat{\mathbf{r}}r^n) = 3r^{n-1} + (n-1)r^{n-1} = (n+2)r^{n-1}. \qquad (26)$$

Vector derivative identities involving the GRADIENT include

$$\nabla(kf) = k\nabla f \qquad (27)$$

$$\nabla(fg) = f\nabla g + g\nabla f \qquad (28)$$

$$\nabla(\mathbf{A} \cdot \mathbf{B}) = \mathbf{A} \times (\nabla \times \mathbf{B}) + \mathbf{B} \times (\nabla \times \mathbf{A}) + (\mathbf{A} \cdot \nabla)\mathbf{B} + (\mathbf{B} \cdot \nabla)\mathbf{A} \qquad (29)$$

$$\nabla(\mathbf{A} \cdot \nabla f) = \mathbf{A} \times (\nabla \times \nabla f) + \nabla f \times (\nabla \times \mathbf{A}) + \mathbf{A} \cdot \nabla(\nabla f) + \nabla f \cdot \nabla \mathbf{A}$$

$$= \nabla f \times (\nabla \times \mathbf{A}) + \mathbf{A} \cdot \nabla(\nabla f) + \nabla f \cdot \nabla \mathbf{A} \qquad (30)$$

$$\nabla\left(\frac{f}{g}\right) = \frac{g\nabla f - f\nabla g}{g^2} \qquad (31)$$

$$\nabla(f + g) = \nabla f + \nabla g \qquad (32)$$

$$\nabla(\mathbf{A} \cdot \mathbf{A}) = 2\mathbf{A} \times (\nabla \times \mathbf{A}) + 2(\mathbf{A} \cdot \nabla)\mathbf{A} \qquad (33)$$

$$(\mathbf{A} \cdot \nabla)\mathbf{A} = \nabla\left(\tfrac{1}{2}\mathbf{A}^2\right) - \mathbf{A} \times (\nabla \times \mathbf{A}). \qquad (34)$$

Vector second derivative identities include

$$\nabla^2 t \equiv \nabla \cdot (\nabla t) = \frac{\partial^2 t}{\partial x^2} + \frac{\partial^2 t}{\partial y^2} + \frac{\partial^2 t}{\partial z^2} \qquad (35)$$

$$\nabla^2 \mathbf{A} = \nabla(\nabla \cdot \mathbf{A}) - \nabla \times (\nabla \times \mathbf{A}). \qquad (36)$$

This very important second derivative is known as the LAPLACIAN.

$$\nabla \times (\nabla t) = \mathbf{0} \qquad (37)$$

$$\nabla(\nabla \cdot \mathbf{A}) = \nabla^2 \mathbf{A} + \nabla \times (\nabla \times \mathbf{A}) \qquad (38)$$

$$\nabla \cdot (\nabla \times \mathbf{A}) = 0 \qquad (39)$$

$$\nabla \times (\nabla \times \mathbf{A}) = \nabla(\nabla \cdot \mathbf{A}) - \nabla^2 \mathbf{A}$$

$$\nabla \times (\nabla^2 \mathbf{A}) = \nabla \times [\nabla(\nabla \cdot \mathbf{A})] - \nabla \times [\nabla \times (\nabla \times \mathbf{A})]$$

$$= -\nabla \times [\nabla \times (\nabla \times \mathbf{A})]$$

$$= -\left\{ \nabla[\nabla \cdot (\nabla \times \mathbf{A})] - \nabla^2(\nabla \times \mathbf{A}) \right\}$$

$$= \nabla^2(\nabla \times \mathbf{A}) \qquad (40)$$

$$\nabla^2(\nabla \cdot \mathbf{A}) = \nabla \cdot [\nabla(\nabla \cdot \mathbf{A})]$$

$$= \nabla \cdot [\nabla^2 \mathbf{A} + \nabla \times (\nabla \times \mathbf{A})] = \nabla \cdot (\nabla^2 \mathbf{A}) \qquad (41)$$

$$\nabla^2[\nabla \times (\nabla \times \mathbf{A})] = \nabla^2[\nabla(\nabla \cdot \mathbf{A}) - \nabla^2 \mathbf{A}]$$

$$= \nabla^2[\nabla(\nabla \cdot \mathbf{A})] - \nabla^4 \mathbf{A} \qquad (42)$$

$$\nabla \times [\nabla^2(\nabla \times \mathbf{A})] = \nabla^2[\nabla(\nabla \cdot \mathbf{A})] - \nabla^4 \mathbf{A} \qquad (43)$$

$$\nabla^4 \mathbf{A} = -\nabla^2[\nabla \times (\nabla \times \mathbf{A})] + \nabla^2[\nabla(\nabla \cdot \mathbf{A})]$$

$$= \nabla \times [\nabla^2(\nabla \times \mathbf{A})] - \nabla^2[\nabla \times (\nabla \times \mathbf{A})]. \qquad (44)$$

Identities involving combinations of vector derivatives include

$$\mathbf{A} \times (\nabla A) = \tfrac{1}{2}\nabla(\mathbf{A} \cdot \mathbf{A}) - (\mathbf{A} \cdot \nabla)\mathbf{A} \qquad (45)$$

$$\nabla \times (\phi\nabla\phi) = \phi\nabla \times (\nabla\phi) + (\nabla\phi) \times (\nabla\phi) = \mathbf{0} \qquad (46)$$

$$(\mathbf{A} \cdot \nabla)\hat{\mathbf{r}} = \frac{\mathbf{A} - \hat{\mathbf{r}}(\mathbf{A} \cdot \hat{\mathbf{r}})}{r} \qquad (47)$$

$$\nabla f \cdot \mathbf{A} = \nabla \cdot (f\mathbf{A}) - f(\nabla \cdot \mathbf{A}) \qquad (48)$$

$$f(\nabla \cdot \mathbf{A}) = \nabla \cdot (f\mathbf{A}) - \mathbf{A}\nabla f, \qquad (49)$$

where (48) and (49) follow from divergence rule (2).

See also CURL, DIVERGENCE, GRADIENT, LAPLACIAN, VECTOR INTEGRAL, VECTOR QUADRUPLE PRODUCT, VECTOR TRIPLE PRODUCT

References
Gradshteyn, I. S. and Ryzhik, I. M. "Vector Field Theorem." Ch. 10 in *Tables of Integrals, Series, and Products, 6th ed.* San Diego, CA: Academic Press, pp. 1081–1092, 2000.
Morse, P. M. and Feshbach, H. "Table of Useful Vector and Dyadic Equations." *Methods of Theoretical Physics, Part I.* New York: McGraw-Hill, pp. 50–54 and 114–115, 1953.

Vector Direct Product

Given VECTORS **u** and **v**, the vector direct product is

$$\mathbf{uv} \equiv \mathbf{u} \otimes \mathbf{v}^{\mathrm{T}}$$

where $\otimes$ is the MATRIX DIRECT PRODUCT and $\mathbf{v}^{\mathrm{T}}$ is the matrix TRANSPOSE. For 3×3 vectors

$$\mathbf{uv} = \begin{bmatrix} u_1 \mathbf{v}^{\mathrm{T}} \\ u_1 \mathbf{v}^{\mathrm{T}} \\ u_1 \mathbf{v}^{\mathrm{T}} \end{bmatrix} = \begin{bmatrix} u_1 v_1 & u_1 v_2 & u_1 v_3 \\ u_2 v_1 & u_2 v_2 & u_2 v_3 \\ u_3 v_1 & u_3 v_2 & u_3 v_3 \end{bmatrix}.$$

Note that if $\mathbf{u} = \hat{\mathbf{x}}_i$, then $u_j = \delta_{ij}$, where δ_{ij} is the KRONECKER DELTA.

See also MATRIX DIRECT PRODUCT, SHERMAN-MORRISON FORMULA, WOODBURY FORMULA

Vector Division

There is no unique solution A to the MATRIX equation $\mathbf{y} = A\mathbf{x}$ unless **x** is PARALLEL to **y**, in which case A is a SCALAR. Therefore, vector division is not defined.

See also MATRIX, SCALAR

Vector Field

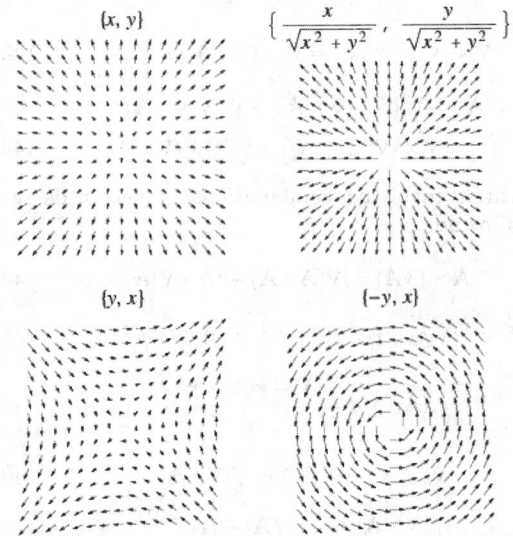

$\{x, y\}$

$\left\{ \dfrac{x}{\sqrt{x^2 + y^2}}, \dfrac{y}{\sqrt{x^2 + y^2}} \right\}$

$\{y, x\}$

$\{-y, x\}$

A MAP $\mathbf{f} : \mathbb{R}^n \mapsto \mathbb{R}^n$ which assigns each **x** a VECTOR FUNCTION $\mathbf{f}(\mathbf{x})$. In French, a vector field is called "un champ." Several vector fields are illustrated above. A vector field is uniquely specified by giving its DIVERGENCE and CURL within a region and its normal component over the boundary, a result known as HELMHOLTZ'S THEOREM (Arfken 1985, p. 79).
FLOWS are generated by vector fields and vice versa. A vector field is a SECTION of its TANGENT BUNDLE.

See also FLOW, SCALAR FIELD, SEIFERT CONJECTURE, TANGENT BUNDLE, VECTOR, WILSON PLUG

References
Arfken, G. "Vector Analysis." Ch. 1 in *Mathematical Methods for Physicists, 3rd ed.* Orlando, FL: Academic Press, pp. 1–84, 1985.
Gray, A. "Vector Fields on $\mathbb{R}^n$" and "Derivatives of Vector Fields on $\mathbb{R}^n$." §11.4 and 11.5 in *Modern Differential Geometry of Curves and Surfaces with Mathematica, 2nd ed.* Boca Raton, FL: CRC Press, pp. 255–258, 1997.
Morse, P. M. and Feshbach, H. "Vector Fields." §1.2 in *Methods of Theoretical Physics, Part I.* New York: McGraw-Hill, pp. 8–21, 1953.

Vector Function

A function of one or more variables whose RANGE is 3-dimensional (or, in general, n-dimensional), as compared to a SCALAR FUNCTION, whose RANGE is 1-dimensional. Vector functions are also called vector-valued functions.

See also COMPLEX FUNCTION, REAL FUNCTION, SCALAR FUNCTION, VECTOR

Vector Harmonic

VECTOR SPHERICAL HARMONIC

Vector Helmholtz Equation

HELMHOLTZ DIFFERENTIAL EQUATION

Vector Integral

The following vector integrals are related to the CURL THEOREM. If

$$\mathbf{F} \equiv \mathbf{c} \times \mathbf{P}(x, y, x), \tag{1}$$

then

$$\int_C d\mathbf{s} \times \mathbf{P} = \int_s (d\mathbf{a} \times \nabla) \times \mathbf{P} \tag{2}$$

If

$$\mathbf{F} \equiv \mathbf{c} F, \tag{3}$$

then

$$\int_C F \, ds = \int_s d\mathbf{a} \times \nabla \mathbf{F} \tag{4}$$

The following are related to the DIVERGENCE THEOREM. If

$$\mathbf{F} \equiv \mathbf{c} \times \mathbf{P}(x, y, x), \tag{5}$$

then

$$\int_V \nabla \times \mathbf{F}\, dV = \int_s d\mathbf{a} \times \mathbf{F}. \tag{6}$$

Finally, if

$$\mathbf{F} \equiv \mathbf{c}F, \tag{7}$$

then

$$\int_V \nabla F\, dV = \int_s F\, d\mathbf{a}. \tag{8}$$

See also CURL THEOREM, DIVERGENCE THEOREM, GRADIENT THEOREM, GREEN'S IDENTITIES, LINE INTEGRAL, SURFACE INTEGRAL, VECTOR DERIVATIVE, VOLUME INTEGRAL

Vector Laplacian

A vector Laplacian can be defined for a VECTOR $\mathbf{A}$ by

$$\nabla^2 \mathbf{A} = \nabla(\nabla \cdot \mathbf{A}) - \nabla \times (\nabla \times \mathbf{A}) \tag{1}$$

in vector notation. The notation $\diamondsuit$ is sometimes also used for a vector Laplacian (Moon and Spencer 1988, p. 3). In tensor notation, $\mathbf{A}$ is written A_μ, and the identity becomes

$$\nabla^2 A_\mu = A_{\mu;\lambda}^{;\lambda} = \left(g^{\lambda\kappa} A_{\mu;\lambda}\right)_{;\kappa}$$

$$= g^{\lambda}{}_{\kappa;\kappa} A_{\mu;\lambda} + g^{\lambda\kappa} A_{\mu;\lambda\kappa}. \tag{2}$$

Similarly, a TENSOR Laplacian can be given by

$$\nabla^2 A_{\alpha\beta} = A_{\alpha\beta;\lambda}^{;\lambda} \tag{3}$$

See also LAPLACIAN, VECTOR POISSON EQUATION

References

Moon, P. and Spencer, D. E. "The Meaning of the Vector Laplacian." *J. Franklin Inst.* **256**, 551–558, 1953.

Moon, P. and Spencer, D. E. *Field Theory Handbook, Including Coordinate Systems, Differential Equations, and Their Solutions, 2nd ed.* New York: Springer-Verlag, 1988.

Vector Multiplication

Although the multiplication of one vector by another is not uniquely defined (cf. SCALAR MULTIPLICATION, which is multiplication of a VECTOR by a SCALAR), several types of useful vector products can be defined, as summarized in the following table.

product name	symbol	result
DOT PRODUCT	$\mathbf{u} \cdot \mathbf{v}$	SCALAR
CROSS PRODUCT	$\mathbf{u} \times \mathbf{v}$	PSEUDOVECTOR
VECTOR DIRECT PRODUCT	$\mathbf{uv}$	TENSOR

Vector multiplication can also be defined for vectors taken three at a time, as summarized in the following table.

product name	symbol	result
VECTOR TRIPLE PRODUCT	$\mathbf{u} \times (\mathbf{v} \times \mathbf{w})$	VECTOR
SCALAR TRIPLE PRODUCT	$[\mathbf{u}, \mathbf{v}, \mathbf{w}]$	PSEUDOSCALAR

A number of VECTOR QUADRUPLE PRODUCTS can also be defined.

See also CROSS PRODUCT, DOT PRODUCT, SCALAR MULTIPLICATION, SCALAR TRIPLE PRODUCT, VECTOR, VECTOR ADDITION, VECTOR DIRECT PRODUCT, VECTOR QUADRUPLE PRODUCT, VECTOR TRIPLE PRODUCT, VECTOR TRIPLE PRODUCT

Vector Norm

Given an n-D VECTOR

$$\mathbf{x} = \begin{bmatrix} x_1 \\ x_2 \\ \vdots \\ x_n \end{bmatrix},$$

a vector norm $\|\mathbf{x}\|$ (sometimes written simply $|\mathbf{x}|$) is a NONNEGATIVE number satisfying

1. $\|\mathbf{x}\| > 0$ when $\mathbf{x} \neq 0$ and $\|\mathbf{x}\| = 0$ IFF $\mathbf{x} = 0$,
2. $\|k\mathbf{x}\| = |k| \|\mathbf{x}\|$ for any SCALAR k,
3. $\|\mathbf{x} + \mathbf{y}\| \leq \|\mathbf{x}\| + \|\mathbf{y}\|$

The vector norm $|\mathbf{x}|_p$ is implemented as Vector-Norm[m, p] in the *Mathematica* add-on package LinearAlgebra`MatrixMultiplication` (which can be loaded with the command <<LinearAlgebra`), where $1 \leq p < \infty$.

See also COMPATIBLE, L1-NORM, L2-NORM, L-INFINITY-NORM, MATRIX NORM, NATURAL NORM, NORM

References

Gradshteyn, I. S. and Ryzhik, I. M. *Tables of Integrals, Series, and Products, 6th ed.* San Diego, CA: Academic Press, p. 1114, 2000.

Vector Ordering

If the first NONZERO component of the vector difference $\mathbf{A} - \mathbf{B}$ is > 0, then $\mathbf{A} \succ \mathbf{B}$. If the first NONZERO component of $\mathbf{A} - \mathbf{B}$ is < 0, then $\mathbf{A} \prec \mathbf{B}$.

See also PRECEDES, SUCCEEDS

Vector Poisson Equation

The PARTIAL DIFFERENTIAL EQUATION

$$\diamond \mathbf{A} = -\nabla \times \mathbf{E},$$

where $\diamond$ is the VECTOR LAPLACIAN.

See also POISSON'S EQUATION, VECTOR LAPLACIAN

References

Moon, P. and Spencer, D. E. "The Meaning of the Vector Laplacian." *J. Franklin Inst.* **256**, 551–558, 1953.
Zwillinger, D. *Handbook of Differential Equations, 3rd ed.* Boston, MA: Academic Press, p. 139, 1997.

Vector Potential

A function $\mathbf{A}$ such that

$$\mathbf{B} \equiv \nabla \times \mathbf{A}.$$

The most common use of a vector potential is the representation of a magnetic field. If a VECTOR FIELD has zero DIVERGENCE, it may be represented by a vector potential.

See also DIVERGENCE, HELMHOLTZ'S THEOREM, PO-TENTIAL FUNCTION, SOLENOIDAL FIELD, VECTOR FIELD

Vector Product

CROSS PRODUCT, SCALAR TRIPLE PRODUCT, VECTOR MULTIPLICATION, VECTOR DIRECT PRODUCT, VECTOR QUADRUPLE PRODUCT, VECTOR TRIPLE PRODUCT

Vector Quadruple Product

There are a number of algebraic identities involving sets of four VECTORS. LAGRANGE'S IDENTITY is given by

$$(\mathbf{A} \times \mathbf{B}) \cdot (\mathbf{C} \times \mathbf{D})$$
$$= (\mathbf{A} \cdot \mathbf{C})(\mathbf{A} \cdot \mathbf{D}) - (\mathbf{A} \cdot \mathbf{D})(\mathbf{B} \cdot \mathbf{C}). \qquad (1)$$

A number of other useful identities include

$$(\mathbf{A} \times \mathbf{B})^2 \equiv (\mathbf{A} \times \mathbf{B}) \cdot (\mathbf{A} \times \mathbf{B})$$
$$= (\mathbf{A} \cdot \mathbf{A})(\mathbf{B} \cdot \mathbf{B}) - (\mathbf{A} \cdot \mathbf{B})(\mathbf{B} \cdot \mathbf{A})$$
$$= \mathbf{A}^2 \mathbf{B}^2 - (\mathbf{A} \cdot \mathbf{B})^2 \qquad (2)$$

$$\mathbf{A} \times (\mathbf{B} \times (\mathbf{C} \times \mathbf{D}))$$
$$= \mathbf{B}(\mathbf{A} \cdot (\mathbf{C} \times \mathbf{D})) - (\mathbf{A} \cdot \mathbf{B})(\mathbf{C} \times \mathbf{D}) \qquad (3)$$
$$(\mathbf{A} \times \mathbf{B}) \times (\mathbf{C} \times \mathbf{D}) = (\mathbf{C} \times \mathbf{D}) \times (\mathbf{B} \times \mathbf{A}) \qquad (4)$$
$$= [\mathbf{A}, \mathbf{B}, \mathbf{D}]\mathbf{C} - [\mathbf{A}, \mathbf{B}, \mathbf{C}]\mathbf{D} \qquad (5)$$
$$[\mathbf{C}, \mathbf{D}, \mathbf{A}]\mathbf{B} - [\mathbf{C}, \mathbf{D}, \mathbf{B}]\mathbf{A}, \qquad (6)$$

where $[\mathbf{A}, \mathbf{B}, \mathbf{C}]$ denotes the SCALAR TRIPLE PRODUCT.

See also LAGRANGE'S IDENTITY, SCALAR TRIPLE PRO-DUCT, VECTOR MULTIPLICATION, VECTOR TRIPLE PRODUCT

Vector Space

A vector space over $\mathbb{R}^n$ is a set of VECTORS for which any VECTORS $\mathbf{X}$, $\mathbf{Y}$, and $\mathbf{Z} \in \mathbb{R}^n$ and any SCALARS r, $s \in \mathbb{R}$ have the following properties:

1. COMMUTATIVITY:
$$\mathbf{X} + \mathbf{Y} = \mathbf{Y} + \mathbf{X}.$$

2. ASSOCIATIVITY of VECTOR ADDITION:
$$(\mathbf{X} + \mathbf{Y}) + \mathbf{Z} = \mathbf{X} + (\mathbf{Y} + \mathbf{Z}).$$

3. Additive identity: For all $\mathbf{X}$,
$$\mathbf{0} + \mathbf{X} = \mathbf{X} + \mathbf{0} = \mathbf{X}.$$

4. Existence of additive inverse: For any $\mathbf{X}$, there exists a $-\mathbf{X}$ such that
$$\mathbf{X} + (-\mathbf{X}) = \mathbf{0}.$$

5. ASSOCIATIVITY of scalar multiplication:
$$r(s\mathbf{X}) = (rs)\mathbf{X}.$$

6. DISTRIBUTIVITY of scalar sums:
$$(r + s)\mathbf{X} = r\mathbf{X} + s\mathbf{X}.$$

7. DISTRIBUTIVITY of vector sums:
$$r(\mathbf{X} + \mathbf{Y}) = r\mathbf{X} + r\mathbf{Y}.$$

8. Scalar multiplication identity:
$$1\mathbf{X} = \mathbf{X}.$$

Let V be a vector space of dimension n over the FIELD of q elements (where q is necessarily a power of a prime number). Then the number of distinct non-singular linear operators on V is

$$M(n, q) = (q^n - q^0)(q^n - q^1)(q^n - q^2) \cdots (q^n - q^{n-1}) \quad (1)$$

and the number of distinct k-dimensional subspaces of V is

$$S(k, n, q) = \frac{(q^n - q^0)(q^n - q^1)(q^n - q^2) \cdots (q^n - q^{k-1})}{M(k, q)}$$
$$\qquad (2)$$
$$= \frac{(q^n - 1)(q^{n-1} - 1)(q^{n-2} - 1) \cdots (q^{n-k+1} - 1)}{(q^k - 1)(q^{k-1} - 1)(q^{k-2} - 1) \cdots (q - 1)}. \quad (3)$$

A consequence of the AXIOM OF CHOICE is that every vector space has a BASIS.

A MODULE is abstractly similar to a vector space, but it uses a RING to define COEFFICIENTS instead of the FIELD used for vector spaces. MODULES have COEFFI-CIENTS in much more general algebraic objects.

See also BANACH SPACE, BASIS (VECTOR SPACE), FIELD, FUNCTION SPACE, HILBERT SPACE, INNER PRODUCT SPACE, MODULE, QUOTIENT VECTOR SPACE,

Ring, Symplectic Space, Topological Vector Space, Vector

References

Arfken, G. *Mathematical Methods for Physicists, 3rd ed.* Orlando, FL: Academic Press, pp. 530–534, 1985.

Vector Spherical Harmonic

The SPHERICAL HARMONICS can be generalized to vector spherical harmonics by looking for a SCALAR FUNCTION ψ and a constant VECTOR $\mathbf{c}$ such that

$$\mathbf{M} \equiv \nabla \times (\mathbf{c}\psi) = \psi(\nabla \times \mathbf{c}) + (\nabla\psi) \times \mathbf{c}$$

$$= (\nabla\psi) \times \mathbf{c} = -\mathbf{c} \times \nabla\psi, \tag{1}$$

so

$$\nabla \cdot \mathbf{M} = 0. \tag{2}$$

Now use the vector identities

$$\nabla^2\mathbf{M} = \nabla^2(\nabla \times \mathbf{M}) = \nabla \times (\nabla^2\mathbf{M})$$

$$= \nabla(\nabla^2\mathbf{c}\psi) = \nabla \times (\mathbf{c}\nabla^2\psi) \tag{3}$$

$$k^2\mathbf{M} = k^2\nabla \times (\mathbf{c}\psi) = \nabla \times (\mathbf{c}\nabla^2\psi) \tag{4}$$

so

$$\nabla^2\mathbf{M} + k^2\mathbf{M} = \nabla \times \left[\mathbf{c}(\nabla^2\psi + \kappa^2\psi)\right], \tag{5}$$

and $\mathbf{M}$ satisfies the vector HELMHOLTZ DIFFERENTIAL EQUATION if ψ satisfies the scalar HELMHOLTZ DIFFERENTIAL EQUATION

$$\nabla^2\psi + k^2\psi = 0. \tag{6}$$

Construct another vector function

$$\mathbf{N} = \frac{\nabla \times \mathbf{M}}{k}, \tag{7}$$

which also satisfies the vector HELMHOLTZ DIFFERENTIAL EQUATION since

$$\nabla^2\mathbf{N} = \frac{1}{k}\nabla^2(\nabla \times \mathbf{M}) = \frac{1}{k}\nabla \times (\nabla^2\mathbf{M})$$

$$= \frac{1}{k}\nabla \times (-k^2\mathbf{M}) = -k\nabla \times \mathbf{M} = -k^2\mathbf{N}, \tag{8}$$

which gives

$$\nabla^2\mathbf{N} + k^2\mathbf{N} = 0. \tag{9}$$

We have the additional identity

$$\nabla \times \mathbf{N} = \frac{1}{k}\nabla \times (\nabla \times \mathbf{M}) = \frac{1}{k}\nabla \times (\nabla \cdot \mathbf{M})$$

$$= \frac{1}{k}\nabla^2\mathbf{M} - \frac{1}{k}\nabla^2\mathbf{M} = \frac{-\nabla^2\mathbf{M}}{k} = k\mathbf{M}. \tag{10}$$

In this formalism, ψ is called the generating function

and $\mathbf{c}$ is called the PILOT VECTOR. The choice of generating function is determined by the symmetry of the scalar equation, i.e., it is chosen to solve the desired scalar differential equation. If $\mathbf{M}$ is taken as

$$\mathbf{M} = \nabla \times (\mathbf{r}\psi), \tag{11}$$

where $\mathbf{r}$ is the radius vector, then $\mathbf{M}$ is a solution to the vector wave equation in spherical coordinates. If we want vector solutions which are tangential to the radius vector,

$$\mathbf{M} \cdot \mathbf{r} = \mathbf{r} \cdot (\nabla\psi \times \mathbf{c}) = (\nabla\psi)(\mathbf{c} \times \mathbf{r}) = 0, \tag{12}$$

so

$$\mathbf{c} \times \mathbf{r} = 0 \tag{13}$$

and we may take

$$\mathbf{c} = \mathbf{r} \tag{14}$$

(Arfken 1985, pp. 707–711; Bohren and Huffman 1983, p. 88).

A number of conventions are in use. Hill (1954) defines

$$V_l^m \equiv -\sqrt{\frac{l+1}{2l+1}}\, Y_l^m\hat{\mathbf{r}} + \frac{1}{\sqrt{(l+1)(2l+1)}}\frac{\partial Y_l^m}{\partial\theta}\hat{\boldsymbol{\theta}}$$

$$+ iM\sqrt{(l+1)(2l+1)}\sin\theta Y_l^m\hat{\boldsymbol{\phi}} \tag{15}$$

$$W_l^m = \sqrt{\frac{l}{2l+1}}\, Y_l^m\hat{\mathbf{r}} + \frac{1}{\sqrt{l(2l+1)}}\frac{\partial Y_l^m}{\partial\theta}\hat{\boldsymbol{\theta}}$$

$$+ \frac{iM}{\sqrt{l(2l+1)}\sin\theta}Y_l^m\hat{\boldsymbol{\phi}} \tag{16}$$

$$X_l^m = -\frac{M}{\sqrt{l(l+1)}\sin\theta}Y_l^m\hat{\boldsymbol{\theta}} - \frac{i}{\sqrt{l(l+1)}}\frac{\partial Y_l^m}{\partial\theta}\hat{\boldsymbol{\phi}} \tag{17}$$

Morse and Feshbach (1953) define vector harmonics called $\mathbf{B}$, $\mathbf{C}$, and $\mathbf{P}$ using rather complicated expressions.

References

Arfken, G. "Vector Spherical Harmonics." §12.11 in *Mathematical Methods for Physicists, 3rd ed.* Orlando, FL: Academic Press, pp. 707–711, 1985.

Blatt, J. M. and Weisskopf, V. "Vector Spherical Harmonics." Appendix B, §1 in *Theoretical Nuclear Physics.* New York: Wiley, pp. 796–799, 1952.

Bohren, C. F. and Huffman, D. R. *Absorption and Scattering of Light by Small Particles.* New York: Wiley, 1983.

Hill, E. H. "The Theory of Vector Spherical Harmonics." *Amer. J. Phys.* **22**, 211–214, 1954.

Jackson, J. D. *Classical Electrodynamics, 2nd ed.* New York: Wiley, pp. 744–755, 1975.

Morse, P. M. and Feshbach, H. *Methods of Theoretical Physics, Part II.* New York: McGraw-Hill, pp. 1898–1901, 1953.

Vector Transformation Law

The set of n quantities v_j are components of an n-D VECTOR $\mathbf{v}$ IFF, under ROTATION,

$$v_i' = a_{ij} v_j$$

for $i = 1, 2, \ldots, n$. The DIRECTION COSINES between x_i' and x_j are

$$a_{ij} \equiv \frac{\partial x_i'}{\partial x_j} = \frac{\partial x_j}{\partial x_i'}.$$

They satisfy the orthogonality condition

$$a_{ij} a_{ik} = \frac{\partial x_j}{\partial x_i'} \frac{\partial x_i'}{\partial x_k} = \frac{\partial x_j}{\partial x_k} = \delta_{jk},$$

where δ_{jk} is the KRONECKER DELTA.

See also TENSOR, VECTOR

Vector Triple Product

The vector triple product identity is also known as the BAC-CAB IDENTITY, and can be written in the form

$$\mathbf{A} \times (\mathbf{B} \times \mathbf{C}) = \mathbf{B}(\mathbf{A} \cdot \mathbf{C}) - \mathbf{C}(\mathbf{A} \cdot \mathbf{B}) \qquad (1)$$

$$(\mathbf{A} \times \mathbf{B}) \times \mathbf{C} = -\mathbf{C} \times (\mathbf{A} \times \mathbf{B})$$

$$= -\mathbf{A}(\mathbf{B} \cdot \mathbf{C}) + \mathbf{B}(\mathbf{A} \cdot \mathbf{C}) \qquad (2)$$

See also BAC-CAB IDENTITY, CROSS PRODUCT, DOT PRODUCT, PERMUTATION SYMBOL, SCALAR TRIPLE PRODUCT, VECTOR MULTIPLICATION, VECTOR QUADRUPLE PRODUCT

References

Arfken, G. "Triple Scalar Product, Triple Vector Product." §1.5 in *Mathematical Methods for Physicists, 3rd ed.* Orlando, FL: Academic Press, pp. 26–33, 1985.

Jeffreys, H. and Jeffreys, B. S. "The Triple Vector Product." §2.092–2.094 in *Methods of Mathematical Physics, 3rd ed.* Cambridge, England: Cambridge University Press, pp. 75–76, 1988.

Vector-Valued Function

VECTOR FUNCTION

Vee

The symbol $\vee$ variously means "disjunction" (i.e., OR in LOGIC) or "join" (for a LATTICE).

See also OR, WEDGE

Velocity

$$\mathbf{v} \equiv \frac{d\mathbf{r}}{dt},$$

where $\mathbf{r}$ is the POSITION VECTOR and d/dt is the derivative with respect to time. Expressed in terms of

the ARC LENGTH,

$$\mathbf{v} = \frac{ds}{dt} \hat{\mathbf{T}},$$

where $\hat{\mathbf{T}}$ is the unit TANGENT VECTOR, so the SPEED (which is the magnitude of the velocity) is

$$v \equiv |\mathbf{v}| = \frac{ds}{dt} = |\mathbf{r}'(t)|.$$

See also ANGULAR VELOCITY, POSITION VECTOR, SPEED

Velocity Vector

The idea of a velocity vector comes from classical physics. By representing the position and motion of a single particle using vectors, the equations for motion are simpler and more intuitive. Suppose the position of a particle at time t is given by the position vector $s(t) = (s_1(t), s_2(t), s_3(t))$. Then the velocity vector $v(t)$ is the derivative of the position,

$$v = \frac{ds}{dt} = \left(\frac{ds_1}{dt}, \frac{ds_2}{dt}, \frac{ds_3}{dt} \right).$$

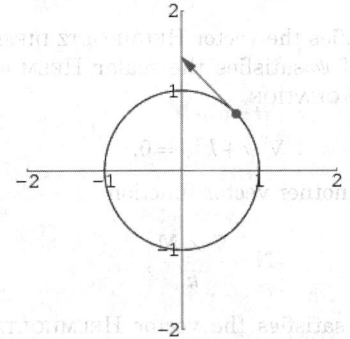

For example, suppose a particle is confined to the plane and its position is given by $s = (\cos t, \sin t)$. Then it travels along the unit circle at constant speed. Its velocity vector is $v = (-\sin t, \cos t)$. In a diagram, it makes sense to translate the velocity vector so it originates at s. In particular, it is drawn as an arrow from s to $s + v$.

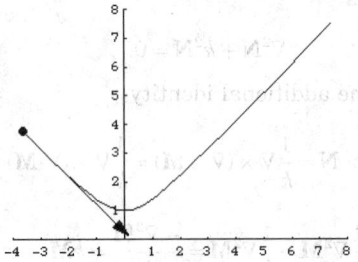

Another example is a particle traveling along a HYPERBOLA specified parametrically by $s(t) = $

($\sinh(t)$, $\cosh(t)$). Its velocity vector is then given by $v = (\cosh(t),\ \sinh(t))$, illustrated above.

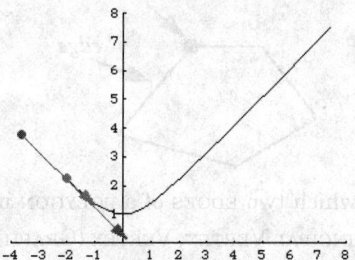

Travel down the same path, but using a different function is called a REPARAMETRIZATION, and the CHAIN RULE describes the change in velocity. For example, the HYPERBOLA can also be parametrized by $r(t) = \left(t, \sqrt{1+t^2}\right)$. Note that $r(\sinh(t)) = s(t)$, and by the CHAIN RULE, $dr/dt(\cosh t) = ds/dt$.

Note that the set of possible velocity vectors forms a VECTOR SPACE. If r and s are two paths through the origin, then so is $r + s$ and the velocity vector of this path is $dr/dt + ds/dt$. Similarly, if α is a scalar, then the path αs has velocity vector αv. It makes sense to distinguish the velocity vectors at different points. In physics, the set of all velocity vectors gives all possible combinations of position and momentum, and is called phase space. In mathematics, the velocity vectors form the tangent space, and the collection of tangent spaces forms the TANGENT BUNDLE.

See also CALCULUS, COORDINATE CHART, DIRECTIONAL DERIVATIVE, EUCLIDEAN SPACE, JACOBIAN, MANIFOLD, TANGENT BUNDLE, TANGENT SPACE, TANGENT VECTOR, VECTOR FIELD, VECTOR SPACE

Venn Diagram

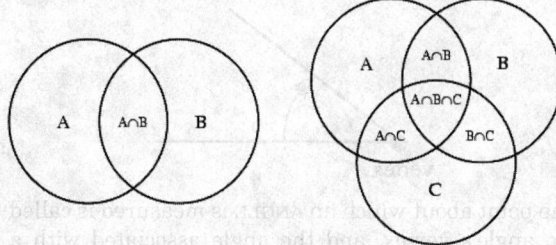

A schematic diagram used in LOGIC theory to depict collections of sets and represent their relationships. The Venn diagrams on two and three sets are illustrated above. The order-two diagram (left) consists of two intersecting circles, producing a total of four regions, A, B, $A \cap B$, and $\varnothing$ (the EMPTY SET, represented by none of the regions occupied). Here, $A \cap B$ denotes the INTERSECTION of sets A and B.

The order-three diagram (right) consists of three symmetrically placed mutually intersecting CIRCLES comprising a total of eight regions. The regions labeled A, B, and C consist of members which are only in one set and no others, the three regions labelled $A \cap B$, $A \cap C$, and $B \cap C$ consist of members which are in two sets but not the third, the region $A \cap B \cap C$ consists of members which are simultaneously in all three, and no regions occupied represents $\varnothing$.

In general, an order-n Venn diagram is a collection of n simple closed curves in the PLANE such that

1. The curves partition the PLANE into 2^n connected regions, and
2. Each SUBSET S of $\{1, 2, \ldots, n\}$ corresponds to a unique region formed by the intersection of the interiors of the curves in S (Ruskey).

Since there are $\binom{n}{k}$ (the BINOMIAL COEFFICIENT) ways to pick k members from a total of n, the number of regions in an order n Venn diagram is

$$N = \sum_{k=0}^{n} \binom{n}{k} = 2^n,$$

(where the region outside the diagram is included in the count).

The region of INTERSECTION of the three CIRCLES $A \cap B \cap C$ in the order three Venn diagram in the special case of the center of each being located at the INTERSECTION of the other two is a geometric shape known as a REULEAUX TRIANGLE.

See also CIRCLE, FLOWER OF LIFE, HARUKI'S THEOREM, INTERSECTION, LENS, MAGIC CIRCLES, REULEAUX TRIANGLE, SEED OF LIFE, Ogilvy, C. S. "Solution to Problem E 1154." *Amer. Math. Monthly* **62**, 584–585, 1955.

References

Cundy, H. and Rollett, A. *Mathematical Models, 3rd ed.* Stradbroke, England: Tarquin Pub., pp. 255–256, 1989.
Ruskey, F. "A Survey of Venn Diagrams." *Elec. J. Combin.* **4**, DS#5, 1997. http://www.combinatorics.org/Surveys/ds5/VennEJC.html.
Ruskey, F. "Venn Diagrams." http://www.theory.csc.uvic.ca/~cos/inf/comb/SubsetInfo.html#Venn.

Verging Construction

NEUSIS CONSTRUCTION

Verhulst Model

LOGISTIC MAP

Verma Module

See also MODULE

References

Huang, J.-S. "Verma Modules." §5.4 in *Lectures on Representation Theory.* Singapore: World Scientific, pp. 52–53, 1999.

Veronese Surface

A smooth 2-D surface given by embedding the PROJECTIVE PLANE into projective 5-space by the homogeneous parametric equations

$$v(x,y,z) = (x^2, y^2, z^2, xy, xz, yz).$$

The surface can be projected smoothly into 4-space, but all 3-D projections have singularities (Coffman). The projections of these surfaces in 3-D are called STEINER SURFACES. The VOLUME of the Veronese surface is $2\pi^2$.

See also STEINER SURFACE

References

Coffman, A. "Steiner Surfaces." http://www.ipfw.edu/math/Coffman/steinersurface.html.

Veronese Variety

VERONESE SURFACE

Versed Sine

VERSINE

Versiera

WITCH OF AGNESI

Versine

$$\text{vers}(z) \equiv 1 - \cos z,$$

where $\cos z$ is the COSINE. Using a trigonometric identity, the versine is equal to

$$\text{vers}(z) = 2\sin^2\left(\tfrac{1}{2}z\right).$$

See also COSINE, COVERSINE, EXSECANT, HAVERSINE

References

Abramowitz, M. and Stegun, C. A. (Eds.). *Handbook of Mathematical Functions with Formulas, Graphs, and Mathematical Tables, 9th printing.* New York: Dover, p. 78, 1972.

Vertex (Graph)

A point of a GRAPH, also called a NODE.

See also EDGE (GRAPH), NULL GRAPH, TAIT COLORING, TAIT CYCLE, TAIT'S HAMILTONIAN GRAPH CONJECTURE, VERTEX (POLYGON)

Vertex (Parabola)

For a PARABOLA oriented vertically and opening upwards, the vertex is the point where the curve reaches a minimum.

Vertex (Polygon)

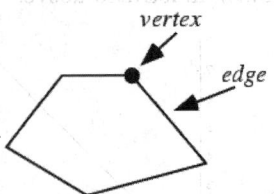

A point at which two EDGES of a POLYGON meet.

See also PRINCIPAL VERTEX, VERTEX (GRAPH), VERTEX (POLYHEDRON)

Vertex (Polyhedron)

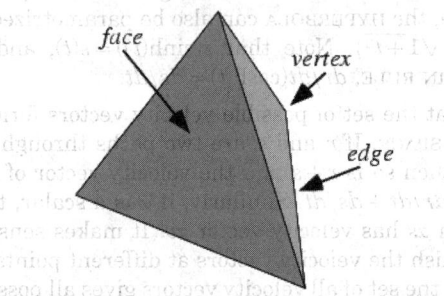

A point at which three of more EDGES of a POLYHEDRON meet. The concept can also be generalized to a POLYTOPE.

See also VERTEX (GRAPH), VERTEX (POLYGON)

Vertex (Polytope)

The vertex of a POLYTOPE is a point where edges of the POLYTOPE meet.

Vertex Angle

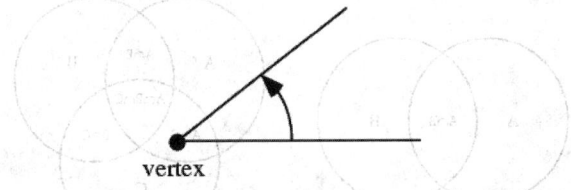

The point about which an ANGLE is measured is called the angle's vertex, and the angle associated with a given vertex is called the vertex angle.

See also ANGLE

Vertex Coloring

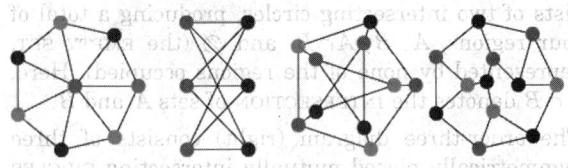

A vertex coloring is an assignment of labels or colors to each vertex of a graph such that no edge connects two identically colored vertices. The most common type of vertex coloring seeks to minimize the number of colors for a given graph. BRELAZ'S HEURISTIC ALGORITHM can be used to find a good, but not necessarily minimal, vertex coloring of a GRAPH. Finding a minimal coloring can be done using brute-force search (Christofides 1971; Wilf 1984; Skiena 1990, p. 214). The minimum number of colors which with the vertices of a graph G may be colored is called the CHROMATIC NUMBER, denoted $\chi(G)$.

The only one-colorable graphs are EMPTY GRAPHS, and two-colorable graphs are exactly BIPARTITE GRAPHS. The FOUR-COLOR THEOREM establishes that all PLANAR GRAPHS are 4-colorable.

See also BRELAZ'S HEURISTIC ALGORITHM, BROOKS' THEOREM, CHROMATIC NUMBER, CHROMATIC POLYNOMIAL, COLORING, EDGE CHROMATIC NUMBER, FOUR-COLOR THEOREM, κ-COLORING

References

Christofides, N. "An Algorithm for the Chromatic Number of a Graph." *Computer J.* **14**, 38–39, 1971.
Gould, R. (Ed.). *Graph Theory.* Menlo Park, CA: Benjamin-Cummings, 1988.
Manvel, B. "Extremely Greedy Coloring Algorithms." In *Graphs and Applications* (Ed. F. Harary and J. Maybee). New York: Wiley, pp. 257–270, 1985.
Matula D. W.; Marble, G.; and Isaacson, J. D. "Graph Coloring Algorithms." In *Graph Theory and Computing* (Ed. R. Read). New York: Academic Press, pp. 109–122, 1972.
Skiena, S. "Finding a Vertex Coloring." §5.5.3 in *Implementing Discrete Mathematics: Combinatorics and Graph Theory with Mathematica.* Reading, MA: Addison-Wesley, pp. 214–215, 1990.
Wilf, H. "Backtrack: An \(1) Expected Time Algorithm for the Graph Coloring Problem." *Info. Proc. Let.* **18**, 119–121, 1984.

Vertex Connectivity

The minimum number of nodes $\kappa(G)$ whose deletion from a GRAPH G disconnects it. Vertex connectivity is sometimes called "point connectivity" or simply "connectivity".

Let $\lambda(G)$ be the EDGE CONNECTIVITY of a graph G and $\delta(G)$ its minimum degree, then for any graph,

$$\kappa(G) \leq \lambda(G) \leq \delta(G)$$

(Whitney 1932, Harary 1994, p. 43).

The vertex connectivity of a graph can be determined with the command VertexConnectivity[g] in the *Mathematica* add-on package DiscreteMath`Combinatorica` (which can be loaded with the command < <DiscreteMath`).

See also DISCONNECTED GRAPH, EDGE CONNECTIVITY, κ-CONNECTED GRAPH, MENGER'S THEOREM

References

Harary, F. *Graph Theory.* Reading, MA: Addison-Wesley, p. 43, 1994.
Skiena, S. *Implementing Discrete Mathematics: Combinatorics and Graph Theory with Mathematica.* Reading, MA: Addison-Wesley, pp. 178–179, 1990.
Whitney, H. "Congruent Graphs and the Connectivity of Graphs." *Amer. J. Math.* **54**, 150–168, 1932.

Vertex Cover

Let S be a collection of subsets of a finite set X. The smallest subset Y of X that meets every member of S is called the vertex cover, or hitting set. However, some authors call any such set a vertex cover, and then refer to the minimum vertex cover (Skiena 1990, p. 218). Finding the hitting set is an NP-COMPLETE PROBLEM.

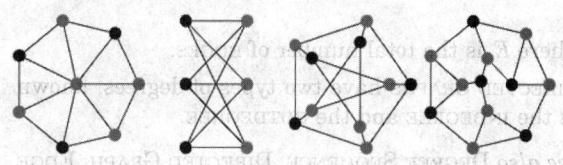

Vertex covers, indicated with red coloring, are shown above for a number of graphs. In a COMPLETE κ-PARTITE GRAPH, and vertex cover contains vertices from at least $k - 1$ stages. The minimum vertex cover of a GRAPH can be computed using MinimumVertexCover[g] in the *Mathematica* add-on package DiscreteMath`Combinatorica` (which can be loaded with the command < <DiscreteMath`).

See also CLIQUE, EDGE COVER, INDEPENDENT SET

References

Skiena, S. "Minimum Vertex Cover." §5.6.2 in *Implementing Discrete Mathematics: Combinatorics and Graph Theory with Mathematica.* Reading, MA: Addison-Wesley, p. 218, 1990.
Skiena, S. S. "Vertex Cover." §8.5.3 in *The Algorithm Design Manual.* New York: Springer-Verlag, pp. 317–318, 1997.

Vertex Degree

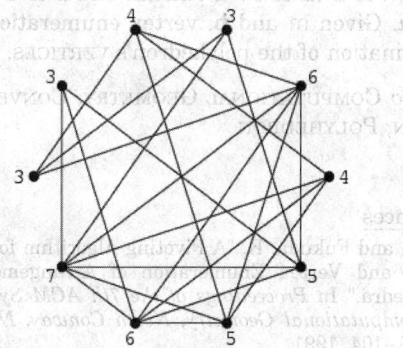

The degree of a VERTEX v of a GRAPH G is the number of EDGES which touch v. The vertex degrees are illustrated above for a random graph. The vertex

degree is also called the local degree or valency. The ordered list of vertex degrees in a given graph is called its DEGREE SEQUENCE. A list of vertex degrees of a graph can be given by VertexDegrees[g] in the *Mathematica* add-on package DiscreteMath'Combinatorica' (which can be loaded with the command < <DiscreteMath').

The minimum vertex degree in a GRAPH G is denoted $\delta(G)$, and the maximum degree is denoted $\Delta(G)$ (Skiena 1990, p. 157).

The VERTEX degree of a point v in a GRAPH, denoted $\rho(v)$, satisfies

$$\sum_{i=1}^{n} \rho(v_i) = 2E,$$

where E is the total number of EDGES.

DIRECTED GRAPHS have two types of degrees, known as the INDEGREE and the OUTDEGREE.

See also DEGREE SEQUENCE, DIRECTED GRAPH, EDGE (GRAPH), EVEN NODE, GRAPH, INDEGREE, LOCAL DEGREE, ODD NODE, OUTDEGREE, PLANTED TREE, VERTEX (GRAPH)

References

Skiena, S. *Implementing Discrete Mathematics: Combinatorics and Graph Theory with Mathematica.* Reading, MA: Addison-Wesley, 1990.

Vertex Enumeration

A CONVEX POLYHEDRON is defined as the set of solutions to a system of linear inequalities

$$m\mathbf{x} \le \mathbf{b},$$

where m is a REAL $s \times d$ MATRIX and $\mathbf{b}$ is a REAL s-VECTOR. Given m and $\mathbf{b}$, vertex enumeration is the determination of the polyhedron's VERTICES.

See also COMPUTATIONAL GEOMETRY, CONVEX POLYHEDRON, POLYHEDRON

References

Avis, D. and Fukuda, K. "A Pivoting Algorithm for Convex Hulls and Vertex Enumeration of Arrangements and Polyhedra." In *Proceedings of the 7th ACM Symposium on Computational Geometry, North Conway, NH, 1991,* pp. 98–104, 1991.
Fukada, K. and Mizukosh, I. "Vertex Enumeration Package for Convex Polytopes and Arrangements, Version 0.41 Beta." http://www.mathsource.com/cgi-bin/msitem?0202-633.

Vertex Figure

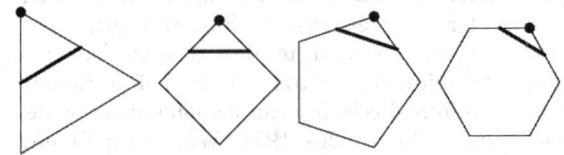

The vertex figure at a vertex V of a POLYGON is the line segment joining the MIDPOINTS of the two adjacent sides meeting at V. For a regular n-gon with side length a, the length v of the vertex figure is

$$v = a \cos\left(\frac{\pi}{n}\right)$$

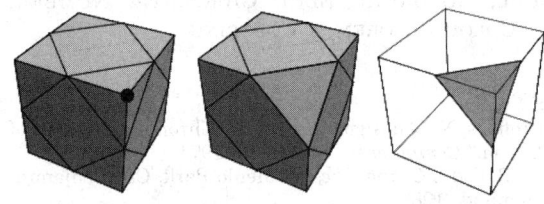

The vertex figure at a vertex V of a POLYHEDRON is the polygon whose sides are the vertex figures of the faces surrounding V. The faces that join at a VERTEX form a SOLID ANGLE whose section by the plane is the vertex figure.

See also MIDPOINT, RECTIFICATION, TRUNCATION

References

Coxeter, H. S. M. "The Polytopes with Regular-Prismatic Vertex Figures." *Phil. Trans. Roy. Soc.* **229**, 330–425, 1930.
Coxeter, H. S. M. *Regular Polytopes, 3rd ed.* New York: Dover, p. 16, 1973.
Cundy, H. and Rollett, A. *Mathematical Models, 3rd ed.* Stradbroke, England: Tarquin Pub., p. 76, 1989.

Vertex Scheme

If K is a SIMPLICIAL COMPLEX, let V be the VERTEX SET of K. Furthermore, let $\Re$ be the collection of all subsets $\{a_0, \ldots, a_n\}$ of V such that the vertices $a_0, \ldots, a_n$ span a SIMPLEX of K. Then the collection $\Re$ is called the vertex scheme of K (Munkres 1993, p. 15).

See also GEOMETRIC REALIZATION, VERTEX SET

References

Munkres, J. R. *Elements of Algebraic Topology.* Perseus Press, 1993.

Vertex Set

The vertex set of a GRAPH is simply a set of all vertices of the graph.

The vertex set V of an ABSTRACT SIMPLICIAL COMPLEX S is the union of one-point elements of S (Munkres 1993, p. 15).

See also DOMINATION NUMBER, EDGE SET, VERTEX SCHEME

References

Munkres, J. R. *Elements of Algebraic Topology.* Perseus Press, 1993.

Vertex-Transitive Graph

A GRAPH such that every pair of vertices is equivalent under some element of its automorphism group. Every nontrivial graph that is EDGE-TRANSITIVE but not vertex-transitive contains at least 20 vertices (Skiena 1990, p. 186). The smallest known CUBIC GRAPH that is EDGE- but not vertex-transitive is the GRAY GRAPH.

See also EDGE-TRANSITIVE GRAPH, FOLKMAN GRAPH, GRAY GRAPH

References

Skiena, S. *Implementing Discrete Mathematics: Combinatorics and Graph Theory with Mathematica.* Reading, MA: Addison-Wesley, 1990.

Vertical

Oriented in an up-down position.

See also HORIZONTAL

Vertical Perspective Projection

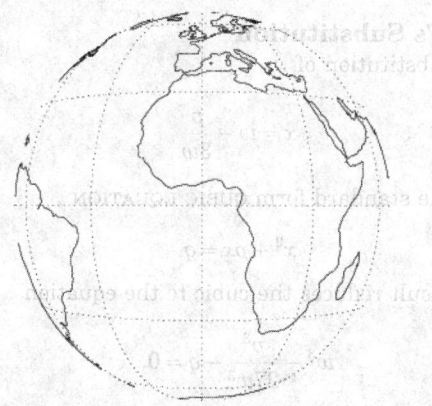

A MAP PROJECTION given by the transformation equations

$$x = k' \cos \phi \sin(\lambda - \lambda_0) \tag{1}$$

$$y = k'[\cos \phi_1 \sin \phi - \sin \phi_1 \cos \phi \cos(\lambda - \lambda_0)], \tag{2}$$

where P is the distance of the point of perspective in

units of SPHERE RADII and

$$k' = \frac{P-1}{P - \cos c} \tag{3}$$

$$\cos c = \sin \phi_1 \sin \phi + \cos \phi_1 \cos \phi \cos(\lambda - \lambda_0) \tag{4}$$

References

Snyder, J. P. *Map Projections--A Working Manual.* U. S. Geological Survey Professional Paper 1395. Washington, DC: U. S. Government Printing Office, pp. 173–178, 1987.

Vertical Rule

BAR, MACRON

Vertical Tangent

A function $f(x)$ has a vertical tangent line at x_0 if f is continuous at x_0 and

$$\lim_{x \to x_0} f'(x) = \pm\infty$$

Vertical-Horizontal Illusion

The HORIZONTAL line segment in the above figure appears to be shorter than the VERTICAL line segment, despite the fact that it has the same length.

See also ILLUSION, MÜLLER-LYER ILLUSION, POGGENDORFF ILLUSION, PONZO'S ILLUSION

References

Fineman, M. *The Nature of Visual Illusion.* New York: Dover, p. 153, 1996.

Vertically Convex Polyomino

COLUMN-CONVEX POLYOMINO

Veryprime

A POSITIVE INTEGER n is a veryprime IFF all primes $p \le \sqrt{n}$ satisfy

$$\begin{cases} |2[n \ (\mathrm{mod}\ p)] - p| \le 1 & \text{very strong} \\ |2[n \ (\mathrm{mod}\ p)] - p| \le \sqrt{p} & \text{strong} \\ |2[n \ (\mathrm{mod}\ p)] - p| \le p/2 & \text{weak} \end{cases}$$

The weak veryprimes are then 2, 3, 5, 7, 11, 13, 17, 19, 23, 37, 43, 47, 53, 67, 73, 103, 107, 137, 157, 173, 227, 347, 487, 773, ... (Sloane's A050264), the strong veryprimes are 2, 3, 5, 7, 11, 13, 17, 19, 23, 37, 43, 47, 53, 67, 73, 137, 227, ..., and the very strong veryprimes are 2, 3, 5, 7, 11, 13, 17, 19, 23, 37, 43, 47, 53,

67, 73, 137, ..., with no others in the first 100,000 primes.

See also QUITEPRIME

References

Ferry, J. "RE: Veryprimes defined." sci.math posting, 09 Sep 1999.
Sloane, N. J. A. Sequences A050264 in "An On-Line Version of the Encyclopedia of Integer Sequences." http://www.re-search.att.com/~njas/sequences/eisonline.html.
Weisstein, E. W. "Integer Sequences." MATHEMATICA NOTEBOOK INTEGERSEQUENCES.M.

Veselov-Novikov Equation

The system of PARTIAL DIFFERENTIAL EQUATIONS

$$\left(\partial_t + \partial_z^3 + \partial_{\bar{z}}^3\right)v + \partial_z(uv) + \partial_{\bar{z}}(vw) \tag{1}$$

$$\partial_{\bar{z}}u = 3\partial_z v \tag{2}$$

$$\partial_z w = 3\partial_{\bar{z}}v \tag{3}$$

where $\bar{z}$ is the COMPLEX CONJUGATE of z.

References

Bogdanov, L. V. "Veselov-Novikov Equation as a Natural Two-Dimensional Generalization of the Korteweg-de Vries Equation." *Theor. Math. Phys.* **70**, 219–233, 1987.
Zwillinger, D. *Handbook of Differential Equations, 3rd ed.* Boston, MA: Academic Press, p. 139, 1997.

Vesica Piscis

LENS

Vibration Problem

Solution of a system of second-order homogeneous ordinary differential equations with constant COEFFICIENTS OF THE FORM

$$\frac{d^2\mathbf{x}}{dt^2} + b\mathbf{x} = 0,$$

where b is a POSITIVE DEFINITE MATRIX. To solve the vibration problem,

1. Solve the CHARACTERISTIC EQUATION of b to get EIGENVALUES $\lambda_1, ..., \lambda_n$. Define $\omega_i \equiv \sqrt{\lambda_i}$.
2. Compute the corresponding EIGENVECTORS $\mathbf{e}_1, ..., \mathbf{e}_n$.
3. The normal modes of oscillation are given by $\mathbf{x}_1 = A_1\sin(\omega_1 t + \alpha_1)\mathbf{e}_1, ..., \mathbf{x}_n = A_n\sin(\omega_n t + \alpha_n)\mathbf{e}_n$, where $A_1, ..., A_n$ and $\alpha_1, ..., \alpha_n$ are arbitrary constants.
4. The general solution is $\mathbf{x} = \Sigma_{i=1}^n \mathbf{x}_i$.

Vickrey Auction

An AUCTION in which the highest bidder wins but pays only the second-highest bid. This variation over the normal bidding procedure is supposed to encourage bidders to bid the largest amount they are willing to pay.

See also AUCTION

References

Vickrey, W. "Counterspeculation, Auctions, and Competitive Sealed Tenders." *J. Finance* **16**, 8–27, 1961. Reprinted in *The Economics of Information, Vol. 1* (Ed. D. K. Levine and S. A. Lippman). Aldershot, Hants, England: Elgar, pp. 8–44, 1995.

Viergruppe

The mathematical group $\mathbb{Z}_2 \times \mathbb{Z}_2$, also denoted D_2. Its multiplication table is

V	I	V_1	V_2	V_3
I	I	V_1	V_2	V_3
V_1	V_1	I	V_3	V_2
V_2	V_2	V_3	I	V_1
V_3	V_3	V_2	V_1	I

See also DIHEDRAL GROUP, FINITE GROUP Z_4

References

Arfken, G. *Mathematical Methods for Physicists, 3rd ed.* Orlando, FL: Academic Press, pp. 184–185 and 239–240, 1985.

Vieta's Substitution

The substitution of

$$x = w - \frac{p}{3w} \tag{1}$$

into the standard form CUBIC EQUATION

$$x^3 + px = q. \tag{2}$$

The result reduces the cubic to the equation

$$w^3 - \frac{p^3}{27w^3} - q = 0, \tag{3}$$

which is easily turned into a QUADRATIC EQUATION in w^3 by multiplying through by w^3 to obtain

$$\left(w^3\right)^2 - q\left(w^3\right) - \frac{1}{27}p^3 = 0 \tag{4}$$

See also CUBIC EQUATION, QUADRATIC EQUATION

Vigesimal

The base-20 notational system for representing REAL NUMBERS. The digits used to represent numbers using vigesimal NOTATION are 0, 1, 2, 3, 4, 5, 6, 7, 8, 9, A, B, C, D, E, F, G, H, I, and J. A base-20 number system was used by the Aztecs and Mayans. The Mayans compiled extensive observations of planetary positions in base-20 notation.

See also BASE (NUMBER), BINARY, DECIMAL, HEXADECIMAL, OCTAL, QUATERNARY, TERNARY

References

Weisstein, E. W. "Bases." MATHEMATICA NOTEBOOK BASES.M.

Vigintillion

In the American system, 10^{63}.

See also LARGE NUMBER

Villarceau Circles

Given an arbitrary point on a TORUS, four CIRCLES can be drawn through it. The first is in the plane of the TORUS and the second is PERPENDICULAR to it. The third and fourth CIRCLES are called Villarceau circles.

See also TORUS

References

Melzak, Z. A. *Invitation to Geometry.* New York: Wiley, pp. 63–72, 1983.
Villarceau, M. "Théorème sur le tore." *Nouv. Ann. Math.* **7**, 345–347, 1848.

Vinculum

A horizontal line placed above multiple quantities to indicate that they form a unit. It is most commonly used to denote

1. A RADICAL ($\sqrt{12345}$),
2. Repeating decimals ($0.\overline{111}$),
3. The distance between two points $\overline{AB}$,
4. The COMPLEX CONJUGATE $\overline{z_1 + z_2}$, or
5. NEGATION of a logical expression, $\overline{A \wedge B} = !(A \wedge B)$.

See also BAR, MACRON, RADICAL, SOLIDUS

References

Bringhurst, R. *The Elements of Typographic Style, 2nd ed.* Point Roberts, WA: Hartley and Marks, p. 286, 1997.

Vinogradov's Theorem

Every sufficiently large ODD number is a sum of three PRIMES (Vinogradov 1937). Ramachandra and Sankaranarayanan (1997) have shown that for sufficiently large n, the error term is $\ll n/(\ln n)^4$. This theorem is closely related to WARING'S PRIME NUMBER CONJECTURE.

See also GOLDBACH CONJECTURE, SCHNIRELMANN'S THEOREM, WARING'S PRIME NUMBER CONJECTURE

References

Ramachandra, K. and Sankaranarayanan, A. "Vinogradov's Three Primes Theorem." *Math. Student* **66**, 1–4 and 27–72, 1997.
Vaughan, R. C. *The Hardy-Littlewood Method.* Cambridge, England: Cambridge University Press, 1981.
Vinogradov, I. M. *The Method of Trigonometrical Sums in the Theory of Numbers* (Russian). Trav. Inst. Math. Stekloff, Vol. 10, 1937.
Vinogradov, I. M. *The Method of Trigonometrical Sums in the Theory of Numbers* (Russian). Trav. Inst. Math. Stekloff, Vol. 23, 1947.
Vinogradov, I. M. *The Method of Trigonometrical Sums in the Theory of Numbers.* London: Interscience, no year given.

Virgule

A diagonal slash resembling the SOLIDUS, but with slightly less slant, used to denote DIVISION for in-line equations such as a/b, $1/(x-1)^2$, etc.

See also SOLIDUS

References

Bringhurst, R. *The Elements of Typographic Style, 2nd ed.* Point Roberts, WA: Hartley and Marks, p. 286, 1997.

Virtual Group

GROUPOID

Visibility

VISIBLE POINT

Visibility Graph

Let S be a set of simple polygonal obstacles in the plane, then the nodes of the visibility graph of S are just the vertices of S, and there is an edge (called a visibility edge) between vertices v and w if these vertices are mutually visible.

References

de Berg, M.; van Kreveld, M.; Overmans, M.; and Schwarzkopf, O. "Visibility Graphs: Finding the Shortest Route." Ch. 15 in *Computational Geometry: Algorithms and Applications, 2nd rev. ed.* Berlin: Springer-Verlag, pp. 307–317, 2000.

Visible Point

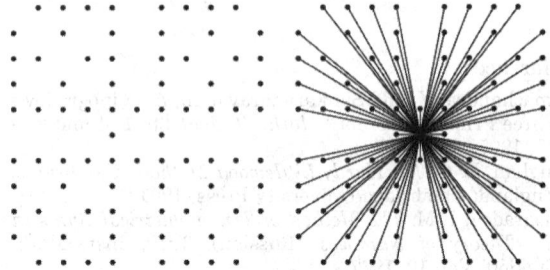

Two LATTICE POINTS (x, y) and (x', y') are mutually visible if the line segment joining them contains no further LATTICE POINTS. This corresponds to the requirement that $(x' - x, y' - y) = 1$, where (m, n) denotes the GREATEST COMMON DIVISOR. The plots above show the first few points visible from the ORIGIN.

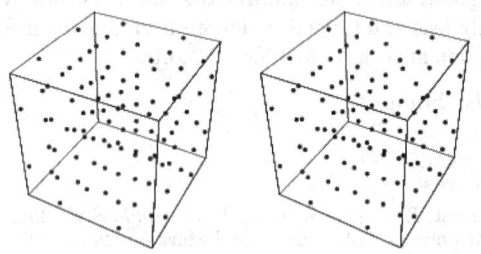

If a LATTICE POINT is selected at random in 2-D, the probability that it is visible from the origin is $6/\pi^2$. This is also the probability that two INTEGERS picked at random are RELATIVELY PRIME. If a LATTICE POINT is picked at random in n-D, the probability that it is visible from the ORIGIN is $1/\zeta(n)$, where $\zeta(n)$ is the RIEMANN ZETA FUNCTION.

An invisible figure is a POLYGON all of whose corners are invisible. There are invisible sets of every finite shape. The lower left-hand corner of the invisible squares with smallest x coordinate of AREAS 2 and 3 are (14, 20) and (104, 6200).

See also LATTICE POINT, ORCHARD VISIBILITY PROBLEM, RIEMANN ZETA FUNCTION

References

Apostol, T. §3.8 in *Introduction to Analytic Number Theory*. New York: Springer-Verlag, 1976.
Asano, T.; Ghosh, S. K.; and Shermer, T. C. "Visibility in the Plane." Ch. 19 in *Handbook of Computational Geometry* (Ed. J.-R. Sack and J. Urrutia). Amsterdam, Netherlands: North-Holland, pp. 829–876, 2000.
Baake, M.; Grimm, U.; and Warrington, D. H. "Some Remarks on the Visible Points of a Lattice." *J. Phys. A: Math. General* **27**, 2669–2674, 1994.
Baake, M.; Moody, R. V.; and Pleasants, P. A. B. Diffraction from Visible Lattice Points and kth Power Free Integers. 19 Jun 1999. http://xxx.lanl.gov/abs/math.MG/9906132/.
Gardner, M. *The Sixth Book of Mathematical Games from Scientific American*. Chicago, IL: University of Chicago Press, pp. 208–210, 1984.
Gosper, R. W. and Schroeppel, R. Item 48 in Beeler, M.; Gosper, R. W.; and Schroeppel, R. *HAKMEM*. Cambridge, MA: MIT Artificial Intelligence Laboratory, Memo AIM-239, p. 17, Feb. 1972.
Herzog, F. and Stewart, B. M. "Patterns of Visible and Nonvisible Lattice Points." *Amer. Math. Monthly* **78**, 487–496, 1971.
Mosseri, R. "Visible Points in a Lattice." *J. Phys. A: Math. Gen.* **25**, L25-L29, 1992.
Schroeder, M. R. "A Simple Function and Its Fourier Transform." *Math. Intell.* **4**, 158–161, 1982.
Schroeder, M. R. *Number Theory in Science and Communication, 2nd ed.* New York: Springer-Verlag, 1990
Steinhaus, H. *Mathematical Snapshots, 3rd ed.* New York: Dover, pp. 100–101, 1999.

Visible Point Vector Identity

A set of identities involving n-D visible lattice points was discovered by Campbell (1994). Examples include

$$\prod_{\substack{(a,b)=1 \\ a \geq 0, b \leq 1}} \left(1 - y^a z^b\right)^{-1/b} = (1-z)^{-1/(1-y)}$$

for $|yz|, |z| < 1$ and

$$\prod_{\substack{(a,b,c)=1 \\ a,b \geq 0, c \leq 1}} \left(1 - x^a y^b z^c\right)^{-1/c} = (1-z)^{-1/[(1-x)(1-y)]}$$

for $|xyz|, |xz|, |yz|, |z| < 1$.

References

Campbell, G. B. "Infinite Products Over Visible Lattice Points." *Internat. J. Math. Math. Sci.* **17**, 637–654, 1994.
Campbell, G. B. "Visible Point Vector Identities." http://www.geocities.com/CapeCanaveral/Launchpad/9416/vpv.html.

Vitali's Convergence Theorem

Let $fn(z)$ be a sequence of functions, each regular in a region D, let $|fn(z)| \leq M$ for every n and z in D, and let $fn(z)$ tend to a limit as $n \to \infty$ at a set of points having a LIMIT POINT inside D. Then $fn(z)$ tends uniformly to a limit in any region bounded by a contour interior to D, the limit therefore being an analytic function of z.

See also MONTEL'S THEOREM

References

Titchmarsh, E. C. *The Theory of Functions, 2nd ed.* Oxford, England: Oxford University Press, p. 168, 1960.

Viviani's Curve

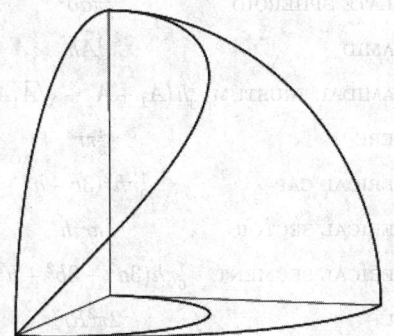

The SPACE CURVE giving the intersection of the CYLINDER

$$(x-a)^2 + y^2 = a^2 \tag{1}$$

and the SPHERE

$$x^2 + y^2 + z^2 = 4a^2. \tag{2}$$

It is given by the PARAMETRIC EQUATIONS

$$x = a(1 + \cos t) \tag{3}$$

$$y = a \sin t \tag{4}$$

$$z = 2a \sin\left(\tfrac{1}{2}t\right). \tag{5}$$

The CURVATURE and TORSION are given by

$$\kappa(t) = \frac{\sqrt{13 + 3\cos t}}{a(3 + \cos t)^{3/2}} \tag{6}$$

$$\tau(t) = \frac{6\cos\left(\tfrac{1}{2}t\right)}{a(13 + 3\cos t)}. \tag{7}$$

See also CYLINDER, CYLINDER-SPHERE INTERSECTION, SPHERE, STEINMETZ SOLID

References
Gray, A. "Viviani's Curve." §8.6 in *Modern Differential Geometry of Curves and Surfaces with Mathematica, 2nd ed.* Boca Raton, FL: CRC Press, pp. 201–202, 1997.

Kenison, E. and Bradley, H. C. *Descriptive Geometry.* New York: Macmillan, p. 284, 1935.

von Seggern, D. *CRC Standard Curves and Surfaces.* Boca Raton, FL: CRC Press, p. 270, 1993.

Viviani's Theorem

For a point P inside an EQUILATERAL TRIANGLE $\triangle ABC$, the sum of the perpendiculars p_i from P to the sides of the TRIANGLE is equal to the ALTITUDE h. This result is simply proved as follows,

$$\triangle ABC = \triangle PBC + \triangle PCA + \triangle PAB. \tag{1}$$

With s the side length,

$$\tfrac{1}{2}sh = \tfrac{1}{2}sp_a + \tfrac{1}{2}sp_b + \tfrac{1}{2}sp_c, \tag{2}$$

so

$$h = p_a + p_b + p_c. \tag{3}$$

See also ALTITUDE, EQUILATERAL TRIANGLE

Vizing Conjecture

Let $\gamma(G)$ denote the DOMINATION NUMBER of a SIMPLE GRAPH G. Then Vizing (1963) conjectured that

$$\gamma(G)\gamma(H) \leq \gamma(G \times H),$$

where $G \times H$ is the GRAPH PRODUCT. While the full conjecture remains open, Clark and Suen (2000) have proved the looser result

$$\gamma(G)\gamma(H) \leq 2\gamma(G \times H).$$

See also DOMINATION NUMBER

References
Clark, W. E. and Suen, S. "An Inequality Related to Vizing's Conjecture." *Electronic J. Combinatorics* **7**, No. 1, N4, 1–3, 2000. http://www.combinatorics.org/Volume_7/v7i1toc.html#N4.

Hartnell, B. and Rall, D. F. "Domination in Cartesian Products: Vizing's Conjecture." In *Domination in Graphs--Advanced Topics* (Ed. T. W. Haynes, S. T. Hedetniemi, and P. J. Slater). New York: Dekker, pp. 163–189, 1998.

Vizing, V. G. "The Cartesian Product of Graphs." *Vycisl. Sistemy* **9**, 30–43, 1963.

Vojta's Conjecture

A conjecture which treats the heights of points relative to a canonical class of a curve defined over the INTEGERS.

References
Cox, D. A. "Introduction to Fermat's Last Theorem." *Amer. Math. Monthly* **101**, 3–14, 1994.

Volterra Integral Equation of the First Kind

An INTEGRAL EQUATION OF THE FORM

$$\phi(x) = \int_a^x k(x,t)\phi(t)\,dt.$$

See also FREDHOLM INTEGRAL EQUATION OF THE FIRST KIND, FREDHOLM INTEGRAL EQUATION OF THE SECOND KIND, INTEGRAL EQUATION, VOLTERRA INTEGRAL EQUATION OF THE SECOND KIND

References

Arfken, G. *Mathematical Methods for Physicists, 3rd ed.* Orlando, FL: Academic Press, p. 865, 1985.

Press, W. H.; Flannery, B. P.; Teukolsky, S. A.; and Vetterling, W. T. "Volterra Equations." §18.2 in *Numerical Recipes in FORTRAN: The Art of Scientific Computing, 2nd ed.* Cambridge, England: Cambridge University Press, pp. 786–788, 1992.

Volterra Integral Equation of the Second Kind

An INTEGRAL EQUATION OF THE FORM

$$\phi(x) = f(x) + \int_a^x k(x,t)\phi(t)\,dt$$

See also FREDHOLM INTEGRAL EQUATION OF THE FIRST KIND, FREDHOLM INTEGRAL EQUATION OF THE SECOND KIND, INTEGRAL EQUATION, VOLTERRA INTEGRAL EQUATION OF THE FIRST KIND

References

Arfken, G. *Mathematical Methods for Physicists, 3rd ed.* Orlando, FL: Academic Press, p. 865, 1985.

Press, W. H.; Flannery, B. P.; Teukolsky, S. A.; and Vetterling, W. T. "Volterra Equations." §18.2 in *Numerical Recipes in FORTRAN: The Art of Scientific Computing, 2nd ed.* Cambridge, England: Cambridge University Press, pp. 786–788, 1992.

Volume

The volume of a solid body is the amount of "space" it occupies. Volume has units of LENGTH cubed (i.e., cm^3, m^3, in^3, etc.) For example, the volume of a box (RECTANGULAR PARALLELEPIPED) of LENGTH L, WIDTH W, and HEIGHT H is given by

$$V = L \times W \times H.$$

The volume can also be computed for irregularly-shaped and curved solids such as the CYLINDER and CUBE. The volume of a SURFACE OF REVOLUTION is particularly simple to compute due to its symmetry.

The following table gives volumes for some common SURFACES. Here r denotes the RADIUS, h the height, and A the base AREA, and, in the case of the TORUS, R the distance from the torus center to the center of the tube (Beyer 1987).

SURFACE	Volume
CONE	$\frac{1}{3}\pi r^2 h$
CONICAL FRUSTUM	$\frac{1}{3}\pi h(R_1^2 + R_2^2 + R_1 R_2)$
CUBE	a^3
CYLINDER	$\pi r^2 h$
ELLIPSOID	$\frac{4}{3}\pi abc$
OBLATE SPHEROID	$\frac{4}{3}\pi a^2 b$
PROLATE SPHEROID	$\frac{4}{3}\pi ab^2$
PYRAMID	$\frac{1}{3}Ah$
PYRAMIDAL FRUSTUM	$\frac{1}{3}h(A_1 + A_2 + \sqrt{A_1 A_2})$
SPHERE	$\frac{4}{3}\pi r^3$
SPHERICAL CAP	$\frac{1}{3}\pi h^2(3r - h)$
SPHERICAL SECTOR	$\frac{2}{3}\pi r^2 h$
SPHERICAL SEGMENT	$\frac{1}{6}\pi h(3a^2 + 3b^2 + h^2)$
TORUS	$2\pi^2 R r^2$

Even simple SURFACES can display surprisingly counterintuitive properties. For instance, the SURFACE OF REVOLUTION of $y = 1/x$ around the x-AXIS for $x \geq 1$ is called GABRIEL'S HORN, and has finite volume, but infinite SURFACE AREA.

The generalization of volume to n DIMENSIONS for $n \geq 4$ is known as CONTENT.

See also ARC LENGTH, AREA, BELLOWS CONJECTURE, CONTENT, HEIGHT, LENGTH (SIZE), SURFACE AREA, SURFACE OF REVOLUTION, VOLUME ELEMENT, VOLUME THEOREM, WIDTH (SIZE)

References

Beyer, W. H. *CRC Standard Mathematical Tables, 28th ed.* Boca Raton, FL: CRC Press, pp. 127–132, 1987.

Volume Element

A volume element is the differential element dV whose VOLUME INTEGRAL over some range in a given coordinate system gives the VOLUME of a solid,

$$V = \iiint_G dx\,dy\,dz. \tag{1}$$

In $\mathbb{R}^n$, the volume of the infinitesimal n-HYPERCUBE bounded by dx_1, ..., dx_n has volume given by the WEDGE PRODUCT

$$dV = dx_1 \wedge \ldots \wedge dx_n \tag{2}$$

(Gray 1997).

The use of the antisymmetric WEDGE PRODUCT instead of the symmetric product $dx_1 \ldots dx_n$ is a technical refinement often omitted in informal usage. Dropping the wedges, the volume element for CURVILINEAR COORDINATES in $\mathbb{R}^3$ is given by

$$dV = |(h_1\hat{\mathbf{u}}_1\,du_1)\cdot(h_2\hat{\mathbf{u}}_2\,du_2) \times (h_3\hat{\mathbf{u}}_3\,du_3)| \tag{3}$$

$$= h_1 h_2 h_3\,du_1\,du_2\,du_3 \tag{4}$$

$$= \left| \frac{\partial \mathbf{r}}{\partial u_1} \cdot \frac{\partial \mathbf{r}}{\partial u_2} \times \frac{\partial \mathbf{r}}{\partial u_3} \right| du_1\,du_2\,du_3 \tag{5}$$

$$= \begin{vmatrix} \dfrac{\partial x}{\partial u_1} & \dfrac{\partial x}{\partial u_2} & \dfrac{\partial x}{\partial u_3} \\[6pt] \dfrac{\partial y}{\partial u_1} & \dfrac{\partial y}{\partial u_2} & \dfrac{\partial y}{\partial u_3} \\[6pt] \dfrac{\partial z}{\partial u_1} & \dfrac{\partial z}{\partial u_2} & \dfrac{\partial z}{\partial u_3} \end{vmatrix} du_1\, du_2\, du_3 \qquad (6)$$

$$= \left| \frac{\partial(x,y,z)}{\partial(u_1,u_2,u_3)} \right| du_1\, du_2\, du_3, \qquad (7)$$

where the latter is the JACOBIAN and the h_i are SCALE FACTORS.

See also AREA ELEMENT, JACOBIAN, LINE ELEMENT, RIEMANNIAN METRIC, SCALE FACTOR, SURFACE AREA, SURFACE INTEGRAL, VOLUME INTEGRAL

References

Gray, A. "Isometries and Conformal Maps of Surfaces." §15.2 in *Modern Differential Geometry of Curves and Surfaces with Mathematica, 2nd ed.* Boca Raton, FL: CRC Press, pp. 346–351, 1997.

Volume Integral

A triple integral over three coordinates giving the VOLUME within some region G,

$$V = \iiint_G dx\, dy\, dz.$$

See also AREA INTEGRAL, INTEGRAL, LINE INTEGRAL, MULTIPLE INTEGRAL, SURFACE INTEGRAL, VOLUME, VOLUME ELEMENT

References

Leathem, J. G. *Volume and Surface Integrals Used in Physics.* 1905.

Volume Theorem

If the top and bottom bases of a solid are equal in area, lie in PARALLEL PLANES, and every SECTION of the solid parallel to the bases is equal in area to that of the base, then the VOLUME of the solid is the product of base and altitude.

See also CAVALIERI'S PRINCIPLE, VOLUME

References

Kern, W. F. and Bland, J. R. "Volume Theorem." §12 in *Solid Mensuration with Proofs, 2nd ed.* New York: Wiley, pp. 27–28, 1948.

von Aubel's Theorem

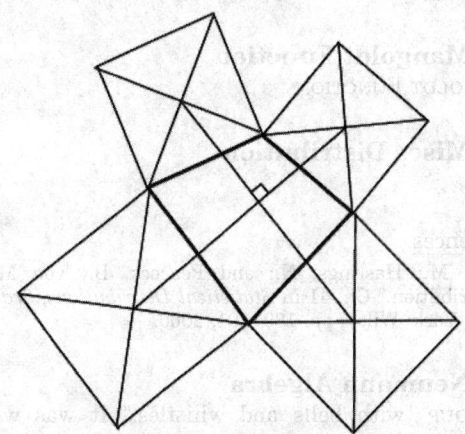

Given an arbitrary QUADRILATERAL, place a SQUARE outwardly on each side, and connect the centers of opposite SQUARES. Then the two lines are of equal length and cross at a RIGHT ANGLE.

See also QUADRILATERAL, RIGHT ANGLE, SQUARE

References

Kitchen, E. "Dörrie Tiles and Related Miniatures." *Math. Mag.* **67**, 128–130, 1994.
Wells, D. *The Penguin Dictionary of Curious and Interesting Geometry.* London: Penguin, p. 11, 1991.

von Dyck's Theorem

Let a GROUP G have a presentation

$$G = (x_1, \ldots, x_n | r_j(x_1, \ldots, x_n), j \in J)$$

so that $G = F/R$, where F is the FREE GROUP with basis $\{x_1, \ldots, x_n\}$ and R is the NORMAL SUBGROUP generated by the r_j. If H is a GROUP with $H = \langle y_1, \ldots, y_n \rangle$ and if $r_j(y_1, \ldots, y_n) = 1$ for all j, then there is a surjective homomorphism $G \to H$ with $x_i \mapsto y_i$ for all i.

See also DYCK'S THEOREM, FREE GROUP, NORMAL SUBGROUP

References

Rotman, J. J. *An Introduction to the Theory of Groups, 4th ed.* New York: Springer-Verlag, p. 346, 1995.

von Kármán Equations

The system of PARTIAL DIFFERENTIAL EQUATIONS

$$\nabla^4 u = E\left(v_{xy}^2 - v_{xx} v_{yy}\right)$$

$$\nabla^4 v = \alpha + \beta\left(u_{yy} v_{xx} + u_{xx} v_{yy} - 2u_{xy} v_{xy}\right),$$

where ∇^4 is the BIHARMONIC OPERATOR.

References

Ames, K. A. and Ames, W. F. "On Group Analysis of the Von Kármán Equation." *Nonlinear Anal.* **6**, 845–853, 1982.

Zwillinger, D. *Handbook of Differential Equations, 3rd ed.* Boston, MA: Academic Press, p. 138, 1997.

von Mangoldt Function

MANGOLDT FUNCTION

von Mises Distribution

References

Evans, M.; Hastings, N.; and Peacock, B. "von Mises Distribution." Ch. 41 in *Statistical Distributions, 3rd ed.* New York: Wiley, pp. 189–191, 2000.

von Neumann Algebra

A GROUP "with bells and whistles." It was while studying von Neumann algebras that Jones discovered the amazing and highly unexpected connections with KNOT THEORY which led to the formulation of the JONES POLYNOMIAL.

References

Iyanaga, S. and Kawada, Y. (Eds.). "Von Neumann Algebras." §430 in *Encyclopedic Dictionary of Mathematics.* Cambridge, MA: MIT Press, pp. 1358–1363, 1980.

von Neumann-Bernays-Gödel Set Theory

This entry contributed by MATTHEW SZUDZIK

von Neumann-Bernays-Gödel set theory (abbreviated "NBG") is a version of SET THEORY which was designed to give the same results as ZERMELO-FRAENKEL SET THEORY, but in a more logically elegant fashion. It can be viewed as a conservative extension of ZERMELO-FRAENKEL SET THEORY in the sense that a statement about sets is provable in NBG if and only if it is provable in ZERMELO-FRAENKEL SET THEORY.

ZERMELO-FRAENKEL SET THEORY is not finitely axiomatized. For example, the AXIOM OF REPLACEMENT is not really a single axiom, but an infinite family of axioms, since it is preceded by the stipulation that it is true "for any set-theoretic formula $A(u,v)$." Montague (1961) proved that ZERMELO-FRAENKEL SET THEORY is not finitely axiomatizable, i.e., there is no finite set of axioms which is logically equivalent to the infinite set of ZERMELO-FRAENKEL AXIOMS. In contrast, von Neumann-Bernays-Gödel set theory has only finitely many axioms, and this was the main motivation in its construction. This was accomplished by extending the language of ZERMELO-FRAENKEL SET THEORY to be capable of talking about CLASSES.

See also CLASS (SET), SET THEORY, ZERMELO-FRAENKEL AXIOMS, ZERMELO-FRAENKEL SET THEORY

References

Itô, K. (Ed.). "Bernays-Gödel Set Theory." §33C in *Encyclopedic Dictionary of Mathematics, 2nd ed., Vol. 1.* Cambridge, MA: MIT Press, p. 148, 1986.

Mendelson, E. *Introduction to Mathematical Logic, 4th ed.* London: Chapman & Hall, 1997.
Montague, R. "Semantic Closure and Non-Finite Axiomatizability. I." In *Infinitistic Methods, Proceedings of the Symposium on Foundations of Mathematics, (Warsaw, 2–9 September 1959).* Oxford, England: Pergamon, pp. 45–69, 1961.

von Staudt Theorem

VON STAUDT-CLAUSEN THEOREM

von Staudt-Clausen Theorem

$$B_{2n} = A_n - \sum_{\substack{p_k \\ (p_k-1)|2n}} \frac{1}{p_k},$$

where B_{2n} is a BERNOULLI NUMBER, A_n is an INTEGER, and the p_ks are the PRIMES satisfying $p_k - 1 | 2k$. For example, for $k = 1$, the primes included in the sum are 2 and 3, since $(2-1)|2$ and $(3-1)|2$. Similarly, for $k = 6$, the included primes are $(2, 3, 5, 7, 13)$, since $(1, 2, 3, 6, 12)$ divide $12 = 2 \cdot 6$. The first few values of A_n for $n = 1, 2, \ldots$ are $1, 1, 1, 1, 1, 1, 2, -6, 56, -528, \ldots$ (Sloane's A000146).

The theorem was rediscovered by Ramanujan (Hardy 1999, p. 11) and can be proved using P-ADIC NUMBERS.

See also BERNOULLI NUMBER, P-ADIC NUMBER

References

Clausen, T. "Theorem." *Astron. Nach.* **17**, 351–352, 1840.
Conway, J. H. and Guy, R. K. *The Book of Numbers.* New York: Springer-Verlag, p. 109, 1996.
Hardy, G. H. *Ramanujan: Twelve Lectures on Subjects Suggested by His Life and Work, 3rd ed.* New York: Chelsea, 1999.
Hardy, G. H. and Wright, E. M. "The Theorem of von Staudt" and "Proof of von Staudt's Theorem." §7.9–7.10 in *An Introduction to the Theory of Numbers, 5th ed.* Oxford, England: Clarendon Press, pp. 90–93, 1979.
Rado, R. "A New Proof of a Theorem of V. Staudt." *J. London Math. Soc.* **9**, 85–88, 1934.
Rado, R. "A Note on the Bernoullian Numbers." *J. London Math. Soc.* **9**, 88–90, 1934.
Sloane, N. J. A. Sequences A000146/M1717 in "An On-Line Version of the Encyclopedia of Integer Sequences." http://www.research.att.com/~njas/sequences/eisonline.html.
Staudt, K. G. C. von. "Beweis eines Lehrsatzes, die Bernoullischen Zahlen betreffend." *J. reine angew. Math.* **21**, 372–374, 1840.

Voronoi Cell

The generalization of a VORONOI POLYGON to n-D, for $n > 2$.

See also DODECAHEDRAL CONJECTURE, VORONOI POLYGON

Voronoi Diagram

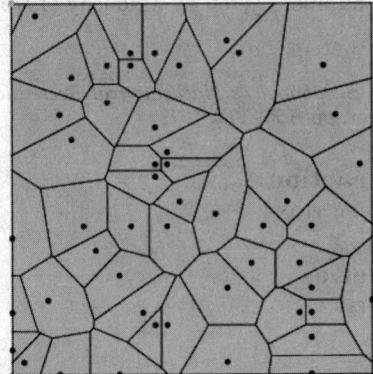

The partitioning of a plane with n points into n convex POLYGONS such that each POLYGON contains exactly one point and every point in a given POLYGON is closer to its central point than to any other. A Voronoi diagram is sometimes also known as a DIRICHLET TESSELLATION. The cells are called DIRICHLET REGIONS, THIESSEN POLYTOPES, or VORONOI POLYGONS. The *Mathematica* command Diagram-Plot[*pts*] in the *Mathematica* add-on package DiscreteMath`ComputationalGeometry` (which can be loaded with the command < <DiscreteMath`) plots the Voronoi diagram of the given list of points.

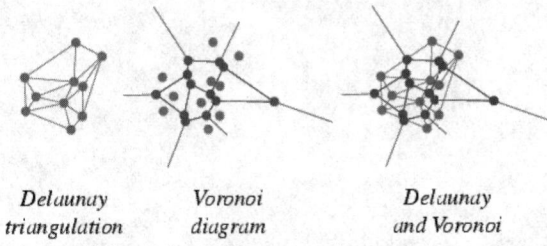

| Delaunay triangulation | Voronoi diagram | Delaunay and Voronoi |

The DELAUNAY TRIANGULATION and Voronoi diagram in $\mathbb{R}^2$ are dual to each other.

See also ART GALLERY THEOREM, COMPUTATIONAL GEOMETRY, DELAUNAY TRIANGULATION, MEDIAL AXIS, TRIANGULATION, VORONOI POLYGON

References

Aurenhammer, F. and Klein, R. "Voronoi Diagrams." Ch. 5 in *Handbook of Computational Geometry* (Ed. J.-R. Sack and J. Urrutia). Amsterdam, Netherlands: North-Holland, pp. 201–290, 2000.
Eppstein, D. "Nearest Neighbors and Voronoi Diagrams." http://www.ics.uci.edu/~eppstein/junkyard/nn.html.
de Berg, M.; van Kreveld, M.; Overmans, M.; and Schwarzkopf, O. "Voronoi Diagrams: The Post Office Problem." Ch. 7 in *Computational Geometry: Algorithms and Applications, 2nd rev. ed.* Berlin: Springer-Verlag, pp. 147–163, 2000.
Guibas, L. and Stolfi, J. "Primitives for the Manipulation of General Subdivisions and the Computations of Voronoi Diagrams." *ACM Trans. Graphics* **4**, 74–123, 1985.
Klee, V. "On the Complexity of d-Dimensional Voronoi Diagrams." *Archiv. Math.* **34**, 75–80, 1980.

Okabe, A.; Boots, B.; and Sugihara, K. *Spatial Tessellations: Concepts and Applications of Voronoi Diagrams, 2nd ed.* New York: Wiley, 2000.
Preparata, F. R. and Shamos, M. I. *Computational Geometry: An Introduction.* New York: Springer-Verlag, 1985.
Skiena, S. S. "Voronoi Diagrams." §8.6.4 in *The Algorithm Design Manual.* New York: Springer-Verlag, pp. 358–360, 1997.

Voronoi Polygon

A POLYGON whose interior consists of all points in the plane which are closer to a particular LATTICE POINT than to any other. The generalization to n-D is called a DIRICHLET REGION, THIESSEN POLYTOPE, or VORONOI CELL.

References

Dirichlet, G. L. "Über die Reduktion der positiven quadratischen Formen mit drei unbestimmten ganzen Zahlen." *J. reine angew. Math.* **40**, 209–227, 1850.
Voronoi, G. "Recherches sur les paralléloèdres Primitives." *J. reine angew. Math.* **134**, 198–287, 1908.
Williams, R. *The Geometrical Foundation of Natural Structure: A Source Book of Design.* New York: Dover, p. 43, 1979.

Voting

The simple process of voting leads to surprisingly counterintuitive paradoxes. For example, if three people vote for three candidates, giving the rankings A, B, C; B, C, A; and C, A, B. A majority prefers A to B, B to C, but also C to A (Gardner 1984, p. 25)! It is also possible to conduct a secret ballot even if the votes are sent in to a central polling station (Lipton and Widgerson, Honsberger 1985).

See also ARROW'S PARADOX, BALLOT PROBLEM, CAKE CUTTING, MAY'S THEOREM, QUOTA SYSTEM, SOCIAL CHOICE THEORY

References

Black, D. *Theory of Committees and Elections.* Cambridge, England: Cambridge University Press, 1958.
Black, D. *A Mathematical Approach to Proportional Representation: Duncan Black on Lewis Carroll.* Boston, MA: Kluwer, 1995.
Gardner, M. *The Last Recreations: Hydras, Eggs, and Other Mathematical Mystifications.* New York: Springer-Verlag, pp. 317–330, 1997.
Gardner, M. *The Sixth Book of Mathematical Games from Scientific American.* Chicago, IL: University of Chicago Press, p. 25, 1984.
Honsberger, R. *Mathematical Gems III.* Washington, DC: Math. Assoc. Amer., pp. 157–162, 1985.
Huntington, E. V. "A Paradox in the Scoring of Completing Teams." *Science* **88**, 287–288, 1938.
Lipton, R. G.; and Widgerson, A. "Multi-Party Cryptographic Protocols."
Niemi, R. G. and Riker, W. H. *Sci. Amer.* **234**, 21–27, Jun. 1976.
Riker, W. H. "Voting and the Summation of Preferences." *Amer. Political Sci. Rev.*, Dec. 1961.
Saari, D. G. *Math. Intell.* **10**, 32, 1988.
Steinhaus, H. *Mathematical Snapshots, 3rd ed.* New York: Dover, pp. 72–74, 1999.

VR Number

A "visual representation" number which is a sum of some simple function of its digits. For example,

$$1233 = 12^2 + 33^2$$

$$2661653 = 1653^2 - 266^2$$

$$221859 = 22^3 + 18^3 + 59^3$$

$$40585 = 4! + 0! + 5! + 8! + 5!$$

$$148349 = !1 + !4 + !8 + !3 + !4 + !9$$

$$4913 = (4 + 9 + 1 + 3)^3$$

are all VR numbers given by Madachy (1979).

References

Madachy, J. S. *Madachy's Mathematical Recreations.* New York: Dover, pp. 165–171, 1979.

Vulgar Fraction

COMMON FRACTION

Vulgar Series

FAREY SERIES

W

W2-Constant

$$W_2 = 1.529954037\ldots.$$

References

Plouffe, S. "W2 Constant." http://www.lacim.uqam.ca/pi-DATA/w2.txt.

Wada Basin

A BASIN OF ATTRACTION in which every point on the common boundary of that basin and another basin is also a boundary of a third basin. In other words, no matter how closely a boundary point is zoomed into, all three basins appear in the picture.

See also BASIN OF ATTRACTION

References

Nusse, H. E. and Yorke, J. A. "Basins of Attraction." *Science* **271**, 1376–380, 1996.

Wadati-Konno-Ichikawa-Shimizu Equation

The PARTIAL DIFFERENTIAL EQUATION

$$iu_t + \left[\left(1 + |u|^2 u\right)^{-1/2} u\right]_{xx} = 0.$$

References

Calogero, F. and Degasperis, A. *Spectral Transform and Solitons: Tools to Solve and Investigate Nonlinear Evolution Equations.* New York: North-Holland, p. 53, 1982.
Zwillinger, D. *Handbook of Differential Equations, 3rd ed.* Boston, MA: Academic Press, p. 135, 1997.

Wagstaff's Conjecture

A modification of the EBERHART'S CONJECTURE proposed by Wagstaff (1983) which proposes that if q_n is the nth prime such that M_{q_n} is a MERSENNE PRIME, then

$$q_n \sim \left(2^{e^{-\gamma}}\right)^n,$$

where γ is the EULER-MASCHERONI CONSTANT.

See also EBERHART'S CONJECTURE

References

Ribenboim, P. *The New Book of Prime Number Records.* New York: Springer-Verlag, p. 412, 1996.
Wagstaff, S. S. "Divisors of Mersenne Numbers." *Math. Comput.* **40**, 385–97, 1983.

Wald's Equation

For a sequence of independent identically distributed random variates $X_1, \ldots, X_N$ and a random positive integer N, the EXPECTATION VALUES satisfy

$$\langle X_1 + \ldots + X_N \rangle = \langle X_1 \rangle \langle N \rangle.$$

See also EXPECTATION VALUE

Walk

A sequence of VERTICES and EDGES such that the VERTICES and EDGES are adjacent. A walk is therefore equivalent to a graph CYCLE, but with the VERTICES along the walk enumerated as well as the EDGES.

See also CIRCUIT, GRAPH CYCLE, PATH, RANDOM WALK

Wallace-Bolyai-Gerwein Theorem

Two POLYGONS are congruent by DISSECTION IFF they have the same AREA. In particular, any POLYGON is congruent by DISSECTION to a SQUARE of the same AREA. Laczkovich (1988) also proved that a CIRCLE is congruent by DISSECTION to a SQUARE (furthermore, the DISSECTION can be accomplished using TRANSLATIONS only).

See also DISSECTION

References

Klee, V. and Wagon, S. *Old and New Unsolved Problems in Plane Geometry and Number Theory.* Washington, DC: Math. Assoc. Amer., pp. 50–1, 1991.
Laczkovich, M. "Von Neumann's Paradox with Translation." *Fund. Math.* **131**, 1–2, 1988.

Wallace-Simson Line

SIMSON LINE

Wallace-Simson Theorem

SIMSON LINE

Wallis Cosine Formula

$$\int_0^{\pi/2} \cos^n x\, dx$$

$$= \begin{cases} \dfrac{\pi}{2} \dfrac{1 \cdot 3 \cdot 5 \cdots (n-1)}{2 \cdot 4 \cdot 6 \cdots n} & \text{for } n = 2, 4, \ldots \\ \dfrac{2 \cdot 4 \cdot 6 \cdots (n-1)}{1 \cdot 3 \cdot 5 \cdots n} & \text{for } n = 3, 5, \ldots. \end{cases}$$

See also WALLIS FORMULA, WALLIS SINE FORMULA

Wallis Formula

The Wallis formula follows from the INFINITE PRODUCT representation of the SINE

$$\sin x = x \prod_{n=1}^{\infty}\left(1 - \frac{x^2}{\pi^2 n^2}\right). \qquad (1)$$

Taking $x = \pi/2$ gives

$$1 = \frac{\pi}{2}\prod_{n=1}^{\infty}\left[1 - \frac{1}{(2n)^2}\right] = \frac{\pi}{2}\prod_{n=1}^{\infty}\left[\frac{(2n)^2 - 1}{(2n)^2}\right], \qquad (2)$$

so

$$\frac{\pi}{2} = \prod_{n=1}^{\infty}\left[\frac{(2n)^2}{(2n-1)(2n+1)}\right] = \frac{2\cdot2}{1\cdot3}\frac{4\cdot4}{3\cdot5}\frac{6\cdot6}{5\cdot7}\cdots. \qquad (3)$$

A derivation due to Y. L. Yung uses the RIEMANN ZETA FUNCTION. Define

$$F(s) \equiv -\mathrm{Li}_s(-1) = \sum_{n=1}^{\infty}\frac{(-1)^n}{n^s}$$

$$= \left(1 - 2^{1-s}\right)\zeta(s) \qquad (4)$$

$$F'(s) = \sum_{n=1}^{\infty}\frac{(-1)^n \ln n}{n^s}, \qquad (5)$$

so

$$F'(0) = \sum_{n=1}^{\infty}(-1)^n \ln n = -\ln 1 + \ln 2 - \ln 3 + \ldots$$

$$= \ln\left(\frac{2\cdot4\cdot6\cdots}{1\cdot3\cdot5\cdots}\right). \qquad (6)$$

Taking the derivative of the zeta function expression gives

$$\frac{d}{ds}\left(1 - 2^{1-s}\right)\zeta(s) = 2^{1-s}(\ln 2)\zeta(s) + \left(1 - 2^{1-s}\right)\zeta'(s) \qquad (7)$$

$$\left[\frac{d}{ds}\left(1 - 2^{1-s}\right)\zeta(s)\right]_{s=0} = -\ln 2 - \zeta'(0)$$

$$= -\ln 2 + \frac{1}{2}\ln(2\pi) = \ln\left(\frac{\sqrt{2\pi}}{2}\right) = \ln\left(\sqrt{\frac{\pi}{2}}\right). \qquad (8)$$

Equating and squaring then gives the Wallis formula, which can also be expressed

$$\frac{\pi}{2} = \left[4^{\zeta(0)}e^{-\zeta'(0)}\right]^2. \qquad (9)$$

The Q-ANALOG of the Wallis formula for $q = 2$ is

$$\prod_{k=1}^{\infty}\left(1 - q^{-k}\right)^{-1} = 3.4627466194\ldots \qquad (10)$$

(Finch).

See also WALLIS COSINE FORMULA, WALLIS SINE FORMULA

References

Abramowitz, M. and Stegun, C. A. (Eds.). *Handbook of Mathematical Functions with Formulas, Graphs, and Mathematical Tables, 9th printing.* New York: Dover, p. 258, 1972.

Finch, S. "Favorite Mathematical Constants." http://www.mathsoft.com/asolve/constant/dig/dig.html.

Jeffreys, H. and Jeffreys, B. S. "Wallis's Formula for π." §15.07 in *Methods of Mathematical Physics, 3rd ed.* Cambridge, England: Cambridge University Press, p. 468, 1988.

Kenney, J. F. and Keeping, E. S. *Mathematics of Statistics, Pt. 2, 2nd ed.* Princeton, NJ: Van Nostrand, pp. 63–4, 1951.

Wallis Sieve

A compact set W_∞ with AREA

$$\mu(W_\infty) = \frac{8}{9}\frac{24}{25}\frac{48}{49}\cdots = \frac{\pi}{4}$$

created by punching a square hole of length $1/3$ in the center of a square. In each of the eight squares remaining, punch out another hole of length $1/(3\cdot5)$, and so on.

Wallis Sine Formula

$$\int_0^{\pi/2}\sin^n x\, dx$$

$$= \begin{cases} \dfrac{\pi}{2}\dfrac{1\cdot3\cdot5\cdots(n-1)}{2\cdot4\cdot6\cdots n} & \text{for } n = 2, 4, \ldots \\[2ex] \dfrac{2\cdot4\cdot6\cdots(n-1)}{1\cdot3\cdot5\cdots n} & \text{for } n = 3, 5, \ldots. \end{cases}$$

See also WALLIS COSINE FORMULA, WALLIS FORMULA

Wallis's Conical Edge

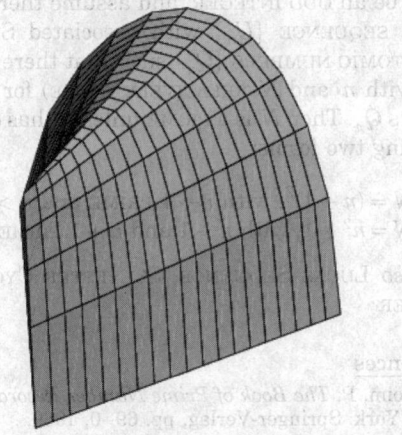

The RIGHT CONOID surface given by the PARAMETRIC EQUATIONS

$$x(u,v) = v \cos u$$
$$y(u,v) = v \sin u$$
$$z(u,v) = c\sqrt{a^2 - b^2 \cos^2 u}.$$

See also RIGHT CONOID

References
Gray, A. *Modern Differential Geometry of Curves and Surfaces with Mathematica, 2nd ed.* Boca Raton, FL: CRC Press, pp. 454–55, 1997.

Wallis's Problem

Find nontrivial solutions to $\sigma(x^2) = \sigma(y^2)$ other than $(x,y) = (4,5)$, where $\sigma(n)$ is the DIVISOR FUNCTION. Nontrivial solutions means that solutions which are multiples of smaller solutions are not considered. For example, multiples m of $(x,y) = (4,5)$ are solutions for $m = 3, 7, 9, 11, 13, 17, 19, 23, 21, \ldots$.

Nontrivial solutions to Wallis's equation include $(x,y) = (4,5)$, $(326, 407)$, $(406, 489)$, $(627, 749)$, $(740, 878)$, $(880, 1451)$, $(888, 1102)$, $(1026, 1208)$, $(1110, 1943)$, $(1284, 1528, 1605)$, $(1510, 1809)$, $(1628, 1630, 2035)$, $(1956, 2030, 2445)$, $(2013, 2557)$, $(2072, 3097)$, $(2508, 2996, 3135, 3745)$, $\ldots$.

See also DIVISOR FUNCTION, FERMAT'S DIVISOR PROBLEM

References
Dickson, L. E. *History of the Theory of Numbers, Vol. 1: Divisibility and Primality.* New York: Chelsea, pp. 54–6, 1952.

Wallpaper Groups

The 17 PLANE SYMMETRY GROUPS. Their symbols are p1, p2, pm, pg, cm, pmm, pmg, pgg, cmm, p4, p4m, p4g, p3, p31m, p3m1, p6, and p6m. For a description of the symmetry elements present in each space group, see Coxeter (1969, p. 413).

References
Coxeter, H. S. M. *Introduction to Geometry, 2nd ed.* New York: Wiley, 1969.
Hilbert, D. and Cohn-Vossen, S. *Geometry and the Imagination.* New York: Chelsea, 1999.
Joyce, D. E. "Wallpaper Groups (Plane Symmetry Groups)." http://aleph0.clarku.edu/~djoyce/wallpaper/.
Schattschneider, D. "The Plane Symmetry Groups: Their Recognition and Notation." *Amer. Math. Monthly* **85**, 439–50, 1978.
Weyl, H. *Symmetry.* Princeton, NJ: Princeton University Press, 1952.
Zwillinger, D. (Ed.). "Crystallographic Groups." §4.2.4 in *CRC Standard Mathematical Tables and Formulae.* Boca Raton, FL: CRC Press, pp. 259–64, 1995.

Walsh Function

Functions consisting of a number of fixed-amplitude square pulses interposed with zeros. Following Harmuth (1969), designate those with EVEN symmetry Cal(k,t) and those with ODD symmetry Sal(k,t). Define the SEQUENCY k as half the number of zero crossings in the time base. Walsh functions with nonidentical SEQUENCIES are ORTHOGONAL, as are the functions Cal(k,t) and Sal(k,t). The product of two Walsh functions is also a Walsh function. The Walsh functions are then given by

$$\mathrm{Wal}(k,t) = \begin{cases} \mathrm{Cal}(k/2, t) & \text{for } k = 0, 2, 4, \ldots \\ \mathrm{Sal}((k+1)/2, t) & \text{for } k = 1, 3, 5, \ldots. \end{cases}$$

The Walsh functions Cal(k,t) for $k = 0, 1, \ldots, n/2 - 1$ and Sal(k,t) for $k = 1, 2, \ldots, n/2$ are given by the rows of the HADAMARD MATRIX $\mathbf{H}_n$.

See also HADAMARD MATRIX, SEQUENCY

References
Beauchamp, K. G. *Walsh Functions and Their Applications.* London: Academic Press, 1975.
Harmuth, H. F. "Applications of Walsh Functions in Communications." *IEEE Spectrum* **6**, 82–1, 1969.
Thompson, A. R.; Moran, J. M.; and Swenson, G. W. Jr. *Interferometry and Synthesis in Radio Astronomy.* New York: Wiley, p. 204, 1986.
Tzafestas, S. G. *Walsh Functions in Signal and Systems Analysis and Design.* New York: Van Nostrand Reinhold, 1985.
Walsh, J. L. "A Closed Set of Normal Orthogonal Functions." *Amer. J. Math.* **45**, 5–4, 1923.

Walsh Index

The statistical INDEX

$$P_{\mathrm{W}} = \frac{\sum \sqrt{q_0 q_n} p_n}{\sum \sqrt{q_0 q_n} p_0},$$

where p_n is the price per unit in period n and q_n is the quantity produced in period n.

See also INDEX

References

Kenney, J. F. and Keeping, E. S. *Mathematics of Statistics, Pt. 1, 3rd ed.* Princeton, NJ: Van Nostrand, p. 66, 1962.

Wangerin Differential Equation

The ORDINARY DIFFERENTIAL EQUATION

$$y'' + \frac{1}{2}\left[\frac{1}{x-a_1} + \frac{1}{x-a_2} + \frac{1}{x-a_3}\right]y'$$

$$+\frac{1}{4}\left[\frac{A_0 + A_1 x + A_2 x^2}{(x-a_1)(x-a_2)(x-a_3)}\right]y = 0.$$

See also LAMÉ'S DIFFERENTIAL EQUATION

References

Moon, P. and Spencer, D. E. *Field Theory for Engineers.* New York: Van Nostrand, p. 157, 1961.
Zwillinger, D. *Handbook of Differential Equations, 3rd ed.* Boston, MA: Academic Press, p. 127, 1997.

Wang's Conjecture

Wang's conjecture states that if a set of tiles can tile the plane, then they can always be arranged to do so periodically (Wang 1961). The CONJECTURE was refuted when Berger (1966) showed that an aperiodic set of tiles existed. Berger used 20,426 tiles, but the number has subsequently been greatly reduced. In fact, Culik (1996) has reduced the number of tiles to 13.

See also TILING

References

Adler, A. and Holroyd, F. C. "Some Results on One-Dimensional Tilings." *Geom. Dedicata* **10**, 49–8, 1981.
Berger, R. "The Undecidability of the Domino Problem." *Mem. Amer. Math. Soc. No.* **66**, 1–2, 1966.
Culik, K. II "An Aperiodic Set of 13 Wang Tiles." *Disc. Math.* **160**, 245–51, 1996.
Grünbaum, B. and Sheppard, G. C. *Tilings and Patterns.* New York: W. H. Freeman, 1986.
Hanf, W. "Nonrecursive Tilings of the Plane. I." *J. Symbolic Logic* **39**, 283–85, 1974.
Kari, J. "A Small Aperiodic Set of Wang Tiles." *Disc. Math.* **160**, 259–64, 1996.
Mozes, S. "Tilings, Substitution Systems, and Dynamical Systems Generated by Them." *J. Analyse Math.* **53**, 139–86, 1989.
Myers, D. "Nonrecursive Tilings of the Plane. II." *J. Symbolic Logic* **39**, 286–94, 1974.
Radin, C. *Miles of Tiles.* Providence, RI: Amer. Math. Soc., pp. 6–, 1999.
Robinson, R. M. "Undecidability and Nonperiodicity for Tilings of the Plane." *Invent. Math.* **12**, 177–09, 1971.
Smith, T. "Penrose Tilings and Wang Tilings." http://www.innerx.net/personal/tsmith/pwtile.html.
Wang, H. "Proving Theorems by Pattern Recognition. II." *Bell Systems Tech. J.* **40**, 1–1, 1961.

Ward's Primality Test

Let N be an ODD INTEGER, and assume there exists a LUCAS SEQUENCE $\{U_n\}$ with associated SYLVESTER CYCLOTOMIC NUMBERS $\{Q_n\}$ such that there is an $n > \sqrt{N}$ (with n and N RELATIVELY PRIME) for which N DIVIDES Q_n. Then N is a PRIME unless it has one of the following two forms:

1. $N = (n-1)^2$, with $n-1$ PRIME and $n > 4$, or
2. $N = n^2 - 1$, with $n-1$ and $n+1$ PRIME.

See also LUCAS SEQUENCE, SYLVESTER CYCLOTOMIC NUMBER

References

Ribenboim, P. *The Book of Prime Number Records, 2nd ed.* New York: Springer-Verlag, pp. 69–0, 1989.

Waring Formula

$$A^n + B^n = \sum_{j=0}^{\lfloor n/2 \rfloor} (-1)^j \frac{n}{n-j}\binom{n-j}{j}(AB)^j(A+B)^{n-2j},$$

where $\lfloor x \rfloor$ is the FLOOR FUNCTION and $\binom{n}{k}$ is a BINOMIAL COEFFICIENT.

See also FERMAT'S LAST THEOREM

Waring's Conjecture

WARING'S PRIME NUMBER CONJECTURE, WARING'S PROBLEM

Waring's Prime Number Conjecture

Every ODD INTEGER n is a PRIME or the sum of three PRIMES. This problem is closely related to VINOGRADOV'S THEOREM.

See also GOLDBACH CONJECTURE, SCHNIRELMANN'S THEOREM, VINOGRADOV'S THEOREM

Waring's Problem

In his *Meditationes algebraicae*, Waring (1770, 1782) proposed a generalization of LAGRANGE'S FOUR-SQUARE THEOREM, stating that every RATIONAL INTEGER is the sum of a fixed number $g(n)$ of nth POWERS of INTEGERS, where n is any given POSITIVE INTEGER and $g(n)$ depends only on n. Waring originally speculated that $g(2) = 4$, $g(3) = 9$, and $g(4) = 19$. In 1909, Hilbert proved the general conjecture using an identity in 25-fold multiple integrals (Rademacher and Toeplitz 1957, pp. 52–1).

In LAGRANGE'S FOUR-SQUARE THEOREM, Lagrange proved that $g(2) = 4$, where 4 may be reduced to 3 except for numbers OF THE FORM $4^n(8k + 7)$ (as proved by Legendre; Hardy 1999, p. 12). In the early twentieth century, Dickson, Pillai, and Niven proved that $g(3) = 9$. Hilbert, Hardy, and Vinogradov proved $g(4) \le 21$, and this was subsequently reduced to $g(4) = 19$ by Balasubramanian *et al.* (1986). Liouville proved (using LAGRANGE'S FOUR-SQUARE THEOREM and LIOUVILLE POLYNOMIAL IDENTITY) that $g(5) \leq 53$, and this was improved to 47, 45, 41, 39, 38, and finally $g(5) \leq 37$ by Wieferich. See Rademacher and Toeplitz (1957, p. 56) for a simple proof. J.-J. Chen (1964) proved that $g(5) = 37$.

Dickson (1936), Pillai (1936), and Niven also conjectured an explicit formula for $g(s)$ for $s > 6$ (Bell 1945, pp. 318 and 602), based on the relationship

$$\left(\frac{3}{2}\right)^n - \left\lfloor\left(\frac{3}{2}\right)^n\right\rfloor = 1 - \left(\frac{1}{2}\right)^n\left\{\left\lfloor\left(\frac{3}{2}\right)^n\right\rfloor + 2\right\}. \quad (1)$$

If the DIOPHANTINE (i.e., n is restricted to being an INTEGER) inequality

$$\operatorname{frac}\left[\left(\frac{3}{2}\right)^n\right] \leq 1 - \left(\frac{3}{4}\right)^n \quad (2)$$

is true, where $\operatorname{frac}(x)$ is the FRACTIONAL PART of x, then

$$g(n) = 2^n + \left\lfloor\left(\frac{3}{2}\right)^n\right\rfloor - 2. \quad (3)$$

This was given as a lower bound by Euler, and has been verified to be correct for $6 \leq n \leq 471{,}600{,}000$ (Kubina and Wunderlich 1990, extending Stemmler 1990). Furthermore, Mahler (1957) proved that at most a FINITE number of n exceed Euler's lower bound. Unfortunately, the proof is nonconstructive.

There is also a related (but more difficult) problem of finding the least INTEGER n such that every POSITIVE INTEGER beyond a certain point (i.e., all but a FINITE number) is the SUM of G_n nth POWERS. From 1920–928, Hardy and Littlewood showed that

$$G(n) \leq (n-2)2^{n-1} + 5 \quad (4)$$

and conjectured that

$$G(k) < \begin{cases} 2k+1 & \text{for } k \text{ not a power of 2} \\ 4k & \text{for } k \text{ a power of 2.} \end{cases} \quad (5)$$

The best currently known bound is

$$G(k) < ck \ln k \quad (6)$$

for some constant c. Heilbronn (1936) improved Vinogradov's results to obtain

$$G(n) \leq 6n \ln n + \left[4 + 3 \ln\left(3 + \frac{2}{n}\right)\right]n + 3. \quad (7)$$

It has long been known that $G(2) = 4$.

Dickson and Landau proved that the only INTEGERS requiring nine CUBES are 23 and 239, thus establishing $G(3) \leq 8$. Wieferich proved that only 15 INTEGERS require eight CUBES: 15, 22, 50, 114, 167, 175, 186, 212, 231, 238, 303, 364, 420, 428, and 454 (Sloane's

A018889), establishing $G(3) \leq 7$ (Wells 1986, p. 70). The largest number known requiring seven CUBES is 8042.

In 1933, Hardy and Littlewood showed that $G(4) \leq 19$, but this was improved in 1936 to 16 or 17, and shown to be exactly 16 by Davenport (1939b). Vaughan (1986) greatly improved on the method of Hardy and Littlewood, obtaining improved results for $n \geq 5$. These results were then further improved by Brüdern (1990), who gave $G(5) \leq 18$, and Wooley (1992), who gave G_n for $n = 6$ to 20. Vaughan and Wooley (1993) showed $G(8) \leq 42$.

Let $G^+(n)$ denote the smallest number such that *almost all* sufficiently large INTEGERS are the sum of $G^+(n)$ nth POWERS. Then $G^+(3) = 4$ (Davenport 1939a), $G^+(4) = 15$ (Hardy and Littlewood 1925), $G^+(8) = 32$ (Vaughan 1986), and $G^+(16) = 64$ (Wooley 1992). If the negatives of POWERS are permitted in addition to the powers themselves, the largest number of nth POWERS needed to represent an arbitrary integer are denoted $eg(n)$ and $EG(n)$ (Wright 1934, Hunter 1941, Gardner 1986). In general, these values are much harder to calculate than are $g(n)$ and G_n.

The following table gives $g(n)$, G_n, $G^+(n)$, $eg(n)$, and $EG(n)$ for $n \leq 20$. The sequence of $g(n)$ is Sloane's A002804.

n	$g(n)$	G_n	$G^+(n)$	$eg(n)$	$EG(n)$
2	4	4		3	3
3	9	≤ 7	≤ 4	[4, 5]	
4	19	16	≤ 15	[9, 10]	
5	37	≤ 18			
6	73	≤ 27			
7	143	≤ 36			
8	279	≤ 42	≤ 32		
9	548	≤ 55			
10	1079	≤ 63			
11	2132	≤ 70			
12	4223	≤ 79			
13	8384	≤ 87			
14	16673	≤ 95			
15	33203	≤ 103			
16	66190	≤ 112	≤ 64		
17	132055	≤ 120			
18	263619	$n \times n$			
19	526502	≤ 138			
20	1051899	≤ 146			

See also EULER'S CONJECTURE, SCHNIRELMANN CONSTANT, SCHNIRELMANN'S THEOREM, VINOGRADOV'S THEOREM

References

Archibald, R. G. "Waring's Problem: Squares." *Scripta Math.* **7**, 33–8, 1940.
Balasubramanian, R.; Deshouillers, J.-M.; and Dress, F. "Problème de Waring pour les bicarrés 1, 2." *C. R. Acad. Sci. Paris Sér. I Math.* **303**, 85–8 and 161–63, 1986.
Bell, E. T. *The Development of Mathematics, 2nd ed.* New York: McGraw-Hill, 1945.
Brüdern, J. "On Waring's Problem for Fifth Powers and Some Related Topics." *Proc. London Math. Soc.* **61**, 457–79, 1990.
Davenport, H. "On Waring's Problem for Cubes." *Acta Math.* **71**, 123–43, 1939a.
Davenport, H. "On Waring's Problem for Fourth Powers." *Ann. Math.* **40**, 731–47, 1939b.
Dickson, L. E. "Waring's Problem and Related Results." Ch. 25 in *History of the Theory of Numbers, Vol. 2: Diophantine Analysis.* New York: Chelsea, pp. 717–29, 1952.
Gardner, M. "Waring's Problems." Ch. 18 in *Knotted Doughnuts and Other Mathematical Entertainments.* New York: W. H. Freeman, pp. 222–31, 1986.
Guy, R. K. "Sums of Squares." §C20 in *Unsolved Problems in Number Theory, 2nd ed.* New York: Springer-Verlag, pp. 136–38, 1994.
Hardy, G. H. *Ramanujan: Twelve Lectures on Subjects Suggested by His Life and Work, 3rd ed.* New York: Chelsea, 1999.
Hardy, G. H. and Littlewood, J. E. "Some Problems of Partitio Numerorum (VI): Further Researches in Waring's Problem." *Math. Z.* **23**, 1–7, 1925.
Hardy, G. H. and Wright, E. M. "The Representation of a Number by Two or Four Squares" and "Representation by Cubes and Higher Powers." Chs. 20–1 in *An Introduction to the Theory of Numbers, 5th ed.* Oxford, England: Clarendon Press, pp. 297–39, 1979.
Hunter, W. "The Representation of Numbers by Sums of Fourth Powers." *J. London Math. Soc.* **16**, 177–79, 1941.
Khinchin, A. Y. "An Elementary Solution of Waring's Problem." Ch. 3 in *Three Pearls of Number Theory.* New York: Dover, pp. 37–4, 1998.
Kubina, J. M. and Wunderlich, M. C. "Extending Waring's Conjecture to 471,600,000." *Math. Comput.* **55**, 815–20, 1990.
Mahler, K. "On the Fractional Parts of the Powers of a Rational Number (II)." *Mathematica* **4**, 122–24, 1957.
Rademacher, H. and Toeplitz, O. *The Enjoyment of Mathematics: Selections from Mathematics for the Amateur.* Princeton, NJ: Princeton University Press, 1957.
Sloane, N. J. A. Sequences A018889 and A002804/M3361 in "An On-Line Version of the Encyclopedia of Integer Sequences." http://www.research.att.com/~njas/sequences/eisonline.html.
Small, C. "Waring's Problem." *Math. Mag.* **50**, 12–6, 1977.
Stemmler, R. M. "The Ideal Waring Theorem for Exponents 401–00,000." *Math. Comput.* **55**, 815–20, 1990.
Stewart, I. "The Waring Experience." *Nature* **323**, 674, 1986.
Vaughan, R. C. "On Waring's Problem for Smaller Exponents." *Proc. London Math. Soc.* **52**, 445–63, 1986.
Vaughan, R. C. and Wooley, T. D. "On Waring's Problem: Some Refinements." *Proc. London Math. Soc.* **63**, 35–8, 1991.
Vaughan, R. C. and Wooley, T. D. "Further Improvements in Waring's Problem." *Phil. Trans. Roy. Soc. London A* **345**, 363–76, 1993a.
Vaughan, R. C. and Wooley, T. D. "Further Improvements in Waring's Problem III. Eighth Powers." *Phil. Trans. Roy. Soc. London A* **345**, 385–96, 1993b.
Waring, E. *Meditationes algebraicae.* Cambridge, England: pp. 204–05, 1770.
Waring, E. *Meditationes algebraicae, 3rd ed.* Cambridge, England: pp. 349–50, 1782.
Waring, E. *Meditationes Algebraicae: An English Translation of the Work of Edward Waring.* Providence, RI: Amer. Math. Soc., 1991.
Wells, D. *The Penguin Dictionary of Curious and Interesting Numbers.* Middlesex, England: Penguin Books, pp. 70 and 75, 1986.
Wooley, T. D. "Large Improvements in Waring's Problem." *Ann. Math.* **135**, 131–64, 1992.
Wright, E. M. "An Easier Waring's Problem." *J. London Math. Soc.* **9**, 267–72, 1934.

Waring's Sum Conjecture

WARING'S PROBLEM

Waring's Theorem

If each of two curves meets the LINE AT INFINITY in distinct, nonsingular points, and if all their intersections are finite, then if to each common point there is attached a weight equal to the number of intersections absorbed therein, the CENTER OF MASS of these points is the center of gravity of the intersections of the asymptotes.

References

Coolidge, J. L. *A Treatise on Algebraic Plane Curves.* New York: Dover, p. 166, 1959.

Wasteful Number

A number n is called wasteful if the number of digits in the prime factorization of n (including powers) uses more digits than the number of digits in n. The first few wasteful numbers are 4, 6, 8, 9, 12, 18, 20, 22, 24 ... (Sloane's A046760). Pinch calls these numbers "frugal" and includes 1 as a frugal number.

See also ECONOMICAL NUMBER, EQUIDIGITAL NUMBER

References

Pinch, R. G. E. "Economical Numbers." http://www.chalcedon.demon.co.uk/publish.html#62.
Rivera, C. "Problems & Puzzles: Puzzle Sequences of Consecutive Economical Numbers.-053." http://www.primepuzzles.net/puzzles/puzz_053.htm.
Santos, B. R. "Problem 2204. Equidigital Representation." *J. Recr. Math.* **27**, 58–9, 1995.
Sloane, N. J. A. Sequences A046760 in "An On-Line Version of the Encyclopedia of Integer Sequences." http://www.research.att.com/~njas/sequences/eisonline.html.
Weisstein, E. W. "Integer Sequences." MATHEMATICA NOTEBOOK INTEGERSEQUENCES.M.

Watchman Theorem

ART GALLERY THEOREM

Watson Identities

Let α, $-\beta$, and $-\gamma^{-1}$ be the roots of the CUBIC EQUATION

$$t^3 + 2t^2 - t - 1 = 0, \tag{1}$$

then the normalized DILOGARITHM $L(x)$ satisfies

$$L(\alpha) - L(\alpha^2) = \frac{1}{7} \tag{2}$$

$$L(\beta) + \frac{1}{2}L(\beta^2) = \frac{5}{7} \tag{3}$$

$$L(\gamma) + \frac{1}{2}L(\gamma^2) = \frac{4}{7}.$$

References

Bytsko, A. G. Two-Term Dilogarithm Identities Related to Conformal Field Theory. 9 Nov 1999. http://xxx.lanl.gov/abs/math-ph/9911012/.
Watson, G. N. *Quart. J. Math. Oxford Ser.* **8**, 39, 1937.

Watson Quintuple Product Identity

QUINTUPLE PRODUCT IDENTITY

Watson-Nicholson Formula

Let $H_\nu^{(i)}(x)$ be a HANKEL FUNCTION OF THE FIRST or SECOND KIND, let $x, \nu > 0$, and define

$$w = \sqrt{\left(\frac{x}{\nu}\right)^2 - 1}.$$

Then

$$H_\nu^{(i)}(x) = 3^{-1/2} w \exp\{(-1)^{i+1} i[\pi/6 + \nu(w - \tfrac{1}{3}w^3 - \tan^{-1} w)]\} H_{1/3}^{(i)}(\tfrac{1}{3}\nu w) + \mathrm{O}\left|\nu^{-1}\right|.$$

References

Iyanaga, S. and Kawada, Y. (Eds.). *Encyclopedic Dictionary of Mathematics.* Cambridge, MA: MIT Press, p. 1475, 1980.

Watson's Formula

Let $J_\nu(z)$ be a BESSEL FUNCTION OF THE FIRST KIND, $Y_\nu(z)$ a BESSEL FUNCTION OF THE SECOND KIND, and $K_\nu(z)$ a MODIFIED BESSEL FUNCTION OF THE FIRST KIND. Also let $\Re[z] > 0$ and require $\Re[\mu - \nu] < 1$. Then

$$J_\mu(z)Y_\nu(z) - J_\nu(z)Y_\mu(z)$$
$$= \frac{4\sin[(\mu - \nu)\pi]}{\pi^2} \int_0^\infty K_{\nu-\mu}(2z \sinh t)e^{-(\mu+\nu)t}\,dt.$$

The fourth edition of Gradshteyn and Ryzhik (2000), Iyanaga and Kawada (1980), and Ito (1987) erroneously give the exponential with a PLUS SIGN. A related integral is given by

$$J_\nu(z)\frac{\partial Y_\nu(z)}{\partial \nu} - Y_\nu(z)\frac{\partial J_\nu(z)}{\partial \nu} = -\frac{4}{\pi} \int_0^\infty K_0(2z \sinh t)e^{-2\nu t}\,dt$$

for $\Re[z] > 0$.

See also DIXON-FERRAR FORMULA, NICHOLSON'S FORMULA

References

Gradshteyn, I. S. and Ryzhik, I. M. Eqns. 6.617.1 and 6.617.2 in *Tables of Integrals, Series, and Products, 6th ed.* San Diego, CA: Academic Press, p. 710, 2000.
Itô, K. (Ed.). *Encyclopedic Dictionary of Mathematics, 2nd ed.* Cambridge, MA: MIT Press, p. 1806, 1987.
Iyanaga, S. and Kawada, Y. (Eds.). *Encyclopedic Dictionary of Mathematics.* Cambridge, MA: MIT Press, p. 1476, 1980.

Watson's Theorem

$$_3F_2\left[\begin{matrix} a, b, c \\ \tfrac{1}{2}(a+b+c),\ c \end{matrix}; 1\right]$$

$$= \frac{\Gamma\left(\tfrac{1}{2}\right)\Gamma\left(\tfrac{1}{2}+c\right)\Gamma\left[\tfrac{1}{2}(1+a+b)\right]\Gamma\left(\tfrac{1}{2}-\tfrac{1}{2}a-\tfrac{1}{2}b+c\right)}{\Gamma\left[\tfrac{1}{2}(1+a)\right]\Gamma\left[\tfrac{1}{2}(1+b)\right]\Gamma\left(\tfrac{1}{2}-\tfrac{1}{2}a+c\right)\Gamma\left(\tfrac{1}{2}-\tfrac{1}{2}b+c\right)},$$

where $_3F_2(a,b,c;d,e;z)$ is a GENERALIZED HYPERGEOMETRIC FUNCTION and $\Gamma(z)$ is the GAMMA FUNCTION (Bailey 1935, p. 16; Koepf 1998, p. 32).

See also GENERALIZED HYPERGEOMETRIC FUNCTION, WATSON-WHIPPLE TRANSFORMATION, WHIPPLE'S IDENTITY

References

Bailey, W. N. "Watson's Theorem." §3.3 in *Generalised Hypergeometric Series.* Cambridge, England: Cambridge University Press, p. 16, 1935.
Koepf, W. *Hypergeometric Summation: An Algorithmic Approach to Summation and Special Function Identities.* Braunschweig, Germany: Vieweg, 1998.

Watson-Whipple Transformation

If at least one of d, e, or f has the form q^{-N} for some nonnegative integer N (in which case both sums terminate after $N+1$ terms), then

$$_8\phi_7\left[\begin{matrix} a, qa^{1/2}, -qa^{1/2}, b, c, d, e, f \\ a^{1/2}, -a^{1/2}, \dfrac{aq}{b}, \dfrac{aq}{c}, \dfrac{aq}{d}, \dfrac{aq}{e}, \dfrac{aq}{f} \end{matrix}; q, \dfrac{a^2q^2}{bcdef}\right]$$

$$= \frac{\left(aq, \dfrac{aq}{de}, \dfrac{aq}{df}, \dfrac{aq}{ef}\right)_\infty}{\left(\dfrac{aq}{d}, \dfrac{aq}{c}, \dfrac{aq}{f}, \dfrac{aq}{def}\right)_\infty} {}_4\phi_3\left[\begin{matrix} \dfrac{aq}{bc}, d, e, f \\ \dfrac{aq}{b}, \dfrac{aq}{c}, \dfrac{def}{a} \end{matrix}; q, q\right],$$

where $(a_1, a_2, \ldots, a_r; q)_\infty$ is a generalized Q-POCHHAMMER SYMBOL

$$(a_1, a_2, \ldots, a_r; q)_\infty = (a_1; q)_\infty (a_2; q)_\infty \ldots (a_r; q)_\infty,$$

and each of $_8\phi_7$ and $_4\phi_3$ is a Q-HYPERGEOMETRIC FUNCTION.

See also Q-HYPERGEOMETRIC FUNCTION, Q-POCHHAMMER SYMBOL, Q-SERIES

References

Gasper, G. and Rahman, M. *Basic Hypergeometric Series.* Cambridge, England: Cambridge University Press, p. 242, 1990.
Gordon, B. and McIntosh, R. J. "Some Eighth Order Mock Theta Functions." To appear in *J. London Math. Soc.* 2000.

Watt's Curve

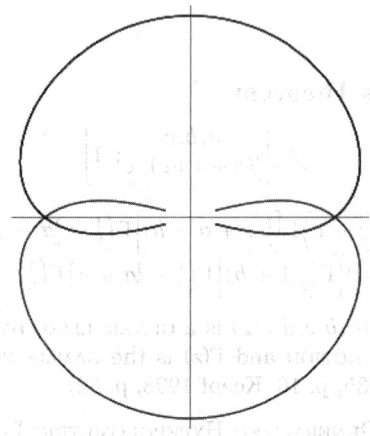

A curve named after James Watt (1736–819), the Scottish engineer who developed the steam engine (MacTutor Archive). The curve is produced by a LINKAGE of rods connecting two wheels of equal diameter. Let the two wheels have RADIUS b and let their centers be located a distance $2a$ apart. Further suppose that a rod of length $2c$ is fixed at each end to the CIRCUMFERENCE of the two wheels. Let P be the MIDPOINT of the rod. Then Watt's curve C is the LOCUS of P.

The POLAR equation of Watt's curve is

$$r^2 = b^2 - \left(a\sin\theta \pm \sqrt{c^2 - a^2\cos^2\theta}\right)^2.$$

If $a = c$, then C is a CIRCLE of RADIUS b with a figure of eight inside it.

See also WATT'S PARALLELOGRAM

References

Lockwood, E. H. *A Book of Curves.* Cambridge, England: Cambridge University Press, p. 162, 1967.
MacTutor History of Mathematics Archive. "Watt's Curve." http://www-groups.dcs.st-and.ac.uk/~history/Curves/Watts.html.

Watt's Parallelogram

A LINKAGE used in the original steam engine to turn back-and-forth motion into approximately straight-line motion.

See also LINKAGE, WATT'S CURVE

References

Rademacher, H. and Toeplitz, O. *The Enjoyment of Mathematics: Selections from Mathematics for the Amateur.* Princeton, NJ: Princeton University Press, pp. 119–21, 1957.

Wave

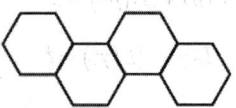

A 4-POLYHEX.

References

Gardner, M. *Mathematical Magic Show: More Puzzles, Games, Diversions, Illusions and Other Mathematical Sleight-of-Mind from Scientific American.* New York: Vintage, p. 147, 1978.

Wave Equation

The wave equation is the important PARTIAL DIFFERENTIAL EQUATION

$$\nabla^2\psi = \frac{1}{v^2}\frac{\partial^2\psi}{\partial t^2}, \tag{1}$$

which can also be written

$$v^2\nabla^2\psi = \psi_{tt}, \tag{2}$$

where ∇^2 is the LAPLACIAN, or

$$\Box^2\psi = 0, \tag{3}$$

where $\Box^2$ is the D'ALEMBERTIAN.

The 1-D wave equation is

$$\frac{\partial^2\psi}{\partial x^2} = \frac{1}{v^2}\frac{\partial^2\psi}{\partial t^2}. \tag{4}$$

In order to specify a wave, the equation is subject to boundary conditions

$$\psi(0, t) = 0 \tag{5}$$

$$\psi(L, t) = 0, \tag{6}$$

and initial conditions

$$\psi(x, 0) = f(x) \tag{7}$$

$$\frac{\partial\psi}{\partial t}(x, 0) = g(x). \tag{8}$$

The wave equation can be solved using the so-called d'Alembert's solution, a FOURIER TRANSFORM method, or SEPARATION OF VARIABLES.

d'Alembert devised his solution in 1746, and Euler subsequently expanded the method in 1748. Let

$$\xi \equiv x - at \tag{9}$$

$$\eta \equiv x + at. \tag{10}$$

By the CHAIN RULE,

$$\frac{\partial^2 \psi}{\partial x^2} = \frac{\partial^2 \psi}{\partial \xi^2} + 2\frac{\partial^2 \psi}{\partial \xi \partial \eta} + \frac{\partial^2 \psi}{\partial \eta^2} \tag{11}$$

$$\frac{1}{v^2}\frac{\partial^2 \psi}{\partial t^2} = \frac{\partial^2 \psi}{\partial \xi^2} - 2\frac{\partial^2 \psi}{\partial \xi \partial \eta} + \frac{\partial^2 \psi}{\partial \eta^2}. \tag{12}$$

The wave equation then becomes

$$\frac{\partial^2 \psi}{\partial \xi \partial \eta} = 0. \tag{13}$$

Any solution of this equation is OF THE FORM

$$\psi(\xi, \eta) = f(\eta) + g(\xi) = f(x+vt) + g(x-vt), \tag{14}$$

where f and g are *any* functions. They represent two waveforms traveling in opposite directions, f in the NEGATIVE x direction and g in the POSITIVE x direction.

The 1-D wave equation can also be solved by applying a FOURIER TRANSFORM to each side,

$$\int_{-\infty}^{\infty} \frac{\partial^2 \psi(x,t)}{\partial x^2} e^{-2\pi ikx} dx = \frac{1}{v^2}\int_{-\infty}^{\infty} \frac{\partial^2 \psi(x,t)}{\partial t^2} e^{-2\pi ikx} dx, \tag{15}$$

which is given, with the help of the FOURIER TRANSFORM DERIVATIVE identity, by

$$(2\pi ik)^2 \Psi(k,t) = \frac{1}{v^2}\frac{\partial^2 \Psi(k,t)}{\partial t^2}, \tag{16}$$

where

$$\Psi(k,t) \equiv F[\psi(x,t)] = \int_{-\infty}^{\infty} \psi(x,t)e^{-2\pi ikx} dx. \tag{17}$$

This has solution

$$\Psi(k,t) \equiv A(k)e^{2\pi ikvt} + B(k)e^{-2\pi ikvt}. \tag{18}$$

Taking the inverse FOURIER TRANSFORM gives

$$\psi(x,t) \equiv \int_{-\infty}^{\infty} \Psi(k,t)e^{2\pi ikx} dx$$

$$= \int_{-\infty}^{\infty} [A(k)e^{2\pi ikvt} + B(k)e^{-2\pi ikvt}]e^{-2\pi ikx} dk$$

$$= \int_{-\infty}^{\infty} A(k)e^{-2\pi ik(x-vt)} dk + \int_{-\infty}^{\infty} B(k)e^{-2\pi ik(x+vt)} dk$$

$$= f_1(x-vt) + f_2(x+vt), \tag{19}$$

where

$$f_1(u) \equiv \mathscr{F}[A(k)] = \int_{-\infty}^{\infty} A(k)e^{-2\pi iku} dk \tag{20}$$

$$f_2(u) \equiv \mathscr{F}[B(k)] = \int_{-\infty}^{\infty} B(k)e^{-2\pi iku} dk. \tag{21}$$

This solution is still subject to all other initial and boundary conditions.

The 1-D wave equation can be solved by SEPARATION OF VARIABLES using a trial solution

$$\psi(x,t) = X(x)T(t). \tag{22}$$

This gives

$$T\frac{d^2 X}{dx^2} = \frac{1}{v^2}X\frac{d^2 T}{dt^2} \tag{23}$$

$$\frac{1}{X}\frac{d^2 X}{dx^2} = \frac{1}{v^2}\frac{1}{T}\frac{d^2 T}{dt^2} = -k^2. \tag{24}$$

So the solution for X is

$$X(x) = C\cos(kx) + D\sin(kx). \tag{25}$$

Rewriting (24) gives

$$\frac{1}{T}\frac{d^2 T}{dt^2} = -v^2 k^2 \equiv -\omega^2, \tag{26}$$

so the solution for T is

$$T(t) = E\cos(\omega t) + F\sin(\omega t), \tag{27}$$

where $v \equiv \omega/k$. Applying the boundary conditions $\psi(0,t) = \psi(L,t) = 0$ to (25) gives

$$C = 0 \quad kL = m\pi, \tag{28}$$

where m is an INTEGER. Plugging (25), (27) and (28) back in for ψ in (23) gives, for a particular value of m,

$$\psi_m(x,t) = [E_m \sin(\omega_m t) + F_m\cos(\omega_m t)]D_m \sin\left(\frac{m\pi x}{L}\right)$$

$$\equiv [A_m \cos(\omega_m t) + B_m \sin(\omega_m t)]\sin\left(\frac{m\pi x}{L}\right). \tag{29}$$

The initial condition $\psi(x,0) = 0$ then gives $B_m = 0$, so (29) becomes

$$\psi_m(x,t) = A_m \cos(\omega_m t)\sin\left(\frac{m\pi x}{L}\right). \tag{30}$$

The general solution is a sum over all possible values of m, so

$$\psi(x,t) = \sum_{m=1}^{\infty} A_m \cos(\omega_m t)\sin\left(\frac{m\pi x}{L}\right). \tag{31}$$

Using ORTHOGONALITY of sines again,

$$\int_0^L \sin\left(\frac{l\pi x}{L}\right)\sin\left(\frac{m\pi x}{L}\right)dx = \tfrac{1}{2}L\delta_{lm}, \tag{32}$$

where δ_{lm} is the KRONECKER DELTA defined by

$$\delta_{mn} \equiv \begin{cases} 1 & m=n \\ 0 & m \neq n \end{cases}, \tag{33}$$

gives

$$\int_0^L \psi(x,0) \sin\left(\frac{m\pi x}{L}\right) dx$$

$$= \sum_{l=1}^{\infty} A_l \sin\left(\frac{l\pi x}{L}\right) \sin\left(\frac{m\pi x}{L}\right) dx$$

$$= \sum_{l=1}^{\infty} A_l \tfrac{1}{2} L \delta_{lm} = \tfrac{1}{2} L A_m, \tag{34}$$

so we have

$$A_m = \frac{2}{L} \int_0^L \psi(x,0) \sin\left(\frac{m\pi x}{L}\right) dx. \tag{35}$$

The computation of A_ms for specific initial distortions is derived in the FOURIER SINE SERIES section. We already have found that $B_m = 0$, so the equation of motion for the string (31), with

$$\omega_m \equiv v k_m = \frac{v m \pi}{L}, \tag{36}$$

is

$$\psi(x,t) = \sum_{m=1}^{\infty} A_m \cos\left(\frac{v m \pi t}{L}\right) \sin\left(\frac{m\pi x}{L}\right), \tag{37}$$

where the A_m COEFFICIENTS are given by (35). A damped 1-D wave

$$\frac{\partial^2 \psi}{\partial x^2} = \frac{1}{v^2} \frac{\partial^2 \psi}{\partial t^2} + b \frac{\partial \psi}{\partial t}, \tag{38}$$

given boundary conditions

$$\psi(0,t) = 0 \tag{39}$$

$$\psi(L,t) = 0, \tag{40}$$

initial conditions

$$\psi(x,0) = f(x) \tag{41}$$

$$\frac{\partial \psi}{\partial t}(x,0) = g(x) \tag{42}$$

and the additional constraint

$$0 < b < \frac{2\pi}{Lv}, \tag{43}$$

can also be solved as a FOURIER SERIES.

$$\psi(x,t) = \sum_{n=1}^{\infty} \sin\left(\frac{n\pi x}{L}\right) e^{-v^2 b t/2}[a_n \sin(\mu_n t) + b_n \cos(\mu_n t)], \tag{44}$$

where

$$\mu_n \equiv \frac{\sqrt{4v^2 n^2 \pi^2 - b^2 L^2 v^4}}{2L} = \frac{v\sqrt{4n^2\pi^2 - b^2 L^2 v^2}}{2L} \tag{45}$$

$$b_n = \frac{2}{L} \int_0^L \sin\left(\frac{n\pi x}{L}\right) f(x)\, dx \tag{46}$$

$$a_n = \frac{2}{L\mu_n} \left\{ \int_0^L \sin\left(\frac{n\pi x}{L}\right)\left[g(x) + \frac{v^2 b}{2} f(x)\right] dx \right\}. \tag{47}$$

To find the motion of a rectangular membrane with sides of length L_x and L_y (in the absence of gravity), use the 2-D wave equation

$$\frac{\partial^2 z}{\partial x^2} + \frac{\partial^2 z}{\partial y^2} = \frac{1}{v^2} \frac{\partial^2 z}{\partial t^2}, \tag{48}$$

where $z(x,y,t)$ is the vertical displacement of a point on the membrane at position (x, y) and time t. Use SEPARATION OF VARIABLES to look for solutions OF THE FORM

$$z(x,y,t) = X(x)Y(y)T(t). \tag{49}$$

Plugging (49) into (48) gives

$$YT \frac{d^2 X}{dx^2} + XT \frac{d^2 Y}{dy^2} = \frac{1}{v^2} XY \frac{d^2 T}{dt^2}, \tag{50}$$

where the partial derivatives have now become complete derivatives. Multiplying (50) by v^2/XYT gives

$$\frac{v^2}{X} \frac{d^2 X}{dx^2} + \frac{v^2}{Y} \frac{d^2 Y}{dy^2} = \frac{1}{T} \frac{d^2 T}{dt^2}. \tag{51}$$

The left and right sides must both be equal to a constant, so we can separate the equation by writing the right side as

$$\frac{1}{T} \frac{d^2 T}{dt^2} = -\omega^2. \tag{52}$$

This has solution

$$T(t) = C_\omega \cos(\omega t) + D_\omega \sin(\omega t). \tag{53}$$

Plugging (52) back into (51),

$$\frac{v^2}{X} \frac{d^2 X}{dx^2} + \frac{v^2}{Y} \frac{d^2 Y}{dy^2} = -\omega^2, \tag{54}$$

which we can rewrite as

$$\frac{1}{X} \frac{d^2 X}{dx^2} = -\frac{1}{Y} \frac{d^2 Y}{dy^2} - \frac{\omega^2}{v^2} = -k_x^2 \tag{55}$$

since the left and right sides again must both be equal to a constant. We can now separate out the $Y(y)$ equation

$$\frac{1}{Y}\frac{d^2Y}{dy^2} = k_x^2 - \frac{\omega^2}{v^2} \equiv -k_y^2, \tag{56}$$

where we have defined a new constant k_y satisfying

$$k_x^2 + k_y^2 = \frac{\omega^2}{v^2}. \tag{57}$$

Equations (55) and (56) have solutions

$$X(x) = E\cos(k_x x) + F\sin(k_x x) \tag{58}$$

$$Y(y) = G\cos(k_y y) + H\sin(k_y y). \tag{59}$$

We now apply the boundary conditions to (58) and (59). The conditions $z(0, y, t) = 0$ and $z(x, 0, t) = 0$ mean that

$$E = 0 \quad G = 0. \tag{60}$$

Similarly, the conditions $z(L_x, y, t) = 0$ and $z(x, L_y, t) = 0$ give $\sin(k_x L_x) = 0$ and $\sin(k_y L_y) = 0$, so $L_x k_x = p\pi$ and $L_y k_y = q\pi$, where p and q are INTEGERS. Solving for the allowed values of k_x and k_y then gives

$$k_x = \frac{p\pi}{L_x} \quad k_y = \frac{q\pi}{L_y}. \tag{61}$$

Plugging (54), (58), (59), (60), and (61) back into (24) gives the solution for particular values of p and q,

$$z_{pq}(x, y, t) = [C_\omega \cos(\omega t) + D_\omega \sin(\omega t)] \left[F_p \sin\left(\frac{p\pi x}{L_x}\right) \right]$$
$$\times \left[H_q \sin\left(\frac{q\pi y}{L_y}\right) \right]. \tag{62}$$

Lumping the constants together by writing $A_{pq} \equiv C_\omega F_p H_q$ (we can do this since ω is a function of p and q, so C_ω can be written as C_{pq}) and $B_{pq} \equiv D_\omega F_p H_q$, we obtain

$$z_{pq}(x, y, t) = [A_{pq}\cos(\omega_{pq}t) + B_{pq}\sin(\omega_{pq}t)]$$
$$\times \sin\left(\frac{p\pi x}{L_x}\right)\sin\left(\frac{q\pi y}{L_y}\right). \tag{63}$$

Plots of the spatial part for modes (1, 1), (1, 2), (2, 1), and (2, 2) follow.

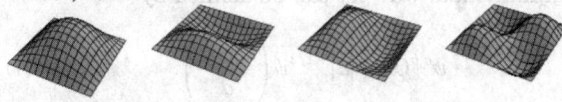

The general solution is a sum over all possible values of p and q, so the final solution is

$$z(x, y, t) = \sum_{p=1}^{\infty}\sum_{q=1}^{\infty}[A_{pq}\cos(\omega_{pq}t)$$

$$+ B_{pq}\sin(\omega_{pq}t)]\sin\left(\frac{p\pi x}{L_x}\right)\sin\left(\frac{q\pi y}{L_y}\right), \tag{64}$$

where ω is defined by combining (57) and (61) to yield

$$\omega_{pq} \equiv \pi v\sqrt{\left(\frac{p}{L_z}\right)^2 + \left(\frac{q}{L_y}\right)^2}. \tag{65}$$

Given the initial conditions $z(x, y, 0)$ and $\frac{\partial z}{\partial t}(x, y, 0)$, we can compute the A_{pq}s and B_{pq}s explicitly. To accomplish this, we make use of the orthogonality of the SINE function in the form

$$I \equiv \int_0^L \sin\left(\frac{m\pi x}{L}\right)\sin\left(\frac{n\pi x}{L}\right)dx = \tfrac{1}{2}L\delta_{mn}, \tag{66}$$

where δ_{mn} is the KRONECKER DELTA. This can be demonstrated by direct INTEGRATION. Let $u \equiv \pi x/L$ so $du = (\pi/L)\,dx$ in (66), then

$$I = \frac{L}{\pi}\int_0^\pi \sin(mu)\sin(nu)\,du. \tag{67}$$

Now use the trigonometric identity

$$\sin\alpha\sin\beta = \tfrac{1}{2}[\cos(\alpha-\beta) - \cos(\alpha+\beta)] \tag{68}$$

to write

$$I = \frac{L}{2\pi}\int_0^\pi \cos[(m-n)u]\,du + \int_0^\pi \cos[(m+n)u]\,du. \tag{69}$$

Note that for an INTEGER $l \neq 0$, the following INTEGRAL vanishes

$$\int_0^\pi \cos(lu)\,du = \frac{1}{l}[\sin(lu)]_0^\pi = \frac{1}{l}[\sin(l\pi) - \sin 0]$$

$$= \frac{1}{l}\sin(l\pi) = 0, \tag{70}$$

since $\sin(l\pi) = 0$ when l is an INTEGER. Therefore, $I = 0$ when $l \equiv m - n \neq 0$. However, I does *not* vanish when $l = 0$, since

$$\int_0^\pi \cos(0\cdot u)\,du = \int_0^\pi du = \pi. \tag{71}$$

We therefore have that $I = L\delta_{mn}/2$, so we have derived (66). Now we multiply $z(x, y, 0)$ by two sine terms and integrate between 0 and L_x and between 0 and L_y,

$$I = \int_0^{L_y}\left[\int_0^{L_x} z(x, y, 0)\sin\left(\frac{p\pi x}{L_x}\right)dx\right]\sin\left(\frac{q\pi y}{L_y}\right)dy. \tag{72}$$

Now plug in $z(x, y, t)$, set $t = 0$, and prime the indices to distinguish them from the p and q in (72),

$$I = \sum_{q'=1}^{\infty} \int_0^{L_y} \left[\sum_{p'=1}^{\infty} A_{p'q'} \int_0^{L_x} \sin\left(\frac{p\pi x}{L_x}\right) \sin\left(\frac{p'\pi x}{L_x}\right) dx \right]$$

$$\times \sin\left(\frac{q\pi y}{L_y}\right) \sin\left(\frac{q'\pi y}{L_y}\right) dy. \qquad (73)$$

Making use of (66) in (73),

$$I = \sum_{q'=1}^{\infty} \int_0^{L_y} \sum_{p'=1}^{\infty} A_{p'q'} \frac{L_x}{2} \delta_{p,p'} \left(\frac{q\pi y}{L_y}\right) \sin\left(\frac{q'\pi y}{L_y}\right) dy, \quad (74)$$

so the sums over p' and q' collapse to a single term

$$I = \frac{L_x}{2} \sum_{p=1}^{\infty} A_{pq} \frac{L_y}{2} \delta_{q,q'} \frac{L_x L_y}{4} A_{pq}. \qquad (75)$$

Equating (74) and (75) and solving for A_{pq} then gives

$$A_{pq} = \frac{4}{L_x L_y} \int_0^{L_y} \left[\int_0^{L_x} z(x,y,0) \sin\left(\frac{p\pi x}{L_x}\right) dx \right] \sin\left(\frac{q\pi x}{L_y}\right) dy. \qquad (76)$$

An analogous derivation gives the B_{pq}s as

$$B_{pq} = \frac{4}{\omega_{pq} L_x L_y} \int_0^{L_y} \left[\int_0^{L_x} \frac{\partial z}{\partial t}(x,y,0) \sin\left(\frac{p\pi x}{L_x}\right) dx \right]$$

$$\times \sin\left(\frac{q\pi x}{L_y}\right) dy. \qquad (77)$$

The equation of motion for a membrane shaped as a RIGHT ISOSCELES TRIANGLE of length c on a side and with the sides oriented along the POSITIVE x and y axes is given by

$$\psi(x,y,t) = \left[C_{pq} \cos(\omega_{pq} t) + D_{pq} \sin(\omega_{pq} t) \right]$$

$$\times \left[\sin\left(\frac{p\pi x}{c}\right) \sin\left(\frac{q\pi y}{c}\right) - \sin\left(\frac{q\pi x}{c}\right) \sin\left(\frac{p\pi y}{c}\right) \right], \qquad (78)$$

where

$$\omega_{pq} = \frac{\pi v}{c} \sqrt{p^2 + q^2} \qquad (79)$$

and p, q INTEGERS with $p > q$. This solution can be obtained by subtracting two wave solutions for a square membrane with the indices reversed. Since points on the diagonal which are equidistant from the center must have the same wave equation solution (by symmetry), this procedure gives a wavefunction which will vanish along the diagonal *as long as p and q are both* EVEN *or* ODD. We must further restrict the modes since those with $p < q$ give wavefunctions which are just the NEGATIVE of (q, p) and (p, p) give an identically zero wavefunction. The following plots show (3, 1), (4, 2), (5, 1), and (5,3).

See also D'ALEMBERTIAN, TELEGRAPH EQUATION

References

Abramowitz, M. and Stegun, C. A. (Eds.). "Wave Equation in Prolate and Oblate Spheroidal Coordinates." §21.5 in *Handbook of Mathematical Functions with Formulas, Graphs, and Mathematical Tables, 9th printing.* New York: Dover, pp. 752–53, 1972.

Morse, P. M. and Feshbach, H. *Methods of Theoretical Physics, Part I.* New York: McGraw-Hill, pp. 124–25 and 271, 1953.

Zwillinger, D. (Ed.). *CRC Standard Mathematical Tables and Formulae.* Boca Raton, FL: CRC Press, p. 417, 1995.

Zwillinger, D. *Handbook of Differential Equations, 3rd ed.* Boston, MA: Academic Press, p. 130, 1997.

Wave Operator

An OPERATOR relating the asymptotic state of a DYNAMICAL SYSTEM governed by the Schrödinger equation

$$i\frac{d}{dt}\psi(t) = H\psi(t)$$

to its original asymptotic state.

See also SCATTERING OPERATOR

Wave Surface

A SURFACE represented parametrically by ELLIPTIC FUNCTIONS.

Wavelet

Wavelets are a class of a functions used to localize a given function in both space and scaling. A family of wavelets can be constructed from a function $\psi(x)$, sometimes known as a "mother wavelet," which is confined in a finite interval. "Daughter wavelets" $\psi^{a,b}(x)$ are then formed by translation (b) and contraction (a). Wavelets are especially useful for compressing image data, since a WAVELET TRANSFORM has properties which are in some ways superior to a conventional FOURIER TRANSFORM.

An individual wavelet can be defined by

$$\psi^{a,b}(x) = |a|^{-1/2} \psi\left(\frac{x-b}{a}\right). \qquad (1)$$

Then

$$W_\psi(f)(a,b) = \frac{1}{\sqrt{a}} \int_{-\infty}^{\infty} f(t)\psi\left(\frac{t-b}{a}\right) dt, \qquad (2)$$

and CALDERÓN'S FORMULA gives

$$f(x) = C_\psi \int_{-\infty}^{\infty} \int_{-\infty}^{\infty} \langle f, \psi^{a,b} \rangle \psi^{a,b}(x) a^{-2} \, da \, db. \quad (3)$$

A common type of wavelet is defined using HAAR FUNCTIONS.

See also FOURIER TRANSFORM, HAAR FUNCTION, LEMARIÉ'S WAVELET, WAVELET TRANSFORM

References

Benedetto, J. J. and Frazier, M. (Eds.). *Wavelets: Mathematics and Applications*. Boca Raton, FL: CRC Press, 1994.

Chui, C. K. *An Introduction to Wavelets*. San Diego, CA: Academic Press, 1992.

Chui, C. K. (Ed.). *Wavelets: A Tutorial in Theory and Applications*. San Diego, CA: Academic Press, 1992.

Chui, C. K.; Montefusco, L.; and Puccio, L. (Eds.). *Wavelets: Theory, Algorithms, and Applications*. San Diego, CA: Academic Press, 1994.

Daubechies, I. *Ten Lectures on Wavelets*. Philadelphia, PA: Society for Industrial and Applied Mathematics, 1992.

Erlebacher, G. H.; Hussaini, M. Y.; and Jameson, L. M. (Eds.). *Wavelets: Theory and Applications*. New York: Oxford University Press, 1996.

Foufoula-Georgiou, E. and Kumar, P. (Eds.). *Wavelets in Geophysics*. San Diego, CA: Academic Press, 1994.

Hernández, E. and Weiss, G. *A First Course on Wavelets*. Boca Raton, FL: CRC Press, 1996.

Hubbard, B. B. *The World According to Wavelets: The Story of a Mathematical Technique in the Making*, 2nd rev. upd. ed. New York: A. K. Peters, 1998.

Jawerth, B. and Sweldens, W. "An Overview of Wavelet Based Multiresolution Analysis." *SIAM Rev.* **36**, 377–12, 1994.

Kaiser, G. *A Friendly Guide to Wavelets*. Cambridge, MA: Birkhäuser, 1994.

Massopust, P. R. *Fractal Functions, Fractal Surfaces, and Wavelets*. San Diego, CA: Academic Press, 1994.

Meyer, Y. *Wavelets: Algorithms and Applications*. Philadelphia, PA: SIAM Press, 1993.

Press, W. H.; Flannery, B. P.; Teukolsky, S. A.; and Vetterling, W. T. "Wavelet Transforms." §13.10 in *Numerical Recipes in FORTRAN: The Art of Scientific Computing*, 2nd ed. Cambridge, England: Cambridge University Press, pp. 584–99, 1992.

Resnikoff, H. L. and Wells, R. O. J. *Wavelet Analysis: The Scalable Structure of Information*. New York: Springer-Verlag, 1998.

Schumaker, L. L. and Webb, G. (Eds.). *Recent Advances in Wavelet Analysis*. San Diego, CA: Academic Press, 1993.

Stollnitz, E. J.; DeRose, T. D.; and Salesin, D. H. "Wavelets for Computer Graphics: A Primer, Part 1." *IEEE Computer Graphics and Appl.* **15**, No. 3, 76–4, 1995.

Stollnitz, E. J.; DeRose, T. D.; and Salesin, D. H. "Wavelets for Computer Graphics: A Primer, Part 2." *IEEE Computer Graphics and Appl.* **15**, No. 4, 75–5, 1995.

Strang, G. "Wavelets and Dilation Equations: A Brief Introduction." *SIAM Rev.* **31**, 614–27, 1989.

Strang, G. "Wavelets." *Amer. Sci.* **82**, 250–55, 1994.

Taswell, C. *Handbook of Wavelet Transform Algorithms*. Boston, MA: Birkhäuser, 1996.

Teolis, A. *Computational Signal Processing with Wavelets*. Boston, MA: Birkhäuser, 1997.

Vidakovic, B. *Statistical Modeling by Wavelets*. New York: Wiley, 1999.

Walker, J. S. *A Primer on Wavelets and their Scientific Applications*. Boca Raton, FL: CRC Press, 1999.

Walter, G. G. *Wavelets and Other Orthogonal Systems with Applications*. Boca Raton, FL: CRC Press, 1994.

"Wavelet Digest." http://www.wavelet.org/wavelet/.

Weisstein, E. W. "Books about Wavelets." http://www.treasure-troves.com/books/Wavelets.html.

Wickerhauser, M. V. *Adapted Wavelet Analysis from Theory to Software*. Wellesley, MA: Peters, 1994.

Wavelet Matrix

Any discrete finite WAVELET TRANSFORM can be REPRESENTED AS a matrix, and such a wavelet matrix can be computed in $\mathcal{O}(n)$ steps, compared to $\mathcal{O}(n \lg n)$ for the FOURIER MATRIX, where $\lg x = \log_2 x$ is the base-2 LOGARITHM. A single wavelet matrix can be built using HAAR FUNCTIONS.

See also FOURIER MATRIX, HAAR FUNCTION, WAVELET, WAVELET TRANSFORM

Wavelet Transform

A transform which localizes a function both in space and scaling and has some desirable properties compared to the FOURIER TRANSFORM. The transform is based on a WAVELET MATRIX, which can be computed more quickly than the analogous FOURIER MATRIX.

See also DAUBECHIES WAVELET FILTER, LEMARIE'S WAVELET, WAVELET MATRIX

References

Blair, D. and MathSoft, Inc. "Wavelet Resources." http://www.mathsoft.com/wavelets.html.

Daubechies, I. *Ten Lectures on Wavelets*. Philadelphia, PA: SIAM, 1992.

DeVore, R.; Jawerth, B.; and Lucier, B. "Images Compression through Wavelet Transform Coding." *IEEE Trans. Information Th.* **38**, 719–46, 1992.

Press, W. H.; Flannery, B. P.; Teukolsky, S. A.; and Vetterling, W. T. "Wavelet Transforms." §13.10 in *Numerical Recipes in FORTRAN: The Art of Scientific Computing*, 2nd ed. Cambridge, England: Cambridge University Press, pp. 584–99, 1992.

Strang, G. "Wavelet Transforms Versus Fourier Transforms." *Bull. Amer. Math. Soc.* **28**, 288–05, 1993.

Weak Convergence

Weak convergence is usually either denoted $x_n \to^w x$ or $x_n \rightharpoonup x$. A SEQUENCE $\{x_n\}$ of VECTORS in an INNER PRODUCT SPACE E is called weakly convergent to a VECTOR in E if

$$\langle x_n, y \rangle \to \langle x, y \rangle \quad as \ n \to \infty, \quad for \ all \ y \in E.$$

Every STRONGLY CONVERGENT sequence is also weakly convergent (but the opposite does not usually hold). This can be seen as follows. Consider the sequence $\{x_n\}$ that converges strongly to x, i.e., $\|x_n - x\| \to 0$ as $n \to \infty$. SCHWARZ'S INEQUALITY now gives

$$|\langle x_n - x, y \rangle| \le \|x_n - x\| \|y\| \quad as \ n \to \infty.$$

The definition of weak convergence is therefore satisfied.

See also INNER PRODUCT SPACE, SCHWARZ'S INEQUALITY, STRONG CONVERGENCE

Weak Law of Large Numbers

A result in probability theory also known as BERNOULLI'S THEOREM or the weak law of large numbers (in contrast to the STRONG LAW OF LARGE NUMBERS). Let $X_1, ..., X_n$ be a sequence of independent and identically distributed random variables, each having a MEAN $\langle X_i \rangle = \mu$ and STANDARD DEVIATION σ. Define a new variable

$$X \equiv \frac{X_1 + \cdots + X_n}{n}. \tag{1}$$

Then, as $n \to \infty$, the sample mean $\langle x \rangle$ equals the population MEAN μ of each variable.

$$\langle X \rangle = \left\langle \frac{X_1 + \cdots + X_n}{n} \right\rangle = \frac{1}{n}(\langle X_1 \rangle + \cdots + \langle X_n \rangle)$$

$$= \frac{n\mu}{n} = \mu. \tag{2}$$

In addition,

$$\mathrm{var}(X) = \mathrm{var}\left(\frac{X_1 + \cdots + X_2}{n}\right)$$

$$= \mathrm{var}\left(\frac{X_1}{n}\right) + \cdots + \mathrm{var}\left(\frac{X_n}{n}\right)$$

$$= \frac{\sigma^2}{n^2} + \cdots + \frac{\sigma^2}{n^2} = \frac{\sigma^2}{n}. \tag{3}$$

Therefore, by the CHEBYSHEV INEQUALITY, for all $\epsilon > 0$,

$$P(|X - \mu| \geq \epsilon) \leq \frac{\mathrm{var}(X)}{\epsilon^2} = \frac{\sigma^2}{n\epsilon^2}. \tag{4}$$

As $n \to \infty$, it then follows that

$$\lim_{n \to \infty} P(|X - \mu| \geq \epsilon) = 0. \tag{5}$$

(Khintchine 1929). Stated another way, the probability that the average $|(X_1 + \cdots + X_n)/n - \mu| < \epsilon$ for ϵ an arbitrary POSITIVE quantity approaches 1 as $n \to \infty$ (Feller 1968, pp. 228–29).

See also ASYMPTOTIC EQUIPARTITION PROPERTY, CENTRAL LIMIT THEOREM, CHEBYSHEV INEQUALITY, FRIVOLOUS THEOREM OF ARITHMETIC, LAW OF TRULY LARGE NUMBERS, STRONG LAW OF LARGE NUMBERS

References

Feller, W. "Laws of Large Numbers." Ch. 10 in *An Introduction to Probability Theory and Its Applications, Vol. 1, 3rd ed.* New York: Wiley, pp. 228–47, 1968.
Feller, W. "Law of Large Numbers for Identically Distributed Variables." §7.7 in *An Introduction to Probability Theory and Its Applications, Vol. 2, 3rd ed.* New York: Wiley, pp. 231–34, 1971.
Khintchine, A. "Sur la loi des grands nombres." *Comptes rendus de l'Académie des Sciences* **189**, 477–79, 1929.
Papoulis, A. *Probability, Random Variables, and Stochastic Processes, 2nd ed.* New York: McGraw-Hill, pp. 69–1, 1984.

Weakly Binary Tree

N.B. A detailed online essay by S. Finch was the starting point for this entry.

A ROOTED TREE for which the ROOT NODE is adjacent to at most two VERTICES, and all nonroot VERTICES are adjacent to at most three VERTICES. Let $b(n)$ be the number of weakly binary trees of order n, then $b(5) = 6$. Let

$$g(z) = \sum_{i=0}^{\infty} g_i z^i, \tag{1}$$

where

$$g_0 = 0 \tag{2}$$

$$g_1 = g_2 = g_3 = 1 \tag{3}$$

$$g_{2i+1} = \sum_{j=1}^{i} g_{2i+1-j} g_j \tag{4}$$

$$g_{2i} = \tfrac{1}{2} g_i(g_i + 1) \sum_{j=1}^{i-1} g_{2i-j} g_j. \tag{5}$$

Otter (Otter 1948, Harary and Palmer 1973, Knuth 1969) showed that

$$\lim_{n \to \infty} \frac{b(n) n^{3/2}}{\zeta^n} = \eta, \tag{6}$$

where

$$\xi = 2.48325\ldots \tag{7}$$

is the unique POSITIVE ROOT of

$$g\left(\frac{1}{x}\right) = 1, \tag{8}$$

and

$$\eta = 0.7916032\ldots. \tag{9}$$

ξ_1 is also given by

$$\xi = \lim_{n \to \infty} (c_n)^{2-n}, \tag{10}$$

where c_n is given by

$$c_0 = 2 \tag{11}$$

$$c_n = (c_{n-1})^2 + 2, \tag{12}$$

giving

$$\eta = \frac{1}{2}\sqrt{\frac{\xi}{\pi}}\sqrt{3 + \frac{1}{c_1} + \frac{1}{c_1 c_2} + \frac{1}{c_1 c_2 c_3} + \ldots} \quad (13)$$

References

Finch, S. "Favorite Mathematical Constants." http://www.mathsoft.com/asolve/constant/otter/otter.html.

Harary, F. *Graph Theory.* Reading, MA: Addison-Wesley, 1969.

Harary, F. and Palmer, E. M. *Graphical Enumeration.* New York: Academic Press, 1973.

Knuth, D. E. *The Art of Computer Programming, Vol. 1: Fundamental Algorithms, 3rd ed.* Reading, MA: Addison-Wesley, 1997.

Otter, R. "The Number of Trees." *Ann. Math.* **49**, 583–99, 1948.

Weakly Complete Sequence

A SEQUENCE of numbers $V = \{v_n\}$ is said to be weakly complete if every POSITIVE INTEGER n beyond a certain point N is the sum of some SUBSEQUENCE of V (Honsberger 1985). Dropping two terms from the FIBONACCI NUMBERS produces a SEQUENCE which is not even weakly complete. However, the SEQUENCE

$$F'_n \equiv F_n - (-1)^n$$

is weakly complete, even with any finite subsequence deleted (Graham 1964).

See also COMPLETE SEQUENCE

References

Graham, R. "A Property of Fibonacci Numbers." *Fib. Quart.* **2**, 1–0, 1964.

Honsberger, R. *Mathematical Gems III.* Washington, DC: Math. Assoc. Amer., p. 128, 1985.

Weakly Connected Component

A weakly connected component is a maximal SUBGRAPH of a DIRECTED GRAPH such that for every pair of vertices u, v in the SUBGRAPH, there is an undirected path from u to v and a directed path from v to u. Weakly connected components can be found using `StronglyConnectedComponents[g]` in the *Mathematica* add-on package `DiscreteMath`-`Combinatorica` (which can be loaded with the command `<<DiscreteMath`) (Skiena 1990, p. 172).

See also WEAKLY CONNECTED DIGRAPH

References

Skiena, S. *Implementing Discrete Mathematics: Combinatorics and Graph Theory with Mathematica.* Reading, MA: Addison-Wesley, 1990.

Weakly Connected Digraph

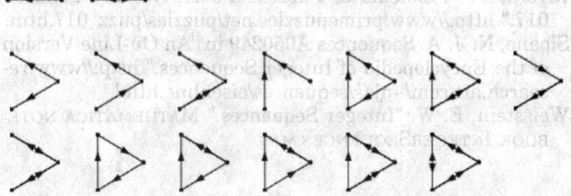

A DIRECTED GRAPH in which it is possible to reach any node starting from any other node by traversing edges in some direction (i.e., not necessarily in the direction they point). The nodes in a strongly connected digraph therefore must all have either OUTDEGREE or INDEGREE of at least 1. The numbers of nonisomorphic simple weakly connected digraphs on $n = 1, 2, \ldots$ nodes are 1, 2, 13, 199, 9364, ... (Sloane's A003085).

See also CONNECTED DIGRAPH, STRONGLY CONNECTED DIGRAPH, WEAKLY CONNECTED COMPONENT

References

Harary, F. and Palmer, E. M. *Graphical Enumeration.* New York: Academic Press, p. 218, 1973.

Skiena, S. "Strong and Weak Connectivity." §5.1.2 in *Implementing Discrete Mathematics: Combinatorics and Graph Theory with Mathematica.* Reading, MA: Addison-Wesley, pp. 172–74, 1990.

Sloane, N. J. A. Sequences A003085/M2067 in "An On-Line Version of the Encyclopedia of Integer Sequences." http://www.research.att.com/~njas/sequences/eisonline.html.

Weakly Differentiable

See also DIFFERENTIABLE

Weakly Independent

An infinite sequence $\{a_i\}$ of POSITIVE INTEGERS is called weakly independent if any relation $\Sigma \epsilon_i a_i$ with $\epsilon_i = 0$ or ± 1 and $\epsilon_i = 0$, except finitely often, IMPLIES $\epsilon_i = 0$ for all i.

See also STRONGLY INDEPENDENT

References

Guy, R. K. *Unsolved Problems in Number Theory, 2nd ed.* New York: Springer-Verlag, p. 136, 1994.

Weakly Prime

A PRIME NUMBER is said to be weakly prime if changing a single digit to every other possible digit produces a COMPOSITE NUMBER when performed on each digit. The first few such numbers are 294001, 505447, 584141, 604171, 971767, 1062599, ... (Sloane's A050249).

See also COMPOSITE NUMBER, PRIME NUMBER

References

--. "Problem #12." http://math.smsu.edu/~les/POW12.html.
Rivera, C. "Problems & Puzzles: Puzzle Weakly Primes.-017." http://www.primepuzzles.net/puzzles/puzz_017.htm.
Sloane, N. J. A. Sequences A050249 in "An On-Line Version of the Encyclopedia of Integer Sequences." http://www.research.att.com/~njas/sequences/eisonline.html.
Weisstein, E. W. "Integer Sequences." MATHEMATICA NOTEBOOK INTEGERSEQUENCES.M.

Weakly Triple-Free Set

TRIPLE-FREE SET

Web Graph

A graph formed by connecting several concentric WHEEL GRAPHS along spokes.

See also WHEEL GRAPH

Weber Differential Equations

Consider the differential equation satisfied by

$$w = z^{-1/2} W_{k,-1/4}\left(\tfrac{1}{2}z^2\right), \tag{1}$$

where W is a WHITTAKER FUNCTION, which is given by

$$\frac{d}{z\,dz}\left[\frac{d\left(wz^{1/2}\right)}{z\,dz}\right] + \left(-\frac{1}{4} + \frac{2k}{z^2} + \frac{3}{4z^4}\right)wz^{1/2} = 0 \tag{2}$$

$$\frac{d^2w}{dz^2} + \left(2k - \tfrac{1}{4}z^2\right)w = 0 \tag{3}$$

(Moon and Spencer 1961, p. 153; Zwillinger 1997, p. 128). This is usually rewritten

$$\frac{d^2 D_n(z)}{dz^2} + \left(n + \tfrac{1}{2} - \tfrac{1}{4}z^2\right)D_n(z) = 0. \tag{4}$$

The solutions are PARABOLIC CYLINDER FUNCTIONS. The equations

$$\frac{d^2 U}{du^2} - \left(c + k^2 u^2\right)U = 0 \tag{5}$$

$$\frac{d^2 V}{du^2} - \left(c - k^2 v^2\right)V = 0, \tag{6}$$

which arise by separating variables in LAPLACE'S EQUATION in PARABOLIC CYLINDRICAL COORDINATES, are also known as the Weber differential equations. As above, the solutions are known as PARABOLIC CYLINDER FUNCTIONS.

Zwillinger (1997, p. 127) calls

$$y'' + \frac{y'}{x} + \left(1 - \frac{v^2}{x^2}\right)y = -\frac{1}{\pi x^2}\left[x + v + (x - v)\cos(v\pi)\right] \tag{7}$$

the Weber differential equation (Gradshteyn and Ryzhik 2000, p. 989).

References

Gradshteyn, I. S. and Ryzhik, I. M. *Tables of Integrals, Series, and Products, 6th ed.* San Diego, CA: Academic Press, p. 989, 2000.
Moon, P. and Spencer, D. E. *Field Theory for Engineers.* New York: Van Nostrand, 1961.
Zwillinger, D. *Handbook of Differential Equations, 3rd ed.* Boston, MA: Academic Press, p. 127, 1997.

Weber Functions

Although BESSEL FUNCTIONS OF THE SECOND KIND are sometimes called Weber functions, Abramowitz and Stegun (1972) define a separate Weber function as

$$\mathscr{E}_v(z) = \frac{1}{\pi}\int_0^\pi \sin(v\theta - z\sin\theta)\,d\theta. \tag{1}$$

Letting $\zeta_n = e^{2\pi i/n}$ be a ROOT OF UNITY, another set of Weber functions is defined as

$$f(z) = \frac{\eta\left(\tfrac{1}{2}(z+1)\right)}{\zeta_{48}\eta(z)} \tag{2}$$

$$f_1(z) = \frac{\eta\left(\tfrac{1}{2}z\right)}{\eta(z)} \tag{3}$$

$$f_2(z) = \sqrt{2}\,\frac{\eta(2z)}{\eta(z)} \tag{4}$$

$$\gamma_2 = \frac{f^{24}(z) - 16}{f^8(z)} \tag{5}$$

$$\gamma_3 = \frac{[f^{24}(z) + 8][f_1^8(z) - f_2^8(z)]}{f^8(z)} \tag{6}$$

(Weber 1902, Atkin and Morain 1993), where $\eta(z)$ is the DEDEKIND ETA FUNCTION. The Weber functions satisfy the identities

$$f(z+1) = \frac{f_1(z)}{\zeta_{48}} \tag{7}$$

$$f_1(z+1) = \frac{f(z)}{\zeta_{48}} \tag{8}$$

$$f_2(z+1) = \zeta_{24}f_2(z) \tag{9}$$

$$f\left(-\frac{1}{z}\right) = f(z) \tag{10}$$

$$f_1\left(-\frac{1}{z}\right) = f_2(z) \tag{11}$$

$$f_2\left(-\frac{1}{z}\right) = f_1(z) \qquad (12)$$

(Weber 1902, Atkin and Morain 1993).

See also ANGER FUNCTION, BESSEL FUNCTION OF THE SECOND KIND, DEDEKIND ETA FUNCTION, *J*-FUNCTION, JACOBI IDENTITIES, JACOBI TRIPLE PRODUCT, MODIFIED STRUVE FUNCTION, *Q*-FUNCTION, STRUVE FUNCTION

References

Abramowitz, M. and Stegun, C. A. (Eds.). "Anger and Weber Functions." §12.3 in *Handbook of Mathematical Functions with Formulas, Graphs, and Mathematical Tables, 9th printing.* New York: Dover, pp. 498–99, 1972.
Atkin, A. O. L. and Morain, F. "Elliptic Curves and Primality Proving." *Math. Comput.* **61**, 29–8, 1993.
Borwein, J. M. and Borwein, P. B. *Pi & the AGM: A Study in Analytic Number Theory and Computational Complexity.* New York: Wiley, pp. 68–9, 1987.
Prudnikov, A. P.; Marichev, O. I.; and Brychkov, Yu. A. "The Anger Function $J_\nu(x)$ and Weber Function $E_\nu(x)$." §1.5 in *Integrals and Series, Vol. 3: More Special Functions.* Newark, NJ: Gordon and Breach, p. 28, 1990.
Weber, H. *Lehrbuch der Algebra, Vols. I-II.* New York: Chelsea, pp. 113–14, 1902.

Weber's Discontinuous Integrals

$$\int_0^\infty J_0(ax)\cos(cx)\,dx = \begin{cases} 0 & a<c \\ \frac{1}{\sqrt{a^2-c^2}} & a>c \end{cases}$$

$$\int_0^\infty J_0(ax)\sin(cx)\,dx = \begin{cases} \frac{1}{\sqrt{c^2-a^2}} & a<c \\ 0 & a>c, \end{cases}$$

where $J_0(z)$ is a zeroth order BESSEL FUNCTION OF THE FIRST KIND.

References

Bowman, F. *Introduction to Bessel Functions.* New York: Dover, pp. 59–0, 1958.

Weber's Formula

$$\frac{1}{2p^2}e^{-(a^2+b^2)/(4p^2)}I_\nu\left(\frac{ab}{2p^2}\right) = \int_0^\infty e^{-p^2t^2}J_\nu(at)J_\nu(bt)t\,dt,$$

where $\Re[\nu] > -1$, $|\arg p| < \pi/4$, and $a, b > 0$, $J_\nu(z)$ is a BESSEL FUNCTION OF THE FIRST KIND, and $I_\nu(z)$ is a MODIFIED BESSEL FUNCTION OF THE FIRST KIND.

See also BESSEL FUNCTION OF THE FIRST KIND, MODIFIED BESSEL FUNCTION OF THE FIRST KIND

References

Iyanaga, S. and Kawada, Y. (Eds.). *Encyclopedic Dictionary of Mathematics.* Cambridge, MA: MIT Press, p. 1476, 1980.

Weber's Theorem

If two curves of the same GENUS (CURVE) > 1 are in rational correspondence, then that correspondence is BIRATIONAL.

References

Coolidge, J. L. *A Treatise on Algebraic Plane Curves.* New York: Dover, p. 135, 1959.

Weber-Sonine Formula

For $\Re[\mu + nu] > 0$, $|\arg p| < \pi/4$, and $a > 0$,

$$\int_0^\infty J_\nu(at)e^{-p^2t^2}t^{\mu-1}dt$$
$$= \left(\frac{a}{2p}\right)^\nu \frac{\Gamma\left[\frac{1}{2}(\nu+\mu)\right]}{2p^\mu\Gamma(\nu+1)}{}_1F_1\left(\tfrac{1}{2}(\nu+\mu);\nu+1;-\frac{a^2}{2p^2}\right),$$

where $J_\nu(z)$ is a BESSEL FUNCTION OF THE FIRST KIND, $\Gamma(z)$ is the GAMMA FUNCTION, and ${}_1F_1(a;b;z)$ is a CONFLUENT HYPERGEOMETRIC FUNCTION.

References

Iyanaga, S. and Kawada, Y. (Eds.). *Encyclopedic Dictionary of Mathematics.* Cambridge, MA: MIT Press, p. 1474, 1980.

Wedderburn's Theorem

A FINITE DIVISION RING is a FIELD.

Weddle's Rule

Let the values of a function $f(x)$ be tabulated at points x_i equally spaced by $h = x_{i+1}-x_i$, so $f_1 = f(x_1)$, $f_2 = f(x_2)$, ..., $f_7 = f(x_7)$. Then Weddle's rule approximating the integral of $f(x)$ is given by the NEWTON-COTES-like formula

$$\int_{x_1}^{x_{6n}} f(x)\,dx = \tfrac{3}{10}h(f_1 + 5f_2 + f_3 + 6f_4 + 5f_5 + f_6$$
$$+ \ldots + 5f_{6n-1} + f_{6n})$$

See also BODE'S RULE, HARDY'S RULE, NEWTON-COTES FORMULAS, SHOVELTON'S RULE, SIMPSON'S 3/8 RULE, SIMPSON'S RULE, TRAPEZOIDAL RULE

References

King, A. E. "Approximate Integration. Note on Quadrature Formulae: Their Construction and Application to Actuarial Functions." *Trans. Faculty of Actuaries* **9**, 218–31, 1923.
Sheppard, W. F. "Some Quadrature-Formulæ." *Proc. London Math. Soc.* **32**, 258–77, 1900.
Whittaker, E. T. and Robinson, G. *The Calculus of Observations: A Treatise on Numerical Mathematics, 4th ed.* New York: Dover, p. 151, 1967.

Wedge

The term "wedge" has a number of meanings in mathematics. It is sometimes used as another name for the CARET symbol, as well as being the notation ($\wedge$) for logical AND.

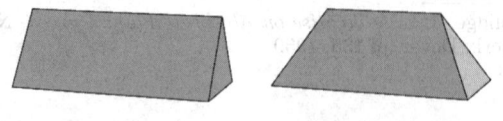

In SOLID GEOMETRY, a wedge is a right triangular PRISM turned so that it rests on one of its lateral rectangular faces (left figure). Harris and Stocker (1998) define a more general type of wedge in which the top edge is symmetrically shortened, causing the end triangles to slant obliquely (right figure).

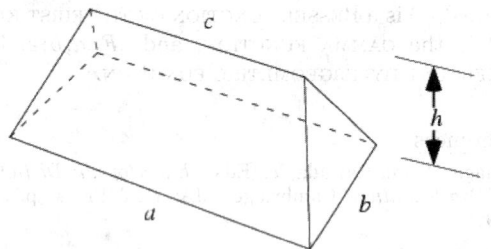

For a wedge of base lengths a and b, height h, and top edge length c, the VOLUME of the wedge is

$$V = \tfrac{1}{6}h(2a + c).$$

In the case $c = a$, this simplifies to $V = ha/2$. The CENTROID is located at a height

$$\bar{z} = \frac{(a + c)h}{2(2a + c)}$$

above the base, which simplifies to h^3 for $c = a$.

See also AND, CARET, CONICAL WEDGE, CYLINDRICAL WEDGE, PRISM, SPHERICAL WEDGE

References

Bringhurst, R. *The Elements of Typographic Style, 2nd ed.* Point Roberts, WA: Hartley and Marks, p. 286, 1997.
Harris, J. W. and Stocker, H. "Wedge." §4.5.2 in *Handbook of Mathematics and Computational Science.* New York: Springer-Verlag, p. 101, 1998.
Weisstein, E. W. "SolidGeometry." MATHEMATICA NOTEBOOK SOLIDGEOMETRY.M.

Wedge Product

The wedge product is the product in an EXTERIOR ALGEBRA. If α and β are DIFFERENTIAL K-FORMS of degrees p and q, respectively, then

$$\alpha \wedge \beta = (-1)^{pq}\beta \wedge \alpha. \tag{1}$$

It is not (in general) COMMUTATIVE, but it is ASSOCIA-TIVE,

$$(\alpha \wedge \beta) \wedge u = \alpha \wedge (\beta \wedge u), \tag{2}$$

and BILINEAR

$$(c_1\alpha_1 + c_2\alpha_2) \wedge \beta = c_1(\alpha_1 \wedge \beta) + c_2(\alpha_2 \wedge \beta) \tag{3}$$

$$\alpha \wedge (c_1\beta_1 + c_2\beta_2) = c_1(\alpha \wedge \beta_1) + c_2(\alpha \wedge \beta_2) \tag{4}$$

(Spivak 1999, p. 203), where c_1 and c_2 are constants. The alternating algebra is generated by elements of degree one, and so the wedge product can be defined using a basis e_i for V:

$$\begin{aligned}\left(e_{i_1} \wedge \ldots \wedge e_{i_p}\right) \wedge \left(e_{j_1} \wedge \ldots \wedge e_{j_q}\right) = e_{i_1} \wedge \ldots \wedge e_{i_p} \\ \wedge \, e_{j_1} \wedge \ldots \wedge e_{j_q}\end{aligned} \tag{5}$$

when the indices $i_1, \ldots, i_p, i_1, \ldots, i_q,$ are distinct, and the product is zero otherwise.

While the formula $\alpha \wedge \alpha = 0$ holds when α has degree one, it does not hold in general. For example, consider $\alpha = e_1 \wedge e_2 + e_3 \wedge e_4$:

$$\alpha \wedge \alpha = (e_1 \wedge e_2) \wedge (e_1 \wedge e_2) + (e_1 \wedge e_2) \wedge (e_3 \wedge e_4)$$

$$+ (e_3 \wedge e_4) \wedge (e_1 \wedge e_2) + (e_3 \wedge e_4) \wedge (e_3 \wedge e_4)$$

$$= 0 + e_1 \wedge e_2 \wedge e_3 \wedge e_4 + e_3 \wedge e_4 \wedge e_1 \wedge e_2 + 0$$

$$= 2e_1 \wedge e_2 \wedge e_3 \wedge e_4 \tag{6}$$

If $\alpha_1, \ldots, \alpha_k$ have degree one, then they are linearly independent IFF $\alpha_1 \wedge \ldots \wedge \alpha_k \neq 0$.

The wedge product is the "correct" type of product to use in computing a VOLUME ELEMENT

$$dV = dx_1 \wedge \ldots \wedge dx_n. \tag{7}$$

The wedge product can therefore be used to calculate DETERMINANTS and volumes of PARALLELEPIPEDS. For example, write $\det A = \det(c_1, \ldots, c_n)$ where c_i are the columns of A. Then

$$c_1 \wedge \ldots \wedge c_n = \det(c_1, \ldots, c_n)e_1 \wedge \ldots \wedge e_n \tag{8}$$

and $|\det(c_1, \ldots, c_n)|$ is the volume of the PARALLELE-PIPED spanned by $c_1, \ldots, c_n$.

In *Mathematica*, a k-form can be written as an ANTISYMMETRIC k-tensor. Using this format, the following *Mathematica* function computes the wedge product. vars.

```
Alt[x_List] := Module[
  {
  p = TensorRank[x],   perms
  },
  perms = Permutations[Range[p]];
    Sum[Signature[perms[[i]]]   Transpose[x,
perms[[i]]],{i, p!}]/p!
 ] Wedge1[a_List, b_List] := Alt[Outer[Times,
 a, b]]
```

It is also possible to use an n-nested binary tree to represent the algebra of differential forms. Using this format, the following *Mathematica* function computes the wedge product recursively.

```
Wedge2[{a_?(! ListQ[#1] &), b_?(! ListQ[#1]
&)}, {c_?(! ListQ[#1] &),    d_?(! ListQ[#1]
&)}] := {a d + b c, b d}  sgn2[a_?ListQ] :=
MapIndexed[(Times[#1, Power[-1, Tr[#2]]] &),
a, {TensorRank[a]}]; Wedge2[{a_List, b_List},
{c_List, d_List}] :=
    {Wedge2[a, d] + Wedge2[sgn2[b], c],
Wedge2[b, d]}
```

See also COHOMOLOGY, CUP PRODUCT, DETERMINANT, DIFFERENTIAL K-FORM, EXTERIOR ALGEBRA, EXTERIOR DERIVATIVE, EXTERIOR POWER, INNER PRODUCT, TENSOR PRODUCT (MODULE), VECTOR SPACE, VOLUME, VOLUME ELEMENT

References

Berger, M. *Differential Geometry.* New York: Springer-Verlag, 1988.
Flanders, H. *Differential Forms with Applications to the Physical Sciences.* New York: Academic Press, 1963.
Spivak, M. *A Comprehensive Introduction to Differential Geometry, Vol. 1, 3rd ed.* Houston, TX: Publish or Perish, pp. 275–80, 1999.
Sternberg, S. *Differential Geometry.* New York: Chelsea, pp. 14–0, 1983.

Weibull Distribution

The Weibull distribution is given by

$$P(x) = \alpha \beta^{-\alpha} x^{\alpha-1} e^{-(x/\beta)^{\alpha}} \tag{1}$$

$$D(x) = 1 - e^{-(x/\beta)^{\alpha}} \tag{2}$$

for $x \in [0, \infty)$, and is implemented in *Mathematica* as `WeibullDistribution[a, b]` in the *Mathematica* add-on package `Statistics`ContinuousDistributions`` (which can be loaded with the command `<<Statistics`). The RAW MOMENTS of the distribution are

$$\mu_1' = b\Gamma(1 + \alpha^{-1}) \tag{3}$$

$$\mu_2' = b^2\Gamma(1 + 2\alpha^{-1}) \tag{4}$$

$$\mu_3' = b^3\Gamma(1 + 3\alpha^{-1}) \tag{5}$$

$$\mu_4' = b^4\Gamma(1 + 4\alpha^{-1}), \tag{6}$$

and the MEAN, VARIANCE, SKEWNESS, and KURTOSIS are

$$\mu = \beta\Gamma(1 + \alpha^{-1}) \tag{7}$$

$$\sigma^2 = \beta^2 \left[\Gamma(1 + 2\alpha^{-1}) - \Gamma^2(1 + \alpha^{-1}) \right] \tag{8}$$

$$\gamma_1 = \frac{2\Gamma^3(1 + \alpha^{-1}) - 3\Gamma(1 + \alpha^{-1})\Gamma(1 + 2\alpha^{-1})}{\left[\Gamma(1 + 2\alpha^{-1}) - \Gamma^2(1 + \alpha^{-1}) \right]^{3/2}}$$

$$+ \frac{\Gamma(1 + 3\alpha^{-1})}{\left[\Gamma(1 + 2\alpha^{-1}) - \Gamma^2(1 + \alpha^{-1}) \right]^{3/2}} \tag{9}$$

$$\gamma_2 = \frac{f(a)}{\left[\Gamma(1 + 2\alpha^{-1}) - \Gamma^2(1 + \alpha^{-1}) \right]^2}, \tag{10}$$

where $\Gamma(z)$ is the GAMMA FUNCTION and

$$f(a) \equiv -6\Gamma^4(1 + \alpha^{-1}) + 12\Gamma^4(1 + \alpha^{-1})\Gamma(1 + 2\alpha^{-1})$$
$$-3\Gamma^2(1 + 2\alpha^{-1}) - 4\Gamma(1 + \alpha^{-1})\Gamma(1 + 3\alpha^{-1})$$
$$+\Gamma(1 + 4\alpha^{-1}). \tag{11}$$

A slightly different form of the distribution is defined by

$$P(x) = \frac{\alpha}{\beta} x^{\alpha-1} e^{-x^{\alpha}/\beta} \tag{12}$$

$$D(x) = 1 - e^{-x^{\alpha}/\beta} \tag{13}$$

(Mendenhall and Sincich 1995). This has RAW MOMENTS

$$\mu_1 = \beta^{1/\alpha}\Gamma(1 + \alpha^{-1}) \tag{14}$$

$$\mu_2 = \beta^{2/\alpha}\Gamma(1 + 2\alpha^{-1}) \tag{15}$$

$$\mu_3 = \beta^{3/\alpha}\Gamma(1 + 3\alpha^{-1}) \tag{16}$$

$$\mu_4 = \beta^{4/\alpha}\Gamma(1 + 4\alpha^{-1}) \tag{17}$$

so the MEAN and VARIANCE for this form are

$$\mu = \beta^{1/\alpha}\Gamma(1 + \alpha^{-1}) \tag{18}$$

$$\sigma^2 = \beta^{2/\alpha}\left[\Gamma(1 + 2\alpha^{-1}) - \Gamma^2(1 + \alpha^{-1}) \right] \tag{19}$$

The Weibull distribution gives the distribution of lifetimes of objects. It was originally proposed to quantify fatigue data, but it is also used in analysis of systems involving a "weakest link."

See also FISHER-TIPPETT DISTRIBUTION

References

Mendenhall, W. and Sincich, T. *Statistics for Engineering and the Sciences, 4th ed.* Englewood Cliffs, NJ: Prentice Hall, 1995.
Spiegel, M. R. *Theory and Problems of Probability and Statistics.* New York: McGraw-Hill, p. 119, 1992.

Weierstrass Approximation Theorem

If f is a continuous real-valued function on $[a, b]$ and if any $\epsilon > 0$ is given, then there exists a POLYNOMIAL p on $[a, b]$ such that

$$|f(x) - P(x)| < \epsilon$$

for all $x \in [a, b]$. In words, any continuous function on a closed and bounded interval can be uniformly approximated on that interval by POLYNOMIALS to any degree of accuracy.

See also MÜNTZ'S THEOREM

References

Jeffreys, H. and Jeffreys, B. S. "Weierstrass's Theorem on Approximation by Polynomials" and "Extension of Weierstrass's Approximation Theory." §14.08–4.081 in *Methods of Mathematical Physics, 3rd ed.* Cambridge, England: Cambridge University Press, pp. 446–48, 1988.

Weierstrass Constant

$$\sigma\left(\tfrac{1}{2}\right) = \tfrac{1}{2} \prod_{\substack{(m,n) \ne \\ (0,0)}} \left[1 - \frac{1}{2(m+ni)}\right] e^{1/[2(m+ni)]+1/[8(m+ni)^2]}$$

$$= \frac{2^{5/4}\sqrt{\pi}e^{\pi/8}}{\Gamma^2\left(\tfrac{1}{4}\right)} = 0.4749493799\ldots.$$

References

Le Lionnais, F. *Les nombres remarquables.* Paris: Hermann, p. 62, 1983.
Plouffe, S. "Weierstrass Constant." http://www.lacim.u-qam.ca/piDATA/weier.txt.
Waldschmidt, M. "Fonctions entières et nombres transcendants." *Cong. Nat. Soc. Sav. Nancy* **5**, 1978.
Waldschmidt, M. "Nombres transcendants et fonctions sigma de Weierstrass." *C. R. Math. Rep. Acad. Sci. Canada* **1**, 111–14, 1978/79.

Weierstrass Elliptic Function

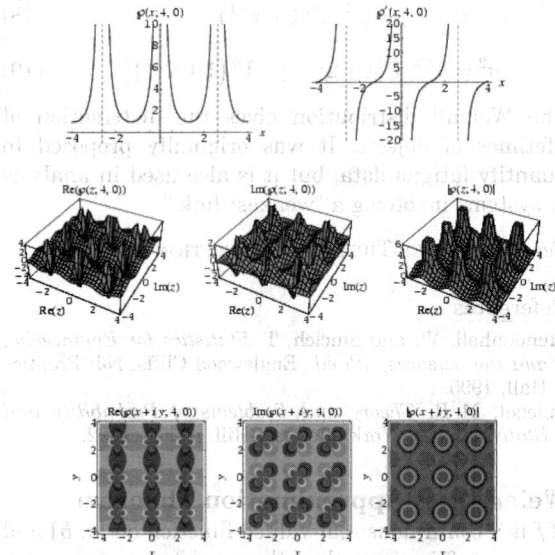

The Weierstrass elliptic functions (or Weierstrass $\wp$-functions, voiced "p-functions") are elliptic functions which, unlike the JACOBI ELLIPTIC FUNCTIONS, have a second-order POLE at $z = 0$. The above plots show the Weierstrass elliptic function $\wp(z)$ and its derivative $\wp'(z)$ for invariants (defined below) of $g_2 = 0$ and $g_3 = 0$. The Weierstrass elliptic function is implemented in

Mathematica as WeierstrassP[u, {$g1$, $g2$}].

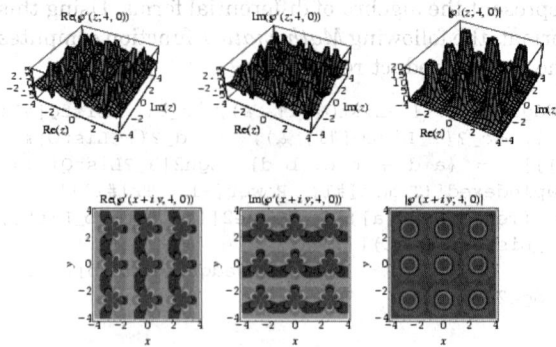

The plots above show the derivatives of the Weierstrass $\wp$-function.

Weierstrass elliptic functions are denoted $\wp(z)$ and can be defined by

$$\wp(z) = \frac{1}{z^2} + \sum_{m,n=-\infty}^{\infty}{}' \left[\frac{1}{(z - 2m\omega_1 - 2n\omega_2)^2} - \frac{1}{(2m\omega_1 + 2n\omega_2)^2}\right] \tag{1}$$

(Whittaker and Watson 1990, p. 434). Write $\Omega_{mn} \equiv 2m\omega_1 + 2n\omega_2$. Then this can be written

$$\wp(z) = z^{-2} + \sum_{m,n}{}' \left[(z - \Omega_{mn})^{-2} - \Omega_{mn}^{-2}\right]. \tag{2}$$

An equivalent definition which converges more rapidly is

$$\wp(z) = \left(\frac{\pi}{2\omega_1}\right)^2 \left[-\frac{1}{3} + \sum_{n=-\infty}^{\infty} \csc^2\left(\frac{z - 2n\omega_2}{2\omega_1}\pi\right)\right.$$

$$\left. - \sum_{n=-\infty}^{\infty}{}' \csc^2\left(\frac{n\omega_2}{\omega_1}\pi\right)\right] \tag{3}$$

(Whittaker and Watson 1990, p. 434). $\wp(z)$ is an EVEN FUNCTION since $\wp(-z)$ gives the same terms in a different order. To specify $\wp$ completely, its periods or invariants, written $\wp(z|\omega_1, \omega_2)$ and $\wp(z; g_1, g_2)$, respectively, must also be specified.

The series expansion of $\wp(z)$ is given by

$$\wp(z) = z^{-2} - \sum_{k=2}^{\infty} c_k z^{2k-2}, \tag{4}$$

where

$$c_2 = \frac{g_2}{20} \tag{5}$$

$$c_3 = \frac{g_3}{28} \tag{6}$$

and

$$c_k = \frac{3}{(2k+1)(k-3)} \sum_{m=2}^{k-2} c_m c_{k-m} \tag{7}$$

for $k \geq 4$ (Abramowitz and Stegun 1972, p. 635). The first few values for c_k for $k \geq 4$ in terms of c_2 and c_3 are given by

$$c_4 = \tfrac{1}{3}c_2^2 \tag{8}$$

$$c_5 = \tfrac{1}{11}(3c_2c_3) \tag{9}$$

$$c_6 = \tfrac{1}{39}\left(2c_2^3 + 3c_3^2\right) \tag{10}$$

$$c_7 = \tfrac{2}{33}2c_2^2c_3 \tag{11}$$

$$c_8 = \tfrac{5}{7293}\left(11c_2^4 + 36c_2c_3^2\right) \tag{12}$$

$$c_9 = \tfrac{29}{2717}\left(c_2^3c_3 + 11c_3^3\right) \tag{13}$$

$$c_{10} = \tfrac{1}{240669}\left(242c_2^5 + 1455c_2^2c_3^2\right) \tag{14}$$

(Abramowitz and Stegun 1972, p. 636).

The Weierstrass elliptic function describes how to get from a TORUS giving the solutions of an ELLIPTIC CURVE to the algebraic form of the ELLIPTIC CURVE.

The differential equation from which Weierstrass elliptic functions arise can be found by expanding about the origin the function $f(z) \equiv \wp(z) - z^{-2}$.

$$\wp(z) - z^{-2} = f(0) + f'(0)z + \frac{1}{2!}f''(0)z^2 + \frac{1}{3!}f'''(0)z^3$$

$$+ \frac{1}{4}f^{(4)}(0)z^4 + \dots. \tag{15}$$

But $f(0) = 0$ and the function is even, so $f'(0) = f'''(0) = 0$ and

$$f(z) = \wp(z) - z^{-2} = \frac{1}{2!}f''(0)z^2 + \frac{1}{4}f^{(4)}(0)z^4 + \dots. \tag{16}$$

Taking the derivatives

$$f' = -2\Sigma'(z - \Omega_{mn})^{-3} \tag{17}$$

$$f'' = 6\Sigma'(z - \Omega_{mn})^{-4} \tag{18}$$

$$f''' = -24\Sigma'(z - \Omega_{mn})^{-5} \tag{19}$$

$$f^{(4)} = 120\Sigma'(z - \Omega_{mn})^{-6}. \tag{20}$$

So

$$f''(0) = 6\Sigma'\Omega_{mn}^{-4} \tag{21}$$

$$f^{(4)}(0) = 120\Sigma'\Omega_{mn}^{-6}. \tag{22}$$

Plugging in,

$$\wp(z) - z^{-2} = 3\Sigma'\Omega_{mn}^{-4}z^2 + 5\Sigma'\Omega_{mn}^{-6}z^4 + \mathcal{O}\left(z^6\right) \tag{23}$$

Define the INVARIANTS

$$g_2 \equiv 60\Sigma'\Omega_{mn}^{-4} \tag{24}$$

$$g_3 \equiv 140\Sigma'\Omega_{mn}^{-6}, \tag{25}$$

then

$$\wp(z) = z^{-2} + \tfrac{1}{20}g_2z^2 + \tfrac{1}{28}g_3z^4 + \mathcal{O}\left(z^6\right) \tag{26}$$

$$\wp'(z) = -2z^{-3} + \tfrac{1}{10}g_2z + \tfrac{1}{7}g_3z^3 + \mathcal{O}\left(z^5\right). \tag{27}$$

Now cube (26) and square (27)

$$\wp^3(z) = z^{-6} + \tfrac{3}{20}g_2z^{-2} + \tfrac{3}{28}g_3 + \mathcal{O}\left(z^2\right) \tag{28}$$

$$\wp'^2(z) = 4z^{-6} - \tfrac{2}{5}g_2z^{-2} - \tfrac{4}{7}g_3 + \mathcal{O}\left(z^2\right). \tag{29}$$

Taking (29) minus $4\times$ (28) cancels out the z^{-6} term, giving

$$\wp'^2(z) - 4\wp^3(z) = \left(-\tfrac{2}{5} - \tfrac{3}{5}\right)g_2z^{-2} + \left(-\tfrac{4}{7} - \tfrac{3}{7}\right)g_3 + \mathcal{O}\left(z^2\right)$$

$$= -g_2z^{-2} - g_3 + \mathcal{O}\left(z^2\right) \tag{30}$$

$$\wp'^2(z) - 4\wp^3(z) + g_2z^{-2} + g_3 = \mathcal{O}\left(z^2\right). \tag{31}$$

But, from (16)

$$\wp(z) = z^{-2} + \tfrac{1}{2!}f''(0)z^2 + \tfrac{1}{4}f^{(4)}(0)z^4 + \dots, \tag{32}$$

so $\wp(z) = z^{-2} + \mathcal{O}(z^2)$ and (31) can be written

$$\wp'^2(z) - 4\wp^3(z) + g_2\wp(z) + g_3 = \mathcal{O}\left(z^2\right). \tag{33}$$

But the Weierstrass elliptic function is analytic at the origin and therefore at all points congruent to the origin. There are no other places where a singularity can occur, so this function is an ELLIPTIC FUNCTION with no SINGULARITIES. By LIOUVILLE'S ELLIPTIC FUNCTION THEOREM, it is therefore a constant. But as $z \to 0$, $\mathcal{O}(z^2) \to 0$, so

$$\wp'^2(z) = 4\wp^3(z) - g_2\wp(z) - g_3 \tag{34}$$

(Whittaker and Watson 1990, pp. 436–37).

The solution to the differential equation

$$y'^2 = 4y^3 - g_2y - g_3 \tag{35}$$

is therefore given by $y = \wp(z + \alpha)$, providing that numbers ω_1 and ω_2 exist which satisfy the equations defining the INVARIANTS. Writing the differential equation in terms of its roots e_1, e_2, and e_3,

$$y'^2 = 4y^3 - g_2y - g_3 = 4(y - e_1)(y - e_2)(y - e_3) \tag{36}$$

(Rainville 1971, p. 312),

$$2\ln(y') = \ln 4 + \sum_{r=1}^{3} \ln(y - e_r) \tag{37}$$

$$\frac{2y''}{y'} = y' \sum_{r=1}^{3}(y - e_r)^{-1} \tag{38}$$

$$\frac{2y''}{y'^2} = \sum_{r=1}^{3}(y-e_r)^{-1} \tag{39}$$

$$2\frac{y'^2y''' - y''(2y'y'')}{y'^4} = -y'\sum_{r=1}^{3}(y-e_r)^{-2} \tag{40}$$

$$\frac{2y'''}{y'^3} - \frac{4y''^2}{y'^4} = -\sum_{r=1}^{3}(y-e_r)^{-2}. \tag{41}$$

Now take (41) divided by 4 plus [(41) divided by 4] quantity squared,

$$\left(\frac{y'''}{2y'^3} - \frac{y''^2}{y'^4}\right) + \left(\frac{y''^2}{4y'^4}\right)$$

$$= -\frac{1}{4}\sum_{r=1}^{3}(y-e_r)^{-2} + \frac{1}{16}\left[\sum_{r=1}^{3}(y-e_r)^{-1}\right]^2 \tag{42}$$

$$\frac{3y''^2}{4y'^4} - \frac{y'''}{2y'^3} = \frac{3}{16}\sum_{r=1}^{3}(y-e_r)^{-2} - \frac{3}{8}y\prod_{r=1}^{3}(y-e_r)^{-1}. \tag{43}$$

The term on the right is half the SCHWARZIAN DERIVATIVE.

The DERIVATIVE of the Weierstrass elliptic function is given by

$$\wp'(z) = \frac{d}{dz}\wp(z) = -2\sum_{m,n}\frac{1}{(z-\Omega_{mn})^3}$$

$$= -2z^{-3} - 2\sum_{m,n}{}'(z-\Omega_{mn})^{-3}. \tag{44}$$

This is an ODD FUNCTION which is itself an elliptic function with pole of order 3 at $z = 0$. The INTEGRAL is given by

$$z = \int_{\wp(z)}^{\infty}(4t^3 - g_2t - g_3)^{-1/2}dt. \tag{45}$$

The second derivative satisfies

$$\wp''\left(\tfrac{1}{2}\omega_1\right) = 2(e_1 - e_2)(e_1 - e_3) \tag{46}$$

(Apostol 1997, p. 23).
A duplication formula is obtained as follows.

$$\wp(2z) = \lim_{y\to z}\wp(y+z)$$

$$= \frac{1}{4}\lim_{y\to z}\left[\frac{\wp'(z) - \wp'(y)}{\wp(z) - \wp(y)}\right]^2 - \wp(z) - \lim_{y\to z}\wp(y)$$

$$= \frac{1}{4}\lim_{h\to 0}\left[\frac{\wp(z) - \wp'(z+h)}{\wp(z) - \wp(z+h)}\right]^2 - 2\wp(z)$$

$$= \frac{1}{4}\left\{\left[\lim_{h\to 0}\frac{\wp'(z) - \wp'(z+h)}{h}\right]\right.$$

$$\times\left.\left[\lim_{h\to 0}\frac{h}{\wp(z) - \wp(z+h)}\right]\right\}^2 - 2\wp(z)$$

$$= \frac{1}{4}\left[\frac{\wp''(z)}{\wp'(z)}\right]^2 - 2\wp(z) \tag{47}$$

(Apostol 1997, p. 24).
A general addition theorem is obtained as follows. Given

$$\wp'(z) = A\wp(z) + B \tag{48}$$

$$\wp'(y) = A\wp(y) + B \tag{49}$$

with zero y and z where $z \not\equiv \pm y \pmod{2\omega_1, 2\omega_2}$, find the third zero ζ. Consider $\wp'(\zeta) - A\wp(\zeta) - B$. This has a pole of order three at $\zeta = 0$, but the sum of zeros ($= 0$) equals the sum of poles for an ELLIPTIC FUNCTION, so $z + y + \zeta = 0$ and $\zeta = -z - y$.

$$\wp'(-z - y) = A\wp(-z - y) + B \tag{50}$$

$$-\wp'(z + y) = A\wp(z + y) + B. \tag{51}$$

Combining (48), (49), and (51) gives

$$\begin{bmatrix} \wp(z) & \wp'(z) & 1 \\ \wp(y) & \wp'(y) & 1 \\ \wp(z+y) & -\wp(z+y) & 1 \end{bmatrix}\begin{bmatrix} A \\ -1 \\ B \end{bmatrix} = \begin{bmatrix} 0 \\ 0 \\ 0 \end{bmatrix}, \tag{52}$$

so

$$\begin{vmatrix} \wp(z) & \wp'(z) & 1 \\ \wp(y) & \wp'(y) & 1 \\ \wp(z+y) & -\wp(z+y) & 1 \end{vmatrix} = 0. \tag{53}$$

Defining $u + v + w = 0$ where $u \equiv z$ and $v \equiv y$ gives the symmetric form

$$\begin{vmatrix} \wp(u) & \wp'(u) & 1 \\ \wp(v) & \wp'(v) & 1 \\ \wp(w) & \wp(w) & 1 \end{vmatrix} = 0 \tag{54}$$

(Whittaker and Watson 1990, p. 440). To get the expression explicitly, start again with

$$\wp'(\zeta) - A\wp(\zeta) - B = 0, \tag{55}$$

where $\zeta = z, y, -z - y$.

$$\wp'^2(\zeta) - [A\wp(\zeta) + B]^2 = 0. \tag{56}$$

But from (34), $\wp'^2(\zeta) = 4\wp^3(\zeta) - g_2\wp(\zeta) - g_3$, so

$$4\wp^3(\zeta) - A^2\wp^2(\zeta) - (2AB + g_2)\wp(\zeta) - (B^2 + g_3) = 0. \tag{57}$$

The solutions $\wp(\zeta) \equiv z$ are given by

$$4z^3 - A^2z^2 - (2AB + g_2)z - (B^2 + g_3) = 0. \tag{58}$$

But the sum of roots equals the COEFFICIENT of the squared term, so

$$\wp(z) + \wp(y) + \wp(z+y) = \tfrac{1}{4}A^2 \tag{59}$$

$$\wp'(z) - \wp'(y) = A[\wp(z) - \wp(y)] \tag{60}$$

$$A = \frac{\wp'(z) - \wp'(y)}{\wp(z) - \wp(y)} \tag{61}$$

$$\wp(z+y) = \frac{1}{4}\left[\frac{\wp'(z) - \wp'(y)}{\wp(z) - \wp(y)}\right]^2 - \wp(z) - \wp(y) \tag{62}$$

(Whittaker and Watson 1990, p. 441).
Half-period identities include

$$x \equiv \wp\left(\tfrac{1}{2}\omega_1\right) = \wp(-h\omega_1 + \omega_1) = e_1 + \frac{(e_1 - e_2)(e_1 - e_3)}{\wp\left(-\tfrac{1}{2}\omega_1\right) - e_1}$$

$$= e_1 + \frac{(e_1 - e_2)(e_1 - e_3)}{x - e_1}. \tag{63}$$

Multiplying through,

$$x^2 - e_1 x = e_1 x - e_1^2 + (e_1 - e_2)(e_1 - e_3) \tag{64}$$

$$x^2 - 2e_1 + [e_1^2 - (e_1 - e_2)(e_1 - e_3)] = 0, \tag{65}$$

which gives

$$\wp\left(\tfrac{1}{2}\omega_1\right) = \tfrac{1}{2}\left\{2e_1 \pm \sqrt{4e_1^2 - 4[e_1^2 - (e_1 - e_2)(e_1 - e_3)]}\right\}$$

$$= e_1 \pm \sqrt{(e_1 - e_2)(e_1 - e_3)}. \tag{66}$$

From Whittaker and Watson (1990, p. 445),

$$\wp'\left(\tfrac{1}{2}\omega_1\right) = -2\sqrt{(e_1 - e_2)(e_1 - e_3)}$$

$$\times \left(\sqrt{e_1 - e_2} + \sqrt{e_1 - e_3}\right). \tag{67}$$

The function is HOMOGENEOUS,

$$\wp(\lambda z | \lambda\omega_1, \lambda\omega_2) = \lambda^{-2}\wp(z|\omega_1, \omega_2) \tag{68}$$

$$\wp(\lambda z; \lambda^{-4}g_2, \lambda^{-6}g_3) = \lambda^{-2}\wp(z; g_2, g_3). \tag{69}$$

To invert the function, find $2\omega_1$ and $2\omega_2$ of $\wp(z|\omega_1, \omega_2)$ when given $\wp(z; g_1, g_2)$. Let e_1, e_2, and e_3 be the roots such that $(e_1 - e_2)/(e_1 - e_3)$ is not a REAL NUMBER > 1 or < 0. Determine the PARAMETER τ from

$$\frac{e_1 - e_2}{e_1 - e_3} = \frac{\vartheta_4^4(0|\tau)}{\vartheta_3^4(0|\tau)}. \tag{70}$$

Now pick

$$A \equiv \frac{\sqrt{e_1 - e_2}}{\vartheta_4^2(0|\tau)}. \tag{71}$$

As long as $g_2^3 \neq 27g_3$, the periods are then

$$2\omega_1 = \pi A \tag{72}$$

$$2\omega_2 = \frac{\pi\tau}{A}. \tag{73}$$

Weierstrass elliptic functions can be expressed in terms of JACOBI ELLIPTIC FUNCTIONS by

$$\wp(u; g_2, g_3) = e_3 + (e_1 - e_3)\,\mathrm{ns}^2\left(u\sqrt{e_1 - e_3}, \sqrt{\frac{e_2 - e_3}{e_1 - e_3}}\right), \tag{74}$$

where

$$\wp(\omega_1) = e_1 \tag{75}$$

$$\wp(\omega_2) = e_2 \tag{76}$$

$$\wp(\omega_3) = -\wp(-\omega_1 - \omega_2) = e_3, \tag{77}$$

and the INVARIANTS are

$$g_2 \equiv 60 \sum_{m,n}{}' \Omega_{mn}^{-4} \tag{78}$$

$$g_3 \equiv 140 \sum_{m,n}{}' \Omega_{mn}^{-6}. \tag{79}$$

Here, $\Omega_{mn} \equiv 2m\omega_1 - 2n\omega_2$.
An addition formula for the Weierstrass elliptic function can be derived as follows.

$$\wp(z + \omega_1) + \wp(z) + \wp(\omega_1)$$

$$= \frac{1}{4}\left[\frac{\wp(z) - \wp'(\omega_1)}{\wp(z) - \wp(\omega_1)}\right]^2 = \frac{1}{4}\frac{\wp'^2(z)}{[\wp(z) - e_1]^2}. \tag{80}$$

Use

$$\wp'(z) = 4\prod_{r=1}^{3}[\wp(z) - e_r], \tag{81}$$

so

$$\wp(z + \omega_1) = -\wp(z) - e_1 + \frac{1}{4}\frac{4\prod_{r=1}^{3}[\wp(z) - e_r]}{[\wp(z) - e_1]^2}$$

$$= -\wp(z) - e_1 + \frac{[\wp(z) - e_2][\wp(z) - e_3]}{\wp(z) - e_1}. \tag{82}$$

Use $\sum_{r=1}^{3} e_r = 0$,

$$\wp(z + \omega_1) = e_1 + \frac{[-2e_1 - \wp(z)][\wp(z) - e_1]}{\wp(z) - e_1}$$

$$+ \frac{\wp^2(z) - \wp(z)(e_2 + e_3) + e_2 e_3}{\wp(z) - e_1}$$

$$= e_1 + \frac{-\wp(z)(e_1 + e_2 + e_3) + e_2 e_3 + 2e_1^2}{\wp(z) - e_1} \tag{83}$$

But $\sum_{r=1}^{3} e_r = 0$ and

$$2e_1^2 + e_2 e_3 = e_1^2 - e_1(e_2 + e_3) + e_2 e_3$$

$$= (e_1 - e_2)(e_1 - e_3), \tag{84}$$

so

$$\wp(z+\omega_1) = e_1 + \frac{(e_1 - e_2)(e_1 - e_3)}{\wp(z) - e_1}. \qquad (85)$$

The periods of the Weierstrass elliptic function are given as follows. When g_2 and g_3 are REAL and $g_2^3 - 27g_3^2 > 0$, then e_1, e_2, and e_3 are REAL and defined such that $e_1 > e_2 > e_3$.

$$\omega_1 = \int_{e_1}^{\infty} \left(4t^3 - g_2 t - g_3\right)^{-1/2} dt \qquad (86)$$

$$\omega_3 = -i \int_{-\infty}^{e_2} \left(g_3 + g_2 t - 4t^3\right)^{-1/2} dt \qquad (87)$$

$$\omega_2 = -\omega_1 - \omega_3. \qquad (88)$$

The roots of the Weierstrass elliptic function satisfy

$$e_1 = \wp(\omega_1) \qquad (89)$$

$$e_2 = \wp(\omega_2) \qquad (90)$$

$$e_3 = \wp(\omega_3), \qquad (91)$$

where $\omega_3 \equiv -\omega_1 - \omega_2$. The e_is are ROOTS of $4t^3 - g_2 t - g_3$ and are unequal so that $e_1 \neq e_2 \neq e_3$.. They can be found from the relationships

$$e_1 + e_2 + e_3 = -a_2 = 0 \qquad (92)$$

$$e_2 e_3 + e_3 e_1 + e_1 e_2 = a_1 = -\tfrac{1}{4}g_2 \qquad (93)$$

$$e_1 e_2 e_3 = -a_0 = \tfrac{1}{4}g_3. \qquad (94)$$

See also ELLIPTIC CURVE, ELLIPTIC FUNCTION, EISENSTEIN SERIES, EQUIANHARMONIC CASE, JACOBI ELLIPTIC FUNCTIONS, LEMNISCATE CASE, PSEUDOLEMNISCATE CASE, WEIERSTRASS SIGMA FUNCTION, WEIERSTRASS ZETA FUNCTION

References

Abramowitz, M. and Stegun, C. A. (Eds.). "Weierstrass Elliptic and Related Functions." Ch. 18 in *Handbook of Mathematical Functions with Formulas, Graphs, and Mathematical Tables, 9th printing.* New York: Dover, pp. 627–71, 1972.

Apostol, T. M. "The Weierstrass $\wp$ Function," "The Laurent Expansion of $\wp$ Near the Origin," "Differential Equation Satisfied by $\wp$," "The Eisenstein Series and the Invariants g_2 and g_3," "The Numbers e_1, e_2, and e_3," and "The Discriminant Δ." §1.6–.11 in *Modular Functions and Dirichlet Series in Number Theory, 2nd ed.* New York: Springer-Verlag, pp. 9–4, 1997.

Eichler, M. and Zagier, D. "On the Zeros of the Weierstrass $\wp$-Function." *Math. Ann.* **258**, 399–07, 1982.

Fischer, G. (Ed.). Plates 129–31 in *Mathematische Modelle/ Mathematical Models, Bildband/Photograph Volume.* Braunschweig, Germany: Vieweg, pp. 126–28, 1986.

Huang, J. "Integral Representation of Harmonic Lattice Sums." *J. Math. Phys.* **40**, 5240–246, 1999.

Rainville, E. D. *Special Functions.* New York: Chelsea, 1971.

Tölke, F. "Spezielle Weierstraßsche $\wp$-Funktionen." Ch. 4 in *Praktische Funktionenlehre, zweiter Band: Theta-Funktionen und spezielle Weierstraßsche Funktionen.* Berlin: Springer-Verlag, pp. 115–44, 1966.

Tölke, F. *Praktische Funktionenlehre, fünfter Band: Allgemeine Weierstraßsche Funktionen und Ableitungen nach dem Parameter. Integrale der Theta-Funktionen und Bilinear-Entwicklungen.* Berlin: Springer-Verlag, 1968.

Whittaker, E. T. and Watson, G. N. *A Course in Modern Analysis, 4th ed.* Cambridge, England: Cambridge University Press, 1990.

Woods, F. S. "The Function $p(u)$." §160 in *Advanced Calculus: A Course Arranged with Special Reference to the Needs of Students of Applied Mathematics.* Boston, MA: Ginn, pp. 381–82, 1926.

Weierstrass Extreme Value Theorem

EXTREME VALUE THEOREM

Weierstrass Factor Theorem

Let any finite or infinite set of points having no finite LIMIT POINT be prescribed, and associate with each of its points a definite positive integer as its order. Then there exists an ENTIRE FUNCTION which has zeros to the prescribed orders at precisely the prescribed points, and is otherwise different from zero. Moreover, this function can be REPRESENTED AS a product from which one can read off again the positions and orders of the zeros. Furthermore, if $G_0(z)$ is one such function, then

$$G(z) = e^{h(z)} G_0(z)$$

is the most general function satisfying the conditions of the problem, where $h(z)$ denotes an arbitrary ENTIRE FUNCTION.

References

Knopp, K. "Weierstrass's Factor-Theorem." §1 in *Theory of Functions Parts I and II, Two Volumes Bound as One, Part II.* New York: Dover, pp. 1–, 1996.

Krantz, S. G. "The Weierstrass Factorization Theorem." §8.2 in *Handbook of Complex Analysis.* Boston, MA: Birkhäuser, pp. 109–10, 1999.

Weierstrass Factorization Theorem

WEIERSTRASS FACTOR THEOREM

Weierstrass Form

A general form into which an ELLIPTIC CURVE over any FIELD K can be transformed is called the Weierstrass form, and is given by

$$y^2 + ay = x^3 + bx^2 + cxy + dx + e,$$

where a, b, c, d, and e are elements of K.

Weierstrass Function

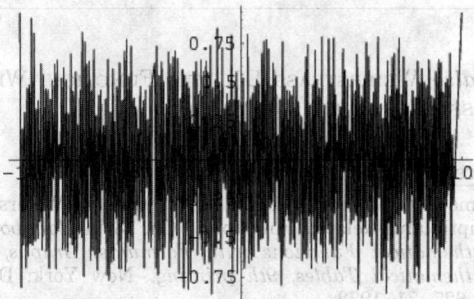

A CONTINUOUS FUNCTION which is nowhere DIFFER-ENTIABLE. It is given by

$$f(x) = \sum_{n=1}^{\infty} b^n \cos(a^n \pi x)$$

where a is an ODD NUMBER, $b \in (0, 1)$, and $ab > 1 + 3\pi/2$. The above plot is for $a = 19$ and $b = 1/2$.

See also BLANCMANGE FUNCTION, CONTINUOUS FUNC-TION, DIFFERENTIABLE

References

Berry, M. V. and Lewis, Z. V. "On the Weierstrass-Mandel-brot Function." *Proc. Roy. Soc. London Ser. A* **370**, 459–84, 1980.
Darboux, G. "Mémoire sur les fonctions discontinues." *Ann. l'École Normale, Ser. 2* **4**, 57–12, 1875.
Darboux, G. "Mémoire sur les fonctions discontinues." *Ann. l'École Normale, Ser. 2* **8**, 195–02, 1879.
du Bois-Reymond, P. "Versuch einer Klassification der will-kürlichen Functionen reeller Argumente nach ihren Än-derungen in den kleinsten Intervallen." *J. für Math.* **79**, 21–7, 1875.
Faber, G. "Einfaches Beispiel einer stetigen nirgends differ-entiierbaren [sic] Funktion." *Jahresber. Deutschen Math. Verein.* **16** 538–40, 1907.
Hardy, G. H. "Weierstrass's Non-Differentiable Function." *Trans. Amer. Math. Soc.* **17**, 301–25, 1916.
Landsberg, G. "Über Differentziierbarkeit stetiger Funktio-nen." *Jahresber. Deutschen Math. Verein.* **17**, 46–1, 1908.
Lerch, M. "Ueber die Nichtdifferentiirbarkeit [sic] gewisser Functionen." *J. reine angew. Math.* **13**, 126–38, 1888.
Mandelbrot, B. B. "Weierstrass Functions and Kin. Ultra-violet and Infrared Catastrophe." *The Fractal Geometry of Nature.* New York: W. H. Freeman, pp. 388–90, 1983.
Pickover, C. A. *Keys to Infinity.* New York: Wiley, p. 190, 1995.
Weierstrass, K. *Abhandlungen aus der Functionenlehre.* Berlin: J. Springer, p. 97, 1886.

Weierstrass Intermediate Value Theorem

If a continuous function defined on an interval is sometimes POSITIVE and sometimes NEGATIVE, it must be 0 at some point.

Weierstrass M-Test

Let $\sum_{k=1}^{\infty} u_n(x)$ be a SERIES of functions all defined for a set E of values of x. If there is a CONVERGENT series of constants

$$\sum_{n=1}^{\infty} M_n,$$

such that

$$|u_n(x)| \le M_n$$

for all $x \in E$, then the series exhibits ABSOLUTE CONVERGENCE for each $x \in E$ as well as UNIFORM CONVERGENCE in E.

See also ABSOLUTE CONVERGENCE, UNIFORM CON-VERGENCE

References

Arfken, G. *Mathematical Methods for Physicists, 3rd ed.* Orlando, FL: Academic Press, pp. 301–03, 1985.
Jeffreys, H. and Jeffreys, B. S. "*M* Test" and "Extension of the *M* Test." §1.1151–.1152 in *Methods of Mathematical Physics, 3rd ed.* Cambridge, England: Cambridge Uni-versity Press, pp. 40–1, 1988.
Knopp, K. *Theory of Functions Parts I and II, Two Volumes Bound as One, Part I.* New York: Dover, p. 73, 1996.

Weierstrass Operator

The operator $e^{vt^2/2}$ which satisfies

$$e^{vt^2/2} p(x) = \frac{1}{\sqrt{2\pi v}} \int_{-\infty}^{\infty} e^{-u^2/(2v)} p(x+u) \, du$$

for $v > 0$.

References

Roman, S. *The Umbral Calculus.* New York: Academic Press, p. 88, 1984.
Rota, G.-C.; Kahaner, D.; Odlyzko, A. "On the Foundations of Combinatorial Theory. VIII: Finite Operator Calculus." *J. Math. Anal. Appl.* **42**, 684–60, 1973.

Weierstrass Point

A POLE of multiplicity less than $p + 1$.

References

Coolidge, J. L. *A Treatise on Algebraic Plane Curves.* New York: Dover, pp. 290–91, 1959.

Weierstrass Product Inequality

If $0 \le a, b, c, d \le 1$, then

$$(1-a)(1-b)(1-c)(1-d) + a + b + c + d \ge 1.$$

References

Honsberger, R. *Mathematical Gems III.* Washington, DC: Math. Assoc. Amer., pp. 244–45, 1985.

Weierstrass Sigma Function

The QUASIPERIODIC FUNCTION defined by

$$\frac{d}{dz} \ln \sigma(z) = \zeta(z), \qquad (1)$$

where $\zeta(z)$ is the WEIERSTRASS ZETA FUNCTION and

$$\lim_{z \to \infty} \frac{\sigma(z)}{z} = 1. \tag{2}$$

Then

$$\sigma(z) = z \prod_{m,n=-\infty}^{\infty}{}' \left[\left(1 - \frac{z}{\Omega_{mn}}\right)\exp\left(\frac{z}{\Omega_{mn}} + \frac{z^2}{2\Omega_{mn}^2}\right)\right], \tag{3}$$

where the term with $m = n = 0$ is omitted from the product. In addition, $\sigma(z)$ satisfies

$$\sigma(z + 2\omega_1) = -e^{2\eta_1(z+\omega_1)}\sigma(z) \tag{4}$$

$$\sigma(z + 2\omega_2) = -e^{2\eta_2(z+\omega_2)}\sigma(z) \tag{5}$$

and

$$\sigma_r(z) = \frac{e^{-\eta_r z}\sigma(z + \omega_r)}{\sigma(\omega_r)} \tag{6}$$

for $r = 1, 2, 3$.

$\sigma(z)$ can be expressed in terms of JACOBI THETA FUNCTIONS using the expression

$$\sigma(z|\omega_1, \omega_2) = \frac{2\omega_1}{\pi\vartheta_1'}\exp\left(-\frac{v^2\vartheta_1'''}{6\vartheta_1'}\right)\vartheta_1\left(v\left|\frac{\omega_2}{\omega_1}\right.\right), \tag{7}$$

where $v \equiv \pi z/(2\omega_1)$, and

$$\eta_1 = -\frac{\pi^2\vartheta_1'''}{12\omega_1\vartheta_1'} \tag{8}$$

$$\eta_2 = -\frac{\pi^2\omega_2\vartheta_1'''}{12\omega_1^2\vartheta_1'} - \frac{\pi i}{2\omega_1}. \tag{9}$$

There is a beautiful series expansion for $\sigma(z)$, given by the DOUBLE SUM

$$\sigma(z) = \sum_{m,n=0}^{\infty} a_{mn}\left(\tfrac{1}{2}g_2\right)^m(2g_3)^n\frac{z^{4m+6n+1}}{(4m + 6n + 1)!}, \tag{10}$$

where $a_{00} = 1$, $a_{mn} = 0$ for either subscript negative, and other values are gives by the RECURRENCE RELATION

$$a_{mn} = 3(m + 1)a_{m+1,n+1} + \tfrac{16}{3}(n + 1)a_{m-2,n+1}$$

$$- \tfrac{1}{3}(2m + 3n - 1)(4m + 6n - 1)a_{m-1,n} \tag{11}$$

(Abramowitz and Stegun 1972, pp. 635–36). The following table gives the values of the a_{mn} coefficients for small m and n.

	$n=0$	$n=1$	$n=2$	$n=3$
a_{0n}	1	-3	-54	14904
a_{1n}	-1	-18	4968	502200
a_{2n}	-9	513	257580	162100440
a_{3n}	69	33588	20019960	-9465715080

a_{4n}	321	2808945	-376375410	-4582619446320
a_{5n}	160839	-41843142	-210469286736	-1028311276281264

See also WEIERSTRASS ELLIPTIC FUNCTION, WEIERSTRASS ZETA FUNCTION

References

Abramowitz, M. and Stegun, C. A. (Eds.). "Weierstrass Elliptic and Related Functions." Ch. 18 in *Handbook of Mathematical Functions with Formulas, Graphs, and Mathematical Tables, 9th printing.* New York: Dover, pp. 627–71, 1972.

Knopp, K. "Example: Weierstrass's σ-Function." §2d in *Theory of Functions Parts I and II, Two Volumes Bound as One, Part II.* New York: Dover, pp. 27–0, 1996.

Tölke, F. "Spezielle Weierstraßsche Sigma-Funktionen." Ch. 9 in *Praktische Funktionenlehre, dritter Band: Jacobische elliptische Funktionen, Legendresche elliptische Normalintegrale und spezielle Weierstraßsche Zeta- und Sigma Funktionen.* Berlin: Springer-Verlag, pp. 164–80, 1967.

Whittaker, E. T. and Watson, G. N. "The Function σ(z)." §20.42 in *A Course in Modern Analysis, 4th ed.* Cambridge, England: Cambridge University Press, pp. 447–48, 450–52, and 458–61, 1990.

Weierstrass Zeta Function

The QUASIPERIODIC FUNCTION defined by

$$\frac{d\zeta(z)}{dz} \equiv -\wp(z) \tag{1}$$

with

$$\lim_{z \to 0}\left|\zeta(z) - z^{-1}\right| = 0. \tag{2}$$

Then

$$\zeta(z) - z^{-1} = -\int_0^z \left[\wp(z) - z^{-2}\right]dz$$

$$= -\Sigma'\int_0^z \left[(z - \Omega_{mn})^{-2} - \Omega_{mn}^{-2}\right]dz \tag{3}$$

$$\zeta(z) = z^{-1} + \sum_{m,n=-\infty}^{\infty}{}' \left[(z - \Omega_{mn})^{-1} + \Omega_{mn}^{-1} + z\Omega_{mn}^{-2}\right] \tag{4}$$

so $\zeta(z)$ is an odd FUNCTION. Integrating $\wp(z + 2\omega_1) = \wp(z)$ gives

$$\zeta(z + 2\omega_1) = \zeta(z) + 2\eta_1. \tag{5}$$

Letting $z = -\omega_1$ gives $\zeta(-\omega_1) + 2\eta_1 = -\zeta(\omega_1) + 2\eta_1$, so $\eta_1 = \zeta(\omega_1)$. Similarly, $\eta_2 = \zeta(\omega_2)$. From Whittaker and Watson (1990),

$$\eta_1\omega_2 - \eta_2\omega_1 = \tfrac{1}{2}\pi i \tag{6}$$

If $x + y + z = 0$, then

$$[\zeta(x) + \zeta(y) + \zeta(z)]^2 + \zeta'(x) + \zeta'(y) + \zeta'(z) = 0 \tag{7}$$

(Whittaker and Watson 1990, p. 446). Also,

$$2 \begin{vmatrix} 1 & \wp(x) & \wp^2(x) \\ 1 & \wp(y) & \wp^2(y) \\ 1 & \wp(z) & \wp^2(z) \\ 1 & \wp(x) & \wp'(x) \\ 1 & \wp(y) & \wp'(y) \\ 1 & \wp(z) & \wp'(z) \end{vmatrix} = \zeta(x+y+z) - \zeta(x) - \zeta(y) - \zeta(z) \quad (8)$$

(Whittaker and Watson 1990, p. 446).

The series expansion of $\zeta(z)$ is given by

$$\zeta(z) = z^{-1} - \sum_{k=2}^{\infty} \frac{c_k z^{2k-1}}{2k-1}, \quad (9)$$

where

$$c_2 = \frac{g_2}{20} \quad (10)$$

$$c_3 = \frac{g_3}{28} \quad (11)$$

and

$$c_k = \frac{3}{(2k+1)(k-3)} \sum_{m=2}^{k-2} c_m c_{k-m} \quad (12)$$

for $k \geq 4$ (Abramowitz and Stegun 1972, p. 635).

See also WEIERSTRASS ELLIPTIC FUNCTION, WEIERSTRASS SIGMA FUNCTION

References

Abramowitz, M. and Stegun, C. A. (Eds.). "Weierstrass Elliptic and Related Functions." Ch. 18 in *Handbook of Mathematical Functions with Formulas, Graphs, and Mathematical Tables, 9th printing.* New York: Dover, pp. 627–71, 1972.
Tölke, F. "Spezielle Weierstraßsche Zeta-Funktionen." Ch. 8 in *Praktische Funktionenlehre, dritter Band: Jacobische elliptische Funktionen, Legendresche elliptische Normalintegrale und spezielle Weierstraßsche Zeta- und Sigma Funktionen.* Berlin: Springer-Verlag, pp. 145–63, 1967.
Whittaker, E. T. and Watson, G. N. "Quasi-Periodic Functions. The Function $\zeta(z)$" and "The Quasi-Periodicity of the Function $\zeta(z)$." §20.4 and 20.41 in *A Course in Modern Analysis, 4th ed.* Cambridge, England: Cambridge University Press, pp. 445–47 and 449–51, 1990.

Weierstrass-Casorati Theorem

An ANALYTIC FUNCTION approaches any given value arbitrarily closely in any ϵ-NEIGHBORHOOD of an ESSENTIAL SINGULARITY.

See also ANALYTIC FUNCTION, ESSENTIAL SINGULARITY

References

Knopp, K. *Theory of Functions Parts I and II, Two Volumes Bound as One, Part I.* New York: Dover, pp. 114–15 and 124–25, 1996.
Krantz, S. G. "The Casorati-Weierstrass Theorem." §4.1.6 in *Handbook of Complex Analysis.* Boston, MA: Birkhäuser, p. 43, 1999.

Weierstrass-Erdman Corner Condition

In the CALCULUS OF VARIATIONS, the condition

$$f_{y'}(x, y, y'(x_-)) = f_{y'}(x, y, y'(x_+))$$

must hold at a corner (x, y) of a minimizing arc E_{12}.

WeierstrassHalfPeriods

WEIERSTRASS ELLIPTIC FUNCTION

WeierstrassInvariants

WEIERSTRASS ELLIPTIC FUNCTION

Weierstrass-Mandelbrot Function

WEIERSTRASS FUNCTION

WeierstrassP

WEIERSTRASS ELLIPTIC FUNCTION

WeierstrassPPrime

WEIERSTRASS ELLIPTIC FUNCTION

Weierstrass's Double Series Theorem

Let all of the functions

$$f_n(z) = \sum_{k=0}^{\infty} a_k^{(n)}(z - z_0)^k$$

with $n = 0, 1, 2, \ldots$, be regular at least for $|z - z_0| < r$, and let

$$F(z) = \sum_{n=0}^{\infty} f_n(z)$$

$$= [a_0^{(0)} + a_1^{(0)}(z - z_0) + \ldots + a_k^{(0)}(z - z_0)^k + \ldots]$$
$$+ [a_0^{(1)} + a_1^{(1)}(z - z_0) + \ldots + a_k^{(1)}(z - z_0)^k + \ldots] + \ldots$$
$$+ [a_0^{(n)} + a_1^{(n)}(z - z_0) + \ldots + a_k^{(n)}(z - z_0)^k + \ldots] + \ldots$$

be uniformly convergent for $z - z_0 \leq \rho < r$ for every $\rho < r$. Then the coefficients in any column form a convergent series. Furthermore, setting

$$a_k^{(0)} + a_k^{(1)} + \ldots + a_k^{(n)} + \ldots = \sum_{n=0}^{\infty} a_k^{(n)} = A_k$$

for $k = 0, 1, 2, \ldots$, it then follows that

$$\sum_{k=0}^{\infty} A_k(z - z_0)^k$$

is the POWER SERIES for $F(z)$, which converges at least for $|z - z_0| < r$.

See also DOUBLE SERIES

References

Knopp, K. *Theory of Functions Parts I and II, Two Volumes Bound as One, Part I.* New York: Dover, p. 83, 1996.

Weierstrass's Gap Theorem

Given a succession of nonsingular points which are on a nonhyperelliptic curve of GENUS p, but are not a group of the canonical series, the number of groups of the first k which cannot constitute the group of simple POLES of a RATIONAL FUNCTION is p. If points next to each other are taken, then the theorem becomes: Given a nonsingular point of a nonhyperelliptic curve of GENUS p, then the orders which it cannot possess as the single pole of a RATIONAL FUNCTION are p in number.

References

Coolidge, J. L. *A Treatise on Algebraic Plane Curves.* New York: Dover, p. 290, 1959.

Weierstrass's Polynomial Theorem

A function, continuous in a finite close interval, can be approximated with a preassigned accuracy by POLYNOMIALS. A function of a REAL variable which is continuous and has period 2π can be approximated by trigonometric POLYNOMIALS.

References

Szego, G. *Orthogonal Polynomials, 4th ed.* Providence, RI: Amer. Math. Soc., p. 5, 1975.

Weierstrass's Theorem

There are at least two theorems known as Weierstrass's theorem. The first states that the only HYPERCOMPLEX NUMBER systems with commutative multiplication and addition are the algebra with one unit such that $e = e^2$ and the GAUSSIAN INTEGERS.

In harmonic analysis, let $U \subseteq \mathbb{C}$ be any OPEN SET, and let a_1, a_2, ..., be a finite or infinite sequence in U (possibly with repetitions) that has no ACCUMULATION POINT in U. There there exists an ANALYTIC FUNCTION f on U whose zero set is precisely $\{a_j\}$ (Krantz 1999, p. 111).

See also GAUSSIAN INTEGER, HYPERCOMPLEX NUMBER, PEIRCE'S THEOREM

References

Krantz, S. G. "Weierstrass's Theorem" §8.3.2 in *Handbook of Complex Analysis.* Boston, MA: Birkhäuser, p. 111, 1999.

WeierstrassSigma

WEIERSTRASS SIGMA FUNCTION

WeierstrassZeta

WEIERSTRASS ZETA FUNCTION

Weighing

n weighings are SUFFICIENT to find a bad COIN among $(3^n - 1)/2$ COINS (Steinhaus 1983, p. 61). vos Savant (1993) gives an algorithm for finding a bad ball among 12 balls in three weighings (which, in addition, determines if the bad ball is heavier or lighter than the other 11), and Steinhaus (1983, pp. 58–1) gives an algorithm for 13 balls.

Bachet's weights problem asks for the *minimum number* of weights (which can be placed in *either* pan of a two-arm balance) required to weigh any integral number of pounds from 1 to 40 (Steinhaus 1983, p. 52). The solution is 1, 3, 9, and 27: 1, $2 = -1 + 3$, 3, $4 = 1 + 3$, $5 = -1 - 3 + 9$, $6 = -3 + 9$, $7 = 1 - 3 + 9$, $8 = -1 + 9$, 9, $10 = 1 + 9$, $11 = -1 + 3 + 9$, $12 = 3 + 9$, $13 = 1 + 3 + 9$, $14 = -1 - 3 - 9 + 27$, $15 = -3 - 9 + 27$, $16 = 1 - 3 - 9 + 27$, $17 = -1 - 9 + 27$, and so on.

See also GOLOMB RULER, PERFECT DIFFERENCE SET, SORTING, THREE JUG PROBLEM

References

Bachet, C. G. Problem 5, Appendix in *Problèmes plaisants et délectables, 2nd ed.* p. 215, 1624.
Ball, W. W. R. and Coxeter, H. S. M. *Mathematical Recreations and Essays, 13th ed.* New York: Dover, pp. 50–2, 1987.
Bellman, R. and Gluss, B. "On Various Versions of the Defective Coin Problem." *Information and Control* **4**, 118–31, 1961.
Descartes, B. *Eureka,* No. 13, Oct. 1950.
Dyson, F. J. "The Problem of the Pennies." *Math. Gaz.* **30**, 231–34, 1946.
Gardner, M. *The Sixth Book of Mathematical Games from Scientific American.* Chicago, IL: University of Chicago Press, pp. 29–3 and 106–09, 1984.
Kraitchik, M. *Mathematical Recreations.* New York: W. W. Norton, pp. 52–5, 1942.
O'Beirne, T. H. Chs. 2 and 3 in *Puzzles and Paradoxes.* Oxford, England: Oxford University Press, 1965.
Pappas, T. "Counterfeit Coin Puzzle." *The Joy of Mathematics.* San Carlos, CA: Wide World Publ./Tetra, p. 181, 1989.
Smith, C. A. B. "The Counterfeit Coin Problem." *Math. Gaz.* **31**, 31–9, 1947.
Steinhaus, H. *Mathematical Snapshots, 3rd ed.* New York: Dover, 1999.
Strong, C. L. "The Amateur Scientist." *Sci. Amer.,* May 1955.
Tartaglia. Book 1, Ch. 16, §32 in *Trattato de' numeri e misure, Vol. 2.* Venice, 1556.
Tweedle, M. C. K. *Math. Gaz.* **23**, 278–82, 1938.
vos Savant, M. *The World's Most Famous Math Problem.* New York: St. Martin's Press, pp. 39–2, 1993.

Weight

The word weight has many uses in mathematics. It can refer to a function $w(x)$ (also called a WEIGHTING FUNCTION or WEIGHT FUNCTION) used to normalize ORTHOGONAL FUNCTIONS. It can also be used to indicate one of a set of a multiplicative constants placed in front of terms in a MOVING AVERAGE, NEWTON-COTES FORMULAS, edge or vertex of a GRAPH or TREE, etc. It also refers to the power k in the multiplicative factor $(c\tau + d)^k$ defining a MODULAR FORM.

The weight of a TREE at a point u is the maximum number of edges in any BRANCH at u (Harary 1994, p. 35).

See also MODULAR FORM, MOVING AVERAGE, NEWTON-COTES FORMULAS, WEIGHTED TREE, WEIGHTING FUNCTION

References

Harary, F. *Graph Theory*. Reading, MA: Addison-Wesley, p. 35, 1994.

Weight (Lie Algebra)

Consider a collection of DIAGONAL MATRICES $H_1, \ldots, H_k$, which SPAN a subspace $\mathfrak{h}$. Then the ith EIGENVALUE, i.e., the ith entry along the diagonal, is a LINEAR FUNCTIONAL on $\mathfrak{h}$, and is called a weight.

The general setting for weights occurs in a REPRESENTATION of a SEMISIMPLE LIE ALGEBRA, in which case the CARTAN SUBALGEBRA $\mathfrak{h}$ is ABELIAN and can be put into diagonal form. For example, consider the standard representation of the SPECIAL LINEAR LIE ALGEBRA $\mathfrak{sl}_3(\mathbb{C})$ on $\mathbb{C}^3$. Then

$$H_1 = \begin{bmatrix} 1 & 0 & 0 \\ 0 & -1 & 0 \\ 0 & 0 & 0 \end{bmatrix} \tag{1}$$

and

$$H_2 = \begin{bmatrix} 1 & 0 & 0 \\ 0 & -1 & 0 \\ 0 & 0 & 1 \end{bmatrix} \tag{2}$$

span the CARTAN SUBALGEBRA $\mathfrak{h}$. There are three weights,

$$\alpha_1(h_{ij}) = h_{11} \tag{3}$$

$$\alpha_2(h_{ij}) = h_{22} \tag{4}$$

and

$$\alpha_3(h_{ij}) = h_{33}, \tag{5}$$

corresponding to the decomposition of

$$\mathbb{C}^3 = \langle e_1 \rangle \oplus \langle e_2 \rangle \oplus \langle e_3 \rangle \tag{6}$$

into its eigenspaces. Note that $\alpha_1 + \alpha_2 + \alpha_3 = 0$, because the matrices have zero TRACE. The eigenvectors e_1, e_2, e_3 are called WEIGHT VECTORS, and the corresponding eigenspaces are called WEIGHT SPACES.

In the important special case of the ADJOINT REPRESENTATION of a SEMISIMPLE LIE ALGEBRA, the weights are called ROOTS and the WEIGHT SPACE is called the ROOT SPACE. The roots generate a DISCRETE LATTICE, called the ROOT LATTICE, in the DUAL SPACE $\mathfrak{h}^*$. The set of all possible weights forms a WEIGHT LATTICE,

which contains the ROOT LATTICE. The REPRESENTATIONS of $\mathfrak{g}$ can be classified using the WEIGHT LATTICE.

See also CARTAN MATRIX, LIE ALGEBRA, ROOT (LIE ALGEBRA), ROOT SYSTEM, SEMISIMPLE LIE ALGEBRA, WEIGHT (LIE ALGEBRA), WEYL CHAMBER, WEYL GROUP

References

Fulton, W. and Harris, J. *Representation Theory*. New York: Springer-Verlag, 1991.
Jacobson, N. *Lie Algebras*. New York: Dover, 1979.
Knapp, A. *Lie Groups Beyond an Introduction*. Boston, MA: Birkhäuser, 1996.

Weight Function

WEIGHTING FUNCTION

Weighted Graph

A TREE in which each branch is given a numerical WEIGHT. A weighted graph is therefore a special type of LABELED GRAPH in which the labels are numbers (which are usually taken to be positive).

See also LABELED GRAPH, TAYLOR'S CONDITION, WEIGHTED TREE

Weighted Inversion Statistic

A STATISTIC w on the SYMMETRIC GROUP S_n is called a weighted inversion statistic if there exists an UPPER TRIANGULAR MATRIX $W = (w_{ij})$ such that

$$w(\sigma) = \sum_{i < j} \chi(\sigma_i > \sigma_j) w_{ij},$$

where χ is the CHARACTERISTIC FUNCTION.

The inversion count ($w_{ij} = 1$ for $i < j$) defined by Cramer (1750) and the major index ($w_{i,i+1} = i$; $w_{ij} = 0$ otherwise) defined by MacMahon (1913) are both weighted inversion statistics (Degenhardt and Milne).

See also INVERSION STATISTIC, SYMMETRIC GROUP

References

Cramer, G. "Intr. à l'analyse de lignes courbes algébriques." Geneva, 657–59, 1750.
Degenhardt, S. L. and Milne, S. C. "Weighted Inversion Statistics and Their Symmetry Groups." Preprint.
MacMahon, P. A. "The Indices of Permutations." *Amer. J. Math.* **35**, 281–22, 1913.

Weighted Tree

A TREE to whose nodes and/or edges labels (usually number) are assigned.

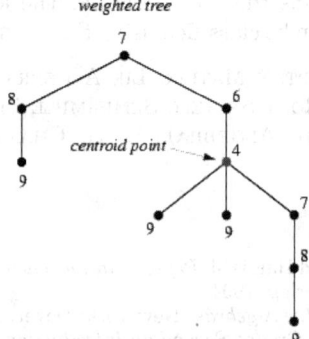

weighted tree

The word "weight" also has a more specific meaning when applied to trees, namely the weight of a TREE at a point u is the maximum number of edges in any BRANCH at u (Harary 1994, p. 35), as illustrated above. A point having minimal weight for the tree is called a CENTROID POINT, and the TREE CENTROID is the set of all CENTROID POINTS.

See also CENTROID POINT, LABELED GRAPH, TAYLOR'S CONDITION, TREE, TREE CENTROID, WEIGHTED GRAPH

References
Harary, F. *Graph Theory*. Reading, MA: Addison-Wesley, 1994.
Weisstein, E. W. "Graphs." MATHEMATICA NOTEBOOK GRAPHS.M.

Weighting Function
A function $w(x)$ used to normalize ORTHONORMAL FUNCTIONS

$$\int [f_n(x)]^2 w(x)\, dx = N_n.$$

See also WEIGHT

Weil-Brezin Map
ZAK TRANSFORM

Weill's Theorem

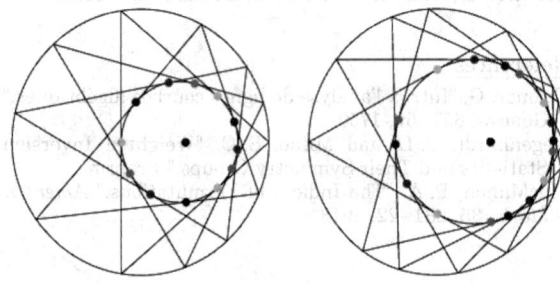

Given the INCIRCLE and CIRCUMCIRCLE of a BICENTRIC POLYGON of n sides, the centroid of the tangent points on the INCIRCLE is a fixed point independent of the particular polygon.

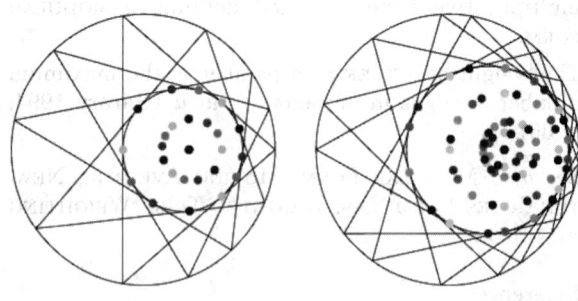

More generally, the LOCUS of the centroid of any number of the n points is a CIRCLE (Casey 1888).

See also BICENTRIC POLYGON, PONCELET'S PORISM

References
Casey, J. *Quart. J. Pure Appl. Math.* **5**, 44, 1862.
Casey, J. *A Sequel to the First Six Books of the Elements of Euclid, Containing an Easy Introduction to Modern Geometry with Numerous Examples,* 5th ed., rev. enl. Dublin: Hodges, Figgis, & Co., p. 164, 1888.
Weill. *Liouville's J. (Ser. 3)* **4**, 270, 1878.

Weingarten Equations
The Weingarten equations express the derivatives of the NORMAL to a surface using derivatives of the position vector. Let $\mathbf{x} : U \to \mathbb{R}^3$ be a REGULAR PATCH, then the SHAPE OPERATOR S of $\mathbf{x}$ is given in terms of the basis $\{\mathbf{x}_u, \mathbf{x}_v\}$ by

$$-S(\mathbf{x}_u) = \mathbf{N}_u = \frac{fF - eG}{EG - F^2}\mathbf{x}_u + \frac{eF - fE}{EG - F^2}\mathbf{x}_v \qquad (1)$$

$$-S(\mathbf{x}_v) = \mathbf{N}_v = \frac{gF - fG}{EG - F^2}\mathbf{x}_u + \frac{fF - gE}{EG - F^2}\mathbf{x}_v, \qquad (2)$$

where $\mathbf{N}$ is the NORMAL VECTOR, E, F, and G the coefficients of the first FUNDAMENTAL FORM

$$ds^2 = E\, du^2 + 2F\, du\, dv + G\, dv^2, \qquad (3)$$

and e, f, and g the coefficients of the second FUNDAMENTAL FORM given by

$$e = -\mathbf{N}_u \cdot \mathbf{x}_u = \mathbf{N} \cdot \mathbf{x}_{uu} \qquad (4)$$

$$f = -\mathbf{N}_v \cdot \mathbf{x}_u = \mathbf{N} \cdot \mathbf{x}_{uv}$$

$$= \mathbf{N}_{vu} \cdot \mathbf{x}_{vu} = -\mathbf{N}_u \cdot \mathbf{x}_v \qquad (5)$$

$$g = -\mathbf{N}_v \cdot \mathbf{x}_v = \mathbf{N} \cdot \mathbf{x}_{vv} \qquad (6)$$

See also FUNDAMENTAL FORMS, SHAPE OPERATOR

References
Gray, A. *Modern Differential Geometry of Curves and Surfaces with Mathematica,* 2nd ed. Boca Raton, FL: CRC Press, pp. 369–71, 1997.

Weingarten Map

SHAPE OPERATOR

Weird Number

A number which is ABUNDANT without being SEMI-PERFECT. (A SEMIPERFECT NUMBER is the sum of any set of its own DIVISORS.) The first few weird numbers are 70, 836, 4030, 5830, 7192, 7912, 9272, 10430, ... (Sloane's A006037). No ODD weird numbers are known, but an infinite number of weird numbers are known to exist. The SEQUENCE of weird numbers has POSITIVE SCHNIRELMANN DENSITY.

See also ABUNDANT NUMBER, SCHNIRELMANN DENSITY, SEMIPERFECT NUMBER

References

Benkoski, S. "Are All Weird Numbers Even?" *Amer. Math. Monthly* **79**, 774, 1972.
Benkoski, S. J. and Erdos, P. "On Weird and Pseudoperfect Numbers." *Math. Comput.* **28**, 617–23, 1974.
Guy, R. K. "Almost Perfect, Quasi-Perfect, Pseudoperfect, Harmonic, Weird, Multiperfect and Hyperperfect Numbers." §B2 in *Unsolved Problems in Number Theory, 2nd ed.* New York: Springer-Verlag, pp. 45–3, 1994.
Sloane, N. J. A. Sequences A006037/M5339 in "An On-Line Version of the Encyclopedia of Integer Sequences." http://www.research.att.com/~njas/sequences/eisonline.html.

Welch Apodization Function

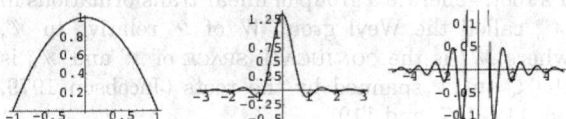

The APODIZATION FUNCTION

$$A(x) = 1 - \frac{x^2}{a^2}.$$

Its FULL WIDTH AT HALF MAXIMUM is $\sqrt{2}a$. Its INSTRUMENT FUNCTION is

$$I(k) = 2a\sqrt{2\pi}\frac{J_{3/2}(2\pi ka)}{(2\pi ka)^{3/2}}$$

$$= a\frac{\sin(2\pi ka) - 2\pi ak\cos(2\pi ak)}{2a^3k^3\pi^3},$$

where $J_\nu(z)$ is a BESSEL FUNCTION OF THE FIRST KIND. It has a width of 1.59044, a maximum of $\frac{4}{3}$, maximum NEGATIVE sidelobe of -0.0861713 times the peak, and maximum POSITIVE sidelobe of 0.356044 times the peak.

See also APODIZATION FUNCTION, INSTRUMENT FUNCTION

References

Press, W. H.; Flannery, B. P.; Teukolsky, S. A.; and Vetterling, W. T. *Numerical Recipes in FORTRAN: The Art of Scientific Computing, 2nd ed.* Cambridge, England: Cambridge University Press, p. 547, 1992.

Well Defined

An expression is called "well defined" (or UNAMBIGUOUS) if its definition assigns it a unique interpretation or value. Otherwise, the expression is said to not be well defined or to be AMBIGUOUS.

For example, the expression abc (the PRODUCT) is well defined if a, b, and c are integers. Because integers are ASSOCIATIVE, abc has the same value whether it is interpreted to mean $(ab)c$ or $a(bc)$. However, if a, b, and c are CAYLEY NUMBERS, then the expression abc is *not* well defined, since CAYLEY NUMBER are not, in general, ASSOCIATIVE, so that the two interpretations $(ab)c$ and $a(bc)$ can be different.

Sometimes, ambiguities are implicitly resolved by notational convention. For example, the conventional interpretation of $a \wedge b \wedge c = a^{b^c}$ is $a^{(b^c)}$, never $\left(a^b\right)^c$, so that the expression $a \wedge b \wedge c$ is well defined even though exponentiation is nonassociative.

The term "well defined" also has a technical meaning in field of PARTIAL DIFFERENTIAL EQUATIONS. A solution to a PARTIAL DIFFERENTIAL EQUATION that is a continuous function of its values on the boundary is said to be well defined. Otherwise, a solution is called ILL DEFINED.

See also AMBIGUOUS, ILL DEFINED, UNDEFINED

Well Order

WELL ORDERED SET

Well Ordered Set

A TOTALLY ORDERED SET $(A, \leq)$ is said to be well ordered IFF every nonempty SUBSET of A has a least element (Ciesielski 1997, p. 38; Moore 1982, p. 2; Rubin 1967, p. 159; Suppes 1972, p. 75). Every finite TOTALLY ORDERED SET is well ordered. The set of integers **Z**, which has no least element, is an example of a set that is *not* well ordered.

An ORDINAL NUMBER is the ORDER TYPE of a well ordered set.

See also AXIOM OF CHOICE, HILBERT'S PROBLEMS, INITIAL SEGMENT, MONOMIAL ORDER, ORDINAL NUMBER, ORDER TYPE, SUBSET, WELL ORDERING PRINCIPLE

References

Ciesielski, K. *Set Theory for the Working Mathematician.* Cambridge, England: Cambridge University Press, 1997.
Ferreirós, J. "Well-Ordered Sets." §8.4 in *Labyrinth of Thought: A History of Set Theory and Its Role in Modern Mathematics.* Basel, Switzerland: Birkhäuser, pp. 274–78, 1999.
Moore, G. H. *Zermelo's Axiom of Choice: Its Origin, Development, and Influence.* New York: Springer-Verlag, 1982.

Rubin, J. E. *Set Theory for the Mathematician.* New York: Holden-Day, 1967.
Séroul, R. *Programming for Mathematicians.* Berlin: Springer-Verlag, pp. 22–3, 2000.
Suppes, P. *Axiomatic Set Theory.* New York: Dover, 1972.

Well Ordering Principle

Every nonempty set of POSITIVE INTEGERS contains a smallest member.

See also AXIOM OF CHOICE, WELL ORDERED SET

References

Apostol, T. M. "The Well-Ordering Principle." §I 4.3 in *Calculus, 2nd ed., Vol. 1: One-Variable Calculus, with an Introduction to Linear Algebra.* Waltham, MA: Blaisdell, pp. 34–5, 1967.
Shanks, D. *Solved and Unsolved Problems in Number Theory, 4th ed.* New York: Chelsea, p. 149, 1993.

Well-Poised

A GENERALIZED HYPERGEOMETRIC FUNCTION

$$_pF_q\begin{bmatrix}\alpha_1,\alpha_2,\ldots,\alpha_p\\\beta_1,\beta_2,\ldots,\beta_q\end{bmatrix};z\end{bmatrix}$$

is said to be well-poised if $p = q + 1$ and

$$1 + \alpha_1 = \beta_1 + \alpha_2 = \ldots = \beta_q + \alpha_{p+1}$$

See also GENERALIZED HYPERGEOMETRIC FUNCTION, K-BALANCED, NEARLY-POISED, SAALSCHÜTZIAN

References

Bailey, W. N. *Generalised Hypergeometric Series.* Cambridge, England: Cambridge University Press, p. 11, 1935.
Koepf, W. *Hypergeometric Summation: An Algorithmic Approach to Summation and Special Function Identities.* Braunschweig, Germany: Vieweg, p. 43, 1998.
Whipple, F. J. W. "On Well-Poised Series, Generalized Hypergeometric Series Having Parameters in Pairs, Each Pair with the Same Sum." *Proc. London Math. Soc.* **24**, 247–63, 1926.
Whipple, F. J. W. "Well-Poised Series and Other Generalized Hypergeometric Series." *Proc. London Math. Soc. Ser. 2* **25**, 525–44, 1926.

Werner Formulas

$$2 \sin\alpha \cos\beta = \sin(\alpha-\beta) + \sin(\alpha+\beta) \tag{1}$$

$$2 \cos\alpha \cos\beta = \cos(\alpha-\beta) + \cos(\alpha+\beta) \tag{2}$$

$$2 \cos\alpha \sin\beta = \sin(\alpha+\beta) - \sin(\alpha-\beta) \tag{3}$$

$$2 \sin\alpha \sin\beta = \cos(\alpha-\beta) - \cos(\alpha+\beta) \tag{4}$$

See also TRIGONOMETRIC ADDITION FORMULAS

Werner Projection

A nonconformal, equal-area projection which is a special case of the BONNE PROJECTION where one of the poles is taken as the standard parallel. Because of its heart shape, this projection is sometimes also called "cordiform."

See also BONNE PROJECTION, MAP PROJECTION

References

MathWorks. "Mapping Toolbox: Bonne Projection." http://www.mathworks.com/access/helpdesk/help/toolbox/map/wernerprojection.shtml.

Weyl Character Formula

References

Hsiang, W. Y. "Weyl Character Formula and the Classification of Complex Irreducible Representations." Lec. 4, §4 in *Lectures on Lie Groups.* Singapore: World Scientific, pp. 74–7, 2000.

Weyl Group

Let $\mathcal{L}$ be a finite-dimensional split SEMISIMPLE LIE ALGEBRA over a FIELD of CHARACTERISTIC 0, $\mathcal{H}$ a splitting CARTAN SUBALGEBRA, and Λ a weight of $\mathcal{H}$ in a representation of $\mathcal{L}$. Then

$$\Lambda' = \Lambda S_\alpha = \lambda - \frac{2(\Lambda,\alpha)}{(\alpha,\alpha)}(\alpha)$$

is also a weight. Furthermore, the reflections S_α with α a root, generate a group of linear transformations in $\mathcal{H}_0^*$ called the Weyl group W of $\mathcal{L}$ relative to $\mathcal{H}$, where $\mathcal{H}^*$ is the CONJUGATE SPACE of $\mathcal{H}$ and $\mathcal{H}_0^*$ is the Q-SPACE spanned by the roots (Jacobson 1979, pp. 112, 117, and 119).

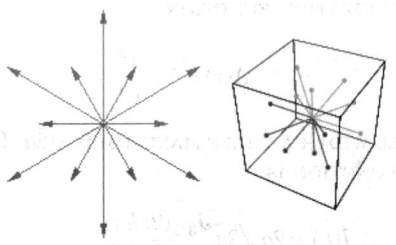

The Weyl group acts on the roots of a semisimple Lie algebra, and it is a finite group. The animations above illustrate this action for Weyl Group acting on the roots of a homotopy from one Weyl matrix to the next one (i.e., it slides the arrows from g to h) in the first two figures, while the third figure shows the Weyl Group acting on the roots of the CARTAN MATRIX of the infinite family of semisimple lie algebras A_3 (cf. DYNKIN DIAGRAM), which is the SPECIAL LINEAR LIE ALGEBRA, $\mathfrak{sl}_4$.

See also CARTAN MATRIX, DYNKIN DIAGRAM, LIE ALGEBRA, LIE GROUP, MACDONALD'S CONSTANT-TERM CONJECTURE, ROOT (LIE ALGEBRA), ROOT SYSTEM, ROOT LATTICE, SEMISIMPLE LIE ALGEBRA,

WEIGHT LATTICE, WEYL CHAMBER

References

Andrews, G. E. "The Macdonald Conjectures." *q*-Series: Their Development and Application in Analysis, Number Theory, Combinatorics, Physics, and Computer Algebra. Providence, RI: Amer. Math. Soc., p. 41, 1986.

Huang, J.-S. "The Weyl Group." §4.5 in *Lectures on Representation Theory*. Singapore: World Scientific, pp. 36–8, 1999.

Jacobson, N. *Lie Algebras.* New York: Dover, pp. 112–19 and 240–43, 1979.

Weyl Reduction

References

Hsiang, W. Y. "Coxeter Groups, Weyl Reduction, and Weyl Formulas." Lec. 4 in *Lectures on Lie Groups.* Singapore: World Scientific, pp. 46–7 and 58–7, 2000.

Weyl Tensor

The TENSOR C_{abcd} defined by

$$R_{abcd} = C_{abcd} + \frac{2}{n-2}\left(g_{a[c}R_{d]b} - g_{b[c}R_{d]a}\right)$$
$$-\frac{2}{(n-1)(n-2)}Rg_{a[c}g_{d]b}, \quad (1)$$

where R_{abcd} is the RIEMANN TENSOR, R is the SCALAR CURVATURE, g_{ab} is the METRIC TENSOR, and $T_{[a_1...a_n]}$ denotes the ANTISYMMETRIC TENSOR part (Wald 1984, p. 40).

The Weyl tensor is defined so that every CONTRACTION between indices gives 0. In particular,

$$C^{\lambda}{}_{\mu\lambda\kappa} = 0 \quad (2)$$

(Weinberg 1972, p. 146). The number of independent components for a Weyl tensor in N-D for $N \geq 3$ is given by

$$C_N = \tfrac{1}{12}N(N+1)(N+2)(N-3) \quad (3)$$

(Weinberg 1972, p. 146). For $N = 3, 4, ...,$ this gives 0, 10, 35, 84, 168, ... (Sloane's A052472).

See also CURVATURE SCALAR, RIEMANN TENSOR

References

Eisenhart, L. P. *Riemannian Geometry.* Princeton, NJ: Princeton University Press, 1964.

Sloane, N. J. A. Sequences A052472 in "An On-Line Version of the Encyclopedia of Integer Sequences." http://www.research.att.com/~njas/sequences/eisonline.html.

Wald, R. M. *General Relativity.* Chicago, IL: University of Chicago Press, 1984.

Weinberg, S. *Gravitation and Cosmology: Principles and Applications of the General Theory of Relativity.* New York: Wiley, 1972.

Weyl, H. "Reine Infinitesimalgeometrie." *Math. Z.* **2**, 384–11, 1918.

Weyl's Criterion

A SEQUENCE $\{x_1, x_2, ...\}$ is EQUIDISTRIBUTED IFF

$$\lim_{N\to\infty} \frac{1}{N}\sum_{n<N} e^{2\pi i m x_n} = 0$$

for each $m = 1, 2, ...$. A consequence of this result is that the sequence $\{\text{frac}(nx)\}$ is dense and EQUIDISTRIBUTED in the interval $[0, 1]$ for irrational x, where $n = 1, 2, ...$ and frac(x) is the FRACTIONAL PART of x (Finch).

See also EQUIDISTRIBUTED SEQUENCE, RAMANUJAN'S SUM

References

Cassels, J. W. S. *An Introduction to Diophantine Analysis.* Cambridge, England: Cambridge University Press, 1965.

Finch, S. "Powers of 3/2 Modulo One." http://www.mathsoft.com/asolve/pwrs32/pwrs32.html.

Kuipers, L. and Niederreiter, H. *Uniform Distribution of Sequences.* New York: Wiley, p. 226, 1974.

Pólya, G. and Szego, G. *Problems and Theorems in Analysis I.* New York: Springer-Verlag, 1972.

Radin, C. *Miles of Tiles.* Providence, RI: Amer. Math. Soc., pp. 79–0, 1999.

Vardi, I. *Computational Recreations in Mathematica.* Redwood City, CA: Addison-Wesley, pp. 155–56 and 254, 1991.

Weyl's Denominator Formula

See also ROOT SYSTEM

References

Simpson, T. "Three Generalizations of Weyl's Denominator Formula." *Electronic J. Combinatorics* **3**, R12 1–1, 1996. http://www.combinatorics.org/Volume_3/volume3.html#R12.

Weyrich's Formula

For r and x real, with $0 \leq \arg\left(\sqrt{k^2 - r^2}\right) < \pi$ and $0 \leq \arg k < \pi$,

$$\tfrac{1}{2}i\int_{-\infty}^{\infty} H_0^{(1)}\left(r\sqrt{k^2 - r^2}\right)e^{irx}dr = \frac{e^{ik\sqrt{r^2 + x^2}}}{\sqrt{r^2 + x^2}},$$

where $H_0^{(1)}(x)$ is a HANKEL FUNCTION OF THE FIRST KIND.

See also HANKEL FUNCTION OF THE FIRST KIND

References

Iyanaga, S. and Kawada, Y. (Eds.). *Encyclopedic Dictionary of Mathematics.* Cambridge, MA: MIT Press, p. 1474, 1980.

W-Function

LAMBERT'S W-FUNCTION

Wheat and Chessboard Problem

Let one grain of wheat be placed on the first square of a CHESSBOARD, two on the second, four on the third, eight on the fourth, etc. How many grains total are placed on an 8×8 CHESSBOARD? Since this is a GEOMETRIC SERIES, the answer for n squares is

$$\sum_{i=0}^{n-1} 2^i = 2^n - 1,$$

a MERSENNE NUMBER. Plugging in $n = 8 \times 8 = 84$ then gives $2^{64} - 1- = 18446744073709551615$.

See also MERSENNE NUMBER

References

Pappas, T. "The Wheat & Chessboard." *The Joy of Mathematics.* San Carlos, CA: Wide World Publ./Tetra, p. 17, 1989.
Steinhaus, H. *Mathematical Snapshots, 3rd ed.* New York: Dover, pp. 23–4, 1999.

Wheel

ARISTOTLE'S WHEEL PARADOX, BENHAM'S WHEEL, WHEEL GRAPH

Wheel Graph

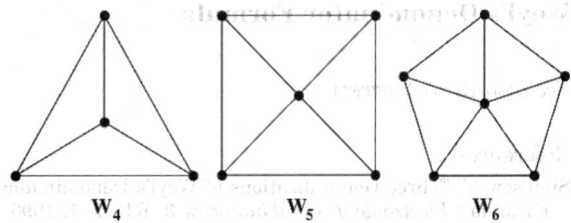

W_4 W_5 W_6

A GRAPH W_n of order n which contains a CYCLE of order $n-1$, and for which every NODE in the cycle is connected to one other NODE (which is known as the HUB). The edges of a wheel which include the HUB are called spokes (Skiena 1990, p. 146). The wheel W_n can be defined as the graph $K_1 + C_{n-1}$, where K_1 is the (trivial) COMPLETE GRAPH on 1 node and C_n is the CYCLE GRAPH. Wheel graphs can be constructed using Wheel[n] in the *Mathematica* add-on package DiscreteMath`Combinatorica` (which can be loaded with the command < <DiscreteMath`).
In a wheel graph, the HUB has DEGREE $n-1$, and other nodes have degree 3. Wheel graphs are 3-connected. $W_4 = K_4$, where K_4 is the COMPLETE GRAPH of order four. The CHROMATIC NUMBER of W_n is

$$\chi(W_n) = \begin{cases} 4 & \text{for } n \text{ odd} \\ 3 & \text{for } n \text{ even.} \end{cases}$$

See also COMPLETE GRAPH, GEAR GRAPH, HUB, WEB GRAPH

References

Harary, F. *Graph Theory.* Reading, MA: Addison-Wesley, p. 46, 1994.
Saaty, T. L. and Kainen, P. C. *The Four-Color Problem: Assaults and Conquest.* New York: Dover, p. 148, 1986.
Skiena, S. "Cycles, Stars, and Wheels." §4.2.3 in *Implementing Discrete Mathematics: Combinatorics and Graph Theory with Mathematica.* Reading, MA: Addison-Wesley, pp. 91 and 144–47, 1990.

Wheel Paradox

ARISTOTLE'S WHEEL PARADOX

Whewell Equation

An INTRINSIC EQUATION which expresses a curve in terms of its ARC LENGTH s and TANGENTIAL ANGLE ϕ.

See also ARC LENGTH, CESÀRO EQUATION, INTRINSIC EQUATION, NATURAL EQUATION, TANGENTIAL ANGLE

References

Yates, R. C. "Intrinsic Equations." *A Handbook on Curves and Their Properties.* Ann Arbor, MI: J. W. Edwards, pp. 123–26, 1952.

Whipple's Identity

Whipple derived a great many identities for GENERALIZED HYPERGEOMETRIC FUNCTIONS, many of which are consequently known as Whipple's identities (transformations, etc.). Among Whipple's identities include

$$_3F_2\left[\begin{matrix} a, 1-a, c \\ e, 1+2c-e \end{matrix}; 1\right] = \frac{2^{1-2c}\pi\Gamma(e)\Gamma(1+2c-e)}{\Gamma\left[\frac{1}{2}(a+e)\right]\Gamma\left[\frac{1}{2}(a+1+2c-e)\right]}$$

$$\times \frac{1}{\Gamma\left[\frac{1}{2}(1-a+e)\right]\Gamma\left[\frac{1}{2}(2+2c-a-e)\right]}$$

(Bailey 1935, p. 15; Koepf 1998, p. 32), where $_3F_2(a,b,c;d,e;z)$ is a GENERALIZED HYPERGEOMETRIC FUNCTION and $\Gamma(z)$ is a GAMMA FUNCTION, and

$$_6F_5\left[\begin{matrix} a, & 1+\frac{1}{2}a, & b, & c, & d, & e \\ & \frac{1}{2}a, & 1+a-b, & 1-a+c, & 1+a-d, & 1+a-e \end{matrix}; 1\right]$$

$$= \frac{\Gamma(1+a-d)\Gamma(1+a-e)}{\Gamma(1+a)\Gamma(1+a-d-e)} {}_3F_2\left[\begin{matrix} 1+a-b-c,d,e; \\ 1+a-b,1+a-c \end{matrix}\right]$$

(Bailey 1935, p. 28).

See also GENERALIZED HYPERGEOMETRIC FUNCTION, WATSON'S THEOREM

References

Bailey, W. N. "Whipple's Theorem on the Sum of a $_3F_2$." §3.4 in *Generalised Hypergeometric Series.* Cambridge, England: Cambridge University Press, p. 16, 1935.
Koepf, W. *Hypergeometric Summation: An Algorithmic Approach to Summation and Special Function Identities.* Braunschweig, Germany: Vieweg, 1998.
Whipple, F. J. W. "Well-Poised Series and Other Generalized Hypergeometric Series." *Proc. London Math. Soc. Ser. 2* **25**, 525–44, 1926.

Whipple's Transformation

$$_7F_6\left[\begin{array}{c} a, 1+\tfrac{1}{2}a, b, c, d, e, -m \\ \tfrac{1}{2}a, 1+a-b, 1+a-c, \\ 1+a-d, 1+a-e, 1+a+m \end{array}\right]$$

$$= \frac{(1+a)_m(1+a-d-e)_m}{(1+a-d)_m(1+a-e)_m}$$

$$\times\, _4F_3\left[\begin{array}{c} 1+a-b-c, d, e, -m \\ 1+a-b, 1+a-c, d+e-a-m \end{array}\right],$$

where $_7F_6$ and $_4F_3$ are GENERALIZED HYPERGEO-
METRIC FUNCTIONS and $\Gamma(z)$ is the GAMMA FUNCTION.

Another transformation due to Whipple (1926) is
given by

$$_4F_3\left[\begin{array}{c} a, b, -z, -n \\ u, v, w \end{array}; 1\right]$$

$$= \frac{\Gamma(u+z+n)\Gamma(w+z+n)\Gamma(v)\Gamma(w)}{\Gamma(v+z)\Gamma(v+n)\Gamma(w+n)\Gamma(w+z)}$$

$$\times\, _4F_3\left[\begin{array}{c} u-a, u-b, -z, -n \\ 1-v-z-n, 1-w-z-n, u \end{array}; 1\right] \quad (1)$$

for one of z and n a NONNEGATIVE INTEGER (Andrews
and Burge 1993).

See also GENERALIZED HYPERGEOMETRIC FUNCTION,
WATSON-WHIPPLE TRANSFORMATION

References

Andrews, G. E. and Burge, W. H. "Determinant Identities."
Pacific J. Math. **158**, 1–4, 1993.
Bailey, W. N. *Generalised Hypergeometric Series.* Cam-
bridge, England: Cambridge University Press, pp. 25
and 29, 1935.
Whipple, F. J. W. "Well-Poised Series and Other General-
ized Hypergeometric Series." *Proc. London Math. Soc. Ser.
2* **25**, 525–44, 1926.
Whipple, F. J. W. "On Well-Poised Series, Generalized
Hypergeometric Series Having Parameters in Pairs,
Each Pair with the Same Sum." *Proc. London Math.
Soc.* **24**, 247–63, 1926.
Whipple, F. J. W. "A Fundamental Relation Between Gen-
eralized Hypergeometric Series." *Proc. London Math. Soc.*
26, 257–72, 1927.

Whirl

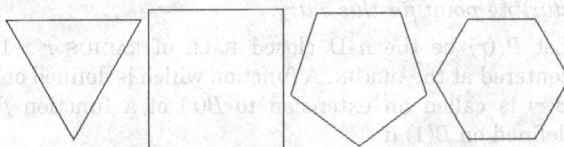

Whirls are figures constructed by nesting a sequence
of polygons (each having the same number of sides),
each slightly smaller and rotated relative to the
previous one. The vertices give the path of the n
mice in the MICE PROBLEM, and form n LOGARITHMIC
SPIRALS.

See also DAISY, DERIVED POLYGON, LOGARITHMIC
SPIRAL, MICE PROBLEM, SWIRL

References

Lauwerier, H. *Fractals: Endlessly Repeated Geometric Fig-
ures.* Princeton, NJ: Princeton University Press, p. 66,
1991.
Pappas, T. "Spider & Spirals." *The Joy of Mathematics.* San
Carlos, CA: Wide World Publ./Tetra, p. 228, 1989.
Weisstein, E. W. "Fractals." MATHEMATICA NOTEBOOK FRAC-
TAL.M.
Weisstein, E. W. "Mice Problem." MATHEMATICA NOTEBOOK
MICEPROBLEM.M.
Wells, D. *The Penguin Dictionary of Curious and Interesting
Geometry.* London: Penguin, pp. 201–02, 1991.

Whisker Plot

BOX-AND-WHISKER PLOT

Whitehead Double

The SATELLITE KNOT of an UNKNOT twisted inside a
TORUS.

See also SATELLITE KNOT, TORUS, UNKNOT

References

Adams, C. C. *The Knot Book: An Elementary Introduction to
the Mathematical Theory of Knots.* New York: W. H.
Freeman, pp. 115–16, 1994.

Whitehead Link

The LINK 05–2–1, illustrated above, with BRAID WORD
$\sigma_1^2\sigma_2^2\sigma_1^{-1}\sigma_2^{-2}$ and JONES POLYNOMIAL

$$V(t) = t^{-3/2}\left(-1 + t - 2t^2 + t^3 - 2t^4 + t^5\right)$$

The Whitehead link has LINKING NUMBER 0. It was
discovered by Whitehead in 1934 (Whitehead 1962,
pp. 21–0) as a counterexample to a piece of an
attempted proof of the POINCARÉ CONJECTURE (Milnor).

See also POINCARÉ CONJECTURE.

References

Milnor, J. "The Poincaré Conjecture." http://www.clay-
math.org/prize_problems/poincare.pdf.
Whitehead, J. H. C. *Mathematical Works, Vol. 2.* London:
Pergamon Press, 1962.

Whitehead Manifold

An open 3-MANIFOLD which is simply connected but is
topologically distinct from Euclidean 3-space.

References

Rolfsen, D. *Knots and Links*. Wilmington, DE: Publish or Perish Press, p. 82, 1976.

Whitehead's Theorem

MAPS between CW-COMPLEXES that induce ISOMORPHISMS on all HOMOTOPY GROUPS are actually HOMOTOPY equivalences.

See also CW-COMPLEX, HOMOTOPY GROUP, ISOMORPHISM

Whitney Singularity

PINCH POINT

Whitney Sum

An operation that takes two VECTOR BUNDLES over a fixed SPACE and produces a new VECTOR BUNDLE over the same SPACE. If E_1 and E_2 are VECTOR BUNDLES over B, then the Whitney sum $E_1 \oplus E_2$ is the VECTOR BUNDLE over B such that each FIBER over B is naturally the DIRECT SUM of the E_1 and E_2 FIBERS over B.

The Whitney sum is therefore the FIBER for FIBER DIRECT SUM of the two BUNDLES E_1 and E_2. An easy formal definition of the Whitney sum is that $E_1 \oplus E_2$ is the pull-back BUNDLE of the diagonal map from B to $B \times B$, where the BUNDLE over $B \times B$ is $E_1 \times E_2$.

See also BUNDLE, FIBER, VECTOR BUNDLE

Whitney Umbrella

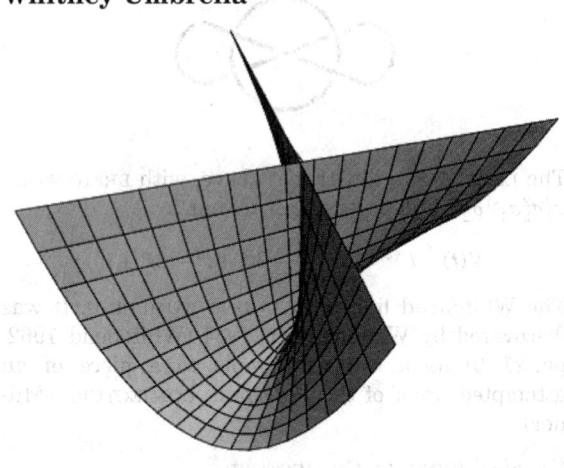

A surface which can be interpreted as a self-intersecting RECTANGLE in 3-D. It is given by the parametric equations

$$x = uv \tag{1}$$

$$y = u \tag{2}$$

$$z = v^2 \tag{3}$$

for $u, v \in [-1, 1]$. The center of the "plus" shape which is the end of the line of self-intersection is a PINCH

POINT. The coefficients of the FIRST FUNDAMENTAL FORM are

$$E = 1 + v^2 \tag{4}$$

$$F = uv \tag{5}$$

$$G = u^2 + 4v^2 \tag{6}$$

and the SECOND FUNDAMENTAL FORM are

$$e = 0 \tag{7}$$

$$f = \frac{2u}{\sqrt{u^2 + 4v^2 + 4v^4}} \tag{8}$$

$$g = -\frac{2u}{\sqrt{u^2 + 4v^2 + 4v^4}} \tag{9}$$

giving AREA ELEMENT

$$dA = \sqrt{u^2 + 4v^2(1 + v^2)} \tag{10}$$

and GAUSSIAN CURVATURE and MEAN CURVATURE

$$K = -\frac{4v^2}{(u^2 + 4v^2 + 4v^4)^2} \tag{11}$$

$$H = -\frac{u(1 + 3v^2)}{(u^2 + 4v^2 + 4v^4)^{3/2}} \tag{12}$$

References

Francis, G. K. *A Topological Picturebook*. New York: Springer-Verlag, pp. 8–, 1987.
Gray, A. "The Whitney Umbrella." *Modern Differential Geometry of Curves and Surfaces with Mathematica, 2nd ed.* Boca Raton, FL: CRC Press, pp. 311 and 401–02, 1997.

Whitney-Graustein Theorem

A 1937 theorem which classified planar regular closed curves up to regular HOMOTOPY by their WINDING NUMBERS. In his thesis, S. Smale generalized this result to regular closed curves on an n-MANIFOLD.

Whitney-Mikhlin Extension Constants

N.B. A detailed online essay by S. Finch was the starting point for this entry.

Let $B_n(r)$ be the n-D closed BALL of RADIUS $r > 1$ centered at the ORIGIN. A function which is defined on $B(r)$ is called an extension to $B(r)$ of a function f defined on $B(1)$ if

$$F(x) = f(x) \ \forall x \in B(1) \tag{1}$$

Given 2 BANACH SPACES of functions defined on $B(1)$ and $B(r)$, find the extension operator from one to the other of minimal norm. Mikhlin (1986) found the best constants χ such that this condition, corresponding to the Sobolev $W(1, 2)$ integral norm, is satisfied,

$$\sqrt{\int_{B(1)}\left[[f(x)]^2+\sum_{j=1}^{n}\left(\frac{\partial f}{\partial x_j}\right)^2\right]dx}$$

$$\le \chi\sqrt{\int_{B(r)}\left[[F(x)]^2+\sum_{j=1}^{n}\left(\frac{\partial F}{\partial x_j}\right)^2\right]dx}. \tag{2}$$

$\chi(1,r)=1$. Let

$$v=\tfrac{1}{2}(n-2), \tag{3}$$

then for $n>2$,

$$\chi(n,r)=\sqrt{1+\frac{I_v(1)}{I_{v+1}(1)}\frac{I_v(r)K_{v+1}(1)+K_v(r)I_{v+1}(1)}{I_v(r)K_v(1)-K_v(r)I_v(1)}}, \tag{4}$$

where $I_v(z)$ is a MODIFIED BESSEL FUNCTION OF THE FIRST KIND and $K_v(z)$ is a MODIFIED BESSEL FUNCTION OF THE SECOND KIND. For $n=2$,

$$\chi(2,r)=\max\left\{\sqrt{1+\frac{I_v(1)}{I_{v+1}(1)}\frac{I_v(r)K_{v+1}(1)+K_v(r)I_{v+1}(1)}{I_v(r)K_v(1)-K_v(r)I_v(1)}},\right.$$

$$\left.\sqrt{1+\frac{I_1(1)}{I_1(1)+I_2(1)}\left[1+\frac{I_1(r)K_0(1)+K_1(r)I_0(1)}{I_1(r)K_1(1)-K_1(r)I_1(1)}\right]}\right\}, \tag{5}$$

For $r\to\infty$,

$$\chi(n,\infty)=\sqrt{1+\frac{I_v(1)}{I_{v+1}(1)}\frac{I_v(1)}{K_v(1)}}, \tag{6}$$

which is bounded by

$$n-1<\chi(n,\infty)<\sqrt{(n-1)^2+4} \tag{7}$$

For ODD n, the RECURRENCE RELATIONS

$$a_{k+1}=a_{k-1}-(2k-1)a_k \tag{8}$$

$$b_{k+1}=b_{k-1}-(2k-1)b_k \tag{9}$$

with

$$a_0=e+e^{-1} \tag{10}$$

$$a_1=e-e^{-1} \tag{11}$$

$$b_0=e^{-1} \tag{12}$$

$$b_1=e^{-1} \tag{13}$$

where E is the constant $2.71828...$, give

$$\chi(2k+1,\infty)=\sqrt{1+\frac{a_k}{a_{k+1}}\frac{b_{k+1}}{b_k}} \tag{14}$$

The first few are

$$\chi(3,\infty)=e \tag{15}$$

$$\chi(5,\infty)=\sqrt{\frac{e^2}{e^2-7}} \tag{16}$$

$$\chi(7,\infty)=\sqrt{\frac{2}{7}}\sqrt{\frac{e^2}{37-5e^2}} \tag{17}$$

$$\chi(9,\infty)=\frac{1}{\sqrt{37}}\sqrt{\frac{e^2}{18e^2-133}} \tag{18}$$

$$\chi(11,\infty)=\frac{1}{\sqrt{133}}\sqrt{\frac{e^2}{2431-329e^2}} \tag{19}$$

$$\chi(13,\infty)=\sqrt{\frac{2}{2431}}\sqrt{\frac{e^2}{3655e^2-27007}} \tag{20}$$

Similar formulas can be given for even n in terms of $I_0(1)$, $I_1(1)$, $K_0(1)$, $K_1(1)$.

References

Finch, S. "Favorite Mathematical Constants." http://www.mathsoft.com/asolve/constant/mkhln/mkhln.html.
Mikhlin, S. G. *Constants in Some Inequalities of Analysis.* New York: Wiley, 1986.

Whittaker Differential Equation

$$\frac{d^2u}{dz^2}+\frac{du}{dz}+\left(\frac{k}{z}+\frac{\frac{1}{4}-m^2}{z^2}\right)u=0 \tag{1}$$

Let $u\equiv e^{-z/2}W_{k,m}(z)$, where $W_{k,m}(z)$ denotes a WHITTAKER FUNCTION. Then (1) becomes

$$\frac{d}{dz}\left(-\tfrac{1}{2}e^{-z/2}W+e^{-z/2}W'\right)+\left(-\tfrac{1}{2}e^{-z/2}W+e^{-z/2}W'\right)$$

$$+\left(\frac{k}{z}+\frac{\frac{1}{4}-m^2}{z^2}\right)e^{-z/2}W=0. \tag{2}$$

Rearranging,

$$\left(\tfrac{1}{4}e^{-z/2}W-\tfrac{1}{2}e^{-z/2}W'-\tfrac{1}{2}e^{-z/2}W'+e^{-z/2}W''\right)p$$

$$+\left(-\tfrac{1}{2}e^{-z/2}W+e^{-z/2}W'\right)+\left(\frac{k}{z}+\frac{\frac{1}{4}-m^2}{z^2}\right)e^{-z/2}W=0 \tag{3}$$

$$-\tfrac{1}{4}e^{-z/2}W+e^{-z/2}W''+\left(\frac{k}{z}+\frac{\frac{1}{4}-m^2}{z^2}\right)e^{-z/2}W=0, \tag{4}$$

so

$$W''+\left(-\frac{1}{4}+\frac{k}{z}+\frac{\frac{1}{4}-m^2}{z^2}\right)W=0, \tag{5}$$

where $W'\equiv dW/dz$ (Abramowitz and Stegun 1972, p. 505; Zwillinger 1997, p. 128). The solutions are known as WHITTAKER FUNCTIONS. Replacing $W(z)$ by $y(x)$, the solutions can also be written in the form

$$y = e^{-x/2}x^{m+1/2}[C_1 U(\tfrac{1}{2} - k + m, 2m + 1, x)$$

$$+C_2 L_{-1/2+k-m}^{2m}(x)], \tag{6}$$

where $U(a, b, z)$ is a CONFLUENT HYPERGEOMETRIC FUNCTION OF THE SECOND KIND and $L_n^a(x)$ is a generalized LAGUERRE POLYNOMIAL.

See also WHITTAKER FUNCTION

References

Abramowitz, M. and Stegun, C. A. (Eds.). *Handbook of Mathematical Functions with Formulas, Graphs, and Mathematical Tables, 9th printing.* New York: Dover, p. 505, 1972.

Zwillinger, D. *Handbook of Differential Equations, 3rd ed.* Boston, MA: Academic Press, p. 128, 1997.

Whittaker Function

Solutions to the WHITTAKER DIFFERENTIAL EQUATION. The linearly independent solutions are

$$M_{k,m}(z) \equiv z^{1/2+m}e^{-z/2} \times \left[1 + \frac{\tfrac{1}{2}+m-k}{1!(2m+1)}z\right.$$

$$\left. + \frac{\left(\tfrac{1}{2}+m-k\right)\left(\tfrac{3}{2}+m-k\right)}{2!(2m+1)(2m+2)}z^2 + \ldots\right],$$

$$\tag{1}$$

and $M_{k,-m}(z)$, where $M_{k,m}(z)$ is a CONFLUENT HYPERGEOMETRIC FUNCTION. In terms of CONFLUENT HYPERGEOMETRIC FUNCTIONS, the Whittaker functions are

$$M_{k,m}(z) = e^{-z/2}z^{m+1/2} \, {}_1F_1\left(\tfrac{1}{2} + m - k, 1 + 2m; z\right) \tag{2}$$

$$W_{k,m}(z) = e^{-z/2}z^{m+1/2} U\left(\tfrac{1}{2} + m - k, 1 + 2m; z\right) \tag{3}$$

(Abramowitz and Stegun 1972, p. 505; Whittaker and Watson 1990, pp. 339–51). However, the CONFLUENT HYPERGEOMETRIC FUNCTION disappears when $2m$ is an INTEGER, so Whittaker functions are often defined instead. The Whittaker functions are related to the PARABOLIC CYLINDER FUNCTIONS. When $|\arg z| < 3\pi/2$ and $2m$ is not an INTEGER,

$$W_{k,m}(z) = \frac{\Gamma(-2m)}{\Gamma\left(\tfrac{1}{2} - m - k\right)} M_{k,m}(z)$$

$$+ \frac{\Gamma(2m)}{\Gamma\left(\tfrac{1}{2} + m - k\right)} M_{k,-m}(z). \tag{4}$$

When $|\arg(-z)| < 3\pi/2$ and $2m$ is not an INTEGER,

$$W_{-k,m}(-z) = \frac{\Gamma(-2m)}{\Gamma\left(\tfrac{1}{2} - m - k\right)} M_{-k,m}(-z)$$

$$+ \frac{\Gamma(2m)}{\Gamma\left(\tfrac{1}{2} + m + k\right)} M_{-k,-m}(-z). \tag{5}$$

Whittaker functions satisfy the RECURRENCE RELATIONS

$$W_{k,m}(z) = z^{1/2}W_{k-1/2,m-1/2}(z)$$

$$+ \left(\tfrac{1}{2} - k + m\right) W_{k-1,m}(z) \tag{6}$$

$$W_{k,m}(z) = z^{1/2}W_{k-1/2,m+1/2}(z)$$

$$+ \left(\tfrac{1}{2} - k - m\right) W_{k-1,m}(z) \tag{7}$$

$$zW'_{k,m}(z) = \left(k - \tfrac{1}{2}z\right) W_{k,m}(z)$$

$$- \left[m^2 - \left(k - \tfrac{1}{2}\right)^2\right] W_{k-1,m}(z). \tag{8}$$

See also CONFLUENT HYPERGEOMETRIC FUNCTION, KUMMER'S FORMULAS, PEARSON-CUNNINGHAM FUNCTION, SCHLÖMILCH'S FUNCTION, SONINE POLYNOMIAL

References

Abramowitz, M. and Stegun, C. A. (Eds.). "Confluent Hypergeometric Functions." Ch. 13 in *Handbook of Mathematical Functions with Formulas, Graphs, and Mathematical Tables, 9th printing.* New York: Dover, pp. 503–15, 1972.

Iyanaga, S. and Kawada, Y. (Eds.). "Whittaker Functions." Appendix A, Table 19.II in *Encyclopedic Dictionary of Mathematics.* Cambridge, MA: MIT Press, pp. 1469–471, 1980.

Meijer, C. S. "Über die Integraldarstellungen der Whittakerschen Funktion $W_{k,m}(z)$ und der Hankelschen und Besselschen Funktionen." *Nieuw Arch. Wisk.* **18**, 35–7, 1936.

Whittaker, E. T. *Bull. Amer. Math. Soc.* **10**, 125–34, 1904.

Whittaker, E. T. and Watson, G. N. *A Course in Modern Analysis, 4th ed.* Cambridge, England: Cambridge University Press, 1990.

Whittaker-Hill Differential Equation

The second-order ORDINARY DIFFERENTIAL EQUATION

$$y'' + [A + B\cos(2x) + C\cos(4x)]y = 0.$$

See also HILL'S DIFFERENTIAL EQUATION, MATHIEU DIFFERENTIAL EQUATION

References

Urwin, K. M. and Arscott, F. M. "Theory of the Whittaker-Hill Equation." *Proc. Roy. Soc. Edinburgh* **69**, 28–4, 1970.

Zwillinger, D. *Handbook of Differential Equations, 3rd ed.* Boston, MA: Academic Press, p. 128, 1997.

Whole Number

One of the numbers 1, 2, 3, ... (Sloane's A000027), also called the COUNTING NUMBERS or NATURAL NUMBERS. 0 is sometimes included in the list of "whole" numbers (Bourbaki 1968, Halmos 1974), but there seems to be no general agreement. Some authors also interpret "whole number" to mean "a number having FRAC-

TIONAL PART of zero," making the whole numbers equivalent to the integers.

Due to lack of standard terminology, the following terms are recommended in preference to "COUNTING NUMBER," "NATURAL NUMBER," and "whole number."

set	name	symbol
..., $-2, -1, 0, 1, 2, ...$	INTEGERS	$\mathbb{Z}$
$1, 2, 3, 4, ...$	POSITIVE INTEGERS	$\mathbb{Z}_+$
$0, 1, 2, 3, 4, ...$	NONNEGATIVE INTEGERS	$\mathbb{Z}^*$
$0, -1, -2, -3, -4, ...$	NONPOSITIVE INTEGERS	
$-1, -2, -3, -4, ...$	NEGATIVE INTEGERS	$\mathbb{Z}_-$

See also COUNTING NUMBER, FRACTIONAL PART, INTEGER, N, NATURAL NUMBER, Z, Z₊, Z₊, Z*

References

Bourbaki, N. *Elements of Mathematics: Theory of Sets.* Paris, France: Hermann, 1968.
Halmos, P. R. *Naive Set Theory.* New York: Springer-Verlag, 1974.
Sloane, N. J. A. Sequences A000027/M0472 in "An On-Line Version of the Encyclopedia of Integer Sequences." http://www.research.att.com/~njas/sequences/eisonline.html.

Width (Partial Order)

For a PARTIAL ORDER, the size of the longest ANTICHAIN is called the width.

See also ANTICHAIN, LENGTH (PARTIAL ORDER), PARTIAL ORDER

Width (Size)

The width of a box is the horizontal distance from side to side (usually defined to be greater than the DEPTH, the horizontal distance from front to back).

See also DEPTH (SIZE), HEIGHT

References

Eppstein, D. "Width, Diameter, and Geometric Inequalities." http://www.ics.uci.edu/~eppstein/junkyard/diam.html.

Wiedersehen Manifold

The only Wiedersehen manifolds are the standard round spheres, as was established by proof of the BLASCHKE CONJECTURE.

See also BLASCHKE CONJECTURE

Wieferich Prime

A Wieferich prime is a PRIME p which is a solution to the CONGRUENCE equation

$$2^{p-1} \equiv 1 \pmod{p^2}.$$

Note the similarity of this expression to the special case of FERMAT'S LITTLE THEOREM

$$2^{p-1} \equiv 1 \pmod{p},$$

which holds for *all* ODD PRIMES. However, the only Wieferich primes less than 4×10^{12} are $p = 1093$ and 3511 (Lehmer 1981, Crandall 1986, Crandall *et al.* 1997). Interestingly, one less than these numbers have suggestive periodic BINARY representations

$$1092 = 10001000100_2$$

$$3510 = 110110110110_2.$$

A PRIME factor p of a MERSENNE NUMBER $M_q = 2^q - 1$ is a Wieferich prime IFF $p^2 | 2^q - 1$. Therefore, MERSENNE PRIMES are *not* Wieferich primes.

If the first case of FERMAT'S LAST THEOREM is false for exponent p, then p must be a Wieferich prime (Wieferich 1909). If $p | 2^n \pm 1$ with p and n RELATIVELY PRIME, then p is a Wieferich prime IFF p^2 also divides $2^n \pm 1$. The CONJECTURE that there are no three POWERFUL NUMBERS implies that there are infinitely many Wieferich primes (Granville 1986, Vardi 1991). In addition, the ABC CONJECTURE implies that there are at least $C \ln x$ Wieferich primes $\leq x$ for some constant C (Silverman 1988, Vardi 1991).

See also ABC CONJECTURE, FERMAT'S LAST THEOREM, FERMAT QUOTIENT, MERSENNE NUMBER, MIRIMANOFF'S CONGRUENCE, POWERFUL NUMBER

References

Brillhart, J.; Tonascia, J.; and Winberger, P. "On the Fermat Quotient." In *Computers and Number Theory* (Ed. A. O. L. Atkin and B. J. Birch). New York: Academic Press, pp. 213–22, 1971.
Crandall, R. *Projects in Scientific Computation.* New York: Springer-Verlag, 1986.
Crandall, R.; Dilcher, K; and Pomerance, C. "A search for Wieferich and Wilson Primes." *Math. Comput.* **66**, 433–49, 1997.
Granville, A. "Powerful Numbers and Fermat's Last Theorem." *C. R. Math. Rep. Acad. Sci. Canada* **8**, 215–18, 1986.
Lehmer, D. H. "On Fermat's Quotient, Base Two." *Math. Comput.* **36**, 289–90, 1981.
Montgomery, P. "New Solutions of $a^{p-1} \equiv 1 \pmod{p^2}$." *Math. Comput.* **61**, 361–63, 1991.
Ribenboim, P. "Wieferich Primes." §5.3 in *The New Book of Prime Number Records.* New York: Springer-Verlag, pp. 333–46, 1996.
Shanks, D. *Solved and Unsolved Problems in Number Theory, 4th ed.* New York: Chelsea, pp. 116 and 157, 1993.
Silverman, J. "Wieferich's Criterion and the abc Conjecture." *J. Number Th.* **30**, 226–37, 1988.
Vardi, I. "Wieferich." §5.4 in *Computational Recreations in Mathematica.* Reading, MA: Addison-Wesley, pp. 59–2 and 96–03, 1991.
Wieferich, A. "Zum letzten Fermat'schen Theorem." *J. reine angew. Math.* **136**, 293–02, 1909.

Wielandt's Theorem

Let the $n \times n$ MATRIX A satisfy the conditions of the PERRON-FROBENIUS THEOREM and the $n \times n$ MATRIX $C = c_{ij}$ satisfy

$$|c_{ij}| \le a_{ij}$$

for $i, j = 1, 2, ..., n$. Then any EIGENVALUE λ_0 of C satisfies the inequality $|\lambda_0| \le R$ with the equality sign holding only when there exists an $n \times n$ MATRIX $D = \delta_{ij}$ (where δ_{ij} is the KRONECKER DELTA) and

$$C = \frac{\lambda_0}{R} DAD^{-1}.$$

References
Gradshteyn, I. S. and Ryzhik, I. M. *Tables of Integrals, Series, and Products, 6th ed.* San Diego, CA: Academic Press, p. 1121, 2000.

Wiener Filter

An optimal FILTER used for the removal of noise from a signal which is corrupted by the measuring process itself.

See also FILTER

References
Press, W. H.; Flannery, B. P.; Teukolsky, S. A.; and Vetterling, W. T. "Optimal (Wiener) Filtering with the FFT." §13.3 in *Numerical Recipes in FORTRAN: The Art of Scientific Computing, 2nd ed.* Cambridge, England: Cambridge University Press, pp. 539–42, 1992.

Wiener Function

BROWN FUNCTION

Wiener Measure

The probability law on the space of continuous functions g with $g(0) = 0$, induced by the WIENER PROCESS.

See also WIENER PROCESS

References
Karatsas, I. and Shreve, S. *Brownian Motion and Stochastic Calculus, 2nd ed.* New York: Springer-Verlag, 1997.

Wiener Numbers

A sequence of UNCORRELATED NUMBERS α_n developed by Wiener (1926–927). The numbers are constructed by beginning with

$$\{1, -1\},$$

then forming the outer product with $\{1, -1\}$ to obtain

$$\{\{\{1, 1\}, \{1, -1\}\}, \{\{-1, 1\}, \{-1, -1\}\}\}.$$

This row is repeated twice, and its outer product is then taken to give

$$\{\{\{1, 1, 1\}, \{1, 1, -1\}, \{1, -1, 1\}, \{1, -1, -1\}\},$$

$$\{\{-1, 1, 1\}, \{-1, 1, -1\}, \{-1, -1, 1\}, \{-1, -1, -1\}\}\}.$$

This is then repeated four times. The procedure is repeated, and the result repeated eight times, and so on. The sequences from each stage are then concatenated to form the sequence $1, -1, 1, 1, 1, -1, -1, 1, -1, -1, 1, 1, 1, -1, -1, 1, -1, -1, \ldots$.

See also UNCORRELATED NUMBERS

References
Papoulis, A. "The Wiener Numbers." *The Fourier Integral and Its Applications.* New York: McGraw-Hill, pp. 258–59, 1962.
Wiener, N. "The Spectrum of an Array and Its Applications to the Study of the Translation Properties of a Simple Class of Arithmetical Functions." *J. Math. Phys.* **6**, 1926–927.

Wiener Process

A continuous-time stochastic process $W(t)$ for $t \ge 0$ with $W(0) = 0$ and such that the increment $W(t) - W(s)$ is Gaussian with mean 0 and variance $t - s$ for any $0 \le s < t$, and increments for nonoverlapping time intervals are independent. Brownian motion (i.e., random walk with random step sizes) is the most common example of a Wiener process.

See also ITÔ'S LEMMA, RANDOM WALK, WIENER PROCESS

References
Karatsas, I. and Shreve, S. *Brownian Motion and Stochastic Calculus, 2nd ed.* New York: Springer-Verlag, 1997.
Papoulis, A. "Wiener-Lévy Process." §15– in *Probability, Random Variables, and Stochastic Processes, 2nd ed.* New York: McGraw-Hill, pp. 292–93, 1984.

Wiener Space

MALLIAVIN CALCULUS, WIENER MEASURE

Wiener-Khintchine Theorem

Recall the definition of the AUTOCORRELATION function $C(t)$ of a function $E(t)$,

$$C(t) \equiv \int_{-\infty}^{\infty} \bar{E}(\tau) E(t + \tau) \, d\tau. \tag{1}$$

Also recall that the FOURIER TRANSFORM of $E(t)$ is defined by

$$E(\tau) \equiv \int_{-\infty}^{\infty} E_v e^{-2\pi i v t} dv, \tag{2}$$

giving a COMPLEX CONJUGATE of

$$\bar{E}(\tau) \equiv \int_{-\infty}^{\infty} \bar{E}_v e^{2\pi i v t} dv \tag{3}$$

Plugging $\bar{E}(\tau)$ and $E(t + \tau)$ into the AUTOCORRELATION function therefore gives

$$C(t) = \int_{-\infty}^{\infty} \left[\int_{-\infty}^{\infty} \bar{E}_\nu e^{2\pi i \nu \tau} d\nu \right] \left[\int_{-\infty}^{\infty} \bar{E}_\nu e^{-2\pi i \nu'(t+\tau)} d\nu' \right] d\tau$$

$$= \int_{-\infty}^{\infty} \int_{-\infty}^{\infty} \int_{-\infty}^{\infty} \bar{E}_\nu E_{\nu'} e^{-2\pi i \tau(\nu'-\nu)} e^{-2\pi i \nu' t} \, d\tau \, d\nu \, d\nu'$$

$$= \int_{-\infty}^{\infty} \int_{-\infty}^{\infty} \bar{E}_\nu E_{\nu'} \delta(\nu'-\nu) e^{-2\pi i \nu' t} d\nu \, d\nu'$$

$$= \int_{\infty}^{\infty} \bar{E}_\nu E_\nu e^{-2\pi i \nu t} d\nu$$

$$= \int_{-\infty}^{\infty} |E_\nu|^2 e^{-2\pi i \nu t} d\nu$$

$$= \mathscr{F} \left[|E_\nu|^2 \right], \tag{4}$$

so, amazingly, the AUTOCORRELATION is simply given by the FOURIER TRANSFORM of the ABSOLUTE SQUARE of $E(\nu)$,

$$C(t) = \mathscr{F} \left[|E(\nu)|^2 \right]. \tag{5}$$

The Wiener-Khintchine theorem is a special case of the CROSS-CORRELATION THEOREM with $f = g$.

See also AUTOCORRELATION, CROSS-CORRELATION THEOREM, FOURIER TRANSFORM

Wiener-Lee Transform

The integral transform obtained by defining

$$\omega \equiv -\tan\left(\tfrac{1}{2}\delta\right), \tag{1}$$

and writing

$$H(\omega) = R(\omega) + iX(\omega), \tag{2}$$

where $R(\omega)$ and $X(\omega)$ are a HILBERT TRANSFORM pair as

$$H(\omega) = \rho(\delta) - i\chi(\delta) \tag{3}$$

(Papoulis 1962, p. 201).

See also HILBERT TRANSFORM, INTEGRAL TRANSFORM

References

Papoulis, A. "Wiener-Lee Transforms." *The Fourier Integral and Its Applications.* New York: McGraw-Hill, pp. 201–03, 1962.

Wiener-Lévy Process

WIENER PROCESS

Wigner 3j-Symbol

The Wigner 3j-symbols are written

$$\begin{pmatrix} j_1 & j_2 & j_3 \\ m_1 & m_2 & m_3 \end{pmatrix} \tag{1}$$

and are sometimes expressed using the related

CLEBSCH-GORDAN COEFFICIENTS

$$C_{m_1 m_2}^j = (j_1 j_2 m_1 m_2 | j_1 j_2 j m) \tag{2}$$

(Condon and Shortley 1951, pp. 74–5; Wigner 1959, p. 206), or RACAH V-COEFFICIENTS

$$V(j_1 j_2 j; m_1 m_2 m). \tag{3}$$

The allowed values of j_1, j_2, j_3, m_1, m_2, and m_3 are given by the constraints placed on CLEBSCH-GORDAN COEFFICIENTS. The Wigner 3j-symbols are returned by the *Mathematica* function ThreeJSymbol[{j1, m1}, {j2, m2}, {j3, m3}].

Connections among the Wigner 3j, Clebsch-Gordan, and Racah V symbols are given by

$$(j_1 j_2 m_1 m_2 | j_1 j_2 j m)$$
$$= (-1)^{m+j_1-j_2} \sqrt{2j+1} \begin{pmatrix} j_1 & j_2 & j \\ m_1 & m_2 & -m \end{pmatrix} \tag{4}$$

$$(j_1 j_2 m_1 m_2 | j_1 j_2 j m)$$
$$= (-1)^{j+m} \sqrt{2j+1} V(j_1 j_2 j; m_1 m_2 - m) \tag{5}$$

$$V(j_1 j_2 j; m_1 m_2 - m) = (-1)^{-j_1+j_2+j} \begin{pmatrix} j_1 & j_2 & j_1 \\ m_2 & m & m_2 \end{pmatrix}. \tag{6}$$

The Wigner 3j-symbols have the symmetries

$$\begin{pmatrix} j_1 & j_2 & j_1 \\ m_1 & m_2 & m \end{pmatrix} = \begin{pmatrix} j_1 & j & j_1 \\ m_2 & m & m_1 \end{pmatrix}$$

$$= \begin{pmatrix} j & j_2 & j_2 \\ m & m_1 & m_2 \end{pmatrix}$$

$$= (-1)^{j_1+j_2+j} \begin{pmatrix} j_2 & j_1 & j \\ m_2 & m_1 & m \end{pmatrix}$$

$$= (-1)^{j_1+j_2+j} \begin{pmatrix} j_1 & j & j_2 \\ m_1 & m & m_2 \end{pmatrix}$$

$$= (-1)^{j_1+j_2+j} \begin{pmatrix} j & j_2 & j_1 \\ m & m_2 & m_1 \end{pmatrix}$$

$$= (-1)^{j_1+j_2+j} \begin{pmatrix} j & j_2 & j \\ -m_1 & -m_2 & -m \end{pmatrix}. \tag{7}$$

The symbols obey the orthogonality relations

$$\sum_{j,m} (2j+1) \begin{pmatrix} j_1 & j_2 & j \\ m_1 & m_2 & m \end{pmatrix} \begin{pmatrix} j_1 & j_2 & j \\ m_1' & m_2' & m \end{pmatrix}$$
$$= \delta_{m_1 m_1'} \delta_{m_2 m_2'} \tag{8}$$

$$\sum_{m_1, m_2} (2j+1) \begin{pmatrix} j_1 & j_2 & j \\ m_1 & m_2 & m \end{pmatrix} \begin{pmatrix} j_1 & j_2 & j' \\ m_1 & m_2 & m' \end{pmatrix}$$
$$= \delta_{jj'} \delta_{mm'}, \tag{9}$$

where δ_{ij} is the KRONECKER DELTA.

General formulas are very complicated, but some specific cases are

$$\begin{pmatrix} j_1 & j_2 & j_1+j_2 \\ m_1 & m_2 & -m_1-m_2 \end{pmatrix} = (-1)^{j_1-j_2+m_1+m_2}$$

$$\times \left[\frac{(2j_1)!(2j_2)!}{(2j_1+2j_2+1)(j_1+m_1)} \right.$$

$$\times \left. \frac{(j_1+j_2+m_1+m_2)!(j_1+j_2-m_1-m_2)!}{(j_1-m_1)(j_2+m_2)(j_2-m_2)!} \right]^{1/2} \quad (10)$$

$$\begin{pmatrix} j_1 & j_2 & j \\ j_1 & -j_1 & -m \end{pmatrix} = (-1)^{-j_1+j_2+m}$$

$$\times \left[\frac{(2j_1)!(-j_1+j_2+j)!}{(j_1+j_2+j+1)!(j_1-j_2+j)!} \right.$$

$$\times \left. \frac{(j_1+j_2+m_1+m_2)!(j_1+j_2-m_1-m_2)!}{(j_1+j_2-j)!(j_1+j_2-j)!(-j_1+j_2-m)!(j+m)!} \right] \quad (11)$$

$$\begin{pmatrix} j_1 & j_2 & j \\ 0 & 0 & 0 \end{pmatrix}$$

$$= \begin{cases} (-1)^g \sqrt{\dfrac{(2g-2j_1)(2g-2j_2)!(2g-2j)!}{(2g+1)!}} \\ \quad \times \dfrac{g!}{(g-j_1)!(g-j_2)!(g-j)!} \\ \qquad \text{if } J = 2g \\ 0 \\ \qquad \text{if } J = 2g+1, \end{cases} \quad (12)$$

for $J \equiv j_1+j_2+j$.

For SPHERICAL HARMONICS $Y_l^m(\theta, \phi)$,

$$Y_{l_1}^{m_1}(\theta, \phi) Y_{l_2}^{m_2}(\theta, \phi)$$

$$= \sum_{l,m} \sqrt{\frac{(2l_1+1)(2l_2+1)(2l+1)}{4\pi}}$$

$$\times \begin{pmatrix} l_1 & l_2 & l \\ m_1 & m_2 & m \end{pmatrix} \bar{Y}_l^m(\theta, \phi) \begin{pmatrix} l_1 & l_2 & l \\ 0 & 0 & 0 \end{pmatrix}. \quad (13)$$

For values of l_3 obeying the TRIANGLE CONDITION $\Delta(l_1 l_2 l_3)$,

$$\int Y_{l_1}^{m_1}(\theta, \phi) Y_{l_2}^{m_2}(\theta, \phi) Y_{l_3}^{m_3}(\theta, \phi) \sin\theta \, d\theta \, d\phi$$

$$= \sqrt{\frac{(2l_1+1)(2l_2+1)(2l_3+1)}{4\pi}} \begin{pmatrix} l_1 & l_2 & l_3 \\ 0 & 0 & 0 \end{pmatrix}$$

$$\times \begin{pmatrix} l_1 & l_2 & l_3 \\ m_1 & m_2 & m_3 \end{pmatrix} \quad (14)$$

and

$$\frac{1}{2} \int Pl_1(\cos\theta) Pl_2(\cos\theta) \sin\theta \, d\theta = \begin{pmatrix} l_1 & l_2 & l_3 \\ 0 & 0 & 0 \end{pmatrix}^2. \quad (15)$$

See also CLEBSCH-GORDAN COEFFICIENT, RACAH *V*-

COEFFICIENT, RACAH *W*-COEFFICIENT, WIGNER 6*J*-SYMBOL, WIGNER 9*J*-SYMBOL

References

Abramowitz, M. and Stegun, C. A. (Eds.). "Vector-Addition Coefficients." §27.9 in *Handbook of Mathematical Functions with Formulas, Graphs, and Mathematical Tables, 9th printing.* New York: Dover, pp. 1006–010, 1972.

Condon, E. U. and Shortley, G. *The Theory of Atomic Spectra.* Cambridge, England: Cambridge University Press, 1951.

de Shalit, A. and Talmi, I. *Nuclear Shell Theory.* New York: Academic Press, 1963.

Gordy, W. and Cook, R. L. *Microwave Molecular Spectra, 3rd ed.* New York: Wiley, pp. 804–11, 1984.

Messiah, A. "Clebsch-Gordan (C.-G.) Coefficients and '3*j*' Symbols." Appendix C.I in *Quantum Mechanics, Vol. 2.* Amsterdam, Netherlands: North-Holland, pp. 1054–060, 1962.

Rose, M. E. *Elementary Theory of Angular Momentum.* New York: Dover, 1995.

Rotenberg, M.; Bivens, R.; Metropolis, N.; and Wooten, J. K. *The 3j and 6j Symbols.* Cambridge, MA: MIT Press, 1959.

Shore, B. W. and Menzel, D. H. *Principles of Atomic Spectra.* New York: Wiley, pp. 275–76, 1968.

Sobel'man, I. I. "Angular Momenta." Ch. 4 in *Atomic Spectra and Radiative Transitions, 2nd ed.* Berlin: Springer-Verlag, 1992.

Wigner, E. P. *Group Theory and Its Application to the Quantum Mechanics of Atomic Spectra, expanded and improved ed.* New York: Academic Press, 1959.

Wigner 6j-Symbol

A generalization of CLEBSCH-GORDAN COEFFICIENTS and WIGNER 3*J*-SYMBOL which arises in the coupling of three angular momenta. The Wigner 6*j*-symbols are returned by the *Mathematica* function SixJSymbol[{*j1*, *j2*, *j3*}, {*j4*, *j5*, *j6*}].

Let tensor operators $T^{(k)}$ and $U^{(k)}$ act, respectively, on subsystems 1 and 2 of a system, with subsystem 1 characterized by angular momentum $\mathbf{j}_1$ and subsystem 2 by the angular momentum $\mathbf{j}_2$. Then the matrix elements of the scalar product of these two tensor operators in the coupled basis $\mathbf{J} = \mathbf{j}_1 + \mathbf{j}_2$ are given by

$$\left(\tau_1' j_1' \tau_2' j_2' J'M' \big| T^{(k)} \cdot U^{(k)} \big| \tau_1 j_1 \tau_2 j_2 JM \right)$$

$$= \delta_{JJ'} \delta_{MM'} (-1)^{j_1+j_2+J} \begin{Bmatrix} J & j_2' & j_1' \\ k & j_1 & j_2 \end{Bmatrix}$$

$$\times \left(\tau_1' j_1' \| T^{(k)} \| \tau_1 j_1 \right) \left(\tau_1' j_2' \| U^{(k)} \| \tau_2 j_2 \right), \quad (1)$$

where

$$\begin{Bmatrix} J & j_2' & j_1' \\ k & j_1 & j_2 \end{Bmatrix}$$

is the Wigner 6*j*-symbol and τ_1 and τ_2 represent additional pertinent quantum numbers characterizing subsystems 1 and 2 (Gordy and Cook 1984).

Edmonds (1968) gives analytic forms of the 6*j*-symbol for simple cases, and Shore and Menzel (1968) and Gordy and Cook (1984) give

$$\begin{Bmatrix} a & b & c \\ 0 & c & b \end{Bmatrix} = \frac{(-1)^s}{\sqrt{(2b+1)(2c+1)}} \qquad (2)$$

$$\begin{Bmatrix} a & b & c \\ 1 & c & b \end{Bmatrix}$$

$$= \frac{2(-1)^{s+1}X}{\sqrt{2b(2b+1)(2b+2)2c(2c+1)(2c+2)}} \qquad (3)$$

$$\begin{Bmatrix} a & b & c \\ 2 & c & b \end{Bmatrix} =$$

$$\frac{2(-1)^s[3X(X-1) - 4b(b+1)c(c+1)]}{\sqrt{(2b-1)2b(2b+1)(2b+2)(2b+3)(2c-1)2c(2c+1)(2c+2)(2c+3)}},$$

$$(4)$$

where

$$s \equiv a + b + c \qquad (5)$$

$$X \equiv b(b+1) + c(c+1) - a(a+1). \qquad (6)$$

See also CLEBSCH-GORDAN COEFFICIENT, RACAH V-COEFFICIENT, RACAH W-COEFFICIENT, WIGNER 3J-SYMBOL, WIGNER 9J-SYMBOL

References

Carter, J. S.; Flath, D. E.; and Saito, M. *The Classical and Quantum 6j-Symbols*. Princeton, NJ: Princeton University Press, 1995.

Edmonds, A. R. *Angular Momentum in Quantum Mechanics, 2nd ed., rev. printing.* Princeton, NJ: Princeton University Press, 1968.

Gordy, W. and Cook, R. L. *Microwave Molecular Spectra, 3rd ed.* New York: Wiley, pp. 807–09, 1984.

Messiah, A. "Racah Coefficients and '6j' Symbols." Appendix C.II in *Quantum Mechanics, Vol. 2.* Amsterdam, Netherlands: North-Holland, pp. 567–69 and 1061–066, 1962.

Rotenberg, M.; Bivens, R.; Metropolis, N.; and Wooten, J. K. *The 3j and 6j Symbols.* Cambridge, MA: MIT Press, 1959.

Shore, B. W. and Menzel, D. H. *Principles of Atomic Spectra.* New York: Wiley, pp. 279–84, 1968.

Wigner 9j-Symbol

A generalization of CLEBSCH-GORDAN COEFFICIENTS and WIGNER 3J- and WIGNER 6J-SYMBOLS which arises in the coupling of four angular momenta and can be written in terms of the WIGNER 3J- and WIGNER 6J-SYMBOLS. Let tensor operators $T^{(k_1)}$ and $U^{(k_2)}$ act, respectively, on subsystems 1 and 2. Then the reduced matrix element of the product $T^{(k_1)} \times U^{(k_2)}$ of these two irreducible operators in the coupled representation is given in terms of the reduced matrix elements of the individual operators in the uncoupled representation by

$$\left(\tau'\tau_1'j_1'\tau_2'j_2'J' \| [T^{(k_1)} \times U^{(k_2)}]^{(k)} \| \tau\tau_1 j_1\tau_2 j_2 J \right)$$

$$= \sqrt{(2J+1)(2J'+1)(2k+1)} \sum_{\tau''} \begin{Bmatrix} j_1' & j_1 & k_1 \\ j_2' & j_2 & k_2 \\ J' & J & k \end{Bmatrix}$$

$$\times (\tau'\tau_1'j_1' \| T^{(k_1)} \| \tau''\tau_1 j_1)(\tau''\tau_2'j_2' \| U^{(k_2)} \| \tau\tau_2 j_2), \qquad (1)$$

where

$$\begin{Bmatrix} j_1' & j_1 & k_1 \\ j_2' & j_2 & k_2 \\ J' & J & k \end{Bmatrix}$$

is a Wigner 9j-symbol (Gordy and Cook 1984). Shore and Menzel (1968) give the explicit formulas

$$\begin{Bmatrix} a & b & C \\ d & e & F \\ G & H & J \end{Bmatrix} = \sum_x (-1)^{2x}(2x+1)$$

$$\times \begin{Bmatrix} a & b & C \\ F & J & x \end{Bmatrix} \begin{Bmatrix} d & e & F \\ b & x & H \end{Bmatrix} \begin{Bmatrix} G & H & J \\ x & a & d \end{Bmatrix} \qquad (2)$$

$$\begin{Bmatrix} a & b & J \\ c & d & J \\ K & K & 0 \end{Bmatrix} = \frac{(-1)^{b+c+J+K}}{\sqrt{(2J+1)(2K+1)}} \begin{Bmatrix} a & b & J \\ d & c & K \end{Bmatrix} \qquad (3)$$

$$\begin{Bmatrix} S & S & 1 \\ L & L & 2 \\ J & J & 1 \end{Bmatrix} = \frac{\begin{Bmatrix} S & L & J \\ L & S & 1 \end{Bmatrix} \begin{Bmatrix} J & L & S \\ L & J & 1 \end{Bmatrix}}{5\begin{Bmatrix} 2 & L & L \\ L & 1 & 1 \end{Bmatrix}}$$

$$+ \frac{(-1)^{S+L+J+1}}{15(2L+1)} \frac{\begin{Bmatrix} S & L & J \\ L & S & 1 \end{Bmatrix}}{\begin{Bmatrix} 2 & L & L \\ L & 1 & 1 \end{Bmatrix}}. \qquad (4)$$

See also CLEBSCH-GORDAN COEFFICIENT, RACAH V-COEFFICIENT, RACAH W-COEFFICIENT, WIGNER 3J-SYMBOL, WIGNER 6J-SYMBOL

References

Gordy, W. and Cook, R. L. *Microwave Molecular Spectra, 3rd ed.* New York: Wiley, pp. 807–09, 1984.

Messiah, A. "'9j' Symbols." Appendix C.III in *Quantum Mechanics, Vol. 2.* Amsterdam, Netherlands: North-Holland, pp. 567–69 and 1066–068, 1962.

Shore, B. W. and Menzel, D. H. *Principles of Atomic Spectra.* New York: Wiley, pp. 279–84, 1968.

Wigner-Eckart Theorem

A theorem of fundamental importance in spectroscopy and angular momentum theory which provides both (1) an explicit form for the dependence of all matrix elements of irreducible tensors on the projection quantum numbers and (2) a formal expression of the conservation laws of angular momentum (Rose 1995).

The theorem states that the dependence of the matrix element $(j'm'|T_{LM}|jm)$ on the projection quantum numbers is entirely contained in the WIGNER 3J-SYMBOL (or, equivalently, the CLEBSCH-GORDAN COEFFICIENT), given by

$$(j'm'|T_{LM}|jm) = C(jLj'; mMm')(j'\|T_L\|j),$$

where $C(jLj'; mMm')$ is a Clebsch-Gordan coeffi-cient and T_{LM} is a set of tensor operators (Rose 1995, p. 85).

See also Clebsch-Gordan Coefficient, Wigner 3J-Symbol

References

Cohen-Tannoudji, C.; Diu, B.; and Laloë, F. "Vector Opera-tors: The Wigner-Eckart Theorem." Complement D_X in *Quantum Mechanics, Vol. 2.* New York: Wiley, pp. 1048–058, 1977.

Eckart, C. "The Application of Group Theory to the Quan-tum Dynamics of Monatomic Systems." *Rev. Mod. Phys.* **2**, 305–80, 1930.

Edmonds, A. R. *Angular Momentum in Quantum Me-chanics, 2nd ed., rev. printing.* Princeton, NJ: Princeton University Press, 1968.

Gordy, W. and Cook, R. L. *Microwave Molecular Spectra, 3rd ed.* New York: Wiley, p. 807, 1984.

Messiah, A. "Representation of Irreducible Tensor Opera-tors: Wigner-Eckart Theorem." §32 in *Quantum Me-chanics, Vol. 2.* Amsterdam, Netherlands: North-Holland, pp. 573–75, 1962.

Rose, M. E. "The Wigner-Eckart Theorem." §19 in *Elemen-tary Theory of Angular Momentum.* New York: Dover, pp. 85–4, 1995.

Shore, B. W. and Menzel, D. H. "Tensor Operators and the Wigner-Eckart Theorem." §6.4 in *Principles of Atomic Spectra.* New York: Wiley, pp. 285–94, 1968.

Wigner, E. P. "Einige Folgerungen aus der Schrödin-gerschen Theorie für die Termstrukturen." *Z. Physik* **43**, 624–52, 1927.

Wigner, E. P. *Group Theory and Its Application to the Quantum Mechanics of Atomic Spectra, expanded and improved ed.* New York: Academic Press, 1959.

Wybourne, B. G. *Symmetry Principles and Atomic Spectro-scopy.* New York: Wiley, pp. 89 and 93–6, 1970.

Wilbraham-Gibbs Constant

N.B. A detailed online essay by S. Finch was the starting point for this entry.

Let a piecewise smooth function f with only finitely many discontinuities (which are all jumps) be defined on $[-\pi, \pi]$ with Fourier series

$$a_k = \frac{1}{\pi} \int_{-\pi}^{\pi} f(t) \cos(kt)\, dt \qquad (1)$$

$$b_k = \frac{1}{\pi} \int_{-\pi}^{\pi} f(t) \sin(kt)\, dt, \qquad (2)$$

$$S_n(f, x) = \frac{1}{2} a_0 + \left\{ \sum_{k=1}^{n} [a_k \cos(kx) + b_k \sin(kx)] \right\}. \qquad (3)$$

Let a discontinuity be at $x = c$, with

$$\lim_{x \to c^-} f(x) > \lim_{x \to c^+} f(x), \qquad (4)$$

so

$$D \equiv \left[\lim_{x \to c^-} f(x) \right] - \left[\lim_{x \to c^+} f(x) \right] > 0. \qquad (5)$$

Define

$$\phi(c) = \frac{1}{2} \left[\lim_{x \to c^-} f(x) + \lim_{x \to c^+} f(x) \right], \qquad (6)$$

and let $x = x_n < c$ be the first local minimum and $x = \xi_n > c$ the first local maximum of $S_n(f, x)$ on either side of x_n. Then

$$\lim_{n \to \infty} S_n(f, x_n) = \phi(c) + \frac{D}{\pi} G', \qquad (7)$$

$$\lim_{n \to \infty} S_n(f, \xi_n) = \phi(c) - \frac{D}{\pi} G', \qquad (8)$$

where

$$G' \equiv \int_0^{\pi} \operatorname{sinc} \theta\, d\theta = 1.851937052\ldots \qquad (9)$$

Here, $\operatorname{sinc} x \equiv \sin x / x$ is the sinc function. The Fourier series of $y = x$ therefore does not converge to $-\pi$ and π at the ends, but to $-2G'$ and $2G'$. This phenomenon was observed by Wilbraham (1848) and Gibbs (1899). Although Wilbraham was the first to note the phenomenon, the constant G' is frequently (and unfairly) credited to Gibbs and known as the Gibbs constant. A related constant sometimes also called the Gibbs constant is

$$G \equiv \frac{2}{\pi} G' = \frac{2}{\pi} \int_0^{\pi} \operatorname{sinc} x\, dx$$

$$= 1.17897974447216727\ldots \qquad (10)$$

(Le Lionnais 1983).

References

Carslaw, H. S. *Introduction to the Theory of Fourier's Series and Integrals, 3rd ed.* New York: Dover, 1930.

Finch, S. "Favorite Mathematical Constants." http://www.mathsoft.com/asolve/constant/gibbs/gibbs.html.

Le Lionnais, F. *Les nombres remarquables.* Paris: Hermann, pp. 36 and 43, 1983.

Zygmund, A. G. *Trigonometric Series 1, 2nd ed.* Cambridge, England: Cambridge University Press, 1959.

Wilcoxon Rank Sum Test

A nonparametric alternative to the two-sample t-test.

See also Paired T-Test, Parametric Test

Wilcoxon Signed Rank Test

A nonparametric alternative to the Paired T-Test which is similar to the Fisher sign test. This test assumes that there is information in the magnitudes of the differences between paired observations, as well as the signs. Take the paired observations, calculate the differences, and rank them from smal-lest to largest by absolute value. Add all the ranks associated with positive differences, giving the T_+ statistic. Finally, the P-value associated with this statistic is found from an appropriate table. The Wilcoxon test is an R-estimate.

See also FISHER SIGN TEST, HYPOTHESIS TESTING, PAIRED T-TEST, PARAMETRIC TEST

Wild Knot

A KNOT which is not a TAME KNOT.

See also TAME KNOT

References

Milnor, J. "Most Knots are Wild." *Fund. Math.* **54**, 335–38, 1964.

Wild Point

For any point P on the boundary of an ordinary BALL, find a NEIGHBORHOOD of P in which the intersection with the BALL's boundary cuts the NEIGHBORHOOD into two parts, each HOMEOMORPHIC to a BALL. A wild point is a point on the boundary that has no such NEIGHBORHOOD.

See also BALL, HOMEOMORPHIC, NEIGHBORHOOD

Wilf Class

Two sets T_1 and T_2 belong to the same Wilf class if $|S_n(T_1)| = |S_n(T_2)|$ for all n, where $S_n(T)$ denotes the set of permutations on $\{1, \ldots, n\}$ that AVOID the pattern T. Two sets having the same Wilf class are said to be WILF EQUIVALENT.

See also AVOIDED PATTERN, WILF EQUIVALENT, PERMUTATION PATTERN

References

Mansour, T. Permutations Avoiding a Pattern from S_k and at Least Two Patterns from S_3. 31 Jul 2000. http://xxx.lanl.gov/abs/math.CO/0007194/.

Wilf Equivalent

Two sets T_1 and T_2 are called Wilf equivalent if they belong to the same WILF CLASS.

See also WILF CLASS, PERMUTATION PATTERN

References

Mansour, T. Permutations Avoiding a Pattern from S_k and at Least Two Patterns from S_3. 31 Jul 2000. http://xxx.lanl.gov/abs/math.CO/0007194/.

Wilf-Zeilberger Pair

A pair of CLOSED FORM functions (F, G) is said to be a Wilf-Zeilberger pair if

$$F(n+1, k) - F(n, k) = G(n, k+1) - G(n, k). \tag{1}$$

The Wilf-Zeilberger formalism provides succinct proofs of *known* identities and allows new identities to be discovered whenever it succeeds in finding a proof certificate for a known identity. However, if the starting point is an unknown hypergeometric sum, then the Wilf-Zeilberger method cannot discover a closed form solution, while ZEILBERGER'S ALGORITHM can.

Wilf-Zeilberger pairs are very useful in proving HYPERGEOMETRIC IDENTITIES OF THE FORM

$$\sum_k t(n, k) = \mathrm{rhs}(n) \tag{2}$$

for which the SUMMAND $t(n, k)$ vanishes for all k outside some finite interval. Now divide by the right-hand side to obtain

$$\sum_k F(n, k) = 1, \tag{3}$$

where

$$F(n, k) \equiv \frac{t(n, k)}{\mathrm{rhs}(n)}. \tag{4}$$

Now use a RATIONAL FUNCTION $R(n, k)$ provided by ZEILBERGER'S ALGORITHM, define

$$G(n, k) \equiv R(n, k) F(n, k). \tag{5}$$

The identity (1) then results. Summing the relation over all integers then telescopes the right side to 0, giving

$$\sum_k F(n+1, k) = \sum_k F(n, k). \tag{6}$$

Therefore, $\sum_k F(n, k)$ is independent of n, and so must be a constant. If F is properly normalized, then it will be true that $\sum_k F(0, k) = 1$.

For example, consider the BINOMIAL COEFFICIENT identity

$$\sum_{k=0}^{n} \binom{n}{k} = 2^n, \tag{7}$$

the function $R(n, k)$ returned by ZEILBERGER'S ALGORITHM is

$$R(n, k) = \frac{k}{2(k - n - 1)}. \tag{8}$$

Therefore,

$$F(n, k) = \binom{n}{k} 2^{-n} \tag{9}$$

and

$$G(n, k) \equiv R(n, k) F(n, k) = \frac{k}{2(k - n - 1)} \binom{n}{k} 2^{-n}$$

$$= -\frac{k n! 2^{-n}}{2(n + 1 - k)! k! (n - k)!} = -\binom{n}{k-1} 2^{-n-1}. \tag{10}$$

Taking

$$F(n + 1, k) - F(n, k) = G(n, k + 1) - G(n, k) \quad (11)$$

then gives the alleged identity

$$\binom{n+1}{k} 2^{-n-1} - \binom{n}{k} 2^{-n}$$

$$= -\binom{n}{k} 2^{-n-1} + \binom{n}{k-1} 2^{-n-1}? \quad (12)$$

Expanding and evaluating shows that the identity does actually hold, and it can also be verified that

$$F(0, k) = \binom{0}{k} = \begin{cases} 1 & \text{for } k = 0 \\ 0 & \text{otherwise,} \end{cases} \quad (13)$$

so $\Sigma_k F(0, k) = 1$ (Petkovsek *et al.* 1996, pp. 25–7).

For any Wilf-Zeilberger pair (*F*, *G*),

$$\sum_{n=0}^{\infty} G(n, 0) = \sum_{n=1}^{\infty} [F(n, n-1) + G(n-1, n-1)] \quad (14)$$

whenever either side converges (Zeilberger 1993). In addition,

$$\sum_{n=0}^{\infty} G(n, 0) = \sum_{n=0}^{\infty} \left[F(s(n+1), n) + \sum_{i=0}^{s-1} G(sn + i, n) \right]$$

$$- \lim_{n \to \infty} \sum_{k=0}^{n-1} F(sn, k), \quad (15)$$

$$\sum_{k=0}^{\infty} F(0, k) = \sum_{n=0}^{\infty} G(n, 0) - \lim_{k \to \infty} \sum_{n=0}^{\infty} G(n, k), \quad (16)$$

and

$$\sum_{n=0}^{\infty} G(n, 0) = \sum_{n=0}^{\infty} \left[\sum_{n=0}^{t-1} F(s(n+1), tn + j) \right.$$

$$\left. + \sum_{n=0}^{s-1} G(sn + i, tn) \right] - \lim_{n \to \infty} \sum_{k=0}^{n-1} F_{s,t}(n, k), \quad (17)$$

where

$$F_{s,t}(n, k) = \sum_{j=0}^{t-1} F(sn, tk + j) \quad (18)$$

$$G_{s,t}(n, k) = \sum_{i=0}^{s-1} G(sn + i, tk) \quad (19)$$

(Amdeberhan and Zeilberger 1997). The latter identity has been used to compute APÉRY'S CONSTANT to a large number of decimal places (Wedeniwski).

See also APÉRY'S CONSTANT, CONVERGENCE IMPROVEMENT, GOSPER'S ALGORITHM, SISTER CELINE'S METHOD, ZEILBERGER'S ALGORITHM

References

Amdeberhan, T. and Zeilberger, D. "Hypergeometric Series Acceleration via the WZ Method." *Electronic J. Combina-*

torics **4**, No. 2, R3, 1–, 1997. http://www.combinatorics.org/Volume_4/wilftoc.html#R03. Also available at http://www.math.temple.edu/~zeilberg/mamarim/mamarimhtml/accel.html.
Cipra, B. A. "How the Grinch Stole Mathematics." *Science* **245**, 595, 1989.
Koepf, W. "Algorithms for *m*-fold Hypergeometric Summation." *J. Symb. Comput.* **20**, 399–17, 1995.
Koepf, W. "The Wilf-Zeilberger Method." Ch. 6 in *Hypergeometric Summation: An Algorithmic Approach to Summation and Special Function Identities.* Braunschweig, Germany: Vieweg, pp. 80–2, 1998.
Petkovsek, M.; Wilf, H. S.; and Zeilberger, D. "The WZ Phenomenon." Ch. 7 in *A = B.* Wellesley, MA: A. K. Peters, pp. 121–40, 1996.
Wilf, H. S. and Zeilberger, D. "Rational Functions Certify Combinatorial Identities." *J. Amer. Math. Soc.* **3**, 147–58, 1990.
Zeilberger, D. "The Method of Creative Telescoping." *J. Symb. Comput.* **11**, 195–04, 1991.
Zeilberger, D. "Closed Form (Pun Intended!)." *Contemporary Math.* **143**, 579–07, 1993.

Wilkie's Theorem

Let $\phi(x_1, \ldots, x_m)$ be an $\mathscr{L}_{\exp}$ formula, where $\mathscr{L}_{\exp} \equiv \mathscr{L} \cup \{e^x\}$ and $\mathscr{L}$ is the language of ordered rings $\mathscr{L} = \{+, -, \cdot, <, 0, 1\}$. Then there exist $n \geq m$ and $f_1, \ldots, f_s \in \mathbb{Z}[x_1, \ldots x_n, e^{x_1}, \ldots e^{x_n}]$ such that $\phi(x_1, \ldots, x_n)$ is equivalent to

$$\exists x_{m+1} \cdots \exists x_n f_1(x_1, \ldots, x_n, e^{x_1}, \ldots, e^{x_n})$$

$$= \ldots = f_s(x_1, \ldots, x_n, e^{x_1}, \ldots, e^{x_n}) = 0$$

(Marker 1996, Wilkie 1996). In other words, every formula is equivalent to an existential formula and every definable set is the projection of an exponential variety (Marker 1996).

References

Marker, D. "Model Theory and Exponentiation." *Not. Amer. Math. Soc.* **43**, 753–59, 1996.
Wilkie, A. J. "Model Completeness Results for Expansions of the Ordered Field of Real Numbers by Restricted Pfaffian Functions and the Exponential Function." *J. Amer. Math. Soc.* **9**, 1051–094, 1996.

Williams p + 1 Factorization Method

A variant of the POLLARD *P*-1 FACTORIZATION METHOD which uses LUCAS SEQUENCES to achieve rapid factorization if some factor *p* of *N* has a decomposition of $p + 1$ in small PRIME FACTORS.

See also LUCAS SEQUENCE, POLLARD P-1 FACTORIZATION METHOD, PRIME FACTORIZATION ALGORITHMS

References

Riesel, H. *Prime Numbers and Computer Methods for Factorization*, 2nd ed. Boston, MA: Birkhäuser, p. 177, 1994.
Williams, H. C. "A *p* + 1 Method of Factoring." *Math. Comput.* **39**, 225–34, 1982.

Wilson Plug

A 3-D surface with constant VECTOR FIELD on its boundary which traps at least one trajectory which enters it.

See also VECTOR FIELD

Wilson Polynomial

The orthogonal polynomial defined by

$$p_n(x;a,b,c,d) = (a+b)_n(a+c)_n(a+d)_n$$

$$\times {}_4F_3\left(\begin{array}{c} -n, a+b+c+d+n-1, a-x, a+x \\ a+b, a+c, a+d \end{array}; 1\right).$$

The first few are

$$p_0(x;a,b,c,d) = 1$$

$$p_1(x;a,b,c,d)$$
$$= abc + abd + acd + bcd + (a+b+c+d)x^2.$$

The Wilson polynomials obey the identity

$$p_n(x;a,b,c,d) = p_n(x;b,a,c,d).$$

References

Koekoek, R. and Swarttouw, R. F. "Wilson." §1.1 in *The Askey-Scheme of Hypergeometric Orthogonal Polynomials and its q-Analogue.* Delft, Netherlands: Technische Universiteit Delft, Faculty of Technical Mathematics and Informatics Report 98–7, pp. 24–6, 1998. ftp://www.twi.tudelft.nl/publications/tech-reports/1998/DUT-TWI-98-7.ps.gz.
Koepf, W. *Hypergeometric Summation: An Algorithmic Approach to Summation and Special Function Identities.* Braunschweig, Germany: Vieweg, p. 116, 1998.
Wilson, J. A. "Some Hypergeometric Orthogonal Polynomials." *SIAM J. Math. Anal.* **11**, 690–01, 1980.

Wilson Prime

A PRIME satisfying

$$W(p) \equiv 0 \pmod{p},$$

where $W(p)$ is the WILSON QUOTIENT, or equivalently,

$$(p-1)! \equiv -1 \pmod{p^2}.$$

5, 13, and 563 (Sloane's A007540) are the only Wilson primes less than 5×10^8 (Crandall *et al.* 1997).

See also BROWN NUMBERS

References

Crandall, R.; Dilcher, K; and Pomerance, C. "A search for Wieferich and Wilson Primes." *Math. Comput.* **66**, 433–49, 1997.
Gonter, R. H. and Kundert, E. G. "All Numbers Up to 18,876,041 Have Been Tested without Finding a New Wilson Prime." Preprint, 1994.
Le Lionnais, F. *Les nombres remarquables.* Paris: Hermann, p. 56, 1983.

Ribenboim, P. "Wilson Primes." §5.4 in *The New Book of Prime Number Records.* New York: Springer-Verlag, pp. 346–50, 1996.
Sloane, N. J. A. Sequences A007540/M3838 in "An On-Line Version of the Encyclopedia of Integer Sequences." http://www.research.att.com/~njas/sequences/eisonline.html.
Vardi, I. *Computational Recreations in Mathematica.* Reading, MA: Addison-Wesley, p. 73, 1991.

Wilson Quotient

$$W(p) \equiv \frac{(p-1)! - 1}{p}.$$

References

Crandall, R.; Dilcher, K; and Pomerance, C. "A search for Wieferich and Wilson Primes." *Math. Comput.* **66**, 433–49, 1997.
Lehmer, E. "On Congruences Involving Bernoulli Numbers and the Quotients of Fermat and Wilson." *Ann. Math.* **39**, 350–60, 1938.

Wilson's Primality Test

WILSON'S THEOREM

Wilson's Theorem

IFF p is a PRIME, then $(p-1)! + 1$ is a multiple of p, that is

$$(p-1)! \equiv -1 \pmod{p}.$$

This theorem was proposed by John Wilson in 1770 (although it was previously known to Leibniz) and proved by Lagrange in 1773. Unlike FERMAT'S LITTLE THEOREM, Wilson's theorem is both NECESSARY and SUFFICIENT for primality. For a COMPOSITE NUMBER, $(n-1)! \equiv 0 \pmod{n}$ except when $n = 4$.

See also FERMAT'S LITTLE THEOREM, WILSON'S THEOREM COROLLARY, WILSON'S THEOREM (GAUSS'S GENERALIZATION)

References

Ball, W. W. R. and Coxeter, H. S. M. *Mathematical Recreations and Essays, 13th ed.* New York: Dover, p. 61, 1987.
Conway, J. H. and Guy, R. K. *The Book of Numbers.* New York: Springer-Verlag, pp. 142–43 and 168–69, 1996.
Hilton, P.; Holton, D.; and Pedersen, J. *Mathematical Reflections in a Room with Many Mirrors.* New York: Springer-Verlag, pp. 41–2, 1997.
Nagell, T. "Wilson's Theorem and Its Generalizations." *Introduction to Number Theory.* New York: Wiley, pp. 99–01, 1951.
Ore, Ø. *Number Theory and Its History.* New York: Dover, pp. 259–61, 1988.
Séroul, R. "Wilson's Theorem." §2.9 in *Programming for Mathematicians.* Berlin: Springer-Verlag, pp. 16–7, 2000.
Shanks, D. *Solved and Unsolved Problems in Number Theory, 4th ed.* New York: Chelsea, pp. 37–8, 1993.

Wilson's Theorem (Gauss's Generalization)

Let $P(n)$ be the product of INTEGERS that are less than or equal to and RELATIVELY PRIME to an integer n. Then

$$P(n) \equiv \prod_{\substack{k=2 \\ k|n}}^{n} k = \begin{cases} -1 \pmod{n} & \text{for } n = 4, p^{\alpha}, 2p^{\alpha} \\ 1 \pmod{n} & \text{otherwise.} \end{cases}$$

When $m = 2$, this reduces to $P \equiv 1 \pmod{2}$ which is equivalent to $P \equiv -1 \pmod{2}$.

See also WILSON'S THEOREM, WILSON'S THEOREM COROLLARY

Wilson's Theorem Corollary

Iff a PRIME p is OF THE FORM $4x + 1$, then

$$[(2x)!]^2 \equiv -1 \pmod{p}.$$

Wimp Transform

The INTEGRAL TRANSFORM defined by

$$(K\phi)(x)$$
$$= \int_{-\infty}^{\infty} G_{p+2,q}^{m,n+2}\left(t \left| \begin{matrix} 1-v+ix, 1-v-ix, (a_p) \\ (b_p) \end{matrix} \right. \right) \phi(t)\, dt,$$

where $G_{c,d}^{a,b}$ is MEIJER'S G-FUNCTION.

References

Samko, S. G.; Kilbas, A. A.; and Marichev, O. I. *Fractional Integrals and Derivatives.* Yverdon, Switzerland: Gordon and Breach, p. 24, 1993.

Winding Number (Contour)

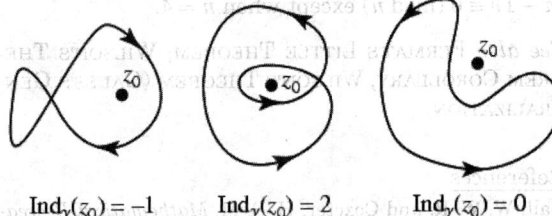

$$\text{Ind}_\gamma(z_0) = -1 \qquad \text{Ind}_\gamma(z_0) = 2 \qquad \text{Ind}_\gamma(z_0) = 0$$

The winding number of a CONTOUR γ about a point z_0, denoted $n(\gamma, z_0)$, is defined by

$$n(\gamma, a) = \frac{1}{2\pi i} \oint_\gamma \frac{dz}{z - z_0}$$

and gives the number of times γ curve passes around a point. The winding number is also called the index, and denoted $\text{Ind}_\gamma(z_0)$.

The contour winding number was part of the inspiration for the idea of the DEGREE of a MAP between two COMPACT, oriented MANIFOLDS of the same DIMEN-

SION. In the language of the DEGREE of a MAP, if γ: $[0,1] \to \mathbb{C}$ is a closed curve (i.e., $\gamma(0) = \gamma(1)$), then it can be considered as a FUNCTION from $\mathbb{S}^1$ to $\mathbb{C}$. In that context, the winding number of γ around a point p in $\mathbb{C}$ is given by the degree of the MAP

$$\frac{\gamma - p}{|\gamma - p|}$$

from the CIRCLE to the CIRCLE.

See also RESIDUE (COMPLEX ANALYSIS)

References

Krantz, S. G. "The Index or Winding Number of a Curve about a Point." §4.4.4 in *Handbook of Complex Analysis.* Boston, MA: Birkhäuser, pp. 49–0, 1999.

Winding Number (Map)

The winding number $W(\theta)$ of a map $f(\theta)$ with initial value θ is defined by

$$W(\theta) \equiv \lim_{n \to \infty} \frac{f^n(\theta) - \theta}{n},$$

which represents the average increase in the angle θ per unit time (average frequency). A system with a RATIONAL winding number $W = p/q$ is MODE-LOCKED, whereas a system with an IRRATIONAL winding number is QUASIPERIODIC. Note that since the RATIONALS are a set of zero MEASURE on any finite interval, almost all winding numbers will be irrational, so almost all maps will be QUASIPERIODIC.

References

Rasband, S. N. *Chaotic Dynamics of Nonlinear Systems.* New York: Wiley, p. 129, 1990.

Windmill

One name for the figure used by Euclid to prove the PYTHAGOREAN THEOREM.

BRIDE'S CHAIR, PEACOCK'S TAIL

Window Function

RECTANGLE FUNCTION

Winkler Conditions

Conditions arising in the study of the ROBBINS AXIOM and its connection with BOOLEAN ALGEBRA. Winkler studied Boolean conditions (such as idempotence or existence of a zero) which would make a ROBBINS ALGEBRA become a BOOLEAN ALGEBRA. Winkler showed that each of the conditions

$$\exists C, \exists D, C \lor D = C$$
$$\exists C, \exists D, !(C \lor D) = !C$$

where $A \lor B$ denotes OR and $!A$ denotes NOT, known as the first and second Winkler conditions, SUFFICES.

A computer proof demonstrated that every Robbins algebra satisfies the second Winkler condition, from which it follows immediately that all Robbins algebras are Boolean.

See also Boolean Algebra, Huntington Axiom, Robbins Algebra, Robbins Axiom

References

McCune, W. "Robbins Algebras are Boolean." http://www-unix.mcs.anl.gov/~mccune/papers/robbins/.

Winkler, S. "Robbins Algebra: Conditions that Make a Near-Boolean Algebra Boolean." *J. Automated Reasoning* **6**, 465–89, 1990.

Winkler, S. "Absorption and Idempotency Criteria for a Problem in Near-Boolean Algebra." *J. Algebra* **153**, 414–23, 1992.

Winograd Transform

A discrete Fast Fourier transform algorithm which can be implemented for $N = 2$, 3, 4, 5, 7, 8, 11, 13, and 16 points.

See also Fast Fourier Transform

Wirtinger's Inequality

If y has period 2π, y' is L^2, and

$$\int_0^{2\pi} y \, dx = 0, \tag{1}$$

then

$$\int_0^{2\pi} y^2 \, dx < \int_0^{2\pi} y'^2 \, dx \tag{2}$$

unless

$$y = A \cos x + B \sin x \tag{3}$$

(Hardy *et al.* 1988).

Another inequality attributed to Wirtinger involves the Kähler form, which in $\mathbb{C}^n$ can be written

$$\omega = -\tfrac{1}{2}i \sum dz_k \wedge d\bar{z}_k. \tag{4}$$

Given $2k$ vectors $X_1, \ldots, X_{2k}$ in $\mathbb{R}^{2n} \simeq \mathbb{C}^n$, let $X = X_1 \wedge \cdots \wedge X_{2k}$ denote the oriented k-dimensional Parallelepiped and $|X|$ its k-dimensional volume. Then

$$\omega^k(X) \leq k! |X|, \tag{5}$$

with equality IFF the vectors span a k-dimensional complex subspace of $\mathbb{C}^n$, and they are positively oriented. Here, ω^k is the kth Exterior power for $1 \leq k \leq n$, and the orientation of a Complex subspace is determined by its Complex structure.

See also Kähler Form

References

Blaschke, W. *Kreis und Kugel.* Leipzig, Germany: p. 105, 1916.

Hardy, G. H.; Littlewood, J. E.; and Pólya, G. "Further Examples: Wirtinger's Inequality." §7.7 in *Inequalities, 2nd ed.* Cambridge, England: Cambridge University Press, pp. 184–87, 1988.

Wirtinger-Sobolev Isoperimetric Constants

Constants γ such that

$$\left[\int_\Omega |f|^q dx \right]^{1/q} \leq \gamma \left[\int_\Omega \sum_{i=1}^N \left| \frac{\partial f}{\partial x_i} \right|^p dx \right]^{1/p},$$

where f is a real-valued smooth function on a region Ω satisfying some Boundary conditions.

References

Finch, S. "Favorite Mathematical Constants." http://www.mathsoft.com/asolve/constant/ws/ws.html.

Wishart Distribution

If X_i for $i = 1, \ldots, m$ has a Gaussian multivariate distribution with mean vector $\mu = 0$ and Covariance matrix Σ, and X denotes the $m \times p$ matrix composed of the row vectors X_i, then the $p \times p$ matrix $\mathsf{X}^\mathsf{T}\mathsf{X}$ has a Wishart distribution with scale matrix Σ and degrees of freedom parameter m. The Wishart distribution is most typically used when describing the Covariance matrix of multinormal samples.

See also F-Distribution, Gaussian Multivariate Distribution, Hotelling T-Squared Distribution

Witch of Agnesi

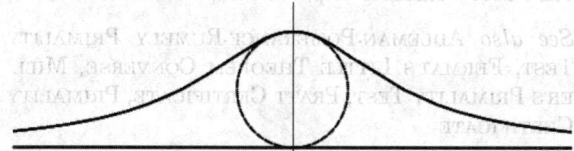

A curve studied and named "versiera" (Italian for "she-devil" or "witch") by Maria Agnesi in 1748 in her book *Istituzioni Analitiche* (MacTutor Archive). It is also known as cubique d'Agnesi or agnésienne. Some suggest that Agnesi confused an old Italian word meaning "free to move" with another meaning "witch." The curve had been studied earlier by Fermat and Guido Grandi in 1703.

It is the curve obtained by drawing a line from the origin through the Circle of radius a (OB), then picking the point with the y coordinate of the intersection with the circle and the x coordinate of the intersection of the extension of line OB with the line $y = 2a$. The curve has Inflection points at $y = 3a/2$. The line $y = 0$ is an Asymptote to the curve.

In parametric form,

$$x = 2a \cot \theta \tag{1}$$

$$y = a[1 - \cos(2\theta)], \qquad (2)$$

or

$$x = 2at \qquad (3)$$

$$y = \frac{2a}{1 + t^2}. \qquad (4)$$

In rectangular coordinates,

$$y = \frac{8a^3}{x^2 + 4a^2}. \qquad (5)$$

See also LAMÉ CURVE

References

Beyer, W. H. *CRC Standard Mathematical Tables, 28th ed.* Boca Raton, FL: CRC Press, p. 226, 1987.

Lawrence, J. D. *A Catalog of Special Plane Curves.* New York: Dover, pp. 90–3, 1972.

MacTutor History of Mathematics Archive. "Witch of Agnesi." http://www-groups.dcs.st-and.ac.uk/~history/Curves/Witch.html.

Yates, R. C. "Witch of Agnesi." *A Handbook on Curves and Their Properties.* Ann Arbor, MI: J. W. Edwards, pp. 237–38, 1952.

Witness

A witness is a number which, as a result of its number theoretic properties, guarantees either the compositeness or primality of a number n. Witnesses are most commonly used in connection with FERMAT'S LITTLE THEOREM CONVERSE. A PRATT CERTIFICATE uses witnesses to prove primality, and MILLER'S PRIMALITY TEST uses witnesses to prove compositeness.

See also ADLEMAN-POMERANCE-RUMELY PRIMALITY TEST, FERMAT'S LITTLE THEOREM CONVERSE, MILLER'S PRIMALITY TEST, PRATT CERTIFICATE, PRIMALITY CERTIFICATE

Witt Geometry

References

Dixon, J. and Mortimer, B. *Permutation Groups.* New York: Springer-Verlag, 1996.

Wittenbauer's Parallelogram

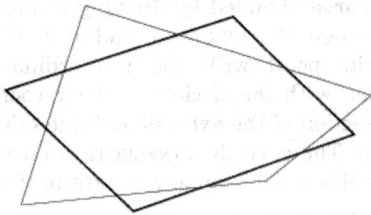

Divide the sides of a QUADRILATERAL into three equal parts. The figure formed by connecting and extending adjacent points on either side of a VERTEX is a PARALLELOGRAM known as Wittenbauer's parallelogram.

See also QUADRILATERAL, WITTENBAUER'S THEOREM

Wittenbauer's Theorem

The CENTROID of a QUADRILATERAL LAMINA is the center of its WITTENBAUER'S PARALLELOGRAM.

See also CENTROID (GEOMETRIC), LAMINA, QUADRILATERAL, WITTENBAUER'S PARALLELOGRAM

Witten's Equations

Also called the SEIBERG-WITTEN INVARIANTS. For a connection A and a POSITIVE SPINOR $\phi \in \Gamma(V_+)$,

$$D_A \phi = 0$$

$$F_+^A = i\sigma(\phi, \phi).$$

The solutions are called monopoles and are the minima of the functional

$$\int_X \left(|F_+^A - i\sigma(\phi, \phi)|^2 + |D_A \phi|^2 \right).$$

See also LICHNEROWICZ FORMULA, LICHNEROWICZ-WEITZENBOCK FORMULA, SEIBERG-WITTEN EQUATIONS

References

Cipra, B. "A Tale of Two Theories." *What's Happening in the Mathematical Sciences, 1995–996, Vol. 3.* Providence, RI: Amer. Math. Soc., pp. 14–5, 1996.

Donaldson, S. K. "The Seiberg-Witten Equations and 4-Manifold Topology." *Bull. Amer. Math. Soc.* **33**, 45–0, 1996.

Kotschick, D. "Gauge Theory is Dead!--Long Live Gauge Theory!" *Not. Amer. Math. Soc.* **42**, 335–38, 1995.

Seiberg, N. and Witten, E. "Monopoles, Duality, and Chiral Symmetry Breaking in $N = 2$ Supersymmetric QCD." *Nucl. Phys. B* **431**, 581–40, 1994.

Witten, E. "Monopoles and 4-Manifolds." *Math. Res. Let.* **1**, 769–96, 1994.

Wolfskehl Prize

A prize of 100,000 German marks offered for the first valid proof of FERMAT'S LAST THEOREM (Ball and Coxeter 1987, p. 72; Barner 1997; Hoffman 1998, pp. 193–94 and 199). The prize was collected by Andrew Wiles after his successful proof of the theorem in the years 1993–995.

See also FERMAT'S LAST THEOREM, MATHEMATICS PRIZES

References

Ball, W. W. R. and Coxeter, H. S. M. *Mathematical Recreations and Essays, 13th ed.* New York: Dover, pp. 69–3, 1987.

Barner, K. "Paul Wolfskehl and the Wolfskehl Prize." *Not. Amer. Math. Soc.* **44**, 1294–303, 1997.

Hoffman, P. *The Man Who Loved Only Numbers: The Story of Paul Erdos and the Search for Mathematical Truth.* New York: Hyperion, pp. 193–99, 1998.

Wolstenholme's Theorem

If p is a PRIME > 3, then the NUMERATOR of

$$1 + \frac{1}{2} + \frac{1}{3} + \ldots + \frac{1}{p-1}$$

is divisible by p^2 and the NUMERATOR of

$$1 + \frac{1}{2^2} + \frac{1}{3^2} + \ldots + \frac{1}{(p-1)^2}$$

is divisible by p. These imply that if $p \geq 5$ is PRIME, then

$$\binom{2p-1}{p-1} \equiv 1 \pmod{p^3}.$$

References

Guy, R. K. *Unsolved Problems in Number Theory, 2nd ed.* New York: Springer-Verlag, p. 85, 1994.
Ribenboim, P. *The Book of Prime Number Records, 2nd ed.* New York: Springer-Verlag, p. 21, 1989.

Woodall Number

Numbers OF THE FORM

$$W_n = 2^n n - 1.$$

The first few are 1, 7, 23, 63, 159, 383, ... (Sloane's A003261). The only Woodall numbers W_n for $n < 100,000$ which are PRIME are for $n = 5312$, 7755, 9531, 12379, 15822, 18885, 22971, 23005, 98726, ... (Sloane's A014617; Ballinger).

See also CULLEN NUMBER, CUNNINGHAM NUMBER, FERMAT NUMBER, MERSENNE NUMBER, SIERPINSKI NUMBER OF THE FIRST KIND

References

Ballinger, R. "Cullen Primes: Definition and Status." http://vamri.xray.ufl.edu/proths/cullen.html.
Caldwell, C. K. "The Top Twenty: Woodall Primes." http://www.utm.edu/research/primes/lists/top20/Woodall.html.
Guy, R. K. "Cullen Numbers." §B20 in *Unsolved Problems in Number Theory, 2nd ed.* New York: Springer-Verlag, p. 77, 1994.
Leyland, P. ftp://sable.ox.ac.uk/pub/math/factors/woodall/.
Ribenboim, P. *The New Book of Prime Number Records.* New York: Springer-Verlag, pp. 360–61, 1996.
Sloane, N. J. A. Sequences A003261/M4379 and A014617 in "An On-Line Version of the Encyclopedia of Integer Sequences." http://www.research.att.com/~njas/sequences/eisonline.html.

Woodbury Formula

$$\left(A + UV^{T}\right)^{-1} = A^{-1} - \left[A^{-1}U\left(1 + V^{T}A^{-1}U\right)^{-1}V^{T}A^{-1}\right].$$

See also SHERMAN-MORRISON FORMULA

References

Golub, G. H. and van Loan, C. F. *Matrix Computations, 3rd ed.* Baltimore, MD: Johns Hopkins, p. 51, 1996.

Woolhouse's Formulas

Let the values of a function $f(x)$ be tabulated at points x_i equally spaced by $h = x_{i+1} - x_i$, so $f_1 = f(x_1)$, $f_2 = f(x_2)$, ..., $f_n = f(x_n)$. Then Woolhouse's formulas approximating the integral of $f(x)$ are given by the NEWTON-COTES-like formulas

$$\int_{x_1}^{x_{11}} f(x)\, dx = 5\left[\frac{223}{3909}(f_1 + f_{11}) + \frac{5875}{18144}(f_2 + f_{10})\right.$$
$$\left. + \frac{4625}{10584}(f_4 + f_8) + \frac{41}{112}f_5\right]$$

$$\int_{x_1}^{x_{29}} f(x)\, dx = 14\left[\frac{7}{195}(f_1 + f_{29}) + \frac{16807}{66690}(f_3 + f_{27})\right.$$
$$\left. + \frac{128}{285}(f_8 + f_{22}) + \frac{71}{135}f_{15}\right].$$

References

King, A. E. "Approximate Integration. Note on Quadrature Formulae: Their Construction and Application to Actuarial Functions." *Trans. Faculty of Actuaries* **9**, 218–31, 1923.
Sheppard, W. F. "Some Quadrature-Formulæ." *Proc. London Math. Soc.* **32**, 258–77, 1900.
Whittaker, E. T. and Robinson, G. "Woolhouse's Formulae." *The Calculus of Observations: A Treatise on Numerical Mathematics, 4th ed.* New York: Dover, p. 158, 1967.
Woolhouse, W. S. B. "On Integration by Means of Selected Values of the Function." *J. Inst. Act.* **27**, 122–55, 1888.

Word

A finite sequence of n letters from some ALPHABET is said to be an n-ary word.

See also CUBEFREE WORD, OVERLAPFREE WORD, SQUAREFREE WORD

Word Sequence

An INTEGER SEQUENCE whose terms are defined in terms of number-related words in some language. For example, the following table gives the sequences of numbers having digits whose English names (zero, one, two, three, four, five, six, seven, eight, nine) are in alphabetical order and also satisfy some other property.

property	Sloane	sequence
ordered	A053432	1, 2, 3, 4, 5, 6, 7, 8, 9, 10, 11, 12, 13, ...
distinct, ordered	A053433	1, 2, 3, 4, 5, 6, 7, 8, 9, 10, 12, 13, 16, ...
prime, ordered	A053434	2, 3, 5, 7, 11, 13, 17, 41, 43, 47, 53, 59, ...
distinct, prime, ordered	A053435	2, 3, 5, 7, 13, 17, 41, 43, 47, 53, 59, 73, ...

See also LOOK AND SAY SEQUENCE

References

Sloane, N. J. A. Sequences A053432, A053433, A053434, and A053435 in "An On-Line Version of the Encyclopedia of Integer Sequences." http://www.research.att.com/~njas/sequences/eisonline.html.

World Line

The path of an object through PHASE SPACE.

Worm

One of the seven 4-POLYHEXES. S. Kim has observed that four worms solve the puzzle of finding a non-three-COLORABLE map with only four congruent countries (as long as no lakes are allowed).

See also COLORABLE

References

Gardner, M. *Mathematical Magic Show: More Puzzles, Games, Diversions, Illusions and Other Mathematical Sleight-of-Mind from Scientific American.* New York: Vintage, p. 147, 1978.
Gosper, R. W. G. "Quattroslabia." http://www.ippi.com/rwg/Quattroslabia.htm.

Worpitzky's Identity

$$x^n = \sum_{k=1}^{n} \left\langle \begin{matrix} n \\ k \end{matrix} \right\rangle \binom{x+k-1}{n},$$

where $\left\langle \begin{matrix} n \\ k \end{matrix} \right\rangle$ is an EULERIAN NUMBER and $\binom{n}{k}$ is a BINOMIAL COEFFICIENT (Worpitzky 1883; Comtet 1974, p. 242).

See also BINOMIAL SUMS, EULERIAN NUMBER

References

Comtet, L. *Advanced Combinatorics: The Art of Finite and Infinite Expansions, rev. enl. ed.* Dordrecht, Netherlands: Reidel, 1974.
Worpitzky. "Studien über die Bernoullischen und Eulerischen Zahlen." *J. reine angew. Math.* **94**, 203–32, 1883.

Wright Function

The ENTIRE FUNCTION

$$\phi(\rho, \beta; z) = \sum_{k=0}^{\infty} \frac{z^k}{k!\,\Gamma(\rho k + \beta)},$$

where $\rho > -1$ and $\beta \in \mathbb{C}$, named after the British mathematician E. M. Wright.

References

Gorenflo, R.; Luchko, Yu.; and Mainardi, F. "Analytical Properties and Applications of the Wright Function." *Fractional Calc. Appl. Anal.* **2**, 383–15, 1999.

Writhe

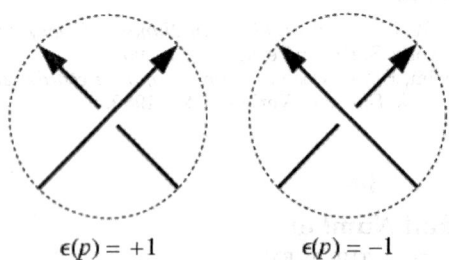

$$\epsilon(p) = +1 \qquad \epsilon(p) = -1$$

Also called the TWIST NUMBER. The sum of crossings p of a LINK L,

$$w(L) = \sum_{p \in C(L)} \epsilon(p), \tag{1}$$

where $\epsilon(p)$ defined to be ± 1 if the overpass slants from top left to bottom right or bottom left to top right and $C(L)$ is the set of crossings of an oriented LINK. The writhe of a minimal knot diagram is *not* a KNOT INVARIANT, as exemplified by the PERKO PAIR, which have differing writhes (Hoste *et al.* 1998). If a KNOT K is AMPHICHIRAL, then $w(K) = 0$ (Thistlethwaite). A formula for the writhe is given by

$$\mathrm{Wr}(K) = \frac{1}{4\pi} \int_K ds \int_K dt \, e^\mu \frac{de^\mu}{ds} \frac{de^\alpha}{dt} \tag{2}$$

where K is parameterized by $x^\mu(s)$ for $0 \le s \le L$ along the length of the knot by parameter s, and the FRAME K_f associated with K is

$$y^\mu = x^\mu(s) + \epsilon n^\mu(s), \tag{3}$$

where ϵ is a small parameter, $n^\mu(s)$ is a unit VECTOR FIELD normal to the curve at s, and the vector field e^μ is given by

$$e^{\mu}(s,t) = \frac{y^{\mu}(t) - x^{\mu}(s)}{|y(t) - x(s)|} \qquad (4)$$

(Kaul 1999).

Letting Lk be the LINKING NUMBER of the two components of a ribbon, Tw be the TWIST, and Wr be the writhe, then the CALUGAREANU THEOREM states that

$$\mathrm{Lk}(K) = \mathrm{Tw}(K) + \mathrm{Wr}(K). \qquad (5)$$

(Adams 1994, p. 187).

See also CALUGAREANU THEOREM, SCREW, TWIST

References

Adams, C. C. *The Knot Book: An Elementary Introduction to the Mathematical Theory of Knots.* New York: W. H. Freeman, 1994.

Hoste, J.; Thistlethwaite, M.; and Weeks, J. "The First 1,701,936 Knots." *Math. Intell.* **20**, 33–8, Fall 1998.

Kaul, R. K. Topological Quantum Field Theories--A Meeting Ground for Physicists and Mathematicians. 15 Jul 1999. http://xxx.lanl.gov/abs/hep-th/9907119/.

Wronskian

$$W(\phi_1, \ldots, \phi_n) \equiv \begin{vmatrix} \phi_1 & \phi_2 & \cdots & \phi_n \\ \phi_1' & \phi_2' & \cdots & \phi_n' \\ \vdots & \vdots & \ddots & \vdots \\ \phi_1^{(n-1)} & \phi_2^{(n-1)} & \cdots & \phi_n^{(n-1)} \end{vmatrix}.$$

If the Wronskian is NONZERO in some region, the functions ϕ_i are LINEARLY INDEPENDENT. If $W = 0$ over some range, the functions are linearly dependent somewhere in the range.

See also ABEL'S DIFFERENTIAL EQUATION IDENTITY, GRAM DETERMINANT, LINEARLY DEPENDENT FUNCTIONS

References

Morse, P. M. and Feshbach, H. *Methods of Theoretical Physics, Part I.* New York: McGraw-Hill, pp. 524–25, 1953.

W-Transform

The W-transform of a function $f(x)$ is defined by the integral

$$(Wf)(x) = \left(W_{pq}^{mn} \begin{vmatrix} v, (\alpha)_p \\ (\beta_q) \end{vmatrix} f(t) \right)(x) \qquad (1)$$

$$= \frac{1}{2\pi i} \int_{\sigma} \Gamma(v - ix - s, v + ix - s)$$

$$\times \Gamma \begin{bmatrix} (\beta_m) + s, & 1 - (\alpha_n) - s \\ (\alpha_p^{n+1}) + s, & 1 - (\beta_q^{m+1}) - s \end{bmatrix} f^*(1-s)\, ds, \qquad (2)$$

where

$$\Gamma \begin{bmatrix} (\beta_m) + s, & 1 - (\alpha_n) - s \\ (\alpha_p^{n+1}) + s, & 1 - (\beta_q^{m+1}) - s \end{bmatrix}$$

$$= \Gamma \begin{bmatrix} \beta_1 + s, & \ldots, & \beta_m + s, & 1 - \alpha_1 - s, & \ldots, & 1 - \alpha_n - s \\ \alpha_{n+1} + s, & \ldots, & \alpha_p + s & 1 - \beta_{m+1} - s, & \ldots & 1 - \beta_q - s \end{bmatrix} \qquad (3)$$

$$= \frac{\prod_{j=1}^{m} \Gamma(\beta_{j+s}) \prod_{j=1}^{n} \Gamma(1 - \alpha_j - s)}{\prod_{j=n+1}^{p} \Gamma(\alpha_{j+s}) \prod_{j=m+1}^{q} \Gamma(1 - \beta_j - s)}, \qquad (4)$$

$\Re[v] > 1/2$, v and the components of the vectors (α_p) and (β_q) are complex numbers satisfying the conditions $\Re[a_p]) \neq 1/2, 3/2, 5/2, \ldots,$ $3/2,$ $5/2,$ $\ldots$ and $\Re[b_q] \neq -1/2, -3/2, -5/2, \ldots,$ $-3/2,$ $-5/2, \ldots,$ $f^*(s)$ is the MELLIN TRANSFORM of a function $f(x)$ and σ is the CONTOUR $\sigma = \{1/2 - i\infty, 1/2 + i\infty\}$.

See also G-TRANSFORM

References

Samko, S. G.; Kilbas, A. A.; and Marichev, O. I. "The W-Transform and Its Inversion." §37.5 in *Fractional Integrals and Derivatives.* Yverdon, Switzerland: Gordon and Breach, pp. 752–58, 1993.

Wulff Shape

An equilibrium MINIMAL SURFACE for a crystal or drop which has the least anisotropic surface free energy for a given volume. It is the anisotropic analog of a SPHERE. In the case of a sessile drop, the Wulff shapes becomes the Winterbottom shape (Dunlop and Magnen 1999, p. 31).

See also SPHERE

References

Dunlop, F. and Magnen, J. "A Wulff Shape from Constructive Field Theory." In *Mathematical Results in Statistical Mechanics, Marseilles, France, July 27–1 1998* (Ed. S. Miracle-Solé, J. Ruis, and V. Zagrebnov). Singapore: World Scientific, pp. 31–2, 1999.

Winterbottom, W. L. "Equilibrium Shape of a Small Particle in Contact with a Foreign Substrate." *Acta Metal.* **15**, 303–10, 1967.

Wulff, G. "Zur Frage der Geschwindigkeit des Wachstums und der Auflösung der Krystallflagen." *Z. Kryst. Mineral.* **34**, 449, 1901.

Wynn's Epsilon Method

A method for numerical evaluation of SUMS and PRODUCTS which samples a number of additional terms in the series and then tries to fit them to a POLYNOMIAL multiplied by a decaying exponential. Wynn's epsilon method can be applied to the terms of a series using the *Mathematica* command `SequenceLimit[l]`.

See also EULER-MACLAURIN INTEGRATION FORMULAS

Wythoff Array

A INTERSPERSION array given by

1	2	3	5	8	13	21	34	55	$\cdots$
4	7	11	18	29	47	76	123	199	$\cdots$
6	10	16	26	42	68	110	178	288	$\cdots$
9	15	24	39	63	102	165	267	432	$\cdots$
12	20	32	52	84	136	220	356	576	$\cdots$
14	23	37	60	97	157	254	411	665	$\cdots$
17	28	45	73	118	191	309	500	809	$\cdots$
19	31	50	81	131	212	343	555	898	$\cdots$
22	36	58	94	152	246	398	644	1042	$\cdots$

the first row of which is the FIBONACCI NUMBERS.

See also BEATTY SEQUENCE, FIBONACCI NUMBER, INTERSPERSION, STOLARSKY ARRAY

References

Kimberling, C. "Fractal Sequences and Interspersions." *Ars Combin.* **45**, 157–68, 1997.
Sloane, N. J. A. "The Wythoff Array and the Para-Fibonacci Sequence." http://www.research.att.com/~njas/sequences/classic.html.

Wythoff Construction

A method of constructing UNIFORM POLYHEDRA.

See also UNIFORM POLYHEDRON

References

Har'El, Z. "Uniform Solution for Uniform Polyhedra." *Geometriae Dedicata* **47**, 57–10, 1993.

Wythoff Symbol

A symbol consisting of three rational numbers that can be used to describe UNIFORM POLYHEDRA based on how a point C in a spherical triangle can be selected so as to trace the vertices of regular polygonal faces. For example, the Wythoff symbol for the TETRAHEDRON is $3|23$. There are four types of Wythoff symbols, $|p\,q\,r$, $p\,|\,q\,r$, $p\,q\,|\,r$ and $p\,q\,r\,|$, and one exceptional symbol, $|\,\frac{3}{2}\,\frac{5}{3}\,3\,\frac{5}{2}$ (which is used for the GREAT DIRHOMBICOSIDODECAHEDRON).

The meaning of the bars | may be summarized as follows (Wenninger 1989, p. 10; Messer). Consider a SPHERICAL TRIANGLE PQR whose angles are π/p, π/q, and π/r.

1. $|\,p\,q\,r$: C is a special point within PQR that traces snub polyhedra by even reflections .
2. $p\,|\,q\,r$ (or $p\,|\,r\,q$) : C is the vertex P .
3. $q\,r\,|\,p$ (or $r\,q\,|\,p$) : C lies on the are PQ and the bisector of the opposite angle R .
4. $pqr|$ (or any permutation of the three letters): C is the incenter of the triangle PQR .

Some special cases in terms of SCHLÄFLI SYMBOLS are

$$p\,|\,q\,2 = p\,|\,2\,q = \{q, p\}$$

$$2\,|\,p\,q = \begin{Bmatrix} p \\ q \end{Bmatrix}$$

$$p\,q\,|\,2 = r \begin{Bmatrix} p \\ q \end{Bmatrix}$$

$$2\,q\,|\,p = t\{p, q\}$$

$$2\,p\,q\,|\,t = \begin{Bmatrix} p \\ q \end{Bmatrix}$$

$$|\,2\,p\,q = s \begin{Bmatrix} p \\ q \end{Bmatrix}$$

See also SCHLÄFLI SYMBOL, SCHWARZ TRIANGLE, UNIFORM POLYHEDRON

References

Har'El, Z. "Uniform Solution for Uniform Polyhedra." *Geometriae Dedicata* **47**, 57–10, 1993.
Messer, P. W. "Closed-Form Expressions for Uniform Polyhedra and Their Duals." Unpublished manuscript.
Wenninger, M. J. *Polyhedron Models.* New York: Cambridge University Press, pp. 8–0, 1989.

Wythoff's Game

A game played with two heaps of counters in which a player may take any number from either heap or the same number from both. The player taking the last counter wins. The rth SAFE combination is $(x, x+r)$, where $x = \lfloor \phi r \rfloor$, with ϕ the GOLDEN RATIO and $\lfloor x \rfloor$ the FLOOR FUNCTION. It is also true that $x + r = \lfloor \phi^2 r \rfloor$. The first few SAFE combinations are $(1, 2)$, $(3, 5)$, $(4, 7)$, $(6, 10)$, ... (Sloane's A000201 and A001950), which are the pairs of elements from the complementary BEATTY SEQUENCES for ϕ and ϕ^2 (Wells 1986, p. 40).

See also BEATTY SEQUENCE, NIM, SAFE

References

Ball, W. W. R. and Coxeter, H. S. M. *Mathematical Recreations and Essays, 13th ed.* New York: Dover, pp. 39–0, 1987.
Coxeter, H. S. M. "The Golden Section, Phyllotaxis, and Wythoff's Game." *Scripta Math.* **19**, 135–43, 1953.
O'Beirne, T. H. *Puzzles and Paradoxes.* Oxford, England: Oxford University Press, pp. 109 and 134–38, 1965.
Sloane, N. J. A. Sequences A000201/M2322 and A001950/M1332 in "An On-Line Version of the Encyclopedia of Integer Sequences." http://www.research.att.com/~njas/sequences/eisonline.html.
Wells, D. *The Penguin Dictionary of Curious and Interesting Numbers.* Middlesex, England: Penguin Books, p. 40, 1986.
Wythoff, W. A. "A Modification of the Game of Nim." *Nieuw Arch. Wiskunde* **8**, 199–02, 1907/1909.

X

x-Axis

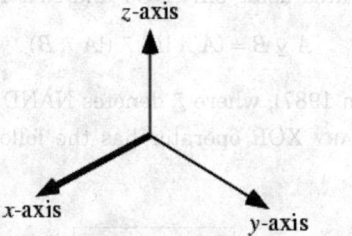

The horizontal axis of a 2-D plot in CARTESIAN COORDINATES. Physicists and astronomers sometimes call this axis the ABSCISSA, although that term is more commonly used to refer to coordinates along the x-AXIS.

See also ABSCISSA, ORDINATE, Y-AXIS, Z-AXIS

Xi Function

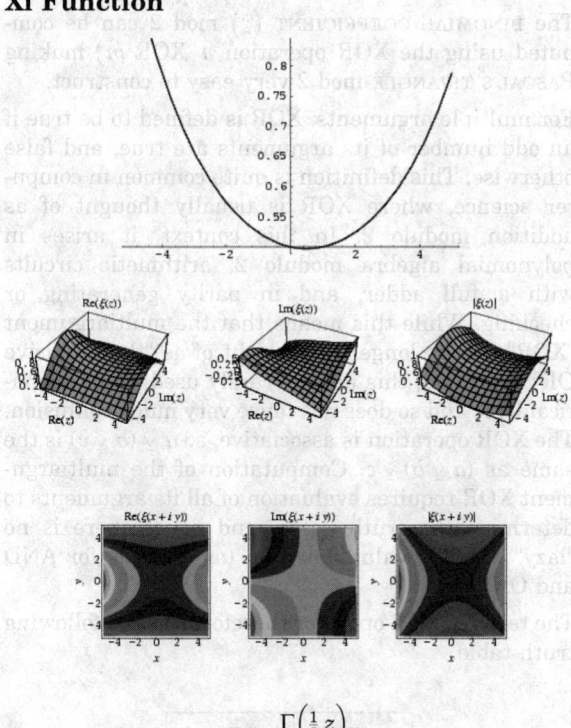

$$\xi(z) \equiv \frac{1}{2} z(z-1) \frac{\Gamma\left(\frac{1}{2} z\right)}{\pi^{z/2}} \zeta(z)$$

$$= \frac{(z-1)\Gamma\left(\frac{1}{2} z + 1\right)\zeta(z)}{\sqrt{\pi^z}}, \tag{1}$$

where $\zeta(z)$ is the RIEMANN ZETA FUNCTION and $\Gamma(z)$ is the GAMMA FUNCTION (Gradshteyn and Ryzhik 2000, p. 1076; Hardy 1999, p. 41). The ξ function satisfies the identity

$$\xi(1-z) = \xi(z). \tag{2}$$

The zeros of $\xi(z)$ and of its DERIVATIVES are all located on the CRITICAL STRIP $z = \sigma + it$, where $0 < \sigma < 1$. Therefore, the nontrivial zeros of the RIEMANN ZETA FUNCTION exactly correspond to those of $\xi(z)$. The function $\xi(z)$ is related to what Gradshteyn and Ryzhik (2000, p. 1074) call $\Xi(t)$ by

$$\Xi(t) \equiv \xi(z), \tag{3}$$

where $z \equiv \frac{1}{2} + it$. This function can also be defined as

$$\Xi(it) \equiv \frac{1}{2}\left(t^2 - \frac{1}{4}\right)\pi^{-t/2-1/4}\Gamma\left(\frac{1}{2} t + \frac{1}{4}\right)\zeta\left(t + \frac{1}{2}\right), \tag{4}$$

giving

$$\Xi(t) = -\frac{1}{2}\left(t^2 + \frac{1}{4}\right)\pi^{it/2-1/4}\Gamma\left(\frac{1}{4} - \frac{1}{2} it\right)\zeta\left(\frac{1}{2} - it\right). \tag{5}$$

The DE BRUIJN-NEWMAN CONSTANT is defined in terms of the $\Xi(t)$ function.

See also DE BRUIJN-NEWMAN CONSTANT, RIEMANN HYPOTHESIS, RIEMANN-SIEGEL FUNCTIONS, RIEMANN ZETA FUNCTION

References

Borwein, J. M.; Bradley, D. M.; and Crandall, R. E. "Computational Strategies for the Riemann Zeta Function." CECM-98:118, 23 Jun 1999. http://www.cecm.sfu.ca/preprints/1999pp.html#98:118.

Brent, R. P. "On the Zeros of the Riemann Zeta Function in the Critical Strip." *Math. Comput.* **33**, 1361–372, 1979.

Brent, R. P.; van de Lune, J.; te Riele, H. J. J.; and Winter, D. T. "On the Zeros of the Riemann Zeta Function in the Critical Strip. II." *Math. Comput.* **39**, 681–88, 1982.

Gradshteyn, I. S. and Ryzhik, I. M. *Tables of Integrals, Series, and Products, corr. enl. 4th ed.* San Diego, CA: Academic Press, 2000.

Hardy, G. H. *Ramanujan: Twelve Lectures on Subjects Suggested by His Life and Work, 3rd ed.* New York: Chelsea, 1999.

Titchmarsh, E. C. and Heath-Brown, D. R. *The Theory of the Riemann Zeta-Function, 2nd ed.* Oxford, England: Oxford University Press, 1986.

x-Intercept

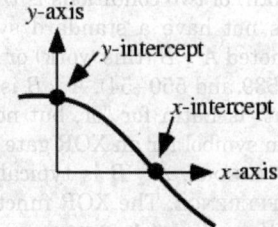

The point at which a curve or function crosses the x-AXIS (i.e., when $y = 0$ in 2-D).

See also LINE, Y-INTERCEPT

XNOR

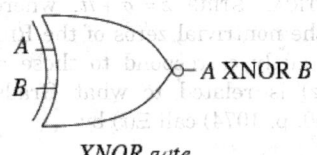

XNOR gate

The CONNECTIVE in logic corresponding to the exclusive nor operation. A XNOR B is equivalent to $(A \wedge B) \vee (!A \wedge !B)$, where $\wedge$ denotes AND, $\vee$ denotes OR, and $!A$ denotes NOT. The circuit diagram symbol for an XNOR gate is illustrated above, and the XNOR TRUTH TABLE is given below.

A	B	A XNOR B
T	T	T
T	F	F
F	T	F
F	F	T

See also AND, BINARY OPERATOR, BOOLEAN ALGEBRA, CONNECTIVE, LOGIC, NAND, NOR, NOT, OR, PASCAL'S TRIANGLE, TRUTH TABLE, XOR

References

Simpson, R. E. "The Exclusive NOR (XNOR) Gate." §12.5.7 in *Introductory Electronics for Scientists and Engineers*, *2nd ed.* Boston, MA: Allyn and Bacon, pp. 539 and 554, 1987.

XOR

Portions of this entry contributed by ROGER GERMUNDSSON

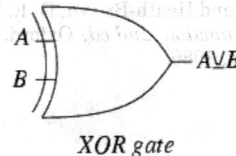

XOR gate

A CONNECTIVE in LOGIC known as the "exclusive or," or EXCLUSIVE DISJUNCTION. It yields true if exactly one (but not both) of two conditions is true. The XOR operation does not have a standard symbol, but is sometimes denoted $A \veebar B$ (this work) or $A \oplus B$ (Simpson 1987, pp. 539 and 550–54). $A \veebar B$ is read "A AUT B," where "aut" is Latin for "or, but not both." The circuit diagram symbol for an XOR gate is illustrated above. In SET THEORY, $A \veebar B$ is typically called the SYMMETRIC DIFFERENCE. The XOR function is implemented in *Mathematica* 4.1 as XOR.

The binary XOR operation $A \veebar B$ is identical to NONEQUIVALENCE $A \not\equiv B$. $A \veebar B$ can be implemented using AND and OR gates as

$$A \veebar B = (A \wedge !B) \vee (!A \wedge B) \qquad (1)$$

$$= (A \vee) \wedge !(A \wedge B), \qquad (2)$$

where $\wedge$ denotes AND and $\vee$ denotes OR, and can be implemented using only NOT and NAND gates as

$$A \veebar B = (A \barwedge !B) \barwedge (!A \barwedge B) \qquad (3)$$

(Simpson 1987), where $\barwedge$ denotes NAND.

The BINARY XOR operator has the following TRUTH TABLE.

A	B	$A \veebar B$
T	T	F
T	F	T
F	T	T
F	F	F

The BINOMIAL COEFFICIENT $\binom{m}{n}$ mod 2 can be computed using the XOR operation n XOR m, making PASCAL'S TRIANGLE mod 2 very easy to construct.

For multiple arguments, XOR is defined to be true if an odd number of its arguments are true, and false otherwise. This definition is quite common in computer science, where XOR is usually thought of as addition modulo 2. In this context, it arises in polynomial algebra modulo 2, arithmetic circuits with a full adder, and in parity generating or checking. While this means that the multiargument "XOR" can no longer be thought of as "the exclusive OR" operation, this form is rarely used in mathematical logic and so does not cause very much confusion. The XOR operation is associative, so $a \veebar (b \veebar c)$ is the same as $(a \veebar b) \veebar c$. Computation of the multiargument XOR requires evaluation of all its arguments to determine the truth value, and hence there is no "lazy" special evaluation form (as there is for AND and OR).

The ternary XOR operator therefore has the following truth table.

A	B	C	$A \veebar B \veebar C$
T	T	T	T
T	T	F	F
T	F	T	F
T	F	F	T
F	T	T	F
F	T	F	T
F	F	T	T
F	F	F	F

See also AND, AUT, BINARY OPERATOR, BOOLEAN ALGEBRA, CONNECTIVE, LOGIC, NAND, NOR, NOT, OR, PASCAL'S TRIANGLE, SYMMETRIC DIFFERENCE, TRUTH TABLE, XNOR

References

Simpson, R. E. "The Exclusive OR (XOR) Gate." §12.5.6 in *Introductory Electronics for Scientists and Engineers, 2nd ed.* Boston, MA: Allyn and Bacon, pp. 550–54, 1987.

Y

Yacht

A 6-POLYIAMOND.

References

Golomb, S. W. *Polyominoes: Puzzles, Patterns, Problems, and Packings, 2nd ed.* Princeton, NJ: Princeton University Press, p. 92, 1994.

Yahtzee

Yahtzee is a game played with five 6-sided DICE. Players take turns rolling the dice, and trying to get certain types of rolls, each with an assigned point value, as summarized in the following table. Players are allowed a total of three rolls, with any subset of dice capable of being set aside at each roll. In addition to runs of a single number, other rolls include 3 of a kind (three of the same number), 4 of a kind (four of the same number), full house (two of one number and three of another), small straight (4 numbers in a row), large straight (5 numbers in a row), Yahtzee (five of the same number), and chance (any roll).

aces	sum of 1s
twos	sum of 2s
threes	sum of 3s
fours	sum of 4s
fives	sum of 5s
sixes	sum of 6s
3 of a kind	sum of all dice
4 of a kind	sum of all dice
full house	25
sm. straight	30
lg. straight	40
Yahtzee	50
chance	sum of all dice

In a variant of the game known as triple Yahtzee, players try to get each type of roll three times over the course of the game instead of just once, with point values for each roll being placed in a single, double, or triple column, whose values are multiplied by the stated weight when scores are totaled. The following tables summarizes the probability of obtaining various rolls. In this table, lower-value rolls are excluded from the results, so, for example, the probability of obtaining a three of a kind excludes rolls that are actually fours of a kind or Yahtzees. Similarly, the three of a kind probability excludes rolls that are full houses, and the two of a kind probability excludes rolls that are small straights.

type	1	2	3	overall
2 of a kind	$\frac{65}{108}$	$\frac{65}{108}$	$\frac{65}{108}$	$\frac{1180205}{1259712}$
3 of a kind	$\frac{25}{162}$			
4 of a kind	$\frac{25}{1296}$			
full house	$\frac{25}{648}$			
sm. straight	$\frac{10}{81}$			
lg. straight	$\frac{5}{162}$			
Yahtzee	$\frac{1}{1296}$	$\frac{83}{6993}$		

type	1	2	3	overall
2 of a kind	60.19%	60.19%	60.19%	93.69%
3 of a kind	15.43%			
4 of a kind	1.93%			
full house	3.86%			
sm. straight	12.35%			
lg. straight	3.09%			
Yahtzee	0.08%	1.19%		

See also DICE

Yanghui Triangle

PASCAL'S TRIANGLE

Yang-Mills Equation

The anti-self-dual Yang-Mills equation is the system of PARTIAL DIFFERENTIAL EQUATIONS

$$\frac{\partial}{\partial \bar{x}_1}\left(\Omega^{-1}\frac{\partial\Omega}{\partial x_1}\right) + \frac{\partial}{\partial \bar{x}_2}\left(\Omega^{-1}\frac{\partial\Omega}{\partial x_2}\right) = 0.$$

References

Ablowitz, M. J.; Costa, D. G.; and Tenenblat, K. "Solutions of Multidimensional Extensions of the Anti-Self-Dual Yang-Mills Equation." *Stud. Appl. Math.* **77**, 37–46 1987.

Zwillinger, D. *Handbook of Differential Equations, 3rd ed.* Boston, MA: Academic Press, p. 139, 1997.

y-Axis

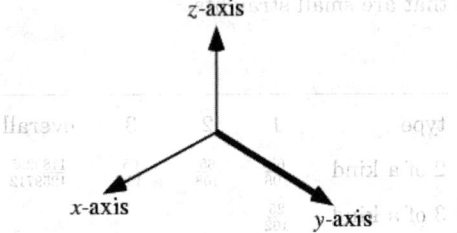

The vertical axis of a 2-D plot in CARTESIAN COORDINATES. Physicists and astronomers sometimes call this axis the ORDINATE, although that term is more commonly used to refer to coordinates along the Y-AXIS.

See also ABSCISSA, ORDINATE, *x*-AXIS, *z*-AXIS

Yff Center of Congruence

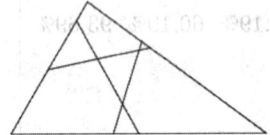

Let three ISOSCELIZERS be constructed on a TRIANGLE, one for each side. Now parallel-displace these ISOSCELIZERS until they concur in a single point. This point is called the Yff center of congruence and has TRIANGLE CENTER FUNCTION

$$\alpha = \sec\left(\tfrac{1}{2} A\right).$$

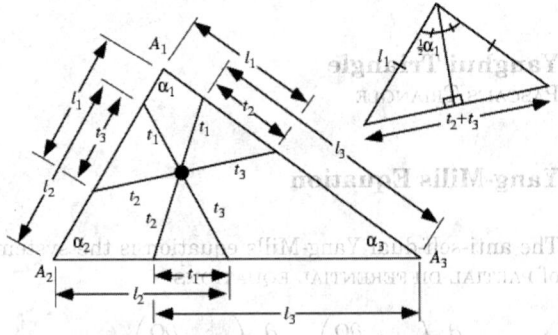

By analogy with the determination of the YFF CENTRAL TRIANGLE, the angle α_1 is related to the isoscelizer distance l_1 and the inner triangle side

lengths t_i are given by

$$\sin\left(\tfrac{1}{2}\alpha_1\right) = \sqrt{\frac{1 - \cos\alpha_1}{2}} = \frac{\tfrac{1}{2}(t_2 + t_3)}{l_1}$$

and so on. Therefore, the length l_i and t_i can be determined by solving the six simultaneous equations

$$l_2 + l_3 - t_1 = s_1$$

$$l_1 + l_3 - t_2 = s_2$$

$$l_1 + l_2 - t_3 = s_3$$

$$\left(\frac{t_2 + t_3}{l_1}\right)^2 = 2\left(1 - \frac{s_2^2 + s_3^2 - s_1^2}{2s_2 s_3}\right)$$

$$\left(\frac{t_1 + t_3}{l_2}\right)^2 = 2\left(1 - \frac{s_1^2 + s_3^2 - s_2^2}{2s_1 s_3}\right)$$

$$\left(\frac{t_1 + t_2}{l_3}\right)^2 = 2\left(1 - \frac{s_1^2 + s_2^2 - s_3^2}{2s_1 s_2}\right).$$

See also CONGRUENT ISOSCELIZERS POINT, ISOSCELIZER, YFF CENTRAL TRIANGLE

References

Kimberling, C. "Yff Center of Congruence." http://cedar.evansville.edu/~ck6/tcenters/recent/yffcc.html.

Yff Central Triangle

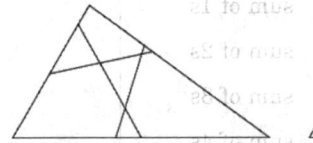 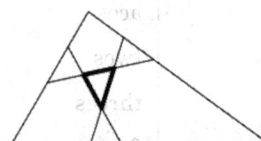

Let three ISOSCELIZERS be constructed on a TRIANGLE, one for each side. This makes all of the inner triangles SIMILAR to each other. However, there is a unique set of three isoscelizers for which the four interior triangles are congruent. The innermost triangle is called the Yff central triangle.

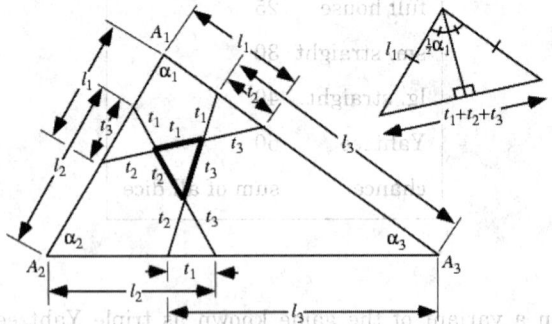

Let the side lengths be denoted s_i, the side lengths of the Yff central triangle t_i, and the distances of the

ISOSCELIZERS from the vertices l_i (for $i = 1, 2, 3$), then the LAW OF COSINES gives

$$\cos \alpha_1 = \frac{s_2^2 + s_3^2 - s_1^2}{2 s_2 s_3}$$

and so on, and trigonometry gives

$$\sin\left(\tfrac{1}{2}\alpha_1\right) = \sqrt{\frac{1 - \cos \alpha_1}{2}} = \frac{\tfrac{1}{2}(t_1 + t_2 + t_3)}{l_1}$$

and so on. Three more equations are obtained by noting that the sums of lengths along each side must sum to that side length. Therefore, the size of the Yff central triangle and the positions of the ISOSCELIZERS can be determined by solving the six simultaneous equations

$$l_2 + l_3 - t_1 = s_1$$

$$l_1 + l_3 - t_2 = s_2$$

$$l_1 + l_2 - t_3 = s_3$$

$$\left(\frac{t_1 + t_2 + t_3}{l_1}\right)^2 = 2\left(1 - \frac{s_2^2 + s_3^2 - s_1^2}{2 s_2 s_3}\right)$$

$$\left(\frac{t_1 + t_2 + t_3}{l_2}\right)^2 = 2\left(1 - \frac{s_1^2 + s_3^2 - s_2^2}{2 s_1 s_3}\right)$$

$$\left(\frac{t_1 + t_2 + t_3}{l_3}\right)^2 = 2\left(1 - \frac{s_1^2 + s_2^2 - s_3^2}{2 s_1 s_2}\right).$$

See also ISOSCELIZER, YFF CENTER OF CONGRUENCE

Yff Points

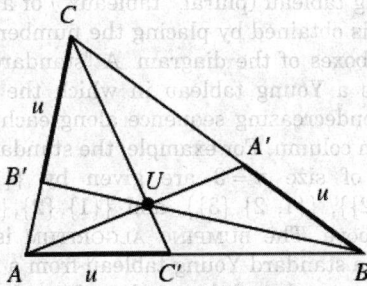

Let points A', B', and C' be marked off some fixed distance x along each of the sides BC, CA, and AB. Then the lines AA', BB', and CC' concur in a point U known as the first Yff point if

$$x^3 = (a - x)(b - x)(c - x). \tag{1}$$

This equation has a single real root u, which can by obtained by solving the CUBIC EQUATION

$$f(x) = 2x^3 - px^2 + qx - r = 0, \tag{2}$$

where

$$p = a + b + c \tag{3}$$

$$q = ab + ac + bc \tag{4}$$

$$r = abc. \tag{5}$$

The ISOTOMIC CONJUGATE POINT U' is called the second Yff point. The TRIANGLE CENTER FUNCTIONS of the first and second points are given by

$$\alpha = \frac{1}{a}\left(\frac{c - u}{b - u}\right)^{1/3} \tag{6}$$

and

$$\alpha' = \frac{1}{a}\left(\frac{b - u}{c - u}\right)^{1/3}, \tag{7}$$

respectively. Analogous to the inequality $\omega \le \pi/6$ for the BROCARD ANGLE ω, $u \le p/6$ holds for the Yff points, with equality in the case of an EQUILATERAL TRIANGLE. Analogous to

$$\omega < \alpha_i < \pi - 3\omega \tag{8}$$

for $i = 1, 2, 3$, the Yff points satisfy

$$u < a_i < p - 3u. \tag{9}$$

Yff (1963) gives a number of other interesting properties. The line UU' is PERPENDICULAR to the line containing the INCENTER I and CIRCUMCENTER O, and its length is given by

$$\overline{UU'} = \frac{4u\overline{IO}\Delta}{u^3 + abc}, \tag{10}$$

where Δ is the AREA of the TRIANGLE.

See also BROCARD POINTS, YFF TRIANGLES

References

Yff, P. "An Analog of the Brocard Points." *Amer. Math. Monthly* **70**, 495–501, 1963.

Yff Triangles

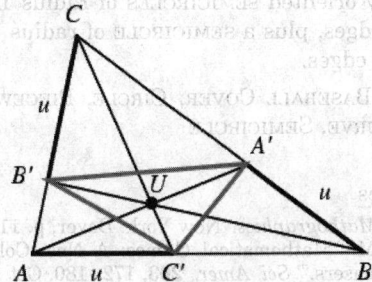

The TRIANGLE $\Delta A'B'C'$ formed by connecting the points used to construct the YFF POINTS is called the first Yff triangle. The AREA of the triangle is

$$\Delta = \frac{u^3}{2R},$$

where R is the CIRCUMRADIUS of the original TRIANGLE ΔABC. The second Yff triangle is formed by connecting the ISOTOMIC CONJUGATE POINTS of A', B', and C'.

See also YFF POINTS

References

Yff, P. "An Analog of the Brocard Points." *Amer. Math. Monthly* **70**, 495–501, 1963.

y-Intercept

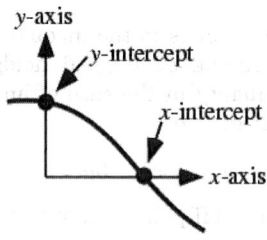

The point at which a curve or function crosses the Y-AXIS (i.e., when $x = 0$ in 2-D).

See also LINE, X-INTERCEPT

Yin-Yang

A figure used in many Asian cultures to symbolize the unity of the two "opposite" male and female elements, the "yin" and "yang." The solid and hollow parts composing the symbol are similar and combine to make a CIRCLE. Each part consists of two equal oppositely oriented SEMICIRCLES of radius 1/2 joined at their edges, plus a SEMICIRCLE of radius 1 joining the other edges.

See also BASEBALL COVER, CIRCLE, PIECEWISE CIRCULAR CURVE, SEMICIRCLE

References

Dixon, R. *Mathographics.* New York: Dover, p. 11, 1991.
Gardner, M. "Mathematical Games: A New Collection of 'Brain-Teasers.'" *Sci. Amer.* **203**, 172–180, Oct. 1960.
Gardner, M. "Mathematical Games: More About the Shapes that Can Be Made with Complex Dominoes." *Sci. Amer.* **203**, 186–198, Nov. 1960.

Young Diagram

FERRERS DIAGRAM, YOUNG TABLEAU

Young Girl-Old Woman Illusion

A perceptual ILLUSION in which the brain switches between seeing a young girl and an old woman.

See also RABBIT-DUCK ILLUSION

References

Pappas, T. *The Joy of Mathematics.* San Carlos, CA: Wide World Publ./Tetra, p. 173, 1989.

Young Tableau

The Young tableau (plural, "tableaux") of a FERRERS DIAGRAM is obtained by placing the numbers 1, ..., n in the n boxes of the diagram. A "standard" Young tableau is a Young tableau in which the numbers form a nondecreasing sequence along each line and along each column. For example, the standard Young tableaux of size $n = 3$ are given by $\{\{1, 2, 3\}\}$, $\{\{1, 3\}, \{2\}\}$, $\{\{1, 2\}, \{3\}\}$, and $\{\{1\}, \{2\}, \{3\}\}$, illustrated above. The BUMPING ALGORITHM is used to construct a standard Young tableau from a permutation of $\{1, \ldots, n\}$, and the number of standard Young tableaux of size 1, 2, 3, ... are 1, 2, 4, 10, 26, 76, 232, 764, 2620, 9496, ... (Sloane's A000085). These numbers can be generated by the RECURRENCE RELATION

$$a(n) = a(n-1) + (n-1)a(n-2)$$

with $a(1) = 1$ and $a(2) = 2$. This is the same as the number of INVOLUTIONS on n elements (Skiena 1990,

p. 32).

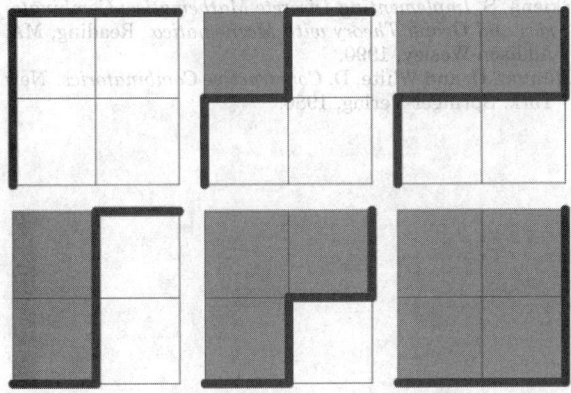

The number of all possible standard Young tableaux of a given shape can also be considered, and can be calculated with the HOOK LENGTH FORMULA. For example, the illustration above shows the 35 standard tableaux of shape $\{3, 2, 1, 1\}$.

The partitions of integers less than or equal to mn in which there are at most n parts and in which no part is larger than m correspond (1) to Young tableaux which fit inside and $m \times n$ rectangle and (2) to lattice paths which travel from the upper right corner of the rectangle to the lower left in $m + n$ leftward and downward steps. The number of Young diagrams fitting inside an $m \times n$ rectangle is given by the BINOMIAL COEFFICIENT $\binom{m+n}{m} = \binom{m+n}{n}$. The above example shows the

$$\binom{2+2}{2} = \binom{4}{2} = \frac{4!}{2!2!} = \frac{24}{4} = 6$$

Young 2×2 diagrams.

There is a correspondence between a PERMUTATION and a pair of Young tableaux, known as the SCHENSTED CORRESPONDENCE.

See also BUMPING ALGORITHM, DURFEE SQUARE, HOOK LENGTH FORMULA, INVOLUTION (PERMUTATION), PARTITION, PARTITION FUNCTION P, RANDOM TABLEAU SCHENSTED CORRESPONDENCE, TABLEAU CLASS

References

Bressoud, D. and Propp, J. "How the Alternating Sign Matrix Conjecture was Solved." *Not. Amer. Math. Soc.* **46**, 637–646.

Comtet, L. "Standard Tableaux." Ch. 2, Exercise 26 in *Advanced Combinatorics: The Art of Finite and Infinite Expansions, rev. enl. ed.* Dordrecht, Netherlands: Reidel, pp. 125–126, 1974.

Fulton, W. *Young Tableaux with Applications to Representation Theory and Geometry.* New York: Cambridge University Press, 1997.

Kreweras, G. "Sur une class de problèmes de dénombrement liés au treillis des partitions d'entiers." *Cahiers Buro* **6**, 2–107, 1965.

Kreweras, G. "Dénombrements de chemins minimaux à sauts imposés." *Comptes rendus* **263**, 1–3, 1966.

Kreweras, G. "Sur une extension du problème dir 'de Simon Newcomb'." *Comptes rendus* **263**, 43–45, 1966.

Kreweras, G. "Traitement simultané du 'problème de Young' et du 'problème de Simon Newcomb'." *Cahiers Buro* **10**, 23–31, 1967.

Messiah, A. Appendix D in *Quantum Mechanics, Vol. 2.* Amsterdam, Netherlands: North-Holland, p. 1113, 1961–62.

Ruskey, F. "Information on Permutations." http://www.theory.csc.uvic.ca/~cos/inf/perm/PermInfo.html#Tableau.

Skiena, S. "Young Tableaux." §2.3 in *Implementing Discrete Mathematics: Combinatorics and Graph Theory with Mathematica.* Reading, MA: Addison-Wesley, pp. 63–76, 1990.

Skiena, S. S. *The Algorithm Design Manual.* New York: Springer-Verlag, pp. 254–255, 1997.

Sloane, N. J. A. Sequences A000085/M1221 in "An On-Line Version of the Encyclopedia of Integer Sequences." http://www.research.att.com/~njas/sequences/eisonline.html.

Stanley, R. P. *Enumerative Combinatorics, Vol. 1.* Cambridge, England: Cambridge University Press, 1999.

Wilf, H. "The Computer-Aided Discovery of a Theorem about Young Tableaux." *J. Symb. Comput.* **20**, 731–735, 1995.

Young's Inequality

Let f be a real-valued, continuous, and strictly increasing function on $[0, c]$ with $c > 0$. If $f(0) = 0$, $a \in [0, c]$, and $b \in [0, f(c)]$, then

$$\int_0^a f(x)\, dx + \int_0^b f^{-1}(x)\, dx \geq ab, \qquad (1)$$

where f^{-1} is the INVERSE FUNCTION of f. Equality holds IFF $b = f(a)$.

Taking the particular function $f(x) = x^{p-1}$ gives the special case

$$\frac{a^p}{p} + \left(\frac{p-1}{p}\right) b^{p/(p-1)} \geq ab, \qquad (2)$$

which is often written in the symmetric form

$$\frac{a^p}{p} + \frac{b^q}{q} \geq ab, \qquad (3)$$

where $a, b \geq 0$, $p > 1$, and

$$\frac{1}{p} + \frac{1}{q} = 1. \tag{4}$$

0 and $b \leq f(a)$, then

$$ab \leq \int_0^a f(x)\,dx + \int_0^b f^{-1}(y)\,dy,$$

where $f^{-1}(y)$ is the INVERSE FUNCTION. Equality holds IFF $b = f(a)$.

References

Cooper, R. "Notes on Certain Inequalities. I." *J. London Math. Soc.* **2**, 17–21, 1927.

Cooper, R. "Notes on Certain Inequalities. II." *J. London Math. Soc.* **2**, 159–163, 1927.

Hardy, G. H.; Littlewood, J. E.; and Pólya, G. "A Theorem of W. H. Young." §8.3 in *Inequalities, 2nd ed.* Cambridge, England: Cambridge University Press, pp. 198–200, 1988.

Mitrinovic, D. S. "Young's Inequality." §2.7 in *Analytic Inequalities.* New York: Springer-Verlag, pp. 48–50, 1970.

Oppenheim, A. "Note on Mr. Cooper's Generalization of Young's Inequality." *J. London Math. Soc.* **2**, 21–23, 1927.

Riesz, F. "Su alcune disuguaglianze." *Boll. Un. Mat. Ital.* **7**, 77–79, 1928.

Takahashi, T. "Remarks on Some Inequalities." *Tôhoku Math. J.* **36**, 99–106, 1932.

Young, W. H. "On Classes of Summable Functions and Their Fourier Series." *Proc. Roy. Soc. London Ser. A* **87**, 225–229, 1912.

Young's Integral

Let $f(x)$ be a REAL continuous monotonic strictly increasing function on the interval $[0, a]$ with $f(0) =$

References

Gradshteyn, I. S. and Ryzhik, I. M. *Tables of Integrals, Series, and Products, 6th ed.* San Diego, CA: Academic Press, p. 1099, 2000.

Young's Lattice

Young's lattice Y_p is the PARTIAL ORDER of partitions CONTAINED within a PARTITION p ordered by containment (Stanton and White 1986; Skiena 1990, p. 77).

See also CONTAINED PARTITION, PARTITION

References

Skiena, S. *Implementing Discrete Mathematics: Combinatorics and Graph Theory with Mathematica.* Reading, MA: Addison-Wesley, 1990.

Stanton, D. and White, D. *Constructive Combinatorics.* New York: Springer-Verlag, 1986.

Z

Z

The DOUBLESTRUCK capital letter Z, $\mathbb{Z}$, denotes the RING of INTEGERS ..., $-2, -1, 0, 1, 2, ...$ The symbol derives from the German word *Zahl*, meaning "number" (Dummit and Foote 1998, p. 1). The RING of integers is sometimes also denoted using the double-struck capital I, $\mathbb{I}$.

See also C, C*, COUNTING NUMBER, I, N, NATURAL NUMBER, Q, R, WHOLE NUMBER, Z-, Z+

References

Dummit, D. S. and Foote, R. M. *Abstract Algebra, 2nd ed.* Englewood Cliffs, NJ: Prentice-Hall, p. 1, 1998.

Z-

The NEGATIVE INTEGERS ..., $-3, -2, -1$.

See also COUNTING NUMBER, NATURAL NUMBER, NEGATIVE, WHOLE NUMBER, Z, Z+, Z*

Z+

The POSITIVE INTEGERS 1, 2, 3, ..., equivalent to N.

See also COUNTING NUMBER, N, NATURAL NUMBER, POSITIVE, WHOLE NUMBER, Z, Z-, Z*

References

Dummit, D. S. and Foote, R. M. *Abstract Algebra, 2nd ed.* Englewood Cliffs, NJ: Prentice-Hall, p. 1, 1998.

Zag Number

An EVEN ALTERNATING PERMUTATION number, more commonly called a TANGENT NUMBER.

See also ALTERNATING PERMUTATION, TANGENT NUMBER, ZIG NUMBER

Zak Transform

This entry contributed by RONALD M. AARTS

The Zak transform is a signal transform relevant to time-continuous signals sampled at a uniform rate and an arbitrary clock phase (Janssen 1988). The Zak transform of a signal can be considered as a mixed time-frequency representation of f

$$(Zf)_T(t, \, v) = T^{1/2} \sum_{k=-\infty}^{\infty} f(t + kT)e^{-2\pi i k v T}$$

for $0 \le t \le T$ and $0 \le v \le T^{-1}$. The Zak transform is sometimes also known as the Weil-Brezin map.

References

Brezin, J. "Function Theory on Metabelian Solvmanifolds." *J. Funct. Analysis* **10**, 33–51, 1972.

Janssen, A. J. E. M. "The Zak Transform: A Signal Transform for Sampled Time-Continuous Signals." *Philips J. Res.* **43**, 23–69, 1988.

Weil, A. "Sur certains groupes d'opérateurs unitaires." *Acta Math.* **111**, 143–211, 1964.

Zak, J. *Phys. Rev. Lett.* **19**, 1385, 1967.

Zak, J. *Phys. Rev.* **168**, 686, 1968.

Zalcman's Lemma

Let f be a family of MEROMORPHIC FUNCTIONS on the UNIT DISK $\mathbb{D}$ which are not normal at 0. Then there exist sequences f_n in F, z_n, ρ_n, and a nonconstant function f meromorphic in the plane such that

$$f_n(z_n + \rho_n z) \to f(z)$$

locally and uniformly (in the spherical sense) in the COMPLEX PLANE $\mathbb{C}$ (Schwick 2000), where $z_n \to 0$ and $\rho_n \to 0$.

References

Schwick, W. "A Note on Zalcman's Lemma." *New Zealand J. Math.* **29**, 71–72, 2000.

Zalcman, L. "A Heuristic Principle in Complex Function Theory." *Amer. Math. Monthly* **82**, 813–817, 1975.

Zarankiewicz's Conjecture

The CROSSING NUMBER for a COMPLETE BIGRAPH is

$$\left\lfloor \frac{n}{2} \right\rfloor \left\lfloor \frac{n-1}{2} \right\rfloor \left\lfloor \frac{m}{2} \right\rfloor \left\lfloor \frac{m-1}{2} \right\rfloor,$$

where $\lfloor x \rfloor$ is the FLOOR FUNCTION. The original proof by Zarankiewicz (1954) contained an error, but was subsequently solved in some special cases by Guy (1969). The conjecture has been shown to be true for all m, $n \le 7$, and Zarankiewicz has shown that in general, the FORMULA provides an upper bound to the actual number.

See also COMPLETE BIGRAPH, CROSSING NUMBER (GRAPH)

References

Guy, R. K. "The Decline and Fall of Zarankiewicz's Theorem." In *Proof Techniques in Graph Theory, Proceedings of the Second Ann Arbor Graph Theory Conference, Ann Arbor, Michigan, 1968.* New York: Academic Press, pp. 63–69, 1969.

Zarankiewicz, K. "On a Problem of P. Turán Concerning Graphs." *Fund. Math.* **41**, 137–145, 1954.

Zariski Topology

A TOPOLOGY of an infinite set whose OPEN SETS have finite complements. The Zariski topology is a TOPOLOGY which is well-suited for the study of polynomial equations in ALGEBRAIC GEOMETRY, since in Zariski topology, there are many fewer OPEN SETS than in the usual METRIC TOPOLOGY. In fact, the only CLOSED SETS are the ALGEBRAIC SETS, which are the zeros of polynomials.

For example, in $\mathbb{C}$, the only nontrivial closed sets are finite collections of points. In $\mathbb{C}^2$, there are also the zeros of polynomials such as lines $ax + by$ and cusps $x^2 + y^3$.

The Zariski topology is not HAUSDORFF. In fact, any two open sets must intersect, and cannot be DISJOINT. Also, the open sets are DENSE, in the Zariski topology as well as in the usual METRIC TOPOLOGY.

Because there are fewer open sets than in the usual topology, it is more difficult for a function to be continuous in Zariski topology. For example, a CONTINUOUS FUNCTION $(\mathbb{C}^n, \text{Zariski}) \to (\mathbb{C}, \text{metric})$ must be a constant function. Conversely, when the range has the Zariski topology, it is easier for a function to be CONTINUOUS. In particular, the polynomials are CONTINUOUS FUNCTIONS $(\mathbb{C}^n, \text{Zariski}) \to (\mathbb{C}, \text{Zariski})$.

See also ALGEBRAIC VARIETY, CATEGORY THEORY, COMMUTATIVE ALGEBRA, CONIC SECTION, IDEAL, PRIME IDEAL, PROJECTIVE VARIETY, SCHEME

References
Bump, D. *Algebraic Geometry.* Singapore: World Scientific, pp. 1–6, 1998.
Hartshorne, R. *Algebraic Geometry.* New York: Springer-Verlag, 1977.

Zaslavskii Map
The 2-D map

$$x_{n+1} = [x_n + \nu(1 + \mu y_n) + \epsilon \nu \mu \cos(2\pi x_n)] \pmod 1$$

$$y_{n+1} = e^{-\Gamma}[y_n + \epsilon \cos(2\pi x_n)],$$

where

$$\mu \equiv \frac{1 - e^{-\Gamma}}{\Gamma}$$

(Zaslavskii 1978). It has CORRELATION EXPONENT $\nu \approx 1.5$ (Grassberger and Procaccia 1983) and CAPACITY DIMENSION 1.39 (Russell *et al.* 1980).

References
Grassberger, P. and Procaccia, I. "Measuring the Strangeness of Strange Attractors." *Physica D* **9**, 189–208, 1983.
Russell, D. A.; Hanson, J. D.; and Ott, E. "Dimension of Strange Attractors." *Phys. Rev. Let.* **45**, 1175–1178, 1980.
Zaslavskii, G. M. "The Simplest Case of a Strange Attractor." *Phys. Let.* **69A**, 145–147, 1978.

Zassenhaus-Berlekamp Algorithm
A method for factoring POLYNOMIALS.

z-Axis

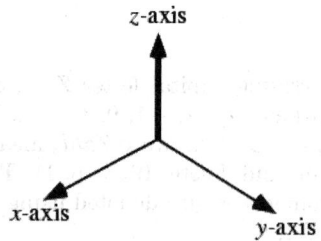

The axis in 3-D CARTESIAN COORDINATES which is usually oriented vertically. CYLINDRICAL COORDINATES are defined such that the z-axis is the axis about which the azimuthal coordinate θ is measured.

See also AXIS, x-AXIS, y-AXIS

z-Distribution
FISHER'S z-DISTRIBUTION, STUDENT'S z-DISTRIBUTION

Zeckendorf Representation
A number written as a sum of nonconsecutive FIBONACCI NUMBERS,

$$n = \sum_{k=0}^{L} \epsilon_k F_k,$$

where ϵ_k are 0 or 1 and

$$\epsilon_k \epsilon_{k+1} = 0.$$

Every POSITIVE INTEGER can be written uniquely in such a form.

See also ZECKENDORF'S THEOREM

References
Grabner, P. J.; Tichy, R. F.; Nemes, I.; and Petho, A. "On the Least Significant Digit of Zeckendorf Expansions." *Fib. Quart.* **34**, 147–151, 1996.
Vardi, I. *Computational Recreations in Mathematica.* Reading, MA: Addison-Wesley, p. 40, 1991.
Zeckendorf, E. "Représentation des nombres naturels par une somme de nombres de Fibonacci ou de nombres de Lucas." *Bull. Soc. Roy. Sci. Liège* **41**, 179–182, 1972.

Zeckendorf's Theorem
The SEQUENCE $\{F_n - 1\}$ is COMPLETE even if restricted to subsequences which contain no two consecutive terms, where F_n is a FIBONACCI NUMBER.

See also FIBONACCI DUAL THEOREM, ZECKENDORF REPRESENTATION

References
Brown, J. L. Jr. "Zeckendorf's Theorem and Some Applications." *Fib. Quart.* **2**, 163–168, 1964.
Keller, T. J. "Generalizations of Zeckendorf's Theorem." *Fib. Quart.* **10**, 95–112, 1972.
Lekkerkerker, C. G. "Voorstelling van natuurlijke getallen door een som van Fibonacci." *Simon Stevin* **29**, 190–195, 1951–52.

Zeeman's Paradox

There is only one point in front of a PERSPECTIVE drawing where its three mutually PERPENDICULAR VANISHING POINTS appear in mutually PERPENDICULAR directions, but such a drawing nonetheless appears realistic from a variety of distances and angles.

See also LEONARDO'S PARADOX, PERSPECTIVE, VANISHING POINT

References

Dixon, R. *Mathographics.* New York: Dover, p. 82, 1991.

Zeilberger-Bressoud Theorem

Dyson (1962abc) conjectured that the constant term in the LAURENT SERIES

$$\prod_{1 \le i \neq j \le n} \left(1 - \frac{x_i}{x_j}\right)^{a_i} \tag{1}$$

is

$$\frac{(a_1 + a_2 + \ldots + a_n)!}{a_1! a_2! \ldots a_n!}, \tag{2}$$

based on a problem in particle physics. The theorem is called DYSON'S CONJECTURE, and was proved by Wilson (1962) and independently by Gunson (1962). A definitive proof was subsequently published by Good (1970).

A q-analog of this theorem (Andrews 1975) states that the coefficient of $x_1^0 x_2^0 \ldots x_n^0$ in

$$\prod_{1 \le i \neq j \le n} \left(\frac{x_i}{x_j} \epsilon_{ij}; q\right)_{a_i} \tag{3}$$

where

$$\epsilon_{ij} \equiv \begin{cases} 1 & \text{for } i < j \\ q & \text{for } i > j \end{cases} \tag{4}$$

is given by

$$\frac{(q; q)_{a_1 + a_2 + \ldots + a_n}}{(q; q)_{a_1}(q; q)_{a_2} \cdots (q; q)_{a_n}}. \tag{5}$$

This can also be stated in the form that the constant term of

$$\prod_{1 \le i < j \le n} \left(1 - x_i/x_j\right)\left(1 - qx_i/q_j\right) \cdots \left(1 - q^{a_i - 1}x_i/x_j\right)$$

$$\times \left(1 - qx_j/x_i\right)\left(1 - q^2 x_j/x_i\right) \cdots \left(1 - q^{a_j} x_j/x_i\right), \tag{6}$$

is the Q-MULTINOMIAL COEFFICIENT

$$\frac{[a_1 + \cdots + a_n]!}{[a_1]! \cdots [a_n]!}, \tag{7}$$

where

$$[n]! \equiv (1)(1+q) \cdots (1+q+\cdots+q^{n-1}). \tag{8}$$

The amazing proof of this theorem was given by Zeilberger and Bressoud (1985).

The full theorem reduces to Dyson's version when $q = 1$. It also gives the Q-ANALOG of DIXON'S THEOREM as

$$\sum_{k=-\infty}^{\infty} (-1)^k q^{k(3k+1)/2} \binom{b+c}{c+k}\binom{c+a}{a+k}\binom{a+b}{b+k}$$

$$= \frac{(q; q)_{a+b+c}}{(q; q)_a (q; q)_b (q; q)_c} \tag{9}$$

(Andrews 1975, 1986). With $q = 1$ and $a = b = c = p$, it gives the beautiful and well-known identity

$$\sum_{k=0}^{2p} (-1)^k \binom{2p}{k}^3 = \frac{(-1)^p (3p)!}{(p!)^3} \tag{10}$$

(Andrews 1986).

See also DIXON'S THEOREM, Q-MULTINOMIAL COEFFICIENT, MACDONALD'S CONSTANT-TERM CONJECTURE, MULTINOMIAL COEFFICIENT

References

Andrews, G. E. "Problems and Prospects for Basic Hypergeometric Functions." In *The Theory and Application of Special Functions* (Ed. R. Askey). New York: Academic Press, pp. 191–224, 1975.

Andrews, G. E. "The Zeilberger-Bressoud Theorem." §4.3 in *q-Series: Their Development and Application in Analysis, Number Theory, Combinatorics, Physics, and Computer Algebra.* Providence, RI: Amer. Math. Soc., pp. 36–38, 1986.

Dyson, F. "Statistical Theory of the Energy Levels of Complex Systems. I." *J. Math. Phys.* **3**, 140–156, 1962a.

Dyson, F. "Statistical Theory of the Energy Levels of Complex Systems. II." *J. Math. Phys.* **3**, 157–165, 1962b.

Dyson, F. "Statistical Theory of the Energy Levels of Complex Systems. III." *J. Math. Phys.* **3**, 166–175, 1962c.

Good, I. J. "Short Proof of a Conjecture by Dyson." *J. Math. Phys.* **11**, 1884, 1970.

Gunson, J. "Proof of a Conjecture of Dyson in the Statistical Theory of Energy Levels." *J. Math. Phys.* **3**, 752–753, 1962.

Wilson, K. G. "Proof of a Conjecture by Dyson." *J. Math. Phys.* **3**, 1040–1043, 1962.

Zeilberger, D. and Bressoud, D. M. "A Proof of Andrews' q-Dyson Conjecture." *Disc. Math.* **54**, 201–224, 1985.

Zeilberger's Algorithm

An ALGORITHM which finds a POLYNOMIAL recurrence for terminating HYPERGEOMETRIC IDENTITIES OF THE FORM

$$\sum_k \binom{n}{k} \frac{\prod_{i=1}^{A}(a_i n + a_i' k + a_i'')!}{\prod_{i=1}^{B}(b_i n + b_i' k + b_i'')!} z^k$$

$$= C \frac{\prod_{i=1}^{\bar{A}}(\bar{a}_i n + \bar{a}_i')!}{\prod_{i=1}^{\bar{B}}(\bar{b}_i n + \bar{b}_i')} \bar{x}^n,$$

where $\binom{n}{k}$ is a BINOMIAL COEFFICIENT, a_i, a_i', $\bar{a}_i$, b_i, b_i',

$\bar{b}_i$ are constant integers and a_i'', $\bar{a}'_i$, b_i'', $\bar{b}'_i$, C, x, and z are complex numbers (Zeilberger 1990). The method was called CREATIVE TELESCOPING by van der Poorten (1979), and led to the development of the amazing machinery of WILF-ZEILBERGER PAIRS.

The also exists a q-analog of the algorithm, called the Q-ZEILBERGER ALGORITHM.

See also BINOMIAL SERIES, BINOMIAL SUMS, GOSPER'S ALGORITHM, HYPERGEOMETRIC IDENTITY, Q-ZEILBERGER ALGORITHM, SISTER CELINE'S METHOD, WILF-ZEILBERGER PAIR

References

Graham, R. L.; Knuth, D. E.; and Patashnik, O. *Concrete Mathematics: A Foundation for Computer Science, 2nd ed.* Reading, MA: Addison-Wesley, 1994.
Koepf, W. "Algorithms for m-fold Hypergeometric Summation." *J. Symb. Comput.* **20**, 399–417, 1995.
Koepf, W. "Zeilberger's Algorithm." Ch. 7 in *Hypergeometric Summation: An Algorithmic Approach to Summation and Special Function Identities.* Braunschweig, Germany: Vieweg, pp. 93–123, 1998.
Krattenthaler, C. "HYP and HYPQ: The Mathematica Package HYP." http://radon.mat.univie.ac.at/People/kratt/hyp_hypq/hyp.html.
Paule, P. and Riese, A. "A *Mathematica* q-Analogue of Zeilberger's Algorithm Based on an Algebraically Motivated Approach to q-Hypergeometric Telescoping." In *Special Functions, q-Series and Related Topics,* Fields Institute Communications **14**, 179–210, 1997.
Paule, P. and Schorn, M. "A Mathematica Version of Zeilberger's Algorithm for Proving Binomial Coefficient Identities." *J. Symb. Comput.* **20**, 673–698, 1995.
Petkovsek, M.; Wilf, H. S.; and Zeilberger, D. "Zeilberger's Algorithm." Ch. 6 in $A = B$. Wellesley, MA: A. K. Peters, pp. 101–119, 1996.
Riese, A. "A Generalization of Gosper's Algorithm to Bibasic Hypergeometric Summation." *Electronic J. Combinatorics* **1**, R19 1–16, 1996. http://www.combinatorics.org/Volume_1/volume1.html#R19.
van der Poorten, A. "A Proof that Euler Missed... Apéry's Proof of the Irrationality of $\zeta(3)$." *Math. Intel.* **1**, 196–203, 1979.
Wegschaider, K. *Computer Generated Proofs of Binomial Multi-Sum Identities.* Diploma Thesis, RISC. Linz, Austria: J. Kepler University, May 1997.
Zeilberger, D. "Doron Zeilberger's Maple Packages and Programs: EKHAD." http://www.math.temple.edu/~zeilberg/programs.html.
Zeilberger, D. "A Fast Algorithm for Proving Terminating Hypergeometric Series Identities." *Discrete Math.* **80**, 207–211, 1990.
Zeilberger, D. "A Holonomic Systems Approach to Special Function Identities." *J. Comput. Appl. Math.* **32**, 321–368, 1990.
Zeilberger, D. "The Method of Creative Telescoping." *J. Symb. Comput.* **11**, 195–204, 1991.

Zeisel Number

A number $N = p_1 p_2 \cdots p_n$ where the p_is are distinct PRIMES and $n \geq 3$ such that

$$p_i = A p_{i-1} + B$$

for $i = 1, 2, ..., n$, p_0 taken as 1, and with A and B some fixed integers. For example, $1885 = 1 \cdot 5 \cdot 13 \cdot$

29 is a Zeisel number with $(A, B) = (2, 3)$ since

$$5 = 2 \cdot 1 + 3; \ 13 = 2 \cdot 5 + 3; \ 29 = 2 \cdot 13 + 3,$$

as is $114985 = 1 \cdot 5 \cdot 13 \cdot 29 \cdot 61$ since

$$5 = 2 \cdot 1 + 3; \ 13 = 2 \cdot 5 + 3; \ 29 = 2 \cdot 13 + 3;$$

$$61 = 2 \cdot 29 + 3.$$

The first few Zeisel numbers are 105, 1419, 1729, 1885, 4505, ... (Sloane's A051015), which correspond to constants $(1, 2)$, $(4, -1)$, $(1, 6)$, $(2, 3)$, $(3, 2)$, ...

References

Brown, K. S. "Zeisel Numbers." http://www.seanet.com/~ksbrown/kmath015.htm.
Sloane, N. J. A. Sequences A051015 in "An On-Line Version of the Encyclopedia of Integer Sequences." http://www.research.att.com/~njas/sequences/eisonline.html.

Zenithal Projection

AZIMUTHAL PROJECTION

Zeno's Paradoxes

A set of four PARADOXES dealing with counterintuitive aspects of continuous space and time.

1. Dichotomy paradox: Before an object can travel a given distance d, it must travel a distance $d/2$. In order to travel $d/2$, it must travel $d/4$, etc. Since this sequence goes on forever, it therefore appears that the distance d cannot be traveled. The resolution of the paradox awaited CALCULUS and the proof that infinite GEOMETRIC SERIES such as $\sum_{i=1}^{\infty}(1/2)^i = 1$ can converge, so that the infinite number of "half-steps" needed is balanced by the increasingly short amount of time needed to traverse the distances.

2. Achilles and the tortoise paradox: A fleet-of-foot Achilles is unable to catch a plodding tortoise which has been given a head start, since during the time it takes Achilles to catch up to a given position, the tortoise has moved forward some distance. But this is obviously fallacious since Achilles will clearly pass the tortoise! The resolution is similar to that of the dichotomy paradox.

3. Arrow paradox: An arrow in flight has an instantaneous position at a given instant of time. At that instant, however, it is indistinguishable

from a motionless arrow in the same position, so how is the motion of the arrow perceived?

4. Stade paradox: A paradox arising from the assumption that space and time can be divided only by a definite amount.

References

Erickson, G. W. and Fossa, J. A. *Dictionary of Paradox.* Lanham, MD: University Press of America, pp. 218–220, 1998.

Gardner, M. *The Sixth Book of Mathematical Games from Scientific American.* Chicago, IL: University of Chicago Press, pp. 163–166, 1984.

Grünbaum, A. *Modern Science and Zeno's Paradoxes.* Middletown, CT: Wesleyan University Press, 1967.

Pappas, T. "Zeno's Paradox--Achilles & the Tortoise." *The Joy of Mathematics.* San Carlos, CA: Wide World Publ./Tetra, pp. 116–117, 1989.

Russell, B. *Our Knowledge and the External World as a Field for Scientific Method in Philosophy.* New York: Routledge, 1993.

Salmon, W. (Ed.). *Zeno's Paradoxes.* New York: Bobs-Merrill, 1970.

Stewart, I. "Objections from Elea." In *From Here to Infinity: A Guide to Today's Mathematics.* Oxford, England: Oxford University Press, p. 72, 1996.

vos Savant, M. *The World's Most Famous Math Problem.* New York: St. Martin's Press, pp. 50–55, 1993.

Zermelo Set Theory

The version of set theory obtained if Axiom 6 of ZERMELO-FRAENKEL SET THEORY is replaced by

6'. Selection axiom (or "axiom of subsets"): for any set-theoretic formula $A(u)$, $\forall x \exists y \forall u(u \in y \equiv u \in x \wedge A(u))$,

which can be deduced from Axiom 6. However, there seems to be some disagreement in the literature about just which axioms of ZERMELO-FRAENKEL SET THEORY constitute "Zermelo Set Theory." Mendelson (1997) does *not* include the AXIOMS OF CHOICE, FOUNDATION, REPLACEMENT In Zermelo set theory, but does includes 6'. However, Enderton (1977) includes the AXIOMS OF CHOICE and FOUNDATION, but does not include the AXIOMS OF REPLACEMENT or Selection.

See also SET THEORY, ZERMELO-FRAENKEL SET THEORY

References

Enderton, H. B. *Elements of Set Theory.* New York: Academic Press, 1977.

Mendelson, E. *Introduction to Mathematical Logic, 4th ed.* London: Chapman & Hall, 1997.

Iyanaga, S. and Kawada, Y. (Eds.). "Zermelo-Fraenkel Set Theory." §35B in *Encyclopedic Dictionary of Mathematics.* Cambridge, MA: MIT Press, p. 135, 1980.

Zermelo, E. "Über Grenzzahlen und Mengenbereiche." *Fund. Math.* **16**, 29–47, 1930.

Zermelo-Fraenkel Axioms

The Zermelo-Fraenkel axioms are the basis for ZERMELO-FRAENKEL SET THEORY. In the following (Iyanaga and Kawada 1980), $\exists$ stands for EXISTS, $\not\exists$ for does not exist, $\in$ for "is an element of," $\varnothing$ for the EMPTY SET, $\forall$ for FOR ALL, $\Rightarrow$ for IMPLIES, ! for NOT (NEGATION), $\wedge$ for AND, $\vee$ for OR, $\equiv$ for "is EQUIVALENT to," and S denotes the union y of all the sets that are the elements of x.

1. AXIOM OF EXTENSIONALITY: $\forall x(x \in a \equiv x \in b) \Rightarrow a = b$.
2. AXIOM OF THE UNORDERED PAIR: $\exists x \forall y(y \in x \equiv y = a \vee y = b)$.
3. AXIOM OF THE SUM SET: $\exists x \forall y(y \in x \equiv \exists z \in a(y \in z))$.
4. AXIOM OF THE POWER SET: $\forall x \exists y y \in x \equiv \forall z \in y(z \in a))$.
5. AXIOM OF THE EMPTY SET: $\exists x \forall y(!y \in x)$.
6. AXIOM OF INFINITY: $\exists x(\varnothing \in x \vee \forall y \in x(y' \in x))$.
7. AXIOM OF SEPARATION: $\exists x \forall y(y \in x \equiv y \in a \wedge A(y))$.
8. AXIOM OF REPLACEMENT (or axiom of comprehension, or axiom of subsets): $\exists x \forall y \in a(\exists z A(y, z) \Rightarrow \exists z \in x A(y, z))$.
9. Axiom of regularity (or AXIOM OF FOUNDATION): $\exists x A(x) \Rightarrow \exists x(A(x) \wedge \forall y \in x(!A(y)))$.
10. AXIOM OF CHOICE: $\forall x \in a \exists A(x, y) \Rightarrow \exists y \forall x \in a A(x, y(x))$.

The system of axioms 1–9 is called ZERMELO-FRAENKEL SET THEORY, denoted "ZF." The system of axioms 1–9 minus the AXIOM OF REPLACEMENT (i.e., axioms 1–7 plus 8) is called ZERMELO SET THEORY, denoted "Z." The set of axioms 1–9 plus the AXIOM OF CHOICE is usually denoted "ZFC."

However, note that there seems to be some disagreement in the literature about just what axioms constitute "ZERMELO SET THEORY." Mendelson (1997) does *not* include the AXIOMS OF CHOICE, FOUNDATION, or REPLACEMENT in Zermelo set theory, but does include the AXIOM OF REPLACEMENT. However, Enderton (1977) includes the AXIOMS OF CHOICE and FOUNDATION, but does not include the AXIOM OF REPLACEMENT.

Abian (1969) proved CONSISTENCY and independence of four of the Zermelo-Fraenkel axioms.

See also AXIOM OF CHOICE, AXIOM OF FOUNDATION, AXIOM OF REPLACEMENT, SET THEORY, VON NEUMANN-BERNAYS-GÖDEL SET THEORY, ZERMELO-FRAENKEL SET THEORY, ZERMELO SET THEORY

References

Abian, A. "On the Independence of Set Theoretical Axioms." *Amer. Math. Monthly* **76**, 787–790, 1969.

Enderton, H. B. *Elements of Set Theory.* New York: Academic Press, 1977.

Itô, K. (Ed.). "Zermelo-Fraenkel Set Theory." §33B in *Encyclopedic Dictionary of Mathematics, 2nd ed., Vol. 1.* Cambridge, MA: MIT Press, pp. 146–148, 1986.

Iyanaga, S. and Kawada, Y. (Eds.). "Zermelo-Fraenkel Set Theory." §35B in *Encyclopedic Dictionary of Mathematics, Vol. 1.* Cambridge, MA: MIT Press, pp. 134–135, 1980.

Mendelson, E. *Introduction to Mathematical Logic, 4th ed.* London: Chapman & Hall, 1997.

Zermelo, E. "Über Grenzzahlen und Mengenbereiche." *Fund. Math.* **16**, 29–47, 1930.

Zermelo-Fraenkel Set Theory

A version of SET THEORY which is a formal system expressed in first-order predicate LOGIC. Zermelo-Fraenkel set theory is based on the ZERMELO-FRAEN-KEL AXIOMS.

ZERMELO-FRAENKEL SET THEORY is not finitely axiomatized. For example, the AXIOM OF REPLACEMENT is not really a single axiom, but an infinite family of axioms, since it is preceded by the stipulation that it is true "For any set-theoretic formula $A(u, v)$." Montague (1961) proved that ZERMELO-FRAENKEL SET THEORY is not finitely axiomatizable, i.e., there is no finite set of axioms which is logically equivalent to the infinite set of ZERMELO-FRAENKEL AXIOMS. VON NEU-MANN-BERNAYS-GÖDEL SET THEORY provides an equivalent finitely axiomized system.

See also LOGIC, SET THEORY, VON NEUMANN-BER-NAYS-GÖDEL SET THEORY, ZERMELO-FRAENKEL AX-IOMS, ZERMELO SET THEORY

References

Montague, R. "Semantic Closure and Non-Finite Axiomatizability. I." In *Infinitistic Methods, Proceedings of the Symposium on Foundations of Mathematics, (Warsaw, 2–9 September 1959).* Oxford, England: Pergamon, pp. 45–69, 1961.

Zermelo, E. "Über Grenzzahlen und Mengenbereiche." *Fund. Math.* **16**, 29–47, 1930.

Zermelo's Axiom of Choice

AXIOM OF CHOICE

Zernike Polynomial

ORTHOGONAL POLYNOMIALS which arise in the expansion of a wavefront function for optical systems with circular pupils. The ODD and EVEN Zernike polynomials are given by

$$
\begin{aligned}
{}^{o}U_n^m(\rho, \phi) \\
{}^{e}U_n^m(\rho, \phi)
\end{aligned} = R_n^m(\rho) \begin{aligned} \sin \\ \cos \end{aligned} (m\phi) \tag{1}
$$

with radial function

$$
R_n^m(\rho) = \sum_{l=0}^{(n-m)/2} \frac{(-1)^l (n-l)!}{l! \left[\frac{1}{2}(n+m) - l\right]! \left[\frac{1}{2}(n-m) - l\right]!} \rho^{n-2l} \tag{2}
$$

for n and m integers with $n \geq m \geq 0$ and $n - m$ EVEN. Otherwise,

$$
R_n^m(\rho) = 0. \tag{3}
$$

Here, ϕ is the azimuthal angle with $0 \leq \phi < 2\pi$ and ρ is the radial distance with $0 \leq \rho \leq 1$ (Prata and Rusch 1989). The radial functions satisfy the orthogonality relation

$$
\int_0^1 R_n^m(\rho) R_{n'}^m(\rho) \rho \, d\rho = \frac{1}{2(n+1)} \delta_{nn'}, \tag{4}
$$

where δ_{ij} is the KRONECKER DELTA, and are related to the BESSEL FUNCTION OF THE FIRST KIND by

$$
\int_0^1 R_n^m(\rho) J_m(v\rho) \rho \, d\rho = (-1)^{(n-m)/2} \frac{J_{n+1}(v)}{v} \tag{5}
$$

(Born and Wolf 1989, p. 466). The radial Zernike polynomials have the GENERATING FUNCTION

$$
\frac{\left[1 + z - \sqrt{1 - 2z(1 - 2\rho^2) + z^2}\right]^m}{(2z\rho)^m \sqrt{1 - 2z(1 - 2\rho^2) + z^2}} = \sum_{s=0}^{\infty} z^s R_{m+2s}^{\pm m}(\rho), \tag{6}
$$

and are normalized so that

$$
R_n^{\pm m}(1) = 1 \tag{7}
$$

(Born and Wolf 1989, p. 465). The first few NONZERO radial polynomials are

$$
R_0^0(\rho) = 1
$$

$$
R_1^1(\rho) = \rho
$$

$$
R_2^0(\rho) = 2\rho^2 - 1
$$

$$
R_2^2(\rho) = \rho^2
$$

$$
R_3^1(\rho) = 3\rho^3 - 2\rho
$$

$$
R_3^3(\rho) = \rho^3
$$

$$
R_4^0(\rho) = 6\rho^4 - 6\rho^2 + 1
$$

$$
R_4^2(\rho) = 4\rho^4 - 3\rho^2
$$

$$
R_4^4(\rho) = \rho^4
$$

(Born and Wolf 1989, p. 465).

The Zernike polynomial is a special case of the JACOBI POLYNOMIAL with

$$
P_{n'}^{(\alpha, \beta)}(x) = (-1)^{n'} \frac{R_n^m(\rho)}{\rho^\alpha} \tag{8}
$$

and

$$
x = 1 - 2\rho^2 \tag{9}
$$

$$
\beta = 0 \tag{10}
$$

$$
\alpha = m \tag{11}
$$

$$
n' = \tfrac{1}{2}(n - m). \tag{12}
$$

The Zernike polynomials also satisfy the RECURRENCE RELATIONS

$$\rho R_n^m(\rho) = \frac{1}{2(n+1)}$$
$$\times \left[(n+m+2)R_{n+1}^{m+1}(\rho) + (n-m)R_{n-1}^{m-1}(\rho) \right] \tag{13}$$

$$R_{n+2}^m(\rho) = \frac{n+2}{(n+2)^2 - m^2} \left\{ \left[4(n+1)\rho^2 - \frac{(n+m)^2}{n} \right] \right.$$
$$\left. - \frac{(n-m+2)^2}{n+2} \right] R_n^m(\rho) - \frac{n^2 - m^2}{n} R_{n-2}^m(\rho) \right\} \tag{14}$$

$$R_n^m(\rho) + R_n^{m+2}(\rho) = \frac{1}{n+1} \frac{d\left[R_{n+1}^{m+1}(\rho) - R_{n-1}^{m+1}(\rho)\right]}{d\rho} \tag{15}$$

(Prata and Rusch 1989). The coefficients A_n^m and B_n^m in the expansion of an arbitrary radial function $F(\rho, \phi)$ in terms of Zernike polynomials

$$F(\rho, \phi) = \sum_{m=0}^{\infty} \sum_{n=m}^{\infty} \left[A_n^m {}^oU_n^m(\rho, \phi) + B_n^m {}^eU_n^m(\rho, \phi) \right] \tag{16}$$

are given by

$$\begin{matrix} A_n^m \\ B_n^m \end{matrix} = \frac{(n+1)}{\epsilon_{mn}^2 \pi} \int_0^1 \int_0^{2\pi} F(\rho, \phi) {}^o_e \begin{matrix} U_n^m(\rho, \phi) \\ U_n^m(\rho, \phi) \end{matrix} \rho \, d\phi \, d\rho, \tag{17}$$

where

$$\epsilon_{mn} \equiv \begin{cases} \epsilon \equiv \dfrac{1}{\sqrt{2}} & \text{for } m=0, \ n \neq 0 \\ 1 & \text{otherwise} \end{cases} \tag{18}$$

Let a "primary" aberration be given by

$$\Phi = a'_{lmn} \bar{Y}_1^{2l+m}(\theta, \phi)\rho^n \cos^m \theta \tag{19}$$

with $2l + m + n = 4$ and where $\bar{Y}$ is the COMPLEX CONJUGATE of Y, and define

$$A'_{lmn} = a'_{lmn} \bar{Y}_1^{2l+m}(\theta, \phi), \tag{20}$$

giving

$$\Phi = \frac{1}{\epsilon_{nm}^2} A_{lmn} R_n^m(\rho) \cos(m\theta). \tag{21}$$

Then the types of primary aberrations are given in the following table (Born and Wolf 1989, p. 470).

Aberration	l	m	n	A	A'
spherical aberration	0	4	0	$A'_{040}\rho^4$	$\epsilon A_{040} R_4^0(\rho)$
coma	0	3	1	$A'_{031}\rho^3 \cos\theta$	$A_{031} R_3^1(\rho) \cos\theta$
astigmatism	0	2	2	$A'_{022}\rho^2 \cos^2\theta$	$A_{022} R_2^2(\rho) \cos(2\theta)$
field curvature	1	2	0	$A'_{120}\rho^2$	$\epsilon A_{120} R_2^0\rho$
distortion	1	1	1	$A'_{111}\rho \cos\theta$	$A_{111} R_1^1(\rho) \cos\theta$

See also JACOBI POLYNOMIAL

References

Bezdidko, S. N. "The Use of Zernike Polynomials in Optics." *Sov. J. Opt. Techn.* **41**, 425, 1974.

Bhatia, A. B. and Wolf, E. "On the Circle Polynomials of Zernike and Related Orthogonal Sets." *Proc. Cambridge Phil. Soc.* **50**, 40, 1954.

Born, M. and Wolf, E. "The Diffraction Theory of Aberrations." Ch. 9 in *Principles of Optics: Electromagnetic Theory of Propagation, Interference, and Diffraction of Light, 6th ed.* New York: Pergamon Press, pp. 459–490, 1989.

Mahajan, V. N. "Zernike Circle Polynomials and Optical Aberrations of Systems with Circular Pupils." In *Engineering and Lab. Notes* **17** (Ed. R. R. Shannon), p. S-21, Aug. 1994.

Prata, A. and Rusch, W. V. T. "Algorithm for Computation of Zernike Polynomials Expansion Coefficients." *Appl. Opt.* **28**, 749–754, 1989.

Wang, J. Y. and Silva, D. E. "Wave-Front Interpretation with Zernike Polynomials." *Appl. Opt.* **19**, 1510–1518, 1980.

Wyant, J. C. "Zernike Polynomials." http://wyant.opt-sci.arizona.edu/zernikes/zernikes.htm.

Zernike, F. "Beugungstheorie des Schneidenverfahrens und seiner verbesserten Form, der Phasenkontrastmethode." *Physica* **1**, 689–704, 1934.

Zhang, S. and Shannon, R. R. "Catalog of Spot Diagrams." Ch. 4 in *Applied Optics and Optical Engineering, Vol. 11.* New York: Academic Press, p. 201, 1992.

Zero

The INTEGER denoted 0 which, when used as a counting number, means that no objects are present. It is the only INTEGER (and, in fact, the only REAL NUMBER) which is neither NEGATIVE nor POSITIVE. A number which is not zero is said to be NONZERO. A ROOT of a function f is also sometimes known as "a zero of f."

Because the number of PERMUTATIONS of 0 elements is 1, 0! (zero FACTORIAL) is defined as 1 (Wells 1986, p. 31). This definition is useful in expressing many mathematical identities in simple form. A number *other than 0* taken to the POWER 0 is defined to be 1, but 0^0 is undefined. Defining $0^0 = 1$ allows some formulas to be expressed simply (Knuth 1997, p. 56), although the same could be said for the alternate definition $0^0 = 0$ (Wells 1986, p. 26). An example of a formula which can be expressed concisely by defining $0^0 = 1$ is the beautiful analytical formula for the integral of the generalized SINC FUNCTION

$$\int_0^{\infty} \frac{\sin^a x}{x^b} \, dx = \frac{\pi^{1-c}(-1)^{\lfloor (a-b)/2 \rfloor}}{2^{a-c}(b-1)!}$$

$$\times \sum_{k=0}^{\lfloor a/2 \rfloor - c} (-1)^k \binom{a}{k} (a - 2k)^{b-1} [\ln(a - 2k)]c$$

given by Kogan, where $a \geq b > c$, $c \equiv a - b \pmod 2$, and $\lfloor x \rfloor$ is the FLOOR FUNCTION.

The following table gives the first few numbers n such that the decimal expansion of k^n contains no zeros, for small k. The largest known n for which 2^n contain no zeros is 86 (Madachy 1979), with no other $n \leq 4.6 \times 10^7$ (M. Cook), improving the 3.0739×10^7 limit obtained by Beeler and Gosper (1972). The values $a(n)$ such that the positions of the right-most zero in $2^{a(n)}$ increases are 10, 20, 30, 40, 46, 68, 93, 95, 129, 176, 229, 700, 1757, 1958, 7931, 57356, 269518, ... (Sloane's A031140). The positions in which the right-most zeros occur are 2, 5, 8, 11, 12, 13, 14, 23, 36, 38, 54, 57, 59, 93, 115, 119, 120, 121, 136, 138, 164, ... (Sloane's A031141). The right-most zero of $2^{781,717,865}$ occurs at the 217th decimal place, the farthest over for powers up to 2.5×10^9.

k	Sloane	n such that k^n contains no 0s
2	Sloane's A007377	1, 2, 3, 4, 5, 6, 7, 8, 9, 13, 14, 15, 16, 18, 19, 24, 25, 27, 28, ...
3	Sloane's A030700	1, 2, 3, 4, 5, 6, 7, 8, 9, 11, 12, 13, 14, 19, 23, 24, 26, 27, 28, ...
4	Sloane's A030701	1, 2, 3, 4, 7, 8, 9, 12, 14, 16, 17, 18, 36, 38, 43, ...
5	Sloane's A008839	1, 2, 3, 4, 5, 6, 7, 9, 10, 11, 17, 18, 30, 33, 58, ...
6	Sloane's A030702	1, 2, 3, 4, 5, 6, 7, 8, 12, 17, 24, 29, 44, ...
7	Sloane's A030703	1, 2, 3, 6, 7, 10, 11, 19, 35
8	Sloane's A030704	1, 2, 3, 5, 6, 8, 9, 11, 12, 13, 17, 24, 27
9	Sloane's A030705	1, 2, 3, 4, 6, 7, 12, 13, 14, 17, 34
11	Sloane's A030706	1, 2, 3, 4, 6, 7, 8, 9, 12, 13, 14, 15, 16, 18, 41, ...

While it has not been proven that the numbers listed above are the only ones without zeros for a given base, the probability that any additional ones exist is vanishingly small. Under this assumption, the sequence of largest n such that k^n contains no zeros for $k = 2, 3, ...$ is then given by 86, 68, 43, 58, 44, 35, 27, 34, 0, 41, ... (Sloane's A020665).

See also 10, APPROXIMATE ZERO, DIVISION BY ZERO, FALLACY, NAUGHT, NEGATIVE, NONNEGATIVE, NONZERO, ONE, POSITIVE, TWO, ZEROFREE

References
Beeler, M. and Gosper, R. W. Item 57 in Beeler, M.; Gosper, R. W.; and Schroeppel, R. *HAKMEM.* Cambridge, MA: MIT Artificial Intelligence Laboratory, Memo AIM-239, p. 22, Feb. 1972.
Knuth, D. E. *The Art of Computer Programming, Vol. 1: Fundamental Algorithms, 3rd ed.* Reading, MA: Addison-Wesley, p. 56, 1997.
Kogan, S. "A Note on Definite Integrals Involving Trigonometric Functions." http://www.mathsoft.com/asolve/constant/pi/sin/sin.html.
Madachy, J. S. *Madachy's Mathematical Recreations.* New York: Dover, pp. 127–128, 1979.
Pappas, T. "Zero-Where & When." *The Joy of Mathematics.* San Carlos, CA: Wide World Publ./Tetra, p. 162, 1989.
Sloane, N. J. A. Sequences A007377/M0485 in "An On-Line Version of the Encyclopedia of Integer Sequences." http://www.research.att.com/~njas/sequences/eisonline.html.
Wells, D. *The Penguin Dictionary of Curious and Interesting Numbers.* Middlesex, England: Penguin Books, pp. 23–26, 1986.

Zero (Root)

ROOT

Zero Divisor

A NONZERO element x of a RING for which $x \cdot y = 0$, where y is some other NONZERO element and the multiplication $x \cdot y$ is the multiplication of the RING. A RING with no zero divisors is known as an INTEGRAL DOMAIN. Let A denote an $\mathbb{R}$-algebra, so that A is a VECTOR SPACE over R and

$$A \times A \rightarrow A$$
$$(x, y) \mapsto x \cdot y.$$

Now define

$$Z \equiv \{x \in A : x \cdot y = 0 \text{ for some nonzero } y \in A\},$$

where $0 \in Z$. A is said to be m-ASSOCIATIVE if there exists an m-dimensional SUBSPACE S of A such that $(y \cdot x) \cdot z = y \cdot (x \cdot z)$ for all y, $z \in A$ and $x \in S$. A is said to be TAME if Z is a finite union of SUBSPACES of A.

References
Finch, S. "Unsolved Mathematics Problems: Zero Structures in Real Algebras." http://www.mathsoft.com/asolve/zero-div/zerodiv.html.

Zero Irrelevancy Proof

CLASSIFICATION THEOREM OF SURFACES

Zero Map

See also IDENTITY MAP

Zero Matrix

A MATRIX consisting of all 0s, denoted 0. The MATRIX EXPONENTIAL of 0 is given by the IDENTITY MATRIX I. An $m \times n$ zero matrix can be generated using `ZeroMatrix[m, n]` in the *Mathematica* add-on package `LinearAlgebra`MatrixMultiplication`

(which can be loaded with the command
`< < LinearAlgebra')`.

See also IDENTITY MATRIX

Zero Section

This entry contributed by RYAN BUDNEY

The zero section of a VECTOR BUNDLE is the SUBMANI-
FOLD of the bundle that consists of all the ZERO
VECTORS.

See also BUNDLE, MANIFOLD, SECTION (BUNDLE),
VECTOR BUNDLE, ZERO VECTOR

Zero Set

If f is a function on an OPEN SET U, then the zero set
of f is the set $Z = \{z \in U : f(z) = 0\}$.

References
Krantz, S. G. *Handbook of Complex Analysis.* Boston, MA:
Birkhäuser, p. 268, 1999.

Zero Vector

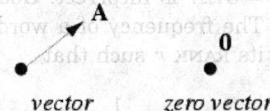

vector   zero vector

A ZERO VECTOR, denoted **0**, is a VECTOR of length 0,
and thus has all components equal to zero.

See also UNIT VECTOR, ZERO VECTOR

Zero-Form

See also DIFFERENTIAL K-FORM, ONE-FORM, TWO-
FORM

Zerofree

An integer whose decimal digits contain no zeros is
said to be zerofree. Zerofree squares are easy to
generate, e.g.,

$$3333333333333334^2$$
$$= 11111111111111115555555555555556. \quad (1)$$

Around 1990, D. Hickerson considered the problem of
finding large zerofree cubes. After some experimentation, he found a formula that generated infinitely
many of them. In March 1998, Bill Gosper asked
about 0-free nth powers, pointing out that heuristi-
cally we should expect there to be infinitely many
zerofree squares, cubes, ..., 21st powers, but only
finitely many 22nd powers, etc. At this point, Hickerson couldn't locate his formula for cubes, and so
came up with the new formula

$$f(n) = \frac{2 \cdot 10^{5n} - 10^{4n} + 17 \cdot 10^{3n-1} + 10^{2n} + 10^{n-2}}{3},$$
$$(2)$$

which is 0-free if $n \equiv 2 \pmod 3$ and $n \geq 5$.

In April 1999, Ed Pegg conjectured on `sci.math` that
there are only finitely many zerofree cubes, so
Hickerson posted his new counterexample, (mista-
kenly claiming that it was the one he had found 10
years ago). A few days later, Lew Baxter posted the
slightly simpler example

$$f(n) = \tfrac{1}{3}(2 \cdot 10^{5n} - 10^{4n} + 2 \cdot 10^{3n} + 10^{2n} + 10^n + 1), \quad (3)$$

known as the BAXTER-HICKERSON FUNCTION.

There is apparently no proof that there exist infinitely many zerofree 4th powers, 5th powers, ..., or
21st powers.

See also BAXTER-HICKERSON FUNCTION, ZERO

Zero-Sum Game

A GAME in which players make payments only to each
other. One player's loss is the other player's gain, so
the total amount of "money" available remains constant.

See also FINITE GAME, GAME

References
Dresher, M. *The Mathematics of Games of Strategy: Theory
and Applications.* New York: Dover, p. 2, 1981.

Zeta

HURWITZ ZETA FUNCTION, RIEMANN ZETA FUNCTION

Zeta Fuchsian

The zeta Fuchsians are class of functions discovered
by Poincaré which are related to the AUTOMORPHIC
FUNCTIONS.

See also AUTOMORPHIC FUNCTION

Zeta Function

A function satisfying certain properties which is
computed as an INFINITE SUM of NEGATIVE POWERS.
The most commonly encountered zeta function is the
RIEMANN ZETA FUNCTION,

$$\zeta(n) \equiv \sum_{k=1}^{\infty} \frac{1}{k^n}.$$

See also DEDEKIND FUNCTION, DIRICHLET BETA
FUNCTION, DIRICHLET ETA FUNCTION, DIRICHLET L-
SERIES, DIRICHLET LAMBDA FUNCTION, EPSTEIN ZETA
FUNCTION, JACOBI ZETA FUNCTION, NINT ZETA FUNC-

TION, PERIODIC ZETA FUNCTION, PRIME ZETA FUNCTION, RIEMANN ZETA FUNCTION, SELBERG ZETA FUNCTION

References

Ireland, K. and Rosen, M. "The Zeta Function." Ch. 11 in *A Classical Introduction to Modern Number Theory, 2nd ed.* New York: Springer-Verlag, pp. 151–171, 1990.

Zeuthen's Rule

On an ALGEBRAIC CURVE, the sum of the number of coincidences at a noncuspidal point C is the sum of the orders of the infinitesimal distances from a nearby point P to the corresponding points when the distance PC is taken as the principal infinitesimal.

References

Coolidge, J. L. *A Treatise on Algebraic Plane Curves.* New York: Dover, p. 131, 1959.

Zeuthen's Theorem

If there is a (v, v') correspondence between two curves of GENUS p and p' and the number of BRANCH POINTS properly counted are β and β', then

$$\beta + 2v'(p-1) = \beta' + 2v(p'-1).$$

See also CHASLES-CAYLEY-BRILL FORMULA

References

Coolidge, J. L. *A Treatise on Algebraic Plane Curves.* New York: Dover, p. 246, 1959.

Zig Number

An ODD ALTERNATING PERMUTATION number, more commonly called an EULER NUMBER or SECANT NUMBER.

See also ALTERNATING PERMUTATION, EULER NUMBER, ZAG NUMBER

Zigzag Permutation

ALTERNATING PERMUTATION

Zig-Zag Triangle

SEIDEL-ENTRINGER-ARNOLD TRIANGLE

Zillion

A generic word for a very LARGE NUMBER. The term has no WELL DEFINED mathematical meaning. Conway and Guy (1996) define the nth zillion as 10^{3n+3} in the American system (million = 10^6, billion = 10^9, trillion = 10^{12}, ...), and 10^{6n} in the British system

(million = 10^6, billion = 10^{12}, trillion = 10^{18}, ...), Conway and Guy (1996) also define the words N-PLEX and N-MINEX for 10^n and 10^{-n}, respectively.

See also LARGE NUMBER

References

Conway, J. H. and Guy, R. K. *The Book of Numbers.* New York: Springer-Verlag, pp. 13–16, 1996.

Zip

Half a ZIP-PAIR.

ZIP Proof

CLASSIFICATION THEOREM OF SURFACES

Zipf's Law

In the English language, the probability of encountering the rth most common word is given roughly by $P(r) = 0.1/r$ for r up to 1000 or so. The law breaks down for less frequent words, since the HARMONIC SERIES diverges. Pierce's (1980, p. 87) statement that $\Sigma P(r) > 1$ for $r = 8727$ is incorrect. Goetz states the law as follows: The frequency of a word is inversely proportional to its RANK r such that

$$P(r) \approx \frac{1}{r \ln(1.78R)},$$

where R is the number of different words.

See also HARMONIC SERIES, RANK (STATISTICS)

References

Bogomolny, A. "Benford's Law and Zipf's Law." http://www.cut-the-knot.com/do_you_know/zipfLaw.html.
Goetz, P. "Phil's Good Enough Complexity Dictionary." http://www.cs.buffalo.edu/~goetz/dict.html.
Li, W. "Zipf's Law." http://linkage.rockefeller.edu/wli/zipf/.
Pierce, J. R. *Introduction to Information Theory: Symbols, Signals, and Noise, 2nd rev. ed.* New York: Dover, pp. 86–87 and 238–239, 1980.

Zip-Pair

A pair of zips, each ZIP being half a zipper, which can be zippered up to close a surface along a curve. The concept of a zip-pair can be extremely useful in topological arguments, and zips can be used to illustrate the construction of the CAP, CROSS-CAP, HANDLE, and CROSS-HANDLE.

See also ZIP

References

Francis, G. K. and Weeks, J. R. "Conway's ZIP Proof." *Amer. Math. Monthly* **106**, 393–399, 1999.

Z-Number

A Z-number is a REAL NUMBER ζ such that

$$0 \le \operatorname{frac}\left[\left(\frac{3}{2}\right)^{k}\xi\right] < \tfrac{1}{2}$$

for all $k = 1, 2, \ldots$, where $\operatorname{frac}(x)$ is the fractional part of x. Mahler (1968) showed that there is at most one Z-number in each interval $[n, n+1]$ for integer n, and therefore concluded that it is unlikely that any Z-numbers exist. The Z-numbers arise in the analysis of the COLLATZ PROBLEM.

See also COLLATZ PROBLEM

References

Flatto, L. "Z-Numbers and β-Transformations." *Symbolic Dynamics and its Applications, Contemporary Math.* **135**, 181–201, 1992.
Guy, R. K. "Mahler's Z-Numbers." §E18 in *Unsolved Problems in Number Theory, 2nd ed.* New York: Springer-Verlag, p. 220, 1994.
Lagarias, J. C. "The $3x+1$ Problem and its Generalizations." *Amer. Math. Monthly* **92**, 3–23, 1985. http://www.cecm.s-fu.ca/organics/papers/lagarias/.
Mahler, K. "An Unsolved Problem on the Powers of 3/2." *Austral. Math. Soc.* **8**, 313–321, 1968.
Tijdman, R. "Note on Mahler's $\frac{3}{2}$-Problem." *Kongel. Norske Vidensk Selsk. Skr.* **16**, 1–4, 1972.

Zöllner's Illusion

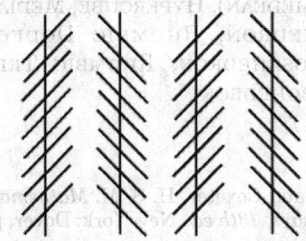

In this ILLUSION, the VERTICAL lines in the above figure are PARALLEL, but appear to be tilted at an angle. In 1860, F. Zöllner sent his discovery in a letter to physicist and scholar J. C. Poggendorff, editor of *Annalen der Physik und Chemie*, who subsequently discovered the related POGGENDORFF ILLUSION.

See also ILLUSION, POGGENDORFF ILLUSION

References

IllusionWorks. "Poggendorf [sic]." http://www.illusion-works.com/html/poggendorf.html.
IllusionWorks. "Zollner." http://www.illusionworks.com/html/zollner.html.
Jablan, S. "Some Visual Illusions Occurring in Interrupted Systems." http://members.tripod.com/~modularity/interr.htm.
Pappas, T. *The Joy of Mathematics.* San Carlos, CA: Wide World Publ./Tetra, p. 172, 1989.

Zome

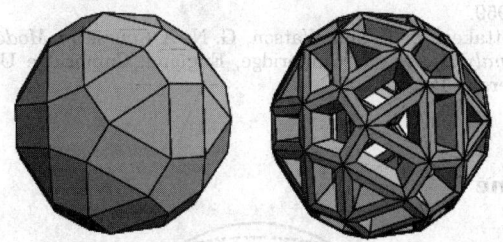

A kit consisting of rods and slotted balls that can be used to construct three-dimensional configurations. The balls into which the rods are placed resembles an "expanded" SMALL RHOMBICOSIDODECAHEDRON, with the squares replaced by rectangles, as illustrated above. The rods come in four colors, and there are three lengths for each color, as summarized in the table below. Here, ϕ is the GOLDEN RATIO.

color	lengths	n
blue	ϕ^{n}	$n = 0, 1, 2$
yellow	$\cos\left(\frac{1}{6}\pi\right)\phi^{n}$	$n = 0, 1, 2$
red	$\cos\left(\frac{1}{10}\pi\right)\phi^{n}$	$n = 0, 1, 2$
green	$\cos\left(\frac{1}{4}\pi\right)\phi^{n}$	$n = -1, 0, 1$

References

Hart, G. W. and Picciotto, H. "Zome Geometry: Hands-on Learning with Zome Models." http://www.georgehart.com/zomebook/zomebook.html.
Zome System. http://www.zometool.com/.

Zonal Harmonic

A SPHERICAL HARMONIC OF THE FORM $P_{l}(\cos\theta)$, i.e., one which reduces to a LEGENDRE POLYNOMIAL (Whittaker and Watson 1990, p. 302). These harmonics are termed "zonal" since the curves on a UNIT SPHERE (with center at the origin) on which $P_{l}(\cos\theta)$ vanishes are l parallels of latitude which divide the surface into zones (Whittaker and Watson 1990, p. 392).

Resolving $P_{l}(\cos\theta)$ into factors linear in $(\cos^{2}\theta)$, multiplied by $(\cos\theta)$ when l is ODD, then replacing $(\cos\theta)$ by z/r allows the zonal harmonic $r^{l}P_{l}(\cos\theta)$ to be expressed as a product of factors linear in x^{2}, y^{2}, and z^{2}, with the product multiplied by z when n is ODD (Whittaker and Watson 1990, p. 1990).

See also LEGENDRE POLYNOMIAL, SECTORIAL HARMONIC, SPHERICAL HARMONIC, TESSERAL HARMONIC

References

Byerly, W. E. "Zonal Harmonics." Ch. 5 in *An Elementary Treatise on Fourier's Series, and Spherical, Cylindrical, and Ellipsoidal Harmonics, with Applications to Problems in Mathematical Physics.* New York: Dover, pp. 144–194, 1959.

Whittaker, E. T. and Watson, G. N. *A Course in Modern Analysis, 4th ed.* Cambridge, England: Cambridge University Press, 1990.

Zone

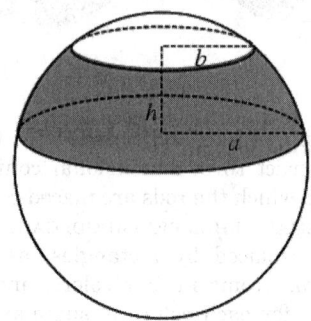

The SURFACE AREA of a SPHERICAL SEGMENT. Call the RADIUS of the SPHERE R, the upper and lower RADII b and a, respectively, and the height of the SPHERICAL SEGMENT h. The zone is a SURFACE OF REVOLUTION about the z-AXIS, so the SURFACE AREA is given by

$$S = 2\pi \int x\sqrt{1 + x'^2}\, dz. \tag{1}$$

In the xz-plane, the equation of the zone is simply that of a CIRCLE,

$$x = \sqrt{R^2 - z^2}, \tag{2}$$

so

$$x' = -z(R^2 - z^2)^{-1/2} \tag{3}$$

$$x'^2 = \frac{z^2}{R^2 - z^2}, \tag{4}$$

and

$$S = 2\pi \int_{\sqrt{R^2 - a^2}}^{\sqrt{R^2 - b^2}} \sqrt{R^2 - z^2} \sqrt{1 + \frac{z^2}{R^2 - z^2}}\, dz$$

$$= 2\pi R \int_{\sqrt{R^2 - a^2}}^{\sqrt{R^2 - b^2}} dz = 2\pi R\left(\sqrt{R^2 - b^2} - \sqrt{R^2 - a^2}\right)$$

$$= 2\pi R h. \tag{5}$$

This result is somewhat surprising since it depends only on the *height* of the zone, not its vertical position with respect to the SPHERE.

See also SPHERE, SPHERICAL CAP, SPHERICAL SEGMENT, ZONOHEDRON

References

Beyer, W. H. (Ed.). *CRC Standard Mathematical Tables, 28th ed.* Boca Raton, FL: CRC Press, p. 130, 1987.

Kern, W. F. and Bland, J. R. "Zone." §35 in *Solid Mensuration with Proofs, 2nd ed.* New York: Wiley, pp. 95–97, 1948.

Zonohedron

A CONVEX POLYHEDRON whose faces all possess a central symmetry (Coxeter 1973, pp. 27–30). Equivalently, a convex polyhedron whose faces are PARALLEL-sided $2m$-gons.

There exist $n(n-1)$ PARALLELOGRAMS in a nonsingular zonohedron, where n is the number of different directions in which EDGES occur (Ball and Coxeter 1987, pp. 141–144). Zonohedra include the CUBE, ENNEACONTAHEDRON, GREAT RHOMBIC TRIACONTAHEDRON, GREAT RHOMBICUBOCTAHEDRON, MEDIAL RHOMBIC TRIACONTAHEDRON, RHOMBIC DODECAHEDRON, RHOMBIC ICOSAHEDRON, RHOMBIC TRIACONTAHEDRON, and RHOMBOHEDRON, as well as the entire class of PARALLELEPIPEDS.

Regular zonohedra have bands of PARALLELOGRAMS which form equators and are called "ZONES." Every convex polyhedron bounded solely by PARALLELOGRAMS is a zonohedron (Coxeter 1973, p. 27). Plate II (following p. 32 of Coxeter 1973) illustrates some equilateral zonohedra. Equilateral zonohedra can be regarded as 3-dimensional projections of n-D HYPERCUBES (Ball and Coxeter 1987).

See also CUBE, ENNEACONTAHEDRON, GREAT RHOMBIC TRIACONTAHEDRON, GREAT RHOMBICUBOCTAHEDRON (ARCHIMEDEAN), HYPERCUBE, MEDIAL RHOMBIC TRIACONTAHEDRON, RHOMBIC DODECAHEDRON, RHOMBIC ICOSAHEDRON, RHOMBIC TRIACONTAHEDRON, RHOMBOHEDRON

References

Ball, W. W. R. and Coxeter, H. S. M. *Mathematical Recreations and Essays, 13th ed.* New York: Dover, pp. 141–144, 1987.

Coxeter, H. S. M. "Zonohedra." §2.8 in *Regular Polytopes, 3rd ed.* New York: Dover, pp. 27–30, 1973.

Coxeter, H. S. M. Ch. 4 in *The Beauty of Geometry: Twelve Essays.* New York: Dover, 1999.

Eppstein, D. "Ukrainian Easter Egg." http://www.ics.uci.edu/~eppstein/junkyard/ukraine/.

Fedorov, E. S. *Zeitschr. Krystallographie und Mineralogie* **21**, 689, 1893.

Fedorov, E. W. *Nachala Ucheniya o Figurakh.* Leningrad, 1953.

Hart, G. "Zonohedra." http://www.georgehart.com/virtualpolyhedra/zonohedra-info.html.

Harp, G. W. "Zonohedrification." *Mathematica J.* **7**, 374–383, 1999.

Kelly, L. M. and Moser, W. O. J. "On the Number of Ordinary Lines Determined by n Points." *Canad. J. Math.* **1**, 210–219, 1958.

Zonotype

The MINKOWSKI SUM of line segments.

Zoomeron Equation

The PARTIAL DIFFERENTIAL EQUATION

$$\left(\frac{\delta^2}{\delta t^2} - \frac{\delta^2}{\delta x^2}\right)\left(\frac{u_{xy}}{u}\right) + 2(u^2)_{xt} = 0.$$

References

Calogero, F. and Degasperis, A. *Spectral Transform and Solitons: Tools to Solve and Investigate Nonlinear Evolution Equations.* New York: North-Holland, p. 58, 1982.
Zwillinger, D. *Handbook of Differential Equations, 3rd ed.* Boston, MA: Academic Press, p. 135, 1997.

Zorn's Lemma

If S is any nonempty PARTIALLY ORDERED SET in which every CHAIN has an upper bound, then S has a maximal element. This statement is equivalent to the AXIOM OF CHOICE.

See also AXIOM OF CHOICE

z-Score

The z-score associated with the ith observation of a random variable x is given by

$$z_i \equiv \frac{x_i - \bar{x}}{\sigma},$$

where $\bar{x}$ is the MEAN and σ the STANDARD DEVIATION of all observations $x_1, \ldots, x_n$.

Zsigmondy Theorem

If $1 \le b < a$ and $(a, b) = 1$ (i.e., a and b are RELATIVELY PRIME), then $a^n - b^n$ has a PRIMITIVE PRIME FACTOR with the following two possible exceptions:

1. $2^6 - 1^6$.
2. $n = 2$ and $a + b$ is a POWER of 2.

Similarly, if $a > b \ge 1$, then $a^n + b^n$ has a PRIMITIVE PRIME FACTOR with the exception $2^3 + 1^3 = 9$.

References

Ribenboim, P. *The Little Book of Big Primes.* New York: Springer-Verlag, p. 27, 1991.

Z-Transform

The Z-transform of $F(t)$ is defined by

$$Z[F(t)] = \mathscr{L}[F^*(t)], \tag{1}$$

where

$$F^*(t) = F(t)\delta_T(t) = \sum_{n=0}^{\infty} F(nT)\delta(t - nT), \tag{2}$$

$\delta(t)$ is the DELTA FUNCTION, T is the sampling period, and $\mathscr{L}[f]$ is the LAPLACE TRANSFORM. An alternative definition is

$$Z[F(t)] = \sum_{\text{residues}} \left(\frac{1}{1 - e^{T_z}z^{-1}}\right)f(z), \tag{3}$$

where

$$f(z) = \sum_{n=0}^{\infty} F(nT)z^{-n}. \tag{4}$$

The inverse Z-transform is

$$Z^{-1}[f(z)] = F^*(t) = \frac{1}{2\pi i} \oint f(z)z^{n-1}\,dz. \tag{5}$$

The GENERATING FUNCTION of $G(t)$ of a sequence of numbers $f(n)$ given by the Z-transform of $f(n)$ in the variable $1/t$ (Germundsson 2000).

It satisfies

$$Z[aF(t) + bG(t)] = aZ[F(t)] + bZ[F(t)] \tag{6}$$

$$Z[F(t + T)] = zZ[F(t)] - zF(0) \tag{7}$$

$$Z[F(t + 2T)] = z^2Z[F(t)] - z^2F(0) - zF(t) \tag{8}$$

$$Z[F(t + mT)] = z^mZ[F(t)] - \sum_{r=0}^{m-1} z^{m-r}F(rt) \tag{9}$$

$$Z[F(t - mT)] = z^{-m}Z[F(t)] \tag{10}$$

$$Z[e^{at}F(t)] = Z[e^{-aT}z] \tag{11}$$

$$Z[e^{-at}F(t)] = Z[e^{aT}z] \tag{12}$$

$$tF(t) = -Tz\frac{d}{dz}Z[F(t)] \tag{13}$$

$$t^{-1}F(t) = -\frac{1}{T}\int_0^z \frac{f(z)}{z}\,dz. \tag{14}$$

Transforms of special functions (Beyer 1987, pp. 426–427) include

$$Z[\delta(t)] = 1 \tag{15}$$

$$Z[\delta(t - mT)] = z^{-m} \tag{16}$$

$$Z[H(t)] = \frac{z}{z - 1} \tag{17}$$

$$Z[H(t - mT)] = \frac{z}{z^m(z - 1)} \tag{18}$$

$$Z[t] = \frac{Tz}{(z - 1)^2} \tag{19}$$

$$Z[t^2] = \frac{T^2z(z + 1)}{(z - 1)^3} \tag{20}$$

$$Z[t^3] = \frac{T^3z(z^2 + 4z + 1)}{(z - 1)^4} \tag{21}$$

$$Z[a^{\omega t}] = \frac{z}{z - a^{\omega T}} \qquad (22)$$

$$Z[\cos(\omega t)] = \frac{z \sin(\omega T)}{z^2 - 2z \cos(\omega T) + 1} \qquad (23)$$

$$Z[\sin(\omega t)] = \frac{z[z - \cos(\omega T)]}{Z^2 - 2z \cos(\omega T) + 1}, \qquad (24)$$

where $H(t)$ is the HEAVISIDE STEP FUNCTION.
In general,

$$Z[t^n] = (-1)^n \lim_{x \to 0} \frac{\delta^n}{\delta x^n} \left(\frac{z}{z - e^{-xT}} \right) \qquad (25)$$

$$\frac{T^n z \sum_{k=1}^{n} \left\langle {n \atop k} \right\rangle z^{k-1}}{(z-1)^{n+1}}, \qquad (26)$$

where the $\left\langle {n \atop k} \right\rangle$ are EULERIAN NUMBERS. Amazingly, the Z-transforms of t^n are therefore generators for EULER'S TRIANGLE.

The discrete z-transform of a sequence $\{a_j\}_{j=-\infty}^{\infty}$ is defined as

$$A(z) = Z[a] = \sum_{k=-\infty}^{\infty} a_k z^{-k} \qquad (27)$$

(Krantz 1999, p. 214). The DISCRETE FOURIER TRANSFORM is therefore a special case of the z-transform with

$$z \equiv e^{-2\pi i/N}. \qquad (28)$$

A z-transform with

$$z \equiv e^{-2\pi i \alpha/N} \qquad (29)$$

for $\alpha \neq \pm 1$ is called a FRACTIONAL FOURIER TRANSFORM.

See also DISCRETE FOURIER TRANSFORM, EULER'S TRIANGLE, EULERIAN NUMBER, FRACTIONAL FOURIER TRANSFORM

References

Arndt, J. "The z-Transform (ZT)." Ch. 3 in "Remarks on FFT Algorithms." http://www.jjj.de/fxt/.
Beyer, W. H. (Ed.). *CRC Standard Mathematical Tables, 28th ed.* Boca Raton, FL: CRC Press, pp. 424–428, 1987.
Bracewell, R. *The Fourier Transform and Its Applications, 3rd ed.* New York: McGraw-Hill, pp. 257–262, 1999.
Germundsson, R. "*Mathematica* Version 4." *Mathematica J.* **7**, 497–524, 2000.

z-Transform (Population)

POPULATION COMPARISON